JUE

OVERD   S

D0712998

LIBRARY
COLLEGE of the REDWOODS
DEL NORTE
883 W. Washington Blvd.
Crescent City, CA 95531

QR 81 .B46 2001

Bergey's manual of
 systematic bacteriology

BERGEY'S MANUAL® OF
Systematic Bacteriology

Second Edition

Volume Two
The *Proteobacteria*

Part C
The *Alpha-, Beta-, Delta-, and Epsilonproteobacteria*

BERGEY'S MANUAL® OF
Systematic Bacteriology
Second Edition

Volume Two
The *Proteobacteria*

Part C
The *Alpha-, Beta-, Delta-, and Epsilonproteobacteria*

Don J. Brenner
Noel R. Krieg
James T. Staley
EDITORS, VOLUME TWO

George M. Garrity
EDITOR-IN-CHIEF

EDITORIAL BOARD
James T. Staley, Chairman, **David R. Boone,** Vice Chairman,
**Don J. Brenner, Paul De Vos, George M. Garrity, Michael Goodfellow,
Noel R. Krieg, Fred A. Rainey, Karl-Heinz Schleifer**

WITH CONTRIBUTIONS FROM 222 COLLEAGUES

 Springer

George M. Garrity, Sc.D.
Bergey's Manual Trust
Department of Microbiology and Molecular Genetics
Michigan State University
East Lansing, MI 48824-4320
USA

Library of Congress Control Number: 2005926296

ISBN-10: 0-387-24145-0
ISBN-13: 978-0-387-24145-6

Printed on acid-free paper.

© 2005, 1984–1989 Bergey's Manual Trust
All rights reserved. This work may not be translated or copied in whole or in part without the written permission of the publisher (Springer Science+Business Media, Inc., 233 Spring Street, New York, NY 10013, USA), except for brief excerpts in connection with reviews or scholarly analysis. Use in connection with any form of information storage and retrieval, electronic adaptation, computer software, or by similar or dissimilar methodology now known or hereafter developed is forbidden.
The use in this publication of trade names, trademarks, service marks, and similar terms, even if they are not identified as such, is not to be taken as an expression of opinion as to whether or not they are subject to proprietary rights.

Printed in the United States of America.

9 8 7 6 5 4

springer.com

*This volume is dedicated to our colleagues,
David R. Boone, Don J. Brenner,
Richard W. Castenholz, and Noel R. Krieg, who
retired from the Board of Trustees of Bergey's Manual
Trust as this edition was in preparation. We deeply
appreciate their efforts as editors and authors; they
have devoted their time and many years in helping
the Trust meet its objectives.*

EDITORIAL BOARD AND TRUSTEES OF BERGEY'S MANUAL TRUST

James T. Staley, *Chairman*
David R. Boone, *Vice Chairman*
George M. Garrity
Paul De Vos
Michael Goodfellow
Fred A. Rainey
Karl-Heinz Schleifer
Don J. Brenner, *Emeritus*
Richard W. Castenholz, *Emeritus*
John G. Holt, *Emeritus*
Noel R. Krieg, *Emeritus*
John Liston, *Emeritus*
James W. Moulder, *Emeritus*
R.G.E. Murray, *Emeritus*
Charles F. Niven, Jr., *Emeritus*
Norbert Pfennig, *Emeritus*
Peter H.A. Sneath, *Emeritus*
Joseph G. Tully, *Emeritus*
Stanley T. Williams, *Emeritus*

Preface to Volume Two of the Second Edition of *Bergey's Manual®of Systematic Bacteriology*

There is a long-standing tradition for the Editors of *Bergey's Manual* to open their respective editions with the observation that the new edition is a departure from the earlier ones. As this volume goes to press, however, we recognize a need to deviate from this practice, by offering a separate preface to each volume within this edition. In part, this departure is necessary because the size and complexity of this edition far exceeded our expectations, as has the amount of time that has elapsed between publication of the first volume of this edition and this volume.

Earlier, we noted that systematic procaryotic biology is a dynamic field, driven by constant theoretical and methodological advances that will ultimately lead to a more perfect and useful classification scheme. Clearly, the pace has been accelerating as evidenced in the super-linear rate at which new taxa are being described. Much of the increase can be attributed to rapid advances in sequencing technology, which has brought about a major shift in how we view the relationships among *Bacteria* and *Archaea*. While the possibility of a universally applicable natural classification was evident as the First Edition was in preparation, it is only recently that the sequence databases became large enough, and the taxonomic coverage broad enough to make such an arrangement feasible. We have relied heavily upon these data in organizing the contents of this edition of *Bergey's Manual of Systematic Bacteriology*, which will follow a phylogenetic framework based on analysis of the nucleotide sequence of the small ribosomal subunit RNA, rather than a phenotypic structure. This departs from the First Edition, as well as the Eighth and Ninth Editions of the *Determinative Manual*. While the rationale for presenting the content of this edition in such a manner should be evident to most readers, they should bear in mind that this edition, as in all preceding ones represents a progress report, rather than a final classification of procaryotes.

The Editors remind the readers that the *Systematics Manual* is a peer-reviewed collection of chapters, contributed by authors who were invited by the Trust to share their knowledge and expertise of specific taxa. Citation should refer to the author, the chapter title, and inclusive pages rather than to the Editors. The Trust is indebted to all of the contributors and reviewers, without whom this work would not be possible. The Editors are grateful for the time and effort that each expended on behalf of the entire scientific community. We also thank the authors for their good grace in accepting comments, criticisms, and editing of their manuscripts. We would also like to thank Drs. Hans Trüper, Brian Tindall, and Jean Euzéby for their assistance on matters of nomenclature and etymology.

We would like to express our thanks to the Department of Microbiology and Molecular Genetics at Michigan State University for housing our headquarters and editorial office and for providing a congenial and supportive environment for microbial systematics. We would also like to thank Connie Williams not only for her expert secretarial assistance, but also for unflagging dedication to the mission of Bergey's Manual Trust and Drs. Julia Bell and Denise Searles for their expert editorial assistance and diligence in verifying countless pieces of critical information and to Dr. Timothy G. Lilburn for constructing many of the phylogenetic trees used in this volume. We also extend our thanks to Alissa Wesche, Matt Chval and Kristen Johnson for their assistance in compilation of the bibliography.

A project such as the *Systematics Manual* also requires the strong and continued support of a dedicated publisher, and we have been most fortunate in this regard. We would also like to express our gratitude to Springer-Verlag for supporting our efforts and for the development of the Bergey's Document Type Definition (DTD). We would especially like to thank our Executive Editor, Dr. William Curtis for his courage, patience, understanding, and support; Catherine Lyons for her expertise in designing and developing our DTD, and Jeri Lambert and Leslie Grossberg of Impressions Book and Journal Services for their efforts during the pre-production and production phases. We would also like to acknowledge the support of ArborText, Inc., for providing us with state-of-the-art SGML development and editing tools at reduced cost. Lastly, I would like to express my personal thanks to my fellow trustees for providing me with the opportunity to participate in this effort, to Drs. Don Brenner, Noel Krieg, and James Staley for their enormous efforts as volume editors and to my wife, Nancy, and daughter, Jane, for their continued patience, tolerance and support.

Comments on this edition are welcomed and should be directed to Bergey's Manual Trust, Department of Microbiology and Molecular Genetics, 6162 Biomedical and Physical Sciences Building, Michigan State University, East Lansing, MI, USA 48824-4320. Email: garrity@msu.edu

George M. Garrity

Preface to the First Edition of *Bergey's Manual® of Systematic Bacteriology*

Many microbiologists advised the Trust that a new edition of the *Manual* was urgently needed. Of great concern to us was the steadily increasing time interval between editions; this interval reached a maximum of 17 years between the seventh and eighth editions. To be useful the *Manual* must reflect relatively recent information; a new edition is soon dated or obsolete in parts because of the nearly exponential rate at which new information accumulates. A new approach to publication was needed, and from this conviction came our plan to publish the *Manual* as a sequence of four subvolumes concerned with systematic bacteriology as it applies to taxonomy. The four subvolumes are divided roughly as follows: (a) the Gram-negatives of general, medical or industrial importance; (b) the Gram-positives other than actinomycetes; (c) the archaeobacteria, cyanobacteria and remaining Gram-negatives; and (d) the actinomycetes. The Trust believed that more attention and care could be given to preparation of the various descriptions within each subvolume, and also that each subvolume could be prepared, published, and revised as the area demanded, more rapidly than could be the case if the *Manual* were to remain as a single, comprehensive volume as in the past. Moreover, microbiologists would have the option of purchasing only that particular subvolume containing the organisms in which they were interested.

The Trust also believed that the scope of the *Manual* needed to be expanded to include more information of importance for systematic bacteriology and bring together information dealing with ecology, enrichment and isolation, descriptions of species and their determinative characters, maintenance and preservation, all focused on the illumination of bacterial taxonomy. To reflect this change in scope, the title of the *Manual* was changed and the primary publication becomes *Bergey's Manual of Systematic Bacteriology*. This contains not only determinative material such as diagnostic keys and tables useful for identification, but also all of the detailed descriptive information and taxonomic comments. Upon completion of each subvolume, the purely determinative information will be assembled for eventual incorporation into a much smaller publication which will continue the original name of the *Manual*, *Bergey's Manual of Determinative Bacteriology*, which will be a similar but improved version of the present *Shorter Bergey's Manual*. So, in the end there will be two publications, one systematic and one determinative in character.

An important task of the Trust was to decide which genera should be covered in the first and subsequent subvolumes. We were assisted in this decision by the recommendations of our Advisory Committees, composed of prominent taxonomic authorities to whom we are most grateful. Authors were chosen on the basis of constant surveillance of the literature of bacterial systematics and by recommendations from our Advisory Committees.

The activation of the 1976 Code had introduced some novel problems. We decided to include not only those genera that had been published in the Approved Lists of Bacterial Names in January 1980 or that had been subsequently validly published, but also certain genera whose names had no current standing in nomenclature. We also decided to include descriptions of certain organisms which had no formal taxonomic nomenclature, such as the endosymbionts of insects. Our goal was to omit no important group of cultivated bacteria and also to stimulate taxonomic research on "neglected" groups and on some groups of undoubted bacteria that have not yet been cultivated and subjected to conventional studies.

The invited authors were provided with instructions and exemplary chapters in June 1980 and, although the intended deadline for receipt of manuscripts was March 1981, all contributions were assembled in January 1982 for the final preparations. The *Manual* was forwarded to the publisher in June 1982.

Some readers will note the consistent use of the stem -var instead of -type in words such as biovar, serovar and pathovar. This is in keeping with the recommendations of the Bacteriological Code and was done against the wishes of some of the authors.

We have deleted much of the synonymy of scientific names which was contained in past editions. The adoption of the new starting date of January 1, 1980 and publication of the Approved Lists of Bacterial Names has made mention of past synonymy obsolete. We have included synonyms of a name only if they have been published since the new starting date, or if they were also on the Approved Lists and, in rare cases with certain pathogens, if the mention of an old name would help readers associate the organism with a clinical problem. If the reader is interested in tracing the history of a name we suggest he or she consult past editions of the *Manual* or the *Index Bergeyana* and its *Supplement*. In citations of names we have used the abbreviation AL to denote the inclusion of the name on the Approved Lists of Bacterial Names and VP to show the name has been validly published.

In the matter of citation of the *Manual* in the scientific literature we again stress the fact that the *Manual* is a collection of authored chapters and the citation should refer to the author, the chapter title and its inclusive pages, not the Editor.

To all contributors, the sincere thanks of the Trust is due; the Editor is especially grateful for the good grace with which the authors accepted comments, criticisms and editing of their manuscripts. It is only because of the voluntary and dedicated efforts of these authors that the *Manual* can continue to serve the science of bacteriology on an international basis.

A number of institutions and individuals deserve special acknowledgment from the Trust for their help in bringing about the publication of this volume. We are grateful to the Department of Biology of the Virginia Polytechnic Institute and State University for providing space, facilities and, above all, tolerance for the diverted time taken by the Editor during the preparation of the book. The Department of Microbiology at Iowa State University of Science and Technology continues to provide a welcome home for the main editorial offices and archives of the Trust and we acknowledge their continued support. A grant (LM-03707) from the National Library of Medicine, National Institutes of Health to assist in the preparation of this and the next volume of the *Manual* is gratefully acknowledged.

A number of individuals deserve special mention and thanks for their help. Professor Thomas O. McAdoo of the Department of Foreign Languages and Literatures at the Virginia Polytechnic Institute and State University has given invaluable advice on the etymology and correctness of scientific names. Those assisting the Editor in the Blacksburg office were R. Martin Roop II, Don D. Lee, Eileen C. Falk and Michael W. Friedman and their help is sincerely appreciated. In the Ames office we were ably assisted by Gretchen Colletti and Diane Triggs during the early period of preparation and by Cynthia Pease during the major portion of the editing process. Mrs. Pease has been responsible for the construction of the List of References and her willingness to handle the cumbersome details of text editing on a big computer is gratefully acknowledged.

John G. Holt

Preface to the First Edition of *Bergey's Manual® of Determinative Bacteriology*

The elaborate system of classification of the bacteria into families, tribes and genera by a Committee on Characterization and Classification of the Society of American Bacteriologists (1911, 1920) has made it very desirable to be able to place in the hands of students a more detailed key for the identification of species than any that is available at present. The valuable book on "Determinative Bacteriology" by Professor F. D. Chester, published in 1901, is now of very little assistance to the student, and all previous classifications are of still less value, especially as earlier systems of classification were based entirely on morphologic characters.

It is hoped that this manual will serve to stimulate efforts to perfect the classification of bacteria, especially by emphasizing the valuable features as well as the weaker points in the new system which the Committee of the Society of American Bacteriologists has promulgated. The Committee does not regard the classification of species offered here as in any sense final, but merely a progress report leading to more satisfactory classification in the future.

The Committee desires to express its appreciation and thanks to those members of the society who gave valuable aid in the compilation of material and the classification of certain species. . . .

The assistance of all bacteriologists is earnestly solicited in the correction of possible errors in the text; in the collection of descriptions of all bacteria that may have been omitted from the text; in supplying more detailed descriptions of such organisms as are described incompletely; and in furnishing complete descriptions of new organisms that may be discovered, or in directing the attention of the Committee to publications of such newly described bacteria.

David H. Bergey, *Chairman*
Francis C. Harrison
Robert S. Breed
Bernard W. Hammer
Frank M. Huntoon
Committee on Manual.
August, 1923.

Contents

Contributors

Wolf-Rainer Abraham
Chemical Microbiology Group, GBF-National Research Centre for Biotechnology, Mascheroder Weg 1, D-38124 Braunschweig, Germany

Paula Aguiar
Portland State University, Portland, OR 97207-0751, USA

Milton J. Allison
Department of Microbiology, Iowa State University, Ames, IA 50011-3211, USA

Rudolf Amann
Nachwuchsgruppe Molekulare Ökologie, Max Planck-Institute für Marine Mikrobiologie, Celsiusstrasse 1, D-28359 Bremen, Germany

Georg Auling
Institute für Mikrobiologie, Universität Hannover, Schneiderberg 50, D-30167 Hannover, Germany

Marcie L. Baer
Biology Department, Shippensburg University, Shippensburg, PA 17257, USA

Simon C. Baker
Birkbeck College, Malet Street, Bloomsbury, London WC1E 7HX, United Kingdom

José Ivo Baldani
Centro Nacional de Pesquisa de Agrobiologia, Empresa Brasileira de Pesquisa Agropecuária, Room 247-23851-970 Seropédica, Caixa Postal 74.505, Rio de Janeiro 465, Brazil

Vera Lúcia Divan Baldani
Centro Nacional de Pesquisa de Agrobiologia, Empresa Brasileira de Pesquisa Agropecuária, Room 247-23851-970 Seropédica, Caixa Postal 74.505, Rio de Janeiro 465, Brazil

David L. Balkwill
Department of Biological Science, Florida State University, Tallahassee, FL 32306-4470, USA

Menachem Banai
Ministry of Agriculture, Veterinary Services & Animal Health, Kimron Veterinary Institute, P.O. Box 12, Bet Dagan 50 250, Israel

Claudio Bandi
Dipartimento di Patologia Animale, Igiene e Sanità Pubblica Veterinaria, Sezione di Patologia Generale e Parassitologia, Università degli Studi di Milano, Via Celoria 10 20133 Milano, Italy

Ellen Jo Baron
Clinical Microbiology/Virology Laboratory, Stanford University Medical Center, Stanford, CA 94305-5250, USA

Janiche Beeder
Section for Biotechnology, Novsk Hydro ASA Research Centre, P. O. Box 2560, N-3901 Porsgruun, Norway

Julia A. Bell
Department of Microbiology and Molecular Genetics, Michigan State University, East Lansing, MI 48824-4320, USA

Nancy M.C. Bleumink-Pluym
Dept. of Bacteriology, Inst. of Infectious Diseases & Immunology, Vet. Medicine, Universität Utrecht, Yalelaan 1, 3584 CL Utrecht, The Netherlands

Eberhard Bock
Inst. für Allgemeine Botanik und Botanischer Garten, Universität Hamburg, Ohnhorststrasse 18, D-22609 Hamburg, Germany

David R. Boone
Department of Environmental Biology, Portland State University, Portland, OR 97207-0751, USA

Edward J. Bottone
Department of Infectious Diseases, The Mount Sinai Hospital, New York, NY 10029-6574, USA

John P. Bowman
School of Agricultural Science, University of Tasmania, Antartic CRC, Private Bag 54, Hobart 7001, Tasmania, Australia

Kristian K. Brandt
Section of Genetics and Microbiology, Department of Ecology, Royal Veterinary and Agricultural University, DK-1871 Frederiksberg, Denmark

Don J. Brenner
Meningitis & Special Pathogens Branch Laboratory Section, Centers for Disease Control & Prevention, Atlanta, GA 30333, USA

Hans-Jürgen Busse
Institut für Bakteriologie, Mykologie und Hygiene, Veterinärmedizinische Universität Wien, Veterinärplatz 1, A-1210 Wien, Austria

Douglas E. Caldwell
Dept. of Applied Microbiology and Food Science, University of Saskatchewan, Saskatoon, 51 Campus Drive, Saskatchewan S7N 5A8 SK, Canada

Wen Xin Chen
Department of Microbiology, Biology College, Beijing Agricultural University, Beijing, P.R. China

John D. Coates
Plant and Microbial Biology, University of California, Berkeley, Berkeley, CA 94720-3102, USA

John D. Coates
Plant and Microbial Biology, University of California, Berkeley, Berkeley, CA 94720-3102. USA

Michael J. Corbel
National Institute for Biol. Standards & Control, Blanche Lane, South Mimms, Potters Bar, Hertfordshire EN6 3QG, United Kingdom

Milton S. da Costa
Departamento de Zoologia, Centro de Neurociências, Universidade de Coimbra, Apartado 3126, P-3004-517 Coimbra, Portugal

Subrata K. Das
Institute of Life Sciences, Nalco square, Bhubaneswar 751 023, India

Gregory A. Dasch
Division of Viral and Rickettsial Diseases, Viral and Rickettsial Zoonoses Branch, National Center for Infectious Diseases, Centers for Disease Control and Prevention, Atlanta, GA 30333, USA

Frank B. Dazzo
Department of Microbiology and Molecular Genetics, Michigan State University. East Lansing, MI 48824-4320, USA

Kim A. DeWeerd
Department of Chemistry, State University of New York, University at Albany, Albany, NY 12222, USA

Ewald B.M. Denner
Abteilung Mikrobiologie und Biotechnologie, Institut für Mikrobiologie und Genetik, Dr. Bohr-Gasse 9, A-1030 Wein, Austria

Richard Devereux
NHEERL, Gulf Ecology Division, U.S.E.P.A., Gulf Breeze, FL 32561, USA

Floyd E. Dewhirst
Department of Molecular Genetics, The Forsyth Institute, 140 The Fenway, Boston, MA 02115-3799, USA

Johanna Döbereiner (Deceased)
Centro Nacional de Pesquisa de Agrobiologia, Empresa Brasiliera de Pesquisa Agropecuária, Room 247, 23851-970 Seropédica, Caixa Postal 74.505, Rio de Janeiro 465, Brazil

Nina V. Doronina
Inst. of Biochemistry & Physiology of Micro-organisms RAS, Laboratory of Methylotrophy, Russian Academy of Sciences, Pushchino-on-the-Oka, Moscow Region 142290, Russia

Galina A. Dubinina
Institute of Microbiology, Russia Academy of Sciences, Prospect 6—let. Oktyabrya 7/2, Moscow, Russia

J. Stephen Dumler
Division of Microbiology, Department of Pathology, The Johns Hopkins Hospital, Univ. School of Medicine, Baltimore, MD 21287-7093, USA

Jürgen Eberspächer
Institut für Mikrobiologie (250), Universität Hohenheim, Garbenstrasse 30, D-70599 Stuttgart, Germany

Thomas W. Egli
Department of Microbiology, EAWAG, Überlandstrasse 133, CH 8600 Düebendorf, Switzerland

Stefanie J.W.H. Oude Elferink
ID TNO Animal Nutrition, P.O. Box 65, 8200 AB Lelystad, The Netherlands

Takayuki T. Ezaki
Department of Microbiology and Bioinformatics, Regeneration and Advanced Medical Science, Gifu University School of Medicine, 40 Tsukasa-machi, Gifu 500 8705, Japan

Mark Fegan
Coop. Research Centre for Tropical Plant Protection, Dept. of Micro. & Parasitology, The University of Queensland, St. Lucia, Brisbane, Queensland 4072, Australia

Andreas Fesefeldt
Geibelallee 12a, 24116 Kiel, Germany

Kai W. Finster
Department of Microbial Ecology, Institute of Biological Sciences, University of Aarhus, Building 540, Ny, Munkegade, DK-8000 Åarhus C, Denmark

James G. Fox
Department of Comparative Medicine, Massachusetts Institute of Technology, Cambridge, MA 02139, USA

Michael Friedrich
Abteilung Biogeochemie, Max Planck-Institut für Terrestrische Mikrobiologie, Karl-von-Frisch-Strasse, D-35043 Marburg, Germany

Georg Fuchs
Mikrobiologie, Institut für Biologie II, Albert-Ludwigs-Universität Freiburg, D-79104 Freiburg, Germany

John A. Fuerst
Center for Bacterial Diversity and Identification, Department of Microbiology, University of Queensland, Brisbane, Queensland 4072, Australia

Jean-Louis Garcia
Laboratoire de Microbiologie, ORSTOM-ESIL-Case 925, Université de Provence, 163, Avenue de Luminy, 13288 Marseille, Cédex 9, France

Monique Garnier (Deceased)
Institut National de la Recherche Agronomique et Université Victor Ségalen, Laboratoire de Biologie Cellulaire et Moléculaire, Bordeaux 2, 33883, BP 81, Villenave d'Ormon Cedex, France

George M. Garrity
Dept. of Microbiology and Molecular Genetics, Michigan State University, East Lansing, MI 48824, USA

Rainer Gebers
Depenweg 12, D-24217 Schönberg/Holstein, Germany

Connie J. Gebhart
Division of Comparative Medicine, University of Minnesota Health Center, Minneapolis, MN 55455, USA

Barbara R. Sharak Genthner
Center for Environmental Diagnostics and Bioremediation, University of West Florida, Pensacola, FL 32514, USA

Peter Gerner-Smidt
Department of Gastrointestinal Infections, Statens Serum Institut, Artillerivej 5, DK-2300 Copenhagen S, Denmark

Monique Gillis
Laboratorium voor Microbiologie Vakgroep WE 10V, Universiteit Gent, K.L. Ledeganckstraat 35, B-9000 Gent, Belgium

Christian Gliesche
Institut für Ökologie, Ernst-Moritz-Arndt-Universität, Greifswald Schwedenhagen 6, D-18565 Kloster/Hiddensee, Germany

José M. González
Departamento de Microbiologia y Biologia Celular, Facultad de Farmacia, Universidad de La Laguna, 38071 La Laguna. Tenerife, Spain

Yvonne E. Goodman
Department of Medical Bacteriology, University of Alberta, Medical Services Building, Edmonton, Alberta, Canada

Vladimir M. Gorlenko
Institute of Microbiology, Russian Academy of Sciences, Prospect 60-letiya, Oktyabrya 7, korpus 2, Moscow 117312, Russia

Hans-Dieter Görtz
Department of Zoology, Biologisches Institut, Universität Stuttgart, Pfaffenwaldring 57, D-70550 Stuttgart, Germany

John J. Gosink
Amgen, Inc., Seattle, WA 98101, USA

Jennifer Gossling
8401 University Drive, St. Louis, MO 63105-3641, USA

Peter N. Green
National Collection of Industrial & Marine Bacteria, 23 St. Machar Drive, Aberdeen AB24 3RY, United Kingdom

Lotta E-L. Hallbeck
Department of Cell and Molecular Biology, Göteborg University, Medicinaregatan 9 C, Box 462, S-405 30 Göteborg, Sweden

Theo A. Hansen
Department of Microbial Physiology, Groningen Biomolecular Sci. & Biotech. Inst., University of Groningen, P. O. Box 14, 9750 AA Haren, The Netherlands

Anton Hartmann
Institute of Soil Ecology, Rhizosphere Biology Division, GSF Research Center, PO Box 1129, D-85764 Neuherberg, München, Germany

Fawzy M. Hashem
Sustainable Agriculture Laboratory, Animal and Natural Resources Institute, Beltsville Agricultural Research Institute,USDA-ARS, Beltsville, MD 20705, USA

Brian P. Hedlund
Department of Biological Sciences, University of Nevada, Las Vegas, Las Vegas, NV 89154-4004, USA

Johann Heider
Mikrobiologie, Institut für Biologie II, Universität Freiburg, Schänzlestrasse 1, D-79104 Freiburg, Germany

Karl-Heinz Hinz
Klinik für Geflügel der Tierärztlichen Hochschule, Bünteweg 17, D-30559 Hannover, Germany

Akira Hiraishi
Department of Ecological Engineering, Toyohashi University of Technology, Tempaku-cho, Toyohashi 441-8580, Japan

Peter Hirsch
Institut für Allgemeine Mikrobiologie der Biozentrum, Universität Kiel, Am Botanischen Garten 1-9, D-24118 Kiel, Germany

Becky Hollen
Department of Biological Sciences, Louisiana State University, Baton Rouge, LA 70803, USA

Barry Holmes
Public Health Laboratory Service, Central Public Health Laboratory, National Collection of Type Cultures, 61 Colindale Avenue, London NW9 5HT, United Kingdom

John Holt
Department of Microbiology and Molecular Genetics, Michigan State University, East Lansing, MI 48824-1101, USA

Philip Hugenholtz
Ecosystem Sciences Division, Department of Environmental Science, Policy, and Management, University of California, Berkeley, Berkeley, CA 94720-3110, USA

Thomas Hurek
Arbeitsgruppe Symbioseforschung, Planck-Institut für Terrestrische Mikrobiologie, Karl-von-Frisch-Strasse, D-35043 Marburg, Germany

Johannes F. Imhoff
Institut für Meereskunde, Abt. Marine Mikrobiologie, Universität Kiel, Düsternbrooker Weg 20, D-24105 Kiel, Germany

Kjeld Ingvorsen
Department of Microbial Ecology, Institute of Biological Sciences, University of Aarhus, Building 540, Ny Munkegade, DK-8000 Aarhus C, Denmark

Francis L. Jackson
Medical Microbiology and Immunology, University of Alberta, 1-41-Medical Sciences Building, Edmonton, Alberta AB T6G 2H7, Canada

Cheryl Jenkins
Department of Microbiology, University of Washington, Seattle, WA 98195-0001, USA

Sibylle Kalmbach
Studienstiftung des Deutschen Volkes, Mirbachstrasse 7, D-53173 Bonn, Germany

Peter Kämpfer
Institut für Angewandte Mikrobiologie, Justus-Liebig-Universität Giessen, Heinrich-Buff-Ring 26-32, IFZ, D-35392 Giessen, Germany

Yoshiaki Kawamura
Department of Microbiology, Gifu University School of Medicine, 40 Tsukasa-machi, Gifu 500 8705, Japan

Donovan P. Kelly
Department of Biological Sciences, University of Warwick, Coventry CV4 7AL, United Kingdom

Suzanne V. Kelly
Professor of Biology, Scottsdale Community College, Scottsdale, AZ 85250, USA

Christina Kennedy
Department of Plant Pathology, College of Agriculture, The University of Arizona, Tucson, AZ 85721-0036, USA

Christina Kennedy
Division of Plant Pathology and Microbiology, Department of Plant Pathology, The University of Arizona, Tucson, AZ 85721-0036, USA

Allen Kerr
Waite Agricultural Research Institute, The University of Adelaide, Glen Osmond 5064, South Australia

Karel Kersters
Lab. voor Microbiologie, Vakgroep Biochemie, Fysiologie en Microbiologie, Rijksuniversiteit Gent, K.L. Ledeganckstraat 35, B-9000 Gent, Belgium

Hans-Peter Klenk
VP Genomics, Epidauros Biotechnology Inc., Am Neuland 1, D-82347 Bernried, Germany

Oliver Klimmek
Biozentrum Niederursel, Institut für Mikrobiologie der Johann Wolfgang Goethe-Universität, Marie-Curie-Strasse 9, D-60439 Frankfurt am Main, Germany

Allan E. Konopka
Department of Biological Science, Purdue University, West Lafayette, IN 47907-2054, USA

Hans-Peter Koops
Abteilung Mikrobiologie, Inst. für Allgemeine Botanik und Botanischer Garten, Universität Hamburg, Ohnhorststrasse 18, D-22609 Hamburg, Germany

Yoshimasa Kosako
The Institute of Physical and Chemical Research, Japan Collection of Microorganisms, RIKEN, Wako-shi, Saitama 351-0198, Japan

Julius P. Kreier
Department of Microbiology, The Ohio State University, Columbus, OH 43201, USA

Noel R. Krieg
Department of Biology, Virginia Polytechnic Institute & State University, Blacksburg, VA 24061-0406, USA

Achim Kröger (Deceased)
Biozentrum Niederursel, Institut für Mikrobiologie der Johann Wolfgang Goethe-Universität, Marie-Curie-Strasse 9, D-60439 Frankfurt am Main, Germany

J. Gijs Kuenen
Faculty of Chemical Tech. & Materials Science, Kluyver Laboratory for Biotechnology, Delft University of Technology, 2628 BC Delft, The Netherlands

Jan Kuever
Department of Microbiology, Institute for Material Testing, Foundation Institute for Materials Science, D-28199 Bremen, Germany

L. David Kuykendall
Molecular Plant Pathology Laboratory, Plant Sciences Institute, United States Department of Agriculture, Beltsville, MD 20705-2350, USA

David P. Labeda
Natl. Ctr. for Agricultural Utilization Research, Microbial Properties Research, U.S. Department of Agriculture, Peoria, IL 61604-3999, USA

Matthias Labrenz
Institut für Allgemeine Mikrobiologie, Biologiezentrum, University of Kiel, Am Botanischen Garten 1-9, 24118 Kiel, Germany

Adrian Lee
School of Microbiology and Immunology, University of New South Wales, Kensington, Sydney, Australia

Werner Liesack
Max Planck-Institut für Terrestrische Mikrobiologie, Karl-von-Frisch-Strasse, D-35043 Marburg, Germany

Timothy Lilburn
ATCC Bioinformatics, Manassas, VA 20110-2209, USA

Niall A. Logan
School of Biological and Biomedical Sciences, Glasgow Caledonian University, Cowcaddens Road, Glasgow G4 0BA, United Kingdom

Derek R. Lovley
Department of Microbiology, University of Massachusetts, Physiology & Ecology of Anaerobic Micro., Amherst, MA 01003, USA

Wolfgang Ludwig
Lehrstuhl für Mikrobiologie, Technische Universität München, Am Hochanger 4, D-85350 Freising, Germany

Barbara J. MacGregor
Max Planck-Institute for Marine Microbiology, Celsiusstrasse 1, D-28359 Bremen, Germany

Michael T. Madigan
Department of Microbiology, Life Science II, Southern Illinois University, Carbondale, IL 62901-6508, USA

Åsa Malmqvist
ANOX AB, Klosterangsvagen 11A, S-226 47 Lund, Sweden

Werner Manz
Section G3, Ecotoxicology and Biochemistry, German Federal Institute of Hydrology, Kaiserin-Augusta-Anlagen 15-17, P. O. Box 20 02 53, D-56002 Koblenz, Germany

Esperanza Martínez-Romero
Centro de Investigación sobre Fijación de Nitrógeno, UNAM, Ap Postal 565–A, Cuernavaca, Morelos, México

Abdul M. Maszenan
Environmental Engineering Research Centre, School of Civil and Structural Engineering, Nanyang Technological University, Block N1, #1a-29, 50 Nanyang Avenue, Singapore 639798

Michael J. McInerney
Department of Botany and Microbiology, The University of Oklahoma, Norman, OK 73019-6131, USA

Steven McOrist
Department of Biomedical Sciences, Tufts University College of Veterinary Medicine, North Grafton, MA 01536, USA

Roy D. Meredith (Deceased)

Joris Mergaert
Laboratorium voor Microbiologie Vakgroep Biochemie, Fysiologie en Microbiol., Universiteit Gent, K.L. Ledeganckstraat 35, B-9000 Gent, Belgium

Ortwin D. Meyer
Lehrstuhl für Mikrobiologie, Universität Bayreuth, Universitätsstrasse 30, D-95440 Bayreuth, Germany

Edward R.B. Moore
Programme of Soil Quality and Protection, The Macaulay Research Institute, Macaulay Dr., Craigiebuckler, AB15 8QH Aberdeen, United Kingdom

R.G.E. Murray
Department of Microbiology and Immunology, The University of Western Ontario, London, Ontario N6A 5C1, Canada

Yasuyoshi Nakagawa
Biological Resource Center (NBRC), Department of Biotechnology, National Institute of Technology and Evaluation, 2-5-8, Kazusakamatari, Kisarazu, Chiba 292-0818, Japan

M. Fernanda Nobre
Departmento de Zoologia, Universidade de Coimbra, Apartado 3126, P-3000 Coimbra, Portugal

Jani L. O'Rourke
School of Microbiology and Immunology, University of New South Wales, Kensington, Sydney, Australia

Bernard Ollivier
Laboratoirede Microbiologie—LMI, ORSTOM, Case 925, Université de Provence, ESIL, 163 Avenue de Luminy, Marseille 13288 Cedex 09, France

Stephen L.W. On
Danish Veterinary Institute, Bülowsvej 27, DK-1790, Copenhagen V, Denmark

Ronald S. Oremland
Water Research Division, U.S. Geological Survey, Menlo Park, CA 94025-3591, USA

Norberto J. Palleroni
Rutgers, North Caldwell, NJ 07006-4146, USA

Bruce J. Paster
Department of Molecular Genetics, The Forsyth Institute, 140 The Fenway, Boston, MA 02115-3799, USA

Bharat K.C. Patel
Microbial Discovery Research Unit, School of Biomolecular Sciences, Griffith University, Nathan Campus, Kessels Road, Brisbane, Queensland 4111, Australia

Dominique Patureau
Laboratoire de Biotechnologie de l'Environnement, INRA Narbonne, avenue des étangs, 11 100 Narbonne, France

Karsten Pedersen
Department of Cell and Molecular Biology, Göteborg University, Medicinaregatan 9 C, Box 462, S-405 30 Göteborg, Sweden

Jeanne S. Poindexter
Department of Biological Sciences, Barnard College, Columbia University, New York, NY 10027-6598, USA

Andreas Pommerening-Röser
Abteilung Mikrobiologie, Inst. für Allgemeine, Botanik und Botanischer Garten, Universität Hamburg, Ohnhorststrasse 18, D-22609 Hamburg, Germany

Bruno Pot
Science Department, Yakult Belgium, Joseph Wybranlaan 40, B-1070 Brussels, Belgium

Fred A. Rainey
Department of Biological Sciences, Louisiana State University, Baton Rouge, LA 70803, USA

Didier Raoult
Faculté de Médecine, CNRS, Unité des Rickettsies, 27 Boulevard Jean Moulin, 13385 Marseille Cedex 05, France

Christopher Rathgeber
Department of Microbiology, The University of Manitoba, Winnipeg, Manitoba R3T 2N2, Canada

Gavin N. Rees
Murray-Darling Freshwater Research Centre, CRC Freshwater Ecology, Ellis Street, Thurgoona, PO Box 921, Albury NSW 2640, Australia

Hans Reichenbach
Arbeitsgruppe Mikrobielle Sekundärstoffe, Gesellschaft für Biotechnologische Forschung mbH, Mascheroder Weg 1, D-38124 Braunschweig, Germany

Barbara Reinhold-Hurek
Universität Bremen, Fachbereich 2, Allgemeine Mikrobiologie, P. O. Box 330440, D-28334 Bremen, Germany

Anna-Louise Reysenbach
Department of Environmental Biology, Portland State University, Portland, OR 97207, USA

Yasuko Rikihisa
Department of Veterinary Biosciences, The Ohio State University, 1925 Coffey Road, Columbus, OH 43210-1093, USA

Lesley A. Robertson
Kluyver Laboratory for Biotechnology, Delft University of Technology, Julianalaan 67, P. O. Box 5057, 2628BC Delft, The Netherlands

Takeshi Sakane
Institute for Fermentation, Osaka, Yodogawa-ku, Osaka 532-8686, Japan

Abigail A. Salyers
Department of Microbiology, University of Illinois-Urbana, Champaign, Urbana, IL 61801-3704, USA

Gary N. Sanden
Epidemic Investigations Laboratory, Meningitis and Special Pathogens Branch, Division of Bacterial and Mycotic Diseases, Centers for Disease Control and Prevention, Atlanta, GA 30333, USA

Hiroyuki Sawada
National Institute of Agro-Environmental Sciences, 3-1-1 Kannondai, Tsukuba, Ibaraki 305-8604, Japan

Bernhard H. Schink
Fakultät für Biologie, Lehrstuhl für Mikrobielle Ökologie, Universität Konstanz, Postfach 55 60, D-78457 Konstanz, Germany

Karl-Heinz Schleifer
Lehrstuhl für Mikrobiologie, Technische Universität München, Am Hochanger 4, Freising D-85350, Germany

Karl-Heinz Schleifer
Lehrstuhl für Mikrobiologie, Technische, Universität München, Am Hochanger 4, Freising D-85350, Germany

Heinz Schlesner
Institut für Allgemeine Mikrobiologie, Universität Kiel, Am Botanischen Garten 1-9, Biologiezentrum, D-24118 Kiel, Germany

Helmut J. Schmidt
Biological Faculty, University of Kaiserslautern, Building 14, Pf 3049, D-67653 Kaiserslautern, Germany

Jean M. Schmidt
Department of Microbiology, Arizona State University, Tempe, AZ 85287-2701, USA

Dirk Schüler
Max Planck-Institute for Marine Microbiology, Celsiusstrasse 1, D-28359 Bremen, Germany

Bernard La Scola
CNRS UMR6020, Unité des Rickettsies, 27 Boulevard Jean Moulin, 13385 Marseille Cedex 05, France

Paul Segers
Lab. voor Microbiologie Vakgroep WE 10V, Universiteit Gent, K.L. Ledeganckstraat 35, B-9000 Gent, Belgium

Robert J. Seviour
Biotechnology Research Centre, La Trobe University, P.O. Box 199, Bendigo VIC 3550, Australia

Richard Sharp
School of Applied Sciences, South Bank University, 103 Borough Road, London SE1 0AA, United Kingdom

Tsuneo Shiba
Shimonoseki University of Fisheries, Dept. of Food Science and Technology, Yoshimi-Nagatahoncho Shimonose, Yamaguchi 759-65, Japan

Martin Sievers
University of Applied Sciences, Department of Biotechnology, Molecular Biology, P. O. Box 335, CH 8820 Wädenswil, Switzerland

Lindsay I. Sly
Centre for Bacterial Diversity and Identification, Department of Microbiology and Parasitology, University of Queensland, St. Lucia, Brisbane, Queensland 4072, Australia

Peter H.A. Sneath
Department of Microbiology and Immunology, School of Medicine, University of Leicester, P.O. Box 138, Leicester LE1 9HN, United Kingdom

Martin Sobieraj
Department of Environmental Biology, Portland State University, P. O. Box 751, Portland, OR 97207-0751, USA

Dimitry Y. Sorokin
Institute of Microbiology, Russian Academy of Sciences, Prospect 60-let. Oktyabrya 7/2, Moscow 117811, Russia

Rob J.M. van Spanning
Department of Molecular Cell Physiology/Molecular Microbial Ecology, Vrije Universiteit, De Boelelaan 1087, NL-1081 HV Amsterdam, The Netherlands

Eva Spieck
Inst. für Allgemeine Botanik und Botanischer Garten, Universität Hamburg, Ohnhorststrasse 18, D-22609 Hamburg, Germany

Georg A. Sprenger
Forschungszentrum Jülich GmbH, Institut für Biotechnologie 1, P. O. Box 1913, D-52425 Jülich, Germany

Stefan Spring
DSMZ-Deutsche Sammlung von Mikroorganismen und Zellkulturen, GmbH, D-38124 Braunschweig, Germany

David A. Stahl
Civil and Environmental Engineering, University of Washington, Seattle, WA 98195-2700, USA

James T. Staley
Department of Microbiology, University of Washington, Seattle, WA 98195-0001, USA

Alfons J.M. Stams
Department of Microbiology, Wageningen Agricultural University, Hesselink Van Suchtelenweg 4, NL-6703 CT Wageningen, The Netherlands

Patricia M. Stanley
Minntech Corporation, North, Minneapolis, MN 55447-4822, USA

John F. Stolz
Department of Biological Sciences, Duquesne University, Pittsburgh, PA 15282-2504, USA

Adriaan H. Stouthamer
Dept. of Molecular Cell Physiology/Molecular Microbial Ecology, Vrije Universiteit, De Boelelaan 1087, NL-1081 HV Amsterdam, The Netherlands

William R. Strohl
Merck & Company, Rahway, NJ 07065-0900, USA

Joseph M. Suflita
Environmental and General Applied Microbiology, Department of Botany & Micro., The University of Oklahoma, Norman, OK 73019-0245, USA

Jörg Süling
Institut für Meereskunde, Abt Marine Mikro-biologie, Universität Kiel, Düsternbrooker Weg 20, D-24105 Kiel, Germany

Jean Swings
Laboratorium voor Microbiologie, Vakgroep WE10V, Fysiologie en Microbiologie, Universiteit of Gent, K.L. Ledeganckstraat 35, B-9000 Gent, Belgium

Ulrich Szewzyk
Department of Microbial Ecology, Technical University Berlin, Franklinstrasse 29, Secr. OE 5, D-10587 Berlin, Germany

Zhiyuan Tan
Department of Microbiology and Molecular Genetics, College of Agronomy, South China Agricultural University, 510642, China

Anders Ternström
ANOX AB, Klosterangsvagen 11A, S-226 47 Lund, Sweden

Tone Tønjum
Institute of Microbiology, Section of Molecular Microbiology A3, Rikshospitalet (National Hospital), Pilestredet 32, N-0027 Olso, Norway

G. Todd Townsend
University of Oklahoma, Norman, OK 73072, USA

Yuri A. Trotsenko
Institute of Biochemistry and Physiology of Microorganisms RAS, Laboratory of Methylotrophy, Prospekt Nauki, 5, Moscow Region 142290, Russia

Hans G. Trüper
Institut für Mikrobiologie und Biotechnologie, Universität Bonn, Mechenheimer Allee 168, W-53115 Bonn, Germany

Richard F. Unz
Department of Civil Engineering, The Pennsylvania State University, University Park, PA 16802-1408, USA

Teizi Urakami
Biochemicals Development Div., Mitsubishi Building, Mitsubishi Gas Chemical Company, 5-2, Marunouchi 2-chome, Chiyoda-ku, Tokyo 100-8324, Japan

Marc Vancanneyt
Laboratorium voor Microbiologie, Universiteit Gent, K.L. Ledeganckstraat 35, B-9000 Gent, Belgium

Peter Vandamme
Lab. voor Microbiologie en Microbiele Genetica, Univeristeit of Gent, Faculteit Wetenschappen, K.L. Ledeganckstraat 35, B-9000 Gent, Belgium

Leana V. Vasilyeva
Institute of Microbiology RAN, 117811, Russian Academy of Sciences, 60-let. Oktyabrya 7 build. 2, Moscow, Russia

Henk W. van Verseveld
Dept. of Molecular Cell Physiology, Molecular and Microbial Ecology, Vrije Universiteit, De Boelelaan 1087, NL-1081 HV Amsterdam, The Netherlands

Paul De Vos
Dept. of Biochem., Physiology and Microbiology (WE 10V), University of Gent, K.L. Ledeganckstraat 35, B-9000 Gent, Belgium

David H. Walker
Department of Pathology, University of Texas Medical Branch, 30l University Boulevard, Galveston, TX 77555-0609, USA

En Tao Wang
Departamento de Microbiologia, Escuela Nacional de Ciencias Biológicas, Instituto Politécnico Nacional, Carpio y Plan de Ayala S/N, México D.F. 11340, México

Naomi L. Ward
The Institute for Genomic Research, Rockville, MD 20850, USA

Richard I. Webb
Department of Microbiology, University of Queensland, Brisbane, Queensland 4072, Australia

Ronald M. Weiner
Cell Biology Cluster, Division of Molecular and Cellular Biosciences, National Science Foundation, Arlington, VA 22230, USA

David F. Welch
Laboratory Corporation of America, Dallas, Texas 75230, USA

Aimin Wen
Food Science and Technology Program, Pacific Agri-Food Research Centre, Summerland BC V0H 1Z0, Canada

Hannah M. Wexler
Department of Veterans Affairs, West Los Angeles Medical Ctr., UCLA School of Medicine, 11301 Wilshire Boulevard, Los Angeles, CA 90073, USA

Robbin S. Weyant
Meningitis & Special Pathogens Branch, Centers for Disease Control and Prevention, Atlanta, GA 30333, USA

Anne M. Whitney
Meningitis & Special Pathogens Branch Lab. Section, MS D-11, Centers for Disease Control & Prevention, Atlanta, GA 30303, USA

Friedrich W. Widdel
Abteilung Mikrobiologie, Max Planck-Institut für Marine Mikrobiologie, Celsiusstrasse 1, D-28359 Bremen, Germany

Jürgen K.W. Wiegel
Department of Microbiology, University of Georgia, Athens, GA 30602-2605, USA

Anne Willems
Laboratorium voor Microbiologie, Universiteit Gent, K.L. Ledeganckstraat 35, B-9000 Gent, Belgium

Henry N. Williams
Department of OCBS, Dental School, University of Maryland at Baltimore, Baltimore, MD 21201-1510, USA

Ann P. Wood
Microbiology Research Group, King's College, London, Div. of Life Sciences, Franklin-Wilkins Building, 150 Stamford Street, London SE1 8WA, United Kingdom

Eiko Yabuuchi
Aichi Medical University, Omiya 4-19-18, Asahi-ku, Osaka 535-0002, Japan

Akira Yokota
Institute of Molecular and Cellular Biosciences, The University of Tokyo, Yayoi 1-1-1, Bunkyo-ku, Tokyo 113-0032, Japan

John M. Young
Mt. Albert Research Centre, Landcare Research New Zealand Ltd., Private Bage 92 170, Auckland, New Zealand

Xue-jie Yu
Department of Pathology, University of Texas Medical Branch, 30l University Boulevard, Galveston, TX 77555-0609

Vladimir V. Yurkov
Department of Microbiology, The University of Manitoba, Winnipeg, Manitoba R3T 2N2, Canada

George A. Zavarzin
Institute of Microbiology, Russian Academy of Sciences, Building 2, Prospect 60-letja Oktyabrya 7a, Moscow 117312, Russia

Bernard A.M. van der Zeijst
National Institute of Public Health and Environ., Antonie van Leeuwenhoeklaan 9, P. O. Box 1, P. O. Box 80.165, 3720 BA Bilthoven, The Netherlands

Tatjana N. Zhilina
Institute of Microbiology, Russian Academy of Sciences, Prospect 60-letja Oktyabrya 7a, Moscow 117312, Russia

Stephen H. Zinder
Department of Microbiology, Cornell University, Ithaca, NY 14853-0001, USA

Class I. **Alphaproteobacteria** *class. nov.*

GEORGE M. GARRITY, JULIA A. BELL AND TIMOTHY LILBURN

Al.pha.pro.te.o.bac.te' ri.a. Gr. n. *alpha* name of first letter of Greek alphabet; Gr. n. *Proteus* ocean god able to change shape; Gr. n. *bakterion* a small rod; M.L. fem. pl. n. *Alphaproteobacteria* class of bacteria having 16S rRNA gene sequences related to those of the members of the order *Caulobacterales.*

The class *Alphaproteobacteria* was circumscribed for this volume on the basis of phylogenetic analysis of 16S rRNA sequences; the class contains the orders *Caulobacterales*, "*Parvularculales*", *Rhizo-* *biales, Rhodobacterales, Rhodospirillales, Rickettsiales*, and *Sphingomon-adales.*

Type order: **Caulobacterales** Henrici and Johnson 1935a, 4.

Order I. **Rhodospirillales** Pfennig and Truper 1971, 17[AL]

GEORGE M. GARRITY, JULIA A. BELL AND TIMOTHY LILBURN

Rho.do.spi.ril.la' les. M.L. neut. n. *Rhodospirillum* type genus of the order; *-ales* suffix to denote order; M.L. fem. n. *Rhodospirillales* the *Rhodospirillum* order.

The order *Rhodospirillales* was circumscribed for this volume on the basis of phylogenetic analysis of 16S rRNA sequences; the order contains the families *Rhodospirillaceae* and *Acetobacteraceae*.

Order is morphologically, metabolically, and ecologically diverse. Includes chemoorganotrophs, chemolithotrophs, and fac- ultative photoheterotrophs; some of the latter are also able to grow photoautotrophically. Other species can grow methylotrophically.

Type genus: **Rhodospirillum** Molisch 1907, 24 emend. Imhoff, Petri and Süling 1998, 796.

Family I. **Rhodospirillaceae** Pfennig and Trüper 1971, 17[AL]

GEORGE M. GARRITY, JULIA A. BELL AND TIMOTHY LILBURN

Rho.do.spi.ril.la' ce.ae. M.L. neut. n. *Rhodospirillum* type genus of the family; *-aceae* ending to denote family; M.L. fem. pl. n. *Rhodospirillaceae* the *Rhodospirillum* family.

The family *Rhodospirillaceae* was circumscribed for this volume on the basis of phylogenetic analysis of 16S rRNA sequences; the family contains the genera *Rhodospirillum* (type genus), *Azospirillum, Inquilinus, Levispirillum, Magnetospirillum, Phaeospirillum, Rhodocista, Rhodospira, Rhodovibrio, Roseospira, Skermanella, Thalassospira*, and *Tistrella* (type genus). *Inquilinus, Thalassospira*, and *Tistrella* were proposed after the cut-off date for inclusion in this volume (June 30, 2001) and are not described here (see Coenye et al., 2002; López-López et al., 2002; and Shi et al., 2002, respectively).

Preferred mode of growth for most genera is photoheterotrophic under anoxic conditions in light. Grow chemotrophically in the dark. *Azospirillum, Magnetospirillum*, and *Skermanella* are chemoorganotrophic. Motile by means of polar flagella; may have lateral flagella.

Type genus: **Rhodospirillum** Molisch 1907, 24 emend. Imhoff, Petri and Süling 1998, 796.

Genus I. **Rhodospirillum** *Molisch 1907, 24[AL] emend. Imhoff, Petri and Süling 1998, 796*

JOHANNES F. IMHOFF

Rho.do.spi.ril' lum. Gr. n. *rhodon* the rose; M.L. neut. n. *Spirillum* a bacterial genus; M.L. neut. n. *Rhodospirillum* the rose *Spirillum.*

Cells are vibrioid to spiral shaped, are motile by means of bipolar flagella, and multiply by binary fission. **Gram negative, belonging to the *Alphaproteobacteria*.**. Internal photosynthetic membranes are present as vesicles or as lamellae forming a sharp angle to the cytoplasmic membrane. Photosynthetic pigments are **bacteriochlorophyll *a*** (esterified with phytol or geranylgeraniol) and **carotenoids of the spirilloxanthin series** with spirilloxanthin itself lacking in some species. **Ubiquinones and rhodoquinones with 8 or 10 isoprene units** are present. **Major cellular fatty acids are** $C_{18:1}$, $C_{16:1}$, and $C_{16:0}$, with $C_{18:1}$ as dominant component (51–55% of total fatty acids).

Grow preferentially photoheterotrophically under anoxic conditions in the light. **Photoautotrophic growth with molecular hydrogen and sulfide** as photosynthetic electron donors may occur. **Chemotrophic growth occurs under microoxic to oxic conditions in the dark.** Some species are very sensitive to oxygen; others grow equally well aerobically in the dark. **Fermentation and oxidant-dependent growth may occur.** Polysaccharides, poly-β-hydroxybutyric acid and polyphosphates may be present as storage products. Growth factors required. **Mesophilic freshwater bacteria with preference for neutral pH.**

The mol% G + C of the DNA is: 62.1–63.5.

Type species: **Rhodospirillum rubrum** (Esmarch 1887) Molisch 1907, 25 (Spirillum rubrum Esmarch 1887, 230.)

FURTHER DESCRIPTIVE INFORMATION

Two species of Rhodospirillum are currently known. Rhodospirillum rubrum, the type species, is one of the most intensively studied of the phototrophic bacteria. Numerous investigations on the physiology and enzymology of this species have focused on its metabolic properties, in particular CO_2 fixation and characterization of ribulose-1,5-bisphosphate carboxylase (Tabita, 1995), ATP generation and coupling-factor ATPase (Gromet-Elhanan, 1995), and nitrogenase and nitrogen fixation (Ludden and Roberts, 1995).

R. rubrum grows very well under photoheterotrophic conditions, but it can also grow under photoautotrophic conditions with molecular hydrogen (Klemme, 1968) or sulfide as the electron donor if supplied at low concentrations (Hansen and van Gemerden, 1972). Autotrophic CO_2 fixation is well documented and occurs via ribulose-1,5-bisphosphate carboxylase (Anderson and Fuller, 1967a, b). This enzyme has been highly purified and is well characterized (e.g., Tabita and McFadden, 1974a, b; Tabita, 1995). Ribulose-1,5-bisphosphate carboxylase is derepressed only at low CO_2 tensions (1.5–2.0% of the atmospheric tension) and can make up to 50% of the total soluble protein of the cells under such conditions (Sarles and Tabita, 1983). In the presence of malate or acetate, however, CO_2 is not assimilated via the reductive pentose phosphate cycle, but by other carboxylating reactions.

Under anoxic dark conditions, R. rubrum is able to ferment sugars and pyruvate (Kohlmiller and Gest, 1951; Gürgün et al., 1976; Gorrell and Uffen, 1977). Pyruvate is cleaved by pyruvate formate lyase, which is specifically induced under these growth conditions (Uffen, 1973; Jungermann and Schön, 1974). From formate, H_2 and CO_2 are formed by a CO-sensitive formic hydrogen lyase (Gorrell and Uffen, 1977). H_2, CO_2, acetate, and eventually propionate are produced as fermentation products from pyruvate. R. rubrum is able to gain energy from the coupling of CO oxidation and H_2 evolution and induces the synthesis of a carbon monoxide dehydrogenase on exposure to CO under anoxic conditions (Bonam et al., 1989; Kerby et al., 1995).

Rhodospirillum rubrum is also able to perform an anaerobic dark metabolism with DMSO and trimethylamine-N-oxide as electron acceptors (Schultz and Weaver, 1982). Under these conditions, growth is possible with succinate, malate, and acetate as substrates, and CO_2 and DMSO or trimethylamine are formed (Schultz and Weaver, 1982).

Rhodospirillum species can grow with ammonia or N_2 as sole nitrogen source (Siefert, 1976; Madigan et al., 1984). The nitrogenase of R. rubrum is subject to post-translational inactivation by ADP-ribosylation under energy-limiting conditions and if fixed nitrogen compounds are available (see Ludden and Roberts, 1995). Ammonia assimilation is mediated by the glutamine synthetase/glutamate synthase reactions, which are NADPH-dependent in R. rubrum (Brown and Herbert, 1977). Although a nitrate reductase is present in R. rubrum, growth with nitrate as sole nitrogen source apparently is not possible (Katoh, 1963; Taniguchi and Kamen, 1963; Ketchum and Sevilla, 1973;

Klemme, 1979). Purines can be used as a nitrogen source under anoxic conditions in the light and under oxic conditions in the dark (Aretz et al., 1978).

Comparably few studies have been performed with Rhodospirillum photometricum (Pfennig et al., 1965; Lehmann, 1976; Sarkar and Banerjee, 1980), which is very sensitive to oxygen and does not grow under oxic conditions in the dark (as does R. rubrum; Pfennig, 1969b), but only under microoxic conditions, provided the oxygen tension is lower than 0.5 kPa (Lehmann, 1976). Like Phaeospirillum species, R. photometricum is presumably unable to induce a second electron transport chain in the presence of oxygen and therefore depends on microoxic conditions in which the internal membrane system and the light-driven electron transport chain are fully expressed. Accordingly, the cells are fully pigmented under these growth conditions (Lehmann, 1976).

Two different types of alcohols are esterified with the bacteriochlorophyll a of Rhodospirillum species. Besides phytol, which is present as an alcohol in the majority of the Purple Nonsulfur Bacteria (PNSB), geranylgeraniol is the major component of R. rubrum and of some strains of R. photometricum (Brockmann and Knobloch, 1972; Künzler and Pfennig, 1973). Carotenoids of the spirilloxanthin series are present in Rhodospirillum species, but R. photometricum is unable to synthesize the end product, spirilloxanthin, and accumulates intermediates of this pathway, such as lycopene, rhodopin, anhydrorhodovibrin, and rhodovibrin (Schmidt, 1978).

Growth of R. photometricum is completely inhibited at penicillin concentrations of only 10 U/ml, whereas in R. rubrum, complete inhibition occurs at penicillin concentrations of more than 1000 U/ml (Weaver et al., 1975a).

ENRICHMENT AND ISOLATION PROCEDURES

Media and growth conditions used for isolation and cultivation of freshwater PNSB in general can also be applied for Rhodospirillum species. Various recipes for appropriate media have been developed in different laboratories (see Biebl and Pfennig, 1981; Imhoff, 1988; Imhoff and Trüper, 1992). One of these, a mineral salts medium that has been used for cultivation of the great majority of PNSB over many years is given in the footnote below.[1] Standard techniques for the isolation of anaerobic bacteria in agar dilution series and on agar plates can be applied for Rhodospirillum species (Biebl and Pfennig, 1981; Imhoff and Trüper, 1992), if care is taken to establish and maintain oxygen-free conditions, especially for oxygen-sensitive species. This can be achieved by addition of 0.05% sodium ascorbate or 0.025% thioglycolate to the growth media in completely filled screw-capped bottles.

---

1. AT medium contains (g/l): KH_2PO_4, 1.0; $MgCl_2 \cdot 6H_2O$, 0.5; $CaCl_2 \cdot 2H_2O$, 0.1; NH_4Cl, 1.0; $NaHCO_3$, 3.0; Na_2SO_4, 0.7; NaCl, 1.0; sulfate-free trace element solution SLA (Imhoff and Trüper, 1977; Imhoff, 1992), 1 ml; and vitamin solution VA (Imhoff and Trüper, 1977; 1992), 1 ml. Organic carbon sources (routinely 10 mM sodium malate, sodium succinate, sodium pyruvate, or sodium acetate) and, for oxygen-sensitive strains, 0.5 g sodium ascorbate or 0.25 g thioglycolate are added separately. The initial pH is adjusted to 6.9. Vitamin solution VA contains in 100 ml of double distilled water: biotin, 10 mg; niacin amide, 35 mg; thiamine dichloride, 30 mg; p-aminobenzoic acid, 20 mg; pyridoxal hydrochloride, 10 mg; calcium pantothenate, 10 mg; and vitamin B_{12}, 5 mg.

The trace element solution SLA has the following composition: $FeCl_2 \cdot 4H_2O$, 1.8 g; $CoCl_2 \cdot 6H_2O$, 250 mg; $NiCl_2 \cdot 6H_2O$, 10 mg; $CuCl_2 \cdot 5H_2O$, 10 mg; $MnCl_2 \cdot 4H_2O$, 70 mg; $ZnCl_2$, 100 mg; H_3BO_3, 500 mg; $Na_2MoO_4 \cdot 2H_2O$, 30 mg; and $Na_2SeO_3 \cdot 5H_2O$, 10 mg. These components are dissolved in 1 liter of double distilled water. The pH of the solution is adjusted with HCl to 2–3.

MAINTENANCE PROCEDURES

Cultures can be preserved by standard techniques in liquid nitrogen or at $-80°C$ in a mechanical freezer.

DIFFERENTIATION OF THE GENUS RHODOSPIRILLUM FROM OTHER GENERA

A number of chemotaxonomic properties distinguish *Rhodospirillum* species from other spiral-shaped Purple Nonsulfur Bacteria (PNSB). They differ from *Phaeospirillum* species, their closest relatives, in major quinone components and cytochrome c structure. Large-type cytochromes c_2 are present in *R. rubrum* and *R. photometricum*, whereas small-type cytochromes c_2 were found in *Phaeospirillum* species (Ambler et al., 1979). Major differentiating properties for phototrophic bacteria in the *Rhodospirillaceae* are shown in Table BXII.α.1. The phylogenetic relationships of these bacteria based on 16S rDNA sequences are shown in Fig. 1 (p. 124) of the introductory chapter "Anoxygenic Phototrophic Purple Bacteria", Volume 2, Part A.

TAXONOMIC COMMENTS

Traditionally all spiral-shaped phototrophic PNSB have been assigned to the genus *Rhodospirillum* (Pfennig and Trüper, 1974). Recognition of the large amount of chemotaxonomic and phylogenetic diversity in the PNSB, and their presence in different groups of the *Proteobacteria*, initially led to the taxonomic reclassification of those species belonging to the *Betaproteobacteria*. "*Rhodospirillum tenue*" was assigned to *Rhodocyclus tenuis* (Imhoff et al., 1984) and this reclassification was later supported by 16S rDNA sequence analyses (Hiraishi et al., 1991a). After this reclassification, all species of the genus *Rhodospirillum* were *Alphaproteobacteria*, though the group remained very heterogeneous in phenotypic properties and genetic relationships.

The description of new species, assigned to the genus *Rhodospirillum*, based merely on their spiral shape, continued until recently, although most of them were quite distinct from *Rhodospirillum rubrum*, the type species of this genus. Four halophilic species were classified together with several freshwater species of the genus *Rhodospirillum*. In addition, *Rhodospirillum centenum* (Favinger et al., 1989) was assigned to this genus, though significant differences from *Rhodospirillum rubrum* had been stated in the species description. More recently, the great genetic distance among the spiral-shaped PNSB has been recognized in several proposals. *Rhodospirillum centenum* was transferred to a new genus as *Rhodocista centenaria* (Kawasaki et al., 1992). Another new spiral-shaped species, *Rhodospira trueperi*, was assigned to a new genus based on significant phenotypic and genotypic differences from *Rhodospirillum rubrum* and other known PNSB (Pfennig et al., 1997).

The anticipated heterogeneity of the genus *Rhodospirillum* became clearly apparent with the 16S rDNA sequence information of most of the known species (Kawasaki et al., 1993b; Imhoff et al., 1998), and these data implied that the spiral-shaped *Alphaproteobacteria* are phylogenetically quite distantly related to each other and do not warrant classification in one and the same genus. Therefore, a reclassification of the spiral-shaped phototrophic *Alphaproteobacteria*, based on distinct phenotypic properties and 16S rDNA sequence similarities, has been proposed (Imhoff et al., 1998; see Table BXII.α.1 and Fig. 1 [p. 124] of the introductory chapter "Anoxygenic Phototrophic Purple Bacteria", Volume 2, Part A).

Major quinone components and fatty acid composition, salt requirements, and phylogenetic relationships based on 16S rDNA sequences were considered of primary importance in defining and differentiating these genera. Several phylogenetic lines of salt dependent species were recognized and the salt-dependence was regarded as a genus-specific property (Imhoff et al., 1998). Four of the genera of spiral-shaped phototrophic *Alphaproteobacteria* were defined as salt-dependent and three are freshwater bacteria. Only *R. rubrum* and *R. photometricum* were maintained as species of the genus *Rhodospirillum*. All other species were transferred to the new genera *Phaeospirillum*, *Rhodovibrio*, *Rhodothalassium*, and *Roseospira* (Imhoff et al., 1998) and are considered in the respective chapters of this volume.

FURTHER READING

Drews, G. and J.F. Imhoff. 1991. Phototrophic purple bacteria. *In* Shively and Barton (Editors), Variations in Autotrophic Life, Academic Press, London. pp. 51–97.

Imhoff, J.F. 1988. Anoxygenic phototrophic bacteria. *In* Austin (Editor), Methods in Aquatic Bacteriology, John Wiley & Sons Ltd., Chichester. pp. 207–240.

Imhoff, J.F. 1992. Taxonomy, phylogeny, and general ecology of anoxygenic phototrophic bacteria. *In* Mann and Carr (Editors), Biotechnology Handbooks: Photosynthetic Prokaryotes, Vol. 6, Plenum Press, New York. pp. 53–92.

Imhoff, J.F. 1999. A phylogenetically oriented taxonomy of anoxygenic phototrophic bacteria. *In* Pescheck, Löffelhardt and Schmetterer (Editors), The Phototrophic Prokaryotes, Plenum Publishing Corporation, New York. pp. 763–774.

Imhoff, J.F., R. Petri and J. Süling. 1998. Reclassification of species of the spiral-shaped phototrophic purple non-sulfur bacteria of the α-Proteobacteria: description of the new genera *Phaeospirillum* gen. nov., *Rhodovibrio* gen. nov., *Rhodothalassium* gen. nov. and *Roseospira* gen. nov. as well as transfer of *Rhodospirillum fulvum* to *Phaeospirillum fulvum* comb. nov., of *Rhodospirillum molischianum* to *Phaeospirillum molischianum* comb. nov., of *Rhodospirillum salinarum* to *Rhodovibrio salinarum* comb. nov., of *Rhodospirillum sodomense* to *Rhodovibrio sodomensis* comb. nov., of *Rhodospirillum salexigens* to *Rhodothalassium salexigens* comb. nov. and of *Rhodospirillum mediosalinum* to *Roseospira mediosalina* comb. nov. Int. J. Syst. Bacteriol. *48*: 793–798.

Imhoff, J.F. and H.G. Trüper. 1992. The genus *Rhodospirillum* and related genera. *In* Balows, Trüper, Dworkin, Harder and Schleifer (Editors), The Prokaryotes: A Handbook on the Biology of Bacteria. Ecophysiology, Isolation, Identification, Applications, 2nd ed., Springer-Verlag, New York. pp. 2141–2155.

Kawasaki, H., Y. Hoshino and K. Yamasato. 1993. Phylogenetic diversity of phototrophic purple non-sulfur bacteria in the *alpha proteobacteria* group. FEMS Microbiol. Lett. *112*: 61–66.

Rodriguez-Valera, F., A. Ventosa, G. Juez and J.F. Imhoff. 1985. Variation of environmental features and microbial populations with salt concentrations in a multi-pond saltern. Microb. Ecol. *11*: 107–116.

DIFFERENTIATION OF THE SPECIES OF THE GENUS RHODOSPIRILLUM

Major differentiating properties between *Rhodospirillum* species are shown in Tables BXII.α.1 and BXII.α.2.

TABLE BXII.α.1. Diagnostic properties of the spiral-shaped phototrophic *Alphaproteobacteria*[a]

Characteristic	*Rhodospirillum rubrum*	*Rhodospirillum photometricum*	*Phaeospirillum fulvum*	*Phaeospirillum molischianum*	*Rhodocista centenaria*	*Rhodospira trueperi*	*Rhodovibrio salinarum*	*Rhodovibrio sodomensis*	*Roseospira mediosalina*	*Roseospirillum parvum*
Cell diameter (μm)	0.8–1.0	1.1–1.5	0.5–0.7	0.7–1.0	1.0–2.0	0.6–0.8	0.8–0.9	0.6–0.7	0.8–1.0	0.4–0.6
Internal membrane system	Vesicles	Stacks	Stacks	Stacks	Lamellae	Vesicles	Vesicles	Vesicles	Vesicles	Lamellae
Motility	+	+	+	+	+	+	+	+	+	+
Color	Red	Brown	Brown	Brown	Pink	Beige	Red	Pink	Pink	Pink
Bacteriochlorophyll	*a*	*a*	*a*	*a*	*a*	*b*	*a*	*a*	*a*	*a*
Growth factors	Biotin	Nicotinamide	*p*-Aminobenzoic acid	Amino acids	Biotin, B$_{12}$	Biotin, thiamine, pantothenate	Cobalamine	Cobalamine	Thiamine, *p*-aminobenzoic acid, nicotinamide	Cobalamine
Aerobic growth	+	−	−	−	+	−	+	+	(+)	(+)
Oxidation of sulfide		−	−	−	nd	+	−	nd	+	+
Salt requirement[b]	None	None	None	None	None	2% (0.5–5)	8–12% (3–24)	12% (6–20)	4–7% (0.5–15)	1–2% (to >6.0%)
Optimal temperature (°C)	30–35	25–30	25–30	30	40–45	25–30	42	35–40	30–35	30
Optimal pH	6.8–7.0	6.5–7.5	7.3	7.3	6.8	7.3–7.5	7.5–8.0	7	7	7.9
Habitat	Fresh water	Fresh water	Fresh water	Fresh water	Fresh water, warm springs	Marine sediments	Saltern	Salt lakes	Salty springs	Marine sediments
Mol% G + C of the DNA	63.8–65.8	64.8–65.8	64.3–65.3	60.5–64.8	69.9	65.7	67.4	66.2–66.6	66.6	71.2
Cytochrome *c* size	Large	Large	Small	Small	nd	nd	None	None	nd	nd
Major quinones	Q-10, RQ-10	Q-8, RQ-8	Q-9, MK-9	Q-9, MK-9	Q-9	Q-7, MK-7	Q-10, MK-10	nd	nd	nd
Major fatty acids:										
C$_{14:0}$	2.1	1.0	0.6	0.7	nd	7.5	1.0	nd	nd	nd
C$_{16:0}$	14.0	25.2	15.1	18.1	nd	27.9	7.4	nd	nd	nd
C$_{16:1}$	27.1	22.2	25.8	36.5	nd	1.2	0.3	nd	nd	nd
C$_{18:0}$	1.3	0.4	1.2	0.7	nd	1.2	23.0	nd	nd	nd
C$_{18:1}$	54.8	51.0	54.5	43.5	nd	60.7	35.2	nd	nd	nd

[a]Symbols: +, positive in most strains; −, negative in most strains; (+), weak growth or microaerobic growth only; nd, not determined; Q-7, ubiquinone 7; Q-8, ubiquinone 8; Q-9, ubiquinone 9; Q-10, ubiquinone 10; Q-9/10, ubiquinones 9 and 10; MK-7, menaquinone 7; MK-9, menaquinone 9; MK-10, menaquinone 10; MK-9/10, menaquinones 9 and 10.

[b]The first set of figures indicates the optimum salt concentration, and the second set– indicates the range of salt concentrations tolerated.

List of species of the genus Rhodospirillum

1. **Rhodospirillum rubrum** (Esmarch 1887) Molisch 1907, 25[AL] (*Spirillum rubrum* Esmarch 1887, 230.)

rub'rum. M.L. neut. adj. *rubrum* red.

Cells are vibrioid shaped to spiral, 0.8–1.0 μm wide; one complete turn of a spiral is 1.5–2.5 μm wide and 7–10 μm long (Fig. BXII.α.1). Internal photosynthetic membranes

are of the vesicular type (Fig. BXII.α2). Anaerobic liquid cultures are pink to deep red, without a brownish tinge, under all conditions. Under oxic conditions, cells are colorless to light pink. Living cells show absorption maxima at 375–377, 510–517, 546–550, 590–595, 807–808, and 881–885 nm. Photosynthetic pigments are bacteriochlorophyll *a* (esterified with geranylgeraniol as major and phytol as minor component) and carotenoids of the spirilloxanthin series with spirilloxanthin as the predominant component.

Cells preferentially grow photoheterotrophically under anoxic conditions in the light with various organic compounds as carbon and electron sources. Photoautotrophic growth occurs with H$_2$ and sulfide as electron donors. Chemotrophic growth occurs under microoxic to oxic conditions in the dark. Fermentative metabolism with pyruvate under anoxic dark conditions and "oxidant-dependent" anaerobic dark metabolism are also possible. The carbon sources utilized are shown in Table BXII.α.2. In addition, alanine and asparagine are used and some strains use propanol. Ammonia, N$_2$, several amino acids, and in some strains, nitrate, adenine, guanine, xanthin, and uric acid may be used as nitrogen source. Sulfate can serve as sole sulfur source. Small amounts of yeast extract may be favorable. Biotin is required as a growth factor.

Mesophilic freshwater bacterium with optimum growth at 30–35°C and pH 6.8–7.0 (pH range: 6.0–8.5). Habitat: stagnant and anoxic freshwater habitats that are exposed to the light. Major quinone components are Q-10 and RQ-10.

The mol% G + C of the DNA is: 63.8–65.8 (Bd).
Type strain: ATCC 11170, DSM 467, NCIB 8255.
GenBank accession number (16S rRNA): D30778, M32020.

2. **Rhodospirillum photometricum** Molisch 1907, 24[AL]

pho.to.me'tri.cum. Gr. n. *phos* light; Gr. adj. *metricus* measuring; M.L. neut. adj. *photometricum* light measuring.

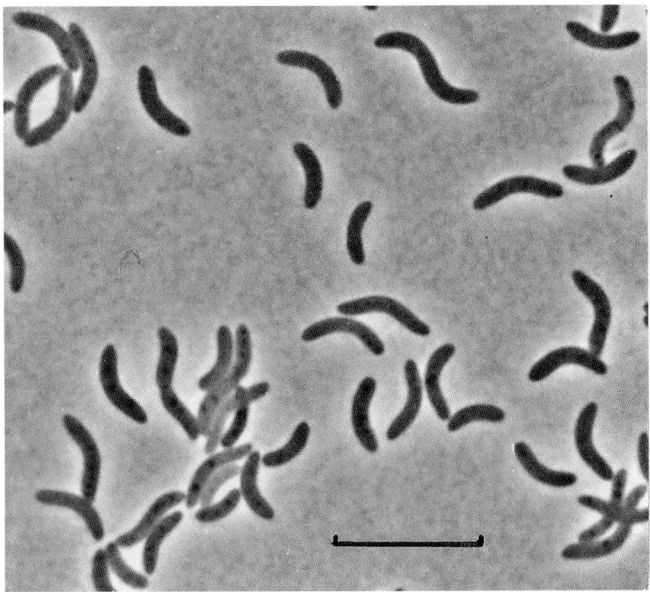

FIGURE BXII.α.1. *Rhodospirillum rubrum* ATCC 11170. Phase-contrast micrograph. Bar = 10 μm. (Courtesy of N. Pfennig).

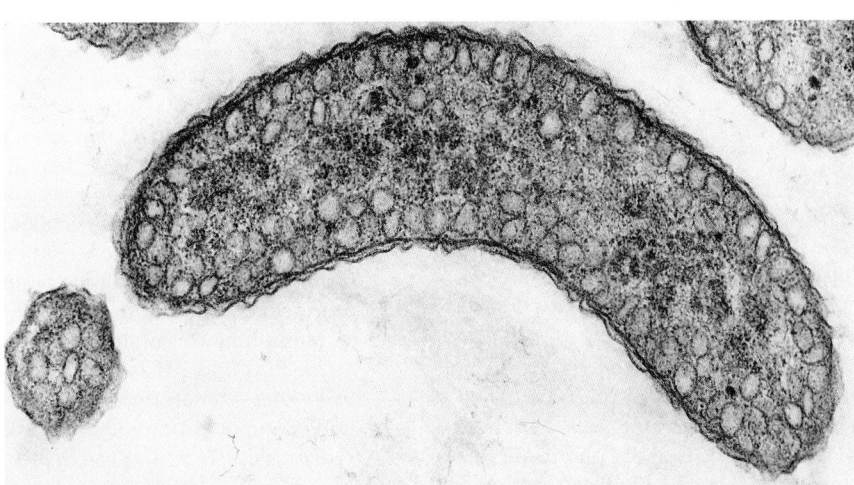

FIGURE BXII.α.2. *Rhodospirillum rubrum* strain FR1 grown anaerobically in the light. Note the vesicular structure of the intracytoplasmic membrane system. × 51,000. (Courtesy of G. Drews and R. Ladwig).

TABLE BXII.α.2. Carbon sources and electron donors used by species of the anoxygenic phototrophic bacteria belonging to the family *Rhodospirillaceae*[a]

Source/donor	*Rhodospirillum rubrum*	*Rhodospirillum photometricum*	*Phaeospirillum fulvum*	*Phaeospirillum molischianum*	*Rhodocista centenaria*	*Rhodospira trueperi*	*Roseospira mediosalina*	*Roseospirillum parvum*
Carbon sources:								
Acetate	+	+	+	+	+	+	+	+
Arginine	+	−	−	−	nd	nd	−	nd
Aspartate	+	+/−	+/−	+/−	+	nd	+	nd
Benzoate	−	−	+	−	nd	nd	−	nd
Butyrate	+	+	+	+	+	+	+	+
Caproate	+	−	+	+	+	nd	nd	nd
Caprylate	nd	−	+	+	+	nd	−	nd
Citrate	−	−	−	−	nd	nd	−	nd
Ethanol	+	+	+	+	nd	-	nd	nd
Formate	−	−	nd	nd	nd	nd	−	nd
Fructose	+/−	+	−	−	−	nd	−	+
Fumarate	+	+	+	+	−	+	+	+
Glucose	−	+	+/−	−	−	nd	−	nd
Glutamate	+	−	nd	nd	+	nd	+	+
Glycerol	−	+	−	−	nd	nd	−	nd
Glycolate	nd	+	nd	−	nd	nd	−	nd
Lactate	+	+	−	+/−	+	+	+	+
Malate	+	+	+	+	−	+	+	+
Mannitol	−	+	−	−	nd	nd	−	nd
Methanol	+/−	−	+/−	−	nd	nd	−	nd
Pelargonate	nd	+/−	+	+	nd	nd	nd	nd
Propionate	+	+/−	+	+	nd	+	+	+
Pyruvate	+	+	+	+	+	+	+	+
Succinate	+	+	+	+	−	+	+	+
Tartrate	−	−	nd	−	nd	nd	−	nd
Valerate	+	+	+	+	+	+	−	+
Electron donors:								
Hydrogen	+	+	nd	nd	nd	nd	nd	nd
Sulfide	+	−	−	−	nd	+	+	+
Sulfur	−	−	−	−	nd	−	−	nd
Thiosulfate	−	−	−	−	nd	nd	−	+

[a]Symbols: +, positive in most strains; −, negative in most strains; +/−, variable in different strains; nd, not determined.

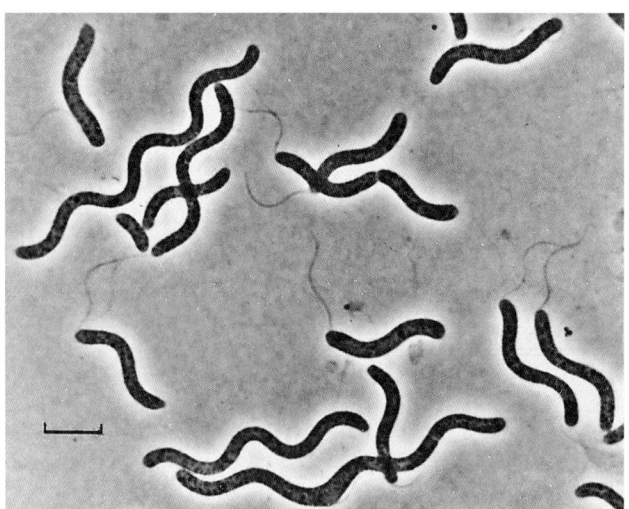

FIGURE BXII.α.3. *Rhodospirillum photometricum* NTHC 132, cultured on malate-yeast extract medium. Phase-contrast micrograph. Bar = 10 μm. (Courtesy or N. Pfennig).

Cells are spirals, 1.1–1.5 μm wide; one complete turn of a spiral is 2.5–4 μm wide and 4–7 μm long; cells 14–30 μm long are common (Fig. BXII.α.3). Internal photosynthetic membranes consist of several lamellar stacks forming a sharp angle with the cytoplasmic membrane. Anaerobic liquid cultures are brown-orange to brown-red or dark brown.

Photosynthetic pigments are bacteriochlorophyll *a* (esterified with phytol and in some strains also with geranylgeraniol) and carotenoids of the spirilloxanthin series. Spirilloxanthin is lacking, but the biosynthetic precursors lycopene and rhodopin are present as major components.

Cells preferentially grow photoheterotrophically under anoxic conditions in the light with various organic compounds as carbon and electron sources. Chemotrophic growth at very low oxygen tensions, but not under oxic growth conditions, is possible. Under microoxic conditions cells are fully pigmented. Carbon sources utilized are shown in Table BXII.α.2. In addition, asparagine, maltose, sucrose, raffinose, adonitol, and dulcitol are used. Not used are cyclohexane carboxylate, mannose, galactose, xylose, and inositol. Nitrogen sources utilized are ammonia, N₂, alanine, glutamate, and asparagine; not utilized are nitrate, urea, and arginine. Sulfate, thiosulfate, cysteine, thioglycolate, and, at low concentrations, sulfide can be used as sulfur sources. Ascorbic acid may be required as a reductant. Nicotinic acid is required as a growth factor.

Mesophilic freshwater bacterium with optimum growth at 25–30°C and pH 6.5–7.5. Habitat: stagnant and anoxic freshwater habitats that are exposed to the light. Major quinone components are Q-8 and RQ-8.

The mol% G + C of the DNA is: 64.8–65.8 (Bd) and 63 (T_m).

Type strain: ATCC 49918, DSM 122, NTHC 132.

GenBank accession number (16S rRNA): AJ222662.

Genus II. *Azospirillum* Tarrand, Krieg and Döbereiner 1979, 79[AL] (Effective publication: Tarrand, Krieg and Döbereiner 1978, 978) emend. Falk, Döbereiner, Johnson and Krieg 1985, 117

José Ivo Baldani, Noel R. Krieg, Vera Lúcia Divan Baldani, Anton Hartmann and Johanna Döbereiner

A.zo.spi.ril'lum. Fr. n. *azote* nitrogen; Gr. n. *spira* a spiral; M.L. dim. neut. n. *spirillum* a small spiral; *Azospirillum* a small nitrogen spiral.

Plump, slightly-curved and straight rods, 0.6–1.7 × 2.1–3.8 μm, often with pointed ends. Intracellular granules of poly-β-hydroxybutyrate are present. Enlarged, pleomorphic forms may occur in old, alkaline cultures, under conditions of excess oxygen or other stress. Gram negative to Gram variable. **Motile in liquid media by a single polar flagellum;** on solid media at 30°C, numerous lateral flagella of shorter wavelength may also be formed. **Nitrogen fixers, exhibiting N_2-dependent growth under microaerobic conditions.** Grow well under an air atmosphere in the presence of a source of fixed nitrogen such as an ammonium or glutamate salts. Cells previously grown in presence of an inorganic nitrogen source may fix nitrogen in air provided that all added nitrogen is exhausted and nitrogenase is derepressed. Possess mainly a respiratory type of metabolism with oxygen and, with some strains, nitrate or nitrite as the terminal electron acceptor. Fermentative metabolism may also occur. **Under severe oxygen limitation, some strains may dissimilate nitrate to nitrite or to nitrous oxide and nitrogen gas.** Optimal temperature for growth varies from 33 to 41°C and pH from 5.5 to 7.5. Some strains may grow and form light or dark pink colonies, often wrinkled and non-slimy, on potato agar. Oxidase positive. Chemoorganotrophic; some strains are facultative hydrogen autotrophs. **Grow well on salts of organic acids such as malate, succinate, lactate or pyruvate. D-fructose and certain carbohydrates may also serve as carbon sources.** Some species require biotin. Growth in presence of 3% NaCl has been observed for some species. **Occur free-living in the soil or associated with the roots, stems, leaves, and seeds** mainly of cereals and forage grasses, although they have also been isolated from coconut plants, vegetables, fruits, legume, and tuber plants. May also be found in freshwater lakes. Root nodules are not induced.

The mol% G + C of the DNA is: 64–71.

Type species: **Azospirillum lipoferum** (Beijerinck 1925) Tarrand, Krieg and Döbereiner 1979, 79 (Effective publication: Tarrand, Krieg and Döbereiner 1978, 978) (*Spirillum lipoferum* Beijerinck 1925, 353.)

FURTHER DESCRIPTIVE INFORMATION

Morphology In complex media such as MPSS broth[1], *A. lipoferum* and *A. brasilense* grow as plump, slightly curved rods and straight cells having a diameter of ~1.0 μm. Many of the cells have pointed ends. In semisolid nitrogen-free malate (NFb) medium[2], *A. lipoferum* develops predominantly into pleomorphic

cells within 48 h (Fig. BXII.α.4), in contrast to *A. brasilense*, which retains mainly the vibrioid form. *A. lipoferum* grows as elongated cells (1.4–1.7 μm × 5 to over 30 μm long), which are nonmotile and have an S shape or helical shape (Fig. BXII.α.4). These forms eventually seem to fragment into shorter ovoid forms, many of which become very large and rounded and may contain several cells filled with phase-refractile granules (probably poly-β-hydroxybutyrate). Alkalinization of the malate medium, due to oxidation of the malate, may be related to development of pleomorphism in *A. lipoferum*. Pleomorphism fails to occur when the organisms are cultured in semi-solid nitrogen-free D-glucose medium, which does not become alkaline (Fig. BXII.α.5).

In semi-solid nitrogen-free malate medium, *A. brasilense* grows mainly as motile, vibrioid cells. Nonmotile, enlarged, pleomorphic forms (C forms) may also occur, especially in older cultures (Eskew et al., 1977; Tarrand et al., 1978), on the surface of nitrogen-free agar media (Berg et al., 1980), in association with plant callus cultures (Berg et al., 1979), or in association with the roots of grass seedlings (Umali-Garcia et al., 1980). A capsule is formed external to the outer wall membrane of C forms (Fig. BXII.α.6) and may be a protective mechanism against unfavorable levels of oxygen under nitrogen-fixing conditions (Berg et al., 1980). Very large, rounded forms containing several cells may be produced (Fig. BXII.α.6). The ultrastructure of the C forms indicates little similarity to the cysts of *Azotobacter*. Lamm and Neyra (1981) have reported that strains of *A. brasilense* and *A. lipoferum* enriched in C forms exhibit a resistance to desiccation and temperature not found in cultures lacking these forms. Cyst forms of *A. brasilense* could be physiologically active in the wheat rhizosphere, as shown by a high level of 16S rRNA-directed oligonucleotide binding (Assmus et al., 1995). Cysts are capable of fixing nitrogen in the absence of an exogenous carbon source (Okon and Itzigsohn, 1992). *A. amazonense* does not develop pleomorphic cells when grown in the semi-solid LGI medium because the pH is maintained around 6. Even when a very poor subsurface pellicle is formed in semi-solid NFb medium (this species is sensitive to alkaline pH), no such cells are observed. In LGI medium, cells remain motile and vibrioid. However, cultures maintained two to three weeks on the bench develop agglomerated cells like cysts due to the dryness of the medium.

Pleomorphic cells have been observed for the species *A. halopraeferens*, *A. irakense*, and *A. largimobile*, depending on the growth conditions. In the case of *A. halopraeferens* this phenomenon is observed in older and very alkaline conditions whereas very long cells (30 μm) are observed when *A. irakense* is grown in nutrient broth (Fig. BXII.α.7). On the other hand, *A. largimobile* forms multicellular conglomerates of nonmotile cells mainly under unfavorable conditions. The most recently described species, *A. doebereinerae*, also develops pleomorphic cells in semi-solid NFb medium.

Encapsulation may be related to resistance to Gram-decolorization exhibited by a small proportion of cells when cultured on MPSS agar at 37°C for 48–72 h. Gram-variability appears to be more pronounced with *A. brasilense* than with *A. lipoferum*.

---

1. MPSS broth (g/l): peptone (Difco), 5.0; succinic acid (free acid), 1.0; $(NH_4)_2SO_4$, 1.0; $MgSO_4 \cdot 7H_2O$, 1.0; $FeCl_3 \cdot 6H_2O$, 0.002; $MnSO_4 \cdot H_2O$, 0.002; pH adjusted to 7.0 with KOH. For solid media add 15.0 g/l agar.

2. NFb medium (g/l): L-malic acid, 5.0; K_2HPO_4, 0.5; $MgSO_4 \cdot 7H_2O$, 0.2; NaCl, 0.02; trace element solution ($Na_2MoO_4 \cdot 2H_2O$, 0.2 g; $MnSO_4 \cdot H_2O$, 0.235 g; H_3BO_3, 0.28 g; $CuSO_4 \cdot 5H_2O$, 0.008 g; $ZnSO_4 \cdot 7H_2O$, 0.024 g; distilled water, 1000 ml), 2.0 ml; bromthymol blue (0.5% aqueous solution [dissolve in 0.2 N KOH]), 2.0 ml; Fe EDTA (1.64% solution), 4.0 ml, vitamin solution (biotin, 0.01 g; pyridoxin, 0.02 g; distilled water, 1000 ml), 1.0 ml; KOH, 4.0; pH adjusted to 6.8 with KOH. For a semi-solid medium add 1.75 g/l agar; for a solid medium add 15.0 g/l agar.

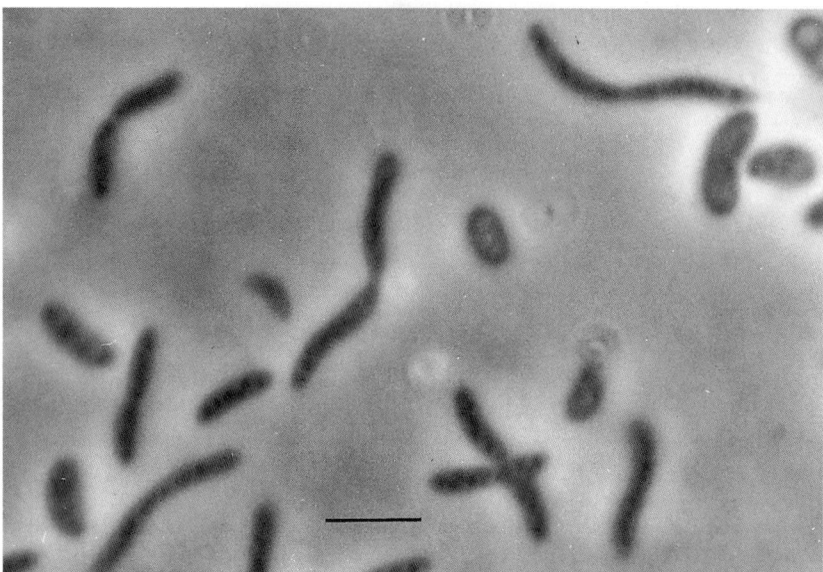

FIGURE BXII.α.4. *Azospirillum lipoferum* ATCC 29707 cultured in semi-solid, nitrogen-free, malate medium at 37°C for 48 h, showing characteristic elongated S-shaped forms and enlarged ovoid forms. Phase contrast microscopy. Bar = 5 μm.

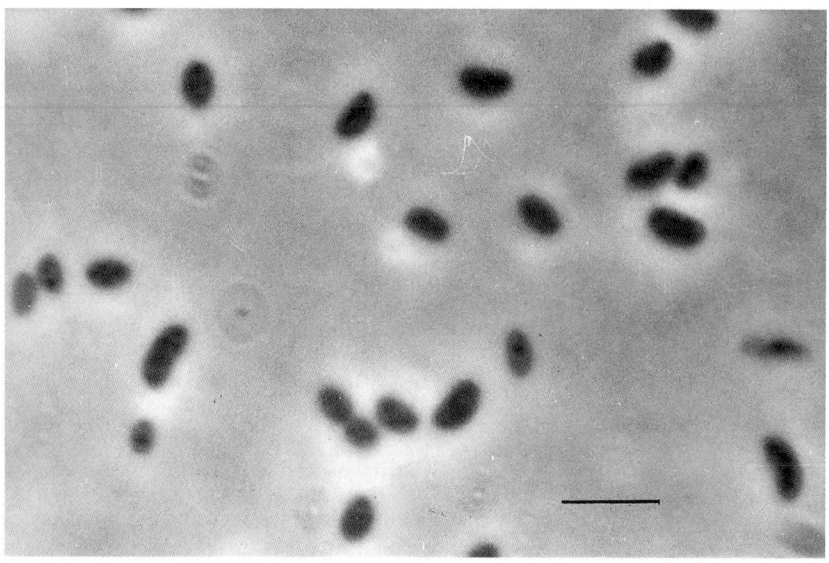

FIGURE BXII.α.5. *Azospirillum lipoferum* ATCC 29707 cultured in semi-solid, nitrogen-free, malate medium at 37°C for 48 h, showing characteristic vibrioid forms. Phase contrast microscopy. Bar = 5 μm.

Cultures grown in MPSS broth appear not to contain encapsulated forms, at least in young cultures, and the cells stain uniformly Gram negative. So far, no Gram variability has been observed for the other *Azospirillum* species.

Poly-β-hydroxybutyrate (PHB) may constitute from 25 to 50% of the dry weight of cells cultured in nitrogen-free media. In cells cultured with an ammonium salt as the nitrogen source, the polymer constitutes only 0.5–1.0% of the cell weight (Okon et al., 1976). However, high amounts of poly-β-hydroxybutyrate are accumulated in cells of *A. brasilense* strain Cd grown in a high C:N medium using D-fructose and ammonium chloride as C and N source (Burdman et al., 1998). The amount of PHB also increases under various stress conditions such as toxic metals and water stress, and this accumulation is correlated with cyst and

floc formation where it accounts for up to 65% of the total dry weight (Olubayi et al., 1998). Cells rich in PHB survive better and promote positive effect on cereals (Assmus et al., 1995; Fallik and Okon, 1996). It has been suggested that inoculants generated from flocculated cells are more advantageous because they seem to be more efficient than those originating from nonflocculated cells, and in addition they can be produced on a large scale and easily separated from the growth medium (Neyra et al., 1995; Burdman et al., 1998). A high amount of PHB has been observed in *A. amazonense* species grown under nitrogen-fixing conditions (Fig. BXII.α.8) but not with an inorganic nitrogen source. Intracellular granules (probably PHB) were also observed in old cells of *A. doebereinerae*.

Azospirilla possess a single polar flagellum when grown in

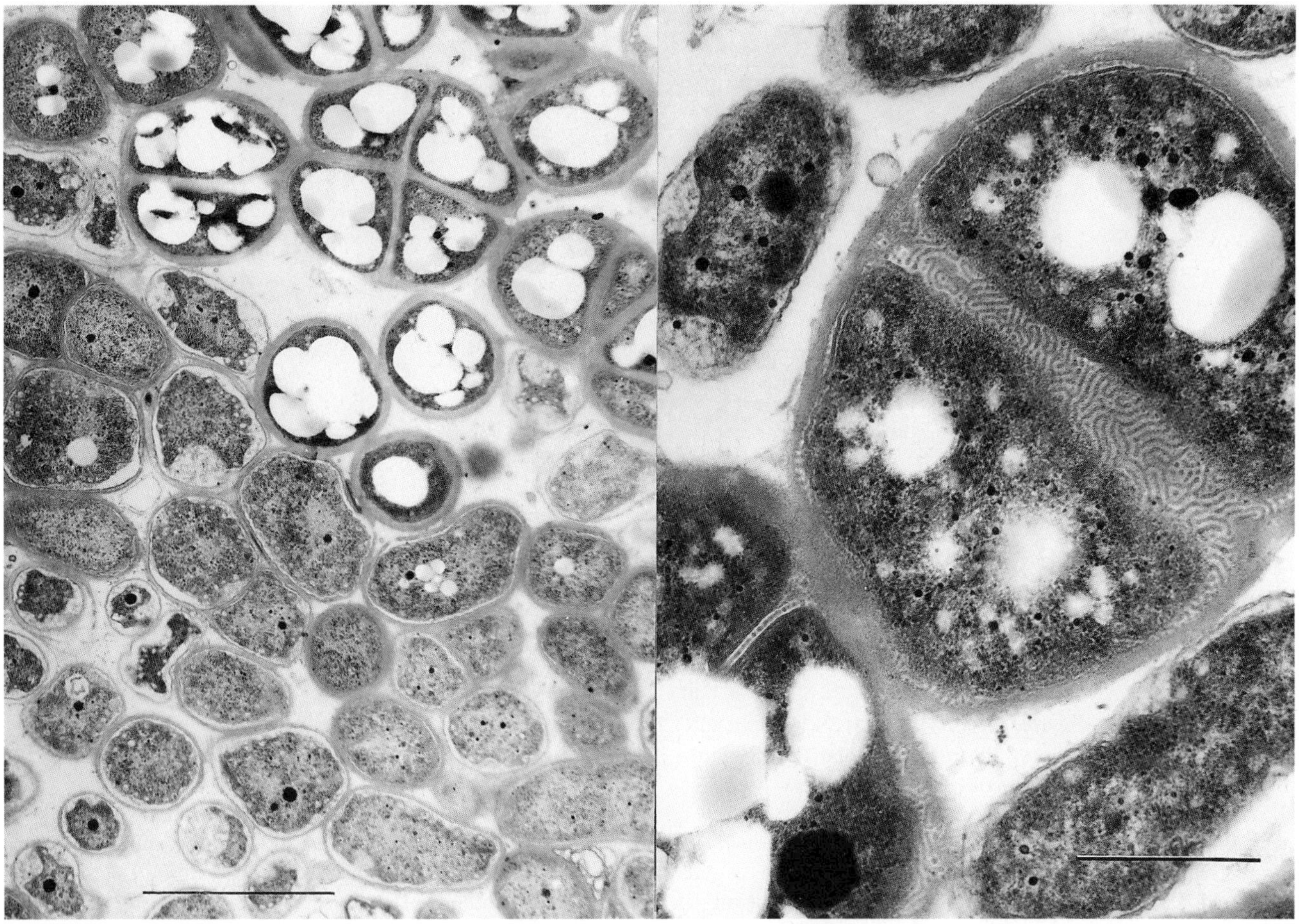

FIGURE BXII.α.6. Ultrastructure of vibrioid and encapsulated (C) forms of *Azospirillum brasilense* ATCC 29145 grown in association with a plant callus (sugarcane). *Left:* gradient of vibrioid forms (*lower*) to C forms (*upper*). Multicellular C forms filled with poly-β-hydroxybutyrate granules (white areas) are seen at the top. Bar = 1.5 μm. *Right:* comparative fine structure of vibrioid forms and C forms. Bar = 0.5 μm. (Reproduced with permission from R.H. Berg et al., Protoplasma *101:* 143–163, 1979, ©Springer-Verlag, Vienna.)

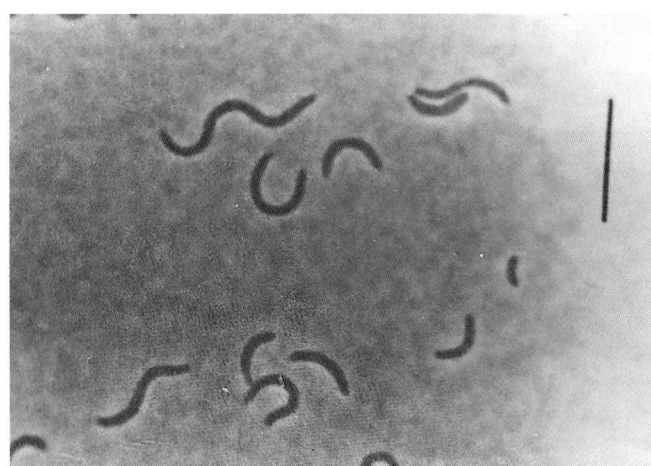

FIGURE BXII.α.7. Phase contrast microscopic examination of *Azospirillum irakense* strain KBC1 grown at 30°C for 20 h in complete nutrient broth showing the curved rods or S-shaped forms. Bar = 10 μm. (Reproduced with permission from K.M. Khammas et al., Research in Microbiology *140:* 679–693, 1989, Editions Scientifiques et Medicales, ©Elsevier, Paris.)

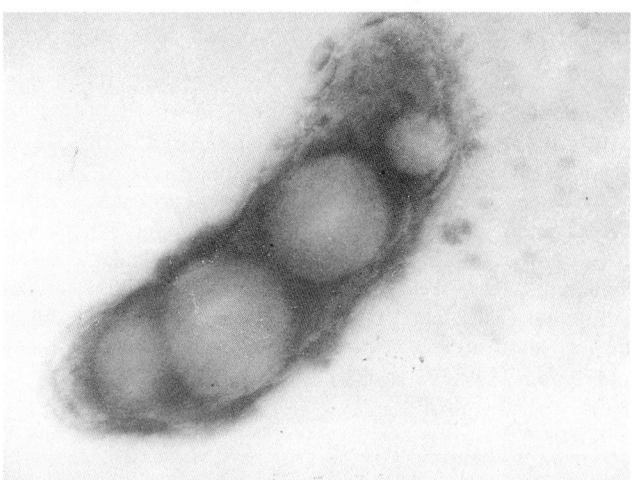

FIGURE BXII.α.8. Transmission electron micrograph of *Azospirillum amazonense* strain ATCC 35119 grown in semi-solid LGI medium for 24 h at 30°C. Note the presence of poly-β-hydroxybutyrate granules inside the cell. (~× 13,000). (Reproduced with permission from F.M. Magalhães et al., Anais da Academia Brasileira de Ciencias *55:* 417–430, 1983, ©Academia Brasileira de Ciencias, Rio de Janeiro.)

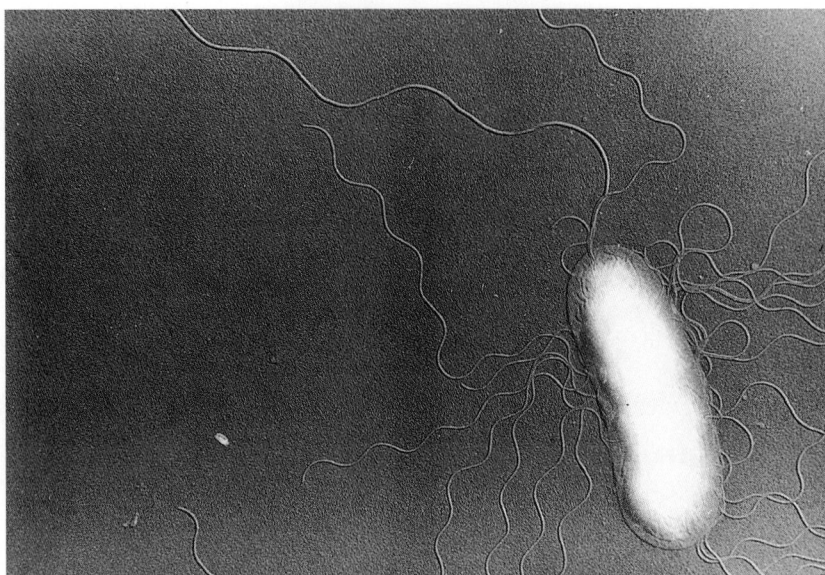

FIGURE BXII.α.9. Electron micrograph of *Azospirillum brasilense* ATCC 29145 cultured on MPSS agar at 30°C for 24 h. Both the single polar flagellum and the numerous lateral flagella can be seen. Shadowed with tungsten oxide. (~× 15,000).

liquid media. However, when cultured on MPSS agar at 30°C numerous lateral flagella in addition to the polar flagellum are formed in the species *A. lipoferum* and *A. brasilense* (see Fig. BXII.α.9). Lateral flagella are also observed in strains of *A. irakense* grown in nutrient broth but not in *A. amazonense*. The polar flagellum appears to be thicker than the lateral flagella and has a longer wavelength. *A. largimobile* cells have a mixed flagellation with a single polar flagellum and 1–10 distinct lateral flagella of different thickness and wavelength when grown in LWA medium. The polar flagellum is not only involved in aero- and chemotaxis (Grishanin et al., 1991), but also in the first adsorption step in the attachment process of *A. brasilense* to wheat roots (Michiels et al., 1991). The function of the lateral flagella is unknown, although a mutant of *A. brasilense* strain Cd lacking both polar and lateral flagella developed a stronger flocculation of cells when grown in a high C:N medium containing D-fructose and ammonium chloride as C and N source (Burdman et al., 1998).

Cultural characteristics On BMS agar[3] after 1–2 weeks of incubation at 33–35°C, colonies of *A. brasilense* and *A. lipoferum* are pink, opaque, irregular or round, often wrinkled, and typically have umbonate elevations (Döbereiner and Baldani, 1979). Pigmentation is best on BMS agar incubated under light. Certain strains and variants of *A. brasilense* form colonies that have a very deep pink color (Eskew et al., 1977; Tarrand et al., 1978). In one such strain (ATCC 29729), this intense color is attributable to the formation of several carotenoid pigments that occur only

under aerobic conditions and may be related to protection of the nitrogenase from oxidative damage (Nur et al., 1981). This hypothesis was supported by the study of Hartmann and Hurek (1988) who compared carotenoid-overproducing mutants of *A. brasilense* strain Sp7 and carotenoid-negative mutants of strain Cd. Carotenoid overproducing mutants exhibited slightly higher oxygen tolerance. On the other hand, no pigment has been observed for strains of the other *Azospirillum* species.

In nitrogen-free media, nitrogen fixation occurs only under microaerobic conditions ($pO_2 \cong 0.003$ atm), due to lack of oxygen protection mechanisms for the nitrogenase. In liquid nitrogen-free media, there is no growth or nitrogen fixation under an air atmosphere except when cells are previously grown in the presence of an exogenous nitrogen source such as glutamate or ammonium chloride (Pedrosa and Yates, 1984; Baldani et al., 1986a). In this case, strains are grown in conical flasks (250 ml) with 100 ml liquid NFbHP[4] medium containing a nitrogen source at 32°C and 120 rpm. Derepressed cells are obtained when the culture reaches an OD_{550} of approximately 1.0. This is the easiest way to obtain nitrogenase-derepressed cells for physiological and biochemical studies, especially when a large number of strains are to be assayed. The fastest growth rates are obtained in continuous or batch cultures when the oxygen input is in equilibrium with bacterial respiration. Depending on the cell density, a gas mixture containing from 1 to 25% oxygen can be bubbled through the culture to achieve this. A simpler way to obtain N_2-dependent growth is by culturing the organisms in semi-solid nitrogen-free media incubated under an air atmosphere. Here, growth is initiated as a thin veil or disc-like pellicle several mil-

---

3. BMS agar: washed, peeled, sliced potatoes, 200 g; L-malic acid, 2.5 g; KOH, 2.0 g; raw cane sugar, 2.6 g; vitamin solution (biotin, 0.01 g; pyridoxin, 0.02 g; distilled water, 1000 ml), 1.0 ml; bromthymol blue (0.5% alcoholic solution), 2 drops; agar, 15.0 g. The potatoes are placed in a gauze bag, boiled in 1 liter of water for 30 min, then filtered through cotton, saving the filtrate. The malic acid is dissolved in 50 ml of water and the bromthymol blue added. KOH is added until the malic solution is green (pH 7.0). This solution, together with the cane sugar, vitamins and agar, is added to the potato filtrate. The final volume is made up to 1 liter with distilled water. The medium is boiled to dissolve the agar, then sterilized by autoclaving.

4. NFbHP medium (g/l): KH_2PO_4, 4.0; K_2HPO_4, 6.0; $MgSO_4 \cdot 7H_2O$, 0.2; NaCl, 0.1; $CaCl_2$, 0.02; Nitrilotriacetic acid 0.056; $FeSO_4 \cdot 7H_2O$, 0.02; sodium lactate, 5.0; biotin, 0.0001; $Na_2MoO_4 \cdot 2H_2O$, 0.002; $MnSO_4 \cdot H_2O$, 0.00235; H_3BO_3, 0.0028; $CuSO_4 \cdot 5H_2O$, 0.00008; $ZnSO_4 \cdot 7H_2O$, 0.00024. Phosphate solutions were autoclaved separately and left to cool before adding to the medium. The volume is completed to 1000 ml and the final pH is 6.8. NH_4Cl is added to a concentration of 20 mM when liquid medium is used. For solid medium add 12 g/l of agar before autoclaving.

limeters or 1 cm below the surface of the medium at a point where the rate of diffusion of oxygen into the medium corresponds to the respiration rate of the organisms so that no excess oxygen remains in solution. As the bacteria multiply, the disc of growth migrates closer to the surface until finally it is just below the surface. The time taken for the veil to move to the surface depends on the *Azospirillum* species. In the case of *A. lipoferum*, *A. brasilense*, and *A. irakense*, the pellicle reaches the surface 30–42 h after incubation, whereas with *A. amazonense*, *A. halopraeferens*, and *A. doebereinerae* it takes around 3–4 days. Okon et al. (1980) have found a similar veil or disc formation even in media containing a source of fixed nitrogen, suggesting that the organisms may prefer microaerobic conditions even when not fixing nitrogen. Under these conditions, the growth is much faster. It has been shown that cells grown in semi-solid NFb or LGI medium containing nitrate can fix nitrogen provided that all accumulated nitrite is reassimilated by the cells and the nitrogenase is derepressed (Magalhães et al., 1983; Baldani et al., 1986a). The optimum level of dissolved oxygen for nitrogenase activity was 0.2 kPa oxygen in *A. brasilense*, *A. lipoferum*, and *A. amazonense* (Hartmann et al., 1985). At transiently increased oxygen levels, a reversible inhibition (switch off) of nitrogenase activity occurred, which was accompanied by a reversible covalent modification of the dinitrogenase reductase of *A. brasilense* and *A. lipoferum* (Hartmann and Burris, 1987). Recently, Vande Broek et al. (1996) demonstrated that *A. brasilense* is slightly more oxygen-sensitive than *A. irakense*, based on the maximum oxygen concentration at which activation of an *A. brasilense* nifH–gusA fusion and acetylene reduction was still observed under the experimental conditions. Comparison of *A. amazonense* and *A. brasilense* grown in nitrogen-free media in batch cultures seemed to show that the first species is much more oxygen sensitive than the latter (Magalhães et al., 1983). However, measurements in a well-mixed chamber where respiration rates, ARA, and dissolved O_2 were controlled showed that nitrogen fixation and derepression of the nitrogenase of *A. amazonense* is tolerant of a somewhat higher pO_2 than those of *A. lipoferum* and *A. brasilense* (Hartmann et al., 1985). This factor seems to be due to the presence of an oxygen-tolerant hydrogenase in *A. amazonense* (Fu and Knowles, 1989).

When supplied with a source of fixed nitrogen, such as an ammonium salt or nitrate, azospirilla can grow under aerobic conditions. Nitrate is assimilated by an assimilatory nitrate reductase (Neyra and van Berkum, 1977). Azospirilla reduce nitrate to nitrite either by an aerobic assimilatory pathway or an anaerobic dissimilatory or respiratory pathway. A group of *A. amazonense* strains (called nir^{++}) assimilates nitrite much faster than most strains of the species (called nir$^+$). Approximately half of the strains of *A. lipoferum* and *A. brasilense* so far isolated can dissimilate nitrite further to nitrous oxide and nitrogen gas (Neyra et al., 1977). Maximum dissimilatory nitrite reductase activity of *A. brasilense* has been observed to occur at high pH values between 7.6 and 8.5 and only under anaerobic conditions (Zimmer et al., 1984). Denitrification has not been observed among strains of *A. amazonense*, *A. irakense*, and *A. largimobile* but it was detected in all strains of *A. halopraeferens*. Nitrite reductase (*nir*) is the key enzyme for denitrification in *Azospirillum* (Magalhães et al., 1978). Plant roots were shown to be mostly infected by non-denitrifying strains (nir$^-$) of *A. lipoferum* and *A. brasilense*. Most of the roots infected by *A. amazonense* strains belong to the nir^{++} group. Chlorate-resistant spontaneous mutants lacking the dissimilatory nitrate reductase (nir$^-$ mutants) and/or the nitrite

reductase (nir$^-$ mutants) have been isolated from *A. lipoferum* and *A. brasilense* species. Nitrogenase activity is retained by such mutants and proceeds even in the presence of 10 mM nitrate. A comparison of several nitrate reductase negative (nir$^-$) mutants of *A. brasilense* strain Sp245 with the wild type carried out in a monoxenic wheat test tube experiment demonstrated the role of nitrate reductase in nitrate assimilation by wheat plants (Ferreira et al., 1987).

Steady-state cultures of denitrifying strains of *A. brasilense* have been obtained under anaerobic conditions in malate containing 100 mM nitrate. When the level of nitrate is lowered to 20 mM, nitrogenase activity occurs but growth is severely decreased (Nelson and Knowles, 1978). Anaerobic suspensions of nitrogenase-containing cells in NFb medium containing 10 mM nitrate exhibit nitrate-dependent nitrogenase activity after 30–60 min, but no growth occurs (Neyra and van Berkum, 1977; Scott et al., 1979); during this period nitrite is accumulated in the medium. A transitory, nitrate-dependent acetylene reduction has been observed in *A. lipoferum* and *A. brasilense* in nitrogen-fixing but not in NO_3^--grown cultures (Bothe et al., 1981).

In peptone-based media, *A. lipoferum* and *A. brasilense* grow abundantly under anaerobic conditions due to the denitrification process, with the nitrate being reduced to nitrite or to nitrous oxide and nitrogen gas. NO_3-dependent anaerobic growth has not yet been shown for *A. amazonense*, *A. halopraeferens*, *A. irakense*, and *A. largimobile*. Other compounds such as nitrite, nitrous oxide, and dimethylsulfoxide (DMSO) can also be used as alternative electron acceptors by *A. lipoferum* and *A. brasilense* strains when NH_4^+ is supplied as a nitrogen source (Döbereiner, 1992a; Hartmann and Zimmer, 1994).

Low concentrations of ammonium completely inhibit nitrogenase activity in *A. lipoferum* and *A. brasilense* (Pedrosa and Yates, 1984; Hartmann et al., 1986). This mechanism, known as "NH_4^+ switch on/off", which involves reversible ADP-ribosylation of the nifH-protein (Fu et al., 1989), has not been observed in *A. amazonense*. In *A. amazonense*, a less effective, noncovalent, reversible inhibition of nitrogenase activity by ammonium chloride occurs. Assimilation of ammonium in azospirilla is mediated by the GS-GOGAT pathway. Glutamate dehydrogenase (GDH), glutamate–oxaloacetate–aminotransferase (GOT), and glutamate–pyruvate–aminotransferase (GTP) also play a role when a high level of ammonium or glutamate is present. *A. lipoferum* and *A. amazonense* exhibit high activities of these enzymes and are able to fix nitrogen using glutamate as carbon and nitrogen source when supplied at low concentrations (1–2 mM), whereas *A. brasilense* and *A. halopraeferens* fix nitrogen in the presence of higher levels of glutamate (20 mM) due to the low rate of glutamate assimilation (Hartmann and Zimmer, 1994). While the amino acids aspartate and histidine showed the same differential effect as glutamate on the nitrogenase activity of azospirilla, the amino acids glutamine, asparagine, arginine, and alanine inhibited nitrogenase effectively (Hartmann et al., 1988).

Azospirillum mutants able to fix nitrogen in the presence of external ammonium sources have been developed. Prototrophic mutants of *A. brasilense* resistant to ethylenediamine (EDA) and able to excrete NH_4^+ were selected. Studies carried out under gnotobiotic conditions showed that these mutants support the growth of wheat plants better than the wild type strains (Christiansen-Weniger and Van Veen, 1991).

Nitrogenase complex The nitrogenase system of *Azospirillum* consists of two components: the Mo–Fe protein and the Fe protein (Ludden et al., 1978). An activating factor for the Fe protein

(Mn^{2+}-dependent activation process) is required for *A. lipoferum* and *A. brasilense* (Ludden et al., 1978; Pedrosa and Yates, 1984) but not for *A. amazonense* (Song et al., 1985). So far, no studies have been carried out for the other *Azospirillum* species. The activating factor is interchangeable with that from *Rhodospirillum rubrum*. An alternative nitrogenase system 3 (Fe-nitrogenase) in addition to the Mo-nitrogenase has been suggested in *A. brasilense* strain Cd (Chakraborty and Samaddar, 1995) but it lacks confirmation. A highly purified dinitrogenase (MoFe-protein) in an active form was obtained from *A. amazonense* (Song et al., 1985) and shown to be a tetramer of 210 kDa with subunits ($\alpha 2\beta 2$) having molecular weights of 50 and 55 kDa, respectively. The dinitrogenase reductase (Fe-protein) is a dimer with subunits of 31 and 35 kDa. Similar results were observed with *A. doebereinerae* strains grown under dinitrogen-dependent conditions, as detected by immunoassay of blotted cell extracts against the Fe-protein of *R. rubrum* and *A. vinelandii* antisera (Eckert et al., 2001). The molecular weights of the nitrogenase component peptides from *A. brasilense* strain Sp7 were determined by immunoprecipitation against antisera from *K. pneumoniae* and were estimated to be 60 kDa and 64 kDa for MoFe protein subunits and 33 kDa and 36 kDa for Fe protein subunits (Nair et al., 1983). The nitrogenase system *in vivo* in *A. lipoferum*, *A. brasilense*, and *A. amazonense* does not liberate H_2 because the latter is recycled by a hydrogenase (Berlier and Lespinat, 1980; Chan et al., 1980; Volpon et al., 1981; Pedrosa and Yates, 1984; Fu and Knowles, 1988). No data are yet available for the other species.

Hydrogenase and autotrophy *A. lipoferum* has the potential for H_2-dependent autotrophic growth under nitrogen-fixing or non-nitrogen-fixing conditions. Strain 208 was able to grow autotrophically under an atmosphere of $H_2/CO_2/O_2/N_2$ (20:5:2:73) in liquid NFb medium (Sampaio et al., 1981). Both ribulose-1,5-bisphosphate carboxylase (RubPcase) and hydrogenase activities were present and were much higher in autotrophically grown cells than in lactate-grown cells. In contrast to *A. lipoferum*, *A. brasilense* ATCC 29145 showed only marginal growth with H_2 under autotrophic conditions. Málek and Schlegel (1981) found that all of the strains of *A. lipoferum* they tested (including the type strain) were capable of hydrogenase activity and autotrophic growth when the O_2 concentration in the gas atmosphere was less than 2% (v/v). RubPcase and hydrogenase activities were present. In the absence of a source of combined nitrogen, the growth rate was very slow under autotrophic conditions and ceased at a cell concentration of 0.2 g/l. With an ammonium salt present, the growth rate was much faster. None of the strains of *A. brasilense* tested was capable of autotrophic growth. In addition, no reports are yet available on autotrophic growth for the other described species of *Azospirillum*.

Methylotrophy Both *A. brasilense* ATCC 29145 and *A. lipoferum* Sp 208 grow well on methane, methanol, or formate as sole energy sources (Sampaio et al., 1981). *A. lipoferum* cells derepressed under CH_4 were devoid of RubPcase and hydrogenase activities; however, RubPcase, but not hydrogenase, could be induced by CO_2 in the presence of methane. These results strongly suggest that *A. brasilense* and *A. lipoferum* have potential for methylotrophic growth and according to Sampaio et al. (1982), these two species could be classified as facultative methylotrophs. However, these authors were not able to demonstrate ^{14}C-incorporation into *A. brasilense* and *A. lipoferum* biomass. These results have not yet been confirmed by other studies, and there are no reports of methylotrophy in other azospirilla species.

Nutritional characteristics Compounds that can serve as sole carbon and energy sources for N_2-dependent growth of *A. brasilense*, *A. lipoferum*, *A. halopraeferens*, *A. irakense*, *A. largimobile*, and *A. doebereinerae* include malate, succinate, pyruvate, and lactate. *A. amazonense* strains can also use these carbon sources, but their sensitivity to alkaline pH values inhibits growth and misleading results can be obtained. A highly buffered medium can circumvent the problem, provided that the initial pH is around 6.0. D-Fructose can also be used as a carbon source by all species except *A. irakense*, in which variable results have been observed. D-Glucose is not used by *A. brasilense* and *A. halopraeferens*. α-Ketoglutarate is used by *A. lipoferum* and by some strains of *A. irakense*. Sucrose has been used as the main carbon source for N_2-dependent growth of *A. amazonense* and *A. irakense*. Studies using ^{14}C as a tracer have been carried out to evaluate the uptake of D-fructose, D-glucose, sucrose, mannitol, and α-ketoglutarate by *A. brasilense*, *A. lipoferum*, and *A. amazonense* (see Döbereiner and Pedrosa, 1987, for more details). The results confirmed the carbon metabolism already known for these species through use of other methodologies. No similar type of study has yet been carried out for the other azospirilla species.

Metabolic characteristics All enzyme activities of the catabolic Embden–Meyerhof–Parnas pathway, the Entner–Doudoroff pathway, and the tricarboxylic acid cycle have been detected in *A. brasilense*, *A. lipoferum*, and *A. amazonense*. However, there are differences among these species, as pointed out by Döbereiner and Pedrosa (1987) and Hartmann and Zimmer (1994) and reviewed here. Unfortunately, no similar studies have been carried out for the other species of *Azospirillum*. The type strains of *A. brasilense* and *A. lipoferum* have been found to phosphorylate fructose to fructose-l-phosphate by means of a phosphoenolpyruvate phosphotransferase system (Goebel and Krieg, 1984). Although both *A. brasilense* and *A. lipoferum* possess glucokinase, only *A. lipoferum* is permeable to D-glucose. Hexokinase activity occurs only in *A. lipoferum*. Sucrose-grown *A. amazonense* cells catabolize D-glucose and D-fructose (produced by the action of β-fructofuranosidase on sucrose) exclusively by the Entner–Doudoroff pathway (Martinez-Drets et al., 1985). *A. amazonense* strains possess fructokinase and glucokinase but not hexokinase. Enzymes of the Embden–Meyerhof–Parnas pathway and the Entner–Doudoroff pathway occur in D-fructose-grown cells of *A. brasilense* and *A. lipoferum*. High levels of NAD(P)-glucose 6-P-dehydrogenase, required for 6-phosphogluconate synthesis, have been detected in strains of *A. lipoferum* and *A. amazonense* but not in *A. brasilense* (Martinez-Drets et al., 1985). On the other hand, the enzyme 1-phosphofructokinase, involved in the phosphorylation of D-fructose, is present in large amounts in *A. lipoferum* and *A. brasilense* but absent in sucrose-grown *A. amazonense*. All three species above lack the key enzyme of the oxidative branch of the hexose monophosphate pathway (HMP). D-Gluconate is metabolized via the Entner–Doudoroff pathway in *A. lipoferum* and *A. brasilense*. Although *A. amazonense* possesses all three enzymes involved in the metabolism of D-gluconate it is unable to grow on this carbon source, probably because it lacks a gluconate transport system (Martinez-Drets et al., 1985). No information is yet available for the other species.

A. lipoferum exhibits weak fermentative ability in media containing D-glucose or D-fructose as a carbon source and with a source of fixed nitrogen. *A. largimobile* possesses fermentative ability in media containing carbohydrate, as determined by the Hugh–Leifson test (Ben Dekhil et al., 1997a). Although strains from all *Azospirillum* species grow far better aerobically than anaerobically, strains of *A. largimobile* have been considered facultative anaerobes. *A. lipoferum* is capable of acidifying D-glucose or D-fructose media anaerobically, of forming very small amounts of gas in Durham vials, of exhibiting slight growth in D-glucose or D-fructose broth under anaerobic conditions, and of forming minute colonies on D-glucose or D-fructose agar anaerobically. Variable results for these tests occur within some strains, and variants with decreased fermentative ability can be selected. Such variants continue to require biotin, use D-glucose as a sole carbon source for N₂-dependent growth, and exhibit the characteristic pleomorphic changes associated with this species. *A. largimobile* also produces acid (but no gas) from D-fructose, D-glucose, D-galactose, D-ribose, and sucrose, whereas *A. halopraeferens* produces acid from D-fructose but only aerobically. No acid is produced from D-glucose or D-fructose in *A. brasilense*, *A. amazonense*, and *A. irakense* either aerobically or anaerobically in a peptone-based medium. Acid is also produced from D-fructose and D-glucose in *A. doebereinerae* strains grown anaerobically (API50 test) (Eckert et al., 2001).

By an auxanographic method using media containing ammonium sulfate as the nitrogen source (Tarrand et al., 1978), the following compounds serve universally as sole carbon sources for all *A. lipoferum*, *A. brasilense*, and *A. halopraeferens* strains: malate, succinate, lactate, pyruvate, fumarate, β-hydroxybutyrate, D-gluconate, glycerol, and D-fructose. *A. irakense* and *A. amazonense* are also able to use these organic acids as carbon source; however, *A. amazonense* strains require a pH of approximately 6.0 in order to use the organic acids for N₂-dependent growth. No data are available for the use of organic acids as carbon source in *A. largimobile*. D-Gluconate and glycerol are not used by *A. amazonense* and *A. irakense*, whereas *A. largimobile* uses glycerol but not D-gluconate. D-Fructose is also catabolized by the other azospirilla species although variable results have been observed for *A. irakense*. Citrate and D-mannitol are not used by strains of *A. amazonense*, *A. irakense*, and *A. brasilense*. A few strains of *A. halopraeferens* do not catabolize citrate. D-Glucose is used by all azospirilla species except *A. halopraeferens* and most strains of *A. brasilense*. However, *A. brasilense* cannot use D-glucose as a sole carbon source for N₂-dependent growth, and it also produces a lower degree of acidification of glucose than does *A. lipoferum* (Tarrand et al., 1978). The use of D-galactose as carbon source is variable among strains of *A. brasilense* and negative for all strains of *A. halopraeferens*, whereas sucrose is catabolized only by strains of *A. amazonense* and *A. irakense*. N-acetylglucosamine is used by *A. largimobile*, *A. lipoferum*, *A. irakense*, variable for *A. amazonense* strains but not used by *A. brasilense* and *A. doebereinerae*. No information is available for *A. halopraeferens*.

The following tests are universally positive for the genus *Azospirillum*: oxidase, phosphatase (weak in *A. largimobile*), urease, and esculin hydrolysis. The following tests are universally negative: starch and gelatin hydrolysis, production of water-soluble pigments, and indole production. Acidification of media containing lactose, sucrose, L-rhamnose, cellobiose, erythritol, dulcitol, or melibiose differs among strains and species of *Azospirillum*. Catalase activity ranges from strong to undetectable. The ability to grow anaerobically with nitrate in peptone-based media is positive for *A. lipoferum*, *A. brasilense*, and *A. halopraeferens* but negative for *A. amazonense*, *A. irakense*, and *A. largimobile*. In addition, the ability to dissimilate nitrate to either nitrite or to nitrous oxide and nitrogen gas differs among strains and species of *Azospirillum*. Most strains of *A. lipoferum*, *A. brasilense*, *A. halopraeferens*, and *A. doebereinerae* are able to denitrify. No denitrification has been observed in strains of *A. amazonense*, *A. irakense*,

and *A. largimobile*, although a report has been published showing a very low rate of denitrification for strain Y1 of *A. amazonense* (Neuer et al., 1985). Hybridization of DNA from five azospirilla species against the *nozZ*-segments containing the nitrous oxide reductase gene from *P. stutzeri* as a probe produced a positive signal with strains of *A. lipoferum*, *A. brasilense*, and *A. halopraeferens* but not for *A. amazonense* and *A. irakense*, confirming the presence of a N₂O reductase in the former three species and its absence in the last two species (Zimmer et al., 1995).

Pectinolytic activity (pectate lyase and pectin methylesterase) has been detected in strains of *A. irakense* (Khammas and Kaiser, 1991). Strains from this species are able to grow and fix nitrogen when pectin is used as the sole carbon source. Pectolytic activity—pectic lyase and endogalacturonase (Umali-Garcia et al., 1980) and polygalacturonic acid transeliminase (Tien et al. 1981)—has also been reported for some strains of *A. lipoferum* and *A. brasilense*. However, the enzymatic activity appears to be weaker than that of *Erwinia* (Tien et al., 1981). In addition, Khammas and Kaiser (1991) failed to demonstrate pectinolytic activity in these two species of *Azospirillum* as well as in strains of *A. amazonense*. Quite recently it was observed that *A. irakense* strain KBC1 also possesses cellulolytic activity—endoglucanase, cellobiohydrolase, and β-glucosidase activities (Vande Broek et al., 1998). No similar activities (pectinolytic and cellulolytic) have been tested for strains of *A. halopraeferens*, *A. largimobile*, and *A. doebereinerae*.

Production of the plant growth substance indoleacetic acid, and also indolelactic acid, gibberellin, and cytokinin-like substances, has been reported in strains of *A. brasilense* (Reynders and Vlassak, 1979; Tien et al., 1979). The production of indole-3-acetic acid by *A. brasilense* and *A. lipoferum* strains has been confirmed by many authors (Hartmann et al., 1983; Fallik et al., 1989; Bar and Okon, 1995) and the pathway for synthesis of IAA as well as the regulatory mechanism involved have been suggested (see Hartmann and Zimmer, 1994; Vande Broek et al., 1998). However, the production of cytokinins and gibberellins is still not clear because these products were detected only via bioassays in very old cultures (see Hartmann and Zimmer, 1994). More recently, the production of abscisic acid in *A. brasilense* (Iosipenko and Ignatov, 1995) and gibberellic acid in strains of *Azospirillum lipoferum* (Bottini et al., 1989) and other azospirilla (Rasul et al., 1998) was demonstrated using the HPLC technique. However, the production was detected with 10-d-old cells. The production of IAA in *Azospirillum irakense* is very low and the gene *trpD*, encoding phosphoribosyl anthranilate transferase, was proposed to be involved in the repression of IAA production (Zimmer et al., 1991). No information about the production of phytohormones by the other *Azospirillum* species is available. Plant response to inoculation with *Azospirillum*, mainly increase in root area, has been reported and attributed to phytohormones such as IAA produced by these bacteria (Bashan and Holguin, 1997).

Other substances such as the vitamins thiamin, niacin, pantothenic acid (Rodelas et al., 1993) and riboflavin (Dahm et al., 1993) have also been detected in *A. brasilense* strains. It has also been reported that *Azospirillum lipoferum* is able to degrade 4-chloronitrobenzene (Russel and Muszynki, 1995) and more recently it was demonstrated that many strains (73 out of 110) of *A. lipoferum* and *A. brasilense* are able to produce cyanide (Gonçalves and Oliveira, 1998). The ability to oxidize phenolic derivatives like syringic aldehyde, acid, or acetosyringone has been detected in *A. lipoferum* strain 4T and is carried out by laccase—an oxidase widespread in fungi and higher plants (Faure et al., 1996).

Azospirillum spp. isolated from saline soils exhibited different degree of osmotolerance. *A. halopraeferens*, isolated from sodic alkaline soils of Pakistan, exhibited the highest osmotolerance, followed by *A. brasilense*, *A. lipoferum*, and *A. amazonense* (Hartmann, 1988); the osmotolerance of *A. irakense* lies between *A. halopraeferens* and *A. brasilense* (Khammas et al., 1989). As compatible solutes in *Azospirillum*, trehalose, glycine betaine, glutamate, and proline were characterized (Hartmann et al., 1991). *A. halopraeferens* is able to take up choline efficiently and to convert it into the most potent osmoprotectant, glycine betaine (Hartmann, 1988; Hartmann et al., 1991). In *A. brasilense* Sp7, a binding protein-dependent, high-affinity uptake of glycine betaine was demonstrated and glycine betaine stimulated growth and nitrogen fixation in *A. brasilense* (Hartmann, 1988; Riou and Rudulier, 1990). In osmotolerant *Azospirillum* spp., choline and glycine betaine, as well as glutamate, proline, and other amino acids, are not or only slightly used as nitrogen or nitrogen and carbon sources (Hartmann et al. 1988; Khammas et al., 1989). Dehydroproline (DHP)-resistant bacteria with improved osmoregulatory properties are available for *A. brasilense* Sp7 (Hartmann et al., 1992). The fact that these mutants appear spontaneously at relatively high rates suggests the ecophysiological importance of this trait.

Under iron-limiting conditions, *A. lipoferum* and *A. brasilense* release the phenolate siderophores 2,3- and 3,5-dihydroxybenzoic acid conjugated with lysine and leucine or ornithine and serine, respectively, which has been called spirillobactin (Saxena et al., 1986; Bachhawat and Ghosh, 1987b). The Fe(III)-spirillobactin complex is taken up via a high-affinity uptake receptor, which involves a specific outer membrane receptor (Bachhawat and Ghosh, 1987a). It has been shown for *A. brasilense* Sp245 that the siderophores ferrichrome, ferrichrysin, and coprogen, as well as the main siderophore of *Streptomyces*, ferrioxamine B, can be used as iron sources too (Hartmann, 1988). However, the iron-scavenging properties vary much among different *Azospirillum* species and strains; e.g., *A. brasilense* strain Sp7 was only a weak siderophore scavenger (Hartmann, 1988). Under severe iron-limiting conditions, which drastically inhibit the growth of *A. brasilense* Sp7 wild type, spontaneous mutants that could readily use coprogen- and ferrichrysin-bound iron, appeared (Hartmann et al., 1992). The high-affinity iron uptake systems of *A. halopraeferens*, *A. irakense*, and *A. doebereinerae* have not yet been studied. Among azospirilla, only *A. irakense* isolates can efficiently hydrolyze desferrioxamine (Winkelmann et al., 1999). The ecological relevance of this unique physiological property has not yet been elucidated. High-affinity-mediated siderophore acquisition could be an important trait for competitiveness in a highly populated habitat such as the rhizosphere.

Genetic characteristics Genetic transformation to generate antibiotic-resistant derivatives has been reported for *A. brasilense* ATCC 29145 (Mishra et al., 1979). Genetic transformation has been also demonstrated for other strains of *A. brasilense* and *A. lipoferum* (Wood et al., 1982; Fani et al., 1986) More recently, DNA transformation of *Azospirillum* using a high-voltage electroporation mechanism was developed (Vande Broek et al., 1989); however, the methodology worked only for *A. brasilense* strains and not for *A. lipoferum* strains. There are no reports of DNA transformation for the other *Azospirillum* species. In contrast to transformation, conjugation appears to be the most efficient method of gene transfer for *Azospirillum* species (see Döbereiner and Pedrosa, 1987). More recently, a *nifH–gusA* fusion plasmid pFAJ21 was transferred by a biparental mating to *A. irakense* (Vande Broek et al., 1996). However, there are no reports of conjugation for the other described *Azospirillum* species.

Genetic recombination has been demonstrated in *A. brasilense* ATCC 29145, using plasmid R68-45 (derived from *Pseudomonas aeruginosa*) to mobilize the chromosome (Franche et al., 1981). The mode of gene transfer promoted by the plasmid appeared to be unpolarized, suggesting the existence of multiple origins of transfer. Similar results were obtained by Bazzicalupo and Gallori (1983) using strains Sp7 and Sp6 of *A. brasilense*. Although much progress has been made, the data obtained so far do not allow a genetic map to be drawn for the chromosome of *A. brasilense*. Several difficulties in constructing the genetic map using this technique have been pointed out by Döbereiner and Pedrosa (1987). Nevertheless, most of the genes involved in the nitrogen fixation process have already been identified in *A. brasilense* and *A. lipoferum* and located on the chromosome (Vande Broek and Vanderleyden, 1995). In addition, the organization and regulation of the *nif* genes as well as the genes involved in nitrogen metabolism are well established in *A. brasilense* (Elmerich et al., 1997). Little information is available so far for the other *Azospirillum* species.

Plasmids and bacteriophages The occurrence of multiple large plasmids in *A. brasilense* and *A. lipoferum* strains was reported by Wood et al. (1982). The authors called these megaplasmids "minichromosomes" that varied in size from 42 to 1850 MDa. Two plasmids of very high molecular weight (1700 and >1800 Kb) have also been observed in several strains (24 out of 28) of *A. brasilense* isolated from sugarcane plants (Caballero-Mellado et al., 1999). These authors also observed that the megaplasmid of 1700 Kb and the plasmid of approximately 600 Kb carry genes homologous to the 16S rDNA, thus reinforcing the hypothesis of the presence of minichromosomes in *Azospirillum brasilense*. More recently, the genome structure of five *Azospirillum* species was analyzed by pulsed-field gel electrophoresis (Martin-Didonet et al., 2000). The authors detected the presence of 8–10 replicons in the species *A. lipoferum* and *A. brasilense* (0.15–2.5 Mbp) and 4–5 in the species *A. amazonense*, *A. irakense*, and *A. halopraeferens* (0.22–2.7 Mbp). They also observed the hybridization of a 16S rDNA probe to some replicons, confirming the existence of multiple chromosomes in the genus. In addition, the authors observed that the *nifHDK* operon is present in the largest replicon. Strains of *A. brasilense* and *A. lipoferum* harboring 1–7 plasmids ranging from 4 to 310 MDa have been described by other authors (Franche and Elmerich, 1981; Plazinski et al., 1983). Plasmids have also been detected in strains of *A. amazonense* and *A. halopraeferens*; however they are present in low numbers and are small in size (Elmerich et al., 1991). No information is yet available for *A. doebereinerae*. Many reports have shown that *Azospirillum* plasmids harbor genes involved in the plant–bacterium interaction and other physiological functions. A plasmid called p90, present in many *A. lipoferum* and *A. brasilense* strains, carries genes involved in chemotaxis, motility, and plant root adsorption (Michiels et al., 1994; Vande Broek and Vanderleyden, 1995). In addition, it also carries *nod* and *exo* genes that are homologous to those frequently found in legume-nodulating rhizobia (Onyeocha et al., 1990). Other functions such as production of melanin in *A. lipoferum* (Givaudan et al., 1991) and IAA biosynthesis in *A. brasilense* (Katzy et al., 1995) have also been detected. The p90 plasmid tagged with Tn5 was mobilized to *Agrobacterium tumefaciens* at a frequency of 10^{-4}, but self-transfer of p90 was not demonstrated (Onyeocha et al., 1990).

Lysogenicity is common in *Azospirillum* (Döbereiner and Pedrosa, 1987). It has been detected in many strains of *A. lipoferum* and *A. brasilense* and the lysis is inducible by mitomycin C (Franche and Elmerich, 1981; Elmerich et al., 1982). However, not all strains of these two species are lysogenic. Temperate bacteriophage Al-1—isolated from soil samples from Brazil—was purified and showed a size and morphology similar to that of coliphage λ (Elmerich, 1983). This phage is lysogenic to some strains of both species (see Döbereiner and Pedrosa, 1987). Phage Ab-1 particles from *A. brasilense* Sp7 were characterized and shown to be infective to strains Sp7 and Cd of *A. brasilense* but not to strains Sp35, Sp59a, RG20a, and Br17 of *A. lipoferum* (Germida, 1984). The presence of bacteriophages in strains of the other species of *Azospirillum* has not been determined.

Bacteriocins Production of bacteriocins has been observed in many strains of *A. lipoferum* and *A. brasilense* in pure culture, but this seems not to be the case for strains living in soil or associated with roots (Oliveira and Drozdowicz, 1981; 1988). Catechol-type siderophores with antimicrobial activity against various bacterial and fungal isolates have been detected in 27 strains of *Azospirillum* (Shah et al., 1992).

Antigenic characteristics Antisera prepared against whole cells have been used in the indirect fluorescent antibody technique to distinguish between *A. lipoferum* and *A. brasilense* and also between groups of strains (Schank et al., 1979; De Polli et al., 1980). Several other reports applying polyclonal antisera to evaluate the colonization of grass plants by *A. lipoferum* and *A. brasilense* have been published (see Kirchhof et al., 1997b, for more details). More recently, with the help of hybridoma technology, several strain-specific monoclonal antibodies were developed for *A. brasilense* (Kirchhof et al., 1997b). Three classes of monoclonal antibodies with high specificity for *Azospirillum brasilense* Sp7 have been characterized by Schloter et al. (1994). One class recognizes a 100 kDa protein subunit of the polar flagellum, and the other two bind to an 85 kDa outer membrane protein and to polysaccharide, respectively. Monoclonal antibodies specific for *A. brasilense* strain Sp245 and *A. brasilense* strain Wa5 were recently characterized (Schloter and Hartmann, 1996; Schloter et al., 1997). As antibody-binding cell surface components, lipopolysaccharides and outer membrane proteins were identified. Studies of wheat root colonization by the above strains using the strain-specific monoclonal antibodies showed that strain Sp245 colonizes the inner root tissue endophytically whereas strain Sp7 colonizes only the root surface (Schloter and Hartmann, 1998). No monoclonal antibodies for other *Azospirillum* species have been produced; therefore, strain-specific root colonization studies involving other species are lacking.

Ecology Azospirilla have a worldwide distribution and occur in large numbers (up to $10^7/g$) in rhizosphere soils and in association with the roots, stems and leaves of a large variety of different plants. Most isolates have been obtained from tropical forage grasses and from cereal plants grown in tropical regions. However, azospirilla have also been frequently isolated from plants grown in temperate regions, including graminoids grown in the Canadian High Arctic sites (Nosko et al., 1994). The idea that *Azospirillum* species were limited to grass plants is no longer tenable. *A. lipoferum* and *A. brasilense* have a widespread occurrence and colonize various plants including legumes, vegetables, and fruits (Bashan and Holguin, 1997; Kirchhof et al., 1997a; Weber et al., 1999). *A. amazonense* has been isolated from roots of maize, sorghum, rice, wheat, forage grasses, and palm trees grown around Brazil. It has also been isolated from roots and stems of sugarcane plants grown in Brazil, from sugarcane roots grown in Hawaii and Thailand, and more recently, from pine-

apple and banana fruits grown in Brazil (Weber et al., 1999). The other *Azospirillum* species appeared to be more restricted: *A. halopraeferens* has so far only been found associated with Kallar grass plants grown in Pakistan. *A. irakense* has so far been isolated from rice plants grown in Iraq (Khammas et al., 1989) and from a freshwater pond in the botanical garden in Tübingen, Germany (Winkelmann et al., 1996). *A. largimobile* has only been isolated from a freshwater lake located in Australia (Ben Dekhil et al., 1997a) while *A. doebereinerae* was so far isolated from roots of the C_4-graminaceous plant *Miscanthus* grown in Germany (Eckert et al., 2001).

In field-grown maize, azospirilla occur on the surface of roots, in the outer cortex, inner cortex, and in the stele (Patriquin and Döbereiner, 1978). Infection of the inner cortex and stele occurs in the absence of significant bacterial colonization or collapse of outlying tissues. Paraxylem vessels can be completely plugged with the bacteria. Infection occurs initially in root branches and spreads longitudinally into main roots. In monoaxenic cultures of pearl millet and guinea grass, azospirilla are found within the mucigel layer of roots and become firmly attached to root hairs. The bacteria enter the roots through lysed root hairs and void spaces created by epithelial desquamation and lateral root emergence (Umali-Garcia et al., 1980). Azospirilla have been observed in intercellular locations within the middle lamella of root tissues; they have also been observed intracellularly, sometimes in very large numbers.

The above observations of root colonization were recently confirmed for *A. brasilense* strains Sp245 and Sp7 inoculated into wheat plants and evaluated using sophisticated techniques such as confocal laser scanning microscopy (SCLM) coupled with fluorescent probes (Assmus et al., 1995), as well as the use of strain-specific monoclonal antibodies (Schloter and Hartmann, 1998) and *nifH–gus* fusion (Vande Broek et al., 1993). There was a marked predominance of *A. brasilense* in the root hair zone as compared with the root tips of wheat plants. *A. brasilense* strain Sp245 was repeatedly detected in the interior of root hairs whereas *A. brasilense* strains Sp7 and Wa3—both isolated from a rhizosphere—colonized only the rhizoplane (Assmus et al., 1995). Strain Sp245, already known to be a colonizer of the root interior (Baldani et al., 1986b), could be found in the intercellular space of the root parenchyma and inside cortex and parenchyma cells, with the point of emergence of lateral roots being the most probable infection site (Assmus et al., 1997). More recently, it was observed that microcolonies are formed in the intercellular space of wheat roots inoculated with strain Sp245, but not with strains Sp7 and Wa3, and that the plant genotype influenced root colonization (Schloter and Hartmann, 1998). Use of strain Sp245 expressing the *gusA* gene showed that the bacteria initially concentrate in the root-hair zones and at sites of lateral root emergence. Proliferation to the other parts of the root was dependent on the status of the nitrogen and carbon sources present in solution (Vande Broek et al., 1993). Although the complete process of association between *Azospirillum* and the host grass plant is still unclear and a model has already been suggested (Del Gallo and Fendrik, 1994), it is known that the colonization requires the attachment of the cell to the root surface. This attachment occurs in two steps: one that is fast, weak, and probably mediated by bacterial proteins, and the second phase (also called anchoring) which is longer and irreversible and probably involves extracellular surface polysaccharides and the polar flagellum (see Vande Broek and Vanderleyden, 1995, for more details). It has been suggested that the *rpoN* gene, involved in nitrate reduction, nitrogen fixation, and cell motility, also controls the colonization of wheat plants by *A. brasilense* Sp7 (Pereg-Gerk et al., 1998). Because the ability of most strains of *Azospirillum* to colonize both the surface and root interior, including the stems and leaves of grass, these bacteria were included in the group called facultative endophytic diazotrophs (Baldani et al., 1997). This term was suggested to distinguish them from the obligate endophytic bacteria that include *Gluconacetobacter diazotrophicus*, a bacterium found colonizing sugarcane plants endophytically (Döbereiner, 1992b).

Several field inoculation experiments applying mainly strains of *A. brasilense* and *A. lipoferum*, and in a few cases *A. amazonense*, have been carried out in the last 20 years at various locations around the world. About 60–70% of all field experiments showed positive effects of the inoculation, with significant yield increases ranging from 5 to 30% (Okon and Labandera-Gonzalez, 1994; Baldani et al., 1997). Few commercial inoculants are available, probably because of the inconsistency and low percentage of response to inoculation. There is still a debate about the main mode of action by which these bacteria contribute to the nitrogen accumulated by the plants. Effects of plant-growth promoting substances, nitrogen fixation *per se*, and the ability of the bacterial nitrate reductase to help in the incorporation of nitrogen assimilated from the soil by the plant (Baldani et al., 1997) have also been mentioned as the major factors involved in the response of the plant to *Azospirillum* inoculation.

ENRICHMENT AND ISOLATION PROCEDURES

Tenfold serial dilutions of roots, stems, or leaf samples are inoculated into 10-ml cotton-plugged serum vials containing 5 ml of semisolid medium and incubated for up to one week at the optimal temperature. Soil samples or smashed pieces (5–8 mm long) of plant tissues can also be used but the incubation time should be only 40–48 h. Nitrogenase activity can be tested by acetylene reduction assay for those vials containing pellicles on the surface.

For vials exhibiting acetylene reduction, a second serial transfer is made. In the case of *A. brasilense*, *A. lipoferum*, and *A. doebereinerae*, the pellicle is transferred to semisolid NFb medium. After 24 h, the growth is streaked out on plates of solid NFb medium containing 0.02 g/l yeast extract. After 1 week, typical small, white, dense, single colonies are sub-cultured to semisolid NFb medium, where subsurface growth in the form of a veil is a presumptive indication of successful enrichment. For final purification, cultures are streaked onto BMS agar and the typical pink, often wrinkled, colonies are transferred for storage and identification (Döbereiner et al., 1995).

For *A. amazonense*, the pellicle is transferred to semi-solid LGI medium[5] and after 3–4 d streaked out on plates of LGI medium containing 0.02% yeast extract. After 5 d, small (2 mm diameter) whitish colonies with firm, dense, but not tenacious consistency and which are partially imbedded into the agar are selected and transferred to new LGI vials and, if growing well, streaked on BMS agar for final purification. On this medium, large, white, flat colonies with raised margin are formed.

For *A. halopraeferens*, roots of *Leptocholoa fusca* plants are collected and the rhizoplane soil separated, diluted, and inoculated into semi-solid modified NFb medium, named SM[6] (containing vitamins, 0.25% NaCl; pH adjusted to 8.5), and incubated at 41°C. Culture from vials showing nitrogenase activity, as determined by the acetylene reduction method, are then transferred to fresh semi-solid SM medium and after 48–72 h of incubation the pellicle is suspended in the same medium containing yeast extract (0.02%). Aliquots are then used to seed molten, cooled (45°C), soft SM (0.8% agar), which is then poured into Petri dishes and incubated for 1 week. Small, white colonies are visible within the agar and are further purified by embedding in the same way. For final purification, colonies are again transferred to semi-solid SM medium and streaked on Tryptic soy agar; cream-colored, circular, flat colonies with an entire margin are formed. No growth occurs on BMS or Congo red agar medium.

For *A. irakense* species, rhizosphere and roots of rice plants are crushed, suspended in sterile distilled water, diluted, and inoculated into semisolid NFb medium. The intermediate purification steps basically follow those for *A. lipoferum*. The final purification is performed on modified semisolid AAM medium[7] (containing 4 ml/l of a 1.64% solution of FeEDTA; pH of medium adjusted to 6.5). Subsurface pellicles are streaked out on the modified AAM agar plates. Colonies formed after 4 d are translucent, glistening, and convex with a regular margin, 1 mm diameter.

In the case of *A. largimobile*, one drop of fresh lake water is spread onto the surface of lake water agar (LWA) plates, dried and incubated for up to 8 d at 28°C. Colonies grown on LWA for 72 h at 28°C are 1–2 mm in diameter, colorless, translucent, low convex, and round, with an entire edge and smooth surface and they readily coalesce. These colonies, containing unicellular or multicellular forms, are further purified on LWA agar where very active motile cells containing refractile granules appear 4–8 h after incubation. These motile cells have a striking resemblance to cells of *Beijerinckia* species. To ensure that the phenomenon of multicellular form occurs, the cells are again inoculated in LWA or lake water salt agar (LWSA). Although *A. largimobile* has shown these peculiar characteristics, strains from this species are able to grow and fix nitrogen in semisolid NFb medium, with many aspects of their growth resembling *A. lipoferum*.

The following method has been used for the specific enrichment of *A. lipoferum*. A nitrogen-free mineral medium is inoculated with soil samples or washed root pieces and incubated under a gas atmosphere of $O_2/CO_2/H_2/N_2$ (5:10:30:55). After one

---

5. LGI medium (g/l): Sucrose, 5.0; K$_2$HPO$_4$, 0.2; KH$_2$PO$_4$, 0.6; MgSO$_4$·7H$_2$O, 0.2; CaCl$_2$·2H$_2$O, 0.02; Na$_2$MoO$_4$·2H$_2$O, 0.002; bromothymol blue 0.5% in 0.2 N KOH, 5ml; FeEDTA (1.64% solution), 4 ml; vitamin solution (biotin, 0.01g; pyridoxal-HCl, 0.02g; distilled water, 100ml), 1 ml; pH adjusted to 6.0–6.2 with H$_2$SO$_4$ and the final volume completed to 1000 ml with distilled water. For semi-solid medium, 1.8 g/l agar was added. For solid medium, add 15 g/l agar and 20–50 mg/l yeast extract.

---

6. SM medium (g/l): DL-malic acid, 5.0; KOH, 4.8; NaCl, 1.2; NaSO$_4$, 2.4; NaHCO$_3$, 0.5; CaCl$_2$, 0.22; MgSO$_4$·7H$_2$O, 0.25; K$_2$SO$_4$, 0.17; Na$_2$CO$_3$, 0.09; Fe-EDTA, 0.077; K$_2$HPO$_4$, 0.13; biotin, 0.0001; MnCl$_2$·4H$_2$O, 0.0002; H$_3$BO$_3$, 0.0002; ZnCl$_2$, 0.00015; CuCl$_2$·2H$_2$O, 0.000002; Na$_2$MoO$_4$·2H$_2$O, 0.002; distilled water, completed to 1000 ml. The final pH of the medium is 8.5 and it does not require further adjustment. For semisolid medium, 2.0 g/l agar were added. Malic acid, KOH, and agar are dissolved in one-half of the total volume and autoclaved. The remaining salts are sterilized by filtration after dissolving them in one-half of the total volume and discarding the precipitate after centrifugation. For solid medium, 8 g/l agar and 20 mg/l yeast extract were added.

7. AAM defined medium (g/l): K$_2$HPO$_4$, 2.0; KH$_2$PO$_4$, 6.0; MgSO$_4$·7H$_2$O, 0.2; NaCl, 0.1; CaCl$_2$·2H$_2$O, 0.026; Na$_2$MoO$_4$·2H$_2$O, 0.002; FeCl$_3$, 0.01; NH$_4$Cl, 1.0; distilled water, 1000 ml. The pH is adjusted to 6.0. For semisolid medium, 1.9 g/l agar were added and the NH$_4$Cl was omitted. For solid medium, 15 g/l agar were added. The carbon sources are sterilized separately (i.e., DL-sodium malate) or filter sterilized (i.e., sucrose) and added aseptically to the minimal medium to give a final concentration of 0.03M.

week, 0.5 ml of the suspension is transferred to fresh medium and incubated as before. The procedure is repeated 2 or 3 times before serial dilutions are plated onto nitrogen-free mineral media containing 1.5% (w/v) agar. Colonies are selected and propagated under autotrophic conditions.

A method has been developed and used to isolate *A. lipoferum* from the rhizosphere of rice plants (Tran Van et al., 1997). Serial dilutions of crushed roots and rhizosphere soil suspensions are inoculated into 18 × 180-mm tubes, each tube containing 8 ml of one of four different liquid N-free media: distilled water, KCl solution (8.5 g/l), NFb liquid medium, and soil extract (obtained by autoclaving a soil suspension in water (50% w/v) at 130°C for 1h and filtering through cellulose filters followed by autoclave treatment). The tubes are incubated at 30–32°C for 15 d in the dark to prevent growth of algae or cyanobacteria. A pellicle is formed 2 mm below the surface of the liquid, and when examined under the microscope normally shows the presence of spiral motile cells. Samples (10 μl) of the pellicle, from the highest dilution, are plated on nutrient agar medium and incubated for 48–72 h. All colony types are then purified. The authors claim that these methods are much simpler than the use of semisolid media; however, many more colonies have to be checked for the presence of nitrogen-fixing ability.

MAINTENANCE PROCEDURES

Stock cultures of *A. brasilense*, *A. lipoferum*, and *A. doebereinerae* may be maintained in semisolid NFb medium at 8–30°C with monthly transfers for *A. brasilense* and biweekly transfers for *A. lipoferum* (Döbereiner, 1992a). Stocks may also be maintained on trypticase soy agar with monthly transfer (Tyler et al., 1979). *A. brasilense* remains viable for several years and *A. lipoferum* for 3–6 months in sterile vermiculite moistened with potato broth; drying should be avoided by tightly sealing the vials with screw caps. A combination of soil and farmyard manure (1:1) can maintain a population of *A. brasilense* at high levels for up to 6 months (Tilak et al., 1979).

Strains can also be kept lyophilized for many years. Cells grown on slanted NFb medium (malic acid is replaced by glucose for *Azospirillum lipoferum* strains) for 48–72 h at 30°C are suspended in 2 ml of a lyophilization solution consisting of 10% sucrose and 5% peptone in 100 ml water. Aliquots (0.2 ml) are then distributed into lyophilization ampules and lyophilization performed as recommended for rhizobia.

For preservation in liquid N_2, heavy suspensions of cells harvested from MPSS broth or MPSS agar plates are prepared in nutrient broth containing 10% (v/v) dimethyl sulfoxide (Tarrand et al., 1978) or in TSS broth (1% trypticase, 0.5% succinate with salts as in NFb medium) in which 15% of the water is replaced by glycerol (Tyler et al., 1979). The suspensions are placed in vials and preserved in liquid nitrogen.

A. amazonense strains may also be maintained in semisolid LGI medium at 30°C with biweekly transfers. Strains from this species have been maintained for more than 15 years in test tubes containing LGI agar medium supplemented with yeast extract and covered with mineral oil. The cultures should never be stored in BMS agar medium even under oil because of the pH sensitivity of the bacteria. Cultures can also be maintained lyophilized after growth for 48–72 h at 30°C in a medium with sucrose as the sole carbon source and following the procedure described above.

A. halopraeferens cells may be maintained at 41°C by biweekly transfer in semi-solid SM medium supplemented with 0.25%

NaCl and adjusted to pH 7.2 (Reinhold et al., 1987). They can also be maintained in liquid nitrogen as described for the other species of *Azospirillum*. *A. irakense* cells are maintained at room temperature on slants of modified AAM agar medium (Khammas et al., 1989) and *A. largimobile* can be maintained on LWA or LWSA agar medium according to the description by Skerman et al. (1983).

Azospirilla can also be maintained for long periods in water. Two colonies grown on the specific agar medium are transferred to micro tubes containing 0.5 ml of distilled sterile water and stored at room temperature. To revive, the cells are homogenized and 20–100 μl are transferred to semi-solid medium and incubated at the appropriate temperature.

PROCEDURES FOR TESTING SPECIAL CHARACTERS

D-Glucose and sucrose used as sole carbon source for growth in semi-solid nitrogen-free medium For D-glucose, a loopful of culture from each species grown in the appropriate semi-solid or liquid medium is inoculated into a tube of semi-solid, nitrogen-free, D-glucose medium (the specific carbon source is replaced by 1.0% D-glucose which has been sterilized by filtration). *A. lipoferum*, *A. amazonense*, and most strains of *A. doebereinerae* form a veil or disc of growth in the depths of the medium; within 3 d at 35–37°C this disc migrates close to the surface of the medium and becomes very dense. *A. brasilense* and *A. halopraeferens* either give no response or form a slight pellicle in the depths of the medium that later disperses. The difference in response between these four species is very pronounced. The other two species, *A. irakense* and *A. largimobile*, also use D-glucose but this characteristic has not been checked under nitrogen-fixing conditions.

For sucrose, a loopful of culture from *A. amazonense* or *A. irakense* grown in liquid medium is inoculated into a vial of semisolid LGI medium containing sucrose as sole carbon source, and the pH is adjusted to 6.0–6.2 for *A. amazonense* and 6.5 for *A. irakense*. *A. amazonense* forms a fine pellicle below the surface very similar to that of the other species in NFb medium except that growth is initially slower; however, a thick surface pellicle is formed after 3–4 d. *A. irakense* forms a white pellicle 2 mm below the surface after 1 d of incubation but reaches the surface within 2–3 d. The other species do not use sucrose as sole carbon source, although some variants of *Azospirillum lipoferum* have been found to use sucrose for growth.

Biotin requirement Glassware should be rinsed copiously with distilled water and subsequently baked in an oven to destroy traces of biotin. A medium of the following composition is used (g/l): K_2HPO_4, 0.5; succinic acid (free acid), 5.0; $FeSO_4 \cdot 7H_2O$, 0.01; $Na_2MoO_4 \cdot 2H_2O$, 0.002; $MgSO_4 \cdot 7H_2O$, 0.2; NaCl, 0.1; $CaCl_2 \cdot 2H_2O$, 0.026; $(NH_4)_2SO_4$, 1.0. The pH is adjusted to 7.0 with KOH solution. Biotin (0.0001 g/l) is added to one portion. The biotin-free and biotin-containing media are sterilized in 5.0-ml amounts in screw-capped tubes by autoclaving. Cultures grown in MPSS broth are inoculated by a loop into 25 ml of one-quarter-strength nutrient broth (Difco) and incubated at 37°C for 24 h. The cells are harvested by centrifugation, washed twice with 10-ml portions of sterile distilled water, and suspended in water to a turbidity of 20 Klett units (blue filter, 16-mm tubes). An aliquot (0.1 ml) of this suspension is used to inoculate each 5.0 ml of medium (with and without biotin). The cultures are incubated for 48 h at the optimal temperature for the species. In cases where growth occurs in the absence of biotin, a second serial transfer is made to media with and without biotin, using 0.1 ml

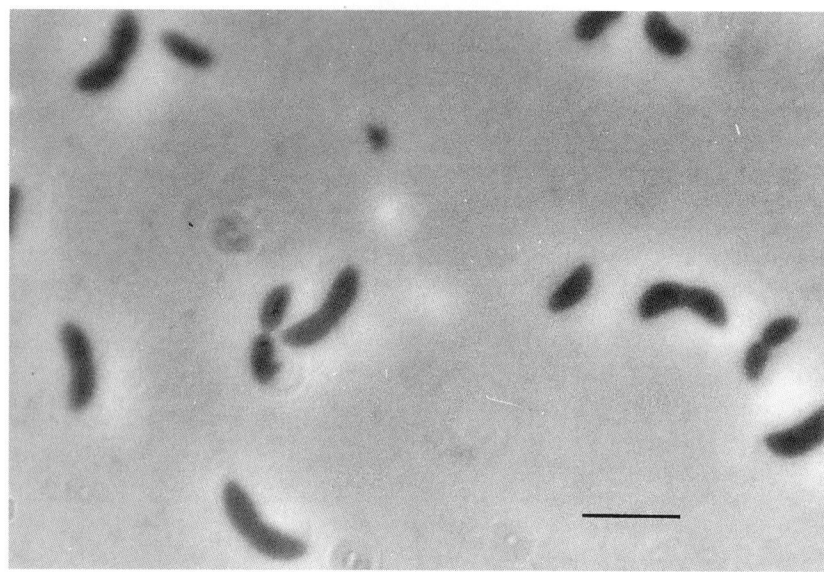

FIGURE BXII.α.10. *Azospirillum brasilense* ATCC 29145 cultured in MPSS broth at 37°C for 24 h. Phase-contrast microscopy. Bar = 5 μm.

from the first culture. This procedure may be repeated at least ten times to make sure that strains are not dependent on biotin.

Development of pleomorphic cells A loopful of growth from a 24 h-old MPSS broth culture is inoculated into a tube of semisolid, nitrogen-free, malate medium (containing 0.05 g/l yeast extract or 0.0001 g/l biotin). The cultures are examined by phase-contrast microscopy after incubation at 37°C for 24–48 h (Fig. BXII.α.10). *A. brasilense* and *A. amazonense* remain mainly vibrioid and motile, whereas *A. lipoferum* becomes wider, longer, nonmotile, and S-shaped or helical. These forms of *A. lipoferum* eventually undergo fragmentation into shorter, ovoid cells, many of which become very large and pleomorphic and are filled with refractile granules. *A. halopraeferens* strains also develop wider, longer and S-shaped cells, which seldom exceed a length of 5 μm. In some cases, very long cells are also observed. Cells of *A. irakense* grown in complete nutrient broth medium are much longer and reach a size of up to 30 μm. *A. largimobile* cells lose motility and form multicellular conglomerates that become optically refractile and reproduce by multi-planar septation. *A. doebereinerae* strains remain motile after growth overnight in liquid medium, but the cells become long and pleomorphic after prolonged growth in semisolid N-free NFb medium.

Auxanographic method for sole carbon sources The following ingredients are dissolved in 50 ml of distilled water passed through a Bantam multibed resin cartridge (Barnstead Co., Boston, MA): $(NH_4)_2SO_4$, 1.0 g; $MgSO_4·7H_2O$, 1.0 g; K_2HPO_4, 2.0 g; $FeCl_3·6H_2O$, 0.0047 g; $MnSO_4·H_2O$, 0.0025 g; $ZnSO_4·7H_2O$, 0.00072 g; $CuSO_4·5H_2O$, 0.000125 g; $CoSO_4·7H_2O$, 0.00014 g; H_3BO_3, 0.000031 g; and $Na_2MoO_4·2H_2O$, 0.000245 g. The pH is adjusted to 2.5 with HCl to dissolve precipitates and then adjusted to 7.0 with KOH. Biotin (0.0001 g) and $CaCO_3$ (0.001 g) are added and the medium is sterilized by autoclaving. The sterile medium is added aseptically to an equal volume of sterile agar solution (15.0 g purified agar (Difco) in 500 ml of distilled water, sterilized by autoclaving) at 45–50°C. Cells are prepared as for the determination of biotin requirement (previous paragraph) except that the final suspension is adjusted to 30 Klett units

instead of 20. Two ml of the cell suspensions are used to spread 20 ml of molten medium at 45–50°C in a Petri dish. After solidification of the medium, sterile 7-mm diameter paper discs (punched from Beckman electrophoresis filter paper, Cat. No. 319328) are dipped into 5% (w/v) aqueous solutions of carbon sources sterilized by filtration. (Solutions of organic acids are adjusted to pH 7.0 with KOH before sterilization.) The saturated discs are then placed near the periphery of the seeded agar plates (3 discs/plate). The plates are incubated at 37°C for 72 h. Any visible zone of turbidity around the discs, as judged by holding the plates against a black background and illuminating them with oblique lighting from the rear, constitutes a positive growth response. New kits for testing these properties are now available on the market (i.e., Biolog, API galleries) and the instructions from the manufacturers should be followed.

Acidification of peptone-based D-glucose medium The following medium is used (g/l): peptone (Difco), 2.0; $MgSO_4·7H_2O$, 1.0; $(NH_4)_2SO_4$, 1.0; $FeCl_3·6H_2O$, 0.002; $MnSO_4·H_2O$, 0.002; bromothymol blue (dissolved in dilute KOH), 0.025. The medium is made up to a volume of 950 ml, adjusted to pH 7.0 and sterilized by autoclaving. After it has cooled, 50 ml of a 20% (w/v) solution of D-glucose (sterilized by filtration) is added aseptically. The development of a yellow color during incubation for 96 h at 37°C indicates acidification.

Fermentative ability Tubes of peptone–D-glucose or D-fructose media, or defined D-glucose or D-fructose broth (Tarrand et al., 1978), are inoculated and placed in GasPak jars (BBL) containing a fresh catalyst. The jars are evacuated, filled once with N_2 and three times with H_2, sealed and incubated at 37°C for up to 2 weeks. Development of a yellow color indicates acidification. Slight growth in the form of a sediment can be detected by agitating the tubes.

Test for autotrophic growth (Málek and Schlegel, 1981) A mineral medium having the following composition (per liter) is used: KH_2PO_4, 2.3 g; $Na_2HPO_4·2H_2O$, 2.9 g; NH_4Cl, 1.0 g; $MgSO_4·7H_2O$, 0.5 g; $CaCl_2·2H_2O$, 0.01 g; ferric ammonium citrate, 0.05 g; and trace element solution[8], 6.0 ml. When necessary, the medium is supplemented with the appropriate growth factors. For testing in liquid culture, 20-ml portions of the medium are distributed into 100- or 250-ml Erlenmeyer flasks. After inoculation, the flasks are placed in desiccator jars or anaerobic jars. The jars are then evacuated and filled with the following gas mixture: $O_2/CO_2/H_2/N_2$ (2:10:60:28). The jars are incubated at 30–37°C with shaking and are periodically evacuated and refilled with fresh gas mixture. For testing growth on agar plates, the mineral medium is solidified with 1.5% agar.

Oligonucleotide probes Probes complementary to a highly variable stretch of helix 55–59 of the 23S rRNA were designed and are available for five species of *Azospirillum* (Kirchhof et al., 1997b). Oligonucleotide probes based on partial 16S rRNA gene sequences have also been designed by Kabir et al. (1995) and were species specific for *A. lipoferum* and *A. amazonense*; another probe recognized both *A. brasilense* and *A. amazonense*. The sequences and dissociation temperatures (°C) of probes used for phylogenetic studies of azospirilla are shown in Table BXII.α.3. Bulk nucleic acids are isolated from strains that were cultivated overnight in nutrient broth or DYGS liquid medium (g/l: dextrose, 2.0; peptone, 1.5; yeast extract, 2.0; K_2HPO_4, 0.5; $MgSO_4$, 0.5; L-glutamic acid 1.5; pH adjusted to 6–6.5), and transferred to a positively charged nylon membrane via spot blotting. Hybridization with radioactive or nonradioactive DIG-labeled probes is performed for 2–12 h (radioactive) or 16 h (nonradioactive) according to the method of Kirchhof et al. (1997a). Signals are detected by autoradiography. Most recently, a comprehensive set of whole-cell-binding 16S rRNA-targeted oligonucleotide probes was developed at genus and species level for all species of the *Azospirillum* cluster (Stoffels et al., 2001). Using fluorescently labeled diagnostic oligonucleotide probes, direct species identification of single cells, as well as *in situ* localization studies, is possible.

DIFFERENTIATION OF THE GENUS *AZOSPIRILLUM* FROM OTHER GENERA

Table BXII.α.4 indicates the characteristics of *Azospirillum* that distinguish it from other diazotrophic bacteria belonging to the *Alphaproteobacteria*.

---

8. Trace element solution SL-6 (Pfennig, 1974) contains (g/l of distilled water): $ZnSO_4·7H_2O$, 0.1; $MnCl_2·4H_2O$, 0.03; H_3BO_3, 0.3; $CoCl_2·5H_2O$, 0.2; $CuCl_2·2H_2O$, 0.01; $NiCl_2·6H_2O$, 0.02; and $Na_2MoO_4·2H_2O$, 0.03.

TAXONOMIC COMMENTS

In 1921–1922, Beijerinck observed extensive development of a spirillum-like bacterium in nitrogen-deficient D-glucose and mannitol solutions that had been inoculated heavily with garden soil or soil from a sand bed. Although the organism grew well at first, it was later displaced by competitive growth of *Azotobacter* and *Clostridium*. However, when calcium malate or lactate was employed as the carbon source instead of the carbohydrates, the organism grew well and was not overgrown by other nitrogen fixers. Beijerinck found that partially purified cultures of the spirillum exhibited increases in nitrogen at the expense of malate, whereas cultures lacking the spirillum failed to show such increases. Pure cultures of the spirillum failed to grow in the absence of a source of fixed nitrogen, and Beijerinck suggested that good growth in partially purified cultures might be attributable to the microaerophilic nature of the spirillum, as suggested by microaerotactic band formation in wet mounts. In general, cells cultured in sugar media were plump, curved rods containing many lipoidal droplets, which sometimes distorted the shape of the cells. On malate or lactate agar, the cells tended to be thinner and straighter, while in dilute bouillon they exhibited a distinct spirillum shape with one or more helical turns. Because of the ease of cultivation on salts of organic acids, as well as the spirillum shape exhibited under certain conditions, Beijerinck considered the organism to be a member of the genus *Spirillum* and to be a bridging organism linking the genus *Spirillum* with the genus *Azotobacter*. He initially named the organism *A. largimobile*, but later renamed it *Spirillum lipoferum* (Beijerinck, 1925). Later studies of *Spirillum lipoferum* by Schröder (1932) also failed to demonstrate N_2 fixation by pure cultures, and the organism was forgotten for many years except for a few scattered reports. However, in 1963 Becking isolated an organism resembling *Spirillum lipoferum* that showed uncontestable nitrogenase activity. Finally, the discovery of the association of such organisms with plant roots (Döbereiner and Day, 1976) caused much interest in *Spirillum lipoferum* and led to detailed taxonomic studies. Tarrand et al. (1978) reserved Beijerinck's specific epithet *lipoferum* for their DNA homology group II (i.e., *A. lipoferum*), because this group seemed to correspond in more ways to Beijerinck's description of *Spirillum lipoferum*, particularly with regard to growth with glucose or mannitol and to formation of spirillum-shaped cells under certain conditions. Since Beijerinck's strains no longer exist, however, their correspondence with homology group II cannot be estimated with certainty.

Analysis of rRNA cistrons indicates that the genus *Azospirillum*

TABLE BXII.α.3. Oligonucleotide sequence probes used for phylogenetic studies of *Azospirillum* species [a]

Name of the probe	Sequence of nucleotides	Species-specific target	TD (°C)[b]
AA 23S rRNA	5′-ACA CCT CCA TGG CAC AC-3′	*A. amazonense*	54
AA/AB 16S rRNA	5′-CGT CCG ATT AGG TAG T-3′	*A. amazonense* and *A. brasilense*	48
AB 23S rRNA	5′-GGG TCC CCA GCC GGG C-3′	*A. brasilense*	60
AH 23S rRNA	5′-TCG CCG CAG CAC GCT-3′	*A. halopraeferens*	52
AI 23S rRNA	5′-GCA TAC TGG TTT TCA G-3′	*A. irakense*	46
AI 16S rRNA	5′-CGT CTG ATT AGG TAG T-3′	*A. irakense*	46
AL 23S rRNA	5′-TAT AAG GCG GGG CTA-3′	*A. lipoferum*	46
AL 16S rRNA	5′-CGT CGG ATT AGG TAG T-3′	*A. lipoferum*	48
Adoeb94	5′-CGT GCG CCA CTG TGC CGA-3′	*A. doebereinerae*	46[c]
Adoeb587	5′-ACT TCC GAC TAA ACA GGC-3′	*A. doebereinerae*	46[c]

[a]Adapted from Kirchhof et al., 1997b.

[b]Temperature of hybridization should be 5°C below the T_D.

[c]It is recommended that formamide (30% v/v) be added to the standard hybridization buffer.

TABLE BXII.α.4. Differential characteristics of the genus *Azospirillum* and other diazotrophic bacteria belonging to the *Alphaproteobacteria*[a]

Characteristic	Azospirillum	Aquaspirillum peregrinum	Rhodospirillum rubrum	Magnetospirillum magnetotacticum	Azorhizobium caulinodans	Xanthobacter autotrophicus	Beijerinckia
Cell shape:							
Vibrioid	+	−	−	−	−	−	−[b]
Helical	−[c]	+	+	+	−	−	−
Cell diameter (μm)	0.6–1.7	0.5–0.7	0.8–1.0	0.2–0.4	0.5–0.6	0.4–0.8	0.5–1.5
Motility	+	+	+	+	+	−[d]	D
Predominant type of flagellation	MP, L	BT	BT	BS	PT, L	±	PT
Nitrogenase activity	+	+	+	+	+	+	+
Nitrogen fixed only under microaerophilic conditions	+	+	+	+	+	+	−
Known to be plant-associated nitrogen fixers	+	−	−	−	+	+	−
Photoautotrophic	−	−	+	−	−	−	−
Root or stem hypertrophies produced	−	−	−	−	+	−	−
Mol% G + C of DNA	64–71	60–64	64–66	65	66–68	65–70	55–61

[a]Symbols: +, typically positive; −, typically negative; ±, differ among strains; D, differs among species; MP, monopolar single flagellum; BT, bipolar tuft; BS, bipolar single flagellum; PT, peritrichous; L, lateral flagella.

[b]*Beijerinckia* may be straight, slightly curved, or pear shaped.

[c]A few helical cells can be observed in *A. halopraeferens*.

[d]Motility depends on the growth conditions (substrate and age).

belongs to the α-subclass of *Proteobacteria* (De Ley, 1992). This subclass, proposed by Woese et al. (1984b), corresponds to the rRNA superfamily IV described by De Ley (1978) and De Smedt et al. (1980). The majority of the nitrogen-fixing bacteria occur as six major groups within the α subclass, with three of them containing most of the known diazotrophic bacteria. The first cluster exhibits 75°C $T_{m(e)}$ and includes the endophytic diazotroph *Gluconacetobacter diazotrophicus* (Yamada et al., 1998a) (formerly *Acetobacter diazotrophicus*) and the first nitrogen-fixing bacterium from the genus *Gluconacetobacter* described (Gillis et al., 1989). The second cluster is formed by *Zymomonas*, *Rhizomonas*, and *Sphingomonas*. A nitrogen-fixing bacterium isolated from rice plants has been identified as *Pseudomonas paucimobilis* (Bally et al., 1983) and is linked to the *Sphingomonas paucimobilis* rRNA branch at $T_{m(e)}$ of 76°C (Gillis and Reinhold-Hurek, 1994). The third cluster comprises the nitrogen-fixing bacteria of the genus *Rhizobium*, *Sinorhizobium*, and *Mesorhizobium*, which are linked at $T_{m(e)}$ of 74.4°C to other non-nitrogen-fixing bacteria of the class. A small cluster is formed by the genera *Rhodobacter* and *Paracoccus* and is linked to the rRNA branch at $T_{m(e)}$ of 67°C. In this cluster the diazotrophic species *Rhodobacter capsulatus* is found. The last two clusters are formed by diazotrophic bacteria that are linked to the main rRNA branch of this class at very similar $T_{m(e)}$ values. The first cluster is formed by *Bradyrhizobium*, *Azorhizobium*, *Beijerinckia*, and *Xanthobacter* in addition to *Rhodopseudomonas*, and is linked to the rRNA branch at 70.4°C $T_{m(e)}$, while the last one— formed by *Azospirillum*, *Rhodospirillum*, *Aquaspirillum peregrinum*, and *Magnetospirillum magnetotacticum*—links at 71°C $T_{m(e)}$. Because the last cluster contains the taxa most closely related to *Azospirillum*, their phenotypic and genetic characteristics should be compared with those of the genus *Azospirillum*.

Aquaspirillum peregrinum exhibits phenotypic characters in common with *Azospirillum*: shows nitrogenase activity under microaerophilic conditions; has cell diameter quite similar to that of some *Azospirillum* species but differs from the latter by having helical cells with flagella in bipolar tufts; a mol% G + C of 60–64; an aquatic habitat rather than an association with plants. A relationship between *Azospirillum* and *Rhodospirillum* is indicated by the 16S rRNA cistron analysis where both genera are linked at 71°C $T_{m(e)}$ within the *Alphaproteobacteria*. Certain phenotypic

similarities are also shared specially by the species *A. lipoferum* and *R. rubrum*. The most striking similarity concerns the activation factor for the Fe protein of nitrogenase. This factor has so far been found only in *Azospirillum* and *Rhodospirillum* (Ludden et al., 1978). It is also required for activation of the Fe protein of *A. brasilense* but not for that of *A. amazonense* (Döbereiner and Pedrosa, 1987). *A. lipoferum* and *R. rubrum* share a growth requirement for biotin. Both *Azospirillum* and *R. rubrum* form a pink or red pigment when grown in the dark under aerobic conditions and form intracellular poly-β-hydroxybutyrate. Despite these similarities, there are some significant differences between the two species. *Azospirillum* is not phototrophic, moreover, *R. rubrum* possesses bipolar tufts of flagella and has cells that are thinner and more helical than those of *Azospirillum*. The mol% G + C of the DNA of *R. rubrum* is 64–66, in contrast to the value of 69–70 for *Azospirillum lipoferum*. The complete 16S rRNA gene sequence of all six *Azospirillum* species has shown that *A. amazonense* and *A. irakense* form a subcluster with *Rhodocista centenaria* (basonym *Rhodospirillum centenum*) (Ben Dekhil et al., 1997a). This *Rhodospirillum* species is also a nitrogen-fixing, nonsulfur, purple phototrophic bacterium, accumulates poly-β-hydroxybutyrate, and has a mol% G + C of the DNA of 68.3, very close to that of *A. amazonense* (67–68); however, it has the ability to form cytoplasmic "R" bodies (Favinger et al., 1989), a characteristic not present in *Azospirillum*.

The species *Magnetospirillum magnetotacticum* (formerly *Aquaspirillum magnetotacticum*), and members of the phototrophic genus *Rhodospirillum* are the nearest relatives of the genus *Azospirillum*, based on 16S rRNA gene sequence analysis (Xia et al., 1994; Ben Dekhil et al., 1997a). Cells from *M. magnetotacticum* are obligately microaerophilic, accumulate poly-β-hydroxybutyrate, fix nitrogen only under microaerophilic conditions, and have a mol% G + C of the DNA of 65 that is in the same range as that of *Azospirillum* (64–71). However, the smaller cell size, type of flagellation, and formation of coccoid bodies 3–4 weeks after growth, as well as the occurrence of magnetite-containing "magnetosomes", excludes it from the genus *Azospirillum*.

The nitrogen fixer *Azorhizobium caulinodans* resembles *Azospirillum* in being a free-living, obligately aerobic, small rod that fixes N_2 microaerobically, as well as being a plant-associated diazotroph. However, the host–bacterium interaction (stem nodule formation) is far more highly specialized than that of *Azospirillum* for symbiotic nitrogen fixation. In addition, *Azorhizobium caulinodans* forms a main cluster with *Xanthobacter autotrophicus*, *Beijerinckia*, and *Bradyrhizobium* which links at 70.4°C $T_{m(e)}$ to the root of the *Alphaproteobacteria*; *Azospirillum* links at 71°C $T_{m(e)}$.

Xanthobacter autotrophicus resembles some species of *Azospirillum* in being a microaerophilic nitrogen fixer, having an association with rice plants, and having a mol% G + C of the DNA (65–70) that is similar to that of *Azospirillum*. However, the cells differ by being smaller (and under certain conditions exhibiting branching), nonmotile, and forming a yellow pigment.

The genus *Beijerinckia* lies within the same *Alphaproteobacteria* group and forms a subcluster together with *Bradyrhizobium*, *Azorhizobium*, and *Xanthobacter* which links to the main RNA branch at 71°C $T_{m(e)}$, according to the rRNA cistron dendrogram for this class (De Ley, 1992). Although *Beijerinckia* cells are straight or slightly curved rods with a diameter similar to that of *Azospirillum*, other characteristics such as the ability to fix nitrogen aerobically, occurrence as a free-living bacterium in soil, and low mol% G + C values (55–61) prevent the inclusion of this genus in *Azospirillum*. Nevertheless, the *Azospirillum largimobile* conglomerate cells show a striking resemblance to cells of *Beijerinckia* species.

By DNA–DNA hybridization, *A. brasilense* strains exhibit a continuum of similarity values with the type strain, ranging from ~70 to 100%. Within the species *A. lipoferum*, strains exhibit similarity values of 70–76% with the type strain. In the case of *A. amazonense*, similarity values range from 56 to 100% with the type strain. A DNA–DNA homology value of 100% was observed for hybridization of one strain of *A. halopraeferens* with the type strain. Within the species *A. irakense*, strains exhibited similarity values of 71–97% with the type strain. For *A. largimobile*, the similarity value was 65% (one isolate) with the type strain.

From DNA–DNA hybridization studies using a membrane-filter competition method, *A. lipoferum* and *A. brasilense* initially were reported to exhibit similarity values of 31–34% (Tarrand et al., 1978). Use of the S1 nuclease method, however, indicated that *A. lipoferum* ATCC 29707 and *A. brasilense* ATCC 29145 exhibited lower similarity values (14–17%) to each other (Falk et al., 1985). A much lower degree of similarity (2–6%) was observed when both species were hybridized against *Azospirillum amazonense* ATCC 35119 (Falk et al., 1985). In the case of *Azospirillum halopraeferens* (LMG 7108), a similarity value of <25% was observed between the three species (Reinhold et al., 1987). A low degree of similarity (4–11%) was also exhibited by the type strain of *Azospirillum irakense* (CIP 103311) to the four above-mentioned species (Khammas et al., 1989). The recently described *Azospirillum largimobile*—formerly *Conglomeromonas largomobilis*—exhibited a very low degree of similarity to *A. amazonense* (5–7%) and *A. brasilense* (13–14%) (Falk et al., 1986); however, it showed a high level of similarity (40–47%) to *Azospirillum lipoferum*. Because of this similarity and other morphological characteristics, Falk et al. (1986) considered strains from this group as belonging to *Azospirillum lipoferum*. Based on the binary sequence similarity values of 16S rRNA genes, Ben Dekhil et al. (1997a) demonstrated that this new species shared 97.1% similarity with *A. lipoferum* but much less with *A. brasilense* (95.2%) and *A. amazonense* (91.7%). Although this difference is small, the authors proposed *A. largimobile* as a new species based on the DNA hybridization data.

An analysis based on the similarity values of the 16S rRNA genes of *Azospirillum* species versus the type species *A. lipoferum* indicates the occurrence of two subgroups within the genus. One includes *A. lipoferum*, *A. brasilense*, *A. halopraeferens*, *A. doebereinerae*, and *A. largimobile*; the other includes *A. amazonense* and *A. irakense* (Ben Dekhil et al., 1997a). In addition, this study showed that *A. brasilense* is most closely related to *A. lipoferum*, followed by *A. halopraeferens* and *A. largimobile*. Fani et al. (1995b) also have observed that *A. lipoferum* and *A. brasilense* are very closely related species, whereas *A. amazonense* was the most divergent. Xia et al. (1994) showed that all type strains of *Azospirillum* formed a deep-branching clade within the *Alphaproteobacteria* when the 16S rDNA sequences were compared. The most recent comprehensive 16S rDNA sequence study of Stoffels et al. (2001) also confirmed a high degree of relatedness among *Azospirillum* species and that they form a cohesive phylogenetic cluster within the *Alphaproteobacteria*. A strain-specific molecular identification of *Azospirillum* strains is possible using RFLP analysis of the whole genome using rarely cutting restriction nucleases and pulsed-field gel electrophoresis (Gündisch et al., 1993).

ACKNOWLEDGMENTS

This chapter is dedicated to the memory of the late Dr. Johanna Döbereiner, the Brazilian soil microbiologist, who, during a long and distinguished career, identified and characterized many novel nitrogen-fixing bacteria.

FURTHER READING

De Ley, J. 1992. The *Proteobacteria*: ribosomal RNA cistron similarities and bacterial taxonomy. *In* Balows, Trüper, Dworkin, Harder and Schleifer (Editors), The Prokaryotes. A Handbook on the Biology of Bacteria: Ecophysiology, Isolation, Identification, Applications., 2nd Ed., Vol. 2, Springer-Verlag, New York. pp. 2111–2140.

Döbereiner, J. 1992. The genera *Azospirillum* and *Herbaspirillum*. *In* Balows, Trüper, Dworkin, Harder and Schleifer (Editors), The Prokaryotes. A Handbook on the Biology of Bacteria: Ecophysiology, Isolation, Identification, Applications, 2nd Ed., Vol. 3, Springer-Verlag, New York. pp. 2236–2253.

Döbereiner, J. 1992. History and new perspective of diazotrophs in association with non-leguminous plants. Symbiosis *13*: 1–13.

Döbereiner, J. and F.O. Pedrosa. 1987. Nitrogen-Fixing Bacteria in Non-Leguminous Crop Plants. Brock/Springer Series in Contemporary Bioscience, Springer-Verlag, New York.

Gillis, M. and B. Reinhold-Hurek. 1994. Taxonomy of *Azospirillum*. *In* Okon (Editor), *Azospirillum*/Plant Associations, CRC Press, Boca Raton. pp. 1–14.

Tarrand, J.J., N.R. Krieg and J. Döbereiner. 1978. A taxonomic study of the *Spirillum lipoferum* group, with descriptions of a new genus, *Azospirillum* gen. nov. and two species, *Azospirillum lipoferum* (Beijerinck) comb. nov. and *Azospirillum brasilense* sp. nov. Can. J. Microbiol. *24*: 967–980.

DIFFERENTIATION OF THE SPECIES OF THE GENUS *AZOSPIRILLUM*

The differential characteristics of the species of *Azospirillum* are indicated in Table BXII.α.5. Other characteristics of the species are presented in Table BXII.α.6.

TABLE BXII.α.5. Characteristics differentiating the species of the genus *Azospirillum*[a,b]

Characteristics	A. lipoferum	A. amazonense	A. brasilense	A. doebereinerae	A. halopraeferens	A. irakense	A. largimobile
Cell width, μm	1.0–1.7	0.8–1.0	1.0–1.2	1.0–1.5	0.7–1.4	0.6–0.9	0.7–1.5
Enlarged, pleomorphic cells develop in alkaline media	+	−	−	+	+	+	+
Growth with 3% NaCl	d	−	−	−	+	+	−
Biotin requirement	+	−	−	−	+	−	+
Pectin hydrolysis	−	−	−	nt	−	+	−
Optimal growth temperature, °C	37	35	37	30	41	33	28
Carbon sources:							
Glucose	+	+	−	d	−	+	+
Mannitol	+	−	−	+	+	−	+
Glycerol	+	−	+	+	+	−	+
Sucrose	−	+	−	−	−	+	−
Hybridization with probes: [c,d]							
AZO 23S rRNA	+	+	+	na	+	+	na
AL 23S/16S rRNA	+	−	−	na	−	−	na
AB 23S rRNA	−	−	+	na	−	−	na
AA 23S rRNA	−	+	−	na	−	−	na
AA/AB 16S rRNA	−	+	+	na	−	−	na
AH 23S rRNA	−	−	−	na	+	−	na
AI 23S/16S rRNA	−	−	−	na	−	+	na
Adoeb94	−	−	−	+	−	+	na
Adoeb587	−	−	−	+	−	−	−

[a]Symbols: +, positive; −, negative; d, differs among strains; na, probes not available; nt, not tested.

[b]Data from Magalhães et al. (1983); Krieg and Döbereiner (1984); Reinhold et al. (1987); Khammas et al. (1989); and Ben Dekhil et al. (1997a).

[c]Data from Kirchhof et al. (1997b).

[d]Data from Eckert et al. (2001).

List of species of the genus Azospirillum

1. **Azospirillum lipoferum** (Beijerinck 1925) Tarrand, Krieg and Döbereiner 1979, 79[AL] (Effective publication: Tarrand, Krieg and Döbereiner 1978, 978 (*Spirillum lipoferum* Beijerinck 1925, 353.)

   li.po′ fe.rum. Gr. n. *lipus* fat; L. v. *fero* to carry; M.L. adj. *lipoferus* fat bearing.

   The characteristics are as described in Tables BXII.α.4, BXII.α.5, and BXII.α.6. The morphological characteristics are depicted in Figs. BXII.α.4 and BXII.α.5. The characteristic pleomorphism seen in malate medium does not occur in D-glucose medium. Cells of certain strains become much thinner and very long in malate medium. Use of D-glucose as the sole carbon source for nitrogen-fixation-dependent growth is more efficient at pH 5.5–6.5 than at pH 7.0–7.5.

   All strains stain uniformly Gram negative when cultured in MPSS broth. Most strains stain uniformly Gram negative when cultured for 48–72 h on MPSS agar but a few strains exhibit a small proportion of cells that are resistant to Gram decolorization.

   Colonies are not slimy on BMS agar. When 0.5% D-glucose is added to the medium, some strains form large, slimy, white colonies. Star-shaped colonies have also been observed for some strains grown on BMS agar plates (Fig. BXII.α.11).

   Aerobic growth in liquid media containing a source of fixed nitrogen often shows extensive clumping, particularly in defined media. No growth occurs anaerobically in the absence of a source of fixed nitrogen. However, nitrogen fixation can be observed in liquid medium containing glu-

   tamate when the cells reach an optical density of approximately 1.0 and all available nitrogen is consumed.

   Strains of *A. lipoferum* that are able to grow lithoautotrophically with H_2 contain an uptake hydrogenase as well as RubPcase. They can be categorized as aerobic hydrogen-oxidizing bacteria.

   Physiological and nutritional characteristics of the species are presented in Tables BXII.α.5 and BXII.α.6. Variants with decreased fermentative ability may arise but continue to require biotin, use D-glucose as a sole carbon source for N_2-dependent growth, and exhibit the characteristic pleomorphism associated with this species in malate medium.

   Strains of this species have been found colonizing several plants including cereals, forage grasses, vegetables, legumes, and the fruits of banana and pineapple plants.

   The mol% G + C of the DNA is: 69–70 (T_m).

   Type strain: BR11080, Sp 59b, ATCC 29707, DSM 1691.

   GenBank accession number (16S rRNA): M59061.

   Additional Remarks: Reference strains: ATCC 29708 (BR11115; Sp Rg 20a), ATCC 29709 (BR11084; Sp Br 17), ATCC 29731 (BR11087; Sp Rg 6xx). The accession numbers of the 16S rDNA sequences from reference strains ATCC 29708 and ATCC 29731 deposited at the EMBL are X79729 and X79730, respectively (Fani et al., 1995a). Other 16S sequences from WO3 and NCIMB 11861 have received the accession numbers X79741 (Fani et al., 1995a) and Z29619 (Xia et al., 1994), respectively.

2. **Azospirillum amazonense** Magalhães, Baldani, Souto, Kuykendall and Döbereiner 1984, 355[VP] (Effective publication: Magalhães, Baldani, Souto, Kuykendall and Döbereiner 1983, 417.)

TABLE BXII.α.6. Other characteristics of the species of the genus *Azospirillum*[a,b]

Characteristics	*A. lipoferum*	*A. brasilense*	*A. amazonense*	*A. halopraeferens*	*A. irakense*	*A. largimobile*	*A. doebereinerae*
Flagella arrangement	MP, L	MP, L	MP	MP	MP, L	MP, L	MP, L
Colony type on:							
BMS agar medium	pink, raised, curled	pink, raised, curled	white, flat raised margin	ng	nd	nd	nd
Congo red medium	scarlet	scarlet	nd	ng	nd	scarlet	scarlet
Dissimilation of:							
NO_3^- to NO_2^-	+	+	d	+	d	+	+
NO_2^- to N_2O	d	d	−	+	−	−	+
NO_3^--dependent anaerobic growth	+	+	−	+	−	−	+
pH range for growth	5.7–6.8	6.0–7.8	5.7–6.5	6.8–8.0	5.5–8.5	nd	6.0–7.0
Acid from:							
Glucose	+	−	−	−	−	+	d
Fructose	+	−	−	+	−	+	+
Acidification of: peptone-based glucose broth	+	−	−	−	−	nd	nd
Sole carbon sources (auxanographic method):							
Succinate, malate, lactate, fumarate	+	+	+	+	+	nd	+
Gluconate	+	+	−	+	−	−	+
Citrate	+	+	−	v	−	+	−
α-Ketoglutarate	+	−	−	nd	v	nd	−
myo-Inositol	v	+	+	−	−	−	+
D-Sorbitol	+	−	−	−	−	+	+
D-Ribose	+	−	+	+	v	+	−
L-Rhamnose	+	v	+	v	+	−	−
Fructose	+	+	+	+	v	+	+
Maltose	−	−	+	nd	+	−	−
Lactose	−	−	v	nd	+	−	−
D-Mannose	v	−	+	+	+	−	−
Habitat:							
Soil and tissues mainly of nonlegumes	+	+	+				
Roots of *Leptochoa fusca* (L.) grown in saline-sodic soil, Pakistan				+			
Soil and roots of rice plants grown in Iraq					+		
Freshwater lake, Australia						+	
Soil and roots of Miscanthus, Germany							+
% similarity in 16S rRNA vs. *A. lipoferum*[c,d]	100	98.1	94	97.2	94	97.1	96.6
Mol% G + C of DNA	69–70	70–71	67–68	69–70	64–67	70	69.6–70.7

[a]Symbols: see standard definitions; nd, not determined; ng, no growth.

[b]Data from Magalhães et al. (1983); Krieg and Döbereiner (1984); Reinhold et al. (1987); Khammas et al. (1989); and Ben Dekhil et al. (1997a).

[c]Based on data from Ben Dekhil et al. (1997a).

[d]Based on data from Eckert et al. (2001).

am.a.zo.nen′se. M.L. adj. *amazonense* pertaining to the Amazon region of Brazil, South America.

The characteristics are as described in Tables BXII.α.4, BXII.α.5, and BXII.α.6, with the following additional features. Colonies on potato agar medium (BMS) containing malate plus sucrose as sole carbon source are 0.5 cm in diameter and are flat with elevated borders. The colonies become bigger and flat with well-delineated borders when both carbon sources are replaced by D-glucose (Fig. BXII.α.12) and become smaller, wet, and smooth when sucrose is omitted from the BMS medium (Fig. BXII.α.12), indicating the sensitivity of this species to alkaline conditions. This is also the reason why only a scant pellicle formation is observed in semi-solid NFb medium containing malate but not with sucrose as carbon source. The structure of the colonies on BMS plates tends to disappear after a long period of incubation. Very dry, white colonies with well-delineated borders are also observed on BMS plates containing sucrose as sole carbon source. Cells remain vibrioid in semi-solid LGI medium with spinning movement. However, long periods of incubation in semi-solid LGI medium allow formation of conglomerate cells similar to cysts observed for *A. lipoferum*. This may be due to the dryness of the medium.

Significant amounts of PHB granules are observed under N_2 growth conditions but much less in the presence of NH_4^+. Nitrogenase activity is more sensitive than in *A. lipoferum* and *A. brasilense*, and *A. amazonense* has a doubling time of 10 hours, in contrast to 5–6 h for *A. lipoferum* and *A. brasilense* at low pO_2.

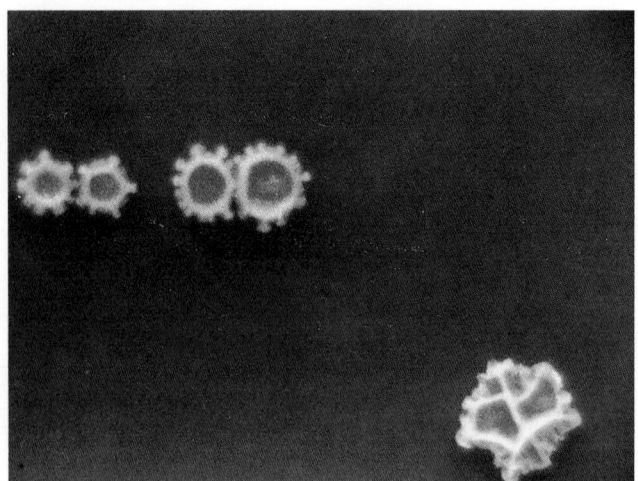

FIGURE BXII.α.11. Colonies of *Azospirillum lipoferum* grown on BMS agar plate for 48 h at 30°C showing starlike forms.

Good aerobic growth occurs in liquid LGI medium containing KNO₃ as nitrogen source but not with NH₄Cl. There is no clumping and the pH remains around 6.0. No growth occurs aerobically in the absence of a source of fixed nitrogen, as observed for the other species. However, nitrogen fixation can be observed in liquid medium containing glutamate when the cells reach optical density around 1.0 and all available nitrogen is consumed.

Fatty acid analysis showed higher amounts of $C_{16:0}$ as compared to the $C_{18:1}$ nonhydroxy fatty acids. This species has been isolated from roots of maize, sorghum, rice and wheat plants, as well as forage grasses grown around Brazil. It was also isolated from roots and stems of sugarcane plants grown in Brazil and from sugarcane roots grown in Hawaii

and Thailand. It was also found associated with roots of palm trees from the Amazon region and more recently from pineapple and banana fruits. The root interior of rice, sorghum, and maize was predominantly colonized by *A. amazonense* strains that do not accumulate nitrite rather than those that produce considerable amounts of nitrite, which afterwards is reassimilated.

The mol% G + C of the DNA is: 67–68 (T_m).

Type strain: BR 11142, Am14, Y1, ATCC 35 119, DSM 2787.

GenBank accession number (16S rRNA): Z29616, X79735.

Additional Remarks: Reference strains: ATCC 35120 (BR 11140; Am18; Y2); ATCC 35121 (BR 11141; Am30; Y6). The accession number for the reference strain ATCC 35120 is X79742 (Fani et al., 1995a).

3. **Azospirillum brasilense** Tarrand, Krieg and Döbereiner 1979, 79[AL] (Effective publication: Tarrand, Krieg and Döbereiner 1978, 979.)

bra.si.len'.se. M.L. adj. *brasilense* pertaining to the country of Brazil, South America.

The characteristics are as described for the genus and as given in Tables BXII.α.4, BXII.α.5, and BXII.α.6, with the following additional features. In semisolid, nitrogen-free, malate medium cells remain mainly vibrioid even when the cultures become alkaline, in contrast to the cells of *A. lipoferum*, *A. irakense*, *A. halopraeferens*, and *A. largimobile*. Some encapsulated (C) forms (Fig. BXII.α.6) may occur, especially in older cultures. C forms occur in abundance in colonies grown on the surface of nitrogen-free media, in association with plant callus cultures, in grass seedlings inoculated with this species, or under stress conditions. Flocculated cells of *A. brasilense* strain Cd grown under conditions of stress accumulate up to 60–65% of total dry weight as PHB.

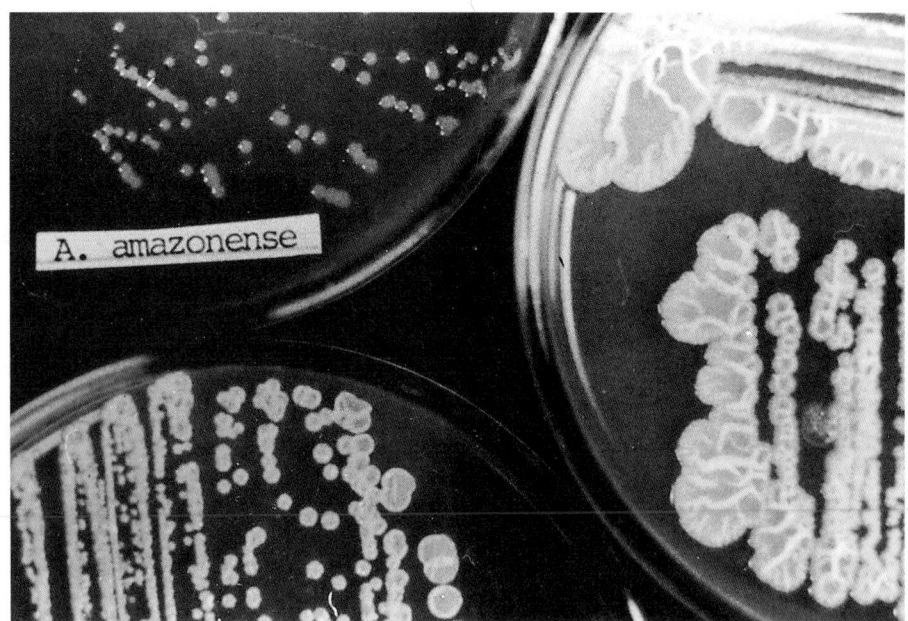

FIGURE BXII.α.12. Colonies of *Azospirillum amazonense* strain ATCC 35119 grown on BMS agar plates containing only malate (*upper left*), malate plus sucrose (*lower left*), or only glucose (*right*) as carbon source for 48 h at 30°C. (Reproduced with permission from J. Döbereiner et al., Brasilia: Embrapa-SPI: Itaguai, RJ: Embrapa-CNPAB, 1995.)

All strains stain uniformly Gram negative when cultured in MPSS broth. Most strains exhibit a small proportion of cells showing resistance to Gram decolorization when cultured on MPSS agar for 48–72 h; this Gram variability may be related to the occurrence of encapsulated forms.

Certain strains and variants of *A. brasilense* form colonies that have a much deeper shade of pink than is usually the case. Carotenoid content has been identified as responsible for the observed characteristic in strain Cd.

Aerobic growth in liquid media containing a source of fixed nitrogen is usually homogeneous and turbid, without clumping. No growth occurs aerobically in the absence of a source of fixed nitrogen. Cells are able to fix nitrogen in liquid aerated medium when pre-grown in the presence of glutamate, up to an optical density around 1.0, and no inorganic nitrogen is available. Strains of this species have been found colonizing several plants including cereals, forage grasses, vegetables, legumes, and banana plants.

The mol% G + C of the DNA is: 70–71 (T_m).

Type strain: BR11001, Sp 7, ATCC 29145, DSM 1690.

GenBank accession number (16S rRNA): X79739.

Additional Remarks: Reference strains: ATCC 29710 (BR11002; Cd); ATCC 29711 (Sp 35). The accession number is Z29617 (Xia et al., 1994) for strain NCIMB 11860.

4. **Azospirillum doebereinerae** Eckert, Weber, Kirchhof, Halbritter, Stoffels and Hartmann 2001, 24^{VP}

doeb' er.ein.er.ae. N.L. gen. fem. n. *doebereinerae* of Döbereiner, in honor of Johanna Döbereiner, who isolated and characterized many *Azospirillum* spp., and other diazotrophic plant-associated bacteria.

The characteristics are as described for the genus and as given in Tables BXII.α.4, BXII.α.5, and BXII.α.6, with the following additional features. Cells are Gram negative, oxidase, catalase, and urease positive, curved rods or S shaped. Long cells are observed, especially in alkaline, semisolid, NFb medium and motile with a winding or snake-like movement. Optimal growth and nitrogen fixation occur in semi-solid NFb medium at 30°C and pH adjusted to 6.5. Production of poly-β-hydroxybutyrate is also observed. No growth or nitrogen fixation occurs in liquid NFb medium. So far, it has been isolated only from the root of *Miscanthus sinensis* cv. "Giganteus" and *Miscanthus sacchariflorus* and also in the rhizosphere soil of these plants grown in Freising, Germany.

The mol% G + C of the DNA is: 70.7 (T_m).

Type strain: GSF71, DSM 13131.

GenBank accession number (16S rRNA): AJ238567.

5. **Azospirillum halopraeferens** Reinhold, Hurek, Fendrik, Pot, Gillis, Kersters, Thielmans and De Ley 1987, 48^{VP}

ha.lo.prae' fe.rens. Gr. n. *hals* salt, the sea L. v. *praeferre* to prefer; M.L. part. adj. *halopraeferens* salt preferring.

The characteristics are as described for the genus and as given in Tables BXII.α.4, BXII.α.5, and BXII.α.6, with the following additional features. Cells are Gram negative, oxidase, and urease positive, vibrioid to S shaped. A few long, helical cells are observed in older, more alkaline medium. Cells are motile in liquid media with rapid corkscrew-like motion. Optimal growth and nitrogen fixation occur in semi-solid SM medium containing 0.25% NaCl at 41°C and pH adjusted to 7.2. Strains are able to grow with 3% NaCl but no nitrogenase activity is observed. No colonies

are formed in BMS or Congo Red medium. Good growth is observed in tryptic soy agar where the colonies are cream-colored, circular and flat with an entire margin.

Production of poly-β-hydroxybutyrate is also observed. No growth or nitrogen fixation occurs in liquid SM medium.

Strains of this species show higher osmotolerance than *A. lipoferum, A. brasilense, A. amazonense, A. doebereinerae,* and *A. largimobile.* Growth and nitrogen fixation is stimulated by glycine betaine, as found in *A. brasilense* but not in *A. amazonense* or *A. lipoferum.*

This species seems to have a very restricted habitat. So far, it has been isolated only from the root surface of Kallar grass (*Leptochloa fusca*) grown in saline-sodic soils in Punjab, Pakistan. Attempts to isolate this species from various plants grown in saline-affected soils around Brazil have been unsuccessful.

The mol% G + C of the DNA is: 69–70 (T_m).

Type strain: Au 4, DSM 3675, LMG 7108.

GenBank accession number (16S rRNA): Z29618, X79731.

6. **Azospirillum irakense** Khammas, Ageron, Grimont and Kaiser 1991, 580^{VP} (Effective publication: Khammas, Ageron, Grimont and Kaiser 1989, 688.)

i.ra.ken' se. L. neut. adj. *irakense* pertaining to the country of Iraq.

The characteristics are as described for the genus and as given in Tables BXII.α.4, BXII.α.5, and BXII.α.6, with the following additional features. The morphological characteristics of the cells in nutrient broth are shown in Fig. BXII.α.7. The cells are Gram-negative, oxidase and catalase positive, curved, vibrioid to S-shaped rods and can reach a length of 30 μm when grown in nutrient broth. Cells are motile with winding or snake-like movements by a single polar flagellum. Colonies are translucent, glistening, convex with regular margin, 1 mm in diameter after 4 d on NFb medium containing yeast extract (20 mg/l) or on AAM agar medium. On tryptic soy agar, colonies are translucent with an opaque center, convex with regular margin, and 3 mm in diameter after 4 d.

Nitrogen fixation occurs only under microaerobic conditions and cells grow well under air atmosphere in the presence of a nitrogen source such as an ammonium salt. Intracellular granules (PHB) are observed.

The physiological and nutritional characteristics are presented in Tables BXII.α.5 and BXII.α.6. Growth occurs in the presence of up to 3% NaCl, but acetylene reduction is optimal when the NaCl concentration is less than 0.01%. All strains hydrolyze esculin and pectin but not hydroxyquinoline-β-glucuronide. Pectin can support growth and nitrogen fixation could be detected in modified, semisolid, AAM medium after 5–11 days of incubation provided that a small amount of fixed nitrogen as starter is present (Khammas and Kaiser, 1991). The species so far has only been isolated from rhizosphere soil and roots of rice plants grown in the region of Diwaniyah in Iraq.

The mol% G + C of the DNA is: 64–67 (T_m).

Type strain: KBC1, CIP 103311, DSM 11586.

GenBank accession number (16S rRNA): Z29583, X 79737.

7. **Azospirillum largimobile** (Skerman, Sly and Williamson 1983) Ben Dekhil, Cahill, Stackebrandt and Sly 1997b, 915^{VP} (Effective publication: Ben Dekhil, Cahill, Stacke-

brandt and Sly 1997a, 74) (*Conglomeromonas largomobilis* subsp. *largomobilis* Skerman, Sly and Williamson 1983, 300.) *lar.gi.mo' bi.le.* L. adj. *largus* a very slow manner (musical); L. v. *mobile* to move; M.L. adj. *largimobile* moving in a very slow manner.

The characteristics are as described for the genus and as given in Tables BXII.α.4, BXII.α.5, and BXII.α.6, with the following additional features. Strains of this species are Gram negative, oxidase, urease, and aminopeptidase positive, and are weakly positive for catalase and phosphatase. Strains exhibit two phases of growth—unicellular and multicellular. Cells are rod shaped with round or tapered ends and a straight or slightly curved axis in the unicellular phase. Motile cells have mixed flagellation. Multicellular conglomerates are formed, with the cells losing motility under unfavorable conditions. The conglomerate dissociates into single motile cells under suitable conditions, forming clear colonies on agar medium. Colonies on lake water agar become buff colored, opaque, low convex, and round with an entire edge and smooth surface after 72 h at 28°C. No encapsulated cells are observed.

N_2 fixation and acetylene reduction were observed only under microaerophilic conditions in semi-solid, nitrogen-free, malate medium with biotin. No nitrogen fixation was detected under aerobic conditions in the presence or absence of a nitrogen source such as ammonium salt.

Other physiological and nutritional characteristics are shown in Tables BXII.α.5 and BXII.α.6. Most of these characteristics of *A. largimobile* are similar to those of *A. lipoferum*, except for acid production from salicin and D-xylose, which is only present in the latter. In addition, *A. largimobile* is DNase positive and shows clearing of egg yolk agar. Strains are able to grow (slowly) in the presence of 2% NaCl. This species is composed of only two strains isolated from fresh lake water in Australia.

The mol% G + C of the DNA is: 69.6–70.8 (T_m).

Type strain: ACM 2041, DSM 9441, UQM 2041.

GenBank accession number (16S rRNA): X90759.

Other Organisms

A noncellulolytic bacterium (strain Mc-2s) having the characteristics of the genus *Azospirillum* was isolated from cellulolytic, nitrogen-fixing, mixed cultures by Wong et al. (1980a). The strain exhibited a biotin requirement, was pleomorphic in nitrogen-deficient malate medium, and could utilize mannitol and α-ketoglutarate. Although these are characteristics of *A. lipoferum*, the strain failed to use D-glucose as a sole carbon source for N_2-dependent growth and did not acidify D-glucose or ribose media. The strain may be a variant of *A. lipoferum*. In another study, Franche and Elmerich (1981) found *Azospirillum* strains K67 and KR77 to be capable of glucose utilization, yet they had no biotin requirement; they were provisionally classified as *A. brasilense*. A third strain, SpN, appeared to be an *A. lipoferum* strain but seemed to require other growth factors in addition to biotin.

Nur et al. (1980) isolated microaerophilic, nitrogen-fixing, vibrioid bacteria having certain characteristics in common with *Azospirillum*. The organisms were associated with grass roots, were vibrioid cells with a single polar flagellum, and could use malate, lactate, arabinose, and galactose, but not mannitol, as sole carbon sources; one isolate could use glucose. Biotin was not required. These strains differed from *Azospirillum* by having a smaller cell diameter (0.5–0.6 μm), by having no lateral flagella, and by forming yellow-pigmented colonies. Their taxonomic placement is uncertain.

Two strains of *Azospirillum* (Am-53 and A1-3) were isolated from soil and plant litter in Göttingen, Germany, as nitrogen-fixing, hydrogen-oxidizing bacteria. These were Gram negative but showed Gram-positive granules and contained lipid droplets. Both had polar flagella and exhibited an active spirillum-like motility. Strain Am-53 (DSM 1727) required biotin, was pleomorphic in nitrogen-deficient malate medium, and in several other characters resembled *A. lipoferum*. Like *A. lipoferum*, it could grow chemolithotrophically on H_2 (Málek and Schlegel, 1981). However, unlike *A. lipoferum*, it could not utilize glucose, malonate, mannitol, or α-ketoglutarate as a sole source of carbon.

Strain Al-3 (DSM 1726) was shorter than *A. lipoferum*. The cells were pleomorphic, but in the presence of 2–3% (w/v) NaCl the cells developed an active spinning motility (around the axis). Morphologically and in several other respects, the strain resembled *A. brasilense*; however, it could grow chemolithotrophically on H_2, could utilize glucose and α-ketoglutarate as a sole carbon source, and could produce acid from glucose aerobically, like *A. lipoferum*. A recent 16S rRNA gene sequence analysis showed that both of these *Azospirillum*-like strains were misclassified. The strain DSM 1726 is closely related to *Agrobacterium tumefaciens*, while DSM 1727 is related to bacteria within the *Gammaproteobacteria* (Xia et al., 1994).

A variant of *Azospirillum halopraeferens* was isolated from roots of rice plants grown in soil with pH 8.0 in India (Balasubramanian and Prabhu, 1989). This isolate grows and fixes nitrogen in semi-solid NFb medium, is highly motile and vibrioid. However, it could grow and fix nitrogen in the presence of 5% NaCl, utilize glucose and sucrose, and acidify peptone-based glucose broth. The ability to use D-glucose and sucrose as carbon sources are characteristics of *Azospirillum irakense*, a diazotroph isolated from rice plants grown in Iraq. Therefore, it seems that the Indian isolate may be more closely related to *A. irakense* than to *A. halopraeferens*.

Two groups of strains showing many morphological and physiological characteristics similar to *A. lipoferum* species were isolated from soil and rice plants grown in North Vietnam (Ngoc Dung et al., 1995). One group (Arm 2-2, GL 1-1) could use D-glucose, sucrose, and α-ketoglutarate, but not mannitol. The other group (strain DA 10-1) also utilized glucose and sucrose, but not mannitol or α-ketoglutarate. Neither group required biotin but formed acid from glucose and scarlet colonies on BMS medium. Additionally, neither group showed positive signals when hybridized with *Azospirillum* probes, and they may represent a new species.

Genus III. **Levispirillum** gen. nov.

BRUNO POT AND MONIQUE GILLIS

Le.vi.spi.ril' lum. L. adj. *laevus* left; M.L. dim. neut. n. *spirillum* small spiral; *Levispirillum* small spiral-shaped bacteria with left-turning helix.

Gram-negative, **rigid, helical cells 0.4–0.7 μm in diameter and 2–22 μm in length.** Helix is left-handed, determined by focusing on the bottom of the cells: the pattern \\\\ indicates a counterclockwise (left-handed) helix. **A polar membrane is present under the cytoplasmic membrane.** Intracellular polyhydroxybutyric acid and coccoid bodies are usually formed. **Bipolar tufts of flagella.** Chemoorganotrophic with an oxidative type of metabolism, using oxygen as a terminal electron acceptor. One species grows anaerobically with nitrate as electron acceptor. Acid is formed from fructose aerobically and anaerobically. Oxidase, catalase, and phosphatase positive; sulfatase negative. Indole negative and esculin positive. Gelatin is not liquefied in 4 d at 30°C or in 7 d at 20°C. One species can liquefy gelatin at 20°C after 28 d. Growth factors not required. Nonphototrophic. No growth on casein, starch, and hippurate. Nitrogenase activity can occur in anaerobic conditions in the presence of ammonium sulfate. No growth at 3% NaCl. Growth on 1% bile and on EMB agar. L-Glutamate, L-asparagine, L-glutamine, and L-proline used as sole C-source; caproate, isocitrate, D-glucose, D-xylose, L-arabinose, and L-tyrosine, L-methionine, L-serine, L-tryptophan, L-isoleucine, and glycine cannot be used as sole C-source. Isolated from pond water and from putrid infusions of freshwater mussels.

The mol% G + C of the DNA is: 60–64.

Type species: **Levispirillum itersonii** comb. nov. (*Aquaspirillum itersonii* (Giesberger 1936) Hylemon, Wells, Krieg and Jannasch 1973b, 370.)

ACKNOWLEDGMENTS

We sincerely acknowledge Dr. N.R. Krieg for kindly providing the template text upon which this chapter was based.

List of species of the genus Levispirillum

1. **Levispirillum itersonii** comb. nov. (*Aquaspirillum itersonii* (Giesberger 1936) Hylemon, Wells, Krieg and Jannasch 1973b, 370.)

   i.ter.so' ni.i. M.L. gen. n. *itersonii* of Iterson; named for G. Van Iterson, a Dutch bacteriologist.

   The morphological characters are shown in Fig. BXII.β.67 and described in Table BXII.β.83. Optimal growth temperature 32–35°C. Chemotaxonomic and physiological characters are as described in Tables BXII.β.80, BXII.β.81, BXII.β.82, and BXII.β.83. Sole carbon sources are listed in Table BXII.β.79. Ammonium salts can serve as sole nitrogen sources. Nitrate supports either no growth or very scanty growth. In 1957, Williams and Rittenberg established a subspecies, subsp. *vulgatum,* to include those strains that use nitrate; however, this could not be confirmed by Terasaki (1979) for the type strain of the subspecies (ATCC 11331), using the methods of Williams and Rittenberg.

   Isolated from pond water and from putrid infusions of freshwater mussels.

   The two subspecies, subsp. *itersonii* and subsp. *nipponicum,* as represented by the type strains IFO 15648 and IFO 13615, respectively, have been shown to be genotypically related at the species level (Kawasaki et al., 1997). The type strain of *L. itersonii* subsp. *nipponicum,* as represented by the subculture ATCC 33333^T, however, could not be located in the *Alphaproteobacteria,* but was found to belong to the *Betaproteobacteria,* where it is a member of the family *Comamonadaceae* (Willems, unpublished results).

   The mol% G + C of the DNA is: 60–66 (T_m).

   Type strain: Giesberger, ATCC 12639, NCIMB 9070, NRRL B-2053.

   GenBank accession number (16S rRNA): Z29620.

   a. **Levispirillum itersonii** subsp. **itersonii** subsp. nov. (*Aquaspirillum itersonii* Hylemon, Wells, Krieg and Jannasch 1973b, 370.)

      Morphology and characteristics as for the species. Differs from the subsp. *nipponicum* by having a cell diameter of 0.4–0.6 μm and rapid formation of coccoid bodies (within 7 d).

   The mol% G + C of the DNA is: 60–64 (T_m).

   Type strain: Giesberger, ATCC 12639, NCIMB 9070, NRRL B-2053.

   GenBank accession number (16S rRNA): Z29620.

   b. **Levispirillum itersonii** subsp. **nipponicum** subsp. nov. (*Spirillum itersonii* subsp. *nipponicum* Terasaki 1973, 58.)

      nip.po' ni.cum. M.L. neut. adj. *nipponicum* pertaining to the country of Japan.

      Morphology and characteristics as for the species. Differs from the subsp. *itersonii* by having a cell diameter of 0.5–0.8 μm and a delayed formation of coccoid bodies (~2 weeks before they become predominant).

      The mol% G + C of the DNA is: 66 (T_m).

      Type strain: KF 8, ATCC 33333, DSM 11590, IFO 13615.

2. **Levispirillum peregrinum** comb. nov. (*Aquaspirillum peregrinum* (Pretorius 1963) Hylemon, Wells, Krieg and Jannasch 1973b, 370.)

   pe.re.gri' num. L. neut. adj. *peregrinum* strange, foreign.

   The morphological characters are shown in Fig. BXII.β.67 and described in Table BXII.β.83. Optimal growth temperature 32°C. Chemotaxonomic and physiological characters are described in Tables BXII.β.80, BXII.β.81, BXII.β.82, and BXII.β.83. Sole carbon sources are listed in Table BXII.β.79. Ammonium salts can be used as sole nitrogen sources. There are conflicting reports concerning the ability to use nitrate as a sole nitrogen source (Terasaki, 1972, 1979; Hylemon et al., 1973b).

   Isolated from a primary oxidation pond and from the putrid infusion of a freshwater mussel.

   The two subspecies, subsp. *peregrinum* and subsp. *integrum,* have been shown to be genotypically related at the species level (Pot, 1996; Kawasaki et al., 1997).

   The mol% G + C of the DNA is: 60–64 (T_m).

   Type strain: ATCC 15387, DSM 1839, NCIB 9435.

   a. **Levispirillum peregrinum** subsp. **peregrinum** subsp. nov. (*Aquaspirillum peregrinum* (Pretorius 1963) Hylemon, Wells, Krieg and Jannasch 1973b, 370.)

Morphology and characteristics as for the species. Differs from the subsp. *integrum* by forming coccoid bodies.

The mol% G + C of the DNA is: 60–62 (T_m).

Type strain: ATCC 15387, DSM 1839, NCIB 9435.

b. **Levispirillum peregrinum** subsp. **integrum** subsp. nov. (*Spirillum peregrinum* subsp. *integrum* Terasaki 1973, 60.) *in'te.grum*. L. neut. adj. *integrum* unchanged (referring here to failure to form coccoid bodies).

Morphology and characteristics as for the species. Differs from the subsp. *peregrinum* by failing to form coccoid bodies as a predominant form even after 28 d of incubation.

The mol% G + C of the DNA is: 64 (T_m).

Type strain: MF 19, ATCC 33334, DSM 11589, IFO 13617.

Genus IV. **Magnetospirillum** *Schleifer, Schüler and Ludwig 1992, 291*[VP] *(Effective publication: Schleifer, Schüler and Ludwig in Schleifer, Schüler, Spring, Weizenegger, Amann, Ludwig and Köhler 1991, 384)*

DIRK SCHÜLER AND KARL-HEINZ SCHLEIFER

Mag.ne'to.spi.ril'lum. Gr. n. *magnes* magnet, comb. form magneto-; Gr. n. *spira* a spiral; M.L. dim. neut. n. *spirillum* a small spiral; *Magnetospirillum* a small magnetic spiral.

Helical (clockwise) **spirilla**, 0.2–0.7 × 3–4 µm. Occur in freshwater sediments. Gram negative. Motile by means of a single flagellum at each pole. Cells **contain variable numbers of magnetosomes**. Intracytoplasmic membranes (magnetosome membrane) are present. Intracellular poly-hydroxy-alkanoate formation. **Magnetotactic. Microaerophilic**, having a respiratory type of metabolism with oxygen as terminal electron acceptor. **Chemoorganotrophic**. Optimum growth temperature is 30°C. No growth in the presence of 1% NaCl. Catalase and oxidase can be absent or present. Denitrify under microaerobic conditions. Growth occurs on various organic acids; carbohydrates are not used. Members form a **monophyletic group within the *Alphaproteobacteria***.

The mol% G + C of the DNA is: 63–64.

Type species: **Magnetospirillum gryphiswaldense** Schleifer, Schüler and Ludwig 1992, 291 (Effective publication: Schleifer, Schüler and Ludwig *in* Schleifer, Schüler, Spring, Weizenegger, Amann, Ludwig and Köhler 1991, 384.)

FURTHER DESCRIPTIVE INFORMATION

Cells of *Magnetospirillum* are helical and 0.2–0.7 × 3–4 µm. Enlarged pleomorphic forms and coccoid bodies are formed in older cultures and under conditions of excess oxygen. Each magnetotactic cell contains a number of electron-dense intracytoplasmic inclusion bodies (magnetosomes), which are aligned in a linear chain (for details of morphology and ultrastructure of the type strain *M. gryphiswaldense* see Fig. BXII.α.13). The magnetosomes consist of membrane-enveloped cubo-octahedral particles of magnetite (Fe$_3$O$_4$) (Balkwill et al., 1980). The size of mature magnetite crystals is 42 nm in diameter. The number of magnetosome particles per cell varies between 0 and 60 depending on growth conditions, oxygen, and iron supply. Magnetite formation requires low oxygen concentrations (<5 kPa) (Blakemore et al., 1985; Schüler and Baeuerlein, 1998). Concentrated suspensions of magnetic cells are dark gray-brown in color, those of nonmagnetic cells are light cream colored. Magnetite formation is coupled to the uptake of bulk amounts of ferric iron at high rates and cells intracellularly accumulate more then 2% iron (dry wt) during magnetite synthesis (Schüler and Baeuerlein, 1998). Production of a hydroxamate-type siderophore has been reported for *M. magnetotacticum* (Paoletti and Blakemore, 1986), but siderophore production could be not detected in *M. gryphiswaldense* (Schüler and Baeuerlein, 1996). Magnetite crystals are enveloped by the magnetosome membrane, which represents an intracytoplasmic membrane compartment with a distinctive protein and lipid composition (Schüler and Baeuerlein, 1997; Gorby et al., 1988). Cells grown under low-iron conditions contain empty magnetosome vesicles. The magnetosome chain is responsible for the tactic response of the cells to magnetic fields that is assumed to have a navigational function by interaction with the Earth's magnetic field (Blakemore and Frankel, 1981). Magnetospirilla exhibit an axial type of magnetotaxis, that is, cells are capable of swimming parallel or antiparallel to the magnetic field by alternately reversing their swimming direction without turning (Frankel et al., 1997).

Magnetospirillum strains are obligately microaerophilic. The level of oxygen tolerance differs between species, and is apparently correlated to the presence or absence of the oxygen-protective enzyme catalase. In semi-solid medium incubated under air atmospheres, initial growth occurs as a thin band some distance from the surface. Colony formation on agar plates is difficult to achieve; surface colonies are formed in *M. magnetotacticum* if catalase is added to the medium (Blakemore et al., 1979) and in *M. gryphiswaldense* on ACA medium (Schultheiss and Schüler, 2003). Best growth occurs in simple media containing short organic acids and minerals (see Tables BXII.α.7 and BXII.α.8). A cytochrome a_1-like hemoprotein (Tamegai et al., 1993), and a cytochrome c_{550} (Yoshimatsu et al., 1995) are present in *M. magnetotacticum*. A ccb-type cytochrome c oxidase functions as terminal oxidase in the microaerobic respiration of *M. magnetotacticum* (Tamegai and Fukumori, 1994). A cytochrome cd_1-type nitrite reductase with Fe(II):nitrite oxidoreductase activity and a speculated function in magnetite synthesis is present in *M. magnetotacticum* (Yamazaki et al., 1995). The purification of a ferric iron reductase that is loosely bound to the cytoplasmic membrane was described for *M. magnetotacticum* (Noguchi et al., 1999).

No growth factors are required. Growth occurs with fumarate, tartrate, malate, succinate, lactate, pyruvate, oxalacetate, malonate, β-hydroxybutyrate, and maleate as sole carbon sources. Carbohydrates are not used. N$_2$, nitrate, and ammonium can serve as sole nitrogen sources (Bazylinski and Blakemore, 1983b; Bazylinski et al., 2000). No growth under strictly anaerobic conditions. However, at low oxygen levels nitrate is dissimilated to nitrous oxide and nitrogen gas in *M. magnetotacticum* (Escalante-Semerena et al., 1980; Bazylinski and Blakemore, 1983a). *Magnetospirillum* species contain C$_{18:1}$, C$_{16:1}$, and C$_{16:0}$ as the major non-polar fatty acids, and C$_{14:0}$, C$_{16:0}$, and C$_{18:0}$ as 3-hydroxy acids; Q-10 is the major ubiquinone (Sakane and Yokota, 1994). *M. gryphiswaldense* and *M. magnetotacticum* contain putrescine and

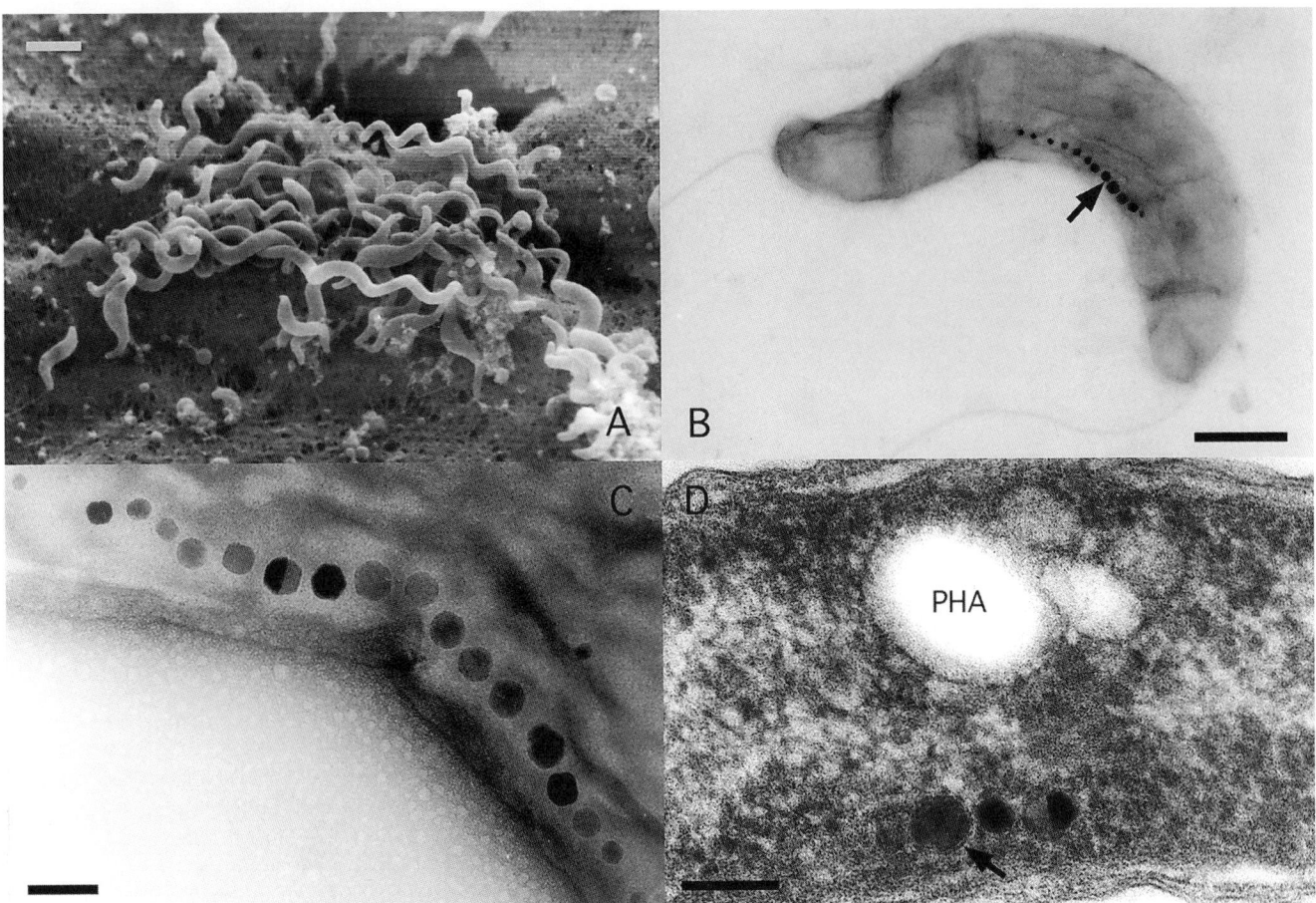

FIGURE BXII.α.13. (*A*) Scanning electron micrograph of cells of *M. gryphiswaldense* showing the characteristic helical morphology of *Magnetospirillum* species (Bar = 1 μm). (*B*) Transmission electron micrograph of a negatively stained cell of *Magnetospirillum gryphiswaldense* showing the bipolar-monotrichous flagellation and the chain of magnetosomes (*arrow*) (Bar = 0.5 μm). (*C*) Transmission electron micrograph of the magnetosome chain of *M. gryphiswaldense*. Cells contain intracellularly up to 60 cubo-octahedral crystals of magnetite (Fe₃O₄) that are 42 nm in size (Bar = 0.1 μm). (*D*) Transmission electron microscopy of a thin-section of a magnetic cell of *M. gryphiswaldense*. Individual magnetite crystals are enveloped by the magnetosome membrane (*arrow*). Cells contain poly-hydroxy-alkanoate granules (PHA) (Bar = 0.1 μm).

spermidine, or putrescine, cadaverine, and spermidine, respectively, as the major polyamines (Hamana et al., 1994).

ENRICHMENT AND ISOLATION PROCEDURES

Magnetospirillum species can be isolated from the oxic–anoxic transition zone of many freshwater sediments. Enrichments of magnetic spirilla can be obtained by the undisturbed incubation of loosely covered jars, filled with 1/3 mud and 2/3 water from the sample origin, during several months at room temperature. As no selective media for the cultivation of magnetic spirilla are

TABLE BXII.α.7. Characteristics differentiating *Magnetospirillum gryphiswaldense* and *M. magnetotacticum*[a]

Characteristics	*M. gryphiswaldense*	*M. magnetotacticum*
Catalase	+	−
Oxidase	+	−[b]
Growth and magnetosome formation in liquid culture exposed to air	+	−

[a]For symbols see standard definitions.

[b]Oxidase test is faintly positive with toluene-treated cells (Maratea and Blakemore, 1981).

known, the effective physical separation of magnetotactic cells from contaminants is crucial in their isolation. Advantage can be taken of the directed, active migration of magnetospirilla in magnetic fields. Large numbers of magnetotactic bacteria can be collected from the sediment-water interface of the enrichments using a bar magnet. Magnetotactic bacteria can then be further separated by a "racetrack" method using semisolid agar or sterile capillaries (Wolfe et al., 1987; Schüler and Köhler, 1992). By application of a magnetic field, magnetotactic bacteria will out-swim contaminating organisms and a highly purified inoculum can be collected at the end of the capillary close to the south pole of a bar magnet. The purified inoculum is then transferred to prereduced semisolid, low-nutrient isolation medium, which is incubated under microaerobic atmospheres containing 0.6–1% oxygen (Blakemore et al., 1979). Alternatively, a two-layer sulfide-oxygen gradient medium has proven useful in the isolation of microaerophilic magnetospirilla by creation of a vertical redox gradient (Schüler et al., 1999).

MAINTENANCE PROCEDURES

Magnetospirilla may be maintained in semisolid growth medium at 30°C with weekly transfer or at room temperature with monthly transfer. Long-term preservation is accomplished by suspending a dense suspension of cells in liquid growth medium containing

TABLE BXII.α.8. Other characteristics of *M. gryphiswaldense* and *M. magnetotacticum*[a]

Characteristics	*M. gryphiswaldense*	*M. magnetotacticum*
Cell size, μm	0.2–0.7 × 3–4	0.2–0.7 × 3–4
Magnetosomes: 0–60 intracellular cubo-octahedral magnetite (Fe_3O_4) crystals, 42 nm	+	+
Magnetic reaction, sensitivity	+	+
Single bipolar flagella	+	+
Intracellular PHA-formation	+	+
Optimal growth temperature, °C	30	30
Growth in the presence of 1% NaCl	−	−
Growth on carbohydrates	−	−
Growth on fumarate, tartrate, malate, succinate, lactate, pyruvate, oxalacetate, malonate, β-hydroxybutyrate, maleate	+	+
Microaerobic reduction of nitrate	+	+
Growth with N_2 as sole nitrogen source, nitrogenase activity	+	+
Hydrolysis of casein, starch, hippurate, esculin, gelatin	−	−
Hydrogen sulfide production from cysteine	−	−
Growth in the presence of 1% glycine	nd	+
Growth in the presence of 1% bile	nd	−
Anaerobic growth with NO_3^-, Fe(III), DMSO, SO_4^- [b]	−	−
Formation of coccoid bodies in older cultures	+	+
Selenite reduction	nd	−
Pigment formation from aromatic amino acids	−	−
Urease, sulfatase, indole	nd	−
Alkaline reaction in litmus milk	nd	−

[a]For symbols see standard definitions; nd, not determined.

[b]Very slow ferric iron-dependent growth in the absence of oxygen has been reported (Guerin and Blakemore, 1992).

10% (v/v) dimethyl sulfoxide, with subsequent freezing in liquid nitrogen and storage at −80°C.

PROCEDURES FOR TESTING SPECIAL CHARACTERS

The most conspicuous trait of *Magnetospirillum* species is their magnetic sensitivity, which can be easily checked in living cells by phase-contrast microscopy in the presence of a permanent magnet, where magnetospirilla will orient and swim parallel to the magnetic field lines. Magnetic reaction of cultures can also be checked by placing a drop of cell suspension onto the top of a magnetic stirrer and watching for the characteristic "flickering", as the light scattering of suspensions of magnetic cells is affected by changing magnetic fields (Schüler et al., 1995). By electron microscopy, intracellular magnetosome crystals are easily revealed in unstained cells as electron-dense inclusions.

A *Magnetospirillum*-specific pair of primers for the PCR-amplification of a 16S rRNA gene fragment has been proposed (Burgess et al., 1993), but these were found to also be homologous to several unrelated, nonmagnetic *Alphaproteobacteria* in a recent database search (D. Schüler, unpublished). Tests for oxygen tolerance and catalase and oxidase activities are further useful diagnostic features.

DIFFERENTIATION OF THE GENUS *MAGNETOSPIRILLUM* FROM OTHER GENERA

The intracellular magnetosome formation and magnetic reaction clearly distinguishes *Magnetospirillum* from other microaerophilic, spiral-shaped bacteria from freshwater.

Magnetospirillum species are most closely related (94–96% 16S rDNA sequence similarity) to several nonmagnetic, photosynthetic, nonsulfur purple *Alphaproteobacteria* of the genus *Phaeospirillum* (*P. fulvum* and *P. molischianum*; Imhoff et al., 1998), with which they share similarities in cell morphology and flagellation, the presence of intracytoplasmic membranes, and the microaerophilic nature of their chemoorganotrophic growth. However, *Magnetospirillum* strains are unable to grow photosynthetically and lack the characteristic photosynthetic pigments (bacteriochlorophyll, carotenoids) of the *Phaeospirillum* species.

Magnetospirillum species are phylogenetically only distantly related to other, mostly uncultured vibrioid, rod-shaped, and coccoid magnetotactic bacteria, and can be distinguished by cell morphology and the shape, size, and arrangement of the magnetosome crystals (Spring and Schleifer, 1995).

TAXONOMIC COMMENTS

The first magnetic spirillum isolated by R. Blakemore was originally described as *Aquaspirillum magnetotacticum* (Maratea and Blakemore, 1981). However, 16S rRNA gene sequence analysis of *A. magnetotacticum* and the newly isolated magnetic strain MSR-1 demonstrated their affiliation with the *Alphaproteobacteria*, while the type species of *Aquaspirillum* falls in the *Betaproteobacteria*. Moreover, magnetic spirilla differ from known *Aquaspirillum* strains in a number of morphological and physiological features such as the presence of magnetic inclusions and the type of flagellation, suggesting classification in a separate genus. Therefore, the new genus *Magnetospirillum* was proposed with the type strain *M. gryphiswaldense* MSR-1, and *A. magnetotacticum* was transferred to *Magnetospirillum* as *M. magnetotacticum* (see Tables BXII.α.7 and BXII.α.8) (Schleifer et al., 1991).

FURTHER READING

Maratea, D. and R.P. Blakemore. 1981. *Aquaspirillum magnetotacticum*, sp. nov., a magnetic spirillum. Int. J. Syst. Bacteriol. *31*: 452–455.

Schleifer, K.H., D. Schüler, S. Spring, M. Weizenegger, R. Amann, W. Lduwig and M. Köhler. 1991. The genus *Magnetospirillum*, gen. nov., description of *Magnetospirillum gryphiswaldense*, sp. nov. and transfer of *Aquaspirillum magnetotacticum* to *Magnetospirillum magnetotacticum*, comb. nov. Syst. Appl. Microbiol. *14*: 379–385.

Schüler, D. 1999. Formation of magnetosomes in magnetotactic bacteria. J. Mol. Microbiol. Biotechnol. *1*: 79–86.

Spring, S. and K.H. Schleifer. 1995. Diversity of magnetotactic bacteria. Syst. Appl. Microbiol. *18*: 147–153.

List of species of the genus Magnetospirillum

1. **Magnetospirillum gryphiswaldense** Schleifer, Schüler and Ludwig 1992, 291[VP] (Effective publication: Schleifer, Schüler and Ludwig *in* Schleifer, Schüler, Spring, Weizenegger, Amann, Ludwig and Köhler 1991, 384.)

   gry.phis.wal.den' se. L. adj. *gryphiswaldense* the Latin name of Greifswald, a town in Germany where the organism was isolated.

   Helical spirilla, 0.7 × 3–4 µm. Catalase and oxidase positive. Microaerophilic, but grows prolifically in agitated liquid medium[1] exposed to air if large inocula (1/10) are used. Growth rates are 0.3–0.1 h[−1]. Isolated by D. Schüler from sediments of a small river (Ryck) near Greifswald, Germany.

   The mol% G + C of the DNA is: 62.7 (HPLC) (Sakane and Yokota, 1994).

   Type strain: MSR-1, DSM 6361.

   GenBank accession number (16S rRNA): Y10109.

2. **Magnetospirillum magnetotacticum** (Maratea and Blakemore 1981) Schleifer, Schüler and Ludwig 1992, 291[VP] (Effective publication: Schleifer, Schüler and Ludwig *in* Schleifer, Schüler, Spring, Weizenegger, Amann, Ludwig and Köhler 1991, 385) (*Aquaspirillum magnetotacticum* Maratea and Blakemore 1981, 454.)

   mag.ne.to.tac.' ti.cum. Gr. n. *magnes* magnet, comb. form *magneto-* Gr. adj. *taktikos* showing orientation or movement directed by a force or agent; *magnetotacticum* capable of orientation with respect to a magnet.

   Helical spirilla, 0.4–0.7 × 3–4 µm. Microaerophilic. No growth of magnetic cells in liquid cultures with free gas exchange to air. Very weak ferric iron-dependent growth in the absence of oxygen has been reported (Guerin and Blakemore, 1992). Nitrate is reduced to N_2 with transient accumulation of nitrous oxide but without nitrite accumulation (Bazylinski and Blakemore, 1983b). Grows in a defined mineral medium[2]. Vitamins are not strictly required for growth, but deletion of vitamins from the growth medium results in a pleomorphic appearance (Blakemore et al., 1979). Catalase, oxidase, urease, sulfatase, and indole are negative. Oxidase test is faintly positive with toluene-treated cells (Maratea and Blakemore, 1981). Isolated by R.P. Blakemore at the University of Illinois from sediments collected in Cedar Swamp, Woods Hole, Massachusetts (USA).

   The mol% G + C of the DNA is: 63 (HPLC) (Sakane and Yokota, 1994).

   Type strain: MS-1, ATCC 31 632, DSM 3856.

   GenBank accession number (16S rRNA): Y10110.

Other Organisms

Several magnetic spirilla isolates have been identified as *Magnetospirillum* species based on cell morphology, ultrastructure, and phylogenetic analysis, and form a monophyletic group including *M. magnetotacticum* and *M. gryphiswaldense* within the *Alphaproteobacteria*. However, their taxonomic characterization is incomplete and they are not yet validly described.

The *Magnetospirillum* strains MGT-1 (GenBank accession number of the 16S rRNA gene sequence: D17515) and AMB-1 (D17514) were isolated from freshwater ponds in Tokyo, Japan (Burgess et al., 1993). Sequence similarities of the 16S rRNA genes of MGT-1 and AMB-1 to that of *M. gryphiswaldense* are 95.0% and 95.8%, respectively, and to that of *M. magnetotacticum* are 99.0% and 99.4% (Schüler et al., 1999). MGT-1 and AMB-1 were reported to have increased oxygen tolerance and to form colonies on agar surfaces exposed to air (Burgess et al., 1993). Seven *Magnetospirillum* strains (MSM-1, 3, 4, 6, 7, 8, 9) were isolated from a freshwater pond in Iowa, USA (EMBL-accession numbers for the sequences of the 16S rRNA genes of the *Magnetospirillum* strains MSM-3, MSM-4, and MSM-6 are Y17389, Y17390, and Y17391, respectively). While five of the isolates were very similar to either *M. gryphiswaldense* or *M. magnetotacticum* by 16S rRNA gene sequence analysis (>99.7%), two new isolates (MSM-3, MSM-4) are likely to represent new species (Schüler et al., 1999). A total of six different sequences was obtained from a single sample site, indicating a significant microdiversity of natural populations of magnetic spirilla.

---

1. *M. gryphiswaldense* medium (mod. after Schüler and Baeuerlein, 1996) contains per liter: KH_2PO_4, 0.5 g; Na-acetate, 1.0 g; soybean peptone (E. Merck), 1.0 g; NH_4Cl, 0.1 g; $MgSO_4 \cdot 7H_2O$, 0.1 g; yeast extract, 0.1 g; agar (for semisolid medium), 2 g; Fe(III) citrate, 50 µM (final concentration). Dissolve and autoclave at 121°C for 15 min. For liquid cultures, inoculate with 1/10 vol and agitate at 100 rpm.

2. *M. magnetotacticum* medium (DSM 380) contains per liter: Vitamin solution (Wolin et al., 1963), 10 ml; trace elements (Wolin et al., 1963), 5 ml; Fe(III) quinate solution, 2 ml; resazurine, 0.5 mg; KH_2PO_4, 0.68 g; $NaNO_3$, 0.12 g; Na-thioglycolate, 0.05 g; L(+)-tartaric acid, 0.3 g; succinic acid, 0.37 g; Na-acetate, 0.05 g; agar (for solid medium), 13.0 g. Dissolve components in the order given, adjust pH to 6.75 with NaOH. Ferric quinate solution, 0.01 M (per liter): $FeCl_3 \cdot 6H_2O$, 4.5 g; quinic acid, 1.9 g, Dissolve and autoclave at 121°C for 15 min. Liquid medium: Purge medium with N_2 gas for 10 min. Under the same atmosphere, anaerobically fill tubes to 1/3 of their volume and seal. Autoclave at 121°C for 15 min. Before inoculation, add sterile air (with hypodermic syringe through the rubber closure) to 1% O_2 concentration in the gas phase.

Genus V. **Phaeospirillum** *Imhoff, Petri, and Süling 1998, 796*[VP]

JOHANNES F. IMHOFF

Phae.o.spi.ril'lum. Gr. adj. *phaeos* brown; M.L. neut. n. *Spirillum* a bacterial genus; M.L. neut. n. *Phaeospirillum* brown *Spirillum*.

Cells are vibrioid to spiral shaped, motile by means of polar flagella and multiply by binary fission. **Gram negative and belong to the *Alphaproteobacteria*.** Internal photosynthetic membranes are lamellar stacks forming a sharp angle with the cytoplasmic membrane. Photosynthetic pigments are **bacteriochlorophyll *a*** (esterified with phytol) and **carotenoids of the spirilloxanthin series**, with spirilloxanthin itself lacking. **Ubiquinones and menaquinones with nine isoprene units (Q-9 and MK-9)** are present. **Major cellular fatty acids are $C_{18:1}$, $C_{16:1}$, and $C_{16:0}$**, with $C_{18:1}$ as dominant component (44–55% of total fatty acids).

Cells grow preferentially **photoheterotrophically** under anoxic conditions in the light. **Chemotrophic growth is possible at very low oxygen tensions in the dark.** Growth factors may be required. **Mesophilic freshwater bacteria with preference for neutral pH.** Habitat: stagnant and anoxic freshwater habitats.

The mol% G + C of the DNA is: 60.5–65.3.

Type species: **Phaeospirillum fulvum** (van Niel 1944) Imhoff, Petri, and Süling 1998, 797 (*Rhodospirillum fulvum* van Niel 1944, 108.)

FURTHER DESCRIPTIVE INFORMATION

Phaeospirillum species are very sensitive to oxygen and do not grow under oxic conditions in the dark (Pfennig, 1969b). They do grow, however, under microoxic conditions in the dark, provided the oxygen tension is lower than 1.5 kPa for *Phaeospirillum fulvum* and 1.0 kPa for *Phaeospirillum molischianum* (Lehmann, 1976). These bacteria are unable to induce a second electron transport chain in the presence of oxygen and depend on microoxic conditions in which the internal membrane system and the light-driven electron transport chain are fully expressed. The cells are fully pigmented under these growth conditions (Lehmann, 1976).

Ammonia assimilation is mediated by the glutamine synthetase/glutamate synthase reactions, which are NADH-dependent in *Phaeospirillum molischianum* and *Phaeospirillum fulvum* (Brown and Herbert, 1977).

ENRICHMENT AND ISOLATION PROCEDURES

Media and growth conditions applied for *Rhodospirillum* species are also suitable for *Phaeospirillum* species, if the oxygen partial pressure is adequately reduced. This can be achieved by addition of 0.05% sodium ascorbate to the growth media in completely filled screw-capped bottles. A recipe for a mineral medium appropriate for cultivation of *Phaeospirillum* species is given with the description of the genus *Rhodospirillum*. Higher fatty acids such as pelargonate and caprylate (concentrations below 0.04% at pH 7.5) provide selective conditions for enrichment of *P. fulvum* and *P. molischianum*. Benzoate may be used for selective enrichment of *P. fulvum*. Standard techniques for isolation of anaerobic bacteria in agar dilution series and on agar plates can be applied to *Phaeospirillum* species (Biebl and Pfennig, 1981; Imhoff, 1988; Imhoff and Trüper, 1992), if care is taken to establish and maintain oxygen-free conditions.

MAINTENANCE PROCEDURES

Cultures can be preserved by standard techniques in liquid nitrogen or at −80°C in a mechanical freezer.

DIFFERENTIATION OF THE GENUS *PHAEOSPIRILLUM* FROM OTHER GENERA

Besides 16S rDNA sequence similarities (Kawasaki et al., 1993b; Imhoff et al., 1998), the size and sequences of the small-size cytochromes c_2 (Ambler et al., 1979), quinone, and fatty acid composition and other properties also distinguish *Phaeospirillum* and *Rhodospirillum* species (see Tables BXII.α.1 and BXII.α.2 in the chapter describing the genus *Rhodospirillum*).

TAXONOMIC COMMENTS

Phaeospirillum species have previously been included in the genus *Rhodospirillum*. Based on significant differences in chemotaxonomic properties and 16S rDNA sequences, they have been transferred into the new genus *Phaeospirillum* (Imhoff et al., 1998). The name *Phaeospirillum* was first proposed by Kluyver and van Niel (1936) for brown-colored, phototrophic, spiral-shaped bacteria but was not included in the Approved List of Names and therefore lost standing in nomenclature. The name was revived (Imhoff et al., 1998). Both *Phaeospirillum* species have highly similar 16S rDNA sequences (greater than 99% similarity) that would allow their treatment as a single species. This was not proposed because of the well-recognized phenotypic differences between the two species. The phylogenetic relationships of these bacteria based on 16S rDNA sequences are shown in Fig. 1 (p. 124) of the introductory chapter "Anoxygenic Phototrophic Purple Bacteria", Volume 2, Part A.

DIFFERENTIATION OF THE SPECIES OF THE GENUS *PHAEOSPIRILLUM*

Phaeospirillum molischianum and *Phaeospirillum fulvum* can be distinguished from each other primarily by cell size, growth factor requirements, and fatty acid composition (Table BXII.α.1 in the description of the genus *Rhodospirillum*). The utilization of benzoate is a characteristic and distinguishing property of *Phaeospirillum fulvum* (Table BXII.α.2 in the description of the genus *Rhodospirillum*).

List of species of the genus Phaeospirillum

1. **Phaeospirillum fulvum** (van Niel 1944) Imhoff, Petri, and Süling 1998, 797[VP] (*Rhodospirillum fulvum* van Niel 1944, 108.)

   ful'vum. M.L. neut. adj. *fulvum* deep or reddish yellow, tawny.

   Cells are vibrioid shaped to spiral, 0.5–0.7 × 3.5 μm; one complete turn of a spiral is 1.0–1.6 μm wide. Internal photosynthetic membranes are present as lamellar stacks forming a sharp angle to the cytoplasmic membrane. Anaerobic liquid cultures are deep brown. Photosynthetic pigments are bacteriochlorophyll *a* (esterified with phytol) and carotenoids of the spirilloxanthin series. Spirilloxanthin is lacking, but the biosynthetic precursors lycopene and rhodopin are present as major components.

Cells grow preferentially photoheterotrophically under anoxic conditions in the light with various organic compounds as carbon and electron sources. Chemotrophic growth at very low oxygen tensions is possible. Unable to adapt to oxic growth conditions. Under microoxic conditions cells are fully pigmented. Carbon sources utilized are shown in Table BXII.α.2 in the description of the genus *Rhodospirillum*. Ammonia and N_2 are used as nitrogen source. Sulfate can be used as sole sulfur source. For optimal development, the addition of ascorbate or thioglycolate as a reductant may be necessary. *p*-Aminobenzoic acid is required as a growth factor.

Mesophilic freshwater bacterium with optimum growth at 25–30°C and pH 7.3 (pH range: 6.0–8.5). Habitat: stagnant and anoxic freshwater habitats that are exposed to the light.

The mol% G + C of the DNA is: 64.3–65.3 (Bd) and 62.1–62.8 (T_m).

Type strain: Pfennig 1360, ATCC 15798, DSM 113.

GenBank accession number (16S rRNA): M14433, M59065.

2. **Phaeospirillum molischianum** (Giesberger 1947) Imhoff, Petri, and Süling 1998, 797[VP] (*Rhodospirillum molischianum* Giesberger 1947, 142.)

mo.li'schi.a' num. M.L. neut. adj. *molischianum* pertaining to Molisch, named for H. Molisch, an Austrian botanist.

Cells are vibrioid to spiral shaped, 0.7–1.0 × 4–6 μm or even longer; one complete turn of a spiral is 1.5–2.5 μm wide. Internal photosynthetic membranes are present as lamellar stacks forming a sharp angle to the cytoplasmic membrane (Fig. BXII.α.14). Anaerobic liquid cultures are brown-orange to brown-red or dark brown. Absorption maxima of living cells are at 375, 465, 488–491, 520–528, 590–595, 803–807, and 850–855 nm. Photosynthetic pigments are bacteriochlorophyll *a* (esterified with phytol) and carotenoids of the spirilloxanthin series. Spirilloxanthin is lacking, but the biosynthetic precursors lycopene and rho-

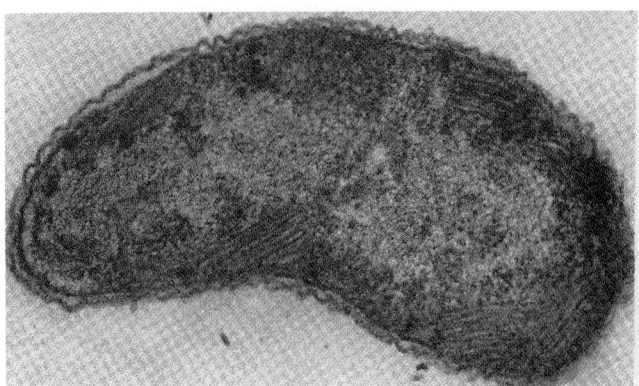

FIGURE BXII.α.14. *Phaeospirillum molischianum* grown anaerobically in the light. Note the position and the lamellar stack type of the intracytoplasmic membrane system. × 90,000. (Courtesy of G. Drews.)

dopin are present as major components. Cells grow preferentially photoheterotrophically under anoxic conditions in the light with various organic compounds as carbon and electron sources. Chemotrophic growth at very low oxygen tensions is possible, but the ability to adapt to oxic growth conditions is lacking. Under microoxic conditions cells are fully pigmented. Carbon sources utilized are shown in Table BXII.α.2 in the description of the genus *Rhodospirillum*. Malonate is not used. Ammonia, N_2, and some amino acids are used as nitrogen sources. For optimal development the addition of ascorbate or thioglycolate as a reductant may be necessary. Complex mixtures of amino acids, such as yeast extract and vitamin-free Casamino acids, stimulate growth considerably and may be required.

Mesophilic freshwater bacterium with optimal growth at 30°C and pH 7.3 (pH range: 6.0–8.5). Habitat: stagnant and anoxic freshwater habitats that are exposed to the light.

The mol% G + C of the DNA is: 60.5–64.8 (Bd) and 62.1–62.6 (T_m).

Type strain: ATCC 14031, DSM 120, NCIB 9957, NTHC 131.

GenBank accession number (16S rRNA): M59067.

Genus VI. **Rhodocista** *Kawasaki, Hoshino, Kuraishi and Yamasato 1994b, 182*[VP] *(Effective publication: Kawasaki, Hoshino, Kuraishi and Yamasato 1992, 548)*

JOHANNES F. IMHOFF

Rho.do.ci' sta. L. fem. n. *rhodon* the rose; L. fem. n. *cista* a basket; M.L. fem. n. *Rhodocista* the rose basket.

Cells are vibrioid to spiral in shape, motile by means of a long polar flagellum. **Gram negative. Belong to the *Alphaproteobacteria*.** Internal photosynthetic membranes occur as lamellae lying parallel to the cytoplasmic membrane. Photosynthetic pigments are **bacteriochlorophyll *a*** and **carotenoids of the spirilloxanthin series. Ubiquinone with nine isoprene units (Q-9) is present.**

Cells grow preferentially photoheterotrophically under anoxic conditions in the light. **Chemotrophic growth under oxic conditions in the dark is possible.**

The mol% G + C of the DNA is: 69.9.

Type species: **Rhodocista centenaria** Kawasaki, Hoshino, Kuraishi and Yamasato 1994b, 182 (Effective publication: Kawasaki, Hoshino, Kuraishi and Yamasato 1992, 548) (*Rhodospirillum centenum* Favinger, Stadtwald and Gest 1994, 182.)

FURTHER DESCRIPTIVE INFORMATION

When grown in liquid culture, *R. centenaria* cells are motile by means of a single polar flagellum (Favinger et al., 1989). On solid media, however, numerous lateral, peritrichous flagella that enable rapid movement on agar surfaces are produced. The kind of swarming motility observed with *R. centenaria* has not yet been observed in other phototrophic purple nonsulfur bacteria, and enables very rapid light-directed motility across the agar surface, which appears to be a true phototactic behavior (Ragatz et al.,

1995). It has also been shown that photo-induced cyclic electron flow is required for the photo-tactical response and that phototaxis is controlled by respiration (Romagnoli et al., 1997). The type strain (ATCC 43720) is not phototactically active (Nickens et al., 1996).

Absorption spectra of cultures grown anaerobically in the light or aerobically in the dark are almost identical, which demonstrates the absence of oxygen repression of photopigment synthesis that is common to most of the anoxygenic phototrophic purple bacteria (Yildiz et al., 1991).

Out of 13 strains under investigation, all were sensitive to rifampicin and streptomycin but highly resistant to neomycin and kanamycin (Nickens et al., 1996).

ENRICHMENT AND ISOLATION PROCEDURES

R. centenaria appears to be widely distributed in warm waters and has been repeatedly isolated from a variety of hot springs with the source water temperature between 34 and 58°C (Nickens et al., 1996). Most of the strains have been isolated with dinitrogen as sole nitrogen source and under anoxic conditions, but after transient exposure to air. Incubation at temperatures between 40 and 45°C under anoxic conditions in the light, in media suitable for freshwater purple nonsulfur bacteria, and with N_2 as the sole nitrogen source, provides selective conditions for the enrichment of *Rhodocista centenaria* (Nickens et al., 1996). Standard techniques for the isolation of anaerobic bacteria in agar dilution series and on agar plates can be applied (Biebl and Pfennig, 1981; Imhoff and Trüper, 1992).

MAINTENANCE PROCEDURES

Cultures are well preserved by standard techniques in liquid nitrogen, by lyophilization or storage at $-80°C$ in a mechanical freezer.

DIFFERENTIATION OF THE GENUS *RHODOCISTA* FROM OTHER GENERA

Differential characteristics and phylogenetic relationships based on 16S rDNA sequences are shown in Table BXII.α.1 and Table BXII.α.2 in the description of the genus *Rhodospirillum* and Fig. 1 (p. 124) of the introductory chapter "Anoxygenic Phototrophic Purple Bacteria", Volume 2, Part A.

TAXONOMIC COMMENTS

Until recently, the genus *Rhodospirillum* represented a heterogeneous assemblage of species that were of spiral shape and capable of phototrophic growth. Four halophilic species were classified together with several freshwater species. Because of its spiral shape and phototrophic capacity, *Rhodospirillum centenum* was also assigned to the genus *Rhodospirillum* (Favinger et al., 1989). The anticipated heterogeneity of the genus *Rhodospirillum* became clearly apparent with the 16S rDNA sequence information of most of the known species (Kawasaki et al., 1993b), and these data implied that the spiral-shaped *Alphaproteobacteria* were phylogenetically quite distantly related to each other and did not warrant classification in one and the same genus. Upon recognition of the great genetic distance between *Rhodospirillum rubrum* and *Rhodospirillum centenum*, the latter was transferred to a new genus as *Rhodocista centenaria* (Kawasaki et al., 1992). The phylogenetic relationships of these bacteria based on 16S rDNA sequences are shown in Fig. 1 (p. 124) of the introductory chapter "Anoxygenic Phototrophic Purple Bacteria", Volume 2, Part A.

FURTHER READING

Favinger, J., R. Stadtwald and H. Gest. 1989. *Rhodospirillum centenum*, sp. nov., a thermotolerant cyst-forming anoxygenic photosynthetic bacterium. Antonie Leeuwenhoek *55*: 291–296.

Imhoff, J.F., R. Petri and J. Süling. 1998. Reclassification of species of the spiral-shaped phototrophic purple non-sulfur bacteria of the α-*Proteobacteria*: description of the new genera *Phaeospirillum* gen. nov., *Rhodovibrio* gen. nov., *Rhodothalassium* gen. nov. and *Roseospira* gen. nov. as well as transfer of *Rhodospirillum fulvum* to *Phaeospirillum fulvum* comb. nov., of *Rhodospirillum molischianum* to *Phaeospirillum molischianum* comb. nov., of *Rhodospirillum salinarum* to *Rhodovibrio salinarum* comb. nov., of *Rhodospirillum sodomense* to *Rhodovibrio sodomensis* comb. nov., of *Rhodospirillum salexigens* to *Rhodothalassium salexigens* comb. nov. and of *Rhodospirillum mediosalinum* to *Roseospira mediosalina* comb. nov. Int. J. Syst. Bacteriol. *48*: 793–798.

Kawasaki, H., Y. Hoshino, H. Kuraishi and K. Yamasato. 1992. *Rhodocista centenaria* gen. nov., sp. nov., a cyst-forming anoxygenic photosynthetic bacterium and its phylogenetic postion in the *Proteobacteria* alpha group. J. Gen. Appl. Microbiol. *38*: 541–551.

Kawasaki, H., Y. Hoshino and K. Yamasato. 1993. Phylogenetic diversity of phototrophic purple non-sulfur bacteria in the *alpha proteobacteria* group. FEMS Microbiol. Lett. *112*: 61–66.

List of species of the genus Rhodocista

1. **Rhodocista centenaria** Kawasaki, Hoshino, Kuraishi and Yamasato 1994b, 182[VP] (Effective publication: Kawasaki, Hoshino, Kuraishi and Yamasato 1992, 548) (*Rhodospirillum centenum* Favinger, Stadtwald and Gest 1994, 182.)

cen.ten.a'ri.a. L. fem. adj. *centenaria* relating to a hundred, to commemorate a century after the publication of the first description of a phototrophic bacterium in 1887.

Cells are vibrioid to spiral in shape, 1–2 × 3 μm. In liquid media they are actively motile by means of a single polar flagellum. When grown on agar, peritrichous flagella are induced, conferring swarming motility. Swarming cells exhibit phototactic behavior. Cells are converted to thick-walled cysts, which are resistant to desiccation and heat, under conditions that also favor the formation of poly-β-hydroxybutyrate, especially in aging conditions on agar. Internal photosynthetic membranes are lamellae lying parallel to the cytoplasmic membrane. Color of anaerobically grown cultures is pink. Living cells show absorption maxima at 475–550, 587, 800, and 875 nm. Photosynthetic pigments are bacteriochlorophyll *a* and carotenoids of the spirilloxanthin series. Photopigment synthesis is not significantly affected by molecular oxygen.

Cells grow preferentially photoheterotrophically under anoxic conditions in the light or chemotrophically under oxic conditions in the dark. Pyruvate, lactate, acetate, alanine, aspartate, glutamate, and glutamine are used as carbon sources and electron donors. Butyrate, valerate, caproate, and caprylate also support phototrophic growth when supplemented with bicarbonate (9 mM). Sugars and C4 dicarboxylic acids, including malate, do not support growth. With butyrate as the sole carbon source, the spiral-shaped cells are converted to cysts, which are desiccation and heat resistant. Filter-fixed cysts kept viability at 55°C for 2 d. Ammonia, N_2, glutamate, glutamine, aspartate, and alanine are utilized as nitrogen sources. Biotin and vitamin B_{12} are required as growth factors.

Freshwater bacterium preferring elevated temperatures, with optimal growth at 40–45°C (upper limit 47°C) and pH 6.8 (pH range 5.7–7.0). Ubiquinone with nine isoprene units (Q-9) is the major quinone. Habitat: neutral to alkaline hot spring effluents, warm soils, effluents of paper factories.

The mol% G + C of the DNA is: 69.9 (T_m).
Type strain: ATCC 43720, DSM 9894, IAM 14193.
GenBank accession number (16S rRNA): D12701.

Genus VII. **Rhodospira** *Pfennig, Lünsdorf, Süling and Imhoff 1998, 328^VP (Effective publication: Pfennig, Lünsdorf, Süling and Imhoff 1997, 44)*

JOHANNES F. IMHOFF

Rho' do.spi' ra. Gr. n. *rhodon* the rose; Gr. n. *spira* the spiral; M.L. fem. n. *Rhodospira* the rose spiral.

Cells are vibrioid to spirilloid in shape, motile by means of flagella and multiply by binary fission. **Gram negative. Belong to the *Alphaproteobacteria.*** Internal photosynthetic membranes are of the vesicular type. Photosynthetic pigments are **bacteriochlorophyll *b*** and **carotenoids. Menaquinones and ubiquinones with seven isoprene units (MK-7 and Q-7) are present. Major cellular fatty acids are C$_{18:1}$ and C$_{16:0}$**, with C$_{18:1}$ as dominant component (~60% of total fatty acids).

Cells grow preferentially photoheterotrophically under anoxic conditions in the light, but poor growth may also be possible under microoxic conditions in the dark. Growth factors are required. **Marine bacteria** requiring salt for growth.

Habitat: salt marsh sediments and laminated microbial mats.

The mol% G + C of the DNA is: 65.7.

Type species: **Rhodospira trueperi** Pfennig, Lünsdorf, Süling and Imhoff 1998, 328 (Effective publication: Pfennig, Lünsdorf, Süling and Imhoff 1997, 44.)

FURTHER DESCRIPTIVE INFORMATION

At present, *Rhodospira trueperi* is the only representative of this genus. *R. trueperi* was isolated from a flat laminated microbial mat in a salt marsh near Woods Hole, Massachusetts, USA. The spiral-shaped bacterium is highly motile, has bipolar tufts of flagella, and internal photosynthetic membranes of the vesicular type. Major photosynthetic pigments are bacteriochlorophyll *b* and the carotenoid tetrahydrospirilloxanthin. *R. trueperi*, a marine organism, shows optimum growth in the presence of salt. It utilizes a number of organic substrates as carbon and energy sources, and requires vitamins and sulfide as a reduced sulfur source for growth. In the presence of sulfide, S⁰ globules are formed outside the cells. S⁰ is not further oxidized to sulfate. *Rhodospira trueperi* has a unique lipid and fatty acid composition (Pfennig et al., 1997). The latter is dominated by C$_{18:1}$ and C$_{16:0}$ and is unique among the spiral-shaped phototrophic *Alphaproteobacteria* (Table BXII.α.1 in the description of the genus *Rhodo-spirillum*). According to 16S rRNA gene sequence analysis, *Rhodospira trueperi* is most similar to *Roseospira mediosalina*. The phylogenetic relationships of these bacteria based on 16S rDNA sequences are shown in Fig. 1 (p. 124) of the introductory chapter "Anoxygenic Phototrophic Purple Bacteria", Volume 2, Part A.

ENRICHMENT AND ISOLATION PROCEDURES

Rhodospira trueperi was isolated from a microbial mat in a salt marsh, requires salt for growth, and depends on the presence of sulfide in growth media (Pfennig et al., 1997). Standard techniques for isolation of anaerobic bacteria in agar dilution series can be applied for *Rhodospira* species (Biebl and Pfennig, 1981; Imhoff, 1988; Imhoff and Trüper, 1992), if care is taken to establish and maintain oxygen-free conditions. *Rhodospira trueperi* can be grown on Pfennig's medium for purple sulfur bacteria (see genus *Chromatium*) with the addition of 2% NaCl and vitamins (or yeast extract).

MAINTENANCE PROCEDURES

Cultures can be preserved by standard techniques in liquid nitrogen or at −80°C in a mechanical freezer.

DIFFERENTIATION OF THE GENUS *RHODOSPIRA* FROM OTHER GENERA

Rhodospira species are characterized by salt requirements typical of marine bacteria, requirement for sulfide as a reduced sulfur source, unusual absorption spectra due to the presence of bacteriochlorophyll *b*, and presence of ubiquinones and menaquinones with seven isoprene units (see Table BXII.α.1). Major differentiating properties between *Rhodospira trueperi* and other phototrophic *Alphaproteobacteria* are shown in Table BXII.α.1 and carbon sources utilized in Table BXII.α.2 in the description of the genus *Rhodospirillum*. The phylogenetic relationships of these bacteria based on 16S rDNA sequences are shown in Fig. 1 (p. 124) of the introductory chapter "Anoxygenic Phototrophic Purple Bacteria", Volume 2, Part A.

List of species of the genus Rhodospira

1. **Rhodospira trueperi** Pfennig, Lünsdorf, Süling and Imhoff 1998, 328^VP (Effective publication: Pfennig, Lünsdorf, Süling and Imhoff 1997, 44.)

    true' pe.ri. M.L. gen. n. *trueperi* of Trüper, named for Hans Georg Trüper, a German microbiologist who contributed significantly to our knowledge of the anoxygenic phototrophic bacteria.

    Cells vibrioid to spirilloid in shape, 0.6–0.8 × 1.5–3.0 μm, and motile by bipolar tufts of flagella (2–5 fibrils). Internal photosynthetic membranes are of the vesicular type. Anaerobically grown cultures are beige to peach-colored. Living cell suspensions show absorption maxima at 397, 458, 490, 600, 689, 801, 889, and 986 nm. Photosynthetic pigments are bacteriochlorophyll *b* and the carotenoid tetrahydrospirilloxanthin.

    Grows photoheterotrophically under strictly anoxic conditions in the light and in the presence of sulfide as a reduced sulfur source. In the presence of sulfide and hydrogen carbonate and under anoxic conditions, acetate, pyruvate, propionate, butyrate, valerate, lactate, fumarate, malate, succinate, and crotonate are used as substrates. Ethanol and cyclohexane carboxylate are not used. Weak growth may occur under microoxic conditions in the dark. Sulfide is oxidized to S^0, which cannot be oxidized further to sulfate. Thiosulfate is not oxidized. Biotin, thiamine, and pantothenate are required as growth factors.

Marine bacterium with optimal growth at 25–30°C, pH 7.3–7.5 and 2% NaCl (in the presence of 0.3% $MgCl_2 \cdot 6H_2O$). Sodium chloride is required for growth, growth range is from 0.5–5.0% NaCl. Habitat: peach-colored layer of a laminated microbial mat of Great Sippewissett Salt Marsh.

The mol% G + C of the DNA is: 65.7 (by chemical analysis).

Type strain: Pfennig 8316, ATCC 700224.

GenBank accession number (16S rRNA): X99671.

Genus VIII. **Rhodovibrio** Imhoff, Petri and Süling 1998, 797[VP]

JOHANNES F. IMHOFF

Rho.do.vi'bri.o. Gr. n. *rhodon* the rose; M.L. masc. n. *Vibrio* a bacterial genus; M.L. masc. n. *Rhodovibrio* the rose *Vibrio*.

Cells are vibrioid to spiral shaped, motile by means of polar flagella and multiply by binary fission. **Gram negative, belonging to the *Alphaproteobacteria*.** Internal photosynthetic membranes are present as vesicles. Photosynthetic pigments are **bacteriochlorophyll *a* and carotenoids of the spirilloxanthin series. Ubiquinones and menaquinones with 10 isoprene units (Q-10 and MK-10) are present. Major cellular fatty acids are $C_{18:1}$ and $C_{18:0}$.**

Cells grow preferentially photoheterotrophically under anoxic conditions in the light. **Chemotrophic growth is possible under microoxic to oxic conditions in the dark.** Complex nutrients are required. **Halophilic bacteria that require NaCl or sea salt for growth** and have salt optima above seawater salinity. Habitat: anoxic zones of hypersaline environments such as salterns and salt lakes that are exposed to the light.

The mol% G + C of the DNA is: 66–67.

Type species: **Rhodovibrio salinarum** (Nissen and Dundas 1985) Imhoff, Petri and Süling 1998, 797 (*Rhodospirillum salinarum* Nissen and Dundas 1985, 224.)

FURTHER DESCRIPTIVE INFORMATION

Growth of *Rhodovibrio* species depends on hypersaline concentrations of salts. Both species grow well at 2 M NaCl. They show best growth in the presence of 0.1 M Mg^{2+} and can grow in media containing 1 M $MgCl_2$ (Mack et al., 1993). In contrast to *R. sodomensis*, which is tolerant to NaBr up to concentrations of 1.5 M, growth of *R. salinarum* is inhibited at much lower concentrations of NaBr (Mack et al., 1993). In response to the external salinity, glycine betaine and ectoine are accumulated as major and minor components, respectively, of the compatible solutes (Severin et al., 1992; Imhoff, 1992).

R. salinarum and *R. sodomensis* are unique among the spiral-shaped phototrophic *Alphaproteobacteria* in that they do not contain a soluble cytochrome c_2 (Moschettini et al., 1997; Bonora et al., 1998). *R. salinarum*, but not *R. sodomensis*, lacks a branched respiratory chain if grown aerobically in the dark (Moschettini et al., 1997). It contains two different high-potential iron–sulfur proteins (HiPIP), whereas *R. sodomensis* lacks HiPIPs (Bonora et al., 1998).

ENRICHMENT AND ISOLATION PROCEDURES

Rhodovibrio species are characterized by their requirement for high salt concentrations and complex nutrients. Therefore, complex media with salt concentrations of 10% and incubation anaerobically in the light provide selective conditions for the enrichment of *Rhodovibrio* species. Standard techniques for isolation

of anaerobic bacteria in agar dilution series and on agar plates can be applied for *Rhodovibrio* species (Imhoff, 1988; Imhoff and Trüper, 1992). The medium of Nissen and Dundas (1984) is suitable for growth of *Rhodovibrio salinarum* and can also be used for *Rhodovibrio sodomensis*. A recipe for this medium is given below.[1]

MAINTENANCE PROCEDURES

Cultures can be preserved by standard techniques in liquid nitrogen, by lyophilization, or storage at $-80°C$ in a mechanical freezer.

DIFFERENTIATION OF THE GENUS *RHODOVIBRIO* FROM OTHER GENERA

Major differentiating properties for *Rhodovibrio* species and other phototrophic spiral-shaped *Alphaproteobacteria* are shown in Table BXII.α.1 in the description of the genus *Rhodospirillum*.

TAXONOMIC COMMENTS

Two species, *Rhodovibrio salinarum* and *Rhodovibrio sodomensis*, are currently recognized (Imhoff et al., 1998). Because of their spiral shape and phototrophic capacity, *Rhodovibrio* species were originally assigned to the genus *Rhodospirillum* (Nissen and Dundas, 1984; Mack et al., 1993). With growing recognition of the high phenotypic and chemotaxonomic diversity of spiral-shaped phototrophic *Alphaproteobacteria* and the great genetic distance between most of these bacteria, a rearrangement of the species of this group became necessary. Several distinct differences between *Rhodovibrio* species and *Rhodospirillum rubrum* led to their reclassification in the new genus *Rhodovibrio* (Imhoff et al., 1998). The genus name *Rhodovibrio* had originally been proposed by Molisch (1907) for bacteria (*Rhodovibrio parvus*) later recognized as belonging to *Rhodopseudomonas palustris*. Because this name was not included in the Approved List of Names, it had no standing in nomenclature and could be revived (Imhoff et al., 1998). The phylogenetic relationships of these bacteria based on 16S rDNA sequences are shown in Fig. 1 (p. 124) of the introductory chapter "Anoxygenic Phototrophic Purple Bacteria", Volume 2, Part A.

---

1. The medium contains (g/l): NaCl, 80; $MgCl_2 \cdot 6H_2O$, 0.4; $CaCl_2 \cdot 2H_2O$, 0.1; $(NH_4)_2SO_4$, 0.5; KH_2PO_4, 2.5; sodium acetate, 1; sodium glutamate 1; and trace element solution SLA, 1 ml. The initial pH is adjusted to 7.0 (see also Imhoff, 1988).

FURTHER READING

Imhoff, J.F., R. Petri and J. Süling. 1998. Reclassification of species of the spiral-shaped phototrophic purple non-sulfur bacteria of the α-*Proteobacteria*: description of the new genera *Phaeospirillum* gen. nov., *Rhodovibrio* gen. nov., *Rhodothalassium* gen. nov. and *Roseospira* gen. nov. as well as transfer of *Rhodospirillum fulvum* to *Phaeospirillum fulvum* comb. nov., of *Rhodospirillum molischianum* to *Phaeospirillum molischianum* comb. nov., of *Rhodospirillum salinarum* to *Rhodovibrio salinarum* comb. nov., of *Rhodospirillum sodomense* to *Rhodovibrio sodomensis* comb. nov., of *Rhodospirillum salexigens* to *Rhodothalassium salexigens* comb. nov. and of *Rhodospirillum mediosalinum* to *Roseospira mediosalina* comb. nov. Int. J. Syst. Bacteriol. *48*: 793–798.

Kawasaki, H., Y. Hoshino and K. Yamasato. 1993. Phylogenetic diversity of phototrophic purple non-sulfur bacteria in the *alpha proteobacteria* group. FEMS Microbiol. Lett. *112*: 61–66.

DIFFERENTIATION OF THE SPECIES OF THE GENUS *RHODOVIBRIO*

Major differentiating properties for *Rhodovibrio* species are shown in Table BXII.α.1 in the description of the genus *Rhodospirillum*.

List of species of the genus Rhodovibrio

1. **Rhodovibrio salinarum** (Nissen and Dundas 1985) Imhoff, Petri and Süling 1998, 797[VP] (*Rhodospirillum salinarum* Nissen and Dundas 1985, 224.)

   sal.i.na' rum. L. fem. pl. n. *salinae* saltern or saltworks; M.L. adj. *salinarum* of a saltern.

   Cells are rod shaped to spiral, 0.8–0.9 × 1.0–3.5 μm, and motile by means of bipolar flagella. Internal photosynthetic membranes are present as vesicles. Color of anaerobically grown cultures is red. Absorption maxima of living cells are at 380, 490, 520, 550, 595, 800, and 870 nm. Photosynthetic pigments are bacteriochlorophyll *a* (esterified with phytol) and carotenoids of the spirilloxanthin series with spirilloxanthin as the major component.

   Growth occurs equally well photoheterotrophically under anoxic conditions in the light and chemoheterotrophically under oxic conditions in the dark. Cells have a complex nutrient requirement and do not grow in the absence of substantial amounts of yeast extract or peptone. In the presence of low amounts of peptone and yeast extract (0.015% each), Casamino acids and lactate support growth, but acetate, citrate, fumarate, glutamate, malate, propionate, pyruvate, succinate, fructose, galactose, glucose, sorbose, sucrose, glycerol, and mannitol do not. Sulfide and S⁰ are not used as electron donors for photosynthesis.

   Obligately and moderately halophilic bacterium with optimum growth at 42°C (temperature range: 20–45°C), pH 7.5–8.0 and 8–12% NaCl (range 3–24%). Habitat: anoxic zones of hypersaline environments such as salterns and salt lakes that are exposed to the light.

   The mol% G + C of the DNA is: 67.4 (Bd) and 68.1 (T_m).
   Type strain: ATCC 35394, DSM 9154.
   GenBank accession number (16S rRNA): M59069.

2. **Rhodovibrio sodomensis** (Mack, Mandelco, Woese and Madigan 1996) Imhoff, Petri and Süling 1998, 797[VP] (*Rhodospirillum sodomense* Mack, Mandelco, Woese and Madigan 1996, 1189.)

   so.do.men' sis. M.L. masc. adj. *sodomensis* pertaining to the Sea of Sodom, the Talmudic name for the Dead Sea.

   Cells vibrio shaped, occasionally true spirilla, 0.6–0.7 × 1.6–2.5 μm and weakly motile, presumably by polar flagella. Internal photosynthetic membranes are of the vesicular type. Photosynthetic cultures appear pink and major absorption maxima of living cells are at 377, 421, 516, 551, 591, and 875–880 nm. Photosynthetic pigments are bacteriochlorophyll *a* and carotenoids of the spirilloxanthin series. A B875 light-harvesting complex (but not a B800/850 complex) is present.

   Growth occurs photoheterotrophically under anoxic conditions in the light only in the presence of yeast extract. At low concentrations of yeast extract, growth is stimulated by acetate, malate, pyruvate, or succinate, but not by propionate, butyrate, and citrate. Chemotrophic growth is possible under oxic conditions in the dark with acetate and succinate in the presence of 0.3–0.5% yeast extract. Nitrogen sources used are ammonia and yeast extract, but evidence for N₂ fixation is lacking. Poly-β-hydroxybutyrate is produced as a storage polymer. Complex nutrients are required, and can be supplied with 0.01% yeast extract.

   Obligately and moderately halophilic bacterium with optimum growth at 35–40°C (range 25–47°C), pH 7.0, and 12% NaCl (salt range: 6–20% NaCl). Extremely bromide tolerant, growing well if 50–75% of NaCl is replaced by NaBr (up to 1.5 M NaBr). Habitat: anoxic parts of sediments and water of the Dead Sea that are exposed to the light.

   The mol% G + C of the DNA is: 66.2–66.6 (T_m).
   Type strain: ATCC 51195.
   GenBank accession number (16S rRNA): M59072.

Genus IX. **Roseospira** *Imhoff, Petri and Süling 1998, 797ᵛᵖ*

JOHANNES F. IMHOFF AND VLADIMIR M. GORLENKO

Ro'se.o.spi' ra. L. adj. *roseus* rosy; Gr. n. *spira* the spiral; M.L. fem. n. *Roseospira* the rosy spiral.

Cells are vibrioid to spiral shaped, motile by means of polar flagella and multiply by binary fission. **Gram negative, belonging to the *Alphaproteobacteria*.** Internal photosynthetic membranes are present as vesicles. Photosynthetic pigments are **bacteriochlorophyll *a* and carotenoids of the spirilloxanthin series**.

Cells grow preferentially photoheterotrophically under anoxic conditions in the light. Chemotrophic growth is possible under microoxic conditions in the dark. Growth factors are required. **Halophilic bacteria that require NaCl or sea salt for growth.**

Habitat: warm sulfur spring with elevated mineral salts concentration.

The mol% G + C of the DNA is: 66.6.

Type species: **Roseospira mediosalina** (Kompantseva and Gorlenko 1984) Imhoff, Petri and Süling 1998, 798 (*"Rhodospirillum mediosalinum"* Kompantseva and Gorlenko 1984, 780.)

ENRICHMENT AND ISOLATION PROCEDURES

Roseospira mediosalina was isolated from microbial mats of a warm sulfur spring (43–51°C) with elevated mineral salts concentration (2% total salts) and alkaline pH (pH 8.2) (Kompantseva and Gorlenko, 1984). Media and growth conditions applied for *Rhodospirillum* species are also suitable for *Roseospira* species, if appropriate salt concentrations are supplied. A recipe for a mineral salts medium, which can be used if 5% NaCl is added, is given with the description of the genus *Rhodospirillum*. Standard techniques for isolation of anaerobic bacteria in agar dilution series and on agar plates can be applied for *Roseospira* species (Imhoff, 1988; Imhoff and Trüper, 1992). *Roseospira mediosalina* can also be grown on Pfennig's medium for purple sulfur bacteria (see Genus *Chromatium*) with the addition of 5% NaCl and vitamins (or yeast extract).

MAINTENANCE PROCEDURES

Cultures can be preserved by standard techniques in liquid nitrogen, by lyophilization, or storage at −80°C in a mechanical freezer.

DIFFERENTIATION OF THE GENUS *ROSEOSPIRA* FROM OTHER GENERA

Major differentiating properties for *Roseospira mediosalina* and other spiral-shaped phototrophic *Alphaproteobacteria* are shown in Table BXII.α.1 and carbon sources utilized in Table BXII.α.2 in the description of the genus *Rhodospirillum*. The phylogenetic relationships of these bacteria based on 16S rDNA sequences are shown in Fig. 1 (p. 124) of the introductory chapter "Anoxygenic Phototrophic Purple Bacteria", Volume 2, Part A.

TAXONOMIC COMMENTS

Because of their spiral shape and phototrophic capacity, *Roseospira* species were originally assigned to the genus *Rhodospirillum*, although DNA homology to *R. rubrum* was insignificant (Kompantseva and Gorlenko, 1984). With growing recognition of the high phenotypic and chemotaxonomic diversity of spiral-shaped phototrophic *Alphaproteobacteria* and the great genetic distance between most of these bacteria (Kawasaki et al., 1993b), a rearrangement of the species of this group became necessary (Imhoff et al., 1998). The great differences between *Roseospira* and *Rhodospirillum* led to their reclassification (Imhoff et al., 1998). The phylogenetic relationships of these bacteria based on 16S rDNA sequences are shown in Fig. 1 (p. 124) of the introductory chapter "Anoxygenic Phototrophic Purple Bacteria", Volume, 2 Part A.

FURTHER READING

Imhoff, J.F., R. Petri and J. Süling. 1998. Reclassification of species of the spiral-shaped phototrophic purple non-sulfur bacteria of the α-*Proteobacteria*: description of the new genera *Phaeospirillum* gen. nov., *Rhodovibrio* gen. nov., *Rhodothalassium* gen. nov. and *Roseospira* gen. nov. as well as transfer of *Rhodospirillum fulvum* to *Phaeospirillum fulvum* comb. nov., of *Rhodospirillum molischianum* to *Phaeospirillum molischianum* comb. nov., of *Rhodospirillum salinarum* to *Rhodovibrio salinarum* comb. nov., of *Rhodospirillum sodomense* to *Rhodovibrio sodomensis* comb. nov., of *Rhodospirillum salexigens* to *Rhodothalassium salexigens* comb. nov. and of *Rhodospirillum mediosalinum* to *Roseospira mediosalina* comb. nov. Int. J. Syst. Bacteriol. *48*: 793–798.

Kompantseva, E.I. and V.M. Gorlenko. 1984. A new species of moderately halophilic purple bacterium, *Rhodospirillum mediosalinum*. Mikrobiologiya *53*: 954-961.

List of species of the genus Roseospira

1. **Roseospira mediosalina** (Kompantseva and Gorlenko 1984) Imhoff, Petri and Süling 1998, 798[VP] (*"Rhodospirillum mediosalinum"* Kompantseva and Gorlenko 1984, 780.)

   me′ di.o.sa.li′ na. L. adj. *medius* medium; L. neut. n. *salinum* saline, salty; M.L. fem. n. *mediosalina* living at a moderate salinity, moderate halophile.

   Cells are vibrioid to spiral shaped, 0.8–1.0 × 2.2–6.0 μm; complete turn of a spiral is 4.0–5.2 μm. They are motile by means of polar flagella and multiply by binary fission. Internal photosynthetic membranes are present as vesicles. Color of cell suspension ranges from pinkish to brownish red. Absorption maxima of living cells occur at 379, 474, 504, 534, 591, 803, and 860 nm. Photosynthetic pigments are bacteriochlorophyll *a* and carotenoids of the normal spirilloxanthin series.

   Growth is preferentially photoheterotrophic in the presence of organic substrates. Photolithoautotrophic growth occurs with sulfide. Sulfide is oxidized to S⁰, which is deposited outside the cell and not further oxidized. Sulfide tolerance is high (up to 2 mM). Thiosulfate does not serve as an electron donor. Chemotrophic growth is possible under microoxic conditions in the dark in the presence of organic substrates. Fatty acids, butyrate, lactate, pyruvate, amino acids (alanine, aspartate, glutamate), intermediates of the tricarboxylic acid cycle (malate, succinate, fumarate), yeast extract, and Casamino acids are used as carbon sources and electron donors. Benzoate, malonate, tartrate, formate, citrate, alcohols, sugars, and sugar alcohols are not used. Ammonium salts and urea can be utilized as nitrogen sources. Capable of assimilatory sulfate reduction. Storage material is poly-β-hydroxybutyrate. Catalase positive. Thiamine, niacin, and *p*-aminobenzoate are required for growth.

   Halophilic bacterium with optimum growth at 30–35°C, pH 7.0, and NaCl concentrations of 4–7% (NaCl range: 0.5–15%). No growth in the absence of salts. No growth above 40°C. Habitat: warm sulfur spring Astara (Azerbaijian) with elevated mineral salts concentration of 2%.

   The mol% G + C of the DNA is: 66.6 (T_m).

   Type strain: BN 280.

   GenBank accession number (16S rRNA): AJ000989.

Genus X. **Roseospirillum** *Glaeser and Overmann 2001, 793^VP (Effective publication: Glaeser and Overmann 1999, 414)*

JOHANNES F. IMHOFF

Ro.se.o.spi.ril' lum. L. adj. *roseus* rosy; M.L. neut. n. *spirillum* the spiral, a bacterial genus; M.L. neut. n. *Roseospirillum* the rosy spiral.

Cells are vibrioid to spiral shaped, motile by means of polar flagella and multiply by binary fission. **Gram negative and belong to the** *Alphaproteobacteria*. Internal photosynthetic membranes are present as lamellar stacks. Photosynthetic pigments are bacteriochlorophyll *a* and carotenoids.

Cells grow photolithoautotrophically and photoheterotrophically under anoxic conditions in the light. Chemotrophic growth is possible under microoxic conditions in the dark. Growth factors are required. Marine bacteria that **require NaCl or sea salt for optimum growth**.

Habitat: marine sediments and microbial mats.

The mol% G + C of the DNA is: 71.2.

Type species: **Roseospirillum parvum** Glaeser and Overmann 2001, 793 (Effective publication: Glaeser and Overmann 1999, 414.)

ENRICHMENT AND ISOLATION PROCEDURES

Media and growth conditions applied for *Rhodospirillum* species are also suitable for *Roseospirillum* species, if appropriate salt concentrations are supplied. A recipe of a mineral salts medium that can be used if 1–2% NaCl is added is given with the description of the genus *Rhodospirillum*. Standard techniques for isolation of anaerobic bacteria in agar dilution series and on agar plates can be applied (Imhoff, 1988; Imhoff and Trüper, 1992).

MAINTENANCE PROCEDURES

Cultures can be preserved by standard techniques in liquid nitrogen, by lyophilization or storage at −80°C in a mechanical freezer.

DIFFERENTIATION OF THE GENUS *ROSEOSPIRILLUM* FROM OTHER GENERA

Major differentiating properties between *Roseospirillum parvum* and other spiral-shaped phototrophic *Alphaproteobacteria* are shown in Table BXII.α.1, carbon sources utilized in Table BXII.α.2. The phylogenetic relationships of these bacteria based on 16S rDNA sequences are shown in Fig. 1 (p.124) of the introductory chapter "Anoxygenic Phototrophic Purple Bacteria", Volume 2, Part A.

List of species of the genus Roseospirillum

1. **Roseospirillum parvum** Glaeser and Overmann 2001, 793^VP (Effective publication: Glaeser and Overmann 1999, 414.) *par' vum.* M.L. neut. adj. *parvum* small.

   Cells are vibrioid to spiral shaped, 0.4–0.6 × 1.8–2.6 µm, motile by means of bipolar flagella and multiply by binary fission. Internal photosynthetic membranes are present as lamellae. Color of a cell suspension is pink to pinkish-red. Absorption maxima of living cells occur at 380, 492, 515, 549, 595, 806, and 911 nm. Photosynthetic pigments are bacteriochlorophyll *a* and the carotenoids spirilloxanthin and lycopenal.

   Photolithoautotrophic growth occurs with sulfide and thiosulfate. Sulfide is oxidized to elemental sulfur, which is deposited outside the cells and not further oxidized to sulfate. In the presence of sulfide and bicarbonate, acetate, propionate, butyrate, valerate, oxoglutarate, pyruvate, lactate, succinate, fumarate, malate, fructose, Casamino acids, yeast extract, L-alanine, and L-glutamate were assimilated. Chemotrophic growth is possible under microoxic conditions in the dark (with thiosulfate or with thiosulfate and acetate). Incapable of assimilatory sulfate reduction. Yeast extract is required to supply growth factors.

   Marine bacterium with optimum growth at 30°C, pH 7.9, and concentrations of 1–2% NaCl (range: up to more than 6% NaCl).

   Habitat: microbial mats of Great Sippewissett Salt Marsh.

   The mol% G + C of the DNA is: 71.2 (HPLC).

   Type strain: 930I, DSM 12498.

   GenBank accession number (16S rRNA): AJ011919.

Genus XI. **Skermanella** *Sly and Stackebrandt 1999, 543^VP*

THE EDITORIAL BOARD

Sker.ma.nel' la. M.L. dim. -ella ending; M.L. fem. dim. n. *Skermanella* named after V.B.D. Skerman, who first isolated this bacterium, and in honor of his contribution to bacterial systematics.

Gram negative, with unicellular and multicellular growth phases. Does not form spores. Unicellular phase cells are rod shaped, with rounded or tapered ends and a straight or curved axis. Motile with single polar flagellum and one or more lateral flagella of different wavelengths. **Multicellular phase conglomerates** arise from cells that lose motility, become optically refractile, and **reproduce by multiplanar septation**. No filamentous structures or buds are formed. Obligately chemoorganotrophic and **facultatively anaerobic**, having fermentative metabolism. Belongs to the *Alphaproteobacteria*.

The mol% G + C of the DNA is: 67.2 ± 0.8.

Type species: **Skermanella parooensis** (Skerman, Sly and Williamson 1983) Sly and Stackebrandt 1999, 543 (*Conglomeromonas largomobilis* subsp. *parooensis* Skerman, Sly and Williamson 1983, 307.)

List of species of the genus Skermanella

1. **Skermanella parooensis** (Skerman, Sly and Williamson 1983) Sly and Stackebrandt 1999, 543[VP] (*Conglomeromonas largomobilis* subsp. *parooensis* Skerman, Sly and Williamson 1983, 307.)

pa.roo.en' sis. M.L. fem. (and masc.) adj. *parooensis* belonging/pertaining to the Paroo, referring to the Paroo Channel in southwest Queensland, Australia, the source of the water from which the organism was isolated.

Characteristics are as described for the genus, with the following additions. Cells are 1.0–1.5 × 1.5 μm (peptone yeast extract agar) or 1.0–1.5 × 2–3 μm (lake water agar). Noncapsulated. Separated colonies on peptone yeast extract agar are in conglomerate form. Colonies are apricot-colored, opaque, round, and raised, with a rough surface and irregular edge after 72 h at 28°C. Conglomerates have dry appearance and are difficult to emulsify. Confluent growth of single cells is mucoid and emulsifies easily. Conglomerates have the same or greater diameter as colonies of the unicellular form. Colonies grown in the presence of Congo red do not turn scarlet. Aging cells contain poly-β-hydroxybutyrate inclusions. Optimal temperature: 28°C; range 10–37°C. Growth occurs on peptone yeast-extract agar, 0.1% peptone agar, glucose-ammonium sulfate agar, citrate-ammonium agar, and lake water agar. Slower growth occurs in the presence of 2% NaCl in the conglomerate form. No growth in the presence of 5% NaCl. Catalase, oxidase, aminopeptidase, phosphatase, deoxyribonuclease, and urease are produced. Nitrate reduced to nitrite. Utilizes citrate and malonate. Hydrolyzes Tween 80, but not esculin, chitin, alginate, cellulose, gelatin, or casein. Acid but not gas from cellobiose, galactose, glucose, inositol, lactose, mannitol, mannose, melibiose, and rhamnose. No acid from adonitol, amygdalin, D-arabinose, arabitol, dextrin, dulcitol, erythritol, ethanol, fructose, glycerol, inulin, maltose, melizitose, raffinose, D-ribose, salicin, sorbitol, L-sorbose, starch, sucrose, trehalose, or xylose. Utilizes the following carbon sources: D-arabinose, L-arabinose, D-fructose, gentiobiose, D-galactose, D-glucose, D-lyxose, D-mannitol, D-mannose, melibiose, L-rhamnose, D-ribose, and D-xylose, but not *N*-acetylglucosamine, D-cellobiose, D-gluconate, glycerol, inositol, lactose, maltose, D-sorbitol, or D-trehalose. The type strain was isolated from fresh water.

The mol% G + C of the DNA is: 67.2 ± 0.8.

Type strain: ACM 2042, CIP 106994, DSM 9527, UQM 2042.

GenBank accession number (16S rRNA): X90760.

Genus Incertae Sedis XII. "**Sporospirillum**" *Delaporte 1964, 257*

BRUNO POT AND MONIQUE GILLIS

Spor.o.spi.ril' lum. Gr. n. *sporos* a seed (spore); Gr. n. *spira* a spiral; M.L. dim. n. *spirillum* a small spiral; M.L. neut. n. *Sporospirillum* a small spore (-forming) spiral.

Rigid, helical bacteria of enormous size, 1.8–4.8 μm in diameter and 40–100 μm in length. Structures that morphologically resemble endospores occur within the cells, but their thermal resistance has not been determined. **The spore-like structures have the ability to rotate and to migrate within the cytoplasm of the bacteria.** They initially develop near the cell poles and subsequently migrate to the center where they are released after the cell ruptures and disintegrates. The Gram reaction has not been reported. The cells are **motile, but no organs of locomotion are evident**. The relationship of the cells to oxygen is unknown. Occur in the intestinal contents of tadpoles. Have not been isolated. A type strain has not been designated.

The mol% G + C of the DNA is: not determined.

Type species: "**Sporospirillum bisporum**" Delaporte 1964, 260.

ACKNOWLEDGMENTS

We sincerely acknowledge Dr. N.R. Krieg for kindly providing the template text upon which this chapter was based.

List of species of the genus "Sporospirillum"

1. "**Sporospirillum bisporum**" Delaporte 1964, 260.
   bi.spo' rum. L. adv. *bis* twice; G. n. *sporos* a seed; M.L. gen. pl. n. *bisporum* of two seeds (spores).

   Cell diameter is 3.5–4.8 μm. Cell length is 50–90 μm. Diameter of helix is 11–15 μm. Wavelength is 27–35 μm. An endospore (2–4 × 10–14 μm) occurs at each pole.

   The mol% G + C of the DNA is: not determined.

   Deposited strain: none designated.

2. "**Sporospirillum gyrini**" Delaporte 1964, 259.
   gy.ri' ni. L. n. *gyrinus* a tadpole; L. gen. n. *gyrini* of a tadpole.

   Cell diameter is 1.8–2.6 μm. Cell length is 40–100 μm. Diameter of helix is 3–6 μm. Wavelength is 13–20 μm. A single endospore is present, 2 × 5–7 μm.

   The mol% G + C of the DNA is: not determined.

   Deposited strain: none designated.

3. "**Sporospirillum praeclarum**" (Collin 1913) Delaporte 1964, 259 (*Spirillum praeclarum* Collin 1913, 62.)
   prae.cla' rum. L. adj. *praeclarum* distinguished, famous.

   Cell diameter is 3.0–4.0 μm. Cell length is 50–100 μm. Diameter of helix is 5–10 μm. Wavelength is 17–23 μm. A single endospore is present, 3–4 × 9–12 μm. A type strain has not been designated.

   The mol% G + C of the DNA is: not determined.

   Deposited strain: none designated.

Family II. **Acetobacteraceae** Gillis and De Ley 1980, 23[VP]

MARTIN SIEVERS AND JEAN SWINGS

A.ce.to.bac.te.ra' ce.ae. M.L. masc. n. *Acetobacter* type genus of the family; *-aceae* ending to denote a family; M.L. fem. pl. n. *Acetobacteraceae* the *Acetobacter* family.

The family *Acetobacteraceae*, the Gram-negative, aerobic, acetic acid bacteria, has undergone many taxonomic changes in respect to their genus status. Several species of acetic acid bacteria were newly described. Relationships among the acetic acid bacteria were studied genotypically by DNA–rRNA hybridization, DNA–DNA hybridization, and ribosomal RNA gene sequences (5S rRNA, 16S rRNA, and 23S rRNA). Changes at the generic level in the classification of some acetic acid bacteria are given in Table BXII.α.9. Phenotypic characteristics of six genera are shown in Table BXII.α.10. *Gluconacetobacter*, *Acidomonas*, and *Asaia* as well as *Gluconobacter* species, possess Q-10 ubiquinone as the major quinone component. The Q-10 equipped species contain in addition minor amounts of ubiquinone of the Q-9 type. The ubiquinone system of *Acetobacter pasteurianus*, *Acetobacter pomorum*, and *Acetobacter aceti* is Q-9 along with minor components of ubiquinone of the Q-8 type.

Type genus: **Acetobacter** Beijerinck 1898, 215.

FURTHER DESCRIPTIVE INFORMATION

Sugar metabolism Acetic acid bacteria catabolize sugars by means of the cytoplasmic hexose monophosphate pathway. Glycolysis is absent due to lack of phosphofructokinase (Leisinger, 1965; Attwood et al., 1991). The Entner–Doudoroff pathway occurs only in cellulose-synthesizing *Gluconacetobacter* strains, where

it appears to be more active than the hexose monophosphate cycle (White and Wang, 1964a, b). The ability to grow on glycerol is due to gluconeogenesis. Specific activities of the key enzymes of the gluconeogenic pathway such as fructose bisphosphatase, fructose bisphosphate aldolase, triosephosphate isomerase, and glycerophosphate dehydrogenase have been described for *Gluconacetobacter diazotrophicus* (Alvarez and Martínez-Drets, 1995). Strains that are able to grow on Hoyer's medium[1] with ethanol as the sole source of carbon and $(NH_4)_2SO_4$ as the only source of nitrogen utilize the glyoxylate bypass (Leisinger, 1965; De Ley et al., 1984).

Alcohol and aldehyde dehydrogenases of acetic acid bacteria The oxidizing systems of acetic acid bacteria are shown in Table BXII.α.11. Acetic acid bacteria oxidize ethanol to acetic acid by two successive catalytic reactions of a membrane-bound alcohol dehydrogenase (ADH) and a membrane-bound aldehyde dehydrogenase (ALDH). The enzymes are bound to the cytoplasmic membrane and face the periplasmic space. The ADH and ALDH complexes are tightly linked to the respiratory chain,

---

1. Hoyer's medium with ethanol consists of (per liter of distilled water): ethanol (99.8 %), 20 ml; $(NH_4)_2SO_4$, 1.0 g; K_2HPO_4, 0.1 g; KH_2PO_4, 0.9 g; $MgSO_4 \cdot 7H_2O$, 0.25 g; and $FeCl_3$, 0.005 g.

TABLE BXII.α.9. Former and current classification of acetic acid bacteria[a]

Former classification	Reference	Current classification	Reference
Acetobacter europaeus	Sievers et al., 1992	*Gluconacetobacter europaeus*	Yamada et al., 1998b
Acetobacter intermedius	Boesch et al., 1998	*Gluconacetobacter intermedius*	Yamada, 2000
Acetobacter oboediens	Sokollek et al., 1998b	*Gluconacetobacter oboediens*	Yamada, 2000
Acetobacter xylinus	Yamada, 1984	*Gluconacetobacter xylinus*	Yamada et al., 1998b
Acetobacter hansenii	De Ley et al., 1984	*Gluconacetobacter hansenii*	Yamada et al., 1998b
		Gluconacetobacter entanii	Schüller et al., 2000
Acetobacter diazotrophicus	Gillis et al., 1989	*Gluconacetobacter diazotrophicus*	Yamada et al., 1998b
		Gluconacetobacter johannae	Fuentes-Ramírez et al., 2001
		Gluconacetobacter azotocaptans	Fuentes-Ramírez et al., 2001
		Gluconacetobacter sacchari	Franke et al., 1999
Acetobacter liquefaciens	Gosselé et al., 1983b	*Gluconacetobacter liquefaciens* (type species)	Yamada et al., 1998b
Acetobacter aceti (type species)	Beijerinck, 1898	*Acetobacter aceti* (type species)	Beijerinck ,1898
Acetobacter pasteurianus	Beijerinck, 1916	*Acetobacter pasteurianus*	Beijerinck, 1916
Acetobacter pomorum	Sokollek et al., 1998b	*Acetobacter pomorum*	Sokollek et al., 1998b
Acetobacter peroxydans	Visser' t Hooft, 1925	*Acetobacter peroxydans*	Lisdiyanti et al., 2000
Acetobacter estunensis	Carr, 1958	*Acetobacter estunensis*	Lisdiyanti et al., 2000
Acetobacter lovaniensis	Frateur, 1950	*Acetobacter lovaniensis*	Lisdiyanti et al., 2000
Acetobacter orleanensis	Henneberg, 1906	*Acetobacter orleanensis*	Lisdiyanti et al., 2000
		Acetobacter indonesiensis	Lisdiyanti et al., 2000
		Acetobacter tropicalis	Lisdiyanti et al., 2000
		Acetobacter orientalis	Lisdiyanti et al., 2001a
		Acetobacter cibinongensis	Lisdiyanti et al., 2001a
		Acetobacter syzygii	Lisdiyanti et al., 2001a
Acetobacter methanolicus	Uhlig et al., 1986	*Acidomonas methanolica* (type species)	Urakami et al., 1989a
		Asaia bogorensis (type species)	Yamada et al., 2000
		Asaia siamensis	Katsura et al., 2001
Gluconobacter oxydans (type species)	De Ley, 1961	*Gluconobacter oxydans* (type species)	De Ley, 1961
Gluconobacter frateurii	Mason and Claus, 1989	*Gluconobacter frateurii*	Mason and Claus, 1989
Gluconobacter cerinus	Yamada and Akita, 1984a	*Gluconobacter cerinus*	Yamada and Akita, 1984a
Gluconobacter asaii	Mason and Claus, 1989	*Gluconobacter asaii*	Mason and Claus, 1989

[a]*Kozakia baliensis* gen. nov., sp. nov. has been published by Lisdiyanti et al., 2002, *International Journal of Systematic and Evolutionary Microbiology* 52: 813–818. *Acetobacter cerevisiae* sp. nov. and *Acetobacter malorum* sp. nov. have been decribed by Cleenwerck et al., 2002, *International Journal of Systematic and Evolutionary Microbiology* 52: 1551–1558.

TABLE BXII.α.10. Differential characteristics of the genera *Acetobacter, Acidomonas, Asaia, Gluconacetobacter, Gluconobacter,* and *Kozakia*[a]

Characteristic	Acetobacter	Acidomonas	Asaia	Gluconacetobacter	Gluconobacter	Kozakia
Flagellation	Peritrichous or nonmotile	Nonmotile	Peritrichous or nonmotile	Peritrichous or nonmotile	Polar or nonmotile	Nonmotile
Oxidation of ethanol to acetic acid	+	+	− or w[b]	+	+	+
Oxidation of acetic acid to CO_2 and H_2O	+	+	+	+[c]	−	w
Oxidation of lactate to CO_2 and H_2O	+	w	+	+ or −	−	w
Growth on 0.35% acetic-acid-containing medium	+	+	−	+	+	+
Growth on methanol	− or w[d]	+	−	−	−	−
Growth on D-mannitol	+ or −	w	+ or −	+ or −	+	+
Growth in the presence of 30% D-glucose	−	−	+	+ or −	− or w	−
Production of cellulose	−	−	−	+ or −	−	−
Production of levan-like mucous substance from sucrose	− or +	−	−	− or +	−	+
Fixation of molecular nitrgen	−	−	−	− or +	−	−
Ketogenesis (dihydroxyacetone) from glycerol	+ or −	w	− or w	+ or −	+	+
Acid production from:						
D-Mannitol	− or +	−	+ or −	+ or −	+	−
Glycerol	− or +	−	+	+	+	+
Raffinose	−	nd	−	−	−	+
Cellular fatty acid type	$C_{18:1\ \omega7}$	$C_{18:1\ \omega7}$	$C_{18:1\ \omega7}$	$C_{18:1\ \omega7}$	$C_{18:1\ \omega7}$	$C_{18:1\ \omega7}$
Ubiquinone type	Q-9	Q-10	Q-10	Q-10	Q-10	Q-10
DNA base composition (mol% G + C)	52–60	63–66	59–61	55–66	54–63	56–57

[a]Symbols: +, 90% or more of the strains positive; w, weakly positive reaction; −, 90% or more of the strains negative.

[b]*Asaia* does not produce acetic acid from ethanol with the exception of one strain producing acid weakly (Yamada et al., 2000).

[c]Overoxidation of acetate to CO_2 and H_2O depends on acetate concentration in the medium.

[d] *A. pomorum* assimilates methanol weakly (Sokollek et al., 1998b).

TABLE BXII.α.11. Oxidizing capacity of acetic acid bacteria

Substrate	Product	Enzyme
Ethanol	Acetic acid	Alcohol dehydrogenase, aldehyde dehydrogenase
iso-Butanol	*iso*-Butyric acid	Alcohol dehydrogenase, aldehyde dehydrogenase
α-Acetolactate	Acetoin	α-Acetolactate decarboxylase
D-Glycerol	Dihydroxyacetone	Glycerol dehydrogenase
D-Mannitol	D-Fructose	D-Mannitol dehydrogenase
D-Sorbitol	L-Sorbose	D-Sorbitol dehydrogenase
D-Glucose	D-Gluconic acid	D-Glucose dehydrogenase
D-Gluconic acid	2-Keto-D-gluconic acid	D-Gluconate dehydrogenase
D-Gluconic acid	5-Keto-D-gluconic acid	D-Gluconate dehydrogenase
2-Keto-D-gluconic acid	2,5-Diketo-D-gluconic acid	2-Keto-D-gluconate dehydrogenase
L-Sorbose	L-Sorbosone	L-Sorbose dehydrogenase
L-Sorbosone	2-Keto-D-gulonic acid	L-Sorbosone dehydrogenase
D-Fructose	5-Keto-D-fructose	D-Fructose dehydrogenase

which transfers electrons via ubiquinone and a terminal ubiquinol oxidase to oxygen as final electron acceptor.

The ADH complex in acetic acid bacteria is composed of two or three subunits. The molecular weight of the subunits estimated by SDS-PAGE are 72, 44, and 15 kDa in *Gluconobacter oxydans* (Kondo and Horinouchi, 1997b), 72, 50, and 15 kDa in *Acetobacter aceti* (Inoue et al., 1989, 1992; Matsushita et al., 1992d), 78, 48, and 20 in *Acetobacter pasteurianus* (Kondo et al., 1995), 80, 54, and 8 kDa in *Acidomonas methanolica* (Frébortová et al., 1997), 72 and 44 kDa in *"Acetobacter polyoxogenes"* (Tayama et al., 1989; Tamaki et al., 1991), and 80 and 49 kDa in *Gluconacetobacter europaeus* (Thurner, 1997). The larger subunit of the ADH possesses a heme C and a pyrroloquinoline quinone (PQQ) as cofactors and requires Ca^{2+} to be active, according to the catalytic mechanism given by Anthony (1996) and Goodwin and Anthony (1998). The second subunit is a cytochrome *c*. The third and smallest subunit of the three-component ADH is discussed to protect the catalytic subunit from proteolysis and to keep the correct conformation of the ADH complex for electron transport on the periplasmic surface (Kondo et al., 1995). The genes encoding the dehydrogenase subunit and the cytochrome *c* sub-

unit of ADH are clustered in the same transcription polarity and were cotranscribed. The cytochrome *c* subunit of the ADH complex of *Gluconobacter oxydans* has an additional function: it is a component of the non-energy-generating cyanide-insensitive bypass terminal oxidase in the respiratory chain (Matsushita et al., 1994). Thus, the cytochrome *c* subunit gene of *Gluconobacter oxydans* is transcribed independently from the dehydrogenase subunit gene (Matsushita et al., 1989). The ADH activity of *Acetobacter pasteurianus* is induced by ethanol. The ADH gene cluster utilizes two different promoters for its expression, in response to the presence or absence of ethanol in the culture medium (Takemura et al., 1993). Since ethanol is the major substrate for energy generation in acetic acid bacteria through the ADH and ALDH reactions, the role of the function of ethanol in regulatory mechanisms is to be elucidated. Transcription of the gene encoding the enzyme arylesterase, which is related in ethylacetate formation from acetic acid and ethanol, is induced by ethanol in *Acetobacter pasteurianus* (Kashima et al., 1999). In vinegar, esters weaken the strong smell of acetic acid (Kashima et al., 1999).

The ALDH complex of acetic acid bacteria, which catalyzes the oxidation of acetaldehyde to acetic acid, is composed of two or three subunits. The molecular masses are 86 and 55 kDa for *Gluconobacter oxydans*, 78, 45, and 14 kDa for *Acetobacter aceti*, and 75 and 19 kDa for "*A. polyoxogenes*" (Tamaki et al., 1989; Matsushita et al., 1994). The ALDH from *Gluconacetobacter europaeus* is composed of three subunits with molecular masses of 79, 49, and 17 kDa. The aldehyde dehydrogenase complex is organized as an operon. The larger catalytic subunit contains heme *b* and a pterin molybdenum instead of PQQ. The middle subunit is a cytochrome *c* with three heme *c* binding sites. The smallest subunit contains two [2Fe-2S] clusters (Thurner, 1997; Thurner et al., 1997). The catalytic subunit of the ALDH of *Gluconacetobacter europaeus* differs by nine amino acids from the corresponding protein of "*A. polyoxogenes*". Each gene of the ADH and ALDH subunits contains a signal sequence encoding for a leader peptide, responsible for translocation of the proteins through the cytoplasmic membrane. The ADH and ALDH of acetic acid bacteria have a broad substrate specificity and oxidize straight and branched chain alcohols and aldehydes into the corresponding carboxylic acids.

Alcohol and aldehyde dehydrogenases dependent on NAD(P)$^+$ have been isolated and characterized from the cytoplasm of acetic acid bacteria. The NAD(P)$^+$-dependent enzymes show an activity optimum at alkaline pH with much lower specific activities than those of quinoprotein enzymes, which show activity optimum at acidic pH (Matsushita et al., 1994).

Aldehydes produced during food processing can react with amino groups to form colored materials. ADH can scavenge aldehydes and carbonyl residues and since carboxylic acids react little with amino acids, the addition of acetic acid bacteria prevents coloration of food and reduces off-flavors (Nomura et al., 1995a). For instance, to reduce the stale flavor of cooked rice, rice grains have been incubated with a freeze-dried powder of acetic acid bacteria at 50°C, which resulted in a decrease of *n*-hexanol in the rice (Nomura et al., 1995b). Another example of the use of acetic acid bacteria to reduce off-flavors in food involves the cloning and expression of the gene that encodes α-acetolactate decarboxylase from *Gluconacetobacter xylinus* in brewer's yeast: during beer fermentation, brewer's yeasts produce α-acetolactate, which is converted to diacetyl by slow nonenzymatic oxidative decarboxylation. Diacetyl causes off-flavors in beer. The α-acetolactate decarboxylase of *Gluconacetobacter xylinus* converts the α-acetolactate directly to acetoin, which has no effect on the flavor of beer (Yamano et al., 1994).

Acetobacter, *Gluconacetobacter*, and *Acidomonas* are, in contrast to *Gluconobacter*, equipped with a complete tricarboxylic acid (TCA) cycle. The amount of TCA cycle enzymes in cells is correlated with the energy requirement of the cells (Rault-Leonardon et al., 1995). Key enzymes of the TCA cycle are influenced by different intermediate metabolites; an example is the citrate synthase of *Gluconacetobacter europaeus*, which is activated by acetate (Sievers et al., 1997). *Gluconacetobacter* and *Acetobacter* species producing high concentrations of acetic acid are not able to keep the internal pH constant and need additional ATP for internal protection against acetate and H$^+$ (Menzel and Gottschalk, 1985). Bacteria have to use acid-resistant systems like genes that are induced by environmental stimuli even if the cytoplasm is acidified by low external pH (Kobayashi et al., 2000). As shown for *Acetobacter aceti* and *Gluconobacter oxydans*, eight acetate-specific stress proteins (Asps) were overexpressed in both species during growth in the presence of acetate. Three Asps were similar and five Asps were different in both species analyzed by two-dimensional gel electrophoresis of total protein extracts (Lasko et al.,

1997). The acetate-activating enzyme, acetyl-CoA synthetase, has been purified from the cytoplasm of *Gluconacetobacter europaeus* and *Gluconobacter oxydans*. The native acetyl-CoA synthetase of *Gluconacetobacter europaeus* and *Gluconobacter oxydans* has a molecular mass near 150 kDa and is composed of two identical subunits with a molecular mass near 75 kDa, suggesting that the enzyme is a dimer. Due to the instability of the enzyme, 15% glycerol was included in the buffer system used. The activity of the acetyl-CoA synthetase was inhibited by AMP, ADP, and Na-citrate (M. Sievers, F. Tschudin, and M. Teuber, unpublished results). The activity of this enzyme is increased significantly in *Acetobacter* cells when acetate is consumed. Addition of glycerol to the culture medium of "*Acetobacter rancens*" increased acetate oxidation (Saeki et al., 1999). The genes responsible for acetic acid resistance in *Acetobacter aceti* are described by Fukaya et al. (1990) and Fukaya et al. (1993).

Respiratory chains Acetic acid bacteria are oxidase negative, with the exception of *Acidomonas methanolica*, which is oxidase positive during growth on methanol. The respiratory chain in *Gluconobacter* consists of cytochrome *c*, ubiquinone, and cytochrome *o* as a terminal ubiquinol oxidase; it also has a cyanide-insensitive alternative terminal ubiquinol oxidase. The respiratory chain in *Acetobacter* and *Gluconacetobacter* has cytochrome *c*, ubiquinone, and a terminal ubiquinol oxidase of the cytochrome a_1, cytochrome *d*, or cytochrome *o* type (Matsushita et al., 1994). The respiratory chain of *Acidomonas methanolica* contains two different terminal oxidases, viz. a cytochrome *c* oxidase and an ubiquinol oxidase (Matsushita et al., 1992c). Under methylotrophic conditions, *Acidomonas methanolica* produces a cytochrome *co* terminal cytochrome *c* oxidase (Chan and Anthony, 1991). When grown under nonmethylotrophic conditions, however, *Acidomonas methanolica* produces cytochrome *o* as the terminal ubiquinol oxidase (Matsushita et al., 1994). Cytochrome *o* from *Acetobacter aceti* and *Acidomonas methanolica* contains heme *o* (Puustinen and Wikström, 1991) and heme *b* in a 1:1 ratio (Matsushita et al., 1992b). One cell type of *Acetobacter aceti* forms smooth-surfaced colonies, while the other cell type of *Acetobacter aceti* forms rough-surfaced colonies. Cells of the former type predominate in shaking cultures, while cells of the latter type predominate in static cultures. There is a difference in the terminal oxidase, which is cytochrome a_1 in cells in shaking culture, but cytochrome *o* in cells grown statically (Matsushita et al., 1992a, 1994). Cytochrome a_1 has a higher affinity for oxygen than cytochrome *o* (Matsushita et al., 1992b), which results in enhanced growth of smooth-surfaced cells of *Acetobacter aceti* in shaking culture. The two different types of colonies of *Acetobacter aceti* with their corresponding types of cytochrome are shown in Fig. BXII.α.15 (Matsushita et al., 1994).

α-Amino acid ester hydrolase The α-amino acid ester hydrolase from "*Acetobacter turbidans*" ATCC 9325 is a beta-lactam antibiotic acylase capable of hydrolyzing beta-lactam antibiotics (cephalexin, ampicillin). The enzyme also catalyzes the transfer of the acyl group from α-amino acid esters to 7-aminocephem and 6-penam compounds. "*Acetobacter turbidans*" ATCC 9325 was described by Takahashi et al. (1972, 1974) as an acetic acid bacterium able to synthesize cephalosporins. The gene *aehA* (AF439262) encodes a dimeric protein with subunits of 70 kDa and an N-terminal leader sequence of 40 amino acids. The α-amino acid ester hydrolase has the motif GXSYXG in the active site, which is found in serine proteases, suggesting that the enzyme is a serine hydrolase (Polderman-Tijmes et al., 2002).

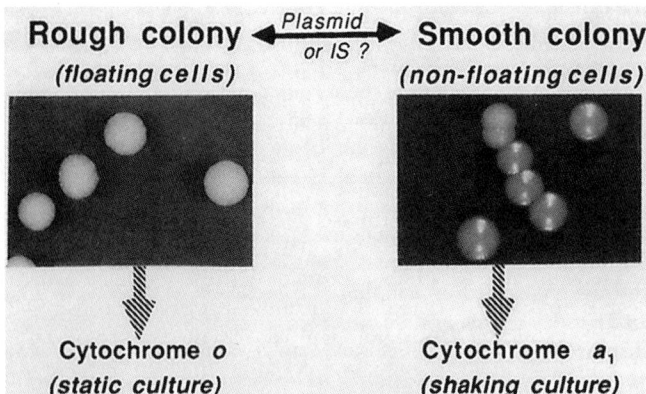

FIGURE BXII.α.15. Two different types of colonies of *Acetobacter aceti* from cells grown in static and shaking cultures. Cells forming rough-surfaced colonies can grow floating on the surface of the culture and thus predominate in static cultures. Cells forming smooth-surfaced colonies are incapable of floating and thus grow only in shaking cultures. The two types of cells produce cytochrome *o* and cytochrome a_1, respectively, as the terminal oxidase. (Reproduced with permission from K. Matsushita et al., Advances in Microbial Physiology, *36:* 247–301, 1994, ©Academic Press Limited, London.)

Production of cellulose and acetan   The production of cellulose by *Gluconacetobacter xylinus* has been regarded as the model system for the study of biochemistry and genetics of cellulose biogenesis and is described by Ross et al. (1991), Cannon and Anderson (1991), and Delmer and Amor (1995). The rate of cellulose production in *Gluconacetobacter xylinus* is proportional to the rate of cell growth, and the yield is dependent on the source of carbon (Embuscado et al., 1994; Oikawa et al., 1995). The cellulose production by *Gluconacetobacter xylinus* strains is enhanced by the addition of small amounts of cellulase (endo-β-1,4-glucanase) from *Bacillus subtilis* to the production culture and the amount of produced acetan is reduced (Tonouchi et al., 1995). Activators for the bacterial cellulose production were compounds like caffeine and related xanthines. The suggested target for these activators was the diguanyl cyclic phosphodiesterase whose inhibition favors the cellulose biosynthesis (Fontana et al., 1991). The produced cellulose is of high purity. Products from bacterial cellulose include wound dressings, skin substitutions, high-quality additives for paper, fiber glass filter sheets, chewing gum, food stabilizers, and acoustic diaphragms for audio instruments (Ross et al., 1991). A mechanical separation method and an alkali treatment for the isolation of bacterial cellulose with removal of bacterial cells were developed (Embuscado et al., 1996).

Gluconacetobacter xylinus secretes long β-1,4 glucan chains in a diameter of 1.5 nm from pores along the longitudinal axis of the cell. These glucans aggregate into 3–4-nm microfibrils by crystallization with subsequent assembly to ribbons in the surrounding medium (Cannon and Anderson, 1991; Ross et al., 1991). The membrane-bound cellulose synthase uses UDP-glucose directly as a substrate and catalyzes the polymerization of the glucose residues to β-1,4 glucan (Saxena et al., 1990). The genes coding for cellulose production are organized in the form of an operon. An operon (*acs*) consisting of three genes is functional for the final steps in cellulose biosynthesis in *Gluconacetobacter xylinus* ATCC 53582. The first gene (*acsAB*) codes for the 168-kDa cellulose synthase, the second gene (*acsC*) codes for a 138-kDa pore-protein, and the third gene (*acsD*) of the operon codes for a 17-kDa protein, which is involved in the crystallization of the microfibrils (Saxena et al., 1994). An additional gene encoding cellulolytic activity (*acsAII*) that is similar to the *acsAB* gene product was detected upstream of the cellulose-synthesizing operon in a *Gluconacetobacter xylinus* strain (Saxena and Brown, 1995). In the *bcs* operon from *Gluconacetobacter xylinus* 1306-3 and BPR 2001 a difference in the numbers of open reading frames coding for cellulose synthase is observed in comparison to the *acs* operon (Wong et al., 1990; Nakai et al., 1998). The first two of the four open reading frames of the *bcs* operon encodes the cellulose synthase subunit A (*bcsA*) and B (*bcsB*). In addition, two open reading frames ORF1 and ORF2 are located upstream of the *bcs* operon. The gene product of ORF1 is identified as cellulase (Standal et al., 1994). The presence of cellulase enhances the production of cellulose, and the integration of IS *1031*A in ORF2 causes the inability to produce cellulose. The promoter of the *bcs* operon upstream from *bcsA* and downstream from ORF1 encoding CMCase is overlapping with the ORF2 (Nakai et al., 1998).

The cellulose synthase is activated by cyclic diguanylic acid (c-di-GMP), which acts as an allosteric effector and stimulates the enzyme reaction rate up to 200-fold (Ross et al., 1987). The enzymes diguanylate cyclase and phosphodiesterase A catalyze the synthesis and degradation, respectively, of c-di-GMP and thus have regulatory effects on cellulose biosynthesis. The genes encoding isoenzymes for phosphodiesterase A and diguanylate cyclase are organized on three unlinked homologous operons (Tal et al., 1998).

A conserved UDP-glucose/UDP-*N*-acetylglucosamine binding motif exists among glycosyltransferases using UDP-glucose or UDP-*N*-acetylglucosamine as a substrate (Delmer and Amor, 1995). Cotton and rice contain amino acid sequences homologous to bacterial sequences of the catalytic part of the cellulose synthase (Pear et al., 1996).

Gluconacetobacter xylinus synthesizes a cellulose mat, which covers the surface of the growth medium in static cultures, whereas round balls of cellulose are formed in shaking cultures. Aeration of cultures of *Gluconacetobacter xylinus* strains by stirring or shaking gives rise to the formation of spontaneous non-cellulose-producing mutants (Ross et al., 1991). This phenomenon is also observed by *Gluconacetobacter europaeus* strains. The Cel⁻ mutants are stable during passages of cultivations in static and shaken cultures.

Acetan, an acidic exopolysaccharide that is related to xanthan, has been isolated from the culture medium of *Gluconacetobacter xylinus* (Couso et al., 1987). Acetan contains glucose, mannose, glucuronic acid, and rhamnose in a molar ratio of 4:1:1:1. The described structure is composed of a pentasaccharide side chain Rha(1-6)Glc (β1-6)Glc(α1-4)GlcA(β1-2)Man, which is α-1,3-linked to every second glucose residue of the β-1,4 glucan chain (Couso et al., 1987). Genes encoding UDP-glucose dehydrogenase, UDP-glycosyl transferase, GDP-mannosyl transferase, and phosphomannose isomerase/GDP-mannose pyrophosphorylase of the acetan biosynthetic pathway have been identified and sequenced from *Gluconacetobacter xylinus* (Griffin et al., 1994, 1996, 1997; Petroni and Ielpi, 1996).

Nitrogen fixation   Diazotrophy, the ability to fix free nitrogen gas (N₂) into cell material by reducing it to ammonium, occurs in species of more than 100 genera of *Bacteria* and *Archaea*. Proteins for nitrogen fixation (*nif*) have common structures and functions, while the degree of linkage and arrangement of specific *nif* and associated genes vary considerably in many diazotrophs (Lee et al., 2000). Among the acetic acid bacteria, *Gluconacetobacter diazotrophicus*, *Gluconacetobacter johannae*, and *Gluconacetobacter azotocaptans* were able to fix molecular nitrogen. The *nifH* gene sequence encoding the Fe protein of the nitrogenase and *nifHDK* genes of *Gluconacetobacter diazotrophicus* are highly similar to those of other nitrogen-fixing bacteria, particularly with other diazotrophic members of the *Alphaproteobacteria* (Franke et al., 1998; Sevilla et al., 1998). The *nifA* and *nifB* genes are adjacent and located upstream from the *nifHDK* genes. The *nifA* gene encoding the transcriptional activator for expression of *nif* genes is repressed at high ammonium concentrations (Teixeira et al., 1999). The cluster of *nif*, *fix*, and associated genes of *Gluconacetobacter diazotrophicus* is about 30.5 kb in size and encodes for 33 proteins with a molecular mass sum up to 1095.5 kDa (Lee et al., 2000). Individual gene products of the cluster of *nif*, *fix*, and associated genes like *mcpA*, which is involved in chemotaxis, are similar to those in species of *Rhizobiaceae* or in *Rhodobacter capsulatus* (Lee et al., 2000).

Insertion sequences Insertion sequences integrated in the genome of acetic acid bacteria are responsible for genetic instability leading to deficiencies in physiological properties like inactivation of alcohol dehydrogenase and loss of cellulose and acetan synthesis. IS elements characterized from acetic acid bacteria are listed in Table BXII.α.12.

The reported IS elements are widely distributed in acetic acid bacteria. Their copy numbers are from 1 to about 16 with the exception of IS*1380*, whose copy number is about 100. The distribution of IS elements in *Gluconacetobacter xylinus*, *Acetobacter pasteurianus*, *Acetobacter aceti*, *Gluconacetobacter europaeus*, *Gluconacetobacter intermedius*, "*A. polyoxogenes*", *Gluconobacter oxydans*, and *Gluconobacter cerinus* indicates frequent horizontal and vertical gene transfer between strains of acetic acid bacteria. The IS *1031* family, consisting of IS*1031*A to IS*1031*D (Coucheron, 1993) and the related IS elements IS *1032* (Iversen et al., 1994) and IS*12528*

(Kondo and Horinouchi, 1997a) are members of the heterogeneous IS*5* group in the higher IS*4* family based on conserved transposase motifs (Rezsöhazy et al., 1993). Additional insertion elements are present in the chromosome of *Acetobacter pasteurianus* NCIB 7214. The strain contains five copies of IS*12528* (Kondo and Horinouchi, 1997a), 10 copies of IS *1452* (Kondo and Horinouchi, 1997c), and 100 copies of IS*1380* (Takemura et al., 1991) with a total insertion element length of 188.5 kb, which is over 6% of the nucleotide sequence of the genome (Kondo and Horinouchi, 1997a). Similarly, *Acetobacter aceti* 1023 contains one copy of IS*12528*, three copies of IS*1452*, and 100 copies of IS*1380* with a total length of 175 kb, corresponding to 6% of the chromosome (Kondo and Horinouchi, 1997a).

Plasmids and phages A variety of plasmids has been reported from strains of *Acetobacter*, *Gluconacetobacter*, and *Gluconobacter*, and genetic transformation systems and plasmid vectors for acetic acid bacteria have been established. An overview of the genetics of *Acetobacter* with respect to acetic acid fermentation has been given by Beppu (1993). Conjugal transfer systems by using broad host-range plasmids based on RK2 replicon as vectors for acetic acid bacteria have been described (Valla et al., 1985; Blatny et al., 1997). For example, plasmid p21R1 (RK2-derivate carrying the gene encoding for levansucrase) was introduced in *Gluconacetobacter diazotrophicus*. In the recombinant bacterium, levansucrase production was increased about fourfold due to the low copy number of p21R1 (Hernández et al., 1999). Electroporation systems for *Gluconobacter oxydans* (Creaven et al., 1994) and *Gluconacetobacter xylinus* (Hall et al., 1992) with heterologous plasmids that gave transformation frequencies of up to 10^5 transformants/μg of DNA have been developed. An efficient electroporation method for transformation of "*A. polyoxogenes*" with plasmid DNA was adapted to the cultivation parameters of this organism (Tayama et al., 1994).

Introduction of the ALDH gene from "*A. polyoxogenes*" in *Gluconacetobacter xylinus* enhanced the production of acetic acid by overexpression of the cloned gene (Fukaya et al., 1989). The

TABLE BXII.α.12. Insertion sequences of acetic acid bacteria

IS element	Strain	GenBank accession number	Mutant deficiency	Insertion	Target site of duplication[a]	bp	Terminal inverted repeat bp	Reference
IS*1380*	*Acetobacter pasteurianus* (NCI1380)	D90424	Alcohol oxidation	Cytochrome *c*	TCGA	1665	15	Takemura et al. (1991)
IS*1031*A[b]	*Gluconacetobacter xylinus* ATCC 23769	M80805	Cellulose production	ORF 2, 500 bp upstream of the operon for cellulose biosynthesis	TGA	930	24	Coucheron (1991), Standal et al. (1994)
IS*1032*	*Gluconacetobacter xylinus* ATCC 23770	U02294	Acetan production	Gene required for acetan production	TCA	916	IR-Left: 14, IR-Right: 16	Iversen et al. (1994)
IS*1452*	*Acetobacter pasteurianus* (NCI1452)	D63923	Alcohol oxidation	Subunit III of the alcohol dehydrogenase complex	CTAR	1411	21	Kondo and Horinouchi (1997c)
IS*12528*	*Gluconobacter oxydans* IFO 12528	D86632	Alcohol oxidation	Dehydrogenase subunit of the alcohol dehydrogenase complex	TMA	904	18	Kondo and Horinouchi (1997a)

[a]Preferred target site M = A,C; R = A,G.

[b]IS*1031*A is 87.1% homolog to IS*1031*C and IS*1031*D from *G. xylinus* ATCC 23769 (Coucheron, 1993).

1440-bp plasmid pAP12875 isolated from *Acetobacter pasteurianus* contains two open reading frames, of which one has similarity to the replicon protein of pVT736-1 from *Actinobacillus actinomycetemcomitans* and the 32-kDa protein of phage Pf3 from *Pseudomonas aeruginosa* (Fomenkov et al., 1995). Plasmid pAH4 was characterized from a cellulose-producing *Acetobacter* (*Gluconacetobacter*) strain; it consisted of 4002 bp, contained an A–T rich region, and encoded four open reading frames. A shuttle vector system of *Escherichia coli* and this strain was constructed by connecting pAH4 to pUC18 (Tonouchi et al., 1994). Mobilization regions of plasmid pAEU601 (3818 bp) from *Gluconacetobacter europaeus* DSM 6160 have DNA similarities to plasmids RSF1010 from *E. coli* and pTF1 from *Thiobacillus ferrooxidans*, contributing to the broad distribution of pAEU601 in *Gluconacetobacter europaeus* strains via conjugation events (Boesch, 1998). Shuttle vectors (Amp^r) for cloning and expression of genes in *Gluconacetobacter europaeus*, *Gluconacetobacter intermedius*, and *E. coli* were constructed by ligation of pJK2-1 from *Gluconacetobacter europaeus* and pUC18 (Trček et al., 2000).

Lysogenic phage Acm1 from *Acidomonas methanolica* has been described and characterized (Wünsche et al., 1983; Mamat et al., 1995). Phage particles from vinegar fermentations, with hexagonal and icosaedric heads, were morphologically characterized and proved to belong to group A and C of Bradley (Bradley, 1967), respectively (Teuber et al., 1987a; Stamm et al., 1989; Defives et al., 1990; Sellmer et al., 1992). Seven morphologically different types of phage particles from vinegar fermentations with hexagonal heads were identified based on head size, tail length, and diameter by Sellmer et al. (1992). Head sizes varied between 60 and 110 nm, the corresponding tail length ranged from 99 to 360 nm, and tail diameter ranged from 14 to 31 nm (Sellmer et al., 1992). *Gluconacetobacter europaeus* does not grow well in soft agar; consequently, the classical plaque test is not suitable to enumerate infective bacteriophages from vinegar fer-

mentations (Sievers and Teuber, 1995). Restriction-modification systems have been detected in *Acetobacter* and *Gluconacetobacter* strains and further characterized (Tagami et al., 1988; Suzuki et al., 1996; Coucheron, 1997; Nwankwo et al., 1997).

TAXONOMIC COMMENTS

Phylogenetic affiliations of *Acetobacter*, *Gluconacetobacter*, *Acidomonas*, *Asaia*, and *Gluconobacter* species and related acidophilic bacteria have been studied on the basis of 5S (Bulygina et al., 1992) and 16S rRNA sequences (Sievers et al., 1994a, 1995a; Kishimoto et al., 1995b; Yamada et al., 1997, 2000; Sokollek et al., 1998b; Franke et al., 1999; Schüller et al., 2000; Lisdiyanti et al., 2001a). All of the 16S rRNA sequences show nucleotide deletions in their loop helices V1, V2, and V3, which are characteristic for the *Alphaproteobacteria*. The clustering of *Gluconacetobacter*, *Acetobacter*, and *Gluconobacter* in the *Alphaproteobacteria* had already been shown by Gillis and De Ley (1980) based on RNA–DNA hybridization studies. All species of the six genera of acetic acid bacteria *Acetobacteraceae*, together with *Acidisphaera rubrifaciens*, *Rhodopila globiformis*, *Acidocella*, *Acidiphilium*, and *Roseococcus thiosulfatophilus* as phylogenetic neighbors, formed a cluster with a distinct line of descent (Fig. BXII.α.16). These organisms were characterized by an acidophilic phenotype with the exception of *Roseococcus thiosulfatophilus*. These acidophilic bacteria became adapted to acidic environments after branching off from the common ancestor of the *Alphaproteobacteria* (Kishimoto et al., 1995b). A phylogenetic tree reflecting the close relationships of *Acetobacter*, *Gluconacetobacter*, *Acidomonas*, *Asaia*, and *Gluconobacter* based on the maximum parsimony analysis is shown in Fig. BXII.α.17. Phylogenetically, the acetic acid bacteria form four major clusters, one containing the *Gluconacetobacter* species with a subcluster comprising *Gluconacetobacter europaeus*, *Gluconacetobacter xylinus*, *Gluconacetobacter intermedius*, *Gluconacetobacter oboediens*, *Gluconacetobacter entanii*, *Gluconacetobacter hansenii*, and one

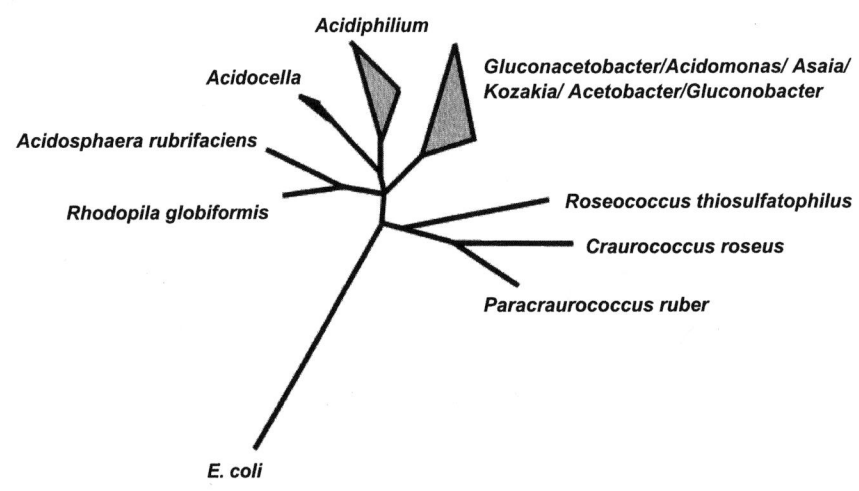

FIGURE BXII.α.16. Phylogenetic tree based on maximum parsimony analysis reflecting the distant relationships of the *Acetobacter–Gluconobacter* cluster and their close relatives among members of the *Alphaproteobacteria*. The topology of the tree was evaluated by applying distance matrix and maximum likelihood analyses. The lengths of the edges indicate the extremes of high and low phylogenetic distances among the members of the particular group. The bar indicates 10% estimated sequence difference. (Courtesy of Wolfgang Ludwig, TU Munich, Germany.)

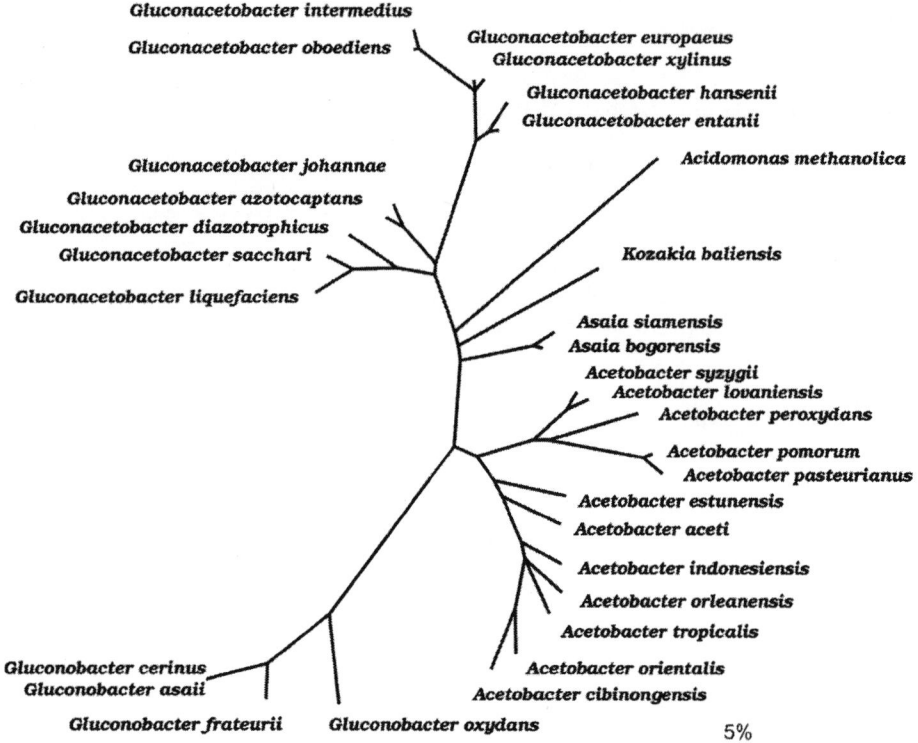

FIGURE BXII.α.17. Phylogenetic tree reflecting the close relationships of the species *Gluconacetobacter, Acidomonas methanolica, Asaia, Acetobacter,* and *Gluconobacter.* The tree is based on the results of a maximum parsimony analysis. The topology of the resulting tree were evaluated by applying the treeing methods (distance matrix, maximum parsimony, and maximum likelihood) to various data sets. The bar indicates 5% estimated sequence difference. (Courtesy of Wolfgang Ludwig, TU Munich, Germany.)

subcluster comprising *Gluconacetobacter sacchari, Gluconacetobacter liquefaciens, Gluconacetobacter diazotrophicus, Gluconacetobacter azotocaptans, Gluconacetobacter johannae,* and a second cluster containing *Acetobacter syzygii, Acetobacter lovaniensis, Acetobacter peroxydans, Acetobacter pomorum, Acetobacter pasteurianus, Acetobacter estunensis, Acetobacter aceti, Acetobacter indonesiensis, Acetobacter orleanensis, Acetobacter tropicalis, Acetobacter orientalis, Acetobacter cibinongensis,* and a third cluster comprising *Gluconobacter oxydans, Gluconobacter asaii, Gluconobacter cerinus,* and *Gluconobacter frateurii. Kozakia baliensis* formed a sublineage separated from *Asaia bogorensis* and *Asaia siamensis. Acidomonas methanolica* represents a distinct phylogenetic line and the separate branching of this organism is supported by the majority of the treeing analyses. The overall 16S rRNA sequence similarity values of the type species of five genera of acetic acid bacteria are given in Table BXII.α.13. The 16S rRNA sequence similarity between the species of the genus *Gluconacetobacter* is above 96.5%. *Gluconacetobacter europaeus, Gluconacetobacter xylinus, Gluconacetobacter intermedius,* and *Gluconacetobacter oboediens* are closely related by sharing over 99% 16S rRNA sequence similarities. Within the genus *Acetobacter,* the overall 16S rRNA sequence similarities are above 95.8% and less than 96,6% with those of other genera. Within the genus *Gluconobacter,* the overall 16S rRNA sequence similarities are 97.0–98.8%. *Gluconobacter* is well separated from the other genera by numerical analyses of protein gel electrophoregrams (Gosselé et al., 1983a, b; Sievers and Teuber, 1995) and by phenotypic features (Gosselé et al., 1983a, b). Two cellulose-producing *Gluconacetobacter* strains (ITDI 2.1 and PA 2.2) involved in nata de coco production were characterized by Bernardo et al. (1998).

Based on 16S rDNA sequence analysis, strain ITDI 2.1 is closely related to *Gluconacetobacter xylinus* and strain PA 2.2 to *Gluconacetobacter hansenii;* these strains are discussed as new subspecies under these species designations (Bernardo et al., 1998).

The 16S–23S rDNA intergenic spacer regions of *Gluconacetobacter europaeus, Gluconacetobacter intermedius,* and *Gluconacetobacter xylinus* contain genes encoding for two tRNA molecules specific for L-isoleucine and L-alanine. The tRNAIle and tRNAAla sequences are identical in these species. *Gluconacetobacter* seems to contain four copies of the *rrn* operons on its chromosome. Downstream from the tRNAAla gene a boxA with the nucleotide sequence TGCTCTTTGATA and a putative boxB consisting of 30 nucleotides with short inverted repeats are present in the spacer regions of these species (Sievers et al., 1996). These antitermination sequences are necessary to prevent premature termination of the precursor rRNA (Condon et al., 1995).

PCR-RFLP of the 16S rDNA with *Taq*I and *Rsa*I allows identification of acetic acid bacteria on genus level and in some cases at species level (Ruiz et al., 2000). 16S–23S rDNA intergenic spacer restriction patterns can be used to identify strains of acetic acid bacteria due to the variability of these sequences. The type strains of *Gluconacetobacter europaeus, Gluconacetobacter intermedius,* and *Gluconacetobacter xylinus* with a high percentage of 16S rDNA similarity have different sequences of their 16S–23S rDNA intergenic spacer regions (Sievers et al., 1996).

For characterization of cultures producing high percentage spirit, wine, and cider vinegar, plasmid profiling has been used to allow definite conclusions regarding origin, stability, and composition of the *Gluconacetobacter europaeus* strains isolated from

TABLE BXII.α.13. Overall 16S rRNA sequence similarity values for *Gluconacetobacter*, *Asaia*, *Acidomonas*, *Acetobacter*, and *Gluconobacter* species

Species	% 16S rRNA sequence similarity														
	Acetobacter aceti	*Acetobacter cibinongensis*	*Acetobacter estunensis*	*Acetobacter indonesiensis*	*Acetobacter lovaniensis*	*Acetobacter orientalis*	*Acetobacter orleanensis*	*Acetobacter pasteurianus*	*Acetobacter peroxydans*	*Acetobacter pomorum*	*Acetobacter syzygii*	*Acetobacter tropicalis*	*Acidomonas methanolica*	*Asaia bogorensis*	*Asaia siamensis*
Acetobacter aceti	100	98	97.9	98	97.2	97.8	97.8	96.8	96.6	97.3	97.5	97.9	95.2	96.5	96.3
Acetobacter cibinongensis	98	100	97.7	98.6	97.4	99	98.1	96.8	96.8	97.2	97.6	98.4	94.7	95.7	95.5
Acetobacter estunensis	97.9	97.7	100	97.5	97.1	97.4	97.5	96.3	97.1	96.6	97.3	97.4	94.9	96.6	96.4
Acetobacter indonesiensis	98	98.6	97.5	100	97.1	97.6	98.3	96.4	96.7	96.7	97.3	98.5	94.7	95.7	95.7
Acetobacter lovaniensis	97.2	97.4	97.1	97.1	100	97.6	97	97.3	97.7	97.6	99.4	97.6	95.1	96.6	96.5
Acetobacter orientalis	97.8	99	97.4	97.6	97.6	100	98.4	96.8	96.5	97.1	97.7	98.7	95.2	96	95.9
Acetobacter orleanensis	97.8	98.1	97.5	98.3	97	98.4	100	95.9	97.5	96.2	96.9	98.7	95.1	95.8	95.7
Acetobacter pasteurianus	96.8	96.8	96.3	96.4	97.3	96.8	95.9	100	97.5	99.4	97.7	96.4	94.7	95.2	95
Acetobacter peroxydans	96.6	96.8	97.1	96.7	97.7	96.5	97.5	97.5	100	97.9	97.8	97	95.2	95.8	95.6
Acetobacter pomorum	97.3	97.2	96.6	96.7	97.6	97.1	96.2	99.4	97.9	100	97.9	96.7	95	95.7	95.5
Acetobacter syzygii	97.5	97.6	97.3	97.3	99.4	97.7	96.9	97.7	97.8	97.9	100	97.7	95.2	96.4	96.2
Acetobacter tropicalis	97.9	98.4	97.4	98.5	97.6	98.7	98.7	96.4	97	96.7	97.7	100	95.1	96.1	95.9
Acidomonas methanolica	95.2	94.7	94.9	94.7	95.1	95.2	95.1	94.7	95.2	95	95.2	95.1	100	96.2	96
Asaia bogorensis	96.5	95.7	96.6	95.7	96.6	96	95.8	95.2	95.8	95.7	96.4	96.1	96.2	100	99.8
Asaia siamensis	96.3	95.5	96.4	95.7	96.5	95.9	95.7	95	95.6	95.5	96.2	95.9	96	99.8	100
Gluconacetobacter liquefaciens	95.9	95.9	95.8	95.6	96.2	96.1	95.6	95.5	95.5	95.7	96.7	96	96.6	96.6	96.4
Gluconacetobacter azotocaptans	95.6	95	95	95	95.6	95.4	95.4	94.9	95	95	96.1	95.6	96	96.4	96.2
Gluconacetobacter diazotrophicus	95.8	95.3	95.4	95.1	95.7	95.3	95.3	95.3	95.2	95.4	96.3	95.3	96	96.2	96
Gluconacetobacter entanii	95.3	94.9	95.3	94.8	95.4	95	95.2	94.8	94.5	95.3	95.6	95.6	95.5	95.9	95.7
Gluconacetobacter europaeus	95.5	95.1	95.5	94.9	95.8	95.1	95.3	94.9	94.5	95.1	95.9	95.8	95.8	96.3	96.1
Gluconacetobacter hansenii	95	94.6	94.9	94.5	95	94.7	94.6	94.3	94.1	94.5	95.3	95.2	95.3	95.5	95.4
Gluconacetobacter intermedius	95.7	95.3	95.8	95.2	96.2	95.4	95.7	95.3	94.9	95.4	96.1	96	95.8	96.6	96.5
Gluconacetobacter johannae	95.4	94.9	95.4	94.8	95.6	95.1	95.4	95.1	95.2	95.3	96	95.4	95.9	96.6	96.1
Gluconacetobacter oboediens	95.7	95.3	95.8	95.2	96.2	95.4	95.7	95.3	94.9	95.4	96.4	96	95.8	96.5	96.3
Gluconacetobacter sacchari	95.7	95.7	26.2	95.6	96.1	96.1	95.7	95.1	95.7	95.3	95.7	96.1	96.4	96.2	96
Gluconacetobacter xylinus	95.3	95	95.4	94.8	95.7	95	95.1	94.6	94.4	94.7	95.4	95.6	95.5	95.7	95.8
Gluconobacter oxydans	95.8	94.9	95.9	95	95.5	95.6	95.3	94.9	95.9	95.3	95.4	95.5	95	95.7	95.8
Gluconobacter asaii	95.3	94.6	95.6	94.9	95.4	95.1	95.1	94.3	95.6	94.8	95.4	94.9	94.6	96.2	96.2
Gluconobacter cerinus	95.3	94.6	95.6	94.9	95.4	95.1	95.1	94.3	95.6	94.8	95.5	94.9	94.6	96.2	96.2
Gluconobacter frateurii	95.5	94.6	95.4	94.8	95.5	95.1	95.1	94.8	96.2	95.2	95.7	94.9	95.2	96.1	96

(continued)

TABLE BXII.α.13. *(cont.)*

% 16S rRNA sequence similarity

Species	*Gluconacetobacter liquefaciens*	*Gluconacetobacter azotocaptans*	*Gluconacetobacter diazotrophicus*	*Gluconacetobacter entanii*	*Gluconacetobacter europaeus*	*Gluconacetobacter hansenii*	*Gluconacetobacter intermedius*	*Gluconacetobacter johannae*	*Gluconacetobacter oboediens*	*Gluconacetobacter sacchari*	*Gluconacetobacter xylinus*	*Gluconobacter oxydans*	*Gluconobacter asaii*	*Gluconobacter cerinus*	*Gluconobacter frateurii*
Acetobacter aceti	95.9	95.6	95.8	95.3	95.5	95	95.7	95.4	95.7	95.7	95.3	95.8	94.6	95.3	95.5
Acetobacter cibinongensis	95.9	95	95.3	94.9	95.1	94.6	95.3	94.9	95.3	95.7	95	94.9	93.7	94.6	94.6
Acetobacter estunensis	95.8	95.4	95.4	95.3	95.5	94.9	95.8	95.4	95.8	96.2	95.4	95.9	94.6	95.6	95.4
Acetobacter indonesiensis	95.6	95	95.1	94.8	94.9	94.5	95.2	94.8	95.2	95.6	94.8	95	94.1	94.9	94.8
Acetobacter lovaniensis	96.2	95.6	95.7	95.4	95.8	95	96.2	95.6	96.2	96.1	95.7	95.5	94.6	95.4	95.5
Acetobacter orientalis	96.1	95.4	95.3	95	95.1	94.7	95.4	95.1	95.4	96.1	95	95.6	94.3	95.1	95.1
Acetobacter orleanensis	95.6	95.4	95.2	95.1	95.3	94.6	95.7	95.4	95.7	95.7	95.1	95.3	94.2	95.1	95.1
Acetobacter pasteurianus	95.5	94.9	95.3	94.8	94.9	94.3	95.3	95.1	95.3	95.1	94.6	94.9	93.6	94.3	94.8
Acetobacter peroxydans	95.5	95	95.2	94.5	94.5	94.1	94.9	95.2	94.9	95.7	94.4	95.9	94.7	95.6	96.2
Acetobacter pomorum	95.7	95	95.4	94.9	95.1	94.5	95.4	95.3	95.4	95.3	94.7	95.3	94	94.8	95.2
Acetobacter syzygii	96.7	96.1	96.3	95.6	95.9	95.3	96.1	96	96.1	96.4	95.7	95.4	94.7	95.5	95.7
Acetobacter tropicalis	96	95.6	95.3	95.6	95.8	95.2	96	95.4	96	96.1	95.6	95.5	94	94.9	94.9
Acidomonas methanolica	96.6	96	96	95.5	95.8	95.3	95.8	95.9	95.8	96.4	96.2	95	93.8	94.6	95.2
Asaia bogorensis	96.6	96.4	96.2	95.9	96.3	95.5	96.6	96.3	96.6	96.5	96.2	95.7	95.2	96.2	96.1
Asaia siamensis	96.4	96.2	96	95.7	96.1	95.4	96.5	96.1	96.5	96.3	96	95.8	95.3	96.2	96
Gluconacetobacter liquefaciens	100	98.1	98.5	97.2	97.4	97	97.2	98	97.2	99.1	97.2	94.3	93.8	94.7	94.9
Gluconacetobacter azotocaptans	98.1	100	98.6	97.7	97.4	97.2	97	99.4	97	97.9	97.2	94.4	93.6	94.5	94.6
Gluconacetobacter diazotrophicus	98.5	98.6	100	97.1	97	97	96.6	98.5	96.6	98.4	96.7	94.5	93.9	94.6	94.9
Gluconacetobacter entanii	97.2	97.7	97.1	100	99	99.3	98.6	97.6	98.6	97	98.8	94.6	93.4	94.2	94.6
Gluconacetobacter europaeus	97.4	97.4	97	99	100	98.6	99.5	97.3	99.5	97.1	99.6	94.8	93.6	94.6	94.9
Gluconacetobacter hansenii	97	97.2	97	99.3	98.6	100	98.2	97.1	98.2	96.8	98.3	94.4	93.2	94	94.2
Gluconacetobacter intermedius	97.2	97	96.6	98.6	99.5	98.2	100	97	100	97	99.2	95.3	93.9	94.9	95.3
Gluconacetobacter johannae	98	99.4	98.5	97.6	97.3	97.1	97	100	97	97.9	97.2	94.5	93.6	94.5	94.6
Gluconacetobacter oboediens	97.2	97	96.6	98.6	99.5	98.2	100	97	100	97	99.2	95.3	93.9	94.9	95.3
Gluconacetobacter sacchari	99.1	97.9	98.4	97	97.1	96.8	97	97.9	97	100	97	94.4	93.6	94.4	94.8
Gluconacetobacter xylinus	97.2	97.2	96.7	98.8	99.6	98.3	99.2	97.2	99.2	97	100	94.6	93.5	94.4	94.7
Gluconobacter oxydans	94.3	94.4	94.5	94.6	94.8	94.4	95.3	94.5	95.3	94.4	94.6	100	97	97.8	98
Gluconobacter asaii	94.7	94.5	94.6	94.2	94.6	94	94.9	94.5	94.9	94.4	94.4	97.9	100	100	98.8
Gluconobacter cerinus	94.7	94.5	94.6	94.2	94.6	94	94.9	94.5	94.9	94.4	94.4	97.9	100	100	98.8
Gluconobacter frateurii	94.9	94.6	94.9	94.6	94.9	94.2	95.3	94.6	95.3	94.8	94.7	98	98	98.8	100

submerged fermentations carried out in acetators and trickle generators (Teuber et al., 1987b; Mariette et al., 1991). Plasmid profile analysis showed that acetators (suspensions of *Gluconacetobacter europaeus*) harbored only one dominant strain, whereas trickle generators (with a microflora in the form of biofilms on mechanical supports such as wooden chips) demonstrated a highly complex strain composition with spirit as substrate (Sievers and Teuber, 1995). Industrial submerged vinegar fermentations are initiated by inoculation with "seed vinegar", a microbiologically undefined fermentation broth from previous fermentations. The lack of defined pure starter cultures is due to the problems in the isolation, culture maintenance, determination of viable counts, and strain preservation of *Gluconacetobacter europaeus* strains responsible for high acid (up to 17% acetic acid) production. Sokollek and Hammes (1997) produced a starter culture of an acetic acid bacterium isolated from an industrial acetator with red wine as substrate for vinegar fermentation. The organism required no acetic acid for growth and is classified as *Gluconacetobacter oboediens* (Sokollek et al., 1998b). The frozen and lyophilized strain LTH 2460 was turned into a starter preparation by revitalization steps and stepwise adjustment of the growing culture to the acetic acid and ethanol composition of the medium used in industrial fermentation (Sokollek and Hammes, 1997). The use of a starter culture leads from spontaneous fermentation to a microbiologically controlled fermentation in vinegar production and reduces the time to start the acetator. RAPD (random amplified polymorphic DNA) analysis with a suitable primer (AGCGGGCGTA or CGCGTGCCCA or GTGGTGGTGGTGGTG) for the PCR was used for characterization of genotypically different strains from spirit vinegar fermentations (Trček et al., 1997). On a strain/species level, electrophoretic protein profiles (Kersters and De Ley, 1975) of total cell protein extracts and their computerized evaluation are very useful for the definition of different phenons within the species *Gluconacetobacter europaeus* and *Gluconacetobacter intermedius* (Sievers and Teuber, 1995; Boesch et al., 1998). Numerical analysis of protein gel electrophoregrams of a wide variety of *Acetobacter* strains has been used for classificatory changes at species level within the genus *Gluconacetobacter* (Gosselé et al., 1983b). Electrophoretic mobilities of metabolic enzymes (multilocus enzyme electrophoresis) have been used to measure genotypic diversity of natural populations (Selander et al., 1986). This technique was applied to *Gluconacetobacter diazotrophicus* strains isolated from sucrose-rich host plants and mealy bugs to determine their degree of genetic relationships (Caballero-Mellado and Martínez-Romero, 1994; Caballero-Mellado et al., 1995). Nitrogen-fixing *Acetobacter* (*Gluconacetobacter*) strains isolated from a new host plant *Eleusine coracana* (finger millet) were characterized upon RAPD profiles grouping into two genetically related clusters (Loganathan et al., 1999). Pulsed-field gel electrophoresis with *Xba*I as restriction enzyme (electrophoresis for 24.5 h at 200 V in 1% agarose with a pulse time ramped from 5 to 40 s) is an appropriate tool for differentiation among strains of acetic acid bacteria.

Enrichment and isolation procedures   A standard medium for enrichment and isolation of acetic acid bacteria, with the exception of *Gluconacetobacter europaeus* and *Gluconacetobacter entanii*, contains (per liter of distilled water) yeast extract, 5.0 g; peptone, 3.0 g; D-glucose, 0.5 g; ethanol (99.8%), 15 ml; cycloheximide, 0.1 g; and 12 g agar. The enrichment medium for the isolation of acetic acid bacteria, especially for *Asaia*, whose growth

is inhibited by 0.35% acetic acid, is composed of (per liter of distilled water): D-sorbitol, 20 g; peptone, 5 g; yeast extract, 3 g; and cycloheximide, 0.1g; adjusted to pH 3.5 with hydrochloric acid (Yamada et al., 2000).

Differentiation of acetic acid bacteria from other genera Acetic acid bacteria (with the exception of *Asaia*) oxidize ethanol aerobically to acetic acid. Acetic acid is accumulated in the medium. After the complete oxidation of ethanol, *Acetobacter*, *Gluconacetobacter*, and *Acidomonas* oxidize acetic acid further to CO_2 and H_2O. This phenomenon provides a rapid phenotypic differentiation of the overoxidation capacity by use of the chalk–ethanol test of Carr and Passmore (1979). All species of acetic acid bacteria with the exception of *Gluconacetobacter europaeus*, *Gluconacetobacter entanii*, *Asaia bogorensis*, and *Asaia siamensis* grow well to moderately on the modified medium.[2] The acetic acid that is initially formed dissolves the calcium carbonate; then further oxidation of the acetic acid by *Acetobacter* and *Gluconacetobacter* leads gradually to a return of the chalk. This medium is used for isolation of acetic acid bacteria from flowers and fruits. A final concentration of 0.01% cycloheximide (stock-solution dissolved in ethanol) is used to prevent growth of most yeast cells.

Frateuria resembles *Acetobacter*, *Gluconacetobacter*, *Acidomonas*, and *Gluconobacter* in its ability to oxidize ethanol to acetic acid. *Frateuria* is not able to overoxidize acetate, but, in contrast to *Gluconobacter* oxidizes lactate to CO_2 and H_2O. *Frateuria* contains ubiquinone of the Q-8 type. Colonies of *Acidiphilium* are light brown, pale pink, red, or violet due to the presence of carotinoids and bacteriochlorophyll *a*. Some strains of *Acidiphilium* produce no bacteriochlorophyll *a*. Growth of *Acidiphilium* is inhibited in contrast to acetic acid bacteria in the presence of 0.25 mM acetate. *Acidiphilium acidophilum* (*Thiobacillus acidophilus*) can be differentiated from acetic acid bacteria by its ability to grow in mineral medium at pH values of 2–4 in the presence of H_2SO_4. *Acidocella* can be differentiated from other genera by its ability to use diaminobutane, 4-aminobutyrate, and 5-aminovalerate as carbon and energy source. *Rhodopila globiformis* grows phototrophically under anaerobic conditions. *R. globiformis* is able to grow photoheterotrophically using ethanol as electron donor and carbon source. The cell shape of *R. globiformis* is spherical. *Roseococcus thiosulfatophilus* contains bacteriochlorophyll *a* and is obligately aerobic. *R. thiosulfatophilus* can be differentiated from acetic acid bacteria by its formation of pink cocci, its oxidation of thiosulfate to sulfate, and its inability to use ethanol, D-mannitol, and D-fructose as carbon sources. *Ketogulonigenium vulgare* is oxidase positive and, in contrast to acetic acid bacteria, not able to grow at pH 4.5. The truncated stem-loop structure in the 16S RNA at position 1241–1296 in the *E. coli* numbering system distinguishes *Ketogulonigenium vulgare* from acetic acid bacteria and some other members of the *Alphaproteobacteria* (Urbance et al., 2001).

In the following chapters, we describe the genera *Gluconacetobacter*, *Acetobacter*, *Acidomonas*, *Asaia*, and *Gluconobacter* of the family *Acetobacteraceae*, which differ in respect to genetic and physiological properties like 16S rRNA sequence similarity, phylogenetic positioning, DNA–DNA similarity, ubiquinone type, oxidation of acetic acid to carbon dioxide and water, and tolerance to acetic acid.

---

2. Medium for the chalk-ethanol test (per liter of distilled water): glucose, 0.5 g; yeast extract, 5.0 g; peptone, 3.0 g; calcium carbonate, 15.0 g; agar, 12.0 g; and ethanol (99.8%), 15 ml (sterilized by filtration and added after sterilization of the basal medium).

Genus I. *Acetobacter* Beijerinck 1898, 215[AL]

MARTIN SIEVERS AND JEAN SWINGS

A.ce.to.bac' ter. L. n. *acetum* vinegar; M.L. n. *bacter* masc. equivalent of Gr. neut. n. *bakterion* rod; M.L.masc. n. *Acetobacter* vinegar rod.

Cells **ellipsoidal to rod shaped**, straight or slightly curved, 0.6–0.9 × 1.0–4.0 μm, occurring singly, in pairs, or in chains. **Motile or nonmotile; if motile, the flagella are peritrichous.** Endospores are not formed. Gram negative. **Obligately aerobic**; metabolism is strictly respiratory with oxygen as the terminal electron acceptor. Never fermentative. Form pale colonies; most strains produce no pigments. A minority of strains produces brown water-soluble pigments or show pink colonies due to the production of porphyrins. Usually catalase positive. Oxidase negative. Absence of gelatin liquefaction, indole production, and H_2S formation. **Oxidizes ethanol to acetic acid. Acetate is oxidized to CO_2 and H_2O.** The best carbon sources for growth are ethanol, glycerol, and glucose. Some strains require *p*-aminobenzoic acid, niacin, thiamin, or pantothenic acid as growth factors. Neither lactose nor starch is hydrolyzed. No production of 2,5-diketo-D-gluconate from D-glucose. Chemoorganotrophs. Optimum temperature is 30°C. Some strains reduce nitrate to nitrite. **The pH optimum for growth is 4.0–6.0. Possesses ubiquinone of the Q-9 type as major quinone.** The predominant fatty acid in *Acetobacter* is the $C_{18:1 \omega 7}$ straight-chain unsaturated acid. ***Acetobacter* species occur in flowers, fruits, palm wine, vinegar, kefir, and fermented foods and can cause infections in grape wine, sake, tequila, cocoa wine, cider, beer, and fermented meat.** *Acetobacter* is not known to have any pathogenic effect toward humans and animals.

The mol% G + C of the DNA is: 50.5–60.3.

Type species: **Acetobacter aceti** (Pasteur 1864) Beijerinck 1898 (*Mycoderma aceti* Pasteur 1864, 125; *Acetobacter aceti* subsp. *aceti* (Pasteur 1864) De Ley and Frateur 1974.)

FURTHER DESCRIPTIVE INFORMATION

Cultivation media *Acetobacter aceti*, *A. pasteurianus*, *A. pomorum*, *A. orleanensis*, *A. peroxydans*, and *A. estunensis* grow well on standard medium (MYP).[1] *Acetobacter indonesiensis* and *Acetobacter tropicalis* are not able to utilize D-mannitol and were cultivated on a basal medium containing glucose, glycerol, and ethanol. Acetic acid bacteria are mesophilic strains with optimal temperature for growth at 28–30°C. Thermotolerant *Acetobacter* spp. capable of producing acetic acid at temperatures of about 37°C have been described (Ohmori et al., 1980; Saeki et al., 1997; Lu et al., 1999).

MAINTENANCE PROCEDURES

Strains of *Acetobacter* can be maintained for 2 weeks at 4°C on agar media used for cultivation. *Acetobacter* strains can be frozen and kept at −75°C in the presence of 24% (v/v) glycerol.

---

1. MYP medium (g/l distilled water): D-mannitol, 25.0; yeast extract, 5.0; peptone, 3.0; agar, 12.0.

DIFFERENTIATION OF THE SPECIES OF THE GENUS *ACETOBACTER*

Characteristics of the species of the genus *Acetobacter* are given in Table BXII.α.14.

TABLE BXII.α.14. Characteristics of the species of the genus *Acetobacter*[a]

Characteristic[b]	*A. aceti*	*A. cibinongensis*	*A. estunensis*	*A. indonesiensis*	*A. lovaniensis*	*A. orientalis*	*A. orleanensis*	*A. pasteurianus*	*A. peroxydans*	*A. pomorum*	*A. syzygii*	*A. tropicalis*
Catalase	+	+	+	+	+	+	+	(+)	−	+	+	+
Ketogenesis from glycerol	+	−	−	−	−	−	d	d	−	+	−	−
Production of acid from D-glucose	+	+	+	+	+	+	+	d	−	(+)	+	+
Production from D-glucose of:												
2-keto-D-gluconate	+	+	+	+	−	+	+	−	−	−	−	+
5-keto-D-gluconate	+	−	−	−	−	−	−	−	−	−	−	−
Nitrate reduction[c]	−	−	+	d	d	−	d	+	−	nd	−	+
Ubiquinone type	Q-9	Q-9	Q-9	Q-9	Q-9	Q-9	Q-9	Q-9	Q-9	Q-9	Q-9	Q-9
Mol% G + C content of DNA	56.2–57.2	53.8–54.5	59.3–59.7	53.3–55.1	57–59	52.0–52.8	56.5–58.7	51.8–53	60.3	50.5	54.3–55.4	55.2–56.6

[a]Symbols: +, 90% or more of the strains positive; (+), weakly positive reaction; d, 11–89% of the strains positive; −, 90% or more of the strains negative; (−), most of the strains negative; nd, not determined.

[b]Data from Sievers et al. (1992); Sokollek et al. (1998b); Lisdiyanti et al. (2001a).

[c]Nitrate reduction was tested from nitrate peptone water (per liter of distilled water, pH 7.0: peptone, 10 g; KNO₃, 2 g) (Franke et al., 1999).

List of species of the genus Acetobacter

1. **Acetobacter aceti** (Pasteur 1864) Beijerinck 1898[AL] (*Mycoderma aceti* Pasteur 1864, 125; *Acetobacter aceti* subsp. *aceti* (Pasteur 1864) De Ley and Frateur 1974.)

a.ce'ti. L. n. *acetum* vinegar; L. gen. n. *aceti* of vinegar.

Cell morphology and colonial characteristics are as described for the genus. All strains are ketogenic toward glycerol and sorbitol and most strains toward D-mannitol. Both 2-keto- and 5-ketogluconic acids are synthesized from D-glucose. 2,5-Diketogluconic acid is not produced. All strains acidify ethanol, *n*-propanol, *n*-butanol, D-xylose, D-mannose, and D-glucose. Ethanol, glycerol, D-mannitol, Na-acetate, and Na-D,L-lactate are good carbon sources for growth. All strains require growth factors in the presence of D-mannitol. Most strains utilize L-alanine and L-proline as a source of nitrogen in the presence of D-mannitol (De Ley et al., 1984). The level of 16S rRNA sequence similarity is 96.6–98.0% between *A. aceti* DSM 3508 and the type strains of the other *Acetobacter* species.

The mol% G + C of the DNA is: 56.2–57.2 (T_m).

Type strain: ATCC 15973, DSM 3508, IMET 10732, JCM 7641, NCIB 8621.

GenBank accession number (16S rRNA): D30768, X74066.

2. **Acetobacter cibinongensis** Lisdiyanti, Kawasaki, Seki, Yamada, Uchimura and Komagata 2002, 3[VP] (Effective publication: Lisdiyanti, Kawasaki, Seki, Yamada, Uchimura and Komagata 2001a, 130.)

ci.bi.non'gen.sis. N.L. adj. *cibinongensis* derived from Cibinong, West Java, Indonesia.

A. cibinongensis is closely related to *Acetobacter orientalis* sharing 99.1–99.3% 16S rRNA sequence similarity by DNA–DNA similarity values of 12–25%. Strains produce D-gluconate and 2-keto-D-gluconate from D-glucose. Acid is produced from D-glucose, ethanol, and *n*-propanol. Acid production from L-arabinose, D-xylose, D-galactose, and D-mannose is strain dependent. *A. cibinongensis* does not produce acid from D-fructose, L-sorbose, D-arabinose, D-mannitol, D-sorbitol, glycerol, and lactose. Weakly positive for growth at 42°C, at pH 3.0 and pH 9.0, in the presence of 20% glucose, and in the presence of 10% ethanol (Lisdiyanti et al., 2001a). The level of 16S rRNA sequence similarity is 96.8–99.0% between *A. cibinongensis* IFO 16605 and the type strains of the other *Acetobacter* species. *A. cibinongensis* was isolated from fruit (mountain soursop) and curd of tufu in Indonesia.

The mol% G + C of the DNA is: 53.8–54.5 (T_m).

Type strain: IFO 16605, JCM 11196, NRIC 0482.

GenBank accession number (16S rRNA): AB052710.

3. **Acetobacter estunensis** (Carr 1958) Lisdiyanti, Kawasaki, Seki, Yamada, Uchimura and Komagata 2001b, 263[VP] (Effective publication: Lisdiyanti, Kawasaki, Seki, Yamada, Uchimura and Komagata 2000, 162) *Acetobacter pasteurianus* subsp. *estunensis* (Carr 1958) De Ley and Frateur 1974, 278; "*Acetobacter estunenses*" (sic) Carr 1958, 157.

es.tun.en'sis. M.L. adj. *estunensis* probably named after the old English name of the village Long Ashton, where J.G. Carr isolated the type strain.

A. estunensis was described by Carr (1958) as "*Acetobacter estunense*". The name, "*A. estunense*", implies that this organism is distinct from any species described by Frateur (Carr, 1958). *A. estunensis* oxidizes glucose to gluconate; 2-ketogluconate is produced but not 5-ketogluconate. Acid is produced from D-mannose and D-xylose but not from D-galactose, D-arabinose, L-arabinose, D-fructose, L-sorbose, D-mannitol, D-sorbitol, glycerol, sucrose, or lactose. *A. estunensis* IFO 13751 contains 89.2% ubiquinone of the Q-9 type and 10.8% of the Q-8 type. *A. estunensis* IFO 13751 was isolated from cider. *A. estunensis* produced pellicles, which are not composed of cellulose since they dissolved after being boiled in 5% NaOH (Lisdiyanti et al., 2000). *A. estunensis* IFO 13751 showed 16S rRNA sequence similarity values of 96.3–97.9% to the type strains of other *Acetobacter* species.

The mol% G + C of the DNA is: 59.3–59.7 (T_m).

Type strain: ATCC 23753, DSM 4493, IFO 13751, LMG 1626, NCIMB 8935.

GenBank accession number (16S rRNA): AB032349.

4. **Acetobacter indonesiensis** Lisdiyanti, Kawasaki, Seki, Yamada, Uchimura and Komagata 2001b, 263[VP] (Effective publication: Lisdiyanti, Kawasaki, Seki, Yamada, Uchimura and Komagata 2000, 162.)

in.do.ne.si.en'sis. M.L. adj. *indonesiensis* referring to Indonesia where most strains studied were isolated.

A. indonesiensis cells are rod shaped, 1.8–2.0 × 0.8–1.0 μm, occurring singly, in pairs, or in chains. Colonies are circular, convex, glistening, and nonpigmented on basal medium containing 1.0% glucose, 1.0% glycerol, 1.0% ethanol, 1.0% peptone, 0.5% yeast extract, and 1.5% agar (Lisdiyanti et al., 2000). No growth on D-mannitol. Oxidizes glucose to gluconate; 2-ketogluconate is produced but not 5-ketogluconate. Does not produce acid from D-arabinose, D-fructose, L-sorbose, D-mannitol, D-sorbitol, glycerol, sucrose, or lactose. Acid production from D-mannose, D-galactose, L-arabinose, and D-xylose is strain dependent. No ketogenesis from glycerol. *A. indonesiensis* IFO 16471 contains 86% ubiquinone of the Q-9 type and 14% of the Q-10 type (Lisdiyanti et al., 2000). The level of 16S rRNA sequence similarity is 96.4–98.6% between *A. indonesiensis* NRIC 0313 and the type strains of other *Acetobacter* species. *A. indonesiensis* was isolated from Indonesian fruits like banana, papaya, zirzak, and mango. *A. indonesiensis* IFO 16471 was obtained from the fruit of zirzak (*Annona muricata*).

The mol% G + C of the DNA is: 53.3–55.1 (T_m).

Type strain: IFO 16471, NRIC 0313.

GenBank accession number (16S rRNA): AB032356.

5. **Acetobacter lovaniensis** (Frateur 1950) Lisdiyanti, Kawasaki, Seki, Yamada, Uchimura and Komagata 2001b, 263[VP] (Effective publication: Lisdiyanti, Kawasaki, Seki, Yamada, Uchimura and Komagata 2000, 162 (*Acetobacter pasteurianus* subsp. *lovaniensis* (Frateur 1950) De Ley and Frateur 1974, 278; "*Acetobacter lovaniense*" (sic) Frateur 1950, 336.)

lo.va.ni.en'sis. M.L. adj. *lovaniensis* referring to the city Louvain (Lovanium), in Belgium, where the type strain was isolated and studied by J. Frateur.

Acetobacter lovaniensis includes *A. pasteurianus* IFO 13753 and IFO 3284, *A. aceti* AJ 2913 and AJ 2914, and several Indonesian isolates from fruits (palm seed and starfruit) and fermented foods (nata de coco, moromi soya, palm wine, and pickle) (Lisdiyanti et al., 2000). *A. lovaniensis* oxidizes glucose to gluconate. 2-ketogluconate and 5-ketogluconate are not produced. Does not produce acid from D-fructose, L-sorbose, D-mannitol, D-sorbitol, glycerol, sucrose, or lactose. Acid production from D-mannose, D-galactose, L-arabinose, and D-xylose is strain dependent. *A. lovaniensis* IFO 13753 contains 84.9% ubiquinone of the Q-9 type, 8.6% of the Q-8 type, and 6.5% of the Q-10 type (Lisdiyanti et al., 2000). The level of 16S rRNA sequence similarity is 97.0–99.4% between *A. lovaniensis* IFO 13753 and the type strains of other *Acetobacter* species.

The mol% $G + C$ of the DNA is: 57–59 (T_m).

Type strain: ATCC 12875, DSM 4491, IFO 13753, LMG 1579, NCIMB 8620.

GenBank accession number (16S rRNA): AB032351.

6. **Acetobacter orientalis** Lisdiyanti, Kawasaki, Seki, Yamada, Uchimura and Komagata 2002, 3[VP] (Effective publication: Lisdiyanti, Kawasaki, Seki, Yamada, Uchimura and Komagata 2001a, 130.)

o.ri.en' ta.lis. M.L. adj. *orientalis* "oriental", referring to the region where the strains were isolated.

A. orientalis cells are rod shaped, 2.0–3.0 × 0.6–0.8 μm. Colonies are circular, convex, glistening, and nonpigmented on basal medium pH 6.8 containing 1.0% glucose, 1.0% glycerol, 1.0% ethanol, 1.0% peptone, 0.5% yeast extract, and 1.5% agar (Lisdiyanti et al., 2001a). Strains produce D-gluconate and 2-keto-D-gluconate from D-glucose. Acid is produced from D-glucose, ethanol, and *n*-propanol. Acid production from L-arabinose, D-xylose, D-galactose, and D-mannose is strain dependent. *A. tropicalis* does not produce acid from D-fructose, L-sorbose, D-arabinose, D-mannitol, D-sorbitol, glycerol, and lactose. Weakly positive for growth at 42°C, at pH 3.0 and pH 9.0, in the presence of 20% glucose, and in the presence of 10% ethanol (Lisdiyanti et al., 2001a). *Acetobacter orientalis* was isolated from canna flower, fruits (starfruit and coconut), and fermented foods (curd of tofu and tempeh) in Indonesia. The level of 16S rRNA sequence similarity is 96.8–99.0% between *A. orientalis* IFO 16606 and the type strains of the other *Acetobacter* species.

The mol% $G + C$ of the DNA is: 52–52.8 (T_m).

Type strain: IFO 16606, JCM 11195, NRIC 0481.

GenBank accession number (16S rRNA): AB052706.

7. **Acetobacter orleanensis** (Henneberg 1906) Lisdiyanti, Kawasaki, Seki, Yamada, Uchimura and Komagata 2001b, 263[VP] (Effective publication: Lisdiyanti, Kawasaki, Seki, Yamada, Uchimura and Komagata 2000, 161) (*Acetobacter aceti* subsp. *orleanensis* (Henneberg 1906) De Ley and Frateur 1974, 278; "*Bacterium orleanense*" Henneberg 1906, 106.)

or.le.an.en' sis. M.L. adj. *orleanensis* referring to the city Orleans and the "Orleans method" for vinegar production.

A. orleanensis includes strains of *A. pasteurianus* IFO 13752, IFO 3170, and IFO 3223, *Gluconacetobacter hansenii* IFO 3296, and six Indonesian isolates from fruits (guava and sapodilla) and fermented foods (nata de coco) (Lisdiyanti et al., 2000). *A orleanensis* oxidizes glucose to gluconate; 2-ketogluconate is produced but not 5-ketogluconate. Does not produce acid from D-arabinose, D-fructose, L-sorbose, D-mannitol, D-sorbitol, glycerol, sucrose, or lactose. Acid production from D-mannose, D-galactose, L-arabinose,

and D-xylose is strain dependent. *A. orleanensis* IFO 13752 contains 81.0% ubiquinone of the Q-9 type, 14.2% of the Q-8 type, and 4.8% of the Q-10 type. The level of 16S rRNA sequence similarity is 95.9–98.7% between *A. orleanensis* IFO 13752 and the type strains of other *Acetobacter* species.

The mol% $G + C$ of the DNA is: 56.5–58.7 (T_m).

Type strain: ATCC 12876, DSM4492, IFO 13752, LMG 1583, NCIMB 8622.

GenBank accession number (16S rRNA): AB032350.

8. **Acetobacter pasteurianus** (Hansen 1879) Beijerinck 1916, 1199[AL] (*Mycoderma pasteurianum* Hansen 1879, 230; *Acetobacter pasteurianum* (sic) Beijerinck 1916, 1199; *Acetobacter pasteurianus* subsp. *ascendens* (Henneberg 1898) De Ley and Frateur 1974, 278; *Acetobacter pasteurianus* subsp. *paradoxus* (Frateur 1950) De Ley and Frateur 1974, 278.)

pas.teur.i.a' nus. M.L. adj. *pasteurianus* of Pasteur; named after Louis Pasteur (1822–1895), French chemist and bacteriologist.

The cell morphology and colonial characteristics are as described for the genus. Not all strains are able to grow on GYC medium (5% D-glucose, 1% yeast extract, 3% CaCO₃, 1.5% agar). Most strains are negative for ketogenesis from glycerol and formation of 5-ketogluconic acid from D-glucose. Some strains are weakly catalase positive. Most strains produce acid from ethanol, *n*-propanol, and *n*-butanol, but not always from D-glucose. All strains require growth factors in the presence of D-mannitol (De Ley et al., 1984). *A. pasteurianus* was found in different kefir grains as milk adapted biotype at numbers varying between 3×10^3 and 3×10^6 CFU/g (Häfliger et al., 1991a, b). *A. pasteurianus* growing in kefir utilizes L-alanine and L-proline as nitrogen and carbon sources (Häfliger et al., 1991b). The level of 16S rRNA sequence similarity is 95.9–99.4% between *A. pasteurianus* DSM 3509 and the type strains of the other *Acetobacter* species.

The mol% $G + C$ of the DNA is: 51.8–53 (T_m).

Type strain: ATCC 33445, DSM 3509, IMET 10733, LMD 22.1.

GenBank accession number (16S rRNA): X71863.

9. **Acetobacter peroxydans** Visser' t Hooft 1925, 225[VP]

per.ox' y.dans. L. pref. *per* very; M.L. part. adj. *oxydans* acid-giving, oxidizing; M.L. part. adj. *peroxydans* strongly oxidizing.

Strain IFO 13755 is the type species and was isolated from ditch water. *A. peroxydans* IFO 13755 oxidizes ethanol, acetate, and lactate but does not produce acid from different carbon sources (D-glucose, D-mannose, D-galactose, D-arabinose, L-arabinose, D-xylose, D-fructose, DL-sorbose, D-mannitol, D-sorbitol, glycerol, sucrose, lactose). Does not produce 2-ketogluconate or 5-ketogluconate. Catalase negative. Peroxidase positive. The level of DNA relatedness of *A. peroxydans* IFO 13755 with other strains of *Acetobacter* is above 46%. The level of 16S rRNA sequence similarity is 96.5–97.9% between *A. peroxydans* IFO 13755 and the type strains of other *Acetobacter* species.

The mol% $G + C$ of the DNA is: 60.3 (T_m).

Type strain: IFO 13755, ATCC 12874, LMG 1635, NCIMB 8618.

GenBank accession number (16S rRNA): AB032352.

10. **Acetobacter pomorum** Sokollek, Hertel and Hammes 1998b, 940[VP]

po.mo' rum. L. n. *pomum* fruit; L. gen. pl. n. *pomorum* of the fruits.

Cells are 0.8–1.2 × 1.3–1.6 µm, nonmotile, and occur mainly in pairs. Colonies are round, regular, convex, soft to liquid, glossy, and beige with a diameter of 0.8–1.5 mm on RAE agar (1a/2e). Plates of RAE agar are prepared by using the double layer technique as described by Entani et al. (1985) with 0.5% agar in the bottom layer and 1% agar in the top layer. RAE agar (1a/2e) is composed of 4% glucose, 1% yeast extract, 1% peptone, 0.338% Na$_2$HPO$_4$·2H$_2$O, 0.15% citric acid·H$_2$O, 1% (v/v) acetic acid, and 2% (v/v) ethanol, in both agar layers (Sokollek and Hammes, 1997). Cells grow on 3% (v/v) ethanol in the presence of acetic acid levels as high as 4% (w/v). Growth occurs in the presence of 30% (w/v) glucose with a weak gluconic acid formation (<10 g/l). 5- and 2-keto-gluconic acid are not formed from glucose. Carbon sources include methanol, ethanol, *n*-propanol, D-glucose, D-fructose, D-maltose, D-ribose, D-xylose, D-sorbitol, D-mannitol, and glycerol. No growth occurs on sucrose, L-lactate, and D-gluconate. Dihydroxyacetone is formed from glycerol. Cellulose is not synthesized. *A. pomorum* DSM 11825 was isolated from a submerged cider vinegar fermentation. The level of 16S rRNA sequence similarity is 96.2–99.4% between *A. pomorum* DSM 11825 and the type strains of the other *Acetobacter* species.

The mol% G + C of the DNA is: 50.5 (T_m).

Type strain: DSM 11825, LTH 2458.

GenBank accession number (16S rRNA): AJ001632.

11. **Acetobacter syzygii** Lisdiyanti, Kawasaki, Seki, Yamada, Uchimura and Komagata 2002, 3[VP] (Effective publication: Lisdiyanti, Kawasaki, Seki, Yamada, Uchimura and Komagata 2001a, 129.)

sy.zy' gi.i. L. gen. *syzygii* derived from N.L. neut. n. *syzygium* referring to the name of the fruit of the Malay rose apple (*Syzygium malaccense*) from which the type strain was isolated.

A. syzygii IFO 16604 is closely related to *A. lovaniensis*

IFO 13753 sharing 99.4% 16S rRNA sequence similarity by DNA–DNA similarity value of 38%. Strains of *A. syzygii* produce acids from D-glucose, ethanol, and *n*-propanol. 2 keto-D-gluconate is not formed. Acid production from L-arabinose, D-xylose, and D-mannose is strain dependent. Weakly positive for growth at 42°C, at pH 3.0 and pH 9.0, in the presence of 20% glucose, and in the presence of 10% ethanol (Lisdiyanti et al., 2001a). The level of 16S rRNA sequence similarity is 96.9–99.4% between *A. syzygii* IFO 16604 and the type strains of the other *Acetobacter* species.

The mol% G + C of the DNA is: 54.3–55.4 (T_m).

Type strain: IFO 16604, JCM 11197, NRIC 0483.

GenBank accession number (16S rRNA): AB052712.

12. **Acetobacter tropicalis** Lisdiyanti, Kawasaki, Seki, Yamada, Uchimura and Komagata 2001b, 263[VP] (Effective publication: Lisdiyanti, Kawasaki, Seki, Yamada, Uchimura and Komagata 2000, 163.)

tro.pi.ca' lis. M.L. adj. *tropicalis* referring to the tropical region where the strains were isolated.

A. tropicalis cells are rod shaped, 1.8–2.0 × 0.5–0.7 µm, occurring singly, in pairs, or in chains. Colonies are circular, convex, glistening, and nonpigmented on basal medium containing 1.0% glucose, 1.0% glycerol, 1.0% ethanol, 1.0% peptone, 0.5% yeast extract, and 1.5% agar (Lisdiyanti et al., 2000). No growth on D-mannitol. Oxidizes glucose to gluconate; 2-ketogluconate is produced but not 5-ketogluconate. Does not produce acid from D-arabinose, D-fructose, L-sorbose, D-mannitol, D-sorbitol, glycerol, sucrose, or lactose. Acid production from D-mannose, D-galactose, L-arabinose, and D-xylose is positive to weak reaction. No ketogenesis from glycerol. The major quinone is Q-9 (86–100%). *A. tropicalis* IFO 16470 showed 16S rRNA sequence similarity values of 96.4–98.7% to the type strains of other *Acetobacter* species. *A. tropicalis* IFO 16470 was isolated from coconut (*Coccos nucifera*).

The mol% G + C of the DNA is: 55.2–56.6 (T_m).

Type strain: IFO 16470, NRIC 0312.

GenBank accession number (16S rRNA): AB032354, AB032355.

Genus II. **Acidiphilium** *Harrison 1981, 331*[VP] *emend. Kishimoto, Kosako, Wakao, Tano and Hiraishi 1995b, 90*

AKIRA HIRAISHI AND JOHANNES F. IMHOFF

A.ci.di.phi' li.um. M.L. n. *acidum* an acid; Gr. adj. *philus* loving; M.L. neut. n. *Acidiphilium* acid lover.

Cells are straight rods, 0.3–1.2 × 4.2 µm, multiply by binary fission and exhibit a pleomorphic tendency at varied pH values and in the presence of different carbon sources. Motile by means of polar, subpolar, and lateral flagella. Do not form spores or capsules. Gram negative. **16S rRNA structures and signatures conform to the *Alphaproteobacteria*.** Cell suspensions and colonies are white to cream, yellow, pink, red, or brown. **Strictly aerobic, chemoorganotrophic and chemolithotrophic bacteria containing photosynthetic pigments**, categorized into the aerobic bacteriochlorophyll-containing bacteria. **Main photosynthetic pigments are zinc-chelated bacteriochlorophyll (Zn-BChl)** *a* **and the carotenoid spirilloxanthin.** Grow with simple organic compounds, such as sugars, as electron donors and carbon sources. **Growth**

is inhibited in the presence of 0.6% yeast extract and low concentrations of acetate (0.25 mM) and lactate (2 mM). One species is a facultative chemolithotroph utilizing elemental sulfur as electron donor. Fe^{2+} does not serve as the electron donor for chemolithotrophic growth but has stimulatory effects on heterotrophic growth. Catalase is positive and oxidase negative or weakly positive. Mesophilic and **obligately acidophilic bacteria** growing in the pH range of 2.0–5.9 (but not at pH 6.1 and above). **Straight-chain monounsaturated C$_{18:1}$ acid is the major component of cellular fatty acids. Ubiquinones with ten isoprene (Q-10) units are present.**

The mol% G + C of the DNA is: 62.9–68.1.

Type species: **Acidiphilium cryptum** Harrison 1981, 331.

FURTHER DESCRIPTIVE INFORMATION

Phylogenetic analyses based on 16S rDNA sequence information have shown that *Acidiphilium* species form a major cluster in the *Alphaproteobacteria*, together with members of the acidophilic chemotrophic bacteria of the genera, *Acidocella*, *Acetobacter*, and *Gluconobacter*, and the acidophilic anoxygenic phototrophic bacterium *Rhodopila* (Sievers et al., 1994a; Kishimoto et al., 1995b). This cluster is deeply branched off from the *Alphaproteobacteria*. (See Fig. 1 p. 136 in the essay "Aerobic Bacteria Containing Bacteriochlorophyll and Belonging to the *Alphaproteobacteria*", Volume 2, Part A.)

Cells of *Acidiphilium* species are straight or slightly curved rods with rounded ends (Figs. BXII.α.18 and BXII.α.19). They vary in size from 0.4–0.8 × 1.0–3.0 μm under optimal growth conditions, but cell size and shape are influenced by the kind and concentration of carbon source and by physico-chemical growth conditions (the range of cell size is shown in the generic description). Swollen cells and filaments (>10 μm long) frequently occur when growing in nutrient-rich media, and this tendency is most pronounced in *Acidiphilium rubrum*. The swollen cells contain discrete, refractive, sudanophilic granules that are possibly poly-β-hydroxybutyrate granules. Electron microscopy of negatively stained cells shows that motile cells have single polar or subpolar flagellum (Fig. BXII.α.20). Two lateral flagella are occasionally found in *Acidiphilium cryptum* (Harrison, 1989) and *Acidiphilium multivorum*. Thin-section electron microscopy shows that any type of intracytoplasmic membrane systems is absent in the cells (Matsuzawa et al., 2000) (Fig. BXII.α.21). Polyphosphate granules as well as poly-β-hydroxybutyrate granules are frequently observed.

Acidiphilium species are strictly aerobic, chemoorganoheterotrophic bacteria (a single species is chemolithotrophic) that require high acidity for growth. They grow in the pH range of 2.5–5.9, some as low as 1.5–2.0, but not at pH 6.1 and above. Growth rates vary remarkably among the species. The type species *A. cryptum* and its closest relatives, *A. multivorum* and *A. organovorum*, exhibit a doubling time of 3–9 h, whereas *A. acidophilum* and *A. rubrum* grow much more slowly with a doubling time of more than 10 h under optimal growth conditions. *Aci-diphilium* species use simple organic compounds as carbon and energy sources for growth. The following sugars and sugar alcohols are good substrates supporting growth of all species: L-arabinose, D-xylose, D-ribose, D-glucose, D-galactose, D-fructose, sucrose, glycerol, arabitol, and mannitol. *Acidiphilium* species utilize the pentose phosphate and Entner–Doudroff pathways for glucose metabolism and lack the Embden–Meyerhoff–Parnas pathway (Shuttleworth et al., 1985). None of the strains uses the following carbon sources: L-rhamnose, cellobiose, trehalose, melibiose, formate, butanol, acetate, propionate, butyrate, valerate, caproate, lactate, pyruvate, glycolate, oxalate, DL-β-hydroxybutyrate, benzoate, *p*-hydroxybenzoate, glycine, cysteine, and lysine. Growth of *Acidiphilium* species in glucose-mineral media is inhibited by low concentrations of acetate or lactate. In addition, succinate and fumarate inhibit growth of some species. Yeast extract at a low concentration of less than 0.1% (w/v) enhances growth of *Acidiphilium* species, but at 0.3% and above it has inhibitory effects on the growth (Harrison, 1981; Kishimoto et al., 1990). This is due to the effects of organic acids contained in the yeast extract (Kishimoto et al., 1990). Trypticase and peptone have similar inhibitory effects.

A. acidophilum is the only species that exhibits chemolithotrophic growth with reduced sulfur compounds as energy sources, but other species may have the ability to oxidize elemental sulfur to sulfate (Harrison, 1981). A number of studies have demonstrated sulfur-dependent chemolithotrophy of and sulfur oxidation by *A. acidophilum* (Guay and Silver, 1975; Harrison, 1981; Pronk et al., 1990; Meulenberg et al., 1992; Hiraishi et al., 1998).

All species are capable of growth on high concentrations of metals, such as 100 mM Fe^{2+} and 100 mM Al^{3+}. *A. rubrum* may accumulate Fe as magnetite (Itoh et al., 1998; Matsuzawa et al., 2000). All species also exhibit high resistance to heavy metals such as copper, nickel, and zinc (Mahapatra and Banerjee, 1996). However, neither Fe^{2+} nor other metal ions can serve as an electron donor for chemolithotrophic growth. However, the addition of ferrous ion stimulates heterotrophic growth significantly. Some strains of *Acidiphilium* species, including *A. cryptum* are capable of coupling the reduction of Fe(III) to the oxidation of a variety of substrates under aerobic or anaerobic conditions (Johnson and McGinness, 1991; Pronk and Johnson, 1992; Johnson, 1998; Küsel et al., 1999). They may co-respire oxygen and Fe(III) under oxic conditions.

A. multivorum has the ability to oxidize arsenite to arsenate, whereas all other species lack this property (Wakao et al., 1994). *A. multivorum* has a number of plasmids of different sizes, one of which is 56 kbp and encodes its arsenic resistance. The arsenic resistance (*ars*) operon from the *A. multivorum* plasmid was cloned and sequenced (Suzuki et al., 1998). This cluster contained five genes in the following order: *arsR*, *arsD*, *arsA*, *arsB*, *arsC*, and the deduced amino acid sequences of all the gene products are homologous to the amino acid sequences of the *ars* gene products of the *Escherichia coli* plasmids. The *ars* operon cloned from *A. multivorum* conferred resistance to arsenate and arsenite upon *E. coli*.

One of the most outstanding characteristics of *Acidiphilium* species is the production of photopigments with Zn-BChl *a* as the major component (for abbreviations for metal-substituted bacteriochlorophylls [BChls] see Takaichi et al., 1999). The chemical structure of Zn-BChl *a* isolated from *Acidiphilium* is the same as Mg-BChl *a* esterified with phytol, except the occurrence of Zn in place of Mg as the central metal (Wakao et al., 1996;

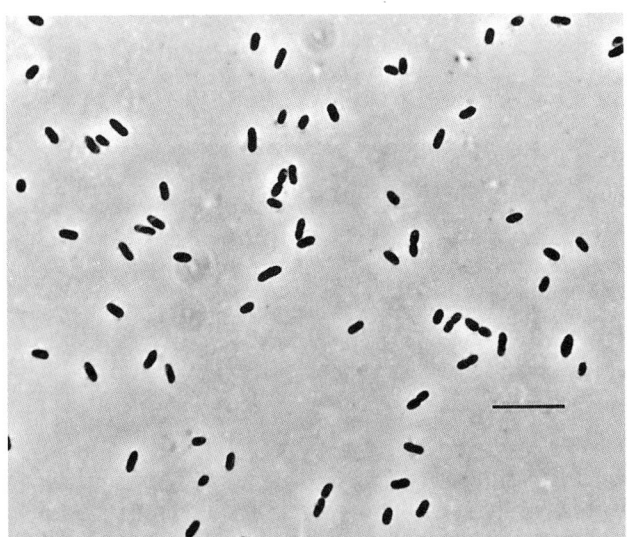

FIGURE BXII.α.18. Phase-contrast photomicrograph showing general cell morphology of *Acidiphilium cryptum* Bar = 5 μm.

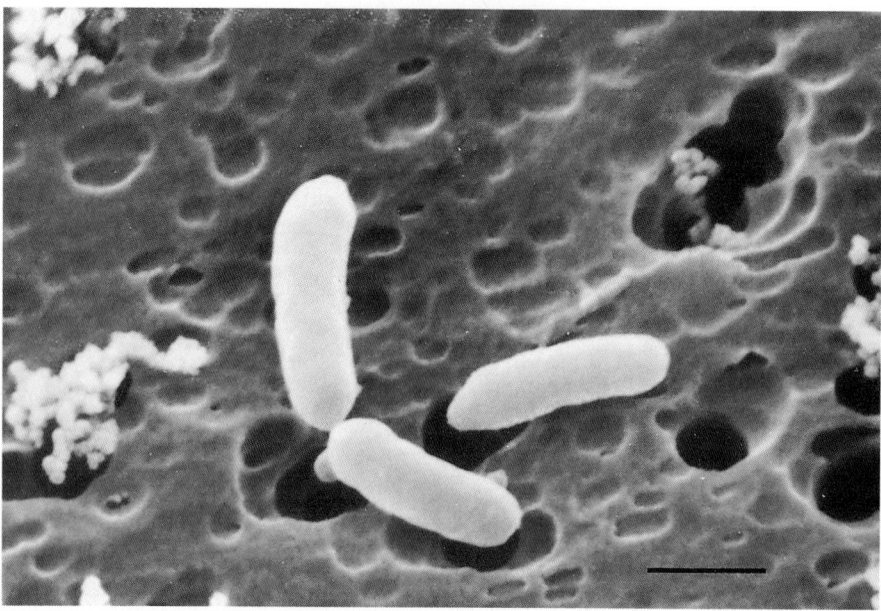

FIGURE BXII.α.19. Scanning electron photomicrograph showing general cell morphology of *Acidiphilium rubrum* ATCC 35905. Bar = 0.9 μm.

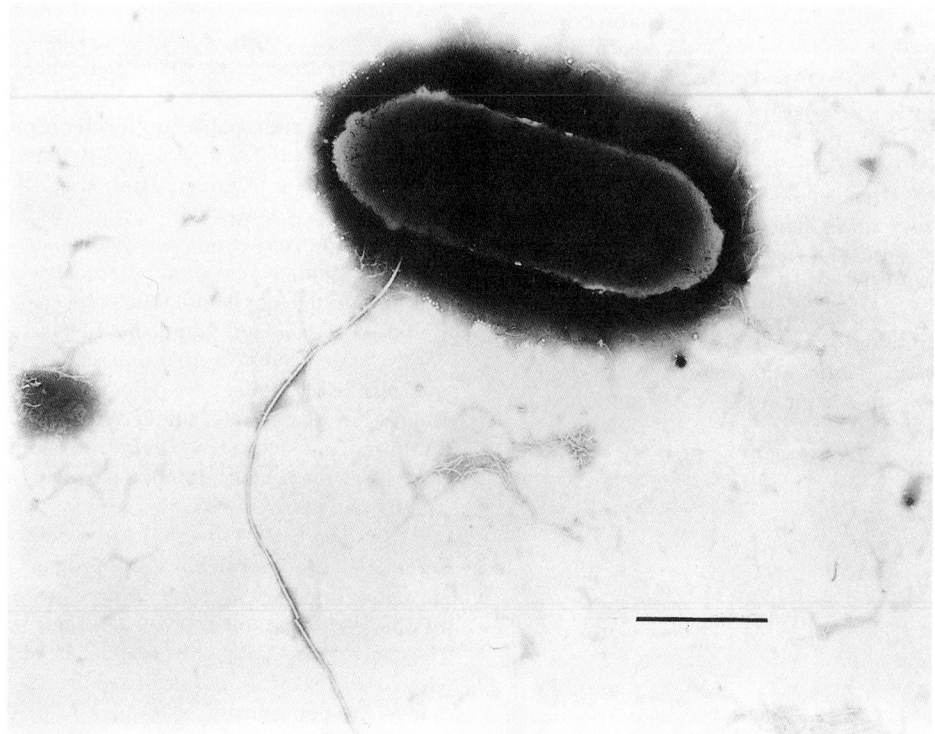

FIGURE BXII.α.20. Electron photomicrograph of a negatively stained cell of *Acidiphilium cryptum* ATCC 33463 having a subpolar flagellum. Bar = 0.5 μm. (Reprinted with permission from A. Hiraishi and K. Shimada, Journal of General Applied Microbiology*47:* 161–180, 2001, ©Japan Society for Bioscience, Biotechnology, and Agrochemistry.)

Kobayashi et al., 1998). The dominant occurrence of Zn-BChl *a* is indicated by an absorption maximum at 763 nm in the acetone-methanol extract, which is blue-shifted by 7 nm compared to the absorption maximum of Mg-BChl *a* in the corresponding near infrared region. Among the *Acidiphilium* species so far estab-

lished, *A. rubrum* is most remarkable for Zn-BChl *a* production and contains 0.8–1.0 nmol of Zn-BChl *a* per mg dry weight of cells under optimal growth conditions. The cellular content of Zn-BChl *a* varies depending on environmental conditions. The production of Zn-BChl *a* occurs only under oxic-dark conditions,

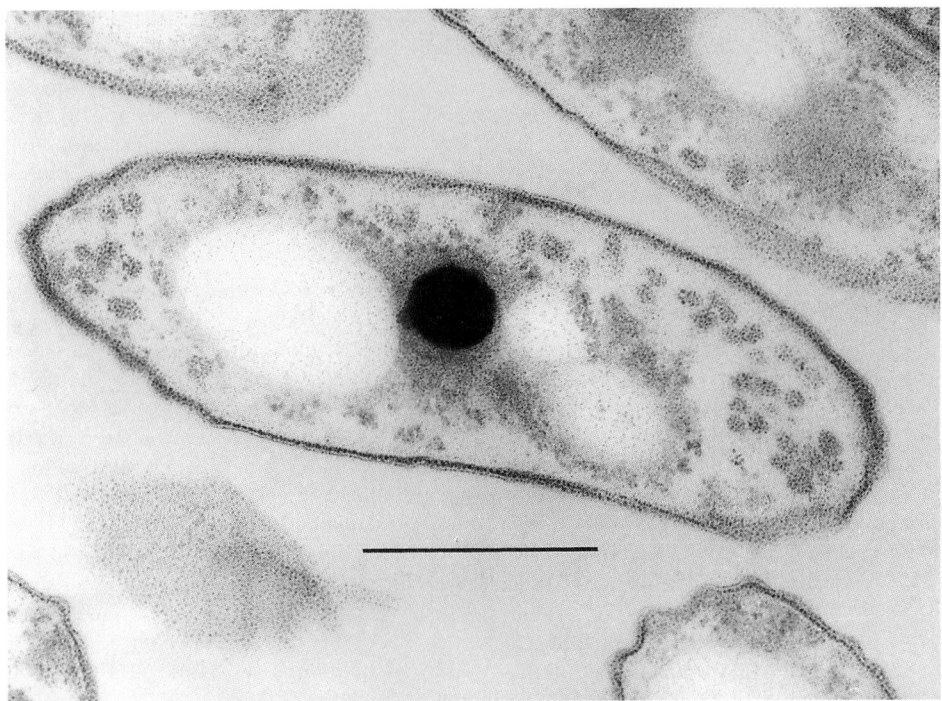

FIGURE BXII.α.21.   Thin-section electron photomicrograph of cells of *Acidiphilium rubrum* ATCC 35905. Bar = 0.5 µm.

indicating the strong requirement of energy produced via aerobic respiratory electron transport for pigment production. Light inhibits photopigment production even under optimal growth conditions, as is the case in other genera of aerobic bacteriochlorophyll-containing bacteria. *Acidiphilium* species produce Mg-BChl *a* and bacteriopheophytin (BPhe) in addition to Zn-BChl *a*, but Zn-BChl *a* always predominates. The molar ratio of Zn-BChl *a*:Mg-BChl *a*:BPhe in the cells is relatively constant among the species; e.g., *A. rubrum* produces them at a molar ratio of 13:2:1 (Wakao et al., 1996; Hiraishi et al., 1998). In *A. rubrum*, a single set of magnesium chelatase homologs catalyzes the insertion of only Mg^{2+} into protoporphyrin IX to yield magnesium protoporphyrin IX monomethyl, and Zn-BChl *a* may be formed by a substitution of Zn^{2+} for Mg^{2+} at a step in the bacteriochlorophyll biosynthesis after formation of magnesium protoporphyrin IX monomethyl ester (Masuda et al., 1999).

Membrane preparations of *A. rubrum* have absorption maxima at 377, 486, 515, 549, 590, 792, and 864 nm. A much lower absorption maximum at around 800 nm compared to the maximum at 864 nm indicates that the cells contain light-harvesting complex I together with the photosynthetic reaction center but lack the peripheral light-harvesting complex LH II, like cells of the anoxygenic phototrophic bacterium *Rhodospirillum rubrum*. Spirilloxanthin is the sole carotenoid component of the antenna complex. Unlike other genera of aerobic bacteriochlorophyll-containing bacteria, such as *Erythrobacter* and *Porphyrobacter*, *Acidiphilium* contains no membrane-free polar carotenoids.

Photosynthetic activities were examined by light-induced absorption changes of the membrane-bound pigments of *A. rubrum* (Wakao et al., 1996). The light-induced difference spectrum showed absorption changes at around 800 and 850 nm, caused by activities of the monomer and special pair BChl molecules in the reaction center complex, respectively. The peak positions assigned to the monomer and special pair BChls are blue-shifted by 7 and 25 nm, respectively, compared to the analogous positions found in *R. rubrum*. This suggests the involvement of Zn-BChl *a*, instead of Mg-BChl *a*, in the photochemical reaction. Although representative species of *Acidiphilium* have been shown to contain a fully active photochemical reaction center, they neither grow anaerobically in the light nor produce photopigments under anoxic or oxygen-limited conditions. The biological significance of the photosynthetic system in *Acidiphilium* has not been fully understood, but it has been shown that light enhances $^{14}CO_2$ incorporation into cells of *A. rubrum* under oxic conditions (Kishimoto et al., 1995a).

Nagashima et al. (1997b) have demonstrated that *Acidiphilium* species contain the *puf* operon, an assemblage of genes coding for α and β polypeptides of the light-harvesting complex LH I, the L and M subunits of photosynthetic reaction center proteins, and the membrane-bound cytochrome *c* subunit. The *puf* operon of *A. rubrum* consists of *pufB, -A, -L, -M*, and *-C* as found in general in anoxygenic phototrophic purple bacteria. A comparative analysis of deduced amino acid sequences showed that His L168, which is highly conserved in the L subunit in the anoxygenic phototrophic bacteria, is replaced by glutamic acid in *Acidiphilium*. This residue was suggested to locate closely to the special pair of BChl molecules and to be involved in the stabilization and function of Zn-BChl *a*, if the three dimensional structures of the *Blastochloris viridis* and *Rhodobacter sphaeroides* reaction centers were taken into account for comparison. A continuous stretch of amino acid sequences of the L and M subunits of *Acidiphilium* species is highly conserved, and a phylogenetic tree based on these amino acid sequences is basically consistent with the 16S rDNA sequence-based tree in the genealogical relationships between *Acidiphilium* and other phototrophic purple bacteria (Hiraishi et al., 1998).

Acidiphilium species contain high amounts (33–43% of the total lipids) of phosphorus-free lipoamino acids in their cell membranes. Two major lipoamino acids were isolated from *A. organovorum* and identified as being α-*N*-3-hydroxystearylornithinyltaurine and α-*N*-3-hydroxystearylornithine, to which $C_{18:1}$ fatty acid is linked by an ester-linkage (Kishimoto et al., 1993a). In addition to these lipids, bisphosphatidylglycerol, phosphatidylglycerol, phosphatidylethanolamine, and phosphatidylcholine are present as the major phospholipids. Straight-chain mono-unsaturated $C_{18:1}$ acid is the major component of whole-cell fatty acids, constituting of 40–90% of the total content (Kishimoto et al., 1993a; Wakao et al., 1994). A considerable amount of cyclopropane $C_{19:0}$ acid (5–21%) is also present.

Acidiphilium species (including new isolates of acidophilic heterotrophic bacteria that most likely belong to this genus) have been isolated from acidic mineral environments, such as pyritic mine drainage, pyritic coal refuse, copper and uranium mine tailings, acidic hot springs, and acidic soil (Tuttle et al., 1968; Manning, 1975; Wichlacz and Unz, 1981; Wichlacz et al., 1986; Kishimoto and Tano, 1987; Harrison, 1989; Wakao et al., 1994; Banerjee et al., 1996; A. Hiraishi, unpublished data). Also, as several strains of *Acidiphilium* species were isolated or found as contaminants from cultures of the iron-oxidizing chemolithotroph *Thiobacillus ferrooxidans* (Guay and Silver, 1975; Harrison et al., 1980; Harrison, 1981, 1989; Johnson and Kelso, 1983; Lobos et al., 1986), they may aid leaching through the consumption of organic matter inhibitory to the growth and activity of *T. ferrooxidans*. Although most *Acidiphilium* species are obligately heterotrophic, their natural habitats seem to be oligotrophic environments with poor organic matter. The ability of *Acidiphilium* to perform photosynthesis with Zn-BChl *a* may be an advantage to grow and survive in such oligotrophic acidic environments. Zn-BChl *a* is much more stable under acidic conditions than Mg-BChl *a* and chlorophylls (Wakao et al., 1996). Therefore, it is likely that the capacity for photosynthesis with Zn-BChl *a* of *Acidiphilium* is a result of its adaptation to acidic mineral environments. In addition, the ability of *Acidiphilium* species to reduce Fe(III), using a variety of organic substrates indicates that they may be of ecological significance in the biogeochemical cycling of iron at oxic-anoxic interfaces in acidic environments.

ENRICHMENT AND ISOLATION PROCEDURES

A suitable medium for growth of *Acidiphilium* contains, per liter of distilled water: 2 g $(NH_4)_2SO_4$, 1 g KH_2PO_4, 0.5 g $MgSO_4 \cdot 7H_2O$, 0.05 g $CaCl_2 \cdot 2H_2O$, and 1 ml trace element solution SL8 (Biebl and Pfennig, 1978), to which are added 0.1% glucose and 0.01% yeast extract (Difco Laboratories). The organic supplements are autoclaved separately. For solid media separately autoclaved 1.5% agar is added; alternatively, 0.6% gellan gum can be used (Kishimoto and Tano, 1987). For isolation and enrichment of *Acidiphilium* from the environment, the medium should be adjusted to pH 2.0–2.5 to minimize growth of fungi and facultatively acidophilic bacteria. Although *Acidocella* strains are frequently recovered on isolation media, it is possible to differentiate colonies of *Acidiphilium* from those of *Acidocella* by pigmentation when the media are incubated for more than one week.

MAINTENANCE PROCEDURES

Cultures are well preserved in liquid nitrogen, by lyophilization, or at −80°C in a mechanical freezer.

DIFFERENTIATION OF THE GENUS *ACIDIPHILIUM* FROM OTHER GENERA

The genus *Acidiphilium* is separated from other genera of aerobic acidophilic heterotrophic bacteria by its phylogenetic position and its outstanding physiological and chemotaxonomic characteristics. Differential characteristics of *Acidiphilium* and phenotypically related genera are indicated in Table BXII.α.15.

TAXONOMIC COMMENTS

The genus *Acidiphilium* was first monotypic with *A. cryptum* as the type species (Harrison, 1983). Later, several species including *A. angustum*, *A. facilis*, *A. organovorum*, and *A. rubrum* (Lobos et al., 1986; Wichlacz et al., 1986) were identified as new members of this genus. A phylogenetic analysis based on 5S rRNA sequences showed a close relationship between *A. cryptum* and *Thiobacillus acidophilus* (Lane et al., 1985). Information about the phylogeny and photosynthetic features of *Acidiphilium* has increasingly been accumulated in the 1990s. Wakao et al. (1993) reported that some members of *Acidiphilium* including the type species *A. cryptum* produced BChls. This report was the first to demonstrate the occurrence of photosynthetic pigments in *Acidiphilium* species. Similar observations were reported by Kishimoto et al. (1995a). Phylogenetic analyses based on 16S rDNA sequences revealed that all species of *Acidiphilium* form a major phylogenetic cluster in the *Alphaproteobacteria* (Lane et al., 1992; Sievers et al., 1994a) and that the BChl-producing species were genetically distant from the non-BChl-producing ones (Kishimoto et al., 1995b). This finding led to the proposal to remove the non-BChl producing species (i.e., *Acidiphilium facilis* and *Acidiphilium aminolytica*) from the genus *Acidiphilium* and to transfer them into the new genus *Acidocella* (Kishimoto et al., 1995b). The emended description of *Acidiphilium* only includes aerobic BChl-containing bacteria.

Acidiphilium angustum and *A. rubrum* were described as distinct species by Wichlacz et al. (1986) based mainly on phenotypic data. However, these two species have been shown to be indistinguishable by 16S rDNA sequencing and genomic DNA–DNA hybridization (Wakao et al., 1994; Kishimoto et al., 1995a; Hiraishi et al., 1998; Table BXII.α.16) and in addition are phenotypically quite similar to each other. These data strongly suggest that the two species names are subjective synonyms. Here we propose to unite *A. angustum* and *A. rubrum* in a single species, retaining the name *A. rubrum*.

Thiobacillus acidophilus (Guay and Silver, 1975) was not included in the Approved List of Bacterial Names (Skerman et al., 1980). After reexamination of the characteristics of *Thiobacillus acidophilus* and confirmation that this bacterium was a facultative chemolithotroph growing equally with elemental sulfur or glucose as the sole energy source, Harrison (1983) revived this name. However, phylogenetic analyses based on small-subunit rRNA sequences showed that *Thiobacillus acidophilus* was more closely related to members of the genus *Acidiphilium* than to any other species of *Thiobacillus* (Lane et al., 1985, 1992; Kishimoto et al., 1995b). Therefore, based on the phylogenetic evidence and the finding that this bacterium contains Zn-BChl *a* and the *puf* genes, as the other *Acidiphilium* species, the transfer of *Thiobacillus acidophilus* to the genus *Acidiphilium* as *Acidiphilium acidophilum* comb. nov. was proposed (Hiraishi et al., 1998).

FURTHER READING

Harrison, Jr., A.P. 1984. The acidophilic thiobacilli and other acidophilic bacteria that share their habitat. Ann. Rev. Microbiol. *38*: 265–292.

TABLE BXII.α.15. Differential characteristics of the genus *Acidiphilium* and phenotypically related genera[a]

Characteristic	*Acidiphilium*	*Acidocella*	*Acidobacterium*
Yellow-orange pigments	D	−	+
Pink-red pigments	D	−	−
Zn-BChl production	+	−	−
Growth inhibition by:			
Acetate (0.25 mM)	+	−	−
Lactate (2 mM)	+	−	−
Chemolithotrophic growth with sulfur	D	−	−
β-Galactosidase activity	−	−	+
Major fatty acid	$C_{18:1}$	$C_{18:1}$	iso-$C_{15:0}$
Major quinone	Q-10	Q-10	MK-8
Bisphosphatidylglycerol	+	−	+
Mol% G + C of DNA	63.2–67.5	58.7–64.4	60.8

[a]Symbols: +, 90% or more of strains positive; −, 90% or more of strains negative; D, different reactions in different species; Q-10, ubiquinone-10; MK-8, menaquinone-8.

TABLE BXII.α.16. Inter- and intraspecies relatedness of genomic DNA in the genus *Acidiphilium*[a]

Species	Level of hybridization (%)				
	A. cryptum	*A. acidophilum*	*A. angustum/A. rubrum*	*A. multivorum*	*A. organovorum*
A. acidophilum	80–100				
A. angustum/A. rubrum	18–22	71–100			
A. cryptum	76–100	8	11–26		
A. multivorum	43–63	10	10–29	83–100	
A. organovorum	64	10	9–26	53–56	100

[a]Data from Wakao et al. (1994) and Hiraishi et al. (1998).

Hiraishi, A. and S. Shimada. 2001. Aerobic anoxygenic photosynthetic bacteria with zinc-bacteriochlorophyll. J. Gen. Appl. Microbiol. *47*: 161–180.

Shimada, K. 1995. Aerobic anoxygenic phototrophs. *In* Blankenship, Madigan and Bauer (Editors), Anoxygenic Photosynthetic Bacteria, Kluwer Academic Publishers, Dordrecht. pp. 105–122.

DIFFERENTIATION OF THE SPECIES OF THE GENUS *ACIDIPHILIUM*

The differential characteristics of the species of *Acidiphilium* are indicated in Table BXII.α.17. Data on genomic DNA relatedness among the species are presented in Table BXII.α.16. Genetically, *Acidiphilium* species can be classified into two subgroups, one of which consists of *A. cryptum*, *A. multivorum*, and *A. organovorum* (*A. cryptum* subgroup) and one of which includes *A. acidophilum* and *A. rubrum* (*A. rubrum* subgroup). The levels of 16S rDNA sequence similarity and DNA–DNA hybridization between the two subgroups are 94–95% and 8–26%, respectively. Species differentiation in the former subgroup is problematic because of only small phenotypic and genetic differences among the species of this subgroup (Tables BXII.α.16 and BXII.α.17). It is impossible to separate *A. cryptum*, *A. multivorum*, and *A. organovorum* by 16S rDNA sequencing (>99.5% interspecies similarity). In addition, genomic DNA–DNA reassociation levels among the three species are relatively high (43–63%). The diagnostic characteristics useful for the identification of the species include growth inhibition by different concentrations of glucose, yeast extract, and trypticase and carbon nutrition (fumarate, succinate, sorbitol, and inositol). However, since the available phenotypic information has been derived from studies of only limited numbers of strains, DNA–DNA hybridization assays are required for the ultimate identification at the species level.

List of species of the genus Acidiphilium

1. **Acidiphilium cryptum** Harrison 1981, 331[VP]

   cryp' tum. Gr. adj. *kryptos* hidden.

   Cells are straight rods, 0.3–0.6 × 0.6–1.5 μm under optimal growth conditions, occurring singly or in pairs. Motile by means of a single polar, or subpolar, flagellum or by two lateral flagella; some strains appear to be nonmotile. Colonies on agar media are circular (1–3 mm in diameter after one week of incubation), smooth, convex, slightly translucent, and white to cream, and turn pink to light brown after continued incubation. Aerobic, chemoorganoheterotrophic bacteria having a respiratory type of metabolism with oxygen as terminal electron acceptor. Some strains may grow with Fe(III) as the terminal electron acceptor under both oxic and anoxic conditions. Carbon sources utilized are listed in Table BXII.α.17. D-Glucose, D-fructose, and some other sugars support good growth. Growth is inhibited by the presence of 1% glucose, 0.3% yeast extract, and 1% trypticase. Growth factors are required and this requirement is satisfied by 0.01% yeast extract. Ammonium salts are used as nitrogen source. Catalase is produced. Oxidase reaction is weakly positive or absent. Urease is negative. Arsenite oxidation is negative. Susceptible to ampicillin, chloramphenicol, rifampin, tetracycline, but resistant to josamycin, lincomycin, penicillin, and streptomycin. Optimal growth occurs at 30–37°C (range: 15–42°C) and at pH 3.0–3.5 (range: pH 2.0–5.9). No growth occurs at 42°C or with 3% NaCl. Habitat: Originally isolated from an iron-oxidizing culture of *Thiobacillus ferrooxidans* and found in strongly acidic environments, including mine water.

TABLE BXII.α.17. General and differential characteristics of the species of the genus *Acidiphilium*[a,b]

Characteristic	A. cryptum	A. acidophilum	A. angustum/A. rubrum	A. multivorum	A. organovorum
Cell width (μm)	0.3–0.6	0.5–0.8	0.7–0.9	0.5–0.9	0.5–0.7
Motility	+	d	d	d	+
Color of colonies:					
Cream/pink/ light brown	+	+	−	+	+
Red/violet	−	−	+	−	−
Zn-BChl content	Low	Trace	High	Low	Trace
Growth factor required	+	−	+	+	+
Chemolithotrophic growth:					
Sulfur	−	+	−	−	−
Thiosulfate	−	+	−	−	−
Growth at 42°C	+	−	−	+	−
Growth with 3% NaCl	+	−	−	+	+
Growth inhibition by:					
1% Glucose	+	+	+	−	−
5% Glucose	+	+	+	−	+
0.3% Yeast extract	+	+	+	−	−
1% Trypticase	+	+	+	−	−
Arsenite oxidation	−	−	−	+	−
Carbon source utilization:					
L-Arabinose	+	+	+	+	+
D-Xylose	+	+	+	+	+
D-Ribose	+	+	+	+	+
L-Sorbose	nd	−	nd	+	nd
D-Fructose	+	+	+	+	+
D-Glucose	+	+	+	+	+
D-Galactose	+	+	+	+	+
D-Mannose	d	+	+	+	+
D-Maltose	+	−	−	−	−
Sucrose	+	+	+	+	+
Lactose	+	−	−	−	−
Raffinose	+	−	−	+	nd
Melibiose	nd	nd	nd	+	nd
Mannitol	+	+	+	+	+
Sorbitol	+	nd	−	+	+
Glycerol	+	+	d	+	+
Inositol	−	nd	−	+	+
Methanol	−	+	−	+	−
Ethanol	d	+	d	+	−
Propanol	−	+	−	+	−
Pyruvate	−	−	−	−	−
Lactate	−	−	−	−	−
Acetate	−	−	−	−	−
Citrate	d	+	+	+	+
Oxoglutarate	d	−	d	+	−
Fumarate (2 mM)	−	−	−	+	+
Succinate (2 mM)	−	−	−	+	+
Malate	d	+	d	+	+
Malonate	nd	−	−	−	nd
Tartrate	+	−	nd	−	−
Gluconate	−	+	nd	+	+
Alanine	−	−	−	+	nd
Leucine	nd	−	nd	−	−
Isoleucine	nd	−	nd	−	−
Proline	nd	+	nd	+	+
Aspartate	nd	+	+	+	nd
Asparagine	+	nd	nd	+	nd
Glutamate	+	+	d	+	+
Arginine	+	nd	−	+	+
Lysine	−	nd	−	+	−
Histidine	nd	+	nd	nd	nd
Phenylalanine	nd	−	nd	+	+
Tryptophan	nd	−	nd	−	−
Sensitivity to:					
Penicilin	−	nd	nd	+	−
Streptomycin	−	nd	nd	+	−
Mol% G + C content of DNA	67.3–68.3	62.9–63.2	63.2–63.4	66.2–68.1	67.4

[a]Symbols: +, 90% or more of strains positive; −, 90% or more of strains negative; d, 11–89% of strains positive; nd, not determined .

[b]Data from Wakao et al. (1994) and Hiraishi et al. (1998).

The mol% G + C of the DNA is: 68–70 (T_m), 69 (Bd), 67.3–67.5 (HPLC).

Type strain: Lhet2, ATCC 33463, DSM 2389.

GenBank accession number (16S rRNA): D30773.

2. **Acidiphilium acidophilum** (Harrison 1983) Hiraishi, Nagashima, Matsuura, Shimada, Takaichi, Wakao and Katayama 1998,1396[VP] (*Thiobacillus acidophilus* (Guay and Silver 1975) Harrison 1983.)

a.ci.do' phi.lum. L. adj. *acidus* sour; M.L. neut. n. *acidum* acid; Gr. adj. *philus* loving; M.L. adj. *acidophilum* acid-loving.

Cells are rod-shaped, 0.5–0.8 × 1.0–1.5 μm , occurring singly, mainly in pairs and rarely chains. Motile by means of single polar flagella or nonmotile. Colonies on agar media are small (1–2 mm in diameter after one week of incubation), round, smooth, convex, slightly translucent, and cream colored; if cells age, they become pink to light brown. Strictly aerobic; facultatively chemolithotroph and mixotroph, growing with elemental sulfur as an energy source. Thiosulfate, trithionate, and tetrathionate also serve as electron donor. Neither sulfite, sulfide, nor ferrous ion is used as electron donor. Polyhedral inclusion bodies (carboxysomes) are present in elemental sulfur-growing cells. Carbon sources utilized are listed in Table BXII.α.17. D-Glucose, D-fructose, and some other sugars support good growth. Growth is inhibited by the presence of 1% glucose, 0.3% yeast extract, or 1% trypticase. No growth factor is required. Ammonium salts are utilized as nitrogen source. Catalase positive. Oxidase negative. Arsenite oxidation is negative. Optimal growth occurs at 25–30°C (range: 15–35°C) and at pH 3.0–3.5 (range: pH 1.5–6.0). No growth occurs in the presence of 3% NaCl. Habitat: Originally isolated from an iron-oxidizing culture of *Thiobacillus ferrooxidans* and found in strongly acidic environments including mine drainage.

The mol% G + C of the DNA is: 62.9–63.2 (T_m).

Type strain: ATCC 27807 (DSM 700).

GenBank accession number (16S rRNA): D86511.

3. **Acidiphilium multivorum** Wakao, Nagasawa, Matsuura, Matsukura, Matsumoto, Hiraishi, Sakurai and Shiota 1995, 197[VP] (Effective publication: Wakao, Nagasawa, Matsuura, Matsukura, Matsumoto, Hiraishi, Sakurai and Shiota 1994, 156.)

mul.ti.vo' rum. L. adj. *multus* many; L. v. *voro* eat, consume; M.L. adj. *multivorum* devouring many kinds of substances.

Cells are rod-shaped, 0.5–0.9 × 1.5–3.8 μm under optimal growth conditions. Motile by means of single polar, subpolar, or lateral flagellum; some strains appear to be nonmotile. Colonies on agar media are circular (1–2 mm in diameter after 3 days of incubation), smooth, slightly convex, opaque, and white to cream, and turn pink to light brown after continued incubation. Strictly aerobic, chemoorganoheterotrophic bacteria. Carbon sources utilized are listed in Table BXII.α.17. D-Glucose, D-fructose, and some other sugars support good growth. Growth is not inhibited by the presence of 1% glucose, 0.3% yeast extract, or 1% trypticase. Growth factors are required and this requirement is satisfied by 0.01% yeast extract. Ammonium salts and nitrate salts are utilized as nitrogen source. Nitrate is reduced to nitrite. No denitrification. Arsenite oxidation is positive. Catalase and urease positive. Oxidase, DNase,

and amino acid carboxylases negative. Hydrogen sulfide produced. Indole not formed. Starch and gelatin are not hydrolyzed. Tween 40 and 60 hydrolyzed. Susceptible to ampicillin, chloramphenicol, penicillin, rifampin, streptomycin, and tetracycline but resistant to josamycin, kanamycin, and lincomycin. Optimal growth occurs at 27–35°C (range: 17–42°C) and at pH 3.2–4.0 (range: pH 1.9–5.6). Growth occurs in the presence of 3% NaCl. Habitat: water and sediments in acidic mine drainage streams.

The mol% G + C of the DNA is: 66.2–68.1 (HPLC).

Type strain: AIU 301, DSM 11245, JCM 8867.

GenBank accession number (16S rRNA): AB006711.

4. **Acidiphilium organovorum** Lobos, Chisolm, Bopp and Holmes 1986, 143[VP]

or.ga.no' vor.um. N.L. n. *organum* organic, compound; L. v. *voro* eat, consume; N.L. adj. *organovorum* devouring organic compounds.

Cells are rod-shaped, 0.5–0.7 × 1.0–1.5 μm under optimal growth conditions, occurring singly, in pairs, and in chains in some cases. Motile by means of polar or subpolar flagella. Colonies on agar media are circular (1–2 mm after 3 days of incubation), smooth, convex, slightly translucent, and white to cream, and turn pink to light brown after continued incubation. Strictly aerobic, chemoorganoheterotrophic bacteria having respiratory type of metabolism with oxygen as terminal electron acceptor. Carbon sources utilized are listed in Table BXII.α.17. D-Glucose, D-fructose, and some other sugars support good growth. Growth is not inhibited by the presence of 1% glucose, 0.3% yeast extract, or 1% trypticase. Growth factors are required, and this requirement is satisfied by 0.01% yeast extract. Ammonium and nitrate salts are utilized as nitrogen source. Catalase and urease positive. Oxidase negative. Arsenite oxidation negative. Susceptible to ampicillin, chloramphenicol, novobiocin, penicillin, rifampin, and tetracycline, but resistant to josamycin, lincomycin, and streptomycin. Optimal growth occurs at 25–30°C (range: 15–42°C) and at pH 3.0–3.5 (range: pH 2.0–6.0). Growth occurs in the presence of 3% NaCl. Habitat: Originally isolated from an iron-oxidizing culture of *Thiobacillus ferrooxidans* and found in acidic mine drainage.

The mol% G + C of the DNA is: is 67.4–67.5 (HPLC).

Type strain: ATCC 43141.

GenBank accession number (16S rRNA): D30775.

5. **Acidiphilium rubrum** Wichlacz, Unz and Langworthy 1986, 200[VP]

ru' brum. L. adj. *rubrum* red colored.

Cells are straight or slightly curved rods, 0.5–0.8 × 1.5–3.2 μm under optimal growth conditions, occurring singly, in pairs, and in chains in some cases. Motile by means of polar or subpolar flagella. Colonies on agar media are small (1 mm in diameter after one week of incubation), round, regular, convex, slightly translucent, and red to violet. Strictly aerobic, chemoorganoheterotrophic bacteria having respiratory type of metabolism. Carbon sources utilized are listed in Table BXII.α.17. D-Glucose, D-fructose, and some other sugars support good growth. Growth is inhibited by the presence of 1% glucose, 0.3% yeast extract, or 1% trypticase. Growth factors are required and this requirement is satisfied by 0.01% yeast extract. Ammonium salts are utilized as nitrogen source. Catalase positive. Oxidase negative. Arsenite oxidation negative. Optimal growth occurs at 25–30°C (range: 15–35°C) and at pH 3.0–3.5 (range: 2.0–5.9). No growth occurs in the presence of 3% NaCl. Habitat: water and sediments in acidic mine drainage streams.

The mol% G + C of the DNA is: 63 (T_m), 63.2–64.0 (HPLC).

Type strain: OP, ATCC 35905.

GenBank accession number (16S rRNA): D30776.

Genus III. **Acidisphaera** *Hiraishi, Matsuzawa, Kanbe, and Wakao 2000b, 1544*[VP]

AKIRA HIRAISHI

A.ci.di.sphae' ra. M.L. n. *acidum* an acid; M.L. fem. n. *sphaera* sphere; M.L. fem. n. *Acidisphaera* acid (-requiring) coccoid microorganism.

Cells are cocci or coccobacilli, 0.7–0.9 × 0.9–1.6 µm. Nonmotile. Spore and capsules are not formed. Multiply by binary fission. Gram negative. Aerobic, having a strictly respiratory type of metabolism with oxygen as the terminal electron acceptor. Chemoorganotrophic. **Produce bacteriochlorophyll *a* esterified with phytol and carotenoids** as photosynthetic pigments only under aerobic growth conditions in the dark. Light stimulates aerobic growth and viability. Do not grow phototrophically under anoxic conditions in the light. Liquid cultures and colonies on agar media are pink or salmon-pink. Growth occurs with a number of simple organic compounds as electron donor and carbon source. Growth on complex media containing peptone is poor. Sulfide, S^0, thiosulfate, and Fe^{2+} do not serve as electron donors for chemolithotrophic growth. Catalase and oxidase positive. Mesophilic. **Obligately acidophilic**, growing in the pH range of 3.5–6.0 (but not at pH 3.0 or 6.5). Freshwater bacteria that do not require NaCl for optimal growth. **Ubiquinones with ten isoprene units** (Q-10) are present. Members of the class *Alphaproteobacteria*. Isolated from acidic mineral environments including surface water and sediment in acidic hot springs (at 42°C and below) and mine drainage streams.

The mol% G + C of the DNA is: 69–70.

Type species: **Acidisphaera rubrifaciens** Hiraishi, Matsuzawa, Kanbe and Wakao 2000b, 1545.

FURTHER DESCRIPTIVE INFORMATION

Cells of *A. rubrifaciens* are cocci, coccobacilli, or short rods and occur singly or in pairs (Fig. BXII.α.22). Motility is not observed at any growth stage. Thin-section electron microscopy shows that cells have typical Gram-negative membranes (Fig. BXII.α.23). Internal photosynthetic membranes are not found. Polyphosphate granules are present.

Colonies grown on agar media are 2 mm in diameter after one week of incubation, and are round, smooth, circular, convex, and salmon pink.

The photosynthetic pigments are magnesium-chelated bacteriochlorophyll (Mg-BChl) *a* esterified with phytol and the carotenoid spirilloxanthin. In addition to this carotenoid component, cells contain much larger amounts of a polar carotenoid that hardly migrates on silica-gel thin-layer chromatography with benzene–acetone (1:1, v/v) as the developing solvent. Membrane preparations from cells grown aerobically in darkness show major absorption maxima at 474–476, 502–503, 545 (shoulder), 590 (shoulder), 801, and 873–874 nm. The very low absorption peak at around 800 nm, compared to the peak at 873–874 nm, suggests that cells contain the core light-harvesting complex (LH I) together with the photosynthetic reaction center, but lack peripheral antenna complexes. The acetone–methanol extract has ab-

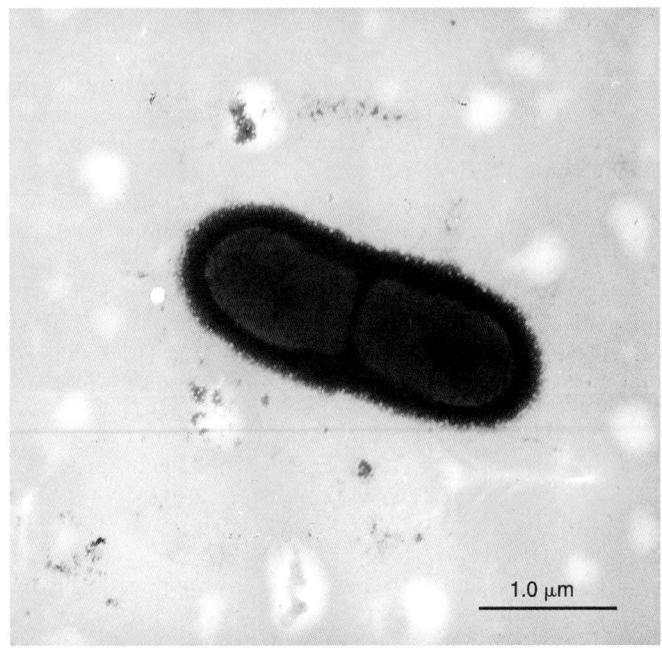

FIGURE BXII.α.22. Electron micrograph showing general cell morphology of *Acidisphaera rubrifaciens* (strain HS-AP3).

sorption maxima at 386, 466–467 (shoulder), 494–495, 525–526, and 770 nm, indicating the presence of Mg-BChl *a* but not Zn-BChl *a* as the major photopigment. However, when cells are grown in the presence of 1 mM zinc sulfate, an additional photopigment, Zn-BChl *a*, occurs in a trace amount (1–2% of the amount of Mg-BChl *a*) (Hiraishi et al., 2000b). As with many other species of aerobic BChl-containing bacteria (Shimada, 1995; Yurkov and Beatty, 1998a), aerobic and dark conditions are most favorable for BChl production by *A. rubrifaciens*. BChl production is repressed completely by continuous illumination. The BChl *a* content of *A. rubrifaciens* strains grown under optimal growth conditions in darkness ranges from 35 to 90 nmol/g dry weight of cells (Hiraishi et al., 2000b). A marked increase in BChl content is found under oligotrophic or starvation conditions (A. Hiraishi, unpublished data).

Strains of *A. rubrifaciens* are aerobic chemoorganotrophic bacteria that grow well under aerobic conditions with shaking but do not exhibit phototrophic growth under anaerobic conditions in the light. They have a doubling time of 11–13 h when grown in gluconate-containing liquid medium in darkness. Continuous incandescent illumination stimulates aerobic growth significantly, provided that the preculture is grown in darkness. The

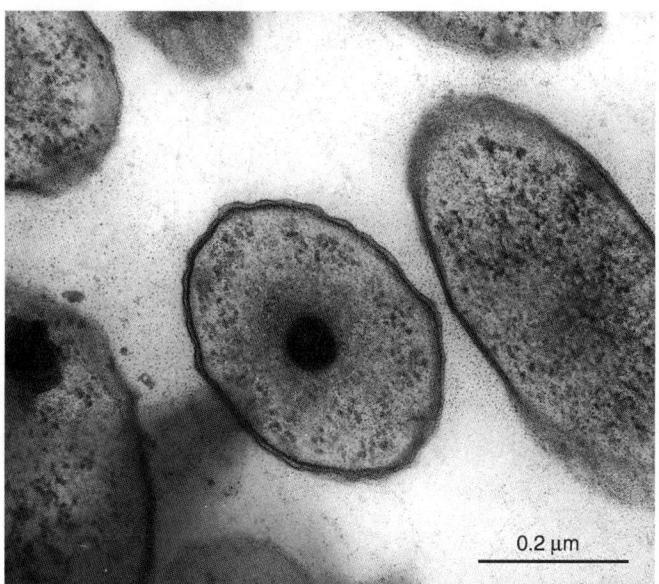

FIGURE BXII.α.23. Thin-section electron micrograph showing ultrastructure of *Acidisphaera rubrifaciens* cells (strain HS-AP3).

growth yield is 1.5–1.8-fold higher in illuminated cultures than in chemotrophic dark cultures. Illumination also enhances the viability of *A. rubrifaciens* cells under starvation conditions, as is the case in some other aerobic BChl-containing bacteria (Shiba, 1984).

The optimum pH for growth is 4.5–5.0. The temperature range for growth is 20–40°C (optimum 30–35°C). Growth is good in the absence of NaCl and is inhibited by adding 3% NaCl.

Catalase and cytochrome oxidase are produced. Hydrolysis of starch, Tween 80, or casein is negative. Nitrate is not reduced to nitrite. Utilization patterns of organic compounds as carbon and energy sources are shown in Tables BXII.α.18 and BXII.α.19. Good carbon sources are gluconate, lactate, malate, and pyru-

vate. Moderate growth is observed with xylose, galactose, glycerol, sorbitol, ethanol, and some intermediates of the tricarboxylic acid cycle. Little or no growth occurs with maltose, lactose, sucrose, lower fatty acids, citrate, benzoate, amino acids, Casamino acids, or peptone.

The major respiratory quinone, Q-10, accounts for 98–99% of the total quinone contents. The remainder detected is Q-9. Other quinone ring groups, such as menaquinones and rhodoquinones, are not found.

The natural habitats and ecological roles of *A. rubrifaciens* have not yet been fully identified. Strains of *A. rubrifaciens* have so far been isolated from acidic mineral environments including surface water and sediment in acidic hot springs (at 42°C and below) and mine drainage streams (Hiraishi et al., 2000b). In mine drainage environments, *A. rubrifaciens* frequently coexists with members of the genus *Acidiphilium*, another group of acidophilic aerobic BChl-containing bacteria (A. Hiraishi, unpublished observations). In view of its characteristic lifestyle as an acidophilic BChl-producing bacterium, *A. rubrifaciens* appears to have an ecological niche similar to that of *Acidiphilium* species, although the former organism differs from the latter in exhibiting weaker acidophily and lacking Zn-BChl as the main photosynthetic pigment. In sunlight-exposed environments, *A. rubrifaciens* can be viable and survive for a long time without any carbon/energy source. Synthesis and expression of the photosynthetic apparatus in *A. rubrifaciens* may be regarded as a result of its adaptation to oligotrophic and mineral environments like hot springs and mine drainage streams.

ENRICHMENT AND ISOLATION PROCEDURES

A suitable medium for growth and purification of *Acidisphaera* contains (per liter distilled water): (NH₄)₂SO₄, 1.0 g; KH₂PO₄, 1.0 g; MgSO₄·7H₂O, 0.5 g; CaCl₂·2H₂O, 0.05 g; and trace element solution SL8 (Biebl and Pfennig, 1978), 1.0 ml. To this medium are added (per liter): gluconate, 2.0–3.0 g, and yeast extract (Difco), 1.0 g. The organic supplements are autoclaved separately and combined with the basal medium before use. The final pH of the medium should be 4.0–4.5. For solid media, separately

TABLE BXII.α.18. Differential characteristics of *Acidisphaera* and phenotypically and phylogenetically related genera of BChl-containing bacteria[a]

Characteristic	*Acidisphaera*	*Acidiphilium*[b]	*Craurococcus*[c]	*Paracraurococcus*[c]	*Roseococcus*[d]	*Rhodopila*[e]
Cell shape	Cocci, short rods	Rods	Cocci	Cocci	Cocci	Cocci
Cell diameter (μm)	0.7–0.9	0.5–0.7	0.8–2.0	0.8–1.5	0.9–1.3	1.6–1.8
Motility	−	+	−	−	+	+
Color of colonies	Salmon-pink	Pink, red	Pink	Red	Pink	Purple-red
Zn-BChl as main pigment	−	+	−	−	−	−
Near IR peak for BChl a (nm)	874	864	872	856	859	865
Anaerobic phototrophy	−	−	−	−	−	+
Optimum pH for growth	4.5–5.0	3.0–3.5	7.5	6.6–6.8	7–8	4.8–5.0
Growth at pH 7	−	−	+	+	+	−
Carbon source:						
Acetate	−	−	−	−	+	−
Succinate	+	D	−	−	+	+
Major quinones	Q-10	Q-10	Q-10	Q-10	Q-10	Q-9(10), MK-9(10), RQ-9(10)
Mol% DNA G + C content	69.1–69.8	62.9–68.3	70.5	70.3–71.0	70.4	66.3
Habitat	Acidic springs and mine water	Acidic mine drainage	Soil	Soil	Cyanobacterial mats in sulfur spring	Acidic sulfur spring

[a]For symbols see standard definitions; Q-10, ubiquinone-10; MK-9, menaquinone-9; RQ-9, rhodoquinone-9. Q-9(10), MK-9(10), and RQ-9(10) mean the presence of those quinones with both 9 and 10 isoprene units in the side chain.

[b]Information from Hiraishi et al. (1998).

[c]Information from Saitoh et al. (1998).

[d]Information from Yurkov and Gorlenko (1992) and Yurkov et al. (1994c).

[e]Information from Pfennig (1974).

autoclaved 1.5% agar or 0.6% gellan gum can be used. Enrichment and isolation of *A. rubrifaciens* from the environment are possible using the above-noted medium, but an alternative use of both gluconate (1.0 g/l) and D-xylose (1.0 g/l) as the carbon source may be more effective in some cases. Cultures are incubated aerobically at 30°C in darkness. Although other aerobic acidophilic bacteria and acid tolerant bacteria are frequently recovered in or on isolation media, enriched cultures or colonies of *A. rubrifaciens* are distinguishable by their pink or salmon-pink color.

MAINTENANCE PROCEDURES

Cultures are preserved in liquid nitrogen or by lyophilization. Preservation in a mechanical freezer at −80°C is also possible.

DIFFERENTIATION OF THE GENUS *ACIDISPHAERA* FROM OTHER GENERA

The genus *Acidisphaera* is differentiated from other genera of aerobic BChl-containing bacteria by a combination of morphological, physiological, and chemotaxonomic characteristics as indicated in Table BXII.α.18.

TAXONOMIC COMMENTS

The description of the genus *Acidisphaera* is based on only one species, *A. rubrifaciens* (Hiraishi et al., 2000b). The 16S rDNA sequences of *A. rubrifaciens* strains place them in the major acidophilic cluster of the class *Alphaproteobacteria*, with members of the genera *Acidiphilium*, *Rhodopila*, *Acetobacter*, and *Gluconobacter* as their phylogenetic relatives. The nearest phylogenetic neighbor is the anaerobic phototrophic bacterium *Rhodopila globiformis* (95% similarity).

Before the description of *A. rubrifaciens*, acidophilic aerobic BChl-containing bacteria had been limited to members of the genus *Acidiphilium*, which contain Zn-BChl *a* rather than Mg-BChl *a* as the major photopigment (Wakao et al., 1996; Hiraishi et al., 1998). Thus, *A. rubrifaciens* is the first description of acidophilic aerobic bacteria containing Mg-BChl *a* as the major photopigment. *A. rubrifaciens* is similar to the moderately acidophilic phototrophic bacterium *Rhodopila globiformis* (Pfennig, 1974) in morphology, physiology, and natural habitats, but *A. rubrifaciens*

TABLE BXII.α.19. Physiological and biochemical characteristics of *Acidisphaera rubrifaciens*[a]

Characteristic	Result/reaction
Optimal temperature (°C) for growth	30–35
Optimal pH for growth (range)	4.5–5.0 (3.5–6.0)
Growth in the presence of:	
0–1% NaCl	+
3% NaCl	−
Vitamin requirement for growth	+
Nitrate reduction to nitrite	−
Catalase	+
Cytochrome oxidase	+
Hydrolysis of starch, casein, or Tween 80	−
Utilization of carbon/energy sources:	
Sugars and sugar alcohols:	
L-Arabinose, D-fructose, and mannitol	d
D-Galactose, D-glucose, glycerol, sorbitol, and D-xylose	+
Inositol, lactose, maltose, D-mannose, and sucrose	−
Alcohols:	
Ethanol	+
Methanol, propanol, and butanol	−
Organic acids:	
Benzoate, citrate, and lower fatty acids	−
Fumarate, gluconate, lactate, malate, pyruvate, and succinate	+
Amino acids and others:	
Alanine, asparagine, aspartate, glutamate, leucine, Casamino Acids, and peptone	−
Yeast extract	+

[a]Symbols: see standard definitions.

differs clearly from the latter by failing to grow anaerobically in the light. The level of 16S rDNA sequence similarity between *A. rubrifaciens* and *R. globiformis* is low enough to warrant different generic allocations.

DIFFERENTIATION OF THE SPECIES OF THE GENUS *ACIDISPHAERA*

The genus *Acidisphaera* is monotypic, with *A. rubrifaciens* as the type species. General characteristics of *A. rubrifaciens* in terms of physiology and biochemistry are shown in Table BXII.α.19.

List of species of the genus Acidisphaera

1. **Acidisphaera rubrifaciens** Hiraishi, Matsuzawa, Kanbe and Wakao 2000b, 1545[VP]

    rub.ri.fac′ i.ens. L. adj. *ruber* red; L. v. *facio* make; M.L. part. adj. *rubrifaciens* red-producing.

    The characteristics are as described for the genus and as listed in Tables BXII.α.18 and BXII.α.19, with the following additional features. Cells are 0.7–0.9 × 0.9–1.6 μm. No intracytoplasmic membrane systems occur. Polyphosphate granules are present. Cell membranes have absorption maxima at 474–476, 502–503, 545, 590, 801, and 873–874 nm. Acetone-methanol extracts from cells have absorp-

    tion maxima at 386, 466–467, 494–495, 525–526, and 770 nm. Sprilloxanthin and a polar pigment are the major carotenoids. No hydrolysis of starch, casein and Tween 80 occurs. Gluconate, lactate, malate, and pyruvate are good carbon sources. Other usable carbon sources are galactose, glucose, D-xylose, glycerol, mannitol, sorbitol, fumarate, malate, succinate, and yeast extract. The habitats are strongly acidic mineral environments, including acidic hot springs (at 42°C and below) and pyritic mine drainage.

    The mol% G + C of the DNA is: 69.1–69.8 (HPLC).
    Type strain: HS-AP3, JCM 10600.
    GenBank accession number (16S rRNA): D86512.

Genus IV. **Acidocella** *Kishimoto, Kosako, Wakao, Tano, and Hiraishi 1996, 362^VP (Effective publication: Kishimoto, Kosako, Wakao, Tano, and Hiraishi 1995b, 90)*

AKIRA HIRAISHI

A.ci.do.cel'la. M.L. n. *acidum* an acid; L. n. *cella* a cell; M.L. n. *Acidocella* acid (-requiring) cell.

Straight or slightly curved rods and coccobacilli with rounded or tapered ends, 0.5–0.8 × 1.0–2.0 μm. Motile by means of polar or lateral flagella; some strains are nonmotile. Nonsporeforming and nonencapsulated. Gram negative. **Aerobic,** having a strictly respiratory type of metabolism with oxygen as the terminal electron acceptor. Cell suspensions and colonies are white, cream, or light brown. Photosynthetic pigments and carotenoids are absent. Mesophilic. **Obligately acidophilic, growing in the pH range of 3.0–6.0;** some strains grow at pH 2.5. No growth occurs at pH ≥6.1. **Chemoorganotrophs,** growing with simple organic compounds as carbon and energy sources. **Growth occurs in the presence of either 0.25 mM acetate, 2 mM lactate, or 4 mM succinate.** Catalase positive. **Oxidase negative or weakly positive. A straight-chain monounsaturated $C_{18:1}$ acid is the major component of cellular fatty acids. Ubiquinones with ten isoprene (Q-10) units are present. 16S rRNA structures and signatures conform to species of the *Alphaproteobacteria.*** Inhabit strongly acidic mineral environments.

The mol% G + C of the DNA is: 58–65.

Type species: **Acidocella facilis** (Wichlacz, Unz, and Langworthy 1986) Kishimoto, Kosako, Wakao, Tano, and Hiraishi 1996, 362 (Effective publication: Kishimoto, Kosako, Wakao, Tano, and Hiraishi 1995b, 90) (*Acidiphilium facilis* Wichlacz, Unz, and Langworthy 1986, 200.)

FURTHER DESCRIPTIVE INFORMATION

Acidocella cells are straight or slightly curved rods with rounded or tapered ends, occurring singly or in pairs (Fig. BXII.α.24). In some cases, they are observed in chains and small flocs. Cells vary from 0.5 to 0.8 μm in width and from 1.0 to 2.0 μm in length under optimal growth conditions. Strains of each species are usually motile, but nonmotile strains are occasionally found. Electron microscopy of negatively stained cells shows that motile cells have polar or lateral flagella (Fig. BXII.α.25). The flagellar arrangement is different from strain to strain; in some cases, cells with polar flagella and with lateral flagella coexist in a single strain.

Colonies on agar media are circular (3–5 mm in diameter after 3 d of incubation), smooth, convex, slightly translucent, and white, cream, or light brown in color. Neither bacteriochlorophylls nor carotenoid pigments are present.

Acidocella species are nutritionally more versatile and exhibit more rapid and luxuriant growth than the phylogenetically and physiologically related *Acidiphilium* species. Most *Acidocella* strains exhibit a doubling time of 2–3 h under optimal growth conditions. The following sugars and sugar alcohols are good substrates supporting growth of all species: L-arabinose, D-xylose, D-ribose, D-glucose, D-galactose, D-fructose, sucrose, glycerol, arabitol, and mannitol. The following compounds are not used as carbon sources: L-rhamnose, maltose, cellobiose, methanol, formate, acetate, propionate, butyrate, valerate, caproate, lactate, and benzoate. As is the case with *Acidiphilium,* acidophilic growth of *Acidocella* strains is inhibited in the presence of high concentrations of organic acids such as acetate, lactate, and succinate when grown in glucose-mineral medium. However, *Acidocella* species are more tolerant to such organic acids at low concentrations.

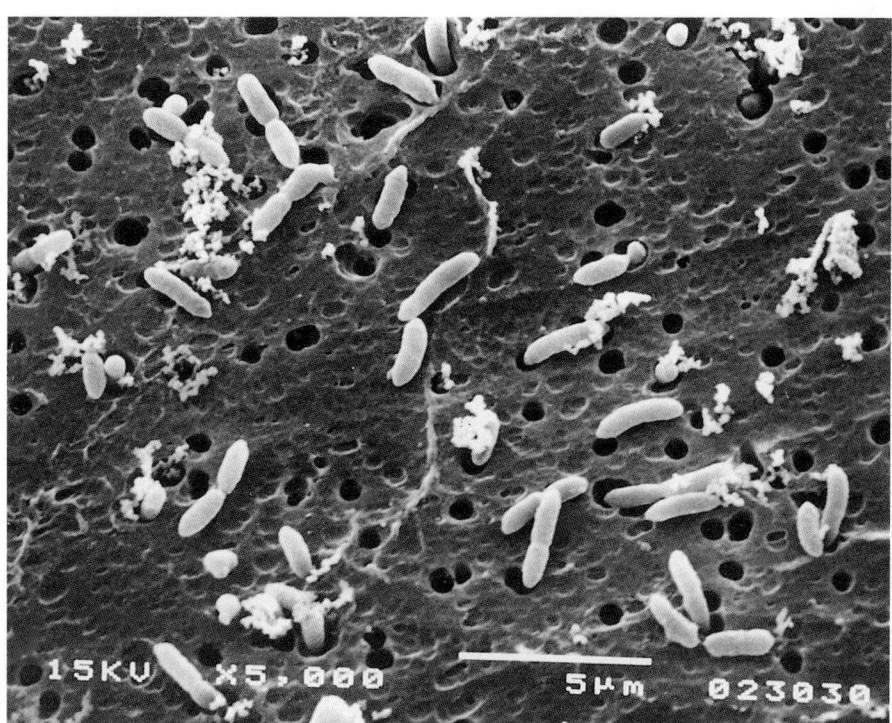

FIGURE BXII.α.24. Scanning electron micrograph showing general cell morphology of *Acidocella facilis* (ATCC 35904^T.)

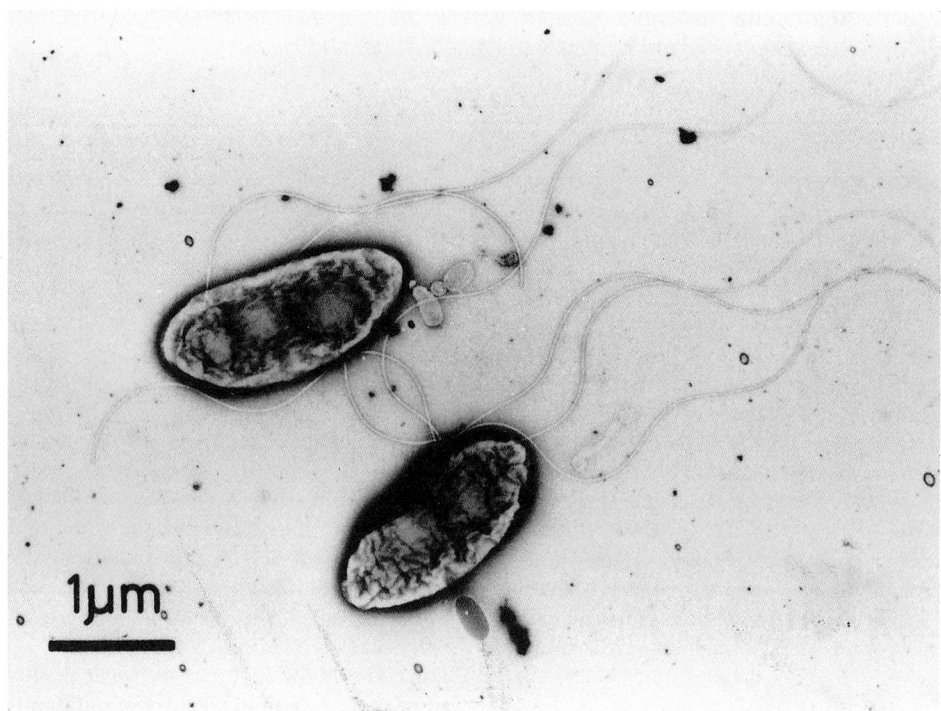

FIGURE BXII.α.25. Electron micrograph of negatively stained cells of *Acidocella facilis* (ATCC 35904[T]) having lateral flagella. (Reprinted with permission from K. Inagaki, Nippon Nogeikagaku Kaishi *71:* 1–8, 1997, ©Japan Society for Bioscience, Biotechnology, and Agrochemistry.)

They are capable of growing in the presence of 0.25 mM acetate, 2 mM lactate, 4 mM succinate, or 0.3% yeast extract, each of which has an inhibitory effect on the growth of *Acidiphilium*.

None of the *Acidocella* species grows chemolithotrophically with S^0 or thiosulfate as the energy source. Neither Fe^{2+} nor other metal ions serve as the electron donor for growth. However, the addition of ferrous or aluminum ions stimulates heterotrophic growth significantly. All species are capable of growth in the presence of 100 mM Fe^{2+} and 100 mM Al^{3+}. Growth yields are much higher in the presence of 100 mM Al^{3+} than in the absence of Al^{3+}, and this growth-promoting effect is more pronounced with aluminum sulfate than with aluminum chloride or aluminum phosphate (A. Hiraishi, unpublished). Some strains also exhibit extremely high resistance to heavy metals such as cadmium, copper, nickel, and zinc; however, these metals extend the lag period and generation time for growth. The heavy-metal resistance of *Acidocella* strains may be plasmid-dependent. When *Acidiphilium multivorum* and *Escherichia coli* were subjected to transformation with a plasmid prepared from an *Acidocella* strain, they became more highly resistant to cadmium and zinc (Ghosh et al., 1997).

Acidocella strains, as well as *Acidiphilium* strains, contain high amounts of phosphorus-free lipoamino acids in their cell membranes. Ornithine amide lipids, $C_{18:1}$ fatty acid esters of α-N-3-hydroxystearylornithine and α-N-3-hydroxystearylornithyltaurine, are the major polar aminolipids (Kishimoto et al., 1993a). In addition to these lipids, phosphatidylglycerol, phosphatidylethanolamine, and phosphatidylcholine are present as the major phospholipids. Bisphosphatidylglycerol is absent. Straight-chain monounsaturated $C_{18:1}$ acid is the major component of whole-cell fatty acids, constituting 57–66% of the total content (Kishimoto et al., 1993b). A considerable amount of cyclopropane $C_{19:0}$ acid (6–21%) is also present. 2-Hydroxymyristic acid is found in these strains. Ubiquinone-10 is the sole respiratory quinone.

The genetic aspects of *Acidocella facilis* have been studied to some degree. The *recA* gene of *A. facilis* has been cloned and sequenced (Inagaki et al., 1993). In the deduced amino acid sequence of the recA protein, *A. facilis* showed a similarity level of 72% to *Thiobacillus ferrooxidans*, 62% to *Escherichia coli*, and 53% to *Bacillus subtilis*.

Like *Acidiphilium* strains, *Acidocella* strains inhabit strongly acidic mineral environments. These organisms have so far been isolated from water and sediment taken from acidic mine drainage streams (Wichlacz et al., 1986; Kishimoto et al., 1993b) and acidic soil from a copper mine (Banerjee et al., 1996). In these environments, *Acidocella* strains frequently coexist with members of the genera *Acidiphilium* and *Acidithiobacillus*. *Acidocella* strains also inhabit acidic hot springs and acidic farm soil, and are the most abundant heterotrophic bacteria isolated from sulfuric acidic tropical soil (I. Nioh, personal communication). Moreover, mud and sewage at slightly acidic pH are possible sources of these organisms (Kishimoto and Tano, 1987; Kishimoto et al., 1993b).

ENRICHMENT AND ISOLATION PROCEDURES

Growth media for *Acidocella* are similar to those for *Acidiphilium*. A suitable medium consists of (per liter of distilled water): $(NH_4)_2SO_4$, 2.0 g; KH_2PO_4, 1.0 g; $MgSO_4 \cdot 7H_2O$, 0.5 g; $CaCl_2 \cdot 2H_2O$, 0.05 g; and trace element solution SL8 of Biebl and Pfennig (1978) (pH 3.0), 1.0 ml. Glucose (3.0 g/l) and yeast extract (Difco, 0.5 g/l) are added to the medium from autoclaved stock solutions to give final concentrations of 3.0 g/l and 0.5 g/l, respectively. When used as a solidified medium, the medium also contains 1.5% agar (separately autoclaved). Alternatively, 0.6% gellan gum can be used for this purpose (Kishimoto and Tano, 1987). For selective isolation and enrichment of *Acidocella* from the environment, the medium is supplemented with 0.2 mM acetate to inhibit co-growth of *Acidiphilium* and is adjusted to pH 2.5–3.0 to minimize growth of fungi and facultatively acidophilic bacteria.

MAINTENANCE PROCEDURES

Cultures can be preserved in liquid nitrogen or by lyophilization. Preservation in a mechanical freezer at −80°C is also satisfactory.

DIFFERENTIATION OF THE GENUS *ACIDOCELLA* FROM OTHER GENERA

The genus *Acidocella* is separated from other genera of chemo-organotrophic proteobacteria by its phylogenetic position and its outstanding physiological and chemotaxonomic characteristics. The differential characteristics of the genera *Acidocella* and *Acidiphilium* are indicated in Table BXII.α.20.

TAXONOMIC COMMENTS

Phylogenetic analyses based on 16S rDNA sequence information have shown that *Acidocella* species form a major cluster in the *Alphaproteobacteria*, together with the acidophilic, chemotrophic genera *Acidiphilium*, *Acetobacter*, and *Gluconobacter*, and the anaerobic phototrophic acidophile *Rhodopila* (Sievers et al., 1994a; Kishimoto et al., 1995b). The genus *Acidiphilium* is the nearest phylogenetic neighbor. The levels of 16S rDNA sequence similarity between members of the genera *Acidocella* and *Acidiphilium* are 92–94%.

The two established species of the genus *Acidocella*, *A. facilis* and *A. aminolytica*, were originally classified in the genus *Acidiphilium* as *Acidiphilium facilis* and *Acidiphilium aminolytica*, respectively. However, these two species were found to differ from other *Acidiphilium* species in the lack of the ability to produce photosynthetic pigments (Wakao et al., 1993; Kishimoto et al., 1995a)

TABLE BXII.α.20. Differential characteristics of the genera *Acidocella* and *Acidiphilium*[a]

Characteristic	*Acidocella*	*Acidiphilium*
Cell width, μm	0.5–0.8	0.3–0.8
Pink or yellow pigment	−	D
Photosynthetic pigments	−	+
Growth inhibition by:		
Acetate, 0.25 mM	−	+
Lactate, 2 mM	−	+
Succinate, 4 mM	−	+
Bisphosphatidylglycerol	−	+
2-Hydroxy fatty acid	$C_{14:0}$	−
Mol% G + C of DNA	58–65	63–68

[a]For symbols see standard definitions.

and in some other phenotypic properties (Kishimoto et al., 1995b). Also, phylogenetic analyses based on 16S rDNA sequences revealed that although *A. facilis* and *A. aminolytica* form a monophyletic cluster with other *Acidiphilium* species in the *Alphaproteobacteria*, there was a significant genetic distance at the generic level between the former two species and the latter (Kishimoto et al., 1995b). These findings led to the conclusion that the non-BChl producing species (i.e., *A. facilis* and *A. aminolytica*) should be removed from the genus *Acidiphilium*.

However, the name *Acidocella* is an orthographic error. The first constituent of the name (Acidi-, from *acidum*, acid) is derived from Latin and therefore the connecting vowel to the second constituent (-cella cell) must be -i-. A correction of the name *Acidocella* to *Acidicella* is recommended so as to be in accord with Rule 57c, Note 1 of the International Code of Nomenclature of Bacteria (Lapage et al., 1992). Likewise, the original spelling of the specific epithet, *A. aminolytica*, should be changed to *aminilytica*.

The levels of 16S rDNA sequence similarity and DNA–DNA hybridization between *A. facilis* and *A. aminolytica* are 98% and 10–15%, respectively (Kishimoto et al., 1993b, 1995b).

ACKNOWLEDGMENTS

The author is indebted to K. Inagaki for providing the electron micrograph and H.G. Trüper for his valuable advice on bacterial nomenclature.

DIFFERENTIATION OF THE SPECIES OF THE GENUS *ACIDOCELLA*

The differential characteristics of the species of *Acidocella* are indicated in Table BXII.α.21.

List of species of the genus Acidocella

1. **Acidocella facilis** (Wichlacz, Unz, and Langworthy 1986) Kishimoto, Kosako, Wakao, Tano, and Hiraishi 1996, 362[VP] (Effective publication: Kishimoto, Kosako, Wakao, Tano, and Hiraishi 1995b, 90) (*Acidiphilium facilis* Wichlacz, Unz, and Langworthy 1986, 200.)

   fa′ci.lis. L. adj. *facilis* ready, quick, with respect to growth.

   The characteristics are as described for the genus and indicated in Tables BXII.α.21 and BXII.α.22. Habitat: acidic mine drainage and acidic soil.

   The mol% G + C of the DNA is: 64.0–64.4 (HPLC).
   Type strain: Strain PW2, ATCC 35904.
   GenBank accession number (16S rRNA): D30774.

2. **Acidocella aminolytica** (Kishimoto, Kosako and Tano 1993b) Kishimoto, Kosako, Wakao, Tano, and Hiraishi 1996, 362[VP] (Effective publication: Kishimoto, Kosako, Wakao, Tano, and Hiraishi 1995b, 90) (*Acidiphilium aminolytica* Kishimoto, Kosako, and Tano 1993b, 135.)

a.mi.no.ly'ti.ca. M.L. n. *aminum* amine; Gr. adj. *lytica* dissolving; M.L. adj. *aminolytica* amine dissolving.

The characteristics are as described for the genus and indicated in Tables BXII.α.21 and BXII.α.22. Habitat: acidic mine drainage and mud.

The mol% G + C of the DNA is: 58.7–59.2 (HPLC).

Type strain: Strain 101, ATCC 51361, DSM 11237, JCM 8796.

GenBank accession number (16S rRNA): D30771.

TABLE BXII.α.21. Differential characteristics of the species of the genus *Acidocella*[a]

Characteristic	A. facilis	A. aminolytica
Growth with 3.5% NaCl	+	−
Hydrolysis of hippurate	−	+
Oxidase	d	−
Carbon source utilization:		
Lactose, ethanol	+	−
Sorbitol, inositol, alanine, lysine, spermine	−	+
Glycerol	+	d
Creatine	−	d
Mol% G + C of DNA	58.7–59.2	64.0–64.4

[a]For symbols see standard definitions.

TABLE BXII.α.22. Other characteristics of the species of the genus *Acidocella*[a]

Characteristic	A. facilis	A. aminolytica
Cell width, μm	0.6–0.8	0.5–0.8
Motility by flagella	+	+
Pigmentation	−	−
pH range for growth	3.0–6.0	3.0–6.0
Growth at 37°C	+	+
Chemolithotrophic growth with Fe^{2+} or S^0	−	−
Hydrolysis of esculin	−	−
Oxidase	d	−
Catalase	+	+
Carbon source utilization:		
L-Arabinose, D-xylose, D-ribose, D-glucose, D-galactose, D-fructose, arabitol, mannitol, succinate, diaminobutane, DL-4-aminobutyrate, DL-5-aminovalerate, arginine	+	+
L-Rhamnose, maltose, cellobiose, starch, methanol, formate, acetate, lactate, glutamate, glycine	−	−
Pyruvate	d	nd
Citrate, cis-aconitate, α-ketoglutarate, fumarate, malate	+	nd
Gluconate	nd	+
Casamino acids, peptone, yeast extract	+	+
Major fatty acid	C$_{18:1}$	C$_{18:1}$
Major quinone	Q-10	Q-10

[a]For symbols see standard definitions; nd, not determined; Q-10, ubiquinone-10.

Genus V. Acidomonas Urakami, Tamaoka, Suzuki and Komagata 1989a, 54[VP]

MARTIN SIEVERS AND JEAN SWINGS

A.ci.do.mo'nas. Gr. adj. *acid* acid; Gr. n. *monas* unit, monad; M.L. fem. n. *Acidomonas* acidophilic monad.

Cells are Gram negative, nonsporeforming, nonmotile, rod shaped, 0.8–1.0 × 1.5–3.0 μm. Cells occur singly, rarely in pairs. Colonies were white to yellow on PYM medium. Strictly aerobic. Facultative methylotroph. Metabolism is respiratory, never fermentative. Optimal temperature for growth is 30°C. Optimal pH for growth is pH 4.0–4.5. Catalase positive. Nitrate is not reduced to nitrite. Oxidizes ethanol to acetic acid. Overoxidizes acetic acid to CO$_2$ and H$_2$O. *Acidomonas* contains C$_{18:1}$ straight-chain unsaturated fatty acid as well as C$_{16:0\ 3OH}$ and C$_{16:0\ 2OH}$ as major components of cellular and hydroxy fatty acids, as do the other genera of the family *Acetobacteraceae* (Yamada et al., 1981; Urakami and Komagata, 1987a; Urakami et al., 1989a). The ubiquinone system is Q-10, along with ubiquinone Q-9 and Q-11 as minor components.

The mol% G + C of the DNA is: 62.

Type species: **Acidomonas methanolica** (Uhlig, Karbaum and Steudel 1986) Urakami, Tamaoka, Suzuki and Komagata 1989a, 54 (*Acetobacter methanolicus* Uhlig, Karbaum and Steudel 1986, 321.)

MAINTENANCE PROCEDURES

For cultivation and maintenance of *Acidomonas* a peptone–yeast extract–malt extract agar (PYM medium) is used. The medium is composed of 0.5% peptone, 0.3% yeast extract, 0.3% malt extract, 1.0% glucose, and 2.0% agar; adjusted to pH 4.5 with 1 N HCl (Urakami et al., 1989a).

Cultures were incubated at 30°C. In addition, the medium 569 is used for cultivation of *Acidomonas methanolica*. It is composed of: KNO$_3$, 1.0 g; MgSO$_4$·7H$_2$O, 0.2 g; CaCl$_2$·2H$_2$O, 0.02 g; Na$_2$HPO$_4$, 0.23 g; NaH$_2$PO$_4$, 0.07 g; FeSO$_4$·7H$_2$O, 1.0 mg; CuSO$_4$·5H$_2$O, 5.0 mg; H$_3$BO$_3$, 10.0 mg; MnSO$_4$·5H$_2$O, 10.0 mg; ZnSO$_4$·7H$_2$O, 70 mg; MoO$_3$, 10 mg; agar, 12 g; distilled H$_2$O, 1000 ml; methanol, 10 ml. Sterile-filtrated methanol is added after sterilization of the basal medium. The pH of the medium is 6.8, growth is aerobically at 30°C.

DIFFERENTIATION OF THE GENUS ACIDOMONAS FROM OTHER GENERA

Acidomonas can be differentiated from *Acetobacter*, *Gluconacetobacter*, *Asaia*, and *Gluconobacter* by growth on methanol and utilization of methanol.

TAXONOMIC COMMENTS

Acetobacter methanolicus was described by Uhlig et al. (1986) as a new *Acetobacter* species and classified as *Acidomonas methanolica* by Urakami et al. (1989a). It was reclassified in the genus *Acetobacter* by Sievers et al. (1994b) based on the 16S rDNA sequence of the type strain. Due to the current classification of acetic acid bacteria (Table BXII.α.9 in *Acetobacteraceae*), the name *Acidomonas methanolica* is used.

List of species of the genus Acidomonas

1. **Acidomonas methanolica** (Uhlig, Karbaum and Steudel 1986) Urakami, Tamaoka, Suzuki and Komagata 1989a, 54[VP] (*Acetobacter methanolicus* Uhlig, Karbaum and Steudel 1986, 321.)

me.tha.no' li.cus. L. n. *methanolum* methanol; L. adj. *methanolicus* methanolic, using methanol as a sole carbon source.

Ellipsoidal to rod-shaped cells 0.6–0.8 × 1.0–1.8 μm, occurring singly, in pairs, or rarely in short chains. Cells are usually motile by peritrichous flagellation. Oxidase negative (but under methylotrophic conditions oxidase positive). *A. methanolica* is a facultatively methylotrophic bacterium and utilizes methanol as the sole source of carbon and energy. Methanol is oxidized to formate and fixed by the ribulose monophosphate cycle for biosynthetic purposes (Steudel et al., 1980; Babel, 1984). *A. methanolica* has two independent respiratory chains for methanol and ethanol (Matsushita et al., 1992c). Two different forms of methanol dehydrogenase have been purified from *A. methanolica* containing PQQ as cofactor (Matsushita et al., 1993). Glu-

cose is oxidized to gluconate, but 2-, 5-ketogluconic acid, and 2,5-diketogluconic acid are not produced. No production of dihydroxyacetone from glycerol; γ-pyrones and brown pigments are not formed. *A. methanolica* utilizes methanol, ethanol, acetic acid, D-glucose, glycerol, and pectin as sole carbon sources. Growth did not occur on D-fructose, D-sorbitol, D,L-lactate, sucrose, and lactose. Growth occurs between pH 2.0 and 5.5. Weak growth occurs on D-mannitol and D-mannose. Acid is produced oxidatively from glucose, ethanol, and methanol, but not from D-mannitol, D-xylose, L-arabinose, D-mannose, D-fructose, D-galactose, D-sorbitol, sucrose, and lactose. A medium composed of 0.3% peptone, 0.5% yeast extract, 0.1% D-glucose, and 1.5% methanol (pH adjusted to 5.5) is used for cultivation of the strains.

The mol% G + C of the DNA is: 62 (T_m).

Type strain: TK 0705, ATCC 43581, DSM 5432, IMET 10945; MB 58.

GenBank accession number (16S rRNA): D30770, X77468.

Genus VI. **Asaia** Yamada, Katsura, Kawasaki, Widyastuti, Saono, Seki, Uchimura and Komagata 2000, 828[VP]

MARTIN SIEVERS AND JEAN SWINGS

A.sa' i.a. M.L. fem. n. *Asaia* derived from Toshinobu Asai, a Japanese bacteriologist who contributed to the systematics of acetic acid bacteria.

Cells are aerobic and rod shaped. Peritrichously flagellated when motile. Does not produce acid from ethanol (with the exception of one strain oxidizing ethanol to acetic acid weakly). Growth is inhibited by 0.35% (v/v) acetic acid. *Asaia* weakly oxidizes acetate and lactate to carbon dioxide and water. No growth on methanol. Growth occurred on mannitol. Produces 2-keto-D-gluconate and

5-keto-D-gluconate but not 2,5-diketo-D-gluconate from D-glucose. Acids are produced from D-glucose, D-fructose, L-sorbose, D-mannitol, and D-sorbitol. The major quinone type is Q-10.

The mol% G + C of the DNA is: 59–61.

Type species: **Asaia bogorensis** Yamada, Katsura, Kawasaki, Widyastuti, Saono, Seki, Uchimura and Komagata 2000, 828.

List of species of the genus Asaia

1. **Asaia bogorensis** Yamada, Katsura, Kawasaki, Widyastuti, Saono, Seki, Uchimura and Komagata 2000, 828[VP]

bo.go.r' en.sis. M.L. adj. *bogorensis* derived from Bogor, Java, Indonesia, where most of the strains were isolated.

Cells are 0.4–1.0 × 0.8–2.0 μm in size. Colonies are pink-yellowish white, shiny, smooth, and raised with an entire margin on AG agar plates (Yamada et al., 2000). AG medium is composed of 0.1% D-glucose, 1.5% glycerol, 0.5% peptone, 0.5% yeast extract, 0.2% malt extract, 0.7% CaCO₃ and 1.5% agar (Yamada et al., 2000). *A. bogorensis* produces acid from D-glucose, D-mannose, D-galactose, D-fructose, L-sorbose, D-xylose, ribitol, *meso*-erythritol, glycerol, and melibiose. No acid production from lactose. *A. bogorensis* produces acid from dulcitol and assimilates dulcitol for growth. Acid production from D-mannitol and D-sorbitol is strain dependent. 16S rRNA sequence similarity value of *A. bogorensis* IFO 16594 with the type strain of *Acidomonas methanolica* is 96.2%.

The mol% G + C of the DNA is: 59–61 (T_m).

Type strain: IFO 16594, JCM 10569, NRIC 0311.

GenBank accession number (16S rRNA): AB025928.

2. **Asaia siamensis** Katsura, Kawasaki, Potacharoen, Saono, Seki, Yamada, Uchimura and Komagata 2001, 562[VP]

si.am' en.sis. M.L. adj. *siamensis* of Siam, old name of Thailand.

Asaia siamensis was isolated from tropical flowers collected in Thailand and Indonesia. Isolates were cultivated for enrichment on sorbitol and dulcitol medium at pH 3.5 (Yamada et al., 2000). Cells are 0.6–1.0 × 1.0–4.5 μm. Colonies are pink, shiny, smooth, and raised with an entire margin on AG agar plates (Katsura et al., 2001). *A. siamensis* produces acid from different sugars (D-glucose, D-mannose, D-fructose, L-sorbose, D-xylose, D-mannitol, D-sorbitol, glycerol, sucrose) but not from lactose and dulcitol. Strains assimilate dulcitol for growth. Acid production from dulcitol is used for differentiation of *A. bogorensis* from *A. siamensis*. 16S rRNA sequence similarity value of *A. siamensis* IFO 16457 with the type strain of *A. bogorensis* is 99.8% by DNA relatedness of 20% and 33%, respectively.

The mol% G + C of the DNA is: 59–60 (T_m).

Type strain: IFO 16457, JCM 10715, NRIC 0323.

GenBank accession number (16S rRNA): AB035416.

Genus VII. *Craurococcus* Saitoh, Suzuki and Nishimura 1998, 1044[VP]

CHRISTOPHER RATHGEBER AND VLADIMIR V. YURKOV

Cra.u.ro.coc′cus. Gr. adj. *crauros* fragile; Gr. n. *coccus* berry; M.L. n. *Craurococcus* fragile coccus.

Cells are coccoid, 0.8–2.0 μm in diameter, nonmotile, Gram negative, and divide by binary fission. Form pink irregular colonies when grown on agar media. **Contain bacteriochlorophyll (Bchl) *a*, spirilloxanthin and carotenoic acids. No growth occurs anaerobically in the light.** Grow heterotrophically under aerobic conditions. Nitrate is reduced to nitrite, dissimilatory denitrification not observed.

The mol% G + C of the DNA is: 70.5.

Type species: **Craurococcus roseus** Saitoh, Suzuki and Nishimura 1998, 1045.

FURTHER DESCRIPTIVE INFORMATION

C. roseus is the only species presently described within the genus *Craurococcus*. Phylogenetically *Craurococcus* forms a cluster with *Paracraurococcus*, *Roseococcus*, *Rhodopila*, and *Acidiphilium* within the *Alphaproteobacteria* (Saitoh et al., 1998).

C. roseus forms irregular light yellowish-red to pink colonies, less than 1 mm in diameter, on agar medium. Cells of *C. roseus* grown in NPGC medium (Saitoh and Nishimura, 1996) adjusted to pH 7.6 were coccoid. The cell size ranges from 0.8 to 2.0 μm in diameter. Cells are nonmotile, occur singly or in pairs, rarely in tetrads. Spore production has not been observed. Cells accumulate poly-β-hydroxybutyrate (PHB) granules when grown under usual culture conditions. Suspension in aqueous solution free of divalent cations causes cells to lyse (Fig. BXII.α.26). Divalent cations such as Ca^{2+} and Mg^{2+} seem to be necessary to maintain cell structure and prevent cell lysis. In liquid culture, divalent cations (0.1–1.0 g/l of $CaCl_2 \cdot 2H_2O$) are required for growth, and optimal growth occurs in BS-XY (Saitoh and Nishimura, 1996) media supplemented with 0.5 g/l $CaCl_2 \cdot 2H_2O$. Supplemental cations are not required for growth on agar media.

In vivo absorption spectra of *C. roseus* show absorption peaks at 800 and 872 nm, indicating the presence of Bchl *a* organized into reaction center and light-harvesting complexes, respectively. Crude pigment extracts in methanol give absorption spectra peaks at 362, 492, and 607 nm due to the presence of carotenoid pigments, as well as a 770 nm peak due to Bchl *a*. Bchl *a* content in cells is estimated to be 0.13 nmol/mg dry cell weight when

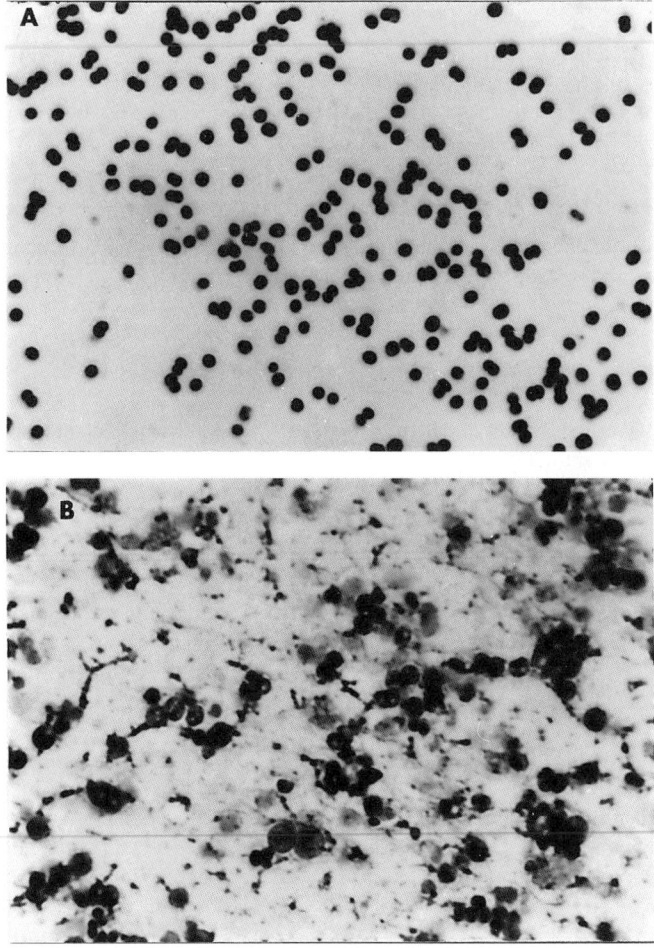

FIGURE BXII.α.26. Light microscopic photographs of *Craurococcus roseus* demonstrating the effect of divalent cations on cell structure when cells are suspended in aqueous solution. (*A*) Cells suspended in sterilized 25 mM $CaCl_2 \cdot 2 H_2O$. (*B*) Cells suspended in sterilized cation free aqueous solution. (Courtesy of Y. Nishimura.)

grown in NPG medium, and 0.04 nmol/mg dry cell weight when grown in NPGC medium (Saitoh and Nishimura, 1996). Absorption difference spectra (illuminated minus unilluminated) show photochemical activity similar to that found in anaerobic and other aerobic anoxygenic phototrophic bacteria. Carotenoids purified by chromatographic procedures include large amounts of carotenoic acids similar to *Pseudomonas radiora*, and smaller amounts of spirilloxanthin.

The cellular fatty acid composition of *C. roseus* consists of large amounts of octadecanoic acid and small amounts of octadecadienoic acid, 2-hydroxyoctadecanoic acid, 3-hydroxytetradecanoic acid, and 3-hydroxyoctadecanoic acid. Ubiquinone is the only quinone type detected. The major ubiquinone detected is Q-10 (89–96%) and ubiquinones Q-7, Q-8, and Q-9 are present in smaller quantities. Cellular lipids consist of phospholipids of phosphotidylcholine, phosphatidylethanolamine, bisphosphatidyl-glycerol, and phosphatidylglycerol.

C. roseus grows heterotrophically under aerobic conditions. Cultures hydrolyze Tweens 20, 40, 60, and 80, and can utilize several substrates as sole carbon and energy sources for heterotrophic growth. Best growth occurs on D-xylose, fumarate, and L-glutamate. Other carbon sources include glycerol, D-fructose, D-fucose, D-galactose, D-glucose, D-arabinose, L-rhamnose, D-ribose, DL-lactate, malonate, pyruvate, gluconate, and L-malate. *C. roseus* does not show gelatinase or amylase activity.

C. roseus is an obligate aerobe, no growth occurs anaerobically under either illuminated or dark conditions. Light at intensity of about 2000 lux has not been shown to stimulate aerobic growth.

Cells do not require salt for growth, and do not grow in media supplemented with 0.4% NaCl. Cells are susceptible to chloramphenicol, penicillin, streptomycin, and tetracycline, and are resistant to polymixin B. Nicotinic acid and pantothenate are required as growth factors.

C. roseus grows at a mesophilic temperature range between 18°C and 37°C, and optimum growth occurs between 28°C and 32°C. The pH range is narrow, with growth occurring only between pH 7.2 and 8.0. Optimal pH for growth is 7.6.

ENRICHMENT AND ISOLATION PROCEDURES

C. roseus was isolated on a 100-fold dilution of nutrient agar media supplemented with 1.5% agar, from soils found in a parking lot in Nerima Ward, Tokyo.

MAINTENANCE PROCEDURES

Cells are maintained on NPG agar medium at pH 8.0. Long-term storage procedures have not been described.

DIFFERENTIATION OF THE GENUS *CRAUROCOCCUS* FROM OTHER GENERA

The genus *Craurococcus* can be differentiated from other known genera of obligately aerobic phototrophic bacteria based on several characteristics. In a phylogenetic tree derived from 16S rRNA gene sequences, *Craurococcus* forms, with the genus *Paracraurococcus*, a separate line of descent within the *Alphaproteobacteria*. This organism can be differentiated from its closest relatives based on color (yellowish-red to pink colonies) as well as specific absorption spectra characteristics. *C. roseus* is capable of growth over very narrow pH, temperature, and salinity ranges (Table BXII.α.23).

Craurococcus has an unusual requirement for divalent cations (Mg^{2+} or Ca^{2+}) when suspended in aqueous solution. Cell lysis occurs if these divalent cations are not present.

TAXONOMIC COMMENTS

The 16S rDNA sequence of *C. roseus* showed that the genus *Craurococcus* is most closely related to the genus *Paracraurococcus* and that this line falls within a cluster formed by the genera *Roseococcus*, *Rhodopila*, and *Acidiphilium*. The position of *Craurococcus* is a branch between the genus *Paracraurococcus* and the other members of the cluster, within the *Alphaproteobacteria*.

FURTHER READING

Shimada, K. 1995. Aerobic anoxygenic phototrophs. *In* Blankenship, Madigan and Bauer (Editors), Anoxygenic Photosynthetic Bacteria, Kluwer Academic Publishers, Dordrecht. pp. 105–122.

Yurkov, V.V. and J.T. Beatty. 1998. Aerobic anoxygenic phototrophic bacteria. Microbiol. Mol. Biol. Rev. *62*: 695–724.

List of species of the genus Craurococcus

1. **Craurococcus roseus** Saitoh, Suzuki and Nishimura 1998, 1045[VP]

   ro.se'us. L. adj. *roseus* rose-colored, pink, pink bacterium.

   Gram negative, forms pink colonies when grown on agar medium. Cells are coccoid, 0.8–2.0 μm in diameter, non-motile. Cells contain Bchl *a* organized into reaction center and light-harvesting complexes, giving *in vivo* absorption spectrum peaks at 800 and 872 nm, respectively. Obligately aerobic chemoorganoheterotroph. The best growth substrates are D-xylose, fumarate and L-glutamate. Other

TABLE BXII.α.23. Distinguishing characteristics of *Craurococcus roseus*, and closely related *Paracraurococcus ruber*[a]

Characteristic	C. roseus	P. ruber
Cell shape and size	Coccoid, 0.8–2.0 μm	Coccoid, 0.8–1.5 μm
Color	Pink	Red
Bchl *a* peaks (nm)	800, 872	802, 856
Growth at 42°C	+	−
Hydrolysis of:		
Tween 80	−	+
Starch	W	−
Gelatin	−	−
Requirement for divalent cations in aqueous suspension	+	+
Requirement for divalent cations in liquid culture	−	+
Growth at:		
pH 6.8	+	−
pH 8.0	+	+

[a]Symbols: +, positive; −, negative, W, weak.

growth substrates are glycerol, D-fructose, D-fucose, D-galactose, D-glucose, D-arabinose, L-rhamnose, D-ribose, DL-lactate, malonate, pyruvate, gluconate, and L-malate. Hydrolyses Tween 20, 40, 60, and 80, but does not hydrolyze starch or gelatin. No growth on D-mannitol, D-sorbitol, methanol, ethanol, 1,2-propanediol, kerosene, acetate, butyrate, citrate, formate, glycolate, propionate, succinate, phthalate, p-hydroxybenzoate, and benzoate.

Optimal growth occurs at pH 7.5 and temperature 28–32°C, NaCl is not required. Incapable of growth at NaCl concentrations of 0.4% and higher. Divalent cations are required for growth. Exhibits oxidase and catalase activity. Reduces nitrate to nitrite. Accumulates poly-β-hydroxybutyrate. Resistant to polymixin B. Susceptible to chloramphenicol, penicillin, streptomycin, and tetracycline. Habitat: soil from a parking lot located in Nerima Ward, Tokyo, Japan.

The mol% G + C of the DNA is: 70.5 (HPLC).
Type strain: NS130, CIP 105707, JCM 9933.
GenBank accession number (16S rRNA): D85828.

Genus VIII. **Gluconacetobacter** Yamada, Hoshino, and Ishikawa 1998b, 32[VP] (Effective publication: Yamada, Hoshino, and Ishikawa 1997, 1249)*

MARTIN SIEVERS AND JEAN SWINGS

Glu.con.a.ce.to.bac' ter. M.L. n. *acidum, gluconicum* gluconic acid; M.L. n. *bacter* masc. equivalent of Gr. neut. n. *bakterion* rod; M.L. masc. n. *Gluconacetobacter* gluconate-vinegar rod.

Cells **ellipsoidal to rod shaped**, straight or slightly curved, 0.6–1.2 × 1.0–3.0 μm, occurring singly, in pairs, or in short chains. **Motile or nonmotile; if motile, the flagella are peritrichous**. Endospores are not formed. Gram negative. **Obligately aerobic**; metabolism is strictly respiratory with oxygen as the terminal electron acceptor. Never fermentative. Some strains produce thick pellicle-forming colonies. Catalase positive and oxidase negative. Absence of gelatin liquefaction, indole production, and H_2S formation. **Oxidize ethanol to acetic acid. Overoxidation of acetate to CO_2 and H_2O depends upon acetate concentration in the medium**. The best carbon sources for growth are ethanol, glucose, and acetate. Acids are formed from glucose and ethanol. Species differ in respect to growth behavior on media with high acetate or glucose concentrations. Optimum temperature is 30°C. **The pH optimum for growth is 2.5–6.0. Possesses ubiquinone of the Q-10 type as major quinone**. The predominant fatty acid in *Gluconacetobacter* is the $C_{18:1 \ \omega7}$ straight-chain unsaturated acid. *Gluconacetobacter* **species occur in vinegar, tea fungus, sugarcane, mealy bug, flowers, and fruits, and can cause infections in beverages and spirituous liquors**. *Gluconacetobacter* is not known to have any pathogenic effect toward humans and animals.

The mol% G + C of the DNA is: 55–67.

Type species: **Gluconacetobacter liquefaciens** (Asai 1935) Yamada, Hoshino and Ishikawa 1998b, 327 (Effective publication: Yamada, Hoshino and Ishikawa 1997, 1250) (*Acetobacter liquefaciens* (Asai 1935) Gosselé, Swings, Kersters, Pauwels, and De Ley 1983c, 896; *Acetobacter aceti* subsp. *liquefaciens* De Ley and Frateur 1974, 277; "*Gluconobacter liquifaciens*" Asai 1935, 610.)

FURTHER DESCRIPTIVE INFORMATION

Based on partial base sequences in positions 1220–1375 (*E. coli* numbering system, Brosius et al., 1981) of 16S rRNA genes, the *Gluconacetobacter* species that have Q-10 clustered remotely from the *Acetobacter* and *Gluconobacter* species (Yamada et al., 1997). The levels of overall 16S rRNA sequence similarity between members within the Q-10 equipped *Gluconacetobacter* species are above 96.5%, whereas those between all of the *Acetobacter* and *Glucon-*

acetobacter species are above 94.0%. The Q-10 equipped *Gluconacetobacter* species formed a cluster closely related to the species of *Acetobacter* in the phylogenetic trees based on complete 16S rRNA sequences (Sievers et al., 1994a). Members of the genus *Acetobacter* have many characteristics in common with *Gluconacetobacter*. *G. europaeus*, *G. intermedius*, *G. oboediens*, *G. entanii*, and *A. pomorum* were isolated from vinegar and share the same ecological niches.

Gluconacetobacter can only be differentiated from *Acetobacter* by the determination of the ubiquinone type. The determination of the ubiquinone type is carried out by the method of reversed-phase paper chromatography (Yamada et al., 1969). Reversed-phase HPLC is used to determine ubiquinones quantitatively (Tamaoka et al., 1983; Franke et al., 1999).

Growth on mannitol agar and acid formation from mannitol occurred in both genera *Gluconacetobacter* and *Acetobacter*. Growth on and acid formation from carbon sources are not necessarily correlated features. Fewer than 10% of the *Acetobacter* strains produce acid from D-mannitol. On the other hand, D-mannitol supported growth for most strains of acetic acid bacteria.

Cultivation media *Gluconacetobacter* species, like *G. diazotrophicus*, *G. hansenii*, and *G. liquefaciens*, grow well on mannitol medium (MYP).[1] *Gluconacetobacter sacchari* strains grow well on GYC medium.[2] *G. xylinus* strains do not develop on MYP medium and should be cultivated on GY medium.[3] *G. europaeus* requires acetic acid for growth and was cultivated on modified AE medium (Entani et al., 1985).[4] *G. intermedius* grows well on MRS medium developed for lactobacilli (de Man et al., 1960), GY medium, or AE medium. *G. johannae* and *G. azotocaptans* were cultivated on GJA medium.[5]

Editorial Note: Yamada et al. (1997) proposed the elevation of the subgenus *Gluconoacetobacter* (Yamada and Kondo, 1984) to the generic level, based on partial base sequences of the 16S rRNA and the ubiquinone system. The name *Gluconoacetobacter* has been corrected to *Gluconacetobacter* in accordance with Rule 61 of the International Code of Nomenclature of Bacteria.

1. MYP medium (g/l distilled water): D-mannitol, 25.0; yeast extract, 5.0; peptone, 3.0; and agar, 12.0.

2. GYC medium (g/l distilled water): D-glucose, 100.0; yeast extract, 10.0; calcium carbonate, 20.0; and agar, 20.0.

3. GY medium (g/l distled water): D-glucose, 100.0; yeast extract, 10.0; and agar, 12.0. (A concentrated glucose solution was autoclaved separately to avoid Maillard reactions.)

4. AE medium (per liter distilled water): yeast extract, 3.0 g; peptone, 4.0 g; D-glucose, 7.5 g; agar, 8.0 g; acetic acid, 30 ml (v/v); and ethanol (99.8%), 30 ml. The ethanol and acetic acid are added after sterilization of the basal medium.

5. GJA medium [medium 920 (per liter of distilled water)]: yeast extract, 2.7; D-glucose, 2.7; D-mannitol, 1.8; MES buffer (Sigma), 4.4; K_2HPO_4, 4.81; KH_2PO_4, 0.65; and agar, 15.0 (final pH 6.7).

ENRICHMENT AND ISOLATION PROCEDURES

LGI medium is used for enrichment of *G. diazotrophicus*. For isolation of *G. diazotrophicus* from sugarcane roots and stems, acetic LGI agar plates (pH 4.5) supplemented with yeast extract (0.05 g/l) and cycloheximide (0.15 g/l) were used (Fuentes-Ramírez et al., 1993). Roots and stems were washed in sterile tap water, macerated in a blender, and for inoculation, dilutions in sugar solution (5% cane sugar in H_2O) were prepared. The acetic LGI medium consists of (quantities per liter): K_2HPO_4, 0.2 g; KH_2PO_4, 0.6 g; $MgSO_4 \cdot 7H_2O$, 0.2 g; $CaCl_2 \cdot 2H_2O$, 0.02 g; $Na_2MoO_4 \cdot 2H_2O$, 0.002 g; $FeCl_3 \cdot 6H_2O$, 0.01 g; bromothymol blue (0.5% solution in 0.2 N KOH), 5 ml; crystallized cane sugar, 100 g; agar, 8.0 g; final pH adjusted to 4.5 with acetic acid (Cavalcante and Döbereiner, 1988).

MAINTENANCE PROCEDURES

Strains of *Gluconacetobacter* can be maintained for 2–3 weeks at 4°C on agar media used for cultivation. *Gluconacetobacter* strains, with the exception of *G. europaeus*, can be frozen and kept at −75°C in the presence of 24% (v/v) glycerol. This maintenance procedure is also possible for *G. intermedius* grown in MRS broth. *Gluconacetobacter* survives lyophilization. Freeze-drying is possible under controlled conditions also for *G. europaeus*, e.g., the type strain of *G. europaeus* DSM 6160. Another reliable method is preservation of *Gluconacetobacter* strains under liquid nitrogen. For preservation of *Gluconacetobacter* isolates from vinegar, cells at the end of the logarithmic growth phase cultivated in modified AE broth were immediately centrifuged at 3000 × g at −10°C. They were resuspended and concentrated in ice-cold 20% malt extract and frozen in liquid nitrogen (Sokollek et al., 1998a).

DIFFERENTIATION OF THE SPECIES OF THE GENUS *GLUCONACETOBACTER*

The differentiation of the species of the genera *Gluconacetobacter*, *Acidomonas*, and *Asaia* is given in Table BXII.α.24.

List of species of the genus Gluconacetobacter

1. **Gluconacetobacter liquefaciens** (Asai 1935) Yamada, Hoshino and Ishikawa 1998b, 327[VP] (Effective publication: Yamada, Hoshino and Ishikawa 1997, 1250) (*Acetobacter liquefaciens* (Asai 1935) Gosselé, Swings, Kersters, Pauwels, and De Ley 1983c, 896; *Acetobacter aceti* subsp. *liquefaciens* De Ley and Frateur 1974, 277; "*Gluconobacter liquifaciens*" Asai 1935, 610.)
li.que.fa' ci.ens. L v. *liquefacio* to liquefy; L. part. adj. *liquefaciens* liquefying.

Colonial and cell morphology are as described for the genus. Motile by peritrichous flagella. 2-Keto-, 2,5-diketo-, and sometimes also 5-ketogluconic acid are synthesized from D-glucose. The majority of the strains produce γ-pyrones from D-glucose and D-fructose. Ketogenesis in *G. liquefaciens* occurs with glycerol, D-mannitol, and sorbitol. Acid is produced from ethanol, *n*-propanol, *n*-butanol, D-mannose, and D-glucose. Most strains are able to grow on *n*-propanol, ethanediol, glycerol, *meso*-erythritol, D-mannitol, D-galactose, D-fructose, D-glucose, Ca-D-gluconate, Ca-D,L-glycerate and Na-D,L-lactate. All strains are able to utilize ammonium as a sole source of nitrogen with ethanol as a carbon source, even without growth factors. Few strains do not require growth factors in the presence of D-mannitol. *G. liquefaciens* causes pink disease in pineapple fruits due to the production of 2,5-diketogluconic acid (Gosselé and Swings, 1986).
The mol% G + C of the DNA is: 62–65 (T_m).
Type strain: ATCC 14835, DSM 5603, IAM 1834, IFO 12388.
GenBank accession number (16S rRNA): X75617.

2. **Gluconacetobacter azotocaptans** Fuentes-Ramírez, Bustillos-Cristales, Tapia-Hernández, Jiménez-Salgado, Wang, Martínez-Romero and Caballero-Mellado 2001, 1312[VP]
a.zo.to.cap' tans. N.L. adj. *azotocaptans* nitrogen-catching.

G. azotocaptans was isolated from the rhizospheres of coffee plants. Cells are straight rods with rounded ends, approximately 1.6–2 × 0.5–0.6 µm, occurring singly or in pairs. Cells are motile by peritrichous flagella. Colonies on potato agar with sugarcane are beige or very light-brownish.

Strains fix molecular nitrogen microaerophilically (Fuentes-Ramírez et al., 2001). Strains grow in 30% D-glucose or sucrose. No growth on D-mannitol, glycerol, and D-xylose. Strains grow on 0.5% ethanol but not with methanol. *G. azotocaptans* can be differentiated from *G. johannae* by RFLP of PCR-amplified 16S rDNA with *Rsa*I generating fragments of 504, 404, 246, 159, and 134 bp, respectively (Fuentes-Ramírez et al., 2001).
The mol% G + C of the DNA is: 64 (T_m).
Type strain: ATCC 700988, DSM 13594.
GenBank accession number (16S rRNA): AF192761.

3. **Gluconacetobacter diazotrophicus** (Gillis, Kersters, Hoste, Janssens, Kroppenstedt, Stephan, Teixeira, Döbereiner and De Ley 1989) Yamada, Hoshino and Ishikawa 1998b, 327[VP] (Effective publication: Yamada, Hoshino and Ishikawa 1997, 1250) (*Acetobacter diazotrophicus* Gillis, Kersters, Hoste, Janssens, Kroppenstedt, Stephan, Teixeira, Döbereiner and De Ley 1989, 362.)
di.a.zo.tro' phi.cus. M.L. masc. adj. *diazotrophicus* one that feeds on dinitrogen.

Straight rods with rounded ends. Motile by lateral or peritrichous flagella. Produces brown water-soluble pigments on D-glucose, yeast extract medium. Dark brown colonies are formed on potato agar supplemented with 10% sucrose. Forms 2-ketogluconic acid and 2,5-diketogluconic acid from glucose. The optimal growth pH is 5.5. High concentrations (10%) of sucrose are used for isolation of *G. diazotrophicus* strains and a medium composed of 2.5% mannitol, 0.5% yeast extract, and 0.3% peptone is suitable for cultivation.

Isolated from roots, stems, and leaves of sugarcane (Gillis et al., 1989), and coffee (Jímnez-Salgado et al., 1997) and, in addition, has been recovered from sweet potato, Cameroon grass (Caballero-Mellado et al., 1995), and mealy bugs, which act as a vector for carrying *G. diazotrophicus* to the leaf tissue of sugarcane (Ashbolt and Inkerman, 1990). *G. diazotrophicus* does not survive in the soil without the presence of host plants and is growing within a low pO_2 environment, which is necessary for the expression and

TABLE BXII.α.24. Differentiating features among species of *Gluconacetobacter*, *Acidomonas*, and *Asaia*[a]

Characteristics	G. liquefaciens	G. azotocaptans	G. diazotrophicus	G. entanii	G. europaeus	G. hansenii	G. intermedius	G. johannae	G. oboediens[b]	G. sacchari	G. xylinus	Acidomonas methanolica	Asaia bogorensis	Asaia siamensis
Growth on 3% (v/v) ethanol in the presence 5–8% acetic acid	−	−	−	+	+	−	+	−	+	−	−	−	−	−
Requirement of acetic acid for growth	−	−	−	+	+	−	−	−	−	−	−	−	−	−
Growth only in the presence of acetic acid and ethanol and glucose[c]	−	−	−	+	−	−	−	−	−	−	−	−	−	−
Growth on the medium of Carr and Passmore	+	nd	+	−	−	+	(+)	nd	nd	+	(+)	+	nd	nd
Growth on methanol	−	−	−	−	−	−	−	−	−	−	−	+	−	−
Formation from D-glucose of:														
5-ketogluconic acid	d	nd	−	−	d	d	−	nd	−	+	+	−	+	+
2,5-diketogluconic acid	+	nd	+	−	−	−	−	nd	−	+	−	−	−	−
Growth on carbon source ethanol	+	+	+	+	+	d	+	+	+	+	d	+	− or w	− or w
Growth in the presence of 30% (w/v) D-glucose	−	+	+	−	−	−	+	+	+	+	−	−	nd	nd
Growth in the presence of 30% (w/v) D-glucose with formation of ≥130 g/l gluconic acid	nd	nd	nd	nd	nd	nd	nd	nd	+	nd	nd	nd	nd	nd
Growth on 0.01% malachite-green agar	+	nd	−	nd	−	nd	nd	nd	nd	+	−	nd	nd	nd
N₂ fixation	−	+	+	−	−	−	−	+	−	−	−	−	−	−
Cellulose formation	−	−	−	−	+ or −	−	+	−	−	−	+	−	−	−
Ubiquinone type	Q-10	Q-10	Q-10	Q-10	Q-10	Q-10	Q-10	Q-10	Q-10	Q-10	Q-10	Q-10	Q-10	Q-10
Mol% G + C content	62–65	64	61–63	58	56–58	58–63	62	58	60	62–67	55–63	62	59–61	59–60
Preferred identification method besides ethanol oxidation and overoxidation of acetic acid:														
Phenotypically[d]				+	+		+			+	+	+	+	+
Genotypically by:														
16S rRNA sequence[e]	+		+			+								
ARDRA[f]		+							+					

[a]Symbols: +, 90% or more of the strains positive; w, weakly positive reaction; d, 11–89% of the strains positive; −, 90 % or more of the strains negative; nd, not determined.

[b]DNAs from the type strains of *G. intermedius* and *G. oboediens* showed species-level similarity of 76% among each other.

[c]Sum of acetic acid and ethanol has to exceed 6% (Schüller et al., 2000).

[d]Data given under the description of the species.

[e]Data from Sievers et al. (1994a).

[f]Amplified rDNA restriction analysis (Fuentes-Ramírez et al., 2001).

functioning of the nitrogenase (James and Olivares, 1997). The mechanisms of infection and colonization of sugarcane by *G. diazotrophicus* are described by James and Olivares (1997).

G. diazotrophicus is able to fix nitrogen in the presence of nitrates and at low pH values (Stephan et al., 1991; Burris, 1994) and seems best adapted to the sugarcane environment, since the nitrogenase activity is stimulated in mixed cultures (Cojho et al., 1993). Strains fix molecular nitrogen under microaerobic conditions. Sucrose concentration of 10% has a positive effect on nitrogenase activity protecting nitrogenase against inhibition by oxygen (Reis and Döbereiner, 1998). During N_2-fixation by *G. diazotrophicus*, activity of glucose dehydrogenase is enhanced and cytochrome a_1 (= cytochrome *ba*) is expressed as terminal oxidase (Flores-Encarnación et al., 1999). Most *G. diazotrophicus* strains produce levan as an exopolysaccharide that is synthesized by levansucrase (Hernández et al., 1995). The bacterium grows in the presence of 30% glucose and exhibits high rates of gluconic acid formation (Attwood et al., 1991). *G. diazotrophicus* grows well on sucrose, D-glucose, D-fructose, D-gluconate, and polyols like D-mannitol, D-sorbitol, and glycerol. The lack of growth on C4-dicarboxylates such as succinate, fumarate, and malate is due to the absence of a transport system for dicarboxylates in *G. diazotrophicus* (Alvarez and Martínez-Drets, 1995; Ureta et al., 1995). The lipopolysaccharide (LPS) of *G. diazotrophicus* was characterized by Fontaine et al. (1995) and contains rhamnose, mannose, and galactose as monosaccharide constituents in all investigated strains.

The mol% G + C of the DNA is: 61–63 (T_m).

Type strain: ATCC 49037, PAL 5, DSM 5601, LMG 7603.

GenBank accession number (16S rRNA): X75618.

4. **Gluconacetobacter entanii** Schüller, Hertel and Hammes 2000, 2019[VP]

en.ta' ni.i. L. gen. *entanii* of Entani, the name honors Etsuzo Entani, a Japanese microbiologist.

Cells are ellipsoidal to rod shaped, straight, or slightly curved, 0.8–1.2 × 1.3–1.6 μm, nonmotile, occurring singly, in chains, and mainly in pairs. Colonies are round, regular, umbonate, soft, glossy, and with a diameter of 1–2 mm on AE agar (Schüller et al., 2000).

G. entanii requires acetic acid in addition to ethanol and glucose for growth. Growth occurs only at total concentrations (sum of acetic acid and ethanol) exceeding 6.0%. At this concentration, *G. entanii* is not able to oxidize acetic acid further to CO_2 and H_2O. Oxidizes ethanol to acetic acid. Growth on 3% (v/v) ethanol in the presence of up to 11% acetic acid (v/v). Gluconate, glycerol, methanol, and lactate are not used as carbon source. The ability to utilize sorbitol and mannitol is strain dependent. No formation of ketogluconic acids from glucose. No formation of cellulose. Good growth on glucose, fructose, and sucrose in AE broth (4a/3e). AE broth (4a/3e) is composed of 0.5% glucose, 0.2% yeast extract, 0.3% peptone, 4% (w/v) acetic acid, and 3% ethanol (v/v) (Entani et al., 1985; Sievers et al., 1992; Sokollek and Hammes, 1997). *Gluconacetobacter entanii* is closely related to *G. hansenii* based on 16S rRNA sequence similarity. *G. entanii* can be distinguished from *G. hansenii* by its ability to grow on 3% (v/v) ethanol in the presence of 4–8% (w/v) acetic acid. *G. entanii* requires, in

contrast to *G. europaeus*, ethanol as well as acetate and glucose for growth.

The mol% G + C of the DNA is: 58 (T_m).

Type strain: DSM 13536, LTH 4560.

GenBank accession number (16S rRNA): AJ251110.

5. **Gluconacetobacter europaeus** (Sievers, Sellmer and Teuber 1992) Yamada, Hoshino and Ishikawa 1998b, 327[VP] (Effective publication: Yamada, Hoshino and Ishikawa 1997, 1250) (*Acetobacter europaeus* Sievers, Sellmer and Teuber 1992, 656.)

eu.ro.pae' us. L. adj. masc. *europaeus* occurring in or coming from Europe.

Cells are rod shaped, straight, 0.7 × 2 μm, occurring singly or in pairs. Mobility and flagella not observed. Pale colonies, no pigments produced. Strains have an absolute requirement of acetic acid for growth. Growth occurs on AE medium in a relative humidity of 92–97%. *G. europaeus* growing on AE medium utilizes glucose and ethanol simultaneously. Cells die rapidly without an oxygen supply (Hitschmann and Stockinger, 1985). Catalase positive, oxidase negative. Acetate is oxidized to CO_2 and H_2O below an acetate concentration of 6%. Glucose is oxidized to 5-ketogluconate or 2-ketogluconate. Some strains produce cellulose/acetan.

Acetic acid bacteria produce hopanoids acting as membrane stabilizers (Rohmer et al., 1984; Ourisson et al., 1987). *G. europaeus* contains novel series of methylhopanoids with an additional methyl group at position C31 (Simonin et al., 1994).

Gluconacetobacter europaeus has been isolated from vinegar fermentations and is used in submerged cultures and trickling generators for industrial production of vinegar.

G. europaeus strains can be freeze-dried under controlled conditions and reactivated. Preservation of strains at −70°C in the presence of 25% glycerol and revival of these strains is still difficult.

The overall 23S rDNA sequence similarity values of *G. europaeus* type strain with the type strains of *G. intermedius* and *G. xylinus* are 99.2% and 98.9%, respectively.

The mol% G + C of the DNA is: 55–58 (T_m).

Type strain: ATCC 51845, DSM 6160.

GenBank accession number (16S rRNA): Z21936.

Additional Remarks: The accession number of the 23S rDNA of *G. europaeus* type strain is X89771 (Boesch et al., 1998).

6. **Gluconacetobacter hansenii** (Gosselé, Swings, Kersters, Pauwels and De Ley 1983c) Yamada, Hoshino and Ishikawa 1998b, 327[VP] (*Acetobacter hansenii* Gosselé, Swings, Kersters, Pauwels and De Ley 1983c, 896.)

han.se' ni.i. M.L. n. *hansenii* of E.C. Hansen, a Danish microbiologist, well known for his study on the acetic acid bacteria.

The cell morphology and colonial characteristics are as described for the genus. Ketogenic toward glycerol, D-mannitol, and sorbitol. Most strains produce 2-ketogluconic acid and 5-ketogluconic acid from D-glucose. Acid is produced from ethanol, *n*-propanol, *n*-butanol, D-mannose, and D-glucose. Strains of *G. hansenii* are unable to grow on ethanol, if ethanol is the sole carbon source and yeast extract is absent. Strains require growth factors.

The mol% G + C of the DNA is: 58–63 (T_m).

Type strain: ATCC 35959, DSM 5602, LMG 1527, NCIMB 8746.

GenBank accession number (16S rRNA): X75620.

7. **Gluconacetobacter intermedius** (Boesch, Trček, Sievers and Teuber 1998) Yamada 2000, 2226[VP] (*Acetobacter intermedius* Boesch, Trček, Sievers and Teuber 1998, 1083.)

in.ter.me' di.us. L. adj. masc. *intermedius* in the middle between.

Cells are rod shaped, straight, 0.7 × 2 μm, occurring singly or in pairs. Motility and flagella not observed. Pale colonies, no pigments produced. Catalase positive, oxidase negative. Acetate is oxidized to CO_2 and H_2O below an acetate concentration of 6%. *G. intermedius* strains grow with or without acetic acid. Growth occurs on AE medium in the presence of 3% (v/v) ethanol and up to 8% (v/v) acetic acid. Eleven of 13 strains of *G. intermedius* grow on GY medium and all strains show strong growth on MRS medium. No oxidation of glucose to 5-ketogluconate. Formation of dihydroxyacetone from glycerol. Most strains do not utilize sucrose as carbon source. Strains produce cellulose/acetan. *Gluconacetobacter intermedius* occurs in tea fungus beverages (Sievers et al., 1995b), spirit vinegar, and cider vinegar fermentations. The type strain TF2 (DSM 11804) was isolated from a commercially available tea fungus beverage (Kombucha) in Switzerland. Kombucha is a refreshing beverage obtained from fermentation of sugared tea with a symbiotic culture of *G. intermedius* and yeasts. The effects of Kombucha on human health are discussed by Dufresne and Farnworth (2000). *G. intermedius* produces a thick cellulose/acetan matrix under static conditions, which covered the surface of the beverage and keeps the cells in close contact with the atmosphere. The overall 23S rDNA sequence similarity values of *G. intermedius* type strain with the type strains of *G. europaeus* and *G. xylinus* are 99.2% and 99.0%, respectively.

An oligonucleotide probe based on the 23S rDNA of *G. europaeus* (23seu: 5′-AATGCGCCAAAAGCCGGAT-3′) developed for *G. europaeus* hybridizes with *G. europaeus* strains and with the following *G. xylinus* strains: LMG 25, ATCC 23768, and ATCC 10245. The probe does not hybridize with *G. intermedius* strains and is thus useful for the differentiation of *G. intermedius* from *G. europaeus* (Boesch et al., 1998).

The mol% G + C of the DNA is: 62 (T_m).

Type strain: TF2, DSM 11804.

GenBank accession number (16S rRNA): Y14694.

Additional Remarks: The accession number of the 23S rDNA sequence of the *G. intermedius* type strain is Y14680.

8. **Gluconacetobacter johannae** Fuentes-Ramírez, Bustillos-Cristales, Tapia-Hernández, Jiménez-Salgado, Wang, Martínez-Romero and Caballero-Mellado 2001, 1312[VP]

jo.han' nae. N.L. gen. n. *johannae* of Johanna, in honor of the Brazilian microbiologist, Johanna Döbereiner, who isolated *G. diazotrophicus* as the first nitrogen-fixing species of the genus *Gluconacetobacter*.

G. johannae was isolated from the rhizospheres of coffee plants. Cells are straight rods with rounded ends, approximately 1.5–1.9 × 0.5–0.6 μm, occurring singly, in pairs, or in short chains. Cells are motile by peritrichous flagella. Colonies on potato agar with sugarcane are brown. Strains fix molecular nitrogen microaerophilically (Fuentes-Ramí-

rez et al., 2001). Strains grow in 30% D-glucose or sucrose. Slight to no growth on D-mannitol and glycerol. Strains grow on D-xylose as carbon source. Strains grow on 0.5% ethanol but not with methanol. *G. johannae* ATCC 700987 showed 45% DNA relatedness to the type strain of *G. azotocaptans*. *G. johannae* can be differentiated from *G. azotocaptans* by RFLP of PCR-amplified 16S rDNA with *Rsa*I generating fragments of 504, 405, 403, and 135 bp, respectively (Fuentes-Ramírez et al., 2001).

The mol% G + C of the DNA is: 58 (T_m).

Type strain: ATCC 700987, DSM 13595.

GenBank accession number (16S rRNA): AF111841.

9. **Gluconacetobacter oboediens** (Sokollek, Hertel, Hammes 1998b) Yamada 2000, 2226[VP] (Sokollek, Hertel, Hammes 1998b, 939.)

ob.oe.di' ens. L. adj. *oboediens* obedient.

G. oboediens DSM 11826 is phenotypically and genotypically similar to *G. intermedius* in respect to growth with and without acetic acid, and 16S rRNA sequence similarity is 100%. The type strains of *G. intermedius* and *G. oboediens* showed 76% DNA relatedness among each other (Sievers and Schumann, unpublished). Based on the DNA–DNA hybridization data and phenotypic similarities, it is obvious that *G. oboediens* is synonymous with *G. intermedius*. *G. oboediens* is able to grow on 3% (v/v) ethanol in the presence of up to 8% (w/w) acetic acid. No absolute requirement of acetic acid for growth. Growth in the presence of 30% (w/v) glucose with a strong formation of gluconic acid (≥130 g/l). Good growth on glucose, fructose, ribose, glycerol, and sucrose on yeast extract peptone agar. Cellulose is not formed.

Gluconacetobacter oboediens, type strain DSM 11826, was isolated from a submerged red wine vinegar fermentation at a factory in the southern part of Germany.

The mol% G + C of the DNA is: 60 (T_m).

Type strain: DSM 11826, LTH 2460.

GenBank accession number (16S rRNA): AJ001631.

10. **Gluconacetobacter sacchari** Franke, Fegan, Hayward, Leonard, Stackebrandt and Sly 1999, 1691[VP]

sacch.ar' i. M.L. gen. n. *sacchari* pertaining to the genus *Saccharum*, sugarcane, from which the organism was isolated.

Cells of *G. sacchari* are ellipsoidal to rod shaped, approximately 0.7–0.9 × 1.3–2.2 μm in size and occur singly, in pairs, or in short chains. Cells are motile by peritrichous flagella. An electron micrograph of the type strain DSM 12717 showing cellular morphology and flagellation is given in Fig. BXII.α.27 (Franke et al., 1999). *Gluconacetobacter sacchari* was isolated from the leaf sheath of sugarcane and from the pink sugarcane mealy bug *Saccharicoccus sacchari*. The occurrence of *G. sacchari* in mealy bug homogenates was detected by *in situ* hybridization with a Cy3-labeled specific primer targeting the 16S rRNA of *G. sacchari* (Franke et al., 2000).

G. sacchari strains grow in the presence of 0.01% malachite-green agar as does *G. liquefaciens*. To test the ability to grow in the presence of the dye malachite green, the agar plates containing (per liter of distilled water): 0.5 g yeast extract, 3.0 g vitamin-free Casamino acids, 1.0 g D-glucose, 1.0 g mannitol, 1.0 g calcium DL-lactate, 25 g agar, and 0.1 g malachite green (Gosselé et al., 1983a). *G. sacchari* strains grow in the presence of 30% D-glucose. Strains grow

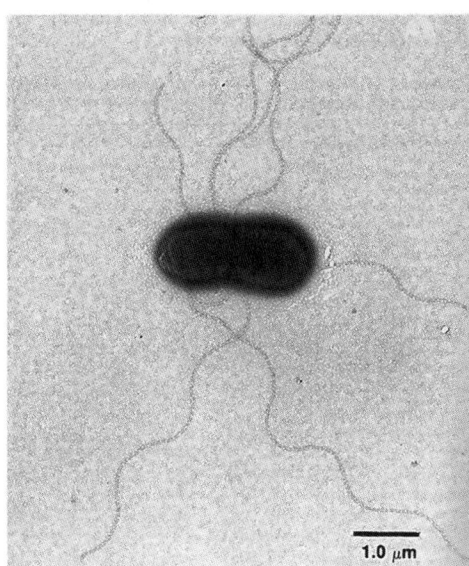

FIGURE BXII.α.27. Electron micrograph of *Gluconacetobacter sacchari* DSM 12717, showing rod-shaped to ellipsoidal cell form and peritrichous flagella. (Reprinted with permission from I.H. Franke et al., International Journal of Systematic and Evolutionary Microbiology *49:* 1681–1693, 1999, ©Society of General Microbiology, UK.)

on mannitol agar. Acid is produced from D-glucose, ethanol, and 1-propanol, but not from D-mannitol or sorbitol.

Strains of *G. sacchari* do not grow on the amino acids L-asparagine, L-glycine, L-glutamine, L-threonine, or L-tryptophan with D-mannitol as the carbon source[6] (Franke et al., 1999). *G. sacchari*, in contrast to *G. liquefaciens*, is not able to utilize these amino acids as the sole nitrogen source with D-mannitol as a carbon source.

Gluconacetobacter sacchari strains are closely related to *G. liquefaciens* by sharing 98.8–99.3% 16S rDNA sequence similarity. The DNA–DNA relatedness of the type strain of *G. sacchari* is 46% with *G. liquefaciens* (Franke et al., 1999). *G. sacchari* does not fix nitrogen in contrast to *G. diazotrophicus*, *G. azotocaptans*, and *G. johannae*.

---

6. The medium for testing the utilization of L-amino acids as sole nitrogen sources contains (per liter of 0.2 M Tris-maleate buffer, pH 5.4): D-mannitol, 30.0 g; L-amino acid, 1.0 g; salt solution A (per liter of distilled water: KH_2PO_4, 100 g and K_2HPO_4, 100 g), 5.0 ml; salt solution B (per liter of distilled water: $MgSO_4 \cdot 7H_2O$, 40.0 g; NaCl, 2.0 g; $FeSO_4 \cdot 7H_2O$, 0.2 g; and $MnSO_4 \cdot H_2O$, 1.5 g), 5.0 ml; D-biotin, calcium pantothenate, thiamine, folic acid, *p*-aminobenzoic acid, and vitamin B_{12}, 0.001 g each; pyridoxal-HCl, niacin and riboflavin, 0.0015 g each.

The mol% G + C of the DNA is: 62–67 T_m).
Type strain: DSM 12717, SRI 1794.
GenBank accession number (16S rRNA): AF127407.

11. **Gluconacetobacter xylinus** (Brown 1886) Yamada, Hoshino and Ishikawa 1998b, 327[VP] (Effective publication: Yamada, Hoshino and Ishikawa 1997, 1250) (*"Bacterium xylinum"* Brown 1886; *Acetobacter xylinus* (Brown 1886) Yamada 1984, 270; *Acetobacter aceti* subsp. *xylinus* (Brown 1886) De Ley and Frateur 1974, 277.)*
xy.li′ nus. L. adj. *xylinus* woody, derived from; Gr. *xylon* wood.

Cells occurring singly, in pairs, in chains, or in small clusters. Peritrichously flagellated when motile. Synthesize cellulose and acetan. No growth on 3% ethanol in the presence of 5–8% acetic acid. *G. xylinus* strains grow well at D-glucose concentration of 100 g per liter medium. Acids are formed from ethanol, D-glucose, and D-xylose.

Acid formation from sucrose, D-galactose, and production of 5-ketogluconic acid from D-glucose is strain dependent. The type strain of *G. xylinus* does not produce acid from sucrose. The DNA from the type strain of *G. xylinus* showed 69–100% similarity to those from seven *G. xylinus* strains that form acid from sucrose (Tanaka et al., 2000b). *G. xylinus* BPR 2001 (*Acetobacter xylinus* subsp. *sucrofermentans*) differs from *G. xylinus* ATCC 23767 by growth on sucrose, by acid formation from sucrose and by oxidation of lactose (Toyosaki et al., 1995). 16S rRNA sequence similarity of *G. xylinus* ATCC 23767 with *G. xylinus* BPR 2001 (*Acetobacter xylinus sucrofermentans* JCM 9730) is 99.6%. Based on DNA–DNA relatedness, *G. xylinus* and *A. xylinus* subsp. *sucrofermentans* are genetically to homologous to maintain a subspecies under this species level (Tanaka et al., 2000b).

The overall 23S rDNA sequence similarity values of *G. xylinus* type strain with the type strains of *G. intermedius* and *G. europaeus* are 99.0% and 98.9%, respectively.

The mol% G + C of the DNA is: 59–63 (T_m).
Type strain: ATCC 23767, DSM 6513, IFO 15237, NCIMB 11664.

GenBank accession number (16S rRNA): X75619.

Additional Remarks: The accession number of the 23S rDNA of *G. xylinus* is X89812 (Boesch et al., 1998).

Editorial Note: The name *Acetobacter xylinum* was corrected to *Acetobacter xylinus* by Euzéby (1997) because the epithet *xylinus* as adjective to *Acetobacter* has to be masculine.

Genus IX. **Gluconobacter** Asai 1935, 689[AL]

MARTIN SIEVERS AND JEAN SWINGS

Glu.co.no.bac′ ter. M.L. n. *acidum, gluconicum* gluconic acid; M.L. n. *bacter* masc. equivalent of Gr. neut. n. *bakterion* rod; M.L. masc. n. *Gluconacetobacter* gluconate rod.

Ellipsoidal to rod-shaped cells, 0.5–1.0 × 2.6–4.2 µm, occurring singly and/or in pairs, rarely in chains. The formation of enlarged, irregular cells may occur. Endospores are not formed. **Motile or nonmotile; if motile, the cells have 3–8 polar flagella**. Gram negative. Obligately aerobic, having a strictly respiratory type of metabolism with oxygen as the terminal electron acceptor.

Metabolism is never fermentative. Optimal temperature for growth is 25–30°C. Optimal pH for growth is 5.0–6.0; most strains will grow at pH 3.5. Catalase positive, oxidase negative. Negative for nitrate reduction, gelatin liquefaction, indole production, and H_2S formation. **Oxidize ethanol to acetic acid. Do not oxidize acetate or lactate to CO_2 and H_2O.** *Gluconobacter* strains generally produce acid during growth on fructose, glucose, xylose, and maltose and tolerate up to 10% glucose. **Ketogenesis occurs from polyalcohols** (for example, dihydroxyacetone from glycerol). **All strains produce 2-ketogluconic acid from D-glucose**, and the majority of strains also form 5-ketogluconic acid. Formation of water-soluble brown pigment is correlated with the production of 2,5-diketogluconic acid and γ-pyrones from D-glucose. Most strains form acid from ethanol, D-mannitol, D-fructose, D-glucose, D-maltose, glycerol, and D-xylose. Few strains produce acid from lactose. The best carbon sources for growth are D-mannitol and D-glucose. **Possesses ubiquinone of the Q-10 type as major quinone**. The predominant fatty acid in *Gluconobacter* is the $C_{18:1\ \omega7}$ straight-chain unsaturated acid. **Isolated from sugar-rich environments such as fruits and flowers, honey bees, grapes and wine, palm sap, cocoa wine, cider, beer, and soft drinks**.

The mol% G + C of the DNA is: 52–64.

Type species: **Gluconobacter oxydans** (Henneberg 1897) De Ley 1961, 304 emend. Mason and Claus 1989, 181.

FURTHER DESCRIPTIVE INFORMATION

Metabolism *Gluconobacter* catabolizes D-glucose by the hexose monophosphate pathway. *Gluconobacter* is not able to overoxidize acetic acid to CO_2 and H_2O due to an incomplete tricarboxylic acid cycle, lacking succinate dehydrogenase (Greenfield and Claus, 1972). *Gluconobacter* strains prefer sugar-enriched environments in contrast to *Acetobacter* and *Gluconacetobacter* strains, which prefer alcohol-enriched environments (Asai, 1968).

Vitamin C production *Gluconobacter* species are used for oxidation of D-sorbitol into L-sorbose in the commercial production of vitamin C (ascorbic acid). Vitamin C is industrially produced by the Reichstein method: D-Glucose → D-Sorbitol → L-Sorbose → Diacetonesorbose → Diacetone-2-keto-L-gulonic acid → L-Ascorbic acid. The process begins with the nonbiological hydrogenation of D-glucose to produce D-sorbitol. The synthesis of L-sorbose from D-sorbitol is carried out biochemically by *Gluconobacter*. The L-sorbose is then converted nonbiologically to L-ascorbic acid. The PQQ-dependent sorbitol dehydrogenase from *G. oxydans* was characterized by Choi et al. (1995a).

Improved methods for vitamin C production that involve the biological formation of 2-keto-L-gulonic acid from D-sorbitol as an intermediate have been developed. The metabolic pathway for biosynthesis of 2-keto-L-gulonic acid from D-sorbitol by *Gluconobacter oxydans* includes L-sorbosone as an intermediate: D-Sorbitol → L-Sorbose → L-Sorbosone → 2-Keto-L-gulonic acid. The 2-keto-L-gulonic acid is then converted nonbiologically to vitamin C (Hoshino et al., 1990; Saito et al., 1997). Mutagenesis of "*Gluconobacter oxydans*" subsp. *melanogenes* IFO 3293 strain resulted in production of 60 g of 2-keto-L-gulonic acid per liter from 100 g D-sorbitol per liter (Sugisawa et al., 1990). Suppression of the formation of L-idonate from 2-keto-L-gulonic acid is a prerequisite for an improved 2-keto-L-gulonic acid production in *Gluconobacter*. Overexpression of the L-sorbose dehydrogenase and L-sorbosone dehydrogenase genes in *G. oxydans* resulted in an improved 2-keto-L-gulonic acid production of the recombinant strain (up to 230%) compared to that of the wild-type strain (Saito et al., 1997). Production of 2-keto-L-gulonic acid from L-sorbose via L-sorbosone is catalyzed in *G. oxydans* IFO 3293 by a membrane-bound L-sorbose dehydrogenase and a L-sorbosone dehydrogenase located in the cytoplasm. Cloning of the gene coding for the membrane-bound L-sorbosone dehydrogenase

from *Gluconacetobacter liquefaciens* IFO 12258 and expression of this gene in *G. oxydans* IFO 3293 revealed a recombinant strain that could carry out the formation of 2-keto-L-gulonic acid by membrane-bound dehydrogenases (Shinjoh et al., 1995). *Ketogulonigenium vulgare* DSM 4025 (former classified as *G. oxydans* DSM 4025) oxidizes L-sorbose to 2-keto-L-gulonic acid (Urbance et al., 2001) and produces vitamin C from L-gulono-γ-lactone. The enzyme L-gulono-γ-lactone dehydrogenase catalyzes this oxidation reaction, has a molecular weight of 110 kDa, and is composed of three subunits (Sugisawa et al., 1995).

Glucose dehydrogenase *Gluconobacter* strains are not known to have any pathogenic effect toward humans or animals but are capable of causing a bacterial rot of apples and pears, which is accompanied by various shades of browning. The bacteria enter the apples through wounds in the cuticula and apple tissue. Strains of *G. oxydans* cause pink disease of pineapple, which is characterized by the formation of pink to deep brown discolorations after heating (e.g., during canning) of the diseased fruit. Glucose dehydrogenase (GDH) is responsible for this fruit discoloration (Cha et al., 1997). GDH catalyzes the oxidation of D-glucose to D-gluconate at the outer surface of the cytoplasmic membrane. D-Gluconate is oxidized by the activity of the D-gluconate-dehydrogenase to 2-ketogluconic acid, which is further converted to 2,5-diketogluconate catalyzed by the 2,5-diketogluconate dehydrogenase (Qazi et al., 1991; Matsushita et al., 1994). 2,5-Diketogluconate may react with natural compounds in the juice to become chromogenic upon heating (Cha et al., 1997). 2,5-Diketogluconate has been also detected in *Gluconacetobacter liquefaciens* strains (Gosselé et al., 1980). Gluconic acid production in *G. oxydans* is mainly the result of the membrane-bound pyrroloquinoline quinone (PQQ)-dependent glucose dehydrogenase, because the activity of the quinoprotein is 30-fold higher than the activity of the cytoplasmic NADP-dependent GDH (Pronk et al., 1989). The PQQ-dependent GDH from *G. oxydans* consists of 808 amino acids encoded from a 2424-bp DNA sequence. A one-point mutation at nucleotide 2359 has replaced the histidine at position 787 with asparagine in the protein of a *G. oxydans* strain. This single amino acid substitution changes the substrate specificity of the quinoprotein GDH, resulting in the conversion of maltose to maltobionic acid in addition to the glucose oxidation (Cleton-Jansen et al., 1991).

Intact cells of *G. oxydans* with glucose oxidation activity in the late exponential phase of 13 μmol of glucose/g dry weight per min were used for construction of a biosensor for determination of glucose. The sensitivity of the glucose sensor was 50 nA/mM glucose and the range of the calibration curve was 0–0.8 mM of glucose concentrations (Švitel et al., 1998).

Gluconobacter strains flourish on sugars and are found in sugary niches such as flowers and fruits. *Gluconobacter* is a typical spoiler of soft drinks causing off-flavors in, e.g., orange juice. *Gluconobacter* strains are also found associated with palm trees, the sap of which is fermented to palm wine. *Gluconobacter oxydans* strains have been isolated from grapes and detected as cider- and beer-spoiling organisms (Swings, 1992).

Plasmids and phages Strains of *G. oxydans* typically cannot utilize lactose as carbon source. The lactose transposon Tn951 was conjugally introduced from *E. coli* (RP1::Tn951) to *G. oxydans*. The heterologous expression level of Tn951-encoded β-galactosidase in a *G. oxydans* strain was less than 5% of the fully induced *E. coli* activity (Condon et al., 1991). Transposon Tn5 was introduced in *G. oxydans* ATCC 9937 by conjugation using plasmid pSUP 2021. Tn5 transposition occurred on the plasmid pVJ1 resulting in a mutant of this strain deficient in glucose dehydrogenase activity (Gupta et al., 1997). The shuttle vector pGE1 (mob, Km^r), which can replicate both in *G. oxydans* and *E. coli*, has been used for cloning the L-sorbosone dehydrogenase gene of *Gluconacetobacter liquefaciens* IFO 12258 in *G. oxydans* IFO 3293 to improve the production of 2-keto-L-gulonic acid in the vitamin C manufacturing process (Shinjoh and Hoshino, 1995).

The *G. oxydans* plasmid pGO128 was cloned in pACYC177 and its complete nucleotide sequence was determined (AJ428837). Plasmid pGO128 from *G. oxydans* DSM 3504 is composed of 4340 bp. The analysis of the nucleotide sequence revealed three open reading frames, located on the same strand. ORF1 encodes for a resolvase. The protein encoded by ORF2 is 38% homologous to a virulence-associated protein vapE from the plant pathogen bacterium *Xylella fastidiosa*.

Phage A-1 (Schocher et al., 1979; Jucker and Ettlinger, 1981; Robakis et al., 1985a) and phages GW6210 and JW2040 (Robakis et al., 1985b) from *G. oxydans* have been isolated and characterized. Phage A-1 caused abnormalities in the microbial process of the oxidation of D-sorbitol to L-sorbose by *G. oxydans* (Schocher et al., 1979).

ENRICHMENT AND ISOLATION PROCEDURES

Enrichment of *Gluconobacter* strains present in flowers, fruits, grapes, bees, beer, palm wine, and soft drinks can be achieved using the following medium (g per liter of distilled water): D-glucose, 100.0; yeast extract, 5.0; peptone, 3.0; acetic acid, 1.0; cycloheximide 0.1. After incubation at 30°C for 3–5 d in flasks with shaking, those cultures showing growth are streaked onto plates of a CaCO$_3$-containing medium, where the colonies dissolve the CaCO$_3$ (Carr and Passmore, 1979). The medium for the chalk-ethanol test is composed of (per liter of distilled water): glucose, 0.5 g; yeast extract, 5.0 g; peptone, 3.0 g; calcium carbonate, 15.0 g; agar, 12.0 g; and ethanol (99.8%), 15 ml (sterile-filtered and added after sterilization of the basal medium). The overoxidation of acetic acid by *Acetobacter*, *Gluconacetobacter*, and *Acidomonas* strains results in a reprecipitation of CaCO$_3$. Colonies that dissolve the CaCO$_3$ without reprecipitation of the CaCO$_3$ are isolated, further purified on GYC medium[1], and characterized for identification and distinction from *Frateuria*.

MAINTENANCE PROCEDURES

Stock cultures of *Gluconobacter* should be grown on GYC medium, MYP medium,[2] or DSYP medium[3] at 28–30°C. The cultures can be maintained for short-term storage in a refrigerator (4–5°C) for 3–4 weeks. For long-term preservation, *Gluconobacter* can be frozen and kept at −75°C in the presence of 24% (v/v) glycerol. *Gluconobacter* strains can be stored as lyophilized cultures.

DIFFERENTIATION OF THE GENUS *GLUCONOBACTER* FROM OTHER GENERA

See Table BXII.α.10 in the chapter on the family *Acetobacteraceae* for differences of *Gluconobacter* from the other genera of acetic acid bacteria. The medium of Carr and Passmore (1979) provides rapid differentiation of *Gluconobacter* from *Acetobacter*. The genus *Frateuria* (*Frateuria* groups in the *Gammaproteobacteria*) most closely resembles the polarly-flagellated *Gluconobacter* but, in contrast to *Gluconobacter*, is able to oxidize lactate to CO$_2$/H$_2$O and does not produce 5-ketogluconic acid. Further information for differentiation of *Gluconobacter* from related genera is given under the section of the family *Acetobacteraceae*.

TAXONOMIC COMMENTS

Within the genus *Gluconobacter*, the species *G. oxydans*, *G. frateurii*, *G. cerinus*, and *G. asaii* were genetically characterized based on 16S rRNA gene sequences (Sievers et al., 1995a) and DNA–DNA hybridization data (Micales et al., 1985; Tanaka et al., 1999). *G. asaii*, *G. frateurii*, and *G. cerinus* showed lower mol% G + C contents of DNA than *G. oxydans*. Phylogenetic trees reflecting the distant and close relationships of *Gluconobacter* based on the results of maximum likelihood analyses are shown in Figs. BXII.α.16 and BXII.α.17 in *Acetobacteraceae*. The overall 16S rRNA gene sequence similarities within the members of the genus *Gluconobacter* are 97.8% to 98.8%, corresponding to 14 to 32 base differences. All four species of *Gluconobacter* form a coherent, closely related cluster that is separated from the other genera of the family *Acetobacteraceae* (Sievers et al., 1994a, 1995a). The ubiquinone composition in *G. oxydans* ATCC 19357 is 3% of Q-9 and 97% of Q-10, in *G. cerinus* IFO 3267, and in *G. asaii* IFO 3276: 4% of Q-9 and 96% of Q-10, respectively, and in *G. frateurii* IFO 3264: 7% Q-9 and 93% Q-10 (Tanaka et al., 1999).

Based on DNA–DNA hybridization data, *G. cerinus* IFO 3262, 3263, 3269 and *G. asaii* IFO 3265 were identified as *G. frateurii* (Tanaka et al., 1999). *G. asaii* IFO 3265 is phenotypically similar to *G. frateurii* (Mason and Claus, 1989). Due to the high values of DNA relatedness (84–96%) between the type strains of *Gluconobacter cerinus* and *Gluconobacter asaii*, *G. asaii* is considered a junior subjective synonym of *G. cerinus* (Yamada et al., 1999; Katsura et al., 2002).

DIFFERENTIATION OF THE SPECIES OF THE GENUS *GLUCONOBACTER*

G. oxydans, *G. frateurii*, *G. cerinus*, and *G. asaii* are differentiated phenotypically by the requirement for nicotinic acid, growth on D-ribitol and L-arabitol (Mason and Claus, 1989), and the formation of acid from D-arabitol, D-ribitol, and L-arabitol (Tanaka et al., 1999) (Table BXII.α.25). Three transfers into nicotinate-deficient media are necessary to demonstrate nicotinate dependence (Mason and Claus, 1989). The nicotinate-deficient medium is composed of 1% vitamin-free Casamino acids (Difco) and 1% mannitol adjusted to pH 6.0, plus a vitamin solution to achieve the final concentrations of 0.00015% pyridoxal hydrochloride, 0.00015% riboflavin, 0.0001% biotin, 0.0001% thiamine, 0.0001% pantothenic acid, and 0.0001% *para*-aminobenzoic acid.

Differential characteristics of the species of the genus *Gluconobacter* are given in Table BXII.α.25.

---

1. GYC medium (g per liter of distilled water): D-glucose, 100.0; yeast extract, 5.0; peptone 3.0; CaCO$_3$, 12.0; agar, 12.0.

2. MYP medium (g per liter of distilled water): D-mannitol, 25.0; yeast extract, 5.0; peptone, 3.0; and agar, 12.0.

3. DSYP medium (g per liter of distilled water): D-sorbitol, 50.0; yeast extract, 5.0; peptone 3.0; and agar 12.0.

TABLE BXII.α.25. Differentiation of *Gluconobacter* species[a]

Characteristic	G. oxydans	G. asaii	G. cerinus	G. frateurii
Growth without nicotinate[b]	−	+	+	+
Growth on D-ribitol[b]	−	−	+[c]	+
Growth on L-arabitol[b]	−	−	+[c]	+
Acid formation from:[d]				
D-Arabitol	−	+	+	+
D-Ribitol	−	−	−	+
L-Arabitol	−	−	−	+

[a]Symbols: see standard definitions.

[b]Data from Mason and Claus (1989) and Yamada and Akita (1984a).

[c]Negative strains are present.

[d]Data from Tanaka et al. (1999). The medium (8 ml) for testing of acid formation contained (g per liter of distilled water): pentitol (D-arabitol, D-ribitol, or L-arabitol), 5.0; peptone, 5.0; yeast extract, 3.0; bromocresol purple, 0.02. Acid formation was judged by change of color from purple to yellow (Tanaka et al., 1999). Formation of acid by the use of this medium was determined by the authors with the type strains of the *Gluconobacter* species and some isolates.

List of species of the genus Gluconobacter

1. **Gluconobacter oxydans** (Henneberg 1897) De Ley 1961, 304[AL] emend. Mason and Claus 1989, 181.

ox′y. dans. M.L. part. adj. *oxydans* acid-giving, oxidizing.

All strains produce 2-ketogluconic acid, 5-ketogluconic acid, and 2,5-diketogluconic acid from D-glucose and dihydroxyacetone from glycerol. *G. oxydans* strains grow to an OD_{620} nm of less than 1.0 after 24 h of incubation and three passages (24 h of incubation each) in media lacking nicotinate. All strains grow to an OD_{620} nm of 0.5 or less in medium containing ribitol or arabitol as the primary carbon source (Mason and Claus, 1989). All strains produce acid from ethanol, *n*-propanol, D-glucose, D-arabinose, D-fructose, D-ribose, D-mannose, D-sorbitol, and D-mannitol, and tolerate up to 10% D-glucose. Carbon sources for growth are D-mannitol, glycerol, or D-sorbitol. Batch cultures of *G. oxydans* have shown that glucose is the preferred substrate of this species rather than lactate (Poget et al., 1994). The type strain of *G. oxydans*, ATCC 19357, does not form acid from D-arabitol, ribitol, L-arabitol, and sucrose (Tanaka et al., 1999).

G. oxydans is not further divided into subspecies, based on cluster analysis of protein profiles from *Gluconobacter* strains (Gosselé et al., 1983a).

The genome sizes of four *G. oxydans* strains were estimated to be between 2240 and 3787 kb by pulsed-field gel electrophoresis upon *Xba*I digestion of their genomes (Verma et al., 1997).

The mol% G + C of the DNA is: 56–64 (T_m).

Type strain: ATCC 19357, IFO 14819, DSM 7145.

GenBank accession number (16S rRNA): X73820.

2. **Gluconobacter asaii** Mason and Claus 1989, 183[VP]

a.sa′i.i. M.L. n. *asaii* of Toshinobu Asai, Japanese microbiologist, who described the genus *Gluconobacter*.

All strains produce acid from ethanol, D-glucose, D-xylose, maltose, and tolerate up to 10% D-glucose. *G. asaii* strains grow to an OD_{620} nm of 1.0 or more after 24 h of incubation and three passages (24 h of incubation each) in media lacking nicotinate. All strains fail to grow beyond an OD_{620} of 0.5 after 24 h when either ribitol or arabitol is used as the primary carbon source. *G. asaii* strain IFO 3297a is characterized by its unique ability to cause alkalinization of polyol media during growth on mannitol, ribitol, or arabitol. The polyol medium is composed of (g per liter of distilled water): vitamin-free Casamino acids (Difco), 3.0; yeast extract, 5.0; and polyol (mannitol, ribitol, or arabitol), 5.0 (Mason and Claus, 1989). The type strain for *G. asaii* (IFO 3276) does not form acid from ribitol, L-arabitol, and sucrose. Since *G. asaii* IFO 3276 showed species-level similarity of 88% by DNA–DNA hybridization with *G. cerinus* IFO 3267 (Tanaka et al., 1999); *G. asaii* IFO 3276 does not represent the species *Gluconobacter asaii* as type strain.

The mol% G + C of the DNA is: 52–55 (T_m).

Type strain: ATCC 49206, IFO 3276, DSM 7148.

GenBank accession number (16S rRNA): AB063287.

3. **Gluconobacter cerinus** Yamada and Akita 1984b, 503[VP] (Effective publication: Yamada and Akita 1984a, 124.)

ce.ri′ nus. L. adj. *cerinus* wax-colored.

Cells are polarly flagellated when motile. Colonies are glossy and smooth. Produces 2-ketogluconate and 5-ketogluconate from glucose. 2,5-Diketogluconate is not produced from D-glucose. Produces dihydroxyacetone from glycerol. Growth occurs on ribitol, xylitol, and L-arabitol. Requires pantothenic acid but not nicotinate for growth. All strains produce acid from ethanol, D-glucose, and D-fructose.

By comparison of electrophoretic patterns from enzymes, the differentiation of *G. cerinus* and *G. oxydans* strains as two separate groups was shown (Yamada and Akita, 1984a).

G. cerinus and *G. frateurii* are phenotypically similar in respect to growth on ribitol and L-arabitol (Yamada and Akita, 1984a; Mason and Claus, 1989). The type strain for *G. cerinus* IFO 3267, in contrast to *G. frateurii* IFO 3264, does not form acid from ribitol and L-arabitol. Weak formation of acid from sucrose is observed by *G. cerinus* IFO 3267 (Tanaka et al., 1999). The type strains of *G. cerinus* and *G. frateurii* have a DNA–DNA similarity value of 23% (Micales et al., 1985) and a 16S rRNA gene sequence similarity value of 98.8%, corresponding to 15 nucleotide differences (Sievers et al., 1995a). The two colony types (glossy and smooth) of the type strain *G. cerinus* IFO 3267 have identical 16S rRNA gene sequences.

The mol% G + C of the DNA is: 54–58 (T_m).

Type strain: ATCC 19441, IFO 3267, DSM 9533.

GenBank accession number (16S rRNA): AB063286.

4. **Gluconobacter frateurii** Mason and Claus 1989, 182[VP]

fra.teur' i.i. M.L. n. *frateurii* of Joseph Frateur (1903–1974), Belgian microbiologist, who is well known for his study of acetic acid bacteria.

All strains produce acid from ethanol, glucose, xylose, maltose, sorbitol, mannitol, lactose, ribitol, and *myo*-inositol, and tolerate up to 15% D-glucose. *G. frateurii* strains grow to an OD_{620} nm of 1.0 or more after 24 h of incubation and three passages (24 h of incubation each) in media lacking nicotinate (except strain IFO 3272). All strains grow to an OD_{620} nm of 1.0 or more after 24 h when either ribitol or arabitol is used as the primary carbon source (Mason and Claus, 1989). The type strain *G. frateurii* IFO 3264 forms acid from D-arabitol, ribitol, L-arabitol, and sucrose (Tanaka et al., 1999).

The mol% *G* + *C* of the DNA is: 53–55 (T_m).

Type strain: ATCC 49207, IFO 3264, DSM 7146.

GenBank accession number (16S rRNA): X82290.

Genus X. **Paracraurococcus** *Saitoh, Suzuki and Nishimura 1998, 1045*[VP]

CHRISTOPHER RATHGEBER AND VLADIMIR V. YURKOV

Pa.ra.cra.u.ro.coc' cus. Gr. prep. *para* like, along side of; Gr. adj. *crauros* fragile; Gr. n. *coccus* berry; M.L. masc. n. *Paracraurococcus* coccus like *Craurococcus.*

Gram-negative, nonmotile cocci, 0.8–1.5 μm in diameter. Cell division occurs by binary fission. Cells grow heterotrophically under aerobic conditions giving rise to red colonies on solid media. **Obligately aerobic, containing bacteriochlorophyll (bchl) *a* organized into reaction center and light-harvesting complexes, as well as carotenoid pigments. No growth occurs anaerobically under illuminated conditions.** Facultatively photoheterotrophic. The habitat is soil.

Type species: **Paracraurococcus ruber** Saitoh, Suzuki and Nishimura 1998, 1046.

FURTHER DESCRIPTIVE INFORMATION

P. ruber is currently the only known representative of the genus *Paracraurococcus*. Phylogenetically this species belongs to the *Alphaproteobacteria*, most closely related to the genera *Craurococcus*, *Roseococcus*, *Rhodopila*, and *Acidiphilium*.

P. ruber forms small (<1mm diameter), irregular, red-to-dark-red colonies on agar media (Fig. BXII.α.28A). Circular, convex, and smooth colonies may appear; these colonies have physiological properties equivalent to strains forming irregular colonies.

Cells of *P. ruber* grown in NPG (Saitoh and Nishimura, 1996) medium adjusted to pH 6.8 appear as cocci. The cells range in size from 0.8 to 1.5 μm in diameter, appear singly or in pairs, and often form aggregates. Cells are nonmotile, spore production has not been observed. *P. ruber* accumulates poly-β-hydroxybutyrate granules when grown under regular culture conditions. Cell division is by way of binary fission (Fig. BXII.α.28B).

Divalent cations appear to be required to maintain proper cell structure in liquid media. Cells lyse when suspended in aqueous solution, however, lysis can be prevented by the addition of divalent cations such as Ca^{2+} or Mg^{2+}. Additions of NaCl, KCl, and sucrose are not sufficient to prevent lysis. The addition of divalent cations to culture media is not required for growth, but has been shown to promote growth in all strains. When cells are grown on solid media, addition of divalent cations is not required and does not affect growth rate.

Intact cells of *P. ruber* show absorption spectrum peaks at 802 nm and 856 nm, corresponding to Bchl *a* organized into reaction center and light-harvesting complexes, respectively. Crude pigment extracts in methanol show absorption spectrum peaks confirming the presence of carotenoid pigments (362 and 492nm) and Bchl *a* (770 nm). Bchl *a* levels have been estimated at 0.3 nmol/g dry cell weight, with slightly higher levels found in strains forming circular, smooth colonies. Cells grown under illuminated conditions have been shown to contain lower levels of Bchl *a*, as compared to cells grown in the dark, indicating that illumination suppresses the synthesis of Bchl *a* in these strains (Saitoh and Nishimura, 1996). Strong inhibition of Bchl *a* synthesis by light has been detected in many aerobic phototrophic bacteria (Yurkov and Beatty, 1998a). Absorption difference spectra (illuminated minus unilluminated) of membrane fractions indicate a photochemically active photosynthetic apparatus, similar to those

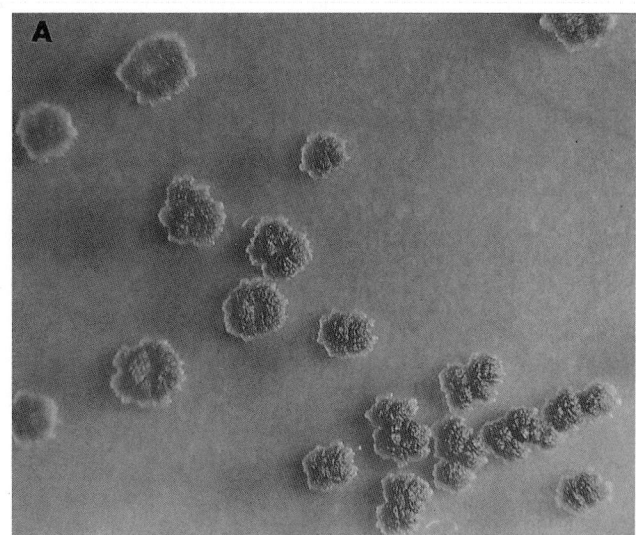

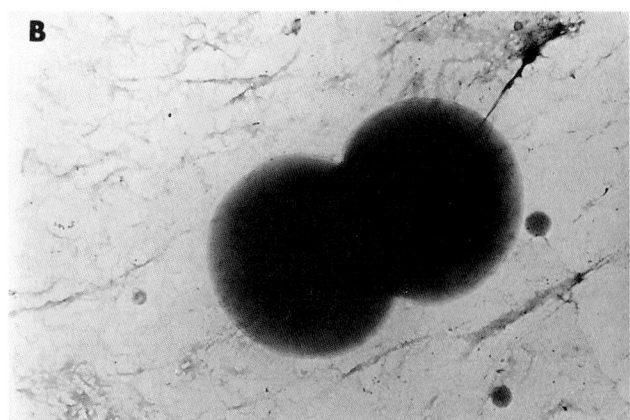

FIGURE BXII.α.28. *Paracraurococcus ruber.* (*A*) Photograph showing red, irregular colonies grown on agar plates. (*B*) Electron microscopic photograph showing division by binary fission. (Courtesy of Y. Nishimura.)

found in other aerobic, as well as typical anaerobic phototrophic bacteria (Saitoh and Nishimura, 1996). *P. ruber* cells contain large amounts of carotenoic acids and smaller amounts of spirilloxanthin.

The cellular fatty acid composition of *P. ruber* consists of octadecanoic acid as well as smaller amounts of octadecadienoic acid, 2-hydroxyoctadecenoic acid, 3-hydroxytetradecanoic acid, and 3-hydroxyoctadecanoic acid. Ubiquinone is the only quinone detected. The major quinone is Q-10 and the quinones Q-7, Q-8, and Q-9 occur in smaller quantities. The cellular lipid profile consists of phospholipids of phosphatidylethanolamine, bisphosphatidylglycerol, and phosphatidylglycerol (Saitoh and Nishimura, 1996).

P. ruber grows heterotrophically under aerobic conditions, and does not grow anaerobically under either illuminated or unilluminated conditions. However light intensities of about 2000 lux stimulate growth under aerobic conditions. Organic substrates utilized are D-xylose, fumarate, L-glutamate, glycerol, D-fructose, D-galactose, D-glucose, D-arabinose, L-rhamnose, D-ribose, D-fucose, DL-lactate, malonate, pyruvate, gluconate, and L-malate. Acids are produced oxidatively from sugars. Hydrolyzes Tween 20, 40, and 60. Tween 80 and gelatin are not hydrolyzed.

NaCl is not required for growth and cells do not grow above 0.4% NaCl. Susceptible to chloramphenicol, streptomycin, and tetracycline, and resistant to penicillin and polymixin B.

Mesophilic, with growth occurring between 16°C and 42°C, with an optimum growth temperature between 30°C and 34°C. The pH range is between 5.5 and 8.1, with optimum growth occurring at pH 6.6–6.8. Nicotinic acid and pantothenate are required as growth factors.

ENRICHMENT AND ISOLATION PROCEDURES

Two strains of *P. ruber* were isolated from soil samples taken from a playground in Yamasaki Noila, of Osakajo Park in Chuoh Ward, Osaka, on a 100-fold dilution of nutrient agar medium supplemented with 1.5% agar.

MAINTENANCE PROCEDURES

Strains can be maintained on a NPG medium (Saitoh and Nishimura, 1996) adjusted to pH 6.8. Long-term storage procedures have not been described.

DIFFERENTIATION OF THE GENUS *PARACRAUROCOCCUS* FROM OTHER GENERA

Paracraurococcus can be differentiated from other genera based on several characteristics. The presence of Bchl *a* organized into reaction center and light-harvesting complexes can be shown by absorption spectrum peaks at 802 and 856 nm, respectively. Light-harvesting complex I, absorbing at 856 nm, is very rarely found among photosynthetic bacteria. The only known species that contains similar light-harvesting complexes is *Roseococcus thiosulfatophilus* (Yurkov and Beatty, 1998a). The presence of Bchl *a* and the inability to grow under illuminated, anaerobic conditions distinguishes *Paracraurococcus* from members of the purple nonsulfur bacteria and places it among the obligately aerobic phototrophic bacteria.

Paracraurococcus can be differentiated from other genera of aerobic phototrophs based on color, red-to-dark-red colonies, as well as its unusual requirement for divalent cations to maintain cell structure in suspension.

On a phylogenetic tree, *Paracraurococcus* forms a separate line of descent with the closely related genus *Craurococcus*. However, *Paracraurococcus* can be distinguished from *Craurococcus* based on differences in absorption spectrum characteristics. *Paracraurococcus* shows absorption spectrum peaks at 802 and 856 nm, whereas *Craurococcus* shows absorption spectrum peaks at 800 and 872 nm, indicating a difference in the structural organization of the light-harvesting complex I between the two genera (see Table BXII.α.23 in Genus *Craurococcus*).

TAXONOMIC COMMENTS

16S rRNA gene sequence analysis shows that *Paracraurococcus* forms a separate line of descent, along with *Craurococcus*, within the *Alphaproteobacteria*. This line falls within a cluster formed by the genera *Roseococcus*, *Rhodopila*, and *Acidiphilium* (Saitoh et al., 1998).

FURTHER READING

Shimada, K. 1995. Aerobic anoxygenic phototrophs. *In* Blankenship, Madigan and Bauer (Editors), Anoxygenic Photosynthetic Bacteria, Kluwer Academic Publishers, Dordrecht. pp. 105–122.

Yurkov, V.V. and J.T. Beatty. 1998. Aerobic anoxygenic phototrophic bacteria. Microbiol. Mol. Biol. Rev. *62*: 695–724.

List of species of the genus Paracraurococcus

1. **Paracraurococcus ruber** Saitoh, Suzuki and Nishimura 1998, 1046[VP]

   ru.ber. L. adj. *ruber* red-colored, red bacterium.

   Gram-negative cocci, 0.8–1.5 μm in diameter, nonmotile, divide by means of binary fission. Cells contain Bchl *a* organized into reaction center and light-harvesting complexes, giving rise to absorption spectrum peaks at 802 and 856 nm, respectively, as well as spirilloxanthin and carotenoic acids. Facultative photoheterotrophs, incapable of anaerobic growth even under illuminated conditions. Aerobic growth is stimulated by light intensities of 2000 lux. Best growth substrates include D-xylose, fumarate and L-glutamate. Glycerol, D-fructose, D-galactose, D-glucose, D-arabinose, L-rhamnose, D-ribose, D-fucose, DL-lactate, malonate, pyruvate, gluconate, and L-malate also serve as the sole carbon source for heterotrophic growth. Incapable of growth on D-mannitol, sorbitol, methanol, ethanol, 1,2-propanediol, kerosene, acetate, butyrate, citrate, formate, glycolate, propionate, succinate, phthalate, *p*-hydroxybenzoate, and benzoate. Hydrolyze Tween 20, 40, and 60, but do not hydrolyze Tween 80. Produce catalase, oxidase, and urease. Reduce nitrate to nitrite. Denitrification not observed. Slight hydrolysis of starch. Do not hydrolyze esculin or gelatin. Do not produce phosphatase, deoxyribonuclease, or indole, but do produce H$_2$S. Voges–Proskauer test and methyl-red test are negative. Accumulate poly-β-hydroxybutyrate granules.

   Optimal growth occurs at pH 6.6–6.8 and 30–34°C. Incapable of growth in NaCl concentrations higher than 0.4%. Require divalent cations to maintain cell structure in suspension. Sensitive to chloramphenicol, streptomycin, and tetracycline. Resistant to penicillin and polymixin B. Nicotinic acid and pantothenate are required as growth factors. Habitat: soil from a playground located in Yamazaki, Noda, of Osakajo Park in Chuoh Ward, Osaka, Japan.

   The mol% G + C of the DNA is: 70.3–71.0 (HPLC).

   Type strain: NS89, CIP 105708, JCM 9931.

   GenBank accession number (16S rRNA): D85827.

Genus XI. **Rhodopila** *Imhoff, Trüper and Pfennig 1984, 341*[VP]

MICHAEL T. MADIGAN AND JOHANNES F. IMHOFF

Rho.do.pi′ la. Gr. n. *rhodon* the rose; M.L. fem. n. *pila* a ball or sphere; M.L. fem. n. *Rhodopila* red sphere.

Cells are spherical to ovoid, motile by means of polar flagella, and divide by binary fission (Fig. BXII.α.29). **Gram negative, belonging to the** *Alphaproteobacteria.* Internal photosynthetic membranes are of the vesicular type. **Photosynthetic pigments are bacteriochlorophyll** *a* **and carotenoids. Major fatty acids are** $C_{18:1}$ **(~75%) and** $C_{16:0}$. Ubiquinones, menaquinones, and rhodoquinones with 9 and 10 isoprene units are present.

Growth occurs preferably photoheterotrophically under anoxic conditions in the light. Cells may be sensitive to oxygen but **grow under microoxic conditions in the dark.** Growth factors are required. Acidophilic freshwater bacteria that live in warm sulfur springs.

The mol% G + C of the DNA is: 66.3.

Type species: **Rhodopila globiformis** (Pfennig 1974) Imhoff, Trüper and Pfennig 1984, 341 (*Rhodopseudomonas globiformis* Pfennig 1974, 205.)

FURTHER DESCRIPTIVE INFORMATION

Rhodopila globiformis is the only known species of this genus and is characterized by a number of peculiar physiological properties. It is an acidophilic bacterium, capable of growth at the lowest pH known for phototrophic purple nonsulfur bacteria (PNSB) (Pfennig, 1969a, 1974). Different pH optima for *Rhodopila globiformis* have been noted during growth with mannitol (pH 4.8–5.0) and fumarate (pH 5.6) as carbon sources (Pfennig, 1974). Defining characteristics are given in Table BXII.α.26; carbon sources and electron donors used are given in Table BXII.α.27. The genus is compared to other phototrophic members of the *Rhodospirillales* in Tables 1 and 2 of the introductory chapter "Anoxygenic Phototrophic Purple Bacteria" (Volume 2, Part A, pp. 121–123). The relationships of these bacteria based on 16S rDNA analysis are shown in Fig. 1 (p. 124) of that chapter.

Rhodopila globiformis is a photoheterotroph, unable to use most of the carbon substrates used by other phototrophic purple nonsulfur bacteria. Ammonia is assimilated via the glutamine synthetase/glutamate synthase reaction (NADH-dependent). Glutamate dehydrogenase is not present, but low activities of an alanine dehydrogenase were found (Madigan and Cox, 1982).

Growth with sulfate as sole sulfur source is possible at low concentrations, but at more than 1 mM sulfate, growth is inhibited. The inhibitory effects of higher sulfate concentrations are thought to be due to misregulation of the enzymes involved (Hensel and Trüper, 1976; Imhoff et al., 1981) and can be overcome by supplementation with a suitable reduced sulfur source. All enzymes necessary for sulfate assimilation are present. Sulfate is reduced via adenosine-5′-phosphosulfate (Imhoff, 1982). Good sulfur sources are cysteine and thiosulfate. Some of the latter is assimilated, but most of it is oxidized to tetrathionate (Then and Trüper, 1981). Sulfite and sulfide are growth-inhibitory even at low concentrations.

Rhodopila globiformis has unique aliphatic ketocarotenoids as major components (Schmidt and Liaaen-Jensen, 1973; Schmidt, 1978) and small amounts of asymmetrical gamma-carotene and neurosporene (S. Takaichi, personal communication). *Rhodopila globiformis* also contains unusual polar lipids (Imhoff et al., 1982), a "small type" cytochrome c_2 (Dickerson, 1980b), and is among the few species of the PNSB that have readily detectable amounts of a high potential iron-sulfur protein (HiPIP).

The type strain of *Rhodopila globiformis* was isolated from a weak red layer present in an acidic sulfur spring along the Gibbon River, Yellowstone National Park (USA). There are several warm springs in this area which are located on the east side of the Grand Loop Road in Yellowstone, about 100 m northeast of Beryl Spring. The springs are warm (~40°C), contain large amounts of S^0 (presumably from the oxidation of sulfide), and can be

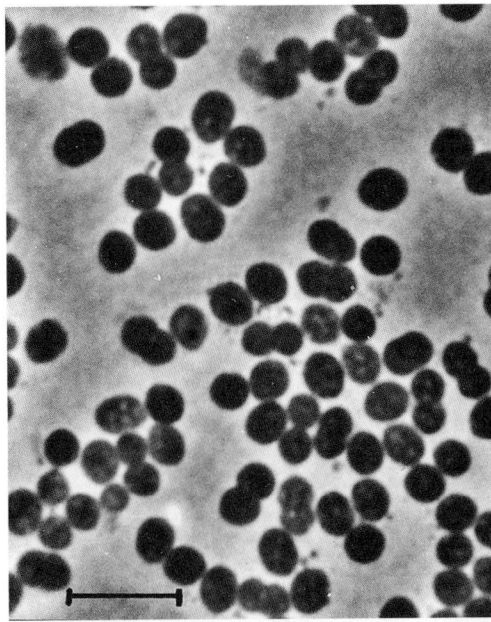

FIGURE BXII.α.29. Phase-contrast photomicrograph of cells of *Rhodopila globiformis* strain DSM 161[T]. Bar = 5 μm. (Courtesy of N. Pfennig.)

TABLE BXII.α.26. Characteristics of *Rhodopila globiformis*[a]

Characteristic	*Rhodopila globiformis*
Cell diameter (μm)	1.6–1.8
Internal membrane system	Vesicles
Motility	+
Color	Purple-red
Bacteriochlorophyll	*a*
Growth factors	Biotin, *p*-aminobenzoic acid
Aerobic growth	(+)
Oxidation of sulfide	−
Salt requirement	None
Optimal temperature (°C)	30–35
Optimal pH	4.8–5.0
Habitat	Fresh water
Mol% G + C of the DNA[c]	66.3
Cytochrome *c* size	Small
Major quinones	Q-9/10, MK-9/10, RQ-9/10
Major fatty acids:	
$C_{14:0}$	5.8
$C_{16:0}$	9.3
$C_{16:1}$	4.7
$C_{18:0}$	1
$C_{18:1}$	74.4

[a]Symbols: +, positive in most strains; −, negative in most strains; (+), weak growth or microaerobic growth only; nd, not determined; Q–9/10, ubiquinones 9 and 10; MK-9/10, menaquinones 9 and 10; RQ-9/10, rhodoquinones 9 and 10.

TABLE BXII.α.27. Carbon sources and electron donors used by *Rhodopila globiformis*[a]

Source/donor	*Rhodopila globiformis*
Carbon sources:	
Acetate	−
Arginine	−
Aspartate	−
Benzoate	−
Butyrate	−
Caproate	−
Caprylate	−
Citrate	−
Ethanol	+
Formate	−
Fructose	+
Fumarate	+
Glucose	+
Glutamate	−
Glycerol	−
Glycolate	−
Lactate	−
Malate	+
Mannitol	+
Methanol	−
Pelargonate	−
Propionate	−
Pyruvate	+
Succinate	+
Tartrate	+
Valerate	−
Electron donors:	
Hydrogen	nd
Sulfide	nd
Sulfur	−
Thiosulfate	−

[a]Symbols: Symbols: +, positive in most strains; −, negative in most strains; +/−, variable in different strains; nd, not determined.

very acidic. Measurements in a series of such springs showed pH values as low as 2.7, with most of them in the range of pH 3–4 (M.T. Madigan, unpublished results). Other locations in Yellowstone where *Rhodopila globiformis* has been observed include warm sulfur springs on the periphery of Nymph Lake; an isolate of *R. globiformis* (strain NL) that matches the type strain in most phenotypic properties was isolated from such a spring (M.T. Madigan, unpublished results). *R. globiformis* coexists in these springs with the extremely acidophilic red alga *Cyanidium caldarum* (Brock, 1978), which supports evidence of the true acidophilic nature of *R. globiformis*.

ENRICHMENT AND ISOLATION PROCEDURES

If samples from suitable natural habitats are available, *Rhodopila globiformis* can be selectively enriched in media with gluconate (or ethanol) as carbon source, thiosulfate as sulfur source, and ammonia or N_2 as nitrogen source at acidic pH (4.8–5.0). Isolation can be achieved by standard procedures for phototrophic purple bacteria under strictly anoxic conditions (Imhoff, 1988; Imhoff and Trüper, 1992). Agar shake cultures established using samples of suspected blooms of *R. globiformis* yield cherry-red colonies within one week at 35°C and 1000 lux incandescent illumination. A suitable medium for *R. globiformis* (after Pfennig, 1974) is given in the footnote below.[1]

1. The medium contains (per liter): KH_2PO_4, 0.5 g; NH_4Cl, 0.4 g; $MgSO_4 \cdot 7H_2O$, 0.4 g; NaCl, 0.4 g; $CaCl_2 \cdot 2H_2O$, 0.05 g; $Na_2S_2O_3 \cdot 5H_2O$, 0.2 g; ferrous citrate, 0.005 g; mannitol, 1.5 g; gluconate, 0.5 g; biotin, 100 µg; *p*-aminobenzoic acid, 200 µg; and trace element solution 6, 10 ml. The initial pH is adjusted to 5.0.

MAINTENANCE PROCEDURES

Cultures of *R. globiformis* are well preserved in liquid nitrogen or at −80°C in a mechanical freezer. Mid-exponential phase cell suspensions should be mixed with glycerol or DMSO to yield final concentrations of 10% and 5%, respectively, and kept at 0°C for 15 min and then frozen immediately.

DIFFERENTIATION OF THE GENUS *RHODOPILA* FROM OTHER GENERA

The genus *Rhodopila* is represented by a single species. *Rhodopila globiformis* differs from all other anoxygenic phototrophic *Alphaproteobacteria* in its outstanding physiological and chemotaxonomic properties and phylogenetic position based on 16S rDNA sequence analysis. First and foremost, acidophily (Pfennig, 1974) and unique carotenoid composition (Schmidt, 1978) characterize *Rhodopila globiformis*. Furthermore, cytochrome c_2 structure (Dickerson, 1980b), quinone and polar lipid composition (Imhoff and Bias-Imhoff, 1995), sulfate assimilation via adenosine-5′-phosphosulfate (Imhoff, 1982), presence of a high potential iron-sulfur protein (HiPIP), and 16S rDNA sequence are characteristics that differentiate *Rhodopila* from other phototrophic *Alphaproteobacteria*. Comparative 16S rDNA sequence analysis of *R. globiformis* shows a clear relationship to both phototrophic and chemotrophic species. A particularly close relationship exists to the acidophilic chemotrophic bacteria including *Acidiphilium* species (Fig. 1 p. 124 in the introductory chapter "Anoxygenic Phototrophic Purple Bacteria", Volume 2, Part A; see also Fig. 1 p. 136 in the introductory chapter "Aerobic Bacteriochlorophyll-Containing Bacteria", Volume 2, Part A). This suggests that the acidophilic lifestyle of *R. globiformis* is a fundamental physiological property rooted in its phylogenetic ties to other acidophilic bacteria. The close relationship to *Acidiphilium* species is of particular interest, because these bacteria contain an extraordinary bacteriochlorophyll *a* with Zn instead of Mg as the chelated metal ion (Wakao et al., 1993, 1996; see Genus *Acidiphilium*).

TAXONOMIC COMMENTS

Several outstanding characteristics of *Rhodopila globiformis* warrant its treatment as a separate genus (Imhoff et al., 1984). This species was originally described as *Rhodopseudomonas globiformis* (Pfennig, 1974) and later reclassified based on unique physiological, chemotaxonomic, and phylogenetic characteristics (Imhoff et al., 1984).

FURTHER READING

Dickerson, R.E. 1980. Evolution and gene transfer in purple photosynthetic bacteria. Nature *283*: 210–212.

Gibson, J., E. Stackebrandt, L.B. Zablen, R. Gupta and C.R. Woese. 1979. A phylogenetic analysis of the purple photosynthetic bacteria. Curr. Microbiol. *3*: 59–64.

Imhoff, J.F. 1982. Occurrence and evolutionary significance of two sulfate assimilation pathways in the *Rhodospirillaceae*. Arch. Microbiol. *132*: 197–203.

Imhoff, J.F. 1992. Taxonomy, phylogeny, and general ecology of anoxygenic phototrophic bacteria. *In* Mann and Carr (Editors), Biotechnology Handbooks: Photosynthetic Prokaryotes, Vol. 6, Plenum Press, New York. pp. 53–92.

Imhoff, J.F., J. Then, F. Hashwa and H.G. Trüper. 1981. Sulfate assimilation in *Rhodopseudomonas globiformis*. Arch. Microbiol. *130*: 234–237.

Imhoff, J.F., H.G. Trüper and N. Pfennig. 1984. Rearrangement of the species and genera of the phototrophic "purple nonsulfur bacteria". Int. J. Syst. Bacteriol. *34*: 340–343.

Madigan, M.T. and S.S. Cox. 1982. Nitrogen metabolism in *Rhodopseudomonas globiformis*. Arch. Microbiol. *133*: 6–10.

Schmidt, K. 1978. Biosynthesis of carotenoids. *In* Clayton and Sistrom (Editors), The Photosynthetic Bacteria, Plenum Press, New York. pp. 729–750.

List of species of the genus Rhodopila

1. **Rhodopila globiformis** (Pfennig 1974) Imhoff, Trüper and Pfennig 1984, 341[VP] (*Rhodopseudomonas globiformis* Pfennig 1974, 205.)

glo.bi.for′ mis. L. n. *globus* sphere; L. n. *forma* shape; M.L. n. *globiformis* of spherical shape.

Cells are spherical to ovoid, diplococcus-shaped before cell division, under optimal growth conditions 1.6–1.8 µm in diameter (depending on the culture conditions diameter ranges from 1.0 to 2.5 µm), and motile by means of polar flagella. Internal photosynthetic membranes are of the vesicular type. Color of cultures grown anaerobically in the light is intensely purple-red; microaerobically grown cells are pink. Absorption spectra of living cells show maxima at 378, 594, 813, and 862 nm and a shoulder at 890 nm. Photosynthetic pigments are bacteriochlorophyll *a* (esterified with phytol) and as major components unusual aliphatic ketocarotenoids.

Cells grow preferably photoheterotrophically under anoxic conditions in the light, but also chemotrophically under microoxic conditions in the dark. Best growth is obtained with gluconate, mannitol, fructose, or ethanol as carbon source. Glucose, tartrate, fumarate, succinate, malate, pyruvate, and yeast extract at low concentration are also assimilated. No growth occurs with fatty acids, lactate, citrate, glycerol, mannose, sorbitol, amino acids, and benzoate as carbon source. Besides ammonia, which is the best nitrogen source, N_2, glutamate, glutamine, aspartate, arginine, urea, asparagine, and alanine are used. Peptone and yeast extract are utilized at low concentrations (0.05%) but are growth-inhibitory at higher concentrations. Nitrate is not assimilated. Thiosulfate, cysteine, or low concentrations of sulfate are the best sulfur sources. Sulfate is growth inhibitory at concentrations above 1 mM. Methionine and tetrathionate can also be used, but sulfide and sulfite inhibit growth even at low concentrations. Biotin and *p*-aminobenzoic acid are required as growth factors.

Acidophilic freshwater bacterium with optimal growth at 30–35°C (no growth at and above 40°C) and at pH 4.8–5.0 with mannitol as carbon source (pH range: 4.2–6.5).

The mol% G + C of the DNA is: 66.3 (Bd).

Type strain: Pfennig 7950, ATCC 35887, DSM 161.

GenBank accession number (16S rRNA): D86513.

Genus XII. **Roseococcus** *Yurkov, Stackebrandt, Holmes, Fuerst, Hugenholtz, Golecki, Gad'on, Gorlenko, Kompantseva and Drews 1994c, 430[VP]*

VLADIMIR V. YURKOV

Ro′ se.o.coc′ cus. M.L. adj. *roseus* rose, pink; Gr. n. *coccus* sphere or spheroidal shape; M.L. masc. n. *Roseococcus* pink spherical bacterium.

Cells are Gram negative, coccoidal, pink and motile by means of polar flagella. Divide by binary fission. **Bacteriochlorophyll *a* and carotenoid pigments** are present. **Obligately aerobic, chemoorganotrophic** (respiratory metabolism) and **facultatively photoheterotrophic. No growth occurs under anaerobic conditions in the light.** May use thiosulfate as an additional energy source. Methanol is not utilized. NaCl not required for growth.

The mol% G + C of the DNA is: 70.4.

Type species: **Roseococcus thiosulfatophilus** Yurkov, Stackebrandt, Holmes, Fuerst, Hugenholtz, Golecki, Gad′ on, Gorlenko, Kompantseva and Drews 1994c, 432.

FURTHER DESCRIPTIVE INFORMATION

The genus *Roseococcus* is phylogenetically related to members of the *Alphaproteobacteria*; this class also includes other aerobic anoxygenic phototrophic bacteria as well as purple nonsulfur photosynthetic and several nonphotosynthetic species. *Roseococcus* is moderately related to *Rhodopila globiformis*, *Thiobacillus acidophilus*, and species of the genus *Acidiphilium* (levels of 16S rDNA sequence homology, >90%) (Yurkov et al., 1994c).

Only one species of the genus is currently isolated and described, *Roseococcus thiosulfatophilus*, represented by pink-red coccoidal cells of 0.9–1.3 × 1.3–1.6 µm in size (Fig. BXII.α.30) and motile by means of a single polar flagellum. The pink-red color of this species is due mainly to two polar red pigments, C_{30} carotene-dioate (4,4′-diapocarotene-4,4′-dioate) and the respective diglucosyl ester (di[β-D-glucopyranosyl]-4,4′-diapocarotene-4,4′-dioate). Together they contribute 95% of the total carotenoid content (Yurkov et al., 1993). Such highly polar C_{30} carotenoid glycosides have never before been observed in other phototrophic bacteria, although the same carotenoid and its diglycosylated form have previously been postulated to exist in *Methylobacterium rhodinum* (formerly *Pseudomonas rhodos*) (Yurkov et al., 1993).

Absorption spectra of the intact cells of *R. thiosulfatophilus* yielded several peaks: at 482, 510, 538, 800, and 859 nm (Yurkov and Gorlenko, 1991). Absorption peaks at 482, 510, and 538 nm are due to the presence of two highly polar red carotenoids described above. Absorption peaks in the infra-red spectrum region indicate the incorporation of bacteriochlorophyll *a* into the reaction center (minor peak at 800 nm) and an unusual type of light-harvesting complex (major peak at 859 nm). Subsequent purification of light-harvesting complexes from these bacteria by detergent treatment of membranes and sucrose density gradient centrifugation revealed the existence of a new *R. thiosulfatophilus* light-harvesting complex I with an absorption peak at 856 nm. The reaction center of *R. thiosulfatophilus* possesses a tightly bound cytochrome *c* (of 44 kDa) that serves as the immediate

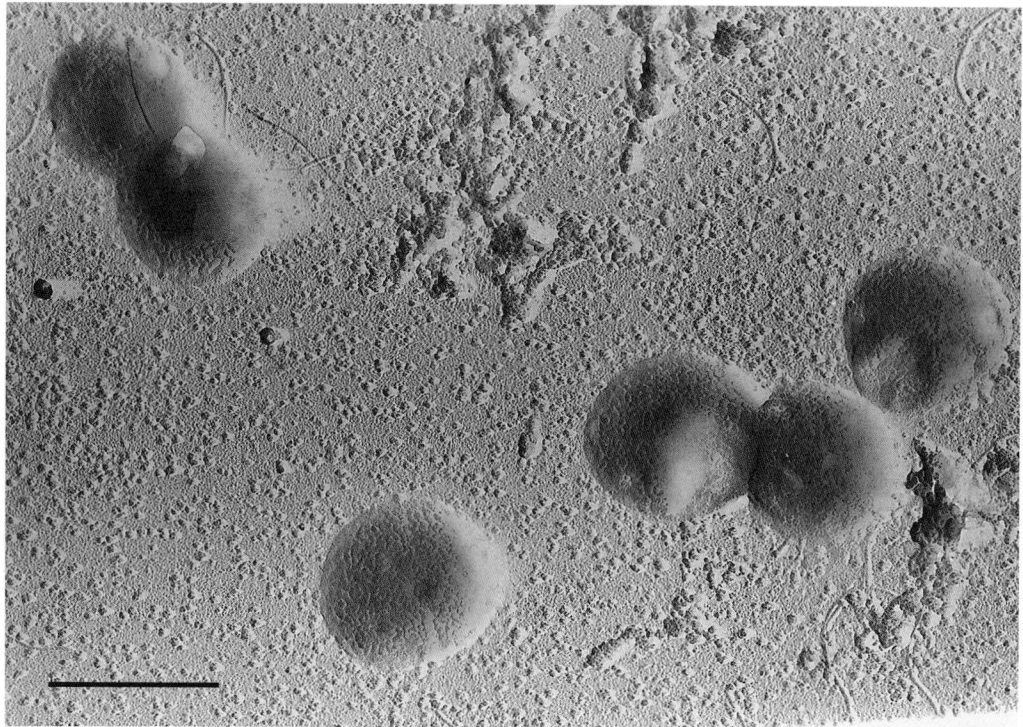

FIGURE BXII.α.30. Coccal cells of *Roseococcus thiosulfatophilus*. Scanning electron micrograph of carbon shadowed cells. Bar = 1.0 μm. (Courtesy of V. Yurkov.)

electron donor to the photooxidized reaction center. The photosynthetic apparatus of *R. thiosulfatophilus* was shown to be functional in terms of a cyclic electron transfer under aerobic conditions. It is not functional anaerobically (Yurkov et al., 1995, 1998a).

Most of the *R. thiosulfatophilus* carotenoids are not bound to the bacteriochlorophyll-protein complexes of the photosynthetic apparatus but to the envelope fraction (cytoplasmic membrane and cell wall) (Yurkov et al., 1993). The excess of carotenoids in the membranes does not contribute to the light-harvesting function of the light-harvesting complexes (Yurkov et al., 1994a).

R. thiosulfatophilus is highly resistant to the toxic heavy metal oxide tellurite (Yurkov et al., 1996). However, the resistance of *R. thiosulfatophilus* to tellurite is not always correlated to its reduction to tellurium. High-level resistance without tellurite reduction was observed for *R. thiosulfatophilus* grown with L-glutamine, succinate, malate, tartrate, or acetate as the organic carbon sources. Tellurite reduction with intracellular deposition of small tellurium crystals in *R. thiosulfatophilus* was shown in rich organic medium (Fig. BXII.α.31). These results are similar to that observed for *Erythromicrobium ezovicum*, implying that tellurite reduction is not essential to confer tellurite resistance and some other important mechanisms could play a role in the resistance character (Yurkov et al., 1996).

ENRICHMENT AND ISOLATION PROCEDURES

R. thiosulfatophilus was isolated in pure culture from a cyanobacterial mat developed in an alkaline sulfide high temperature spring situated in the Bol' shaya River valley. Isolation procedures are similar to that described for the genus *Erythromicrobium* (see the chapter on the *Erythromicrobium*).

MAINTENANCE PROCEDURES

Maintenance of *R. thiosulfatophilus* in liquid culture, on agar plates and long-term preservation are the same as described for the genus *Erythromicrobium*.

DIFFERENTIATION OF THE GENUS *ROSEOCOCCUS* FROM OTHER GENERA

The genus *Roseococcus* has many specific features that differentiate this genus from other genera of aerobic anoxygenic phototrophic bacteria: very high mol% G + C DNA content (70.4) and low DNA homology determined by DNA–DNA hybridization, morphological peculiarities (cell shape and color), carotenoid composition, photosynthetic apparatus components, ability to oxidize thiosulfate to sulfate, and resist high concentrations of toxic tellurite in a special manner (See above).

Roseococcus thiosulfatophilus shares high 16S rDNA sequence similarity with the purple nonsulfur bacterium *Rhodopila globiformis*, which is the closest phylogenetic relative (Yurkov et al., 1994c). Both *Roseococcus thiosulfatophilus* and *R. globiformis* contain bacteriochlorophyll *a* and are able to oxidize thiosulfate. Both these microorganisms also form pink colonies and their mol% G + C contents are high (66.3 and 70.4). There are, however, considerable differences in the growth conditions which they prefer; *Rhodopila globiformis* prefers to grow anaerobically under photoheterotrophic conditions, while *Roseococcus thiosulfatophilus* is obligately aerobic. The antenna systems of these two organisms are also different. *Rhodopila globiformis* seems to have two light-harvesting complexes (with absorption peaks at 813, 862, and 890 nm), whereas *Roseococcus thiosulfatophilus* has only one (with an absorption peak at 856 nm).

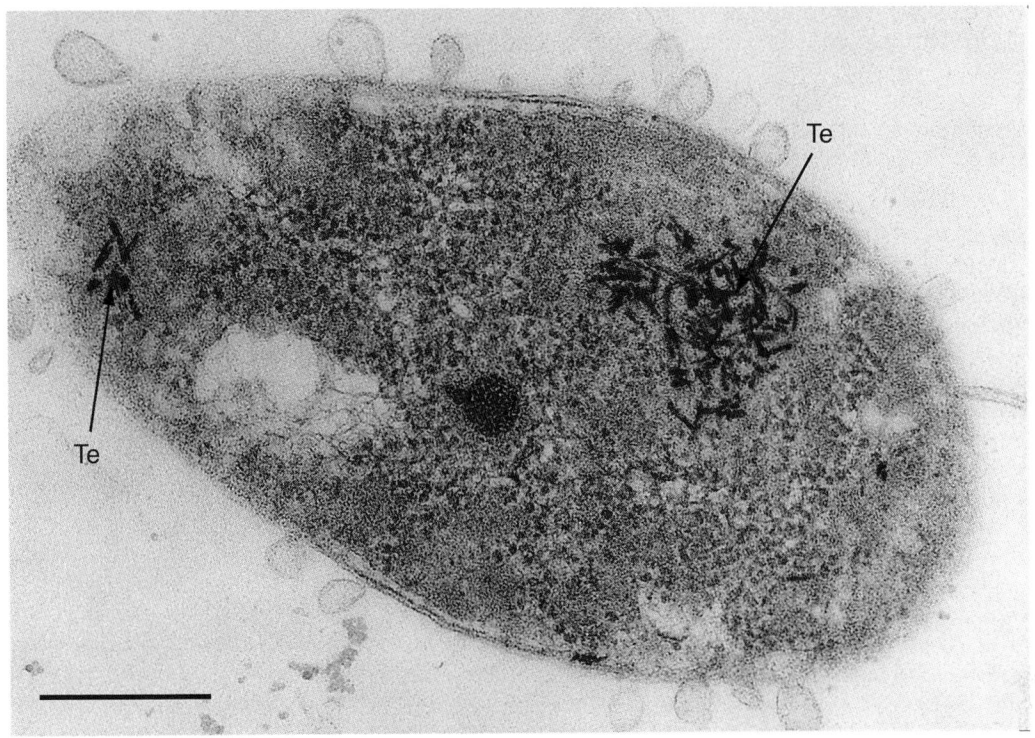

FIGUREBXII.α.31. *Roseococcus thiosulfatophilus*, electron microscopy of ultrathin sections. Intracellular localization of tellurium (*Te; indicated by arrows*) as a product of tellurite reduction. Bar = 0.25 μm. (Courtesy of V. Yurkov.)

TAXONOMIC COMMENTS

The genus *Roseococcus* was created by Yurkov and Gorlenko in 1991 for bacterial strains isolated from a sulfide high temperature spring, based on their morphological, physiological, and DNA–DNA hybridization analyses (Yurkov and Gorlenko, 1991). A valid description of this genus was supplemented by additional results on the photosynthetic apparatus organization, carotenoid composition, and phylogeny (Yurkov et al., 1994c).

FURTHER READING

Shimada, K. 1995. Aerobic anoxygenic phototrophs. *In* Blankenship, Madigan and Bauer (Editors), Anoxygenic Photosynthetic Bacteria, Kluwer Academic Publishers, Dordrecht. pp. 105–122.

Yurkov, V. and J.T. Beatty. 1998. Aerobic anoxygenic phototrophic bacteria. Microbiol. Mol. Biol. Rev. *62*: 695–724.

List of species of the genus Roseococcus

1. **Roseococcus thiosulfatophilus** Yurkov, Stackebrandt, Holmes, Fuerst, Hugenholtz, Golecki, Gad'on, Gorlenko, Kompantseva and Drews 1994c, 432[VP]

    thi' o.sul.fa.to' phi.lus. M.L. adj. *thiosulfatophilus* thiosulfate liking.

    Gram-negative, pink cocci that are 0.9–1.3 × 1.3–1.6 μm and are motile by means of single polar flagellum. Cells contain bacteriochlorophyll *a* and carotenoids. Membranes isolated from cells grown semiaerobically in the dark have absorption maxima at 478, 501, and 505 nm (carotenoids) and at 800 and 855 nm (bacteriochlorophyll). The major carotenoid is C_{30} carotene-dioate. Aerobic, chemoorganotrophic, and facultatively photoheterotrophic.

    Growth occurs on yeast extract and when succinate, acetate, pyruvate, citrate, lactate, malate, or glutamate are used as sole carbon source; weak growth occurs with glucose, maltose, and glycerol. Fructose, sucrose, ribose, arabinose, butyrate, formate, fumarate, propionate, benzoate, tartrate, methanol, ethanol, mannitol, and glycolate are not used. Gelatin, starch, and Tween-80 are not hydrolyzed.

    The cells can oxidize thiosulfate to sulfate in the presence of organic compounds. Ribulose diphosphate carboxylase activity is not detected. The cells are susceptible to tetracycline, streptomycin, polymyxin-B, erythromycin, amikacin, kanamycin, neomycin, aureomycin, vancomycin, chloramphenicol, and fusidic acid, but resistant to gentamicin, lincomycin, nystatin, bacitracin, and penicillin. Oxidase and catalase positive.

    Resistant to the heavy metal oxide tellurite. Reduction of tellurite results in the accumulation of elemental tellurium inside the cells. Tellurite resistance and reduction depend on supplemented organic carbon source. The storage material consists of polysaccharides, poly-β-hydroxybutyrate, and polyphosphate.

    Habitat: freshwater cyanobacterial mat of a thermal alkaline sulfide spring (54°C; pH 9.3; 7.4 mg/l sulfide; 1.6 mg/l oxygen).

    The mol% G + C of the DNA is: 70.4 (T_m).

    Type strain: Strain RB3, ATCC 700004, DSM 8511.

    GenBank accession number (16S rRNA): X72908.

Genus XIII. **Roseomonas** Rihs, Brenner, Weaver, Steigerwalt, Hollis, and Yu 1998, 627[VP]
(Effective publication: Rihs, Brenner, Weaver, Steigerwalt, Hollis, and Yu 1993, 3282)

ROBBIN S. WEYANT AND ANNE M. WHITNEY

Ro.se.o.mo′nas. M.L. adj. *roseus* rosy, rose-colored, or pink; Gr. n. *monas* a unit; M.L. n. *Roseomonas* a pink-pigmented bacterium.

Plump cocci, coccobacilli, or short rods. Arranged singly and less frequently in pairs or short chains. Gram negative. **Aerobic.** Growth occurs on trypticase soy agar with and without 5% sheep blood, heart infusion agar with and without 5% rabbit blood, chocolate agar, and buffered charcoal yeast extract (BCYE) agar. Most strains grow on MacConkey agar. Growth occurs at 25°C, 30°C, 35°C, and for most strains, 42°C. Optimal growth temperature is 35°C. No growth in media containing greater than 6% NaCl. Pinpoint colonies appear after 48–72 h incubation on BCYE agar. Colonies are raised, entire, glistening, and often mucoid. **Pale pink growth pigment is produced. Catalase and urease positive.** Oxidase variable. Motility variable. Indole is not produced. Associated with human infections, usually as a secondary or opportunistic pathogen. Interrelatedness of species by DNA–DNA hybridization is 7–53%.

The mol% G + C of the DNA is: 65–71.

Type species: **Roseomonas gilardii** Rihs, Brenner, Weaver, Steigerwalt, Hollis, and Yu 1998, 627 (Effective publication: Rihs, Brenner, Weaver, Steigerwalt, Hollis, and Yu 1993, 3282.)

FURTHER DESCRIPTIVE INFORMATION

Phylogenetic treatment The genus *Roseomonas* contains three named species and three currently unnamed genomospecies: *R. gilardii*, *R. cervicalis*, *R. fauriae*, and *Roseomonas* genomospecies 4, 5, and 6. Fig. BXII.α.32 illustrates the phylogenetic relatedness among the species of *Roseomonas* based on 16S rDNA sequence similarity. As currently defined, *Roseomonas* is a dichotomous genus with one cluster of species, including *R. gilardii*, *R. cervicalis*, and *Roseomonas* genomospecies 4 and 5 in the family *Acetobacteraceae* and another cluster, including *R. fauriae* and *Roseomonas* genomospecies 6 in the family *Rhodospirillaceae*. Both *Roseomonas* clusters are in the *Alphaproteobacteria*.

Cell and flagellum morphology *Roseomonas* cells can be visualized using the Gram stain and other common staining techniques. Gram-stained preparations of 48-hour cultures show weakly staining Gram-negative plump coccoid rods, occasionally appearing in pairs or short chains. Cells of *R. fauriae* and *Roseomonas* genomospecies 6 have a more rod-like appearance than cells of the other species. Motility varies between species and between strains within species. Motile strains are found in all species except *Roseomonas* genomospecies 5. When present, motility is provided by a single polar flagellum per cell (Weyant et al., 1996).

Cellular fatty acid (CFA) composition *Roseomonas* species produce two characteristic CFA profiles that correlate with the dichotomy of the genus. The profile produced by *R. gilardii*, *R. cervicalis*, and *Roseomonas* genomospecies 4 and 5 is characterized by large amounts (43–53%) of *cis*-11-octadecenoic, δ-*cis*-11,12-methyleneoctadecanoic (10–25%), 2-hydroxyoctadecanoic (2–5%), and 2-hydroxy-δ-*cis*-11,12-methyleneoctadecanoic (3–11%) acids. The profile produced by *R. fauriae* and *Roseomonas* genomospecies 6 is characterized by very large amounts (63–90%) of *cis*-11-octadecenoic and small amounts of 3-hydroxytetradecanoic (2–6%), *cis*-9-hexadecenoic (3–12%), *cis*-9,12-octadeca-

dienoic, *cis*-9-octadecenoic, and octadecanoic acids (Wallace et al., 1990; Weyant et al., 1996).

Colonial and cultural characteristics Growth is achieved in aerobic conditions on most nonselective media, including trypticase soy agar with and without 5% sheep blood, heart infusion agar with and without 5% rabbit blood, chocolate agar, nutrient agar, Sabouraud dextrose agar, and buffered charcoal yeast extract (BCYE) agar. Most strains also grow on MacConkey and Thayer–Martin media. When cultured under optimal conditions, pinpoint colonies are observed after 48 hours of incubation at 35°C. Colonies are typically shiny, raised, entire, and often mucoid. Young colonies have a pale pink color that darkens with age.

Genetics The organization of the genus *Roseomonas* was determined in a DNA–DNA hybridization-based study of 42 strains by Rihs et al. (1993). Using the guidelines of the *ad hoc* committee on reconciliation of approaches to bacterial systematics (Wayne et al., 1987), six genomospecies, three named and three currently unnamed, were identified. The DNA mol% G + C contents of the six genomospecies range from 65.0 for *Roseomonas* genomospecies 6 to 70.4 for *R. cervicalis*. Lewis et al. (1997) have applied molecular subtyping approaches, including pulsed-field gel electrophoresis of chromosomal digests (PFGE) and random amplified polymorphic DNA (RAPD) analysis, to *R. gilardii* clinical isolates. RAPD analysis was reported to be more discriminatory than PFGE and less susceptible to interference by mucoid extracellular material produced by the isolates (Lewis et al., 1997).

Antibiotic sensitivity A significant proportion of the current *Roseomonas* literature contains information related to antibiotic susceptibility of *R. gilardii* strains (Gilardi and Faur, 1984; Korvick et al., 1989; Rihs et al., 1993; Lewis et al., 1997; Sandoe et al.,1997). Antibiotics to which all *R. gilardii* strains are susceptible include the aminoglycosides (gentamicin, amikacin, and tobramycin), tetracycline, and imipenem. One strain tested by Sandoe et al. (1997) was sensitive to netilmicin. All *R. gilardii* strains are resistant to piperacillin and erythromycin. The majority of strains tested were resistant to first- and second-generation cephalosporins. Susceptibility to third-generation cephalosporins varied within and between studies. The eight strains studied by Lewis et al. (1997) and two strains studied by Korvick et al. (1989) were susceptible to ceftriaxone, but only 1 of 23 strains studied by Rihs et al. (1993) was susceptible to this antibiotic. Likewise, the two strains studied by Korvick et al. (1989) were susceptible to cefotaxime but only 1 of 23 strains studied by Rihs et al. (1993) was susceptible. All strains studied were resistant to ceftazidime. Quinolone susceptibility also varied between studies. All eight strains studied by Lewis et al. (1997) and the strain studied by Sandoe et al. (1997) were susceptible to ciprofloxacin, but only 15 of 23 strains studied by Rihs et al. (1993) were susceptible. The study by Rihs et al. (1993) also included norfloxacin, to which 19 of 23 strains tested were susceptible. Susceptibility to sulfamethoxazole/trimethoprim and to penicillins (including natural, extended spectrum, and combinations with β-lactamase inhibitors) was variable in all studies that included more than

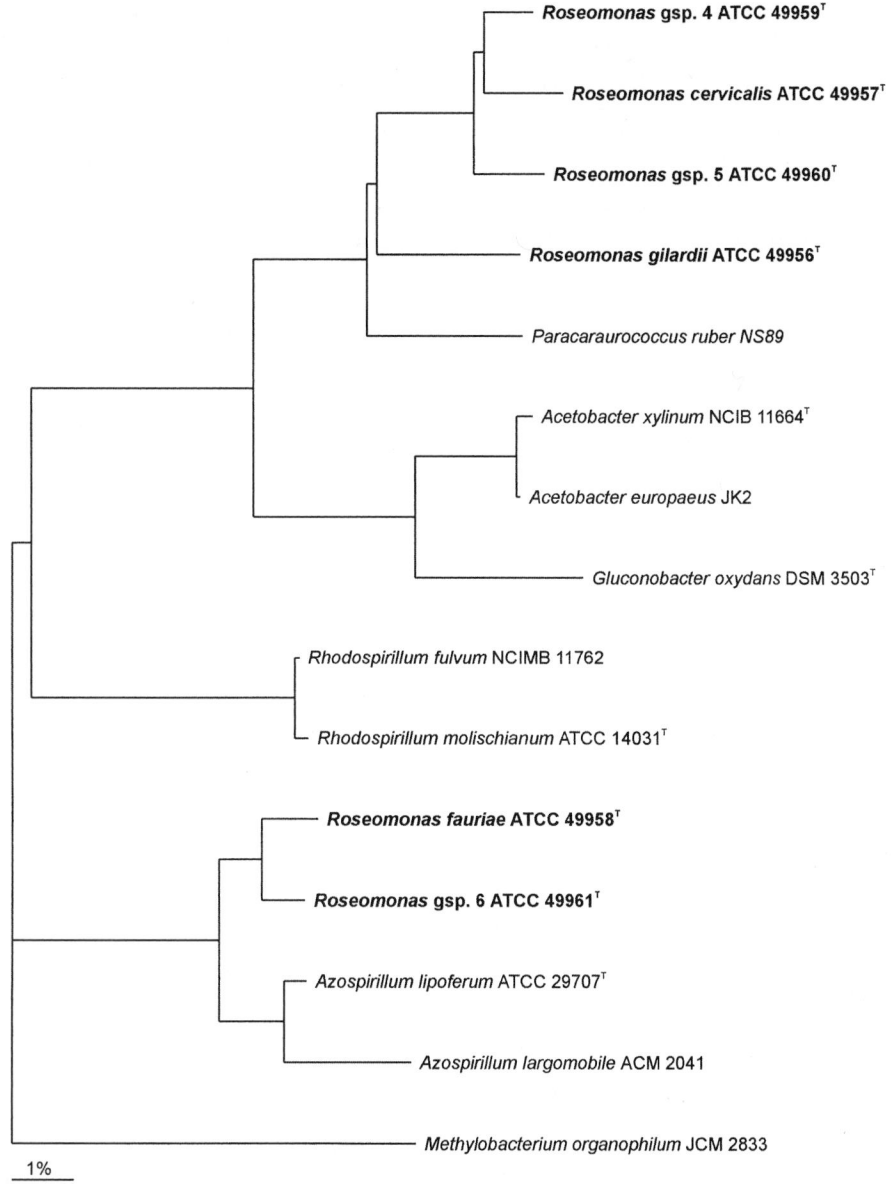

FIGURE BXII.α.32. Phylogenetic tree based on 16S rDNA sequence homologies between the type strains of *Roseomonas* species and type or reference strains of closely related taxa. *Acetobacter xylinum* and *Acetobacter europaeus* have been proposed as *Gluconacetobacter xylinus* and *Gluconacetobacter europaeus* (Yamada et al., 1997, 1998b); *Rhodospirillum fulvum* and *Rhodospirillum molischianum* have been proposed as *Phaeospirillum fulvum* and *Phaeospirillum molischianum* (Imhoff et al., 1998).

one strain. Two strains tested by Korvick et al. (1989) were sensitive to chloramphenicol but resistant to clindamicin, vancomycin, and rifampin.

Rihs et al. (1993) also provided *in vitro* antibiotic susceptibility profiles for strains of the other *Roseomonas* species. All strains tested were sensitive to aminoglycosides, quinolones, tetracycline, imipenem, and penicillins in combination with β-lactamase inhibitors. Most strains were resistant to natural and extended spectrum penicillins, aztreonam, nitrofurantoin, and cephalosporins. Variable results were obtained for the sulfonamides.

Pathogenicity *Roseomonas* species have been associated with human infections, usually as secondary or opportunistic patho-

gens. There is currently no evidence that indicates *Roseomonas* species cause disease in plants or animals other than humans. Most *Roseomonas* infections have been reported in patients with significant underlying disease and, in many cases, *Roseomonas* strains were isolated in mixed culture. In a series of 35 cases reviewed by Struthers et al. (1996), *Roseomonas* strains were most commonly isolated from middle-aged women with one of several underlying conditions, including cancer and diabetes; blood was the most common source of isolation; and only 75% of strains were isolated in pure culture. Similar findings were reported by Lewis et al. (1997), who reviewed a series of eight cases of *R. gilardii* infection. All eight patients had underlying conditions, including ovarian cancer, AIDS, multiple myeloma, non-Hodgkin's lymphoma, and breast cancer. *R. gilardii* was isolated from blood cultures of seven patients; all of whom had central venous catheters and four of whom had polymicrobic infections. Based on the cases reported in the literature, *R. gilardii* is more likely to be associated with human infections than are the other *Roseomonas* species.

Ecology The ecology of *Roseomonas* has not been studied directly, with the great majority of known strains having been isolated from human clinical specimens. There are some suggestions in the literature that these organisms may reside in potable water. Of 42 *Roseomonas* strains studied by Rihs et al. (1993), three *R. gilardii* strains were isolated from potable water. Wallace et al. (1990) reported the source of 2 of the 156 strains that they studied to be a saline contaminant and ice. Another possible ecological niche for these organisms is human skin or mucosal surfaces. The type strain of *R. cervicalis* was isolated from a human cervix culture and *Roseomonas* genomospecies 5 has been isolated as a commensal from young adults attending a sexually transmitted diseases clinic (Rihs et al., 1993; Struthers et al., 1996).

ENRICHMENT AND ISOLATION PROCEDURES

Roseomonas species grow on most commonly used bacteriological media, including trypticase soy agar with and without 5% sheep blood, heart infusion agar with and without 5% rabbit blood, Sabouraud dextrose agar, nutrient agar, buffered charcoal yeast extract agar, and chocolate agar. Selective media that may be used to enrich for these organisms include Thayer–Martin agar and MacConkey agar, although a few MacConkey-negative strains have been described (Odugbemi et al., 1988; Weyant et al., 1996). Growth is inhibited in media containing more than 6% NaCl. Optimal growth temperature is 35°C and aerobic atmospheric conditions are satisfactory.

MAINTENANCE PROCEDURES

Strains can be maintained for short periods (up to 3 months) as cultures in motility deeps. Motility medium deeps in screw cap tubes are inoculated with a sterile inoculation needle by stabbing fresh growth approximately 5 cm into the agar with a single stroke. The tubes are incubated for 1–2 d at 35°C with the screw cap opened slightly to allow for air exchange. After growth is observed in the tube, the cap is tightened to prevent desiccation and the tube is stored at room temperature. Successful long-term storage can be achieved by suspending fresh growth in defibrinated rabbit blood and freezing in liquid nitrogen (Weyant et al., 1996).

PROCEDURES FOR TESTING SPECIAL CHARACTERS

The oxidation–fermentation (OF) medium of King (Weyant et al., 1996) is more sensitive than the Hugh–Leifson formulation (Hugh and Leifson, 1953) in detecting acid production from carbohydrates with *Roseomonas* strains (Rihs et al., 1993). Long-wave absorbance of UV light is a relatively simple procedure for differentiating *Roseomonas* from *Methylobacterium* strains. Colonies grown on BCYE agar are exposed to a long wave UV lamp (365 nm) at a distance of approximately 5 inches (15 cm) in a dark room. *Methylobacterium* colonies will appear dark due to the absorbance of the light, whereas *Roseomonas* species will be easily visible due to the reflection of the light (Rihs et al., 1993). Interlab variability for the oxidase test with *R. gilardii* strains has been reported (Rihs et al., 1993). When tested by the Kovacs

method (Kovács, 1956), some strains may test negative at 10 seconds but will produce a weak reaction after 30 seconds.

DIFFERENTIATION OF THE GENUS *ROSEOMONAS* FROM OTHER GENERA

Roseomonas shares many phenotypic characteristics with *Methylobacterium*, a genus of aerobic, pink-pigmented, oxidative, Gram-negative rods. *Methylobacterium* strains have also been isolated from human clinical specimens (Weyant et al., 1996). Characteristics useful in differentiating these two genera are given in Table BXII.α.28. *Roseomonas* strains fail to produce acid from methanol, assimilate acetamide, and absorb long-wave (365 nm) UV light (Rihs et al., 1993). Most *Roseomonas* strains grow on MacConkey agar, although detectable growth may require 3–7 d of incubation. Esculin hydrolysis and motility are useful in differentiating strains of some *Roseomonas* species from *Methylobacterium*. Strains of *R. fauriae* and *Roseomonas* genomospecies 6 are esculin-positive, and strains of *Roseomonas* genomospecies 5 are nonmotile, which differentiate them from *Methylobacterium* strains.

TAXONOMIC COMMENTS

Strains subsequently shown to be *Roseomonas* were first described in a study of pink-pigmented clinical isolates by Gilardi and Faur. They designated their strains as an "unnamed pink-pigmented taxon" that could be differentiated phenotypically from *Methylobacterium* (Gilardi and Faur, 1984). This term was used by Odugbemi et al. and Korvick, et al. to describe additional isolates from human blood cultures (Odugbemi et al., 1988; Korvick et al., 1989). In 1990 Wallace, et al. published an extensive phenotypic and cellular fatty acid analysis of 156 pink-pigmented strains submitted to the U.S. Centers for Disease Control and Prevention for identification. Included within the study group were the strains described by Odugbemi et al. and by Gilardi and Faur. Four phenotypic groups (Pink Coccoid Groups I, II, III, and IV) that were differentiated by esculin hydrolysis, oxidation of D-xylose and D-mannitol, and cellular fatty acid profiles were described (Wallace et al., 1990). In 1993, Rhis et al., published a polyphasic classification study of 42 strains, including the strains from Gilardi and Faur, Odugbemi et al., Korvick et al., and representatives of all four Pink Coccoid Groups from Wallace et al. Using DNA–DNA hybridization, in conjunction with morphological, biochemical, and antibiotic susceptibility profile analysis, the genus containing six new species (*Roseomonas gilardii*, *R. cervicalis*, *R. fauriae*, and *Roseomonas* genomospecies 4, 5, and 6) was described (Rihs et al., 1993). Recently obtained 16S rDNA sequence information now suggests phylogenetic dichotomy within the genus *Roseomonas* (Fig. BXII.α.32). The type species, *R. gi-*

TABLE BXII.α.28. Differentiation of *Roseomonas* species from *Methylobacterium* species[a,b]

Characteristic	*Roseomonas* species	*Methylobacterium* species
Acid from methanol	−	+
Assimilation of acetamide	−	+
Absorption of long-range UV light	−	+
Growth on MacConkey agar	+[c]	−
Esculin hydrolysis	d	−
Motility	d	+

[a]For symbols see standard definitions.

[b]Data from Rihs et al. (1993) and Weyant et al. (1996).

[c]Some strains require 3–7 days for detection of growth.

lardii, along with *R. cervicalis* and *Roseomonas* genomospecies 4 and 5 are mostly closely related to taxa in the family *Acetobacteraceae*. *R. fauriae* and *Roseomonas* genomospecies 6, however, are more closely related to taxa in the family *Rhodospirillaceae*. Cellular morphology (more rod-like for *R. fauriae* and *Roseomonas* genomospecies 6), CFA profile analysis, and intraspecies DNA–DNA hybridization findings also suggest a dichotomy between the species clustered around *R. gilardii* and those clustered around *R. fauriae* (Rihs et al., 1993; Weyant et al., 1996).

FURTHER READING

Gilardi, G.L. and Y.C. Faur. 1984. *Pseudomonas mesophilica* and an unnamed taxon, clinical isolates of pink-pigmented oxidative bacteria. J. Clin. Microbiol. *20*: 626–629.

Rihs, J.D., D.J. Brenner, R.E. Weaver, A.G. Steigerwalt, D.G. Hollis and V.L. Yu. 1993. *Roseomonas*, a new genus associated with bacteremia and other human infections. J. Clin. Microbiol. *31*: 3275–3283.

Wallace, P.L., D.G. Hollis, R.E. Weaver and C.W. Moss. 1990. Biochemical and chemical characterization of pink-pigmented oxidative bacteria. J. Clin. Microbiol. *28*: 689-693.

DIFFERENTIATION OF THE SPECIES OF THE GENUS *ROSEOMONAS*

Biochemical tests useful in the differentiation of *Roseomonas* species are given in Table BXII.α.29. *R. fauriae* and *Roseomonas* genomospecies 6 are the only *Roseomonas* species that contain strains which test positive for esculin hydrolysis. The ability to produce acid from L-arabinose, D-galactose, and D-xylose differentiates *R. fauriae* from *Roseomonas* genomospecies 6. Acidification of D-mannose differentiates *Roseomonas* genomospecies 4 from the other esculin-negative species. Motility and glycerol oxidation are useful in differentiating between *R. gilardii*, *R. cervicalis*, and *Roseomonas* genomospecies 5. All *Roseomonas* genomospecies 5 strains studied thus far have been negative for either citrate alkalinization or acid production from glycerol, which differentiates this species from *R. gilardii* (Rihs et al., 1993).

List of species of the genus Roseomonas

1. **Roseomonas gilardii** Rihs, Brenner, Weaver, Steigerwalt, Hollis, and Yu 1998, 627^{VP} (Effective publication: Rihs, Brenner, Weaver, Steigerwalt, Hollis, and Yu 1993, 3282.) *gi.lar' di.i.* M.L. gen. n. *gilardii* named after Gerald L. Gilardi, who first described these organisms in 1978.

   The morphological and cultural characteristics are as described for the genus. Other characteristics of this species are given in Table BXII.α.30.

   This species has been associated with opportunistic infections of humans, including bacteremia. Although a few strains have been isolated from potable water, the natural reservoir of this species is unknown.

   The mol% G + C of the DNA is: 67.6–71.2 (T_m).
   Type strain: 5424, ATCC 49956, CIP 104026.

2. **Roseomonas cervicalis** Rihs, Brenner, Weaver, Steigerwalt, Hollis, and Yu 1998, 627^{VP} (Effective publication: Rihs, Brenner, Weaver, Steigerwalt, Hollis, and Yu 1993, 3282.) *cer.vi.ca' lis.* M.L. adj. *cervicalis* from the cervix.

   The morphological and cultural characteristics are as described for the genus. Other characteristics of this species are given in Table BXII.α.30.

   This species has been associated with urogenital and eye infections of humans.

   The mol% G + C of the DNA is: 70.4 (T_m).
   Type strain: E7107, ATCC 49957, CIP 104027.

3. **Roseomonas fauriae** Rihs, Brenner, Weaver, Steigerwalt, Hollis, and Yu 1998, 627^{VP} (Effective publication: Rihs, Brenner, Weaver, Steigerwalt, Hollis, and Yu 1993, 3282.) *faur' i.a.e.* M.L. n. *fauriae* named after Yvonne Faur, who first described *Roseomonas* isolates in 1978.

   The morphological and cultural characteristics are as described for the genus. Other characteristics of this species are given in Table BXII.α.30.

   This species has been associated with bacteremia and wound infections of humans.

   The mol% G + C of the DNA is: 68 (T_m).
   Type strain: C610, ATCC 49958, CIP 104028.

TABLE BXII.α.29. Characteristics useful in differentiating *Roseomonas* species[a,b]

Characteristic	R. gilardii	R. cervicalis	R. fauriae	Roseomonas genomospecies 4	Roseomonas genomospecies 5	Roseomonas genomospecies 6
Esculin hydrolysis	−	−	+	−	−	+
Citrate alkalinization	+	d	d	−	d[c]	+
Motility	d	+	+	d	−	+
Acid production from:						
L-Arabinose	d	d	+	d	d	−
D-Galactose	d	−	+	d	d	−
Glycerol	+	−	+	−	d[b]	+
D-Mannose	−	−	−	+	−	−
D-Xylose	d	d	+	+	d	−

[a]For symbols see standard definitions.

[b]Data from Rihs et al. (1993) and Weyant et al. (1996).

[c]All *Roseomonas* genomospecies 5 strains studied thus far have been negative for either citrate alkalinization or acid production from glycerol (Rihs et al., 1993).

TABLE BXII.α.30. Biochemical characteristics of *Roseomonas* species[a,b]

Characteristic	R. gilardii	R. cervicalis	R. fauriae	Roseomonas genomospecies 4	Roseomonas genomospecies 5	Roseomonas genomospecies 6
Pink growth pigment	+	+	+	+	+	+
Absorption of UV light (365 nm)	–	–	–	–	–	–
Motility	d	+	+	d	–	+
Growth at:						
25°C	+	+	+	+	+	+
35°C	+	+	+	+	+	+
42°C	d	+	+	+	d	+
Growth on:						
MacConkey agar[c]	+	+	+	+	+	+
Salmonella–Shigella agar	–	–	d	–	–	–
Cetrimide agar	–	–	–	–	–	–
Growth in nutrient broth	+	+	+	+	+	+
Growth in nutrient broth with 6% NaCl	d	–	d	d	–	+
Catalase	+	+	+	+	+	+
Oxidase	v[d]	+	+	+	+	+
Simmons citrate	+	+	d	–	d	+
Christensen's urea	+	+	+	+	+	+
Nitrate reduction	–	–	+	+	–	+
Gas from nitrate	–	–	d	–	–	–
Indole	–	–	–	–	–	–
Esculin hydrolysis	–	–	+	–	–	+
H$_2$S production (TSI butt)[e]	–	–	–	–	–	–
o-Nitrophenyl-β-D-galactopyranoside	–	–	–	–	–	–
Phenylalanine deaminase	–	–	–	–	–	–
L-Lysine decarboxylase	–	–	–	–	–	–
L-Arginine dihydrolase	–	–	–	–	–	–
L-Ornithine decarboxylase	–	–	–	–	–	–
Acid production from:						
D-Glucose	d	–	d	–	–	–
D-Xylose	d	d	+	+	d	–
D-Mannitol	d	–	–	–	–	–
Lactose	–	–	–	–	–	–
Sucrose	–	–	–	–	–	–
Maltose	–	–	–	–	–	–
D-Mannose	d	–	–	+	–	–
L-Arabinose	d	d	+	d	d	–
Fructose	+	+	+	d	+	+
D-Galactose	d	–	+	d	d	–
Salicin	–	d	d	–	–	+
Dulcitol	–	–	–	–	–	–
Methanol	–	–	–	–	–	–
Raffinose	–	–	–	–	–	–
L-Rhamnose	–	–	–	–	–	–

[a]For symbols see standard definitions.

[b]Data from Rihs et al. (1993) and Weyant et al. (1996).

[c]Some strains require 3–7 days for detection of growth.

[d]Interlab variability has been reported for testing *R. gilardii* strains by the method of Kovacs (Kovács, 1956); some strains may produce a weak and extremely delayed reaction (Rihs et al., 1993).

[e]TSI, triple sugar iron agar.

Other Organisms

1. *Roseomonas* genomospecies 4 Rihs, Brenner, Weaver, Steigerwalt, Hollis, and Yu 1993, 3282.

   The morphological and cultural characteristics are as described for the genus. Other characteristics of this species are given in Table BXII.α.30.

   This genomospecies has been isolated from human wound, ear, and cervix specimens.

   The mol% G + C of the DNA is: 67.8 (T_m).

   Deposited strain: E7832, ATCC 49959.

2. *Roseomonas* genomospecies 5 Rihs, Brenner, Weaver, Steigerwalt, Hollis, and Yu 1993, 3282.

   The morphological and cultural characteristics are as described for the genus. Other characteristics of this species are given in Table BXII.α.30.

   This genomospecies has been isolated from human blood, bone tissue, and breast tissue.

   The mol% G + C of the DNA is: 65 (T_m).

   Deposited strain: F4700, ATCC 49960.

3. *Roseomonas* genomospecies 6 Rihs, Brenner, Weaver, Steigerwalt, Hollis, and Yu 1993, 3282.

   The morphological and cultural characteristics are as described for the genus. Other characteristics of this species are given in Table BXII.α.30.

   This genomospecies has been isolated from human wound material.

   The mol% G + C of the DNA is: 65.4 (T_m).

   Deposited strain: F4626, ATCC 49961.

Genus XIV. **Stella** Vasilyeva 1985, 520[VP]

LEANA V. VASILYEVA

Stel'la. M.L. fem. n. *Stella* star, to denote star-shaped morphology of cells.

Cells are flat, **six-pronged stars**, **radially symmetrical**, and 0.7–3.0 μm. Occur singly or in pairs. Some strains possess **gas vesicles**. Spores are not formed. **Gram negative**. Aerobic. Reproduction is by **symmetrical cell division**.

Chemoorganotrophic, using a variety of amino acids or organic acids. Aerobic and oxidative. **Oligocarbophilic**.

Occur in soil, freshwaters, and artificial ecosystems where complete decomposition of organic matter is underway; typical representative of the microflora of dispersal systems.

The mol% G + C of the DNA is: 69.3–73.5 (T_m).

Type species: **Stella humosa** Vasilyeva 1985, 520.

FURTHER DESCRIPTIVE INFORMATION

Flat, star-shaped bacteria were first observed by Nikitin and colleagues (Nikitin et al., 1966) and shortly thereafter by Staley (Staley, 1968). Since that time, these prosthecobacteria have been discussed in a number of publications (Vasilyeva, 1970, 1972b, 1980, 1984, 1986; Vasilyeva et al., 1974; Hirsch et al., 1977; Stanley et al., 1979; Hirsch and Schlesner, 1981; Staley et al., 1981; Terekhova et al., 1981; Schlesner, 1983; Reimer and Schlesner, 1989).

The characteristic morphology of *Stella* species is that of a flat, six-pointed star ("Star of David"). There are also intermediate forms among the flat prosthecobacteria, such as the flat triangles observed in organisms assigned to the genus *Labrys* (Vasilyeva and Semenov, 1984), and the genus *Angulomicrobium* (Vasilyeva et al., 1979).

The reproduction of star-shaped bacteria is quite similar in all strains studied. Cell division proceeds by cross-wall formation along a line where the cell has the smallest diameter and between opposite pairs of prongs. Both daughter cells retain three prongs each from the mother cell and then form three new additional prongs (Vasilyeva, 1972b). The cell cytoplasm appears to be similar to that of other Gram-negative bacteria, and so far, no obvious intracellular membrane component that might account for the prong structures that occur on the flat cells of the organism has been observed. Poly-β-hydroxybutyrate storage granules are apparent within some cells. Morphology of the star-shaped forms can vary with changes in concentration of nutrient in the broth, as well as within different strains.

The phylogenetic position of the genus *Stella* indicates that the star-shaped organism belongs to the *Alphaproteobacteria* (Fischer et al., 1985; Stackebrandt et al., 1988a, b; Stackebrandt, et al., 1988a).

Strains of the genus *Stella* are able to utilize a variety of organic acids or amino acids as substrates. The majority of strains utilize pyruvate, citrate, α-ketoglutarate, succinate, malate, acetate, and glutamate. Some strains additionally utilize lysine, glutamine, cysteine, cystine, L-alanine, asparagine, aspartate, and gluconate, and only one or a few strains utilize fumarate, butyrate, valerate, D-alanine, arginine, proline, threonine, or histidine. None of the strains utilizes propionate, benzoate, urea, ethanol, methanol, monomethylamine, leucine, isoleucine, methionine, valine, tryptophan, phenylalanine, glycine, or oxyproline. Complex carbohydrate or protein polymers, cellulose, starch, and gelatin are

not catabolized. In addition to the compounds listed, star-shaped bacteria require 0.01% yeast extract or Casamino acids in the growth medium. A mixture containing 0.1% L-glutamic acid and various B vitamins can be substituted for the yeast extract. Star-shaped bacteria are obligately aerobic, and their metabolism is oxidative (Vasilyeva et al., 1974).

The strains of *Stella* were obtained from a variety of soil, aquatic, and animal fecal materials (Hirsch and Schlesner, 1981; Vasilyeva, 1985; Schlesner, 1992).

From ecological observations performed with direct microscopic techniques and enrichment broth cultures, it is apparent that these organisms are found in habitats where active degradation of organic substances is occurring. Thus, these prosthecobacteria are typical representatives of the aerobic oligotrophic microflora of dispersal (Zavarzin, 1970), now designated as dissipotrophs (Vasilyeva and Zavarzin, 1995), which are defined as an ecological group of bacteria that utilize low molecular weight compounds formed by other microorganisms during hydrolysis of particulate organic matter.

ENRICHMENT AND ISOLATION PROCEDURES

The most successful method for recovery of *Stella* species involves the use of enrichment cultures containing prosthecobacteria growing in association with aerobic cellulose-decomposing bacteria on Hutchinson medium. The unique morphology of these star-shaped bacteria is usually obvious under phase-contrast microscopy. Colonies on agar plates are usually mixed, and a long series of subculturing and reisolation of single colonies is required for selection of pure cultures.

Pure cultures of organisms can be maintained either in dilute meat-peptone media (1:5), such as nutrient broth with beef extract and peptone added and solidified with 2% (w/v) agar, or in media described by Staley (1989). This defined medium supplemented with 0.01% yeast extract can be used for determination of carbon source utilization.

MAINTENANCE PROCEDURES

Strains can be maintained on slants in the refrigerator for at least 6 months. Lyophilization can be used for long-term preservation.

DIFFERENTIATION OF THE GENUS *STELLA* FROM OTHER GENERA

The cluster of star-shaped bacteria appears to represent a clear and distinct group within the collection of budding and/or appendaged bacteria. The mode of cell division, occurring as a cross-wall separating two equal daughter cells with a profile of two three-pronged flat crowns, provides a major means to separate organisms in the genus *Stella* from other prosthecobacteria, such as *Hyphomicrobium, Hyphomonas, Pedomicrobium, Rhodomicrobium, Ancalomicrobium, Ancalochloris, Prosthecomicrobium,* and *Prosthecochloris*. The flat cellular morphology of *Stella* is shared with the genus *Labrys,* but the latter multiplies by budding. DNA–DNA homology between *Stella* species and *Labrys* is quite low.

DIFFERENTIATION OF THE SPECIES OF THE GENUS *STELLA*

The two species, *S. humosa* and *S. vacuolata,* of the genus *Stella* are differentiated by the presence or absence of gas vesicles, which correlates with a number of other characters, such as substrate utilization and a slightly different mol% G + C content of their DNA. All vacuolate strains had DNA–DNA homologies in the range of 74–100%; however, some nonvacuolated strains had DNA–DNA homologies in the range of 60–72% while others had homologies as low as 3%, indicating that stellas are heter-ogeneous. Two additional species, "*S. aquatica*" and "*S. pusilla*", were described by Schlesner (Schlesner, 1983); differentiating characters, however, remain unclear.

List of species of the genus Stella

1. **Stella humosa** Vasilyeva 1985, 520^{VP}

   hu.mo′sa. M.L. fem. adj. *humosa* soil, earth.

   Cells are flat, six-pronged stars 0.7–3.0 µm in diameter and occur singly or in pairs (Fig. BXII.α.33). No clusters or other aggregates are formed. Gram negative with typical three-layered cell envelope. Nonmotile. Multiplication by symmetrical cell division. Aerobic. Poly-β-hydroxybutyrate granules are formed as the reserve substance.

   Colonies are grayish white, circular, compact, and up to 2.5 mm in diameter after incubation for 14 d at 28°C.

   Chemoorganotrophic, utilizing as sole energy source a limited number of organic acids of the tricarboxylic acid cycle and amino acids for respiration. Carbohydrates are not fermented or utilized. Yeast extract is required by all strains but may be replaced by glutamic acid and B vitamins in some strains. No hydrolytic activity. Polysaccharides and proteins not degraded. Oligocarbophilic.

   Optimal temperature for growth is 28–30°C. Optimal pH is neutral for the type strain and is slightly alkaline for others. Growth of some strains is stimulated by the addition of up to 1% NaCl. Catalase and oxidase positive.

   Sensitive to neomycin (0.5–1.0 µg/ml); moderately sensitive to penicillin (8 µg/ml for type strain) and monomycin (2–10 µg/ml) as well as to most other antibacterial antibiotics (Terekhova et al., 1981).

   Plasmids have so far not been detected.

   The mol% G + C of the DNA is: 69.3–72.9 (T_m).

   Type strain: AUCM B-1137, ATCC 43930, DSM 5900.

2. **Stella vacuolata** Vasilyeva 1985, 521^{VP}

   va′cu.o.la.ta. L. adj. *vacuus* empty, void; L. adj. *latus* broad, wide; M.L. adj. *vacuolata* large areas in cytoplasm that appear empty due to gas vesicle formation.

   Cells are flat, six-pronged stars 1.9–2.5 µm in diameter. Occur singly or in pairs. Clusters or aggregates are not formed. Gram negative, nonmotile. Multiplication by symmetrical division into two initially three-pronged daughter cells. Distinctive morphological feature is the presence of gas vesicles throughout cell cytoplasm (Fig. BXII.α.34). Cells exhibit considerable buoyancy.

   Colonies on agar prepared with dilute meat-peptone media (1:5) are milky white, circular and viscous and may reach 2.5 mm in diameter.

   Growth in liquid broth: A pellicle is formed and sediment occurs on the bottom of tube. Growth is inhibited by shaking cultures and by 1% NaCl.

   Chemoorganotrophic, aerobic. Selected amino acids and organic acids of the tricarboxylic acid cycle are utilized as energy sources. Yeast extract is required for growth, although Casamino acids or L-glutamic acid and a B vitamin mixture can substitute for yeast extract. Carbohydrates are

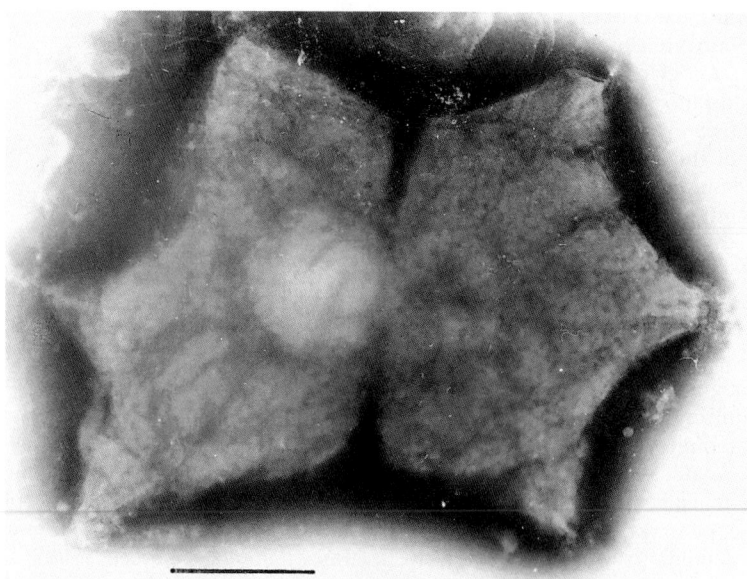

FIGURE BXII.α.33. *Stella humosa* type strain AUCMB-1137. Phosphotungstic acid-negative stain. Bar = 0.5 µm. (Reprinted with permission from E.N. Mischustin, Izvestiya Akademii Nauk SSSR. Seriya Biologicheskaya *5:* 730, 1980, ©Izdatel′stvo Nauka, Moscow, Russia.)

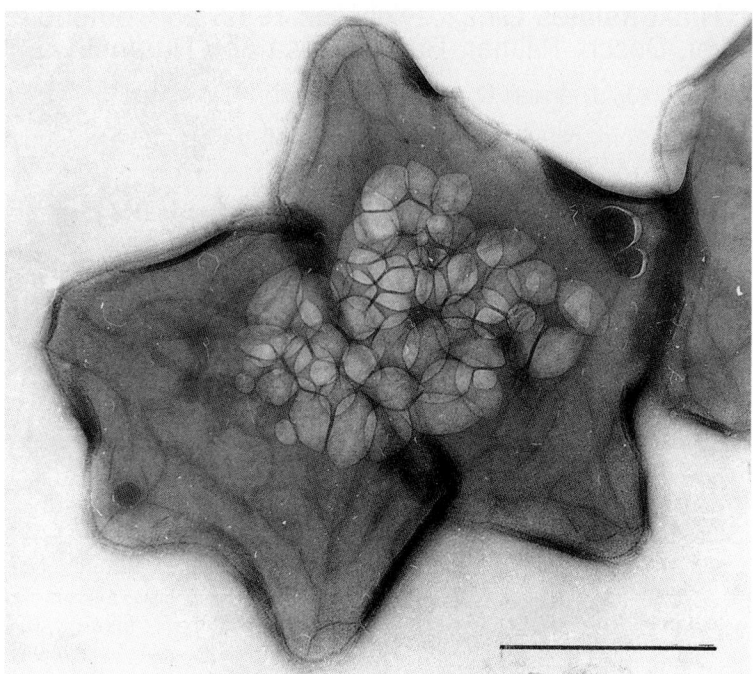

FIGUREBXII.α.34. *Stella vacuolata* Bright areas in cytoplasm are gas vesicles. Uranylacetate-negative stain. Bar = 1.0 μm.

not utilized but do not inhibit growth. Do not possess the ability to utilize polymers.

Optimal temperature for growth: 28°C. Optimal pH for growth: neutral or slightly alkaline. Plasmids were not detected.

Habitat: isolated from horse manure and sewage sludge from a piggery.

The mol% G + C of the DNA is: 70–73.5 (T_m).

Type strain: AUCM B-1552, ATCC 43931, DSM 5901.

Genus XV. **Zavarzinia** *Meyer, Stackebrandt and Auling 1994, 182[VP] (Effective publication: Meyer, Stackebrandt and Auling 1993, 393)*

THE EDITORIAL BOARD

Za.var.zi' ni.a. M.L. n. *Zavarzinia* named for Georgi Alexandrovich Zavarzin, the Russian microbiologist who with his co-workers made the first thorough investigation of mesophilic, Gram-negative carboxidotrophic strains.

The genus description is the same as the description for *Zavarzinia compransoris*.

The mol% G + C of the DNA is: 66.1.

Type species: **Zavarzinia compransoris** (ex Nozhevnikova and Zavarzin 1974) Meyer, Stackebrandt and Auling 1994, 182 (Effective publication: Meyer, Stackebrandt and Auling 1993, 393.)

List of species of the genus Zavarzinia

1. **Zavarzinia compransoris** (ex Nozhevnikova and Zavarzin 1974) Meyer, Stackebrandt and Auling 1994, 182[VP] (Effective publication: Meyer, Stackebrandt and Auling 1993, 393.)

com.pran' so.ris. L. *compransoris* of a dinner companion.

Gram-negative, curved rod. Motile by means of a polar flagellum, 6–8.5 μm in length. Colonies are thin and translucent. Cells swarm on wet agar surfaces. Catalase positive. Thiamine required for growth. "*Pseudomonas gazotropha*" provides thiamine in mixed cultures. Utilizes organic acids (except citrate) and amino acids as sole carbon and energy sources. No growth on sugars. Under aerobic conditions, utilizes ammonia, nitrate, nitrite, hydroxylamine, and urea as nitrogen sources. Heterotrophic denitrification occurs with pyruvate as the electron donor and forming N_2O. Chemolithoautotrophic growth with CO, with a doubling time of 18 h, and incorporation of 4% of the CO oxidized into cell carbon. Possesses cytochromes *a*, *b*, and *c*, but not b_{563}.

The mol% G + C of the DNA is: 66.1.

Type strain: ATCC 51430, DSM 1231, LMG 5821, LMG 8357, Z-1155.

Order II. **Rickettsiales** Gieszczykiewicz 1939, 25^{AL} emend. Dumler, Barbet, Bekker, Dasch, Palmer, Ray, Rikihisa and Rurangirwa 2001, 2156

J. Stephen Dumler and David H. Walker

Rick.ett.si.a' les. M.L. fem. n. *Rickettsia* type genus of the order; *-ales* ending to denote order; M.L. fem. pl. n. *Rickettsiales* the *Rickettsia* order

Rod-shaped, coccoid or irregularly shaped bacteria with typical Gram-negative cell walls and no flagella. Multiply only inside host cells. Can be cultivated in living tissues such as those of embryonated chicken eggs and metazoan cell cultures. All are regarded as parasitic or mutualistic. The bacteria are parasitic forms associated with host cells of the mononuclear phagocyte system, the hematopoietic system, or the vascular endothelium of vertebrates; with various organs and tissues of helminths; or with tissues of arthropods, which may act as vectors or primary hosts. May cause disease in man or in other vertebrate or invertebrate hosts. Mutualistic forms in insects and helminths may be required for development and reproduction of the host under some circumstances.

The mol% G + C of the DNA is: 28–56.

Type genus: **Rickettsia** da Rocha-Lima 1916, 567.

Taxonomic Comments

In the ninth edition of *Bergey's Manual of Determinative Bacteriology,* Section 9, the order *Rickettsiales* was divided into three families, *Rickettsiaceae, Bartonellaceae,* and *Anaplasmataceae.* The recent removal of the family *Bartonellaceae* as well as some genera and species in the family *Anaplasmataceae* from the order also removes all species that could be propagated extracellularly—either on artificial media or on the surface of host cells. In addition, modern molecular classification methods and the recognition of shared biological and morphological features have allowed a reclassification of some *Rickettsiaceae* genera and species into the family *Anaplasmataceae.* These methods have also allowed the removal of some genera and species from *Rickettsiaceae* and *Anaplasmataceae* as well as the removal of the tribe structure of the family *Rickettsiaceae.* Currently, the order contains only the families *Rickettsiaceae* and *Anaplasmataceae.*

Family I. **Rickettsiaceae** Pinkerton 1936, 186^{AL} emend. Dumler, Barbet, Bekker, Dasch, Palmer, Ray, Rikihisa and Rurangirwa 2001, 2156

Xue-jie Yu and David H. Walker

Rick.ett.si.a' ce.ae. M.L. fem. n. *Rickettsia* type genus of the family; *-aceae* ending to denote family; M.L. fem. pl. n. *Rickettsiaceae* the *Rickettsia* family.

Mainly **diplococcus-shaped, but can also be rod shaped or coccoid.** Gram negative. **Obligately intracellular. Intimately associated with arthropod hosts.** No flagella or endospores occur. No member of the family has yet been cultivated in cell-free media. Some species can be parasitic in man and other vertebrates, causing diseases (e.g., typhus and related illnesses in man) that are transmitted by arthropods (lice, fleas, ticks and mites). Some are confined to the invertebrate host as pathogens or symbionts.

The mol% G + C of the DNA is: 29–33.

Type genus: **Rickettsia** da Rocha-Lima 1916, 567.

Taxonomic Comments

Historically, the designation "rickettsia" has been used indiscriminately for many small rods that could not be cultivated and were not otherwise identified. Most frequently, but not always, these organisms were seen in association with arthropods. The last edition of the *Manual of Systematic Bacteriology* (1st edition, 1984) reflects the considerable progress that has been made in establishing a more precise definition of the family *Rickettsiaceae* and in eliminating species that do not fit the definition. The process continues with this edition. In the last edition, the family *Rickettsiaceae* included three tribes: *Rickettsieae, Ehrlichieae,* and *Wolbachieae.* In this edition, the tribes are eliminated. The tribe *Rickettsieae* contained three genera: *Rickettsia, Rochalimaea,* and *Coxiella.* The genus *Rickettsia* is retained in the family *Rickettsiaceae.* Neither *Rochalimaea* nor *Coxiella* are genotypically related to *Rickettsia* except for a superficial phenotypic similarity between *Rickettsia* and *Coxiella. Rochalimaea* has been combined with *Bartonella* and removed from the family *Rickettsiaceae* (Brenner et al., 1993). *Coxiella* has been removed from the family *Rickettsiaceae* and placed in the order *"Legionellales".* Organisms in tribes *Ehrlichieae* and *Wolbachieae* have been removed from the family *Rickettsiaceae* and combined into the family *Anaplasmataceae. Rickettsia tsutsugamushi* was placed in a new genus *Orientia* as *Orientia tsutsugamushi* (Tamura et al., 1995). At present the family *Rickettsiaceae* consists of two genera: *Rickettsia* and *Orientia.* All organisms in the family are phylogenetically related (Fig. BXII.α.35).

Genus I. *Rickettsia* da Rocha-Lima 1916, 567^{AL}

Xue-jie Yu and David H. Walker

Rick.ett' si.a. M.L. fem. n. *Rickettsia* named after Howard Taylor Ricketts, who first associated organisms of this description with spotted fever and typhus and who died of typhus contracted in the course of his studies.

Short, often paired rods, 0.3–0.5 × 0.8–2.0 μm. The rickettsial envelope has a typical Gram-negative structure with a bilayer inner membrane, a peptidoglycan layer, and a bilayer outer membrane. The cells are often surrounded by a protein microcapsular layer and slime layer. Rickettsiae retain basic fuchsin when stained by the method of Giménez (1964). The organisms are

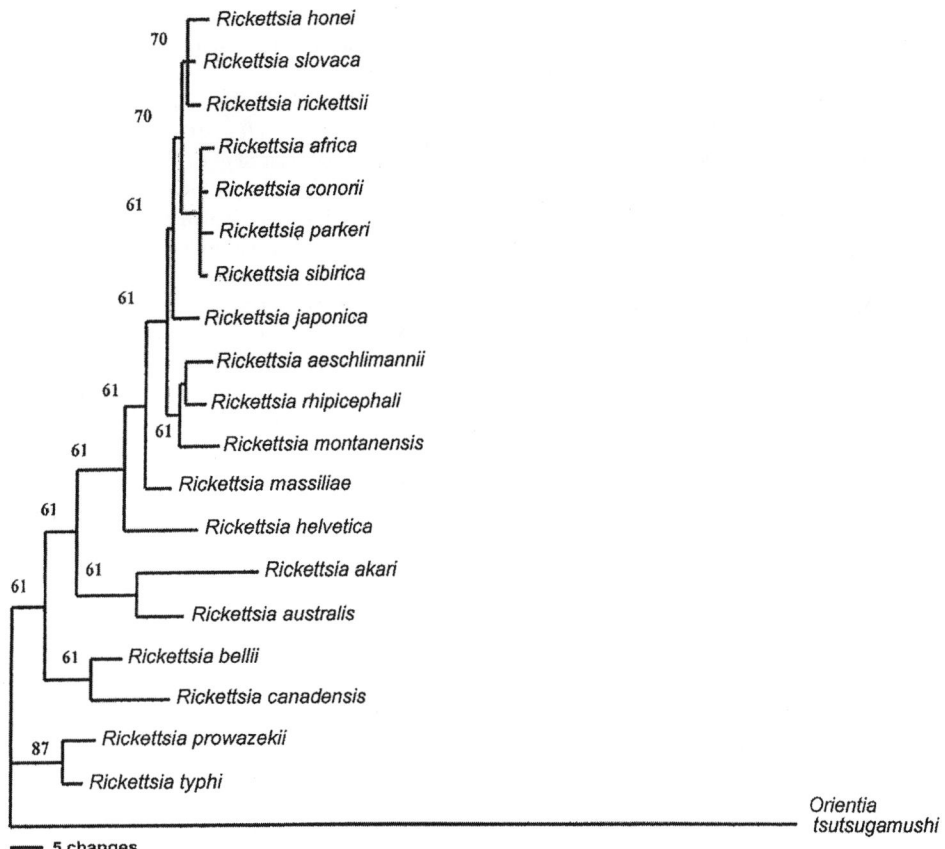

FIGURE BXII.α.35. Phylogenetic relationships of the organisms in the family *Rickettsiaceae* based on the DNA sequences of the 16S rRNA genes (GenBank accession numbers: *R. aeschlimannii*, RAU74757; *R. africae*, RIRRGDA; *R. akari*, RAU12458; *R. australis*, RAU17644; *R. bellii*, RBU11014; *R. canadensis*, RCU15162; *R. conorii*, RIRRGDH; *R. helvetica*, RIRRGDK; *R. honei*, AF060705; *R. japonica*, RIRRGDL; *R. massiliae*, RIRRGDI; *R. montanensis*, RIRRGDN; *R. parkeri*, RIRRRDA; *R. prowazekii* RIRGGSA; *R. rhipicephali*, RIRRGDO; *R. rickettsii*, RIRRGDP; *R. sibirica*, RIRRS16SRG; *R. slovaca*, RIRRGDX; *R. typhi*, RIRRGDU; *O. tsutsugamushi*, RIRRTKP16B). The length of each pair of branches represents the distance between sequence pairs. The numbers on the branch indicate the bootstrap values.

obligately intracellular and reside free in the cytoplasm of the eucaryotic host cell, where they divide by binary fission. Rickettsiae of the spotted fever group (SFG) may also reside in the nucleus of the eucaryotic host cells. Rickettsiae are closely associated with arthropods (ticks, mites, fleas, lice, and other insects) for their maintenance in nature. Their natural cycle usually involves both a vertebrate and an invertebrate host. For some, the arthropod host is both a reservoir and a vector. Transovarian transmission of the agent from the infected female to the next generation is the essential mechanism for the maintenance of many species (Burgdorfer, 1988). Rickettsial cells are usually unstable when separated from host components, except for highly stable forms found in the feces of arthropod hosts; stability can be enhanced by certain proteins, sucrose, and reagents that tend to maintain the integrity of outer membranes, osmolarity, and ATP level. Rickettsiae are best preserved by rapid freezing and storage below −50°C. The cells are rapidly inactivated at 56°C. Rickettsiae derive energy from the metabolism of glutamate via the citric acid cycle, but do not utilize glucose. They transport and metabolize phosphorylated compounds but do not synthesize or degrade nucleoside monophosphates. Rickettsiae are etiological agents of typhus and spotted fevers in humans. There are 21 recognized species.

The mol% G + C of the DNA is: 29–33.

Type species: **Rickettsia prowazekii** da Rocha-Lima 1916, 567 (Nom. Cons. Opin. 19, Jud. Comm. 1958, 158.)

FURTHER DESCRIPTIVE INFORMATION

Phylogeny In the 1st edition of *Bergey's Manual of Systematic Bacteriology*, the organisms in the genus *Rickettsia* were divided into the typhus group (TG) and the spotted fever group (SFG). *R. prowazekii*, *R. typhi*, and *R. canadensis* were included in the TG; these species stimulate antibodies that react more strongly with *Proteus vulgaris* OX19 antigen than OX2 antigen and share LPS antigens. The other rickettsiae were included in the SFG; they stimulate antibodies that react more strongly with antibody specific for *Proteus vulgaris* OX2 antigen and share their own LPS antigens.

In the present edition, *Rickettsia* species are divided into three groups based on phylogeny (Fig. BXII.α.35), as follows. The TG includes *R. prowazekii* and *R. typhi*. The SFG includes *R. aeschlimannii*, *R. africae*, *R. akari*, *R. australis*, *R. conorii*, *R. felis*, *R. helvetica*, *R. honei*, *R. japonica*, *R. massiliae*, *R. montanensis*, *R. parkeri*, *R. peacockii*, *R. rhipicephali*, *R. rickettsii*, *R. sibirica*, and *R. slovaca* (Bouyer et al., 2001). The third group, the ancestral group, includes *R. bellii*, *R. canadensis*, and the AB male-killing bacterium;

this group represents early divergent lineages within the genus (Stothard et al., 1994).

The numbers of named SFG species have increased markedly since the last edition owing to the improvement of isolation approaches and molecular methods for identification. However, there is no consensus concerning the criteria used to define a species of *Rickettsia*. The 16S rRNA gene sequences show 97.2–99.9% similarity among the SFG rickettsiae. The 16S rRNA gene sequences are ≥99.7% similar among the following *Rickettsia* species: *R. africae*, *R. conorii*, *R. massiliae*, *R. parkeri*, *R. sibirica*, *R. slovaca*, and *R. rickettsii*. Thus, many designated species of SFG rickettsiae actually might be clones or strains of a single species according to the standards used for classifying other bacteria. In this chapter, we have compiled information about all *Rickettsia* species that were either included in the Approved Lists of Bacterial Names validly published after January 1980 regardless of where they stand in phylogeny. Rickettsiae need to be classified in the future by an authorized committee according to defined criteria.

Cell structure   Although rickettsiae are relatively small bacteria, they closely resemble other Gram-negative bacteria. In smears from yolk sac, tissue, or cell culture, rickettsiae are best visualized by the Giménez (1964) stain. By this procedure, rickettsiae stain bright red with basic fuchsin, while the background is decolorized and stains a pale greenish blue with the malachite green counterstain (Fig. BXII.α.36A and B).

The rickettsial cell wall contains peptidoglycan, which consists of glutamic acid, alanine, and diaminopimelic acid in a molar ratio of 1.0:2.3:1.0. The small amount of lysine found in the peptidoglycan preparation suggests that a peptidoglycan-linked lipoprotein(s) might be present (Pang and Winkler, 1994).

In ultrathin sections viewed by electron microscopy, rickettsiae are surrounded by a typical Gram-negative envelope (Fig. BXII.α.36E–G). The rickettsial cytoplasm contains ribosomes and strands of DNA. Organisms of the genus *Rickettsia* are typically surrounded in the host cell by an electron-lucent zone that has been proposed to represent a slime layer, which is stabilized by the presence of antibodies (Fig. BXII.α.36E). During the course of infection, rickettsial morphology can change: older cells can become smaller and more electron dense. In *R. prowazekii*, cells may stop dividing after prolonged cultivation and become 3–4 μm long. They also may contain translucent vacuole-like structures, which in some cases appear to occupy as much as 25% of the cytoplasm (Wisseman and Waddell, 1975; Silverman et al., 1980). Intracytoplasmic crystalline structures may also be formed, presumably by a DNA-binding protein homologous to the Dps protein of *E. coli* (Frenkiel-Krispin and Minsky, 2002; personal communication, Dr. Vsevolod Popov). SFG rickettsiae usually do not form vacuoles and crystalline structures.

Cultivation   Rickettsiae must be cultivated in tissue culture or yolk sac of developing chicken embryos (Cox, 1941). L929 and Vero cells are used most frequently, but a great variety of other cells has been used, including chicken embryo fibroblasts, golden hamster BHK-21 cells, HEL cells, monocytes and polymorphonuclear leukocytes.

In irradiated chicken embryo cells at 34°C, *R. prowazekii* multiplies with a generation time of about 9 h until a very high rickettsial density is reached (Fig. BXII.α.36C). If the inoculum is derived from rickettsiae harvested past their logarithmic growth phase, exponential growth is preceded by a lag phase. The nucleus is not invaded. The rickettsiae are released by the disruption of massively infected cells (Wisseman and Waddell, 1975). Growth in other cell types is qualitatively comparable.

Plaques have been produced on a variety of cell monolayers by procedures similar to, but somewhat more difficult than, those used in virology, since monolayers must be maintained for 5–11 days and the introduction of antibiotics must be carefully avoided (Wike and Burgdorfer, 1972). Plaques appear early and are usually larger for most SFG rickettsiae than with TG rickettsiae. SFG rickettsiae form plaques after 5–8 days with a diameter of 2–3 mm. TG rickettsiae form smaller plaques (1 mm) between 8–10 days. TG rickettsiae produce visible plaques on primary chick embryo fibroblasts and on low-passage mouse embryo fibroblasts but do not form reproducible plaques on continuous cell culture lines. TG rickettsiae may produce visible plaques on continuous cell lines if modified procedures are used. One procedure involves a primary overlay with a medium at pH 6.8, which is followed 2–3 days later with a secondary overlay at neutral pH that contains 1 μg of emetine per ml and 20 μg of NaF per ml. Another procedure involves overlay with a medium containing 50 ng of dextran sulfate per ml (Policastro et al., 1996).

Nutrition and metabolism   The nutritional requirements of the rickettsiae, as distinct from those of their host cells, are not known. Rickettsiae grow in heavily irradiated cells (Weiss and Dressler, 1958) that have lost the ability to divide, and they grow in the presence of a low level of cycloheximide that inhibits host protein synthesis. Under these conditions, rickettsiae incorporate exogenous amino acids and adenine, but not thymidine (Weiss et al., 1972). Thus, rickettsial growth occurs independently of host cell protein synthesis, host cell division, and DNA or RNA synthesis. Rickettsiae require a CO_2-enriched atmosphere to grow in chicken embryo cells when an organic buffer is substituted for sodium bicarbonate.

Evolutionarily, rickettsiae are thought to be derived from a free-living bacterial ancestor by the following sequence of events. A free-living aerobic bacterium entered and established an intracellular parasitic relationship with the pre-eucaryote (i.e., an ancestral eucaryote lacking organelles). This made it possible for many chemical substrates to be readily obtained by the bacterium from the host and used without further metabolic modification by the bacterium. The genes of the bacterium involved in metabolic pathways such as glycolysis, fermentation and biosynthesis of small molecules were then redundant with the host cell genes. These redundant genes then had the freedom to mutate, and the mutations might result either in genes with new functions or in pseudogenes. Pseudogenes eventually would be deleted from the genome to make the bacterium more efficient in its use of energy. Gene reduction eventually resulted in a small genome in intracellular bacteria. The loss of genes essential to a free-living mode made the bacterium further dependent on the host cell to supply the substrates, and eventually it lost its ability to live outside the host cell.

Much evidence can be adduced for the above scheme. For instance, no genes required for *de novo* synthesis of nucleotides have been found in the genome of *R. prowazekii*. However, the *R. prowazekii* genome encodes all of the enzymes required for the interconversion of nucleoside monophosphates into all of the other required nucleotides (Andersson et al., 1998). Thus, rickettsiae depend on the monophosphate compounds of the host cell for their nutrition, but they can also utilize their di- and triphosphates. AMP is neither synthesized nor degraded, a situation that seems to be true of the other nucleoside monophosphates as well (Williams and Peterson, 1976). Rickettsiae possibly acquire most amino acids from host cells directly, because only the genes associated with biosynthesis of lysine and serine are found in the genome of *R. prowazekii*. *R. prowazekii*

FIGUREBXII.α.36. Interaction with host cells and fine structure of rickettsiae. (*A* and *B*) Giménez stained preparations. (*C-G*) Transmission electron micrographs of ultrathin sections stained with uranyl acetate and lead citrate. *A*, Human F-1000 fibroblast infected with the Breinl strain of *Rickettsia prowazekii*. A large mass of rickettsiae comprises the entire left portion of the cell with some perinuclear organisms evident. No nuclear involvement is apparent (× 1200). *B*, Secondary chicken embryo fibroblast infected with the Sheila Smith strain of *Rickettsia rickettsii*. Note the sparse, diffusely distributed cytoplasmic organisms and the two distinct compact masses of the organisms within the nuclear region (× 1200). *C*, *R. prowazekii*-infected secondary chicken embryo fibroblast late in infection, showing large numbers of free cytoplasmic rickettsiae without vacuolar membrane, most of which contain vacuole-like structures characteristic of typhus organisms in the stationary phase of growth (× 13,600). *D R. rickettsii* (Sheila Smith strain) in a human endothelial cell (HUVEC) 15 minutes after infection is in the process of escaping from the phagosome. Breaks in the phagosomal membrane (a), and formation of an F-actin tail (b) are apparent (× 31,350). *E*, *R. prowazekii* (Breinl strain) released from an infected host cell and treated with specific human immune serum to demonstrate the slime layer on the surface of the organism (× 67,000). *F*, Cell envelope of *R. prowazekii* including microcapsular layer. The envelope of *R. rickettsii* is morphologically similar (× 196,000). *G*, Cell envelope of *Orientia tsutsugamushi*. Note the thickened outer leaflet of the envelope compared with that of *R. prowazekii* (*above*) (× 171,000). (From the collection of Charles L. Wisseman, Jr., and David J. Silverman. *A*, Reproduced with permission from C.L. Wisseman, Jr. and A.D. Waddell, Infection and Immunity *11:* 1391–1401, 1975, ©American Society for Microbiology; *B*, Reproduced with permission from C.L. Wisseman, Jr. et al., Infection and Immunity *14:* 1052–1064, 1976, ©American Society for Microbiology; *C*, Reproduced with permission from D.J. Silverman et al., Infection and Immunity *29:* 778–790, 1980, ©American Society for Microbiology; *D*, Courtesy of M.E. Eremeeva; *E*, Reproduced with permission from D.J. Silverman et al., Infection and Immunity *22:* 233–246, 1978, ©American Society for Microbiology; *F* and *G*, Reproduced with permission from D.J. Silverman and C.L. Wisseman, Jr. Infection and Immunity *21:* 1020–1023, 1978, ©American Society for Microbiology.

requires host cell proline. Maximal rickettsial growth occurs only in host cells with an intracellular proline pool of 1.0 mM or greater (Austin and Winkler, 1988a). *R. prowazekii* also requires host cell serine or glycine for growth (Austin et al., 1987). Interconversion of serine and glycine is catalyzed by a rickettsial serine hydroxymethyltransferase. Rickettsiae have a membrane-bound ATP/ADP translocase that mediates exchange of ATP and ADP. Rickettsiae exchange extracellular ATP for intracellular ADP without a change in the total adenylate pool. The concentration of phosphate acts as a regulatory signal for the rate of

ADP-ATP exchange. The influx of ATP is greatly favored over ADP under conditions of low phosphate in the host cell cytoplasm. Upon transport into the cell, ATP is hydrolyzed by ATPase to generate a membrane potential in the absence of electron transport. Unlike *Chlamydia* cells, which depend totally on the host cell for energy, rickettsiae also synthesize their own ATP (see below). Thus, we do not know to what extent rickettsiae depend on the host cell for ATP. The fact that both *R. prowazekii* and *R. conorii* have five copies of an ATP/ADP translocase gene indicates that a large amount of ATP may be required by rickettsiae from host cells. The general notion is that rickettsiae acquire host cell ATP early in the infectious cycle via the ATP/ADP translocase and that rickettsiae may synthesize their own ATP to compensate for the depletion of cytosolic ATP late in the infection. The ATP/ADP translocase gene was possibly acquired from the host cell, because it is found only in rickettsiae and chlamydiae (Wolf et al., 1999)

Genes encoding all enzymes of the tricarboxylic acid (TCA) cycle are present in *R. prowazekii*. The primary source of energy is oxidation of glutamate (Bovarnick and Snyder, 1949). Glutamine and pyruvate are also utilized, but to a lesser extent (Weiss, 1973). Glutamine is converted to glutamate by a deamidase. The amino group of the glutamate is transferred to oxaloacetate with the formation of aspartate and α-ketoglutarate by a reversible glutamate-oxaloacetate transaminase. In addition, glutamate is also converted to NH_3 and α-ketoglutarate by a NAD(P)-dependent glutamate dehydrogenase (Weiss, 1973). Glutamate oxidation drives electron transport coupled to oxidative phosphorylation as well as active transport of at least two amino acids, lysine and proline. In *R. prowazekii* the incorporation of P_i into ATP accompanying glutamate oxidation is catalyzed by a membrane-bound ATPase and is sensitive to dicyclohexylcarbodiimide, cyanide, arsenite, and 2,4-dinitrophenol.

Glucose and glucose-6-phosphate are not utilized by rickettsiae. *R. prowazekii* transports uridine 5'-diphosphoglucose (UDPG) but not glucose. *R. prowazekii* takes up glucose phosphates to a much lesser extent than UDPG. *R. prowazekii* does not have hexokinase and phosphoglucomutase, enzymes required for the metabolism of glucose and glucose-6-phosphate (Winkler and Daugherty, 1986). The inability to use glucose as a substrate indicates that rickettsiae may use UDPG as precursors for synthesis of the slime layer, peptidoglycan, and LPS. Rickettsiae must also obtain glycolytic intermediates and products such as acetyl coenzyme A, which are required in the citrate synthase reaction of the TCA cycle. The genes encoding three components (E1–E3) of the pyruvate dehydrogenase complex are found in *R. prowazekii*, indicating that this organism may utilize cytosolic pyruvate. Rickettsiae may use host cell pyruvate by the same mechanism as mitochondria. Pyruvate is imported into mitochondria directly from the cytoplasm, and subsequent conversion into acetyl coenzyme A and CO_2 is catalyzed by pyruvate dehydrogenase. In *R. prowazekii*, pyruvate dehydrogenase activity is dependent on coenzyme A, NAD, and thiamine pyrophosphate and is inhibited by NADH, but not adenylates.

Antigenic structure   The major antigens of *Rickettsia* are lipopolysaccharide, lipoprotein, outer membrane proteins, and heat shock proteins. The Weil-Felix reaction has been used as a presumptive diagnostic test for rickettsial diseases. It is based on the cross-reaction of antibodies to rickettsial antigens from primary rickettsial infections with the somatic antigens of three strains: *Proteus vulgaris* strains OX19 and OX2, and *Proteus mirabilis* strain OXK. The rickettsial antigens do not elicit an anamnestic antibody response to these *Proteus* antigens in recrudescent typhus. The antibodies of Japanese spotted fever patients are reactive with the LPS of OX2 as well as OX19. Thus, the cross-reactive antigens between *Rickettsia* and *Proteus* are most likely present in the LPS. The O-polysaccharides of the LPS of typhus rickettsiae are composed of glucose, glucosamine, quinovosamine, and phosphorylated hexosamine, which are also found in the O-polysaccharide of the LPS from *P. vulgaris* OX19 used in the Weil-Felix test. These findings suggest that these O-polysaccharides may represent the antigens common to LPSs from TG rickettsiae and *P. vulgaris* OX19 (Amano et al., 1998).

The antibodies of scrub typhus patients recognize the antigens of OXK only (Amano et al., 1996).

A 17-kDa protein is a genus-common protein of rickettsiae that has been identified in all *Rickettsia* species examined (Anderson, 1990). The sequences of the gene encoding the 17-kDa protein gene are conserved among rickettsial species, indicating the importance of it to the survival of the rickettsiae. The protein is predicted to be a lipoprotein, and part of the protein is surface exposed. Thus, it has been speculated that the 17-kDa protein may play a scaffolding and protective role in the rickettsiae (Anderson, 1990).

All rickettsiae have a 135-kDa outer membrane protein B (OmpB) that has been identified as an S-layer protein of rickettsiae (Ching et al., 1990). OmpB is the most abundant rickettsial surface protein, and it contains species-, group- and genus-specific epitopes (Anacker et al., 1987).

SFG rickettsiae have an additional outer membrane protein, OmpA. *R. felis* and *R. peacockii* have an OmpA that is truncated by the presence of premature stop codons. OmpA contains a hydrophilic region of tandem repeat units, which consists of over 40% of the amino acid sequences (Anderson et al., 1990). The repeat units are not identical and thus are divided into three types, with type I composed of 75 amino acids, type II of 72 amino acids, and type III of 85 amino acids. The type II repeats are less conserved and are further divided into two subtypes (IIa and IIb) (Anderson et al., 1990). The *ompA* sequences up-stream and down-stream of the repeat region are conserved among SFG rickettsial species. Although the sequences of the repeats are conserved among *R. rickettsii*, *R. conorii* and *R. akari*, the number and the order of arrangement of repeat units vary among SFG rickettsiae (Gilmore, 1993). However, the third type of repeat unit found in *R. australis* differs greatly from other SFG rickettsial repeat units. *R. australis* has only type III repeat units, which have only 21% identity to the type I repeat of *R. rickettsii* (Stenos and Walker, 2000). The antigenic diversity of SFG rickettsiae is determined in large part by the number, order, and type of repeat units.

Both OmpA and OmpB have been demonstrated to stimulate protective immunity against rickettsial challenge of vaccinated animals (Sumner et al., 1995; Dasch et al., 1999).

A striking feature of rickettsiae is that all the identified rickettsial outer membrane proteins belong to the autotransporter family. These proteins include OmpA, OmpB, and four ORFs encoding the hypothetical proteins Sca1 (Accession RP018), Sca2 (RP081), Sca3 (RP451), and Sca5 (RP704) (Andersson et al., 1998). Unlike other transport systems, autotransporter proteins are transported across the cytoplasmic or inner membrane and the outer membrane without the assistance of accessory proteins. An autotransporter protein consists of three functional domains: the amino-terminal leader sequence, the secreted mature (α) protein and a carboxy-terminal (β) domain. The signal peptide inserts itself into the inner membrane, directing the export of the precursor molecule by action of the general secretion pathway into the periplasm. The precursor is cleaved by the signal peptidase to release the mature polypeptide into the periplasm. Once in the periplasm, the β-domain of the protein is inserted into the outer membrane to form a β-barrel pore, and the passenger domain is translocated to the cell surface through the pore. The precursor of OmpB is a 168 kDa-protein that is post-translationally processed to the 135-kDa mature OmpB by cleavage of a 32-kDa β-peptide (Hackstadt et al., 1992).

The amino acid sequences of the β-peptides of all five rickettsial autotransporters are highly homologous and share homology with autotransporters from distantly related bacteria (Henderson et al., 1998). The diversity among the passenger domains that are coupled with the conserved β-domain suggests that the autotransporters are either derived originally from a single gene or resulted from recombination of unrelated proteins with the β-domains (Henderson et al., 1998). Sca 1 and Sca 2 consist of only the β-domain without the passenger peptide. Thus, Sca 1 and Sca 2 may be involved in the transport of other rickettsial proteins. Most of the autotransporter proteins are adhesins or proteases of bacteria, indicating that OmpA, OmpB, Sca3, and Sca5 might be important rickettsial virulence factors.

Drug and antibiotic susceptibility Rickettsiae are naturally resistant to most classes of antimicrobial agents. Contemporary methods determine rickettsial susceptibility to antimicrobial drugs in cell culture. Even the most effective antimicrobial activities against *Rickettsia* are bacteriostatic rather than bactericidal. The antimicrobial drugs that are most effective, i.e., that have the lowest minimal inhibitory concentrations, belong to the following classes: tetracyclines, chloramphenicol, rifampin, fluoroquinolones (pefloxacin, ofloxacin, ciprofloxacin) and some, but not all, macrolides (josamycin, azithromycin, clarithromycin). Doxycycline is the drug of choice in most clinical settings, except for patients who are pregnant or hypersensitive to tetracyclines (Raoult et al., 1990; Walker and Sexton, 1999). There has been long empiric experience with the use of chloramphenicol for the successful treatment of rickettsioses, but the outcome is less satisfactory for Rocky Mountain spotted fever than treatment with doxycycline (Holman et al., 2001). Josamycin has been shown to be an effective treatment of boutonneuse fever during pregnancy. Ciprofloxacin, ofloxacin, pefloxacin, clarithromycin, and azithromycin have been reported to ameliorate the course of boutonneuse fever (Bella et al., 1990; Cascio et al., 2001). The results of treatment with rifampin have been less conclusive, and *R. massiliae*, *R. aeschlimannii*, *R. montanensis*, and *R. rhipicephali* are relatively resistant to rifampin. *Rickettsia prowazekii* is susceptible to erythromycin *in vitro*; however, rapid selection of antimicrobial resistant mutants occurs. *Rickettsia* species are resistant to aminoglycosides and to penicillin and other β-lactam antimicrobial agents.

Sulfonamides are not only ineffective, they actually exacerbate rickettsial infections. In contrast, an analog of sulfonamide, *p*-aminobenzoic acid, is rickettsiostatic and was used successfully to treat Rocky Mountain spotted fever in the 1940s prior to the availability of broad spectrum antibiotics.

Pathogenicity A number of *Rickettsia* species cause severe disease in humans. Before the advent of the broad-spectrum antibiotics, epidemic typhus and Rocky Mountain spotted fever—caused by *R. prowazekii* and *R. rickettsii*, respectively—had a very high case fatality rate. Epidemic typhus has changed the course of history on many occasions (Zinsser, 1935). Rocky Mountain spotted fever and epidemic typhus have exacted a heavy toll on their early investigators.

Other named rickettsiae and their diseases include *R. typhi* (murine typhus), *R. conorii* (boutonneuse fever, Mediterranean spotted fever, Astrakhan fever, Israeli spotted fever), *R. sibirica* (North Asian tick typhus), *R. australis* (Queensland tick typhus), *R. akari* (rickettsialpox), *R. japonica* (Japanese spotted fever), *R. africae* (African tick bite fever), *R. honei* (Flinders Island spotted fever), and *R. felis* (flea-borne spotted fever).

Rickettsiae are inoculated into the skin in the saliva during feeding by an infected tick, mite, or flea or by scratching of rickettsia-laden feces deposited by an infected louse or flea. Rick-

ettsiae are distributed throughout the body via the bloodstream where they enter their principal target, endothelial cells.

In the interaction of rickettsiae with host cells, the entry process involves three steps: attachment, internalization and escape from the phagosome. Rickettsiae adhere to the host cell by means of a rickettsial adhesin and a host cell receptor. Experimental evidence indicates that a ligand and receptor interaction mediate rickettsial attachment. Adsorption of *R. prowazekii* to L-cells and human endothelial cells can be saturated and is dependent on time and multiplicity of infection (Walker and Winkler, 1978; Walker, 1984). Pure *R. prowazekii* is more efficient than a crude suspension in entry into chicken embryo cells (Wisseman et al., 1976). The difference is due to competition between rickettsial receptors on host cell membrane fragments and on intact host cells. The receptors on host cells have yet to be identified, although experimental evidence suggests the host receptor for rickettsiae contains cholesterol as a binding moiety. *R. prowazekii* binds to the cholesterol-containing fraction of chloroform-methanol extracts of erythrocyte membranes (Ramm and Winkler, 1976). Digitonin and amphotericin B, which bind to cholesterol, inhibit the adsorption of *R. prowazekii* to erythrocyte ghosts and reduce plaque formation by *R. rickettsii* in chicken embryo cell monolayers (Walker et al., 1983). However, there is no effect of these agents on the interaction of *R. prowazekii* and L929 cells (Austin and Winkler, 1988b).

OmpA has been implicated as an adhesin of *R. rickettsii* because monoclonal antibodies to OmpA block *R. rickettsii* attachment (Li and Walker, 1998). There is evidence that both OmpA and OmpB are adhesins of *R. japonica* (Uchiyama, 1999).

Rickettsiae enter the host cell via induced phagocytosis that requires undefined rickettsial activity and host cell cytoskeletal rearrangements. Adsorption and internalization require metabolically competent rickettsiae and active participation of the host cell. Inactivation of *R. prowazekii* results in decreased adsorption to L929 cells and endothelial cells (Walker, 1984). *R. prowazekii* adheres to, but is unable to enter, cells treated with a cytoskeletal actin polymerization inhibitor such as cytochalasin B (Walker and Winkler, 1978). Owing to rickettsial infection of nonphagocytic cells, the internalization process is usually called invasion.

Once phagocytosed by the host cell, rickettsiae rapidly escape from the phagosome prior to its fusion with a lysosome (Fig. BXII.α.36D). A rickettsial enzyme is hypothesized to lyse the phagosome membrane to release rickettsiae. The enzyme is most likely a phospholipase because the incubation of *R. prowazekii* with erythrocytes results in hemolysis and formation of free fatty acids and lysophosphatides (Winkler and Miller, 1981; Walker et al., 2001). The phospholipase is of rickettsial origin rather than host cell origin (Winkler, 1990). The nature of the rickettsial phospholipase has not been determined directly. Although phospholipase A2 (PLA2) has been suggested as the rickettsial enzyme that lyses the phagosomal membrane, rickettsial genome sequencing has not revealed a gene encoding a PLA2. A rickettsial protein (Accession RP534) has a calcium-independent PLA2 motif (Yu et al., 2000b), which is homologous to the gene for a cytotoxin (ExoU) of *Pseudomonas aeruginosa* (Vallis et al., 1999). Within the cytosol, rickettsiae acquire nutrients, ATP, amino acids, and nucleic acid precursors from the host by active transport mechanisms and replicate slowly (Winkler, 1990).

SFG rickettsiae stimulate actin-based mobility, resulting in intercellular spread. TG rickettsiae accumulate to massive quantities intracellularly until the cell bursts, releasing the rickettsiae. Observations in cell culture systems suggest that the mechanisms of intracellular movement and destruction of the host cell differ among SFG and TG rickettsiae (Silverman and Wisseman, 1979; Silverman et al., 1980). TG rickettsiae are released from host cells by lysis of the cells. After infection with *R. prowazekii* or *R. typhi*, the rickettsiae continue to multiply until the cell is packed with organisms and then bursts. Cell death possibly results from apoptosis or from membranolytic activity that previously has been hypothesized to be phospholipase A2 (Winkler and Miller, 1982; Walker et al., 1984; Winkler and Daugherty, 1989; Silverman et al., 1992; Manor et al., 1994; Winkler et al., 1994; Ojcius et al., 1995). Before lysis, TG rickettsia-infected host cells have a normal ultrastructural appearance. SFG rickettsiae seldom accumulate in large numbers and do not burst the host cells. They escape from the cell by stimulating polymerization of host cell-derived F-actin tails (Fig. BXII.α.37), which propel them through the cytoplasm and into filopodia, from the tip of which they emerge (Schaechter et al., 1957; Teysseire et al., 1992; Heinzen et al., 1993, 1999). The rickettsial protein responsible for the actin-based movement of SFG rickettsiae has yet to be identified. The genome sequence of *R. conorii* has not revealed a rickettsial protein homologous to ActA or Ics, the proteins responsible for actin polymerization by *Listeria monocytogenes* and *Shigella flexneri*, respectively; however, a hypothetical protein of 520 residues (RC0909) exhibits an overall organization similar to that of ActA. Both proteins are highly charged at the N-terminus and have a central proline-rich region. RC0909 has a weak similarity to the WASP (Wiskott-Aldrich Syndrome Protein) homology domain 2, which regulates the formation of the actin filaments (Ogata et al., 2001).

Rickettsia rickettsii stimulates the production of reactive oxygen species by infected endothelial cells, resulting in oxidative stress and lipid peroxidation-mediated injury to host cell membranes (Walker et al., 1983, 1984; Silverman and Santucci, 1988; 1990;

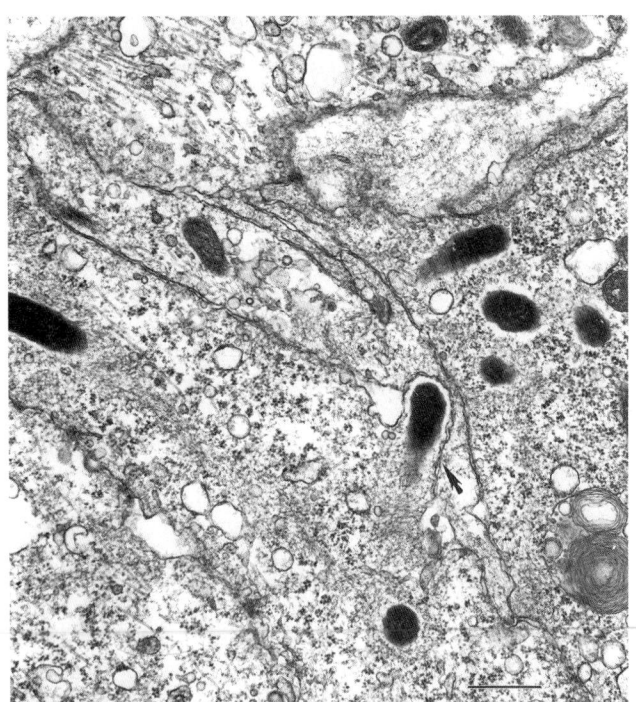

FIGURE BXII.α.37. Electron photomicrograph of *Rickettsia conorii*-infected Vero cells. A rickettsial cell with an actin tail protrudes into an adjacent cell (*arrowhead*). Bar represents 1 μm. Photomicrograph courtesy of Vsevolod Popov.

Eremeeva et al., 2001; Walker et al., 2001). *R. rickettsii* also directly stimulates activation of transcription factor NF-κB by a proteosome-independent mechanism. The consequences include inhibition of endothelial cell apoptosis as well as proinflammatory effects (Sporn et al., 1997; Sahni et al., 1998). The key pathophysiologic effect of disseminated rickettsial infection of endothelium is increased vascular permeability—a life-threatening event in the brain and lungs in Rocky Mountain spotted fever and epidemic typhus and in a smaller proportion of patients with boutonneuse fever and murine typhus. A hallmark of most rickettsioses is a rash that represents multiple focal networks of rickettsia-infected microvascular endothelium in the skin. Except in Rocky Mountain spotted fever and epidemic and murine typhus, another characteristic visible sign of rickettsiosis is an eschar—the site of cutaneous necrosis where the tick or mite inoculated the rickettsiae.

Antirickettsial immune mechanisms are dominated by T-lymphocyte-mediated cellular immunity. Cytokines, particularly gamma interferon and tumor necrosis factor-α, secreted by T-lymphocytes, macrophages, and natural killer cells, activate infected endothelial cells and other cells to kill intracellular rickettsiae by nitric oxide-dependent mechanisms and other mechanisms. Cytotoxic CD8 T-lymphocytes are critical effectors of the clearance of rickettsiae. Antibodies directed against OmpA and OmpB, but not lipopolysaccharide, also contribute to protective immunity. The potential pathologic effects of cytokines and cytotoxic T-lymphocyte activity have not been determined.

Ecology A consistent characteristic of *Rickettsia* species is their residence in an arthropod host as at least a part of their ecological niche. Transovarian maintenance from one generation of tick, mite, or flea to the next via infected ova that hatch into infected larvae is a factor in the maintenance in nature of all SFG rickettsiae (e.g., *R. akari*, *R. australis*, *R. conorii*, *R. felis*, *R. honei*, *R. peacockii*, *R. rickettsii*, and *R. sibirica*) as well as *R. typhi*. For some *Rickettsia* species, such as *R. peacockii*, transovarian transmission appears to be the only mechanism of survival (Niebylski et al., 1996, 1997b). For others, such as *R. rickettsii*, the organism exerts a pathologic effect on the tick that would eventually result in extinction of the rickettsia (Niebylski et al., 1999). Thus, it appears that its virulence comprises an additional survival strategy, the ability to invade and grow in a vertebrate host (e.g., cotton rats) to a sufficient level and duration of rickettsemia to establish infection in uninfected feeding larval and nymphal ticks, with the subsequent transovarian maintenance of the rickettsiae. In this way the death of infected ticks that is brought about by the pathogenic effects of their rickettsial infection is balanced by replacement by new infected tick lines. Ticks infected with highly virulent rickettsiae (e.g., *R. rickettsii*) are infected by fewer rickettsiae than ticks infected with rickettsiae of no apparent pathogenicity (e.g., *R. montanensis*). For *R. typhi*, transovarian maintenance is less important than horizontal spread via rickettsemic *Rattus*.

In the human-body louse (*Pediculus humanus corporis*) cycle of *R. prowazekii*, the lice do not transmit the rickettsiae vertically to the next generation. Indeed, *R. prowazekii* kills 100% of infected *P. humanus corporis* lice. However, the likely natural cycle of *R. prowazekii* in the flying squirrel (*Glaucomys volans*) involves its own flea (*Orchopeas howardii*) and louse (*Neohematopinus sciuropteri*) without severe pathologic effects on the vertebrate and invertebrate hosts (Bozeman et al., 1975, 1981). Epidemics of louse-borne typhus depend upon the establishment of latent human infection with *R. prowazekii*. Years later, recrudescence of the infection occurs, accompanied by rickettsemia. Human body lice acquire the rickettsiae by feeding upon the patient's infected blood and then move to nonimmune persons. It is postulated that this human-louse cycle was first established in North America after accidental human infection originating from contact with infected flying squirrel ectoparasites in the pre-Columbian era. Human body lice were present in the New World at that time.

Knowledge of rickettsial ecology is preponderantly based on investigations of medical epidemiology and hematophagous arthropods. The discovery of rickettsial strains closely related to *R. bellii* in herbivorous insects and in association with plant pathology (papaya bunchy top disease) indicates that our view of the diversity and evolutionary origin of *Rickettsia* is likely overly anthropocentric.

ENRICHMENT AND ISOLATION PROCEDURES

Isolation of rickettsiae from ecologic survey specimens or from clinical material requires strict adherence to safety precautions. A biological safety level 3 facility and an approved safety cabinet in which negative air pressure is maintained are recommended.

Rickettsiae are most commonly isolated in cell culture, although historically the isolation of rickettsiae employed inoculation of adult male guinea pigs or embryonated chicken eggs (Walker, 1996). Vero, L-929, HEL, and MRC5 are cell lines commonly used for isolation of rickettsiae. Heparin-anticoagulated plasma or buffy coat ideally is collected before the patient is given antirickettsial treatment. Isolation of rickettsiae is enhanced by centrifugation of the inoculum onto monolayers in shell vials (La Scola and Raoult, 1996a). Rickettsiae are detected by examining monolayers stained by Giemsa, Giménez, or immunofluorescence methods with 82% of positive samples identified after 48 hours incubation. Rickettsiae are susceptible to some classes of antibiotics. Thus, the culture media ideally should not contain antibiotics although penicillin, streptomycin, and sulfonamides have been used to suppress the growth of bacteria from contaminated specimens.

If the samples are contaminated with bacteria or if cell culture is not available, rickettsiae can be also isolated using animals. The guinea pig is the animal of choice for rickettsiae of the typhus and spotted fever groups. Burgdorfer et al. (1975) found the meadow vole (*Microtus pennsylvanicus*) to be particularly susceptible and valuable for the isolation of spotted fever rickettsiae. The mouse is the animal of choice for *R. australis* and *R. akari*. Rickettsial species vary considerably in virulence for the guinea pig and mouse (Table BXII.α.31). Spleen and other organs are collected from guinea pigs during the 2nd and 3rd day of fever or from moribund mice and are passed into cell culture or embryonated eggs. When a rickettsial species of low virulence, such as *R. montanensis* is expected, the spleen is passed on the 10th–12th day after injection, even in the absence of fever or other signs of illness.

Chicken embryos are generally used for the production of stocks of rickettsiae isolated in laboratory animals. Embryonated eggs can also be used for primary isolation if the inoculum is free from adventitious agents (blood or ticks whose exterior surfaces have been decontaminated by immersion in Merthiolate and by repeated washing). Chicken embryos must be obtained from flocks maintained on a rigorous antibiotic-free diet. Yolk sac inoculation during the 5th–7th d of embryo development is the only satisfactory route for propagation of rickettsiae in eggs (Cox, 1941). The optimal conditions for cultivation and harvest vary with the rickettsial group (Table BXII.α.31).

Rickettsiae have also been isolated from the hemolymph of ticks by the shell vial centrifugation-enhanced method of cell culture. A drop of hemolymph is obtained by amputating the distal portion of one or more legs and is inoculated into a shell vial containing a L929 cell monolayer. The vials are centrifuged at 700 × g for 1 h to increase rickettsial attachment and entry into the cells.

Purification of rickettsiae is accomplished by the separation of rickettsiae from host cell components by differential centrifugation or density gradient centrifugation.

Release of rickettsiae from host cells is achieved as follows. If rickettsiae are grown in yolk sacs, the yolk sacs should be ground using a tissue homogenizer and diluted in 10 volumes of sucrose phosphate glutamate (SPG) buffer (0.22 M sucrose, 0.01 M potassium phosphate and 0.005 M potassium glutamate [pH 7.0]) (Bovarnick et al., 1950). If cell culture is used to propagate rickettsiae, the cells should be harvested in SPG buffer and sonicated for 10-sec periods several times on ice until most of cells are broken.

To remove large particles of debris, the rickettsial suspension is centrifuged at 200 × g for 10 min. The supernatant is then centrifuged at 10,000 × g for 20 min to pellet the rickettsiae. The aqueous phase (and fat layer if yolk sacs are involved) is discarded, and the pellet is resuspended in SPG buffer in an appropriate volume for the Renografin centrifugation.

Renografin is the common gradient material used for purification of rickettsiae by density gradient centrifugation (Weiss et al., 1975). The gradient can be either linear or discontinuous. A linear gradient is made using two concentrations of Renografin, 32% and 42%, by a gradient mixer. A discontinuous gradient consists of Renografin concentrations of 32, 36, and 42%. A discontinuous gradient may be made by adding the solutions in ascending or descending order. If the solution is added in ascending order, the 42% solution is added to the bottom of the centrifuge, then the 36% solution is added slowly on top of the 42% solution using a pipette or a syringe along the edge of the tube. The tip of the pipette or the syringe should just touch the surface of the 42% solution to avoid disturbing the solution. Finally, the 32% solution is added on top of the 36% solution by same method. If the solution is added in descending order, the 32% solution is added into the bottom of the centrifuge tube, then the 36% solution is added slowly by inserting the tip of the pipette or syringe to the bottom of the tube. Finally, the 42% solution is added slowly by inserting the tip of the pipette or syringe to the bottom of the tube. A rickettsial suspension prepared by differential centrifugation is added on top of the Renografin solution. Five ml of rickettsial suspension may be added to a 30 ml Renografin gradient column. The rickettsiae are centrifuged at 87,275 × g for 60 min. Two bands (the heavy and light bands) are usually visible at the interfaces between the 32% and 36% Renografin layers and between the 36% and 42% Renografin layers. The bands are collected by aspiration using a pipette or a syringe inserted into the middle of a band or by puncture of the centrifuge tube beneath the band. The suspension is diluted using SPG to two volumes minimum and mixed well by inverting the tube. The suspension is centrifuged at 11,400 × g for 20 min. The pellet contains the purified rickettsiae, which are then suspended in SPG buffer for storage.

The organisms in the heavy band are defective in their infective and metabolic activities, as compared to organisms from the light band. The greater density of heavy-banding organisms is due to their lack of an intact permeability barrier to Renografin. The proportion of heavy-banding organisms in a rickett-

TABLE BXII.α.31. Differentiation of the species of the family *Rickettsiaceae*[a]

Species	Geographic distribution	Arthropod host	Intracellular location	Peak titer in chicken embryo occurs	Optimal temperature for growth in chicken embryo	Hemolytic activity	Susceptibility to infection	OmpA	LPS[b]
Typhus group									
R. prowazekii	Worldwide	Human louse, flying squirrel flea and louse	Cytoplasm	Prior to death	35°C	+	Guinea pig	−	T
R. typhi	Worldwide	Flea	Cytoplasm	Prior to death	35°C	+	Guinea pig, mouse	−	T
Spotted fever group									
R. aeschlimannii	Africa	Tick	Cytoplasm[c]	24–72 h after death	32–34°C			+	S
R. africae	Africa, Caribbean	Tick	Cytoplasm[c]	24–72 h after death	32–34°C	−	Guinea pig, mouse	+	S
R. akari	Worldwide	Mite	Cytoplasm, nucleus	24–72 h after death	32–34°C			+	S
R. australis	Australia	Tick	Cytoplasm, nucleus	24–72 h after death	32–34°C	−	Mouse	+	S
R. conorii	Eurasia, Africa	Tick	Cytoplasm, nucleus	24–72 h after death	32–34°C	−	Guinea pig	+	S
R. felis	Americas, Europe	Flea	Cytoplasm[c]					+[d]	
R. helvetica	Europe, Asia	Tick	Cytoplasm[c]	24–72 h after death	32–34°C			+	S
R. honei	Australia, Asia	Tick	Cytoplasm, nucleus	24–72 h after death	32–34°C			+	S
R. japonica	East Asia	Tick	Cytoplasm, nucleus					+	
R. massiliae	Europe	Tick	Cytoplasm[c]					+	
R. montanensis	America	Tick	Cytoplasm, nucleus	24–72 h after death	32–34°C	−	Guinea pig	+	S
R. parkeri	America	Tick	Cytoplasm, nucleus	24–72 h after death	32–34°C	−		+	S
R. peacockii	America	Tick	Cytoplasm					−[e]	
R. rhipicephali	America	Tick	Cytoplasm, nucleus					+	S
R. rickettsii	Americas	Tick	Cytoplasm, nucleus	24–72 h after death	32–34°C	−	Guinea pig	+	S
R. sibirica	Asia	Tick	Cytoplasm, nucleus	24–72 h after death	32–34°C	−	Guinea pig	+	S
R. slovaca	Europe	Tick	Cytoplasm, nucleus	24–72 h after death	32–34°C			+	S
Ancestral group									
R. bellii	America	Tick	Cytoplasm, nucleus	Prior to death	32–34°C	−			B
R. canadensis	America	Tick	Cytoplasm, nucleus	Prior to death	35°C	+		+	T
Scrub typhus group									
O. tsutsugamushi	Asia, Australia	Mite	Cytoplasm, nucleus	Prior to death	35°C	−	Mouse	−	−

[a]For symbols see standard definitions.

[b]LPS type: T typhus type, S Spotted fever type, B = *R. bellii* type.

[c]Nucleus location was not reported.

[d]*R. felis* has a truncated *ompA*.

[e]*R. peacockii* has an *ompA* pseudogene.

sial suspension is influenced by the growth phase when they are harvested from infected yolk sacs, as well as by the conditions and media to which they subsequently are exposed (Hanson et al., 1981).

MAINTENANCE PROCEDURES

Rickettsial stocks are usually prepared from infected cell culture or yolk sacs. Infected cells should be harvested before all the cells have detached from the flask. Portions of the infected cells are stored as stock. Infected yolk sacs are ground and diluted to 10 volumes in SPG buffer. Stocks of rickettsiae have been maintained as crude yolk sac suspensions for decades at $-70°C$. When diluted or separated from host constituents, viability declines at rates that depend on the degree of purification and on the choice of suspending fluid. The diluent most frequently used as the maintenance medium is SPG buffer. Stability is further enhanced by the addition of Mg^{2+} and a substance such as brain-heart infusion broth (BHI), bovine serum albumin, or Renografin. Storage in liquid nitrogen is recommended for specimens of unusual value.

Rickettsiae lose viability when they are repeatedly frozen and thawed, maintained at 4°C for a few days, or kept at room temperature for several hours. However, they may be quite stable in desiccated arthropods or louse feces, which occasionally have been a source of unexpected infection (see Smadel, 1965; Snyder, 1965; Woodward and Jackson, 1965).

DIFFERENTIATION OF THE GENUS *RICKETTSIA* FROM OTHER GENERA

Rickettsiae can be easily differentiated from free-living bacteria by their obligately intracellular growth feature. The genus *Rickettsia* can be differentiated from *Anaplasma*, *Ehrlichia*, and *Coxiella* by its intracellular location. Rickettsiae grow free in the cytoplasm or nucleus as compared to *Anaplasma*, *Ehrlichia*, and *Coxiella*, which grow in the cytoplasm within a vacuole. The characteristics for differentiation between *Rickettsia* and *Orientia* are summarized in Table BXII.α.32.

ACKNOWLEDGMENTS

We are indebted to Drs. David J. Silverman Department of Microbiology, School of Medicine, University of Maryland, Baltimore, Vsevolod Popov, Department of Pathology, University of Texas Medical Branch, Galveston, Akira Tamura and Hiroshi Urakami, Department of Microbiology, Niigata College of Pharmacy, Niigata, for providing the illustrations.

DIFFERENTIATION OF THE SPECIES OF THE GENUS *RICKETTSIA*

The differential characteristics of the three groups of *Rickettsia* and some of the species characteristics are presented in Table BXII.α.31. Although most species display some obvious differential characteristics, final identification requires genetic and antigenic analysis.

List of species of the genus Rickettsia

1. **Rickettsia prowazekii** da Rocha-Lima 1916, 567[AL] (Nom. Cons. Opin. 19, Jud. Comm. 1958, 158.)

   pro.wa.ze'ki.i. M.L. gen. n. *prowazekii* of Prowazek; named after Stanislav von Prowazek, an early investigator of the etiology of typhus who died of typhus contracted in the course of his studies.

   The description of the genus and, in particular, of the typhus group within the genus is based largely on studies of this species. Although cells are generally small, variation in size and morphology is most pronounced in this species. Unusually long cells (4 μm, single or in short chains) and cells with prominent vacuoles appear with moderate frequency in stationary cultures.

   Highly infectious for chicken embryos, which die within 4–13 days after inoculation, depending on the quantity of infectious organisms in the inoculum. Optimal yields of rickettsiae are obtained by inoculating the embryos with small inocula (10 viable rickettsiae per egg) and harvesting the yolk sacs 10–14 days later, just before the embryos die. Highly infectious also for monolayers of chick embryo fibroblasts, L-929 cells, and Vero cells. Plaques are usually small (0.5–1.5 mm in diameter) and turbid (see Table BXII.α.31). The virulent Breinl strain, but not the avirulent E strain, multiplies to a moderate extent in human macrophage cultures (Gambrill and Wisseman, 1973).

   R. prowazekii is the etiologic agent of epidemic typhus fever, which is acquired through contact with lice. It is also the etiologic agent of recrudescent typhus (called Brill–Zinsser disease) and occasionally of sporadic typhus in individuals who have been in contact with flying squirrels or their ectoparasites (McDade et al., 1980; Duma et al., 1981). The most effective arthropod vector of epidemic typhus is the human body louse, *Pediculus humanus corporis*, possibly because it takes frequent large blood meals and because it tends to desert febrile hosts to seek new ones. The head louse is equally susceptible to infection but has not been implicated in typhus fever transmission, possibly because it imbibes only very small amounts of blood (Murray and Torrey, 1975). When the rickettsiae are ingested in an infectious blood meal, they enter in the midgut epithelial cells of the louse and undergo massive multiplication. The rickettsiae grow profusely in the cells of the gut epithelium of the louse, even when the ingested human blood contains high levels of antibodies (Wisseman and Waddell, 1975). Heavily infected epithelial cells are released into the lumen, are discharged with the feces of the louse, and are the source of human infection when the louse feces are inoculated into the skin by scratching. Lice invariably succumb to infection, usually within 1–2 weeks, rarely surviving longer than 3 weeks. When louse infestation is very heavy or when large volumes of rickettsial suspensions are processed in the laboratory, infection may occur by inhalation of aerosols.

TABLE BXII.α.32. Differentiation of the genera of the family *Rickettsiaceae*[a]

	Rickettsia	Orientia
Cell wall components		
LPS	+	−
Peptidoglycan	+	−
OmpB	+	−
17 kDa predicted lipoprotein	+	−
56 kDa protein	−	+
Mol% G + C of the DNA	29–33	28.1–30.5
Electron lucent zone	+	−

[a]For symbols see standard definitions.

The guinea pig is highly susceptible to infection, but develops a mild disease, usually manifested only as fever lasting approximately a week. The cotton rat, *Sigmodon hispidus*, is also highly susceptible to infection, but signs of disease and death are produced only by doses in excess of 3×10^5 viable rickettsiae. Inapparent infection elicits solid immunity to the homologous species and to *R. typhi*. True infection does not occur in the mouse, but acute toxic death is produced by the intravenous injection of at least 10^6 viable rickettsiae.

Of the laboratory-induced variants of *R. prowazekii*, the most notable is strain E, isolated from a typhus patient in Madrid in 1941 and passed in rapid succession in eggs 255 times. This strain has limited virulence for the guinea pig and low virulence for man. It has been used as a living vaccine (Fox, 1956). Other laboratory-induced variants have been obtained from strain E. They include strains unaffected by erythromycin and strains of increased resistance to *p*-aminobenzoic (PABA) and to chloramphenicol (Weiss, 1960; Weiss and Dressler, 1962). Strain E easily reverts to virulence during passage in guinea pigs. The mechanisms of attenuation and reversion to virulence are not known, although virulence is associated with differences in methylation of lysine residues of OmpB (Ching et al., 1993).

Strains isolated from various geographic locations are remarkably similar to each other in biological properties. Differences in % DNA–DNA hybridization were negligible (range 92–100; mean = 97) among strains isolated in Poland, Spain, and Burundi from human sources and from flying squirrels from the United States (Myers and Wisseman, 1980). Minor differences in protein migration patterns obtained by isoelectric focusing in polyacrylamide gels can distinguish strains derived from Eastern Europe, Spain or Africa, or from American flying squirrels. Another minor difference between the attenuated E strain and the virulent strains in protein migration patterns has been noted (Dasch et al., 1978). The genome size of *R. prowazekii* is 1,111,523 bp (Andersson et al., 1998). No resident plasmid or bacteriophage has been identified in *R. prowazekii*. The *R. prowazekii* genome has 834 open reading frames.

The mol% G + C of the DNA is: 29.0 (T_m).

Type strain: Breinl strain, ATCC VR-142.

GenBank accession number (16S rRNA): M21789.

2. **Rickettsia aeschlimannii** Beati, Meskini, Thiers and Raoult 1997, 553[VP]

ae.schli.man' ni.i. L. gen. n. *aeschlimannii* of Aeschlimann, named after Andre Aeschlimann, a Swiss zoologist.

The organisms are 0.7–1.1 µm in length, with a mean diameter of 0.3 µm. *R. aeschlimannii* grows in Vero, L929, and HEL cell lines. Mouse antisera against the organism show cross-reactivity with antigens of other SFG rickettsiae. 16S rRNA gene sequence analysis can distinguish *R. aeschlimannii* from other SFG rickettsiae. The most closely related species are *R. massiliae* and *R. rhipicephali*. The sole available isolate was obtained from *Hyalomma marginatum* collected in Morocco (Beati et al., 1997). Two human patients have been reported as infected with *R. aeschlimannii* (Raoult et al., 2002; Pretorius and Birtles, 2002).

The mol% G + C of the DNA is: not determined.

Type strain: MC16 (Reference Center for Rickettsioses, Marseille, France).

3. **Rickettsia africae** Kelly, Beati, Mason, Matthewman, Roux and Raoult 1996, 612[VP]

a' fri.cae. L. gen. n. *africae* pertaining to Africa, the continent where the organism was isolated.

The dimensions of the organisms are 0.3–0.5×0.9–1.6 µm. Growth occurs in the yolk sacs of chicken embryos, L929, Vero, and human embryonic lung fibroblast cells. Plaque formation does not occur in infected Vero cells. Electron microscopy shows that the organisms grow free in the cytoplasm with an outer electron lucent zone and a trilaminar cell wall. The organism can be differentiated from closely related rickettsiae such as *R. parkeri*, *R. sibirica*, and *R. rickettsii* by pulsed field gel electrophoresis after digestion of the DNA with *Eag*I, *Sam*I, and *Bss*HII. The 16S rRNA gene sequences of *R. africae* have 99.9% similarity to those of *R. conorii*, and 99.7% similarity to those of *R. sibirica*. The genome size of *R. africae* is 1.248 kb.

The reservoirs of *R. africae* are *Amblyomma hebraeum* and *A. variegatum*. Rickettsiae can be maintained transstadially and transovarially in *A. hebraeum*; 72% of *A. hebraeum* in Zimbabwe carry these rickettsiae (Kelly et al., 1996).

R. africae causes African tick bite fever. The disease has been reported from sub-Saharan Africa where boutonneuse fever is also endemic. African tick bite fever is a mild disease, whereas boutonneuse fever may be more severe. The skin rash may be sparse, vesicular, or absent. Many patients have multiple eschars and associated regional lymphadenopathy (Kelly et al., 1996).

The mol% G + C of the DNA is: not determined.

Type strain: Z9-Hu (Reference Center for Rickettsioses, Marseille, France).

4. **Rickettsia akari** Huebner, Jellison and Pomerantz 1946, 1682[AL]

a.ka' ri. Gr. neut. n. *akari* a mite.

Among SFG rickettsiae, *R. akari* is relatively distantly related to *R. rickettsii*. It grows somewhat more profusely in the yolk sac of chicken embryos than does *R. rickettsii* (Shepard and Lunceford, 1976) and is somewhat more cytotoxic when grown in cell cultures (Weiss et al., 1972; Kokorin et al., 1978). The mouse, however, is highly susceptible to infection and is the animal of choice for isolation. Differences in virulence between strains of this species and differences in susceptibility between inbred strains of mice have been demonstrated (Anderson and Osterman, 1980). The moderate virulence for the guinea pig is about the same as that of *R. conorii*.

In man, *R. akari* causes rickettsialpox. This disease is characterized by a generalized, typically vesicular rash, fever, malaise, and lymphadenitis, which resolve in about a week. The disease in man was first observed in 1946 in New York City (Huebner et al., 1946). It has been reported mainly from urban areas along the Atlantic coast of the United States, in the Crimea and southern Ukraine and Croatia (Radulovic et al., 1996). In the mid-1940s, about 180 cases were reported annually in the United States, but only sporadic cases have been recognized in recent years (Wong et al., 1979).

This species is readily distinguished from the other members of the spotted fever group by DNA sequence analysis and immunologic evaluation. The main vector is the mite, *Liponyssoides sanguineus*, which maintains the rickettsia transovarially. The nymph and adult stages of the mite feed on the house mouse, *Mus musculus*, but they may attack other animals and man. This *Rickettsia* has also been isolated from rats and from a wild Korean rodent, *Microtus fortis pelliceus* (Jackson et al., 1957).

The genome has a size of 1,231,204 bp and contains 1677 open reading frames and 33 tRNA genes (Eremeeva et al., 2002). Differences in electrophoretic migration patterns of the solubilized proteins between *R. akari* and the other species of the spotted fever group are easily visualized (Obijeski et al., 1974; Pedersen and Walters, 1978).

The mol% G + C of the DNA is: 32.3 (determined from the genome sequence).

Type strain: Kaplan, MK, ATCC VR-148.

This strain was isolated from the blood of a patient in New York City (Huebner et al., 1946). Strain 29 isolated from a mouse by Fuller et al. (1951) and a strain isolated from a mite in the U.S.S.R. (received from Zhdanov [Zhdanov and Korenblit, 1950]) are frequently used as reference strains.

5. **Rickettsia australis** Philip 1950, 786[AL]

aus.tra' lis. L. fem. adj. *australis* southern.

R. australis is a member of a clade that also includes *R. akari* and *R. felis* and is somewhat separate from the core of the SFG2. *R. australis* is not highly virulent for the guinea pig and produces a relatively mild disease in humans. It is highly virulent for the newborn mouse, which has been successfully used for isolation (Campbell and Domrow, 1974). It causes severe, highly invasive disease in several inbred strains of mice but manifests low toxicity for the adult mouse (Bell and Pickens, 1953). Growth characteristics are not appreciably different from those of related species. Identification, as with other species, is based on DNA sequence analysis or immunologic testing.

The disease in man is called Queensland tick typhus. *R. australis* was first isolated in 1944–1945 from the blood of two military patients on field exercises in belts of dense forest interspersed in grassy savannah in North Queensland, Australia (Andrew et al., 1946). Cases have also been recognized along the southeastern coast of Queensland and near Sydney. The scrub tick, *Ixodes holocyclus*, has been implicated in these infections. A strain was also isolated from *I. tasmani* collected from a wild rat (Campbell and Domrow, 1974).

Protein migration patterns are not appreciably different from those of related species (Pedersen and Walters, 1978). The genome size of *R. australis* is 1256–1276 kbp as determined by pulsed-field gel electrophoresis (Roux et al., 1992).

The mol% G + C of the DNA is: not known.

Type strain: NIAID Phillips 32.

GenBank accession number (16S rRNA): L36101, U12459.

The type strain was isolated from a patient by Andrew et al. (1946).

6. **Rickettsia bellii** Philip, Casper, Anacker, Cory, Hayes, Burgdorfer and Yunker 1983, 105[VP]

be.lli' i. M.L. gen. n. *bellii* of Bell, named in honor of a rickettsiologist E. John Bell, who first isolated the organism.

The organisms are longer than most other rickettsiae with dimensions of 0.3–0.4 × 2.0–3.0 μm during the log phase of growth and sometimes 10–15 μm under suboptimal conditions in tissue culture. Like other rickettsiae, *R. bellii* possess an outer slime layer; however, *R. bellii* can be differentiated from other rickettsiae because it has no discernible microcapsular layer. Most strains of *R. bellii* grow

poorly in yolk sacs although the type strain was originally isolated in embryonated eggs. Vero cell culture is a satisfactory system for primary isolation. Other cells such as *Xenopus laevis* (South African clawed toad) and tick cell lines have been used to isolate *R. bellii*. The cytopathic effect varies among isolates of *R. bellii*. Some isolates form characteristic and well defined but faint ring-shaped plaques, and others do not form plaques in Vero cells. Although *R. bellii* possesses antigens that cross-react with antibodies to *R. rickettsii*, *R. prowazekii*, and *R. typhi*, the protein composition of *R. bellii* differs from that of TG and SFG rickettsiae. Since 1974, 263 *R. bellii* isolates have been identified by immunofluorescence.

R. bellii has been detected in ticks from all geographic regions of the United States. In many locations, *R. bellii* is the predominant *Rickettsia* sp. in ticks. *R. bellii* comprises 39% (41/106) of isolates from *D. andersoni* in Montana, 82% (59/72) of isolates from North Carolina, and 83% (128/154) of isolates from Ohio. *R. bellii* has been identified in six species of ixodid ticks (*Dermacentor variabilis*, *D. andersoni*, *D. occidentalis*, *D. parumapertus*, *D. albipictus*, and *Haemaphysalis leporispalustris*) and two species of argasid ticks (*Argas cooleyi* and *Ornithodoros concanensis*) (Philip et al., 1983).

A symbiont of a pea aphid (*Acyrthosiphon pisum*)—a plant-feeding insect—and an organism associated with a plant disease called papaya bunch top disease are phylogenetically placed in a clade with *R. bellii* (Chen et al., 1996; Davis et al., 1998).

Rickettsia bellii is nonpathogenic for mice and guinea pigs. Limited replication may occur in voles for some strains, but these strains cannot be maintained by serial passage in voles. There is no evidence that *R. bellii* causes human illness.

The mol% G + C of the DNA is: 30 (T_m). This value is similar to that of the TG rickettsiae but lower than that of the SFG rickettsiae.

Type strain: RML 369-C (Rocky Mountain Laboratories Collection).

GenBank accession number (16S rRNA): U11014.

RML 369-C (Rocky Mountain Laboratories Collection). This strain was isolated in embryonated chicken eggs from a triturated pool of unfed adult *Dermacentor variabilis* ticks collected from vegetation near Fayetteville, Arkansas, in June of 1966. After the fifth passage in yolk sacs, this organism regularly kills chicken embryos 4–5 days after inoculation (Philip et al., 1983).

7. **Rickettsia canadensis** McKiel, Bell and Lackman 1967, 509[AL] (*Rickettsia canada* (sic) McKiel, Bell and Lackman 1967, 509.)

ca' na.den.sis. M.L. n. *canadensis* referring to Canada, the country where the organism was first isolated.

R. canadensis was previously known as *R. canada*. The name was corrected to *R. canadensis* because *canadensis* is the Latin adjective referring to Canada (Roux and Raoult, 2000).

R. canadensis can be grown in laboratory-reared ticks (Burgdorfer and Brinton, 1975), but grows sparingly in the louse (Weyer and Reiss-Gutfreund, 1973). Unlike some of the species of the spotted fever group, *R. canadensis* is not highly cytopathic and is cultivated in chicken embryos by the procedure used for the typhus group (Table BXII.α.31). Plaques on monolayers are small as are those of the typhus group (Woodman et al., 1977). Virulence for the guinea pig and mouse is quite low. At least 10^6 viable cells are required to induce fever in the guinea pig, and 10^5 viable cells are required for seroconversion in mice (Ormsbee et al., 1978). The hemolytic activity as measured with sheep erythrocytes and the rate of CO_2 production from glutamate are approximately the same as those obtained with *R. prowazekii* and *R. typhi* (Woodman et al., 1977). *R. canadensis* reacts strongly with guinea pig and rabbit sera prepared against the typhus group rickettsiae but weakly with sera against spotted fever group rickettsiae; this weak reaction is due to the presence of epitopes in its LPS that are shared with TG rickettsiae prepared according to a particular immunization protocol (McKiel et al., 1967). Specific antigens can readily be demonstrated with mouse antisera (Philip et al., 1978). As with *R. prowazekii* and *R. typhi*, *R. canadensis* has antigens in common with *Proteus vulgaris* OX19, whereas SFG rickettsiae share antigens with the *Proteus vulgaris* OX2 antigen. Thus, *R. canadensis* was previously classified in the typhus group. However, *R. canadensis* resembles SFG rickettsiae in its other features, such as its ecological niche in the tick and its ability to grow in the nucleus of host cells. Moreover, *R. canadensis* has OmpA (Vishwanath, 1991), which is a common feature of SFG rickettsiae but not typhus rickettsiae. Thus, *R. canadensis* does not fit into either the TG or SFG of rickettsiae.

Two strains were isolated from pools of engorged rabbit ticks (*Haemaphysalis leporispalustris*) collected near Richmond, Ontario; one strain was isolated from an indicator rabbit, the other from a wild snowshoe hare (McKiel et al., 1967). Only one of these two strains is still available. Another strain was isolated in California from a *H. leporispalustris* tick feeding on a California jackrabbit (*Lepus californicus*) (Lane et al., 1981). There is serological evidence that human infection may have occurred (Bozeman et al., 1970).

Although the mol% G + C of the DNA of *R. canadensis* is identical to the values for *R. prowazekii* and *R. typhi*, the genome size is distinctly larger (1.49×10^9 Da). Thus, the extent of DNA–DNA hybridization between *R. canadensis* and the other two species of the typhus group varies somewhat owing to differences in the size of the genomes (38% and 46% hybridization with *R. prowazekii* and *R. typhi*, respectively). A degree of DNA–DNA hybridization occurs between *R. canadensis* and *R. rickettsii* that is similar to that between *R. canadensis* and *R. typhi* (Myers and Wisseman, Jr., 1981). When compared to *R. prowazekii* and *R. typhi* on the basis of migration patterns in polyacrylamide gel electrophoresis of solubilized proteins, *R. canadensis* displays a number of differences, although a basic similarity can still be recognized (Dasch et al., 1978).

The mol% G + C of the DNA is: 29 (T_m).

Type strain: 2678, ATCC VR-610.

GenBank accession number (16S rRNA): L36104.

8. **Rickettsia conorii** Brumpt 1932, 1199[AL]

co.no′ri.i. M.L. gen. n. *conorii* of Conor; named after A. Conor who, in collaboration with A. Bruch, provided the first description of boutonneuse fever.

R. conorii resembles *R. rickettsii*, but it is antigenically distinct and less virulent for guinea pigs and humans; however, the Malish 7 strain is more virulent for C3H/HeN mice. Growth in cell cultures has the basic features of other rickettsiae of the spotted fever group, including early spread between host cells (Oaks and Osterman, 1979), and it is cultivated by procedures identical to those used for *R. rickettsii*. This species can be differentiated from related species of the spotted fever group by appropriate DNA sequence analysis and serological tests (Bell and Stoenner, 1960; Bozeman et al., 1960). Differences among strains of *R. conorii* include different numbers, type, and order of tandem repeat units of OmpA. Genetic analyses reveal clustering of some isolates (e.g., those similar to Israeli isolates, those similar to Astrakhan isolates, and those similar to the Malish 7 South African isolate).

The human disease varies in severity and is fatal in 1–5% of hospitalized cases. It is called boutonneuse fever, Mediterranean spotted fever, Israeli tick typhus, Astrakhan spotted fever, Kenya tick typhus, Indian tick typhus, or other names that designate the locality of occurrence. It is normally transmitted by the bite of the tick, but may also be acquired through the skin or conjunctiva when the ticks are crushed. After a period of apparent low incidence, there was a marked rise in number of cases diagnosed during the 1970s.

R. conorii is the most geographically dispersed rickettsia of the spotted fever group. It has been recognized in most of the regions bordering on the Mediterranean Sea and Black Sea, Israel, Kenya and other parts of North, Central, and South Africa, and India. The brown dog tick, *Rhipicephalus sanguineus* is the prevailing vector. The involvement of rabbits in maintaining an infected tick population was suggested by association of the disease with large rabbit populations on endemic islands and by the drop in human disease following the myxomatosis epizootic that reduced the rabbit population in Europe (Hoogstraal, 1967, 1981).

The genome of *R. conorii* contains 1,268,755 bp (Ogata et al., 2001). No resident plasmid or bacteriophage has been identified in *R. conorii*. The electrophoretic profiles of the soluble proteins are similar but not identical to those of *R. rickettsii* and *R. sibirica* (Anacker et al., 1980).

The mol% G + C of the DNA is: 32.4 (determined from the genome sequence).

Type strain: Malish 7.

GenBank accession number (16S rRNA): AE008647 (genome segment), L36105 (Moroccan strain).

The type strain is a South African patient strain, isolated by J.H. Gear in 1946. This strain and the Moroccan strain (ATCC VR-141) are used most frequently in laboratory studies.

9. **Rickettsia felis** Bouyer, Stenos, Crocquet-Valdes, Moron, Popov, Zavala-Velazquez, Foil, Stothard, Azad and Walker 2001, 346[VP]

fe′ lis. L. gen. n. *felis* of the cat; named after the cat flea, *Ctenocephalides felis*, in which the organism was first observed by electron microscopy.

In 1990 during a study investigating potential vectors for *Neorickettsia risticii*, rickettsia-like organisms were observed in adult cat fleas obtained from EL Labs. Thus, the organism was originally designated as the ELB agent. The organisms are 0.25–0.45 µm × 1.5 µm and are found in the midgut, tracheal matrix, muscles and reproductive tissues of the fleas. The organism contains a trilaminar cell wall that is characteristic of rickettsiae with a well-defined inner cell membrane and outer membrane. The dimensions of the microcapsular layer, outer and inner leaflets of the outer membrane, and the periplasmic space strongly resemble those in other *Rickettsia* species.

Phylogenetic analysis of the 17 kDa antigen gene, *ompA*, and *ompB* and other genes places *R. felis* as a member of the SFG.

R. felis has been cultivated in the XTC-2 cell line derived from *Xenopus laevis*. *R. felis* grows most rapidly at 28°C in XTC-2 cells, which die at a temperature of 32°C. The organism also grows to a lesser extent in Vero cells at 28°C or 32°C. L-929, and MRC-5 cell lines are unable to support the permanent growth of *R. felis*. Electron microscopy shows that *R. felis* is free in the cytoplasm but not in the nucleus of the cells.

The organism is maintained in the flea by transovarial transmission. Horizontal transmission or the acquisition of *R. felis* by fleas feeding on cats or artificially infected meals has not been demonstrated. *R. felis* has been associated with opossums and their fleas in Texas and California. Serological evidence and PCR detection indicate that *R. felis* infection occurs in humans in the United States, Mexico, Brazil, and Germany. The organism has also been detected in cat fleas from Spain and Ethiopia; thus, *R. felis* is likely distributed worldwide.

The mol% G + C of the DNA is: not determined.

Type strain: CNCM I-2363, Marseille-URRWFXCal2 (Reference Center for Rickettsioses, Marseille, France).

10. **Rickettsia helvetica** Beati, Peter, Burgdorfer, Aeschlimann and Raoult 1993, 524[VP]

hel.ve' ti.ca. N.L. adj. *helvetica* pertaining to Helvetica, the Latin name of Switzerland where the organism was originally isolated.

R. helvetica, formerly called the Swiss agent, was first isolated from *Ixodes ricinus* ticks from Switzerland (Burgdorfer et al., 1979). Recently, *R. helvetica* has been isolated from *Ixodes ricinus* ticks from France (Parola et al., 1998). Rickettsial isolates from *Ixodes ovatus*, *I. persulcatus*, and *I. monospinosus* from Japan are identical or closely related to *R. helvetica* (Fournier et al., 2002). In serological tests, *R. helvetica* antigens are weakly cross-reactive with antisera to other SFG rickettsiae.

R. helvetica grows well in yolk sacs of embryonated eggs, chicken embryo fibroblasts, L929, and Vero cells. Human embryonic lung fibroblasts do not support the growth of *R. helvetica* in primarily isolation. *R. helvetica* does not cause a cytopathic effect in Vero cells, and it can be continuously subcultured by trypsinization of the monolayers. Unlike other SFG rickettsiae, *R. helvetica* does not stimulate actin tail formation when it grows inside the host cell. *R. helvetica* exhibits moderate pathogenicity in meadow voles; however, rabbits, guinea pigs, and mice are not susceptible (Fournier et al., 2000). A serological survey suggests that a portion of tick-exposed humans have antibodies reactive with *R. helvetica* in France (Nilsson et al., 1999). However, the pathogen has not been isolated from ill humans.

The mol% G + C of the DNA is: not determined.

Type strain: C3 (Reference Center for Rickettsioses, Marseille, France), ATCC VR-1375.

GenBank accession number (16S rRNA): L36214.

11. **Rickettsia honei** Stenos, Roux, Walker and Raoult 1998, 1403[VP]

ho.ne.i. L. gen. n. *honei* of Hone, named after Frank Sandland Hone, an early pioneer in Australian rickettsiology.

An SFG rickettsiosis-like ailment was identified in 1991 on Flinders Island of Australia and was named Flinders Island spotted fever (FISF). The causative agent of FISF was isolated from buffy coat preparations from the blood of two patients (Baird et al., 1992) and named *R. honei* (Stenos et al., 1998).

R. honei and the Thai tick typhus rickettsia (TT-118) are identical in DNA sequences of the gene encoding the 17 kDa protein, *gltA*, and *ompA*, and they have only 1 nucleotide difference in the 16S rRNA gene. Thus, *R. honei* and TT-118 are considered a single species (Stenos et al., 1998).

TT-118 was isolated from a pool of immature *Rhipicephalus* and *Ixodes* of unknown species collected in Thailand (Robertson and Wisseman, 1973). A *Rickettsia* identical to TT-118 has recently been detected by PCR from an *Ixodes granulatus* tick from Thailand (Kollars et al., 2001). Migratory birds have been postulated to spread *R. honei* widely. A rickettsia genetically identical to *R. honei* has been detected by PCR in *Aponoma hydrosauri* ticks from Flinders Island and Tasmania. This tick is a reptile-associated tick, but it also bites humans (Whitworth et al., 2003).

The mol% G + C of the DNA is: not determined.

Type strain: RB, ATCC VR-1472.

GenBank accession number (16S rRNA): U17645, AF060705.

The type strain of *R. honei* was isolated from buffy coat of a patient's blood by inoculation of buffalo green monkey kidney cells.

12. **Rickettsia japonica** Uchida, Uchiyama, Kumano and Walker 1992, 303[VP]

ja.po' ni.ca. N.L. adj. *japonica* pertaining to Japan, the country from which the first isolates were identified.

The dimensions of the organisms are 0.4–0.5 × 0.8–1.5 μm. Unlike some other SFG rickettsiae, *R. japonica* can persistently infect Vero cells. *R. japonica* does not cause a cytopathic effect in Vero cells but does cause cytopathic effects in L929, BHK21/13, and primary chicken embryo fibroblast cells. The rickettsiae grow primarily in cytoplasm and rarely in nuclei. The plaque morphology produced by *R. japonica* differs from that produced by most other SFG rickettsiae except *R. rickettsii*. *R. japonica* forms target like plaques (a dye-stained inner focus of viable cells surrounded by a clear zone of necrotic cells) after 9–11 days of incubation in Vero cells. Other SFG rickettsiae form clear plaques. A variant that forms clear plaques was isolated from a guinea pig inoculated with *R. japonica*. About 0.01 to 0.001% of cells carrying strain YH produced clear plaques on Vero cells (Uchida et al., 1992).

The organisms are pathogenic for guinea pigs. Guinea pigs develop fever and scrotal swelling. C3H/He strain of inbred mice and conventional strain ddY mice are resistant to *R. japonica* infection. Chicken embryos die 5–7 days after yolk sac inoculation. Like other SFG rickettsiae, *R. japonica* does not exhibit hemolytic activity.

R. japonica is presumably transmitted by ticks. It has been cultivated from *Dermacentor taiwanensis* and *Haemaphysalis flava* (Fournier et al., 2002).

The disease caused by *R. japonica* infection is Japanese spotted fever. The disease was first found in Japan in 1984. The disease is mainly distributed along the coast of southwestern and central Japan where the climate is warm, and most cases occur between April and October. The clinical features of the disease include fever (100%), headache (80%), shaking chills (87%), skin eruption (100%), and tick bite eschars (90%) (Mahara, 1997). Serologic evidence indicates that disease might occur in eastern Asia and adjacent islands (Feng et al., 1991; Camer et al., 2000). However, the antibodies against *R. japonica* might have been stimulated by other SFG rickettsiae in these areas.

"*R. heilongjiangii*", a rickettsial isolate from both ticks and humans in northeastern China, is closely related to *R. japonica*. The 16S rRNA gene sequence of "*R. heilongjiangii*" has 99.6% similarity to that of *R. japonica* (Zhang et al., 2000b). Further studies need to be performed to determine whether "*R. heilongjiangii*" should be designated a strain of *R. japonica*.

The mol% G + C of the DNA is: 31.2 (T_m).

Type strain: YH; ATCC VR-1363.

GenBank accession number (16S rRNA): L36213 (strain YM).

The YH strain was isolated from the blood of a patient in Japan by Uchida et al. (1985).

13. **Rickettsia massiliae** Beati and Raoult 1993, 839[VP]

mas.si' li.ae. L. gen. n. *massiliae* from Massilia, the Latin name of Marseille, where the organism was first isolated.

The dimensions of the organisms are 0.3–0.4 × 0.6–1 μm. Growth occurs in the cytoplasm of eucaryotic cells. *R. massiliae* can be cultivated in L929, Vero, or human embryonic lung fibroblast cells. *R. massiliae* has been isolated from *Rhipicephalus turanicus* and *R. sanguineus* ticks in France. *R. massiliae* has also been detected in ticks from Greece, Portugal, Spain, and the Central African Republic where *Rhipicephalus lunulatus*, *R. mushamae*, and *R. sulcatus* are hosts (Beati and Raoult, 1993; Babalis et al., 1994; Beati et al., 1996). Pathogenicity of *R. massiliae* for humans and animals has not been demonstrated.

The mol% G + C of the DNA is: not determined.

Type strain: Mtu1, ATCC VR-1376.

GenBank accession number (16S rRNA): L36214.

Mtu1 was isolated from the hemolymph of a *Rhipicephalus turanicus* tick collected in Marseille, France (Reference Center for Rickettsioses, Marseille, France).

14. **Rickettsia montanensis** corrig. (ex Lackman, Bell, Stoenner and Pickens 1965) *Rickettsia montana* (sic) Weiss and Moulder 1984b, 356[VP] (Effective publication: Weiss and Moulder 1984a, 697.)

mon.tan' en.sis. M.L. gen. n. *montanensis* from Montana, the state where the organism was first isolated.

Resembles other members of the spotted fever group in its growth characteristics in the chicken embryo and antigenic composition, but it is avirulent for the guinea pig and mouse. Although antibodies reactive with this species have been detected in dogs, *R. montanensis* is not associated with signs or symptoms of spotted fever in humans or experimentally inoculated dogs. It is distinguished from other rickettsiae by DNA sequence analysis and immunologic testing (Philip et al., 1978). *R. montanensis* was first isolated in eastern Montana, repeatedly from *D. variabilis*, occasionally from *D. andersoni*. It has been also isolated with notable frequency from rodents (genera *Microtus* and *Peromyscus*) and from *D. variabilis* in various parts of the U.S. (Lackman et al., 1965; Philip et al., 1978; Linnemann et al., 1980).

The mol% G + C of the DNA is: not determined.

Type strain: ATCC VR-611.

GenBank accession number (16S rRNA): L36099.

ATCC VR-611 is the tick strain of Bell et al. (1963).

15. **Rickettsia parkeri** Lackman, Bell, Stoenner and Pickens 1965, 137[AL]

par' ke.ri. M.L. gen. n. *parkeri* of Parker; named after Ralph R. Parker, a founder of the Rocky Mountain Laboratory.

This species is cultivated in eggs by the same procedure used for the other species of the spotted fever group. *R. parkeri* is identified by DNA sequence analysis or immunologic evaluation. *R. parkeri* produces a nonfatal disease in the guinea pig that is characterized by fever and reddening of the scrotum; the most abundant growth takes place in the testicular tissue (Parker et al., 1939). Human infection has not been reported. The species has been isolated from *Amblyomma maculatum* ticks collected from domestic animals in Texas, Georgia and Mississippi (Lackman et al., 1949).

The mol% G + C of the DNA is: not determined.

Type strain: NIAID maculatum 20.

GenBank accession number (16S rRNA): U12461.

The type strain was isolated in 1948 from *A. maculatum* ticks collected from sheep in Mississippi (Bell and Pickens, 1953).

16. **Rickettsia peacockii** Niebylski, Schrumpf, Burgdorfer, Fischer, Gage and Schwann 1997b, 451[VP]

pea.cock' i.i. M.L. gen. n. *peacockii* of Peacock, named after M.G. Peacock, a well-respected rickettsiologist.

R. peacockii is a endosymbiont of wood ticks (*Dermacentor andersoni*) collected from the Bitterroot Valley in the western Montana. *R. peacockii* is transstadially and transovarially transmitted in ticks. Massive infections of *R. peacockii* are observed only in *D. andersoni* ovarian tissues. The organisms are not found in tick hemocytes, salivary glands, midgut, Malpighian tubules, or hypodermal tissues. Thus, *R. peacockii* is likely unable to be transmitted to mammals.

The 16S rRNA gene sequence is most similar to *R. rickettsii* R strain and *R. slovaca* (99.7% for both species). The sequence of the *ompA* gene is most similar to *R. slovaca* (93.0%). The *ompA* gene of *R. peacockii* has a single nucleotide deletion at position 334. The deletion results in a prematurely truncated OmpA of *R. peacockii* with 110 amino acids only. Thus, presumably *R. peacockii* does not have a functional OmpA (Niebylski et al., 1997b).

R. peacockii is postulated to interfere with the stable maintenance of virulent rickettsiae in nature (Burgdorfer et al., 1981) because the distribution of *R. peacockii* coincides with a low incidence of Rocky Mountain Spotted Fever (RMSF) on the east side of the Bitterroot Valley, where early cases of RMSF were recognized on the west side of the valley. *R. peacockii* infects approximately 70% of wood ticks from the east side but is uncommon in the ticks from the west side of the valley. A plausible explanation for the *R. peacockii* interference is that infections of *R. peacockii* in tick ovarian tissues might inhibit transovarial transmission of virulent *R. rickettsii* (Niebylski et al., 1997b).

The mol% G + C of the DNA is: not determined.

Type strain: Skalkaho.

GenBank accession number (16S rRNA): U55820.

This strain was characterized directly from an infected *D. andersoni* tick (SK-594) that was used to generate the 16S rRNA gene sequence and *ompA* gene sequence. The organism has been recently co-cultured from the *D. andersoni* cells (Simser et al., 2001).

17. **Rickettsia rhipicephali** (ex Burgdorfer, Brinton, Krynski and Philip 1978) Weiss and Moulder 1988, 221[VP] (Effective publication: Weiss and Moulder 1984a, 698.)

rhi.pi.ce'pha.li. M.L. gen. n. *rhipicephali* of rhipicephalus; named after its natural tick host *Rhipicephalus sanguineus.*

Although this species is isolated with difficulty in the chicken embryo and the guinea pig, it is readily recovered in male meadow voles (*Microtus pennsylvanicus*), in which it produces massive infection in the tissues of the tunica vaginalis. This species can also be cultivated on monolayers of chick embryo fibroblasts, Vero, and mouse L cells. Although growth is profuse, damage to the cells is limited, and plaques are small and turbid, reflecting a mixture of viable and dead host cells. There is no evidence that *R. rhipicephali* is pathogenic for humans. A serosurvey of acute and convalescent serum samples from 80 dogs in which Rocky Mountain spotted fever had been considered as a differential diagnosis identified one dog with a fourfold increase in antibody titer to *R. rhipicephali* (Breitschwerdt et al., 1995). *R. rhipicephali* has been detected in about 19% of the brown dog ticks (*R. sanguineus*) removed from dogs in central and northern Mississippi and from ticks collected in Texas and North Carolina (Burgdorfer and Brinton, 1975; Burgdorfer et al., 1978). It has also been isolated with high frequency from *Dermacentor andersoni* ticks collected in western Montana (Philip and Casper, 1981). The genome size of *R. rhipicephali* is 1252–1258 kb as determined by pulsed-field gel electrophoresis (Roux et al., 1992).

The mol% G + C of the DNA is: 32.2 (T_m) (Anacker et al., 1980).

Type strain: Burgdorfer 3-7-female 6.

GenBank accession number (16S rRNA): L36216.

This strain was isolated in 1973 in Mississippi by Burgdorfer et al. (Burgdorfer et al., 1975).

18. **Rickettsia rickettsii** (Wolbach 1919) Brumpt 1922, 757[AL] (*Dermacentroxenus rickettsii* Wolbach 1919, 87.)

rick.ett'si.i. M.L. gen. *rickettsii* of Ricketts; named after Howard Taylor Ricketts for his classic studies of the etiology of Rocky Mountain spotted fever.

R. rickettsii is the most widely studied species of the spotted fever group. The cells are slightly smaller and more uniform in size than those of *R. prowazekii.*

Because of the high virulence of *R. rickettsii* for chicken embryos, the embryos die before extensive growth has taken place, and the procedure for optimal harvest from yolk sacs differs considerably from the one employed for the typhus group (Table BXII.α.31). It is performed as follows (Stoenner et al., 1962). Embryos, 4–5 d old, are inoculated with a sufficient number of viable cells to kill most embryos within 4–5 d. The eggs are incubated at 33.5°C (lowest temperature compatible with survival of most embryos) and maintained in an incubator at 32°C for 2 d after death of the embryos. Even under the best conditions, yields of *R. rickettsii* cells from yolk sac are smaller than those of *R. prowazekii* or *R. typhi*, but they are somewhat higher than yields obtainable from cell cultures.

Multiplication occurs primarily in the cytoplasm, but intranuclear growth is sufficiently prominent to have stimulated early investigators to use it as a criterion for the classification of the spotted fever group. Because of cell-to-cell spread via actin-based mobility and greater cytopathogenicity, plaques on monolayers occur earlier and are larger than those formed by the typhus group; however, sheep or rabbit erythrocytes are not hemolyzed (Table BXII.α.31).

Strains of this species vary considerably in virulence for the guinea pig. Virulence for humans and guinea pigs appears to vary independently.

The guinea pig is highly susceptible to infection. The more virulent strains induce fever and scrotal necrosis, and the infection is often fatal. In humans, *R. rickettsii* is the causative agent of Rocky Mountain spotted fever, the most severe disease of the spotted fever group. The disease is characterized by high fever and widespread damage to the small blood vessels; this damage results in a skin rash, increased vascular permeability associated with non-cardiogenic pulmonary edema and vascular lesions in various organs including encephalitis. Before the era of the tetracycline and chloramphenicol antibiotics, the case fatality rates were as high as 90% in the Bitterroot Valley of Montana and as low as 5% in the Snake River Valley of Idaho and 25% in Long Island. Because the disease responds well to tetracyclines, present mortality rates are relatively low, provided the disease is recognized and treated promptly. Dogs are susceptible to natural infection and develop clinical signs (Lissman and Benach, 1980). The mouse is quite resistant to infection, and doses in excess of 10^6 viable cells are required to produce a significant antibody response.

The metabolism of *R. rickettsii* is similar to that of other rickettsiae. Biochemical investigations have focused on loss and restoration of viability, and metabolic activity is best maintained under microaerophilic conditions or in the presence of reduced glutathione or protein. Reversible changes in rickettsial activity occur in the tick: prolonged refrigeration reduces the virulence of the rickettsiae for the guinea pig, but virulence is restored when the ticks have a blood meal or are incubated at 37°C for 24–48 h. Spencer and Parker (1923) postulated that virulence of *R. rickettsii* in the tick vector is linked directly to the physiological state of the tick and defined this phenomenon as "reactivation." The reactivation may result from growth of rickettsiae or differential expression of rickettsial virulence factors at the elevated temperature or after stimulation by components of the blood meal. *R. rickettsii* numbers increase 100-fold in the hemolymph of partially engorged ticks compared to unfed infected ticks (Wike and Burgdorfer, 1972). Differential expression of rickettsial proteins and ultrastructural changes has been confirmed by immunoblot and electron microscopy. Electron microscopy reveals that reactivated *R. rickettsii* in ticks incubated at 37°C or in ticks fed on animals has a discrete microcapsular layer and a discrete electron-lucent slime layer outside the microcapsular layer. In starved ticks, the microcapsular layer and the slime layer of rickettsiae are inconspicuous or ragged (Hayes and Burgdorfer, 1982). Immunoblot analysis indicates that rickettsial proteins of 42, 43, 48, 75 and 100 kDa are induced in a tick cell line when the incubation temperature is raised from 28°C to 34°C (Policastro et al., 1997).

In a series of brilliant experiments conducted in 1906 and 1907, Ricketts (1911) clearly established the basic features of the ecology of the agent of Rocky Mountain spotted fever in the tick. *R. rickettsii* is confined to the Western Hemisphere (Table BXII.α.31), but at present, it is encountered much more frequently in the southeastern and central United States, particularly in North Carolina and Oklahoma, than in the Rocky Mountains. In the western

United States, the most common human vector is the wood tick, *Dermacentor andersoni*. In the eastern two thirds of the United States and parts of the Far West, the American dog tick, *Dermacentor variabilis*, is the chief vector. The brown dog tick, *Rhipicephalus sanguineus*, and the cayenne tick *Amblyomma cajennense* have been implicated in transmission of human infections in Mexico and South America.

Although there is considerable serological evidence of widespread exposure to SFG rickettsiae among wild vertebrates and domestic dogs, demonstration of infection by recovery of the microorganisms has been difficult, presumably because sufficient numbers of rickettsiae are present in the vertebrate host only during brief periods. Natural infection among vertebrates was first demonstrated in the 1930s in Brazil in domestic and wild dogs (*Canis brasiliensis*), the opossum (*Didelphis marsupialis*), the wild rabbit (*Sylvilagus minensis*), and the Brazilian capybara (*Cavia aperea*) (Moreira and de Magalhaes, 1937). Recovery of *R. rickettsii* has been reported in the United States (Gould and Miesse, 1954) from the meadow vole (*Microtus pennsylvanicus*), opossum, cotton rat (*Sigmodon hispidus*), cottontail rabbit (*Sylvilagus floridanus*), whitefooted mouse (*Peromyscus* sp.) and pine vole (*Pitymis pinetorum*) (Bozeman et al., 1967). In the western United States, *R. rickettsii* has been isolated from the chipmunk (*Eutamias amoenus*), snowshoe hare (*Lepus americanus*) and the golden-mantled ground squirrel (*Citellus lateratis tescorum*) (Burgdorfer et al., 1962). Rickettsemic cotton rats have been demonstrated experimentally as a source of *R. rickettsii* for *D. variabilis* ticks, and cotton rats are susceptible to the infection when transmitted by tick bite.

Transmission to man occurs through the bite of an infected tick that remains attached to the skin for several hours. Although man is only an incidental host, human infection reflects a changing ecology. From 1910 to 1930 most of the cases, about 100–600 per year, were diagnosed within the area of distribution of *D. andersoni* in the Rocky Mountain region. Since 1930, there has been a decrease in incidence in the Rocky Mountain region along with increased recognition of the disease in the southeastern parts of the United States associated with transmission by *D. variabilis*. From 1948 to 1959, the number of cases decreased from an annual rate of 500 to an annual rate of 200. After 1959, the disease increased to an annual rate greater than 1000 cases by 1977. A decrease in reported cases occurred during the 1980s and early 1990s, but there were hints of an increase in the late 1990s. A strain of *R. rickettsii* has been isolated from patients in Costa Rica (Fuentes, 1979).

The genome is 1,257,710 bp in size and contains 1486 open reading frames and 33 tRNA genes (Eremeeva et al., 2002). The proteins of various strains of *R. rickettsii* have shown no differences except for minor epitope differences of the HLP strain, an isolate that has been identified only in ticks (Anacker et al., 1980).

The mol% G + C of the DNA is: 32.4 (determined from the genome sequence).

Type strain: ATCC VR-149.

GenBank accession number (16S rRNA): L36217.

Strain Sheila Smith was isolated from a patient in Montana (Bell and Pickens, 1953). Reference strain: strain R, was isolated from *Dermacentor andersoni*.

19. **Rickettsia sibirica** Zdrodovskii 1949, 20[AL]

si.bi′ri.ca. M.L. fem. adj. *sibirica* pertaining to Siberia.

This species closely resembles *R. rickettsii*, although most strains are less virulent for animals and humans. Cinematographic observations by Kokorin et al. (1978) have revealed intense mobility of the rickettsiae within their host cells, and some movement from cell to cell, but only limited cytotoxicity occurs. The species is cultivated by procedures identical to those used for *R. rickettsii*. Virulence for the guinea pig is quite variable, and in some cases, only fever of short duration is produced. An occasional strain kills mice and hamsters (Bazlikova and Brezina, 1978), and inbred strains of mice vary in susceptibility to this organism (Kekcheeva et al., 1978). The disease in humans resembles moderately severe or mild Rocky Mountain spotted fever. It is called Siberian or North Asian tick typhus.

This species is differentiated from the other strains of the spotted fever group by DNA sequence analysis of *ompA*, *ompB*, the 17 kDa lipoprotein gene, the citrate synthase gene, or the 120 kDa cytoplasmic antigen gene. Differentiation can also be achieved by immunologic analysis using IFA titration with sera prepared according to a particular protocol in mice (Philip et al., 1978). Toxicity neutralization tests in mice demonstrate the close relationship of *R. sibirica* with *R. rickettsii* (Lackman et al., 1965).

The habitat of *R. sibirica* consists of foci extending from the Pacific Coast of Russia and China in the east to Armenia in the west and from Siberia in the north to northern China in the south. The foci are usually associated with steppe landscapes with low rainfall close to foothills and mountain ranges, and they may extend to the dry slopes of the mountains. *R. sibirica* has been detected in ticks of the following genera: *Demacentor* (*D. nuttalli*, *D. marginatus*, and *D. silvarum*), *Haemaphysalis* (*Ha. punctata*, *Ha. concinna*, and *Ha. japonica*), *Hyalomma* (*Hy. dromedarii*, *Hy. asiaticum*, *Hy. detritum*, *Hy. anatolicum*, *Hy. plumbeum*, *Hy. marginatus*), *Ixodes* (*I. ricinus*, *I. persulcatus*, *I. apronophorus*, and *I. plumbeus*), and *Rhipicephalus* (*R. sanguineus*, *R. turanicus*, and *R. schulzei*). These ticks feed on birds, numerous small wild rodents, and domestic animals. Antibodies to *R. sibirica* have been detected from at least 18 kinds of mammals, including Siberian squirrels or susliks (genus *Citellus*), chipmunks (*Eutamias*), hamsters (*Cricetus*), lemmings (*Lagurus*), hares (*Lepus*), domestic and field mice, and voles (Hoogstraal et al., 1967). *R. sibirica* has been isolated from wild rodents and birds but not from domestic animals. Rodents are the main seasonal reservoir of *R. sibirica* in its natural focus, and birds may serve as the vehicle to establish new foci over long distances. Humans are infected through the bite of an infected tick.

The genome size is 1,255,665 bp in size and contains 1316 open reading frames and 33 tRNA genes (Eremeeva et al., 2002). The electrophoretic migration patterns in polyacrylamide gel of the solubilized proteins are similar but not identical to those of *R. rickettsii* (Pedersen and Walters, 1978).

The mol% G + C of the DNA is: 32.4 (determined from the genome sequence).

Type strain: 246, ATCC VR-151.

GenBank accession number (16S rRNA): L36218.

Strain 246 was isolated from *Dermacentor nuttalli* in the U.S.S.R. about 1949 (Bell and Stoenner, 1960).

20. **Rickettsia slovaca** Sekeyova, Roux, Xu, Rehácek and Raoult 1998, 1458[VP]

slo.va.ca′. L. gen. n. *slovaca* from Slovakia, the country where the organism was first isolated.

The dimensions of the organisms are $0.37–0.45 \times 0.8–1.2$ μm. *R. slovaca* can be propagated in yolk sacs of chicken embryos, HEL, Vero, L929, and tick cells. No cytopathologic effect is observed in any of the infected cell lines. Electron microscopy reveals that the organisms are free in the host cell cytoplasm and are surrounded by an electron-lucent zone. The organism has very low pathogenicity for mice and guinea pigs. The only observed change in inoculated mice is a slightly enlarged spleen. Guinea pigs develop brief and mild fever after inoculation of the organisms. *R. slovaca* causes asymptomatic infection in *Lepus europaeus* and hamsters. *R. slovaca* has been associated with human infections that are characterized as usually afebrile with an eschar and regional lymphadenopathy after the bite of a *D. marginatus* or *D. reticulatum* tick. *R. slovaca* was isolated from a pool of male and female *Dermacentor marginatus* ticks collected in Slovakia. Rickettsial strains closely related to *R. slovaca* have been isolated from Armenia, Crimea, France, Portugal, Switzerland, Ukraine, and Yugoslavia, in the geographic distribution of *D. marginatus* and *D. reticulatum* (Sekeyova et al., 1998).

The mol% G + C of the DNA is: not determined.

Type strain: B (Reference Center for Rickettsioses, Marseille, France).

21. **Rickettsia typhi** (Wolbach and Todd 1920) Philip 1943, 304[AL] (*Dermacentroxenus typhi* (Wolbach and Todd 1920) *Rickettsia mooseri* Monteiro 1931, 97.)

ty′phi. Gr. n. *typhus* cloud, hence stupor associated with rickettsial encephalitis; M.L. n. *typhus* fever, typhus; M.L. gen. n. *typhi* of typhus.

R. typhi is the valid name for the species, appearing on the Approved Lists of Bacterial Names (Skerman et al., 1980). However, the species has also been called "*R. mooseri*" in honor of Herman Mooser, who clearly differentiated the species from *R. prowazekii* based on its virulence for the guinea pig (Mooser, 1928).

R. typhi is more virulent than *R. prowazekii* for guinea pigs and mice (Table BXII.α.31). Unlike *R. prowazekii*, it can be passed indefinitely in the rat and may persist for months in the rat brain. Rats (*Rattus norvegicus* and *R. rattus*, especially those located in urban areas) and other rodents are the primary reservoirs. *R. typhi* has a worldwide distribution and has been reported from nearly all countries where investigators have competently searched for it. The rat louse, *Polyplax spinulosus*, and the rat flea, *Xenopsylla cheopis*, are the chief transmitters of the organism from rat to rat. The human flea, *Pulex irritans*, and the human body louse are highly susceptible to infection and may play roles in transmission in populations with high ectoparasitic infestation. The rat-rat flea cycle has been interrupted in the United States by control measures, including killing of fleas with DDT. Currently the principal *R. typhi* maintenance cycle in the United States involves the cat flea (*Ctenocephalides felis*) and opossums (*Didelphis virginianus*), and the cat flea is the apparent vector for human infections.

The human disease is a mild form of typhus and is called murine or endemic typhus, but it is also known by names that reflect the geographic location. "Tabardillo" in Mexico might be murine typhus transmitted by lice. Murine typhus is reported with variable frequency, undoubtedly largely because of DDT interruption of the transmission by fleas; for instance, in the United States, a peak of 5400 cases occurred in 1944, but the incidence was less than 100 cases in 1958. Many cases are not properly documented (Traub et al., 1978), due in part to effective treatment of the disease with antibiotics.

Some of the antigens of *R. typhi* demonstrable in serological tests are very similar, if not identical, to those of *R. prowazekii*, while others are species-specific. The antigens containing common epitopes are numerous and include LPS, OmpB, the 17kDa proteins, GroES and GroEL. OmpB-containing species-specific epitopes as well as 10–15% of the total cellular protein are released as a soluble fraction when the cells are suspended in a hypotonic solution that lacks Mg^{2+} and is incubated at 45°C for 20 min (Dasch, 1981). The two species can also be distinguished immunologically by species-specific monoclonal antibodies; cross-absorption of defined antisera and subsequent IFA titration or immunoblotting; cross-challenge of guinea pigs vaccinated with formalin-inactivated antigens; or mouse toxicity neutralization tests.

There have been no extensive attempts to isolate mutant strains of *R. typhi* in the laboratory. Differences between *R. typhi* and *R. prowazekii* in cell morphology and in the mechanism of interaction with eucaryotic cells undoubtedly exist, such as the relatively ineffective actin-based mobility of *R. typhi* as compared to absence of actin-based mobility of *R. prowazekii*; however, objective criteria of differentiation based on such characteristics have not been developed. Differences in metabolic activities have not been described, and the two species have been used interchangeably for metabolic investigations.

The mol% G + C of the DNA of *R. typhi* is similar to that of *R. prowazekii*. The genome size of *R. typhi* is $1,133 \pm 44$ kb (Eremeeva et al., 1993). Although no differences in the degree of DNA–DNA hybridization have been found between strains of *R. typhi*, the degree of hybridization between *R. typhi* and *R. prowazekii* is 70–79%, (Myers and Wisseman, 1980). Similarly, no differences in electrophoretic protein migration patterns have been noted between strains of *R. typhi*, but consistent differences in the migration patterns of malate dehydrogenase and of several unidentified proteins have been found between *R. typhi* and *R. prowazekii* (Dasch et al., 1978).

The mol% G + C of the DNA is: 29 (T_m).

Type strain: Wilmington, ATCC VR-144.

GenBank accession number (16S rRNA): M21789.

Strain ATCC VR-144 is that of Maxcy (1929).

Other Organisms

1. *"Rickettsia amblyommii"*

A rickettsial agent that was first described by Burgdorfer et al. (1974) was later characterized as a new SFG rickettsia and referred to as WB-8-2 (Burgdorfer et al., 1981). WB-82-2 was isolated from an *Amblyomma americanum* tick collected from vegetation in Tennessee (Burgdorfer et al., 1974). The organism invades all tick tissues but, in general, produces a mild to moderate infection except in the ovary where it may be abundant and from where it is passed via eggs to as many as 100% of progeny. The organism grows well in avian cell cultures but very poorly in embryonated hen's eggs, which die 5–7 days after inoculation. WB-8-2 is nonpathogenic for guinea pigs. In male meadow voles (*Microtus pennsylvanicus*), it produces microscopically detectable mild and transient infections in the tunica vaginalis only after inoculation of heavily infected tissue culture suspensions. Epidemiological evidence suggests that WB-8-2 is nonpathogenic for humans (Burgdorfer et al., 1981). Weller et al. (1998) isolated another strain of this rickettsia (MOAa) from an adult female *A. americanum* tick collected in Bolling County, Missouri. The MOAa organism is highly cytopathogenic for RAE25 cells, a tick cell line. Phylogenetic analysis indicates that strains MOAa and WB-8-2 are closely related. They represent different strains of "*Rickettsia amblyommii*". Stains MOAa and WB-8-2 are most closely related to *R. montanensis* (Weller et al., 1998).

2. "*Rickettsia cooleyi*"

A rickettsial agent exists in *Ixodes scapularis* ticks collected from Anderson County in eastern Texas. The organism has not yet been cultivated. It was identified by amplification of rickettsial genes by the polymerase chain reaction. DNA sequencing showed the highest nucleotide sequence similarity with *R. australis* for the 17-kDa protein gene, with *R. helvetica* for *gltA*, and with *R. montanensis* for *ompA*. The organism was provisionally designated as the Cooleyi agent after Dr. Robert A. Cooley, who served the Montana State Board of Entomology from 1899 to 1944 (Billings et al., 1998). Weller et al. (1998) have also sequenced *ompA* of a symbiont in ovarian tissues of *I. scapularis*. The *ompA* sequences of the *I. scapularis* symbiont and "*Rickettsia cooleyi*" are identical.

Species Incertae Sedis

1. "**Rickettsia monacensis**" Simser, Palmer, Fingerle, Wilske, Kurtti and Munderloh 2002.

    mo.na.cen'sis. M.L. n. *Monacum* Munich, a German city; M.L. adj. *monacensis* from/of Munich.

The name has been proposed for a rickettsial agent that exists in *Ixodes ricinus* collected in the English Garden in Munich, Germany (Simser et al., 2002). The cell size is 1.0–1.5 × 0.3–0.4 μm. The organisms resemble SFG rickettsiae in their ultrastructure, and they reside free in the cytoplasm and occasionally within the nuclei of host cells. The slime layer is thinner (<30 nm) than that of *R. rickettsii* (30–60 nm). This rickettsia was originally isolated by using a tick cell line (ISE6) and subsequently cultivated in L-929 cells. The organisms are cytopathic for ISE6 cells and cause cell lysis after the third passage. The rickettsia induces polymerization of host actin in both tick and mammalian cells. The immunodominant proteins are OmpA and OmpB. Phylogenetically the organism is distantly related to other SFG rickettsiae. Its pathogenicity for animals and humans is unknown.

The mol% G + C of the DNA is: not determined.

Deposited strain: Type strain: IrR/Munich.

Strain IrR/Munich was isolated from a female *I. ricinus* tick (deposited by U. Munderloh at the Rickettsial and Ehrlichial Diseases Research Laboratory, University of Texas Medical Branch, Galveston, Texas, USA).

Genus II. **Orientia** *Tamura, Ohashi, Urakami and Miyamura 1995, 590*[VP]

XUE-JIE YU AND DAVID H. WALKER

O'ri.en'ti.a. M.L. fem. n. *Orientia* pertaining to the Orient, the area where the organisms are widely distributed.

Short rods 0.5–0.8 × 1.2–3.0 μm. **Obligately intracellular. Grow free in the cytosol of the host cell**; rarely invade the nucleus. Bacterial cells are released surrounded by host cell membrane. **Cell walls lack lipopolysaccharide and peptidoglycan.** The outer leaflet of the cell wall is considerably thicker than the inner leaflet, unlike the wall of rickettsiae. The genome of *Orientia* is 1.1–1.5 × 10⁹ Da, larger than that of *Rickettsia*. Maintained by transovarian transmission in its various **trombiculid mite hosts. Etiological agents of scrub typhus in humans.** Transmitted to humans by mite larvae. Found in Asia and Australia.

The mol% G + C of the DNA is: 28.1–30.5 (HPLC).

Type species: **Orientia tsutsugamushi** (Hayashi 1920) Tamura, Ohashi, Urakami and Miyamura 1995, 590 (*Rickettsia tsutsugamushi* (Hayashi 1920) Ogata 1931, 252; "*Theileria tsutsugamushi*" Hayashi 1920, 63.)

FURTHER DESCRIPTIVE INFORMATION

16S rRNA gene analysis indicates that the organisms are separate from the *Rickettsia* species, with a similarity value of 90.2–90.6%. This genus contains a single species, *O. tsutsugamushi*.

Cell structure *O. tsutsugamushi* is stained by a modification of the Giménez procedure used for the other rickettsiae; this modified procedure requires preliminary destaining with ferric nitrate and counterstaining with fast green. This species is also satisfactorily stained by Giemsa stain after Carnoy's fixation.

In ultrathin sections, *O. tsutsugamushi* is surrounded by a cytoplasmic membrane and a cell wall, with a clear periplasmic space between the cytoplasmic membrane and cell wall. In the cytoplasm, the electron-dense ribosome-rich area and the less dense network area with DNA fibers are distinctive. Orientiae possess an unusual Gram-negative cell wall that lacks lipopolysaccharide and peptidoglycan. The outer leaflet of the cell wall is considerably thicker than the inner leaflet, while the opposite is true of *Rickettsia* (Fig. BXII.α.36, F and G) (Silverman and Wisseman, 1978). Possibly as a result, *O. tsutsugamushi* manifests much greater tenacity of adherence to host cell components, which renders purification much more difficult. In contrast to *Rickettsia*, *O. tsutsugamushi* does not have an electron-lucent zone.

Cultivation *O. tsutsugamushi* is cultivated well in the yolk sac of 5–7 d old chicken embryos if the inoculum is relatively large

$(10^4–10^6$ viable cells per egg) and if the rickettsiae are harvested just before the death of the embryos. It is also cultivated well in various cell lines, including HeLa, BHK, Vero, and L929 cells. *O. tsutsugamushi* produces small plaques on cell monolayers after 11–17 d of incubation. CO_2 enrichment is not necessary for intracellular growth (Kopmans-Gargantiel and Wisseman, 1981).

Mice are highly sensitive to infection with *O. tsutsugamushi.* Mice have been used for primary isolation and passage of orientiae. Mice are more sensitive to orientiae by intraperitoneal inoculation than by subcutaneous inoculation.

Multiplication cycle in host cells   Orientiae reside free in the cytosol of the host cell. They enter the host cell by attachment and induce phagocytosis. The adhesin has not been identified, but host cell surface heparan sulfate glucosaminoglycan plays a role as a receptor (Ihn et al., 2000). To escape the phagosome, orientiae have to be metabolically active (Rikihisa and Ito, 1982). Movement of orientiae within the cytoplasm to the perinuclear microtubule organizing center, where they replicate, is mediated by microtubules through interaction with dynein—the minus end-directed microtubule associated motor protein—rather than via F-actin tails (Kim et al., 2001). In mouse peritoneal mesothelial cells, *O. tsutsugamushi* multiplies in the cytoplasm, moves to the cell periphery, and separates from the cell surface surrounded by a host cell membrane (Tsuruhara et al., 1982) (Fig. BXII.α.38). Orientiae that are enveloped by the host membrane enter other mesothelial cells, apparently by a phagocytic mechanism. The organisms escape from the phagocytic vacuole as the vacuole membrane and host cell membrane coat disintegrate (Ewing et al., 1978). Cells heavily infected with orientiae undergo apoptotic cell death in association with decreased content of focal adhesion kinase and paxillin, decreased actin stress fiber polymerization, and decreased expression of anti-apoptotic Bcl-2 (Kee et al., 1999). *O. tsutsugamushi* achieves a high intracytoplasmic density, especially in the perinuclear region, and only rarely invades the nucleus.

Nutrition and metabolism   There is little information about the nutrition and metabolism of *O. tsutsugamushi*, a topic that has been neglected in the past 20–30 years. *O. tsutsugamushi*-infected cells incorporate ^{14}C labeled amino acids and adenine at a much higher level than uninfected cells during the period from 3–6 days postinoculation (Weiss et al., 1973). *O. tsutsugamushi* incorporates radiolabeled proline into rickettsial protein (Tamura et al., 1982). In contrast to *Rickettsia*, orientiae do not require CO_2 for growth (Kopmans-Gargantiel and Wisseman, 1981).

Antigenic structure   The major proteins of *O. tsutsugamushi* have the following sizes: 110, 80, 70, 60, 56, 47, 42, 35, 28, and 25 kDa (Tamura et al., 1985). Except for the 70- and 60-kDa proteins, all of these are surface proteins. The 56, 28, and 25 kDa proteins are heat labile; they migrate at molecular sizes of 43, 25, and 21 kDa, respectively (Urakami et al., 1986) when they are not heat denatured. The most abundant antigens of *O. tsutsugamushi* are the 60-kDa heat shock protein and the 56-kDa surface protein. The 60-kDa protein is the homolog of the GroEL protein family (Stover et al., 1990). Genetic analysis of the DNA sequence of *groESL* of isolates from several geographic foci has revealed a limited number of variants within each focus and a diversity of strains from focus to focus. In spite of its cytoplasmic location, the 60-kDa protein stimulates antibodies in the sera of some acutely ill patients (Ohashi et al., 1988). The 56-kDa protein varies among geographic isolates of *O. tsutsugamushi* and thus is

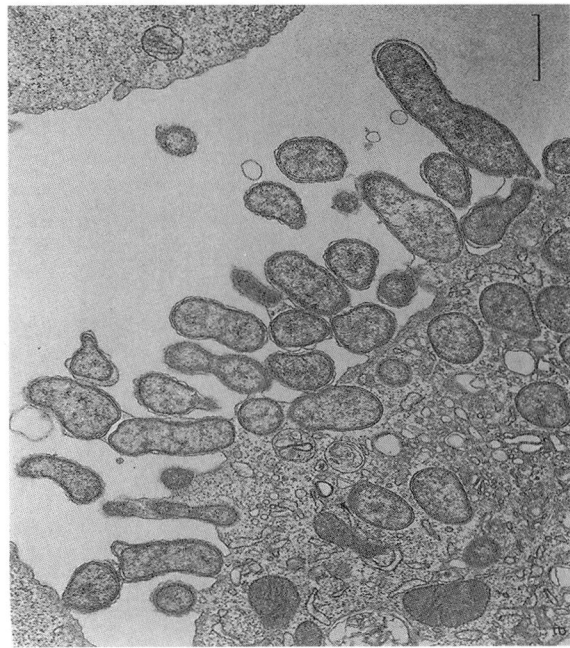

FIGURE BXII.α.38.   Electron photomicrograph of *Orientia tsutsugamushi* budding from a host cell at 72 h post infection. Photograph courtesy of A. Tamura and H. Urakami.

called type-specific antigen (TSA; Ohashi et al., 1992). The TSAs of six *Orientia* antigenic variants range from 55,308 to 56,745-daltons with 521–532 amino acids. TSA has the feature of transmembrane proteins with alternating hydrophobic and hydrophilic regions. Analysis of the TSAs of *Orientia* variants has revealed four variable domains with spans of 16–40 amino acids. The variable domains are located in the hydrophilic regions of the molecule that are likely surface-exposed and show different amino acid sequences among the strains (Ohashi et al., 1992). Early studies of *Orientia* isolates demonstrated three antigenic types: Gilliam, Karp, and Kato. Phylogenetic analyses based on homologies of 56-kDa type-specific antigen genes have classified *Orientia* isolates from China, Korea, Japan, and southeast Asia into seven genotypes designated Gilliam, Karp, Kato, Kawasaki, Kuroki, Shimokoshi and LX-1. All isolates originating in southeast Asia—including the prototype Gilliam and Karp strains isolated in Burma and New Guinea, respectively—are distantly located in the phylogenetic tree from the isolates originating in Japan, Korea, and China; this finding indicates that strains of *O. tsutsugamushi* distributed in northeastern and southeastern Asia are of different types (Enatsu et al., 1999). The antigenic variability of *Orientia* poses a challenge for vaccine design. Immunization with *Orientia* confers relatively strong protection that lasts only 1–3 years against challenge by the homologous strain, whereas protection is very weak and short-lived against heterologous strains (Seong et al., 2001).

The mol% G+C of the DNA of *Orientia* strains is similar to that of TG rickettsiae but lower than that of SFG rickettsiae. The genome of *O. tsutsugamushi* is $1.1–1.5 \times 10^9$ Da and consists of 2,400–2,700 kbp, as determined by pulse field gel electrophoresis. The genome of *Orientia* is larger than that of *Rickettsia*. The 16S rRNA gene sequence similarity among the antigenic variants of *O. tsutsugamushi*, including the Gilliam, Kato, Karp, Kawasaki, Kuroki, and Shiimokoshi strains, is $\geq 98.5\%$.

When injected intraperitoneally into mice, virulent strains of *O. tsutsugamushi* cause peritonitis, splenomegaly and death in 10–24 days. However, strains can vary greatly in virulence. The Karp strain, for example, is more virulent than the Gilliam strain for most outbred mice. Certain inbred mouse strains are highly resistant to the Gilliam strain; this resistance is controlled by a single, autosomal, dominant gene and does not involve susceptibility to the Karp strain (Groves et al., 1980).

Ecology *Orientia tsutsugamushi* is maintained by transovarian transmission in its various trombiculid mite hosts. Scrub typhus cases occur when larval mites (chiggers) encounter humans as a source of a tissue fluid meal taken from the skin. Each mite host species has its own geographic distribution and seasonal activity pattern that determine the occurrence of scrub typhus (Audy, 1968). Because mites feed on their host only once, the role of the rodents in the maintenance of the natural cycle of rickettsial infection appears even more remote than in the case of the spotted fever rickettsiae. Although wild rats become infected and chiggers acquire rickettsiae when feeding on rats, the chiggers do not transmit the organisms to the next generation (Traub and Wisseman, 1974). *Orientia* requires no other host than the mite in its life cycle.

Orientia tsutsugamushi is encountered in an area of the Orient that extends from India and Pakistan in the west to Japan, the northern portions of Australia, and the intervening islands in the Pacific Ocean in the east and includes southeastern Siberia, Korea, southeast Asia, southern China, the Philippines and Indonesia. These rickettsiae are usually found in circumscribed foci or "ecological islands," which have the proper vegetation and proper concentration of mites and their wild rat hosts. The habitats are usually characterized by the presence of changing ecological conditions, wrought by man or nature, and expressed by transitional types of vegetation (Traub and Wisseman, 1974).

The mite most commonly associated with scrub typhus is *Leptotrombidium deliense*, but several other trombiculid mites, including *L. fletcheri*, *L. akamushi*, *L. arenicola*, and *L. scutellare*, have been shown to be naturally infected and to transmit the rickettsia transovarially. Shortly after its emergence from the egg, the six-legged larva, or chigger, either remains in the soil or travels upward a few cm on debris or dead vegetation until it can attach and burrow into the skin of any animal it happens to contact. Following a meal of tissue juices, it returns to the soil to resume a free-living existence. The vertebrates most commonly infected are rodents of the genus *Rattus*, although isolations from temperate zone rodents, including *Apodemus* and *Microtus*, have been reported. The wide dissemination of *O. tsutsugamushi* on islands that are separated from each other and from the mainland by large bodies of water can best be explained by assuming that migrant birds play a role in the transport of the chiggers (Traub and Wisseman, 1974).

ENRICHMENT AND ISOLATION PROCEDURES

Orientia tsutsugamushi is isolated by inoculation of mice, cell culture, or embryonated chicken eggs, as described above. *O. tsutsugamushi* can be identified by polymerase chain reaction amplification of the DNA of a species-specific gene, DNA sequencing of genes that are available in sequence databases, and reactivity with species-specific antibodies.

The cell wall of *Orientia* is fragile and is easily disrupted by mechanical treatment, such as strong homogenization and osmotic shock. Orientiae are also very adherent to one another, and it is difficult to resuspend them once they are pelleted. Renografin density gradient centrifugation is not suitable for the purification of orientae. Percoll density gradient centrifugation is the choice for their purification because Percoll has no osmotic effect, and the purification procedure does not involve the pelleting of the organisms until the final stage (Tamura et al., 1982).

Orientia-infected cells are harvested by centrifugation at 300 × g for 10 min and suspended in 1% of the original volume in buffer containing 0.033 M Tris hydrochloride, pH 7.4, and 0.25 M sucrose (TS). The cells are homogenized with a tissue homogenizer with 20–30 strokes. The homogenate is centrifuged at 200 × g for 10 min. The supernatant is mixed with 40% Percoll in TS buffer and centrifuged at 25,000 × g for 60 min. Two bands are formed in the centrifuge tube. The upper band (near the top) consists of cell debris and is removed by suction. The lower band (near the bottom) consists of orientiae and is collected with a capillary pipette. The orientiae are washed by centrifugation at 6,000 × g for 20 min in TS buffer.

MAINTENANCE PROCEDURES

Stocks of *O. tsutsugamushi* are usually preserved frozen at −70° to −80° or in the vapor phase of liquid nitrogen. They may also be preserved in a lyophilized state. The diluent most frequently used is SPG (0.22 M sucrose, 0.01 M potassium phosphate (pH 7.0) and 0.005 M potassium glutamate) (Bovarnick et al., 1950).

DIFFERENTIATION OF THE GENUS *ORIENTIA* FROM OTHER GENERA

Orientiae grow free in the cytoplasm or nucleus as compared to *Anaplasma*, *Ehrlichia* and *Coxiella*, which grow in the cytoplasm within a vacuole. The characteristics for differentiation between *Rickettsia* and *Orientia* are summarized in Table BXII.α.32 in the chapter on the genus *Rickettsia*.

TAXONOMIC COMMENTS

O. tsutsugamushi, formerly called *Rickettsia tsutsugamushi*, was removed from the genus *Rickettsia* based on the phylogenetic differences between the two genera (Tamura et al., 1995). Phylogenetic analysis of 16S rRNA gene sequences reveals that *O. tsutsugamushi* is located apart from the *Rickettsia* species with a similarity value of 90.2–90.6% (Tamura et al., 1995) (Fig. BXII.α.35).

List of species of the genus Orientia

1. **Orientia tsutsugamushi** (Hayashi 1920) Tamura, Ohashi, Urakami and Miyamura 1995, 590 (*Rickettsia tsutsugamushi* (Hayashi 1920) Ogata 1931, 252; *"Theileria tsutsugamushi"* Hayashi 1920, 63.)

   tsu.tsu.ga.mu.shi. M.L. n. *tsutsugamushi* popular name of the disease caused by this species, generally interpreted to mean mite disease.

The description of the species is the same as that of the genus.

The mol% G + C of the DNA is: 28.1–30.5 (HPLC).

Type strain: Karp, ATCC VR-150.

GenBank accession number (16S rRNA): D38623, U17257.

The type strain is that of Derrick and Brown (1949). Reference strains include Gilliam, ATCC VR-312 (Bennett et al., 1949), and Kato, ATCC VR-609 (Shishido et al., 1958).

Family II. **Anaplasmataceae** Philip 1957, 980[AL] emend. Dumler, Barbet, Bekker, Dasch, Palmer, Ray, Rikihisa and Rurangirwa 2001, 2156

J. STEPHEN DUMLER, YASUKO RIKIHISA AND GREGORY A. DASCH

A.na.plas.ma.ta'ce.ae. M.L. *Anaplasma* type genus of the family; *-aceae* ending to denote a family; M.L. fem. pl. n. *Anaplasmataceae* type genus of the family.

Rickettsial organisms pathogenic for certain mammals, including man, and birds; non-pathogenic for some arthropods, insects, and helminths. **The predominant mammalian host cells are of bone marrow or hematopoietic origin,** including erythrocytes, monocytes or macrophages, neutrophils, and platelets. Also may grow within tick or other invertebrate cells. Members of the family share a high degree of nucleotide sequence similarity in regard to 16S rDNA and to the *groESL* operon. The **organisms grow within a cytoplasmic vacuole, but not in the nucleus;** they appear as compact to loose inclusions containing as few as one to many individual organisms. Infected cells may contain more than one inclusion per cell. The mulberry-like appearance of the inclusions has led to the use of the term "morulae" for them. **Organisms may have two distinct morphological forms: dense-core and reticulate bodies.** Gram negative. Nonmotile. **Certain species are adapted to existence in ticks.** Some differential characteristics of the genera of *Anaplasmataceae* are shown in Table BXII.α.33.

Type genus: **Anaplasma** Theiler 1910, 7 emend. Dumler, Barbet, Bekker, Dasch, Palmer, Ray, Rikihisa and Rurangirwa 2001, 2157.

TABLE BXII.α.33. Key to the genera of the family *Anaplasmataceae.*

	Anaplasma	*Ehrlichia*	*Neorickettsia*	*Wolbachia*
Host cell infected	Erythrocytes, granulocytes	Mononuclear phagocytes, granulocytes	Mononuclear phagocytes, trematode cells	Arthropod ovaries
Ultrastructural morphology	Multiple bacteria in single vacuole; no intravacuolar fibrillar matrix ; vacuoles do not contact mitochondria and endoplasmic reticulum	Multiple bacteria in single vacuole; intravacuolar fibrils present; vacuoles contact mitochondria and endoplasmic reticulum	Small clusters or individuals cells within vacuoles; vacuoles divide with bacterial division	Single bacteria within vacuoles; vacuoles contact endoplasmic reticulum
Mol% G + C of the DNA	43–56	32–46	30–43	Not known

Genus I. **Anaplasma** *Theiler 1910, 7*[AL] *emend. Dumler, Barbet, Bekker, Dasch, Palmer, Ray, Rikihisa and Rurangirwa 2001, 2157*

J. STEPHEN DUMLER, YASUKO RIKIHISA AND GREGORY A. DASCH

A.na.plas'ma. Gr. pref. *an* without; Gr. n. *plasma* anything formed or molded; M.L. neut. n. *Anaplasma* a thing without form.

Small, often pleomorphic, coccoid to ellipsoidal cells 0.3–0.4 μm in diameter found in cytoplasmic vacuoles in mammalian host cells, often in inclusion bodies (morulae) in mature or immature hematopoietic cells, in peripheral blood, or in tissues, usually organs containing mononuclear phagocytes (spleen, liver, bone marrow) of mammalian hosts. Morulae are suspended in a nonfibrillar matrix. Two morphologic forms: larger reticulate cells and smaller "dense core" forms. In blood smears stained by Romanowsky methods, the organisms appear as dense, homogeneous, bluish-purple, round inclusions 0.3–2.5 μm in diameter. Gram negative. Nonmotile. No spores or resistant stages. **Ticks are the only known biological vectors;** mechanical vectors include biting flies or other fomites. The organisms can be propagated in tick spp. Some anaplasmas can be propagated in tick cell lines or mammalian cells of hematopoietic origin. **Cause disease in canids, humans, and ruminants.** Variably pathogenic in cattle, goats, sheep, deer, horses, and rodents. Placed in the *Alphaproteobacteria* by **16S ribosomal RNA sequence analysis; organisms possess not less than 96% 16S rRNA gene similarity**. Where studied, these bacteria possess major surface protein genes with a high degree of sequence similarity nontandemly dispersed throughout the chromosome.

The mol% G + C of the DNA is: 43 and 56.
Type species: **Anaplasma marginale** Theiler 1910, 7.

FURTHER DESCRIPTIVE INFORMATION

Phylogeny By 16S rRNA gene and *groESL* operon sequence analysis, the genus *Anaplasma* forms a clade distinct from *Ehrlichia, Wolbachia, Neorickettsia, Rickettsia,* and *Orientia* in the *Alphaproteobacteria,* Order *Rickettsiales.* When 16S rRNA gene and *groESL* operon sequences are aligned, all species in the genus are more than 96% and 74% similar, respectively. Where investigated, organisms possess multiple similar genes that encode major surface proteins that vary in molecular size from approximately 36 to 49 kDa. The bacteria are only distantly related to other obligate intracellular bacteria in genera such as *Chlamydia* and *Coxiella,* and unrelated to bacteria lacking a cell wall, such as *Eperythrozoon* and *Haemobartonella.*

Cell morphology *Anaplasma* species stain bluish-purple with Romanowsky methods. They occur in membrane-bound vacuoles in the cytoplasm of cells of hematopoietic origin, forming inclusions that contain variable numbers of organisms (Simpson, 1972, 1974; Hildebrandt et al., 1973) (See Fig. BXII.α.39). Various spe-

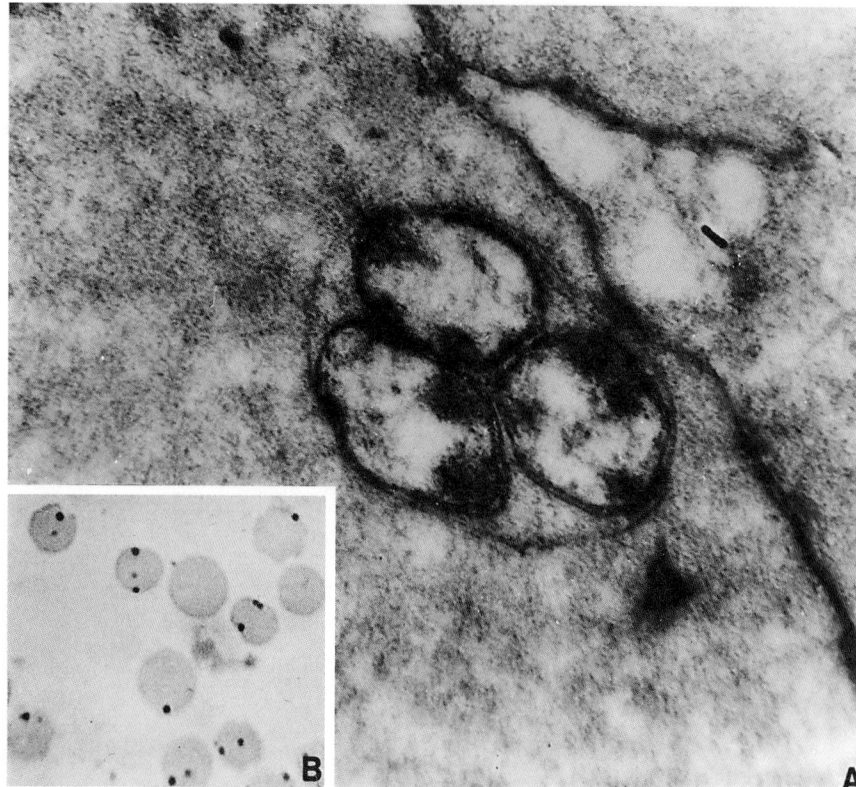

FIGURE BXII.α.39. *Anaplasma marginale.* A, Transmission electron micrograph of an infected erythrocyte. Three individual bacteria bound by cell wall and plasma membrane and lacking intravacuolar fibrils and tubules are shown within the vacuole (× 51,000). *B,* Intraerythrocytic inclusion bodies stained by a Romanowsky method (Giemsa stain, × 815).

cies infect predominantly erythrocytes, neutrophils, or mononuclear phagocytes *in vivo.* The individual organisms are approximately 0.3 μm in diameter but may vary considerably in size and shape. Inclusions (morulae) range in size up to 4.0 μm in diameter. The organisms are weakly Gram-negative. Anaplasmas are best stained with Romanowsky stains, where the organisms develop a dark blue appearance as opposed to the violet color of the eucaryotic cell nucleus. The organisms stain poorly with the Giménez method but easily with acridine orange.

Cell wall composition Little is known of the components of the cell wall of *Anaplasma* spp. Lipopolysaccharide and lipooligosaccharide have not yet been detected, and it is not known if the cell wall contains peptidoglycan.

Fine structure By electron microscopy, anaplasmas reside within membrane-bound vacuoles that are very early endosomes for *Anaplasma phagocytophilum* (Webster et al., 1998; Mott et al., 1999). The ultrastructure of the cell envelope reveals simple inner and outer leaflets similar to those of Gram-negative bacteria (Popov et al., 1998). The internal structures consisting of chromatin strands and ribosomes are readily visualized. Two distinct morphologic forms may be detected: dense-core forms that contain a relatively dense central or eccentrically-placed condensation of chromatin strands, and reticulate cell forms that contain a homogeneous loose matrix of chromatin strands, among which ribosomes are also spread. Reticulate forms are typically identified *in vivo,* and dense core forms are identified predominantly during *in vitro* propagation. Both forms undergo binary fission, suggesting that a developmental cycle associated with these mor-

phologies is unlikely. *Anaplasma* species may produce an abundant membrane, which on occasion wraps around individual cells or invaginates into the cell. Unlike the genus *Ehrlichia, Anaplasma* species lack a fibrillar matrix within the vacuolar space.

Cultural characteristics *A. phagocytophilum* has been successfully propagated in primary human neutrophil cultures and in a variety of human (Goodman et al., 1996) and non-human myeloid or myelogenous lineage cell lines (HL-60, THP-1, M1). *A. phagocytophilum* and *A. marginale* have also been propagated in tick cell tissue culture (Munderloh et al., 1996a, b). Limited propagation of *A. marginale* in erythrocyte culture has also been achieved. No *Anaplasma* species has been cultivated on cell-free media. In cell culture, infection is observed at first with one or few bacteria within the vacuole, previously referred to as an initial body. With continued binary fission of the bacteria, increasingly larger intravacuolar microcolonies are formed, which are called intermediate bodies or morulae, depending upon size. Ultrastructural examination has shown that all vacuoles may contain either reticulate cells, dense-cored cells, or both, and that each of these forms undergoes binary fission; these findings argue against a well-defined intracellular life cycle, as observed for *Chlamydia* spp. Multiple infected vacuoles may exist in a single cell, presumably owing to the lack of fusion of infected endosomes.

For *A. marginale,* recent data suggest that a restricted set of major surface protein genes (*msp2*) may be transcribed during tick infection. However, these results were not corroborated when different strains were examined, which again argues against distinct life cycle-specific bacterial components.

Antigenic structure All *Anaplasma* species contain a multigene family encoding major surface proteins that are the immunodominant antigens of the genus. The genes are generally characterized by the presence of highly conserved 5′ and 3′ sequences that flank a hypervariable core region (Barbet, 1995; Zhi et al., 1998; Viseshakul et al., 2000). One or several of these genes may be transcriptionally active to produce protein antigens of various molecular sizes ranging from 36 to 49 kDa (Asanovich et al., 1997). Other antigenic components may be conserved among species, such as the heat shock proteins (GroESL), or they may be species-specific.

Antibiotic sensitivity All *Anaplasma* species are susceptible to tetracycline antibiotics. Some fluoroquinolone antibiotics, such as enrofloxacin and trovofloxacin, have *in vivo* and *in vitro* efficacy for some species (Guglielmone et al., 1996; Klein et al., 1997). Anaplasmas are not susceptible to β-lactam antibiotics, aminoglycosides, macrolides, chloramphenicol, or sulfa-containing drugs.

Pathogenicity Little is known about the mechanisms by which *Anaplasma* species cause disease. *In vitro*, infected cells undergo both necrosis and apoptosis, and erythrocytes may lyse (Goodman et al., 1996; Waghela et al., 1997). The *in vivo* infection may be associated with febrile hemolytic anemia for *A. marginale*, and with non-specific febrile illness with leukopenia, mild anemia, thrombocytopenia, and mild hepatic injury for *A. phagocytophilum*. While the anemia associated with *A. marginale* may result from direct erythrocyte lysis, it is highly unlikely that the degree of cytopenia observed with *A. phagocytophilum* infections results from direct cytolysis of infected cells (Bakken et al., 1996). "*A. platys*" is associated with febrile cyclical thrombocytopenia associated with the appearance of morulae in platelets in peripheral blood. It is likely that most infections are asymptomatic. Clinically apparent infection may be observed in a range of animals including cattle, goats, sheep, deer, llamas, dogs, horses, and humans.

Anaplasma species gain access to their host cells by adherence to surface components that generally are glycosylated scaffold proteins (McGarey and Allred, 1994; Herron et al., 2000). After adherence, the bacteria are internalized and remain within vacuoles that are early endosomes for *A. phagocytophilum* (Mott et al., 1999). Infection of granulocytic cells *in vitro* results in the secretion of chemokines (Klein et al., 2000), whereas infection of neutrophils *in vitro* results in delayed apoptosis (Yoshiie et al., 2000).

Ecology Ticks are vectors for all *Anaplasma* species that have been studied. *Dermacentor andersoni*, *Boophilus microplus*, *Ixodes scapularis*, *I. pacificus*, and *I. ricinus*, among other ticks, are important biological vectors. Mechanical vectors, such as male ticks and biting flies are also important for transmission of *A. marginale*. *Anaplasma* species are ingested by adult and nymphal stage ixodid ticks when feeding on an infected host. The bacteria are maintained in the tick by transstadial, but not transovarial transmission; thus, emergent larval ticks are not infectious. Domestic and wild ruminants—especially cervids—may be important reservoirs for all species of the genus. Other important reservoirs for *A. phagocytophilum* include small mammals such as mice and other rodents.

ENRICHMENT AND ISOLATION PROCEDURES

Both *A. marginale* and *A. phagocytophilum* have been isolated by inoculation of whole blood in tick cell cultures (Munderloh et al., 1996a, b), whereas *A. phagocytophilum* is easily isolated in granulocyte cell cultures using whole blood or leukocyte preparations (Goodman et al., 1996). *Anaplasma* species are obligately intracellular bacteria that cannot be isolated by axenic methods. Isolation has also been achieved by inoculation of fresh or frozen blood from infected animals into naïve susceptible animals, such as cows for *A. marginale* and horses or mice for *A. phagocytophilum*.

MAINTENANCE PROCEDURES

A. marginale can be maintained for long periods in tick cell culture, and for shorter periods in erythrocyte cultures (Munderloh et al., 1996b; Waghela et al., 1997). *A. phagocytophilum* can be maintained for long periods (>1 year) by serial propagation in granulocyte cell lines such as HL-60 cells, although *in vitro* passage may alter the propensity of the bacteria to elicit clinical manifestations in susceptible animals (Goodman et al., 1996; Pusterla et al., 2000a). Cultured cells containing the bacteria may be stored in medium containing 30% fetal bovine serum and 10% dimethyl sulfoxide at −80°C for months to years, and in liquid nitrogen for years. Both *A. marginale* and *A. phagocytophilum* can be maintained by serial passage in susceptible animals hosts. Neither *A. bovis* nor "*A. platys*" have been cultivated *in vitro*.

DIFFERENTIATION OF THE GENUS *ANAPLASMA* FROM OTHER GENERA

Anaplasma species may be morphologically difficult to distinguish from *Ehrlichia* and *Neorickettsia* species that grow within bone marrow-derived cells. The genus is separated from related genera by its obligate intracellular growth in bone marrow-derived cells or tick cells, by its growth within a vacuole of the infected host cell, and by finding similarities of >97% in the 16S rRNA gene sequence or >74% in the *groESL* operon sequences compared to sequences of established strains.

TAXONOMIC COMMENTS

Phylogenetic comparisons comprise the most objective comparators and are used as the major tool to evaluate genetic similarities and to establish objective and reproducible taxonomic criteria that cannot be achieved reliably with phenotypic or clinical data. *A. marginale* is the most distinct species in the genus in terms of phenotypic and genotypic characteristics; in fact, some authorities consider this taxon to qualify for a separate genus position.

A. marginale, *A. ovis*, and *A. centrale* are all recognized as valid species in the genus *Anaplasma*. However, the 16S rRNA gene sequences of strains of each, with the exception of a single Japanese strain, are at least 99.1% similar. Thus, it is likely that these represent single species variants, as initially suggested by Theiler (1911). Moreover, the tight clustering of erythrocytic anaplasmas with *A. marginale* is also supported by other genotypic and phenotypic characters, including shared 19, 36, and 105 kDa protein antigens (McGuire et al., 1984, Palmer et al., 1988a, Visser et al., 1992, Palmer et al., 1998). Although *A. caudatum* is proposed and listed as a unique species, some authorities believe that it is a species synonomous with *A. marginale* that represents an artifact of experimental, rather than natural, tick-transmitted infection. A strain of *A. centrale* exists that has 1.8% 16S rRNA gene nucleotide divergence from other *A. centrale* and *A. marginale* strains that were phenotypically characterized, casting significant doubt on morphological methods of taxonomy for this species and genus and suggesting that other erythrocytic *Anaplasma* species exist. Additional study will be required to delineate the precise taxonomic relationships among these closely related bacteria.

The name *Anaplasma* was selected by Dumler et al. (2001) for the emended, newly combined genus because of historical precedence (Theiler, 1910). The exact identity of the previously described *"Cytoecetes microti"* (Tyzzer, 1938) is not certain; however, there is a clear phenotypic similarity to *A. phagocytophilum*. The name *"Cytoecetes"* is not recognized on the Approved Lists of Bacterial Names, but if sufficient data become available that support the separation of *Anaplasma marginale* (and other erythrocytic anaplasmas such as *A. ovis*) from *A. phagocytophilum*, *A. bovis*, and *"A. platys"*, the designation *"Cytoecetes"* would have historical precedence and would be validly published.

FURTHER READING

Barbet, A.F. 1995. Recent developments in the molecular biology of anaplasmosis. Vet. Parasitol. *57*: 43–49.

Dumler, J.S., A.F. Barbet, C.P.J. Baker, G.A Dasch, G.H. Palmer, S.C. Ray, Y. Rikihisa and F.R. Rurangirwa. 2001. Reorganization of the genera in the families *Rickettsiaceae* and *Anaplasmataceae* in the order *Rickettsiales*. Unification of some species of *Ehrlichia* with *Anaplasma*, *Cowdria* with *Ehrlichia* and *Ehrlichia* with *Neorickettsia*, description of six new species combinations and designation of *Ehrlichia equi* and "HGE agent" as subjective synonyms of *Ehrlichia phagocytophila*. Int. J. Syst. Evol. Microbiol. *51*: 2145–2165.

Palmer, G.H., W.C. Brown and F.R. Rurangirwa. 2000. Antigenic variation in the persistence and transmission of the ehrlichia *Anaplasma marginale*. Microbes Infect. *2*: 167–176.

Rikihisa, Y. 1991. The tribe *Ehrlichieae* and ehrlichial diseases. Clin. Microbiol. Rev. *4*: 286–308.

DIFFERENTIATION OF THE SPECIES OF THE GENUS *ANAPLASMA*

The preferred method for definitive identification of *Anaplasma* species depends upon sequence analysis of all or part of the 16S rRNA gene or the *groESL* operon or by PCR amplification using species-specific oligonucleotide primers. The full 16S rRNA gene sequence may be identified by many different methods, including commercially available systems such as the MicroSeq™ Full Gene 16S rDNA Bacterial Sequencing Kit (Applied Biosystems, Foster City, CA). Species-specific primers for *A. marginale*, *A. phagocytophilum* and *"A. platys"* are shown in Table BXII.α.34. Various methods and protocols for diagnostic identification in clinical samples have been described; the diagnostic sensitivity and specificity should be determined by each laboratory performing the clinical assay.

See Table BXII.α.35 for key features of Anaplasma species.

List of species of the genus Anaplasma

1. **Anaplasma marginale** Theiler 1910, 7[AL]

mar.gi.na′ le. L. n. *margo, marginis* edge, margin; M.L. neut. adj. *marginale* marginal, referring to location of the organism within the erythrocytes.

The characteristics are as described for the genus and as listed in Tables BXII.α.35 and BXII.α.36. See also Fig. BXII.α.39.

A. marginale is the causative agent of bovine anaplasmosis, a severe febrile hemolytic anemia of cattle that occurs after tick bites or mechanical transmission by other arthropods. The disease is worldwide in distribution.

A. marginale possesses polymorphous major surface proteins (MSPs), several of which are similar to proteins in *A. phagocytophilum* and some of which are encoded by multigene families (Barbet, 1995). A low level, cyclical, persistent infection is established and is accompanied by emergence of antigenic variants in MSP2. MSP1a, MSP1b, and MSP2 are associated with hemagglutinating activity and may represent adhesins of *A. marginale* (McGarey and Allred, 1994). Epitopes of MSP2 and MSP5 are conserved among *A. marginale*, *Anaplasma centrale*, and *Anaplasma ovis*, and both the *msp2* and *msp3* genes of *A. marginale* are present in the genome of *A. ovis*. No tick-specific antigen expression has been proven (Rurangirwa et al., 1999, 2000).

Tetracycline compounds and dithiosemicarbazone inhibit replication and ameliorate clinical manifestations. Penicillin, streptomycin, sulfonamides, and arsenicals are inactive (Barrett et al., 1965; Ristic, 1981).

Since infection results in low-level persistence, natural immunity is incomplete. Immunity that results after elimination of persistently infected state may persist for at least 8 mo. Vaccination using whole intact bacteria or subcomponents has resulted in variable levels of protective immunity (Palmer et al., 1988a; Montenegro-James et al., 1991, Tebele et al., 1991). Cross-protection between *A. marginale* and *A. centrale* has been reported (Palmer et al., 1988b).

The relationship of *A. marginale* to erythrocytic anaplasmas other than *A. centrale* is not certain, although distinct sequences of the 16S rRNA genes have been reported.

A. marginale causes clinical signs with infection in cattle, occasionally sheep or goats, and giraffes (Kuttler, 1984). Inapparent infection may occur in zebu, water buffalo (*Babalus babalis*), bison (*Bison bison*), African antelopes, black wildebeest (*Connochaetes gnou*), blesbuck (*Damaliscus albifrons*), and duiker (*Sylvicapra grimmi grimmi*), American deer (southern black-tailed, Rocky Mountain mule deer, Virginia white-tailed), pronghorn (*Antilocapra americana americana*), elk, bighorn sheep (*Ovis canadensis canadensis*), and camel (*Camelus bactrianus*). *A. marginale* has not been demonstrated to be infectious for rodents or other small mammals. Persistent infection leading to a carrier "reservoir" state is best documented for cattle; other domestic and wild ungulates and ruminants may also play a role (Kuttler, 1984). Vectors for *A. marginale* include *Dermacentor andersoni* ticks in North America and *Boophilus microplus* ticks in Africa. Transstadial but not transovarial transmission occurs (Stich et al., 1989); thus, the predominant mechanism for natural maintenance involves a mammalian reservoir that is expanded by tick vectors and mechanical or fomite vectors.

Isolation is best achieved by serial passage of blood from infected to susceptible animals. The bacterium may be maintained in stable form when frozen ($\leq -70°C$) in anticoagulated blood containing dimethyl sulfoxide; such preparations can be later transfused into susceptible hosts. The bacteria may also be propagated *in vitro* by cultivation in IDE8 tick cells (Munderloh et al., 1996b). Isolated bacteria may be identified by polymerase chain reaction amplification of conserved *A. marginale* genes, by hybridization of bacterial DNA with *A. marginale* DNA probes, or with *A. marginale* monoclonal or polyclonal antibodies (McGuire et al., 1984; Barbet, 1995; Torioni et al., 1998).

A. marginale persists at low levels in the blood of infected

TABLE BXII.α.34. Species-specific oligonucleotide primers for PCR identification of *Anaplasma* species.

Anaplasma species	Targeted gene	Forward primer	Reverse primer	Hybridization probe
A. marginale[a]	msp1	5'-GTATGGCACGTAGTCTTGGGATCA-3'	5'-CAGCAGCAGCAAGACCTTCA-3'	
A. phagocytophilum[b]	16S rRNA	5'-AACGGATTAGGCTTTATAGCTTGCT-3'	5'-TTCCGTTAAGAAGGATCTAATCTCC-3'	
"A. platys"[c]	16S rRNA	5'-TGTCGTAGCTTGCTATG-3'	5'-CGTTTTGTCTCTGTGTTG-3'	5'-GAAGATAATGACGGTACCC-3'

[a]From Stich et al., 1993.

[b]From Edelman and Dumler, 1996.

[c]From Chang and Pan, 1996.

cattle (Torioni et al., 1998). During the initial infection, antigenic variants emerge because of expression of new MSP2 genes, perhaps by recombinational events at the level of the chromosome (French et al., 1998). The nucleic acid sequences of the multigenes in the genome of a single *A. marginale* isolate may be as little as 25% identical. Several of the MSPs of *A. marginale* form complexes within the membrane of the bacterium.

A. marginale is the most phylogenetically distinct species in the genus *Anaplasma*, differing from other species by between 2.3 and 3.5% identity in the 16S rRNA gene sequence and by 25% in the *groESL* operon sequence.

The mol% G + C of the DNA is: 56 (spectral analysis; Alleman et al., 1992).

Type strain: no culture isolated.

GenBank accession number (16S rRNA): M60313.

2. **Anaplasma bovis** (Donatien and Lestoquard 1936) Dumler, Barbet, Bekker, Dasch, Palmer, Ray, Rikihisa and Rurangirwa 2001, 2158[VP] (*Rickettsia bovis* Donatien and Lestoquard 1936, 1061; *"Ehrlichia bovis"* Moshkovski 1945, 18.) bo'vis. L. n. *bos* the ox; L. gen. n. *bovis* of the ox.

The characteristics are as described for the genus and as listed in Tables BXII.α.35 and BXII.α.36. Cells are coccoid to ellipsoidal and are often pleomorphic. They infect mononuclear phagocytes, mostly monocytes, of cattle. The morphologic and ultrastructural appearance is not different from other members of *Anaplasma*.

A. bovis is the causative agent of bovine ehrlichiosis of Africa, the Middle East, India, and Sri Lanka, a disease that is clinically characterized by fluctuating fever, lymphadenopathy, depression, and, occasionally, death (Rioche, 1966). Infection is usually inapparent in endemic regions, except when exacerbation occurs during periods of stress, with splenectomy, or with other infections such as rinderpest.

A. bovis has not yet been cultivated *in vitro*, but may be propagated by serial passage in susceptible cattle. Identification may be achieved by sequence analysis of polymerase chain reaction-amplified DNA from animal blood or ticks.

The principal clinical manifestations include fever, anorexia and diarrhea, and—infrequently—involvement of the central nervous system with drowsiness and convulsions. Leukopenia and thrombocytopenia may occur.

A. bovis is transmitted by ticks, including *Rhipicephalus appendiculatus* and *Amblyomma variegatum*, in Africa (Rioche, 1966; Matson, 1967) and possibly by *Amblyomma cajennense* ticks in Guadaloupe and Brazil. *A. bovis*-like 16S rRNA gene sequences have been detected in *Haemaphysalis leporispalustris* and *Ixodes scapularis* ("dammini") ticks on Nantucket Island in the northeastern U.S. (Goethert and Telford, 2000). This latter observation was also associated with detection in cottontail rabbits, suggesting the existence of a non-ruminant ecological cycle. *A. bovis* infections have been described in North America, Africa, the Middle East, and Asia (India and Sri Lanka).

The 16S rRNA gene sequences are represented by two submissions in GenBank, one from South Africa and one from North America, which are ≥99.5% identical.

The mol% G + C of the DNA is: not known.

Type strain: no strain isolated.

GenBank accession number (16S rRNA): U03775.

TABLE BXII.α.35. List of diagnostic features of *Anaplasma* species.

	A. bovis	A. marginale	A. phagocytophilum	"A. platys"
Usual host species	Cattle	Cattle	Ruminants, horses, dogs, humans	Dogs
Usual infected host cell type	Monocytes	Erythrocytes	Neutrophils	Platelets
Present in peripheral blood[a]	±	+ +	+ +	+
Serological reactions with:[b]				
A. bovis	+ + +	−	−	?
A. marginale	?	+ + +	−	?
A. phagocytophilum	?	−	+ + +	−
"A. platys"	?	?	−	+ + +
Ehrlichia canis	+	−	+	−

[a]Symbols: ±, rarely present or difficult to identify; +, present infrequently; + +, present moderately often.

[b]Symbols: −, none; +, weak; + + +, strong; ?, unknown or not reported.

TABLE BXII.α.36. Descriptions of the species of the genus *Anaplasma*.

	A. marginale	A. bovis	A. phagocytophilum	"A. platys"
Geographic distribution	Worldwide	Africa, Asia, North America?	Europe, Asia, North America, South America	North America, Europe, Taiwan
Infected host(s) with clinical signs	Cattle, ruminants	Cattle	Ruminants, horses, dogs, humans	Dogs
Reservoir host(s)	Cattle, wild ruminants and cervids (deer, elk, water buffalo, sheep)	Cattle?	Small rodents, wild ruminants and cervids, felids (mountain lions), black bear	Dogs
Vectors	Boophilus microplus, Dermacentor andersoni, mechanical vectors (biting flies, fomites)	Rhipicephalus appendiculatus, Amblyomma variegatum, Amblyomma cajennense, Ixodes scapularis (dammini)?, Haemaphysalis leporispalustris	Ixodes scapularis, I. pacificus, I. ricinus, I. persulcatus, I. spinipalpis	Rhipicephalus sanguineus?
Host cell	Erythrocytes	Monocytes	Granulocytes (neutrophils, eosinophils, basophils)	Platelets
In vitro cultivation	Tick cells, erythrocytes, endothelial cells	−	Granulocyte cell lines, tick cells, endothelial cells	−

3. **Anaplasma caudatum** (Kreier and Ristic 1963) Ristic and Kreier 1984c, 355[VP] (Effective publication: Ristic and Kreier 1984a, 722) (*Paranaplasma caudatum* Kreier and Ristic 1963, 701)

cau'da.tum. L. neut. n. *cauda* tail; *caudatum* tailed, with a tail.

Anaplasma caudatum is similar to *A. marginale* except that each cell possesses an appendage, usually in the form of a tapering tail, a loop, a disk, or ring, that can only be visualized through immunologic or ultrastructural techniques. By electron microscopy, the tails are not directly attached to the bacterium, but contain some bacterial-specific antigens and are predominantly composed of polymerized F-actin filaments of unknown origin (Kocan et al., 1978; Stich et al., 1997). The exact relationship of *A. caudatum* to *A. marginale* is uncertain.

The mol% G + C of the DNA is: unknown.

Type strain: no culture isolated.

4. **Anaplasma centrale** (ex Theiler 1911) Ristic and Kreier 1984c, 355[VP] (Effective publication: Ristic and Kreier 1984a, 722.)

cen.tra'le. L. neut. adj. *centrale* central, referring to the location of the organism within erythrocytes.

Anaplasma centrale is similar to *A. marginale*, except that the disease produced in cattle is usually mild, and the lo-

cation of the bacterium within the erythrocyte cytoplasm is central and not peripheral.

This species shares many genes and antigens with *A. marginale* and *A. ovis* (Palmer et al., 1988b, 1998; Visser et al., 1992). Cross-protection between *A. marginale* and *A. centrale* has been reported (Abdala et al., 1990; Turton et al., 1998). Strains of *A. centrale* possess both specific and genus antigens; however strains of *A. marginale* differ from each other as much as they do from *A. centrale* (McGuire et al., 1984; Palmer et al., 1988b; Visser et al., 1992).

The exact relationship of *A. centrale* to *A. marginale* is uncertain, but the 16S rRNA genes are at least 99.2% identical.

The mol% G + C of the DNA is: not determined.

Type strain: Israel.

GenBank accession number (16S rRNA): AF309869.

5. **Anaplasma ovis** Lestoquard 1924, 784[AL]

o'vis. L. gen. n. *ovis* of the sheep.

Anaplasma ovis is similar to *A. marginale* except that the hosts with clinical signs are usually restricted to sheep and goats.

This species contains multiple genes that are also present in *A. marginale*, including *msp5* and multiple *msp2* and *msp3* genes (Visser et al., 1992; Palmer et al., 1998).

Cross-protection between *A. marginale* and *A. ovis* has

not been reported, although T cells from *A. marginale* immunized animals proliferate when exposed to *A. ovis* antigens (Brown et al., 1998).

The exact relationship of *A. ovis* to *A. marginale* is uncertain, but the sequences of the 16S rRNA genes are at least 99.6% identical.

The mol% G + C of the DNA is: not determined.

Type strain: Idaho.

GenBank accession number (16S rRNA): AF309865.

6. **Anaplasma phagocytophilum** (Foggie 1951) Dumler, Barbet, Bekker, Dasch, Palmer, Ray, Rikihisa and Rurangirwa 2001, 2158[VP] (*Rickettsia phagocytophila* Foggie 1951, 4; *Ehrlichia phagocytophila* Philip 1962, 42.)

pha.go.cy.to′phi.lum. Gr. inf. *phagein* to eat, devour; Gr. n. *kytos* a vessel, enclosure; Gr. inf. *philein* to love; M.L. adj. *phagocytophilum* fond of devouring cells (in microbiology, attractive to phagocytes).

The characteristics are as described for the genus and as listed in Tables BXII.α.35 and BXII.α.36. See also Figs. BXII.α.40 and BXII.α.41. Cell morphology is as described for the genus. Morulae can be up to 6 μm in diameter (Popov et al., 1998). Dense-core and reticulate cells occur together in vacuoles; both forms divide by binary fission. There is no fibrillar matrix; the vacuolar space may contain empty vesicles. The bacterial cytoplasmic membrane may protrude into the periplasmic space or invaginate into the interior of the cell.

A. phagocytophilum causes tick-borne fever of ruminants, mainly in Western Europe (Gordon et al., 1932; Hudson, 1950; Foggie, 1951). Strains previously known as *Ehrlichia equi* and HGE agent cause equine granulocytic ehrlichiosis (Gribble, 1969; Stannard et al., 1969; Madigan and Gribble, 1987), a form of canine granulocytic ehrlichiosis (Greig et al., 1996; Pusterla et al., 1997), and human granulocytic ehrlichiosis in northern Europe and the Western Hemisphere (Bakken et al., 1994; Chen et al., 1994; Goodman et al., 1996). Serological evidence indicates that asymptomatic infection occurs in humans and animals. Clinical signs in humans and animals include fever accompanied by leukopenia and thrombocytopenia; opportunistic infections may occur in humans and animals (Walker and Dum-

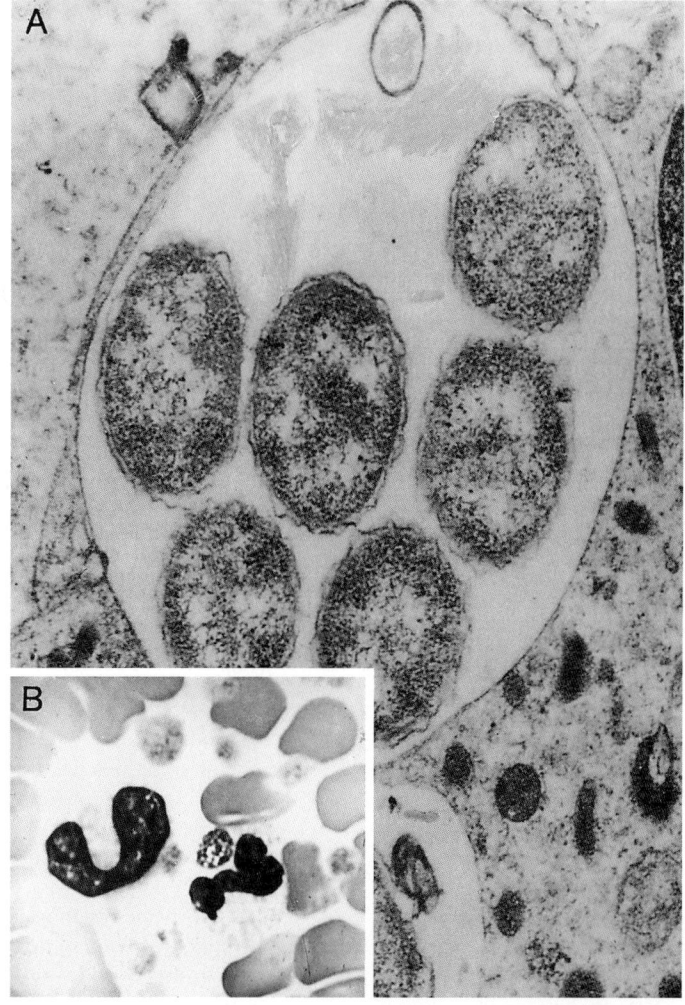

FIGURE BXII.α.40. *Anaplasma phagocytophilum A,* Transmission electron micrograph of an infected equine blood granulocyte with an intracytoplasmic inclusion body (morula). Several single bacteria bound by a rippled cell wall and plasma membrane are evident (× 51,000). (Reproduced by permission from D.M. Sells et al., Infection and Immunity, *13:* 273–280, 1976, ©American Society for Microbiology, Washington, D.C.). *B,* An intragranulocytic inclusion body stained by a Romanowsky method (Giemsa stain, × 1,020) (Courtesy of Cynthia Holland).

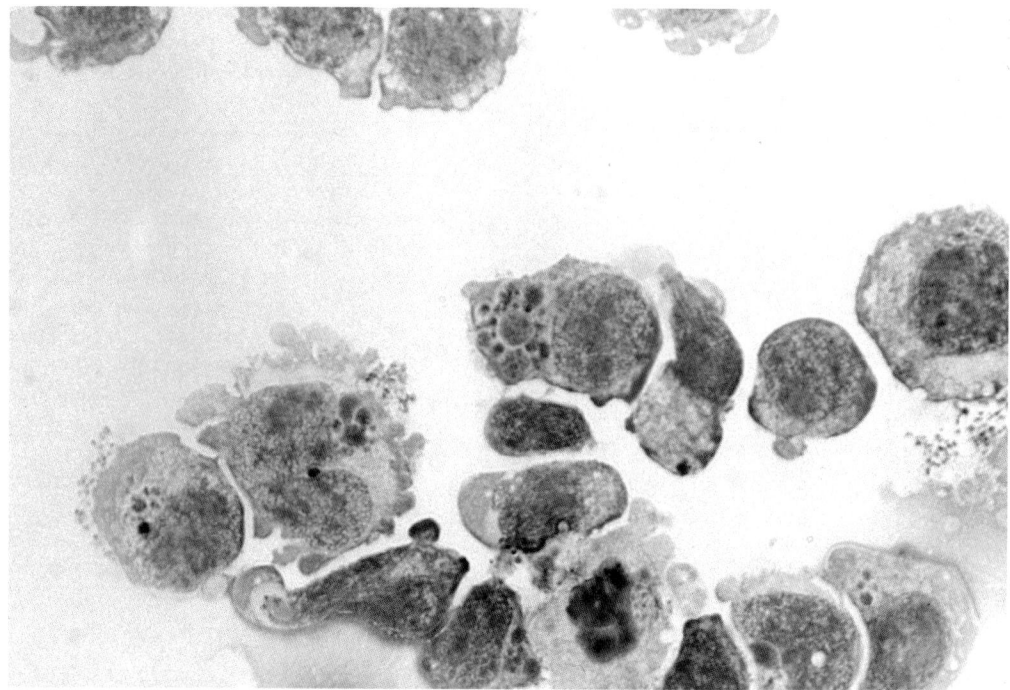

FIGURE BXII.α.41. *Anaplasma phagocytophilum* intracytoplasmic inclusions (morulae) that contain many individual bacteria in cultured human HL-60 promyelocytic leukemia cells, stained with a Romanowsky stain (LeukoStat stain, × 360).

ler, 1996). Other signs include depression and anorexia in horses and dogs and limb edema and ataxia in horses. Humans may suffer liver damage (Aguero-Rosenfeld et al., 1996; Bakken et al., 1996). The organism is widely distributed geographically.

Tick vectors include species of the *Ixodes persulcatus* complex (MacLeod and Gordon, 1933; Foggie, 1951; Richter et al., 1996; Telford et al., 1996). Organisms are maintained in ticks by transstadial but not transovarial transmission. Important reservoirs include rodents (*Peromyscus leucopus*) and potentially large mammals such as cattle and wild ruminants, including cervids and deer.

All isolates cross-react serologically and share antigens with *Ehrlichia canis*, *E. chaffeensis*, and *E. ruminantium*. Major antigens are 42–49 kDa outer membrane proteins encoded by a multigene family (Dumler et al., 1995; Asanovich et al., 1997; Zhi et al., 1997, 1998; Murphy et al., 1998). The primary structures of these antigens are similar to corresponding antigens of *A. marginale*, *E. ruminantium*, *E. canis*, *E. chaffeensis*, and *Wolbachia*.

A. phagocytophilum is susceptible to low concentrations of tetracycline antibiotics, rifamycins, and some fluoroquinolones such as trovafloxacin and resistant to moderate or high concentrations of β-lactams, aminoglycosides, macrolides, and chloramphenicol (Klein et al., 1997).

A. phagocytophilum may be isolated from infected animals by passage of fresh or dimethyl sulfoxide-treated frozen anticoagulated blood specimens or by *in vitro* cultivation in granulocyte cultures, such as HL-60 cells, or in tick cells such as IDE8 cells. Infected cells may be stored for several months at −70°C or for years in liquid N₂ if highly infected. The storage medium is RPMI 1640 medium supplemented with 10% dimethyl sulfoxide. Identification is best achieved

by polymerase chain reaction amplification of *A. phagocytophilum*-specific gene targets or by immunocytochemical identification using specific polyclonal or monoclonal antibodies. The genome size is approximately 1500 kbp (Rydkina et al., 1999).

Within the species, *A. phagocytophilum* 16S rRNA gene sequences exhibit >99.5% identity, and *groESL* sequences exhibit ≥99.0% identity.

The mol% G + C of the DNA is: 43.8 (by gene sampling).
Type strain: Webster.
GenBank accession number (16S rRNA): U02521.

7. **"Anaplasma platys"** (French and Harvey 1983) Dumler, Barbet, Bekker, Dasch, Palmer, Ray, Rikihisa and Rurangirwa 2001, 2159 (**"Ehrlichia platys"** French and Harvey 1983, 2410.)

pla'tys. Gr. adj. *platys* flat, the word from which platelet is derived.*

The characteristics are as described for the genus and as listed in Tables BXII.α.35 and BXII.α.36. The cells are coccoid to ellipsoidal, often pleomorphic, and infect platelets of dogs. Except for the presence in platelets, the morphologic and ultrastructural appearance is not different from that of other members of *Anaplasma* (Mathew et al., 1997).

This species is the causative agent of infectious cyclic thrombocytopenia of dogs, a disease that results in decreasingly severe episodes of fever and thrombocytopenia. The bacterium may also cause inapparent infection in rumi-

---

Editorial Note: The name of this species has not been validly published, because the etymology of *platys* was missing from the descriptions.

nants (Allsopp et al., 1997). The principal clinical manifestation is thrombocytopenia in the absence of fever and clinical signs. Hemorrhage occurs rarely if associated with surgery or accidents.

The organisms are likely transmitted by ticks, perhaps including *Rhipicephalus sanguineus* or *Amblyomma* spp. The observation of infection in sheep and impalas suggests a potential role for ruminants as reservoirs (Allsopp et al., 1997; Du Plessis et al., 1997).

This species has not yet been cultivated *in vitro*, but it can be propagated by serial intravenous passage in susceptible dogs. Infected blood can be stored in liquid N_2 when diluted 1:2 in phosphate buffered saline containing a final concentration of 7.5% glycerol.

Identification is best achieved by polymerase chain reaction amplification of "*A. platys*"-specific or white-tailed deer *Ehrlichia*-specific gene targets (Little et al., 1997; Mathew et al., 1997) or by immunocytochemical identification using specific polyclonal antibodies (Simpson and Gaunt, 1991).

"*A. platys*" is serologically distinct from *Ehrlichia canis*. The relapsing nature of the infection is similar to that seen in bovine anaplasmosis and may similarly be associated with changes in surface protein expression.

The 16S rRNA gene sequences among three strains are ≥99.5% identical, including one strain originating from South African sheep.

16S rRNA gene sequences that have been amplified from the blood of white-tailed deer and an *Amblyomma ameri-*

canum tick in North America are most similar to those of "*A. platys*", with a nucleic acid sequence identity between 96.5 and 98.1% (Dawson et al., 1996; Brandsma et al., 1999). The 16S rRNA gene sequences indicate that the "white-tailed deer" group forms a reproducible and unique clade that is between 96.9% and 98.6% identical to other members of the *Anaplasma* genus, but is not more than 92.3% identical to any member of the genera *Ehrlichia*, *Wolbachia*, *Neorickettsia*, *Orientia*, or *Rickettsia*. The "white-tailed deer" bacteria have never been morphologically identified or cultivated *in vitro*. "*A. platys*" has not yet been cultivated *in vitro*.

The mol% G + C of the DNA is: not known.

Deposited strain: no strain isolated.

GenBank accession number (16S rRNA): M82801.

Species Incertae Sedis

1. **"Anaplasma mesaeterum"** Uilenberg, van Vorstenbosch and Perie 1979, 21.

    mes.ae.ter′um. L.

The taxonomic relationship of this organism to the genus *Anaplasma* is not known. The organisms were identified in sheep in The Netherlands. The description is similar to that of *A. ovis* except for an increased pathogenicity for sheep and a decreased proportion of peripherally located bacteria in the infected erythrocytes.

Transmission is by either *Ixodes ricinus* or *Haemaphysalis punctata* ticks.

Genus II. **Ehrlichia** *Moshkovski 1945, 18*[AL] *emend. Dumler, Barbet, Bekker, Dasch, Palmer, Ray, Rikihisa and Rurangirwa 2001, 2157*

J. Stephen Dumler, Yasuko Rikihisa and Gregory A. Dasch

Ehr.lich′i.a. M.L. fem. n. *Ehrlichia* named after Paul Ehrlich, a German bacteriologist.

Coccoid to ellipsoidal cells. Found in cytoplasmic vacuoles in endothelial or hemopoietic mammalian host cells, including macrophages, monocytes, neutrophils, and cells of the bone marrow, liver, spleen, or lymph nodes. Often occur in clusters (morulae). Pleomorphic; occur as **reticulate cells and "dense core forms" with condensed cytoplasm** (Popov et al., 1998). Gram negative. Nonmotile. Some species grow in cultured tick cells or in cultured mammalian monocytes, macrophages, or endothelial cells. **Cause disease in ruminants, canids, rodents, and humans. Ticks are the primary vectors and hosts.** Belong to the *Alphaproteobacteria* by 16S rDNA gene sequence analysis; the organisms exhibit ≥ 97.6% 16S rRNA gene sequence identity with one another. Where studied, these bacteria possess surface protein antigen genes that are tandemly arranged in the chromosome and have a high degree of sequence similarity (Ohashi et al., 1998a, b; Reddy et al., 1998; McBride et al., 2000; Yu et al., 2000a).

The mol% G + C of the DNA is: 30–56.

Type species: **Ehrlichia canis** (Donatien and Lestoquard 1935) Moshkovski 1945, 18 (*Rickettsia canis* Donatien and Lestoquard 1935, 419) emend. Dumler, Barbet, Bekker, Dasch, Palmer, Ray, Rikihisa and Rurangirwa 2001, 2159.

FURTHER DESCRIPTIVE INFORMATION

Phylogeny By 16S rRNA gene and *groESL* operon sequence analysis, the genus *Ehrlichia* forms a clade distinct from *Ana-*

plasma, Wolbachia, Neorickettsia, Rickettsia, and *Orientia* in the *Alphaproteobacteria*, Order *Rickettsiales*. All species in the genus show more than 97.6% and 86.3% similarity in the 16S rRNA gene and *groESL* operon sequences, respectively. Where investigated, the organisms possess multiple similar genes that encode surface protein antigens varying in molecular size from approximately 28–32 kDa (Ohashi et al., 1998a, b; Reddy et al., 1998; McBride et al., 2000; Yu et al., 2000a). The bacteria are only distantly related to other obligate intracellular bacteria in genera such as *Chlamydia* and *Coxiella* and are unrelated to cell-wall-free bacteria such as *Eperythrozoon* and *Haemobartonella*.

Cell morphology Ehrlichiae occur in membrane-bound vacuoles in the cytoplasm of host cells of hematopoietic origin and, for some species, in endothelial cells. The bacteria form inclusions (morulae) that contain variable numbers of organisms (Simpson, 1972, 1974; Hildebrandt et al., 1973) (See Figs. BXII.α.42 and BXII.α.43). The individual organisms are approximately 0.5 μm in diameter, but size may vary considerably. The morulae range in size up to 4.0 μm in diameter. The organisms are weakly Gram negative. Ehrlichiae are best stained with Romanowsky stains, where the organisms develop a dark blue appearance as opposed to the violet color of the host cell nucleus. Organisms stain poorly with the Giménez method but easily with acridine orange.

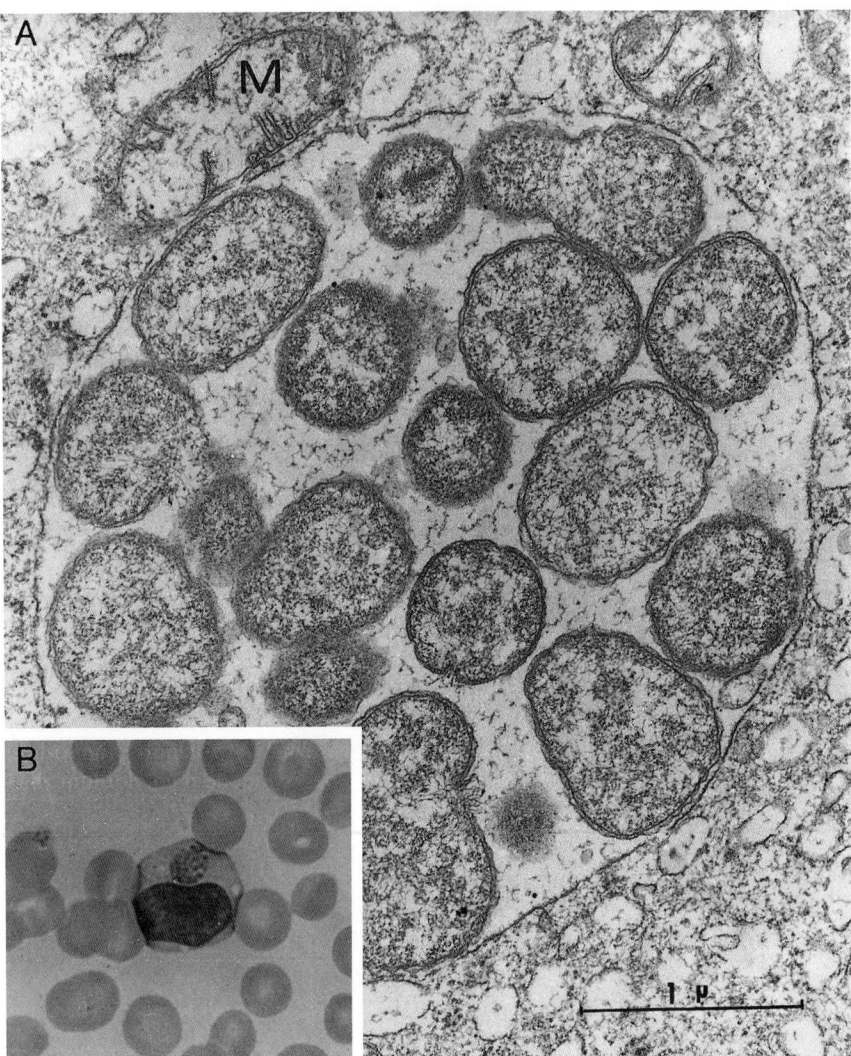

FIGURE BXII.α.42. *Ehrlichia canis. A,* Transmission electron micrograph of an infected canine blood monocyte with an intracytoplasmic inclusion body (morula). Numerous bacteria with distinct plasma membranes and rippled outer cell walls are shown. Note the presence of fibrils and occasional tubules within the intravacuolar, extrabacterial space. *M* = mitochondrion (× 60,000). (Reproduced with permission from P.K. Hildebrandt et al., Infection and Immunity, 7: 265–271, 1973, ©American Society for Microbiology, Washington, D.C. *B,* An intramonocytic inclusion body stained by a Romanowsky method (Giemsa × 680) (Courtesy of S.A. Ewing, Veterinary Medicine, Oklahoma State University, Stillwater).

Cell wall composition   Little is known of the components of the cell wall in *Ehrlichia*. Lipopolysaccharide and lipooligosaccharide have not yet been detected, and it is not known if the cell wall contains peptidoglycan.

Fine structure   By electron microscopy, ehrlichiae reside within membrane-bound vacuoles that are early endosomes for *E. chaffeensis* (Mott et al., 1999). The ultrastructure of the cell envelope reveals simple inner and outer leaflets similar to those of Gram-negative bacteria (Popov et al., 1998). Internal structures consisting of chromatin strands and ribosomes are readily visualized. Two distinct morphologic forms may be detected, dense-core forms that contain a relatively dense central or eccentrically located condensation of chromatin strands, and reticulate cell forms that contain a homogeneous loose matrix of chromatin strands among which ribosomes are spread. Reticulate cells are typically identified *in vivo*, and dense core cells are

identified predominantly during *in vitro* propagation. Both forms undergo binary fission, suggesting that a developmental cycle associated with these morphologies is unlikely. *Ehrlichia* species may produce an abundant membrane that on occasion wraps around individual ehrlichiae or invaginates into the ehrlichial cell and occasionally forms tubule and vesicle profiles in the vacuolar space. Unlike the genus *Anaplasma*, *Ehrlichia* species possess a fibrillar matrix within the vacuolar space.

Cultural characteristics   *E. canis, E. chaffeensis, E. ruminantium,* and *E. muris* have been successfully propagated *in vitro*. Most species may be propagated in primary monocyte cultures established from blood or in macrophage/histiocyte cell lines derived from humans, dogs, and mice (e.g. THP-1, DH82, and P388D1, among others). *E. canis* has been successfully cultivated in primary canine monocyte cultures and in canine-mouse hybrid cells (Nyindo et al., 1971; Stephenson and Osterman, 1977; Hemelt

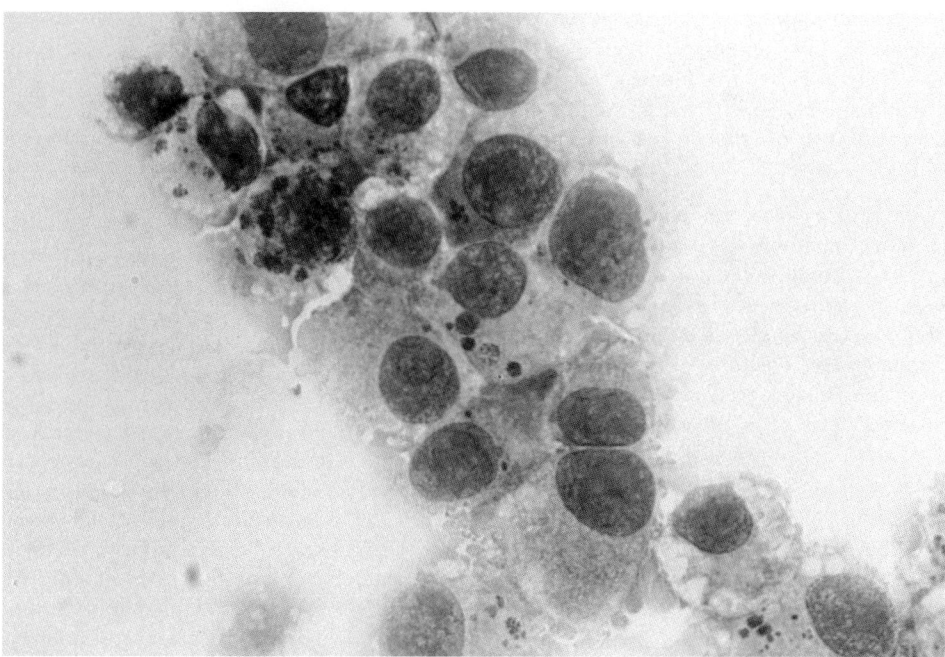

FIGURE BXII.α.43. *Ehrlichia chaffeensis.* Intracytoplasmic inclusions that contain many individual bacteria in cultured canine DH82 histiocytes, stained with a Romanowsky stain (LeukoStat stain, × 340).

et al., 1980; Dawson et al., 1991). *E. chaffeensis* has also been propagated in fibroblast cell lines (Vero, HEL, BGM) and in endothelial cell lines (Dawson et al., 1993; Brouqui et al., 1994; Chen et al., 1995b). *E. ruminantium* is often propagated in primary bovine aortic endothelial cell cultures (Martinez et al., 1993; Yunker, 1995; Perez et al., 1997). *E. canis* has also been propagated in tick cell tissue culture (Ewing et al., 1995). No *Ehrlichia* species has been cultivated on cell-free medium. In cell culture, infection is observed at first with one or few bacteria within the vacuole, previously referred to as an "initial body". With continued binary fission of the bacteria, increasingly larger intravacuolar microcolonies are formed, which are called intermediate bodies or morulae, depending upon size. Ultrastructural examination has shown that all vacuoles may contain either reticulate or dense-cored cells or both and that each of these forms undergoes binary fission; these findings argue against a well-defined intracellular life cycle as observed for *Chlamydia* spp. Multiple infected vacuoles may exist in a single cell, presumably owing to the lack of fusion of infected endosomes.

Antigenic structure All *Ehrlichia* species contain a multigene family that encodes major surface proteins. These proteins are the immunodominant antigens of the genus, and the genes encoding them have a relatively high degree of similarity to corresponding genes in the genera *Anaplasma* and *Wolbachia*. The genes are usually characterized by the presence of several highly conserved sequences that flank several variable sequences. B cell epitopes map to the amino acids encoded by the variable regions, suggesting a role in immune evasion. *In vitro*, from one to many of these genes are transcriptionally active and yield protein antigens of various molecular sizes ranging from 28–32 kDa. Other antigenic components may be conserved, such as the heat shock proteins (GroESL), or they may be species specific.

Antibiotic susceptibility Ehrlichiae are most strongly inhibited during *in vitro* propagation by rifampin and tetracycline antibiotics, including tetracycline hydrochloride, oxytetracycline, and doxycycline. Other antibiotics are ineffective, including aminoglycosides, macrolides, fluoroquinolones, chloramphenicol, and β-lactams. Patients with human monocytic ehrlichiosis have a shorter duration of fever and hospitalization when treated with either a tetracycline or chloramphenicol than with other antibiotics.

Pathogenicity The mechanisms by which *Ehrlichia* species cause disease are poorly understood. During *in vitro* propagation, infected mononuclear phagocytes undergo both necrosis and apoptosis. The *in vivo* infection presents as a non-specific febrile illness with leukopenia, mild anemia, thrombocytopenia, and mild hepatic injury for most species. By contrast, *E. canis* infection in some dog breeds results in a persistent infection that terminates with recrudescence months to years later and is often fatal. Infections and ehrlichial loads appear to be more severe in immunocompromised hosts (splenectomized, corticosteroid therapy, HIV infection, transplant recipients, severe combined immunodeficiency); however, under most other circumstances, the degree of cytopenia observed is out of proportion to the quantity of ehrlichiae detected. An exception may occur in *E. ruminantium* infections, where a significant proportion of the pathogenesis may be related directly to vascular injury after endothelial cell infection, particularly in the cerebral microvascular beds. The proportion of infections that result in clinical signs is still not known, but among those with clinical signs, the proportion with severe morbidity may be as high as 10–50%. Clinically evident infection may be observed in a range of animals including cattle, goats, sheep, deer, other domestic and wild ruminants, dogs and other canidae, mice and other small rodents, and humans.

Ehrlichia species gain access to host cells by adherence to surface proteins that are usually glycosylated. The bacteria are then internalized within early endosomes; for *E. chaffeensis* these endosomes selectively accumulate transferrin receptors, effectively sequestering the infected vacuole into a receptor salvage pathway that precludes lysosome fusion. Infection of macrophages *in vitro* results in low-level secretion of proinflammatory cytokines, whereas binding of immunoglobulin-opsonized *E. chaffeensis* to the macrophage surface results in a high-level release of proinflammatory cytokines.

Ecology Ticks are the primary vectors and hosts for *Ehrlichia* species. Although the microorganisms are passed transstadially in ticks, definite transovarial transmission of identified ehrlichiae has not been demonstrated. The range of potential vertebrate hosts for *Ehrlichia* species is not completely defined owing to a lack of clinical signs in many reservoir hosts, including deer and small mammals. Some mammalian hosts, such as white-tailed deer, some ruminants, and dogs, maintain infectivity for long intervals (months to years) in the absence of clinical signs, whereas other mammals, such as some dog breeds and humans, develop sterile immunity after clinically apparent primary infection. Thus, natural maintenance of ehrlichiae is dependent upon horizontal transmission involving acutely and persistently infected mammals and ticks.

ENRICHMENT AND ISOLATION PROCEDURES

Some species of *Ehrlichia* can be isolated and propagated in cell culture. For isolation, blood is the most appropriate specimen for inoculation. Blood is fractionated into component leukocyte fractions or inoculated directly into appropriate cell cultures, including primary monocyte cultures or cell lines of macrophages (canine DH82, human UL37 or differentiated HL-60, murine P388D1), myelomonocytic cells (human THP-1), fibroblasts (Vero, L929), or endothelial cells. Cultured cells are examined by Romanowsky staining, immunofluorescent staining, or by nucleic acid detection methods to establish the presence of the infectious agent. Primary isolation may require intervals as short as several days or longer than 30 d. Partially purified ehrlichiae may be prepared by density gradient centrifugation using diatrizoate meglumine (Renografin), sucrose, or Percoll gradients.

MAINTENANCE PROCEDURES

E. canis, *E. ewingii*, *E. ruminantium*, and *E. muris* can be serially passed by animal inoculation using blood, spleen, or other tissues. Infectious blood can be stored by suspension with glycerol or dimethyl sulfoxide at 4°C for several days to one week, at −80°C for weeks or months, and in liquid nitrogen for years. Cultured cells containing the bacteria can be stored in media containing 30% fetal bovine serum and 10% dimethyl sulfoxide at −80°C for months to years and in liquid nitrogen for years.

DIFFERENTIATION OF THE GENUS *EHRLICHIA* FROM OTHER GENERA

Characteristics useful for the differentiation of the genus *Ehrlichia* from the other genera of the family *Anaplasmataceae* are listed in Table BXII.α.33 of the section describing the family. *Ehrlichia* species may be morphologically difficult to distinguish from *Anaplasma* and *Neorickettsia* species that grow within bone marrow-derived cells. By ultrastructural examinations, *Ehrlichia* spp. tend to form large morulae containing many cells suspended in an intravacuolar, extrabacterial fibrillar matrix, whereas *Anaplasma* spp. tend to lack additional intravacuolar structures and form smaller morulae containing low numbers of bacteria. Differen-

tiation is best achieved by using DNA sequence analysis of the 16S rRNA gene or the *groESL* operon.

TAXONOMIC COMMENTS

Historically, *Ehrlichia* species were detected as intracytoplasmic clusters of bacteria (morulae) within peripheral blood leukocytes. An essential characteristic was the presence of the bacteria within membrane-bound vacuoles. Other important characteristics included the host mammalian species infected, the geographic distribution of the infectious agent, and a minor degree of antigenic cross-reactivity among proposed species. Most species were recognized as transmissible by the bite of a tick vector.

With the advent of molecular biological methods, a more accurate understanding of the phylogenetic and potential taxonomic positions of *Ehrlichia* species has emerged. Taxonomic placements in the genus *Ehrlichia* are now largely determined by nucleotide sequence analysis of the 16S rRNA genes and nucleotide and amino acid sequence analyses of the *groESL* operon and GroEL proteins. This information is used in conjunction with an assessment of the degree of antigenic cross-reactivity, the presence of similar outer membrane protein antigens, the tropism for mammalian cell hosts derived from bone marrow precursors, and the range of mammalian and tick hosts.

Members of the genus *Ehrlichia* share greater than 97.6% 16S rRNA gene nucleotide sequence similarity. The branching order established in dendrograms and phylogenetic analyses mimics similar arrangements predicted by the sequence analysis of the *groESL* operon and by antigenic analysis of the GroEL proteins. The specific host cell type, the infected host mammalian species, and the geographic distribution do not correlate with predicted genetic relationships; however, residence within an intracytoplasmic endosome and transmissibility by tick vectors are characteristic of all species. Although clear differences in the types of clinical manifestations are observed in appropriate infected mammalian hosts, the weight of objective genetic, antigenic, and pathogenic features provides unequivocal evidence of an evolutionary linkage. The current genetic taxonomic system has yielded several significant changes from previous methods of classification; these changes include the removal of several species from the genus *Ehrlichia* into the genera *Anaplasma* or *Neorickettsia*, and the integration of the genus *Cowdria* into the genus *Ehrlichia* (Dumler et al., 2001). It is anticipated that whole genome sequences may further still refine the genetic classification and yield new combinations not currently considered.

FURTHER READING

Dame, J.B., S.M. Mahan and C.A. Yowell. 1992. Phylogenetic relationship of *Cowdria ruminantium*, agent of heartwater, to *Anaplasma marginale* and other members of the order *Rickettsiales* determined on the basis of 16S rRNA sequence. Int. J. Syst. Bacteriol. *42*: 270–274.

Popov, V.L., V.C. Han, S.M. Chen, J.S. Dumler, H.M. Feng, T.G. Andreadis, R.B. Tesh and D.H. Walker. 1998. Ultrastructural differentiation of the genogroups in the genus *Ehrlichia*. J. Med. Microbiol. *47*: 235–251.

Rikihisa, Y. 1991. The tribe *Ehrlichieae* and ehrlichial diseases. Clin. Microbiol. Rev. *4*: 286–308.

Wen, B., Y. Rikihisa, J. Mott, P.A. Fuerst, M. Kawahara and C. Suto. 1995. *Ehrlichia muris* sp. nov., identified on the basis of 16S rRNA base sequences and serological, morphological, and biological characteristics. Int. J. Syst. Bacteriol. *45*: 250–254.

Yu, X., J.W. McBride, X. Zhang and D.H. Walker. 2000. Characterization of the complete transcriptionally active *Ehrlichia chaffeensis* 28 kDa outer membrane protein multigene family. Gene *248*: 59–68.

DIFFERENTIATION OF THE SPECIES OF THE GENUS *EHRLICHIA*

The preferred method for definitive identification of *Ehrlichia* species depends upon sequence analysis of all or part of the 16S rRNA gene or the *groESL* operon or upon PCR amplification using species-specific oligonucleotide primers. The full 16S rRNA gene sequence may be identified by many different methods, including commercially available systems such as the MicroSeq™ Full Gene 16S rDNA Bacterial Sequencing Kit (Applied Biosys-

tems, Foster City, CA). Species-specific primers for *E. canis, E. chaffeensis, E. muris,* and *E. ruminantium* are shown in Table BXII.α.37. Various methods and protocols for diagnostic identification in clinical samples have been described; the diagnostic sensitivity and specificity should be determined by each laboratory performing the clinical assay.

List of species of the genus Ehrlichia

1. **Ehrlichia canis** (Donatien and Lestoquard 1935) Moshkovski 1945, 18[VP] (*Rickettsia canis* Donatien and Lestoquard 1935, 419) emend. Dumler, Barbet, Bekker, Dasch, Palmer, Ray, Rikihisa and Rurangirwa 2001, 2159.
ca' nis. M.L. gen. n. *canis* of the dog.

The characteristics are as given for the genus and as listed in Tables BXII.α.38 and BXII.α.39. See also Fig.

BXII.α.42. *E. canis* forms a phylogenetically distinct clade in the genus *Ehrlichia*, differing from other species by 1.3–2.4% in the 16S rRNA gene sequence and by 8.7–14.5% in the *groESL* operon sequence. *E. canis* possesses polymorphic major surface proteins encoded by multigene families similar to those in other *Ehrlichia* species and more distantly related to those in *Anaplasma* and *Wolbachia* species. The

TABLE BXII.α.37. Species-specific oligonucleotide primers for PCR identification of *Ehrlichia* species.

Species	Targeted gene	Forward primer	Reverse primer
E. canis	16S rRNA[a]	5′-CAATTGCTTATAACCTTTTGGTTATAAAT-3′	5′-ATAGGGAAGATAATGACGGTACCTATA-3′
E. chaffeensis	16S rRNA[a]	5′-CAATTATTTATAGCCTCTGGTTATAGGA-3′	5′-ATAGGGAAGATAATGACGGTACCTATA-3′
E. ewingii	16S rRNA[a]	5′-CAATTCCTAAATAGTCTCTGACTATTTGA-3′	5′-ATAGGGAAGATAATGACGGTACCTATA-3′
E. muris	16S rRNA[a]	5′-TAGCTACCCATAGCTTTTTTAGCTATAGG-3′	5′-ATAGGGAAGATAATGACGGTACCTATA-3′
E. ruminantium	16S rRNA[a]	5′-CAGTTATTTATAGCTTCGGCTATGAGTATC-3′	5′-ATAGGGAAGATAATGACGGTACCTATA-3′
E. ruminantium	*map1*[b]	5′-GATGTAATACAGGAAGAG-3′	5′-CTATTCTTGGTCCATTC-3′

[a]Based upon unique and shared regions as in Anderson et al., 1992b.

[b]From Kock et al., 1995.

TABLE BXII.α.38. Diagnostic features of *Ehrlichia* species.[a]

Characteristic	*E. canis*	*E. chaffeensis*	*E. ewingii*	*E. muris*	*E. ruminantium*
Usual host species	Dogs	Dogs, humans, deer	Dogs, humans	Mice	Cattle
Usual infected host cell type	Monocytes	Monocytes	Neutrophils	Monocytes	Endothelial cells, neutrophils
Present in peripheral blood[a]	+ +	+ +	+	−	±
Serological reactions with:[b]					
E. canis					+ +
E. chaffeensis	+ + +				
E. ewingii	+ + +	+ + +			
E. muris	+ + +	+ + +	+ + +		
E. ruminantium	+ +	+ +	+ +	?	
Anaplasma marginale	−	−	−	−	±
Anaplasma phagocytophilum	±	±	±	?	±

[a]Symbols: ±, rarely present or difficult to identify; +, present infrequently; + +, present moderately often.

[b]Symbols: −, none; +, weak; + +, moderate; + + +, strong; ?, unknown or not reported.

TABLE BXII.α.39. Other characteristics of the species of the genus *Ehrlichia*.

	E. canis	*E. chaffeensis*	*E. ewingii*	*E. muris*	*E. ruminantium*
Geographic distribution	Worldwide	North America	North America	Japan	Africa, Caribbean
Host(s) with clinical signs	Dogs, other canids	Humans, dogs	Dogs, humans	Mice	Cattle, sheep, goats, other ruminants
Reservoir host(s)	Canids	Cervids, canids	Dogs	Mice	Ruminants
Vectors	*Rhipicephalus sanguineus*	*Amblyomma americanum, Dermacentor variabilis?*	*Amblyomma americanum*	*Haemaphysalis flava*	*Amblyomma variegatum, Amblyomma cajennense*
Host cell	Monocytes and macrophages	Monocytes and macrophages	Neutrophils	Macrophages	Endothelial cells and neutrophils
In vitro cultivation	Tick cells, macrophages	Tick cells, macrophages, endothelial cells, fibroblasts	Not cultivated *in vitro*	Macrophages	Endothelial cells, neutrophils, tick cells

function of the polymorphic proteins that vary from 28–32 kDa is not known, but owing to the presence of multiple copies, it is speculated that these proteins might play a role in immune evasion.

Tetracycline antibiotics are effective at eliminating infection *in vitro* but may have some limitations *in vivo*. Other antibiotics, including imidocarb dipropionate, are not effective.

Infection by *E. canis* stimulates the production of antibodies that also strongly react with *E. chaffeensis*, *E. muris*, *E. ruminantium*, and *E. ewingii* and react more weakly with *Anaplasma phagocytophilum* and *Neorickettsia* spp. Since infection results in low-level persistence, natural immunity is often incomplete, and recrudescence can occur months to years after primary infection. Immunity resulting in elimination of a persistent infection may occasionally occur. Different strains of *E. canis* may not induce homologous or heterologous cross-protection. Protection against other *Ehrlichia* species, including *E. chaffeensis*, is not induced by infection with *E. canis*.

E. canis is the causative agent of canine monocytic ehrlichiosis, a severe febrile infection of canids that occurs after tick bites and is associated with pancytopenia. The disease is worldwide in distribution. Asymptomatic human infection has been documented in one case.

Clinical signs of infection occur in canids—including dogs, coyotes, foxes, and wolves, among others. The organisms have not been demonstrated to be infectious for rodents or other small mammals. Persistent infection leading to a carrier "reservoir" state may develop in canids. The major worldwide vector for *E. canis* is *Rhipicephalus sanguineus*, the brown dog tick. Transstadial but not transovarial transmission occurs; thus, the predominant mechanism for natural maintenance involves a mammalian reservoir that is expanded by tick vectors.

Isolation can be achieved by inoculation of leukocytes from infected dogs onto monolayers of canine DH82 histiocyte cells. Infected cells may be stored for months or years frozen at −80°C or for years in liquid nitrogen if maintained in a medium with at least 30% serum and 10% dimethyl sulfoxide. Isolated bacteria may be identified by reaction with polyclonal or monoclonal antibodies or by polymerase chain reaction amplification of conserved *E. canis* genes.

The mol% G + C of the DNA is: 35.3 (by gene sampling).

Type strain: Oklahoma.

GenBank accession number (16S rRNA): M73221.

2. **Ehrlichia chaffeensis** Anderson, Dawson, Jones and Wilson 1992a, 327[VP] (Effective publication: Anderson, Dawson, Jones and Wilson 1991, 2841.)

chaf fe.en.sis. N.L. fem. adj. *chaffeensis* of Chaffee; pertaining to Fort Chaffee in western Arkansas, where the patient from whom the first isolate was prepared was identified; L. adj. *-ensis* derived from; M.L. adj. *chaffeensis* from Chaffee.

The characteristics are as given for the genus and as listed in Tables BXII.α.38 and BXII.α.39. See also Fig. BXII.α.43. *E. chaffeensis* forms a phylogenetically distinct clade in the genus *Ehrlichia*; it differs from other species by 1.0–1.9% in 16S rRNA gene sequence and by 6.4–17.7% in the *groESL* operon sequence. Possesses polymorphic major surface proteins of approximately 28 kDa. These proteins

are encoded by 21 distinct genes in a multigene family. These genes are similar to those in other ehrlichiae and are more distantly related to those of *Anaplasma* and *Wolbachia* spp. The nucleic acid sequences of the genes encoding the proteins vary between strains. The function of these polymorphic proteins is not known, but it is speculated that they play a role in immune system evasion. The organisms also possess genes that encode major membrane glycoproteins ranging from 120–140 kDa.

Tetracycline and rifamycin are effective at eliminating infection *in vitro*. Aminoglycosides, fluoroquinolones, macrolides, sulfa-containing, and β-lactam antibiotics are not effective *in vitro*.

E. chaffeensis induces strong cross-reactive serologic responses to *E. canis*, *E. ewingii*, *E. muris*, and *E. ruminantium*, and weaker responses to *Anaplasma phagocytophilum*, and possibly to *Rickettsia rickettsii* and *R. typhi*. Serologic responses can be differentiated by protein immunoblotting.

Infection in deer, goats, and dogs may be persistent, and these animals may be important natural reservoirs. Infection in humans is generally self-limited, and persistence is exceedingly rare. Second infections and recrudescence have not been well investigated in animals, but are probably rare in humans.

E. chaffeensis is a causative agent of monocytic ehrlichiosis in humans and dogs. Infection can result in a severe illness with fever, pancytopenia, and occasionally other manifestations after tick bites. The disease appears to be limited to North America although serologic evidence of infection has been reported in humans in Europe and Africa. Asymptomatic and symptomatic canine infections and asymptomatic human infection are well documented. Persistent infection leading to a carrier "reservoir" state may develop in canids and cervids. The major vector for *E. chaffeensis* is *Amblyomma americanum*, the Lone Star tick, although limited data suggests that *Dermacentor variabilis*, the American dog tick, may also play a role in transmission. Transstadial transmission occurs, but transovarial passage has not been documented; thus, the predominant mechanism for natural maintenance involves a mammalian reservoir that is expanded by tick vectors.

Isolation can be achieved by inoculation of infected leukocytes onto monolayers of canine DH82 histiocyte cells. Infected cells can be stored for months or years frozen at −80°C or in liquid nitrogen for years if maintained in medium with at least 30% serum and 10% dimethyl sulfoxide. Isolated bacteria can be identified by reaction with polyclonal or monoclonal antibodies or by polymerase chain reaction amplification of conserved *E. chaffeensis* genes.

The mol% G + C of the DNA is: 33.9 (by gene sampling).

Type strain: Arkansas, ATCC CRL-10679.

GenBank accession number (16S rRNA): M73222.

3. **Ehrlichia ewingii** Anderson, Greene, Jones, and Dawson 1992b, 301[VP] emend. Dumler, Barbet, Bekker, Dasch, Palmer, Ray, Rikihisa and Rurangirwa 2001, 2159.

ew.in' gi.i. N.L. gen. n. *ewingii* of Ewing, named in honor of Sidney A. Ewing for his pioneering work with this agent.

The characteristics are as given for the genus and as listed in Tables BXII.α.38 and BXII.α.39. *E. ewingii* occupies a distinct clade in the genus *Ehrlichia*, with at least 99.7% 16S rRNA gene sequence similarity among strains. *E. ewingii* differs from other species by 1.0–1.4% in the 16S rRNA gene sequence and by 9.9–15.1% in the *groESL* operon sequence. *E. ewingii* also contains at least one copy of a homolog of the p28 gene family of *E. chaffeensis*.

E. ewingii induces antibodies that strongly cross-react with *E. canis* and *E. chaffeensis* by indirect fluorescent antibody tests but can be differentiated from those species by protein immunoblot analysis.

E. ewingii is a causative agent of granulocytic ehrlichiosis in humans and dogs; infections can result in a mild to moderate illness with fever, pancytopenia, and occasionally other manifestations after tick bites. Dogs develop polyarthritis more often than with *E. canis* infection. The disease has been described only in North America. *E. ewingii* may be transmitted by *Amblyomma americanum* (Lone Star tick) under experimental conditions.

No *in vitro* isolation has been accomplished, although limited *in vitro* replication of *ex vivo*-infected neutrophils has been described. The organism may be maintained by intravenous inoculation of dog blood that contains infected neutrophils or from "carrier" dogs.

The mol% G + C of the DNA is: 40.6 (by gene sampling).

Type strain: Stillwater.

GenBank accession number (16S rRNA): M73227.

4. **Ehrlichia muris** Wen, Rikihisa, Mott, Fuerst, Kawahara and Suto 1995b, 254[VP] emend. Dumler, Barbet, Bekker, Dasch, Palmer, Ray, Rikihisa and Rurangirwa 2001, 2159.

mu' ris. L. gen. n. *muris* of the mouse; the species was first isolated from a mouse.

The characteristics are as given for the genus and as listed in Tables BXII.α.38 and BXII.α.39. *E. muris* occupies a distinct clade in the genus *Ehrlichia*, with at least 99.7% 16S rRNA gene sequence similarity among different strains. *E. muris* differs from other species by 1.0–1.9% in the 16S rRNA gene sequence and by 6.4–15.3% in the *groESL* operon sequence. The organisms possess polymorphic major surface proteins of approximately 28 kDa that are encoded by at least two distinct genes; these genes are similar to those in other *Ehrlichia* and more distantly related to those of *Anaplasma* and *Wolbachia* spp.

Tetracyclines effectively prevent infection of mice, but penicillin, aminoglycosides, and sulfonamides do not.

E. muris induces antibodies that cross-react strongly with *E. canis* and *E. chaffeensis* by indirect fluorescent antibody tests and weakly with *Neorickettsia sennetsu*. Antibodies stimulated by *E. muris* infection may be differentiated by protein immunoblot analysis from those elicited by other *Ehrlichia* spp.

E. muris is a pathogen that causes mild to severe clinical signs—including splenomegaly and lymphadenopathy—in wild mice of Japan and perhaps other regions in Asia. The organisms can be detected in splenic and peritoneal macrophages of infected mice. *E. muris* can be transmitted by *Haemaphysalis flava* ticks under natural conditions.

E. muris can be cultivated in the canine histiocyte cell line DH82 maintained in minimal essential medium supplemented with 10% fetal bovine serum and 2 mM L-glutamine; cultures are incubated at 37°C under an air atmosphere enriched with 5% CO_2. No growth occurs in chicken embryo yolk sacs.

The mol% G + C of the DNA is: 38.5 (by gene sampling).

Type strain: AS145, ATCC VR-1411.

GenBank accession number (16S rRNA): U15527.

5. **Ehrlichia ruminantium** (Cowdry 1925) Dumler, Barbet, Bekker, Dasch, Palmer, Ray, Rikihisa and Rurangirwa 2001, 2158[VP] (*Rickettsia ruminantium* Cowdry 1925, 231; *Cowdria ruminantium* Moshkovski 1947, 62.)

ru.mi.nan' ti.um. M.L. gen. pl. n. *ruminantium* of *Ruminantia*, formerly a common name for cud-chewing animals.

The characteristics are as given for the genus and as listed in Tables BXII.α.38 and BXII.α.39. *E. ruminantium* forms a distinct clade in the genus *Ehrlichia*, sharing at least 99.1% nucleotide sequence similarity in the 16S rRNA gene among strains. *E. ruminantium* differs from other species by 1.3–1.6% in the 16S rRNA gene sequence and by 14.5–17.7% in the *groESL* operon sequence. The organisms possess polymorphic major surface proteins encoded by multigene families similar to those in other *Ehrlichia* species and more distantly related to those in *Anaplasma* and *Wolbachia* spp. The function of the polymorphic proteins, which average approximately 32 kDa, is not known, but it is speculated that they might play a role in immune evasion.

Tetracycline antibiotics are effective at eliminating infection *in vivo*.

Infection by *E. ruminantium* stimulates the production of antibodies that also strongly react with *E. chaffeensis* and *E. canis* but react more weakly with *Anaplasma phagocytophilum*, *A. marginale*, *Rickettsia* spp., *Coxiella burnetii*, and *Neorickettsia* spp. Since infection results in low-level persistence, natural immunity is often incomplete and may result in recrudescence months or years after primary infection. Different strains of *E. ruminantium* may not induce homologous or heterologous cross-protection potentially owing to antigenic diversity in this species.

E. ruminantium is the causative agent of heartwater or cowdriosis, a severe febrile infection of ruminants that occurs after tick bites. The disease is associated with fever, organ dysfunction, and severe neurological signs. The disease is limited to sub-Saharan Africa and the Caribbean. Infection of exotic and domestic ruminants is generally severe, although some native ruminant species are refractory to clinical signs or develop only mild illness. Some strains are pathogenic for mice. Persistent infection leading to a carrier "reservoir" state often develops.

The major vectors for *E. ruminantium* are *Amblyomma* ticks, in particular *A. variegatum* and *A. habreum*. Transstadial but not transovarial transmission occurs; thus, the predominant mechanism for natural maintenance involves horizontal transmission through vertebrates.

Isolation can be achieved by inoculation of blood from infected animals onto monolayers of cultured endothelial cells. Infected cells can be stored for months or years when frozen at −80°C or for years in liquid nitrogen. Isolated bacteria can be identified by reaction with antibodies or by polymerase chain reaction amplification of conserved *E. ruminantium* genes.

The mol% G + C of the DNA is: 32.2 (by gene sampling).

Type strain: Welgevonden.

GenBank accession number (16S rRNA): X61659 (Crystal Springs strain).

Genus III. **Neorickettsia** Philip, Hadlow and Hughes 1953, 257[AL] emend. Dumler, Barbet, Bekker, Dasch, Palmer, Ray, Rikihisa and Rurangirwa 2001, 2157

YASUKO RIKIHISA, J. STEPHEN DUMLER AND GREGORY A. DASCH

Ne.o.rick.ett′si.a. Gr. pref. *neo-* new; M.L. fem. n. *Rickettsia* type genus of the family, *Rickettsiaceae*; M.L. fem. n. *Neorickettsia* the new *Rickettsia*

Coccoid or pleomorphic cells that reside in cytoplasmic vacuoles within monocytes and macrophages of dogs, horses, bats, and humans. Small (0.2–0.4 µm) electron dense forms and relatively large (0.8–1.5 µm) lighter forms may occur. Tissues of adult **trematode (fluke) vectors** and all other fluke stages—eggs, miracidia, rediae, sporocysts, cercariae, and metacercariae—produce infection when injected into susceptible hosts, indicating transovarial transmission in the vector. Transstadial transmission also occurs. Gram negative. Nonmotile. Most species are cultivable in peripheral blood monocytes, myelomonocytic cell lines, or promyelocytic cell lines. Some species are pathogenic to laboratory mice. **Etiologic agents of diseases of dogs and other canids, horses, and humans.** Where studied, these bacteria possess an antigenically cross-reactive group-specific protein gene with a high degree of sequence identity. Sensitive to tetracycline antibiotics. The estimated genome size is 860–900 kb.

The mol% G + C of the DNA is: 42.

Type species: **Neorickettsia helminthoeca** Philip, Hadlow and Hughes 1953, 257.

FURTHER DESCRIPTIVE INFORMATION

Phylogeny By 16S rRNA gene and *groESL* sequence analysis, the genus *Neorickettsia* forms a clade distinct from *Anaplasma*, *Ehrlichia*, *Wolbachia*, *Rickettsia*, and *Orientia* in the *Alphaproteobacteria*. When 16S rRNA gene sequences are aligned, all species in the genus are more than 95% similar. The amino acid sequence identity between GroEL of *N. sennetsu* and *N. risticii* is 97.6%; however, the identity of this gene between *Neorickettsia* spp. and *Ehrlichia* spp. or *Anaplasma* spp. is no greater than 59.2%.

Cell morphology *Neorickettsia* species are found in the cytoplasm of monocytes in the blood and in macrophages of lymphoid tissues of infected *Canidae*, humans, bats, and horses. In Romanowsky-stained preparations the organisms are dark blue to purple. Individual round forms are most common, but small clusters (microcolonies) of bacteria in inclusions called morulae are also present. These may be mistaken for large bacteria. The morulae are less compact and are smaller than those of *Ehrlichia* spp. in DH82 cells (Rikihisa et al., 1991).

Ultrastructure By electron microscopy small (0.2–0.4 µm) dense forms and relatively large (0.8–1.5 µm) light forms have been recognized in several species of *Neorickettsia* (Rikihisa et al., 1985, 1991; Rikihisa, 1990b). Ribosomes, fine DNA strands, and two layers (outer and inner) of membrane, are present in the organisms (Rikihisa, et al., 1991) (see Fig. BXII.α.44). Neorickettsiae, like the members of the genera *Ehrlichia* and *Anaplasma*, show no thickening of the inner leaflet of the outer membrane or the outer leaflet of the inner membrane (Rikihisa, 1990b; Rikihisa et al., 1985, 1991). *Neorickettsia* species multiply by binary fission and are found in early endosomes enriched with transferrin receptors in the cytoplasm (Barnewall et al., 1999). Individual organisms or groups of organisms are tightly enveloped by the endosomal membrane. Genes encoding lipid A biosynthesis were not detected, and most genes required for biosynthesis of peptidoglycan were absent (Lin and Rikihisa, 2003).

Cultural characteristics *Neorickettsia* species have not been grown in ordinary bacteriological media or in yolk sacs. Where tested most species are cultivable in peripheral blood monocytes, peritoneal macrophages, myelomonocytic cell lines, or promyelocytic cell lines such as P338D, DH82, and U937 (Rikihisa, 1991b).

Metabolism Members of the genus *Neorickettsia* possess an aerobic type of metabolism and are asaccharolytic. The metabolic activities of *N. risticii* and *N. sennetsu* have been investigated. Neorickettsiae can utilize glutamine and glutamate and can generate ATP (Weiss et al., 1988, 1989) as rickettsiae do, but unlike rickettsiae, neorickettsiae prefer to use glutamine rather than glutamate. They exhibit this preference because the organisms are enveloped by the host membrane and because glutamine enters endosomes more readily than glutamate. Like the genus *Rickettsia*, the genus *Neorickettsia* cannot utilize glucose-6-phosphate or glucose. The greatest metabolic activity of the *Neorickettsia* is observed at pH 7.2 to 8.0 and declines rapidly below pH 7 (Weiss et al., 1988); thus the genus *Neorickettsia* is not acidophilic.

Genetics Genome sizes of *N. sennetsu* and *N. risticii* are approximately 880 and 870 kb, respectively (Rydkina et al., 1996). The sequences of the 16S rRNA genes of *N. helminthoeca*, *N. sennetsu*, and *N. risticii* have been determined. *GroESL* has been cloned from *N. sennetsu* (Zhang et al., 1997), *N. risticii* (Sumner et al., 1997), and from SF agent and *N. helminthoeca* (Rikihisa, et al., unpublished data). The unique feature of the *groESL* operon of *Neorickettsia* spp. is that there is no spacer between *groES* and *groEL* (*groES* and *groEL* overlap by one base). The HSP70 gene of *N. sennetsu* has been cloned (Zhang et al., 1998).

The genes encoding the 51-kDa protein antigen have been identified in various strains of *N. risticii* from horses and trematodes (Barlough et al., 1998; Dutta et al., 1998; Reubel et al., 1998; Kanter et al., 2000) and in other *Neorickettsia* spp. (Rikihisa et al., unpublished data). Amino acid sequence identities of the 51-kDa protein among *N. risticii* strains are greater than 77% (Kanter et al., 2000). Another antigenic protein, 50/85 kDa in size, shows strain-dependent polymorphism, being 50 kDa in strain 25-D and 85 kDa in strain 90-12 by SDS-PAGE, the higher value is due to the presence of tandem repeat nucleotide sequences which vary in size, number and type of repeats (Biswas et al., 1998). Genes encoding a type IV secretion apparatus and two-component regulatory systems were identified (Rikihisa, unpublished data).

Antigenic structure Indirect fluorescent antibody (IFA) and Western immunoblot analyses of purified cells of *N. helminthoeca*, *N. sennetsu* and *N. risticii* using antisera from infected animals have shown that the surface antigens of these organisms are highly cross-reactive (Rikihisa et al., 1991). The cross-reactive antigens between *N. risticii* and *N. sennetsu*, as detected by Western immunoblot analysis, are proteins 51–55 kDa in size (Rikihisa, 1996). The 55-kDa GroEL homolog, the 50/85-kDa strain-specific antigen, and the 51-kDa antigen are in this range (Vemulapalli et al., 1998). The 51- and 50/85-kDa proteins are apparently unrelated. The 51-kDa protein has a putative signal peptide sequence and is thus a potential outer membrane protein (Vemulapalli et al., 1998), whereas the 50/85-kDa protein does not. GroEL is also a dominant antigen in *Neorickettsia* spp. and has been shown to cross-react with the GroEL of *Ehrlichia* spp., *Orientia* spp. and *Rickettsia* spp. but not with the GroEL of *Escherichia coli* (Zhang et al., 1997).

Pathogenicity *N. helminthoeca* is the causative agent of salmon poisoning disease of the *Canidae*, *N. sennetsu* is the agent of human Sennetsu ehrlichiosis, and *N. risticii* is the causative agent of Potomac horse fever or equine monocytic ehrlichiosis. When examined with an *in vitro* cultivation assay system, professional phagocytes—monocytes and macrophages—of host animal species have no intrinsic resistance to infection with *Neorickettsia* spp. *Neorickettsia* spp. also cause systemic febrile illness with thrombocytopenia and leukopenia, with or without lymphadenopathy. *Neorickettsia* spp. induce intestinal manifestations, which are especially severe with *N. risticii*, which causes watery diarrhea. Evidence of a severe inflammatory reaction is usually absent at necropsy. Pathogenicity has been studied most extensively for *N. risticii*. *N. risticii* specifically binds to macrophages, and both the host cell receptor and the *N. risticii* ligand appear to be proteins (Messick and Rikihisa, 1993). Neorickettsiae enter macrophages by receptor-mediated endocytosis, not by phagocytosis, and reside in the early endosomes that do not fuse with lysosomes (Wells and Rikihisa, 1988; Messick and Rikihisa, 1993, Rikihisa et al., 1994). Neorickettsial internalization and proliferation are dependent on Ca^{2+}-calmodulin-signaling and protein tyrosine phosphorylation (Rikihisa et al., 1995; Zhang and Rikihisa, 1997). *N. sennetsu* resides in early endosomes enriched with transferrin receptor and upregulates transferrin receptor mRNA by activating iron-responsive protein 1 (Barnewall et al., 1999). This activation might be a mechanism to facilitate iron uptake by neorickettsiae.

Neorickettsial release appears to occur not only by cell lysis but also through exocytosis by fusion of the inclusion membrane with the plasma membrane. In the case of intestinal epithelial cells, which make a monolayer tightly connected by circumferential zones of intercellular junctions, neorickettsiae appear to be transmitted between adjacent cells *in vitro* by a coupled exocytosis in one cell and endocytosis in an adjacent cell (Rikihisa, 1990a). *N. risticii* can be consistently reisolated from equine peripheral blood monocytes from days 1–28 after infection and 8 days after spontaneous resolution of clinical signs (Mott et al., 1997). *N. risticii* may persist much longer, since in a pregnant mare-fetus system, *N. risticii* has been isolated from the aborted fetus of a mare that was experimentally infected up to 3.5 months prior to the abortion (Dawson et al., 1987; Long et al., 1995).

Mouse pathogenicity of *Neorickettsia* spp. varies. *N. sennetsu* is pathogenic for mice. Clinical signs are ruffled fur, inactivity, and mild diarrhea followed by death in about 3–4 weeks after inoculation (Misao and Kobayashi, 1954). Necropsy of mice reveals marked enlargement of the spleen, the liver, and lymph nodes throughout the body. All human isolates have been obtained from the peritoneal macrophages of the mouse after intraperitoneal inoculation of patient's blood. *N. helminthoeca* is not pathogenic for mice. *N. risticii* can establish infection in mice and horses with or without causing apparent clinical illness. Based on a murine model for Potomac horse fever, the disease is apparently dose-dependent, i.e., with a low dose of *N. risticii*, innate defense mechanisms appear to contain *Neorickettsia* cells below a threshold level. Only at higher doses can the organism cause

disease and pathological changes (Rikihisa et al., 1987). Horses that recover are immune to Potomac horse fever for up to 20 months after infection (Palmer et al., 1990). Exposure to killed organisms can protect mice from *N. risticii* infection (Rikihisa, 1991a). Both humoral and cell-mediated immune responses appear to play significant roles in this protection. The presence of neutralizing antibodies and antibody-dependent cellular toxicity has been demonstrated using a cell culture system (Messick and Rikihisa, 1992a; Rikihisa et al., 1993, 1994) and a murine model (Rikihisa et al., 1993). Polyclonal antiserum to *N. risticii* does not inhibit binding of *N. risticii* cells to P388D macrophages or their internalization, but internalized antibody-coated *N. risticii* fails to survive (Messick and Rikihisa, 1994). Because macrophages bear the Fc receptor, the Fab fragment of the antibody was used to analyze the inhibitory mechanism. Since the Fab fragment of the anti-*N. risticii* IgG blocks the binding of *N. risticii* to macrophages, antibody-coated *N. risticii* most likely enters macrophages via the Fc receptor. Fc-receptor-mediated uptake may deliver *N. risticii* to the intracellular compartment that is susceptible to lysosomal fusion. Alternatively, because ^{14}C-L-glutamine metabolism by purified host cell-free *N. risticii* is blocked by polyclonal antibody, direct metabolic inhibition may make *N. risticii* unable to survive in the macrophage (Messick and Rikihisa, 1994). *N. risticii* is very sensitive to nitric oxide, which is generated by macrophage cytoplasmic nitric oxide synthase induced by interferon gamma (IFN-γ) (Park and Rikihisa, 1991, 1992). Activities of T cells that generate IFN-γ are, however, generally severely depressed in infected mice in a time- and dose-related manner (Rikihisa et al., 1987). One cause for depressed T-cell response was found in macrophages. Class II histocompatibility antigen (Ia antigen) induction on the surface of *N. risticii*-infected macrophages (antigen-presenting cells) is suppressed *in vitro* (Messick and Rikihisa, 1992b), suggesting inhibition of antigen-specific T-cell activation in *N. risticii* infection. *N. risticii* induces only low levels of tissue necrosis factor alpha (TNF-α) and prostaglandin E2 production in murine macrophages *in vitro* (van Heeckeren et al., 1993). In contrast, infected macrophages produce significant amounts of interleukin 1 (IL-1).

Ecology Where known, *Neorickettsia* spp. are transmitted and maintained in trematodes found in *Pleuroceridae* snails (Nyberg et al., 1967; Millemann and Knapp, 1970). Infection of mammals is required for completion of the life cycle of the trematodes, but it is not directly required for maintenance of *Neorickettsia* spp. Dogs, horses, bats, and humans accidentally acquire *Neorickettsia* spp. by ingestion of the metacercaria stage of trematodes in fish or adult aquatic insects. The life cycle of *N. helminthoeca* includes growth in eggs of a digenetic microcercous trematode, *Nanophyetus salmincola*, in the redia stage of the trematode in the snail *Oxytrema silicula*, in the metacercaria stage of the trematode in salmonid fish, and in monocytes and macrophages in the dog. *N. helminthoeca* infects neither the fish nor the snail carrying the trematode. The disease has been reported from the Pacific Coast of Northern California, Oregon, and Washington. Additionally, several salmon poisoning cases have been reported in Vancouver Island, Canada (Booth et al., 1984).

N. risticii has been found in digenetic and virgulate trematodes of the family *Lecithodendoriidae*, which infect snails in the *Pleuroceridae* family, namely, *Elimia* and *Juga* spp., in the midwestern and northeastern U.S. and in California, respectively (Barlough et al., 1998; Reubel et al., 1998; Kanter et al., 2000). *N. risticii*-infected trematodes then penetrate and develop into metacer-

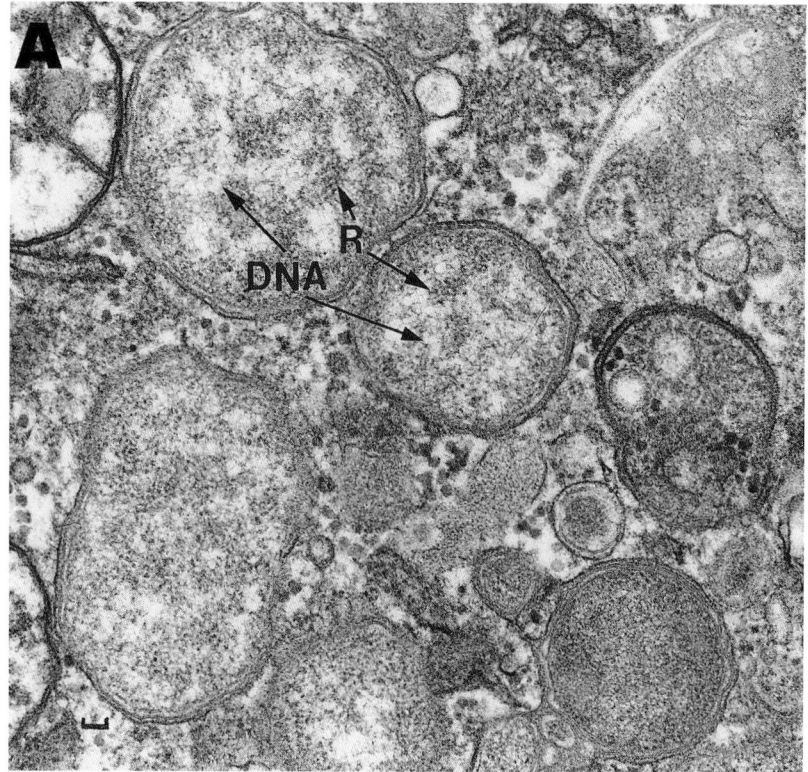

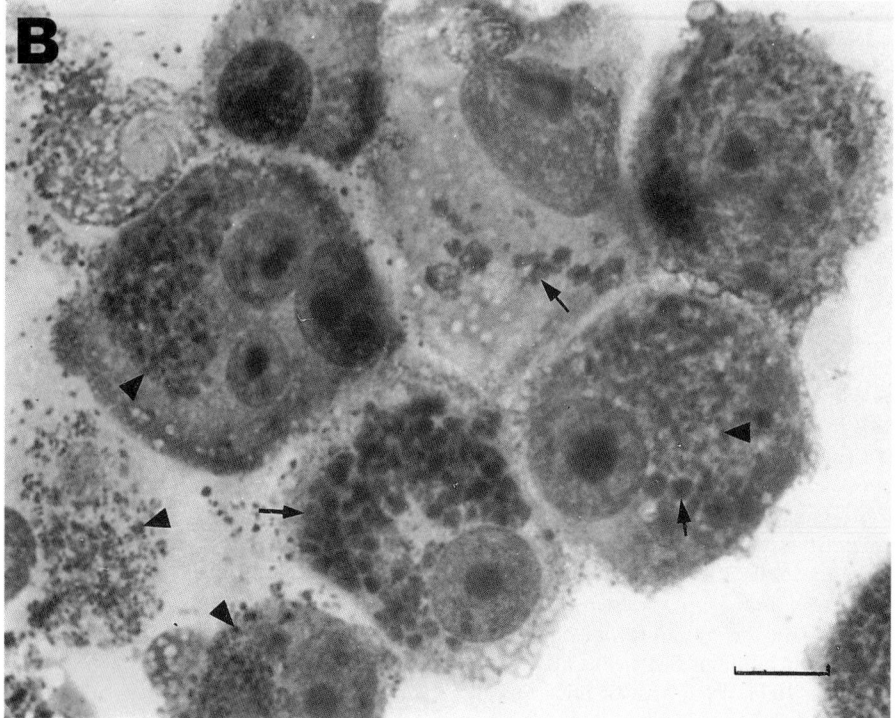

FIGURE BXII.α.44. *Neorickettsia helminthoeca.* *A*, Ultrathin section of an infected, cultured, canine monocytic cell (cell line DH82). *N. helminthoeca* organisms are tightly enveloped by a host cell membrane (*arrowheads*). Each bacterium is surrounded by a distinct plasma membrane and an outer membrane. Fine DNA strands and ribosomes (*R*) are evident in the organisms. (× 37,000). *B*, *N. helminthoeca* are seen in the cytoplasm of DH82 cells as small inclusions containing several organisms (*arrows*), as individual organisms (*arrowheads*), and as extracellular organisms. Preparation stained by Diff-Quik stain (× 1,275) (Reproduced with permission from Y. Rikihisa, et al., Journal of Clinical Microbiology *29*: 1928–1933, 1991, American Society for Microbiology, Washington, D.C.)

cariae in nymphs of a variety of aquatic insect including mayflies, caddis flies, and stone-flies (Chae et al., 2000). *N. risticii* survives through metamorphosis of insects and persists in adult insects; horses can acquire Potomac horse fever by ingestion of infected adult caddis flies (Madigan et al., 2000a; Mott et al., 2002). Despite epidemiological evidence that patients acquire Sennetsu fever by eating raw gray mullet fish, *N. sennetsu* has not been isolated from trematodes. However, the *Stellantochasmus falcatus* (SF) agent, which is more closely related to *N. risticii* than to *N. sennetsu*, has been isolated directly from metacercariae of the *S. falcatus* trematode in the gray mullet fish in a Sennetsu fever-endemic region in Japan (Fukuda et al., 1973; Yamamoto, 1978; Wen et al., 1996).

Antibiotic sensitivity All *Neorickettsia* spp. are sensitive to tetracycline antibiotics; however, they are not susceptible to β-lactam antibiotics, aminoglycosides, macrolides, chloramphenicol, or sulfonamides (Rikihisa and Jiang, 1988, 1989).

ENRICHMENT AND ISOLATION PROCEDURES

N. helminthoeca has been isolated from the blood, spleen, and liver of dogs that have been infected by feeding them metacercariae-infected salmon; the organism was then propagated in the canine myelocytic leukemia cell line DH82 (Rikihisa et al., 1991). *N. risticii* has been isolated from the blood of horses infected with canine peripheral blood monocytes, U937 cells, or P388D cells (Rikihisa and Perry, 1984, 1985; Dutta et al., 1985; Holland et al., 1985). *N. risticii* has also been isolated from the blood of horses infected by oral administration of adult caddis flies or by subcutaneous inoculation of cercaria or sporocysts obtained from snails collected in an endemic area (Madigan et al., 2000a; Pusterla, et al., 2000b). *N. sennetsu* was isolated by intraperitoneal inoculation of patients' blood into laboratory mice (Fukuda et al., 1954; Misao and Kobayashi, 1954). The SF agent has been directly isolated by oral or intraperitoneal inoculation of mice with metacercariae excised from gray mullet fish (Fukuda et al., 1973).

MAINTENANCE PROCEDURES

The organism can be preserved by storing infected primary blood monocyte cultures or infected DH82 or P388D cell cultures in liquid nitrogen. The organism can also be preserved by storage at −80°C of a 20% suspension of homogenized infected lymph node or spleen tissue in RPMI 1640 medium containing 10% dimethyl sulfoxide and 15% fetal bovine serum (Rikihisa et al., 1991).

DIFFERENTIATION OF THE GENUS *NEORICKETTSIA* FROM OTHER GENERA

Neorickettsiae are distinguished from *Ehrlichia*, *Anaplasma*, and *Wolbachia* spp. by their 16S rRNA gene sequences or *groESL* gene sequences, by the presence of a 51-kDa protein gene unique to *N. risticii*, *N. sennetsu*, and SF agent, by the lack of *msp2* or *p44* homologs of *Anaplasma* spp. (Zhi et al., 1999) or OMP-1 homologs of *Ehrlichia* spp. (Ohashi et al., 1998a), by not being tickborne, and by their life cycle in trematodes.

TAXONOMIC COMMENTS

Sequence comparisons of almost complete 16S rRNA genes revealed that the sequence similarity of a horse isolate of *N. risticii*—isolate 081—to *N. sennetsu* Miyayama is 99.7%—slightly greater than the similarity to either the *N. risticii* type strain (99.6%) or to *N. risticii* Kentucky. The 16S rRNA gene sequence similarities between *N. helminthoeca* and *N. sennetsu* strains or *N. risticii* strains are 95.4–95.8% (Wen et al., 1996). According to Stackebrandt and Goebel (1994), a 16S rRNA gene sequence similarity of less than 97% between strains indicates that they represent different species, but at 97% or higher 16S rRNA gene sequence similarity, DNA relatedness must be used to determine whether strains belong to different species. Therefore, *N. sennetsu* and *N. risticii* belong to the same species based on 16S rRNA gene sequences.

The 16S rRNA gene sequence similarity between *N. risticii*, the agent of Rocky Mountain spotted fever, and *Neorickettsia* spp. is low (80%) (Pretzman et al., 1995).

FURTHER READING

Madigan, J.E., N. Pusterla, E. Johnson, J.S. Chae, J.B. Pusterla, E. Derock and S.P. Lawler. 2000. Transmission of *Ehrlichia risticii*, the agent of Potomac horse fever, using naturally infected aquatic insects and helminth vectors: preliminary report. Equine Vet. J. *32*: 275–279.

Millemann, R.E. and S.E. Knapp. 1970. Biology of *Nanophyetus salmincola* and "salmon poisoning" disease. Adv. Parasitol. *8*: 1–41.

Rikihisa, Y. 1991. The tribe *Ehrlichieae* and ehrlichial diseases. Clin. Microbiol. Rev. *4*: 286–308.

Rikihisa, Y. 1997. Rickettsial diseases in horses. *In* Bayly and Reed (Editors), Equine Internal Medicine, W.B. Saunders, Philadelphia. pp. 112–123.

Rikihisa, Y., H. Stills and G. Zimmerman. 1991. Isolation and continuous culture of *Neorickettsia helminthoeca* in a macrophage cell line. J. Clin. Microbiol. *29*: 1928–1933.

Rikihisa, Y. and G. Zimmerman. 1995. Salmon poisoning disease. *In* Kirk and Bonagura (Editors), Current Veterinary Therapy XII Small Animal Practice, W.B. Saunders Co., Philadelphia. pp. 297–300.

DIFFERENTIATION OF THE SPECIES OF THE GENUS *NEORICKETTSIA*

Diagnostic features of *Neorickettsia* spp. are shown in Table BXII.α.40. The preferred methods for definitive identification of *Neorickettsia* species are sequence analysis of all or part of 16S rRNA gene, the *groESL* operon, or the *gltA* gene, or alternatively, amplification of portions of these genes by the polymerase chain reaction using species-specific oligonucleotide primers. Species-specific primers for *N. helminthoeca* and *N. risticii* and *N. sennetsu* are shown in Table BXII.α.41.

List of species of the genus Neorickettsia

1. **Neorickettsia helminthoeca** Philip, Hadlow and Hughes 1953, 257[AL]

hel.minth′oe.ca. Gr. n. *helmins, helminthis* worm; Gr. n. *oikos* house; M.L. fem. adj. *helminthoeca* worm-dwelling.

TABLE BXII.α.40. Descriptive and diagnostic features of *Neorickettsia* species[a]

Characteristic	*N. helminthoeca*	*N. risticii*	*N. sennetsu*
Geographic distribution	U.S.A., Canada	North and South America, India, Europe	Japan, Malaysia
Reservoir host(s) and vectors	*Nanophyetus salmincola*	*Lecithodendriidae* trematodes	unknown trematode
Usual host species and infected host(s) that show clinical signs	Canidae	Horses	Humans
Usual infected host cell type	Monocytes	Monocytes, intestinal epithelial cells, mast cells	Monocytes
Presence in peripheral bood	+	+	+
Serological reaction with:			
N. helminthoeca antiserum	+ + +	+ +	+ +
N. risticii antiserum	+ +	+ + +	+ + +
N. sennetsu antiserum	+ +	+ + +	+ + +

[a]Symbols: +, low; + +, medium; + + + high.

TABLE BXII.α.41. Species-specific oligonucleotide primers for PCR identification of *Neorickettsia* species

Species	Target gene	Forward primer	Reverse primer
N. helminthoeca	16S RNA[a]	5'-GGACTTTTGACTGCTTGCCAG-3'	5'-TGGGTACCGTCATTATCTTCC-3'
N. risticii	16S rRNA[a]	5'-GGAATCAGGGCTGCTTGCAGC-3'	5'-TGGGTACCGTCATTATCTTCC-3'
	groESL[b]	5'-GGTTACAAGGTAATTAACAAC-3'	5'-CGGCAATCTTGTTACCGATT-3'
N. sennetsu	16S RNA[a]	5'-GGAATCAAAGCTGCTTGCCAG-3'	5'-TGGGTACCGTCATTATCTTCC-3'
	groESL[b]	5'-GGTTATAAGGTGATGAATCAG-3'	5'-CGGCAATCTTGCCACCAATC-3'

[a]Anderson et al. (1992c); Pretzman et al. (1995); Kanter et al. (2000).

[b]Sumner et al. (1997); Zhang et al. (1997).

The characteristics are as described for the genus and as listed in Table BXII.α.40. *N. helminthoeca* is the most phylogenetically distinct species in the genus *Neorickettsia*, differing from other species by approximately 4.5% in the 16S rRNA gene sequence. *N. helminthoeca* causes salmon poisoning disease, an acute and highly fatal disease of domestic and wild canidae. Clinical signs are fever, anorexia, inactivity, depression, vomiting, and diarrhea. In infected dogs, the major lymph nodes are enlarged. Although the trematode vector infects a wide variety of species of animals, salmon poisoning disease occurs chiefly in *Canidae* (dogs, coyotes, foxes) and is occasionally seen in immunosuppressed individuals of other animal species. Venereal transmission from an infected male to a female has also been reported. Rectal and aerosol transmission is also possible. Blood, feces, and lymph node aspirates are infectious; thus infected dogs must be isolated and care used so as not to induce iatrogenic transmission.

Ingested metacercariae in salmonid fish mature in 5–6 days in the gut, and the adult stage attaches deep in the interior of intestinal mucosa, inducing inflammation, i.e., hyperemia, inflammatory cell migration, and edema around the parasites. By an unknown mechanism, *N. helminthoeca* is transferred to monocytes and macrophages, which migrate through blood and lymphatic vessels and lodge in somatic and visceral lymph nodes. *N. helminthoeca* circulates in the blood of orally infected dogs starting day 8–12 after infection, as evidenced by the successful reisolation of *N. helminthoeca* from the blood from day 8–11 after infection until death (Rikihisa et al., 1991). Lymph nodes appear to be the primary site of multiplication of neorickettsiae. Like *N. risticii*, *N. helminthoeca* appears to exhibit intestinal tissue tropism, since not only the oral route of infection but also the intravenous inoculation of cell-cultured organisms induces severe hemorrhage throughout the small intestine. Nonsuppurative meningitis or meningoencephalitis and brain lesions have also been noted in

both natural and experimental *N. helminthoeca* infections. The severe central nervous system depression seen in salmon poisoning disease is probably related to these brain lesions. Dogs experimentally infected with *N. helminthoeca* P.O. develop IgG antibody 13–15 days after infection as detected by IFA testing (Rikihisa et al., 1991).

Clinical signs are characterized by anorexia, inactivity, depression, and sudden onset of fever above 39.8°C and up to 42.7°C, followed by hypothermia over the next 4–8 days. Vomiting may precede the typical watery yellowish diarrhea that is sometimes tinged with blood. Rapid weight loss is evident and the lack of ingestion of water, coupled with fluid loss through vomiting and diarrhea, results in severe dehydration with both electrolyte and acid-base imbalances. Serous nasal and ocular discharges are sometimes seen. Lymphadenopathy of the major nodes is easily palpable.

Microscopically, lesions are generally mild and proliferative in character, with very little evidence of degeneration and necrosis in the various organs including the brain, although extensive necrosis may accompany the proliferative response in lymphoid tissue. Lesions consist of mononuclear cell infiltration. Generalized lymph node enlargement is due to marked infiltration of macrophages, accompanied by severe depletion of small lymphocytes and a loss of germinal centers.

Elokomin fluke fever agent was reported to occur along the Elokomin River in the state of Washington. The agent has a wider host range than *N. helminthoeca* and is reported to infect bears, raccoons, and ferrets in addition to the Canidae. Clinical signs of Elokomin fluke fever are milder than those of salmon poisoning disease. The agent has not been cultured in a continuous cell line. It is currently considered a strain of *N. helminthoeca* but much information is lacking (Rikihisa and Zimmerman, 1995).

The mol% G + C of the DNA is: 42 (by gene sampling).

Type strain: no culture isolated.

GenBank accession number (16S rRNA): U12457.

2. **Neorickettsia risticii** (Holland, Weiss, Burgdorfer, Cole and Kakoma 1985) Dumler, Barbet, Bekker, Dasch, Palmer, Ray, Rikihisa and Rurangirwa 2001, 2159^{VP} (*Ehrlichia risticii* Holland, Weiss, Burgdorfer, Cole and Kakoma 1985, 524.)

ris.ti' cii. M.L. n. *ristic* from Miodrag Ristic, meaning Ristic's.

The characteristics are as described for the genus and as listed in Table BXII.α.40. *N. risticii* cells are generally round, sometimes pleomorphic, and may be elongated, especially in tissue culture. The cells divide by binary fission. As determined by electron microscopy and immunoperoxidase staining, the organisms are found in membrane-lined vacuoles within the cytoplasm of infected eucaryotic host cells, primarily macrophages, mast cells, and glandular epithelial cells in the intestine of the horse (Rikihisa et al., 1984, 1985; Steele et al., 1986). *N. risticii* cells are shed into the intestinal lumen (Rikihisa et al., 1985) and are found in feces (Mott et al., 1997). Morulae and individually enveloped forms appear to transform into one another. An intermediate stage has been described (Rikihisa et al., 1985); it appears similar to a moderately electron-dense ehrlichia that is tightly enveloped by a host membrane. The membrane is continuous with a membrane surrounding a morula. *N. risticii* cells are primarily seen as individual forms in the T-84, P388D, and U-937 cell lines, in primary equine monocyte culture, and in infected equine tissues. The individual form is especially common in intestinal epithelial cells.

N. risticii is the causative agent of Potomac horse fever, which is also called equine monocytic ehrlichiosis. The disease may also be called equine ehrlichial colitis and acute equine diarrhea syndrome, depending on the clinical signs displayed in a given horse. Serological data indicate that Potomac horse fever occurs in 43 of the United States, in two provinces in Canada (Ontario and Saskatchewan), and in France, Italy, Venezuela, India, Brazil, and Australia (Rikihisa et al., 1990; Ristic, 1990).

Clinical signs are an acute onset of fever (up to 107°F), depression, anorexia, decreased borborygmi in all abdominal quadrants, subcutaneous edema of the legs and ventral abdomen, dehydration, and diarrhea (Rikihisa, 1997). Laminitis and severe abdominal pain occur in 15–25% and 5–10% of cases, respectively. Development of these signs is the most common reason for euthanizing infected horses. Laminitis may progress, even when other clinical signs resolve. Diarrhea may be mild to severe (pipestream), and occurs in 10–30% of cases. In some horses diarrhea is transient; in others, it persists for several days. Some horses may have no diarrhea. The case fatality rate ranges from 5–30%. Transplacental transmission of *N. risticii* occurs and the organism may induce abortion or resorption of the fetus or infect foals, which then require extensive neonatal care. Leukopenia and rebound leukocytosis are prominent hematological changes. Anemia, plasma protein concentration, and packed cell volume may increase. Thrombocytopenia may also be observed. Pathogenicity in laboratory mice varies among strains.

Watery diarrhea is caused by a reduction in electrolyte transport (Na$^+$, Cl$^-$); thus, there is lack of water resorption, mainly in the large and small colons (Rikihisa et al., 1992). Infected intestinal epithelial cells lose microvilli, which may

contribute to the reduced electrolyte transport and water resorption. An increase in intracellular cyclic AMP (cAMP) is found in both infected mouse macrophages and infected horse intestinal tissues. This change in cAMP content may also contribute to the reduced luminal absorption of Na$^+$ and Cl$^-$ in the colon, and thus to the lack of water absorption and diarrhea.

Genetic and antigenic divergence among strains has been examined (Chaichanasiriwithaya et al., 1994). The Ohio 081 strain was significantly different from Maryland, Virginia, Kentucky, or other Ohio strains in protein composition and 16S rRNA gene sequence. In agreement with these findings, the identity of the P51 amino acid sequences between the type strain and the Ohio 081 strain is 80.6%, the lowest among all P51s examined (Kanter et al., 2000). The greatest 16S rRNA gene sequence difference was between the type strain of *N. risticii* and the Bunn strain, followed by the Ohio 081 strain (14 and 10 bases, respectively) (Wen et al., 1995a). Genetic comparisons of strains do not show segregation of strains among those from the blood of horses, those from snails or trematodes, or those from aquatic insects. Thus, it appears that strains that affect horses are also present in trematodes in snails and insects. The genome size is 880 kb (Rydkina et al., 1996).

The mol% G + C of the DNA is: 42 (by gene sampling).

Type strain: HRC-IL, Illinois, ATCC VR-986.

GenBank accession number (16S rRNA): M21290.

3. **Neorickettsia sennetsu** (Misao and Kobayashi 1956) Dumler, Barbet, Bekker, Dasch, Palmer, Ray, Rikihisa and Rurangirwa 2001, 2159^{VP} (*Rickettsia sennetsu* Misao and Kobayashi 1956) *Ehrlichia sennetsu* Ristic and Huxsoll 1984, 355.

sen' ne' tsu. M.L. n. *sennetsu* from Japanese, meaning glandular fever.

The characteristics are as described for the genus and as listed in Table BXII.α.40. The morphology is similar to that of *N. risticii*. *N. sennetsu* is the causative agent of human sennetsu rickettsiosis fever, also called Hyuga fever and Kagami fever. The clinical aspects may vary from mild headache, slight back pain, and a low-grade fever to a severe form of the disease characterized by persistent high fever, anorexia, lethargy, lymphadenopathy, and prominent hematologic abnormalities. The disease is limited to Western Japan and Malaysia (Ristic, 1990). *N. sennetsu* is not pathogenic for horses, but horses infected with it are protected from developing Potomac horse fever when challenged with *N. risticii* (Rikihisa et al., 1988). The 16S rRNA gene sequences and the *groESL* gene sequences (Zhang et al., 1997) are similar in *N. risticii* and *N. sennetsu*. The mode of transmission of *N. sennetsu* is unknown, although metacercariae that parasitize the gray mullet fish are suspected as the vector because of the association of the disease with the consumption of a raw fish of this type (Fukuda et al., 1973). Mice are highly susceptible to *N. sennetsu* and are used to isolate the organism from human specimens (Fukuda et al., 1954; Misao and Kobayashi, 1954). The genome size is 859 kb based on whole genome sequencing (Rikihisa, unpublished data).

The mol% G + C of the DNA is: 42 (by genome sequencing).

Type strain: Miyayama, ATCC VR-367.

GenBank accession number (16S rRNA): M73219.

Genus IV. **Wolbachia** *Hertig 1936, 472*[AL]

BERNARD LA SCOLA, CLAUDIO BANDI AND DIDIER RAOULT

Wol.ba' chia. M.L. fem. n. *Wolbachia* named after S. Burt Wolbach, who described the rickettsial agent of Rocky Mountain spotted fever and, in collaboration with Marshall Hertig, studied the rickettsia-like microorganisms of insects.

Pleomorphic bacteria that appear as small rods (0.5–1.3 μm in length) and coccoid forms (0.25–1 μm); large forms (1–1.8 μm in diameter) may be observed. Grow in vacuoles of host cells. Do not grow outside of cells. **Associated with arthropods and filarial nematodes.** Responsible for reproductive alterations in arthropods. With few exceptions, not pathogenic for the host. **Strains associated with nematodes responsible for indirect pathogenesis in humans. Susceptibility to doxycycline and rifampin demonstrated** *in vitro* **and** *in vivo.*

The mol% G + C of the DNA is: not determined.

Type species: **Wolbachia pipientis** Hertig 1936, 472.

FURTHER DESCRIPTIVE INFORMATION

Discovered in 1924 and described in 1936 by Hertig (Hertig and Wolbach, 1924; Hertig, 1936), *Wolbachia pipientis* is currently regarded as the sole species of the genus *Wolbachia* (Dumler et al., 2001). *Wolbachia* is a member of the *Alphaproteobacteria.*

Older descriptions of the genus *Wolbachia* included two species, *Wolbachia persica* and *Wolbachia melophagi*, that have characteristics distinct from *W. pipientis* (Fig. BXII.α.45). *Wolbachia persica* belongs to the *Gammaproteobacteria* (Weisburg et al., 1989; Forsman et al., 1994) and should be reclassified as *"Francisella persica"* (Niebylski et al., 1997a). The exact taxonomic position of *Wolbachia melophagi* has yet to be determined, but its description as an extracellular bacterium in the lumen of its host, *Melophagus ovinus*—a wingless fly commonly called sheep ked (Weiss et al., 1984)—is not in accordance with the strictly intracellular location of bacteria of the family *Anaplasmataceae.* In addition, a 16S rDNA gene sequence obtained from *W. melophagi* shows high homology with the 16S rRNA gene of species in the genus *Bartonella* (R.J. Birtles and D.H. Molyneux, unpublished GenBank accession X89110). A formal proposal to remove *W. persica* and *W. melophagi* from the genus *Wolbachia* has recently been published (Dumler et al., 2001). Descriptions of *Wolbachia persica* and

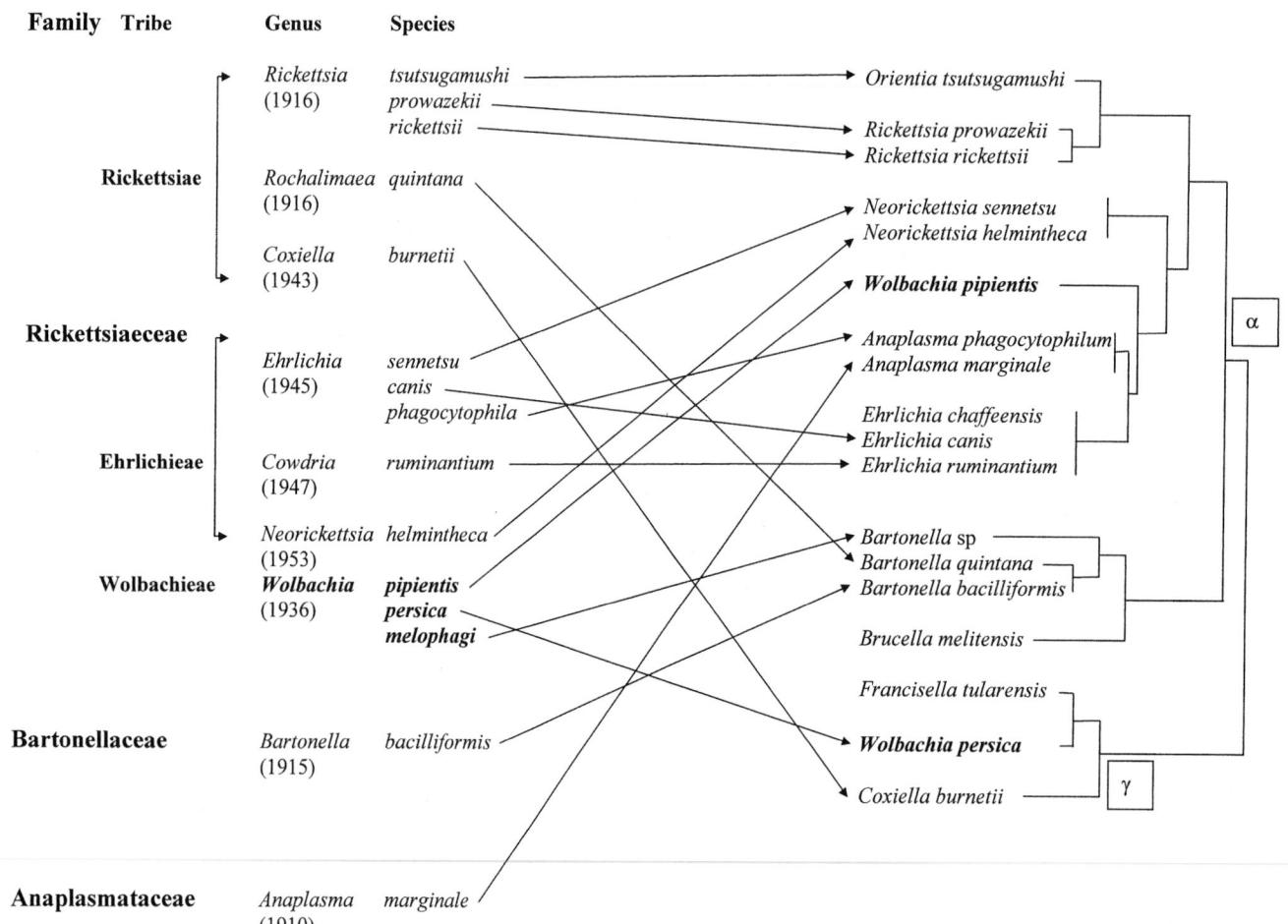

FIGURE BXII.α.45. The left-hand side of the figure indicates the classification of the *Rickettsiae* in *Bergey's Manual.* (Reproduced with permission from E. Weiss et al., *In* Krieg and Holt (editors), *Bergey's Manual of Systematic Bacteriology,* ©Williams & Wilkins Co., 1984, pp. 711–713). The classification on the right-hand side is based on a comparison of 16s rRNA gene sequences including the recent reorganization in the families *Rickettsiaceae* and *Anaplasmataceae* (Dumler et al., 2001).

Wolbachia melophagi can be found in the chapters on *Francisella* and *Bartonella*, respectively.

Additional *Wolbachia*-like organisms have been described, including *"Wolbachia postica"*, *"Wolbachia trichogrammae"*, and *"Wolbachia popcon"* (Hsiao and Hsioa, 1985; Louis and Nigro, 1989; Min and Benzer, 1997).

W. pipientis is not visible by Gram staining in spite of its Gram-negative cell wall structure (Fig. BXII.α.46). The organisms can be stained using conventional Giemsa's stain or rapid staining such as Diff-Quick (Fig. BXII.α.47). With the Gimenez stain, the bacteria can be visualized, but they appear as dark blue structures within a blue-green cytoplasm and not as pink-red structures as do rickettsiae. A peptidoglycan layer has not been observed and the cells display some plasticity (Yen and Barr, 1974; Wright et al., 1978; Wright and Barr, 1980). Association with bacteriophage-like particles has been described previously (Wright et al., 1978).

More recently, a bacteriophage-like genetic element, named bacteriophage WO, has been identified in *Wolbachia* (Masui et al., 2000).

W. pipientis multiplies by binary fission in the vacuoles of host cells and is surrounded by a membrane of host origin (Fig. BXII.α.46). In arthropods, the bacteria are present mostly in the cytoplasm of cells in the reproductive organs, but they can also be observed in other tissues, including nervous tissue and hemocytes (Louis and Nigro, 1989; Rigaud et al., 1991). In filarial nematodes, *Wolbachia* is present in the lateral cords and in the female reproductive apparatus (Kozek, 1977; Kozek and Marroquin, 1977). In both arthropods and nematodes, *Wolbachia* is transovarially transmitted to the offspring (Werren, 1997; Bandi et al., 2001).

A strain of *W. pipientis* has recently been established in an *Aedes albopictus*-derived cell line that allows production of signif-

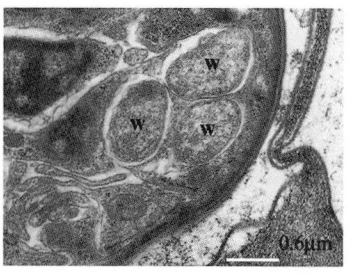 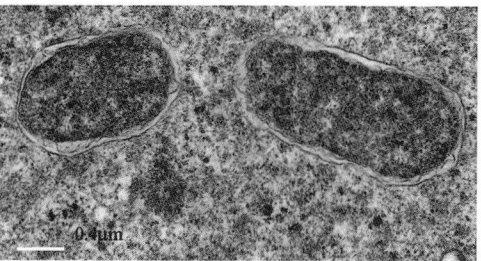

a **b**

FIGURE BXII.α.46. *Wolbachia pipientis* in filarial nematodes. (*a*) *Wolbachia* (*W*) inside an embryo of the dog heartworm *D. immitis*. (*b*) Details of *Wolbachia* endosymbionts inside an oocyte of the lymphatic filarial worm *Brugia pahangi* (photographs courtesy of Luciano Sacchi).

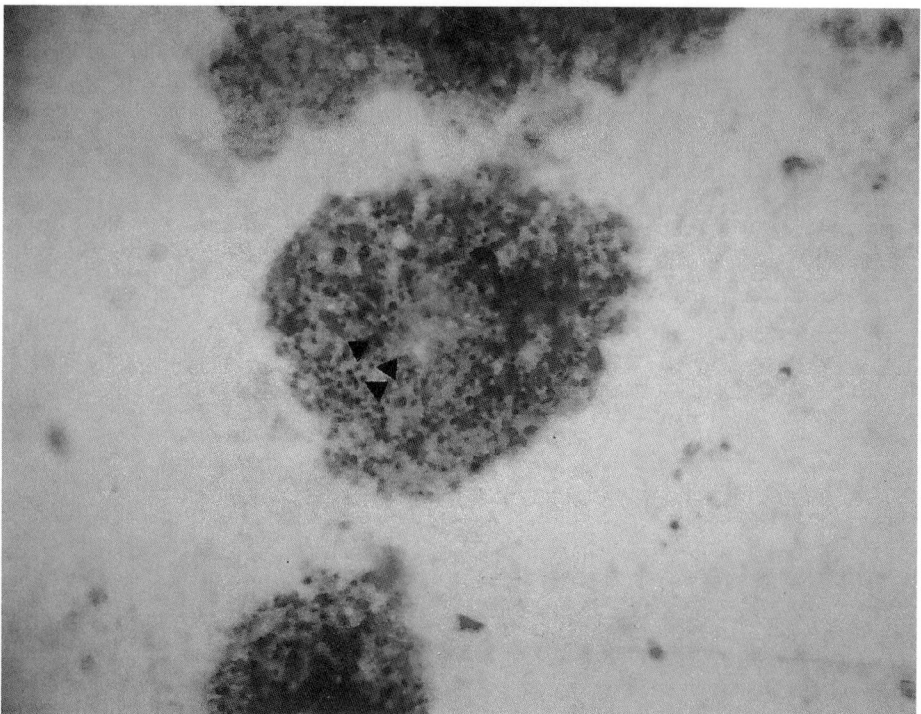

FIGURE BXII.α.47. O'Neill strain of *Wolbachia pipientis* in heavily infected Aa23 cells of *Aedes albopictus* using Giemsa's stain. Three bacterial cells are indicated by arrows.

icant amounts of the bacteria (O'Neill et al., 1997). This strain has been deposited at the American Type Culture Collection (ATCC VR-1529).

W. pipientis was originally observed in the gonads of the mosquito *Culex pipiens* (Hertig and Wolbach, 1924; Hertig, 1936). Later surveys showed *Wolbachia* to be present in at least 16% of sampled insect species (Werren et al., 1995a). The organism has also been found in representatives of other classes of arthropods (Rousset et al., 1992) and more recently in filarial nematodes of the family *Onchocercidae* (Sironi et al., 1995). Several species of filarial nematodes are known to be infected by *Wolbachia* (Table BXII.α.42). *Wolbachia*-infected taxa and the prevalence of infection have been reviewed recently (Taylor and Hoerauf, 1999; Bandi et al., 2001; Stevens et al., 2001) (Table BXII.α.42). In arthropods, *Wolbachia* is patchily distributed among insect populations and species (e.g., both infected and uninfected populations can be observed in the same species; Werren, 1997). In infected nematode species, all individuals and populations thus far examined are infected (Bandi et al., 2001). *Wolbachia* strains that infect arthropods can tolerate the cellular environment of diverse hosts, as demonstrated by studies with microinjection in various host cells (Boyle et al., 1993; Braig et al., 1994). In addition, there is evidence for horizontal transmission of *Wolbachia* in arthropods, and the same host may be infected by different strains of *Wolbachia* belonging to different genomic groups (Werren, 1997; Jamnongluk et al., 2002). Horizontal transmission and multiple infections have not been documented for nematode-infecting wolbachiae (Bandi et al., 1999, 2001; Casiraghi et al., 2001a).

Wolbachia does not usually cause obvious damage to its host cell. Nevertheless, a variant strain detected in *Drosophila melanogaster* showed massive proliferation in adult cells leading to widespread degeneration of tissues followed by early death of the insect (Min and Benzer, 1997). In addition, the male-killing *Wolbachia* strains present in arthropods (see below) are obviously pathogenic to male embryos (Hurst et al., 1999).

In arthropods, *Wolbachia* infection is usually associated with alterations in host reproduction, which include killing of male embryos, induction of parthenogenesis, feminization of genetic males, and cytoplasmic incompatibility (CI) (reviewed in Werren, 1997; Stouthamer et al., 1999; Stevens et al., 2001). It must be noted that *Wolbachia* is transmitted to the offspring by females, whereas males are usually not involved in transmission. All the reproductive alterations effected by *Wolbachia* in arthropods are interpreted as having the overall effect of increasing the transmission rate of *Wolbachia* (Werren, 1997). For example, parthenogenesis and feminization cause a *Wolbachia*-infected female to generate more female offspring, which will in turn transmit the bacterium. In CI, embryonic death is observed after mating between males that are infected by certain strains of *Wolbachia* and females that are either uninfected or infected with an incompatible *Wolbachia* strain. i.e., males infected by *Wolbachia* do not

TABLE BXII.α.42. Prevalence of *Wolbachia* in arthropods and filarial nematodes[a]

Host taxa	*Wolbachia* prevalence	Reference
Arthropods:		
Arthropoda	76%	Jeyaprakash and Hoy, 2000
Arachnida	(1/2)	
Insecta	(47/61)	
Arthropoda	18.4%	Werren and Windsor, 2000
Arachnida	(1/12)	
Insecta	(28/145)	
Arthropoda	16.5%	Werren et al., 1995a
Arachnida	(0/3)	
Insecta	(26/154)	
Insecta (*Hymenoptera* and *Diptera*)	26%	Cook and Butcher, 1999
Insecta (*Hymenoptera*)	59.30%	Plantard et al., 1999
Insecta (*Hymenoptera*)	50%	Wenseleers et al., 1998
Crustacea (*Isopoda*)	30%	Bouchon et al., 1998
Insecta	21.70%	West et al., 1998
Insecta (*Dictyoptera*)	89%	Bandi et al., 1997
Filarial nematodes:		
Acanthocheilonema viteae	No *Wolbachia* observed	McLaren et al., 1975; Bandi et al., 1998; Hoerauf et al., 1999
Brugia malayi	*Wolbachia* observed	Bandi et al., 1998; Taylor et al., 1999
Brugia pahangi	*Wolbachia* observed	Bandi et al., 1998; Taylor et al., 1999
Dirofilaria immitis	*Wolbachia* observed	Sironi et al., 1995
Dirofilaria repens	*Wolbachia* observed	Bandi et al., 1998
Litomosoides sigmodontis	*Wolbachia* observed	Bandi et al., 1998; Hoerauf et al., 1999
Mansonella ozzardi	*Wolbachia* observed	Casiraghi et al., 2001a
Onchocerca flexuosa	No *Wolbachia* observed	Henkle-Dührsen et al., 1998; Plenge-Bönig et al., 1995
Onchocerca gibsoni	*Wolbachia* observed	Bandi et al., 1998
Onchocerca gutturosa	*Wolbachia* observed	Bandi et al., 1998
Onchocerca lupi	*Wolbachia* observed	Egyed et al., 2002
Onchocerca ochengi	*Wolbachia* observed	Bandi et al., 1998
Wuchereria bancrofti	*Wolbachia* observed	Bandi et al., 1998; Taylor et al., 1999

[a]Data from Taylor and Hoerauf (1999), Bandi et al. (2001), and Stevens et al. (2001). For filarial nematodes, the *Wolbachia*-infected species included in the table are those for which positive *Wolbachia* PCR and *Wolbachia* gene sequences have been obtained; the *Wolbachia*-negative filarial species included in the table (*A. viteae* and *O. flexuosa*) are those for which the evidence for the absence of intracellular bacteria or *Wolbachia* has been reported in independent experimental studies based on different approaches (e.g., electron microscopy and PCR); for other filarial species that may be either infected or uninfected by *Wolbachia*, see the tables published by Taylor and Hoerauf (1999) and Bandi et al. (2001).

transmit it, but they sterilize those females that do not carry *Wolbachia* or those that carry a different compatibility type of *Wolbachia*. This reduction in the fitness of uninfected females implies an increase of the fitness of infected ones, thereby favoring the spread of the CI-inducing *Wolbachia* in the host population.

The molecular mechanisms of the reproductive alterations caused by *Wolbachia* are not known, even though the events leading to phenomena like CI have been described at the cytological level (e.g., Tram and Sullivan, 2002). In addition to the effects on arthropod reproduction, there is one case of a wasp (*Asobara tabida*) in which *Wolbachia* is required for oogenesis (Dedeine et al., 2001). *Wolbachia* has also been shown to rescue a deleterious genetic mutation that affects oogenesis in *Drosophila melanogaster* (Starr and Cline, 2002). Moreover, the number of *Wolbachia* cells in filarial nematodes has been shown to be positively associated with longevity (Taylor, personal communication).

There are different lines of evidence implicating *Wolbachia* in the pathogenesis of filariasis and in the side effects of anti-filarial therapy. In particular, extracts of filariae harboring wolbachiae stimulate human and mouse monocytes to produce pro-inflammatory cytokines and other mediators of inflammation (Brattig et al., 2000; Taylor et al., 2000), and they also attract granulocytes to the cornea in an animal model of river blindness, with thickening and opacity of the corneal stroma (Saint Andrè et al., 2002). Both corneal alterations and monocyte stimulation have been shown to be dependent on receptors required for the host cell response to lipopolysaccharide (LPS). In addition, monocyte stimulation is inhibited by an LPS antagonist/inhibitor, but not by heat treatment of the filarial extract (Taylor et al., 2000). Thus, there is overall evidence that filariae that harbor *Wolbachia* strains also contain LPS, which might play an important role in the immunopathogenesis of filariasis. The presence of LPS in these filariae is supported by a positive *Limulus* amebocyte lysate assay on filarial extracts (Brattig et al., 2000; Taylor et al., 2000). However, the LPS of *Wolbachia* has not yet been purified. The fact that *Wolbachia* actually interacts with the host immune system is further documented by the presence of antibodies against the *Wolbachia* surface protein (WSP) in animals infected with filarial nematodes (Bazzocchi et al., 2000b; Punkosdy et al., 2001). In addition, real-time PCR has allowed quantification of the release of *Wolbachia* DNA after microfilaricidal treatment of patients affected by onchocerciasis. There is a correlation between the PCR signal for *Wolbachia*, the release of mediators of inflammation, and the side effects of therapy (Keiser et al., 2002). It must be noted that post-treatment reactions in filariasis partially resemble those observed in acute bacteremia, both at the clinical level and in terms of markers of inflammation (Haarbrink et al., 2000; Keiser et al., 2002).

Susceptibility of *W. pipientis* to tetracycline was first demonstrated by addition of 17-50 µg/ml of tetracycline to the diet of larvae, resulting in the elimination of *Wolbachia* from the insects (Weiss et al., 1984). Studies of antibiotic susceptibility of *W. pipientis* were extended by studies of the susceptibility of the O'Neill strain in the Aa23 cell line to five antibiotic agents (Hermans et al., 2001) (Table BXII.α.43). *In vivo* studies in filarial nematode models have also demonstrated the activity of rifampicin and doxycycline and the inactivity of ciprofloxacin against *Wolbachia* (Hoerauf et al., 2000a; Towson et al., 2000). Treatment with tetracycline and tetracycline derivatives has also been shown to have detrimental effects on filariae that harbor *Wolbachia*

TABLE BXII.α.43. Effects of antibiotics on the O'Neill strain of *Wolbachia pipientis*[a, b]

Antibiotic	MIC (mg/l)	MBC (mg/l)
Doxycycline	0.0625	0.25
Oxytetracycline	4	1
Rifampicin	0.0625	2
Ciprofloxacin	>8	>8
Penicillin	>256	>256

[a]Data from Hermans et al. (2001).

[b]MIC, minimal inhibitory concentration; MBC, minimal bacteriocidal concentration.

(Bandi et al., 1999; Hoerauf et al., 1999, 2000a, b; McCall et al., 1999; Langworthy et al., 2000; Towson et al., 2000; Casiraghi et al., 2002). These detrimental effects include inhibition of worm development, infertility, and adulticidal effects (reviewed in Bandi et al., 2001). Since tetracycline treatment has no detrimental effect on the *Wolbachia*-free filarial worm A. *viteae*, it is assumed that, in infected species, *Wolbachia* is needed by the host nematode (Hoerauf et al., 1999; McCall et al., 1999). The discovery that filarial nematodes are susceptible to antibiotics has opened new prospects for the control of filariasis: clinical trials have already demonstrated the potential utility of doxycycline for the control of river blindness (Hoerauf et al., 2000b).

ENRICHMENT AND ISOLATION PROCEDURES

The sole isolation of *W. pipientis* reported to date is that of cells grown successfully in an Aa23 cell line from *Aedes albopictus* mosquitoes. The cell line was established as described by Tesh and Modi (1983). These cells (both infected ones and ones freed of *W. pipientis* by antibiotic treatment) are routinely grown at 28°C in a mixture (1:1 v/v) of Mitsuhashi-Maramorosh insect medium and Schneider's insect medium supplemented with 10–15% bovine fetal serum. Cells are grown in 25-cm³ flasks containing 5 ml of medium. The cells are passaged every week by shaking the flask and centrifuging the resulting cell suspension at low speed. The cell pellet is resuspended in the same amount of medium, and a new flask is seeded with 20% of the resuspended cells. For production of large amounts of bacteria, infected cells from one flask may be harvested every five days and inoculated into three cell culture flasks with fresh medium. The O'Neill strain can be propagated on human embryonic lung (HEL) fibroblast monolayers (CCL-137, American Type Culture Collection, Rockville, MD) at 28 and 37°C under previously described conditions (Raoult et al., 2001). The strain can also be propagated in C6/36—(CRL-1660, ATCC)—another mosquito cell line (from A. *albopictus*) grown in Leibowitz-15 medium with L-glutamine and L-amino acids, 5% (v/v) fetal bovine serum and 2% (v/v) tryptose phosphate at 28°C (Fenollar, La Scola, Taylor, and Raoult, unpublished data).

MAINTENANCE PROCEDURES

The O'Neill strain is preserved by rapid freezing and storage at −80°C.

DIFFERENTIATION OF THE GENUS *WOLBACHIA* FROM OTHER GENERA

Wolbachia are differentiated from *Anaplasma*, *Ehrlichia*, and *Neorickettsia* because they do not form morulae and because they infect only arthropods and filarial nematodes, not mammals. They are also distinguished by their 16S rRNA gene sequence.

TAXONOMIC COMMENTS

A phylogenetic analysis of the 16S rRNA gene clearly assigned *W. pipientis* to the *Alphaproteobacteria* and revealed a close relationship to the genera *Ehrlichia*, *Anaplasma* and *Neorickettsia* (Fig. BXII.α.45) (O'Neill et al., 1992). This study showed that *W. pipientis sensu stricto* (i.e., the bacterium present in the ovaries of *C. pipiens*) formed a monophyletic clade with other insect-associated microorganisms, and the authors suggested that this analysis supported the provisional "classification of these bacteria as members of the same species" (O'Neill et al., 1992). Further analyses based on the 16S rRNA gene and other genes identified other bacteria found in arthropods and in filarial nematodes as close relatives of *W. pipientis* (Rousset et al., 1992; Stouthamer et al., 1993; Werren et al., 1995b; Bandi et al., 1998; Vandekerckhove et al., 1999; Lo et al., 2002). All these bacteria form a coherent, monophyletic clade composed of at least six different main clusters (Fig. BXII.α.48), which have been indicated as supergroups A–F (Lo et al., 2002). In agreement with the opinion expressed by O'Neill et al. (1992), it has been suggested that the six supergroups be provisionally regarded as belonging to the sole valid species of the genus, *W. pipientis*, until formal proposals are made to elevate these clusters at the species rank (Bandi et al., 2001; see also discussion in Dumler et al., 2001).

At a finer taxonomic scale, a system based on the level of similarity in the *wsp* gene sequence has been proposed for strain grouping (Zhou et al., 1998). However, evidence for recombination between *Wolbachia* strains of supergroups A and B has raised questions about the usefulness of strain assemblages based

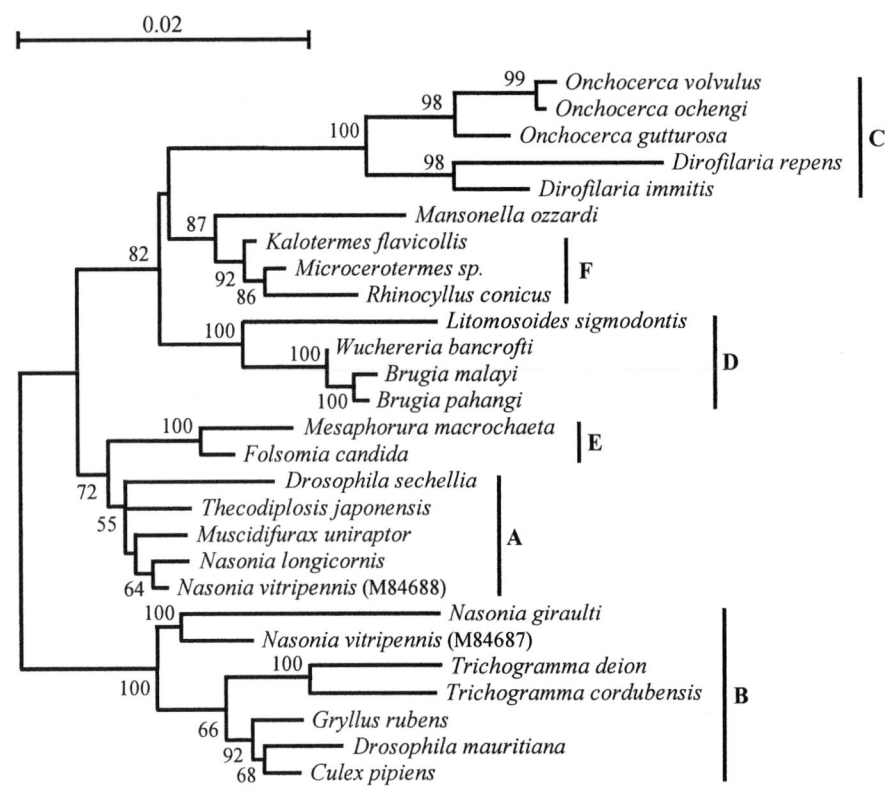

FIGURE BXII.α.48. Unrooted phylogenetic tree showing the six supergroups of *Wolbachia pipientis* (A–F) Lo et al., 2002. Names at the terminal nodes are those of the host species. The scale bar indicates distance in substitutions per nucleotide. Numbers at the nodes are the bootstrap confidence values based on 100 replicates (values below 50% are not indicated). Supergroups A, B, E, and F are found in arthropods. Supergroups C and D are found in filarial nematodes. The positioning of the *Wolbachia* found in the filarial nematode *Mansonella ozzardi* as a sister group of F wolbachiae is still to be evaluated, since only the 16S rRNA gene is available for this symbiont. The tree (neighbor-joining method; Kimura correction) is based on the gene coding for the 16S rRNA. This tree includes *Wolbachia* sequences derived from: *Nasonia vitripennis*, GenBank accession number M84688; *Mesaphorura macrochaeta*, GenBank accession number AJ422184; *Grillus rubens*, GenBank accession number U83092; *Muscifidurax uniraptor*, GenBank accession number L02882; *Rhinocyllus conicus*, GenBank accession number M85267; *Nasonia longicornis*, GenBank accession number M84691; *Nasonia vitripennis*, GenBank accession number M84687; *Nasonia giraulti*, GenBank accession number M84689; *Trichogramma cordubensis*, GenBank accession number L02883; *Trichogramma deion*, GenBank accession number L02884; *Culex pipiens*, GenBank accession number X61768; *Drosophila mauritiana*, GenBank accession number U17060; *Drosophila sechellia*, GenBank accession number U17059; *Folsomia candida*, GenBank accession number AF179630; *Thecodiplosis japonensis*, GenBank accession number AF220604; *Kalotermes flavicollis* Y11377; *Microcerotermes* sp., GenBank accession number AJ292347; *Brugia pahangi*, GenBank accession number AJ012646; *Brugia malayi*, GenBank accession number AJ010275; *Wuchereria bancrofti*, GenBank accession number AF093510; *Onchocerca ochengi*, GenBank accession number AJ010276; *Dirofilaria immitis*, GenBank accession number Z49261; *Mansonella ozzardi*, GenBank accession number AJ279034; *Dirofilaria repens*, GenBank accession number AJ276500; *Litomosoides sigmodontis*, GenBank accession number AF069068; *Onchocerca volvulus*, GenBank accession number AF069069; *Onchocerca gutturosa*, GenBank accession number AJ276498.

on single gene sequences (Jiggins et al., 2001; Werren and Bartos, 2001).

The inference of phylogenetic relationships among strains of *Wolbachia* and the description of the six supergroups have been based on the analyses of 16S rRNA and *ftsZ* genes (Werren et al., 1995b; Bandi et al., 1998; Lo et al., 2002). At the level of supergroup definition, analyses of other genes (*wsp, groESL*) have led to results that are generally congruent with those based on 16S rDNA and *ftsZ* genes (Masui et al., 1997; Bazzocchi et al. 2000a). However, most of the phylogenetic trees thus far published for *Wolbachia* are actually unrooted (e.g., Werren et al., 1995a; Bandi et al., 1998; Zhou et al., 1998). In the absence of a reliable rooting for the overall phylogeny of *Wolbachia* and of suitable outgroup sequences, the relationships among the six supergroups are indeed not yet established. The possibility that one of the supergroups is paraphyletic cannot be excluded.

Supergroups A, B, E, and F contain the wolbachiae found in arthropods, whereas supergroups C and D contain the wolbachiae of filarial nematodes. One possible exception is the *Wolbachia* strain harbored by the filarial nematode *Mansonella ozzardi*; this strain appears to be more closely related to members of supergroup F (Lo et al., 2002). In supergroups A and B the phylogeny of *Wolbachia* is not always congruent with the host phylogeny (O'Neill et al., 1992; Werren et al., 1995b); this find-

ing suggests that *Wolbachia* can be horizontally transmitted among arthropod hosts (Werren, 1997). In supergroups C and D the phylogeny of *Wolbachia* is consistent with the phylogeny of the host nematodes, which suggests strict vertical transmission of the symbionts (Casiraghi et al., 2001b). In a single infected nematode species, all the individuals thus far examined have been shown to harbor *Wolbachia* strains with identical *wsp* gene sequences, while *Wolbachia* strains from different nematode species possess differences in this gene (Bazzocchi et al., 2000b). This result supports the hypothesis of strict vertical transmission and suggests that the association with *Wolbachia* is species-specific in filarial nematodes.

FURTHER READING

Bandi, C., A.J. Trees and N.W. Brattig. 2001. *Wolbachia* in filarial nematodes: evolutionary aspects and implications for the pathogenesis and treatment of filarial diseases. Vet. Parasitol. *98*: 215–238.

Stevens, L.A., R. Giordano and R.F. Fialho. 2001. Male-killing, nematode infections, bacteriophage infection and virulence of cytoplasmic bacteria in the genus *Wolbachia*. Annu. Rev. Ecol. Syst. *32*: 519–545.

Stouthamer, R., J.A.J. Breeuwer and G.D.D. Hurst. 1999. *Wolbachia pipientis*: microbial manipulator of arthropod reproduction. Annu. Rev. Microbiol. *53*: 71–102.

Werren, J.H. 1997. Biology of *Wolbachia*. Annu. Rev. Entomol. *42*: 587–609.

List of species of the genus Wolbachia

1. **Wolbachia pipientis** Hertig 1936, 472[AL]

pi.pi.en'tis. M.L. n. *pipiens* specific epithet of the host mosquito, *Culex pipiens*; M.L. gen. n. *pipientis* of pipiens.

The characteristics are as described for the genus.

The mol% G + C of the DNA is: unknown.

Type strain: no type strain.

Additional Remarks: The sole available strain for *in vitro* growth is O'Neill strain (see above).

Genus Incertae Sedis V. **Aegyptianella** Carpano 1929, 12[AL]

YASUKO RIKIHISA AND JULIUS P. KREIER

Ae.gyp'ti.a.nel'la. dim. ending *-ella* M.L. fem. dim. n. *Aegyptianella* named after Egypt where the organism was described in 1929.

In blood smears stained by Romanowsky methods, the inclusions **appear in erythrocytes as purple intracytoplasmic bodies** of 0.3–4 μm in size. By electron microscopy, each inclusion contains between 1 and 26 pleomorphic cocci with trilaminar outer membranes. They are obligate parasites of domestic and wild birds. Transmitted by ticks.

The mol% G + C of the DNA is: not known.

Type species: **Aegyptianella pullorum** Carpano 1929, 12.

FURTHER DESCRIPTIVE INFORMATION

Phylogeny By 16S rRNA gene and *groEL* operon sequence analysis, *Aegyptianella pullorum* belongs to the family *Anaplasmataceae* (Rikihisa et al., 2003). Identities of 16S rRNA genes and *groEL* operon sequences of *A. pullorum* to those of *Anaplasma* species were 92.7–93.3% and 71.4–73.3%, respectively. Identities of 16S rRNA gene and *groEL* operon sequences of *A. pullorum* to those of *Ehrlichia* species were 86.0–88.2 and 68.3–69.2%, respectively. Identities of 16S rRNA gene and *groEL* operon sequences of *A. pullorum* to those of *Neorickettsia* species were 78.2–79.2% and 40.0–41.5%, respectively.

Cell morphology In blood smears stained with Giemsa, the inclusions appear in the host's erythrocytes in a variety of forms: compact, round or oval, ring- or horseshoe-shaped, polygonal or polymorphic. They are violet-reddish in color, with a diameter of 0.3–4.0 μm. In larger inclusions, clearly defined small cocci of 0.25–0.4 μm resembling *Anaplasma* sp. can be distinguished by electron microscopy. The organisms reproduce by binary fission. These bacteria may also be found free in the plasma and in phagocytic cells. Organisms in phagocytic cells are probably phagocytosed and not growing, but undergoing digestion (Gothe, 1967b, 1971).

Fine structure In erythrocytes, the inclusions are separated from the erythrocyte cytoplasm by a single membrane presumably of erythrocyte origin. The bacterial cytoplasm contains ribosomes and DNA strands. The organisms are enveloped in a trilaminar outer membrane (Gothe, 1967a, 1971; Bird and Garnham, 1969; Castle and Christensen, 1985) (Fig. BXII.α.49). Scanning electron microscopy of infected erythrocytes revealed that the mode of entrance of organisms into erythrocytes may be endocytosis followed by formation of an erythrocytic vesicle. The exit of the organisms from parasitized erythrocytes may be the reverse of the invasive mechanism, an exocytosis. Generally, however, the affected erythrocytes are injured by the parasites, resulting in release of the parasites into the plasma by host cell lysis (Gothe and Burkhardt, 1979).

Cultural characteristics and antibiotic susceptibility Multiplication of *A. pullorum* has not been observed to occur in cell-free media or in tissue cultures. Attempts at continuous propagation of the organism in chicken embryos have not been successful (Gothe, 1971). Only broad-spectrum antibiotics of the tetracycline series, dithiosemicarbazones, and pleuromutilins have a bacteriocidal efficacy with a significant chemotherapeutic influence on the course of the infection in chickens (Gothe, 1971; Gothe and Mieth, 1979).

Ecology Chickens are naturally infected with *A. pullorum* by the ticks *Argas (Persicargas) persicus, A. (P.) walkerae, A. (P.) sanchezi,* and *A. (P.) radiates.* The tick is a biological vector. Experimental infection can be achieved by subcutaneous, intramuscular, intravenous, and intraperitoneal inoculation of infected blood or by scarification followed by application of the infected blood to the scarified area. In addition to infection of chickens, natural infections have been described in geese, ducks, quail, and ostriches. Wild birds that have been experimentally infected are *Turtur erythrophrys* and *Balearica pavonina* (Curasson and Andrjesky,

1929), *Turtur senegalensis, Milvus aegyptiacus,* and *Vidua principalis* (Curasson, 1938).

Infection is transstadial in ticks (Gothe, 1967c, 1971; Hadani and Dinur, 1968). Transovarial transmission has also been observed (Hadani and Dinur, 1968; Gothe, 1971).

MAINTENANCE PROCEDURES

The infectivity of *A. pullorum* in chicken blood can be preserved up to nearly 7 years by storage in liquid nitrogen (Raether and Seidenath, 1977). Cryopreservation does not affect the ability of the parasites to propagate in the vector tick *Argas (Persicargas) walkerae* (Gothe and Hartmann, 1979).

DIFFERENTIATION OF THE GENUS *AEGYPTIANELLA* FROM OTHER GENERA

Characteristics useful for differentiating *Aegyptianella* from the other members of the family *Anaplasmataceae* are provided in the key to this family (see Table BXII.α.33 of the chapter describing the family *Anaplasmataceae*).

TAXONOMIC COMMENTS

At present several species have been described in the genus *Aegyptianella* including those bacteria infecting erythrocytes of poikilotherms (Gothe, 1978). Detailed morphological and biochemical investigations have only been carried out with *A. pullorum.*

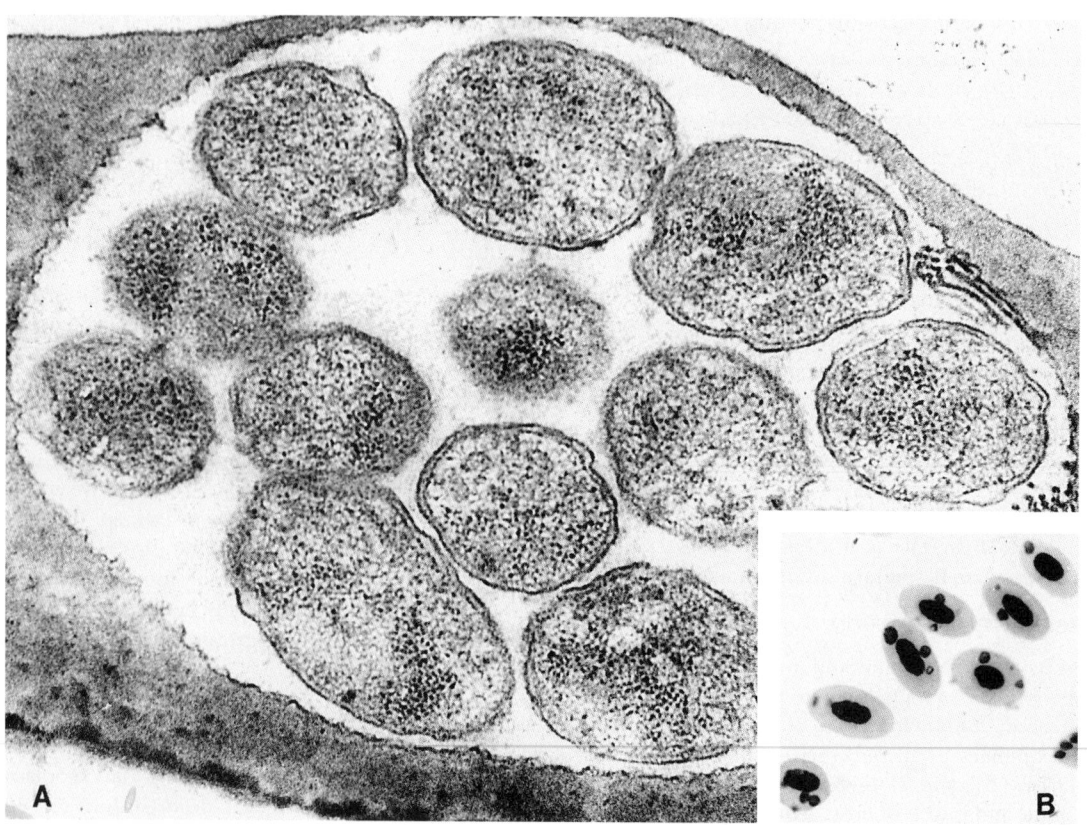

FIGURE BXII.α.49. *Aegyptianella pullorum. A,* ultrathin section of an infected erythrocyte, showing an inclusion body containing 12 parasites (× 90,000). *B,* intraerythrocytic inclusion bodies stained by the Giemsa method. (Reproduced with permission from R. Gothe, Zeitschrift für Parasitenkunde, *29:* 119–129, 1967, ©Springer-Verlag, Berlin.)

Electron microscopy studies have been performed for *"Aegyptianella botuliformis"*, *"Aegyptianella ranarum"*, and *"Aegyptianella bacterifera"*, which show that these three spp. are different from *A. pullorum*. Other species have been described but, except for *"A. ranarum"*, neither their true identify nor their relationship to *A. pullorum* has been established (see *Species incertae sedis*).

FURTHER READING

Bird, R.G. and P.C. Garnham. 1969. *Aegyptianella pullorum* Carpano 1928- -fine structure and taxonomy. Parasitology. *59*: 745–752.

Gothe, R. and J.P. Kreier. 1977. *Aegyptianella, Eperythrozoon,* and *Haemobartonella. In* Kreier (Editor), Parasitic Protozoa, Vol. IV, Academic Press, New York. pp. 251–294.

DIFFERENTIATION OF THE SPECIES OF THE GENUS *AEGYPTIANELLA*

The following species are tentatively classified in the genus *Aegyptianella*; genetic studies are required to learn how similar they are. Except for *A. pullorum*, these species have not been formally described and therefore do not presently have standing in nomenclature. Characteristics are listed in Table BXII.α.44.

List of species of the genus Aegyptianella

1. **Aegyptianella pullorum** Carpano 1929, 12[AL]
   pul.lo' rum. L. gen. pl. n. *pullorum* of young fowls.

   The characteristics are as described for the genus.
   The mol% G + C of the DNA is: unknown.
   Type strain: No culture isolated.
   Additional Remarks: At the time of publication, the only available 16S rRNA gene sequence for *A. pullorum* was AY125087.

Species Incertae Sedis

1. **"Aegyptianella bacterifera"** Barta, Boulard and Desser 1989, 14 ("Cytamoeba bacterifera" Labbé 1894, 104.)
   bac.te.ri' fe.ra. Gr. dim. n. *bacterium* a small rod; L. suffix *-fer (-ferus),* from L. v. *fero* to bear; M.L. adj. *bacterifera* bearing small rods.

   Up to 12 rod-shaped organisms that are 2.3–4.9 × 0.5 µm are found in membrane-bound inclusions of erythrocytes of *Rana esculata* in Corsica and *R. nigromaculata* in China.

2. **"Aegyptianella botuliformis"** Huchzermeyer, Horak, Putterill and Earle 1992, 100.
   botuli' formis. L. gen. n. *Botulus* sausage; L. n. *forma* shape, form; M.L. adj. *botuliformis* sausage-shaped.

   In small inclusions it resembles *A. pullorum*, but in larger inclusions organisms are sausage-shaped with up to eight organisms tightly packed in an inclusion. It does not infect chickens but produces a high and long-lasting parasitemia in guinea fowl. Infection is associated with *Amblyomma hebraeum* or *A. marmoreum* infestation.
   Type strain: no culture isolated.
   The mol% G + C of the DNA is: unknown.

3. **"Aegyptianella carpani"** Battelli 1947, 212.
   Found in the snake *Naia nigricollis.*

4. **"Aegyptianella elgonensis"** Mutinga and Dipeolu 1989, 410.
   Found in the lizards collected near Mt. Elgon.

5. **"Aegyptianella ranarum"** Desser 1987, 53.
   rana' rum. L. n. *rana* frog; M.L. adj. *ranarum* of frog.

   Up to 120 rod-shaped organisms that are 150–200 nm × 1–1.7 µm are found in membrane-bound inclusions of erythrocytes of bullfrogs (*Rana catesbeiana*), green frogs (*R. clamitans*), and mink frogs (*R. septentrionalis*).

   Identity of the 16S rRNA gene sequence of *"A. ranarum"* (AY208995) to that of *A. pullorum* (AY125087) is 60.9%, showing that *"A. ranarum"* does not belong to the genus *Aegyptianella* or even to the family *Anaplasmataceae* (Rikihisa, unpublished results).

6. **"Sogdianella moshkovskii"** (Laird and Lari 1957) Schurenkova 1938, 936 (*"Babesia moshkovskii"* Laird and Lari 1957, 794.)
   Found in various wild birds.

7. **"Tunetella emydis"** Brumpt and Lavier 1935, 548.
   Found in the tortoise *Emys leprosa.*

TABLE BXII.α.44. Differential characteristics of the species of the genus *Aegyptianella*

Characteristic	*A. pullorum*[a]	*"A. bacterifera"*[b]	*"A. botuliformis"*[c]	*"A. ranarum"*[d]
Morphology	Cocci	Rods	Rods	Rods
Host	Chicken, goose, duck, ostrich, quail, wild turkey	Frog (*Rana nigromaculata, R. esculata*)	Helmeted guinea fowl (*Numida meleagris*)	Frog (*Rana catesbeiana, R. septentrionalis, R. clamitans*)
Location	North Africa, North America	China, Corsica	South Africa	Canada
Possible vector	*Argas walkerae, A. sanchezi, A. radiatus*	Unknown	*Amblyomma hebraeum, A. marmoreaun*	Leech (*Batracobdella picta*)

[a]Gothe and Kreier, 1984; Castle and Christensen, 1985.

[b]Werner, 1993; Desser and Bartha, 1989.

[c]Huchzermeyer et al., 1992.

[d]Desser, 1987.

Family III. **Holosporaceae** *fam. nov.*

Hans-Dieter Görtz and Helmut J. Schmidt

Ho.lo' spo.ra.ce.ae. Gr. fem. adj. *holos* whole, complete; Gr. n. *sporus* seed; M.L. n. *spora* spore; M.L. fem. n. *Holospora* whole spore; *-ceae* ending to denote family; *Holosporaceae* family of *Holospora* bacteria.

Bacteria are present specifically in the **micronucleus or macronucleus** of *Paramecium* species. Exists in two forms: the **reproductive form** is a short rod 1.0–3.0 × 0.5 µm. It undergoes binary fission and may give rise to the **long infectious form**, measuring 5.0–20.0 µm with rounded or tapered ends, that can infect *Paramecium* and become established in the nucleus. The infectious form has a voluminous periplasm filled with fine, granular material with a less electron-dense pale tip, called the special tip. The condensed-looking bacterial protoplasm of the infectious form contains ribosomes and mesosome-like structures. The cytoplasm is polarly positioned opposite to the special tip in one half of the cell (Fig. BXII.α.50). The cytoplasm stains with DNA-specific dyes. Gram-negative, nonmotile, obligate symbiont. No toxic effects of *Holospora*-bearing paramecia on paramecia lacking the symbiont were observed. Paramecia can outgrow *Holospora* species, but reinfection occurs readily; mass cultures show up to 100% infection.

Type genus: **Holospora** (ex Hafkine 1890) Gromov and Ossipov 1981, 351.[VP]

Further descriptive information

The family currently contains only the single genus *Holospora*. In 1890 the genus *Holospora*, containing three species, was described by Hafkine (1890). Nine species have now been identified, but only four of them were validly published (see list of the species of the genus *Holospora*). Holosporas share a number of unique features. All invade the nuclei of *Paramecium* (Fig. BXII.α.51) (Ciliophora, Protozoa) (Hafkine, 1890; Gromov and Ossipov, 1981; Görtz, 1983, 1986; Fokin, 1989). They are host specific, infecting only certain species of the genus *Paramecium*. All species show a developmental cycle with an infectious and a reproductive form (Fig. BXII.α.52). Only the infectious form is infectious and even short infectious forms were found to be fully infectious (Kawai and Fujishima, 2000). The infectious form shows a unique cellular organization (Fig. BXII.α.50).

Holosporas are truly parasitic. Infectious forms of the bacteria are regularly released by the host cells into the surrounding medium and may infect new host cells. If the macronucleus of a paramecium (see Fig. BXII.α.53 for the organization of the ciliate cell) is infected, this may affect its growth rate and under unfavorable conditions even kill the cell. Under favorable conditions, host cells are not damaged, and some natural populations have been found stably infected for years with infection rates of close to 100%.

Micronuclear-specific infections are different. In an infection of the micronucleus with *Holospora elegans*, host cell conjugation (sexual propagation) was no longer successful (Görtz and Fujishima, 1983) and exconjugant cells were not viable. Apparently, new macronuclei were not functional. The inhibition of successful conjugation renders the host cells genetically dead.

The individual host cell is not necessarily killed by infection with *Holospora*. Rather, bacteria and host appear to be well adapted. In starving *Paramecium* or after inhibition of host protein synthesis, most bacteria differentiate into the infectious form (Fujishima, 1993), which is a resting stage and does not multiply. In addition, paramecia seem to have mechanisms to cure themselves of endonuclear symbionts. Skoblo et al. (1990), Ossipov et al. (1993), and Fokin and Skovorodkin (1991, 1997) have discovered that *Holospora* may be synchronously and completely lysed in the host nuclei of certain strains of *Paramecium* after an infection. This lysis may be due to an unknown defense mech-

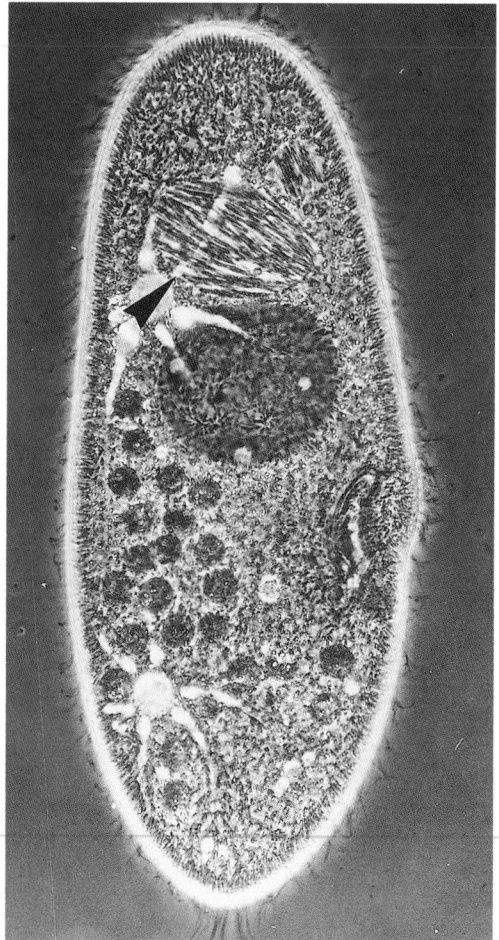

FIGURE BXII.α.51. *Holospora elegans* in the micronucleus (*arrow*) of *Paramecium caudatum*. Due to the infection, the micronucleus has almost reached the size of the macronucleus.

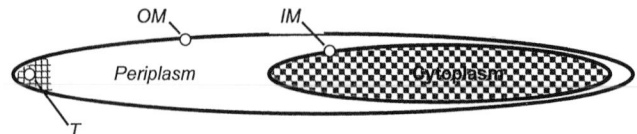

FIGURE BXII.α.50. Diagram of the infectious form of *H. obtusa*. The cytoplasm is condensed and shifted towards one pole. The electron-dense material of the periplasm fills more than half of the cell, with less electron-dense material (*T*) at the pole opposite to the cytoplasm. *IM*, inner membrane; *OM*, outer membrane.

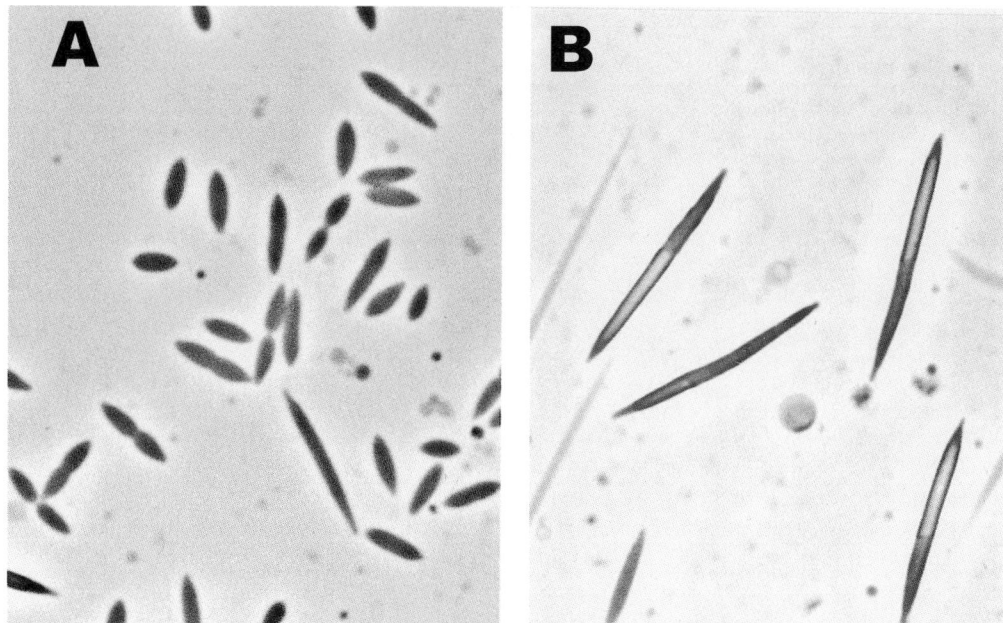

FIGURE BXII.α.52. Reproductive (*A*) and infectious (*B*) forms of *Holospora elegans* released from the micronucleus of *Paramecium caudatum.*

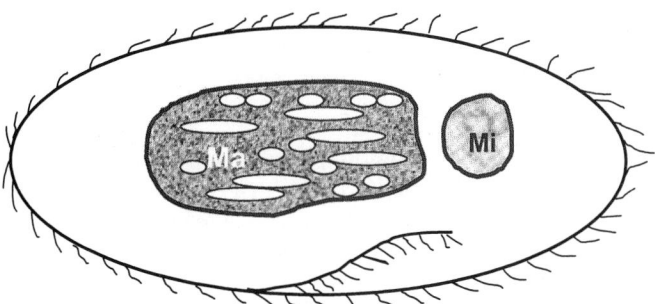

FIGURE BXII.α.53. Diagram of a ciliate cell infected by a *Holospora* species in the macronucleus (schematic representation). *Mi*, micronucleus; *Ma*, macronucleus. The micronucleus is the generative nucleus and transcriptionally inactive. The macronucleus is the somatic nucleus and transcriptionally active. After conjugation, a new macronucleus is developed from the new micronucleus, both deriving from the syncaryon.

anism against intracellular infections evolved in ciliates. After lysis of the bacteria, the host cells are cured and remain viable.

Developmental cycle *Holospora* show a developmental cycle with a specialized infectious form (Figs. BXII.α.50, BXII.α.52). After division of the host cell, only infectious forms are released into the medium, while the reproductive forms remain in the host nuclei. Some reproductive forms may then develop into further infectious forms. The developmental cycle is described in Fig. BXII.α.54.

A new infection may begin with the ingestion of an infectious form. The bacterium leaves the phagosome, is transported through the cytoplasm, and is taken into the host nucleus by fusion of the two membranes of the transport vesicle with those of the nuclear envelope (Görtz and Wiemann, 1989). Transport to the nucleus, and within the nucleus during its division, appears to be mediated by the cytoskeleton of the host cell (Ossipov and Podlipaev, 1977; Görtz and Wiemann, 1989).

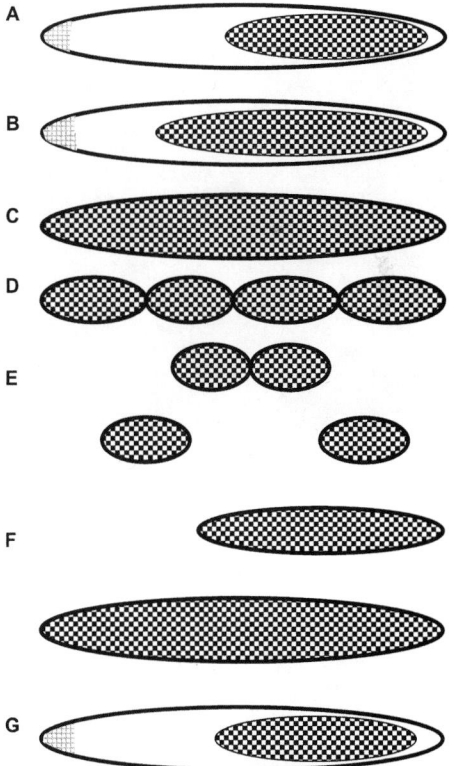

FIGURE BXII.α.54. Diagram of the development of *H. obtusa*. In early stages of invasion, the cytoplasm of the infectious form extends into the periplasmic space, while the volume of periplasm decreases (*A*, *B*); after the bacterium arrives in the host nucleus, the whole volume of the cell is occupied by the cytoplasm (*C*); the cell constricts at several points and divides into small cells (*D*); by this point, the reproductive form is established and multiplies by binary division (*E*); reproductive forms may grow out in the host nucleus (*F*). The resulting long cells develop into the infectious forms by depositing periplasmic material and condensing the cytoplasm (*G*).

TABLE BXII.α.45. Diagnostic table of *Holospora*[a]

Characteristics	*H. undulata*	*H. elegans*	*H. caryophila*	*H. obtusa*	*"H. acuminata"*	*"H. recta"*	*"H. curviuscula"*	*"H. bacillata"*	*"H. curvata"*
Cell morphology of infectious form	Spiral, ends tapered	Straight, ends tapered	Spiral, ends tapered	Straight, ends rounded	Straight rod, ends tapered	Straight, one end tapered	Curved, ends tapered	Straight, ends rounded	Curved
Cell size (μm)	0.8–1 × 10–25	0.8–1 × 10–20	0.2–0.3 × 5–8	0.8–1 × 10–25	0.5–0.6 × 5–8	0.7–1 × 10–15	0.4–0.5 × 4–10	0.7–0.8 × 5–17	0.7–0.9 × 12–20
Host:									
Paramecium biaurelia			+						
Paramecium bursaria					+		+		
Paramecium caudatum	+	+	+	+		+			
Paramecium calkinsi								+	+
Paramecium novaurelia			+						
Paramecium woodruffi								+	
Nucleus:									
Macronucleus			+	+			+	+	+
Micronucleus	+	+			+	+			
Holospora "species group"[b]	I	I	II	I	I	I		II	II

[a]For symbols see standard definitions.

[b]*Holospora* species group: I, infectious forms are collected in the connecting piece of the dividing host nucleus and released by the host cell; those species tested were labeled by *in situ* hybridization using a *Holospora*-specific probe (Fokin et al., 1996); II, infectious forms are not collected in the connecting piece of the dividing host nucleus but remain distributed in the dividing nucleus; those species were not labeled by *in situ* hybridization using a *Holospora*-specific probe.

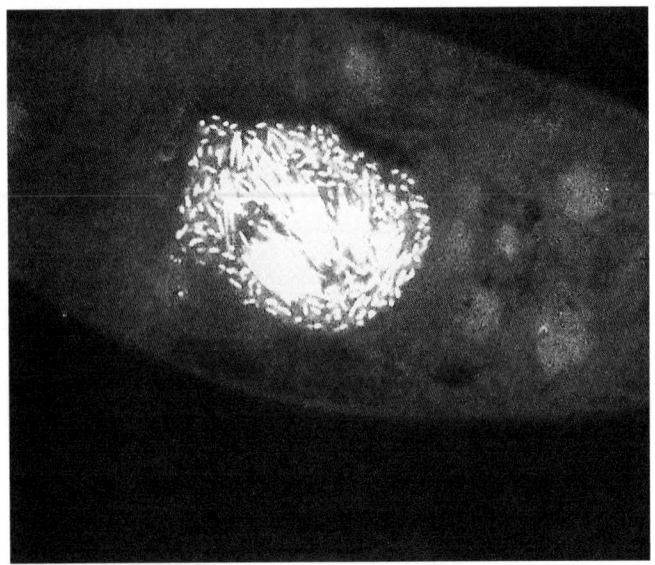

FIGURE BXII.α.55. *Holospora obtusa* in the macronucleus of *Paramecium caudatum*. As a result of hybridization with an *H. obtusa*-specific probe, the bacteria are labeled (Reproduced with permission from S.I. Fokin, European Journal of Protistology Supplement 1, *32:* 19–24, 1996, ©Urban & Fischer Verlag GmbH & Co.)

The infectious form has a voluminous periplasm that contains a number of stage-specific proteins, some of which appear to be released during the infection process (Görtz and Wiemann, 1989; Fujishima et al., 1990, 1997; Wiemann and Görtz, 1991; Dohra et al., 1994). Some proteins that were immuno-localized in the periplasm of the infectious form were found on the surface of the bacteria after their ingestion into the phagosome, or associated with the phagosome membrane. This is what would be expected if such proteins were used for communication with host membranes. Released periplasmic proteins could also protect the bacteria against lysosomal enzymes of the host or inactivate such enzymes.

The gene of a small periplasmic protein of 5.4 kDa was sequenced by Dohra et al. (1997). Northern blot hybridization showed that the gene is highly expressed in the intermediate form, a transitional stage in the development from the reproductive into the infectious form of the bacterium. Amino acid sequence similarities with other polypeptides have not been found (Dohra et al., 1997). It has been suggested that the protein may function in the recognition process in the early phase of infection. Dohra et al. (1998) have also identified a GroEL-like protein in *H. obtusa*. The gene is selectively expressed in the reproductive form.

The infectious form is polarly organized (Figs. BXII.α.50, BXII.α.52) (Dohra and Fujishima, 1999). Half of the cell is occupied by the voluminous periplasm that contains a number of stage-specific proteins. The proteins appear to be released during the infection process. Some proteins were immunolocalized in the periplasm of the infectious form and during the invasion process were found on the surface of the bacteria after their ingestion into the phagosome, associated with the phagosome membrane (Fujishima et al., 1990, 1997; Görtz et al., 1990; Wiemann and Görtz, 1991; Dohra et al., 1994; and unpublished results).

Phylogeny of the *Holosporaceae* *Holospora* species were the first intracellular bacteria in *Paramecium* for which the phylogenetic position was determined (Amann et al., 1991). *H. obtusa*, *H. elegans*, and *H. undulata* belong to the *Alphaproteobacteria*. The closest relative among other symbionts in ciliates was found to be *Caedibacter caryophilus*, and the closest relatives among other bacteria found to date were *Rickettsia* and *Ehrlichia* species (Amann et al., 1991; Springer et al., 1993). It is tempting to regard the striking biology (developmental cycle; host specificity for *Paramecium*, etc.) and the unique morphology of the infectious form as homologous features, proving the close relationship and monophyletic origin of these bacteria. New observations, however, cast doubt on this possibility (Fokin et al., 1996).

The behavior of the infectious forms of certain *Holospora* species, assembling as they do in the connecting piece of the dividing host nucleus, is certainly highly advanced and must be regarded as an apomorphic feature. This behavior ensures that the infectious forms are specifically collected and released by the host cell. *H. caryophila*, *H. bacillata*, and *H. curvata* do not share this feature with the other holosporas. It is not known how the infectious forms of these species leave their host nuclei. They are either more primitive than the other holosporas, as a quantitative separation of infectious forms and reproductive forms is not observed, or they are not closely related phylogenetically. Two species groups may presently be distinguished in the genus *Holospora* (Table BXII.α.45). It has been hypothesized that the unique behavior of the infectious form to polarly deposit enormous amounts of periplasmic materials could be encoded on a plasmid or phage genome. However, no plasmid or phage genome has been found (Rautian et al., unpublished results). It is consistent with this observation that the "more advanced" species tested (in which the infectious forms are collected in the separation spindle) gave a positive result after *in situ* hybridization (Fig. BXII.α.55) using an oligonucleotide probe designed for *H. obtusa* (Amann et al., 1991; Fokin et al., 1996).

Genus I. **Holospora** *(ex Hafkine 1890) Gromov and Ossipov 1981, 351*[VP]*

HANS-DIETER GÖRTZ AND HELMUT J. SCHMIDT

Ho.lo'spo.ra. Gr. fem. adj. *holos* whole, complete; Gr. n. *sporus* seed; M.L. n. *spora* spore; M.L. fem. n. *Holospora* whole spore.

The description of the genus is that of the family. Diagnostic characteristics are given in Table BXII.α.45.

Type species: **Holospora undulata** (ex Hafkine 1890) Gromov and Ossipov 1981, 351.

TAXONOMIC COMMENTS

The species listed under "Other Organisms" have not been validly published. They show the typical features of the genus, namely the expression of a life cycle with infectious and reproductive forms and a strong infectivity for the host nuclei. They share a number of biological features with the validated *Holospora* species and are therefore listed in this chapter. Except for "*H. recta*", where this has been questioned (see below), the bacteria appear to be good species with a distinct host specificity, nucleus-specificity, and clear morphology.

List of species of the genus Holospora

1. **Holospora undulata** (ex Hafkine 1890) Gromov and Ossipov 1981, 351[VP]

   un.du.la'ta. L. fem. adj. *undulatus* undulated, with waves.

   Lives in the micronucleus of *P. caudatum*. Reproductive form short, spindle shaped; infectious form long, spiral shaped. Diagnostic characteristics are given in Table BXII.α.45.

   The mol% G + C of the DNA is: not determined.

   Type strain: original description and illustration of Hafkine (1890).

   Additional Remarks: Clone M1-48 of *P. caudatum* containing *H. undulata* has been deposited in the culture collection of the Laboratory of Invertebrate Zoology, Biological Research Institute, State University of St. Petersburg, Russia.

2. **Holospora caryophila** (ex Preer, Preer and Jurand 1974) Preer and Preer 1982, 141[VP] (*Cytophaga caryophila* Preer, Preer and Jurand 1974, 156.)

   ca.ry.o'phi.la. G. n. *caryum* nut, kernel, nucleus; Gr. adj. *philus* loving, M.L. fem. adj. *caryophila* nucleus loving.

   Lives in the macronucleus (Figs. BXII.α.56 and BXII.α.57) of *P. biaurelia* or *P. caudatum*. Reproductive form thin, fusiform rod, 0.3–0.5 × 1.0–3.0 μm; spiral infectious form, 0.2–0.3 × 5–6 μm, tapered ends. Is highly infective to a few stocks of *Paramecium biaurelia* and *P. caudatum*. Known previously as alpha (Preer, 1969).

   Recent observations may cast some doubt upon the inclusion of *H. caryophila* in the genus *Holospora*. It was observed that the behavior of *H. caryophila* in dividing host nuclei is different from that of *H. undulata*, *H. elegans*, *H. obtusa*, and "*H. acuminata*". While in those species the in-fectious forms are collected in the connecting piece of the dividing nucleus and are then released from the nucleus, infectious forms of *H. caryophila* leave the host nucleus by an unknown mode (Fokin et al., 1996). Moreover, in contrast to *H. undulata*, *H. elegans*, and *H. obtusa*, *H. caryophila* is not recognized by the *Holospora* probe in *in situ* hybridization ("*H. acuminata*" not tested; Fokin et al., 1996). Because of these differences to *H. undulata*, *H. obtusa*, and *H. elegans*, *H. caryophila* has been considered to belong to a second group of *Holospora* (Table BXII.α.45).

   The mol% G + C of the DNA is: not determined.

   Type strain: ATCC 30694.

   Additional Remarks: The type strain of *H. caryophila* is in stock 562 of *P. tetraurelia* (ATCC 30694) (Preer et al., 1974).

3. **Holospora elegans** (ex Hafkine 1890) Preer and Preer 1982, 141[VP]

   e'le.gans. L. adj. *elegans* choice, elegant.

   Lives in the micronucleus of *P. caudatum*. Reproductive form short, spindle-shaped; infectious form long rod with tapered ends. Diagnostic characteristics are given in Table BXII.α.45.

   The mol% G + C of the DNA is: not determined.

   Type strain: ATCC 50008.

   Additional Remarks: The type strain of *H. elegans* is in stock C101 of *P. caudatum* (ATCC 50008) (Görtz and Dieckmann 1980).

4. **Holospora obtusa** (ex Hafkine 1890) Gromov and Ossipov 1981, 351[VP]

   ob.tu'sa. L. fem. adj. *obtusa* obtuse, blunt.

   Short, fusiform, reproductive rod about 3.0 μm long that undergoes binary fission and grows in infective form up to 20 μm long. Ends of rod rounded (Fig. BXII.α.58). Found in *P. caudatum* in the macronucleus. Diagnostic characteristics are given in Table BXII.α.45.

   The mol% G + C of the DNA is: not determined.

Editorial Note: This genus was described by J. R. Preer, Jr. and L. B. Preer (1984) in the first edition of *Bergey's Manual of Systematic Bacteriology*. We have used their descriptions and figures and have added newer information.

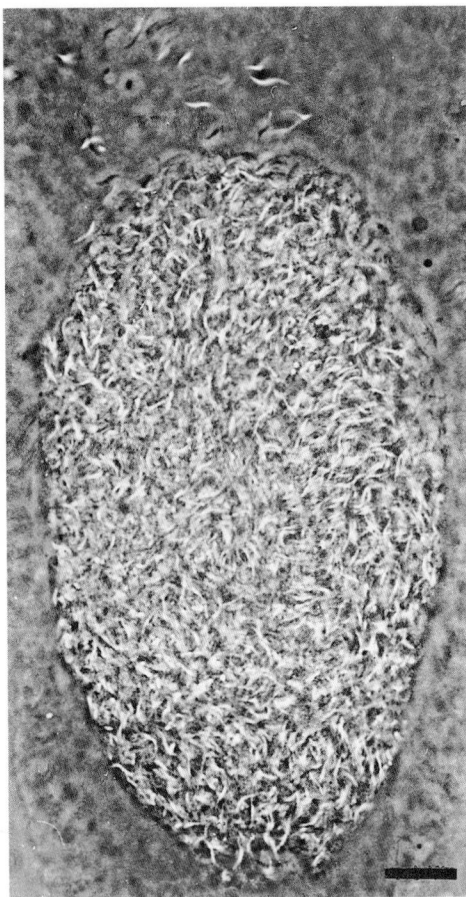

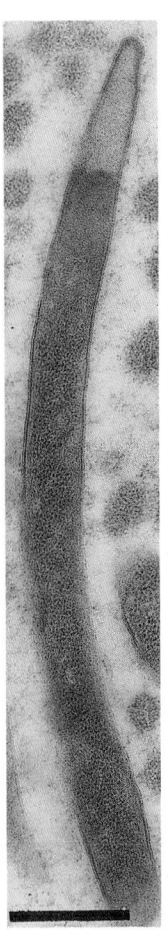

FIGURE BXII.α.56. Macronucleus of *Paramecium biaurelia* stock 562. The spiral endosymbiont filling the macronucleus is *Holospora caryophila*. Osmium–lacto–orcein preparation, whole mount, bright phase-contrast. Bar = 10 μm. (Reproduced with permission from L.B. Preer, Journal of Protozoology *16*: 570–578, 1969, ©Society of Protozoologists.)

FIGURE BXII.α.57. *Holospora caryophila* of *Paramecium biaurelia* stock 562, spiral form. Longitudinal section. Bar = 0.5 μm. (Reproduced with permission from L.B. Preer, Journal of Protozoology *16*: 570–578, 1969, ©Society of Protozoologists.)

Type strain: The original description and illustration was given by Hafkine (1890).

Additional Remarks: Clone M.115 of *P. caudatum* containing *H. obtusa* in its macronuclei has been deposited in the culture collection of the Laboratory of Invertebrate Zoology, Biological Research Institute, St. Petersburg State University (Gromov and Ossipov, 1981).

5. **"Holospora acuminata"** Ossipov, Skoblo, Borchsenius, Rautian and Podlipaev 1980, 927.

Lives in the micronucleus of *Paramecium bursaria* (Ossipov et al., 1980). Reproductive form short fusiform rod; infectious form 5.0–8.0 × 0.5–0.6 μm, straight, both ends

tapered. The species name has not been validly published. There is, however, little doubt that it belongs to the genus *Holospora*, because of many biological features shared with the three species *H. undulata, H. obtusa,* and *H. elegans.* In addition, they are recognized with an oligonucleotide probe designed for *H. obtusa*.

The mol% G + C of the DNA is: not determined.

Deposited strain: AC61-10.

Additional Remarks: In stock AC61-10 of *P. bursaria* (deposited in the culture collection of the Laboratory of Invertebrate Zoology, Biological Research Institute, State University of St. Petersburg, Russia).

Other Organisms

1. *"Holospora bacillata"* Fokin and Sabaneyeva 1993, 393.

Lives in the macronucleus of *Paramecium calkinsi* (Fokin and Sabaneyeva, 1993) and *Paramecium woodruffi* (Fokin et al., 1996; Fokin and Sabaneyeva, 1997). Infectious form 5–17.0 × 0.7–0.8 μm, straight, both ends rounded. Diagnostic characteristics as given in Table BXII.α.45.

The mol% G + C of the DNA is: not determined.

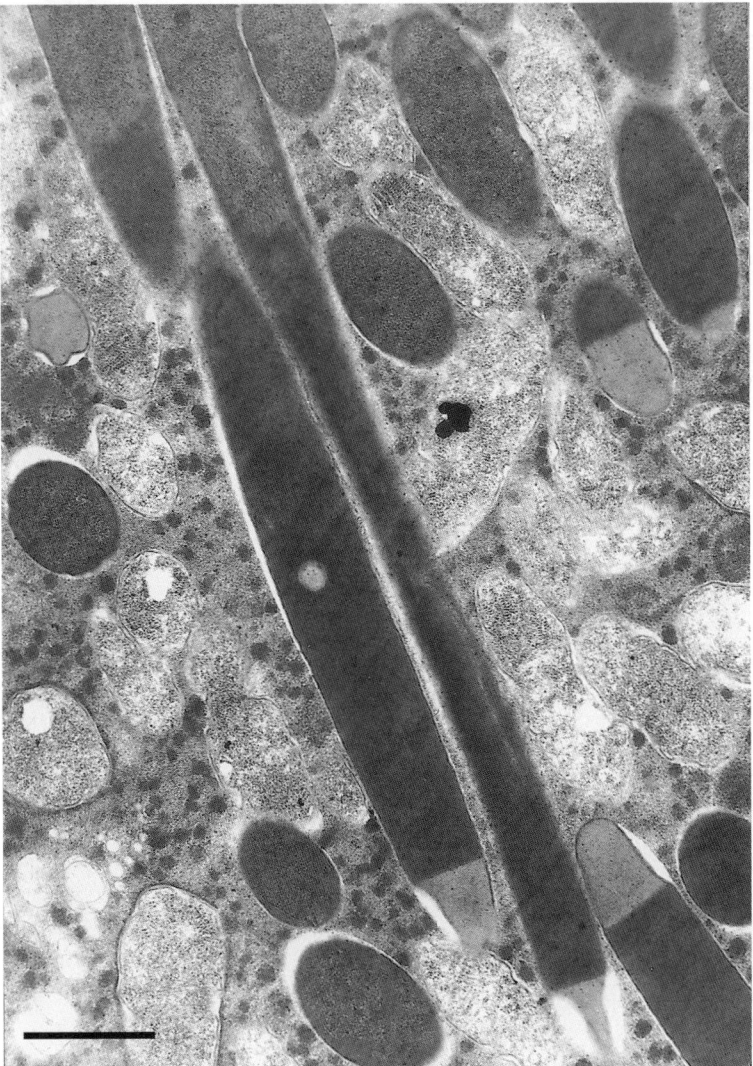

FIGURE BXII.α.58. *Holospora obtusa,* macronuclear symbiont of *Paramecium caudatum.* Infectious form stains darkly; the noninfectious form is light. Bar = 4.3 μm (Reproduced with permission from H.-D. Görtz, 1980. *In* Schwemmler and Schenk (Editors), Endocytobiology: Endosymbiosis and Cell Biology, Vol. 1, ©Walter deGruyter and Co, pp. 381–392.)

2. *"Holospora curvata"* Fokin and Sabaneyeva 1993, 393.

Lives in the macronucleus of *Paramecium calkinsi* (Fokin and Sabaneyeva, 1993). Infectious form 12–20 × 0.7–0.9 μm, curved, both ends rounded. Diagnostic characteristics as given in Table BXII.α.45.

The mol% G + C of the DNA is: not determined.

3. *"Holospora curviuscula"* Borchsenius, Skoblo and Ossipov 1983, 96.

Lives in the macronucleus (sometimes also micronucleus) of *P. bursaria.* Reproductive form short, spindle shaped; infectious form 4.0–10.0 × 0.4–0.5 μm, slightly curved rod with tapered ends. Diagnostic characteristics as given in Table BXII.α.45.

The mol% G + C of the DNA is: not determined.

4. *"Holospora recta"* Fokin 1991, 139.

rec.ta. L. adj. *rectus* straight; M.L. fem. adj. *recta* straight.

Lives in the micronucleus of *Paramecium caudatum* (Fokin, 1991). Infectious form 10–15 × 0.7–1.0 μm, straight, one end rounded, one end tapered. Diagnostic characteristics as given in Table BXII.α.45. Whereas all other species listed in this chapter appear to be taxonomically sound, this has been questioned for *"H. recta"* by Rautian and Ossipov (personal communication), who found a strain of *H. elegans* with a number of individuals exhibiting *"H. recta"*-like features. The question may finally be resolved after molecular data have been obtained from the different strains.

The mol% G + C of the DNA is: not determined.

Genus *Incertae Sedis* II. **Caedibacter** *(ex Preer, Preer and Jurand 1974) Preer and Preer 1982, 140^{VP}**

HANS-DIETER GÖRTZ AND HELMUT J. SCHMIDT

Cae' di.bac.ter. L. n. *caedes* act of killing; M.L. masc. n. *bacter* the masculine equivalent of the Gr. neut. n. *bactrum* a rod; masc. n. *caedibacter* the bacterium which kills.

Straight rods or coccobacilli 0.4–1.0 × 1.0–4.0 μm. Gram negative, nonflagellated, nonmotile. Intracellular (endosymbiotic) bacteria in *Paramecium*. Culture free of *Paramecium* impossible or exceedingly difficult. The ability to produce refractile inclusion (R) bodies was seen as a unique feature of the genus. R bodies are proteinaceous ribbons, approximately 10 μm long, coiled inside the bacterial cell (Figs. BXII.α.59, BXII.α.60, and BXII.α.61). The hollow cylindrical structure formed may be 0.4–0.8 μm long and about 0.4 μm in diameter. R bodies unroll when ingested into a phagosome and under certain *in vitro* conditions (for details see Preer et al., 1974; Quackenbush, 1988; Pond et al., 1989). R bodies are associated with spherical phage-like structures or covalently closed circular DNA plasmids. All species are toxic to certain sensitive strains of paramecia; bacteria confer

killer-trait or mate killer-trait upon their host cells. An exception is *"C. macronucleorum"* from *Paramecium duboscqui*, where no toxic effects were observed. Cells containing R bodies are usually larger than cells that do not contain R bodies, and contain many spherical phage-like structures of covalently closed, circular, DNA plasmids. Recently, a symbiotic bacterium from *Acanthamoeba* was included in the genus *Caedibacter* because of its sequence similarity to *Caedibacter caryophilus* (Horn et al., 1999). This new *Caedibacter* species, *Candidatus* Caedibacter acanthamoebae, does not establish R bodies. Also interesting is the relatively close relationship of *Caedibacter* to the genus *Holospora* and to the "NHP bacterium", etiologic agent of necrotizing hepatopancreatitis in shrimp (Loy et al., 1996; Horn et al., 1999).

The mol% G + C of the DNA is: 34–44.

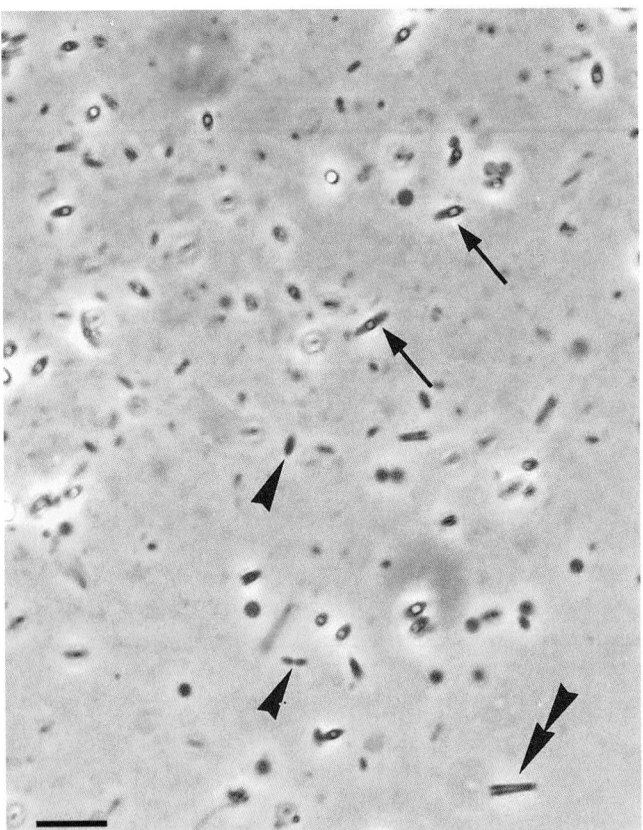

FIGURE BXII.α.59. Fresh squash of *Paramecium novaurelia,* washed free of bacteria, viewed by bright phase-contrast. Note *Caedibacter caryophilus* present in two forms: small rods (*arrowheads*) and large spindle-shaped cells containing refractile bodies (*arrows*) characteristic of the genus. Free R bodies, partly stretched, (*double arrowhead*) are also visible. Bar = 2.4 μm.

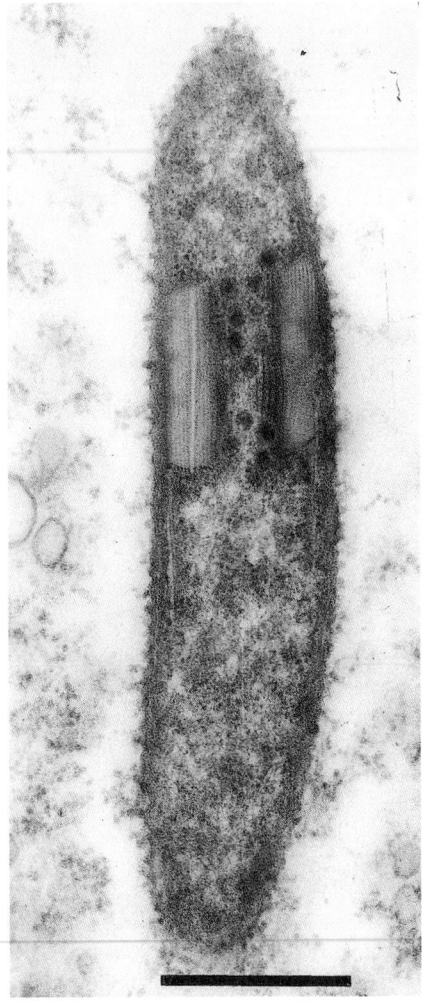

FIGURE BXII.α.60. *Caedibacter varicaedens,* endosymbiont of *Paramecium biaurelia* stock 7. Note spherical phages inside the coiled R body. Longitudinal section. Bar = 0.5 μm. (Reproduced with permission from J.R. Preer, Jr. and A. Jurand, Genetical Research *12:* 331–340, 1968, ©Cambridge University Press.)

---

*Editorial Note: This genus was described by J.R. Preer, Jr. and L.B. Preer (1984) in the first edition of *Bergey's Manual of Systematic Bacteriology.* The authors have used their descriptions and added new information.

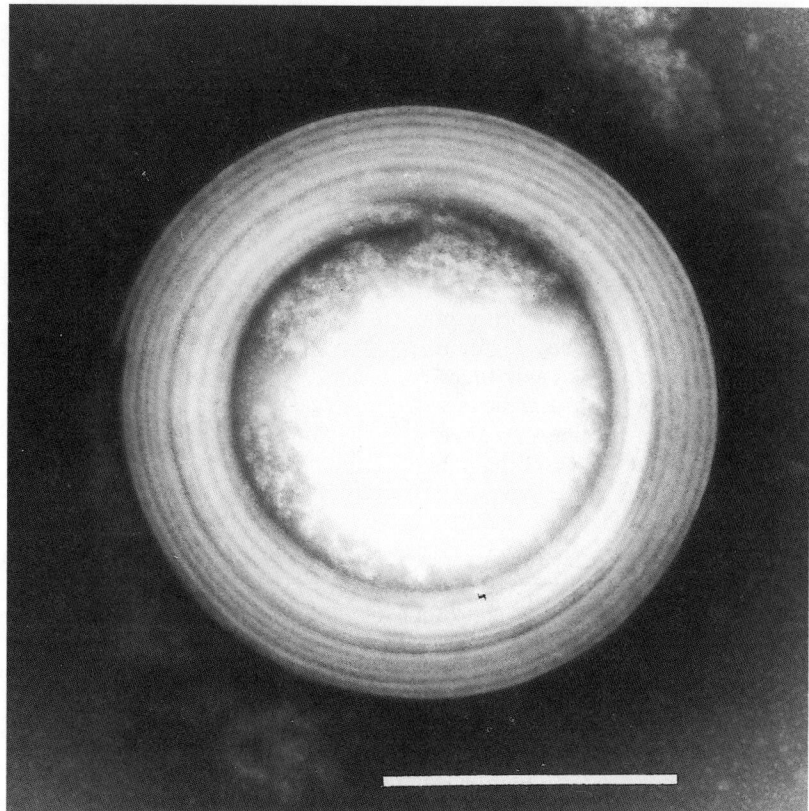

FIGURE BXII.α.61. Intact R body from *Caedibacter varicaedens* of *Paramecium biaurelia* stock 511. Phosphotungstic acid. Bar = 2.5 μm. (Reproduced with permission from J.R. Preer, Jr. et al., Bacteriological Reviews *38:* 113–163, 1974, ©American Society for Microbiology.)

Type species: **Caedibacter taeniospiralis** (ex Preer, Preer and Jurand 1974) Preer and Preer 1982, 140 (*Caedobacter taeniospiralis* [sic] Preer, Preer and Jurand 1974, 157.)

FURTHER DESCRIPTIVE INFORMATION

The distinguishing characteristic of *Caedibacter* species is the presence of refractile (R) bodies. Between less than 10% and up to 50% of the bacteria in a host cell contain R bodies. Cells containing R bodies are called "brights"; cells without R bodies are called "non-brights". Only a very weak infectivity, if any, is observed for non-brights. The coiled R body, a distinctive structure, is seen with bright phase-contrast microscopy either as doughnut-shaped or as a pair of parallel rods, depending upon its orientation (Fig. BXII.α.62). Although the structure is readily resolved in bright phase-contrast, it is often obscure in dark phase-contrast, because dark phase-contrast optics are usually of such high contrast that phase reversal occurs in most aqueous media. R bodies themselves appear to result from the induction of phage-like or plasmid-like extrachromosomal DNAs (Quackenbush, 1988; Pond et al., 1989). The induction appears to be lethal: R body-containing cells do not have the capacity to reproduce. It is interesting that R bodies have been observed in free-living bacteria belonging to the genus *Pseudomonas* (Lalucat et al., 1979). The toxicity of *C. taeniospiralis* and *C. pseudomutans*, formerly known as kappa, is produced by the ingestion of R body-containing kappas by sensitive strains of paramecia.

R bodies can be induced to unroll (Fig. BXII.α.63) and in some cases, reroll; they have been shown to unroll in the food vacuoles of sensitive paramecia. In some species of *Caedibacter*, the R body unrolls from the inside (Fig. BXII.α.64); in others, from the outside. Rupture of the membrane of the food vacuole occurs, and the contents of the food vacuole, including the R bodies, pass into the cytoplasm of the paramecium. The toxins themselves have never been obtained in soluble form and their nature is unknown. Ingestion of the fourth species, *C. paraconjugatus*, does not produce toxic effects on sensitives. Instead it is a mate-killer, and sensitive paramecia die only after contact with paramecia bearing *C. paraconjugatus* during conjugation. The endosymbiont discovered by Estève (1978) in the macronucleus of a strain of *Paramecium caudatum* was later found again and described as *C. caryophilus* (Schmidt et al., 1987a; Euzéby, 1997).

Note added in proof: According to an analysis of 16S rRNA sequences, *C. taeniospiralis* (isolated from stock 51k of *Paramecium tetraurelia*) belongs to the *Gammaproteobacteria*, whereas *C. caryophilus* belongs to the *Alphaproteobacteria* (Beier et al., 2002).

FURTHER READING

Heckmann, K. and H.D. Görtz. 1991. Procaryotic symbionts of ciliates. *In* Balows, Trüper, Dworkin, Harder and Schleifer (Editors), The Prokaryotes a Handbook on the Biology of Bacteria: Ecophysiology, Isolation, Identification, Applications, Springer Verlag, Berlin, Heidelberg, New York. pp. 3865–3890.

Pond, F.R., I. Gibson, J. Lalucat and R.L. Quackenbush. 1989. R-body producing bacteria. Microbiol. Rev. *53*: 25–67.

Preer, J.R., Jr., L.B. Preer and A. Jurand. 1974. Kappa and other endosymbionts in *Paramecium aurelia*. Bacteriol. Rev. *38*: 113–163.

Quackenbush, R.L. 1988. Endosymbionts of killer paramecia. *In* Görtz (Editor), Paramecium, Springer Verlag, Heidelberg, New York. pp. 406–418.

Soldo, A.T. 1974. Intracellular particles in *Paramecium aurelia*. *In* Wagtendonk, v. (Editor), *Paramecium*: A Current Survey, Elsevier, Amsterdam. pp. 375–442.

List of species of the genus Caedibacter

1. **Caedibacter taeniospiralis** (ex Preer, Preer and Jurand 1974) Preer and Preer 1982, 140[VP]

   taen.i.o.spi.ral' is. L. n. *taenia* ribbon; L. adj. *spiralis* coiled; M.L. masc. adj. *taeniospiralis* coiled ribbon.

   Description of the species as for the genus. Rods, 0.4–0.7 × 1.0–2.5 μm. Found exclusively in the cytoplasm of *Paramecium tetraurelia*. Contain plasmids (Dilts, 1976). R bodies unroll from the inside and contain plasmids. Ingestion of R body-containing symbionts by sensitive paramecia causes development of small blisters or humps on their surface within 2–3 hours preceding the death of the paramecium.

   The mol% G + C of the DNA is: 41 (Bd).

   Type strain: ATCC 30632.

   Additional Remarks: The type strain of *Caedibacter taeniospiralis* was isolated from stock 51 of *Paramecium tetraurelia* (ATCC 30632) (Preer et al., 1974)

2. **Caedibacter caryophilus** Schmidt, Görtz and Quackenbush 1987a, 461[VP]

   ca.ry.o' phi.lus. Gr. n. *caryum* nucleus; Gr. adj. *philus* loving; M.L. adj. *caryophilus* nucleus loving.

In macronucleus of *P. caudatum*. Cells without R bodies 1–1.5 × 0.4 μm, cells with R bodies 1.5–2.5 × 0.7 μm. R bodies unrolling from the inside. Outer terminus of unrolled R bodies blunt, inner terminus acute. R bodies associated with phages. Width of R bodies 0.8 μm. Sensitive strains of *Paramecium* are killed by paralysis. Co-infections with *Holospora* species in natural populations.

The mol% G + C of the DNA is: 35 (T_m).

Type strain: ATCC 50168.

GenBank accession number (16S rRNA): X71837.

Additional Remarks: The type strain of *Caedibacter caryophilus* was isolated from stock C221 of *P. caudatum* (ATCC 50168) (Schmidt et al., 1987a).

3. **Caedibacter paraconjugatus** Quackenbush 1982, 266[VP] (Effective publication: Quackenbush 1978, 186.)

   par.a.con.ju.ga' tus. Gr. prep. *para* alike; L. part. adj. *conjugatus* conjugated; also the specific epithet of a mate killer (*Pseudocaedibacter conjugatus*); M.L. masc. part. adj. *paraconjugatus* similar to mate killers.

   Small rods. Fewer than 1% of the cells contain R bodies, which are smaller than those found in the other species of

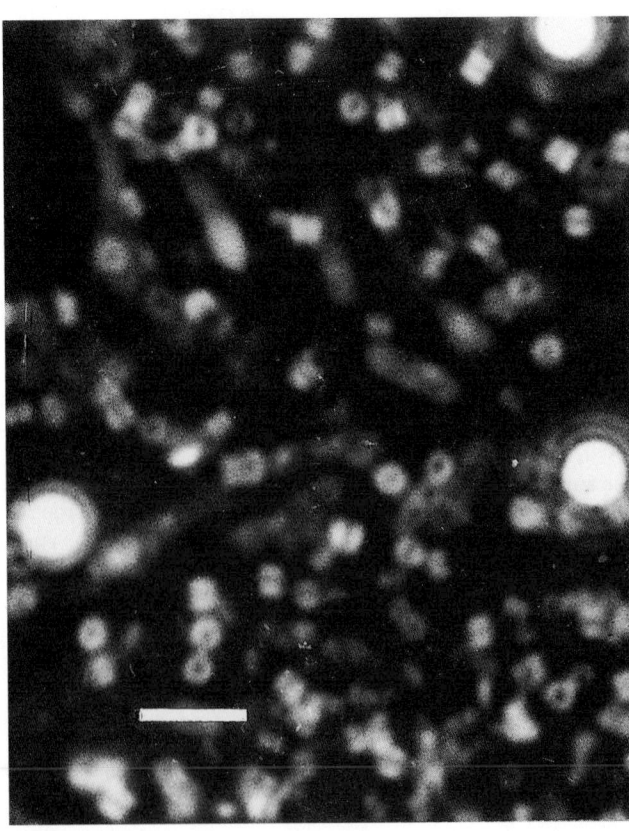

FIGURE BXII.α.62. Isolated R bodies from *Caedibacter varicaedens* of *Paramecium biaurelia* stock 7, viewed with bright phase-contrast. R bodies appear doughnut-shaped when viewed on end, and as two parallel rods when viewed from the side. The large bright spheres are latex particles. Bar = 2 μm.

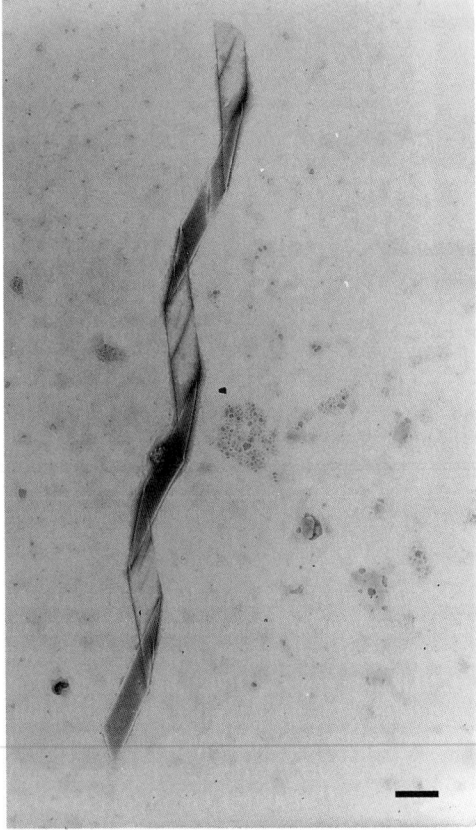

FIGURE BXII.α.63. Unrolled R body isolated from *Caedibacter varicaedens* of *Paramecium biaurelia* stock 1039. Phosphotungstic acid. Bar = 1 μm. (Reproduced with permission from J.R. Preer Jr. et al., Bacteriological Reviews *38*: 113–163, 1974, ©American Society for Microbiology.)

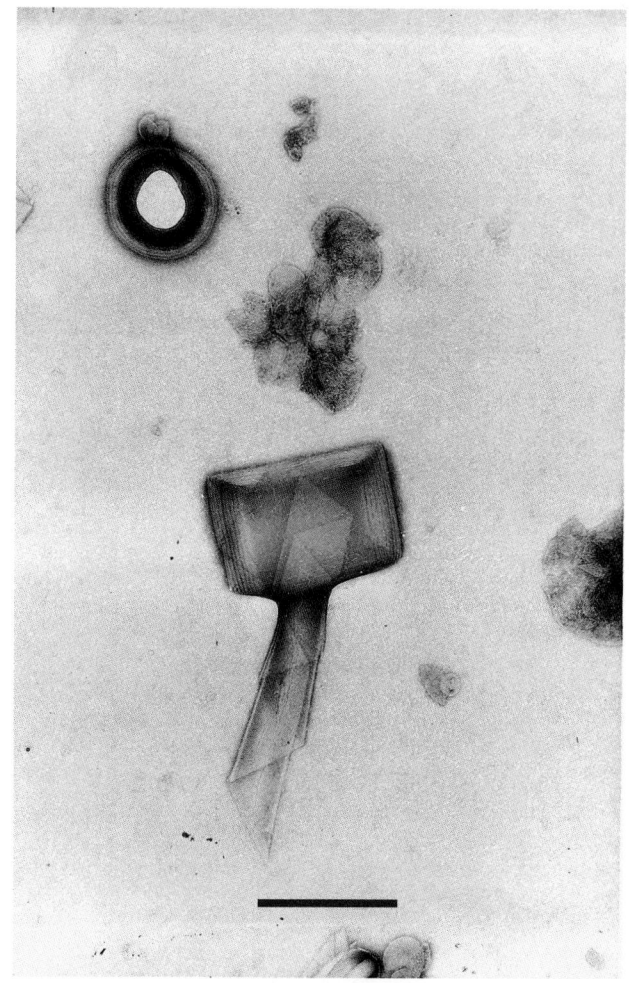

FIGURE BXII.α.64. R bodies isolated from *Caedibacter taeniospiralis* of *Paramecium tetraurelia* stock 51. Note intact doughnut-shaped R body and, below, an R body unrolling from inside. Phosphotungstic acid. Bar = 0.5 μm. (Reproduced with permission from L.B. Preer, Journal of Cell Science, *11:* 581–600, 1972, ©Company of Biologists, Ltd.)

Caedibacter. Ingestion of cells by sensitive paramecia does not cause any observable effects. Cell-to-cell contact between host and sensitive paramecia is required for toxic effects (mate killing) to be observed in the sensitive paramecia. Found in *P. biaurelia.* Contain phage-like structures.

The mol% G + C of the DNA is: not determined.

Type strain: ATCC 30638.

Additional Remarks: The type strain of *Caedibacter paraconjugatus* was isolated from stock 570 of *Paramecium biaurelia* (ATCC 30638) (Preer et al., 1974).

4. **Caedibacter pseudomutans** Quackenbush 1982, 266^{VP} (Effective publication: Quackenbush 1978, 186.)

pseu.do.mu′ tans. Gr. adj. *pseudo* false; L. part. adj. *mutans* changing; M.L. part. adj. *pseudomutans* false changing, referring to the fact that it was once thought to be a mutant of *C. taeniospiralis* (Dippell, 1950).

Cigar-shaped rods approximately 0.5 × 1.5 μm. Found in *P. tetraurelia.*

The mol% G + C of the DNA is: 44 (Bd).

Type strain: ATCC 30633.

Additional Remarks: The type strain of *Caedibacter pseudomutans* was isolated from stock 51ml of *Paramecium tetraurelia* (ATCC 30633) (Preer et al., 1974).

5. **Caedibacter varicaedens** Quackenbush 1982, 266^{VP} (Effective publication: Quackenbush 1978, 186.)

var.i.cae′ dens. L. adj. *varis* different; L. v. *caedo* to kill; M.L. part. adj. *varicaedens* killing in different ways.

Rods 0.4–1.9 × 2–4 μm. Different strains cause either vacuolization or paralysis or rapid reverse rotation while swimming (spin killing) in sensitive paramecia. R bodies unroll from the outside. One of the commonest killers of *P. biaurelia.* Most strains contain spherical phage-like structures (Fig. BXII.α.60).

The mol% G + C of the DNA is: 40–41 (Bd).

Type strain: ATCC 30637.

Additional Remarks: The type strain of *Caedibacter varicaedens* was isolated from stock 7 of *Paramecium biaurelia* (ATCC 30637) (Preer et al., 1974).

6. *Candidatus* Caedibacter acanthamoebae Horn, Fritsche, Gautom, Schleifer and Wagner 1999, 364.

a′ canth.a.moe.bae. L. gen. sing. n. *acanthamoebae* of *Acanthamoeba*, genus name of host protozoa.

Phylogenetic position, *Alphaproteobacteria;* not cultivated on cell-free media; Gram reaction, negative; rod-shaped ~0.7–3.3 μm in length, 0.22–0.33 μm in diameter; basis of assignment, 16S rDNA sequence (accession number AF1321138) and nucleotide S-S-CaeAc-998-a-A-18 (5′-TCTTGTCTCCGCGATCCC-3′); association and host, intracellular symbiont of *Acanthamoeba polyphaga* HN-3; mesophilic (Horn et al., 1999).

Other Organisms

1. "*Caedibacter macronucleorum*" Fokin and Görtz 1993, 322.

ma.cro.nu.cle.o′ rum. Gr. adj. *macro* big; L. n. *nucleus* nucleus; *macronucleorum* of the macronucleus.

In macronucleus of *Paramecium duboscqui.* Cells without R bodies 1–1.5 × 0.4 μm, cells with R bodies 1.5–2.5 × 0.7 μm). R bodies unrolling from the inside. Outer terminus of unrolled R bodies blunt, inner terminus acute. R bodies associated with phages. Width of R bodies 0.8 μm. Sensitive strains of *Paramecium* are killed by paralysis. Co-infections with *Holospora* species in natural populations. As was the

rule for intracellular bacteria in protozoa before the advent of molecular sequencing, this organism was named according to morphology and biological behavior. "*Caedibacter macronucleorum*" had been grouped into the genus *Caedibacter*, because it was found as an intracellular bacterium in a protozoon and because of its ability to form R bodies like the valid species of the genus. Since no molecular data have been obtained to date, the true phylogenetic position of the bacterium is not known.

Genus Incertae Sedis III. **Lyticum** (ex Preer, Preer and Jurand 1974) Preer and Preer 1982, 141^{VP}*

HANS-DIETER GÖRTZ AND HELMUT J. SCHMIDT

Ly' ti.cum. L. adj. *lyticus* dissolving; M.L. neut. n. *Lyticum* dissolver.

Large rods 0.6–0.8 μm in diameter, straight, curved, or spiral. Length of single forms 3.0–5.0 μm. Numerous **peritrichous flagella**. Although cultivation independently from *Paramecium* has been reported, it has not been confirmed. Produce labile **toxins which kill** sensitive strains of paramecia **very quickly by lysis**. Gram negative. Nonmotile or almost so, in spite of numerous, well-developed flagella (Fig. BXII.α.65). Occurs in *Paramecium biaurelia*, *P. tetraurelia*, and *P. octaurelia* (Fig. BXII.α.66).

The mol% G + C of the DNA is: 27–49.

Type species: **Lyticum flagellatum** (ex Preer, Preer and Jurand 1974) Preer and Preer 1982, 140.

FURTHER DESCRIPTIVE INFORMATION

These very large endosymbionts are unique due to their conspicuous flagella and rapidly acting toxins (less than 30 minutes at room temperature). The nature of the toxin has not been revealed. *Paramecium triaurelia*, *P. pentaurelia*, and *P. novaurelia* are particularly sensitive to the toxins. The substantially different values for mol% G + C of DNA reported by Behme and by Soldo (see Preer et al., 1974 for a discussion) are unresolved.

DIFFERENTIATION OF THE GENUS *LYTICUM* FROM OTHER GENERA

Lyticum flagellatum is a straight rod found in *Paramecium tetraurelia* and *P. octaurelia*, while *L. sinuosum* is a strikingly curved rod found in *Paramecium biaurelia*.

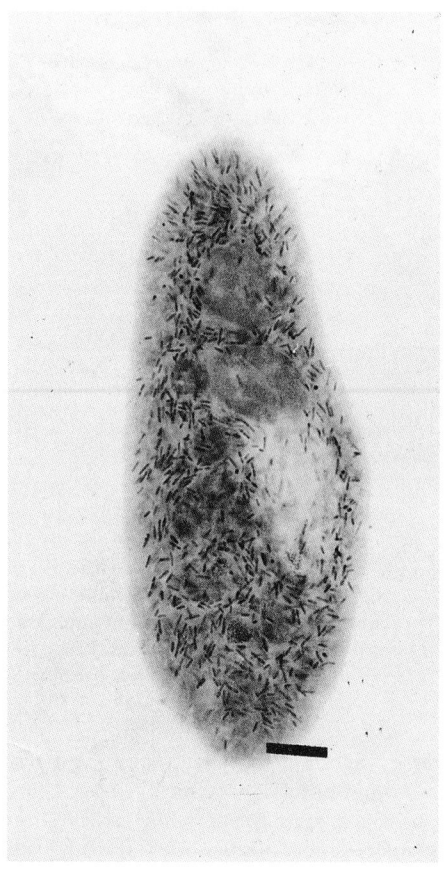

FIGURE BXII.α.66. *Paramecium tetraurelia* stock 239 bearing endosymbiont *Lyticum flagellatum*. The numerous black rods throughout the cytoplasm are the endosymbionts. Osmium–lacto–orcein preparation, whole mount, dark phase-contrast. Bar = 20 μm. (Reproduced with permission J.R. Preer, Jr. et al., Bacteriological Reviews *38*: 113–163, 1974, ©American Society for Microbiology.)

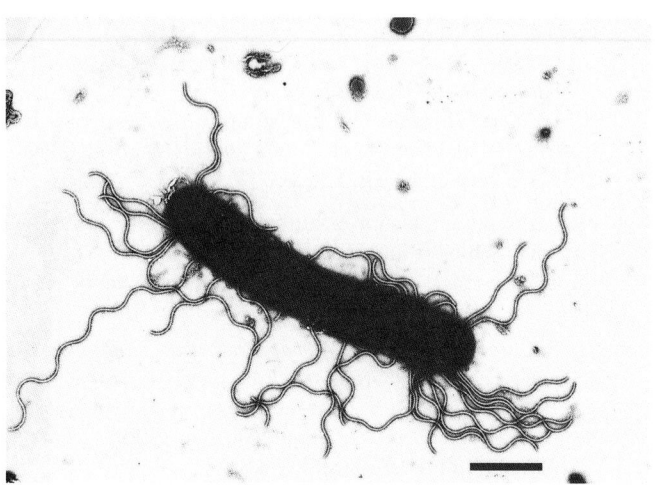

FIGURE BXII.α.65. *Lyticum flagellatum* isolated from *Paramecium octaurelia* stock 327. Phosphotungstic acid. Bar = 1 μm. (Reproduced with permission J. R. Preer, Jr. et al., Bacteriological Reviews *38*: 113–163, 1974, ©American Society for Microbiology.)

List of species of the genus Lyticum

1. **Lyticum flagellatum** (ex Preer, Preer and Jurand 1974) Preer and Preer 1982, 140^{VP}

   fla.gel.la' tum. L. neut. n. part. adj. *flagellatum* flagellated.

   Straight rods 0.6–0.8 × 2.0–4.0 μm, with many peritrichous flagella (Fig. BXII.α.65). Originally called lambda. Found within the cytoplasm of *Paramecium tetraurelia* and *P. octaurelia*. Stock 299 of *P. octaurelia* containing lambda does not require folic acid, whereas symbiont-free lines of 299 do (Soldo and Godoy, 1973).

   The mol% G + C of the DNA is: 27 (Bd) in one strain (Soldo) and 49 (Bd) in another (Behme; Preer et al. 1974).

   Type strain: 299, ATCC 30700.

Editorial Note: This genus was described by J.R. Preer, Jr. and L.B. Preer (1984) in the first edition of *Bergey's Manual of Systematic Bacteriology*. The authors have used their descriptions and figures and have added newer information.

2. **Lyticum sinuosum** (ex Preer, Preer and Jurand 1974) Preer and Preer 1982, 140[VP]

sin′u.o.sum. L. neut. adj. *sinuosum* winding, sinuous.

Curved or spiral rods 0.7–0.9 × 2.0.–10.0 μm, sometimes forming chains of 2–3 cells. Originally called sigma. Found within the cytoplasm of *Paramecium biaurelia.*

The mol% G + C of the DNA is: 45 (Bd).

Type strain: 114, ATCC 30696.

Genus Incertae Sedis IV. *Candidatus* Odyssella Birtles, Rowbotham, Michel, Pitcher, Lascola, Alexou-Daniel and Raoult 2000, 71

THE EDITORIAL BOARD

Od.ys.sel′la. Gr. dim. fem. n. *Odyssella* pertaining to Odysseus.

Motile, rod-shaped (0.2–0.5 × 0.7–1.0 μm) **intracytoplasmic parasite of** *Acanthamoeba* spp. Does not grow axenically. Most virulent at temperatures above 30°C.

The mol% G + C of the DNA is: 41 ± 1 (T_m).

Type species: *Candidatus* Odyssella thessalonicensis Birtles, Rowbotham, Michel, Pitcher, Lascola, Alexou-Daniel and Raoult 2000, 71.

FURTHER DESCRIPTIVE INFORMATION

Electron micrographs show that the organisms are present in the cytoplasm of infected *Acanthamoeba* cells and that at lower growth temperatures the bacterial cells are sometimes surrounded by a clear zone containing vesicles that appear to have budded from the bacterial cell membrane.

The organism infects the following *Acanthamoeba* species: *A. polyphaga, A. lenticulata, A. castellanii, A. quinaludunensis,* and *A. comandoni.*

Comparisons of 16S rDNA sequences place the organism in the *Alphaproteobacteria*; however, its 16S rDNA sequence was not closely related to those of any of the bacteria to which it was compared. It was most similar to that of two other intracellular parasites: *Holospora obtusa* and *Caedibacter caryophilus.*

ENRICHMENT AND ISOLATION PROCEDURES

Candidatus Odyssella thessalonicensis was first isolated from an *Acanthamoeba* sp. obtained from an air-conditioning system. It was propagated on *A. polyphaga* Linc Ap-1 using the methods of Birtles et al. (1996).

MAINTENANCE PROCEDURES

Suspensions of infected *Acanthamoeba* cells can be stored at −70°C (Birtles et al., 1996).

List of species of the genus Odyssella

1. *Candidatus* Odyssella thessalonicensis *sp. nov.* Birtles, Rowbotham, Michel, Pitcher, Lascola, Alexou-Daniel and Raoult 2000, 71.

   thess.al.lon′i.cen′sis. M.L. masc. n. adj. *thessalonicensis* pertaining to the Greek city Thessalonika, where the organism was isolated.

The genus description and the species description are identical.

The mol% G + C of the DNA is: 41 ± 1 (T_m).

Type strain: L13 (only described strain).

GenBank accession number (16S rRNA): AF069496.

Genus Incertae Sedis V. *Candidatus* Paracaedibacter Horn, Fritsche, Gautom, Schleifer and Wagner 1999, 364

HANS-DIETER GÖRTZ AND HELMUT J. SCHMIDT

For a description see the list of species.

TAXONOMIC COMMENTS

A formal description and diagnosis has not yet been published. Because of the relatively low overall 16S rDNA sequence simi-larities (87–88%) of the two new species *Candidatus* Paracaedibacter acanthamoebae and *Candidatus* Paracaedibacter symbiosus, Horn et al. (1999) propose the provisional classification of these symbionts in the new genus *Candidatus* Paracaedibacter.

List of species of the genus Candidatus Paracaedibacter

1. *Candidatus* Paracaedibacter acanthamoebae Horn, Fritsche, Gautom, Schleifer and Wagner 1999, 364.

   Not cultivated on cell-free media; Gram negative; rod shaped, ~1.3–1.7 × 0.22–0.24 μm. Based on 16S rDNA sequence and nucleotide probe S-S-PcaeC9-217-a-A-18 (5′-GGGCTGCTCAATTGGCGA-3′), *Candidatus* Paracaedibacter acanthamoebae is assigned to the *Alphaproteobacteria*. Intracellular symbiont of *Acanthamoeba* sp. UWC9. Mesophilic.

   GenBank accession number (16S rRNA): AF132137.

2. *Candidatus* Paracaedibacter symbiosus Horn, Fritsche, Gautom, Schleifer and Wagner 1999, 364.

Not cultivated on cell-free media; Gram negative; rod shaped, ~1.3–1.7 × 0.22–0.24 μm. Based on 16S rDNA sequence and nucleotide probe S-S-PcaeE39-217-a-A-18 (5′-GGGCTGTTCCTTTAGCGA-3′), *Candidatus* Paracaedibac-

ter symbiosus is assigned to the *Alphaproteobacteria*. Intracellular symbiont of *Acanthamoeba* sp. UWE39. Mesophilic. *GenBank accession number (16S rRNA)*: AF132139.

Genus Incertae Sedis VI. **Pseudocaedibacter** Quackenbush 1982, 267^VP (Effective publication Quackenbush 1978, 186)*

HANS-DIETER GÖRTZ AND HELMUT J. SCHMIDT

Pseu.do.cae′di.bac.ter. Gr. adj. *pseudo* false; M.L. n. *Caedibacter* genus of endosymbionts that include organisms commonly known as kappa; M.L. masc. n. *Pseudocaedibacter* false kappa particles.

Rods 0.25–0.7 × 0.5–4 μm. Gram negative. Nonmotile. **Do not produce R-body-containing cells.** May or may not confer a killer trait upon their host paramecia. Occurs in *Paramecium primaurelia*, *P. biaurelia*, *P. pentaurelia*, and *P. octaurelia*. Whether lacking the ability to produce R bodies justifies separation of the genus *Pseudocaedibacter* from the genus *Caedibacter* may be questioned because the newly found endosymbiont from *Acanthamoeba polyphaga*, *Candidatus* Caedibacter acanthamoebae, does not establish R bodies but showed 93% sequence similarity (16S rRNA) to *Caedibacter caryophilus* and was therefore included in the genus *Caedibacter* (Horn et al., 1999).

The mol% G + C of the DNA is: 35–39.

Type species: **Pseudocaedibacter conjugatus** (ex Preer, Preer and Jurand 1974) Quackenbush 1982, 267 (Effective publication: Quackenbush 1978, 187) (*Caedibacter conjugatus* Preer, Preer and Jurand 1974, 157.)

FURTHER DESCRIPTIVE INFORMATION

Pseudocaedibacter is much like *Caedibacter*, except that no R bodies are found. It includes the typical mate-killing forms (*P. conjugatus*, called mu), the small killing forms of *Paramecium octaurelia* (*Pseudocaedibacter minuta*, gamma), and the nondescript forms with no killing action (*P. falsus*, nu).

TAXONOMIC COMMENTS

For formal reasons, *"Pseudocaedibacter glomeratus"* is listed under Other Organisms, because the species name has not been validated. However, it should be stated that the validated species, too, were only included in the genus because of a few biological features. As no molecular or physiological data are available for any of the species, the whole genus and its species appear arbitrary.

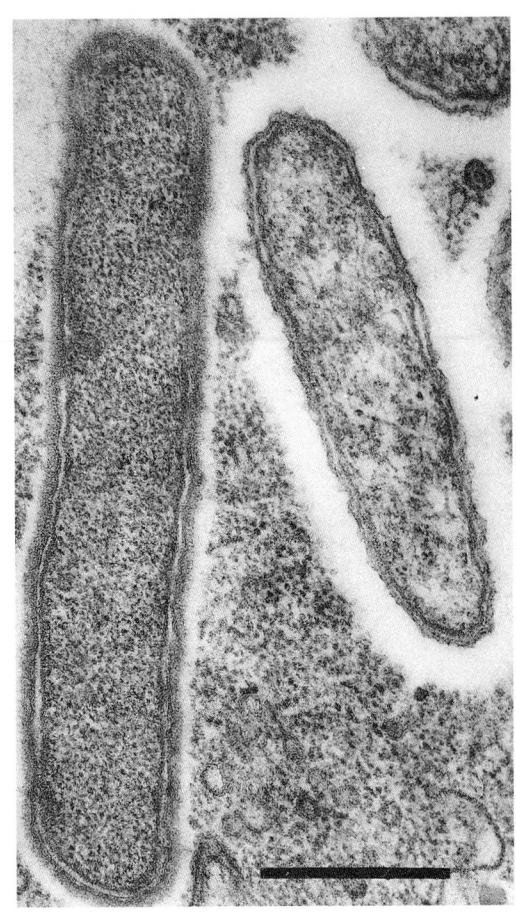

FIGURE BXII.α.67. Two endosymbionts present in a single cell of *Paramecium octaurelia* stock 131. *Left, Tectibacter vulgaris* (note electron-dense material surrounding outer membrane); *right, Pseudocaedibacter conjugatus.* Longitudinal section. Bar = 0.5 μm. (Reproduced with permission from J.R. Preer, Jr. et al., Bacteriological Reviews *38:* 113–163, 1974, ©American Society for Microbiology.)

---

*Editorial Note: This genus was described by J.R. Preer, Jr. and L.B. Preer (1984) in the first edition of *Bergey's Manual of Systematic Bacteriology.* The authors have used their descriptions and figures and have added newer information.

List of species of the genus Pseudocaedibacter

1. **Pseudocaedibacter conjugatus** (ex Preer, Preer and Jurand 1974) Quackenbush 1982, 267^VP (Effective publication: Quackenbush 1978, 187) (*Caedibacter conjugatus* Preer, Preer and Jurand 1974, 157.)
   con.ju.ga′ tus. L. masc. part. adj. *conjugatus* conjugated.

Rods 0.3–0.5 × 1–4 μm. The species is called mu and (except for *Caedibacter paraconjugatus*) is the only symbiont responsible for the mate killer phenotype in the *Paramecium aurelia* complex (Fig. BXII.α.67). It produces the toxin capable of killing sensitive strains of *Paramecium* only after

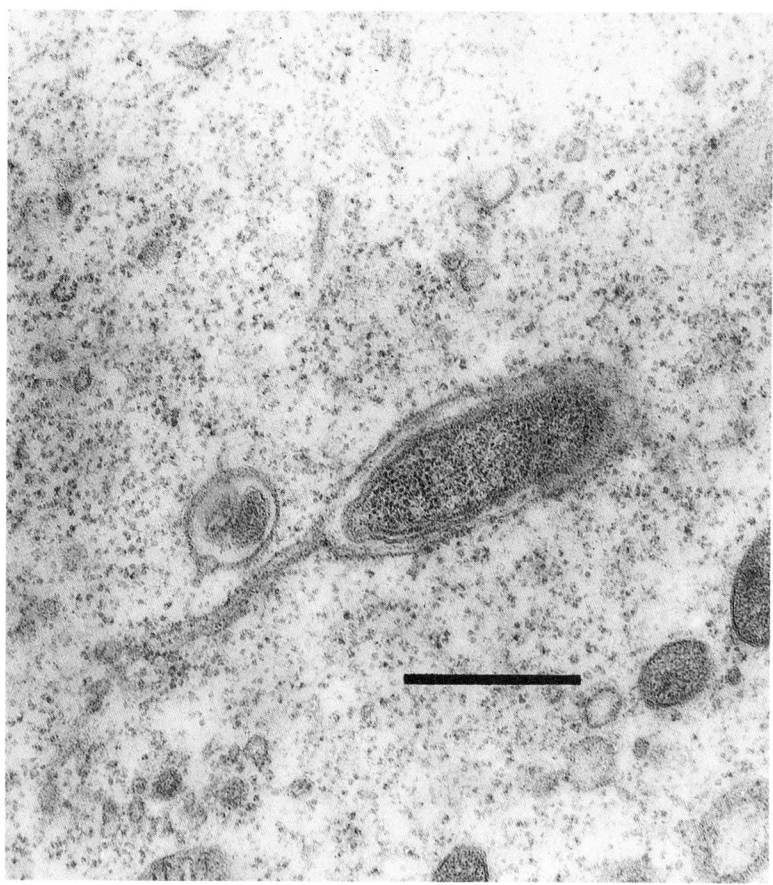

FIGURE BXII.α.68. *Pseudocaedibacter minutus,* endosymbiont of *Paramecium octaurelia* stock 565. No additional outer membrane. Longitudinal section. Bar = 0.5 μm. (Reproduced with permission from A. Jurand.)

cell-to-cell contact between killers and sensitives at conjugation. Cultivation outside the cytoplasm of *Paramecium* has been reported to occur on a very complex medium (Williams, 1971). Found in the cytoplasm of *Paramecium primaurelia* and *P. octaurelia*.

The mol% G + C of the DNA is: 35–37 (Bd).

Type strain: ATCC 30796.

Additional Remarks: The type strain of *P. conjugatus* is isolated from stock 540 of *Paramecium primaurelia* (ATCC 30796) (Preer et al., 1974).

2. **Pseudocaedibacter minutus** (Preer, Preer and Jurand 1974) Quackenbush 1982, 267[VP] (Effective publication: Quackenbush 1978, 187) (*Caedibacter minutus* Preer, Preer and Jurand 1974, 157.)

mi.nu' tus. L. masc. adj. *minutus* small.

Rods, often double, 0.25–0.35 × 0.5–1.0 μm (singles). This very small cell is unique among the endosymbionts of *Paramecium* in being surrounded by an extra set of membranes, apparently continuous with the endoplasmic reticulum of its host (Fig. BXII.α.68). Although it never rises to high concentrations in the cytoplasm, the paramecia that bear it are nevertheless very strong killers. Found only in

the cytoplasm of *Paramecium octaurelia*. Originally called gamma.

The mol% G + C of the DNA is: 38 (Bd).

Type strain: ATCC 30699.

Additional Remarks: Isolated from stock 214 of *Paramecium octaurelia* (ATCC 30699) (Preer et al., 1974).

3. **Pseudocaedibacter falsus** (Preer, Preer and Jurand 1974) Quackenbush 1982, 267[VP] (Effective publication: Quackenbush 1978, 187) (*Caedibacter falsus* Preer, Preer and Jurand 1974, 157.)

fal' sus. L. masc. adj. *falsus* false.

Rods 0.4–0.7 × 1.0–1.5 μm. No toxic action known, although the forms (called nu) found in *Paramecium pentaurelia* are said to increase the resistance of their hosts to the toxin produced by *Lyticum flagellatum* (Holtzman, 1959). The strains found in *Paramecium tetraurelia* were once regarded as mutants of *Caedibacter taeniospiralis* and called pi. Found also in the cytoplasm of *Paramecium biaurelia*.

The mol% G + C of the DNA is: 36 (Bd).

Type strain: ATCC 30640.

Additional Remarks: Isolated from stock 1010 of *P. biaurelia* (ATCC 30640) (Preer et al., 1974).

Other Organisms

1. *"Pseudocaedibacter glomeratus"* Fokin and Ossipov 1986, 1003.

Rods, 1.0–1.2 × 0.3 µm. No flagella. Gram negative, no central nucleoid. Bacteria are encircled by host membranes in a symbiontophoral vacuole. These vacuoles are surrounded by lacunae of endoplasmic reticulum with numerous ribosomes. No killing or toxic actions known.

The mol% G + C of the DNA is: not determined.

The type strain is in *Paramecium pentaurelia* strain Bp171, culture collection of the Laboratory of Protozoan Karyology of the Biological Research Institute of St.Petersburg State University, Russia.

Genus Incertae Sedis VII. "Pseudolyticum" Boss, Borchsenius and Ossipov 1987, 98*

HANS-DIETER GÖRTZ AND HELMUT J. SCHMIDT

pseu'do.ly.ti.cum. Gr. adj. *pseudo* false; L. adj. *lyticus* dissolving; M.L. neut. n. *Lyticum* genus of endosymbionts producing labile toxins which kill sensitive strains of paramecia very quickly by lysis; M.L. n. *Pseudolyticum* false *Lyticum*.

Straight rods, 2.0 × 3.5–14.0 µm. Bacteria are not infective and do not confer a killer trait on their host. Gram negative. Nonmotile in spite of numerous flagella. Occur in *Paramecium caudatum*.

The mol% G + C of the DNA is: not determined.

Type species: **"Pseudolyticum multiflagellatum"** Boss, Borchsenius and Ossipov 1987, 98.

List of species of the genus Pseudolyticum

1. **"Pseudolyticum multiflagellatum"** Boss, Borchsenius and Ossipov 1987, 98.

mul.ti.fla.gel.la'tum. L. neut. adj. *multus* many; L. neut. part. adj. *flagellatum* flagellated; *multiflagellatum* with many flagella.

As the only species, its characteristics are those of the genus. Straight rods 2.0 × 3.5–14.0 µm. Bacteria are not infective and do not confer a killer trait on their host. Occurs in *Paramecium caudatum*. Similar endosymbiotic bacteria in *P. caudatum* have been described by Dieckmann (1977). The bacteria described by Dieckmann were, however, infectious and motile when outside their host cells.

The mol% G + C of the DNA is: not determined.

Deposited strain: in stocks E39-3 and E39-10 (deposited in the culture collection of the Laboratory of Invertebrate Zoology, Biological Research Institute, State University of St. Petersburg, Russia).

*Editorial Note: The genus has not been validly published, due to the custom of Russian protozoologists at the time of its description to publish new symbiont names in Russian journals. However named, species were only included in the genus because of a few biological features. As no molecular or physiological data are available for any of the species, the whole genus and its species appear arbitrary.

Genus Incertae Sedis VIII. Tectibacter (ex Preer, Preer and Jurand 1974) Preer and Preer 1982, 140^VP*

HANS-DIETER GÖRTZ AND HELMUT J. SCHMIDT

Tec.ti.bac'ter. L. masc. n. *tectum* covering; M.L. masc. n. *bacter* the equivalent of Gr. neut. n. *bactrum* a rod; M.L masc. n. *Tectibacter* the bacterium with a covering.

Straight rods 0.4–0.7 × 1–2 µm. Distinguished by **outer covering around its cell wall**, visible in sections with the transmission electron microscope. Sparsely flagellated, peritrichous. No known strains are toxic to protozoa. Gram negative. Often observed to be motile. Occur widely among strains of *Paramecium primaurelia*, *P. biaurelia*, *P. tetraurelia*, *P. sexaurelia*, and *P. octaurelia*, often with other symbionts (Fig. BXII.α.67).

Type species: **Tectibacter vulgaris** (ex Preer, Preer and Jurand 1974) Preer and Preer 1982, 140.

*Editorial Note: This genus was described by J.R. Preer, Jr. and L.B. Preer (1984) in the first edition of *Bergey's Manual of Systematic Bacteriology*. We have used their descriptions and figures and have added newer information.

List of species of the genus Tectibacter

1. **Tectibacter vulgaris** (ex Preer, Preer and Jurand 1974) Preer and Preer 1982, 140^VP

vulgar'is. L. masc. adj. *vulgaris* common.

The characteristics are as described for the genus.

The mol% G + C of the DNA is: not determined.

Type strain: ATCC 30697.

Additional Remarks: The type strain of *Tectibacter vulgaris* was isolated from stock 225 of *Paramecium sexaurelia* (ATCC 30697) (Preer et al., 1974).

Order III. **Rhodobacterales** *ord. nov.*

GEORGE M. GARRITY, JULIA A. BELL AND TIMOTHY LILBURN

Rho.do.bac.ter.a' les. M.L. masc. n. *Rhodobacter* type genus of the order; *-ales* suffix to denote order; M.L. fem. n. *Rhodobacterales* the *Rhodobacter* order.

The order *Rhodobacterales* was circumscribed for this volume on the basis of phylogenetic analysis of 16S rRNA sequences; the order contains the family *Rhodobacteraceae.*

Description is the same as for the family *Rhodobacteraceae.*

Type genus: **Rhodobacter** Imhoff, Trüper and Pfennig 1984, 342.

Family I. **Rhodobacteraceae** *fam. nov.*

GEORGE M. GARRITY, JULIA A. BELL AND TIMOTHY LILBURN

Rho.do.bac.ter.a' ce.ae. M.L. masc. n. *Rhodobacter* type genus of the family; *-aceae* ending to denote family; M.L. fem. pl. n. *Rhodobacteraceae* the *Rhodobacter* family.

The family *Rhodobacteraceae* was circumscribed for this volume on the basis of phylogenetic analysis of 16S rRNA sequences; the family contains the genera *Rhodobacter* (type genus), *Ahrensia, Albidovulum, Amaricoccus, Antarctobacter, Gemmobacter, Hirschia, Hyphomonas, Jannaschia, Ketogulonicigenium, Leisingera, Maricaulis, Methylarcula, Octadecabacter, Pannonibacter, Paracoccus, Pseudorhodobacter, Rhodobaca, Rhodothalassium, Rhodovulum, Roseibium, Roseinatronobacter, Roseivivax, Roseobacter, Roseovarius, Rubrimonas, Ruegeria, Sagittula, Staleya, Stappia,* and *Sulfitobacter. Albidovulum, Leisingera, Pseudorhodobacter, Pannonibacter,* and *Jannaschia* were proposed after the cut-off date for inclusion in this volume (June 30, 2001) and are not described here (see Albuquerque et al.,

2002; Schaefer et al., 2002; Uchino et al., 2002; Borsodi et al., 2003, and Wagner-Döbler et al., 2003, respectively).

Family is phenotypically, metabolically, and ecologically diverse. Includes photoheterotrophs that can also grow photoautotrophically or chemotrophically under appropriate environmental conditions, chemoorganotrophs with either strictly aerobic or facultatively anaerobic respiratory metabolism, facultatively fermentative organisms, and facultative methylotrophs. Some denitrify. Many are aquatic. Many require NaCl for growth.

Type genus: **Rhodobacter** Imhoff, Trüper and Pfennig 1984, 342.

Genus I. **Rhodobacter** *Imhoff, Trüper and Pfennig 1984, 342*[VP]

JOHANNES F. IMHOFF

Rho.do.bac' ter. Gr. n. *rhodon* the rose; M.L. masc. n. *bacter* equivalent of Gr. neut. n. *bakterion* a rod; M.L. masc. n. *Rhodobacter* red rod.

Cells are ovoid or rod-shaped, motile by polar flagella or nonmotile, divide by binary fission or budding, may produce capsules and slime, and may form chains of cells. **Gram negative; belong to the *Alphaproteobacteria.*** Phototrophically grown cells **form vesicular or lamellar internal photosynthetic membranes.** The color of phototrophic cultures is yellow-green to yellow-brown, while aerobic cultures are pink to red. **Photosynthetic pigments are bacteriochlorophyll *a* (esterified with phytol) and carotenoids of the spheroidene series. Ubiquinone 10 is the major quinone. Major fatty acids are C_{18} and C_{16} saturated and monounsaturated fatty acids with $C_{18:1}$ as predominant component.**

Photoheterotrophic growth occurs under anoxic conditions in the light with a variety of organic compounds as carbon and electron sources. **Photoautotrophic growth may be possible with hydrogen, sulfide, or thiosulfate as an electron donor. Chemotrophic growth may occur by aerobic respiration and under anoxic dark conditions, also by denitrification, fermentation, and oxidant-dependent metabolism.** Polysaccharides, poly-β-hydroxybutyric acid and polyphosphates may be present as storage products. Growth factors are required.

Mesophilic freshwater bacteria with optimal growth at neutral pH and in the absence of NaCl.

Habitats: freshwater environments providing organic substrates and reduced oxygen concentrations; sewage ponds, eutrophic lakes and the like.

The mol% G + C of the DNA is: 64.4–73.2.

Type species: **Rhodobacter capsulatus** (Molisch 1907) Imhoff, Trüper and Pfennig 1984, 342 (*Rhodonostoc capsulatum* Molisch 1907, 23; *Rhodopseudomonas capsulata* (Molisch 1907) van Niel 1944, 92.)

FURTHER DESCRIPTIVE INFORMATION

Cells of all *Rhodobacter* species have similar shape and size but different tendencies to form capsules, slime, or chains. Chain formation is also dependent on the growth conditions. Characteristic for *R. capsulatus* is the formation of zigzag chains, which has not been observed in the other species. However, not all strains of this species form zigzag chains, and this phenomenon is best observed in mineral media. Most of the strains that form zigzag chains in mineral media form straight chains in complex media, in which a marked tendency of spheroplast formation was observed, especially if the concentrations of yeast extract were higher than 0.7% (Weaver et al., 1975a).

Physiologically, *Rhodobacter* species are among the most versatile species of the purple nonsulfur bacteria. They can perform a number of different growth modes. Most of the biochemical and genetic research with purple nonsulfur bacteria has been performed with either of the two species, *Rhodobacter capsulatus* and *Rhodobacter sphaeroides.* All species grow well photoheterotrophically under anoxic conditions in the light with a variety of carbon sources and electron donors. Most species grow photoautotrophically with hydrogen (except *R. veldkampii*). Molecular hydrogen is an excellent electron donor for *R. capsulatus*, while *R. sphaeroides* grows only slowly with hydrogen (Klemme, 1968; Gest et al., 1983). Sulfide is used as a photosynthetic electron donor in all species but *R. blasticus*. The oxidation product of sulfide is elemental sulfur in most species; only in *R. veldkampii* is sulfate the final oxidation product, and elemental sulfur appears as an intermediate outside the cells (Hansen et al., 1975). Thiosulfate and elemental sulfur are used only by *R. veldkampii*. Sulfate is assimilated via 3′-phosphoadenosine-5′-phosphosulfate (PAPS) in those species which are able to use sulfate as sole sulfur source (Imhoff, 1982). *R. veldkampii* lacks this capability and depends on reduced sulfur compounds for growth.

R. capsulatus can grow well under chemoautotrophic conditions aerobically in the dark with molecular hydrogen as electron donor (Madigan and Gest, 1979; Siefert and Pfennig, 1979). Under oxic dark conditions, it can also use sulfide as electron donor (Kompantseva, 1981). Under anoxic dark conditions, pyruvate and sugars can be fermented (Gürgün et al., 1976; Schultz and Weaver, 1982). Sugars, as well as succinate, malate, or acetate, can support anaerobic dark growth of *R. capsulatus* if DMSO or trimethylamine-*N*-oxide is present as electron acceptor (Yen and Marrs, 1977; Madigan and Gest, 1978; Schultz and Weaver, 1982). Molar growth yields from cultures grown anaerobically in the dark with fructose and DMSO were about 60% of that obtained from aerobic respiratory growth with fructose (Schultz and Weaver, 1982) and about 4–5 times higher than those from cultures grown under anoxic dark conditions without DMSO.

Nitrate reductase-linked electron transport and membrane potential generation accompanied by nitrate reduction to nitrite have been found in strains of *R. capsulatus* (McEwan et al., 1983). Nitrite is accumulated and not reduced further. Some strains of *R. sphaeroides* have the ability to denitrify under anoxic dark conditions (Satoh et al., 1974, 1976; Pellerin and Gest, 1983). Denitrification is also found in *R. azotoformans* (Hiraishi et al., 1996).

The abilities to assimilate and dissimilate nitrate are not linked to each other. Denitrifying strains of *R. sphaeroides* are unable to use nitrate as an assimilatory nitrogen source (Satoh et al., 1976) but have the capacity to fix dinitrogen derived from denitrification (Kelley et al., 1982). Nitrate is assimilated under anoxic conditions in the light by strains of *R. capsulatus*, *R. sphaeroides* (Klemme, 1979), and *R. veldkampii* (Hansen et al., 1975).

Ammonia is certainly the preferred nitrogen source. It is assimilated via the glutamine synthetase/glutamate synthase (NADPH-dependent) reactions (Brown and Herbert, 1977) in *R. capsulatus* and *R. sphaeroides*. A nucleotide-unspecific glutamate dehydrogenase is present in *R. sphaeroides* (Engelhardt and Klemme, 1978), but absent in *R. capsulatus*, in which an alanine dehydrogenase, which acts primarily in the catabolism of alanine, was found (Johannson and Gest, 1976; Tolxdorff-Neutzling and Klemme, 1982).

Besides ammonia, dinitrogen is commonly used as a nitrogen source by *Rhodobacter* species (Siefert, 1976; Madigan et al., 1984). Nitrogen fixation is possible not only under anoxic conditions in the light but also under microoxic to oxic conditions in the dark (Madigan et al., 1979; Siefert and Pfennig, 1980) and under anoxic denitrifying conditions in strains of *R. sphaeroides* (Kelley et al., 1982). A number of organic nitrogen compounds also serve as sources for cellular nitrogen, including purines (Aretz et al., 1978) and pyrimidines (Kaspari, 1979).

Most species are sensitive to penicillin (Weaver et al., 1975a;

de Bont et al., 1981a; Hansen and Imhoff, 1985). With the exception of *R. sphaeroides*, growth inhibition is almost complete at 0.1 U/ml penicillin. Growth of *R. sphaeroides* is completely inhibited at 1–100 U/ml of penicillin.

ENRICHMENT AND ISOLATION PROCEDURES

Standard media and techniques for enrichment and isolation of PNSB are suitable for *Rhodobacter* species (see Genus *Rhodospirillum*). In enrichment cultures set up for purple nonsulfur bacteria, members of this genus will usually grow faster and outcompete other purple nonsulfur bacteria. Media for the enrichment of *Rhodobacter* species should not contain NaCl in order to specifically exclude *Rhodovulum* species. Media with organic substrates and for most species, the provision of autotrophic conditions with sulfide, thiosulfate, or hydrogen as electron donors are equally effective for enrichment of *Rhodobacter* species (Biebl and Pfennig, 1981; Imhoff, 1988; Imhoff and Trüper, 1992). Photoautotrophic conditions with hydrogen and dinitrogen as electron and nitrogen sources, respectively, appear highly selective for *R. capsulatus*.

MAINTENANCE PROCEDURES

Cultures are well preserved by use of standard techniques with liquid nitrogen, by lyophilization, or storage at −80°C in a mechanical freezer.

DIFFERENTIATION OF THE GENUS *RHODOBACTER* FROM OTHER GENERA

Characteristic properties of *Rhodobacter* species are the ovoid to rod-shaped cell morphology, the presence of vesicular internal membranes (except in *R. blasticus*), and carotenoids of the spheroidene series. All these properties are shared with *Rhodovulum* species. In addition, *Rubrivivax gelatinosus* (*Betaproteobacteria*) forms carotenoids of the spheroidene series. *Rhodobacter* species can be distinguished from *Rhodovulum* species by the lack of a substantial NaCl requirement for optimal growth, i.e., they show the typical response of freshwater bacteria. This does not exclude, however, very minor requirements for the sodium ion that have been demonstrated e.g., for *R. sphaeroides*, which has a growth optimal at 4 mM sodium chloride (Sistrom, 1960).

Rhodobacter species have a number of characteristic chemotaxonomic properties that enable their distinction from other genera. All investigated species have a large-type cytochrome c_2 (Ambler et al., 1979) and as sole quinone component Q-10 (Imhoff, 1984a). Those species that are able to assimilate sulfate use the pathway via 3′-phosphoadenosine-5′-phosphosulfate (PAPS; Imhoff, 1982). The lipopolysaccharides of investigated species (*R. capsulatus*, *R. sphaeroides*, *R. blasticus*, *R. veldkampii*) contain glucosamine as sole amino sugar in their lipid A moieties, have phosphate, amide-linked 3-OH-14:0 and/or 3-oxo-14:0 and ester-linked 3-OH-10:0 (Weckesser et al., 1995). All of these properties are shared with *Rhodovulum* species. Differentiation of the genera and species of *Rhodobacter*, *Rhodobaca* and *Rhodovulum* is possible based on 16S rDNA sequences (see Fig. 3 [p. 129] in the introductory chapter "Anoxygenic Phototrophic Purple Bacteria", Volume 2, Part A), by phenotypic characteristics (see Tables BXII.α.46, BXII.α.47, and BXII.α.48), and by DNA–DNA hybridization.

TAXONOMIC COMMENTS

Species of this genus (*R. capsulatus*, *R. sphaeroides*, *R. blasticus*) were formerly included in the genus *Rhodopseudomonas*. The rec-

TABLE BXII.α.46. Differentiating characteristics of the genera *Rhodobacter*, *Rhodobaca*, and *Rhodovulum*[a]

Characteristic	*Rhodobacter*	*Rhodobaca*	*Rhodovulum*
Salt required for optimal growth	−	+	+
Optimal pH	6.5–7.5	9.0	6.5–7.5
Final oxidation product of sulfide	S^0/SO_4^{2-}	S^0	SO_4^{2-}
Utilization of:			
Formate	+/−	−	+
Thiosulfate	+/−	nd	+
Polar lipid composition:			
Phosphatidylcholine	+/−	nd	−
Sulfolipid	+/−	nd	+
Mol% G + C of genomic DNA	64–70	58.8	62–69
Light-harvesting complexes	LHI and LHII	LHI	LHI and LHII
Natural habitat	Freshwater and terrestrial environments	Soda lakes	Hypersaline and marine environments
16S rRNA signature(s) at position(s):[b]			
359	G	G	A
408	C	C	C
578	A	G	G
1311	G	G	C
1353–1355	CGT	CGG	CGT
1365–1367	ACG	CCG	ACG
1473	G	G	A
1449–1452	GCAA	CAAT	TTC/AG

[a]Symbols: +, positive in most strains; −, negative in most strains; +/−, variable in different strains; nd, not determined.

[b]Nucleotide position numbers are those of the *E. coli* numbering system.

TABLE BXII.α.47. Differentiating characteristics of *Rhodobacter* species[a]

Characteristic	*Rhodobacter capsulatus*	*Rhodobacter azotoformans*	*Rhodobacter blasticus*	*Rhodobacter sphaeroides*	*Rhodobacter veldkampii*
Cell diameter (μm)	0.5–1.2	0.6–1.0	0.6–0.8	2.0–2.5	0.6–0.8
Motility	+	+	−	+	−
Internal membrane system:					
Vesicle	+	+		+	+
Lamellae			+		
Cell division:					
Binary fission	+	+		+	+
Budding			+		
Slime production	±	+	−	±	−
NaCl required	−[b]	−[b]	−	−[b]	−
Sulfate assimilated	+	+	−	+	−
Oxidation products of sulfide	S^0	S^0	−	S^0	Sulfate
Aerobic dark growth	+	+	+	+	+
Anaerobic growth with:					
Nitrate (denitrification)	−	+	−	±	−
Dimethylsulfoxide	+	−	nd	+	nd
Trimethylamine-*N*-oxide	+	−	nd	+	nd
Vitamins required[c]	t (b, n)	b, n, t	b, n, t, B_{12}	b, t, n	b, *p*-ABA, t
Utilization of:					
Formate	+	+	−	−	−
Citrate	−	nd	+	+	−
Tartrate	−	−	−	+	−
Mannitol	±	+	+	+	−
Glycerol	−	+	+	+	−
Ethanol	−	nd	−	+	−
Hydrogen	+	nd	+	+	−
Thiosulfate	−	−	−	−	+
Mol% G + C of DNA	65.5–66.8 (Bd), 68.1–69.6 (T_m)	69.5–70.2 (HPLC)	65.3 (Bd)	68.4–69.9 (Bd), 70.8–73.2 (T_m)	64.4–67.5 (T_m)

[a]Symbols +, positive; −, negative; ±, variable in different strains, nd, not determined.

[b]Optimal growth in the absence of NaCl but able to grow at 3% NaCl.

[c]b, biotin; B_{12}, vitamin B_{12}; n, niacin; *p*-ABA, *p*-aminobenzoic acid; t, thiamine; (b, n), a few strains require biotin and/or niacin.

ognition of morphologically and chemotaxonomically distinct characteristics and early phylogenetic analyses led to their separation from *Rhodopseudomonas* and classification in the genus *Rhodobacter* (Imhoff et al., 1984; Kawasaki et al., 1993a). *Rhodobacter blasticus* is the only species of this genus that forms buds during cell division and contains internal photosynthetic membranes of the lamellar type (Eckersley and Dow, 1980). Because these were characteristic properties of the genus *Rhodopseudomonas* at that time, this bacterium was described as a *Rhodopseudomonas* species (Eckersley and Dow, 1980) and retained within this genus in a

TABLE BXII.α.48. Photosynthetic electron donors and carbon sources of *Rhodobacter* species[a]

Donor/source	R. capsulatus	R. azotoformans	R. blasticus	R. sphaeroides	R. veldkampii
Formate	+	+	−	−	−
Acetate	+	+	+	+	+
Propionate	+	+	+	+	+
Butyrate	+	+	+	+	+
Valerate	+	nd	nd	+	+
Caproate	+	nd	nd	+	+
Caprylate	+	−	nd	+	+
Pelargonate	−	nd	nd	+	+
Pyruvate	+	+	+	+	+
Lactate	+	+	+	+	+
Malate	+	+	+	+	+
Succinate	+	+	+	+	+
Fumarate	+	+	+	+	+
Tartrate	−	−	−	+	−
Citrate	±	nd	+	+	−
Aspartate	±	nd	nd	nd	+
Arginine	nd	nd	nd	nd	nd
Glutamate	+	+	+	+	+
Benzoate	−	−	−	−	−
Gluconate	−	nd	nd	+	nd
Glucose	+	+	+	+	+
Fructose	+	+	+	+	−
Mannose	−	+	nd	+	nd
Mannitol	±	+	+	+	−
Sorbitol	±	+	+	+	−
Glycerol	−	+	+	+	−
Methanol	−	−	−	±	−
Ethanol	−	nd	−	+	−
Propanol	+	nd	nd	nd	−
Hydrogen	+	nd	+	+	−
Sulfide	+	+	−	+	+
Thiosulfate	−	−	−	−	+
Sulfur	−	−	−	−	+

[a]Symbols: +, positive in most strains; −, negative in most strains; ±, variable in different strains; (+) weak growth or microaerobic growth only; nd, not determined.

reclassification of the purple nonsulfurbacteria because of insufficient supporting data (Imhoff et al., 1984). With the availability of 16S rDNA sequences and supporting chemotaxonomic data (e.g., lipopolysaccharide analyses; see Weckesser et al., 1995), the great similarity and phylogenetic affiliation of this species to *Rhodobacter* became apparent and led to its reclassification as *Rhodobacter blasticus* (Kawasaki et al., 1993a). These authors correctly proposed *Rhodobacter blasticus* comb. nov. as the new name but did not give a formal description of the new combination. Unfortunately, this name appeared incorrectly in validation list no. 51 of the IJSB as "*Rhodobacter blastica*" and therefore a revision of this name was proposed by Euzéby (1997). As the genus name *Rhodobacter* is masculine, the correct name is *R. blasticus*.

Strains of *R. sphaeroides* that were confirmed as this species by DNA–DNA hybridization studies (de Bont et al., 1981a) are able to denitrify (Satoh et al., 1976; Pellerin and Gest, 1983). The denitrifying *Rhodobacter azotoformans* is clearly distinct from *R. sphaeroides* by DNA–DNA hybridization (Hiraishi et al., 1996).

DIFFERENTIATION OF THE SPECIES OF THE GENUS *RHODOBACTER*

The apparent morphological similarity, highly similar pigmentation, and great intraspecies variability regarding the utilization of carbon sources demands careful analyses to identify and differentiate *Rhodobacter* species. Problems in species identification that may arise due to high metabolic versatility and flexibility of the *Rhodobacter* species have been discussed in regard to the differentiation of *R. sphaeroides* and *R. capsulatus* (Imhoff, 1989). Characteristic properties to differentiate species of *Rhodobacter* are summarized in Tables BXII.α.47 and BXII.α.48. The utilization of formate, citrate, tartrate, mannitol, glycerol, ethanol, hydrogen, and thiosulfate is of diagnostic value (Table BXII.α.48).

Rhodobacter species are well characterized and clearly distinct based on 16S rDNA sequence comparison (Fig. 3 (p. 129) of the introductory chapter "Anoxygenic Phototrophic Purple Bacteria", Volume 2, Part A) and by DNA–DNA hybridization (de Bont et al., 1981a; Ivanova et al., 1988; Hiraishi et al., 1996). In addition, all species of *Rhodobacter* (and *Rhodovulum*) that were included in a study of fatty acid and polar lipid composition can be differentiated on this basis (Imhoff, 1991; Imhoff and Bias-Imhoff, 1995).

List of species of the genus Rhodobacter

1. **Rhodobacter capsulatus** (Molisch 1907) Imhoff, Trüper and Pfennig 1984, 342[VP] (*Rhodonostoc capsulatum* Molisch 1907, 23; *Rhodopseudomonas capsulata* (Molisch 1907) van Niel 1944, 92.)

cap.su.la' tus. L. dim. n. *capsula* a small chest, capsule; M.L. masc. adj. *capsulatus* capsuled.

Cells are ovoid to rod-shaped, 0.5–1.2 × 2.0–2.5 μm, sometimes even longer. Spherical cells occur in media below pH 7.0 and are often irregularly arranged in chains resembling streptococci. Ovoid and rod-shaped cells are characteristic in media above pH 7.0, but pleomorphic cells appear and media become mucoid above pH 8.0. Chains of cells in zigzag arrangement are typical, but in some strains, these chains may be straight. Most strains that show zigzag arrangement in mineral media form straight chains in complex media. Capsules of varying thickness are formed. Cells are motile by means of polar flagella. Internal photosynthetic membranes appear as vesicles.

Cultures grown anaerobically in the light are yellowish brown or greenish to deep brown. When grown in the presence of oxygen, they are red. Anaerobically grown cells change their color to a distinct red when shaken with air for a few hours; light enhances the color change. Absorption spectra of living cells show maxima at 376–378, 450–455, 478–480, 508–513, 590–592, 802–805, and 860–863 nm. Photosynthetic pigments are bacteriochlorophyll *a* (esterified with phytol) and carotenoids of the spheroidene series, including spheroidene and hydroxyspheroidene. These two carotenoids are converted to the corresponding ketocarotenoids under oxic conditions, which causes the color change to red.

Photoautotrophic growth with molecular hydrogen as electron donor is excellent, but is also possible with sulfide, which is oxidized to elemental sulfur only; thiosulfate and elemental sulfur are not used. Photoheterotrophic growth occurs anaerobically in the light with a variety of organic compounds. Chemotrophic growth aerobically in the dark occurs heterotrophically with organic substrates or autotrophically with molecular hydrogen at the full oxygen tension of air. With sugars, anaerobic dark growth occurs in the presence of DMSO or trimethylamine-*N*-oxide as oxidant. Some strains may use nitrate as electron acceptor under similar conditions, reducing it to nitrite. Marginal growth is possible during fermentation of pyruvate or sugars. Carbon sources utilized are listed in Table BXII.α.48.

Ammonia, dinitrogen, and a number of amino acids are used as nitrogen sources; nitrate only by some strains; guanine, xanthine, uric acid, cytidine, uracil, thymine, and adenine only under oxic conditions. Sulfate assimilation occurs. Thiamine is required as a growth factor; some strains also require biotin or biotin and niacin.

All strains of *R. capsulatus* are susceptible to lysis by at least 1 of 16 bacteriophages, which are species specific and do not attack other purple nonsulfur bacteria (Wall et al., 1975).

Mesophilic freshwater bacteria with optimal growth at pH 7.0 (range: 6.5–7.5), 30–35°C (temperature maximum for most strains is above 36°C) and in the absence of added salt (NaCl).

Habitat: stagnant waters exposed to the light and with reduced oxygen concentrations, sewage plants, eutrophic ponds and lakes, and the like.

The mol% G + C of the DNA is: 65.5–66.8 (Bd) and 68.1–69.6 (T_m).

Type strain: ATCC 11166, ATH 2.3.1, DSM 1710, NCIB 8286.

GenBank accession number (16S rRNA): D16428, D134.

2. **Rhodobacter azotoformans** Hiraishi, Muramatsu and Ueda 1997a, 601[VP] (Effective publication: Hiraishi, Muramatsu and Ueda 1996, 175.)

a.zo.to.for' mans. Fr. n. *azote* nitrogen; L. part. adj. *formans* forming; M.L. adj. *azotoformans* nitrogen forming.

Cells are ovoid to rod-shaped, 0.6–1.0 × 0.9–1.5 μm, motile by means of single polar flagella and divide by binary fission. Pleomorphic and swollen cells occur when growing in the presence of peptone or yeast extract at a concentration of 0.1% or more. Slime is produced under phototrophic growth conditions. Neither zigzag arrangement of cells in chains nor rosette formation is found. Internal photosynthetic membranes appear as vesicles. The color of phototrophic cultures is yellow-green to yellow-brown, while aerobic cultures are pink to red. Phototrophically grown cells have absorption maxima at 376, 449, 476, 510, 589, 800, and 850 nm. Photosynthetic pigments are bacteriochlorophyll *a* (esterified with phytol) and carotenoids of the spheroidene series.

Growth occurs under anoxic conditions in the light and under oxic conditions in the dark at full atmospheric oxygen tension. Sulfide at low concentrations (less than 0.5 mM), but not thiosulfate, is utilized under phototrophic conditions and oxidized to elemental sulfur as the final oxidation product. Anaerobic growth in darkness occurs with nitrate (by denitrification) but not with dimethylsulfoxide or trimethylamine *N*-oxide as a terminal electron acceptor. Photoheterotrophy with various organic compounds is the preferred mode of growth. Good carbon sources are acetate, pyruvate, lactate, succinate, fumarate, malate, D-xylose, D-fructose, D-glucose, D-mannose, D-mannitol, L-alanine, L-asparagine, L-glutamate, peptone, Casamino acids, and yeast extract. Growth also occurs with formate, propionate, butyrate, malonate, D-dulcitol, D-sorbitol, glycerol, and L-leucine. Not utilized are caprylate, tartrate, glycolate, benzoate, and methanol. Nitrogen sources are ammonia, dinitrogen, and glutamate. Nitrogenase-dependent photoevolution of hydrogen gas is found under ammonium-limited conditions. Sulfate is assimilated. Biotin, niacin, and thiamine are required as growth factors.

Mesophilic freshwater bacterium with optimal growth at pH 7.0–7.5, 30–35°C and in the absence of sodium chloride. Growth is possible at 3% NaCl.

Habitat: sewage and activated sludge of wastewater treatment plants.

The mol% G + C of the DNA is: 69.5–70.2 (HPLC).

Type strain: KA25, JCM 9340.

GenBank accession number (16S rRNA): D70846.

3. **Rhodobacter blasticus** (Eckersley and Dow 1981) Kawasaki, Hoshino, Hirata and Yamasato 1994a, 852[VP] (Effective publication: Kawasaki, Hoshino, Hirata and Yamasato 1993a, 362.) (*Rhodopseudomonas blastica* Eckersley and Dow 1981, 216.)

blas' ti.cus. Gr. adj. *blasticos* to bud; M.L. masc. adj. *blasticus* apt to bud.

Cells are ovoid to rod-shaped, 0.6–0.8 × 1.0–2.5 μm, nonmotile, reproduce by budding, and form sessile buds. Although mother and daughter cells appear morphologically similar at cell division, they differ with regard to the time taken to reach the next division. The mother cell initiates cell growth immediately after division, whereas the daughter cell has to undergo an obligate period of maturation. In media containing yeast extract, abnormally swollen cells and spheroplasts are formed. Internal photosynthetic membranes are present as lamellae underlying and parallel to the cytoplasmic membrane.

Color of cell suspensions is orange-brown if grown under anoxic conditions in the light, but changes to red in the presence of oxygen. Cells grown under oxic conditions in the dark are colorless to faint pink. Absorption spectra of living cells have maxima at 378, 418, 476, 506, 590, 795, and 862 nm. Photosynthetic pigments are bacteriochlorophyll *a* and carotenoids of the spheroidene series.

Preferred growth is photoheterotrophically under anoxic conditions in the light. Photoautotrophic growth occurs with hydrogen as electron donor; sulfide and thiosulfate are not used. Chemotrophic growth is possible under oxic conditions in the dark. Ammonia, dinitrogen, glutamate, aspartate, glutamine, alanine, ornithine, tyrosine, thymine, and urea serve as nitrogen sources; nitrate and nitrite are not used. Sulfate is assimilated. In addition to thiamine and niacin, biotin and vitamin B_{12} are required as growth factors (Schmidt and Bowien, 1983).

Mesophilic freshwater bacterium with optimal growth at pH 6.5–7.5, 30–35°C and in the absence of salt (NaCl).

Habitat: stagnant waters exposed to the light and with reduced oxygen concentrations, sewage plants, eutrophic ponds, and lakes.

The mol% G + C of the DNA is: 65.3 (Bd).

Type strain: NCIB 11576, ATCC 33485.

GenBank accession number (16S rRNA): D16429, D138.

4. **Rhodobacter sphaeroides** (van Niel 1944) Imhoff, Trüper and Pfennig 1984, 342[VP] (*Rhodopseudomonas sphaeroides* van Niel 1944, 95.)

sphae.ro'i.des. Gr. adj. *sphaeroides* spherical.

Cells have highly variable morphology, especially in media containing complex nutrients; morphology is more uniform in mineral salts media (Fig. BXII.α.69). Cells are spherical to ovoid, 2.0–2.5 × 2.5–3.5 μm in sugar-containing media, 0.7–4.0 μm wide under other conditions. Frequently cells occur in pairs or as a chain of beads which in many instances are connected by a thin filament and slightly unequal in size. As cultures age, they become viscous due to slime production, except when sugars serve as carbon source. Slime formation is enhanced in complex media. In young cultures, cells are motile by polar flagella; motility ceases in alkaline media. Internal photosynthetic membranes appear as vesicles.

Cultures grown under anoxic conditions in the light are greenish brown to dark brown, if grown in the presence of oxygen they are red. The brown color of cells grown anaerobically in the light changes to red if cells are shaken with air; light stimulates this change. Absorption spectra of living cells have maxima at 372–375, 446–450, 474–481, 507–513, 588–590, 800–805, 850–852, and 870–880 nm. The relative absorbances at 850–880 nm vary greatly with the culture conditions. Photosynthetic pigments are bacteriochlorophyll *a* (esterified with phytol) and carotenoids of the spheroidene series, including spheroidene and hydroxyspheroidene, which are converted to their corre-

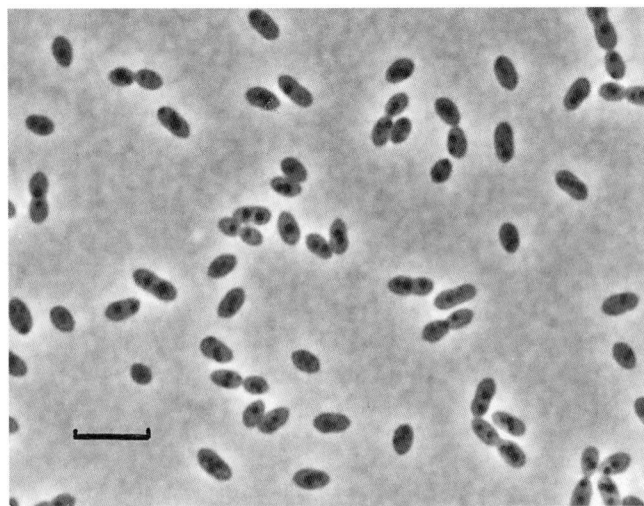

FIGURE BXII.α.69. *Rhodobacter sphaeroides* ATCC 17032 grown in mineral medium with 0.2% succinate and 0.05% yeast extract at pH 6.8. Phase-contrast micrograph. Bar = 5 μm. (Courtesy of N. Pfennig.)

sponding ketocarotenoids under oxic conditions and thereby cause the color change to red.

Preferred growth is photoheterotrophically under anoxic conditions. Photoautotrophic growth with molecular hydrogen or sulfide as electron donor is slow; thiosulfate and elemental sulfur are not used; sulfide is oxidized to elemental sulfur only. Good growth is observed under oxic conditions in the dark with a number of organic carbon sources. In addition, oxidant-dependent anaerobic dark growth with DMSO and TMAO as electron acceptors and several organic substrates is possible. Anaerobic dark fermentative growth with pyruvate and sugars is only marginal. Some strains are able to denitrify under anoxic conditions in the dark.

Organic substrates utilized are shown in Table BXII.α.48. Lower fatty acids are used at low concentrations. Higher fatty acids are toxic. Growth with glycerol depends on the presence of CO_2. Acids are produced in sugar-containing media under all conditions but disappear later. Nitrogen sources are ammonia, dinitrogen, alanine, glutamate, and aspartate; dinitrogen is also used under microoxic to oxic growth conditions in the dark. Some strains use uric acid, guanine, xanthine, cytidine, uracil, thymine, adenine (only under oxic conditions), and nitrate. Sulfate is assimilated. Thiamine, biotin, and niacin are required as growth factors.

Mesophilic freshwater bacterium with optimal growth at pH 7.0 (range: 6.0–8.5), 30–34°C, and in the absence of NaCl. Growth is possible at 3% NaCl.

Habitat: stagnant waters exposed to the light and with reduced oxygen concentrations, sewage plants, eutrophic ponds, and lakes.

The mol% G + C of the DNA is: 68.4–69.9 (Bd) and 70.8–73.2 (T_m).

Type strain: ATCC 17023, ATH 2.4.1, DSM 158, NCIB 8253.

GenBank accession number (16S rRNA): X53853, X53854, X53855.

5. **Rhodobacter veldkampii** Hansen and Imhoff 1985, 115[VP]

veld.kamp' i.i. M.L. gen. n. *veldkampii* of Veldkamp; named for H. Veldkamp, a Dutch microbiologist.

Cells are ovoid to rod-shaped, 0.6–0.8 × 1.0–1.3 µm, nonmotile, have a pronounced tendency to form chains of cells under certain growth conditions, and do not produce slime. Internal photosynthetic membranes appear as vesicles. Color of photosynthetically grown cells varies from yellowish brown to dark brown and red. Aerobically grown cells are red. Absorption spectra of photosynthetically grown cells have maxima at 373, 448, 477, 510, 589, 803, and 855 nm. Photosynthetic pigments are bacteriochlorophyll *a* and carotenoids of the spheroidene series.

Photoheterotrophic growth occurs under anoxic conditions in the light with a variety of carbon compounds, if a reduced sulfur source is provided. Sulfide, thiosulfate, and elemental sulfur, but not hydrogen, serve as electron donors for photoautotrophic growth. Chemotrophic growth occurs aerobically in the dark. During phototrophic growth with sulfide, elemental sulfur is deposited outside the cells and oxidized further to sulfate after sulfide depletion in batch cultures. In sulfide-limited chemostat cultures, sulfate is the major oxidation product. High concentrations of sulfide (5 mM) are tolerated. The carbon compounds utilized are shown in Table BXII.α.48.

Nitrogen sources are ammonia, dinitrogen, and nitrate. Sulfate cannot be assimilated and reduced sulfur sources are required: sulfide, cysteine, and cystine are used, but not sulfite, thiosulfate, and methionine. Biotin, thiamine, and *p*-aminobenzoic acid are required as growth factors; 0.01% yeast extract can replace the vitamin requirement.

Mesophilic freshwater bacterium with optimal growth (with succinate as substrate and in the presence of cysteine) at pH 7.5, 30–35°C, and in the absence of NaCl.

Habitat: stagnant waters exposed to the light and with reduced oxygen concentrations, sewage plants, eutrophic ponds, and lakes.

The mol% G + C of the DNA is: 64.4–67.5 (T_m).

Type strain: Hansen 51, ATCC 35703.

GenBank accession number (16S rRNA): D137, D16421.

Genus II. **Ahrensia** Uchino, Hirata, Yokota and Sugiyama 1999, 1[VP] (Effective publication: Uchino, Hirata, Yokota and Sugiyama 1998, 208

NOEL R. KRIEG

Ahrens.i.a. M.L. dim ending *-ia* ; M.L. fem. n. *Ahrensia* named after R. Ahrens, a German microbiologist, who contributed to the taxonomy of marine species of *Agrobacterium*.

Rods, 0.6–1.0 × 2.0–4.0 µm. Motile. Do not form spores. Gram negative. **Aerobic,** having a strictly respiratory type of metabolism with oxygen as the terminal electron acceptor. **Na⁺ is required for growth.** Optimal growth occurs between 20 and 30°C; growth occurs at 5°C but not at 37°C. **Catalase and oxidase positive.** Nitrate is not reduced to nitrite or to gas. Gelatin, starch, chitin, and alginate are not hydrolyzed. Indole is not produced. The major fatty acid is $C_{18:1}$. The 3-hydroxy fatty acid is $C_{12:0\ 3OH}$. 2-Hydroxy fatty acids are absent. The major quinone is ubiquinone 10. The genus belongs to the class *Alphaproteobacteria*. Isolated from seawater.

The mol% G + C of the DNA is: 48.

Type species: **Ahrensia kieliensis** (ex Ahrens 1968) Uchino, Hirata, Yokota and Sugiyama 1999, 1 (Effective publication: Uchino, Hirata, Yokota and Sugiyama 1998, 208) (*Agrobacterium kieliense* Ahrens 1968, 156.)

FURTHER DESCRIPTIVE INFORMATION

Some discrepancies regarding the phenotypic characteristics of *Ahrensia* have become evident. Although Uchino et al. (1998) defined the genus *Ahrensia* as having polar flagella, they also indicated in a table that *A. kieliensis* possessed peritrichous flagella. Rüger and Höfle (1992) reported *A. kieliensis* to have peritrichous or degenerately peritrichous flagella.

It is also not clear whether *A. kieliensis* can use carbohydrates. Rüger and Höfle (1992) reported that none of the 34 compounds tested could be used as sole sources of carbon and energy after incubation at 20°C for 4 weeks. These compounds included glucose and 11 other sugars, the salts of eight organic acids, glycerol, mannitol, sorbitol, and 7 amino acids. Acid production, however, was reported from glucose and xylose after 4–6 weeks of incubation at 20°C.

DIFFERENTIATION OF THE GENUS *AHRENSIA* FROM OTHER GENERA

It is not yet clear how the genus can be readily differentiated phenotypically from other aerobic marine rods such as *Alteromonas, Deleya, Marinomonas, Marinobacter, Neptunomonas, Oceanospirillum kriegii,* and *Oceanospirillum jannaschii*. It differs from the genus *Stappia* by having $C_{12:0\ 3OH}$ as its major 3-OH fatty acid rather than $C_{14:0\ 3OH}$.

TAXONOMIC COMMENTS

Ahrens (1968) classified a new marine species in the genus *Agrobacterium* as *Agrobacterium kieliense* but later withdrew this proposal (Allen and Holding, 1974). DNA–rRNA hybridization studies by De Smedt and De Ley (1977) showed that *Agrobacterium kieliense* was not an agrobacterium, although with a $T_{m(e)}$ of 72°C it did belong on the *Agrobacterium–Rhizobium–Phyllobacterium* branch. 16S rDNA sequence analysis by Uchino et al. (1998) showed that *Agrobacterium kieliense* was separated from other genera in the *Proteobacteria*, and the species was assigned to a new genus, *Ahrensia*, as *Ahrensia kieliense* (sic). In this edition of the *Manual*, the species is placed in the class *Alphaproteobacteria*, the order *Rhodobacterales*, and the family *Rhodobacteraceae*.

List of species of the genus Ahrensia

1. **Ahrensia kieliensis** (ex Ahrens 1968) Uchino, Hirata, Yokota and Sugiyama 1999, 1[VP] (Effective publication: Uchino, Hirata, Yokota and Sugiyama 1998, 208) (*Agrobacterium kieliense* Ahrens 1968, 156.)

kiel.i.ensis. M.L. fem. adj. *kieliensis* pertaining to Kiel, Germany.

The characteristics are as described for the genus. Isolated from brackish seawater of the Baltic Sea.

The mol% G + C of the DNA is: 48.

Type strain: B9, ATCC 25656, DSM 5980, IAM 12618, IFO 15762, NCMB 2205.

GenBank accession number (16S rRNA): D88524.

Genus III. **Amaricoccus** *Maszenan, Seviour, Patel, Rees and McDougall 1997, 732[VP]*

ABDUL M. MASZENAN, ROBERT J. SEVIOUR AND BHARAT K.C. PATEL

A.ma' ri.coc.cus. Gr. n. *amara* sewage duct; Gr. n. *coccus* grain; L. n. *Amaricoccus* spherical cells from sewage ducts.

Cocci, 1.3–1.8 μm in diameter, usually arranged in tetrads. Gram negative. Nonmotile. Do not form spores or store polyphosphate granules. Oxidase positive. Aerobic, having a strictly respiratory type of metabolism with oxygen as the terminal electron acceptor. Growth occurs at 20–37°C and at pH 5.5–9.0. Chemoheterotrophic. A variety of organic compounds can be used as carbon sources, including D-fructose, glucose, mannitol, L-glutamic acid, D-mannose, D-sorbitol, sucrose, L-rhamnose, L-arabinose, L-leucine, and L-histidine. Isolated from activated sludge and sewage.

The mol% G + C of the DNA is: 51–63.

Type species: **Amaricoccus kaplicensis** Maszenan, Seviour, Patel, Rees and McDougall 1997, 733.

FURTHER DESCRIPTIVE INFORMATION

The following enzymes are produced: alkaline phosphatase, esterase, esterase-lipase, lipase, leucine arylamidase, valine arylamidase, acid phosphatase, naphthol-AS-BI-phosphohydrolase, and α-glucosidase. Do not produce cystine arylamidase, trypsin, chymotrypsin, α-galactosidase, β-galactosidase, β-glucuronidase, β-glucosidase, *N*-acetyl-β-glucosaminidase, α-mannosidase, α-fucosidase, lysine decarboxylase, ornithine decarboxylase, and arginine dihydrolase.

The following compounds are used as carbon sources: dextrin, L-arabinose, D-arabitol, cellobiose, D-fructose, L-fucose, D-galactose, α-D-glucose, *m*-inositol, maltose, D-mannitol, D-mannose, D-psicose, L-rhamnose, D-sorbitol, sucrose, D-trehalose, turanose, xylitol, methyl pyruvate, monomethyl succinate, α-hydroxybutyric acid, β-hydroxybutyric acid, α-ketobutyric acid, DL-lactic acid, succinic acid, succinamic acid, D-alanine, L-alanine, L-asparagine, L-glutamic acid, maltotriose, palatinose, D-ribose, salicin, D-tagatose, D-xylose, D-lactic acid methyl ester, L-lactic acid, D-malic acid, methyl succinate, pyruvic acid, *N*-acetyl-L-glutamic acid, and adenosine. The following compounds are not used: α-cyclodextrin, *N*-acetyl-D-galactosamine, *i*-erythritol, α-D-lactose, lactulose, D-raffinose, D-glucosaminic acid, D-saccharic acid, sebacic acid, L-alanyl-glycine, glycyl-L-aspartic acid, L-histidine, L-leucine, L-ornithine, L-pyroglutamic acid, uridine, inulin, amygdalin, α-methyl-D-galactoside, β-methyl-D-mannoside, sedoheptulosan, stachyose, and UMP.

Cech and Hartman (1993) suggested that "G-Bacteria" (*Amaricoccus* species) grew well in activated sludge systems operating with alternating aerobic/anaerobic zones, and were dominant when glucose was the carbon source rather than acetate. They hypothesized that "G-Bacteria" were detrimental to the process of biological phosphate removal as they outcompeted polyphosphate-accumulating bacteria (PAB) by anaerobic assimilation of glucose to the storage compound poly-β-hydroxybutyrate, which was subsequently used under aerobic conditions to form glycogen instead of polyphosphate. However, pure cultures of *Amaricoccus* have failed to demonstrate any ability for anaerobic substrate utilization (Seviour et al., 2000).

Amaricoccus species have so far been isolated only from activated sludge. Data from whole-cell hybridization studies with a fluorescently labeled rRNA-targeted oligonucleotide probe suggest that *Amaricoccus* is widespread in the activated sludge environment (Maszenan et al., 2000).

ENRICHMENT AND ISOLATION PROCEDURES

Cech and Hartman (1990, 1993) initially isolated a pure culture of a bacterium described as "G-Bacteria" from a laboratory-scale activated sludge reactor sample. The isolate was subsequently shown to belong to a new genus *Amaricoccus* and described as *Amaricoccus kaplicensis* (Maszenan et al., 1997). A pure culture of "G-Bacteria" was also isolated using a Skerman micromanipulator and called "*Tetracoccus cechi*", but because its 16S rRNA gene sequence was identical to that of *A. kaplicensis*, it was not taxonomically validated (Blackall et al., 1997). Maszenan et al. (1997, 1998) used a micromanipulator to isolate additional strains from activated sludge biomass samples onto a range of different media and found the Glucose Sulfide (GS) agar medium of Williams and Unz (1985)[1] to be the best for isolation purposes. Once isolated, pure culture cells were able to grow well on a range of media (Maszenan et al., 1997).

DIFFERENTIATION OF THE GENUS *AMARICOCCUS* FROM OTHER GENERA

Members of the order *Rhodobacterales* are heterogenous, comprising different morphologies and physiologies and varying in the mol% G + C value of the DNA. The genus *Amaricoccus* can be differentiated from phylogenetically closely related members of *Rhodobacterales* by the phenotypic criteria listed in Table BXII.α.49.

TAXONOMIC COMMENTS

Phylogenetic analyses based on 16S rDNA sequencing indicate that *Amaricoccus* species belong to the order *Rhodobacterales* in the class *Alphaproteobacteria* within the phylum *Proteobacteria* (Fig. BXII.α.70). The genus appears to be almost equidistant from *Paracoccus*, *Rhodobacter*, and *Rhodovulum* (91% similarity), and less closely related to *Roseobacter* (88% similarity).

The type strains of the four *Amaricoccus* species exhibit a high

---

1. The medium contains (g/l distilled water): glucose, 0.15; (NH₄)₂SO₄, 0.50; CaCO₃, 0.10; Ca(NO₃)₂, 0.10; KCl, 0.05; K₂HPO₄, 0.05; MgSO₄·7H₂O, 0.05; Na₂S·9H₂O, 0.187; 1.0 ml of vitamin solution (Eikelboom, 1975); agar technical No. 3 (Oxoid), 15; pH adjusted to 7.3 (± 0.2). The vitamin solution contains (μg/l): biotin, 5; calcium pantothenate, 100; cocarboxylase, 100; cyanocobalamin (vitamin B₁₂), 5; folic acid, 5; inositol, 100; niacin, 100; *p*-aminobenzoic acid, 100; pyridoxine, 100; riboflavin, 100; thiamine, 100.

TABLE BXII.α.49. Differential characteristics of the genus *Amaricoccus* and some closely related genera[a,b]

Characteristic	*Amaricoccus*	*Paracoccus*	*Rhodobacter*	*Rhodovulum*	*Rubrimonas*
Cell size (μm)	1.3–1.8	1.0–1.3	0.5–1.2	0.6–1.0	1–1.5 × 1.2–2.0
Cell shape:					
Cocci	+[c]				
Coccoid to short rods		+			
Ovoid to rod-shaped			+	+	
Small rods					+
Motile	−	−	+[d]	+[d]	+
Mode of respiration:					
Aerobic	+	+		+	+
Anaerobic			+	+	
Mol% G + C content of DNA	51–63	64–67	64–70	62–69	74.0–74.8
Habitat:					
Activated sludge	+				
Freshwater and terrestrial environments			+		
Marine or hypersaline environments				+	
Saline lake					+
Sewage	+	+			
Soil		+			
Growth physiology:					
Autotroph		+			
Chemoheterotroph	+	+			
Photoautotroph			+		
Photoheterotroph			+	+	+
Carbon sources:					
D-Fructose	+	+	V	+	+
Glucose	+	+	+	V	+
Mannitol	+	+	V	−	
L-Glutamic acid	+	+	+[e]	−	
D-Mannose	+	+		−	
D-Sorbitol	+	+		−	
Sucrose	+	+		−	−
L-Leucine	−	+		+	
L-Histidine	−	+			
L-Rhamnose	+	−			
L-Arabinose	+			w	+

[a]Symbols: +, positive; −, negative; W, weakly positive; V, variable.

[b]Modified from Maszenan et al. (1997) and Suzuki et al. (1999a).

[c]Arranged in tetrads.

[d]However, nonmotile species have been reported.

[e]*Rhodobacter adriaticus* does not utilize L-glutamic acid.

level of 16S rDNA similarity (96.6%) to one another, but are deemed sufficiently different in their phenotypic characteristics to justify placement in separate species. A numerical taxonomic analysis of the phenotypic characteristics indicates that none of the type strains are linked to the others by an S_{sm} value greater than 0.85.

FURTHER READING

Cech, J.S. and P. Hartman. 1993. Competition between polyphosphate and polysaccharide accumulating bacteria in enhanced biological phosphate removal systems. Water Res. *27*: 1219–1225.

Maszenan, A.M., R.J. Seviour, B.K.C. Patel, G.N. Rees and B.M. McDougall. 1997. *Amaricoccus* gen. nov., a gram-negative coccus occurring in regular packages or tetrads, isolated from activated sludge biomass, and descriptions of *Amaricoccus veronensis* sp. nov., *Amaricoccus tamworthensis* sp. nov., *Amaricoccus macauensis* sp. nov., and *Amaricoccus kaplicensis* sp. nov. Int. J. Syst. Bacteriol. *47*: 727–734.

Maszenan, A.M., R.J. Seviour, B.K.C. Patel, G.N. Rees and B. McDougall. 1998. The hunt for the G-bacteria in activated sludge biomass. Water Sci. Technol. *37*: 65–69.

Maszenan, A.M., R.J. Seviour, B.K.C. Patel and J. Wanner. 2000. A fluorescently-labelled rRNA targeted oligonucleotide probe for the *in situ* detection of G-bacteria of the genus *Amaricoccus* in activated sludge. J. Appl. Microbiol. *88*: 826–835.

Seviour, R.J., A.M. Maszenan, J.A. Soddell, V. Tandoi, B.K. Patel, Y. Kong and P. Schumann. 2000. Microbiology of the "G-bacteria" in activated sludge. Environ. Microbiol. *2*: 581–593.

Suzuki, T., Y. Muroga, M. Takahama, T. Shiba and Y. Nishimura. 1999. *Rubrimonas cliftonensis* gen. nov., sp. nov., an aerobic bacteriochlorophyll-containing bacterium isolated from a saline lake. Int. J. Syst. Bacteriol. *49*: 201–205.

Williams, T.M. and R.F. Unz. 1985. Isolation and characterization of filamentous bacteria present in bulking activated-sludge. Appl. Microbiol. Biotechnol. *22*: 273–282.

DIFFERENTIATION OF THE SPECIES OF THE GENUS *AMARICOCCUS*

Table BXII.α.50 lists the features that differentiate the four species of *Amaricoccus*.

List of species of the genus Amaricoccus

1. **Amaricoccus kaplicensis** Maszenan, Seviour, Patel, Rees and McDougall 1997, 733[VP]

ka.pli.cen' sis. M.L. adj. *kaplicensis* referring to Kaplice, Czech Republic, the source of the type strain.

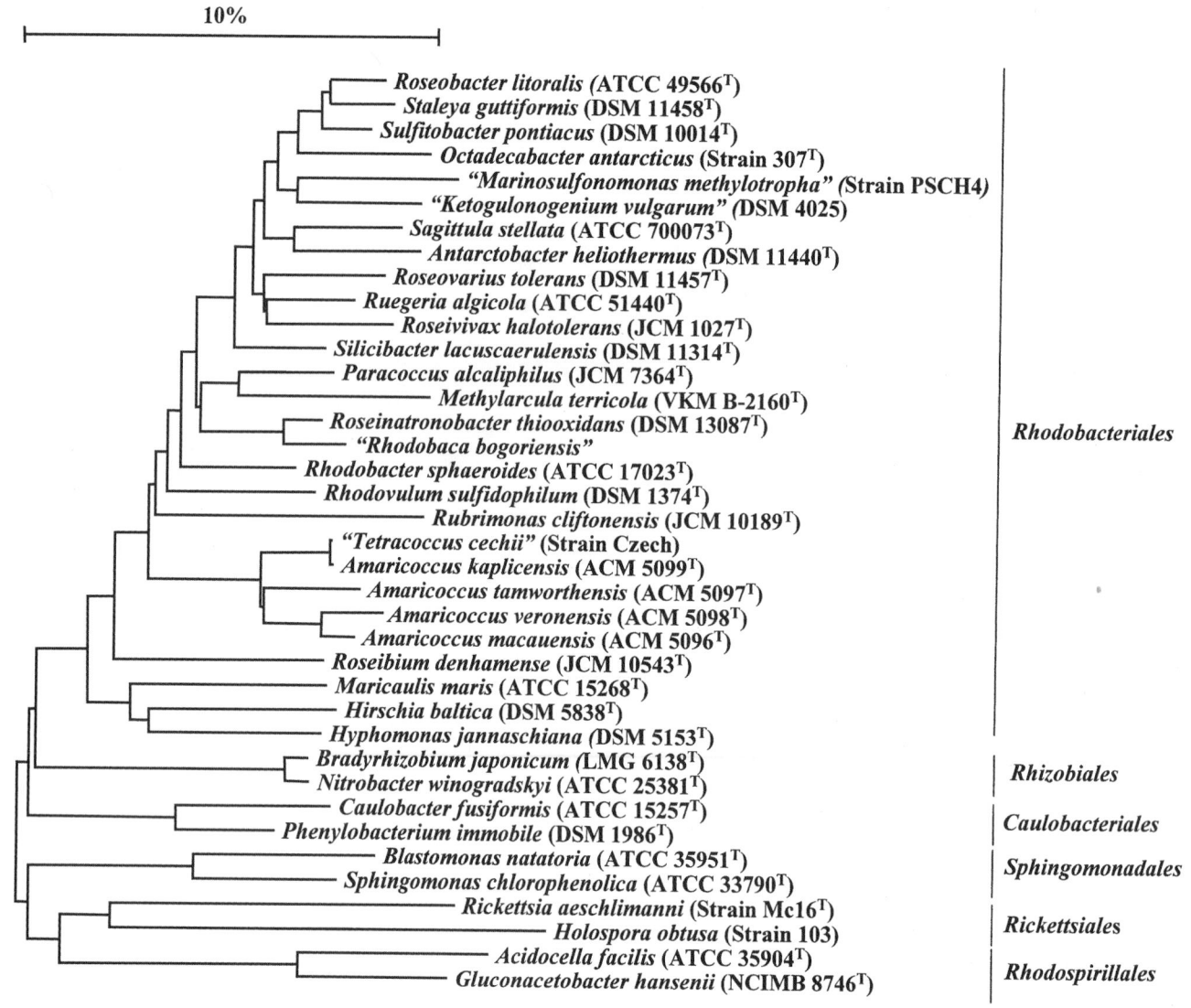

FIGURE BXII.α.70. The phylogenetic position of genus *Amaricoccus* as a member of the order *Rhodobacterales* is shown. The order *Rhodobacterales*, together with the orders *Rhizobiales, Caulobacteriales, Sphingomonadales, Rickettsiales,* and *Rhodospirillales,* constitute the class *Alphaproteobacteria.* The strains used in the analysis, their culture collection numbers and corresponding 16S rRNA gene sequences extracted from GenBank/EMBL are listed below. *Roseobacter litoralis* (X78312, ATCC 49566^T), *Staleya guttiformis* (SY16427, DSM 11458^T), *Sulfitobacter pontiacus* (Y13155, DSM 10014^T), *Octadecabacter antarcticus* (U14583, strain 307^T), "*Marinosulfonomonas methylotrophus*" (U62894, strain PSCH4), "*Ketogulonogenium vulgare*" (AF136849, DSM 4025), *Sagittula stellata* (U58356, ATCC 700073^T), *Antarctobacter heliothermus* (Y11552, DSM 11445^T), *Roseovarius tolerans* (Y11551, DSM 11457^T), *Ruegeria algicola* (X78315, ATCC 51440^T), *Roseivivax halotolerans* (D85831, JCM 1027^T), *Silicibacter lacuscaerulensis* (U77644, DSM 11314^T), *Paracoccus alcaliphilus* (Y17512, JCM 7364^T), *Methylarcula terricola* (AF030437, VKM B-2160^T), *Roseinatronobacter thiooxidans* (AF249749, DSM 13087^T), *Rhodobaca bogoriensis* (AF248638^T), *Rhodobacter sphaeroides* (X53854, ATCC 17023^T), *Rhodovulum sulfidophilum* (U55277, DSM 1374^T), *Rubrimonas cliftonensis* (D85834, JCM 10189^T), "*Tetracoccus cechi*" (Y09610^T, strain Czech), *Amaricoccus kaplicensis* (U88041, ACM 5099^T), *Amaricoccus tamworthensis* (U88044, ACM 5097^T), *Amaricoccus veronensis* (U88043, ACM 5098^T), *Amaricoccus macauensis* (U88042, ACM 5096^T), *Roseibium denhamense* (D85832, JCM 10543^T), *Maricaulis maris* (AJ227802, ATCC 15268^T), *Hirschia baltica* (X52909, DSM 5838^T), *Hyphomonas jannaschiana* (AF082789, DSM 5153^T), *Bradyrhizobium japonicum* (S46916, LMG 6138^T), *Nitrobacter winogradskyi* (L35506, ATCC 25381^T), *Caulobacter fusiformis* (AJ007803, ATCC 15257^T), *Phenylobacterium immobile* (Y18216, DSM 1986^T), *Blastomonas natatoria* (X73043, ATCC 35951^T), *Sphingomonas chlorophenolica* (X87161, ATCC 33790^T), *Rickettsia aeschlimannii* (U74757, Strain Mc16^T), *Holospora obtusa* (endosymbiont), *Acidocella facilis* (D30774, ATCC 35904^T) and *Gluconacetobacter hansenii* (X75620, NCIMB 8746^T). Phylogenetic analysis was performed on 1230 unambiguous nucleotides using dnadist and neighbor-joining programs which form part of the PHYLIP suite of software. Bar = 10 nucleotide changes per 100 nucleotides. The following abbreviations have been used: ^{T}, Type culture; ATCC, American Type Culture Collection; DSM, Deutsche Sammlung von Mikroorganismen und Zellkulturen GmbH; ACM, Australian Collection of Microorganisms; JCM, Japan Collection of Microorganisms; NCIMB, National Collection of Industrial and Marine Bacteria; VKM, All-Russian Collection of Microorganisms.

Pleomorphic cells, 1.6 μm in diameter. Colonies are mucoid on GS agar. D-galactosidic acid, lactone, and α-keto-valeric acid can be used as carbon sources. Does not utilize gentiobiose, β-methyl-D-glucoside, inositol, or lactamide. Nitrate reduced to nitrite.

The mol% G + C of the DNA is: 56 (T_m).
Type strain: Ben 101, ACM 5099.
GenBank accession number (16S rRNA): U88041.

TABLE BXII.α.50. Characteristics differentiating the species of the genus *Amaricoccus*[a]

Characteristic	*A. kaplicensis*	*A. macauensis*	*A. tamworthensis*	*A. veronensis*
Average cell size (μm)	1.6	1.3	1.8	1.8
Mol% G + C content of DNA	56	63	51	55
Growth on:				
Glycogen	+	+	−	+
Tween 40	−	−	+	−
Tween 80	−	−	+	+
N-Acetyl-D-glucosamine	+	+	+	−
Adonitol	−	−	−	+
Gentiobiose	−	+	+	+
D-Melibiose	−	−	+	−
β-Methyl-D-glucoside	−	+	+	+
D-Glucuronic acid	−	+	−	−
Acetic acid	+	+	−	+
cis-Aconitic acid	+	+	−	−
Citric acid	−	+	−	−
L-Phenylalanine	−	+	−	−
L-Proline	−	+	−	+
α-Ketoglutaric acid	−	−	+	−
D-Serine	+	+	−	−
α-Ketobutyric acid	+	+	−	+
L-Threonine	−	+	−	+
D,L-Carnitine	−	+	−	+
γ-Aminobutyric acid	−	+	−	−
Urocanic acid	−	+	+	−
Inosine	+	+	−	+
Thymidine	+	+	−	+
Phenyl ethylamine	+	+	−	+
Formic acid	+	+	−	+
D-Galactonic acid	+	−	−	−
D-Galacturonic acid	−	−	+	−
D-Gluconic acid	+	+	+	−
γ-Hydroxybutyric acid	+	+	+	−
p-Hydroxyphenyl acetic	−	+	−	−
Itaconic acid	−	+	−	−
α-Ketovaleric acid	−	−	−	−
Malonic acid	−	+	−	−
Propionic acid	+	+	−	−
Quinic acid	−	+	−	−
Bromosuccinic acid	+	+	−	+
Glucuronamide	−	+	−	+
Alaninamide	+	+	−	+
L-Aspartic acid	+	+	−	+
Glycyl-L-glutamic acid	−	−	−	+
Putrescine	−	+	−	−
2-Aminoethanol	−	+	−	−
2,3-Butanediol	−	−	−	+
Glycerol	+	+	+	−
D,L-α-Glycerol phosphate	−	+	−	−
Glucose-L-phosphate	−	+	−	+
Glucose-6-phosphate	−	+	−	+
α-Cyclodextrin	−	−	+	+
Mannan	−	−	+	+
N-acetylmannosamine	−	−	−	+
Arbutin	+	+	−	+
D-Melezitose	−	+	−	+
3-Methyl glucose	−	−	+	+
α-Methyl-D-glucoside	−	+	−	+
Lactamide	−	+	+	+
2-Deoxyadenosine	+	+	−	+
β-Methyl-D-galactoside	−	+	−	−
Fructose-6-phosphate	−	+	−	+
Hydroxy-L-proline	−	+	−	−
AMP	−	+	−	+
TMP	−	+	−	+

[a]Symbols: see standard definitions.

2. **Amaricoccus macauensis** Maszenan, Seviour, Patel, Rees and McDougall 1997, 733[VP]

ma.cau.en′sis. M.L. adj. *macauensis* referring to Macau, the source of the type strain.

Cells are 1.3 μm in diameter. D-Glucuronic acid, citrate, *p*-hydroxyphenylacetate, itaconate, malonate, quinic acid, glucuronamide, β-methyl-D-galactoside, hydroxy-L-proline, L-phenylalanine, γ-aminobutyric acid, putrescine, 2-aminoethanol, and DL-α-glycerol phosphate are used as carbon sources. Lysine decarboxylase activity is present.

The mol% G + C of the DNA is: 63 (T_m).

Type strain: BEN 104, ACM 5096.

GenBank accession number (16S rRNA): U88042.

3. **Amaricoccus tamworthensis** Maszenan, Seviour, Patel, Rees and McDougall 1997, 733[VP]

tam.worth.en′sis. M.L. adj. *tamworthensis* referring to Tamworth, Australia, the source of the type strain.

Cells are 1.3 μm in diameter. Individual tetrad cells are joined by an interconnecting fibril network. Tween 40, D-melibiose, D-galacturonic acid, and α-ketoglutaric acid can be used as carbon sources. Glycogen, acetic acid, formic acid, bromosuccinic acid, alaninamide, L-aspartic acid, inosine, thymidine, phenylethylamine, arbutin, and 2-deoxyadenosine are not used.

The mol% G + C of the DNA is: 51 (T_m).

Type strain: BEN 103, ACM 5097.

GenBank accession number (16S rRNA): U88044.

4. **Amaricoccus veronensis** Maszenan, Seviour, Patel, Rees and McDougall 1997, 733[VP]

ve.ron.en′sis. M.L. adj. *veronensis* referring to Verona, Italy, the source of the type strain.

Cells are 1.8 μm in diameter. Adonitol, glycyl-L-glutamic acid, 2,3-butanediol, *N*-acetylmannosamine, and α-methyl-D-glucoside can be used as carbon sources. *N*-Acetyl-D-galactosamine, D-gluconic acid, γ-hydroxybutyric acid, and glycerol are not used.

The mol% G + C of the DNA is: 58 (T_m).

Type strain: Ben 102, ACM 5098.

GenBank accession number (16S rRNA): U88043.

Genus IV. **Antarctobacter** *Labrenz, Collins, Lawson, Tindall, Braker and Hirsch 1998, 1369*[VP]

MATTHIAS LABRENZ AND PETER HIRSCH

Ant.arc′to.bac.ter. M.L. *Antarctica* name of the South Pole continent; derived from Gr. pref. *anti* against (on the other side); Gr. adj. *arcticos* northern; M.L. masc. n. *bacter* Gr. n. *baktron* a rod or staff; *Antarctobacter* a rod-shaped bacterium from Antarctica.

Rods, one or both cell poles narrower, sizes vary due to age and growth conditions: 0.8–1.2 × 2.0–33.6 μm. Rosettes are formed. Gram negative. **Daughter cells may be motile.** No resting stages are known. **Aerobic**, with a strictly respiratory type of metabolism. **Intracellular granules of poly-β-hydroxybutyrate are present.** Cells have an absolute requirement for Na⁺. Temperature range <3–43.5°C, salinity range <1.0 to <10.0% NaCl and <10 to >150‰ of artificial seawater (ASW), pH range 5.3 to >9. **Oxidase and catalase positive.** Do not grow photoautotrophically with H_2/CO_2 (80:20) or photoorganotrophically with acetate or glutamate. Do not contain bacteriochlorophyll *a* (bchl *a*). Occur in marine habitats.

The mol% G + C of the DNA is: 62–63.

Type species: **Antarctobacter heliothermus** Labrenz, Collins, Lawson, Tindall, Braker and Hirsch 1998, 1371.

FURTHER DESCRIPTIVE INFORMATION

Visible aerobic growth of *A. heliothermus* appears after 3–5 d on medium PYGV (Staley, 1968) + 25‰ (or 40‰) artificial seawater (ASW; Lyman and Fleming, 1940) at 20°C. Colonies are circular, smooth, convex, 1–4 mm in diameter and brownish yellow. On R2A agar (Difco, Detroit, USA) + ASW the colonies are brownish red. Cell growth of *A. heliothermus* appears to be monopolar since one cell end is usually narrower and shorter, possibly indicating a budding process. In such cases, the "daughter cell" is often positioned at an angle to the "mother cell". The cell sizes of the strains EL-54, EL-165, EL-185, and EL-219[T] are very similar; a peculiar characteristic of laboratory-cultured cells, as well as those from the source (Ekho Lake, Vestfold Hills, East Antarctica), is the frequent formation of elongated cells up to 33 μm in length. Motility is observed in some strains. Electron microscopy reveals 1–3 subpolar flagella in strains EL-54 and EL-185.

A. heliothermus has an absolute requirement for Na⁺, but none for K⁺, Mg²⁺, Ca²⁺, Cl⁻ or SO₄²⁻. This species is susceptible to chloramphenicol, penicillin G, streptomycin, tetracycline, and polymyxin B. Nitrate is reduced to nitrite. Nitrite is not reduced aerobically, anaerobic reduction to N₂ is variable. Poly-β-hydroxybutyrate is formed intracellularly.

In the presence of available nitrogen, *A. heliothermus* uses succinate, butyrate, glutamate, acetate, pyruvate, malate, citrate, and α-D-glucose, but not methanol or methanesulfonic acid. In the absence of other N-sources, glutamate is utilized as a C- and N-source. Without any added nitrogen source, there occurs a slight growth with acetate, butyrate, pyruvate, malate, or succinate. Metabolism of carbon sources tested with the Biolog system was described by Labrenz et al. (1998).

Two strains of *A. heliothermus* (EL-54 and EL-165) grow microaerophilically, but they do not grow anaerobically in a photolithoautotrophic mode with H_2/CO_2 (80:20) in the gas phase or photoorganotrophically. Bacteriochlorophyll *a* is never found in cell suspensions nor in methanolic extracts. Absorption spectra of methanolic extracts are very similar to those of *Ruegeria algicola* (Lafay et al., 1995; Uchino et al., 1998) when grown on PYGV + 25‰ ASW.

The peptidoglycan of *A. heliothermus* contains *m*-diaminopimelic acid, and the peptidoglycan of strain EL-185 is of the directly cross-linked type A1γ (Schleifer and Kandler, 1972). Respiratory quinone is Q-10. Phosphatidylglycerol and phosphatidylcholine are present as well as unknown phospholipids and an aminolipid. The dominant fatty acid is C₁₈:₁ (81.4–84.0%); other fatty acids are C₁₂:₁ 3OH (3.0–3.6%), C₁₆:₀ (0.6–1.1%), C₁₆:₁ (2.3–2.7%), C₁₈:₀ (0.7–1.4%), and C₁₉:₀ cyclo (1.9–2.7%).

Comparative 16S rDNA sequence analysis shows that *A. heliothermus* is a member of the *Alphaproteobacteria*, specifically associated with the *Roseobacter* cluster of organisms, which includes *Sagittula*, *Octadecabacter*, *Silicibacter*, *Prionitis lanceolata* gall symbiont, *Roseobacter*, *Roseovarius*, *Ruegeria*, *Staleya*, and *Sulfitobacter*.

ENRICHMENT AND ISOLATION PROCEDURES

For *A. heliothermus*, three enrichment procedures may be used: (1) 5 ml of the original water sample are inoculated into 10 ml of sterile medium "398" (Labrenz et al., 1998) and incubated for four weeks at 15°C under dim light (4.1 mmol photons m^{-2} s^{-1}). (2) Original water samples (50 ml) are amended with 1 ml of a filter-sterilized solution of 12.5 mg/ml Bacto Yeast Extract in original sample water and incubated at 15°C and at 4.1 mmol photons m^{-2} s^{-1}. (3) Original water samples (0.5 ml) are spread directly on agar plates of medium PYGV that have been prepared with original water of 10‰ salinity. Incubation at 15°C and in the dark yields the best results. Pure cultures are isolated by several dilution transfers on the corresponding agar media.

MAINTENANCE PROCEDURES

Strains of *A. heliothermus* are initially cultured on PYGV or R2A agar with 25 or 40‰ ASW. After incubation at 16–26°C to allow abundant growth, the cultures may be maintained in a refrigerator (4°C) for 6 months. They can be preserved indefinitely by lyophilization or in 15% glycerol (frozen at −72°C).

PROCEDURES FOR TESTING SPECIAL CHARACTERS

For the study of 16S rRNA gene fragments, PCR amplification as described by Hudson et al. (1993) and Labrenz et al. (1998) is recommended.

DIFFERENTIATION OF THE GENUS *ANTARCTOBACTER* FROM OTHER GENERA

Table BXII.α.51 indicates those characteristics of *Antarctobacter* that differentiate it from other morphologically, physiologically, or chemotaxonomically similar organisms. Fig. BXII.α.71 presents an unrooted tree showing phylogenetic relationships of *Antarctobacter heliothermus* and closely related *Alphaproteobacteria*.

TAXONOMIC COMMENTS

The sum of respiratory lipoquinones, fatty acids, and polar lipid data indicates that *A. heliothermus* belongs to a group of organisms (at the genus or family rank) within the *Proteobacteria*.

Comparative 16S rDNA sequence analysis also shows that it is a member of the *Alphaproteobacteria* and specifically associated with the *Roseobacter* supercluster of organisms, which includes *Sagittula, Octadecabacter, Silicibacter, Prionitis lanceolata* gall symbiont, *Roseovarius, Ruegeria, Staleya,* and *Sulfitobacter*. Sequence divergence values of >5% show this genus is genetically distinct from all currently recognized members of the *Proteobacteria*. Furthermore, bootstrap resampling shows that *A. heliothermus* does not possess a particularly significant phylogenetic affinity with any individual species within the above-mentioned genera.

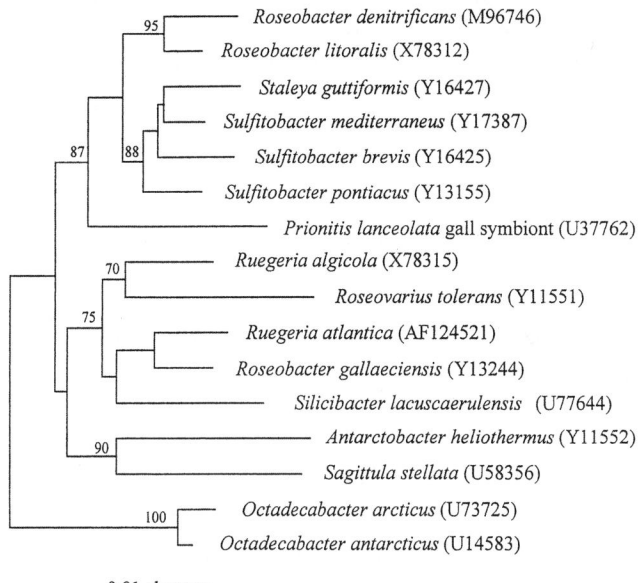

FIGURE BXII.α.71. Unrooted tree showing phylogenetic relationships of *Antarctobacter heliothermus* and closely related *Proteobacteria*. The tree was constructed using the neighbor-joining method; it was based on a comparison of approximately 1300 nucleotides. Bootstrap values, expressed as a percentage of 500 replications, are given at branching points; only those above 70% are shown.

TABLE BXII.α.51. Differential characteristics of the genus *Antarctobacter* and other morphologically, physiologically, or chemotaxonomically similar organisms[a]

Characteristic	Antarctobacter	Octadecabacter arcticus	Roseobacter litoralis	Roseovarius tolerans	Sagittula stellata	Staleya guttiformis	Sulfitobacter pontiacus
Rosettes formed	+	−	−	−	+	+	+
Gas vesicles	−	+	−	−	−	−	−
Motile	d	−	+	d	+	+	+
Maximum length 10.9–33.6 μm	+	−	−	−	−	−	−
Growth at <8.5°C	+	+	+	+	+	+	+
Utilize methanol as carbon source	−	−	−	−	+	−	−
Oxidase activity	+	−	+	+	+	+	+
Cellulase activity	−	nd	nd	−	+	nd	nd
Bacteriochlorophyll *a*	−	−	+	d	−	V	−
Phosphatidylcholine	+	nd	−	+	nd	+	nd
Diphosphatidylglycerol	−	nd	+	+	nd	−	nd
Phosphatidylethanolamine	−	nd	−	+	nd	+	nd
C$_{18:2}$ fatty acid	−	−	+	+	−	−	nd
C$_{12:1\ 3OH}$ fatty acid	+	−	−	−	+	−	nd
Colony color brown-yellow	+	−	−	−	−[b]	−	−
Mol% G + C of DNA	62–63	57	56–58	63	65	55–56	62

[a]For symbols see standard definitions; nd, not determined; V, variable.

[b]Cream (González et al., 1997b).

List of species of the genus Antarctobacter

1. **Antarctobacter heliothermus** Labrenz, Collins, Lawson, Tindall, Braker and Hirsch 1998, 1371[VP]

he.li.o.ther′ mus. Gr. n. *helios* sun; Gr. adj. *thermos* hot; M.L. masc. n. *heliothermus* heated by the sun, referring to the heliothermal water layers of Ekho Lake, the source.

Rods with an average length of 3.8–7.7 μm. Further morphological description as for the genus. Physiological and nutritional characteristics of the species are presented in Table BXII.α.52. Optimal growth occurs at 16–26°C with salt concentrations of 20–60‰ NaCl or 10–70‰ ASW. Optimal pH for growth is 6.9–7.8. Colonies on medium PYGV + 25‰ (or 40‰) ASW are circular, smooth, convex, 1–4 mm in diameter and brownish yellow. Isolated from hypersaline, meromictic, and heliothermal Ekho Lake, East Antarctica.

The mol% G + C of the DNA is: 62–63 (HPLC).

Type strain: EL-219, DSM 11445.

GenBank accession number (16S rRNA): Y11552.

Additional Remarks: Reference strain: EL-165, DSM 11440.

TABLE BXII.α.52. Other characteristics of *Antarctobacter heliothermus*

Characteristic	*A. heliothermus*
Oxidase and catalase activity	+
NO_3^- reduced to NO_2^-	+
NO_2^- reduced to N_2	v
Requirement of thiamine or nicotinic acid	
Requirement of pantothenate or vitamin B_{12}	−
Stimulation by biotin	+
Hydrolysis of gelatin or DNA	+
Hydrolysis of Tween 80 or alginate	−
Hydrolysis of starch	d
Indole test	−
H_2S production	−
Voges–Proskauer test	−
Acid from glucose	−
Growth with succinate, glutamate, butyrate, acetate, citrate, pyruvate, malate, α-D-glucose	+
Growth with methanesulfonic acid	−
Susceptible to chloramphenicol, penicillin G, streptomycin, polymyxin B or tetracycline	+

Genus V. **Gemmobacter** *Rothe, Fischer, Hirsch, Sittig and Stackebrandt 1988, 328*[VP]
(Effective publication: Rothe, Fischer, Hirsch, Sittig and Stackebrandt 1987, 97)

PETER HIRSCH AND HEINZ SCHLESNER

Gem.mo.bac′ ter. L. n. *gemma* a bud; M.L. masc. n. *bacter* equivalent; Gr. neut. n. *bactrum* a rod; M.L. masc. n. *Gemmobacter* a budding rod.

Rod-shaped cells with rounded poles or ovoid, 1.0–1.2 × 1.25–2.7 μm. New cell formation by budding from one cell pole or laterally (Fig. BXII.α.72). Daughter cells may stay connected with the mother cell resulting in formation of short chains. Not acid-fast, do not form spores or cysts; without prosthecae or other appendages. Nonmotile. **Gram negative.** Colonies colorless, round with entire margins. Temperature range 17–39°C, optimal temperature for growth 31°C.

Aerobic or facultatively anaerobic, chemoorganotrophic. **Sugars and ethanol are fermented.** Do not produce gas, D- or L-lactate, acetate or ethanol. **Dissimilatory nitrate reduction positive.** The respiratory ubiquinone is **Q-10.**

The mol% G + C of the DNA is: 63.4.

Type species: **Gemmobacter aquatilis** Rothe, Fischer, Hirsch, Sittig and Stackebrandt 1988, 328 (Effective publication: Rothe, Fischer, Hirsch, Sittig and Stackebrandt 1987, 98.)

FURTHER DESCRIPTIVE INFORMATION

Cells of *Gemmobacter* are morphologically and physiologically similar to those of members of the *Blastobacter* genus and *Sphingomonas natatoria* (Table BXII.α.53). Analysis by the 16S rRNA cataloging approach indicated that *Gemmobacter* is a member of the *Alphaproteobacteria* (Rothe et al., 1987).

ENRICHMENT AND ISOLATION PROCEDURES

The strain was isolated from water of a forest pond near Augusta (Michigan, USA). A 100-ml water sample was placed in a 250-ml Erlenmeyer flask and Bacto-peptone (Difco) was added to a final concentration of 0.005%. After plugging the flask with cotton,

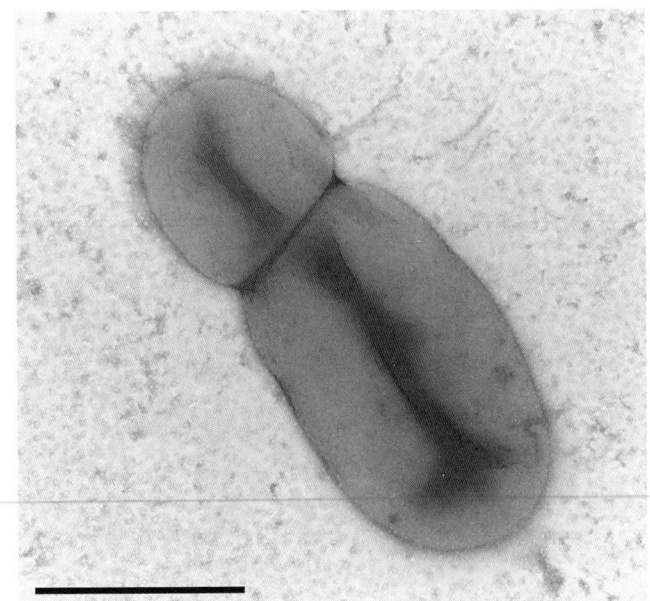

FIGURE BXII.α.72. Budding cell of *Gemmobacter aquatilis* IFAM 1031[T]. Transmission electron micrograph of negatively stained cell. Bar = 1 μm.

TABLE BXII.α.53. Properties of diagnostic value for *Gemmobacter aquatilis*, *Blastobacter* spp., and *Sphingomonas natatoria*[a]

Characteristic	Gemmobacter aquatilis	Blastobacter aggregatus	Blastobacter capsulatus	Blastobacter denitrificans	Sphingomonas natatoria
Cell shape	Ovoid to rods, short chains	Ovoid to rods	Rods, often bent	Rods with rounded poles	Rods with rounded end, slightly curved or wedge-shaped
Initial bud shape	Spherical or ovoid	Rods	Ovoid	Rods	Spherical or ovoid
Bud origin	Polar to subpolar	Distant pole	Distant pole or lateral	Slightly subpolar	Polar
Capsule formation	−	−	+	−	−
Motility	−	+	−	+	+
Rosette formation	−	+	−	−	+
Colony pigmentation	Colorless	Colorless	Colorless	Yellow	Yellow or pale pink
Utilization of sucrose	+	+	+	−	nd
N-Acetylglucosamine	−	−	−	+	nd
Succinate, malate, tartrate	+	−	−	+	nd
L-Arginine	+	+	−	−	nd
L-Proline	+	+	+	−	nd
Methanol (0.5% v/v)	−	−	−	+	−
Ethanol (0.4% v/v)	+	+	−	+	nd
Aerobically produce acid from glucose	+	+	+	−	+
Mol% G + C	63.4	60.4	58.9	64.5	65
Nicked 23S rRNA	+	−	+	−	nd
Hydroxy fatty acids, (rel. %):[b]					
$C_{10:0\ 3OH}$	23.7	−	−	−	−
$C_{14:1\ 2OH}$	−	−	−	−	0.6
$C_{14:0\ iso}$	−	−	−	−	52.2
$C_{14:0\ 3OH}$	−	1.2	0.4	nd	−
$C_{14:0\ 3OH}$	−	81.7	85	nd	−
$C_{15:0\ 2OH}$	−	−	−	−	3.9
$C_{16:1\ \omega7\ 2OH}$	−	−	−	−	1.2
$C_{16:1\ \omega5\ 2OH}$	−	−	−	−	16.2
$C_{16:0\ 2OH}$	−	−	−	−	25.7
$C_{17:1\ \omega6\ 2OH}$	−	−	−	−	0.4
$C_{18:1\ 3OH}$	21.7	5.1	3.5	nd	−
$C_{18:0\ 3OH}$	53.4	5.7	7.6	nd	−
$C_{20:1\ 3OH}$	−	5.4	−	nd	−

[a]For symbols see standard definitions; nd, not determined.

[b]Data from Sittig and Hirsch (1992).

the enrichment was incubated for 12 d in the dark at 20–25°C. Subsamples were then streaked onto medium PYGV[1] (Staley, 1968). Single colonies were further purified on the same medium.

MAINTENANCE PROCEDURES

The strain can be kept on slants at 4–5°C; it should be subcultured every three months. The strain is easily revived from lyophilized cultures.

PROCEDURES FOR TESTING SPECIAL CHARACTERS

Light microscopy with phase contrast and agar-coated glass slides (Pfennig and Wagener, 1986).

DIFFERENTIATION OF THE GENUS *GEMMOBACTER* FROM OTHER GENERA

Table BXII.α.53 lists the major features that differentiate *Gemmobacter aquatilis* from closely related taxa.

List of species of the genus Gemmobacter

1. **Gemmobacter aquatilis** Rothe, Fischer, Hirsch, Sittig and Stackebrandt 1988, 328[VP] (Effective publication: Rothe, Fischer, Hirsch, Sittig and Stackebrandt 1987, 98.)

   a.qua′ti.lis. L. neut. adj. *aquatilis* belonging to the water, describing its biotope.

   Description as for the genus. Further characteristics are given in Tables BXII.α.53 and BXII.α.54.

   The mol% G + C of the DNA is: 63.4 (T_m).

   Type strain: ATCC 49971, DSM 3857, IFAM 1031.

TABLE BXII.α.54. Further characteristics of *Gemmobacter aquatilis*[a]

Characteristic	Result/Reaction
Carbon source utilization (1 g/l):	
Glucose	+
Maltose	+
Lactose	+
Sucrose	+
Mannitol	+
Methanol (0.5% v/v)	−
Ethanol (0.4% v/v)	+
L-Arginine	+
L-Glutamate	+
Malate	+
Succinate	+

(*continued*)

---

1. PYGV consists of (per liter distilled water): peptone, 0.25 g; yeast extract, 0.25 g; glucose, 0.25 g; Hutner's basal salts medium (Cohen-Bazire et al., 1957), 20 ml; vitamin solution no. 6 (Staley, 1968), 10 ml.

TABLE BXII.α.54.   (*cont.*)

Characteristic	Result/Reaction
Tartrate	+
N-Acetylglucosamine	−
Fermentation of (2 g/l):	
Glucose	+
Maltose	+
Lactose	+
Sucrose	+
Mannitol	+
Ethanol	+
Utilization of nitrogen sources (0.5 g/l):	
Ammonia	+
Nitrate	+
Urea (4.6 g/l)	−
Hydrolysis of (2.5 g/l):	
Casein	−
Gelatin	−

(*continued*)

TABLE BXII.α.54.   (*cont.*)

Characteristic	Result/Reaction
Starch (3 g/l)	−
Reaction in litmus milk	−
Catalase	+
Cytochrome oxidase	+
Peroxidase	+
Major fatty acids:	
$C_{18:1\ \omega7}$	+
$C_{18:2\ \omega7,\ 13}$	+
$C_{19:0\ \omega cyclo\ 7-8}$	+
Phospholipids:	
Phosphatidylglycerol	+
Phosphatidylethanolamine	−
Phosphatidylmonomethylethanolamine	−
Phosphatidyldimethylethanolamine	+
Phosphatidylcholine	−
Biphosphatidylglycerol	+

[a]For symbols see standard definitions.

Genus VI. **Hirschia** *Schlesner, Bartels, Sittig, Dorsch and Stackebrandt 1990, 449*[VP]

HEINZ SCHLESNER

Hirsch' i.a. M.L. fem. n. *Hirschia* honoring Peter Hirsch, a German microbiologist, who is an expert on budding and hyphal bacteria.

Rod-shaped or oval cells, 0.5–1.0 × 0.5–6.0 μm with one or two polar hyphae (prosthecae) with a diameter of 0.2 μm. Buds are formed at the tips of the hyphae. Daughter cells are motile by a single polar flagellum. **Gram negative. Aerobic** and **chemoorganotrophic. PHB is not stored.** Utilize various sugars and organic acids. **C_1-compounds are not used as a carbon source.** Ammonia and amino acids are nitrogen sources for growth. The ubiquinone system is a **Q-10** system. Hydroxy fatty acids are of the **3-OH** type.

The mol% G + C of the DNA is: 45–47 (T_m).

Type species: **Hirschia baltica** Schlesner, Bartels, Sittig, Dorsch and Stackebrandt 1990, 449.

FURTHER DESCRIPTIVE INFORMATION

Cells of *Hirschia* spp. are morphologically very similar to members of the genera *Hyphomicrobium* and *Hyphomonas* (Fig. BXII.α.73). Even the life cycle resembles that of the above genera, i.e., a motile swarmer cell loses the flagellum, the cell produces a hypha at one pole and a bud is formed at the tip of the hypha. A flagellum is produced at the distal pole of the bud. The mature bud separates from the mother cell by fission of the hyphal tip. Additional daughter cells may be produced by the mother cell.

Analysis of the 16S rRNA gene sequence of the type strain indicated that the genus *Hirschia* is a member of the *Alphaproteobacteria* and is distinctly but remotely related to members of the genera *Hyphomicrobium*, *Filomicrobium*, *Pedomicrobium*, and *Dichotomicrobium*. The 23S rRNA is not nicked (Schlesner et al., 1990).

ENRICHMENT AND ISOLATION PROCEDURES

The strains were obtained from brackish surface water (Kiel Fjord) after either enrichment or direct plating of sample on medium 13 (M13): peptone, 0.25 g/l; yeast extract, 0.25 g/l; glucose, 0.25 g/l; artificial seawater, 250 ml/l; vitamin solution, 10 ml/l; Hutners basal salts medium, 20 ml/l; 100 mM Tris/HCl (pH 7.5), 50 ml/l; agar, 18 g/l (Schlesner, 1986).

Three enrichment procedures are described by Schlesner et al., 1990:

1. Add glucose (0.05% final concentration) to the sample and incubate in a cotton stoppered Erlenmeyer flask at room temperature.
2. Add *N*-acetylglucosamine (0.1% final concentration) to the sample and incubate in a cotton-stoppered Erlenmeyer flask at room temperature.
3. The Petri dish procedure: A layer (about 1 cm) of water agar (1.8% agar in distilled water) on the bottom of a Petri dish (diameter, 25 cm) is covered with the sample to a depth of about 2 cm. The Petri dish is incubated at room temperature.

The cultures are examined by phase-contrast microscopy for the presence of hyphal bacteria. Samples are then streaked on M13 agar. After three weeks of incubation at 25°C, colonies are screened by phase-contrast microscopy. Pure cultures are obtained by repeatedly streaking on M13 agar.

MAINTENANCE PROCEDURES

Hirschia strains grown on slants can be kept at 4–5°C. They should be subcultured every three months. They are easily revived from lyophilized cultures and can be stored at −70°C in a solution of 50% glycerol in M13.

PROCEDURES FOR TESTING SPECIAL CHARACTERS

This genus is easily distinguished from other hyphal bacteria by the yellow colonies formed on M13 agar.

DIFFERENTIATION OF THE GENUS *HIRSCHIA* FROM OTHER GENERA

Table BXII.α.55 lists the major features that differentiate *Hirschia* from other genera of budding prosthecate bacteria.

FURTHER READING

Schlesner, H., C. Bartels, M. Sittig, M. Dorsch and E. Stackebrandt. 1990. Taxonomic and phylogenetic studies on a new taxon of budding, hyphal *Proteobacteria*, *Hirschia baltica* gen. nov., sp. nov. Int. J. Syst. Bacteriol. *40*: 443–451.

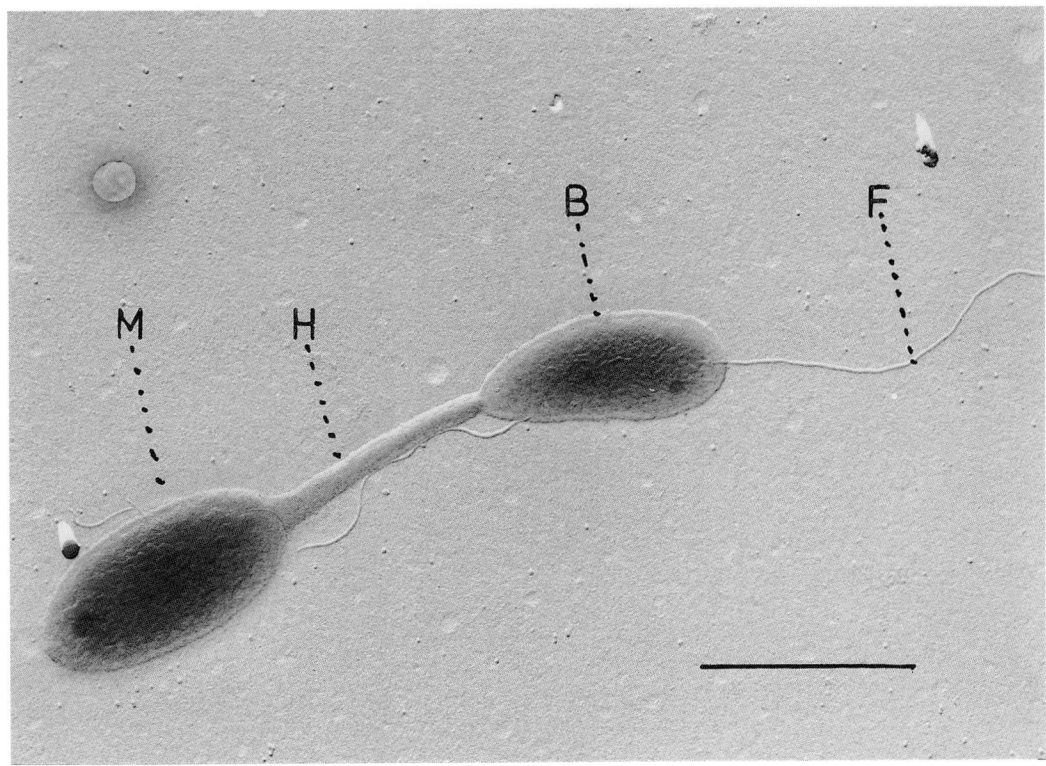

FIGURE BXII.α.73. *Hirschia baltica*, IFAM 1415, showing a mother cell (*M*), a hypha (*H*), and a bud (*B*) with polarly inserted flagellum (*F*). Platinum shadowing. Bar = 1 μm.

List of species of the genus Hirschia

1. **Hirschia baltica** Schlesner, Bartels, Sittig, Dorsch and Stackebrandt 1990, 449

   bal′ti.ca. M.L. fem. adj. *baltica* pertaining to the Baltic Sea.

   Cells are rods, 0.5–1.0 × 0.5–6.0 μm, elliptical or ovoid. Colonies are mucoid or dry and yellow. Optimal growth occurs between 22 and 28°C. Seawater is required for growth. Further characteristics are listed in Table BXII.α.56.

   The habitat is brackish water. The isolates were obtained from the Baltic Sea (Kiel Fjord).

   The mol% G + C of the DNA is: 45–47 (T_m); type strain 46 (T_m).

   Type strain: SH 149, ATCC 49814, DSM 5838, IFAM 1418.

   GenBank accession number (16S rRNA): X52909.

   Additional Remarks: The 16S rRNA gene sequence of *Hirschia baltica* IFAM 1418 has been deposited at the European Molecular Biology Laboratory (EMBL), Heidelberg, Germany.

TABLE BXII.α.55. Differentiation of *Hirschia* from other hyphal budding bacteria

Characteristic	*Hirschia*[a]	*Hyphomicrobium*[b,c]	*Hyphomonas*[c,d]	*Filomicrobium*[c,e]	*Dichotomicrobium*[c,f]	*Pedomicrobium*[c,g]	*Rhodomicrobium*[h,i]
Cells ovoid, rod-, pear-, or bean-shaped	+	+	+	−	−	+	+
Cells nearly spherical	+	−	+	−	−	+	+
Cells nearly tetrahedral	−	−	−	−	+	−	−
Cells fusiform	−	−	−	+	−	−	−
Bud elongates with long axis of hyphae	+	+	+	+	+	−	+
Bud elongates perpendicular to hyphal long axis	−	−	−	−	−	+	−
Photosynthetic	−	−	−	−	−	−	+
Color of colonies[j]	Y	b	C or G	r	Rb	B	R

(continued)

TABLE BXII.α.55. *(cont.)*

Characteristic	*Hirschia*[a]	*Hyphomicrobium*[b,c]	*Hyphomonas*[c,d]	*Filomicrobium*[c,e]	*Dichotomicrobium*[c,f]	*Pedomicrobium*[c,g]	*Rhodomicrobium*[h,i]
Able to use C₁-compounds as the sole source of carbon	−	+	−	−	−	−	−
Able to use amino acids as carbon source	+	−	+	−	−	−	−
Able to use sugars as carbon source	+	−	−	−	d	d	−
Able to use organic acids as carbon source	+	+	−	+	+	+	+
Fe and/or Mn are oxidized	−	v	−	−	−	+	−
Number of hyphae	1 or 2	1 or 2	1 or 2	2 or 3	up to 4	1 to 5	1 or 2
Flagellation monotrichous	+	+	+	−	−	+	−
Flagellation peritrichous	−	−	−	−	−	−	+
Main quinone component:							
Q9		+		+			
Q10	+		+		+	+	+
Q11			+				
Mol% G + C of DNA (T_m)	45–47	59–65	57–62	62	62–64	62–67	62–64

[a]Data from Schlesner et al. (1990).

[b]Data from Hirsch (1989).

[c]Data from Sittig and Hirsch (1992).

[d]Data from Weiner et al. (1985).

[e]Data from Schlesner (1987).

[f]Data from Hirsch and Hoffmann (1989a).

[g]Data from Gebers (1981).

[h]Data from Duchow and Douglas (1949).

[i]Data from Collins and Jones (1981).

[j]For symbols see standard definitions. Y, yellow; B, dark brown; b, brown; C, colorless; G, gray; R, red; r, light red; Rb, reddish brown.

Other Organisms

Besides the three strains of *H. baltica* (IFAM 1408, 1415, and 1418[T]), a fourth strain (IFAM 1538) was isolated and investigated. Morphological characteristics are similar to *H. baltica*, the base composition of the DNA is 47 mol% G + C (T_m). Differences are found in the utilization of carbon sources, i.e., the ability to utilize lactose and glycerol and the inability to utilize acetate,

propionate, pyruvate, lactate, or alanine. Furthermore, with the LRA ZYM Osidase System (API bioMérieux, Montalieu, France) the enzymes α-D-galactosidase, β-D-galactosidase, phospho-β-D-galactosidase, β-D-fucosidase, and β-D-lactosidase were found in IFAM 1538, but not in the strains of *H. baltica*, while only the strains of *H. baltica* tested positive for α-D-arabinosidase.

TABLE BXII.α.56. Characteristics of *Hirschia baltica*[a]

Characteristic	Reaction
Carbon source utilization:	
Cellobiose	+
Fructose	d
Glucose	+
Maltose	d
Lactose	−
Xylose	−
Raffinose	−
Rhamnose	d
Ribose	−
Trehalose	−
Adonitol	−

(continued)

TABLE BXII.α.56. *(cont.)*

Characteristic	Reaction
Ethanol	−
Glycerol	−
Mannitol	−
Methanol	−
Acetate	+
Adipate	−
Butyrate	+
Caproate	+
Citrate	−
Formate	−
Fumarate	−
Lactate	+
Malate	−

(continued)

TABLE BXII.α.56. *(cont.)*

Characteristic	Reaction
Propionate	+
Pyruvate	+
Succinate	d
Gluconate	+
Glucuronate	d
Alanine	+
Arginine	d
Asparagine	+
Aspartic acid	+
Glutamic acid	+
Glutamine	+
Glycine	−
Histidine	−
Isoleucine	+
Leucine	d
Lysine	−
Proline	+
Serine	d
Valine	−
Methylammonium chloride	−
Formamide	−
N-Acetylglucosamine	d
Amygdaline	d
Utilization of nitrogen sources:	
Ammonia	+
Glutamic acid	+
Nitrate	+
Urea	+
Acetamide	−
N-Acetylglucosamine	−
Formamide	−

(continued)

TABLE BXII.α.56. *(cont.)*

Characteristic	Reaction
Nicotinic acid	−
Hydrolysis of:	
Alginate	+
DNA	d
Gelatin	+
Starch	d
Tween 80	+
Production of NH_3 from peptone	+
Production of H_2S from thiosulfate	d
Sensitive to:	
Ampicillin	+
Oxytetracycline	+
Polymyxin B	+
Streptomycin	+
Tetracycline	+
Bacitracin	−
Nalidixic acid	−
Oxacillin	−
Fatty acids:	
$C_{16:0}$	+
$C_{18:1\ \omega7}$	+
$C_{18:2\ \omega7,\ 13}$	+
Cyclopropane fatty acids	(+)[b]
Branched fatty acids	−
Phospholipids:	
Phosphatidylglycerol	+
Glycolipids	−
Mesodiaminopimelic acid	+

[a]For symbols see standard definitions.

[b]Present in minor amounts (up to 10% of the total fatty acids).

Genus VII. **Hyphomonas** *(ex Pongratz 1957) Moore, Weiner and Gebers 1984, 71[VP] emend. Weiner, Melick, O'Neill and Quintero 2000, 466*

RONALD M. WEINER

Hy.pho.mo'nas. Gr. n. *hyphos* filament; Gr. n. *monas* a unit, monad; M.L. fem. n. *Hyphomonas* hypha-bearing unit.

Rod-shaped to oval mature cells measure $0.5–1.0 \times 1.0–3.0$ μm and may become larger and rounder just prior to bud formation. Unicellular. Cell division occurs by budding. **Buds are produced at tips of single polar prosthecae**, which are 0.2–0.3 μm in diameter and 1–5 times the length of the cell body. Prosthecae are nonseptate and rarely branch under normal growth conditions. Pleomorphic. *Hyphomonas* spp. have a **biphasic life cycle** and normally generate only a single polar prostheca (hypha). Young daughter cells (i.e., newly formed buds) are oval to pear shaped, lack prosthecae, and are smaller than the mother cell. Motile by means of a **single polar to lateral flagellum located on developing buds** or young daughter cells. **Gram negative**. Not acid-fast. Aerobic. Nonsporeforming. Chemoorganotrophic. *Hyphomonas* spp. **catabolize amino acids or tricarboxylic acid cycle intermediates** for energy and growth. All strains investigated thus far are catalase positive, oxidase positive, nonproteolytic, nonsaccharolytic, and nonpathogenic. Gelatin and starch not hydrolyzed; no indole from tryptophan; no DNase, ornithine decarboxylase, lysine decarboxylase, or coagulase activity. With one exception, all species denitrify. Amino acids are required for heterotrophic growth. Optimal temperature for growth ranges from 22 to 37°C at one atmosphere of pressure. Prefer slightly alkaline conditions for growth. **All strains were isolated from marine sources.**

The mol% G + C of the DNA is: 57–64.

Type species: **Hyphomonas polymorpha** (ex Pongratz 1957) Moore, Weiner and Gebers 1984, 71.

FURTHER DESCRIPTIVE INFORMATION

Normal-appearing cells of the type strain of *Hyphomonas polymorpha* (ATCC 33881, IFAM PS 728) are shown in Fig. BXII.α.74. All species include strains that have capsules and/or holdfasts (Langille and Weiner, 1998; Quintero, et al., 1998). Old cultures or poor growth conditions produce a large number of aberrant cell forms. These include giant cells, spindle-shaped or triangular cells, cells with unusually long or branched prosthecae, cells with intercalary buds and cells with hyphae originating from locations other than the poles. Daughter cells may be half their normal dimensions. Polyphosphate and poly-β-hydroxybutyrate granules, which are normally present (Weiner et al., 1985), may become especially pronounced under poor growth conditions. A number of traits set *Hyphomonas* spp. apart from other bacteria that bud through a prostheca (Table BXII.α.57).

The appearance of the cell wall in thin sections is typical of other Gram-negative bacteria. The cytoplasmic membrane is continuous with the cell wall, and no organized membrane structures are observed. The composition of the cell wall of *Hyphomonas*

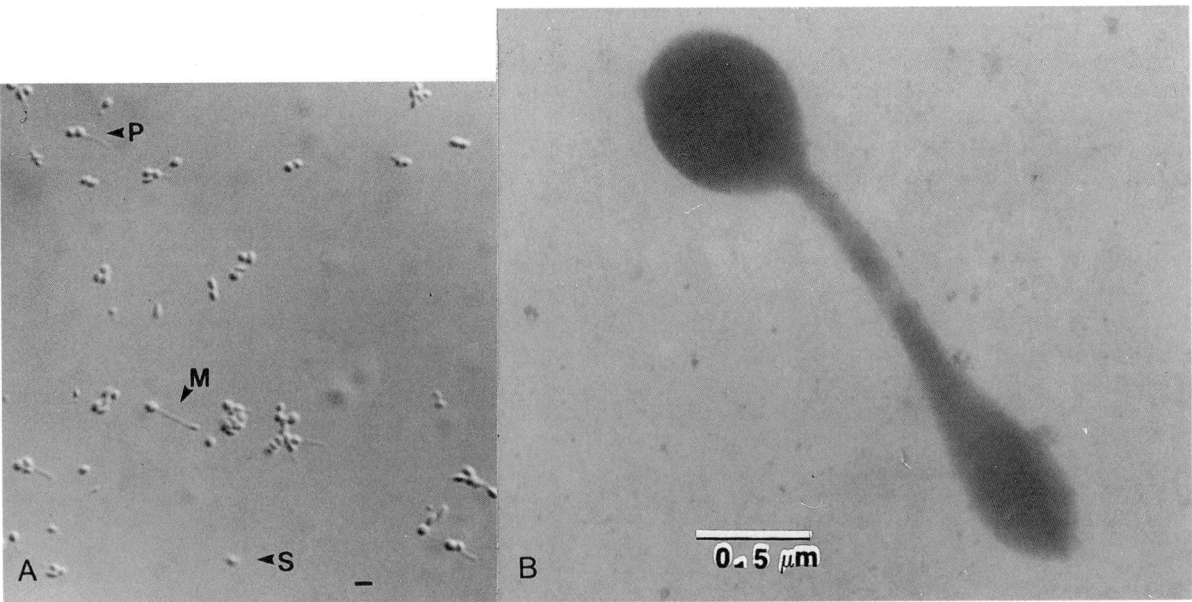

FIGURE BXII.α.74. *H. polymorpha*, type species. *A*, type strain S (ATCC 33881; IFAM PS 728). Nomarski photomicrograph showing budding mother cell (*M*), cell with prostheca (*P*), and swarmer cells (*S*). Flagella are not visible. Bar = 1 μm. *B*, strain R (ATCC 33880; IFAM PR 727). Scanning electron micrograph of mother cell showing bud formation. (Courtesy of Sorando and R. Weiner).

TABLE BXII.α.57. Differential characteristics of the genera of bacteria that reproduce by budding from the tip of the prostheca [a]

Characteristic	*Hyphomonas*	*Hirschia*[b]	*Hyphomicrobium*	*Pedomicrobium*	*Rhodomicrobium*
Photosynthetic	−	−	−	−	+
Able to use C_1 compounds as a source of carbon	−	−	+	D	−
Requires amino acids as a source of carbon	+	−	−	−	−
Able to use a variety of organic acids and alcohols as a source of carbon	−	−	−	+	+
Produces spores or cysts	−	−	−	−	+
Capable of depositing a heavy layer of iron or manganese salts on the cell surface	−	−	−	+	−
Main cell body of mother cell retains size and shape during cell cycle	−	−	+	−	+
Number of hyphae	1[c]	1–2	1[c]	1–5	2
Position of hyphae	Polar	Polar	Polar		Polar
PHB storage granules	+	−	+	+	+
Colony color	G/b	Y	b	B	R
Mol% G + C	57–64	45–47	59–65	62–67	62–64

[a]Symbols: +, 90% or more of strains are positive; −, 90% or more of strains are negative; D, differs, depending on species of genus; G, gray; b, brown; B, dark brown; R, red; Y, yellow (Moore et al., 1984).

[b]Schlesner et al., 1990.

[c]Occasionally two hyphae are present.

neptunium was found to be much more similar to that of *Escherichia coli* than to a strain of *Hyphomicrobium* (Jones and Hirsch, 1968). The cells are sensitive to detergents (see Table BXII.α.58) and are readily lysed in a solution of ethylenediaminetetraacetate with 1% sodium lauryl sulfate. Cells stain well with aniline dyes. The use of Lugol's iodine as a mordant facilitates the staining of the narrow hyphae. Best results are obtained with silver impregnation.

The life cycle of these organisms is generally similar to those of other prosthecate, budding bacteria (Hirsch, 1974a) and has been investigated in some detail (Wali et al., 1980; Moore, 1981a; Quintero, et al., 1998; Weiner, 1998). At 36°C, *Hyphomonas nep-*

tunium requires about 265 min to complete a full cycle. During the first 85 min, the cell exists as a motile swarmer cell. The flagellum becomes detached, and the cell produces a prostheca at one, usually the narrower, of the cell poles. Some species produce polar fimbriae at the other pole (Quintero et al., 1998). The major outer membrane protein profile undergoes a marked change (Shen et al., 1989). This prosthecal growth stage requires approximately 95 min of the total cycle. At the end of this stage, the size of the main cell body may increase by up to 2–4 times. The bud development stage occupies the remaining 85 min of the life cycle. The bud is produced at the very tip of the prostheca and synthesizes a single flagellum (Quintero et al., 1998). The

TABLE BXII.α.58. Characteristics of species of the genus *Hyphomonas*[a]

Characteristic	*H. polymorpha* PS 728/PR 727[T]	*H. adhaerens* MHS-3	*H. hirschiana* VP5[T]	*H. jannaschiana* VP2[T]/VP4	*H. jannaschiana* VP1/VP3	*H. johnsonii* MHS-2	*H. neptunium* LE617[T]/H13	*H. oceanitis* SCH 1325[T]	*H. rosenbergii* VP-6
Type of growth in broth:									
Granular		+			+	+			+
Pellicle			+						
Ropy		+							
Turbid	+	+		+					+
Optimal temperature (°C)	30–37	25–37	25–31	37	37	25–37	30–37	20–30	25–45
% NaCl for growth, optimal	0.5–3.5	3.5–7.5	3.5–7.5	3.5–7.5	3.5–7.5	2.5–5.0	2.5–5.0	2.5–5.0	3.0–5.0
% NaCl for growth, range	0.5–5.0	1.5–12.0	2.0–15.0	2.0–5.0	2.0–15.0	1.5–6.0	1.0–7.5	1.0–7.5	1.0–12.0
Optimal pH	7.0–7.4	7.6	7.6	7.6	7.6	7.4	8.0	7.6	8.0
Microaerophilic growth	−	Slight	−	Slight	Slight	Slight	−	Slight	−
Requires methionine	+[b]	−	+	−	−	−	+	+	nd
Requires biotin	−	−	−	+	+	−	−	+	nd
Film formation on growth vessel	− or slight[c]	+++	Slight	Slight	+	±	Slight	−	++
Rosette formation	−	−	−	−	+	+	−	−	+
Brown pigment at 31–41°C	−	±	−	+	+	−	−	−	+
Resistance to mechanical lysis[d]	2–5	nd	9	9	9	nd	2	6–8	nd
Hemolysis on sheep blood agar	χ	χ	χ	α	α	α	α	χ	α
Growth on TSI agar	−	−	−	−	−	−	−	−	−
Growth on MacConkey agar	−	−	−	−	−	−	−	−	−
Mannitol fermentation	−	−	−	−	−	−	−	−	−
Acid from glucose, lactose	−	−	−	−	−	+	−	−	−
Gelatin hydrolysis	−	−	−	−	−	−	−	−	−
Starch hydrolysis	−	−	−	−	−	−	−	−	−
Indole from tryptophan	−	−	−	−	−	−	−	−	−
$NO_3^- \rightarrow NO_2^-$	−	+	+	+	+	+	+	+	+
$NO_2^- \rightarrow N_2, NH_3$	−	+	+	+	+	+	+	+	+
DNase	−	+	−	−	−	−	−	−	+
Ornithine decarboxylase	−	−	−	−	−	−	−	−	−
Lysine decarboxylase	−	−	−	−	−	−	−	−	−
Catalase	+	+	+	+	+	+	+	+	+
Coagulase	−	−	−	−	−	−	−	−	−
Sensitivity to:									
Tellurite (0.5%)	S	S	S	S	S	S	S	S	S
Crystal violet (0.2%)	S	S	S	S	S	S	S	S	R
Neutral red (0.4%)	R	S	R	R	R	R	R	R	R
Brilliant green (0.2%)	S	S	S	S	S	S	S	S	S
Methylene blue (10%)	S	S	S	S	S	R	S	S	S
Sodium lauryl sulfate (1%)	S	S	S	S	S	S	S	S	S
Teepol 610 (80%)	S	nd	S	S	S	nd	S	S	nd
Methyl violet (0.04%)	R	S	R	R	R	R	R	R	S
Pyronin (0.4%)	S	nd	S	S	S	nd	S	S	nd
NP-40 (0.01%)	I	nd	S	S	S	nd	S	S	nd
Tween 80 (0.01–1.0%%)	R	R	R	R	R	R	R	S	R
Erythromycin (15 µg)	S	nd	S	S	S	nd	S	S	nd
Rifampin (5 µg)	S	nd	S	S	S	nd	S	S	nd
Cephalothin (30 µg)	S	nd	S	S	S	nd	S	S	nd

(continued)

TABLE BXII.α.58. *(cont.)*

Characteristic	*H. polymorpha* PS 728/PR 727[T]	*H. adhaerens* MHS-3	*H. hirschiana* VP5[T]	*H. jannaschiana* VP2[T]/VP4	*H. jannaschiana* VP1/VP3	*H. johnsonii* MHS-2	*H. neptunium* LE617[T]/H13	*H. oceanitis* SCH 1325[T]	*H. rosenbergii* VP-6
Novobiocin (30 μg)	S	S	S	S	S	S	S	R	S
Kanamycin (30 μg)	S	S	S	S	S	S	S	S	S
Chloramphenicol (30 μg)	S	S	S	S	S	S	S	S	S
Penicillin (10 IU)[e]	S	R	I	S	S	S	I	R	S
Ampicillin (10 μg)	S	R	S	S	S	S	S	R	R
Streptomycin (10 μg)	I	S	S	S	S	S	S	I	S
Mol% G + C of DNA	60–61	60	57	60	60	64	62–60	59	61

[a]Symbols: +, 90% or more of strains are positive; −, 90% or more of strains are negative; S, sensitive; I, intermediate; R, resistant; nd, not determined. Data from Weiner et al. (1985) and Weiner et al. (2000).

[b]*H. polymorpha* PR 727[T] does not require methionine, but growth is stimulated by methionine.

[c]+/−, patchy film; ++, contiguous, clearly visable film; +++, thick, uniform film.

[d]Minutes required for 99% lysis by Brownwill Biosonik IV sonicator; low probe at full power.

[e]Sensitivities were assessed on 50% concentrations of marine broth (Zobell, 1941). Sensitivity on other medium may be different (Hirsch, 1974a).

mother cell is pushed through the medium by the motile bud, in contrast to *Hyphomicrobium* species, which are pulled by the motile bud (Moore, 1981a). Whether this constitutes a serious diagnostic feature, however, is uncertain. The mature bud eventually separates from the mother cell by fission of the prosthecal tip. After a short period of prosthecal growth, further daughter cells may be produced by the mother cell. Altering the temperature of growth not only influences the time course of the life cycle but also results in a shift in the relative proportion of the various cell types present in a given culture. DNA synthesis is discontinuous and occurs just prior to bud formation. Membrane structures are involved in DNA segregation (Zerfas et al., 1997).

Good growth occurs on Casamino acids, blood agar, and other rich media. The addition of sea salts is required by nearly all strains and is stimulatory to the others (Havenner et al., 1979; Weiner et al., 2000). Growth in a defined medium supplemented with a mixture of glutamic acid, aspartic acid, methionine, and serine as substrates has been reported for *Hyphomonas neptunium* (Havenner et al., 1979). *H. hirschiana* and *H. polymorpha* have the same requirements as *H. neptunium; H. jannaschiana* requires biotin but not methionine; *H. oceanitis* requires the four amino acids plus biotin, folic acid, pyridoxine, riboflavin, and *p*-aminobenzoic acid. *H. johnsonii* differs from the other species of *Hyphomonas*, being able to utilize glucose as a carbon source in the presence of amino acids. *H. adhaerens* does not have auxotrophic requirements for amino acids but requires them (or peptone or protein) for carbon and energy.

CO$_2$ is the primary end product of metabolism, and the medium becomes alkaline, probably due to the evolution of ammonia (Leifson, 1964; Havenner et al., 1979). Deamination and tricarboxylic acid cycle oxidation of amino acids is the major catabolic pathway (Havenner et al., 1979).

No growth on mineral salts media with CaCO$_3$, acetate, propionate, lactate, ethanol, glycerol, glycine, or C$_1$ compounds as the sole source of carbon. Growth is not influenced by light.

Under certain conditions, some strains may produce a dark brown, acid-insoluble, base-soluble pigment (see Table BXII.α.58), identified as a pyomelanin, the product of *p*-hydroxyphenylpyruvate hydroxylase via homogentisic acid (Kotob et al., 1995).

H. neptunium, H. polymorpha, H. oceanitis, and *H. hirschiana* have a Q-11 ubiquinone type along with a significant amount of Q-10 (approximately 10% of total ubiquinones) and minor amounts of Q-9 and Q-12. *H. jannaschiana* has Q-10 as its major quinone with trace amounts of Q-9 and Q-11 (Urakami and Komagata, 1987b).

Hyphomonas spp. synthesize high percentages of octadecenoic acid, (C$_{18:1}$; Table BXII.α.59), albeit in varying amounts. These range from 11% in *H. jannaschiana* to 80% in MHS-3, a value exceeding that found in all extant *Hyphomonas* spp. but correlating well with most other budding/prosthecate bacteria (i.e., *Rhodomicrobium, Pedomicrobium,* and *Hyphomicrobium*) (Urakami and Komagata, 1987b; Sittig and Hirsch, 1992). *Hyphomonas* also contains novel lipids (Batrakov et al., 1996).

The membrane protein profile similarities of the eight species range from 30 to 80% (Dagasan and Weiner, 1986; Weiner et al., 2000), and all but one species, *H. johnsonii*, synthesize large amounts of a 47 kDa protein. All species are serologically related, based upon exposed determinants (Weiner et al., 2000).

Proteases (Shi et al., 1997) and S-layer proteins (Shen and Weiner, 1998) have been identified and characterized.

Table BXII.α.60 shows the % overall DNA–DNA homologies among the eight extant *Hyphomonas* species.

ENRICHMENT AND ISOLATION PROCEDURES

Most of the currently available isolates have been obtained by direct plating of samples onto commonly used media such as blood agar, medium 383 (casitone, 2 g/l; yeast extract, 1 g/l; MgCl$_2$, 1 g/l; [Leifson, 1964]) and marine agar (Difco 2216; [Zobell, 1941]). However, a simple procedure has been reported for their isolation from water samples (Moore, 1981b). The sample is brought to 0.005% (w/v) peptone and 0.005% (w/v) yeast extract. The cultures are incubated aerobically, and after times ranging from about one day to a week or more, are plated onto solid medium. The colonies, which arise relatively slowly, are screened microscopically for cells with the typical morphology of *Hyphomonas* species and are replated for purification and identification. Three of the eight species have been isolated from warm water hydrothermal vents, either from shellfish beds near the Galapagos Islands, at a depth of 2600 m (Jannasch and Wir-

TABLE BXII.α.59. Fatty acid profiles of *Hyphomonas*[a]

Strain	Reference	% Total fatty acids[b]						
		$C_{15:0}$	$C_{16:0}$	$C_{16:1}$	$C_{17:0}$	$C_{17:0\ cyclo}$	$C_{17:1}$	$C_{18:1}$
H. polymorpha PS728[T]	Sittig and Hirsch, 1992[c]	9	0	1	25	nd[d]	14	12
	Weiner et al., 2000[e]	6	2	1	23	nd	24	15
	Urakami and Komagata, 1987b[f]	10	2	2	19	44	nd	10
H. adhaerens MHS-3[T]	Weiner et al., 2000[e]	0	10	2	0	nd	0	80
H. hirschiana VP-5[T]	Sittig and Hirsch, 1992[c]	0	30	2	0	nd	0	54
	Weiner et al., 2000[e]	7	9	3	11	nd	13	39
	Urakami and Komagata, 1987b[f]	1	9	7	2	11	nd	67
H. jannaschiana VP-2[T]	Sittig and Hirsch, 1992[c]	5	4	1	18	nd	14	22
	Weiner et al., 2000[e]	5	2	0	31	nd	22	11
	Urakami and Komagata, 1987b[f]	1	7.2	3	6	10	nd	65
H. johnsonii MHS-2[T]	Weiner et al., 2000[e]	0	21	3	0	nd	0	64
H. neptunium Le670[T]	Sittig and Hirsch, 1992[c]	10	5	1	19	nd	18	19
	Weiner et al., 2000[e]	8	8	4	10	nd	14	37
	Urakami and Komagata, 1987b[f]	8	2	2	16	55	nd	10
H. oceanitis SCH89[T]	Sittig and Hirsch, 1992[c]	0	34	6	0	nd	0	53
	Weiner et al., 2000[e]	0	27	6	0	nd	0	49
	Urakami and Komagata, 1987b[f]	0	22	13	1	1	nd	60
H. rosenbergii VP-6[T]	Weiner et al., 2000[e]	2	32	6	0	nd	3	43

[a]Data from Weiner et al., 2000.

[b]Rounded to nearest 1% of total detected fatty acids.

[c]Cells harvested in late logarithmic phase at 25°C.

[d]nd; no data.

[e]Cells harvested in late logarithmic phase at 27°C.

[f]Cells harvested during stationary phase at 30°C.

TABLE BXII.α.60. DNA–DNA reassociation values of the eight *Hyphomonas* species[a]

Strains	*H. polymorpha* PS-728[T]	*H. adhaerens* MHS-3[T]	*H. neptunium* LE-670[T]	*H. johnsonii* MHS-2[T]	*H. hirschiana* VP-5[T]	*H. oceanitis* Sch89[T]	*H. jannaschiana* VP-1	*H. rosenbergii* VP-6[T]
PS-728	100							
MHS-3	32	100						
LE-670	27	66	100					
MHS-2	13	25	25	100				
VP-5	34	19	54	12	100			
Sch89	28	11	12	23	20	100		
VP-1	8	41	32	18	23	26	100	
VP-6	5	30	11	21	19	2	4	100

[a]Results calculated as the relationship of at least three separate experiments, averaging six values including the reciprocal. All the values listed are combined averages from the reciprocal hybridizations, which are within 10% of one another. Data from Weiner et al., 2000.

sen, 1981), or from the Guyamas Basin at a depth of 2000 m. Members of the genus have been identified as primary colonizers of submerged surfaces in marine waters (Baier et al.,1983; Frolund et al., 1996). Due to their relatively slow growth, they are not readily isolated, though they can be well represented on submerged surfaces in marine waters (Weidner et al., 1996).

MAINTENANCE PROCEDURES

The members of this genus are generally quite hardy and will survive up to several months in liquid or solid growth medium at 4°C. They are easily revived from lyophilized cultures and can be stored at −20°C over a period of several years, or for a longer period of time at −70°C, in a solution of 60% glycerol in 0.05 M KH_2PO_4 at pH 7.

DIFFERENTIATION OF THE GENUS *HYPHOMONAS* FROM OTHER GENERA

The eight species have signature 16S rDNA sequences (Weiner et al., 2000) that place them in the *Alphaproteobacteria* (Stackebrandt et al., 1988a). Each is related to the other above the 96% similarity level. A member of the most closely related genus, *Hirschia*, (Schlesner et al., 1990), is related to each of the strains

of *Hyphomonas* at the 90 ± 1% level. 16S rDNA sequence analyses support the notion that *Hyphomonas*, a genus of marine bacteria, is more closely related to the marine line of *Caulobacter* (Stahl et al., 1992) than to terrestrial, budding, prosthecate genera (e.g., *Hyphomicrobium*; Fig. BXII.α.75).

Table BXII.α.57 lists other major features which differentiate *Hyphomonas* from other genera of prosthecate, budding bacteria. Support for the separation of this genus is also indicated by the low level of intergeneric cross-reactions seen in serological studies (Powell et al., 1980; Weiner et al., 2000), DNA–DNA homology studies (Moore and Hirsch, 1972; Moore and Staley, 1976; Gebers et al., 1984, Weiner et al., 2000), rRNA–DNA studies (Moore, 1977), fatty acid analyses (Urakami and Komagata, 1987b; Sittig and Hirsch, 1992; Weiner, et al., 2000), rRNA cistron similarities (Roggentin and Hirsch, 1989), comparisons of membrane protein profiles (Dagasan and Weiner, 1986; Weiner et al., 2000), and other criteria (Gebers et al., 1985; Köbel-Boelke et al., 1985; Nikitin et al., 1990).

ACKNOWLEDGMENTS

The authors thank J. Poindexter, R. Gebers, J. Smit, and H. Jannasch for providing isolates and helpful discussion. Colleagues and members of my

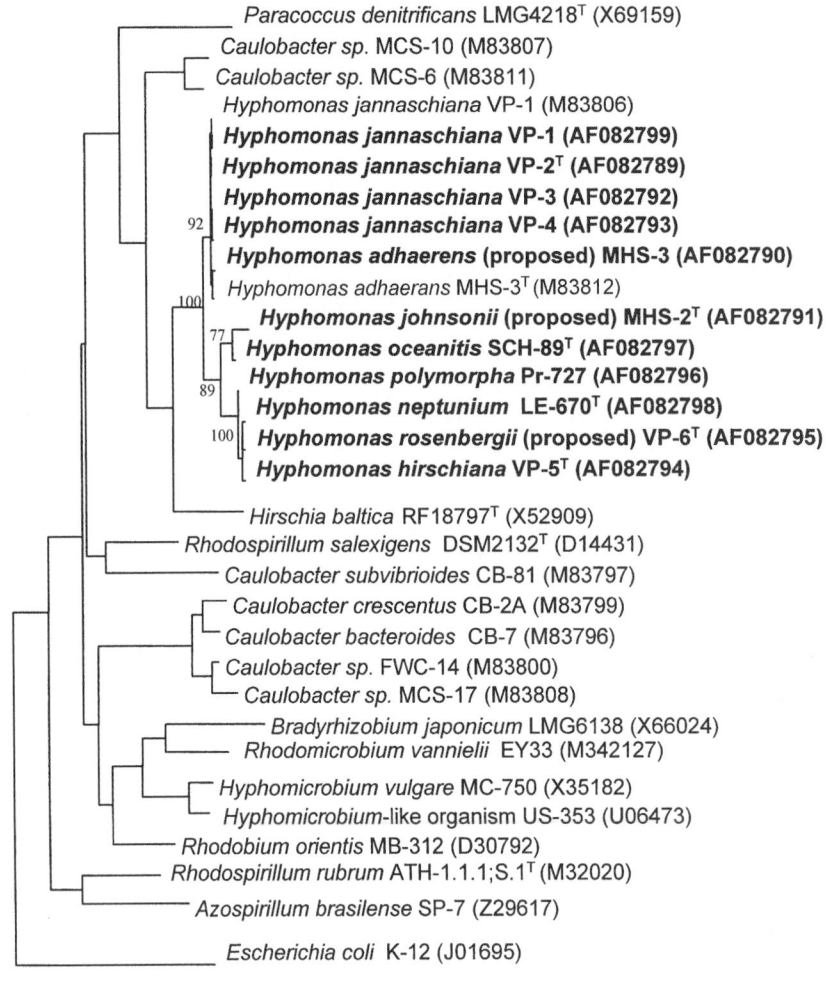

FIGURE BXII.α.75. Phylogenetic analysis of *Hyphomonas*. Tree constructed using the fastDNA maximum likelihood program from the Ribosomal RNA Database Project (RDP) (Larsen et al., 1993). Sequences obtained in this study are in boldface type. Sequences for comparison were obtained from the RDP (Larsen et al., 1993). Bootstrap values (expressed as percentages of 100 replications), pertinent to *Hyphomonas* species, are shown at the appropriate branch points. *Escherichia coli* MRE600 was used as an outgroup to establish the root of the tree. Data are reproduced with permission from R.M. Weiner et al., International Journal of Systematic and Evolutionary Microbiology: *50:* 459–469, 2000, ©International Union of Microbiological Societies.

laboratory who have contributed to the manuscript are also acknowledged. These include K. Busch, R. Devine, L. Dagasan, R. Gherna, K. Guthrie, G. Lacy, J. Johnson, M. Melick, K. O'Neill, E. Quintero, F. Singleton, and H. G. Trüper. A portion of this work was supported by Maryland Industrial Partnerships (MIPS), FDA–JIFSAN, and Maryland Sea Grant.

FURTHER READING

Moore, R.L. 1981. The biology of *Hyphomicrobium* and other prosthecate, budding bacteria. Annu. Rev. Microbiol. *35:* 567–594.

Moore, R.L., R.M. Weiner and R. Gebers. 1984. Genus *Hyphomonas* Pongratz 1957 nom. rev. emend., *Hyphomonas polymorpha* Pongratz 1957 nom. rev. emend., and *Hyphomonas neptunium* (Leifson 1964) comb. nov. emend., (*Hyphomicrobium neptunium*). Int. J. Syst. Bacteriol. *34:* 71–73.

Quintero, E.J., K. Busch and R.M. Weiner. 1998. Spatial and temporal deposition of adhesive extracellular polysaccharide capsule and fimbriae by *Hyphomonas* strain MHS-3. Appl. Environ. Microbiol. *64:* 1246–1255.

Weiner, R.M. 1998. Plasticity of bacteria. Survival mechanisms. *In* Rosenberg (Editor), Recent Advances in Microbial Ecology and Infectious Disease, ASM Press, Washington D.C. pp. 17–29.

Weiner, R.M., R.A. Devine, D.M. Powell, L. Dagasan and R.L. Moore. 1985. *Hyphomonas oceanitis* sp. nov., *Hyphomonas hirschiana* sp. nov., and *Hyphomonas jannaschiana* sp. nov. Int. J. Syst. Bacteriol. *35:* 237–243.

Weiner, R.M., M. Melick, K. O'Neill and E.J. Quintero. 2000. *Hyphomonas adhaerens* sp. nov., *Hyphomonas johnsonii* sp. nov. and *Hyphomonas rosenbergii* sp. nov., marine budding and prosthecate bacteria. Int. J. Syst. Evol. Microbiol. *50:* 459–469.

DIFFERENTIATION OF THE SPECIES OF THE GENUS *HYPHOMONAS*

Most criteria, other than 16S rDNA sequence analysis, support the placement of *Hyphomonas* as a discrete genus but do not clearly establish relationships between all species. These criteria include serological identity based upon surface antigen (Powell et al., 1980; Weiner et al., 2000), some DNA–DNA homology results (Moore and Hirsch, 1972; Moore and Staley, 1976; Gebers et al., 1986, Weiner et al., 2000), rRNA–DNA studies (Moore, 1977), fatty acid analyses (Weiner et al., 2000), and comparisons of membrane protein profiles (Dagasan and Weiner, 1986; Weiner et al., 2000).

The best criterion to ascertain the interrelationship of *Hyphomonas* species is genetic similarity. The phylogenetic tree (Fig. BXII.α.75) shows the relatedness of each of the species of *Hyphomonas* according to 16S rDNA sequence analysis. *H. rosenbergii* (nov.), *H. hirschiana, H. polymorpha*, and *H. neptunium* cluster at the 99.4% similarity level. *H. adhaerens* (nov.) and *H. jannaschiana* share sequence similarities of 99.3% and, as expected, each of the strains of *H. jannaschiana* is related a high level (99.3%) as well. *H. johnsonii* (nov.) is most closely related to *H. oceanitis* at the 98.7% level, and neither of these species is closely related to other species of *Hyphomonas*.

Thus, 16S rDNA sequence analyses divide the eight species of *Hyphomonas* into: 1) a tight group of four species, *H. rosenbergii, H. hirschiana, H. polymorpha*, and *H. neptunium*; 2) a group with two species, *H. adhaerens* and *H. jannaschiana*; and 3) a less tight group of two species, *H. johnsonii* and *H. oceanitis*. Likewise, the DNA–DNA hybridization data (Table BXII.α.60) show *H. neptunium* to be closely related to *H. hirschiana* and supports both of the two species groupings, i.e., *H. adhaerens* and *H. jannaschiana* plus *H. johnsonii* and *H. oceanitis*. (DNA–DNA hybridization data also suggest that *H. neptunium* forms a tighter relationship with *H. adhaerens* based upon DNA–DNA reassociation values (Table BXII.α.60) than by 16S rDNA sequence comparisons.

However, DNA–DNA reassociation values can be inflated by shared plasmids and other horizontal genetic exchange).

Phenotypic characteristics tend to support these placements, as can be ascertained from an examination of Table BXII.α.58. Rosette formation is a result of holdfast (polar adhesive capsule) production. *H. rosenbergii* synthesizes both a capsule that surrounds the entire cell, including main body, prostheca and bud, and a holdfast (Langille and Weiner, 1998). *H. adhaerens* produces a capsule that surrounds only the main body of the mother cell, mediating side/side adhesion as well as pole/pole adhesion (Quintero and Weiner, 1995; Quintero et al., 1998). The holdfast of *H. rosenbergii* and capsule of *H. adhaerens* are chemically different, while sharing certain properties such as an abundance of amino sugars (Quintero and Weiner, 1995). The LPS of *H. adhaerens* is antigenically unique from that of any other surface antigen of any other *Hyphomonas* species (Weiner et al., 2000).

H. johnsonii is clearly the outlier species of *Hyphomonas* based on almost all criteria. So far, it is the only species shown to be able to utilize sugars for carbon and energy. Additionally, it has little DNA–DNA hybridization similarity with other *Hyphomonas* species (Table BXII.α.60) and lacks some genus-specific phenotypic characteristics (Jones and Krieg, 1984), including outer membrane proteins (Weiner et al., 2000), Furthermore, it has the lowest serological identity with any of the other *Hyphomonas* species (Weiner et al., 2000).

The site of isolation (e.g., hydrothermal vent, benthos, coastal) does not appear to have a bearing on the relationships between *Hyphomonas* species.

List of species of the genus Hyphomonas

1. **Hyphomonas polymorpha** (ex Pongratz 1957) Moore, Weiner and Gebers 1984, 71[VP]

po.ly.mor′pha. Gr. adj. *poly* many; Gr. n. *morphe* shape, body; M.L. adj. *polymorpha* many shapes.

Pongratz (1957) reported that upon isolation on solid media, colonies appear as smooth or rough types. The smooth colonies are round, convex, watery, and translucent and can be emulsified easily. A capsule covers the main body of the reproductive cell (not prostheca). Rough colonies are rare. They are smaller and dry and form a central crater after several days. The colonies are not readily emulsified, and suspensions remain granular. They lack capsules.

No growth on inulin, dextrin, glycogen, esculin, glycerol, erythritol, mannitol, sorbitol, or urea. Reduces neutral red, methylene blue and Janus green. Produces H_2S. Other characteristics are given in Table BXII.α.58.

Contrary to the original description (Pongratz, 1957), NO_3^- and NO_2^- are not reduced, indole is not formed from tryptophan, and the rough strain, PR 727, is motile. It is not known whether these differences are due to the continuous subculturing of these strains.

Not virulent for mice (5 ml), rats (1 ml), guinea pigs (1 ml), or rabbits (1 ml) if a suspension which contains 5 × 10^9 cells/ml is injected subcutaneously or intraperitoneally. The organisms cannot be recovered from treated animals. The major fatty acids are $C_{17:0}$, $C_{17:1}$, and octadecanoate ($C_{18:1}$) at 24, 19, and 14%, respectively, of the total fatty acids (Table BXII.α.59). Isolated only once from a patient (deep sea diver) with infectious sinusitis. The overall DNA–DNA similarities with other *Hyphomonas* species are reported in Table BXII.α.60.

The mol% G + C of the DNA is: 60 (Bd) (Mandel et al., 1972); rough strain (IFAM PR 727, ATCC 33880), 61 (Bd). *Type strain*: PS728, ATCC 33881; IFAM PS728. *GenBank accession number (16S rRNA)*: AJ227813.

2. **Hyphomonas adhaerens** Weiner, Melick, O′ Neill and Quintero 2000,467[VP]

ad.hae′ rens. L. part. adj. *adhaerens* hanging on, sticking to.

Main body of the mother cell is prolate spheroid, ~1–2 μm in diameter, and has one prostheca, 0.2 × 1–5 μm. Buds are motile by a single flagellum. The main body of the reproductive cell, but not the prostheca, is surrounded by capsular polysaccharide. Gram negative. Not acid fast. No endospores. Aerobic. Additional phenotypic characteristics are reported in Table BXII.α.58.

Colonies are round, undulate, about 1.5 mm in diameter after 3 days at 30°C on marine agar. In liquid media, there is granular turbidity due to adhering cell masses, and thick biofilm forms on the surface of the growth vessel. Does not form rosettes. The optimal temperature range for growth is 25–37°C. The ocean salts growth range is 1.5–12%. The optimal pH growth range is 5.7–8.7. Nitrate is reduced. Sheep erythrocytes are not hemolyzed. Susceptible to novobiocin, streptomycin, tellurite, crystal violet, brilliant green, and methylene blue, and resistant to 1.0% Tween 80, penicillin, and ampicillin.

The major fatty acid is octadecanoate ($C_{18:1}$) at 80% of the total fatty acids (Table BXII.α.59). Isolated in 1982 by J. Smit from an inshore slough in Puget Sound, Pacific Ocean. The overall DNA–DNA similarities with other *Hyphomonas* species are reported in Table BXII.α.60.

The mol% G + C of the DNA is: 60 (Bd).
Type strain: MHS-3, ATCC 43965.
GenBank accession number (16S rRNA): AF082790.

3. **Hyphomonas hirschiana** Weiner, Devine, Powell, Dagasan and Moore 1985, 242[VP]

hir' schi.an.a. L. fem. adj. *hirschiana* pertaining to Hirsch; named for P. Hirsch for his contributions to the study of prosthecate, budding bacteria.

Cellular morphology is like that of *H. polymorpha*. Colonies are small (1.0 mm in diameter after 48 h), dull gray, dry, and circular to irregular, with dimpled elevations and lobate margins. Growth in broth produces a pellicle at the surface and a light film of growth on the sides and bottom of the culture vessel.

A mixture of L-glutamic acid, L-aspartic acid, L-serine, and L-methionine is required for growth. The temperature range for optimal growth is 25–31°C. The NaCl concentration for growth is 2.0–15.0%. Optimal pH for growth: 7.6.

Nitrate is reduced. Sheep erythrocytes are not hemolyzed. Susceptible to 0.01% NP-40, novobiocin, and ampicillin. Resistant to 1.0% Tween 80. Intermediate susceptibility to penicillin and streptomycin. Treatment for 9 min or more with a Brownwill Biosonik IV sonicator (low probe at full power) is required to produce 99% lysis of cells.

A major fatty acid is octadecanoate ($C_{18:1}$) at up to 67% of the total fatty acids (Table BXII.α.59). The type strain was isolated in 1979 from shellfish beds near hydrothermal vents on the floor of the mid-Pacific Ocean by H. Jannasch. The overall DNA–DNA similarities with other *Hyphomonas* species are reported in Table BXII.α.60.

The mol% G + C of the DNA is: 57 (Bd).
Type strain: VP5, ATCC 33886.
GenBank accession number (16S rRNA): AF082794.

4. **Hyphomonas jannaschiana** Weiner, Devine, Powell, Dagasan and Moore 1985, 240[VP]

jan' nasch.i.an.a. L. fem. adj. *jannaschiana* pertaining to Jannasch; named for H. Jannasch for his contributions to marine microbiology.

Mature mother cells are 0.5–0.8 × 1.0–3.0 μm, and the main body of the cell tends to be elongated and bullet- or spindle-shaped. Young swarmer cells are motile and pear-shaped and resemble those of other members of the genus. On Zobell marine agar supplemented with 5% sheep blood, colonies are small, dull gray, dry, and circular to irregular with dimpled elevations and lobate margins (strains VP1 and VP3) or gray, circular, convex, entire, glistening, and mucoid (strains VP2[T] and VP4), resembling *H. neptunium* colonies, which have similar shape and are off-white. In broth, growth appears granular, and a film of tan cells covers the sides and bottom of the culture vessel (strains VP1 and VP3), or growth may appear ropy and produce only a slight film (strains VP2[T] and VP4). Strains VP1 and VP3 tend to form rosette-like cell aggregates.

A brown pigment, identified as a pyomelanin, is produced in type 2216 marine broth at 31–37°C. A mixture containing glutamic acid, aspartic acid, and serine as substrates and the cofactor biotin is required for chemoorganotrophic growth. The temperature for optimal growth is 37°C. The NaCl concentration for optimal growth is 3.5–7.5%. Optimal pH for growth: 7.6.

Nitrate is reduced. Produces α-hemolysis on Zobell marine medium supplemented with 5% sheep blood. Susceptible to NP-40 (0.01%), novobiocin, penicillin, ampicillin, and streptomycin. Resistant to Tween 80 (1.0%). Very resistant to breakage by sonication.

A major fatty acid is octadecanoate ($C_{18:1}$) at up to 65% of the total fatty acids (Table BXII.α.59). Four strains, ATCC 33882 (VP1), ATCC 33883[T] (VP2[T]), ATCC 33884 (VP3), and ATCC 33885 (VP4), of *H. jannaschiana* were isolated in 1979 from shellfish beds near hydrothermal vents on the floor of the mid-Pacific Ocean. The appearance is typical of Gram-negative bacteria. The only notable fine-structural difference among the strains is the consistent presence of large dark granules in strains VP1 and VP3. The occurrence of these granules is independent of the medium used for culturing. Strains cannot be differentiated on the basis of morphological, serological, nutritional or biochemical tests. The overall DNA–DNA similarities with other *Hyphomonas* species are reported in Table BXII.α.60.

The mol% G + C of the DNA is: 60 (Bd).
Type strain: VP2, ATCC 33883.
GenBank accession number (16S rRNA): AF082789.

5. **Hyphomonas johnsonii** Weiner, Melick, O' Neill and Quintero 2000, 467[VP]

john.so' ni.i. M.L. gen. n. *johnsonii* of Johnson, named after the (American) molecular taxonomist John Johnson.

Main body of the mother cell is prolate spheroid, ~1 μm in diameter and has one prostheca, 0.2 × 1 μm. Buds are motile by a single flagellum. Gram negative. Not acid fast. No endospores. Aerobic. Additional phenotypic characteristics are reported in Table BXII.α.58.

Colonies are round, convex, about 1.5 mm in diameter after 3 days at 30°C on marine agar. In liquid medium, growth is granular and/or ropy. May form thin film on growth vessel. Can utilize sugars, in the presence of amino acids, for carbon and energy. The optimal temperature range for growth is 25–37°C. The ocean salts growth range is 1.5–6.0. The pH growth range is 5.7–8.0. Nitrate is reduced. α-Hemolysis in the presence of sheep erythrocytes. Susceptible to novobiocin, penicillin, ampicillin, streptomycin, tellurite, crystal violet, brilliant green; resistant to 1.0% Tween 80, methylene blue, and neutral red.

The major fatty acid is octadecanoate ($C_{18:1}$) at 64% total fatty acids (Table BXII.α.59). The outer membrane protein profiles, DNA homology, and serology have low but significant similarity to other strains of *Hyphomonas*. Isolated in 1982 by J. Smit from an inshore slough in Puget Sound, Pacific Ocean. The overall DNA–DNA similarities with other *Hyphomonas* species are reported in Table BXII.α.60.

The mol% G + C of the DNA is: 64 (Bd).
Type strain: MHS-2, ATCC 43964.
GenBank accession number (16S rRNA): AF082791.

6. **Hyphomonas neptunium** (Leifson 1964) Moore, Weiner and Gebers 1984, 71[VP] (*Hyphomicrobium neptunium* Leifson 1964, 249.)

nep.tu' ni.um. L. n. *neptunus* god of the sea, probably should be L. gen. n. *neptuni*.

See Table BXII.α.58 for characteristics. A major fatty acid is octadecanoate ($C_{18:1}$) at up to 37% of the total fatty acids (Table BXII.α.59). Isolated from a sample of stored seawater or obtained from the harbor at Barcelona, Spain. The overall DNA–DNA similarities with other *Hyphomonas* species are reported in Table BXII.α.60.

> *The mol% G + C of the DNA is*: 61.7 (Bd).
> *Type strain*: LE 670, IFAM LE 670, ATCC 15444.
> *GenBank accession number (16S rRNA)*: AF082798.

7. **Hyphomonas oceanitis** Weiner, Devine, Powell, Dagasan and Moore 1985, 240[VP]

o.cean.i' tis. N.L. fem. n. *oceanitis* daughter of the ocean.

Cells are round to oval and ~0.9 μm in diameter and usually have one hypha, which is up to 3 times the length of the mother cell. Under favorable growth conditions, a small proportion of cells produce two hyphae, one at each of the poles. Both hyphae are capable of producing buds concurrently. Buds (daughter cells) are pear-shaped and become larger and rounder when they are mature (mother cells) and producing buds. Intercalary buds (i.e., two or more daughter cells remaining attached to the distal pole(s) of the stalk(s) of the mother cell) are common. Buds are motile until hyphae form. Gram negative. Not acid-fast. No endospores. Aerobic or microaerophilic.

On agar, the colonies are colorless, raised, semitranslucent or opaque and, after 3 days, up to 1.5 mm in diameter. In liquid media, growth produces uniform turbidity without a pellicle or sediment. The best growth occurs in media containing sea salts. Does not grow on the mineral salts medium used for *Hyphomicrobium vulgare*. A mixture of L-glutamic acid, L-aspartic acid, L-serine, and L-methionine is required for growth. Requires biotin.

The temperature range for optimal growth is 20–30°C. The NaCl concentration for optimal growth is 1.0–7.5%. Optimal pH for growth is: 7.6. Nitrate is reduced. Sheep erythrocytes are not hemolyzed. Under optimal conditions in marine broth, growth is slower than the growth of other species of *Hyphomonas*. Susceptible to NP-40 (0.01%) and Tween 80 (0.1–1.0%). Resistant to novobiocin, penicillin, ampicillin, and streptomycin. Requires 6–8 min of treatment with a Brownwill Biosonik IV sonicator (low probe at full power) to produce 99% lysis of cells.

A major fatty acid is octadecanoate ($C_{18:1}$) at up to 60%

of the total fatty acids (Table BXII.α.59). Isolated from the Baltic Sea in 1979 by H. Schlesner. The overall DNA–DNA similarities with other *Hyphomonas* species are reported in Table BXII.α.60.

> *The mol% G + C of the DNA is*: 59 (Bd).
> *Type strain*: SCH 1325, IFAM 1325, ATCC 33879.
> *GenBank accession number (16S rRNA)*: AF082797.

8. **Hyphomonas rosenbergii** Weiner, Melick, O' Neill and Qintero 2000, 467[VP]

ro.sen.ber' gi.i. M.L. gen. n. *rosenbergii* of Rosenberg, named after the (Israeli) microbial ecologist Eugene Rosenberg.

Main body of the mother cell is prolate spheroid, ~1 μm in diameter and has one prostheca, 0.2 × 1 μm. Buds are motile by a single flagellum. Synthesizes a capsule, which surrounds the entire cell at all growth stages, and a polar holdfast, which is temporally synthesized. Gram negative. Not acid fast. No endospores. Aerobic. Additional phenotypic characteristics are reported in Table BXII.α.58.

Colonies are round, convex, about 1.5 mm in diameter after 3 days at 37°C on marine agar. In liquid medium, growth is granular and/or ropy. Biofilm forms on walls of growth vessel. Cells form rosettes. The optimal temperature range is 25–45°C. The ocean salts growth range is 1.0–12.0. The pH growth range is 5.7–8.9. Nitrate is reduced. Sheep erythrocytes are not hemolyzed.

Susceptible to novobiocin, penicillin, ampicillin, streptomycin, tellurite, crystal violet, brilliant green, and methylene blue; resistant to 1.0% Tween 80 and neutral red. The major fatty acid is octadecanoate ($C_{18:1}$) at 43% total fatty acids. The outer membrane protein profiles, DNA homology, and serology have low but significant similarity to other strains of *Hyphomonas*. Isolated in 1985 from the Guaymas Basin thermal vent region (2000 m deep), Gulf of Mexico by H. Jannasch. The overall DNA–DNA similarities with other *Hyphomonas* species are reported in Table BXII.α.60.

> *The mol% G + C of the DNA is*: 61 (Bd).
> *Type strain*: VP-6, ATCC 43869.
> *GenBank accession number (16S rRNA)*: AF082795.

Genus VIII. **Ketogulonicigenium** *corrig.* Urbance, Bratina, Stoddard and Schmidt 2001, 1068[VP] *(Ketogulonigenium (sic) Urbance, Bratina, Stoddard and Schmidt 2001, 1068)*

THE EDITORIAL BOARD

Ke.to.gu.lo.ni.ci.ge' ni.um. N.L. n. *acidum ketogulonicum* ketogulonic acid; N.L. neut. suff. *genium* genium that which produces; L. neut. n. *Ketogulonicigenium* that which produces ketogulonic acid.

Gram-negative chemoorganotrophic facultative anaerobes. Cells ovoid or rod-shaped (0.5–0.7 × 0.8–1.3 μm). Oxidase and catalase positive. **Convert L-sorbose to 2-keto-L-gulonic acid. Major fatty acids $C_{16:0}$, and $C_{18:1}$ ω7c/ω9t/ω12t.** Temperature optimal 27–31°C; pH optimal 7.2–8.5. Grow best on arabinose, glycerol, inositol, lactose, mannitol, and sorbitol.

The mol% G + C of the DNA is: 52.1–54.

Type species: **Ketogulonicigenium vulgare** Urbance, Bratina, Stoddard and Schmidt 2001, 1069 (*Ketogulonigenium vulgare* (sic) Urbance, Bratina, Stoddard and Schmidt 2001, 1069.)

FURTHER DESCRIPTIVE INFORMATION

Analysis of 16S rDNA sequences placed the two species of the genus *Ketogulonicigenium* in the family *Rhodobacteraceae* of the *Alphaproteobacteria*, distinct from a cluster containing the genera *Gluconobacter* and *Acetobacter*. The five strains of *Ketogulonicigenium* were phenotypically similar but could be separated into two species on the basis of DNA/DNA hybridization; the five strains include four new isolates and DSM 4025, which was previously classified as *Gluconobacter oxydans* (Urbance et al., 2001).

ENRICHMENT AND ISOLATION PROCEDURES

Enrichments were carried out in liquid medium with L-sorbose as the predominant carbon source and with $CaCO_3$ (6 g/l) added to buffer the medium against acidic metabolic products (Urbance et al., 2001). Culture supernatants were examined for 2-keto-L-gulonic acid production by TLC and HPLC; positive cultures were serially diluted and spread on agar of the same composition as the enrichment medium. Individual colonies were purified by streaking and then tested for the ability to produce 2-keto-L-gulonic acid by TLC and HPLC (Urbance et al., 2001).

MAINTENANCE PROCEDURES

Isolates were maintained on trypticase soy agar or another complex medium, SYM agar, described in Urbance et al. (2001). Long-term storage was achieved by freezing at $-70°C$ in peptone-yeast extract-mannitol medium amended with 20% glycerol (Urbance et al., 2001).

DIFFERENTIATION OF THE SPECIES OF THE GENUS *KETOGULONICIGENIUM*

Ketogulonicigenium vulgare is nonmotile, does not grow at 37°C, has a pH optimum of 7.2–8.0, and grows best at a sodium ion concentration of 31.2 mM. *Ketogulonicigenium robustum* is motile, grows at 37°C, has a pH optimum of 8.0–8.5, and grows best at sodium ion concentrations between 117–459 mM (Urbance et al., 2001).

List of species of the genus Ketogulonicigenium

1. **Ketogulonicigenium vulgare** Urbance, Bratina, Stoddard and Schmidt 2001, 1069[VP] (*Ketogulonigenium vulgare* (sic) Urbance, Bratina, Stoddard and Schmidt 2001, 1069.)
   vul.ga′ re. L. neut. adj. *vulgare* common.

   Description as for the genus with the following additional characteristics. Nonmotile. No growth at 37°C. pH optimum 7.2–8.0. Optimum Na^+ concentration for growth 31.2 mM.
   The mol% G + C of the DNA is: 53.4–54.0 (T_m).
   Type strain: : DSM 4025 (patent strain).
   GenBank accession number (16S rRNA): AF136849.

2. **Ketogulonicigenium robustum** Urbance, Bratina, Stoddard and Schmidt 2001, 1069[VP] (*Ketogulonigenium robustum* (sic) Urbance, Bratina, Stoddard and Schmidt 2001, 1069.)
   ro.bus′ tum. L. adj. *robustum* strong.

   Description as for the genus with the following additional characteristics. Motile. Grow at 37°C. pH optimum 8.0–8.5. Optimum Na^+ concentration for growth 117–459 mM.
   The mol% G + C of the DNA is: 52.1 (T_m).
   Type strain: X6L, KCTC 0858BP, NRRL B-21267.
   GenBank accession number (16S rRNA): AF136850.

Genus IX. **Maricaulis** *Abraham, Strömpl, Meyer, Lindholst, Moore, Christ, Vancanneyt, Tindall, Bennasar, Smit and Tesar 1999, 1071[VP]*

JEANNE S. POINDEXTER

Ma′ ri.cau′ lis. L. neut. n. *mare* the sea; L. masc. n. *caulis* stalk; M.L. masc. n. *Maricaulis* stalk from the sea.

Cells single, unbranched, with poles gently rounded or slightly tapered; long axis typically straight and cell shape **bacteroid**, but may be slightly curved in some cells and some isolates and cell shape **subvibrioid**; 0.4–0.6 × 1–2 μm during growth in most media. The younger pole of the dividing cell bears a **single flagellum**, and the older pole bears a prostheca (the **stalk**) developed by outgrowth of the cell envelope. Stalk diameter is constant along its length, varying from 0.11 to 0.18 μm among isolates. Binary **fission is asymmetric**: prior to fission, one pole bears a stalk and the other a single flagellum, and fission results in the separation of a non-motile stalked cell and a motile, monoflagellate swarmer cell. At the base of the flagellum and at the outer tip of the stalk is a small mass of adhesive material, the **holdfast**, which confers adhesiveness on each of the progeny.

Gram negative. All isolates grow aerobically with O_2 as terminal electron acceptor; principal respiratory quinone is ubiquinone-10. Nitrate may be reduced to nitrite anaerobically; N_2 is not produced. Some isolates (12 of 25 tested) can grow anaerobically on peptone-yeast extract medium without additional carbohydrates. Colonies are circular, convex and glistening, with a smooth margin, butyrous in texture, and colorless. In unagitated liquid cultures, cells accumulate as a **surface film** or pellicle and develop as a ring of growth on the vessel wall at or just below the air-liquid interface. Growth in agitated liquid cultures is evenly dispersed.

Chemoorganotrophic and **oligotrophic**; grow optimally in media such as peptone-yeast extract containing 0.05–0.3% (w/v) organic solutes prepared in sea water or artificial sea salts solution, but not in undiluted standard marine broth such as Zobell's medium 2216, in which growth is inhibited or cells are deformed and viability is low. Optimal temperature range for growth 20–25°C; tolerated range for growth 10–35°C. Optimal pH for growth near neutrality; pH 6–8 tolerated. Growth in peptone or Casamino acids media requires the addition of sea salts or NaCl; optimal NaCl concentration for growth 2–6% (w/v); tolerated range 0.5–8% (w/v). Maximum specific rates of exponential growth 0.17–0.23 h^{-1}. Do not grow in defined media, and growth is not stimulated by B vitamins; almost all isolates require unidentified growth factors available in peptone. Do not use alcohols as sole sources of carbon.

Comparative analysis of 16S rDNA sequences is consistent with placement of *Maricaulis* among the *Alphaproteobacteria*. The positions of the 15 sequences grouped as *Maricaulis* in Fig. BXII.α.111 of the chapter describing the genus *Caulobacter* (see the family *Caulobacteraceae*) are represented here in Fig. BXII.α.76 by the seven nonredundant sequences with their strain identities and GenBank accession numbers shown. These seven sequences were determined for isolates known to exhibit *Maricaulis* phenotypes. The positions of the sequences for *Hirschia baltica*, "marine caulobacter" isolate MCS23, and the six nonredundant se-

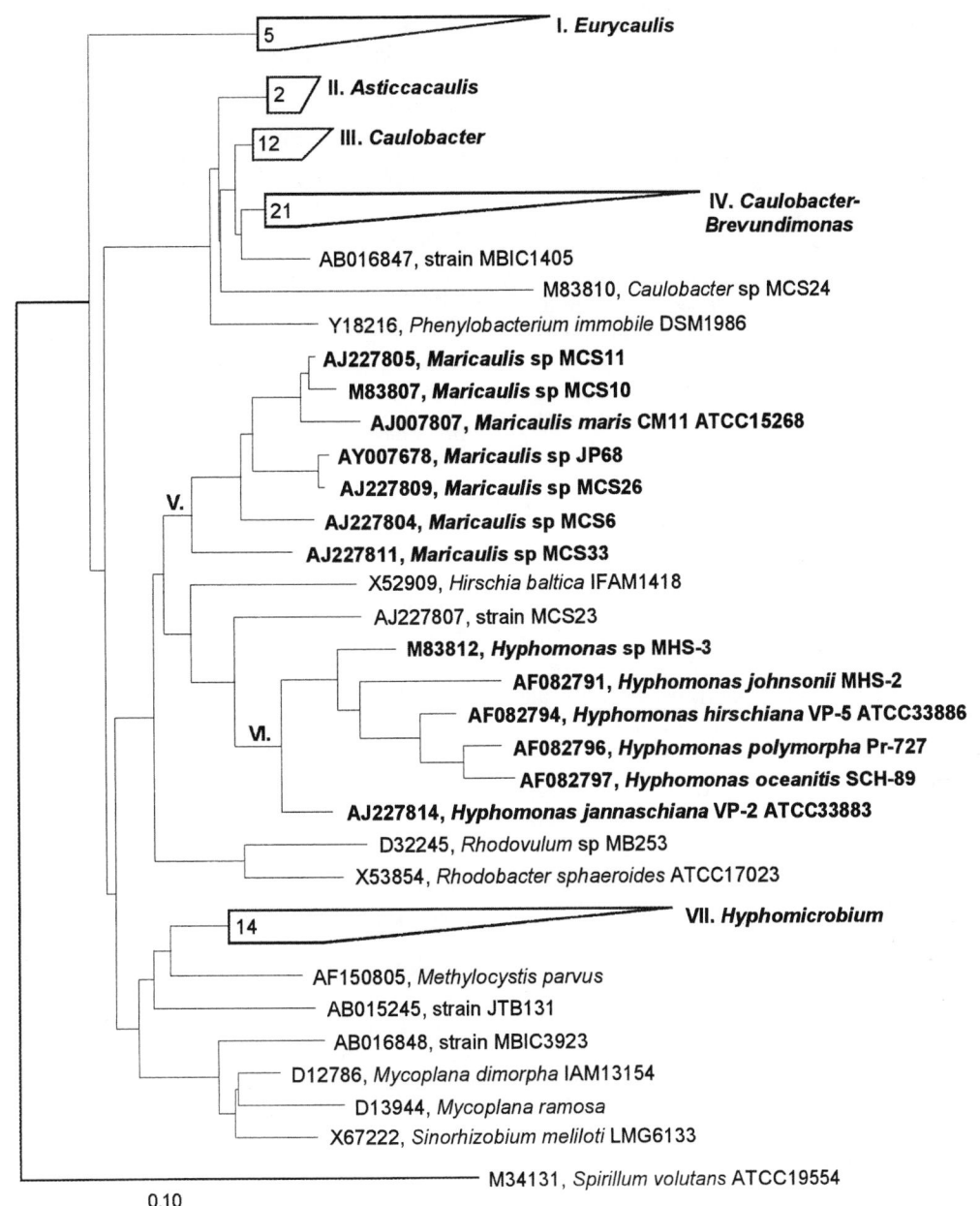

FIGURE BXII.α.76. Phylogenetic tree of 81 16S rDNA sequences, with group V. *Maricaulis* and group VI. *Hyphomonas* expanded to display GenBank accession numbers and strain identities (see Fig. BXII.α.114 in the chapter on the genus *Caulobacter*). Also represented are the positions of *Maricaulis* sequences AJ227810 (by AJ227805); AJ007804, AJ227802, and AJ227803 (by AJ007807); AJ227808 (by AJ227809); and AJ227806, M83809, and M83811 (by AJ227811). Represented *Hyphomonas* sequences are AF082795 and AF082798 (by AF082794); AJ227813 (by AF082796); AF082789, AF082790, AF082792, AF082793, and AF082799 (by AJ227814); and M83806 (by M83812). Bar indicates evolutionary distance.

quences among the 15 reported for *Hyphomonas* spp. are also displayed to illustrate their similarity to *Maricaulis* sequences. (See "Taxonomic Comments" in the chapter in which the genus *Caulobacter* is described.)

Sources: isolated from filtered, stored seawater and from natural bodies of seawater. Not encountered as clinical isolates, and not known to be pathogenic for plants or animals.

The mol% G + C of the DNA is: 62.5–64.0.

Type species: **Maricaulis maris** (Poindexter 1964) Abraham, Strömpl, Meyer, Lindholst, Moore, Christ, Vancanneyt, Tindall, Bennasar, Smit and Tesar 1999, 1071 (*Caulobacter maris* Poindexter 1964, 289.)

FURTHER DESCRIPTIVE INFORMATION

Morphology and fine structure Known isolates of *Maricaulis* are either bacteroid or subvibrioid, i.e., cell poles are gently rounded or tapered, and the long axis is not strongly curved. However, vibrioid caulobacters are seen in samples and enrichment cultures of seawater, and the present description should not prevent an isolate from being identified provisionally as *Mari-*

caulis. Vibrioid marine isolates have been obtained from deep-sea vent communities (Poindexter, unpublished), but further characterizations are required to determine whether they are otherwise significantly different from isolates on which the present description is based.

As in freshwater caulobacters, a holdfast is present at the distal tip of the *Maricaulis* stalk; the holdfasts of all adhesive marine strains tested bound Wheat Germ Agglutinin (Merker and Smit, 1988; Abraham et al., 1999; 2001) and so must contain some *N*-acetylglucosamine. This is a higher proportion of WGA-binding strains than was found for freshwater caulobacters, suggesting that the composition of holdfast material may be constrained by the ionic environment of the marine habitat.

Maricaulis isolates have not been examined in ultrathin sections to determine the ultrastructure of the stalk. Although these marine bacteria are susceptible to disintegration during preparation for electron microscopy of whole cells, the slenderness of the stalks on the cells that do remain intact during specimen preparation and the absence of budding from the tip of the prostheca are features consistent with an internal structure comparable to that of freshwater caulobacters (see Morphology in the chapter on the genus *Caulobacter*). In contrast to freshwater caulobacters, however, the stalks of *Maricaulis* isolates do not contain stalk bands. An S-layer coat has not been discerned in *Maricaulis* cells, but their cell surface is poorly preserved by the procedures that have been used to study this feature in *Caulobacter* cells.

Cellular composition The principal phospholipid in *Maricaulis* cells is phosphatidylglycerol, and the most abundant fatty acid species are $C_{18:1}$ and $C_{16:0}$. All isolates tested also contain lower, but significant, amounts of $C_{18:1\ \omega7c\ 11CH_3}$ and species of $C_{17:1}$ and $C_{16:1}$, $C_{17:0}$, and—in contrast to freshwater caulobacters—$C_{18:0}$ and $C_{11:0\ iso\ 3OH}$ (Abraham et al., 1999; Abraham et al., 2001). Phospholipids are significantly less abundant than in freshwater caulobacters and most other bacteria, and phosphoglucolipids are not detectable. All seven isolates so far analyzed possess sulfoquinovosyl diacylglycerols, one of which was identified as 1-O-(α-6′,6-deoxy-aldohexopyranosyl-6′-sulfonic acid)-3-O-diacylglycerol, and four also formed at least small amounts of taurineamide diacylglycerols (Abraham et al., 1999). This pattern of cellular lipids is similar to that found by the same laboratory for *Hyphomonas* isolates, but differs noticeably by the absence of sulfoquinovosyl diacylglycerides from the cells of these budding, hyphal bacteria and of one isolate (MCS23) regarded as a caulobacter throughout its isolation. This is lipid pattern "M" of Abraham et al. (1997).

Genome structure Native plasmids were detected in two marine isolates by Anast and Smit (1988), but not in the majority of strains. More significantly for potential employment of these organisms in marine habitats, every isolate tested was able to propagate and express genes (for antibiotic resistance) carried on plasmids transferred to the marine caulobacters by conjugation with *Escherichia coli* donors. Ribosomal DNA sequences have been determined for a large number of *Maricaulis* and *Maricaulis*-like isolates. *Maricaulis* sequences constitute one of five coherent groups among caulobacterial sequences. The minimum similarity value among sequences grouped within V. *Maricaulis* in Fig. BXII.α.76 is 92.2%. The highest similarities outside the genus are to sequences for organisms identified as *Hyphomonas jannaschiana* (90.6–92.0%). Two sequences are similar to *Maricaulis* and *H. jannaschiana* sequences: the sequence for *Hirschia baltica* is up to 89.8% similar to *Maricaulis* sequences, and the

sequence for isolate MCS23, described as a caulobacter, is up to 89.5% similar. Bacteriophages lytic for *Maricaulis* isolates have not been reported. (See "Taxonomic Comments" below and the chapter on the genus *Caulobacter* for further discussion of 16S rDNA sequences and other bacteriophages.)

Antibiotic and metal susceptibilities *Maricaulis* isolates are typically susceptible to most antibiotics tested. However, nine of 20 isolates tested were found resistant to 10 µg/ml or more of $HgCl_2$. Mercury resistance was not correlated with the presence of plasmids, and a probe could not detect the mercury reductase gene.

ENRICHMENT AND ISOLATION PROCEDURES

Maricaulis isolates are obtained by the basic procedure described for *Caulobacter* (see the chapter on the genus *Caulobacter*), beginning with a sample of estuarine or seawater. Deep-sea *Maricaulis* isolates were obtained by allowing glass cover slips to reside in an open box for several weeks near a deep-sea vent; caulobacters attached to the glass surfaces, which were used as inocula for enrichment cultures in dilute peptone-seawater medium (Poindexter, unpublished). Once isolated, *Maricaulis* strains can be maintained on slants of dilute peptone (0.05%, w/v)–Casamino acids (0.05%, w/v) media prepared with 75–80% seawater or artificial sea salts solution, transferred every two months, incubated for 2–3 d at room temperature, then refrigerated. Lyophilization is not a suitable means of preservation.

PROCEDURES FOR TESTING SPECIAL CHARACTERS

The distinctive features of these bacteria are morphological—the asymmetry of the dividing cell, the single polar flagellum, the holdfast, and the absence of bands in the stalks. However, intact cells are difficult to preserve for the electron microscopical examinations needed for close examination of these features. Aldehyde preservation is effective with some isolates, and acetic acid (2%, v/v) for others, but many disintegrate during specimen preparation. Close observation by phase-contrast microscopy of wet cells is, however, adequate to determine the asymmetry of dividing cells, the presence of stalks, and adhesiveness and motility of cells.

DIFFERENTIATION OF THE GENUS *MARICAULIS* FROM OTHER GENERA

Four groups of bacteria other than caulobacters produce long cellular appendages of the envelope that are of constant diameter; these bacteria also tend to occur in habitats with caulobacters. They are distinguishable from caulobacters, and from each other, by the morphology of their reproductive stages. The five groups are distinguished as in Table BXII.α.101, and the five species groups of caulobacters are distinguished as shown in Table BXII.α.102, both in the chapter in which the *Caulobacter* is described. Relationships to nonprosthecate bacteria implied by 16S rDNA sequence analysis are illustrated and discussed further in "Taxonomic Comments" in the chapter describing the genus *Caulobacter*.

TAXONOMIC COMMENTS

Among the 16S rDNA sequences that were found similar to those of *Maricaulis* isolates, the entire set determined for *Hyphomonas* isolates is clearly the most similar (more than 90% similarity), while the sequences of freshwater caulobacters are less similar to *Maricaulis* sequences than are those of phototrophic bacteria such as *Rhodobacter* and *Rhodopseudomonas*. This phylogenetic relationship could reflect a common marine ancestry of *Maricaulis* and *Hyphomonas* in the ocean, an ancestry further implied by the similarity of lipid composition of these two groups. In each morphological type, the marine isolates are more like their co-habitants than like their freshwater/soil morphological twins. It is not so easy to distinguish these two kinds of dimorphic, prosthecate bacteria upon first observing them, even in pure cultures. For example, the "VP" designation on several *Hyphomonas* strains was used to label deep-sea Vent Prosthecate bacteria until morphology could be established unequivocally for pure populations (Poindexter, unpublished). Among the marine caulobacters collected by Anast and Smit (1988), one isolate (MCS23) appears closer, by several criteria, to *Hyphomonas* than to any isolates within the coherent *Maricaulis* group. This isolate might reflect

a shift in reproductive function between prostheca and cell, but the direction in which it might be evolving cannot be inferred.

Two species are recognized, distinguished by physiological traits as shown in Tables BXII.α.61 and BXII.α62..

TABLE BXII.α.61. Differential characteristics of the species of the genus *Maricaulis*[a]

Characteristic	*M. maris*	"*M. halobacteroides*"
Growth with 4% NaCl	+	−
Carbon sources generally used:		
Carbohydrates	+	+
Amino acids	−	+
Other organic acids	−	+
Starch hydrolyzed	−	+
Nitrate reduced to nitrite anaerobically	+	−

[a]Symbols: +, 90% or more of strains are positive; −, 90% or more of strains are negative.

TABLE BXII.α.62. Other characteristics of the species of the genus *Maricaulis*[a]

Characteristic	*M. maris*	"*M. halobacteroides*"
Carbon-source utilization:[b]		
Arabinose	−	+
Ribose	−	+
Xylose	+	+
Glucose	+	+
Galactose	−	+
Mannose	−	+
Lactose	−	+
Maltose	+	+
Sucrose	+	+
Proline	−	+
Acetate	−	+
Butyrate	−	+
Pyruvate	−	+
Starch hydrolysis	−	+

[a]Symbols: +, 90% or more of strains are positive; −, 90% or more of strains are negative. Neither species utilizes any of fructose, alanine, aspartate, glutamate, tyrosine, pimelate, malate, fumarate, or succinate when provided as sole source of carbon in dilute (0.005%, w/v, of each of peptone and yeast extract) media prepared with mineral nutrients and 3% NaCl.

[b]Carbon-source utilizations were determined with D-isomers of sugars, with L-isomers of amino acids, or (when single isomers were not available) with racemic mixtures.

List of species of the genus Maricaulis

1. **Maricaulis maris** (Poindexter 1964) Abraham, Strömpl, Meyer, Lindholst, Moore, Christ, Vancanneyt, Tindall, Bennasar, Smit and Tesar 1999, 1071[VP] (*Caulobacter maris* Poindexter 1964, 289.)

ma'ris. L. gen. n. *maris* of the sea.

Rod-shaped cells slender. Colonies colorless. Organic growth factor requirements not satisfied by mixtures of B vitamins and amino acids. Growth requires NaCl (1–4%, w/v). Known isolates utilize few sugars and do not hydrolyze starch; they do not utilize individual amino acids as carbon sources, but grow on peptone-Casamino acids media.

Strain CM11 was isolated from filtered, stored seawater in Pacific Grove, CA.

The mol% G + C of the DNA is: 62.5 (HPLC).

Type strain: CM11, ACM 5106, ATCC 15268, DSM 4734.

GenBank accession number (16S rRNA): AJ227802.

Additional Remarks: One of two marine species included in this list. Vibrioid marine and estuarine isolates, also dependent on NaCl supplementation for growth in complex organic media, are known; their characteristics have not been reported. At least some of the isolates of marine caulobacters reported by Anast and Smit (1988) could be assigned here, but others may require accommodation in another species. Some of those isolates appear more subvibrioid than bacteroid, but whether cell morphology will serve as a suitable subgeneric criterion among *Maricaulis* isolates has yet to be determined.

Other Organisms

1. "*Maricaulis halobacteroides*" (*Caulobacter maris* Poindexter 1964, 289.)

hal.o.bac.ter.oi'des. Gr. n. *hals* salt; M.L. masc. *bacter* Gr. neut. n. *bactrum* rod; Gr. n. *eidus* form, shape; M.L. adj. *halobacteroides* salt (-needing) and rod shaped.

Rod-shaped cells slender. Colonies colorless. Organic growth factor requirements not satisfied by mixtures of B vitamins and amino acids. Starch is hydrolyzed, and most

sugars, but few individual amino acids, can be utilized as sole sources of carbon for growth. Growth requires NaCl (0.5–3%, w/v) and is inhibited at 4% (w/v) NaCl.

Strain CM13 was isolated from filtered, stored seawater in Pacific Grove, CA.

The mol% G + C of the DNA is: not determined.

Deposited strain: CM13, ATCC 15269.

GenBank accession number (16S rRNA): AJ007804.

Additional Remarks: The original type strain deposited with the ATCC appears to have been, very early, overgrown by strain CM11 (*M. maris*), resulting in identity of the 16S rDNA sequences that have been determined for the two strains (see Fig. BXII.α.76) and DNA–DNA hybridization >90% (Moore et al., 1978; Abraham et al., 1997). Other copies of the ATCC strain (DSM 4734 and ACM 5106) may also be strain CM11. The problem probably arose from attempts to maintain the strains as lyophilized suspensions, a state from which it is difficult to revive these marine bacteria. At present, it seems wiser to continue to recognize the species and thereby possibly provide a taxon suitable for recent isolates of this type of marine caulobacter.

Genus X. **Methylarcula** Doronina, Trotsenko, and Tourova 2000a, 1857[VP]

NINA V. DORONINA AND YURI A. TROTSENKO

Me.thyl.ar.cu′la. Fr. *méthyle* the methyl radical; L. fem. n. *arcula* small box; M.L. fem. n. *Methylarcula* methyl-using small box.

Rods 0.5–0.8 × 0.8–2.0 μm. Nonmotile. Gram negative. **Poly-β-hydroxybutyrate granules are formed.** Endospores and prosthecae are not formed. Colonies are white or pale pink. Do not produce pyocyanin and fluorescein. **Aerobic**, having a strictly respiratory type of metabolism with oxygen as the terminal electron acceptor. **Moderately halophilic**; NaCl is required for growth. Ectoine is accumulated intracellularly as the main osmoprotectant. Growth does not occur on peptone/yeast extract medium, with or without NaCl. Nitrate is not reduced to nitrite. Chemoorganotrophic. **Facultatively methylotrophic. Assimilate C_1 compounds via the isocitrate lyase-negative (icl⁻) serine pathway. Methylamine, sugars, and some organic acids are used as carbon and energy sources.** Ammonium salts, some amino acids, and methylamine are used as nitrogen sources. Growth factors are not required. **Oxidase positive.** Acids are produced from sugars oxidatively. Acetoin, hydrogen sulfide, and ammonium are not formed. Indole is formed from L-tryptophan. The major ubiquinone is Q-10. The dominant phospholipids are phosphatidylethanolamine and phosphatidylcholine. The predominant cellular fatty acids are straight-chain unsaturated ($C_{18:1}$), saturated ($C_{18:0}$), and cyclopropane ($C_{19:0}$) acids. Belongs to the *Alphaproteobacteria*.

The mol% G + C of the DNA is: 57–60.

Type species: **Methylarcula marina** Doronina, Trotsenko and Tourova 2000a, 1858.

FURTHER DESCRIPTIVE INFORMATION

Cells are lysed when transferred into distilled water after NaCl treatment. The divalent cations Mg^{2+} and Ca^{2+} added to the 0.5 M NaCl washing solution at concentrations as low as 50 mM prevent cell lysis. The ectoine pool increases when the NaCl concentration in the growth medium is increased.

Methylamine dehydrogenase and amine oxidase are absent in methylamine-grown cells. Alternatively, they contain an inducible γ-glutamylmethylamide synthetase/lyase and *N*-methylglutamate synthase/lyase. Both enzyme systems produce formaldehyde, which is further oxidized by glutathione-independent formaldehyde dehydrogenase to formate. The latter is ultimately oxidized to CO_2 by phenazine methosulfate-linked and NAD-dependent formate dehydrogenases. Formaldehyde assimilation occurs via the serine pathway as confirmed by the presence of the appropriate specific enzymes: hydroxypyruvate reductase, serine-glyoxylate aminotransferase, and malate lyase. *Methylarcula* strains either lack or have very low isocitrate lyase (icl⁻) activity and during growth on C_1 compounds consequently implement the icl⁻ variant of the serine pathway. Primary assimilation of ammonia occurs both by reductive amination of α-ketoglutarate and via the glutamate cycle.

ENRICHMENT AND ISOLATION PROCEDURES

Successful enrichment has been achieved by inoculating water or soil samples with a salinity of 6–11% (pH 6.5–7.5) into mineral medium MK[1]. Enrichment cultures were grown for 1 week at 29°C in 750-ml Erlenmeyer flasks containing 100 ml of the medium, with shaking at 120 rpm; they were then diluted 1:10³ and plated on solidified MK medium containing 2.0% Bacto agar (Difco).

MAINTENANCE PROCEDURES

Methylarcula cultures can be stored for 2 months on MK agar slants at 4°C. For long-term preservation, *Methylarcula* strains can be lyophilized in skim milk by using common procedures for aerobes.

DIFFERENTIATION OF THE GENUS *METHYLARCULA* FROM OTHER GENERA

The major characteristics differentiating the genus *Methylarcula* from other related genera are summarized in Table BXII.α.63.

TAXONOMIC COMMENTS

In the phylogenetic tree derived from 16S rRNA sequences, *Methylarcula* is located in the order *Rhodobacterales* in the class *Alphaproteobacteria*, whereas most other genera of methylobacteria having the serine pathway belong to the order *Rhizobiales* in the class *Alphaproteobacteria*. The moderately halophilic marine genus *Methylophaga*, which uses the ribulose monophosphate pathway, belongs to the class *Gammaproteobacteria*. The similarity values for species *Methylarcula* and other members of the order *Rhodobacterales* in the class *Alphaproteobacteria*, including the marine methylotroph "*Marinosulfonomonas*" (Holmes et al., 1997), fell into the range 87.1–92.6%.

Methylarcula has less than 10% DNA–DNA similarity with members of *Methylorhabdus*, *Methylobacterium*, *Aminobacter*, and *Methylopila*.

---

1. MK medium contains (g/l of distilled water): $(NH_4)_2SO_4$, 2.0; KH_2PO_4, 2.0; NaCl, 60.0; $MgSO_4 \cdot 7H_2O$, 0.2 g; $FeSO_4 \cdot 7H_2O$, 0.002 (Doronina et al., 2000a). The pH of the medium was adjusted to 7.2 prior to autoclaving. Methylamine was added to a final concentration of 3.0 g/l.

TABLE BXII.α.63. Major differentiating characteristics of facultative serine pathway methylobacteria belonging to various genera[a]

Characteristic	*Methylarcula*	*Aminobacter*	*Hyphomicrobium*	*"Marinosulfonomonas"*	*Methylobacterium*	*Methylopila*	*Methylorhabdus*	*"Methylosulfonomonas"*
Morphology (flagella)	−	+	+	−	+	+	−	+
Reproduction by:								
Budding	−	+	+	−	−	−	−	−
Division	+	−	−	+	+	+	+	+
Hyphae formation	−	−	+	−	−	−	−	−
Oxidase	+	+	−	+	+	+	−	−
Catalase	±	+	+	+	+	−	+	+
Carotenoids	−	−	−	−	+	−	−	−
Reduction of NO₃ to NO₂	−	+	+	nd	+	+	+	nd
Methylamine metabolism:								
Amine dehydrogenase or	−	−	−	nd	+	+	+	nd
N-methylglutamate derivatives	+	+	+	nd	+	−	−	nd
γ-Glutamylmethylamide lyase	+	−	−	nd	−	−	−	nd
Isocitrate lyase	−	+	+	nd	−	−	−	nd
Cyclopropane acid, $C_{19:0\ cyclo}$	+	+	+	+	Trace	+	+	+
Methyloctadecenoic acid, $C_{18:1\ \omega7\ CH_3}$	−	−	+	nd	−	+	+	nd
Major ubiquinone	Q-10	Q-10	Q-9	nd	Q-10	Q-10	Q-10	nd
Tolerance to NaCl (%)	12	2.5	3	3.5	2.5	2	2	0.5
Growth at pH 10.0	+	−	−	−	−	−	−	−
Utilization of methanol	−	−	+	+	+	+	+	+
Mol% G + C of DNA	57–61	62–64	61–65	57	60–70	66–70	66–67	61

[a]Symbols: +, present; −, absent; ±, variable; nd, not determined.

List of species of the genus Methylarcula

1. **Methylarcula marina** Doronina, Trotsenko and Tourova 2000a, 1858[VP]

ma.ri′ na. L. fem. n. *mare* sea; L. fem. adj. *marina* of the sea.

Fig. BXII.α.77 illustrates the morphological features. Colonies on methylamine agar medium are white or pale pink, round, convex, 1–2 mm in diameter. Oxidase positive. Urease, lipase, and catalase negative. Nitrate is not utilized. Able to grow at 10–42°C, at pH 5.0–10.5, and in the presence of 0.05–12% NaCl. Optimal conditions for growth are 29–35°C, pH 7.5–8.5, and 3–8% NaCl. Utilizable carbon sources are methylamine, fructose, glucose, maltose, lactose, mannose, ribose, trehalose, galactose, xylose, sucrose, succinate, pyruvate, and acetate. Methylamine is oxidized to formaldehyde by the N-methylglutamate pathway enzymes γ-glutamylmethylamide synthetase/lyase and N-methylglutamate synthase/lyase.

The type strain was isolated from Azov Sea estuary water.

The mol% G + C of the DNA is: 60.4 (T_m).

Type strain: h1, VKM B-2159.

GenBank accession number (16S rRNA): AF030437.

2. **Methylarcula terricola** Doronina, Trotsenko and Tourova 2000a, 1858[VP]

ter.ri′ co.la. Gr. adj. *terricola* of the soil.

Fig. BXII.α.77 illustrates the morphological features. Colonies on methylamine agar medium are white, 2 mm in diameter. Oxidase and catalase positive. Urease and lipase negative. Nitrate is not utilized. Able to grow at 10–40°C, pH 5.5–10.0, and in the presence of 0.05–14% NaCl. Optimal conditions for growth are 29–32°C, pH 7.5–8.5, and 3–6% NaCl. Utilizable carbon sources are mono- and dimethylamine, fructose, glucose, maltose, lactose, mannose, ribose, trehalose, galactose, xylose, sucrose, succinate, pyruvate, and acetate. Methylamine is oxidized to formaldehyde by the N-methylglutamate pathway enzymes γ-glutamylmethylamide synthetase/lyase and N-methylglutamate synthase/lyase.

The level of DNA–DNA similarity with the type species *Methylarcula marina* is ~25–30% (membrane method).

The type strain was isolated from coastal salty soil of the Black Sea (Crimea).

The mol% G + C of the DNA is: 57.1 (T_m).

Type strain: h37, VKM B-2160.

GenBank accession number (16S rRNA): AF030436.

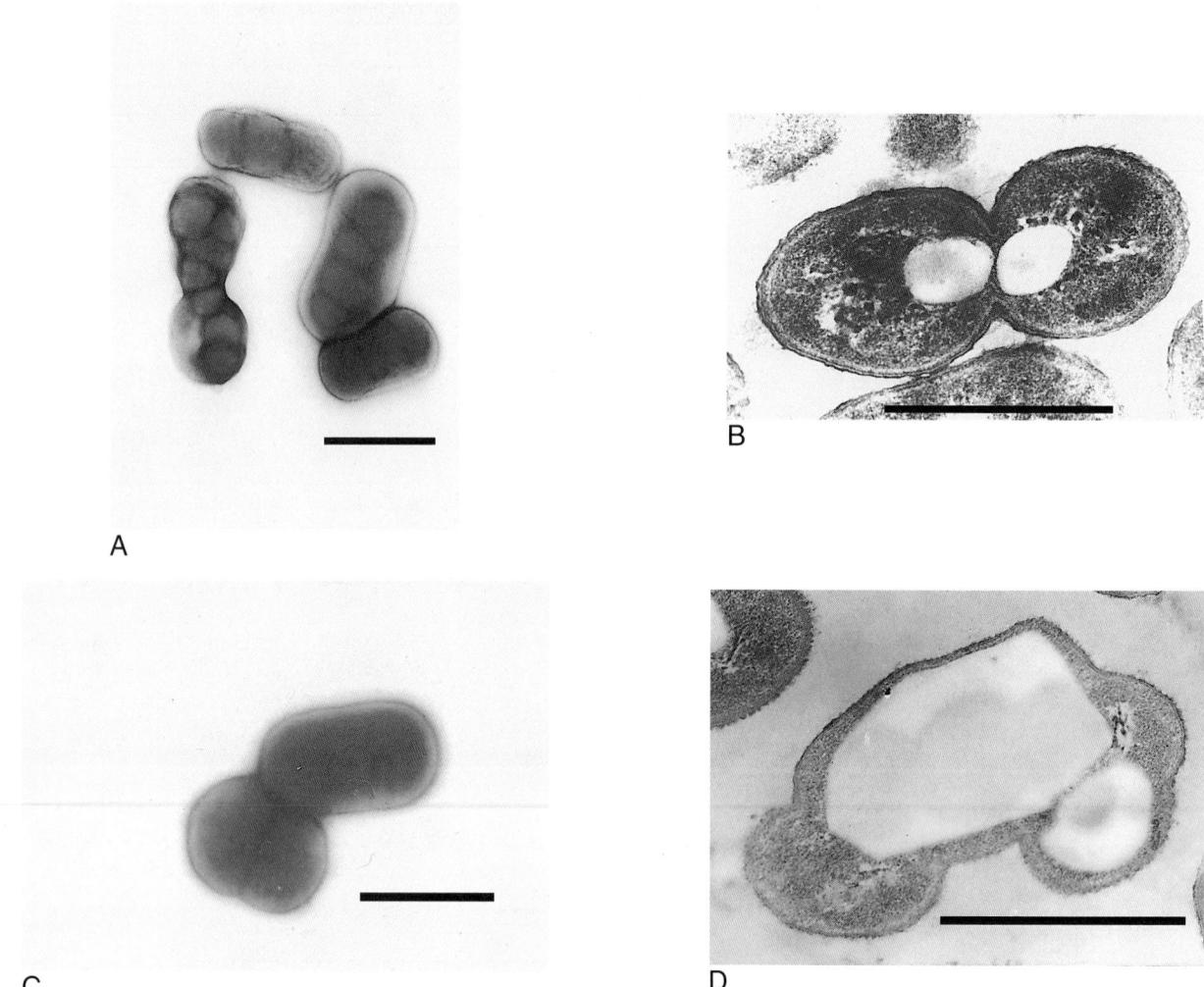

FIGURE BXII.α.77. *Methylarcula marina* (*A, B*) and *Methylarcula terricola* (*C, D*). *A* and *C*, negatively stained cells; *B* and *D*, ultrathin sections showing cell wall structure and granules of poly-β-hydroxybutyrate. Bars = 1 μm.

Genus XI. **Octadecabacter** *Gosink, Herwig and Staley 1998, 327*[VP] *(Effective publication: Gosink, Herwig and Staley 1997, 363)*

JOHN J. GOSINK

Oc.ta.dec' a.bac.ter. Gr. pref. *okto* eight; Gr. pref. *deca* ten; Gr. neut. n. *bakterion* a rod; M.L. masc. n. *Octadecabacter* an 18-carbon fatty acid–containing rod.

Straight or pleomorphic rods 0.6–0.8 × 1.6–4.8 μm. Nonmotile. All known strains produce gas vesicles. Gram negative. Aerobic or microaerophilic, possessing a strictly respiratory type of metabolism, with oxygen as the terminal electron acceptor. Do not reduce NO_3^- to NO_2^-. Do not grow anaerobically. Do not produce bacteriochlorophyll *a*. Form white, circular, convex, entire colonies on SWCm agar.[1] Growth occurs at pH 6.5–8.5 and at temperatures as low as 4°C in media having 17–70‰ salinity (4.8% NaCl). **Catalase positive. Oxidase negative.** Growth occurs on few (if any) organic compounds as sole carbon sources at 0.2% concentration. Growth occurs sparingly on selected carbon sources, including L-glutamate, glycerol, and mixed amino acids in dilute yeast extract. **Octadecenoic acid ($C_{18:1}$) is the major fatty acid.** Member of the class *Alphaproteobacteria.* Isolated from polar marine sea ice.

The mol% G + C of the DNA is: 56–57 (HPLC).

Type species: **Octadecabacter arcticus** Gosink, Herwig and

---

1. SWCm agar (Irgens et al., 1989) has the following composition (g per l of distilled water): NaCl, 12.0; MgSO₄·7H₂0, 7.0; MgCl₂·6H₂0, 5.2; CaCl₂·2H₂0, 1.1; KCl, 0.7; KH₂P0₄, 0.01; ferric citrate, 0.001; NH₄Cl, 0.4; yeast extract, 0.4; beef extract, 0.4; tryptone, 0.4; vitamins (mg per l of distilled water): pyridoxine·HCl, 10; calcium pantothentate, 5; nicotinamide, 5; *p*-aminobenzoic acid, 5; riboflavin, 5; thiamine·HCl, 5; biotin, 2; folic acid, 2; cyanocobalamine (B₁₂), 0.1), 10 ml; trace elements solution ((g per l of distilled water): H₃BO₃, 0.2; CaCl₂·2H₂O, 0.2;

ZnSO₄·7H₂O, 0.1; MnCl·4H₂O, 0.03; Na₂MoO₄·2H₂O, 0.03; NiCl₂·6H₂O, 0.02; CuCl₂·2H₂O, 0.01; adjust pH to 3.5.), 2.0 ml; agar, 15; adjust pH to 7.0 before sterilizing.

Staley 1998, 327 (Effective publication: Gosink, Herwig and Staley 1997, 363.)

FURTHER DESCRIPTIVE INFORMATION

Octadecabacter strains appear as rods or pleomorphic rods that form small, white, circular, convex, entire colonies when grown on SWCm agar plates at 4°C for several weeks. These strains do not produce bacteriochlorophyll *a* or any associated pigments.

All known *Octadecabacter* strains produce gas vesicles. The role of gas vesicles in this genus is unknown. Gas vesicles may serve to buoy cells in the water column up to the sea ice microbial community. Alternatively, gas vesicles may serve as a mechanism for cell dispersal during spring breakup of the sea ice.

Octadecabacter (and some related genera) are distinguished by a high level of octadecenoic acid: 70–80% of the total cellular fatty acids (Gosink and Staley, 1995). The location and *cis* or *trans* nature of the double bond in the octadecenoic acid (either $C_{18:1\ \omega7c}$, $C_{18:1\ \omega9t}$, or $C_{18:1\ \omega12t}$) of *Octadecabacter* has not been determined. The remainder of the fatty acids are mostly $C_{16:1\ \omega7c}$, $C_{16:0}$, and $C_{10:0\ 3OH}$ and several other fatty acids, each of which constitutes 3% or less of the total.

The strains grow on few (if any) organic compounds as sole carbon sources at 0.2% concentration. However, limited growth occurs when 0.2 g/l yeast extract is added to the basal medium. All of the polar gas vacuolate strains grow on L-glutamate, glycerol, and mixed amino acids in the presence of 0.2 g/l yeast extract.

Phylogenetically closely related taxa have been isolated from a number of marine environments around the world. The physiological capabilities and ecological roles of these bacteria are quite diverse (Sorokin, 1995; Bowman et al., 1997a; Gonzáles and Moran, 1997; Labrenz et al., 1999).

ENRICHMENT AND ISOLATION PROCEDURES

All currently known strains have been obtained from polar marine sea ice. Strains can be collected by melting ice fragments at room temperature several hours to overnight. The meltwater is then plated on suitable media and incubated at 4°C. *Octadecabacter* has only been cultivated from natural environments by plating on SWCm medium; however, media using several other carbon sources may also work. In general, only dilute carbon sources should be used. *Octadecabacter* colonies are usually only seen as small white colonies after several weeks of growth. The presence of gas vesicles gives the colonies a chalky opaque coloration. The occurrence of gas vesicles can be confirmed by phase and electron microscopy. *Octadecabacter* strains are readily distinguished from other gas vacuolate strains by their color, slow growth, and fatty acid composition.

MAINTENANCE PROCEDURES

Stocks are maintained for general work on SWCm at 4°C. Care must be taken when transferring or working with the strains to keep them and the growth media at 4°C. Cultures can be killed by leaving them out at room temperature for several hours or overnight. Long-term storage is best done by transferring cells to SWCm broth with 25% glycerol or 10% DMSO and freezing at −80°C.

DIFFERENTIATION OF THE GENUS *OCTADECABACTER* FROM OTHER GENERA

In terms of phenotypic characteristics, *Octadecabacter* can be differentiated from *Roseobacter* and *Sulfitobacter* primarily on the basis of fatty acid composition, lack of pigmentation, lack of bacteriochlorophyll *a*, and gas vesicles (Table BXII.α.64).

Octadecenoic acid ($C_{18:1}$) comprises a reproducibly high proportion (70–80%) of the total cellular fatty acids of *Octadecabacter* (Gosink and Staley, 1995). However, this fatty acid is found in relative abundance in both pigmented and nonpigmented, bacteriochlorophyll *a*–producing and nonbacteriochlorophyll *a*–producing, psychrotrophic, and mesophilic, acidophilic and nonacidophilic species of the *Alphaproteobacteria* (Fuerst et al., 1993; Kishimoto et al., 1995b).

The known members of *Octadecabacter*, unlike *Roseobacter*, do not produce bacteriochlorophyll *a* or pigments.

Another difference between *Octadecabacter* vs. *Roseobacter* and *Sulfitobacter* is the presence of gas vesicles. Phase contrast microscopy of cultures of *R. denitrificans* and *Ruegeria algicola* under different growth conditions at different stages of growth never reveals the presence of gas vesicles (Gosink et al., 1997). Likewise, there are no reports of closely related genera, including *Sulfitobacter*, that produce gas vesicles.

TABLE BXII.α.64. Features that differentiate *Octadecabacter* from *Roseobacter* and *Sulfitobacter*[a,b]

Characteristic	*Octadecabacter*	*Roseobacter denitrificans*	*Sulfitobacter pontiacus*
Motility	−	+	+
Poly-β-hydroxybutyrate granule production	−	−	+
Colony color	White	Brick red	Colorless
Bacteriochlorophyll *a*	−	+	
Growth at 19°C	−	+	+
Utilization as a carbon source:			
N-acetyl-glucosamine	D	−	−
Citrate	D	+	−
Fructose	−	+	+
Glycerol	D	−	+
Leucine	−	w	
Pyruvate, propionate	D	+	+
Ribose	D	+	−
Gelatin hydrolysis	−	+	−
Oxidase	−	+	+
NO₃⁻ reduction	−	+	
Mol % G + C of DNA	56–57	59.6	62.1

[a]Symbols: see standard definitions; w, weak reaction; blank space, not determined or not applicable.

[b]Data from Shiba, 1991a; Sorokin and Lysenko, 1993; Sorokin, 1995; Gosink et al., 1997.

TAXONOMIC COMMENTS

Eighteen strains of bacteria have been associated with the genus *Octadecabacter*. Of these, four strains (238, 307, 308, and 309) have been examined closely, and two strains (238 and 307) have been officially recognized (Gosink et al., 1997).

Phylogenetic analysis of the 16S rDNA of strains 238 and 307 shows that *Octadecabacter* is a member of the *Alphaproteobacteria* and that it is related to the genera *Roseobacter* (Shiba, 1991a; Gosink et al., 1997) and *Sulfitobacter* (Sorokin, 1995; Gonzáles and Moran, 1997). Several different methods were employed in analyzing the 16S rDNA sequences, including distance, parsimony, and likelihood methods, using various parameters for each type of analysis. Some of the trees obtained by these various methods differed from each other depending on which evolutionary framework (distance, parsimony, or likelihood) and what specific assumptions (transition/transversion ratio, weighting sets, step matrices, etc.) were applied. Regardless of the method employed, *Octadecabacter* is rooted deeply within the *Rhodobacter* group (Maidak et al., 1999) but not within the genera *Paracoccus* or *Rhodobacter*. The most likely tree, as evaluated under a likelihood framework by the Kishino–Hasegawa test (Kishino and Hasegawa, 1989), is shown in Fig. BXII.α.78. Other phylogenetic relationships can be seen in Bowman et al. (1997a), Gonzáles and Moran (1997), and Labrenz et al. (1999).

Octadecabacter arcticus and *R. denitrificans* show DNA–DNA hybridization values of 35% ± 9%. *Octadecabacter antarcticus* and *R. denitrificans* have DNA–DNA hybridization values of 42% ± 15%. These values are below 70%, the minimum value required to be considered different isolates of the same species (Wayne et al., 1987).

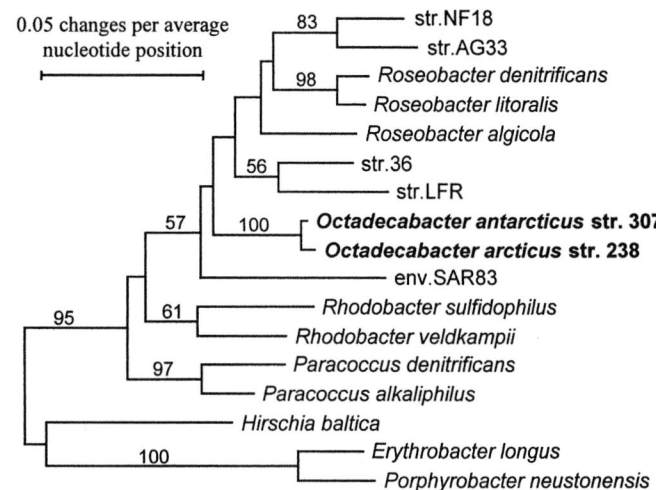

FIGURE BXII.α.78. Phylogenetic position of *Octadecabacter* in relation to *Roseobacter* and closely related species. This tree is depicted under a likelihood framework with a transition to transversion ratio of 1.15. The scale bar represents approximately 0.05 changes per average nucleotide position. Numbers above the branches indicate percent bootstrap support for that branch out of 100 likelihood bootstrap resamplings. (Reprinted with permission from J.J. Gosink, et al., Systematic and Applied Microbiology, 1997, *20:* 356–365.)

DIFFERENTIATION OF THE SPECIES OF THE GENUS *OCTADECABACTER*

Characteristics differentiating *Octadecabacter arcticus* from *Octadecabacter antarcticus* are listed in Table BXII.α.65. Other characteristics are listed in Table BXII.α.66.

TABLE BXII.α.65. Characteristics differentiating between the species of the genus *Octadecabacter* and two unnamed *Octadecabacter* strains[a]

Characteristic	*Octadecabacter arcticus* strain 238	*Octadecabacter antarcticus* strain. 307	*Octadecabacter* sp. strain 308	*Octadecabacter* sp. strain 309
Colony color	White	White	White	White
Growth at pH values of:				
8.5	w	+	+	+
9.5	−	+	+	+
Growth at 15°C	w	−	−	−
Growth in 4.8% NaCl	w	+	+	+
Growth with carbon sources:				
N-Acetyl-glucosamine	−	w	−	+
DL-Aspartate	w	−	−	+
Citrate, D-ribose	w	−	−	−
D-Glucose, succinate	w	w	−	w
L-Glutamate	w	+	w	+
Glycerol	+	w	+	+
Glycolic acid	w	w	−	+
Propionate	w	−	+	+
Pyruvate	w	−	w	w
Requirement for vitamins:				
Biotin	−	−[b]	+	+
Thiamine, nicotinic acid, pantothenic acid	+	−[b]	+	+

[a]Symbols: +, positive; −, negative; w, weakly positive.

[b]A cofactor or vitamin is required other than biotin, thiamine, nicotinic acid, or pantothenic acid.

List of species of the genus Octadecabacter

1. **Octadecabacter arcticus** Gosink, Herwig and Staley 1998, 327[VP] (Effective publication: Gosink, Herwig and Staley 1997, 363.)

arc' ti.cus. M.L. gen. *arcticus* of the arctic.

The characteristics are as described for the genus and listed in Tables BXII.α.65 and BXII.α.66, with the following additional information. Long irregular or pleomorphic rods from 0.6–0.8 × 2.4–4.0 μm. Known strains produce gas vesicles. The organisms require thiamine, nicotinic acid, and pantothenic acid for growth. Grow well on glycerol and

TABLE BXII.α.66. Other characteristics of the genus *Octadecabacter*[a]

Characteristic	Reaction or result
Growth at pH:	
5.5	−
6.5–7.6	+
Growth at a temperature of:	
4–10 °C	+
19°C or greater	−
Salinity growth range::	
0–0.24% NaCl	−
1.2–4.8% NaCl	+
Growth with carbon sources:	
Casamino acids	+
Acetate, butyrate, ethanol, D-fructose, L-leucine, methanol, L-proline, sucrose	−
Hydrolysis of gelatin, starch	−

[a]Symbols: +, positive; −, negative.

mixed amino acids only in the presence of small amounts of yeast extract. Grow at temperatures from 4–15°C. Predominant fatty acids when grown on Marine medium 2216 at 10°C are $C_{18:1\ \omega7c}$, $C_{18:1\ \omega9t}$, or $C_{18:1\ \omega12t}$ (75%), $C_{16:1\ \omega7c}$ (8%), $C_{16:0}$ (6%), and $C_{10:0\ 3OH}$ (4%).

The mol% G + C of the DNA is: 57 (HPLC).
Type strain: 238.
GenBank accession number (16S rRNA): U73725.

2. **Octadecabacter antarcticus** Gosink, Herwig and Staley 1998, 327[VP] (Effective publication: Gosink, Herwig and Staley 1997, 363.)

ant.arc' ti.cus. M.L. gen. *antarcticus* of Antarctica

The characteristics are as described for the genus and listed in Tables BXII.α.65 and BXII.α.66, with the following additional information. Long irregular or pleomorphic rods 0.6–0.8 × 1.6–4.8 μm. Known strains produce gas vesicles. Grow well on L-glutamate and mixed amino acids in the presence of small amounts of yeast extract. Do not grow on vitamin-free media supplemented with only biotin, thiamine, nicotinic acid, and pantothenic acid. Growth temperature range is 4–10°C. On Marine Medium 2216 at 10°C the predominant fatty acids are $C_{18:1\ \omega7c}$, $C_{18:1\ \omega9t}$, or $C_{18:1\ \omega12t}$ (77%), $C_{16:1\ \omega7c}$ (12%), $C_{16:0}$ (6%), and $C_{10:0\ 3OH}$ (2%).

The mol% G + C of the DNA is: 56 (HPLC).
Type strain: 307, CIP 106732.
GenBank accession number (16S rRNA): U14583.

Genus XII. **Paracoccus** Davis 1969, 384[AL] emend. Ludwig, Mittenhuber and Friedrich 1993, 366

ROB J.M. VAN SPANNING, ADRIAAN H. STOUTHAMER, SIMON C. BAKER AND HENK W. VAN VERSEVELD

Pa.ra.coc' cus. Gr. prep. *para* like, alongside of; Gr. n. *coccus* a grain, berry; M.L. masc. n. *Paracoccus* like a coccus.

Spherical cells 0.5–0.9 μm in diameter **or very short rod-shaped cells** 1.1–1.3 μm in diameter. **Occur singly, in pairs, or in small clusters.** Usually not encapsulated. **Gram negative.** Nonsporeforming and nonmotile. **Aerobic with a strictly respiratory metabolism.** *N*-oxides such as nitrate, nitrite, and nitrous oxide **can be used as terminal electron acceptors** for respiration under anaerobic conditions. **Nitrate is reduced to N_2** via nitrite, nitric oxide, and nitrous oxide (denitrification). Oxidase and catalase positive. **Chemoorganotrophic growth** occurs with a wide variety of organic compounds as carbon source. **Chemolithoautotrophic growth** occurs with CO_2 as carbon source, and H_2, methanol, methylamine, or thiosulfate as the electron-donating free-energy source. **Occur in water, soil, sewage, and sludge.**

The mol% G + C of the DNA is: 64–67.

Type species: **Paracoccus denitrificans** (Beijerinck and Minkman 1910) Davis 1969, 384 emend. Rainey, Kelly, Stackebrandt, Burghardt, Hiraishi, Kaayama and Wood 1999, 650 (*Micrococcus denitrificans* Beijerinck and Minkman 1910, 53.)

FURTHER DESCRIPTIVE INFORMATION

Paracoccus strains grow readily on ordinary media. No growth factors are required during heterotrophic growth. Colonies that are present on nutrient agar after 3 days incubation at 30°C are usually 2–4 mm in diameter, circular, smooth, low convex, moist,

and opaque. The cells do not produce carotenoid pigments. The cell wall is of type II or III (Walther-Mauruschat et al., 1977) and its composition is typical for Gram-negative bacteria. It has a thickness of 25–55 nm. No special structures have been observed in response to changes in growth conditions (Kocur et al., 1968; Sleytr and Kocur, 1973; Walther-Mauruschat et al., 1977; Nokhal and Mayer, 1979). During growth with an excess of carbon or energy source, intracellular granules of poly-β-hydroxybutyrate (PHB) are set aside as reserve for carbon and energy. PHB enhances the viability of resting cells. No extracellular hydrolysis of PHB occurs.

Quinones The presence of ubiquinone-10 is a very characteristic feature of all bacteria belonging to this group (for references see Stouthamer, 1992). In general, ubiquinone-8 is characteristic for the *Betaproteobacteria* and ubiquinone-9 for the *Gammaproteobacteria*. The nature of the respiratory quinone is therefore an important chemotaxonomic trait.

Fatty acid analyses Fatty acid methyl ester analyses of members of the genus have shown that strain LMD 22.21 and relatives cluster together, while *P. denitrificans* GB17 and strain LMD 52.44 are well separated (Baker et al., 1995). Approximately 80% of the total fatty acids is oleic acid (Wilkinson et al., 1972). The predominant phospholipid is phosphatidylglycerol (52%) The remaining phospholipids are mostly phosphatidylcholine (31%), phosphatidylethanolamine (5.8%), and cardiolipin (3.2%). Lipid A analyses confirm not only the present taxa of the purple nonsulfur bacteria (formerly *Rhodospirillaceae*), but also the phylogenetic relatedness of distinctly phototrophic to distinctly non-phototrophic bacteria, as was suggested by cataloging of 16S rRNA. For example, lipid A in lipopolysaccharides with ester-bound $C_{10:0\ 3OH}$ and the rare amide-linked $C_{14:0\ 3\ OXO}$ is common to the phototrophic *Rhodobacter capsulatus* and *Rhodobacter sphaeroides* and also to *P. denitrificans* and *P. versutus* (Weckesser and Mayer, 1988).

Metabolism   Members of *Paracoccus* are versatile organisms that can adapt their metabolism to the prevailing environmental conditions (Stouthamer, 1980; Van Verseveld and Stouthamer, 1992; Van Spanning et al., 1995). They are nonfermentative, and they do not produce extracellular xenobiotics or metabolites. A wide variety of organic compounds may serve as sole carbon and free-energy sources during heterotrophic growth, including D-glucose, D-fructose, sucrose, trehalose, mannose, succinate, mannitol, choline, acetate, propionate, lactate, pyruvate, malonate, butyrate, malate, citrate, gluconate, *p*-hydroxybenzoate, histidine, leucine, proline, serine, asparagine, glutamine, ethanol, propanol, butanol, glycerol, and sorbitol (Vogt, 1965; Davis et al., 1969; Pichinoty et al., 1977c). No growth occurs on xylose, rhamnose, starch, gelatin, glycogen, lactose, cellulose, benzoate, phenol, tryptophan, arginine, threonine, ethanolamine, or urea. All enzymes of the tricarboxylic acid cycle are present in *Paracoccus*. Glucose is metabolized by the Entner–Douderoff pathway, by the hexose monophosphate pathway (Forget and Pichinoty, 1965), or by a combination of the two. The glycolytic pathway is absent (Slabas and Whatley, 1977).

During autotrophic growth, H_2, methanol, methylamine, carbon disulfide (Jordan et al., 1997), or thiosulfate (Friedrich and Mitrenga, 1981) can be used as free-energy sources. Oxidation of methanol is catalyzed by methanol dehydrogenase, a periplasmic enzyme encoded by the *mxa* gene cluster. The enzyme contains pyrroloquinoline quinone (PQQ) as a prosthetic group. Methylamine is oxidized by methylamine dehydrogenase, a periplasmic enzyme with a tryptophan tryptophyl quinone (TTQ) as prosthetic group. The enzyme is encoded by the *mau* gene cluster. The oxidation product of methanol and methylamine is formaldehyde, which is further oxidized via formate to CO_2 in the cytoplasm by formaldehyde and formate dehydrogenases. *Paracoccus* uses the Calvin cycle to fix CO_2 during autotrophic growth (Kornberg et al., 1960), and during carbon-limited mixed autotrophic/heterotrophic growth (Van Verseveld et al., 1979). Population analysis in a denitrifying methanol-fed sand filter in a municipal wastewater treatment plant revealed that 3.5% of the total cell counts corresponded to a *Paracoccus* sp., whereas in a parallel sand filter without methanol feed and no detectable denitrification only very few *Paracoccus* spp. could be detected (Neef et al., 1996). These observations indicate that autotrophic denitrifying conditions could provide the natural environmental niche for *P. denitrificans*, even though it has been difficult to grow axenic cultures under these conditions. *P. denitrificans* GB17 (formerly *Thiosphaera pantotropha*) differs from other strains by lacking the capacity to grow with methanol. However, methanol-utilizing mutants of GB17 have been isolated (Egert et al., 1993; Moir and Ferguson, 1993; Ras et al., 1995). In these GB17 mutants, methanol is oxidized by an altered ethanol dehydrogenase (Ras et al., 1995).

Two major pathways for the oxidation of thiosulfate exist in microorganisms (Kelly et al., 1997; Friedrich, 1998). In the first pathway, called the *Paracoccus* sulfur oxidation pathway, tetrathionate is not an intermediate. In the other pathway, tetrathionate is a free intermediate in the oxidation process. The first pathway is present in *P. versutus* and *P. denitrificans*. *Rhodobacter capsulatus* and *Rhodobacter sphaeroides* cannot oxidize thiosulfate (Imhoff and Trüper, 1992). Other *Rhodobacter* spp. and *Rhodopseudomonas palustris* can oxidize thiosulfate and possibly use the same type of enzyme as *P. denitrificans*. Recently the gene for sulfite dehydrogenase was cloned from *P. denitrificans* GB17 and was shown to be essential for thiosulfate oxidation (Wodara et al., 1997).

P. denitrificans is sensitive to potassium cyanide. Three cytochrome oxidases, ba_3, cbb_3, and aa_3 (de Gier et al., 1994) have been detected in *P. denitrificans*. The reduction of oxygen is coupled to the translocation of protons in the aa_3-type (van Verseveld et al., 1981) and ba_3-type (Puustinen et al., 1989; Richter et al., 1994) oxidases. The proton-pumping capacity of the cbb_3-type is still a matter of debate (de Gier et al., 1994; Raitio and Wikström, 1994).

Gases other than CO_2 produced by *Paracoccus* are nitric oxide, nitrous oxide, and dinitrogen during denitrification.

In the absence of oxygen, nitrate can serve as terminal electron acceptor for respiration. Nitrate is then reduced via nitrite, nitric oxide, and nitrous oxide, by the corresponding reductases (Boogerd et al., 1981; Timkovich et al., 1982; Snyder and Hollocher, 1987).

P. denitrificans is able to express two different types of nitrate reductase. One is membrane bound, encoded by the *narGHJI* gene cluster, and expressed during the shift from aerobic to anaerobic growth. The enzyme has its catalytic center facing the cytoplasm and has the ability to reduce chlorate as well. *In vivo*, both methyl and benzylviologens may act as electron donors to this enzyme. The other nitrate reductase is located in the periplasm, and is encoded by the *nap* operon. It is expressed independently of the oxygen concentration. This enzyme is unable to reduce chlorate. Both nitrate reductases carry molybdenum in their catalytic pocket.

Nitrite reductase is a periplasmic enzyme of the cd_1-type (Timkovich et al., 1982). It consists of two identical subunits of about 63 kDa, each subunit harboring one heme c and one heme d_1 (Moir et al., 1993). The X-ray structure shows that the two types of heme are located in separate domains (Fulop et al., 1993).

Nitric oxide reductase is a membrane-bound enzyme consisting of two subunits. The prosthetic groups are heme c, found in the small subunit, and two hemes b and a non-heme iron, found in the large subunit (Girsch and De Vries, 1997). With respect to its architecture and putative metal ligands, the latter subunit resembles subunit I of terminal heme-copper oxidases (Saraste and Castresana, 1994; Van der Oost et al., 1994).

Nitrous oxide reductase is a copper-containing dimeric enzyme found in the periplasm (Snyder and Hollocher, 1987). Studies on its counterpart in *Pseudomonas stutzeri* suggest that each monomer contains approximately four copper ions. Two of these form a binuclear copper center structurally similar to that of Cu_A in cytochrome c oxidases (Farrar et al., 1994).

The optimal pH for denitrification is between 7.0 and 8.5 and the optimal temperature is between 30 and 37°C. The rate of denitrification can be enhanced by adding a small amount of yeast extract.

P. denitrificans is susceptible to ampicillin, tetracycline, streptomycin, kanamycin, erythromycin, chloramphenicol, rifampicin, and gentamicin. *P. denitrificans* is resistant to trimethoprim. Unable to fix nitrogen. There is no antigenic scheme available for *Paracoccus* species.

Paracoccus can use ammonium or nitrate as nitrogen sources. During oxidation of methylamine, the cell can use the released ammonia as nitrogen source. Two pathways serve for assimilation of ammonia in *P. denitrificans*, one with glutamate dehydrogenase operating at high ammonia concentrations, the other with glutamate-ammonia ligase and glutamate synthase at low concentrations (Mikes et al., 1991).

The distribution of metabolic systems among various selected species of the *Alphaproteobacteria* has been reviewed by Stouthamer (1992). Table BXII.α.67 is an update of the results from this publication. The division of the *Alphaproteobacteria* into subgroups is as given by Stouthamer (1992). When a property occurs only in some of the strains of a species, this is indicated in Table BXII.α.67 as "+". It is evident from Table BXII.α.67 that several pathways are very common in these bacteria. Denitrification, nitrogen fixation, hydrogenase formation, autotrophic CO_2 fixation by ribulose bisphosphate carboxylase, and methylotrophy are widespread, but others are very characteristic for but not limited to *Paracoccus*, such as heterotrophic nitrification, aerobic denitrification, and thiosulfate oxidation.

The physiological differences between *P. denitrificans* and *P. pantotrophus* are less fundamental than previously thought (Stouthamer et al., 1997). Heterotrophic nitrification was originally thought to be present only in strain GB17 but was later also detected in other *P. denitrificans* strains (Crossman et al., 1995). The key enzyme in this process, ammonia monooxygenase (AMO), has recently been purified from *P. denitrificans* (Moir et al., 1996a). The enzyme has a number of similarities with the AMO of the chemolithotrophic bacterium *Nitrosomonas europaea*. The hydroxylamine oxidase of *P. denitrificans* is completely different from that of *N. europaea* (Wehrfritz et al., 1993; Moir et al., 1996b) and donates electrons to the cytochrome bc_1 complex. The process of heterotrophic nitrification is thus a process that differs in a large number of aspects from chemolithotrophic nitrification.

Recently the influence of carbon source on aerobic denitrification has been studied with chemostat cultures of *P. denitrificans*. The occurrence of aerobic denitrification was found to be strongly stimulated by growth with reduced substrates, e.g., bu-

TABLE BXII.α.67. Characteristics differentiating two denitrifying species of the genus *Paracoccus* from selected species of other genera of the class *Alphaproteobacteria* of the Phylum *Proteobacteria*[a,b]

Characteristic	*Paracoccus denitrificans*	*Paracoccus versutus*	*Azospirillum lipoferum*	*Bradyrhizobium japonicum*	*Bradyrhizobium strain BTAi1*	*Erythrobacter longus*	*Methylobacterium extorquens*	*Methylobacterium organophilum*	*Methylosinus trichosporium*	*Nitrobacter winogradskyi*	*Oligotropha carboxidovorans*	*Rhizobium leguminosarum*	*Rhodobacter capsulatum*	*Rhodobacter sphaeroides*	*Rhodopseudomonas acidophila*	*Rhodopseudomonas palustris*	*Roseobacter denitrificans*	*Xanthobacter tagetidis*
Order:[c]																		
Rhizobiales				+	+		+	+	+	+	+	+			+	+		+
Rhodospirillales			+															
Rhodobacterales	+	+											+	+			+	
Sphingomonadales						+												
Family:																		
Bradyrhizobiaceae				+	+					+	+				+	+	+	
Hyphomicrobiaceae																		+
Methylobacteriaceae							+	+										
Methylocystaceae									+									
Rhizobiaceae												+						
Rhodobacteraceae	+	+											+	+				
Rhodospirillaceae			+															
Sphingomonadaceae						+												
Denitrification	+	+	+	+		+	d	−[d]	d	+	−	+	b[e]	+	−	+	+	d
Photosynthesis	−	−	−	−	−	−	−	−	−	−	−	−	+	+	+	+	−	−
Aerobic photosynthesis	−	−	−	−	+	+	+	−	−	−	−	−	−	−	−	−	+	−
Methylotrophy	+	+	+	+		−	+	+		−	+		+		+			+
Hydrogenase	+	−	+	+		−				+	+		+	+	+	+		+
Ribulose bisphosphate carboxylase	+	+	+	+						+	+		+	+	+	+	−	+
Serine pathway	−	−				−	+	+	+									
Methanotrophy[f]	−	−	+	−	−	−	−	−	+									
N$_2$ fixation	−	+	+	+	+		−		+			+	+		+	+		+
Heterotrophic nitrification	+		c				−		c	−								
Thiosulfate oxidation	+	+											−	−				+

[a]For symbols see standard definitions.

[b]Data from Stouthamer (1992).

[c]See taxonomic outline.

[d]Some methylobacteria reduce nitrate to nitrite.

[e]*R. capsulatum* is able to reduce nitrate to nitrite and nitrous oxide to nitrogen.

[f]All methanotrophs also oxidize ammonia to nitrite.

tyrate (Sears et al., 1997), suggesting that the process is used to increase the capacity for electron transfer.

Genetic aspects The molecular genetics of *P. denitrificans* have been studied in greater detail in the last decade and these efforts have been summarized in detail in Baker et al. (1998).

The *P. denitrificans* Pd1222 genome harbors at least five copies of an insertion sequence IS1248. In addition, two copies of an element referred to as an integration region have been reported (Van Spanning et al., 1995).

In three strains of *P. denitrificans*, DSM 65, DSM 413, and DSM 415, large plasmids of molecular mass greater than 300×10^6 Da were resolved (Gerstenberg et al., 1982). *P. denitrificans* DSM 413 has three chromosomes of 1.83, 1.16, and 0.67 Mbp (Winterstein and Ludwig, 1998). The two smaller chromosomes were suggested to be linear molecules. Essential genes coding for respiratory enzymes were found distributed over all three chromosomes without any obvious clustering on any of the three molecules. The total genomic size of *P. denitrificans* of 3.66 Mbp is in agreement with an earlier estimate using a different approach (Nokhal and Schlegel, 1983). Mutations have been introduced randomly by chemical mutagenesis methods (Harms et al., 1996) or by locus-specific gene exchange techniques (Van Spanning et al., 1991). Strain Pd1222, a derivative of DSM 413, was selected from the results of a mutagenesis approach for enhanced frequencies of conjugation. This mutant had defects in two restriction modification systems and is widely used as host in gene cloning and transposon mutagenesis experiments.

Enrichment and Isolation Procedures

The majority of *Paracoccus* strains can be enriched by culturing soil or mud samples in media containing CO_2 as carbon source and hydrogen, methanol, or methylamine as the free-energy source. The preferred electron acceptor under these growth conditions is oxygen. With nitrate, growth is very slow or absent (Vogt, 1965). Alternatively, *Paracoccus* can be enriched in anaerobic media supplemented with a heterotrophic carbon and energy source and nitrate as terminal electron acceptor, or in liquid minimal medium containing tartrate or succinate incubated under a nitrous oxide atmosphere (Pichinoty et al., 1977a, b, c). These approaches may not always be successful for the isolation of *P. denitrificans* because it is not one of the major denitrifiers in all types of soil (Gamble et al., 1977). *P. denitrificans* GB-17 can be isolated from wastewater and sulfide-rich ecosystems.

Maintenance Procedures

Stock cultures of *Paracoccus* strains can be maintained at room temperature in any rich agar-containing medium with Brain Heart Infusion (BHI, Difco) or Nutrient Broth (Difco) and 0.3% agar, pH 7.0. The cultures remain viable up to a year without subculturing if they are sealed with a rubber stopper or a cork that has been soaked in hot paraffin wax. Cultures survive for at least five years when frozen at $-70°C$ in media to which 10–20% (w/v) glycerol has been added. Strains may also be preserved indefinitely by lyophilization.

Procedures for Testing Special Characters

In general, cultivation of *Paracoccus denitrificans* is possible in any size of bottle or glass fermenter. The organism has been studied exhaustively during continuous cultivation in the chemostat and the retentostat (chemostat with 100% biomass retention) (Jones et al., 1977; Meijer et al., 1977; Van Verseveld and Stouthamer, 1978; Van Verseveld et al., 1979, 1986). A medium widely used for large-scale cultivation is that described by Chang and Morris (1962). Sometimes it is necessary to add a trace elements solution (Boogerd et al., 1983). During autotrophic growth, 0.01% yeast extract and 0.05% sodium bicarbonate are added to initiate growth (Van Verseveld and Stouthamer, 1978).

Differentiation of the Genus *Paracoccus* from Other Genera

Table BXII.α.67 indicates the differential characteristics of *Paracoccus*.

Taxonomic Comments

The type strain *Paracoccus denitrificans* ATCC 17741 (Davis et al., 1969), formerly designated *Micrococcus denitrificans*, was first isolated by Beijerinck (Beijerinck and Minkman, 1910; Nokhal and Schlegel, 1980). The following species have been regarded as close relatives of *Paracoccus denitrificans* and are placed in the same genus on the basis of 16S rRNA gene sequence analyses and DNA–DNA hybridization: *P. pantotrophus* (formerly designated *Thiosphaera pantotropha*) and *P. versutus* (formerly designated *Thiobacillus versutus*; Katayama et al., 1995). The recently isolated species *P. kocurii* (Ohara et al., 1990), *P. alcaliphilus* (Urakami et al., 1989b), *P. aminophilus* (Urakami et al., 1990b), *P. aminovorans* (Urakami et al., 1990b), *P. thiocyanatus* (Katayama et al., 1995), *P. solventivorans* (Siller et al., 1996), *P. marcusii* (Harker et al., 1998), *P. alkenifer* (Lipski et al., 1998), and *P. carotinifaciens* (Tsubokura et al., 1999) have also been placed in the genus *Paracoccus*. *P. halodenitrificans* has been moved from the genus *Paracoccus* into the genus *Halomonas* (Dobson and Franzmann, 1996).

The *c*-type cytochrome and protein profiles were compared for a number of cultures of *P. denitrificans* obtained from a range of culture collections. The cultures fell into two groups corresponding to the two original isolates of this bacterial species. Members of one group, which included NCIMB 8944, ATCC 13543, ATCC 17741, ATCC 19367, Pd 1222, and DSM 413, were similar or identical to LMD 22.21. The second group, including DSM 65 and LMG 4218, were similar or identical to LMD 52.44 (Goodhew et al., 1996). These groupings were not compatible with the recorded history of culture deposition. Mass spectrometry and amino acid sequence comparisons showed that the cytochrome c_{550} of the LMD 52.44 culture group differed by 16% from that of the LMD 22.21 group, and yet there was only 1% difference from the cytochrome c_{550} of *Thiosphaera pantotropha*. These results suggest that consideration should be given to creation of a new species of *P. pantotrophus*, which would include *Thiosphaera pantotropha* and *P. denitrificans* LMD 52.44. At the moment, *P. denitrificans* and *T. pantotropha* are combined in the *Paracoccus* genus (Ludwig et al., 1993), since the nucleotide sequence of the 16S rRNA genes of these latter two organisms was found to be the same.

Additional strains have been isolated in independent studies. Eleven strains were isolated by (Nokhal and Schlegel, 1983) and analyzed based on 235 characteristics. Together with DSM 413 and DSM 415, these strains were placed into four subgroups with similarities ranging from 74 to 88%. This grouping was in accordance with that deduced from DNA–DNA reassociation kinetics (Auling et al., 1980).

The genus *Paracoccus* belongs to the α-subdivision of nonsulfur purple bacteria, the so-called *Alphaproteobacteria* (Woese, 1987), and encompasses 13 defined species that have been included in the Validation Lists. Members of the genus *Paracoccus* fall in the

Rhodobacter group of the *Alphaproteobacteria*. Nearest neighbors are *Rhodobacter*, *Amaricoccus*, *Octadecabacter*, *Rhodovulum*, *Roseobacter*, *Sagittula* and "*Tetracoccus*". They share a number of properties with *Paracoccus*, including the ability to grow anaerobically with nitrate. The resemblance of the respiratory networks is striking, and most components of the pathways show structural homology to one another. Analyses of 23S and 16S rRNA gene sequences indicated a close relationship to *Rhodobacter capsulatus* (Fox et al., 1980), suggesting that this organism and *Paracoccus denitrificans* arose from a common ancestor, after which it lost the genes encoding the photosynthetic apparatus. In addition, the amino acid sequence of cytochrome c_{550} from *P. denitrificans* most closely resembles that of cytochrome c_2 from *R. sphaeroides* and *R. capsulatus* (Dickerson, 1980a).

DIFFERENTIATION OF THE SPECIES OF THE GENUS *PARACOCCUS*

The differential and other characteristics of the species of *Paracoccus* are indicated in Table BXII.α.68.

TABLE BXII.α.68. Differential features of the species of the genus *Paracoccus*[a,b]

Characteristic	P. denitrificans	P. alcaliphilus	P. alkenifer	P. aminophilus	P. aminovorans	P. carotinifaciens	P. kocurii	P. marcusii	P. methylutens	P. pantotrophus	P. thiocyanatus	P. versutus	P. solventivorans
Motility	−	−	−	−	−	+	−	−	−	−	−	+	−
Nitrate respiration	+	−	+	−	−	−	+	−	+	+	+	+	−
Vitamin requirement:													
Biotin	−	+	−	−	−	−	−	−	−	−	−	−	−
Thiamin	−	−	−	+	+	−	+	−	−	−	+	−	−
None	+	−	+	−	−	+	−	+	−[c]	+	−	+	+
Energy sources:													
Methanol	+	+	+	−	−		−	−	+	−	−	−	−
Formate	+						+	+		−		+	
Ethanol	+	+	+		−				+			+	−
Butanol	+			−	−						+		
Methylamine	+	+	−	+	+		+	−	+	−		+	
Dimethylamine	+	−	−	+	+		+	−	−				−
Trimethylamine	+	−	−	+	+		+	−	−				
Glycerol	+	+		+	+	+	−	+	+		−	+	−
Glucose	+	−		+	+	+	−	+	+	+	+	+	
Fructose	+	+		−	+			+	+	+	+	+	
Galactose	+	+	−	+	+	+		+		−	+	+	
Ribose	−		−	+	+		+		+		+		+
Maltose	+	−	−	−	−	+	−	+	+	+	−	+	
Sucrose	+	−	−	−	−	+	−	+	+		−	+	−
Lactose		−	−	−	−	+						−	
Mannitol	+	+	−		+	+	−	+	+	+	+	+	−
Inositol	+	+		−	−		−	+	+				
Xylose	+	+	−	+	−	+		−	−				
Mannose		+	−	−	+	+		+	+	+			
Sorbitol	+	+	−	−	+	+	−	+	+	+			
Arabinose	+	+	−	+	+	+	−	+	+	−	+	+	
Trehalose	+	−	−		−	+	−	+	−				
Benzoate	−				−					+	+		−
Gluconate	+		−	+		+		+		+	+	+	+
Acetate	+	+		+	+			−	+	+	+	+	
Lactate								+		+	+	+	
Citrate	−	+		−	−	−		+	+	+	−		
Succinate	+	+		+	+	−		+	+	+	+	+	+
Malate	+					+		−	+	+	−	+	
Pyruvate	+		+		+		+		+	+	+	+	+
Trimethyl-*N*-oxide	−			+	+	+	+						
Formamide		−	−	+	+								
N-methylformamide		−	−	+	+								
N,N-dimethylformamide		−	−	+	+								
Thiosulfate	−				+		−		−		+		+
Mol% G + C of DNA	66.5	64.6		63.8	66.8	67	70.2	66	67	66	66.5	66.8	68.5

[a]For symbols see standard definitions.

[b]Recently, three additional species of *Paracoccus* have been described: *Paracoccus kondratievae* (Doronina and Trotsenko, 2000, 2001a), *Paracoccus seriniphilus* (Pukall et al., 2003), and *Paracoccus zeaxanthinifaciens* (Berry et al., 2003).

[c]Requires vitamin B_{12} (Doronina et al., 1998b).

List of species of the genus Paracoccus

1. **Paracoccus denitrificans** (Beijerinck and Minkman 1910) Davis 1969, 384 emend. Rainey, Kelly, Stackebrandt, Burghardt, Hiraishi, Kaayama and Wood 1999, 650 (*Micrococcus denitrificans* Beijerinck and Minkman 1910, 53.)

de.ni.tri' fi.cans. L. prep. *de* away from; L. n. *nitrum* soda; M.L. *nitrum* nitrate; M.L. v. *denitrifico* denitrify; M.L. part. adj. *denitrificans* denitrifying.

The description is as given for the genus and in Table BXII.α.68, with the following additional features. Cells are rod shaped, 0.4–0.5 × 1.1–1.7 μm, occurring singly or in pairs. Sometimes motile with a tuft of flagella. The optimal growth temperature is 30–37°C; the optimal pH for growth is 7.5–8. Electron acceptors are oxygen and nitrate. N_2 is produced from nitrate from respiration. S^0 is oxidized slowly. Tetrathionate and thiocyanate do not support growth. In addition to those mentioned in Table BXII.α.68, glycine, L-alanine, L-valine, L-leucine, L-isoleucine, L-threonine, L-cysteine, L-phenylalanine, L-tyrosine, L-histidine, L-aspartate, L-asparagine, L-glutamate, L-glutamine, L-lysine, L-arginine, L-ornithine, L-cellobiose, adipate, tartrate, sarcosine, creatine, *m*-hydroxybenzoate, *p*-hydroxybenzoate, glycerate, formaldehyde, lactate, and butyrate can serve as energy and carbon sources, but cyclohexanol, L-methionine, L-tryptophan, *p*-aminobenzoate, and DL-mandelate do not. Some strains can grow with carbon disulfide as sole source of carbon and energy. Some if not all strains contain megaplasmids of at least 450 kb in size. For other variations between strains, see Nokhal and Schlegel (1983), Urakami et al. (1989b), and Rainey et al. (1999).

The mol% G + C of the DNA is: 64–67 (T_m).

Type strain: ATCC 17741, DSM 65, DSM 413, LMD 22.21, LMG 4218.

GenBank accession number (16S rRNA): Y16927, Y16935, D13480, Y16928, X69159.

2. **Paracoccus alcaliphilus** Urakami, Tamaoka, Suzuki and Komagata 1989b, 118[VP]

al.ca.li.phi' lus. M.L. n. *alcali* from Arabic *al* end; *qaliy* soda ash; Gr. adj. *philum* loving; M.L. adj. *alcaliphilus* liking alkaline media.

The description is as given for the genus and in Table BXII.α.68, with the following additional features. Cells are coccoid or rod like, 0.5–0.9 × 0.9–2.0 μm occurring singly, in pairs, or in clusters. Light-scattering PHB inclusions are evident. The preferred electron acceptor is oxygen. Nitrite is produced from nitrate, but nitrate respiration is not performed. See Table BXII.α.68 for use of energy sources. Biotin is required for growth on minimal media. Optimal growth temperature, 30°C ; optimal pH for growth, 8–9. NaCl is inhibitory at concentrations >3%. Isolated from wastewater (activated sludge).

The mol% G + C of the DNA is: 64–66 (T_m).

Type strain: TK 1015, ATCC 51199, DSM 8512, JCM 7364.

GenBank accession number (16S rRNA): D32238.

3. **Paracoccus alkenifer** Lipski, Reichert, Reuter, Spröer and Altendorf 1998, 535[VP]

al.ke' ni.fer. M.L. n. *alkenum* unsaturated hydrocarbons; L. suff. *-fer* carrying; M.L. adj. *alkenifer* referring to the occurrence of unusual monounsaturated fatty acids in whole-cell hydrolysates.

The description is as given for the genus and in Table BXII.α.68, with the following additional features. Cells are coccoid or rod like, 0.4–0.6 × 0.9–1.7 μm. Nitrate is reduced to N_2. See Table BXII.α.68 for use of energy sources. Carbohydrates and amines are not utilized. Lactate, butyrate, hydroxybutyrate, asparagine, and acetone are utilized. Optimal growth temperature, 30°C; optimal pH for growth, 6–9. Isolated from biofilters used in treatment of waste gas from an animal rendering plant.

The mol% G + C of the DNA is: not reported.

Type strain: A901/1, DSM 11593.

GenBank accession number (16S rRNA): Y13827.

4. **Paracoccus aminophilus** Urakami, Araki, Oyanagi, Suzuki and Komagata 1990b, 289[VP]

a.mi.no' phi.lus. M.L. n. *aminum* amine; Gr. adj. *phila* loving; M.L. adj. *aminophilus* amine loving.

The description is as given for the genus and in Table BXII.α.68, with the following additional features. Cells are coccoid or rod like, 0.5–0.9 × 0.9–2.0 μm, occurring singly, in pairs, or in clusters. Light-scattering PHB inclusions are evident. The preferred electron acceptor is oxygen. Nitrite is produced from nitrate, but nitrate respiration is not performed. See Table BXII.α.68 for use of energy sources. Thiamine is required for growth on minimal media. NaCl is inhibitory at concentrations >3%. Optimal growth temperature, 30°C; optimal pH for growth, 6.5–8.0. Isolated from soil.

The mol% G + C of the DNA is: 68.5 (HPLC) or 70 (T_m).

Type strain: DM-15, ATCC 49673, DSM 8538, JCM 7686.

GenBank accession number (16S rRNA): D32239.

5. **Paracoccus aminovorans** Urakami, Araki, Oyanagi, Suzuki and Komagata 1990b, 289[VP]

a.mi.no' vo.rans. L. n. *aminum* amine; L. v. *voro* to eat, devour; L. part. adj. *vorans* eating; *aminovorans* amine eating.

The description is as given for the genus and in Table BXII.α.68, with the following additional features. Cells are coccoid or rod like, 0.5–0.9 × 0.9–2.0 μm, occurring singly, in pairs, or in clusters. Light-scattering PHB inclusions are evident. The preferred electron acceptor is oxygen. Nitrite is produced from nitrate, but nitrate respiration is not performed. See Table BXII.α.68 for use of energy sources. Thiamine is required for growth on minimal media. NaCl is inhibitory at concentrations >3%. Optimal growth temperature is 30–37°C; optimal pH for growth, 7–8. Isolated from soil.

The mol% G + C of the DNA is: 67-68 (T_m).

Type strain: DM-82, ATCC 49632, DSM 8537, JCM 7685.

GenBank accession number (16S rRNA): D32240.

6. **Paracoccus carotinifaciens** Tsubokura, Yoneda and Mizuta 1999, 281[VP]

ca.ro.ti.ni.fa' ci.ens. M.L. neut. n. *carotinum* carotene; L. part. adj. *faciens* making, producing; M.L. part. adj. *carotinifaciens* carotene/carotenoid-producing.

The description is as given for the genus and in Table BXII.α.68, with the following additional features. Cells are rod shaped, 0.3–1.0 × 1.0–5.0 μm. Motile by means of peritrichous flagella. Cells are elongated or swollen at the early stage of growth with 1% yeast extract present in liquid medium. Nitrate is not reduced. Produces astaxanthin. Colonies are orange to red. Optimal growth temperature, 25–30°C, no growth at 37°C; optimal pH for growth, 8–9. Isolated from soil.

The mol% G + C of the DNA is: 67 (T_m).

Type strain: E-396, IFO 16121.

7. **Paracoccus kocurii** Ohara, Katayama, Tsuzaki, Nakamoto and Kuraishi 1990, 293[VP]

ko.cu'ri.i. M.L gen. n. *kocurii* of Kocur; named after Miloslav Kocur, Czechoslovakian bacteriologist.

The description is as given for the genus and in Table BXII.α.68, with the following additional features. Cells are rod shaped, 0.5–0.8 × 0.7–1.1 μm, occurring singly or in pairs. Light-scattering PHB inclusions are evident. The preferred electron acceptor is oxygen. Nitrogen is produced from nitrate. In addition to those mentioned in Table BXII.α.68, *n*-butyrate and lactate can serve as carbon and energy sources, but cyclohexanol, glycine, L-alanine, L-valine, L-leucine, L-isoleucine, L-threonine, L-cysteine, L-methionine, L-phenylalanine, L-tyrosine, L-histidine, L-tryptophan, L-aspartate, L-asparagine, L-glutamate, L-glutamine, L-lysine, L-arginine, L-ornithine, L-cellobiose, adipate, tartrate, *p*-hydroxybenzoate, *p*-aminobenzoate, and DL-mandelate do not. Thiamine is required for growth on minimal media. NaCl is inhibitory in concentrations at or above 3%. Optimal growth temperature, 25–30°C; optimal pH for growth, 6.6–8.2. Isolated from wastewater (activated sludge).

The mol% G + C of the DNA is: 71 (T_m).

Type strain: B, ATCC 49631, DSM 8536, JCM 7684.

GenBank accession number (16S rRNA): D32241.

8. **Paracoccus marcusii** Harker, Hirschberg and Oren 1998, 547[VP]

mar.cu'si.i. N.L. adj. *marcusii* referring to the late Menashe Marcus, an Israeli geneticist.

The description is as given for the genus and in Table BXII.α.68, with the following additional features. Cocci to short rods, 1–2 × 1.0–1.5 μm, occurring in pairs or short chains. Colonies are bright orange, due to the accumulation of carotenoids, including astaxanthin. Grows on a wide range of carbon sources. In addition to those mentioned in Table BXII.α.68, cellobiose, lactose, melibiose, gentiobiose, glucuronic acid, galacturonic acid, erythritol, xylitol, adonitol, arabitol, propionate, *cis*-aconitate, malonate, and alanine can serve as carbon and energy sources, but fucose, rhamnose, raffinose, starch, glycogen, *N*-acetylglucosamine, *N*-acetylgalactosamine, α-glycerolphosphate, glucose-1-phosphate, glucose-6-phosphate, 2,3-butanediol, Tween 40 and 80, asparagines, aspartate, glutamate, histidine, leucine, ornithine, phenylalanine, proline, serine, threonine, inosine, uridine, and thymidine do not. Weak growth in 6% NaCl. No growth at NaCl >8%. Optimal growth temperature, 25–30°C, weak growth at 35°C; optimal pH for growth, 7. Isolated as a contaminant on agar plates.

The mol% G + C of the DNA is: 66 (HPLC).

Type strain: MH1, DSM 11574.

9. **Paracoccus methylutens** Doronina, Trotsenko, Krausova and Suzina 1998c, 1083[VP] (Effective publication: Doronina, Trotsenko, Krausova and Suzina 1998b, 235.)

me.thyl.u'tens. M.L. neut. n. *methylum* methyl compound; L. part. adj. *utens* using; *methylutens* using methyl groups.

The description is as given for the genus with the following additional features. It is a facultative methylotrophic species capable of growth on dichloromethane, methanol, methylamine, and formate, but not formaldehyde. Nitrate is reduced to nitrite. Grows on a wide range of organic substrates, but not on acetamide, rhamnose, raffinose, or trehalose. Its major fatty acid is $C_{18:1 \omega 7}$.

The mol% G + C of the DNA is: 67 (T_m).

Type strain: DM12, VKM B-2164.

10. **Paracoccus pantotrophus** (Robertson and Kuenen 1984) Rainey, Kelly, Stackebrandt, Burghardt, Hiraishi, Katayama and Wood 1999, 650[VP] (*Thiosphaera pantotropha* Robertson and Kuenen 1984, 91.)

pan.to'troph.us. Gr. pre. *panto* all; Gr. n. *trophos* feeder; M.L. adj. *pantotrophus* omnivorous.

The description is as given for the genus and in Table BXII.α.68, with the following additional features. Cocci occurring singly, in pairs, or short chains. In addition to those mentioned in Table BXII.α.68, fumarate can be used as electron acceptor, acetoin is produced, grows on adipate, propionate, acetone, propane-1,2-diol, isopropanol, glutamate, isoleucine, leucine, alanine, histidine, proline, propanol, acetol, and propionaldehyde, and does not grow on pimelate, methylacetate, methyl ethyl ketone, propylene oxide, dulcitol, and glycogen. Autotrophic growth on sulfide, thiosulfate, and hydrogen under both aerobic and denitrifying conditions by the type strain. All three substrates can be used for mixotrophic growth. Some strains are capable of aerobic denitrification. Some strains are capable of aerobic growth on methanol and formate. Some strains contain plasmids of 85–110 kb in size and megaplasmids greater than 450 kb in size. Optimal growth temperature, 37°C (range 15–42°C); optimal pH for growth, 8 (range 6.5–10.5). Isolated from a sulfide-oxidizing, effluent treatment pilot plant.

The mol% G + C of the DNA is: 66 (T_m).

Type strain: ATCC 35512, DSM 2944 , LMD 82.5.

GenBank accession number (16S rRNA): Y16933.

11. **Paracoccus solventivorans** Siller, Rainey, Stackebrandt and Winter 1996, 1129[VP] emend. Lipski, Reichert, Reuter, Spröer and Altendorf 1998, 535.

sol.ven.ti.vo'rans. L. v. *solvo* melt, free; M.L. neut. n. *solventum* solvent; L. v. *voro* to eat, devour; L. part. adj. *vorans* eating; *solventivorans* solvent-eating.

The description is as given for the genus and in Table BXII.α.68, with the following additional features. This organism has a Gram-negative cell wall architecture, but the Gram stain reaction is variable. Cells are coccoid or rod like, 0.4–0.5 × 0.9–1.5 μm. Light-scattering PHB inclusions are evident. NaCl is inhibitory at concentrations >0.2%. Optimal growth temperature, 30–37°C; optimal pH for growth, 7–8. No vitamin addition is required. The electron acceptors are oxygen and nitrate. N_2 is produced from nitrate from respiration. Acetone is oxidized to CO_2 and metabolized to PHB. Besides acetone and in addition to Table BXII.α.68, 2-butanone, 2-propanol, fumarate, glutamate, aspartate, asparagine, α-ketoglutarate, isoleucine, glycine, propionate, *n*-butyrate, 3-hydroxybutyrate, acetoacetate, and Casamino Acids (Difco) can serve as carbon and energy sources. Many other substrates are not utilized. Acetone is degraded via acetoacetate.

The mol% G + C of the DNA is: 68.5 (HPLC) or 70 (T_m).

Type strain: L1, DSM 6637.

GenBank accession number (16S rRNA): Y07705.

12. **Paracoccus thiocyanatus** Katayama, Hiraishi and Kuraishi 1996, 625[VP] (Effective publication: Katayama, Hiraishi and Kuraishi 1995, 1475.)

thi.o.cy.a' na.tus. Gr. n. *thium* sulfur; Gr. n. *kyanos* dark blue; M.L. adj. *thiocyanatus* referring to the ability to use thiocyanate.

The description is as given for the genus and in Table BXII.α.68, with the following additional features. Cells are rod shaped, 0.5–0.7 × 0.8–1.3 μm, occurring singly or in pairs. Polyhedral inclusion bodies (carboxysomes) are absent in cells. Optimal growth temperature, 30–35°C; optimal pH for growth, 7.0–8.5. The addition of thiamine is required for growth in minimal media. The electron acceptors are oxygen and nitrate. N_2 is produced from nitrate from respiration. Thiocyanate is oxidized to CO_2. S^0 is oxidized slowly. Thiosulfate is oxidized to tetrathionate by chemoorganotrophically grown cells or to sulfate by thiocyanate-grown cells. Tetrathionate does not support growth. The organism is catalase and oxidase positive. Besides thiocyanate and in addition to those mentioned in Table BXII.α.68, L-alanine, L-serine, L-leucine, L-isoleucine, L-glutamate, L-proline, L-histidine, L-phenylalanine, propionate, *n*-butyrate, glutarate, *m*-hydroxybenzoate, and *p*-hydroxybenzoate serve as carbon and energy sources, but L-aspartate, L-tryptophan, L-cysteine, cellobiose, cyclohexanol, oxalate, adipate, pimelate, *o*-hydroxybenzoate, *p*-aminobenzoate, and DL-mandelate do not.

The mol% G + C of the DNA is: 66.5 (HPLC) or 67.6 (T_m).

Type strain: THI 011, IAM 12816, IFO 14569.

13. **Paracoccus versutus** (Harrison 1983) Katayama, Hiraishi and Kuraishi 1996, 625[VP] (Effective publication: Katayama, Hiraishi and Kuraishi 1995, 1476) (*Thiobacillus versutus* Harrison 1983, 216.)

ver.su' tus. L. adj. *versutus* versatile.

Cells are rod shaped, 0.4–0.5 × 1.1–1.7 μm, occurring singly or in pairs. Nonsporeforming. Motile with a tuft of flagella. Optimal growth temperature, 30–37°C; optimal pH for growth, 7.5–8.0. No vitamin addition is required. The electron acceptors are oxygen and nitrate. N_2 is produced from nitrate from respiration. S^0 is oxidized slowly. Tetrathionate and thiocyanate do not support growth. The organism is catalase and oxidase positive. In addition to those mentioned in Table BXII.α.68, L-alanine, L-serine, L-leucine, L-isoleucine, L-aspartate, L-glutamate, L-proline, L-histidine, L-phenylalanine, propionate, *n*-butyrate, 2-oxoglutarate, glutarate, *m*-hydroxybenzoate, and *p*-hydroxybenzoate serve as carbon and energy sources, but L-tryptophan, L-cysteine, cellobiose, cyclohexanol, oxalate, pimelate, *o*-hydroxybenzoate, *p*-aminobenzoate, and DL-mandelate do not.

The mol% G + C of the DNA is: 67–68 (T_m).

Type strain: A2, ATCC 25364, CCM 2505, DSM 582, IAM 12814.

GenBank accession number (16S rRNA): Y16932.

Genus XIII. **Rhodobaca** *Milford, Achenbach, Jung and Madigan 2001, 793*[VP] *(Effective publication: Milford, Achenbach, Jung and Madigan 2000, 25)*

JOHANNES F. IMHOFF

Rho.do.bac' ca. Gr. n. *rhodon* the rose; L. fem. n. *baca* berry; M.L. fem. n. *Rhodobaca* red (rose) berry.

Cells are ovoid or rod shaped, are motile by polar flagella, and divide by binary fission. **Gram negative; belong to the *Alphaproteobacteria*.** Phototrophically grown cells **form vesicular internal photosynthetic membranes.** The color of phototrophic cultures is yellow to yellow-brown, while aerobic cultures are pink to red. **Photosynthetic pigments are bacteriochlorophyll *a* and carotenoids of the spheroidene series.**

Photoheterotrophic growth occurs under anoxic conditions in the light with a variety of organic compounds as carbon and electron sources. **Chemotrophic growth occurs by aerobic respiration.** Growth factors are required.

Mesophilic bacterium with an elevated temperature optimal and best growth at alkaline pH and in the presence of low concentrations of mineral salts.

Habitats: alkaline soda lakes of low to moderate salinity.

The mol% G + C of the DNA is: 58.8.

Type species: **Rhodobaca bogoriensis** Milford, Achenbach, Jung and Madigan 2001, 793 (Effective publication: Milford, Achenbach, Jung and Madigan 2000, 25.)

FURTHER DESCRIPTIVE INFORMATION

In many phenotypic properties, *Rhodobaca bogoriensis* resembles *Rhodobacter* and *Rhodovulum* species. Characteristic properties that reflect adaptation to the natural environment, African soda lakes (Lake Bogoria, Kenya), are the high pH-optimal at pH 9 and the temperature optimal at 39°C (Milford et al., 2000). No growth occurs below 30°C. Good growth is from 0–6% NaCl with an optimum from 1–2%. *Rhodobaca bogoriensis* apparently is unable to fix dinitrogen and does not grow photoautotrophically, either with hydrogen or with sulfide as electron donors. Sulfide is oxidized to extracellular elemental sulfur (Milford et al., 2000). Spectral properties of whole cells indicate the presence of only one light-harvesting pigment complex (LHI, core antenna complex) unlike *Rhodobacter* and *Rhodovulum* species, but similar to bacteria such as *Rhodospirillum rubrum*, *Rhodobium marinum*, *Rhodocista centenaria*, *Rhodovibrio sodomense*, *Rhodospira trueperi*, and *Roseospirillum parvum*. The sequence of the 16S rDNA clearly distinguishes *Rhodobaca* from *Rhodobacter* and *Rhodovulum* species but also shows the association of this bacterium with these two genera in the *Alphaproteobacteria* (see Fig. 3 [p. 129] in the introductory chapter "Anoxygenic Phototrophic Purple Bacteria", Volume 2, Part A. Differentiating phenotypic characteristics are given in Table BXII.α.46 of the chapter on the genus *Rhodobacter*.)

ENRICHMENT AND ISOLATION PROCEDURES

Standard media and techniques for enrichment and isolation of purple nonsulfur bacteria are suitable for *Rhodobaca* species (see chapter Genus *Rhodospirillum*). Media for the enrichment of *Rhodobaca* species should have alkaline pH and contain 1–6% mineral salts.

MAINTENANCE PROCEDURES

Cultures are well preserved by standard techniques in liquid nitrogen, by lyophilization or storage at −80°C in a mechanical freezer.

DIFFERENTIATION OF THE GENUS *RHODOBACA* FROM OTHER GENERA

Characteristic properties of *Rhodobaca* are the ovoid to rod-shaped cell morphology, the presence of vesicular internal membranes, and the content of carotenoids of the spheroidene series. All these properties are shared with *Rhodovulum* and *Rhodobacter* species. A clear differentiation of *Rhodobaca* from *Rhodobacter* and *Rhodovulum* species is possible based on 16S rDNA sequences (see Fig. 3 [p. 129] in the introductory chapter "Anoxygenic Phototrophic Purple Bacteria"., Volume 2, Part A. Differentiating phenotypic characteristics are given in Table BXII.α.46 of the chapter on the genus *Rhodobacter*.)

List of species of the genus Rhodobaca

1. **Rhodobaca bogoriensis** Milford, Achenbach, Jung and Madigan 2001, 793[VP] (Effective publication: Milford, Achenbach, Jung and Madigan 2000, 25.)

   bo.go.ri.en'sis. M.L. fem. adj. *bogoriensis* pertaining to Lake Bogoria, a soda lake in Kenya, Africa.

   Cells are ovoid to rod shaped, 0.8–1.0 × 0.8–1.5 μm. Cells are motile by means of polar flagella. Internal photosynthetic membranes appear as vesicles.

   Cultures grown anaerobically in the light are yellow to yellowish brown. When grown in the presence of oxygen, the color turns to red. Absorption spectra of living cells show maxima at 376–378, 450–455, 478–480, 508–513, 590–592, 802–805, and 860–863 nm. Photosynthetic pigments are bacteriochlorophyll *a* and carotenoids of the spheroidene series, with demethylspheroidenone as major component.

   Photoheterotrophic growth occurs anaerobically in the light with a variety of organic compounds. No evidence for photoautotrophic growth with H_2 or sulfide as electron donors. Chemoheterotrophic growth aerobically in the dark occurs at the full oxygen tension of air. Carbon sources supporting excellent phototrophic growth include acetate, pyruvate, malate, succinate, fumarate, butyrate, valerate, mannitol, heptanoate, xylose, glucose, fructose, and sucrose. Ammonia, aspartate, glutamate, asparagine, glutamine, and Casamino acids are used as nitrogen sources; N_2 fixation is absent. Sulfate assimilation is present. Biotin and thiamine are required as growth factors; vitamin B_{12} is growth stimulatory.

   Mesophilic and alkaliphilic bacterium with optimal growth at pH 9.0 (range: 7.5–10.0), 39°C (range: 30–45°C), and in the presence of 1–2% NaCl.

   Habitat: alkaline and saline soda lakes exposed to the light and with reduced oxygen concentrations.

   The mol% G + C of the DNA is: 58.8 (T_m).

   Type strain: LBB1, ATCC 700920.

   GenBank accession number (16S rRNA): AF248638.

Genus XIV. **Rhodovulum** *Hiraishi and Ueda 1994a, 21[VP]*

JOHANNES F. IMHOFF

Rho.do'vu.lum. Gr. n. *rhodon* rose; L. dim. n. *ovulum* small egg; M.L. neut. n. *Rhodovulum* small red egg.

Cells are ovoid to rod shaped, are motile by means of flagella or nonmotile, and divide by binary fission. **Gram negative; belong to the *Alphaproteobacteria*.** Phototrophically grown cells **form vesicular internal photosynthetic membranes. Photosynthetic pigments are bacteriochlorophyll *a* (esterified with phytol) and carotenoids of the spheroidene series.** The color of phototrophic cultures is yellow-green to yellow-brown, while aerobic cultures are pink to red. **Ubiquinone 10 is the major quinone. Major fatty acids are $C_{18:1}$ (predominant), $C_{18:0}$, and $C_{16:0}$.**

Cells grow preferably photoheterotrophically under anoxic conditions in the light. **Photoautotrophic growth may occur with sulfide, thiosulfate, hydrogen, and ferrous iron.** The final oxidation product of sulfide is sulfate. **Chemotrophic growth in the dark occurs under microoxic to oxic conditions and under anoxic conditions by fermentation and oxidant-dependent metabolism.** Polysaccharides, poly-β-hydroxybutyric acid, and polyphosphates may be present as storage products. Growth factors are required.

Mesophilic marine bacteria that require sodium chloride for growth.

Habitats: marine and hypersaline environments rich in organic matter, containing hydrogen sulfide, and exposed to the light.

The mol% G + C of the DNA is: 62.1–73.2.

Type species: **Rhodovulum sulfidophilum** (Hansen and Veldkamp 1973) Hiraishi and Ueda 1994a, 21 (*Rhodopseudomonas sulfidophila* Hansen and Veldkamp 1973, 55; *Rhodobacter sulfidophilus* Imhoff, Trüper, and Pfennig 1984, 342.)

ENRICHMENT AND ISOLATION PROCEDURES

Standard techniques and media for the isolation of phototrophic bacteria in agar dilution series and on agar plates can be applied for *Rhodovulum* species, if appropriate concentrations of salts are included (see chapter Genus *Rhodospirillum*; Biebl and Pfennig, 1981; Imhoff, 1988; Imhoff and Trüper, 1992). Media containing 2–3% NaCl will be selective for *Rhodovulum* species. Under these conditions, enrichment cultures set up for purple nonsulfur bacteria will often yield *Rhodovulum* species as dominant members. Media for enrichment and isolation with organic substrates, or, for most species, autotrophic conditions with reduced sulfur compounds or hydrogen, may be chosen. Vitamins must be supplied. For *R. iodosum* and *R. robiginosum*, media with ferrous iron and containing 2.5–5% NaCl provide selective conditions.

MAINTENANCE PROCEDURES

Cultures are well preserved by standard techniques in liquid nitrogen, by lyophilization, or storage at $-80°C$ in a mechanical freezer.

DIFFERENTIATION OF THE GENUS *RHODOVULUM* FROM OTHER GENERA

Differentiation of the genera and species of *Rhodobacter*, *Rhodobaca*, and *Rhodovulum* is possible based on 16S rDNA sequences (see Fig. 3 [p. 129] in the introductory chapter "Anoxygenic Phototrophic Purple Bacteria", Volume 2, Part A, and by DNA–DNA hybridization. Differentiating phenotypic characteristics are given in Table BXII.α.46 of the chapter on the genus *Rhodobacter*.)

TAXONOMIC COMMENTS

Species of this genus (*R. sulfidophilum, R. adriaticum*) have formerly been included in the genus *Rhodopseudomonas*. The recognition of morphologically and chemotaxonomically distinct characteristics and early phylogenetic analyses led to their separation from *Rhodopseudomonas* and classification in the genus *Rhodobacter* (Imhoff et al., 1984). With the availability of 16S rDNA sequences and additional species and isolates, it became apparent that the marine and halophilic species that depend on NaCl form a group phylogenetically separate from the freshwater species. Both groups were taxonomically separated and the marine species transferred to the new genus *Rhodovulum* (Hiraishi and Ueda, 1994a). The newly described *Rhodobacter euryhalinus* (Kompantseva, 1985) was also included in this new genus as *Rhodovulum euryhalinum*.

DIFFERENTIATION OF THE SPECIES OF THE GENUS *RHODOVULUM*

Characteristic properties to differentiate species of *Rhodovulum* are summarized in Tables BXII.α.69 and BXII.α.70.

List of species of the genus Rhodovulum

1. **Rhodovulum sulfidophilum** (Hansen and Veldkamp 1973) Hiraishi and Ueda 1994a, 21[VP] (*Rhodopseudomonas sulfidophila* Hansen and Veldkamp 1973, 55; *Rhodobacter sulfidophilus* Imhoff, Trüper, and Pfennig 1984, 342.)

   sul.fi.do′phi.lum. M.L. n. *sulfidum* sulfide; Gr. adj. *philos* loving; M.L. adj. *sulfidophilum* sulfide loving.

   Cells are ovoid to rod shaped, $0.6–0.9 \times 0.9–2.0$ μm, motile by means of polar flagella. Eventually, slime may be produced and short straight chains may be formed. Internal photosynthetic membranes appear as vesicles. Color of cell suspensions depends on the redox state of the culture and is from yellowish green through yellowish brown to dark brown and brown-red. In the presence of oxygen, cultures are red. Absorption spectra of living cells show maxima at 374–378, 451–455, 480–489, 508–512, 588–592, 800–805,

and 850–855 nm. Photosynthetic pigments are bacteriochlorophyll *a* (esterified with phytol) and carotenoids of the spheroidene series with spheroidene and hydroxyspheroidene, which are converted to their corresponding ketocarotenoids under oxic conditions and thereby cause the color change to red.

Preferentially grow photoheterotrophically under anoxic conditions in the light using a variety of organic compounds as carbon sources and electron donors. Photoautotrophic growth occurs with sulfide, thiosulfate, and molecular hydrogen as electron donors. Sulfide and thiosulfate are oxidized to sulfate without intermediary accumulation of elemental sulfur; elemental sulfur is oxidized slowly. In mineral media sulfide is tolerated up to 5–6 mM; in the presence of complex nutrients such as 0.01% yeast extract,

TABLE BXII.α.69. Differentiating characteristics of *Rhodovulum* species[a]

Characteristic	R. sulfidophilum	R. adriaticum	R. euryhalinum	R. iodosum	R. robiginosum	R. strictum
Cell diameter (μm)	0.6–0.9	0.5–0.8	0.7–1.0	0.5–0.8	0.5–0.8	0.6–1.0
Motility	+	−	+	−	−	+
NaCl optimal (%)	1–6	2.5–7.5	0.5–12	2.5–5	2.5–5	0.8–1.0
pH optimal	6.5–8.0	6.5–7.0	7.0–8.0	6.5; 7.0–7.3[b]	6.5; 7.3–7.7[b]	8.0–8.5
Aerobic dark growth	+	−	−	+	+	+
Sulfide tolerance	High[c]	High	High	nd	nd	High
Sulfate assimilated	+	−	−	+	+	+
Vitamins required[d]	b, n, *p*-ABA, t	b, t	b, n, *p*-ABA, t	b, n	b, n	b, *p*-ABA, t
Utilization of:						
Citrate	−	−	−	−	−	(+)
Tartrate	−	−	−	nd	nd	(+)
Mannitol	±	−	±	+	+	−
Glycerol	+	+	+	−	+	−
Ethanol	±	+	±	−	−	−
Gluconate	nd	+	nd	nd	nd	nd
Hydrogen	+	−	+	+	+	−
Thiosulfate	+	+	+	+	+	+
Ferrous iron	−	−	−	+	+	nd
Mol% G + C of the DNA	66.3–66.6 (HPLC), 68.9–73.2 (T_m)	64.9–66.7 (T_m)	62.1–68.6 (T_m)	66 (HPLC)	69 (HPLC)	67.3–67.7 (HPLC)

[a]Symbols: +, positive in most strains; −, negative in most strains; ± variable in different strains; (+) weak growth or microaerobic growth only; nd, not determined.

[b]The first value is with ferrous iron; the second value is with acetate.

[c]Good growth at concentrations ≥ 1–2 mM.

[d]b, biotin, n, niacin, *p*-ABA, *p*-aminobenzoic acid, t, thiamine.

TABLE BXII.α.70. Photosynthetic electron donors and carbon sources of *Rhodovulum* species[a]

Donor/source	R. sulfidophilum	R. adriaticum	R. euryhalinum	R. iodosum	R. robiginosum	R. strictum
Formate	+	+	±	−	−	+
Acetate	+	+	+	+	+	+
Propionate	+	+	+	+	+	+
Butyrate	+	−	+	+	−	+
Valerate	+	+	nd	+	−	+
Caproate	+	+	nd	−	−	+
Caprylate	+	nd	+	−	−	−
Pelargonate	+	nd	−	nd	nd	−
Pyruvate	+	+	+	+	+	+
Lactate	+	+	+	+	+	+
Malate	+	+	+	+	+	+
Succinate	+	+	+	+	+	+
Fumarate	+	+	+	+	−	+
Tartrate	−	−	−	nd	nd	(+)
Citrate	−	−	−	−	−	+
Aspartate	±	nd	+	−	−	+
Arginine	−	nd	−	nd	nd	nd
Glutamate	+	nd	±	+	+	−
Benzoate	−	−	−	−	−	−
Gluconate	±	nd	+	nd	nd	nd
Glucose	+	+	+	−	−	+
Fructose	±	−	+	−	−	+
Mannose	±	nd	nd	nd	nd	−
Mannitol	±	−	±	+	+	−
Sorbitol	±	nd	nd	nd	nd	−
Glycerol	+	+	+	−	+	−
Methanol	−	−	−	−	−	−
Ethanol	±	+	±	−	−	−
Propanol	±	−	nd	nd	nd	−
Hydrogen	+	−	nd	+	+	nd
Sulfide	+	+	+	+	+	+
Thiosulfate	+	+	+	+	+	+
Sulfur	+	+	+	+	+	nd
Ferrous iron	−	−	−	+	+	nd

[a]Symbols: +, positive in most strains; −, negative in most strains; ± variable in different strains; (+) weak growth or microaerobic growth only; nd, not determined.

the tolerance is 7–8 mM. Chemotrophic growth is possible under oxic conditions in the dark. Carbon sources utilized are shown in Table BXII.α.70.

Ammonia and, in most strains, dinitrogen but not nitrate are used as nitrogen sources. Sulfate is assimilated and can serve as sole sulfur source under photoheterotrophic growth conditions and during growth with molecular hydrogen. In addition, sulfite, thiosulfate, cysteine and reduced glutathione are used as sulfur sources. Biotin, niacin, thiamine, and *p*-aminobenzoic acid are required as growth factors.

Mesophilic marine bacterium with optimal growth at pH 6.5–8.0 (with sulfide as electron donor) or at pH 5.0–7.5 (if grown with malate), 30–35°C, and 1–6% NaCl.

Habitats: marine sediments and coastal marine waters containing hydrogen sulfide and rich in organic matter.

The mol% G + C of the DNA is: 68.9–73.2 (T_m) and 66.3–66.6 (HPLC).

Type strain: Hansen W4, ATCC 35886, DSM 1374.

GenBank accession number (16S rRNA): D16423, D13475.

2. **Rhodovulum adriaticum** (Neutzling, Imhoff, and Trüper 1984) Hiraishi and Ueda 1994a, 22[VP] (*Rhodopseudomonas adriatica* Neutzling, Imhoff and Trüper 1984, 503; *Rhodobacter adriaticus* Imhoff, Trüper and Pfennig 1984, 342.)

a.dri.a' ti.cum. M.L. adj. *adriaticum* pertaining to the Adriatic Sea.

Cells are ovoid to rod shaped, 0.5–0.8 × 1.3–1.8 μm, nonmotile, often occurring in short straight chains, form capsules, and produce slime. Internal photosynthetic membranes appear as vesicles. Color of cell suspensions is yellowish brown to dark brown. Absorption spectra of living cells show maxima at 374–378, 447–450, 475–480, 508–512, 588–590, 802–805, and about 869 nm. Photosynthetic pigments are bacteriochlorophyll *a* and carotenoids of the spheroidene series.

Photoheterotrophic growth occurs under anoxic conditions in the light with a variety of organic compounds. Photoautotrophic growth is possible with sulfide, elemental sulfur, and thiosulfate as electron donor. During oxidation of sulfide to sulfate, elemental sulfur is intermediately formed outside the cells. Molecular hydrogen is not used. Sensitive to oxygen, but microaerobic growth in the dark is possible. In the presence of ascorbate as a reductant, phototrophic growth is stimulated. Carbon sources utilized are shown in Table BXII.α.70.

Good nitrogen sources are ammonia, dinitrogen and some amino acids; growth with urea is poor. Nitrate is not assimilated but is reduced to nitrite. Sulfate is not assimilated. Growth depends on reduced sulfur sources such as sulfide, thiosulfate, cysteine, and elemental sulfur. Biotin and thiamine are required; niacin stimulates growth.

Mesophilic marine bacterium with optimal growth at pH 6.5–7.0 (pH range: 6.9–8.5), 25–30°C, and 2.5–7.5% NaCl.

Habitats: marine sediments and coastal marine waters containing hydrogen sulfide and rich in organic matter.

The mol% G + C of the DNA is: 64.9-66.7 (T_m).

Type strain: Imhoff 6II, ATCC 35885, DSM 2781.

GenBank accession number (16S rRNA): D16418, D13476.

3. **Rhodovulum euryhalinum** (Kompantseva 1989b) Hiraishi and Ueda 1994a, 22[VP] (*Rhodobacter euryhalinus* Kompantseva 1989b, 205.)

eu.ry.ha.li' num. Gr. adj. *eurys* wide; Gr. n. *hals* salt; M.L. adj. *euryhalinum* living in a wide range of salinity.

Cells are ovoid to rod shaped, $0.7–1.0 \times 1.5–3.0$ μm, motile by polar flagella, divide by binary fission and often occur in short straight chains. Internal photosynthetic membranes appear as vesicles. Color of anaerobic cultures is greenish yellow, yellowish brown, to dark brown. Aerobic cultures are red. Absorption spectra of living cells show maxima at 378, 450–460, 479–486, 509–517, 592, 805, and 855 nm. Photosynthetic pigments are bacteriochlorophyll *a* and carotenoids of the spheroidene series.

Photoheterotrophic growth occurs under anoxic conditions in the light with a variety of organic compounds. Photoautotrophic growth is possible with sulfide, thiosulfate, and hydrogen as electron donor. During oxidation of sulfide to sulfate, elemental sulfur is intermediately formed outside the cells. Chemotrophic growth in the dark occurs with organic substrates and under autotrophic conditions with sulfide at microoxic conditions. Carbon sources utilized are shown in Table BXII.α.70. Ammonia is used as nitrogen source. Most strains do not assimilate sulfate, but depend on reduced sulfur sources; sulfide, cysteine, and cystine are used. Biotin, thiamine, niacin, and *p*-aminobenzoate are required as growth factors.

Mesophilic marine bacterium with optimal growth at pH 7.0–8.0, 25–35°C, and 0.5–12% NaCl. No growth in the absence of NaCl.

Habitats: marine sediments, coastal marine waters, and continental salt waters containing hydrogen sulfide and rich in organic matter.

The mol% G + C of the DNA is: 62.1–68.6 (T_m); type strain: 66.3 (T_m) and 65.5 (HPLC).

Type strain: Kompantseva KA-65, DSM 4868.

GenBank accession number (16S rRNA): D13479, D16426.

4. **Rhodovulum iodosum** Straub, Rainey and Widdel 1999, 734[VP]

i.o.do' sum. Gr. adj. *iodes* violet, rusty; M.L. adj. *iodosum* indicating the formation of rusty ferric iron deposits.

Cells are ovoid to rod shaped, $0.5–0.8 \times 2.4–3.8$ μm, nonmotile and divide by binary fission. Color of strictly anaerobically grown cell suspensions is yellowish to brown. Photosynthetic pigments are bacteriochlorophyll *a* and carotenoids of the spheroidene series. Photoheterotrophic growth occurs under anoxic conditions in the light using a variety of organic compounds.

Photoautotrophic growth occurs with sulfide, sulfur, thiosulfate, and molecular hydrogen as electron donors. In addition, ferrous iron serves as photosynthetic electron donor. During growth on iron sulfide, ferric iron and sulfate are formed. Chemotrophic growth under oxic conditions in the dark is possible (with acetate as carbon source). Under anoxic dark conditions, no growth occurs by fermentation or with nitrate, ferric iron, DMSO, and TMAO as electron acceptors. Carbon sources utilized are shown in Table BXII.α.70.

Ammonia is used as a nitrogen source. Sulfate is not assimilated, but reduced sulfur sources are required. Thiosulfate, elemental sulfur, sulfide, and cysteine can be used, but not sulfate and sulfite. Biotin and niacin are required as growth factors.

Mesophilic marine bacterium with optimal growth at pH 7.0–7.3 (with acetate as substrate) or pH 6.5 (pH range: 6.3–6.8, with ferrous iron as substrate), 20–25°C, and 2.5–5% NaCl (range: 2–7%).

Habitats: intertidal mud flats.

The mol% G + C of the DNA is: 66 (HPLC).

Type strain: N1, DSM 12328.

GenBank accession number (16S rRNA): Y15011.

5. **Rhodovulum robiginosum** Straub, Rainey and Widdel 1999, 734[VP]

ro.bi.gi.no' sum. M.L. adj. *robiginosum* rusty, indicating the formation of rusty ferric iron deposits.

Cells are ovoid to rod shaped, $0.5–0.8 \times 1.6–3.2$ μm, nonmotile, and multiply by binary fission. Color of strictly anaerobically grown cell suspensions is yellowish to brown. Photosynthetic pigments are bacteriochlorophyll *a* and carotenoids of the spheroidene series.

Grow photoheterotrophically under anoxic conditions in the light using a variety of organic compounds, and photoautotrophically with sulfide, sulfur, thiosulfate, and molecular hydrogen as electron donors. In addition, ferrous iron serves as photosynthetic electron donor. During growth on iron sulfide, ferric iron and sulfate are formed. Chemotrophic growth under oxic conditions in the dark is possible (with acetate as carbon source). Under anoxic dark conditions, no growth occurs by fermentation or with nitrate, ferric iron, DMSO, and TMAO as electron acceptors. Carbon sources utilized are shown in Table BXII.α.70.

Ammonia is used as a nitrogen source. Sulfate is not assimilated, but reduced sulfur sources are required. Thiosulfate, elemental sulfur, sulfide, and cysteine can be used, but not sulfate and sulfite. Biotin, niacin, and vitamin B_{12} are required as growth factors.

Mesophilic marine bacterium with optimal growth at pH 7.3–7.7 (with acetate as substrate) or pH 6.5 (pH range: 6.3–6.8, with ferrous iron as substrate), 25–28°C, and 2.5–5% NaCl (range: 1–7%).

Habitats: intertidal mud flats.

The mol% G + C of the DNA is: 69 (HPLC).

Type strain: N2, DSM 12329.

GenBank accession number (16S rRNA): Y15012.

6. **Rhodovulum strictum** Hiraishi and Ueda 1995, 325[VP]

stric' tum. M.L. part. n. adj. *strictum* strict, accurate, referring to the fact that the cells require strict growth conditions.

Cells are ovoid to rod shaped, $0.6–1.0 \times 1.0–2.5$ μm, motile by means of polar flagella and divide by binary fission. Phototrophically grown cells contain internal photosynthetic membranes of the vesicular type. Photosynthetic cultures are yellow-green to yellow-brown, while aerobic cultures grown in the dark are pink to red. Colonies that develop aerobically on agar media in the dark are circular, convex, with entire margins, and red in color. Photosynthetic pigments are bacteriochlorophyll *a* and carotenoids of the spheroidene series.

Facultative photoheterotrophic bacteria that can grow anaerobically in the light or aerobically in the dark at full atmospheric oxygen tension. Sulfide and thiosulfate are used as electron donors for phototrophic growth, with sulfate as the final oxidation product.

Good carbon sources for phototrophic growth are formate, acetate, propionate, butyrate, lactate, pyruvate, fumarate, malate, succinate, glycolate, fructose, alanine, leucine, Casamino acids, peptone, and yeast extract. Weak or slow growth occurs with valerate, caproate, tartrate, citrate, arabinose, and glucose. No or little growth occurs with caprylate, pelargonate, malonate, sucrose, galactose, mannose, adonitol, mannitol, sorbitol, glycerol, methanol, ethanol, propanol, asparagine, aspartate, or glutamate. Sulfate is assimilated as sole sulfur source. Biotin, *p*-aminobenzoic acid, and thiamine are required as growth factors.

Mesophilic, marine bacterium with optimal growth at pH 8.0–8.5 (pH range: 7.5–9.0), 30–35°C, and in the presence of 0.8–1% NaCl (range: 0.25–3.0%).

Habitat: tidal and seawater pools and similar marine environments.

The mol% G + C of the DNA is: 67.3–67.7 (HPLC).

Type strain: MB-G2, JCM 9220.

GenBank accession number (16S rRNA): D16419.

Genus XV. **Roseibium** Suzuki, Muroga, Takahama and Nishimura 2000, 2155[VP]

THE EDITORIAL BOARD

Ro.sei' bi.um. M.L. adj. *roseus* rose/pink; Gr. n. *bios* life; M.L. neut. n. *Roseibium* pink life.

Gram-negative aerobic chemoheterotrophic motile rods (0.5–0.8 × 1.0–4.0 µm for most strains). Peritrichous flagella. **Produce bacteriochlorophyll *a* aerobically. Do not grow as anaerobic phototrophs.** Catalase and oxidase positive. Produce phosphatase and nitrate reductase. Major cellular fatty acid $C_{18:1}$.

The mol% G + C of the DNA is: 57.6–63.4 (HPLC).

Type species: **Roseibium denhamense** Suzuki, Muroga, Takahama and Nishimura 2000, 2155.

FURTHER DESCRIPTIVE INFORMATION

Roseibium strains were isolated from marine environments including sand and surfaces of red algae (Shiba et al., 1991; Suzuki et al., 2000).

All *Roseibium* strains tested were ONPG positive and produced indole. They hydrolyzed gelatin and utilized L-aspartate, butyrate, L-glutamate, and pyruvate. They produced acid from D-fructose,

D-glucose, maltose, and D-ribose. All strains were Voges–Proskauer negative and did not produce H_2S. They were unable to utilize ethanol, glycolate, or methanol and did not produce acid from lactose or L-arabinose. None hydrolyzed Tween 80, starch, or alginate. All were streptomycin and chloramphenicol sensitive and tetracycline and penicillin resistant (Suzuki et al., 2000).

Analysis of 16S rDNA sequences placed the *Roseibium denhamense* and *R. hamelinense* type strains in the *Alphaproteobacteria* (Suzuki et al., 2000).

ENRICHMENT AND ISOLATION PROCEDURES

These bacteria were isolated from seawater, sand, and the surfaces of marine organisms from the coasts of Australia (Shiba et al., 1991; Suzuki et al., 2000). Procedures and media used are given by Shiba et al. (1991).

DIFFERENTIATION OF THE SPECIES OF THE GENUS *ROSEIBIUM*

R. denhamense utilizes acetate, fumarate, and DL-malate; *R. hamelinense* does not. *R. hamelinense* utilizes DL-lactate; *R. denhamense* does not. *R. denhamense* produces acid from D-galactose and D-

xylose; *R. hamelinense* does not. *R. hamelinense* grows in 0% NaCl; *R. denhamense* does not.

List of species of the genus Roseibium

1. **Roseibium denhamense** Suzuki, Muroga, Takahama and Nishimura 2000, 2155[VP]

den.ha.men' se. M.L. adj. *denhamense* referring to Denham, Australia, the source of the type strain.

Description as for the genus with the following additional characteristics: utilizes acetate, fumarate, and DL-malate; produces acid from D-galactose and D-xylose; does not grow in 0% NaCl.

The mol% G + C of the DNA is: 57.6–60.4 (HPLC).

Type strain: OCh 254, JCM 19543.

GenBank accession number (16S rRNA): D85832.

2. **Roseibium hamelinense** Suzuki, Muroga, Takahama and Nishimura 2000, 2155[VP]

ha.me.li.nen' se. M.L. adj. *hamelinense* referring to Hamelin Pool, Australia, the source of the type strain.

Description as for the genus with the following additional characteristics: utilizes DL-lactate and grows in 0% NaCl.

The mol% G + C of the DNA is: 59.2–63.4 (HPLC).

Type strain: OCh 368, JCM 10544.

GenBank accession number (16S rRNA): D85836.

Genus XVI. **Roseinatronobacter** Sorokin, Tourova, Kuznetsov, Bryantseva and Gorlenko 2000b, 1415[VP] (Effective publication: Sorokin, Tourova, Kuznetsov, Bryantseva and Gorlenko 2000a, 81)

THE EDITORIAL BOARD

Ro.se.i.nat.ro.no.bac' ter. M.L. adj. *roseus* pink; M.L. n. *natron* soda; M.L. masc. n. *bacter* rod; M.L. n. *Roseinatronobacter* pink rod from soda lake.

Gram-negative, strictly aerobic, heterotrophic, nonmotile, lemon-shaped rods (0.5–0.8 × 0.8–2.2 µm); cells single or in chains. **pH optimal 10.** [Na^+] optimal 0.4–0.6 M. **Produce bacterio-**

chlorophyll *a*. Oxidize thiosulfate to sulfate. Do not grow autotrophically.

The mol% G + C of the DNA is: 61.5 (T_m).

Type species: **Roseinatronobacter thiooxidans** Sorokin, Tourova, Kuznetsov, Bryantseva and Gorlenko 2000b, 1415 (Effective publication: Sorokin, Tourova, Kuznetsov, Bryantseva and Gorlenko 2000a, 82.)

FURTHER DESCRIPTIVE INFORMATION

The organism was isolated from an alkaline lake (Sorokin et al., 2000a).

Roseinatronobacter thiooxidans ALG 1 oxidizes thiosulfate to sulfate; addition of thiosulfate to cultures growing aerobically on organic substrates increases the amount of biomass produced; thus, it is able to grow lithoheterotrophically with thiosulfate. *Roseinatronobacter thiooxidans* ALG 1 also produces bacteriochlorophyll *a* but fails to grow autotrophically in the light in the absence of oxygen. Substrates used include acetate, aspartate, benzoate, caproate, caprylate, citrate, fructose, fumarate, glucose, glutamate, glycerol, glycolate, glyoxalate, lactate, maleate, mannitol, propionate, pyruvate, sorbitol, succinate, and valerate. Nitrate can be reduced to nitrite; nitrate, nitrite, ammonium ion, and amino acids can be used as sources of nitrogen (Sorokin et al., 2000a).

Analysis of 16S rDNA sequences placed *Roseinatronobacter* in the *Alphaproteobacteria* (Sorokin et al., 2000a).

ENRICHMENT AND ISOLATION PROCEDURES

This organism was obtained from an alkaline liquid enrichment containing acetate and thiosulfate (Sorokin et al., 2000a).

List of species of the genus Roseinatronobacter

1. **Roseinatronobacter thiooxidans** Sorokin, Tourova, Kuznetsov, Bryantseva and Gorlenko 2000b, 1415[VP] (Effective publication: Sorokin, Tourova, Kuznetsov, Bryantseva and Gorlenko 2000a, 82.)

    thi.o.oxi.dans'. Gr. n. *thios* sulfur; M.L. part. adj. *thiooxidans* sulfur-oxidizing.

    Description: same as given for the genus.
    The mol% G + C of the DNA is: 61.5 (T_m).
    Type strain: ALG 1, DSM 13087.
    GenBank accession number (16S rRNA): AF249749.

Genus XVII. **Roseivivax** Suzuki, Muroga, Takahama and Nishimura 1999a, 632[VP]

CHRISTOPHER RATHGEBER AND VLADIMIR V. YURKOV

Ro.se.i.vi'vax. M.L. adj. *roseus* rose colored, pink; L. adj. *vivax* living; M.L. masc. n. *Roseivivax* pink living organism.

Cells are Gram-negative, slender rods 0.5–1.0 × 1.0–5.0 μm. Motile by means of subpolar flagella. Form pink colonies when grown on agar media due to the presence of carotenoid pigments. **Grow heterotrophically and produce bacteriochlorophyll (Bchl) *a* under aerobic conditions. Reproduction occurs by binary fission. Do not grow anaerobically under illuminated conditions.** The habitat is saline lakes.

The mol% G + C of the DNA is: 59.7–64.4.

Type species: **Roseivivax halodurans** Suzuki, Muroga, Takahama and Nishimura 1999a, 632.

FURTHER DESCRIPTIVE INFORMATION

Two species of *Roseivivax* are presently described, *Roseivivax halodurans* and *Roseivivax halotolerans*. Phylogenetically these species belong to the *Alphaproteobacteria*, and are most closely related to members of the genera *Roseobacter*, *Sagittula*, *Octadecabacter*, and *Sulfitobacter*. Both species form circular, smooth, slightly convex, entire, glistening, opaque, pink colonies on agar media. Although all isolated strains produce photosynthetic pigments, they grow only under aerobic conditions, and do not grow anaerobically even in the presence of light. This peculiarity is common to all representatives of the so-called aerobic phototrophic bacteria.

Cells are slender rods 0.5–1.0 × 1.0–5.0 μm, and reproduce by binary fission. Cells are motile by means of subpolar flagella.

Optimal growth occurs at pH between 7.5 and 8.0, and at temperatures between 27°C and 30°C. *Roseivivax* species show a broad tolerance for saline conditions, with growth occurring between 0% and 20% NaCl. A variety of organic substrates can be utilized for growth, although the specific substrates utilized vary between species. Members of this genus are catalase and oxidase positive, produce indole, but do not produce H₂S. Voges–Proskauer test is negative, ONPG reaction is positive.

The major cellular fatty acid present is $C_{18:1}$ and the major quinone is ubiquinone Q-10, with smaller amounts of ubiquinones Q-7, Q-8, and Q-9.

ENRICHMENT AND ISOLATION PROCEDURES

Both *R. halodurans* and *R. halotolerans* were isolated from the saline lake, Lake Clifton, on the west coast of Australia. *R. halodurans* was isolated from a sample containing charophytes, whereas *R. halotolerans* was isolated from a sample containing epiphytes, which had developed on living stromatolites. Strains were grown on PPES-II medium (Suzuki et al., 1999a) and pure cultures were obtained by replating of separate colonies.

MAINTENANCE PROCEDURES

Both *Roseivivax* spp. can be maintained on PPES-II slant agar culture medium (Shiba et al., 1991).

DIFFERENTIATION OF THE GENUS *ROSEIVIVAX* FROM OTHER GENERA

Genus *Roseivivax* can be differentiated from other known genera based on 16S rRNA gene sequence analysis, as the two known members form a distinct cluster within the *Alphaproteobacteria*. Members of this genus are related to members of the aerobic phototrophic genus *Roseobacter*, but can be differentiated from them by differences in absorption spectrum characteristics. *Roseivivax* shows an absorption spectrum similar to that of *Erythrobacter longus* (Nishimura et al., 1994), with a major peak at 871–873 nm and a smaller peak at 803–805 nm, which correspond to Bchl *a* incorporated into light-harvesting complex I and reaction centers, respectively. In contrast, the absorption spectrum of *Roseobacter* shows a major peak at about 806 nm due to the presence of an unusual bacteriochlorophyll-protein complex, B806, as well as a smaller peak at about 870 nm, due to the light-harvesting complex I.

Roseivivax forms circular pink colonies when grown on agar media. Growth occurs heterotrophically under aerobic conditions, but not under anaerobic conditions even in the light. This peculiarity differentiates *Roseivivax* from all known genera of the purple nonsulfur bacteria, which possess Bchl *a* and can grow under both conditions, aerobic dark and anaerobic light.

Members of the genus *Roseivivax* can be differentiated from each other based on differences in 16S rRNA gene sequence, homology of the DNA, and absorption spectrum characteristics, as well as physiological and biochemical properties, including salt requirement, carbon sources utilized, presence of urease, phosphatase, gelatinase, and the ability to reduce nitrate (Table BXII.α.71).

TAXONOMIC COMMENTS

Shiba et al. (1991) described several strains of aerobic bacterio-chlorophyll *a*-containing bacteria isolated from different saline locations on the east and west coasts of Australia. These strains were divided into four groups (GI, GII, GIII, and GIV), based on cell color, type of absorption spectrum, and cell morphology. *R. halotolerans* and *R. halodurans* were placed into the GII group along with 17 other similar strains.

However, based on DNA–DNA hybridization studies, this group was further divided into four subgroups based on DNA homology. *R. halotolerans* and *R. halodurans* fell outside of these four subgroups, indicating that the genus *Roseivivax* is not closely related to any of the other strains in the GII group (Nishimura et al., 1994).

16S rRNA gene sequence analysis shows that *Roseivivax* forms a separate cluster within the *Alphaproteobacteria*. The closest relatives are members of the genera *Roseobacter*, *Sagittula*, *Octadecabacter*, and *Sulfitobacter*.

FURTHER READING

Shimada, K. 1995. Aerobic anoxygenic phototrophs. *In* Blankenship, Madigan and Bauer (Editors), Anoxygenic Photosynthetic Bacteria, Kluwer Academic Publishers, Dordrecht. pp. 105–122.

Yurkov, V.V. and J.T. Beatty. 1998. Anoxygenic aerobic phototrophic bacteria. Microbiol. Mol. Biol. Rev. *62*: 695–724.

List of species of the genus Roseivivax

1. **Roseivivax halodurans** Suzuki, Muroga, Takahama and Nishimura 1999a, 632[VP]

ha.lo.du' rans. Gr. n. *hals* salt; L. pres. part. *durans* enduring; M.L. part. adj. *halodurans* salt enduring.

Gram-negative rods, 0.5–1.0 × 1.0–5.0 μm, motile by subpolar flagella. Form circular pink colonies on solid media. Cells contain Bchl *a* organized into reaction center and light harvesting complexes, giving absorption spectrum peaks at 803 and 873 nm, respectively. Tolerate high salt concentrations with growth occurring from 0 to 20.0% NaCl. Obligately aerobic heterotrophs, do not grow anaerobically under illuminated conditions. Cells utilize D-glu-cose, L-arabinose, D-fructose, D-galactose, lactose, maltose, D-ribose, sucrose, acetate, butyrate, citrate, DL-lactate, DL-malate, pyruvate, succinate, L-aspartate, and L-glutamate as the sole carbon source. Cells do not utilize fumarate, glycolate, ethanol, methanol, or D-xylose. Do not hydrolyze gelatin, Tween 80, starch, or alginate. Produce catalase, oxidase, phosphatase, and nitrate reductase; do not produce urease. Voges–Proskauer test is negative, ONPG reaction is positive, produce indole but not H_2S.

Optimal growth occurs at pH 7.5–8.0 and at temperatures between 27 and 30°C. Resistant to penicillin and tetracycline, but susceptible to chloramphenicol and strep-

---

TABLE BXII.α.71. Distinguishing characteristics of species *Roseivivax halodurans*, *Roseivivax halotolerans*, and the species *Rubrimonas cliftonensis*[a]

Characteristic	Roseivivax halodurans	Roseivivax halotolerans	Rubrimonas cliftonensis
Nitrate reductase	+	−	+
Phosphatase	+	−	+
Urease	−	+	−
Hydrolysis of gelatin	−	+	−
Utilization of:			
Acetate	+	+	−
Butyrate	+	+	W
Glycolate	−	−	+
DL-Lactate	+	+	−
DL-Malate	+	+	−
Succinate	+	+	−
Ethanol	−	−	+
D-Fructose	+	W	+
D-Glucose	+	W	+
D-Galactose	+	−	+
Maltose	+	−	W
D-Ribose	+	−	+
D-Xylose	−	W	+
Lactose	+	−	−
Sucrose	+	−	−
Growth in presence of:			
0% NaCl	W	+	−
20% NaCl	W	+	−
Near infrared Bchl *a* peaks	803, 873	805, 871	806, 871

[a]Symbols: +, positive; −, negative; W, weak.

tomycin. Habitat: charophytes in a saline lake, Lake Clifton, on the west coast of Australia.

The mol% G + C of the DNA is: 64.4 (HPLC).

Type strain: Och 239, ATCC 700843, CIP 105983, JCM 10272, NBRC 16685.

GenBank accession number (16S rRNA): D85829.

2. **Roseivivax halotolerans** Suzuki, Muroga, Takahama and Nishimura 1999a, 633[VP]

ha.lo.to' le.rans. Gr. n. *hals* salt; L. pres. part. *tolerans* tolerating; M.L. part. adj. *halotolerans* salt tolerating.

Gram-negative rods, 0.5–1.0 × 1.0–5.0 μm, motile by subpolar flagella, form circular pink colonies on agar media. Cells contain Bchl *a* organized into reaction center and light harvesting complexes, exhibiting absorption spectrum peaks at 805 and 871 nm, respectively. Require salt for growth, able to tolerate high salinity with growth occurring between 0.5 and 20.0% NaCl. Aerobic heterotrophs, do not grow anaerobically under illuminated conditions. Sub-

strates utilized as sole carbon source are D-glucose, D-fructose, D-xylose, acetate, butyrate, citrate, DL-lactate, DL-malate, pyruvate, succinate, L-aspartate, and L-glutamate. Does not grow on fumarate, glycolate, ethanol, methanol, L-arabinose, D-galactose, lactose, maltose, D-ribose, or sucrose. Hydrolyze gelatin, do not hydrolyze Tween 80, starch, or alginate. Produce catalase, oxidase, and urease. Do not have phosphatase or nitrate reductase activity. Produce indole but not H_2S, Voges–Proskauer test is negative, ONPG reaction is positive.

Optimal growth occurs at pH 7.5–8.0, and temperature 27–30°C. Resistant to penicillin and tetracycline, but susceptible to chloramphenicol and streptomycin. Habitat: epiphytes growing on living stromatolites in Lake Clifton, a saline lake on the west coast of Australia.

The mol% G + C of the DNA is: 59.7 (HPLC).

Type strain: Och 210, ATCC 700842, CIP 105984, JCM 10271, NBRC 16686.

GenBank accession number (16S rRNA): D85831.

Genus XVIII. **Roseobacter** *Shiba 1991b, 331[VP] (Effective publication: Shiba 1991a, 144)*

TSUNEO SHIBA AND JOHANNES F. IMHOFF

Ro.se.o.bac' ter. M.L. adj. *roseus* rose-colored, pink; M.L. masc. n. *bacter* equivalent of Gr. neut. n. *bacterion* a rod; M.L. masc. n. *Roseobacter* pink rod-shaped bacterium.

Cells are ovoid or rod shaped, motile by subpolar flagella, divide by binary fission. They are **Gram negative** and are members of the *Alphaproteobacteria*. Colonies are circular, smooth, slightly convex, and pink-to-red in color. Absorption spectra of cell suspensions have major maxima in the near infrared region at 805–807 nm and a smaller one at 868–873 nm (Fig. BXII.α.79). **Bacteriochlorophyll (BChl)** *a* is present. The major carotenoid is spheroidenone. Phosphatidylglycerol and diphosphatidylglycerol are present. The main cellular fatty acid is $C_{18:1}$.

Aerobic chemoorganotrophic bacteria. Do not grow phototrophically and do not produce photosynthetic pigments under anoxic conditions in the light. Carbon sources and electron donors supporting growth are simple organic compounds. Gelatin and Tween 80 are hydrolyzed. Catalase and oxidase are present. Biotin, thiamine and nicotinic acid are required as growth factors.

The major respiratory quinone is ubiquinone-10.

Mesophilic and neutrophilic bacteria from the marine environment. Optimal growth is at pH 7.0–8.0 and 20–30°C. Low concentrations of sodium ions are required.

The mol% G + C of the DNA is: 56–60.

Type species: **Roseobacter litoralis** Shiba 1991b, 331 (Effective publication: Shiba 1991a, 144.)

FURTHER DESCRIPTIVE INFORMATION

Roseobacter is an aerobic chemoheterotrophic bacterium containing bacteriochlorophyll (Shiba and Harashima, 1986). According to 16S rDNA sequence similarity, it is closely related to *Sulfitobacter, Staleya, Sagittula, Octadecabacter, Ruegeria, Antarctobacter, Roseovarius, Roseivivax,* and *Marinosulfonomonas,* which form a cluster of bacteria in the *Alphaproteobacteria* that have been isolated from marine environments. Of these bacteria only *Roseovarius, Roseivivax,* and *Staleya* contain BChl *a*.

The requirement of oxygen for photosynthesis and BChl synthesis in *R. denitrificans* is in contrast to the inhibitory effect of oxygen on photosynthesis and bacteriochlorophyll synthesis in

anoxygenic phototrophic purple bacteria (Harashima et al., 1982; Shiba, 1984, 1987; Okamura et al., 1985; Nishimura et al., 1996; Porra et al., 1996; Kortlüke et al., 1997).

BChl synthesis in *Roseobacter* requires oxygen and dark conditions (Shioi and Doi, 1988; Takamiya et al., 1992). Even though the bacteria grow anaerobically with nitrate or trimethylamine-*N*-oxides (TMAO) as terminal electron acceptor, BChl is not synthesized under these conditions (Arata et al., 1988). The 13'-oxo group of the isocyclic ring E of BChl is derived from molecular oxygen via an oxygenase, whereas the same biosynthetic step in *Rhodobacter sphaeroides* is an anaerobic process (Porra et al., 1996).

The photosynthetic apparatus of *Roseobacter* is similar to that found in the anaerobic phototrophic bacteria of the *Rhodobacteraceae*. In *R. denitrificans*, BChl *a* is esterified with phytol as in

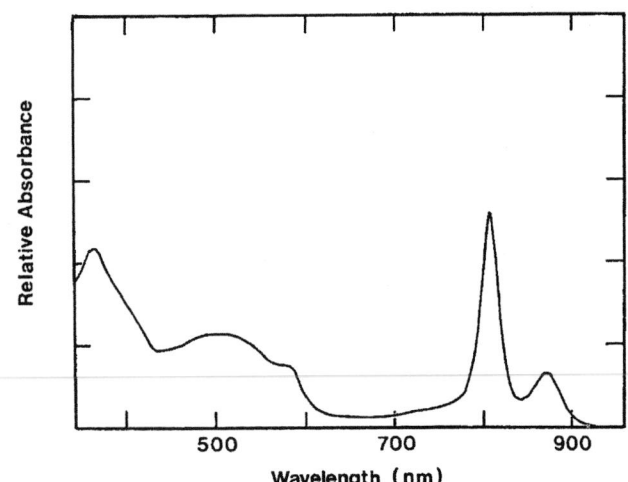

FIGURE BXII.α.79. Absorption spectrum of a membrane fraction of *Roseobacter litoralis* (Reproduced with permission from T. Shiba, Systematic and Applied Microbiology *14*: 140–145, 1991, ©Urban & Fischer Verlag.)

Rhodobacter and many anoxygenic phototrophic purple bacteria (Shiba, 1991a). The major carotenoid is spheroidenone, an oxo-derivative of spheroidene present in *Rhodobacter* species (Harashima and Nakada, 1983). Two BChl-protein complexes, RC-870 and B806, have been isolated from *R. denitrificans* (Shimada et al., 1985). RC-870 corresponds to a reaction center-light harvesting complex (RC-LH1 complex) of the anoxygenic phototrophic purple bacteria with which the complex purified from *R. denitrificans* shares almost identical absorption spectra. One mole of the RC contains 4 mol of BChl *a*, 2 mol of bacteriopheophytin *a*, 4 mol of cytochrome c_{554}, 2 mol of ubiquinone-10 and an unknown amount of carotenoids. The B806 complex corresponds to the B800-850 (LH2) of anoxygenic phototrophic purple bacteria, but lacks the absorption maximum at approx. 850 nm. The BChl-proteins of *R. denitrificans* seem to be integrated in vesicular internal membrane structures. The diameter of the vesicles, 60–120 nm, is larger than that of the anaerobic photosynthetic bacteria (40–60 nm) (Iba et al., 1988). The outer layer of *R. denitrificans* is characterized by a lipopolysaccharide containing no heptose, as in *Rhodobacter* (Neumann et al., 1995).

Roseobacter spp. are aerobic chemoheterotrophic bacteria that are unable to grow phototrophically under anoxic conditions. Light is utilized by *R. denitrificans* as an auxiliary energy source for ATP synthesis only under oxic conditions (Shiba, 1984; Okamura et al., 1985). A photosynthetic electron transfer system operates under anaerobic conditions, if the bacteria can utilize nitrate, nitrite, or TMAO as the terminal oxidant (Arata et al., 1988; Takamiya et al., 1988). Since this photosynthetic system competes with the respiratory system for the same electron transfer component, light suppresses the respiratory activity (Harashima et al., 1982). Therefore, there is no difference in exponential growth in the light and in the dark. A positive effect of light on growth is observed only when respiratory activity is suppressed under certain environmental conditions. For example, growth yield is enhanced only when cultures are illuminated at stationary growth phase (Shioi, 1986; Harashima et al., 1987). The viability of *R. denitrificans* is also enhanced by light in starvation survival media lacking energy-yielding substrates (Shiba, 1984).

Roseobacter has been isolated from green seaweed (Shiba et al., 1979; Shiba, 1991a). Although *Roseobacter* species have not been isolated from seawater, a rDNA clone SAR83, which is phylogenetically closely related to *Roseobacter denitrificans*, is predominant in the bacterial plankton population of Sargasso Sea (Britschgi and Giovannoni, 1991; Fuhrman et al., 1993). Many aerobic bacteriochlorophyll-containing bacteria, the relationship of which to *Roseobacter* is not yet clarified, have been isolated from cyanobacterial patches, seaweeds, sea grasses, sea sands, and surface seawater. The population of aerobic bacteriochlorophyll-containing bacteria may comprise as much as 20–50% of the total aerobic chemoheterotrophic bacteria (Shiba et al., 1991; Shiba, 1995).

ENRICHMENT AND ISOLATION PROCEDURES

Roseobacter strains have been isolated on agar plates with a modified PPES-II medium (Shiba, 1989) which contains per liter: 2 g polypeptone (Nihon Pharmaceutical Co.), 1 g proteose–peptone no. 3 (Difco), 1 g Bacto soytone (Difco), 1 g Bacto yeast extract (Difco), 0.1 g ferric citrate, 700 ml artificial seawater (Lyman and Fleming, 1940), and 300 ml distilled water. Inclusion of iron strongly enhances BChl synthesis. The pH is adjusted to 7.6–7.8. The agar plates are incubated at 20–25°C. Pink- or red-pigmented colonies are formed on agar plates.

Shioi (1986) used a medium which contains in 1 liter of distilled water: 20 g NaCl, 5.0 g $MgCl_2 \cdot 6H_2O$, 2.0 g Na_2SO_4, 0.5 g KCl, 0.5 g $CaCl_2 \cdot 2H_2O$, 0.2 g $NaHCO_3$, 0.1 g ferric citrate, 2.0 g yeast extract (Difco), 1.0 g polypeptone, 1.0 g Casamino acids, and 1.0 ml glycerol. The growth yield in this medium is approximately twice as high as that in PPES-II medium. Elimination of the trace elements Br, Sr, B, and F from the medium of Shioi (1986) has no effect on growth.

MAINTENANCE PROCEDURES

Roseobacter can be maintained at 20°C as stab cultures in media containing 0.4% agar (Difco). Cultures should be transferred monthly. Maintenance on agar slant cultures at room temperature is not recommended. The bacteria remain viable for 6 months when slant cultures are kept at −80°C. Cultures may also be preserved in liquid nitrogen or by lyophilization.

DIFFERENTIATION OF THE GENUS *ROSEOBACTER* FROM OTHER GENERA

Roseobacter differs from anoxygenic phototrophic bacteria in the *Rhodobacteraceae* in its inability to perform photosynthesis under anoxic conditions. Based on 16S rDNA sequence analysis, the genus is most closely related to *Sulfitobacter, Staleya, Sagittula, Octadecabacter, Ruegeria, Antarctobacter, Roseovarius, Roseivivax*, and *Marinosulfonomonas* (Sorokin, 1995; González et al., 1997b; Gosink et al., 1997; Holmes et al., 1997; Labrenz et al., 1998, 1999; Uchino et al., 1998; Pukall et al., 1999; Suzuki et al., 1999a; Labrenz et al., 2000). It can be differentiated from these bacteria by pigment and phospholipid composition, capabilities for degradation of organic substances, and other properties. Unlike *Roseobacter*, cells of *Sulfitobacter, Sagittula, Octadecabacter, Ruegeria*, and *Antarctobacter* do not contain bacteriochlorophyll. Although *Staleya, Roseivivax*, and *Roseovarius* contain BChl *a*, their absorption spectra do not show the large peak at ~800 nm. Phosphatidylethanolamine is present in the cells of *Staleya* and *Roseovarius* but not in *Roseobacter*. Tween 80 is hydrolyzed by *Roseobacter, Staleya*, and *Roseovarius tolerans* strain EL-52, but not by species of the other genera. *Staleya, Roseovarius*, and *Antarctobacter* can grow at 150% seawater but *Roseobacter* cannot. *Roseobacter* differs from *Marinosulfonomonas* and *Sagittula* in lacking the capacity for methylotrophic growth (Holmes, et al., 1997). *Roseobacter* can grow at 30°C but the gas-vacuolated bacterium *Octadecabacter* cannot (Gosink et al., 1997). Differential characteristics of the genus *Roseobacter* and related genera are shown in Table BXII.α.72.

TAXONOMIC COMMENTS

The genus *Roseobacter* was proposed for aerobic chemoheterotrophic bacteria which contain BChl *a* and have a strict requirement of oxygen for synthesis of bacteriochlorophyll (Shiba, 1991a).

Subsequently, *Roseobacter algicola* and *Roseobacter gallaeciensis* were included in the genus *Roseobacter* based only on 16S rDNA sequence information, although their phenotypic characteristics were not coincident with the definition of the genus *Roseobacter* (Lafay et al., 1995; Ruiz-Ponte et al., 1998). Neither species contains BChl *a*. Data on levels of DNA–DNA reassociation to the type species of *Roseobacter litoralis* are lacking (Lafay et al., 1995). In addition, the phylogenetic position of *Roseobacter algicola* is more distant from *Roseobacter litoralis* and *R. denitrificans* than are *Sagittula, Sulfitobacter*, and *Octadecabacter* (González et al., 1997b; Labrenz et al., 1999).

Roseobacter algicola has been reassigned to a new genus as *Rue-*

TABLE BXII.α.72. Comparison of different genera of bacteria related to *Roseobacter*[a]

	Roseobacter	*Roseobacter gallaeciensis*	*Ruegeria*	*Roseivivax*	*Roseovarius*	*Antarctobacter*	*Octadecabacter*	*Sulfitobacter*	*Staleya*	*Sagittula*	*Marinosulfonomonas*
Bacteriochlorophyll	+	−	−	+	+	−	−	−	V	−	−
Bchl protein (nm)[b]	805/873			873	877–879				800-2/861-2		
Motility	+	+	V	+	+	V	−	+	+	+[c]	−
Degradation of:											
Tween 80	+	−	−	−	V	−	nd	nd	+	−	nd
Gelatin	+	−	+	V	−	+	−	V	−	−	nd
Requirement of:											
Biotin	+	−	+	nd	W	W	V	nd	W	enhance	nd
Niacin	+	−	−	nd	W	+	V	nd	W	enhance	nd
Oxidase	+	+	+	+	W	+	−	+	+	+	+
Diphosphatidylglycerol	+	nd	+	nd	+	−	nd	+	−	nd	nd
Phosphatidylethanolamine	−	nd	+	nd	+	−	nd	+	+	nd	nd
Gas vacuole	−	nd	−	nd	nd	−	+	−	+	−	−
Growth at 150% seawater	−	−	−	nd	+	+	−	−	+	−	−
Growth at:											
30°C	+	+	+	+	+	+	+	+	+	+	+
4°C	+	−	−	nd	+	+	+	+	+	+	−
Utilization of:											
Methanol	−	nd	−	−	−	−	−	−	−	+	+
Butyrate	−	+	V	+	+	+	−	+	−	+	nd

[a]Symbols: +, positive; −, negative; nd, not described; V, variable; W, weak; enhance, growth enhancement.

[b]Cell suspension spectra.

[c]Motility of *Sagittula* was assumed by the presence of flagella in the suspension, but motile cells were not seen.

geria algicola (Uchino et al., 1998) and according to 16S rDNA analysis constitutes a robust monophyletic cluster with *R. gallae-* *ciensis* (Suzuki et al., 1999a). Neither species is considered in this chapter.

List of species of the genus Roseobacter

1. **Roseobacter litoralis** Shiba 1991b, 331[VP] (Effective publication: Shiba 1991a, 144.)

li.to.ra' lis. L. gen. n. *litoris* of the seashore; M.L. masc. adj. *litoralis* belonging or pertaining to the seashore.

Cells are rod-to-ovoid shaped, 0.6–0.9 × 1.2–2.0 μm, motile by subpolar flagella. Cell suspensions and colonies are pink. Cell suspensions show a major absorption band at 805–807 nm and a minor one at 868–873 nm. Bacteriochlorophyll *a* esterified with phytol is present. The major carotenoid is spheroidenone. Phosphatidylglycerol and diphosphatidylglycerol are present. The main cellular fatty acid is $C_{18:1}$.

Strictly aerobic chemoheterotrophic bacteria that grow on several organic substrates under oxic conditions. They produce photosynthetic pigments under aerobic conditions, but are unable to grow photosynthetically in the absence of oxygen. Under oxic conditions, photosynthetic energy generation is possible. Methanol is not utilized. Dissimilatory nitrate reducing activity is absent. Chemoautotrophic growth with H_2 is not possible. Biotin, thiamin, and nicotinic acid are required as growth factors.

Optimal growth is at pH 7–8 and 20–30°C. Sodium ions are required.

Habitat: surface of high-tidal seaweeds.

The mol% G + C of the DNA is: 57.2 ± 0.9 (HPLC).

Type strain: OCh149, ATCC 49566, DSM 6996, IFO 15278.

GenBank accession number (16S rRNA): X78312.

2. **Roseobacter denitrificans** Shiba 1991a, 331[VP] (Effective publication: Shiba 1991a, 144.)

de.ni.tri' fi.cans. M.L. part. adj. *denitrificans* denitrifying.

Cells are rod-to-ovoid shaped, 0.6–0.9 × 1.0–2.0 μm, motile by subpolar flagella. Internal photosynthetic membranes are of vesicular type. Cell suspensions and colonies are pink. Cell suspensions show a major absorption band at 805–807 nm and a minor one at 868–873 nm. Bacteriochlorophyll *a* esterified with phytol is present. The major carotenoid is spheroidenone. The major quinone is ubiquinone-10. Phosphatidylglycerol, diphosphatidylglycerol, and phosphatidylcholine are present. The main cellular fatty acid is $C_{18:1}$.

Growth occurs chemoheterotrophically under oxic conditions with several organic substrates. They produce photosynthetic pigments under aerobic conditions, but are incapable of photosynthetic growth in the absence of oxygen. Under oxic conditions, photosynthetic energy generation is possible. Methanol is not utilized. Photosynthetic activity is also found in the presence of nitrate or trimethylamine-*N*-oxide as electron acceptors. Nitrate is reduced to N_2O.

Chemoautotrophic growth with H_2 is not possible. Biotin, thiamin, and nicotinic acid are required as growth factors.

Optimal growth is at pH 7–8 and 20–30°C. Sodium ions are required.

Habitat: surface of high-tidal seaweeds.

The mol% G + C of the DNA is: 59.6 ± 0.5 (HPLC).

Type strain: OCh114, ATCC 33942, DSM 7001, IFO 15277.

GenBank accession number (16S rRNA): L01784, M59063.

Genus XIX. **Roseovarius** Labrenz, Collins, Lawson, Tindall, Schumann and Hirsch 1999, 145[VP]

MATTHIAS LABRENZ AND PETER HIRSCH

Ro' se.o.va' ri.us. L. adj. *roseus* rosy; L. adj. *varius* diverse, varied; M.L. masc. n. *Roseovarius* a variably rosy bacterium.

Rods, one or both cell poles pointed, multiplying by monopolar growth, i.e., by a budding process. Gram negative. Daughter cells may be motile. No resting stages are known. **Aerobic**, with a strictly respiratory type of metabolism. **Intracellular granules of poly-β-hydroxybutyrate (PHB) are present. Bacteriochlorophyll *a* (bchl *a*) may be produced.** Cells have an absolute requirement for Na$^+$. Temperature range for growth <3–43.5°C, salinity range for growth <10 to >150‰ of artificial seawater (ASW), NaCl tolerance range <1.0 to 10.0% and pH range for growth 5.3 to >9. **Catalase and oxidase weakly positive.** Do not grow photoautotrophically with H_2/CO_2 (80:20) or photoorganotrophically with acetate or glutamate. Occur in marine habitats.

The mol% G + C of the DNA is: 62–64.

Type species: **Roseovarius tolerans** Labrenz, Collins, Lawson, Tindall, Schumann and Hirsch 1999, 145.

FURTHER DESCRIPTIVE INFORMATION

Visible growth of *R. tolerans* appears at 20°C after 3–5 d on medium PYGV (Staley, 1968) prepared with 25‰ (or 40‰) artificial seawater (ASW; Lyman and Fleming, 1940) or on R2A agar (Difco, Detroit) with the appropriate ASW concentration. Colonies are circular, smooth, convex, 1–2 mm in diameter and red, pink, beige to light red, beige, or whitish beige. Cell sizes: 0.7–1.0 × 1.1–2.2 μm. Cell growth appears to be monopolar since one cell end is usually more narrow and shorter, which indicates a budding process (Hirsch, 1974a). In PHBA medium (Labrenz et al., 1998), motility can be observed. Small daughter cells show predominantly tumbling motion and only rarely directed movements, but flagella were not observed.

Bchl *a* is found in strains EL-78, EL-83, EL-171, and EL-172^T in cell suspensions grown in the dark (Labrenz et al., 1999). Absorbances characteristic of bchl *a* with a large peak at 877–879 nm and smaller ones at 799–802 nm and 589–591 nm are similar to maxima found in bchl *a*-containing anoxygenic phototrophs (Biebl and Drews, 1969), but they differ from the maxima of bchl *a*-containing *Roseobacter denitrificans* and *Roseobacter litoralis* (Shiba, 1991a) or *Staleya guttiformis* (Labrenz et al., 2000). Under identical growth conditions, bchl *a* could not be detected in suspensions or methanolic extracts of *R. tolerans* strains EL-52, EL-90, EL-164, or EL-222. Even concentrated extracts of these strains lacked bchl *a* when thin layer chromatography was applied. The *in vitro* absorption spectra of methanolic *R. tolerans* extracts show a large peak at 767–769 nm and smaller ones at 605–607, 699–700 nm, and around 350 nm. Unlike *Roseobacter denitrificans* or *Staleya guttiformis*, the production of bchl *a* by these EL-strains is repressed in constant dim light.

Colony colors of strains EL-83 and EL-171 were white-beige to beige during the first six years of cultivation; at that time these strains did not produce bchl *a*. Then the colony colors changed to light red and *in vivo* bchl *a* spectra had a small peak at 868–871 nm. Five months later, colonies of EL-83 became pinkish and since then this strain produces bchl *a* just as EL-78 and EL-172^T do. Methanolic extracts of all four bchl-producing EL-strains have identical absorption spectra. Unlike in *Roseobacter denitrificans*, vesicular intracytoplasmatic membrane systems (Harashima et al., 1982) are not found in ultrathin sections of aerobically and dark grown cells of EL-172^T. *R. tolerans* has an absolute requirement for Na$^+$, but not for K$^+$, Mg^{2+}, Ca^{2+}, Cl$^-$, or SO$_4^{2-}$.

In the presence of available nitrogen, *Roseovarius tolerans* uses acetate, pyruvate, malate, succinate, butyrate, or glutamate, but not citrate, methanesulfonic acid, methanol, or α-D-glucose. Without any added nitrogen compounds, this species grows weakly on acetate, pyruvate, succinate, malate, or butyrate. Glutamate is used with and without an additional source of combined nitrogen. No growth on glucose anaerobically in the absence of nitrate. Metabolism of carbon sources tested with the Biolog system is described in Labrenz et al. (1999). Strain EL-222 grows microaerophilically.

The peptidoglycan of *R. tolerans* contains *m*-diaminopimelic acid. Respiratory quinone is Q10. Phospholipids present are: diphosphatidylglycerol, phosphatidylglycerol, phosphatidylcholine, and phosphatidylethanolamine as well as an unknown phospholipid and an aminolipid. The dominant fatty acid is $C_{18:1}$ (70.2%); other characteristic fatty acids are $C_{18:2}$ (10.6%), $C_{12:0\ 2OH}$ (2.4%), $C_{12:1\ 3OH}$ (3.6%), $C_{16:1}$ (0.8%), $C_{16:0}$ (6.2%), and $C_{18:0}$ (0.8%).

Comparative 16S rRNA gene sequence analysis shows *R. tolerans* to be a member of the *Alphaproteobacteria*. Highest 16S rRNA gene sequence relatedness (93–95%) is displayed with species of the genus *Ruegeria* (Uchino et al., 1998), viz. *R. algicola* (Lafay et al., 1995) as well as with *Roseobacter*, *Antarctobacter*, *Sagittula*, *Staleya*, *Octadecabacter*, and *Sulfitobacter*. (See Fig. BXII.α.71 of the chapter describing the genus *Antarctobacter*.)

ENRICHMENT AND ISOLATION PROCEDURES

Two enrichment procedures may be used for *R. tolerans*: (1) Original water samples (0.5 ml) are spread directly on agar plates of medium PYGV prepared with water of 72‰ salinity or with 67‰ or 130‰ ASW salinity. The same procedure may be carried out with Sabouraud–Dextrose–Agar prepared with 72‰ salinity water. Incubation is at 4°C or 15°C in the dark or at 4.1 μmol photons m^{-2} s^{-1}. Pure cultures are isolated by several dilution transfers on the corresponding agar media. (2) Original water samples (50 ml) are amended with 1 ml of a filter-sterilized solution of 2.5 mg/ml Bacto Yeast Extract prepared in ASW (72‰); incubation at 15°C and in dim light of 4.1 μmol photons m^{-2} s^{-1}.

MAINTENANCE PROCEDURES

Strains of *Roseovarius tolerans* can initially be cultivated on PYGV or R2A agar with 25‰ or 40‰ ASW. After incubation at 8–33°C to allow abundant growth, cultures may be maintained at 4°C for 6 months. They can also be preserved indefinitely by lyophilization or frozen at −72°C in 15% glycerol (v/v).

PROCEDURES FOR TESTING SPECIAL CHARACTERS

For the study of 16S rRNA gene fragments, PCR amplification is recommended as described by Hudson et al. (1993) and Labrenz et al. (1999).

DIFFERENTIATION OF THE GENUS *ROSEOVARIUS* FROM OTHER GENERA

Table BXII.α.73 lists characteristics of *Roseovarius* that differentiate it from other morphologically, physiologically, or chemotaxonomically similar organisms. Fig. BXII.α.71 of the chapter describing the genus *Antarctobacter* presents an unrooted tree showing phylogenetic relationships of *Roseovarius tolerans* with closely related *Alphaproteobacteria*.

TAXONOMIC COMMENTS

The combination of respiratory lipoquinone, fatty acid, and polar lipid data indicates that *R. tolerans* belongs with organisms (at the genus or family rank) within the *Alphaproteobacteria*. $C_{18:1}$ is the characteristic fatty acid for the *Alphaproteobacteria* whereas $C_{18:2}$ occurs in most cases in *R. tolerans* and *Roseobacter* spp. These two groups are differentiated by fatty acids in lower proportions, such as $C_{12:0\ 2OH}$ and $C_{12:1\ 3OH}$ (*R. tolerans*), or $C_{10:0\ 2OH}$ and $C_{14:1\ 3OH}$ (*Roseobacter denitrificans* and *Roseobacter litoralis*) (Labrenz et al., 2000). When grown on Bacto Marine Broth (Difco), *Ruegeria algicola* lacks these characteristic fatty acids. However, the polar lipid patterns of *Roseovarius tolerans* and *Ruegeria algicola* are nearly identical (Labrenz et al., 1999) and they differ from those of *Roseobacter denitrificans*, *Roseobacter litoralis*, or *Antarctobacter heliothermus*. In addition, the mol% G + C ratio of *Ruegeria algicola* (64–65) resembles that of *R. tolerans* (62–64) but differs from that of other *Roseobacter* species (56.3–60.1). Comparative 16S rRNA gene sequence analysis, chemotaxonomic, biochemical, and physiological studies clearly show the close relationship of *Roseobacter denitrificans* with *Roseobacter litoralis*. On the other hand, *Ruegeria algicola* appears to be more related to, albeit different from, *Roseovarius tolerans*, as shown by physiological and fatty acid data (Labrenz et al., 1999).

16S rRNA gene sequencing data indicate the affiliation of *R. tolerans* with the *Alphaproteobacteria*; specifically, there is an association with the *Roseobacter* cluster, which also includes *Ruegeria, Staleya, Sulfitobacter, Antarctobacter, Sagittula, Octadecabacter, Silicibacter*, and a *Prionitis lanceolata* gall symbiont. Sequence divergence values of >5% (4.7% to *Ruegeria algicola*) show *R. tolerans* to be phylogenetically distinct from all currently recognized members of the *Proteobacteria*. Furthermore, bootstrap resampling shows *R. tolerans* does not possess a particularly significant phylogenetic affinity with any individual species within the above-mentioned *Roseobacter* cluster.

List of species of the genus Roseovarius

1. **Roseovarius tolerans** Labrenz, Collins, Lawson, Tindall, Schumann and Hirsch 1999, 145[VP]

   to'le.rans. L. part. adj. *tolerans* enduring stress conditions.

   Rods with a size of 0.74–0.83 × 1.34–1.94 μm. Further morphological descriptions as for the genus. Physiological and nutritional characteristics are presented in Table BXII.α.74. Optimal growth occurs between 8°C and 33.5°C with salt concentrations of 1.0–8.0% NaCl or with 10–130‰ ASW. Optimal pH for growth is 5.9 to >9.0. Colonies on medium PYGV or R2A + 25 (or 40‰) ASW are smooth, convex, and red, pinkish, beige to red, beige, or whitish beige. Isolated from hypersaline, meromictic, and heliothermal Ekho Lake, East Antarctica.

   The mol% G + C of the DNA is: 62–64 (HPLC).
   Type strain: EL-172, DSM 11457.
   GenBank accession number (16S rRNA): Y11551.
   Additional Remarks: Reference strains include EL-222, DSM 11463.

TABLE BXII.α.73. Differential characteristics of the genus *Roseovarius* and other morphologically, physiologically, or chemotaxonomically similar organisms[a]

Characteristic	Roseovarius tolerans	Antarctobacter heliothermus	Octadecabacter arcticus	Roseobacter litoralis	Ruegeria algicola	Sagittula stellata	Staleya guttiformis	Sulfitobacter pontiacus
Rosettes formed	−	+	−	−	−	+	+	+
Bud formation	+	+	−	−	−	nd	+	−
Bacteriochlorophyll *a*	d	−	−	+	−	−	v	−
Motility	d	d	−	+	+	+	+	+
Maximum length 10.9–33.6 μm	−	+	−	−	−	−	−	−
Growth at <8.5°C	+	+	+	+	−	+	+	+
Gelatin hydrolysis	−	+	−	+	+	−	−	−
Utilize methanol as a carbon source	−	−	−	−	−	+	−	−
Utilize citrate as a carbon source	−	+	d	+	+	+	−	−
Oxidase	W	+	−	+	+	+	+	+
Phosphatidylcholine	+	+	nd	−	+	nd	+	nd
Diphosphatidylglycerol	+	−	nd	+	+	nd	−	nd
Phosphatidylethanolamine	+	−	nd	−	+	nd	+	nd
$C_{18:2}$ fatty acid	+	−	−	+	+	−	+	nd
$C_{12:1\ 3OH}$ fatty acid	+	+	−	−	−	+	−	nd
$C_{12:0\ 2OH}$ fatty acid	+	−	−	−	−	−	−	nd
Mol% G + C of DNA	62–64	62–63	57	56–58	64–65	65	55–56	62

[a]For symbols see standard definitions; nd, not determined; V, variable; W, weak reaction.

TABLE BXII.α.74. Other characteristics of *Roseovarius tolerans*[a]

Characteristic	R. tolerans
Catalase activity	+
Oxidase activity	w
Bchl *a* absorption bands *in vivo* at 589–591 nm, 799–802 nm, and 877–879 nm	d
Bchl *a* absorption bands *in vitro*[b] at 605–607 nm, 699–700 nm, and 767–769 nm	d
Reduction of NO_3^-	–
Requirement of vitamin B_{12}	d
Requirement of thiamin or nicotinic acid	w
Requirement of pantothenate	–
Stimulation by biotin	+
Hydrolysis of gelatin, starch, or alginate	–

(*continued*)

TABLE BXII.α.74. (*cont.*)

Characteristic	R. tolerans
Hydrolysis of Tween 80 or DNA	d
Indole test	–
H_2S production	–
Voges–Proskauer test	–
Acid from glucose	–
Growth with succinate, glutamate, butyrate acetate, pyruvate, malate	+
Growth with citrate, α-D-glucose methanesulfonic acid	–
Susceptible to chloramphenicol and streptomycin	+
Susceptible to polymyxin B	–
Susceptible to penicillin G or tetracycline	d

[a]For symbols see standard definitions; w, weak reaction.

[b]In methanolic extracts.

Genus XX. **Rubrimonas** Suzuki, Muroga, Takahama, Shiba and Nishimura 1999b, 204[VP]

CHRISTOPHER RATHGEBER AND VLADIMIR V. YURKOV

Ru.bri.mo'nas. L. adj. *ruber* reddish; Gr. n. *monas* unit; M.L. fem. n. *Rubrimonas* reddish monad.

Rubrimonas **cells are Gram-negative, short to ovoid rods, 1.0–1.5 × 1.2–2.0 μm. Motile by polar flagella. Division occurs by binary fission.** Forms pink circular colonies when grown on agar media. **Pink coloration is due to the presence of carotenoid pigments. Cells grow heterotrophically under aerobic conditions and produce bacteriochlorophyll (Bchl) *a*. No growth occurs under anaerobic conditions even in the light.** Habitat is saline lakes.

The mol% G + C of the DNA is: 70.4–74.8.

Type species: **Rubrimonas cliftonensis** Suzuki, Muroga, Takahama, Shiba and Nishimura 1999b, 204.

FURTHER DESCRIPTIVE INFORMATION

Rubrimonas cliftonensis is the only species presently described. Phylogenetically, *Rubrimonas* falls within the *Alphaproteobacteria* forming a separate branch. The closest relative is *Rhodobacter veldkampii* with a 16S rDNA sequence similarity of 89.8%.

R. cliftonensis forms circular, smooth, slightly convex, opaque, pink colonies when grown on agar medium. Cells grown in PPES-II media (Suzuki et al., 1999b) are ovoid rods 1.0–1.5 × 1.2–2.0 μm. Optimal growth occurs at pH 7.5–8.0 and temperature 27–30°C. NaCl is required, with growth occurring in media supplemented with 0.5–7.5% NaCl.

R. cliftonensis is an obligately aerobic heterotroph that produces Bchl *a* under aerobic conditions. Incapable of anaerobic growth even under illumination. A wide variety of organic substrates can be used to support heterotrophic growth, which is typical of other aerobic phototrophic bacteria. Cells utilize D-glucose, L-arabinose, D-fructose, D-galactose, D-ribose, D-xylose, citrate, glycolate, pyruvate, and ethanol. Acids are produced from sugars. Urease, phosphatase, and nitrate reductase are produced (See genus Table BXII.α.71 in Genus *Roseivivax*).

The absorption spectrum of membrane fractions shows a major absorbance peak at 806 and a smaller peak at 871 nm, indicating the presence of Bchl *a* organized into light-harvesting II and light-harvesting I complexes, respectively. This absorption spectrum is similar to that found among members of the genus *Roseobacter*, indicating a similar organization of the photosynthetic apparatus (Nishimura et al., 1994).

The major cellular fatty acid is $C_{18:1}$, making up 68–70% of the cellular fatty acid profile. $C_{16:0}$, $C_{18:0}$, $C_{20:0}$, $C_{19:1}$, and $C_{14:0\ 3OH}$ are found in smaller quantities. The principal ubiquinone is Q-10, making up 98–99% of the quinone content; trace amounts of the quinones Q-8 and Q-9 are also present.

ENRICHMENT AND ISOLATION PROCEDURES

R. cliftonensis was isolated from water samples taken from Lake Clifton, an isolated saline lake on the west coast of Australia. Lake Clifton is a ground water-fed lake, which exhibits a large range in salinity from 14.5 kg/m³ to 31.5 kg/m³ (~1.5–3.2%) (Rosen et al., 1996). Samples were plated on PPES-II agar medium (Suzuki et al., 1999b). Pure cultures were obtained by replating of separate colonies.

MAINTENANCE PROCEDURES

R. cliftonensis can be maintained on PPES-II slant agar culture medium. Long-term storage procedures have not been described.

DIFFERENTIATION OF THE GENUS *RUBRIMONAS* FROM OTHER GENERA

Genus *Rubrimonas* can be differentiated from other genera based on 16S rRNA gene sequence, as well as a relatively high mol% G + C content of 74.8. *R. cliftonensis* forms a distinct branch only distantly related to other phototrophic and nonphototrophic genera. The nearest relative is *Rhodobacter veldkampii*, a purple nonsulfur bacterium. *Rubrimonas* is an obligately aerobic phototrophic bacterium, which forms circular pink colonies on agar media due to the presence of carotenoid pigments. It can be distinguished from nonphototrophic genera by absorption spectrum peaks at 806 and 871 nm, corresponding to the presence of Bchl *a*. Aerobic phototrophic bacteria are distinguished from closely related purple nonsulfur bacteria by their inability to utilize light for anaerobic photosynthetic growth.

TAXONOMIC COMMENTS

Shiba et al. (1991) described several strains of aerobic bacteriochlorophyll *a*-containing bacteria from saline locations on the east and west coasts of Australia. Thirty-seven strains were divided into four groups (GI, GII, GIII, and GIV) based on cell color, type of absorption spectrum, and cell morphology. DNA–DNA hybridization studies have shown that the two strains making up the GIII group were not closely related to members of the other

three groups (Nishimura et al. 1994), and they were subsequently described as *Rubrimonas cliftonensis*.

16S rDNA sequence analysis shows that *Rubrimonas* forms a separate branch within the *Alphaproteobacteria*, with no close relatives. The nearest relative is *Rhodobacter veldkampii* with 89.8% sequence similarity.

FURTHER READING

Shimada, K. 1995. Aerobic anoxygenic phototrophs. *In* Blankenship, Madigan and Bauer (Editors), Anoxygenic Photosynthetic Bacteria, Kluwer Academic Publishers, Dordrecht. pp. 105–122.

Yurkov, V.V. and J.T. Beatty. 1998. Anoxygenic aerobic phototrophic bacteria. Microbiol. Mol. Biol. Rev. *62*: 695–724.

List of species of the genus Rubrimonas

1. **Rubrimonas cliftonensis** Suzuki, Muroga, Takahama, Shiba and Nishimura 1999b, 204^VP

   clif.to.nen′sis. M.L. adj. *cliftonensis* referring to Lake Clifton, Australia, the source of the type strain.

   Gram-negative short to ovoid rods (1.0–1.5 × 1.2–2.0 µm), motile possessing polar flagella, divide by means of binary fission. Form circular, pink colonies when grown on agar media. Cells produce Bchl *a* under aerobic conditions, giving rise to absorption spectrum peaks at 806 and 871 nm, corresponding to Bchl *a* incorporated into light-harvesting II and light-harvesting I complexes respectively. Produce carotenoid pigments. Obligately aerobic heterotroph, incapable of anaerobic growth even under illuminated conditions. Utilize D-glucose, L-arabinose, D-fructose, D-galactose, D-ribose, D-xylose, citrate glycolate, pyruvate,

   and ethanol. Do not utilize acetate, fumarate, DL-lactate, DL-malate, succinate, methanol, L-glutamate, lactose, or sucrose. Acids are produced from sugars. Do not hydrolyze starch, gelatin, alginate, or Tween 80. Cells produce catalase, oxidase, nitrate reductase, phosphatase, and urease. Voges–Proskauer test and ONPG reaction are negative. Produce indole but not H$_2$S.

   NaCl is required, with growth occurring at concentrations ranging from 0.5 to 7.5%. Optimal growth occurs at pH 7.5–8.0 and temperature 27–30°C. Cells are resistant to penicillin and are susceptible to chloramphenicol, streptomycin, and tetracycline. Habitat: saline lake water from Lake Clifton, Australia.

   The mol% G + C of the DNA is: 74.0–74.8 (HPLC).
   Type strain: Och317, CIP 105913, JCM 10189.
   GenBank accession number (16S rRNA): D85834.

Genus XXI. **Ruegeria** Uchino, Hirata, Yokota and Sugiyama 1999, 1^VP (Effective publication: Uchino, Hirata, Yokota and Sugiyama 1998, 208)

THE EDITORIAL BOARD

Rue.ger′ia. M.L. *-ia* ending; M.L. fem. n. *Ruegeria* honoring Rueger, a German microbiologist, for his contribution to the taxonomy of marine species of *Agrobacterium*.

Gram-negative, **ovoid to rod-shaped cells**, 0.6–1.6 × 1.0–4.0 µm. Motile by polar flagella, or nonmotile. Do not form spores. **Aerobic.** Oxidase and catalase positive. No photosynthetic growth. Bacteriochlorophyll *a* is absent. **Major quinone is ubiquinone 10.**

The mol% G + C of the DNA is: 55–59.

Type species: **Ruegeria atlantica** (Rüger and Höfle 1992) Uchino, Hirata, Yokota and Sugiyama 1999, 1 (Effective publication: Uchino, Hirata, Yokota and Sugiyama 1998, 208) (*Agrobacterium atlanticum* Rüger and Höfle 1992, 141.)

List of species of the genus Ruegeria

1. **Ruegeria atlantica** (Rüger and Höfle 1992) Uchino, Hirata, Yokota and Sugiyama 1999, 1^VP (Effective publication: Uchino, Hirata, Yokota and Sugiyama 1998, 208) (*Agrobacterium atlanticum* Rüger and Höfle 1992, 141.)

   at.lan′ti.ca. M.L. adj. *atlantica* pertaining to the Atlantic Ocean as the locality.

   Gram-negative, ovoid to rod-shaped cells, 0.6–1.6 × 1.0–4.0 µm. Motile by polar flagella or nonmotile. Do not form spores. Aerobic. Oxidase and catalase positive. Non-photosynthetic. Bacteriochlorophyll *a* is absent. Nonmotile. Nitrate reduced to nitrite. Seawater or Na$^+$ required for growth. Major fatty acid is C$_{18:1}$. 3-Hydroxy fatty acids are C$_{12:0\ 3OH}$ and C$_{14:1\ 3OH}$. 2-Hydroxy fatty acid is C$_{16:0\ 2OH}$. Major quinone is ubiquinone 10.

   The mol% G + C of the DNA is: 55–58.

   Type strain: 1480, ATCC 700000, CIP 105975, DSM 5823, IAM 14463, IFO 15792.
   GenBank accession number (16S rRNA): D88526.

2. **Ruegeria algicola** (Lafay, Ruimy, Rausch de Traubenberg, Breittmayer, Gauthier and Christen 1995) Uchino, Hirata, Yokota and Sugiyama 1999, 1^VP (Effective publication: Uchino, Hirata, Yokota and Sugiyama 1998, 209) (*Roseobacter algicola* Lafay, Ruimy, Rausch de Traubenberg, Breittmayer, Gauthier and Christen 1995, 295.)

   al.gi′co.la. L. n. *alga* algae; L. subst. *cola* dweller; M.L. n. *algicola* algae dweller.

   Gram-negative, ovoid to rod-shaped cells, 0.6–1.6 × 1.0–4.0 µm. Motile by means of one or two subpolar flagella. Do not form spores. Aerobic. Oxidase and catalase positive.

Not photosynthetic. Bacteriochlorophyll *a* is absent. Cells are ovoid during exponential growth. Colonies on salt-containing agar are beige when young, pinkish beige after 96 h. Optimal temperature 25–30°C. No denitrification. The following tests are positive: oxidase, catalase, gelatinase, esculinase, β-galactosidase, and amylase. Do not accumulate polyhydroxybutyrate. Isolated from a culture of the toxin-producing dinoflagellate *Prorocentrum lima* PLV2. Require Na$^+$ for growth. Major fatty acid is C$_{18:1}$. 3-Hydroxy fatty acids are C$_{12:0\ 3OH}$, C$_{10:0\ 3OH}$, and C$_{14:1\ 3OH}$. 2-Hydroxy fatty acids are absent. Major quinone is ubiquinone 10.

The mol% G + C of the DNA is: unknown.

Type strain: FF3, ATCC 51440, DSM 10251, IAM 14591.

GenBank accession number (16S rRNA): X78315.

3. **Ruegeria gelatinovorans** (Rüger and Höfle 1992) Uchino, Hirata, Yokota and Sugiyama 1999, 1^{VP} (Effective publica-

tion: Uchino, Hirata, Yokota and Sugiyama 1998, 209) (*Agrobacterium gelatinovorum* Rüger and Höfle 1992, 141.)

ge.la.ti.no′vor.ans. M.L. gelatin; L. v. *voro* to devour; M.L. adj. *gelatinovorans* gelatin-devouring.

Gram-negative, ovoid to rod-shaped cells, 0.6–1.6 × 1.0–4.0 μm. Motile by means of polar flagella. Do not form spores. Aerobic. Oxidase and catalase positive. Not photosynthetic. Bacteriochlorophyll *a* is absent. Nitrate reduced to nitrite. Acids produced from glycerol but not glucose, fructose, maltose, or xylose. Isolated from seawater of the Baltic Sea. Require seawater or Na$^+$ for growth. Major fatty acids are C$_{18:1}$ and C$_{18:0}$. 3-Hydroxy fatty acid is C$_{12:0\ 3OH}$. 2-Hydroxy fatty acids are absent. Major quinone is ubiquinone 10.

The mol% G + C of the DNA is: 59.

Type strain: B6, ATCC 25655, DSM 5887, IAM 12617.

GenBank accession number (16S rRNA): D88523.

Genus XXII. **Sagittula** *González, Mayer, Moran, Hodson and Whitman 1997b, 778*^{VP}

JOSÉ M. GONZÁLEZ

Sa.git′tu.la. L. fem. n. *sagittula* small arrow, referring to the shape of the bacterium.

Rod-shaped cells, 0.9 × 2.3 μm. Gram negative. Cells attach by one pole to particles of cellulose and lignocellulose by means of a polar holdfast structure that can be seen by electron microscopy. Endospores and cysts are not formed. Capsules are produced. **Intracellular granules of polyhydroxybutyrate are formed. Aerobic, having a strictly aerobic type of metabolism with oxygen as the terminal electron acceptor.** Do not denitrify. Colonies are nonpigmented. Bacteriochlorophyll *a* is not present. **Oxidase and catalase positive. Sea-salt-based medium is required for growth.** Chemoorganotrophic. Methanol, various carbohydrates and amino acids, and some aromatic compounds such as *p*-coumarate, cinnamate, ferulate, and vanillate are utilized. Organic nitrogen compounds, ammonium, and nitrate serve as nitrogen sources. **Cellulose is hydrolyzed. Synthetic lignin preparations are partially solubilized and mineralized in the presence of glucose.** 16S rRNA gene sequence analysis positions this genus in a group of marine bacteria (except for the moderate halophile *Silicibacter*) within the family *Rhodobacteraceae*. Isolated from a salt marsh at 2% salinity on the coast of Georgia, USA, by enrichment with lignin-rich pulp mill effluent.

The mol% G + C of the DNA is: 65.0.

Type species: **Sagittula stellata** González, Mayer, Moran, Hodson and Whitman 1997b, 778.

FURTHER DESCRIPTIVE INFORMATION

As seen by electron microscopy, each cell has a holdfast structure at one pole and the cell envelope has numerous surface vesicles derived from the outer membrane (Fig. BXII.α.80). Cells form rosettes and aggregates, especially at the stationary phase of growth.

Sagittula contains the following major fatty acids: C$_{16:0}$, C$_{18:0}$, C$_{12:1\ 3OH}$, C$_{19:0\ cyclo\ \omega 8c}$, C$_{18:1\ \omega 7c}$, C$_{18:1\ \omega 9t}$, and C$_{18:1\ \omega 12t}$.

The only species in the genus was isolated from a salt marsh at 2% salinity on the coast of Georgia, USA, by enrichment with the high-molecular-weight fraction of pulp mill effluent. Although the species so far includes only one strain, closely related organisms were detected by molecular techniques in unpolluted

coastal water, which receives high inputs of lignocellulosic material from aquatic plants.

ENRICHMENT AND ISOLATION PROCEDURES

The type strain was isolated on YTSS^1 agar plates from a marine enrichment community growing on the high-molecular-weight fraction of a black liquor sample from pulp mill effluent, which is rich in lignin, lignin byproducts, and other plant polymers in a smaller proportion (González et al., 1997b). The original inoculum was from a salt marsh on the coast of Georgia, USA. The enrichment medium consisted of filter-sterilized seawater containing 5 mM NH$_4$NO$_3$, 1 mM KH$_2$PO$_4$, and the liquor fraction at a concentration of 20 mg C per liter. This medium was inoculated with cloudy, brown-green seawater from the salt marsh. Flasks were incubated aerobically with shaking. After enrichment, the organism was isolated on YTSS agar plates incubated at room temperature.

MAINTENANCE PROCEDURES

The type strain is maintained in YTSS broth with 15% glycerol and 15% DMSO at −70°C. Lyophilized cultures are also used.

DIFFERENTIATION OF THE GENUS *SAGITTULA* FROM OTHER GENERA

Table BXII.α.75 lists characteristics differentiating *Sagittula* from other aerobic marine bacteria. *Sagittula* does not oxidize sulfite, and this characteristic distinguishes it from *Sulfitobacter*. The absence of bacteriochlorophyll *a* production, major fatty acids, and the mol% G + C content of the DNA distinguish it from *Roseobacter denitrificans*, *Roseobacter litoralis*, and *Octadecabacter*. Temperature tolerance is different from *Octadecabacter* and *Silicibacter* (Table BXII.α.75). 16S rRNA gene sequences of *Sagittula* and "*Marinosulfonomonas*" show a low level of homology (90%).

---

1. YTSS agar (g/l distilled water): yeast extract, 4.0; tryptone (Difco), 2.5; sea salts (Sigma Chemical Co., St. Louis, MO), 20; and agar, 18. YTSS broth is the same except that agar is omitted.

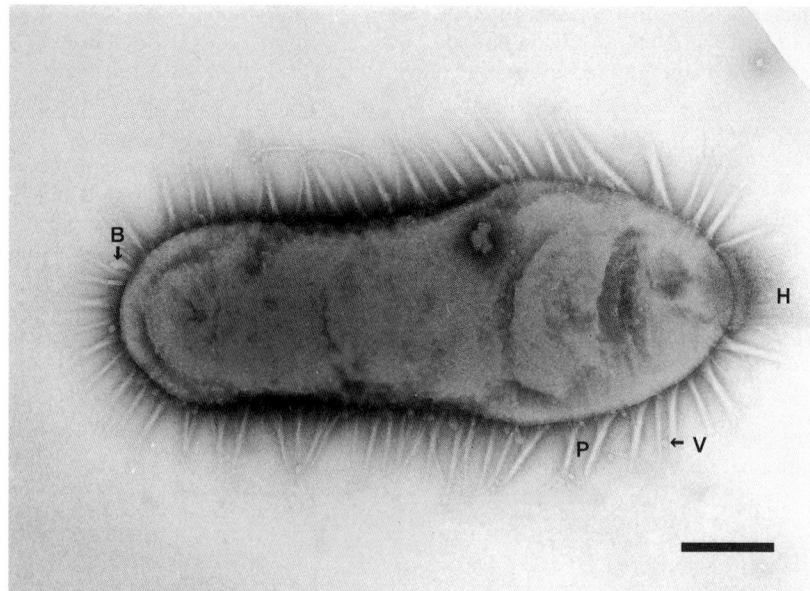

FIGURE BXII.α.80. Electron micrograph of *Sagittula stellata*. Negatively stained. Cells are covered by polysaccharides fibrils (*P*) and blebs (*B*). Vesicles (*V*) are also seen unattached. Holdfast structure (*H*) at the wider cell pole is indicated. The cells attach to cellulose and lignocellulose particles by the wider pole. Bar = 0.2 μm. (Courtesy of F. Mayer).

TABLE BXII.α.75. Differential characteristics of *Sagittula* and some phylogenetically closely related bacteria[a]

Characteristic	Sagittula stellata	"Marinosulfonomonas"	Octadecabacter species	Roseobacter denitrificans	Ruegeria algicola	Silicibacter	Sulfitobacter pontiacus
Cells arranged in rosettes	+	+	−	−	−	−	+
Gas vacuoles	−	−	+	−	−	+	−
Oxidase	+	+	−	+	+	+	+
Nitrate red	−	−	−	+/−	−	+	−
Bchl *a*	−		−	+	−	−	−
Growth temperature, °C	10–41	20–37	4–15	2–30	10–37	22–50	4–35
Optimal temperature, °C	30	30			25–30	45	22–25
Utilization of:							
DMSP	+	−					
Formate	+	+					
Methanol	+	+	−		−		
Sulfite oxidation	−			−			+
Mol% G + C DNA	65		56–57	56–60	55–59	66	58–62

[a]For symbols see standard definitions.

TAXONOMIC COMMENTS

Based on 16S rRNA gene sequence analysis, the closest related genera are *Sulfitobacter, Roseobacter, Ruegeria, Octadecabacter, Silici-bacter*, and "*Marinosulfonomonas*". The genus has a signature secondary structure since a deletion corresponding to Helix 11 of the 16S rRNA occurs. This deletion spans 11 bases starting at position 200 (*E. coli* numbering system).

List of species of the genus Sagittula

1. **Sagittula stellata** González, Mayer, Moran, Hodson and Whitman 1997b, 778[VP]

stel.la′ ta. L. adj. *stellata* starry, here referring to the cell arrangement.

Colonies on marine agar 2216 are light cream colored.

The characteristics are as described for the genus with the following additional information. The cells show polarity, the width of one half is greater than that of the other. The holdfast structure is located at the thicker pole. Vesicles are not only seen on the cell surface but also are free in the medium. The cells occur singly and arranged as rosettes, especially at stationary phase, and form short chains in marine broth 2216.

Growth occurs between pH 5.5 and 8.5; optimal, 7.5. The temperature range for growth is 4–41°C; optimal, 30°C. Sea-salt-based medium is required for growth. Vitamins are required for optimal growth. Does not grow in the absence of a source of fixed nitrogen.

Growth occurs on methanol, the salts of various organic acids, monosaccharides, disaccharides, amino acids, and aromatic compounds. Among the latter are lignin-related compounds such as *p*-coumarate, cinnamate, ferulate, and vanillate. Cells also utilize organic sulfur compounds and similar compounds that are produced by aquatic plants and algae, such as dimethylsulfoniopropionate, 3-mercaptopropionate, methanethiol, and glycine betaine. Strain E-37 quickly oxidizes dimethylsulfide to dimethylsulfoxide in aer-obic conditions without further degradation of dimethyl-sulfoxide (González et al., 1999). Strain E-37 was able to solubilize and partially mineralize artificial lignin in the presence of glucose.

The mol% G + C of the DNA is: 65.0 (HPLC).

Type strain: E-37, ATCC 700073; CIP 105237, DSM 11524.

GenBank accession number (16S rRNA): U58356.

Genus XXIII. **Staleya** Labrenz, Tindall, Lawson, Collins, Schumann and Hirsch 2000, 310[VP]

MATTHIAS LABRENZ AND PETER HIRSCH

Sta'ley.a. M.L. fem. n. *Staleya* named after the American microbiologist J.T. Staley in recognition of his work on budding and appendaged bacteria and his contributions to polar microbiology.

Rods, one or both cell poles narrower, multiplying by monopolar growth, i.e., by a budding process. Gram negative. Cells may be motile. No resting stages known. **Aerobic**, with a strictly respi-ratory type of metabolism. **Intracellular granules of poly-β-hy-droxybutyrate (PHB) are present.** Cells have a weak requirement for Na$^+$. Temperature range for growth <4–32°C, salinity range <10 to <150‰ of artificial seawater (ASW), NaCl tolerance range <1.0–4.0% and pH tolerance range 5.3–6.5 to >9. Oxidase and weakly catalase positive. Do not grow photoautotrophically with H$_2$/CO$_2$ (80:20) or photoorganotrophically with acetate or glutamate. Bacteriochlorophyll *a* (BChl *a*) may be produced. Occur in marine habitats.

The mol% G + C of the DNA is: 55–56.

Type species: **Staleya guttiformis** Labrenz, Tindall, Lawson, Collins, Schumann and Hirsch 2000, 310.

FURTHER DESCRIPTIVE INFORMATION

Visible growth of *S. guttiformis* appears at 20°C after 3–5 d on medium PYGV (Staley, 1968) prepared with 25‰ artificial sea-water (ASW; Lyman and Fleming, 1940). Colonies are circular, smooth, convex, 1–3 mm in diameter and beige, older colonies also pink. Cell sizes: 1.0–1.5 × 1.5–8.9 μm. Cell growth appears to be monopolar since one cell end is usually more narrow and shorter, which indicates a budding process (Hirsch, 1974a). Ro-settes are formed frequently. Cells are motile with one flagellum.

Bacteriochlorophyll *a* is found in older, dark-grown cell sus-pensions of *Staleya guttiformis*. Absorbance values characteristic of BChl *a* have a large peak at 861–862 nm, and smaller ones at 800–802 nm and 590–592 nm. They differ from maxima of the BChl *a*-containing *Roseobacter denitrificans* and *Roseobacter litoralis* (Shiba, 1991a) or *Roseovarius tolerans* (Labrenz et al., 1999). Un-like *Roseovarius tolerans*, the production of BChl *a* by *S. guttiformis* is not totally repressed in constant dim light.

S. guttiformis has a weak requirement for Na$^+$, but not for K$^+$, Mg^{2+}, Ca^{2+}, Cl$^-$ or SO$_4^{2-}$.

Cells do not grow on glucose anaerobically in the absence of nitrate. Metabolism of carbon sources tested with the Biolog sys-tem is described by Labrenz et al. (2000). Acetoin or acids are not produced from glucose. Neither sulfide nor indole is pro-duced. Do not reduce nitrate anaerobically.

The peptidoglycan of *S. guttiformis* contains *m*-diaminopimelic acid. Respiratory quinone is Q-10. Phospholipids present are: phosphatidylglycerol, phosphatidylcholine, and phosphatidyleth-anolamine as well as an unknown aminolipid. The predominant fatty acid is C$_{18:1\ \omega 7c}$ (79.7%); other characteristic fatty acids are C$_{10:0\ 3OH}$ (5.9%), C$_{14:1\ 3OH}$ (2.1%), C$_{16:0}$ (3.9%), C$_{18:0}$ (0.7%), C$_{18:2}$ (5.3%), and C$_{19:1}$ (1.4%).

Comparative 16S rRNA gene sequence analysis shows that *S. guttiformis* is a member of the *Alphaproteobacteria*. Highest 16S rRNA gene sequence relatedness (98%) is displayed with *Sulfi-tobacter mediterraneus* (Pukall et al., 1999), a new species described simultaneously with *Staleya*. In addition, high sequence related-ness is observed with *Sulfitobacter pontiacus* (Sorokin, 1995), *S. brevis* (Labrenz et al., 2000), *Roseobacter denitrificans* and *Roseobacter litoralis* (Shiba, 1991a). Other species belonging to the *Alphapro-teobacteria* show lower levels of relatedness. An unrooted tree de-picting the phylogenetic relationships of *S. guttiformis* and its closest relatives is shown in Fig. BXII.α.71 of the chapter de-scribing the genus *Antarctobacter*. The results of treeing analyses showed that this species phylogenetically clusters with *Sulfitobacter* spp.

ENRICHMENT AND ISOLATION PROCEDURES

An enrichment procedure used for *S. guttiformis*: Original water samples (1.0 ml) are spread directly on medium PYGV agar plates prepared with water of 15‰ salinity. Incubation is at 15°C at 4.1 μmol photons m^{-2} s^{-1}.

MAINTENANCE PROCEDURES

Staleya guttiformis can initially be cultivated on PYGV agar with 10–40‰ ASW. After incubation at 12–20°C to allow abundant growth, cultures may be maintained in a refrigerator (4°C) for 6 months. They can also be preserved by lyophilization or frozen at −72°C in 15% glycerol (v/v).

PROCEDURES FOR TESTING SPECIAL CHARACTERS

For the study of 16S rRNA gene fragments, PCR amplification is recommended as described by Hudson et al. (1993) and La-brenz et al. (1998).

DIFFERENTIATION OF THE GENUS *STALEYA* FROM OTHER GENERA

Table BXII.α.76 lists characteristics of *Staleya* that differentiate it from other morphologically, physiologically, or chemotaxonom-ically similar organisms. Fig. BXII.α.71 of the chapter describing the genus *Antarctobacter* presents an unrooted tree showing phy-logenetic relationships of *Staleya guttiformis* with closely related *Alphaproteobacteria*.

TAXONOMIC COMMENTS

Sequencing of 16S rRNA genes confirmed *S. guttiformis* is a member of the *Alphaproteobacteria*. The combination of respiratory lipoquinone, fatty acid, and polar lipid data also indicates that it belongs with bacteria (above the species rank) of the *Alphaproteobacteria* (Labrenz et al., 2000). Comparative 16S rRNA gene sequence analysis shows that *S. guttiformis* is specifically associated with the *Sulfitobacter–Roseobacter* cluster, but not with *Ruegeria algicola* or *Roseobacter gallaeciensis*. It is evident from treeing analyses that the genus *Roseobacter*, as currently recognized, is interspersed with several other taxa. This is also reflected by the affiliation of the former *Roseobacter algicola* with the genus *Ruegeria* (Uchino et al., 1998).

The chemotaxonomic data indicate that members of the genus *Roseobacter* (i.e., *R. litoralis* and *R. denitrificans*) have a distinctive polar lipid composition, in which both species have phosphatidylglycerol, diphosphatidylglycerol, and an unidentified amino lipid as the major components. In contrast, *S. guttiformis*, *Sulfitobacter brevis*, and *S. pontiacus* all synthesized phosphatidylglycerol, phosphatidylethanolamine, and phosphatidylcholine, together with the same unidentified amino lipid. This polar lipid composition serves to distinguish these two groups from one another.

Phylogenetically, *S. guttiformis* is closely related to *Sulfitobacter* spp. Morphologically, it is also more similar to *Sulfitobacter* than to *Roseobacter* strains: its cells multiply by monopolar growth (a budding process) and form rosettes. The ability to oxidize sulfite and thereby to increase growth characterizes *S. pontiacus* (Sorokin, 1995). *S. mediterraneus* grows also to higher cell densities with 10 mM sulfite in an acetate-supplemented medium (Pukall et al., 1999). *S. brevis* tolerates lower initial sulfite concentrations in an acetate-supplemented medium than *S. pontiacus*, but not so low as *Staleya guttiformis* (Labrenz et al., 2000).

Especially, chemotaxonomic data indicate differences between *Staleya* and *Sulfitobacter*. *Staleya* does not synthesize the fatty acids $C_{16:1}$, $C_{17:1}$, $C_{17:0}$, or $C_{18:1\ 10CH_3}$ as well as the phospholipid diphosphatidylglycerol. Examination of the mol% G + C base content showed that the value for *Staleya* was about 6% lower than that for the *Sulfitobacter* type species, *S. pontiacus*.

Other differences concerned BChl *a*, which was present in *Staleya guttiformis*, as shown by *in vivo* absorption bands at 800–802 and 861–862 nm. These peaks were clearly absent from *Roseobacter* spp. and *Sulfitobacter* cells. The importance of BChl *a* for a taxonomic separation of these aerobic bacteria has been questioned (Labrenz et al., 1999).

In conclusion, bacteria of the *Staleya–Sulfitobacter–Roseobacter* cluster appear to be phylogenetically quite closely related, but are phenotypically very diverse, which makes their taxonomic separation difficult. 16S rDNA sequence data and similarities in chemotaxonomic data make it appear likely that the genera *Staleya*, *Sulfitobacter*, and *Roseobacter* share a common ancestor (Labrenz et al., 1999). Presently, *Staleya* is placed between *Roseobacter* (Shiba, 1991a) and *Sulfitobacter* (Sorokin, 1995).

TABLE BXII.α.76. Differential characteristics of the genus *Staleya* and other morphologically, physiologically, or chemotaxonomically similar organisms[a]

Characteristic	*Staleya*	*Antarctobacter heliothermus*	*Octadecabacter arcticus*	*Roseobacter litoralis*	*Roseovarius tolerans*	*Ruegeria algicola*	*Sagittula stellata*	*Sulfitobacter mediterraneus*	*Sulfitobacter pontiacus*
Rosettes formed	+	+	−	−	−	−	+	+	+
Budding cell division	+	+	−	−	+	−	nd	+	+
Oxidase	+	+	−	+	+	+	+	+	+
BChl *a*	v	−	−	+	+	−	−	−	−
Growth at 4°C	+	+	+	+	+	−	+	+	+
Growth at 37°C	−	+	−	−	+	+	+	−	−
Tween 80 hydrolysis	+	−	nd	+	−	−	−	nd	−
Gelatin hydrolysis	−	+	−	+	−	+	−	+	
Citrate utilized as carbon source	−	+	v	+	−	+	+	+	−
Butyrate utilized as carbon source	−	+	−	−	w	−	+	+	+
Fatty acids:									
$C_{10:0\ 3OH}$	+	−	+	+	−	−	−	+	+
$C_{12:1\ 3OH}$	−	+	−	−	+	−	+	−	−
$C_{12:0\ 2OH}$	−	−	−	−	+	−	−	−	−
$C_{14:1\ 3OH}$	+	−	−	+	−	−	−	−	+
$C_{16:1}$	−	+	+	−	+	−	−	+	+
$C_{17:1}$	−	−	−	−	−	−	−	+	+
$C_{17:0}$	−	−	−	−	−	−	−	+	+
$C_{18:2}$	+	−	−	+	+	+	−	−	−
$C_{18:1\ \omega9c}$	−	−	−	−	−	−	−	+	−
$C_{18:1\ 10CH_3}$	−	−	−	−	−	−	−	+	−
$C_{18:1\ 18CH_3}$	−	−	−	−	−	−	−	+	−
$C_{19:0\ cyclo}$	−	+	−	−	−	−	+	−	−
$C_{19:1}$	+	−	−	−	−	−	−	−	−
Diphosphatidylglycerol	−	−	nd	+	+	+	nd	nd	+
Phosphatidylethanolamine	+	−	nd	−	+	+	nd	nd	+
Phosphatidylcholine	+	+	nd	−	+	+	nd	nd	+
Mol% G + C content	55–56	62–64	57	56–59	63	64–65	65	59	62–63

[a]For symbols see standard definitions; nd, not determined; w, weak reaction.

List of species of the genus Staleya

1. **Staleya guttiformis** Labrenz, Tindall, Lawson, Collins, Schumann and Hirsch 2000, 310[VP]

gut.ti.for' mis. L. fem. n. *gutta* the drop; M.L. *guttiformis* drop shaped.

Rods with average size of 1.1 × 1.8 µm. Further morphological description as for the genus. Physiological and nutritional characteristics are presented in Table BXII.α.77. Optimal growth occurs between 12 and 20°C with 1.0% NaCl or 10–40‰ ASW. Optimal pH for growth is 7.0–8.5. Colonies on medium PYGV are smooth, convex, and beige, later pink. Isolated from hypersaline, meromictic, and heliothermal Ekho Lake, East Antarctica.

The mol% G + C of the DNA is: 55–56 (HPLC).

Type strain: EL-38, DSM 11458.

GenBank accession number (16S rRNA): Y16427.

TABLE BXII.α.77. Other characteristics of *Staleya guttiformis*[a]

Characteristic	S. guttiformis
Catalase activity	w
BChl *a* absorption bands *in vivo* at 590–592 nm, 800–802 nm and 861–862 nm	v
Aerobic reduction of NO_3^- to NO_2^-	+
Requirement of vitamin B_{12}	−
Requirement of thiamin, biotin, or nicotinic acid	w
Requirement of pantothenate	+
Hydrolysis of starch or alginate	−
Hydrolysis of DNA	+
Growth with succinate, glutamate, acetate, pyruvate, malate	+
Growth with α-D-glucose	w
Growth with methanesulfonic acid or methanol	−
Susceptible to chloramphenicol, penicillin G, tetracycline, polymyxin B, nalidixic acid or streptomycin	+

[a]For symbols see standard definitions; w, weak reaction.

Genus XXIV. **Stappia** Uchino, Hirata, Yokota and Sugiyama 1999, 1[VP] (Effective publication: Uchino, Hirata, Yokota and Sugiyama 1998, 208)

THE EDITORIAL BOARD

Stap.pi.a. M.L. dim. *-ia* ending; M.L. fem. n. *Stappia* honoring Stapp, a Belgian microbiologist, for his contribution to the taxonomy of marine species of *Agrobacterium*.

Gram-negative **rods**, 0.6–1.0 × 2.0–4.0 µm. Motile by means of polar flagella. Do not form spores. **Aerobic.** Nonphototrophic. Oxidase and catalase positive. Seawater or Na⁺ required for growth. Major fatty acids are $C_{18:1}$ and $C_{18:0}$. 3-Hydroxy fatty acid is $C_{14:0\ 3OH}$. 2-Hydroxy fatty acids are absent. **Major quinone is ubiquinone 10.** Isolated from marine sediment and seawater.

The mol% G + C of the DNA is: 59.

Type species: **Stappia stellulata** (Rüger and Höfle 1992) Uchino, Hirata, Yokota and Sugiyama 1999, 1 (Effective publication: Uchino, Hirata, Yokota and Sugiyama 1998, 208) (*Agrobacterium stellulatum* Rüger and Höfle 1992, 141.)

List of species of the genus Stappia

1. **Stappia stellulata** (Rüger and Höfle 1992) Uchino, Hirata, Yokota and Sugiyama 1999, 1[VP] (Effective publication: Uchino, Hirata, Yokota and Sugiyama 1998, 208) (*Agrobacterium stellulatum* Rüger and Höfle 1992, 141.)

stel' lu.la.ta. M.L. fem. n. *stella* star-shaped; M.L. adj. *stellulata* star-shaped.

Gram-negative rods, 0.6–1.0 x 2.0–4.0 µm. Motile by means of polar flagella. Do not form spores. Aerobic. Nonphototrophic. Oxidase and catalase positive. Seawater or Na⁺ required for growth. Nitrate is not reduced to nitrate but is reduced to gas. No acid from glucose, maltose, or mannitol after 4–6 weeks. Major fatty acids are $C_{18:1}$ and $C_{18:0}$. 3-Hydroxy fatty acid is $C_{14:0\ 3OH}$. 2-Hydroxy fatty acids are absent. Major quinone is ubiquinone 10. Isolated from marine sediment and seawater.

The mol% G + C of the DNA is: 59.

Type strain: ATCC 15215, CIP 105977, DSM 5886, IAM 12621, IFO 15764.

GenBank accession number (16S rRNA): D88525.

2. **Stappia aggregata** (ex Ahrens 1968) Uchino, Hirata, Yokota and Sugiyama 1999, 1[VP] (Effective publication: Uchino, Hirata, Yokota and Sugiyama 1998, 208.)

ag.gre.ga' ta. L. adj. *aggregatus* joined together, referring to the frequent formation of rosettes.

Gram-negative rods, 0.6–1.0 × 2.0–4.0 µm. Motile by means of polar flagella. Do not form spores. Aerobic. No photosynthetic growth. Oxidase and catalase positive. Seawater or Na⁺ required for growth. Nitrate is not reduced to nitrate but is reduced to gas. No acid from glucose or xylose. Weak or variable acid production from fructose, glycerol, and maltose after 4–6 weeks. Major fatty acids are $C_{18:1}$ and $C_{18:0}$. 3-Hydroxy fatty acid is $C_{14:0\ 3OH}$. 2-Hydroxy fatty acids are absent. Major quinone is ubiquinone 10. Isolated from marine sediment and seawater.

The mol% G + C of the DNA is: 59.

Type strain: B1, ATCC 25650, IAM 12614, NCMB 2208.

GenBank accession number (16S rRNA): D88520.

Genus XXV. *Sulfitobacter* Sorokin 1996, 362[VP] (Effective publication: Sorokin 1995, 304)

DIMITRY Y. SOROKIN, FRED A. RAINEY, RICHARD I. WEBB AND JOHN A. FUERST

Sul.fi.to.bac' ter. M.L. n. *sulfitum* sulfite; M.L. masc. form *bacter* Gr. neut. n. *bactrum* rod; M.L. masc. n. *Sulfitobacter* sulfite rod.

Rod-shaped cells 0.5–1.0 × 1.0–3.0 μm. Often dumbbell shaped, with a tendency for polar growth. **Motile by means of 1–5 subpolar flagella.** Gram negative. Inside the cells, the nucleoid and polar region are often in separate compartments. **Obligately aerobic, having a strictly respiratory type of metabolism with oxygen as the terminal electron acceptor. Obligately chemoorganotrophic, but can use energy from the oxidation of thiosulfate and sulfite during growth in the presence of an organic carbon source. Cells cannot grow autotrophically with sulfur compounds or hydrogen as electron donors, but are able to oxidize thiosulfate, sulfur, and sulfite to sulfate.** Can grow in presence of at least 10 mM sulfite; some strains can grow with up to 60 mM sulfite. **NaCl is required for growth.** Mesophilic. Neutrophilic.

The mol% G + C of the DNA is: 59–62.

Type species: **Sulfitobacter pontiacus** Sorokin 1996, 362 (Effective publication: Sorokin 1995, 304.)

FURTHER DESCRIPTIVE INFORMATION

Ultrastructure Many cells of *Sulfitobacter pontiacus*, especially taken from batch culture, display an unusual compartmentalized cell ultrastructure (Fig. BXII.α.81). At least some cells appear to possess a major cell compartment bounded by a single membrane and containing the cell's nucleoid as well as electron-dense ribosome-like particles. A less electron-dense cytoplasm surrounds the compartment, with a larger region of this material at one pole comprising a polar cap. Such organization appears in thin sections whether cells are either chemically fixed or cryofixed and processed via cryosubstitution. An indication of such a structure may also be apparent from negatively stained whole cells (Sorokin, 1995). This type of cell organization has been referred to as a pirellulosome, an organelle reported previously only in the genus *Pirellula* (Lindsay et al., 1997), a member of the order *Planctomycetales*, which is a distinct phylogenetic lineage within the domain *Bacteria*. Members of at least two genera of planctomycetes (*Pirellula* and *Gemmata*) display unusual compartmented cell organization (Fuerst and Webb, 1991; Fuerst, 1995). Pirellulosomes of *S. pontiacus* differ from those in *Pirellula* in that in cells of the latter a nucleoid is always clearly visible as a condensed fibrillar region even after cryosubstitution, whereas in *Sulfitobacter* a condensed nucleoid in the pirellulosome is not clearly apparent in cryosubstituted cells, but is revealed after chemical fixation. This absence of distinct nucleoid in cryosubstituted cells is typical of the limited model bacteria so far examined by cryosubstitution. The occurrence of membrane-bounded nucleoid-containing cell compartments of analogous organization in two phylogenetically widely separated members of separate divisions or phyla of the domain *Bacteria* suggests either retention of an ancient characteristic of cell organization within deeply divergent lineages or convergent evolution of an analogous structure with similar function.

Metabolism Although *Sulfitobacter* species are obligate heterotrophs, *S. pontiacus* possesses a unique ability to derive additional metabolic energy from sulfite oxidation under substrate-limiting conditions, thus belonging to a specific type of lithoheterotrophs. The high tolerance to sulfite has also been demonstrated for other *Sulfitobacter* species but whether or not they are capable of sulfite oxidation and sulfite-dependent lithoheterotrophy remains unclear.

Habitat Sulfur-oxidizing chemolithoheterotrophic bacteria of the type represented by *S. pontiacus* rather than the autotrophic thiobacilli were found to dominate the sulfate-forming population in the Black Sea. This can be explained by better adaptation of these versatile heterotrophs to the conditions of the redox layer of the Black Sea, where relatively diluted upward fluxes of reduced sulfur compounds meet downward fluxes of organic matter from the euphotic surface layer. Recently, it was reported that bacteria closely related to *Sulfitobacter* represent a substantial part of heterotrophic population in littoral marine environments (González and Moran, 1997; Pukall et al., 1999).

Sulfitobacter brevis was isolated from the hypersaline lake, Ekho Lake, in East Antarctica (Vestfold Hills). The salinity of this lake increases with depth, with the formation of interface layers. *Sulfitobacter mediterraneus* was isolated from a microcosm prepared with natural seawater from the Mediterranean Sea.

ENRICHMENT AND ISOLATION PROCEDURES

Sulfitobacter pontiacus, together with other strains of sulfate-producing chemolithoheterotrophic bacteria from the Black Sea, were enriched on mineral medium containing seawater and thiosulfate. Although pure cultures of such bacteria cannot be grown on synthetic mineral media and demand an organic carbon source, the low amount of organic matter present in seawater, in combination with inorganic energy source allows selection of chemolithoheterotrophs. Pure cultures can be successfully grown on A medium with the following composition (per liter): NaCl, 20.0 g; HEPES (N-[2-hydroxyethyl]piperazine-N'-[ethane-sulfonic acid]), 7 g ; K_2HPO_4, 1.0 g; NH_4Cl, 0.5 g; $MgSO_4 \cdot 7H_2O$, 1.0 g; $CaCl_2 \cdot 2H_2O$, 0.2 g; sodium acetate, 2.8 g; sodium thiosulfate, 2.5 g; yeast extract, 0.5 g; trace elements solution (Pfennig and Lippert, 1966), 1 ml; final pH 7.8. Short-term preservation of the active cultures (<1 month) is possible by storage of agar slopes at 4°C. Best survival during long-term storage (6–12 months) was registered when concentrated cultures were kept at −70°C with glycerol (15% v/v final concentration).

For the enrichment of *Sulfitobacter brevis*, 50 ml of saline Ekho Lake water (salinity 15‰) was amended with a filter-sterilized solution of 12.5 mg/ml. Oxoid yeast extract and incubated at 15°C under dim light. Colonies were obtained on PYGV agar[1] containing 25‰ (v/v) artificial seawater.

Enrichment of *Sulfitobacter mediterraneus* was accomplished by preparing a microcosm filled with 300 liters of natural seawater collected from a depth of 1 m at a station (42° 31′ N, 03° 11′ E) located about 2400 m from Banyuls-sur-Mer , France (Mediterranean Sea) and filtered through a 200-μm nylon mesh. At various times, isolation was performed by spreading 0.1 ml of 1:10, 1:100, and 1:1000 dilutions of the seawater onto plates of Marine Agar 2216 (Difco). Between 30 and 50 colonies were selected from each sampling time. Isolates were preserved by freezing at −80°C in 40% glycerol or by storing at 20°C on marine agar.

---

1. PYGV medium consists of (per liter): Bacto peptone (Difco), 0.25; Bacto yeast extract (Difco), 0.25; Hutner's basal mineral salt solution HBM (Cohen-Bazire et al., 1957), 20 ml; and vitamin solution No. 6 (van Ert and Staley, 1971), 10 ml. The medium is solidified with 18 g agar. For salt-requiring organisms, the medium is supplemented with 25‰ artificial sea water (ASW; Lyman and Fleming, 1940.

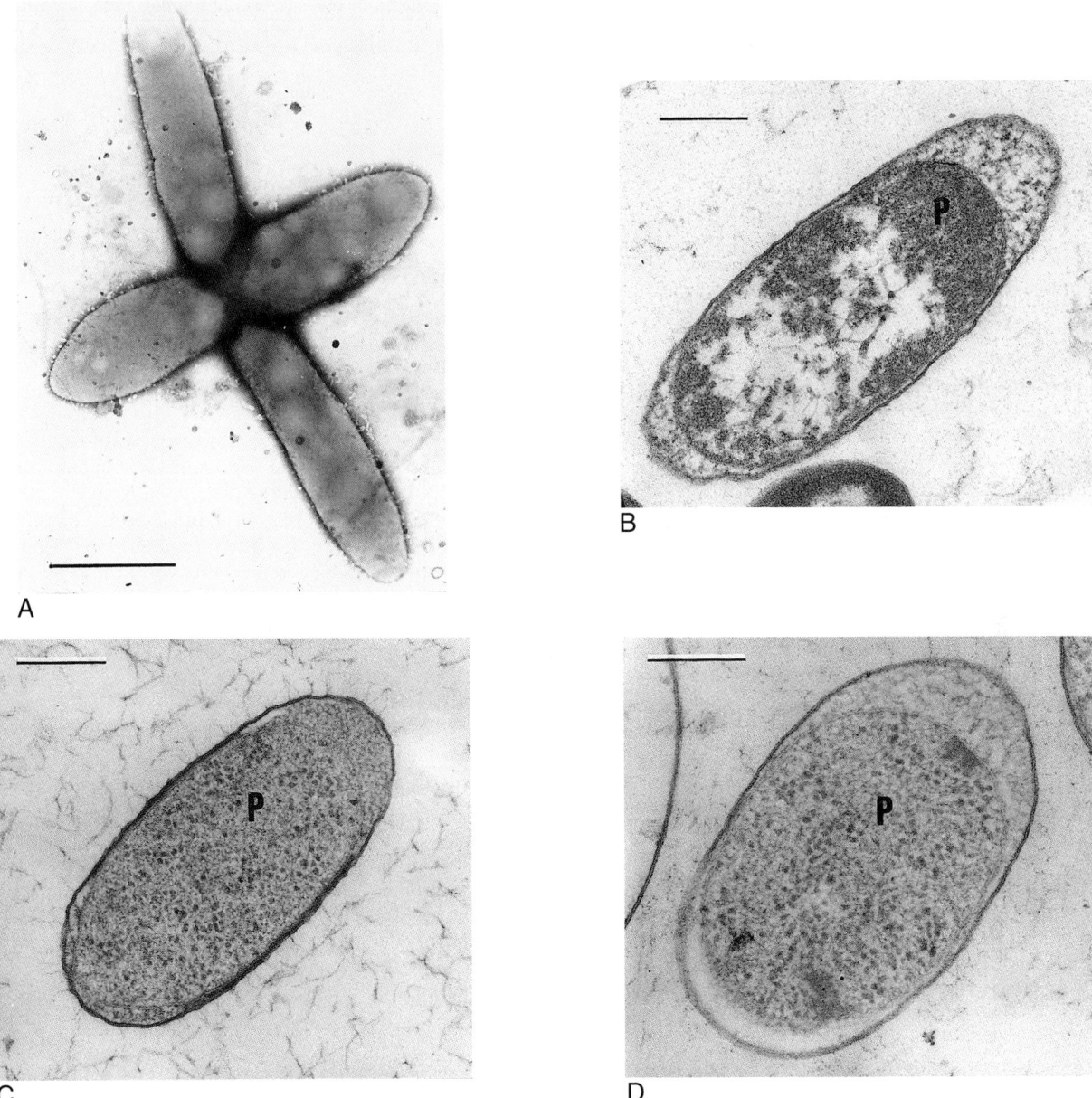

FIGURE BXII.α.81. Cell morphology of *Sulfitobacter pontiacus* (transmission electron micrographs). *A–C*, Strain CHLG 10. *D*, Strain CHLG 5. *A*, Total preparation stained with phosphotungstic acid. Bar = 1 μm. *B*, Thin-sectioned chemically fixed cell showing a major pirellulosome cell compartment (*P*) bounded by a single membrane and containing fibrillar nucleoid and electron-dense particulate cytoplasm and less electron-dense polar regions continuous with cytoplasm surrounding the membrane-bounded compartment. Bar = 0.1 μm. *C–D*, Thin-sectioned cryosubstituted cells, showing a major pirellulosome compartment (*P*) containing relatively large electron-dense particles bounded by a single membrane and a polar cap continuous with cytoplasm surrounding the pirellulosome. Bar = 0.1 μm.

DIFFERENTIATION OF THE GENUS *SULFITOBACTER* FROM OTHER GENERA

Table BXII.α.78 provides some characteristics that can be used to differentiate this genus from the closely related genera of the *Roseobacter* cluster.

TAXONOMIC COMMENTS

The original description of *Sulfitobacter pontiacus* was based on two closely related strains, which had a DNA–DNA reassociation value of about 90%: strain ChLG 10 (DSM 10014[T]) and ChLG 5 (DSM 10015), isolated from the Black Sea (Sorokin, 1995). 16S rRNA gene sequence analysis revealed another representa-

tive of this species, strain SED3, isolated from the littoral waters near Australia (Ward-Rainey et al., 1996). Although different morphologically, this strain possesses a highly active sulfite dehydrogenase, typical for *Sulfitobacter*, and it showed a high percentage of DNA–DNA reassociation (>70%) with the type strain ChLG 10.

Phylogenetic analyses based on 16S rDNA sequence comparison demonstrate that *S. pontiacus* belongs to the *Alphaproteobacteria*, with its closest relatives being two of the three species of the aerobic bacteriochlorophyll-containing erythrobacteria of the genus *Roseobacter*, namely *Roseobacter litoralis* and *Roseobacter denitrificans* (Fig. BXII.α.82). The 16S rDNA sequence of *S. pontiacus* strain ChLG 10 has 96.7% and 97.5% similarity to the 16S

TABLE BXII.α.78. Differential characteristics of the genus *Sulfitobacter* and other phylogenetically related genera from the *Alphaproteobacteria*[a]

Characteristics	*Sulfitobacter*	*Antarctobacter*	*Octadecabacter*	*Roseobacter*[b]	*Sagittula*	*Staleya*
Morphology:						
Rosettes formation	+	+	−	−	+	+
Polar growth, pointed poles	+	+	−	−	+	+
Gas vesicles	−	−	+	−	−	−
Physiology:						
Sulfite oxidation	+	nd	nd	−	−	nd
Bchl *a*	−	−	−	+	−	+
Psychrophily	−	−	+	−	−	−
Denitrification	−	+	−	±	−	−
Gelatinase	−	+	−	+	−	−
Hydrolysis of cellulose and lignin	−	−	−	−	+	−
Biotin requirement	+	+	−	+	−	+
Mol% G + C of DNA	58–62	62.5	56–57	56–60	65	55–56.3

[a]For symbols see standard definitions; ±, variable; nd, not determined.

[b]Based on Bchl *a*–containing species.

rDNA sequences of *Roseobacter denitrificans* and *Roseobacter litoralis*, respectively. The only other complete 16S rDNA sequence with a high similarity to the *S. pontiacus* strain ChLG 10 sequence is that of strain S34, a marine bacterium isolated from the Sargasso Sea (Suzuki et al., 1997). With a 16S rDNA similarity value of 99.2% between the sequences of strain S34 and strain ChLG 10, it seems clear that strain S34 is a member of the genus *Sulfitobacter* and most probably an additional strain of the species *S. pontiacus*.

In the 16S rDNA sequence databases, there are additional partial (<500 nucleotides) sequences to which that of *S. pontiacus* (ChLG 10) shows greater than 98% sequence similarity. These sequences include those of the isolates EE-36 (99.7%), GAI-21 (98.8%) (González and Moran, 1997), and SED3 (99.3%) (Ward-Rainey et al., 1996). The high similarity between these sequences demonstrates the wide geographical distribution of strains of the genus *Sulfitobacter*.

The relatedness of *Sulfitobacter* to members of the genus *Roseobacter* is not surprising, considering that the class *Alphaproteobacteria* contains many species able to oxidize sulfur compounds, both photo- and chemotrophically. They are likely descendents of the purple nonsulfur bacteria, in particular of *Rhodobacter* spe-cies. It could be speculated that colorless sulfate-forming litho-heterotrophs, like *S. pontiacus*, descended from the purple non-sulfur bacteria via erythrobacteria. In fact, some of the latter are capable of oxidizing reduced sulfur compounds (Yurkov et al., 1994b). Experiments with representatives of the genus *Roseobacter* demonstrated that the atypical *Roseobacter* species, *R. algicola*, was able to oxidize thiosulfate and sulfite to sulfate and possessed moderate sulfite dehydrogenase activity.

Analysis of nearly complete 16S rRNA gene sequences showed that the type strain of *S. brevis* was 96.5–98% similar to that of *S. pontiacus*. A high degree of sequence similarity occurred with *Roseobacter denitrificans* and *Roseobacter litoralis*. No significant DNA–DNA reassociation occurred between *S. brevis* and *S. pontiacus* (Labrenz et al., 2000).

16S rDNA sequence analysis of the type strain of *Sulfitobacter mediterraneus* indicated 98.2% similarity to the type strain of *S. pontiacus*. DNA–DNA reassociation experiments indicated a 46% hybridization value between the two species. Except for a quantitative difference in fatty acid composition, there are no distinct phenotypic differences between the two species (Pukall et al., 1999).

DIFFERENTIATION OF THE SPECIES OF THE GENUS *SULFITOBACTER*

No clear-cut differences between the three species were found. The real differentiation was achieved based on DNA–DNA hybridization and 16S rDNA analysis. The presence of sulfite de-hydrogenase and the ability to grow lithoheterotrophically with sulfur compounds remains to be tested for *S. brevis* and *S. mediterraneus*.

List of species of the genus Sulfitobacter

1. **Sulfitobacter pontiacus** Sorokin 1996, 362[VP] (Effective publication: Sorokin 1995, 304.)

   pon.ti.a′ cus. Gr. n. *Pontus* Black Sea; M.L. adj. *pontiacus* from the Black Sea.

   Cells grown in batch cultures vary in shape and size from regular rods (0.5–0.8 × 1.0–1.5 µm) to swollen filaments (up to 20 µm) and are very rarely motile. Bud-like minicells often appear in such cultures, especially when grown on solid medium. Cells grown in continuous culture are more regular, mostly motile rods with sharp edges, often arranged in rosettes (Fig. BXII.α.81a).

   Sulfitobacter pontiacus benefits from the oxidation of reduced sulfur compounds such as thiosulfate and sulfite, being able to oxidize these to sulfate. This ability was proven in experiments with acetate-limited continuous culture (thiosulfate- or sulfite-dependent increase of specific growth yield) and with washed cells (sulfite-dependent ATP production). An extremely active, AMP-independent, soluble sulfite dehydrogenase allows *S. pontiacus* to grow in the presence of very high concentrations of sulfite in acetate-limited continuous culture, which was not demonstrated previously even for chemolithoautotrophic sulfur-oxidizing bacteria. During cultivation, the bacterium gradually adapts to increasing sulfite concentrations and, at the same time, the specific biomass yield and the efficiency of acetate utilization increases.

   Sole carbon and energy sources include acetate, propionate, butyrate, pyruvate, lactate, gluconate, acids of the Krebs cycle (except citrate and *cis*-aconitate), glycerol, glutamate, proline, aspartate, asparagine, serine, L-α-alanine, ornithine, and arginine as the sole carbon and energy sources. Ammonium salts and some amino acids are used

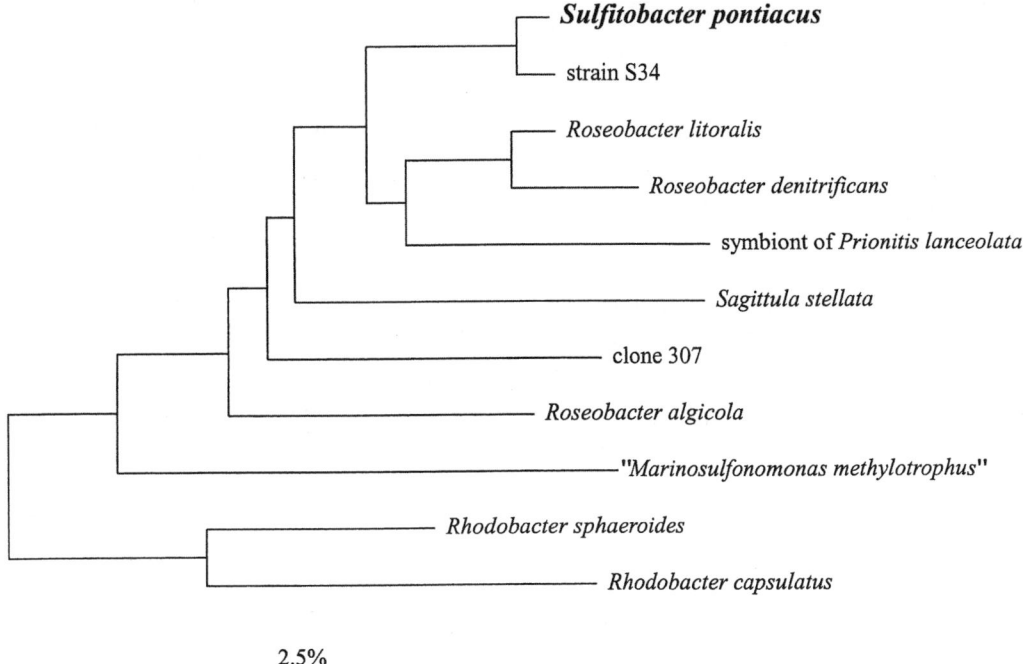

FIGURE BXII.α.82. Phylogenetic dendrogram based on comparison of 16S rDNA sequences indicating the position of *S. pontiacus* and its closest relatives within the *Alphaproteobacteria*. The dendrogram was reconstructed from evolutionary distances (Jukes and Cantor, 1969) by the neighbor-joining method (Saitou and Nei, 1987). Bar = 2.5 inferred nucleotide changes per 100 nucleotides. The accession numbers of the sequences used in these analyses are as follows: *Sulfitobacter pontiacus* (Y13155), strain S34 (U87407), *Roseobacter litoralis* (X78312), *Roseobacter denitrificans* (M59063), symbiont of *Prionitis lanceolata* (U37762), *Sagittula stellata* (U58356), clone 307 (U14583), *Roseobacter algicola* (X78315), "*Marinosulfonomonas methylotrophus*" (U62894), *Rhodobacter sphaeroides* (X53854), and *Rhodobacter capsulatus* (D16428).

as nitrogen sources. Growth on a mineral medium with acetate is stimulated by the presence of yeast extract and biotin. NaCl is required for growth; range, 5–80 g/l; optimal, 20–25 g/l). Temperature range for growth is 4–35°C (optimal, 22–25°C). The pH range for growth is 6.5–8.5 (optimal, 7.5–7.8).

No hydrolytic activities are present. Catalase and oxidase positive. Cytochromes c_{551}, b_{558}, and cytochrome *c* oxidase aa_3 are detectable in cell-free extracts. Isolated from the sulfide-oxygen interface water of the Black Sea.

The mol% G + C of the DNA is: 61.7–62.5 (T_m).

Type strain: ChLG 10, DSM 10014, VKM B-2022.

GenBank accession number (16S rRNA): Y13155.

2. **Sulfitobacter brevis** Labrenz, Tindall, Lawson, Collins, Schumann and Hirsch 2000, 311[VP]

bre' vis. L. adj. *brevis* short, referring to the short cells.

Rods with one or both cell poles pointed, multiplying by monopolar growth, i.e., by a budding process. Cell sizes vary: 0.8–1.0 × 1.1–1.5 µm, with a mean size of 0.9 × 1.3 µm. Rosettes may be formed. Cells may be motile. Poly-β-hydroxybutyrate granules may be present. Endospores are not formed. On PYGV + ASW the colonies are smooth, convex, and yellowish brown. Do not grow photoautotrophically with H_2/CO_2 (80:20) or photoorganotrophically with acetate or glutamate. Peroxidase, catalase, and oxidase positive. Bacteriochlorophyll *a* is not produced.

Optimal growth temperature, 3–26°C. NaCl requirement, 1.0–2.0% NaCl or 10–80‰ ASW. Optimal pH for growth, 7.5–8.0. Cells have a requirement for pantothenate

and thiamin, a weak requirement for biotin and nicotinic acid, and none for vitamin B_{12}. Tween 80 is hydrolyzed, but not alginate, gelatin, DNA, or starch. Growth occurs on acetate, pyruvate, malate, succinate, citrate, butyrate, or glutamate, but not on methanesulfonic acid or methanol. α-D-Glucose is utilized weakly. Cells are susceptible to chloramphenicol, streptomycin, penicillin G, polymyxin B, and tetracycline, but not to nalidixic acid. Nitrate is not reduced. H_2S and indole are not produced. Methyl red and Voges–Proskauer negative.

Polar lipids present are diphosphatidylglycerol, phosphatidylglycerol, phosphatidylcholine, phosphatidylethanolamine, and an unknown aminolipid. Dominant fatty acids are $C_{18:1\ \omega7c}$ and $C_{18:1\ \omega9c}$, and $C_{18:2}$ is present as two isomers. Other characteristic fatty acids are $C_{10:0\ 3OH}$, $C_{14:1\ 3OH}$, $C_{16:0}$, and $C_{18:0}$. Major respiratory quinone is Q-10. Source: water sample from Ekho Lake, Antarctica.

The mol% G + C of the DNA is: 57.9–58.1 (HPLC).

Type strain: EL-162, DSM 11443.

GenBank accession number (16S rRNA): Y16425.

3. **Sulfitobacter mediterraneus** Pukall, Buntefuß, Frühling, Rohde, Kroppenstedt, Burghardt, Lebaron, Bernard and Stackebrandt 1999, 518[VP]

me.di.ter.ra.ne' us. N.L. adj. *mediterraneus* pertaining to the Mediterranean Sea.

Cells grown on marine agar at 25°C are rod shaped, 1–3 × 0.5–0.8 µm. Cells are motile by means of 1–5 subpolar flagella. Bacteria grown on marine agar, supplemented with acetate, tend to form rosettes and contain poly-β-hydroxybutyrate granules. Colonies on Marine agar are 1.2–1.4 mm in diameter, circular, convex, with entire or undulate margin, translucent and cream-colored. Temperature range for growth, 4–35°C; optimal, 17–28°C. pH range for growth, 6.5–8.5; optimal, 7.0–7.5. NaCl (2–80 g/l, optimal 15–20 g/l) is required for growth. Catalase and oxidase positive. Bacteriochlorophyll *a* is not present.

In minimal media with or without the addition of yeast extract or vitamins, growth occurs at 28°C with glucose, mannitol, gluconate, adipate, acetate, malate, pyruvate, lactate, propionate, butyrate, serine, proline, ornithine, alanine, asparagine, glutamate, and glycerol. Glutarate, *cis*-aconitate, tryptophan, caprate, urea, arabinose, and mannose are not utilized. Weak growth is detected with aspartate. Nitrate is not reduced. Utilization of citrate and maltose is stimulated in media containing yeast extract or biotin. Utilization of *N*-acetylglucosamine and activity of arginine

hydrolase, β-glucosidase, and gelatin hydrolase are induced after a longer incubation period (>3 d). Phenylacetate is used after 24 h incubation at 25°C. Growth occurs on 10 mM sulfite in acetate (10 mM)-supplemented HEPES (*N*-[2-hydroxyethyl]piperazine-*N*'-[ethane-sulfonic acid]) medium. In TSB[2] medium containing 3% NaCl, the main fatty acid is *cis*-11 octadecenoic acid. Hexadecanoic acid occurs in smaller amounts. Signature nucleotides for 16S rDNA are located at positions 418–425 (C-G), 591–648 (C-G), 592–647 (A-U), and 599–639 (A-U). Isolated from coastal waters of the west Mediterranean Sea.

The mol% G + C of the DNA is: 59 (HPLC).

Type strain: CH-B427, DSM 12244.

GenBank accession number (16S rRNA): Y17387.

---

2. Tryptone Soya Broth was used for FAME analysis of *Sulfitobacter mediterraneus* and is available from Oxoid.

Genus Incertae Sedis XXVI. Rhodothalassium Imhoff, Petri and Süling 1998, 797[VP]

JOHANNES F. IMHOFF

Rho'do.tha.las'si.um. Gr. n. *rhodon* the rose; Gr. adj. *thalassios* belonging to the sea; M.L. neut. n. *Rhodothalassium* the rose belonging to the sea.

Cells are vibrioid to spiral-shaped, motile by means of polar flagella and multiply by binary fission. **Gram-negative members of the *Alphaproteobacteria*.** Internal photosynthetic membranes are present as lamellar stacks lying parallel to the cytoplasmic membrane. Photosynthetic pigments are **bacteriochlorophyll *a*** (esterified with phytol) and **carotenoids of the spirilloxanthin series.** **Ubiquinones and menaquinones with 10 isoprene units (Q-10 and MK-10) are present. Major cellular fatty acids are C$_{18:1}$, C$_{18:0}$, and C$_{16:0}$**, with C$_{18:1}$ as dominant component (~60% of total fatty acids). **Cells grow preferably photoheterotrophically** under anoxic conditions in the light. **Chemotrophic growth is possible under oxic conditions in the dark.** Amino acids may be required as growth factors. **Obligately halophilic bacteria that require NaCl or sea salt for growth**, have salt optima above seawater salinity and an extended salt tolerance. Habitat: anoxic zones of hypersaline environments such as salterns and salt lakes that are exposed to the light.

The mol% G + C of the DNA is: 64.0.

Type species: **Rhodothalassium salexigens** (Drews 1982) Imhoff, Petri and Süling 1998, 797 (*Rhodospirillum salexigens* Drews 1982, 384.)

FURTHER DESCRIPTIVE INFORMATION

Rhodothalassium salexigens at present is the only representative of this genus. It is a moderately halophilic bacterium that requires elevated concentrations of salt for growth and is among the few extremely halotolerant species (up to 20% of salts) of the purple nonsulfur bacteria. Glycine betaine is accumulated as osmoticum and compatible solute in response to the external salt concentrations (Imhoff, 1992; Severin et al., 1992). Growth depends on glutamate as a growth factor or nitrogen source (Madigan et al., 1984). Cultures grown in the presence of 4 mM glutamate and dinitrogen form readily detectable amounts of nitrogenase (Madigan et al., 1984). The cell wall of *Rhodothalassium salexigens* contains peptidoglycan and proteins but lacks glycolipids and lipopolysaccharides (Golecki and Drews, 1980; Weckesser et al., 1995). Growth of *Rhodothalassium salexigens* is inhibited by a number of antibiotics such as tetracycline, chloramphenicol, penicil-

lin, ampicillin, cycloserine, nisin, vancomycin, and bacitracin, but not by oxacillin (Drews, 1981). Additional properties are given in Table BXII.α.79.

ENRICHMENT AND ISOLATION PROCEDURES

Rhodothalassium salexigens appears to be widely distributed in hypersaline environments such as salterns and salt lakes. Several strains of this species have been isolated from a saltern at the Mediterranean shore of Spain (Rodriguez-Valera et al., 1985). Standard techniques for isolation of purple nonsulfur bacteria in agar dilution series and on agar plates can be applied to *Rhodothalassium* species (Imhoff, 1988; Imhoff and Trüper, 1992). Media with neutral pH and elevated salt concentrations containing complex nutrients are suitable for their enrichment. A modification of the medium of Golecki and Drews (1980) has been successfully used and a recipe is given below (see also Imhoff, 1988).[1] *Rhodothalassium salexigens* also grows well on the complex medium of Nissen and Dundas (1984).

MAINTENANCE PROCEDURES

Cultures are well preserved by standard techniques in liquid nitrogen, by lyophilization or storage at −80°C in a mechanical freezer.

DIFFERENTIATION OF THE GENUS *RHODOTHALASSIUM* FROM OTHER GENERA

With respect to the natural environment, the obligate requirement for salt, cell morphology, and internal membrane structure, *Rhodothalassium* is clearly distinguished from other spiral-shaped phototrophic *Alphaproteobacteria*. In addition, it has a characteristic fatty acid and quinone composition and is clearly separated from other phototrophic purple bacteria by its 16S rDNA sequence. Major differentiating properties between *Rhodothalassium salexigens* and other phototrophic *Alphaproteobacteria* can be

---

1. The medium contains (per liter): 100 g NaCl, 3.5 g MgCl$_2$·6H$_2$O, 0.3 g KH$_2$PO$_4$, 10 mM sodium malate, 1.5 g yeast extract, 1.5 g protease peptone, and 1 ml trace element solution SLA. The initial pH is adjusted to 7.0.

TABLE BXII.α.79. Properties of *Rhodothalassium salexigens*[a]

Characteristic	*Rhodothalassium salexigens*
Cell diameter (μm)	0.6–0.7
Internal membrane system	Lamellae
Motility	+
Color	Red
Bacteriochlorophyll	*a*
Growth factors	Glutamic acid
Aerobic growth	+
Oxidation of sulfide	−
Salt requirement	6–8% (5–20%)
Optimal temperature (°C)	40
Optimal pH	6.6–7.4
Habitat	Saltern
Mol% G + C of the DNA[b]	64.0
Carbon sources utilized:	
Acetate	+
Arginine	−
Citrate	+
Fructose	−
Glucose	+
Glutmate	+
Glycerol	+
Lactate	−
Pyruvate	+
Succinate	+
Electron donors utilized:	
Sulfide	−
Sulfur	−
Thiosulfate	−
Major quinones	Q-10, MK-10
Major fatty acids:	
$C_{14:0}$	3.8
$C_{16:0}$	16.1
$C_{16:1}$	1.5
$C_{18:0}$	17.8
$C_{18:1}$	59.9

[a]Symbols: +, positive in most strains; −, negative in most strains; Q-10, ubiquinone 10; MK-10, menaquinone 10.

[b]Determined by buoyant density centrifugation.

seen by comparing Table BXII.α.1 in the chapter for the genus *Rhodospirillum* with Table BXII.α.79. Carbon sources used can be compared between Table BXII.α.2 in the chapter for the genus *Rhodospirillum* and Table BXII.α.79. The phylogenetic relationship of these bacteria based on 16S rDNA sequences is presented in Fig. 1 (p. 124) in the introductory chapter on "Anoxygenic Phototrophic Purple Bacteria", Volume 2, Part A.

TAXONOMIC COMMENTS

Because of its spiral shape and phototrophic capacity, *Rhodothalassium salexigens* was originally assigned to the genus *Rhodospirillum*, which comprised all spiral-shaped purple nonsulfur bacteria at that time (Drews, 1981). High diversity among these bacteria was revealed by their chemical composition in regard to fatty acids and quinones, but also in respect to the structure of internal membrane systems, to growth factor requirement and salt dependency. Because of outstanding phenotypic differences and the great genetic distance from the type species of this genus, *Rhodospirillum rubrum*, and to other purple nonsulfur bacteria, it was reclassified as *Rhodothalassium salexigens* (Imhoff et al., 1998).*

FURTHER READING

Imhoff, J.F., R. Petri and J. Süling. 1998. Reclassification of species of the spiral-shaped phototrophic purple non-sulfur bacteria of the *Alphaproteobacteria*: description of the new genera *Phaeospirillum* gen. nov., *Rhodovibrio* gen. nov., *Rhodothalassium* gen. nov. and *Roseospira* gen. nov. as well as transfer of *Rhodospirillum fulvum* to *Phaeospirillum fulvum* comb. nov., of *Rhodospirillum molischianum* to *Phaeospirillum molischianum* comb. nov., of *Rhodospirillum salinarum* to *Rhodovibrio salinarum* comb. nov., of *Rhodospirillum sodomense* to *Rhodovibrio sodomensis* comb. nov., of *Rhodospirillum salexigens* to *Rhodothalassium salexigens* comb. nov. and of *Rhodospirillum mediosalinum* to *Roseospira mediosalina* comb. nov. Int. J. Syst. Bacteriol. *48*: 793–798.

Imhoff, J.F. and H.G. Trüper.. 1992. The genus *Rhodospirillum* and related genera. *In* Balows, Trüper, Dworkin, Harder and Schleifer (Editors), The Prokaryotes: A Handbook on the Biology of Bacteria. Ecophysiology, Isolation, Identification, Applications, Springer-Verlag, New York. pp. 2141–2155.

Kawasaki, H., Y. Hoshino and K. Yamasato. 1993. Phylogenetic diversity of phototrophic purple non-sulfur bacteria in the alpha proteobacteria group. FEMS Microbiol. Lett. *112*: 61–66.

Editorial Note: Based on 16S rDNA sequence information, *Rhodothalassium salexigens* forms a distinct phylogenetic lineage, separate from other so-called "purple nonsulfur bacteria." This separation is so significant that it precludes inclusion of the genus *Rhodothalassium* in any of the families of the *Alphaproteobacteria*. Though its phylogenetic position would allow *Rhodothalassium* to be considered a separate family, 16S rDNA sequence evidence is only available from a single strain. Therefore, further data should be obtained prior to assigning this genus to a family. At present, the genus *Rhodothalassium* is considered a *genus incertae sedis* with distant affiliations to the families *Rhodospirillaceae*, *Rhodobacteraceae*, and *Rhizobiaceae*.

List of species of the genus Rhodothalassium

1. **Rhodothalassium salexigens** (Drews 1982) Imhoff, Petri and Süling 1998, 797[VP] (*Rhodospirillum salexigens* Drews 1982, 384.)

   sal.ex' i.gens. L. n. *sal* salt; L. part. adj. *exigens* demanding; M.L. adj. *salexigens* salt-demanding.

   Cells are rod-shaped to spiral, 0.6–0.7 × 1–6 μm; one complete turn of a spiral is 0.8–0.9 μm wide. Internal photosynthetic membranes occur as lamellae lying parallel to the cytoplasmic membrane. Color of anaerobically grown liquid cultures is red. Absorption maxima of living cells are at 375, 485, 515, 550, 590, 800, 840, and 875 nm. Photosynthetic pigments are bacteriochlorophyll *a* (esterified with phytol) and carotenoids of the spirilloxanthin series with spirilloxanthin as the predominant component. Cells grow photoheterotrophically under anoxic conditions in the light with various organic compounds as carbon and electron sources or under oxic conditions in the dark. Carbon compounds assimilated are shown in Table BXII.α.79. Photoautotrophic growth with molecular hydrogen, sulfide, or thiosulfate as electron donor is not possible. Ammonia or dinitrogen cannot serve as sole nitrogen source; casein hydrolysate is used as a carbon and nitrogen source. Glutamate is required for growth, but no other growth factors are required. Obligately and moderately halophilic bacterium with optimal growth at 40°C (range 20–45°C), pH 6.6–7.4, and 6–8% NaCl (range: 5–20% NaCl). No growth occurs in the absence of salt. Habitat: anoxic zones of hypersaline environments such as salterns and partially evaporated pools of seawater with decaying plants that are exposed to the light.

   The mol% G + C of the DNA is: 64.0 (Bd, T_m).

   Type strain: WS 68, ATCC 35888, DSM 2132.

   GenBank accession number (16S rRNA): D14431.

Order IV. **Sphingomonadales** *ord. nov.*

Eiko Yabuuchi and Yoshimasa Kosako

Sphing.o.mon.a.da' les. M.L. fem. n. *Sphingomonas* type genus of the order; *-ales* ending
to denote an order; M.L. fem. pl. n. *Sphingomonadales* the order of *Sphingomonas.*

Gram-negative, nonsporeforming rod-shaped, ovoid, or pleo-morphic cells. Reproduction is usually by binary fission, but in some members, polar growth or budding has been observed by electron microscopy. Motile or nonmotile. **Instead of lipopolysaccharide (LPS) in cell wall, glucuronosyl ceramide (SGL-1) and 2-OH myristic acid are present, without 3-OH acids.** Instead of $C_{14:0\ 2OH}$ nonhydroxy myristic acid in *Zymomonas mobilis.* Aerobic, having a strictly respiratory type of metabolism with oxygen as the terminal electron acceptor, with the exception of the genus *Zymomonas,* which is **facultatively anaerobic and has a strictly fermentative type of metabolism.** Some species synthesize bacteriochlorophyll *a* and are **facultative phototrophs. Chemoheterotrophic. The major respiratory quinone is ubiquinone Q-10** in those genera that have been tested. Belong to the class *Alphaproteobacteria.* Free living and widely distributed in nature.

TABLE BXII.α.80. 16S rDNA sequence homology ratios of 49 type strains of 48 species in 13 genera of *Alphaproteobacteria* to *Sphingomonas paucimobilis* EY 2395[T]

Order[a]	Family	Genus	Species	Type strains used	Sequence data accession number	% Similarity to GIFU 2395[T]
IV	I	I	Sphingomonas paucimobilis	GIFU[b] 2395[T]	D16144	100
IV	I	I	S. parapaucimobilis	JCM 7510[T]	X72721	97.7
IV	I	I	S. sanguinis	IFO 13937[T]	D13726	97.7
IV	I	I	S. roseiflava	MK341[T]	D84520	96.8
IV	I	I	S. adhaesiva	GIFU 11458[T]	D16146	96
IV	I	I	S. pituitosa	EDIV[T]	AJ243751	95.8
IV	I	I	S. trueperi	LMG 2142[T]	X97776	95.5
IV	I	I	S. aquatilis	KCTC 2881[T]	AF131295	95.4
IV	I	I	S. koreensis	KCTC 2882[T]	AF131296	95.4
IV	I	I	S. echinoides	ATCC 14820[T]	AB021370	94.7
IV	I	I	S. asaccharolytica	IFO 10564[T]	Y09639	94.3
IV	I	I	S. mali	IFO 10550[T]	Y09638	94.1
IV	I	I	S. melonis	DAPP PG-244[T]	AB055563	94
IV	I	I	S. pruni	IFO 15498[T]	Y09637	93.9
IV	I	I	S. wittichii	RW1[T]	AB021492	93.6
IV	I	I	S. rosa	IFO 15208[T]	D13945	92.5
IV	I	I	S. aromaticivorans	IFO 16084[T]	AB25014	92.3
IV	I	I	S. yanoikuyae	GIFU 9882[T]	D16145	92.1
IV	I	I	S. cloacae	S-3[T]	AB040739	92.1
IV	I	I	S. suberifaciens	IFO 15211[T]	D13737	91.9
IV	I	I	S. xenophaga	BN6[T]	X94098	91.9
IV	I	I	S. subterranea	IFO 16086[T]	AB25012	91.8
IV	I	I	S. ursincola	KR-99[T]	Y10677	91.8
IV	I	I	S. subarctica	KF-3[T]	X94103	91.7
IV	I	I	S. taejonensis	KCTC 2884R	AF131297	91.7
IV	I	I	S. chlorophenolica	ATCC 33790[T]	X87161	91.4
IV	I	I	S. macrogoltabidus	IFO 15033[T]	D13723	91.4
IV	I	I	S. stygia	IFO 16085[T]	AB25013	91.4
IV	I	I	S. capsulata	GIFU 11526	D16147	91.3
IV	I	I	S. herbicidovorans	Not recorded	AB042233	91.2
IV	I	I	S. natatoria	DSM 3183[T]	Y13774	90.7
IV	I	I	S. terrae	IFO 15098[T]	D13727	90.2
IV	I	I	S. alaskensis	RB2256[T]	Z73631	90.9
IV	I	I	S. chungbukensis	DJ77[T]	AF129257	90.8
IV	I	VI	Zymomonas mobilis subsp. mobilis	ATCC 10988[T]	RDP[c]	90.4
IV	I	VI	Zymomonas mobilis subsp. pomaceae	ATCC 29192[T]	RDP	90.4
IV	I	III	Erythromicrobium ramosum	E5[T]	X72909	92.1
IV	I	II	Erythrobacter litoralis	T4[T]	X72962	91.9
IV	I	V	Sandaracinobacter sibiricus	RB16-17[T]	Y10678	91.8
IV	I	IV	Porphyrobacter tepidarius	OT3[T]	D84429	91.3
IV	I	II	Erythrobacter longus	OCH 101[T]	M96744	90.7
IV	I	IV	Porphyrobacter neustonensis	ACM 2844[T]	L01785	90.6
I	I	I	Rhodospirillum rubrum	ATCC 11170[T]	D30778	85.2
I	I	II	Acetobacter aceti	DSM 3508[T]	X94066	84.3
II	I	I	Rickettsia prowazekii	Not recorded	M21789	83.4
III	I	I	Rhodobacter capsulatus	ATCC 11166[T]	D16428	84.8
V	I	I	Caulobacter subvibrioides	ATCC 15264[T]	AB008392	85.6
V	I	III	Brevundimonas diminuta	DSM 1635[T]	X87274	87
VI	I	I	Rhizobium leguminosarum	IAM 12609[T]	D14513	87.6

[a]Order I. *Rhodospirillales*[AL]; II. *Rickettsiales*[AL]; III. *Rhodobacterales*; IV. *Sphingomonadales*; V. *Caulobacterales*[AL]; VI. *Rhizobiales.*

[b]Same strain as EY 2395[T].

[c]RDP, Ribosomal Data Project.

The mol% G + C of the DNA is: 59–68.5; (*Z. mobilis*, 49.1).

Type genus: **Sphingomonas** Yabuuchi, Yano, Oyaizu, Hashimoto, Ezaki and Yamamoto 1990b, 321 (Effective publication: Yabuuchi, Yano, Oyaizu, Hashimoto, Ezaki and Yamamoto 1990a, 116) emend. Takeuchi, Hamana and Hiraishi 2001, 1414; emend. Yabuuchi, Kosako, Fujiwara, Naka, Matsunaga, Ogura and Kobayashi 2002, 1489.

FURTHER DESCRIPTIVE INFORMATION

If motile, motility is by means of polar or subpolar-monotrichous, polar-multitrichous, or peritrichous flagella.

In addition to SGL-1, galacturonosyl ceramide (SGL-1′) is detected in several species. SGL-1 is commercially used to moisturize human skin.

No data are available concerning the quinones of *Zymomonas mobilis*.

Strains of some species biodegrade dibenzo-*p*-dioxin and related compounds. Cells that are aggregated or attached to goethite (α-FeOOH) particles inactivate chlorine (Gauthier et al., 1999). Some extracellular polysaccharides, e.g., sphingan (Pollock, 1993), are of industrial use.

Strains of the genera have been isolated from a variety of environments, including chemically contaminated soil, subsurface sediment, river water, seawater, alkaline spring, hot springs, distilled water, sewage, plants, hospitals, or found in association with diseased plant and human hosts. The wide variety of isolation sources is reflected in the nutritional diversity of the organisms.

Circumscription of the Order Yabuuchi et al. (1990a) transferred the species *Pseudomonas paucimobilis* Holmes et al. 1977a

(previously known as CDC group IIK-2) to a new genus, *Sphingomonas*, as the type species, *Sphingomonas paucimobilis*. Eventually, the genus was expanded to include 34 species. These were later included in a new family, *Sphingomonadaceae*, by Kosako et al. (2000a). This family is presently the only family in the order *Sphingomonadales* and contains nine genera.

The percent similarity of the 16S rDNA oligonucleotide sequences of the various members of *Sphingomonadales* to that of the type species of the genus *Sphingomonas* ranges from 90.4% to 97.7%. The percent similarity of the 16S rDNA of five other orders in the *Alphaproteobacteria* ranges from 83.4% to 87.6% (Table BXII.α.80). Analysis of the 16S rDNA sequences indicates a close relationship of *Zymomonas mobilis* to the genus *Sphingomonas* (Fig. BXII.α.83). Signature nucleotide sequences of the 16S rDNA (Table BXII.α.81) and the dendrogram of six type species of *Sphingomonadaceae* (Fig. BXII.α.84) indicate the base of construction of this order. The mol% G + C of the DNA ranges from 59 to 68.5, with the exception of *Z. mobilis*, which has a value of 49.1.

It is unlikely that levels of 16S rDNA sequence similarity or the presence or absence of specific cellular components can define the order. It is now apparent that phenotypic characteristics such as flagellar morphology, fermentation vs. nonfermentation of glucose, fermentative vs. respiratory metabolism, and (as mentioned by Kondratieva et al., 1992) the synthesis of bacteriochlorophyll *a* cannot be used as a key characteristic to exclude any genus or family from the order. It is possible that development of new, easy-to-use identification methods based on molecular techniques such as DNA-tips might be able to accurately differentiate the taxa within the order.

TABLE BXII.α.81. Signature sequences of 16S rDNA oligonucleotides among the members of Class *Alphaproteobacteria*

Order	Species	Position						
		108–116	433–442	750–754	822–825	855–858	1334–1336	1385
IV	*S. paucimobilis* and 32 *Sphingomonas* sp.	CACGGGTGC	GCTCTTTTAC	TTGAC	ATAA	GGCG	GGC	A
	S. stygia	CACGGGTGC	GCTCTTTTAC	[CTGAC][a]	ATAA	GGCG	GGC	A
	Erythrobacter longus	CACGGGTGC	GCTCTTTTAC	TTGAC	ATAA	GGCG	GGC	A
	Erythrobacter litoralis	CACGGGTGC	GCTCTTTTAC	TTGAC	ATAA	GGCG	* * *[b]	*
	Erythromicrobium ramosum	CACGGGTGC	GCTCTTTTAC	TTGAC	ATAA	GGCG	* * *	*
	Sandaracinobacter sibiricus	CACGGGTGC	GCTCTTTTAC	TTGAC	[ACGA]	GGCG	GGC	A
	Z. mobilis	CACGGGTGC	GCTCTTTTAC	TTGAC	ATAA	GGCG	GGC	A
I	*Rhodospirillum rubrum*	CACGGGTGA[c]	GCTCTTTCGG	TTGAC	AGTG	GTCG	GTC	G
I	*Acetobacter aceti*	GACGGGTGA	GCACTTTCGG	CTGAC	TGTG	GTCG	GGT	G
II	*Rickettsia prowazekii*	GACGGGTGA	GCTCTTTTAG	CTGAC	AGTG	TTCG	GTT	G
III	*Rhodobacter capsulatus*	GACGGGTGA	GCTCTTTCAG	CTGAC	ATGC	GACA	GTT	G
V	*Caulobacter subvibrioides*	* * * * GGTGA	ATACTTTCAC	CTGAC	ATTG	GACG	GTT	G
V	*Brevundimonas diminuta*	GACGGGTGA	ATTCTTTCAC	CTGAC	ATTG	GACG	GTT	G
VI	*Rhizobium leguminosarum*	GACGGGTGA	GCTCTTTCAC	CTGAC	AATG	GGCG	GTT	G
	Escherichia coli JO-1	GACGGGTGA	GTACTTTCAG	CTGAC	TCGA	TCCG	GTC	G

[a][ ], Sequence containing nucleotide(s) different from those of *S. paucimobilis*.

[b]*, Defect of nucleotide.

[c]Gothic letter: nucleotide different from that of organism in Order IV.

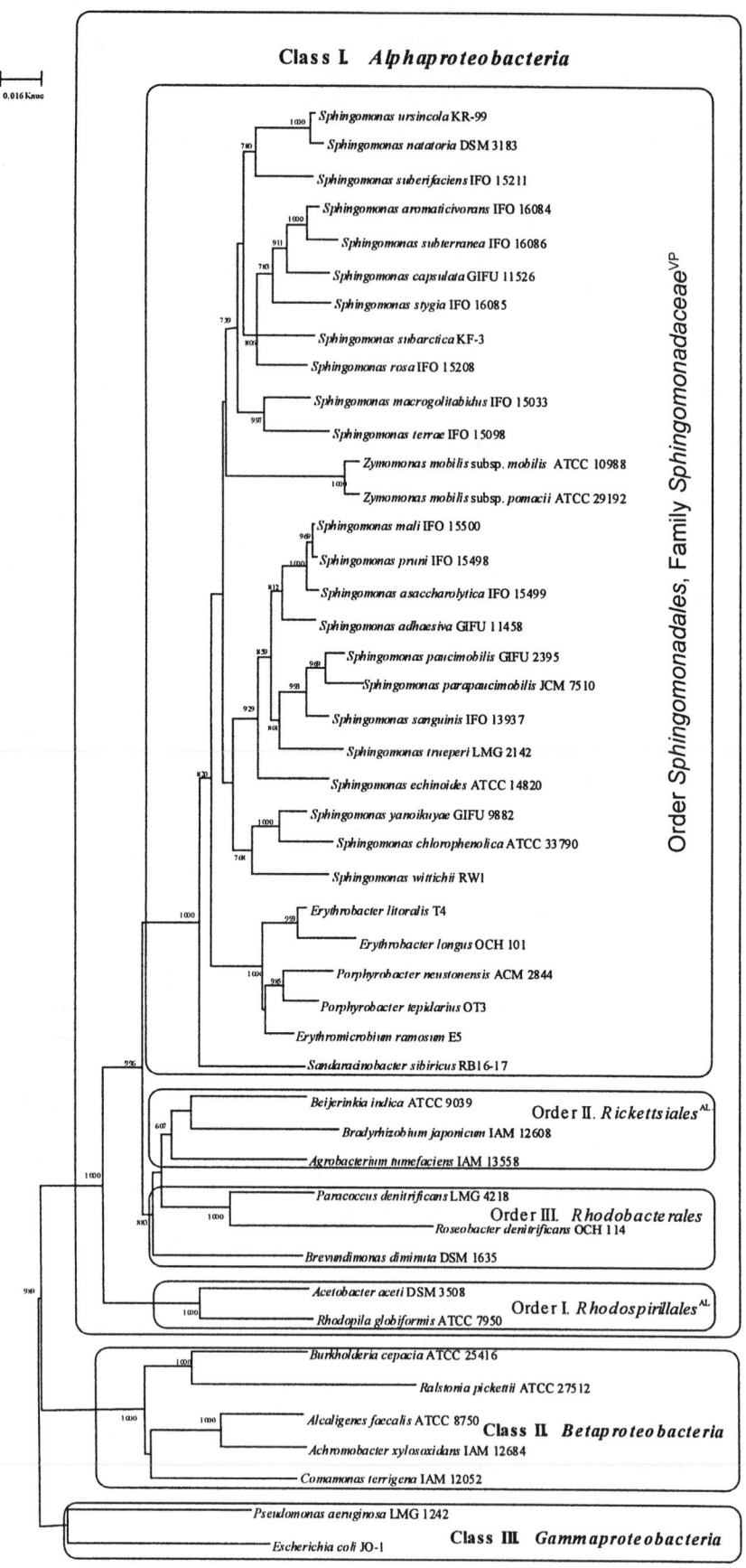

FIGURE BXII.α.83. Phylogenetic position of *Sphingomonadales*, *Sphingomonadaceae*, and *Sphingomonas* among the members of *Proteobacteria* obtained by the neighbor-joining analysis (Saitou and Nei, 1987) of 16S rDNA. Scale bar = 16 nucleotide substitutions per 1000 nucleotides of the 16S rDNA sequence. Boot strap values from 1000 analyses are shown at the branch points. Class I. *Alphaproteobacteria*. Order I. *Rhodospirillales*. Order II. *Rickettsiales*. Order III. *Rhodobacterales*. Order IV. *Sphingomonadales*. Class II. *Betaproteobacteria*. Class III. *Gammaproteobacteria*.

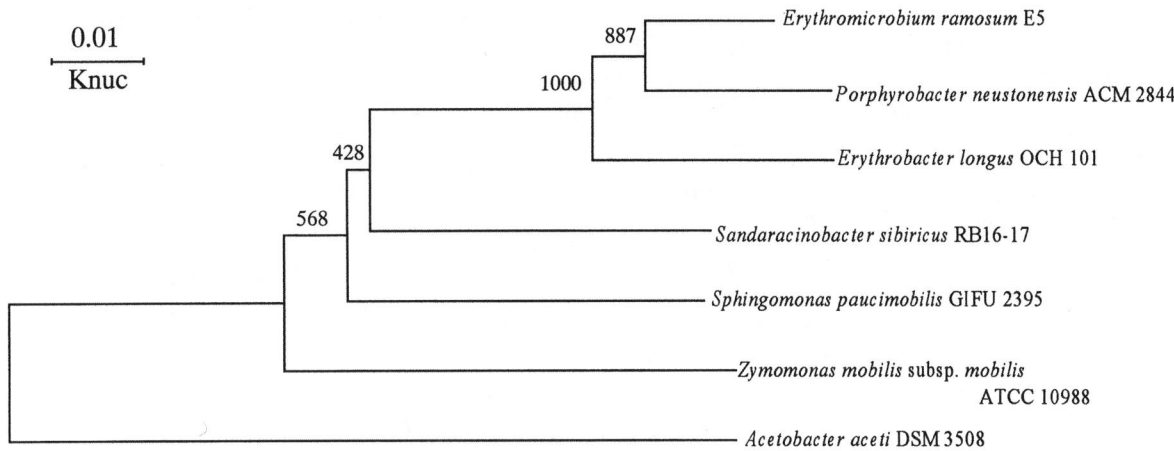

FIGURE BXII.α.84. Phylogenetic position of type species of six genera of family *Sphingomonadaceae* obtained by the neighbor-joining analysis of 16S rDNA. *Acetobacter aceti* was used as an out group species.

Family I. **Sphingomonadaceae** Kosako, Yabuuchi, Naka, Fijiwara and Kobayashi 2000b, 1953[VP] (Effective publication: Kosako, Yabuuchi, Naka, Fijiwara and Kobayashi 2000a, 563)

EIKO YABUUCHI AND YOSHIMASA KOSAKO

Sphing.o.mon.a.da' ceae. M.L. fem. n. *Sphingomonas* type genus of the family; *-aceae* ending to denote an family; M.L. fem. pl. n. *Sphingomonadaceae* the family of *Sphingomonas*.

The characteristics are as given for the order. Differential characteristics of the genera in the family *Sphingomonadaceae* are listed in Table BXII.α.82.

The mol% G + C of the DNA is: 59–68.5 (*Z. mobilis*, 49.1).

Type genus: **Sphingomonas** Yabuuchi, Yano, Oyaizu, Hashi-moto, Ezaki and Yamamoto 1990b, 321 (Effective publication: Yabuuchi, Yano, Oyaizu, Hashimoto, Ezaki and Yamamoto 1990a, 116) emend. Takeuchi, Hamana and Hiraishi 2001, 1414; emend. Yabuuchi, Kosako, Fujiwara, Naka, Matsunaga, Ogura and Kobayashi 2002, 1489.

TABLE BXII.α.82. Differential characteristics of the genera in the family *Sphingomonadaceae*[a]

Characteristics	*Sphingomonas*	*Erythrobacter*	*Erythromicrobium*	*Porphyrobacter*	*Sandaracinobacter*	*Zymomonas*
Cell diameter, μm	0.2–1.4	0.5–0.7	0.6–1.0	0.5–0.7	0.3–0.5	1.0–1.2
Cell length, μm	0.5–4.0	0.7–5.0	1.3–2.5	0.8–1.4	1.5–2.5	2.0–6.0
Strictly respiratory type of metabolism	+	+	+	+	+	−
Glucuronosyl ceramide present	+	+	+	+	nd	+
Galacturonosyl ceramide present	D	−	−	−	nd	−
2-OH myristic acid present	+	+	+	+	nd	−
Bacteriochlorophyll *a*	D	+	+	+	+	−
Facultative photorophic	D	+	+	+	+	−
Major carotenoid	Nostoxanthin[b]	Zeaxanthin[c]	Zeaxanthin[c]	Carotenoid sulfate	Carotenoids[d]	−
Oxidase	D	+	+	−	+	−
Catalase	+	+	+	+	−	+
Growth factor requirement	D	+	+	−	nd	+
Fermentation of glucose	−	−	−	−	−	+
Colony color	Orange, yellow, white, colorless	Brown to red	Orange	Orange to red	Yellow to orange	−
Major respiratory quinone	Q-10	Q-10	Q-10	Q-10	Q-9, Q-10	nd
NaCl requirement	−	+	−	−	−	−
Mol% G + C of DNA	59–67.2	63.6–66	63.6–64.2	66.4–65.0	68.5	47.7–48.5

[a]Symbols: +, postive; −, negative; D, different reaction in different species; nd, not determined.

[b]Data from Jenkins et al. (1979).

[c]Data from Hanada et al. (1997).

[d]Data from Yurkov et al. (1997).

Genus I. **Sphingomonas** Yabuuchi, Yano, Oyaizu, Hashimoto, Ezaki and Yamamoto 1990b, 321[VP] (Effective publication: Yabuuchi, Yano, Oyaizu, Hashimoto, Ezaki and Yamamoto 1990a, 116) emend. Takeuchi, Hamana and Hiraishi 2001, 1414; emend. Yabuuchi, Kosako, Fujiwara, Naka, Matsunaga, Ogura and Kobayashi 2002, 1489

EIKO YABUUCHI AND YOSHIMASA KOSAKO

Sphin.go.mo′ nas. Gr. gen. n. *sphingos* of sphinx; Gr. n. *monad* unit, monad; M.L. fem. n. *Sphingomonas* a sphingosine-containing monad.

Straight or slightly curved rods or ovoid cells 0.2–1.4 × 0.5–4.0 μm. Gram negative. Asporogenous. Reproduction in most species is by binary fission; budding or asymmetric division as visualized by electron microscopy occurs in two species. Motile or nonmotile. A rosette-like aggregation caused by polar fimbriae occurs in some species. **Aerobic**, having a strictly respiratory type of metabolism with oxygen as the terminal electron acceptor. Anaerobic nitrate respiration does not occur. Esculin is hydrolyzed. **Two species with bacteriochlorophyll *a* are facultative photoorganotrophs.** Colony color varies from orange or yellow to white to nonpigmented. Catalase positive. Oxidase positive or negative. **Glucuronosyl-(1→1)-ceramide (SGL-1), galacturonosyl-β(1→1)-ceramide (in several species), and 2-hydroxymyristic acid occur, but not 3-hydroxy fatty acids** (Table BXII.α.83, Figs. BXII.α.85, BXII.α.86, and BXII.α.87). The lipopolysaccharide (LPS) of the cell wall is replaced by sphingolipids (Figs. BXII.α.88 and BXII.α.89). $C_{18:1\ \omega 9t}$ and $C_{18:1\ \omega 7c}$ are the major nonpolar fatty acids and **2-hydroxymyristic acid** is the major 2-hydroxy acid (Table BXII.α.84). Some species are opportunistic pathogens, causing meningitis, septicemia, peritonitis, and neonatal infections in intensive care units. Pathogenicity toward animals is not known. **Free living** in natural and man-made environments.

The mol% G + C of the DNA is: 59–68.

Type species: **Sphingomonas paucimobilis** (Holmes, Owen, Evans, Malnick and Wilcox 1977a) Yabuuchi, Yano, Oyaizu, Hashimoto, Ezaki and Yamamoto 1990b, 321 (Effective publication: Yabuuchi, Yano, Oyaizu, Hashimoto, Ezaki and Yamamoto 1990a, 116) (*Pseudomonas paucimobilis* Holmes, Owen, Evans, Malnick and Wilcox 1977a, 133.)

FURTHER DESCRIPTIVE INFORMATION

Cellular and flagellar morphology Cell sizes range from 0.4 × 1.4 μm in *S. paucimobilis* to 0.7–1.2 × 1.3–4.0 μm in *S. xenophaga.* By Gram staining, the cells of the type strains of most species are straight or slightly curved rods with rounded ends, whereas those of *S. echinoides* have sharp ends (Stolz et al., 2000). Electron micrographs of *S. natatoria* and *S. ursincola* have revealed a budding type of reproduction, but this has not been seen in photomicrographs of Gram-stained preparations. A rosette-like arrangement was reported in *S. paucimobilis* (Yabuuchi et al., 1979), *S. natatoria* (*B. natatorius* Sly 1985), and *S. echinoides* Heumann and Marx (1964) described the star (rosette)-formation of the organism and illustrated the phenomenon by electron micrographs. Cells aggregated or attached to goethite (α-FeOOH) particle inactivate chlorine (Gauthier et al., 1999). The aggregation is caused by attachment and irreversible contraction of the monopolar fimbriae of two or more cells.

Diffuse spreading growth does not occur on 0.3% semisolid agar plates in *S. capsulata* and 13 other species. These are recognized as nonmotile. Peritrichous flagellation of *S. trueperi* EY4218[T] is shown in Fig. BXII.α.90.

Colony characteristics Colonies on agar plating media suitable for each species are pinpoint in size after 48 h or more incubation at 25–30°C. Colonies are approximately 1 mm in diameter when fully grown, circular, domed, and smooth with entire margins. Colonies are pigmented at the beginning of their appearance on agar media. Members of the genus, except the type strain of *S. herbicidovorans*, do not produce water-soluble pigments. Pigmentation of colonies differs from species to species

TABLE BXII.α.83. Distribution of the five molecular species of sphingoglycolipid (SGL) among the members of *Sphingomonadaceae*[a]

SGL on TLC	Species	EY No.	Long-chain base(s)	Carbohydrate moiety(ies)	Hydroxy fatty acid(s)
1	*S. paucimobilis*	2395[T]	$C_{18:0}$, $C_{20:1}$, $C_{21\ cyclo}$	Glucuronic acid	$C_{14:0\ 2OH}$
	S. capsulata	4216[T]	$C_{18:0}$, $C_{20:1}$, $C_{21\ cyclo}$	Glucuronic acid	$C_{14:0\ 2OH}$
	S. yanoikuyae	4208[T]	$C_{20:1}$, $C_{21\ cyclo}$	Glucuronic acid	$C_{14:0\ 2OH}$
	Erythrobacter longus	4203[T]	$C_{20:1}$	Glucuronic acid	$C_{14:0\ 2OH}$, $C_{15:0\ 2OH}$, $C_{16:0\ 2OH}$
	S. natatoria	4220[T]	$C_{19:1}$, $C_{20:1}$	Glucuronic acid	$C_{14:0\ 2OH}$, $C_{15:0\ 2OH}$, $C_{16:0\ 2OH}$
	S. cloacae	4361[T]	nd	Glucuronic acid	nd
	Zymomonas mobilis	4209[T]	$C_{16:0}$	Glucuronic acid	$(C_{14:0}, C_{16:0})^{b}$
1′	*S. yanoikuyae*	4208[T]	$C_{20:1}$, $C_{21\ cyclo}$	Galacturonic acid	$C_{14:0\ 2OH}$
	S. terrae	4207[T]	nd	Galacturonic acid	nd
	S. macrogoltabidus	4304[T]	nd	Galacturonic acid	nd
	S. wittichii	4224[T]	nd	Galacturonic acid	nd
	S. cloacae	4361[T]	nd	Galacturonic acid	nd
2	*S. paucimobilis*	2395[T]	nd	Glucosamine, glucuronic acid	nd
3	*S. capsulata*	4216[T]	$C_{18:0}$, $C_{20:1}$, $C_{21\ cyclo}$	Galactose, glucosamine, glucuronic acid	$C_{14:0\ 2OH}$
4	*S. paucimobilis*	2395[T]	$C_{18:0}$, $C_{20:1}$, $C_{21\ cyclo}$	Mannose, galactose, glucosamine, glucuronic acid	$C_{14:0\ 2OH}$
	S. sanguinis	2397[T]	nd	nd	nd
	S. parapaucimobilis	4213[T]	nd	nd	nd

[a]Abbreviations: cyclo, cyclopropanoic acid; nd, not determnined.

[b]Nonhydroxy acid.

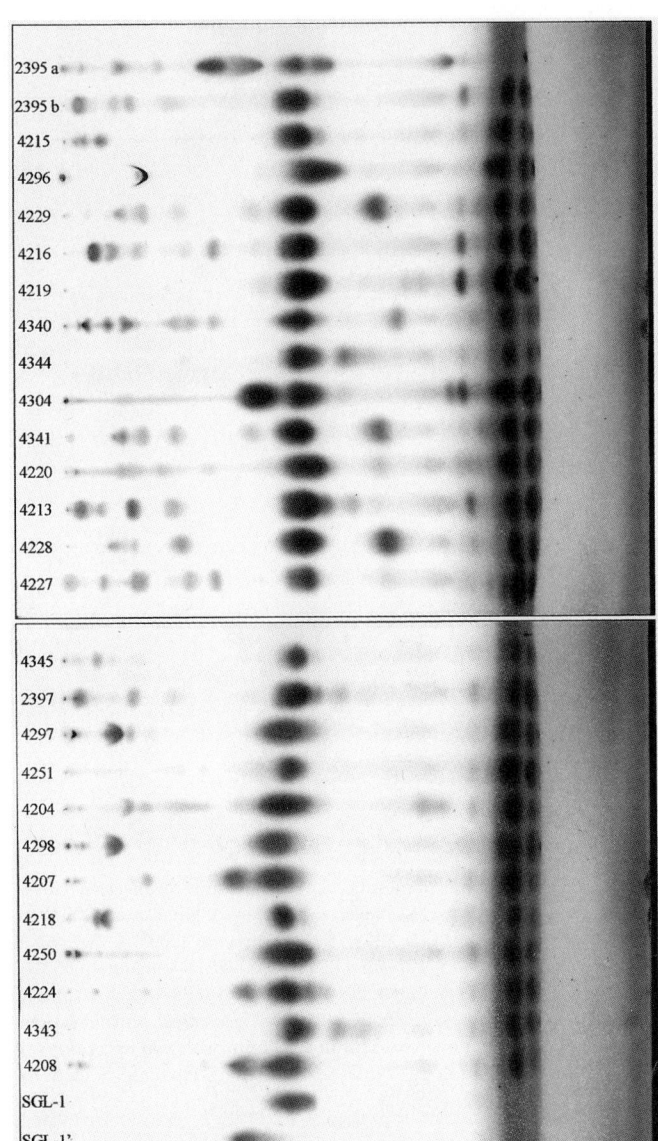

FIGURE BXII.α.85. Thin-layer chromatograms of alkali-stable lipids of type strains of 26 *Sphingomonas* species. EY strain number and species, from top to bottom: 2395^T, *S. paucimobilis* (*a* = crude lipids; *b* = alkali-stable lipids); 4215^T, *S. adhaesiva*; 4296^T, *S. aromaticivorans*; 4229^T, *S. asaccharolytica*; 4216^T, *S. capsulata*; 4219^T, *S. chlorophenolica*; 4340^T, *S. echinoides*; 4344^T, *S. herbicidovorans*; 4304^T, *S. macrogoltabidus*; 4341^T, *S. mali*; 4220^T, *S. natatoria*; 4213^T, *S. parapaucimobilis*; 4228^T, *S. pruni*; 4227^T, *S. rosa*; 4345, *S. roseiflava*; 2397, *S. sanguinis*; 4297^T, *S. stygia*; 4251^T, *S. subarctica*; 4204^T, *S. suberifaciens*; 4298^T, *S. subterranea*; 4207^T, *S. terrae*, 4218^T, *S. trueperi*; 4250^T, *S. ursincola*; 4224^T, *S. wittichii*; 4343^T, *S. xenophaga*; 4208^T, *S. yanoikuyae*, purified SGL-1, purified SGL-1'. Solvent system: chloroform/methanol/acetic acid/H$_2$O (100:20:12:5, by vol).

and may be influenced by the medium and time of incubation. A characteristic deep-yellow color is due to the carotenoid nostoxanthin (Jenkins et al., 1979), rather than the brominated arylpolyamine xanthomonadin that is produced by *Xanthomonas* sp. (Starr et al., 1977; Jenkins and Starr, 1985). Colonies of some initially nonpigmented strains such as *S. yanoikuyae* become lemon yellow after 3 days incubation at room temperature. The type strain of *S. herbicidovorans* produces the water-soluble brown pigment alcapton by accumulation of homogentisic acid in pep-

tone media (Table BXII.α.85), as observed in the pyomelanogenic strain of *Pseudomonas aeruginosa* (Mann, 1969; Yabuuchi and Ohyama, 1972; Gessard, 1981). The type strain of *S. herbicidovorans* represents the second confirmed bacterium with a defective homogentisicase, which results in accumulation of homogentisic acid. The mechanism is similar to that in the human inborn error, alcaptonuria.

Sphingoglycolipids The presence of sphingolipids among the cells of eucaryotes and procaryotes is well known. Sphingolipids are divided into sphingophospholipids and sphingoglycolipids. The skeleton of sphingolipids is ceramide, which is composed of sphingosine (long chain base coupled with a fatty acid molecule by acidic amide linkage). Sphingomyelin is one of the most common mammalian sphingophospholipids. Sphingophospholipids are the characteristic component of the cellular lipids of *Sphingobacterium* spp. and *Prevotella* spp. Molecular species of sphingoglycolipids (SGLs) are differentiated according to their number (1 to 4) and kinds of carbohydrates (glucuronic acid, glucosamine, galactose, and mannose). These carbohydrates are linked to the alcoholic hydroxic base by a glycosidic bond. Both glucuronosyl-ceramide (SGL-1) (Yamamoto et al., 1978) and galacuturonosyl-ceramide (SGL-1') (Naka et al., 2000) are novel sphingoglycolipids among procaryotes and eucaryotes. The chemical structure, function, and distribution of SGL of *Sphingomonas* species in *Proteobacteria* have been reported to replace lipopolysaccharide in the cell wall of Gram-negative rods (Kawahara et al., 1990, 1991, 1999, 2000; Kawasaki et al., 1994c).

A key enzyme in sphingolipid biosynthesis, serine palmitoyltransferase (SPT, EC 2.3.1.50), has been successfully purified. When the gene encoding 420 amino acid residues was overexpressed in *Escherichia coli*, the recombinant SPT was indistinguishable from the native enzyme, both catalytically and spectrophotometrically (Ikushiro et al., 2001). *Sphingomonas* SPT is recognized as a prototype of the eucaryotic enzyme, and an investigation of the sphingolipid biosynthetic enzymes of *Sphingomonas* would help to clarify sphingolipid biosynthesis of both procaryotes and higher organisms (Ikushiro et al., 2001).

Localization of SGL-4 of the type strain of *S. paucimobilis* has been visualized by gold-labeled immunoelectron microscopy (Fig. BXII.α.88), and a model of the outer membrane of *S. paucimobilis* is shown in Fig. BXII.α.89. Stimulation of phagocytosis and phagosome-lysosome fusion in human neutrophiles was reported (Miyazaki et al., 1995). Induction and release of monokine by SGL of type strain of *S. paucimobilis* by different mechanisms from those of LPS were reported (Tahara and Kawazu, 1994; Krziwon et al., 1995).

Cellular fatty acid composition The presence of 2-hydroxymyristic acid and the lack of 3-hydroxy acid of any kind in cellular lipid are important characteristics of members of *Sphingomonadaceae* (except the anaerobic species *Zymomonas mobilis*) (Fig. BXII.α.86). Minor components may change their concentration because of culture conditions and are not useful as differential characters among species.

The fatty acid α-hydroxylase of the type strain of *S. paucimobilis* was partially purified, and direct involvement of hydrogen peroxide in the α-hydroxylation was confirmed (Fig. BXII.α.91). Interestingly, molecular oxygen was not required for α-hydroxylation if hydrogen peroxide was present (Matsunaga et al., 1996). The nucleotide sequence of the fatty acid α-hydroxylase gene was deposited in DDBJ/EMBLE/GenBank with accession number AB006957 (Matsunaga et al., 1998), and the deduced amino acid sequence of the enzyme is shown in Fig. BXII.α.92. The

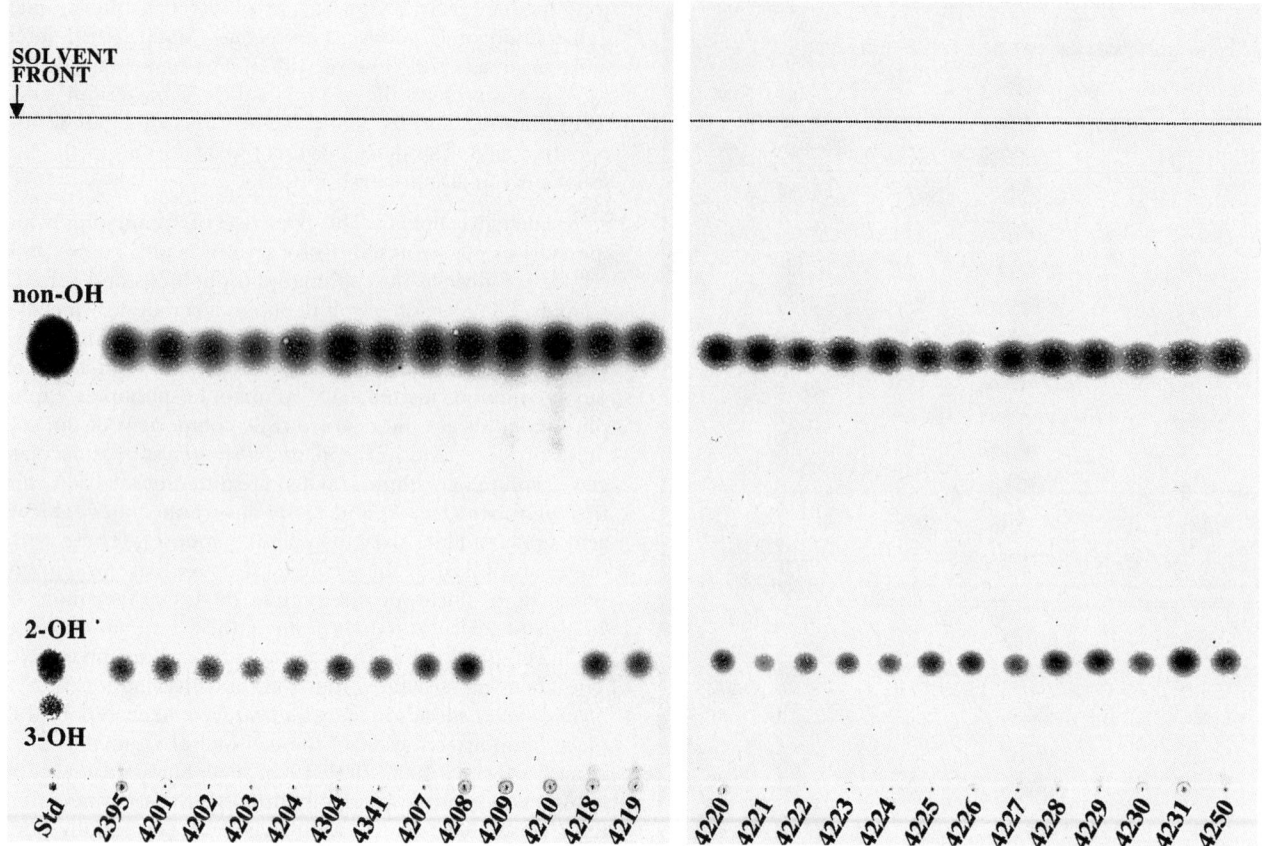

FIGURE BXII.α.86. Thin-layer chromatograms of cellular fatty acids of 21 type and 5 reference strains of member species of family *Sphingomonadaceae*. *Std* = standard, EY strain number and species, from left to right: 2395[T] = *S. paucimobilis*; 4201 = *S. yanoikuyae*; 4202 = *S. yanoikuyae*; 4203[T] = *Erythrobacter longus*; 4204[T] = *S. suberifaciens*; 4304[T] = *S. macrogoltabidus*; 4341[T] = *S. mali*; 4207[T] = *S. terrae*; 4208[T] = *S. yanoikuyae*; 4209[T] = *Zymomonas mobilis*; 4210[T] = *Z. mobilis* subsp. *pomacii*; 4218[T] = *S. trueperi*; 4219[T] = *S. chlorophenolica*; 4220[T] = *S. natatoria*; 4221 = *S. chlorophenolica*; 4222[T] = *Erythrobacter litoralis*; 4223[T] = *Erythromicrobium ramosum*; 4224[T] = *S. wittichii*; 4225 = *Sphingomonas* sp. SS3; 4226 = *S. yanoikuyae* B1; 4227[T] = *S. rosa*; 4228[T] = *S. pruni*; 4229[T] = *S. asaccharolytica*; 4230[T] = *Porphyrobacter neustonensis*; 4231[T] = *Porphyrobacter tepidarius*; 4250[T] = *S. ursincola*. Solvent system: *n*-hexane/diethyl ether (80:20, by vol). Nonhydroxy acid and no 3-hydroxyacid in all 26 strains. 2-hydroxy acid in 24 strains with two exceptions of *Zymomonas* strains. (Chromatogram reproduced from Y. Kosako et al., Microbiology and Immunology, *44*: 563–575, 2000, ©Japanese Society for Bacteriology, Tokyo.)

fatty acid α-hydroxylase gene was defined as a cytochrome P450 enzyme and overexpressed in *Escherichia coli* (Matsunaga et al., 1997).

Biodegradation of dibenzo-*p*-dioxin and related compounds Soil and water pollution by polychlorinated derivatives of dibenzo-*p*-dioxin (DD) has become a problem worldwide. In addition to the most toxic compound 2,3,7,8-tetrachloro-dibenzo-*p*-dioxin (TCDD), there are many possible congeners, as shown in Fig. BXII.α.93. Accordingly, numerous articles have been published concerning the aerobic degradation pathways (Fig. BXII.α.94) used by strain RW1 of *Sphingomonas wittichii*—the best-known metabolizer of DD and DF (dibenzofuran) (Wittich et al., 1992; Bünz and Cook, 1993; Bünz et al., 1993; Happe et al., 1993; Moore et al., 1993; Bertini et al., 1995; Thakur, 1996; Arfmann et al., 1997; Armengaud and Timmis, 1997, 1998; Megharaj et al., 1997; Armengaud et al., 1998, 1999; Keim et al., 1999; Leung et al., 1999; Vuilleumier et al., 2001). Bioremediation of environments polluted with halogenated DDs and DFs seems difficult because of inefficient attack on the highly halogenated congeners by the initial dioxygenase of RW1 and because of the lack of catabolic pathways for the mineralization of halogenated intermediates.

Other species of *Sphingomonas* are capable of degrading a huge range of recalcitrant compounds of environmental concern. *S. paucimobilis* can degrade biphenyl (Davison et al., 1996, 1999) and utilize the herbicide diclofop-methyl (Smith-Greenier and Adkins, 1996), γ-hexachlorocyclohexane (Miyauchi et al., 1998), and aromatic hydrocarbons (Shuttleworth et al., 2000). *S. yanoikuyae* can degrade (polycyclic) aromatic hydrocarbons (Klecka and Gibson, 1980; Eaton et al., 1996b; Khan et al., 1996a; Kim et al., 1997; Lloyd-Jones and Lau, 1997; Kim and Zylstra, 1999; Kazunga and Aitken, 2000; Shuttleworth et al., 2000). *S. subarctica* degrades chlorophenols (Nohynek et al., 1996a); *S. aromaticivorans*, *S. subterranea*, and *S. stygia* degrade aromatic hydrocarbons (Balkwill et al., 1997); *S. herbicidovorans* degrades Mecoprop (Zipper et al., 1996); *S. chlorophenolica* degrades pentachlorophenol (Nohynek et al., 1995, 1996a; McCarthy et al., 1997; Ohtsubo et al., 1999; Xu et al., 1999; Xun et al., 1999); and *S. xenophaga* degrades naphthalene sulfonic acids and *N,N*-dimethylaniline (Stolz et al., 2000).

Numerous *Sphingomonas* isolates are capable of degrading diphenyl ethers (Schmidt et al., 1992a, b, 1993), dibenzofuran, and substituted dibenzofurans (Fortnagel et al., 1990; Harms and Zehnder, 1994; Harms et al., 1995; Wittich et al., 1999), 2,4-D

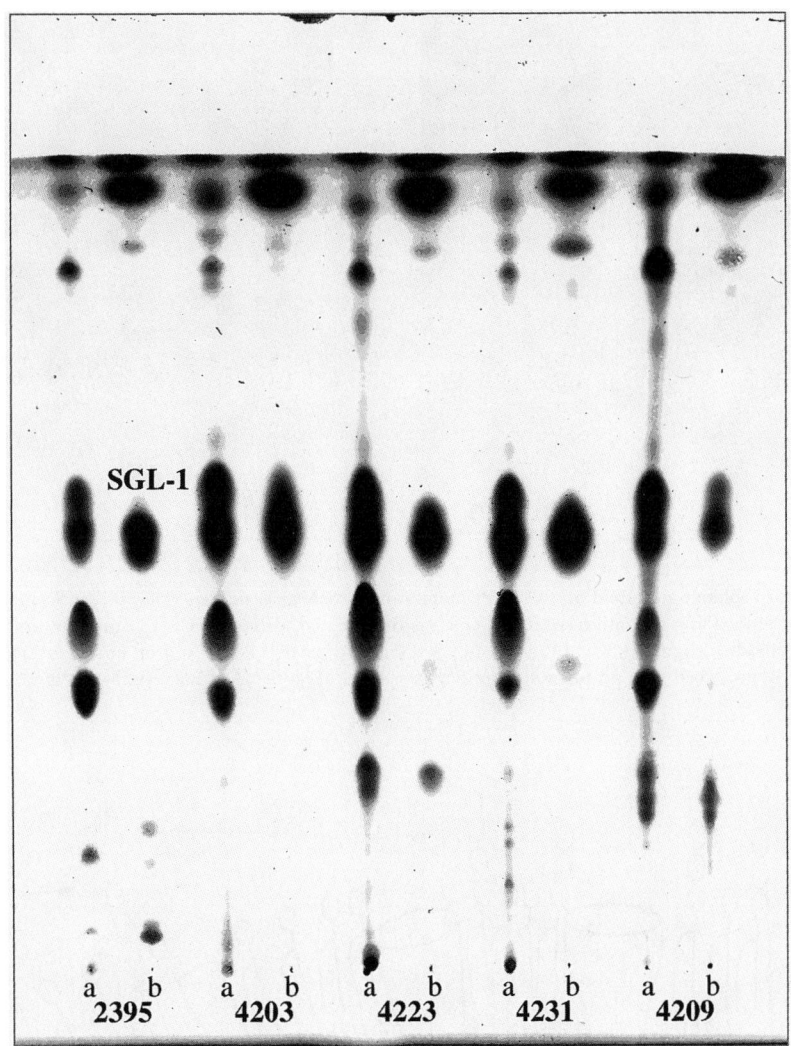

FIGURE BXII.α.87. Thin-layer chromatograms of cellular lipids and alkali-stable lipids of type strains in the family *Sphingomonadaceae.* EY strain number and species, from left to right: 2395^T = *S. paucimobilis;* 4203^T = *Erythrobacter longus;* 4223^T = *Erythromicrobium ramosum;* 4209 ^{T} = *Zymomonas mobilis.* Solvent system: chloroform/methanol/ acetic acid/H$_2$O (100:20:12:5, by vol).

(Ka et al., 1994a, b, c), (polycyclic) aromatic hydrocarbons (Fredrickson et al., 1995; Dagher et al., 1997; Ederer et al., 1997; Harayama, 1997; Mueller et al., 1997), aromatics and chloroaromatics (Yrjala et al., 1998), pentachlorophenol (Karlson et al., 1995), gentisate (Werwath et al., 1998), 4-methylquinoline (Pfaller et al., 1999), and carbofuran (Ogram et al., 2000).

Strain RW1 previously adapted to DD- and DF-contaminated soil was found to survive better in DD- and DF-amended soil and biodegrade DD and DF more efficiently than bacteria that had not been pre-adapted (Megharaj et al., 1997). There are some other interesting reports on biodegradation of compounds of environmental concern in polluted soil by strains of this genus (Thomas et al., 2000) and utilization of herbicide dichlorofopmethyl by soil microorganism as the sole source of carbon and energy (Smith-Greenier and Adkins, 1996). The use of pentachlorophenol-degrading sphingomonads for environmental clean up was reported (Colores et al., 1995) and may be exploited in the future.

Extracellular heteropolysaccharide Production of agar-like polysaccharide by *Pseudomonas* species was reported by Kang et al. (1982). Gellan S-60 (Kang and Veeder, 1982a), welan S-130 (Kang and Veeder, 1982b), rhamsan, S-88 (Kang and Veder, 1985), S-198, and S-657 are members of a family of microbial polysaccharides (Moorhouse, 1987). The structure of these Gellan-related heteropolysaccharides was called "sphingan" by Pollock (1993), who recognized the producers as members of the genus *Sphingomonas*. The biochemical functions of the glycosyl transferase genes essential for biosynthesis of exopolysaccharides have been reported by Pollock et al. (1998). A heteropolysaccharide isolated from a culture of a *"Pseudomonas elodea"* strain is a repeating linear tetrasaccharide composed of D-glucose, D-glucuronic acid, and L-rhamnose in a ratio 2:1:1 (Jansson et al., 1983); side chains were not present. An α-rhamnosidase of *Sphingomonas* sp. R1 has been reported (Hashimoto and Murato, 1998). X-ray crystallographic analysis and genetic studies were reported for the alginate lyases of a strain of *Sphingomonas* sp.

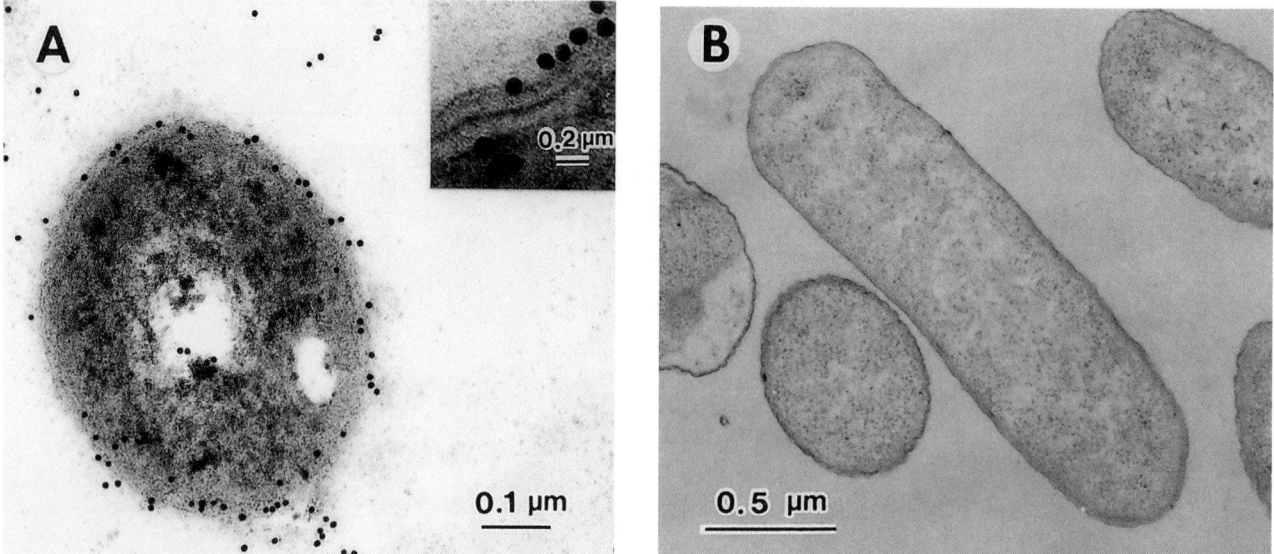

FIGURE BXII.α.88. Localization of sphingoglycolipid on the outer membrane of *Sphingomonas paucimobilis*. (A) *Sphingomonas paucimobilis* IAM 12576[T], immunogold-labeled soma after embedding. Ultrathin-sectioned cell was first treated with anti-SGL-4A antibody, and then stained with gold-labeled secondary antibody. Primary antibody recognized SGL-4A alone, not SGL-1. Gold particles observed on entire surface of the soma, but none on (B) *Escherichia coli* used as control. (Reproduced with permission from S. Kawasaki et al., Journal of Bacteriology *176*: 284–290, 1994, ©American Society for Microbiology.)

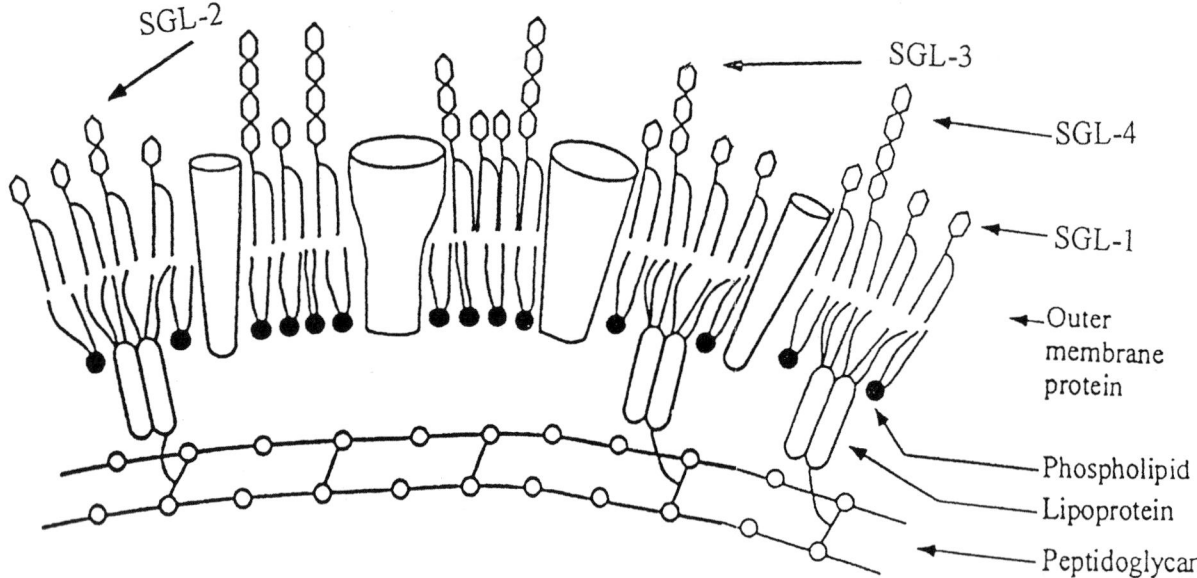

FIGURE BXII.α.89. Outer membrane model of *Sphingomonas paucimobilis*. SGL-1 glucuronic acid–Ceramide; SGL-2 glucosamine–glucuronic acid–Ceramide; SGL-3 galactose–glucosamine–glucuronic acid–Ceramide; SGL-4 = mannose–galactose–glucosamine–glucuronic acid–Ceramide. (Modified with permission from K. Kawahara et al., Journal of Industrial Microbiology and Biotechnology *23*: 408–413, 1999, ©Specialist Journals Division, Nature Publishing Group.)

(Yoon et al., 2000a, b). A *pgmG* gene, which encodes a bifunctional enzyme in Gellan gum-producing *S. paucimobilis*, was reported by Videira et al. (2000).

Industrial usage of the heteropolysaccharides has been reported (Peik et al., 1983, 1985), and modified heteropolysaccharides might possibly be used for oil recovery (Robinson and Stipanovic, 1989).

Ecological features Sphingomonads are free living in natural and man-made environments. Yellow-pigmented aerobic Gram-negative rods have often been isolated from polluted and un-

polluted environments and identified as members of the genus *Sphingomonas*. Furthermore, many yellow-colored isolates formerly assigned to the genus *Flavobacterium*, and others earlier classified as *Pseudomonas* and possibly *Moraxella*, might belong to the genus *Sphingomonas*, but they have not been reinvestigated to date.

Metabolic features The presence of either spermidine or homospermidine differs among species. Assimilation tests for selected organic compounds can serve to differentiate *Sphingomonas* species (Table BXII.α.86). Esculin is hydrolyzed, although

TABLE BXII.α.84. Cellular fatty acid composition (%) of the type strains of 39 species and 2 subspecies in the family *Sphingomonadaceae*[a,b]

Fatty acids, % of total	*Sphingomonas paucimobilis* EY 2395[T]	*Sphingomonas adhaesiva* EY 4215[T]	*Sphingomonas alaskensis* EY 4374[T]	*Sphingomonas aquatilis* KCTC 2881[Tc]	*Sphingomonas aromaticivorans* EY 4296[T]	*Sphingomonas asaccharolytica* EY 4929[T]	*Sphingomonas capsulata* EY 4216[T]	*Sphingomonas chlorophenolica* EY 4219[T]	*Sphingomonas changbukensis* EY 4375[T]	*Sphingomonas cloacae* EY 4361[T]	*Sphingomonas echinoides* EY 4340[T]	*Sphingomonas herbicidovorans* EY 4344[T]	*Sphingomonas koreensis* EY 4376[T]	*Sphingomonas macrogoltabidus* EY 4304[T]	*Sphingomonas mali* EY 4341[T]	*Sphingomonas melonis* EY 4350[T]	*Sphingomonas natatoria* EY 4220[T]	*Sphingomonas parapaucimobilis* EY 4213[T]	*Sphingomonas pituitosa* EY 4370[T]	*Sphingomonas pruni* EY 4228[T]	*Sphingomonas rosa* EY 4227[T]
$C_{14:1}$																					
$C_{14:0}$	1	1		1	tr		tr		tr	tr		tr	tr	tr	tr	1		2	tr		tr
$C_{15:0\ iso}$					2						14										
$C_{15:0}$		1	3	nd	tr	tr	tr						tr		1			tr	2		
$C_{16:1\ \omega7c}$	2	4	3		16		2	7	5	3	1	5	1	29	tr	4	5	3	1		17
$C_{16:1}$		1	1	1	1		1	1	3	2		1	3	3		1		tr	2		tr
$C_{16:0}$	17	13	3	20.6	3	4	7	9	8	4	8	6	7	10	14	9	6	7	12	10	9
$C_{17:1}$			9	nd	5	3									3						
$C_{17:0\ cyclo}$		4	40		12	32	2	tr	4			1	4	tr	20	2	1	2	9	3	tr
$C_{17:0}$		1	3	nd	1	9	tr	tr	tr		4		tr		6	tr		tr	tr		tr
$C_{18:2}$																					
$C_{18:1\ \omega9v}$	60	55	11	61	43	19	66	66	65	69	41	67	61	43	45	51	65	61	49	55	60
$C_{18:1\ \omega7c}$																					
$C_{18:1}$		tr			tr	tr	2	1		4		2	3	tr	2	tr	1	3	tr	2	tr
$C_{18:0}$	1	tr	3	1	2	tr	2	tr	tr	4	2	tr	tr	tr	tr	5	1	2	4	1	
$C_{19:1}$						tr															
$C_{19:0\ cyclo}$	trb					1															
$C_{19:0}$						tr					2										
$C_{14:0\ 2OH}$	18	18	2	10.9	10	3	16	10	14	13	23	14	17	7	5	26	8	17	20	14	11
$C_{15:0\ 2OH}$			21	nd	4	15	tr											tr	2		
$C_{16:1\ 2OH}$																	6				
$C_{16:0\ 2OH}$				nd				tr	1			3	3	4	3		4	tr		2	
$C_{14:0\ 3OH}$																					
$C_{16:0\ 3OH}$																					
Other	tr	tr	1	5		11		3		tr	5			3		1	2	tr	tr	11	1
Total	99	98	100	101	99	97	98	98	99	99	100	99	99	99	99	99	99	100	97	100	98

[a]Hydrolysis was carried out in HCl–methanol(1:5,v/v), 100°C, 3 h.

[b]Abbreviations: tr, <1.0%; cyclo, cyclopropanoic acid; nd, not determined.

[c]Data from Lee et al. (2001).

prolonged incubation may be needed for a positive reaction by some species. Additional information can be found in Nohynek et al. (1996a), Kämpfer et al. (1997), Denner et al. (1999), Stolz et al. (2000), and Denner et al. (2001).

Antibiotic sensitivity Of 33 type strains tested, 91–100% were susceptible to tetracyclines, amikacin, gentamicin, panipenem, and imipenem (Table BXII.α.87). Differences in susceptibilities to amoxicillin (67%) and amoxicillin/clavulanic acid (91%) suggest the presence of penicillinase in these species (Table BXII.α.87).

Pathogenicity *Sphingomonas* species can act as opportunistic pathogens (Decker et al., 1992) and have caused meningitis (Sakai et al., 1978; Hajiroussou et al., 1979), bacteremia (Calubiran et al., 1990), septicemia (Casadevall et al., 1992), peritonitis (Glupczynski et al., 1984), and neonatal infections in intensive care units. Their pathogenicity toward animals is not known. Induction of monokine production in human mononuclear cells (Flad and Ulmer, 1995) and stimulation of phagocytosis and phagosome–lysosome fusion in human neutrophils (Miyazaki et al., 1995) have been reported.

DIFFERENTIATION OF THE GENUS *SPHINGOMONAS* FROM OTHER GENERA

Together with the results of phylogenetic analysis of 16S rDNA nucleotide sequence, the presence of glucuronosyl ceramide is the key characteristic that defines the genus *Sphingomonas* and the family *Sphingomonadaceae* (Figs. BXII.α.85, BXII.α.86, BXII.α.87, and BXII.α.88). Thus it is not enough to refer to the presence of sphingoglycolipid (correctly glucuronosyl ceramide) as "sphingolipid" or even "long chain bases" to define an organism as a member of the genus *Sphingomonas*. The presence of 2-OH myristic acid as the major hydroxy fatty acid and the absence of any kind of 3-hydroxy acid are also important characteristics throughout the family *Sphingomonadaceae*; however, nonpolar myristic acids are present instead of a 2-OH myristic acid in the facultative anaerobic species *Zymomonas mobilis* (Fig. BXII.α.86).

TABLE BXII.α.84. *(cont.)*

Species and Strains

Fatty acids, % of total	*Sphingomonas roseiflava* EY 4345[T]	*Sphingomonas sanguinis* EY 2397[T]	*Sphingomonas stygia* EY 4297[T]	*Sphingomonas subarctica* EY 4251[T]	*Sphingomonas suberifaciens* EY 4204[T]	*Sphingomonas subterranea* EY 4298[T]	*Sphingomonas taejonensis* EY 4377[T]	*Sphingomonas terrae* EY 4207[T]	*Sphingomonas trueperi* EY 4218[T]	*Sphingomonas ursincola* EY 4250[T]	*Sphingomonas wittichii* EY 4224[T]	*Sphingomonas xenophaga* EY 4343[T]	*Sphingomonas yanoikuyae* EY 4208[T]	*Erythrobacter longus* EY 4203[T]	*Erythrobacter litoralis* EY 4222[T]	*Erythromicrobium ramosum* EY 4223[T]	*Porphyrobacter tepidarius* EY 4231[T]	*Porphyrobacter neustonensis* EY 4230[T]	*Zymomonas mobilis* subsp. *mobilis* EY 4209[T]	*Zymomonas mobilis* subsp. *pomaceae* EY 4210[T]
$C_{14:1}$																			tr	tr
$C_{14:0}$	tr	2	1		2	tr	tr		tr		tr	tr	tr						8	6
$C_{15:0\ iso}$																				
$C_{15:0}$			tr		tr	6	2							tr					tr	
$C_{16:1\ \omega7c}$	13	3	18	6	13	11	3	4	1	9	6	22	14	tr	2	2	tr		2	tr
$C_{16:1}$	tr	tr	tr	2	1	tr	tr	tr		2	2	2	2							
$C_{16:0}$	10	12	3	8	4	7	4	2	8	5	12	7	9	4	tr	11	2	5	6	7
$C_{17:1}$		1			tr	2	7	8						4						
$C_{17:0\ cyclo}$	3	tr	4	6	11	45	45			10	tr	tr	tr	2	1	1	tr	7		
$C_{17:0}$		tr	tr	1	2	6	4			10			tr	3	6	2	tr			
$C_{18:2}$														19						
$C_{18:1\ \omega9v}$	57	51	49	57	51	52	6	13	77	50	54	57	52	50	73	68	73	58	80	68
$\quad C_{18:1\ \omega7c}$																				
$C_{18:1}$		1	1	tr		tr		tr		1	tr	1	2		tr	tr			2	2
$C_{18:0}$		2	1	tr	tr	tr	tr		tr	tr		tr	tr	tr			tr		1	1
$C_{19:1}$									tr											tr
$C_{19:0\ cyclo}$					tr				1		3									tr
$C_{19:0}$																tr				
$C_{14:0\ 2OH}$	16	28	18	17	13	10	1	2	13	9	10	10	14	6	7	11	16	8		
$C_{15:0\ 2OH}$			3		2	3	17	13							7		tr			
$C_{16:1\ 2OH}$					tr					tr			tr		2	5	1			
$C_{16:0\ 2OH}$				7	tr				tr				2		tr	3	2	3	4	
$C_{14:0\ 3OH}$																				
$C_{16:0\ 3OH}$																				
Other				2	3		4	2		tr	11		3		2	tr	tr	14	3	16
Total	99	99	99	99	96	98	99	96	99	98	98	99	98	98	99	98	97	99	99	97

[a] Hydrolysis was carried out in HCl–methanol(1:5,v/v), 100°C, 3 h.

[b] Abbreviations: tr, <1.0%; cyclo, cyclopropanoic acid; nd, not determined.

[c] Data from Lee et al. (2001).

TAXONOMIC COMMENTS

The genus *Sphingomonas** was so named by Yabuuchi et al. (1990a, b) because of the presence of a specific sphingoglycolipid (SGL-1) containing glucuronic acid as the carbohydrate moiety of its molecule (Yamamoto et al., 1978). SGL-1 was first found in the cellular lipids of the type strain of *Flavobacterium devorans* ATCC 10829. The species *Flavobacterium devorans* (Zimmermann 1890) Bergey et al. 1923b was cited in the Approved Lists of Bacterial Names (Skerman et al., 1980) and maintained its nomenclatural standing. However, ATCC strain 10829 was the only strain preserved in a culture collection and no other living cultures of the species were available. Furthermore, the history of the species revealed that ATCC 10829 was a misidentified later isolate and not the holotype strain of the species (Yabuuchi et al., 1979). Thus, when the strain was later classified as *Pseudomonas paucimobilis*, the priority of specific epithet *devorans* over *paucimobilis* could not be accepted (Yabuuchi et al., 1979). Owen and Jackman (1982) revealed the close relationship between the type strains of *Pseudomonas paucimobilis* and *F. devorans* by a high DNA–DNA hybridization ratio (93 ± 6%) and protein pattern similarity (87%).

Based on the results of a phylogenetic analysis of 16S rDNA and the presence of sphingoglycolipid in cellular lipids of the type strains, the following genera have been placed into the genus *Sphingomonas* by Yabuuchi et al. (1999b): *"Rhizomonas"* van Bruggen et al. 1990, 186; *Blastomonas* Sly and Cahill 1997, 567 emend. Hiraishi et al. 2000a; and *Erythromonas* Yurkov et al. 1997, 1177.

The 16S rDNA sequence similarity of type strains of 34 *Sphin-*

*Editorial Note: Readers should be advised that there are conflicting opinions regarding the taxonomic status of the genus *Sphingomonas*. At this time, there is no strong phenotypic support for the genera *Novosphingobium*, *Sphingobium*, and *Sphingopyxis*, which are well defined in phylogenetic models based on 16S rDNA sequences. Therefore, Yabuuchi et al. (2002) regard these genera as junior objective synonyms of *Sphingomonas*.

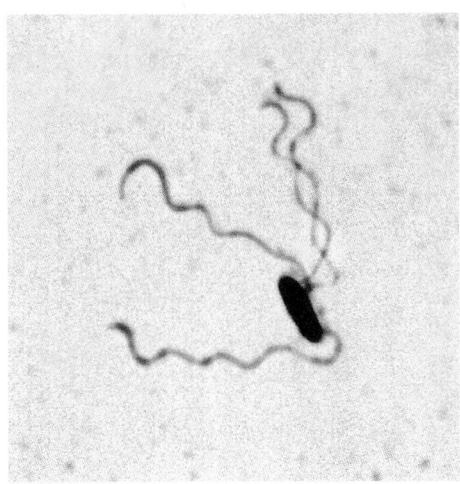

FIGURE BXII.α.90. Flagellar morphology of *Sphingomonas trueperi* EY 4218^T. Four flagella around the soma. Leifson flagella stain.

be distinguished by a combination of phenotypic tests, which are easily performed in every microbiological laboratory. The phylogenetic tree derived from 16S rDNA sequence analysis seems likely to support splitting the genus, but an examination of the phenotypic characteristics of 27 type strains among the 33 species currently assigned to the genus *Sphingomonas* in relation to their phylogeny indicates that—as shown in Fig. BXII.α.95—there is no phenotypic evidence to support this proposal (Yabuuchi et al., 2002). We therefore concluded that *Sphingobium*, *Novosphingobium*, and *Sphingopyxis* are junior objective synonyms of *Sphingomonas*, and it is unreasonable to exclude *S. yanoikuyae*, *S. herbicidovorans*, *S. chlorophenolica*, *S. rosa*, *S. subarctica*, *S. stygia*, *S. subterranea*, *S. aromaticivorans*, *S. capsulata*, *S. macrogoltabidus*, and *S. terrae* from the genus *Sphingomonas*. A dendrogram showing the relationships among the 16S rDNA sequences of *Sphingomonas* species is given in Fig. BXII.α.96. Furthermore, Takeuchi et al. (2001) utilized the rejected generic name "*Rhizomonas*" in combination with the specific epithet *suberifaciens*. It is an obvious violation of the Code. "*Rhizomonas suberifaciens*" was transferred to the genus *Sphingomonas*, as *Sphingomonas suberifaciens* (Yabuuchi et al., 1999a).

Because of the detection of bacteriochlorophyll *a* and *puf* genes encoding the proteins related to photosynthetic activities from both *Blastomonas natatoria* and *Erythromonas ursincola*, together with phenotypic, chemoanalytic, and phylogenetic data, Hiraishi et al. (2000a) published an emended description of *Blastomonas natatoria* (Sly 1985) Sly and Cahill 1997 and a proposal of *Blastomonas ursincola* comb. nov. These two species had been assigned to the genus *Sphingomonas* as *Sphingomonas natatoria* (Sly 1985) Yabuuchi et al. 1999b and *Sphingomonas ursincola* (Yurkov et al. 1997) Yabuuchi et al. 1999b, based on the presence of glucuronosyl seramide (SGL-1) in their cellular lipids and the results of phylogenetic analysis of 16S rDNA sequences. In spite of the report and proposal by Hiraishi et al. (2000a), phototrophic ability is widely distributed among procaryotes, and its presence does not necessarily indicate a close phylogenetic relationship among these taxa (Kondratieva et al., 1992).

gomonas species to the type strain of *S. paucimobilis*—the type species of the genus—ranges from 98.1% to 91.0% (see Table BXII.α.80 in the description of the order *Sphingomonadales*). Because of the rapid increase in the number of species, the diversity of metabolic ability, and the wide distribution of organisms in various ecological niches, the definition of the genus *Sphingomonas* has been subjected to much taxonomic discussion chiefly based on the results of phylogenetic analysis of 16S rDNA nucleotide sequences (Takeuchi et al., 1994, 2001; Balkwill et al., 1997; Kämpfer et al., 1997; Hiraishi et al., 2000a; Stolz et al., 2000). Although there has been a proposal to split the genus *Sphingomonas* into four genera, *Sphingomonas*, *Sphingobium*, *Novosphingobium*, and *Sphingopyxis* (Takeuchi et al., 2001), on the basis of phylogenetic analysis of 16S rDNA sequences and polyamine profiles, from the practical aspect, it is essential that any genus not be delineated by phylogenetic analysis alone, but also

DIFFERENTIATION OF THE SPECIES OF THE GENUS *SPHINGOMONAS*

The differential features and other characteristics of the species of *Sphingomonas* are listed in Tables BXII.α.84, BXII.α.85, BXII.α.86, BXII.α.87, BXII.α.88, and BXII.α.89. Data on members of the genera *Erythrobacter*, *Porphyrobacter*, *Sandaracinobacter*, and *Zymomonas* are provided for comparison in Tables BXII.α.84 (fatty acid composition), BXII.α.85 (morphological and biochemical characteristics), and BXII.α.88 (oxidative acid production).

List of species of the genus Sphingomonas

1. **Sphingomonas paucimobilis** (Holmes, Owen, Evans, Malnick and Wilcox 1977a) Yabuuchi, Yano, Oyaizu, Hashimoto, Ezaki and Yamamoto 1990b, 321^{VP} (Effective publication: Yabuuchi, Yano, Oyaizu, Hashimoto, Ezaki and Yamamoto 1990a, 116) (*Pseudomonas paucimobilis* Holmes, Owen, Evans, Malnick and Wilcox 1977a, 133.)
pau.ci.mo' bi.lis. L. adj. *paucis* few; L. adj. *mobilis* mobile; M.L. fem. adj. *paucimobilis* intended to mean few motile cells.

The characteristics are as described for the genus and in Tables BXII.α.84, BXII.α.85, BXII.α.86, BXII.α.87, BXII.α.88, and BXII.α.89, with the following additional information. Colonies are convex and deep yellow on ordinary peptone media. Soma size is 0.7 × 1.4 µm (Holmes et al., 1977a). A few cells are motile with single polar flagellum. Some cells form a rosette-like arrangement. In addition to SGL-1, SGLs-2, 3, and 4 are detected on TLC (Table BXII.α.83, Figs. BXII.α.85 and BXII.α.87). SGL induces monokine production in human mononuclear cells (Krziwon et al., 1995) and stimulates phagocytosis and phagosome-lysosome fusion in human neutrophil cells (Miyazaki et al., 1995). Major polyamine is homospermidine (Busse and Auling, 1988; Hamana and Matsuzaki, 1993; Segers et al., 1994; Takeuchi et al., 1995; Takeuchi et al., 2001); one strain was reported to produce trace amounts of spermidine (Segers et al., 1994). The structure and function of sphingan (Pollock, 1993; Lobas et al., 1994; Pollock et al., 1994; Sutherland, 1994) and lyases (Sutherland and Kennedy, 1996) have been reported. Human clinical cases of infection have been reported (Crane et al., 1981; Glupczynski et al., 1984; Calubiran et al., 1990; Casadevall et al., 1992; Decker et al., 1992; Lemaitre et al., 1996). Type strain was isolated from hospital respirators.

TABLE BXII.α.85. Morphological, physiological, and biochemical characteristics of 42 type strains of 40 species and 2 subspecies in *Sphingomonadaceae*[a]

Substrate or test	*Sphingomonas paucimobilis* EY 2395[T]	*Sphingomonas adhaesiva* EY 4215[T]	*Sphingomonas alaskensis* EY 4374[T]	*Sphingomonas aquatilis* KCTC 2881[Tb]	*Sphingomonas aromaticivorans* EY 4296[T]	*Sphingomonas chlorophenolica* EY 4219[T]	*Sphingomonas capsulata* EY 4216[T]	*Sphingomonas asaccharolytica* EY 4229[T]	*Sphingomonas chungbukensis* EY 4375[T]	*Sphingomonas cloacae* EY 4361[T]	*Sphingomonas echinoides* EY 4340[T]	*Sphingomonas herbicidovorans* EY 4344[T]	*Sphingomonas koreensis* EY 4376[T]	*Sphingomonas macrogoltabidus* EY 4304[T]	*Sphingomonas mali* EY 4341[T]	*Sphingomonas melonis* EY 4350[T]	*Sphingomonas natatoria* EY 4920[T]	*Sphingomonas parapaucimobilis* EY 4213[T]	*Sphingomonas pituitosa* EY 4370[T]	*Sphingomonas pruni* EY 4228[T]	*Sphingomonas rosa* EY 4227[T]
Gram-negative rod shaped	+	+	+	+	+	+	+	+	Ov	+	+	+	+	+	+	+	+	+	+	+	+
Aerobic growth	+	+	+	+	+	+	+	+	+	+	+	+	+	+	+	+	+	+	+	+	+
Fermentation of glucose	−	−	−	−	−	−	−	−	−	−	−	−	−	−	−	−	−	−	−	−	−
Growth in the presence of 3% NaCl	+	+	+	−	+	−	+	−	+	−	+	−	−	+	+	−	+	+	+	+	+
Alkapton production	−	−	−	−	−	−	−	−	−	−	−	+	−	−	−	−	−	−	−	−	−
Pigmentation of colonies	DY	DY	Y	Y	DY	LY	DY	LY	Y	GW	LY	Y	Y	GW	LY	Y	O	DY	DY	GW	GW
Bacteriochlorophyll *a*	−	−	−	−	−	−	−	−	−	−	−	−	−	−	−	−	+	−	−	−	−
Oxidase	−	−	+	+	+	−	+	−	+	+	+	+	+	−	+	w	−	−	−	+	+
Catalase	+	+	+	+	+	+	+	+	+	+	+	+	+	+	+	+	+	+	+	+	+
Fermentation of glucose	−	−	−	−	−	−	−	−	−	−	−	−	−	−	−	−	−	−	−	−	−
Hydrolysis of:																					
Esculin	+	+	+	−	+	+	+	+	−	+	+	+	+	+	+	+	+	+	+	+	+
Gelatin	+	+	+	−	+	−	−	+	−	−	+	−	−	+	−	−	−	−	−	−	−
Starch	+	+	+	−	+	+	+	−	+	−	−	+	+	+	−	+	+	+	+	−	−
Tween 80	+	+	+	−	+	−	+	−	−	−	+	+	+	+	−	+	+	+	+	−	−
Citrate, Simmons	+	−	−	−	−	−	+	−	−	−	+	−	−	+	−	−	−	−	−	−	−
Malonate	−	−	−	−	−	−	+	−	−	−	−	−	−	+	−	−	NG	−	−	−	+
Arginine, Moller	−	−	−	−	−	−	−	−	−	−	−	−	−	−	−	−	−	−	−	−	−
Lysine, ornithine, Moller	−	−	−	−	−	−	−	−	−	−	−	−	−	−	−	−	−	−	−	−	−
Gas from nitrate	−	−	−	−	−	−	−	−	−	−	−	−	−	−	−	−	−	−	−	−	−
Nitrite from nitrate	−	−	−	−	−	−	−	−	−	−	−	+	−	−	−	−	−	−	+	+	−
Zn test on negative NO₂ test	+	+	+	−	+	+	+	+	+	+	nd	+	+	+	+	+	+	−	nd	+	+
Phenylalanine deaminase	−	−	−	−	−	−	+	−	−	−	−	−	−	−	−	−	−	−	−	−	−
Urease	−	−	+	−	−	−	−	−	−	−	−	−	−	−	−	−	−	−	−	−	−
Acylamidase	−	−	−	−	−	−	+	−	−	−	−	−	−	−	−	−	−	−	−	NG	−
DNase	−	−	−	+	NG	−	+	−	−	−	−	−	−	−	−	−	+	−	−	NG	+

[a]Symbols: +, positive reactions; −, negative reactions; Ov, ovoid; Y, yellow; LY, light yellow; GW, grayish white; O, orange; PY, pale yellow; OB, orange brown; B, brown; RO, reddish orange; YO, yellowish orange; NG, no growth; nd, not determined.

[b]Data from Lee et al. (2001). Blank space, no data available.

[c]Data from Yurkov et al. (1997). Blank space, no data available.

The mol% G + C of the DNA is: 62–64 (HPLC).

Type strain: ATCC 29837, DSM 1098, EY 2395, GIFU 2395, IAM 12576, IFO 11385, JCM 7516, LMG 1227, NCTC 11030.

GenBank accession number (16S rRNA): U37337, X72722.

2. **Sphingomonas adhaesiva** Yabuuchi, Yano, Oyaizu, Hashimoto, Ezaki and Yamamoto 1990b, 321[VP] (Effective publication: Yabuuchi, Yano, Oyaizu, Hashimoto, Ezaki and Yamamoto 1990a, 116.)

ad.hae.si.va. L. *adhaerere* adhere; *adhaesiva* intended to mean sticking to agar medium.

The characteristics are as described for the genus and in Tables BXII.α.84, BXII.α.85, BXII.α.86, BXII.α.87, BXII.α.88, and BXII.α.89, with the following additional information. Cells have polar monotrichous flagella. Circular and yellow-pigmented colonies stick to agar medium, but corrosion of the agar has not been observed. No growth occurs on Simmons' citrate medium. Acid production in OF medium from carbohydrates takes 4 or more days of incubation. Major long-chain bases are $C_{18:0}$ (39%) and $C_{21:1}$ (26%); major fatty acids are $C_{18:1\,cis}$ (46%) and $C_{14:0\,2OH}$ (23%). Major polyamine is homospermidine (Hamana and Matsuzaki, 1993; Segers et al., 1994; Takeuchi et al., 1995). The type strain was isolated from ultraviolet-irradiated water at a surgical operation theater.

The mol% G + C of the DNA is: 67.2 (HPLC).

Type strain: ATCC 51229, CCUG 27290, DSM 7418, EY 4215, GIFU 11458, IFO 5099, JCM 7370, LMG 0922.

GenBank accession number (16S rRNA): D16146, X72720.

3. **Sphingomonas alaskensis** Vancanneyt, Schut, Snuwaert, Goris, Swings and Gottschal 2001, 78[VP]

TABLE BXII.α.85. *(cont.)*

Substrate or test	*Sphingomonas roseiflava* EY 4345[T]	*Sphingomonas sanguinis* EY 2397[T]	*Sphingomonas stygia* EY 4297[T]	*Sphingomonas subarctica* EY 4251[T]	*Sphingomonas suberifaciens* EY 4204[T]	*Sphingomonas subterranea* EY 4298[T]	*Sphingomonas taejonensis* EY 7377[T]	*Sphingomonas terrae* EY 4207[T]	*Sphingomonas trueperi* EY 4218[T]	*Sphingomonas ursincola* EY 4250[T]	*Sphingomonas wittichii* EY 4224[T]	*Sphingomonas xenophaga* EY 4343[T]	*Sphingomonas yanoikuyae* EY 4208[T]	*Erythrobacter longus* EY 4203[T]	*Erythrobacter litoralis* EY 4222[T]	*Erythromicrobium ramosum* EY 4223[T]	*Porphyrobacter neustonensis* EY 4230[T]	*Porphyrobacter tepidarius* EY 4231[T]	*Sandaracinobacter sibiricus* RB 16-17[Tc]	*Z. mobilis* subsp. *mobilis* EY 4209[T]	*Z. mobilis* subsp. *pomacii* EY 4209[T]
Gram-negative rod shaped	+	+	+	+	+	+	+	+	+	Ov	+	+	+	+	+	+	+	Ov	+	+	+
Aerobic growth	+	+	+	+	+	+	+	+	+	+	+	+	+	+	+	+	+	+	+	−	−
Fermentation of glucose	−	−	−	−	−	−	−	−	−	−	−	−	−	−	−	−	−	−	−	+	+
Growth in the presence of 3% NaCl	−	+	−	+	−	−	−	+	+	−	+	+	−	+	+	−	−	+	+	+	+
Alkapton production	−	−	−	−	−	−	−	−	−	−	−	−	−	−	−	−	−	−	−	−	−
Pigmentation of colonies	O	DY	LY	LY	Y	DY	PY	DY	LY	OB	GW	DY	GW	B	RO	O	RO	O	YO	GW	GW
Bacteriochlorophyll *a*	−	−	−	−	−	−	−	−	−	+	−	−	−	+	+	+	+	+	+	−	−
Oxidase	−	−	+	+	−	+	+	+	+	+	+	+	−	+	−	+	+	+	+	−	−
Catalase	+	+	+	+	+	+	+	+	+	+	+	+	+	+	+	+	+	+	nd	−	+
Fermentation of glucose	−	−	−	−	−	−	−	−	−	−	−	−	−	−	−	−	−	−	−	+	+
Hydrolysis of:																					
Esculin	+	+	+	+	+	+	+	+	+	+	+	+	+	+	−	−	−	−	−	nd	nd
Gelatin	+	+	+	+	−	+	−	+	−	+	−	+	−	+	−	−	−	+	−	−	−
Starch	+	+	−	−	−	−	−	−	−	−	+	−	−	−	−	+	+	+	−	−	−
Tween 80	+	+	−	−	−	+	−	−	−	−	+	−	−	w	+	+	+	+	−	−	−
Citrate, Simmons	−	−	+	−	−	−	−	−	−	−	+	−	+	nd	nd	+	−	+	−	−	−
Malonate	−	−	−	−	−	+	−	−	−	−	−	−	−	nd	nd	−	−	−	−	nd	nd
Arginine, Moller	−	−	−	−	−	−	−	−	−	−	−	−	−	nd	nd	−	−	−	−	−	−
Lysine, ornithine, Moller	−	−	−	−	−	−	−	−	−	−	−	−	−	nd	nd	−	−	+	−	nd	nd
Gas from nitrate	+	−	−	−	−	−	−	−	−	−	−	−	−	nd	nd	+	−	−	−	−	−
Nitrite from nitrate	+	−	−	−	−	−	−	+	−	−	−	−	−	nd	nd	+	−	+	−	−	−
Zn test on negative NO₂ test	−	+	+	+	+	+	+	+	+	+	+	+	+	nd	nd	−	+	+	−	+	+
Phenylalanine deaminase	−	−	−	−	−	−	−	−	−	−	−	+	−	nd	nd	−	−	+	−	nd	nd
Urease	−	−	−	−	−	−	−	−	−	−	−	−	−	nd	nd	nd	nd	nd	nd	nd	nd
Acylamidase	−	−	+	−	−	−	−	−	−	−	−	+	−	nd	nd	+	−	−	−	nd	nd
DNase	−	−	+	+	−	+	−	+	−	−	−	−	−	nd	nd	−	NG	−	−	nd	nd

[a]Symbols: +, positive reactions; −, negative reactions; Ov, ovoid; Y, yellow; LY, light yellow; GW, grayish white; O, orange; PY, pale yellow; OB, orange brown; B, brown; RO, reddish orange; YO, yellowish orange; NG, no growth; nd, not determined.

[b]Data from Lee et al. (2001). Blank space, no data available.

[c]Data from Yurkov et al. (1997). Blank space, no data available.

a.las.ken'sis. M.L. adj. *alaskensis* pertaining to Alaska, the source of the type strain.

The characteristics are as described for the genus and in Tables BXII.α.84, BXII.α.85, BXII.α.86, BXII.α.87, BXII.α.88, and BXII.α.89, with the following additional information. Motile small rod. Soma size 0.2–0.5 × 0.5–3.0 μm. Although isolated at low temperatures (4–8°C), the optimal growth temperature for the type strain is approximately 37°C. Growth occurs at 44–48°C. Does not grow in Bacto-OF medium. Acid is produced in Bacto-CTA medium from maltose and trehalose, but not from D-glucose (Table BXII.α.88). For results with the API 20NE and API 50CH systems, see Vancanneyt et al. (2001). Isolated from seawater from Alaska.

The mol% G + C of the DNA is: 65 (HPLC).

Type strain: DSM 13593, EY 4374, LMG 18877, RB 2256.

GenBank accession number (16S rRNA): AF145754.

4. **Sphingomonas aquatilis** Lee, Shin, Yoon, Takeuchi, Pyun and Park 2001, 1495[VP]

a.qua.til'is. L. fem. adj. *aquatilis* aquatic, growing in water.

The characteristics are as described for the genus and in Tables BXII.α.84, BXII.α.85, BXII.α.88, and BXII.α.89, with the following additional information. Motile by a single polar flagellum. Colonies are circular, entire, low convex, smooth, opaque and yellow. Gelatin is not liquefied. Esculin is not hydrolyzed. D-Melibiose, D-sucrose, glucose, malate, maltose, N-acetylglucosamine, rhamnose, and salicin are assimilated, but 2-ketogluconate, 5-ketogluconate, adipate, caprate, citrate, D-ribose, D-sorbitol, gluconate, glycogen, L-arabinose, mannose, and phenylacetate are not assimilated. Acid is weakly produced oxidatively from glucose and rhamnose, but not from glycerol. Isolated from natural mineral water in Taejon City, Korea.

The mol% G + C of the DNA is: 63 (HPLC).

Type strain: JSS7, KCTC 2881, KCCM 41067.

GenBank accession number (16S rRNA): AF131295.

Additional Remarks: Reference strain: EY 4383, JSS28, KCTC 2883, KCCM 41066.

5. **Sphingomonas aromaticivorans** Balkwill, Drake, Reeves,

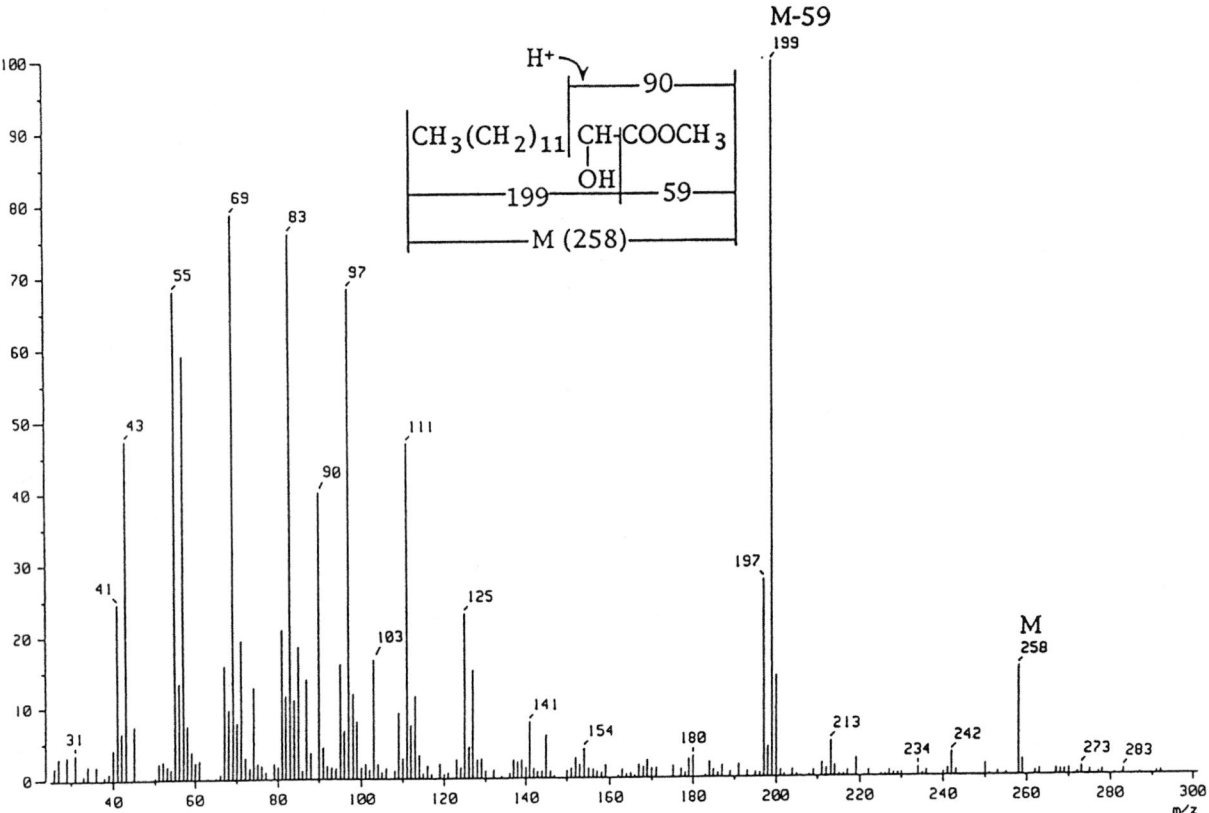

FIGURE BXII.α.91. Mass spectrum of the reaction product of H_2O_2-dependent α-hydroxylation of myristic acid. The mass spectrum of the methylated product gave ions of m/z 258 and 199, which were identical to those of methyl-2-hydroxcymyristic acid (*M*) and its characteristic fragment (*M-59*), respectively. (Reproduced with permission from I. Matsunaga et al., Federation of European Biochemical Societies Letters, *386:* 252–254, 1996, ©Elsevier Science B.V., Amsterdam.)

Fredreckson, White, Ringelberg, Chandler, Romine, Kennedy and Spadoni 1997, 199[VP]*

a.ro.ma.ti.ci' vo.rans. N.L. n. *aromaticus* aromatic compound; L. part. *vorans* eating; *aromaticivorans* eating aromatic compounds.

The characteristics are as described for the genus and in Tables BXII.α.84, BXII.α.85, BXII.α.86, BXII.α.87, BXII.α.88, and BXII.α.89, with the following additional information. Nonmotile. Colonies are circular entire low convex smooth opaque and deep yellow. SGL-1 is present but not SGL-1'. In addition to $C_{14:0\ 2OH}$, $C_{15:0\ 2OH}$ is detected as a minor component. Major polyamine is spermidine (Busse et al., 1999; Takeuchi et al., 2001). Isolated from saturated Atlantic coastal plain terrestrial subsurface sediments.

The mol% G + C of the DNA is: 62.9–65.0 (HPLC).

Type strain: F199, DSM 12444, EY 4296, JCM 16084, SMCC F199.

GenBank accession number (16S rRNA): U20756.

Additional Remarks: Reference strains include strains B0522 (SMCC B0522) and B0695 (SMCC B0695).

6. **Sphingomonas asaccharolytica** Takeuchi, Sakane, Yanagi, Yamasato, Hamana and Yokota 1995, 341[VP]

a.sac.cha.ro.ly' ti.ca. Gr. pref. *a* not; Gr. n. *sacchar* sugar; Gr. adj. *lytica* able to loosen; M.L. fem. adj. *asaccharolytica* not digesting sugar.

The characteristics are as described for the genus and in Tables BXII.α.84, BXII.α.85, BXII.α.86, BXII.α.87, BXII.α.88, and BXII.α.89, with the following additional information. Nonmotile because neither actively motile in wet mount preparation nor demonstrates any spreading growth on semisolid motility agar plate. Colonies are light yellow. In spite of the specific epithet, acid is produced oxidatively in OF basal medium from eight kinds of carbohydrates, including glucose and maltose (Table BXII.α.88). β-Galactosidase positive. 3-Ketolactose is produced and the major polyamine is homospermidine (Takeuchi et al., 1995; Takeuchi et al., 2001); Kämpfer et al. (1997) reported the production of minor amounts of spermidine in addition to homospermidine. Isolated from the roots of *Malus* spp. (apple) in Tsukuba City, Japan.

The mol% G + C of the DNA is: 64.8 (HPLC).

Type strain: Y-345, ATCC 51839, DSM 10564, EY 4229, IFO 15499, JCM 10279, LMG 17539.

GenBank accession number (16S rRNA): Y09639.

7. **Sphingomonas capsulata** (Leifson 1962) Yabuuchi, Yano, Oyaizu, Hashimoto, Ezaki and Yamamoto 1990b, 321[VP] (Effective publication: Yabuuchi, Yano, Oyaizu, Hashimoto,

---

Editorial Note: Novosphingobium aromaticivorans (Balkwill et al. 1997) Takeuchi et al. 2001, 1415 is a junior objective synonym of *S. aromaticivorans* (Yabuuchi et al., 2002).

FIGURE BXII.α.92. Nucleotide sequence of 1.3-kb fragment containing the fatty acid α-hydroxylase gene and deduced amino acid sequence. The deduced amino acid sequence is represented under the nucleotide sequence by single-letter code. The asterisk and double underline indicate the termination codon and putative Shine-Dalgarno sequence, respectively. Helix-1, helix-K, aromatic region, and heme-binding region are underlined. Heme-binding cysteine is boxed. (Reproduced from I. Matsunaga et al., Journal of Biological Chemistry, *272:* 23592–23596, 1997, ©American Society for Biochemistry and Molecular Biology, Inc.)

Ezaki and Yamamoto 1990a, 117) (*Flavobacterium capsulatum* Leifson 1962, 161.)*

cap.su.la′ ta. L. n. *capsula* a small chest, capsule; M.L. fem. adj. *capsulata* encapsulated.

The characteristics are as described for the genus and in Tables BXII.α.84, BXII.α.85, BXII.α.86, BXII.α.87, BXII.α.88, and BXII.α.89, with the following additional information. Nonmotile without flagellum. Produces definitely yellow colonies. In spite of the specific epithet (Leifson, 1962), the organism lost its ability to produce capsule. Phenylalanine deaminase positive. Major polyamine reported as spermidine (Busse and Auling, 1988; Hamana and Matsuzaki, 1993; Segers et al., 1994; Takeuchi et al., 1995; Takeuchi et al., 2001). Isolated from distilled water.

The mol% G + C of the DNA is: 63.1 (HPLC).

Type strain: ATCC 14666, DSM 30916, EY 4216, GIFU 11526, IFO 12533, JCM 7508, LMG 2830, NCIMB 9890.

GenBank accession number (16S rRNA): D16147.

8. **Sphingomonas chlorophenolica** Nohynek, Suhonen, Nurmiaho-Lassila, Hantula and Salkinoja-Salonen 1996b, 625[VP]

Editorial Note: Novosphingobium capsulatum (Leifson 1962) Takeuchi et al. 2001, 1415 is a junior objective synonym of *S. capsulata* (Yabuuchi et al., 2002).

FIGURE BXII.α.93. Structure of polychlorinated dibenzo-*p*-dioxins (PCDDs), dibenzofurans (PCDFs), and diphenyl ethers (PCDEs). The number of possible congeners is given in brackets. (Reproduced from R.-M. Wittich. 1998. *In* Wittich (Editor) Biodegradation of Dioxins and Furans. Springer-Verlag, Berlin, and R.G. Landes Co., pp. 1–28.)

(Effective publication: Nohynek, Suhonen, Nurmiaho-Lassila, Hantula and Salkinoja 1995, 536.)*

chlor.o.phen.o' li.ca. L. fem. adj. *chloro* containing chlorine; N.L. n. *pheno* phenol; L. fem. adj. *chlorophenolica* relating to chlorophenols.

The characteristics are as described for the genus and in Tables BXII.α.84, BXII.α.85, BXII.α.86, BXII.α.87, BXII.α.88, and BXII.α.89, with the following additional information. Rods vary in size, 0.3–0.7 × 1.0–3.5 μm. Acylamidase and DNase are positive, but hydrolysis of starch and Tween 80 are negative. Fimbriae or filaments may occur. Major polyamine is spermidine (Busse et al., 1999; Takeuchi et al., 2001).

Isolated from soil contaminated with wood preserving chlorophenols.

The mol% G + C of the DNA is: 63–67 (T_m).

Type strain: ATCC 33790, DSM 6284, EY 4219, IFO 16172, JCM 10275, LMG 17554.

GenBank accession number (16S rRNA): U60171, X87161.

9. **Sphingomonas chungbukensis** Kim, Chun, Bae and Kim 2000a, 1646[VP]

chung.bu.ken' sis. M.L. adj. *chungbukensis* pertaining to Chungbuk National University.

The characteristics are as described for the genus and in Tables BXII.α.84, BXII.α.85, BXII.α.86, BXII.α.87, BXII.α.88, and BXII.α.89, with the following additional information. Curved rods, 0.5–1 × 0.5–3.0 μm. Nonmotile. Colonies are smooth and yellow. Produce a great quantity of extracellular polysaccharide. Esculin and Tween 80 are not hydrolyzed. The major respiratory quinone is ubiqui-

none 10. Isolated from chemically contaminated freshwater sediment in Taejon, Republic of Korea.

The mol% G + C of the DNA is: 63 (T_m).

Type strain: DJ77, EY 4375, IMSNU 11152, JCM 11454, KCTC 2955.

GenBank accession number (16S rRNA): AF159257.

10. **Sphingomonas cloacae** Fujii, Urano, Ushio, Satomi and Kimura 2001, 608[VP]

clo.a' cae. L. n. *cloaca* sewer, the source of the organism.

The characteristics are as described for the genus and in Tables BXII.α.84, BXII.α.85, BXII.α.86, BXII.α.87, BXII.α.88, and BXII.α.89, with the following additional information. Cells are 1.1–1.4 × 2.0–3.1 μm. Motile. Colonies are opaque and white. Unable to grow at 4° or 42°C. Both SGL-1 and SGL-1′ are present. The major isoprenoid quinone is ubiquinone Q-10. Isolated from wastewater of a sewage-treatment plant in Tokyo.

The mol% G + C of the DNA is: 63 (HPLC).

Type strain: S-3, EY 4361, IAM 14885, JCM 10874.

GenBank accession number (16S rRNA): AB040739.

11. **Sphingomonas echinoides** (Heumann 1962) Denner, Kämpfer, Busse, and Moore 1999, 1108[VP] (*Pseudomonas echinoides* Heumann 1962.)

e.chi.noi' des. Gr. adj. *echinos* spiny appearance; Gr. n. *eidus* form, shape; M.L. adj. *echinoides* spiny shaped.

The characteristics are as described for the genus and in Tables BXII.α.84, BXII.α.85, BXII.α.86, BXII.α.87, BXII.α.88, and BXII.α.89, with the following additional information. Curved rods 0.8 × 1.9 μm, with sharp ends. Motile by polar flagella of 1.9 μm wavelength. Colonies are light yellow. Form cell aggregates (rosettes) both on solid and in liquid media. β-Galactosidase positive. For chromatogram of lipids, see Fig. BXII.α.85. Major polyamine is homospermidine (Denner et al., 1999; Takeuchi et al., 2001). Denner et al. (1999) also reported production of spermidine. Isolated as a laboratory contaminant on a nutrient agar plate.

The mol% G + C of the DNA is: 65.8 (T_m) (Owen and Jackman, 1982).

Type strain: ATCC 14820, DSM 1805, EY 4340, ICBP 2835, IFO 15742, JCM 10637, NCIMB 9420.

GenBank accession number (16S rRNA): AB021370, AJ012461.

12. **Sphingomonas herbicidovorans** Zipper, Nickel, Angst, and Kohler 1997, 601[VP] (Effective publication: Zipper, Nickel, Angst and Kohler 1996, 4319.)**

herb.i.ci.do' vo.rans. L. n. *herba* an herb; *cido-cide* joint word to mean killing; L. v. *voro* to devour; M.L. fem. adj. *herbicidovorans* herbicide-devouring, referring to its ability to utilize herbicides as a sole source of carbon and energy.

The characteristics are as described for the genus and in Tables BXII.α.84, BXII.α.85, BXII.α.86, BXII.α.87, BXII.α.88, and BXII.α.89, with the following additional information. The type strain, strain MH (previously designated *Flavobacterium* sp. strain MH) was able to utilize the chiral herbicide (RS)-2-(4-chloro-2-methylphenoxy)propionic acid (mecoprop) as the sole carbon end en-

---

Editorial Note: Sphingobium chlorophenolicum (Nohynek et al. 1996a) Takeuchi et al. 2001, 1415 is a junior objective synonym of *S. chlorophenolica* (Yabuuchi et al., 2002).

**Editorial Note: (Zipper et al. 1997) Takeuchi et al. 2001, 1415 is a junior objective synonym of *S. herbicidovorans* (Yabuuchi et al., 2002).

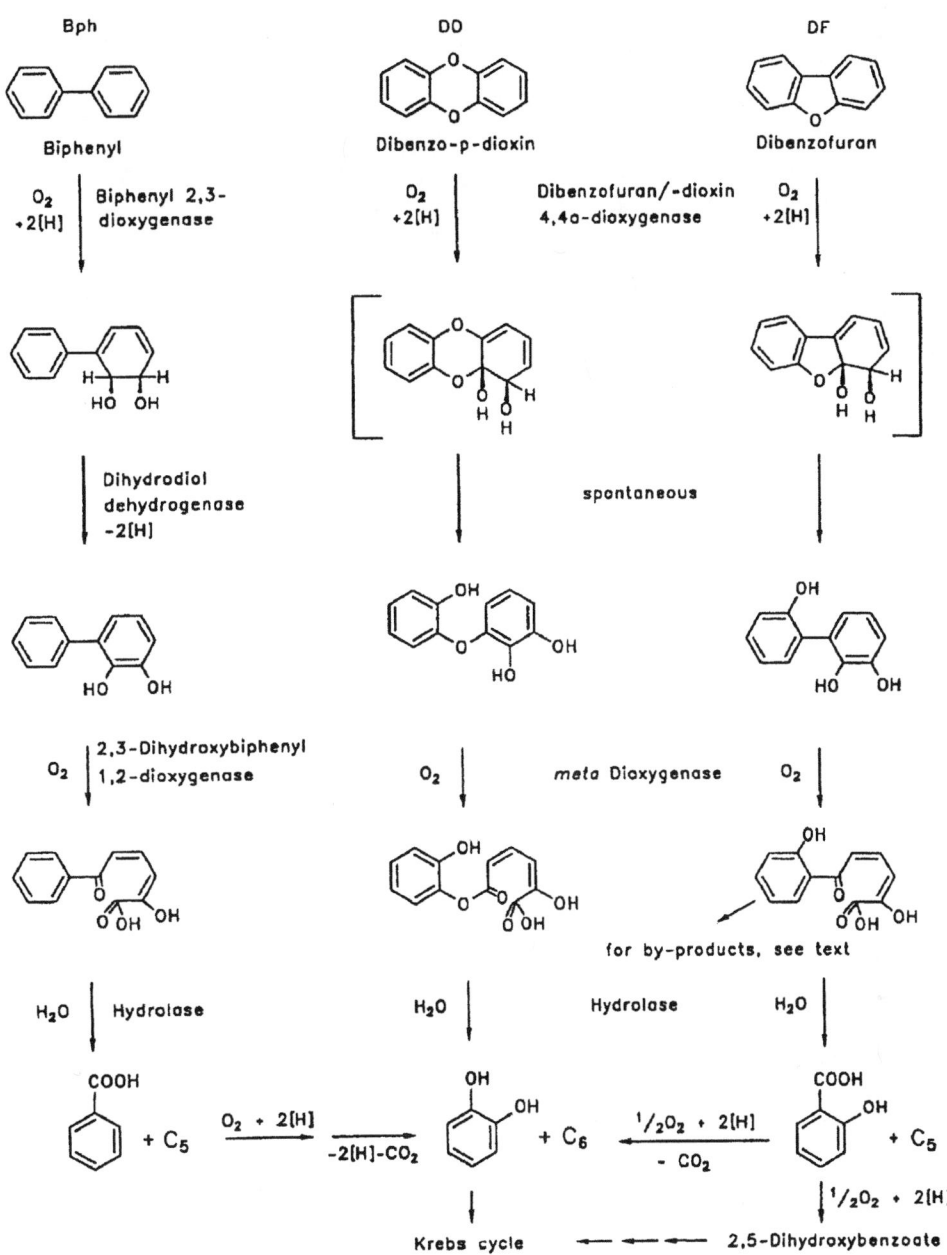

FIGURE BXII.α.94. Pathways for the complete aerobic degradation of biphenyl, dibenzo-*p*-dioxin, and dibenzo-furan. (Reproduced from R.-M. Wittich. 1998. *In* Wittich (Editor) Biodegradation of Dioxins and Furans. Springer-Verlag, Berlin, and R.G. Landes Co., pp. 1–28.)

ergy source. Colonies are yellow. Produce the water-soluble brown pigment alkaptone due to accumulation of homogentisic acid by metabolic failure of phenylalanine or tyrosine. A chromatogram of SGL-1 is shown in Fig. BXII.α.85. Major polyamine is spermidine (Busse et al., 1999; Takeuchi et al., 2001). Isolated from soil.

The mol% G + C of the DNA is: 62.5 (HPLC).

Type strain: MH, ATCC 700291, CIP 106705, DSM 11019, EY 4344, LMG 18315.

GenBank accession number (16S rRNA): AB042233.

13. **Sphingomonas koreensis** Lee, Kook Shin, Yoon, Takeuchi, Pyun and Park 2001, 1496[VP]

ko.re.en′ sis. N.L. fem. adj. *koreensis* pertaining to Korea, where the new organisms were isolated.

The characteristics are as described for the genus and in Tables BXII.α.84, BXII.α.85, BXII.α.86, BXII.α.87, BXII.α.88, and BXII.α.89, with the following additional information. Motile rods with a single polar flagellum. Colonies are yellow. Acid is produced oxidatively in OF medium from 6 of 26 carbohydrates. The major isoprenoid quinone is ubiquinone Q-10. Isolated from natural mineral water in Taejon City, Korea.

The mol% G + C of the DNA is: 66 (HPLC).

Type strain: JSS26, EY 4376, JCM 11456, KCTC 2882, KCCM 41069.

GenBank accession number (16S rRNA): AF131296.

14. **Sphingomonas macrogoltabidus** Takeuchi, Kawai, Shimada, and Yokota 1993b, 864[VP] (Effective publication: Takeuchi, Kawai, Shimada and Yokota 1993a, 236.)*

TABLE BXII.α.86. Selected results of assimilation tests (Biotype 100) by 31 *Sphingomonas* species[a,b]

Biotype 100, Substrates or tests	*S. paucimobilis* EY 2395	*S. adhaesiva* EY 4215	*S. alaskensis* EY 4374	*S. aromaticivorans* EY 4296	*S. asaccharolytica* EY 4229	*S. capsulata* EY 4216	*S. chlorophenolica* EY 4219	*S. chungbukensis* EY 4375	*S. cloacae* EY 4361	*S. echinoides* EY 4340	*S. herbicidovorans* EY 4344	*S. koreensis* EY 4376	*S. macrogoltabidus* EY 4304	*S. mali* EY 4341	*S. natatoria* EY 4220	*S. parapaucimobilis* EY 4213
Betaine	+	−	−	−	−	−	−	−	−	−	−	−	−	−	−	−
α-Ketoglutarate	−	+	−	−	−	−	−	−	−	−	−	−	−	−	−	+
D(+)Trehalose	+	+	+	−	+	+	−	−	−	+	−	+	+	+	+	+
D-Galacturonate	−	−	−	+	−	−	−	−	−	−	−	−	−	−	−	−
D-Saccharate	−	−	−	−	+	−	−	−	−	−	−	−	−	−	−	−
(−)-Quinate	−	−	−	+	−	+	−	−	−	−	−	−	−	−	−	−
D(−)Ribose	−	−	−	−	−	−	+	−	−	−	−	−	−	−	−	−
m-Hydroxybenzoate	−	−	−	−	−	−	−	+	−	−	−	−	−	−	−	−
trans-Aconitate	+	−	−	−	−	−	−	−	−	+	−	−	−	−	−	+
L(−)-Malate	+	+	−	−	+	+	−	−	−	+	−	+	+	−	−	+
α-L-Rhamnose	−	−	−	+	+	+	−	−	−	−	−	+	−	−	+	−
L-Serine	+	−	−	−	+	−	−	−	−	−	−	−	+	−	−	−
D(−)-Tartrate	−	−	−	−	−	−	−	−	−	−	−	−	+	−	−	−
α-D(+)-Melibiose	+	+	−	+	+	−	−	−	−	−	−	−	−	−	+	−
Protocatechuate	+	−	−	−	−	+	+	+	−	−	−	−	+	−	+	−
5-Keto-L-Gluconate	−	−	−	−	−	−	−	−	−	−	−	−	−	−	−	+
meso-Tartrate, D(+)-malate, caprylate,	−	−	−	−	+	−	−	−	−	−	−	−	−	−	−	−
D(+)-Galactose	+	−	−	−	+	+	−	−	−	−	−	−	−	+	−	+
L-Tryptophan	−	−	−	−	−	−	−	−	−	−	−	−	−	−	−	−
cis-Aconitate	+	−	−	−	+	−	−	−	−	+	−	−	−	−	−	+
D-Glucosamine	−	−	−	−	−	−	−	−	−	−	−	−	−	−	−	−
D-Lyxose, i-erythritol, 2-keto-D-gluconate	−	−	−	−	−	−	−	−	−	−	−	−	−	−	−	−
d-Alanine	−	−	−	−	+	−	−	+	+	−	−	−	−	−	−	−
Phenylacetate	−	−	−	−	−	−	−	−	−	−	−	−	−	−	−	−
L-Aspartate	+	+	−	−	+	+	+	−	−	−	−	+	+	−	+	−
Succinate	+	+	−	−	+	+	+	−	−	−	+	+	−	−	−	+
1-O-Methyl-α-galactopyranoside	+	−	−	+	−	−	−	−	−	−	−	−	−	−	−	+
D-Tagatose	−	−	−	−	−	−	−	−	−	−	−	−	−	−	−	−
DL-Glycerate	−	−	−	−	+	−	−	−	−	−	−	−	−	−	−	−
p-Hydroxybenzoate	+	−	−	+	−	+	+	+	+	−	−	−	−	−	−	−
Mucate	−	−	−	−	+	−	−	−	−	−	−	−	−	−	−	−

[a]For symbols see standard definitions.

[b]The 17 substrates not assimilated by the 31 strains are dulcitol, L(+)-sorbose, D(+)-arabitol, L(−)-arabitol, D-sorbitol, adonitol, hydroxyquinoline-β-glucuronide, L(+)-tartrate, 3-phenylpropionate, m-coumarate, trigonelline, histamine, histidine, DL-α-amino-n-valerate, ethanolamine, malonate, and 3-O-methyl-D-glucopyranose.

mac.ro.gol.ta' bi.dus. M.L. fem. n. *macrogol* a trade name for a polyethylene glycol product; L. adj. *tabidus* dissolving; M.L. fem. adj. *macrogoltabida* polyethylene glycol dissolving.

The characteristics are as described for the genus and in Tables BXII.α.84, BXII.α.85, BXII.α.86, BXII.α.87, BXII.α.88, and BXII.α.89, with the following additional information. Motile by a single polar flagellum. Colonies are grayish white. Polyethylene glycol 4000 is assimilated. Major polyamine is spermidine (Takeuchi et al., 1995; Takeuchi et al., 2001). Isolated from soil.

The mol% G + C of the DNA is: 63.2–65.0 (HPLC).

Type strain: CIP 104196, DSM 8826, EY 4304, IFO 15033, JCM 10192, LMG 17324.

GenBank accession number (16S rRNA): D13723.

15. **Sphingomonas mali** Takeuchi, Sakane, Yanagi, Yamasato, Hamana and Yokota 1995, 341[VP]

mal' i. M.L. gen. n. *mali* of *Malus*, the apple genus, the source of the organism.

The characteristics are as described for the genus and in Tables BXII.α.84, BXII.α.85, BXII.α.86, BXII.α.87, BXII.α.88, and BXII.α.89, with the following additional information. Cells are motile. Colonies are light yellow. DNase positive, but does not hydrolyze gelatin and starch. Major polyamine is homospermidine (Takeuchi et al., 1995; Takeuchi et al., 2001). Isolated from the roots of *Malus* spp. (apple) in Tsukuba City, Japan.

The mol% G + C of the DNA is: 65.4–65.9 (HPLC).

Type strain: Y-547, ATCC 51480, DSM 10565, EY 4341, IFO 15500, JCM 10193, LMG 17331.

GenBank accession number (16S rRNA): Y09638.

16. **Sphingomonas melonis** Buonaurio, Stravato, Kosako, Fujiwara, Naka, Kobayashi, Cappelli and Yabuuchi 2002, 2086.

me.lo' nis. M.L. n. *melo* melon; L. gen. n. *melonis* of melon (*Cucumis melo* var. *inodorus*, Spanish melon), referring to the fruit of the plant for which the organism was pathogenic.

Editorial Note: Sphingopyxis macrogoltabida (Takeuchi et al. 1993a) Takeuchi et al. 2001, 1416 is a junior objective synonym of *S. macrogoltabidus* (Yabuuchi et al., 2002).

TABLE BXII.α.86. *(cont.)*

Biotype 100, Substrates or tests	*S. pituitosa* EY 4370	*S. pruni* EY 4228	*S. rosa* EY 4227	*S. roseiflava* EY 4345	*S. sanguinis* EY 2397	*S. stygia* EY 4297	*S. subarctica* EY 4251	*S. subterranea* EY 4298	*S. taejonensis* EY 4377	*S. terrae* EY 4207	*S. trueperi* EY 4218	*S. ursincola* EY 4250	*S. wittichii* EY 4224	*S. xenophaga* EY 4343	*S. yanoikuyae* EY 4208	No. of positive EY strains	No. of negative EY strains
Betaine	−	−	−	−	−	−	−	−	−	−	−	−	−	−	−	1	30
α-Ketoglutarate	−	−	−	−	−	−	−	−	−	−	−	−	−	−	−	2	29
D(+)Trehalose	+	+	+	−	+	−	+	−	−	+	+	−	+	−	+	19	12
D-Galacturonate	−	−	−	−	−	−	+	−	−	−	+	−	−	−	−	3	28
D-Saccharate	−	−	−	−	−	+	−	−	−	−	−	−	−	−	+	3	28
(−)-Quinate	+	−	+	−	−	−	−	+	−	−	−	+	−	−	+	7	24
D(−)Ribose	−	−	−	−	−	−	−	−	−	−	−	−	+	−	−	2	29
m-Hydroxybenzoate	−	−	−	−	+	−	−	−	−	−	−	−	−	−	−	2	29
trans-Aconitate	−	−	−	+	−	+	−	−	−	−	−	−	+	−	−	6	25
L(−)-Malate	+	−	−	+	−	−	+	−	−	+	+	−	+	+	+	17	14
α-L-Rhamnose	−	+	+	−	−	−	+	+	−	−	+	−	−	−	+	12	19
L-Serine	−	−	−	−	+	−	+	−	−	−	+	−	+	−	−	7	24
D(−)-Tartrate	−	−	−	−	−	−	−	−	+	−	−	−	−	−	−	2	29
α-D(+)-Melibiose	+	+	−	+	−	−	−	−	−	−	+	+	−	−	+	13	18
Protocatechuate	+	−	+	−	−	−	+	+	−	−	+	−	−	+	+	13	18
5-Keto-L-Gluconate	−	−	−	−	−	−	−	−	−	−	−	−	−	−	−	1	30
meso-Tartrate, D(+)-malate, caprylate,	+	−	−	−	−	−	−	−	−	−	−	−	−	−	−	2	29
D(+)-Galactose	+	+	+	−	+	−	+	+	−	−	+	−	+	+	+	17	14
L-Tryptophan	−	−	+	−	−	−	−	−	−	−	−	−	−	−	−	1	30
cis-Aconitate	−	−	−	+	−	−	+	−	−	−	−	−	+	−	−	7	24
D-Glucosamine	−	−	−	−	+	−	−	−	−	−	−	−	+	−	−	2	29
D-Lyxose, *i*-erythritol, 2-keto-D-gluconate	−	−	−	−	−	+	−	−	−	−	−	−	−	−	−	1	30
d-Alanine	−	−	−	−	−	−	+	−	+	−	−	−	−	−	−	5	26
Phenylacetate	−	−	−	−	−	−	−	+	−	−	−	−	+	−	−	3	28
L-Aspartate	+	−	−	+	−	−	+	−	+	−	−	−	+	−	+	14	17
Succinate	+	−	−	+	+	−	+	−	−	+	+	−	+	+	+	17	14
1-*O*-Methyl-α-galactopyranoside	−	−	−	+	−	−	−	−	−	−	+	−	−	−	−	5	26
D-Tagatose	−	−	−	−	−	−	−	−	−	−	−	+	+	−	−	2	29
DL-Glycerate	−	−	−	−	−	−	−	−	−	−	−	−	+	−	−	2	29
p-Hydroxybenzoate	+	−	+	−	−	−	+	+	−	−	−	−	−	+	+	11	20
Mucate	−	−	−	−	−	−	−	−	−	−	−	−	−	−	+	2	29

[a]For symbols see standard definitions.

[b]The 17 substrates not assimilated by the 31 strains are dulcitol, L(+)-sorbose, D(+)-arabitol, L(−)-arabitol, D-sorbitol, adonitol, hydroxyquinoline-β-glucuronide, L(+)-tartrate, 3-phenylpropionate, *m*-coumarate, trigonelline, histamine, histidine, DL-α-amino-*n*-valerate, ethanolamine, malonate, and 3-*O*-methyl-D-glucopyranose.

The characteristics are as described for the genus and in Tables BXII.α.84, BXII.α.85, BXII.α.87, BXII.α.88, and BXII.α.89, with the following additional information. Non-motile. Colonies are deep yellow. Isolated from fruits of yellow Spanish melons (*Cucumis melo* var. *inodorus*) in Almeria (Spain). Causal agent of brown spot of melon fruits. Organisms identified as strains of *Sphingomonas* sp., had previously been reported as a causative agent of melon fruit disease (Buonaurio et al., 2001).

The mol% G + C of the DNA is: 65 (HPLC).

Type strain: DAPP-PG 224, EY 4350, LMG 19484.

GenBank accession number (16S rRNA): AB055863.

17. **Sphingomonas natatoria** (Sly 1985) Yabuuchi, Kosako, Naka, Suzuki and Yano 1999b, 935[VP] (Effective publication: Yabuuchi, Kosako, Naka, Suzuki and Yano 1999a, 347) (*Blastobacter natatorius* Sly 1985, 40; *Blastomonas natatoria* (Sly 1985) Sly and Cahill 1997, 568.)*

Editorial Note: Blastomonas natatoria (Sly 1985) Sly and Cahill 1997, 568 emend. Hiraishi, Kuraishi and Kawahara 2000a, 1117 is a junior objective synonym of S. natatoria (Yabuuchi et al., 2002).

na.ta.to′ri.a. M.L. fem. adj. *natatoria* of a swimming place (pool).

The characteristics are as described for the genus and in Tables BXII.α.84, BXII.α.85, BXII.α.86, BXII.α.87, BXII.α.88, and BXII.α.89, with the following additional information. Straight rod. Rosette-like arrangements of cells are seen. Unable to visualize either budding or asymmetric division of cells by optical microscopy. Though single polar flagellum was observed by electron microscopy (Sly, 1985), neither active motility in wet mount preparation nor diffuse spreading growth on semisolid motility agar plate was demonstrated. Colonies are orange pigmented. Aerobic and facultative photoorganotrophs, possessing bacteriochlorophyll *a*. Acid produced oxidatively from fructose and maltose but not from glucose. Isolated from a fresh water swimming pool. Major polyamine is spermidine (Busse et al., 1999; Takeuchi et al., 2001).

The mol% G + C of the DNA is: 64.5 (HPLC).

Type strain: ACM 2507, ATCC 35951, DSM 3183, EY 4220, IFO15649, JCM 10396, LMG 17322, NCIMB 12085.

GenBank accession number (16S rRNA): Y13774.

18. **Sphingomonas parapaucimobilis** Yabuuchi, Yano, Oyaizu,

TABLE BXII.α.87. Susceptibilities of type strain of 33 *Sphingomonas* species against 36 antibacterial agents determined by Kirby–Bauer method[a]

Antibiotic (mg/disk)	Clinical, hospital, laboratory							PP		Plant					Soil and sediments		
	S. paucimobilis EY 2395[T]	*S. adhaesiva* EY 4215[T]	*S. capsulata* EY 4216[T]	*S. echinoides* EY4340[T]	*S. parapaucimobilis* EY 4213[T]	*S. sanguinis* EY 2397[T]	*S. yanoikuyae* EY 4208[T]	*S. melonis* EY 4350[T]	*S. suberifaciens* EY 4204[T]	*S. asaccharolytica* EY 4229[T]	*S. mali* EY 4341[T]	*S. pruni* EY 4228[T]	*S. rosa* EY 4227[T]	*S. roseiflava* EY4345[T]	*S. macrogoltabidus* EY4304[T]	*S. trueperi* EY 4218[T]	*S. herbicidovorans* EY4344[T]
Doxycycline (10)	S	S	S	S	S	S	S	S	S	S	S	S	S	S	S	S	S
Tetracycline (30)	S	S	S	S	S	S	S	S	S	S	S	S	S	S	S	S	S
Minocycline (30)	S	S	S	S	S	S	S	S	S	S	R	S	S	S	S	S	S
Amikacin (30)	S	S	S	S	S	S	S	S	S	S	R	S	S	S	IM	S	S
AMPC/CVA (20/10)	S	S	S	S	S	S	S	S	S	S	R	S	S	S	IM	S	S
Gentamicin (10)	S	S	S	S	S	S	S	R	S	S	S	S	R	S	S	IM	S
Panipenem (10)	S	S	S	S	S	S	S	S	S	S	S	S	R	S	S	IM	S
Imipenem (10)	S	S	S	S	S	IM	S	S	S	S	S	S	S	S	S	IM	S
Dibekacin (30)	S	S	S	S	S	S	S	S	S	S	S	S	S	S	S	S	S
Sparfloxacin (5)	S	S	IM	R	S	S	S	S	S	IM	S	S	S	R	S	S	S
Clarithromycin (15)	S	S	S	S	S	S	S	S	S	IM	S	R	S	S	S	S	S
Levofloxacin (5)	S	S	S	S	S	S	S	S	S	R	S	S	S	S	S	S	S
Cefotaxime (30)	S	S	S	S	R	R	S	R	S	IM	S	S	R	S	R	S	S
Meropenem (10)	S	S	S	S	S	S	S	S	S	IM	S	S	R	S	R	S	S
Erythromycin (15)	S	R	IM	R	S	S	IM	IM	S	S	R	R	IM	S	S	IM	S
Cefaclor (30)	S	S	S	S	R	R	S	R	S	S	S	R	S	S	R	S	R
Ofloxacin (5)	S	IM	R	S	S	S	S	S	S	R	S	R	S	S	S	IM	S
Tosufloxacin (5)	S	IM	R	R	S	S	IM	S	S	IM	IM	S	R	S	S	S	S
Ceftazidime (30)	S	S	S	S	R	S	S	R	S	S	R	S	R	S	R	S	R
Amoxicillin (25)	S	S	S	S	R	R	S	R	S	S	S	R	S	S	R	S	S
Ciprofloxacin (5)	S	IM	R	S	S	S	S	S	S	R	R	IM	S	S	R	IM	S
Ampicillin (10)	S	S	S	S	R	R	S	R	S	S	S	R	S	S	R	S	R
Polymyxin B (300)	IM	R	IM	R	IM	R	S	R	S	R	R	R	S	S	S	R	S
Sulfamethoxazole-trimethoprim (23.75/1.25)	S	R	S	IM	S	S	S	S	S	S	S	R	R	S	S	R	R
Flomoxef (30)	S	R	S	R	S	R	R	R	S	R	R	IM	R	S	R	R	S
Roxithromycin (15)	S	R	R	R	S	S	R	IM	IM	IM	R	R	IM	S	IM	R	S
Cefmetazole (30)	R	IM	S	R	R	R	R	R	S	S	R	R	R	S	R	R	R
Cefoperazone (75)	S	S	IM	S	R	R	R	R	S	R	R	R	R	R	R	R	R
Penicillin (10)	S	R	S	S	R	R	S	R	S	S	R	R	R	R	R	R	R
Carumonam (30)	S	R	R	R	R	R	R	R	S	S	R	S	S	IM	R	R	R
Piperacillin (100)	R	IM	IM	S	R	R	R	R	S	R	R	R	R	R	R	R	S
Norfloxacin (10)	IM	R	R	R	R	S	IM	IM	R	S	R	R	R	S	R	R	IM
Aztreonam (30)	IM	R	R	R	R	R	R	R	S	R	R	IM	R	R	R	R	R
Moxalactam (30)	R	R	IM	S	R	R	R	R	S	IM	R	IM	R	R	R	R	R
Cefazolin (30)	R	R	S	S	R	R	R	R	S	R	R	R	R	S	R	R	R
Trimethoprim (5)	R	R	R	R	R	R	R	R	R	R	R	R	R	S	S	R	R
Summary:																	
R	5	12	8	9	16	16	9	17	3	9	15	19	21	5	16	14	10
%R	14	33	22	25	44	44	25	47	8	25	42	53	58	14	44	39	28
S + IM	31	24	28	27	20	20	27	19	33	27	21	17	15	31	20	22	26

[a]Abbreviations: PP, plant pathogenic; AMPC, amoxicillin; CVA, clavulanic acid; S, susceptible; IM, intermediate; R, resistant.

Hashimoto, Ezaki and Yamamoto 1990b, 321[VP] (Effective publication: Yabuuchi, Yano, Oyaizu, Hashimoto, Ezaki and Yamamoto 1990a, 116.)

pa.ra.pau.ci.mo.bi' lis. Gr. prep. *para* alongside of, resembling; M.L. *paucimobilis* specific epithet of *Sphingomonas paucimobilis*; M.L. gen. n. *parapaucimobilis* intended to mean like the species of *S. paucimobilis*.

The characteristics are as described for the genus and in Tables BXII.α.84, BXII.α.85, BXII.α.86, BXII.α.87, BXII.α.88, and BXII.α.89, with the following additional information. Cellular and colonial characteristics are similar to those of *S. paucimobilis*. Glucuronosyl ceramide (SGL-1) is present, but not galacturonosyl ceramide (SGL-1′) (Fig. BXII.α.85). Major component of polyamine is homospermidine (Hamana and Matsuzuki, 1993; Takeuchi et al., 1995; Takeuchi et al., 2001); Segers et al. (1994) reported production of spermidine. Isolated from human urine.

The mol% G + C of the DNA is: 64–65 (HPLC).

Type strain: ATCC 51231, CCUG 27291, DSM 7463, EY 4213, GIFU 11387, IFO 15100, JCM 7510, LMG 10923.

GenBank accession number (16S rRNA): X72721.

19. **Sphingomonas pituitosa** Denner, Paukner, Kämpfer, Moore, Abraham, Busse, Wanner and Lubitz 2001, 837[VP]

pi.tu.i.to' sa. L. fem. adj. *pituitosa* slimy.

TABLE BXII.α.87.　*(cont.)*

Antibiotic (mg/disk)	S. terrae EY 4207[T]	S. aromaticivorans EY4296[T]	S. stygia EY4297[T]	S. subterranea EY4298[T]	S. chungbukensis EY 4875[T]	S. chlorophenolica EY 4219[T]	S. cloacae EY 4361[T]	S. subarctica EY4251[T]	S. wittichii EY 4224[T]	S. pituitosa EY 4370[T]	S. xenophaga EY4343[T]	S. ursincola EY4250[T]	S. natatoria EY 4220[T]	S. alaskensis EY 4374[T]	S. koreensis EY 4376[T]	S. taejonensis EY 4877[T]	No. of resistant strains
	Soil and sediments				Sludge				Fresh water					Sea or mineral water			
Doxicycline (10)	S	S	S	S	S	S	S	S	S	S	S	S	S	S	S	S	0
Tetracycline (30)	S	S	S	S	S	S	S	S	IM	S	IM	S	S	S	S	S	0
Minocycline (30)	S	S	S	S	S	S	S	S	S	R	S	S	S	S	S	S	2
Amikacin (30)	S	S	S	S	S	S	S	S	S	S	S	S	S	R	R	IM	2
AMPC/CVA (20/10)	S	S	S	S	S	S	IM	S	S	S	S	S	S	IM	R	R	3
Gentamicin (10)	S	S	S	S	S	S	S	S	S	S	S	R	S	R	R	S	3
Panipenem (10)	S	S	S	S	S	S	S	IM	S	S	S	S	S	S	R	IM	3
Imipenem (10)	S	S	S	S	S	S	R	S	S	R	S	S	S	S	R	S	3
Dibekacin (30)	S	S	S	S	S	S	R	S	S	S	S	R	S	R	R	IM	4
Sparfloxacin (5)	S	R	S	IM	S	S	S	R	R	S	S	IM	S	S	IM	S	5
Clarithromycin (15)	R	S	S	S	S	S	R	R	S	R	IM	S	S	S	R	R	6
Levofloxacin (5)	S	IM	R	R	S	S	S	R	R	S	S	IM	S	S	S	S	6
Cefotaxime (30)	S	S	S	S	S	S	S	S	S	S	S	S	S	R	R	R	7
Meropenem (10)	IM	S	S	S	S	S	R	IM	S	R	S	S	S	R	R	R	7
Erythromycin (15)	IM	IM	S	IM	S	R	S	IM	R	S	IM	IM	S	IM	R	R	8
Cefaclor (30)	IM	M	S	IM	S	S	S	R	S	IM	S	S	S	R	R	R	10
Ofloxacin (5)	S	R	R	R	S	S	S	S	R	S	S	S	S	S	S	IM	10
Tosufloxacin (5)	IM	R	R	R	S	S	S	S	R	S	IM	IM	R	R	R	IM	10
Ceftazidime (30)	S	S	S	S	S	S	R	IM	S	R	S	S	S	R	R	R	11
Amoxicillin (25)	S	S	S	S	S	S	R	S	S	R	R	S	S	IM	R	IM	11
Ciprofloxacin (5)	S	R	R	R	S	S	S	R	R	IM	S	S	S	S	R	IM	11
Ampicillin (10)	S	S	S	S	R	S	R	S	S	R	R	S	S	R	R	R	14
Polymyxin B (300)	R	S	R	R	S	S	S	S	IM	R	S	IM	IM	S	R	S	14
Sulfamethoxazole-trimethoprim (23.75/1.25)	R	IM	R	R	R	S	S	R	R	IM	R	R	S	R	R	R	A6
Flomoxef (30)	R	S	S	S	S	S	R	IM	S	R	S	IM	IM	R	R	R	16
Roxithromycin (15)	R	IM	S	S	S	R	R	R	R	IM	IM	R	S	R	R	R	16
Cefmetazole (30)	R	R	S	R	S	S	S	R	S	S	S	IM	IM	R	R	R	17
Cefoperazone (75)	S	S	S	S	R	S	R	S	IM	R	R	S	S	IM	R	R	17
Penicillin (10)	S	S	S	R	S	R	R	S	S	S	R	S	S	R	R	R	18
Carumonam (30)	R	S	S	S	R	S	R	IM	S	S	IM	R	S	R	R	R	18
Piperacillin (100)	S	S	S	S	S	S	R	S	S	S	R	R	S	R	R	R	19
Norfloxacin (10)	IM	R	R	R	S	S	R	R	R	R	S	IM	S	S	R	R	19
Aztreonam (30)	R	IM	S	S	R	R	S	S	R	S	IM	R	IM	R	R	R	21
Moxalactam (30)	R	R	S	R	R	R	S	R	R	S	R	R	R	R	R	R	25
Cefazolin (30)	S	R	S	R	R	R	R	R	S	R	R	R	S	R	R	R	25
Trimethoprim (5)	R	R	R	R	R	R	S	R	R	S	R	R	R	R	R	R	29
Summary:																	
R	10	9	8	12	10	8	14	13	13	12	9	9	4	18	31	21	
%R	28	25	22	33	28	22	39	36	36	33	25	25	11	50	86	58	
S + IM	26	27	28	24	26	28	22	23	23	24	16	27	32	18	5	15	

[a]Abbreviations: PP, plant pathogenic; AMPC, amoxicillin; CVA, clavulanic acid; S, susceptible; IM, intermediate; R, resistant.

The characteristics are as described for the genus and in Tables BXII.α.84, BXII.α.85, BXII.α.86, BXII.α.87, BXII.α.88, and BXII.α.89, with the following additional information. Cells are 0.4–0.75 × 1.0–3.0 μm. Motile by a single polar flagellum. Strictly aerobic. Colonies are deep yellow. Poly-β-hydroxybutyrate is accumulated (Denner et al., 2001). Glucuronosyl ceramide (SGL-1) is present. Major polyamine is homospermidine (Denner et al., 2001). Isolated from the water of a eutrophic fountain in Vienna, Austria, in which an algal bloom was observed.

The mol% G + C of the DNA is: 64.5 (HPLC).
Type strain: EDIV, CIP 106154, DSM 13101, EY 4370.
GenBank accession number (16S rRNA): AJ243751.

20. **Sphingomonas pruni** Takeuchi, Sakane, Yanagi, Yamasato, Hamana and Yokota 1995, 340[VP]
pru′ni. M.L. gen. n. *Prunus* genus of peach; L, gen. n. *pruni* of peach (*Prunus persica*), the source of the organism.

The characteristics are as described for the genus and in Tables BXII.α.84, BXII.α.85, BXII.α.86, BXII.α.87, BXII.α.88, and BXII.α.89, with the following additional information. Neither active motility in wet mount preparation nor diffuse spreading growth on semisolid motility agar plate was observed. Colonies are grayish white. Glucuronosyl ceramide (SGL-1) is present, but not SGL-1′ (Fig. BXII.α.85). Major polyamine is homospermidine (Takeuchi

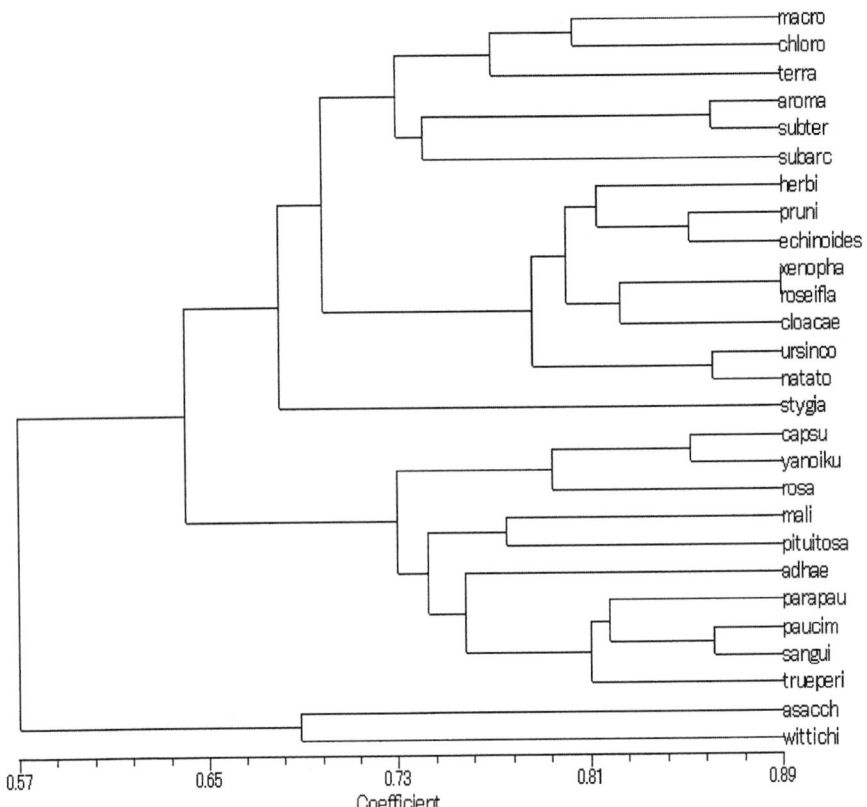

FIGURE BXII.α.95. Numerical analysis of 106 features of type strains of 27 *Sphingomonas* species. Numerical analysis was done by using NTSYS-pc Version 2 program package (Applied Biostatistics Inc. Setauket, NY). Coefficient was calculated by Simple Matching, and the dendrogram was drawn by UPGMA (Sneath and Sokal, 1973). Name of *Sphingomonas* species (new genus by Takeuchi et al., 2001): macro = *S. macrogoltabidus* (*Sphingopyxis*); chloro = "*S. chlororaphis*" (*Sphingobium*); terra = *S. terrae* (*Sphingopyxis*); aroma = *S. aromaticivorans* (*Sphingomonas*); subter = *S. subterranea* (*Sphingomonas*); subarc = *S. subarctica* (*Sphingomonas*); herbi = *S. herbicidovorans* (*Sphingobium*); pruni = *S. pruni* (*Sphingomonas*); echinoides = *S. echinoides* (*Sphingomonas*); xenopha = *S. xenophaga*; roseifla = *S. roseiflava* (*Sphingomonas*); cloacae = *S. cloacae*; ursincola = *S. ursincola* (*Blastomonas*); natato = *S. natatoria* (*Blastomonas*); stygia = *Sphingomonas stygia* (*Novosphingobium*); capsu = *S. capsulata* (*Novosphingobium*); yanoiku = *S. yanoikuyae* (*Sphingobium*); rosa = *S. rosa* (*Novosphingobium*); mali = *S. mali* (*Sphingomonas*); pituitosa = *S. pituitosa*; adhae = *S. adhaesiva* (*Sphingomonas*); parapau = *S. parapaucimobilis* (*Sphingomonas*); sangui = *S. sanguinis* (*Sphingomonas*); trueperi = *S. trueperi*; assach = *S. asaccharolytica* (*Sphingomonas*); wittichi = *S. wittichii*(*Sphingomonas*).

et al., 1995; Takeuchi et al., 2001). Isolated from the roots of *Prunus persica* (peach) in Tsukuba City, Japan.

The mol% G + C of the DNA is: 65.4 (HPLC).

Type strain: Y-250, DSM 10566, EY 4228, IFO 15498, JCM 10277, LMG 18380.

GenBank accession number (16S rRNA): D28568, Y09637.

21. **Sphingomonas rosa** Takeuchi, Sakane, Yanagi, Yamasato, Hamana and Yokota 1995, 340^VP*

ro' sa. M.L. n. *rosa* rose, the source of the organism.

The characteristics are as described for the genus and in Tables BXII.α.84, BXII.α.85, BXII.α.86, BXII.α.87, BXII.α.88, and BXII.α.89, with the following additional information. Neither active motility in wet mount preparation nor diffuse spreading growth on semisolid motility agar plate was observed. Colonies are light yellow. Glucuronosyl ceramide (SGL-1) is present, but not galacturonosyl ceramide (SGL-1′) (Fig. BXII.α.85). Major polyamine is sper-

midine (Takeuchi et al., 1995; Takeuchi et al., 2001). Isolated from the hairy roots of *Rosa* spp. (rose).

The mol% G + C of the DNA is: 64.7–65.0 (HPLC).

Type strain: ATCC 51837, DSM 7285, EY 4227, IFO 15208, JCM 10276, LMG 17328, NCPPB 2661.

GenBank accession number (16S rRNA): D13945.

22. **Sphingomonas roseiflava** Yun, Shin, Hwang, Kuraishi, Sugiyama and Kawahara 2000b, 1415^VP (Effective publication: Yun, Shin, Hwang, Kuraishi, Sugiyama and Kawahara 2000a, 17) (*Sphingomonas roseoflava* (sic) Yun, Shin, Hwang, Kuraishi, Sugiyama and Kawahara 2000b, 1415.)

ro.se.i.fla' va. L. adj. *roseus* rose colored; L. adj. *flavus* yellow; N.L. fem. adj. *roseiflava* rose-yellow.

The characteristics are as described for the genus and in Tables BXII.α.84, BXII.α.85, BXII.α.86, BXII.α.87, BXII.α.88, and BXII.α.89, with the following additional information. Nonmotile rod. Colonies are orange. Glucuronosyl ceramide (SGL-1) is present but not galacturonosyl ceramide (SGL-1′) (Fig. BXII.α.85). Two types of sphingoglycolipid are present: one contains glucuronic acid, and the other contains glucuronic acid, glucosamine, and hex-

Editorial Note: Novosphingobium rosa (Takeuchi et al. 1995) Takeuchi et al. 2001, 1415 is a junior objective synonym of *S. rosa* (Yabuuchi et al., 2002).

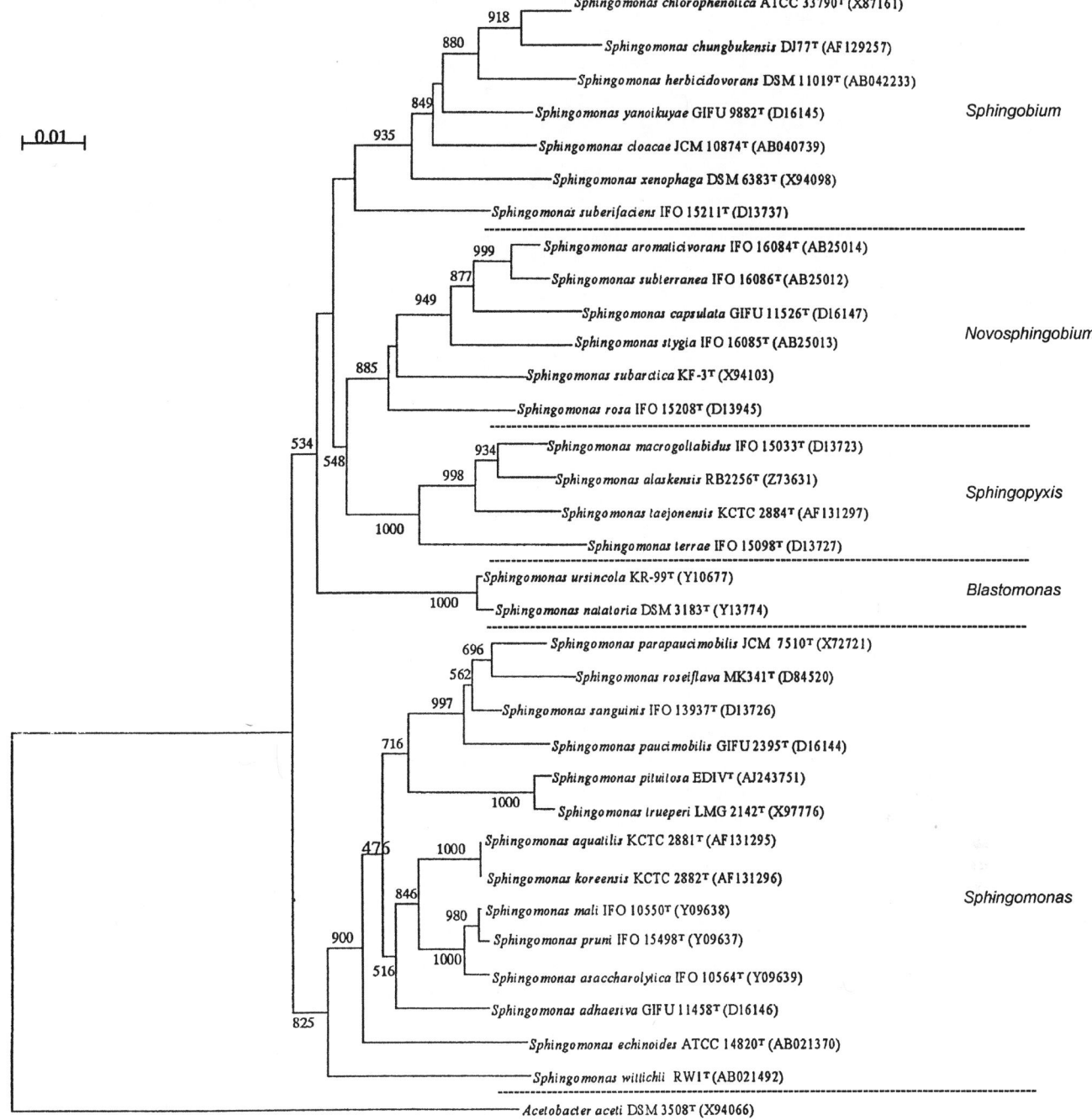

FIGURE BXII.α.96. Dendrogram based on the results of phylogenetic analysis of 16S rDNA sequence of type strains of 33 *Sphingomonas* species. Nucleotide substitution rates (K_{nuc}) (Kimura, 1980) were determined by using the CLUSTAL W program (Thompson et al., 1994). Neighbor-joining method (Saitou and Nei, 1987) was used to reconstruct a phylogenetic tree from the distance matrices by the NJPLOT written by Manolo Gouy (Laboratoire de Biometrie, Univ. Lyon, Villeurbanne, France). Alignment gaps and unidentified base positions were not taken into consideration for the calculations. To evaluate the topology of the phylogenetic tree, a bootstrap analysis was performed with 1000 bootstrapped trials. Bootstrap values are indicated at the branching points. Clusters containing *Sphingomonas* species assigned by Takeuchi et al., 2001 to the genera *Sphingobium*, *Sphingopyxis*, and *Novosphingobium* as well as the closely related *Blastomonas* species are indicated on the right.

oses (Yun et al., 2000a). Frequently isolated from the ear of the plant *Setaria viridis*.

The mol% G + C of the DNA is: 66–68 (HPLC).

Type strain: MK3412, EY 4345, IAM 14823.

GenBank accession number (16S rRNA): D84520.

23. **Sphingomonas sanguinis** Takeuchi, Kawai, Shimada and

Yokota 1993b, 864[VP] (Effective publication: Takeuchi, Kawai, Shimada and Yokota 1993a, 237) (*Sphingomonas sanguis* (sic) Takeuchi, Kawai, Shimada, and Yokota 1993b, 864.)

san'gui.nis. L. gen. n. *sanguinis* of the blood.

The characteristics are as described for the genus and in Tables BXII.α.84, BXII.α.85, BXII.α.86, BXII.α.87,

TABLE BXII.α.88. Oxidative acid production of the type strains of 38 species of *Sphingomonadaceae* in Bacto-Oxidation–Fermentation (OF), Cystine–tryptone agar, or Marine OF media[a,b]

Carbohydrates	*S. paucimobilis* EY 2395[T] (a)[c,d]	*S. alaskensis* EY 4374[T] (b)	*S. adhaesiva* EY 4215[T] (a)	*S. aquatilis* KCTC 2881[T] (a)[e]	*S. aromaticivorans* EY 4296[T] (b)	*S. asaccharolytica* EY 4229[T] (a)	*S. capsulata* EY 4216[T] (b)	*S. chlorophenolica* EY 4219[T] (a)	*S. chungbukensis* EY 4375[T] (a)	*S. cloacae* EY 4361[T] (a)	*S. echinoides* EY 4340[T] (a)	*S. herbicidovorans* EY 4344[T] (b)	*S. koreensis* EY 4376[T] (a)	*S. macrogoltabidus* EY 4304[T] (b)	*S. mali* EY 4341[T] (a)	*S. natatoria* EY 4220[T] (b)	*S. parapaucimobilis* EY 4213[T] (a)	*S. pituitosa* EY 4370[T] (a)	*S. pruni* EY 4228[T] (a)
Adonitol	−	−	−		−	−	−	−	−	−	−	−	−	−	−	−	−	−	−
D-Arabinose	−	−	−		−	+	+	−	−	−	+	−	−	−	−	−	−	−	−
L-Arabinose	+	−	+		−	+	+	−	+	−	+	+	+	+	−	+	+	+	+
Cellobiose	−	−	−		+	+	−	+	−	−	−	+	+	−	−	+	+	+	+
Dulcitol	−	−	−		+	+	−	+	−	−	−	−	−	−	+	+	+		
Ethanol (3%)	+	−	+		−	NG	+	−	−	−	−	−	NG	−	+	−	+	−	+
Fructose	+	−	+		+	−	+	−	−	−	+	−	+	−	+	+	+	−	+
Galactose	+	−	+		−	−	+	−	+	−	+	+	−	−	+	−	+	+	+
Glucose	−	−	+		+	+	+	+	−	−	+	+	+	−	+	−	+	+	+
Glycerol	−	−	+	−	−	−	−	−	−	−	−	+	−	−	+	−	−	−	−
Inositol	−	−	−		−	−	−	−	−	−	−	−	−	−	−	−	−	−	−
Inulin	−	−	−		−	−	−	−	−	−	−	−	−	−	−	−	−	−	−
Lactose	−	−	+		−	−	+	+	−	−	−	−	−	−	−	−	+	+	+
Maltose	−	−	−		+	+	+	+	+	+	+	−	−	−	+	−	+	+	+
Mannitol	−	−	−		+	−	+	−	−	−	−	−	−	−	+	−	+		
Mannose	+	−	+		+	+	+	+	−	−	+	−	−	−	+	−	+	+	+
Melezitose	−	−	+		−	+	+	−	−	−	−	−	−	−	−	−	+		
Melibiose	−	−	−		−	−	−	−	−	−	−	−	−	+	−	−	+		
Raffinose	−	−	−		−	−	−	+	−	−	−	−	−	−	−	−	+		
Rhamnose	−	−	−		+	+	−	+	−	−	−	−	−	−	+	−	+		
D-Ribose	−	−	−		−	−	−	+	−	−	−	−	−	−	−	−	−		
Salicin	−	−	−		−	−	−	+	−	−	−	−	−	−	−	−	−		
Sorbitol	−	−	−		−	−	−	−	−	−	−	−	−	−	−	−	−	−	−
Sucrose	−	+	+		+	−	+	+	−	−	+	−	+	+	+	−	+	+	+
Trehalose	−	+	+		−	−	−	−	−	−	+	−	+	+	+	−	+	+	−
Xylose	+	−	+		−	+	+	−	−	+	+	+	−	+	−	+	+	+	−
Total No. of carbohydrates oxidized	6	2	12		10	8	12	14	3	1	10	5	6	7	12	3	14	9	7

[a] *Sphingomonas suberifaciens* EY 4202[T] was omitted because of no growth on any of three media.

[b] Symbols: +, positive reaction within 4 d; −, negative reaction for 4 weeks; NG, no growth.

[c] EY, Eiko Yabuuchi.

[d] (a) Bacto-Oxidation–Fermentation (OF) basal medium; (b) Bacto-Cystine–tryptone agar (CTA) medium; (c) Bacto-Marine Oxidation–Fermentation (MOF) medium.

[e] Data from Lee et al. (2001).

BXII.α.88, and BXII.α.89, with the following additional information. Rods. Motile by a single polar flagellum. Colonies are deep yellow. A thin-layer chromatogram of the alkali-stable lipids is shown in Fig. BXII.α.85. Major component of polyamine is homospermidine (Hamana and Matsuzaki, 1993; Segers et al., 1994; Takeuchi et al., 1995; Takeuchi et al., 2001); Segers et al. (1994) reported trace amounts of spermidine. PEG 4000 and PEG 6000 are not assimilated (Takeuchi et al., 1993a). Isolated from blood.

The mol% G + C of the DNA is: 61.8 (HPLC).

Type strain: ATCC 51382, CDC B4562, CIP 104197, DSM 13885, EY 2397, GIFU 2397, IFO 13937, JCM 7514, LMG 17325, NCTC 11032.

GenBank accession number (16S rRNA): D13726.

24. **Sphingomonas stygia** Balkwill, Drake, Reeves, Fredreckson, White, Ringelberg, Chandler, Romine, Kennedy and Spadoni 1997, 199[VP]*

sty′gi.a. L. masc. n. *Styx* underworld river in classical Greek mythology; L. fem. adj. *stygia* pertaining to the underworld, subterranean.

The characteristics are as described for the genus and in Tables BXII.α.84, BXII.α.85, BXII.α.86, BXII.α.87, BXII.α.88, and BXII.α.89, with the following additional information. Nonmotile rods. Colonies are light yellow. Glucuronosyl ceramide (SGL-1) is present, but not galacturonosyl ceramide (SGL-1′) (Fig. BXII.α.85). Major component of polyamine is spermidine (Busse et al., 1999; Takeuchi et

---

Editorial Note: Novosphingobium stygium (Balkwill et al. 1997) Takeuchi et al. 2001, 1415 is a junior objective synonym of S. stygia (Yabuuchi et al., 2002).

TABLE BXII.α.88. *(cont.)*

Carbohydrates	Sphingomonas rosa EY 4227[T] (a)	Sphingomonas roseiflava EY 4345[T] (b)	Sphingomonas sanguinis EY 2397[T] (a)	Sphingomonas stygia EY 4297[T] (a)	Sphingomonas subarctica EY 4251[T] (a)	Sphingomonas subterranea EY 4298[T] (b)	Sphingomonas taejonensis EY 4377[T] (a)	Sphingomonas terrae EY 4207[T] (b)	Sphingomonas trueperi EY 4218[T] (a)	Sphingomonas ursincola EY 4250[T] (a)	Sphingomonas wittichii EY 4224[T] (a)	Sphingomonas xenophaga EY 4343[T] (b)	Sphingomonas yanoikuyae EY 4208[T] (a)	Sphingomonas melonis EY 4350[T] (a)	Erythrobacter longus EY 4203[T] (a)	Erythrobacter litoralis EY 4222[T] (a)	Erythromicrobium ramosum EY 4223[T] (a)	Porphyrobacter neustonensis EY 4349[T] (b)	Porphyrobacter tepidarius EY 4231[T] (a)
Adonitol	−	−	−	−	−	−	−	−	−	−	−	−	−	−	−	−	−	−	−
D-Arabinose	−	+	−	−	−	−	−	−	−	+	−	−	−	−	−	−	−	+	−
L-Arabinose	+	−	−	−	+	+	−	−	+	−	−	+	+	+	−	−	−	−	−
Cellobiose	+	+	−	+	−	−	−	+	+	+	−	+	−	+	+	−	+	−	+
Dulcitol	+	−	−	−	+	−	−	+	−	+	−	−	−	−	−	−	−	−	+
Ethanol (3%)	NG	−	+	NG	−	NG	NG	−	−	−	−	−	−	−	−	+	−	−	−
Fructose	+	+	−	−	+	+	−	−	−	+	−	−	−	+	−	−	−	+	+
Galactose	+	+	+	−	+	−	+	−	+	−	+	−	−	+	+	−	−	−	+
Glucose	+	+	−	−	+	−	+	−	+	−	+	−	−	+	+	+	−	−	−
Glycerol	−	−	+	−	−	−	−	−	−	−	+	−	−	+	−	−	−	+	+
Inositol	−	−	−	+	−	−	−	−	−	−	−	−	−	−	−	+	−	−	−
Inulin	−	−	−	−	−	−	−	−	−	−	−	−	−	−	−	+	−	−	−
Lactose	+	+	+	+	+	−	−	−	−	−	−	−	−	+	−	−	−	−	−
Maltose	+	+	−	+	+	−	+	+	+	+	+	−	−	+	+	−	−	+	+
Mannitol	−	−	−	+	+	−	−	+	−	+	−	−	−	−	−	−	−	−	+
Mannose	+	+	+	−	+	−	+	+	+	+	−	−	−	+	−	−	−	+	+
Melezitose	−	−	−	+	+	−	−	−	+	−	−	−	−	−	−	−	−	−	−
Melibiose	−	−	−	−	−	+	−	−	−	−	−	−	−	−	−	−	−	−	−
Raffinose	+	−	+	−	−	−	−	−	−	−	−	−	−	−	−	−	+	−	−
Rhamnose	+	−	−	−	+	−	−	+	−	+	−	−	−	−	−	−	−	−	+
D-Ribose	−	−	−	+	+	−	−	+	−	+	+	−	−	−	−	−	−	+	+
Salicin	+	−	−	−	−	−	−	+	−	+	+	−	−	−	+	−	−	−	−
Sorbitol	−	−	−	−	−	−	−	−	−	+	−	−	−	−	−	−	−	−	+
Sucrose	+	+	−	+	+	−	−	−	+	+	−	−	−	+	+	−	−	+	−
Trehalose	+	+	−	−	+	−	+	+	−	−	−	+	+	+	−	−	−	−	−
Xylose	+	+	+	−	+	+	−	−	+	−	−	−	+	+	+	−	−	−	−
Total No. of carbohydrates oxidized	15	11	8	8	15	4	4	9	11	7	11	4	3	12	7	4	1	7	11

[a] *Sphingomonas suberifaciens* EY 4202[T] was omitted because of no growth on any of three media.

[b] Symbols: +, positive reaction within 4 d; −, negative reaction for 4 weeks; NG, no growth.

[c] EY, Eiko Yabuuchi.

[d] (a) Bacto-Oxidation–Fermentation (OF) basal medium; (b) Bacto-Cystine–tryptone agar (CTA) medium; (c) Bacto-Marine Oxidation–Fermentation (MOF) medium.

[e] Data from Lee et al. (2001).

al., 2001). Isolated from saturated Atlantic coastal plain terrestrial subsurface sediments.

The mol% G + C of the DNA is: 65.4 (HPLC).

Type strain: B0712, ATCC 700280, DSM 12425, EY 4297, JCM 16085, SMCC B0712.

GenBank accession number (16S rRNA): U20775.

25. **Sphingomonas subarctica** Nohynek, Nurmiaho-Lassila, Suhonen, Busse, Mohammadi, Hantula, Rainey and Salkinoja-Salonen 1996a, 1053[VP]*

sub.arc′ti.ca. M.L. adj. *subarcticus* below the arctic, because the organism was isolated from a subarctic area, Finland.

The characteristics are as described for the genus and in Tables BXII.α.84, BXII.α.85, BXII.α.86, BXII.α.87, BXII.α.88, and BXII.α.89, with the following additional in-

formation. Though an electron micrograph of a cell with four flagella arranged peritrichously appeared (Nohynek et al., 1996a), neither active motility in wet mount preparation nor diffuse spreading growth on semisolid motility agar plate were observed. Colonies are light yellow. Glucuronosyl ceramide (SGL-1) is present, but not galacturonosyl ceramide (SGL-1′) (Fig. BXII.α.85). Strain KF1 degrades 2,4,6-tri- and 2,3,4,6-tetrachlorophenols but not pentachlorophenol. Major polyamine is spermidine (Nohynek et al., 1996a; Takeuchi et al., 2001). Isolated from the biofilm of an activated sludge reactor.

The mol% G + C of the DNA is: 65–67 (T_m).

Type strain: KF1, DSM 10700, EY 4251, HAMBI 2110, JCM 10398.

GenBank accession number (16S rRNA): X94102.

26. **Sphingomonas suberifaciens** (van Bruggen, Jochimsen and Brown 1990) Yabuuchi, Kosako, Naka, Suzuki and Yano

---

Editorial Note: Novosphingobium subarcticum (Balkwill et al. 1997) Takeuchi et al. 2001, 1415 is a junior objective synonym of *S. subarctica* (Yabuuchi et al., 2002).

TABLE BXII.α.89. Results of API 20NE of type strains of 34 *Sphingomonas* species[a]

Substrate or test[b]	*S. paucimobilis* EY 2395[T]	*S. adhaesiva* EY 4215[T]	*S. alaskensis* EY 4374[T]	*S. aquatilis* KCTC 2881[Tc]	*S. aromaticivorans* EY 4296[T]	*S. asaccharolytica* EY 4229[T]	*S. capsulata* EY 4216[T]	*S. chlorophenolica* EY 4219[T]	*S. chungbukensis* EY 4375[T]	*S. cloacae* EY 4361[T]	*S. echinoides* EY 4340[T]	*S. herbicidovorans* EY 4344[T]	*S. koreensis* EY 4376[T]	*S. macrogoltabidus* EY 4304[T]	*S. mali* EY 4341[T]	*S. melonis* EY 4150[T]	*S. natatoria* EY 4920[T]
p-Nitro-β-D-galactopyranoside	+	+	+		+	+	+	+	−	−	−	+	+	+	+	−	−
Esculin	+	+	+	+	+	+	+	−	+[d]	+	+	+	+	+	+	+	+
Assimilation of:																	
Glucose	+	+	+		+	+	+	+	+	+	+	+	+	+	+	−	+
L-Arabinose	+	−	−	−	+	+	+	−	+	+	+	−	−	−	+	−	+
D-Mannose	+	+	−	−	+	+	+	+	+	+	+	−	−	−	+	+	−
D-Mannitol	−	−	−		−	−	−	−	−	−	−	−	−	−	−	−	−
N-acetyl-D-glucosamine	+	+	−	+	+	+	+	−	−	+	+	+	−	+	−	+	−
Maltose	+	+	+	+	+	+	+	−	+	+	+	−	+	−	+	−	+
Potassium gluconate	+	−	−	−	−	+	−	+	+	+	+	−	−	−	+	−	−
n-Caprate	−	−	−	+	−	−	−	+	−	+	+	−	−	−	−	−	−
Adipate	+	−	−	−	−	−	−	−	−	−	+	−	−	−	−	−	+
DL-Malate	+	+	−	−	−	−	−	−	+	+	+	−	+	−	−	−	−
Sodium citrate	+	−	−	−	−	−	−	−	−	−	−	−	−	−	−	−	−
Phenylacetate	−	−	−	−	−	−	−	−	−	−	−	−	−	−	−	−	−

[a]For symbols see standard definitions.

[b]All 26 strains positive for hydrolysis of esculin and Zinc dust test in negative NO₂ test. All 26 strains negative for NO$_2$ from NO$_3$, indole from tryptophan, acid from glucose, arginine dihydrolase activity, hydrolysis of urea and gelatin.

[c]Data from Lee et al. (2001).

[d]Positive reaction after more than 5 d.

1999b, 935[VP] (Effective publication: Yabuuchi, Kosako, Naka, Suzuki and Yano 1999a, 347) (*Rhizomonas suberifaciens* van Bruggen, Jochimsen and Brown 1990, 186.)*

su.be.ri.fa′ci.ens. L. gen. n. *suberus* of cork, corky; L. part. adj. *faciens* making, producing; M.L. part. adj. *suberifaciens* corky making.

The characteristics are as described for the genus and in Tables BXII.α.84, BXII.α.85, BXII.α.87, and BXII.α.89, with the following additional information. Cells of the type strain are thin and short straight rods. Actively motile in wet mount preparation. Spreading growth occurs on semisolid (0.3%) agar medium. Colonies on agar plate of ATCC medium 1700 (*Rhizobium* medium) after 24 h at 26°C are colorless and translucent. Unable to determine acid production from carbohydrates, because neither OF nor CTA medium supports the growth of the organism. Glucuronosyl ceramide (SGL-1) is present but not galacturonosyl ceramide (SGL-1′) (Fig. BXII.α.85). Major polyamine is spermidine (Takeuchi et al., 1995; Takeuchi et al., 2001). Isolated from the corky roots of lettuce (*Lactuca sativa*).

The mol% G + C of the DNA is: 59.0 (HPLC).

Type strain: Ca1, ATCC 49355, DSM 7465, EY 4204, IFO 15211, JCM 8521, NCPPB 3629.

GenBank accession number (16S rRNA): D13737.

27. **Sphingomonas subterranea** Balkwill, Drake, Reeves, Fredreckson, White, Ringelberg, Chandler, Romine, Kennedy and Spadoni 1997, 199[VP]**

sub.ter.ra′ne.a. L. fem., adj. *subterranea* underground, subterranean.

The characteristics are as described for the genus and in Tables BXII.α.84, BXII.α.85, BXII.α.86, BXII.α.87, BXII.α.88, and BXII.α.89, with the following additional information. Colonies are deep yellow. Glucuronosyl ceramide (SGL-1) is present, but not galacturonosyl ceramide (SGL-1′) (Fig. BXII.α.85). Major component of polyamine is spermidine (Takeuchi et al., 2001). Isolated from saturated Atlantic coastal plain terrestrial subsurface sediments.

The mol% G + C of the DNA is: 60.0 (HPLC).

Type strain: B0478, ATCC 700279, DSM 12447, EY 4298, JCM 16086, SMCC B0478.

GenBank accession number (16S rRNA): U20773.

28. **Sphingomonas taejonensis** Lee, Kook Shin, Yoon, Takeuchi, Pyun and Park 2001, 1497[VP]

tae.jon.en′sis. N.L. fem. adj. *taejonensis* of Taejon, Korea, the geographical origin of the species.

The characteristics are as described for the genus and in Tables BXII.α.84, BXII.α.85, BXII.α.86, BXII.α.87, BXII.α.88, and BXII.α.89, with the following additional information. Motile by a single polar flagellum. Colonies are pale yellow. Sphingolipid is present, though no details have been reported. Isolated from natural mineral water from Taejon City, Korea.

The mol% G + C of the DNA is: 63 (HPLC).

Type strain: JSS54, EY 4377, JCM 11457, KCTC 2884, KCM 41068.

GenBank accession number (16S rRNA): AF131297.

Editorial Note: The generic name *Rhizomonas* Orla-Jensen 1909 has been placed on the List of Rejected Generic Names, Opinion 14 of the Judicial Commission, Lapage et al. (1992).

**Editorial Note: *Novosphingobium subterraneum* (Balkwill et al. 1997) Takeuchi et al. 2001, 1415 is a junior objective synonym of *S. subterranea* (Yabuuchi et al., 2002).

TABLE BXII.α.89. *(cont.)*

Substrate or test[b]	S. parapaucimobilis EY 4213[T]	S. pituitosa EY 4370[T]	S. pruni EY 4228[T]	S. rosa EY 4227[T]	S. roseiflava EY 4345[T]	S. sanguinis EY 2397[T]	S. stygia EY 4297[T]	S. subarctica EY 4251[T]	S. suberifaciens EY 4204[T]	S. subterranea EY 4298[T]	S. taejonensis EY 7377[T]	S. terrae EY 4377[T]	S. trueperi EY 4218[T]	S. ursincola EY 4250[T]	S. wittichii EY 4224[T]	S. xenophaga EY 4343[T]	S. yanoikuyae EY 4208[T]
p-Nitro-β-D-galactopyranoside	+	+	+	+	+	+	+	+	+	+	−	−	+	−	−	+	+
Esculin	+	+	+	+	+	+	+	+	+	+	+	+	+	+[d]	+	+	+
Assimilation of:																	
Glucose	+	+	+	+	+	+	+	+	−	+	+	−	+	−	−	+	+
L-Arabinose	+	+	+	+	+	+	+	+	−	+	−	−	+	−	−	+	+
D-Mannose	+	+	−	−	−	+	−	+	−	+	−	+	−	−	−	−	−
D-Mannitol	−	−	−	−	−	−	−	−	−	−	−	−	−	−	−	−	−
N-acetyl-D-glucosamine	+	+	+	+	−	+	−	+	−	−	+	−	+	−	−	−	+
Maltose	+	+	+	+	+	+	+	+	+	−	+	+	+	+	−	+	+
Potassium gluconate	+	+	−	−	−	+	+	−	−	−	−	−	−	−	−	−	+
n-Caprate	−	+	−	−	−	−	−	−	−	−	−	−	−	−	−	−	−
Adipate	−	−	−	−	−	−	−	+	−	−	−	−	−	−	−	−	−
DL-Malate	+	+	−	−	+	+	−	−	−	+	+	+	−	−	+	+	+
Sodium citrate	+	−	−	−	−	−	−	−	−	−	−	−	−	−	−	−	+
Phenylacetate	−	−	−	−	−	−	−	−	−	−	−	−	−	−	−	−	−

[a]For symbols see standard definitions.

[b]All 26 strains positive for hydrolysis of esculin and Zinc dust test in negative NO₂ test. All 26 strains negative for NO₂ from NO₃, indole from tryptophan, acid from glucose, arginine dihydrolase activity, hydrolysis of urea and gelatin.

[c]Data from Lee et al. (2001).

[d]Positive reaction after more than 5 d.

29. **Sphingomonas terrae** Takeuchi, Kawai, Shimada and Yokota 1993b, 864[VP] (Effective publication: Takeuchi, Kawai, Shimada and Yokota 1993a, 236.)*

ter′rae. L. n. *terra* earth; M.L. gen. n. *terrae* of the earth.

The characteristics are as described for the genus and in Tables BXII.α.84, BXII.α.85, BXII.α.86, BXII.α.87, BXII.α.88, and BXII.α.89, with the following additional information. Rods with round ends. Motile by a single polar flagellum. Colonies are deep yellow. PEG 6000 is assimilated in a mixed culture. Both glucuronosyl ceramide (SGL-1) and galacturonosyl ceramide (SGL-1′) are present (Fig. BXII.α.85). Polyamines reported include spermidine alone (Takeuchi et al., 1995; Takeuchi et al., 2001) and homospermidine together with a trace of spermidine (Segers et al., 1994). Isolated from active sludge.

The mol% G + C of the DNA is: 63.0–64.9 (HPLC).

Type strain: E-1-A, ATCC 51381, CIP 104198, DSM 8831, EY 4207; IFO 15098, JCM 10195, LMG 17326.

GenBank accession number (16S rRNA): D13727.

30. **Sphingomonas trueperi** Kämpfer, Denner, Meyer, Moore and Busse 1997, 579[VP]

true′per.i. L. gen. n. *truepari* of Trüper; named in honor of the German microbiologist Hans G. Trüper in recognition of his numerous contributions to the taxonomy of the *Proteobacteria.*

The characteristics are as described for the genus and in Tables BXII.α.84, BXII.α.85, BXII.α.86, BXII.α.87, BXII.α.88, and BXII.α.89, with the following additional in-

formation. Motile with peritrichous flagella. Colonies are light yellow. Able to grow in the presence of the 10% sodium chloride and 0.02% sodium azide (Kämpfer et al., 1997). Reported polyamines include homospermidine together with spermidine (Busse and Auling, 1988) or homospermidine alone (Takeuchi et al., 2001). Glucuronosyl ceramide (SGL-1) present but not galacturonosyl ceramide (SGL-1′) (Fig. BXII.α.85). Isolated from soil.

The mol% G + C of the DNA is: 65.6 (T_m).

Type strain: ATCC 12417, DSM 7225, EY 4218, JCM 10278, LMG 2141, NCIMB 9391.

GenBank accession number (16S rRNA): X97776.

31. **Sphingomonas ursincola** (Yurkov, Stackebrandt, Buss, Vermeglio, Gorlenko and Beatty 1997) Yabuuchi, Kosako Naka, Suzuki and Yano 1999b, 935[VP] (Effective publication: Yabuuchi, Kosako, Naka, Suzuki and Yano 1999a, 347) (*Erythromonas ursincola* Yurkov, Stackebrandt, Buss, Vermeglio, Gorlenko and Beatty 1997, 1177.)**

ur.sin′co.la. M.L. adj. *ursincola* neighbor or compatriot of bears.

The characteristics are as described for the genus and in Tables BXII.α.84, BXII.α.85, BXII.α.86, BXII.α.87, BXII.α.88, and BXII.α.89, with the following additional information. Short rod or ovoid cells. Rosettes are not formed. In wet mount preparation, cells seem nonmotile and coccoid. No diffuse spreading growth on semisolid motility agar plate. Unable to visualize either budding or asymmetric division of cells by optical microscopy of Gram-

*Editorial Note: Sphingopyxis terrae (Takeuchi et al. 1993a) Takeuchi et al. 2001, 1416 is a junior objective synonym of S. terrae (Yabuuchi et al., 2002).

**Editorial Note: Blastomonas ursincola (Yurkov et al. 1997) Hiraishi et al. 2000a, 1117 is a junior objective synonym of S. ursincola (Yabuuchi et al., 2002).

stained preparations. Aerobic and facultative photoorgano-trophic. Colonies are orange brown. Needs 5 d for esculin hydrolysis. Glucuronosyl ceramide (SGL-1) present but not galacturonosyl ceramide (SGL-1′). (Fig. BXII.α.85). Isolated from a freshwater cyanobacterial mat.

The mol% G + C of the DNA is: 64.8 (HPLC).

Type strain: KR-99, DSM 9006, EY 4250, JCM 10397.

GenBank accession number (16S rRNA): Yl0677.

32. **Sphingomonas wittichii** Yabuuchi, Yamamoto Terakubo, Okamura, Naka, Fujiwara, Kobayashi, Kosako and Hiraishi 2001, 289^{VP}

wi.tti′ chi.i. M.L. gen. n. wittichii of Wittich, referring to Rolf-Michael Wittich, the German bacteriologist who first isolated this potent metabolizer of dibenzo-p-dioxin from the water of the river Elbe and described the metabolism of the compound by this organism.

The characteristics are as described for the genus and in Tables BXII.α.84, BXII.α.85, BXII.α.86, BXII.α.87, BXII.α.88, and BXII.α.89, with the following additional information. Rods with rounded ends. Actively motile, with a single polar or subpolar flagellum. Colonies are grayish white and become faintly yellow after incubation for 3 d at 30°C. Both glucuronosyl ceramide (SGL-1) and galacturonosyl ceramide (SGL-1′) are present (Fig. BXII.α.85). Polyamines include homospermidine and smaller amounts of spermidine (Busse et al., 1999). Isolated from water of the River Elbe.

The mol% G + C of the DNA is: 67 (T_m).

Type strain: Wittich RW1, DSM 6014, EY 4224, JCM 10273, SMUM 2128.

GenBank accession number (16S rRNA): AB021492.

33. **Sphingomonas xenophaga** Stolz, Schmidt-Maag, Denner, Busse, Egli and Kämpfer 2000, 40^{VP}

xe.no′ pha.ga. Gr. adj. xenos foreign; Gr. n. phagos eater; M.L. fem. adj. xenophaga eating foreign (xenobiotic) compounds.

The characteristics are as described for the genus and in Tables BXII.α.84, BXII.α.85, BXII.α.86, BXII.α.87,

BXII.α.88, and BXII.α.89, with the following additional information. Motile by a polar flagellum. Colonies are deep yellow. Glucuronosyl ceramide (SGL-1) present, but not galacturonosyl ceramide (SGL-1′) (Fig. BXII.α.85). Major polyamine is spermidine (Stolz et al., 2000). The type strain was isolated from river water (River Elbe, Germany).

The mol% G + C of the DNA is: 62.1–63.3 (T_m).

Type strain: BN6, DSM 6383, EY 4343.

GenBank accession number (16S rRNA): X94098.

34. **Sphingomonas yanoikuyae** Yabuuchi, Yano, Oyaizu, Hashimoto, Ezaki and Yamamoto 1990b, 321^{VP} (Effective publication: Yabuuchi, Yano, Oyaizu, Hashimoto, Ezaki and Yamamoto 1990a, 116.)*

ya.no.i.ku′yae. N.L. adj. yanoikuyae named in honor of Professor Ikuya Yano, the Japanese bacteriologist who first recognized the second major spot of alkaline-stable glycolipid (SGL-1′, now known as galacturonosyl ceramide) on TLC.

The characteristics are as described for the genus and in Tables BXII.α.84, BXII.α.85, BXII.α.86, BXII.α.87, BXII.α.88, and BXII.α.89, with the following additional information. Actively motile with a single polar or subpolar flagellum. Colonies are grayish white, never became definitely yellow after prolonged incubation at room temperature. Both glucuronosyl ceramide (SGL-1) and galacturonosyl ceramide (SGL-1′) are present (Fig. BXII.α.85). Major component of polyamine reported as spermidine (Hamana and Matsuzaki, 1993; Segers et al., 1994; Takeuchi et al., 1995; Busse et al., 1999; Takeuchi et al., 2001). Type strain isolated from a hospital specimen (Bruun, 1982); other strains from plant roots.

The mol% G + C of the DNA is: 61.7 (HPLC).

Type strain: ATCC 51230, CCUG 28380, DSM 7462, EY4208, GIFU 9882, IFO 15102, JCM 7371, LMG 11252.

GenBank accession number (16S rRNA): D16145.

---

*Editorial Note: Sphingobium yanoikuyae (Yabuuchi et al. 1990b) Takeuchi et al. 2001, 1415 is a junior objective synonym of S. yanoikuyae (Yabuuchi et al., 2002).

Genus II. *Blastomonas* Sly and Cahill 1997, 567^{VP} emend. Hiraishi, Kuraishi and Kawahara 2000a, 1117

LINDSAY I. SLY AND PHILIP HUGENHOLTZ

Blas.to.mo′ nas. Gr. n. blastos bud shoot; Gr. n. monas a unit, monad; M.L. fem. n. Blastomonas a budding monad.

Cells are ovoid or rod-shaped and reproduce by budding or asymmetric cell division. They occur singly or in pairs and **may form rosette-like aggregates.** No stalks and prosthecae are found. **Gram negative.** Nonsporeforming. **Motile by means of polar flagella. Strictly aerobic. Chemoorganotroph** and **facultative photoorganoheterotroph.** No growth occurs under anaerobic conditions in the light. **Produces BChl a.** Colonies and cell suspensions are **yellow to orange due to carotenoids. Mesophilic** and **neutrophilic. Catalase and oxidase positive.** Nitrate is not reduced. Acid is not produced in Hugh and Leifson's OF medium. The **major whole-cell fatty acid is $C_{18:1 \, \omega 9}$. The major hydroxy-fatty acid is $C_{14:0 \, 2OH}$.** $C_{15:0 \, 2OH}$ and $C_{16:0 \, 2OH}$ are present as minor components. 3-OH fatty acids are absent. Monosaccharide-type glycosphingolipids are present. **Ubiquinone-10 is the major respiratory quinone.**

Belongs to the *Alphaproteobacteria*. Habitat is fresh water.

The mol% G + C of the DNA is: 65.

Type species: **Blastomonas natatoria** (Sly 1985) Sly and Cahill 1997, 568 emend. Hiraishi, Kuraishi and Kawahara 2000a, 1117 (Blastobacter natatorius Sly 1985, 43.)

FURTHER DESCRIPTIVE INFORMATION

The cellular morphology of *Blastomonas* is shown in Fig. BXII.α.97. Some variation in cell shape and arrangement occurs. Cells are usually rod-shaped with straight sides, but may be swollen or wedge-shaped, or have a slightly curved axis (Sly, 1985). In *B. natatoria* each cell has a simple mucilaginous holdfast at its non-reproductive pole, by which it attaches to solid surfaces or other cells to form rosettes (Sly, 1985). Rosette formation in *B. ursincola* has not been reported. Further information on morphological and phenotypic characters may be found in publi-

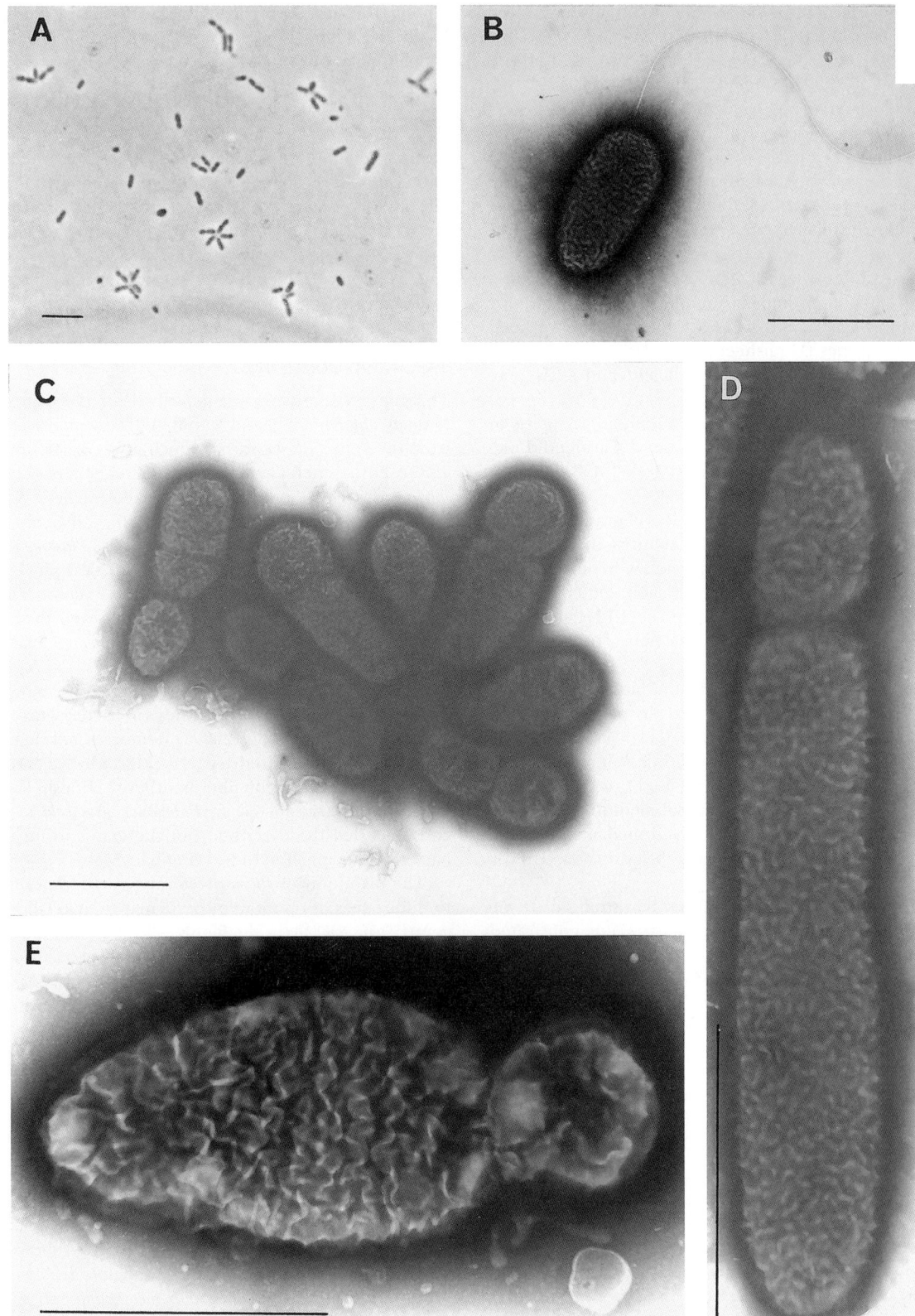

FIGURE BXII.α.97. Morphology of *Blastomonas natatoria* ACM 2057. (*A*) Photomicrograph showing individual cells and budding cells in rosette formation. Bar = 5 μm. (*B*) Electron micrograph of a daughter cell (bud) with a single polar flagellum. Bar = 1 μm. (*C*) Electron micrograph of budding cells in rosette formation. Bar = 1 μm. (*D*) and (*E*) Electron micrographs demonstrating different cell morphologies. Bar = 1 μm. (Reprinted with permission from L.I. Sly, International Journal of Systematic Bacteriology *35:* 40–45, 1985, ©International Union of Microbiological Societies.)

cations by Sly and Hargreaves (1984), Sly (1985), Sly and Cahill (1997), Yurkov et al. (1997), and Hiraishi et al. (2000a). Chemotaxonomic characteristics are described in Sittig and Hirsch (1992), Yurkov et al. (1997), and Hiraishi et al. (2000a).

Considerable research has been undertaken on the photosynthesis characteristics. Bacteriochlorophyll a has been detected in the type strains of both *B. natatoria* and *B. ursincola*, but the latter strain produces the photosynthetic pigment in higher amounts (Hiraishi et al., 2000a). *B. ursincola* was shown to have bacteriochlorophyll a incorporated into the photochemically active photosynthetic reaction center and core light harvesting (LH1) complexes (Yurkov et al., 1997, 1998a, b). The *puf* genes coding for proteins of the L and M subunits of the photosynthetic reaction center complex and of the LH1 complex have been detected in both species (Hiraishi et al., 2000a). *B. natatoria* and *B. ursincola* show 4–5% difference in their *pufL* and *pufM* gene sequences (Hiraishi et al., 2000a).

B. natatoria and *B. ursincola* contain carotenoid pigments in addition to the bacteriochlorophyll a pigment. Carotenoid pigments give three absorption peaks at 425–430, 453–458, and 481–485 nm (Yurkov et al., 1997; Hiraishi et al., 2000a) and together with bacteriochlorophyll a produce the colors of orange in young cultures and dark orange-brown in older cultures. In *B. ursincola* bacteriochlorophyll a is present in membrane-bound protein-pigment complexes, consisting of the reaction center with absorption peaks at 751, 801, and 853 nm and core LH1 absorbing at 867 nm (Yurkov et al., 1997). The reaction center contains tightly bound tetraheme cytochrome c.

Some information is available about the biochemistry of *B. ursincola*. The tricarboxylic acid cycle and the glyoxylate shunt are present, but the key enzyme of the Calvin cycle, ribulose biphosphate carboxylase, is not present (Yurkov et al., 1997). *B. ursincola* is very resistant to tellurite, growing in the presence of 2700 µg/ml in acetate-containing minimal medium (Yurkov et al., 1997). Tellurite is reduced and transformed into metallic tellurium accumulated as metal crystals in the cytoplasm (Yurkov et al., 1996).

Recently, Rickard et al. (2000) reported that strains of *B. natatoria* isolated from biofilms (Buswell et al., 1997) demonstrated intraspecies coaggregation between strains and intergeneric coaggregation with *Micrococcus luteus*. Coaggregation was maximal in the stationary phase of growth, and each member of a coaggregating pair carried either a heat- and protease-sensitive protein (lectin) adhesin or a saccharide receptor, as coaggregation was reversed by sugars.

ENRICHMENT AND ISOLATION PROCEDURES

No enrichment media have been designed for the isolation of *Blastomonas*. The type strain of *B. natatoria* was isolated from a swimming pool using a Millipore *Pseudomonas* Count Water Tester (Sly and Hargreaves, 1984). However, *B. natatoria* may be grown on dilute media such as Staley's peptone yeast glucose medium (Staley, 1981b) on which the colonies appear pink (Sly, 1985). On nutrient agar and R2A agar (Reasoner and Geldreich, 1985) colonies are yellow (Sly, 1985; Rickard et al., 2000).

MAINTENANCE PROCEDURES

Cultures of *Blastomonas* grow well on dilute nutrient media. *B. natatoria* grows well on peptone, yeast extract, glucose (PYG) medium (Staley, 1981b), or R2A medium (Reasoner and Geldreich, 1985). Cultures may be preserved by cryogenic storage in liquid nitrogen when suspended in sucrose peptone broth con-

taining 10% glycerol, and by freeze-drying in glucose peptone broth containing horse serum. Media for *B. ursincola* requires supplementation with 20 µg/l Vitamin B_{12} (Yurkov et al., 1997).

DIFFERENTIATION OF THE GENUS *BLASTOMONAS* FROM OTHER GENERA

Budding of bacteria is widespread throughout the *Proteobacteria* and occurs in phylogenetically diverse genera with widely different physiological characteristics (Rothe et al., 1987; Hugenholtz et al., 1994; Sly and Cahill, 1997) including *Agromonas*, *Blastobacter*, *Blastomonas*, *Gemmobacter*, *Rhodopseudomonas*, *Methylosinus*, and *Nitrobacter*. Tables BXII.α.90 and BXII.α.91 give the characteristics that are useful in differentiating the species of heterotrophic Gram-negative rod-shaped budding bacteria.

TAXONOMIC COMMENTS

The genus *Blastomonas* was described for the type species *Blastomonas natatoria* (Sly and Cahill, 1997), which was transferred from the genus *Blastobacter* in which it was originally placed (Sly, 1985). Phylogenetic analysis of the members of the genus *Blastobacter* revealed that the genus was polyphyletic and required taxonomic revision (Hugenholtz et al., 1994; Sly and Cahill, 1997). The second species of the genus, *Blastomonas ursincola*, was first described as an aerobic bacteriochlorophyll a containing organism and assigned to a new genus *Erythromonas*, as *Erythromonas ursincola*, by Yurkov et al. (1997), even though the type strains of *Blastomonas natatoria* and *Erythromonas ursincola* showed 99.8% 16S rRNA sequence similarity. This taxonomic decision was made with regard to their different physiologies and the important position of photosynthesis in bacterial taxonomy. However later, Hiraishi et al. (2000a) demonstrated that *Blastomonas natatoria* also synthesized bacteriochlorophyll a aerobically and contained *puf* genes for photosynthesis. Hiraishi et al. (2000a) consequently transferred *Erythromonas ursincola* to *Blastomonas* and emended the description of the genus to include aerobic bacteriochlorophyll-a-synthesizing bacteria.

The taxonomy of *Blastomonas* species continues to be subject to differences of opinion, in particular their relationship to the genus *Sphingomonas* in the family *Sphingomonadaceae*. Takeuchi et al. (2001) proposed that the genus *Sphingomonas* be split into four separate genera (*Sphingomonas sensu stricto*, *Sphingobium*, *Novosphingobium*, and *Sphingopyxis*), primarily on phylogenetic and chemotaxonomic evidence, and proposed that the genera *Blastomonas* and "*Rhizomonas*" be retained to avoid taxonomic confusion. Yabuuchi et al. (1999a), on the other hand, argue that all species in the *Sphingomonadaceae*, including the species of *Blastomonas* be combined into the genus *Sphingomonas*.

Phylogenetically, the genus *Blastomonas* belongs to the *Sphingomonadales*, a monophyletic group in the *Alphaproteobacteria* (Fig. BXII.α.98). *Blastomonas* comprises a monophyletic lineage, not robustly affiliated with any other lineage within the *Sphingomonadales* (including "*Rhizomonas*", which is often shown branching with *Blastomonas*). There is no reproducible (robust) branching order of lineages within the *Sphingomonadales*, so it could be argued that all taxa in the *Sphingomonadales* shown in Fig. BXII.α.98 could be reclassified into the genus *Sphingomonas* to preserve phylogenetic coherence of the group if a "lumping" strategy was adopted. However, it could also be argued that all independent monophyletic lineages within the *Sphingomonadales* (currently there are 10) could be classified as separate genera, thus retaining *Blastomonas* as a distinct genus. Applying this criterion, several species of *Sphingomonas* in Fig. BXII.α.98 would need to be re-

TABLE BXII.α.90. Differential characteristics of the genus *Blastomonas* species and other heterotrophic Gram-negative rod-shaped budding bacteria[a]

Characteristic[b]	*Blastomonas natatoria*	*Blastomonas ursincola*	*Blastobacter henricii*	*Blastobacter aggregatus*	*Blastobacter capsulatus*	*Blastobacter denitrificans*	"*Blastobacter aminooxidans*"	"*Blastobacter viscosus*"	*Agromonas oligotrophica*	*Gemmobacter aquatilis*
Cellular morphology:										
Shape	Rods, slightly curved, or wedge-shaped	Rods, slightly curved, or wedge-shaped	Rods, wedge- or club-shaped, often curved	Ovoid to rod-shaped	Rods, often bent and tapering on budding pole, older cells Y-shaped	Rods with rounded cell poles	Pleomorphic rods with minute appendages	Pleomorphic rods, often bent and branched	Bent, branched rods	Ovoid to rod-shaped, short chains
Cell length (μm)	1.0–3.0	1.3–2.6	2.0–4.5	1.5–2.3	1.5–2.3	1.5–2.3	1.5–3.0	1.0–3.2	2.0–7.0	1.2–2.7
Cell width (μm)	0.5–0.8	0.8–1.0	0.7–1.0	0.6–0.8	0.7–0.9	0.6–0.8	0.8–1.0	0.5–0.9	0.6–1.0	1.0–1.2
Initial bud shape	Spherical or ovoid	Spherical or ovoid	Spherical or oblong	Rods	Ovoid	Rods	Ovoid	Ovoid	Spherical	Spherical or ovoid
Bud origin	Polar	Polar	Polar	Narrow pole	Narrow pole or lateral	Slightly subpolar	Polar or lateral	Polar or subpolar	Polar	Polar or subpolar
Capsule formation	−	−	ND	−	+	−	+	+	ND	−
Motility	+	+	−	+	−	+	−	−	+	−
Rosette formation	+	−	+	+	−	−	−	−	+	−
Colony pigmentation	Yellow, orange, brown	Yellow orange, brown	ND	Colorless, slightly brownish when older	Colorless	Colorless, brownish when older	Yellow (orange)	Yellow	Colorless	Colorless
Bacteriochlorophyll *a* produced aerobically	+	+	−	−	−	−	−	−	−	−
Origin	Freshwater	Freshwater	Forest brook water	Lake water	Eutrophic pond water	Lake water	Activated sludge	Activated sludge	Paddy soil	Forest pond
Mol% G + C	65	65	ND	60	59	65	69	66	66	63

[a]For symbols see standard definitions.

[b]Data from Doronina et al. (1983), Hirsch and Müller (1985), Hiraishi et al. (2000a), Loginova and Trotsenko (1979), Rothe et al. (1987), Saito et al. (1998), Sly (1985), Sly and Cahill (1997), Trotsenko et al. (1989), Yurkov et al. (1997).

TABLE BXII.α.91. Differential biochemical characteristics of the species of the genus *Blastomonas*[a]

Characteristic	Blastomonas natatoria	Blastomonas ursincola
Growth with 3% NaCl[b]	+	−
Hydrolysis of:		
Casein	−	+
DNA	(+)	+
Esculin	+	−
Gelatin	(+)	−
Carbon source utilization:[c]		
L-Arabinose	+	−
D-Glucose	+	(+)
Malate	−	(+)
D-Sorbitol	−	(+)
D-Xylose	+ +	+

[a]Data from Hiraishi et al. (2000a).

[b]Symbols for physiological and biochemical tests: +, positive; (+), weakly positive; −, negative.

[c]Symbols for carbon source utilization tests: + +, good growth; +, moderate growth; −, little or no growth.

classified according to their generic affiliation, and *Erythrobacter*, *Erythromicrobium*, and *Porphyrobacter* may need to be reclassified into a single genus. Papers by Sly and Cahill (1997), Takeuchi et al. (2001), and Yurkov et al. (1997) argue on phylogenetic evidence that *Blastomonas* belongs to a deep separate lineage without a close relationship to the genus *Sphingomonas*. Hiraishi et al. (2000a) proposes that *Blastomonas* should be retained on phylogenetic and phenotypic evidence, particularly its budding morphology and photosynthesis capacity, which clearly separate it from *Sphingomonas*. On the other hand, Yabuuchi et al. (1999a) have proposed the transfer of *B. natatoria* and *B. ursincola* to the genus *Sphingomonas*.

On balance, the authors support the retention of the genus *Blastomonas* at this time to avoid taxonomic confusion. Based on budding morphology and aerobic bacteriochlorophyll *a* synthesis, *Blastomonas* can easily be differentiated from *Sphingomonas*. Hiraishi et al. (2000a) support the retention of the genus *Blastomonas* and note that whereas most species of the genus *Sphingomonas* contain monosaccharide- and oligosaccharide-type glycosphingolipids, *B. natatoria* and *B. ursincola* contain monosaccharide-type glycosphingolipids only.

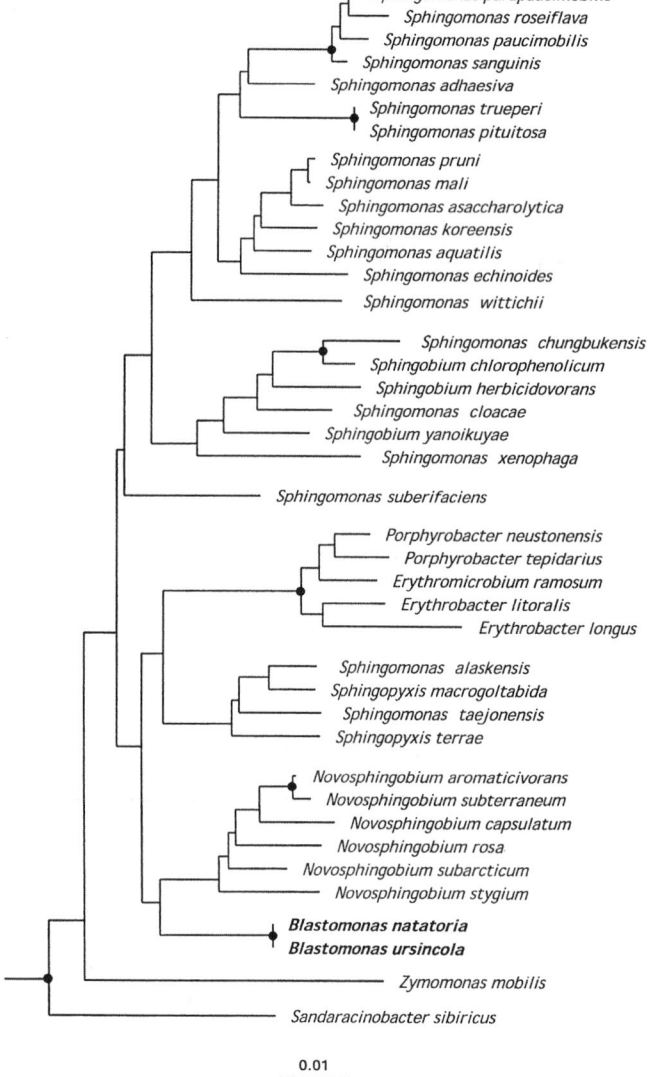

FIGURE BXII.α.98. 16S rRNA evolutionary distance tree of the *Sphingomonadales*. The outgroups used were *Rhodospirillum rubrum*, *Rhodobacter capsulatus*, and *Agrobacterium tumefaciens*. Robust internal nodes (>90 bootstrap confidence) are indicated by a solid circle on the node. The intra-order sequence divergence within the *Sphingomonadales* is 9%.

List of species of the genus *Blastomonas*

1. **Blastomonas natatoria** (Sly 1985) Sly and Cahill 1997, 568[VP] emend. Hiraishi, Kuraishi and Kawahara 2000a, 1117 (*Blastobacter natatorius* Sly 1985, 43.)

   na.ta.to′ri.a. M.L. masc. adj. *natatorius* of a swimming place [pool], the source of the water from which the organism was first isolated.

   Cells are ovoid or rod-shaped, 0.6–0.9 × 1–2.5 μm, and reproduce by budding or asymmetric cell division. Colonies on complex media containing peptone and beef extract or yeast extract are circular, convex, smooth, and opaque and grow to 2 mm within 1 week of incubation. Color of colonies is yellow, orange, or brown, depending upon the composition of growth media and age of culture. Aerobic chemoorganotrophy is the preferred mode of growth. No chemolithotrophic growth with H₂, sulfide, or thiosulfate is found. Optimal temperature for growth is 30–35°C. Optimal pH is 7.0–7.5. Growth occurs in the presence of 3% NaCl.

   No growth factors are required. Esculin, starch, gelatin, Tween 80, and DNA are hydrolyzed. Chitin, cellulose, and casein are not hydrolyzed. Urease, phenylalanine deaminase, indole, and H₂S are not produced. Good growth occurs with D-xylose, maltose, pyruvate, glutamate, peptone, or yeast extract as sole carbon sources. Other usable carbon sources are L-arabinose, D-glucose, acetate, propionate, butyrate, succinate, fumarate, and Casamino acids. No or little growth occurs with D-fructose, D-mannose, cellobiose, lactose, mannitol, sorbitol, lactate, methanol, propanol, formate, citrate, malate, phenylacetate, benzoate, dichlorophenol, dibenzofuran, dibenzo-*p*-dioxin, or naphthalene. The major phospholipids are phosphatidylglycerol, phosphatidylethanolamine, phosphatidyldimethylethanolamine, and phosphatidylcholine. Inhabits freshwater environments. Isolated from a freshwater swimming pool.

The mol% G + C of the DNA is: 65 (HPLC).

Type strain: ACM 2507, ATCC 35951, DSM 3183, JCM 10396, NCIMB 12085, UQM 2507.

GenBank accession number (16S rRNA): AB024288.

2. **Blastomonas ursincola** (Yurkov, Stackebrandt, Buss, Vermeglio, Gorlenko and Beatty 1997) Hiraishi, Kuraishi and Kawahara 2000a, 1117[VP] (*Erythromonas ursincola* Yurkov, Stackebrandt, Buss, Vermeglio, Gorlenko and Beatty 1997, 1177.)

ur.sin' co.la. M.L. adj. *ursincola* neighbor or compatriot of bears.

Cells are 0.8–1.0 × 1.3–2.6 μm in size. No growth occurs in the presence of 3% NaCl. Growth factors are not required, but growth is stimulated significantly by vitamins. Starch, casein, Tween 80, and DNA are hydrolyzed. Esculin, chitin, cellulose, and gelatin are not hydrolyzed. Urease, phenylalanine deaminase, indole, and H_2S are not produced. Good growth occurs with maltose, pyruvate, glutamate, peptone, or yeast extract as sole carbon source. Other usable carbon sources are D-xylose, D-glucose, D-sorbitol, acetate, propionate, butyrate, succinate, fumarate, malate, and Casamino acids. No or little growth occurs with L-arabinose, D-fructose, D-mannose, cellobiose, lactose, D-mannitol, lactate, methanol, ethanol, propanol, formate, citrate, phenylacetate, benzoate, dichlorophenol, dibenzofuran, dibenzo-*p*-dioxin, or naphthalene. Inhabits freshwater environments. Isolated from a cyanobacterial mat of a thermal spring on Kamchatka Island (Russia).

The mol% G + C of the DNA is: 65 (HPLC).

Type strain: DSM 9006, KR-99.

GenBank accession number (16S rRNA): AB024289.

Genus III. "**Citromicrobium**" Yurkov, Krieger, Stackebrandt and Beatty 1999, 4523

VLADIMIR V. YURKOV

Ci.tro.mi.cro' bi.um. Gr. n. *citron* citron; Gr. adj. *micros* small; Gr. n. *bios* life; M.L. n. *Citromicrobium* citron-colored microbe.

Cells are pleomorphic, depending on the growth phase of cultures, coccoid to ovoid rods, often forming Y-cells. Motile by one polar or subpolar flagellum. Gram negative, highly variable in its mode of multiplication. Cultures are **intensely lemon-yellow colored because of carotenoid pigments, and contain bacteriochlorophyll a. Photosynthetic apparatus contains reaction center (RC) and light harvesting (LH) complexes. No growth occurs anaerobically in the light.** Incapable of autotrophic growth. **Obligately aerobic.** No dissimilatory denitrification activity detected. The habitat is marine.

The mol% G + C of the DNA is: 67.5.

Type species: "**Citromicrobium bathyomarinum**" Yurkov, Krieger, Stackebrandt and Beatty 1999, 4523.

FURTHER DESCRIPTIVE INFORMATION

"*Citromicrobium bathyomarinum*" is the only "*Citromicrobium*" species currently described. Phylogenetically, this species is closely related to other aerobic phototrophic bacteria and belongs to the *Alphaproteobacteria*.

On agar plates, "*C. bathyomarinum*" forms small (2–4 mm) lemon-yellow colonies with a smooth surface. With age, the color of the colonies turns intensely yellow. After growth aerobically in liquid media aerated by shaking, cultures are slightly aggregative, and bright yellow. The culture does not grow anaerobically in the light or under dark conditions. In agar (0.7%) deeps, growth occurs only at the surface (the aerobic and semiaerobic zones) of agar tubes. Light is not required for growth.

Cells of strain "*C. bathyomarinum*" grown in rich organic (RO) liquid medium that contains 0, 0.5, 1.0, 2.0, 3.0, or 5.0% NaCl, are similar morphologically and unusually pleomorphic. Depending on the growth phase of cultures, the morphology of cells ranges from almost coccoid (0.4–0.5 × 0.5–0.8 μm), through ovoid rods (0.4–0.5 × 1.0–1.2 μm), to thread-like formations of up to 5 cells. In RO medium with higher NaCl content (7% or 10%), bean shaped or wavy cells as well as long thread-like formations of up to 10 cells are found. Coccoid cells from young cultures are motile by one polar or subpolar flagellum (Yurkov and Beatty, 1998b; Yurkov et al., 1999).

"*C. bathyomarinum*" is variable in its mode of cell division. Budding, ternary fission, binary division, and symmetric and asymmetric constrictions were observed. This bacterium forms Y-cells, a rare type of bacterial multiplication, which results in the possibility of three daughter cells produced from one mother cell (Fig. BXII.α.99). Cells often remain attached after division, apparently by means of a membranous connective material of unknown nature, and are surrounded by a thin "bubbly" substance (Fig. BXII.α.99). Therefore, cells in liquid culture are slightly aggregative, such that many individual cells remain in contact after division.

Electron microscopic thin sections showed that "*C. bathyomarinum*" has a Gram-negative cell wall. The cytoplasmic membrane was visible, but no obvious intracytoplasmic membranes (ICMs) were detected. No inclusions indicative of storage materials were seen in cells harvested from RO medium.

The *in vivo* absorption spectrum of "*C. bathyomarinum*" has a major peak at 867 nm, indicating the presence of bacteriochlorophyll *a* incorporated into LH complex I (LHI). The small peak at 800 nm indicates the presence of the photosynthetic RC. The photosynthetic apparatus organization of an LH system associated with the RC (photosynthetic unit) in this species was indicated by isolation and purification of LHI-RC particles after lysis and treatment of cells with detergent, and sucrose gradient fractionation. In addition to the LHI-RC-enriched fraction, another fraction that contained a low amount of apparently an LHII complex (absorption peaks at 799 and 849 nm) was reported (Yurkov et al., 1999).

The above data indicate the presence of LHI, LHII, and RC complexes in cells of "*C. bathyomarinum*", and the total number of photosynthetic units per cell is similar to the number of units determined from the spectra of other aerobic phototrophic bacteria. Cells contain 0.4–0.6 nmol of BChl/mg of protein. "*C. bathyomarinum*" produces BChl when grown aerobically in the dark, but production of this pigment is significantly repressed on rich media such as RO, or media containing yeast extract or Casamino acids. The most pronounced BChl synthesis was detected in a minimal medium containing acetate, glutamate, or butyrate as the sole source of carbon.

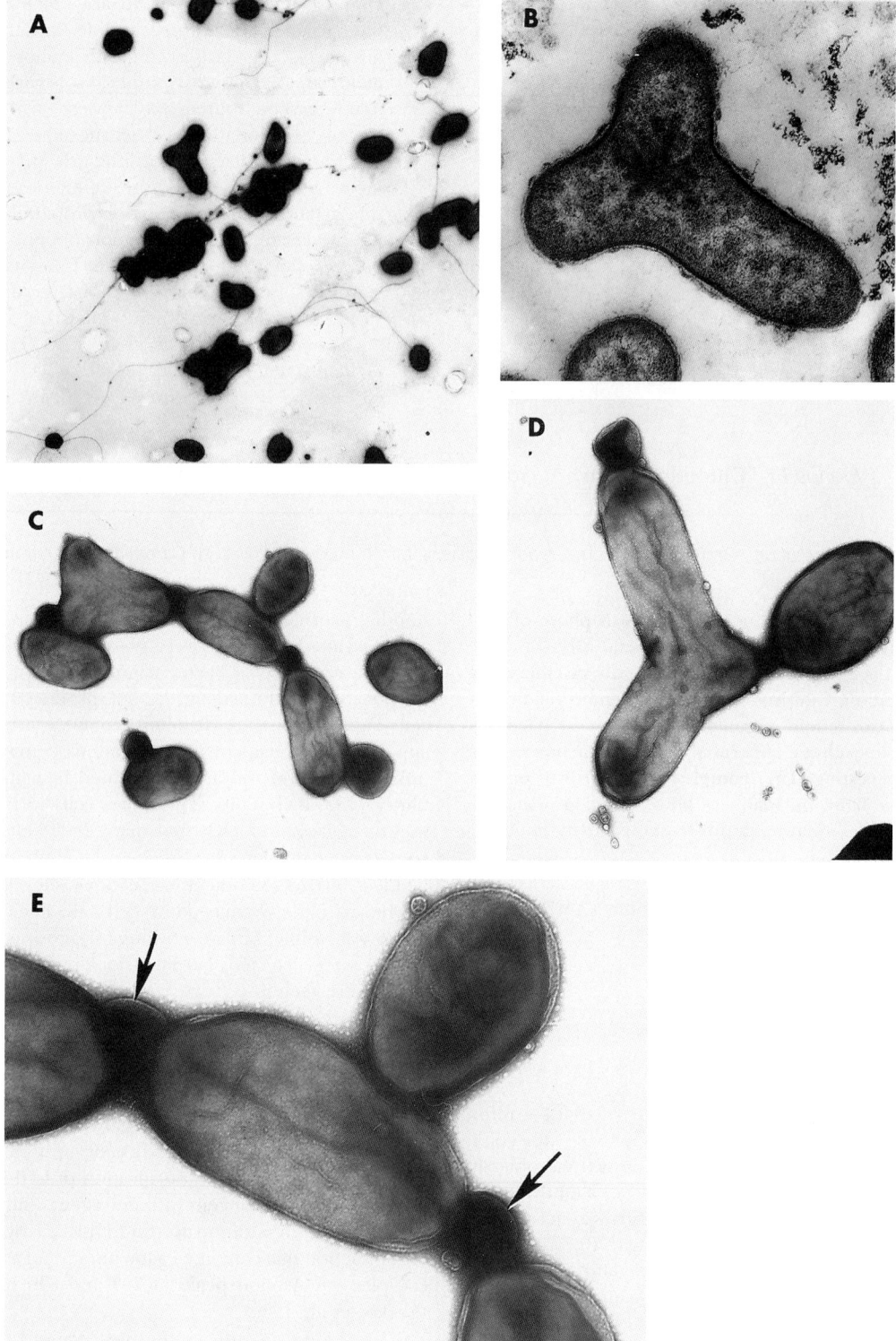

FIGURE BXII.α.99. Electron micrographs of negatively stained (A, C, D, E) and thin-sectioned (B) cells of *"Citromicrobium bathyomarinum"*. (*A*) Pleomorphism of cells, (*B*) Y-shaped form during early stage of cell division. The nucleoid is distributed in three directions, (*C* and *D*) Cells of different morphologies connected by an unknown material, (*E*) Enlarged fragment of panel *C*, showing membranous connective material (*arrows*).

The yellow color of cells and the three peaks at 433, 457, and 487 nm indicate the presence of carotenoids, apparently of the carotene type. The ratio of the absorbance at the LHI BChl absorption peak (867 nm) to that at the main carotenoid absorption peak (457 nm) is about 1:8. However, the ratio of carotenoids:BChl was much higher in intact cells than in partially purified photosynthetic complexes. It seems that most of these carotenoids were not intimately bound to the components of the photosynthetic apparatus (LH and RC). As noted above, ICM invaginations were not observed in the electron microscopic thin sections. Possibly the photosynthetic apparatus of "*C. bathyomarinum*" is restricted to the cytoplasmic membrane.

"*C. bathyomarinum*" is broadly tolerant to culture conditions such as salinity, temperature, and pH. Thus, growth occurs in a freshwater medium and a medium supplemented with 10% NaCl, at temperatures ranging from 4° to 45°C, and at pH values of 5.5–10.0. Therefore, "*C. bathyomarinum*" is a salt-, pH-, and thermotolerant strain.

"*C. bathyomarinum*" utilizes an unusually low number of substrates compared to other aerobic phototrophic bacteria. Glutamate, butyrate, and yeast extract are the best carbon sources, and acetate and glucose support weak growth. The substitution of nitrate for ammonia as a nitrogen source, and an increase or decrease in aeration did not affect the variety of organic substrates that supported growth. Gelatin is not hydrolyzed, but Tween 20, Tween 80, and starch are hydrolyzed, indicating the absence of gelatinase and the presence of lipolytic and amylolytic activities (Table BXII.α.92).

TABLE BXII.α.92. Growth and physiological properties of "*Citromicrobium bathyomarinum*" strain JF-1 and phylogenetically related species[a]

Characteristic	"*Citromicrobium bathyomarinum*" JF-1	*Blastomonas ursincola*	*Erythrobacter longus*	*Erythromicrobium ramosum*	*Porphyrobacter neustonensis*
Growth at pH:					
5	−	−	−	−	−
5.5	+	−	−	−	−
6	+	+	−	+	−
7	+	+	+	+	+
8	+	+	+	+	+
9	+	+	+	+	+
9.5	+	+	+	+	−
10	w	−	−	−	−
Growth at (°C):					
4	+	−	w	−	−
15	+	+	+	+	w
20	+	+	+	+	+
30	+	+	+	+	+
37	+	w	w	−	w
40	+	−	−	−	−
42	+	−	−	−	−
45	+	−	−	−	−
50	−	−	−	−	−
Utilization of:					
Acetate	w	+	−	+	−
Pyruvate	−	+	+	+	+
L-Glutamate	+	+	+	+	−
Butyrate	+	+	+	+	−
Citrate	−	+	+	+	−
Malate	−	+	−	+	−
Succinate	−	+	−	+	+
Lactate	−	+	nd	+	−
Formate	−	−	nd	−	−
D-Glucose	w	+	+	+	+
D-Fructose	−	+	+	+	+
Methanol	−	−	−	−	−
Ethanol	−	−	nd	+	−
Yeast extract	+	+	+	+	+
Hydrolysis of:					
Gelatin	−	−	+	−	−
Tween 20	+	+	nd	−	+
Tween 80	+	+	+	−	+
Starch	+	−	−	−	−
Reduction of:					
NO_3^- to NO_2^-	−	−	+	−	nd
NO_2^- to N_2	−	−	−	−	−
TeO_3^{-2} to Te	+	+	+	+	nd
Oxidase	+	+	+	+	+
Catalase	+	+	+	+	+
Susceptibility to:					
Chloramphenicol (100 µg)	+	+	+	+	+
Penicillin (20 µg)	−	−	−	−	+
Streptomycin (50 µg)	−	−	−	−	−
Fusidic acid (0.5 µg)	+	+	nd	+	nd
Polymixin B (100 µg)	+	+	−	+	nd

[a]For symbols see standard definitions; nd, not determined; w, weak growth.

As a good representative of the aerobic phototrophic bacteria, "*C. bathyomarinum*" is highly resistant to tellurite and reduces this compound to tellurium apparent as intracellular crystals with the minimal inhibitory concentration of tellurite shown to be 2000 μg/ml (Yurkov et al., 1999).

ENRICHMENT AND ISOLATION PROCEDURES

"*C. bathyomarinum*" strain JF-1 was isolated from samples obtained from the vicinity of nonbuoyant regions of plumes emitted from hydrothermal vents on the Juan de Fuca Ridge (Northeastern Pacific Ocean; ~47°57′N, 129°05′W; about 2000 m beneath the ocean surface). Descriptions of the samples and bacterial heterotrophic population enumerated on the medium used are given in Yurkov and Beatty (1998b). The techniques used to isolate and cultivate "*C. bathyomarinum*" are similar to those described for *Erythromicrobium*.

MAINTENANCE PROCEDURES

"*Citromicrobium*" can be preserved by standard procedures in liquid nitrogen, by freezing at −70°C or below, or lyophilization (see chapter on *Erythromicrobium* for details).

DIFFERENTIATION OF THE GENUS *CITROMICROBIUM* FROM OTHER GENERA

Genus "*Citromicrobium*" can be easily differentiated from other known genera of the aerobic phototrophic bacteria. On a phylogenetic tree based on 16S rDNA sequences, "*C. bathyomarinum*" represents an independent branch within the *Alphaproteobacteria*. This organism has a very characteristic lemon yellow color, which has been used to compose a taxonomic name for this species. "*C. bathyomarinum*" is very pleomorphic (Fig. BXII.α.99) and variable in its mode of reproduction. The Y-cell type of cellular division, a type rarely observed among bacteria, has been found as one of the ways "*C. bathyomarinum*" reproduces itself (Fig. BXII.α.99). Cells often remain attached after division, apparently by means of membranous connective material of unknown nature. This species is capable of growth over a broad range of NaCl concentrations, pH values, and temperatures. However, it

seems to have quite limited metabolic possibilities, being able to grow on an unusually low (for this physiological group) number of substrates (Table BXII.α.92).

TAXONOMIC COMMENTS

The 16S rDNA sequences of strain JF-1 showed that the phylogenetically closest relatives of the genus "*Citromicrobium*" were members of the genera *Sphingomonas*, *Sandaracinobacter*, *Erythrobacter*, *Erythromonas* (*Sphingomonas*), and *Porphyrobacter* (Fig. BXII.α.100). The phylogenetic position of "*C. bathyomarinum*" is as a branch between the genus *Erythromonas* (*Sphingomonas*) and the *Erythromicrobium*–*Porphyrobacter*– *Erythrobacter* cluster within the *Alphaproteobacteria*.

FURTHER READING

Shimada, K. 1995. Aerobic anoxygenic phototrophs. *In* Blankenship, Madigan and Bauer (Editors), Anoxygenic Photosynthetic Bacteria, Kluwer Academic Publishers, Dordrecht. pp. 105–122.

Yurkov, V.V. and J.T. Beatty. 1998. Aerobic anoxygenic phototrophic bacteria. Microbiol. Mol. Biol. Rev. *62*: 695–724.

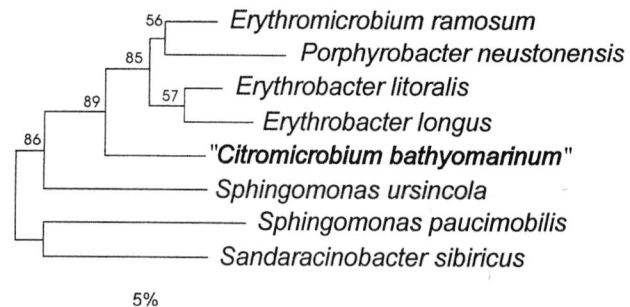

FIGURE BXII.α.100. Dendrogram showing the phylogenetic position of "*Citromicrobium bathyomarinum*" among members of the genera *Sphingomonas*, *Erythrobacter*, *Sandaracinobacter*, and *Porphyrobacter*. Note: *Blastomonas ursincola* (Yurkov et al. 1997) Hiraishi et al. 2000a, 1117 is a junior objective synonym of *Sphingomonas ursincola* (Yabuuchi et al., 2002). Bar = 5% sequence divergence.

List of species of the genus "Citromicrobium"

1. **"Citromicrobium bathyomarinum"** Yurkov, Krieger, Stackebrandt and Beatty 1999, 4524.

   ba.thy.o.ma.ri′ num. Gr. adj. *bathys* deep; L. adj. *marinum* oceanic; M.L. adj. *bathyomarinum* deeply oceanic.

   Gram negative, lemon-yellow pigmented, pleomorphic. Cells may be almost coccoid (0.4–0.5 × 0.5–0.8 μm), ovoid rods (0.4–0.5 × 1.0–1.2 μm), or thread-like structures. Coccoid cells are motile by means of one polar or subpolar flagellum. Cells contain BChl *a* and carotenoid pigments. BChl *a* bound to proteins gives *in vivo* absorption peaks at 800 and 867 nm. Photosynthetic apparatus is organized in RC, LHI, and LHII complexes, as evidenced by partially purified preparations in which the LHI complex yields a peak at 866 nm and the LHII complex at 799 and 849 nm.

   Aerobic, chemoorganotroph, and facultative photoheterotroph. The best growth substrates are L-glutamate, butyrate, and yeast extract; weak growth on minimal media containing acetate or D-glucose. No growth on pyruvate, citrate, malate, succinate, lactate, formate, D-fructose, meth-

anol, or ethanol. Optimal temperature for growth 20–42°C. Capable of growth over a salinity range from 0% to 10% NaCl in RO medium with an optimum of 1–5%. The pH optimum is from 6.0 to 8.0. Exhibits oxidase, catalase, lipase, and amylase, but not gelatinase activities. No dissimilatory denitrification, or anaerobic growth in the presence of trimethylamine oxide (TMAO). Glucose is fermented to acid products without gas generation. Demonstrates a high level of resistance to tellurite (up to 2000 μg/ml in an acetate-glutamate minimal medium). Tellurite is reduced and transformed to metallic tellurium causing blackening of the culture. Resistant to penicillin and streptomycin; sensitive to chloramphenicol, fusidic acid, and polymyxin B.

Habitat: the vicinity of nonbuoyant regions of plumes emitted from hydrothermal vents on the Juan de Fuca Ridge (northeastern Pacific Ocean).

The mol% G + C of the DNA is: 67.5 (HPLC).

Deposited strain: JF-1.

GenBank accession number (16S rRNA): Y16267.

Genus IV. Erythrobacter Shiba and Simidu 1982, 215[VP]

TSUNEO SHIBA AND JOHANNES F. IMHOFF

E.ryth'ro.bac.ter. Gr. adj. erythros red; M.L. masc. n. bacter rod; M.L. masc. n. Erythrobacter red rod.

Cells are ovoid to rod-shaped, $0.2–0.4 \times 1.0–5.0$ μm, multiply by binary fission, and are motile by means of polar or subpolar flagella. Gram negative and belong to the *Alphaproteobacteria*. Cell suspensions and colonies are orange or pink. **Bacteriochlorophyll *a* is present. The major respiratory quinone is ubiquinone-10.** The main cellular fatty acid is $C_{18:1}$.

Aerobic chemoorganoheterotrophic bacteria. Do not grow chemoautotrophically or under anoxic conditions in the light. Metabolism is predominantly respiratory. Under microoxic conditions, small amounts of acid are produced from carbohydrates. Oxidase and catalase positive. Tween 80 is hydrolyzed.

Mesophilic and neutrophilic bacteria from the marine environment. Optimum growth is at pH 7.0–8.0 and 25–30°C. Grow well at salt concentrations of 0.1–9.6%.

The mol% G + C of the DNA is: 57–67.

Type species: **Erythrobacter longus** Shiba and Simidu 1982, 216.

FURTHER DESCRIPTIVE INFORMATION

Erythrobacter belongs to the *Alphaproteobacteria*, together with *Porphyrobacter*, *Erythromicrobium*, "*Citromicrobium*," *Erythromonas*, *Sandaracinobacter*, *Sphingomonas*, and *Blastomonas*. Although *Erythrobacter* contains bacteriochlorophyll *a* (Bchl *a*), internal membrane systems are not present. The cellular level of Bchl is only around 2 nmol/mg of dry weight for *E. longus*, or 0.9–3.9 nmol/mg of protein for *E. litoralis*, and comparable to the level in heterotrophically grown cells of most purple nonsulfur bacteria (Harashima et al., 1978; Yurkov et al., 1994c). A reaction center–light-harvesting complex I (RC-LH I) is present, but not a light-harvesting complex II. In the near infrared region, the *in vivo* absorption spectrum of Bchl *a* reveals a minor maximum at 800–807 nm and a major one at 866–868 nm. *Erythrobacter* contains a large amount of polar carotenoids not included in the RC-LH l complex (Shimada et al., 1985) and not engaged in photosynthetic activities (Noguchi et al., 1992). Among about 20 different polar carotenoids, the most abundant are erythroxanthin sulfate and caloxanthin sulfate (Takaichi et al., 1991). The RC-LH I complex contains bacteriorubixanthinal, zeaxanthin, and hydroxyderivatives of β-carotene (Takaichi et al., 1988). Terminal oxidase in the respiratory system of *E. longus* is aa_3-type cytochrome oxidase (Fukumori et al., 1987).

The amine composition of *E. longus* is characteristic, in that only spermidine, but no appreciable amounts of putrescine, is found (Hamana et al., 1985). A high level of melatonin, *N*-acetyl-5-methoxytryptamine, which is known to offer protection against oxygen radicals, is found in *E. longus*. Cell homogenates of dark-grown, but not of light-grown, cells retain melatonin-producing enzyme activity (Tilden et al., 1997).

Erythrobacter grows as an aerobic chemoorganoheterotroph (Shiba and Simidu, 1982). Neither phototrophic growth nor light-dependent ATP synthesis are found, though an operative light-driven cyclic electron transfer system occurs (Harashima et al., 1982). The physiological function of Bchl *a* is not yet known. In contrast to anoxygenic phototrophic bacteria, Bchl synthesis is inhibited by light, but not suppressed at atmospheric oxygen tension (Harashima et al., 1980, 1987; Shiba, 1987). *E. litoralis* is quite resistant to tellurite (Yurkov et al., 1996).

E. litoralis was isolated from a cyanobacterial mat in a supralittoral zone of the North Sea, which is flooded approximately

twice a month. Due to rainfall and evaporation induced by wind and sunlight, the salinity in the pore water may fluctuate greatly and reach high levels. Growth of *E. litoralis* at salt concentrations of 0.5–9.6% may reflect adaptation to these conditions (Yurkov et al., 1994c).

E. longus was isolated from the green seaweed *Enteromorpha linza* distributed in a high intertidal zone. This species can also grow in a wide salinity range of 0.1–7% (Sato et al., 1989; Shiba et al., 1991; Shiba, 1995).

ENRICHMENT AND ISOLATION PROCEDURES

Enrichment of *Erythrobacter* species has been achieved from marine beach sand, surface seawater, and several seaweeds. Samples were diluted 10-fold with seawater and spread on agar plates. The agar plates were incubated at 20°C. Characteristic pink or orange colonies were streaked out on further agar plates until pure cultures were obtained.

The culture medium used for *E. litoralis* contained per liter of distilled water: 1.0 g sodium acetate, 1.0 g yeast extract (Difco), 1.0 g Bacto Peptone (Difco), 15.0 g NaCl, 0.3 g KCl, 0.5 g of $MgSO_4 \cdot 7H_2O$, 0.05 g $CaCl_2 \cdot 2H_2O$, 0.3 g NH_4Cl, 0.3 g K_2HPO_4, 20 μg vitamin B_{12}, and 1 ml of a trace element solution (Drews, 1983). The pH is adjusted to 7.6–7.8.

For *E. longus*, the modified PPES-II medium (Shiba, 1989) was used, containing per liter: 2 g polypeptone (Nihon Pharmaceutical Co.), 1 g proteose–peptone (Difco no. 3), 1 g Bacto soytone (Difco), 1 g Bacto yeast extract (Difco), 0.1 g ferric citrate, 300 ml of distilled water, 700 ml artificial seawater (Lyman and Fleming, 1940), and 15 g agar.

MAINTENANCE PROCEDURES

Erythrobacter can be maintained at 20°C as stab cultures in media containing 0.4% agar (Difco). Cultures should be transferred monthly. Maintenance on agar slant cultures at room temperature is not recommended. The bacteria remain viable for 6 months when slant cultures are kept at -80°C. Cultures may also be preserved in liquid nitrogen or by lyophilization.

DIFFERENTIATION OF THE GENUS ERYTHROBACTER FROM OTHER GENERA

Erythrobacter differs from the genus *Sphingomonas* by the presence of Bchl *a* (Takeuchi et al., 1994). The color of cell suspensions and colonies of *Erythrobacter* is orange or pink, while those of *Sphingomonas* are yellow or white. The difference in colony color suggests differences in carotenoid composition. *Erythrobacter* also differs from the two genera in the presence of 3-hydroxy fatty acids (Urakami and Komagata, 1988; Takeuchi et al., 1993a).

Erythrobacter can be distinguished from *Erythromicrobium* by the composition of Bchl-protein complexes. *Erythrobacter* contains only a RC-LH I complex, while *Erythromicrobium* contains RC-LH I and LH II complexes. In addition to near infrared absorption maxima at 799 and 870 nm, *Erythromicrobium* has a maximum at 838 nm (Shimada et al., 1985; Yurkov et al., 1997).

Cell form and type of fission distinguish *Erythrobacter* from *Blastomonas*. *Erythrobacter* is rod shaped and reproduces by binary fission, while *Blastomonas* is ovoid and reproduces by budding. It does not contain 3-hydroxy fatty acids. *Erythrobacter* and *Sandaracinobacter* are distinguished by their color: colonies of *Erythro-*

bacter are orange or pink and show an *in vivo* absorption maximum at 470 nm, while those of *Sandaracinobacter* are yellow-orange and have maxima at 424, 450, and 474 nm (Shiba and Simidu, 1982; Yurkov et al., 1997).

Erythrobacter differ from *Porphyrobacter* in carotenoid composition. *Erythrobacter* contain erythroxanthin sulfate as a major polar carotenoid, while *Porphyrobacter* contain a carotenoid sulfate, the structure of which is not yet identified, but different from erythroxanthin sulfate. Zeaxanthin is a major non-polar carotenoid in *Erythrobacter*, whereas it is minor in *Porphyrobacter* (Takaichi et al., 1990 1991; Hanada et al., 1997). *Erythrobacter* species were isolated from marine environments and show a slight requirement for sodium ion, while *Porphyrobacter* species were isolated from terrestrial environments and do not have a requirement for sodium (Shiba and Simidu, 1982; Yurkov et al., 1994c). *Erythrobacter* are different from "*Citromicrobium*" in colony color and cell form. "*Citromicrobium*" is yellow-pigmented and pleomorphic.

TAXONOMIC COMMENTS

The genus *Erythrobacter* contains two species. The strain OChll4, assigned to the genus *Erythrobacter*, was originally transferred to the genus *Roseobacter* as *Roseobacter denitrificans* (Shiba, 1991a). Based on 16S rDNA sequence analysis, "*Erythrobacter sibiricus*" has been removed from the genus (Yurkov and Gorlenko, 1990). After tentative assignment to the genus *Erythromicrobium*, this species is now included in a new genus as *Sandaracinobacter sibiricus* (Yurkov et al., 1997).

List of species of the genus Erythrobacter

1. **Erythrobacter longus** Shiba and Simidu 1982, 216[VP]
lon' gus. L. adj. *longus* long.

Cells are long rods, 0.3–0.4 × 1.0–5.0 μm, motile by means of subpolar flagella. The cells do not form internal membranes. Cell suspensions and colonies are orange, having *in vivo* absorption maxima at 807 (minor) and at 470 nm and 866–867 nm (major). Bacteriochlorophyll *a* esterified with phytol is present. Contain erythroxanthin sulfate and bacteriorubixanthinal as major carotenoids. Main cellular fatty acid is $C_{18:1}$. $C_{15:0\ 2OH}$ and $C_{14:0\ 2OH}$ are present.

Aerobic, chemoorganoheterotrophic, predominantly respiratory metabolism. Utilize glucose, acetate, pyruvate, glutamate, and butyrate as sole carbon sources. Methanol is not utilized. Some strains reduce nitrate to nitrite. Oxidase, catalase, and phosphatase positive. The Voges–Proskauer and methyl-red tests are negative. H_2S is not produced. Indole is produced. Tween 80 and gelatin are hydrolyzed. Some strains hydrolyze alginate. Biotin is required as growth factor.

Mesophilic and neutrophilic bacterium with optimal growth at pH 7.0–8.0 and 25–30°C. Good growth at salt concentrations of 0.1–7.0%.

Habitat: oxic marine environments, especially on seaweeds.

The mol% G + C of the DNA is: 57 (HPLC).

Type strain: OCh101, ATCC 33941, DSM 6997, IFO 14126.

GenBank accession number (16S rRNA): L01786, M96744, M59062.

2. **Erythrobacter litoralis** Yurkov, Stackebrandt, Holmes, Fuerst, Hugenholtz, Golecki, Gad' on, Gorlenko, Kompantseva and Drews 1994c, 432[VP]
li.to.ra' lis. L. gen. n. *litoris* of the seashore; M.L. masc. adj. *litoralis* belonging or pertaining to the seashore.

Cells are rod shaped, 0.2–0.3 × 1.0–1.3 μm, motile by means of polar or subpolar flagella. Short chains of up to five cells may occur. Internal membranes are not formed. Color of cell suspensions and colonies is pink or orange, but become red or brown in older cultures. Near-infrared absorption maxima are at 800 and 868 nm. Bchl *a* is present. The main carotenoids are bacteriorubixanthinal and erythroxanthin sulfate.

Aerobic chemoorganoheterotrophic metabolism. Growth occurs on glucose, fructose, butyrate, glutamate, acetate, and lactate, and weak growth on succinate. Methanol is not utilized. Nitrate is not reduced. Tween 80 is hydrolyzed; gelatin and starch are not. Catalase and oxidase positive. Susceptible to chloramphenicol, tetracycline, and fusidic acid. Resistant to penicillin, streptomycin, and polymyxin B.

Mesophilic and neutrophilic bacterium with optimal growth at pH 7.0–8.0 and 25–30°C. Good growth at salt concentrations of 0.5–9.6%.

Habitat: marine cyanobacterial mat in supralittoral zones.

The mol% G + C of the DNA is: 67 (T_m).

Type strain: T4, DSM 8509, IAM 14332.

GenBank accession number (16S rRNA): ABO13354.

Genus V. **Erythromicrobium** *Yurkov, Stackebrandt, Holmes, Fuerst, Hugenholtz, Golecki, Gad'on, Gorlenko, Kompantseva and Drews 1994c, 432[VP]*

VLADIMIR V. YURKOV

E.ry' thro.mi.cro' bi.um. Gr. adj. *erythrus* red; Gr. adj. *micros* small; Gr. n. *bios* life; M.L. n. *Erythromicrobium* red microbe.

Pleomorphic bacteria, Gram negative, and usually motile by means of flagella. Divide by binary or ternary fission. **Real branching may occur.** Colonies are red-orange. Cells **contain bacteriochlorophyll *a* and carotenoids.**

Aerobic. Cultures do not grow anaerobically in the light and do not grow chemoautotrophically. Ribulose diphosphate carboxylase is not detected. No fermentation and no denitrification activities occur.

The mol% G + C of the DNA is: 62.5–68.5.

Type species: **Erythromicrobium ramosum** Yurkov, Stackebrandt, Holmes, Fuerst, Hugenholtz, Golecki, Gad' on, Gorlenko, Kompantseva and Drews 1994c, 432.

FURTHER DESCRIPTIVE INFORMATION

Phylogenetically, the genus *Erythromicrobium* is associated with members of the *Alphaproteobacteria* and is closely related to *Ery-*

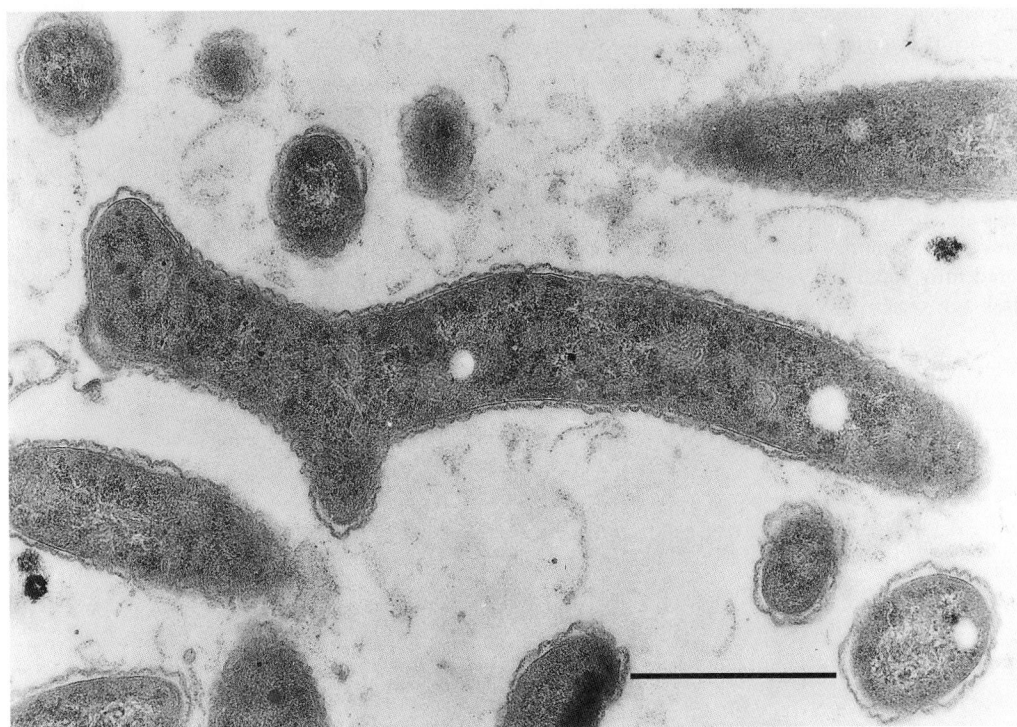

FIGURE BXII.α.101. *Erythromicrobium ramosum.* Ramifying thread-like cell showing true branching. Electron micrograph, ultrathin section. Bar = 0.5 μm. (Printed with permission from V. Yurkov.)

throbacter (level of 16S rDNA sequence similarity, 98%) (Yurkov et al., 1994c). The genera *Erythromicrobium*, *Erythrobacter*, and *Porphyrobacter* form a phylogenetic cluster in the *Alphaproteobacteria*. The genus *Sphingomonas* was shown to be the most closely related nonphotosynthetic genus within the *Alphaproteobacteria* (Yurkov et al., 1994c).

Three described species of *Erythromicrobium*—*E. ramosum*, "*E. ezovicum*", and "*E. hydrolyticum*"—are very long rods and produce characteristic thread-like cells, dividing by symmetric or asymmetric constriction. For *E. ramosum* and "*E. hydrolyticum*" ternary fission and branching were demonstrated (Fig. BXII.α.101) (Yurkov and Gorlenko, 1992).

All *Erythromicrobium* species synthesize a large amount of carotenoid pigments that determine the color of the organism. *E. ramosum* was shown to produce at least 10 different kinds of carotenoids (Yurkov et al., 1993). The two predominant carotenoids, the orange erythroxanthin sulfate and the red bacteriorubixanthinal, are very polar. The *in vivo* absorption spectra of "*E. ezovicum*", "*E. hydrolyticum*", and *E. ramosum* all have major carotenoid peaks at 466 and 478 nm, indicating a similar carotenoid composition. The carotenoids impact an intense color in liquid cultures, red-orange (Yurkov et al., 1997).

The photosynthetic apparatus of *Erythromicrobium* contains a reaction center and two types of antenna complexes, LHI and LHII. Reaction centers do not possess reaction center-bound cytochrome *c*, and the photooxidized special pair (P$^+$) of the reaction center is directly reduced by a soluble cytochrome *c* (Yurkov et al., 1995, 1998a). LHI with absorption maximum at 871 nm is similar to that measured in many anaerobic phototrophic and other aerobic phototrophic bacteria. Isolated LHII showed bacteriochlorophyll absorption maxima at 798–800 and 832–833 nm, indicating the presence of new types of LHII in *Erythromicrobium* cells. The long-wavelength band of LHII at 832–

833 nm is about 20 nm less than that usually observed in purple bacteria (Yurkov et al., 1997). Although it has some peculiarities, the photosynthetic apparatus of *Erythromicrobium* is functional in terms of a cyclic electron transfer system. The photoinduced cyclic electron transfer occurs only under relatively oxidized (aerobic) conditions, as elucidated by light-induced absorbance changes in intact cells. Under anaerobic conditions, no light-induced reaction center absorbance changes were observed. The lack of photochemistry under anaerobic conditions is consistent with the inability of these bacteria to grow by light-dependent photophosphorylation in the absence of oxygen (Yurkov et al., 1995, 1998a).

Erythromicrobium species possess ubiquinone Q-10 as the major quinone. No menaquinones or rhodoquinones have been detected. The ubiquinone Q-9 was detected as a minor quinone in addition to Q-10 in *E. ramosum*, "*E. ezovicum*", and "*E. hydrolyticum*". The quinone Q-10 of "*E. hydrolyticum*" and *E. ramosum* seems to exist as a methylated form (Gogotov and Gorlenko, 1995).

The genus *Erythromicrobium* demonstrated high-level resistance to tellurite and accumulation of metallic tellurium crystals due to tellurite reduction (Fig. BXII.α.102) (Yurkov et al., 1996). Tellurite resistance and tellurium accumulation depend on medium composition, particularly on organic carbon source (Table BXII.α.93).

The major determinative characteristics of the genus *Erythromicrobium* are summarized in Table BXII.α.94.

ENRICHMENT AND ISOLATION PROCEDURES

Erythromicrobium species have been isolated from the samples of freshwater cyanobacterial mats that developed in alkaline warm springs (pH 9.5, 25°C) of the Bol' shaya River valley (Baykal Lake region in Russia). It was isolated by direct inoculation of a homogenized mat sample with dilutions on agar plates of rich organic (RO) medium (Yurkov et al., 1994c) containing (g/l): yeast extract, 1.0; Bacto peptone, 1.0; sodium acetate, 1.0; KCl, 0.3; $MgSO_4 \cdot 7H_2O$, 0.5; $CaCl \cdot 2H_2O$, 0.05; NH_4Cl, 0.3; K_2HPO_4, 0.3. The medium was supplemented with 20 mg/l of vitamin B_{12} and 1.0 ml/l of a trace element solution (Drews, 1983). The plates were cultivated in the dark at 30°C and pH 7.6–7.8. Characteristic red-orange colonies are streaked out on agar plates by the usual aerobic techniques. When a pure culture is obtained, a single colony is transferred into liquid RO medium and cultivated in an Erlenmeyer flask aerobically in the dark.

MAINTENANCE PROCEDURES

Erythromicrobium liquid cultures (taken from late logarithmic growth phase) and agar surface cultures remained viable after storage at 4°C for at least 2 months. Long-term preservation of *Erythromicrobium* species is possible by storage in liquid nitrogen or freezing at −70°C. For this purpose, heavy cell suspensions of liquid cultures (mid-logarithmic growth phase) are supplemented with glycerol (final concentration, 30%) as a protective agent. These cell suspensions are filled into 2-ml plastic screw-cap tubes and freeze-stored. Lyophilization can also be used as a method for *Erythromicrobium* preservation.

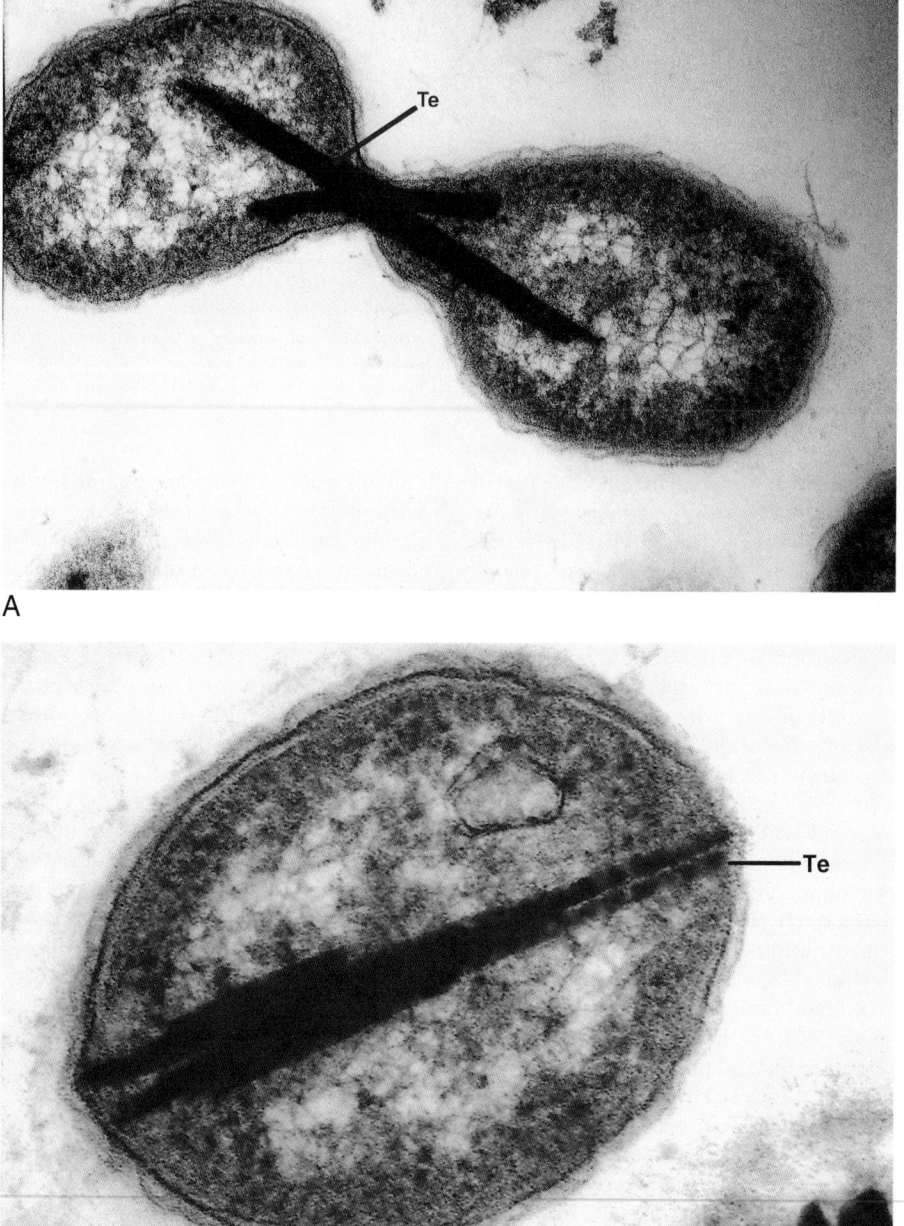

FIGURE BXII.α.102. Intracellular localization of accumulated tellurium (*Te*, indicated by *arrows*) by *Erythromicrobium ramosum.* (Printed with permission from V. Yurkov.) *A*, Two daughter cells with tellurium crystals apparently interfering with cell division. *B*, Long tellurium crystals extending across cell. Bars = 0.25 μm.

TABLE BXII.α.93. Reduction of K_2TeO_3 by *Erythromicrobium* species depending on organic carbon source[a]

Carbon source	*E. ramosum*[b]			"*E. ezovicum*"[b]			"*E. hydrolyticum*"[b]		
	A	B	C	A	B	C	A	B	C
R.O.[c]	+	50	750	+	500	1200	+	n.r.	5
Yeast extract	+	250	1500	+	500	2000	+	n.r.	5
L-Glutamine	+	250	1200	+	750	1200	+	2000	2000
Succinate	+	500	1200	+	750	1200	+	2000	2000
Malate	+	250	1200	+	750	1200	+	1200	1200
Tartrate	±	100	500	±	n.r.	5	±	5	5
Acetate	+	1000	2300	+	1000	2500	+	500	500
Ethanol	+	250	1000	+	100	250	+	5	5

[a]Symbols: +, good growth; ±, weak growth; n.r., not reduced.

[b]Column A: growth without tellurite; column B: highest rate and completeness of K_2TeO_3 reduction (μg/ml); column C: minimal inhibitory concentration (μg/ml).

[c]R.O., rich organic medium.

TABLE BXII.α.94. Major characteristics of *Erythromicrobium* representatives[a]

Characteristic	*E. ramosum*	"*E. ezovicum*"	"*E. hydrolyticum*"
Cell shape	Bacilli, branched	Long bacilli	Bacilli, branched
Cell size (μm)	0.7–1.0 × 1.6–2.5	0.6–0.8 × 2.7–2.8	0.7–1.1 ×1.8–2.5
Color	Red-orange	Red-orange	Red-orange
Major carotenoid *in vivo* peaks, nm	466, 478	466, 478	466, 478
Utilization of:			
Glucose	+	+	+
Maltose	+	+	+
Acetate	+	+	+
Pyruvate	+	−	+
Butyrate	+	+	−
Malate	+	+	+
Citrate	+	+	+
Succinate	+	+	+
Lactate	+	+	+
Ethanol	+	+	+
Methanol	−	−	−
Hydrolysis of:			
Starch	−	−	+
Gelatin	−	−	+
Tween-80	−	−	+
Mol% DNA G + C content	64.2	62.5	65.2

[a]Symbols: +, substrate is utilized; −, substrate is not utilized.

DIFFERENTIATION OF THE GENUS *ERYTHROMICROBIUM* FROM OTHER GENERA

Species of the genus *Erythromicrobium* are freshwater bacteria inhabiting cyanobacterial mats of warm springs. The cells of certain representatives exhibit morphological pleomorphism and the ability to divide by ternary fission because of real branching of the cells. *E. ramosum*, "*E. ezovicum*", and "*E. hydrolyticum*" are very long rods that often produce characteristic thread-like cells (Figs. BXII.α.101 and BXII.α.103B). The specific carotenoid composition determines the major carotenoid absorption peaks of the intact cell absorption spectra at 466 and 478 nm, apparent as an intense red-orange color in liquid cultures (Yurkov et al., 1997). The photosynthetic apparatus of all *Erythromicrobium* species described so far is unique. In addition to the reaction center and light-harvesting complex I, it contains a new type of light-harvesting complex II, with absorption maxima at 798–800 nm and 832–833 nm. The presence of this light-harvesting complex II in *Erythromicrobium* distinguishes its *in vivo* absorption characteristics from the absorption spectra of other aerobic and anaerobic anoxygenic phototrophic genera.

TAXONOMIC COMMENTS

The genus *Erythromicrobium* was introduced by Yurkov et al. (1992a) as a genus of freshwater obligately aerobic, facultatively photoheterotrophic bacteria that included five species: *E. ramosum*, "*E. ezovicum*", "*E. hydrolyticum*", "*E. sibiricum*", and "*E. ursincola*". The genus was validly published in 1994 with *E. ramosum* as the type strain (Yurkov et al., 1994c). In 1997, the genus *Erythromicrobium* was taxonomically reorganized, resulting in the exclusion of "*E. sibiricum*" and "*E. ursincola*" from the genus transfer to two new genera, *Sandaracinobacter* and *Erythromonas*, respectively (Yurkov et al., 1997).

FURTHER READING

Shimada, K. 1995. Aerobic anoxygenic phototrophs. *In* Blankenship, Madigan and Bauer (Editors), Anoxygenic Photosynthetic Bacteria, Kluwer Academic Publishers, Dordrecht. pp. 105–122.

Yurkov, V. and J.T. Beatty. 1998. Aerobic anoxygenic phototrophic bacteria. Microbiol. Mol. Biol. Rev. *62*: 695–724.

DIFFERENTIATION OF THE SPECIES OF THE GENUS *ERYTHROMICROBIUM*

Although they resemble each other physiologically and morphologically, *Erythromicrobium* species have some distinguishing features (Table BXII.α.94). All described species differing from each other by the mol% DNA G + C content and lower levels of DNA similarity, determined using DNA–DNA hybridization (Yurkov et al., 1991a). "*E. hydrolyticum*" demonstrates specifically high hy-

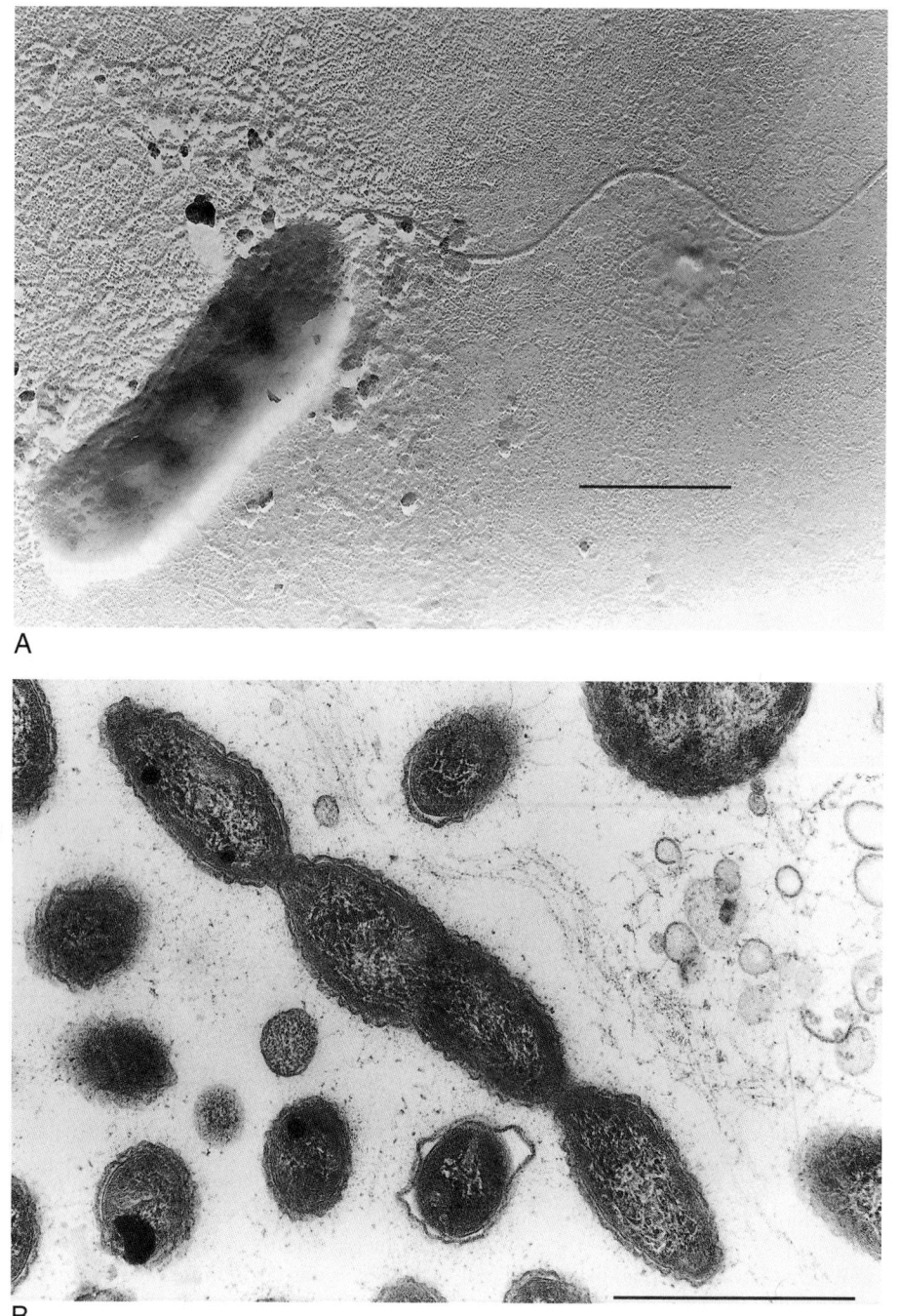

FIGURE BXII.α.103. *"Erythromicrobium ezovicum"*. Electron micrographs. (Printed with permission from V. Yurkov.) *A*, Scanning micrograph showing polar flagellum. *B*, Long thread-like formation of cells. Bars = 1.0 μm.

drolytic activities utilizing starch, gelatin, and Tween-80 as sole carbon sources. This species can oxidize thiosulfate to tetrathionate in aerobic energy metabolism as an addition to using organic energy sources. Representatives of *Erythromicrobium* are resistant to the toxic heavy metal oxide tellurite, reduce tellurite to tellurium, and accumulate metallic tellurium crystals inside the cells (Yurkov et al., 1996). However, the MIC (minimal inhibitory concentration) level of tellurite is species specific (Table BXII.α.93).

List of species of the genus Erythromicrobium

1. **Erythromicrobium ramosum** Yurkov, Stackebrandt, Holmes, Fuerst, Hugenholtz, Golecki, Gad' on, Gorlenko, Kompantseva and Drews 1994c, 432[VP]

ra.mo′ sum. L. adj. *ramosum* ramifying, referring to the morphology of the cells.

Gram-negative, red-orange rods that are 0.6–1.0 × 1.3–

2.5 μm. Cells may branch (Fig. BXII.α.101). Multiplication occurs by binary or ternary fission. Motile by means of polar flagella. Bacteriochlorophyll *a* and carotenoids are present. The cytoplasmic membrane contains a reaction center and two light-harvesting complexes, LHI with an absorption maximum at 870 nm and unusual "blue shifted" LHII with absorption maxima at 798 and 832 nm. The major carotenoids are the very polar compound erythroxanthin sulfate and bacteriorubixanthinal. Optimal growth occurs at temperatures between 25 and 30°C and at pH values between 7.0 and 8.5.

The optimal substrates for growth are glucose, sucrose, maltose, acetate, pyruvate, butyrate, malate, succinate, fumarate, propionate, glutamate, casein hydrolysate, and yeast extract. Growth also occurs on fructose, citrate, lactate, tartrate, and ethanol. No growth occurs on ribose, arabinose, formate, benzoate, mannitol, glycerol, and glycolate. Methanol is not utilized. Starch, gelatin, and Tween-80 are not hydrolyzed. Oxidase and catalase positive. The tricarboxylic acid cycle operates. The glyoxylate shunt has been observed in strain E4(2), but not in strain E5.

Susceptible to the following antibiotics: tetracycline, polymyxin B, erythromycin, nalidixic acid, amikacin, gentamicin, nystatin, bacitracin, kanamycin, neomycin, aureomycin, vancomycin, novobiocin, chloramphenicol, and fusidic acid. Resistant to penicillin, ampicillin, streptomycin, and nystatin. The storage compounds are polysaccharides, poly-β-hydroxybutyric acid, and polyphosphates.

Resistant to heavy metal oxide tellurite. Reduction of tellurite results in the accumulation of elemental tellurium inside cells. Tellurite resistance and reduction depend on supplemented organic carbon source (Fig. BXII.α.102; Table BXII.α.93).

Habitat: cyanobacterial mat from an alkaline spring (pH 9.5; 25°C).

The mol% G + C of the DNA is: 64.2 (T_m).

Type strain: E5, ATCC 700003, CIP 106927, DSM 8510, JCM 10282.

GenBank accession number (16S rRNA): AB013355 and X72909.

2. **"Erythromicrobium ezovicum"** Yurkov and Gorlenko 1992, 248.

e.zo' vi.cum. M.L. adj. *ezovicum* referring to the Russian river name Ezovca, site of isolation.

Gram-negative, red-orange colored rods, 0.6–0.8 × 2.7–2.8 μm. Divide by symmetric or asymmetric constrictions. Nonbranching thread-like cells are characteristic. Motile by means of a polar flagellum (Fig. BXII.α.103). Cells produce bacteriochlorophyll *a* and carotenoids. *In vivo* absorption spectrum of bacteriochlorophyll *a* has maxima at 868 (light-harvesting complex I) and 798 nm, and a shoulder at 836 nm (light-harvesting complex II). Carotenoid acetone extracts have two peaks, at 466 and 478 nm.

Aerobic, chemoorganotrophic, facultative photoorganotrophic. No growth in the light under anaerobic conditions. Fermentation and denitrification activities not observed. The optimal temperature and pH are 25–30°C and 7.0–8.0, respectively.

Best growth in media containing glucose, sucrose, maltose, malate, succinate, casein hydrolyzate, and yeast extract. Good growth obtained on fructose, acetate, butyrate, cit-

rate, lactate, fumarate, propionate, tartrate, ethanol, and glutamate. Ribose, arabinose, pyruvate, formate, benzoate, methanol, mannitol, glycerol, and glycolate are not utilized. Gelatin, starch, and Tween-80 are not hydrolyzed.

A closed tricarboxylic acid cycle and the glyoxalate shunt operate. Oxidase and catalase positive.

Sensitive to the following antibiotics: tetracycline, polymixin-B, amikacin, gentamicin, bacitracin, kanamycin, neomycin, aureomycin, vancomycin, chloramphenicol, and fusidic acid. Resistant to penicillin, ampicillin, streptomycin, erythromycin, nalidixic acid, lincomycin, mycostatin, and novobiocin.

High-level resistance to tellurite was established. However, unlike other species of *Erythromicrobium*, tellurite resistance of *"E. ezovicum"* is not always correlated with tellurite reduction to metallic tellurium (Table BXII.α.93). Other mechanisms, such as continuous tellurite efflux or tellurite complexing or methylation may play an important role in the resistance character. Tellurite is biotransformed to tellurium in glutamate-, succinate-, or malate-containing media (Yurkov et al., 1996).

Habitat: freshwater cyanobacterial mat that developed in alkaline spring (20–25°C; pH 9.5).

The mol% G + C of the DNA is: 62.5 (T_m).

Deposited strain: E-1.

3. **"Erythromicrobium hydrolyticum"** Yurkov and Gorlenko 1992, 249.

hyd.ro.ly' ti.cum. L. adj. *hydrolyticum* splitting water.

Gram-negative, red-orange rods 0.7–1.1 × 1.8–2.5 μm or more. Cells may branch. Multiply by binary or ternary fission. Motility not observed. Cells contain bacteriochlorophyll *a* and carotenoids. Bacteriochlorophyll *a* in intact cells has major peak at 868 nm (indicating the presence of light-harvesting complex I), a minor peak at 798 nm, and a shoulder at 836 nm (indicating the presence of "blue shifted" light-harvesting complex II). Carotenoid acetone extracts have peaks at 466 and 478 nm.

Aerobic, chemoorganotrophic, facultative photoorganotrophic. Thiosulfate may be utilized as an additional source of energy, being oxidized to tetrathionate. Anaerobic growth in the light does not occur. Fermentation and denitrification activities are not detected. Favorable growth temperature, 25–30°C. Optimal pH, 7.0–8.0.

Best growth occurs in the presence of glucose, maltose, pyruvate, malate, succinate, fumarate, casein hydrolyzate, and yeast extract. Fructose, acetate, citrate, tartrate, and ethanol support a good growth. Lactate supports weak growth. Ribose, arabinose, butyrate, formate, propionate, benzoate, methanol, mannitol, glycerol, and glycollate cannot be used as organic carbon sources. Starch, gelatin, and Tween 80 are hydrolyzed.

Tellurite resistant. Transform tellurite to metallic tellurium (Table BXII.α.93).

The closed tricarboxylic acid cycle and the glyoxalate shunt operate. Oxidase and catalase positive. Storage compounds: polysaccharides and polyphosphates.

Habitat: cyanobacterial mat from an alkaline spring (20–25°C; pH 9.5).

The mol% G + C of the DNA is: 65.2 (T_m).

Deposited strain: E4(1).

Genus VI. **Erythromonas*** *Yurkov, Stackebrandt, Buss, Vermeglio, Gorlenko and Beatty 1997, 1177VP*

VLADIMIR V. YURKOV

E.ry.thro.mo' nas. Gr. adj. *erythrus* red; Gr. n. *monas* a unit, monad; M.L. fem. n. *Erythromonas* red monad.

Ovoid, Gram negative, motile by means of polarly flagellated cells. Do not form chains. Reproduce by **budding** or asymmetric division. The cells are **orange-brown due to carotenoid pigments. Contain bacteriochlorophyll *a*. Aerobic, chemoorganotrophic, and facultatively photoheterotrophic. No photosynthetic growth occurs under anaerobic conditions.** NaCl is not required for growth.

The mol% G + C of the DNA is: 65.4.

Type species: **Erythromonas ursincola** Yurkov, Stackebrandt, Buss, Vermeglio, Gorlenko and Beatty 1997, 1177.

FURTHER DESCRIPTIVE INFORMATION

The monospecific genus *E. ursincola* is phylogenetically related to members of the *Alphaproteobacteria* that form a separate subline on the evolutionary tree of *Sphingomonas* with a 16S rDNA homology of 92.5% (Yurkov et al., 1997).

E. ursincola cells are ovoid (0.8–1.0 × 1.3–2.6 μm) and reproduce by budding or asymmetric division. The carotenoid composition of the species, apparent as absorption peaks at 430, 458, and 485 nm of the intact cell absorption spectra, produces the characteristic orange-brown color of *E. ursincola*. Bacteriochlorophyll *a* (BChl *a*) is incorporated into the photosynthetic apparatus that is organized in the reaction center with tightly bound tetraheme cytochrome *c* and light-harvesting complex I absorbing at 867 nm. Under anaerobic conditions, no light-induced absorbance changes can be observed in the photosynthetic apparatus of *E. ursincola*. In contrast, when cells were grown aerobically, typical photochemical reactions were revealed confirming aerobic photosynthetic functionality of the reaction center (Yurkov et al., 1995, 1998a).

E. ursincola has a complex soluble (three cytochromes *c* of 6.5, 9.0, and 14.0 kDa) and membrane-bound (four cytochromes *c* of 14.3, 21.0, 24.0, and 40.0 kDa) cytochrome population (Yurkov et al., 1997). Interestingly, *E. ursincola* contains an unusually small soluble cytochrome *c* of 6.5 kDa. Such a small cytochrome *c* is rare in bacteria and is present in *Hydrogenobacter thermophilus* (6.0 kDa), *Methylomonas* strain A4 (4.0 kDa), and an aerobic phototrophic species, *Roseococcus thiosulfatophilus* (4.0 and 6.5 kDa). The quinone composition of *E. ursincola* is represented by the only quinone Q-10 (Gogotov and Gorlenko, 1995).

Similarly to other aerobic phototrophic bacteria, *E. ursincola* is highly resistant to tellurite and reduces this compound to tellurium apparent as intracellular crystals. However, the minimal inhibitory concentration of tellurite (2700 μg/ml) for *E. ursincola* in acetate medium is significantly higher than that found in other species (Yurkov et al., 1996).

ENRICHMENT AND ISOLATION PROCEDURES

Strain KR99 designated as *E. ursincola* was isolated from samples of a cyanobacterial mat that developed in freshwater thermal springs (pH 6.7–7.0 and temperature, 34–40°C). The Neskuchninskii spring is situated on Kunashir Island (Southern Kurily,

Russia). The techniques used to isolate and cultivate *E. ursincola* are similar to those described for *Erythromicrobium*.

MAINTENANCE PROCEDURES

Erythromonas can be preserved by standard procedures in liquid nitrogen, by freezing at −70 to −80°C or by lyophilization.

DIFFERENTIATION OF THE GENUS *ERYTHROMONAS* FROM OTHER GENERA

The genus *Erythromonas* is separated from other genera of the aerobic anoxygenic phototrophic bacteria by its distinguishing morphological, physiological, as well as genetic properties. Its comparatively large cell size, ovoid form, budding, and characteristic absorption spectrum of the intact cells are distinctive phenotypic properties.

The major morphological and physiological differences between *Erythromonas* and its most closely related genera of freshwater aerobic phototrophic genera are shown in Table BXII.α.95.

TAXONOMIC COMMENTS

Newly isolated in 1991, strain KR99 was initially assigned to the genus *Erythromicrobium* under the species name *"E. ursincola"* (Yurkov et al., 1991a). However, detailed phylogenetic investigation of 16S rDNA sequences and precise chemotaxonomic research performed on this bacterium later clearly revealed that a new genus named *Erythromonas* with *E. ursincola* as a type species should be created (Yurkov et al., 1997).

Phylogenetic comparisons of 16S rDNA sequences determined that *E. ursincola* clusters with *Blastomonas natatoria* (99.8% sequence identity) (Yurkov et al., 1997). However, significant phenotypic and physiological differences exist between *E. ursincola* and *B. natatoria* that preclude their assignment to the same genus. *B. natatoria* is a nonphotosynthetic species containing carotenoid pigments and lacking bacteriochlorophyll, whereas *E. ursincola* is physiologically related to obligately aerobic anoxygenic phototrophic bacteria producing carotenoids and BChl *a*. *E. ursincola* contains BChl *a* incorporated into a photochemically active reaction center and light-harvesting complexes, and contains specific electron transfer carriers of a cyclic photosynthetic pathway (such as cytochrome *c* bound to the reaction center, soluble cytochrome c_2, the reaction center Q_A primary electron acceptor, and the "special pair" P of the reaction center). High 16S rDNA sequence similarity between these two species indicates a close phylogenetic relationship and that they share a common ancestor; conceivably one is an evolutionary progenitor of the other. However, due to the occurrence of significant physiological differences (photosynthesis is a restricted mode of energy generation), they are not designated as members of the same genus for taxonomic purposes.

FURTHER READING

Shimada, K. 1995. Aerobic anoxygenic phototrophs. *In* Blankenship, Madigan and Bauer (Editors), Anoxygenic Photosynthetic Bacteria, Kluwer Academic Publishers, Dordrecht. pp. 105–122.

Yurkov, V.V. and J.T. Beatty. 1998. Aerobic anoxygenic phototrophic bacteria. Microbiol. Mol. Biol. Rev. *62:* 695–724.

Editorial Note: The type and only species of the genus *Erythromonas* has been transferred to the genus *Sphingomonas*.

TABLE BXII.α.95. Determinative characteristics of taxonomically closely related freshwater aerobic anoxygenic phototrophic genera

Characteristics	Erythromonas	Sandaracinobacter	Erythromicrobium	Porphyrobacter
Cell shape	Ovoid	Thin, long rods	Rods, branched	Pleomorphic
Cell size, μm	0.8–1.0 × 1.3–2.6	0.3–0.5 × 1.5–2.5	0.7–1.0 × 1.6–2.5	0.4–0.8 × 1.1–2.0
Color	Orange-brown	Yellow-orange	Red-orange	Orange-red
Carotenoid *in vivo* peaks, nm	430, 458, 485	424, 450, 474	466, 478	464, 491
BChl *a in vivo* peaks, nm	800, 867	800, 867	798, 832, 868	799, 869
Mol% DNA G + C content	65.4	68.5	64.2	65–66

List of species of the genus Erythromonas

1. **Erythromonas ursincola** Yurkov, Stackebrandt, Buss, Vermeglio, Gorlenko and Beatty 1997, 1177[VP]

ur.sin′co.la. M.L. adj. *ursincola* neighbor or compatriot of bears.

Cells are Gram-negative and ovoid 0.8–1.0 × 1.3–2.6 μm. Long chains are not formed. Reproduce by budding or asymmetric division. Cells motile by means of a unique polar flagellum. Cells contain BChl *a* and carotenoid pigments. Carotenoids have three main absorption peaks at 430, 458, 485 nm *in vivo*, and in combination with BChl *a*, determine the culture color: orange in young culture and dark orange-brown in older liquid or young agar cultures. BChl *a* is present in membrane-bound, protein-pigment complexes, consisting of the reaction center with absorption peaks at 751, 801, and 853 nm and core light-harvesting complex I absorbing at 867 nm. Reaction center contains tightly bound tetraheme cytochrome *c* with a molecular weight of 40.0 kDa. Total cytochrome *c* composition of cells growing in the dark is very abundant and represented by soluble cytochrome *c* of 6.5, 9.0, and 14.0 kDa, and membrane-bound cytochrome *c* of 14.3, 21.0, and 24.0 kDa. In cells growing in the dark, only quinone Q-10 was determined.

Aerobic, chemoorganotrophic, facultatively photoorganoheterotrophic. Best growth is on media containing glucose, fructose, sucrose, maltose, acetate, glutamate, propionate, casein hydrolyzate, or yeast extract. Good growth on pyruvate, butyrate, malate, or succinate. Poor growth on media supplemented with arabinose, citrate, lactate, glycerol or mannitol. No growth detected in media containing ribose, formate, benzoate, tartrate, methanol, ethanol, or glycolate. The tricarboxylic acid cycle and glyoxylate shunt operate during growth on acetate containing medium. The key enzyme of the Calvin cycle, ribulose diphosphate carboxylase, is not found. No anaerobic growth in the light, fermentation, or denitrification is found.

Optimal growth temperature is 25–30°C. Freshwater organism, does not require NaCl for growth. Optimal pH is 7.0–8.0. Yeast extract and vitamin B$_{12}$ satisfy the requirement for growth factors. Oxidase and catalase positive. Tween 80 is hydrolyzed. Lipase activity of 64.9 U/g of biomass. Starch and gelatin are not hydrolyzed. Sensitive to tetracycline, polymyxin B, amikacin, gentamicin, neomycin, aureomycin, vancomycin, novobiocin, chloramphenicol, and fusidic acid. Resistant to penicillin, ampicillin, streptomycin, erythromycin, nalidixic acid, lincomycin, mycostatin, bacitracin, and kanamycin. Very resistant to tellurite. Can grow in the presence of tellurite up to 2700 μg/ml in acetate-containing minimal medium. Resistance to tellurite depends on organic carbon source in the medium. Tellurite can be reduced and transformed into metallic tellurium, which is accumulated as metal crystals in the cell cytoplasm.

Storage compounds: polyphosphates.

Habitat: freshwater cyanobacterial mat developing in the thermal springs at pH 6.7–7.0 and temperature 34–40°C.

The mol% G + C of the DNA is: 65.4 (*T$_m$*).

Type strain: KR-99, EY 4250, CIP 106843, DSM 9006, JCM 10397.

GenBank accession number (16S rRNA): Y10677, AB024289.

Genus VII. **Porphyrobacter** *Fuerst, Hawkins, Holms, Sly, Moore and Stackebrandt 1993, 132*[VP]

Akira Hiraishi and Johannes F. Imhoff

Por.phy.ro.bac′ter. Gr. adj. *porphyreos* purple; M.L. masc. n. *bacter* equivalent of Gr. neut. n. *bacterion* a rod; M.L. masc. n. *Porphyrobacter* porphyrin-producing rod.

Cells are straight, pleomorphic, ovoid-to-short rods, or cocci, 0.5–1.0 × 0.8–2.0 μm. Spores and capsules are not formed. Multiply by budding or binary fission. Gram negative. Members of the *Alphaproteobacteria*. Motile or nonmotile. Do not form any type of internal membranes. **Synthesize bacteriochlorophyll *a* esterified with phytol and carotenoids** as photosynthetic pigments only under aerobic conditions in the dark.

Strictly aerobic, chemoorganoheterotrophic bacteria containing bacteriochlorophyll. Do not grow phototrophically under anoxic conditions in the light. Simple organic compounds, peptone, and yeast extract used as electron donors and carbon sources. Produce acid but not gas from glucose. Catalase positive and oxidase negative. Some strains may require vitamins for growth.

Mesophilic to moderately thermophilic, neutrophilic freshwater bacteria.

Straight-chain octadecenoic acid (C$_{18:1}$) is the major cellular fatty acid. 2-Hydroxy fatty acids and sphingoglycolipids are present. 3-Hydroxy fatty acids are absent. Ubiquinone-10 is the major quinone.

The mol% G + C of the DNA is: 65.0–66.4.

Type species: **Porphyrobacter neustonensis** Fuerst, Hawkins, Holms, Sly, Moore and Stackebrandt 1993, 132.

FURTHER DESCRIPTIVE INFORMATION

Based on 16S rDNA sequence analysis, *Porphyrobacter* species belong to the *Alphaproteobacteria*, with *Erythrobacter* and *Erythromicrobium* species as phylogenetic neighbors. The nearest phylogenetic neighbor is *Erythromicrobium ramosum*, with 98% 16S rDNA sequence identity to *Porphyrobacter neustonensis*.

Cell morphology of *Porphyrobacter* varies depending upon the species. The type species, *P. neustonensis*, contains rod-shaped and pleomorphic cells and in some cases exhibits coccoid forms (Fig. BXII.α.104). The cells reproduce by polar growth and budding, and this mode of cell division is similar to a division type found in *Planctomycetales*, in which the daughter cell appears as a small spherical bud. Cells produce multifibrillar stalk-like fascicle structures and crateriform structures on the cell surface. Cells of *P. neustonensis* are motile by means of subpolar or peritrichous flagella (Fig. BXII.α.105). Cells of *P. tepidarius* are ovoid-to-short rods, nonmotile, and multiply by binary fission (Fig. BXII.α.106). Thin-section electron microscopy shows that *Porphyrobacter* species do not contain internal membranes (Fig. BXII.α.107). Thin-sections of *P. neustonensis* indicated the presence of an electron-dense cell wall layer in the mother cell but the absence of such a layer in the bud (Fig. BXII.α.108). Analyses of crude cell wall material of *P. neustonensis* showed that muramic acid and diaminopimelic acid were present in the cell wall (Fuerst et al., 1993).

A group of highly polar carotenoids, including carotenoid sulfates and bacteriorubixanthin, are major components (Hanada et al., 1997). Nostaxanthin is also present in *P. tepidarius*. Unlike *Erythromicrobium* and *Erythrobacter* species, *Porphyrobacter* species lack zeaxanthin. Absorption spectra of living cells or membrane preparations have a low absorption maximum at

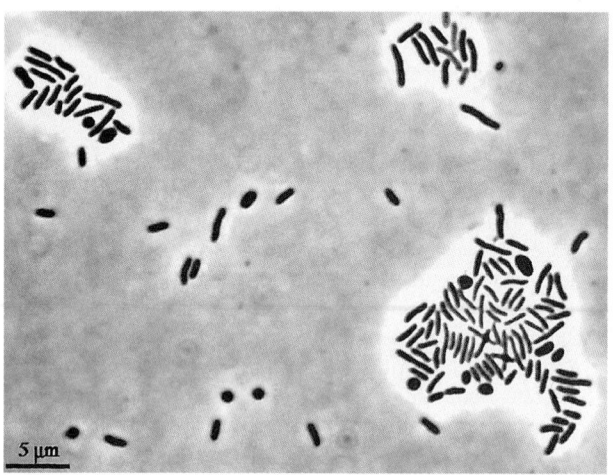

FIGURE BXII.α.104. Phase-contrast photomicrograph showing general cell morphology of *Porphyrobacter neustonensis*. Rod-shaped and coccoid cells coexist. Bar = 5 μm.

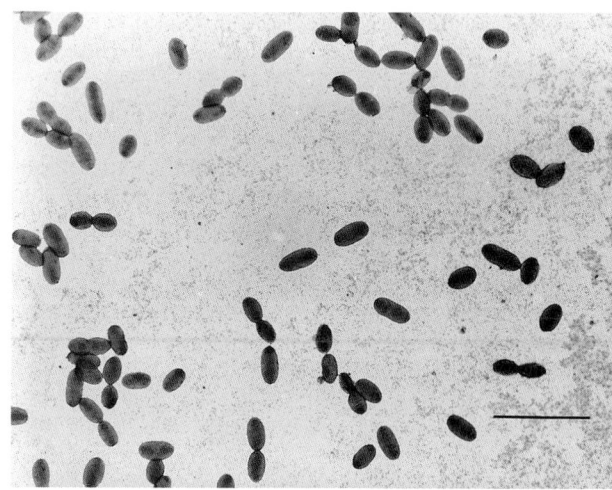

FIGURE BXII.α.106. Electron photomicrograph of negatively stained cells of *Porphyrobacter tepidarius*. Scale: 2.4 cm = 5 μm. (Reproduced with permission from S. Hanada et al., International Journal of Systematic Bacteriology, *47*: 408–413, 1997 ©International Union of Microbiological Societies.)

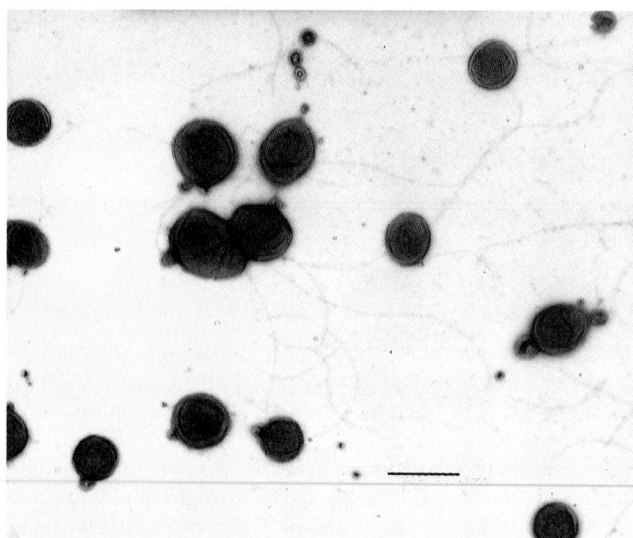

FIGURE BXII.α.105. Electron photomicrograph of negatively stained cells of *Porphyrobacter neustonensis*. Bar = 1 μm. (Reproduced with permission from J.A. Fuerst et al., International Journal of Systematic Bacteriology, *43*: 125–134, 1993 ©International Union of Microbiological Societies.)

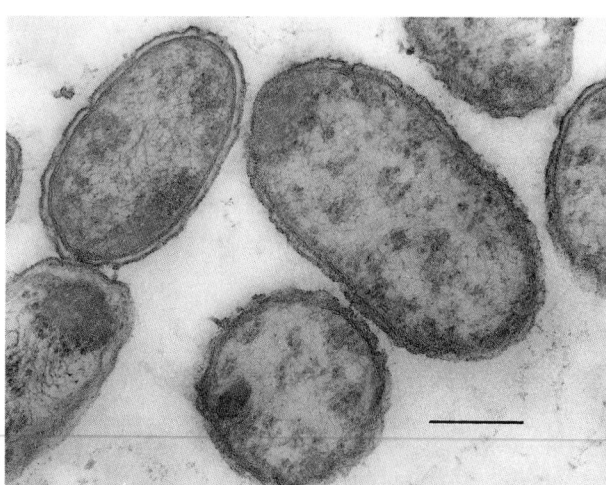

FIGURE BXII.α.107. Thin-section electron photomicrograph showing the ultrastructure of *Porphyrobacter tepidarius*. Scale 2.3 cm = 0.5 μm. (Reproduced with permission from S. Hanada et al., International Journal of Systematic Bacteriology, *47*: 408–413, 1997 ©International Union of Microbiological Societies.)

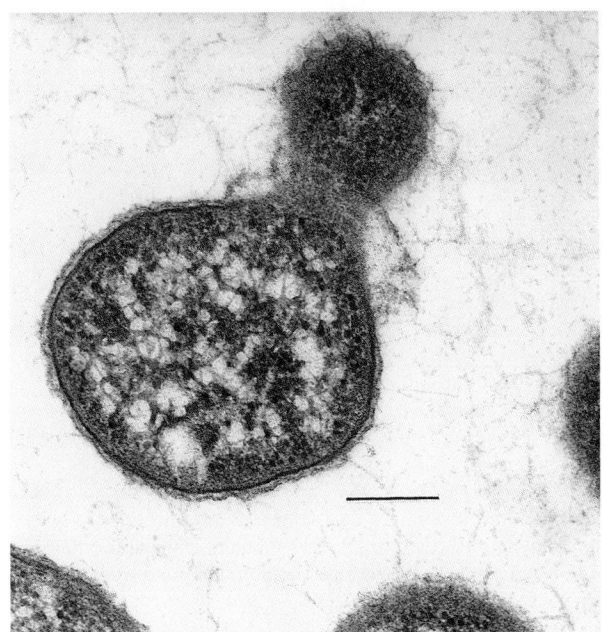

FIGURE BXII.α.108. Thin-section electron photomicrograph showing the ultrastructure of *Porphyrobacter neustonensis*. Bar = 0.2 μm. (Reproduced with permission from J.A. Fuerst et al., International Journal of Systematic Bacteriology, *43*: 125–134, 1993 ©International Union of Microbiological Societies.)

around 800 nm and a high absorption maximum at 868–871 nm. These spectral patterns in the near-infrared region indicate that the cells contain the core light-harvesting (LH) complex (B870), together with the photosynthetic reaction center, and lack the peripheral light-harvesting complex LHII, as is the case in other aerobic bacteriochlorophyll-containing bacteria of the *Alphaproteobacteria*.

Like many other aerobic photosynthetic bacteria, *Porphyrobacter* produces BChl *a* only under aerobic growth conditions in darkness. The amount of BChl *a* produced by *P. neustonensis* ACM 2844 is maximal in gently stirred dark broth cultures (J.A. Fuerst, personal communication). Standing broth cultures produce 0.08 ± 0.05 nmol/mg dry wt of BChl *a*, whereas stirred cultures produce 1.04 ± 0.09 nmol/mg dry wt and shaken cultures 0.65 ± 0.10 nmol/mg dry wt. The maximum BChl *a* content of *P. neustonensis* is somewhat lower than, or comparable to, the BChl *a* content reported for the phylogenetic neighbor *Erythrobacter longus* (Harashima et al., 1982). Due to the presence of carotenoids, liquid cultures and colonies on agar media are orange to red. The intensity of pigmentation depends upon the growth medium. For *P. neustonensis*, casitone–yeast extract agar is best for pigment production (Fuerst et al., 1993).

Porphyrobacter species are aerobic chemoheterotrophic bacteria that grow well under aerobic dark conditions but not anaerobically in the light. They do not grow by anaerobic respiration with nitrate, dimethyl sulfoxide, or trimethylamine *N*-oxide as the terminal electron acceptor. Most strains exhibit a doubling time of 2–4 h under optimal growth conditions. Simple organic compounds such as glucose are used as electron donors and carbon sources for growth. Peptone and yeast extract are also good substrates. Hydrolytic activities against starch, chitin, gelatin, and DNA are absent, whereas Tween 80 hydrolysis is positive in all strains.

The lipid composition of *Porphyrobacter* has not yet been studied in detail, but some information on fatty acid profiles is available. The major cellular fatty acid is straight-chain octadecenoic acid ($C_{18:1}$). Some strains of *P. neustonensis* contain either $C_{17:1}$ or $C_{18:2}$ as a second major component. Characteristic for *Porphyrobacter* is the production of 2-OH fatty acid components and glycosphingolipids (A. Hiraishi, unpublished data). These properties are found in other genera of the *Alphaproteobacteria*, such as *Erythrobacter* and the obligately chemotrophic relative *Sphingomonas*. All *Porphyrobacter* species contain ubiquinone-10 as their sole respiratory quinones.

The natural habitats of *Porphyrobacter* are freshwater environments. *P. neustonensis* strains have been isolated from the air-water interface of freshwater bodies (Fuerst et al., 1993). *P. tepidarius* was isolated from cyanobacterial mats developing in the Usami hot spring (Shizuoka Prefecture, Japan), which contains brackish water (Hanada et al., 1997).

ENRICHMENT AND ISOLATION PROCEDURES

Growth media and cultural conditions commonly used for the isolation of freshwater aerobic bacteria can be used for the isolation of *P. neustonensis*. The surface water of freshwater environments is a possible source of this organism. As an isolation medium, casitone–yeast extract agar medium can be used (Fuerst et al., 1993).

The medium used for isolation of *P. tepidarius* is PE agar medium (Hanada et al., 1995, 1997), which includes acetate, succinate, glutamate, yeast extract, and Casamino acids as carbon sources. The medium is inoculated with water samples from thermal environments at 40–50°C and incubated at pH 5–8 and 45°C in darkness. After several days of incubation, orange colonies may appear on the agar medium.

MAINTENANCE PROCEDURES

Cultures are well preserved in liquid nitrogen or by lyophilization. Preservation in an electric freezer at −80°C is also possible.

DIFFERENTIATION OF THE GENUS *PORPHYROBACTER* FROM OTHER GENERA

Differential characteristics of the genus *Porphyrobacter* and the phylogenetically and phenotypically related genera *Erythromicrobium* and *Erythrobacter* are shown in Table BXII.α.96.

TAXONOMIC COMMENTS

Porphyrobacter neustonensis is the first organism described as a budding species of aerobic bacteriochlorophyll-containing bacteria (Fuerst et al., 1993). Phylogenetically and physiologically, *P. neustonensis* is closely related to *Erythromicrobium* and *Erythrobacter*, but differences in morphological properties and in natural habitats led to the proposal of a new genus for this species. Based on phylogenetic evidence, *P. tepidarius* and *P. neustonensis* are closely related species, although cell morphology and temperature relations distinguish the two (Hanada et al., 1997).

Intensive studies on the isolation and phylogenetic analysis of aerobic BChl-containing bacteria in freshwater and marine environments have yielded new budding strains in both the genera *Porphyrobacter* and *Erythrobacter* (A. Hiraishi, unpublished data; M. Suzuki, personal communication). In addition, the marine bacterium "*Agrobacterium sanguineum*" (Ahrens and Rheinheimer, 1967) has been shown to belong to the genus *Porphyrobacter* phylogenetically (A. Hiraishi, unpublished). In view of these observations and the close phylogenetic relationship of *Porphyrobacter*

TABLE BXII.α.96. Differential characteristics of the genus *Porphyrobacter* and related aerobic bacteria containing bacteriochlorophyll[a]

Characteristic	*Porphyrobacter*	*Erythromicrobium*	*Erythrobacter*
Pleomorphic rods/cocci	D	−	−
Cell width (μm)	0.5–1.2	0.6–1.0	0.3–0.4
Motility by flagella	D	+	+
Carotenoid:			
Carotenoid sulfate	+	−	−
Erythroxanthin sulfate	−	+	+
Zeaxanthin	−	+	+
NaCl required for growth	−	−	+
Tween 80 hydrolysis	+	−	+
Habitat	Freshwater	Freshwater	Marine
Mol% G + C of DNA	65–67	64	57–67

[a]Symbols: +, 90% or more of strains positive; −, 90% or more of strains negative; D, different reactions in different species.

species to *Erythromicrobium* (>98% similarity in 16S rDNA sequence) and *Erythrobacter* (>95% similarity), a taxonomic rearrangement of these genera may result from future studies.

ACKNOWLEDGMENTS

The authors are indebted to J.A. Fuerst and S. Hanada for their contribution of figures, unpublished information, and helpful discussion.

DIFFERENTIATION OF THE SPECIES OF THE GENUS *PORPHYROBACTER*

General and differential characteristics of the two species *P. neustonensis* and *P. tepidarius* are included in Table BXII.α.97. The average level of genomic DNA–DNA hybridization between *P.* *neustonensis* and *P. tepidarius* is 30%. The two species also show 99% 16S rDNA sequence similarity.

TABLE BXII.α.97. General and differential characteristics of the species of the genus *Porphyrobacter*[a,b]

Characteristic	*P. neustonensis*	*P. tepidarius*
Cell shape	pleomorphic rods or cocci	ovoid-to-short rods
Cell division	budding	binary fission
Motility by flagella	+	−
Color of colonies	Orange, red	Orange
Optimal growth temperature (°C)	28–30	40–48
Growth with 3% NaCl	−	−
Vitamin required for growth	−	Biotin
Catalase activity	+	+
Oxidase activity	−	−
Nitrate reduction to nitrite	−	nd
Antibiotic susceptibility:		
Chloramphenicol (100 μg/ml)	+	+
Penicillin (20 U)	+	+
Streptomycin (50 μg/ml)	−	nd
Major quinone	Q-10	Q-10
Mol% G + C of DNA	65.7–66.4	65
Hydrolysis of:		
Starch	−	+
Chitin	−	nd
Esculin	d	nd
Casein	d	−
Gelatin	−	−
Tween 80	+	+
DNA	−	nd
Electron donor/carbon source:		
D-Xylose	+	nd
D-Glucose	+	+
D-Galactose	+	nd
D-Mannose	+	nd
D-Fructose	d	+

(*continued*)

TABLE BXII.α.97. (*cont.*)

Characteristic	*P. neustonensis*	*P. tepidarius*
Maltose	+	nd
Cellobiose	d	nd
Lactose	−	−
Sucrose	+	nd
Raffinose	d	nd
Trehalose	d	nd
Mannitol	d	nd
Methanol	−	−
Ethanol	−	−
Formate	−	−
Acetate	−	+
Butyrate	−	+
Pyruvate	+	−
Lactate	−	−
Citrate	−	−
Succinate	d	−
Fumarate	d	−
Malate	−	−
Benzoate	−	−
Alanine	d	nd
Asparagine	d	nd
Glutamate	−	+
Histidine	d	nd
Isoleucine	d	nd
Ornithine	d	nd
Phenylalanine	−	nd
Proline	+	nd
Yeast extract	+	+

[a]Symbols: +, 90% or more of strains positive; −, 90% or more of strains negative; d, 11–89% of strains positive; nd, infomation unavailable; Q-10, ubiquinone-10.

[b]Substrates tested but not utilized by *P. neustonensis*: L-arabinose, D-ribose, L-rhamnose, inulin, adonitol, dulcitol, erythritol, sorbitol, caproate, caprylate, glycolate, tartrate, phtalate, leucine, lysine, methionine, phenylalanine, tyrosine, and valine.

List of species of the genus Porphyrobacter

1. **Porphyrobacter neustonensis** Fuerst, Hawkins, Holms, Sly, Moore and Stackebrandt 1993, 132[VP]

neu.sto.nen' sis. Gr. n. *neustos* swimming, floating; M.L. masc. adj. *neustonensis* occurring at the air–water interface layer.

Cells are pleomorphic rods or cocci, 0.5–1.0 × 0.8–2 μm and multiply by polar growth or budding. Produce multifibrillar, stalk-like, fascicle structures and crater-form structures on the cell surface. Motile by means of subpolar or peritrichous flagella. The absorption spectrum of living cells has maxima at 460, 494, 799–806, and 868–871 nm. Contain a group of highly polar carotenoids and bacteriorubixanthinal as the major components.

Obligately aerobic, chemoorganoheterotrophic bacterium. No phototrophic growth anaerobically in the light. Good growth occurs with xylose, glucose, galactose, mannose, maltose, sucrose, pyruvate, proline, and yeast extract. Methanol, ethanol, formate, acetate, lactate, and malate are not used. Tween 80, but not starch or gelatin, is hydrolyzed. Susceptible to penicillin G and chloramphenicol, but resistant to streptomycin. No vitamins required as growth factors.

Mesophilic freshwater bacterium. Optimal growth is at 28–30°C; growth range is 10–37°C. No growth in the presence of 1.3% NaCl.

Habitat: air–water interface of freshwater subtropical ponds.

The mol% G + C of the DNA is: 65.0–66.4 (HPLC).

Type strain: ACM 2844, DSM 9434.

GenBank accession number (16S rRNA): AB03327, L01785, M96745.

2. **Porphyrobacter tepidarius** Hanada, Kawase, Hiraishi, Takaichi, Matsuura, Shimada and Nagashima 1997, 413[VP]

tep.i.dar' ius. L. n. *tepidarium* a warm bath; M.L. adj. *tepidarius* warm bathing.

Cells are ovoid to rod-shaped, 0.5–0.7 × 0.8–1.4 μm, nonmotile and multiply by binary fission. Internal membranes may be absent. Colonies and liquid cultures are orange due to the presence of carotenoids and bacteriochlorophyll *a*. The *in vivo* absorption spectra show maxima at 460, 494, 596, 800, and 870 nm. The main carotenoids are OH-β-carotene sulfate derivatives, nostoxanthin, and bacteriorubixanthinal.

Obligately aerobic, chemoorganoheterotrophic bacterium. No phototrophic growth anaerobically in the light. Good growth occurs with glucose, acetate, butyrate, glutamate, Casamino acids, and yeast extract as sole energy sources. Methanol, ethanol, pyruvate, malate, and succinate are not used. Starch and Tween 80, but not gelatin, are hydrolyzed. Susceptible to penicillin and chloramphenicol, but resistant to streptomycin. Biotin required as growth factor.

Moderately thermophilic freshwater bacterium, able to tolerate up to 1.3% NaCl. Optimal growth is at 40–48°C; growth range is 30–50°C. No growth in the presence of 2% NaCl.

Habitat: cyanobacterial mats in brackish water of hot springs (Shizuoka, Japan).

The mol% G + C of the DNA is: 65.0 (HPLC).

Type strain: OT3, DSM 10594.

GenBank accession number (16S rRNA): AB033328, D84429.

Genus VIII. **Sandaracinobacter** *Yurkov, Stackebrandt, Buss, Vermeglio, Gorlenko and Beatty* 1997, 1177*[VP]

VLADIMIR V. YURKOV

San.da.ra.ci' no.bac' ter. Gr. adj. *sandaracinos* orange-colored; Gr. n. *bacter* rod; M.L. masc. n. *Sandaracinobacter* orange-colored rod.

Cells are thin, long rods, forming chains. Motile by means of subpolar flagella. **Gram negative.** Divide by binary division. Cultures are **intensely yellow-orange because of carotenoid pigments. Contain bacteriochlorophyll *a*.**

Aerobic chemoorganotrophic and facultative photoheterotrophic metabolisms. No growth occurs anaerobically in light. Ribulose diphosphate carboxylase is not detected. No fermentation or denitrification activities observed. The habitat is fresh water. Not halophilic.

The mol% G + C of the DNA is: 68.5.

Type species: **Sandaracinobacter sibiricus** Yurkov, Stackebrandt, Buss, Vermeglio, Gorlenko and Beatty 1997, 1177.

FURTHER DESCRIPTIVE INFORMATION

Currently the genus *Sandaracinobacter* is represented by a single species, *S. sibiricus*. On the 16S rDNA tree, *Sandaracinobacter* forms a single subline of descent within the *Alphaproteobacteria* (Yurkov et al., 1997).

Thin, long (0.3–0.5 × 1.5–2.5 μm), rod-shaped cells of *S. sibiricus* are intensely yellow-orange colored due to their high carotenoid content. The carotenoid composition, or at least the major pigments, of *S. sibiricus* (absorption at 424, 450, and 474 nm) are different from those detected for other aerobic phototrophic species and determine its unusual color. The cells of *S. sibiricus* often produce long chains (sometimes up to 10 cells) and reproduce by binary division. Budding and branching have not been observed. The cells are enclosed in a thin capsule. *S. sibiricus* is a nonhalophilic species, salt concentrations above 1% NaCl strongly inhibit bacterial growth.

S. sibiricus utilizes a variety of sugars such as glucose, fructose, sucrose, and maltose (Yurkov and Gorlenko, 1990). The catabolism of glucose by *S. sibiricus* occurs mainly via the Entner–Doudoroff pathway. The species possesses glucose-6-phosphate dehydrogenase and 2-keto-3-deoxygluconate-aldolase, two main enzymes of this pathway. Low activity of the key enzyme of the Embden–Meyerhof pathway, fructose-diphosphate-aldolase, was also detected in the cells growing in glucose-supplemented medium. It was concluded that this enzyme functions in biosynthesis. No 6-phosphogluconate dehydrogenase was detected in *S. sibiri-*

---

*Editorial Note: Readers are advised that *Sandaracinobacter* and *Sandaracinobacter sibiricus* may be illegitimate names as the type material was not available from any public culture collection at the time this volume went to press.

cus, suggesting that the pentose monophosphate pathway does not function in this bacterium (Yurkov et al., 1992b).

The photosynthetic apparatus of *Sandaracinobacter* consists of a reaction center and one type of light-harvesting complex I (LHI), which absorbs maximally at 867 nm (Yurkov et al., 1997). The very fast cytochrome photooxidation observed after flash excitation in the cells of *S. sibiricus* indicates that the immediate electron donor to the reaction center in this species is a reaction center-bound cytochrome *c* (44.0 kDa) (Yurkov et al., 1995). Recently, reaction centers were purified from *S. sibiricus* membranes, confirming a typical overall organization of the reaction center (Yurkov et al., 1998a). The photochemical activity of the *S. sibiricus* photosynthetic apparatus was found to exist only under aerobic conditions (Yurkov et al., 1995, 1998a). Quinones Q-9 and Q-10 are the quinones of *S. sibiricus* (Gogotov and Gorlenko, 1995).

S. sibiricus demonstrated high-level resistance to toxic tellurite and accumulation of metallic tellurium crystals inside the cells. The level of tellurite resistance is strongly dependent on the organic carbon source used for growth. The highest resistance was found in yeast extract (sole carbon source)-containing medium and reached 1200 μg of tellurite per ml of medium (Yurkov et al., 1996).

S. sibiricus accumulates polyphosphates that provide a reserve of inorganic phosphate (Yurkov et al., 1991b). Osmium-stained granules of polyphosphates are found under nearly all experimental conditions studied—in the light and the dark, with high and low aeration. The highest amount of polyphosphate was accumulated in a growth medium supplemented with sucrose. Under such conditions, polyphosphate granules occupied about 30–40% of the total cell volume (Fig. BXII.α.109). Another in-

clusion body found in *S. sibiricus* consists of polyhydroxyalkanoate compound (a lipid-like compound that is formed from β-hydroxybutyric acid units) revealed as electron-transparent granules on electron micrographs (Yurkov et al., 1991b). Polyhydroxyalkanoate formation occurs when cells grow in media unbalanced for nitrogen (urea as a nitrogen source), as well as during incubation in a medium lacking fixed nitrogen. Replacement of ammonium with nitrate as the source of nitrogen also results in pronounced formation of polyhydroxyalkanoates with large granules occupying 40–50% of the total cell volume (Fig. BXII.α.110).

ENRICHMENT AND ISOLATION PROCEDURES

S. sibiricus was isolated from thin microbial mats that developed around the underwater hydrothermal vents of the Bol'shaya river bottom. The mats were formed by cyanobacteria *Oscillatoria subcapitata*, diatoms, and the purple bacteria *Thiocapsa roseopersicina* and *Rhodopseudomonas palustris*. The mats were situated at the boundary of the anaerobic and aerobic zones at pH 8.9–9.4, temperature 9–33°C, and hydrogen sulfide and oxygen concentrations of 0.6–4.5 mg/l and 10.4–4.1 mg/l, respectively. The water redox potential was high, $E_h = 230$–330 mV. Thus, despite the presence of sulfide, the mat surface was washed by water with a high oxygen content (Yurkov and Gorlenko, 1990). Isolation procedures are identical to those described for the genus *Erythromicrobium*.

MAINTENANCE PROCEDURES

S. sibiricus can be readily maintained under short- and long-term storage following procedures described for the genus *Erythromicrobium*.

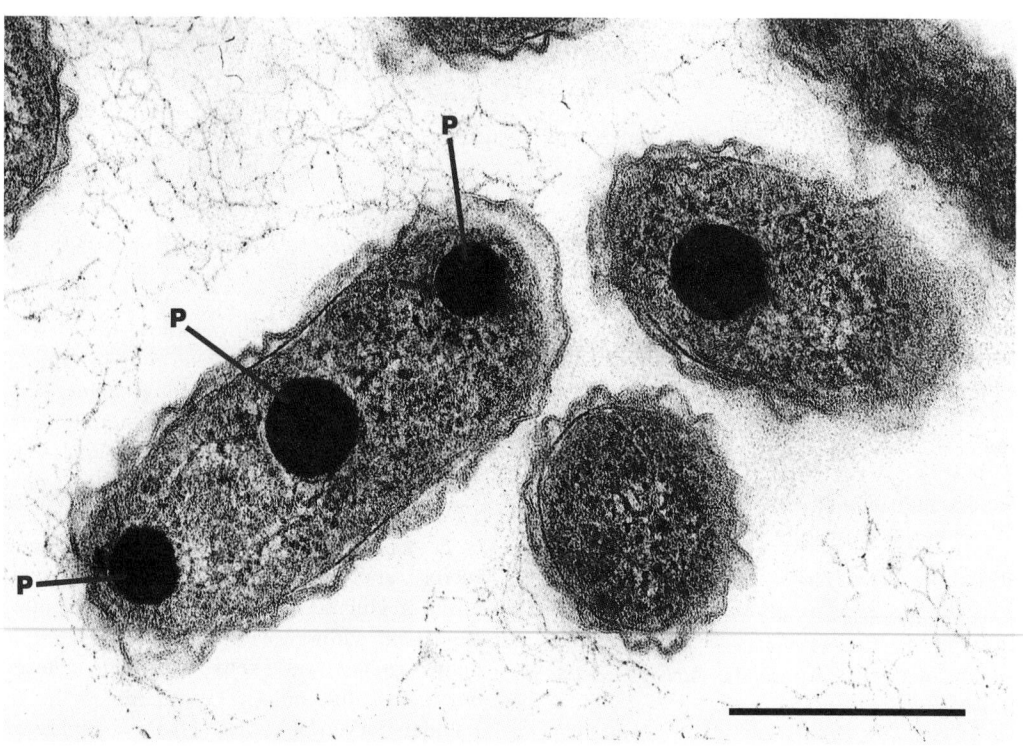

FIGURE BXII.α.109. Intracytoplasmic components of *S. sibiricus* revealed by electron microscopy of ultrathin sections. Electron dense granules of polyphosphates (*P* indicated by arrows). Bar = 0.5 μm. (Printed with permission from V. Yurkov).

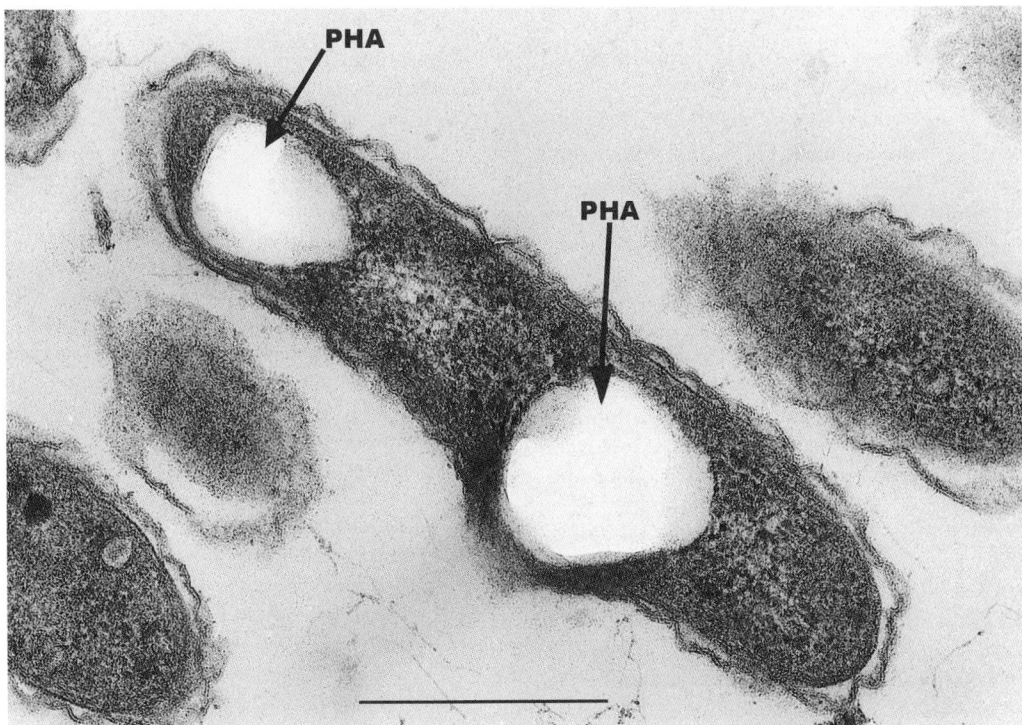

FIGURE BXII.α.110. *S. sibiricus.* Big electron clear granules of polyhydroxyalkanoate (*PHA*; indicated by *arrows*) often occupied up to 40–50% of the total cell volume. Note cell shape distortion by some granules. Bar = 0.5 μm. (Printed with permission from V. Yurkov.)

DIFFERENTIATION OF THE GENUS *SANDARACINOBACTER* FROM OTHER GENERA

Genomic difference of *Sandaracinobacter* from other genera of aerobic anoxygenic phototrophic bacteria was revealed by the DNA composition, DNA–DNA hybridization, and 16S rDNA sequence analyses. As determined by thermal denaturation, the mol% G + C content of *S. sibiricus* is 68.5. According to DNA–DNA hybridization, this species has a low level of DNA similarity with other species investigated. The highest level of DNA similarity (20–27%) was found with *Erythromonas ursincola* (formerly described as *"Erythromicrobium ursincola"*) (Yurkov et al., 1991a). Based on the 16S rDNA sequence data, *Sandaracinobacter* branches independently from its most closely related aerobic phototrophic genus *Erythromonas*, and nonphotosynthetic genera *Blastomonas*, *"Rhizomonas"*, and *Sphingomonas* (Yurkov et al., 1997).

The major morphological and physiological features of *Sandaracinobacter* that distinguish it from its closest relatives of freshwater aerobic phototrophic bacteria are shown in Table BXII.α.95 of the chapter describing genus *Erythromonas*.

TAXONOMIC COMMENTS

The first representative of a freshwater aerobic bacteriochlorophyll *a* (BChl *a*)-containing bacterium (strain RB16-17), isolated from thin microbial mats, was ascribed to the marine genus *Erythrobacter* and named *"Erythrobacter sibiricus"* (Yurkov and Gorlenko, 1990). This assignment was made based on bacteriochlorophyll *a* production and strict aerobiosis. However, since that time *"E. sibiricus"* was reclassified into the genus *Erythromi-crobium*, a new genus for freshwater aerobic anoxygenic phototrophic bacteria (Yurkov and Gorlenko, 1992; Yurkov et al., 1991a, 1992a). Based on phenotypic similarities, the five species—*"E. sibiricus"*, *"E. ursincola"*, *"E. ezovicum"*, *"E. hydrolyticum"*, and *E. ramosum*—were included in the same genus, *Erythromicrobium* (Yurkov et al., 1991a). However, DNA–DNA hybridization data showed that DNA from the species *"E. sibiricus"* and *"E. ursincola"* had very low relatedness (11–27%) to the other three species of this genus. It was proposed that additional physiological, biochemical, and DNA sequence analyses might demonstrate clear differences between *"E. sibiricus"* and *"E. ursincola"* on the one hand, and the other orange freshwater bacteria, on the other hand (Yurkov et al., 1991a). In agreement with this proposal, an analysis of 5S rRNA sequences confirmed the genetic heterogeneity of the genus *Erythromicrobium* (Turova et al., 1995).

Recent results on the morphology, physiology, pigment composition, light-harvesting antenna, reaction center organization, and electron carriers of five *Erythromicrobium* representatives led to excluding *"E. sibiricus"* from the genus *Erythromicrobium* and, based on its unique phylogenetic position, describing this microorganism as the type species of a new genus, *Sandaracinobacter* (Yurkov et al., 1997).

FURTHER READING

Shimada, K. 1995. Aerobic anoxygenic phototrophs. *In* Blankenship, Madigan and Bauer (Editors), Anoxygenic Photosynthetic Bacteria, Kluwer Academic Publishers, Dordrecht. pp. 105–122.

Yurkov, V.V. and J.T. Beatty. 1998. Aerobic anoxygenic phototrophic bacteria. Microbiol. Mol. Biol. Rev. *62*: 695–724.

List of species of the genus Sandaracinobacter

1. **Sandaracinobacter sibiricus** Yurkov, Stackebrandt, Buss, Vermeglio, Gorlenko and Beatty 1997, 1177[VP]

si.bi' ri.cus. L. adj. *sibiricus* isolated in Siberia.

Gram-negative, yellow-orange pigmented thin, long rods, 0.3–0.5 × 1.5–2.5 μm or more. Motile by means of subpolar flagella (up to three). Cells contain BChl *a* and carotenoid pigments. Carotenoids have major absorption peaks at 424, 450, and 474 nm *in vivo*. The cytoplasmic membranes contain a reaction center (RC) and light-harvesting complex I (LHI) with absorption peaks at 750, 799, 857 nm for RC, and 867 nm for LHI. Tetraheme cytochrome *c* of 44.0 kDa is tightly bound to RC and is its immediate electron donor. Additionally contain soluble cytochrome *c* of 14.0 kDa and membrane-bound cytochrome *c* of 30.0 kDa (cytochrome of bc_1 complex). Contain quinones Q-9 and Q-10. Menaquinone is not found.

Aerobic chemoorganotroph and facultative photoheterotroph. The best growth substrates are butyrate, sucrose, casein hydrolysate, and yeast extract. Good growth is observed on acetate and maltose; weak growth on media containing glucose, fructose, pyruvate, propionate, or glycerol. No utilization of ribose, sorbitol, benzoate, fumarate, formate, succinate, citrate, malate, methanol, or ethanol.

Optimal temperature for growth is 25–30°C. Freshwater bacteria; salinity above 1.0% of NaCl in media strongly inhibits growth. The pH optimum is 7.5–8.5. The bacteria exhibit oxidase activity and lack catalase activity. Hydrolyze Tween 60 and do not hydrolyze gelatin or starch.

Resistant to chloramphenicol, fusidic acid, streptomycin, amikacin, bacitracin, kanamycin, neomycin, and novobiocin. Sensitive to penicillin, ampicillin, tetracycline, polymyxin-B, erythromycin, nalidixic acid, lincomycin, mycostatin, aureomycin, and vancomycin.

Demonstrate a high level of resistance to tellurite. Tellurite resistance depends on medium composition, particularly on organic carbon source. The highest tellurite concentration tolerated is 1200 μg/ml in media containing acetate or yeast extract as a sole organic source. Tellurite can be reduced and transformed into metallic tellurium accumulated inside of cells.

Storage compounds: polyphosphates, polysaccharide, and polyhydroxyalkanoate.

Habitat: freshwater algobacterial mat near hydrothermal sulfide-containing vents along the river bottom.

The mol% G + C of the DNA is: 68.5 (T_m).

Type strain: RB16-17.

GenBank accession number (16S rRNA): Y10678.

Genus IX. **Zymomonas** *Kluyver and van Niel 1936, 399*[AL]

GEORG A. SPRENGER AND JEAN SWINGS

Zy.mo' mo.nas or Zy.mo.mo' nas. Gr. n. *zyme* leaven, ferment; Gr. n. *monas* a unit, monad; M.L. fem. n. *Zymomonas* fermenting monad.

Rod-shaped cells with rounded ends, occasionally ellipsoidal, usually in pairs, 2–6 × 1.0–1.4 μm. Gram negative. **Usually nonmotile; if motile, they possess one to four polar flagella.** Motility may be lost spontaneously. **Facultatively anaerobic; some strains are obligately anaerobic.** Optimum temperature 25–30°C. Colonies on the standard medium[1] are glistening, regularly edged, white to cream colored, 1–2 mm in diameter after 2 d at 30°C. Oxidase negative. Chemoorganotrophic, growing on and **fermenting 1 mol of glucose or fructose to almost 2 mol of ethanol, 2 mol of CO_2, and some organic acids such as lactic acid.** Some strains may also utilize sucrose, but other carbon sources are not used. Gluconate can be degraded but does not serve as sole carbon or energy source (Strohdeicher et al., 1988). Sorbitol and gluconolactone are formed when grown on sucrose or mixtures of glucose and fructose by a so far unique enzyme, glucose-fructose oxidoreductase. **Membranes contain pentacyclic triterpenoids of the hopane series** (Sahm et al., 1993). Gelatinase negative. Nitrates are not reduced and indole is not produced. **Zymomonas tolerates 5% ethanol and is acid tolerant, growing at pH 3.5–7.5.**

Good growth is obtained only when a mixture of amino acids is present in the medium, but no one amino acid is essential. All strains require biotin and pantothenate. **Zymomonas occurs as a spoiler in beers, ciders, and perries; as fermenting agents in** *Agave* sap, palm sap, and sugarcane juice; and on honeybees and in ripening honey.

The mol% G + C of the DNA is: 47.5–49.5.

Type species: **Zymomonas mobilis** (Lindner 1928a) Kluyver and van Niel 1936, 399 (*Thermobacterium mobile* Lindner 1928b, 253; *Zymomonas anaerobia* Kluyver 1957, 199.)

FURTHER DESCRIPTIVE INFORMATION

Morphology *Zymomonas* cells are mostly straight rods with rounded or ovoid ends, occurring singly or in pairs. *Zymomonas* cells form no spores, and contain no detectable intracellular lipids, glycogen, or poly-β-hydroxybutyrate. Some individual strains form rosette-like cell aggregations, cell chains, curved or U-shaped cells, or filamentous cells. Most strains are nonmotile.

Oxygen relationships Although *Zymomonas* has a fermentative type of metabolism, it is able to grow aerobically, and therefore should be qualified as a facultative anaerobe. Strains deficient in alcohol dehydrogenase are obligate aerobes (Wecker and Zall, 1987). An electron transport chain is present in the cytoplasmic membrane (Strohdeicher et al., 1990; Kalnenieks et al., 1993; Kim et al., 1995) but no oxidative phosphorylation occurs in the presence of glucose (Kalnenieks et al., 1993).

Cultural characteristics Deep colonies in solid standard me-

---

1. The standard medium (SM) has the following composition (g/l distilled water): D-glucose, 20; and yeast extract, 5. Another medium (RM) was introduced by Bringer et al. (1985) (g/l distilled water): D-glucose, 20; yeast extract, 10; KH_2PO_4, 1.0; $(NH_4)_2SO_4$, 1.0; and $MgSO_4 \cdot 7H_2O$, 0.5. A defined medium (Fein et al., 1983) consists of (g/l distilled water): glucose, 20; KH_2PO_4, 3.5; $(NH_4)_2SO_4$, 2.0; $MgSO_4 \cdot 7H_2O$, 1.0; $FeSO_4 \cdot 7H_2O$, 0.01; and 2-[N-Morpholino]ethanesulfonate (MES), 19.52. The medium is adjusted to pH 5.5 and autoclaved. A sterile solution of biotin and Ca-pantothenate is added after autoclaving to a final concentration of 0.001 g/l each.

dium are lenticular, regular, entire edged, butyrous, white or cream colored, and 1–2 mm in diameter after 2–4 d at 30°C. Anaerobic surface colonies are spreading, entire edged, convex or umbonate, and 1–4 mm in diameter after 2–7 d at 30°C (Swings and De Ley, 1977). When incubated aerobically, colonies reach a maximum diameter of 1.5 mm or appear as microcolonies (Swings et al., 1977).

Most (90%) of the *Zymomonas* strains are able to grow between pH 3.85 and pH 7.55. At pH 3.5, 43% of the strains develop, indicating a high acid tolerance (Swings and De Ley, 1977). This feature is not at all surprising as the natural niche of the genus is in acid palm wines, ciders, and beers at pH 4 or below. *Zymomonas* cannot grow in liquid standard medium at pH 3.05.

Zymomonas grows best between 25°C and 30°C. At 38°C, 74% of the strains grow, but at 40°C, growth is rare (De Ley and Swings, 1976; Swings and De Ley, 1977). *Zymomonas* slowly develops at 15°C (Millis, 1951; Dadds et al., 1973) but not at 4°C. Growth at 36°C is the best phenotypic test to differentiate *Z. mobilis* subsp. *mobilis* (+) from *Z. mobilis* subsp. *pomacii* (−) (Swings et al., 1977). *Zymomonas* is killed by exposure to 60°C for 5 min.

Glucose metabolism *Zymomonas* grows easily in liquid media containing either D-glucose or D-fructose; a dense turbidity accompanied by abundant CO_2 formation develops after 1–2 d at 30°C. The final pH in the standard medium after 3 d at 30°C is 4.8–5.2. The acidification of the medium is more pronounced upon incubation at higher temperatures. Strain-specific flocculent or compact cell deposits are formed. Half of the strains grow in glucose concentrations up to 40% (Swings and De Ley, 1977).

Zymomonas is an unusual bacterium in that it ferments glucose anaerobically by the Entner–Doudoroff pathway (for references, see Conway, 1992), followed by a pyruvate decarboxylation and reduction of acetaldehyde, according to the following general fermentation balance:

1 glucose → (1.58–1.93) ethanol + (1.7–1.9) CO_2 + (0.02–0.2) lactate or other organic acid + (0.011) cell material $[CH_2O]$.

Small amounts of acetaldehyde, acetyl methyl carbinol, and glycerol are also formed. During the dissimilation of glucose under aerobic conditions, ethanol, acetaldehyde, and acetate are formed. In the presence of oxygen, NADH (from fermentation) is used by a membrane-bound NADH oxidase to reduce oxygen to water. Consequently, less NADH is available for the reduction of acetaldehyde to ethanol, and the accumulating acetaldehyde is growth inhibiting (Bringer et al., 1984; Pankova et al., 1985). Some acetate is formed from acetaldehyde through an NADP-dependent acetaldehyde dehydrogenase, with highest activity after aerobic growth (Barthel et al., 1989). Strains deficient in alcohol dehydrogenase are obligate aerobes (Wecker and Zall, 1987). A membrane-bound glucose dehydrogenase containing pyrroloquinoline-quinone as a prosthetic group (Strohdeicher et al., 1988), feeds the electrons derived from glucose into the respiratory chain, which contains ubiquinone-10 (Strohdeicher et al., 1990; Kalnenieks et al., 1993). A membrane-associated F_0F_1-ATPase has been purified (Reyes and Scopes, 1991), and ATP yields via oxidative phosphorylation have been shown with aerated cells using ethanol or acetaldehyde as substrates, but not with glucose (Kalnenieks et al., 1993). All enzymes from the Entner–Doudoroff pathway, including pyruvate decarboxylase and alcohol dehydrogenases, of *Z. mobilis* have been purified, the respective genes have been cloned, and the sequences thereof

have been established (Scopes et al., 1985; Pawluk et al., 1986; Aldrich et al., 1992; Conway, 1992; Sprenger et al., 1993)

Only a truncated tricarboxylic acid cycle is functional in *Zymomonas*. Malate and oxaloacetate are provided by action of a phosphoenolpyruvate carboxylase and a malic enzyme (Dawes et al., 1970; Bringer-Meyer and Sahm, 1989). From the pentose-phosphate pathway enzymes, transaldolase is missing (Feldmann et al., 1992; Zhang et al., 1995). Acetyl-CoA for anabolism is provided by an anaerobically active pyruvate dehydrogenase complex (Bringer-Meyer and Sahm, 1993; Neveling et al., 1998).

Only 2% of the glucose is incorporated in the cells, producing 48% of the cellular carbon (Belaïch and Senez, 1965); the rest of the carbon is derived from yeast extract components. The growth yield coefficients ($Y_{glucose}$ = 3.88–9.32, as determined by several authors) indicate that the growth of these organisms is not very efficient.

Sucrose metabolism Sucrose is fermented and used for growth by many *Zymomonas* strains (for references, see Preziosi et al., 1990; Sprenger, 1996). This property is lost occasionally upon subculturing on D-glucose (Shimwell, 1950). Sucrose fermentation is inducible (Kluyver and Hoppenbrouwers, 1931; Dadds et al. 1973; Richards and Corbey, 1974) and can lead to capsule formation (Kirk et al., 1994): 35g/l of levan is formed from 150 g/l of sucrose (Yoshida et al., 1990). Other by-products of sucrose fermentation are fructo-oligosaccharides, sorbitol, and gluconate (Leigh et al., 1984; Viikari, 1984). Sorbitol and gluconolactone are formed by an intermolecular oxidation-reduction catalyzed by a periplasmic NADP-containing glucose-fructose oxidoreductase that so far has been detected only in *Zymomonas* (Zachariou and Scopes, 1986; Loos et al. 1991; G. A. Sprenger, unpublished results). Sorbitol is used as compatible solute and is accumulated intracellularly at high osmotic stress conditions (Loos et al., 1994). Sucrose is split extracellularly by action of levansucrase or invertase.

Glucose and fructose are taken up by facilitated diffusion through a transport protein that prefers glucose to fructose. *Zymomonas* appears to be the only known bacterium that relies solely on such a uniport type for sugar uptake (DiMarco and Romano, 1985 Parker et al. 1995; Weisser et al. 1995, 1996). Some strains form mannitol from fructose (Viikari, 1984). Dihydroxyacetone and glycerol are excreted after growth under suboptimal conditions with high fructose in the medium (Viikari, 1984) by the subsequent action of a dihydroxyacetone phosphate phosphatase and dihydroxyacetone reductase (Horbach et al., 1994).

Nitrogen sources The nitrogen source for growth can be supplied as peptone, yeast extract, nutrient broth, beer, palm juice, or apple juice or a mixture of 20 amino acids. (Groups of amino acids, individual amino acids, NH_4Cl, or $(NH_4)_2SO_4$ can also serve as nitrogen sources, but this has not been verified for every *Zymomonas* strain.) In synthetic media, the growth yield is lower and the generation time longer than in complex media, but the ethanol yield remains constant (Belaïch and Senez, 1965).

Growth factors *Zymomonas* strains require biotin and pantothenate as growth factors. No strain requires nicotinic acid. Only six strains need additional growth factors. The most exacting strain (VP3) requires additional vitamin B_2, lipoic acid, riboflavin, and folic acid (Van Pee et al., 1974).

Ethanol tolerance *Zymomonas* grows in the presence of 5% ethanol, and many strains grow at even higher concentrations

(Swings and De Ley, 1977; Bringer-Meyer and Sahm, 1988). The organism can produce high concentrations of ethanol, up to 13% (w/v) in batch cultures (Rogers et al., 1982), making it an attractive organism for industrial ethanol production. The main targets for detrimental action of ethanol are the cell membrane (Carey and Ingram, 1983) and fermentative activity; at an ethanol concentration of 10%, 3-phosphoglycerate is accumulated intracellularly concomitant with a decrease in fermentative activity (Strohhäcker et al., 1993). High ethanol levels or heat stress induces a characteristic protein pattern in the cells (Michel and Starka, 1986). A major protective function against ethanol can be ascribed to the hopanoids, pentacyclic triterpenoids, which occur in large amounts (up to 3% of membrane dry weight or 40–50% of total lipid content) in the membrane (Bringer et al., 1985; Flesch and Rohmer, 1989; Hermans et al., 1991; Sahm et al., 1993). Isoprenoids are synthesized exclusively via the pyruvate/glyceraldehyde-3-phosphate pathway (Rohmer et al., 1993). Other major membrane constituents are vaccenic acid (up to 70% of fatty acids; Tornabene et al., 1982; Carey and Ingram, 1983), sphingolipids as the precursor 3-oxosphinganine (Sutter, 1991) or free ceramide (N-palmitoyl-dihydroxyhexadecane; Tahara and Kawazu, 1990), and the newly discovered dialkylcyclohexadiene-carbinol (Koukkou et al., 1998). No evidence for ketodeoxyoctulosonic acid, heptoses, or hydroxy fatty acids has been found (Tornabene et al., 1982).

Other physiological features   The following physiological tests are positive for the genus: catalase and superoxide dismutase (Bringer et al., 1984; Pankova et al., 1985; Shvinka et al., 1989); reduction of methylene blue, thionin, and 2,3,5-triphenyltetrazolium chloride; formation of traces of acetyl methyl carbinol and the production of a characteristic fruity odor. The following tests are negative: growth in 0.5% yeast extract, nutrient broth, or 1% peptone broth, in liquid standard medium + 2% NaCl; indole production, nitrate reduction, hydrolysis of gelatin, hydrolysis of Tween 60 and Tween 80, and oxidase.

Pathogenicity   *Zymomonas* is not known to be pathogenic for humans, animals, or plants. Lindner (1929, 1931) recommended the use of *Zymomonas* in human nutrition as a kind of yogurt. Antagonistic effects of *Zymomonas* against bacteria and fungi *in vitro*, and the therapeutic use of *Zymomonas* in cases of chronic enteric and gynecological infections, have been reported (Swings and De Ley, 1977; Falcao de Morais et al., 1993).

Ecology   In sweet English ciders, *Zymomonas* is the causative agent of a secondary fermentation, known as "cider sickness". The first description of this phenomenon was given by Barker and Hillier (1912). Cider sickness is recognized by frothing and abundant gas formation, a typical change in the aroma and flavor, reduction of sweetness, and marked turbidity forming a heavy deposit afterward (Barker, 1948; Millis, 1951; Carr and Passmore, 1971). A cider disorder known as framboise, or "framboisement" in France, is also attributed to *Zymomonas*. Millis (1951) also isolated *Zymomonas* from "sick" perries. Lindner (1928b) discovered that *Zymomonas* is the fermentative agent that transforms the sugary *Agave* sap (aguamiel) to pulque in Mexico. *Zymomonas* is a serious beer contaminant, particularly in English cask beers, producing a heavy turbidity and an unpleasant odor due to acetaldehyde and H_2S. Dennis and Young (1982) isolated five strains of *Zymomonas* from an ale brewery using an isolation medium containing actidione and Schiff's reagent as a dye to detect acetaldehyde production. *Zymomonas* has not been reported in lager beers. Palm wines are prepared in the Far East and in Africa from the sap of *Arenga*, *Raphia*, and *Elaeis* palms and are known to harbor *Zymomonas* as a fermentative agent. *Zymomonas* is also present in fermenting sugarcane juice in Brazil, and on bees and ripening honey in Spain.

Plasmids   Indigenous plasmids in a size range of 1.5 kb to >40 kb of DNA have been described. Some antibiotic or heavy metal resistance features have been ascribed to the presence of these plasmids (for references, see Sprenger et al., 1993). Indigenous plasmids have been used to construct shuttle vectors for the expression of homologous and heterologous genes in *Z. mobilis* for metabolic engineering (Conway et al. 1987; Sprenger et al., 1993).

Potential for industrial applications and metabolic engineering   *Zymomonas* may be important as an industrial ethanol producer and it offers advantages over traditional yeast fermentation: it has higher specific rates of glucose uptake and ethanol production; it gives higher ethanol yield and lower biomass; it grows anaerobically; and it has a high ethanol tolerance. Some authors have studied extensively the kinetics of ethanol fermentation in both batch and continuous cultures at high glucose concentrations (Lee et al., 1979; Rogers et al., 1982). Attempts to broaden the narrow substrate and product spectrum of *Z. mobilis* have yielded strains that can utilize carbon sources such as mannitol (Buchholz et al., 1988) after chemical mutagenesis. Introduction of heterologous genes from various microorganisms has led to strains that can utilize D-xylose (Feldmann et al., 1992; Zhang et al., 1995), D-mannose (Weisser et al., 1996), L-arabinose (Deanda et al., 1996), β-glucosides, or lactose (Sprenger, 1993). Similarly, strains that produce L-alanine (Uhlenbusch et al., 1991) or carotenoids (Misawa et al., 1991) have been described. Several reviews discuss the use of *Zymomonas* in the production of bioethanol from renewable carbon sources (Buchholz et al., 1987; Ingram et al., 1989; Johns et al., 1992; Sahm et al., 1992; Doelle et al., 1993; Sprenger, 1993, 1996; Sprenger et al., 1993).

ENRICHMENT AND ISOLATION PROCEDURES

The following medium, originally designed as a detection medium for *Zymomonas* in breweries (Dadds, 1971), can be recommended for enrichment and has the following composition (g/l distilled water): malt extract, 3; yeast extract, 3; D-glucose, 20; peptone, 5; and actidione, 0.02. The pH is adjusted to 4.0. Ethanol is added to a final concentration of 3% (v/v). The presence of *Zymomonas* is indicated by abundant gas production after 2–6 d incubation at 30°C.

MAINTENANCE PROCEDURES

Zymomonas cultures held in the standard medium at room temperature are transferred every 2–3 weeks. *Zymomonas* survives the ordinary lyophilization procedure for many years. Cell viability is rapidly lost in ionic dilution buffers as sodium citrate, sodium chloride, or sodium phosphate. A 10 mM Tris buffer (pH 8) containing 33 mM $MgCl_2$ should be used for serial dilutions at low temperatures (Buchholz and Eveleigh, 1989).

DIFFERENTIATION OF THE GENUS *ZYMOMONAS* FROM OTHER GENERA

Table BXII.α.98 indicates the most salient features that differentiate *Zymomonas* from other genera. *Zymomonas* can be differentiated from *Sphingomonas* by the absence of the fatty acid

$C_{14:0\ 2OH}$, which is typically present in all *Sphingomonas* species (Yabuuchi et al., 1990a; Takeuchi et al., 1993a; Balkwill et al., 1997), and also by the much lower mol% G + C of its DNA. *Zymomonas* is phenotypically and genotypically well defined and is easily recognized. Its most outstanding feature is the quantitative fermentation of glucose or fructose (and to a lesser extent of sucrose)—but no other sugars—to equimolar amounts of ethanol and CO_2. This feature makes *Zymomonas* a unique ethanol-producing bacterium.

Zymomonas is excluded from the *Enterobacteriaceae* based on polar flagellation, inability to reduce nitrates, growth at pH 4, and growth in the presence of 5% ethanol.

Genetically, phenotypically, and ecologically, *Zymomonas* is distantly related to the acetic acid bacteria: they both occur in acid, sugary, and alcoholized niches such as tropical plant juices and beer. They are ecologically complementary in that *Zymomonas* produces ethanol, which is further oxidized by the acetic acid and bacteria. The acetic acid bacteria are differentiated from *Zymomonas* by their strictly aerobic growth requirements and the mol% G + C values of their DNA.

TAXONOMIC COMMENTS

The genus *Zymomonas* belongs to the family *Sphingomonadaceae* based on phylogenetic analysis of 16S rRNA gene sequences (Balkwill et al., 1997). The genus *Zymomonas* contains only one species: *Zymomonas mobilis*. Strains ATCC 29192, NCIB 8777, and 10565 are almost identical and are united in *Zymomonas mobilis* subsp. *pomacii* (Swings et al., 1977). All the other strains belong in *Zymomonas mobilis* subsp. *mobilis*.

All *Zymomonas* strains have mol% G + C values for their DNA within the narrow range of 47.5–49.5. The tightness of the genus *Zymomonas* is also reflected in the high phenotypic similarity (S_{sm} ≥88%) between the strains (De Ley and Swings, 1976). DNA–DNA hybridizations show a nucleotide sequence similarity of >76%; only one strain, ATCC 29192, is aberrant with less than 32% DNA duplexing (Swings and De Ley, 1975). Because of the high phenotypic similarity, the authors refrained from proposing the subspecies *Zymomonas mobilis* subsp. *pomacii* as a separate species, although the DNA–DNA hybridization with reference strain 5.3 was very low (i.e., 32%), so low a value that it would have allowed the creation of two separate species. However, no additional strains and hybridization values were available to further support the separation. The homogeneity of the genus *Zymomonas* is further demonstrated by the computer-assisted comparison of electropherograms of the soluble cell proteins. The genome size has been determined to be in the range of 1.4–1.5 × 10^9 Da (Swings and De Ley, 1975; Kang and Kang, 1998).

Plasmid profiles from both *Z. mobilis* subspecies have been described and have been used to discriminate laboratory strains (Tonomura et al., 1982; Stokes et al., 1983; Skotnicki et al., 1984; Walia et al., 1984; Scordaki and Drainas, 1987; Yablonsky et al., 1988; Misawa and Nakamura, 1989; Degli-Innocenti et al. 1990).

List of species of the genus Zymomonas

1. **Zymomonas mobilis** (Lindner 1928a) Kluyver and van Niel 1936, 399[AL] (*Thermobacterium mobile* Lindner 1928b, 253; *Zymomonas anaerobia* Kluyver 1957, 199.)

mo'bi.lis. L. adj. *mobilis* movable, motile.

The description of the species is as for the genus and as listed in Tables BXII.α.98 and BXII.α.99. The differentiation of its two subspecies is indicated in Table BXII.α.100.

The mol% G + C of the DNA is: 47.5–49.5 (T_m).

Type strain: ATCC 10988, DSMZ 424.

Additional Remarks: phenotypic centrotype ATCC 29191.

a. **Zymomonas mobilis** *subsp.* **mobilis** (Lindner 1928a) De Ley and Swings 1976, 156[AL] (*Zymomonas mobilis* biovar anaerobia Richards and Corbey 1974, 243; *Zymomonas mobilis* biovar recifensis Gonçalves de Lima, De Araújo, Schumacher and Cavalcanti Da Silva 1970, 3; *Zymomonas anaerobia* biovar anaerobia (Shimwell 1937) Carr 1974, 353; *Saccharomonas anaerobia* biovar immobilis Shimwell 1950, 182; *Zymomonas anaerobia* biovar immobilis (Shimwell 1937) Carr 1974, 353.)

mo'bi.lis. L. adj. *mobilis* movable, motile.

TABLE BXII.α.98. Differential characteristics of the genus *Zymomonas* and other genera[a]

Characteristics	Zymomonas	Acetobacter	Aeromonas	Gluconobacter	Sphingomonas	Vibrio
Gram variability occurs	−	+	−	+	−	−
Flagellar arrangement:						
Polar only	+	−	+	+	+	+
Peritrichous	−	+	−	−	−	−
Oxygen tolerance:						
Growth under both aerobic and anaerobic conditions	+	−	+	−	−	+
Growth under aerobic conditions only	−	+	−	+	+	−
Oxidase	−	−	+	−	+	+[b]
Carbohydrate metabolism:						
Fermentative and respiratory	+	−	+	−	−	+
Respiratory only	−	+	−	+	+	−
Gas from D-glucose	+	−	D	−	−	−
1 mol of glucose fermented to 2 mol of ethanol and 2 mol of CO_2	+	−	−	−	−	−
Nitrate reduction	−	−	+	−	−	+
Growth factors required	+[c]	+[d]		−		
Growth at pH 4.0	+	+	−	+		−
Inhibited by novobiocin	+	D	−	+		d
Mol% G + C of DNA	47.5–49.5	51–65	57–62	56–64	60–65.4	38–51

[a]Symbols: +, typically positive; −, typically negative; D, differs among species.

[b]Some species exhibit a negative or weak oxidase reaction.

[c]Pantothenate and biotin.

[d]Pantothenate and/or niacin.

TABLE BXII.α.99. Other characteristics of *Zymomonas mobilis*[a]

Characteristics	Reaction or result
Occurrence of "fruity" odor when cultured in standard medium	+
Growth in 0.5% yeast extract broth, in 0.5% peptone broth, or in beer	−
Growth in beer containing 2% glucose or on malt agar	+
Salt tolerance, growth in SM in presence of:	
0.5% NaCl	+
1.0% NaCl	d
2.0% NaCl	−
pH range, growth in SM at pH:	
3.05	−
3.50	d
4.0–7.0	+
7.5	d
8.0	−
Temperature range, growth in SM at:	
30–36°C	+
38°C	d
40°C	−
Ethanol tolerance, growth in SM containing:	
5.5% Ethanol	+
7.7% Ethanol	d
Glucose tolerance, growth in SM containing:	
20% Glucose	+
40% Glucose	d
Growth in SM containing 0.1% neutral red	+
Vitamin requirements:	
Pantothenate and biotin	+
Lipoic acid, folic acid, niacin, *p*-aminobenzoic acid, riboflavin, cyanocobalamin	−
Catalase	+
Oxidase	−
Acetyl methyl carbinol formed (Voges–Proskauer)	W
Reduction of nitrate	−
H$_2$S produced	d
Survival at 60°C for 5 min	−
Final pH in SM at 30°C	4.8–5.2
Carbon sources:	
D-glucose, D-fructose	+
Sucrose	d
D-Mannose, L-sorbose, D- and L-arabinose, L-rhamnose, D-xylose, D-ribose, D-sorbitol, salicin, dulcitol, D-mannitol, adonitol, erythritol, glycerol, ethanol, D-galacturonate, DL-malate, succinate, pyruvate, DL-lactate, tartrate, citrate, starch, dextrin, raffinose, D-trehalose, maltose, lactose, D-cellobiose	−
Urease	d
Methylene blue reduction	+
Thionin reduction	+
Triphenyltetrazolium reduction	+
Indole	−
Hydrolysis of gelatin, Tween 60, Tween 80	−
L-ornithine, L-arginine, and L-lysine decarboxylases	d
Antimicrobial agents (amount per disk):	
Chloramphenicol, 30 µg; fusidic acid, 10 µg; sulfafurazole, 500 µg; tetracycline, 10 µg	S
Ampicillin, 10 µg; cephaloridine, 10 µg; erythromycin, 10 µg; vancomycin, 10 µg	d
Bacitracin, 5 U; gentamicin, 10 µg; kanamycin, 10 µg; lincomycin, 10 µg; methicillin, 10 µg; nalidixic acid, 30 µg; neomycin, 10 µg; novobiocin, 30 µg; penicillin, 5 U; polymyxin, 300 U; streptomycin, 10 µg	R
Actidione, 0.01%	R

[a]For symbols see standard definitions; R, resistant; S, susceptible; and W, weak.

See Table BXII.α.100 for differentiation of this subspecies from the subsp. *pomacii*.

Isolated from bees, from ripening honey in Spain, from the fermenting sap of *Agave americana* (*atrovirens*) in Mexico, from fermenting palm juice (*Arenga pinnata*) in Indonesia, from sugarcane juice in Queensland, Australia, and the Fiji Islands (Warr et al., 1984), from *Elaeis guineensis* and *Raphia uinifera* in Zaire and Nigeria, and from fermenting sugarcane juice or molasses in Brazil. It has also been isolated in England from beer, from the surface of brewery yards, and from the brushes of cask-washing machines.

The mol% G + C of the DNA is: 47.5–49.5 (T_m).

Type strain: ATCC 10988, DSMZ 424.

Additional Remarks: phenotypic centrotype ATCC 29191.

b. **Zymomonas mobilis** *subsp.* **pomacii** (Millis 1956) De Ley and Swings 1976, 156^{AL} (*Zymomonas anaerobia* subsp. *pomaceae* (sic) Millis 1956, 527; *Zymomonas mobilis* subsp. *pomaceae* (sic) (Millis 1956); De Ley and Swings 1976, 156.) *pom.a' ci.i.* V.L. n. *pomacium* cider; M.V.L. gen. n. *pomacii* of cider.

See Table BXII.α.100 for differentiation of this subspecies from the subspecies *mobilis*.

Isolated in England from sick cider and from apple pulp.

The mol% G + C of the DNA is: unknown.

Type strain: ATCC 29192, NCIB 11200.

Reference strains NCIB 8777 and NCIB 10565.

TABLE BXII.α.100. Differentiation between subspecies of *Zymomonas mobilis*[a]

Characteristics	*Z. mobilis* subsp. *mobilis*	*Z. mobilis* subsp. *pomacii*
Colony diameter after aerobic growth on SM for 7 d at 30°C	*1.5 mm*	*<1.0 mm*
Growth in SM at 36°C	+	−
Percent DNA–DNA similarity with strain 5.3[b]	76–100	<32
Clustering level of protein electropherograms[c]	Cluster together above r = 0.88	Cluster at r = 0.75 with subsp. *mobilis*
Infrared spectra of intact cells:[d]		
Distinct peak at 960 cm^{-1}	+	−
Shoulder only, 960 cm^{-1}	−	+

[a]Symbols: +, typically positive; −, typically negative.

[b]Data from Swings and De Ley (1975).

[c]Data from Swings et al. (1976).

[d]Data from Swings and Van Pee (1977).

Order V. **Caulobacterales** Henrici and Johnson 1935b, 4^{AL}

GEORGE M. GARRITY, JULIA A. BELL AND TIMOTHY LILBURN

Cau'lo.bac.ter.a' les. M.L. masc. n. *Caulobacter* type genus of the order; *-ales* ending to denote order; M.L. fem. n. *Caulobacterales* the *Caulobacter* order.

The order *Caulobacterales* was circumscribed for this volume on the basis of phylogenetic analysis of 16S rRNA gene sequences; the order contains the family *Caulobacteraceae*.

Description is the same as for the family *Caulobacteraceae*.

Type genus: **Caulobacter** Henrici and Johnson 1935b, 83 emend. Abraham, Strömpl, Meyer, Lindholst, Moore, Christ, Vancanneyt, Tindall, Bennasar, Smit and Tesar 1999, 1070.

Family I. **Caulobacteraceae** Henrici and Johnson 1935b, 4^{AL}

GEORGE M. GARRITY, JULIA A. BELL AND TIMOTHY LILBURN

Cau'lo.bac.ter.a' ce.ae. M.L. masc. n. *Caulobacter* type genus of the family; *-aceae* ending to denote family; M.L. fem. pl. n. *Caulobacteraceae* the *Caulobacter* family.

The family *Caulobacteraceae* was circumscribed for this volume on the basis of phylogenetic analysis of 16S rRNA gene sequences; the family contains the genera *Caulobacter* (type genus), *Asticcacaulis*, *Brevundimonas*, and *Phenylobacterium*.

Oligotrophic and chemoorganotrophic, with a strictly aerobic respiratory metabolism. *Caulobacter*, *Asticcacaulis*, and some spe-

cies of *Brevundimonas* produce prosthecae; *Phenylobacterium* does not.

Type genus: **Caulobacter** Henrici and Johnson 1935b, 83 emend. Abraham, Strömpl, Meyer, Lindholst, Moore, Christ, Vancanneyt, Tindall, Bennasar, Smit and Tesar 1999, 1070.

Genus I. **Caulobacter** *Henrici and Johnson 1935b, 83^{AL} emend. Abraham, Strömpl, Meyer, Lindholst, Moore, Christ, Vancanneyt, Tindall, Bennasar, Smit and Tesar 1999, 1070*

JEANNE S. POINDEXTER

Cau'lo.bac' ter. L. n. *caulis* stalk; M.L. masc. *bacter* form of Gr. neut. n. *bactrum* rod; M.L. masc. n. *Caulobacter* stalk(ed) rod.

Cells single, unbranched, with tapered poles; long axis typically distinctly curved, conferring a **vibrioid** shape, but may be straight in some isolates and the cell shape **fusiform**; 0.4–0.6×1–2 µm during growth in most media. The younger pole of the dividing cell bears a **single flagellum**, and the older pole bears a prostheca (the **stalk**) developed by outgrowth of the cell envelope. The stalk includes outer membrane, peptidoglycan, cell membrane, and a core sometimes observed to be occupied in part by membranes but not in any case by any other discernible cytoplasmic components. Stalk diameter is constant along its length, varying from 0.11 to 0.18 µm among isolates; the stalk possesses at least one **stalk band** as a substructure after the cell has matured and completed at least one cell cycle. Binary fission is constrictive and is completed without formation of a septum. The stalk-bearing progeny cell grows and eventually repeats the **asymmetric cell division**. The flagellum-bearing progeny cell, after a period of motility, releases the flagellum and develops its stalk at the previously flagellated site as it grows and proceeds to its **asymmetric cell division**. At the base of the flagellum and at the outer tip of the stalk is a small mass of adhesive material, the **holdfast**, which confers adhesiveness on each of the progeny.

Gram negative. Strictly **respiratory** and **aerobic**; O_2 serves as the only terminal electron acceptor for growth, although nitrate may be reduced to nitrite. The principal respiratory quinone is ubiquinone-10. Colonies are circular, convex, and glistening, with smooth margins; butyrous in texture; may be colorless or yellow; upon aging, colorless colonies tend to become light brown. In static liquid cultures, cells accumulate as a **surface film** or pellicle and develop as a ring of growth on the vessel wall at or just below

the air-liquid interface. Growth in agitated liquid cultures is evenly dispersed.

Chemoorganotrophic and **oligotrophic**; grow optimally in media such as peptone–yeast extract containing 0.05–0.3% (w/v) organic solutes, but not in standard nutrient broth with 0.8% (w/v) organic solutes or richer media such as tryptic soy broth. Do not produce acid or gas from sugars during growth. Optimal temperature for growth 25–30°C; tolerated range for growth 10–35°C. Optimal pH near neutrality; pH 6–9 tolerated. Typically do not grow in media containing 2% (w/v) NaCl. Maximum specific rates of exponential growth 0.17–0.46 h^{-1}. Isolates are diverse in their nutritional requirements. Isolates of one species grow in defined media with glucose or glutamic acid and minerals, other isolates are distinctly stimulated by riboflavin or cyanocobalamin, and some require unidentified growth factors available in peptone. Glucose and glutamic acid are the most widely utilized carbon sources.

Sequences of 16S rDNA are consistent with placement of *Caulobacter* among the *Alphaproteobacteria*. The positions of the 44 sequences grouped as *Caulobacter* in Fig. BXII.α.111 (see "Taxonomic Comments") are represented in Fig. BXII.α.112 by 12 nonredundant sequences which are grouped in Fig. BXII.α.113 and in Fig. BXII.α.114; their strain identities and GenBank accession numbers are displayed in Fig. BXII.α.112. These 12 sequences represent 11 isolates known to exhibit *Caulobacter* phenotypes as well as *Mycoplana segnis*, which is phenotypically distinguishable from *Caulobacter* species. (See "Taxonomic Comments" for descriptions of these trees and for further comments.)

Sources: isolated from distilled, tap and commercial bottled

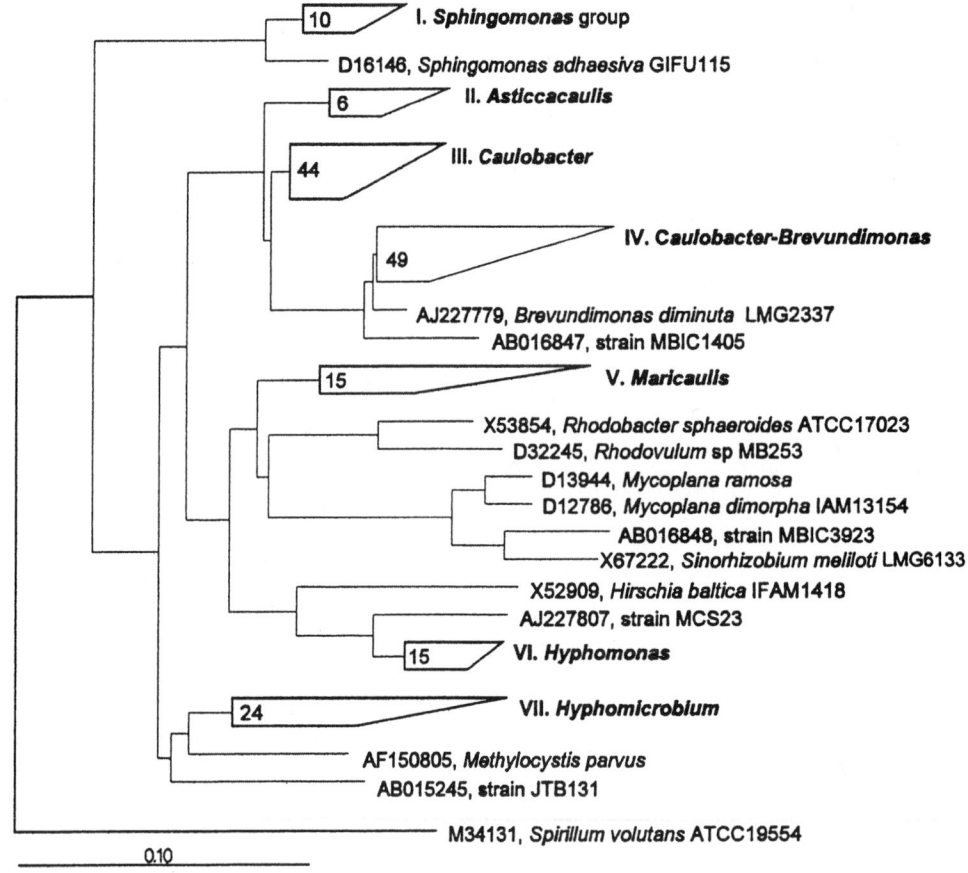

FIGURE BXII.α.111. Phylogenetic tree of 177 16S rDNA sequences built by neighbor-joining analysis, using all available sequences for caulobacters and nonphototrophic hyphal-budding bacteria available as of January 2001. The set includes many sequences that are redundant (determined for the same strain and identical within experimental error), as well as four whose labels are suspect (see text). Within the tree are 99 sequences for caulobacters, 40 for non-phototrophic hyphal-budding bacteria, 13 included by virtue of sequence similarity without phenotypic descriptions of the isolates, and 25 for isolates that are phenotypically distinguishable from prosthecate bacteria. The seven groups accommodate all but 14 of the sequences; excluded are one caulobacter (strain MCS23), one hyphal-budding bacterium (*Hirschia baltica* IFAM1418), three sequences for undescribed organisms, and nine for isolates known to be phenotypically distinguishable from caulobacters and hyphal-budding bacteria. Bar indicates evolutionary distance. Each enclosed numeral indicates the number of sequences grouped. The minimum similarities within each of the caulobacterial genus groups are: I. *Sphingomonas* group (*C. leidyi*), 95.7%; II. *Asticcacaulis*, 95.9%; III. *Caulobacter*, 94.9%; IV. Group IV caulobacters, 94.8%; and V. *Maricaulis*, 94.7%. See Fig. BXII.α.114 for a more stringent (maximum-likelihood) analysis of these sequences.

drinking water, natural bodies of fresh water, and sewage (principally from activated sludge during secondary treatment). Not encountered as clinical isolates and not known to be pathogenic for plants or animals.

The mol% G + C of the DNA is: 62–65.

Type species: **Caulobacter vibrioides** Henrici and Johnson 1935b, 84.

FURTHER DESCRIPTIVE INFORMATION

Morphology Morphological details of *Caulobacter* cells are illustrated in Fig. BXII.α.115, Fig. BXII.α.116, and Fig. BXII.α.117. Cell shape and life cycle are major characteristics in the description of caulobacters, and the redistribution of these bacteria among five genera (see "Taxonomic Comments") increases the usefulness of morphology in the description of each genus. Cells of *Caulobacter* isolates have distinctly tapered poles; only isolates of the species *C. fusiformis*, atypical in several ways, yet more reasonably accommodated here than elsewhere in the

family, lack the distinct curvature of the others, all of which are vibrioid. Neither the sharp tapering nor the cell curvature varies with cultivation in liquid media under conditions that support exponential growth of morphologically uniform populations. Morphology of these, as of most bacteria, becomes aberrant when cells are subjected to conditions unfavorable for growth. In unfavorable conditions (e.g., being buried under millions of other cells in a colony on an agar surface, or in media rich in soluble organic nutrients, ammonium ions, or NaCl), cells may become irregular in many ways, which include branching of cells or of stalks, uneven bloating, failure of stalk elongation, development of supernumerary stalks, loss of motility, heavy shedding of the S-layer and other cell surface components, frequent lysis and, typically, a decrease in ability to form colonies, i.e., loss of viability. Such populations are not suitable for description of the morphology of an isolate.

Morphology is uniform, swarmers are motile, cells are adhesive, stalks elongate, and bands are added through successive

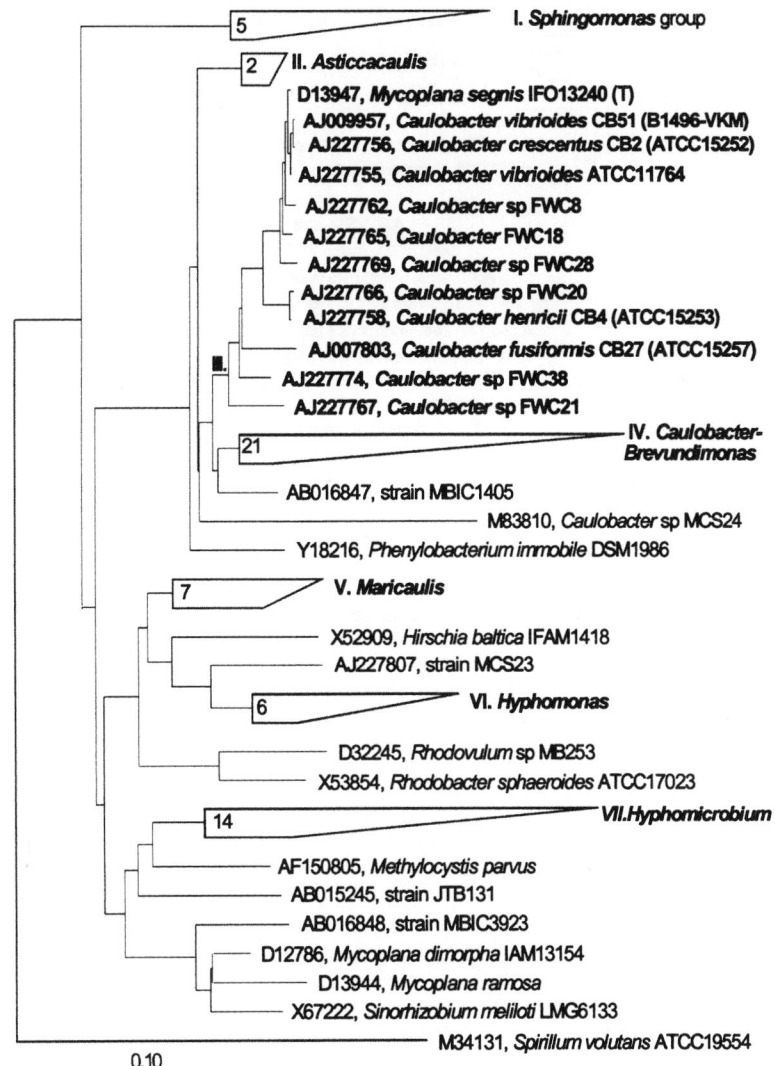

FIGURE BXII.α.112. Phylogenetic tree of 81 16S rDNA sequences, with group III. *Caulobacter* expanded to display GenBank accession numbers and strain identities. (See Fig. BXII.α.114.) Also represented are the positions of sequences AB023427 (by D13947); AB008531, AF125194, AJ227754, AJ227757, AJ227768, M83798, M83799, M83801, M83802, M83803, M83804, and X52281 (by AJ009957 and AJ227756); AJ227761, AJ227763, AJ227764, AJ227773, AJ227776, and AJ227777 (by AJ227755); AJ227770 (by AJ227762); AJ227772 and AJ227775 (by AJ227765); AJ227760 and AJ227771 (by AJ227769); AB008532 and AJ007805 (by AJ227758 and AJ227766); AB008533, AJ227759, and M83796 (misidentified as *Caulobacter bacteroides* strain CB7) (by AJ007803); and M83805 (by AJ227774). Bar indicates evolutionary distance. Each enclosed numeral indicates the number of sequences grouped.

cell cycles when *Caulobacter* isolates are cultivated aerobically in dilute media at neutral pH and at or below 30°C. Isolates identifiable as *Caulobacter* species by additional, non-morphological criteria fail to thrive and typically die in rich media, apparently due to inhibition of one or more steps in their morphogenetic sequence. "Rich" may apply to a single excess nutrient, as demonstrated for ammonium by Felzenberg et al. (1996). Generally, for any isolate, mean stalk length decreases with increasing concentration of nutrients in peptone–yeast extract media and with increasing phosphate concentration in both complex and defined media (Schmidt and Stanier, 1966; Haars and Schmidt, 1974; Poindexter, 1984; Felzenberg et al., 1996; Poindexter and Staley, 1996). Growth of the majority of isolates is adversely affected by phosphate concentrations in excess of 5 mM.

The flagella of most species contain three flagellins of slightly different mobilities (29, 27, and 25 kDa) in denaturing electrophoretic gels. The periodic synthesis and assembly of the flagellum of *C. crescentus* and of chemotaxis proteins are influenced by a hierarchy of expression of dozens of genes that is coordinated with the regulation of many other growth and morphogenetic genes (reviewed in, *inter alia*, Wu and Newton, 1997). The basal body has five rings, rather than the four typical of Gram-negative cells; it is possible that the fifth ring participates in some way in the regular shedding of the *Caulobacter* flagellum that precedes the onset of stalk development and general cellular growth.

Fine structure Stalk bands (see Figs. BXII.α.115D, BXII.α.116C, and BXII.α.117C) occur in all *Caulobacter* isolates. Each band is a set of concentric rings that are more electron-opaque and structurally more rigid than the remainder of the

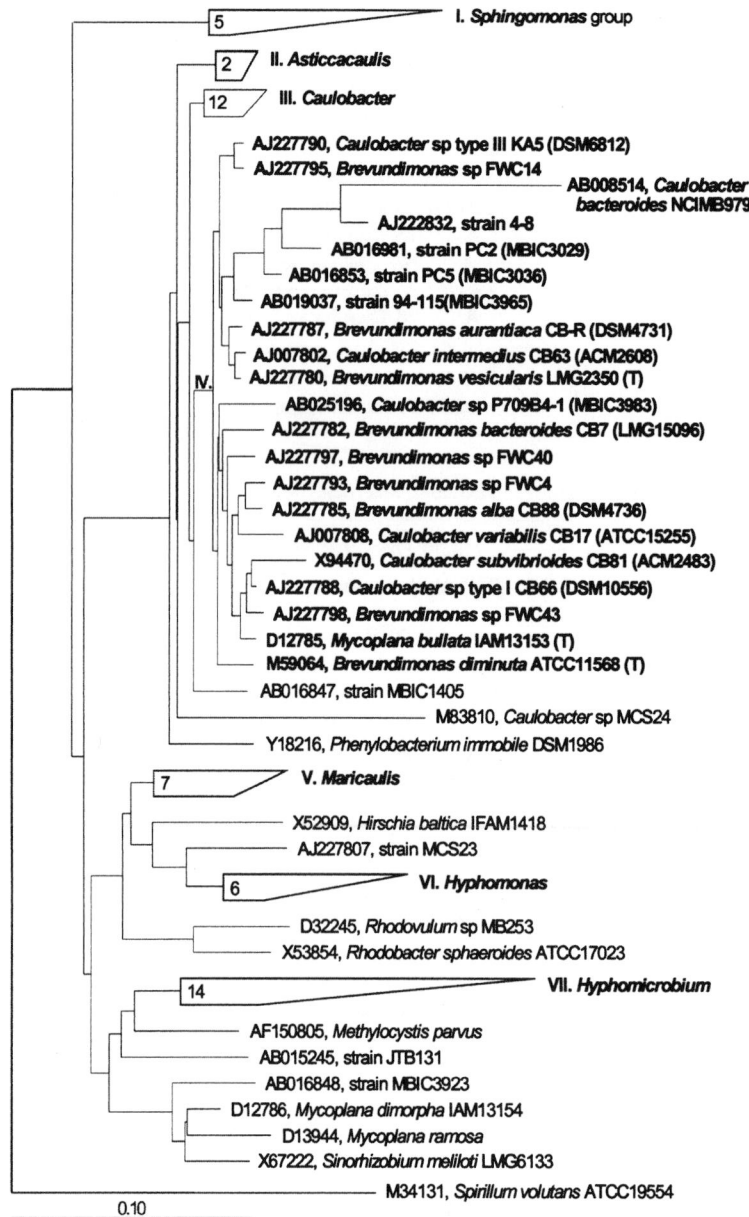

FIGURE BXII.α.113. Phylogenetic tree of 81 16S rDNA sequences, with group IV caulobacters expanded to display GenBank accession numbers and strain identities. (See Fig. BXII.α.114) Also represented are the positions of sequences M83800 (by AJ227795); AB016980 (by AB016981); AB016846, AB016849, AB016850, AB016851, AB016852, and AJ227796 (by AJ227787); AB023784 (by AJ007802); AB021414, AJ007801, and AJ227781 (by AJ227780); AJ227799 and M83808 (by AB025196) (neighbor-joining analysis groups M83810 with these sequences, although maximum-likelihood analysis places it in a unique position); AB008513 and AJ227792 (by AJ227782); AJ227794 (by AJ227793); AB008393 (by AJ227785); AB0083392 and AJ227784 (by X94470); AJ227789 and AJ227791 (by AJ227788). The position of the *Mycoplana bullata* sequence AB023428 is represented by D12785, same species. The positions of *Brevundimonas diminuta* sequences AB021415, AJ227778, and X87274 are represented by M59064, same species. Two additional sequences available for *B. diminuta* are not represented: D49422 falls within this group, but at a unique position, and AJ227779 falls outside the group (see Fig. BXII.α.111). Bar indicates evolutionary distance. Each enclosed numeral indicates the number of sequences grouped.

stalk (Jones and Schmidt, 1973). Earlier evidence that one band is added to the growing stalk during each cell cycle completed by the mature stalked cell of *C. crescentus* (Staley and Jordan, 1973) has been confirmed (Poindexter and Staley, 1996) and this trait exploited in the determination of the reproductive rate of diverse caulobacters *in situ* in a freshwater lake (Poindexter et al., 2000). The stalk band appears to be one more cellular

feature to be added to the growing list of cell cycle-regulated events in the life of *C. crescentus*, and presumably of other species as well.

The cell and stalk of *C. crescentus* have long been known to be coated with an S-layer (Poindexter et al., 1967), a cellular feature of a wide variety of both Gram-negative and -positive bacteria (Sleytr et al., 1988). J. Smit and his coworkers have dem-

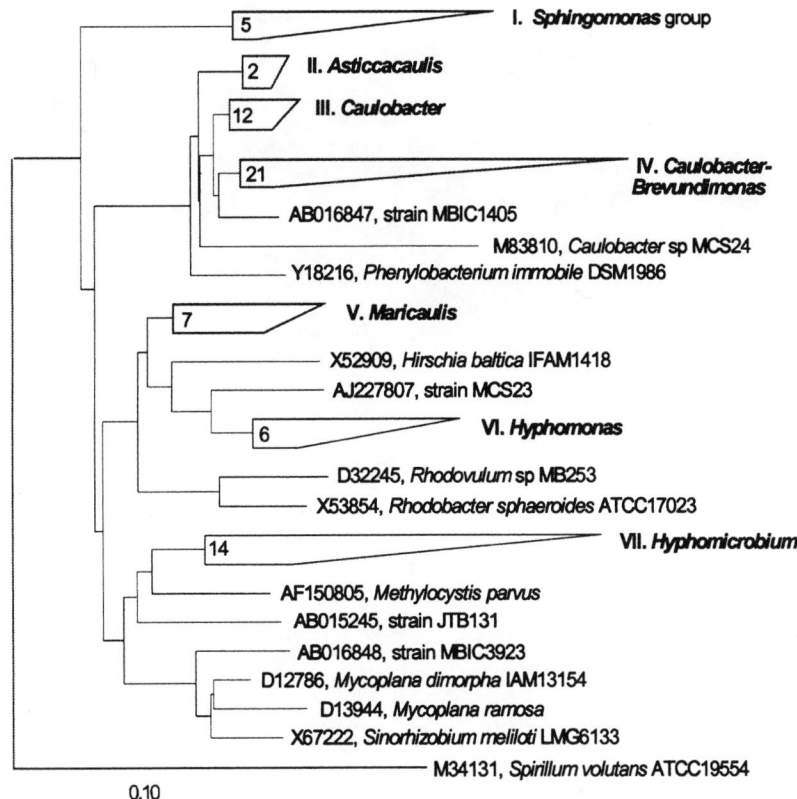

FIGURE BXII.α.114. Phylogenetic tree of 98 16S rDNA sequences built by maximum likelihood (DeSoete) analysis. After removal of redundant sequences (determined to be for the same strain and identical within experimental error), the four sequences whose labels are suspect, and one sequence that appeared to be only very distantly related (X97079, strain TM16; not described phenotypically), all positions of the complete tree are here represented by 81 sequences: 37 for caulobacters, 20 for hyphal-budding bacteria, 14 for bacteria phenotypically distinguishable as nonprosthecate bacteria, and 10 included by virtue of sequence similarity without phenotypic descriptions of the isolates. Relevant branches are expanded in Fig. BXII.α.112 and BXII.α.113 and in other chapters (*Asticcacaulis*, Fig. BXII.α.119 and *Maricaulis* Fig. BXII.α.76), where GenBank accession numbers and strain identities are shown, and sequences represented but not shown are listed in the legends. Bar indicates evolutionary distance. Each enclosed numeral indicates the number of sequences grouped.

onstrated that this structural feature is typical of *Caulobacter* isolates; exceptions occur only among isolates that are atypical in other respects as well. Attachment of the S-layer to the cell surface is mediated by calcium and requires an anchoring oligosaccharide. Both the S-layer protein and the oligosaccharide are antigenic, and serological cross-reaction occurs among the S-layers and their anchors (Walker et al., 1992). An S-layer gene probe prepared from *C. crescentus* detects similar genes in most *Caulobacter* isolates, but not in Group IV caulobacters or random nonprosthecate isolates from wastewater samples that yielded *Caulobacter* spp., suggesting one cultivation-independent means for enumeration of *Caulobacter* spp. in environmental samples (MacRae and Smit, 1991). Although an S-layer may provide some defense against *Bdellovibrio* predation (Koval and Hynes, 1991), the function of this layer is as uncertain in *Caulobacter* species as it is in other bacteria.

A third surface structure of all caulobacters is the holdfast. Again, it is J. Smit and his coworkers who have most thoroughly examined this structure, both in regard to its chemistry (Merker and Smit, 1988; MacRae and Smit, 1991; Abraham et al., 1999) and (in *C. crescentus*) its genetics (Mitchell and Smit, 1990). As noted by Merker and Smit (1988), the function of this organelle is uncomplicated in caulobacters; in nature, it serves to mediate

adhesion of the cell to surfaces. It does not coat the entire cell surface, trap molecules, or join cells in biofilms. In dense laboratory cultures, cells stick to each other in rosettes; while that is helpful in recognition and differentiation of caulobacter isolates by microscopy, it is an artifact of cultivation (Poindexter, 1964). Some *Caulobacter* species also bear pili, which typically arise around the base of the flagellum and disappear as stalk development is initiated. Pili probably also participate in adhesion of cells to surfaces, and are demonstrably able to bind RNA bacteriophages (Schmidt, 1966).

Assays based on the binding of plant lectins by holdfasts, the susceptibility of adhesiveness to competition by monomers, dimers and trimers of *N*-acetylglucosamine, and the stability of adhesion in the presence of various hydrolytic enzymes have consistently implied that the holdfasts of most *Caulobacter*, Group IV caulobacters, and *Maricaulis* (but not *C. leidyi*) isolates consist of or contain some oligo-*N*-acetylglucosamine that participates in adhesion. In contrast to the marine caulobacters, all of which bind wheat-germ agglutinin (WGA) on their holdfasts, some isolates of each of the freshwater genera did not bind WGA; some of those also failed to bind any of six other lectins tested, indicating some diversity of holdfast composition in the freshwater genera (Merker and Smit, 1988).

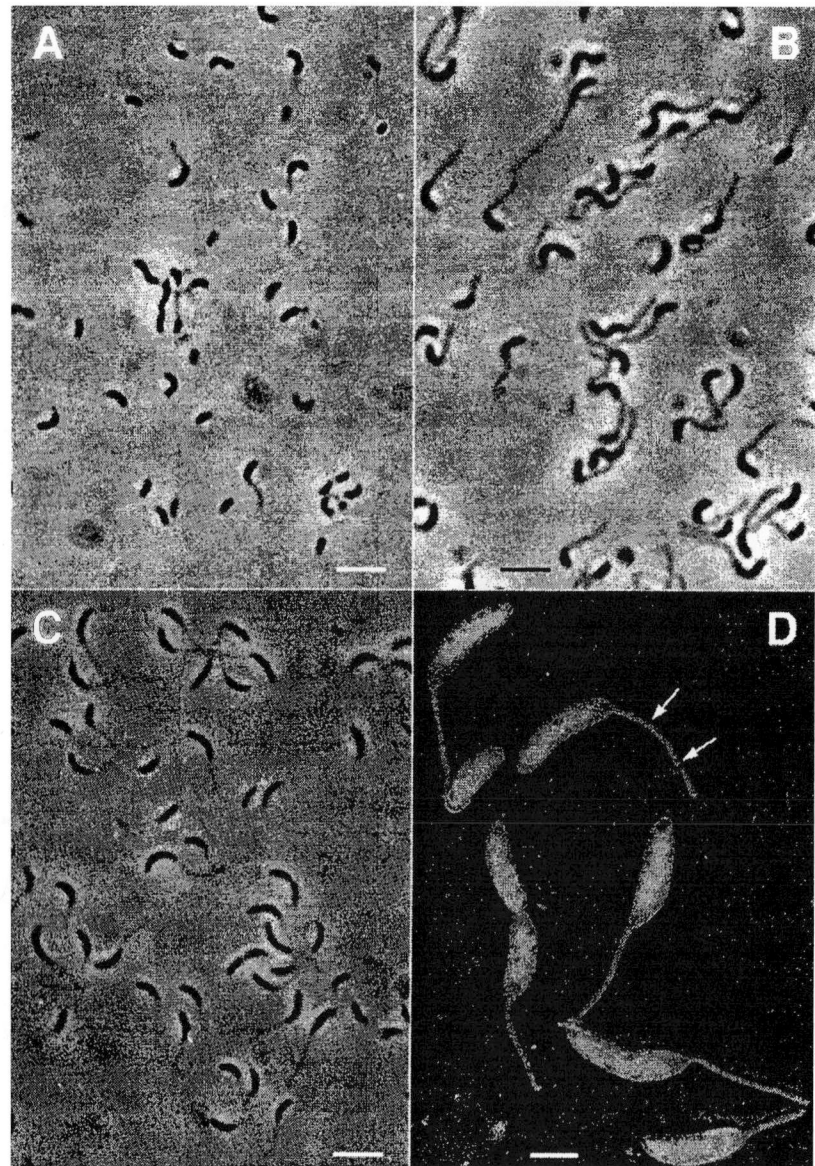

FIGURE BXII.α.115. *C. crescentus* grown in peptone–yeast extract medium (*A*) and in glucose–glutamate minimal medium (*B–D*). *A–B*, strain CB15 (ATCC19089), phase-contrast microscopy of dried smears to which Gray's flagella stain mordant and cover slips were added. *C–D*, strain CB2 (ATCC15252). *C*, phase-contrast microscopy, wet mount. *D*, electron microscopy, shadowed specimen, negative image; arrows indicate stalk bands. Bars = 5 μm (*A–C*) and 1 μm (*D*).

Cellular composition All isolates examined possess a thin peptidoglycan sacculus that extends continuously through the cell pole into the stalk. The peptidoglycan is composed of essentially the same directly cross-linked, diaminopimelic acid-containing polymer that is typical of Gram-negative bacteria. However, while whole-cell peptidoglycan exhibits a 50% excess of muramic acid over glucosamine (Poindexter and Hagenzieker, 1982), the relative proportions of the two sugars appear to be reversed in the stalk peptidoglycan, which has been assayed separately only for *C. crescentus*. This implies that stalk outgrowth involves a modification of the peptidoglycan or of its synthesis. The composition of the peptide side chains of the peptidoglycan may vary with medium composition; glycine is a component in some media (Markiewicz et al., 1983), but it is not invariably

detected in the peptidoglycan. Major membrane proteins unique to the stalk have not been discerned.

The principal phospholipid in *Caulobacter* and group IV caulobacter cells is phosphatidylglycerol, and the most abundant fatty acid species are $C_{18:1}$ and $C_{16:0}$. All isolates tested also contain lower, but significant amounts of $C_{18:1\ \omega7c\ 11CH_3}$ and of species of $C_{17:1}$, $C_{16:1}$, $C_{17:0}$, $C_{15:0}$, $C_{14:0}$. *Caulobacter* isolates contain $C_{12:1\ 3OH}$ and an irresolvable mixture of C_{12} fatty acids designated ECL 11.798 by Abraham et al. (1999). A single type of phosphoglucolipid (PGL), 3-O-[6'-(*sn*-glycero-3'-phosphoryl)-α-D-glycopyranosyl]-*sn*-glycerol, is identifiable in this genus. This is lipid pattern C of Abraham et al. (1997), found in 34 of 37 *Caulobacter* isolates, but in only one of 19 Group IV caulobacter isolates tested. The three exceptional *Caulobacter* strains contained PGLs

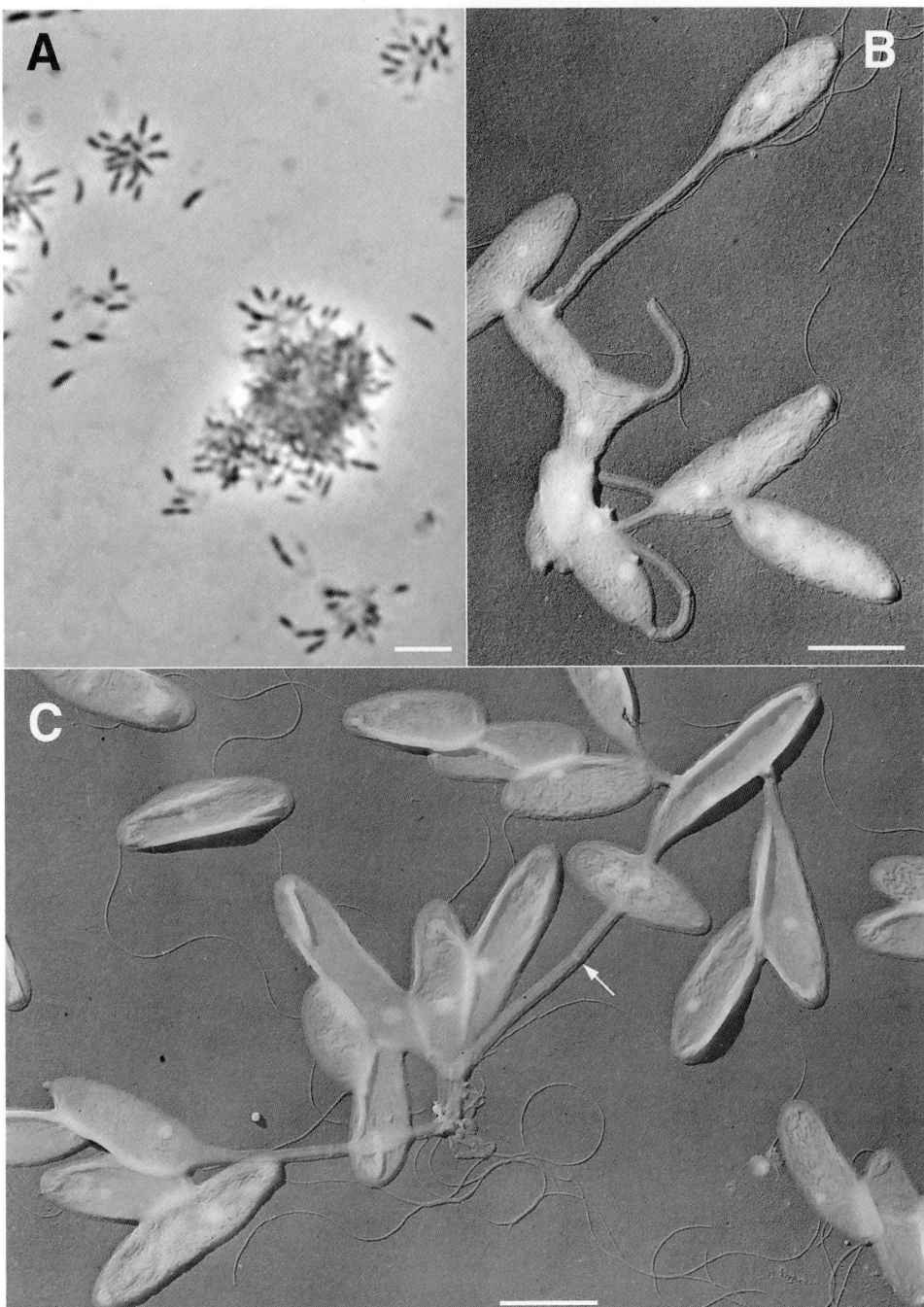

FIGURE BXII.α.116. *C. fusiformis* strain CB27 (ATCC15257) grown in peptone–yeast extract medium. *A*, phase-contrast microscopy, wet mount. Bar = 5 μm. *B–C*, electron microscopy, shadowed specimen, negative image. *Arrow* in (*C*) indicates a stalk band. Y-shaped cells, like the one visible here, are not common. Bars = 1 μm.

that also occur in some Group IV caulobacter isolates. Typically, Group IV caulobacters are distinguished from *Caulobacter* by the presence of $C_{12:0\ 3OH}$ and the absence of $C_{12:1\ 3OH}$ and ECL 11.798. Eight of 12 Group IV caulobacter isolates analyzed contain the PGL typical of *Caulobacter*, and all 12 also contain at least two other PGL types resolvable by the chromatographic methods used in these analyses. This is lipid pattern B of Abraham et al. (1997), found in 18 of 19 Group IV caulobacter isolates, but in only two of 37 *Caulobacter* isolates tested. Five of 12 Group IV caulobacter isolates also contain sulfoquinovosyl diacylglycerols,

one of which was identified as 1-O(α-6′,6-deoxy-aldohexopyra-nosyl-6'-sulfonic acid)-3-O-diacylglycerol. Such compounds are uncommon in nonphototrophic bacteria, but are typical of *Maricaulis* isolates.

Enzymes   As far as tested, *Caulobacter* and Group IV caulo-bacters employ the Entner–Doudoroff pathway for hexose ca-tabolism. Enzymes for catabolism of some carbohydrates are in-duced by exposure to substrate, but induced levels are not much more than 10-fold higher than noninduced levels (Poindexter,

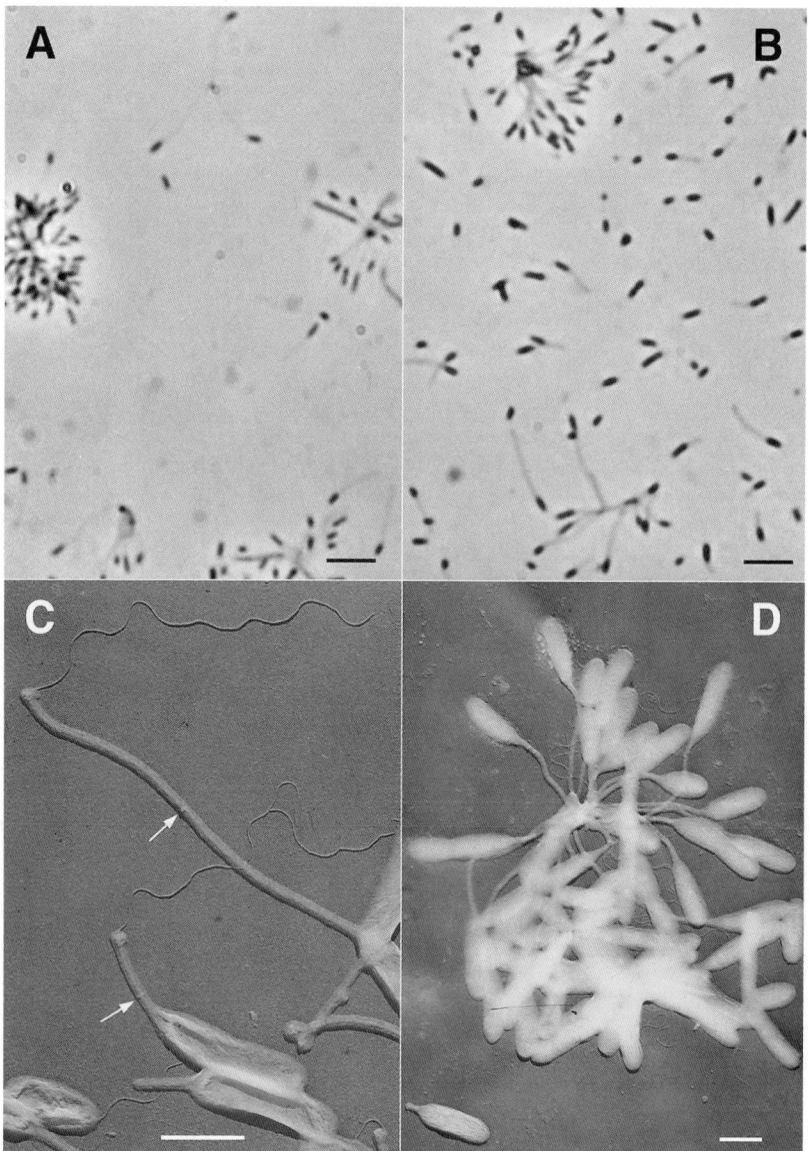

FIGURE BXII.α.117. Group IV caulobacter isolates of the bacteroides cell shape grown in peptone–yeast extract medium. *A–C*, strain 25 (VKM v-1183); *D*, strain CB11a (ATCC19090). *A*, phase-contrast microscopy, wet mount. *B*, phase-contrast microscopy of dried smear to which Gray's flagella stain mordant and cover slip were added. *C–D*, electron microscopy, shadowed specimens, negative image. *Arrows* in *C* indicate stalk bands. *A–B*, Bars = 5 μm; *C–D*, Bars = 1 μm.

1981). Ammonium assimilation has been examined only in *C. crescentus*, which lacks glutamic dehydrogenase and employs the glutamine synthetase–glutamate synthase system only (Ely et al., 1978). This may in part account for susceptibility of growth to excessive (relative to glutamate) availability of ammonium (Felzenberg et al., 1996). *C. crescentus* cells have more than one enzymatic defense against toxic oxygen species. Two superoxide dismutases (SOD) are made, a cytoplasmic Fe-SOD and a Cu-Zn-SOD that is localized in the periplasm (Steinman and Ely, 1990); the ability of the periplasmic enzyme to protect cells against extracellular superoxide has been demonstrated (Schnell and Steinmann, 1995). *C. crescentus* also produces a catalase/peroxidase whose activity increases significantly in the stationary phase of laboratory cultures (Schnell and Steinmann, 1995). The pres-

ence, location, and inducibility of these enzymes may contribute to two characteristics of caulobacter ecology: their tolerance of close association with O_2 generators such as algae, and their slow deterioration during periods of nutrient deprivation.

Genome structure The complete genome sequence of *Caulobacter crescentus* strain CB15 has been determined (Nierman et al., 2001) and is available through GenBank as accession number AE005673. *C. crescentus* was the first free-living alphaproteobacterium to be analyzed this thoroughly. In the report an interpretation of the genome structure and probable roles of the 3767 genes recognized is presented. The report by Nierman et al., (2001), in particular, gives considerable attention to the characteristics of this genome that are consistent with the oligotrophic, aquatic existence of the organism.

DNA–DNA hybridization has not been used extensively with caulobacters since the study by Moore et al. (1978), where only *C. vibrioides* and *C. crescentus* showed hybridization values greater than 58%, and some pairs of strains were more than 90% homologous. Analysis of restriction endonuclease digests of the 16S–23S rDNA interspacer regions (ISR) of *C. vibrioides* and *C. crescentus* strains and of other *Caulobacter* isolates revealed two patterns, but neither was indicative of either species (Abraham et al., 1999). More such analyses—both hybridization and ISR analysis—would be helpful in determining the extent of genomic overlap between isolates that grow vigorously, rapidly, and independently of riboflavin (*C. crescentus*) and those that grow at lower rates, with lower yields, and require both riboflavin and other unidentified organic micronutrients (*C. vibrioides*).

Plasmids, often quite large (>50 kb), have been detected in some *Caulobacter* and Group IV caulobacter isolates (Gregory and Staley, 1982; Schoenlein and Ely, 1983; MacRae and Smit, 1991). The identities of genes on the native plasmids have not yet been determined. However, expression of several foreign plasmid-borne genes has been demonstrated in *C. crescentus*. These include genes for mosquitocidal toxins from Gram-positive bacteria (Thanabalu et al., 1992), and a phosphate-regulated porin gene from *Pseudomonas aeruginosa* (Walker et al., 1991). Ribosomal rDNA sequences have been determined for a large number of caulobacter isolates. *Caulobacter* sequences constitute one of five coherent groups among caulobacterial sequences. The minimum similarity among sequences grouped within "III. *Caulobacter*" in Fig. BXII.α.111 is 94.9%. All known sequences for 16S rDNA are less similar except sequences for organisms identified as *Mycoplana segnis* (95.9–99.9%) and *Phenylobacterium immobile* (95.5–96.4%). See "Taxonomic Comments," below, for further discussion of 16S rDNA sequences and of bacteriophages.

Antibiotic susceptibilities *Caulobacter* and Group IV caulobacter isolates obtained from natural fresh water and from drinking water have usually exhibited susceptibility to both cell wall- and protein synthesis-inhibiting antibiotics (Poindexter, 1964; Nikitin et al., 1990; MacRae and Smit, 1991), but resistance has been encountered in the majority of isolates from wastewater treatment systems. Some of those isolates appear resistant to antibiotics of various chemical types, targets, and mechanisms of antibacterial action, implying that sewage treatment facilities—like hospitals—may provide environments that encourage the propagation, and possibly the transfer, of genes for antibiotic resistance (MacRae and Smit, 1991).

Enrichment and Isolation Procedures

Because of their unique morphology, caulobacter cells can be recognized in natural materials by phase-contrast and electron microscopy. Generally, they are most readily observed in water samples with low soluble organic nutrient content and among the bacteria attached to the surfaces of algal thalli. They are commonly found among the bacterial contaminants of unialgal cultures and may influence the course of development in the algae (Klaveness, 1982). They are also frequently detected in moist soils, either attached to diatoms or among members of rhizosphere communities, and in activated sludge. In samples from such environments, particularly aqueous ones, caulobacters may account for more than 40 of every 100 viable aerobic, oligotrophic, chemoheterotrophic bacteria. When they occur in such relative abundance, caulobacters can be isolated by direct plating, and their colonies identified and selected for purification as described below for colonies from enrichment cultures.

The lack of identification of any physiologic peculiarity within this group has prevented the development of a strongly selective enrichment procedure. Enrichment depends on a mechanical process: the accumulation of stalked cells in the air–water interface of a water sample or soil extract that has been protected during incubation and examination against turbulence, which would disturb the air–water interface. The most dependable, although tedious, procedure for enrichment and isolation of caulobacters has not changed since it was introduced by Houwink (1955). A sample of water or a water extract of soil is enriched with at most 0.01% (w/v) peptone and allowed to stand at room temperature. Within a few days, a sample of the surface film is taken with a bacteriological loop or a cover slip and is examined by phase-contrast microscopy for the presence of stalked cells. Some cells may possess a stalk at each pole and so be indistinguishable from *Prosthecobacter* (*q.v.*) until clones have been isolated and examined for the motility characteristic of caulobacters. When caulobacter cells account for 10–50% of the population, a sample (again, taken with a loop, not a pipette) is streaked on a dilute (0.05%, maximum 0.1%, w/v) peptone or peptone–yeast extract medium prepared with 1.0% or 1.5% agar. By the third or fourth day of incubation, small, hyaline or crystalline, noniridescent colonies of caulobacters begin to appear. It is helpful to remove these colonies while they are very small, by using sterile toothpicks, to small sites ("patches") on secondary plates, for two reasons. First, samples of oligotrophic populations often include bacteria that swarm as a continuous film on the agar surface and invade other colonies. Second, preparation of a wet mount to screen the initial, small colony may consume the entire colony. It is preferable to begin microscopic screening only after the patches of growth on the secondary plate have developed.

Identification requires phase-contrast microscopy because of the small diameter of the stalk (<0.2 μm), which is below the resolution afforded by ordinary light microscopy. The addition of a stain to the wet mount or the use of a mordanted stain such as is designed for flagella may be helpful. The mordant alone applied to dried smears clearly reveals stalks by phase-contrast microscopy. However, examination of wet mounts of living cells is necessary because motility is characteristic of caulobacters, and its detection aids particularly in distinguishing caulobacter and Group IV caulobacters from *Prosthecobacter*. In the dense populations developed on an agar surface, caulobacter cells will adhere to each other's holdfasts, and the characteristic rosettes with cells peripheral to the stalks, which are united in a common holdfast, are often the most dependable way to detect the presence of stalks. In enrichment samples and in pure populations, stalked cells become trapped within air bubbles in wet mounts, and stalks not clearly discernible in the suspended population are more obvious in such regions.

Typical caulobacters multiply less rapidly than do bacteria such as *Pseudomonas* and *Flavobacterium* species that are often present in the same samples. It has been suggested that serial dilutions of the sample might allow the caulobacters to develop without becoming outnumbered by such other types. However, this procedure is suitable only when the caulobacters are initially predominant and are not attached to other cells, two conditions not commonly met. Particularly in dilute media, the bacteria that develop first exhaust the nutrients and preclude multiplication of slower organisms, even if those were more numerous in the initial sample (Belyaev, 1967; Poindexter, unpublished).

It is also possible to enrich a water sample *in situ* by "baiting" the caulobacters with cover slips, to which they will attach. The submerged cover slip can be removed, rinsed well with sterile water, then used as inoculum for a stationary culture as described above. Caulobacter attachment is tenacious; the cells cannot be efficiently scraped from the cover slip but should be allowed to release swarmers by cell division. Some success with thin plastic foil carefully laid on the surface of quiet water has also been reported.

In any of these procedures, the most helpful factor important to successful isolation is that the sample be low in soluble organic nutrient content. Caulobacters are not necessarily absent from richer samples, but their proportion is typically low, and their reproductive rate is not competitive in nutrient-rich cultures. Nevertheless, they can be isolated from soil (e.g., Poindexter, 1964; Belyaev, 1969) and from wastewater treatment facilities, particularly from secondary-stage activated sludge (MacRae and Smit, 1991).

MAINTENANCE PROCEDURES

Once isolated, *Caulobacter* species can be dependably maintained in at least four ways. First, vegetative stocks should be maintained on 1% agar slants of dilute (0.05–0.3% organic material) complex medium or defined medium (if possible, with a given isolate) that contains at least 100 mg organic carbon/mg phosphate-phosphorus and 1 mM each of $MgSO_4$ and $CaCl_2$. Such a medium promotes the storage of carbon reserves and prolongs the viability of the population during storage. Isolates should be transferred every 8 or 9 weeks, incubated 2 or 3 days at 20–25°C (higher incubation temperatures reduce stability during storage), then refrigerated. Second, cells grown in dilute complex medium can be stored in small volumes frozen at −70°C without cryoprotectant. Survival varies among isolates, ranging from 10 to 50%. Such frozen cultures can be thawed at room temperature and transferred to growth medium to resume vegetative cultivation. This is the most dependable method of maintenance of freshwater and soil caulobacters. Third, washed cells diluted to 10^3–10^4 cells/ml of sterile water and sealed in ampules can be stored at room temperature for 2–3 years. This procedure is more dependable than lyophilization but requires periodic recultivation. The fourth method, lyophilization, is the least dependable means of maintenance. If a protectant such as milk solids is used, the rehydrated specimen should be diluted immediately in order to avoid the inhibitory effect of the solids. Lyophilization on strips of sterile filter paper, without any additive, has proven to be the most dependable means of lyophilization.

PROCEDURES FOR TESTING SPECIAL CHARACTERS

All caulobacter species groups are distinguished by the morphology of the dividing cell—specifically, by the asymmetry of its appendages. Both appendages are best characterized by electron microscopy. Shadowed specimens provide the clearest image, but negatively stained specimens need not be washed and so preserve a higher proportion of attached flagella. The holdfast is best detected by its adhesive function, although it, too, is discernible by electron microscopy.

DIFFERENTIATION OF THE GENUS *CAULOBACTER* FROM OTHER GENERA

Three groups of bacteria other than caulobacters produce long cellular appendages of the envelope that are of constant diameter. These bacteria may occur in habitats with caulobacters, and a motile stage is produced by one type (Poindexter, 1992). Nev-

ertheless, they are distinguishable from caulobacters, and from each other, by the morphology of their reproductive stages, as shown in Table BXII.α.101. Relationships to non-prosthecate bacteria implied by 16S rDNA analysis are illustrated and discussed under "Taxonomic Comments," below. The five species groups of caulobacters are distinguished as shown in Table BXII.α.102.

TAXONOMIC COMMENTS

The genus *Caulobacter* was created by Henrici and Johnson (1935b) on the basis of their observations of bacterial populations attached to slides submerged in a freshwater lake. They did not obtain isolates or study pure cultures. Nevertheless, they correctly inferred the developmental sequence of the life cycle and recognized that this type of bacterium was unique. The laboratories of several investigators, notably Shapiro, Newton, Brün, Mullins, Gomes, Marczynski, Gober, and Ely, continue to explore that life cycle and the intricate regulatory genetic system that guides it (Gober and Marques, 1995; Marczynski and Shapiro, 1995; Wu and Newton, 1997). Those genetic and developmental studies have exploited the ability of the species *C. crescentus* to grow in unsupplemented, chemically defined media and thereby to present itself as the caulobacter most suitable for detectable expression of genetic change. *C. crescentus* has been found to be amenable to various means of genetic manipulation and analysis and is now one of the most thoroughly characterized bacterial genomic systems. Focus on that species has yielded a vast amount of information with considerable significance for prokaryotic developmental biology and molecular genetics. The major taxonomic import of those studies is the finding that the type of asymmetric cell division unique to the caulobacters reflects a complex of genes, not a single gene or operon or cassette, whose expression has been interwoven into the information that guides reproduction and on which viability depends. Whatever the non-prosthecate ancestry of this group, evolution of the caulobacter type of life cycle has clearly involved steps that led to a nearly irreversible commitment to it. The rarest morphogenetic mutant of *C. crescentus* is a stalkless but otherwise unaltered clone, although loss of motility or of adhesiveness can occur through mutations that do not affect other aspects of morphogenesis or viability. Nature may yet provide a recognizable stalkless caulobacter, but at first isolation, it is certain to be assigned to a non-prosthecate taxon. Meanwhile, nature does provide a diversity of bacterial types that exhibit the life cycle inferred by Henrici and Johnson. That diversity is the focus of this and two other chapters (*Asticcacaulis* and *Maricaulis*) in this *Manual*. These comments are related to all three chapters and to all four species groups. (*C. leidyi* is not included in this edition of the *Manual*.)

The initial system for classification of species of caulobacters (Poindexter, 1964; Poindexter and Lewis, 1966), continued in previous editions of this *Manual* (Poindexter, 1974, 1989), employed cell morphology as the principal subdividing criterion, in part because of the limited availability of other significant taxonomic criteria. More compelling, however, was the recognition that cell form under standard conditions is a complex and highly conserved feature of unicellular organisms; a feature such as cell pole contour is intimately associated with the reproductive mechanism, and mutations that affect cell shape can be lethal when expressed and often can be studied only in conditional mutants.

More than a decade of chemotaxonomic, ultrastructural, and genomic studies of the caulobacters has yielded taxonomically useful information that can now be added to morphological and

TABLE BXII.α.101. Differential morphology among genera that produce cylindrical prosthecae[a]

Characteristic	Caulobacters[b]	Prosthecobacter	Ancalomicrobium	Budding hyphal bacteria[c]
Reproduction:				
Asymmetric binary fission	+			
Symmetric binary fission		+		
Budding from cell surface			+	
Budding from hyphal tip				+
Prosthecae per cell	1 (or 2[d])	1	3–4	1
Holdfast	+	+	−	+
Flagellum	1	0	0	1

[a]For symbols see standard definitions.

[b]*Caulobacter, Asticcacaulis, Maricaulis C. leidyi,* and Group IV caulobacters.

[c]*Hyphomonas, Hyphomicrobium, Pedomicrobium,* and *Filomicrobium.*

[d]One species of *Asticcacaulis* only.

TABLE BXII.α.102. Differential characteristics among five species groups of caulobacters[a]

Characteristic	C. leidyi	Caulobacter	Group IV caulobacters	Asticcacaulis	Maricaulis
16S rRNA pattern[b]	I	III	IV	II	V
Morphology:					
Cells poles tapered	−	+	d	−	−
Cells distinctly curved	−	+	d	−	−
Stalk bands	−	+	+	+	−
S-layer	nd	+	−	nd	−
Holdfast location on:					
Stalk tip	+	+	+		+
Cell pole				+	
Cellular pigments:[c]					
None	+	+	+	+	+
Yellow	+	+	+	−	−
Gold, orange, red	−	−	+	−	−
Cellular lipid pattern[d]	nd	C	B	A	M
WGA-binding holdfast	−	+	+	+	+
Vitamins known to stimulate growth of some isolates:					
Riboflavin	−	+	−	−	−
Cyanocobalamin	−	+	−	−	−
Biotin	−	−	+	+	−
Growth in media containing:					
NaCl, 0% (w/v)	+	+	+	+	−
NaCl, 2% (w/v)	+	−	+	d	d
Sea salts, 1×	nd	−	d	−	+

[a]Symbols: +, 90% or more of strains are positive; −, 90% or more of strains are negative; d, 11–89% of strains are positive; nd, not determined.

[b]See Figs. BXII.α.111, BXII.α.114, and BXII.α.118.

[c]+, some isolates positive; −, no known isolates positive.

[d]See generic descriptions of *Caulobacter* for lipid patterns B and C, of *Maricaulis* for lipid pattern M, and of *Asticcacaulis* for lipid pattern A.

nutritional traits in the continuing effort to recognize relationships among caulobacters and propose a phylogenetic classification.

As of 1 January 2001, 153 16S rDNA sequences of at least 1200 nucleotides in length, representing 71 caulobacter isolates and 39 isolates of hyphal budding bacteria, had been deposited with data banks. All 153 sequences, plus 24 sequences for other bacteria, were used to build a series of trees to guide this reconsideration of caulobacter classification. Neighbor-joining, maximum likelihood, and parsimony analyses repeatedly generated four major groups, which were further divisible into seven subgroups—five of caulobacters and two of hyphal budding bacteria. The tree shown in Fig. BXII.α.118 exemplifies the core topology of those trees. The tree displays the identity and location of 58 sequences, as follows: 1) all 37 sequences available for type strains of species and subspecies of caulobacters and hyphal budding bacteria that have been named or proposed; 2) 20 for strains for which sufficient phenotypic characterization is available to regard them as dimorphic prosthecate bacteria and as representative of other unnamed isolates with similar descriptions and similar sequences; 3) a sequence for *Spirillum volutans*, a genus of *Alphaproteobacteria* that occurs in caulobacter-laden waters. A neighbor-joining tree built using all 177 sequences is shown as groups of sequences in Fig. BXII.α.111. A DeSoete maximum-likelihood tree built using 98 sequences (trimmed here to 81: 57 nonredundant sequences for caulobacters and hyphal-budding bacteria, the 13 most similar sequences that could be found among nonprosthecate bacteria, 10 similar sequences for organisms of unknown phenotype, and the *S. volutans* sequence) is shown as groups of sequences in Fig. BXII.α.114. The relevant branches of this tree are expanded to display strain identity and sequence accession numbers for *Caulobacter* (Fig. BXII.α.112), Group IV caulobacters (Fig. BXII.α.113), *Asticcacaulis* (Fig. BXII.α.119), and *Maricaulis* (Fig. BXII.α.76 in the *Maricaulis* chapter), along with the positions of other similar sequences.

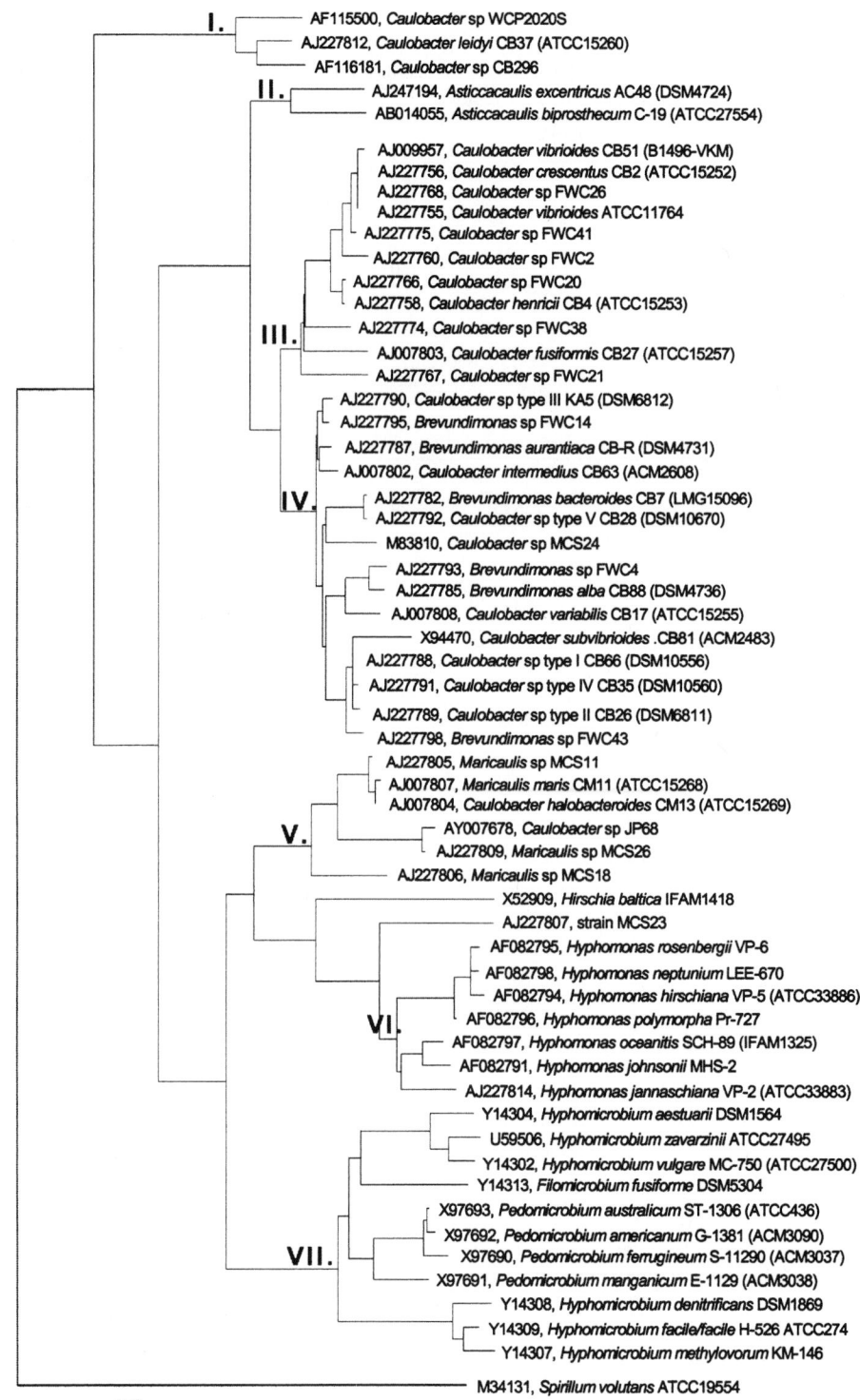

FIGURE BXII.α.118. Phylogenetic tree of 58 16S rDNA sequences of caulobacters (38), nonphototrophic hyphal-budding bacteria (19), and *Spirillum volutans* (1) built using maximum likelihood (PAUP 4.0) analysis. The correlations between 16S rDNA pattern and genus are as follows. I. *Sphingomonas* group (*C. leidyi*), II. *Asticcacaulis*, III. *Caulobacter*, IV. Group IV caulobacters, V. *Maricaulis*, VI. *Hyphomonas*, VII. *Hyphomicrobium*. Bar indicates evolutionary distance.

The four major groups of caulobacters and hyphal budding bacteria reflect the natural distribution of these organisms: a pair of genera (*Maricaulis* and *Hyphomonas*) characteristically isolated from marine habitats; three species groups (*Caulobacter*, *Asticca-*caulis, and the group IV caulobacters) encountered principally in freshwater environments; a diverse group (which includes *C. leidyi*) isolated from soil, arthropod guts, desert soils, mine tailings, and fresh water that exhibit exceptional tolerance toward

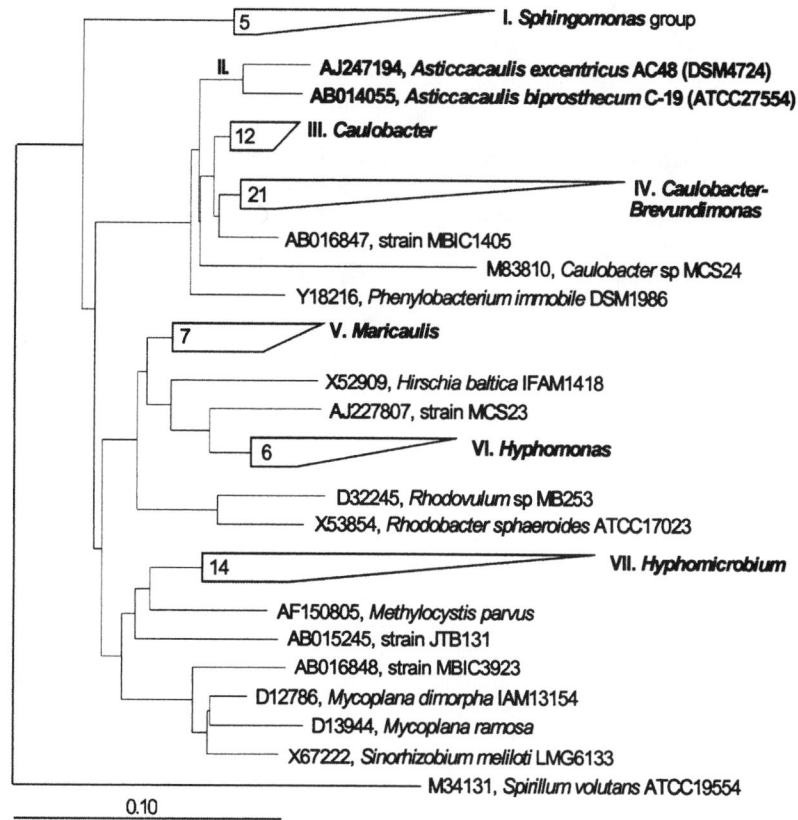

FIGURE BXII.α.119. Phylogenetic tree of 81 16S rDNA sequences, with group II. *Asticcacaulis* to display GenBank accession numbers and strain identities (see Fig. BXII.α.114 in the *Caulobacter* chapter). Also represented are the positions of sequences AB016610 and AF115499 (by AJ247194), and AF115501 and AJ247193 (by AB014055). Bar indicates evolutionary distance. Each enclosed numeral indicates the number of sequences grouped.

changes in environmental conditions; and oligocarbophilic, CO_2-dependent methylotrophs (*Hyphomicrobium*) that are widely distributed in soil, water, and sewage systems. Natural distribution and 16S rDNA sequence similarities correlate with differences in morphology, life cycle, certain nutritional and physiological properties, and details of ultrastructure and cell composition. Accordingly, they seem to constitute well-defined taxa, and the seven subgroups should be genera.

BLAST searches among GenBank sequences revealed different relationships outside the seven groups. Sequences like those of *C. leidyi* are more similar to sequences of *Sphingomonas* species than to those of any of the other caulobacters, whereas the sequences of the two species of *Asticcacaulis* appear to be meaningfully similar only to sequences of *Caulobacter* and Group IV caulobacters. The sequence of one species of Gram-negative branching bacteria, *Mycoplana segnis*, is similar to some *Caulobacter* sequences. At least one strain of *M. segnis* contains cellular lipids similar to those found in *Caulobacter* isolates, but this organism is readily distinguished from caulobacters by its morphology, its lack of adhesiveness, its reproduction by multiple fragmentation, its peritrichous flagella, and much of its physiology. The sequences of other *Mycoplana* spp. group near other prosthecate bacteria: *M. bullata* near Group IV caulobacters and *M. dimorpha* and *M. ramosa* near *Hyphomicrobium*. Whether these sequence similarities imply an evolutionary relationship between branching and prostheca development cannot be assessed at present.

Some Group IV caulobacter sequences are noticeably similar to sequences of the two nonprosthecate species of *Brevundimonas*

(*B. diminuta* and *B. vesicularis*); *B. vesicularis*, in particular, also shares similarities of lipid composition with Group IV caulobacters. As suggested by Abraham et al. (1999), it is conceivable that the *Brevundimonas* species are of prosthecate ancestry, but are now "locked in the motile phase" (Abraham et al., 1999, p. 1070). However, general phenotypic similarity is low, and without the prosthecate species, *Brevundimonas* comfortably accommodates nonstalked, small rods that bear short-wavelength, polar flagella, divide symmetrically, and grow abundantly on rich media designed for the isolation and cultivation of clinically encountered microorganisms.

The greater similarity of sequences of the two marine genera to each other than to any of the nonmarine groups strongly suggests an independent evolutionary origin of the appendages in the marine organisms. The absence of bands from the stalks of marine caulobacters (Anast and Smit, 1988), as from all budding hyphae, supports the inference that the stalks of marine and freshwater caulobacters are less likely to be homologous structures than are the stalks and hyphae of *Maricaulis* and *Hyphomonas*, respectively. Unlike freshwater caulobacters, *Maricaulis* and *Hyphomonas* isolates share detectable 16S rDNA similarity with some purple photoheterotrophic bacteria (*Rhodopseudomonas*, *Rhodovulum*) and with members of the *Rhizobium* group. The sequences of nonmarine hyphal budding bacteria (the *Hyphomicrobium* group) are not significantly similar to any caulobacter sequences or to those of their marine morphological twins; they are more similar to sequences of nonhyphal methylotrophic bacteria (*Methylosinus* and *Methylocystis*). The 16S rDNA sequence similarities between the two types of hyphal budding bacteria are of the same degree as similarities between each group and rhizobial sequences.

The 16S rDNA sequences imply that isolates previously assigned to *Caulobacter* represent three independent origins of the dimorphic life cycle characterized by asymmetric binary fission that produces a stalked cell and a monoflagellate swarmer cell—informally, "caulobacters." To accommodate this implication, it is appropriate to redistribute former *Caulobacter* species and isolates among three groups. (1) *Maricaulis* (Abraham et al. 1999) has been proposed to accommodate the marine caulobacters. (2) A new genus will be proposed to accommodate *C. leidyi* and other nonmarine caulobacters that appear to be more closely related to *Sphingomonas* than to any of the other caulobacters (Kurtz and Poindexter, unpublished). (3) The remaining caulobacter species and unassigned isolates appear to comprise a phylogenetically coherent group that also includes *Asticcacaulis*. On the bases of relative similarities of 16S rDNA sequences and of phenotypic characteristics as shown in Table BXII.α.102, these non-marine caulobacters can be appropriately accommodated in three species groups: *Caulobacter* (rDNA group III) for classical vibrioid, colorless or lightly pigmented caulobacters as described (without isolates) by Henrici and Johnson (1935b), as well as one fusiform type; *Asticcacaulis* (rDNA group II), which continues to accommodate colorless, rod-shaped caulobacters with stalks that are banded (as in the other two species groups), but do not bear the cell's holdfast; and a new genus that will be proposed (Kurtz and Poindexter, unpublished) to accommodate caulobacters of varied but rarely classically vibrioid cell shape that are often deeply pigmented and seem adapted to various levels of environmental salinity (rDNA Group IV).

This redistribution of caulobacters has been guided by investigations of 16S rDNA sequences and chemotaxonomic properties reported since the previous edition of this *Manual*, principally by the laboratories of W.R. Abraham, D.I. Nikitin, L.I. Sly, J. Smit, E. Stackebrandt, and D. Stahl (Anast and Smit, 1988; Merker and Smit, 1988; Stackebrandt et al., 1988a; Nikitin et al., 1990; MacRae and Smit, 1991; Stahl et al., 1992; Walker et al., 1992; Abraham et al., 1997, 1999; Sly et al., 1997, 1999). It is offered as a means of reducing heterogeneity within the genera used for the classification of these prosthecate bacteria and of resolving to some extent the question of the significance of cell morphology and life cycle as taxonomically useful traits. The accumulated work of several laboratories now implies a polyphyletic history of both caulobacter stalks and budding bacterial hyphae. Nevertheless, it is reasonable to propose that the detection of both nonmotile prosthecate and motile (mono-flagellated cells) in a pure population should be regarded as taxonomically helpful in assigning such an isolate to a taxon of phenotypically and genomically similar organisms.

Bacterial classification has become heavily dependent on genomic sequences, particularly of 16S rDNA, as one standard of taxonomic acceptability. Although genes descend and function within a genomic and cytoplasmic context, and one gene does not a genus make, sequences of 16S rRNA genes change very slowly and alert the systematist to spurious similarities of other characteristics that may well be instances of convergent evolution. They also aid in the recognition of phenotypic characteristics that are probably of phylogenetic import, and so help in interpreting the selective pressures that have influenced the course of evolution of a kind of organism. Without doubt, 16S rDNA sequences are sound and valuable guides to probable natural relationships, but they are neither all-powerful (Stackebrandt

and Goebel, 1994) nor—in practice—totally dependable. As this author delved into the sequences available for motile, prosthecate bacteria, several problems were discovered that have already led to confusion and occasional misinterpretation in caulobacter taxonomy. First, it was not at all uncommon to encounter sequences that had been determined for the same isolate (strain), but are currently listed and labeled in sequence banks as though that strain's sequence represents more than one species, even more than one genus! A second impediment to exploitation of this promising approach to microbial taxonomy was the difficulty of distinguishing between high similarity of phylogenetic significance and high similarity of techniques employed with the same strain obtained from different culture collections, labeled according to its shelf position in each collection rather than by the original label, which should accompany and identify the strain wherever it is used for any purpose. The current presence of unrecognizable strain redundancy encourages the discovery of "groups" created by unacknowledged label changes, not by natural selection.

The worst problem was the persistence in data banks of sequences that are mislabeled, probably because the cultures from which the DNA was extracted were either mislabeled or contaminated (see, e.g., comments in Sly et al. (1997), and in Abraham et al. (2001)). When such a sequence is detected, it would benefit every systematist who employs sequences if the label were corrected or the sequence withdrawn.

In preparation for this review of caulobacter classification, an effort was made to identify redundant sequences (those determined for the same strain and essentially identical) and sequences from probably mislabeled cultures. Of the 153 sequences deposited, 43 were regarded as recognizably redundant, and four were suspected to be mislabeled (M83797, recognized by Sly et al., 1997; M83796, recognized by Abraham et al., 1999; AJ007799 and AJ007800, recognized by Abraham et al. (2001)). While the overall approach and ultimate validity of 16S rDNA sequence analyses are undeniably valuable in systematics, there are pitfalls for the unwary attempting to use the expanding collection of data. It would be a great service to bacteriological (and probably all biological) systematics if someone regularly reviewed, and—with depositors' assent—removed or corrected the labels of sequences from misidentified cultures, cross-referenced sequences that purportedly display the same genomic region from the same strain, and otherwise edited, corrected, and updated taxonomic assignations with which sequences are labeled.

Lipid composition has become a taxonomically useful criterion in many bacterial groups, and the caulobacters are not an exception. Among these bacteria, there is a strong correlation between lipid composition and 16S rDNA sequence, providing two strong legs under the table on which to place three new (since the previous edition of this *Manual*) genera to accommodate isolates formerly or tentatively assigned to *Caulobacter*. Studies of caulobacterial lipids prior to the application of automated methods for lipid analysis (Chow and Schmidt, 1974; Carter and Schmidt, 1976; DeSiervo and Homola, 1980; DeSiervo, 1985) found evidence of two features of taxonomic importance: 1) fatty acids were not clearly distinctive among (then) *Caulobacter* species, and 2) marine caulobacters possessed significantly lower levels of phospholipids than did freshwater isolates. Differences in lipid composition were also detected between growing and stationary-phase cells.

More recently, technological developments have provided more detailed analysis of bacterial lipids, and the greatest detail regarding caulobacterial lipids has been reported by the laboratory of W.R. Abraham (Abraham et al. 1997, 1999, 2001). Those studies reported fatty acid and polar lipid composition for isolates of four groups of caulobacters (*Caulobacter*, Group IV caulobacters, *Asticcacaulis*, and *Maricaulis*) and one of budding hyphal bacteria (*Hyphomonas*) in comparative studies that also included isolates of two nonprosthecate genera (*Brevundimonas* and *Mycoplana*). *C. leidyi* was included, but seemed to be characterized mainly by the absence of identifiable polar lipids and the presence of $C_{14:0\ 2OH}$, an uncommon bacterial fatty acid that is, however, common among *Sphingomonas* spp. The two nonprosthecate types share some similarities with caulobacters in both lipid composition and 16S rDNA sequence. The major distinguishing features of the lipids detected are presented in both procedural and chemical detail in the reports from Abraham's group, which should be consulted. A similar composition was reported for one freshwater isolate by Batrakov et al. (1997). A brief summary of these reports is presented here.

Three patterns of lipid composition, designated "C", "B", and "M" by Abraham et al. (1997), appear to distinguish the three groups *Caulobacter*, Group IV caulobacters, and *Maricaulis*, respectively (see Table BXII.α.102), and a fourth pattern, designated "H", is found in *Hyphomonas*. The lipids of *Asticcacaulis* isolates appear to be somewhat more like those of *Caulobacter* than of the other genera. All groups, even *C. leidyi*, share five fatty acids. The greatest differences, as anticipated by DeSiervo's studies, were detected between freshwater and marine caulobacters. The marine groups possess significant amounts of six fatty acids that are absent (five) or very low (the sixth) in freshwater organisms; the reverse was observed for three fatty acids—significant in freshwater, absent or scant in marine isolates.

Glucolipid patterns were also found to be distinctive. One type of phosphoglucolipid (PGL), 3-O-[6'-(*sn*-glycero-3'-phosphoryl)-α-D-glycopyranosyl]-*sn*-glycerols was the only such lipid found in the two *Asticcacaulis* and in 34 of 37 *Caulobacter* isolates tested. The exceptional *Caulobacter* isolates possess that PGL plus at least two others, as do 9 of the 13 Group IV caulobacter isolates and 1 of the 4 *Brevundimonas* strains. PGLs, at least one of which is present in every freshwater isolate, are scant or undetectable in marine isolates. Sulfonolipids were detected in many strains. All seven *Maricaulis* strains, 10 of 19 Group IV caulobacter strains, and two *Caulobacter* strains possess sulfoquinovosyl diacyglycerols (SQ), as do two of the four *Brevundimonas* strains and two marine isolates tentatively assigned to *Brevundimonas*. Sulfur-containing lipids were not found in *Asticcacaulis* isolates (Abraham et al., 2001). The *Hyphomonas* isolates lack SQ, but contain taurineamide diacylglycerides, found also in four *Maricaulis*, one *Caulobacter*, and one *Brevundimonas* strain. SQ are uncommon among nonphototrophic organisms, and it is remarkable and probably taxonomically useful that they occur in both freshwater and marine caulobacters.

Seven kinds of bacteriophages have been isolated for caulobacters, two of which are not commonly encountered among noncaulobacters (Schmidt and Stanier, 1965; reviewed in Poindexter, 1981). The first of these is the most frequently reported type of caulophage; it is produced by strains of *Caulobacter*, Group IV caulobacters, and *Asticcacaulis*. The virions have large (50–65 × 170–260 nm), prolate, cylindrical heads and long (200–320 nm), flexible, noncontractile tails, and genomes of 2S DNA; they appear most like *Siphoviridae* among formal phage families. These phages exhibit wide host ranges among caulobacters; an individual phage may infect not only more than one species, but also more than one of the three freshwater species groups. The second, uncommon type consists of small icosahedral RNA phages (family *Leviridae*) produced by both *Caulobacter* and Group IV caulobacter isolates. Unlike the DNA-containing caulophages, each of the RNA phages is host species-specific, for *C. crescentus*, *C. fusiformis*, or *Brevundimonas bacteroides* (a Group IV caulobacter). Phages have not been reported to be produced by *Maricaulis* or *C. leidyi*. None of the caulophages tested is lytic for non-caulobacters, and phages lytic for other genera are not lytic for caulobacters. Like most events in the life of caulobacters, the ability to adsorb and allow infection by some of the phages are traits expressed only intermittently during the cell cycle (Schmidt, 1966; Bendis and Shapiro, 1970; Lagenaur et al., 1977; Bender et al., 1989), usually during the swarmer stage. This is a further indication that asymmetric fission, with its intricate regulatory background and multiplicity of cell surface participants, continues to be the distinctive trait of caulobacters.

Four species of *Caulobacter* are recognized: *C. vibrioides*, *C. crescentus*, *C. henricii*, and *C. fusiformis*. Distinguishing and other traits are shown in Tables BXII.α.103 and BXII.α.104, respectively. The other nine species included in the previous edition of this *Manual* have been removed to other assignments: *C. bacteroides*, *C. henricii* subsp. *aurantiacus*, *C. intermedius*, *C. subvibrioides*, *C subvibrioides* subsp. *albus*, and *C. variabilis* to *Brevundimonas* as six species; *C. halobacteroides* and *C. maris* to *Maricaulis*; and *C. glutinosus*, *C. kusnezovii*, and *C. leidyi* to limbo.

ACKNOWLEDGMENTS

The author is deeply grateful to Dr. Thomas M. Schmidt of the Department of Microbiology and Molecular Genetics and Center for Microbial Ecology, Michigan State University, for his guidance in the analysis of 16S rDNA sequences and preparation of phylogenetic trees. His patience and generosity as a tutor are inexhaustible. The taxonomic conclusions, of course, are the responsibility of the author alone. The analysis and writing were supported in part by a Barnard College Faculty Grant during the author's sabbatical leave at Michigan State University in 2001.

List of species of the genus Caulobacter

1. **Caulobacter vibrioides** Henrici and Johnson 1935b, 84[AL]

    vib.ri.oi' des. M.L. n. *Vibrio* name of a genus; Gr. n. *eidus* form, shape; M.L. adj. *vibrioides* resembling a vibrio.

    Vibrioid cells slender or nearly ovoid. Colonies colorless or pale yellow. Vitamin B_2 essential for growth, but additional unidentified growth factors available in peptone–yeast extract also required by most strains.

    Strain CB51 was isolated from a freshwater pond in Berkeley, CA.

    The mol% G + C of the DNA is: 64–65 (Bd, determined for two isolates).

    Type strain: CB51, DSM 9893, VKM B-1496.

    GenBank accession number (16S rRNA): AJ009957, AJ227754.

    Additional Remarks: It is not clear whether any of the copies of strain CB51 currently available from culture collections is identical with the original isolate (Poindexter, 1964), which was not deposited with the ATCC. CB51 appears to have been or to have very early become overgrown by a strain of *C. crescentus*, resulting in identity of the 16S rDNA sequences that have been determined for CB51 and *C. crescentus* strain CB2, and DNA–DNA hybridization >90% (Moore et al., 1978). Response to riboflavin, the most easily tested distinction between the two species, has not been

TABLE BXII.α.103. Differential characteristics of the species of the genus *Caulobacter*[a]

Characteristic	C. vibrioides	C. crescentus	C. fusiformis	C. henricii
Long axis distinctly curved	+	+	−	+
Colonies:[b]				
Colorless	+	+	−	−
Pale yellow	+	−	−	−
Bright yellow	−	−	−	+
Dark yellow	−	−	+	−
Carbon sources generally used:				
Carbohydrates	+	+	−	+
Amino acids	d	+	+	+
Other organic acids	d	+	−	d
Primary alcohols	−	+	−	−
Organic growth factors required:				
Riboflavin	+	−	−	−
Cyanocobalamin	−	−	−	+
Amino acids	−	−	−	d
Other, unidentified	+	−	+	−
Maximum specific rate of exponential growth >0.35 h^{-1}	−	+	−	−

[a]Symbols: +, 90% or more of strains are positive; −, 90% or more of strains are negative; and d, 11–89% of strains are positive.

[b]+, some isolates positive; −, no known isolates positive.

TABLE BXII.α.104. Other characteristics of the species of the genus *Caulobacter*[a]

Characteristic	C. vibrioides	C. crescentus	C. fusiformis	C. henricii
Carbon-source utilization:[b]				
Arabinose	+	d	−	d
Ribose	+	−	−	d
Xylose	+	+	−	+
Glucose	+	+	−	+
Galactose	+	+	−	+
Mannose	+	+	−	d
Fructose	+	−	−	d
Lactose	+	+	−	d
Maltose	+	+	−	+
Sucrose	+	+	−	+
Alanine	+	+	−	+
Aspartate	+	+	−	+
Glutamate	+	+	+	+
Proline	+	+	−	+
Tyrosine	d	+	−	d
Acetate	−	+	d	+
Butyrate	d	+	−	+
Pimelate	nd	+	−	+
Pyruvate	d	d	−	+
Malate	d	d	−	+
Fumarate	+	d	−	+
Succinate	+	−	−	+
Methanol	−	+	−	−
Ethanol	−	+	−	d
Propanol	−	+	−	−
Butanol	+	+	−	+
Pentanol	−	+	−	−
Starch hydrolysis	+	d	−	+
Sensitivity to bacteriophages:[c]				
Type I:				
Phages 1,3	−	d	−	−
Others	+	+	+	+
Type II	+	+	+	−
Type III	+	+	−	+
RNA phages	+	+	+	−

[a]Symbols: +, 90% or more of strains are positive; −, 90% or more of strains are negative; d, 11–89% of strains are positive; and nd, not determined.

[b]Carbon-source utilizations were determined with D-isomers of sugars, with L-isomers of amino acids, or (when single isomers were not available) with racemic mixtures.

[c]2sDNA phage types: I, head prolate cylinder 50–65 × 170–260 nm, tail noncontractile 200–320 nm long; II, head elongated polyhedron 65–70 × 100–105 nm, tail noncontractile 260–300 nm long; III, head icosahedron 50–80 nm diameter, tail noncontractile 150–200 nm long. ssRNA phages: icosahedron 20–29 nm diameter, no tail.

reported for current copies of "CB51." As suggested by Sly et al. (1997), a search for original CB51 DNA should be made, or a substitute neotype strain matching the phenotype of *C. vibrioides* designated. Krasil'nikov and Belyaev (1973) reported 20 isolates of *C. vibrioides*. Babinchak and Gerencser (1976) reported 28 vibrioid isolates; however, characteristics other than morphology and phage sensitivities have not been described for their isolates. MacRae and Smit (1991) reported the isolation of 25 colorless vibrioid caulobacters from wastewater facilities; although they noted that many of the isolates were distinctly stimulated by supplementation of complex media with riboflavin, they did not indicate which of their isolates responded. Those isolates would be accommodated by *C. vibrioides*, whereas the riboflavin-indifferent isolates should be assigned to *C. crescentus*. Further colorless, vibrioid freshwater isolates were obtained from other sources and used by Merker and Smit (1988) and by Abraham et al. (1997, 1999), but these were not significantly different from each other or the wastewater isolates in lipid composition. Different strains in the collection did, however, bind different plant lectins, a trait that might eventually prove to be an additional aid in distinguishing between the two colorless vibrioid species, *C. vibrioides* and *C. crescentus*.

2. **Caulobacter crescentus** Poindexter 1964, 288[AL]

cres'cen.tus. L. adj. *crescentus* of the moon in its first quarter, crescent.

Vibrioid cells slender (see Fig. BXII.α.115). Colonies colorless, with centers becoming tan or dark pink upon aging. Lack of growth factor requirements is unique among *Caulobacter* species, as is the ability to utilize primary alcohols as sole sources of carbon and energy in defined media; other species may utilize butanol, but not methanol, ethanol, propanol or pentanol, as do *C. crescentus* isolates. Some of the wastewater caulobacters isolated by MacRae and Smit (1991) may be *C. crescentus*, but this species cannot be distinguished from *C. vibrioides* by the traits (lipid composition, 16S rDNA sequences, lectin binding) so far reported for these isolates.

Strain CB2 was isolated from tap water in Berkeley, CA.

The mol% G + C of the DNA is: 62–67 (Bd, determined for one isolate; T_m, determined for three isolates; paper electrophoresis of hydrolysate, determined for one isolate; complete sequence for one isolate).

Type strain: CB2, ATCC 15252.

GenBank accession number (16S rRNA): AJ227756.

3. **Caulobacter fusiformis** Poindexter 1964, 289[AL]

fus.i.form'is. L. n. *fusus* spindle; L. n. *forma* shape, form; M.L. adj. *fusiformis* spindle-shaped.

Fusiform cells slender (see Fig. BXII.α.116). Colonies bright yellow. Sugars not utilized as carbon sources. Organic growth factor requirements not satisfied by mixtures of B vitamins, amino acids, and purine and pyrimidine bases.

Strain CB27 was isolated from a freshwater pond in Berkeley, CA.

The mol% G + C of the DNA is: not determined.

Type strain: CB27, ACM 5108, ATCC 15257, DSM 4728.

GenBank accession number (16S rRNA): AJ227759, AJ007803, AB008533.

Additional Remarks: Krasil'nikov and Belyaev's (1973) strain 25 (VKM v-1183) was assigned by them to this species. However, based on its morphology, pigmentation, and ability to utilize a variety of sugars, it is assignable to Group IV caulobacters. Another fusiform isolate, strain 1 (VKM v-1189), designated *"Caulobacter rossii"* sp. nov. by Krasil'nikov and Belyaev (1973), is not assignable to *C. fusiformis*; because only one such isolate is known and its physiological characteristics are predominantly inabilities and appear variable, *"C. rossii"* is not included in this list. Babinchak and Gerencser (1976) reported five fusiform isolates; however, characteristics other than morphology and phage sensitivities have not been described for their isolates. Yellow fusiform isolates were not obtained from wastewater by MacRae and Smit (1991).

4. **Caulobacter henricii** Poindexter 1964, 288[AL]

hen.ric'i.i. M.L. gen. n. *henricii* of Henrici; named for A.T. Henrici, who observed stalked bacteria on slides that had been submerged in freshwater.

Vibrioid cells slender. Colonies bright yellow and may be somewhat translucent. Vitamin B_{12} typically required as growth factor in peptone–yeast extract media.

Strain CB4 was isolated from a freshwater pond in Berkeley, CA.

The mol% G + C of the DNA is: 62–65 (Bd, determined for two isolates).

Type strain: CB4, ATCC 15253, DSM4730, ACM5105.

GenBank accession number (16S rRNA): AJ227758, AJ007805, AB008532.

Additional Remarks: Krasil'nikov and Belyaev's (1973) strain 44 (VKM v-1190), designated *"Caulobacter rutilus"* sp. nov. is not distinguishable from *C. henricii*. As described here, *C. henricii* accommodates all known *Caulobacter* isolates of yellow-pigmented, riboflavin-indifferent vibrioid types. Only two wastewater isolates (FWC20 and FWC23) reported by MacRae and Smit (1991) were yellow and vibrioid; the 16S rDNA sequence of one (FWC20) has been determined and found to be more similar (99.8%) to the type strain of *C. henricii* than to any other caulobacter sequence.

Genus II. **Asticcacaulis** *Poindexter 1964, 282*[AL]

JEANNE S. POINDEXTER

A'stic.ca.cau'lis. Gr. *alpha* privative without; Anglo-Saxon n. *sticca* stick; L. n. *caulis* stalk; L. masc. n. *Asticcacaulis* stalk that does not stick.

Cells single, unbranched, **rod-shaped**, poles blunt or gently rounded; 0.5–0.7 × 1–3 μm during growth in most media. Some cells in any growing population have one subpolar or one or two lateral **prosthecae**. Each prostheca includes outer membrane, peptidoglycan, cell membrane, and a core sometimes observed to be occupied in part by membranes, but other cytoplasmic components cannot be discerned through most of its length. Beyond the cell–prostheca juncture, prostheca diameter is constant, 0.10–0.15 μm; each prostheca possesses at least one **stalk band** as a substructure after the cell has matured and completed at least one cell cycle. Other cells in the same population bear a **single, subpolar flagellum**. Each type of cell bears a small mass of adhesive material, the **holdfast**, at one pole; the holdfast site is not coincident with the site of the flagellum or of the prostheca(e). Binary fission occurs by **septation**, typically resulting in the production of a longer, prosthecate cell and a shorter, flagellated cell. Fission may occur in cells lacking prosthecae. In both instances, **cell division** is **unequal**.

Gram negative. Strictly **respiratory** and **aerobic** but may be somewhat O_2 sensitive; O_2 serves as the only terminal electron acceptor for growth, although nitrate may be reduced to nitrite. Colonies circular, convex, glistening, with smooth margins, butyrous in texture, and colorless. In standing liquid cultures, cells accumulate as a **surface film** or heavier pellicle and develop as a ring of growth on the vessel wall at or just below the air–liquid interface. Growth in agitated liquid cultures is evenly dispersed.

Chemoorganotrophic and **oligotrophic**; grow readily in media such as peptone–yeast extract containing 0.05–0.3% (w/v) organic solutes, but not in standard nutrient broth with 0.8% (w/v) organic solutes. During growth, may produce acid from sugars but do not produce gas. Optimal temperature for growth: 25–30°C; tolerated range for growth: 15–35°C. Optimal pH near neutrality; pH 6–9 tolerated. Typically do not grow in media containing 2% (w/v) NaCl. Maximum specific rates of exponential growth: 0.23–0.57 h^{-1}. All isolates require biotin as the only organic micronutrient. Glucose, fructose, maltose, or lactose may be utilized as the sole carbon source.

Sequences of 16S rDNA are consistent with placement of *Asticcacaulis* among the *Alphaproteobacteria*. The positions of the six sequences grouped as *Asticcacaulis* in Fig. BXII.α.111 in the chapter describing the genus *Caulobacter* are represented in Fig. BXII.α.119 (in the *Caulobacter* chapter) by the two nonredundant sequences grouped in Fig. BXII.α.114 in *Caulobacter*; their strain identities and GenBank accession numbers are displayed. These sequences were determined for the type strain of each species. (See "Taxonomic Comments" in the chapter describing the genus *Caulobacter* for a description of this tree.)

Sources: isolated from tap water and natural bodies of fresh water. Not encountered as clinical isolates, and not known to be pathogenic for plants or animals. Rarely encountered; only four isolates of *Asticcacaulis excentricus* are known, and only one of *Asticcacaulis biprosthecum* has been described.

The mol% G + C of the DNA is: 55–61.

Type species: **Asticcacaulis excentricus** Poindexter 1964, 292.

FURTHER DESCRIPTIVE INFORMATION

Morphology and fine structure The prosthecae of all isolates are banded (see Figs. BXII.α.120B and BXII.α.121C). Each band is a set of concentric rings that are more electron-opaque and structurally more rigid than is the remainder of the prostheca (Schmidt and Swafford, 1975). Earlier evidence that one band is added to each growing prostheca during each cell cycle of *Caulobacter crescentus* (Staley and Jordan, 1973) has been confirmed for both that species and for *Asticcacaulis biprosthecum* (Poindexter and Staley, 1996). However, the function of the bands is unknown. They occur in the prosthecae of only three species groups: *Caulobacter*, *Asticcacaulis*, and "other caulobacters."

The pattern of prostheca development in *Asticcacaulis* species as it relates to the cell cycle appears to vary among isolates and, in a given isolate, with composition of the growth medium (Pate et al., 1973; Larson and Pate, 1975). In some cultures, development occurs only toward the end of exponential growth; in others, practically all dividing cells are prosthecate in all growth phases of the culture. As far as is known, only one progeny cell of each reproductive event is flagellated, whether or not the other progeny cell bears a prostheca. Prosthecae that arise laterally are usually inherited by the nonmotile progeny cell, but may also be inherited by the motile cell. However, the capacity for development is not lost by a clone arising from a nonmotile, nonprosthecate cell; clones without prosthecae have not been encountered as spontaneous variants. As with other caulobacters, prostheca length is greatly increased in dilute media; prostheca length is also promoted in defined media, especially when phosphate concentration is growth limiting (Schmidt and Stanier, 1966), but also when the available carbon source is only slowly utilized (Larson and Pate, 1975).

Cellular composition The thin peptidoglycan layer of the *Asticcacaulis* cell body continues without interruption into the prostheca. Peptidoglycan of both known species is similar with respect to the glycan component, in which muramic acid is present in a 50% excess over glucosamine (Poindexter and Hagenzieker, 1982). The species, however, differ in peptide composition. *A. biprosthecum* contains a significantly higher proportion of glutamic acid than is found in *A. excentricus* and other Gram-negative bacteria. Prostheca peptidoglycan has not been separately assayed in either species.

The principal respiratory quinone in *Asticcacaulis* cells is ubiquinone-10, and the principal phospholipid is phosphatidylglycerol. The most abundant fatty acid species are $C_{18:1}$ and $C_{16:0}$. Both isolates tested (one of each species) also contain lower, but significant amounts of $C_{18:1\ \omega7c\ 11CH_3}$ and of species of $C_{17:1}$ and $C_{15:0}$. As in *Caulobacter*, the principal hydroxy fatty acid detected by Abraham et al. (2001) was $C_{12:1\ 3OH}$, although a study of the lipid composition of an isolate described as "*A. biprosthecum*-like" (Sittig and Schlesner, 1993) reported detection of $C_{12:0\ 3OH}$, as well. Neither isolate studied by Abraham et al. (2001) contained the PGLs found in *Caulobacter* and "other caulobacters."

Enzymes Intermediary metabolism has been examined in only a preliminary fashion. The Entner–Doudoroff pathway for carbohydrate dissimilation is present. Enzymes for catabolism of some carbohydrates are induced by exposure to substrate, but induced levels are only 1.5–10-fold higher than uninduced levels.

Genome structure DNA–DNA hybridization values were reported by Moore et al. (1978) to be in the range of 33–88% among isolates of *A. excentricus*, but to be undetectable between the type strains of the two species in this genus. Ribosomal rDNA sequences have been determined for both species of *Asticcacaulis*, which appear to comprise one of five coherent groups among caulobacterial sequences. The minimum similarity between the sequences that represent the two isolates grouped within "II. *Asticcacaulis*" in Fig. BXII.α.111 in the *Caulobacter* chapter is 95.9%. All known sequences for 16S rDNA are less similar; the highest similarities outside the genus are with sequences for organisms identified as *Caulobacter henricii* (93.7%) or *Brevundimonas aurantiaca* (formerly *C. henricii* subsp. *aurantiacus* (94.7%).

Three morphological types of bacteriophages that are lytic for *Asticcacaulis* species have been isolated from fresh water and sewage (Pate et al., 1979; reviewed in Poindexter, 1981). All are two-stranded DNA phages with long, flexible, noncontractile tails. Heads may be icosahedral, prolate cylindrical, or elongated polyhedral. Of more than 40 phage isolates known from *Asticcacaulis*, only two (of the prolate cylindrical type) are lytic for a few *Caulobacter* and "other caulobacter" isolates; all others are lytic only

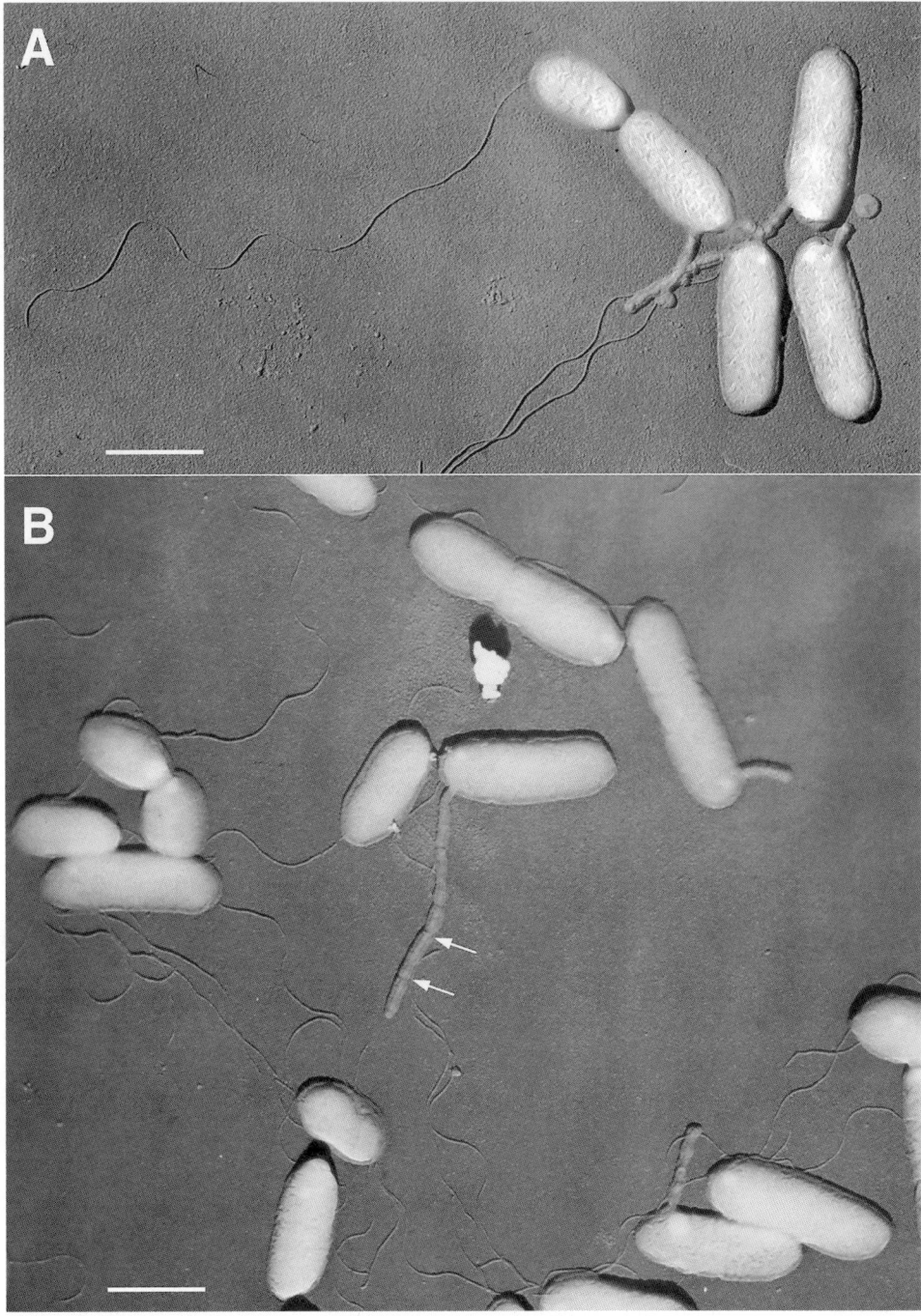

FIGURE BXII.α.120. *A. excentricus* strain S-3 grown in peptone–yeast extract medium. Electron microscopy, shadowed specimen, negative image. *Arrows* in (*B*) indicate stalk bands. Bars = 1 μm.

for *Asticcacaulis* isolates. No phages that are lytic for any other genus (including other caulobacters) are also lytic for *Asticcacaulis* isolates. Generally, this genus appears to be isolated from other genera with respect to phage propagation. Genetic studies have not been done for this genus, but *A. excentricus* has been found to be capable of accepting and expressing plasmid-borne genes, including genes for insect larvicide production (Liu et al., 1996).

See "Taxonomic Comments" below and in the chapter describing the genus *Caulobacter* for further discussion of 16S rDNA sequences and other bacteriophages.

ENRICHMENT AND ISOLATION PROCEDURES

Asticcacaulis is not frequently observed in natural samples and is rarely isolated. Isolation has been achieved by the same procedure as described for *Caulobacter* (*q.v.*). A sample of clean water is enriched with at most 0.01% (w/v) peptone and allowed to stand at room temperature. Within a few days, a sample of the surface film, taken with a bacteriological loop or a cover slip, is examined by phase-contrast microscopy for the presence of prosthecate cells or rosettes of rod-shaped cells. When such cells account for 10–50% of the population, a sample (again, taken with a loop, not a pipette) is streaked on a dilute (0.05%; max-

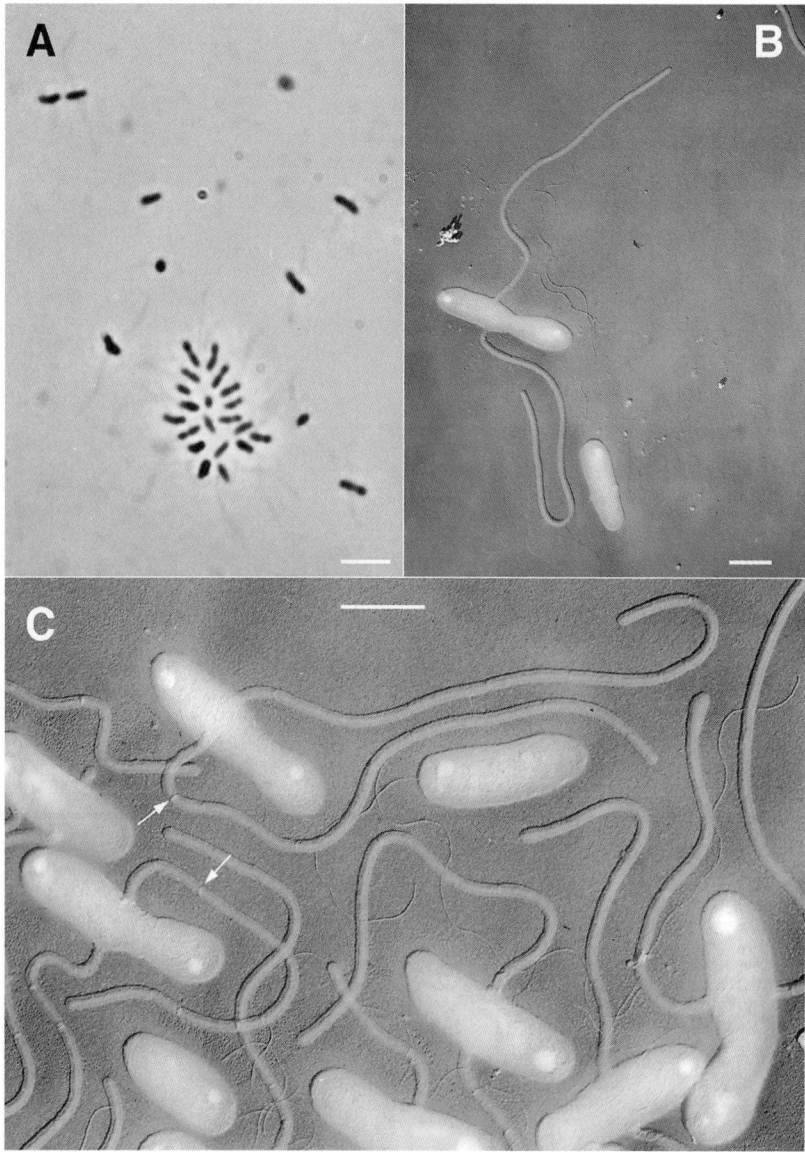

FIGURE BXII.α.121. *A. biprosthecum* strain C-19 grown in peptone–yeast extract medium. *A*, phase-contrast microscopy, wet mount. *B–C*, electron microscopy, shadowed specimen, negative image. *Arrows* in (*C*) indicate stalk bands. Bars = 5 μm (*A*) and 1 μm (*B–C*).

imum 0.1%, w/v) peptone or peptone–yeast extract medium prepared with 1.0 or 1.5% agar. By the third or fourth day of incubation of the plates, small, hyaline or crystalline, noniridescent colonies of prosthecate bacteria begin to appear. These colonies, or patches of growth inoculated from them, are then screened by phase-contrast microscopy. Clones of *Asticcacaulis* are often difficult to recognize during the microscopic screening. However, any clone that includes frequent rosettes should be purified, cultivated in dilute (e.g., 0.1% peptone, 0.05% yeast extract) broth, and examined by electron microscopy for the presence of prosthecate and flagellated cells in whole-cell mounts, such as shadowed or negatively stained specimens.

A. biprosthecum has been isolated only once, by a modification of this procedure in which a loopful of the enrichment culture was streaked on a dilute medium containing 0.02% beef extract, 0.05% tryptone, 0.05% yeast extract, and 0.02% sodium acetate, prepared with 1.5% agar. After several days of incubation at 30°C,

very small colonies were transferred as deep cultures to the same medium prepared with 0.4% agar. *A. biprosthecum* was detected by microscopic examination of the submerged colonies (Pate et al., 1973).

MAINTENANCE PROCEDURES

Once isolated, *Asticcacaulis* species can be maintained as vegetative, frozen, or lyophilized populations. Vegetative stocks should be maintained on 1% agar slants of dilute (0.05–0.3% organic material) complex medium, transferred every 8 or 9 weeks, incubated 2 or 3 days at 20–25°C, and then refrigerated. Cells grown in dilute complex medium can be stored in small volumes frozen at −70°C without cryoprotectant. Such frozen cultures can be thawed at room temperature and transferred to growth medium to resume vegetative growth. Lyophilization is a dependable means for maintenance, either in milk solids or on strips of filter paper.

Procedures for Testing Special Characters

This genus is distinguished by the number, positions, and substructures of its appendages, particularly the prosthecae. The appendages are best examined by electron microscopy. Shadowed specimens provide the clearest image, but negatively stained specimens need not be washed and so preserve a higher proportion of attached flagella. The holdfast is best detected by its adhesive function, although it, too, is discernible by electron microscopy.

Differentiation of the Genus Asticcacaulis from Other Genera

Four groups of bacteria other than caulobacters produce long cellular appendages of the envelope that are of constant diameter; these bacteria also tend to occur in the same habitats as caulobacters. They are distinguishable from caulobacters and from each other by the morphology of their reproductive stages. The five groups are distinguished in Table BXII.α.101 in the *Caulobacter* chapter, and the five genera of caulobacters are distinguished in Table BXII.α.102 in the *Caulobacter* chapter. Relationships to nonprosthecate bacteria implied by 16S rDNA analysis are illustrated and discussed further in "Taxonomic Comments" in the chapter describing the genus *Caulobacter*.

Taxonomic Comments

The genus *Asticcacaulis* was created (Poindexter, 1964) to accommodate isolates originally regarded as *Caulobacter* species, but whose prosthecae were not adhesive and consequently not equivalent to *Caulobacter* stalks as adhesive organelles. Subsequent investigations of a greater variety of prosthecate bacteria have revealed that only caulobacterial appendages lack cytoplasm along most or all of their length and that stalk bands occur only in freshwater caulobacters. The unique structure of the freshwater caulobacter appendages suggests that they are homologous. Although the mol% G + C of *Asticcacaulis* DNA is lower than that known for other freshwater caulobacters, and only two of more than 100 caulophage isolates are lytic for both *Asticcacaulis* and other freshwater caulobacters, 16S rDNA sequence analysis places *Asticcacaulis* well within the caulobacter group (*Caulobacter*,

Asticcacaulis, and "other caulobacters"). Its closest nonprosthecate relative is *Brevundimonas*, whose sequences are more similar to "other caulobacter" sequences than to those of *Asticcacaulis*.

Two species are recognized and are distinguished by the number and position of the prosthecae (Table BXII.α.105); other characteristics are listed in Table BXII.α.106. Individual strains of the icosahedral-head DNA phages are lytic for both species, but cellular genomic DNA–DNA similarity has not been detected (Moore et al., 1978).

TABLE BXII.α.106. Other characteristics of the species of the genus *Asticcacaulis*[a]

Characteristic	A. excentricus	A. biprosthecum
Carbon-source utilization:[b]		
Arabinose	d	−
Ribose	−	nd
Xylose	+	+
Glucose	+	+
Galactose	+	+
Mannose	+	−
Fructose	+	+
Lactose	+	+
Maltose	+	+
Sucrose	+	−
Alanine	+	+
Aspartate	+	+
Glutamate	+	+
Proline	+	+
Tyrosine	d	−
Acetate	+	−
Butyrate	−	nd
Pimelate	−	nd
Pyruvate	+	+
Malate	+	−
Fumarate	+	−
Succinate	+	−
Methanol	d	−
Ethanol	+	+
Propanol	d	−
Butanol	d	−
Pentanol	d	−
Starch hydrolysis	+	+
Sensitivity to bacteriophages:[c]		
Type I:		
Phages 1, 3	−	−
Others	+	+
Type II	−	+
Type III	+	+
RNA phages	−	−

[a]Symbols: +, 90% or more of strains are positive; −, 90% or more of strains are negative; d, 11–89% of strains are positive; nd, not determined.

[b]Carbon-source utilizations were determined with D-isomers of sugars, with L-isomers of amino acids, or (when single isomers were not available) with racemic mixtures.

[c]2sDNA phage types: I, head prolate cylinder 50–65 × 170–260 nm, tail noncontractile 200–320 nm long; II, head elongated polyhedron 65–70 × 100–105 nm, tail noncontractile 260–300 nm long; III, head icosahedron 50–80 nm diameter, tail noncontractile 150–200 nm long. ssRNA phages: icosahedron 20–29 nm diameter, no tail.

TABLE BXII.α.105. Differential characteristics of the species of the genus *Asticcacaulis*[a]

Characteristic	A. excentricus	A. biprosthecum
Typical number of prosthecae per cell:		
One	+	−
Two	−	+
Position of prostheca:		
Subpolar	+	−
Lateral	−	+

[a]Symbols: +, 90% or more of strains are positive; and −, 90% or more of strains are negative.

List of species of the genus Asticcacaulis

1. **Asticcacaulis excentricus** Poindexter 1964, 292[AL]

   ex.cen' tri.cus. L. pref. *ex* out, beyond; Gr. n. *centron* center of circle; M.L. adj. *excentricus* out from the center.

   Rod-shaped cells thick (see Fig. BXII.α.120). A single prostheca arises from a subpolar site previously occupied by a single flagellum. Colonies colorless. Biotin is the only growth factor required in glucose–ammonium salts me-

dium. Sugars are the preferred carbon sources; all isolates can also use ethanol, but use of other primary alcohols is variable among isolates.

   Strain AC48 was isolated from a freshwater pond in Berkeley, CA.

   The mol% G + C of the DNA is: 55–60 (Bd, T_m).

   Type strain: AC48, ACM 1263, ATCC 15261, DSM 4724.

GenBank accession number (16S rRNA): AJ247194.

2. **Asticcacaulis biprosthecum** Pate, Porter and Jordan 1973, 582[AL]

bi.pros.thec′ um. L. pref. *bi, bis* twice; Gr. fem. n. *prosthece* appendage; M.L. adj. *biprosthecum* twice-appendage(d).

Rod-shaped cells thick (see Fig. BXII.α.121). One or, more often, two prosthecae arise near the equator of the cell at roughly diametrically opposite positions from each other; neither site is coincidental with the site occupied by the single, subpolar flagellum. Colonies colorless. Biotin is the only growth factor required in glucose–ammonium salts medium, but growth is markedly stimulated by mixtures of amino acids; supplementation with single amino acids can be inhibitory. Somewhat sensitive to dissolved O_2, growing faster and more efficiently (as measured by yield) when allowed to reduce medium prior to aeration. Pili as well as holdfast material participate in adhesion.

Strain C-19 was isolated from a freshwater pond in Madison, WI.

The mol% G + C of the DNA is: 61 (T_m).

Type strain: C-19, ACM 2498, ATCC 27554, DSM 4723, MBIC 3411.

GenBank accession number (16S rRNA): AJ247193.

Additional Remarks: There is a single known isolate of this species; strain C-19 was the original strain designation, but this isolate also appears in the literature as AC-2. One "*A. biprosthecum*-like" isolate has been reported (Sittig and Schlesner, 1993). (Note: The species epithet was misspelled "biprosthecium" in the 1989 Approved List; the spelling originally proposed by Pate et al., 1973, should be used.)

Genus III. **Brevundimonas** *Segers, Vancanneyt, Pot, Torck, Hoste, Dewettinck, Falsen, Kersters and De Vos 1994, 507[VP] emend. Abraham, Strömpl, Meyer, Lindholst, Moore, Christ, Vancanneyt, Tindall, Bennasar, Smit and Tesar 1999, 1070*

MARC VANCANNEYT, PAUL SEGERS, WOLF-RAINER ABRAHAM AND PAUL DE VOS

Brev.un.di′ mo.nas. L. adj. *brevis* short; L. fem. n. *unda* wave; Gr. *monas* a unit, monad; M.L. fem. n. *Brevundimonas* bacteria with short wavelength flagella.

Gram-negative, rod-shaped, subvibrioid or vibrioid cells, 0.4– 0.5 × 1–2 μm. **Cells of some species can form prosthecae (stalks).** These species are characterized by an asymmetric cell division whereby fission results in a prosthecate (nonmotile) and a flagellated (motile) cell. Motility by means of single polar flagella. Nonsporeforming. A yellow or orange carotenoid pigment may be formed.

Aerobic, having a strictly respiratory type of metabolism with oxygen as the terminal electron acceptor. Catalase positive. Nitrate is rarely reduced.

Chemoorganotrophic and oligotrophic. Good growth occurs on most common media at a pH range of 6.0–8.0 and at 25–30°C. **Growth factors are required by all strains.** Growth occurs without NaCl, but is optimal at 0.5–2% NaCl (w/v). All species can utilize pyruvate, and most isolates are also positive for growth on the organic acids acetate, butyrate, fumarate, and succinate and on the amino acids glutamate and proline. Glucose, galactose, maltose, and starch can be utilized by all species except the type species, which has a restricted nutritional spectrum. Growth on primary alcohols is usually negative. No or weak acid production from sugars.

Dominant fatty acids are $C_{12:0\ 3OH}$, $C_{14:0}$, $C_{15:0}$, $C_{16:0}$, $C_{16:1}$, $C_{17:0}$, $C_{17:1\ \omega6c}$, and $C_{17:1\ \omega8c}$, and $C_{18:1}$. Polar lipids are α-D-glucopyranosyl diacylglycerol, α-D-glucopyranuronosyl diacylglycerol, 1,2-di-*O*-acyl-3-*O*-[D-glucopyranosyl-(1→4)-α-D-glucopyranuronosyl] glycerol, 6-phosphatidyl-α-D-glucopyranosyl diacylglycerol (main mass numbers 1413 and 1439 Da), and phosphatidylglycerol. Most strains contain sulfoquinovosyl diacylglycerol. Ubiquinone Q-10 and the polyamines spermidine and homospermidine are present in the species examined. Cells accumulate poly-β-hydroxybutyrate as a reserve material but are, so far as known, not able to hydrolyze this polymer. The genus belongs to the class *Alphaproteobacteria*. Strains are isolated from water, soil, and clinical specimens.

The mol% G + C of the DNA is: 62–68.

Type species: **Brevundimonas diminuta** (Leifson and Hugh 1954b) Segers, Vancanneyt, Pot, Torck, Hoste, Dewettinck, Falsen, Kersters and De Vos 1994, 507 (*Pseudomonas diminuta* Leifson and Hugh 1954b, 68.)

FURTHER DESCRIPTIVE INFORMATION

Phylogenetic treatment Within the class *Alphaproteobacteria*, *Brevundimonas* belongs to the family *Caulobacteraceae*, which also includes the genera *Asticcacaulis*, *Caulobacter*, and *Phenylobacterium*. *Brevundimonas* is most closely linked to *Caulobacter*, with 16S rDNA sequence similarities of 95–96% (Fig. BXII.α.122; Abraham et al., 1999). Lower similarities of 93–94% and 92–93% are obtained with *Asticcacaulis* and *Phenylobacterium*, respectively. Within *Brevundimonas*, the interspecies 16S rDNA sequence similarities are higher than 96%. Remarkably, the misclassified species, "*Mycoplana bullata*", groups among the *Brevundimonas* species, with similarities above 97% (see below). No phylogenetic subgroups are recognized that reflect the distinct morphological characteristics of the species (see below; Fig. BXII.α.122; Abraham et al., 1999).

Cell morphology Cell morphology may be a primary criterion for distinguishing between the dimorphic prosthecate and nonprosthecate *Brevundimonas* species. Reproduction of six prosthecate species, *Brevundimonas alba*, *Brevundimonas aurantiaca*, *Brevundimonas bacteroides*, *Brevundimonas intermedia*, *Brevundimonas subvibrioides*, and *Brevundimonas variabilis*, results in the separation of two cells that are morphologically and behaviorally different from each other. Cells divide by binary transverse fission (which is constrictive without formation of a septum), in which the younger pole bears a single flagellum, and the older pole bears a prostheca derived from the cell envelope. The stalk includes membranes and peptidoglycan, but not cytoplasmic components. The stalk-bearing progeny cell grows and eventually repeats the asymmetric cell division. The flagellum-bearing progeny cell, after a period of motility, releases the flagellum and develops its stalk at the previously flagellated site as it grows and proceeds to its asymmetric cell division (Poindexter, 1989). At the base of the flagellum and at the outer tip of the stalk is a

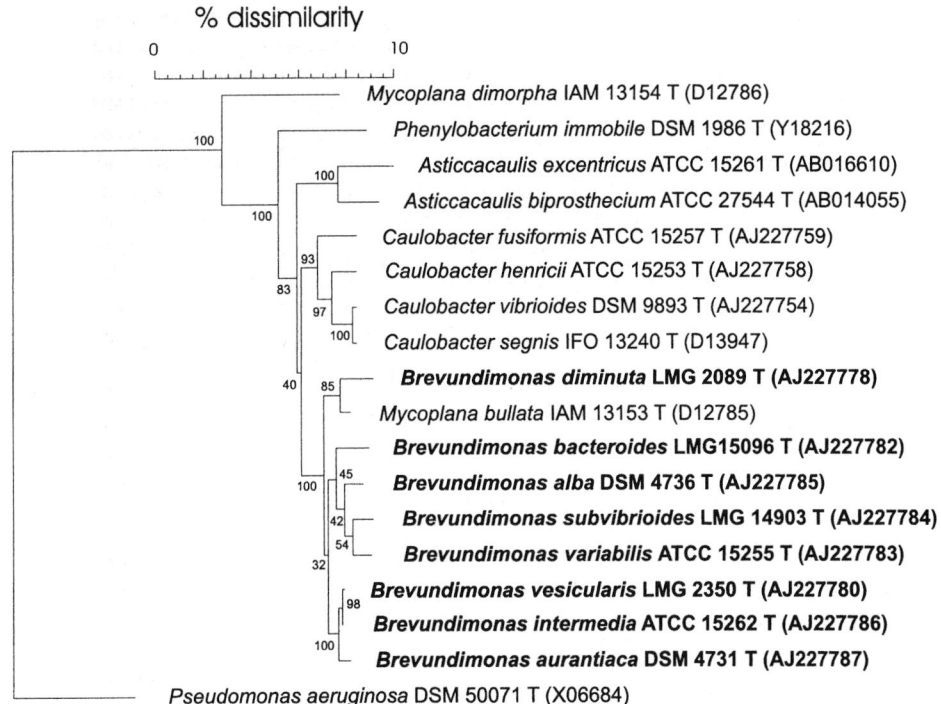

% dissimilarity

0   10

Mycoplana dimorpha IAM 13154 T (D12786)

Phenylobacterium immobile DSM 1986 T (Y18216)

Asticcacaulis excentricus ATCC 15261 T (AB016610)

Asticcacaulis biprosthecium ATCC 27544 T (AB014055)

Caulobacter fusiformis ATCC 15257 T (AJ227759)

Caulobacter henricii ATCC 15253 T (AJ227758)

Caulobacter vibrioides DSM 9893 T (AJ227754)

Caulobacter segnis IFO 13240 T (D13947)

Brevundimonas diminuta LMG 2089 T (AJ227778)

Mycoplana bullata IAM 13153 T (D12785)

Brevundimonas bacteroides LMG15096 T (AJ227782)

Brevundimonas alba DSM 4736 T (AJ227785)

Brevundimonas subvibrioides LMG 14903 T (AJ227784)

Brevundimonas variabilis ATCC 15255 T (AJ227783)

Brevundimonas vesicularis LMG 2350 T (AJ227780)

Brevundimonas intermedia ATCC 15262 T (AJ227786)

Brevundimonas aurantiaca DSM 4731 T (AJ227787)

Pseudomonas aeruginosa DSM 50071 T (X06684)

FIGURE BXII.α.122. Phylogenetic tree based on a comparison of 16S rDNA sequences of strains belonging to the genera *Brevundimonas* and related taxa (accession numbers are given in brackets). *T*, indicates type strain.

small mass of adhesive material, the holdfast, which confers adhesiveness on each of the progeny (Poindexter, 1989). Stalk position is species-specific or, for one species, may vary among strains, but the stalk length varies greatly among isolates (0.11–0.18 µm; Poindexter, 1989). The cell is rod-shaped, vibrioid, or subvibrioid, depending on the species, but cell shape may be influenced by medium composition (e.g., phosphate concentration) or culture age (Poindexter, 1989).

Strains of the nonprosthecate species, *Brevundimonas diminuta* and *Brevundimonas vesicularis,* are short rods with an unusual short-wavelength flagellar morphology (average wavelength, 0.62 µm). The flagellum originates for most cells from the periphery of the pole rather than from the center of the pole (Leifson and Hugh, 1954b; Clark et al., 1984).

Cultural characteristics Colonies are usually circular, convex, and glistening with smooth margins. Colony diameter varies with strain and medium, but usually reaches 3–5 mm after several days of incubation. Colonies are colorless or may have a pink, yellow, orange, or dark-red carotenoid pigment (Table BXII.α.107; Poindexter, 1964, 1989; Ballard et al., 1968).

The prosthecate *Brevundimonas* species grow well in peptone–yeast extract media supplemented with salts (Poindexter, 1981, 1989). Although in the literature it is indicated that they generally require a low concentration of organic material for growth, they are able to grow on common, nutrient-rich culture media, like tryptic soy agar or nutrient agar (M. Vancanneyt, personal communication). In standing broth cultures, they may form a pellicle that develops at the surface of the medium and adheres firmly to the wall of the culture vessel (Poindexter, 1989).

The nonprosthecate species, *B. diminuta* and *B. vesicularis,* grow on blood agar, infusion agar, tryptic soy and nutrient agar, Mueller–Hinton medium, and King A medium. After 3 d, a brownish-black, diffusible pigment is produced by most *B. dimi-*

nuta strains. Contrary to Richard and Kiredjian (1995), Gilardi (1978b) has indicated that *B. vesicularis* might occasionally produce a brown discoloration of the agar medium surrounding the colony. *B. diminuta* produces a water-soluble brown pigment. On heart infusion agar with tyrosine, more than 60% of the *B. vesicularis* strains produce a tan-brown, water-soluble pigment, and about 15% of the strains produce a yellow pigment (Gilardi, 1978b; Clark et al., 1984). Most *B. diminuta* strains and only about 25% of the *B. vesicularis* strains grow on MacConkey medium. No growth occurs on *Salmonella–Shigella* agar or on mineral base medium with acetate (Gilardi, 1978b; Gilligan, 1995). Hemolysis is not observed, and the egg-yolk reaction for lecithinase is negative (Gilardi, 1978b).

The optimal growth temperature is 28–30°C. Maximum growth for most species occurs at around 37°C, although nonprosthecate species grow well at this temperature, and about 50% of *B. diminuta* strains are able to grow at 42°C. Suboptimal growth is observed at 25°C, and no growth is observed at 4°C (Ballard et al., 1968; Gilardi, 1978b; Poindexter 1981).

Optimal pH is around neutrality. For the species examined, growth is observed in a range of pH 6–8 (Poindexter, 1964), although strains have been isolated from water and soil with a pH of about 5–9 (Poindexter, 1981).

Although *Brevundimonas* strains can grow without sodium chloride, optimal growth is obtained in a range of 0.5–2% NaCl for all species. Depending on the species, reduced growth may occur with salt concentrations of 2–6% and no growth is obtained at concentrations of 6–8% NaCl (Abraham et al., 1999).

Nutrition All species require growth factors in chemically defined media (Poindexter 1964; Ballard et al., 1968). For the prosthecate species, growth factors (available in peptone or yeast extract) remain undetermined, except for *B. intermedia,* for which growth is supported by biotin (Poindexter, 1981). Both non-

TABLE BXII.α.107. Differentiation of the *Brevundimonas* species from phylogenetically or phenotypically related taxa[a,b]

Characteristic	Brevundimonas Nonprosthecate species[c]	Brevundimonas Prosthecate species[d]	Mycoplana bullata	Caulobacter Prosthecate species[e]	Caulobacter segnis	Asticcacaulis	Mycoplana sensu stricto[f]	Pseudomonas (rRNA group I[g])
Cell morphology:								
Rods	+							+
Dimorphic rods		+		+		+		
Irregular rods (branching filaments)			+		+		+	
Prosthecae:								
Polar		+		+				
Subpolar/lateral						+		
None	+		+		+		+	+
Flagellation:								
Peritrichous			+		+		+	
Polar	+[h]	+[i]		+[j]				+[l]
Subpolar						+[k]		
Pigmentation	C, O, Y	C, P, Y, O	W	C, P, Y, O	W	C	W	C, Fl, Yo
Growth factors:								
Biotin	+	+				+	+	
Cyanocobalamin	+							
Cystine	+							
Panthothenate	+		+					
Riboflavin					+			
Thiamine							+	
Vitamin B₂ or B₁₂				+				
Other unidentified growth factors		+	+	+				
Nitrate reduction	Rarely	– or d	Weak or –	– or d	–	d	d	d
Carbon assimilation:								
Normal					+	+	+	+
Normal to restricted		+		+				
Restricted	+		+					
Major polar lipids	PG, PGL, GL, GOL, DGL, SQD	PG, PGL, GL, GOL, DGL, SQD	PG, PGL, GL, GOL, DGL, SQD	PG, PGL, GL, GOL	PG, PGL, GL, GOL	PG, GL, GOL	nd	DPG, PE, PG, PC
Major hydroxy fatty acids	$C_{12:0\ 3OH}$	$C_{12:0\ 3OH}$	$C_{12:0\ 3OH}$	$C_{12:1\ 3OH}$	$C_{12:0\ 3OH}$, $C_{12:1\ 3OH}$, $C_{16:1\ 2OH}$	$C_{12:1\ 3OH}$	$C_{14:0\ 3OH}$	$C_{10:0\ 3OH}$, $C_{12:0\ 3OH}$
No growth above NaCl (%, w/v)	6–8	6–8[m]	6–8	1–2[m]	1–2	2–3	nd	nd
Major polyamines	SPD, HOMOSPD	nd	nd	nd	nd	nd	nd	PUT, SPD
Ubiquinone	Q-10	Q-10	Q-10	Q-10	Q-10		Q-10	Q-9
Mol% G + C of DNA	65–68	62–67	66–68	62–67	66–68	55–61	64–65	55–65

[a]For symbols, see standard definitions; nd, not determined; C, colorless; O, orange; Y, yellow; Yo, yellow-orange; P, pink; W, white; Fl, fluorescent; DGL, 1,2-di-*O*-acyl-3-O-[D-glucopyranosyl-(1→4)-α-D-glucopyranuronosyl]glycerol; DPG, diphosphatidylglycerol; GL, glucosyldiacylglycerol; GOL, glucuronosyldiacylglycerol; PC, phosphatidylcholine; PE, phosphatidylethylamine; PG, phosphatidyl glycerol; PGL, phosphatidyl glucosyl diacylglycerol; SQD, sulfoquinovosyl diacylglycerol; SPD, spermidine; HOMOSPD, homospermidine; PUT, putrescine; Q-8–Q-10, ubiquinone with 8–10 isoprene units, respectively.

[b]Data from references Yamada et al. (1982); Oyaizu and Komagata (1983), Palleroni (1984), Poindexter (1989), Urakami et al. (1990d), Yabuuchi et al. (1990a), Stead (1992), Takeuchi et al. (1993a), and Abraham et al. (1999)

[c] *B. alba, B. aurantiaca, B. bacteroides, B. intermedia, B. subvibrioides,* and *B. variabilis.*

[d] *B. diminuta* and *B. vesicularis.*

[e] *C. fusiformis, C. henricii,* and *C. vibrioides.*

[f] *Mycoplana dimorpha* and *Mycoplana ramosa.*

[g]According to Palleroni (1984).

[h]Single flagellum with short wavelength.

[i]Single flagellum.

[j]Single flagellum.

[k]Single flagellum.

[l]Flagella with long wavelength.

[m]Varies depending on species.

prosthecate species require pantothenate, biotin, and cyanocobalamin to support growth. *B. diminuta* requires the amino acid cystine, for which methionine may be substituted, resulting in less effective growth (Ballard et al., 1968).

The nutritional spectrum of the genus *Brevundimonas* is very variable and can be used for species differentiation (Table BXII.α.108). Only a few carbon sources are assimilated by the majority of the isolates: the organic acids acetate, butyrate, fumarate, pyruvate, and succinate; the amino acids glutamate and proline; and the carbohydrates glucose, galactose, maltose, and starch. The latter carbohydrates are not utilized by *B. diminuta*, which is regarded as an alkali-producing species (Gilardi, 1978b)

TABLE BXII.α.108. Differential features among species of the genus *Brevundimonas*[a,b]

Characteristic	B. diminuta[c,d,e]	B. alba	B. aurantiaca	B. bacteroides	B. intermedia	B. subvibrioides	B. variabilis	B. vesicularis[c,d,e]
Morphology:								
Cell type:								
Bacteroid	+			+			+	+
Subvibrioid		+				+		
Vibrioid			+		+			
Prostheca (stalk)	−	+	+	+	+	+	+	−
Nonstalked pole tapered				−	+		−	
Stalk invariably central				+	+	+	−	
Pigmentation	C	C	R, G	C, Y, O	C	O, G	C, Ro	P, Y, O, C
Required growth factors:								
Biotin	+			−	+	−	−	+
Cyanocobalamin	+							+
Cystine	+							−
Pantothenate	+	−	−	−	−	−	−	+
Amino acids				d	−	−	−	
Others, unidentified		+		+	+	+	+	
Assimilation of:								
DL-Arabinose[f]	−	−	−	+	+	−	−	−
D-Cellobiose	−							d(+)
D-Fructose[f]	−	−	−	+	−	−	−	−
D-Galactose[f]	−	+	+	+	+	+	+	+
D-Glucose[f]	−	+	+	+	+	+	+	+
Lactose	−	−	−	d	+	−	−	−
Maltose	−	+	+	d	+	+	+	+
D-Mannose[f]	−	d	−	+	+	−	−	
L-Rhamnose[f]	−							d(−)
D-Ribose	−	d	−	+	−	−	+	−
Starch	−	+	+	+	+	d	+	d(+)
Sucrose	−	+	+	+	+	+	d	
DL-Xylose[f]	−	−	−	+	+	+	d	−
D-α-Alanine and L-Alanine[f]	d(−)	+	+	d	+	+	d	d(−)
Casamino acids		+	+	d	+	+	+	
L-Arginine[f]	d(−)	−	−	−	−	−	−	
L-Aspartate[f]	+	+	+	d	+	+	−	d(−)
L-Glutamate[f]	+	+	+	d	+	+	+	+
L-Histidine[f]	+	−	−	−	−	−	−	−
L-Isoleucine[f]	d(+)	+	−	−	−	−	−	
L-Leucine[f]	+	d	−	d	−	d	+	d(−)
L-Proline[f]	+	d	+	d	+	d	+	+
L-Norleucine[f]	d(−)							
DL-Norvaline[f]	d(+)							d(−)
L-Serine[f]	+	−	+	−	+	d	−	d(−)
L-Threonine[f]	d(+)							d(−)
L-Tyrosine[f]	d(−)	−	−	−	−	−	−	−
L-Valine[f]	d(−)	+	−	−	−	−	−	
Acetate	+	+	+	+	+	+	d	d(−)
Aconitate	−							d(−)
Butyrate	+	d	+	d	+	d	+	d
Fumarate	d(−)	+	+	+	+	+	+	d(+)
Isovalerate	d(+)							−
2-Ketoglutarate	d(−)							−
DL-Lactate[f]	−	−	+	d	−	−	−	d(+)
L-Malate[f]	d(−)	d	+	d	+	d	−	d(+)
Pimelate	−	−	−	−	+	−	−	
Succinate	d(−)	+	+	+	+	+	+	d(+)
n-Valerate	d(+)							−
Butanol	−	d	−	−	+	−	d	−
Ethanol		d	+	d	d	−	−	d
Methanol	−	−	−	−	−	−	−	
Pentanol	−	d	−	−	−	−	−	
Propanol		d	−	−	−	−	−	
Enzyme activity:								
Nitrate reduction	−	−	−	d	−	−	−	−
Hydrolysis of esculin	−							+
Starch hydrolysis		−	+	(+)	+	+	+	+
α-Glucosidase	−							+
Valine arylamidase	−							d(−)
Chymotrypsin	d(+)							−

[a]For symbols see standard definitions; C, colorless; R, red; Ro, red-orange; G, golden; Y, Yellow; O, orange; P, pink.

[b]Data from Poindexter (1964, 1981, 1989), Palleroni (1984), Urakami et al. (1990d), Segers et al. (1994), and Abraham et al. (1999).

[c]Reactions of the type strain are given between brackets for *B. diminuta* and *B. vesicularis*.

[d]Additional nutritional features for *B. diminuta* and *B. vesicularis* as determined by API 50CH, 50AO, and 50AA galleries. Substrates used by fewer than 10% of the strains are: adonitol, amygdalin, D-arabitol, L-arabitol, arbutin, dulcitol, erythritol, D-fucose, L-fucose, β-gentiobiose, gluconate, 2-ketogluconate, 5-ketogluconate, N-acetyl-glucosamine, methyl-α-D-glucose, glycerol, glycogen, inositol, inulin, D-lyxose, mannitol, methyl-α-D-mannoside, D-melezitose, D-melibiose, D-raffinose, salicin, sorbitol, L-sorbose, D-tagatose, trehalose, D-turanose, xylitol, methyl-β-xyloside, adipate, azelate, benzoate, m-hydroxybenzoate, o-hydroxybenzoate, p-hydroxybenzoate, caprate, n-caproate, caprylate, citraconate, citrate, isobutyrate, glutarate, DL-glycerate, glycolate, heptanoate, itaconate, levulinate, maleate, D-malate, malonate, D-mandelate, L-mandelate, mesaconate, oxalate, phenylacetate, phthalate, isophthalate, terephthalate, pelargonate, propionate, sebacate, suberate, D-tartrate, L-tartrate, *meso*-tartrate, acetamide, amylamine, 2-aminobenzoate, 3-aminobenzoate, 4-aminobenzoate, benzylamine, betaine, diaminobutane, butylamine, DL-2-aminobutyrate, DL-3-aminobutyrate, DL-4-aminobutyrate, L-citrulline, creatine, L-cysteine, ethanolamine, ethylamine, glucosamine, glycine, histamine, DL-kynurenine, L-norleucine, L-lysine, L-methionine, L-ornithine, L-phenylalanine, sarcosine, spermine, trigonelline, tryptamine, D-tryptophan, L-tryptophan, urea, and DL-5-aminovalerate. DL-β-hydroxybutyrate is used by all strains of both species. Also, the following substrates are not used by either species (according to conventional test results; Segers et al., 1994): anthranilate, benzoylformate, 2,3-butyleneglycol, n-dodecane, eicosanedioate, ethyleneglycol, geraniol, n-hexadecane, hippurate, isobutanol, isopropanol, kynurenate, methylamine, mucate, naphthalene, nicotinate, phenol, phenylethanediol, propyleneglycol, quinate, saccharate, testosterone and DL-α-aminovalerate. A negative reaction is also obtained by conventional tests (Segers et al., 1994) for gelatin liquefaction, indole formation, urease, lysine and ornithine decarboxylases, arginine dihydrolase, phenylalanine deaminase, lecithinase (egg yolk), and lipase (Tween 80).

[e]Activity of the following enzymes as determined for *B. diminuta* and *B. vesicularis* by API ZYM. Always absent: lipase (C_{14}), cystine arylamidase, α-fucosidase, α-galactosidase, β-galactosidase, β-glucosidase, N-acetyl β-glucosidase, and α-mannosidase; present: alkaline and acid phosphates, ester lipase (C_8), leucine arylamidase, phosphoamidase, trypsin, and indophenol oxidase.

[f]Isomer not specified for *B. alba*, *B. aurantiaca*, *B. bacteroides*, *B. intermedia*, *B. subvibrioides*, and *B. variabilis*.

because of its inability to utilize most carbohydrates. Characteristic for both nonprosthecate species is also the utilization of DL-β-hydroxybutyrate, a feature that has not been tested for the prosthecate species. It is clear from Table BXII.α.108 that several phenotypic characteristics are available solely for either the prosthecate or the nonprosthecate species, which is a result from their—until recently—different generic classification. No comprehensive phenotypic studies are available in which all species of the genus are compared in a consistent and reproducible way.

The nonprosthecate species are oxidase positive and urease and gelatinase negative, and they do not produce indole. The prosthecate species have, as far as is known, not been examined for these features. Metabolic studies using *B. bacteroides* have indicated the presence of 2-keto-3-deoxy-6-phosphogluconate aldolase which suggests that glucose is catabolized via the Entner–Doudoroff pathway (Poindexter, 1981). *Brevundimonas* strains also have the ability to degrade or detoxify aromatic compounds and organophosphates (Poindexter, 1981; Weissenfels et al., 1990; Caldwell et al., 1991; Labuzek et al., 1994; Hoskin et al., 1995; Hong and Raushel, 1996; Davis et al., 1997), a characteristic that has recently been used to develop an amperometric enzyme biosensor for the direct measurement of parathion (Sacks et al., 2000). *B. diminuta* strains are also characterized by a specific control mechanism for DAHP synthetase (3-deoxy-D-arabino-heptulosonate 7-phosphate synthetase), an activity that is not always recovered for *B. vesicularis* (Whitaker et al., 1981). Byng et al. (1980) have demonstrated the presence of an NADP-dependent prephenate dehydrogenase (tyrosine biosynthesis) in nonprosthecate *Brevundimonas* species, and West (1992) has studied their ribonucleoside catabolic enzyme activities. Polyvinyl alcohol oxidase activity has been demonstrated for a particular *B. vesicularis* strain (Kawagoshi and Fujita, 1997; Kawagoshi et al., 1997). EDTA resistance is demonstrated for *B. diminuta* (Wilkinson, 1968). Finally, the unusual metabolic capacities of some *Brevundimonas* members have recently been investigated for bioconversion of intermediate components of cephalosporin production (e.g., Kim et al., 2000b).

Lipid composition The polar lipids are α-D-glucopyranosyl diacylglycerol, α-D-glucopyranuronosyl diacylglycerol, 6-phosphatidyl-α-D-glucopyranosyl diacylglycerol, phosphatidylglycerol, 1,2-di-O-acyl-3-O-[D-glucopyranosyl- (1→4) -α-D-glucopyranuronosyl] glycerol, and sulfoquinovosyl diacylglycerol (Abraham et al., 1997).

Data on the lipopolysaccharides are available only for the two nonprosthecate species and have some unusual features (Wilkinson and Taylor, 1978). The lipid A contains 2,3-diamino-2,3-dideoxy-D-glucose instead of 2-amino-2-deoxy-D-glucose (the backbone of lipid A in lipopolysaccharides of most Gram-negative bacteria). The component 9-hydroxy-delta-tetradecalactone has been detected in lipid A of both species (Arata et al., 1989). Acetylated mannan and D-threo-pent-2-ulose have been isolated from the lipopolysaccharides of one *B. diminuta* strain NCTC 8545 (Wilkinson 1981b, a). Purified lipid A preparations of *B. vesicularis* may activate macrophages to resist infection by an opportunistic bacterium in mice (Arata et al., 1994).

Brevundimonas strains are characterized by the following dominant fatty acids $C_{12:0\ 3OH}$ (1–3%), $C_{14:0}$ (tr–5%), $C_{15:0}$ (tr–8%), $C_{16:0}$ (10–24%), $C_{16:1}$ (1–11%), $C_{17:0}$ (tr–8%), $C_{17:1\ \omega6c}$ (tr–6%), $C_{17:1\ \omega8c}$ (1–11%), and $C_{18:1}$ (39–69%) (Abraham et al., 1999). Species may be differentiated on the basis of quantitative differences in these fatty acids. *B. diminuta* and *B. alba* are further differentiated from the other *Brevundimonas* species by the presence of a significant amount of $C_{19:0\ cyclo\ \omega8c}$ (methyleneoctadecanoic acid; Yamada et al., 1982; Segers et al., 1994; Abraham et al., 1997, 1999).

Polyamines Prosthecate species have not been examined for the presence of polyamines. *B. diminuta* strains produce homospermidine as the major polyamine component (1.7–3.4 μmol/g protein) as well as trace amounts of spermidine. *B. vesicularis* strains (except one strain) produce spermidine as the major component (1.6–2.5 μmol/g protein), homospermidine as a minor component (0.1–0.5 μmol/g protein), and, in some cases, trace amounts of norspermidine, putrescine, and spermine. Two currently unnamed *Brevundimonas* strains yield a similar polyamine pattern to the *B. vesicularis* strains, and another strain produces only spermidine in high amounts (Segers et al., 1994).

Genetic features Rather small genomes (3–3.5 Mb) are observed for the nonprosthecate species (Grothues and Tümmler, 1991). No information on genome size is available for the prosthecate species. Plasmids have so far been detected only for *B. bacteroides* (Poindexter, 1989). DNA–DNA hybridization studies among all currently described species of *Brevundimonas* (M. Vancanneyt, personal communication) have shown relatedness values below 70%, thus confirming their status as separate species. Intermediate levels of DNA–DNA relatedness of 50% and 41% are observed between the nonprosthecate species *B. vesicularis* and the prosthecate species *B. aurantiaca* and *B. intermedia*, respectively. The latter two species show 65% DNA–DNA relatedness, indicating their close relationship.

Antimicrobial agents All species are sensitive to penicillin G and streptomycin. Significantly more data on antibiotic resistance are available for the two nonprosthecate species, which may be explained by their occurrence in clinical sources. *B. diminuta* is sensitive to aminoglycosides and β-lactam antibiotics and is highly sensitive to tetracyclines and rifampicin. Susceptibility to carbenicillin, sulfonamides, and chloramphenicol differs among strains of the species. All strains are resistant to nalidixic acid (a diagnostic feature), furans, trimethoprim, colistin, ampicillin, cephalothin, polymyxin, and gentamicin. *B. vesicularis* is usually more sensitive to antibiotics than is *B. diminuta*, especially to cephalothin (Gilardi, 1976; Richard and Kiredjian, 1995); however, treatment with tobramycin and ceftazidime of a case of *B. vesicularis* bacteremia in a patient was not successful (Planes et al., 1992).

B. subvibrioides and *B. bacteroides* have been found to be sensitive to some *Caulobacter* phages (Poindexter, 1989).

Source or habitat The organisms are typical aquatic bacteria. At least some species have been isolated from mineral water (Jayasekara et al., 1999). Prosthecate species may be isolated from soil (e.g., *B. alba*). Nonprosthecate species have also been identified in clinical specimens, such as blood cultures, infected biological fluids, urine, wounds, vagina, eye, and tissue cultures (Aspinall and Graham, 1989; Morais and da Costa, 1990; Gilardi, 1991; Planes et al., 1992; Abraham et al., 1997; Giardini et al., 1997). In particular, *B. vesicularis* has been reported as an important nosocomial agent (Gilad et al., 2000).

ENRICHMENT AND ISOLATION PROCEDURES

There is no strong selective enrichment procedure described for the isolation of *Brevundimonas* strains. Nevertheless, the tolerance of prolonged nutrient scarcity by prosthecate members of the genus provides a dependable physiological basis for their enrichment. Successful enrichments from water and from soil suspended in water are obtained when the sample is allowed to stand undisturbed for one to several weeks (Poindexter, 1992). Enrichment/isolation procedures are also described based on the accumulation of the stalked cells at the air–water interface or on a cover slip of a liquid culture with low nutrient content that is carefully protected against turbulence (Houwink, 1955; Poindexter, 1989). Methods and media for enrichment and further differentiation from other organisms are described in more detail in this chapter and in the literature (Leifson and Hugh, 1954b; Poindexter, 1964, 1981, 1989, 1992; Ballard et al., 1968; Gilardi, 1978b; Clark et al., 1984; Palleroni, 1984; Segers et al., 1994; Gilligan, 1995; Grimont et al., 1996).

MAINTENANCE PROCEDURES

Strains can be easily cultured aerobically on most common agar media containing peptone or yeast extract at 30°C. These media support growth without growth-factor supplements, which are necessary in defined media and enhance growth in complex media (Table BXII.α.107). Nonprosthecate cultures can be maintained for at least 1 week at room temperature or in a refrigerator before being transferred. Prosthecate isolates streaked on dilute complex media (and incubated for 2 or 3 d at 20–30°C) should be transferred every 8–9 weeks when refrigerated (Poindexter, 1989).

Cultures may be stored for many years by lyophilization, by freezing at −80°C, or in liquid nitrogen. In general, cryoprotective agents such as 10% glycerol or dimethyl sulfoxide are added to cultures before freezing, although prosthecate species can be frozen at −70°C without cryoprotectant (Poindexter, 1989).

DIFFERENTIATION OF THE GENUS *BREVUNDIMONAS* FROM OTHER GENERA

Characteristics that differentiate *Brevundimonas* from phylogenetically and phenotypically related genera and species are listed in Table BXII.α.107. *Caulobacter* is the closest phylogenetic neighbor with 16S rDNA sequence similarities of 95–96%. Based on morphological features, both nonprosthecate species of *Brevundimonas* are easily distinguished from *Caulobacter* species of which all, except *Caulobacter segnis*, are stalk-forming organisms. Within the latter genus, *C. segnis* has a unique morphology with irregular, peritrichously flagellated rods and branching filaments. Morphology does not allow differentiation among prosthecate species of both taxa. There are also no straightforward nutritional properties that unequivocally distinguish both genera, although particular tests can be used to characterize individual species (Table BXII.α.108; Poindexter, 1989). Some useful cultural and chemotaxonomic differentiating features between the genera *Brevundimonas* and *Caulobacter* are their salt tolerances and polar lipid and fatty acid contents (Table BXII.α.107; Stahl et al., 1992; Abraham et al., 1999). Strains of the genus *Caulobacter* show optimal growth at 0.5% NaCl and reduced or no growth at salt concentrations of 1–2% (depending on the species). *Brevundimonas* strains have a broader optimal salt concentration of 0.5–2% NaCl. The polar lipids 1,2-di-*O*-acyl-3-*O*-[D-glucopyranosyl-(1→4)-α-D-glucopyranuronosyl]glycerol, and sulfoquinovosyl diacylglycerol are present in most *Brevundimonas* strains (>80%) and absent in

Caulobacter. *Brevundimonas* is furthermore differentiated from *Caulobacter* by the absence of significant amounts of $C_{12:0\ 3OH}$ and of an unknown fatty acid with ECL 11.789, and by the presence of at least traces of an unknown fatty acid with ECL 17.897 and higher amounts of $C_{12:0\ 3OH}$ and $C_{18:1}$ (Abraham et al., 1999).

Differentiation among *Brevundimonas* and *Caulobacter* strains is also obtained based on the reactivity of antisera. A 43-kDa protein exhibits genus-specific epitopes (Abraham et al., 1999). These authors have also demonstrated that desorption chemical ionization (DCI) mass spectrometry of glycolipids and subsequent multivariate analysis of the relative intensities of the $[M+NH_4]^+$ ions between m/z 740–840 has allowed differentiation between the genera *Brevundimonas* and *Caulobacter*.

Asticcacaulis is, at the generic level, the closest phylogenetic neighbor of *Brevundimonas* and *Caulobacter*, but clearly represents a separate branch within the family *Caulobacteraceae* (16S rDNA sequence similarity of 93–94%; Abraham et al., 2001). The genus *Asticcacaulis* was created by Poindexter (1964) to accommodate prosthecate isolates originally regarded as *Caulobacter* species. They are morphologically distinguished from prosthecate *Brevundimonas* and *Caulobacter* species by the site of adhesion of sessile cells to substrata; *Brevundimonas* and *Caulobacter* cells adhere by the distal tip of the prostheca, while *Asticcacaulis* cells adhere by the cell pole, and the prostheca is not adhesive (Poindexter, 1989, 1992). With a mol% G + C content between 55 and 61, *Asticcacaulis* can be differentiated from *Brevundimonas* and *Caulobacter* by its mol% G + C content of 62–68. The phylogenetic branches are further differentiated by the lack of 1,2-diacyl-3-*O*-[6′-phosphatidyl-α-D-glucopyranosyl]glycerol (PGL) in *Asticcacaulis*. Strains of this genus also do not contain 1,2-di-*O*-acyl-3-*O*[D-glucopyranosyl-(1→4)-α-D-glucopyranuronosyl]glycerol (DGL) found in most *Brevundimonas* strains but not in strains of the genus *Caulobacter* (Abraham et al., 2001). These authors have demonstrated, furthermore, that *Asticcacaulis* species are characterized by the presence of $C_{12:1\ 3OH}$ in their fatty acid profile, a component that has also been observed to be characteristic for *Caulobacter* species, but not for *Brevundimonas* taxa.

A deeper branching in the *Caulobacteraceae* is found for *Phenylobacterium*, a taxon characterized by nonmotile rods or cocci that do not produce prosthecae. Although there are some common features with the other members of the family, such as their growth factor requirements and carbohydrates for growth, the genus is easily distinguished by its high nutritional specialization. Good growth occurs on chloradizon, antipyrin, and L-phenylalanine (Lingens et al., 1985).

The two nonprosthecate species of *Brevundimonas* were previously classified in *Pseudomonas* (Palleroni, 1984). Both species can easily be differentiated from members of the present genus *Pseudomonas* (in general terms restricted to rRNA group 1; Palleroni, 1984) not only by their separate phylogenetic position, but also by one or more of the following characteristics: cell morphology, short wavelength of the flagella, polyamines and quinone content, a restricted nutritional spectrum, and their salt tolerance (Table BXII.α.107; Ballard et al., 1968; Gilardi, 1978b; Palleroni, 1984; Segers et al., 1994; Abraham et al., 1997, 1999). Both species are further distinguished by the absence of phosphatidylethanolamine and by the presence of phosphatidylglycerol, phosphatidylglucosyldiacylglycerol, and some major glycosyldiacylglycerols in their polar lipids (Wilkinson and Bell, 1971; Wilkinson and Taylor, 1978; Wilkinson and Galbraith, 1979;

Barnes et al., 1989). The presence or absence of specific 2-OH and 3-OH fatty acids can be used as diagnostic features to differentiate *Brevundimonas* taxa from other members of the former genus *Pseudomonas* (Table BXII.α.107; Oyaizu and Komagata, 1983; Palleroni, 1984; Segers et al., 1994). Also, genomic differences are found, like the small genomes of *Brevundimonas* spp. (3–3.5 Mb) and the absence of host factor for coliphage Qβ RNA replication (DuBow and Ryan, 1977; Grothues and Tümmler, 1991).

TAXONOMIC COMMENTS

The genus *Brevundimonas* currently contains eight species and belongs to the class *Alphaproteobacteria* as a member of the family *Caulobacteraceae*. Originally, *Brevundimonas* contained only *B. diminuta* and *B. vesicularis*, which were reclassified members of the genus *Pseudomonas sensu lato* (Segers et al., 1994). Techniques based on rRNA similarities (Palleroni et al., 1973; De Vos and De Ley, 1983) demonstrated the phylogenetic heterogeneity of the latter genus and showed that the rRNA Group IV members, *Pseudomonas diminuta* and *Pseudomonas vesicularis*—often called the "diminuta group" (Ballard et al., 1968; Gilardi, 1978b)— were found to occupy a unique position within the *Alphaproteobacteria*. It was also shown by Segers et al. (1994) that some strains that were preliminarily classified into EF group 21, a heterogenous group containing organisms that show phenotypic similarities to *Sphingomonas paucimobilis* (formerly *Pseudomonas paucimobilis*), are members of *Brevundimonas*.

In 1999, 16S rRNA gene sequencing studies indicated that *B. vesicularis* and *B. diminuta* are phylogenetically interrelated with *Caulobacter intermedius* and *Caulobacter variabilis* (Abraham et al., 1999; Sly et al., 1999) and with *Caulobacter subvibrioides* and *Caulobacter bacteroides*, as well as with the misclassified *Mycoplana bullata* (Abraham et al., 1999). Other validly described species of the genus *Caulobacter*, including the type species *Caulobacter vibrioides* and *Caulobacter segnis*, constitute a separate phylogenetic branch within the family *Caulobacteraceae* (Abraham et al., 1999; Fig. BXII.α.122).

A formal reclassification has been made based on phylogenetic data and the genus extended with six new species, *B. bacteroides*, *B. intermedia*, *B. subvibrioides*, *B. variabilis*, *B. alba*, and *B. aurantiaca*. The latter two were originally described as *C. subvibrioides* subsp. *alba* and *C. henricii* subsp. *aurantiaca*, respectively. Based on their distinct phylogenetic position in the genus, they have been elevated to the species rank (Abraham et al., 1999).

Abraham et al. (1999) have also emended the genus description of *Caulobacter*, which currently contains, in addition to the type species *Caulobacter vibrioides*, the species *Caulobacter fusiformis*, *Caulobacter henricii*, *Caulobacter leidyi*, and *Caulobacter segnis*. Cau-

lobacter segnis was originally named *Mycoplana segnis*, a misclassified member of *Mycoplana* that belongs to a different lineage in the *Alphaproteobacteria* within the family *Rhizobiaceae*. Furthermore, Abraham et al. (1999) have proposed that *Caulobacter crescentus* is a subjective synonym of *C. vibrioides*.

As shown in Fig. BXII.α.122, not only *Caulobacter segnis* (*Mycoplana segnis*) was misclassified, but also *Mycoplana bullata*, which clearly belongs on phylogenetic arguments to *Brevundimonas*. Additionally, 39% DNA relatedness has been found to the type strain of *B. diminuta* by DNA–DNA hybridizations (M. Vancanneyt, personal communication). Up to now, however, no formal reclassification has been proposed for the misclassified *Mycoplana bullata*. Except for the very distinct morphologic characteristics, i.e., peritrichously flagellated, rod-shaped cells forming filaments before fragmentation, most genomic, chemotaxonomic and nutritional features of *Mycoplana bullata* are analogous to those of related species of *Brevundimonas* (Table BXII.α.107; Yanagi and Yamasato, 1993; Abraham et al., 1999). If a formal reclassification is proposed to include *M. bullata* as a member of *Brevundimonas*, the emended genus description of the latter will have to encompass the features of *Mycoplana bullata*, as indicated below.

FURTHER READING

Abraham, W.R., C. Strömpl, H. Meyer, S. Lindholst, E.R. Moore, R. Christ, M. Vancanneyt, B.J. Tindall, A. Bennasar, J. Smit and M. Tesar. 1999. Phylogeny and polyphasic taxonomy of *Caulobacter* species. Proposal of *Maricaulis* gen. nov. with *Maricaulis maris* (Poindexter) comb. nov. as the type species, and emended description of the genera *Brevundimonas* and *Caulobacter*. Int. J. Syst. Bacteriol. *49*: 1053–1073.

Ballard, R.W., M. Doudoroff, R.Y. Stanier and M. Mandel. 1968. Taxonomy of the aerobic pseudomonads: *Pseudomonas diminuta* and *P. vesiculare*. J. Gen. Microbiol. *53*: 349–361.

Gilardi, G.L. 1978. Identification of *Pseudomonas* and related bacteria. *In* Gilardi (Editor), Glucose Nonfermenting Gram-Negative Bacteria in Clinical Microbiology, CRC Press, West Palm Beach, Florida. pp. 15–44.

Palleroni, N.J. 1984. Genus I *Pseudomonas*. *In* Krieg and Holt (Editors), Bergey's Manual of Systematic Bacteriology, 1st Ed., Vol. 1, The Williams & Wilkins Co., Baltimore. pp. 141–199.

Poindexter, J.S. 1964. Biological properties and classification of the *Caulobacter* group. Bacteriol. Rev. *28*: 231–295.

Poindexter, J.S. 1989. The genus *Caulobacter* and the genus *Asticcacaulis*. *In* Staley, Byrant, Pfenning and Holt (Editors), Bergey's Manual of Systematic Bacteriology, 1st Ed., Vol. 3, The Williams & Wilkins Co., Baltimore. pp. 1924–1942.

Segers, P., M. Vancanneyt, B. Pot, U. Torck, B. Hoste, D. Dewettinck, E. Falsen, K. Kersters and P. De Vos. 1994. Classification of *Pseudomonas diminuta* Leifson and Hugh 1954 and *Pseudomonas vesicularis* Busing, Doll, and Freytag 1953 in *Brevundimonas* gen. nov. as *Brevundimonas diminuta* comb. nov. and *Brevundimonas vesicularis* comb. nov., respectively. Int. J. Syst. Bacteriol. *44*: 499–510.

DIFFERENTIATION OF THE SPECIES OF THE GENUS *BREVUNDIMONAS*

Features for differentiating *Brevundimonas* species are the cell morphology and cell division, stalk formation, pigmentation, requirement for growth factors, nutritional spectrum, and other characteristics as shown in Table BXII.α.108.

List of species of the genus Brevundimonas

1. **Brevundimonas diminuta** (Leifson and Hugh 1954b) Segers, Vancanneyt, Pot, Torck, Hoste, Dewettinck, Falsen, Kersters and De Vos 1994, 507[VP] (*Pseudomonas diminuta* Leifson and Hugh 1954b, 68.)
   di.mi.nu′ta. L. adj. *minutus* small; M.L. fem. adj. *diminuta* defective, minute.

   Short rods. Motile by means of one polar flagellum that has a short wavelength (0.6–1 μm). No prosthecae or special cell division. No pigmentation. The description of *B. diminuta* is the same as for the genus, with the additional characters given in Table BXII.α.107 and BXII.α.108.

Isolated from water, aqueous solutions and diverse clinical specimens in man and animals.

The mol% G + C of the DNA is: 66–68 (T_m).

Type strain: ATCC 11568, DSM 7234, IMET 10409, LMG 2089.

GenBank accession number (16S rRNA): AJ227778.

2. **Brevundimonas alba** (ex Poindexter 1964) Abraham, Strömpl, Meyer, Lindholst, Moore, Christ, Vancanneyt, Tindall, Bennasar, Smit and Tesar 1999, 1070[VP]

al′ba. L. fem. adj. *alba* white.

Cell morphology varies within a clone; some poles distinctly tapered, others rounded; long axis gently curved. Colonies colorless. The description of *B. alba* is the same as that for the genus, with the additional characters given in Table BXII.α.107 and BXII.α.108.

Strains are characterized by the presence of detectable amounts of $C_{19:0\ cyclo\ \omega 8c}$.

The type strain was isolated from soil.

The mol% G + C of the DNA is: 67 (Bd); 68 (HPLC).

Type strain: CB88, DSM 4736, LMG 18360.

GenBank accession number (16S rRNA): AJ227785.

3. **Brevundimonas aurantiaca** (ex Poindexter 1964) Abraham, Strömpl, Meyer, Lindholst, Moore, Christ, Vancanneyt, Tindall, Bennasar, Smit and Tesar 1999, 1071[VP]

au.ran.ti.a′ ca. M.L. fem. adj. *aurantiaca* orange.

Vibrioid, unusually small cells. Dark golden colonies. The description of *B. aurantiaca* is the same as that for the genus, with the additional characters given in Table BXII.α.107 and BXII.α.108.

The type strain was isolated from a contaminated *Chlorella* culture.

The mol% G + C of the DNA is: 67 (HPLC).

Type strain: CB-R, DSM 4731, LMG 18359.

GenBank accession number (16S rRNA): AJ227787.

4. **Brevundimonas bacteroides** (Poindexter 1964) Abraham, Strömpl, Meyer, Lindholst, Moore, Christ, Vancanneyt, Tindall, Bennasar, Smit and Tesar 1999, 1071[VP] (*Caulobacter bacteroides* Poindexter 1964, 272.)

bac.ter.oi′ des. M.L. *bacter* masc. form of Gr. neut. n. *bactrum* rod; Gr. n. *eidus* form, shape; M.L. adj. *bacteroides* rod shaped.

Slender, rod-shaped cells. Stalk typically has a bulbous distal tip. Colonies colorless or brightly pigmented yellow or orange. The description of *B. bacteroides* is the same as that for the genus, with the additional characters given in Table BXII.α.107 and BXII.α.108.

Isolated from water and soil.

The mol% G + C of the DNA is: 66 (Bd); 68 (HPLC).

Type strain: CB7, ATCC 15254 , DSM 4726, LMG 15096.

GenBank accession number (16S rRNA): AJ227782.

5. **Brevundimonas intermedia** (Poindexter 1964) Abraham, Strömpl, Meyer, Lindholst, Moore, Christ, Vancanneyt, Tindall, Bennasar, Smit and Tesar 1999, 1071[VP] (*Caulobacter intermedius* Poindexter 1964, 272.)

in.ter.med′ i.a. L. adj. *intermedia* in the middle degree, between extremes.

Short, vibrioid cells. Colorless colonies. The description of *B. intermedia* is the same as that for the genus, with the

additional characters given in Table BXII.α.107 and BXII.α.108.

Isolates reported from pond water.

The mol% G + C of the DNA is: 66 (HPLC).

Type strain: CB63, ATCC 15262, DSM 4732, LMG 18361.

GenBank accession number (16S rRNA): AJ227786.

6. **Brevundimonas subvibrioides** (Poindexter 1964) Abraham, Strömpl, Meyer, Lindholst, Moore, Christ, Vancanneyt, Tindall, Bennasar, Smit and Tesar 1999, 1071[VP] (*Caulobacter subvibrioides* Poindexter 1964, 272.)

sub.vib.ri.oi′ des. L. pref. *sub* almost, somewhat, near; M.L. n. *vibrio* name of a genus; Gr, n. *eidus* resembling; M.L. adj. *subvibrioides* somewhat like a vibrio.

Cell morphology variable within a clone: some poles distinctly tapered, others rounded; long axis gently curved or not curved. Colonies orange, golden, or colorless. The description of *B. subvibrioides* is the same as that for the genus, with the additional characters given in Table BXII.α.107 and BXII.α.108.

Isolated from water and soil.

The mol% G + C of the DNA is: 67 (Bd); 68 (HPLC).

Type strain: CB81, ATCC 15264, LMG 14903.

GenBank accession number (16S rRNA): AJ227784.

7. **Brevundimonas variabilis** (Poindexter 1989) Abraham, Strömpl, Meyer, Lindholst, Moore, Christ, Vancanneyt, Tindall, Bennasar, Smit and Tesar 1999, 1071[VP] (*Caulobacter variabilis* Poindexter 1989, 495.)

var.i.a′ bil.is. M.L. adj. *variabilis* variable.

Short, thick, rod-shaped cells; position of stalk variable, arising from the center of the cell pole or from a subpolar (eccentric) site. Colonies red-orange or colorless. The description of *B. variabilis* is the same as that for the genus, with the additional characters given in Table BXII.α.107 and BXII.α.108.

Isolated from water and soil.

The mol% G + C of the DNA is: 65 (HPLC).

Type strain: CB17, ATCC 15255, DSM 4737, LMG 18362.

GenBank accession number (16S rRNA): AJ227783.

8. **Brevundimonas vesicularis** (Büsing, Döll and Freytag 1953) Segers, Vancanneyt, Pot, Torck, Hoste, Dewettinck, Falsen, Kersters and De Vos 1994, 508[VP] (*Pseudomonas vesicularis* (Büsing, Döll and Freytag 1953) Galarneault and Leifson 1964, 167; "*Corynebacterium vesiculare*" Büsing, Döll and Freytag 1953, 76.)

ve.si.cu.la′ ris. M.L. adj. *vesicularis* pertaining to a vesicle.

Short rods. Motile by means of one polar flagellum that has a short wavelength (0.6–1 μm). No prosthecae or special cell division. An intracellular carotenoid pigment (pink, yellow, or orange) is usually produced. The description of *B. vesicularis* is the same as that for the genus with the additional characters given in Table BXII.α.107 and BXII.α.108.

Isolated from water, aqueous solutions, and diverse clinical specimens from humans and animals.

The mol% G + C of the DNA is: 65–66 (T_m).

Type strain: ATCC 11426, CCUG 2032, DSM 7226, LMG 2350.

GenBank accession number (16S rRNA): AJ227780.

Other Organisms

1. *Mycoplana bullata* Gray and Thornton 1928, 83[AL]

bul.la′ ta. L. adj. *bullata* with a knob.

As shown above and reported in the literature (Fig. BXII.α.122; Abraham et al., 1999), *M. bullata* can also be phylogenetically considered a member of the genus *Brevundimonas*.

Cells are rod-shaped with round ends, occurring singly or rarely in pairs, motile by means of peritrichous flagella. They form branching filaments prior to fragmentation. No stalk formation. Calcium pantothenate and an unidentified compound are required for growth. Nitrate reduction and catalase activity are weak or negative. The Voges–Proskauer test is positive. Acid is produced oxidatively from D-glucose but is not produced from D-arabinose, D-xylose, D-mannose, D-fructose, D-galactose, maltose, sucrose, lactose, trehalose, D-sorbitol, D-mannitol, inositol, glycerol, or soluble starch. Other features are given in Table BXII.α.107 and in Urakami et al. (1990d).

Isolated from soil.

The mol% G + C of the DNA is: 66–68 (T_m).

Deposited strain: ATCC 4278, DSM 7126, IFO 13290, LMG 17157.

GenBank accession number (16S rRNA): D12785.

Genus IV. **Phenylobacterium** Lingens, Blecher, Blecher, Blobel, Eberspächer, Fröhner, Görisch, Görisch and Layh 1985, 38[VP]

JÜRGEN EBERSPÄCHER

Phe.ny.lo.bac.te′ ri.um. Gr. n. *phen* benzene; Gr. n. *bacterion* a small rod; M.L. neut. n. *Phenylobacterium* benzene bacterium, so named because of its unique preference for phenyl moieties from heterocyclic compounds as carbon sources.

Rods, coccobacilli, or cocci; 0.7–1.0 × 1.0–2.0 μm occurring singly, in pairs, or in short chains. Some strains tend to clump. In old cultures, pleomorphic forms, such as long rods (1.0 × 2.0–4.0 μm), long chains (10–50 μm) connected by filaments, and elliptical forms, may occur. Capsule stain negative; flexible capsule present. **Nonmotile.** Nonsporeforming. Non-acid-fast. Sheaths and prosthecae not produced. Nonpigmented. **Gram negative.** Growth on chloridazon–mineral salts agar is slow and colonies are small (1–2 mm) after 2–3 weeks. Colonies may be either smooth, convex, moist with shiny surfaces and entire edges, and easily emulsified in saline, or rough, dry, and not emulsifiable in saline. In liquid media, there is slight production of a greenish-yellow nonfluorescent pigment; in media containing L-phenylalanine a yellowish-green fluorescent pigment is produced. **Strictly aerobic. Catalase positive, weakly oxidase positive. Chemoorganotrophic, having a strictly respiratory type of metabolism.** Optimal temperature, 28–30°C. No growth at 4°C or 37°C; at 37°C cultures die within several days. Optimal pH 6.8–7.0; growth occurs between pH 6.5 and 8; no growth at pH 4 or 9. **Osmotically sensitive. Vitamin B_{12} is a growth factor. High nutritional specialization; grows well on chloridazon, antipyrin, and L-phenylalanine (most strains only after long lag phase).** Slow growth on L-glutamate, pyruvate, fumarate, succinate, and malate as well as on diluted complex media (0.5–2 g/l peptone). **Most sugars, alcohols, amino acids, carboxylic acids, and ordinary complex media are not utilized.** Does not denitrify; does not produce nitrite from nitrate. NH_4^+ and NO_3^- are used as sole sources of nitrogen; no growth with N_2. Gelatin, casein, starch, and esculin not hydrolyzed. Urease negative. Litmus milk negative. Weak H_2S production from thiosulfate or cysteine. Methyl red and Voges–Proskauer negative. Indole negative. No acid or gas produced from sugars or alcohols. Not pathogenic for rats or rabbits. **Isolated from soil after enrichment in mineral salts media containing chloridazon, antipyrin, or pyramidon** (synthetic heterocyclic compounds).

The mol% G + C of the DNA is: 65–68.5.

Type species: **Phenylobacterium immobile** Lingens, Blecher, Blecher, Blobel, Eberspächer, Fröhner, Görisch, Görisch and Layh 1985, 38.

FURTHER DESCRIPTIVE INFORMATION

Two strains of *Phenylobacterium immobile*, namely strains E and A13 (see Fig. BXII.α.123, parts a and b) form exclusively coccal rods, which tend to clump, and growth in liquid cultures is sometimes flocculent. All other strains have short rod-shaped cells, which occur singly, in pairs, or in short chains. These strains never clump in liquid cultures. Chains are found mainly in cultures of strains that form smooth colonies on agar plates (e.g., strain K_2, see Fig. BXII.α.123c). In old cultures, especially when the bacteria are cultivated in a dilute complex medium that allows only poor growth, pleomorphic forms, such as long rods, long chains of cells connected by small filaments, or club-shaped and elliptical forms, sometimes occur.

Capsule staining and Gram staining are negative. Electron microscopy of thin sections reveal the typical Gram-negative cell wall pattern. Ultrathin sections of ruthenium red-treated cells of strain K_2, which forms smooth colonies, show a microcapsule surrounded by a slime layer of acidic polysaccharides (Fig. BXII.α.124a). No slime layer is detected in ruthenium-red-stained cells of strain E, which forms rough colonies on agar (Fig. BXII.α.124b). According to Costerton et al. (1981), *Phenylobacterium* cells possess flexible capsules, which can be made visible only by electron microscopy, but not rigid capsules, which represent the typical India ink-excluding capsule type.

The murein main component from the type strain is identical to the C_6 muropeptide of *E. coli*. The carbohydrate moiety from the lipopolysaccharide of the type strain consists of heptose, 3-deoxyoctulosonic acid, and D-glucose in a molar ratio of 1:2:2.3 (Weisshaar and Lingens, 1983). Lipid A is composed of 1 mol 2,3-diamino-2,3-dideoxy-D-glucose, 2 mol amide-bound fatty acids, and 2.6 mol ester-bound fatty acids per mol lipid A. 2,3-diamino-2,3-dideoxy-D-glucose is typical of *Alphaproteobacteria* but absent in other Gram-negative bacteria, which contain glucosamine as the main lipid A component. Amide-bound fatty acids are 3-hydroxydodecanoic acid and 3-hydroxyhexadecanoic acid; ester-linked fatty acids are dodecanoic acid and 3-hydroxy-5-*cis*-dodecenoic acid. The detection of the latter, unusual fatty acid, not found in nature before, can be used to demonstrate the presence of *Phenylobacterium immobile* in soil samples (Bellmann and Lingens, 1985).

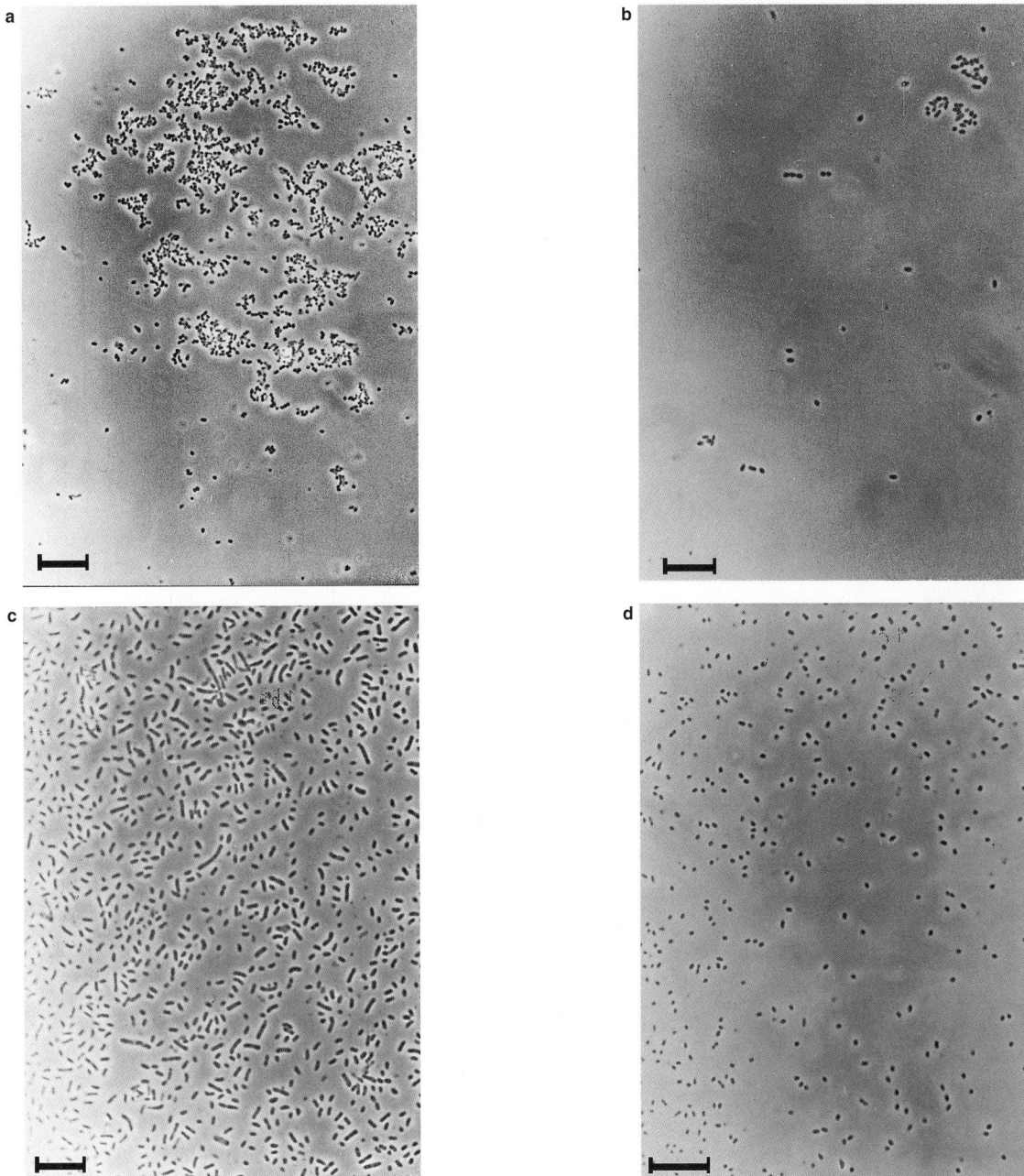

FIGURE BXII.α.123. Phase contrast photomicrographs of cells of *Phenylobacterium immobile*. *a*) Strain A₁₃; *b*) strain E, the type strain; *c*) strain K₂; and *d*) strain N. Bars = 10 μm.

In liquid cultures, during growth on mineral salts medium with phenylalanine, especially at higher concentrations (3–5 g/l), a yellowish-green fluorescent pigment is produced. On chloridazon or antipyrin mineral salts media, a greenish-yellow nonfluorescent pigment is formed.

Optimal growth on agar plates occurs on mineral salts medium containing chloridazon, antipyrin, or L-phenylalanine as the carbon source. Single colonies are not visible before 4–7 days, and after 2–3 weeks the colonies are 1–2 mm in diameter, circular with entire edges, slightly raised, and not adherent to the agar. Nearly half of the strains form smooth and shiny colonies that can be readily emulsified in water. The other strains have rough and dry colonies that clump when suspended in water.

Each different strain of *Phenylobacterium immobile* does not utilize all three xenobiotic substrates (formula, Fig. BXII.α.125). Whereas chloridazon and antipyrin are well utilized by most strains, only 7 of the 22 isolates use pyramidon as a growth substrate (Table BXII.α.109). When pyramidon is added to media containing chloridazon or antipyrin, the growth of all isolates is inhibited.

The pathway for the degradation of the three xenobiotics follows the well-known route for the oxidative dissimilation of aromatic compounds. From these xenobiotic compounds, only the phenyl moiety is used as a carbon source; the heterocyclic moiety remains unchanged. In the first step, O_2 is incorporated in the benzene nucleus by the action of a dioxygenase. The

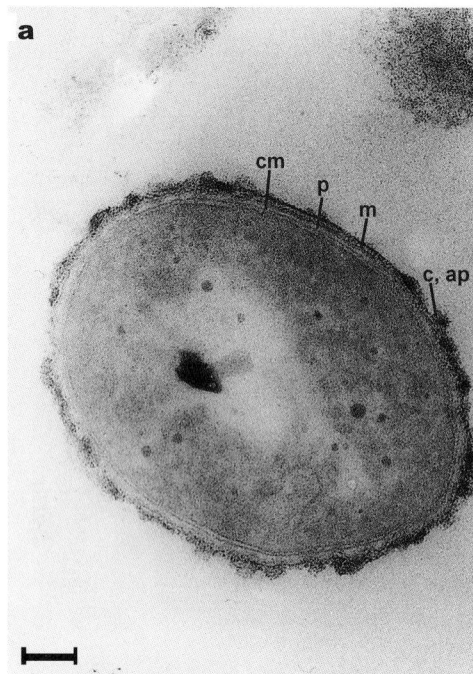

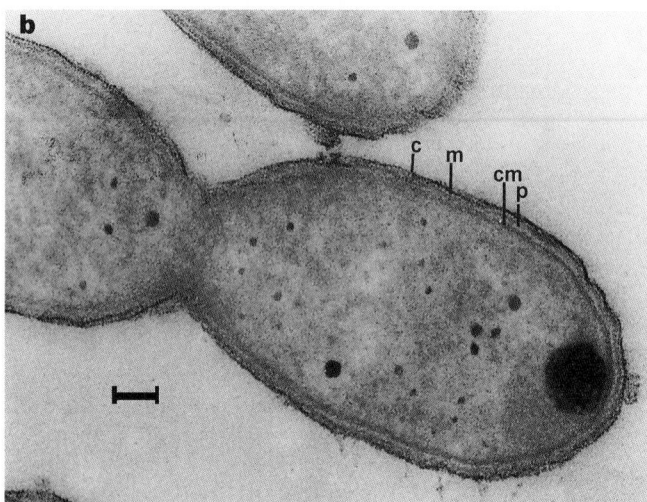

FIGURE BXII.α.124. Ultrathin sections of ruthenium red-treated cells of *Phenylobacterium immobile. a*) strain K₂; *b*) strain Eᵀ. *cm*, cytoplasmic membrane; *p*, periplasm; *m*, outer membrane; *c*, capsule (not stained); *c,ap*, capsule consisting of acid polysaccharides (stained with ruthenium red).

FIGURE BXII.α.125. The herbicide chloridazon and the structurally related analgesics antipyrin and pyramidon were used for the enrichment of *Phenylobacterium immobile* from soil. These xenobiotics are also the best growth substrates for the bacteria.

resulting *cis*-dihydro-dihydroxy compound is oxidized by a dehydrogenase, yielding a catechol derivative, which leads to a dephenylated heterocycle and 2-hydroxymuconate by *meta*-cleavage and the subsequent action of an amidase. Meta-cleaving enzymes from *P. immobile* form a distinct group among nonheme-iron dioxygenases (Schmitt et al., 1984) 2-Hydroxymuconate is further converted via 2-oxo-4-hydroxyvalerate to pyruvate and acetaldehyde, two compounds of intermediary metabolism that are potential carbon sources for the bacterium.

Citric acid cycle enzymes and enzymes catalyzing anaplerotic routes are present in *P. immobile* strain E. Tyrosine is synthesized exclusively via the arogenate pathway, not via prephenate (Keller et al., 1982).

More than 20 different heterocyclic or aromatic compounds that are structurally related to chloridazon or antipyrin have been tested as possible carbon sources. Most chloridazon analogs with an altered heterocyclic moiety are good growth substrates. A substitution at the aromatic nucleus, as for *o-*, *m-*, or *p*-methylchloridazon, makes the compound nondegradable. Of the aniline derivatives tested, *N*-methylacetanilide and *N*-methylformanilide are poor growth substrates. The following compounds fail to support growth: benzene, toluene, phenol, catechol, benzaldehyde, benzoate, and a number of mono- and dihydroxylated benzoates. One strain (strain N) grows well on L-phenylalanine with a normal lag phase of 1 day; all other strains have lag phases of 2–3 weeks but then grow well on phenylalanine. The long lag phases are observed only for the first transfer of the strains from chloridazon or antipyrin to phenylalanine; thereafter the bacteria grow immediately. Phenylalanine-induced cells also grow well on phenylpropionate, phenylpyruvate, and phenyllactate.

All strains tested are susceptible to tetracycline, bacitracin, chloramphenicol, kanamycin, and streptomycin. The majority (18 of 22) are susceptible to penicillin G and 15 of 22 strains are susceptible to novobiocin. All are resistant to cephaloridine, cloxacillin, fusidic acid, lincomycin, methicillin, and sulfafurazol.

P. immobile strains were found to be harmless when tested in rats and rabbits (Kaiser et al., 1981). No adverse reactions were observed when cells of strains E and N, respectively, were orally administered to rats over a period of 7 d. Exposure via air in inhalation experiments did not lead to specific pulmonary changes. Intracutaneous injection did not cause adverse skin reactions, and intraperitoneal injections did not kill the rats, although bacteria entered the blood. Intravenous injections of living and formalin-inactivated bacteria into rabbits during immunization did not lead to toxic effects.

As shown in Table BXII.α.109, depending on the strain, between one and six different plasmids have been found, which vary in size from 8–300 mDa (Kreis et al., 1981).

Agglutination and immunofluorescence tests with antisera against 4 strains of *Phenylobacterium immobile* have revealed the serological uniformity of the different strains (Layh et al., 1983). Slight differences in immune reactions allow a classification of the strains into 5 serological subgroups. No relationship was found between *Phenylobacterium immobile* and 40 representative Gram-negative bacteria, including *Acinetobacter calcoaceticus, Azospirillum brasilense, Caulobacter* spp., *Paracoccus denitrificans*, some *Pseudomonas* spp., *Rhizobium* spp., *Rhodomicrobium vannielii*, and *Rhodopseudomonas capsulata*. A slight but significant immunofluorescence reaction has been observed with *Brevundimonas vesicularis, Gluconobacter oxydans, Aquaspirillum itersonii*, and *Rhodospirillum rubrum* (Dorfer et al., 1985). Crossed immunoelectrophoresis reveals a serological relationship between *Phenylobacterium immobile* and *Brevundimonas diminuta*.

TABLE BXII.α.109. Characteristics of the different strains of *Phenylobacterium immobile*[a]

Characteristic	Strain (laboratory designation)[b]																					
	A_6	A_{11}	A_{12}[c]	A_{13}	A_{14}	C_2[d]	ET[e]	J_1[f]	J_2	K_2[g]	K_3	K_5	L	M_{11}	M_{13}	M_{15}	N[h]	R[i]	Z_5	Z_6[j]	Z_7	Z_8
Carbon source for enrichment[k]	A	A	A	A	A	C	C	C	C	C	C	C	C	P	P	P	C	C	C	C	C	C
Growth on:																						
Chloridazon	+	–	+	–	–	+	+	+	+	+	+	+	+	–	–	–	+	+	+	+	+	+
Antipyrin	+	+	+	+	+	+	+	+	+	+	+	+	+	+	+	+	+	+	+	+	+	+
Pyramidon	–	+	–	–	–	+	–	–	+	–	–	–	+	+	+	+	–	–	–	–	–	+
Colonies[l]	R	R	S	R	R	R	R	S	S	S	S	S	S	S	R	R	S	S	R	R	R	R
Cells tend to clump	+	+	–	+	+	+	+	+	+	–	–	–	–	–	+	+	–	–	+	+	+	–
Serological subgroup[m]	V	III	V	V	V	II (m)	V (m)	IV	IV	I	II	II	III	I (m)	III	III	IV	III	V	V	V	I/II
Number of plasmids	3	2	2	4	2	1	3	3	4	2	2	2	4	nd	nd	3	3	3	4	6	4	3
Plasmid size (mDa)	8.6, 11.7, 14.0	10.3, 18.8	16.3, 32.8	10.3, 15.2, 15.7, 21.0	10.3, 15.2	170	8.6, 11.7, 14.0	11.7, 14.0, 19.1	10.6, 22.6, 133, 304	14.0, 21.5	14.0, 20.2	14.0, 21.5	11.7, 14.0, 17.3, 213	nd	nd	11.0, 17.0, 22.6	9.0, 17.7, 262	8.6, 11.7, 14.0	6.4, 11.7, 14.0, 19.8	7.7, 8.6, 11.7, 13.4, 14.0, 19.1	11.7, 13.4, 14.0, 19.8	11.7, 14.0, 19.1

[a]For symbols see standard definitions; nd, not determined.

[b]Strains were isolated from soils in the following locations: A_6, A_{14}, M_{15}, Z_5, Z_6, Z_7, Z_8 - Hohenheim, Germany; A_{11}, A_{12}, A_{13}, M_{11}, M_{13} - Wimpfen, Germany; K_2, K_3, K_5 - Eldoret-Nakaru, Kenya; L - Limburgerhof, Germany; N - Lincoln, Nebraska USA; R - Rothschwaige, Germany; C_2 - Cardwell, Australia; ET - Ecuador; J_1, J_2 - Lyngby, Denmark;

[c]ATCC 35972, DSM 2115.

[d]CCM 3864.

[e]ATCC 35973, DSM 1986.

[f]ATCC 35974, DSM 2116.

[g]ATCC 35975, DSM 2117.

[h]ATCC 35976, DSM 2113.

[i]CCM 3865.

[j]ATCC 35977, DSM 2114.

[k]A, antipyrin; C, chloridazon; P, pyramidon.

[l]R, rough; S, smooth.

[m](m), marginal position.

The detection of *Phenylobacterium immobile* as a new genus was the unintentional result of studies on the breakdown of the herbicide chloridazon. Chloridazon is the active ingredient of the herbicide Pyramin®. This compound has been used since about 1960 for the control of weeds in sugar beet and beet root cultures. The decomposition of chloridazon (formerly named pyrazon) was demonstrated to be a microbial process (Drescher and Otto, 1969; Frank and Switzer, 1969). The isolation of bacteria with the ability to grow on chloridazon as the sole carbon source was described by Engvild and Jensen (1969) and by Fröhner et al. (1970). Since 1969, more than 20 different strains that grow with the herbicide chloridazon or the structurally related analgesics antipyrin and pyramidon have been isolated from soils from various places over the world (Table BXII.α.109).

All attempts to identify the different isolates of chloridazon-degrading bacteria on the basis of routine characters have led to unsatisfactory results. The high nutritional specialization of these bacteria, which grow optimally on man-made compounds and utilize poorly only a few normal carbon sources, is their most distinguishing feature, along with the fact that nearly all biochemical tests give negative results.

P. immobile seems to be a typical inhabitant of the upper aerobic part of the soil. Different strains have been isolated from soil samples from various locations all over the world. All efforts to isolate chloridazon-degrading bacteria which grow at 37°C have failed. For this purpose, soil and water samples from hot springs, from near-volcanic regions, and from regions with tropical climates were subjected to chloridazon enrichment. Attempts to demonstrate the breakdown of chloridazon in soil or in mud samples under anaerobic conditions failed. In one case, a slow degradation of chloridazon in river water was observed, but efforts to isolate chloridazon-degrading bacteria from this specific water sample failed. However, one cannot rule out the possibility that phenylobacteria may occur in aquatic habitats.

The technique for the enrichment of chloridazon-degrading bacteria leads to the isolation of organisms that are able to utilize synthetic molecules not normally encountered in nature. Obviously, the chemical character of the selective agent (heterocyclic plus a phenyl moiety) is a special challenge for microbial cells, for which only a special sort of organism (*Phenylobacterium immobile*) seems to have the adequate response. This view is supported by the fact that in many different soil samples, the same sort of bacteria are always isolated. Furthermore, efforts to isolate the same bacteria with L-phenylalanine as the selective agent are unsuccessful. Enrichment techniques and nutritional characteristics raise the question of which substrates are used by these bacteria in their natural environments. Using an immunofluorescence membrane filter technique, evidence has been obtained that *Phenylobacterium immobile*, or serologically closely related organisms, also occur in soils which have never been treated with the herbicide. This suggests that these bacteria are able to survive in nature without their optimal growth substrate by utilizing other substrates—perhaps L-phenylalanine or mixtures of organic compounds such as pyruvate, succinate, malate, fumarate, and L-glutamate, which allow moderate growth.

Enrichment and Isolation Procedures

The following procedure leads to the isolation of bacteria that are able to utilize as a sole carbon source man-made compounds not normally encountered in nature. As a synthetic substrate for selective enrichment, either the herbicide chloridazon, or the analgesics antipyrin or pyramidon (Fig. BXII.α.125) can be ap-

plied. Chloridazon, the active ingredient of the herbicide Pyramin®, can be obtained from Riedel de Haen, D-30918 Seelze, Germany; antipyrin and pyramidon (4-dimethylaminoantipyrin) are available from Sigma-Aldrich Chemie, D-89555 Steinheim, Germany. About 300 g soil or compost is mixed with 0.5 g chloridazon, antipyrin, or pyramidon, and the preparation is incubated at 30°C or room temperature in a flower pot and regularly moistened with water. Degradation of the xenobiotic compound can be observed chromatographically (TLC or HPLC) by testing the excess water that drains from the flower pot for the presence of the xenobiotic. Decomposition is complete when the xenobiotic is no longer detectable, and a new compound corresponding to the dephenylated heterocyclic moiety of the xenobiotic appears, usually after one to several weeks, depending on the soil. A 5 g sample of the active soil is then placed into an Erlenmeyer flask containing 50 ml of mineral salts medium[1] supplemented with the xenobiotic as the carbon source at a concentration of 0.04– 0.1%.

This culture is incubated on a rotary shaker at 30°C, and degradation is monitored chromatographically. When the decomposition of the xenobiotic is complete, 1 ml of the culture fluid is transferred into a new Erlenmeyer flask. After 5–10 transfers, a sample of the liquid culture is streaked onto agar plates containing the same medium. Single colonies, which normally appear after 1–3 weeks, are picked and again streaked onto agar. After 5–10 transfers, pure cultures can usually be obtained. Growth on chloridazon mineral salts agar allows the removal of chloridazon to be perceived as a clearance zone around the bacterial colonies (see Fig. BXII.α.126). Since none of the *P. immobile* strains grows on ordinary complex media, the inoculation of a complex agar medium can be used for testing purity. This medium contains per liter deionized water: peptone, 5 g; meat extract, 3 g; yeast extract, 5 g, NaCl, 5 g. Growth on this medium indicates contamination of the culture.

Maintenance Procedures

For short-term preservation, the bacteria are transferred onto agar at intervals of 2–3 weeks. For long-term preservation, a bacterial suspension in skim milk is dropped onto silica gel grains and stored at 4°C. Good results are obtained with this method when transfer is repeated every 2–3 years; some of the strains have been viable even after a period of 10 years. Storage at −80°C of a concentrated bacterial suspension in fresh chloridazon-mineral salts medium supplemented with 15% glycerol has resulted in good viability after a storage of more than 8 years.

Procedures for Testing Special Characters

Nutritional specialization is a main feature of *Phenylobacterium immobile*. Therefore, tests for growth on chloridazon or antipyrin are of special importance. Optimal growth and maximum cell yield are achieved at 30°C in mineral salts medium with chloridazon or antipyrin at 0.4–1 g/l. Under these conditions, a dou-

---

1. The mineral salts medium has the following composition per liter deionized water: $Na_2HPO_4 \cdot 12H_2O$, 0.7 g; KH_2PO_4, 0.3 g; $(NH_4)_2HPO_4$, 0.7 g; $(NH_4)H_2PO_4$, 0.3 g; $(NH_4)_2SO_4$, 0.1 g; trace element solution (see text) 1 ml; vitamin B_{12} solution (0.03 mg/ml), 1 ml, $MgSO_4 \cdot 7H_2O$, 0.25 g; $CaCl_2 \cdot 6H_2O$, 0.05 g. To avoid precipitates, the magnesium and calcium salts are each dissolved separately. Trace element solution per liter deionized water: $MnSO_4 \cdot 4H_2O$, 400 mg; $ZnSO_4 \cdot 7H_2O$, 400 mg; $FeCl_3 \cdot 6H_2O$, 200 mg; $CuSO_4 \cdot 5H_2O$, 40 mg; H_3BO_3, 500 mg; $(NH_4)_2MoO_4 \cdot 4H_2O$, 200 mg; KI, 100 mg; biotin, 100 mg. Either chloridazon, antipyrin, or pyramidon is added as the carbon source at a concentration of 0.4–1 g/l. The pH of the medium is 7.0.

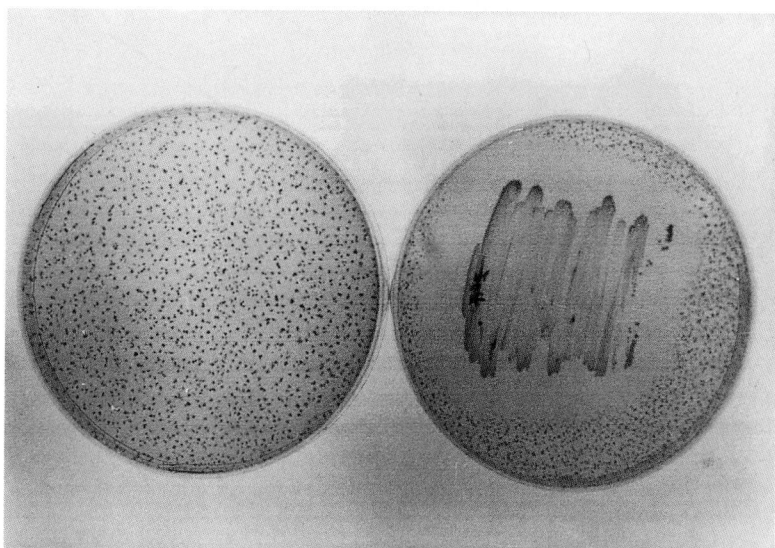

FIGURE BXII.α.126. Agar plates containing mineral salts medium with 2 g/l chloridazon. Fine crystals of chloridazon have precipitated in the agar. On the right-hand plate, a clearance zone around the bacterial smear of *Phenylobacterium immobile* has formed by chloridazon degradation. At the left margin of the bacterial smear, a large crystal is visible, which corresponds to the dephenylated heterocyclic moiety of chloridazon, the main metabolite of chloridazon degradation.

bling time of 7–8 h is observed, and, depending on the strain, a yield of about 0.4–1.0 g/l bacteria (wet weight) of culture fluid is obtained. Mineral salts medium with either 0.2% chloridazon or 0.1% antipyrin as the carbon source, supplemented with 15 g/l agar, allows good growth of *Phenylobacterium immobile*. Growth on agar may be made visible as follows: Agar plates containing mineral salts medium with 2 g/l chloridazon show a fine precipitate of chloridazon crystals in the agar. During growth on this agar, *Phenylobacterium immobile* removes the crystals by degradation, and a clearance zone around the bacterial smear develops (Fig. BXII.α.126). In agar cultures 4 or more weeks old, a new, and in most cases, relatively large sort of crystals forms within the bacterial smear. These crystals have been identified as the dephenylated heterocyclic moiety of chloridazon, which is a dead-end metabolite of chloridazon degradation.

The osmotic sensitivity and nutritional specialization of *Phenylobacterium immobile* do not allow the use of routine media for biochemical characterization. Therefore, tests have to be performed in modified media on which the bacteria are able to multiply. Mineral salts medium with 0.5 g chloridazon or 1 g antipyrin as the carbon source can be applied for testing catalase, oxidase, antibiotic susceptibility, and arginine-, lysine- and ornithine-decarboxylase. Mineral salts medium containing 1 g peptone, 1 g yeast extract, and 0.5 g antipyrin or chloridazon can be used for testing urease, indole reaction, Voges–Proskauer, methyl red, H_2S production, production of nitrite from nitrate, gelatin, starch, cellulose, and esculin hydrolysis.

DIFFERENTIATION OF THE GENUS *PHENYLOBACTERIUM* FROM OTHER GENERA

Small subunit rRNA sequence analysis has shown sequence similarities between *P. immobile* and several *Caulobacter* and *Mycoplana* species, as well as one *Afipia* genospecies. A phylogenetic tree reflecting the relationships of *P. immobile* and selected members of the *Caulobacteraceae* group is shown in Fig. BXII.α.127. The sequence similarity values suggest a close phylogenetic relation between these members of the *Alphaproteobacteria*, despite great differences in their phenotypes and ways of life. *P. immobile* is not prosthecate like *Caulobacter*; however, it shares several properties, such as an oligotrophic nature with ready growth on dilute, complex medium (below 0.1% organic material), a strictly respiratory and aerobic metabolism, and a similar mol% G + C content of the DNA with *Caulobacter*. *Phenylobacterium* is Gram-negative like *Mycoplana* but, unlike *Mycoplana*, it does not exhibit branching filaments that fragment into motile, irregular rods—a trait that has led several workers to place *Mycoplana* in the order *Actinomycetales*. From DNA–RNA hybridization studies *M. dimorpha* and *M. bullata* seem to be remote relatives of the family *Rhizobiaceae*, a family whose members were also found to show some sequence similarity with *Phenylobacterium*; however, *Phenylobacterium* is neither a nitrogen fixer nor associated with plants. *Afipia* is the causal organism of cat scratch disease, whereas *Phenylobacterium* seems to be nonpathogenic. Comparative sequence analysis of the small subunit rRNA has demonstrated that the root-nodulating *Bradyrhizobium japonicum* and the budding bacteria *Blastobacter denitrificans* form a tight phylogenetic group with the human pathogens *Afipia felis* and *Afipia clevelandensis* (Willems and Collins, 1992).

TAXONOMIC COMMENTS

Analysis of the 16S rDNA sequence[2] (1466 nucleotides) of the type strain (strain E), obtained by direct sequencing of the PCR

2. The 16S rDNA of the type strain was amplified by polymerase chain reaction using primers 27f and 1522rN from the 5′- and 3′-ends of the gene. Direct sequencing of the amplified product was performed using primers to conserved regions of the rRNA. The complete sequence of the 16S rDNA of the type strain, comprising 1466 nucleotides, was deposited at the EMBL Nucleotide Sequence Database under the accession number Y18216. Potential diagnostic targets for *Phenylobacterium*-specific probes were identified by analyzing the complete small subunit rRNA sequence data set (ARB data base; approximately 13,000 sequences) using the probe design and probe match tools of the ARB program package. The most promising diagnostic probes are: 5′-CCUGAUCGCCGGAGAGAU-3′ (*E. coli* pos. 1000–1017) and 5′-AAUGGUACUGCCGAGGUU-3′ (*E. coli* pos. 1151–1168).

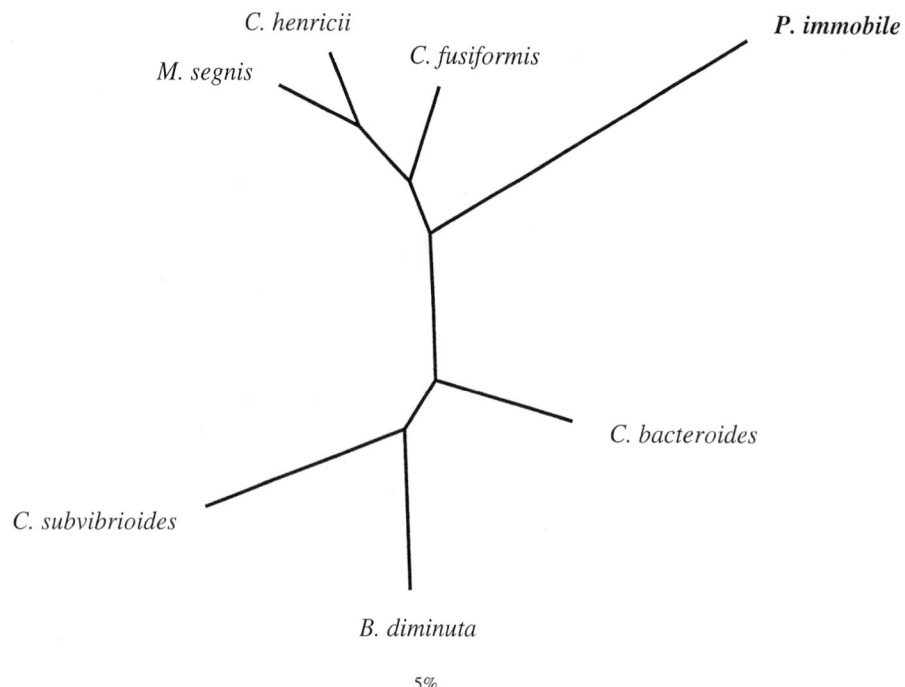

FIGURE BXII.α.127. Phylogenetic tree reflecting the relationships of *Phenylobacterium immobile* and selected members of the *Caulobacteraceae* group (*Caulobacter* spp., *Mycoplana* spp., and *Brevundimonas diminuta*). Only type strains for which 16S rRNA sequences are available are shown. The bar indicates 5% estimated sequence divergence. The tree was constructed on the basis of the complete small subunit rRNA sequence data (ARB data base, approximately 13,000 sequences. Compare: Ludwig, W. and Strunk, O. 1996 ARB: a software environment for sequence data.). The tree topology was evaluated by performing maximum parsimony and distance matrix analyses of the complete small subunit rRNA data set or a subset comprising all almost complete proteobacterial sequences together with reference data from selected representatives of the other bacterial lines of descent, respectively. Multiple data sets which varied with respect to the selection of alignment positions included were analyzed applying the alternative treeing methods. Alignment positions were included or excluded according to their degree of conservation. The majority of results support the depicted tree topology, which is based upon a data set comprising alignment positions that share identical residues in at least 50% of all sequences from members of the *Caulobacteraceae* group.

amplified gene product, reveals that *Phenylobacterium immobile* is a member of the *Caulobacteraceae*. Sequence similarities of 91.6–94.5% have been found for various *Caulobacter* species (*C. henricii*, *C. fusiformis*, *C. vibrioides*, *C. crescentus*, *C. bacteroides*, and *C. subvibrioides*). In addition, 95.9% similarity is obtained with *Afipia* genosp. 14, 94.3% with *Caulobacter segnis* (*Mycoplana segnis*), 93.2% with *M. bullata*, and 93.1% with *Brevundimonas diminuta*. Significantly, lower similarities of 90% and below are obtained with species of *Rhizobium*, *Paracoccus*, *Agrobacterium*, and other members of the *Alphaproteobacteria*. These results are in accordance with partial sequence analysis of 16S rRNA data of the type strain (Ludwig et al., 1984). Based on the formerly applied nucleotide cataloging method, *Phenylobacterium immobile* is a member of the *Alphaproteobacteria* with highest similarity (S_{AB} values of 0.51) to *Brevundimonas diminuta* (formerly *Pseudomonas diminuta*) and *Rhizobium leguminosarum*.

Several other features support the phylogenetic position of *Phenylobacterium* that has been established by small subunit rRNA sequence analysis. The presence of 2,3-diamino-2,3-dideoxy-D-glucose as a lipid A constituent of *Phenylobacterium immobile* indicates its relationship to the *Alphaproteobacteria*. This unusual sugar has also been detected in *Rhodopseudomonas viridis*, *R. palustris*, *R. sulfoviridis*, *Brevundimonas diminuta*, *B. vesicularis*, and *Nitrobacter winogradskyi* (Weckesser and Mayer, 1987). Busse and Auling (1988) have shown that polyamines may serve as a useful chemotaxonomic marker within the *Proteobacteria*. Like other species of the *Alphaproteobacteria*, *P. immobile* contains *syn*-homospermidine exclusively, as well as ubiquinone Q-10, composed of 10 isoprenoid units (R.M. Kroppenstedt, J. Eberspächer, and F. Lingens, unpublished). Weak serological reactions of *Phenylobacterium immobile* occur only with members of the *Alphaproteobacteria* (Dorfer et al., 1985) and not with any other Gram-negative bacteria.

DNA hybridization tests reveal 100% similarity of DNA preparations among four different strains with DNA of strain R. No similarity has been found using DNA from *Agrobacterium tumefaciens*, *Bacillus subtilis*, *Escherichia coli*, *Pseudomonas fluorescens*, or calf thymus. Morphological, physiological, biochemical, and nutritional studies, and serological and enzymological data have also demonstrated the high degree of similarity among the more than 20 different isolates. Therefore, the isolates have been grouped together in one single species. On the other hand, minor, but significant, differences exist in cell and colony morphology, in plasmid pattern, in the utilization of the xenobiotics chloridazon, antipyrin and pyramidon, and in serological properties that allow a classification into 5 serological subgroups (Table BXII.α.109).

The construction of *Phenylobacterium*-specific probes using 16S rDNA sequence analysis data should allow the detection of *Phenylobacterium immobile* and closely related bacteria in environmental samples without prior enrichment with xenobiotics. This technique using diagnostic target probes seems to be more specific than the previously applied serological method. Based on the probe examination of environmental samples, it would seem to be feasible to isolate *P. immobile* or *P. immobile*-like organisms without using the selective enrichment technique for growth on the xenobiotics chloridazon, antipyrin, or pyramidon.

ACKNOWLEDGMENTS

The author is greatly indebted to Dr. Wolfgang Ludwig, Institute of Microbiology, Technical University of Munich, for processing 16S rDNA data and constructing a phylogenetic tree, and to Prof. Dr. Michael Schlömann and his coworkers from the Institute of Microbiology, University of Stuttgart, for his introduction into the 16S-rDNA sequencing method.

FURTHER READING

Lingens, F., R. Blecher, H. Blecher, F. Blobel, J. Eberspächer, C. Fröhner, H. Görisch, H. Görisch and G. Layh. 1985. *Phenylobacterium immobile* gen. nov., sp. nov., a Gram-negative bacterium that degrades the herbicide chloridazon. Int. J. Syst. Bacteriol. *35*: 26–39.

List of species of the genus Phenylobacterium

1. **Phenylobacterium immobile** Lingens, Blecher, Blecher, Blobel, Eberspächer, Fröhner, Görisch, Görisch and Layh 1985, 38[VP]

im.mo'bi.le. L. adj. *immobile* nonmotile.

The characteristics are as described for the genus and in Tables BXII.α.109 and BXII.α.110. *P. immobile* uses relatively few simple carbon sources. Growth occurs on L-glutamate, pyruvate, fumarate, succinate, and malate, but more poorly than on chloridazon or L-phenylalanine. A mixture of these compounds, e.g., pyruvate plus L-glutamate, yields better growth than do single compounds, although the growth rate and cell yield are still lower than with chloridazon. No growth occurs on more than 40 compounds tested as possible carbon sources, including various sugars, alcohols, carboxylic acids, and amino acids. A similar spectrum of compounds is negative for acid or gas production. Complex media with 10–20 g/l peptone or meat extract plus yeast extract does not support growth, whereas with 0.5–2 g/l peptone plus yeast extract, growth occurs, but more slowly than on chloridazon. *Phenylobacterium immobile* is osmotically sensitive; the addition of 5–7 g NaCl per liter chloridazon mineral salts medium result in considerable growth inhibition, 10 g NaCl leads to total growth inhibition.

The mol% G + C of the DNA is: 65–66.5 (Bd); strain C_2 has a somewhat higher value of 68.5 (Bd).

Type strain: E, ATTC 35973, DSMZ 1986.

GenBank accession number (16S rRNA): Y18216.

TABLE BXII.α.110. General characteristics of *Phenylobacterium immobile*[a]

Characteristic	Phenylobacterium immobile
Cell diameter, μm	0.7–1.0
Cell length, μm	1.0–2.0
Cells single, in pairs or short chains	+
Motility	−
Spores, sheaths, prosthecae	−
Pigmented	−
Long rods or chains or elliptical cells in old cultures	d
Gram-stain, Ziehl–Neelsen-stain, capsule-stain	−
Colonies smooth	d
Colonies rough	d
Strictly aerobic	+
Catalase	+
Oxidase	w
Growth at 4°C and 37°C	−
Optimal growth temperature, °C	28–30
Growth at pH 4 and 9	−
Optimal pH	6.8–7.0
Growth inhibition by 0.5–0.7% NaCl	+
Growth factor vitamin B_{12}	+
Optimal growth on chloridazon, antipyrin, and phenylalanine (after long lag phase)	+
Fluorescent pigment on phenylalanine medium	+

(continued)

TABLE BXII.α.110. *(cont.)*

Characteristic	Phenylobacterium immobile
Non-fluorescent pigment on chloridazon medium	+
Moderate growth on glutamate, succinate, pyruvate, fumarate, malate	+
Growth on sugars, alcohols, amino acids (40 compounds tested)	−
Growth on aromatic compounds (18 compounds tested)	−
Acid or gas from sugars and alcohols (34 compounds tested)	−
Growth on complex media	−
Growth on dilute complex media (0.5–2 g/l organic material)	+
NH_4^+ and NO_3^- used as N-sources	+
Denitrification	−
N_2-fixation	−
Gelatine, casein, starch, esculin hydrolyzed	−
Litmus milk	−
H_2S from cysteine or thiosulfate	w
Methyl red, Voges–Proskauer, indole	−
Mol% G + C of DNA (Bd)	65–68.5

[a]For symbols see standard definitions.

Order VI. **Rhizobiales** *ord. nov.*

L. DAVID KUYKENDALL

Rhi.zo.bi.a' les. M.L. n. *Rhizobium* type genus of the family *Rhizobiaceae*; *-ales* ending to denote an order; M.L. fem. n. *Rhizobiales* the *Rhizobium* order.

The order *Rhizobiales* is a phenotypically heterogeneous assemblage of Gram-negative bacteria and is based solely on 16S rRNA gene sequence analysis.

Type genus: **Rhizobium** Frank 1889, 338.

Families of the Order Rhizobiales

    I. Family *Rhizobiaceae*
    II. Family *Bartonellaceae*
    III. Family *Brucellaceae*
    IV. Family *Phyllobacteriaceae*
    V. Family *Methylocystaceae*
    VI. Family *Beijerinckiaceae*
    VII. Family *Bradyrhizobiaceae*
    VIII. Family *Hyphomicrobiaceae*
    IX. Family *Methylobacteriaceae*
    X. Family *Rhodobiaceae*

Family I. **Rhizobiaceae** Conn 1938, 321[AL]

L. DAVID KUYKENDALL

Rhi.zo.bi.a' ce.ae. M.L. n. *Rhizobium* type genus of the family; *-aceae* ending to denote a family; M.L. fem. n. *Rhizobiaceae* the *Rhizobium* family.

The family *Rhizobiaceae* is a phenotypically heterogeneous assemblage of aerobic, Gram-negative rod-shaped bacteria and is based solely on 16S rRNA gene sequence analysis. (Table BXII.α.111)

Type genus: **Rhizobium** Frank 1889, 338.

Genera of the Family Rhizobiaceae

    I. Genus *Rhizobium*
    II. Genus *Agrobacterium*
    III. Genus *Carbophilus*
    IV. Genus *Chelatobacter*
    V. Genus *Ensifer*
    VI. Genus *Sinorhizobium*

TABLE BXII.α.111.   Some determinative phenotypic features of the genera of the family *Rhizobiaceae*[a]

Characteristic	*Rhizobium, Sinorhizobium, Allorhizobium*	*Agrobacterium*	*Carbophilus*	*Chelatobacter*	*Ensifer*
Reproduce by budding at one pole of the cell	−	−	nd	d	+
Attach endwise to various host bacteria and may cause lysis of the host cells	−	nd	−	−	+
Some strains induce hypertrophisms in plants as root nodules with or without symbiotic nitrogen fixation	+[b]	−[b]	nd	nd	nd
Some strains induce tumorous galls in plants	−[b]	+[b]	nd	nd	nd
Aminopolycarboxylic acid, nitrilotriacetic acid can be utilized as a sole source of carbon/energy and nitrogen	−	−	nd	+	nd
Can grow chemolithoautotrophically with carbon monoxide (CO) as a sole carbon and energy source	−	−	+	nd	nd

[a]For symbols see standard definitions; nd, not determined.

[b]Traditionally, *Agrobacterium* has been associated with the incitement of tumorous galls and/or hairy roots in plants, whereas *Rhizobium*, *Sinorhizobium*, and *Allorhizobium* have been associated with the symbiotic induction of root nodules. However, because these features are plasmid-mediated, this is not a reliable basis for classification of these genera.

Genus I. Rhizobium *Frank 1889, 338*[AL]

L. DAVID KUYKENDALL, JOHN M. YOUNG, ESPERANZA MARTÍNEZ-ROMERO, ALLEN KERR AND HIROYUKI SAWADA

Rhi.zo'bi.um. Gr. n. *rhiza* a root; Gr. n. *bios* life; M.L. neut. n. *Rhizobium* that which lives in a root.

Rods 0.5–1.0 × 1.2–3.0 μm. Nonsporeforming. Gram negative. **Motile by 1–6 peritrichous flagella.** Fimbriae have been described on some strains. **Aerobic**, possessing a respiratory type of metabolism with oxygen as the terminal electron acceptor. Optimal temperature for growth, 25–30°C; some species can grow at temperatures >40°C. Optimal pH for growth, 6–7; range pH 4–10. Generation times of *Rhizobium* strains are 1.5–5.0 h. **Colonies** are usually white or beige, circular, convex, semi-translucent or opaque, raised and mucilaginous, **usually 2–4 mm in diameter within 3–5 days on yeast-mannitol-mineral salts agar (YMA). Growth on carbohydrate media is usually accompanied by copious amounts of extracellular polysaccharide.** Pronounced turbidity develops after 2 or 3 days in aerated or agitated broth. Chemoorganotrophic, utilizing a wide range of carbohydrates and salts of organic acids as sole carbon sources, without gas formation. Cellulose and starch are not utilized. **Produce an acidic reaction in mineral-salts medium containing mannitol or other carbohydrates.** Ammonium salts, nitrate, nitrite, and most amino acids can serve as nitrogen sources. Strains of some species will grow in a simple mineral salts medium with vitamin-free casein hydrolysate as the sole source of both carbon and nitrogen, but strains of many species require one or more growth factors such as biotin, pantothenate, or nicotinic acid. Peptone is poorly utilized. Casein, starch, chitin, and agar are not hydrolyzed. **All known *Rhizobium* species include strains which induce hypertrophisms in plants as root nodules with or without symbiotic nitrogen fixation.** Some cells of symbiotic bacterial species enter root hair cells of leguminous plants (Family Leguminosae) via invagination or by wounds ("crack entry") and elicit the production of root nodules wherein the bacteria engage as intracellular symbionts, usually fixing nitrogen. Many well-defined nodulation (*nod*) and nitrogen fixation (*nif*) genes are clustered on large plasmids or megaplasmids (pSyms). Plasmid transfer between species results in the expression and stable inheritance of the particular plant-interactive properties of the plasmid-donor species. Plant host specificity is usually for a few legume genera but may, in some strains, include a wide variety of legume genera and is to some extent determined by the chemical structure of the lipochito-oligosaccharide *Nod* factors produced. These chitin-like molecules induce nodule organogenesis in the absence of bacteria. **In root nodules the bacteria occur as endophytes that exhibit pleomorphic forms, termed "bacteroids", which reduce or fix gaseous atmospheric nitrogen into a combined form utilizable by the host plant.**

The mol% G + C of the DNA is: 57–66.

Type species: **Rhizobium leguminosarum** (Frank 1879) Frank 1889, 338 (*Schinzia leguminosarum* Frank 1879, 397.)

FURTHER DESCRIPTIVE INFORMATION

Introductory Note This treatment of the genus *Rhizobium* also includes information on the closely related genera *Agrobacterium*, and *Sinorhizobium*. The rationale for this treatment is provided under Taxonomic Comments.

Following Bradbury (1986), strains in past literature called *Agrobacterium* biovar 1 are herein referred to as *Agrobacterium tumefaciens*. *Agrobacterium* biovar 2 strains are referred to as *Agrobacterium rhizogenes*. *Agrobacterium* biovar 3 strains are referred to

as *Agrobacterium vitis* (See Table BXII.α.114 in the chapter on the genus *Agrobacterium*). Where necessary, the ability of pathogenic strains to cause crown gall tumors or the hairy root condition, previously attributed to strains using the names "*Agrobacterium tumefaciens*" and "*Agrobacterium rhizogenes*", is indicated by reference to the tumorigenic or rhizogenic ability, respectively, of strains in species of *Agrobacterium*. Nonpathogenic strains previously named *Agrobacterium radiobacter* are referred to either as nonpathogenic strains of *Agrobacterium tumefaciens* and *Agrobacterium rhizogenes* or merely as nonpathogenic *Agrobacterium* if the species designation has not been identified. Ti or Ri plasmids determine the pathogenic status of strains (see below). Species comprising pathogenic or nonpathogenic strains can be reported as tumorigenic as a (Ti strain) or (Ti), as rhizogenic as a (Ri strain) or (Ri), or as nonpathogenic strains of the species where relevant.

Morphology The formation of star- or rosette-shaped aggregates of cells by several *Agrobacterium* strains has been described by Beijerinck and van Delden (1902), Stapp and Knösel (1956) and Knösel (1962). Granules of poly-β-hydroxybutyrate are common in older cells, so that upon simple staining the rods appear banded. Strains of *R. leguminosarum* often contain metachromatic granules, demonstrated by staining with methylene blue, washing with dilute iodine, and staining with neutral red (Graham and Parker, 1964). Fig. BXII.α.128 shows a cell of *R. leguminosarum* biovar trifolii.

All species are motile by one to six flagella. For most species examined, insertion is peritrichous. Strains in species with a single flagellum (*R. galegae*, *R. mongolense*, *Sinorhizobium fredii*, *Sinorhizobium saheli*, *Sinorhizobium terangae*, and *Sinorhizobium xinjiangense*) also appear to have peritrichous organization expressed as polar or sub-polar insertion. Only *R. hainanense* is reported as being unambiguously polarly flagellated.

Cell wall composition The cell wall structure of *Rhizobium* is generally similar to that of other Gram-negative bacteria. The peptidoglycan consists of glutamic acid, alanine, diaminopimelic acid, and amino sugars. In addition, leucine, phenylalanine, serine, and aspartic acid have been detected in relatively large amounts in the peptidoglycan layer of several pathogenic strains. Lipopolysaccharide (LPS) cell wall composition varies from strain to strain but consistently contains 2-keto-3-deoxyoctanoic acid (KDO), uronic acids, glucosamine, glucose, mannose, rhamnose, fucose, and galactose (Carlson, 1982). The structures of LPS from a number of species have been determined, and all of them contain the unusually long 27-hydroxyoctacosanoic acid (Jeyaretnam et al., 2002; Sharypova et al., 2003). Rhizobia also have an unusually complex composition of membrane phospholipids, among them phosphatidylcholine, and under conditions of phosphorus limitation, phospholipids can be replaced by membrane lipids that do not contain phosphorus (López-Lara et al., 2003).

Fine structure As revealed by electron microscopy and biochemical analyses, cellulose-containing fibrils are formed by pathogenic *Agrobacterium* strains during their attachment to plant cells *in vitro*. These fibrils anchor the bacteria to the plant cell surface

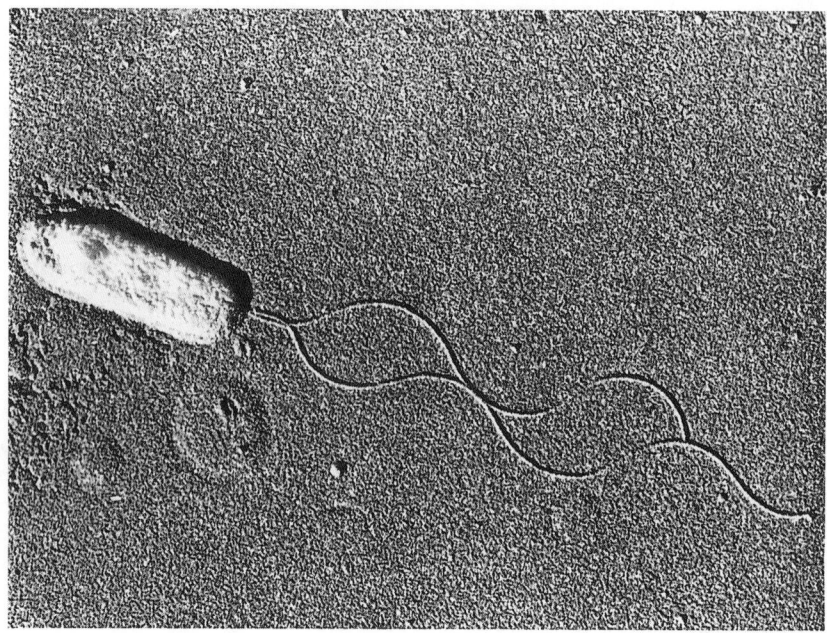

FIGURE BXII.α.128. Cell of *Rhizobium leguminosarum* biovar trifolii showing two polar flagella (× 14,000).

(Matthysse et al., 1981). Lipopolysaccharides of the outer membrane of cell envelopes play a role in the attachment of the bacteria to the wound site of the plant (Whatley et al., 1976). Smit et al. (1989) purified an adhesin that appears to mediate the first step in attachment of nodulating *Rhizobium* and pathogenic *Agrobacterium* bacterial cells to plant root hair tips.

Colonial and cultural characteristics On carbohydrate-containing solid media, the majority of *Rhizobium* strains produce circular, low convex to convex, mucous, glistening, opaque, white to beige-colored colonies, with an entire edge and a diameter of 2–4 mm after 5–6 days of incubation at 28°C.

Most strains grow rapidly on a mineral salts medium containing yeast extract and any one of a wide variety of carbohydrates (Vincent et al., 1979). Acid is usually produced to a moderate degree from carbohydrates (Norris, 1965).

Growth on carbohydrate-containing solid media may be opaque, clear or translucent (Fig. BXII.α.129) but may also exhibit small opaque areas within a clear slime. Variants that produce small colonies often fix little nitrogen symbiotically. Growth of *Agrobacterium* on nutrient agar is moderate, whereas abundant growth is obtained on media containing yeast extract and a suitable carbohydrate such as glucose, sucrose, or lactose (see Maintenance Procedures).

Almost all strains of *Rhizobium* species—but not *Agrobacterium* species—form only white colonies on yeast extract-mannitol-mineral-salts medium containing 0.0025% Congo red.

Some strains are encapsulated. All produce abundant water-soluble extracellular polysaccharide, the principal constituent of which is acidic heteropolysaccharide (80–90%). The remaining constituents, in most strains, are neutral, unbranched, β-2-linked glucans, of which some cyclics are important for nodule development (York et al., 1980; Breedveld and Miller, 1994). Certain chromosomal genes termed *chv* or *ndv* are responsible for the production of cyclic glucans essential for either virulence or nodule development in all species of *Agrobacterium* or *Rhizobium*. All strains within a species produce the same acidic heteropolysac-

charide except for *R. leguminosarum* biovar phaseoli, which possesses a unique heteropolysaccharide. Curdlan production has been reported in several isolates of *R. leguminosarum* biovar trifolii (Ghai et al., 1981). The formation of water-insoluble β-1,3-glucans has been reported in some strains of pathogenic *Agrobacterium* (Nakanishi et al., 1976). The synthesis of glycogen by *Agrobacterium tumefaciens* strain B6 is regulated at the level of ADP-glucose synthesis (Eidels et al., 1970).

Nutrition and growth conditions The temperature range for growth, which is highly strain dependent, is 4–40°C; however, growth at 4°C is rare, and only certain species can grow at 40°C. The temperature maximum for *R. leguminosarum* is 38°C. All of the strains grow between 20°C and 28°C. *Agrobacterium rhizogenes* cannot grow above 30°C, whereas *Sinorhizobium saheli* and *Sinorhizobium terangae* can grow at 44°C (de Lajudie et al., 1994). The pH range for growth for the entire genus *Rhizobium* is 4–10.

Intermediates of the tricarboxylic acid cycle and several amino acids can be utilized as sole sources of carbon. The majority of *Agrobacterium tumefaciens* strains can grow on a minimal medium with nitrate or ammonium salts as the nitrogen source. *Agrobacterium rhizogenes* strains do not utilize nitrate unless biotin is supplied; some strains require both L-glutamic acid and biotin. Strains belonging to *Agrobacterium rubi* require L-glutamic acid and yeast extract (Starr, 1946; Lippincott and Lippincott, 1969; Keane et al., 1970).

Metabolism and metabolic pathways The principal mechanisms of glucose catabolism in *Rhizobium* are the Entner-Doudoroff pathway and the pentose cycle (Katznelson and Zagallo, 1957; Vardanis and Hochster, 1961; Martínez-De Drets and Arias, 1972; Arthur et al., 1973, 1975; Ronson and Primrose, 1979). It is unlikely that the Embden-Meyerhof-Parnas pathway operates in *Rhizobium* spp. because activities of fructose-1,6 diphosphate aldolase and 6-phosphofructokinase are low. Polyols are substrates for an inducible dehydrogenase that converts mannitol to fructose and arabitol to xylulose (Martínez-De Drets and Arias,

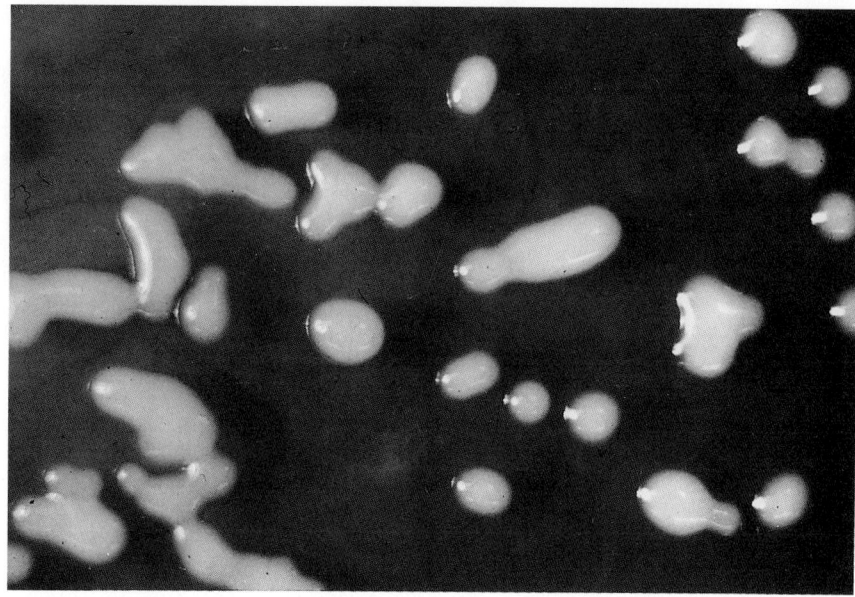

FIGURE BXII.α.129. Colonies of *Rhizobium leguminosarum* on mineral-salts mannitol agar (× 1.5).

1970). L-Arabinose is metabolized to α-ketoglutarate (Duncan, 1979).

The tricarboxylic acid cycle is operative, and the enzymes of the glyoxylate bypass are present (Johnson et al., 1966; Arthur et al., 1973; Chern et al., 1976a). Pyruvate carboxylase is an important anaplerotic enzyme (Chern et al., 1976b; Ronson and Primrose, 1979).

Glucuronic acid and glucaric acid are metabolized via 2-keto-3-deoxy-D-glucaric acid to α-ketoglutaric acid (Chang and Feingold, 1970). The initial step in the catabolism of L-sorbose by some *Agrobacterium tumefaciens* strains is the reduction to sorbitol, followed by oxidation of the latter compound to D-fructose (Van Keer et al., 1976). The majority of *Agrobacterium* strains characteristically oxidize a large number of carbohydrates (disaccharides, bionic acid, and several monosaccharides) to the corresponding 3-uloses (Bernaerts and De Ley, 1960a, b; Fukui et al., 1963; De Ley et al., 1966). The vigorous and unusual oxidation of lactose to 3-ketolactose is, so far, unique to *Agrobacterium tumefaciens* and is the basis of a simple and specific diagnostic test for the rapid differentiation of this species from other *Agrobacterium* and *Rhizobium* spp. that have been tested (Bernaerts and De Ley 1963; de Lajudie et al., 1994). These specific oxidations are catalyzed by an inducible hexopyranoside:cytochrome *c* oxidoreductase (D-glucoside 3-dehydrogenase), containing flavin adenine dinucleotide as cofactor (Hayano and Fukui, 1967; Van Beeumen and De Ley, 1968; Nakamura and Tyler, 1977). Although an "alpha-3"-ketoglucosidase was detected in a strain of *Rhizobium* (Hayano and Fukui, 1970; Hayano et al., 1973), 3-ketosucrose and 3-ketolactose are probably not involved as essential intermediates in the metabolism of sucrose and lactose, respectively (Kurowski and Pirt, 1971). Conditions have been worked out for increasing the yield of 3-ketoglycosides (Tyler and Nakamura, 1971; Fensom et al., 1974; Kurowski et al., 1975).

Anaerobic growth by nitrate reduction has been reported for *Sinorhizobium fredii* (Hynes et al., 1985) and *Sinorhizobium meliloti* (Daniel et al., 1982). *Sinorhizobium fredii* produces N_2O and *Sinorhizobium meliloti* produces N_2 as the end products of denitrification.

Rhizobium contains at least two soluble cytochromes *c*: a cytochrome c_{552} and a cytochrome c_{556}. Cytochrome c_{552} has been sequenced (Van Beeumen et al., 1980). It belongs to the cytochrome *c* sequence class IB (*sensu* Ambler, 1973) and, of all known procaryotic cytochromes *c*, shows the highest amino acid sequence homology with mitochondrial cytochrome *c* of tuna fish (Van Beeumen et al., 1980). Cytochrome c_{556} from *Rhizobium* belongs to the cytochrome *c* sequence class II (*sensu* Ambler, 1973), because its single heme group is bound near the C-terminus (Van Beeumen et al., 1980).

Auxotrophic mutants of *Rhizobium* are sometimes symbiotically defective and the isolation and study of such mutants has provided insights into the biochemical prerequisites of symbiosis (Kuykendall, 1981). For example, auxotrophy toward adenine, uracil, and leucine is often associated with symbiotic ineffectiveness in *Sinorhizobium meliloti* (Dénarié et al., 1976).

Fatty acids Fatty acid profiles have been studied for some species. Using principal components analysis of whole cell fatty acid methyl esters (FAME), Jarvis et al. (1996) reported three clusters: 1) *R. leguminosarum, R. etli, R. tropici, Agrobacterium rhizogenes, Sinorhizobium fredii,* and *Sinorhizobium meliloti;* 2) *R. galegae, Agrobacterium tumefaciens, Agrobacterium rubi,* and *Agrobacterium vitis;* and 3) *R. huakuii* and *R. loti,* (now allocated to *Mesorhizobium*). Sawada et al. (1992d) could differentiate *Agrobacterium tumefaciens, Agrobacterium rhizogenes,* and *Agrobacterium vitis* using fatty acid profiles. *Agrobacterium tumefaciens* could be separated from *A. vitis* based on quantitative differences in $C_{16:1}$ and $C_{18:1}$ acids, and the presence of $C_{17:0 \text{ cyclo}}$ acids. Quantitative differences in $C_{19:0 \text{ cyclo}}$, and the presence of $C_{15:0 \text{ iso 3OH}}$ acids and the absence of $C_{18:1 \text{ 3OH}}$ acids differentiated *Agrobacterium rhizogenes* from the other two species (Sawada et al., 1992d). Tighe et al. (2000) studied 600 strains and found that the fatty acid composition of all the fast-growing species differed from that of *Bradyrhizobium* and *Mesorhizobium,* but there were no clear differences among fast-growing genera except for the relative concentration of $C_{16:0 \text{ 3OH}}$ fatty acid, which was quantitatively lower in *Sinorhizobium* than in most *Rhizobium* strains. Species-level identification was, however, possible in many cases, and thus the data of Tighe et al. (2000) is a valuable resource.

Genetics A circular linkage map of the *R. leguminosarum* chromosome was first constructed by Beringer and Hopwood (1976). The genome of *Sinorhizobium meliloti* has also been mapped (Kondorosi et al., 1977; Meade and Singer, 1977). These two species have similar chromosomal gene arrangements (Beringer et al., 1987). *Sinorhizobium meliloti* has a chromosome size of 3.7×10^6 bp, and its sequence has been reported (Galibert et al., 2001).

There are only incomplete comparisons of *Rhizobium* spp. by DNA–DNA reassociation. DNA–DNA reassociation data are available for the following species: *R. galegae, R. leguminosarum*, and *R. tropici* (Martínez-Romero et al., 1991); *Sinorhizobium fredii* and *Sinorhizobium meliloti* (de Lajudie et al., 1994); and *Sinorhizobium saheli* and *Sinorhizobium terangae* (de Lajudie et al. 1994). An early study was made of *Agrobacterium tumefaciens, Agrobacterium rhizogenes*, and *Agrobacterium rubi* (De Ley 1972, 1974).

Easily amplified DNA regions called amplicons, apparently controlling both adaptability and biological interactions, have been documented in *Rhizobium* genomes (Palacios et al., 1998). As these become better defined through mapping and DNA sequencing, their structure can be expected to be of particular interest.

Plasmids Plasmids and megaplasmids as large as 1600 kb can constitute as much as 50% of the total genome. Sequences of plasmids have been reported (Freiberg et al., 1997; Barnett et al., 2001; Finan et al., 2001).

Hypertrophic activity and plasmids All species (though not all strains) of *Rhizobium* and *Agrobacterium* are associated with plant hypertrophisms and, with a few exceptions, are associated with root hypertrophisms. Hypertrophy can involve formation of either nitrogen-fixing nodules in *Rhizobium* or of oncogenous (tumorous) galls or hairy roots in *Agrobacterium*. These activities are confined to particular species. Genes (pSym, pTi, and pRi) associated with hypertrophisms are usually carried on large plasmids. These genes and the associated hypertrophying activity play a central role in the ecology of these genera. Plant transformations using the agrobacterial tumorigenic system are now routine (see reviews by Kado, 1991; Zambryski, 1992; Hooykaas and Beijersbergen, 1994). Transformations in filamentous fungi have also been reported (De Groot et al., 1998).

Plasmid-regulated nitrogen fixation Some members of the *Rhizobiaceae* and *Phyllobacteriaceae* are characterized by their ability to incorporate functional genetic elements as plasmids or symbiotic islands (Sullivan and Ronson 1998) which permit them to establish pathogenic (oncogenic) or symbiotic nitrogen-fixing relationships with plants. Nitrogen-fixing symbioses involving members of the *Rhizobiaceae* are restricted to plants of the family Leguminosae (van Rhijn and Vanderleyden, 1995), with one exception. *Rhizobium* cells contain as many as 10 naturally occurring plasmids ranging in size from less than 100 kb to megaplasmids of more than 1000 kb. In some instances, the combined length of plasmids and megaplasmids approximates that of the chromosome, meaning that up to 50% of the *Rhizobium* genome is not in the chromosome. Both *nod* and *nif* genes controlling nodulating and nitrogen-fixing ability have always been found clustered together on one or more *Rhizobium* plasmids or megaplasmids called pSym. Numerous genetics studies have delineated *Rhizobium* genes essential for legume nodulation and symbiotic nitrogen fixation (Rossen et al., 1984; Török et al., 1984; Egelhoff and Long, 1985; Egelhoff et al., 1985; Jacobs et al., 1985; Debellé et al., 1986; Evans and Downie, 1986; Göttfert et al., 1986; Horvath et al., 1986; Rostas et al., 1986; Shearman et al., 1986; Aguilar et al., 1987; Fisher et al., 1987; Honma and Ausubel, 1987; Cremers et al., 1988; Davis et al., 1988; Surin and Downie, 1988; Cervantes et al., 1989; De Maagd et al., 1989; Schwedock and Long 1989, 1994; Barnett and Long, 1990; Economou et al., 1990; Honma et al., 1990; Surin et al., 1990; Baev et al., 1991, 1992; Kondorosi et al., 1991a, b; Rushing et al., 1991; Baev and Kondorosi, 1992). The nodulation-controlling genes are organized into several coordinately regulated operons. For example, the common *nodABC* operon is present in all legume symbiotic strains and can complement strains in different genera. For *R. etli, nodA* is separated from *nodBC* (Vázquez et al., 1991; Vázquez et al., 1993), and for *Mesorhizobium loti, nodB* is independent of *nodAC* (Scott et al., 1996). Others may be present in different species as allelic variants, such as *nodEF*, and these are host-specific and hence not interchangeable. Such *nod* or *hsn* genes are sometimes present only in certain strains.

pSyms are not essential for survival of the *Rhizobium* strains in soil. Non-nodulating soil bacteria identified as *Rhizobium* species have been isolated which can only form nodules after transconjugation with related symbiotic strains. Transfer of symbiotic plasmids among *Rhizobium* species has been reported under laboratory conditions and has been demonstrated by sequence comparison of natural isolates. Martínez et al. (1987) demonstrated nitrogen-fixing nodules formed by *Agrobacterium tumefaciens* carrying a conjugally transferred pSym from *R. tropici* (as *R. phaseoli* type II). Similar results were obtained for transconjugants of *Agrobacterium tumefaciens* containing pSym from *Rhizobium* strains that nodulate *Phaseolus vulgaris* (Brom et al., 1988). A pSym of *R. leguminosarum* biovar trifolii has been introduced into 15 non-nodulating bacterial isolates identified as *M. loti, R. leguminosarum, R. tropici, Sinorhizobium meliloti*, and four isolates related to *R. leguminosarum* (Sivakumaran et al., 1997). By comparison of *nifH* sequence type to 16S rRNA gene sequence type, Haukka et al. (1998) demonstrated that similar symbiotic genes could be found in different 16S rRNA gene backgrounds, indicating horizontal transfer across species boundaries.

There is considerable variation in the nodulating and nitrogen-fixing capacity of individual strains. In *Rhizobium* strains with the capacity for symbiotic activity, most if not all genes that specify and regulate nodulating and nitrogen-fixing abilities are carried on one or more plasmids. In this respect, members of the genus differ from *Bradyrhizobium* (see the chapter on the genus *Bradyrhizobium*), in which all symbiosis-controlling genes have been shown to be carried on the chromosome, and *Mesorhizobium* (see the chapter on the genus *Mesorhizobium*), in which many strains have symbiosis-controlling genes located on the chromosome. The systematics of these plasmid-determined symbiotic associations is reviewed in Young and Johnston (1989).

The expression of nodulation genes is controlled by the presence of flavonoids excreted by the host plant. For, example, the regulatory *nodD* gene controls *nodABC* expression. Flavonoids produced by various legumes seem to interact specifically with particular NodD proteins, which vary in structure according to *Rhizobium* species, as do the Nod Factors (NF). Compatibility between flavonoid and NodD protein is thought to be a major factor in host specificity. (For detailed reviews of nodulation and nitrogen fixation genetics and biochemistry, see Schultze et al., 1994; van Rhijn and Vanderleyden, 1995; Dénarié et al., 1996.)

Another regulation gene, *nolR*, has been reported to be common in symbiotic species of *Rhizobium* and *Sinorhizobium*. This gene was not found in species of *Azorhizobium*, *Bradyrhizobium*, *Mesorhizobium*, or *Agrobacterium* (Kiss et al., 1998). This gene repressed the expression of both *nodABCIJ* and *nodD* genes, resulting in decreased Nod factor production. Recently, a novel family of *nod* gene inducers, aldonic acids, was reported for *Sinorhizobium meliloti*, *Mesorhizobium loti*, and *Bradyrhizobium* (as *B. lupini*) strains (Gagnon and Ibrahim 1998).

Root nodules induced by some strains of symbiotic *Rhizobium* contain substances called rhizopines (Murphy et al., 1987, 1995), analogous to the opines found in pathogenic *Agrobacterium* strains (see later discussion). Those described so far are substituted scyllo-inosamines. Both synthetic and catabolic genes are located on pSym. Genes involved in synthesis are active in the symbiotic bacteroid state and catabolic genes are active in the free-living cells (Murphy et al., 1988; Dessaux et al., 1998). The likely ecological benefit of rhizopine activity for producing strains has been reported (Gordon et al., 1996). Rhizopine production has been reported for *R. leguminosarum* biovar viciae and *Sinorhizobium meliloti*. No rhizopine production was detected from isolates of *R. etli*, *R. tropici*, or *R. leguminosarum* biovar trifolii and biovar phaseoli (Wexler et al., 1995). Rhizopine synthesis (*mos*) and catabolism (*moc*) genes from *R. leguminosarum* and *Sinorhizobium meliloti* have been sequenced (Wexler et al., 1996b). Further study showed that some non-symbiotic soil bacteria, including *Arthrobacter*, *Aeromonas*, *Alcaligenes*, and *Pseudomonas*, could catabolize rhizopine. No DNA sequences homologous to *nodB* and *nodC* and no effective nodulation on *Medicago sativa* were detected in two strains that were related to *Sinorhizobium meliloti* based on partial 16S rDNA sequence analysis (Gardener and Bruijn, 1998); this result indicated that the gene(s) related to rhizopine metabolism may be carried on the chromosome or on plasmids other than pSym.

Two megaplasmids in *Sinorhizobium meliloti* are involved in the effective nodulation of alfalfa (Hynes et al., 1986). In *R. leguminosarum* (Hynes and McGregor, 1990), as in *R. etli* CFN42, plasmids other than the pSym (*nod-nif* plasmid) are required for an effective symbiosis and may carry *lps* genes (García-de los Santos and Brom, 1997; Vinuesa et al., 1999).

Plant specificity in nitrogen fixation The legume-nodulating ability of *Rhizobium* species appears to be specific to a few plant species or genera. An exception is *Sinorhizobium* sp. strain NGR234, which nodulates 112 legume genera (Pueppke and Broughton, 1999). In such instances, nodulation can occur without nitrogen fixation. Since Lerouge et al. (1990) established the chemical structure of the nodule-inducing compound produced by *Sinorhizobium meliloti*, a general hypothesis has been proposed: host specificity in all legume microsymbionts is related to the chemical structure of specific lipochitooligosaccharide Nod factors (Dénarié et al., 1992, 1996). Nodulation factors of strain NGR234 include variants of Nod factors (Price et al., 1992). Structures of Nod factors produced by several species have been described (Poupot et al., 1993, 1995; Lorquin et al., 1997; Yang et al., 1999; Snoeck et al., 2001; Pacios-Bras et al., 2002).

Plasmid-mediated plant-pathogenic (oncogenic) activity in *Agrobacterium* spp. The early literature on plasmid-mediated plant pathogenic (oncogenic) activity was reviewed by Nester et al. (1984). A recent review is given by Binns and Costantino (1998). Oncogenic (tumorigenic or rhizogenic [hairy root]) activity in the four plant-pathogenic *Agrobacterium* species is mediated by genes that are largely or wholly borne on one or more large (>150 kb) plasmids. Tumorigenic activity is conferred by Ti plasmids and rhizogenic activity is conferred by Ri plasmids. Tumorigenic genes on the Ti plasmid comprise (a) T-DNA genes, and (b) virulence (*vir*) genes. Wounding of susceptible plant tissue activates *vir* genes that facilitate the transfer of a component of the Ti (or Ri) plasmid, the T-DNA (8–22 kb). The T-DNA fragment is integrated into the plant nucleus apparently at random (Chyi et al., 1986), in one or more copies (Chilton et al., 1980; Lemmers et al., 1980; Willmitzer et al., 1980; Zambryski et al., 1980). T-DNA carries all necessary genes for tumor growth, the most important for tumorigenesis being those associated with auxin and cytokinin synthesis, which are expressed in the plant.

Comparative analysis indicates a correlation between sequence structure of the 16S–23S rRNA intergenic spacer region and the type of Ti plasmid (nopaline, vitopine, or octopine/cucumopine) present in strains of *Agrobacterium vitis* (Otten et al., 1996).

Another large Ri plasmid is involved in the hairy root disease of plants caused by rhizogenic *Agrobacterium* strains (Moore et al., 1979; White and Nester, 1980a). Little overall sequence similarity to other Ti plasmids has been detected (White and Nester, 1980b). There is one small region of conserved similarity between the Ri plasmid and an octopine Ti plasmid (pTi-B6806), but the former shows no similarity to the T-DNA region of the latter plasmid. The Ri plasmid is compatible with other Ti plasmids and thus represents a new incompatibility class of plasmids (White and Nester, 1980b).

Large plasmids that are not involved in pathogenic activity have been found in nonpathogenic strains of *Agrobacterium tumefaciens* and *Agrobacterium rhizogenes* (Merlo and Nester, 1977; Sheikholeslam et al., 1979). In addition, large plasmids have been discovered in addition to the Ti plasmid in several tumorigenic *Agrobacterium* strains. Some strains possess an additional 2.1×10^6 base-pair linear chromosome in addition to two or more very large plasmids (Allardet-Servent et al., 1993). The sequence of the *A. tumefaciens* C58 genome was reported (Goodner et al., 2001).

Opines are unusual amino acid derivatives produced in tumor or hairy root tissues induced by pathogenic strains of *Agrobacterium* species. Opines are synthesized from common plant compounds but are not utilized by plants or by most microorganisms. They are utilized by *Agrobacterium* strains that carry the specific opine-inducing plasmids associated with oncogenicity. The opine concept as originally proposed was that specific genes associated with T-DNA permitted the synthesis of opines from plant photosynthetic products in a form whose availability as nutrients was restricted to tumorigenic strains bearing the relevant plasmid (Schell et al., 1979). At least eleven opines have been identified: octopine, lysopine, nopaline, succinamopine, leucinopine, cucumopine, heliopine, chrysopine, mikimopine, agropine, and agrocinopines (Chang et al., 1989; Dessaux et al., 1992; Chilton et al., 1995). In addition, imino acids (Moore et al., 1997), mannopine (Petit et al., 1983), and vitopine (Szegedi et al., 1988) have been reported. It is unlikely that this list is exhaustive (Moore et al., 1997). Opine catabolism by pathogenic *Agrobacterium* strains is mediated by genes, located on one or more plasmids including the Ti plasmid, that are not transferred to the plant cell nucleus (Montoya et al., 1977). Pathogenic strains can utilize more than one opine (Moore et al., 1997). Some opines can also induce conjugal transfer of the Ti plasmid to nontumorigenic strains and may contribute to the dissemination of the infectious plasmid (Gelvin 1992; Guyon et al., 1993). Analysis of the distribution of Ti plasmids in terms of their opine genes showed that the ecology of plasmid-bearing strains is highly complex (Moore et al., 1997). In only a few samples could field outbreaks of crown gall be traced to a clonal origin of infection. In most collections, field tumors were induced by Ti plasmids of more than one opine type. Field tumors of some hosts yielded no detectable opines, even though opine-utilizing bacteria were present. Bacterial isolates from other hosts (plum and cherry) showed the best correspondence between the opine in tumors (nopaline) and the presence of bacteria that catabolized that opine. However, several unusual opine catabolic combinations were identified, including isolates that catabolized a variety of opines but were nonpathogenic. There are indications of some specificity between pathogenic *Agrobacterium* species, opine type, and host plant (Lopez et al., 1988; Sawada et al., 1992a). The opine concept (Schell et al., 1979) assumed specific utilization of opines by tumorigenic *Agrobacterium* strains bearing relevant plasmids, and that these compounds could not be utilized by other soil organisms. Since the original proposal, however, it has become clear that a wide range of soil organisms have the capacity to metabolize opines. These include *Pseudomonas* species (Beaulieu et al., 1983; Tremblay et al., 1987b) and Gram-positive coryneform bacteria (Tremblay et al., 1987a). These data collectively suggest that mechanisms explaining the involvement of opines may be more complex than the original model.

At present, oncogenic activity is associated with *Agrobacterium tumefaciens*, *Agrobacterium rhizogenes*, *Agrobacterium rubi*, and *Agrobacterium vitis*. There may be several additional pathogenic bacterial populations that also merit classification as species. Bouzar et al. (1995) records a pathogen associated with aerial infections of *Ficus benjamina*, and Sawada and Ieki (1992b) report phenotypically distinct strains isolated from affected plants.

Pathogenic host range and pathogenicity   The host range of tumorigenic *Agrobacterium* strains is reported to be very wide. De Cleene and De Ley (1976) described at least 640 plant species belonging to 331 genera in 93 families of dicotyledon and gymnosperm plants as susceptible to transformation by *Agrobacterium* (Ti strains) (such as *Agrobacterium tumefaciens*). De Cleene and De Ley (1981) reported 37 plant species belonging to 30 genera in 15 families of dicotyledonous plants as susceptible to transformation by *Agrobacterium* (Ri strains) (such as *Agrobacterium rhizogenes*). None of the 250 monocotyledonous species investigated was susceptible to the disease, except some members of the orders Liliales and Arales. Bradbury (1986) listed almost 400 plant species affected by tumorigenic strains and over 50 plant species affected by rhizogenic strains.

The inference that crown gall- and hairy-root-inducing strains of *Agrobacterium tumefaciens* and *Agrobacterium rhizogenes* have wide host ranges is based on reports of the many hosts from which the pathogens have been isolated and on the many reports of tumor induction in experimental inoculations. However, there have also been indications of host specificity within some populations (Panagopoulos and Psallidas, 1973; Anderson and Moore 1979; Moore and Cooksey 1981; Paulus et al., 1991a, b; Palumbo et al., 1998). In some instances, strains are pathogenic to a relatively narrow range of host plants (Unger et al., 1985). Furthermore, although crown gall has been reported on some host species in some countries, it is not necessarily found in these same hosts in other countries where crown gall is known. The family of Ti plasmids may have restricted host ranges in varying degrees and chromosomal background may affect specificity. It

seems clear that some strains naturally infect host plants from several unrelated genera whereas others are more specific (Paulus et al., 1991b). The nature of specificity, whether it is a function of the bacterial strain or of the plasmid (Loper and Kado 1979; Close et al., 1985; D'Souza et al., 1993), has not been generally confirmed, although modifications to the Ti-plasmid have been implicated (Paulus et al., 1991a). There is no indication of host range specificity as it occurs in the pathogenic species and pathovars of *Pseudomonas* or *Xanthomonas*. Although *Agrobacterium rubi* is identified as a species isolated from galls on the canes of *Rubus* spp., the specificity of this pathogen to *Rubus* is in doubt because it has been shown to have a wide host range (Sawada et al., 1992a). *Agrobacterium vitis* is found as the predominant tumorigenic species specific to *Vitis* spp. (Thies et al., 1991), and *Agrobacterium vitis* strains have occasionally been isolated from other hosts, such as *Actinidia* (Sawada and Ieki 1992a). *Agrobacterium vitis* appears to be unique among pathogenic *Agrobacterium* species in being associated with a root decay symptom (Burr et al., 1987).

Upon infection of wounded plant tissues, tumorigenic *Agrobacterium* strains can transform plant cells into autonomously proliferating cells. In nature, the swellings mostly occur at the transition zone between the stem and the root system of the host plant, hence the name "crown gall disease." Small spherical growths or elongated ridges can occur on the stems of *Rubus* spp. such as raspberry and bramble bushes. Some oncogenic strains cause hairy root on susceptible plants (such as apple trees and roses) but cause crown gall on other plants. On some plants (e.g., *Kalanchoë*), oncogenic strains of *Agrobacterium* spp. can induce the formation of teratomata, characterized by the development of aberrant shoots, leaves, or roots developing from the tumor tissue. The type of disease produced (differentiated or undifferentiated tumors) is probably determined by both the bacterial Ti plasmid and the host plant (Gresshoff et al., 1979).

A prerequisite for tumorigenesis is the wounding of the host. Infection can occur during various stages of the life of a plant via wounds caused by growth, germination (e.g., peaches and almond), subterranean insects, or mechanical injuries (e.g., pruning, grafting, and replanting of trees in nurseries).

Tumorigenic and rhizogenic activity was initiated by *Agrobacterium tumefaciens* and *Agrobacterium rhizogenes* at 20°C and 27°C, respectively. The latter temperature, however, was not conducive to tumorigenesis by *Agrobacterium vitis*. Temperature effects were mediated by the choice of host plant (Charest and Dijon, 1985).

Crown gall disease seldom kills plants, but growth is often impaired and stunted. Significant damage and economic loss can occur on stone fruit and grape (De Cleene, 1979).

Plasmid exchange between *Rhizobium* and *Agrobacterium* species   Intergeneric transmissibility of Ti and nodulating plasmids has been demonstrated from nodulating *Rhizobium* spp. to tumorigenic *Agrobacterium* spp. (Martínez et al., 1987; Brom et al., 1988; Abe et al., 1998), and from tumorigenic *Agrobacterium* spp. to nodulating *Rhizobium* spp. (Hooykaas et al., 1977) and to *Phyllobacterium myrsinacearum* (van Veen et al., 1988); this supports a close relationship for these genera and points to promiscuous plasmid exchange between taxa. Novikova and Safronova (1992) reported transconjugants of *Agrobacterium tumefaciens* harboring the pSym genes of *R. galegae* that formed an effective symbiosis with *Medicago sativa*. The finding by Nesme et al. (1987) that crown gall that occurred in a poplar nursery was caused by naturally occurring resident mixed populations of both *Agrobacte-*

rium tumefaciens (Ti strains) and *Agrobacterium rhizogenes* (Ti strains) supports the idea that Ti plasmids may be promiscuous in the resident tumorigenic *Agrobacterium* species. Plasmid homologies do not correlate with any numerical classification of pathogenic *Agrobacterium* spp. (Currier and Nester, 1976), hence it is generally assumed that plasmid-borne Ti and Ri genes are readily transmitted within and between strains of *Agrobacterium tumefaciens* and *Agrobacterium rhizogenes*.

Although plasmids conferring pathogenicity and nodulation are transmissible between pathogenic and nodulating genera in the laboratory, in nature these characteristics appear to be specific to the particular species. The relative specificity between particular nodulating *Rhizobium* spp. and their legume symbionts, and the pathogenic specificity indicated for *Agrobacterium vitis* and, perhaps, *Agrobacterium rubi* suggests that some plasmid incompatibilities exist, causing a restriction on transmission or subsequent gene expression. Furthermore, Bouzar et al. (1993) and Otten et al. (1996) demonstrated a correlation between the form of resident plasmids and host chromosome, suggesting possible restrictions on exchange in nature.

Agrocin and trifolitoxin activity Agrocin 84 from the non-pathogenic *Agrobacterium rhizogenes* strain 84 (ICMP 3379; NCPPB 2407) (New and Kerr, 1972; Kerr and Htay, 1974) is plasmid-encoded (Ellis et al., 1979). It is a toxic analog of an adenine nucleotide (Roberts et al., 1977) and selectively inhibits pathogenic *Agrobacterium* strains harboring a nopaline plasmid. It is effective against *Agrobacterium tumefaciens* and *Agrobacterium rhizogenes* but not *Agrobacterium vitis* or *Agrobacterium rubi* (Ma et al., 1985; van Zyl et al., 1986; Psallidas 1988; Sawada et al., 1992a). Sensitivity towards agrocin 84 is determined by the Ti plasmid. Dipping seeds, roots, or wounded plant surfaces in suspensions of strain 84 has been used with success worldwide for the biological control of crown gall disease (Moore and Warren, 1979; Kerr, 1980). Strain 84 is available in commercial preparations as a biological control agent and has found wide application. *Agrobacterium vitis* strains are insensitive to agrocin 84. Other agrocin-producing strains, effective against *Agrobacterium vitis*, have been isolated (Staphorst et al., 1985; Chen and Xiang 1986; Webster et al., 1986; Webster and Thompson 1988; Xie et al., 1993).

Trifolitoxin (TFX) is a post-translationally modified peptide antibiotic produced by *R. leguminosarum* biovar trifolii T24 (Breil et al., 1993). TFX is toxic to non-producing strains within a distinct taxonomic group of the *Alphaproteobacteria*, and it appears to give an ecological advantage for nodulation by the producing strain. Eight genes have been identified for the production of this toxin (Triplett et al., 1994; Breil et al., 1996).

Bacteriophages Lysogeny for either active plaque-forming or defective bacteriophages is widespread in tumorigenic *Agrobacterium* spp. Morphological, biological, and physicochemical properties and genetic relationships of several of the isolated phages or phage-like particles have been determined (Beardsley, 1955; Zimmerer et al., 1966; Stonier et al., 1967; De Ley et al., 1972; Manasse et al., 1972; Vervliet et al., 1975). Virulent bacteriophages of *Rhizobium* were the subject of numerous studies published prior to 1950 (Allen and Allen, 1950), but *Agrobacterium* phages, although not as extensively studied in earlier times, have more recently been isolated from sewage and soil (Roslycky et al., 1963; Boyd et al., 1970a, b).

The range of hosts susceptible to a particular *Rhizobium* bacteriophage is highly variable. In some instances, it is limited to relatively few strains within a single host species; in others, it may cross taxonomic boundaries. Cross-infection studies of *Sinorhizobium meliloti* and *Sinorhizobium fredii* (Hashem et al., 1996) indicate that bacteriophage lysis is not usually sufficiently specific to identify species or individual strains. Some strains of *Rhizobium* are lysogenic. Bacteriocins have been reported (Roslycky, 1967; Venter et al., 2001), as well as a parasitic *Bdellovibrio*.

Antigenic structure Early serological studies indicated that strains belonging to *Agrobacterium tumefaciens* and *Agrobacterium rhizogenes* could be distinguished from each other by serological reactions (Keane et al., 1970; Lopez, 1978). Alarcón et al. (1987) and Sawada et al. (1992c) found serological heterogeneity between these species and with *Agrobacterium vitis*. *Rhizobium* spp. show extensive cross-reaction with *Agrobacterium* strains (Graham, 1971).

Most serological reactions (agglutination, gel diffusion, precipitation, and fluorescent antibody) show strain specificity and have traditionally been of great value in identifying particular strains of *Rhizobium* in nodules of field plants or in laboratory investigations. The agglutination reaction, using crushed nodule extracts, is the most widely used for field work because of its simplicity, although it is complicated by cross-reactions and autoagglutination. Surface antigens, although useful for strain recognition, are limited in their usefulness for species identification. Early work, such as that of Vincent and Humphrey (1970), who reported on the antigen structure in the "biovars" of *R. leguminosarum*, should now be reinterpreted in the light of modern taxonomic revisions of the species. Sawada et al. (1992b) differentiated several serogroups in *Agrobacterium vitis* using a slide agglutination test.

Ecology *Rhizobium* occurs worldwide in soils and especially in the rhizosphere of plants. As many as 10^6–10^7 cells/g soil of symbiotic *Rhizobium* have been reported. *Rhizobium* strains capable of degrading 2-sulfonato-fatty-acid-methyl-esters (Masuda et al., 1995) and *Agrobacterium* strains capable of utilizing phthalate (Nomura et al., 1989) were reported as common soil inhabitants in contaminated soils. The identification of these strains as authentic rhizobia and agrobacteria needs confirmation.

Natural interactions between *Rhizobium* strains within the rhizosphere are complex, as indicated by the extent to which different *Rhizobium* populations are shown to compete in the infection processes. Agrocin-producing strains have a proven role in competing with tumorigenic *Agrobacterium* strains to inhibit infection. A similar competitive process occurs when attempts are made to nodulate seedling legumes with effective *Rhizobium* strains. Naturally occurring strains that are nonefficient in nitogen fixation can be more effective in infecting and nodulating plants, thereby limiting plant growth (Triplett and Sadowsky, 1992). Strains of *Rhizobium* that are non-nodulating and occurring naturally in soils have been well documented (Soberón-Chávez and Nájera, 1989; Segovia et al., 1991). Symbiotic *Rhizobium leguminosarum* biovar trifolii strains have been reported as natural endophytes in the roots of rice (Yanni et al., 1997), and *R. etli* strains, as endophytes of maize (Gutiérrez-Zamora and Martínez-Romero, 2001). Plant-pathogenic *Agrobacterium* species have also been isolated from a crown gall tumor on alfalfa (Palumbo et al., 1998).

Although bacteria within the genus *Rhizobium* have been shown to have a worldwide distribution, unique species may be isolated from limited geographic regions, normally related to the distribution of their hosts (Martínez-Romero and Caballero-Mellado, 1996). *Rhizobium* species may have been spread internationally with inoculated legume plants, with seed (Pérez-Ramírez et al., 1998), or soil. Caballero-Mellado and Martínez-Romero (1999) reported that soil fertilization limited the genetic diversity of *Rhizobium* in bean nodules. Vance (1998) reviewed agronomic aspects of commercial inoculants that are used to enhance legume crop cultivation.

Agrobacterium strains have also been reported in a variety of human clinical specimens (CDC group Vd-3) (Lautrop, 1967; Riley and Weaver, 1977; Gilardi, 1978a; Rubin et al., 1980). They are usually 3-ketolactose-positive and nonpathogenic to tomato. It is believed that these clinical isolates occur either as incidental inhabitants in the patient or as contaminants introduced during sample manipulation. The authenticity of these strains as true *Agrobacterium* spp. needs to be confirmed.

Antibiotic sensitivity *Rhizobium* strains are resistant to a variable spectrum of antibiotics (Davis, 1962). Most are susceptible to tetracycline. Although there is wide strain-to-strain variation in resistance, *Rhizobium* strains are intrinsically more sensitive than *Bradyrhizobium* to tetracycline, penicillin G, viomycin, vancomycin, and streptomycin. Streptomycin-resistant mutants, which are usually effective as nodulating strains, are important in ecological field studies on strain competition. In general, wild-type *Agrobacterium* tumorigenic species have been reported to be sensitive to chlorotetracycline, gentamicin, neomycin, novobiocin, oxytetracycline, and tetracycline (Kersters et al., 1973) but are commonly resistant to nalidixic acid. Growth is inhibited by low concentrations (3–780 µg/ml of medium) of metacycline, doxycycline, sigmamycin (tetracycline + oleandomycin), and triacetyloleandomycin (Goedert, 1973).

ENRICHMENT AND ISOLATION PROCEDURES

Although *Rhizobium* strains are common soil inhabitants, they are best isolated from freshly excised legume root nodules. Identification is relatively easy if strains are isolated from host plant nodules. Isolation is difficult if strains are isolated directly from the soil or if strains are non-infective, unless they have unique genetic markers. Isolation of symbiotic species from soil generally requires the use of trap hosts, which are leguminous plants grown in the soil and from which nodules are selected for subsequent isolation of rhizobia. Nodules collected in the field can be temporarily stored in small vials containing silica gel held under a cotton plug.

In order to isolate symbionts, healthy root nodules—with a small portion of root attached if they are very small (<1.0 mm)—are surface sterilized by exposure to commercial 3% H_2O_2 solution or 5% commercial hypochlorite (3% available chlorine) solution for 5–60 min depending on their size. This treatment is followed by a wash in sterile water. The nodules are crushed in a small drop of sterile 0.05% peptone or 0.1M phosphate and a loopful of the suspension is streaked onto surface-dried plates of yeast extract-mannitol agar (YMA)[1] prepared without $CaCO_3$. Alternatively, a small loopful of the crushed nodule suspension can be streaked onto successive plates of agar medium. Large nodules can be sliced with a sterile scalpel blade and portions of the interior removed with a needle. Bacteria can readily be isolated from young galls and nodules on different parts of plants.

Isolation from older hypertrophying tissue is more difficult. Incubation is at 28°C for 3 or more days. Well-isolated, white, mucoid, glistening, hemispheric colonies are restreaked onto fresh plates for subsequent confirmation, which requires reinfection of the original host under careful aseptic conditions where uninoculated controls are devoid of nodules. If heavy fungal contamination is expected, the agar medium used for initial isolation should contain 0.002% actidione. *Rhizobium* strains grow poorly on 0.04% peptone/1% glucose mineral salts agar and show little pH change. This medium can serve as a useful contamination check: colony formation in 2 days at 28°C and a marked pH change are not characteristic of *Rhizobium*.

MAINTENANCE PROCEDURES

YMA slant cultures in sealed containers can be stored for 2–3 years or longer at 2°C. Cultures usually survive for around 2 months at 15°C. Long-term storage at −20°C or −80°C is recommended: turbid suspensions from fresh broth cultures are mixed with equal volumes of sterile 80% glycerol in water and allowed to stand for 1 hr at room temperature before storing in small aliquots in the freezer. Individual aliquots are thawed as required. Norris (1963) described a preservation method using small porcelain beads which, after inoculation and drying over silica gel, can be used individually to inoculate YMB for subsequent recovery of the bacteria.

Stock cultures of pathogenic species may be routinely maintained on agar slants in screw-capped vials at 4°C for 2 months on YMA or on either of the following media (in g/l of tap water): (a) glucose, 20; yeast extract, 10; $CaCO_3$, 20; and agar, 20; or (b) glucose, 10; yeast extract, 10; $(NH_4)_2SO_4$, 1.0; KH_2PO_4, 0.25; and agar, 20.

Lyophilized cultures stored at 4°C remain viable for at least 25 years.

PROCEDURES FOR TESTING SPECIAL CHARACTERS

Carbohydrate and organic acid utilization have been determined using inoculated plates of the medium of Elkan and Kwik (1968). On the dried surface of these plates are placed absorbent paper discs previously saturated with a 10% solution of the organic compound and slowly dried. During incubation at 28°C the plates are examined daily using indirect lighting against a black, nonreflecting background.

API Biotype galleries (BioMerieux, La Balme-les-Grottes, France) and similar standardized systems to test for the utilization of standard ranges of substrates are increasingly used to obtain reproducible biochemical data.

Details on the methods used to assess nodulation response under greenhouse or growth room conditions are given by Vincent (1970), and Somasegaran and Hoben (1994) have produced an excellent methods book for the novice researcher of symbiotic species. Moore et al. (1988) have provided useful advice for performing inoculation tests for pathogenicity studies and summarize recipes for diagnostic media and tests.

Sawada et al. (1995) have reported a method for detecting the presence of Ti and Ri plasmids by specific amplification of components using the polymerase chain reaction.

DIFFERENTIATION OF THE GENUS *RHIZOBIUM* FROM OTHER GENERA

Members of *Rhizobium* are distinguished from those in the related genera *Mesorhizobium* and *Phyllobacterium* by differences in growth rate, fatty acid profiles, and 16S rDNA sequence. Members of

---

1. Yeast extract-mannitol agar (YMA) contains (g/l of distilled water): D-mannitol, 10.0; KH_2PO_4, 0.5; $MgSO_4 \cdot 7H_2O$, 0.2; NaCl, 0.1; $CaCO_3$, 4.0; yeast extract (Difco), 0.4; agar, 15.0; pH 6.8–7.0. Sterilize at 121°C for 15 or 30 min depending on the volume. The $CaCO_3$ is omitted for the preparation of pour plates or for liquid medium.

Rhizobium are not distinguished from those in the related genera *Allorhizobium* or *Sinorhizobium* by any phenotypic characters except those that form the individual species circumscriptions. The genus *Agrobacterium* is distinguished from the genera containing nitrogen-fixing species, including *Rhizobium*, only because its members have oncogenic capabilities. *Rhizobium* is distinguished from *Sinorhizobium* not only based on differences in their 16S rDNA sequences but also on the basis of other gene sequences (Gaunt et al., 2001) and by *nolR* gene hybridization (Toledo et al., 2003).

The features that differentiate nodulating strains of *Rhizobium* species from morphologically and physiologically similar organisms are given in Tables BXII.α.112 and BXII.α.113. Strains of nonpathogenic agrobacteria and non-nodulating rhizobia can be isolated from soils and are difficult to allocate to genera or species based on characteristics reported in Table BXII.α.112. It is necessary to resort to specific molecular probes for reliable identification.

TAXONOMIC COMMENTS

Overviews of the relationships of bacterial nitrogen-fixing genera are given in Young (1992, 1994), Martínez-Romero (1994), Lindström et al. (1995, 1998), Martínez-Romero and Caballero-Mellado (1996), and Young and Haukka (1996).

At that time of publication of the first edition of *Bergey's Manual of Systematic Bacteriology*, the family *Rhizobiaceae* comprised *Rhizobium* (Jordan, 1984a), *Bradyrhizobium* (Jordan, 1984a), *Phyllobacterium* (Knösel, 1984a), and *Agrobacterium* (Kersters and De Ley, 1984a).

Since then, the relationships of nodulating, nitrogen-fixing species have been investigated by comparative analysis of 16S rDNA sequence data. Sequences of the type strains, obtained from international databases, have been subjected to various forms of algorithmic and parsimonious analysis in order to establish their phylogenetic relationships (Sawada et al., 1993; Willems and Collins 1993; de Lajudie et al., 1994, 1998b, 1998a; Nour et al., 1995; Rome et al., 1996; Young and Haukka 1996; Amarger et al., 1997; Tan et al., 1997; Lindström et al., 1998). Comparison of these analyses with others that have been published (Rome et al., 1996; Jarvis et al., 1997; de Lajudie et al., 1998a, b; van Berkum et al., 1998; Wang et al., 1998), and with Fig. BXII.α.130 of this chapter, shows how much results can vary depending on the selection of sequences and the form of analysis. Notwithstanding, all data clearly support the separation of *Bradyrhizobium* Jordan 1982 (now assigned to the family *Bradyrhizobiaceae*, Kuykendall, this volume) and *Azorhizobium* Dreyfus et al., 1988 (now assigned to the family *Hyphomicrobiaceae*, Kuykendall, this volume) as distantly related to all the other pathogenic and nodulating species. These latter species are allocated to the closely related families, *Rhizobiaceae* and *Phyllobacteriaceae*. Nitrogen-fixing and oncogenic (hypertrophying) species in *Agrobacterium*, *Rhizobium*, and *Sinorhizobium* are assigned to the *Rhizobiaceae* and species of *Mesorhizobium* (Jarvis et al., 1997; Chen and Kuykendall, this volume) are assigned to the *Phyllobacteriaceae* (Chen and Kuykendall, this volume).

The family *Rhizobiaceae* includes symbiotic and pathogenic species in the genera *Agrobacterium*, *Rhizobium*, and *Sinorhizobium*. Species allocated to these genera are found in two or three clusters. One cluster comprises *Sinorhizobium*: *Sinorhizobium fredii* (the type species), *Sinorhizobium arboris*, *Sinorhizobium kostiense*, *Sinorhizobium medicae*, *Sinorhizobium meliloti*, *Sinorhizobium saheli*, *Sinorhizobium terangae*, and *Sinorhizobium xinjiangense*. *S. arboris* and *S.*

kostiense were only recently described (Nick et al., 1999). The second cluster is more heterogeneous and may be considered to be represented by two subgroups. Subgroup 2a includes *R. leguminosarum* (the type species), *R. etli*, *R. gallicum*, *R. giardinii*, *R. hainanense*, *R. mongolense*, *R. tropici*, and *Agrobacterium rhizogenes*. Subgroup 2b includes *R. galegae*, *R. huautlense*, *Agrobacterium tumefaciens* (the type species), *Agrobacterium rubi*, *Agrobacterium vitis*, and *Allorhizobium undicola* (the type species). All these species have base differences amounting to less than 7% of the total 16S rDNA sequence. The extent of statistical support for individual branches and their relative positions depend on the form of phylogenetic analysis and the selection of sequences. Eardly et al. (1996), Martínez-Romero and Caballero-Mellado (1996), and Young and Haukka (1996) note anomalies in sequence analyses that are attributable to recombination events between species, a conclusion supported but qualified by Wernegreen and Riley (1999). As yet the significance and implications of recombination on the inference of phylogenetic relationships are unclear.

The family also contains, as outliers to the rhizobial species, strains named *Blastobacter* spp. and "*Liberibacter*" spp., which do not have symbiotic or pathogenic characteristics. Other strains of *Blastobacter* spp. are to be found in the families *Methylobacteriaceae* (four strains), *Bradyrhizobiaceae* (one strain), and *Sphingomonadaceae* (one strain). *Blastobacter aggregatus* ATCC 43293 and *Blastobacter capsulatus* ATCC 43294 in the *Rhizobiaceae* are therefore perhaps incorrectly named. The new genus "*Liberibacter*" represents strains of the fastidious organism that is the pathogen of citrus greening disease. This organism appears to be relatively distantly related to *Rhizobium*.

When first proposed, *Sinorhizobium* (Chen et al., 1988b) was based on only a small number of nutritional and biochemical tests, and its validity was questioned by Jarvis et al. (1992) on the basis of partial 16S rDNA sequence analysis and on the interpretation of numerical data. This genus has since been examined in greater detail and an emended circumscription of the genus has been produced (de Lajudie et al., 1994). However, this circumscription does not delineate a taxon distinct from *Rhizobium*, and the polyphasic data reported (PAGE of total proteins and carbon source utilization tests) do not support a coherent taxon. Moreover, the protein data show *Sinorhizobium fredii* (the type strain) as an outlier to the other species, and carbon source utilization data show *Sinorhizobium* species intermingled with *Azorhizobium*, *Bradyrhizobium*, and *Rhizobium*. Support for this genus, distinct from *Rhizobium*, is based on comparative 16S rDNA sequence data alone (de Lajudie et al., 1994).

Allorhizobium, is a monospecific genus established because—as indicated by its name—comparative analysis of 16S rDNA sequence data for *Allorhizobium undicola* indicated that this organism belonged to an outlying branch (de Lajudie et al., 1998b). The closest neighboring species in their analysis was *Agrobacterium vitis*. The species *Allorhizobium undicola* is well defined based on DNA–DNA reassociation, PAGE of total proteins, and carbon source utilization tests. As with *Sinorhizobium*, the circumscription of the genus does not delineate a taxon distinct from *Rhizobium*, and the polyphasic data do not support a close relationship between *Allorhizobium undicola* and *Agrobacterium vitis* and other agrobacteria. In proposing a new genus, rather than either nominating the species as a nitrogen-fixing member of *Agrobacterium ex tempore* or allocating it to *Rhizobium ex tempore*, the authors are committed to a nomenclature in which either *Agrobacterium vitis* is renamed *Allorhizobium vitis* or is recognized in its own monospecific genus. While monospecific genera can sensibly be named

TABLE BXII.α.112. Characteristics of *Rhizobium*, *Agrobacterium*, *Allorhizobium*, and *Sinorhizobium* species[a,b]

Characteristics	*Rhizobium leguminosarum*	*Rhizobium etli*	*Rhizobium galegae*	*Rhizobium gallicum*	*Rhizobium giardinii*	*Rhizobium hainanense*	*Rhizobium huautlense*	*Rhizobium mongolense*	*Rhizobium tropici*	*Agrobacterium tumefaciens*	*Agrobacterium rhizogenes*	*Agrobacterium rubi*	*Agrobacterium vitis*	*Allorhizobium undicola*	*Sinorhizobium fredii*	*Sinorhizobium medicae*	*Sinorhizobium meliloti*	*Sinorhizobium saheli*	*Sinorhizobium terangae*	*Sinorhizobium xinjiangense*
Polar flagella			1–2						yes					yes						1
Peritrichous flagella or one subpolar flagellum	2–6		1–2			1		1		1–4	1–4	1–4	1–4		1–3		2–6	1	1	1–3
3-ketolactose produced	−	−	+	−	−	−			−	+	−	+	+	−	−	−	+	−	−	
Growth factors required:																				
Biotin	+	−	+	−					−	−	+	+	+	−	−		+			
Pantothenate	d	−	−	−					−	−	+	+	+	−		d	d			
Thiamine	+	−	−	−			+		−	−				−		+	−			
pH range	4–9		5–9.5[c]	>4 to <8	4–8.5	5–10	5–9	4–10	4–10		5–9				5–10.5	5–10	4.5–9.5			5–10.5
Grows at 28°C	+	+	+	+	+	+	+	+	+	+	+	+	+	+	+	+	+	+	+	+
Grows at 35°C	+	+	+	+	+	+	+	+	+	+	+	+	+		+	+	+	+	+	+
Grows at 40°C	−	−	d[c]	−	d	+	+	−	+	+	−	−	−	−	−	d	d	+	+	+
Grows in 1% NaCl	−	−	−	−	−	+	−	−	−	+	−	−	+	−	+	+	+	+	+	+
Grows in 2% NaCl	−	−	−	−	−	+	−	−	−	+	−	−	+	−	d	+	d	+	+	(+)
Growth in Luria–Bertani medium	−	−	−	−	−	+	−	−	+	+				−	−	−	+	−	−	−
Oncogenicity to few or many plant genera[d]	−	−	−	−	−	−	−	−	−	many[d]	many[d]	few[d]	few[d]	−	−	−	−	−	−	−
Symbiotic nodulating/nitrogen-fixing ability[d]	+	+	+	+	+	+	+	+	+	−	−	−	−	+	+	+	+	+	+	+

[a]For symbols see standard definitions.

[b]Data are from Graham and Parker (1964), Jordan (1984a), Kersters and De Ley (1984a), Kerr (1992) and original descriptions of species.

[c]Unpublished data of E.T. Wang (personal communication) using the methods of Wang et al. (1998).

[d]Because oncogenicity and nitrogen-fixing symbioses are plasmid-mediated and the stability of resulting host specificity is also uncertain, these characteristics are not reliable bases for classification or identification of these species.

TABLE BXII.α.113. Carbon source utilization tests which differentiate *Rhizobium*, *Agrobacterium*, *Allorhizobium*, and *Sinorhizobium* species[a,b]

Substrate[c,d]	Rhizobium leguminosarum	Rhizobium galegae	Rhizobium tropici	Agrobacterium tumefaciens	Agrobacterium rhizogenes	Agrobacterium vitis	Allorhizobium undicola	Sinorhizobium fredii	Sinorhizobium meliloti	Sinorhizobium saheli	Sinorhizobium terangae
Number of strains[e]	37	2	3	7	3	2	6	2	3	4	20
Acetate	+	+	d	+	–	+	d	–	d	+	+
N-acetylglucosamine	+	+	+	+	+	+	–	+	+	+	+
Aconitate	–	–	+	+	–	+	–	d	–	+	d
Adonitol	+	+	+	+	+	+	+	+	+	+	+
L-(alpha)-Alanine	+	+	+	+	d	+	d	–	d	d	d
beta-Alanine	–	–	–	–	–	–	d	–	+	–	–
DL-3-Amino butyrate	–	–	d	d	–	–	d	–	+	d	+
DL-4-Amino butyrate	–	–	–	–	–	+	d	+	+	d	d
DL-5-Aminovalerate	–	d	d	–	d	d	–	–	d	–	+
Amygdalin	–	–	+	+	+	d	–	d	+	+	d
D-Arabinose	+	+	+	+	+	d	–	+	+	+	+
L-Arabitol	–	–	+	–	+	–	+	–	+	–	–
Arbutin	+	+	+	+	+	+	+	+	+	+	+
L-Arginine	+	d	–	–	+	–	d	d	+	+	d
L-Aspartate	–	–	+	+	+	+	d	–	+	+	d
Butyrate	–	–	–	–	–	+	–	–	–	–	d
Citrate	–	d	d	–	+	+	d	d	–	–	+
L-Citrulline	+	–	d	–	d	–	d	–	–	+	–
L-Cysteine	+	–	–	–	d	–	–	–	+	–	–
Dulcitol	+	–	–	+	d	–	–	–	+	+	+
Erythritol	+	–	d	–	+	+	d	d	+	+	+
Ethanolamine	–	d	–	–	+	–	–	+	d	+	d
D-Fucose	+	+	d	+	+	+	d	d	d	d	–
Gluconate	+	+	+	+	+	–	–	–	–	–	–
L-Glutamate	+	–	+	+	+	d	d	+	+	+	+
Glutarate	+	–	d	d	d	+	–	–	–	–	–
DL-Glycerate	–	d	+	+	+	+	d	d	d	+	d
Glycolate	–	–	–	–	–	–	–	–	–	+	–
m-Hydroxybenzoate	+	+	–	d	+	+	d	–	–	+	d
p-Hydroxybenzoate	+	+	d	d	d	+	d	–	d	+	d
DL-3-Hydroxybutyrate	+	+	–	+	d	+	d	–	d	+	d
Isobutyrate	–	–	–	–	–	–	–	–	d	d	d
L-Isoleucine	+	–	–	–	d	–	d	–	d	d	+
2-Ketogluconate	+	+	+	+	+	d	d	–	+	d	–
5-Ketogluconate	–	–	–	+	+	–	–	–	–	–	–
2-Ketoglutarate	d	d	+	–	d	–	d	d	–	–	–
DL-Lactate	–	+	+	+	+	+	+	+	+	+	+

(continued)

TABLE BXII.α.113. *(cont.)*

Substrate[c,d]	*Rhizobium leguminosarum*	*Rhizobium galegae*	*Rhizobium tropici*	*Agrobacterium tumefaciens*	*Agrobacterium rhizogenes*	*Agrobacterium vitis*	*Allorhizobium undicola*	*Sinorhizobium fredii*	*Sinorhizobium meliloti*	*Sinorhizobium saheli*	*Sinorhizobium terangae*
L-Leucine	+	d	d	d	d	−	−	d	d	d	+
L-Lysine	+	d	−	+	+	−	−	−	+	+	+
D-Lyxose		+	+	+	+	+	+	−	+	d	d
Malonate		−	d	−	+	−	−	−	−	−	−
D-Mandelate		−	+	−	+	−	−	−	d	−	−
L-Mandelate		−	+	−	+	−	−	−	−	+	+
D-Melibiose	+	+	+	+	+	+	−	+	+	d	d
D-Melezitose		−	−	+	+	−	−	d	+	−	−
Methyl-D-glycoside		d	+	−	+	−	−	−	+	−	+
Methyl-D-xyloside		−	d	+	+	−	+	+	+	−	−
L-Ornithine		d	d	−	+	−	+	+	+	+	+
L-Phenylalanine	+	+	−	−	+	−	−	d	−	−	−
Propionate	−	+	+	+	+	+	(+)	d	d	+	d
Pyruvate		+	+	+	+	+	(+)	d	+	+	(+)
D-Raffinose		+	+	+	+	+	d	+	+	+	+
Salicin		−	d	−	+	+	d	+	+	d	d
Sarcosine		+	d	−	+	−	−	−	−	−	−
L-Serine	+	+	d	d	+	d	+	−	d	−	d
L-Sorbose		−	−	+	+	−	−	−	+	−	−
D-Tagatose		−	−	+	+	d	−	−	−	−	d
D-Tartrate		−	+	+	+	+	−	−	+	+	d
L-Tartrate		−	+	+	−	+	−	−	−	−	−
meso-Tartrate		−	−	+	−	d	−	−	−	+	d
L-Threonine	+	d	d	−	+	d	+	−	d	+	d
Trigonelline		−	+	−	d	+	d	d	+	d	−
L-Tyrosine	+	+	−	−	−	−	−	−	d	d	d
L-Valine	+	+	+	+	+	−	−	−	+	+	+
Xylitol		+	+	+	+	−	−	−	d	−	d
L-Xylose		−	+	d	+	d	−	−	d	d	d

[a]For symbols see standard definitions; (+), weak reaction.

[b]Carbon source utilization data is from de Lajudie et al. (1994) and de Lajudie et al. (1998a). Data for *R. leguminosarum* are from Amarger et al. (1997).

[c]Substrates which gave positive reactions by strains of all species: L-arabinose, D-cellobiose, D-fructose, D-galactose, D-glucose, L-histidine, inositol, D-mannose, rhamnose, ribose, sorbitol, D-turanose, and D-xylose. Fumarate, glycerol, lactose, L-malate, maltose, mannitol, sucrose, succinate, and trehalose were substrates for which only one species expressed variable negative reactions.

[d]Substrates which gave negative reactions by strains of all species: adipate, D-("alpha")-alanine, DL-2-aminobutyrate, amylamine, azelate, benzoate, benzylamine, butylamine, caprate, *n*-caprate, caprylate, citraconate, diaminobutane, esculin, ethylamine, glycine, glycogen, heptanoate, histamine, *o*-hydroxybenzoate, inulin, isophthalate, isovalerate, itaconate, DL-kynurenine, levulinate, maleate, mesaconate, L-methionine, DL-norvaline, oxalate, pelargonate, phenylacetate, phthalate, pimelate, sebacate, spermine, starch, suberate, terephthalate, D-tryptophan, tryptamine, urea, and *n*-valerate. 2-Aminobenzoate, 3-aminobenzoate, 4-aminobenzoate, L-cysteine, fumarate, methyl-D-mannoside, L-norleucine, and sucrose were substrates for which only one species expressed variable positive reactions.

[e]Data are included only where information is available for two or more strains.

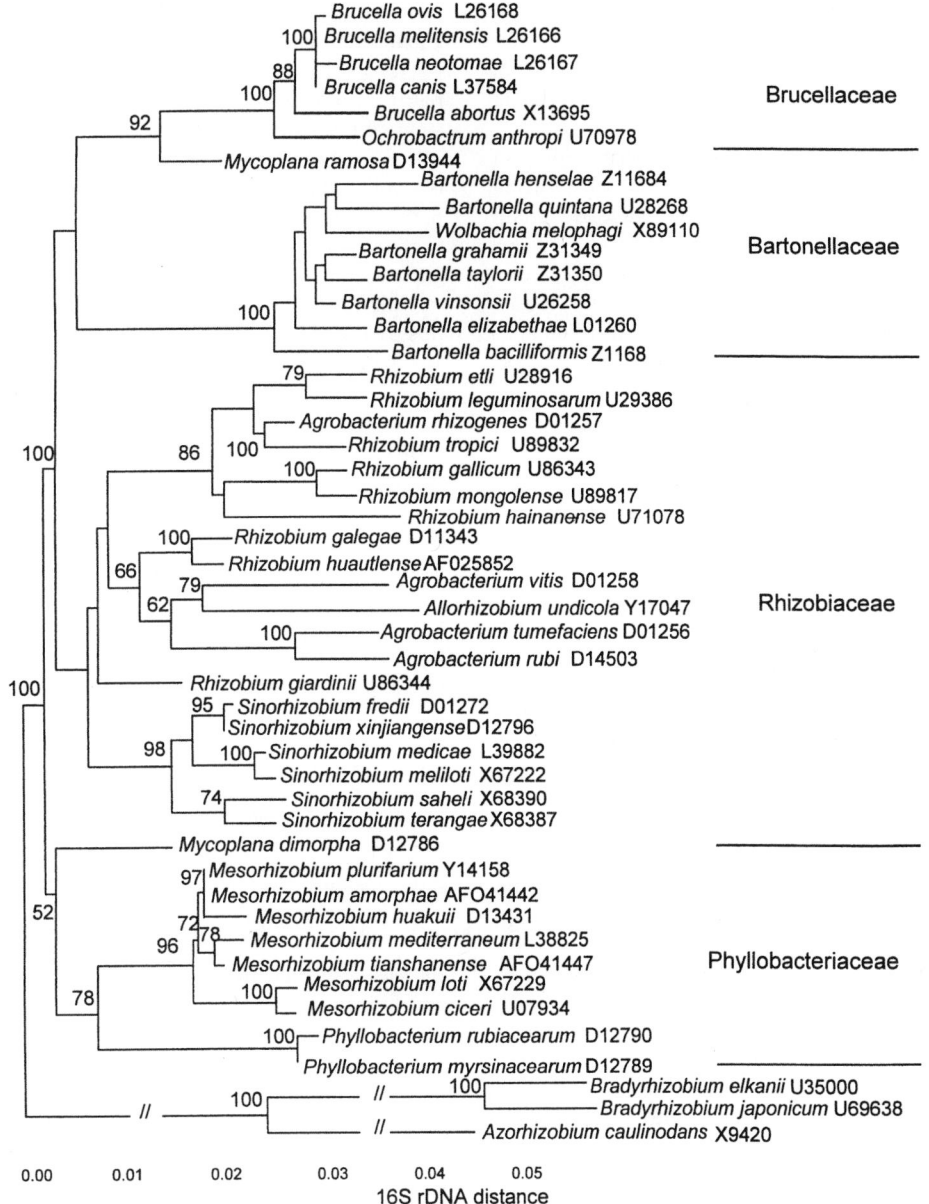

FIGURE BXII.α.130. Neighbor-joining tree expressing the relationships of genera within the *Rhizobiaceae* (*Agrobacterium, Allorhizobium, Rhizobium,* and *Sinorhizobium* and related families) based on 16 S rDNA sequences. Bootstrap probabilities (as percentages) are determined from 1000 resamplings.

when there is clear justification based on a unique circumscription of the taxon, the practice of naming taxa when they are merely outlying members of larger clades must be questioned. In such cases, where the description of the genus is indistinguishable from adjacent genera and where the species description forms the basis of the generic description, there is no basis for forestalling an on-going process of naming monospecific genera across the clade. This approach to classification implies the creation of many genera containing small numbers of species with circumscriptions indistinguishable from *Rhizobium.*

In regard to "rhizobia," legume plants are represented by more than 14,000 species (Jordan 1984a; Lindström et al., 1998), of which fewer than 1% have been investigated to establish the identity of their associated nitrogen-fixing bacterial species. Considering the relatively close relationships of taxa within the *Rhi-*

zobiaceae, it is highly probable that new taxa will be characterized that will be intermediate between the present named species and genera (*Agrobacterium, Rhizobium,* and *Sinorhizobium*). Intermediate taxa can be expected to obscure the apparent deep branches between currently named genera (Martínez-Romero and Caballero-Mellado, 1996).

Murray et al. (1990) have made clear the expectation that at the generic level, taxa should be supported with phenotypic descriptions. Proposed minimal standards for *Agrobacterium* and *Rhizobium* require that generic as well as species names should be based on both phenotypic and phylogenetic data (Graham et al., 1991). Phylogenetic divisions can only be understood as generating distinct genera where these have evolved into discrete phenotypic groups. Genera differentiated solely on sequence data have the same nomenclatural status as taxa erected as *Candidatus,* as described by Murray and Stackebrandt (1995). Furthermore, 16S rDNA sequence data cannot be accepted uncritically where there is evidence of recombination within sequences (Eardly et al., 1996; Martínez-Romero and Caballero-Mellado, 1996; Young and Haukka, 1996).

With the exception of the 16S rDNA-based discrimination of *Sinorhizobium* as a distinct clade, the generic circumscriptions of *Agrobacterium*, *Allorhizobium*, *Rhizobium*, and *Sinorhizobium* contain no demonstrated characteristics by which these genera can be supported as distinct taxa. Multiple character analysis towards a phenetic (Goodfellow and O'Donnell, 1993; Goodfellow et al., 1997) or polyphasic (Vandamme et al., 1996b) classification aims to produce coherent taxa with relevant circumscriptions. Electrophoretic protein patterns, and numerical analysis of nutritional and biochemical data (de Lajudie et al., 1994) gave no support for segregation of species into the separate genera *Agrobacterium*, *Rhizobium*, and *Sinorhizobium*. Terefework et al. (1998) and de Lajudie et al. (1998a) provide data showing that various pathogenic and nodulating, symbiotic bacteria belonging to the genera *Agrobacterium* and *Rhizobium* are interspersed on sub-branches of 16S or 23S phylogenetic trees. Analysis of fatty acid profiles showed that these three genera were closely related (Jarvis et al., 1996), but that species of *Mesorhizobium* formed a distinct group. More recently, Tighe et al. (2000) showed that the concentration of $C_{16:0\ 3OH}$ fatty acid was generally somewhat lower in *Sinorhizobium* than *Rhizobium*. 16S rDNA sequence data clearly show that *Agrobacterium* spp. (Willems and Collins, 1993) are closely related to *Rhizobium* spp. such as *Rhizobium galegae* (Lindström, 1989), and the recently described *Rhizobium huautlense* (Wang et al., 1998). *Agrobacterium rhizogenes* is always found among authentic *Rhizobium* spp. Many past studies have suggested the

need for amalgamation of *Agrobacterium* and *Rhizobium* (Graham 1964, 1976; Heberlein et al., 1967; De Ley, 1968; White, 1972; Kerr, 1992; Sawada et al., 1993). It has long been clear that pathogenic (*Agrobacterium*) and nitrogen-fixing (*Rhizobium*) species are interspersed, and there seems to be a paucity of justification for the on-going separation of these genera. *Agrobacterium* is a polyphyletic genus that is an artificial amalgamation of plant-pathogenic species (Young et al., 2001).

The four named genera—*Agrobacterium*, *Allorhizobium*, *Rhizobium*, and *Sinorhizobium*—are closely related in genomic and phenotypic terms. It has recently been proposed by the authors of this treatment that *Agrobacterium*, *Allorhizobium*, and *Rhizobium* be amalgamated into a single genus, *Rhizobium*, based on their close genomic and phenotypic similarity (Young et al., 2001). We maintain that *Allorhizobium* is an artificial genus and in phylogenetic terms is part of the *Agrobacterium/Rhizobium* cluster. Indeed it seems clear that its creation was only justified based on the anomalous state of *Agrobacterium* nomenclature at the time it was proposed by de Lajudie et al., (1998a). The branches between *Agrobacterium* and *Rhizobium* on the one hand and *Sinorhizobium* on the other suggest an evolutionary divergence that could be interpreted as preliminary to the formation of new genera. At present these two clades are probably incipient genera and merit no more than recognition as subgenera.[2]

---

2. Since the completion of this manuscript, a proposal to include species of *Agrobacterium*, *A. radiobacter*, *A. rhizogenes*, *A rubi*, *A. vitis*, and *Allorhizobium undicola* in *Rhizobium* with an emended description of the genus, has been made (Young et al., 2001), as a more natural polyphasic, interpretation of the the taxonomy of the family *Rhizobiaceae*. The status of the genus *Sinorhizobium* is considered to need further evaluation.

DIFFERENTIATION OF THE SPECIES OF THE GENUS *RHIZOBIUM*

Morphological, biochemical, and nutritional data are given in Tables BXII.α.112 and BXII.α.113. Nutritional data have been reported for only about half of *Rhizobium* species (Table BXII.α.113). Data are included here only if more than one strain for each species has been reported. Some recently named species have been established largely using comparative 16S rDNA sequence analyses. Phenotypic descriptions of these species, as well as a comprehensive comparative investigation of the whole genus, are needed.

The phenotypic descriptions of the symbiotic nodule-forming

Rhizobium species are based on D.C. Jordan's treatment in the first edition of *Bergey's Manual of Systematic Bacteriology* in 1984 and on original descriptions of new species. The phenotypic description of *Agrobacterium* is based on studies by De Ley et al. (1966), Lippincott and Lippincott (1969), Keane et al. (1970), White (1972), Kersters et al. (1973), Panagopoulos and Psallidas (1973), Kerr and Panagopoulos (1977), Süle (1978), Panagopoulos et al. (1978), Holmes and Roberts (1981), Sawada and Ieki (1992b), de Lajudie et al. (1994), and Amarger et al. (1997).

List of species of the genus Rhizobium

1. **Rhizobium leguminosarum** (Frank 1879) Frank 1889, 338[AL] (*Schinzia leguminosarum* Frank 1879, 397.)
   le.gu.mi.no.sa' rum. M.L. fem. n. *Leguminosae* old family name of the legumes; M.L. gen. pl. n. *leguminosarum* of legumes.

   The characteristics are as given for the genus and listed in Tables BXII.α.112 and BXII.α.113. The cells have either 1 or 2 polar flagella or 2–6 peritrichous flagella. Growth does not occur at 39–40°C. The pH range for growth is 4.5–9.0. Growth does not occur in YMA containing 2% NaCl. Pantothenate and, for some strains, thiamine are required as growth factors. Well characterized genetically. Three biovars of *R. leguminosarum* have been defined, biovar trifolii (previously classified as *Rhizobium trifolii* Dangeard 1926[AL]), biovar phaseoli, and biovar viceae, based on nodulating specificity. Selected strains of biovar phaseoli have been elevated to the status of species as *R. etli*, *R. gallicum*, *R.*

giardinii, *R. mongolense*, and *R. tropici*. Residual strains representing *R. leguminosarum* need to be re-examined and the description of the species needs to be emended.

   R. leguminosarum nodulates with some, but not necessarily all, *Lathyrus* spp., *Lens* spp., temperate species of *Phaseolus* (*P. vulgaris*, *P. angustifolius*, *P. multiflorus*), *Pisum* spp., *Trifolium* spp., and *Vicia* spp.

   The mol% G + C of the DNA is: 59–63 (T_m).

   Type strain: ATCC 10004, DSM 30132, NCIB 11478, USDA 2370.

   GenBank accession number (16S rRNA): U29386.

2. **Rhizobium etli** Segovia, Young, and Martínez-Romero 1993, 376[VP]
   et' li. L. n. *etl* bean; N.L. gen. n. *etli* of bean.

   The characteristics are as given for the genus and listed

in Table BXII.α.112. Fast growing: colonies are 2–4 mm in diameter after 2–4 d on peptone–yeast extract agar. No growth occurs on Luria broth medium or on peptone yeast-extract medium lacking calcium. Growth occurs on a minimal medium containing malate as a carbon source. Maximum temperature for growth is 37°C. Selected strains of *R. leguminosarum* biovar phaseoli were differentiated as this new species on the basis of differences in protein profiles, antibiotic resistance profiles, serological types, DNA–DNA reassociation data, plasmid profiles, exopolysaccharide structures, and multilocus enzyme electrophoresis.

The species contains two named biovars: *Rhizobium etli* biovar phaseoli and biovar mimosae (Wang et al., 1999a). The species nodulates and fixes nitrogen in association with *Phaseolus vulgaris* and some other legumes, such as *Mimosa affinis*. Nonsymbiotic strains are included in the species.

The mol% G + C of the DNA is: 59-63 (T_m).

Type strain: CFN 42, ATCC 51251, ICMP 13642, USDA 9032.

GenBank accession number (16S rRNA): U28916.

3. **Rhizobium galegae** Lindström 1989, 365[VP]

ga.le′ gae. M.L. fem. gen. n. *galegae* of *Galega*, a genus of leguminous plants.

The characteristics are as given for the genus and listed in Tables BXII.α.112 and BXII.α.113. Motile by 1–2 polar or subpolar flagella. Relatively slow growing. Colonies on YMA are more than 1.0 mm in diameter after 7 d at 28°C. Growth does not occur on YMA containing 2% NaCl. Maximum temperature for growth is 33–37°C. Most strains form a serum zone and give an alkaline reaction in litmus milk. Hydrolyzes urea but does not precipitate calcium glycerophosphate or reduce nitrate. Requires pantothenate as a vitamin supplement or growth factor, but not thiamine. Utilizes relatively few organic substrates as sole sources of carbon (Table BXII.α.113). A preliminary report gave information on this species (Lindström and Lehtomäki, 1988). Nodulates *Galega orientalis* and *Galega officinalis* and is reported to be specific to these species.

The mol% G + C of the DNA is: 63 (T_m).

Type strain: HAMBI 540, ATCC 43677, DSM 11542, ICMP 13643, LMG 6214.

GenBank accession number (16S rRNA): D11343, X67226.

4. **Rhizobium gallicum** Amarger, Macheret, and Laguerre 1997, 1005[VP]

gal′ li.cum. L. adj. *gallicum* pertaining to Gallia; the country of origin, France.

The characteristics are as given for the genus and listed in Table BXII.α.112. Relatively fast growing. Colonies are 2–4 mm in diameter within 2–3 d at 28°C. Growth does not occur on YMA containing 1% NaCl. Resistant to nalidixic acid. Separated from other *Rhizobium* species by cluster analysis of phenotypic data. The status of *R. gallicum* as an authentic species is supported by amplified 16S rDNA restriction analysis, comparative 16S rDNA sequence analysis, DNA–DNA reassociation, and nutritional data.

Two biovars, *R. gallicum* biovar gallicum and *R. gallicum* biovar phaseoli, are established based on nodulating specificity. *R. gallicum* biovar gallicum nodulates and fixes nitrogen in association with *Leucaena leucocephala*, *Macroptilium atropurpureum*, *Onobrychis viciifolia*, and *Phaseolus* spp.

whereas *R. gallicum* biovar phaseoli nodulates *Phaseolus* spp. only.

The mol% G + C of the DNA is: not available.

Type strain: R602sp, MSDJ1109.

GenBank accession number (16S rRNA): AF008130, U86343.

5. **Rhizobium giardinii** Amarger, Macheret, and Laguerre 1997, 1005[VP]

giar.di′ ni.i. N.L. gen. n. *giardinii* of Giardini, a Brazilian microbiologist who isolated the organism.

The characteristics are as given for the genus and listed in Table BXII.α.112. Relatively fast growing. Colonies are 2–4 mm in diameter within 2–3 d at 28°C. Growth does not occur on YMA containing 2% NaCl. Resistant to nalidixic acid. The status of *R. giardinii* as an authentic species is supported by amplified 16S rDNA restriction analysis, DNA–DNA hybridization, and comparative 16S rDNA sequence analysis.

Two biovars, *R. giardinii* biovar giardinii and *R. giardinii* biovar phaseoli, are established based on nodulating specificity. *R. giardinii* biovar giardinii nodulates *Phaseolus* spp., *Leucaena leucocephala*, and *Macroptilium atropurpureum*, but does not fix nitrogen with *Phaseolus vulgaris*. *R. giardinii* biovar phaseoli nodulates *Phaseolus* spp. and is weakly efficient in fixing nitrogen in association with that host.

The mol% G + C of the DNA is: not available.

Type strain: H152, MSDJ0144.

GenBank accession number (16S rRNA): U86344.

6. **Rhizobium hainanense** Chen, Tan, Gao, Li, and Wang 1997b, 872[VP]

hai.na.nen′ se. M.L. neut. adj. *hainanense* pertaining to Hainan Province in China.

The characteristics are as given for the genus and listed in Table BXII.α.112. Motile by a single polar flagellum. Relatively fast growing, with a generation time of 2–4 h. Colonies are 2–4 mm in diameter after 3 d growth on yeast extract mannitol agar. Temperature optimal for growth, 25–30°C. Strains can grow at 40°C. Optimal pH for growth, 6–8; pH range, 5–10. Grows on YMA containing 2% NaCl.

The status of *R. hainanense* as an authentic species is supported by cluster analysis of phenotypic features, DNA–DNA reassociation data, and comparative 16S rDNA sequence analysis.

Nodulates *Acacia sinicus*, *Arachis hypogaea*, *Centrosema pubescens*, *Desmodium gyroides*, *D. sinuatum*, *D. triquetrum*, *D. heterophyllum*, *Macroptilium lathyroides*, *Stylosanthes guianensis*, *Tephrosia candida*, *Uraria crinita*, and *Zornia diphylla*.

The mol% G + C of the DNA is: 59–63 (T_m).

Type strain: I66, CCBAU 57015, DSM 11917, ICMP 13690.

GenBank accession number (16S rRNA): U71078.

7. **Rhizobium huautlense** Wang, van Berkum, Beyene, Sui, Dorado, Chen and Martínez-Romero 1998, 696[VP]

hu.aut.len′ se. N.L. adj. *huautlense* of Huautla, the region in Mexico where the organisms were isolated.

The characteristics are as given for the genus and listed in Table BXII.α.112. Relatively fast growing, with a generation time of 2.0–2.2 h. Colonies are 2–4 mm in diameter within 2–3 d at 28°C. Grows at 40°C. Growth does not occur on YMA containing 2% NaCl. Requires thiamine as a growth factor. Comparative sequence analysis of 16S rDNA indicates that this species is closely related to *R. galegae*. These two species are differentiated based on multilocus enzyme electrophoresis, DNA–DNA reassociation, size of compatible Sym plasmids, and a small number of other features. Nodulates *Sesbania herbacea*, *S. rostrata*, and *Leucaena leucocephala*.

The mol% G + C of the DNA is: 57–59 (T_m).
Type strain: S02, ICMP 13551, USDA 4900.
GenBank accession number (16S rRNA): AF025852.

8. **Rhizobium lupini** (Schroeter 1886) Eckhardt, Baldwin and Fred 1931, 273[AL] (*Phytomyxa lupini* Schroeter 1886, 135.)
lu.pi′ni. M.L. masc. n. *Lupinius* generic name of lupine; M.L. gen. n. *lupini* of *Lupinus*.

A limited description is given in Jordan and Allen (1974). Jordan (1984a) recognized the affinities of this species with *Bradyrhizobium* species but felt that evidence was lacking to propose the transfer of *R. lupini* to the genus. Since then comparative sequence analyses of 16S rDNA from nodulating strains of *Lupinus* support transfer of the species; however, the 16S rDNA sequence of the type strain, which has unfortunately been contaminated with *B. japonicum* but still contains another distinct species (van Berkum et al., 1998), appears not to have been analyzed. The status of this species clearly needs work.

Nodulates *Lupinus* spp. and *Ornithopus* spp. Limited nodulation of *Glycine* spp. and the cowpea miscellany.
The mol% G + C of the DNA is: not available.
Type strain: ATCC 10319, DSM 30140.

9. **Rhizobium mongolense** van Berkum, Beyene, Bao, Campbell and Eardly 1998, 21[VP]
mon.go.len′se. L. neut. adj. *mongolense* pertaining to Inner Mongolia, the region where the bacteria were isolated.

The characteristics are as given for the genus and listed in Table BXII.α.112. Motile by a single polar or subpolar flagellum. Relatively fast growing: colonies 1–3 mm in diameter within 3–5 days on arabinose-gluconate agar. No growth occurs on YMA containing 1% NaCl or at pH values below 4.0. Resistant to bacitracin, cefuperazone, and pen-

icillin G. *R. mongolense* shares 99.2% similarity in its 16S rDNA sequence with *R. gallicum* (van Berkum et al., 1998), and it may therefore be a junior synonym of *R. gallicum*. Isolated from *Medicago ruthenica*. Nodulates *Medicago ruthenica* and *Phaseolus vulgaris*.
The mol% G + C of the DNA is: not available.
Type strain: USDA 1844 (ICMP 13688).
GenBank accession number (16S rRNA): U89817

10. **Rhizobium tropici** Martínez-Romero, Segovia, Mercante, Franco, Graham and Pardo 1991, 424[VP]
tro′pi.ci. M.L. gen. n. *tropici* of the tropic (of Cancer).

The characteristics are as given for the genus and listed in Tables BXII.α.112 and BXII.α.113. Motile by peritrichous flagella. Relatively fast growing, with a doubling time of 1.6–2.0 h. Colonies are 2–4 mm in diameter within 2–4 days at 30°C on mannitol–yeast extract agar and peptone yeast-extract agar. Can grow at 40°C. Growth occurs at pH 4.5–7. Does not produce 3-ketolactose. Utilizes a range of organic substrates as sole sources of carbon (Table BXII.α.113).

Previously classified as Type II strains of *Rhizobium leguminosarum* biovar phaseoli. This species comprises Type A and Type B strains, which may represent two distinct species. Distinguished from *Rhizobium leguminosarum* by host range, *nif* gene organization, high temperature tolerance, and extreme acid tolerance, and are said to be more symbiotically stable. Distinguished from other *Rhizobium* species by DNA–DNA reassociation, multilocus enzyme electrophoresis profiles, in biochemical tests, and 16S rDNA sequence comparison.

Forms nodules on *Phaseolus vulgaris*, *Leucaena* spp., and with other legume species. Type A strain CFN299 also nodulates *Amorpha fruticosa*.
The mol% G + C of the DNA is: 60–62 (T_m).
Type strain: ATCC 49672, ICMP 13646, IFO 15427, LMG 9503, USDA 9030.
GenBank accession number (16S rRNA): U89832, X77125.

Genus II. Agrobacterium Conn 1942, 359[AL]*

JOHN M. YOUNG, ALLEN KERR AND HIROYUKI SAWADA

A.gro.bac.te′ri.um. Gr. n. *agros* a field; Gr. dim. neut. n. *bakterion* a small rod; M.L. neut. n. *Agrobacterium* a small field rod.

Rods 0.6–1.0 × 1.5–3.0 μm, occurring singly or in pairs. Nonsporeforming. Gram negative. **Motile by 1–4 peritrichous flagella. Aerobic**, possessing a strictly respiratory type of metabolism with oxygen as the terminal electron acceptor. Some strains are capable of anaerobic respiration in the presence of nitrate. Most strains are able to grow under reduced oxygen tensions in plant tissues. Optimal temperature for growth: 25–28°C. Colonies are usually convex, circular, smooth, nonpigmented to light beige. **Growth on carbohydrate-containing media is usually accompa-**

nied by copious extracellular polysaccharide slime. Catalase positive. **Usually oxidase positive** and urease positive. Indole is not produced. Chemoorganotrophs, utilizing a wide range of carbohydrates, salts of organic acids, and amino acids as carbon sources, but not cellulose, starch, agar, or chitin. **Produce an acid reaction in mineral salts media containing mannitol and other carbohydrates.** Ammonium salts and nitrates can serve as nitrogen sources for strains of some species; others require amino acids and additional growth factors. 3-ketoglycosides are produced by the majority of strains belonging to *A. tumefaciens*. **Strains of some species in this genus invade the crown, roots, and stems of a great variety of dicotyledonous and some gymnospermous plants via wounds, causing transformation of the plant cells into autonomously proliferating tumor cells.** Onco-

Editorial Note: Young et al. (2001) have proposed the transfer of the type species of the genus *Agrobacterium*, *A. tumefaciens*, as well as the species *A. rhizogenes*, *A. rubi*, and *A. vitis* to the genus *Rhizobium*.

genicity is correlated with the presence of a large tumor-inducing plasmid. Habitat: soil. Oncogenic strains occur mainly in soils previously contaminated with diseased plant material. Some non-oncogenic *Agrobacterium* strains have been isolated from human clinical specimens.

The mol% G + C of the DNA is: 57–63.

Type species: **Agrobacterium tumefaciens** (Smith and Townsend 1907) Conn 1942, 359[AL] (*Bacterium tumefaciens* Smith and Townsend 1907, 672; *Agrobacterium radiobacter* (Beijerinck and van Delden 1902) Conn 1942, 359; *Agrobacterium radiobacter* biovar radiobacter (Beijerinck and van Delden 1902) Keane, Kerr and New 1970, 594; *Agrobacterium radiobacter* biovar tumefaciens (Smith and Townsend 1907) Keane, Kerr and New 1970, 594; *Agrobacterium radiobacter* pathovar *tumefaciens* (Smith and Townsend 1907) Young, Dye, Bradbury, Panagopoulos and Robbs 1978, 156.)

FURTHER DESCRIPTIVE INFORMATION

Introductory note on nomenclature used in this chapter Following Bradbury (1986), strains in past literature called *Agrobacterium* biovar 1 are herein referred to as *A. tumefaciens*. *Agrobacterium* biovar 2 strains are referred to as *A. rhizogenes*. *Agrobacterium* biovar 3 strains are referred to as *A. vitis*. Where necessary, the ability of pathogenic strains to cause crown gall tumors or the hairy root condition, previously attributed to strains using the names "*Agrobacterium tumefaciens*" and *A. rhizogenes*, is indicated by reference to the tumorigenic or rhizogenic capacity respectively of strains in species of *Agrobacterium*. Nonpathogenic strains previously named *A. radiobacter* are referred to as nonpathogenic strains of *A. tumefaciens* and *A. rhizogenes* or as nonpathogenic *Agrobacterium* if the species designation has not been identified. Ti or Ri plasmids determine the pathogenic status of strains (See below). Species comprising pathogenic or nonpathogenic strains are reported as tumorigenic as a (Ti strain) or (Ti); as rhizogenic as a (Ri strain) or (Ri); or as nonpathogenic strains of the species where relevant. Table BXII.α.114 explains the relationships of names in the literature.

Acid is produced in mineral salts media from L-arabinose, cellobiose, D-fructose, D-glucose, lactose, maltose, melezitose, L-rhamnose, trehalose, D-xylose, adonitol, arabitol, ethanol, mannitol, salicin, and other carbohydrates.

Plant diseases associated with agrobacteria are commonly known as crown gall, hairy root, cane gall, and grapevine gall. Strains of some species possess a wide host range (*A. rhizogenes* and *A. tumefaciens*), whereas others (*A. rubi* and *A. vitis*) possess a limited host range, perhaps confined to single plant genera. The tumors are self-proliferating and can be transmitted by grafting.

Tumor induction is correlated with the presence of a large tumor-inducing plasmid (Ti-plasmid) in the bacterium. Hairy root induction is associated with the presence of an Ri plasmid. Both plasmids are involved in similar mechanisms of symptom production. Some strains lack oncogenic plasmids and are nonpathogenic.

The molecular size of the *Agrobacterium* genome ranges from 3.0 to 3.6 × 10⁹ bp.

Additional information about agrobacteria can be found in the chapter on the genus *Rhizobium*.

ENRICHMENT AND ISOLATION PROCEDURES

Several selective media have been described for the isolation of *Agrobacterium* species from soil and crown gall tissues (Moore et al., 1988). The medium of Schroth et al. (1965)[1] is a general medium for the isolation of *A. tumefaciens*. Medium 1A[2] of Brisbane and Kerr (1983) can be used for most *Agrobacterium tumefaciens* strains. Medium 2E of Brisbane and Kerr (1983)[3] is used for *Agrobacterium rhizogenes*; erythritol was selected as the sole carbon source in this medium because *Agrobacterium tumefaciens* strains cannot utilize it (see Table BXII.α.113 in the chapter on the genus *Rhizobium*). Agrobacteria can be isolated from soil and from young crown gall tissues by spreading 0.1 ml of the appropriate dilution of soil or extracts of gall tissue with an L-shaped glass rod over one of the surface-dried media in Petri dishes, which are subsequently incubated at 27°C. Typical convex, glistening, and circular colonies with an entire edge are transferred for storage and identification.

Other selective media were described by Clark (1969), Kado and Heskett (1970), and Moore et al. (1980). A selective medium[4] has been developed for *Agrobacterium vitis* (Roy and Sasser 1983).

PROCEDURES FOR TESTING SPECIAL CHARACTERS

Pathogenicity No single host plant will serve for the assay of virulence of tumorigenic strains of *Agrobacterium* because the host range of some strains is restricted (Panagopoulos and Psallidas, 1973; Anderson and Moore, 1979) (see above). Sunflower (*Helianthus annuus*) may be host to the widest range of tumorigenic strains. Young vigorously growing plants are recommended. Stems of 1–2 week-old sunflower plants are inoculated with sterile needles or sharpened sterile toothpicks dipped in colonies or heavy (>10⁸ cfu/ml) aqueous bacterial suspensions prepared from 2-d-old cultures on any medium producing satisfactory growth. Pathogenicity can usually be recorded within a week after inoculation. The young stems of various varieties of the following plants can also be used: tobacco (*Nicotiana tabacum*), tomato (*Lycopersicon esculentum*) and *Kalanchoë daigremontiana* (Anderson and Moore, 1979). Tumor formation can be rather slow, taking from 2–4 weeks. As positive and negative controls, wounded plants should be inoculated with known pathogenic and nonpathogenic *Agrobacterium* strains. Inoculated plants should be

---

1. The medium of Schroth et al. (1965) consists of (g/l of distilled water): agar, 20.0; mannitol, 10.0; NaNO₃, 4.0; MgCl₂, 2.0; calcium propionate, 1.2; MgHPO₄·3H₂O, 0.2; MgSO₄·7H₂O, 0.1; NaHCO₃, 0.075; and magnesium carbonate, 0.075. The pH is adjusted to 7.1 with 1 N HCl. After the medium is autoclaved and cooled to 50–55°C, the following compounds are added aseptically to give final concentrations of (mg/l): berberine, 275; sodium selenite, 100; penicillin G (1625 U/mg), 60; streptomycin sulfate, 30; cycloheximide, 250; tyrothricin, 1.0; and bacitracin (65 U/mg), 100.

2. Medium 1A of Brisbane and Kerr (1983) contains (g/l of distilled water): L (−) arabitol, 3.04; NH₄NO₃, 0.16; KH₂PO₄, 0.54; K₂HPO₄, 1.04; MgSO₄·7H₂O, 0.25; sodium taurocholate, 0.29; crystal violet (0.1% w/v aqueous), 2 ml; and agar, 15.0. Autoclave; cool to 50°C, then add 1 ml of a filter-sterilized 2% solution of cycloheximide and 6.6 ml of a filter-sterilized 1% solution of Na₂SeO₃.

3. Medium 2E for *Agrobacterium rhizogenes* strains (Brisbane and Kerr, 1983) consists of (g/l of distilled water): NH₄NO₃, 0.16; erythritol, 3.05; KH₂PO₄, 0.54; K₂HPO₄, 1.04; MgSO₄·7H₂O, 0.25; sodium taurocholate, 0.29; yeast extract (1% w/v aqueous), 1 ml; malachite green (0.1% w/v aqueous), 5 ml; agar, 15. Autoclave; cool to 50°C, then add 1 ml of a filter-sterilized 2% solution of cycloheximide and 6.6 ml of a filter-sterilized 1% solution of Na₂SeO₃.

4. The medium for *Agrobacterium vitis* of Roy and Sasser (in Moore et al., 1988) contains (g/l): adonitol, 4.0; KH₂PO₄, 0.7; K₂HPO₄, 0.9; MgSO₄·7H₂O, 0.2; NaCl, 0.2; H₃BO₃, 1.0; yeast extract, 0.14; agar, 15; chloranthanil (Bravo 500; 4% aqueous), 0.5 ml. Adjust to pH 7.2, autoclave; cool to 50°C and add the following as filter-sterilized solutions: triphenyltetrazolium chloride, 80 mg; D-cycloserine, 20 mg; trimethroprim (in acidified water), 20 mg. Colonies of *Agrobacterium vitis* have a dark red center with white margins. Comparison with an authentic strain is recommended.

TABLE BXII.α.114. Relationships between different proposed nomenclature for the genus *Agrobacterium*

Species names based on natural classification		Species names based on pathogenicity	
Names used in this text (after Holmes and Roberts, 1981; Bradbury, 1986; Holmes, 1988)	After Keane et al. (1970); New and Kerr (1972); Kerr and Panagopoulos (1977); Panagopoulos et al. (1978)	Allen and Holding, 1974; Approved Lists (Skerman et al., 1980)	After Kersters and De Ley, 1984a
A. tumefaciens (Ti strain or Ti)	*A. radiobacter*[a] biovar tumefaciens (biotype 1)	*A. tumefaciens*	*A. tumefaciens* (biovar 1)
A. tumefaciens (Ri strain or Ri)	*A. radiobacter* biovar rhizogenes (biotype 1)	*A. rhizogenes*	*A. rhizogenes* (biovar 1)
A. tumefaciens (nonpathogenic)	*A. radiobacter* biovar radiobacter (biotype 1)	*A. radiobacter*	*A. radiobacter* (biovar 1)
A. rhizogenes (Ti strain or Ti)	*A. radiobacter* biovar tumefaciens (biotype 2)	*A. tumefaciens*	*A. tumefaciens* (biovar 2)
A. rhizogenes (Ri strain or Ri)	*A. radiobacter* biovar rhizogenes (biotype 2)	*A. rhizogenes*	*A. rhizogenes* (biovar 2)
A. rhizogenes (nonpathogenic)	*A. radiobacter* biovar radiobacter (biotype 2)	*A. radiobacter*	*A. radiobacter* (biovar 2)
A. rubi (Ti strain or Ti)[b]	*A. radiobacter* biovar tumefaciens (biotype 2)	*A. rubi*	*A. rubi*
A. vitis (Ti strain or Ti)[b]	*A. radiobacter* biovar tumefaciens (biotype 3)	*A. vitis*	*A. tumefaciens* (biovar 3)
A. vitis (nonpathogenic)	NR[c]	NR	NR

[a]Use of the species epithet *radiobacter* in place of *tumefaciens* is now not considered acceptable in terms of the Code (Sawada et al., 1993; Bouzar, 1994).

[b]Only tumorigenic (Ti) capability has been reported for this species.

[c]NR, not recorded.

kept in a greenhouse at 20–27°C. When no greenhouse facilities are available, disks of carrot roots (*Daucus carota*) are also useful (Klein and Tenebaum, 1955; Lippincott and Lippincott, 1969). This is a convenient procedure, provided that at least 10 slices from different carrot roots are inoculated per strain and that proper controls are included, because false positive responses (cambial swellings) occasionally occur. Because pathogenic strains belonging to *A. vitis* and isolated from grapevines usually display a restricted host specificity, pathogenicity tests for these strains should be performed on the green tender shoots of grapevines (Panagopoulos and Psallidas, 1973; Panagopoulos et al., 1978).

The root-inducing ability of *A. rhizogenes* (Ri strains) is usually tested by the carrot disk assay (Lippincott and Lippincott, 1969; Moore et al., 1979) or on *Kalanchoë daigremontiana* (White and Nester, 1980b). The interpretation of such experiments is sometimes difficult because some tumorigenic strains are known to induce typical hairy root symptoms on *Kalanchoë* plants (De Cleene and De Ley, 1981).

DIFFERENTIATION OF THE GENUS *AGROBACTERIUM* FROM OTHER GENERA

See Table BXII.α.112 in the chapter on the genus *Rhizobium*. Apart from phenotypic characters that distinguish them as individual species, there are no common distinct characters that differentiate these species as separate from members of the genus *Rhizobium*.

TAXONOMIC COMMENTS

Inferred phylogenies based on comparative analyses of 16S rDNA sequence data show that plant-pathogenic *Agrobacterium* spp. are intermingled with fast-growing *Rhizobium* spp., together with *Allorhizobium undicola*. The genus itself is established only based on pathogenicity characteristics of its species. Apart from pathogenicity tests and nomenclatural problems posed by the confusing species epithets, the generic taxonomic relationships of *Agrobac-*

terium spp. are discussed fully and further descriptive information is given in the chapter on *Rhizobium*.

Nomenclatural problems *Agrobacterium tumefaciens* (Smith and Townsend 1907) Conn 1942 (type species) was the name given to strains of *Agrobacterium* capable of inducing tumorigenic reactions in many host genera (although some of these strains, isolated from *Vitis* spp., appeared to be specific to grape). *Agrobacterium rhizogenes* (Riker et al., 1930) Conn 1942 is comprised of strains capable of inducing a hairy-root (rhizogenic) reaction in host plants. *Agrobacterium radiobacter* (Beijerinck and van Delden 1902) Conn 1942 consisted of nonpathogenic *Agrobacterium* strains, and *A. rubi* (Hildebrand 1940) Starr and Weiss 1943 contained strains capable of inducing tumorigenic reactions in *Rubus* spp. Although *A. rhizogenes* and *A. tumefaciens* comprised populations from soil and from a large number of dicotyledonous hosts, *A. rubi* was represented by isolates from *Rubus* which appeared to be specific to that host in nature. This classification was predominantly, and for some species solely, based on the phytopathogenic behavior of these bacteria. Though of practical use and widely supported as a special purpose classification, it bears no relationship to, and is inconsistent with, natural classifications of *Agrobacterium* as now understood.

When Smith and Townsend and Riker et al. first proposed the names *A. tumefaciens* and *A. rhizogenes*, they followed the custom of giving names which reported a distinctive character of the species—in this instance their pathogenic symptoms. However, the tumorigenic and rhizogenic characters are carried on different plasmids, with the consequence that the pathogenic characters are separately transmissible between species. The result is that both species *A. tumefaciens* and *A. rhizogenes* are represented by strains which may be either tumorigenic, rhizogenic, or—if lacking either kind of plasmid—nonpathogenic. Notwithstanding customary naming, in this extreme example apparently distinctive features are not characteristic of all members of the taxon, are found in these related taxa, and are mobile.

Although it was believed in the past that the different pathogenic reactions would be supported by physiological reactions that would justify species discriminations (Burkholder and Starr, 1948), this assumption has long been shown to be false (Young et al., 1992). Although the establishment of these species (*A. rhizogenes* and *A. tumefaciens*) based on distinct pathogenic characters was supported by Kersters and De Ley (1984a) and by subsequent workers, others have supported an alternative nomenclature based on natural classification. Holmes and Roberts (1981) and Holmes (1988) proposed a rational and consistent nomenclature for these taxonomic groups, which is used here. Bradbury (1986) also supported this classification but it has not found widespread acceptance for reasons that will be discussed below.

The classification of *Agrobacterium* species has been thoroughly studied using the following techniques: (a) numerical analysis of phenotypic characteristics (White, 1972; Kersters et al., 1973; Holmes and Roberts, 1981); (b) biochemical and physiological tests (Keane et al., 1970; Kersters et al., 1973; Kerr and Panagopoulos, 1977; Süle, 1978; Holmes and Roberts, 1981); (c) fatty acid methyl ester profiles (Sawada et al., 1992d; Jarvis et al., 1996); (d) DNA–DNA reassociation (De Ley, 1972 ,1974); (e) measurements of the thermal stability of DNA–DNA hybrids (De Ley et al., 1973); and (f) comparison of electrophoretograms of soluble proteins (Kersters and De Ley, 1975). The results obtained by the above mentioned methods corroborated each other and indicated that the genus *Agrobacterium* consisted of at least three genetically and phenotypically different groups or clusters. These groups corresponded to biovars 1, 2, and 3 of Keane et al. (1970). Recently, this number has been increased to four, possibly five, groups now recognized as species (see below).

Species recognized based on their overall phenotypic and genomic relatedness are given below under the List of the species of the genus *Rhizobium* according to the type strains allocated to each species population.

Bradbury (1986), Holmes and Roberts (1981), Holmes (1988), and Young et al. (1992) have supported use of the names *A. tumefaciens* and *A. rhizogenes*, in accord with natural classification, recognizing pathogenic strains according to their tumorigenic, rhizogenic, and nonpathogenic states. However, the epithet *tumefaciens* has become so entrenched as the name for pathogenic populations of *Agrobacterium* spp. with tumorigenic capabilities in a special purpose classification, that it is difficult for it also to be used unambiguously for its proper purpose in natural classification. Attempts to resolve the difficulty by recognizing *A. radiobacter* in place of *A. tumefaciens* (Young et al., 1978; Kersters

and De Ley, 1984a; Sawada et al., 1993) cannot be adopted easily because *A. tumefaciens*, the type species, has been conserved over *A. radiobacter* (Judicial Commission, 1970). Bouzar (1994) sought clarification of the proposal of Sawada et al. (1993) which, with the response of Sawada et al. (Bouzar 1994), the Judicial Commission of the ICSB deemed to have resolved the matter (L.G. Wayne, personal communication). The practical and usual solution to this nomenclatural confusion has been to use an artificial classification and informal nomenclature in which the species names *A. tumefaciens* or *A. rhizogenes* are applied to the pathogenic plasmid-borne states, and the terms biotype or biovar are applied to the natural species groups.

If the genus *Agrobacterium* continues to be differentiated from *Rhizobium* in the future, then the use of a natural classification that recognizes *Agrobacterium* species will probably require a radical change of nomenclature by application of the Code (Lapage et al., 1992). For instance, the name *A. tumefaciens* could be rejected (Rule 23a) as a *nomen ambiguum*; a name that has been used with different meanings and has thus become a source of error (Rule 56a). Rejection would also require the designation of a new type species. The obvious candidate is *A. radiobacter*. A more radical option could involve the application of new names (and a new type species), extending the proposal of Kersters and De Ley (1984a). For both these proposals, it would be necessary to make a Request for an Opinion to the Judicial Commission of the ICSB. The four named genera—*Agrobacterium*, *Allorhizobium*, *Rhizobium*, and *Sinorhizobium* —are closely related in genomic and phenotypic terms. It has recently been proposed that *Agrobacterium*, *Allorhizobium*, and *Rhizobium* be amalgamated into a single genus, *Rhizobium*, based on their close genomic and phenotypic similarity (Young et al., 2001) (see *Rhizobium* chapter in this volume).

Elsewhere, the application of pathovars in terms of the Standards for Naming Pathovars (Dye et al., 1980) has been proposed (Young et al., 1978; Kersters and De Ley, 1984a). However, the fact that most pathogenicity genes are carried on a plasmid means that the pathogenic character of any strain is unstable. This lack of stability would make uncertain the application of pathovar names to particular strains, most notably to pathotype strains. For pathogenic strains in *Agrobacterium* therefore, this formal special purpose nomenclature (Dye et al., 1980) seems inappropriate. Species comprising pathogenic or non-pathogenic strains could be reported as tumorigenic, as a Ti strain or Ti, as rhizogenic, as a Ri strain or Ri, or as nonpathogenic strains of the species, where relevant.

List of species of the genus Agrobacterium

1. **Agrobacterium tumefaciens** (Smith and Townsend 1907) Conn 1942, 359[AL] (*Bacterium tumefaciens* Smith and Townsend 1907, 672; *Agrobacterium radiobacter* (Beijerinck and van Delden 1902) Conn 1942, 359; *Agrobacterium radiobacter* biovar radiobacter (Beijerinck and van Delden 1902) Keane, Kerr and New 1970, 594; *Agrobacterium radiobacter* biovar tumefaciens (Smith and Townsend 1907) Keane, Kerr and New 1970, 594; *Agrobacterium radiobacter* pathovar *tumefaciens* (Smith and Townsend 1907) Young, Dye, Bradbury, Panagopoulos and Robbs 1978, 156.)

   tu.me.fa' ci.ens. L. n. *tumor* a swelling tumor; L. v. *facere* to make, to produce; M.L. part. adj. *tumefaciens* tumor producing.

   The characteristics are as described for the genus and

as listed in Tables BXII.α.112 and BXII.α.113 in the chapter on the genus *Rhizobium*. Temperature optimal for growth 25–28°C and grows at 37°C. Grows in media containing 2% NaCl. Produces 3-ketolactose. Has no growth factor requirements. Utilizes a relatively wide range of organic substrates as sole sources of carbon (see Table BXII.α.113 in the chapter on the genus *Rhizobium*). The status of *A. tumefaciens* as an authentic species with the *Rhizobium* clade is supported by comparative 16S rDNA sequence analysis.

   This species corresponds to biotype 1 of Keane et al. (1970), group I of White (1972), cluster 1 of Kersters et al. (1973), biovar 1 of Willems and Collins (1993) and of Sawada et al. (1993). *A. tumefaciens* comprises 3-ketolactose-positive, tumorigenic strains, as well as hairy root-forming strains and nonpathogenic strains. It includes the type strain of *A. tumefaciens*, a tumorigenic strain, as well as the type strain of *A. radiobacter*, a nonpathogenic strain.

Ti or Ri plasmids determine the pathogenic status of strains. The species comprises pathogenic or nonpathogenic strains, which can be reported as tumorigenic, as a Ti strain or Ti, as rhizogenic as a Ri strain or Ri, or as nonpathogenic strains of the species where relevant. Pathogenic strains have a wide, and perhaps complex, host range.

An alternative special purpose nomenclature involves naming strains according to their pathogenic character, as *A. tumefaciens*, *A. rhizogenes*, or *A. radiobacter* for tumorigenic, rhizogenic and nonpathogenic strains, respectively, and classifying them as *Agrobacterium* biovar 1.

The epithet *tumefaciens* takes precedence in *Agrobacterium* because it is the designated type species of this genus.

The mol% G + C of the DNA is: 57–63 (T_m).

Type strain: ATCC 23308, DSM 30205, ICMP 5856, LMG 187, NCPPB 2437.

GenBank accession number (16S rRNA): D01256, D14500, M11223.

2. **Agrobacterium rhizogenes** (Riker, Banfield, Wright, Keitt and Sagen 1930) Conn 1942, 359[AL] emend. Sawada, Ieki, Oyaizu and Matsumoto 1993, 701 (*Bacterium rhizogenes* Riker, Banfield, Wright, Keitt and Sagen 1930, 536; *Agrobacterium radiobacter* biovar rhizogenes (Riker, Banfield, Wright, Keitt and Sagen 1930) Keane, Kerr and New 1970, 594; *Agrobacterium radiobacter* pathovar *rhizogenes* (Riker, Banfield, Wright, Keitt and Sagen 1930) Young, Dye, Bradbury, Panagopoulos and Robbs 1978, 156.)

rhi.zo'ge.nes. Gr. n. *rhiza* a root; Gr. v. *gennao* to make, to produce; M.L. adj. *rhizogenes* root-producing.

The characteristics are as described for the genus and as listed in Tables BXII.α.112 and BXII.α.113 in the chapter on the genus *Rhizobium*. Temperature optimal for growth 25–28°C. Does not grow at 35°C. Does not grow in media containing 2% NaCl. Does not produce 3-ketolactose. Requires biotin as a growth factor. Utilizes a relatively wide range of organic substrates as sole sources of carbon (Table BXII.α.113). The status of *A. rhizogenes* as an authentic species in the genus *Rhizobium* is supported by numerical analysis of nutritional and biochemical data and by comparative 16S rDNA sequence analysis.

This species corresponds to biotype 2 of Keane et al. (1970), group III of White (1972), and cluster 2 of Kersters et al. (1973). It comprises 3-ketolactose-negative, tumorigenic, rhizogenic, and nonpathogenic strains.

Ti or Ri plasmids determine the pathogenic status of strains. The species comprises pathogenic or nonpathogenic strains, which can be reported as tumorigenic, as a Ti strain or Ti, as rhizogenic as a Ri strain or Ri, or as nonpathogenic strains of the species where relevant. Pathogenic strains have a wide, and perhaps complex, host range.

An alternative special purpose nomenclature involves naming strains according to their pathogenic character, as *A. tumefaciens*, *A. rhizogenes*, or *A. radiobacter* for tumorigenic, rhizogenic, and nonpathogenic strains, respectively, and classifying them as *Agrobacterium* biovar 2.

The mol% G + C of the DNA is: 59–63 (T_m).

Type strain: ATCC 11325, DSM 30148, ICMP 5794, IFO 13257, LMG 150, NCPPB 2991.

GenBank accession number (16S rRNA): D01257, D14501.

3. **Agrobacterium rubi** (Hildebrand 1940) Starr and Weiss 1943, 316[AL] (*Phytomonas rubi* Hildebrand 1940, 694.)

ru'bi. L. n. *Rubus* generic name of blackberry; L. gen. n. *rubi* of *Rubus*

The characteristics are as described for the genus and as listed in Table BXII.α.112 in the chapter on the genus *Rhizobium*. Temperature optimal for growth 25–28°C. Does not grow at 35°C (Sawada and Ieki, 1992b); grows at 37°C (Kerr, 1992). Does not grow in media containing 2% NaCl. Does not produce 3-ketolactose. Growth rate on ordinary media is characteristically slower than for the other species. Requires L-glutamic acid, biotin, pantothenate, and nicotinic acid (present in yeast extract) as growth factors. The status of *A. rubi* as an authentic species is supported by numerical analysis of nutritional and biochemical data and by comparative 16S rDNA sequence analysis.

So far, pathogenic strains of this species have been reported only from *Rubus* spp.

Tumorigenic strains bearing Ti plasmids determine the pathogenic status of strains. Isolated from aboveground cane galls on *Rubus* spp. (black raspberry, boysenberry). The natural host range is apparently limited to *Rubus* spp., but artificial inoculations indicate a wider host range (Anderson and Moore, 1979).

The mol% G + C of the DNA is: 57.6–58.8 (T_m).

Type strain: ATCC 13335, CFBP 1317, ICMP 6428, IFO 13261, LMG 156, NCPPB 1854.

GenBank accession number (16S rRNA): D14503, X67228.

4. **Agrobacterium vitis** Ophel and Kerr 1990, 240[VP]

vi'tis. L. fem. n. *vitis* wine plant; L. gen. fem. n. *vitis* of the wine plant.

The characteristics are as described for the genus and as listed in Tables BXII.α.112 and BXII.α.113 in the chapter on the genus *Rhizobium*. Temperature optimal for growth 25–28°C. Does not grow at 35°C. Grows in medium containing 2% NaCl. Does not produce 3-ketolactose. Strains require biotin for growth. Produces acid from adonitol and mannitol, but not from alpha-methyl-D-glucoside, erythritol, dulcitol, melezitose, or arabitol. Utilizes relatively few organic substrates as sole sources of carbon (Table BXII.α.113).

The status of *A. vitis* as an authentic species is supported by numerical analysis of nutritional and biochemical data and by comparative 16S rDNA sequence analysis. The species comprises strains previously referred to as *Agrobacterium* biovar 3 *sensu* Kerr and Panagopoulos (1977) as described by Süle (1978) and Panagopoulos et al. (1978). *A. vitis* strains have been isolated from grapevines in many countries including Greece and Hungary (Panagopoulos et al., 1978), People's Republic of China (Ma et al., 1985), Australia (Ophel and Kerr, 1990), Japan (Sawada et al., 1990), France, Germany, and the United States (Otten et al., 1996), Korea, and, to date, the geographic distribution of the pathogen has reflected that of the host. Many strains of this species display a limited host range. So far, pathogenic strains of this species have been reported largely from *Vitis* spp. although Sawada and Ieki (1992a) report an *A. vitis* strain (as biovar 3) from *Actinidia*.

Ti plasmids determine the pathogenic status of strains. The species comprises pathogenic or nonpathogenic strains, which can be reported as tumorigenic, as Ti strain or Ti, or nonpathogenic strains of the species where relevant. Strains are generally isolated from *Vitis* spp. grape, but they have occasionally been isolated from other dicotyledonous plant species.

The mol% G + C of the DNA is: 59 (T_m).

Type strain: ATCC 49767, ICMP 10752, LMG 8750, NCPPB 3554.

GenBank accession number (16S rRNA): D01258, D14502, U45329, X67225.

Other Organisms

A small group of yellow-pigmented, 3-ketolactose-positive bacteria, including *"Chromobacterium folium"* (*"Chromobacterium lividum"*) strains NCTC 10590 and 10591, that were considered to constitute another distinct group of agrobacteria (Holmes and Roberts, 1981) have since been classified as *Sphingomonas yanoikuyae* (Takeuchi et al., 1995).

Marine, star-shaped, aggregate-forming bacteria named *A. ferrugineum* Ahrens and Rheinheimer 1967, *A. gelatinovorans* Ahrens 1968 (as *A. gelatinovorum*), and *A. stellulatum* Stapp and Knösel 1954 were not included in the Approved Lists. However, Rüger and Höfle (1992) reported phenotypic and genotypic data which did "not definitely support exclusion of marine strains from *Agrobacterium*." As an *ex tempore* measure they proposed these species, together with *A. atlanticum* Rüger and Höfle 1992 and *A. meteori* Rüger and Höfle 1992, as members of the genus *Agrobacterium*.

A comparative analysis of 16S rDNA sequence data has shown the marine *Agrobacterium* spp. to cluster in two subdivisions in the *Alphaproteobacteria*. *A. stellulatum* and *"A. kieliense"* were in the "alpha-2" subdivision, but were more distant from *Rhizobium* than genera of the *Bartonellaceae*, *Brucellaceae*, and *Phyllobacteriaceae*. *A. meteori*, *A. atlanticum*, *A. ferrugineum*, and *A. gelatinovorans* were in the "alpha-3" subdivision (Uchino et al., 1997). Following a reinvestigation of phenotypic characters and of inferred phylogenies based on comparative analyses of 16S rDNA sequences, Uchino et al. (1998) allocated the marine agrobacteria as follows: *Ruegeria atlantica* (Rüger and Höfle 1992) Uchino et al. 1999[VP] (syn: *Agrobacterium meteori* Rüger and Höfle 1992), *Ruegeria gelatinovorans* (Rüger and Höfle 1992) Uchino et al. 1999 (as *R. gelatinovora*[VP]), *Stappia stellulata* (Rüger and Höfle 1992) Uchino et al. 1999[VP]. *Agrobacterium ferrugineum* is a member of an unidentified genus in the alpha-3 sub-group of the *Proteobacteria*.

An unnamed *Agrobacterium* species has been reported from *Ficus benjamina* (Bouzar et al., 1995). Identity as a new species is based on the presence of a unique 16S rRNA domain, on differences in nutritional profiles, and on an unusual opine metabolism. Tumorigenic and nonpathogenic strains have been isolated.

Genus III. **Allorhizobium** de Lajudie, Laurent-Fulele, Willems, Torck, Coopman, Collins, Kersters, Dreyfus and Gillis 1998a, 1288[VP]*

L. David Kuykendall and Frank B. Dazzo

Al.lo.rhi.zo'bi.um. Gr. adj. *allos* other; M.L. neut. n. *Rhizobium* a bacterial generic name; M.L. neut. n. *Allorhizobium* the other *Rhizobium*, to refer to the fact that it is phylogenetically separate from other *Rhizobium* species.

Rods 0.5–0.7 × 2.0–4.0 μm. Nonsporeforming. Gram negative. Aerobic, having a strictly respiratory type of metabolism with oxygen as the terminal electron acceptor. **Relatively fast growing**; colonies 0.5–3.0 mm develop in 1–2 days on yeast–mannitol–mineral salts agar, and pronounced turbidity develops after 1–2 days in agitated broth media. Chemoorganotrophic. **A wide range of carbohydrates, organic acids, and amino acids are used as sole carbon sources for growth. Ketolactose is not produced from lactose. Growth on carbohydrate media is usually accompanied by extracellular polysaccharide production.** At the molecular level, the genus can be distinguished from related species in other genera by SDS-PAGE whole cell protein analysis, ITS PCR-RFLP, and 16S rRNA gene sequencing. **Initiates the production of root nodules, in which the bacteria occur as nitrogen-fixing intracellular symbionts.** Exhibits specificity to certain temperate-zone and tropical-zone leguminous plants (family Leguminoseae).

The mol% G + C of the DNA is: 60.1.

Type species: **Allorhizobium undicola** De Lajudie, Laurent-Fulele, Willems, Torck, Coopman, Collins, Kersters, Dreyfus and Gillis 1998a, 1288.

Further descriptive information

Like most legume-nodulating bacteria, these organisms are Gram-negative rods but, compared with most of the fast-growing *Rhizobium*, bacteria isolated from nodules of *Neptunia natans* growing in India are even faster growing, developing approximately 2 mm colonies in 2 d on yeast mannitol agar (Vincent, 1970). Colonies are beige, round, creamy, and smooth; like most *Rhizobium* species, this organism, motile in liquid medium, also produces copious slime, which is composed of extracellular polysaccharides (de Lajudie et al., 1998a). Utilizes a wide range of carbohydrates, organic acids, and amino acids for carbon and energy sources (see Table BXII.α.113 in the chapter on the genus *Rhizobium*).

In the aquatic environment, symbiotic infection of *Neptunia* occurs by wound or "crack" entry, rather than via root hairs, followed by formation of infection threads that penetrate host nodule cells and release the endosymbionts (Subbao-Rao et al., 1995). Thus, nodules are formed at the base of lateral and adventitious roots, wherein the bacteria occur as intracellular symbionts that fix N_2.

Editorial Note: Young et al. (2001) have proposed the transfer of the type and only species of *Allorhizobium*, *A. undicola*, to the genus *Rhizobium*.

This species exhibits host specificity, since no isolates have been found which can nodulate *Sesbania rostrata*, *Sesbania pubescens*, *Sesbania grandiflora*, *Vigna ungiuculata*, or *Macroptilium atrpurpureum* (de Lajudie et al., 1998a). This species forms nodules on the temperate-zone *Medicago sativa* (alfalfa) and a number of tropical-zone (*Neptunia natans*, *Acacia senegal*, *Acacia seyal*, *Acacia tortilis* subsp. *radiana*, *Lotus arabicus*, *Faidherbia albida*) leguminous plants of the family Leguminoseae (Subbao-Rao et al., 1995; de Lajudie et al., 1998a).

ENRICHMENT AND ISOLATION PROCEDURES

This organism can be isolated from fresh nodules. A sample of water or soil can be used to inoculate seeds of the host plant *Neptunia natans* that have been surface-sterilized and scarified with concentrated sulfuric acid for about 5 min, followed by several rinses with sterile water. Plants are grown aseptically, so that uninoculated controls are devoid of nodules. Fahraeus plant growth medium is recommended (Fåhraeus, 1957). Nodules formed are surface sterilized in 50% Clorox for a few minutes, followed by several sterile water rinses. The contents of squashed nodules can be streaked on YEM for isolation.

MAINTENANCE PROCEDURES

Lyophilized cultures stored at 4°C are recommended.

DIFFERENTIATION OF THE GENUS *ALLORHIZOBIUM* FROM OTHER GENERA

See Table BXII.α.112 in the chapter on the genus *Rhizobium*. Apart from phenotypic characters that distinguish *Allorhizobium*

undicola as an individual species, there are no common distinct characters that differentiate *Allorhizobium* from members of the genus *Rhizobium*, *Agrobacterium*, or *Sinorhizobium*.

TAXONOMIC COMMENTS

Inferred phylogenies based on comparative analyses of 16S rDNA sequence data show that *Allorhizobium undicola* is intermingled with fast-growing *Rhizobium* spp. and plant pathogenic *Agrobacterium* spp. (Young et al., 2001). Further, the circumscription of the genus does not delineate a taxon distinct from *Rhizobium* (Young et al., 2001). *Allorhizobium* is a monospecific genus established because—as indicated by its name—the data for *Allorhizobium undicola* represented an outlying branch in the particular comparative analysis of 16S rDNA sequence data (de Lajudie et al., 1998a). The closest neighboring species in this analysis is *Rhizobium vitis* (formerly *Agrobacterium vitis*). Comparison of this analysis with others that have been published (Sawada et al., 1993; Willems and Collins, 1993; de Lajudie et al., 1994, 1998b, a; Nour et al., 1995; Rome et al., 1996; Amarger et al., 1997; Chen et al., 1997b; Jarvis et al., 1997; Tan et al., 1997; van Berkum et al., 1998; Wang et al., 1998) and with Fig. BXII.α.130 of the chapter on the genus *Rhizobium* shows how much results can vary depending on the selection of sequences and upon the form of analysis. The species *R. undicola* is well defined based on DNA–DNA reassociation, PAGE of total proteins, and carbon source utilization tests. Further taxonomic comments are given in the chapter on the genus *Rhizobium*.

List of species of the genus Allorhizobium

1. **Allorhizobium undicola** De Lajudie, Laurent-Fulele, Willems, Torck, Coopman, Collins, Kersters, Dreyfus and Gillis 1998a, 1288.
   un.di'co.la. L. n. *unda* water; L. n. *cola* dweller; M.L. n. *undicola* water dweller, referring to the isolation of these strains from nodules of the aquatic plant *Neptunia natans*.

   The characteristics are as described for the genus and listed in Tables BXII.α.112 and BXII.α.113 in the chapter on the genus *Rhizobium*. Exhibits specificity to some temperate-zone (*Medicago sativa*) and some tropical-zone (*Neptunia natans*, *Acacia senegal*, *Acacia seyal*, *Acacia tortilis* subsp.

raddiana, *Lotus arabicus*, *Faidherbia albida*) leguminous plants (family Leguminoseae) and incites the production of root nodules wherein the bacteria occur as intracellular symbionts. No strain has been found to nodulate *Sesbania rostrata*, *Sesbania pubescens*, *Sesbania grandiflora*, *Vigna unguiculata*, or *Macroptilium atropurpureum*. Wound or crack-entry infection of *Neptunia* occurs, rather than infection via root hairs in the aquatic environment. The closest relative is *Agrobacterium vitis* (96.2% 16S ribosomal RNA similarity).

The mol% G + C of the DNA is: 60.1 (T_m).
Type strain: ORS 992, LMG 11875.
GenBank accession number (16S rRNA): Y17047.

Genus IV. **Carbophilus** *Meyer, Stackebrandt and Auling 1994, 182*[VP] *(Effective publication: Meyer, Stackebrandt and Auling 1993, 393)*

ORTWIN O. MEYER

car.bo.phi'lus. L. masc. *carbo* carbon; M.L. part. *philus* loving, loving carbohydrates and other carbonaceous substrates.

Rod-shaped cells with slightly tapered ends, 0.7–0.9 × 0.9–1.7 μm. Motility variable; when present, it is by means of up to five peritrichous flagella. Flagellation is maximal in the late exponential growth phase, when more than 80% of the cells are flagellated. Formation of aggregates is rare. **Gram negative.** Colonies are white to cream colored. Organic growth factors are not required. **Aerobic**, having a strictly respiratory type of metabolism, with oxygen as the terminal electron acceptor. Neither denitrification nor reduction of nitrate to nitrite has yet been observed. **Facultatively chemolithoautotrophic.** CO is utilized as a

substrate under aerobic chemolithoautrophic conditions. Chemoorganoheterotrophic substrates encompass a wide variety of the salts of organic acids (except citrate), amino acids, sugars, and sugar alcohols (see Table 1 of Meyer et al., 1993). Suitable nitrogen sources are ammonia, hydroxylamine, nitrate, and urea (Frunzke and Meyer, 1990). Dinitrogen is not fixed. Not phototrophic. Not methanogenic. Isolated from soil at the city of Moscow (Russia).

The mol% G + C of the DNA is: 62.8 (T_m).
Type species: **Carbophilus carboxidus** (ex Nozhevnikova and

Zavarzin 1974) Meyer, Stackebrandt and Auling 1994, 182 (Effective publication: Meyer, Stackebrandt and Auling 1993, 393.)

FURTHER DESCRIPTIVE INFORMATION

CO-oxidizing *Carbophilus carboxidus* grows with CO as energy source and CO_2 as carbon source under aerobic chemolithoautotrophic conditions, with a generation time of about 40 h (Cypionka et al., 1980). *C. carboxidus* is an exception among the aerobic CO-oxidizing bacteria in that strains capable of growing at the expense of H_2 plus CO_2 are not known. The electron transport system of *C. carboxidus* contains *b*-, *c*-, and *a*-type cytochromes at levels similar to those of other aerobic, respiratory bacteria (Cypionka and Meyer, 1983b). The *o*-type cytochrome b_{536} serves as alternative CO-insensitive terminal oxidase.

C. carboxidus contains the megaplasmids pHCG1-a (558 kb), pHCG1-b (428 kb), and pHCG1-c (129 kb) (Kraut and Meyer, 1988). The structural genes for CO dehydrogenase (*coxLMS*) and ribulosebisphophshate carboxylase (*cbbL*) are plasmid encoded (Kraut et al., 1989).

For appropriate media, cultivation and growth conditions, type of CO dehydrogenase, and enrichment and isolation procedures, refer to the genus *Oligotropha*.

DIFFERENTIATION OF THE GENUS *CARBOPHILUS* FROM OTHER GENERA

The ability to utilize CO under aerobic chemolithoautotrophic conditions occurs only in the genera *Carbophilus*, *Oligotropha*, *Bradyrhizobium*, and *Zavarzinia*. A property that separates *Carbophilus carboxidus* from all other aerobic carboxidotrophic bacteria is its inability to grow chemolithoautotrophically on H_2 and CO_2.

The genus *Carbophilus* can be differentiated from *Oligotropha* and *Zavarzinia* by its peritrichous flagellation, the wide variety of organic substrates utilized, particularly sugars, and a different profile of fatty acids and polyamines (see Table 1 of Meyer et al., 1993). *C. carboxidus* does not fix N_2. *Carbophilus* is differentiated from *Nitrobacter* by the inability to utilize ammonium as an energy source for chemolithoautotrophic growth. It is differentiated from *Agrobacterium* by not being infectious to plants.

TAXONOMIC COMMENTS

The genus *Carbophilus* with the single species *C. carboxidus* has been allocated to the class *Alphaproteobacteria* in the phylum *Proteobacteria* (Fig. BXII.α.185 in the genus *Oligotropha*) on the basis of 16S ribosomal RNA cataloging, the presence of signature oligonucleotides in 16S rRNA catalogues (see Table 5 of Auling et al., 1988), the presence of ubiquinone Q-10 as the major quinone (see Table 3 of Auling et al., 1988), the presence of *sym*-homospermidine and putrescine as the major polyamines (see Table 4 of Auling et al., 1988), and the presence of *cis*-11,12-octadecenoic acid as the main fatty acid and 11-methyl-*cis*-11,12-octadecenoic acid as diagnostic fatty acid (see Table 3 of Auling et al., 1988). The position of *Carbophilus* is in an individual line of descent within the *Alphaproteobacteria* together with the nitrilotriacetic acid utilizing members of the genus *Chelatobacter* (Meyer et al., 1993). The 16S rRNA similarity coefficient of 0.78 separating these two taxa is lower than those found to delineate phenotypically well-described and phylogenetically coherent genera, such as *Bradyrhizobium*, *Rhizobium*, or *Nitrobacter* (Meyer et al., 1993).

List of species of the genus Carbophilus

1. **Carbophilus carboxidus** (ex Nozhevnikova and Zavarzin 1974) Meyer, Stackebrandt and Auling 1994, 182[VP] (Effective publication: Meyer, Stackebrandt and Auling 1993, 390.)

   car.box'idus. L. n. connected with carbon oxides.

   The description is as given for the genus. The strain carboxydobacterium Z-1171 was isolated from soil near a stream at Neskuchny Garden, city of Moscow, Russia; it was originally described as "*Achromobacter carboxydus*" and later as "*Alcaligenes carboxydus*" (Nozhevnikova and Zavarzin, 1974; Zavarzin and Nozhevnikova, 1976, 1977; Zavarzin, 1978; Cypionka et al., 1980; Meyer and Schlegel, 1983).

   The mol% G + C of the DNA is: 62.8 (T_m).

   Type strain: ATCC 51424, CIP 105722, DSM 1086, Z-1171.

Genus V. **Chelatobacter** *Auling, Busse, Egli, El-Banna and Stackebrandt 1993b, 624[VP]*
(Effective publication: Auling, Busse, Egli, El-Banna and Stackebrandt 1993a, 109)

THOMAS W. EGLI AND GEORG AULING

Che.la' to.bac.ter. Gr. n. *chele* claw; M.L. v. *chelato* to form claw-like complexes with divalent cations, i.e., to chelate; M.L. masc. n. *bacter* equivalent of Gr. neut. n. *bakterion* rod or staff; M.L. masc. n. *Chelatobacter* chelating rod.

Rods 0.7–0.9 × 1–2 μm, often pleomorphic and budding. L-, Y-, and X-shaped forms occur. Gram negative. **Motile, usually by two or three subpolar flagella. Obligately aerobic.** Optimal growth temperature, 28–30°C; no growth at 41°C and 4°C. Slow growers with all substrates tested so far (μ_{max} <0.25 h^{-1}). Poly-β-hydroxybutyrate is accumulated. All strains are sensitive to β-lactam antibiotics but resistant to nalidixic acid. **The metal-chelating aminopolycarboxylic acid nitrilotriacetic acid (NTA) can be utilized as a sole source of carbon/energy and nitrogen.** A great variety of sugars, acids, alcohols, and methylated amines are utilized. Vitamins are not required. Ubiquinone Q-10 is present. The main polyamine present is *sym*-homospermidine; pu-

trescine and spermidine occur as major polyamines and spermine as a minor polyamine.

The mol% G + C of the DNA is: 62–63.

Type species: **Chelatobacter heintzii** Auling, Busse, Egli, El-Banna and Stackebrandt 1993b, 624 (Effective publication: Auling, Busse, Egli, El-Banna and Stackebrandt 1993a, 110.)

FURTHER DESCRIPTIVE INFORMATION

Cell morphology All the strains isolated so far are of similar morphology, although they differ slightly with respect to their physiological properties (Egli et al., 1988). Generally the cells are motile, usually by two to three subpolar flagella (Fig. BXII.α.131a); however, for one strain (TE 10) peritrichously flag-

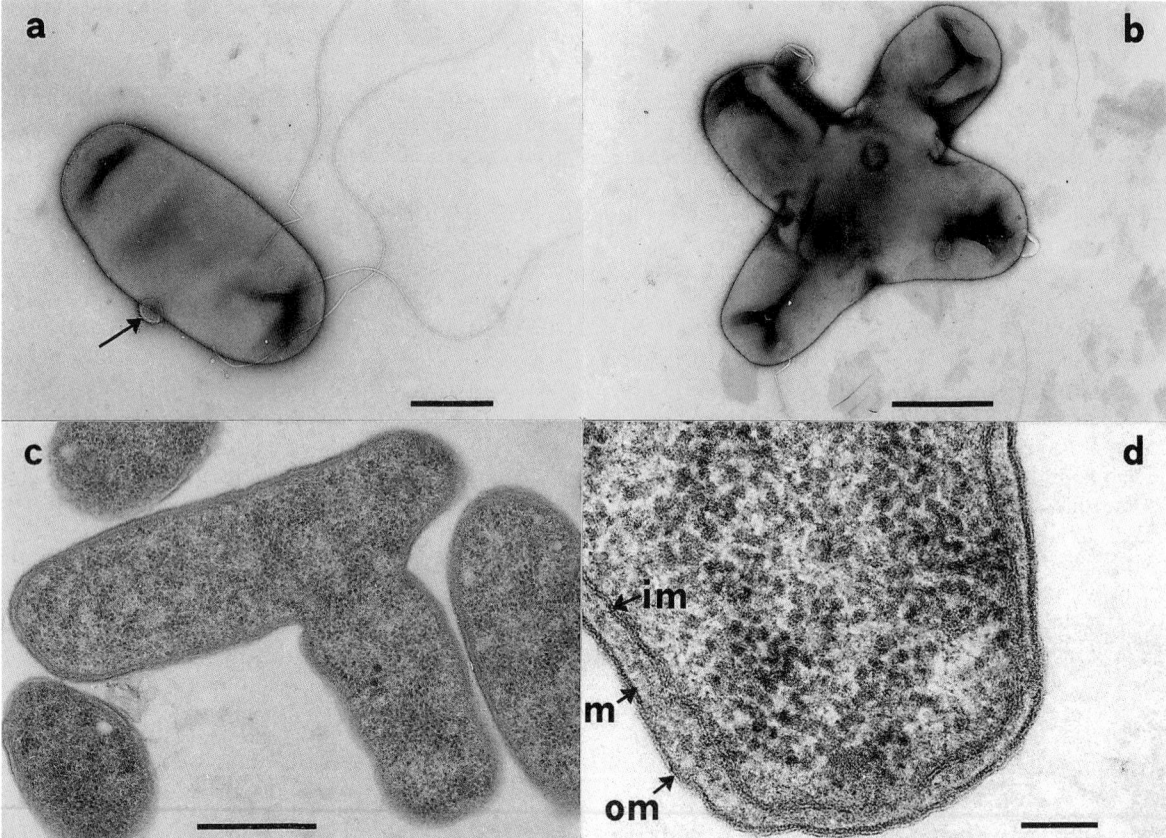

FIGURE BXII.α.131. Electron micrographs of *Chelatobacter heintzii* strains. (*a*), negatively stained cell of isolate TE 8 with the subpolar location of flagella and the initiation of cell budding (*arrow*). Bar = 0.5 μm. (*b*), negatively stained cell of isolate TE 8 demonstrating pleomorphism. Bar = 0.5 μm. (*c*), thin-sectioned cells of isolate TE 7 exhibiting pleomorphism. Bar = 0.5 μm. (*d*), higher magnification of thin-sectioned cell of isolate TE 6 demonstrating the Gram-negative cell wall with inner membrane (*im*), murein layer (*m*), and outer membrane (*om*), Bar = 0.1 μm.

ellated cells were also observed (Wehrli and Egli, 1988). The proportion of cells that are motile depends on the growth phase and substrate. In addition to cells dividing normally by binary fission, L-, Y-, and X-shaped cells are frequently detected in thin-sectioned, freeze-fractured, and negatively stained preparations (Fig. BXII.α.131b, c). In addition, bud-like structures are often observed where growth of daughter cells is obviously initiated (Fig. BXII.α.131a, b). The significance of this pleomorphism and whether or not it is dependent on nutritional or particular environmental conditions is still unknown; however, it does not result from growth with the metal-chelating compound NTA because pleomorphic cells are also observed during growth in complex media.

All isolates exhibit a peptidoglycan layer of 4–6 nm thickness (Wehrli and Egli, 1988) and an outer membrane (Fig. BXII.α.131d). The outer membrane is always closely associated with the underlying murein layer, and this is probably the reason why in freeze-fractured cells breakage occurs exclusively at the inner membrane and not at the outer membrane as well, as has been observed with most other Gram-negative bacteria (Wehrli and Egli, 1988).

Colony morphology   Strains that grew well on Plate Count Agar (PCA, from Difco, Detroit, MI) exhibited round and smooth, beige-brown colonies that had a diameter of 2–3 mm after 4–5 days of growth. However, most of the strains grew faster and reliably on 0.2 × PCA broth solidified with 1.5% agar (see

below), and on this medium the colonies of all isolates were similar: round, light-beige, slightly translucent and small in size (1.0–1.5 mm in diameter after 4–5 days). The colonies were round and volcano- or sometimes fried-egg-like, frequently with one or two concentric rings (Fig. BXII.α.132a, b, c, e). Colonies grown on NTA-containing media were white, tough, and pinpoint in size at the early stages of development (Fig. BXII.α.132b, viewed with a stereo-microscope with incident light); later a brown center developed and finally (4–5 days after streaking out) the whole colony turned black and polygonal as shown in Fig. BXII.α.132d (viewed with transmitted light).

Cultural characteristics   The chemically defined medium (SM) used for isolation of *Chelatobacter heintzii* strains in both batch and chemostat culture (Egli et al., 1988; Egli and Weilenmann, 1989) contained (g/l distilled water): $MgSO_4 \cdot 7H_2O$, 1.0 g; $CaCl_2 \cdot 2H_2O$, 0.2 g; $Na_2HPO_4 \cdot 2H_2O$, 0.41 g; KH_2PO_4, 0.26 g; 1 ml of the trace element stock solution described by Pfennig et al. (1981) with NTA as the chelating agent (5.2 g/l); and 1 ml of a vitamin stock solution. The vitamin solution contains (per liter): pyridoxine·HCl, 100 mg; thiamine·HCl, riboflavin, nicotinic acid, D-calcium pantothenate, *p*-amino benzoic acid, lipoic acid, nicotinamide, and vitamin B_{12}, 50 mg each; biotin, 20 mg; and folic acid, 20 mg. This medium was supplemented with either NTA or other carbon and/or nitrogen sources (up to a maximum of 5 g/l carbon). For growth with carbon sources that contained no nitrogen, the SM was supplemented with NH_4Cl, 0.54 g/l.

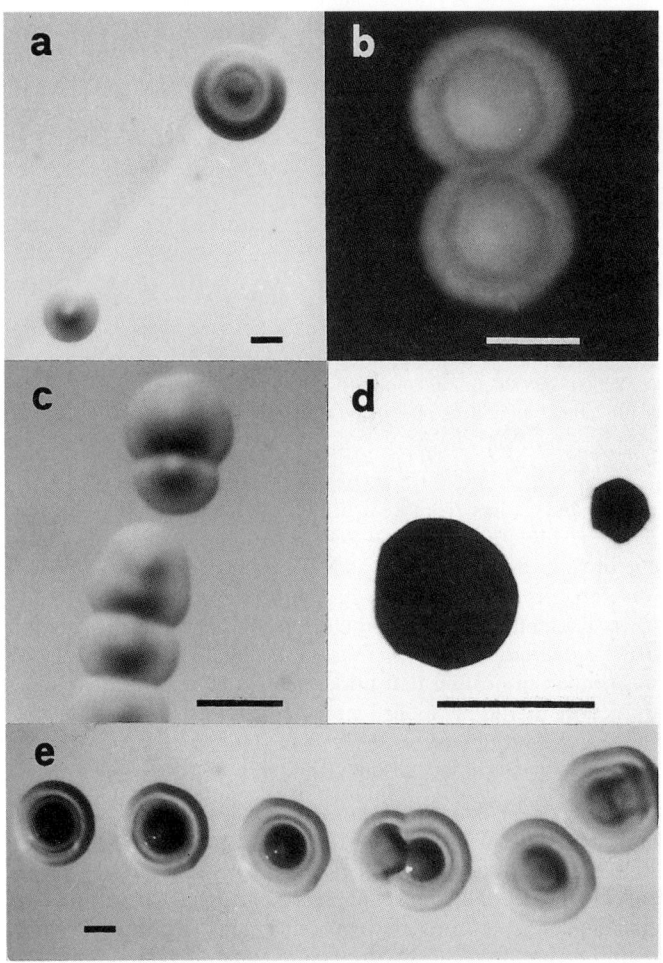

FIGURE BXII.α.132. Morphology of *Chelatobacter heintzii* colonies. (*a*), PCA-grown strain TE 8. (*b*), NTA-grown strain TE 8. (*c*), PCA-grown strain TE 9. (*d*), colony of NTA-grown strain TE 9 after 7 days, note the polygonal shape of the colonies. (*e*), colonies of PCA-grown strain TE 4. Bars = 0.5 mm.

For isolation and growth on agar plates this medium is supplemented with 1.5% agar plus NTA (0.5 g/l). For isolation, and in early experiments, vitamins were included in the medium, but later it was found that for growth of all strains of *Chelatobacter heintzii*, vitamins can be omitted.

Growth characteristics Isolates differed with respect to their growth on media containing high nutrient concentrations. Most of the strains grew poorly and only after heavy inoculation on PCA plates of the original concentration. However, faster growth occurred on diluted (10–20% of the original concentration) PCA

medium broth amended with 1.5% agar, although the colonies obtained were smaller.

Growth in batch culture with NTA as the only carbon/energy and nitrogen source results in the excretion of ammonia, causing an increase in pH of the culture medium. Under such conditions growth ceases at approximately pH 9 and, after transfer into new medium, the cells exhibit extended lag phases (often 5–10 d) until growth resumes. Growth with mixtures of utilizable carbon sources plus NTA resulted in a simultaneous utilization of the two substrates and in faster growth than with NTA alone. For example, during simultaneous utilization of glucose plus NTA, a maximum specific growth rate of 0.17 h^{-1} was reported for strain TE 8 compared to 0.11 h^{-1} during growth with NTA alone (Table BXII.α.115).

Nutritional and physiological characteristics The nutritional and physiological characteristics of different strains of *Chelatobacter heintzii* are listed in Table BXII.α.116. All strains utilized acetate, benzoate, citrate, glycollate, lactate, malate, propionate, pyruvate, succinate, *n*-butanol, ethanol, glycerol, *iso*-propanol, *n*-propanol; the sugars *N*-acetylglucosamine, arabinose, D-cellobiose, esculin, D-fructose, galactose, gentiobiose, D-glucose, inositol, D-maltose, mannitol, D-mannose, rhamnose, D-ribose, D-sorbitol, sucrose, D-xylose, L-alanine, L-arginine, L-aspartate, L-glutamate, L-glycine, L-lysine, L-phenylalanine, L-proline, L-serine, and acetamide. No strain grew with: adonitol, aniline, arabitol, dimethyl formamide, dimethyl sulfoxide, erythritol, [S,S]-ethylene diamine disuccinate, formate, gluconate, N-(2-acetamido)-iminodiacetate, D-lactose, malonate, methane, methanol, methyl acetamide, methyl diethyl amine, *n*-decane, oxalate, phenol, tetraethyl ammonium chloride, triethanol amine, tris-hydroxymethyl amino methane, urea, xylene, and H$_2$/CO$_2$.

In addition, none of the strains produced gas from glucose, indole or H$_2$S, was able to ferment sugars, to hydrolyze DNA, to fix N$_2$, to denitrify with NTA or with glucose, to produce acid from glucose or ethanol, to grow on mannitol (*Rhizobium*) agar, or was acid fast. For all strains, growth was slightly affected by the presence of 1% NaCl and no growth occurred with 10% NaCl.

Metabolism and metabolic pathways A key nutritional feature of members of the genus *Chelatobacter* is their ability to grow with the chelating agent NTA as the only source of carbon, energy and nitrogen. The biochemistry of NTA utilization has been studied mainly in *Chelatobacter heintzii* ATCC 29600 (Egli, 1994) and is shown in Fig. BXII.α.133. After transport into the cell, two enzymatic steps are sufficient to transform NTA into the central metabolites glycine and glyoxylate. Transport of NTA is most probably energy-dependent (summarized in Bucheli-Witschel and Egli, 2001). Recently, it was suggested that free NTA (and not a metal-NTA complex) is the form that is transported (Bolton et al., 1996). This implies that the equilibrium between free NTA and metal-complexed NTA will be crucial for the transport of

TABLE BXII.α.115. Maximum specific growth rate of three NTA-utilizing strains during batch growth with various substrate combinations[a,b]

Substrate	*Chelatobacter heintzii* TE 8	*Chelatobacter heintzii* TE 9	*Chelatococcus asaccharovorans* TE 1
NTA	0.11	0.1	0.07
NTA/acetate	0.15	0.13	0.13
NTA/glucose	0.17	0.13	0.07
NH$_4^+$/acetate	0.19	0.23	0.12

[a]Adapted from Egli et al. (1988).

[b]Growth rates (h^{-1}) measured at 30°C, pH 6.8–7.0, and substrate concentrations of 250 mg/l.

FIGURE BXII.α.133. Metabolic pathway for NTA in *Chelatobacter heintzii* ATCC 29600. *IDA*, membrane-bound iminodiacetate dehydrogenase; *NTA-Mo*, NTA monooxygenase, component A (*cA*) and B (*cB*), respectively. Note that the function of cB can be taken over by other FMN-reducing oxidoreductases.

this compound into the cell and that labile metal-NTA complexes should be degraded preferentially.

NTA monooxygenase The first intracellular step is catalyzed by a monooxygenase (NTA-Mo). NTA-Mo was originally isolated as a two component system, consisting of component cA and a FMN-containing protein cB with both components being present as dimers (Uetz et al., 1992). Immunological studies indicated that both cA and cB are present in all *Chelatobacter heintzii* strains tested so far (Table BXII.α.116; Uetz, 1992). It was demonstrated recently that cA acts as an independent monooxygenase whereas cB is supplying the enzyme with $FMNH_2$ and that FMN oxidoreductases from distantly related organisms and enzyme systems can supply cA with $FMNH_2$(Witschel et al., 1997). The substrate specificity of NTA-Mo is very narrow, so far NTA complexed with Mg^{2+} is the only compound that is known to be accepted (Uetz et al., 1992). Neither free NTA nor Ca^{2+}-complexed NTA is accepted as a substrate. Both components have been cloned and sequenced from *Chelatobacter heintzii* ATCC 29600 (see below).

Iminodiacetate dehydrogenase A membrane-bound iminodiacetate dehydrogenase (IDA-DH) which is present in all presently known NTA-degrading strains, was found to be the second enzyme in the catabolism of NTA (Uetz and Egli, 1993). The enzyme differs from succinate dehydrogenase and experiments with IDA-DH extracted from membranes and reconstituted in artificial membrane vesicles indicated that it feeds electrons abstracted from IDA into the electron transfer chain at the level of ubiquinone.

Regulation of NTA-degrading enzymes Recently, regulation of the ability to degrade NTA and the expression of NTA-Mo was studied under different growth conditions in carbon-limited continuous culture (Bally, 1994; Bally et al., 1994; Egli, 1995; Bally and Egli, 1996). NTA was consumed simultaneously with easily degradable carbon substrates such as, e.g., glucose and the extent of expression of NTA-Mo was dependent on the ratio of NTA:glucose in the feed medium (and, hence, the ratio of the two carbon sources consumed by the cell). Significant expression of NTA-Mo was only observed when the contribution of NTA to the total carbon consumed by the culture was higher than 1%. The experiments indicate that in wastewater treatment plants or the natural environment, NTA-degrading bacterial cells are probably not growing with NTA as a single carbon source but with mixtures of NTA plus naturally available carbon substrates, and that NTA-degrading enzymes are probably rarely fully induced under environmental growth conditions. These results are sup-

ported by studies from real and model wastewater treatment plants (Bally, 1994) showing that in plants receiving more than 1% of their organic carbon as NTA, both components of NTA-Mo were expressed, whereas in plants receiving only little NTA it was not detectable. Nevertheless, in both systems the fraction of NTA-degrading cells (as assessed with cell surface antibodies) was similar, indicating that under such conditions probably the induction of the catabolic enzymes was more responsible for successful degradation of NTA than the enrichment of NTA-degrading bacterial cells (Egli, 1995).

Genetics With regard to catabolic genes, in *Chelatobacter heintzii* ATCC 29600 the genes for the components cA and cB of NTA-Mo (see above) have been cloned and sequenced recently (Knobel et al., 1996). Two open reading frames (most probably chromosomally encoded) oriented divergently with an intergenic region of 307 bp could be assigned to the two components. The deduced gene products of *ntaA* showed significant homology only to SoxA (involved in dibenzothiphene degradation) and SnaA (involved in pristamycin synthesis), whereas that of *ntaB* shared weak homology in one domain with other NADH:FMN oxidoreductases. Unfortunately, these homologies give no conclusive answer yet with respect to the possible evolutionary origin of NTA-Mo. Additionally, an open reading frame was found downstream of *ntaA* which shares considerable homology in the N-terminal region with the GntR class of bacterial regulator proteins and, therefore, may encode a regulatory protein involved in the regulation of *ntaA* and *ntaB* expression. This information was recently confirmed (Xu et al., 1997). Hybridization experiments with different gene probes from *Chelatobacter heintzii* ATCC 29600 indicate that *ntaA* is highly conserved in the different strains investigated so far (even in *Chelatococcus asaccharovorans*), whereas *ntaB* was not detected in all strains. This confirms the recent observation that probably any FMN-reducing oxidoreductase can supply FMNH to NTA-Mo, i.e., component A (Witschel et al., 1997; Xu et al., 1997).

Antigenic structure Polyclonal antibodies (α-*Cb* and α-*Cc*) have been raised against whole cells (El-Banna, 1989; Wilberg et al., 1993) and isolated cell walls (Bally, 1994) of NTA-utilizing *Chelatobacter* and *Chelatococcus* strains. In Ouchterlony double diffusion or indirect immunofluorescence tests, the antisera raised against the two central NTA-utilizing *Chelatobacter* strains, i.e., *Chelatobacter heintzii* strains TE 6 and ATCC 29600, did not cross-react with cell homogenates from both *Chelatococcus asaccharovorans* strains. Similarly, the antiserum raised against *Chelatococcus asaccharovorans* strain TE 2 did not cross-react with homogenates

TABLE BXII.α.116. Physiological properties of NTA-utilizing, obligately aerobic strains of *Chelatobacter heintzii* and *Chelatococcus asaccharovorans*[a,b,c]

Characteristic	*Chelatobacter heintzii*[d]	*Chelatococcus asaccharovorans*[d]
Hydrolysis of:		
Gelatin[e,f]	d	+
ONPG[f,g]	d	+
Urea[f]	d	+
Starch	−[h]	−
Protein	−[i]	
Tween 80	−[j]	−
Production of:		
Nitrite[e,f]	d	+
Acetoin[f,k]	+[l]	−
Pigment on King's medium	−	−
Presence of:		
Oxidase[f,m]	+	+
Catalase[f]	+[n]	+
NTA-monooxygenase cA[o]	+	+
NTA-monooxygenase cB[o]	+	−
PHB	+[p]	+
Growth characteristics:		
41°C	−	+
4°C	−[q]	+
KCN	+[r]	+
Vitamin requirement	−	+
Tellurite[s]	−[t]	−
Carbon substrates used:		
Methylamine	+[u]	−
Dimethylamine	+[u]	−
Trimethylamine	+[u]	−
Butyrate	+[v]	−
Fucose	+[w]	−
Glycyl-glycine	d	−
Glyoxylate	+[x]	+
Iminodiacetate	+[y]	−
D-(+)-Raffinose	d	−
Sarcosine	d	−
Skim milk	−[z]	−
Xylitol	−	−

[a]Symbols: +, positive result obtained with seven or more of the nine strains; (+), weakly positive result; −, negative result obtained with seven or more of the nine strains; d, three to six strains out of nine strains tested differed in their reaction (results of individual strains in Egli et al., 1988).

[b]Data from Egli et al. (1988) and Egli and Weilenmann (1989).

[c]Many of the tests were performed with both NTA-grown cells and cells grown on PCA plates.

[d] *Chelatobacter heintzii* strains TE 4–TE 10 and ATCC 27109 and ATCC 29600; *Chelatococcus asaccharovorans* strains TE 1 and TE 2. All strains were enriched in batch culture, except for *Chelatobacter heintzii* strain TE 4.

[e]Only cells grown on NTA-agar plates were positive; cells grown on PCA plates were negative.

[f]Test performed with API-20B.

[g]As in e, except for strain ATCC 27109 that gave positive results on NTA as well as PCA plates.

[h]TE 6 and ATCC 29600 were slightly positive.

[i]ATCC 29600 was positive, TE 4 and 9 were slightly positive.

[j]ATCC 29600 was positive.

[k]PCA-grown cells of strain TE 5 were negative.

[l]ATCC 27109 was negative.

[m]PCA-grown cells of strains TE 4–8 and 10 were weakly positive, those of strain TE 9 were negative.

[n]CA grown cells of TE 4 and 9 were negative.

[o]Results from Uetz (1992).

[p]TE 10 and ATCC 27109 were slightly positive.

[q]ATCC 27109 was positive.

[r]TE 4–8 were only slightly positive.

[s]+, formation of black colonies; −, no growth.

[t]TE 9 and ATCC 27109 were positive.

[u]TE 4 and TE 10 were negative.

[v]TE 10 and ATCC 29600 were negative.

[w]TE 9 and 10, ATCC 27109, and ATCC 29600 were weakly positive.

[x]TE 4 and 10 were negative.

[y]TE 4 was negative.

[z]ATCC 29600 was positive; TE 4 and TE 9 were slightly positive.

of the two central strains of *Chelatobacter heintzii* (Wilberg et al., 1993).

Cross-reaction of α-*Cb* was also tested with a variety of bacteria that were expected to co-exist with members of *Chelatobacter* in the same habitat. No cross reaction was observed for *Agrobacterium tumefaciens*, *Alcaligenes faecalis*, *Azotobacter chroococcum*, *Bradyrhizobium japonicum*, *Comamonas acidovorans*, *Comamonas testosteroni*, *Escherichia coli* ML30, *Klebsiella pneumoniae*, NTA-utilizing strain TE 11 (denitrifying), *Paracoccus denitrificans*, *Pseudomonas fluorescens*, *Pseudomonas putida*, *Rhizobium leguminosarum*, *Rhizobium meliloti* and *Rhodobacter* sp. (Wilberg et al., 1993). A slight cross-reaction was observed with *Carbophilus carboxidus* DSM 1086, *Aminobacter aminovorans* NCIB 9039, *Chelatobacter heintzii* strains TE 4 to TE 10, and *Ralstonia eutropha*. Highly specific antibodies were prepared from outer membrane fractions of *Chelatobacter heintzii* ATCC 29600 using a depletion method (Bally, 1994). The only bacteria still exhibiting significant cross-reaction with the purified antibodies were *Rhizobium phaseoli* and *Carbophilus carboxidus*.

Antibiotics and drug resistance The susceptibility of NTA-utilizing *Chelatobacter heintzii* and *Chelatococcus asaccharovorans* strains towards a range of antibiotics was tested (Auling et al., 1993a) and a summary of the results is provided in Table BXII.α.117. All *Chelatobacter heintzii* strains were resistant to nalidixic acid and sensitive towards penicillin G, ampicillin, carbenicillin, deoxycycline, vancomycin, and novobiocin. It has not yet been determined whether this extreme sensitivity towards β-lactam antibiotics is a property that allows the differentiation of *Chelatobacter* spp. from close neighbors. However, the resistance of all strains towards nalidixic acid might be used to isolate such strains with a medium amended with this antibiotic.

TABLE BXII.α.117. Resistance of *Chelatobacter heintzii* and *Chelatococcus asaccharovorans* towards a range of different antibiotics[a,b]

Antibiotics[c]	*Chelatobacter heintzii*[d]	*Chelatococcus asaccharovorans*[e]
β-Lactams:		
Penicillin G, 10U	S	S
Ampicillin, 10 µg	S	d
Carbenicillin, 100 µg	S	S
Cephaloridine, 30 µg	S	R
Cephalexin, 30 µg	d	R
Aminoglycosides:		
Streptomycin, 10 µg	d	R
Neomycin, 30 µg	S	S
Kanamycin, 30 µg	d	S
Gentamycin, 10 µg	S	S
Tobramycin, 10 µg	d	S
Other antibiotics:		
Tetracycline, 30 µg	S	S
Doxycycline, 30 µg	S	S
Erythromycin, 15 µg	S	S
Chloramphenicol, 30 µg	d	d
Rifampicin, 30 µg	S	S
Polymyxin B, 300 U	d	S
Novobiocin, 30 µg	S	S
Vancomycin, 30 µg	S	R
Nalidixic acid, 30 µg	R	R
Trimethoprim, 1.25 µg; + sulfamethoxazole, 23.75 µg	d	R

[a]Symbols: S, at least eight or all strains sensitive; R, at least eight or all strains resistant; d, strains differed in their reaction.

[b]Data from Auling et al. (1993a).

[c]Absolute amount of antibiotic used in test.

[d]Nine *Chelatobacter heintzii* strains were tested (TE 4 to 10, ATCC 27109, ATCC 29600).

[e]Two *Chelatococcus asaccharovorans* strain were tested (TE 1 and TE 2).

Ecology Data collected so far with surface antibodies (Wilberg et al., 1993; Bally, 1994) and 16S rRNA probes (Neef, 1997) indicate that members of the genus *Chelatobacter* are ubiquitously distributed in the environment. This is supported by the ready isolation of *Chelatobacter* strains from soil, sediment, surface and wastewater. Even in an environment with little chance of having ever been exposed to NTA, i.e., a small pristine alpine stream that contained no detectable NTA (<0.2 µg/l), cells cross-reacting with antibodies raised against *Chelatobacter heintzii* ATCC 29600 were detected at 1.2×10^5 cells per liter (corresponding to 0.27% of the total bacterial population) (Bally, 1994). In nutrient-rich environments such as eutrophic surface waters or wastewater treatment plants, the total number of cross-reacting cells generally increased by several orders of magnitude. For example, both Wilberg et al. (1993) and Bally (1994) reported numbers of NTA-degrading *Chelatobacter* cells in aerated tanks of several Swiss wastewater treatment plants in the range of 10^{10} cells/l (corresponding to 0.1–1% of the total bacterial population). Similar figures were recently reported using a molecular probe specific for the 16S rRNA of *Chelatobacter heintzii* ATCC 29600 (Neef, 1997).

Polyamines and ubiquinones All *Chelatobacter heintzii* strains contained *sym*-homospermidine as the main, putrescine and spermidine as major, and spermine as minor polyamine compounds (Auling et al., 1993a). Ubiquinone Q-10 was present in all strains. As it has been frequently stated that NTA is utilized by specialized pseudomonads, this polyamine and quinone profile allows *Chelatobacter heintzii* to be clearly distinguished from the true (fluorescent) *Pseudomonas* spp., which contain putrescine and spermidine as the main polyamines and ubiquinone Q-9 (Auling et al., 1991, 1993a).

Soluble protein patterns The soluble protein pattern for *Chelatobacter heintzii* has been compared to that of *Chelatococcus asaccharovorans* (Auling et al., 1993a). *Chelatobacter* and *Chelatococcus* strains exhibited clearly different patterns. Among the different *Chelatobacter* strains, the protein patterns were highly similar, except for two isolates (strains TE 4 and 7). Although the soluble protein patterns of strains TE 4 and 7 were virtually identical, the two strains differed considerably with respect to their nutritional and physiological properties (Egli et al., 1988).

Miscellaneous comments In the 1970s, eutrophication of surface waters by phosphorus from detergents and agriculture stimulated discussion about banning sodium tripolyphosphate (STPP) from laundry detergents and replacing it with non-phosphorus-containing metal-chelating agents such as aminopolycarboxylic acids (Mottola, 1974; Tiedje, 1980; Anderson et al., 1985). Aminopolycarboxylic acids are metal-sequestering compounds used in large amounts in a variety of domestic and industrial applications including cleaning agents, agricultural fertilizers, for water treatment and descaling of boilers, in the photographic industry, the dying of textiles, during pulp and paper production, for metal finishing and rubber processing, or in food, pharmaceuticals, and cosmetics (McCrary and Howard, 1979; Egli, 1988; Egli et al., 1990). Aminopolycarboxylic acids include both man-made (e.g., ethylenediaminetetraacetic acid [EDTA] and nitrilotriacetic acid [NTA]) and microbially synthesized (e.g., *S,S*-ethylenediaminedisuccinic acid) compounds. The most extensively used synthetic chelating agents are EDTA and NTA. All applications are water based and therefore their susceptibility to biodegradation during wastewater treatment and in the aquatic environment is an important criterion for assessing their environmental impact and toxicity (Anderson et al., 1985).

Of all the synthetic chelating agents, NTA and EDTA have received the most attention, NTA because of its controversial application in laundry detergents as a substitute for sodium tripolyphosphate, and EDTA because of its slow biodegradability and its ubiquitous presence in aqueous systems (Tiedje, 1980; Wolf and Gilbert, 1992).

It was soon established that elimination of NTA, one of the most promising substitutes for STPP, from wastewater and natural waters was exclusively brought about by microbial action. This was the motivation for setting up the first enrichment cultures to isolate NTA-degrading microorganisms (Focht and Joseph, 1971; Cripps and Noble, 1973; Enfors and Molin, 1973a, b; Tiedje et al., 1973; Kakii et al., 1986). Most of these isolates were identified based on rather superficial characterization and assigned to the genus *Pseudomonas*, although the isolation of NTA-utilizing nonpseudomonads was also reported (Egli et al., 1990). However, with the exception of the two strains deposited at the ATCC (strains ATCC 29600 and 27109), none of these isolates seems to be available anymore. Despite the poor characterization of these isolates, the (probably false) conclusion was drawn that the biodegradation of NTA was primarily a trait inherent to specialized *Pseudomonas* strains (Tiedje, 1980; Anderson et al., 1985). From the description given in the original publications and the fact that the two ATCC strains were described as similar, it seems highly likely that most of these isolates were the first *Chelatobacter* strains isolated from the environment. Today the ubiquitous occurrence of members of the genus *Chelatobacter* and *Chelatococcus* (Wilberg et al., 1993; Bally, 1994) gives a strong indication that these bacteria are a major component of the NTA-degrading microbial community in wastewater treatment plants and aerobic environments.

Other strains reported to grow at the expense of NTA have included *Bacillus*, *Listeria*, *Rhodococcus*, and yeast species (Egli, 1994). However, most of these strains have been poorly characterized and were lost. Nevertheless, these reports suggest that the ability to utilize NTA is probably not restricted to *Chelatobacter* and *Chelatococcus* species. This is supported by the recent finding of an EDTA-degrading isolate (clearly different from *Chelatobacter* and *Chelatococcus*) that was also able to metabolize NTA (Witschel et al., 1997).

ENRICHMENT AND ISOLATION PROCEDURES

From soil, wastewater, or surface waters, strains belonging to the genus *Chelatobacter* can be easily enriched in either batch or chemostat cultures using NTA as the only source of carbon, energy, and nitrogen (Egli et al., 1988; Egli and Weilenmann, 1989). All of the presently known strains of *Chelatobacter* are resistant towards nalidixic acid (Table BXII.α.117). This might be used to isolate such strains with a medium containing this antibiotic.

MAINTENANCE PROCEDURES

All strains can be maintained in a freeze-dried condition. Alternatively, cultures grown in SM with NTA and amended with either glycerol (15%, v/v) or DMSO (50%, v/v) can be stored in liquid nitrogen. However, with either method revival of cultures may take a while. Best results have been obtained when streaking cultures from liquid nitrogen directly onto agar plates with SM containing NTA as the only carbon and nitrogen source. When maintaining cultures on selective NTA agar plates, it is recommended to transfer them to new plates at least every week because growth leads to an increased pH in and around the colonies. Cells stored under such conditions exhibit long lag phases (up to several weeks) before they restart growing. Liquid cultures behave similarly. Long lag times and irreproducible growth can be avoided by transferring the cells to fresh medium whilst they are still growing exponentially. Note that concentrations of NTA that are too high, combined with low buffering capacity of the medium or no pH control, may lead to an early cessation of growth.

PROCEDURES FOR TESTING SPECIAL CHARACTERS

Growth with NTA as the only source of carbon, energy, and nitrogen is a key property of *Chelatobacter*. Consumption of NTA can be measured either by the disappearance of dissolved organic carbon (DOC) paralleled by the excretion of ammonia, or via the consumption of NTA. A HPLC method specially developed for the analysis of NTA in culture media and cell extracts has been described by Schneider et al. (1988).

DIFFERENTIATION OF THE GENUS *CHELATOBACTER* FROM OTHER GENERA

The presently available information indicates that members of the genus *Chelatobacter* are most likely closely related to members of the genus *Aminobacter*. The nutritional and physiological properties that allow separation of the two genera are listed in Table BXII.α.118.

TAXONOMIC COMMENTS

The presence of ubiquinone Q-10 (Egli et al., 1988), and the characteristic polyamine pattern with *sym*-homospermidine as the main, and putrescine and spermidine as major components (Auling et al., 1993a), clearly allocates members of the genus *Chelatobacter* to the *Alphaproteobacteria*, and to the family *Rhizobiaceae* (D. Kuykendall, personal communication). The polyamine pattern indicates a relationship to *Phyllobacterium* (Auling et al., 1991) but distinguishes *Chelatobacter* from *Ochrobactrum*. A weak serological cross-reaction with *Aminobacter aminovorans* (Green and Gillis, 1989; Urakami et al., 1992) and even less with *Carbophilus carboxidus* (Meyer et al., 1993), formerly *"Alcaligenes carboxydus"* (Auling et al., 1988), was observed (Wilberg et al., 1993). Oligonucleotide catalogs indicated the latter as the nearest neighbor of *Chelatobacter* (Auling et al., 1993a). Meanwhile, 16S rDNA sequence data allow grouping of *Chelatobacter* with *Mesorhizobium* (Jarvis et al., 1997), 4-chloro-2-methylphenol-degrading (Lechner et al., 1995) members of *Defluvibacter* (Fritsche et al., 1999a), two species of *Pseudaminobacter* (Katayama-Fujimura et al., 1983, 1984b; Kämpfer et al., 1999), and *Aminobacter* (Urakami et al., 1992).

The genus presently comprises nine strains and contains only

TABLE BXII.α.118. Nutritional and physiological properties that allow differentiation between the genera *Chelatobacter* and *Aminobacter*[a,b]

Characteristics	*Chelatobacter*	*Aminobacter aminovorans*[c]	*A. aganoensis, A. niigataensis*[d]
Morphology	many pleomorphic cells	rods, pleomorphic forms	rods, budding, no pleomorphic forms reported
Substrate utilization:			
NTA[e]	+	−	−
Aspartate	+	−	+
Citrate	+[f]	−	−
Dimethylamine	+[g]	−	+
Ethanol	+	−	−
Malate	+[f]	−	−/+
Phenylalanine	+	−	−
Fucose	−	d	nd
Sarcosine	d[f]	−	nd
Succinate	+	−[c] (+)[d]	+
Glutamate	+	d	nd
Growth on mannitol (*Rhizobium*) agar	−	+	+
Growth on standard complex medium[h]	weak	abundant	weak
Growth at 4°C	−	nd	nd
Growth at 5°C	nd	+	nd
Urease	−[i]	+ (−)[d]	−
Nitrate to nitrite	−[j]	d	nd
Gelatin liquification	+[k]	−	−
Acid from arabinose	+	+	−
Voges–Proskauer	+[l]	−	−
16S rRNA gene sequence to *C. heintzii* DSM6450[Tm]	100%	99.8%	99.9%
Serological relationship (cross reaction with ATCC 29600)	+	−	nd

[a]Data collected from Egli et al. (1988), Egli and Weilenmann (1989), Green and Gillis (1989), and Urakami et al. (1992).

[b]Symbols: +, positive; −, negative; d, individual strains differ in their properties; nd, not reported.

[c]Most data from Green and Gillis (1989) (including the strains NCIB 9039, 11590, and 11591, later assigned to this species by Urakami et al., 1992).

[d]Data from Urakami et al. (1992).

[e]Strains tested in our laboratory were *A. aminovorans* NCIB 9039, *A. aganoensis* DSM 7051, *A. niigataensis* DSM 7050.

[f]From Egli et al. (1988), differing from Kämpfer et al. (1999).

[g]TE 4 and 10 are negative.

[h]Plate count agar, nutrient broth, peptone water, or PYG broth.

[i]TE 5 and 9, ATCC 27109, and 29600 are positive.

[j]TE 7 and 8 are positive, TE 4 and 5 slightly positive.

[k]TE 9 and 10 and ATCC 29600 are negative.

[l]ATCC 27109 negative.

[m]From Kämpfer et al. (1999). Strains used were *Chelatobacter heintzii* DSM 6450[T], *Aminobacter aminovorans* DSM 7048[T], *A. aganoensis* DSM 7051[T], and *A. niigataensis* DSM 7050[T].

a single species, *Chelatobacter heintzii.* Based on DNA–DNA hybridization studies (Auling et al., 1993a), the genus is divided into two groups, A1 and A2 (Fig. BXII.α.134). However, this division does not reflect the nutritional and physiological properties of the different *Chelatobacter* isolates (see Table BXII.α.116). The reference strain for DNA homology cluster A1 is *C. heintzii* strain ATCC 29600, which is also the type strain of the species. Strain *C. heintzii* TE 6 is the central strain of the DNA homology cluster A2.

The increasing availability of new relatives of *Chelatobacter* may require a change in its nomenclatural status in the future; a clustering of *Chelatobacter* with *Aminobacter* was recently proposed by Kämpfer et al. (2002). However, any revision based on phylogenetic considerations will meet with problems, since the related taxa are located within the notoriously "shallow" branch formerly known as the "alpha-2" branch of the *Proteobacteria.*

ACKNOWLEDGMENTS

We are indebted to Ernst Wehrli for supplying us with excellent electron micrographs. In addition, we thank D. Kuykendall for allowing us to include unpublished information. TE would like to thank H.U. Weilenmann and all the students that have contributed to the study of NTA-utilizing microorganisms. Furthermore, the generous financial support of research on NTA in the laboratory of TE by grants from the Swiss National Science Foundation, Lever AG Switzerland and Unilever Port Sunlight, the Research Commission of ETH Zürich, and by EAWAG is gratefully acknowledged.

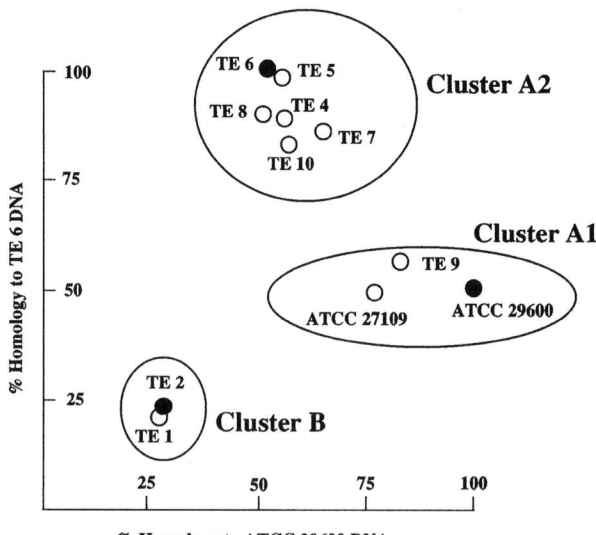

FIGURE BXII.α.134. DNA similarity groups of Gram-negative, obligately aerobic, NTA-utilizing isolates belonging to the genera *Chelatobacter* and *Chelatococcus* (Reproduced with permission from G. Auling et al., Systematic and Applied Microbiology *16:* 104–112, 1993, ©Urban & Fischer Verlag, Jena.)

List of species of the genus Chelatobacter

1. **Chelatobacter heintzii** Auling, Busse, Egli, El-Banna and Stackebrandt 1993b, 624[VP] (Effective publication: Auling, Busse, Egli, El-Banna and Stackebrandt 1993a, 110.)

    heintz'i.i. M.L. gen. n. *heintzii* of Heintz; named after the chemist W. Heintz who was the first to synthesize the chelating agent NTA (Heintz 1862, 1865) and to describe some of its properties.

    The characteristics are as described for the genus and listed in Tables BXII.α.115, BXII.α.116, and BXII.α.117.
    The mol% G + C of the DNA is: 62–63 (T_m).
    Type strain: ATCC 29600, DSM 10368.

    Additional Remarks: All NTA-utilizing *Chelatobacter heintzii* strains described here have been deposited at the German Collection of Microorganisms and Cell Cultures (DSM), Braunschweig, Germany, where they have been assigned the following numbers: TE 4, DSM 6463; TE 5, DSM 6464; TE 6, DSM 6465; TE 7, DSM 6466; TE 8, DSM 6449; TE 9, DSM 6450; TE 10, DSM 6451. The other strains are available from the American Type Culture Collection, Rockville, MD, U.S.A., i.e., the strain isolated by Tiedje et al. (1973), namely *Chelatobacter heintzii* ATCC 29600 (also available from DSM under accession number DSM 10368) and ATCC 27109 (originally isolated by Focht and Joseph, 1971).

Genus VI. **Ensifer** Casida 1982, 343[VP]

DAVID L. BALKWILL

En'si.fer. L. adj. *ensifer* sword-bearing; M.L. masc. *Ensifer* sword bearer.

Rods 0.7–1.1 × 1.0–1.9 μm, occurring singly or in pairs. **Reproduction by budding at one pole of the cell,** with the bud then elongating to produce **asymmetric polar growth. Attaches endwise to various living Gram-positive and Gram-negative host bacteria and may cause lysis of the host cells. Host cells are not required for growth.** Gram negative, but may stain poorly. **Motile by a tuft of three to five subpolar flagella.** Aerobic; does not grow anaerobically in the presence of light. Optimal growth occurs at 27°C; good growth occurs at 20°C and 37°C. Not heat resistant. Weakly catalase positive. Nitrate and nitrite are reduced. Nitrification is negative for nitrite and ammonia. The **metabolism of glucose and galactose is oxidative.** Growth is inhibited by 4% NaCl but not by 2.5% NaCl. Grows well on most media. Definite but slow growth on soil extract agar and on 1.5% Noble agar in distilled water. Agar is not hydrolyzed. Utilizes a variety of organic carbon sources. A segment of the *Ensifer* 16S rDNA sequence (TACGGAGACGTTT, corresponding to positions 1009 through

1021 in the *Escherichia coli* 16S rDNA sequence; Brosius et al., 1978) might represent a signature sequence for this genus, as it did not match any other bacterial sequences available in public databases as of June, 2003.

The mol% G + C of the DNA is: 67 (T_m) and 63 (Bd).
Type species: **Ensifer adhaerens** Casida 1982, 343.

FURTHER DESCRIPTIVE INFORMATION

E. adhaerens was originally designated as strain A (Casida, 1980). Growth of *E. adhaerens* is initiated by budding at one pole of the cell, after which the bud elongates to produce asymmetric polar growth (Fig. BXII.α.135). This growth usually widens to equal the diameter of the mother cell. The daughter cell eventually separates from the mother cell by binary fission. Either the mother and daughter cells are the same size, or the daughter cell is smaller.

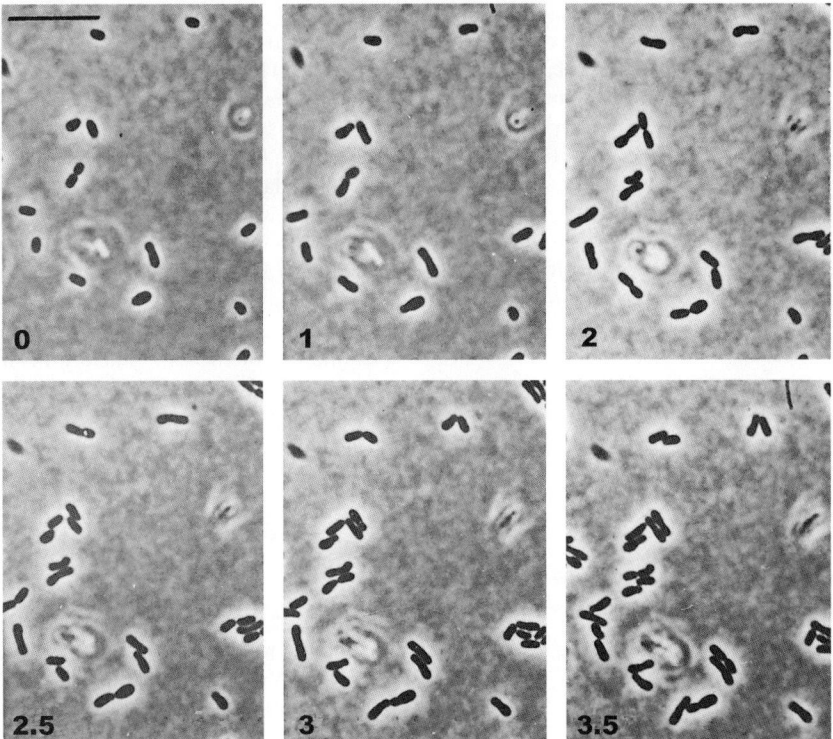

FIGURE BXII.α.135. Growth of *Ensifer adhaerens* on full-strength heart infusion agar. Incubation was for 0, 1, 2, 2.5, 3, and 3.5 h. Phase-contrast microscopy of a slide culture. Bar = 10 μm. (Reproduced with permission from L.E. Casida, Jr., International Journal of Systematic Bacteriology *32:* 339–345, 1982, ©International Union of Microbiological Societies.)

Measurements from electron micrographs indicate that some of the nonattached daughter cells are 20–30% smaller (length or width or both) than the nondividing mother cells. There is no tube or filament between mother and daughter cells during binary fission. After fission, growth resumes as new buds at the newly formed poles of both the mother and daughter cells. There does not appear to be any requirement for a rest or maturation period before growth resumes. Bud formation does not always occur at the pole of the cell; buds sometimes form on the side of a mother cell in old cultures.

The budding and division processes of *E. adhaerens* are similar to those of *Rhodopseudomonas blastica* (Eckersley and Dow, 1980). However, the latter organism is photosynthetic, grows anaerobically in the light, and divides only when the mother and daughter cells are of equal size. Moreover, the daughter cell must undergo an obligate period of maturation before it initiates growth.

A bar of darkly stained material is frequently visible along the side of the cell in whole-cell preparations of *E. adhaerens* that are stained with uranyl acetate and observed by electron microscopy. This bar becomes wider near one or both poles of the cell. When cells of *E. adhaerens* are attached to host cells, such as those of *Micrococcus luteus*, this bar often extends between the *E. adhaerens* cell and the host cell. The chemical composition of the bar is not known. In addition, it is not readily apparent in whole-cell preparations negatively stained with phosphotungstic acid instead of uranyl acetate, and it has not been located unequivocally in thin-sectioned material.

When *E. adhaerens* is grown in the presence of *M. luteus* (or some other sensitive bacterium) on agar medium, large numbers of *E. adhaerens* cells attach endwise to *M. luteus* cells. The *E. adhaerens* cells are situated side by side, closely packed, in a picket fence arrangement. When the prevailing nutrient and pH con-

ditions (in soil or on laboratory media; pH 6 is optimal) are satisfactory, this arrangement eventually results in lysis of the *M. luteus* cells. There is no specialized morphological structure on the *E. adhaerens* cell for attachment purposes. Rosettes are not formed.

E. adhaerens is not an obligate predator. It grows well on most media in the absence of potential host cells. In fact, the overall growth of *E. adhaerens* usually does not increase as a result of tracking (Casida, 1980) or lysis of *M. luteus* cells. However, some increased growth of *E. adhaerens* does occur when both organisms are placed on Noble agar with or without 0.1% glucose. Thus *E. adhaerens* probably benefits from this interaction only if the nutritive value of the medium is very low, a situation that is comparable to the presumed low availability of nutrients occurring in soils that have not recently received organic matter. In soil, *E. adhaerens* cells attach directly to host cells, so that even the small amounts of lytic factor produced under low-nutrient conditions can be used effectively in lysing host cells. On agar media, this lytic factor is diffusible and can act on host cells at a distance from the *E. adhaerens* cells.

E. adhaerens produces colonies 10–15 mm in diameter after 6 d on heart infusion agar prepared at one-tenth of the recommended strength (but containing 1.5% agar). These colonies are grayish white, circular with undulate margins, convex, slimy, moist, and opaque (but may appear almost translucent because of excessive slime production). Growth on agar slants is abundant, opaque, grayish white, smooth, flat, slimy, and moist. Pellicle and sediment (with little turbidity) are produced in a broth medium containing 0.5% peptone, 0.1% yeast extract, and 0.1% glucose.

Definite but slow growth occurs during sequential transfers on 1.5% Noble agar in distilled water. Host cells are not required, and the agar is not hydrolyzed. This growth is equivalent to that obtained on soil extract agar. Good growth is obtained on a synthetic medium containing 0.1% glucose, 0.1% NH_4NO_3, 0.1% KH_2PO_4, 0.1% NaCl, 0.1% $MgSO_4 \cdot 7H_2O$, 0.1% L-glutamic acid, and 1.5% agar (pH 7.0). *E. adhaerens* utilizes a variety of organic carbon sources including glucose, galactose, mannose, rhamnose, xylose, mannitol, sorbitol, glycerol, L-glutamic acid, L-alanine, L-asparagine, and L-glutamine. Acetate is used only slowly and does not inhibit glucose utilization. *E. adhaerens* grows on pure gelatin without hydrolysis. Starch is not hydrolyzed. Good growth occurs on blood agar, with no hemolysis. Good growth also occurs on desoxycholate agar, initially appearing whitish purple and then changing to buff.

The metabolism (Hugh and Leifson, 1953) of glucose and galactose is oxidative. No growth occurs under petrolatum. Gas is not produced in the absence of petrolatum; only trace amounts of acid, if any, are produced.

E. adhaerens cells do not survive 30 s of heating in tap water at 71°C.

Thirteen lytic phages for *E. adhaerens* were isolated from soil by Germida and Casida (1983) and used for phage typing of eight *E. adhaerens* host strains. The phages were also used in assays to monitor *E. adhaerens* predatory activity against other bacteria in soil (see below).

The habitat for *E. adhaerens* is soil. The numbers of *E. adhaerens* cells in soils can be estimated (Casida, 1980) by gently applying dilutions of soil to the surfaces of pregrown lawns of *M. luteus* cells (one-tenth-strength heart infusion agar, with 1.5% agar). After continued incubation, *E. adhaerens* produces small, thin, transparent colonies that appear as small moist areas that expand and coalesce. These colonies are not visible by transmitted light. The plates must be viewed from above, using light arriving at an oblique angle. The *E. adhaerens* colonies are on the surface of the *M. luteus* lawn. Under these conditions, they neither penetrate the lawn nor lyse the *M. luteus* cells. The cell numbers of specific *E. adhaerens* strains in soil can also be estimated with a most probable number (MPN) method that combines phage analysis with a dilution frequency procedure (Makkar and Casida, 1987b). A small (but known) number of lytic phage specific for the strain to be enumerated is added to each soil dilution. A tenfold or greater increase in PFU after a suitable incubation period is then taken to indicate that *E. adhaerens* host cells were present in a dilution. This method is capable of detecting *E. adhaerens* strains in soils even when they are present in very low numbers.

E. adhaerens has been isolated from a range of soil types. However, the strains from different soils do not necessarily cross-react when tested by phage typing (Germida and Casida, 1983). The predatory activity of *E. adhaerens* and its host range can be detected and followed in soils with the indirect phage analysis procedure of Germida and Casida (1983). Some bacteria, such as a myxobacterium and a *Streptomyces* predatory bacterium, succumb to *E. adhaerens* attack in soil but do not do so in laboratory cultures.

Indigenous *E. adhaerens* cells in soil attack only certain species of bacteria that are added to the soil. Studies using indirect phage analysis (Germida and Casida, 1983; Zeph, 1986; Zeph and Casida, 1986) have shown that *E. adhaerens* attacks *Agromyces ramosus*, *Micrococcus luteus*, a *Myxococcus* species soil isolate, *Staphylococcus aureus*, and two *Streptomyces* soil strains (C2 and 34). *E. adhaerens* does not attack *Actinomyces humiferus*, *Agrobacterium tumefaciens*, *Arthrobacter globiformis*, *Azotobacter vinelandii*, *Bacillus stearothermophilus*, *Bacillus subtilis*, *Bacillus thuringiensis*, *Escherichia coli*, *Pseudomonas aeruginosa*, *Rhizobium leguminosarum*, *Salmonella typhi*, or *Sinorhizobium meliloti*. However, in a few instances *E. adhaerens* will

attack another bacterial predator, such as the *Myxococcus* soil isolate that is itself attacking certain of the above bacteria (Germida and Casida, 1983).

E. adhaerens is a component of a sequence of three predatory bacteria in soil that respond naturally when host cells, such as those of *M. luteus*, are added to the soil (Casida, 1980). A *Streptomyces* predator responds quickly with growth and attack on *M. luteus*. This is followed by growth of *E. adhaerens*, which attaches to and lyses both *M. luteus* and the streptomycete. Finally, a myxobacterium (*Myxococcus*) multiplies and attacks residual *M. luteus* cells but is, in turn, attacked by *E. adhaerens*. *E. coli* is not attacked directly by *E. adhaerens* in soil. Instead, the *E. coli* cells activate germination of the myxobacterium microcysts so that *E. coli* is destroyed, and this is followed by *E. adhaerens* attack on the myxobacterium (Germida and Casida, 1983; Liu and Casida, 1983).

E. adhaerens can also activate *Bacillus* spores (Mormak and Casida, 1985). In laboratory cultures, *E. adhaerens* cells promote a reversible germination of *B. subtilis* spores that does not proceed beyond the activation stage. Indigenous *E. adhaerens*-like bacteria promote the same type of limited spore germination in unamended soil. However, spores associated with *E. adhaerens*-like bacteria in soils that have been nutritionally enriched by incubation with ground alfalfa progress further in the germination process, and some of them may germinate fully, as implied by the eventual appearance of ghosted vegetative cells. There is no direct evidence that the indigenous *E. adhaerens*-like bacteria attack the germinated spores in nutritionally enriched soil, and cells of *E. adhaerens* do not attack vegetative cells of *B. subtilis* in laboratory cultures. Mormak and Casida (1985) have suggested that because small quantities of dipicolinic acid, calcium, and certain amino acids are released during spore activation, *E. adhaerens* and other soil bacteria may utilize these compounds for small amounts of growth or for maintaining cell viability.

E. adhaerens attaches to other bacteria in soil, but it also attaches to other bacteria (or does not become detached) during the preparation of dilutions of soil for plating of other soil organisms. Therefore, it may be in the colonies of the other soil organisms, and it may not be apparent that this is the case. *E. adhaerens* is difficult to remove from cultures of other bacteria unless the other organisms will grow on media containing 4% NaCl. *E. adhaerens* does not grow in the presence of 4% NaCl.

Germida et al. (1985) have reported that *E. adhaerens* can be used to bioassay for plant-available manganese in soils. The assay is specific for manganese, and the results correlate significantly with plant yields.

ENRICHMENT AND ISOLATION PROCEDURES

Variations of two procedures can be used for enrichment and isolation of *E. adhaerens*. In the first procedure, washed *M. luteus* cells, or cells of some other host organism, are added to natural soil, followed by the addition of enough water to adjust the soil water content to 50–60% of the soil's moisture-holding capacity. Alternatively, the cells are mixed with soil and sand in a soil percolation device (Germida and Casida, 1983), and water is percolated through the column of soil. Incubation is for 4–6 days at 25–27°C. After incubation, dilutions of soil are plated on desoxycholate agar or MacConkey agar. The dilutions of soil may also be spread carefully over the surfaces of pregrown lawns of *M. luteus* (on one-tenth strength heart infusion agar, with 1.5% agar), after which the *M. luteus* plates are incubated further. (See previous description for the appearance of the resulting *E. adhaerens* colonies.)

In the second isolation procedure, *M. luteus* cells, or cells of some other host organism, are applied as a smear to a sterile glass slide. The cells are allowed to become just dry at room temperature. They are then immediately placed in contact with natural soil. The slides can be partially buried in soil, either outdoors or in the laboratory. They are placed vertically in the soil, so that only the top 2 cm (approximately) protrudes. Water is added to the soil at intervals to maintain a range of 40–65% of moisture-holding capacity. Alternatively, a sterile glass ring 11 mm high × 25 mm in diameter (e.g., as cut from a test tube) may be placed on the smear and filled with soil. The soil is tamped lightly to ensure direct contact with the smear. Moisture-holding capacity is adjusted to 65%, and incubation is at 27°C in sterile Petri plates. Additional water is added as needed. The incubation time for the *M. luteus* host cells is 3–4 d. After completion of the incubation of these slides, or of buried slides, the soil is gently removed to expose the area of the slide surface where the *M. luteus* smear was placed. An inoculating loop that has been heated and plunged into agar while still hot is touched against this area of the slide. The loop is then streaked through a lawn of *M. luteus* cells or onto the surfaces of plates of desoxycholate or MacConkey agar without bacterial lawns.

MAINTENANCE PROCEDURES

E. adhaerens survives well on refrigerated slants of heart infusion agar made up at one-tenth strength (with 1.5% agar). It can also be lyophilized by common procedures used for aerobes.

DIFFERENTIATION OF THE GENUS *ENSIFER* FROM OTHER GENERA

The genus *Ensifer* is separated from other genera of aerobic, motile, nonphotosynthetic, Gram-negative rods by its method of multiplication. It reproduces by budding at one pole of the cell, with the bud then elongating to produce asymmetric polar growth. Separation of the cells occurs by binary fission, after which growth resumes immediately as new buds at the newly formed poles of both the mother and daughter cells.

In addition to the above, *E. adhaerens* attaches with one of its poles to various species of Gram-positive and Gram-negative bacteria. If enough *E. adhaerens* cells are present, they will position themselves side by side in a picket fence arrangement around the host bacterium. *M. luteus* is a good host organism for demonstrating this behavior. Depending on the species of the host involved in the interaction, *E. adhaerens* may proceed to kill and lyse the host if the pH and background nutritional level are suitable.

E. adhaerens can be differentiated, or even removed, from some other genera of bacteria by using a medium containing 4% NaCl. *E. adhaerens* does not grow in the presence of 4% NaCl but can grow in the presence of 2% NaCl.

The 16S rDNA sequence for *E. adhaerens* differs from those of all species of *Sinorhizobium* (the most closely related genus according to phylogenetic analyses) at positions 658, 659/746, 747, and 1012/1017, and from those of all but one species of *Sinorhizobium* at positions 1010/1019 and 1011/1018 (see below for details). The segment of the *E. adhaerens* 16S rDNA sequence from positions 1009 through 1021 did not match any bacterial sequences in public databases (as of June, 2003) and, therefore, may be unique to *Ensifer*.

TAXONOMIC COMMENTS

E. adhaerens-like bacteria have been isolated from several soils. These bacteria have the characteristics of *E. adhaerens*, but do not demonstrate a cross-reaction with strains A or 7A when they

are examined by phage typing. These *E. adhaerens*-like bacteria are considered to be strains of *E. adhaerens*, with the lack of cross-reaction being due to strain specificity of the particular bacteriophages that have been used.

16S rDNA sequences have been determined for two strains of *E. adhaerens*: ATCC 33212[T] and ATCC 33499 (Balkwill, unpublished data; GenBank accession numbers AF191739 and AF191738, respectively). Phylogenetic analyses of these sequences with distance matrix (Fig. BXII.α.136), parsimony, and maximum likelihood methods indicated that *E. adhaerens* should be placed in the *Rhizobiaceae*. Within this family, *E. adhaerens* is most closely related to the genus *Sinorhizobium*. Similarities between the sequences for *E. adhaerens* and those for validly published species of *Sinorhizobium* are high (98.2% to 99.0%). On the other hand, the *Sinorhizobium* and *Ensifer* strains are always assigned to distinct (but closely related) clusters within the phylogenetic tree (see Fig. BXII.α.136), and the node at which the branching between these two clusters occurs is resolved consistently regardless of the method used to analyze the sequences. A number of consistent differences between the *Ensifer* and all available *Sinorhizobium* 16S rDNA sequences were also noted. These were located at *E. coli* positions (Brosius et al., 1978) 658

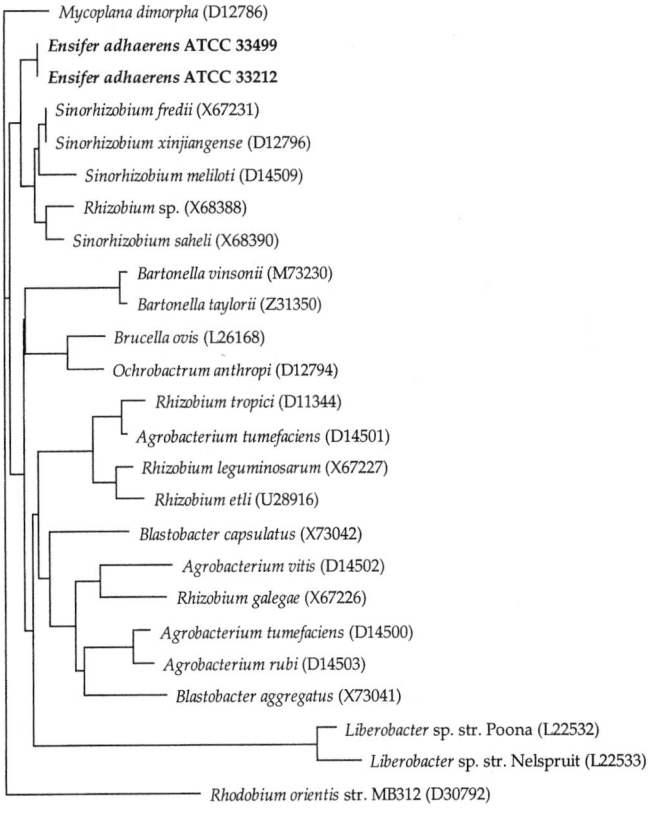

FIGURE BXII.α.136. Phylogenetic tree for two strains of *Ensifer adhaerens* and selected strains of *Bacteria*, based on a distance matrix analysis. Bar = 2 substitutions per 100 bases. *Rhodobium orientis* was used as the outgroup. Parsimony and maximum likelihood analyses yielded virtually identical trees with respect to the positions of *Ensifer* and *Sinorhizobium* and the branching patterns within the cluster containing strains of these two genera. The numbers in parentheses are the GenBank/EMBL accession numbers for the sequences used in the analysis. The GenBank accession numbers for the *Ensifer* sequences are as follows: *E. adhaerens* ATCC 33212[T], AF191739; *E. adhaerens* ATCC 33499, AF191738.

(A in *Ensifer* vs. C in *Sinorhizobium*), 659/746 (T–A pair vs. C–G), 747 (T vs. G), and 1012/1017 (G–C vs. A–T). Additional differences between the *Ensifer* sequences and those for all *Sinorhizobium* species except *Sinorhizobium terangae* were detected at positions 1010/1019 (A–T in *Ensifer* vs. C–G in *Sinorhizobium*) and 1011/1018 (C–G vs. G–C).

The close relatedness of *Ensifer* and *Sinorhizobium* detected by phylogenetic analyses of 16S rDNA sequences, and the high similarities between the 16S rDNA sequences of these organisms, could have implications for the status of *Ensifer* as a separate genus. *Ensifer* is not known to invade root hairs of plants and cause formation of root nodules, but this has not been tested. Similarly, *Sinorhizobium* species are not known to prey on other species of bacteria, but this has not been tested either. In addition, it is not known whether any of the unique traits of *Ensifer* are plasmid-coded and, thus, should not be included in the genus

definition. Designation of *Ensifer* as a distinct genus appears to be justified based on current information. However, it is recommended that a detailed direct comparison of *Ensifer* and *Sinorhizobium* strains be carried out to more accurately assess this situation.

ACKNOWLEDGMENTS

For the most part, this chapter is a minor revision of the original description of *Ensifer* written for the previous edition of the *Manual* by L.E. Casida, Jr. The principal changes include the addition of information from analysis of 16S rDNA sequences and the addition of data from several more recent publications. The current author thanks L.E. Casida Jr. for providing reprints and other information related to his research on *Ensifer adhaerens*. The two strains of *E. adhaerens* for which 16S rDNA sequences were determined were obtained from the American Type Culture Collection.

List of species of the genus Ensifer

1. **Ensifer adhaerens** Casida 1982, 343[VP]
   ad.haer' ens. L. adj. *adhaerens* adherent.

   The characteristics are as described for the genus. Two variants have been isolated from cultures of the type strains. These variants produce less slime, which results in smaller colonies and a drier and slightly whiter appearance of the colonies.

   A strain of *E. adhaerens* that differs from the type strain in its rate of attack on *M. luteus* cells but not in most other characteristics, including phage typing, was isolated from the same soil as strain A. However, the soil had been in-

cubated with added *B. subtilis* spores instead of *M. luteus* cells. This strain (designated strain 7A) has been deposited with the American Type Culture Collection under the number ATCC 33499. Strain 7A can be differentiated from strain A by its growth on desoxycholate agar; strain 7A growth remains purple throughout 10 d and then becomes whitish purple.

   The habitat is soil.

   The mol% G + C of the DNA is: 67 (T_m) and 63 (Bd).

   Type strain: A, ATCC 33212.

   GenBank accession number (16S rRNA): AF191739.

Genus VII. **Sinorhizobium** *Chen, Yan and Li 1988b, 396*[VP] *emend. de Lajudie, Willems, Pot, Dewettinck, Maestrojuan, Neyra, Collins, Dreyfus, Kersters and Gillis 1994, 732*

L. David Kuykendall, Fawzy M. Hashem and En Tao Wang

Si.no.rhi.zo' bi.um. L. n. *sinae* China; Gr. n. *rhiza* a root; Gr. n. *bios* life; M.L. neut. n. *Sinorhizobium* which lives in a root in China; *Rhizobium* isolated in China.

Rods 0.5–1.0 × 1.2–3.0 µm. Commonly pleomorphic under adverse growth conditions. Usually contain granules of polyhydroxybutyrate, which are refractile by phase-contrast microscopy. Nonsporeforming. Gram negative. **Motile** by one polar or subpolar flagellum or one to six peritrichous flagella. Fimbriae have been described on a few strains. **Aerobic, possessing a respiratory type of metabolism with oxygen as the terminal electron acceptor.** Optimal growth temperature, 25–30°C, but can tolerate a wide range of temperatures. Most strains can grow at 35°C; some strains can grow at 10°C, and others (*S. saheli* and *S. terangae*) can grow at temperatures up to 42–44°C (de Lajudie et al., 1994). Tolerate 1.0% NaCl, and some strains grow well on Yeast Mannitol (YM) medium containing 4.5% NaCl. Optimal pH for growth is near neutral (6–8), but some strains can grow at pH 5.0, and others tolerate pH 10.5. Colonies are circular, convex, usually opaque, sometimes translucent, raised, and mucilaginous. **Colonies are usually 2–4 mm in diameter within 3–5 days of growth on yeast mannitol-mineral salts agar** and pronounced turbidity develops after 2–3 days in aerated or agitated broth. Generation times are 3–6 h. Chemoorganotrophic, **utilizing a wide**

range of carbohydrates and salts of organic acids as carbon sources, producing acid without gas formation. Cellulose and starch are not utilized. Produce an acidic reaction in mineral-salts medium containing mannitol. Growth on carbohydrate media is often accompanied by **copious extracellular polysaccharide slime production**. Ammonium salts, nitrate, nitrite, and most amino acids can serve as nitrogen sources. **All strains require pantothenate and nicotinic acid.** Peptone is poorly utilized. Casein and agar are not hydrolyzed. **Oxidase positive.** Catalase positive. 3-Ketolactose not produced from lactose. **The organisms are characteristically able to invade the root hairs of temperate-zone and tropical-zone leguminous plants (family Leguminoseae) and incite production of root nodules wherein the bacteria occur as intracellular nitrogen-fixing microsymbionts.** Strains exhibit varying degrees of host specificity, determined by the type of Nod Factors produced. In root nodules the bacteria occur as endophytes exhibiting pleomorphic forms, termed "bacteroids", which reduce or fix gaseous atmospheric nitrogen into a combined form utilizable by the host plant.

The mol% G + C of the DNA is: 57–66.

Type species: **Sinorhizobium fredii** (Scholla and Elkan 1984) Chen, Yan and Li 1988b, 396 (*Rhizobium fredii* Scholla and Elkan 1984, 484.)

FURTHER DESCRIPTIVE INFORMATION

Granules of poly-β-hydroxybutyrate are common in older cells, so that upon simple staining the rods appear banded. As in the genus *Rhizobium*, all species are motile by one or more flagella. For most species examined that have multiple flagella, the insertion of the flagella is peritrichous, with random placement. Strains in those species with a single flagellum (*Sinorhizobium fredii*, *S. saheli*, *S. terangae*, and *S. xinjiangense*) are monotrichous with polar or subpolar insertion.

Cell wall composition The cell wall structure of *Sinorhizobium* is generally similar to that of other Gram-negative bacteria. The peptidoglycan consists of glutamic acid, alanine, diaminopimelic acid, and amino sugars. In addition, leucine, phenylalanine, serine, and aspartic acid have been detected in relatively large amounts in the peptidoglycan layer of several pathogenic strains. Lipopolysaccharide cell wall composition varies from strain to strain but consistently contains 2-keto-3-deoxyoctanoic acid (KDO), uronic acid, glucosamine, glucose, and galactose (Carlson, 1982).

Other genetic and physiological features of *Sinorhizobium* are generally the same as for *Rhizobium*. See the chapter on *Rhizobium*. In fact, as everyone in this area of specialization knows, much of what is known about *Rhizobium* was learned from studies on *S. meliloti*.

ENRICHMENT AND ISOLATION PROCEDURES

Like other legume-nodulating bacteria, members of the genus *Sinorhizobium* live in soils where legumes occur and may persist for years without host plants. They can be most suitably isolated from the root nodules formed on suitable host plants. Surface sterilization of root nodules is followed by sterile water rinses until the contents can be released into sterile physiological saline and streaked onto a medium with Congo red. Like *Rhizobium*, *Sinorhizobium* colonies remain white as they do not take up the dye.

MAINTENANCE PROCEDURES

Strains of *Sinorhizobium* are usually maintained on YEM medium, containing 20–50% glycerol at either −20°C or −70°C when available. Freeze-drying is also an accepted practice.

DIFFERENTIATION OF THE GENUS *SINORHIZOBIUM* FROM OTHER GENERA

See Tables BXII.α.112 and BXII.α.113 in the chapter on the genus *Rhizobium*. Apart from phenotypic characters that distinguish them as individual species, there are no common distinct characters that differentiate these species as separate from members of the genus *Rhizobium*. The closely related genus *Ensifer* has not been tested for a variety of features unique to legume-nodulating and symbiotic nitrogen-fixing genera including these very traits and its failure to absorb Congo Red. These closely related genera can now be recognized by 16S rDNA sequencing and other molecular methods. *Sinorhizobium* can be differentiated from unrelated genera by cellular fatty acid analysis. Although *Sinorhizobium* and *Rhizobium* generally have qualitatively identical fatty acid composition, there appears to be a quantitative difference in $C_{16:0\ 3OH}$ fatty acid between *Sinorhizobium* and *Rhizobium*, in that the concentration is generally lower in *Sinorhizobium* than in *Rhizobium* (Tighe et al., 2000).

TAXONOMIC COMMENTS

The original proposal of *Sinorhizobium* Chen et al., 1988b, was questioned on the basis that there was little to discriminate this genus from *Rhizobium* (Graham et al., 1991; Jarvis et al., 1992). Subsequently, de Lajudie et al. (1994) proposed an emended description of the genus based on a comparative analysis of 16S rDNA sequences, which showed that *Rhizobium meliloti* and *Sinorhizobium fredii*, together with two new species *Sinorhizobium terangae* and *Sinorhizobium saheli*, formed a separate phylogenetic branch distinct from *Rhizobium* species (*R. leguminosarum*, *R. tropici*). Rome et al. (1996) named a sixth species, *Sinorhizobium medicae*, for strains isolated from various *Medicago* species, together with strains previously recognized as *Rhizobium meliloti* type B (Eardly et al., 1990). Two new species, *S. arboris* and *S. kostiense*, were described recently from legume trees in Sudan and Kenya (Nick et al., 1999). Wang et al. (1999a) reported three new groups within this genus for isolates from *Leucaena leucocephala* in Mexico.

Inferred phylogenies based on comparative analyses of 16S rDNA sequence data show that *Sinorhizobium* species, and *Ensifer adhaerens* (Casida, 1982), a bacterial predator of bacteria in soil, form a clade(s) distinct from *Rhizobium* and *Agrobacterium* species. Apart from phenotypic characters which distinguish them as individual species, and except for the quantitative difference in $C_{16:0\ 3OH}$ fatty acid content described above, there are no common distinct characters which differentiate these species as separate from members of the genus *Rhizobium*. At the time of this writing, perhaps only about one-half of the investigators in the field accept the taxonomic validity of *Sinorhizobium*. This new genus, together with *Ensifer*, will probably not stand if supporting data that provide evidence of distinct taxa cannot be gathered. The present fragile nature of these genera will be further undermined when strains from different soils and hitherto untested legume/*Rhizobium* combinations are found to be intermediate between the genera.

The taxonomic relationships of *Sinorhizobium* are discussed fully and further descriptive information is given in the chapter on *Rhizobium*. Perhaps the ultimate tool of taxonomy, the complete genome sequence for *S. meliloti*, was recently reported by Galibert et al. (2001).

DIFFERENTIATION OF THE SPECIES OF THE GENUS *SINORHIZOBIUM*

See Tables BXII.α.112 and BXII.α.113 in the chapter on the genus *Rhizobium*. Reports of specific phenotypic differences are needed for some species. The species of *Sinorhizobium* can also be differentiated by comparative sequence analysis of 16S rRNA gene sequence data. Fatty acid analysis can also be useful to discriminate *Sinorhizobium* from other related genera (Jarvis et al., 1996).

List of species of the genus Sinorhizobium

1. **Sinorhizobium fredii** (Scholla and Elkan 1984) Chen, Yan and Li 1988b, 396^VP (*Rhizobium fredii* Scholla and Elkan 1984, 484.)

fred' i.i. M.L. gen. n. *fredii* of E.B. Fred.

The characteristics are as described for the genus and as listed in Tables BXII.α.112 and BXII.α.113 in the chapter on the genus *Rhizobium*. Motile by one polar flagellum or by 1–3 peritrichous flagella. Relatively fast growing: colonies are 2–4 mm in diameter after 3–5 days growth on yeast extract mannitol agar. Temperature growth optimum is 25–30°C. Most strains grow at 35°C, and some grow at 10°C. Optimal pH for growth, 6–8; range pH 5.0–10.5. Most strains grow on YMA containing 1.0% NaCl, but a few strains tolerate >3.0% NaCl. Require pantothenate and nicotinate as growth factors. Utilize relatively few organic substrates as sole sources of carbon. Two chemovars were proposed—fredii and siensis—that were differentiated by the ability of siensis to reduce litmus and produce a final pH in growth media greater than 6.1, and its resistance to kanamycin (20 ppm). Later revisions have led to a consolidation of these groups. *S. fredii* can be differentiated from the other *Sinorhizobium* species by auxanographic characteristics, protein SDS-PAGE profile, DNA–DNA reassociation, and comparative analysis of 16S rDNA sequence data.

This species effectively nodulates *Cajanus cajan*, *Glycine max* cv. Peking, *Glycine soja*, and *Vigna unguiculata*. Dowdle and Bohlool (1985) reported new strains that are symbiotically competent with North American cultivars of soybean. Some original strains, including the type strain, were also found to nodulate *Medicago sativa* (Hashem et al., 1997) with efficient symbiotic nitrogen fixation (Kuykendall et al., 1999).

The mol% G + C of the DNA is: 60–64 (T_m).

Type strain: PRC 205, ATCC 35423, ICMP 11139, NCIB 12104, USDA 205.

GenBank accession number (16S rRNA): D01272, X67231.

2. **Sinorhizobium arboris** Nick, de Lajudie, Eardly, Suomalainen, Paulin, Zhang, Gillis and Lindström 1999, 1366^VP

ar' bo.ris. L. fem. n. *arbor* tree; n. *arboris* of the tree.

Short, aerobic, Gram-negative rods that are motile by one or two polar or subpolar flagella. Maximum growth temperature is about 42°C. On YEM medium, most strains produce circular, cream-colored, semi-translucent colonies that become very mucilaginous and often spread over the entire plate within 2–4 d. Utilize a wide range of carbon sources and amino acids for growth. Grow on L-isoleucine and tolerate 3% (w/v) NaCl, heavy metals (but not Al), and antibiotics. Produce melanin and grow at pH 8.5. However this species cannot be identified by physiological features and biochemical traits alone. Nodulates *Acacia sengal* and *Prosopis chilensis*. SDS-PAGE whole protein pattern, MLEE, rep-PCR genomic fingerprinting, RFLP analysis of amplified 16S rRNA, total DNA–DNA hybridization, and 16S rDNA sequence comparison are useful for distinguishing closely related species at the molecular level.

The mol% G + C of the DNA is: 60.6–61.8 (T_m).

Type strain: HAMBI 1552, LMG 14919.

GenBank accession number (16S rRNA): Z78204.

3. **Sinorhizobium kostiense** Nick, de Lajudie, Eardly, Suomalainen, Paulin, Zhang, Gillis and Lindström 1999, 1366^VP

kos.ti.en' se. L. neut. adj. *kostiense* pertaining to Kosti, the region in Sudan where most of these organisms have been isolated.

Short, aerobic, Gram-negative rods that are motile by one or two polar or subpolar flagella. Maximum growth temperature is about 38–40°C for most strains. On YEM medium, most strains produce circular, cream-colored, semi-translucent colonies that become very mucilaginous and often spread over the entire plate within 2–4 d. Utilize a relatively narrow range of carbon sources and amino acids for growth. Tolerate 1% (w/v) NaCl; sensitive to heavy metals except lead and copper. Sensitive to most antibiotics. Produce melanin but fail to grow at pH 5.5 or 8.5. This species cannot be identified by physiological features and biochemical traits alone. Nodulates *Acacia sengal* and *Prosopis chilensis*. SDS-PAGE whole protein pattern, MLEE, rep-PCR genomic fingerprinting, RFLP analysis of amplified 16S rRNA, total DNA–DNA hybridization, and 16S rDNA sequence comparison are useful for distinguishing closely related species at the molecular level.

The mol% G + C of the DNA is: 57.9–61.6 (T_m).

Type strain: HAMBI 1489, LMG 15613.

GenBank accession number (16S rRNA): Z78203.

4. **Sinorhizobium medicae** Rome, Fernandez, Brunel, Normand and Cleyet-Marel 1996, 979^VP

me' di.cae. L. fem. n. *medica* from *medica lucerne* (plant belonging to the genus *Medicago*); L. gen. n. *medicae* of *medica*.

The characteristics are as described for the genus and as listed in Table BXII.α.112 in the chapter on the genus *Rhizobium*. Fast growing: semi-translucent, circular, and mucoid colonies spread over the plate in 3–5 days on yeast extract mannitol agar. Temperature growth optimum, 25–30°C. Most strains grow at 35°C but not at 42°C; 40°C is the maximum temperature for growth. Optimal pH for growth, 6–8; range pH 5.0–10. Grows in media containing 2% NaCl but does not grow well in media containing 3% NaCl. Resistant to nalidixic acid. Differentiated from other *Sinorhizobium* and *Rhizobium* spp. based on differences in 16S rDNA sequence data. Sequence similarities indicate a very close relationship to *S. meliloti* (99.7%).

Strains have been isolated from various *Medicago* species in different geographical sites including southern France and the eastern Mediterranean basin, which are in the center of origin of the genus *Medicago*. Isolated from *Medicago orbicularis*, *M. polymorpha*, *M. rugosa*, and *M. truncatula*. *Sinorhizobium medicae* strains fix nitrogen with *M. polymorpha*, whereas *S. meliloti* forms ineffective nodules.

The mol% G + C of the DNA is: 61.0–63.0 (T_m).

Type strain: A 321, USDA 1037.

GenBank accession number (16S rRNA): L39882.

5. **Sinorhizobium meliloti** (Dangeard 1926) de Lajudie, Willems, Pot, Dewettinck, Maestrojuan, Neyra, Collins, Dreyfus, Kersters, and Gillis 1994, 731^VP (*Rhizobium meliloti* Dangeard 1926, 194.)

me.li.lo' ti. M.L. masc. n. *Melilotus* generic name of sweet clover; M.L. gen. n. *meliloti* of *Melilotus*.

The characteristics are as described for the genus and as listed in Tables BXII.α.112 and BXII.α.113 in the chapter on the genus *Rhizobium*. Motile by 2–6 peritrichous flagella. Variable growth on YMA containing 2% NaCl. Variable growth at 39–40°C. Does not require thiamine or pantothenate for growth; some strains require biotin. pH range for growth, 5.0–9.5. Utilizes a relatively wide range of organic substrates as sole sources of carbon. Distinguished from other *Sinorhizobium* and *Rhizobium* species by DNA–DNA reassociation data, differences in protein profiles obtained by SDS-PAGE, and 16S rDNA sequence comparison. Sequence similarities indicate a close relationship to *S. medicae*. Two subpopulations have been distinguished by DNA hybridization and multilocus enzyme electrophoresis.

Forms nitrogen-fixing nodules on *Melilotus*, *Medicago*, and *Trigonella*. Gao and Yang (1995) reported a Chinese *Sinorhizobium meliloti* strain that nodulated and fixed nitrogen in association with both alfalfa and soybean. Well characterized genetically. Symbiosis-controlling genes are carried on megaplasmids. In addition, strains harbor various numbers (0–4) of large pRme plasmids (90–500 kb) (Boivin et al., 1997).

The mol% G + C of the DNA is: 62–63 (T_m).

Type strain: ATCC 9930, ICMP 12623, LMG 6133, USDA 1002.

GenBank accession number (16S rRNA): X67222.

6. **Sinorhizobium saheli** de Lajudie, Willems, Pot, Dewettinck, Maestrojuan, Neyra, Collins, Dreyfus, Kersters, and Gillis 1994, 732[VP]

sa'hel.i. N.L. gen. n. *saheli* of the Sahel, the region in Africa from which they were isolated.

The characteristics are as described for the genus and as listed in Tables BXII.α.112 and BXII.α.113 in the chapter on the genus *Rhizobium*. Motile by one or more polar or subpolar flagella. Colonies can be mucoid and spreading on yeast extract mannitol agar. Old colonies become brown. Can grow at 44°C. Utilizes a range of organic substrates as sole sources of carbon (Table BXII.α.113 in the chapter on the genus *Rhizobium*). Distinguished from other *Sinorhizobium* and *Rhizobium* species by DNA–DNA reassociation data, differences in protein profiles obtained by SDS-PAGE, and 16S rDNA sequence comparison. Haukka et al. (1998) reported two different sequences for a 230-nucleotide segment of the 16S ribosomal RNA gene in the type strain of *S. saheli*.

Nodulates different *Sesbania* species (*S. cannabina*, *S. grandiflora*, *S. rostrata*, *S. pachycarpa*) growing in the Sahel; in addition to these species, strains can nodulate *Acacia seyal*, *Leucaena leucocephala*, and *Neptunia natans*. Can form effective stem nodules on *Sesbania rostrata* when plants are not previously nodulated on their roots (Boivin et al., 1997).

The mol% G + C of the DNA is: 65–66 (T_m).

Type strain: ORS 609, DSM 11273, ICMP 13648, LMG7837.

GenBank accession number (16S rRNA): X68390.

7. **Sinorhizobium terangae** de Lajudie, Willems, Pot, Dewettinck, Maestrojuan, Neyra, Collins, Dreyfus, Kersters, and Gillis 1994, 732[VP] (*Sinorhizobium teranga* (sic) de Lajudie, Willems, Pot, Dewettinck, Maestrojuan, Neyra, Collins, Dreyfus, Kersters, and Gillis 1994, 732.)

te'ran.gae. N.L. gen. n. *terangae* of *teranga* hospitality, in the language of West African Wolof people; referring to the isolation of this species from different host plants.

The characteristics are as described for the genus and as listed in Tables BXII.α.112 and BXII.α.113 in the chapter on the genus *Rhizobium*. Motile by one or several polar or subpolar flagella. Colonies are mucoid and spreading on yeast extract mannitol agar. Old colonies become brown. Can grow at 44°C. Utilizes a range of organic substrates as sole sources of carbon (Table BXII.α.113 in the chapter on the genus *Rhizobium*). Differentiated from other *Sinorhizobium* and *Rhizobium* species by DNA–DNA reassociation data, differences in protein profiles obtained by SDS-PAGE, and 16S rDNA sequence comparison.

Nodulates *Acacia* species (*A. senegal*, *A. laeta*, *A. tortilis* subsp. *raddiana*, *A. horrida*, *A. mollissima*) and *Sesbania* species (*S. rostrata*, *S. cannabina*, *S. aculeata*, *S. sesban*). Two biovars were distinguished based on the host range of strains (Lortet et al., 1996): *S. terangae* biovar sesbaniae for strains able to nodulate *Sesbania* species (*S. grandiflora*, *S. pubescens*, and *S. rostrata*) but not *Acacia* species (*A. senegal*, *A. tortilis* subsp. *raddiana*, and *A. nilotica*) nor *Leucaena leucocephala*; *S. terangae* biovar acaciae for strains able to nodulate *Acacia* species (*A. senegal*, *A. tortilis* subsp. *raddiana*, and *A. nilotica*) and *Leucaena leucocephala* but not *Sesbania* species (*S. grandiflora*, *S. pubescens*, and *S. rostrata*).

The mol% G + C of the DNA is: 60.8–61.6 (T_m).

Type strain: ORS 1009, DSM 11282, ICMP 13649, LMG 7834.

GenBank accession number (16S rRNA): X68391.

8. **Sinorhizobium xinjiangense** Chen, Yan and Li 1988b, 396[VP] (*Sinorhizobium xinjiangensis* Chen, Yan and Li 1988b, 396.)

xin.jian.gen'se. M.L. adj. *xinjiangense* pertaining to the suburbs of Xinjiang, China.

The characteristics are as described for the genus and as listed in Table BXII.α.112 in the chapter on the genus *Rhizobium*. Motile by one polar flagellum or by 1–3 peritrichous flagella. Relatively fast growing: colonies are 2–4 mm in diameter after 3–5 days growth on yeast extract mannitol agar. Temperature growth optimum, 25–30°C. Most strains grow at 35°C and some grow at 10°C. Optimal pH for growth, 6.0–8.0; range, 5.0–10.5. Does not grow on YMA containing 1.5% NaCl, but some strains tolerate 4.5% NaCl.

The species was proposed based on nutritional and biochemical differences from *R. leguminosarum*, *S. fredii*, *S. meliloti*, *Agrobacterium tumefaciens*, and *A. rhizogenes*. 16S rDNA sequence similarities indicate a close relationship to *S. fredii*. The species differs from *S. fredii* in that, unlike *S. fredii*, *S. xinjiangense* cannot grow at low and high pH (<5.5 and >8.5 respectively). All strains of *S. xinjiangense* produce acid in litmus milk, and are sensitive to a number of antibiotics such as vancomycin (25 μg/ml), chloramphenicol (125 μg/ml), penicillin (25 μg/ml), and streptomycin (5 μg/ml). Isolated from *Glycine max*.

The mol% G + C of the DNA is: 60–64 (T_m).

Type strain: ATCC 49357, CCBAU 110, IAM 14142, ICMP 11141.

GenBank accession number (16S rRNA): D12796.

Family II. **Bartonellaceae** Gieszczykiewicz 1939, 25[AL]

DAVID F. WELCH

Bar.to.nel.la' ce.ae. M.L. fem. n. *Bartonella* type genus of the family; *-aceae* ending to denote family; M.L. fem. pl. n. *Bartonellaceae* the *Bartonella* family.

Small coccobacilli, may be beaded or filamentous and <3 μm in their greatest diameter. Erythrocytic forms stain lightly with many aniline dyes but distinctly with Giemsa's stain after methanol fixation. Gram negative, not acid-fast. Some species have polar flagella. Cultivable, but highly fastidious, *in vitro*, on blood-enriched bacteriological media. Arthropod transmission has been established. Etiological agents demonstrable in diverse clinical material from humans and erythrocyte-associated bacteria of other vertebrates. Cause bartonellosis, cat scratch disease, bacillary angiomatosis, endocarditis, trench fever, and a spectrum of other human inflammatory lesions.

On the basis of 16S rRNA gene sequence analysis, the family *Bartonellaceae* is in the *Alphaproteobacteria*. The genus *Bartonella* was removed from the order *Rickettsiales* and the family *Rickettsiaceae*, resulting in a single genus within the family *Bartonellaceae* (Brenner et al., 1993). The genus *Rochalimaea*, then contained within the family *Rickettsiaceae*, was at the same time combined with *Bartonella*, and subsequently *Grahamella* was merged with *Bartonella* (Birtles et al., 1995).

Type genus: **Bartonella** Strong, Tyzzer and Sellards 1915, 808 emend. Birtles, Harrison, Saunders and Molyneux 1995, 7.

Genus I. **Bartonella** Strong, Tyzzer and Sellards 1915, 808[VP] emend. Birtles, Harrison, Saunders and Molyneux 1995, 7

DAVID F. WELCH

Bar.to.nel' la. M.L. dim. ending *-ella*; M.L. fem. dim. n. *Bartonella* named after Alberto L. Barton, who described these organisms in 1909, after studying the agent of Carrion's disease.

Morphology is small (0.5–0.6 × 1.0 μm), **slightly curved bacillus**. Not acid-fast. **Faintly stains Gram negative** (stains poorly or not at all with many aniline dyes) but satisfactorily with Romanowsky's or Giemsa's stain. Warthin-Starry silver staining of fixed tissue sections reveals bacilli in clusters. May be seen in stained blood films appearing as rounded or ellipsoidal forms or as slender, straight, curved, or bent rods, occurring singly or in groups. They characteristically occur in chains of several segmenting organisms, sometimes swollen at one or both ends and frequently beaded. In the tissues, they are situated **within the cytoplasm of endothelial cells** as isolated elements or are grouped in rounded masses. **Intraerythrocytic forms occur** in the blood of felines, small rodents, birds, fish, and other animals. In cultures, the cells may be very **autoadherent**. Some species possess unipolar flagella. **The presence of pili** is associated with the marked adherence and may mediate specific interaction with host endothelial cells and erythrocytes leading to intracellular localization.

Aerobic but **highly fastidious**. May be cultivated on media enriched with blood components in the presence of air or 5% CO_2. **Growth occurs at 20–37°C after prolonged incubation** (7–21 d). No growth on MacConkey or nutrient agars. Cocultivation with an endothelial cell line can also be performed and this method may be more successful in recovering organisms from specimens such as tissue. There are **numerous reservoirs and vectors** for *Bartonella* spp. The organisms are transmitted by arthropod vectors (*Lutzomyia verrucarum*, *Pediculus humanus*, *Ctenocephalides felis* [*Siphonaptera pulicidae*], and possibly ticks); *B. bacilliformis* is found only in the Andes region of South America. Etiological agents of human bartonellosis, cat scratch disease, bacillary angiomatosis, peliosis hepatis, trench fever, endocarditis, and neuroretinitis. **Six species cause *Bartonella*-associated infectious diseases in humans.** The organisms are usually catalase and oxidase negative, aerobic, and they do not produce acid from carbohydrates. *Bartonella* spp. have relatively simple gas–liquid chromatography profiles consisting mainly of $C_{18:1}$, $C_{18:0}$, and $C_{16:0}$ acids. A portion of the 16S–23S rRNA intergenic spacer region can be targeted with primers that can distinguish *Barto-*

nella from other genera including those closely related within the *Alphaproteobacteria*.

The mol% G + C of the DNA is: 37–41.

Type species: **Bartonella bacilliformis** (Strong, Tyzzer, Brues, Sellards and Gastiaburu 1913) Strong, Tyzzer and Sellards 1915, 808 (*Bartonia bacilliformis* Strong, Tyzzer, Brues, Sellards and Gastiaburu 1913, 1715.)

FURTHER DESCRIPTIVE INFORMATION

Stained blood films have been used to detect intraerythrocytic *B. bacilliformis* in patients with Oroya fever. Bacilli may be demonstrated by use of the Warthin-Starry silver stain during the early stages of lymphadenopathy in cat scratch disease due to *B. henselae*, but typically not during the later granulomatous stage of inflammation. When seen in blood smears, *B. bacilliformis* consists of small, polymorphic forms. The maximum morphological range is seen in the blood of man, where the organisms appear as red–violet rod or coccal forms situated on or in the red cells when stained with Giemsa's stain. Bacilliform bodies are the most typical, measuring 0.25–0.5 × 1.0–3.0 μm. The cells are often curved and may show polar enlargement and granules at one or both ends. Rounded organisms measure ~0.75 μm in diameter and a ring-like variety is sometimes abundant. By light microscopy or by "stripping" in the pseudoreplica technique for electron microscopy (Peters and Wigand, 1955), the organisms appear to be situated on the surface of erythrocytes; however, they have also been reported to occur within erythrocytes in thin sections observed by electron microscopy (Cuadra and Takano, 1969). The Gram stain of a colony from solid media reveals small, Gram-negative, slightly curved rods resembling *Campylobacter*, *Helicobacter*, or *Haemophilus*. Cells, especially of *B. henselae*, are very autoadherent, as can be demonstrated by attempting to scrape colonies off a culture plate with a loop. In semisolid media, a mixture of rods and granules appears. The organisms may occur singly or in large and small, irregular, dense collections measuring up to 25 μm or more in length. Punctiform, spindle-shaped, and ellipsoidal forms that vary in size from 0.2–0.5 $\times$ 0.3–3.0 μm also occur.

The organisms have a trilaminar cell wall, the formation of which can be inhibited by penicillin. The cellular fatty acid composition among *Bartonella* spp. is relatively simple, compared to that of many other Gram-negative bacteria, in that they have gas–liquid chromatography profiles consisting mainly of $C_{18:1}$, $C_{18:0}$, and $C_{16:0}$ acids. *B. elizabethae* and *B. vinsonii* contain a greater amount of $C_{17:0}$ than the other species. *B. bacilliformis* contains a significant amount of $C_{16:1}$. An unusual branched-chain fatty acid (11-methyloctadec-12-enoic acid) is found in *Afipia* spp. but not in *Bartonella* spp., and *Brucella* spp. contain a relatively large amount of $C_{19:0}$ cyclopropane acids. In culture, the cells of two species (*B. bacilliformis* and *B. clarridgeiae*) possess a tuft of 1–10 unipolar flagella (Peters and Wigand, 1955; Lawson and Collins, 1996a). Other species, especially *B. henselae*, *B. alsatica*, and *B. tribocorum*, have pili. Phenotypically similar to type 4 pili, they mediate adherence to and entry into human epithelial cells (Batterman et al., 1995). Flagella have not been demonstrated in tissues, but the pili have been.

Bartonella spp. demonstrate various colonial morphologic types, from clear (*B. bacilliformis*) to smooth white (*B. quintana* and *B. alsatica*) to rough tan (*B. henselae* and *B. tribocorum*). Several species embed in the agar. Colonies of *B. henselae* are typically of two types: (i) irregular, raised, whitish, rough (cauliflower or molar tooth or verrucous), and dry in appearance or (ii) smaller, circular, tan, and moist in appearance, tending to pit and adhere to the agar. Both types are usually present in the same culture (Fig. BXII.α.137). The degree of colonial heterogeneity varies by species and by strain, and is probably related to the degree of piliation. Repeated subcultures of *B. henselae* tend to have increasing proportions of smooth colonies. *B. quintana* may appear as uniformly smooth colonies in primary cultures. Cultures of *B. henselae* on blood agar may produce an odor similar to the caramel odor (diacetyl) produced by *Streptococcus milleri*.

The majority of isolates require more than 7 days of incubation before they can be detected by culture. A source of blood or hemoglobin in the medium is necessary in most cases. *B. bacilliformis* may be cultivated in semisolid agar containing fresh rabbit serum and rabbit hemoglobin or containing the blood of human, horse, or rabbit, with and without the addition of fresh tissue and carbohydrates. It may also be cultivated in other culture media containing blood, serum, or plasma, in Huntoon's hormone agar, in semisolid gelatin media, and in blood–glucose–cysteine agar. It can also be grown in certain tissue cultures and in the chorioallantoic fluid and yolk sac of the chicken embryo. The other species can be recovered on blood or chocolate agar media as well.

Bartonella spp. are typically inert biochemically. In conventional carbohydrate utilization tests no acid or gas production occurs from amygdalin, L-arabinose, dextrin, dulcitol, fructose, D-galactose, D-glucose, inulin, lactose, maltose, D-mannitol, D-mannose, raffinose, L-rhamnose, salicin, sucrose, trehalose, or D-xylose. Gelatin is not liquefied. Esculin is not hydrolyzed. H_2S is not detected with lead acetate.

Bartonella is a monophyletic genus within the *Alphaproteobacteria*. Species of *Bartonella* are animal cell associated but closely related to the plant cell associated genera, *Agrobacterium* and *Rhizobium*. Common pathogenic domains necessary for invasion and survival in association with cells have been preserved in the chromosomes of both the animal- and plant-associated *Alphaproteobacteria*. These have evolved through reductions of the larger genomes, including a second chromosome in some instances, of chemoautotrophic ancestors (Moreno, 1998). There is also a high level (~95%) of similarity between *Bartonella* and *Brucella* according to 16S rRNA gene sequence analysis (Brenner et al., 1993). Phylogenetic analysis of *Bartonella* species can be done based on either 16S rRNA or citrate synthase (*gltA*) gene sequencing. The latter has been compared to 16S rRNA gene sequence analysis and some investigators (Birtles and Raoult, 1996) believe the *gltA*-derived phylogeny is more useful than the phylogeny derived from 16S rDNA sequence data for investigating the evolutionary relationships of *Bartonella* species. The *gltA* method may result in amplification of two products of differing

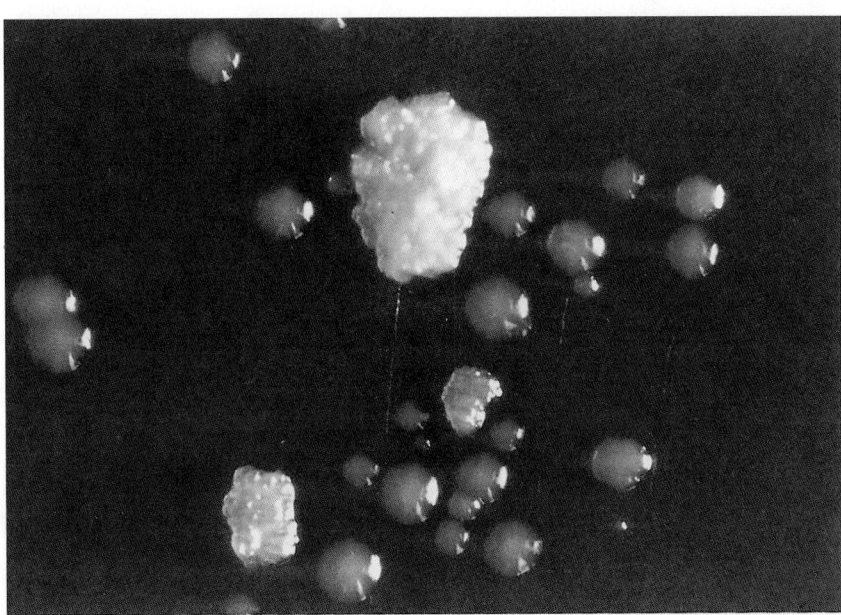

FIGURE BXII.α.137. Colonies of *Bartonella henselae* (40× magnification), showing heterogeneous mixture of smooth and rough types.

sizes from *B. bacilliformis* (Birtles, 1995). Limited work with heat shock protein (groEL) genes of some *Bartonella* species has also been conducted, leading to similar phylogenetic conclusions (Haake et al., 1997).

A portion of the 16S rDNA consisting of 241 nucleotides can be amplified using primers p24E and p12B (Relman et al., 1990a). Signature sequences, e.g., a 3 base pair difference between *B. quintana* and *B. henselae* (Koehler et al., 1992), occur within this fragment.

A genus-specific primer set targeting the internal portions of the 16S–23S intergenic spacer region can distinguish *Bartonella* species from other *Alphaproteobacteria*, including *Brucella abortus*, *Agrobacterium tumefaciens*, and *Rhizobium meliloti* (Minnick and Barbian 1997). These investigators also developed species-specific primers. Distinguishing among isolates of the same species using DNA fingerprinting methods has been applied mainly with *B. henselae*. Pulsed-field gel electrophoresis (PFGE), 16S rRNA type-specific PCR, enterobacterial repetitive intergenic consensus (ERIC)-PCR, repetitive extragenic palindromic (REP) PCR, and arbitrarily primed (AP)-PCR have all been used to clearly distinguish several subtypes (Sander et al., 1998). These methods provide useful epidemiological tools and demonstrate a high degree of genetic heterogeneity.

FtsZ proteins (the highly conserved cell division protein) of *B. bacilliformis*, *B. henselae*, and *B. quintana* are about twice as large as the FtsZ proteins reported in most other organisms. The amino acid sequences of *B. henselae* and *B. quintana* ftsZ are 81–83% identical to the corresponding protein in *B. bacilliformis*, suggesting that localized differences within the sequence of the *Bartonella* ftsZ genes may be used as the basis for species-specific identification (Kelly et al., 1998b). *B. henselae* harbors a bacteriophage that has been cloned and sequenced (Anderson et al., 1997; Bowers et al., 1998). Associated with the bacteriophage is a 31-kDa major protein (Pap31) that is targeted to the cell membrane. *B. vinsonii* subsp. *berkhoffii* contains a 12-base insertion sequence in the 16S rRNA gene that is not present in other *Bartonella* species (Kordick et al., 1996).

The FtsZ protein is antigenic in *B. bacilliformis* and in *B. henselae* due to its partial exposure at the cell surface (Padmalayam et al., 1997). Antigenicity is species and strain specific as indicated by the lack of cross-protection between *B. henselae* and *B. clarridgeiae* and the lack of protection between *B. henselae* types I, II, and a wildlife strain (Yamamoto et al., 1998). On the other hand, serological cross-reactions do occur between species and with the genera *Chlamydia* and *Bartonella*. Maurin et al. (1997) reported eight patients originally diagnosed as having *Chlamydia pneumoniae* endocarditis whose sera reacted with *B. quintana* antigens in a microimmunofluorescence technique. Adsorption of sera with *B. quintana* or *C. pneumoniae* antigens removed anti-*C. pneumoniae* antibodies, whereas adsorption with *C. pneumoniae* antigens did not change antibody titers to *B. quintana*. There is also cross reactivity between *B. henselae* and *Coxiella*. La Scola and Raoult (1996b) demonstrated that >50% of chronic Q fever patients they studied had antibodies that reacted significantly against *B. henselae* antigen.

Agglutination of suspensions of *B. bacilliformis* by sera from convalescent patients has been reported. Immune sera fix complement in the presence of the organisms. When various isolates have been employed, no significant titer differences have been found in quantitative tests. Immune rabbit sera do not agglutinate *Proteus* strains OX19, OX2, or OXK at titers above 1:20. Serodiagnosis of cat scratch disease by indirect immunofluorescence or enzyme immunoassay is a practical technique and carries a sensitivity of approximately 85% and a specificity of approximately 95%. IgM reactivity is directed to an 8-kDa band and IgG to 209-, 208.5-, 208-, 116-, and 80-kDa bands by Western blot analysis of *B. henselae* serum samples (Litwin et al., 1997).

Bartonella spp. generally display high levels of *in vitro* susceptibility to antibacterial agents. Agar dilution testing shows resistance only to fosfomycin, colistin, and vancomycin (Maurin et al., 1995). *B. bacilliformis* is resistant *in vivo* to neosalvarsan and to other arsenical compounds in general. It is sensitive to penicillin, streptomycin, chloramphenicol, and oxytetracycline. Oral chloramphenicol is the standard treatment for *B. bacilliformis* infection. When grown with penicillin, the organism produces L forms (Sharp, 1968). *B. henselae*, *B. quintana*, and *B. elizabethae* are susceptible to the macrolides azithromycin, clarithromycin, dirithromycin, erythromycin, and roxithromycin based on testing in Vero cell cultures (Ives et al., 1997). Agents that should be considered in therapy of cat scratch disease include azithromycin or rifampin. In the initial antimicrobial treatment of hepatosplenic cat-scratch disease in children, rifampin therapy, initiated alone or in combination, results in improvement within 1–5 days (Arisoy et al., 1999). A placebo-controlled clinical study showed that azithromycin caused an 80% decrease in initial lymph node volume in 7 of 14 azithromycin-treated cat scratch disease patients during the first 30 days of observation (Bass et al., 1998). Erythromycin, doxycyline, or ciprofloxacin is recommended for therapy of bacillary angiomatosis.

B. bacilliformis, *B. quintana*, *B. henselae*, and *B. elizabethae* are the primary pathogenic species in humans. *B. clarridgeiae*, originally isolated from a domestic cat, may cause lymphadenopathy in humans (Kordick et al., 1997). *B. vinsonii* is potentially pathogenic, as documented by a human isolate of *B. vinsonii* subsp. *arupensis* (Welch et al., 1999). *B. henselae* and *B. clarridgeiae* were discovered during the 1990s, largely in the course of testing for opportunistic infections in human immunodeficiency virus (HIV)-infected individuals. In recent years, *B. quintana* bacteremic infection (trench fever) in the absence of HIV has been identified sporadically, mainly in homeless persons in North America and Europe, but it was a major cause of morbidity during WW I. There is archeological evidence for pre-Colombian existence of bartonellosis due to *B. bacilliformis* in South America. The spectrum of diseases caused by *Bartonella* spp. other than *B. bacilliformis* includes trench fever, cat scratch disease, bacillary angiomatosis, bacillary peliosis, relapsing bacteremia, endocarditis, lebers neuroretinitis, and aseptic meningitis (Wong et al., 1995a; Raoult et al., 1996).The mechanism of pathogenesis by which *Bartonella* spp. cause disease is not fully understood. A central feature is an effect that causes proliferation of microvascular endothelial cells and neovascularization (angiogenesis). Surface-exposed determinants such as adhesins and other outer membrane proteins are of substantial importance in the pathogenesis (Burgess and Anderson, 1998). A process of adherence, followed by invasion of feline erythrocytes, is involved in establishing long-term bacteremia in cats (Mehock et al., 1998). Intracellular localization of *B. quintana* has been demonstrated *in vivo* and within endothelial cell cultures (Brouqui and Raoult, 1996). Upon engulfment by human endothelial cells, the organisms appear in vacuoles similar to the morulae seen in ehrlichiae- or chlamydiae-infected cells. *B. vinsonii* subsp. *vinsonii*, *B. grahamii*, *B. taylorii*, and *B. doshiae* were isolated from small wild mammals. *B. vinsonii* subsp. *berkhoffii* was isolated from a dog with endocarditis and *B. vinsonii* subsp. *arupensis* was isolated from naturally infected mice and a man with a febrile illness. Human bartonellosis is a classically biphasic disease caused by *B. bacilliformis* and manifested as a progressive anemia (Oroya fever) followed by a cutaneous eruption (Verruga Peruana). While typical bartonellosis has remained endemic for the past century in highland provinces near the Peruvian border, there have recently been an increasing number of atypical cases in which monophasic verrucous cutaneous disease is the only clinical manifestation (Amano et al., 1997). These cases consist of mild clinical disease, possibly associated with less virulent bacterial strains.

The transmission of Oroya fever is dependent on the ecology of the sand fly (*Phlebotominae*: *Lutzomyia verrucarum*) vector and is therefore typically confined to an elevation of 2500–8000 feet above sea level in a band less than 100 miles wide and ~1000 miles long on the western slopes of the Andes mountains in Peru, Ecuador, and Colombia. However, the atypical cases tend to now be disseminating or re-emerging in previously disease-free areas.

Experimental Oroya fever has not been successfully produced in animals, except rarely in an atypical form in monkeys. Experimental Verruga Peruana has been produced in man and in a number of species of monkeys. The other species of *Bartonella* are less geographically confined. The vector of *B. quintana* is *Pediculus humanus*, the human body louse. Outbreaks of trench fever (also known as Volhynia fever, Meuse fever, His-Werner disease, shinbone fever, shank fever, and quintan or five-day fever) therefore occur focally, often associated with conditions of poor sanitation and personal hygiene that may predispose to exposure to lice. Nonhuman vertebrate reservoirs have not been identified for *B. bacilliformis*, *B. quintana*, or *B. elizabethae*. No arthropod vector has been identified for *B. elizabethae*. *B. henselae* or *B. clarridgeiae* can induce chronic infection in specific-pathogen-free cats (Guptill et al., 1998, Kordick et al., 1999). There may be minimal clinical signs but histological changes occur and *Bartonella* DNA can be detected in tissues, supporting an etiologic role for *Bartonella* species in idiopathic diseases of cats. *B. henselae* also causes reproductive failure in female cats, but has not been found to be vertically transmitted. In contrast, evidence does suggest the possibility of vertical transmission of *Bartonella* spp. among natural rodent hosts (Kosoy et al., 1998). Stray cats are a major reservoir of *Bartonella* spp., which can be transmitted to pet cats and, consequently, to humans (Heller et al., 1997). It has been demonstrated that the cat flea readily transmits *B. henselae* to cats (Chomel et al., 1996). Cats are also the reservoir for *B. clarridgeiae* and *B. koehlerae*. Mice, voles, and rats carry subspecies of *B. vinsonii*. *Bartonella* can be detected in various rodent populations, most of which are also reservoirs of tick-borne pathogens such as *Borrelia burgdorferi*, *Ehrlichia* spp., and *Babesia microti* (Kosoy et al., 1997; Hofmeister et al., 1998). These observations, along with the case reports of bacteremia due to *B. henselae* in men who had sustained tick bites prior to their illnesses (Lucey et al., 1992), suggest that *Bartonella* spp. may also be transmitted by ticks. Schouls et al. (1999) have also found *Bartonella* DNA in a high proportion of an *Ixodid* tick population.

ENRICHMENT AND ISOLATION PROCEDURES

The specimen source of most isolates of *Bartonella* is blood or tissue. The fastidious nature of these organisms requires that precautions be taken to minimize delayed attempts at isolation. If storage of specimens is necessary, they should be kept frozen. Blood collected in tubes containing EDTA can be plated after 26 d at $-65°C$ with no loss of sensitivity (Brenner et al., 1997). Blood-lysis tubes (Isolator; Wampole, Cranbury, NJ) yield good recovery from freshly collected specimens (Welch et al., 1992). *Bartonella* spp. have been isolated using agar, in semisolid media, and in broth. Broth-based systems generally tend to have lower sensitivity. If a broth-based blood culture system that relies on CO_2 detection to indicate growth is used, it should be combined with acridine orange staining at the end of a 7-d incubation and subculture to solid media (Spach et al., 1995). Combining the subculture of blood culture broth into shell vials has been reported to produce sensitivity of 71% for recovery of *B. quintana* or *B. henselae* from patients with endocarditis, bacillary angiomatosis, and lymph nodes of cat scratch disease and no prior antibiotic therapy (La Scola and Raoult, 1999b). The sensitivity of culture was still low when compared with that of PCR-based detection or serological methods of diagnosis. A modified RPMI 1640 medium has also been described by Wong et al. (1995b) for recovery of *B. henselae* from both tissue and blood. Isolator-processed blood should be plated on enriched (chocolate- or blood-containing) medium incubated at 35–37°C (30°C for *B. bacilliformis*) under conditions of 5–10% CO_2 and >40% humidity. For optimal recovery, the medium should be as freshly prepared as possible. Plates should be sealed after the first 24 h of incubation to preserve the moisture content of the medium. Isolates have been obtained from liver, spleen, lymph node, and skin after homogenization either by direct plating or by cocultivation with an endothelial cell line (Koehler et al., 1992). The cocultivation method may be more successful in recovering organisms from tissue specimens. Selective culture techniques have not been developed, so recovery of isolates from contaminated specimens may not be possible. During Oroya fever, the organisms can be isolated from blood and from endothelial cells of lymph nodes, spleen, and liver. In cases of Verruga Peruana, they are found in the blood and in the eruptive lesions. The organisms can also be isolated from the sand fly vector (*Phlebotomus* spp.). Other species have also been isolated from their respective arthropod vectors, and *B. henselae* is readily isolated from the blood of cats. In human cases of cat scratch disease, attempts at isolating the organism are rarely successful.

MAINTENANCE PROCEDURES

Viability of cultures can usually be maintained by passage on blood or chocolate agar at biweekly intervals. During serial transfers, the greatest longevity of *B. bacilliformis* is achieved at a temperature of 28°C. Cultures of this and the other species are best maintained during storage at $-70°C$ in a blood- or hemoglobin-containing medium. Long-term storage is best accomplished by lyophilization. Retrieval of viable cultures that have been lyophilized may be enhanced by suspending dried cells in a blood–broth mixture followed by culture on fresh blood agar media.

DIFFERENTIATION OF THE GENUS *BARTONELLA* FROM OTHER GENERA

The organism originally described as the cat scratch bacillus (*Afipia felis*) is distinct genotypically and phenotypically from *Bartonella* spp. It is ~90% related to *Bartonella* based on 16S rRNA gene sequence similarity. See Table BXII.α.119 for additional characteristics useful in differentiating the *Bartonella* spp. from the genus *Afipia*. Another organism found in the blood of cats and similarly confused in the past with *Bartonella*, *Haemobartonella felis*, can now be clearly identified using a set of *H. felis*-specific primers that selectively amplify a 1316-bp DNA fragment of the 16S rRNA gene of *H. felis* (Messick et al., 1998).

TABLE BXII.α.119. Differential characteristics of the *Bartonella* species, *Afipia felis*, and *Brucella melitensis*[a]

Characteristic	*B. bacilliformis*	*B. alsatica*	*B. clarridgeiae*	*B. doshiae*	*B. elizabethae*	*B. grahamii*	*B. henselae*	*B. koehlerae*	*B. peromysci*
Optimal temp, °C	25–30	35	35–37	35–37	35–37	35–37	35–37	35	20–28
Growth in nutrient broth	−	−	−		−	−	−		
Growth on heart infusion agar with X factor					+		−	−	
Hemolysis	−	−	−		v		−		−
Growth in <10 d	+	−	+	v	+	+	v	−	+
Oxidase	−	−	−		−		−		
Catalase	+	−	−		−		v		
Nitrate reduction	−	−	−		−		−		
Indole	−	−	−		−		−		
Urease	−	−	−		−		−		
Glucose oxidation or fermentation	−	−		−	−		−	−	−
Voges–Proskauer	−	−	−	+		+	−		
p-Nitrophenyl-β-D-galactopyranosidase	−	−	−		−		−	−	
p-Nitrophenyl-α-D-galactopyranosidase	−	−	−	−	−		−	−	
bis *p*-Nitrophenyl-phosphatase	+	−			+		+	+	
p-Nitrophenyl-*N*-acetyl-β-D-glucosaminidase	−	−	−	−	−	−	−		−
p-Nitrophenyl-α-D-glucopyranosidase	−	−	−	−	−		−		
p-Nitrophenyl-β-D-glucopyranosidase	−	−	−	−	−		−		
p-Nitrophenyl phosphatase	−								−
p-Nitrophenyl-α-L-fucopyranosidase	−	−		−	−		−		
p-Nitrophenyl-α-D-mannopyranosidase	−	−	−	−	−		−		
L-Leucine-β-naphthyl-amidase	+	+		+	+	+	+	+	
DL-Methionine-β-naphthyl-amidase	+	+			+		+	+	
L-Lysine-β-naphthylamidase (alkaline)	+	+			+		+	+	
L-Lysine-β-naphthyl-amidase (acidic)	+	+			w		+	+	
Glycylglycine-β-naphthylamidase	+	+			+		+	+	
Glycine-β-naphthylamidase	+	+	+		+		+	+	
L-Proline-β-naphthylamidase	−	+	+	−	−	+	+	−	
L-Arginine-β-naphthyl-amidase	+	+	+		+		+	+	
L-Pyrrolindonyl-β-naphthylamidase	−	−		−	−	−	−	−	
L-Tryptophan-β-naphthylamidase	+	+			+		+	+	
3-Indoxyl phosphatase	−	−			−		−	−	
Flagella	+	−	+				−	−	−
Twitching motility					−		+	−	
Major cellular fatty acids:									
C16:0	+		+		+		+		
C16:1 ω7c	+								
C17:0					+				
C18:0			+				+		
C18:1 ω7c	+		+		+		+		
C19:0 cyclo									
CBr-19:1									
Reactivity with fluorescent antibody to:									
B. bacilliformis	+			−		−	−		
B. doshiae	−			+	−	w	−		
B. elizabethae	−			−	+	−	−		
B. grahamii	−			w	w	+	w		
B. henselae	−		−		−		+		
B. quintana	−		−	−	−	−	−		
B. taylorii	−			w	−	−	w		
B. vinsonii	−			−	−	−	−		

[a]Symbols: +, positive; −, negative; w, weakly positive; v, variable (only some isolates have the characteristic).

(*continued*)

TAXONOMIC COMMENTS

Bartonella is the only genus of the family *Bartonellaceae* within the *Alphaproteobacteria*. *B. bacilliformis* was the original and only member of the genus until 1993. The genus *Rochalimaea* was then united with the genus *Bartonella* in the family *Bartonellaceae* based on a proposal by Brenner et al. (1993). 16S rRNA gene sequence data and DNA hybridization data revealed high levels of relatedness between *Bartonella bacilliformis* and the four *Rochalimaea* species, indicating that these species were members of a single genus. The name *Bartonella* was retained as the genus name since it had nomenclatural priority over the name *Rochalimaea*. More recently, the genera of *Grahamella* and *Bartonella* were merged, and at present, 16 species/subspecies belong to the genus. The former *Rochalimaea* and, subsequently, *Grahamella* spp. were uni-

TABLE BXII.α.119. *(cont.)*

Characteristic	B. quintana	B. talpae	B. taylorii	B. tribocorum	B. vinsonii subsp. vinsonii	B. vinsonii subsp. arupensis	B. vinsonii subsp. berkhoffii	Afipia felis	Brucella melitensis
Optimal temp, °C	35–37		35–37	35	35–37	35–37	35–37	25–30	36–38
Growth in nutrient broth	−	−	−	−	−	−	−	+	−
Growth on heart infusion agar with X factor	v				v	+	−	+	+
Hemolysis	−	−			−	−	−	−	−
Growth in <10 d	+	+	+	−	+	+	+	+	+
Oxidase	v			−	v	−	−	+	+
Catalase	−			−	v	−	v	−	+
Nitrate reduction	−			−	−	−	−	+	+
Indole	−			−	−	−	−	−	−
Urease	−			−	−	−	−	+	+
Glucose oxidation or fermentation	−	−		−	−	−	−	−	
Voges–Proskauer	−		+		−				−
p-Nitrophenyl-β-D-galactopyranosidase	−			−	−	−	−		−
p-Nitrophenyl-α-D-galactopyranosidase	−		−	−	−	−	−		−
bis p-Nitrophenyl-phosphatase	v			−	+	+	+		−
p-Nitrophenyl-N-acetyl-β-D-glucosaminidase	−	−	−	−	−	−	−		+
p-Nitrophenyl-α-D-glucopyranosidase	−			−	−	−	−		+
p-Nitrophenyl-β-D-glucopyranosidase	−			−	−	−	−		−
p-Nitrophenyl phosphatase	−	−		−	−	−	−		+
p-Nitrophenyl-α-L-fucopyranosidase	−		−	−	−	−	−		−
p-Nitrophenyl-α-D-mannopyranosidase	−		−	−	−	−	−		−
L-Leucine-β-naphthyl-amidase	+		+	+	+	+	+		−
DL-Methionine-β-naphthyl-amidase	+			+	+	w	+		+
L-Lysine-β-naphthylamidase (alkaline)	+			+	+	+	+		+
L-Lysine-β-naphthyl-amidase (acidic)	−			+	−	−	+		−
Glycylglycine-β-naphthylamidase	+			+	+	+	+		+
Glycine-β-naphthylamidase	+			+	+	+	+		+
L-Proline-β-naphthylamidase	+		+	−	v	−	+		+
L-Arginine-β-naphthyl-amidase	+			+	+	+	+		+
L-Pyrrolindonyl-β-naphthylamidase	−		−		−	−	−		−
L-Tryptophan-β-naphthylamidase	+			+	+	+	+		−
3-Indoxyl phosphatase	−			−	−	−	−		−
Flagella	−	−		−	−	−	−	+	−
Twitching motility	+				−	−	−		
Major cellular fatty acids:									
C$_{16:0}$	+				+	+	+		+
C$_{16:1\ \omega7c}$									
C$_{17:0}$					+	+			+
C$_{18:0}$	+				+	+	+		+
C$_{18:1\ \omega7c}$	+				+	+	+	+	
C$_{19:0\ cyclo}$								+	+
C$_{Br\text{-}19:1}$								+	
Reactivity with fluorescent antibody to:									
B. bacilliformis	−		−		−				
B. doshiae	w		w		w				
B. elizabethae	−		−		−				
B. grahamii	w		w		w				
B. henselae	v		−		v	−		−	
B. quintana	+		−		−	−			
B. taylorii	w		+		w				
B. vinsonii	w		−		+				

[a]Symbols: +, positive; −, negative; w, weakly positive; v, variable (only some isolates have the characteristic).

fied as *Bartonella* based on DNA–DNA hybridization data showing that they were not as closely related as previously thought to members of the order *Rickettsiales* (Brenner et al., 1993).

There are a number of unclassified isolates, mostly recovered from rodents, presumed to be *Bartonella* spp. based on 16S rRNA gene sequence data. Further growth in the number of species and subspecies in the genus is thus expected to occur in the future.

ACKNOWLEDGMENTS

This description of the genus *Bartonella* contains information presented by Miodrag Ristic and Julius P. Kreier in the first edition of this *Manual.*

FURTHER READING

Anderson, B.E. and M.A. Neuman. 1997. *Bartonella* spp. as emerging human pathogens. Clin. Microbiol. Rev. *10:* 203–219.

Brenner, D.J., S.P. O'Connor, D.G. Hollis, R.E. Weaver and A.G. Steigerwalt. 1991. Molecular characterization and proposal of a neotype strain for *Bartonella bacilliformis.* J. Clin. Microbiol. *29:* 1299–1302.

Brenner, D.J., S.P. O'Connor, H.H. Winkler and A.G. Steigerwalt. 1993.

Proposals to unify the genera *Bartonella* and *Rochalimaea*, with descriptions of *Bartonella quintana* comb. nov., *Bartonella vinsonii* comb. nov., *Bartonella henselae* comb. nov., and *Bartonella elizabethae* comb. nov., and to remove the family *Bartonellaceae* from the order *Rickettsiales.* Int. J. Syst. Bacteriol. *43:* 777–786.

Maurin, M., R. Birtles and D. Raoult. 1997. Current knowledge of *Bartonella* species. Eur. J. Clin. Microbiol. Infect. Dis. *16:* 487–506.

Relman, D.A., P.W. Lepp, K.N. Sadler and T.M. Schmidt. 1992. Phylogenetic relationships among the agent of bacillary angiomatosis, *Bartonella bacilliformis*, and other alpha-proteobacteria. Mol. Microbiol. *6:* 1801–1807.

List of species of the genus Bartonella

1. **Bartonella bacilliformis** (Strong, Tyzzer, Brues, Sellards and Gastiaburú 1913) Strong, Tyzzer and Sellards 1915, 808[AL] (*Bartonia bacilliformis* Strong, Tyzzer, Brues, Sellards and Gastiaburú 1913, 1715.)

ba.cil.li.for′mis. L. dim. n. *bacillus* a small staff, rodlet; L. n. *forma* shape, form; M.L. adj. *bacilliformis* rod-shaped.

Displays characteristics of the genus *Bartonella*. Optimal growth occurs at 25–28°C on enriched media. Cells have 1–10 polar flagella; some may have subpolar or lateral flagella. Etiologic agent of Carrion's disease (Oroya fever, Verruga Peruana). Other characteristics useful in identification are shown in Table BXII.α.119.

Genome size is 4×10^8 Da.

The mol% G + C of the DNA is: 39 (T_m).

Type strain: ATCC 35685.

GenBank accession number (16S rRNA): Z11683.

Additional Remarks: The neotype strain ATCC 35685 was proposed by Brenner et al. (1991b).

2. **Bartonella alsatica** Heller, Kubina, Mariet, Riegel, Desacour, Dehio, LaMarque, Kasten, Boulouis, Mounteil, Chomel and Piémont 1999, 287[VP]

al.sa′ti.ca. L. adj. *alsatica* from Alsace, the region in eastern France near the Rhine River where wild rabbits, from which strains of the species were isolated and identified, were trapped.

Displays characteristics of the genus *Bartonella*. Colonies grown on blood agar appeared after 10 d as small, white, smooth, regular colonies (diameter ~1 mm). Additional characteristics useful in Identification are shown in Table BXII.α.119.

The mol% G + C of the DNA is: 37 (capillary electrophoresis).

Type strain: IBS 382, CIP 105477.

GenBank accession number (16S rRNA): AJ002139.

3. **Bartonella birtlesii** Bermond, Heller, Barrat, Delacour, Dehio, Alliot, Monteil, Chomel, Boulouis and Piémont 2000, 1978[VP]

birt.les′i.i. M.L. gen. n. *birtlesii* of Richard J. Birtles, whose studies have contributed to an improved understanding of the taxonomy of the genus.

The mol% G + C of the DNA is: not determined.

Type strain: IBS 325, CIP 106294, CCUG 44360.

GenBank accession number (16S rRNA): AF204274.

4. **Bartonella clarridgeiae** Lawson and Collins 1996b, 836[VP] (Effective publication: Lawson and Collins 1996a, 71.)

clar.ridge′i.a.e. M.L. fem. adj. *clarridgeiae* named in honor of Jill E. Clarridge III, the microbiologist who first isolated the organism, in Houston, Texas.

Displays characteristics of the genus *Bartonella*. Possesses polar flagella. Additional characteristics useful in identification are shown in Table BXII.α.119.

The mol% G + C of the DNA is: not determined.

Type strain: Houston-2 cat, ATCC 51734.

GenBank accession number (16S rRNA): X89208.

5. **Bartonella doshiae** Birtles, Harrison, Saunders and Molyneux 1995, 7[VP]

do′shi.ae. M.L. gen. n. *doshiae* named in honor of Nivedita Doshi, who was technically responsible for work with *Legionella* and *Bartonella* at the Central Public Health Laboratory in London.

Displays characteristics of the genus *Bartonella*. Additional characteristics useful in identification are shown in Table BXII.α.119.

The mol% G + C of the DNA is: 41 (T_m).

Type strain: R18, ATCC 700133, NCTC 12862.

GenBank accession number (16S rRNA): Z31351.

6. **Bartonella elizabethae** (Daly, Worthington, Brenner, Moss, Hollis, Weyant, Steigerwalt, Weaver, Daneshvar and O'Connor 1993) Brenner, O'Connor, Winkler and Steigerwalt 1993, 785[VP] (*Rochalimaea elizabethae* Daly, Worthington, Brenner, Moss, Hollis, Weyant, Steigerwalt, Weaver, Daneshvar and O'Connor 1993, 880.)

e.liz′a.beth.a.e. M.L. fem. adj. *elizabethae* named after St. Elizabeth's Hospital in Brighton, Massachusetts, where the organism was isolated.

Displays characteristics of the genus *Bartonella*. There is incomplete hemolysis on rabbit blood agar. Additional characteristics useful in identification are shown in Table BXII.α.119. Etiologic agent of endocarditis and neuroretinitis.

The mol% G + C of the DNA is: 41 (T_m).

Type strain: F9251, B91-002005, ATCC 49927.

GenBank accession number (16S rRNA): L01260.

7. **Bartonella grahamii** Birtles, Harrison, Saunders and Molyneux 1995, 7[VP]

gra.ham′i.i. M.L. gen. n. *grahamii* of Graham, named in honor of G.S. Graham-Smith who observed the organisms subsequently named *Grahamella* in the blood of moles.

Displays characteristics of the genus *Bartonella*. Additional characteristics useful in identification are shown in Table BXII.α.119.

The mol% G + C of the DNA is: 40 (T_m).

Type strain: V2, ATCC 700132, NCTC 12860.

GenBank accession number (16S rRNA): Z31349.

8. **Bartonella henselae** (Regnery, Anderson, Clarridge, Rodriguez-Barradas and Jones 1992) Brenner, O'Connor, Winkler and Steigerwalt 1993, 785[VP] (*Rochalimaea henselae* Regnery, Anderson, Clarridge, Rodriguez-Barradas and Jones 1992, 272.)

hen' sel.a.e. M.L. gen. n. *henselae* named in honor of Diane M. Hensel, who isolated many of the original strains detected in bacteremic patients from Oklahoma.

Displays characteristics of the genus *Bartonella*. Additional characteristics useful in identification are shown in Table BXII.α.119. Etiologic agent of septicemia in immunocompromised hosts, endocarditis, bacillary angiomatosis, peliosis hepatis, cat scratch disease, and HIV-associated neurological syndromes.

The mol% G + C of the DNA is: 41 (T_m).

Type strain: Houston-1, G5436, ATCC 49882.

GenBank accession number (16S rRNA): M73229.

9. **Bartonella koehlerae** Droz, Chi, Horn, Steigerwalt, Whitney and Brenner 2000, 423[VP] (Effective publication: Droz, Chi, Horn, Steigerwalt, Whitney and Brenner 1999, 1122.)

koeh' ler. ae. M.L. fem. adj. *koehlerae* named in honor of Jane E. Koehler, who was the first to isolate *Bartonella* species from bacillary angiomatosis lesions and whose studies of *B. quintana* and *B. henselae* isolates from human immunodeficiency virus-infected patients have contributed to an improved understanding of *Bartonella*-associated disease in humans.

Displays characteristics of the genus *Bartonella*. Growth is optimal on chocolate agar, and primary colonies are observed after 14 d of incubation at 35°C in a CO_2-enriched environment. Additional characteristics useful in identification are shown in Table BXII.α.119.

The mol% G + C of the DNA is: not determined.

Type strain: C-29, ATCC 700693.

GenBank accession number (16S rRNA): AF076237.

10. **Bartonella peromysci** (Ristic and Kreier 1984b) Birtles, Harrison, Saunders and Molyneux 1995, 7[VP] (*Grahamella peromysci* (ex Tyzzer 1942) Ristic and Kreier 1984b, 719.)

pe.ro.mys' ci. M.L. gen. n. *peromysci*, of *Peromyscus* a genus of mice.

Displays characteristics of the genus *Bartonella*. Additional characteristics useful in identification are shown in Table BXII.α.119. Distinguished from *B. talpae* by the host specificity.

The mol% G + C of the DNA is: unknown.

Type strain: No type strain available.

11. **Bartonella quintana** (Schmincke 1917) Brenner, O'Connor, Winkler and Steigerwalt 1993, 784[VP] (*Rochalimaea quintana* (Schmincke 1917) Krieg 1961, 163; *Rickettsia quintana* Schmincke 1917, 961.)

quin.ta' na. M.L. fem. adj. *quintana* fifth, referring to 5-day fever and the clinical disease produced by the species.

Displays characteristics of the genus *Bartonella*. Etiologic agent of trench fever, bacillary angiomatosis, and septicemia. Additional characteristics useful in identification are shown in Table BXII.α.119.

The mol% G + C of the DNA is: 40 (T_m).

Type strain: ATCC VR-358.

GenBank accession number (16S rRNA): M11927, M73228.

12. **Bartonella talpae** (Ristic and Kreier 1984b) Birtles, Harrison, Saunders and Molyneux 1995, 7[VP] (*Grahamella talpae* (ex Brumpt 1911) Ristic and Kreier 1984b, 719.)

tal' pae. M.L. gen. n. *talpae* of *Talpa*; M.L. fem. n. a genus of moles.

Displays characteristics of the genus *Bartonella*. Additional characteristics useful in identification are shown in Table BXII.α.119. Distinguished from *B. peromysci* by the host specificity.

The mol% G + C of the DNA is: unknown.

Type strain: No type strain is available.

13. **Bartonella taylorii** Birtles, Harrison, Saunders and Molyneux 1995, 7[VP]

tay.lor' i.i. M.L. gen. n. *taylorii* of Taylor, named in honor of A.G. Taylor who led various microbiologic studies at the Central Public Health Laboratory in London.

Displays characteristics of the genus *Bartonella*. Additional characteristics useful in identification are shown in Table BXII.α.119.

The mol% G + C of the DNA is: 41 (T_m).

Type strain: M6, NCTC 12861.

GenBank accession number (16S rRNA): Z31350.

14. **Bartonella tribocorum** Heller, Riegel, Hansmann, Delacour, Bermond, Dehio, Lamarque, Monteil, Chomel and Piémont 1998, 1338[VP]

tri.bo.co' rum. L. n. gen. pl. *Triboci* the tribes, mentioned by Caesar (51 BC) in his *Commentarii de Bello Gallico*, which were living in the region near the Rhine River in eastern France. Wild rats, from which two strains of the species were isolated, were trapped there.

Displays characteristics of the genus *Bartonella*. Colonies grown on blood agar appear as small, white, smooth, regular colonies (~1 mm diameter) after 10 d. Electron microscopic examination shows small bacilli without flagella, approximately 1–2 × 0.5 µm. Additional characteristics useful in identification are shown in Table BXII.α.119.

The mol% G + C of the DNA is: 38 (capillary electrophoresis).

Type strain: IBS 506, CIP 105476.

GenBank accession number (16S rRNA): AJ003070.

15. **Bartonella vinsonii** (Weiss and Dasch 1982) Brenner, O'Connor, Winkler and Steigerwalt 1993, 785[VP] emend. Kordick, Swaminathan, Greene, Wilson, Whitney, O'Connor, Hollis, Matar, Steigerwalt, Malcolm, Hayes, Hadfield, Breitschwerdt and Brenner 1996, 708 (*Rochalimaea vinsonii* Weiss and Dasch 1982, 313.)

vin.so' ni.i. N.L. gen. n. *vinsonii* named in honor of J. William Vinson who, with Henry S. Fuller, originally demonstrated that *Bartonella vinsonii* subsp. *vinsonii* (*Rochalimaea vinsonii*) could be grown on blood agar.

The mol% G + C of the DNA is: See subspecies data below.

Type strain: See subspecies below.

a. **Bartonella vinsonii** *subsp.* **vinsonii** (Weiss and Dasch 1982) Brenner, O'Connor, Winkler and Steigerwalt 1993, 785[VP] emend. Kordick, Swaminathan, Greene, Wilson, Whitney, O'Connor, Hollis, Matar, Steigerwalt, Malcolm, Hayes, Hadfield, Breitschwerdt and Brenner 1996, 708 (*Bartonella vinsonii* (Weiss and Dasch 1982) Brenner, O'Connor, Winkler and Steigerwalt 1993, 785; *Rochalimaea vinsonii* Weiss and Dasch 1982, 313.)

Displays characteristics of the genus *Bartonella*. Additional characteristics useful in identification are shown in Table BXII.α.119. Vole is the animal host (also known as the "Canadian vole agent").

The mol% G + C of the DNA is: 41 (T_m) (Daly et al., 1993).

Type strain: ATCC VR-152.

GenBank accession number (16S rRNA): L01259, Z31352.

b. **Bartonella vinsonii** *subsp.* **arupensis** Welch, Carroll, Hofmeister, Persing, Robison, Steigerwalt and Brenner 2000, 3[VP] (Effective publication: Welch, Carroll, Hofmeister, Persing, Robison, Steigerwalt and Brenner 1999, 2601.)

a.rup.en'sis. N.L. fem. adj. *arupensis* coming from ARUP (Associated Regional and University Pathologists, Inc.) in Salt Lake City, Utah, the laboratory where the type strain was initially characterized.

Exhibits characteristics of the species *Bartonella vinsonii*. Grows on heart infusion agar in the presence of X factor. Other characteristics useful in identification are shown in Table BXII.α.119. Isolated from a human and mice (*Peromyscus leucopus*). Presumptively pathogenic for humans.

The mol% G + C of the DNA is: not determined.

Type strain: OK 94-513, ATCC 700727, isolated from a 62-year-old bacteremic man.

GenBank accession number (16S rRNA): U71322.

Additional Remarks: Other reference strain, UMB (Hofmeister et al., 1998, 413).

c. **Bartonella vinsonii** *subsp.* **berkhoffii** Kordick, Swaminathan, Greene, Wilson, Whitney, O'Connor, Hollis, Matar, Steigerwalt, Malcolm, Hayes, Hadfield, Breitschwerdt and Brenner 1996, 908[VP]

berk.hof'fi.i. M.L. gen. n. *berkhoffii* named in honor of Herman A. Berkhoff, a veterinary microbiologist whose research contributed to the understanding of *Bartonella*

infections in domestic animals and recognition of this subspecies.

Exhibits characteristics of the species *Bartonella vinsonii*. Can be distinguished from *Bartonella vinsonii* subsp. *vinsonii* by DNA methods. 16S rRNA gene contains an insertion sequence designated I1-I12 in the variable region V2. Other characteristics useful in identification are shown in Table BXII.α.119. Etiologic agent of canine endocarditis.

The mol% G + C of the DNA is: not determined.

Type strain: 93-CO1, ATCC 51672.

GenBank accession number (16S rRNA): L35052.

Species Incertae Sedis

1. **Wolbachia melophagi** (Nöller 1917) Philip 1956, 267[AL] (Nöller 1917, 70.)*

me.lo.pha'gi. M.L. gen. n. *melophagi* of *Melophagus*, named after the genus of its natural host, *Melophagus ovinus*, a wingless fly commonly called sheep ked.

The characteristics are as described for the genus *Wolbachia* with the following additional characteristics. Coccoid cells 0.4–0.6 μm or short rods 0.3–1.0 μm. Grow on blood-glucose-bouillon agar and chicken embryo yolk sacs but not in cultured eucaryotic cells. Natural hosts are a wingless fly, *Melophagus ovinus*, and *Diptera* that infest goats, pigs, and horses. Grows extracellularly in the host insect's gut, where the wolbachiae form rows of short rods associated with the intestinal epithelium. Apparently nonpathogenic for hosts.

The mol% G + C of the DNA is: not determined.

Type strain: no strain isolated.

GenBank accession number (16S rRNA): X89110 (strain MO6).

Editorial Note: Dumler et al. (2001) have indicated that 16S rDNA sequence analysis places *Wolbachia melophagi* in the genus *Bartonella*.

Family III. **Brucellaceae** Breed, Murray and Smith 1957, 394[AL]

GEORGE M. GARRITY, JULIA A. BELL AND TIMOTHY LILBURN

Bru.cel.la'ce.ae. M.L. fem. n. *Brucella* type genus of the family; *-aceae* ending to denote family; M.L. fem. pl. n. *Brucellaceae* the *Brucella* family.

The family *Brucellaceae* was circumscribed for this volume on the basis of phylogenetic analysis of 16S rRNA gene sequences; the family contains the genera *Brucella* (type genus), *Mycoplana*, and *Ochrobactrum*.

Chemoorganotrophs with an aerobic, respiratory metabolism. Family includes pathogens and soil organisms.

Type genus: **Brucella** Meyer and Shaw 1920, 173[AL]

Genus I. **Brucella** Meyer and Shaw 1920, 173[AL]

MICHAEL J. CORBEL AND MENACHEM BANAI

Bru. cel'la. L. dim. ending *-ella*; M.L. fem. n. *Brucella* named after Sir David Bruce, who first recognized the organism causing undulant (Malta) fever.

Cocci, coccobacilli, or short rods, 0.5–0.7 × 0.6–1.5 μm. Arranged singly and, less frequently, in pairs, short chains, or small groups. True capsules are not produced. Do not usually show true bipolar staining. Resting stages are not known. Gram negative. **Nonmotile;** do not produce flagella. **Aerobic,** possessing a respiratory type of metabolism and having a **cytochrome-based electron transport system** with oxygen or nitrate as the terminal electron acceptor. Nitrate reductase is produced. **Many strains**

require supplementary CO$_2$ for growth, especially on primary isolation. Colonies on serum-dextrose agar or other clear media are transparent, raised, convex, with an entire edge and a smooth, shiny surface. They appear a **pale honey color** by transmitted light. Smooth strains produce perosamine synthetase and a distinctive lipopolysaccharide (LPS). Nonsmooth variants of the smooth species occur, but there are also stable nonsmooth nomenspecies with a distinctive host range. Optimal temperature for growth 37°C. Growth occurs between 20 and 40°C. Optimal pH for growth 6.6–7.4. **Catalase positive.** Usually oxidase positive, but negative strains occur. Chemoorganotrophic. Most strains require complex media containing several amino acids, thiamin, nicotinamide, iron, and magnesium ions; some strains may be induced to grow on minimal media containing an ammonium salt as the sole nitrogen source. **Growth is improved by serum or blood, but hemin (X-factor) and nicotinamide adenine dinucleotide (NAD: V-factor) are not essential. Acid production does not occur from carbohydrates in conventional media,** except for *B. neotomae.* Do not produce indole. **Do not liquefy gelatin** or inspissated serum. Do not lyse erythrocytes. Do not produce **acetyl methyl carbinol** (Voges–Proskauer test.) **The methyl red test is negative.** Intracellular parasites, transmissible to a wide range of animal species including man. The genome typically comprises two circular chromosomes, but a single large chromosome is present in *B. suis* biovar 3.

The genus is essentially genetically **monospecific,** nomenspecies and biovars reflecting deletions, insertions, or rearrangements in a largely conserved genome. The 16S rRNA gene sequence data, distinctive lipid composition, shared proteins and sensitivity to trifolitoxin indicate affiliation to the order *Rhizobiales* in the *Alphaproteobacteria.* The closest relationships are to the genus *Ochrobactrum* and to a lesser extent to the genus *Mycoplana,* and the three genera comprise the known membership of the family *Brucellaceae.*

The mol% G + C of the DNA is: 57.9–59.

Type species: **Brucella melitensis** (Hughes 1893) Meyer and Shaw 1920, 179 (*Streptococcus melitensis* Hughes 1893, 235.)

FURTHER DESCRIPTIVE INFORMATION

Cell morphology When grown in nutritionally adequate liquid or solid media such as serum-dextrose broth (SDB)[1] or serum-dextrose agar (SDA)[2], *Brucella* cells are coccoid, coccobacilli, or short rods with slightly convex sides and rounded ends. Freshly isolated strains tend to be more coccoid than laboratory-adapted cultures; this is also true of organisms growing *in vivo*. In general, the morphology of *Brucella* strains is constant and pleomorphic forms are rare, except in old cultures growing under adverse conditions.

Brucella cells stain readily by conventional methods. Although not truly acid fast, they do tend to resist decolorization by weak acids and thus stain red by Macchiavello's stain or by the modified Ziehl–Neelsen technique used by Stamp et al. (1950). They are usually stained red by the modified Köster method (Christofferson and Ottosen, 1941), but *B. ovis* is an exception. True capsules do not occur although capsule-like structures have been reported.

Fine structure The ultrastructure of the *Brucella* cell is broadly similar to that of other Gram-negative bacteria but shows a number of significant differences from that of cells of the *Enterobacteriaceae,* as typified by *Escherichia coli* (De Petris et al., 1964; Dubray, 1972, 1976; Dubray and Plommet, 1976).

The *Brucella* cell wall is composed of an outer membrane comprising an external layer of lipopolysaccharide (LPS), a range of outer membrane proteins including some with porin activity, lipoproteins, and phospholipids. In negatively stained thin sections, this layer appears about 9 nm thick (Dubray, 1976). It is the location of the major surface antigens in both smooth and nonsmooth cells. It is supported by an electron-dense inner layer 3.5 nm thick corresponding to cross-linked muramic-acid-containing peptidoglycan. Some of the outer membrane proteins extend through this layer to the periplasmic region, which appears as a zone of low electron density 3–6 nm thick in smooth-phase cells but up to 30 nm thick in nonsmooth cells (Dubray and Plommet, 1976). As in other Gram-negative bacteria, it is the site of periplasmic enzymes involved in cell wall biosynthesis and various metabolic functions.

The cytoplasmic inner membrane has a typical three-layered lipoprotein structure (De Petris et al., 1964; Dubray and Plommet, 1976). Granular aggregations adjacent to the cytoplasmic membrane mark the location of polyribosomal complexes (Dubray, 1972, 1976).

The *Brucella* cytoplasm is homogeneous and is interspersed with small vacuoles and polysaccharide-containing granules (Dubray, 1972). The nuclear apparatus comprises an osmiophobic mass intersected by osmiophilic filamentous structures (De Petris et al., 1964; Peschkov and Feodorov, 1978).

Cell wall composition The detailed chemical structure of the *Brucella* cell wall has not been fully determined. Gross analyses have indicated that the cell wall accounts for about 21% of the total bacterial dry weight in smooth cultures and 14% in nonsmooth strains. The walls of smooth *Brucella* cells contain approximately 37% protein, 14% carbohydrate, 18% lipid, 0.4% muramic acid, and 0.1% 2-keto-3-deoxyoctulosonic acid (KDO). For nonsmooth *Brucella* cells walls the corresponding values are approximately 47.5% protein, 13% carbohydrate, 17% lipid, 0.4% muramic acid, and 0.1% KDO (Kreutzer et al., 1977).

The external layer of the outer membrane comprises mainly LPS interspersed with a variety of proteins and lipids. The LPS of smooth *Brucella* cell walls is unusual in that it partitions into the phenol layer on phenol–water extraction (Baker and Wilson, 1965). The LPS of the nonsmooth strains is extremely hydrophobic and cannot be extracted with phenol–water but is soluble in phenol–hexane–chloroform (Moreno et al., 1979).

The LPS of both forms consists of a lipid A containing both 2,3-diamino-2,3-dideoxyglucose and 2-amino-2-deoxyglucose in the glycose backbone, a feature of some other members of the *Proteobacteria* (Weckesser and Mayer, 1988). Both amide- and ester-linked fatty acids are attached to the aminoglycose skeleton.

---

1. Serum-dextrose broth (SDB): tryptone-soya broth (Oxoid), 30 g; distilled water, 1000 ml; sterile horse serum (inactivated at 56°C for 30 min), 50 or 100 ml; D-glucose (25%, w/v, solution autoclaved at 105°C for 20 min), 40 ml. The medium is prepared by dissolving the tryptone-soya broth powder in the water and autoclaving at 115°C for 15 min. After cooling, the sterile horse serum and glucose are added aseptically. Note: Any good quality peptone medium such as Tryptose broth (Difco), or *Brucella* broth (Gibco) may be used as alternatives to tryptone-soya broth. Serum-dextrose agar (SDA): blood agar base No. 2 (Oxoid) 40 g; distilled water, 1000 ml; sterile horse serum (inactivated at 56°C for 30 min), 50 or 100 ml; D-glucose ((25% w/v) solution autoclaved at 105°C for 20 min), 40 ml. The blood agar base is dissolved in the water with the aid of gentle heating and then autoclaved at 121°C for 15 min. After cooling, the horse serum and glucose are added aseptically and plates or slopes poured immediately. Note: Tryptose agar (Difco) or *Brucella* agar (Gibco) are satisfactory alternative basal media.

The amide-linked acids include 3-O-$C_{(16:0)\ 12:0}$ (25%), 3-O-$C_{(16:0)\ 13:0}$ (4%), 3-O-$C_{(16:0)\ 14:0}$ (64%), and 3-O-$C_{(18:0)\ 14:0}$ (7%) as diesters with $C_{16:0\ 3OH}$ as the unsubstituted fatty acid. The ester-linked acids comprise $C_{16:0}$, $C_{16:0\ 3OH}$, $C_{18:0\ 3OH}$, and $C_{18:0}$ acids, which account for 37%, 12.5%, 3.5%, and 4.5%, respectively, of the total fatty acids. Lactobacillic and unsaturated fatty acids are absent from the lipid A of *Brucella* although represented in many other lipid components (Cherwonogrodzky et al., 1990).

The lipid A is linked through KDO to the core polysaccharide composed of D-glucose, D-mannose, and 6-amino 6-deoxyglucose (quinovosamine) (Bowser et al., 1974). Heptose is absent from *Brucella* LPS. In the case of some rough strains, a short homopolymer of 4,6-dideoxy-4-formamido-D-mannose (*N*-formyl-D-perosamine) may be attached to the core polysaccharides (Perry and Bundle, 1990), but in many strains this is absent and some may also be deficient in quinovosamine (Moreno et al., 1984). D-Glucose and D-mannose, and possibly quinovosamine, are components of the major epitope of nonsmooth *Brucella* strains.

In the case of smooth strains, the LPS carries an O chain linked to the core polysaccharide. The O chain comprises a homopolymer of about 100 residues of *N*-formyl-D-perosamine. In the case of A-dominant *B. abortus* LPS, the O chain consists of a majority of glycose residues linked alpha-1,2, but with a very small proportion linked alpha-1,3 (Caroff et al., 1984a, b; Bundle et al., 1987).

In the case of M-dominant *B. melitensis* LPS, the O chains consist of unbranched linear polymers of pentasaccharide units, comprising four residues linked alpha-1,3. The difference in linkage produces penta- or hexasaccharide units with different preferred conformations. The common presence of nonterminal tetrasaccharide units of alpha-1,2 linked *N*-formyl-D-perosamine explains the crossreactivity observed between the LPS of all smooth *Brucella* strains (C epitope) (Cloeckaert et al., 1998).

The A- and M-specific epitopes are actually present as minority structures in both types of LPS. In the case of strains typified by *B. abortus* biovar 1, the O chains contain predominantly A epitopes with a very small proportion of M epitope, attributable to the few alpha-1,3 linked residues. In strains typified by *B. melitensis* biovar 1, the reverse situation applies. It should be noted that strains of *B. abortus*, *B. melitensis*, or *B. suis* can be A-, M-, or A- and M-antigen positive (see Table BXII.α.120 and Dubray and Limet, 1987). Strains that are both A- and M-antigen positive

synthesize LPS with O chains that contain both A and M structural features in relatively high proportion (Perry and Bundle, 1990).

N-Acylated-D-perosamine also occurs in the O chains of the LPS complexes of *E. coli* 0157, *Escherichia hermannii*, *Salmonella* O30 (Bundle et al., 1987), *Stenotrophomonas maltophilia* strain 555, *Vibrio cholerae* (Redmond, 1979; Kenne et al., 1982), and *Yersinia enterocolitica* 09 (Caroff et al., 1984b), all of which crossreact serologically with smooth *Brucella* strains (Corbel, 1985). Crossreactions also occur between *Brucella* and *Francisella tularensis* (Francis and Evans, 1926; Ohara et al., 1974), although the basis of this has yet to be determined.

The outer membrane proteins include the 88–94-kDa high molecular weight group 1, the 43 kDa and 36–38-kDa porin proteins of group 2 (Douglas et al., 1984), and the 25–27-kDa proteins of group 3, as well as minor proteins of 15–31 kDa. A lipoprotein of 8 kDa is also covalently linked to the peptidoglycan skeleton. These components are present in all the nomenspecies but with quantitative differences (Santos et al., 1984; Verstreate and Winter, 1984). In addition to the above, *B. melitensis* contains a 31 kDa outer-membrane protein and another of 30–40-kDa.

In all *Brucella* strains, the group 2 proteins are by far the most abundant. The genes for the 36-kDa porins (*omp2a* and *omp2b*) have been cloned, and the peptide sequences determined (Ficht et al., 1990). The function of the group 3 proteins, formerly assumed to be analogues of Omp A, has yet to be determined. The *omp2* genes and those encoding the Group 3 proteins demonstrate polymorphisms, which permit grouping into clusters corresponding approximately to nomenspecies and biovars (Cloeckaert et al., 1995). The 8 kDa lipoprotein resembles the Braun lipoprotein of *E. coli* in molecular weight, isoelectric point, and amino acid composition but differs from it in being surface exposed (Gomez-Miguel et al., 1987). The Omp 10, Omp 16, and Omp 19 outer membrane proteins are also reported to be lipoproteins (Tibor et al., 1999).

The *Brucella* outer membrane is unusual in being particularly rich in myristic, palmitic, and stearic acids, in containing moderate quantities of *cis*-vaccenic and arachidonic acids, low quantities of C_{17} and C_{19} cyclopropane fatty acids and no hydroxy fatty acids. This unusual fatty acid composition may contribute to hydrophobic interactions in the *Brucella* outer membrane and enhance its stability (Cherwonogrodzky et al., 1990).

TABLE BXII.α.120. Differentiation of the species and biovars of the genus *Brucella*[a]

Characteristics	B. melitensis biovars			B. abortus biovars								B. suis biovars					B. ovis	B. neotomae	B. canis
	1	2	3	1	2	3[b]	4	5	6[b]	7	9	1	2	3	4	5			
CO_2 requirement	−	−	−	[+]	[+]	[+]	[+]	−	−	−	−	−	−	−	−	−	+	−	−
H_2S production	−	−	−	+	+	+	+	−	[−]	[+]	+	+	−	−	−	−	−	+	−
Growth on media containing:[c]																			
Thionine	+	+	+	−	−	+	−	+	+	+	+	+	+	+	+	+	+	−[d]	+
Basic fuchsin	+	+	+	+	−	+	+	+	+	+	+	[−]	−	+	[−]	−	[−]	−	[−]
Agglutination with monospecific antisera:																			
A	−	+	+	+	+	+	−	−	+	+	−	+	+	+	+	−	−	+	−
M	+	−	+	−	−	−	+	+	−	+	+	−	−	−	+	+	−	−	−
R	−	−	−	−	−	−	−	−	−	−	−	−	−	−	−	−	+	−	+

[a]Symbols: +, positive for all strains; [+], positive for most strains; [−], negative for most strains; −, negative for all strains.

[b]For more certain differentiation of biovar 3 and 6, thionine at 1:25,000 (w/v) is used; biovar 3 gives a positive growth response, biovar 6 is negative.

[c]Dye concentration, 1:50,000 (w/v).

[d]Growth will occur in the presence of thionine at a concentration of 1:150,000 (w/v).

The peptidoglycan framework of the cell wall comprises a glycan skeleton of D-glucosamine and muramic acid linked by short chains of alanine, glutamic acid, and alpha, epsilon-diaminopimelic acid. However, it also contains covalently linked proteins and lipid and is unusually resistant to lysozyme, even in the presence of detergents and chelating agents.

The lipid composition of *Brucella* cells has a number of distinctive features, which reinforce the taxonomic affiliation of the genus to the group of bacteria known as the alpha 2 subdivision of the *Proteobacteria.*

The bound lipids are mainly associated with the LPS, lipoprotein, and glycolipid fractions of the outer membrane. The free lipid fraction comprises mainly phospholipids and neutral lipids (Wober et al., 1964; Thiele and Kehr, 1969; Thiele and Schwinn, 1969). Phosphatidylcholine is the principal phospholipid, in contrast to most bacterial species (Thiele and Schwinn, 1973). It may also form an antigenic determinant (Casao et al., 1998). Its presence confers distinctive properties on the outer membrane and may account for its reduced susceptibility to phospholipases and lysozyme. Phosphatidylglycerol and diphosphatidylglycerol are also present as major lipids. Phosphatidyl ethanolamine, phosphatidyl serine, and cardiolipin are represented as minor components. The fatty acids associated with the phospholipid fraction are unusual in that lactobacillic ($C_{19:\ cyclo}$) acid, typical of Gram-positive but not Gram-negative species, is usually the major cyclopropane fatty acid, together with its metabolic precursor *cis*-vaccenic ($C_{18:1\ cyclo}$) acid. *B. canis* is an exception, and has *cis*-vaccenic acid as the major fatty acid, with lactobacillic acid in only trace amounts. This largely accounts for the distinct position of *B. canis* vis-à-vis the other nomenspecies when fatty acid composition is used for taxonomic analysis (Tanaka et al., 1977).

The neutral lipids include unusual wax-like esters containing large quantities of myristic ($C_{14:0}$), palmitoleic ($C_{16:1}$), stearic ($C_{18:0}$) and *cis*-vaccenic ($C_{18:1}$) acids. Lactobacillic acid is absent from these compounds.

Other distinctive free lipid components of *Brucella* include ubiquinone Q-10 and ornithine-containing lipids (Thiele and Schwinn, 1973). The latter make up 32% of the total neutral lipid. They contain lactobacillic, *cis*-vaccenic, and palmitoleic acids in ester linkage and palmitic and stearic acids in amide linkage, but no hydroxy fatty acid. The function of the ornithine lipids is unknown but they are structural components of the outer membrane. It has been suggested that they are implicated in the attachment of *Brucella* cells to the surface of macrophages and lymphocytes (Cherwonogrodzky et al., 1990). Minor free lipid components include 1,3 and 1,2 diesters of glycerol and monoesters of ethylene glycol.

Cultural characteristics Most *Brucella* strains behave as slow-growing, fastidious organisms on primary isolation. Although laboratory-adapted strains may be induced to grow in synthetic media containing an ammonium salt as the sole nitrogen source, the majority of fresh isolates have complex nutritional requirements and grow poorly on ordinary nutrient media unless these are supplemented with blood, serum, or tissue extracts. Liver infusion agar was at one time widely employed for the cultivation of brucellae, but better-defined media of more consistent composition are now preferred.

For most purposes, SDA is the medium recommended and will support the growth of all species and most strains (Jones and Morgan, 1958). Tryptose agar (Difco), *Brucella* agar

(Gibco),[3] and Tryptone-soya agar (Oxoid) or equivalent media will support the growth of most strains without serum supplementation and the growth of nearly all, if heated equine serum is added to a final concentration of 5–10% (v/v). The function of the serum is not simply nutritional but is also reported to neutralize inhibitors present in the peptone component of ordinary culture media, and may be replaced by other colloids, including Tween 40. Strains of *B. abortus* biovar 2 and some of biovar 4, and *B. ovis* are the most fastidious and grow best on media containing 10% (v/v) serum.

Potato infusion agar[4] supports the growth of many *Brucella* strains and is often employed as the medium of choice for antigen or vaccine production, as it does not favor dissociation (Alton et al., 1988). For the most fastidious strains, it may be necessary to supplement this medium with horse serum to achieve satisfactory growth. Although wide variations occur between strains within nomen species, in general the most rapid growth and largest colony size are achieved by isolates of *B. suis* biovar 1 and 3 and *B. canis*. Growth is least vigorous for *B. ovis*, followed by many strains of *B. melitensis* biovar 1 and *B. suis* biovar 2. Strains of the other species and biovars usually occupy an intermediate position.

On primary isolation, colonies of any *Brucella* strain are rarely visible before 48 h. At this stage, colonies on SDA are usually 0.5–1.0 mm in diameter, raised, convex with a circular outline and an entire edge. In transmitted light, the colonies of smooth strains have a shiny surface and appear a clear pale yellow. In reflected light, the colonies have a smooth glistening surface but are slightly opalescent and bluish gray. The colonies of non-smooth strains are of similar size and shape to smooth colonies but vary considerably in color, consistency, and surface texture. They range from smooth-intermediate (SI) variants, which are morphologically indistinguishable from smooth (S) colonies but may differ in antigenic properties and phage susceptibility, through intermediate (I) forms to rough (R) and mucoid (M) variants.

R colonies are usually much less transparent than S colonies, with a more granular, dull surface, and range in color from matte white, yellowish white, or buff, to brown. Unlike S colonies, which are soft and easily emulsifiable to form stable suspensions in saline solutions, R colonies are often friable or viscous and difficult to detach cleanly from the agar surface. They will not form uniform suspensions in saline solutions but produce granular aggregates, threads, or clumps. M colonies are similar to R colonies in color and opacity but have a sticky glutinous texture.

The colonial variants of *Brucella* are best studied after four days' growth on glycerol-dextrose agar (GDA),[5] under oblique illumination as described by Henry (1933). Differentiation of the various colonial types is greatly facilitated by staining with ammonium oxalate-crystal violet before examination in reflected

---

3. Albimi *Brucella* agar is no longer available, but an equivalent product is obtainable as *Brucella* agar from Gibco Laboratories, Grand Island, New York, NY 14072, U.S.A. High quality peptone based media from other suppliers are also satisfactory.

4. Potato infusion agar: Bacto potato infusion agar (Difco), 49.0 g; glycerol, 20.0 g; distilled water, 1000 ml. The glycerol is dissolved in the water and the dehydrated medium is suspended in this solution and heated to the boiling point until dissolved. The medium is sterilized at 121°C for 15 min. It should be prepared immediately before use.

5. Glycerol dextrose agar: to blood agar base No. 2 (Oxoid) or Tryptose agar (Difco) that has been autoclaved and cooled to 56°C, sterile solutions of glycerol and D-glucose are added to give final concentrations of 2% (w/v) and 1% (w/v), respectively.

light (White and Wilson, 1951). Under these conditions, S colonies appear pale yellow, R colonies are stained red with a coarse granular appearance, and other dissociated colonies are stained various shades of pink, purple, or blue. Apart from S, R, and M colonies, numerous transitional phases may occur in cultures undergoing dissociation to the nonsmooth state.

Nonsmooth variants may arise as a result of a genetic deletion leading to synthesis of an incomplete LPS structure. This can result in mutation or deletion of genes in the *rfb* operon, of which the perosamine synthetase gene is particularly critical (Godfroid et al., 1998). However, in some nonsmooth strains other changes may accompany the LPS modification and involve deeper structures within the cell wall (Dubray and Plommet, 1976; Kreutzer and Robertson, 1979).

Most *Brucella* strains grow moderately well on sheep blood agar but the colonial appearance is not distinctive. The organisms are nonhemolytic, but a greenish brown discoloration may develop around the colonies. This is most apparent in old cultures and is probably attributable to alkali production.

The more vigorous strains of *B. abortus*, *B. melitensis*, and *B. suis* will grow on MacConkey agar, producing small lactose-negative colonies. In general, the growth of the more fastidious *Brucella* strains is inhibited on media containing bile salts, tellurite, or selenite. Tolerance to synthetic dyes varies considerably between strains and is employed as the basis for differentiation of biovars (see below and Table BXII.α.120).

Nutrition and growth conditions   Growth in simple nutrient liquid media is usually poor unless these are supplemented with blood, serum, or tissue extracts. Most strains will grow fairly well on unsupplemented, high-quality, enriched, peptone-based media such as *Brucella* broth (Gibco) or Tryptone broth (Difco).

Good growth is obtained in SDB or other media supplemented with serum. It is essential to maintain adequate aeration if satisfactory growth is to be obtained. After static incubation for 7 days at 37°C in SDB, smooth strains produce a slight to moderate uniform turbidity with a light, powdery deposit. Nonsmooth strains may produce a granular or slimy deposit, variable turbidity, and pellicle formation, sometimes accompanied by a "stalactite" appearance. Growth in static liquid media favors dissociation of S-phase cultures to nonsmooth forms. Vigorous aeration will prevent this if the medium remains adequately buffered near neutral pH.

In semisolid media, CO_2-requiring cultures of *B. abortus* and *B. ovis* produce a disk of growth a few millimeters below the surface of the medium. CO_2-independent *Brucella* species produce a uniform turbidity from the surface down to a depth of a few millimeters. There is no growth under the anaerobic conditions prevailing in static deep cultures.

The optimal growth temperature for all *Brucella* strains is 36–38°C, but growth of most strains will occur in the range 20–40°C. All strains lose viability at 56°C, although temperatures as high as 85°C may be necessary to ensure sterilization of dense suspensions (Swann et al., 1981).

Metabolism and metabolic pathways   The organisms are aerobic, although many strains grow best under microaerophilic conditions. The electron transport of *B. abortus* (the only species closely studied in this regard) consists of a branched system involving cytochromes $a + a_3$, b, c, and o, flavoproteins, and ubiquinone. It is unusually resistant to respiratory inhibitors (Rest and Robertson, 1975). Energy-yielding processes are essentially oxidative, and *Brucella* cultures show little ability to acidify car-

bohydrate media in conventional tests. They have been reported to lack phosphofructokinase (Robertson and McCullough, 1968a, b), although this enzyme was reported in extracts of *B. suis* by Roessler et al. (1952). The inability to acidify carbohydrate media has also been attributed to inhibition by peptone constituents, and acidic reactions have been demonstrated in peptone-free media (Pickett and Nelson, 1955). *B. neotomae* is exceptional in that it is reported to produce acid (but not gas) from D-glucose, D-galactose, L-arabinose, and D-xylose in conventional peptone-water sugar media (Stoenner and Lackman, 1957).

Glucose catabolism occurs via the hexose monophosphate pathway in conjunction with the tricarboxylic acid cycle (Robertson and McCullough, 1968a, b; Rest and Robertson, 1974). *meso*-Erythritol is metabolized by many *Brucella* strains in preference to glucose (Anderson and Smith, 1965), and D-erythritol-1-phosphate and other intermediates in the erythritol pathway will reduce the entire electron transport system (Rest and Robertson, 1975).

Although some laboratory-adapted strains will grow in minimal medium with an ammonium salt as the sole nitrogen source (McCullough and Dick, 1943), the nutritional requirements of *Brucella* cultures in general are complex. Multiple amino acids, thiamin, biotin, nicotinamide, iron, and magnesium ions are essential for growth; iron, manganese, and other trace elements exert a regulatory action. Iron acquisition is dependent on 2,3-dihydroxybenzoic acid siderophore production (Bellaire et al., 1999). Surplus iron is sequestered and stored by a bacterioferritin, which may also facilitate *in vivo* survival by blocking the catalytic role of iron in the formation of reactive oxygen intermediates (Denoel et al., 1995). The growth of many strains is stimulated by calcium pantothenate and *meso*-erythritol. Very few strains will grow with citrate as the sole carbon source.

Sulfur-containing amino acids and proteins are degraded and may be reduced to H_2S, but this varies with species and biovar. Indole is never produced from tryptophan or its proteins, and acetyl methyl carbinol is not produced from glucose (Voges–Proskauer reaction). Hydrolytic activity towards proteins in general is very limited and gelatin, inspissated serum, and litmus milk are not digested. *Brucella* strains either render the latter alkaline or produce no visible change.

The supplementary CO_2 required by strains of *B. abortus* and *B. ovis* for growth is used as a nutritional factor and not simply to lower oxygen tension or pH. It is incorporated directly into pyrimidines, glycine, and alanine (Newton et al., 1954).

The optimal pH for growth is between 6.6 and 7.4. Cultures die rapidly at pH 3.5 or below. Most *Brucella* strains produce alkali on protein or peptone-containing media, and this may act as a growth-limiting factor. Culture media should be adequately buffered near pH 7 for optimal growth. The optimal osmotic pressure is between 2 and 6 atmospheres, equivalent to between 0.05 and 0.15 M NaCl.

A wide range of carbohydrate and amino acid substrates is oxidized (McCullough and Beal, 1951). Manometric measurement of oxidation rates with selected substrates produces metabolic patterns that are characteristic of each species and some biovars (Meyer, 1961; Meyer and Cameron, 1961a, b). The oxidative metabolic patterns show a close correlation with the phage-sensitivity pattern and the preferred natural hosts of the species (Meyer and Morgan, 1962). As indicated in Table BXII.α.121, they are of primary importance in defining the species of *Brucella* and are of additional value in classifying the biovars of *B. suis*

TABLE BXII.α.121. Differentiation of the species of the genus *Brucella*[a]

Characteristic	B. melitensis	B. abortus	B. suis biovar					B. canis	B. neotomae	B. ovis
			1	2	3	4	5			
Lysis by Phage at RTD:										
Tb	NL	L	NL	NL	NL	NL	NL	NL	PL	NL
Wb	NL	L	L	L	L	L	L	NL	L	NL
Fi	NL	L	PL	PL	PL	L	L	NL	L	NL
BK₂	L	L	L	L	L	L	L	NL	L	NL
R/O	NL	PL	NL	NL	NL	NL	NL	NL	NL	L
R/C	NL	NL	NL	NL	NL	NL	NL	L	NL	L
Oxidation of Substrate:										
L-Alanine	+	+	d	−	d	−	−	d	d	d
L-Asparagine	+	+	−	d	−	−	+	−	+	+
L-Glutamic acid	+	+	−	d	d	d	+	+	+	+
L-Arabinose	−	+	+	+	−	−	−	d	+	−
D-Galactose	−	+	d	d	−	−	−	d	+	−
D-Ribose	−	+	+	+	+	+	+	+	d	−
L-Glucose	+	+	+	+	+	+	+	+	+	−
D-Xylose	−	d	+	+	+	+	+	−	−	−
L-Arginine	−	−	+	+	+	+	+	+	−	−
DL-Citrulline	−	−	+	+	+	+	+	+	−	−
DL-Ornithine	−	−	+	+	+	+	+	+	−	−
L-Lysine	−	−	+	−	+	+	+	+	−	−
meso-Erythritol	+	+	+	+	+	+	+	d	+	−
Preferred host:										
Cattle		+								
Desert wood rats									+	
Dogs								+		
Goats	+									
Hares				+						
Reindeer						+				
Rodents							+			
Sheep	+									+
Swine			+	+	+					

[a]Symbols: +, positive; −, negative; d, doubtful; NL, no lysis; PL, partial lysis.

(Stableforth and Jones, 1963; Jones, 1967; Jones and Wundt, 1971).

Procedures for the determination of oxidative metabolic patterns by manometric methods are described by Morgan and Gower (1966), Corbel et al. (1979), and Alton et al., (1988). A nonquantitative technique, using thin layer chromatography to detect substrate utilization, has also been employed (Balke et al., 1977; Corbel et al., 1978). More recently, a simplified colorimetric method that uses chromogenic tetrazolium substrates, and has the advantages of speed and reduced hazard, has been developed (Broughton and Jahans, 1997). Irrespective of technique, the range of substrates used includes L-arabinose, D-glucose, D-ribose, D-xylose, *meso*-erythritol, L-alanine, L-asparagine, L-glutamic acid, L-arginine, L-lysine, DL-citrulline, and DL-ornithine. For some purposes, other substrates including adonitol, L-histidine, L-serine, D-amino acids, and urocanic acid may also be used.

All strains are catalase positive and superoxide dismutases are produced (Bricker et al., 1990). The species *B. abortus*, *B. melitensis*, *B. suis*, and *B. canis* are usually oxidase positive in tests with tetramethyl-*p*-phenylenediamine, but some strains are oxidase negative and this can be a useful epidemiological marker. Most *Brucella* strains produce nitrate reductase and reduce nitrates to nitrites. Nitrites may be further reduced (Zobell and Meyer, 1932). *B. canis*, *B. neotomae*, and *B. suis* strains normally show very strong urease activity. Most strains of *B. abortus* and *B. melitensis* also produce urease, but a few do not. This may be useful as an epidemiological marker. *B. ovis* does not usually hydrolyze urea, but some strains may do so on prolonged incubation.

Genetics

Genome composition The members of the genus form a close-knit group. The DNA base composition is in the range 57.9–59 mol% G + C. DNA–DNA hybridization studies have indicated that all nomenspecies show >90% similarity to the type strain (Hoyer and McCullough, 1968a, b; Verger et al., 1985; De Ley et al., 1987). This is consistent with the concept of a single species, and Verger et al. (1985) have proposed a system of classification in which all of the current nomenspecies would be classified as biovars of *B. melitensis*. However, recent advances in knowledge of the molecular genetics of *Brucella* have provided a clearer insight into the phylogenetic position of the genus and the evolution of the distinct types within it. Unexpectedly, the molecular genetic data have provided evidence in support of the classical differentiation into nomenspecies and biovars, which were defined on the basis of patterns of phenotypic characteristics that aligned with the preferred host specificity (Michaux-Charachon et al., 1997).

Restriction endonuclease analysis using low-cleavage-frequency enzymes such as *Xba*I or *Not*I, combined with pulsed field gel electrophoresis, produced patterns which were characteristic of each nomenspecies but showed little difference between biovars within these taxa (Allardet-Servent et al., 1988). This study also suggested a genome size of about 2.6×10^6 bp for *B. melitensis* 16M and *B. suis* 1330[T]. However, development of a partial physical map of the *B. melitensis* biovar 1 strain 16M genome suggested an apparent size of about 3.3×10^6 bp (Allardet-Servent et al., 1991). An extension of this study led to the conclusion that the genome consisted of two chromosomes of 2.1

and 1.5×10^6 bp, respectively (Michaux et al., 1993). This was confirmed by insertion into each replicon of the I-*Sce*-I unique restriction site, which is a nonsymmetrical double-stranded sequence extending over 18 bp, followed by restriction fragment analysis with *Pac*I and *Spe*I (Jumas-Bilak et al., 1995). Two rRNA operons and the gene for the DnaK heat shock protein are located on the large chromosome, and a third rRNA operon and the gene encoding the equivalent of the GroEL chaperonin are located on the small chromosome. As both replicons encode functions essential for survival and replication, they are clearly chromosomes rather than plasmids (Michaux et al., 1993).

Application of these methods to the full range of nomenspecies and biovars has produced a rather more complex picture, which has confirmed genomic differences between phenotypically distinguishable types (Michaux-Charachon et al., 1997; Jumas-Bilak et al., 1998). *B. melitensis* biovars 2 and 3 have a similar genomic structure to biovar 1. *B. abortus* strains have a large chromosome of about 2.1×10^6 bp and a small replicon of about 1.15×10^6 bp. However, in biovars 1–4 the small chromosome contains a 640×10^3 bp inversion, whereas in biovars 5–9 it resembles that of *B. melitensis*. The *B. suis* biovar 1 genome is grossly similar in size and structure to that of *B. melitensis* biovar 1, except that its small chromosome is about 50×10^3 bp longer. In contrast, *B. suis* biovars 2 and 4 each have a large chromosome of 1.85×10^6 bp and a small chromosome of 1.35×10^6 bp. *B. suis* biovar 3 appears to be unique in that it contains a single chromosome of 3.2×10^6 bp. This molecule incorporates the sequences present in the large and small chromosomes of the other types. The genome structure of *B. canis* is indistinguishable from that of *B. suis* biovar 1, reinforcing the view that it is a stable R mutant of that organism. On the other hand, the large chromosome of *B. neotomae* resembles that of *B. melitensis* biovar 1, whereas its small chromosome resembles that of *B. suis* biovar 1. The *B. ovis* genome is broadly similar to that of *B. melitensis* biovar 1 but contains 30 copies of the insertion sequence IS*6501*, compared with the 5–10 copies found in the other nomenspecies. The complete genome of *B. melitensis* 16M has been published (DelVecchio et al., 2002). Analysis of this genome has indicated the absence of Type I, II, and III secretion systems but the presence of Type IV and V secretion systems and genes encoding flagellin, hemolysin, and fusion proteins.

Within the genomes of all strains studied, numerous small insertions and deletions, ranging from $1–32 \times 10^3$ bp, and three *rrn* sites have been found. These may contribute to observed differences between nomenspecies and biovars (Mercier et al., 1996; Michaux-Charachon et al., 1997). Insertion and repeated sequences have been widely documented in *Brucella* strains (Halling and Zehr, 1990). These may influence the expression of specific genes. For example, the BCSP31 protein is not expressed by *B. ovis*, possibly because of the presence of multiple copies of a repeated sequence downstream of the gene. The insertion sequence IS*6501* is universally present and has proved to be useful for identification of the genus (Ouahrani et al., 1993).

RNA composition   Analysis of ribosomal RNA gene sequences has disclosed the unexpected phylogenetic position of the genus *Brucella* (De Ley et al., 1987; Dorsch et al., 1989). 5S rRNA gene sequence analysis indicates that *Brucella* is highly homogeneous, with about 95% relatedness to the *Agrobacterium* group but with a more distant relationship to *Bartonella*. The 16S rRNA gene sequence is typical of the *Alphaproteobacteria*. 16S rRNA/DNA binding has confirmed the close relationship to *Ochrobactrum* and

a rather more distant one to *Agrobacterium*, *Mycoplana*, and *Phyllobacterium*.

Extrachromosomal genetic elements   Hitherto, there have been conflicting reports on the natural presence of plasmids or other forms of extrachromosomal DNA, such as transposons, in *Brucella*. Indirect evidence for the occurrence of plasmids, such as infectious antibiotic resistance, is also lacking. However, a small cryptic plasmid has been reported to be present in many strains. The existence of bacteriocins specific for the genus *Brucella* has not been confirmed, although bacteriocin-like effects have been occasionally reported (Pickett and Nelson, 1950; Todorov and Koleva-Todorova, 1971). Many bacteriophages active upon members of the genus *Brucella* have been described. These phages have not been shown to lyse bacteria of other genera and thus are of taxonomic value for identification at both genus and species level.

Phages and phage typing   The phage-susceptibility pattern is of major importance in definition of the species of *Brucella*. For culture identification, it is convenient to use phage preparations standardized at the routine test dilution (RTD). This is defined as the minimum concentration that produces complete lysis of the propagating strain for that phage. It is essential that phage strains be maintained on the approved propagating strain, as the host range may be modified by adaptation to new strains.

Based on their host range, the phages may be classified into seven distinct groups. These are summarized in Table BXII.α.122.

The phages in group 1, typified by the Tbilisi (Tb) strain (Popkhadze and Abashidze, 1957), are capable of efficient replication only in cells of *B. abortus* in the S, SI, or I phases. Limited replication also occurs in S, SI, or I cultures of *B. neotomae*, but the efficiency of plating is low. At high concentrations, these phages produce lysis of S, SI, or I cultures of *B. suis* by a bacteriocin-like effect. Cultures of *B. melitensis*, *B. canis*, and *B. ovis* are not lysed, nor are cultures of *B. abortus*, *B. neotomae*, or *B. suis* in the M or R phases.

Group 2 comprises those phages typified by the Firenze (Fi) strain 75/13 (Corbel and Thomas, 1976). These replicate in, and form plaques on, S, SI, and I cultures of *B. abortus*, *B. neotomae*, and *B. suis* but not on those of other *Brucella* species. They are also inactive on M and R strains of *Brucella*. The efficiency of plating is somewhat higher on *B. abortus* strains than on those of *B. neotomae* or *B. suis* biovar 4, and very much higher on these than on *B. suis* biovars 1, 2, and 3.

Group 3 includes those phages typified by the Weybridge (Wb) strain (Morris and Corbel, 1973). All of these replicate and form plaques on S, SI, and I cultures of *B. abortus*, *B. neotomae*, and *B. suis*. The efficiency of plating on the three species varies between the phages of this group, within a range of about 2 $\log_{10}$ units. Some of these phages have been reported to form plaques on cultures of *B. melitensis*, but the efficiency of plating is very low (Moreira-Jacob, 1968; Douglas and Elberg, 1976). None of the phages of group 3 will lyse M or R strains of any *Brucella* species, including *B. canis* and *B. ovis*.

Group 4 comprises the Berkeley phages Bk_0, Bk_1, and Bk_2 (Douglas and Elberg, 1976, 1978). The most useful of these, Bk_2, replicates and causes lysis in S-phase cultures of *B. abortus*, *B. melitensis*, *B. neotomae*, and *B. suis*. It shows no lytic activity towards *B. canis*, *B. ovis*, or other nonsmooth *Brucella* strains, even at high concentrations. Its lytic activity towards the smooth strains is drastically reduced as these undergo dissociation to nonsmooth forms. This is particularly the case with *B. melitensis* cultures, and

TABLE BXII.α.122. Lytic activity of phages for smooth (S) and rough (R) *Brucella* species[a]

Phage Group	Phage strain	Titer	*B. melitensis*		*B. abortus*		*B. canis*	*B. neotomae*		*B. ovis*	*B. suis*	
			S	R	S	R	R	S	R	R	S	R
1	Tb	RTD	NL	NL	L	NL	NL	NL or PL	NL	NL	NL	NL
		RTD × 10⁴	NL	NL	L	NL	NL	L	NL	NL	L	NL
2	Fi 75/13	RTD	NL	NL	L	NL	NL	L	NL	NL	PL	NL
3	Wb	RTD	V	NL	L	NL	NL	L	NL	NL	L	NL
4	BK₂	RTD	L or PL	NL	L	NL	NL	L	NL	NL	L	NL
5	R	RTD	NL	NL	V	L	NL	NL	NL	NL	NL	NL
	R/O	RTD	NL	NL	V	L	NL	V	NL	L	V	NL
	R/C	RTD	NL	NL	NL	L	L	NL	NL	L	NL	NL
6	Iz₁	RTD	L or PL	V	L	NL	NL	L	PL	NL	L[b]	V[b]
											PL[c]	NL[c]
7	Np	RTD	NL	NL	L	NL	NL	NL	NL	NL	NL	NL
		RTD × 10⁴	NL	NL	L	NL	NL	L	NL	NL	NL	NL

[a] L, confluent lysis; PL, partial lysis, single plaques, or growth inhibition; NL, no lysis; V, variable, some strains lysed; RTD, routine test dilution.

[b] Biogroups 1 and 4.

[c] Biogroups 2, 3, and 5.

the efficiency of plating of Bk₂ phage on this species may vary considerably between strains.

Group 5 includes those phages lytic for nonsmooth *Brucella* strains. All are derived from phage R, which was developed as a mutant selected from a mixture of phages active on smooth *Brucella* strains (Corbel, 1977b, 1979). Phage R is lytic for nonsmooth cultures of *B. abortus* but not for other species. The strain is genetically unstable, however, and produces smooth-specific phage mutants at high frequency during replication. Phage R/O is lytic for *B. ovis* and some S-phase *B. abortus* and *B. suis* strains, but not for *B. melitensis* or nonsmooth cultures of *B. abortus*, *B. melitensis*, *B. suis*, *B. neotomae*, or *B. canis*. It is genetically unstable and its host range is subject to variation between successive batches. Phage R/M is lytic for some R-phase cultures of *B. melitensis* and, like phages R and R/O, is genetically unstable. Phage R/C was developed by passage of phage R/O on cultures of *B. canis* RM 6/66. Unlike the other phages in this group, it is relatively stable in host range and is therefore the most useful for taxonomic purposes. It is lytic for *B. canis*, *B. ovis*, and nonsmooth strains of *B. abortus*. Some nonsmooth strains of *B. suis* and *B. melitensis* are inhibited by it at high concentrations. It is not lytic for completely smooth strains of *B. abortus*, *B. melitensis*, *B. suis*, or *B. neotomae*, although it will produce plaques on strains of *B. abortus* in the SI, I, M, or R phases.

Group 6 is typified by the Izatnagar (Iz₁) phage. This is lytic for smooth cultures of all *Brucella* nomenspecies and produces patchy incomplete lysis on rough cultures of *B. melitensis*, *B. neotomae*, and *B. suis* (Corbel et al., 1988). The lytic activity towards nonsmooth cultures is rapidly lost if the phage is propagated on smooth cultures.

Group 7 is represented by the Nepean (Np) phages. These have a lytic spectrum similar to the Group 1 phages but produce complete lysis of *B. neotomae* at RTD (routine test dilution) and do not lyse *B. suis* at 10⁴ RTD (Rigby, 1990).

The brucellaphages are morphologically similar, with a hexagonal head 55–80 nm in width and a tubular, apparently noncontractile tail 14–33 nm long. Most of the phage strains have a mean head diameter of about 60 nm and a tail length of about 25 nm. Attachment of the phages is mediated via short fibrous structures linked to the distal end of the tail, which interact with protein or glycoprotein components forming part of the outer membrane of the bacteria (Corbel, 1977a). All of the brucellaphages are relatively stable to organic solvents and nonionic and anionic detergents. They are inactivated by heat, cationic detergents, and oxidizing agents, show a variable stability to proteolytic enzymes and reducing agents, and do not require divalent cations for their interaction with *Brucella* cells. All contain DNA as their genetic material and, for the phages of group 1, this has a mol% G + C of 45.3–46.7(T_m). In spite of their isolation from different sources and geographical areas over a wide range of time, their general properties and the results of restriction nuclease analysis indicate that the brucellaphages hitherto described probably comprise host range variants of a single phage (Rigby, 1990).

Although phage typing is used primarily for identification at the nomenspecies level and usually gives a clear-cut result, some *Brucella* strains may show deviation from the standard pattern of lysis. This is particularly true for *B. melitensis* strains, which can be divided into subtypes according to pattern of susceptibility to Bk₂, Iz₁, and Wb phages (Corbel, 1987, 1989b).

Genetic exchange Chromosomal DNA exchange by conjugation has not been demonstrated between *Brucella* strains. This has impeded the application of classical methods of genetic analysis to the genus. However, transfer of plasmids by conjugation has been demonstrated between *Brucella* strains and *E. coli* under laboratory conditions. Tn5-associated kanamycin resistance has been transferred using a triparental cross protocol (Smith and Heffron, 1987). The IncP broad-host-range plasmid R751 encoding trimethoprim resistance has also been transferred from *E. coli* to reference strains of the six *Brucella* nomenspecies (Verger et al., 1993). Similar results were obtained with the broad-host-range plasmids pTH10 (IncP, Tc^R, Km^R), pSa (IncW, Tc^R, Km^R), and pR751 (IncP, Tp^R). The three plasmids were transferred successfully by conjugation from an *E. coli* donor to *B. abortus* S19, and were maintained in the transconjugant strain with no effects on biotyping properties. Moreover, plasmid transfer was also demonstrated from *B. abortus* to a recipient *E. coli* strain. In contrast, electroporation of narrow-range ColE1-derived plasmids failed to manifest a replicating plasmid, but rather facilitated transposon integration into the chromosome (Rigby and Fraser, 1989). Most recent molecular transformations of *B. abortus* have used the plasmid pBR1MCS (Kovach et al., 1994).

Variation Spontaneous variation in some of the properties of *Brucella* cultures is not uncommon. This usually results from mutations involving the modification of individual characteristics. Probably the most frequent mutation is the production of nonsmooth variants from S-phase organisms. Inactivation of the perosamine synthetase gene results in production of stable rough mutants (Godfroid et al., 1998). Other characteristics are also subject to variation. CO_2-independent variants are rapidly selected from CO_2-dependent cultures of *B. abortus* and, much less frequently, from *B. ovis*. Loss of H_2S production and urease production occurs occasionally. Changes in resistance to various dyes, including basic fuchsin, thionine, thionine blue, safranin O, and to *meso*-erythritol, and various antibiotics, occur with varying frequency. Modification of such characteristics may result in alteration of biovars. For example, *B. abortus* biovar 2 can convert to biovar 1 by acquiring resistance to basic fuchsin; the mutation rate for resistance to this dye is 6.0×10^{-10}/cell division (Shibata et al., 1962). Variations in surface antigens or phage sensitivity in S-phase cultures occur infrequently; nevertheless, smooth phage-resistant variants of *B. abortus* have been isolated under both laboratory and field conditions (Corbel and Morris, 1974, 1975; Harrington et al., 1977).

In general, *Brucella* strains maintain their nomenspecies identity, which is closely related to their preferred host specificity.

Cell wall-defective variants of *Brucella* may be induced by penicillin or glycine (Hines et al., 1964; Roux and Sassine, 1971; Hatten, 1973), hormones (Meyer, 1976), or cell cultures of immune macrophages (McGhee and Freeman, 1970a). They may also be recovered from the blood or tissues of animals or humans infected with *Brucella* strains (Nelson and Pickett, 1951; Ross and Corbel, 1980). Artificially induced spheroplasts are osmotically unstable and require hypertonic media for survival and growth. Naturally occurring *Brucella* L-forms are less osmotically sensitive but have more exacting growth requirements than the parent *Brucella* strains. Thus, they may require specially enriched media containing reducing agents (Nelson and Pickett, 1951) or high concentrations of horse serum (Corbel et al., 1980).

Cell wall-defective *Brucella* variants are highly pleomorphic and, if induced to grow on solid media, they produce bizarre colonies. These may vary from tenacious, granular, fried egg colonies of the *Mycoplasma* type, which may or may not be surrounded by a lipid film, to colonies resembling those of normal smooth or intermediate *Brucella* cultures, but very much smaller (Nelson and Picket, 1951; Corbel et al., 1980). *Brucella* L-forms and spheroplasts normally show partial or complete loss of surface antigens but may also express surface antigens characteristic of nonsmooth cultures. They usually show a reduced sensitivity to brucellaphages but may be susceptible to growth inhibition or lysis. Their pathogenicity towards experimental animals is slight or absent, unless they revert to the parent form.

Antigenic structure   Smooth species of *Brucella* show complete crossreaction in agglutination tests with unabsorbed antisera to smooth *Brucella* organisms. This crossreaction does not extend to nonsmooth variants in the M or R phases. Crossreactions between nonsmooth strains can be demonstrated by agglutination tests with unabsorbed antiserum to a rough *Brucella* strain or to the R antigen of *B. ovis*. By using crossabsorbed antisera, different quantitative distribution of two major surface epitopes, A and M, can be demonstrated in smooth strains (Wilson and Miles, 1932) and is of value in differentiating biovars of the major species (see Table BXII.α.120). It should be emphasized that this procedure should only be performed with high-quality monoclonal or polyclonal reagents of validated specificity. These antigenic determinants, together with the common C epi-

tope, are present on the O chain of the LPS, the immunodominant surface antigen of smooth *Brucella* species. The structure of this, and the related R LPS, is discussed under "Cellular composition".

An antigenic relationship, attributable to the presence of *N*-acylated-D-perosamine, has been confirmed between the LPS antigens of smooth *Brucella* species and *Yersinia enterocolitica* 0:9, *Francisella tularensis*, *Salmonella* 0:30 serotypes, *Vibrio cholerae*, and the 0:157 antigen of *E. coli* and *E. hermannii* (Corbel, 1985). The 8-kDa Braun lipoprotein also crossreacts with the homologous protein of *E. coli* (Gomez-Miguel et al., 1987).

Serological crossreactions between nonsmooth *Brucella* strains and organisms of other genera have received relatively little attention. Evidence for crossreactions between *B. canis* and *Actinobacillus equuli*, mucoid strains of *Pseudomonas aeruginosa*, and some serotypes of *Pasteurella multocida* has been presented (Weber, 1976; Carmichael et al., 1980). The outer membrane lipoproteins share antigenic determinants with *Agrobacterium* and other members of the former *Rhizobiaceae* (Cloeckaert et al., 1999).

Minor surface or subsurface antigens common to smooth and certain nonsmooth *Brucella* strains have been described. These include the "native hapten", also referred to as component 1 or second polysaccharide, present in phenol-water, ether-water, or trichloroacetic acid extracts of smooth strains and the rough strain *B. melitensis* B115 (Diaz et al., 1968; Moreno et al., 1981), which corresponds to free O chain (Cloeckaert et al., 1992). This is present in crude preparations of the polysaccharide B, which was formerly thought to be an antigen but is now known to be a nonantigenic cyclic beta-glucan (Perry and Bundle, 1990).

Many of the minor surface or subsurface antigens previously reported in smooth and some nonsmooth *Brucella* strains and designated as χ, β, and γ antigens, and the f antigen (Freeman et al., 1970; McGhee and Freeman, 1970b) have subsequently been identified with outer membrane proteins, bacterioferritin, or enzymes. Some of these proteins, including those with molecular masses of 10-, 16.5-, 19-, 25–27-, 36–38-, and 89-kDa, stimulate significant immune responses in infected individuals. Specific functions have been attached to some, such as the lumazine synthetase (Goldbaum et al., 1999), bacterioferritin (Denoel et al., 1995), and stress proteins (Robertson and Roop, 1999).

The soluble internal antigens released on disruption of *Brucella* cells can be demonstrated by immunodiffusion, immunoelectrophoresis, or immunoblotting. They are, in many cases, common to both smooth and nonsmooth strains. Some appear to be unique to the *Brucella* genus, but several at least are shared with other genera in the *Rhizobiales*, particularly *Ochrobactrum* (Velasco et al., 1998). Of particular importance is the L7/L12 ribosomal protein. This plays a key role in the induction of cell-mediated immunity and is a major protective antigen of *Brucella* (Bachrach et al., 1994a, b; Oliveira and Splitter, 1996).

Antibiotic sensitivity   *Brucella* strains show sensitivity to a wide range of antimicrobial agents, but there is limited correlation between activity *in vitro* and therapeutic efficacy, mainly because the latter is influenced by pharmacodynamics, and sustained high intracellular levels are essential for eradication of the infection. Variations in susceptibility between nomenspecies, and even between biovars and strains within species, occur. Antibiogram resistance patterns have been used for characterization at the strain level and have suggested a clonal structure, at least in *B. abortus* biovar 1 (Corbel, 1989a).

Sensitivity to beta-lactams is variable; some strains are sensitive to benzylpenicillin, ampicillin, and amoxicillin but most are resistant to methicillin, nafcillin, piperacillin, and ticarcillin. Similarly, sensitivity to cephalosporins tends to be limited except that for some third generation agents such as cefotaxime, ceftizoxime, and ceftriaxone, MICs are in the range 0.25–2 mg/l. Sensitivity to most macrolides is low but clarithromycin and azithromycin are active in the range 2–8 mg/l. The MIC for chloramphenicol is 2–3 mg/l for most strains. For rifampicin and rifapentine, MICs are in the range 0.1–2 mg/l. Similarly, sensitivity to tetracyclines is generally about 0.1 mg/l.

Nearly all strains are resistant to nalidixic acid, but MICs for fluoroquinolones are in the range 0.5–1 mg/l. Sensitivity to cotrimoxazole is borderline. Most *Brucella* strains are resistant to polymyxins, amphotericin B, bacitracin, cycloheximide, clindamycin, lincomycin, nystatin, and vancomycin at concentrations that inhibit many other organisms, and this fact has allowed the formulation of selective culture media (Farrell, 1974). *Brucella* strains are sensitive to the rhizobial peptide trifolitoxin, and on this basis have been classified in the bacteria known as Group 1 of the alpha-2 *Proteobacteria* (Triplett et al., 1994).

Pathogenicity *Brucella* species are pathogenic for a wide variety of animals, frequently producing generalized infections with a bacteremic phase followed by localization in the reproductive organs and the reticuloendothelial system. Infection in the pregnant animal often results in placental and fetal infection and this frequently causes abortion. The organisms may localize in mammary tissue and can be excreted in the milk. Because all the main species of meat- and milk-producing domesticated animals are susceptible to brucellosis and act as sources of human infection, the economic impact of the disease is enormous.

Typically, growth *in vivo* is intracellular and the organisms can survive within both granulocytes and monocytes. Infections in the natural host are rarely lethal and often mild, with clinical manifestations occurring mainly in the pregnant animal. Nevertheless, localization can occur in a wide range of organs with production of a variety of lesions.

All the *Brucella* species may produce infection in laboratory animals including guinea pigs, mice, and rabbits, but the severity of the infection varies considerably with the virulence of the infecting strain. For the guinea pig, the order of pathogenicity is *B. melitensis* > *B. suis* > *B. abortus* > *B. neotomae* > *B. canis* > *B. ovis* (Braude, 1951; Isayama et al., 1977).

The more pathogenic strains usually produce a local abscess at the site of inoculation, followed by bacteremia of varying duration. The regional lymph nodes may become enlarged and granulomatous changes develop. Similar changes occur in the liver and spleen and frequently in other organs, particularly the testes and epididymides. *B. melitensis* and *B. suis* biovars sometimes produce fatal infections (Braude, 1951). The other species rarely produce severe disease, and infection is usually self-limiting within a period varying from a few weeks to more than 6 months. The mouse is more susceptible to persistent infection with *B. neotomae*, *B. canis*, and *B. ovis* than the guinea pig (Isayama et al., 1977). Pathogenic effects of these nomenspecies are limited to slight-to-moderate splenic enlargement.

Infection in humans arises from direct or indirect contact with infected animals or by consumption of milk or meat products derived from them. Person-to-person transmission is extremely rare and plays no part in the natural history of the disease. Entry

of the organisms may be via the respiratory or gastrointestinal mucosa, or the percutaneous route. The subsequent pathogenesis is believed to follow a similar pattern to that observed in experimental animals, with proliferation in lymphoid tissue succeeded by a bacteremic phase of variable duration with, in some cases, localization in specific organs. The infection may be completely subclinical, or it can produce a subacute or acute febrile illness. In the absence of adequate antibiotic therapy, this can persist for many months and may be accompanied by the development of severe complications such as endocarditis, meningoencephalitis, arthritis, spondylitis, and orchitis. A postinfectious, chronic, debilitating syndrome may also result. *B. melitensis* accounts for the majority of severe infections, followed by *B. suis*. *B. abortus* and *B. canis* are usually associated with milder disease. Infection elicits both antibodies and cell-mediated immunity. The LPS is the dominant antigen in the serological response but antibodies and delayed hypersensitivity to a variety of proteins can develop. Further details are given by Young and Corbel (1989) and Madkour (2000).

Ecology Although easily cultivated *in vitro*, under natural conditions *Brucella* species behave as obligate parasites and do not pursue an existence independent of their animal hosts. However, under suitable conditions of temperature and humidity they can persist in the environment, for example in soil or surface water, for long periods. Their distribution is worldwide, apart from the few countries from which they have been successfully eradicated. Considerable local variations in the prevalence of particular nomenspecies and biovars occur. This is of epidemiological importance (Corbel, 1989a).

ENRICHMENT AND ISOLATION PROCEDURES

The isolation of *Brucella* may be attempted from any tissue or secretion. Those most likely to yield positive cultures include abortion material (placental cotyledon, amniotic fluid, vaginal discharge, fetal gastric contents, fetal lung, and fetal liver), lymph nodes, bone marrow, mammary gland, uterus, seminal vesicles and accessory glands, testes, and epididymides, or other organs with local lesions, milk, colostrum, semen, and blood.

Uncontaminated materials may be inoculated directly on to SDA plates. Where contaminating organisms are likely to be present, a selective medium should be used. SDA supplemented with the antibiotic formulation of Farrell (1974)[6] is satisfactory for the isolation of most smooth strains of *Brucella*, but may be too inhibitory for *B. canis* and *B. ovis*. For the isolation of these species from contaminated material, modified Thayer–Martin medium (Brown et al., 1971), or SDA containing 10% (v/v) heated horse serum and VCN-F inhibitor (BBL),[7] may be used. These media are less selective than Farrell's SDA, and it is advisable to dilute heavily contaminated material, such as semen, in 5–10 volumes

---

6. Farrell's selective medium: prepared from SDA. After addition of horse serum and glucose to the molten medium cooled to 56°C, antibiotics are added to give the following final concentrations: bacitracin, 25 U/ml; polymyxin B, 5 U/ml; cyclohexamide, 100 μg/ml; vancomycin, 20 μg/ml; nalidixic acid, 5 μg/ml; and nystatin, 100 U/ml.

7. Modified SDA antibiotic medium: SDA is prepared as previously described, but with horse serum added to a final concentration of 10% (v/v) and No. 1 agar (Oxoid) added to give a final concentration of 3% (w/v). To each liter of molten medium cooled to 56°C, one vial of VCN inhibitor (Difco; reconstituted in 10 ml of sterile distilled water) is added. This is followed by addition of 1 ml furadantin solution (10 mg/ml of 0.1 N NaOH), and the plates are poured immediately. The medium contains vancomycin (3 μg/ml), sodium colistimethate (7.5 μg/ml), nystatin (12.5 U/ml), and furadantin (10 μg/ml) (final concentrations).

of sterile isotonic solution and filter it through a membrane filter (0.8 μm pore size) before plating it out. All cultures should be incubated at 37°C under air supplemented with 5–10% (v/v) CO_2—unless isolation of a CO_2-independent species of *Brucella* is being attempted.

Direct inoculation of selective media is usually satisfactory for isolation of *Brucella* from heavily infected materials, even in the presence of contaminating organisms. However, when the *Brucella* concentration is likely to be low, as in blood, milk, or semen samples, enrichment procedures should be used. Incubation of milk samples at 4°C concentrates *Brucella* cells in the cream layer, which should then be cultured.

Enrichment may be achieved either by intramuscular inoculation of the samples into guinea pigs, followed by culture of the spleen tissue four weeks later, or by growth in a liquid enrichment medium. Guinea pig inoculation is only likely to succeed with virulent strains of *B. melitensis*, *B. abortus*, and *B. suis*.

Enrichment cultures may be performed by the two-phase system of Castañeda (1947). As liquid phase, SDB supplemented with 25% (w/v) trisodium citrate is used. The solid phase is SDA containing 2.5% (w/v) agar. The medium is made selective by adding the antibiotic formulation of Farrell (1974) to the liquid phase.

For cultures of *B. ovis* and *B. canis* the antibiotic formulation may be replaced by VCN-F inhibitor. The sample is mixed with the liquid phase and the bottles incubated at 37°C under air (with supplementary CO_2 if required). At intervals of 2–3 days, the liquid phase is tipped over the solid phase and the bottles re-incubated. Incubation is continued for up to 6 weeks or until colonies appear on the solid phase, whichever is sooner.

Blood or tissue samples may be enriched for *B. canis* by inoculation of the yolk sacs of 6–8-day-old chick embryos. Yolk from eggs with dead embryos is plated onto SDA or other suitable medium. Culture of cell-wall-defective forms of *Brucella* may be attempted by culturing samples of blood, synovial fluid, and solid tissue on Farrell's selective medium enriched with 20% (v/v) horse serum. Incubation should be continued for at least 14 days before discarding the plates. The plates should be examined for microcolonies under a stereomicroscope.

MAINTENANCE PROCEDURES

Cultures may be maintained for short periods by streaking onto SDA slopes, incubating for 72 h under air (+ 10% CO_2, v/v, if required) and then sealing the slope and storing at 4°C. This procedure needs to be repeated every 6–8 weeks. It is unsatisfactory for long-term maintenance of strains, as these are liable to change their characteristics on repeated subculturing.

Brucella cultures may be preserved satisfactorily by vacuum drying. The strains are grown on SDA slopes incubated at 37°C under air (+ 10% CO_2, if required) for 72 h. The growth is washed off the slopes and suspended in sterile rabbit serum to give a suspension of about 10^{10} organisms/ml. This is distributed in 0.1-ml volumes into sterile tubes sealed with cotton-wool plugs. The tubes are stored over phosphorus pentoxide in a desiccator kept at 4°C. The desiccator is evacuated daily until a pressure of 0.05-mm Hg can be obtained. The tubes are then sealed under vacuum in glass ampoules containing a small quantity of silica gel. Under these conditions, *Brucella* cultures may remain viable for many years.

Brucella strains may also be preserved satisfactorily by freeze-drying. General directions for this have been given by Lapage

et al. (1970) and details of the technique by Boyce and Edgar (1966).

Storage in liquid nitrogen will maintain *Brucella* cultures with a smaller decrease in viability than is produced by either vacuum drying or freeze-drying (Davies et al., 1973). Cultures are grown for 72 h on SDA slopes at 37°C in air (+ 10% CO_2, if required) and the growth suspended in single-strength Bacto-glutamate medium[8] to form a dense suspension. This is left undisturbed at 4°C for 7 days, after which volumes of up to 1 ml are placed in sterile glass screw-capped vials. The vials are allowed to equilibrate in the vapor phase or immersed in the liquid.

PROCEDURES FOR TESTING SPECIAL CHARACTERS

Safety precautions   All *Brucella* isolates should be regarded as potentially pathogenic for man. Adequate precautions against accidental infection should be taken at all stages of work involving live cultures. It is strongly recommended that all such procedures be performed by suitably trained staff using efficient exhaust protective cabinets. Particularly hazardous techniques, such as the determination of oxidative metabolism rates by respirometric methods, are best left to reference laboratories experienced in their use. Where possible, hazardous procedures should be replaced by methods that do not require viable bacteria or else are easily contained. Specific indications of safety precautions to be adopted during work with *Brucella* cultures have been given by Corbel et al. (1979) and Alton et al., (1988).

Oxidative metabolism rates   For respirometric methods, cells are grown in Roux flasks containing SDA for 48 h. They are harvested in Sorensen's phosphate buffer (pH 7.0), washed 3 times by centrifugation (10,000 × g for 15 min), and standardized turbidimetrically to a cell concentration equivalent to 0.9 mg N/ml, as determined by micro-Kjeldahl analysis. Measurements of oxygen uptake are performed using either a Warburg constant volume respirometer or a Gilson differential respirometer (see *Manometric Techniques* by Umbreit et al., 1972). Oxygen uptake with various substrates is expressed as μl/mg of cell nitrogen per hour ($Q_{O_2}N$). The following substrates are normally used for *Brucella* studies: group 1: L-alanine, L-asparagine, and L-glutamic acid; group 2: L-arginine, DL-citrulline, DL-ornithine, and L-lysine; group 3: L-arabinose, D-galactose, D-glucose, D-ribose, D-xylose, and *meso*-erythritol. Other substrates are sometimes used. The substrates are prepared as 10% (w/v) solutions in Sorensen's phosphate buffer (pH 7.0), sterilized by filtration, and stored at −20°C until needed. The DL-citrulline solution should be sterilized by autoclaving at 121°C for 15 min.

The $Q_{O_2}N$ values obtained by either the Gilson or the Warburg method are corrected for endogenous O_2 uptake (by subtracting the $Q_{O_2}N$ value for controls lacking substrate) and are compared with expected values for each species (Meyer and Cameron, 1961a, b; Meyer, 1969; Verger and Grayon, 1977). The results may be conveniently expressed in the form of a three-level metabolic profile (Verger and Grayon, 1977) or as a pattern of + and − signs (see Table BXII.α.121).

Thin-layer chromatography   The oxidative metabolic pattern may be determined semi-quantitatively by thin-layer chromatog-

---

8. Bacto-glutamate medium: Bacto-casitone (Difco) or equivalent, 25 g, is dissolved in 100 ml of distilled water and autoclaved at 115°C for 20 min. To this solution, 50 g of sucrose and 10 g of monosodium glutamate are added and dissolved by steaming for 10 min. The medium is filtered under positive pressure through Seitz EKS filters into sterile containers and finally autoclaved at 106°C for 15 min.

raphy. The results do not always coincide precisely with those obtained by respirometry, possibly because nonoxidative degradation of substrates may occur; however, the overall pattern is generally satisfactory for culture identification. The method used is a modification of that described by Balke et al. (1977). The amino acid and carbohydrate substrates are incubated with the *Brucella* suspension, and the supernatant fluids are then tested for the presence of the substrate. It is essential to include *Brucella* reference strains as standards during such testing. Details of the technique are given by Corbel and Brinley-Morgan (1984).

Substrate-specific tetrazolium reduction The method of Broughton and Jahans (1997) can be performed under containment conditions and is much more convenient and safer than respirometry. The strains are cultured on Trypticase soy agar (or equivalent) layers for 48–72h at 37°C in 10% CO_2. The bacteria are then harvested in phosphate-buffered saline, pH 7.2 and adjusted turbidimetrically to a concentration equivalent to 10^{10} organisms/ml. The substrates L-alanine, L-asparagine, L-glutamic acid, L-arginine, DL-ornithine, L-lysine, D-galactose, D-ribose, D-xylose, *meso*-erythritol, and urocanic acid are prepared as separate solutions in PBS at 20 g/l, adjusted to pH 7.2 if required, and sterilized by membrane filtration. These stock solutions are stored at 4°C and diluted to 1.25g/l in sterile PBS immediately before use. Each column in a flat-bottomed microtiter plate is loaded with 100 µl of substrate. Volumes of 50 µl of each cell suspension are added to four rows of wells, and the plates incubated at 37°C in 10% CO_2 for 18 h. After incubation, 50 µl of a sterile solution of 1 g/l of 3-[4,5-dimethylthiazol-2yl] 2,5 tetrazolium bromide (MTT) in PBS is added to each well. After 1 h at room temperature (18–22°C), 50 µl of 40% formaldehyde is added to each well and after a further 2–4 h, the optical density of each well is determined using a microplate reader at 630 nm. The values for the four replicates of each substrate are calculated. Comparisons are performed with reference strains included in each run. The pattern of utilization of substrates is similar to that obtained with respirometry, but the oxidation of urocanate distinguishes *B. suis* from *B. canis*.

Determination of phage sensitivity The colonial phase of the test culture must first be determined by streaking GDA plates and incubating under a suitable gas atmosphere at 37°C for up to 5 days. The colonial morphology is then examined in obliquely transmitted light according to Henry (1933) or after ammonium oxalate-crystal violet staining according to White and Wilson (1951). If the culture has colonies of only a single phase, representatives of these are selected for phage typing. If a mixture of smooth and nonsmooth colonial forms is present, then smooth colonies are selected for subculturing and phage typing.

For routine phage typing, a suspension of cells is prepared as described for the slide agglutination test (see below). Plates of SDA or TSA are inoculated with this suspension so as to produce confluent growth. Typing may be performed using the Tb phage standardized at the routine test dilution (RTD) and at 10,000 × RTD, or by using phages representing groups 1–5 of Corbel and Thomas (1980) standardized at RTD. For typing smooth cultures, Tb phage alone, when used at both concentrations, is usually adequate, but for nonsmooth or atypical cultures, the additional phages are useful. Discrete drops of phage preparation (~25 µl) are applied to the surface of the inoculated plate and allowed to soak in the agar. The plates are then inverted and incubated for 37°C under the appropriate gas atmosphere. They are in-

spected for growth and phage lysis after 24 h incubation, and at further 24 h intervals if required. The results of the phage sensitivity test can only be interpreted accurately if the colonial phase of the culture is known. In addition to the initial examination of the colonial morphology of the culture, a further check should be done by testing a sample of growth from each phage plate for agglutinability in 0.1% aqueous acriflavine (clumping indicates nonsmooth cells).

CO_2 requirement The culture is streaked onto duplicate SDA slopes. One of these is incubated at 37°C under an air atmosphere, the other under air supplemented with 5–10% CO_2. Growth on both slopes indicates an absence of CO_2-dependence, whereas growth only with CO_2 supplementation indicates a CO_2 requirement. The test should be performed on freshly isolated cultures as CO_2-dependence can be rapidly lost by *Brucella* strains.

H_2S production The culture is inoculated onto an SDA or TSA slope, and a strip of lead acetate-impregnated paper is placed in the mouth of the tube. This must not come into contact with the medium or with condensation on the wall of the tube. The culture is incubated under the appropriate atmosphere for 4–5 days. The paper strip is examined and changed daily. Blackening of the paper indicates H_2S production. Slight blackening of the tip of the paper for the first day only is not considered positive.

Agglutination with monospecific antisera The growth from an SDA slope incubated for 48 h is suspended in ~0.5 ml of sterile saline to give an opacity equivalent to ~10^{10} cells/ml. One drop of this suspension is mixed with an equal volume of 0.1% aqueous acriflavine on a glass slide and examined for agglutination. The absence of agglutination indicates a smooth culture. In this case, one drop of the suspension is added to each of a series of drops of A and M monospecific antisera, and to negative control serum, on a glass slide. After agitation for 1 min, each drop is examined for agglutination. Smooth *Brucella* cultures will produce agglutination with either A and /or M sera, but not with the negative control serum.

If the culture is nonsmooth (agglutination occurs with acriflavine), one drop of culture suspension is added to a drop of R antiserum and to a drop of negative control serum, with subsequent agitation for 1 min. The absence of agglutination in either serum will exclude a nonsmooth *Brucella* strain. Agglutination with only the R antiserum is strongly suggestive of a nonsmooth *Brucella*. Agglutination with both sera may also indicate nonsmooth *Brucella*, although rough organisms of other genera occasionally react in this way.

Urease activity Christensen's agar slopes are each inoculated with a single loopful of culture suspension prepared as described for the slide agglutination tests. The slopes are examined immediately and after 15 min, 1 h, 2 h, and 24 h incubation at 37°C. Cultures of *B. suis*, *B. canis*, and *B. neotomae* almost invariably produce an immediate positive reaction, turning the medium magenta within 15 min. Most strains of *B. abortus* and *B. melitensis* will give positive reactions after 1–2 h and nearly all after 18–24 h. The reference strain *B. abortus* 544 is an exception and is urease-negative even at 24 h. *B. ovis* cultures are also urease-negative under these conditions, although some strains will give positive reactions if incubated for up to 7 days.

Growth in the presence of dyes Cell suspensions prepared as for the slide agglutination tests are used to inoculate plates of SDA or other basal media containing the dyes basic fuchsin or thionine[9]. The plates are divided into four quadrants and one loopful of suspension is applied to each quadrant and streaked five times in succession without recharging the loop. The plates are incubated at 37°C under the appropriate gas atmosphere for up to four days. Growth on three or more streaks is considered to indicate resistance to the dye. Growth on only one or two streaks is not considered significant. Other dyes, such as safranin, methyl violet, pyronine Y, thionine blue, or malachite green, are occasionally used in addition to basic fuchsin and thionine. In each case, it is essential that plates inoculated with the biovar 1 reference strains for each *Brucella* species are incubated and examined in parallel with the test cultures.

Nucleic acid amplification procedures Procedures using the polymerase chain reaction (PCR) are being used increasingly for diagnostic purposes. Because of their speed, sensitivity, and safety they are likely to replace culturing methods in many instances. This means that in the future, fewer strains are likely to be available for conventional typing. A useful degree of typing is achievable by molecular methods if attention is paid to the selection of suitable primers. Use of the rRNA gene spacer region (*rrs-rrl*) and IS*6501* allows identification at genus level. Selection of primers for the *omp2a, 2b,* and *25* genes will permit putative identification to nomenspecies and, in many cases, biovar level. The procedures used are those described in the original reports (Halling and Zehr, 1990; Cloeckaert et al., 1995; Ficht et al., 1996; Fox et al., 1998c).

DIFFERENTIATION OF THE GENUS *BRUCELLA* FROM OTHER GENERA

Most of the bacterial species known to cross-react serologically with *Brucella* are unlikely to be misidentified as members of this genus, as they are easily distinguishable by their morphological, cultural, and biochemical characteristics. A possible exception is *F. tularensis*; however, this organism is unlikely to grow on ordinary brucella culture media, is not susceptible to any known brucella phages, and is usually very much smaller than *Brucella* cells. It also shows fermentative activity towards glucose and a variety of other carbohydrates, is catalase negative, has a low cytochrome *c* content, produces rapidly lethal infections in mice, and does not share internal antigens with *Brucella* strains (although unabsorbed antisera will cross-react in agglutination and immunofluorescence tests). Moreover, its DNA base composition (33–36 mol% G + C) is quite distinct from that of *Brucella*, and there is no relatedness of polynucleotide sequences via DNA–DNA hybridization (Hoyer and McCullough, 1968b).

Confusion is most likely to occur with other small, Gram-negative, nonfermentative bacteria. These include *Bordetella* species, particularly *Bordetella bronchiseptica*, which is often isolated from animal sources, and species of *Achromobacter, Acinetobacter, Branhamella, Kingella, Moraxella,* and *Neisseria*. Some of the less frequently identified *Pseudomonas* species have also been con-

fused with *Brucella*, as have occasional strains of *Haemophilus* and *Pasteurella*.

Differentiation of these taxa is achieved in the first instance by careful examination of Gram-stained smears. The morphology of many of these organisms is usually sufficiently distinctive to differentiate them from *Brucella*. Examination of growth and colonial morphology on brucella culture media incubated at 20°C and 37°C, and motility tests at both these temperatures, will enable many other isolates to be eliminated.

At this stage the colonial phase of a suspected *Brucella* strain should be determined. Slide agglutination tests with unabsorbed antiserum to smooth and nonsmooth *Brucella* strains will then permit recognition of true members of the genus in nearly every case. Occasionally, nonsmooth organisms other than *Brucella* are agglutinated by antisera to nonsmooth *Brucella* strains. Subsequent tests with phage R/C will permit *B. canis, B. ovis,* and nonsmooth *B. abortus* strains to be definitely identified as members of the genus. The identification of nonsmooth strains of the other species can occasionally present difficulties, particularly if they give atypical results in one or more of the routine typing procedures. In such cases it is necessary to resort to further tests for genus identification. These may include pulsed-field gel electrophoresis of DNA extracts (Allardet-Servent et al., 1988), PCR using genus-specific primers (Halling and Zehr, 1990; Cloeckaert et al., 1995; Fox et al. 1998c), gas-liquid chromatography of the fatty acid methyl esters (Tanaka et al., 1977), and serological demonstration of genus-specific antigens by immunodiffusion or immunoblotting (Velasco et al., 1998).

TAXONOMIC COMMENTS

Molecular genetic studies, particularly those involving sequencing and hybridization of the ribosomal RNA fractions, have clearly demonstrated that the genus *Brucella* forms a discrete homogeneous group, which, of known taxa, is most closely related to the genus *Ochrobactrum*. It quite clearly belongs in the *Alphaproteobacteria*. Sequencing of 5S and 16S rRNA and competitive binding studies of these molecules with DNA (De Ley et al., 1987; Dorsch et al., 1989; Relman et al., 1992), lipid composition (Weckesser and Mayer, 1988), antigenic relatedness (Velasco et al., 1998; Cloeckaert et al., 1999), and multi-locus enzyme electrophoresis have confirmed a close genetic relationship to *Ochrobactrum* and, to a lesser extent, *Agrobacterium, Mycoplana,* and *Phyllobacterium*, and a somewhat more distant relationship to *Bartonella*. *Brucella* also shares gene sequences essential for pathogenicity with *Agrobacterium* (Sola-Landa et al., 1998; O'Callaghan et al., 1999). The cyclic glucan synthetase gene is also shared with other species within the *Rhizobiales* (De Iannino et al., 1998). The phylogenetic relationships summarized by Yanagi and Yamasato (1993) suggested an affiliation to the *Rhizobiaceae*. However, a more recent analysis of this family indicates that *Brucella* and *Ochrobactrum* form a discrete group outside it (Fig. BXII.α.138). This suggests that until evidence to the contrary is presented, these two genera and *Mycoplana* should represent the only known members of the *Brucellaceae*.

The issue of subdivision within the genus is less clear-cut. DNA–DNA binding studies within the genus have shown all nomenspecies to share similar polynucleotide sequences; competitive binding assays have indicated a relatedness of >90% (Hoyer and McCullough, 1968a, b; Verger et al., 1985). This degree of DNA relatedness is indicative of a very close genetic similarity between all members of the genus and is consistent with the concept of a single species. This conclusion has been supported

---

9. Dye sensitivity test media: a 0.1% solution of basic fuchsin or thionine is prepared in distilled water and heated at 100°C for 1 h. Portions of this stock solution are then added to molten basal media such as SDA or TSA so as to produce a final dye concentration between 10 and 40 μg/ml. The exact concentration required to produce satisfactory differentiation of biovars must be determined for each batch of dye medium by use of *Brucella* reference strains. The plates should be incubated overnight at 37°C before use, to reveal any contaminants. Dyes from National Aniline Division, Allied Chemical and Dye Co., New York, NY, USA were originally recommended but are no longer available. Dyes from other sources may be used but each batch must be validated before use. Basic fuchsin is now difficult to obtain and has been replaced by *p*-rosaniline.

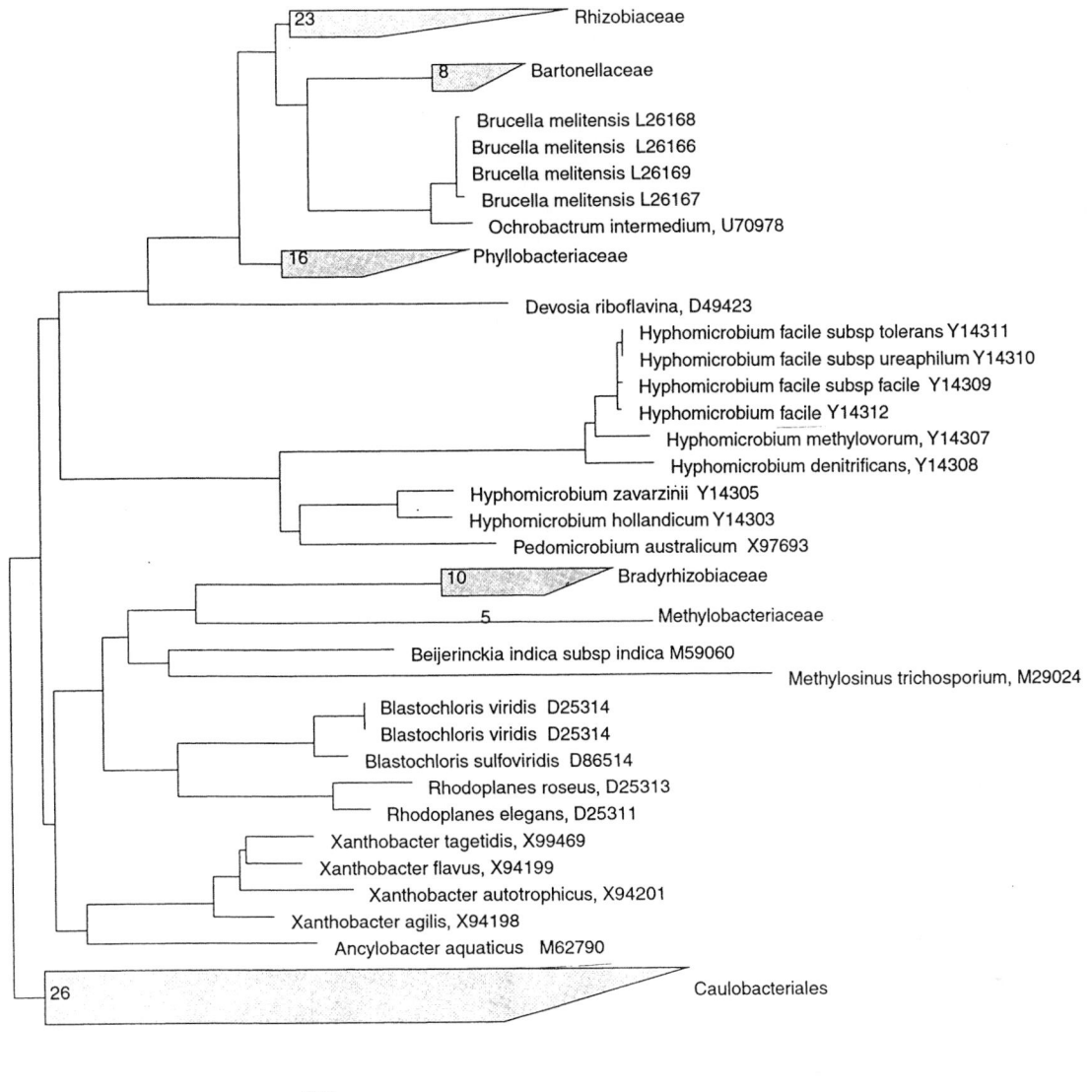

FIGURE BXII.α.138. Phylogenetic tree showing the relationships between the families of *Rhizobiaceae, Bartonellaceae, Bradyrhizobiaceae, Methylobacteriaceae,* and the order *Caulobacterales,* prepared using the weighted least squares method implemented by PAUP, with inverse-squared weighting. Distances were calculated using 1182 positions and corrected according to the HKY85 model (Courtesy of the Ribosomal Database Project).

by gene sequencing and genome mapping as indicated above. These have shown that the various nomenspecies share an essentially similar genomic structure, the main differences being in arrangement rather than content. Current evidence suggests that the presently recognizable subdivisions evolved from a precursor monochromosomal strain resembling *B. suis* biovar 3. The double chromosomes present in the other nomenspecies and biovars contain very similar genes to those represented in the single chromosome, suggesting that they arose through scission of the latter. This emphasizes the essential unity of the genus. Nevertheless significant differences in the genetic, biochemical, cultural, and pathogenic properties of the strains corresponding to its recognized subdivisions do exist, and these are correlated with host specificity (Fig. BXII.α.139). While the concept of separate species is not tenable, it is clear that subdivision at a level higher than biovar should be maintained. To minimize the confusion that would be likely to result from the adoption of a completely novel nomenclature, it is proposed that the current division into nomenspecies should be retained until an adequate genetically based classification is devised.

The present system of classification was devised by the Subcommittee on Taxonomy of the genus *Brucella* at its first meeting (Stableforth and Jones, 1963), with amendments introduced at its subsequent meetings (Jones, 1967; Jones and Wundt, 1971; Wundt and Morgan, 1975). Although a number of studies have produced suggestions for modifications of this scheme, none of the proposed changes has yet emerged as entirely satisfactory. Any improved system needs to take into account the genetic evidence indicating the monospecific nature of the genus, but at the same time acknowledge the significant divisions which exist below species level, particularly in relation to genomic arrangements and host specificity. At this level, the main problem is one of terminology, as clearly differences are discernible within the genus at a higher level than that of the conventionally defined biovars. The application of genetic typing methods has not resolved this problem; rather it has added a further dimension of

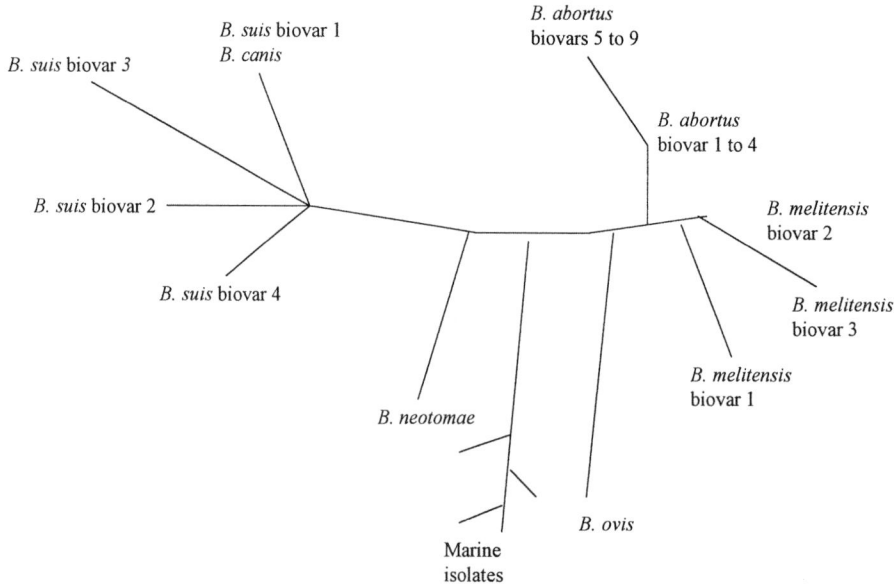

FIGURE BXII.α.139. Phylogenetic tree of the *Brucella* species.

complexity as the level of subdivision can be increased progressively by increasing the number of gene polymorphisms examined.

The biovars defined by conventional biotyping tests correspond to epidemiologically significant variants and this scheme is still useful, although it is evidently incomplete. It is also internally inconsistent, as the number of differential characteristics necessary to define a biovar is variable. For example, the conventional scheme identifies only three biovars of *B. melitensis* based on serology. In fact, strains of this organism may show a similar range of sensitivity to basic fuchsin and thionine to that shown by *B. abortus* strains and this extends the number of possible biovars to twelve, all of which have been detected at some time (Banai et al., 1990; Corbel, 1991; Bardenstein et al., 2002). Similarly, strains of the other nomenspecies that do not fit into the current biotyping scheme have also been described but not accorded separate status.

The conventional scheme also needs to be extended to ac-commodate the strains isolated from marine mammals. These differ from the currently recognized nomenspecies by as much as the latter differ from each other. It is also clear that distinct groups exist within the marine strains, corresponding to isolations from seals, porpoises, dolphins, or whales (Ross et al., 1996; Clavareau et al., 1998; Miller et al., 1999). These distinctive types may need to be regarded as separate nomenspecies. They show some divergence from the "mainstream" nomenspecies and quite clearly form a distinct branch of the genus (Fig. BXII.α.139).

At one point, the designation "*B. maris*" was suggested for the marine isolates. However, both conventional and genotyping methods indicate the heterogeneity of the group. The marine strains break down into four clear subgroups corresponding to oxidative metabolism pattern and preferred host species. These are distributed across seven marine-specific genotypes according to *omp2* and *omp25* PCR-RFLP typing. The official status of these isolates has yet to be decided but they undoubtedly comprise a novel subdivision within the genus *Brucella*.

List of species of the genus Brucella

1. **Brucella melitensis** (Hughes 1893) Meyer and Shaw 1920, 179[AL] (*Streptococcus melitensis* Hughes 1893, 235).
   me.li.ten′sis. L. adj. *Melitensis* of or pertaining to the Island of Malta (Melita) where first isolated (Bruce, 1893).

   The morphological and cultural characteristics are as described for the genus. Other characteristics of the species and its biovars are indicated in Tables BXII.α.120 and BXII.α.121. The usual natural hosts are sheep and goats, but other species, including cattle, pigs, and humans, may be infected. Smooth cultures are usually pathogenic for the guinea pig and the mouse. In general, smooth cultures of *B. melitensis* tend to be more virulent for laboratory animals than *B. abortus* and may cause fatal infections. Nonsmooth cultures are usually avirulent for both laboratory animals and the natural hosts.
   The mol% G + C of the DNA is: 58 (T_m).
   Type strain: 16M, ATCC 23456, NCTC 10094.
   GenBank accession number (16S rRNA): L26166.

   Additional Remarks: The complete genome of strain 16M has been published by DelVecchio et al. (2002). Strain 16M is also the reference strain for biovar 1. The reference strain for biovar 2 is 63/9 (ATCC 23457; NCTC 10508), and that for biovar 3 is Ether (ATCC 23458; NCTC 10509).

2. **Brucella abortus** (Schmidt 1901) Meyer and Shaw 1920, 176[AL] (*Bacterium abortus* Schmidt *in* Schmidt and Weiss 1901, 266.)
   a.bor′tus. L. gen. n. *abortus* of abortion, miscarriage; first isolated by Bang (1897).

   The morphological and cultural characteristics are as described for the genus. Other characteristics of the species and its biovars are indicated in Tables BXII.α.120 and BXII.α.121. The usual natural hosts are cattle and other bovidae. Horses, camels, sheep, deer, dogs, man, and other species may also be infected. Placentitis and abortion are usually produced in the pregnant animal. Pathogenic for laboratory animals including rabbits, guinea pigs, and mice; the guinea pig is probably the most susceptible. Rough strains are usually avirulent, but mice retain the organisms in the spleen for some time after inoculation.

The mol% G + C of the DNA is: 56 (Bd) or 57 (T_m).

Type strain: 544, ATCC 23448, NCTC 10093.

Additional Remarks: ATCC 23448 is also the reference strain for biovar 1. For biovar 2 (referred to as "dye-sensitive *B. abortus*" by Wilson, 1933) the reference strain is 86/8/59 (ATCC 23449; NCTC 10501). For biovar 3 (referred to as "Rhodesian abortus", "non-CO$_2$-requiring, thionine-resistant *B. abortus*" (Bevan, 1930) or "CO$_2$-requiring, thionine-resistant *B. abortus*" (Van der Schaaf and Rosa, 1940), the reference strain is Tulya (ATCC 23450; NCTC 10502). The reference strain for biovar 4 is 292 (ATCC 23451; NCTC 10503). For biovar 5 (referred to as British *melitensis* [Stableforth, 1959]) the reference strain is B3196 (ATCC 23452; NCTC 10504). The reference strain for biovar 6 is 870 (ATCC 23453; NCTC 10505). The reference strain for biovar 7, 63/75 (ATCC 23454; NCTC 10506), was reported to be a mixture of both A- and M-dominant strains and this biovar has been suspended. For biovar 9 (referred to as "H$_2$S-producing *B. melitensis*" by Taylor et al., 1932) the reference strain is C68 (ATCC 23455; NCTC 10507). Biovar 8 was originally described as "CO$_2$-requiring *B. melitensis*" by Taylor et al. (1932), but as no cultures of this biovar are known to exist, the status of this biovar was suspended by the Subcommittee on the Taxonomy of the genus *Brucella* in 1978 (Corbel, 1982).

3. **Brucella canis** Carmichael and Bruner 1968, 579[AL]

ca' nis. L. gen. n. *canis* of the dog.

The morphological and cultural characteristics are as described for the genus. Other characteristics are indicated in Tables BXII.α.120 and BXII.α.121. No biovars are recognized. On incubation for more than a few days, the colonies become very tenacious and viscous. The growth is almost impossible to emulsify and forms a ropy agglutinate in physiological saline. In *Brucella* broth incubated at 37°C for seven days, a moderate turbidity is produced, with a ropy or viscous sediment that cannot be uniformly resuspended. Older cultures may form a fine surface pellicle, which is easily disrupted. In serum-dextrose broth or well-buffered media, the ropy sediment or pellicle is often not produced. Cultures are always in the rough or mucoid phase on primary isolation. A smooth phase is not known. Cultures are not agglutinated by antisera monospecific for the A and M antigens, but do agglutinate with antiserum to the R antigen of *B. ovis*. Cross-reactions with the surface antigens of nonsmooth strains of other *Brucella* species also occur. Internal antigens are shared with smooth and nonsmooth strains of all *Brucella* strains. Metabolic activity is very similar to *B. suis* biovar 1, but urocanic acid is oxidized.

Pathogenic for the dog, producing chronic bacteremia and localizing granulomatous lesions. Epididymo-orchitis and prostatitis are produced in the male, and metritis, placentitis, and abortion are produced in the pregnant female. The infection is occasionally transmitted to humans; natural infections in other species have not been authenticated. Infection in guinea pigs and mice may be established by inoculation of large doses of organisms. The mouse retains the infection longer than the guinea pig.

The mol% G + C of the DNA is: 56 (T_m).

Type strain: RM6/66, ATCC 23365, NCTC 10854.

Additional Remarks: Genetic and other evidence suggests *B. canis* is a stable rough variant of *B. suis* biovar 1.

4. **Brucella neotomae** Stoenner and Lackman 1957, 947[AL]

ne.o.to' mae. M.L. fem. n. *Neotoma*, *Neotoma lepida* generic name of the desert wood rat of the Western U.S.A (Thomas); M.L. fem. gen. n. *neotomae* of the desert wood rat, the host from which the organism was first isolated.

The morphological and cultural characteristics are as described for the genus. Other characteristics are indicated in Tables BXII.α.120 and BXII.α.121. Nonpathogenic for cattle, sheep, goats, and pigs. Not proven to be pathogenic for humans. Does not apparently produce disease in its natural host (the desert wood rat) and shows minimal pathogenicity for laboratory animals. Guinea pigs develop slight splenomegaly, and sometimes epididymo-orchitis or testicular abscesses following intraperitoneal inoculation. Small granulomatous lesions develop in the liver and spleen. Mice are more susceptible to infection, and lesions may be produced by frequently passaged strains.

The mol% G + C of the DNA is: 56–57 (T_m).

Type strain: 5K33, ATCC 23459; NCTC 10084.

5. **Brucella ovis** Buddle 1956, 351[AL]

o' vis. L. gen. n. *ovis* of the sheep.

The morphological and cultural characteristics are as described for the genus. Although other species of *Brucella* stain red by modified Köster' s stain, *B. ovis* stain blue. Other characteristics are listed in Tables BXII.α.120 and BXII.α.121; no biovars are recognized. L-alanine, L-aspartic acid, L-glutamic acid, adonitol, and DL-serine are oxidized. Cultures are known to exist only in the nonsmooth colonial phase. They do not agglutinate with antisera monospecific for A and M surface antigens, but are agglutinated by antisera to the R surface antigen of *B. ovis*. Cross-reactions occur with the surface antigens of nonsmooth strains of other *Brucella* species. Many of the internal antigens are shared with other *Brucella* species, irrespective of colonial phase.

Pathogenic for sheep, producing epididymo-orchitis in the male and placentitis and abortion in the pregnant female. Goats may be infected experimentally and subclinical infections may be produced in cattle, guinea pigs, rabbits, mice, and gerbils.

The mol% G + C of the DNA is: 57–58 (T_m).

Type strain: 63/290, ATCC 25840, NCTC 10512.

GenBank accession number (16S rRNA): L26168.

Additional Remarks: The *B. ovis* genome contains many more copies of the insertion sequence IS*6501* than the genomes of the other nomenspecies.

6. **Brucella suis** Huddleson 1929, 12[AL]

su' is. L. gen. n. *suis* of the pig.

The morphological and cultural characteristics are as described for the genus. Other characteristics of the species and its biovars are indicated in Tables BXII.α.120 and BXII.α.121. Biovars 1, 2, and 3 are naturally pathogenic for pigs. Biovar 2 also naturally infects hares. Strains resembling biovar 3 have also been isolated from various species of rodents. Biovar 4 is naturally pathogenic for reindeer (Davydov, 1961). Biovar 5 has only been isolated from rodents

(Corbel, 1982; 1988). All biovars, with the possible exception of biovar 2, are pathogenic for humans. Other species, including dogs, horses, and many species of rodents, may also be infected. In the natural hosts, generalized infections are produced, with localizing lesions, particularly in the genitalia. The testes, epididymides, and seminal vesicles are usually severely affected in the male. Metritis, placentitis, and abortion are produced in the pregnant female.

Pathogenic for laboratory animals including rabbits, guinea pigs, and mice. Splenomegaly and widespread granulomatous and suppurative lesions are produced. Biovars 1 and 3 are usually the most virulent, and heavy inocula may produce fatal infections in guinea pigs.

The mol% G + C of the DNA is: 56–57 (T_m).

Type strain: 1330, ATCC 23444, NCTC 10316.

GenBank accession number (16S rRNA): L26169.

Additional Remarks: ATCC 23444 is also the reference strain for biovar 1 (referred to as "American *suis*"). For biovar 2 (referred to as "Danish or European *B. suis*" by Thomsen, 1929), the reference strain is Thomsen (ATCC 23445; NCTC 10510). For biovar 3 (referred to as "American *melitensis*" or "dye-resistant *B. suis*" [Huddleson, 1957]), the reference strain is 686 (ATCC 23446; NCTC 10511). This biovar is unique in having a single chromosome of 3.2×10^6 base pairs. For biovar 4 (formerly "*Brucella rangiferi tarandi*" [Davydov, 1961]), the reference strain is 40 (ATCC 23447; NCTC 11364). For biovar 5 the reference strain is 513 (NCTC 11996). Biovar 3 is probably closest to the ancestral prototype of the genus. It is believed that the double chromosomes of the other nomenspecies and biovars arose because of rearrangements during replication.

Other Organisms

"Brucella maris"

The designation "*Brucella maris*" has been suggested for strains of *Brucella* isolated from marine mammals. However, it is now clear that these comprise at least four distinct groups, according to conventional typing and host range, and at least seven types based on restriction fragment polymorphism of the *omp2a, 2b,* and *omp25* genes. Three new nomenspecies have been proposed to cover the known variants. These are "*B. phocae*" with two biovars, associated with seals, "*B. phoecoenae*" and "*B. delphini*" with one biovar each, isolated from porpoises and dolphins, respectively. All strains oxidize D-glutamic acid, ribose, and xylose, whereas none oxidize L-alanine, L-aspartic acid, L-arginine, L-ornithine, or L-lysine. Strains of "*B. phocae*" biovars 1 and 2 oxidize *i*-erythritol but only biovar 2 oxidizes urocanic acid. Both "*B. phocoenae*" and "*B. delphini*" oxidize D-galactose; the former oxidizes *i*-erythritol but not urocanic acid, whereas the latter shows the reverse pattern. Nearly all strains are lysed by Bk₂ phage and none are lysed by phage R/Cat-RTD. Strains of "*B. phocae*"

biovar 2 and "*B. delphini*" are lysed by Tb and Fi phages at RTD, whereas "*B. phocae*" biovar 1 and "*B. phocoenae*" are not. Only strains of "*B. phocae*" biovar 2 are usually lysed by Wb phage. There is some variation in phage sensitivity between strains within putative nomenspecies, suggesting that phage typing could be applicable for subgrouping.

Strains of both "*B. phocae*" biovars require CO_2, but the other nomenspecies do not. None produce H_2S, but all are urease positive and grow in the presence of basic fuchsin and thionine. European isolates obtained to date have been A-antigen dominant, but North American strains dominant for M antigen have also been described. The latter may correspond to an additional nomenspecies. Pathogenicity for natural host species has not been clearly defined; subclinical infection seems the usual outcome. Potentially pathogenic for humans (Brew et al., 1999).

The status of this group requires resolution. It is clear that it represents a novel subdivision within the genus, but with considerable internal diversity.

Genus II. **Mycoplana** *Gray and Thornton 1928, 82*[AL] *emend. Urakami, Oyanagi, Araki, Suzuki and Komagata 1990d, 439*

Teizi Urakami and Paul Segers

My.co.pla′na. Gr. *mykes* fungus; Gr. *plane* a wandering; M.L. fem. n. *Mycoplana* fungus wanderer.

Cells are **curved or irregular rods** with rounded ends, 0.5–0.8 × 2.0–3.0 µm. Occur singly or rarely in pairs. The cells form **branching filaments** prior to fragmentation into irregular rods. Gram negative and nonsporeforming. **Motile by means of peritrichous flagella.**

Colonies are white to light yellow.

Aerobic, with a **strictly respiratory** type of metabolism; not fermentative. Grow well on standard culture media, between pH 6 and 8, at 30°C, but not in medium containing 3% NaCl and not at 42°C. The methyl red test is negative. Indole and hydrogen sulfide are not produced. Hydrolysis of gelatin and starch does not occur. Ammonia is produced. Denitrification is negative and litmus milk is not changed. Acid is produced only oxidatively from sugars. Urease and oxidase are produced. A water-soluble fluorescent pigment is not produced on King A or King B media. The cells **accumulate granules of poly-β-hydroxybutyrate. Occur**

in soil. The major fatty acids are octadecenoic acid ($C_{18:1}$), in large amounts, hexadecanoic acid ($C_{16:0}$), and hexadecenoic acid ($C_{16:1}$). The major hydroxy fatty acid is $C_{14:0\ 3OH}$. They contain Q-10 as major ubiquinone.

The mol% G + C of the DNA is: 63–65.

Type species: **Mycoplana dimorpha** Gray and Thornton 1928, 82 emend. Urakami, Oyanagi, Araki, Suzuki and Komagata 1990d, 440.

FURTHER DESCRIPTIVE INFORMATION

The genus *Mycoplana* as emended and described by Urakami et al. (1990d) is paraphyletic, as shown by rRNA gene sequence analysis (Abraham et al., 1999, see also the chapter on *Brevundimonas*). *Mycoplana dimorpha* and *Mycoplana ramosa* belong to a separate lineage within the family *Brucellaceae*, whereas the type strains of *Mycoplana bullata* and *Mycoplana segnis* belong to the family *Caulobacteraceae*. Furthermore, *M. bullata* was shown to belong to the genus *Brevundimonas* and *M. segnis* was reclassified as *Caulobacter segnis* (Abraham et al., 1999). Since the last two species belong to other genera, the authentic genus *Mycoplana* contains only the species *M. dimorpha* (the type species) and *M. ramosa*. Therefore, in this chapter, a corresponding restrictive description is given for the genus. *M. bullata* and *Caulobacter segnis* (*M. segnis*) are described in the chapters on the genera *Brevundimonas* and *Caulobacter*, respectively, the genera in which they are presently classified.

Colonies of the genus *Mycoplana* are not mucous. Voges–Proskauer and catalase tests are positive. Ammonium salts, urea, peptone, Casamino acids, and, for most strains, nitrate are utilized as nitrogen and or carbon sources. Biotin or thiamine is required for growth. They utilize as sole source of carbon L-arabinose, D-xylose, D-glucose, D-mannose, D-fructose, D-galactose, D-sorbitol, D-mannitol, glycerol, L-glutamic acid, L-aspartic acid, acetic acid, putrescine, spermidine, and spermine. Lactose, trehalose, inositol, soluble starch, citric acid, formic acid, methanol, ethanol, butanol, monomethylamine, methane, and butane are not utilized.

In an extensive comparative study between members of the *Rhizobiaceae*, assimilation of 147 carbohydrates by the type strains of *M. dimorpha* and *M. ramosa* (NCIB 9440) was shown (de Lajudie et al., 1994). In this study, the assimilation of maltose, sucrose and succinate was shown to be negative for *M. dimorpha*, and it was demonstrated that the assimilation of ribose, L-sorbose, inositol, 2-ketogluconate, butyrate, DL-3-hydroxybutyrate, L-valine, betaine, spermine, and glucosamine can be used to distinguish the two species *M. dimorpha* and *M. ramosa*.

In the emended description of *Mycoplana* (Urakami et al., 1990d), strains of *M. dimorpha* and *M. ramosa* were shown to produce oxidatively acid from L-arabinose, D-xylose, D-glucose, D-mannose, D-fructose, D-galactose, D-sorbitol, D-mannitol, and glycerol, but not from maltose, sucrose, lactose, trehalose, inositol, and soluble starch. Hydrolysis of gelatin and starch was not observed. Contradicting the results of Gray and Thornton (1928), *Mycoplana* strains produced no acid from carbohydrates, and hydrolysis of starch and gelatin liquefaction were variable features.

The ability to decompose aromatic compounds, such as phenol and *m*-cresol, is included in the original description (Gray and Thornton, 1928).

Mycoplana strains that are diazotrophic under microaerobic conditions have been described. Several *M. dimorpha* and *M. bullata* strains showed nitrogenase activity and ^{15}N enrichment when grown on N-free media (Pearson et al., 1982).

Q-10 is the major ubiquinone; also detected were small amounts of Q-9 and traces of Q-11.

ENRICHMENT AND ISOLATION PROCEDURES

Isolation of *Mycoplana* strains was originally described by Gray and Thornton (1928): inoculate 100 ml of a solution of mineral salts, to which 0.05–0.1% phenol or 0.05% *m*-cresol is added, with soil (0.5–1 g). If growth occurs, make transfers into fresh flasks of the same medium. After further plating and purification on an agar medium, *Mycoplana* strains can be isolated.

MAINTENANCE PROCEDURES

Cultures can be maintained on most common media. Abundant growth occurs in nutrient broth, PYG broth, and peptone water, and on these solid media, when grown at 30°C in air. Cultures

can be stored at room temperature or in the refrigerator, but should be transferred weekly, although longer storage in a refrigerator is possible.

Cultures may be stored for many years by lyophilization, freezing at −80°C, or in liquid nitrogen. Cryoprotective agents such as 10% glycerol or DMSO should be added to cultures before freezing, and heavy cell concentrations should be used.

DIFFERENTIATION OF THE GENUS *MYCOPLANA* FROM OTHER GENERA

Mycoplana belongs phylogenetically to the family *Brucellaceae*, which also contains the genera *Brucella* and *Ochrobactrum*. *Mycoplana* can easily be differentiated from the other genera by the ability to form branching filaments (Fig. BXII.α.140), motility (*Brucella* is nonmotile), accumulation of granules of poly-β-hydroxybutyrate, utilization and acid production from several carbohydrates (Holmes et al., 1988; de Lajudie et al., 1994; see Table BXII.α.123 in the chapter on *Ochrobactrum*), and the hydrolysis of L- isoleucyl-β-naphthylamide and L-prolyl-β-naphthylamide hydroxychloride (API-ZYM test, Holmes et al., 1988).

The former *Mycoplana* species *M. bullata* and *C. segnis* can be distinguished from the authentic *Mycoplana* species *M. dimorpha* and *M. ramosa* by their phylogenetic position, mucoid colonies, requirement for growth of special unknown growth factors or peptone, no production of acid from several carbohydrates, occurrence of $C_{12:0\ 3OH}$, and only small amounts of $C_{14:0\ 3OH}$ and larger amounts of hexadecanoic acid ($C_{16:0}$), and the higher mol% G + C content (66–68), (Urakami et al., 1990d).

Mycoplana strains resemble *Methylobacterium*, *Xanthobacter*, and *Oerskovia* strains based on branching filament formation and/or a few chemotaxonomic properties. One can easily differentiate those taxa by their different phylogenetic position, Gram stain, the fine structure of the cells, colony color, utilization of methanol, and hydroxy fatty acid composition (Urakami et al., 1990d).

TAXONOMIC COMMENTS

In 1928, Gray and Thornton created the genus *Mycoplana* for branching bacteria isolated from soil and able to decompose aromatic compounds. They classified this new genus with two species, *M. dimorpha* (the type species) and *M. bullata*, in the family *Mycobacteriaceae*.

The ability of the strains to form branching filaments prior to breaking up into motile rods has led several workers to place the genus *Mycoplana* in the order *Actinomycetales*, although they have a cell wall of the Gram-negative type, containing *meso*-DAP and numerous amino acids (Sukapure et al., 1970; Cross and Goodfellow, 1973; Lechevalier and Lechevalier, 1981a, c). In different editions of *Bergey's Manual of Determinative Bacteriology*, the genus was placed in the family *Mycobacteriaceae* or in the family *Pseudomonadaceae* (Breed, 1957) or even, in the eighth edition, left out (Buchanan and Gibbons, 1974). In the first edition of *Bergey's Manual of Systematic Bacteriology*, it was mentioned in the section "Nocardioform Actinomycetes", as a comment, supplementary to the genus *Oerskovia* (Lechevalier and Lechevalier, 1981b).

Mycoplana, *M. dimorpha*, and *M. bullata* are valid names since they appear on the Approved Lists of Bacterial Names (Skerman et al., 1980).

The genus description was emended in 1990, and two new species, *M. ramosa* and *M. segnis*, were proposed by Urakami et

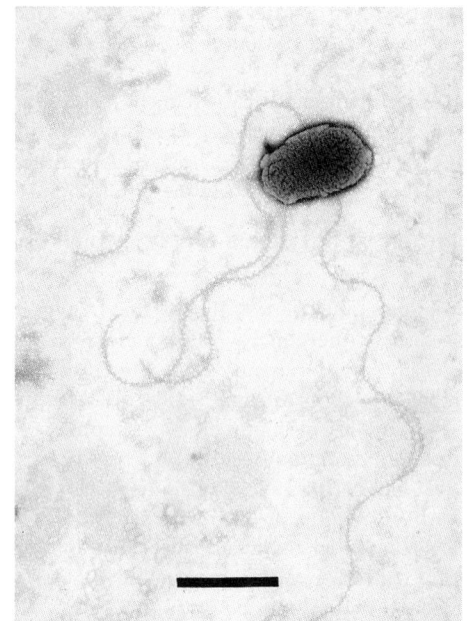

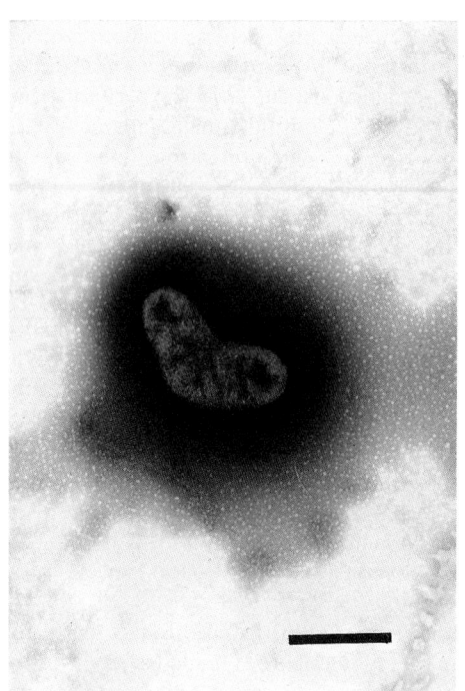

FIGURE BXII.α.140. Negatively stained cells of *Mycoplana dimorpha* ATCC 4279[T] (Bar = 1 μm). (*A*) peritrichous flagella; (*B*) exhibiting branching.

on strain authenticity investigations, identified as the original type strain for *M. bullata*. *M. bullata* strain NCIB 9440, supposedly originating from the same source, differs from this species. A new species, *M. ramosa*, was proposed for this strain. *M. segnis* was proposed as a new taxon with two strains.

Recent rRNA similarity studies showed that the genus *Mycoplana* is paraphyletic (Yanagi and Yamasato, 1993; Abraham et al., 1999; see Fig. BXII.α.122 in the chapter on *Brevundimonas*). With two species, *M. dimorpha* and *M. ramosa*, *Mycoplana* constituted a separate branch, closely related to *Brucella* and *Ochrobactrum* in the family *Brucellaceae*. 16S rRNA gene sequencing studies showed *M. bullata* to cluster in the *Caulobacteraceae* as a separate branch within the emended genus *Brevundimonas* (Abraham et al., 1999; see chapter *Brevundimonas*). It can be considered as a *Brevundimonas* species. *M. segnis*, a member of the same rRNA family, is more closely related to the type species of *Caulobacter* and has been reclassified as *Caulobacter segnis* (see Fig.BXII.α.122 and Table BXII.α.107 in the chapter *Brevundimonas*; Abraham et al., 1999). The unexpectedly high level of interrelationship of both former *Mycoplana* species with the emended genera *Brevundimonas* and *Caulobacter* was also confirmed by polar lipid and fatty acid analysis, mol% G + C content, and phenotypic characterization.

Sensitivity to the rhizobial peptide antibiotic trifolitoxin was found in *Mycoplana* species and in organisms of related genera (Triplett et al., 1994).

Lipopolysaccharides were isolated and characterized from *M. dimorpha*, *M. ramosa*, *M. bullata*, and *C. segnis* (Tharanathan et al., 1993). Similarities in chemical composition between related organisms could be shown. *M. ramosa* and *M. dimorpha* expressed ladder-like patterns on DOC-PAGE, albeit weakly, indicating some S-type lipopolysaccharides and lipid A of the D-glucosamine type. *M. bullata* and *M. segnis* showed an R-type on DOC-PAGE and had lipid As of the lipid A-DAG-type, which exclusively contained 2,3-diamino-2,3-dedeoxy-D-glucose as lipid A sugar (Tharanathan et al., 1993).

"*Mycoplana rubra*" (Devries and Derx, 1953), a pink methylamine-utilizing bacterium was deposited in the National Collection of Industrial Bacteria (Aberdeen, UK). This organism is now classified in the facultative methylotrophic genus *Methylobacterium* (Urakami and Komagata, 1984; Bousfield and Green, 1985).

FURTHER READING

Abraham, W.R., C. Strömpl, H. Meyer, S. Lindholst, E.R. Moore, R. Christ, M. Vancanneyt, B.J. Tindall, A. Bennasar, J. Smit and M. Tesar. 1999. Phylogeny and polyphasic taxonomy of *Caulobacter* species. Proposal of *Maricaulis* gen. nov. with *Maricaulis maris* (Poindexter) comb. nov. as the type species, and emended description of the genera *Brevundimonas* and *Caulobacter*. Int. J. Syst. Bacteriol. *49*: 1053–1073.

Gray, P.H.H. and H.G. Thornton. 1928. Soil bacteria that decompose certain aromatic compounds. Centralbl. Bakteriol. Parasitenkd. Infektionskr. 2. Abt. *73*: 74–96.

Urakami, T., H. Oyanagi, H. Araki, K.I. Suzuki and K. Komagata. 1990. Recharacterization and emended description of the genus *Mycoplana* and description of two new species, *Mycoplana ramosa* and *Mycoplana segnis*. Int. J. Syst. Bacteriol. *40*: 434–442.

al. (1990d). In a polyphasic approach, they recharacterized and clarified the authenticity of the type strains of *M. dimorpha* and *M. bullata* from different origins as well as a few other *Mycoplana* strains. From the strains of *M. bullata*, received from different culture collections, only ATCC 4278[T] (IFO 13290[T]) was, based

DIFFERENTIATION OF THE SPECIES OF THE GENUS *MYCOPLANA*

Characteristics useful in distinguishing the two authentic *Mycoplana* species *M. dimorpha* and *M. ramosa* are the ability of the first species to assimilate maltose, sucrose, and succinic acid, and its requirement of biotin for growth. *M. ramosa* requires thiamine

for growth (Urakami et al., 1990d). Differences in the utilization and acid production from other carbohydrates between the two species have also been shown (Holmes et al., 1988; de Lajudie et al., 1994).

List of species of the genus Mycoplana

1. **Mycoplana dimorpha** Gray and Thornton 1928, 98[AL] emend. Urakami, Oyanagi, Araki, Suzuki and Komagata 1990d, 440.

di.mor' pha. Gr. adj. *dimorpha* two forms.

Morphology and characteristics are as given for the genus.

Reduction of nitrate is variable. Biotin is required for growth. Maltose, sucrose, and succinic acid are also assimilated.

The mol% G + C of the DNA is: 64–65 (HPLC).

Type strain: ATCC 4279, IFO 13291, NCIB 9439, LMG 4061.

2. **Mycoplana ramosa** Urakami, Oyanagi, Araki, Suzuki and Komagata 1990d, 441[VP]

ra.mo' sa. M.L. *ramosa* ramous, ramose, branched, branchlike.

Morphology and characteristics are as given for the genus.

Reduction of nitrate and the Voges–Proskauer test are weakly positive. Thiamine is required for growth. Maltose, sucrose, and succinic acid are not assimilated.

The mol% G + C of the DNA is: 64 (HPLC).

Type strain: ATCC 49678, NCIB 9440, IFO 15249, LMG 3026.

Genus III. **Ochrobactrum** Holmes, Popoff, Kiredjian and Kersters 1988, 412[VP]

BARRY HOLMES

O.chro.bac' trum. Gr. adj. *ochros* pale, colorless; Gr. neut. n. *baktron* a staff or stick; M.L. neut. n. *Ochrobactrum* a colorless rod.

Rods with parallel sides and rounded ends, typically 1.0 × 1.5–2.0 μm; shorter oval forms, 1.0–1.5 μm in length, may occur. Cells usually occur singly. **Gram negative. Motile** by means of peritrichous flagella. **Obligately aerobic,** having a strictly **respiratory** type of metabolism. Optimal growth temperature in the range 20–37°C. Colonies on nutrient agar are smooth, low convex, and translucent (opaque and mucoid, quickly becoming confluent in *O. tritici*). Chemoorganotrophic, using a variety of amino acids, organic acids, and carbohydrates as carbon sources. Acid is produced from glucose, arabinose, ethanol, fructose, rhamnose, and xylose. **Oxidase positive. Catalase positive.** Indole is not formed. **Nitrates usually reduced.** Growth occurs on MacConkey agar. Urease usually positive. Ornithine decarboxylase (ODC) negative. Occur in human clinical specimens, also in soil samples and wheat roots. Parameters of DNA–rRNA hybrids indicate that *Ochrobactrum* belongs to the *Brucella* branch within rRNA superfamily IV. 16S rRNA gene sequence analysis places *Ochrobactrum* in the group of bacteria known as the α-2 subclass of the *Proteobacteria.* At the suprageneric level, *Ochrobactrum* is related to *Brucella, Phyllobacterium, Rhizobium,* and *Agrobacterium.*

The mol% G + C of the DNA is: 56–59.

Type species: **Ochrobactrum anthropi** Holmes, Popoff, Kiredjian and Kersters 1988, 412.

FURTHER DESCRIPTIVE INFORMATION

Cells of *Ochrobactrum* grown on nutrient agar are monomorphic straight rods and have rounded ends. The usual cell dimensions are 1.0 × 1.5–2.0 μm. Shorter oval forms, 1.0–1.5 μm in length, may occur.

Sudanophilic bodies have not been detected with the use of Sudan Black B.

After incubation for 24 h at 37°C on nutrient agar, colonies are <0.5 mm in diameter, circular, low convex, smooth, moist, glistening, translucent, and butyrous, and have an entire edge; pigment is not produced.

Growth at 20°C and 37°C is equally good. Few strains grow at 42°C, and while clinical strains have not been reported to grow at 5°C, environmental isolates will grow at 4°C.

Immunological cross-reactions between *O. anthropi* and *Bru-cella* spp. are apparent in animals infected by *Brucella* spp. Pulsed-field gel electrophoresis and polymerase chain reaction genome fingerprinting based on repetitive chromosomal sequences (rep-PCR) have been applied to the typing of strains of *O. anthropi* (van Dijck et al., 1995). Two chromosomes are present.

Clinical strains have been reported as resistant to ampicillin, amoxicillin + clavulanic acid, aztreonam, cefamandole, cefonicid, cefoperazone, cefoxitin, cefsulodin, ceftazidime, cefuroxime, cephalothin, chloramphenicol, erythromycin, fosfomycin, kanamycin, mezlocillin, pipemidic acid, piperacillin, pristinamycin, streptomycin, and ticarcillin. Strains are said to be sensitive to amikacin, cefoperazone, ceftriaxone, ciprofloxacin, gentamicin, imipenem, netilmicin, nalidixic acid, pefloxacin, rifampicin, tetracycline, and vancomycin; most are also susceptible to moxalactam. However, different authors find different results with cefotaxime, colistin, newer fluoroquinolones, and trimethoprim-sulfamethoxazole. Monotherapy with an appropriate aminoglycoside or an appropriate β-lactam (such as ceftriaxone) has yielded a good clinical response in the treatment of bacteremia. However, despite determining *in vitro* susceptibility to imipenem in initial isolates, treatment of two patients with this agent failed to eradicate the organism (Kern et al., 1993).

In humans, *O. anthropi* has been reported from various clinical specimens, mostly as a cause of bacteremia but also of endophthalmitis, meningitis, necrotizing fasciitis, pancreatic abscess, and puncture wound osteochondritis of the foot. *O. anthropi* is apparently often community acquired and can be pathogenic in critically ill or immunocompromised patients, with or without indwelling catheters; although it can produce clinically significant infections, it appears to be of relatively low virulence. The organism has also been reported from activated sludge, biofilm in water supply lines, soil, and the termite gut. *O. intermedium* has also been isolated from human blood and soil. *O. grignonense* and *O. tritici* have so far been reported predominantly from soil and wheat roots, respectively.

ENRICHMENT AND ISOLATION PROCEDURES

Special procedures are not required for the isolation of *Ochrobactrum* strains.

MAINTENANCE PROCEDURES

Stock cultures can be maintained for several months on a slant of Dorset egg medium (Barrow and Feltham, 1993) in a metal screw-capped bijou bottle stored at 4°C. Cultures can be maintained for longer periods by suspending growth from an 18-h-old agar slant culture in defibrinated rabbit blood, transferring to a small tube (capped or plugged), and freezing in a mixture of dry ice and alcohol prior to storage at −50°C. Strains of *Ochrobactrum* may also be preserved by lyophilization.

PROCEDURES FOR TESTING SPECIAL CHARACTERS

A major feature of *Ochrobactrum* strains is their motility by means of from two to several subpolar peritrichous flagella. Motility should be determined in hanging-drop preparations after overnight growth in nutrient broth, incubated at either 37°C or room temperature (18–22°C). The peritrichous nature of the flagella and their lateral attachment are best demonstrated by electron microscopy.

The saccharolytic ability of strains must be determined in a medium with either a low peptone content, such as Hugh and Leifson medium, or with no peptone, such as ammonium salt sugars. In high peptone-content media, alkali production from the peptone will mask the acid produced by oxidation of the carbohydrate.

DIFFERENTIATION OF THE GENUS *OCHROBACTRUM* FROM OTHER GENERA

Characteristics for the differentiation of *Ochrobactrum* from closely related and phenotypically similar genera are given in Table BXII.α.123.

TAXONOMIC COMMENTS

Holmes et al. (1988) found two DNA hybridization groups among the 56 strains for which they proposed the name *O. anthropi*. The majority of strains tested grouped around the proposed type strain of the species (NCTC 12168 LMG 3331), with an average relative binding ratio of 80.5% ± 7.2%. A second, smaller group of strains formed around reference strain NCTC 12171 (= LMG 3301), with an average relative binding ratio of 87% ± 11%. Two further reference strains occupied a somewhat separate position. The two DNA hybridization groups hybridized at an average degree of relative binding of 45% ± 4%. The major hybridization group of 32 strains could be divided into biovars A and D (corresponding to the *Achromobacter* Groups A and D of Holmes and Dawson 1983). The DNA–DNA hybridization data, however, clearly demonstrated that the three strains constituting

biovar D showed a high degree of DNA relatedness (≥90%) to the strains of biovar A, despite having phenotypic differences. The second DNA hybridization group of five strains contained two which formed biovar C (corresponding to the *Achromobacter* Group C of Holmes and Dawson 1983). However, the remaining three strains of the second DNA hybridization group (including LMG 3301) corresponded to biovar A, as did the two strains which occupied an intermediate hybridization position.

In summary, Holmes et al. (1988) found *O. anthropi* to be phenotypically homogeneous except for five strains; three of these represented a separate biovar but were genotypically indistinguishable from the major group of strains. *O. anthropi* was also genotypically homogeneous, except for seven strains, but only two of these could be distinguished phenotypically from the major group of strains. Holmes et al. (1988) proposed recognition of their group of strains as a single species. This view was not accepted by Velasco et al. (1998), who re-examined LMG 3301 (= NCTC 12171) and some additional strains, and found them more closely related to *Brucella* than to the group containing the type strain of *O. anthropi*, NCTC 12168 (= LMG 3331). The two groups could be differentiated by 16S rRNA gene sequence analysis and by 16S rRNA PCR with *Brucella*-specific primers. It was specifically recommended that distinct genomic groups not be proposed as named species until adequate phenotypic characters for differentiation from related genomic groups (Wayne et al., 1987) are available. Despite the fact that the two groups differed only in susceptibility to colistin, Velasco et al. (1998) considered this sufficient to propose a separate species, *O. intermedium*, for the group containing LMG 3301 (= NCTC 12171).

Subsequently, Lebuhn et al. (2000) studied isolates of *Ochrobactrum* from soil samples and wheat roots. They described two new species, *O. grignonense* and *O. tritici*, and found additional phenotypic characters to distinguish *O. intermedium* from *O. anthropi*. They also countenanced the future transfer of *O. intermedium* to a new genus, in order to yield monophyletic lineages that correspond to results from genotyping.

FURTHER READING

Holmes, B., M. Popoff, M. Kiredjian and K. Kersters. 1988. *Ochrobactrum anthropi* gen. nov., sp. nov. from human clinical specimens and previously known as group Vd. Int. J. Syst. Bacteriol. *38*: 406–416.

Lebuhn, M., W. Achouak, M. Schloter, O. Berge, H. Meier, M. Barakat, A. Hartmann and T. Heulin. 2000. Taxonomic characterization of *Ochrobactrum* sp. isolates from soil samples and wheat roots, and description of *Ochrobactrum tritici* sp. nov. and *Ochrobactrum grignonense* sp. nov. Int. J. Syst. Evol. Microbiol. *50*: 2207–2223.

TABLE BXII.α.123. Differentiation of *Ochrobactrum* from closely related and phenotypically similar genera[a]

Characteristic	Ochrobactrum	Agrobacterium	Alcaligenes	Brucella	Mycoplana	Phyllobacterium	Rhizobium
Acid from:							
Ethanol	+	6/7	14/19	−	−	−	−
Fructose	+	+	−	−	+	+	+
Raffinose	−	+	−	−	−	−	1/3
Utilization of:							
Glycine	+	−	13/19	−	−	−	−
Succinate	+	+	+	−	−	+	+
Rhamnose	+	+	−	−	−	+	+
Mannitol	+	+	−	−	+	+	+
Sorbitol	+	+	−	−	+	+	+
D-Arabitol	+	+	−	−	+	+	+
Hydrolysis of:							
L-Isoleucyl-β-naphthylamide	−	−	−	−	+	−	−

[a]Symbols: +, all strains tested were positive; −, all strains tested were negative. x/x, number of strains positive/number tested. Data from Holmes et al. (1988).

DIFFERENTIATION OF THE SPECIES OF THE GENUS *OCHROBACTRUM*

Phenotypic differentiation among the four named species of *Ochrobactrum* is given in Table BXII.α.124.

List of species of the genus Ochrobactrum

1. **Ochrobactrum anthropi** Holmes, Popoff, Kiredjian and Kersters 1988, 412[VP]

an.thro' pi. Gr. n. *anthropos* a human being; N.L. gen. n. *anthropi* of a human being.

The characteristics are as described for the genus. Additional characteristics are listed in Table BXII.α.125. The species was originally divided into three biovars A, C, and D, corresponding to the *Achromobacter* groups A, C, and D, respectively, of Holmes and Dawson (1983). On genomic grounds, however, only biovars A and D can be retained for *O. anthropi*; strains of biovar A differ from those of biovar D in producing acid from adonitol and sucrose, also in failing to utilize ʟ-arabitol.

The mol% G + C of the DNA is: 56–59 (T_m).

Type strain: NCTC 12168, CIP 82.11, CIP 14970, DSM 6882, IAM 14119, LMG 3331.

GenBank accession number (16S rRNA): D12794.

2. **Ochrobactrum grignonense** Lebuhn, Achouak, Schloter, Berge, Meier, Barakat, Hartmann and Heulin 2000, 2221[VP]

gri.gno.nen' se. Fr. n. *Grignon* a location in France; L. neut. suff. *-ense* indicating provenance; N.L. neut. adj. *grignonense* pertaining to Grignon, region from which the strains were isolated.

The characteristics are as described for the genus. Additional characteristics are listed in Table BXII.α.125. The principal diagnostic phenotypic characteristics of *O. grignonense* are production of acid from arabinose and melibiose, assimilation of mannose but not maltose, and utilization of malonic acid but not adonitol or ᴅ-glucosaminic acid.

The mol% G + C of the DNA is: 58 (T_m).

Type strain: OgA9a, LMG 18954, DSM 13338.

GenBank accession number (16S rRNA): AJ242581.

3. **Ochrobactrum intermedium** Velasco, Romero, López-Goñi, Leiva, Díaz and Moriyón 1998, 767[VP]

in' ter.me' dium. L. neut. adj. *intermedium* of intermediate position.

The characteristics are as described for the genus. Additional characteristics are listed in Table BXII.α.125. There are few phenotypic tests to differentiate *O. intermedium* from *O. anthropi*, but strains of the former are said to differ from those of the latter in being resistant to both colistin and polymyxin B, in failing to produce urease, and in failing to utilize quinic acid.

The mol% G + C of the DNA is: 58 (T_m).

Type strain: NCTC 12171, LMG 3301.

GenBank accession number (16S rRNA): U70978.

4. **Ochrobactrum tritici** Lebuhn, Achouak, Schloter, Berge, Meier, Barakat, Hartmann and Heulin 2000, 2222[VP]

tri' ti.ci. M.L. gen. n. *tritici*, from *Triticum*, generic name for wheat, from which the strains were isolated.

The characteristics are as described for the genus. Additional characteristics are listed in Table BXII.α.125. The principal diagnostic phenotypic characteristics of *O. tritici* are assimilation of gluconate, and utilization of sebacic acid but not γ-hydroxybutyric acid.

The mol% G + C of the DNA is: 59 (T_m).

Type strain: SCII24, LMG 18957, DSM 13340.

GenBank accession number (16S rRNA): AJ242584.

TABLE BXII.α.124. Differentiation of *Ochrobactrum* species[a]

Characteristics	*O. anthropi*	*O. grignonense*	*O. intermedium*	*O. tritici*
Acid from:				
Arabinose	−	± to +	−	−
Melibiose	−	+	−	−
Antimicrobial susceptibility:				
Colistin	S/I	R	R	S
Polymyxin B	S	R	R	S
Assimilation of:				
Maltose (48 h)	+	−	+	+
Mannose (24 h)	−	+	−	−
Urease production (24 h)	+	−	−	+
Utilization of:				
Adonitol	+	−	+	+
Cellobiose	+	− to ±	+	−
Gentiobiose	+	− to ±	+	−
Malonic acid	−	+	−	−
Quinic acid	+	−	−	−
ᴅ-Glucosaminic acid	+	−	+	+
ᴅ-Trehalose	+	−	+	+
γ-Hydroxybutyric acid	+	+	+	−

[a]For symbols see standard definitions; ±, borderline reaction; I, intermediate; R, resistant; S, susceptible. Data from Lebuhn et al., 2000.

TABLE BXII.α.125.   Additional characteristics of the species of the genus *Ochrobactrum*[a]

Characteristics	*O. anthropi*	*O. grignonense*	*O. intermedium*	*O. tritici*
Arginine dihydrolase	−	−	−	−
Assimilation of:				
Adipate	−	−	−	−
Arabinose (48 h)	+	+	+	+
Citrate (24 h)	−	+	+	− to ±
Gluconate (48 h)	− to ±	− to ±	− to ±	+
Glucose (48 h)	+	+	+	+
Mannitol	−	−	−	−
Phenylacetate	−	−	−	−
Chloramphenicol susceptibility	R	I/S	R	I
Denitrification	+	+	+	+
Fermentation of glucose	−	−	−	−
Gelatin hydrolysis	−	−	−	−
H$_2$S production	−	−	−	−
Indole production	−	−	−	−
Lysine decarboxylase	−	−	−	−
Ornithine decarboxylase	−	−	−	−
Production of β-galactosidase (ONPG and PNPG)	−	−	−	−
Tryptophan deaminase	−	−	−	−
Utilization of:				
Citrate (24 h)	−	−	−	−
Citrate (48 h)	+	+	+	+
Bromosuccinic acid	+	+	+	+
Glycerol	± to +	+	−	−
Glycyl-L-aspartic acid	+	+	+	+
Glycyl-L-glutamic acid	+	+	+	+
Hydroxy-L-proline	+	+	+	+
Inosine	+	+	+	+
Maltose	+	+	+	+
Monomethyl succinate	+	+	+	+
Phenylethylamine	−	−	−	−
Psicose	+	+	+	+
Sebacic acid	− to ±	− to ±	−	+
Succinic acid	+	+	+	+
Turanose	+	+	+	+
Tween 40	+	+	+	+
Tween 80	+	+	+	+
Uridine	± to +	+	− to ±	− to ±
Xylitol	− to ±	±	+	−
α-D-Glucose	+	+	+	+
α-Ketobutyric acid	+	+	+	+
α-Ketoglutaric acid	+	+	+	+
β-Hydroxybutyric acid	+	+	+	+
D-Alanine	+	+	+	+
D-Arabitol	+	+	+	+
D-Fructose	+	+	+	+
D-Galactose	+	+	+	+
D-Mannose	+	+	+	+
DL-α-Glycerol phosphate	± to +	± to +	−	− to ±
DL-Carnitine	+	− to ±	+	− to ±
DL-Lactic acid	+	+	+	+
i-Erythritol	+	+	+	+
L-Alanine	+	+	+	+
L-Alanyl-glycine	+	+	+	+
L-Arabinose	+	+	+	+
L-Asparagine	+	+	+	+
L-Aspartic acid	+	+	+	+
L-Fucose	+	+	+	+
L-Glutamic acid	+	+	+	+
L-Histidine	+	+	+	+
L-Leucine	+	+	+	+
L-Ornithine	+	+	+	+
L-Phenylalanine	−	−	−	−
L-Proline	+	+	+	+
L-Rhamnose	+	+	+	+
L-Serine	+	+	+	+
L-Threonine	+	+	+	+
m-Inositol	+	+	+	+
γ-Aminobutyric acid	+	+	+	+

[a]For symbols see standard definitions; ±, borderline reaction; I, intermediate; R, resistant; S, susceptible. Data from Lebuhn et al., 2000.

Family IV. **Phyllobacteriaceae** *fam. nov.*

JORIS MERGAERT AND JEAN SWINGS

Phyl.lo.bac.teri.a' ce.ae. M.L. neut. n. *Phyllobacterium* type genus of the family; *-aceae* ending to denote family; M.L. fem. pl. n. *Phyllobacteriaceae* the *Phyllobacterium* family.

Rod-shaped, ovoid, or reniform cells when cultured *in vitro.* **Nonsporeforming. Gram negative. Aerobic.** Cells cultured *in vitro* are motile by means of polar, subpolar, or lateral flagella. Strains grow well on complex solid media at 28°C. **Occur in leaf nodules and the rhizosphere of higher plants.**

The mol% G + C of the DNA is: 60–62 (De Smedt and De Ley, 1977).

Type genus: **Phyllobacterium** (ex Knösel 1962) Knösel 1984b, 356 (Effective publication: Knösel 1984a, 254.)

TAXONOMIC COMMENTS

The family presently contains seven genera. Based on 16S rRNA gene sequence analysis the type genus, *Phyllobacterium*, is closely related phylogenetically to the genus *Mesorhizobium*. Together these genera form a separate branch in the *Alphaproteobacteria* and are phylogenetically related to the genera *Rochalimaea, Mycoplana, Bartonella, Rhizobium, Agrobacterium, Sinorhizobium, Allorhizobium,** *Ochrobactrum,* and *Brucella* (Yanagi and Yamasato, 1993; de Lajudie et al., 1994, 1998a, b; Jarvis et al., 1997). Apart from 16S rRNA gene similarities, there are as yet no other arguments available to unite the genera *Mesorhizobium* and *Phyllobacterium* in a single family. Both genera are very different, not only on the basis of habitat, plant interactions, and biochemical characteristics, but also chemotaxonomically. By analysis of total cellular fatty acid composition, the genera *Phyllobacterium, Mesorhizobium,* and members of the *Agrobacterium–Rhizobium–Sinorhi-*

**Editorial Note: Allorhizobium* has been moved to the family *Rhizobiaceae* and is no longer listed in *Phyllobacteriaceae*.

zobium group form clearly separated entities, as exemplified in Fig. BXII.α.141 (see also Table BXII.α.126, and Jarvis et al., 1996). Unlike *Mesorhizobium* and other rhizobia, *Phyllobacterium* shows no evidence of nitrogen fixation (Lersten and Horner, 1976; Van Hove, 1976), and the genus lacks the *nif* HFK-like genes (Lambert et al., 1990), which occur in all the classical nitrogen-fixing bacteria.

Some differential features permitting differentiation of *Phyllobacterium* and *Mesorhizobium* are given in Table BXII.α.126.

ACKNOWLEDGMENTS

The authors are indebted to the Bijzonder Onderzoeksfonds (Belgium) for personnel grants, and to Margo Cnockaert for technical assistance.

TABLE BXII.α.126. Characteristics differentiating *Phyllobacterium* and *Mesorhizobium*[a]

Characteristic	*Phyllobacterium*	*Mesorhizobium*
Assimilation of:[b]		
Lactose	−	+
n-Valerate, aconitate, *p*-hydroxybenzoate, isovalerate	+	−
Nitrogen fixation	−	+
Occurs in root nodules	−	+
Occurs in leaf nodules	+	−
Fatty acids in whole cell extracts: [c]		
$C_{13:0 \text{ iso } 3OH}$, $C_{17:0 \text{ iso}}$	−	+
$C_{16:0 \text{ 3OH}}$, $C_{18:1 \text{ 2OH}}$, $C_{20:0}$	+	−

[a]For symbols, see standard definitions.

[b]Data from de Lajudie et al. (1994).

[c]For strains investigated and methods used: see Figure BXII.α.141; Methods used according to Jarvis et al. (1996).

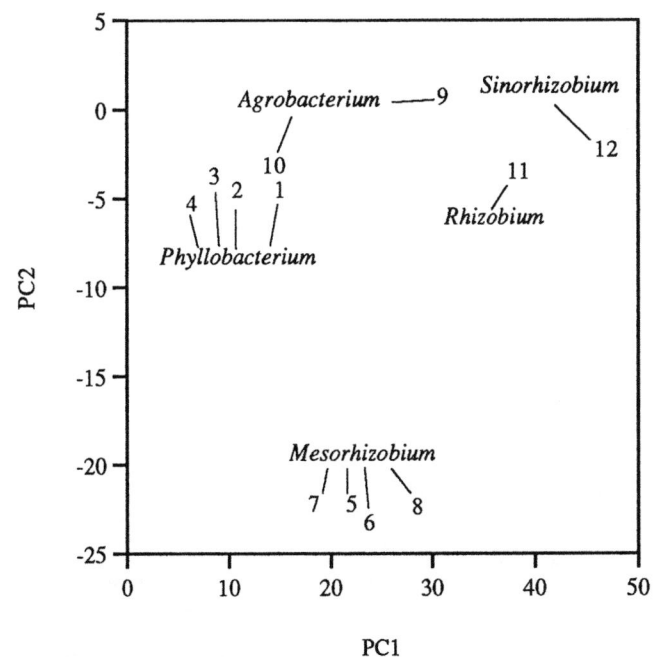
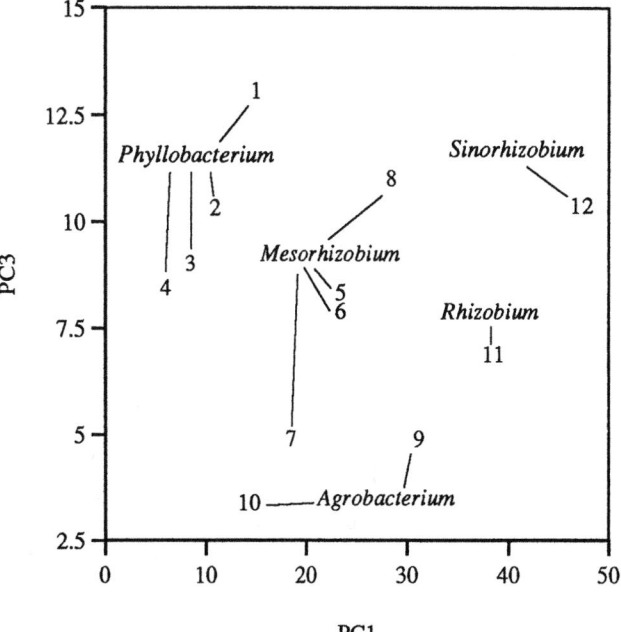

FIGURE BXII.α.141. Principal component analysis of fatty acid compositions of *Phyllobacterium* and phylogenetically related genera. Methods used are according to Jarvis et al. (1996). The following strains were investigated: *Phyllobacterium myrsinacearum* LMG 1t1[T], LMG 2t2, LMG 8225, and LMG 8229; *Mesorhizobium loti* LMG 6123, LMG 6125[T], and LMG 6126; *Mesorhizobium huakuii* LMG 14107[T], *Agrobacterium rhizogenes* LMG 150[T]; *Agrobacterium tumefaciens* LMG 187[T]; *Rhizobium leguminosarum* LMG 8820; and *Sinorhizobium fredii* LMG 6217[T].

Genus I. **Phyllobacterium** *(ex Knösel 1962) Knösel 1984b, 356[VP] (Effective publication: Knösel 1984a, 254)*

JORIS MERGAERT AND JEAN SWINGS

Phyl.lo.bac.te' ri.um. Gr. n. *phyllos* leaf; Gr. dim. neut. n. *bakterion* a small rod; M.L. neut. n. *Phyllobacterium* leaf bacterium.

Cells cultured *in vitro* are **rod-shaped, ovoid, or reniform**. Clumping and/or **star formation** occurs when cells are grown in carrot juice medium. **Gram negative. Aerobic**, having a strictly respiratory type of metabolism with oxygen as the terminal electron acceptor. Cultures are **motile** by means of polar, subpolar, or lateral flagella. Optimal temperature for growth, 28–34°C. After 1–2 days of incubation on nutrient agar, *Phyllobacterium* colonies are whitish-gray, punctiform or circular, with a colony diameter of <1 mm, and regularly edged. Most cultures are mucoid and confluent. After 4 d, colonies reach a maximum diameter of 4 mm. **Oxidase and catalase positive. Chemoorganotrophic**, using a variety of sugars or salts of organic acids as carbon sources. Do not hydrolyze starch, pectin or cellulose. **Occur in leaf nodules and the rhizosphere of higher plants.**

The mol% G + C of the DNA is: 60.3–61.3 (De Smedt and De Ley, 1977).

Type species: **Phyllobacterium myrsinacearum** (ex Knösel 1962) Knösel 1984b, 356 (Effective publication: Knösel 1984a, 254) (*Phyllobacterium rubiacearum* Knösel 1984a, 254.)

FURTHER DESCRIPTIVE INFORMATION

Phyllobacterium within leaf nodules of plants develop into pleomorphic cells that are rod shaped, ellipsoidal, or branched (Fig. BXII.α.142). In liquid media, especially in carrot juice medium[1], the cells appear as motile, straight rods, which form characteristic star clusters (Fig. BXII.α.143) after an initial phase of intensive swimming.

The maximum temperature at which growth occurs is 36°C; the minimum is 4–5°C. Cells heated in saline are killed within 10 min at 52°C.

Whole-cell fatty acid extracts of cells grown on modified TY medium[2] are composed of mainly octadecenoic acids ($C_{18:1}$) and cyclopropaneoctadecanoic acid ($C_{19:0\ cyclo}$), lower amounts of hydroxylated straight-chain fatty acids ($C_{16:0\ 3OH}$ and $C_{18:1\ 2OH}$) and hexadecanoic acid ($C_{16:0}$), and minor amounts of other fatty acids.

Acid is produced from a wide range of carbohydrates without gas formation: pentoses, hexoses, maltose, inulin, glycerol, adonitol, sorbitol, and dulcitol. Starch, pectin, and cellulose are not hydrolyzed. Ammonium salts, nitrates, and most amino acids can serve as nitrogen sources. Strains may or may not reduce nitrate to nitrite.

Growth of *Phyllobacterium* is inhibited by deoxycycline, novobiocin, framycetin, and tetracycline (Lambert et al., 1990). Sensitivity towards the rhizobial peptide antibiotic trifolitoxin has been reported (Triplett et al., 1994).

Although over 400 plant species of three genera of Rubiaceae and one genus of Myrsinaceae have been reported to have bacterial leaf nodules (see Lersten and Horner, 1976), *Phyllobacterium* has been isolated by Knösel (1962) only from *Pavetta zimmermanniana* (Rubiaceae), *Ardisia crispa*, and *Ardisia crenata* (Myrsinaceae). Although Knösel (1984a) has stated that *Phyllobacterium* is able to induce nodules on leaves, there is still no proof for the leaf nodulation capacity of *Phyllobacterium* (Swings et al., 1992), despite more than a century of research on leaf nodulation (reviewed by Lersten and Horner, 1976).

It has been suggested that leaf-nodule bacteria produce plant growth hormones, particularly cytokinins, which are necessary for the normal function of the plant (Rodrigues-Pereira et al., 1972; Fletcher and Rhodes-Roberts, 1976). The auxin indolyl-3-acetic acid has been isolated from the growth medium of a *Phyllobacterium* strain (Lambert et al., 1990).

It has long been claimed that leaf-nodule bacteria can fix nitrogen, but this has been contested by Van Hove (1976) and Lersten and Horner (1976). The lack of the ability to fix nitrogen is also indicated by the absence of the *nif* HFK-like genes (Lam-

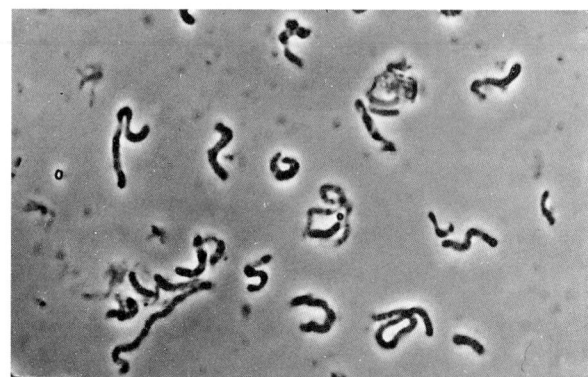

FIGURE BXII.α.142. Phase-contrast photomicrograph showing bacteroids of *Phyllobacterium myrsinacearum* (synonym *Phyllobacterium rubiacearum*) from leaf nodules of *Pavetta zimmermanniana* (× 2000).

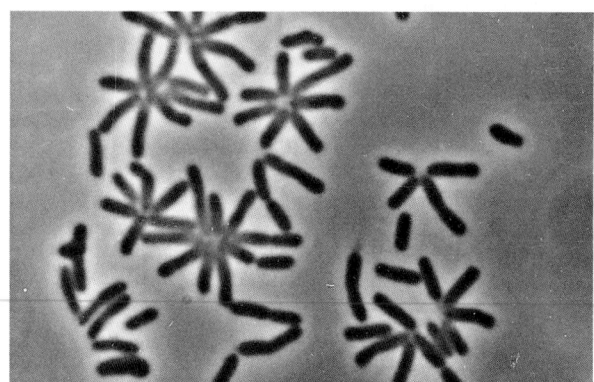

FIGURE BXII.α.143. Phase-contrast photomicrograph showing star clusters of cells of *Phyllobacterium myrsinacearum* (synonym *Phyllobacterium rubiacearum*) in carrot juice medium after incubation at 28°C for 24 h (× 2000)

---

1. Carrot juice medium has the following composition: fresh carrot juice, 500 ml; water, 500 ml; $FeSO_4 \cdot 7H_2O$, 0.1 g; $MnSO_4 \cdot H_2O$, 0.1 g; pH 7.2. The medium is sterilized by fractional sterilization. For a solid medium, add 15.0 g/l agar. (Knösel, 1984a).

2. According to the methods described by Jarvis et al. (1996) and Sasser (1990a).

bert et al., 1990), which occur in all the classical nitrogen-fixing bacteria.

Phyllobacterium is able to interact with plant tissues, as demonstrated by the tumor induction on *Kalanchoe* plants by a Ti plasmid-carrying *Phyllobacterium* strain (Lambert et al., 1990). Tumor induction involves the attachment of bacterial cell wall sites to plant cell wall sites prior to the induction of T-DNA transfer. The chromosomal genes *chv*A, *chv*B, *exo*C, and *att* are the only ones found to be involved in attachment, but these cannot be detected in *Phyllobacterium* strains, suggesting that other genes with similar functions are present. There are no data indicating that *Phyllobacterium* is pathogenic or deleterious to plants.

A number of *Phyllobacterium* isolates show antifungal and antibacterial activities (Lambert et al., 1990). The fact that *Phyllobacterium* is a predominant bacterium on the root surface of sugar beet plant tissues, its capacity to communicate with plant tissues, and its nonpathogenic status make this bacterium an interesting new candidate for use in plant-growth promotion or biological control of soil-borne diseases. Indeed, some of the sugar beet isolates exert broad antibacterial and antifungal activity (Lambert et al., 1990). It has also been shown that a *Phyllobacterium* strain is able to reduce the symptoms of vascular *Fusarium* wilt when artificially introduced in seedlings of cotton plants prior to inoculation with *Fusarium oxysporum* f. sp. vasinfectum (Chen et al., 1995a).

ENRICHMENT AND ISOLATION PROCEDURES

To isolate strains from leaf nodules, Knösel (1984a) used the following procedure. Washed leaf pieces showing nodules are macerated by rubbing and placed in saline. After shaking, dilutions are plated onto carrot juice agar containing yeast extract. After incubation at 28°C, typical colonies are transferred into liquid carrot juice medium for micromorphological characterization. The cultures should be observed by phase-contrast microscopy after 24–48 h to confirm the presence of star clusters (Fig. BXII.α.143).

Lambert et al. (1990) described a procedure for isolating strains from the rhizosphere of sugar beets. The entire root system of the plant is carefully washed to remove adhering soil, then vigorously shaken for 15 minutes using a flask shaker in a phosphate-buffered saline solution containing 0.025% Tween 20. Serial dilutions of the resulting suspensions are plated on 10% trypticase soy agar. After 2 d incubation at 28°C, *Phyllobacterium* colonies are beige-colored, ~5 mm large, convex with entire margins, and very mucoid and glistening.

To extract vascular sap from cotton plants for the subsequent isolation of *Phyllobacterium*, the use of the Scholander pressure bomb has been shown to be a successful technique (Hallmann et al., 1997).

MAINTENANCE PROCEDURES

Stock cultures on trypticase soy agar or nutrient agar may be maintained at 4°C for 2–3 months. Long-term preservation can be achieved by lyophilization or by storing cell suspensions in 25% glycerol at −70°C (Swings et al., 1992).

TABLE BXII.α.127. Differential characteristics of *Phyllobacterium* and *Ochrobactrum*[a,b]

Characteristic	*Phyllobacterium*	*Ochrobactrum*
Utilization of:[c]		
L-citruline, glutarate, *meso*-erythritol, glycine	−	+
L-tryptophan	+	−
Hydrolysis of:		
p-nitrophenyl-α-maltoside and *p*-nitrophenyl-α-xylopyranoside[d]	+	−

[a]For symbols, see standard definitions.
[b]Data from Swings et al. (1992).
[c]Using API 50CH, 50AO and 50AA auxanography strips (Biomérieux).
[d]Using API ZYM enzymic strips (Biomérieux).

DIFFERENTIATION OF THE GENUS *PHYLLOBACTERIUM* FROM OTHER GENERA

The genus *Phyllobacterium* is phenotypically most similar to the genus *Ochrobactrum* (Swings et al., 1992). *Phyllobacterium* is isolated from plants or the rhizosphere, whereas *Ochrobactrum* is typically of clinical origin. Table BXII.α.127 shows useful characteristics for the differentiation of the two genera from each other.

TAXONOMIC COMMENTS

The genus *Phyllobacterium* was originally restricted to bacteria that develop within leaf nodules of the tropical ornamental plants. Two species, *Phyllobacterium myrsinacearum* (from *Ardisia* leaf nodules) and *Phyllobacterium rubiacearum* (from *Pavetta* leaf nodules), have been defined on the basis of plant source, nitrate reduction, and flagellar characteristics. However, conflicting results have been obtained with regard to the nitrate-reduction capability by Lambert et al. (1990). It has become evident that the natural habitat of these bacteria is not limited to leaf nodules, and that these microorganisms are also very common on the rhizoplane and phylloplane of other plants (including sugar beet roots and *Polymnia* leaves) and in roots of cotton plants (Lambert et al., 1990; Hashidoko et al., 1994; Chen et al., 1995a; Hallmann et al., 1997). The phyllobacteria from *Ardisia*, *Pavetta*, and sugar beet roots are phenotypically very similar (Lambert et al., 1990) and the 16S rRNA sequences of the type strains of *P. myrsinacearum* and *P. rubiacearum* differ from each other by only 2 nucleotides (Yanagi and Yamasato, 1993). The sugar beet isolates show whole-cell protein-electrophoretic patterns almost identical to those of the type strain of *P. rubiacearum* and very similar to those of *P. myrsinacearum* strains (Lambert et al., 1990). Strains from these three different habitats also show very similar fatty acid compositions (see Fig. BXII.α.141 in *Phyllobacteriaceae*). DNA pairing data are not available to draw definite conclusions about speciation within the genus. Based upon their phenotypic and chemotaxonomic resemblance and for practical reasons of identification, all the phyllobacteria are united here within a single species, *Phyllobacterium myrsinacearum* (subjective synonym *Phyllobacterium rubiacearum*) (Mergaert et al., 2002).

ACKNOWLEDGMENTS

The authors are indebted to the Bijzonder Onderzoeksfonds (Belgium) for personnel grants.

List of species of the genus Phyllobacterium

1. **Phyllobacterium myrsinacearum** (ex Knösel 1962) Knösel 1984b, 356 (Effective publication: Knösel 1984a, 254) (*Phyllobacterium rubiacearum* Knösel 1984a, 254.)

   myr.si.na.ce.a′rum. M.L. fem. pl. n. *Myrsinaceae* family of plants; M.L. fem. gen. pl. n. *myrsinacearum* of the myrsine family.

Cells cultured *in vitro* have a maximum size of 1.1 × 2.2 μm. The bacteroids in leaf nodules are rod-shaped or ellipsoidal, with some branched forms. A pellicle is formed in liquid media. Other characteristics are listed in Table BXII.α.128. Found in leaf nodules of tropical ornamental plants (species of Myrsinaceae and Rubiaceae) (see Figs. BXII.α.144 and BXII.α.145), on the surface of sugar beet (*Beta vulgaris*) roots (Lambert et al., 1990), in internal tissues of cotton plants (*Gossypium*) (Chen et al., 1995a; Hallmann et al., 1997), and from damaged leaf surfaces of *Polymnia sonchifolia* (Hashidoko et al., 1994).

The mol% G + C of the DNA is: 60.3–61.3 (T_m) (De Smedt and De Ley, 1977).

Type strain: ATCC 43590, DSM 5892, IAM 13584, LMG 2t2 (holotype).

GenBank accession number (16S rRNA): Strain LMG 2t2, D12789.

Additional Remarks: Reference strains are LMG 1t1 (subjective synonym *P. rubiacearum*) and LMG 8225 (from sugar beet roots). The GenBank accession number (16S rRNA) for LMG 1t1 is D12790.

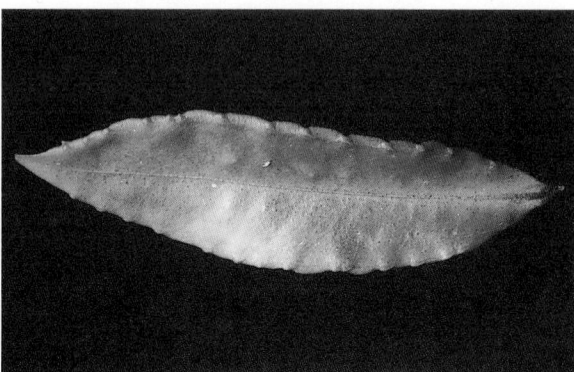

FIGURE BXII.α.144. Photograph showing a leaf of *Ardisia crispa* with nodules located at the leaf margin.

FIGURE BXII.α.145. Photograph showing a leaf of *Pavetta zimmermanniana* with nodules distributed all over the leaf blade.

TABLE BXII.α.128. Physiological and biochemical features of *Phyllobacterium myrsinacearum*[a,b]

Characteristic	Reaction
Glucose fermentation, Voges–Proskauer, methyl red reaction, 3-ketolactose test, reducing compounds from gluconate	−
Oxidase, catalase	+
Arginine dihydrolase; arginine, ornithine, and lysine decarboxylases; urease	−
H₂S production from thiosulfate	+
Indole formation from tryptophan	−
Pectate breakdown, gelatin liquefaction, starch hydrolysis, Tween 20 and Tween 40 hydrolysis, deoxyribonuclease	−
Growth at pH 4.2	−
Growth at pH 5.3 and pH 7.8	+
Growth at 36°C	+
Growth in the presence of 3% NaCl	±
Growth in the presence of 10% glucose	+
Growth in litmus milk	−
Growth on NH₄Cl, KNO₃, or sodium glutamate as sole source of nitrogen	+
Utilization of the following carbohydrates as carbon source (API 50CH):[c]	
Glycerol, D-arabinose, L-arabinose, D-ribose, D-xylose, adonitol, D-galactose, D-glucose, D-fructose, D-mannose, L-rhamnose, dulcitol, *meso*-inositol, D-mannitol, sorbitol, methyl-α-D-glucoside, N-acetylglucosamine, D-cellobiose, maltose, sucrose, trehalose, xylitol, β-gentiobiose, D-turanose, D-lyxose, D-tagatose, D-fucose, L-fucose, D-arabitol, D-gluconate, 2-ketogluconate	+
Utilization of the following organic acids (sodium salts) as carbon source (API 50AO):[c]	
Acetate, propionate, butyrate, isobutyrate, *n*-valerate, isovalerate, succinate, fumarate, glycolate, D,L-lactate, D,L-glycerate, D,L-3-hydroxybutyrate, D-malate, L-malate, pyruvate, aconitate, citrate, and *p*-hydroxybenzoate	+
Utilization of the following amino acids and amines as carbon source (API 50AA):[c]	
L-α-Alanine, L-leucine, L-isoleucine, L-serine, L-tryptophan, trigonelline, L-threonine, L-aspartate, L-glutamate, L-proline, betaine, D,L-4-aminobutyrate, ethanolamine, and glucosamine	+
Growth inhibition by deoxycycline (30 μg), novobiocin (30 μg), framycetin (100 μg), and tetracycline (30 μg)	+

[a]Symbols: +, all strains positive; −, all strains negative; ±, all strains weakly positive.

[b]Data from Lambert et al. (1990).

[c]Using the API 50CH, 50AO, and 50AA auxanography strips (Biomérieux).

Genus II. *Aminobacter* Urakami, Araki, Oyanagi, Suzuki and Komagata 1992, 90^VP

TEIZI URAKAMI

Am.i.no.bac' ter. M.L. n. *aminum* amine; M.L. n. *bacter* rod, staff; M.L. masc. n. *Aminobacter* amine rod.

Rods, 0.5–0.9 × 1.0–3.0 μm, with rounded ends. Occur singly (rarely in pairs). Motile by means of subpolar flagella. Nonspore-forming. Gram negative. **Reproduction occurs by budding.** Poly-β-hydroxybutyrate granules are accumulated in the cells. Colonies are white to light yellow. Aerobic, having a strictly respiratory type of metabolism. Cells grow abundantly in nutrient broth and PYG broth. No water-soluble florescent pigment is produced. Methyl red and Voges–Proskauer negative. Indole and H_2S are not produced. Gelatin and starch are not hydrolyzed. **Ammonia is produced.** Denitrification does not occur. Litmus milk is not changed. Acids are produced from sugars oxidatively but not fermentatively. Monomethylamine, trimethylamine, trimethyl-amine-*N*-oxide, and sugars are utilized as carbon sources. Methanol, methane, and hydrogen are not utilized. Growth factors are not required. Ammonia, nitrate, urea, peptone, and methylamine are utilized as nitrogen sources. Oxidase and catalase positive. Urease negative. Good growth occurs between pH 6.0 and 8.0. No growth above pH 9.0 or below pH 5.0. Good growth occurs at 30 and 37°C; no growth at 42°C. No growth in the presence of 3% NaCl. The cellular fatty acids include a large amount of straight-chain, unsaturated $C_{18:1}$ acid. The hydroxy acids include a large amount of $C_{12:0\ 3OH}$. The ubiquinone system is ubiquinone Q-10.

The mol% G + C of the DNA is: 62–64.

Type species: **Aminobacter aminovorans** (den Dooren de Jong 1926) Urakami, Araki, Oyanagi, Suzuki and Komagata 1992, 90 (*Pseudomonas aminovorans* den Dooren de Jong 1926, 161.)

ENRICHMENT AND ISOLATION PROCEDURES

Aminobacter strains are isolated from soils at 30°C by the enrichment culture technique, using methylamine compounds or methylformamide compounds (Urakami et al., 1990a, c). Monomethylamine, dimethylamine, trimethylamine, trimethylamine-*N*-oxide, and tetramethylammonium (TMAH) are used as methylamine compounds and *N*-methylformamide and *N,N*-dimethylformamide (DMF) are used as methylformamide compounds.

MAINTENANCE PROCEDURES

Working stock cultures should be transferred every 7 d to ensure viability. This requirement for frequent transfer appears to be independent of the type of maintenance medium or temperature of storage. Stock strains may be preserved indefinitely by lyophilization in 10% skim milk. Strains may also be preserved for at least 5 years by storage at −70°C in glycerol–PYG broth (PYG broth supplemented with 15% glycerol). Organisms are cultivated approximately 24 h in PYG broth medium, and glycerol is added to culture broth (final glycerol concentration, 15%). 1.0 ml of dense suspension of organisms is transferred to an autoclaved half-dram vial and stored in a −70°C freezer.

PROCEDURES FOR TESTING SPECIAL CHARACTERS

Acid production from carbohydrates Hugh and Leifson oxidation–fermentation basal medium (Difco) is used with each of the carbohydrates. The carbohydrates (5 g/l) are added aseptically to the autoclaved, cooled basal medium from a 10% stock solution that has been sterilized by filtration.

The GasPak™ anaerobic system is used for anaerobic conditions. Acidification of the media is investigated up to 4 weeks.

Urease, oxidase, and catalase test Urease activity is determined on Christensen medium incubated for 7 d. The oxidase test is performed with cytochrome oxidase test paper (Nissui Seiyaku, Tokyo, Japan). Catalase activity is detected by the production of bubbles upon addition of a 3% hydrogen peroxide solution to cultures grown on agar slants of PYG medium, which consists of 0.5% peptone, 0.5% yeast extract, and 0.5% glucose (pH 7.0).

Utilization of carbon compounds Utilization of carbon compounds is carried out at a concentration of 0.15% (w/v). Utilization of methane is tested under an atmosphere containing $CH_4/O_2/CO_2$ (5:4:1) with a rotary shaker in a tightly stopped conical flask containing carbon compound-free medium. Utilization of hydrogen is tested in the same manner, but in an atmosphere containing $H_2/O_2/CO_2$ (8:1:1).

DIFFERENTIATION OF THE GENUS *AMINOBACTER* FROM OTHER GENERA

The minimal characteristics for differentiating the genus *Aminobacter* from related genera are shown in Table BXII.α.129.

Aminobacter strains resemble *Blastobacter aggregatus* and *Blastobacter denitrificans* in motility and colony pigmentation, but are differentiated from *B. aggregatus* on the basis of acid production from mannitol under aerobic conditions, utilization of ethanol and succinic acid, tolerance of NaCl, and DNA base composition, and from *B. denitrificans* on the basis of acid production from mannitol under aerobic conditions, utilization of methanol, ethanol, and sucrose, tolerance of NaCl, and denitrification (Hirsch and Muller, 1985; Urakami et al., 1992).

Aminobacter strains resemble *Hyphomicrobium* strains in that they are nonpigmented, Gram-negative, methylamine-utilizing bacteria that metabolize C_1 compounds via the serine pathway (Bellion and Hersh, 1972; Wagner and Lecvitch, 1975). However, *Aminobacter* strains are distinguished from *Hyphomicrobium* species based on formation of hyphae, utilization of methanol and carbohydrates, quinone system, and hydroxy acid composition (Hirsch, 1984; Urakami and Komagata, 1986b, 1987a, b).

TAXONOMIC COMMENTS

In his classical work, den Dooren de Jong (1926) isolated from soil enrichments with various amines (methylamine, trimethylamine, tetramethylammonium (TMAH), ethylamine, and ethylurea) a group of organisms, among which seven strains were placed in the genus *Pseudomonas*. den Dooren de Jong created the name *Pseudomonas aminovorans* for the group because of the ability of the strains to utilize various amines as sole carbon and energy sources. *Pseudomonas aminovorans*, which is a typical species of methylamine-utilizing bacteria, was included in Section V in *Bergey's Manual of Systematic Bacteriology* (Palleroni, 1984) with other *Pseudomonas* species whose natural relationships were unknown. These methylamine-utilizing bacteria were distinguished clearly from genus *Pseudomonas sensu stricto*, i.e., from the *Pseudomonas fluorescens* rRNA branch (De Vos et al., 1989), by the following characteristics: motility by subpolar flagella, multipli-

TABLE BXII.α.129. Differential characteristics of the genus *Aminobacter* and related organisms [a]

Characteristic	Aminobacter	Blastobacter aggregatus	Blastobacter denitrificans	Brevundimonas diminuta	Brevundimonas vesicularis	Hyphomicrobium	Rhizobium	Sphingomonas
Colony color:								
Colorless	+	+	+	+	+	+	+	−
Yellow	−	−	−	−	−	−	−	+
Colonies have viscosity	−	−	−	−	−	−	+	−
Hyphae formed	−	−	−	−	−	+	−	−
Denitrification	−	−	+	−	−	D	nd	nd
Growth with 3% NaCl	−	+	+	−	−	−	−	−
Acid from mannitol (aerobic)	+	−	−	nd	nd	−	nd	−
Carbon sources:								
Methanol	−	−	+	−	−	+	−	−
Monomethylamine	+	nd	nd	−	−	+	−	−
Ethanol	−	+	+	D	+	D	nd	nd
Sucrose	+	+	−	−	−	−	nd	D
Succinic acid	+	−	+	D	+	−	nd	nd
Ubiquinone system:								
Q-9	−	nd	nd	−	−	+	nd	−
Q-10	+	nd	nd	+	+	−	nd	+
Major hydroxy acids:								
$C_{12:0\ 3OH}$	+	nd	nd	+	+	−	−	−
$C_{14:0\ 3OH}$, $C_{16:0}$	−	nd	nd	−	−	+	−	+
$C_{14:0\ 3OH}$	−	nd	nd	−	−	−	+	+
Mol% G + C of DNA	62–64	60	65	66	66	59–65	59–64	64–66

[a]For symbols, see standard definitions; nd, no data.

cation by budding (Green and Gillis, 1989), utilization of methylamine, quinone system, cellular fatty acid composition, hydroxy fatty acid composition, DNA–DNA hybridization, and rRNA–DNA hybridization (De Ley et al., 1987; De Vos et al., 1989).

De Ley et al. (1987) studied the rRNA cistron similarities of *Pseudomonas* and *Pseudomonas*-like strains and reported that the type strain of *Pseudomonas aminovorans* was actually a member of the *Rhizobium loti* rRNA branch, not of the genus *Pseudomonas*. Furthermore, Green and Gillis (1989) pointed out the phenotypic resemblance between *Pseudomonas aminovorans* strains and the genus *Blastobacter*, which contains heterotrophic, rod-shaped, budding bacteria that do not fix nitrogen.

In 1990, the tetramethylammonium (THAH)-utilizing strain TH-3 of Urakami et al. (1990a) and the *N,N*-dimethylformamide (DMF)-utilizing strain DM-81 of Urakami et al. (1990c) were isolated and found to resemble *P. aminovorans*. Urakami et al. (1992) studied the chemotaxonomic characteristics of five methylamine-utilizing bacteria, including strains TH-3 and DM-81, and proposed the transfer of these strains and *Pseudomonas aminovorans* to a new genus, *Aminobacter*. *Pseudomonas aminovorans* became *Aminobacter aminovorans*, and two additional species were created: *Aminobacter aganoensis* for strain TH-3[T] and *Aminobacter niigataensis* for strain DM-81[T].

Although *Aminobacter* is actually a member of the *Rhizobium loti* rRNA branch as determined by rRNA–DNA hybridization (Jarvis et al., 1986; De Ley et al., 1987), it can be differentiated from *R. loti sensu stricto* on the basis of the viscosity of the colonies, utilization of methylamine, production of extracellular gum (Jordan, 1984b), hydroxy fatty acid composition (Yokota, 1989), and rRNA–DNA hybridization data (De Ley et al., 1987; De Vos et al., 1989). In addition, the failure of representative *Aminobacter* strains to reduce acetylene has been described by Green and Gillis (1989). Furthermore, these methylamine-utilizing bacteria have been isolated from soils enriched with various amines, which serve as sole carbon and energy sources.

DIFFERENTIATION OF THE SPECIES OF THE GENUS *AMINOBACTER*

Characteristics useful in differentiation of the species of *Aminobacter* are listed in Table BXII.α.130.

TABLE BXII.α.130. Differential characteristics of the species of the genus *Aminobacter*[a]

Characteristic	A. aminovorans	A. aganoensis	A. niigataensis
Nitrate reduction	−	w	w
Growth in peptone water	−	w or −	w or −
Oxidative production of acid from L-arabinose	+	−	−
Carbon sources:			
Dimethylamine	−	+	+
Formamide	−	−	w
N-methylformamide, *N,N*-dimethylformamide	−	−	+
Tetramethylammonium hydroxide	−	+	−

[a]For symbols, see standard definitions.

List of species of the genus Aminobacter

1. **Aminobacter aminovorans** (den Dooren de Jong 1926) Urakami, Araki, Oyanagi, Suzuki and Komagata 1992, 90[VP] (*Pseudomonas aminovorans* den Dooren de Jong 1926, 161.) *am.i.no′ vo.rans.* M.L. n. *aminum* amine; L. v. *voro* to devour; L. part. adj. *aminovorans* amine-devouring, digesting.

The characteristics are as described for the genus and as given in Tables BXII.α.129 and BXII.α.130. In addition, it has the following characteristics. Growth occurs in peptone water. Nitrate is not reduced to nitrite. Acids are produced oxidatively from L-arabinose, D-xylose, D-glucose, D-mannose, D-fructose, D-galactose, maltose, sucrose, trehalose, D-sorbitol, D-mannitol, inositol, and glycerol. No acid production from lactose or soluble starch. Carbon sources for growth include L-arabinose, D-xylose, D-glucose, D-mannose, D-fructose, D-galactose, maltose, sucrose, trehalose, D-sorbitol, D-mannitol, inositol, glycerol, succinic acid, acetic acid, monomethylamine, trimethylamine, and trimethylamine-*N*-oxide. Lactose, soluble starch, citric acid, formic acid, ethanol, methanol, dimethylamine, formamide, *N*-methylformamide, DMF, TMAH, methane, and H$_2$ are not utilized.

The type strain was isolated from soil enrichments containing various amines by de Dooren de Jong in 1926.

The mol% G + C of the DNA is: 62.5 (HPLC).

Type strain: ATCC 23314, DSM 7048, JCM 7852, NCIB 9039, NCTC 10684.

GenBank accession number (16S rRNA): AJ011759.

Additional Remarks: Reference strain MS, ATCC 23819, NCIB 11591.

2. **Aminobacter aganoensis** Urakami, Araki, Oyanagi, Suzuki and Komagata 1992, 91[VP]

a.ga.no.en′ sis. M.L. masc. adj. *aganoensis* coming from the Agano River, Niigata, Japan.

The characteristics are as described for the genus and as given in Tables BXII.α.129 and BXII.α.130. In addition, it has the following characteristics. Poor growth occurs in peptone water. Nitrate is weakly reduced to nitrite. Acids are produced oxidatively from D-xylose, D-glucose, D-mannose, D-fructose, D-galactose, maltose, sucrose, trehalose, D-sorbitol, D-mannitol, inositol, and glycerol. Weak acid production from D-glucose, D-mannose, maltose, sucrose, trehalose, D-sorbitol, D-mannitol, and inositol. No acid production from L-arabinose, lactose, or soluble starch. L-Arabinose, D-xylose, D-glucose, D-mannose, D-fructose, D-galactose, maltose, sucrose, trehalose, D-sorbitol, D-mannitol, inositol, glycerol, succinic acid, acetic acid, monomethylamine, dimethylamine, trimethylamine, trimethylamine-*N*-oxide, and TMAH are utilized as carbon sources. Acetic acid is utilized weakly. Lactose, soluble starch, citric acid, formic acid, ethanol, methanol, formamide, *N*-methylformamide, DMF, methane, and H$_2$ are not utilized.

The mol% G + C of the DNA is: 63.8 (HPLC).

Type strain: Strain TH-3, DSM 7051, JCM 7854.

GenBank accession number (16S rRNA): AJ011760.

3. **Aminobacter niigataensis** Urakami, Araki, Oyanagi, Suzuki and Komagata 1992, 91[VP]

ni.i.ga.ta.en′ sis. M.L. masc. adj. *niigataensis* coming from the Niigata region of Japan.

The characteristics are as described for the genus and as given in Tables BXII.α.129 and BXII.α.130. In addition, it has the following characteristics. Poor growth occurs in peptone water. Nitrate is weakly reduced to nitrite. Acid is produced oxidatively from D-xylose, D-glucose, D-mannose, D-fructose, D-galactose, maltose, sucrose, trehalose, D-sorbitol, D-mannitol, inositol, and glycerol. No acid production from L-arabinose, lactose, or soluble starch. L-arabinose, D-xylose, D-glucose, D-mannose, D-fructose, D-galactose, maltose, sucrose, trehalose, D-sorbitol, D-mannitol, inositol, glycerol, succinic acid, acetic acid, monomethylamine, dimethylamine, trimethylamine, trimethylamine-*N*-oxide, *N*-methylformamide, and DMF are utilized as carbon sources. Formamide is utilized weakly. Lactose, soluble starch, citric acid, formic acid, ethanol, methanol, TMAH, methane, and H$_2$ are not utilized.

The type strain was isolated as a DMF-utilizing bacterium from soil by Urakami et al. (1990c).

The mol% G + C of the DNA is: 63.2 (HPLC).

Type strain: Strain DM-81, DSM 7050, JCM 7853.

GenBank accession number (16S rRNA): AJ011761.

Genus III. **Aquamicrobium** *Bambauer, Rainey, Stackebrandt and Winter 1998b, 631[VP]*
(Effective publication: Bambauer, Rainey, Stackebrandt and Winter 1998a, 300)

THE EDITORIAL BOARD

A.qua.mi.cro′ bi.um. L. n. *aqua* water; L. n. *microbium* a microbe; *Aquamicrobium* a bacterium living in water/wastewater.

Gram negative, pleomorphic or regularly formed short **rods**. Motile or nonmotile. Respiratory metabolism using either O$_2$ or nitrate under anoxic conditions. **Nitrate reduced to nitrite or N$_2$.** Mesophilic. Optimal pH 6–9. Salt tolerant, up to 3% NaCl. Utilize sugars, fatty acids, and heterocyclic aromatic compounds as carbon sources. Utilize nitrate or O$_2$ as electron acceptors.

The mol% G + C of the DNA is: 59–65.

Type species: **Aquamicrobium defluvii** Bambauer, Rainey, Stackebrandt and Winter 1998b, 631 (Effective publication: Bambauer, Rainey, Stackebrandt and Winter 1998a, 300.)

ENRICHMENT AND ISOLATION PROCEDURES

Aquamicrobium defluvii strain NKK was isolated from activated sewage sludge, using thiophene-2-carboxylate as the sole carbon source and nitrate as the electron acceptor, as described in Bambauer et al. (1998a). Sludge (10% v/v) was added to mineral salts medium containing 2 mM thiophene-2-carboxylate and 10 mM nitrate. The enrichment culture was incubated at 37°C for 3 weeks, and 1 ml was transferred to fresh medium. Pure cultures were obtained after several transfers.

List of species of the genus Aquamicrobium

1. **Aquamicrobium defluvii** Bambauer, Rainey, Stackebrandt and Winter 1998a, 631[VP] (Effective publication: Bambauer, Rainey, Stackebrandt and Winter 1998a, 300.)
de.flu' vi.i. M.L. neut. n. *defluvium* wastewater; *defluvii* from waste water.

Gram-negative rods, 0.5–0.8 × 1.5–2.5 μm. Motile. Do not form spores. Catalase and oxidase positive. Colonies are white, circular, convex, 2–4 mm in diameter after 3–5 d. Optimal temperature 30–37°C. Optimal pH 7.5–8.5. Growth occurs at 41°C and with 2% NaCl. Vitamins required for growth. Respiratory metabolism with O_2 or nitrate as electron acceptors. Nitrate reduced to nitrite. Chemoorganotrophic, utilizing thiophene-2-carboxylate, acetate, propionate, butyrate, crotonate, glucose, fructose, mannose, xylose, mannitol, and sorbitol as carbon sources.

The mol% G + C of the DNA is: 61.7.

Type strain: NKK, CIP 105610, DSM 11603.

GenBank accession number (16S rRNA): Y15403.

Genus IV. **Defluvibacter** *Fritsche, Auling, Andreesen and Lechner 1999b, 1325[VP] (Effective publication: Fritsche, Auling, Andreesen and Lechner 1999a, 202)*

THE EDITORIAL BOARD

De.flu.vi.bac' ter. M.L. n. *defluvium* waste water; Gr. hyp. masc. n. *bacter* rod; M.L. masc. n. *Defluvibacter* referring to its origin from activated sludge of a waste water treatment plant.

Gram-negative **rods**, occurring as single cells. Motile. **Strictly aerobic.** Oxidase and catalase positive. Does not reduce nitrate. Utilizes the following carbohydrates: D-glucose, D-fructose, D-mannose, D-ribose, D-xylose, and L-lyxose. Some amines and amino acids used as carbon sources. **Major ubiquinone is ubiquinone 10.** Major fatty acid is $C_{18:1}$. $C_{12:0\ 3OH}$ is present in small amounts.

Spermidine is the major polyamine. The genus belongs in the *Alphaproteobacteria* and is most closely related to *Pseudaminobacter*, *Mesorhizobium*, and *Phyllobacterium*.

Type species: **Defluvibacter lusatiensis** Fritsche, Auling, Andreesen and Lechner 1999b, 1325 (Effective publication: Fritsche, Auling, Andreesen and Lechner 1999a, 202)

List of species of the genus Defluvibacter

1. **Defluvibacter lusatiensis** Fritsche, Auling, Andreesen and Lechner 1999b, 1325[VP] (Effective publication: Fritsche, Auling, Andreesen and Lechner 1999a, 202)
lu.sa.ti.en.sis. M.L. gen. n. *lusatiae* referring to the German province of Lausitz (Latin name *Lusatia*), where the organism was isolated.

Short rods, 0.6–0.8 × 1.5–3 μm, occurring singly. Motile by means of a single polar flagellum. Do not form spores. Colonies are grayish white, circular, and mucoid, with a diameter of 2 mm after 2 d on nutrient agar. Can reach 4 mm after prolonged incubation. Optimal temperature 30–37°C (range 12–44°C). Optimal pH 7.0–7.7 (range 6.0–9.2). Grows in peptone medium. Does not form indole. Does not hydrolyze urea, starch, gelatin, casein, DNA, Tween 80, or esculin. Utilization of carbohydrates requires yeast extract or growth factors. Carbohydrates utilized are DL-α-amino-*n*-valerate, putrescine, L-tryptophan, malonate, glutarate, and 2-ketoglutarate. Can utilize the following as carbon sources: 4-chloro-2-methylphenol, 2,4-dichlorophenol, and 4-chlorophenol. Phenol is degraded by a different set of enzymes. Does not degrade benzoate, *m*- and *p*-hydroxybenzoate, 3,4-dihydroxybenzoate, gentisate (2,5-dihydroxybenzoate), and phenylacetate. Isolated from activated sludge of an industrial waste water plant at Schwarzheide.

The mol% G + C of the DNA is: 61.4.

Type strain: S1, CIP 106844, DSM 11099.

GenBank accession number (16S rRNA): AJ132378.

Genus V. **Candidatus** *Liberibacter Jagoueix, Bové and Garnier 1994, 385 (Candidatus Liberobacter (sic) Jagoueix, Bové and Garnier 1994, 385)*

MONIQUE GARNIER

Li.be.ri' bac.ter. L. n. *liber* phloem; Gr. n. *bakterion* a small rod; N.L. masc. n. *Liberibacter* rod in the phloem.

Filamentous bacteria occurring in the phloem sieve tubes of plants (Fig. BXII.α.146). Round forms are also observed but have been shown to correspond to degenerating bacteria. The bacteria are also present in the hemolymph and salivary glands of insect vectors (Psyllidae) responsible for the transmission. Like most phloem-restricted bacteria, *Candidatus* Liberibacter spp. have resisted *in vitro* cultivation (Garnier and Bové, 1993). The characterized species originated from sweet orange (*Citrus sinensis*). The original description is based on bacteria present in the phloem sieve tubes of sweet orange trees affected by huanglongbing (ex. greening disease) in Poona (India) and Nelspruit (South Africa).

Type species: *Candidatus* Liberibacter asiaticus Jagoueix, Bové and Garnier 1994, 385 (*Candidatus* Liberobacter asiaticum (sic) Jagoueix, Bové and Garnier 1994, 385.)

FURTHER DESCRIPTIVE INFORMATION

Phylogenetic treatment Phylogenetic analyses of the 16S rRNA gene indicated *Candidatus* Liberibacter spp. belong to the *Alphaproteobacteria*; their closest cultivated relatives are members of the group of bacteria known as the alpha-2 subgroup belonging to the genera *Bartonella*, *Bradyrhizobium*, *Agrobacterium*, *Brucella*, and *Afipia* (Fig. BXII.α.147) (Jagoueix et al., 1994).

Based on sequence comparisons of ribosomal protein genes of the beta operon, and on biological properties, two species have been recognized: *Candidatus* Liberibacter asiaticus and *Candidatus* Liberibacter africanus.

Strain morphology Electron microscopy measurements on thin sections show that the filamentous forms of the liberibacteria have a diameter of 0.2–0.3 μm. Variations in diameter occur between organisms and sometimes within a single organism. Round forms are larger (0.5 μm) with a less dense cytoplasm and often show plasmolysis (Figs. BXII.α.146 and BXII.α.148). The *Candidatus* Liberibacter spp. envelope is a membranous wall characteristic of Gram-negative bacteria (Fig. BXII.α.149), but the peptidoglycan layer is hardly visible (Garnier et al., 1984). There is no evidence for flagella or pili. *Candidatus* L. asiaticus and *Candidatus* L. africanus cannot be distinguished on the basis of morphology.

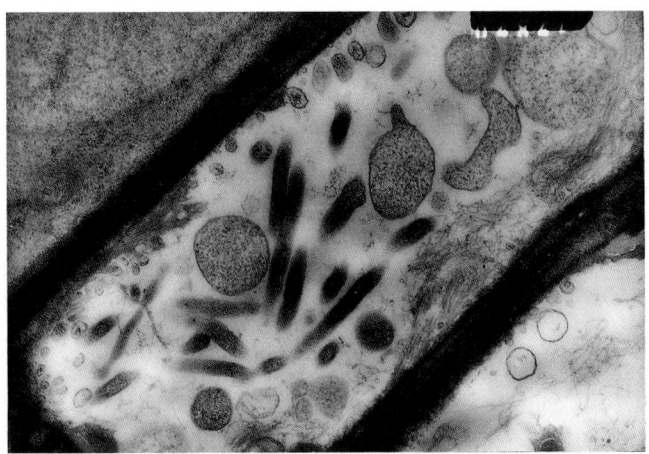

FIGURE BXII.α.146. *Candidatus* Liberibacter cells in a phloem sieve tube of huanglongbing-affected sweet orange leaves (× 12,000).

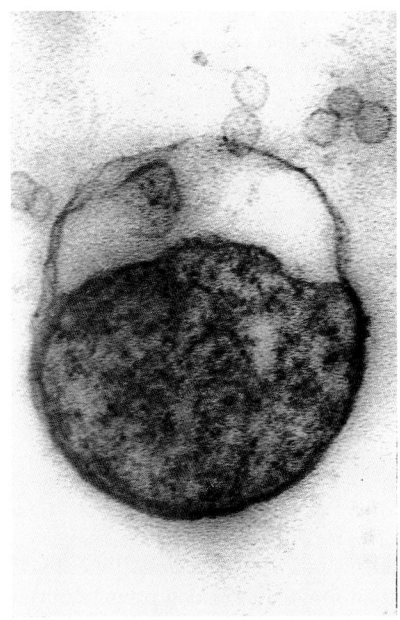

FIGURE BXII.α.148. Round plasmolyzed *Candidatus* Liberibacter (× 100,000).

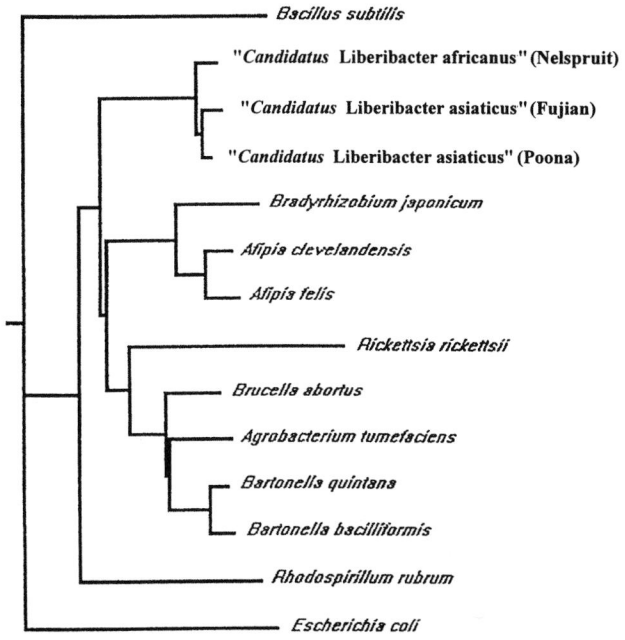

FIGURE BXII.α.147. Phylogenetic tree constructed with 16S rDNA sequences from GenBank.

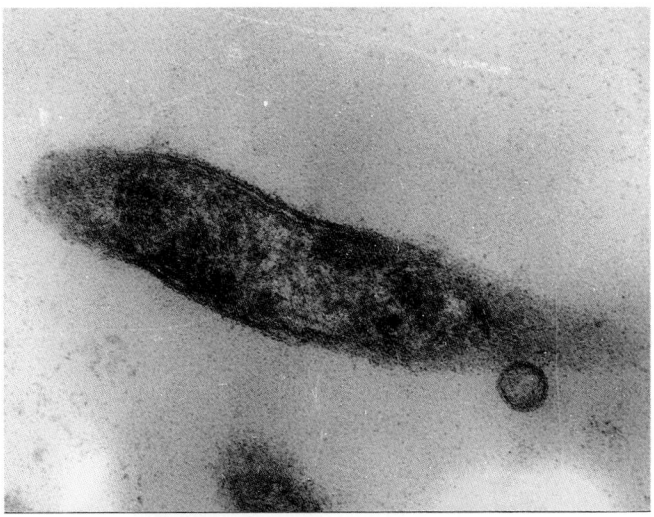

FIGURE BXII.α.149. Filamentous *Candidatus* Liberibacter cell showing the membranous cell wall (× 80,000).

Ecological data, host range *Candidatus* Liberibacter spp. are phytopathogenic bacteria affecting most, if not all, *Citrus* species and some other genera and species in the Rutaceae family. They are responsible for huanglongbing (ex. greening disease). Natural infections are limited to members of the Rutaceae plant family, but experimental transmissions by dodder (*Cuscuta* sp.) to periwinkle (*Catharanthus roseus*, Apocynaceae) and tobacco (*Nicotiana* sp., Solanaceae) have been achieved (Garnier and Bové, 1983; Bové and Garnier, 1992). *Candidatus* Liberibacter asiaticus is present in Asia (from Pakistan to China) and induces symptoms in a temperature range of 25–40°C, while *Candidatus* Liberibacter africanus is present in South and Eastern Africa as well as Cameroon and does not induce symptoms when temperatures are above 30°C (Bové et al., 1974; Garnier and Bové, 1996).

Psyllids become contaminated by feeding on infected plants; they remain infective all their life. *Candidatus* Liberibacter cells can cross the insect gut membranes and multiply into the hemolymph, from which they reach salivary glands. From there on, they can be reinoculated into plants. There is no evidence of transovarial transmission of the liberibacteria. Two psyllids, *Diaphorina citri* in Asia (Capoor et al., 1967) and *Trioza erytreae* in Africa (McLean and Oberholzer, 1965), are vectors of the liberibacteria.

Diaphorina citri has also been present in South and Central America for many years; it has recently reached Florida and the West Indies. Liberibacteria, however, are not present in the American continent and the West Indies. Recently *D. citri* has also been reported from Iran. In the Arabian Peninsula (South Western Saudi Arabia, Yemen), Mauritius, and Reunion islands, both psyllid vectors and the two *Candidatus* Liberibacter spp. are present (Garnier and Bové, 1996).

ENRICHMENT AND ISOLATION PROCEDURES

To date, all attempts to grow liberibacteria in axenic cultures have failed.

In periwinkle plants, the *Candidatus* Liberibacter cells reach higher titers than in citrus. Purified preparations of phloem tissue from infected periwinkle plants are enriched in *Candidatus* Liberibacter cells.

MAINTENANCE PROCEDURES

Strains are maintained in citrus or periwinkle plants by repeated graft-inoculations.

PROCEDURES FOR TESTING SPECIAL CHARACTERS

Specific identification Cells belonging to *Candidatus* Liberibacter can be identified by amplification and sequencing of the 16S rDNA. The complete sequence of the 16S rRNA gene is deposited in GenBank under the accession number L22532 for *Candidatus* Liberibacter asiaticus and L22533 for *Candidatus* Liberibacter africanus. Primers for specific amplification of the *Candidatus* Liberibacter 16S rDNA have been developed and yield a 1160 bp amplicon with both species (Jagoueix et al., 1996). The oligonucleotide sequences complementary to unique regions of the 16S rDNA are 5′-GCGCGTATGCAATACGAGCGGCA-3′ for *Candidatus* Liberibacter asiaticus and 5′-GCGCGTATTTTATAC-GAGCGGCA-3′ for *Candidatus* Liberibacter africanus. Digestion of the amplified DNA with *Xba*I is required for species identification.

Primers A2 (5′-TATAAAGGTTGACCTTTCGAGTTT-3′) and J5 (5′-ACAAAAGCAGAAATAGCACGAACAA-3′) defined in the ribosomal protein genes *rplA* and *rplJ* of the *Candidatus* Liberibacter spp. yield amplicons of 703 bp and 669 bp for *Candidatus* Liberibacter asiaticus and *Candidatus* Liberibacter africanus, respectively (Hocquellet et al., 1999).

DNA probes specific for each species or for strains within species have also been obtained (Villechanoux et al., 1992; Hocquellet et al., 1997).

Seven serovars have been demonstrated in the *Candidatus* Liberibacter spp. with the 12 monoclonal antibodies available so far (Gao et al., 1993).

ACKNOWLEDGMENTS

Monique Garnier-Semancik died suddenly in May, 2003. She was Director of Research at INRA Bordeaux and served as director of a "Joined Research Unit" in which INRA researchers and University teachers work together. Most of her scientific contributions were to the field of phloem- and xylem-restricted plant pathogenic bacteria, which won her respect both in France and internationally.

List of species of the genus Candidatus Liberibacter

1. *Candidatus* Liberibacter asiaticus Jagoueix, Bové and Garnier 1994, 385 (*Candidatus* Liberobacter asiaticum (sic) Jagoueix, Bové and Garnier 1994, 385.)
   as.iat′i.cus. M.L. adj. *asiaticus* from Asia.

   The type strain Poona was isolated from a sweet orange tree in India.
   The mol% G + C of the DNA is: not determined.
   GenBank accession number (16S rRNA): AY192576, L22532.

2. *Candidatus* Liberibacter africanus Jagoueix, Bové and Garnier 1994, 379.
   afr.ic′a.nus. M.L. adj. *africanus* from Africa.

   The type strain Nelspruit was isolated from a sweet orange tree in South Africa.

   The mol% G + C of the DNA is: not determined.
   GenBank accession number (16S rRNA): L22533.

   a. *Candidatus* Liberibacter africanus subsp. capensis Garnier, Jagoueix-Eveillard, Cronje, le Roux and Bové 2000, 2124.

   Detected in an ornamental rutaceous tree (*Calodendrum capense*) in South Africa. 16S rDNA and ribosomal protein gene sequence analysis revealed that it is different from the two previously described species but more closely related to *Candidatus* L. africanus.

   The mol% G + C of the DNA is: not determined.
   GenBank accession number (16S rRNA): AF137368.

Genus VI. Mesorhizobium Jarvis, van Berkum, Chen, Nour, Fernandez, Cleyet-Marel and Gillis 1997, 897[VP]

WEN XIN CHEN, EN TAO WANG AND L. DAVID KUYKENDALL

Me.so.rhi.zo'bi.um. Gr. adj. mesos middle; M.L. neut. n. Rhizobium bacterial generic name; M.L. neut. n. Mesorhizobium the meso-growing rhizobium, referring to the growth rate intermediate between those of the genera Rhizobium and Bradyrhizobium.

Rods 0.4–0.9 × 1.2–3.0 μm. Commonly pleomorphic under adverse growth condition or in the root nodules as bacteroids. Usually contain granules of poly-β-hydroxybutyrate, which are refractile by phase-contrast microscopy. Nonsporeforming. Gram negative. **Motile** by one polar or subpolar flagellum or by peritrichous flagella. Aerobic, possessing a respiratory type of metabolism with oxygen as the terminal electron acceptor. Optimal temperature 25–30°C. Optimal pH 6–8. Growth in yeast extract–mannitol agar (YMA)[1] produces colonies that are circular, convex, semitranslucent, raised, and mucilaginous, 2–4 mm diameter within 5–6 d for some species or 1 mm after a 7-d incubation for other species. **The generation times of Mesorhizobium strains range from 4–15 h.** Chemoorganotrophic, utilizing a wide range of carbohydrates and salts of organic acids as carbon sources without gas production. Cellulose and starch are not utilized. **Produces an acidic reaction in YMA.** Ammonium salts, nitrates, urea, and most amino acids are utilized as nitrogen sources. Peptone is poorly utilized. Mesorhizobium strains are only weakly proteolytic, but can produce a slow digestion in litmus milk; some strains form a clear serum zone. Some strains require thiamin, nicotinamide, and riboflavin for growth. **The organisms are characteristically able to invade the root hairs of a wide range of temperate, subtropical, and tropical leguminous plants, inciting production of root nodules where the bacteria reduce atmospheric nitrogen into a combined form available for the host plants.** All strains exhibit host specificity.

The mol% G + C of the DNA is: 59–64.

Type species: **Mesorhizobium loti** (Jarvis, Pankhurst, and Patel 1982) Jarvis, van Berkum, Chen, Nour, Fernandez, Cleyet-Marel and Gillis 1997, 898 (Rhizobium loti Jarvis, Pankhurst, and Patel 1982, 378.)

FURTHER DESCRIPTIVE INFORMATION

In young cultures, the cells are short rods of various cell sizes. The cell width of some strains is 0.25 μm, and the cell length of some strains is 4.0 μm (de Lajudie et al., 1998b). In old cultures, or under adverse environmental conditions, the cells are usually pleomorphic, swollen, club-shaped, or branched.

In host nodules, the bacteroids of M. huakuii are mostly club-shaped and approximately 40–50 times larger than the rod-shaped cells in the infection threads or in vitro culture. The invaded nodule cell contains many swollen bacteroids and a few rod-shaped bacteria. Microculture of bacteroids and bacteria released from bacterial nodule protoplasm reveals that only the rod-shaped cells multiply; the club-shaped bacteroids do not (Fig. BXII.α.150). (Cao et al., 1984). The ratio of viable (may multiply) counts and total counts of bacteroids released from protoplasts of Astragalus sinicus is 0.27:100 for M. huakuii, which is between the values reported for Sinorhizobium meliloti in Medicago sativa (0.11:100) and for Bradyrhizobium japonicum in Glycine max (81.08:100) (Cao et al., 1984).

The unique phenotypic characters of this genus are the moderately slow or slow growth rate of the strains and their acid production in YMA. The generation times (4–15 h) of Mesorhizobium strains (Table BXII.α.131) are generally slower than those of fast growers classified in the genera Rhizobium and Sinorhizobium (<6 h) and overlap those of Bradyrhizobium species (>6 h). Mesorhizobium species produce acid in mineral medium, such as YMA, which is a characteristic of Rhizobium and Sinorhizobium; they do not produce an alkaline reaction like that characteristic of Bradyrhizobium (Chen et al., 1995c; Wang et al., 1998). The temperature ranges are highly strain dependent and vary between 4–42°C. Growth at 4°C and 10°C has been observed only with strains of M. tianshanense, and growth at 42°C is documented only for strains of M. plurifarium, a bacterium from tropical regions. The maximum temperature is 37–40°C for strains from temperate regions. The pH range for the genus is 4.0–10.0, except for M. tianshanense, which originated from an alkaline environment and can grow only in the narrow range of pH 6.0–8.0. M. loti is the most acid-tolerant species and grows at pH 4.0.

Mesorhizobium strains tested produce a slow digestion in litmus milk with a slightly alkaline reaction.

The bacteria in this genus fix nitrogen only in root nodules. No stem-nodule formation nor nitrogen-fixation in free-living conditions has been reported for any Mesorhizobium strain.

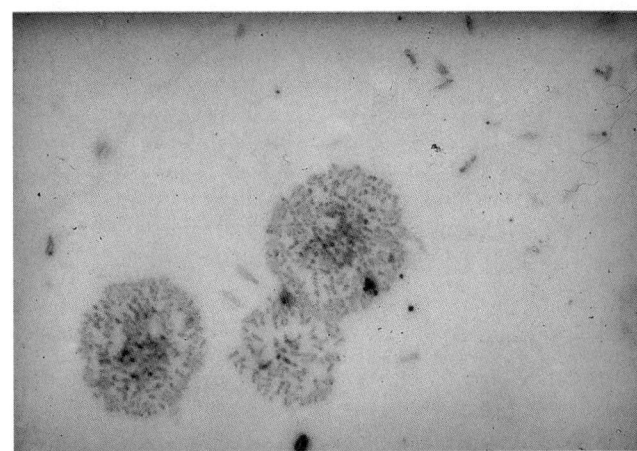

FIGURE BXII.α.150. Photomicrograph showing a microculture of bacterioids and bacteria released from bacteroidal host protoplast. There are 3 microcolonies in the lower left corner, which were developed from regular rods after a 3–4-d incubation in wet chamber. The club-shaped bacteroids elsewhere in the picture could not grow. (Reproduced with permission from Y.Z. Cao et al., Scientia Sinica (Series B) 27: 593–600, 1984, ©China Science Press.)

---

1. Yeast extract–mannitol agar (YMA): mannitol, 10.0 g; KH₂PO₄, 0.5 g; MgSO₄·7H₂O, 0.2 g; NaCl, 0.1g; CaCO₃, 4.0 g; yeast water, 100 ml (or yeast extract [Difco], 0.4 g); agar, 15.0 g; and distilled water, 1 liter; pH 6.8–7.0. Sterilize at 121°C for 15 min. The CaCO₃ is omitted in the preparation of pour plates or of liquid medium (YMB) used for turbidimetric measurement of growth. The yeast water is prepared by mixing 100 g of baker's compressed yeast with 1 liter of cold water, allowing the mixture to stand at room temperature for 1–2 h, and then steaming for 40–60 min. After the mixture has been centrifuged or allowed to settle, the clear supernatant fluid is autoclaved at 121°C for 15 min (Kleczkowska et al., 1968).

TABLE BXII.α.131. Characteristics differentiating the species of the genus *Mesorhizobium*[a]

Characteristic	M. loti[b]	M. amorphae[e]	M. ciceri[d]	M. huakuii[e]	M. mediterraneum[d]	M. plurifarium[f]	M. tianshanense[e]
Monotrichous flagellation	+	nd	nd	+	nd	nd	d
Colony diameter, mm (incubation time)	1 (7 d)	1 (7 d)	2–4 (3–5 d)	2–4 (5–6 d)	2 (5–7 d)	0.5–2 (2–3 d)	1–2 (5–7 d)
Generation time (h)	nd	5–13	<6	4–6	>5	nd	5–15
Maximum growth temperature (°C)	39	37	40	37	40	42	37
Maximum NaCl tolerated for growth (%, w/v)	2.0	<1.0	2.0	1.0	2.0	1.0	1.0
pH range for growth	4.0–10.0	5.0–9.0	5.0–10.0	5.0–9.5	9.5	nd	6.0–8.0
Sole carbon sources:							
D- and L-Arabinose/maltose	+	nd	+	+	+	+	d
D-Fructose/sucrose	+	+	+	+	+	+	d
D- and L-Fucose/L-proline	+	nd	+	−	+	+	d
Fumarate	+	nd	+	+	+	+	−
Inositol	+	+	+	−	+	+	−
D- and L-Malate	+	+	+	+	+	+	−
D-Raffinose	d	+	−	+	−	d	d
Erythritol	d	nd	−	−	−	−	+
L-Aspartate	−	nd	+	+	−	d	d
L-Ornithine	d	+	+	+	−	nd	+
L-Xylose	+	nd	+	+	−	+	+
β-Alanine	d	+	+	−	−	nd	d
Mol% G+C of DNA (T_m)	59–64	64	63–64	59–64	63–64	63–64	59–63

[a]For symbols, see standard definitions.

[b]Data from Jarvis et al. (1982).

[c]Data from Wang et al. (1999b).

[d]Data from Nour et al. (1994a, 1995).

[e]Data from Chen et al. (1991b, 1995c).

[f]Data from de Lajudie et al. (1994, 1998b).

Strains contain 1–5 plasmids, with molecular sizes ranging from 60–940 kb, depending on the species and strain. Symbiotic plasmids carrying the *nod* and *nif* genes have been reported in *M. huakuii* (Chan et al., 1988; Xu and Murooka, 1995; Zou et al., 1997) and *M. amorphae* (Wang et al., 1999b). In other species, the symbiotic genes might be carried on the chromosome, as in *Bradyrhizobium*, since no symbiosis-controlling plasmids have been identified (Wang et al., 1998). Novel and complex chromosomal arrangements have been reported for the nodulation genes of *M. loti* (Scott et al., 1996). The gene *nodB* is separated from *nodACIJ* in *M. loti*, but they are in the same operon in other legume-nodulating bacteria. Lateral transfer of symbiotic genes in the field has been reported between *M. loti* and other related bacteria (Sullivan et al., 1995), and a transferable chromosomal element carrying the symbiotic genes has been identified for *M. loti* (Sullivan and Ronson, 1998).

M. loti strains produce extracellular polysaccharides, which contain uronic acid, galactose, and glucose (Jarvis et al., 1982). Somatic antigens are highly strain specific, but internal antigens show a reaction of identity among strains and can be used to distinguish *M. loti* from other *Rhizobium* species (Jarvis et al., 1982). No serological crossreactions (immunofluorescence) have been observed among chickpea rhizobia and other rhizobia (Kingsley and Bohlool, 1983). As analyzed by the method of Vincent (1982), antisera to strains of *M. huakuii* either do not react or do so only weakly with strains of *Astragalus hamosus* and *A. adsurgens* (Chan et al., 1988).

Most strains are resistant to streptomycin, erythromycin, ampicillin, kanamycin, polymyxin, and nystatin at concentrations of 10–30 μg/ml (Chan et al., 1988; Chen et al., 1991b, 1995c; Wang et al., 1998). Two species isolated from chickpea plants, *M. mediterraneum* and *M. ciceri* (Nour et al., 1994a, 1995), can be differentiated by antibiotic tests: in contrast to *M. ciceri*, *M. mediterraneum* is susceptible to carbenicillin and resistant to chloram-

phenicol, nalidixic acid, and trimethoprim–sulfamethoxazole (Nour et al., 1995).

Strains of *Mesorhizobium* occur in vast areas from arctic to tropic regions, according to their hosts. The host range of the species within genus *Mesorhizobium* varies from a single legume species to several species in different genera. Some *Mesorhizobium* species share hosts with each other or with bacteria from other rhizobial genera. Four restriction-fragment-length-polymorphism (RFLP) groups of *Cicer*-nodulating bacteria have been identified by Kuykendall et al. (1993a). *Cicer arietinum* is a host for *M. ciceri*, *M. mediterraneum*, and two other genomic species (Nour et al., 1994a, 1995). *Acacia* species have been reported to nodulate with *M. plurifarium*, *Sinorhizobium terangae*, and some as yet unnamed species (de Lajudie et al., 1994). Among *Mesorhizobium* species, *M. tianshanense* has the broadest host range.

ENRICHMENT AND ISOLATION PROCEDURES

Isolation from root nodules and soil is by the techniques described for the genus *Rhizobium*.

MAINTENANCE PROCEDURES

A suitable medium is YMA. Storage recommendations are the same as given for the genus *Rhizobium*.

PROCEDURES FOR TESTING SPECIAL CHARACTERS

Procedures for testing special characters are the same as given for the genus *Rhizobium*.

DIFFERENTIATION OF THE GENUS *MESORHIZOBIUM* FROM OTHER GENERA

Phenotypically, the moderately slow growth rate and acid production in YMA can differentiate *Mesorhizobium* strains from other related bacteria, such as *Agrobacterium*, *Rhizobium*, *Sinorhizobium*, *Allorhizobium*, *Bradyrhizobium*, and *Azorhizobium*.

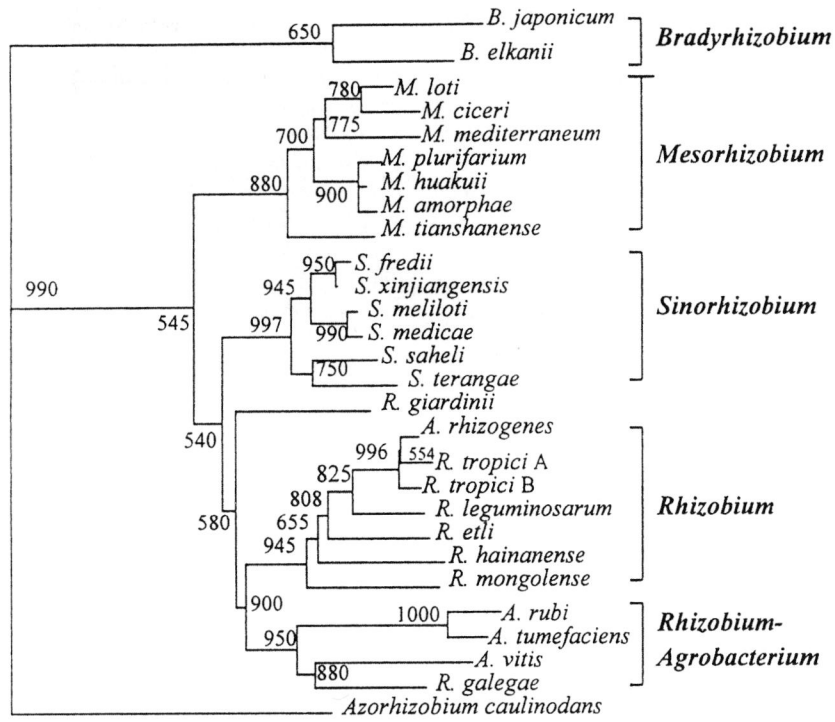

FIGURE BXII.α.151. Dendrogram showing the phylogenetic relationships of species of *Mesorhizobium* and related genera. 16S rRNA gene sequences were obtained from GenBank, DDBJ, and EMBL (analyzed by Z.Y. Tan). The similarity values were calculated with Jukes–Cantor coefficient and the neighbor-joining method was used for construction of the phylogenetic tree and 1000 subsamples were generated for bootstrap analysis using the package of PHYLIP version 3.572c (Felsenstein, 1993).

Analyses of rRNA sequences can differentiate *Mesorhizobium* species from other related bacteria on the phylogenetic basis (Fig. BXII.α.151). Ribotyping or PCR-based RFLP of 16S rRNA genes can serve as a simple and quick method for identification of *Mesorhizobium* strains (Laguerre et al., 1994; Wang et al., 1998). *Mesorhizobium* species have fingerprinting patterns different from other related bacteria and form a group distinct from other bacteria according to cluster analysis of RFLP patterns.

Other differences among the genus *Mesorhizobium* and related bacteria have been demonstrated by numerical taxonomy, SDS-PAGE patterns of total cell proteins, MLEE patterns, PCR-based fingerprinting (ribotyping, ARDRA, RAPD, AFLP), and total DNA–DNA relatedness.

Taxonomic Comments

The genus *Mesorhizobium*, together with some other genera, such as *Agrobacterium*, *Allorhizobium*, *Rhizobium*, and *Sinorhizobium*, constitutes an rRNA cluster in the *Alphaproteobacteria* class. The 16S rRNA sequence similarities between *M. loti* (type species of genus *Mesorhizobium*) and representatives of related genera, specifically *Rhizobium leguminosarum*, *Agrobacterium tumefaciens*, *Sinorhizobium fredii*, *Bradyrhizobium japonicum*, and *Azorhizobium caulinodans*, are 93.2, 93.1, 95.1, 87.7, and 89.4, respectively.

In the first edition of *Bergey's Manual of Systematic Bacteriology*, all of the legume root nodule bacteria were divided into two genera: *Rhizobium* and *Bradyrhizobium* (Jordan, 1984a). *M. loti*, the type species of the genus *Mesorhizobium*, was included in the genus *Rhizobium* as *R. loti* (Jarvis et al., 1982). Several new species phylogenetically related to *R. loti* were later described, including

R. huakuii from *Astragalus sinicus* (Chen et al., 1991b), *R. ciceri*, and *R. mediterraneum* from *Cicer arietinum* (Nour et al., 1994a, 1995), and *R. tianshanense* from several leguminous plants (Chen et al., 1995c). It was recognized that *R. loti* and its relatives were quite different from both the fast-growing symbiotic species, such as *R. leguminosarum* and *S. meliloti*, and the slow growing bacteria in the genus *Bradyrhizobium*, now in a new family, the *Bradyrhizobiaceae*. Comparative analyses of the complete 16S rRNA sequences from representative strains for the species indicated that all these species clustered in a close branch distinct from related bacteria (Nour et al., 1995; Tan et al., 1997; de Lajudie et al., 1998b). The strains of *Mesorhizobium* produce acid in YMA, and they generally exhibit growth rates slower than those of *R. leguminosarum* and *S. meliloti*. Consequently, there is good agreement that these organisms warrant the designation of a new genus in the family *Rhizobiaceae* (Lindström et al., 1995; Young and Haukka, 1996). The name *Mesorhizobium*, referring to the moderately slow growth rate, was first suggested by W.X. Chen and coworkers and afterwards validly proposed (Jarvis et al., 1997) to include five species: *M. ciceri*, *M. huakuii*, *M. loti*, *M. mediterraneum*, and *M. tianshanense*. Subsequently, three new species have been described: *M. plurifarium*, from the tropical trees *Acacia senegal*, *Leucaena leucocephala*, and *Prosopis juliflora* (de Lajudie et al., 1998b); *M. amorphae*, from *Amorpha fruticosa*, originating from Chinese soils (Wang et al., 1999b); and *M. chacoense* (Velázquez et al., 2001).

Although 16S rRNA gene analysis can differentiate *Mesorhizobium* from other related genera, it is not as useful for defining the species within this genus. For example, *M. loti* is not coherent and consists of strains with differing phylogenies, since there are more differences among the strains within this species than there are among different *Mesorhizobium* species (Laguerre et al., 1997). Thus, the definition of species within this genus is based mainly upon DNA–DNA hybridization, associated with distinctive features.

Diversity has been investigated for some *Mesorhizobium* species using different methods. Two genomic groups have been distinguished within *M. loti* by DNA–DNA hybridization and by 16S rRNA analyses (de Lajudie et al., 1998b). *M. plurifarium* has been reported to contain two subgroups divided by protein SDS-PAGE and DNA–DNA hybridization (de Lajudie et al., 1998b). *M. amorphae* has been reported as one of the three *Mesorhizobium* groups from *A. fruticosa*, each of which contains strains with different enzyme electrophoretic types (ET) and different plasmid con-

tents. These three groups have different phylogenies within the *Mesorhizobium* branch but share the same symbiotic plasmid (Wang et al., 1998). *M. huakuii* has been described for nine strains isolated from *A. sinicus*, grown in the southern part of China. Recently 200 strains from the same host species growing in seven provinces of China have been divided into four genotypic groups (including *M. huakuii*) by PCR-RFLP of 16S rRNA analysis (Zhong, personal communication). Further taxonomic work would be desirable to clarify the relationships among the defined species and new or unnamed groups.

ACKNOWLEDGMENTS

The authors are indebted to Dr. M. Gillis, Dr. E. Martínez-Romero, Dr. K. Lindström, and other members in the International Subcommittee for the Taxonomy of *Rhizobium/Agrobacterium* for their valuable advice.

DIFFERENTIATION OF THE SPECIES OF THE GENUS *MESORHIZOBIUM*

The differential characteristics of the species of *Mesorhizobium* are indicated in Table BXII.α.131.

List of species of the genus Mesorhizobium

1. **Mesorhizobium loti** (Jarvis Pankhurst, and Patel 1982) Jarvis, van Berkum, Chen, Nour, Fernandez, Cleyet-Marel and Gillis 1997, 898[VP] (*Rhizobium loti* Jarvis Pankhurst, and Patel 1982, 378.)

   lo' ti. M.L. masc. n. *Lotus* generic name of leguminous plants; M.L. gen. n. *loti* of *Lotus*.

   The characteristics are as described for the genus and as indicated in Table BXII.α.131. Cells are motile, predominantly by one polar or subpolar flagellum. Bacteroids in nodules are club-shaped and branched and contain inclusion bodies composed of poly-β-hydroxybutyrate (Jarvis et al., 1982). Strains usually lack plasmids, but a plasmid has recently been found in strain NZP 2213 (Wang et al., 1998). All strains utilize one or more of the following carbohydrates as sole carbon sources and form acidic products from them: glucose, galactose, fructose, arabinose, xylose, rhamnose, maltose, sucrose, lactose, trehalose, raffinose, mannitol, and dulcitol (Jarvis et al., 1982).

   Strains normally form nitrogen-fixing root nodules on the following hosts: *Lotus corniculatus*, *Lotus tenuis*, *Lotus japonicum*, *Lotus krylovii*, *Lotus filicalius*, and *Lotus schoelleri*. Ineffective nodules are formed on *L. pedunculatus*, *L. hispidus*, and *L. angustissiumum*. Information concerning additional plant hosts is presented by Jarvis et al. (1982).

   The mol% G + C of the DNA is: 59–64 (T_m).

   Type strain: ATCC 33669, DSM 2626, LMG 6125, NZP 2213.

   GenBank accession number (16S rRNA): D14514, X67229.

2. **Mesorhizobium amorphae** Wang, van Berkum, Sui, Beyene, Chen and Martínez-Romero 1999b, 63[VP]

   a.mor' phae. M.L. gen. n. *amorphae* of *Amorpha*, a genus of plant with which the species forms a nitrogen-fixing symbiosis.

   The characteristics are as described for the genus and as indicated in Table BXII.α.131. The cell size is 0.41–0.65 × 0.47–1.68 μm. Produces acid in mineral-salt medium containing mannitol and alkaline in litmus milk. The generation times are from 6–13 h in YM broth. The species can use L-arabinose, D-fructose, D-glucose, rhamnose, sucrose,

L-xylose, inositol, fumarate, alanine, and ornithine as sole carbon sources and ammonium salts and tyrosine as sole nitrogen sources. Strains are resistant to cefoperazone (75 μg/ml) and tetracycline (30 μg/ml). Usually, strains harbor one to three plasmids (150–930 kb). The genes for symbiosis reside on the 930-kb plasmids. There is only one *nifH* gene copy. Strains of this species have been isolated from nodules of *Amorpha fruticosa* growing in north China. No nodulation by the reference strain, ATCC 19665, has been observed on any selected host plant of other described microsymbiont species, except its original one, and *Amorpha fruticosa* does not nodulate with any type strains of other described species within the genera *Rhizobium*, *Sinorhizobium*, *Mesorhizobium*, *Bradyrhizobium*, and *Azorhizobium* (Wang et al., 1999b).

   The species can be differentiated at the molecular level from other *Mesorhizobium* species and related genera by PCR-RFLP of the 16S rRNA gene, multilocus enzyme electrophoresis (MLEE) patterns, DNA–DNA relatedness, and 16S rRNA sequence.

   The DNA–DNA similarity between *M. amorphae* and *M. loti* is 35.0% (Southern blotting method), and sequence similarity of 16S rRNA gene between them is 98.2%.

   The mol% G + C of the DNA is: 63–64 (T_m).

   Type strain: ATCC 19665.

   GenBank accession number (16S rRNA): AF041442.

3. **Mesorhizobium chacoense** Velázquez, Igual, Willems, Fernández, Muñoz, Mateos, Abril, Toro, Normand, Cervantes, Gillis and Martínez-Molina 2001, 1019[VP]

   cha.co.en' se. N.L. neut. adj. *chacoense* from El Chaco, Argentina, where the type strain was isolated.

   Motile Gram-negative rods. Colonies on yeast mannitol agar opaque, white, and convex. Aerobic. Acid produced from adonitol, L-arabinose, galactose, lactose, maltose, melibiose, rhamnose, sucrose, and trehalose in NH₄NO₃-containing media. *N*-acetylglucosaminidases, α-maltosidases, and β-xylosidases produced. Resistant to ciprofloxacin, cloxacillin, and erythromycin. Isolated from trees of the genus *Prosopsis*. Form root nodules and fix nitrogen in association with *P. alba*, *P. chilensis*, and *P. flexuosa*.

The mol% G + C of the DNA is: 62 (HPLC).

Type strain: LMG 19008.

GenBank accession number (16S rRNA): AJ278249.

4. **Mesorhizobium ciceri** (Nour, Fernandez, Normand and Cleyet-Marel 1994a) Jarvis, van Berkum, Chen, Nour, Fernandez, Cleyet-Marel and Gillis 1997, 897[VP] (*Rhizobium ciceri* Nour, Fernandez, Normand and Cleyet-Marel 1994a, 520.)

ci' ce.ri. L. gen. n. *ciceri* of *Cicer.*

The characteristics are as described for the genus and as indicated in Table BXII.α.131. Sole carbon sources are glycerol, arabinose, ribose, D-xylose, D-glucose, D-fructose, D-mannose, rhamnose, inositol, mannitol, *N*-acetylglucosamine, maltose, saccharose, trehalose, D-turanose, L-fucose, D-arabitol, succinate, fumarate, malate, pyruvate, L-histidine, tryptophan, L-glutamate, D,L-kynurenine, L-proline, betaine, D,L-4-aminobutyrate, 2-aminobenzoate, 3-aminobenzoate, 4-aminobenzoate, glucosamine, galactose, D,L-lactate, L-tyrosine, D-lyxose, glycerate, L-asparate, L-ornithine, β-alanine, L-xylose, L-alanine, and L-leucine (Nour et al., 1994b). No serological crossreaction is observed between this organism and 55 strains of *R. leguminosarum, B. japonicum,* and others (Kingsley and Bohlool, 1983). Strains of this species are isolated from chickpeas grown in uninoculated fields over a wide geographic range, including Spain, the United States of America, India, Russia, Turkey, Morocco, and Syria. Of 71 chickpea isolates tested, 70 nodulate only the original host plant and do not nodulate the 88 species belonging to Fabaceae and Mimosaceae (Gaur and Sen, 1979). Consequently, this species may be considered a monospecific cross-nodulation system.

This species can be distinguished at the molecular level from other *Mesorhizobium* species and related genera by DNA–DNA relatedness, PCR-RFLP of 16S rRNA, and 16S rRNA sequencing.

The DNA similarity between *M. ciceri* and *M. loti* is 7.5% (S$_1$ nuclease method). The 16S rRNA sequence similarity between them is 99.8%.

The mol% G + C of the DNA is: 63–64 (T_m).

Type strain: UPM-Ca7, ATCC 51585, DSM 11540, LMG 14898.

GenBank accession number (16S rRNA): U07934.

5. **Mesorhizobium huakuii** (Chen, Li, Qi, Wang, Yuan and Li 1991b) Jarvis, van Berkum, Chen, Nour, Fernandez, Cleyet-Marel, and Gillis 1997, 897[VP] (*Rhizobium huakuii* Chen, Li, Qi, Wang, Yuan and Li 1991b, 278.)

hua.kui' i. N.L. gen. n. *huakuii* in honor of Huakui Chen, a Chinese professor of soil microbiology, who is the pioneer in investigating the microsymbionts isolated from *Astragalus sinicus.*

The characteristics are as described for the genus and as indicated in Table BXII.α.131. Motile by means of a single polar or subpolar flagellum. Cells usually contain granules of poly-β-hydroxybutyrate. The bacteroids in nodules are mostly club-shaped, in vast amounts and approximately 40–50 times larger than the unswollen rod-shaped bacterium (Cao et al., 1984). The generation time of the strains in the species varies from 4–6 h. Strains use D-arabinose, cellobiose, D-fructose, D-glucose, lactose, sucrose, maltose, D-raffinose, L(+)-rhamnose, malate, fumarate, oxalate, or succinate as a sole carbon source. Ammonium nitrate, urea,

and many amino acids, including arginine, citrulline, lysine, methionine, ornithine, threonine, tryptophan, and tyrosine, are utilized as sole nitrogen sources. DL-Leucine and valine are not utilized. Some strains require thiamin, nicotinamide, and riboflavin for growth (Chen et al., 1991b). Strains are resistant to the following antibiotics (μg/ml): streptomycin, 10; tetracycline, 10; erythromycin, 20; penicillin, 10; kanamycin, 5; polymyxin, 10; nystatin, 30; neomycin, 10. Most strains contain 1–5 plasmids (66–942 kb). Usually, the *nod* and *nif* genes are carried on the largest plasmid (942 kb) as a gene cluster (Xu and Murooka, 1995; Zou et al., 1997). The other plasmids are also related to symbiosis. A derivative strain cured of the middle-sized plasmid does not nodulate the host plant, has an altered lipopolysaccharide, and grows more slowly than the parent strain. Curing the strain of the smallest plasmid results in delayed nodulation and loss of nitrogen-fixation ability. Curing the strain of each of these plasmids reduces the acid tolerance (Zou et al., 1997). The original host of the species is *Astragalus sinicus,* a green manure crop grown in rice fields in the southern parts of China, Japan, and Korea. This species either cannot nodulate other leguminosarum species or can do so only weakly, and isolates from nodules of 24 species of 8 genera of Leguminoseae cannot nodulate *A. sinicus,* except for isolates from *Desmodium hoterotyllum* (Chen and Shu, 1944). Consequently, the *M. huakuii* symbiosis may be a monospecific cross-nodulation system (Chen et al., 1992).

This species can be differentiated at the molecular level from other *Mesorhizobium* species and related genera by SDS-PAGE of whole-cell proteins, DNA–DNA relatedness, and 16S rRNA sequence.

The DNA–DNA similarity between type strains of *M. loti* and *M. huakuii* is 22.3%. The 16S rRNA sequence similarity between them is 98.3%.

The mol% G + C of the DNA is: 59–64 (T_m).

Type strain: 103, ATCC 51122, CCBAU 2609, DSM 6573, IAM 14158.

GenBank accession number (16S rRNA): D12797.

6. **Mesorhizobium mediterraneum** (Nour, Normand, Cleyet-Marel and Fernandez 1995) Jarvis, van Berkum, Chen, Nour, Fernandez, Cleyet-Marel and Gillis 1997, 898[VP] (*Rhizobium mediterraneum* Nour, Normand, Cleyet-Marel and Fernandez 1995, 647.)

me.di.ter.ra' ne.um. L. adj. *mediterraneum* midland, inland.

Characteristics are as described for the genus and as indicated in Table BXII.α.131. Strains that have been tested can grow at 4°C and cannot grow at pH 5.0 and 10.0; these differentiate them from *M. ciceri.* Sole carbon sources include 39 compounds that are the same as those used by *M. ciceri.* D-Lyxose, DL-glycerate, L-asparate, L-ornithine, β-alanine, L-xylose, L-alanine, and L-leucine are not used (Nour et al., 1994b). Strains comprise a single serological group (Kingsley and Bohlool, 1983). This species may be a monospecific cross-nodulation system (Gaur and Sen, 1979). It can be distinguished at the molecular level from other *Mesorhizobium* species and related genera by DNA–DNA relatedness, PCR-RFLP of the intergenic spacer (IGS) between the 16S and 23S rRNA genes (the average length of the IGS DNA is 1000 bp), and total 16S rRNA sequence analysis.

The DNA–DNA similarity between *M. mediterraneum* and

M. loti is 31%, and the 16S rRNA sequence similarity between them is 97.2%.

> *The mol% G + C of the DNA is*: 63–64 (T_m).
> *Type strain*: ATCC 51670, UPM-Ca36, DSM 11555.
> *GenBank accession number (16S rRNA)*: L38825.

7. **Mesorhizobium plurifarium** de Lajudie, Willems, Nick, Moreira, Molouba, Hoste, Torck, Neyra, Collins, Lindström, Dreyfus and Gillis 1998b, 380[VP]

plu.ri.fa'ri.um. M.L. adj. *plurifarus* from adv. *plurifarium* in a different place, referring to the fact that this species contains strains isolated from several places in East Africa, West Africa, and South America.

The characteristics are described as for the genus and as indicated in Table BXII.α.131. Cells are motile in liquid media. Colonies on YMA are beige, round, convex to droplike, 0.5–2.0 mm in diameter within 2–3 d at 30°C. Strains utilize glycerol, arabinose, ribose, L-xylose, L-sorbose, D-mannose, rhamnose, inositol, mannitol, sorbitol, *N*-acetylglucosamine, D-cellobiose, maltose, lactose, D-melibiose, D-raffinose, xylitol, D-turanose, D-lyxose, succinate, fumarate, glycolate, D,L-lactate, D-malate, L-malate, pyruvate, L-histidine, L-glutamate, L-ornithine, L-arginine, L-proline, betaine, and glucosamine as sole carbon sources. Erythritol, methyl-L-xyloside, amygdalin, glutarate, D-tartrate, D-mandelate, L-mandelate, glycine, D-(−)-alanine, L-norleucine, D,L-norvaline, L-serine, L-cysteine, L-citrulline, creatine, diaminobutane, and spermine are utilized (de Lajudie et al., 1994).

Most of the strains can nodulate *Acacia senegal*, *A. tortilis*, *A. nilotica*, *A. seyal*, *Leucaena leucocephala*, and *Neptunia oleracea*, but not *Sesbania rostrata*, *S. pubescens*, *S. grandiflora*, *Ononis repens*, and *Lotus corniculatus* (Lortet et al., 1996; de Lajudie et al., 1998b). This species can be differentiated at the molecular level from other *Mesorhizobium* species and related genera by SDS-PAGE of whole-cell proteins, DNA–DNA hybridization, and 16S rRNA sequence analysis.

The DNA–DNA similarity between this species and *M. loti* is 10–13%. The 16S rRNA sequence similarity between this species and *M. loti* is 98.0%.

> *The mol% G + C of the DNA is*: 62.6–64.4 (T_m).
> *Type strain*: LMG 11892, ORS 1032.
> *GenBank accession number (16S rRNA)*: Y14158.

8. **Mesorhizobium tianshanense** Jarvis, van Berkum, Chen, Nour, Fernandez, Cleyet-Marel and Gillis 1997, 898[VP] (*Rhizobium tianshanense* Chen, Wang, Wang, Li, Chen and Li 1995c, 158.)

tian.shan.en'se. M.L. adj. *tianshanense* referring to the Tianshan Mountains in the Xingjiang region of the People's Republic of China, where strains were isolated.

The characteristics are described as for the genus and as indicated in Table BXII.α.131. Cells of some strains move by means of peritrichous flagella, others by means of polar and subpolar flagella. The generation time of strains varies from 5–15 h in YM broth. Even the strains with a slow growth rate (generation time >8 h) produce an acid reaction in YMA. All tested strains use D-glucose, mannitol, glycerol, D-xylose, D-fucose, erythritol, D,L-proline, L-(+)-glutamic acid, and rhamnose; some strains can use lactose, maltose, sucrose, pyruvate, raffinose, sorbitol, fructose, mannose, arabinose, citrate, and dulcitol. Tagatose, galactose, melibiose, sorbose, inulin, benzoate, malate, vanillic acid, inositol, salicytol, fumarate, pyrocatechol, starch, and dextrin are not utilized. Sole nitrogen sources are ammonium salts, nitrate, and many kinds of amino acids, including glutamate, glycyl-L-leucine, valine, threonine, arginine, proline, and glycine. No megaplasmid has been detected, but some strains harbor three plasmids (166–468 kb) (Zou and Chen, unpublished). Strains have been isolated from *Glycyrrhiza pallidiflora*, *G. uralensis*, *Glycine max*, *Sophora alopecuroides*, *Swainsonia salsula*, *Caragara polourensis*, and *Halimodendron holodendron* growing in Xinjiang Region of China. Most of these plants are wild and indigenous to that region, except *Glycine max*, a cultivated crop that originated in northeastern China. The type strain A-1BS nodulates all of the original host species of *M. tianshanense* strains but not *Pisum sativum*, *Medicago sativa*, *Phaseolus vulgaris*, *Trifolium repens*, *Lotus corniculatus*, *Leucaena leucocephala*, *Macroptilium atropurpureum*, *Vigna unguiculata*, and *Astragalus sinicus*. Therefore, the hosts of *M. tianshanense* form a single cross-nodulation group (Chen et al., 1995c).

The species can be differentiated from other *Mesorhizobium* species and related genera by phenotypic features, DNA–DNA relatedness, and 16S rRNA sequence.

The DNA–DNA similarity between the type strain of this species and that of *M. loti* is 4.4% (Southern blotting method), and the 16S rRNA gene sequence similarity between them is 97.6%.

> *The mol% G + C of the DNA is*: 59–63 (T_m).
> *Type strain*: A-1BS, CCBAU 3306, DSM 11417.
> *GenBank accession number (16S rRNA)*: AF041447, U71079.

Other Organisms

Four unnamed species within the genus *Mesorhizobium* have been reported for some symbiotic or nonsymbiotic strains that were isolated from the rhizosphere of *Lotus corniculatus* (Sullivan et al., 1995, 1996). These isolates have been identified by RFLP of different genes from *M. loti*, including exopolysaccharide genes and symbiotic genes, total DNA–DNA hybridization, MLEE, and 16S rRNA gene sequencing. They are considered to be potential recipients that can acquire and express symbiotic-controlling genes from related microsymbiotic strains.

Two other *Mesorhizobium* groups were also identified when the species *M. amorphae* was described (Wang et al., 1999b). These two groups were also isolated from *Amorpha fruticosa*, and they share moderate DNA–DNA similarity (35–50%) and the same symbiotic plasmid with *M. amorphae* but have distinct phylogenies of their 16S rRNA genes. They also have high phenotypic similarities to *M. amorphae*. These might be included in *M. amorphae* when further evidence is obtained.

Genus VII. **Pseudaminobacter** Kämpfer, Müller, Mau, Neef, Auling, Busse, Osborn and Stolz 1999, 894[VP]

PETER KÄMPFER

Pseud.ami.no.bac′ter. Gr. adj. *pseudos* false; M.L. *aminobacter* generic name of a bacterium; *Pseudaminobacter* false aminobacters.

Rods 0.5–0.8 × 1.0–2.0 µm, with rounded ends. **Motile**. Gram negative. **Aerobic**, having a strictly respiratory type of metabolism with oxygen as the terminal electron acceptor. **Oxidase and catalase positive**. Colonies on nutrient agar at 25°C are circular, entire, slightly convex and smooth, glistening, and pale beige. Contains ubiquinone Q-10. Major polyamines are spermidine, *sym*-homospermidine, and putrescine. Polar lipid pattern characterized by nearly equal amounts of phosphatidylcholine, phosphatidylglycerol, phosphatidyldimethylethanolamne, phosphatidylmonomethylethanolamine, phosphatidylethanolamine, and diphosphatidylglycerol. The major cellular fatty acid is $C_{18:1}$. The hydroxy acids include the hydroxylated fatty acid $C_{15:0 \text{ iso } 3OH}$. Isolated from wastewater and river water.

The mol% G + C of the DNA is: 62.9–63.9.

Type species: **Pseudaminobacter salicylatoxidans** Kämpfer, Müller, Mau, Neef, Auling, Busse, Osborn and Stolz 1999, 894.

FURTHER DESCRIPTIVE INFORMATION

P. salicylatoxidans BN12[T] was described as an aerobic bacterium, able to degrade substituted naphthalenesulfonates and substituted salicylates and originated from a 6-aminonaphthalene-2-sulfonate degrading microbial consortium from the German river Elbe (Nörtemann, 1987). *P. defluvii* THI 051[T] was isolated during a study on thiobacilli (Katayama-Fujimura et al., 1983,

1984b), but it became soon obvious that the organism was not an authentic *Thiobacillus* species.

Growth occurs on nutrient agar (Oxoid), tryptone soya agar (Oxoid), trypticase soy broth (BBL) supplemented with 1.5% (w/v) agar (BBL), and R2A agar (Oxoid).

DIFFERENTIATION OF THE GENUS *PSEUDAMINOBACTER* FROM OTHER GENERA

Table BXII.α.132 lists the main differential characteristics of the genera belonging to this rRNA group.

TAXONOMIC COMMENTS

16S rDNA sequence analyses of the type strains of the two *Pseudaminobacter* species clearly place the genus in the family within the *Alphaproteobacteria*. In phylogenetic trees, the highest similarities occur with *Aminobacter* species (>96.7%). Together with the genera *Defluvibacter*, *Mesorhizobium*, *Phyllobacterium*, *Sinorhizobium*, *Ochrobactrum*, and *Aminobacter* (sequence similarities between 16S rRNA genes are above 93%), they form a branch referred to as the group of bacteria known as the alpha-2-subgroup of the *Proteobacteria*. In Fig. BXII.α.152 a phylogenetic tree is presented.

At present, each species is represented by only one strain (Kämpfer et al., 1999).

DIFFERENTIATION OF THE SPECIES OF THE GENUS *PSEUDAMINOBACTER*

The two species can be differentiated based on several biochemical tests as indicated in Table BXII.α.133.

TABLE BXII.α.132. Features that differentiate the genus *Pseudaminobacter* from related genera of the *Alphaproteobacteria*[a,b]

Characteristic	Pseudaminobacter	Aminobacter[c]	Defluvibacter	Mesorhizobium	Ochrobactrum	Phyllobacterium
Major polyamines:						
Putrescine	+	+		+		+
Spermidine	+		+		+	
sym-Homospermidine	+	+		+		+
Presence of fatty acid $C_{12:0 \text{ 3OH}}$	−	+	+	+	−	−
Assimilation of:						
Malonate			+	−	−	−
Glutarate	D	D	+	−	+	−
Norvaline			+	−	+	−
L-Tryptophan	−	−	+	D	−	D
Putrescine	−	−	+	−	−	−
Nitrate reduction			−	−	+	+

[a]Symbols: see standard definitions; blank space, not determined.

[b]Data from Fritsche et al. (1999a) and Kämpfer et al. (1999).

[c]The genus *Chelatobacter* is not listed separately, because it is considered as a later synonym of *Aminobacter* (Kämpfer et al., 2002)

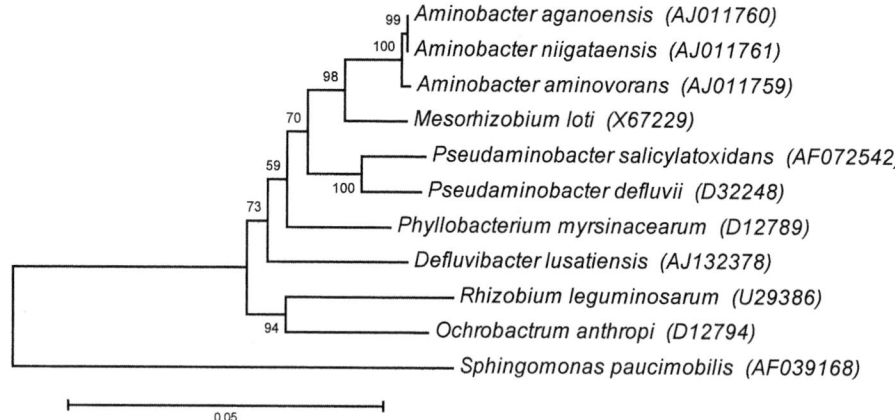

FIGURE BXII.α.152. Phylogenetic tree based on 16S rRNA sequences of *Pseudaminobacter* species and close relatives. The underlying distance matrix has been calculated with the neighbor joining tool of the ARB package using the Jukes-Cantor correction. All gaps in the sequences were excluded. *S. terrae* was used as the outgroup. Accession numbers of the analyzed sequences are indicated. Bar = 10% sequence divergence.

List of species of the genus Pseudaminobacter

1. **Pseudaminobacter salicylatoxidans** Kämpfer, Müller, Mau, Neef, Auling, Busse, Osborn and Stolz 1999, 894[VP]

sa.li.cy.lat.oxi' dans. N.L. *salicylatoxidans* oxidizing salicylate, because the organism oxidizes salicylate in an unusual manner.

The characteristics are as described for the genus and as listed in Tables BXII.α.132 and BXII.α.133, with the following additional features. Cell size, 0.5–0.8 × 1.0–1.5 μm. Colonies on nutrient agar are circular, entire, slightly convex and smooth, glistening, and beige, 1–3 mm in diameter. Growth occurs at 20–40°C. The major polyamines are spermidine, *sym*-homospermidine, and putrescine. Utilize adonitol, *cis*-aconitate, D-maltose, D-trehalose, glutarate, and L-malate but not L-arabinose, D-cellobiose, D-galactose, D-mannose, sucrose, L-ornithine, and L-serine. Do not hydrolyze *p*-nitrophenyl α-D-glucopyranoside and *p*-nitrophenyl-β-D-glucopyranoside. Isolated from a microbial consortium originating from the river Elbe, Germany.

The mol% G + C of the DNA is: 63.9 (HPLC).
Type strain: BN12, CIP 106963, DSM 6986.
GenBank accession number (16S rRNA): AF072542.

2. **Pseudaminobacter defluvii** Kämpfer, Müller, Mau, Neef, Auling, Busse, Osborn and Stolz 1999, 895[VP]

de.flu' vi.i. L. neut. gen. n. *fluvius* wastewater, because the organism was isolated from activated sludge.

The characteristics are as described for the genus and as listed in Tables BXII.α.132 and BXII.α.133, with the following additional features. Coccoid to rod-shaped cells, 0.5–0.8 × 0.8–1.2 μm. Colonies on nutrient agar are circular, entire, slightly convex and smooth, glistening, and beige, 1–3 mm in diameter. Growth occurs between 10° and 40°C. No growth occurs at 5° or 45°C. Utilize D-glucose, ribose, D-xylose, acetate, and propionate but not D-maltose, D-trehalose, adonitol, mannitol, sorbitol, L-malate, and L-aspartate. Isolated from a Japanese activated sludge enriched with thiocyanate.

The mol% G + C of the DNA is: 62.9 (HPLC).
Type strain: THI 051, CIP 107185, IFO 14570.
GenBank accession number (16S rRNA): D32248.

TABLE BXII.α.133. Differential characteristics of the type strains of *Pseudaminobacter* species[a]

Characteristic[b]	*P. salicylatoxidans* strain BN12	*P. defluvii* strain THI 051
Acid produced from:[c]		
D-Mannitol	w	−
Dulcitol	w	−
Melibiose	w	−
Utilization of:[d]		
D-Maltose	+	−
D-Trehalose	+	−
Adonitol	+	−
D-Mannitol	+	−
D-Sorbitol	+	−
cis-Aconitate	+	−
Glutarate	+	−
L-Malate	+	−
L-Aspartate	+	−
L-Ornithine	−	+
L-Serine	−	+
4-Hydroxybenzoate	+	−

[a]Symbols: +, positive; −, negative; w, weakly positive. Test results given in the table were read after 72 h of incubation at 30°C.

[b]Tests for hydrolysis of various substrates (not listed in this table) failed to differentiate the two strains. Both type strains hydrolyzed bis-*p*-nitrophenyl-phosphate, *p*-nitrophenyl-phosphorycholine, and L-alanine-*p*-nitroanilide. Neither strain hydrolyzed esculin, *p*-nitrophenyl-β-L-galactopyranoside, *p*-nitrophenyl-β-D-glucuronide, *p*-nitrophenyl-α-D-glucopyranoside, *p*-nitrophenyl-β-D-glucopyranoside, *p*-nitrophenyl-phosphoryl-choline, 2-deoxythymidine-5′-*p*-nitrophenyl-phosphate, and L-glutamate-γ-3-carboxy-*p*-nitroanilide.

[c]Acid formation from carbohydrates in most cases was very weak (even after prolonged incubation). Both strains showed weak acid formation from D-glucose. Neither strain formed acid from lactose, sucrose, salicin, adonitol, inositol, sorbitol, L-arabinose, raffinose, rhamnose, maltose, D-xylose, trehalose, cellobiose, methyl-D-glucoside, and erythritol.

[d]Both strains utilized *N*-acetyl-D-glucosamine, D-glucose, D-ribose, D-xylose, acetate, propionate, 4-aminobutyrate, DL-3-hydroxybutyrate, DL-lactate, oxoglutarate, pyruvate, L-alanine, β-alanine, L-histidine, L-leucine, and L-proline. Neither strain utilized L-arabinose, arbutin, D-cellobiose, D-fructose, D-galactose, gluconate, α-D-melibiose, L-rhamnose, sucrose, salicin, *i*-inositol, maltitol, putrescine, *trans*-aconitate, adipate, azelate, citrate, fumarate, itaconate, mesaconate, suberate, L-phenylalanine, L-tryptophan, 3-hydroxybenzoate, and L-phenylacetate.

Family V. **Methylocystaceae** *fam. nov.*

JOHN P. BOWMAN

Me.thy.lo.cyst.a′ce.ae. M.L. masc. n. *Methylocystis* type genus of the family; *-aceae* ending to denote a family; M.L. fem. pl. n. *Methylocystaceae* the *Methylocystis* family.

Cells are **pyriform, vibrioid, reniform, or rod-like** in shape, sometimes arranged in rosettes. **Reproduce by budding or binary division. Produce heat- and desiccation-resistant exospores or lipoidal cysts.** Motility varies; if present, cells are propelled by a polar and/or subpolar flagellar tuft. Cells contain **type II intracytoplasmic membranes,** which are arranged as layers in parallel to the periphery of the cell wall. Possess an **aerobic, strictly respiratory metabolism. Obligately methanotrophic, utilizing only methane and methanol as sole carbon and energy sources.** C_{2+} and other C_1 compounds are not utilized. Fix formaldehyde for cell carbon via the **serine pathway.** Enzymes for the Benson–Calvin cycle pathway are absent. The tricarboxylic acid pathway is complete. **Fix atmospheric nitrogen** by means of an oxygen-sensitive nitrogenase. Mesophilic. Nonhalophilic. Major habitats include soils, freshwater sediments, and groundwater. Primary fatty acids are $C_{18:1\ \omega 8c}$ and $C_{18:1\ \omega 7c}$. Primary quinone is ubiquinone-8 (Q-8).

The mol% G + C of the DNA is: 61–67 (T_m).

Type genus: **Methylocystis** (ex Romanovskaya, Malashenko and Bogachenko 1978) Bowman, Sly, Nichols and Hayward 1993c, 751.

FURTHER DESCRIPTIVE INFORMATION

Circumscription of the family Bacteria that have the capability to oxidize and utilize methane as a sole carbon and energy source are referred to as methanotrophic bacteria or methanotrophs (Whittenbury et al., 1970b). The original description of the family *Methylococcaceae* (Whittenbury and Krieg, 1984) included two distinct groups of methanotrophs referred to in the literature as "type I" and "type II" methanotrophs. In addition, there is a third group called the "type X" methanotrophs (Hanson and Hanson, 1996), which represents a subgroup of type I methanotrophs. Type I and II methanotrophs possess extensive biological differences, which are morphological, biochemical, chemotaxonomic, and phylogenetic in nature. The order *"Methylococcales"*, which includes the type I methanotrophs, is classified in the *Gammaproteobacteria* and is thus dealt with in another part of this volume. The Family *Methylocystaceae* represents all type II methanotrophs and includes the genera *Methylocystis* and *Methylosinus*. Phenotypic properties and fatty acid components differentiating these genera are shown in Tables BXII.α.134 and BXII.α.135, respectively. The facultative methylotroph *Methylobacterium* represents a distinct taxonomic group and does not have close enough taxonomic association with *Methylocystis* or *Methylosinus* to be included in the family *Methylocystaceae*.

Methane oxidation The major characteristic that separates methanotrophs from other procaryotes is their ability to utilize methane as a sole carbon and energy source. Methanotrophs oxidize methane to CO_2 in a dissimilatory pathway that generates energy and allows access to metabolizable carbon units in the form of formaldehyde. The first step of this pathway is the oxidation of methane to methanol by action of methane monooxygenase (MMO). Methanol is then oxidized to formaldehyde by a pyrrolquinol quinone-containing methanol dehydrogenase. Formaldehyde is then either used for cell carbon (see below) or oxidized to formate and then CO_2, these final steps forming reducing equivalents needed to drive methane oxidation.

TABLE BXII.α.134. Differentiation of the major types of methanotrophs[a]

Characteristic	Type I[b]	Type II[c]	Type X[d]
Resting stages:			
Azotobacter-type cysts	+[e]	−	+
Exospores	−	+	−
Lipoidal cysts	−	+	−
Intracytoplasmic membranes:			
Type I	+	−	+
Type II	−	+	−
Soluble methane monooxygenase (sMMO)	−	+	−
Carbon assimilation pathway:			
RuMP	+	−	+
Serine	−	+	−
Benson-Calvin cycle enzymes	−	−	+
Fatty acid carbon chain length	16	18	16
Major quinone:			
MQ-8	+	−	+
Q-8	+	+	−
Mol% G + C (T_m)	43–60	61–67	56–65

[a]For symbols see standard definitions.

[b]Genera include *Methylosphaera, Methylobacter, Methylomicrobium,* and *Methylomonas*; family *Methylococcaceae,* class *Gammaproteobacteria.*

[c]Genera include *Methylosinus* and *Methylocystis*; family *Methylocystaceae,* class *Alphaproteobacteria.*

[d]Genera include *Methylococcus*; family *Methylococcaceae,* class *Gammaproteobacteria.*

[e]May have no resting stage.

TABLE BXII.α.135. Phospholipid fatty acid profiles of the genera *Methylosinus* and *Methylocystis*[a]

Fatty acids	Percent composition	
	Methylocystis	*Methylosinus*
$C_{16:0}$	0.9–5.0	0.7–2.2
$C_{16:1\ \omega 7c}$	0.3–2.6	9.3–14.2
$C_{18:1\ \omega 13c}$	0–4.9	0
$C_{18:1\ \omega 9c}$	0–0.4	0–0.2
$C_{18:1\ \omega 8c}$	61.0–74.6	65.9–70.5
$C_{18:1\ \omega 7c}$	15.3–18.6	12.8–17.7
$C_{18:0}$	2.1–5.8	0–0.2
$C_{18:1\ cyclo}$	1.0–4.7	0–2.2
$C_{19:0\ cyclo}$	1.5–4.9	0–3.0

[a]Data from Bowman et al. (1993c).

In all type II methanotrophs, MMO is present in a membrane-bound form and is referred to as particulate MMO (pMMO) (Semrau et al., 1995). Synthesis of a cytoplasmic version of MMO, referred to as soluble MMO (sMMO) (Dalton, 1992; Murrell, 1992; Lipscomb, 1994), is common in type II methanotrophs but is relatively rare in type I methanotrophs (Bowman et al., 1993a; Bowman and Sayler, 1994; Miguez et al., 1997). Although pMMO and sMMO are genetically distinct (Martin and Murrell, 1995), they are controlled by a common copper-inducible regulatory pathway (Nielsen et al., 1997). Soluble MMO is copper-repressible and contains three components. The first component is a hydroxylase that contains an unusual non-heme iron–oxygen linked active site. The second component is a regulatory protein, and the third is a ferredoxin-based reductase. Particulate MMO has a high affinity to methane, allowing for more efficient growth yields than with sMMO. sMMO appears to be an adaptation to copper-limiting growth conditions (Hanson and Hanson, 1996). However, sMMO has an exceptionally broad substrate specificity and can co-metabolize an extraordinarily wide range of aliphatic, aromatic, and heterocyclic compounds (Haber et al., 1984). This property has been exploited biotechnologically, with methanotrophs being proposed for organic transformation applications in industrial applications (Lidstrom and Stirling, 1990) and as bioremediation agents (Hanson and Hanson, 1996).

The amino acid sequences of the polypeptides and the nucleotide sequences of the genes (mmoA, mmoB, mmoC and mmoX) of sMMO of type I and II methanotrophs show a high degree of conservation (Stainthorpe et al., 1991; Murrell, 1992). These genes have been used to develop general oligonucleotide probes that are specific for different types of methanotrophs (Holmes et al., 1995b; McDonald and Murrell, 1997a, b; Miguez et al., 1997; Dedysh et al., 1998b). Several physiological comparisons have been drawn between methanotrophs and ammonia-oxidizing bacteria because both groups are capable of oxidizing methane and ammonia by similar means (Hanson and Hanson, 1996). For example, type II methanotrophs are able to oxidize ammonia to nitrite and nitrous oxide, although the oxidation rates are lower than those in ammonia-oxidizing bacteria (Yoshinari, 1985). An evolutionary link has been drawn between type I methanotrophs and ammonia-oxidizers (such as *Nitrosococcus oceani*) owing to the similarity of pMMO (*pmo*A gene) and ammonia monooxygenase genes (Holmes et al., 1995a). However, the *pmo*A sequences from type II methanotrophs are comparatively more divergent compared to *pmo*A gene sequences from type I methanotrophs (Holmes et al., 1995a; Bodrossy et al., 1997).

C$_1$ carbon assimilation Type II methanotrophs are specialized to utilize C$_1$ compounds, including methane, methanol, and formaldehyde. C$_{2+}$ (acetate, D-glucose, etc.) and other C$_1$ compounds (formate, methylamine, etc.) are not utilized as energy sources, but they may be used as supplementary carbon sources when cells are grown in the presence of methane or methanol (Hanson et al., 1992). Formaldehyde is toxic to methanotrophic cells at relatively low concentrations and so is often not tested as a carbon source. In addition, freshly isolated methanotrophs often prove sensitive to methanol (due to accumulation of formaldehyde); consequently, methanol should be tested only at low concentrations (10–30 mM) or provided as a vapor in a sealed container. Methanotrophs can be "trained" to tolerate higher concentrations of methanol (Lidstrom, 1988). Type II methanotrophs use the serine pathway to incorporate formaldehyde into synthetic pathways (Large and Quayle, 1963), in contrast to type I methanotrophs, which utilize the ribulose monophosphate (RuMP) pathway to fix carbon (Strom et al., 1974). The serine pathway has been described in detail by Hou (1984). The enzyme α-hydroxypyruvate reductase is often used to detect the serine pathway and thus can be used as a direct way to distinguish type I and II methanotrophs; however, some type I methanotrophs may possess a low level of activity for α-hydroxypyruvate reductase. Type II methanotrophs may also be distinguished by their lack of key enzymes (i.e., hexose phosphate synthase) for the RuMP pathway (Colby et al., 1979).

Intracytoplasmic membranes When methanotrophs are grown in the presence of methane and methanol, characteristic intracytoplasmic membranes (ICMs) are formed from convolutions of the cytoplasmic membrane. ICMs are about 80–90 nm thick, appear as typical lipid bilayers, and in type II methanotrophs are arranged in pairs parallel to the cytoplasmic membrane (type II ICM) (Best and Higgins, 1981) (Fig. BXII.α.153). The relative ICM content in cells is enhanced under reduced methane tensions, at elevated copper ion concentrations, and when a stable and rapid rate of methane transfer is available (Best and Higgins, 1981; Scott et al., 1981).

Ecology Methane is an important greenhouse gas and the most abundant organic gas in the atmosphere (Crutzen, 1991). Methanotrophs are the largest global methane sink and as a result are ubiquitous in nature. They produce the highest and most active populations in environments with a stable gas exchange in which both oxygen and methane are readily available (Reeburgh et al., 1993). Recent evidence suggests methanotrophs make up a high proportion of the total bacterial biomass (up to 40%) in many aquatic environments and surface sediments (Ross et al., 1997; P.I. Boon, personal communication). Stable isotopic analyses (methane is highly ^{13}C-depleted) indicate that a considerable proportion of carbon found in aquatic life at different trophic levels has its origin from methanotrophic bacteria (Boschker et al., 1998; Boon et al., unpublished).

Many studies have used indirect, culture-independent means to study the ecology of type II methanotrophs. These include indirect immunofluorescence (Reed and Dugan, 1978; Smirnova and Archipova, 1981), lipid signatures (Nichols et al., 1985, 1987; Bowman et al., 1991a) and probes or PCR to detect pMMO and sMMO (Tsien and Hanson, 1992; Hanson et al., 1993; Brusseau et al., 1994; McDonald et al., 1995, 1996; McDonald and Murrell, 1997b; Ross et al., 1997; Dedysh et al., 1998b), methanol dehydrogenase (McDonald and Murrell, 1997a), and 16S rRNA genes (Brusseau et al., 1994; Holmes et al., 1995a, 1996; McDonald et al., 1996). This use of culture-independent means is perhaps due to the effort and time necessary to obtain pure cultures. In only a few instances have studies coupled culture-independent and -dependent investigations, thereby permitting a definitive identification of the type II methanotrophs present in environmental samples. In any case, type II methanotrophs are ubiquitous in

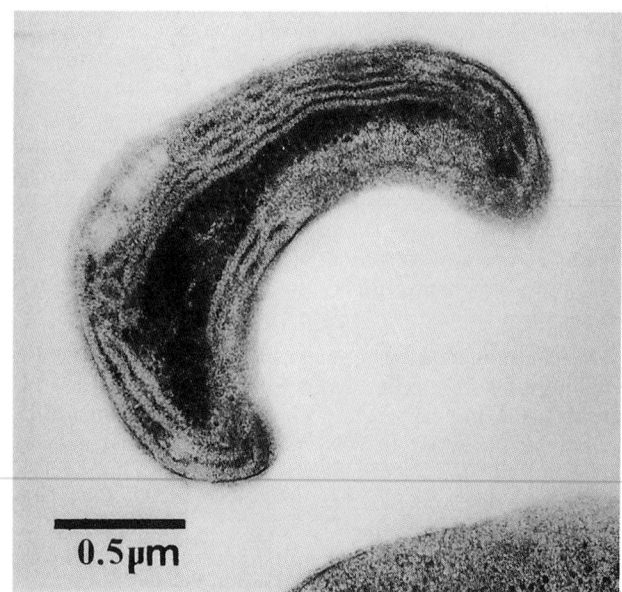

FIGURE BXII.α.153. Electron micrograph of a thin section of *Methylocystis parvus* showing the typical ultrastructure of type II intracytoplasmic membranes.

aquatic and terrestrial ecosystems and have been successfully isolated or enriched from marshes and swamps; roots of aquatic macrophytes; sediments of rivers and streams; rice paddies; ponds, lakes; soils of meadows, forests, peat bogs, and tundra; groundwater; and sewage sludge (Whittenbury et al., 1970b; Gal' chenko, 1977, 1994; Haubold, 1978; Heyer et al., 1984; Nichols et al., 1985; Bowman et al., 1993a, c; Hanson et al., 1993; Vecherskaya et al., 1993; Boon et al., 1996; Calhoun and King, 1998; Dedysh et al., 1998b). In general, type II methanotrophs appear to have a narrower distribution than do type I methanotrophs because they have not been isolated from marine habitats or low and high temperature ecosystems and have not been found to form endosymbioses.

Type II methanotrophs outcompete type I methanotrophs in copper- and nitrogen-limiting oligotrophic environments such as groundwater and soil (Bowman et al., 1993a; Graham et al., 1993) and in environments with low oxygen tensions where methane is not limiting for growth (Amaral et al., 1995a, b). Though as common as type I methanotrophs in various freshwater environments (Saralov et al., 1984; Gal' chenko, 1994; Ross et al., 1997), type II methanotrophs preferentially concentrate in surface layers of sediment (Reed and Dugan, 1978; Hanson et al., 1993)

that have high methane fluxes. When present together, type I and II methanotrophs can produce either mutually beneficial or antagonistic effects depending on the strains compared (Starostina et al., 1995; Pashkova et al., 1997).

FURTHER READING

Bowman, J.P., L.I. Sly, P.D. Nichols and A.C. Hayward. 1993. Revised taxonomy of the methanotrophs: description of *Methylobacter* gen. nov., emendation of *Methylococcus*, validation of *Methylosinus* and *Methylocystis* species, and a proposal that the family *Methylococcaceae* includes only the group I methanotrophs. Int. J. Syst. Bacteriol. *43*: 735–753.

Hanson, R.S. and T.E. Hanson. 1996. Methanotrophic bacteria. Microbiol. Rev. *60*: 439–471.

Hanson, R.S., A.I. Netrusov and K. Tsuji. 1992. The obligate methanotrophic bacteria *Methylococcus*, *Methylomonas*, and *Methylosinus*. *In* Balows, Trüper, Dworkin, Harder and Schleifer (Editors), The Prokaryotes. A Handbook of Bacteria: Ecophysiology, Isolation, Identification, Applications, 2nd Ed., Vol. 3, Springer-Verlag, New York. pp. 2350–2364.

Whittenbury, R.A., K.C. Phillips and J.F. Wilkinson. 1970. Enrichment, isolation and some properties of methane-utilizing bacteria. J. Gen. Microbiol. *61*: 205–218.

Genus I. **Methylocystis** *(ex Romanovskaya, Malashenko and Bogachenko 1978) Bowman, Sly, Nichols and Hayward 1993c, 751*[VP]

JOHN P. BOWMAN

Me.thyl.o.cys' tis. Fr. *methyle* the methyl radical; *cystis* bag; M.L. n. *Methylocystis* methyl bag.

Cells **are small, rod-like to reniform** in shape, 0.3–0.5 × 0.5–1.5 µm, usually arranged singly. **Reproduce by binary division.** Nonmotile. Cells contain **type II intracytoplasmic membranes,** which are arranged as multiple layers along the periphery of the cell wall. May form **cylindrical spinae.** Cells may contain a **desiccation-resistant lipoidal cyst resting stage.** Form inclusions of poly-β-hydroxybutyrate. **Aerobic,** possessing a strictly respiratory type of metabolism with oxygen as the terminal electron acceptor. **Obligately methanotrophic,** utilizing only methane and methanol as sole carbon and energy sources. C_{2+} and other C_1 compounds are not utilized. Fixes formaldehyde for cell carbon via the serine pathway. Enzymes for the Benson-Calvin cycle pathway are absent and the tricarboxylic acid pathway is complete. **Fixes atmospheric nitrogen.** Mesophilic, neutrophilic, and nonhalophilic ecophysiology with optimal growth at about 25–30°C and pH 7.0. Major habitats include rice paddy soils, sewage, and freshwater sediments. Primary fatty acids are $C_{18:1 \omega 8c}$ and $C_{18:1 \omega 7c}$. Primary quinone is ubiquinone-8 (Q-8). Member of the family *Methylocystaceae* in the *Alphaproteobacteria.*

The mol% G + C of the DNA is: 61–67 (T_m).

Type species: **Methylocystis parvus** (ex Romanovskaya, Malashenko and Bogachenko 1978) Bowman, Sly, Nichols and Hayward 1993c, 751.

FURTHER DESCRIPTIVE INFORMATION

Methylocystis strains grow well in nitrate mineral salts (NMS) medium[1] under a methane/air headspace (usually provided in a

1:1 ratio). Crimped sealed serum vials, anaerobic jars, or desiccators with an inlet tap provide useful vessels for cultivating strains. Colonies on NMS agar or silica gels appear initially white but become pale pink following extended incubation, probably due to accumulated cytochromes. Colonies have a convex and circular shape with an entire edge and butyrous consistency. *Methylocystis echinoides* colonies are pinpoint on standard agar nitrate mineral salts media but develop larger colonies on silica gel solidified media. Over extended incubation times, many *Methylocystis* strains form a diffusible tan pigment, the intensity of which varies from strain to strain but is thought to be linked to cyst formation (Whittenbury et al., 1970a).

Fatty acid profiles *Methylocystis* species possess a fatty acid profile consisting of mainly $C_{18:1 \omega 8c}$, $C_{18:1 \omega 7c}$, and $C_{18:0}$ with smaller amounts of cyclopropane fatty acids (Table BXII.α.135). The profile is very similar to that of *Methylosinus*, differing only in a greater abundance of $C_{18:0}$ and a lower level of $C_{16:1 \omega 7c}$ (Bowman et al., 1991a). The fatty acid $C_{18:1 \omega 8c}$ is an unusual feature that is almost unique to the Family *Methylocystaceae* and is very rarely encountered in other procaryotes (Nichols et al., 1985; Bowman et al., 1991a). In ecological studies $C_{18:1 \omega 8c}$ has become a very useful signature for the detection and quantification of type II methanotroph populations in environmental samples (Nichols et al., 1985, 1987; Guckert et al., 1991; Boon et al., 1996; Guezennec and Fiala-Medioni, 1996; Sundh et al., 1997). The major lipopolysaccharide-derived hydroxy fatty acids

---

1. Nitrate mineral salts (NMS) medium consists of the following: 100 ml of a 10× NMS salt stock solution (10 g KNO₃, 10 g MgSO₄·7H₂O, and 2 g anhydrous CaCl₂, dissolved in 700 ml distilled water) is diluted in distilled water to approximately 1 liter. To this is added 0.1 ml of iron EDTA solution (3.8 g FeEDTA in 100 ml distilled water), 1 ml molybdate solution (0.26 g Na₂MoO₄·2H₂O per liter distilled

water), 1 ml of a trace element solution (1 g CuSO₄·5H₂O, 2.5 g FeSO₄·7H₂O, 2 g ZnSO₄·H₂O, 0.075 g H₃BO₃, 0.25 g CoCl₂·6H₂O, 1.25 g EDTA disodium salt, 0.1 g MnCl₂·4H₂O and 0.05 g NiCl₂·6H₂O, per liter distilled water), and (optional) 1.5% agar. The solution is sterilized by autoclaving. After cooling to approximately 50°C, 10 ml of phosphate buffer (71.6 g Na₂HPO₄·12H₂O and 26 g KH₂PO₄ per liter distilled water) is added.

include $C_{14:0\ 3OH}$ and $C_{18:0\ 3OH}$. In addition, *Methylocystis* species also possess unusual ω-1 hydroxy fatty acids with carbon chain lengths of 26 and 28 (Skerratt et al., 1992) that make up about 15% of the total hydroxy fatty acids. These hydroxy acids have been previously detected in freshwater sediments (Mendoza et al., 1987) and *Methylocystis* and *Methylosinus* represent the first recognized biological source of these particular lipids. *Methylocystis* species contain a suite of polar lipids that vary slightly between strains, possibly due to cultivation conditions (Andreev and Galchenko, 1983). Polar lipids present include phosphatidylethanolamine, phosphatidylglycerol, diphosphatidylglycerol, phosphatidylmethylethanolamine, phosphatidyl-*N*,*N*-dimethylethanolamine, and lysophosphatidylglycerol. The major quinone in *Methylocystis* is Q-8 (Collins and Green, 1985).

ENRICHMENT AND ISOLATION PROCEDURES

Methylocystis species can be enriched and isolated from the microaerobic and aerobic sediments of freshwater lakes and rivers, soil, and sewage samples. A small amount of sample is added to a liquid NMS medium in serum vials or in cotton wool-stoppered flasks placed within airtight containers. Methane is added directly to the vials and containers, usually by first removing a portion of the headspace. The best methane/air ratio to use is equivocal but should be in a range of 1:10–1:1. The methane should be of high purity. Natural gas may contain acetylene, which is a suicide substrate of pMMO and sMMO and will hinder or prevent growth even at low concentrations. Static incubation should then proceed at 25–30°C. Growth from the enrichments can then be plated onto mineral salts agar plates which are then incubated under 1:1 methane/air at 25°C.

One of the most problematic areas of methanotroph study lies in obtaining pure cultures. In practically all situations, methanotroph enrichments are severely contaminated by nonmethanotrophic (often methylotrophic) bacteria, which can easily overgrow and/or predate the cultures. Due to the relatively slow growth of type II methanotrophs such as *Methylocystis* and the humid conditions in which plates are incubated, fungal contamination is frequent unless containers are thoroughly cleaned with ethanol before each use. Addition of fungicides such as cycloheximide or nystatin to the medium is usually effective in reducing this problem.

A straightforward method for the isolation of *Methylocystis* and other methanotrophs uses NMS agar media containing a small amount of yeast extract (0.05% w/v) and methanol (0.05% v/v) (Malashenko et al., 1975), with incubation still occurring under a methane/air atmosphere. The enrichment culture is then serially diluted onto the media to the point of extinction. Single colonies on spread plates are then transferred to liquid media. A number of passages from liquid media to spread plates and back to liquid media are often required. The purity of the culture can be assessed by assuring that strains show no growth on nutrient media in the presence or absence of methane and that cells possess a reasonably consistent morphology. More details on enrichment, isolation, and the potential problems have been detailed by Whittenbury et al. (1970b) and Hanson et al. (1992).

Some *Methylocystis* strains, especially those of *M. echinoides*, grow poorly on agar-solidified media. Highly purified agars at lower concentrations may lead to improved growth. Alternatively, silica gel can be employed (Gal'chenko et al., 1975, 1977); however, silica gel is often difficult and time-consuming to prepare. For direct purification of these strains, an alternative approach involves serially diluting enrichments in NMS liquid media to extinction in 96-well plastic titer trays (Bowman et al., 1997b). Several strains can be purified simultaneously in the same tray. After sufficient incubation, the wells of the highest dilutions showing growth are examined by microscopy. A number of separate transfers may be required to obtain morphologically homogeneous cultures.

MAINTENANCE PROCEDURES

Plate or slant cultures can be stored for several months at 4°C when stored under an atmosphere of 1:1 methane/air. Takeda (1988) found that the shelf life could be increased to over 12 months if cultures were kept under a 100% nitrogen atmosphere. *Methylocystis* strains are also amenable to cryopreservation (with 20% dimethylsulfoxide as a cryoprotectant) and freeze drying (with 20% skim milk or 10% horse serum as the cryoprotectant).

PROCEDURES FOR TESTING SPECIAL CHARACTERS

The acetylene reduction assay traditionally used for detecting nitrogenase activity cannot be used for methanotrophs growing on methane because of co-oxidation of ethylene by MMO and because of the toxicity of the acetylene. Nitrogenase can be detected in methanotrophs grown at low oxygen partial pressures (in semi-solid agar) on methanol or in pre-grown cells supplied with a suitable source of reducing power such as formate, hydrogen, or ethanol (Murrell and Dalton, 1983b; Toukdarian and Lidstrom, 1984a, b; Takeda, 1988).

Putative soluble methane monooxygenase activity in methanotrophs can be rapidly assessed using the naphthalene oxidation technique (Brusseau et al., 1990). The methanotroph to be tested is cultured in copper-free NMS liquid media. Glassware used for cultivation should be washed with 1% (v/v) nitric acid to remove contaminating copper ions. An equal volume of a saturated naphthalene solution (approximately 30 mg/l at 25°C) is added to a suspension of methanotrophic cells that have been degassed briefly with nitrogen to remove residual methane. The mixture is then incubated at room temperature for approximately 1 h. 20 µl of a freshly prepared solution of 1% (w/v) tetraotized *o*-dianisidine (Fast Blue salt B) is then added and mixed. The presumptive presence of soluble methane monooxygenase is indicated by the immediate formation of a reddish semi-soluble diazo compound formed from α-naphthol, which can be measured by spectrophotometry at 540 nm. An adaptation of this procedure to test colonies on agar plates is also available (Graham et al., 1992).

DIFFERENTIATION OF THE GENUS *METHYLOCYSTIS* FROM OTHER GENERA

Characteristics differentiating *Methylosinus* from *Methylocystis* are given in Table BXII.α.136.

TAXONOMIC COMMENTS

Phylogenetic analysis based on 16S rRNA sequencing indicates that *Methylocystis* strains should be classified in the *Alphaproteobacteria* (Bratina et al., 1992; Brusseau et al., 1994) and that they are closely related to strains of the genus *Methylosinus*. Due to the poor quality of some sequences, it is difficult to make a clear delineation of *Methylocystis* species from *Methylosinus* species. The close phylogenetic relationship of the genera is supported by the similarity of nucleotide distribution width and asymmetry of DNA melting curves across the genera (De Ley, 1969; Gebers et al., 1985; Bowman et al., 1991b). However, *Methylocystis* strains have

TABLE BXII.α.136. Differential characteristics of the genera *Methylocystis* and *Methylosinus*[a]

Characteristics	*Methylocystis*	*Methylosinus*
Morphology:		
Reniform to rod-like	+	−
Pyriform or vibrioid	−	+
Motility	−	+
Exospores	−	+
Lipid cysts, spinae	D	−

[a]For symbols see standard definitions.

genomes 30–50% larger than those of *Methylosinus* strains; *Methylocystis* genomes range from 4.4–5.6 Mb in size (Bowman et al., 1991b). The DNA base composition of *Methylocystis parvus* is 63–67 mol% G + C (T_m), while the values for *Methylocystis echinoides* are slightly lower (61–62 mol% G + C [T_m]) (Meyer et al., 1986; Lysenko et al., 1988; Bowman et al., 1991a, b). DNA–DNA hybridization studies indicate *M. parvus* and *M. echinoides* are genetically distinct, sharing only low levels of DNA hybridization, and that no significant levels of hybridization occur between *Methylocystis* and *Methylosinus* strains (Lysenko et al., 1988; Bowman et al., 1991a, b).

Although *Methylocystis* was originally described by Romanovskaya et al. (1978) based on the phenotypic groupings created by Whittenbury et al. (1970b), the genus was not included in the Approved List of Bacterial Names (Skerman et al., 1980). Only *M. parvus* was included in the genus at the time, and several other species including *"Methylocystis methanolicus"* (Gal' chenko et al., 1977), *"Methylocystis pyreformis"*, *"Methylocystis fuscus"* (Gal' chenko, 1977), and *"Methylocystis fistulosa"* (Meyer, 1977) were ignored owing to lack of data, lack of a type strain, and/or the possibility that some were still mixed cultures (D. Prauser, personal communication). The genus was subsequently revived and described in 1993 (Bowman et al., 1993c). Studies using protein electrophoresis (Gal' chenko and Nesterov, 1981) and serotyping (Bezrukova et al., 1983) still suggest that the diversity in *Methylocystis* is greater than what is currently recognized.

Strains designated *M. echinoides* by Gal' chenko et al. (1977) were very similar to the strain IC 493S/5 (IMET 10491) of Haubold (1978) according to protein electrophoretic patterns (Gal' chenko and Nesterov, 1981), and strain IMET 10491 subsequently became the type strain of *M. echinoides*. *"Methylocystis minimus"* IMET 10519 (Whittenbury et al., 1970b; Romanovskaya et al., 1978) is a subjective synonym of *M. parvus* (Bowman et al., 1993c), as it shares about 68% similarity by DNA hybridization with *M. parvus* strain OBBP. Strain A of *"Methylovibrio soehngenii"* (Hazeu and Steenis, 1970) probably should be described as a strain of *M. parvus* (Anthony, 1982).

DIFFERENTIATION OF THE SPECIES OF THE GENUS *METHYLOCYSTIS*

Phenotypic characteristics differentiating the species *M. parvus* and *M. echinoides* are presented in Table BXII.α.137. Additional phenotypic characteristics for these species are presented in Table BXII.α.138.

TABLE BXII.α.137. Differential characteristics of the species of the genus *Methylocystis*[a]

Characteristic	*M. parvus*	*M. echinoides*
Morphology:		
Reniform	+	−
Short rods	−	+
Spinae	−	+
Lipid cysts	+	−
Growth on agar	good	poor
Desiccation resistant	+	−
Growth at 37°C	+	−
Mol% G+C (T_m)	63–67	61–62

[a]For symbols see standard definitions.

TABLE BXII.α.138. Other characteristics of the species of the genus *Methylocystis*[a]

Characteristics	*M. parvus*	*M. echinoides*
Polyphosphate	d	−
Poly-β-hydroxybutyrate inclusions	+	+
Melanin-like pigments	d	+
pH range for growth	5.5–8.5	5.5–8.5
Growth in the presence of:		
0.02% NaN₃	d	+
0.0075% KCN	+	+
0.001% crystal violet	−	−
0.001% malachite green	d	−
0.01% SDS	+	+
Lysed by 2% SDS	−	−
Oxidase/catalase activity	+	+
Phosphatase, urease	d	−
Triphenyltetrazolium chloride (TTC) reduction	−	−
Nitrogen fixation	+	+

[a]For symbols see standard definitions.

List of species of the genus Methylocystis

1. **Methylocystis parvus** (ex Romanovskaya, Malashenko and Bogachenko 1978) Bowman, Sly, Nichols and Hayward 1993c, 751[VP]

   par' vus. L. adj. *parvus* small.

   Phenotypic characteristics are presented in the genus description, in the text, and in Tables BXII.α.137 and BXII.α.138.

   Isolated from soil and fresh water sediments.

   The mol% G + C of the DNA is: 63–67 (T_m).

   Type strain: OBBP, ACM 3309, ATCC 35066, IMET 10483, NCIMB 11129.

   GenBank accession number (16S rRNA): M29026, Y18945.

2. **Methylocystis echinoides** (ex Gal' chenko, Shishkina, Suzina and Trotsenko 1977) Bowman, Sly, Nichols and Hayward 1993c, 751[VP]

   ech.i.noi' des. Gr. adj. *echino* spiny, hedgehog-like; Gr. suff. *ides* similar to; M.L. neut. adj. *echinoides* spiny like a hedgehog.

   Phenotypic characteristics are presented in the genus description, in the text, and in Tables BXII.α.137 and BXII.α.138.

   Isolated from sewage and fresh water sediments.

   The mol% G + C of the DNA is: 61–62 (T_m).

   Type strain: IMET 10491, NCIMB 13100.

   GenBank accession number (16S rRNA): L20848.

   Additional Remarks: This sequence is not from the type strain.

Species Incertae Sedis

1. **"Methylocystis fistulosa"** Meyer 1977, 19.

2. **"Methylocystis fuscus"** Gal'chenko 1977, 15.

3. **"Methylocystis methanolicus"** Gal'chenko, Shishkina, Suzina and Trotsenko 1977, 726.

Deposited strain: 36, NCIMB 13101, IMET 10597.

4. **"Methylocystis pyreformis"** Gal'chenko 1977, 15.
   Deposited strain: NCIMB 13102.
   GenBank accession number (16S rRNA): L20803.

Genus II. **Albibacter** Doronina, Trotsenko, Tourova, Kuznetsov and Leisinger 2001, 1056[VP]

NINA V. DORONINA AND YURI A. TROTSENKO

Al.bi.bac'ter. L. adj. *albus* white; M.L. masc. n. *bacter* equivalent of Gr. neut. n. *baktron* a small rod; M.L. masc. n. *Albibacter* small, white rod.

Rods 0.9–1.0 × 1.2–1.8 µm, singly, in pairs, or in clusters. **Nonmotile**. Gram negative. Multiplication is by binary fission. Endospores and prosthecae are not formed. Colonies are white. Obligately **aerobic**, having a strictly respiratory type of metabolism with oxygen as the terminal electron acceptor. Nitrate is reduced to nitrite. Growth occurs on nutrient agar and peptone–yeast extract–glucose (PYG) agar. **Facultatively methylotrophic and chemolithotrophic. Assimilate C$_1$ compounds via the ribulose bisphosphate pathway.** Able to grow on a wide spectrum of polycarbon substrates. Sugars are utilized aerobically but not fermented. Ammonium salts, nitrate, urea, peptone, some amino acids, and methylated amines are utilized as nitrogen sources. Growth factors are not required. Methyl red and Voges–Proskauer negative. Catalase and urease positive. **Oxidase activity is very low.** Indole is formed from L-tryptophan in the mineral medium with methanol as a sole carbon and energy source and with KNO$_3$ as the nitrogen source. The major ubiquinone is Q-10. The predominant cellular fatty acids are *cis*-vaccenic (C$_{18:1\ \omega7}$) and palmitic (C$_{16:0}$). The dominant phospholipids are phosphatidylethanolamine, phosphatidylglycerol, phosphatidylcholine, and phosphoserine. Belongs to the *Alphaproteobacteria*.

The mol% G + C of the DNA is: 66.7.

Type species: **Albibacter methylovorans** Doronina, Trotsenko, Tourova, Kuznetsov and Leisinger 2001, 1056.

FURTHER DESCRIPTIVE INFORMATION

Albibacter species are facultatively autotrophic and methylotrophic. Good growth occurs on mineral salt medium with gas mixture of CO$_2$/H$_2$/O$_2$, dichloromethane (0.05–0.1%, v/v), methanol (0.5%, v/v), methylamine (0.3%, w/v), or formate (0.05%, w/v) as the carbon and energy sources. Dichloromethane-grown cells contain an inducible, reduced glutathione (GSH)-dependent dichloromethane dehalogenase, whereas methylamine- or methanol-grown cells possess the appropriate pyrroloquinoline quinone-linked dehydrogenases. Formaldehyde is further oxidized by a GSH-dependent formaldehyde dehydrogenase to formate. The latter is finally oxidized to CO$_2$ by formate dehydrogenase. The cells assimilate CO$_2$ via the ribulose bisphosphate (RuBP) pathway and show phosphoribulokinase and ribulosebisphosphate carboxylase activity. Neither the serine nor the ribulose monophosphate pathway of C$_1$ assimilation is operative, due to the absence of the appropriate specific enzymes, namely hydroxypyruvate reductase (NADH), serine-glyoxylate aminotransferase, ATP malate lyase, and hexulosephosphate synthase, respectively.

The cells have no 2-keto-3-deoxy-6-phosphogluconate aldolase; therefore, the Entner–Doudoroff pathway is not functioning. In addition, the enzymes of the pentose phosphate pathway (glucose-6-phosphate and 6-phosphogluconate dehydrogenases) possess low activities. Hence, fructose-1,6-bisphosphate aldolase plays an important role in metabolic conversions of phosphotrioses. The cells contain pyruvate dehydrogenase as well as a complete set of the citric acid cycle and glyoxylate shunt enzymes.

The primary assimilation of ammonia occurs by both reductive amination of α-ketoglutarate and by the glutamate cycle.

ENRICHMENT AND ISOLATION PROCEDURES

Albibacter methylovorans was isolated on dichloromethane agar from enrichment culture that had been inoculated with a groundwater sample (Doronina et al., 2001).

MAINTENANCE PROCEDURES

The bacteria can be stored in liquid mineral medium K, which has the following composition (g/l): KH$_2$PO$_4$, 2.0; (NH$_4$)$_2$SO$_4$, 2.0; NaCl, 0.5; MgSO$_4$·7H$_2$O, 0.125; FeSO$_4$·7H$_2$O, 0.002; pH 7.2. The cultures can be stored in the liquid medium at 4°C for at least one month and on agar slants at 4°C for 2–3 months. Cells can also be frozen in medium K containing 40% glycerol and stored at −70°C or lyophilized with a cryoprotectant (skim milk).

DIFFERENTIATION OF THE GENUS *ALBIBACTER* FROM OTHER GENERA

Albibacter methylovorans differs from the facultatively chemolithotrophic and methylotrophic members of the genera *Xanthobacter*, *Blastobacter*, *Angulomicrobium*, *Ancylobacter*, and *Ralstonia* by morphological features. The representatives of *Xanthobacter* form pleomorphic cells, and their colonies are slimy and are yellow due to the water-insoluble carotenoid pigment, zeaxanthin dirhamnozide. Species of the genera *Blastobacter* (irregular, ovoid, pleomorphic rods) and *Angulomicrobium* (tetrahedron-shaped cells) multiply by budding. The vibrioid cells of the genus *Ancylobacter* have a characteristic morphology and rings are formed occasionally prior to cell division. *Ralstonia* (formerly *Alcaligenes*) species are motile by one to eight peritrichous flagella.

TAXONOMIC COMMENTS

The levels of DNA similarity with representatives of *Xanthobacter*, *Blastobacter*, *Angulomicrobium*, *Ancylobacter*, and *Ralstonia* are less than 7%. Phylogenetic analysis by 16S rDNA sequencing (Doronina et al., 2001) confirmed the absence of close relatedness between *A. methylovorans* and the species of the genera *Xanthobacter* and *Blastobacter*. Although *A. methylovorans* resembles *Paracoccus* species in morphology, ubiquinone system, C$_1$-assimilation pathway, and range of mol% G + C content, it has very low DNA–DNA similarities (5–7%) with the type species of this genus and thus belongs to another subgroup of the *Alphaproteobacteria*. Despite rather high 16S rDNA sequence similarity to *Methylopila capsulata*, *A. methylovorans* is clearly distinct from this serine pathway methylotroph by morphology, cellular phospholipids and fatty acids, autotrophic growth, and RuBP pathway operation, as well as by a very low DNA–DNA similarity. In the phylogenetic tree derived from 16S rDNA sequences, the genus *Albibacter* forms a distinct branch within the *Alphaproteobacteria* (Doronina et al., 2001).

List of species of the genus Albibacter

1. **Albibacter methylovorans** Doronina, Trotsenko, Tourova, Kuznetsov and Leisinger 2001, 1056[VP]

   me.thy.lo.vor' ans. Fr. *méthyle* the methyl group; L. part. adj. *vorans* devouring, digesting; *methylovorans* digesting methyl groups.

   The characteristics are as described for the genus, with the following additional features.

   Fig. BXII.α.154 illustrates the morphological features.

   Colonies on methanol or peptone agar are circular and 1–2 mm in diameter, white, convex, translucent to opaque, and mucoid. Growth occurs at 10–35°C and pH 6.0–9.0. Optimal conditions for growth are 28–30°C, pH 7.5–8.0. No growth occurs in the presence of 3% NaCl. Starch is hydrolyzed but not gelatin or cellulose.

   Utilizable carbon sources are dichloromethane, methanol, methylamine, formate, CO_2/H_2, D-glucose, D-fructose, D-mannose, L-arabinose, D-xylose, sorbose, ribose, maltose, sucrose, D-sorbitol, D-mannitol, inositol, glycerol, dulcitol, adonitol, ethanol, acetate, α-ketoglutarate, fumarate, pyruvate, succinate, malate, citrate, oxaloacetate, propionate, *cis*-aconitate, L-glutamate, L-alanine, and acetamide. Not utilizable are methane, chloromethane, formaldehyde, di- and trimethylamines, thiocyanate, thiosulfate, dimethylsulfoxide, and dimethylacetamide. Yeast extract (0.01%), or a mixture of biotin and thiamin (both at 20 μg/l), stimulates growth.

   Resistant to penicillin and rifampicin, but sensitive to gentamicin, kanamycin, ampicillin, neomycin, novobiocin, nalidixic acid, tetracycline, and lincomycin.

   The type (and only) strain was isolated from groundwater in Switzerland.

   The mol% G + C of the DNA is: 66.7 (T_m).

   Type strain: DM10, DSM 13819, VKM B-2236.

   GenBank accession number (16S rRNA): AF273213.

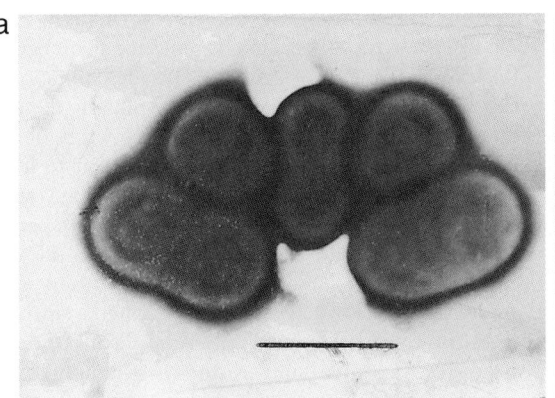 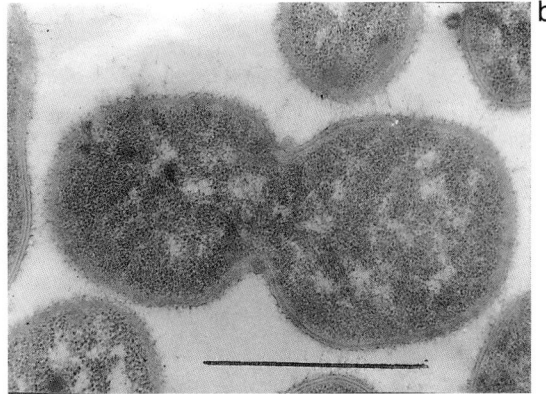

FIGURE BXII.α.154. *Albibacter methylovorans* (*a*) Negatively stained cells. (*b*) Ultrathin sections showing cell wall structure. Bars = 1 μm.

Genus III. **Methylosinus** (ex Romanovskaya, Malashenko and Bogachenko 1978) Bowman, Sly, Nichols and Hayward 1993c, 751[VP]

JOHN P. BOWMAN

Me.thyl.o.si' nus. Fr. *methyle* the methyl radical; *sinus* bend; M.L. n. *Methylosinus* methyl bender.

Cells are **pyriform or vibrioid** in shape, 0.5–1.5 × 1.5–3.0 μm, usually arranged in rosettes of 4–6 cells. Reproduce by **binary and budding division**. In budding division, the bud contains a **heat and desiccation-resistant exospore**, which germinates into a vegetative daughter cell; this **daughter cell is motile**, propelled by a tuft of 3–10 polar and/or subpolar flagella. Cells contain **type II intracytoplasmic membranes** that are arranged as multiple layers along the periphery of the cell wall. Strongly accumulate inclusions of poly-β-hydroxybutyrate. Aerobic, possessing a **strictly respiratory type of metabolism with oxygen as the terminal electron acceptor. Obligately methanotrophic**, utilizing only methane and methanol as sole carbon and energy sources. C_{2+} and other C_1 compounds are not utilized. Fix formaldehyde for cell carbon via the serine pathway. Enzymes for the Benson–Calvin cycle pathway are absent. The tricarboxylic acid pathway is complete. **Fix atmospheric nitrogen by means of an oxygen-sensitive nitrogenase.** Mesophilic. Nonhalophilic. Optimal temperature, 25–30°C; optimal pH, 6.5–7.0. Major habitats include soil, freshwater sediments and groundwater. **Primary fatty acids are $C_{18:1\ \omega 8c}$ and $C_{18:1\ \omega 7c}$. Primary quinone is ubiquinone-8 (Q-8).** Member of the Family *Methylocystaceae* found in the *Alphaproteobacteria*.

The mol% G + C of the DNA is: 62–67 (T_m).

Type species: **Methylosinus trichosporium** (ex Romanovskaya, Malashenko and Bogachenko 1978) Bowman, Sly, Nichols and Hayward 1993c, 751.

FURTHER DESCRIPTIVE INFORMATION

Morphology and exospore formation Cells of *Methylosinus trichosporium* have a pear-shaped (pyriform) morphology with a width of 0.5–1.5 μm and a length of 1.5–3 μm. *Methylosinus sporium* cells have a similar size but are vibrioid. When grown in liquid media, both species form rosettes consisting of 4–6 cells. Photomicrographs of the morphology of *Methylosinus* species have been published by Whittenbury et al. (1970a, b). In logarithmic phase, the cells divide by standard binary division. As cultures enter the stationary growth phase, an increasing proportion of cells begin to sporulate and reproduce by budding, either when the cells are arranged in rosettes or singly. The bud that is formed consists of an exospore surrounded by a part of the cell wall. A sporulating cell ceases to divide or bud, becomes granulated, and eventually lyses. It will often produce an extensive fibrillar capsule, which is not formed by vegetative cells (Whittenbury et al., 1970a). Exospore formation by *M. trichosporium* is stimulated by oxygen limitation (Reed and Dugan, 1978). When spores appear, they are initially non-refractile, Gram-negative, coccoid bodies at the end of sporulating cells. As the spores mature, they become increasing refractile and acid-fast. The malachite green spore stain (Doetsch, 1981) can be used to visualize mature spores. Thin sections show that the exospore consists of an electron-dense outer coat (exosporium) surrounded by a cell wall derived from the parent cell. The *M. trichosporium* exospores possess a capsular coat that is attached to the exosporium but is distinct from the parent cell's capsular layer (Reed et al., 1980). Within the exospore, there is a laminated inner coat and a poorly defined cortex (Reed et al., 1980; Titus et al., 1982). Dipicolinic acid and respiratory activity are not detectable in the mature exospores. The exospores are resistant to desiccation and can survive 18 months of drying. Exospores can also withstand heating at 85°C for 10 min and 10 min of ultrasonication. The exospores are structurally similar to those formed by *Rhodomicrobium vannielii* (Titus et al., 1982). When freshly formed, exospores take ~2–3 d to germinate, whereas heat- or desiccation-treated exospores may take weeks to recover. During the germination process, which lasts 5–7 h, the exospores lose their refractility and shed the capsular coats (if any). The exosporium then opens to release a short rod-like daughter cell, which rapidly becomes motile by forming a tuft of 3–10 unsheathed flagella. This cell usually undergoes a few generations of binary division before settling into a rosette to commence exospore production.

Cell walls *Methylosinus* strains produce poly-β-hydroxybutyrate as an internal carbon reserve (Weaver et al., 1975b; Best and Higgins, 1981). The cell walls are of the usual Gram-negative type; however, unlike most other Gram-negative bacteria, *Methylosinus* strains are very resistant to lysis by detergents (such as sodium dodecyl sulfate) and to lytic bacteria (Bowman et al., 1993c; Starostina and Pashkova, 1993). Cell lysis can be caused by a lysozyme pretreatment (1 mg/ml lysozyme at 37°C, for 30–60 min), followed by addition of 2% sodium dodecyl sulfate.

Cultural characteristics *Methylosinus* strains grow abundantly in nitrate mineral salts broth and on media solidified by all brands of agar (see *Methylocystis* for medium formula). The colonies of *M. trichosporium* and *M. sporium* are similar: circular, convex, with an entire edge and creamy consistency. Colonies typically reach 1–2 mm in diameter on NMS agar after 7 d incubation. One strain (IMV B-3037), identified as a strain of *M. sporium*, forms brilliant red nondiffusible pigments that have a slight green metallic sheen. Spectral analysis suggests they are similar to the prodigiosin pigments of *Serratia marcescens*. Two different pigment types were detected and are referred to as methylosin A and B (Strauss and Berger, 1983).

Methylosinus strains are mesophilic and neutrophilic with growth occurring from 10 to 40°C and pH 5.5–9.0. Optimal growth occurs at approximately 30°C and at pH 6.5–7.0. Most strains will either not grow or grow only poorly in the presence of NaCl concentrations greater than 0.3 M. No strains have been found to possess growth factor requirements.

Nutrition and metabolism *Methylosinus* strains are strictly aerobic and obligately methanotrophic. The only carbon and energy sources supporting growth are methane and methanol (Meyers, 1982; Bowman et al., 1993c). Formaldehyde is assimilated by cells for cell carbon via the serine pathway (Wolfe and Higgins, 1979). All strains possess a complete tricarboxylic acid cycle containing a NADP-dependent isocitrate dehydrogenase (Colby et al., 1979). Strains can grow over a wide range of oxygen concentrations (<0.5–60%, v/v) and are not microaerophilic as has been suggested in some previous studies. Only when oxygen levels drop below 0.5% does growth become limiting (Ren et al., 1997). Interestingly, *M. trichosporium* OB3b can survive anoxia for several months (presumably due to exospore formation) and can rapidly respond when methane and oxygen once again become available (Roslev and King, 1994, 1995). Oxygen limitation results in a reduction in the amount of intracytoplasmic membrane (ICM) formation (Scott et al., 1981). When grown under either methane or nitrogen limitation, cells also appear to lack the typical paired type II ICMs, instead having membrane vesicles (Best and Higgins, 1981; Scott et al., 1981). All strains produce catalase and cytochrome *c* oxidase. Nitrate and ammonia salts, amino acids, yeast extract and Casamino acids (Difco) can be used as sources of nitrogen (Warner et al., 1983; Toukdarian and Lidstrom, 1984b; Bowman et al., 1993c). Ammonia and nitrate are metabolized by the glutamine synthetase-glutamine 2-oxoglutarate aminotransferase system (Shishkina and Trotsenko, 1979; Murrell and Dalton, 1983a; Toukdarian and Lidstrom, 1984b). In addition, *Methylosinus* species are capable of fixing atmospheric nitrogen by means of an oxygen-sensitive nitrogenase (Murrell and Dalton, 1983b), which in some *Methylosinus* strains shows homology with the nifH genes of *Klebsiella pneumoniae* (Toukdarian and Lidstrom, 1984a; Oakley and Murrell, 1993).

To date, all strains tested form soluble methane monooxygenase (sMMO), which can be detected using the naphthalene oxidation assay (Brusseau et al., 1990), by a gene probe (Stainthorpe et al., 1991), or by PCR detection of sMMO genes by means of specific primers (Holmes et al., 1995b; McDonald et al., 1995). Many studies have focused on the capability of *M. trichosporium* OB3b to co-oxidize a wide range of carbon substrates by its production of sMMO. The compounds oxidized are too many to list but include a wide range of aliphatic, heterocyclic, and aromatic compounds (Burrows et al., 1984). Several studies have focused on the industrial applications of this biocatalytic ability, particularly in the production of epoxides for plastics manufacture (Hou, 1984). Another focus of research has been in the bioremediation field. *M. trichosporium* OB3b and other similar strains can co-metabolize a wide range of chlorinated aliphatic compounds, including major groundwater pollutants such as trichloroethylene, chloroform, and tetrachloroethylene (Oldenhuis et al., 1989; Tsien et al., 1989; Castro et al., 1996; Hamamura et al., 1997; Moran and Hickey, 1997; Tartakovsky et al., 1998). However, the industrial application of methanotrophs has been hampered by their relatively slow growth and their requirement for methane, which is a potentially explosive substrate that also competitively inhibits cometabolic reactions. Another problem is that trace copper levels can suppress sMMO activity, thereby eliminating or reducing transformation rates (Oldenhuis et al., 1989; Lontoh and Semrau, 1998). To overcome this, constitutive mutants lacking sMMO activity have been created by chemical mutagenesis and by marker exchange (Phelps et al., 1992; Fitch et al., 1993; Martin and Murrell, 1995; Tellez et al., 1998). These mutants can cometabolize trichloroethylene in the presence of high levels of copper. Some of these copper-tolerant mutants have been used for the development of treatment strategies for chlorinated aliphatic pollutants (Tschantz et al., 1995). In addition the sMMO gene cluster has been successfully cloned into *Pseudomonas putida* strain F1, which is not only able to degrade trichloroethylene but also possesses the ability to grow rapidly; moreover, it lacks the problem associated with methane competitive inhibition (Jahng and Wood, 1994; Jahng et al., 1996).

Plasmids Strains of *M. trichosporium* possess three large plasmids of cryptic function, which are 147 kb, 150 kb, and 160 kb in size, respectively. *M. sporium* ATCC 35069 has two cryptic plasmids, each around 100 kb in size. Restriction endonuclease analysis has indicated that different *M. trichosporium* strains possess identical plasmids; however, Southern blot hybridization analysis indicated that these plasmids share negligible homology with the plasmids of *M. sporium* (Lidstrom and Stirling, 1990).

DNA composition *Methylosinus* strains possess DNA base composition values of 62–67 mol% G + C (T_m). *M. trichosporium* strains possess values of 62–63 (T_m), while the values for *M. sporium* are slightly higher, at 65–67 (T_m) (Table BXII.α.139). The characteristics of the DNA melting curves support a close relationship between *Methylosinus* and *Methylocystis*, as they possess similar nucleotide distribution widths and curve asymmetries (De Ley, 1969; Gebers et al., 1985; Bowman et al., 1991b). The genome size of various *Methylosinus* strains, as calculated by renaturation kinetics, is in the range of 3.0–3.5 Mb, somewhat smaller than that of *Methylocystis* species (Bowman et al., 1991b).

Protein analysis Analysis of protein banding patterns has shown *M. sporium* and *M. trichosporium* to be similar, sharing about 60% of the bands (Gal'chenko and Nesterov, 1981). Only low levels of serological cross reactivity have been detected between antisera raised from *M. trichosporium* and *M. sporium* (Bezrukova et al., 1983).

TABLE BXII.α.139. Differentiation of the species of the genus *Methylosinus*[a]

Characteristics	M. trichosporium	M. sporium
Morphology:		
Pyriform	+	−
Vibrioid	−	+
Melanin-like pigments	−	d
Phosphatase, urease	−	+
Mol% G + C of the DNA	62–63	65–67

[a]For symbols see standard definitions.

Lipids *Methylosinus* species possess phospholipid fatty acids that consist mainly of $C_{18:1\ \omega 8c}$, $C_{18:1\ \omega 7c}$, and $C_{16:1\ \omega 7c}$, with smaller amounts of $C_{18:0}$ and cyclopropane fatty acids (Weaver et al., 1975b; Makula, 1978; Nichols et al., 1985; Bowman et al., 1991b; Guckert et al., 1991) (Table BXII.α.136). The profile is very similar to that of *Methylocystis* and differs only in possessing a greater abundance of $C_{16:1\ \omega 7c}$ and a lower level of $C_{18:0}$. The fatty acid $C_{18:1\ \omega 8c}$ is an unusual feature that is practically unique to the family *Methylocystaceae* (Nichols et al., 1985; Bowman et al., 1991b). In ecological studies, $C_{18:1\ \omega 8c}$ has become a very useful signature for the detection and quantification of type II methanotroph populations in environmental samples (Nichols et al., 1985, 1987; Bowman et al., 1991a, b; Guckert et al., 1991; Ross et al., 1997; Sundh et al., 1997). In addition, major lipopolysaccharide-derived hydroxy fatty acids include $C_{14:0\ 3OH}$, $C_{16:0\ 3OH}$, and $C_{18:0\ 3OH}$ (Bowman et al., 1991a; Skerratt et al., 1992). In addition, *Methylosinus sporium* possesses unusual ω-1 hydroxy fatty acids with carbon chain lengths of 26 and 28 (Skerratt et al., 1992) that make up about 15% of the total hydroxy fatty acids. These hydroxy acids have been previously detected in freshwater sediments (Mendoza et al., 1987), and *Methylocystis* and *Methylosinus* represent the first recognized biological source of these particular lipids.

The neutral sugar components of the LPS oligosaccharide core in *M. trichosporium* OB3b include mostly rhamnose, glucose, and heptose (Sutherland and Kennedy, 1986). *Methylosinus* species contain a suite of phospholipids (Makula, 1978; Andreev and Galchenko, 1983), including (with relative % levels): phosphatidyldimethylethanolamine (50%), phosphatidylglycerol (13%), phosphatidylmethylethanolamine (20%), phosphatidylcholine (10%), and lysophosphatidylglycerol (5%). The major quinone in *Methylosinus* is Q-8 (Collins and Green, 1985). Neunlist and Rohmer (1985) found that *M. trichosporium* OB3b contained two free triterpenoids of the hopane family, including an aminotriol (35-aminobacteriohopane-32,33,34-triol) and a novel tetrol (35-aminobacteriohopane-31,32,33,34-tetrol). The aminotriol has been observed previously in *Rhodomicrobium vannielii* (Neunlist and Rohmer, 1985).

Ecology *Methylosinus* strains have been isolated from a variety of soil and freshwater sediment habitats with *M. sporium* being more frequently isolated than is *M. trichosporium* (Bowman, 1992). *Methylosinus* dominates the culturable methanotroph population from groundwater (Bowman et al., 1993a) and has been isolated from the root systems of various aquatic macrophytes (Calhoun and King, 1998). Indirect immunofluorescence studies give a rough indication of the presence of *Methylosinus* species and indicate that *Methylosinus* species congregate at high populations in surface sediments of various freshwater and brackish water bodies (Reed and Dugan, 1978; Abramochkina et al., 1987; Malashenko et al., 1987; Gal'chenko et al., 1988; Gal'chenko, 1994). *Methylosinus* also occurs in rice paddy soils (Saralov and Babnazarov, 1982) and has been detected by gene probe in blanket peat bogs (McDonald et al., 1996).

ENRICHMENT AND ISOLATION PROCEDURES

Procedures for enrichment and isolation are the same as described for the genus *Methylocystis*.

MAINTENANCE PROCEDURES

Methods for preservation and storage of strains are the same as described for the genus *Methylocystis*.

DIFFERENTIATION OF THE GENUS *METHYLOSINUS* FROM OTHER GENERA

Characteristics differentiating *Methylosinus* from *Methylocystis* are given in Table BXII.α.136.

TAXONOMIC COMMENTS

Phylogenetic studies place *Methylosinus* within the *Alphaproteobacteria*. The genus is very closely related to *Methylocystis* and is more distantly related to an assemblage of facultatively methylotrophic, nitrogen-fixing, and phototrophic bacteria, including *Azorhizo-* *bium, Xanthobacter, Ancylobacter, Blastochloris, Rhodoplanes, Rhodopseudomonas viridis, Rhodopseudomonas rosea, Rhodopseudomonas acidophila,* and *Beijerinckia*. Owing to the relatively poor quality of some of the 16S rRNA sequences, the distinction between *Methylosinus* and *Methylocystis* based on phylogeny alone is ambiguous.

No significant DNA–DNA hybridization (<25%) exists between *Methylosinus trichosporium* and *Methylosinus sporium* strains and with various strains of *Methylocystis* (Lysenko et al., 1988; Bowman et al., 1993c).

DIFFERENTIATION OF THE SPECIES OF THE GENUS *METHYLOSINUS*

Characteristics differentiating *M. trichosporium* and *M. sporium* are shown in Table BXII.α.139. Additional phenotypic characteristics for these species are presented in Table BXII.α.140.

List of species of the genus Methylosinus

1. **Methylosinus trichosporium** (ex Romanovskaya, Malashenko and Bogachenko 1978) Bowman, Sly, Nichols and Hayward 1993c, 751[VP]

   tri.cho.spo' ri.um. Gr. n. *trix* hair; L. n. *spora* spore; *trichosporium* hair spore-former.

   Phenotypic characteristics are presented in Tables BXII.α.139 and BXII.α.140.

   Isolated from soil, fresh water sediments, and groundwater.

   The mol% G + C of the DNA is: 62–63 (T_m).

   Type strain: OB3b, ACM 3311, ATCC 35070, IMET 10543, NCIMB 11131.

   GenBank accession number (16S rRNA): M29024, Y18947.

2. **Methylosinus sporium** (ex Romanovskaya, Malashenko and Bogachenko 1978) Bowman, Sly, Nichols and Hayward 1993c, 751[VP]

   spo' ri.um. M.L. n. *sporium* spore-former.

   Phenotypic characteristics are presented in Tables BXII.α.139 and BXII.α.140.

   Isolated from soil and fresh water sediments.

   The mol% G + C of the DNA is: 65–67 (T_m).

   Type strain: 5, ACM 3306, ATCC 35069, IMET 10545, NCIMB 11126.

   GenBank accession number (16S rRNA): Y18946.

TABLE BXII.α.140. Other characteristics of the species of the genus *Methylosinus*[a]

Characteristics	M. trichosporium	M. sporium
Poly-β-hydroxybutyrate inclusions	+	+
Heat and desiccation resistance	+	+
pH range for growth	5.5–9.0	5.5–9.0
Growth at 37°C	+	+
Growth in the presence of:		
0.02% NaN₃	+	+
0.0075% KCN	+	+
0.001% crystal violet	−	d
0.001% malachite green	+	+
0.01% SDS	+	+
3% NaCl	−	−
Lysed by 2% SDS	−	−
Oxidase/catalase activity	+	+
Triphenyltetrazolium chloride (TTC) reduction	−	−
Nitrogen fixation	+	+

[a]For symbols see standard definitions.

Genus Incertae Sedis IV. **Methylopila** *Doronina, Trotsenko, Krausova, Boulygina and Tourova 1998a, 1319[VP]*

NINA V. DORONINA AND YURI A. TROTSENKO

Me.thy.lo.pi' la. Fr. *méthyle* the methyl radical; Gr. adj. *pila* ball or sphere; M.L. *Methylopila* methyl-using sphere.

Rods, 0.5–0.7 × 1.0–1.3 μm, occurring singly or in pairs. **Motile or nonmotile.** If motile, the cells have a single lateral flagellum. Gram negative. **Poly-β-hydroxybutyrate granules are formed.** Prosthecae are not formed. Colonies are white. Do not produce pyocyanine and fluorescein. Methyl red and Voges–Proskauer negative. **Oxidase and urease positive.** Indole is formed from L-tryptophan on the mineral medium with methanol as the sole carbon and energy source and KNO₃ as the sole nitrogen source. Produce acid from sugars oxidatively. **Aerobic**, having a strictly respiratory type of metabolism with oxygen as the terminal electron acceptor. Chemoorganotrophic. **Facultatively methylotrophic. Assimilate C₁ compounds by the serine pathway (isocitrate lyase⁻ variant).** Growth factors are not required. Optimal pH 6.5–7.5; no growth above pH 9 or below pH 5. Optimal temperature, 28–35°C; no growth at 42°C. No growth with 7% NaCl. NaCl is not required for growth. Cellular fatty acid profile is characterized by the presence of 45–70% *cis*-vaccenic acid (C₁₈:₁ ω7); main quinone is Q-10. Dominant cell phospholipids are phosphatidylethanolamine and phosphatidylcholine. The genus belongs to the order *Rhizobiales* in the class *Alphaproteobacteria*.

The mol% G + C of the DNA is: 66–70.

Type species: **Methylopila capsulata** Doronina, Trotsenko, Krausova, Boulygina and Tourova 1998a, 1320.

FURTHER DESCRIPTIVE INFORMATION

After division, the cells remain connected by a constriction that appears to be formed by the outer membrane.

The serine-pathway-specific enzymes (hydroxypyruvate reductase, serine–glyoxylate aminotransferase, malate lyase) are present. Isocitrate lyase is absent.

Primary assimilation of ammonia occurs both by reductive amination of α-ketoglutarate and via the glutamate cycle.

ENRICHMENT AND ISOLATION PROCEDURES

Methylopila capsulata strains can be isolated from soil samples as colonies on the mineral base medium K[1] with 2% (w/v) of Difco agar and 1% methanol (Doronina et al., 1998a). *Methylopila helvetica* strains can be isolated from groundwater and active sludge as dichloromethane-utilizing bacteria on the medium DM[2]. Wa-

---

1. K medium contains (g/l of distilled water): $(NH_4)_2SO_4$, 2.0; KH_2PO_4, 2.0; NaCl, 0.5; $MgSO_4 \cdot 7H_2O$, 0.125; $FeSO_4 \cdot 7H_2O$, 0.002. The pH of the medium is adjusted to 7.2 prior to autoclaving. Methanol is added to a final concentration of 1% (v/v).

2. DM medium contains (g/l of distilled water): KH_2PO_4, 1.36; Na_2HPO_4, 2.13; $(NH_4)_2SO_4$, 2.0; $MgSO_4 \cdot 7H_2O$, 0.2. Prior to autoclaving, the pH of the medium is adjusted to 7.2. After sterilization, 1 liter of the medium is supplemented with 1 ml of a trace element solution containing (g/l): $MnSO_4 \cdot 5H_2O$, 1.0; H_3BO_3, 1.0; $CaCl_2$, 0.25; $ZnCl_2$, 0.25; NH_4VO_3, 0.1; $CoCl_2 \cdot 6H_2O$, 0.25 and $Ca(NO_3)_2$, 25.

ter and sludge samples (0.5 g) (Gälli and Leisinger, 1985) are used for direct inoculation of 10 ml DM medium in closed 50 ml flasks containing 100 μmol CH_2Cl_2. Enrichment cultures are grown at 30°C for 5 d on a rotary shaker and those acidifying the medium are then transferred to fresh medium. After three transfers, the enrichment cultures are then plated into Petri dishes with solidified medium containing 2% (w/v) of Difco agar and 0.01% (w/v) of bromothymol blue, and incubated in a desiccator with 0.05% (v/v) CH_2Cl_2 in the gas phase; pure cultures can then be isolated from individual colonies that change the medium color.

MAINTENANCE PROCEDURES

Strains may be maintained on agar slants at 4°C for at least 2 months. Lyophilization may be used for long-term preservation, with skim milk as a cryoprotectant.

DIFFERENTIATION OF THE GENUS *METHYLOPILA* FROM OTHER GENERA

The major characteristics differentiating the genus *Methylopila* from other facultative serine pathway methylobacteria are summarized in Table BXII.α.63 in Genus *Methylarcula*.

TAXONOMIC COMMENTS

In the phylogenetic tree derived from 16S rRNA sequences, the genus *Methylopila* forms a distinct branch within the order *Rhizobiales* of the class *Alphaproteobacteria*.

Methylopila species share less than 10% DNA–DNA similarity with members of *Methylorhabdus, Methylobacterium, Aminobacter, Hyphomicrobium, Methylarcula,* and *Pseudomonas*.

List of species of the genus Methylopila

1. **Methylopila capsulata** Doronina, Trotsenko, Krausova, Boulygina and Tourova 1998a, 1320[VP]

cap.su.la' ta. L. n. *capsule* a small chest, capsule; M.L. adj. *capsulata* having a capsule.

Pleomorphic rods or cocci, occurring singly or in pairs, having capsules covered in spikes. Motile by means of a single lateral flagellum. Fig. BXII.α.155a and b illustrate the morphological features. Colonies on nutrient agar or glucose–potato agar are round, viscous, semitransparent, convex, with even edges, 2 mm in diameter. Nitrates are reduced to nitrites slowly. Catalase activity is weak. H_2S is produced during growth in nutrient broth. The following carbon sources are used: methanol, mono-, di-, and trimethylamine, butanol, ethanol, glycerol, maltose, sucrose, L-arabinose, D-fructose, D-glucose, succinate, fumarate, and pyruvate. Weak hydrolysis of gelatin and starch occurs. The following nitrogen sources are used: ammonium salts, nitrate, urea, peptone, some amino acids, and methylated amines. Methylamine is oxidized to formaldehyde by amine dehydrogenase.

The type strain was isolated from soil of Tashkent City, Uzbekistan.

The mol% G + C of the DNA is: 67.2 (T_m).

Type strain: IM1, ATCC 700716, VKM B-1606.

GenBank accession number (16S rRNA): AF004844.

2. **Methylopila helvetica** Doronina, Trotsenko, Tourova, Kuznetsov and Leisinger 2000c, 1953[VP] (Effective publication: Doronina, Trotsenko, Tourova, Kuznetsov and Leisinger 2000b, 216.)

hel.ve' ti.ca. M.L. neut. adj. *helvetica* from Helvetia, an old name of Switzerland.

Short rods or cocci, occurring singly or rarely in pairs. Motile by means of a single lateral flagellum. Fig. BXII.α.155c and d illustrate the morphological features. No growth occurs on nutrient agar or peptone–yeast extract–glucose agar. Colonies on methanol–salt agar are white, round, viscous, semitransparent, convex, 1–3 mm in diameter. Nitrate is reduced to nitrite. H_2S and NH_3 are not formed. Catalase positive. The following carbon sources are used: dichloromethane, methanol, formate, mono- and trimethylamine, ethanol, glycerol, pyruvate, fumarate, and succinate. Gelatin and starch are not hydrolyzed. Ammonium salts, nitrate, methylated amines, and some amino acids are utilized as nitrogen sources. Methylamine is oxidized to formaldehyde by the *N*-methylglutamate pathway enzymes γ-glutamylmethylamide synthetase and *N*-methylglutamate synthase/lyase.

The type strain was isolated from groundwater in Switzerland.

The mol% G + C of the DNA is: 68.5 (T_m).
Type strain: DM9, CIP 106788, VKM B-2189.
GenBank accession number (16S rRNA): AF227126.

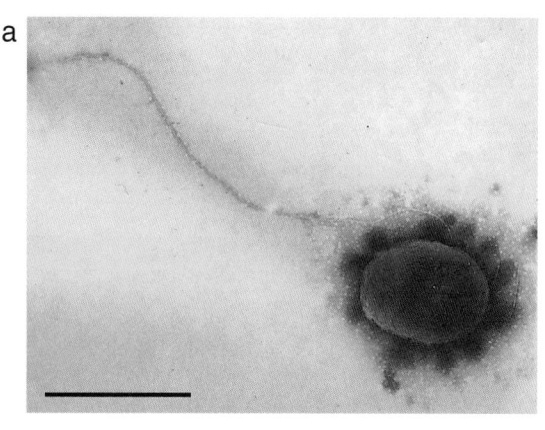
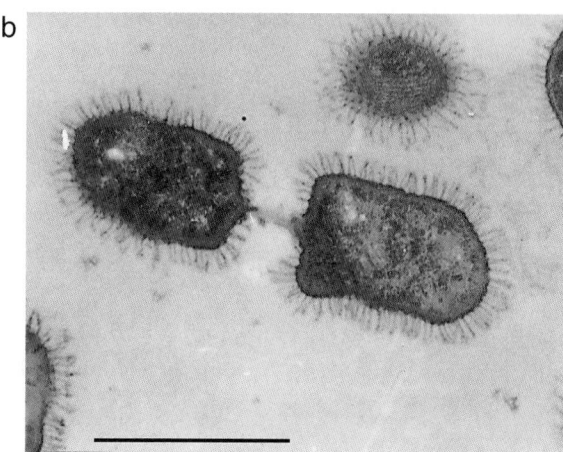
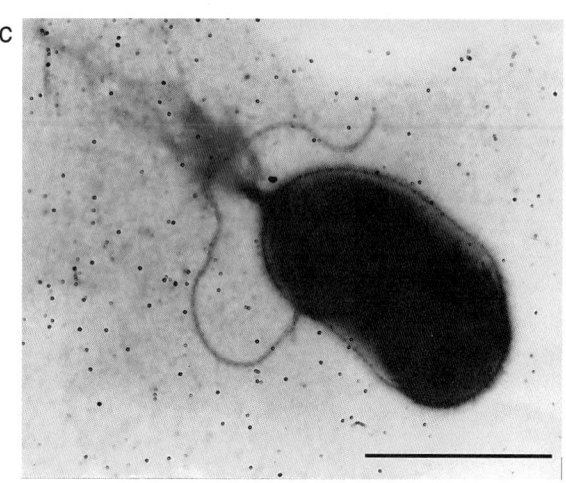
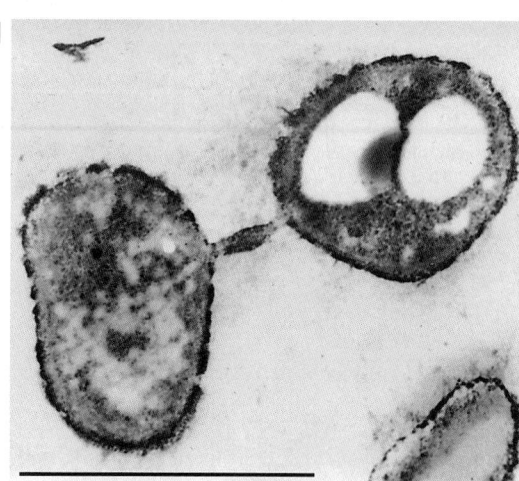

FIGURE BXII.α.155. *Methylopila capsulata* (*a, b*) and *Methylopila helvetica* (*c, d*). *a* and *c*, negatively stained cells, note the subpolar location of the single flagellum; *a*, the large amount of capsular material is evident. *b* and *d*, ultrathin sections; constriction in the final stage of cell division and granules of PHB. Bars = 1 μm.

Family VI. **Beijerinckiaceae** *fam. nov.*

GEORGE M. GARRITY, JULIA A. BELL AND TIMOTHY LILBURN

Beij.e.rinck.i.a' ce.ae. M.L. fem. n. *Beijerinckia* type genus of the family; *-aceae* ending to denote family; M.L. fem. pl. n. *Beijerinckiaceae* the *Beijerinckia* family.

The family *Beijerinckiaceae* was circumscribed for this volume on the basis of phylogenetic analysis of 16S rRNA sequences; the family contains the genera *Beijerinckia* (type genus), *Chelatococcus*, *Methylocapsa*, and *Methylocella*. *Methylocapsa* was proposed after the

cut-off date for inclusion in this volume (June 30, 2001) and is not described here (see Dedysh et al. (2002)).

Aerobic. Form poly-β-hydroxybutyrate granules. Family is metabolically diverse. *Beijerinckia* and *Methylocella* fix nitrogen.

Type genus: **Beijerinckia** Derx 1950a, 145.

Genus I. **Beijerinckia** *Derx 1950a, 145*[AL]

CHRISTINA KENNEDY

Beij.e.rinck' i.a. M.L. fem. n. *Beijerinckia* named after M.W. Beijerinck, the Dutch microbiologist (1851–1931).

Straight or slightly curved rods, ∼**0.5–1.5 × 7–4.5 µm,** with rounded ends. Cells occur singly or appear as dividing pairs. Sometimes large, misshapen cells 3.0 × 5.0–6.0 µm occur; these are occasionally branched or forked. Intracellular granules of **poly-β-hydroxybutyrate (PHB)** are formed, generally one at each pole. Cysts (enclosing one cell) and capsules (enclosing several cells) may occur in some species. Gram negative. Motile by peritrichous flagella or nonmotile. **Aerobic,** having a strictly respiratory type of metabolism. **N$_2$ is fixed under aerobic or microaerobic conditions.** Optimal temperature for growth, 20–30°C; no growth occurs at 37°C. **Growth occurs between pH 3.0 and pH 9.5–10.0. Liquid cultures can become a highly viscous, semitransparent mass;** in some species the whole medium becomes opalescent and turbid, and adhering slime is not produced. **On agar media, especially under N$_2$-fixing conditions, copious slime is produced** and giant colonies with a smooth, folded, or plicate surface develop; some strains form slime having a more granular consistency. **Catalase positive. Glucose, fructose, and sucrose are utilized** by all strains and oxidized to CO_2. No growth occurs on peptone medium. Glutamate is utilized poorly or not at all. Species are found in soils, particularly **tropical regions**.

The mol% G + C of the DNA is: 54.7–60.7.

Type species: **Beijerinckia indica** (Starkey and De 1939) Derx 1950a, 146 (*Azotobacter indicum* (sic) Starkey and De 1939, 337.)

FURTHER DESCRIPTIVE INFORMATION

Cell morphology The typical microscopic appearance of *Beijerinckia* cells is shown in Figs. BXII.α.156 and BXII.α.157. Cells may be bicellular due to crosswall formation in the middle of the longitudinal direction of the cell (Figs. BXII.α.158 and

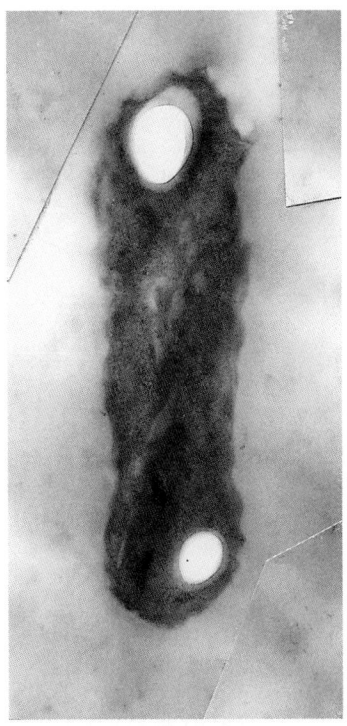

FIGURE BXII.α.157. *Beijerinckia indica* electron micrograph of a thin section showing the two polar lipoid bodies, which are surrounded by a membrane (× 33,300).

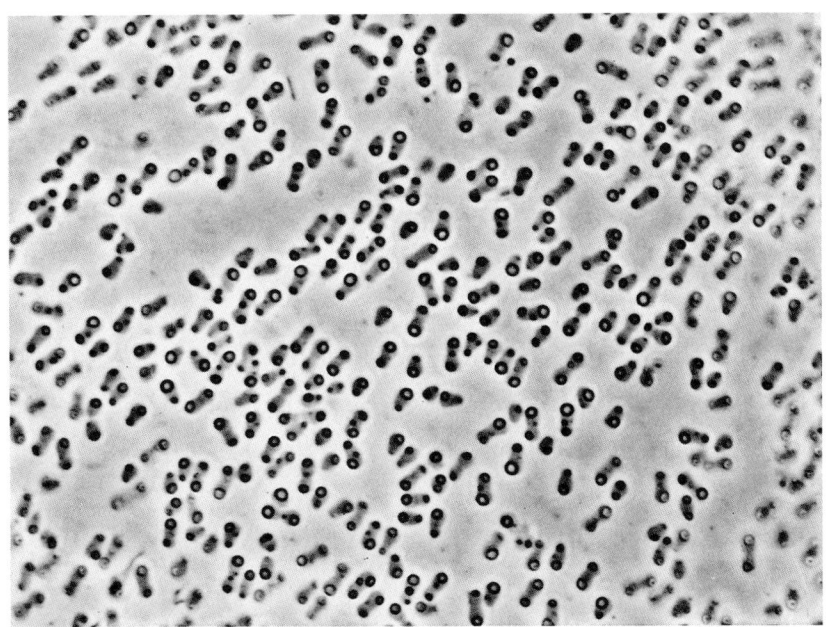

FIGURE BXII.α.156. *Beijerinckia indica* cells cultured on nitrogen-free glucose mineral agar (pH 5.0). The typical appearance of the cells and their intracellular polar lipid bodies is illustrated. Living preparation, phase-contrast microscopy (× 1500).

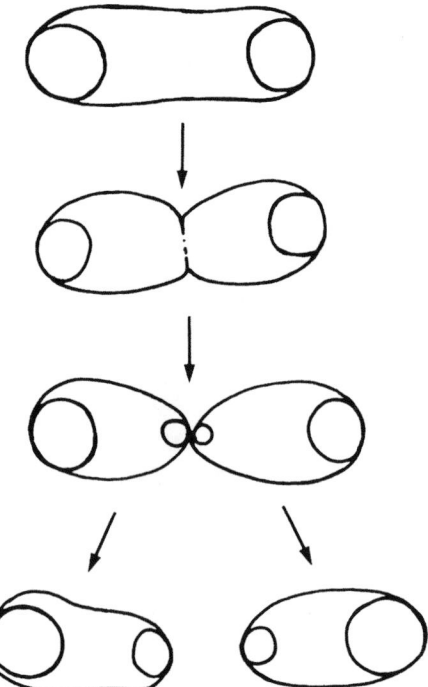

FIGURE BXII.α.158. Diagram of the life cycle of a *Beijerinckia* cell. A cell in division forms a crosswall in the middle of the longitudinal axis of the cell. In actively dividing cells, intermediate stages can often be seen.

BXII.α.159). Under certain cultural conditions, some *Beijerinckia* strains show coccoid cells without terminal lipoid globules (Fig. BXII.α.160). Sometimes, especially in *B. mobilis* strains, more than two lipoid globules per cell occur (Fig. BXII.α.162). The lipoid material is poly-β-hydroxybutyrate (PHB) (Becking, 1974, 1984b). Cyst and capsule formation occur in some species (*B. fluminensis*, *B. mobilis*, and *B. indica*) (Figs. BXII.α.161, BXII.α.162, and BXII.α.163).

The flagella of motile cells are peritrichous in those strains studied (Thompson and Skerman, 1979). Flagella appear to originate from one-half of the often dumbbell-shaped cells. The wave pattern is normal or curly, the wavelength has an average value of ~1.1–1.3 μm, and the amplitude is 0.26–0.35 μm. The amplitude of the waves in *B. fluminensis* and *B. derxii* strains is usually somewhat larger (0.26–0.35 μm) than in strains of *B. indica* (0.26 μm) (Thompson and Skerman, 1979).

Colonial and cultural characteristics On agar media, especially under N_2-fixing conditions, *Beijerinckia* strains may produce giant colonies containing copious amounts of exopolysaccharide. This slime is often extremely tough, tenacious, or elastic, which makes it difficult to remove part of a colony with a loop. In *B. mobilis* and *B. fluminensis* the exopolysaccharide has a more granular consistency (Becking, 1984a).

On nitrogen-free mineral agar medium[1], *Beijerinckia* species exhibit various kinds of colonial characteristics and pigmentation, and in liquid cultures they show differences in viscosity and pellicle formation; see the individual species descriptions for details.

---

1. N-free mineral medium, g/l: glucose, 20.0; K_2HPO_4, 0.8; KH_2PO_4, 0.2; $MgSO_4 \cdot 7H_2O$, 0.5; $FeCl_3 \cdot 6H_2O$, 0.025 or 0.05; $Na_2MoO_4 \cdot 2H_2O$, 0.005; $CaCl_2$, 0.05; and agar, 15.0. The pH is adjusted to 6.9. The $CaCl_2$ may be omitted to obtain a calcium-free medium.

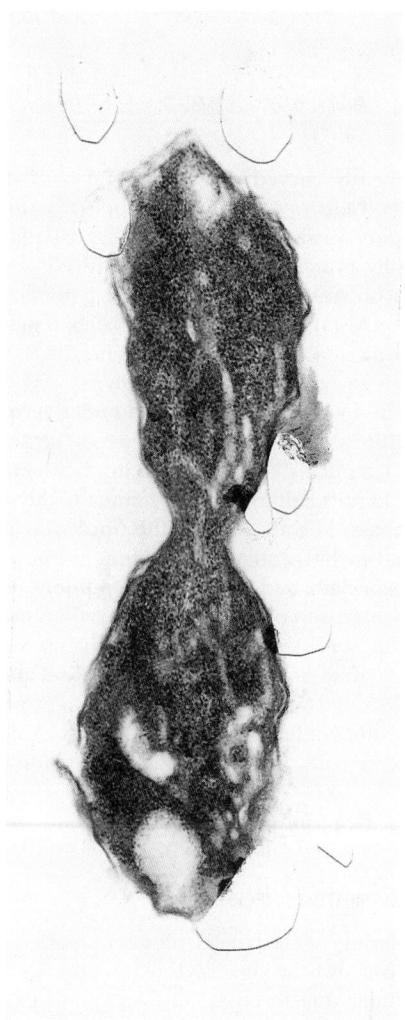

FIGURE BXII.α.159. Electron micrograph of a *Beijerinckia* cell in the process of division. The constriction in the middle of the cell is clearly visible and the two terminal lipoid bodies of the original cell can be seen (× 33,000).

The temperature range for growth of *Beijerinckia* species is from 10 to 35°C. Cells are resistant to freezing: no reduction of viability occurs when stored for 3–4 months at −4°C (Becking, 1961).

Metabolism and metabolic pathways Carbon sources utilized include many sugars, organic alcohols, and organic acids (see Table BXII.α.141 and Table BXII.α.142). Since the compilation of strain characteristics by Thompson and Skerman (1979) and Becking (1984a), few reports of new features have appeared. One report describes the ability of *B. mobilis* to utilize several aromatic compounds as carbon sources and as energy sources for nitrogen fixation (Chen et al., 1993); these include benzoate, catechol, 4-hydroxybenzoate, naphthalene, protocatechuate, and 4-toluate. Most *Beijerinckia* species hydrolyze starch.

In alkaline, nitrogen-free, glucose minimal medium, *Beijerinckia* strains decrease the pH to 4.0–5.0. The acids produced are mainly acetic and a small amount of lactic (Kauffmann and Toussaint, 1951; Becking, 1961).

As with other N_2 fixers, *Beijerinckia* species require molybdenum for optimal growth and N_2 fixation. The molybdenum requirement—4.0–35.0 mg/l (Becking, 1962)—is notably higher

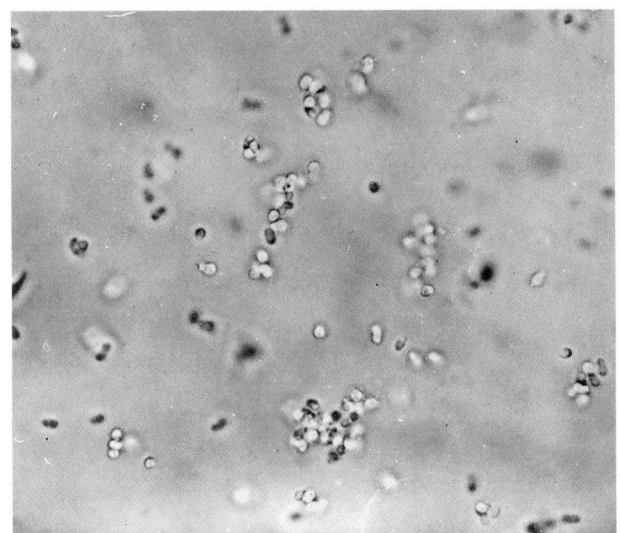

FIGURE BXII.α.160. *Beijerinckia mobilis* from an aged culture. The individual cells often lack the characteristic polar lipoid bodies and are more rounded in form, resembling certain *Azotobacter* species (× 1000).

than that of *Azotobacter* and *Azomonas*. Moreover, unlike *Azotobacter*, the molybdenum cannot be replaced by vanadium (Becking, 1962). This finding suggests that alternative nitrogenases are not present in *Beijerinckia*, at least not the nitrogenase that contains vanadium (nitrogenase-2) (for further explanation, see the genus *Azotobacter*, this volume). Growth of *Beijerinckia* species under N_2-fixing conditions does not require calcium, in contrast to most (but not all) *Azotobacter* species; indeed, $CaCO_3$ is even slightly inhibitory because it tends to prolong the lag phase of growth. The efficiency of N_2 fixation in *Beijerinckia* strains varies from 6 to 17 mg N_2 per g glucose consumed and is inversely proportional to growth (Becking, 1984a); the efficiency is greater at lower carbohydrate and/or oxygen supply levels (Spiff and Odu, 1973; Becking, 1978). The exocellular polysaccharide produced in copious amounts by species of *Beijerinckia* may protect nitrogenase from oxygen damage, as was shown for *B. derxii* subsp. *venezuelae* (Barbosa and Altherthum, 1992). Removal of the exopolysaccharide resulted in a drastic loss of nitrogenase activity.

Strains of *Beijerinckia indica* and *Beijerinckia mobilis* have been reported to contain large amounts of triterpenoids of the hopane series. In the former species, both bacteriohopanetetrol and aminobacteriohopanetriol were present, whereas in the latter, amino-bacteriohopanetriol was the only C35 hopanoid (Vilcheze et al., 1994). Whether this difference represents a distinguishing feature for all, or several, members of each species was not indicated.

Occurrence and transfer of plasmids   No studies involving isolation of mutants or genetic analysis of metabolism have been reported. The broad-host-range IncP plasmid RP4 could be mobilized into *B. indica* by conjugation and stably maintained (Naik et al., 1994). A tributyltin-sensitive strain, *Beijerinckia* sp. MC-27, isolated from freshwater sediment became resistant to this compound after transfer of a plasmid associated with chromium resistance from *Pseudomonas aeruginosa* PAO1 (Miller et al., 1995). The presence of indigenous plasmids in species of *Beijerinckia* was reported (Murai et al., 1990) but no function could be assigned.

Ecology   *Beijerinckia* strains were originally isolated from a quartzite soil (pH 4.5) of Malaysia (Altson, 1936) and later from acid soils of Dacca, Bangladesh (pH 4.9), and Insein in Burma (pH 5.2) by Starkey and De (1939). They were later observed to be widely distributed in acidic tropical soils of Africa, Southeast Asia, and South America (Kluyver and Becking, 1955; Becking, 1961). *B. indica* is the most commonly encountered species and has been isolated from tropical soils of all continents and sometimes from nontropical regions. *Beijerinckia* strains can also be recovered from waterlogged soils of wet rice fields (Becking 1961, 1978). In a survey of 392 soils of worldwide distribution, *Beijerinckia* strains were found infrequently in some temperate and subtropical soils (Becking, 1959, 1961; see also the review by Becking, 1981). Nevertheless, their occurrence in temperate habitats might be more widespread than originally thought, based on reports of *Beijerinckia* spp. being isolated from European white fir (Streichan and Schink, 1986). Moreover, acidophilic, methanotrophic organisms with a soluble methane monooxygenase were isolated from three boreal forests in acidic northern wetlands of West Siberia and European north Russia (Dedysh et al., 1998a), and analysis of 16S rDNA sequence from these organisms placed them close to *Beijerinckia indica*. However, more recent studies show that these organisms—unlike *Beijerinckia* species—are unable to grow on sugars and other multicarbon compounds. In addition, *Beijerinckia* cannot grow on methane, and genes coding for methane monooxygenase have not been detected in *B. indica* (Dedysh, personal communication). These bacteria are now named *Methylocella palustris* (Dedysh et al., 2000).

B. mobilis occurs mainly in very acid soils (pH 4.0–4.5) of tropical Southeast Asia, Africa, and South America, and this is in accordance with its ability to fix nitrogen optimally at pH 3.9 (Becking, 1961). This high degree of acid tolerance might be useful for the specific enrichment and isolation of this species. *B. fluminensis* was originally isolated from a "Baixada Fluminensis" (pH 4.2–5.2) of Rio de Janeiro and from some other Brazilian soils (Döbereiner and Ruschel, 1958). It was also isolated from several African and Asian soils (Becking, 1961). *B. derxii* was originally isolated from an Australian soil (Tchan, 1957) and then found in several South American and Asian locations, as well as in further locations in Australia (Thompson and Skerman, 1979).

In addition to soil or water habitats, *Beijerinckia* species are also found in plant rhizosphere and phyllosphere habitats. Associations with roots of sugarcane in Brazil and with a salt-tolerant grass *Leptochloa fusca* in Pakistan have been reported (Zafar et al., 1987; Baldani et al., 1997). *Beijerinckia* species have also been isolated from the leaves of plants such as coffee, cocoa, and cotton (Ruinen, 1956, 1961; Murty, 1984). There is no evidence that nitrogen fixation by *Beijerinckia* species provides an amount of fixed nitrogen sufficient to benefit plant growth.

ENRICHMENT AND ISOLATION PROCEDURES

An acidic, nitrogen-free medium[2] can be used for selective isolation of *Beijerinckia* from soil. The low pH of this medium favors development of beijerinckias, which are acid tolerant, and inhibits growth of other organisms, especially *Azotobacter* species. The requirement for trace elements (iron, molybdenum) in this medium is provided by the soil used as the inoculum. The medium is poured as thin layers (2–3 mm deep) into Petri dishes to allow good aeration. This partially inhibits the development

---

2. Enrichment medium (g/l of distilled water): glucose, 20.0; KH_2PO_4, 1.0; and $MgSO_4 \cdot 7H_2O$, 0.5. The pH is adjusted to 5.0.

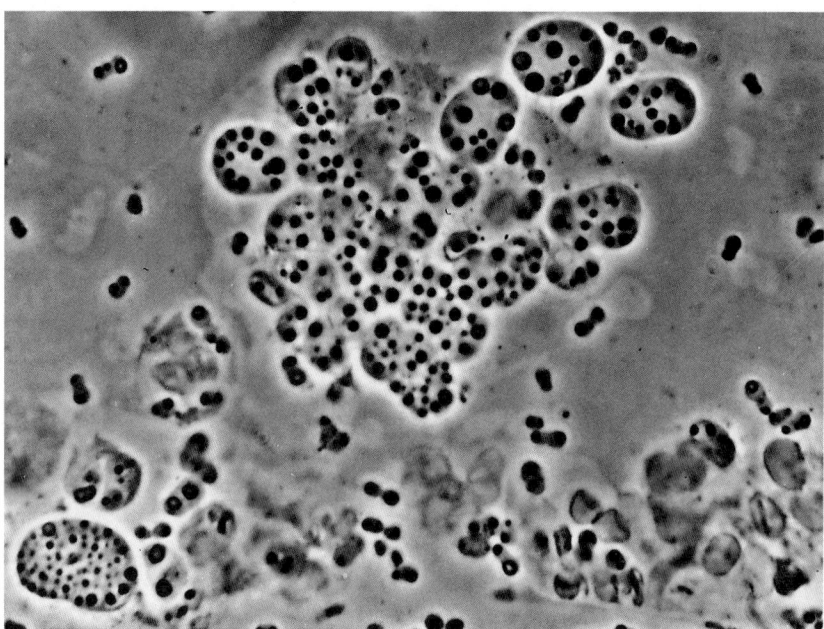

FIGURE BXII.α.161. *Beijerinckia fluminensis* cultured in nitrogen-free glucose mineral agar (pH 5.0), showing distinct capsule formation. The capsules enclose a large number of individual cells. Living preparation, phase-contrast microscopy (× 1500).

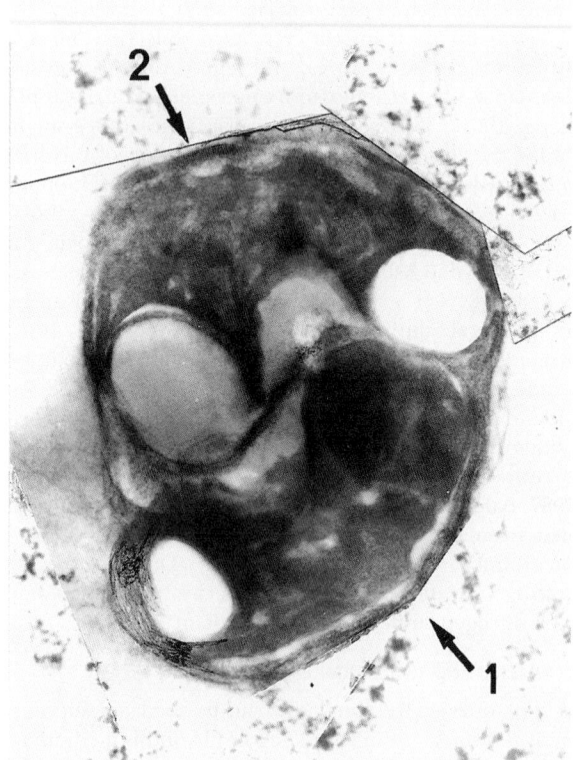

FIGURE BXII.α.162. Electron micrograph of a capsule of *Beijerinckia mobilis* containing two cells (see *arrows 1* and *2*). The terminal lipoid bodies of each cell are visible and also the distinct capsular wall (see *arrow 1*) (× 33,000).

or more weeks (*Beijerinckia* strains grow slowly). The entire culture may eventually change into a viscous mass due to slime production. The cultures are examined microscopically at various times for the presence of characteristic *Beijerinckia*-like cells. When such cells occur, the enrichment culture is plated onto a nitrogen-free, mineral agar medium (see Further Descriptive Information). Although an acidic agar medium can be used for isolation, it is not recommended because the agar may be hydrolyzed.

The sieved-soil plate method of Winogradsky (1932), using nitrogen-free mineral agar (pH 4.5–5.0) with glucose or sucrose (10 or 20 g/l), can also be applied. *Beijerinckia* colonies develop after 2–3 weeks around the soil particles on the plates. The type strain of *B. indica* was obtained by this method. In general, however, it is less satisfactory than the use of liquid enrichment media because purification is more difficult (the slime is more tenacious) when one starts with solid media. If the plating method is chosen, a drop inoculation is recommended rather than spreading, because of the higher number of colonies obtained with the former approach (Barbosa et al., 1995).

On agar media, *Beijerinckia* species form characteristic, highly raised, glistening colonies containing a tough, elastic slime. For further purification, a similar medium is used, but it is made neutral or alkaline by the addition of $CaCO_3$[3]. This is because the slime is more soluble under alkaline conditions, and it is easier to suspend the cells in sterile tap water or liquid medium for further streaking. On the alkaline agar medium, *B. indica* strains usually form highly raised colonies, whereas *B. mobilis* colonies are flatter and produce a uniform reddish brown or amber-brown color on aging.

of anaerobic or facultative N_2 fixers. Approximately 0.5 g of soil, 10 ml of water sample, or detached leaves can be used as the inoculum. The enrichment cultures are incubated at 30°C for 2

---

3. N-free mineral agar medium with $CaCO_3$: composition is similar to that given in Further Descriptive Information, but the $CaCl_2$ is replaced by $CaCO_3$ (10–20 g/l).

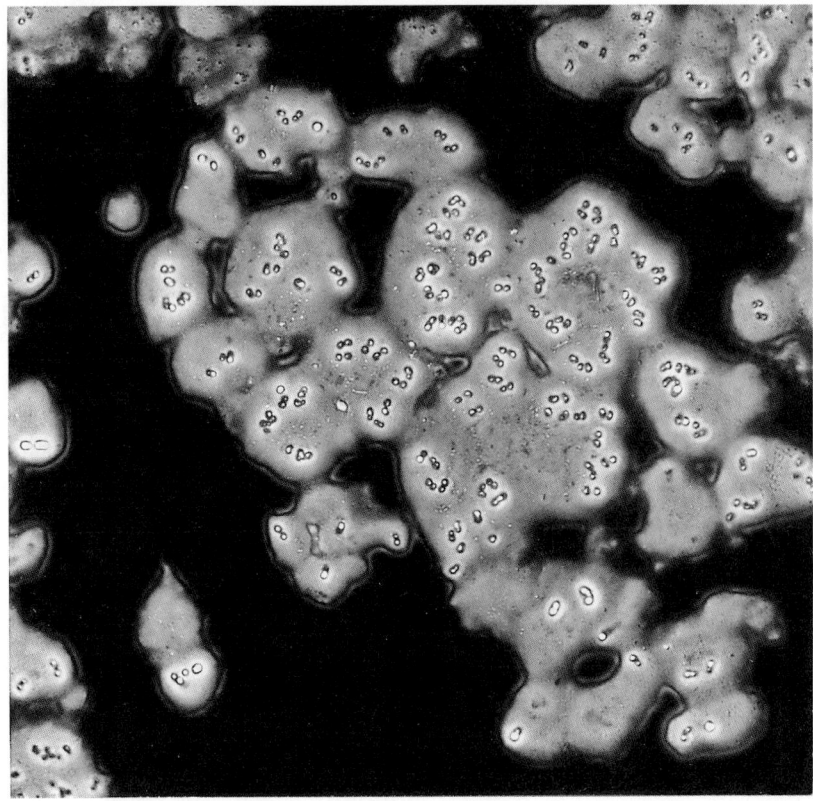

FIGURE BXII.α.163. Cells of *Beijerinckia indica* suspended in India ink, showing the polysaccharide formation around the cells. Living preparation, phase-contrast microscopy. (× 1350)

TABLE BXII.α.141. Differential characteristics of the species of the genus *Beijerinckia* and the genus *Derxia*[a]

Characteristics	B. indica	B. derxii	B. fluminensis	B. mobilis	Derxia gummosa
Water-soluble, green fluorescent pigment	−	+	−	−	−
Colony color after aging	P	B	P	AB	Br
Motility	[−][b]	−	[−][b]	+	+
Resistant to 1% peptone	d	−	−	d	+
Starch hydrolyzed	−	d	−	−	−
Growth on asparagine as C and N source	−	−	−	d	+
Urea hydrolyzed	+	+	−	+	+
H₂S production from cysteine	−	−	+	−	−
Tween 20 hydrolyzed	−	−	−	−	+
Indole produced	−	−	−	−	+
Antagonism to Gram positive organisms	−	d	−	−	+

[a]For symbols, see standard definitions; P, pink; B, buff; AB, amber-brown; Br, brown. For additional distinguishing characteristics, see Table BXII.α.142, utilization of carbon compounds.

[b]If positive, motility occurs mostly in young stages and the cells are usually only weakly motile.

Specific enrichment procedures No specific procedure is known to select for a particular *Beijerinckia* species, and all existing strains are random isolates obtained from soil using one of the general procedures outlined above. In general, carbon source utilization is not distinctive for particular species, although certain substrates might be useful for enrichment of one or two particular species because of their preferential use: e.g., benzoate for enrichment of *B. mobilis* and mannose for *B. indica* or *B. fluminensis* (see Table BXII.α.142). Thompson and Skerman (1979) suggested that certain substrates or inhibitors might be useful for enrichment of various species, although this has not yet been experimentally tested. From the properties of the strains studied, the following compounds have potential value for enrichment or selection:

1. *B. indica*: L-arabinose, D-mannose, glycerol, caprylate, and *trans*-aconitate. Glycerol might be useful for *B. indica*, and for *Derxia gummosa* and some *B. fluminensis* strains. Nitrilotriacetate might also be useful for *B. indica*.
2. *B. mobilis*: pentan-2-ol, 1,3-butylene glycol, asparagine, and *n*-valerate might be useful. Phenol (0.05%) might be used to inhibit the growth of other *Beijerinckia* species.
3. *B. indica* and *B. mobilis*: propan-1-ol, butan-1-ol, or 1,3-propylene glycol for enrichment.

4. *B. indica* subsp. *lacticogenes* and *B. mobilis*: caproate, *p*-hydroxybenzoate, or phenol for enrichment.

5. *B. derxii* and *B. fluminensis*: α-methyl-D-glucoside, maltose, melibiose, or melezitose for enrichment. Where both species are present, *B. derxii* would likely outgrow the extremely slow-growing *B. fluminensis*.

6. *B. derxii* subsp. *venezuelae*: L-arabitol for enrichment.

Because the colony morphology and chromogenesis of growth on solid media differs among the various *Beijerinckia* species, primary selection of colonies from the enrichment cultures is done mainly on the basis of these colonial characteristics. A more precise identification is performed later using additional differential characteristics.

MAINTENANCE PROCEDURES

Beijerinckia strains can be lyophilized in skim milk or dextran-sodium glutaminate solution on filter paper and stored in the dark at room temperature (Becking, 1984a).

Storage has also been achieved on the usual agar media in tubes plugged with sterile rubber seals, with storage in the dark at room temperature (Antheunisse, 1972, 1973); after 10 years, 33% of the cultures retained viability. *Beijerinckia* cultures stored under a seal of sterile liquid paraffin or mineral oil generally survive for at least 3–5 years (Becking, 1984a).

Strains may also be preserved indefinitely in liquid nitrogen. At the type culture collection in Delft, The Netherlands, DMSO (10%, v/v) is added to cultures in the log phase or end of log phase, and the cultures are frozen as rapidly as possible in liquid nitrogen. For recovery, the vials are thawed rapidly in a waterbath at 37°C (Becking, 1984a).

Cryopreservation in 10% glycerol was also successful (Thompson, 1987). See the genus *Azotobacter* for details.

DIFFERENTIATION OF THE GENUS *BEIJERINCKIA* FROM OTHER GENERA

Beijerinckia strains can be distinguished from other aerobic N₂ fixers by their great acid tolerance (which allows them to grow well at pH 4.0 or 5.0), by their failure to form a pellicle on the surface of liquid media, and by their ability to make a liquid medium viscous by slime production (Fig. BXII.α.164). Moreover, on solid media they produce characteristic large, slimy colonies having a tough, tenacious, and sometimes elastic slime. Because of this exopolysaccharide production, it is often difficult to subculture portions of a colony for purification. For other features distinguishing *Beijerinckia* strains from *Derxia**, the most closely related genus, see Tables BXII.α.141 and BXII.α.142.

Beijerinckia cells can be distinguished from those of *Azotobacter* and *Azomonas* by their generally smaller size, by their more rod-shaped or sometimes pear- or dumbbell-shaped appearance, and by the characteristic presence of a lipoid body at each pole (Fig. BXII.α.156). Many strains of *Beijerinckia* utilize nitrate poorly or not at all and, in this respect, differ from strains of *Azotobacter*.

Although both *Beijerinckia* and *Derxia* produce slimy colonies on agar and viscosity in broth, *Beijerinckia* strains can be distinguished by (a) failure to produce dark mahogany-brown colonies with aging, (b) cells that contain bipolar lipoid bodies rather than numerous lipoid bodies throughout the whole cell, (c) failure to form a pellicle at the surface of liquid media, and (d) a positive catalase reaction. In addition, a number of C sources should distinguish the two genera; for instance, *Beijerinckia* species, but not *Derxia*, can utilize sorbose and raffinose, and *Derxia*, but not *Beijerinckia*, utilizes aspartate, glutamate, or ethylamine (Table BXII.α.142).

TAXONOMIC COMMENTS

When numerical analysis methods are applied to species of *Beijerinckia*, *Azotobacter*, and *Azomonas*, the *Beijerinckia* species fuse into a single, apparently coherent group (Thompson and Skerman, 1979). Using a wide range of attributes, and considering all strains of named and unnamed *Beijerinckia* species, numerical analysis supports the concept of a separate genus for these bacteria. In addition, the experiments reported by De Smedt et al. (1980), in which ^{14}C-labeled rRNA from *Beijerinckia indica* was hybridized with filter-fixed DNA from a wide variety of Gram-negative bacteria, indicated the genus *Beijerinckia* to be a heterogeneous but coherent group. From rRNA cistron similarities, it was concluded that *Beijerinckia* and *Azotobacter*/*Azomonas* belong to different classes. Analysis of the 16S rDNA sequence of *Beijerinckia indica* places this genus within the phylum *Proteobacteria*. The family *Beijerinckiaceae*, including the genera *Beijerinckia*, *Derxia*, and *Chelatococcus*, is in the class *Alphaproteobacteria* and the order *Rhizobiales*. This family is most closely related to the families *Methylocystaceae* and *Bradyrhizobiaceae*.

Thompson and Skerman (1979) used many phenotypic characters of *Beijerinckia* strains for numerical analysis, yielding a hierarchical classification. These authors confirmed the presence of four species, in two main groups within the genus. Group 1294 (*B. fluminensis* + *B. derxii*) fuses with group 1289 (*B. indica* subsp. *indica* + *B. indica* subsp. *lacticogenes* + *B. mobilis*) to produce the group 1297. Group 1289 strains generally differed from group 1294 strains in having thinner cells and using caprylate, propan-1-ol, 1,3-propylene glycol, *trans*-aconitate, nitrilotriacetate, and L-arabinose, but not maltose and α-methyl-D-glycoside, as sole carbon sources. Moreover, strains of group 1289 produced acid from L-arabinose and glycerol and were resistant to 0.5% NaCl, 0.05% phenol, and 1.0% sodium benzoate. One widely studied member of the genus, *Beijerinckia* sp. strain B1, was reclassified as a strain of *Sphingomonas yanoikuyae* on the basis of 16S rDNA analysis (Khan et al., 1996a).

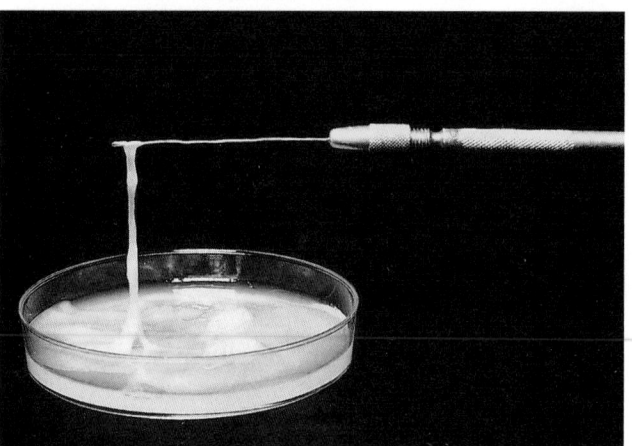

FIGURE BXII.α.164. Enrichment culture of *Beijerinckia* inoculated with tropical soil, demonstrating the highly viscous consistency of the medium after 3 weeks (× 0.8).

Editorial Note: Because of the 16S rDNA placement, the genus *Derxia* has been moved to the family *Alcaligenaceae* and is no longer listed in *Beijerinckiaceae*.

TABLE BXII.α.142. Utilization of carbon compounds by *Beijerinckia* and *Derxia* species[a]

Carbon compounds utilized[b,c]	*B. indica* (14)[d]	*B. derxii* (21)	*B. fluminensis* (10)	*B. mobilis* (2)	*Derxia gummosa* (6)
Arabinose	+	−	d	+	−
Galactose	+	+	+	−	−
Fructose	+	+	+	−	+
Melibiose	d	d	+	−	−
Maltose	−	+	d	−	−
Mannose	+	−	d	−	−
Sorbose	+	d	+	+	−
Raffinose	+	+	d	d	−
Xylose	−	−	+	−	−
Butanol	+	−	d	+	+
Propanol	+	−	−	+	+
Glycerol	+	−	+	+	+
Sorbitol	+	+	+	−	d
Mannitol	+	+	d	d	d
Acetate	d	+	−	+	+
Citrate	+	d	−	d	+
Oxaloacetate	+	d	−	−	+
Fumarate	+	d	d	d	+
Malate	+	d	−	d	−
Malonate	−	−	−	−	+
Glycolate	+	+	−	+	+
Benzoate	−	−	−	+	−
L-Ascorbate	+	d	+	−	−
Aspartate	−	−	−	−	+
Glutamate	−	−	−	−	+
Ethylamine	−	−	−	−	+

[a]For symbols, see standard definitions.

[b]Data represent a merger of information in Thompson and Skerman (1979) and Becking (1984a) in the 1st edition of *Bergey's Manual of Systematic Bacteriology*. Not all compounds tested by Thompson and Skerman (1979) are included.

[c]All species can utilize glucose, sucrose, lactate, pyruvate, succinate, gluconate, and fumarate. All species fail to utilize ribose, fucose, cellobiose, lactose, trehalose, glutarate, and oxoglutarate.

[d]Numbers in parentheses represent the number of strains tested.

ACKNOWLEDGMENTS

The description of the genus as given by J.-H. Becking in the 1st edition of the *Manual* remains largely unchanged; information has been reorganized, reevaluated, and updated.

FURTHER READING

Becking, J.H. 1992. The Genus *Beijerinckia*. *In* Balows (Editor), The Prokaryotes. A Handbook on the Biology Of Bacteria: Ecophysiology, Isolation, Identification, Applications, 2nd Ed., Vol. 3, Springer-Verlag, New York. pp. 2254–2267.

List of species of the genus Beijerinckia

1. **Beijerinckia indica** (Starkey and De 1939) Derx 1950a, 146[AL] (*Azotobacter indicum* (sic) Starkey and De 1939, 337.) *in' di.ca.* L. fem. adj. *indica* of India.

The characteristics are as described for the genus and as given in Tables BXII.α.141 and BXII.α.142, with the following additional information. Straight or slightly curved rods 0.5–1.2 × 1.6–3.0 μm. PHB granules persist in aged cultures. No resting stages occur; neither cyst nor spore formation is observed.

Agar colonies are raised. At first, they are semitransparent but soon become uniformly turbid or opaque white. On aging, the colonies develop a light reddish pink, cinnamon, or fawn color on neutral or alkaline media; on acid media, they remain colorless. On acid media, the slime is more tenacious, tough, and elastic than on alkaline media. Giant colonies may develop, first with a smooth surface (Fig. BXII.α.165), but later with a folded or wrinkled surface (Fig. BXII.α.166).

Liquid media become viscous as cell density increases.

Color may be produced, but is less prominent than on agar.

Grows between pH 3.0 and 10.0 (optimum is 4.0–10.0). Temperature range for growth, 10–35°C; no growth at 37°C.

Growth on, and utilization of, nitrate is poor, and N$_2$ is fixed in preference to utilization of nitrate in the medium (Becking, 1962). Weak growth occurs on malt agar, no growth in plain broth or peptone agar.

Widely distributed in acid tropical soils.

The mol% G + C of the DNA is: 54.7–58.5 (T_m) (De Ley and Park, 1966; De Smedt et al., 1980).

Type strain: Delft E.II.12.1.1, ATCC 9039, DSM 1715, NCIB 8712, WR-119.

GenBank accession number (16S rRNA): M59060.

Derx (1950a) described "*B. indica* biovar alba", which was distinguished by its lack of pigmentation on aging; however, under extreme (alkaline) conditions it produced a pink pigment. In the hierarchical classification of Thompson and Skerman (1979), one co-type (strain WR-236) of *B. indica* biovar alba was placed in *B. indica* subsp. *lacti-*

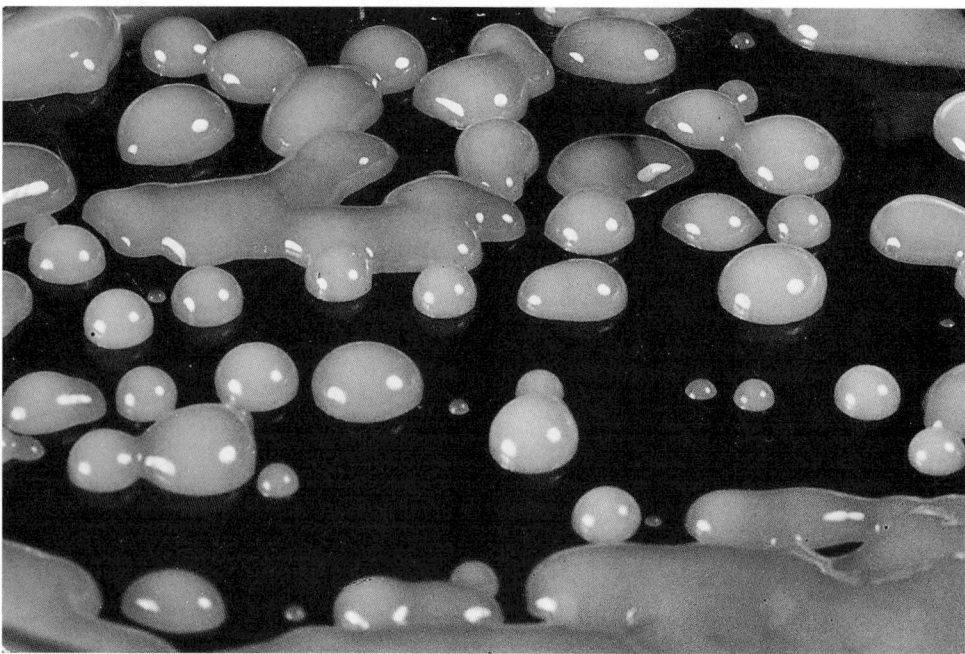

FIGURE BXII.α.165. Typical colony type of *Beijerinckia indica* on nitrogen-free glucose mineral agar. The colonies are highly raised and have a very tough, elastic slime. In young cultures these colonies are colorless and transparent (× 2).

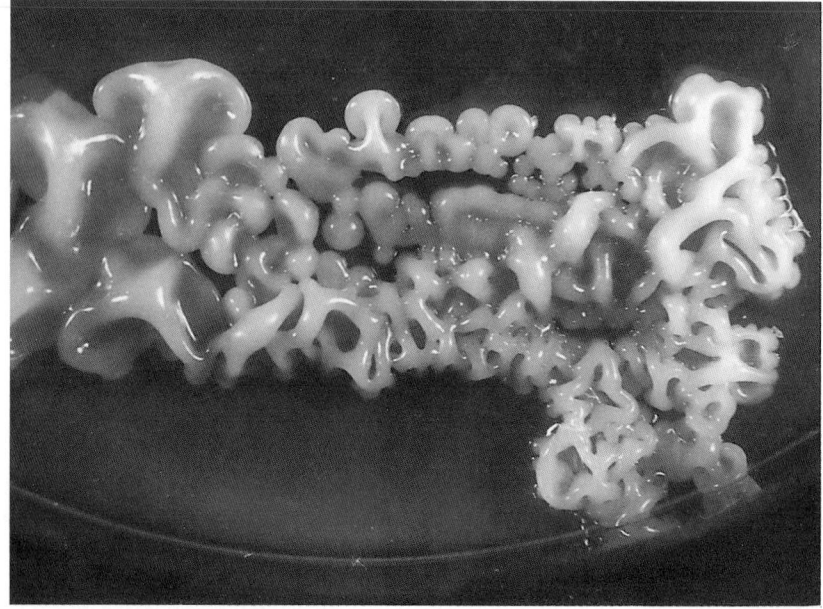

FIGURE BXII.α.166. Typical colony of an aged culture of *Beijerinckia indica* on nitrogen-free glucose mineral agar. The colonies increase greatly in size due to copious slime production. The colonies become massive and opaque, with a plicate surface. In this stage they often attain a light reddish, pink, or cinnamon color, especially on neutral or alkaline media (× 2).

cogenes (group 1270) and the other (strain WR-235) was placed in *B. derxii* subsp. *venezuelae* (group 1271).

a. **Beijerinckia indica** *subsp.* **indica** (Starkey and De 1939) Derx 1950a, 146[AL] (*Azotobacter indicum* (sic) Starkey and De 1939, 337.)

Thompson and Skerman (1979) distinguished *B. indica* subsp. *indica* from subsp. *lacticogenes* by differences in organic carbon utilization, nitrate reduction, and resistance to peptone-nitrogen. Based on their study of nine strains of subsp. *indica* and four or five strains of subsp. *lacticogenes*, the only absolute character that differentiated the two subspecies was the failure of subsp. *indica* to utilize *p*-hydroxybenzoate as a sole carbon source. Nitrate was reduced to nitrite by eight of nine strains of subsp. *indica* but not by four of four strains of subsp. *lacticogenes*. The differences in utilization of sole carbon sources by subsp. *indica* vs. subsp. *lacticogenes* were as follows: propan-2-ol, 7/9 vs. 5/5; butan-2-ol, 0/9 vs. 3/5; D-arabitol, 4/9 vs. 1/5; phenol, 0/9 vs. 4/5; caproate, 0/9 vs. 4/5; adipate, pimelate, suberate, azelate, and sebacate, 0/9 vs. 1/5 or 1/4; α-oxybutyrate, 0/9 vs. 1/5; fumarate, DL-malate, tartrate (D, L, and *meso*), oxaloacetate, mucate, and *trans*-aconitate, 9/9 vs. 3–4/5. None of the strains of subsp. *indica* lacked flagella, whereas half the strains of subsp. *lacticogenes* did lack flagella.

The mol% $G + C$ of the DNA is: 54.7–58.5 (T_m) (De Ley and Park, 1966; De Smedt et al., 1980).

Type strain: Delft E.II.12.1.1, ATCC 9039, DSM 1715, NCIB 8712, WR-119.

GenBank accession number (16S rRNA): M59060.

b. **Beijerinckia indica** *subsp.* **lacticogenes** (Kauffmann and Toussaint 1951) Thompson and Skerman 1981, 215^{VP} (Effective publication: Thompson and Skerman 1979, 332) (*Azotobacter lacticogenes* Kauffmann and Toussaint 1951, 710)

lac.ti.co'ge.nes. M.L. n. *acidum lacticum* lactic acid; Gr. v. *gennaio* to produce; M.L. adj. *lacticogenes* lactic acid producing (which is an error: acetic acid is produced).

This subspecies is distinguished from *B. indica* subsp. *indica* by the characteristics described above, and by the consistency of the colonies (less elastic and rubbery (cartilaginous) and more butyrous and brittle than those of subsp. *indica*). Moreover, subsp. *lacticogenes* is somewhat less resistant to diamond fuchsin, brilliant green, sodium fluoride, and streptomycin. In contrast to subsp. *indica*, four of five strains of subsp. *lacticogenes* grown on agar containing *p*-hydroxybenzoate could metabolize protocatechuate via the *ortho*-cleavage pathway (Thompson and Skerman, 1979).

The mol% $G + C$ of the DNA is: unknown.

Type strain: WR-119, ATCC 19361.

2. **Beijerinckia derxii** Tchan 1957, 315^{AL}

derx'i.i. M.L. gen. n. *derxii* of Derx; named after H.G. Derx, the Dutch microbiologist (1894–1953).

The characteristics are as described for the genus and as given in Tables BXII.α.141 and BXII.α.142, with the following additional information. Single straight or curved rods, or rods with clavate extremities, 1.5–2.0 × 3.5–4.5 μm. Polar lipoid bodies are very large and conspicuous. No cyst or capsule formation occurs. Nonmotile.

Colonies are highly raised, slimy, and smooth. The chemical composition of the slime has not yet been examined. Colonies are at first semitransparent or opaque white, but after 2–3 weeks, a yellow-green, water-soluble, fluorescent pigment is produced, particularly on iron-deficient media. When the pigment first appears it remains within the colony, but later it diffuses into the agar medium. Under certain conditions, pigment production on agar media may be very poor or absent.

Liquid cultures become uniformly turbid and pigment production is usually less than on solid media.

Growth occurs between pH 4.0 and 9.0 (optimum, 6.0–7.0). There is no growth at pH 3.0 or 11.0. Temperature range for growth, 10–35°C; no growth at 37°C.

Isolated from soils from Queensland, Northern Australia, and neutral and alkaline soils of South Africa.

The mol% $G + C$ of the DNA is: 59.1 ± 1.6 (T_m) (De Ley and Park, 1966).

Type strain: Q13 of Tchan, ATCC 49361, DSM 2328, UQM 1968.

a. **Beijerinckia derxii** *subsp.* **derxii** Tchan 1957, 315^{AL}

In the hierarchical classification of Thompson and Skerman (1979), the strains within the species *B. derxii* can be divided into two main groups: group 1263 and group 1271. Group 1263 contains a co-type of *B. derxii* Tchan 1957 and three co-type strains of "*Beijerinckia congensis*" Hilger 1965, as well as three Australian isolates. This group was named by Thompson and Skerman (1979) as *B. derxii* subsp. *derxii*. Differences between this subspecies and *B. derxii* subsp. *venezuelae* (group 1271) are not markedly consistent, apart from utilization of nitrate, as outlined below in the description of the subsp. *venezuelae*.

The mol% $G + C$ of the DNA is: not available.

Type strain: Q13 of Tchan, ATCC 49361, DSM 2328, UQM 1968.

b. **Beijerinckia derxii** *subsp.* **venezuelae** (Materassi, Florenzano, Balloni and Favilli 1966) Thompson and Skerman 1981, 215^{VP} (Effective publication: Thompson and Skerman 1979, 343) (*Beijerinckia venezuelae* Materassi, Florenzano, Balloni and Favilli 1966, 210.)

ven.e.zue'lae. M.L. gen. n. *venezuelae* of Venezuela, South America.

In the hierarchical classification of Thompson and Skerman (1979), group 1271 is classified as *B. derxii* subsp. *venezuelae*. This group consists of six co-type strains of "*B. venezuelae*" Materassi et al. 1966, one co-type of "*B. indica* biovar alba" Derx 1950a, and three Australian isolates.

Differences between group 1263 (subsp. *derxii*) and group 1271 (subsp. *venezuelae*) are not markedly consistent. In general, strains of group 1271 are distinguished by not utilizing nitrate as a sole source of nitrogen, being nonmotile, not hydrolyzing glycogen, growing over a slightly wider pH range (not more than 0.5 pH unit at each end of the range), and by differences in utilization of several organic compounds as sole sources of carbon. Of 14 strains of subsp. *venezuelae* and 7 strains of subsp. *derxii* tested by Thompson and Skerman (1979), the utilization of these substrates was as follows: melibiose, 5/14 vs. 7/7; propan-2-ol, 6/14 vs. 0/7; L-arabitol, 6/14 vs. 0/7; propionate, 7/14 vs. 0/7; fumarate, 8/14 vs. 1/7; and L-ascorbate, 6/14 vs. 5/7. Eleven of 14 strains of subsp. *venezuelae* used nitrate as a nitrogen source, whereas only 1 of 7 strains of subsp. *derxii* could do so.

The mol% $G + C$ of the DNA is: not available.

Type strain: 2 of Materassi, DSM 2329, WR-222.

Of the six strains of subsp. *venezuelae* provided by G. Florenzano (i.e., strains WR-221 to WR-226), it is not clear how many independent isolates are represented (Becking, 1984a).

3. **Beijerinckia fluminensis** Döbereiner and Ruschel 1958, 269[AL]

flu.mi.nen' sis. M.L. adj. *fluminensis* named after the locality "Baixada Fluminense", State Rio de Janeiro, Brazil, from which soil it was first isolated.

The characteristics are as described for the genus and as given in Tables BXII.α.141 and BXII.α.142, with the following additional information. Straight or slightly curved rods, 1.0–1.5 × 3.0–3.5 µm. Older cultures show characteristic large capsules enclosing 2–10 or more individual cells. Division of cells within the capsules has been observed. Motility is slow or absent, especially in older cells.

Colonies are typically small and granular, moderately raised, with an irregular rough surface (Fig. BXII.α.167). The slime is not liquid, tenacious, or elastic, but more granular and stiff; its chemical composition has not yet been examined. Colonies are at first opaque white, becoming pink, reddish brown, or fulvous (like *B. indica*) after 1–2 weeks on neutral or alkaline media.

Slime production in liquid media is reduced. No pellicle or viscosity occurs, but a bluish white turbidity develops.

Grows between pH 3.5 and 9.2. Temperature range for growth, 10–35°C (optimum, 26–33°C); no growth at 37°C.

Found in acidic soils of South America, Africa, and Asia (China, Indonesia).

The mol% G + C of the DNA is: 56.2 ± 1.8 (T_m) (type strain) (De Ley and Park, 1966; De Smedt et al., 1980).

Type strain: CD10 of Döbereiner and Ruschel, DSM 2327, UQM 1685.

4. **Beijerinckia mobilis** Derx 1950b, 10[AL] (*Beijerinckia mobile* (sic) Derx 1950b, 10.)

mo' bi.lis. L. fem. adj. *mobilis* movable, motile.

The characteristics are as described for the genus and as given in Tables BXII.α.141 and BXII.α.142, with the following additional information. Straight, curved, or pear-shaped rods, 0.6–1.0 × 1.6–3.0 µm. Misshaped or forked cells sometimes occur. "*Ascococcus*"-like clusters of cells are often visible in older cultures. The typical polar lipoid bodies may disappear in aging cells, and the cells are then more rounded and resemble *Azotobacter* cells (Fig. BXII.α.160). Motility is conspicuous.

Agar colonies are not as raised as those of *B. indica*, and slime production is less (Fig. BXII.α.168). The slime is neither elastic nor sticky; its chemical composition has not yet been examined. Older cultures on neutral or alkaline agar media show a typical dark amber or deep reddish brown color.

Broth cultures do not become viscous. There is a tendency to form a pellicle at the surface.

Grows between pH 3.0 and 10.0. Optimal growth and N_2 fixation occur at pH 4.0–5.0 and decrease sharply at the more alkaline values (Becking, 1961). Temperature range for growth, 10–35°C; no growth at 37°C.

All strains tested have grown well on nitrate or ammonium salts as the nitrogen source (in contrast to *B. indica*). Weak growth or no growth occurs on urea, glycine, glutamate, or tyrosine. All strains grow on leucine and casein agar. Moderate growth occurs on malt agar. The differences in levan production from sucrose observed by Derx (1950b) are variable and cannot be used for differentiation of this species.

Common in Indonesian (Java) soils; also isolated from soils of South America (Surinam) and tropical Africa.

The mol% G + C of the DNA is: 57.3 (T_m) (De Smedt et al., 1980).

Type strain: Delft E.III.12.2, ATCC 35011, DSM 2326, UQM 1969.

FIGURE BXII.α.167. Typical colonies of *Beijerinckia fluminensis* on nitrogen-free glucose mineral agar. This species forms rather small, raised colonies with a highly plicate surface. In this species the slime has a granular consistency (× 4).

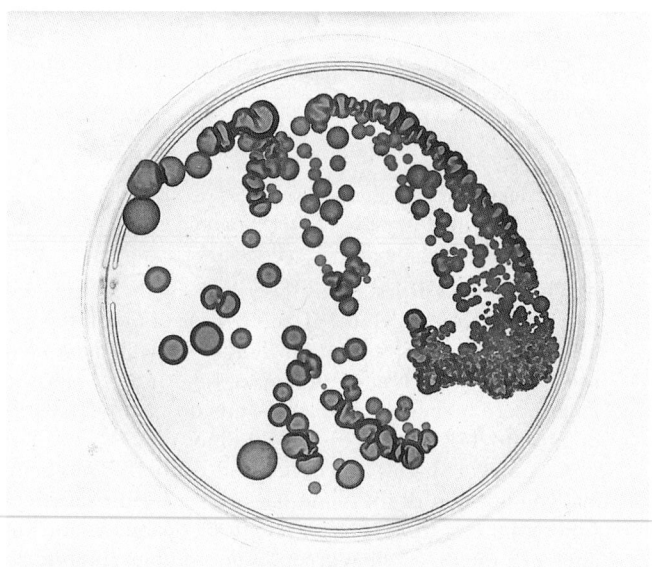

FIGURE BXII.α.168. Colonies of *Beijerinckia mobilis* on a nitrogen-free glucose mineral agar containing CaCl₂. On this transparent medium, the species forms only small raised colonies having a typical amber-brown color on aging (× 0.7).

Genus II. **Chelatococcus** Auling, Busse, Egli, El-Banna and Stackebrandt 1993b, 624[VP]
(Effective publication: Auling, Busse, Egli, El-Banna and Stackebrandt 1993a, 109)

THOMAS W. EGLI AND GEORG AULING

Che.la' to.coc.cus. Gr. n. *chele* claw; M.L. v. *chelato* to form claw-like complexes with divalent cations, i.e., to chelate; Gr. n. *coccus* berry, used for a bacterium of roughly spherical shape; M.L. masc. n. *Chelatococcus* chelating coccus.

Short, stout, almost coccoid rods, 0.5–1 × 1–2 µm; resemble diplococci during exponential growth. Gram negative. **Nonmotile. Obligately aerobic.** Optimal growth temperature, 35–37°C; temperature range for growth, 4–41°C. Slow growth with all substrates tested so far (μ_{max} <0.2 h^{-1}). Poly-β-hydroxybutyrate is accumulated within the cells. **Utilize the metal-chelating aminopolycarboxylic acids nitrilotriacetic acid (NTA) and S,S-ethylenediaminedisuccinic acid as sole sources of carbon, energy, and nitrogen.** Generally, sugars are not utilized (except for glycerol). **Require one or more vitamins.** All strains are sensitive to β-lactam antibiotics but resistant to nalidixic acid. Ubiquinone Q-10 is present. The main polyamine present is *sym*-homospermidine; putrescine and spermidine occur as major polyamines, and spermine as a minor polyamine.

The mol% G + C of the DNA is: 63.

Type species: **Chelatococcus asaccharovorans** Auling, Busse, Egli, El-Banna and Stackebrandt 1993b, 624 (Effective publication: Auling, Busse, Egli, El-Banna and Stackebrandt 1993a, 110.)

FURTHER DESCRIPTIVE INFORMATION

Morphological features The typical morphology of *Chelatococcus asaccharovorans* is shown in Fig. BXII.α.169a and b. Ultrathin sections reveal a Gram-negative cell envelope with a cytoplasmic membrane, a murein layer 6–8 nm wide, an outer membrane, and an additional proteinaceous surface layer (Fig. BXII.α.169c, d). The murein layer is always in close contact with the outer membrane. The outer proteinaceous crystalline surface layer (S-layer) is approximately 15 nm wide. Its outer side appears smooth, whereas, depending on the plane of sectioning, the inward-facing side of the S-layer exhibits a regular arrangement of protrusions (Fig. BXII.α.169d). This S-layer is a rather brittle structure. Easy detachment of the S-layer may explain the patchy surface structure of deep-etched cells (Fig. BXII.α.169c). It is not yet known whether this is an intrinsic property of the S-layer or an artifact resulting from excessive mechanical stress during preparation of the cells for electron microscopy. Detached patches with a hexagonal arrangement are usually found in the culture medium (Wehrli and Egli, 1988). The repeating units are most probably hexamers, with a central hole and a center-to-center spacing of approximately 16 nm. Since smooth cell surfaces were occasionally observed in deep-etched samples of both strains, the possibility that the S-layer is covered by a further outer protein layer remains. Although such S-layers are common in both prokaryotic domains (*Bacteria* and *Archaea*), their value for taxonomic purposes has not yet been shown (Messner and Sleytr, 1992).

Colony morphology Colonies of *Chelatococcus* are similar in appearance to those of *Chelatobacter* (see Fig. BXII.α.132 in the genus *Chelatobacter*). After 4–5 d incubation at 30°C on complex medium (PCA, Plate Count Agar, Difco), *Chelatococcus asaccharovorans* strain TE1 develops round and smooth colonies with a diameter of 3–4 mm, which appear white-to-beige and mucoid. On selective NTA mineral medium (see below), a different colony morphology is observed with a stereo-microscope, 25-fold magnification. Even well-developed colonies never exceed 0.5 mm in diameter. The smaller colony size on NTA agar (compared to growth on PCA) is probably due to a rise in pH resulting from the excretion of ammonia during growth with NTA as the only source of carbon, energy, and nitrogen. Colonies appear white (incident light) and translucent (transmitted light) when young (1–2 d). With prolonged incubation, the colonies are still white (incident light), but under the stereo-microscope (transmitted light) they appear brown and finally black, and their shape becomes "polygonal" (3–5 d) as shown in Fig. BXII.α.132. *Chelatococcus asaccharovorans* strain TE2 may be a nonmucoid mutant of strain TE1. Colonies are smaller (1 mm after 4–5 d) on both complex and mineral agar. They appear firm and raised on NTA agar.

Cultural characteristics The chemically-defined medium used for isolation, maintenance, and growth of *Chelatococcus asaccharovorans* strains in both batch and chemostat culture contains (g/l distilled water): $MgSO_4 \cdot 7H_2O$, 1.0; $CaCl_2 \cdot 2H_2O$, 0.2; $Na_2HPO_4 \cdot 2H_2O$, 0.41; KH_2PO_4, 0.26; 1 ml of the trace element stock solution described by Pfennig et al. (1981) with NTA as the chelating agent (5.2 g/l), and 1 ml of a vitamin stock solution. The vitamin solution contains (per liter): pyridoxine-HCl, 100 mg; thiamine-HCl, riboflavin, nicotinic acid, D-calcium pantothenate, *p*-aminobenzoic acid, lipoic acid, nicotinamide and vitamin B_{12}, 50 mg each; biotin, 20 mg; and folic acid, 20 mg. This medium is supplemented with either NTA or other carbon and/or nitrogen sources (up to a maximum of 5 g/l carbon). For growth with carbon sources containing no nitrogen, the medium is supplemented with NH_4Cl (0.54 g/l). For isolation and growth on agar plates, this medium was supplemented with 1.5% agar plus NTA (0.5 g/l). Growth is dependent on addition of the above vitamin mixture (Egli et al., 1988) or yeast extract (0.1 g/l).

Growth characteristics Growth in batch culture with NTA as the only carbon/energy and nitrogen source results in the excretion of ammonia, which causes an increase in pH of the culture medium. Under such conditions, growth ceases at approximately pH 9 and after transfer into new medium, the cells exhibit extended lag times (often 5–10 d) until growth resumes. Growth with mixtures of utilizable carbon sources plus NTA results in simultaneous utilization of the two substrates, and in faster growth than with NTA alone. For example, during simultaneous utilization of acetate plus NTA a maximum specific growth rate of 0.13 h^{-1} was reported, compared to 0.07 h^{-1} during growth with NTA alone (see Table BXII.α.115 in the genus *Chelatobacter*).

Nutritional and physiological characteristics For the nutritional and physiological characteristics of *Chelatococcus asaccharovorans* strains and NTA-utilizing *Chelatobacter heintzii* strains, the reader is referred to Table BXII.α.116 in the chapter on the genus *Chelatobacter*. Additional properties of *Chelatococcus asaccharovorans* are as follows. Substrates utilized for growth include acetate, alanine, aspartate, *n*-butanol, citrate, ethanol, S,S-ethylenediaminedisuccinate, glutamate, glycerol, glycine, lactate, malate, phenylalanine, propionate, proline, *n*-propanol, isopro-

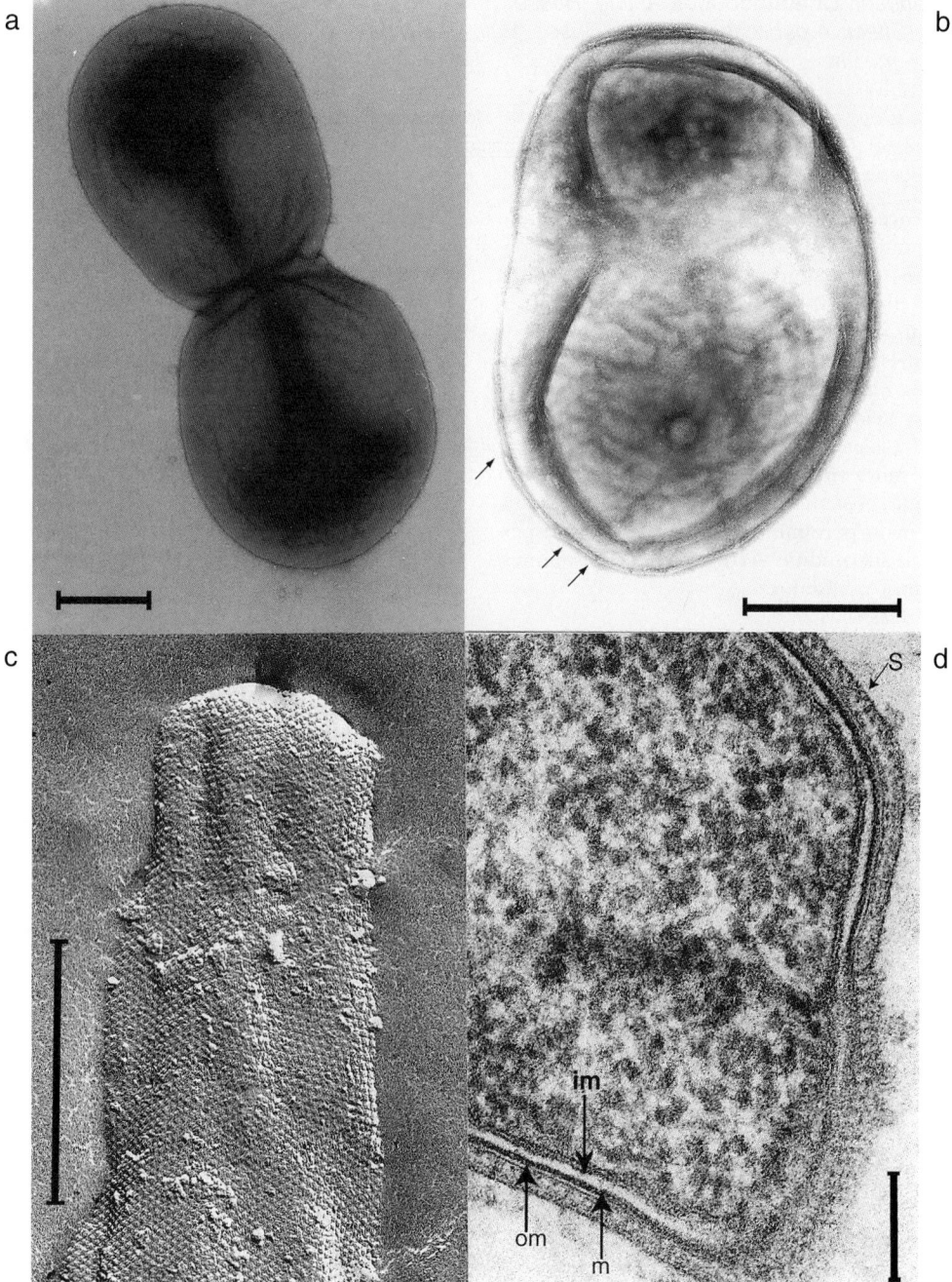

FIGURE BXII.α.169. Morphology of *Chelatococcus asaccharovorans* strains. (*a*), negatively stained cell of isolate TE 1 from the exponential growth phase, bar = 0.5 μm; (*b*), negatively stained cell of isolate TE 2 demonstrating the proteinaceous surface layer; the cracks in the S-layer suggest that it is brittle in nature (*arrows*), bar = 0.5 μm; (*c*), surface layer of strain TE 2 as revealed by deep-etching, note the paracrystalline arrangement of subunits, with a center-to-center distance of 16 nm, and the patchy appearance of the surface layer, bar = 0.5 μm; (*d*), higher magnification of thin-sectioned cell of isolate TE 1 demonstrating the fine structure of the cell wall with inner membrane (*im*), murein layer (*m*), outer membrane (*om*), and proteinaceous surface layer (*s*), bar = 0.1 μm.

panol, pyruvate, succinate, and serine. Substrates not utilized for growth include acetamide, adonitol, aniline, arabinose, L-arabitol, arginine, benzoate, *n*-decane, butyrate, D-(+)-cellobiose, cellulose, erythritol, dimethylformamide, dimethylsulfoxide, esculin, formate, fucose, D-(−)-fructose, galactose, gentiobiose, D-(+)-glucose, glycollate, glycyl-glycine, H_2/CO_2, inositol, lactose, lysine, mannitol, D-(+)-mannose, malate, D-(+)-maltose, methane, methanol, methylacetamide, methyldiethylamine, *N*-(2-acetamido)iminodiacetate, *N*-acetylglucosamine, oxalate, phenol, D-(+)-raffinose, rhamnose, D-(+)-ribose, D-(−)-sorbitol, sucrose, tetraethylammoniumchloride, triethanolamine, tris(hydroxymethyl)aminomethane, urea, xylitol, xylene, and D-(+)-xylose.

Other properties of *Chelatococcus asaccharovorans* strains include the following: no production of indole, H_2S, or acetoin (VP negative); no fermentation of glucose; no hydrolysis of starch or DNA and only weak hydrolysis of protein (skim milk); no dinitrogen fixation; no denitrification with NTA, glucose, or acetate; no acid production from glucose or ethanol; not acid fast; no growth on tellurite agar plates; no growth on mannitol (*Rhizobium*) agar; no gas from glucose; weak growth on gluconate. Growth occurs in the presence of 1% NaCl but not with 10% NaCl.

Metabolism and metabolic pathways A key nutritional feature of members of *Chelatococcus* is their ability to grow with the chelating agent NTA as the only source of carbon, energy, and nitrogen. (For a short overview on the use and importance of aminopolycarboxylates as metal-chelating agents, readers are referred to the chapter on *Chelatobacter*.) The current knowledge is summarized in Fig. BXII.α.133 in the chapter on the genus *Chelatobacter*. The data strongly support an identical catabolic route in both genera, i.e., initially, transport of NTA into the cell, a step which is most probably energy-dependent, and subsequently two enzymatic steps, catalyzed by a soluble NTA-monooxygenase complex and a membrane-bound dehydrogenase. These enzymes are sufficient to degrade NTA to the central metabolites glycine and glyoxylate.

NTA monooxygenase The first intracellular step is catalyzed by a monooxygenase (NTA-Mo). NTA-Mo was originally isolated from *Chelatobacter heintzii* ATCC 29600 as a two-component system consisting of component A (cA) and a FMN-containing protein component B (cB), with both components being present as dimers (Uetz et al., 1992). Using antibodies raised against cA isolated from *Chelatobacter heintzii* ATCC 29600, the presence of cA was confirmed in *Chelatococcus asaccharovorans* strains, whereas no cross reaction was observed in cell-free extracts of *Chelatococcus* strains with antibodies raised against the flavin-containing cB (Uetz et al., 1992). Later, cA purified from *Chelatococcus asaccharovorans* TE 2 was found to be virtually identical to the protein present in *Chelatobacter heintzii* ATCC 29600 (Uetz, 1992), i.e., only 1 out of the first 17 amino acids of cA was different (in position 3 Asn instead of Asp was present in cA purified from *Chelatococcus*). A hybrid enzyme composed of cA purified from *Chelatococcus* and cB from *Chelatobacter heintzii* ATCC 29600 exhibits the typical NTA-Mo activity known from strain ATCC 29600. According to Witschel et al. (1997), cA of *Chelatococcus* acts as a monooxygenase supplied with $FMNH_2$ by an independent FMN oxidoreductase. Recent data suggest that any FMN oxidoreductase can supply reduced FMN to the NTA-Mo (Witschel et al., 1997; Xu et al., 1997). Although not yet investigated in detail, the substrate specificity of NTA-Mo from *Chelatococcus* is probably as narrow as that found for NTA-Mo isolated from *Chelatobacter*, with NTA being the only substrate presently known (Uetz, 1992).

Iminodiacetate dehydrogenase A membrane-bound iminodiacetate dehydrogenase (IDA-DH) catalyses the second enzymatic step during biodegradation of NTA in *Chelatococcus asaccharovorans* strain TE 2, as also occurs in members of *Chelatobacter* (Uetz and Egli, 1993).

Regulation of NTA-degrading enzymes The enzyme system for NTA utilization is inducible (Uetz, 1992) but no detailed information is presently available. Nevertheless, other carbon sources do not seem to repress the synthesis of NTA-utilizing enzymes, as growth in batch culture with mixtures of NTA plus acetate results in the simultaneous utilization of the two substrates concomitant with an enhanced growth rate (Egli et al., 1988; see also Table BXII.α.115 in the chapter describing the genus *Chelatobacter*.).

Genetics Hybridization experiments with different gene probes from *Chelatobacter heintzii* ATCC 29600 strongly indicate that both the genes encoding *ntaA* and the ORF1—which probably encodes a regulatory protein—are present in *Chelatococcus asaccharovorans* strain TE 2 in a similar arrangement as that reported for the *Chelatobacter* strain (Knobel, 1997). This supports the data reported by Uetz (1992) on the N-terminal sequence of NTA-Mo purified from *Chelatococcus* strain TE 2. In contrast, only weak hybridization was found with probes directed towards *ntaB*, confirming the conclusion drawn by Uetz (1992), based on immunological evidence, that quite a different FMN oxidoreductase is supplying $FMNH_2$ to NTA-Mo in this strain.

Antigenic structure Polyclonal antibodies have been raised against whole cells (Wilberg et al., 1993) or isolated cell walls (Bally, 1994) of NTA-utilizing *Chelatobacter* and *Chelatococcus* strains. In Ouchterlony double diffusion or indirect immunofluorescence tests, the antisera raised against the two central NTA-utilizing *Chelatobacter* strains, i.e., *Chelatobacter heintzii* strains TE 6 and ATCC 29600, did not crossreact with cell homogenates from either *Chelatococcus asaccharovorans* strain. Similarly, the antiserum raised against *Chelatococcus asaccharovorans* strain TE 2 did not crossreact with homogenates of the two central strains of *Chelatobacter heintzii* (Wilberg et al., 1993). Crossreaction of *Chelatococcus* antibodies was also tested with a variety of bacteria that were expected to co-exist with *Chelatococcus* in the same habitat, and no crossreaction was found with any of the strains tested (for the strains tested, see the section on antigenic structure in the chapter on the genus *Chelatobacter*).

Antibiotics and drug resistance The susceptibility of *Chelatococcus asaccharovorans* strains TE 1 and TE 2 to a range of antibiotics was tested (Auling et al., 1993a; see Table BXII.α.117 in the chapter on the genus *Chelatobacter*. Both *Chelatococcus asaccharovorans* strains were resistant to cephaloridine, cephalexin, nalidixic acid, vancomycin, and trimethoprim-sulfamethoxazole, but sensitive to penicillin G, carbenicillin, most of the aminoglycosides tested (neomycin, kanamycin, gentamicin, tobramycine, except for streptomycin), and to tetracycline, doxycycline, erythromycin, rifampicin, polymyxin B, and novobiocin. The resistance of *Chelatococcus* strains to nalidixic acid, in combination with their resistance to vancomycin, might be used to isolate such strains with a medium containing NTA as the sole source of carbon and energy, as these two antibiotics would suppress growth of the faster-growing *Chelatobacter* strains.

Ecology Data reported so far indicate that cells of *Chelatococcus* are ubiquitously distributed in the aquatic environment, and that their number increases with both increasing eutrophication and temperature. Cells crossreacting with antibodies raised against *Chelatococcus asaccharovorans* TE 1 (α-*Cc*) have been detected in surface waters and wastewater (Wilberg et al., 1993; Bally, 1994). In surface waters, their numbers were in the order of 1×10^3 to 5×10^5 cells/l (0.03–3‰ of the total bacterial population). Cells of *Chelatococcus* have even been detected in mountain streams with no detectable NTA present and little chance of having been ever exposed to NTA; only in a cold pristine alpine stream were cells crossreacting with α-*Cc* not observed (Bally, 1994). Generally, the number of crossreacting cells increased by several orders of magnitude in nutrient-rich environments such as polluted rivers or wastewater treatment plants. For example, in aerated reactors of two Swiss wastewater treatment plants (with NTA concentrations between 10 and 100 μg/l) numbers of α-Cc-crossreacting cells up to 2×10^{11} per liter were observed (in the range of one to a few percent of the total bacterial counts); many of the cells exhibited the typical diplococcoid shape (Bally, 1994). In this environment, members of *Chelatococcus* seem as numerous as strains of *Chelatobacter* or even more so (although these numbers have to be treated with caution considering possible crossreactivity of the antibodies used).

Polyamines and ubiquinones The two *Chelatococcus asaccharovorans* strains studied contained *sym*-homospermidine as the main polyamine compound, putrescine and spermidine as major components, and spermine as a minor component (Auling et al., 1993a). Ubiquinone Q-10 was present in both strains of *Chelatococcus asaccharovorans*.

Soluble protein patterns The soluble protein patterns of *Chelatococcus asaccharovorans* strains have been compared to those of *Chelatobacter heintzii* strains (Auling et al., 1993a). *Chelatobacter* and *Chelatococcus* strains exhibited clearly different patterns, whereas the soluble protein pattern of the two *Chelatococcus* strains tested were virtually identical.

ENRICHMENT AND ISOLATION PROCEDURES

Only two strains of *Chelatococcus* have been isolated to date. Both were isolated with a batch enrichment procedure using NTA as the only source of carbon, energy, and nitrogen (Egli, 1988). However, they were not as easily enriched as members of the genus *Chelatobacter*. The fact that the optimal growth temperature of *Chelatococcus* strains (35–37°C) is considerably higher than that of *Chelatobacter* strains, together with their resistance to vancomycin, might be a way to preferentially enrich members of this genus when using NTA as the only source of carbon and nitrogen for growth.

MAINTENANCE PROCEDURES

Strains can be maintained indefinitely in a freeze-dried condition. Alternatively, cultures grown in SM with NTA and amended with either glycerol (15%, v/v) or DMSO (50%, v/v) can be stored in liquid nitrogen. With either method, revival of cultures may take a while. Best results have been obtained by streaking cultures from liquid nitrogen directly onto SM agar plates containing NTA as the only carbon and nitrogen source. When maintaining cultures on selective NTA agar plates, it is recommended to transfer them to new plates at least every week, because growth leads to an increased pH in and around the colonies. Cells stored under such conditions exhibit long lag phases (up to several weeks) before they restart growing. Cultures grown in liquid media behave similarly. To avoid long lag times and erratic growth responses, it is best to transfer the cells to fresh medium when they are still growing exponentially. Excessive concentrations of NTA, combined with the low buffering capacity of the medium or lack of pH control, might lead to an early cessation of growth.

PROCEDURES FOR TESTING SPECIAL CHARACTERS

Growth with NTA as the only source of carbon, energy, and nitrogen is a key property of *Chelatococcus*. Consumption of NTA can be measured either by the disappearance of dissolved organic carbon (DOC) paralleled by the excretion of ammonia, or via the consumption of NTA. A HPLC method specially developed for the analysis of NTA in culture media and cell extracts has been described by Schneider et al. (1988).

DIFFERENTIATION OF THE GENUS *CHELATOCOCCUS* FROM OTHER GENERA

Differentiation of *Chelatococcus* from the closely related genospecies *Methylobacterium organophilum* and *Bradyrhizobium japonicum* (see taxonomic comments) can be easily done, based on the ability to grow with NTA as the sole carbon, nitrogen, and energy source, and the coccobacillary shape of the cells, their nonmotility, and the inability to grow on sugars (Green, 1992; Auling et al., 1993a, or see *Methylobacterium* in this *Manual*).

TAXONOMIC COMMENTS

The presence of ubiquinone Q-10 (Egli et al., 1988) and the characteristic polyamine pattern, with *sym*-homospermidine as the main polyamine and putrescine plus spermidine as major polyamines (Auling et al., 1993a), clearly allocate the members of the genus *Chelatococcus* to the *Alphaproteobacteria*. Sequencing of a 150-nucleotide 16S rRNA gene fragment (position 1220–1377) revealed nucleotides characteristic of the *Alphaproteobacteria* at specific positions, but a difference of 13 nucleotides between *Chelatococcus* and *Chelatobacter* indicated only a low degree of relationship between these genera (Auling et al., 1993a). This conclusion is supported by the absence of serological crossreaction between both genera of NTA-utilizing bacteria (Wilberg et al., 1993). Recently, analysis of complete 16S rRNA gene sequences revealed that *Methylobacterium organophilum* and *Bradyrhizobium japonicum* are more closely related to *Chelatococcus asaccharovorans* than to *Rhodopseudomonas acidophila* (C. Strömpl, personal communication). The next relative in this group for which 16S rRNA gene sequence data are available is *Beijerinckia indica*, with 94.5% sequence homology (C. Strömpl, personal communication).

A comparison of the complete 16S rDNA sequence of the type strain *Chelatococcus asaccharovorans* strain TE 2 with a partial sequence of strain TE 1 has so far shown identity of the two 16S rRNA genes (C. Strömpl, personal communication).

ACKNOWLEDGMENTS

We are indebted to Ernst Wehrli for supplying us with excellent electron micrographs. Also we thank C. Strömpl for supplying us with unpublished information on the phylogenetic position of *Chelatococcus asaccharovorans*. T.E. would like to thank H.U. Weilenmann and the students that have contributed to the study of NTA-utilizing microorganisms. Furthermore, the generous financial support of research on NTA in the laboratory of T.E. by grants from the Swiss National Science Foundation, Lever AG Switzerland and Unilever Port Sunlight, the Research Commission of ETH Zürich, and by EAWAG is gratefully acknowledged.

List of species of the genus Chelatococcus

1. **Chelatococcus asaccharovorans** Auling, Busse, Egli, El-Banna and Stackebrandt 1993b, 624[VP] (Effective publication: Auling, Busse, Egli, El-Banna and Stackebrandt 1993a, 110.)

   a.sac.cha.ro.vo′ rans. Gr. pref. *a* not; Gr. n. *sacchar* sugar; L. v. *voro* to devour; M.L. part. adj. *asaccharovorans* not devouring sugars.

The characteristics are as described for the genus. To date, only two strains (TE1, DSMZ 6461; TE2, DSMZ 6462) represent the species.

The mol% G + C of the DNA is: 63 (T_m).

Type strain: TE 2, DSM 6462.

GenBank accession number (16S rRNA): AJ294349.

Genus III. **Methylocella** *Dedysh, Liesack, Khmelenina, Suzina, Trotsenko, Semrau, Bares, Panikov and Tiedje 2000, 967[VP]*

THE EDITORIAL BOARD

Me.thyl.o.cel′la. M.L. n. *methyl* the methyl group; L. n. *cella* a cell; M.L. n. *Methylocella* methyl-using cell.

Encapsulated, nonmotile straight or curved Gram-negative rods (0.6–1.0 × 1.0–2.5 μm). One large **poly-β-hydroxybutyrate granule** at each pole. **Form exospores.** Contain **intracytoplasmic membranes. Grow only on C₁ compounds using serine pathway.** Complete TCA cycle. **Fixes N₂.** Major fatty acids are $C_{18:1}$ forms.

The mol% G + C of the DNA is: 61.2.

Type species: **Methylocella palustris** Dedysh, Liesack, Khmelenina, Suzina, Trotsenko, Semrau, Bares, Panikov and Tiedje 2000, 967.

FURTHER DESCRIPTIVE INFORMATION

A single species, *Methylocella palustris*, has been described; three independent isolates have been examined in detail. *Methylocella palustris* inhabits acidic peat bogs in northern Russia and Siberia (Dedysh et al., 1998a, b). Both the temperature range (10–20°C) and the pH range (4.5–7.0) for growth, as well as a requirement for dilute environments, reflect adaptation to this habitat (Dedysh et al., 2000).

Methylocella palustris is able to grow on methane and methanol as sole source of energy and carbon; the serine pathway is the route of carbon fixation. The organisms are able to fix N₂ in microaerobic environments and are able to use nitrate, ammonium ion, and organic nitrogen (Dedysh et al., 2000).

Intracytoplasmic membranes consist of spherical vesicles formed from the cytoplasmic membrane and located next to it. These membranes do not resemble either the vesicular discs found in type I methanotrophs or the laminar membranes found in type II methanotrophs (Dedysh et al., 2000).

Analysis of 16S rDNA sequences showed that *Methylocella palustris* is most closely related to *Beijerinckia indica* and *Rhodopseudomonas acidophila*; it is related to but clearly different from other type II methylotrophs (*Methylocystis* and *Methylosinus* spp.) in the *Alphaproteobacteria* (Dedysh et al., 1998a, 2000). Attempts to PCR-amplify portions of the *pmoA* gene from *Methylocella palustris* isolates using two sets of primers that yielded specific products from *Methylococcus capsulatus* (type X methylotroph), *Methylosinus trichosporium* (type II methylotroph), and *Methylomicrobium album* (type I methylotroph) were unsuccessful. Southern hybridization of *Methylocella palustris* genomic DNA digests with a probe specific for the *Methylococcus capsulatus pmoA* gene produced only weak hybridization at low stringency, further indicating substantial genetic distance between *Methylocella palustris* and other methylotrophs in the *Alphaproteobacteria* (Dedysh et al., 1998a, 2000). An analysis of the fatty acid profiles of methylotrophs of all known metabolic types of methylotrophs (I, II, and X) also showed that *Methylocella palustris* was distinct from all other strains examined (Fang et al., 2000).

ENRICHMENT AND ISOLATION PROCEDURES

Enrichment and isolation procedures have been described by Dedysh et al. (1998a, b). Enrichment required multiple passages and was achieved in dilute liquid mineral medium under a 50% methane:air atmosphere with or without KNO₃ as a nitrogen source.

MAINTENANCE PROCEDURES

Strains can be maintained by subculturing monthly (Dedysh et al., 1998a, b).

DIFFERENTIATION OF THE GENUS *METHYLOCELLA* FROM OTHER GENERA

Dedysh et al. (2001) have developed FISH (fluorescence *in situ* hybridization) probes based on 16S rDNA sequences that can be used to distinguish *Methylocella palustris* from other methylotrophs in enrichment cultures and environmental samples.

List of species of the genus Methylocella

1. **Methylocella palustris** Dedysh, Liesack, Khmelenina, Suzina, Trotsenko, Semrau, Bares, Panikov and Tiedje 2000, 967[VP]

   pa.lus′tris. L. n. *palus* a bog; M.L. adj. *palustris* bog-inhabiting.

   Description as for the genus with the following additional characteristics. Utilize methane and methanol. Growth optima at pH 5.5 and 20°C. Inhibited by 0.5% NaCl.

   The mol% G + C of the DNA is: 61.2 (T_m).

   Type strain: K, ATCC 700799.

   GenBank accession number (16S rRNA): Y17144.

Family VII. **Bradyrhizobiaceae** *fam. nov.*

GEORGE M. GARRITY, JULIA A. BELL AND TIMOTHY LILBURN

Bra.dy.rhi.zo.bi.a' ce.ae. M.L. neut. n. *Bradyrhizobium* type genus of the family; *-aceae* ending to denote family; M.L. fem. pl. n. *Bradyrhizobiaceae* the *Bradyrhizobium* family.

The family *Bradyrhizobiaceae* was circumscribed for this volume on the basis of phylogenetic analysis of 16S rRNA sequences; the family contains the genera *Bradyrhizobium* (type genus), *Afipia*, *Agromonas*, *Blastobacter*, *Bosea*, *Nitrobacter*, *Oligotropha*, *Rhodoblastus*, and *Rhodopseudomonas*.

The family is phenotypically, metabolically, and ecologically diverse. It includes organisms that fix N_2, photosynthetic organisms, organisms capable of aerobic and/or anaerobic respiration, and human pathogens.

Type genus: **Bradyrhizobium** Jordan 1982, 137.

Genus I. **Bradyrhizobium** *Jordan 1982, 137*[VP]

L. DAVID KUYKENDALL

Bra.dy.rhi.zo' bi.um. Gr. adj. *bradus* slow; M.L. neut. n. *Rhizobium* a bacterial generic name; M.L. neut. n. *Bradyrhizobium* the slow (growing) rhizobium.

Rods $0.5–0.9 \times 1.2–3.0$ μm. **Commonly pleomorphic under adverse growth conditions.** Usually contain granules of poly-β-hydroxybutyrate that are refractile by phase-contrast microscopy. **Nonsporeforming. Gram negative. Motile by one polar or subpolar flagellum.** Fimbriae have not been described. **Aerobic, possessing a respiratory type of metabolism with oxygen as the terminal electron acceptor. Optimal temperature 25–30°C.** Optimal pH, 6–7, although lower optima may be exhibited by strains from acid soils. Colonies are circular, opaque, rarely translucent, white, and convex, and tend to be granular in texture; they do not exceed 1.0 mm in diameter in less than 5–6 days incubation on A1EG medium. **Turbidity develops only after 3–4 days in agitated broth. Generation times are 9–18 h.** Chemoorganotrophic, utilizing a range of carbohydrates and salts of organic acids as carbon sources, without gas formation; arabinose and other pentoses are preferred carbon sources. Cellulose and starch are not utilized. Produce an alkaline reaction in mineral salts medium containing mannitol and/or many other carbohydrates. **Growth on carbohydrate media is usually accompanied by extracellular polysaccharide slime production particularly with glycerol, gluconate, or mannitol.** Some strains can grow chemolithotrophically in the presence of H_2, CO_2, and low levels of O_2. Ammonium salts, usually nitrates, and some amino acids, can serve as nitrogen sources. Peptone is poorly utilized (except for strains isolated from *Lotononis*). Casein and agar are not hydrolyzed. There is usually no requirement for vitamins with the rare exception of biotin, which also may be inhibitory to some strains. 3-Ketoglycosides are not produced (Bernaerts and De Ley, 1963). **The organisms are characteristically able to enter the root hairs of tropical-zone and some temperate-zone leguminous plants (family Leguminosae) and incite the production of root nodules, in which the bacteria occur as intracellular nitrogen-fixing symbionts.** Some strains, especially *B. elkanii*, fix nitrogen in the free-living state when examined under special conditions.

The mol% G + C of the DNA is: 61–65.

Type species: **Bradyrhizobium japonicum** (Kirchner 1896) Jordan 1982, 137 (*"Rhizobacterium japonicum"* Kirchner 1896, 221; *Rhizobium japonicum* (Kirchner 1896) Buchanan 1926, 90.)

FURTHER DESCRIPTIVE INFORMATION

In young cultures, cells are short rods but in older cultures or under adverse growth conditions, such as low concentrations of calcium or magnesium, cells are commonly pleomorphic (swollen and elongated). Older cells stain to give a banded appearance

because of large accumulations of poly-β-hydroxybutyrate. Within the root nodules the bacteroids are rod shaped and slightly swollen but not branched or highly distorted, and they contain polyphosphate inclusions as well as poly-β-hydroxybutyrate accumulations.

All strains produce large amounts of water-soluble extracellular polysaccharide. The main constituent is acidic heteropolysaccharide (80–90%); the remainder is neutral β-1,2-, β-1-3-, and α-1,2-glucans (Breedveld and Miller, 1994). About six different types of acidic heteropolysaccharides have been noted in an examination of 30 or more strains of *Bradyrhizobium* and possess three distinctive features: (a) heterogeneity in composition and structure, (b) the frequent presence of methylated sugars (Dudman, 1976, 1978; Kennedy and Bailey, 1976), and (c) the presence of D-galacturonic acid (Dudman, 1976) in some strains.

Young cells are motile with one polar or subpolar flagellum. Growth on carbohydrate-containing solid media is clear, translucent, or opaque white with a tendency to become tan on prolonged incubation. Colonies of bacteria from *Lotononis bainesii* are pink to deep red but those from *Lotononis angolensis* are colorless.

Most strains grow on a chemically defined minimal medium with arabinose as a carbon source and NH_4Cl as the sole source of nitrogen (Kuykendall, 1987). All strains grow on a buffered mineral salts medium containing yeast extract and either glucose, galactose, gluconate, glycerol, fructose, or arabinose. Disaccharides such as lactose and sucrose are usually not used. Colony-type derivatives differing in symbiotic nitrogen fixation ability and carbon source utilization occur (Kuykendall and Elkan, 1976) (Fig. BXII.α.170).

As with *Rhizobium*, *Bradyrhizobium* strains lack the ability to absorb Congo red from a yeast extract–mannitol–mineral salts medium containing this dye (0.0025% final concentration). The colonies are colorless white or very faintly pink colonies, whereas most other soil bacteria produce colonies that are red on this differential medium.

The optimal temperature for most strains is 25–30°C. The maximum growth temperature ranges from 33–35°C, with many strains failing to grow above 34°C. Usually acid-tolerant with most strains growing at pH 4.5. Over 30% of the strains will grow at pH 4.0 and a few as low as pH 3.5. Growth usually does not occur above pH 9.0.

An alkaline reaction is produced in litmus milk without the production of a clear, upper serum zone.

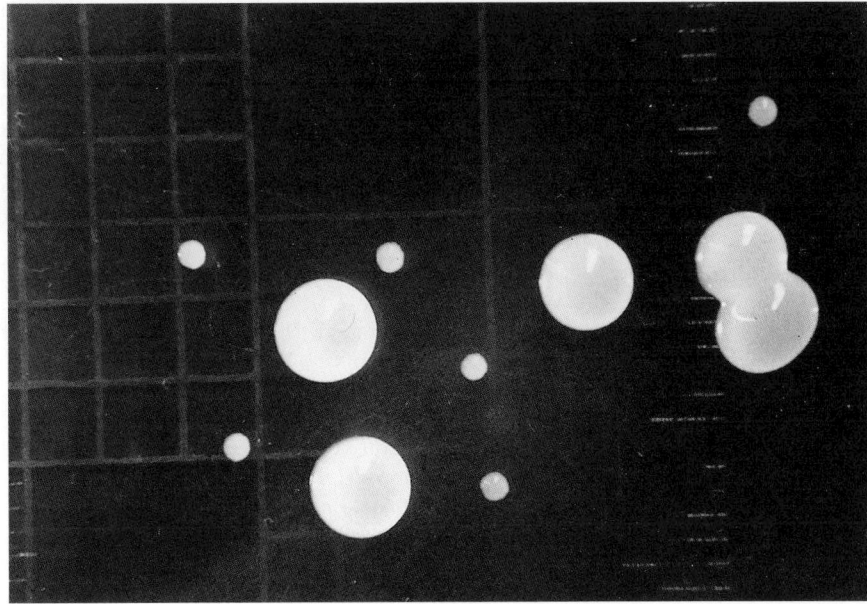

FIGURE BXII.α.170. *Bradyrhizobium japonicum* strain USDA 110 derivatives I-110 (smaller colonies) and L1-110 (larger colonies) after 14 d incubation at 30°C grown on HM salts agar (Cole and Elkan, 1973) with yeast extract (0.25 g/l), L-arabinose (0.5 g/l), and D-mannitol (5.0 g/l).

Bradyrhizobium strains fail to grow in media containing 2% NaCl, do not produce H_2S, and do not form a precipitate in calcium glycerophosphate medium (Hofer, 1941). Penicillinase production is common.

The Entner–Doudoroff pathway is employed in carbohydrate degradation (Martínez-De Drets and Arias, 1972; Mulongoy and Elkan, 1977) with the simultaneous operation of the Embden–Meyerhof–Parnas pathway (Mulongoy and Elkan, 1977). Low levels of fructose-1,6-bisphosphate aldolase are present, however. $NADP^+$-dependent 6-phosphogluconate dehydrogenase is not present but an NAD-linked 6-phosphogluconate dehydrogenase occurs, suggesting operation of a new pathway. The tricarboxylic acid cycle operates in both free-living cells (Keele et al., 1969; Mulongoy and Elkan, 1977) and in symbiotic bacteroids from root nodules (Stovall and Cole, 1978). A pathway for the direct oxidation of gluconate by the tricarboxylic acid cycle, via 2-keto- and 2,5-diketogluconate and α-ketoglutarate, has been described by Keele et al. (1970). Many differences in carbon metabolism between fast-growing *Rhizobium* and slow-growing *Bradyrhizobium* were described by Elkan and Kuykendall (1982); they seem especially significant since these genera are now in separate families.

The ability to take up H_2 was shown by the laboratory of Dr. Harold J. Evans to be a key determining factor in the efficiency of symbiotic nitrogen fixation, because without this capability the nodules evolve large quantities of H_2 as a byproduct of the nitrogenase enzyme action. Some strains do not evolve any H_2; such strains possess an active uptake hydrogenase, which also enables them to grow chemolithotrophically in an atmosphere of $H_2/CO_2/O_2$ (10:15:1) (the balance being N_2). Under such conditions, ribulose bisphosphate carboxylase is primarily responsible for the fixation of CO_2 (Hanus et al., 1979; Lepo et al., 1980). Mutants unable to grow chemolithotrophically fall into several classes, including those impaired in H_2 uptake and those deficient in CO_2 uptake (Maier, 1981).

Nitrogenase activity by free living cells occurs in certain strains but only in media containing selected carbon sources and under a low level of oxygen (Keister, 1975; Kurtz and LaRue, 1975; McComb et al., 1975; Pagan et al., 1975).

Auxotrophic mutants (Kuykendall, 1981) were characterized by Wells and Kuykendall (1983) and Kummer and Kuykendall (1989). Most tryptophan-requiring mutants do not nodulate soybeans but those that specifically lack only tryptophan synthetase do form nodules. An indole glycerol phosphate-requiring mutant, strain TA-11, was discovered to produce phenotypic revertants having the capacity to quantitatively form more nodules (Hunter and Kuykendall, 1990; Kuykendall and Hunter, 1991).

Host specificity, nitrogen fixation (*nif*) and nodulation (*nod*) genes have been analyzed (Kaluza and Hennecke, 1984; Lamb and Hennecke, 1986; Nieuwkoop et al., 1987; Banfalvi et al., 1988; Gottfert et al., 1990a, b, 1992; Sadowsky et al., 1991; Wang and Stacey, 1991; Barbour et al., 1992; Dockendorff et al., 1994; and Stacey et al., 1994). Unlike the fast-growing *Rhizobiaceae*, *Bradyrhizobium* strains have only one rRNA operon (Kundig et al., 1995). A correlated physical and genetic map is available and all *nif* and *nod* genes are contained in a large but discrete region of the 8700 kb chromosome (Kundig et al., 1993). Kaneko et al. (2002) recently reported the complete polynucleotide sequence of *B. japonicum* USDA 110. Much is known about the function, organization and sequence of symbiosis-controlling genes (see reviews by Schultze et al., 1994; Stacey et al., 1995; van Rhijn and Vanderleyden, 1995). Kaluza et al. (1985) demonstrated the presence of a repeated DNA sequence, RSα, clustered around the symbiosis-controlling region of the genome, and Minamisawa et al. (1998) recently described some new strains with an unusually high RSα copy number.

Bacteriophages active against *Bradyrhizobium* have been isolated from soil (Hashem et al., 1986). Bacteriocins have been reported (Roslycky, 1967).

Bradyrhizobium strains are often resistant to a number of antibiotics (tetracyclines, streptomycin, penicillin G, viomycin, vancomycin), but are said to be more sensitive to growth inhibitors such as D-alanine or ethidium bromide (Jordan, 1984c).

Bradyrhizobium strains cause nodule production on *Glycine* (soybean), *Vigna* (cowpea), *Macroptilium* (siratro), certain species of *Lotus*, and a wide variety of leguminous plants that are also nodulated by *Mesorhizobium loti*, such as *Acacia*. In fact there appear to be three distinct groups of *Acacia* species: one group nodulated by *M. loti*, one group nodulated by *Bradyrhizobium* sp. (*Acacia*) and one group nodulated by both of these species (Dreyfus and Dommergues, 1981). As with *Rhizobium*, *Bradyrhizobium* strains produce specific Nod Factors thought to control host specificity (Dénarié et al., 1992; Carlson et al., 1993; Stacey et al., 1995).

A highly specific nodulation of the nonleguminous plant *Parasponia* (*Trema*) is caused by a strain of *Bradyrhizobium* (Trinick, 1973, 1976). This is the only completely validated instance of plant nodulation that occurs outside of the Leguminosae.

ENRICHMENT AND ISOLATION PROCEDURES

Isolation from root nodules and soil is by the techniques described for the genus *Rhizobium* of the *Rhizobiaceae*.

MAINTENANCE PROCEDURES

A suitable maintenance medium is A1EG (Kuykendall, 1987) or arabinose enriched (0.1% yeast extract) gluconate, as given by Kuykendall et al. (1988). Storage recommendations are the same as given for *Rhizobium*, except *Bradyrhizobium* persists on agar slants in tightly closed screw-capped tubes for more than five years at 4°C.

PROCEDURES FOR TESTING SPECIAL CHARACTERS

For antibiotic resistance determinations, Petri dishes containing about 25 ml of A1EG agar medium containing a specific concentration of antibiotic are streaked for isolation with about 10 µl of a fresh 3–5 day-old broth culture. Growth of isolated colonies is determined visually after 10 d incubation. Resistance to a particular concentration is defined as the ability of a strain to form colonies at that concentration.

DIFFERENTIATION OF THE GENUS *BRADYRHIZOBIUM* FROM OTHER GENERA

The ability to fix nitrogen symbiotically in association with legumes is the primary characteristic that distinguishes *Bradyrhizobium* from other closely related genera in the new family *Bradyrhizobiaceae*. It is easily differentiated from the not so closely-related, relatively fast-growing, legume-nodulating microsymbionts in *Mesorhizobium*, *Rhizobium*, and *Azorhizobium* (Table BXII.α.143).

TAXONOMIC COMMENTS

In the first edition of *Bergey's Manual of Systematic Bacteriology*, *Bradyrhizobium* was still in the family *Rhizobiaceae*. The genus *Rhizobium* had previously been subdivided into two groups, distin-

guished by their growth rate in yeast extract–mannitol–mineral salts medium, flagellar arrangement, DNA base composition and the genera of host plants nodulated (Jordan and Allen, 1974). Jordan (1982) transferred the slow-growing root nodule bacteria into a new genus, *Bradyrhizobium*, separate from the fast growing, acid-producing, nodulating bacteria, which subsequently were placed in genera of their own. This revision resulted from numerous studies involving numerical taxonomy (Graham, 1964; Moffett and Colwell, 1968), DNA base ratios (Wagenbreth, 1961; De Ley and Rassel, 1965), nucleic acid hybridization (Heberlein et al., 1967; Gibbins and Gregory, 1972), cistron similarities (De Smedt and De Ley, 1977), serology (Graham, 1963; Vincent and Humphrey, 1970; Humphrey et al., 1973; Vincent, 1977), composition of extracellular gum (Dudman, 1976; Kennedy, 1976; Kennedy and Bailey, 1976; Dudman, 1978), carbohydrate metabolism (Martínez-De Drets and Arias, 1972), bacteriophage sensitivity (Napoli et al., 1980), antibiotic sensitivity (Strzelcowa, 1968), protein composition (Roberts et al., 1980), and type of bacteroid inclusion bodies (Craig et al., 1973). Based on 16S rRNA sequences analyses, *Bradyrhizobium* is now grouped with phylogenetically related organisms in a new separate family, the *Bradyrhizobiaceae*.

Three species of *Bradyrhizobium* are presently recognized; at least 16 genomospecies have also been isolated (Lafay and Burdon, 1998).

In the first edition of *Bergey's Manual of Systematic Bacteriology*, the genus *Bradyrhizobium* had only one recognized species, *B. japonicum*. However, DNA–DNA hybridization studies by Hollis et al. (1981) had earlier suggested that a subset of strains in *B. japonicum* could perhaps be separated into a distinct species, and that two DNA homology groups of bona fide *B. japonicum* existed as well. *B. elkanii* (Kuykendall et al., 1992, 1993b; Devine and Kuykendall, 1996) was named to recognize a species of *Bradyrhizobium* bacteria clearly distinct from *B. japonicum*. Extra-slow-growing strains from alkaline soils were then named *Bradyrhizobium liaoningense* Xu et al. (1995).

The taxonomic position of nodule bacteria from the pasture legume *Lotononis* is uncertain (Norris, 1958). Isolates from *L. bainesii* are red because of an intracellular red carotenoid pigment, although isolates from *L. angolensis* are nonpigmented. As with other strains of *Bradyrhizobium*, these strains are monotrichous, grow slowly, and produce an alkaline reaction and no serum zone in litmus milk; however, cultured cells appear as enlarged, banded ovoids. Resistance of these strains to ultraviolet light is greater than that exhibited by other strains of *Bradyrhizobium* or of *Rhizobium* (Law, 1979). They also utilize peptone, are not recognized by antisera that detect other slow growers or species of *Rhizobium*, and the mol% G + C of the DNA is 68–69 (T_m) (Godfrey, 1972). Nodulation by these strains is extremely specialized. Nodulation by *Lotononis* spp. and *Macroptilium atropurpureum* is effective, whereas by selected species of *Aeschynomene* and *Crotolaria* is ineffective.

Comments concerning *Sinorhizobium fredii* (Scholla and Elkan, 1984; Chen et al., 1988b) are presented under Taxonomic Com-

TABLE BXII.α.143. Phenotypic differentiation of *Bradyrhizobium* from other genera of nodule-forming, nitrogen-fixing legume microsymbionts

Characteristic	*Bradyrhizobium*	*Azorhizobium*	*Mesorhizobium*
Grows *in vitro* on fixed N	−	+	−
Fixes nitrogen *ex planta*	+	+	−
Slow doubling times, >8h	+	−	−

ments for the genus *Rhizobium*. These fast growing root-nodule bacteria are capable of producing effective nodules on *Glycine soja* and *G. max* cv Peking, but ineffective nodules on commercial lines of soybean.

ACKNOWLEDGMENTS

The author appreciates Dr. Bob Davis, Research Leader of the Molecular Plant Pathology Laboratory, for support and encouragement, thanks Dr. Thomas E. Devine whose collaboration and friendship made a real difference, and was indebted to Professors Gerald Elkan and Hauke Hennecke who were mentors. Drs. Fawzy Hashem, Babita Saxena, Jim Hunter, Greg Upchurch, Mulongoy Kalemani, Gail Hollowell, Robin Kummer, Margaret Roy, Ketan Shah, Matthias Hahn, Bill Gillette, Susan Wells, and Tom Wacek all helped investigate the properties of *Bradyrhizobium*.

DIFFERENTIATION OF THE SPECIES OF THE GENUS *BRADYRHIZOBIUM*

While *B. elkanii* and *B. japonicum* are readily distinguished by a variety of phenotypic analyses, there is a paucity of data available on qualitative differences between *Bradyrhizobium liaoningense* and the two previously named species (Table BXII.α.144). This highlights the need for phenotypic characterization of the third species.

List of species of the genus Bradyrhizobium

1. **Bradyrhizobium japonicum** (Kirchner 1896) Jordan 1982, 137[VP] (*"Rhizobacterium japonicum"* Kirchner 1896, 221; *Rhizobium japonicum* (Kirchner 1896) Buchanan 1926, 90.) *ja.po' ni.cum.* M.L. adj. *japonicum* pertaining to Japan.

   The characteristics are as described for the genus and as indicated in Table BXII.α.144. Whole cell fatty acid content is 1.3% $C_{16:1\,\omega7c}$, 3.6% $C_{16:1}$, 8.8% $C_{16:0}$, 1.2% $C_{19:0\,cyclo}$, and 81.2% $C_{18:1}$, when grown on A1EG agar medium at 30°C for 6 days. Resistant to trimethoprim at 50 mg/l and to vancomycin at 100 mg/l but sensitive to nalidixic acid (50 mg/l), tetracycline (100 mg/l), streptomycin (100 mg/l), erythromycin (250 mg/l), chloramphenicol (500 mg/l), rifampicin (500 mg/l), and carbenicillin (500 mg/l). Does not grow on Beringer's TY Medium that is used to grow *Rhizobium*. The exocellular polysaccharide contains glucose, mannose, galacturonic acid, and galactose. Denitrification positive. Bacteroids in root nodules are usually viable when plated onto media. Cells of *B. japonicum* have one polar or subpolar flagellum. The species normally causes the formation of root nodules on species of *Glycine* (soybean) (see Fig BXII.α.171) and on *Macroptilium atropurpureum* (siratro). Some strains express hydrogenase activity with the soybean host and thus are more efficient in symbiotic nitrogen fixation.

   The mol% G + C of the DNA is: 61–65 (T_m).
   Type strain: USDA 6, ATCC 10324, DSMZ 30131.
   GenBank accession number (16S rRNA): U69638, X87272.

2. **Bradyrhizobium elkanii** Kuykendall, Saxena, Devine, and Udell 1993b, 398[VP] (Effective publication: Kuykendall, Saxena, Devine, and Udell 1992, 504.)

el.ka' ni.i. M.L. n. *elkanii* of G.H. Elkan.

Most characteristics are as described for the genus and as indicated in Table BXII.α.144. Unlike *B. japonicum*, *B. elkanii* readily fixes nitrogen *ex planta*, strains generally produce indole acetic acid in liquid or broth cultures, they produce at least some growth on TY medium, and they generally grow at 34°C or higher temperatures. Strains are better established in the southeastern United States than the rest of the country. Whole cell fatty acid content is 0.5% $C_{16:1}$, 11.1% $C_{16:0}$, 0.8% $C_{17:0\,cyclo}$, 24.7% $C_{19:0\,cyclo}$, and 62.3% $C_{18:1}$, when grown on A1EG agar medium at 30°C for 6 days. Like *B. japonicum*, it is resistant to trimethoprim at 50 mg/l and to vancomycin at 100 mg/l. Unlike *B. japonicum*, *B. elkanii* is resistant to nalidixic acid (50 mg/l), tetracycline (100 mg/l), streptomycin (100 mg/l), erythromycin (250 mg/l), chloramphenicol (500 mg/l), rifampicin (500 mg/l), and carbenicillin (500 mg/l). The exocellular polysaccharide contains rhamnose and 4-*o*-methyl-glucuronic acid. Cells lack nitrite reductase activity but can possess the ability to reduce nitrate to nitrite. The species normally causes the formation of root nodules on species of *Glycine* (soybean), non-nodulating *rj1rj1* soybean, black-eyed peas (*Vigna*), mungbean, and on *Macroptilium atropurpureum* (siratro). Unlike *B. japonicum*, *B. elkanii* often produces rhizobitoxine-induced chlorosis on sensitive soybean cultivars (Devine et al., 1988; 1990). Strains often are hydrogenase positive on *Vigna* but not *Glycine*, suggesting more symbiotic affinity or compatibility with the former than the latter. *B. elkanii* may have originated on a continent other than Asia, where *B. japonicum* originates. Bottomley et al. (1994) used

TABLE BXII.α.144. The differential characterisitics of the species of *Bradyrhizobium*[a]

Characteristic	Bradyrhizobium japonicum	Bradyrhizobium elkanii	Bradyrhizobium liaoningense
Extra-slow growth (>18h)	−	−	+
Antibiotic resistance	−	+	−
Broad host range	−	+	NR
EPS has rhamnose	−	+	NR
ex planta Nitrogenase	L	H	NR
$C_{19:0\,cyclo}$	H	L	NR
Nitrite reduction	+	−	NR
Indole-3-acetic acid produced in broth	−	+	NR
L-Rhamnose utilization	−	+	NR
D-Glucosamine utilization	−	+	NR
D-Alanine utilization	−	+	NR
Malonic acid utilization	+	−	NR

[a]For symbols, see standard definition; L, low; H, high; NR, not reported.

multilocus enzyme electrophoresis to confirm the genetically distinct lineages of *B. japonicum* and *B. elkanii*. The data of Barrera et al. (1997) indicate that many Mexican *Lupinus*-nodulating bacteria belong to *B. elkanii* but some are *B. japonicum*.

> *The mol% G + C of the DNA is*: 61–65 (T_m).
> *Type strain*: USDA 76, ATCC 49852, DSMZ 11554.
> *GenBank accession number (16S rRNA)*: U35000.

3. **Bradyrhizobium liaoningense** Xu, Ge, Cui, Li and Fan 1995, 706[VP]

> *li.a.o.ning.en' se*. M.L. adj. *liaoningense* referring to a province in China where they originated.

This species consists of extra-slow-growing bradyrhizobia from alkaline soils. Colonies are only 0.2–1.0 mm on YMA medium (recipe given in *Rhizobium* chapter) after 7–14 days. Unlike *B. elkanii*, this species is more susceptible to antibiotics at low concentrations. Numerical taxonomy and serotyping were also used to establish an identity for this species apart from the two previously described species. A representative strain, 2062, had less than 25% DNA homology with strains of either *B. japonicum* or *B. elkanii*.

> *The mol% G + C of the DNA is*: 61–65 (T_m).
> *Type strain*: 2281.
> *GenBank accession number (16S rRNA)*: X86065.
> *Additional Remarks*: This sequence is not from the type strain.

Other Organisms

Ladha and So (1994) used numerical taxonomy to show that *Bradyrhizobium* sp. (*Aeschynomene*) strains may constitute a separate taxon from *B. japonicum* and *B. elkanii* but FAME and 16S rRNA sequence data did not support this hypothesis (So et al., 1994). Wong et al. (1994) demonstrated that *Bradyrhizobium* spp. (*Aeschynomene*) are closely related to *B. japonicum*. Other *Bradyrhizobium* species are known to occur but have not yet been named. For example, the fatty acid data of Graham et al. (1995) clearly document distinct microorganisms. Dupuy et al. (1994) established that *Acacia*-nodulating *Bradyrhizobium* strains in Africa corresponded to either *B. japonicum*, *B. elkanii*, or one or more as yet unnamed *Bradyrhizobium* spp. Related organisms cause nodule production on certain species of *Lotus* (*L. uliginosus* and *L. pedunculatus*) and also on *Vigna* and species of *Lupinus*, *Ornithopus*, *Cicer*, *Sesbania*, *Leucaena*, *Mimosa*, *Lablab*, and *Acacia*, which are also nodulated by the relatively fast-growing species *Mesorhizobium loti*. Such strains also usually nodulate *Macroptilium* and, more rarely, *Glycine*. It is suggested that until such time as further species or biovars are created within the genus *Bradyrhizobium*, these organisms (other than the species described above) be desig-

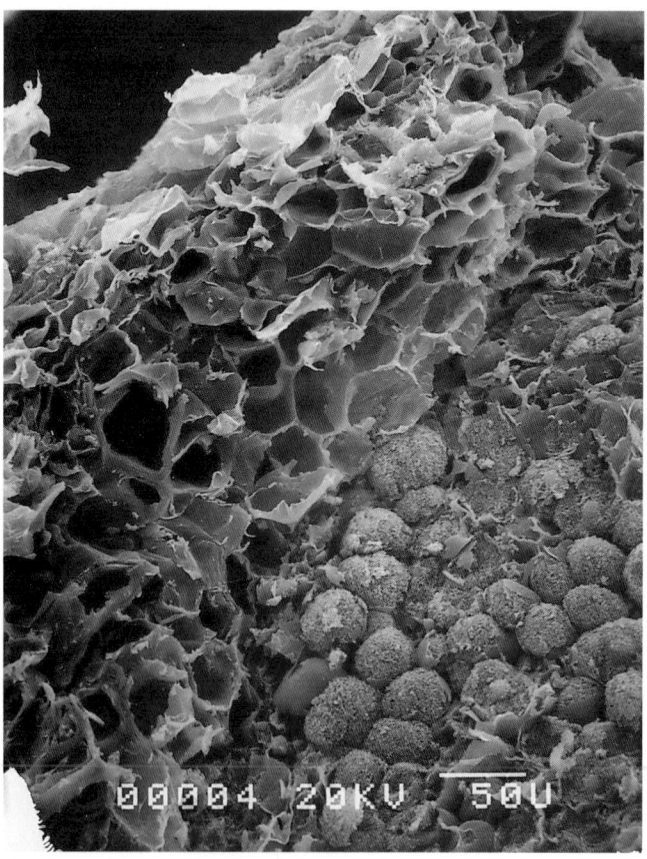
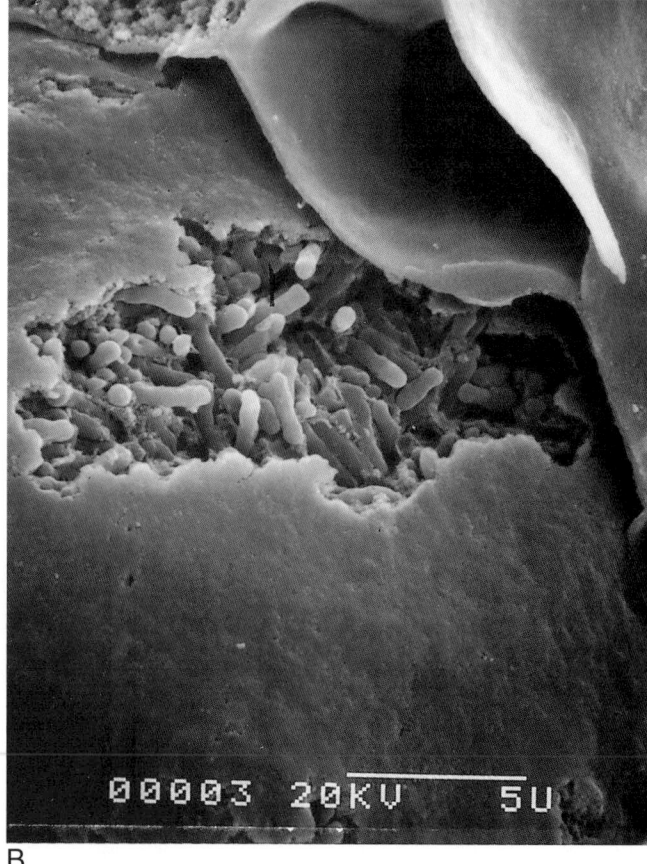

A   B

FIGURE BXII.α.171. Scanning EM of the freeze-fractured interior of a root nodule formed by *Bradyrhizobium japonicum* strain I-110 on soybean. (*A*) Plant cells filled with viable bacteria are in the center portion. (*B*) The cell wall of a bacteria-filled cell was partially torn away, showing the rod-shaped bacteria within. (EM micrographs were taken in the laboratory of Dr. Bill Wergin at Beltsville, MD.)

nated as *Bradyrhizobium* sp. with the name of the appropriate host plant given in parenthesis immediately following; e.g., *Bradyrhizobium* sp. (*Acacia*). A number of *Bradyrhizobium* sp. (*Arachis*)

strains were found to be closely related to *B. japonicum* (Van Rossum et al., 1995).

Genus II. **Afipia** Brenner, Hollis, Moss, English, Hall, Vincent, Radosevic, Birkness, Bibb, Quinn, Swaminathan, Weaver, Reeves, O'Connor, Hayes, Tenover, Steigerwalt, Perkins, Daneshvar, Hill, Washington, Woods, Hunter, Hadfield, Ajello, Kaufmann, Wear and Wenger 1992, 327^VP (Effective publication: Brenner, Hollis, Moss, English, Hall, Vincent, Radosevic, Birkness, Bibb, Quinn, Swaminathan, Weaver, Reeves, O'Connor, Hayes, Tenover, Steigerwalt, Perkins, Daneshvar, Hill, Washington, Woods, Hunter, Hadfield, Ajello, Kaufmann, Wear and Wenger, 1991, 2457.)

ROBBIN S. WEYANT AND ANNE M. WHITNEY

A.fip′i.a. L. fem. n. *Afipia* derived from the abbreviation AFIP, for the Armed Forces Institute of Pathology, where the type strain of the type species was isolated.

Gram-negative rods in the *Alphaproteobacteria*. **Motile** by means of one or two polar, subpolar, or lateral flagella. **Optimal growth is obtained at 30°C** and growth does not occur at or above 42°C. Grows well on buffered charcoal yeast extract (BCYE) agar and in nutrient broth with 0% added NaCl. Growth does not occur in nutrient broth with 6% added NaCl and is not enhanced by increased CO_2. Colonies are gray-white, glistening, convex and opaque, 0.5–1.5 mm in diameter at 72 hours of incubation at 30°C on BCYE agar. Nonfermentative metabolism; acid is not produced from D-glucose, lactose, maltose, or sucrose. **Positive for oxidase, urease, and litmus milk alkalinization**. Negative for hemolysis, gas production from nitrate, production of indole and H_2S (triple sugar iron agar method), and hydrolysis of gelatin and esculin. Isolated from potable water. Associated with human infection. Interrelatedness of species by DNA–DNA hybridization is 12–69%.

The mol% G + C of the DNA is: 61.5–69.

Type species: **Afipia felis** Brenner, Hollis, Moss, English, Hall, Vincent, Radosevic, Birkness, Bibb, Quinn, Swaminathan, Weaver, Reeves, O'Connor, Hayes, Tenover, Steigerwalt, Perkins, Daneshvar, Hill, Washington, Woods, Hunter, Hadfield, Ajello, Kaufmann, Wear and Wenger 1992, 327 (Effective publication: Brenner, Hollis, Moss, English, Hall, Vincent, Radosevic, Birkness, Bibb, Quinn, Swaminathan, Weaver, Reeves, O'Connor, Hayes, Tenover, Steigerwalt, Perkins, Daneshvar, Hill, Washington, Woods, Hunter, Hadfield, Ajello, Kaufmann, Wear and Wenger, 1991, 2458.)

FURTHER DESCRIPTIVE INFORMATION

Phylogenetic treatment On the basis of DNA hybridization and phenotypic characterization, the genus *Afipia* currently contains 3 named species, *Afipia felis*, *A. broomeae*, and *A. clevelandensis* as well as three unnamed genomospecies (Brenner et al., 1991a). When compared with other members of the *Alphaproteobacteria* by 16S rDNA sequence analysis, *Afipia* species are intermingled with species of *Rhodopseudomonas*, *Bradyrhizobium*, *Nitrobacter*, and *Blastobacter* as shown in the phylogenetic tree in Fig. BXII.α.172. The 16S rRNA genes of *A. felis*, *A. broomeae*, and *A. clevelandensis* share 97.6–98.5% similarity to each other, and 97.0–98.5% to *Afipia* genomospecies 3, *Rhodopseudomonas palustris*, *Nitrobacter winogradskyi*, *Nitrobacter hamburgensis*, "*Bradyrhizobium lupini*", *Bradyrhizobium japonicum*, and *Blastobacter denitrificans*. The high degree of sequence similarity supports Willems and Collins' proposal that these species could be classified as members of a single genus (Willems and Collins, 1992). The 16S rRNA sequences for *Afipia*

genomospecies 1 and 2, however, are 95–95.9% similar to those of the named *Afipia* species and *Afipia* genomospecies 3.

Eleven additional genomospecies have been tentatively included in the genus *Afipia* based on phenotypic characteristics, chemotaxonomy, and DNA relatedness (Weyant et al., unpublished data). These strains are described in the section Other Organisms.

Cell and flagella morphology *Afipia* cells can be visualized using the Gram stain and other common staining techniques. Gram-stained preparations of 48-hr BCYE cultures show short to medium-length (>1.0 μm) Gram-negative rods of medium width (>0.5 μm). Cells are randomly distributed and do not form chains or clusters. Motility is achieved by one or two polar or subpolar flagella. Flagella may be easily visualized using the Ryu method (Ryu, 1937; Kodaka et al., 1982). Characteristic Gram and flagella stains of *Afipia* species have been illustrated by English et al. (1988) and Weyant et al. (1995).

Cellular fatty acid (CFA) composition and ubiquinone content All *Afipia* species share a unique CFA profile of predominately branched fatty acids ($C_{16:1 \, \omega7c}$, $C_{18:1 \, \omega7c}$, and $C_{Br-19:1}$). The high amount of $C_{Br-19:1}$ (10–23%) is unique to *Afipia*. All *Afipia* species also contain $C_{17:0 \, cyclo}$ and/or $C_{19:0 \, cyclo \, \omega7c}$ in proportions ranging from 1 to 15% (Moss et al., 1990a; Weyant et al., 1995). Ubiquinone 10 is the major isoprenoid ubiquinone of *Afipia* (Moss et al., 1991).

Colonial and cultural characteristics *Afipia* strains grow in various common growth media, including heart infusion agar with 5% rabbit blood, trypticase soy agar with 5% sheep blood, tryptone glucose yeast extract agar, and BCYE. Broth media that support *Afipia* growth include heart infusion broth and nutrient broth. Growth occurs at 25, 30, and 35°C, but not at or above 42°C. Optimal growth is at pH 6.8 (Müller, 1995). Growth occurs under aerobic conditions and is not enhanced by increased CO_2. Growth does not occur in nutrient broth containing 6% NaCl and only rarely occurs on MacConkey agar. Growth is weakly enhanced by ferric pyrophosphate, cysteine-HCl, and α-ketoglutarate (Müller, 1995).

Optimal growth is observed on BCYE agar under aerobic conditions at an incubation temperature of 30–32°C. Growth is relatively slow, with colonies appearing between 48 and 72 hours of incubation under optimal conditions. Colonies are gray-white, glistening, convex, and opaque, with smooth edges. Colony diameters range from 0.5 to 1.5 mm. No hemolysis is observed on blood agar.

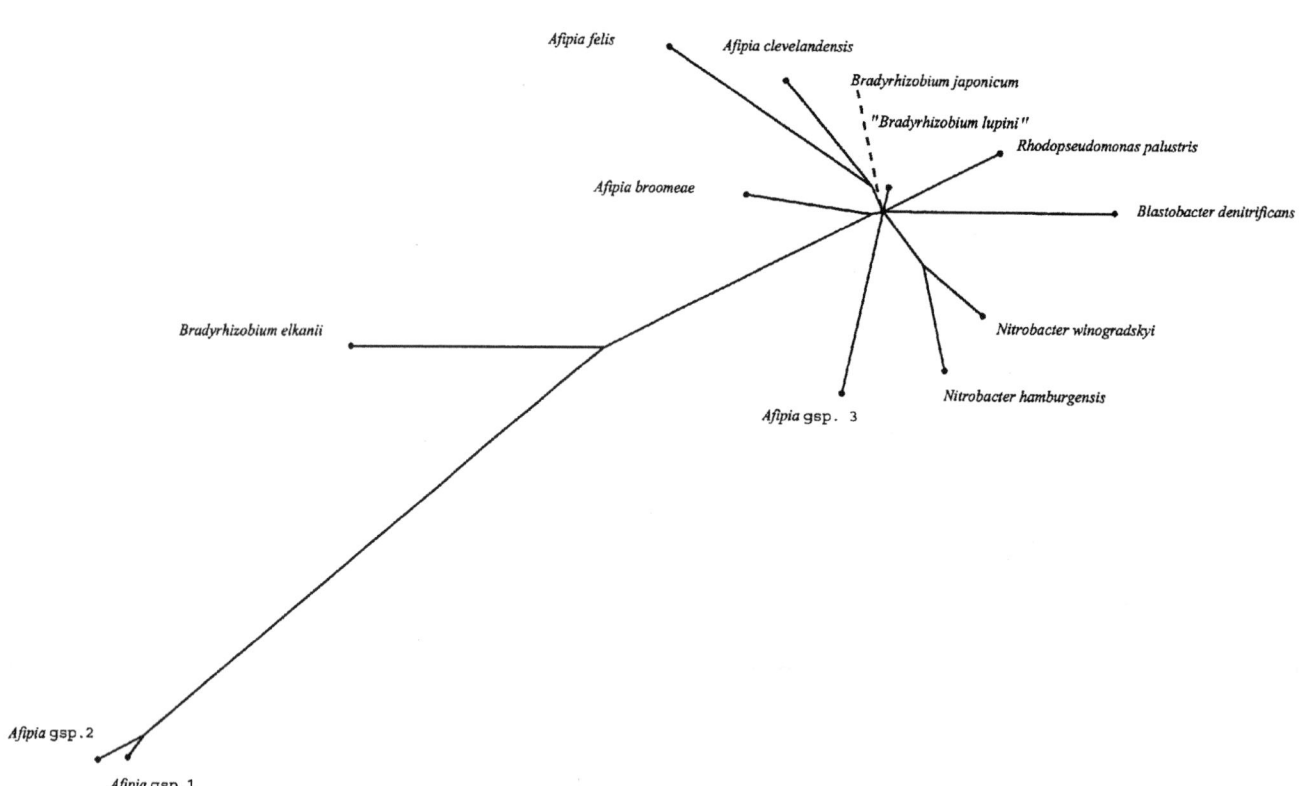

FIGURE BXII.α.172. Phylogenetic tree based on 16S rDNA sequence homologies between the type strains of *Afipia* species and type or reference strains of closely related taxa.

Genetics Mol% G + C content of the DNA ranges from 61.5 in *A. broomeae* to 69 in *Afipia* genomospecies 1 (Brenner et al., 1991a). All *A. felis* strains described by Brenner et al. (1991a) contained a 44-kb plasmid, while the other 5 species or genomospecies did not contain a plasmid.

Antibiotic sensitivity *In vitro* antibiotic sensitivity information for *Afipia* is subject to two limitations. The incubation temperature (30–32°C) differs from the standard 37°C used in the development of the National Committee for Clinical Laboratory Standards (1990); therefore, these standards cannot be used to describe resistance without further clarification. The second limitation is the small number of strains from which these data have been derived. The total published susceptibility literature for this genus consists of the original and five additional *A. felis* strains (Brenner et al., 1991a; Maurin et al., 1993), one *A. clevelandensis* strain, three *A. broomeae* strains, and one strain each of *Afipia* genomospecies 1, 2, and 3 (Brenner et al., 1991a). These studies report *A. felis* to be resistant to a wide variety of antimicrobial agents, including most beta-lactams, ciprofloxacin, and tetracycline, but susceptible to aminoglycosides and imipenem. *A. clevelandensis* and *A. broomeae* share similar susceptibility profiles with *A. felis*, except for slightly higher MICs (4–16 µg/ml) to aminoglycosides. *Afipia* genomospecies 1 and 2 are resistant to all antibiotics tested thus far including beta-lactams, ciprofloxacin, tetracycline, imipenem, and aminoglycosides. Conversely, *Afipia* genomospecies 3 is sensitive to all of the above antibiotic classes

except cefoperazone, cefoxitin, and ceftazidime. These generalizations may change as more *Afipia* strains are isolated and tested.

Pathogenicity *A. felis* has been detected in lymph nodes of patients with cat scratch disease (CSD) by culture and PCR on at least four occasions (English et al., 1988; Birkness et al., 1992; Alkan et al., 1995; Giladi et al., 1998); however, the vast majority of CSD cases appear to be caused by *Bartonella henselae*. One of the criteria for differential diagnosis of CSD is a bite or scratch from a cat. *A. felis*, however, has not been isolated from a feline source. Although English reported that antibodies made to *A. felis* reacted with specimens from CSD patients (English et al., 1988), other investigators have reported that antibody titers to *A. felis* in CSD patients were similar to titers in control groups (Amerein et al., 1996).

A. felis is capable of intracellular growth in amoebae (La Scola and Raoult, 1999a) and in tissue culture with human monocytes, macrophages, HeLa cells, and HMEC-1 cells (Birkness et al., 1992). In human macrophages *in vitro*, *A. felis* survives in phagosomes apparently by inhibiting phagosome–lysosome fusion (Brouqui and Raoult, 1993).

The isolation history of the other *Afipia* species suggests that they may act as opportunistic pathogens in humans. *A. clevelandensis* ATCC 49720 was isolated from a tibial biopsy specimen of a patient with necrotizing pancreatitis (Hall et al., 1991). *A. broomeae* ATCC 49717 was isolated from human sputum while *A.*

broomeae ATCC 49719 grew in a culture of the synovial fluid of a diabetic man with arteriosclerosis (Brenner et al., 1991a). *Afipia* genomospecies 1 ATCC 49721 was isolated from pleural fluid and *Afipia* genomospecies 2 ATCC 49722 was cultured from a bronchial lavage (Brenner et al., 1991a). *Afipia* genomospecies 3 ATCC 49723 was cultured from a water sample from Indiana, and its pathogenicity is unknown (Brenner et al., 1991a).

Ecology The majority of confirmed strains of *Afipia* have been isolated from human clinical specimens, with insufficient exposure information to identify a primary source. Multiple lines of evidence indicate that these organisms reside in fresh water. The type strain of *Afipia* genomospecies 3 was isolated from fresh water (Brenner et al., 1991a), and *A. felis* has been isolated from potable water in a hospital (La Scola and Raoult, 1999a). Many of the *in vitro* characteristics of these organisms, including their preferred growth temperature (30–32°C), lack of growth enhancement by blood or blood products, and lack of growth in elevated NaCl concentrations are also consistent with a fresh water habitat.

ENRICHMENT AND ISOLATION PROCEDURES

When incubated at 30–32°C under aerobic conditions, *Afipia* strains grow on a variety of nonselective media, including brain-heart infusion, heart infusion (with and without 5% rabbit blood), chocolate, potato dextrose, trypticase-soy (with and without 5% sheep blood), and BCYE agars. Liquid media that support *Afipia* growth include heart infusion, brain-heart infusion, and nutrient broth. Although no selective media have yet been described for these organisms, incubation of cultures at 25–30°C favors the growth of *Afipia* over many other taxa that prefer higher temperatures.

Primary isolation of *Afipia felis* strains from lymph nodes of patients with cat scratch disease is achieved by grinding tissues in equal volumes of phosphate buffered saline (pH 7.6) and sea sand, removing the sand by centrifugation at 500 × g for five minutes, then centrifuging the supernatant at 10,000 × g for ten minutes to obtain pelleted material. The pelleted material is then subcultured in biphasic brain-heart infusion media at 30–32°C and growth is observed between 1 and 6 days of incubation (English et al., 1988). Primary isolation is also achieved by cultivation of sonicated lymph node tissues on HeLa cell monolayers (Brenner et al., 1991a).

MAINTENANCE PROCEDURES

Long-term preservation of cultures may be achieved by suspending freshly grown cells in trypticase soy broth with 20% glycerol or defibrinated rabbit blood and freezing at −70°C or in vapor nitrogen. For short-term maintenance (less than one year), strains should be inoculated as stabs into semisolid motility medium deeps in screw-capped tubes, incubated at 30°C until growth is observed, then placed at room temperature with the caps tightly closed (Weyant et al., 1995).

DIFFERENTIATION OF THE GENUS *AFIPIA* FROM OTHER GENERA

Other taxa that are isolated from clinical specimens and share similar phenotypic characteristics with *Afipia* are given in Table BXII.α.145. *Legionella* species, *Francisella tularensis*, and *Bordetella pertussis* may all be isolated from clinical specimens by culture on BCYE agar. The lower preferred growth temperature, the ability to grow in nutrient broth, and the hydrolysis of urea differentiates *Afipia* from these organisms. *Bartonella* species are closely related to *Afipia* phylogenetically and may also be isolated from lymph node specimens of individuals with cat-scratch disease. Unlike *Afipia*, these organisms do not grow well on BCYE agar or nutrient broth. With the exception of *B. bacilliformis*, these species prefer warmer growth temperatures and are nonmotile. The presence of $C_{Br-19:1}$ as a predominating cellular fatty acid is also useful in differentiating *Afipia* from these other organisms.

Detailed phenotypic comparison studies including *Afipia* and its phylogenetic neighbors *Rhodopseudomonas*, *Bradyrhizobium*, *Rhizobium*, *Nitrobacter*, and *Blastobacter* have not yet been described. A limited analysis, including only type strains, indicates that *Bradyrhizobium* and *Blastobacter* acidify a wider range of carbohydrates (D-glucose, adonitol, D-galactose, D-mannose, and L-rhamnose) than *Afipia* (Weyant et al., unpublished findings). Neither the extracellular polysaccharide slime produced by *Bradyrhizobium* strains when grown on carbohydrate-containing media nor the branching or budding characteristic of *Blastobacter* cells have been observed with *Afipia* strains. *Rhodopseudomonas*, unlike *Afipia* and the other taxa in this phylogenetic cluster, is a phototroph.

FURTHER READING

Brenner, D.J., D.G. Hollis, C.W. Moss, C.K. English, G.S. Hall, J. Vincent, J. Radosevic, K.A. Birkness, W.F. Bibb, F.D. Quinn, B. Swaminathan, R.E. Weaver, M.W. Reeves, S. O'Connor, P. Hayes, F. Tenover, A.G. Steigerwalt, B. Perkins, M.I. Daneshvar, B.C. Hill, J.A. Washington, T. Woods, S. Hunter, D.J. Wear and J. Wenger. 1991. Proposal of *Afipia*, gen. nov., with *Afipia felis*, sp. nov. (formerly the cat scratch disease bacillus), *Afipia clevelandensis*, sp. nov. (formerly the Cleveland Clinic Foundation strain), *Afipia broomeae*, sp. nov., and three unnamed genospecies. J. Clin. Microbiol. *29*: 2450–2460.

English, C.K., D.J. Wear, A.M. Margileth, C.R. Lissner and G.P. Walsh. 1988. Cat-scratch disease: isolation and culture of the bacterial agent. JAMA (J. Am. Med. Assoc.). *259*: 1347–1352.

Müller, H.E. 1995. Investigations of culture and properties of *Afipia* spp. Zentbl. Bakteriol. *282*: 18–23.

DIFFERENTIATION OF THE SPECIES OF THE GENUS *AFIPIA*

A biochemical characterization of 6 *Afipia* species is given in Table BXII.α.146. Characteristics useful in differentiating these species are given in Table BXII.α.147. The ability to reduce nitrate is a unique characteristic that differentiates *A. felis* from the other species. Likewise, the ability to alkalinize citrate is unique to *Afipia* genomospecies 1. *A. clevelandensis* is characterized by its inability to produce acid from carbohydrates. Demonstration of specific protease activity is useful in differentiating the remaining species.

List of species of the genus Afipia

1. **Afipia felis** Brenner, Hollis, Moss, English, Hall, Vincent, Radosevic, Birkness, Bibb, Quinn, Swaminathan, Weaver, Reeves, O'Connor, Hayes, Tenover, Steigerwalt, Perkins, Daneshvar, Hill, Washington, Woods, Hunter, Hadfield, Ajello, Kaufmann, Wear and Wenger 1992, 327^VP (Effective publication: Brenner, Hollis, Moss, English, Hall, Vincent, Radosevic, Birkness, Bibb, Quinn, Swaminathan, Weaver, Reeves, O'Connor, Hayes, Tenover, Steigerwalt, Perkins, Daneshvar, Hill, Washington, Woods, Hunter, Hadfield, Ajello, Kaufmann, Wear and Wenger, 1991, 2458.)

   fe'lis. L. gen. n. *felis* of the cat.

TABLE BXII.α.145. Differentiation of *Afipia* from similar taxa[a]

Characteristic	Afipia	Bartonella	Legionella	Francisella tularensis	Bordetella pertussis
Cellular morphology:					
Straight rods	+				
Thin, slightly curved rods		+			
Thin rods			+		
Minute coccobacilli				+	+
Cell size (µm)	0.5–1.0 × 1–3	0.5–0.6 × 1–2	0.5–0.6 × 1–2	0.5–0.6 × 1–2	0.2–0.5 × 0.5–2.0
Optimal growth temperature (°C)	25–30	25–37	35–37	35–37	35–37
Growth required/enhanced by:					
Cysteine	+	−	+	+	−
Rabbit blood	−	+	−	+	−
Growth in nutrient broth	+	−	−	−	−
Acid production from D-glucose	−	−	−	+	
Hydrolysis of:					
Gelatin	−	nd	v	−	−
Urea	+ or (+)	−	−	−	−
Nitrate reduction	v	−	−	−	−
Oxidase	+	v	+	v	+
Motility	v	v	+	−	−
Predominating cellular fatty acids:					
$C_{10:0}$				+	
$C_{14:0}$				+	
$C_{I14:0}$			+		
$C_{15:0\ anteiso}$			+		
$C_{16:0}$		+		+	+
$C_{16:1\ \omega7c}$		+			+
$C_{16:0\ iso}$			+		
$C_{17:0}$		+			
$C_{17:0\ anteiso}$			+		
$C_{17:0\ cyclo}$	+				
$C_{18:0}$	+			+	
$C_{18:1\ \omega7c}$	+	+			
$C_{24:1\ \omega15c}$				+	
$C_{Br-19:1}$	+				
$C_{18\ 3OH}$				+	

[a]Symbols: +, >90% species positive; (+), delayed positive; −, <10% species positive; v, 11–89% species positive; nd, not determined; for *B. pertussis* the symbols refer to strains. Data from Brenner, et al. (1991a) and Müller (1995).

The morphological and cultural characteristics are as described for the genus. Other characteristics for this species are given in Table BXII.α.146. Rare cases of cat scratch disease have been associated with this species. The natural reservoir of this species is unknown. Most confirmed isolates have been recovered from human clinical specimens. The type strain was isolated from a lymph node of a person with cat scratch disease.

The mol% G + C of the DNA is: 62.5 (T_m).

Type strain: BV, ATCC 53690, CDC B91-007352, DSM 7326.

2. **Afipia broomeae** Brenner, Hollis, Moss, English, Hall, Vincent, Radosevic, Birkness, Bibb, Quinn, Swaminathan, Weaver, Reeves, O'Connor, Hayes, Tenover, Steigerwalt, Perkins, Daneshvar, Hill, Washington, Woods, Hunter, Hadfield, Ajello, Kaufmann, Wear and Wenger 1992, 327[VP] (Effective publication: Brenner, Hollis, Moss, English, Hall, Vincent, Radosevic, Birkness, Bibb, Quinn, Swaminathan, Weaver, Reeves, O'Connor, Hayes, Tenover, Steigerwalt, Perkins, Daneshvar, Hill, Washington, Woods, Hunter, Hadfield, Ajello, Kaufmann, Wear and Wenger, 1991, 2458.)

broome' a.e. N.L. gen. n. *broomeae* named after Dr. Claire V. Broome in recognition of her contributions to the epidemiology and microbiology of cat scratch disease, legionellosis, listeriosis, toxic shock syndrome, Brazilian purpuric fever, and many other diseases.

The morphological and cultural characteristics are as described for the genus. Other characteristics of the species are given in Table BXII.α.146. The natural reservoir of this species is unknown. All confirmed strains have been isolated from human clinical material, including blood, lung tissue, and bone marrow. Presumptively pathogenic for humans. The type strain was isolated from human sputum in New Zealand in 1981.

The mol% G + C of the DNA is: 61.5 (T_m).

Type strain: ATCC 49717, CDC B91-007286, DSM 7327.

GenBank accession number (16S rRNA): U87759.

3. **Afipia clevelandensis** Brenner, Hollis, Moss, English, Hall, Vincent, Radosevic, Birkness, Bibb, Quinn, Swaminathan, Weaver, Reeves, O'Connor, Hayes, Tenover, Steigerwalt, Perkins, Daneshvar, Hill, Washington, Woods, Hunter, Hadfield, Ajello, Kaufmann, Wear and Wenger 1992, 327[VP] (Effective publication: Brenner, Hollis, Moss, English, Hall, Vincent, Radosevic, Birkness, Bibb, Quinn, Swaminathan, Weaver, Reeves, O'Connor, Hayes, Tenover, Steigerwalt, Perkins, Daneshvar, Hill, Washington, Woods, Hunter, Hadfield, Ajello, Kaufmann, Wear and Wenger, 1991, 2457.)

cleve.land.en' sis. N.L. fem. adj. *clevelandensis* coming from Cleveland, Ohio, USA, where the type strain was isolated.

The morphological and cultural characteristics are as described for the genus. Other characteristics of the species are given in Table BXII.α.146. The natural reservoir of this species is unknown. The only known strain of this species was isolated from a tibial biopsy specimen of a human. Presumptively pathogenic for humans. The type strain was

TABLE BXII.α.146. Biochemical and growth characteristics of *Afipia* species[a]

Characteristic	A. felis	A. broomeae	A. clevelandensis	Afipia genomospecies 1	Afipia genomospecies 2	Afipia genomospecies 3
Motility	+	+	+	+	+	+
Growth at:						
25°C	+	+	+	+	+	+
35°C	+	v	+	+	+	+
42°C	−	−	−	−	−	−
Growth on:						
MacConkey agar	v	−	−	−	−	−
Salmonella–Shigella agar	−	−	−	−	−	−
Cetrimide agar	−	−	−	−	−	−
Growth in nutrient broth	+	+	+	+	+	+
Growth in nutrient broth with 6% NaCl	−	−	−	−	−	−
Catalase	v	+	−	v	+	−
Oxidase	+	+	+	+	+	+
Alkalinization of:						
Citrate	−	−	−	+	−	−
Sodium acetate	+ or (+)	V	(+)	−	(+)	(+)
Acetamide	−	−	−	−	−	−
Serine	−	−	−	−	−	−
Tartrate	v	−	−	(+)	−	−
Litmus milk	+ or (+)	(+)	(+)	(+)	(+)	(+)
Hydrolysis of:						
Urea	+ or (+)	+ or (+)	(+)	+	+	(+)
Esculin	−	−	−	−	−	−
Gelatin	−	−	−	−	−	−
Nitrate reduction	+	−	−	−	−	−
Gas from nitrate	−	−	−	−	−	−
Nitrite reduction	−	−	−	−	−	−
Indole production	−	−	−	−	−	−
H₂S production (TSI butt)	−	−	−	−	−	−
Phenylalanine deaminase	+	−	−	+	−	+
L-Lysine decarboxylase	−	−	−	−	nd	−
L-Arginine dihydrolase	−	−	−	−	nd	−
L-Ornithine decarboxylase	−	−	−	−	nd	−
Acid phosphatase	+	+	+	+	+	+
Alkaline phosphatase	+	+	+	+	+	+
Phosphodiamidase	+	+	+	+	+	+
Sulfatase	+w	+w	+w	+w	+w	+w
Glycine aminopeptidase	+	+	+	+	+	+
L-Lysine aminopeptidase	+	+	+	+	+	+
Amylase	−	−	−	−	−	−
N-Acetyl-β-D-glucosaminidase	−	−	−	−	−	−
Galactosidase (α and β)	−	−	−	−	−	−
Glucosidase (α and β)	−	−	−	−	−	−
β-Glucuronidase	−	−	−	−	−	−
Fucosidase	−	−	−	−	−	−
Mannosidase	−	−	−	−	−	−
Xylosidase	−	−	−	−	−	−
Chymotrypsin	−	+w	+w	+w	+w	+w
L-Phenylalanine aminopeptidase	−	+w	+w	+	+	+
Trypsin	−	+	−	+w	+	+w
L-Histidine aminopeptidase	−	+w	−	+	+	+
L-Asparagine aminopeptidase	+	−	−	+	+	+
L-Cystine aminopeptidase	−	−	−	+	+	+
L-Isoleucine aminopeptidase	−	−	−	+	+	+
L-Leucine aminopeptidase	−	−	−	+	+	+
DL-Methionine aminopeptidase	−	−	−	+	+	+
L-Proline aminopeptidase	−	−	−	+	+	+
L-Tryptophan aminopeptidase	−	−	−	+	+	+
L-Tyrosine aminopeptidase	−	−	−	+	−	−
L-Valine aminopeptidase	−	−	−	+	−	−
Acid production from:						
D-Glucose	−	−	−	−	−	−
D-Xylose	+w	(+)	−	(+)	(+)	(+)
D-Mannitol	−	−	−	(+)	−	−
Lactose	−	−	−	−	−	−
Sucrose	−	−	−	−	−	−
Maltose	−	−	−	−	−	−

[a]Symbols are as follows: +, >90% strains positive; (+), delayed positive, may require 3–7 days incubation; +w, weak positive; −, <10% strains positive; v, 11–89% strains positive; nd, not determined. Data from Brenner, et al. (1991a) and Müller (1995).

TABLE BXII.α.147. Characteristics useful in differentiating *Afipia* species[a]

Characteristic	A. felis	A. broomeae	A. clevelandensis	Afipia genomospecies 1	Afipia genomospecies 2	Afipia genomospecies 3
Growth on MacConkey agar	v	−	−	−	−	−
Catalase	v	+	−	v	+	−
Alkalinization of:						
Citrate	−	−	−	+	−	−
Sodium acetate	+ or (+)	v	(+)	−	(+)	(+)
Tartrate	v	−	−	(+)	−	−
Nitrate reduction	+	−	−	−	−	−
Phenylalanine deaminase	+	−	−	+	−	+
L-Phenylalanine aminopeptidase	−	+w	+w	+	+	+
Trypsin	−	+	−	+w	+	+w
L-Asparagine aminopeptidase	+	−	−	+	+	+
L-Tyrosine aminopeptidase	−	−	−	+	−	−
L-Valine aminopeptidase	−	−	−	+	−	−
Acid production from:						
D-Xylose	+w	(+)	−	(+)	(+)	(+)
D-Mannitol	−	−	−	(+)	−	−

[a]Symbols are as follows: +, >90% species positive; (+), delayed positive, requires 3–7 days incubation; +w, weak positive; −, <10% strains positive; v, 11–89% strains positive. Data from Brenner et al. (1991a) and Müller (1995).

isolated in 1988 from a tibial biopsy specimen taken from a 69-year-old man in Ohio, USA.

The mol% G + C of the DNA is: 64 (T_m).

Type strain: ATCC 49720, CDC B91-007353, DSM 7315.

4. *Afipia* genomospecies 1

The morphological and cultural characteristics are as described for the genus. Other characteristics for the species are given in Table BXII.α.146. The natural reservoir of this species is unknown. All confirmed strains have been isolated from human clinical material. Presumptively pathogenic for humans. ATCC 49721 was isolated at Oklahoma, USA, from human pleural fluid in 1981.

The mol% G + C of the DNA is: 69 (T_m).

Deposited strain: ATCC 49721, CDC B91-007287.

5. *Afipia* genomospecies 2

The morphological and cultural characteristics are as

described for the genus. Other characteristics for the species are given in Table BXII.α.146. The natural reservoir of this species is unknown. All confirmed strains have been isolated from human clinical material. Presumptively pathogenic for humans. ATCC 49722 was isolated at Indiana, USA, from a human bronchial wash specimen in 1989.

The mol% G + C of the DNA is: 67 (T_m).

Deposited strain: ATCC 49722, CDC B91-007290.

6. *Afipia* genomospecies 3

The morphological and cultural characteristics are as described for the genus. Other characteristics for the species are given in Table BXII.α.146. The natural reservoir of this species appears to be fresh water. ATCC 49723 was isolated at Indiana, USA, from water in 1990.

The mol% G + C of the DNA is: 65.5 (T_m).

Deposited strain: ATCC 49723, CDC B91-00729.

Other Organisms

We have studied multiple clinical and non-clinical isolates that are phenotypically similar to *Afipia*, but fall into different DNA hybridization groups. These isolates have been provisionally designated *Afipia* genomospecies 4, 5, 6, 7, 8, 9, 10, 11, 12, 13, and 14 (Weyant, unpublished findings). Phylogenetic analysis, using 16S rRNA sequence data indicate that genomospecies 4, 5, and 10 fall into the same *Rhodopseudomonas–Blastobacter–Nitrobacter* cluster as the 6 described species. Genomospecies 11, 13, and 14 associate most closely with *Caulobacter* and *Sphingomonas*, whereas the remaining genomospecies represent novel taxa. Additional phenotypic and chemotaxonomic studies are underway to clarify the appropriate taxonomic designation of these "*Afipia*" strains.

Genus III. **Agromonas** *Ohta and Hattori 1985, 223*[VP] *(Effective publication: Ohta and Hattori 1983, 43)*

CHRISTINA KENNEDY

Ag.ro.mon' as. Gr. n. *agros* a field; Gr. n. *monas* a unit, monad; M.L. fem. n. *Agromonas* field monad.

Bent, branched, and/or budding rods, usually 0.6–1.0 × 2–7 μm, when grown in diluted nutrient broth (NB/100). Motile by a polar flagellum. Gram negative. Colonies are colorless to white. No spores or microcysts formed. Aerobic. **Oligotrophic**; grows under conditions of low organic carbon supply (<1 mg/ml). **Fixes N₂ under microaerobic conditions. Catalase and oxidase positive.** Cellulose and starch are not hydrolyzed. Casein and gelatin are

not hydrolyzed. Cellular fatty acids consist mainly of a straight-chain unsaturated acid $C_{18:1}$, with smaller amounts of $C_{16:0}$ and $C_{19:1}$. Ubiquinone Q-10 is present. Isolated from rice paddy soils.

The mol% G + C of the DNA is: 65.1–66.0.

Type species: **Agromonas oligotrophica** Ohta and Hattori 1985, 223 (Effective publication: Ohta and Hattori 1983, 43.)

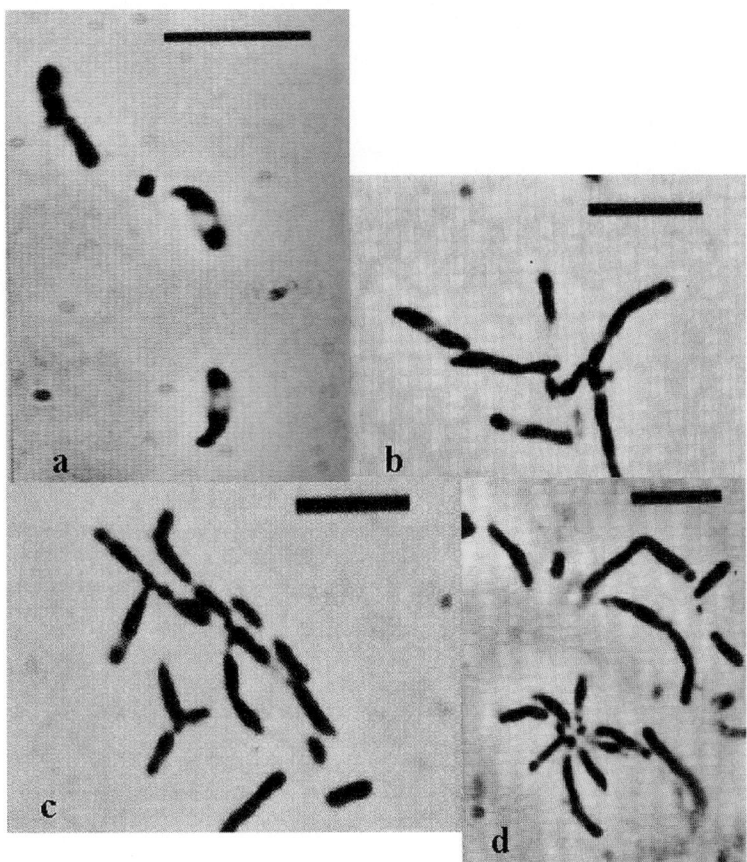

FIGURE BXII.α.173. Phase-contrast photomicrographs of 3-d-old (*a*, *b*) and 8-d-old (*c*, *d*) cultures of *A. oligotrophica* grown in NB/100 (*a*) and NB/10 (*b*, *c*, *d*) medium; (*a* and *b*) bent and budding rods in which both ends were darker than the rest of the cell; (*c*) branched rods; (*d*) a cell aggregate (rosette). Bar = 5 μm. (Reproduced with permission from H. Ohta and T. Hattori, *Antonie van Leeuwenhoek 49:* 429–446, 1983, ©Kluwer Academic Publishers, Dordrecht, Netherlands.)

FURTHER DESCRIPTIVE INFORMATION

Cell morphology Cells have an irregular rod-shaped morphology of dimensions 0.6–1.0 × 2–7 μm (Ohta and Hattori, 1983). The irregular shape of the cells is shown in Fig. BXII.α.173. Cell division is by irregular budding and/or elongation and pinching off, not by the usual septum formation (Hattori et al., 1995).

Cell envelope Outer layers are typical for *Proteobacteria*. Electron microscopy shows a multilayered envelope structure consisting of an outer layer connected to a dark thin peptidoglycan layer, a double-layered cytoplasmic membrane, and a light space between outer and inner layers (Hattori et al., 1995) (Fig. BXII.α.174a).

Fine structure The cytoplasm is unusually divided into several compartments surrounded by the cytoplasmic membrane (Hattori et al., 1995) (Fig. BXII.α.174b). Compartmentalization develops through invagination and growth of the cytoplasmic membrane. Compartments are often connected with each other and spaces between are frequently filled with electron-dense material (Fig. BXII.α.174c). Formation of compartments is inhibited by NaCl in PM/100 medium[1] and does not occur in PM/10, although growth occurs normally in the latter medium. These findings suggest that the developed cytoplasmic membrane and the compartmentalization of cytoplasm are not required for rapid growth but rather for growth in highly diluted nutrients. Another unusual feature is that electron-dense cores are visible in the cytoplasm, numbering from three to five. They probably represent nucleoids, as indicated by staining with the DNA-specific stain DAPI (4,6-diamidino-2-phenylindole) (Hattori et al., 1995).

Colonial or cultural characteristics: nutrition and growth Colonies are colorless/white, punctiform, pulvinate, and small (<0.5 mm diameter) (Ohta and Hattori, 1983). Growth is very sensitive to supply of organic compounds and occurs only in dilute media. Growth can occur in 1/10,000 nutrient broth (NB/10,000),[2] with cells reaching a density of >10^5 cells/ml. Growth also occurs on NB/1000, NB/100, NB/10, but not on NB. The linear relationship between the dilution of NB and cell yield indicates that growth is dependent on supplied organic carbon and is not autotrophic. Cultures reach a density of 1–5 × 10^8 cells/ml after 7–8 d growth on NB/100 or NB/10 (Ohta and Taniguchi, 1988a). The optimal growth temperature is 25–27°C.

Utilizable carbon sources are shown in Table BXII.α.148. The phenolic acids ferulic, *p*-coumaric, and *p*-anisic, but not benzoic,

1. Full-strength PM medium consists (g/l): peptone, 10.0; and meat extract. 10.0.

2. Full-strength NB contains (g/l of tap water): peptone, 10.0 g; meat extract, 10.0 g; and NaCl, 5.0 g; pH 7.0–7.2.

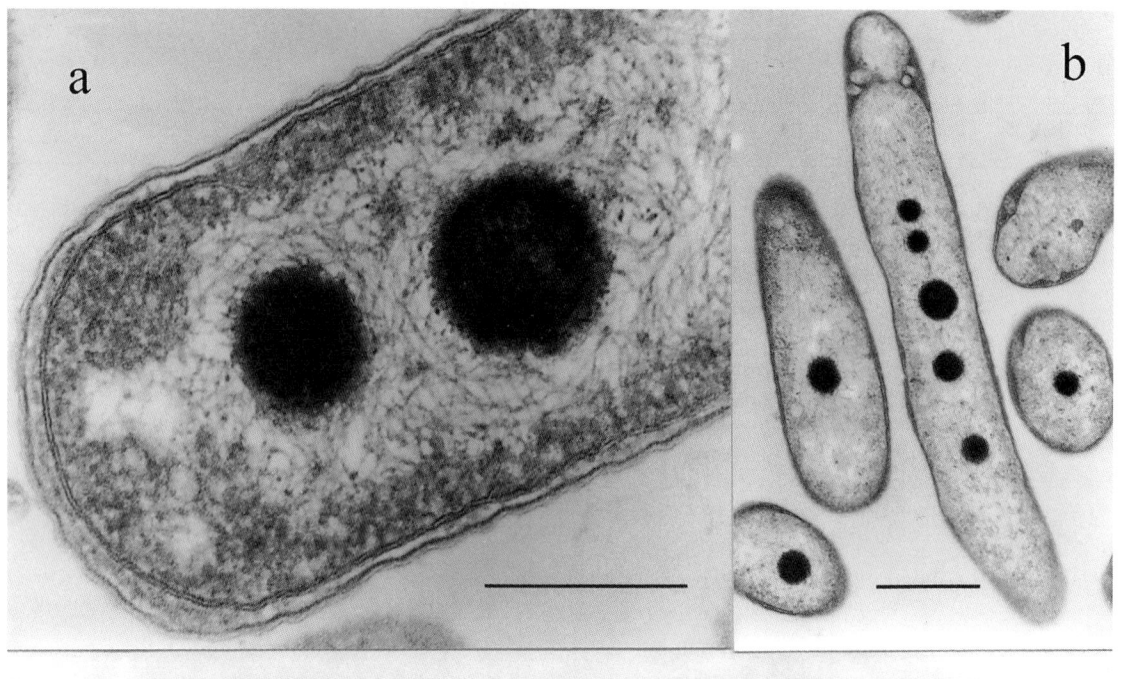

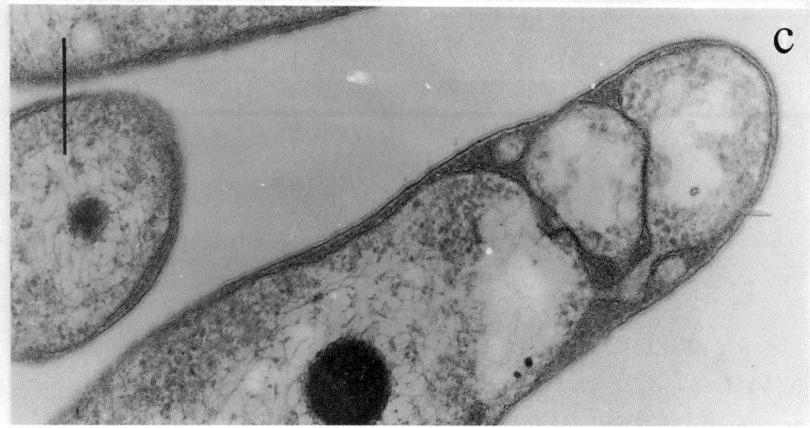

FIGURE BXII.α.174. Electron micrograph of *Agromonas oligotrophica* grown in PM/100. Sections were stained with 2% (w/v) aqueous uranyl acetate and saturated lead citrate. (*a*) The multilayered structure of cell envelope and the nucleoid-like structures in cytoplasm. Bar = 0.2 µm. (*b*) Growing cells; note that the cytoplasm is divided into compartments with electron-dense cores in nuclear zones. Bar = 1 µm. (*c*) Compartments can be separate or connected. Bar = 0.2 µm. (Reproduced with permission from T. Hattori et al., Journal of General and Applied Microbiology *41:* 23–30, 1995, ©Microbiology Research Foundation, Tokyo, Japan.)

support growth. Ferulic acid degradation requires organic nutrients to be present in the growth medium. The pathway includes conversion of ferulic acid to vanillate and then probably to protocatechuate (Ohta, 2000).

Growth occurs microaerobically on nitrogen-free medium (NF).[3] On NF semi-solid medium, cells grow at a depth of 18–25 mm below the surface. On media supplemented with 50 mg/l $(NH_4)_2SO_4$, growth occurs at a depth of 8–12 mm (Ohta and Hattori, 1983), indicating use of ammonium as the N source at a higher pO_2 than for utilization of N_2. Acetylene reduction occurs, verifying nitrogen fixation by *A. oligotrophica*; the specific activity is ~10 nmoles/min/mg protein, similar to the activity in many other free-living diazotrophs. Nitrate is reduced to nitrite.

Metabolism and metabolic pathways *A. oligotrophica* is an aerobic/microaerobic organism not extensively characterized in terms of metabolic features. In one study of respiratory characteristics, growth of *A. oligotrophica* on 1% trypticase peptone plus 0.1% yeast extract was biphasic (Ohta and Taniguchi, 1988b). The molar growth yield on oxygen consumed in the first phase increased 1.7-fold in the second phase. Cytochromes of the *a*, *b*, and *c* type are present, as well as a CO-binding *b*-type cytochrome, thought to be cytochrome *o*. The pattern was not detectably changed by medium dilution.

Ecology Strains of *Agromonas oligotrophica* have been isolated from rice paddy soils. They may play a role in decomposing organic matter such as the phenolic acids, ferulic, and coumaric,

---

3. NF medium contains (per liter of distilled water): K_2HPO_4, 50 mg; $MgSO_4 \cdot 7H_2O$, 10 mg; NaCl, 5 mg; $CaCl_2$, 0.1 mg; $FeCl_3$, 0.1mg; $NaMoO_4 \cdot 2H_2O$, 0.2 mg; 2-oxoglutaric acid, 0.3 or 1 g; pH 6.8–7.5. The carbon source—2-oxoglutaric acid—is sterilized by filtration and added aseptically to the autoclaved medium.

TABLE BXII.α.148. Characteristics of *Agromonas oligotrophica*[a]

Characteristic	Reaction
Growth at:	
4°C	−
10–37°C	+
42°C	d
Hydrolysis of gelatin, casein, starch, and cellulose	−
Fluorescent pigment production	+
Polyhydroxybutyrate production	+
Catalase, oxidase	+
Growth on:	
NB	−
NB/10–NB/10,000	+
Utilization of:	
Glucose, galactose, mannose, xylose, L-arabinose, acetic acid, lactic acid, gluconate, pyruvate, citrate, 2-oxo-glutarate, succinate, L-malate, ferulic, acid, p-coumaric acid, p-anisic acid	+
Cellobiose, maltose, lactose, raffinose, benzoate, methanol	−

[a]Symbols: see standard definitions.

and recycling other nutrients in environments low in organic substrates. Nitrogen fixation may be of ecological significance, allowing occupation of this niche.

Ohta and Hattori (1983) reported that irregular rod-shaped, oligotrophic bacteria, many of which were identical to *A. oligotrophica*, were abundant in the roots of rice plants numbering 10^8–10^9 cells/g dry matter. Whether fixed nitrogen can be provided for plant nutrition is not known.

Fatty acid composition The predominant fatty acid in five strains tested is an 18-carbon straight-chain unsaturated acid with one double bond ($C_{18:1}$) (Ohta and Hattori, 1983). Present also among the isolates was a $C_{16:0}$ fatty acid (at 9–14% of total) and $C_{19:1}$ fatty acid (6–18% of total). Ubiquinone Q-10 was found in all isolates.

ENRICHMENT AND ISOLATION PROCEDURES

Oligotrophic bacteria are widely distributed in nature and share the ability to grow on low nutrient media. At microsites of soil, selective growth of oligotrophic bacteria occurs at higher probabilities with lower concentrations of nutrients and at lower probabilities with higher concentrations of nutrients (Hattori, 1981). Consequently, the isolation of *Agromonas* is based on this feature. Soil suspensions are diluted 10^{-3} and 10^{-4}; 1 ml of each dilution is plated onto NB/100 medium containing 1% agar. After 1 week to 1 month, colonies are tested for sensitivity to nutrients by plating on 1× NB. NB-sensitive colonies are then examined for other characteristics as described.

MAINTENANCE PROCEDURES

The medium for maintenance is NB/100 containing 0.3–0.4% agar, dispensed in screw-cap tubes. Cells are kept as a stabbed culture at room temperature. Cells remain viable for several years; transfer every 3–4 years is recommended. An alternative method is the suspension of cells in 20% glycerol followed by storage at −80°C. Cells stored this way are less viable in the long

term than those described above (H. Ohta, personal communication).

DIFFERENTIATION OF THE GENUS *AGROMONAS* FROM OTHER GENERA

The organisms most closely related to *Agromonas*, as indicated by 16S rDNA analysis, are species of *Blastobacter*, *Bradyrhizobium*, and the unclassified DNB strains described by Saito et al. (1998) (see Taxonomic Comments). Table BXII.α.149 summarizes the differential features of these organisms.

TAXONOMIC COMMENTS

Based on comparison of morphological and physiological characteristics, Ohta and Hattori (1983) placed five paddy soil isolates in a new genus, *Agromonas*, and a new species, *Agromonas oligotrophica*. The morphological and physiological characteristics were most similar to those of the genera *Xanthobacter* and *Azospirillum*, but 16S rDNA analysis (Saito et al., 1998) indicated a close relationship between *Agromonas oligotrophica* and *Blastobacter denitrificans*, which form a group (Fig. BXII.α.175). *B. denitrificans* does not fix nitrogen and, unlike *A. oligotrophica*, it can utilize methanol and denitrify (Hirsch and Müller, 1985). However, *B. denitrificans* is oligotrophic and the rod-shaped cells are bent and budding like those of *A. oligotrophica*. Other oligotrophs, called DNB (dilute nutrient broth) bacteria, are also placed in this group (Saito et al., 1998). These DNB strains were isolated from grassland soil by Ohta and Hattori (1980), and only 1 of 11 was able to fix N_2. The DNB isolates were similar to each other in a number of ways but were not assigned a genus or species name. The group containing *B. denitrificans*, *A. oligotrophica*, and the unclassified DNB bacteria is most closely related to a *Bradyrhizobium* sp. Unlike the latter species, neither *A. oligotrophica* nor the DNB bacteria can nodulate siratro (*Macroptilium atropurpureum*). In DNA–DNA hybridization tests, the DNB bacteria appeared more closely related to the *Bradyrhizobium* sp. than to *A. oligotrophica*. The taxonomic cluster carrying *A. oligotrophica*, *B. denitrificans*, the DNB strains, and the *Bradyrhizobium* sp. is related most closely with a group composed of *Rhodopseudomonas palustris*, two *Nitrobacter* species, and the genus *Afipia* (together named the BANA domain; Saito et al., 1998). *Agromonas oligotrophica* is a member of the phylum *Proteobacteria*, order *Rhizobiales*, and the family *Bradyrhizobiaceae*.

ACKNOWLEDGMENTS

I am grateful to Drs. T. Hattori, K. Minamisawa, R. Hattori, and H. Ohta for their cooperation in providing clarifying information, reprints, and electron micrographs, and for their helpful comments concerning the manuscript. Their work allowed *Agromonas oligotrophica* to be known to the world.

FURTHER READING

Hattori, T. and R. Hattori. 2000. The plate count method: an attempt to delineate the bacterial life in the microhabitat of soil. *In* Bollag and Stotzsky (Editors), Soil Biochemistry, Vol. 10, Marcel Dekker Inc., New York, NY. pp. 271–302.

Saito, A., H. Mitsui, R. Hattori, K. Minamisawa and T. Hattori. 1998. Slow-growing and oligotrophic soil bacteria phylogenetically close to *Bradyrhizobium japonicum*. FEMS Microbiol. Ecol. *25*: 277–286.

List of species of the genus Agromonas

1. **Agromonas oligotrophica** Ohta and Hattori 1985, 223[VP] (Effective publication: Ohta and Hattori 1983, 43.)
   ol.i.go.tro′phi.ca. Gr. adj. *oligo* few, low; Gr. n. *tropheia* nourishment; M.L. adj. *oligotrophica* low nutrients.

The characteristics are as described for the genus and in Tables BXII.α.148 and BXII.α.149. Additional characteristics are as follows. Several cells may adhere to each

TABLE BXII.α.149. Characteristics distinguishing *Agromonas oligotrophica* from related taxa[a]

Characteristic	*Agromonas oligotrophica*	DNB bacteria[b]	*Blastobacter denitrificans*	*Bradyrhizobium japonicum* (USDA Strain 110)	*Azospirillum brasilense*	*Xanthobacter* sp.
Budding	+	+	+	−	−	+
Flagellar arrangement:						
Polar	+	−	−	+	+	−
Subpolar	−	+	+	−	+	−
Acid from glucose	−	−	+	−	+	−
Denitrification	−	−	+	−	d	−
Methanol utilization	−	−	+	−	−	−
Nodulation of siratro	−	−	−	+	−	−
N₂ fixation in culture	+	−	−	±[c]	+	+

[a]Symbols: see standard definitions.

[b]Dilute nutrient broth.

[c]Low activity, only in certain media (Keister, 1975).

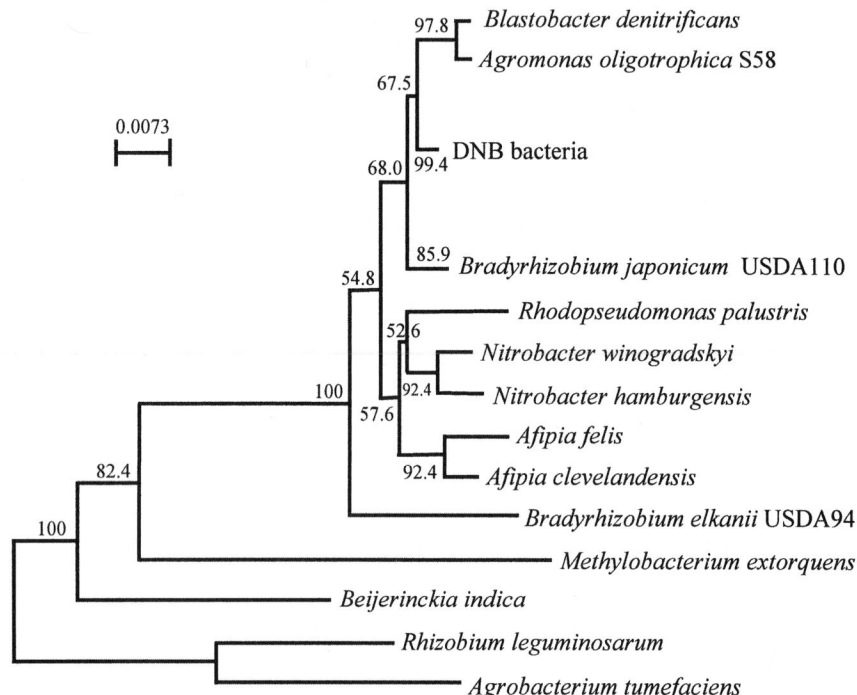

FIGURE BXII.α.175. Phylogenetic relationships of *Agromonas oligotrophica* to other members of the *Alphaproteobacteria* based on near full-length sequence similarities of 16S rRNA genes. The phylogenetic tree was constructed using the neighbor-joining method. Bootstrap values are shown at nodes. The scale bar indicates substitutions per site. The group DNB are unclassified oligotrophs, mostly non-nitrogen fixing, isolated from grassland soil (Saito et al., 1998).

other, forming a rosette. Colonies are colorless/white, punctiform, pulvinate, and entire.

NaCl, KCl, Casamino acids, peptone, and meat extract inhibit growth at 0.5–1.0%. Several sugars and organic acids are utilized. The aromatic acids ferulic, *p*-coumaric, and *p*-anisic can be utilized, but not benzoic acid. Neither acid

nor gas is produced from glucose. Cellulose and starch are not hydrolyzed.

The mol% G + C of the DNA is: 65.1–66.0 (T_m).

Type strain: ATCC 43045, DSM 12412, JCM1494.

GenBank accession number (16S rRNA): D78366.

Genus IV. **Blastobacter** *Zavarzin 1961, 962[AL] emend. Sly 1985, 44*

LINDSAY I. SLY AND PHILIP HUGENHOLTZ

Blas.to.bac' ter. Gr. n. *blastos* bud, shoot; M.L. n. *bacter* equivalent of Gr. masc. n. *bactrum* rod; M.L. masc. n. *Blastobacter* a budding rod.

Cells **ovoid rods, wedge or club shaped**, or pleomorphic, **often slightly curved and occasionally branched**. Cell poles are rounded or slightly tapering on one pole. Cell size range is 0.5–1.0 × 1.0–

4.5 μm. Cells **may form rosettes. Gram negative. New cell formation and multiplication occur by budding** on the free cell pole, subpolarly or laterally. Young **buds initially rod shaped, ovoid,** or spherical to oblong. Buds may be released or remain attached. **Motile, flagellated swarmer cells may occur** in some species. Do not form spores or cysts. Some strains produce large amounts of exopolymer. Grow in liquids as turbidity, pellicle, or precipitate. Colony pigmentation may be yellow, or colorless, becoming brown in older cultures.

Aerobic. Heterotrophic. Oxidase, catalase, and peroxidase positive. Some species grow chemolithotrophically with hydrogen. Some species may fix CO_2 reductively when grown on methanol or methylated amines. Optimal pH between 6.8 and 7.8. Temperature range for growth is 10–46°C. Carbon and energy sources may be alcohols, sugars, organic acids, or some amino acids. Ammonium, nitrate, urea, peptone, yeast extract, or casein hydrolysate may be utilized as nitrogen sources. May reduce nitrate to nitrite, or denitrify.

Isolated from freshwater lakes, ponds, and groundwater, as well as activated sludge.

The mol% G + C of the DNA is: 59–69.

Type species: **Blastobacter henricii** Zavarzin 1961, 962.

FURTHER DESCRIPTIVE INFORMATION

Cells of *Blastobacter* generally have a rod-shaped morphology and reproduce by budding at the poles or laterally, with cells often attaching at the nonreproductive pole to form rosettes (Fig. BXII.α.176). The characteristics of the species of *Blastobacter* are mainly based on the study of the type strains of the five species and little is known of the diversity within each species. All strains have yet to be characterized over a constant range of tests in a single polyphasic study, making it difficult at this time to compare the species. A continuing difficulty is the absence of a culture for the type species *B. henricii*, whose description is based solely on morphological characteristics obtained from examination of an enrichment culture (Zavarzin, 1961). 16S rRNA sequences are available for the three validly published species *B. aggregatus*, *B. capsulatus*, and *B. denitrificans*, but not for the two species "*B. aminooxidans*" and "*B. viscosus*" that have not been validated.

Hirsch and co-workers have searched for *Blastobacter* cultures from various freshwater habitats and obtained pure cultures from lakes and ponds (Hirsch and Müller, 1985), resulting in the description of three new species—*B. aggregatus*, *B. capsulatus*, and *B. denitrificans*—which together with *B. henricii* are the only validly described species. In the absence of a type culture for *B. henricii*, the

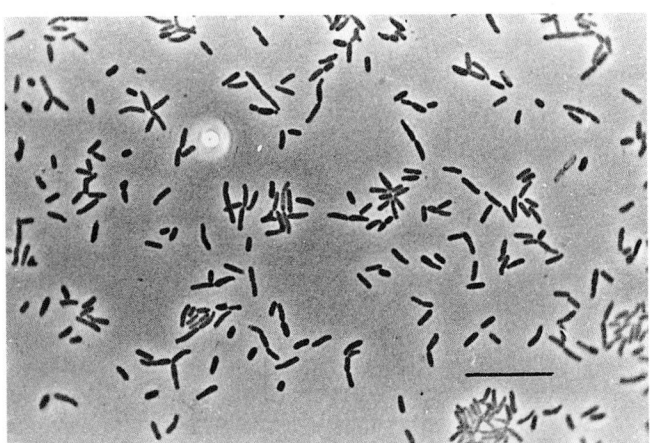

FIGURE BXII.α.176. Photomicrograph of actively growing culture of *Blastobacter aggregatus* on PYG medium, showing budding, rod-shaped cells, and rosette formation. Bar = 10 μm.

Hirsch and Müller (1985) argued that their new species belonged to the genus *Blastobacter* based on cellular morphological similarity, and physiological differences with other budding bacteria such as *Rhodopseudomonas*, *Nitrobacter*, and *Methylosinus trichosporium*.

Morphologically, cells of *B. aggregatus*, *B. capsulatus*, and *B. denitrificans* exhibit budding, and no cross-septation occurs without prior budding (Hirsch and Müller, 1985). Cells of these species are essentially rod shaped and of varying length; shorter cells are often ovoid. Cells of *B. capsulatus* IFAM 1004 have a tendency to bend or even twist helically (Hirsch and Müller, 1985). Of these, only *B. aggregatus* IFAM 1003 forms rosettes, and is therefore most morphologically similar to the description of the type species *B. henricii*. Two species, *B. aggregatus* and *B. denitrificans*, produce motile swarmer cells with flagella while *B. capsulatus* and *B. henricii* are nonmotile.

B. aggregatus, *B. capsulatus*, and *B. denitrificans* utilize D-glucose, D-fructose, lactose, D-ribose, D-galactose, mannitol, glycerol, pyruvate, α-oxoglutarate, acetate, L-glutamate, and L-histidine as carbon compounds, but are unable to utilize maltose, starch, dextrin, inulin, fumarate, lactate, caproate, citrate, propionate, indole, phthalate, D,L-phenylalanine, L-aspartate, leucine, L-serine, glycine, and alanine. These species are reported to produce acid but no gas anaerobically from D-fructose, and D-galactose in the Hugh and Leifson fermentation test (Hirsch and Müller, 1985). Production of acid aerobically from D-glucose and mannitol varies among species (Table BXII.α.150). In addition to the generic characters, these species do not decarboxylate arginine or lysine, deaminate lysine, or phenylalanine, hydrolyze gelatin or cellulose, or produce acetoin, indole, or H_2S. Ammonium, nitrate, and methane are not oxidized. Litmus milk is not acidified, peptonized, or coagulated (Hirsch and Müller, 1985).

Two nomenclaturally invalid species, "*B. aminooxidans*" and "*B. viscosus*", were isolated from activated sludge and studied extensively for their morphological and physiological characteristics (Loginova and Trotsenko, 1979; Doronina et al., 1983).* These species grow on C_1 compounds as carbon and energy sources and are facultative autotrophs capable of chemolithotrophic growth with hydrogen. Strains of "*B. viscosus*" (Loginova and Trotsenko, 1979) were isolated from activated sludge of the purifying installation of the Baikal Pulp and Paper Combine in Russia. Studies on the formation of the bud in "*B. viscosus*" show that budding begins with a centripetal ingrowth of the plasma membrane and the cell wall with full delimitation of the bud from the maternal cell. Incomplete separation and multiple budding leads to the formation of pleomorphic cells. Crystalloid formations consisting of parallel arrangements of electron dense and electron transparent filamentous structures are observed in the region of bud formation (Loginova and Trotsenko, 1979).

When grown on a medium with methanol as carbon source, "*B. viscosus*" produces a heteropolysaccharide consisting of galactose, glucose, rhamnose, xylose, and glucuronic acid (Loginova and Trotsenko, 1979).

"*B. viscosus*" grows weakly with methylamine, dimethylamine, formate, formamide, and dimethylformamide, but not trimethylamine, formaldehyde, dimethyl sulfoxide, or dimethylsulfone as carbon sources (Loginova and Trotsenko, 1979).

Ribulose phosphate carboxylase is induced in "*B. viscosus*" when C_1 compounds are utilized, but no activities of specific

---

Editorial Note: Recently "*Blastobacter viscosus*" and "*Blastobacter aminooxidans*" have been reclassified as *Xanthobacter viscosus* and *Xanthobacter aminoxidans* (Doronina and Trotsenko, 2003).

TABLE BXII.α.150. Differential characteristics of the genus *Blastobacter* species and other heterotrophic, Gram-negative, rod-shaped budding bacteria[a]

Characteristic	*Blastobacter henricii*	*Blastobacter aggregatus*	*Blastobacter capsulatus*	*Blastobacter denitrificans*	"*Blastobacter aminooxidans*"	"*Blastobacter viscosus*"	*Blastomonas natatoria*	*Agromonas oligotrophica*	*Gemmobacter aquatilis*
Cellular morphology:									
Shape	Rods, wedge- or club-shaped, often curved	Ovoid to rod-shaped	Rods, often bent and tapering on budding pole, older cells Y-shaped	Rods with rounded cell poles	Pleomorphic rods with minute appendages	Pleomorphic rods, often bent and branched	Rods, slightly curved, or wedge-shaped	Bent, branched rods	Ovoid to rod-shaped, short chains
Cell length (μm)	2.0–4.5	1.5–2.3	1.5–2.3	1.5–2.3	1.5–3.0	1.0–3.2	1.0–3.0	2.0–7.0	1.2–2.7
Cell width (μm)	0.7–1.0	0.6–0.8	0.7–0.9	0.6–0.8	0.8–1.0	0.5–0.9	0.5–0.8	0.6–1.0	1.0–1.2
Initial bud shape:									
Ovoid									
Rods			+		+	+			
Spherical				+					
Spherical or ovoid		+					+	+	+
Spherical or oblong	+								
Bud origin	Polar	Narrow pole	Narrow pole or lateral	Slightly subpolar	Polar or lateral	Polar or subpolar	Polar	Polar	Polar or subpolar
Capsule formation	nd[b]	–	+	–	+	+	–	nd	–
Motility	–	+	–	+	–	–	+	+	–
Rosette formation	+	+	–	–	–	–	+	+	–
Colony pigmentation	nd	Colorless, slightly brownish when older	Colorless	Colorless, brownish when older	Yellow (orange)	Yellow	Yellow or pale pink	Colorless	Colorless
Origin	Forest brook water	Lake water	Eutrophic pond water	Lake water	Activated sludge	Activated sludge	Swimming pool	Paddy soil	Forest pond
Mol% G + C	nd	60	59	65	69	66	65	66	63

(*continued*)

TABLE BXII.α.150. *(cont.)*

Characteristic	*Blastobacter henricii*	*Blastobacter aggregatus*	*Blastobacter capsulatus*	*Blastobacter denitrificans*	*"Blastobacter aminooxidans"*	*"Blastobacter viscosus"*	*Blastomonas natatoria*	*Agromonas oligotrophica*	*Gemmobacter aquatilis*
Utilization of carbon sources:									
Sucrose	nd	+	+	−	+	+	nd	nd	+
Cellobiose	nd	+	+	−	+	+	nd	−	nd
Salicin	nd	−	+	−	nd	nd	nd	nd	nd
Glucuronic acid lactone	nd	−	−	+	nd	nd	nd	nd	nd
N-Acetyl-glucosamine	nd	−	−	+	nd	nd	nd	nd	−
Succinate, malate	nd	−	−	+	+	+	nd	+	+
Formate, formamide	nd	−	−	+	+	+	nd	nd	nd
Tartrate	nd	nd	−	+	nd	−	nd	nd	+
Glutamate	nd	+	nd	nd	nd	nd	nd	nd	+
L-Arginine	nd	+	−	−	−	−	nd	nd	+
L-Proline	nd	−	+	−	nd	−	−	−	nd
Methanol	nd	−	−	+	−	+	nd	nd	−
Ethanol	nd	+	−	+	+	+	nd	nd	+
Acid production from:									
Glucose	nd	+	+	−	−	+	+	−	+
Mannitol	nd	−	+	−	nd	nd	−	nd	+
Utilization of urea as N source	nd	+	−	nd	nd	−			
Nitrate reduction:									
Assimilatory	nd	−	+	−	+	+	−	nd	+
Dissimilatory	nd	−	+	+	+	+	−	nd	+
Hydrogen autotrophy	nd	nd	nd	nd	+	+	−	nd	
Temperature range for growth (°C)	nd	13–43	14–35	13–46	10–34	10–34	11–39	nd	17–39
Temperature optimum (°C)	nd	36	27	41	29–32	28–30	25–30	nd	31
pH optimum for growth	nd	6.9	7.3–7.8	6.8–7.2	7.2–7.8	6.8–7.2	nd	nd	nd

[a]Data from Loginova and Trotsenko (1979), Doronina et al. (1983), Hirsch and Müller (1985), Rothe et al. (1987), Trotsenko et al. (1989), and Saito et al. (1998).

[b]nd, not determined.

enzymes of the serine and hexulose monophosphate pathways characteristic of methylotrophs were observed. "*B. viscosus*" is apparently able to utilize the carbon from methanol autotrophically after its oxidation to CO_2 (Loginova and Trotsenko, 1979).

Like "*B. viscosus*", "*B. aminooxidans*" was also isolated from activated sludge. "*B. aminooxidans*" was the first organism in which autotrophic assimilation of methylated amines was shown. As well as the ability to utilize methylated amines, "*B. aminooxidans*" is able to grow chemolithotrophically with hydrogen (Doronina et al., 1983). Trimethylamine is oxidized to dimethylamine and formaldehyde by dehydrogenase (phenazine methosulfate). Methylamine is further oxidized to formaldehyde and ammonia, and the formaldehyde is oxidized via formate to CO_2. The CO_2 is assimilated by the ribulose bisphosphate pathway. The serine and hexose monophosphate pathways are not present. The cells possess enzymes of the tricarboxylic acid cycle and the glyoxylate shunt. Ammonium is assimilated via reductive amination of α-glutarate, pyruvate, and glyoxylate, and via the glutamate cycle (Doronina et al., 1983).

In "*B. aminooxidans*", trimethylamine and methylamine are oxidized by a monooxygenase (NADH or NADPH). Formaldehyde is oxidized via formate to CO_2 by dehydrogenases: the CO_2 is refixed into 3-phosphoglyceric acid by using the autotrophic ribulose-1,5-bisphosphate pathway (Trotsenko et al., 1989). Phosphoribulose kinase is also present and active. In "*B. viscosus*" autotrophic growth occurs in an atmosphere of $H_2/O_2/CO_2$ or with methanol. In both cases, cells assimilate CO_2 via the ribulose bisphosphate pathway and show active phosphoribulokinase (PRK) and ribulose bisphosphate carboxylase (RBPC). In contrast to PRK, the RBPC is completely repressed in glucose-grown cells. The primary CO_2 acceptor is regenerated by transaldolase and transketolase activity (Trotsenko et al., 1989).

Cells of "*B. viscosus*" possess dehydrogenase activity, catalyzing methanol oxidation via formaldehyde and formate to CO_2. The serine and hexulose-phosphate pathways of C_1 metabolism do not operate due to absence of hydroxypyruvate reductase, serine-glyoxylate aminotransferase, ATP malate lyase, and hexulose-phosphate synthase. Fructose-1,6-bisphosphate aldolase and glyceraldehyde-phosphate dehydrogenase (NAD) play an important role in metabolic conversions of phosphotrioses. The cells contain all enzymes of the citric acid cycle with lower levels in methanol-grown cells than in glucose-grown cells. C_4 compounds are resynthesized mainly by carboxylation of pyruvate and phosphoenolpyruvate (Trotsenko et al., 1989).

ENRICHMENT AND ISOLATION PROCEDURES

A low-nutrient medium containing glucose, peptone, and yeast extract such as that described by Staley (1981b) is suitable for the enrichment and isolation of *Blastobacter*. Cultures may be isolated by direct plating of environmental samples, or after enrichment. Alternatively, water samples may be enriched with 0.005% peptone and incubated aerobically at 18–23°C for 1–3 weeks (Hirsch and Müller, 1985; Trotsenko et al., 1989). "*Blastobacter viscosus*" was isolated from activated sludge after enrichment in mineral salts medium containing 1% methanol at 30°C for 7 d (Loginova and Trotsenko, 1979).

MAINTENANCE PROCEDURES

Cultures of *B. aggregatus*, *B. capsulatus*, and *B. denitrificans* are reported (Hirsch and Müller, 1985) to grow well on dilute peptone–yeast extract–glucose (PYG) medium (Staley, 1981b). "*B. viscosus*" grows well on media with methanol as the carbon source

(Loginova and Trotsenko, 1979), and "*B. aminooxidans*" may be grown on media with trimethylamine as the carbon source or on glucose potato agar (Doronina et al., 1983). Cultures may be preserved by cryogenic storage in liquid nitrogen when suspended in sucrose peptone broth containing 10% glycerol, and by freeze-drying in glucose peptone broth containing horse serum.

DIFFERENTIATION OF THE GENUS *BLASTOBACTER* FROM OTHER GENERA

Budding of bacteria is widespread throughout the *Proteobacteria* and occurs in phylogenetically diverse genera with widely different physiological characteristics (Rothe et al., 1987; Hugenholtz et al., 1994; Sly and Cahill, 1997) including *Agromonas*, *Blastobacter*, *Blastomonas*, *Gemmobacter*, *Rhodopseudomonas*, *Methylosinus*, and *Nitrobacter*. Table BXII.α.150 gives the characteristics that are useful in differentiating the species of Gram-negative, heterotrophic rod-shaped budding bacteria.

TAXONOMIC COMMENTS

The taxonomy of the genus *Blastobacter* Zavarzin 1961 is in need of revision. The polyphyletic nature of the genus is well established (Hugenholtz et al., 1994; Sly and Cahill, 1997), and emphasizes that morphological characteristics such as budding cell division, which define the genus, are not phylogenetically useful features at the genus level and may group distantly related species in the genus as currently defined (Zavarzin, 1961; Sly, 1985; Trotsenko et al., 1989). Cell division by nonprosthecate budding is confined to the *Alphaproteobacteria* but is widely distributed in physiologically diverse genera within the class. As reported previously (Hugenholtz et al., 1994; Sly and Cahill, 1997), resolution of the taxonomic confusion is impeded by the lack of a type strain for the type species, *Blastobacter henricii* (Zavarzin, 1961; Skerman et al., 1989), which was never obtained in pure culture.

The genus *Blastobacter* was proposed by Zavarzin (1961) to include rosette-forming, budding, rod-shaped or wedge-shaped bacteria that were observed in a filter paper enrichment of reduced iron-containing water from a northern Russian forest brook. Zavarzin was unable to isolate the cells in pure culture, and the description of *Blastobacter henricii* is based on drawings and observations.

Several additional *Blastobacter* species have been validly described since 1961 (Moore and Moore, 1992). These include *Blastobacter aggregatus*, *Blastobacter capsulatus*, *Blastobacter denitrificans* (Hirsch and Müller, 1985), and *Blastobacter natatorius* (Sly and Hargreaves, 1984; Sly, 1985). *Blastobacter natatorius* was later transferred and became the type species of the genus *Blastomonas* in the first step to clarify the taxonomy of the genus *Blastobacter* (Sly and Cahill, 1997). Other taxonomically invalid species include "*Blastobacter aminooxidans*" (Doronina et al., 1983), "*Blastobacter viscosus*" (Loginova and Trotsenko, 1979), and "*Blastobacter novus*" (Rezanka et al., 1991). Several authors have demonstrated that there is a high degree of heterogeneity in the genus *Blastobacter* with respect to phenotype (Trotsenko et al., 1989), cellular fatty acids and phospholipids (Sittig and Hirsch, 1992), and molecular phylogeny (Rothe et al., 1987; Green and Gillis, 1989; Hugenholtz et al., 1994; Sly and Cahill, 1997; Willems et al., 2001).

Fig. BXII.α.177 shows the phylogenetic relationships of the species of *Blastobacter* for which 16S rRNA sequences are available. An analysis of the 16S rRNA sequences of the validated species shows that species of *Blastobacter* are polyphyletic and belong to

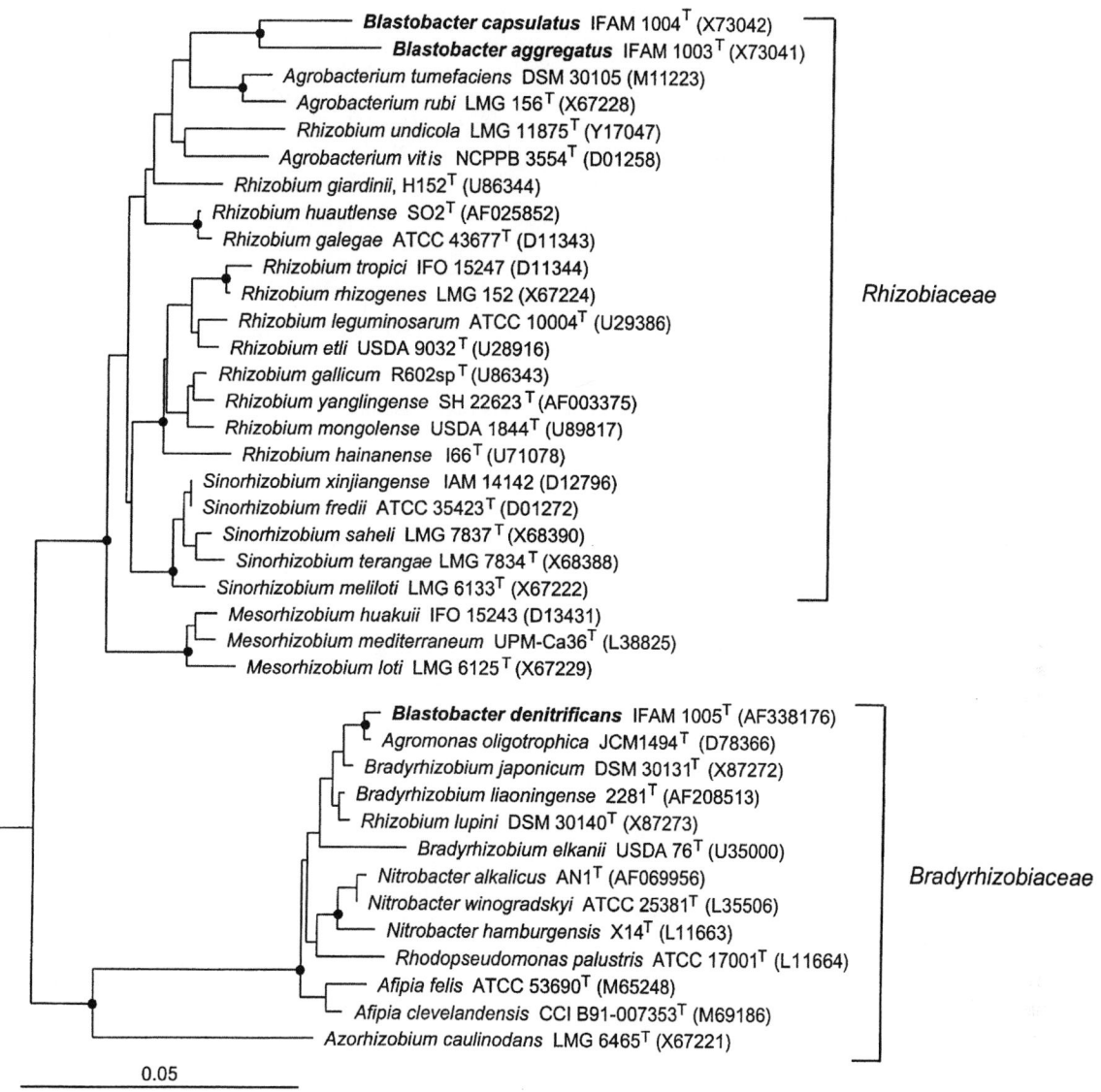

FIGURE BXII.α.177. Phylogenetic tree showing the polyphyletic positions of *Blastobacter* species in the *Alphaproteobacteria* based on 16S rRNA sequence similarities.

separate lineages in the *Alphaproteobacteria*. *Blastobacter aggregatus* and *Blastobacter capsulatus* are close phylogenetic relatives belonging to a strongly supported branch of the family *Rhizobiaceae*, whereas *Blastobacter denitrificans* belongs to a cluster containing the genus *Bradyrhizobium* and *Agromonas oligotrophica* in the family *Bradyrhizobiaceae*. *Blastobacter aggregatus* and *Blastobacter capsulatus* are closely related to each other (96.6% sequence similarity) and to *Agrobacterium tumefaciens* (96.0–96.9% sequence similarity). This confirms the previous finding of Rothe et al. (1987) concerning the close relationship between *Blastobacter aggregatus* and *A. tumefaciens* determined by rRNA oligonucleotide catalogue analysis. *Blastobacter denitrificans* on the other hand is most closely related to the type species of the nitrogen-fixing genus *Agromonas* from paddy soil, *Agromonas oligotrophica*, and to *Bradyrhizobium japonicum*. The relationship of *Blastobacter denitrificans* to *Bradyrhizobium japonicum* was first observed by Willems and Collins (1992) based on 16S rRNA sequence similarities and by Green and Gillis (1989) based on rRNA cistron similarities. Willems et al. (2001) showed that the 16S rRNA sequences of photosynthetic

Bradyrhizobium strains, *Agromonas oligotrophica*, and *Blastobacter denitrificans* belong to a well-supported cluster and may represent a separate genus. Further data on the DNA–DNA hybridization levels between the species in this cluster are required to determine the relationships between the members of this phylotype.

Correction of the taxonomic problems within *Blastobacter* is not straightforward because a culture of the type species *Blastobacter henricii* was never isolated. Hugenholtz et al. (1994) proposed that a new type species for the genus be designated. In the absence of physiological and phylogenetic information about the type species, there is no way of knowing to which phylogenetic lineage the true blastobacters as described by Zavarzin (1961) belong. One solution is to reserve the genus *Blastobacter* at this time for *Blastobacter henricii* in case a culture matching the description can be isolated from the same habitat in the future, and to describe new genera for the other species. Given that the description of *B. henricii* is too limited for reliable assignment of isolates to the species, a better solution is to retain the genus for one of the phylotypes of extant species. Circumscription of the genus to include *B. henricii*, *B. aggregatus*, and *B. capsulatus* would be straightforward. The taxonomic position of *Blastobacter denitrificans*, on the other hand, is more problematic given its close relationship with the budding bacterium *Agromonas oligotrophica* and photosynthetic *Bradyrhizobium* strains. A reasonable taxonomic solution, therefore, will be to retain *Blastobacter aggregatus* and *Blastobacter capsulatus* in the genus *Blastobacter* on the grounds that they belong to a well-supported phylogenetic group sufficiently distant from related taxa at the genus level and that *Blastobacter aggregatus* appears to have characteristics closest to the original description of *Blastobacter* (Hirsch and Müller, 1985). We propose that the species *B. henricii*, *B. aggregatus*, and *B. capsulatus* be retained in the genus *Blastobacter* and that *Blastobacter denitri-*

ficans be transferred to the genus *Agromonas*. DNA–DNA hybridization between *Blastobacter denitrificans* and *Agromonas oligotrophica* will be required to determine species relationships and nomenclatural priority within the genus *Agromonas*. The budding bacterium *Gemmobacter aquatilis* was shown to be most closely related to *Rhodobacter capsulatus* by rRNA catalogue analysis (Rothe et al., 1987). A complete 16S rRNA sequence is required to determine the exact phylogenetic position of this phenotypically and phylogenetically related species. Determination of the phylogenetic positions of the invalid species "*B. aminooxidans*" and "*B. viscosus*" will be required to resolve the remaining taxonomic uncertainty in the genus.

List of species of the genus Blastobacter

1. **Blastobacter henricii** Zavarzin 1961, 962[AL]

hen.ri′ci.i. M.L. gen. n. *henricii* of Henrici; named for A. Henrici, an American microbiologist who may have been the first to see bacteria belonging to the genus *Blastobacter*.

Cells are rod-, wedge-, or club-shaped, 0.7–1.0 × 2.0–4.5 µm. Cells form rosettes by attaching to each other with the nonreproductive, frequently tapered pole. A glistening corpuscle was seen in the center of the rosette. Single buds (spherical to oblong) are formed terminally on the blunt cell pole. Released buds are 0.3 µm wide and nonmotile. Originally found in a cylinder containing iron-rich forest brook water from Northern Russia and to which shreds of filter paper had been added. Growth was best in a zone of reduced iron at a pH of 6.2. A pure culture was not obtained.

The mol% G + C of the DNA is: unknown.

Type strain: No culture has been isolated.

2. **Blastobacter aggregatus** Hirsch and Müller 1986, 354[VP] (Effective publication: Hirsch and Müller 1985, 284.)

ag.gre.ga′ tus. L. adj. *aggregatus* joined together, referring to the frequent formation of rosettes.

Cells are 0.6–0.8 × 1.5–2.3 µm, ovoid to rod shaped. Multiplication by rod-shaped buds formed at the narrow and unattached cell pole. Bud cells are motile and attach to each other to form rosettes. Colonies colorless to slightly beige or brownish, round with entire edges, dull to shiny. Liquid cultures turbid. Grow well in dilute media containing peptone, yeast extract, and glucose. Vitamins not required. Grow in presence of up to 36 g/l NaCl. Further characteristics are given in Table BXII.α.150. Isolated from surface water of Lake Höftsee (Holstein, Germany).

The mol% G + C of the DNA is: 60 (T_m).

Type strain: Müller 161, ATCC 43293, DSM 1111, IFAM 1003.

GenBank accession number (16S rRNA): X73041.

3. **Blastobacter capsulatus** Hirsch and Müller 1986, 354[VP] (Effective publication: Hirsch and Müller 1985, 285.)

cap.su.la′ tus. L. n. *capsula* a small chest, capsule; M.L. neut. n. *capsulatus* encapsulated.

Cells are 0.7–0.9 × 1.5–2.3 µm, rod shaped to short ovoid, often bent and narrowing on one cell pole. Older cells frequently Y shaped. Buds are produced terminally on the narrow cell pole or occasionally laterally. Cells form capsules. Nonmotile. Rosettes not formed. Grow in liquids

as turbidity or pellicle. Exopolymer produced. Nitrate is reduced without gas formation. Growth occurs in the presence of up to 27 g/l NaCl. Grow well in dilute media containing peptone, yeast extract, and glucose. Vitamins not required. Further characteristics are given in Table BXII.α.150. Isolated from a shallow eutrophic pond near Westensee (Kiel, Germany).

The mol% G + C of the DNA is: 59 (T_m).

Type strain: Müller 216, ATCC 43294, DSM 1112, IFAM 1004.

GenBank accession number (16S rRNA): X73042.

4. **Blastobacter denitrificans** Hirsch and Müller 1986, 354[VP] (Effective publication: Hirsch and Müller 1985, 285.)

de.ni.tri′ fi.cans. L. prep. *de* away from; L. n. *nitrum* soda; M.L. n. *nitrum* nitrate; M.L. v. *denitrifico* denitrify; M.L. part. adj. *denitrificans* denitrifying.

Cells are 0.6–0.8 × 1.5–2.3 µm, rod shaped with rounded poles. Rod-shaped buds formed subpolarly, motile with 1–3 subpolar flagella. Do not produce capsules. Rosettes not formed. Growth in liquid media is turbid. Colonies glistening, round with entire edges, initially colorless, later beige to brownish in transmitted light. Grow well in dilute media with peptone, yeast extract, and glucose. Grow well with C_1 compounds such as methanol, formate, or formamide. Denitrification occurs with nitrogen gas formed from nitrate anaerobically. Growth occurs in the presence of up to 27g/l NaCl. Further characteristics are given in Table BXII.α.150. Isolated from surface water of Lake Plussee (Holstein, Germany).

The mol% G + C of the DNA is: 65 (T_m).

Type strain: Müller 222, ATCC 43295, DSM 1113, IFAM 1005.

GenBank accession number (16S rRNA): X66025.

5. **"Blastobacter aminooxidans"** Doronina, Govorukhina and Trotsenko 1983, 552.

a.mi.no.ox′ i.dans. M.L. n. *aminum* amine; M.L. v. *oxido* make acid, oxidize; M.L.part. adj. *aminooxidans* oxidizing amines.

Cells are 0.8–1.0 × 1.5–3.0 µm, rod shaped, often pleomorphic, forming Y-shaped cells. Cells frequently possess minute tube-like appendages on one pole. Multiplication is by nonmotile, oval buds formed terminally or laterally. Rosettes not formed. Colonies on agar media with trimethylamine or glucose are yellow, 2 mm in diameter, convex, round, glistening, and opaque. Colonies have a smooth surface, an entire edge, uniform consistency, and are viscous. Exopolymer not produced. Vitamins are not required. Gelatin and starch are hydrolyzed; milk is alkalinized, but not peptonized or coagulated. Nitrates are reduced to nitrite. Acid but no gas produced from glucose. Methyl red and Voges–Proskauer tests are negative. Carbon sources utilized include monomethylamine, dimethylamine, or trimethylamine, ethanol, butanol, mannitol, xylose, glucose, raffinose, fumarate, malate, and lactate. Nitrogen sources utilized are ammonium, nitrate and peptone, methylated amines, and certain amino acids, but not nitrite. Autotrophic growth occurs in an atmosphere of $H_2/CO_2/O_2$. The main fatty acids of whole cells are $C_{18:1}$ (48%) and $C_{19:0}$ (27%). Further characteristics are given in Table BXII.α.150. Isolated from activated sludge of a sewage purification system at a pulp and paper mill.

The mol% G + C of the DNA is: 69 (T_m).

Deposited strain: 14a (Culture Collection Institute of Microbial Biochemistry and Physiology, Academy of Sciences, Russia.)

6. **"Blastobacter viscosus"** Loginova and Trotsenko 1979, 650. *vis.co'sus.* L. adj. *viscosus* sticky.

Cells are rod shaped to pleomorphic, often bent, and occasionally branched. Microcapsules may be present. Cell size range is 0.5–0.9 × 1.0–3.2 μm. Buds are produced polarly and laterally and are ovoid and nonmotile. Rosettes not formed. Colonies on agar containing peptone and glucose are 3–4 mm in diameter, round, convex, shiny, and opaque, with an even edge, smooth surface, and a slimy consistency. Colony pigmentation is yellow. In liquid medium with methanol and glucose, growth occurs as turbidity and a slimy sediment. Vitamins not required. Exopolysaccharide containing galactose, glucose, rhamnose, xylose, and glucuronic acid is formed. Carbon sources utilized are galactose and sucrose but not methane, alanine, or glycine. Autotrophic growth occurs in an atmosphere of $H_2/O_2/CO_2$ or with methanol. Utilizes ammonium, peptone, and amino acids as nitrogen sources. Gelatin, starch, and cellulose are not hydrolyzed. Acid but no gas produced from glucose. Milk alkalinized but not peptonized or coagulated. Acetoin, indole, or H_2S not produced. The main fatty acids of PYG agar-grown cells are $C_{18:1}$ (63%) and $C_{16:0}$ (14%). Isolated from activated sludge of the drainage system of the Baikal paper mill in Russia.

The mol% G + C of the DNA is: 66 (T_m).

Deposited strain: 7d, UCMV-1439D (Culture Collection, Institute Microbial Biochemistry and Physiology, Academy of Sciences, Pushchino, Russia.)

Genus V. **Bosea** Das, Mishra, Tindall, Rainey and Stackebrandt 1996, 985[VP]

SUBRATA K. DAS

Bos'e.a. M.L. gen. n. *Bosea* named after Sir J.C. Bose, the founder of Bose Institute, Calcutta, India.

Rod shaped cells, 0.85 × 1.4-1.6 μm. Gram negative. **Aerobic.** Occurring singly. **Motile by a single polar flagellum.** Optimal temperature for growth: 30–32°C, range; 20–37°C. Optimal pH: 7.5–8.0; range, 6.0–9.0. **Chemolithoheterotrophic**, able to obtain energy from the oxidation of reduced sulfur compounds in the presence of organic carbon. No autotrophic growth occurs. Catalase- and oxidase-positive. Found in cultivated soil.

The mol% G + C of the DNA is: 68.2.

Type species: **Bosea thiooxidans** Das, Mishra, Tindall, Rainey and Stackebrandt 1996, 985.

FURTHER DESCRIPTIVE INFORMATION

Colonies on GYM medium[1] supplemented with $Na_2S_2O_3 \cdot 5H_2O$ and an organic substrate (Das et al., 1996) are round, circular, smooth, mucoid, and cream colored. *Bosea thiooxidans* has a single polar flagellum, which is observed with electron microscopy (Fig. BXII.α.178). Pigments are not produced.

The phospholipid composition consists of phosphatidyl glycerol, phosphatidyl ethanolamine, phosphatidyl choline, diphosphatidyl glycerol, and an amino lipid. The cellular fatty acid composition of *B. thiooxidans* strain BI-42 is presented in Table BXII.α.151.

Neither ammonia nor nitrate serves as the sole nitrogen source for growth. *B. thiooxidans* lacks glutamate synthase activity and therefore is a glutamate auxotroph.

In batch culture, growth and thiosulfate oxidation by this organism in GYM medium without any extra carbon source is gra-

tuitous, since it does not increase the growth yield. The low cell yield indicates that, like other chemolithoheterotrophic organisms such as *Thiobacillus* Q (Gommers and Kuenen, 1988) and *Catenococcus thiocycli* (Sorokin, 1992), *B. thiooxidans* cannot assimilate carbon dioxide autotrophically; a failure to detect ribulose 1,5-bisphosphate carboxylase is consistent with this conclusion. A marked stimulation of growth does occur when GYM medium is supplemented with sodium succinate in addition to thiosulfate. The growth yield varies with the final concentration of thiosul-

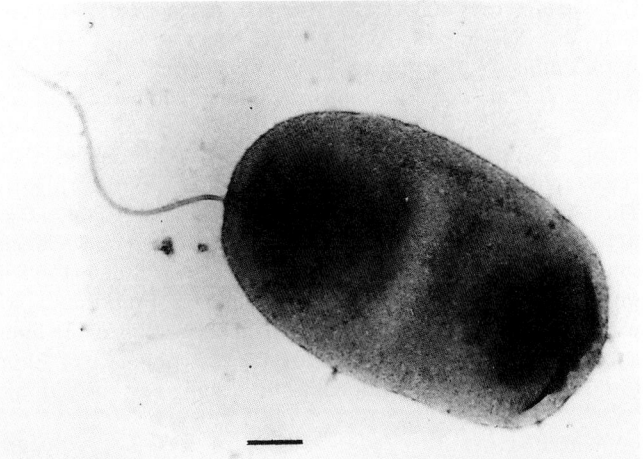

FIGURE BXII.α.178. Transmission electron micrograph of *Bosea thiooxidans* showing the single, polar flagellum. Bar = 200 nm. (Reprinted with permission from S.K. Das et al, International Journal of Systematic Bacteriology *46*: 981–987, 1996, ©Society for General Microbiology.)

---

1. GYM-medium (Glutamate–yeast extract–mineral salts medium), g/l distilled water: Na_2HPO_4, 4.0; KH_2PO_4, 1.5; $MgCl_2 \cdot 5H_2O$, 0.1; sodium glutamate, 0.5; and yeast extract powder, 0.1. Final pH, 8.0.

TABLE BXII.α.151. Cellular fatty acid composition of *Bosea thiooxidans* (strain BI-42)

Fatty acids	Designation	%[a]
Nonhydroxylated acids:		
Pentadecanoic acid	$C_{15:0}$	2.33
cis-Hexadec-9-enoic acid	$C_{16:1\ \omega7c}$	3.7
Hexadecanoic acid	$C_{16:0}$	7.58
cis-Hepta-9-enoic acid	$C_{17:1\ \omega8c}$	3.56
cis-9,10-Methylene hexadecanoic acid	$C_{17:0\ cyclo}$	2.08
Heptadecanoic acid	$C_{17:0}$	4.82
cis-Octadec-9-enoic acid	$C_{18:1\ \omega7c}$	61.55
Octadecanoic acid	$C_{18:0}$	0.67
cis-Cyclo-10,11-methylene octadecanoic acid	$C_{19:0\ cyclo\ \omega8c}$	8.35
10-Methyl octadecanoic acid	$C_{19:0\ 10CH_3}$	0.71
Hydroxylated acids:		
3-Hydroxypentadecanoic acid	$C_{15:0\ 3OH}$	0.49
3-Hydroxyhexadecanoic acid	$C_{16:0\ 3OH}$	3.25
3-Hydroxyheptadecanoic acid	$C_{17:0\ 3OH}$	0.9

[a]Percentages of the total fatty acids. (Adapted by permission of the Society for General Microbiology and with permission from S.K. Das et al., International Journal of Systematic Bacteriology *46:* 981–987, 1996.)

fate; maximum growth (A$_{660}$, 1.2–1.3) occurs at a concentration of 5 g/l. Under these growth conditions there is a stoichiometric conversion of thiosulfate to sulfate with a concurrent decrease in the pH of the medium from 8.0–6.6. The thiosulfate is almost consumed within 36 h and its oxidation enhances heterotrophic carbon assimilation. Cytochrome *c*, *b*, and *aa$_3$* are present. Cytochrome *c* occurs in both the soluble and membrane fractions whereas cytochrome *b* is membrane bound only. The genetic regulation of thiosulfate metabolism by *B. thiooxidans* has been studied by transposon insertion mutagenesis (Das and Mishra, 1996). Several thiosulfate metabolism defective mutants were isolated and a comparative enzymatic study with the wild type strain suggested that the enzymes thiosulfate oxidase, sulfite oxidase, and cytochrome *c$_{550-552}$* are mainly responsible for thiosulfate oxidation.

The MICs (minimum inhibitory concentrations) of several antibiotics for the type strain are (μg/ml): streptomycin, 120; tetracycline, 10; neomycin, 40; chloramphenicol, 20; and rifampicin, 15. Ampicillin fails to inhibit growth even at 200 μg/ml.

ENRICHMENT AND ISOLATION PROCEDURES

For isolation, use cultivated soil from different agricultural fields. Place the soil samples (25 g) in Petri plates and moisten with sterile distilled water. Enrich by adding sodium thiosulfate, sodium sulfite, and S⁰. After thorough mixing, incubate the plates at 30°C for 10 d. Further enrich in mineral salt broth (pH 8.0) supplemented with phenol red (0.02 g/l) and either Na$_2$S$_2$O$_3$·5H$_2$O (5 g/l) or thiosulfate plus yeast extract (5 g/l). The mineral salts medium contains (per liter): Na$_2$HPO$_4$, 4.0 g; KH$_2$PO$_4$, 1.5 g; MgCl$_2$·5H$_2$O, 0.1 g; NH$_4$Cl, 1.0 g; and trace metal solution (Vishniac and Santer, 1957), 2 ml. Adjust the pH to 8.0 with 4 N NaOH. Inoculate 1 g of enriched soil from each Petri plate into a 250-ml flask containing 50 ml of enrichment medium and incubate in a water bath shaker at 220 rpm at 30°C. After two days, if a yellow color develops (indicating acid formation),

plate serial dilutions onto mineral salts-thiosulfate agar and mineral salts-thiosulfate-yeast extract agar. After 4 days at 30°C a few colonies surrounded by yellow halos should appear on mineral salts-thiosulfate-yeast extract agar (but not on the mineral salts-thiosulfate agar). Select these colonies individually, streak and restreak onto similar media until a pure culture is obtained.

MAINTENANCE PROCEDURES

For short-term preservation, streak cultures onto mixed substrate agar or Luria agar plates, and incubate until growth becomes visible. Store the plates at 4°C and transfer at two-month intervals. Alternatively, grow cultures to the early stationary growth phase, freeze them in the growth medium with 15% glycerol, and store at −20°C. The cultures should remain viable for several months to a year or more. Lyophilization is the method of choice for long-term preservation.

DIFFERENTIATION OF THE GENUS *BOSEA* FROM OTHER GENERA

Bosea is similar to the genus *Thiobacillus* in regard to mol% G + C content of its DNA, respiratory quinones, and ability to oxidize inorganic sulfur compounds. However, it possesses hydroxy fatty acids in combination with 10-methyl and cyclic fatty acids—a feature that is not found in autotrophic and facultatively chemolithoautotrophic sulfur oxidizers or any other member of the alpha group of the *Proteobacteria*. The presence of ubiquinone (Q-10), phosphatidyl choline and a high content of unsaturated nonhydroxy fatty acid C$_{18:1\ \omega7c}$ is similar to that of *Methylobacterium*; however, the latter is not able to oxidize inorganic sulfur compounds. Unlike *Thiobacillus* sp., *Bosea* grown on mixed substrate agar does not produce sulfur deposits on colony surfaces even though both have a chemolithoheterotrophic mode of thiosulfate oxidation. Although both *Bosea* and *Thiobacillus versutus* denitrify under heterotrophic growth conditions, the absence of denitrification under mixed substrate growth conditions differentiates *Bosea* from *Thiosphaera pantotropha* and *T. versutus*.

TAXONOMIC COMMENTS

16S rDNA sequence analysis of the type strain indicates that *Bosea* constitutes a new genus of the *Alphaproteobacteria*. The 16S rDNA sequence has been deposited in the EMBL database under accession number X81044. A total of 1377 nucleotides of the 16S rDNA of the type strain were amplified and sequenced. The sequence was most similar to that of *Beijerinckia indica* (level of similarity, 92.8%) and to those of *Rhodopseudomonas palustris*, *Nitrobacter winogradskyi*, *Blastobacter denitrificans*, and *Bradyrhizobium japonicum* (levels of similarity 92.0–92.5%) of the *Alphaproteobacteria*. The distance matrix phylogenetic tree based on dissimilarity values of *B. thiooxidans* as compared to those of nineteen reference strains of the *Alphaproteobacteria* indicated a new lineage located between the methylotrophs, the genus *Beijerinckia*, and the *R. palustris* group (Fig. BXII.α.179). No close relationship was found between *B. thiooxidans* and other sulfur oxidizing bacteria such as *Thiobacillus acidophilus* and *Acidiphilium* species (levels of 16S rDNA similarity were less than 88%).

List of species of the genus Bosea

1. **Bosea thiooxidans** Das, Mishra, Tindall, Rainey and Stackebrandt 1996, 985[VP]

   thi.o.ox′ i.dans. Gr. n. *thion* sulfur; M.L. v. *oxido* make acid, oxidize; M.L. part. adj. *thiooxidans* oxidizing sulfur.

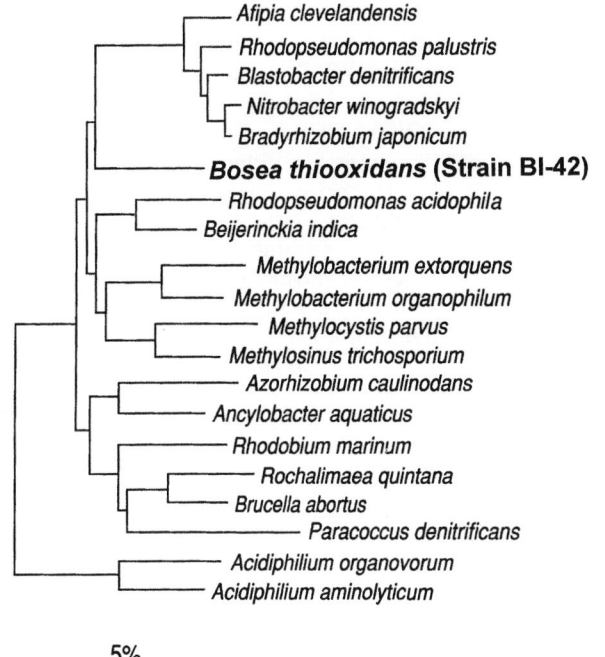

Afipia clevelandensis
Rhodopseudomonas palustris
Blastobacter denitrificans
Nitrobacter winogradskyi
Bradyrhizobium japonicum
Bosea thiooxidans (Strain BI-42)
Rhodopseudomonas acidophila
Beijerinckia indica
Methylobacterium extorquens
Methylobacterium organophilum
Methylocystis parvus
Methylosinus trichosporium
Azorhizobium caulinodans
Ancylobacter aquaticus
Rhodobium marinum
Rochalimaea quintana
Brucella abortus
Paracoccus denitrificans
Acidiphilium organovorum
Acidiphilium aminolyticum

5%

FIGURE BXII.α.179. Phylogenetic tree based on dissimilarity values, showing the relationships between *Bosea thiooxidans* (strain BI-42) and related reference organisms. Note that *Rochalimaea quintana* has been transferred to *Bartonella* and is now *Bartonella quintana*. Bar = 5% nucleotide difference. (Reprinted with permission from S.K. Das et al, International Journal of Systematic Bacteriology 46: 981–987, 1996, ©Society for General Microbiology.)

Characteristics are as described for the genus. Glutamate, glutamine, and aspartate, but not ammonium salts, nitrate, or urea, can serve as nitrogen sources. Glutamate auxotrophy occurs and the organisms require yeast extract powder (0.1 g/l) for growth. There is no anaerobic growth in the presence or absence of nitrate; however, under microaerobic conditions denitrification occurs with gas production in GYM medium containing malate, succinate, glucose, or sucrose. Gas is not formed after the addition of thiosulfate to the above growth media.

Organic compounds supporting heterotrophic growth include glucose, fructose, sorbose, xylose, pyruvate, rhamnose, ribose, arabinose, galactose, citrate, gluconate, succinate, malate, acetate, glutamine, proline, aspartic acid, cysteine, serine, asparagine, alanine, and lysine. The following do not support growth: formate, glycerol, mannitol, raffinose, leucine, glycine, isoleucine, methionine, tyrosine, tryptophan, phenylalanine, glyoxylate, lactate, propionate, salicylate, butyrate, cyclohexanol, *p*-amino benzoate, or methanol.

Growth occurs on Simmons citrate agar and MacConkey agar. Tests using methyl red, urease, starch hydrolysis, indole production, H_2S production, Voges–Proskauer, gelatin hydrolysis, and pigment production are negative.

Tetrathionate is oxidized slowly. Sulfite, thiocyanate and S^0 do not support growth.

Ubiquinone 10 is the major ubiquinone.

Isolated from agricultural field soil near Calcutta, India.

The mol% G + C of the DNA is: 68.2 (T_m).

Type strain: BI 42, ATCC 700366, DSM 9653.

GenBank accession number (16S rRNA): X81044.

Genus VI. **Nitrobacter** *Winogradsky 1892, 127[AL] Nom. Cons. Opin. 23 Jud. Comm. 1958, 169*

EVA SPIECK AND EBERHARD BOCK

Ni.tro.bac' ter. L. n. *nitrum* nitrate; M.L. n. *bacter* the masc. form of Gr. neut. n. *bactrum* a rod; M.L. masc. n. *Nitrobacter* nitrate rod.

Pleomorphic rod- or pear-shaped cells 0.5–0.9 × 1.0–2.0 µm. Cells reproduce by budding or binary fission. **Intracytoplasmic membranes occur as a polar cap of flattened vesicles** in the cell periphery. Gram negative. Cells may be motile by means of a single polar to subpolar flagellum. **Grows lithoautotrophically and chemoorganotrophically. Under oxic conditions, nitrite is the preferred energy source and carbon dioxide is the main source of carbon. Aerobic, but also capable of anaerobic respiration with nitrate.** Under anoxic conditions nitrate is reduced to nitrite, nitric oxide, and nitrous oxide. Occurs in aerobic and microaerophilic habitats where organic matter is mineralized.

The mol% G + C of the DNA is: 59–62.

Type species: **Nitrobacter winogradskyi** Winslow, Broadhurst, Buchanan, Krumwiede, Rogers and Smith 1917, 552 emended mut. char. Watson 1971, 264.

FURTHER DESCRIPTIVE INFORMATION

Nitrobacter cells are facultative lithoautotrophs that obtain energy from the oxidation of nitrite to nitrate during lithoautotrophic growth. The main source of carbon is carbon dioxide. Alternatively, pyruvate, formate and acetate can serve as energy and carbon sources even in the absence of nitrite. The optimal pH

range for growth is 7.5–8.0; the temperature range for growth is 5–37°C with an optimum between 28 and 30°C. Heterotrophic growth is often unbalanced and is accompanied by the formation of large quantities of poly-β-hydroxybutyrate granules. Glycogen and polyphosphates granules are also found as cytoplasmic inclusions. Carboxysomes are present in most but not all species of *Nitrobacter*.

Additional details and a comparison of the biochemical properties of *Nitrobacter* to those of other nitrite-oxidizing genera can be found in the introductory chapter "Lithoautotrophic Nitrite-Oxidizing Bacteria." Details about the ecology of nitrite-oxidizing bacteria and of the phylogeny of these organisms are described in the introductory chapter, Nitrifying Bacteria.

Nitrobacter cells may be either rod-shaped (Fig. BXII.α.180) or pear-shaped. A pear-shaped cell of *Nitrobacter* with a polar cap of intracytoplasmic membranes is shown in Fig. BXII.α.181. The cell wall differs from that found in other Gram-negative bacteria in that the inner side of the wall is more electron dense than the outer one. A similar asymmetry occurs in the cytoplasmic and intracytoplasmic membranes (Fig. BXII.α.181); carboxysomes are formed (Fig. BXII.α.181). Freeze-etched or negatively stained preparations reveal that the inner surface of these mem-

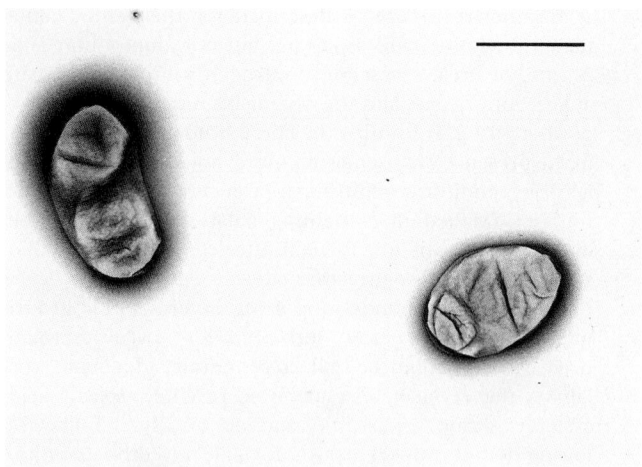

FIGURE BXII.α.180. Negatively stained short rods of *Nitrobacter winogradskyi* Engel. Bar = 1 μm.

branes is covered with densely packed particles with a size of 8–10 nm (Tsien et al., 1968, Remsen and Watson, 1972, Sundermeyer and Bock, 1981, Sundermeyer-Klinger et al., 1984). These particles are composed of the nitrite oxidoreductase enzyme (NOR). Electron microscopy of the isolated enzyme reveals uniform particles with a size of 8 nm (Meincke et al., 1992). Particulate membranes have been labeled with monoclonal antibodies recognizing the α- and β-subunits of the NOR (Spieck et al., 1996a). The particle locations correlate with immunogold-labeling of the α- and β-subunits of nitrite oxidoreductase; labeling of the α-subunit in *Nitrobacter hamburgensis* is shown in Fig. BXII.α.182. The enzyme forms a periodic arrangement in paired rows (Fig.BXII.α.183); a digital image processing analysis of this two-dimensional structure is shown in Fig. BXII.α.184. The molecular weight of a single particle is 186 kDa, suggesting that it is most likely composed of an αβ-heterodimer (Spieck et al., 1996b); the α-subunit has a mass of 115–130 kDa and the β-subunit a mass of 65 kDa.

Lithotrophic growth is slow. The generation time varies from 8 h to several days.

In *Nitrobacter* the concentration of NOR varies with growth conditions. Synthesis of the enzyme is induced by nitrite or nitrate, and the enzyme is the major constituent of nitrite-oxidizing membranes (Bock et al., 1991). In *Nitrobacter hamburgensis* the nitrite-oxidizing activity of mixotrophically grown cells is higher than that of autotrophically grown cells (Milde and Bock, 1985). During heterotrophic growth the NOR is repressed by a factor of more than 90%. When O_2 is absent, NOR shows nitrate reductase activity (Sundermeyer-Klinger et al., 1984). Furthermore, a membrane-bound nitrite reductase which transformed nitrite to NO in low oxygen conditions has been copurified with the NOR (Ahlers et al., 1990). Depending upon the isolation procedure, the NOR is obtained as complexes containing 2–3 subunits (Tanaka et al., 1983; Sundermeyer-Klinger et al., 1984). The membrane-associated α-subunit (115–130 kDa) and β-subunit (65 kDa) can be solubilized by heat treatment at 55°C. Investigations by electron paramagnetic resonance spectroscopy (EPR) showed that the catalytically active enzyme includes molybdenum and iron-sulfur centers which are involved in the transformation of nitrite to nitrate (Meincke et al., 1992). The cytochromes a_1 and c_1 have been coisolated with NOR in procedures

employing different detergents. Additional major proteins of the intracellular membranes had molecular masses of 14, 28, and 32 kDa. The 32-kDa protein is the γ-subunit of the NOR (Sundermeyer-Klinger et al., 1984). Tanaka et al. (1983) suggested that the subunit structure of the NOR (cytochrome a_1c_1) of *Nitrobacter winogradskyi* (254 kDa) consists of proteins with molecular masses of 55, 29 and 19 kDa. The pH optimum of the isolated enzyme was 8.0, and the Km value for nitrite was determined to be 0.5–2.6 mM by Tanaka et al. (1983) and 3.6 mM by Sundermeyer-Klinger et al. (1984). A Km value of about 0.9 mM was calculated for the reduction of nitrate. Kirstein and Bock (1993) identified the genes of the NOR and obtained the amino acid sequence of the NOR β-subunit of *Nitrobacter hamburgensis* X14. This protein contained four cysteine clusters (probably three [4Fe-4S] and one [3Fe-4S]) and shared significant sequence similarities with the β-subunits of the dissimilatory nitrate reductases (NRA, NRZ) of *Escherichia coli*. The β-NOR is thought to function as an electron-channeling protein between the nitrite-oxidizing α-subunit and the membrane-integrated electron transport chain. Recently, the genes encoding the α-subunit of the NOR were sequenced in the four described species of *Nitrobacter* (Degrange, personal communication). As in the case of the β-NOR, the *Nitrobacter* α-NORs possess significant sequence similarities to the α-subunits of several dissimilatory nitrate reductases (e.g., *Escherichia coli* or *Pseudomonas* species).

ENRICHMENT AND ISOLATION PROCEDURES

Nitrite oxidizers can be isolated using a mineral medium containing nitrite; the compositions of media for lithotrophic, mixotrophic, and heterotrophic growth are given in Table BXII.α.152. Serial dilutions of enrichment cultures must be incubated for one to several months in the dark. Since nitrite oxidizers are sensitive to high partial pressures of oxygen, cell growth on agar surfaces is limited. Pure cultures of *Nitrobacter alkalicus* were obtained by multiple passages in liquid medium of colonies from nitrite agar (Sorokin et al., 1998). Nitrite oxidizers can be separated from heterotrophic contaminants by Percoll gradient centrifugation and subsequent serial dilution (Ehrich et al., 1995).

MAINTENANCE PROCEDURES

Nitrifying organisms can survive starvation for more than one year when kept at 17°C in liquid medium. Nevertheless, cells should be transferred to fresh media every four months. Table BXII.α.152 lists three different growth media for nitrite oxidizers. Freezing in liquid nitrogen is a suitable technique for maintenance of stock cultures suspended in a cryoprotective buffer containing sucrose and histidine. When freeze-dried on lavalite or polyurethane, about 0.5% of *Nitrobacter* cells survive for one year (L. Lin, personal communication). Another possibility for the storage of *Nitrobacter* for several years is cultivation in 1l-bottles filled to the top with complex medium and closed by a screw top. Glycerol should be used instead of pyruvate to keep the pH-value stable for a long period. Since the bacteria are able to oxidize nitrite to nitrate aerobically and subsequently able to reduce the nitrate anaerobically, a high cell yield can be obtained using this method (Freitag et al., 1987).

DIFFERENTIATION OF THE GENUS *NITROBACTER* FROM OTHER GENERA

The nitrite oxidizers are a diverse group of long or short rods, cocci, and spirilla. The genera *Nitrobacter* and *Nitrococcus* possess

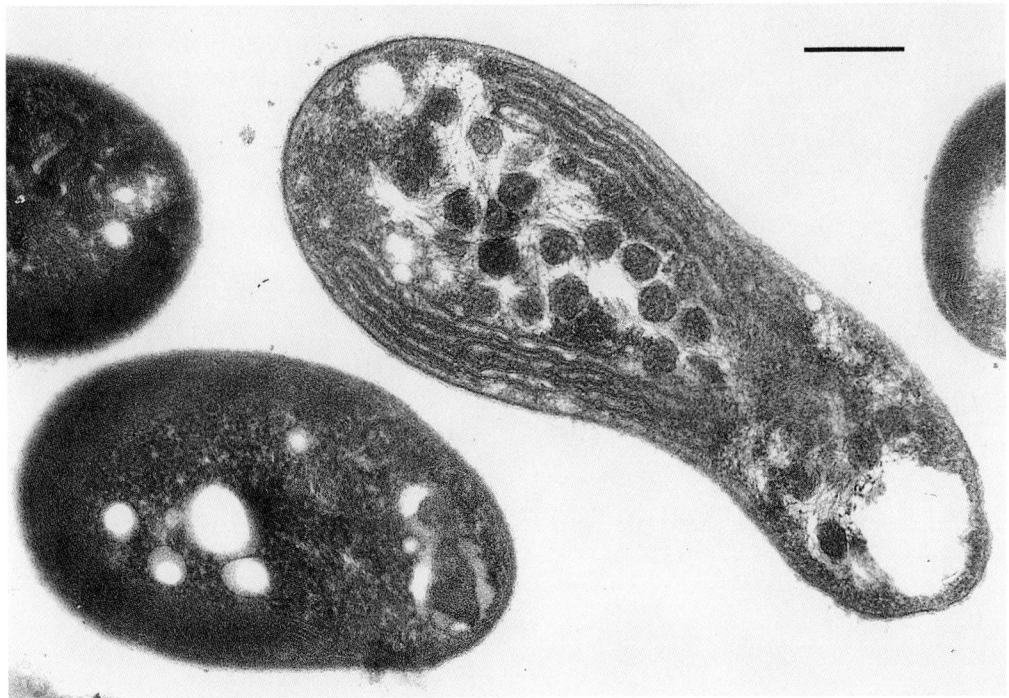

FIGURE BXII.α.181. Ultrathin section of *Nitrobacter vulgaris* strain nevada with intracytoplasmic membranes and carboxysomes. Bar = 250 nm.

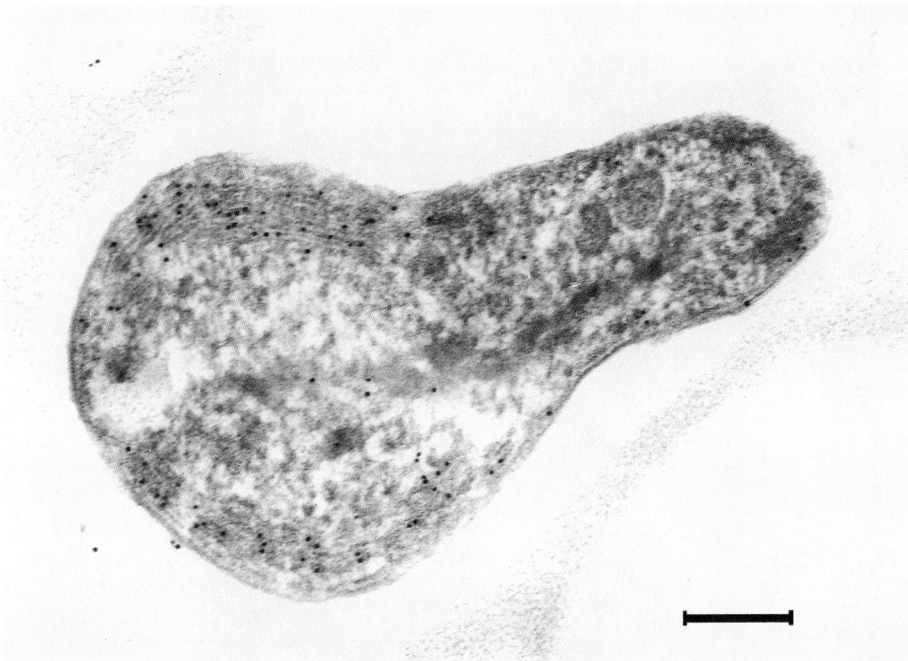

FIGURE BXII.α.182. Ultrathin section of *Nitrobacter hamburgensis* × 14. The nitrite oxidoreductase (NOR) was localized by immunogold-labeling at the cytoplasmic and intracytoplasmic membranes with monoclonal antibodies (MAbs 153-2) recognizing the α-subunit. (Reproduced with permission from E. Spieck et al., FEMS Microbiology Letters *139:* 71–76, 1996, ©Elsevier Science B.V., Amsterdam.) Bar = 250 nm.

a complex arrangement of intracytoplasmic membranes in the form of flattened vesicles or tubes. The taxonomic categorization is based on the work of Sergei and Helene Winogradsky (Winogradsky, 1892). Traditionally, the classification of genera is performed primarily on cell shape and arrangement of intracytoplasmic membranes. So far, four morphologically distinct genera (*Nitrobacter, Nitrococcus, Nitrospina,* and *Nitrospira*) have been described; the four genera contain a total of eight species (Watson

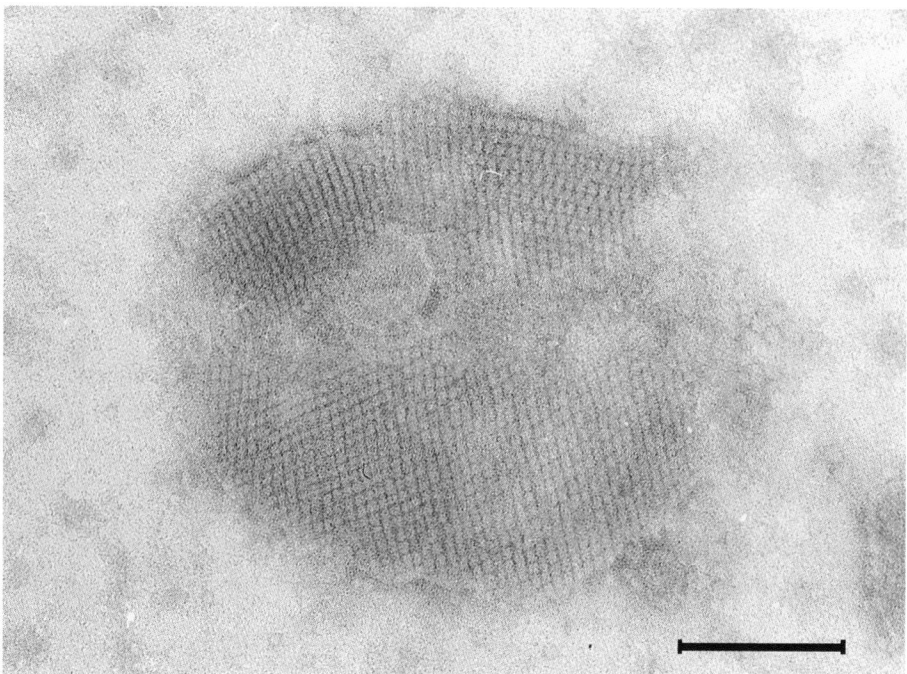

FIGURE BXII.α.183. Isolated negatively stained nitrite-oxidizing membrane of *Nitrobacter hamburgensis* × 14 with crystalline arrays of NOR particles. Bar = 100 nm..(Reproduced with permission from E. Spieck et al., Journal of Structural Biology *117:* 117–123, 1996, Academic Press Inc., Orlando.)

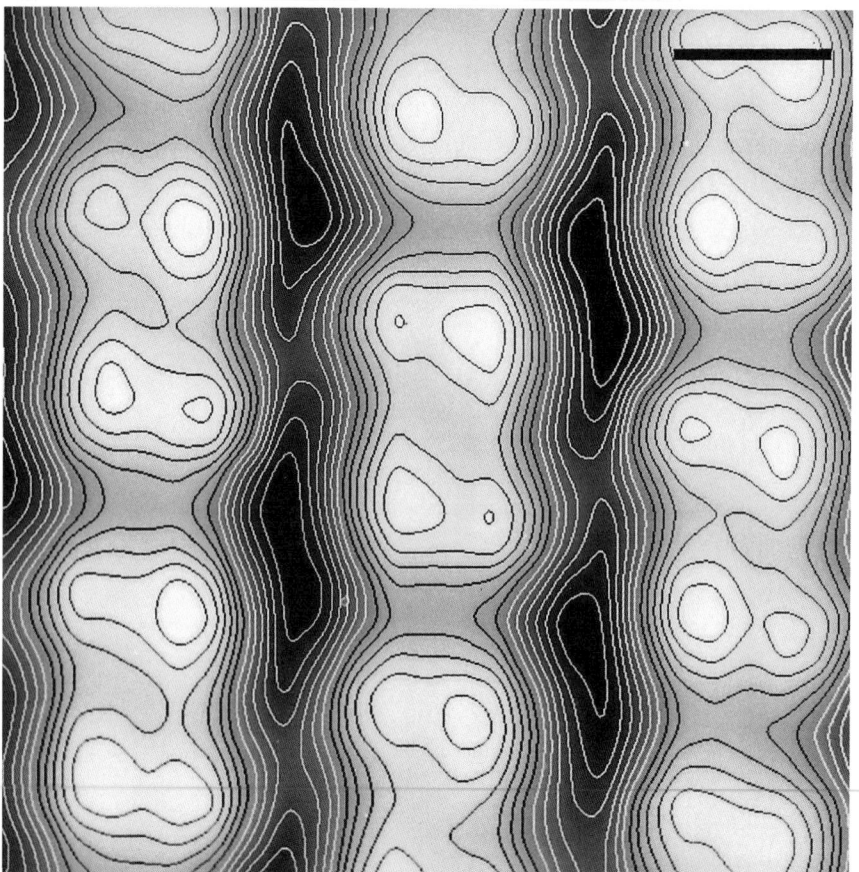

FIGURE BXII.α.184. Image reconstruction of the NOR of *Nitrobacter hamburgensis* × 14 by correlation averaging. The two-dimensional lattice is composed of particle-dimers, arranged in rows. Protein appears bright. Bar = 5nm. (Reproduced with permission from E. Spieck et al., Journal of Structural Biology *117:* 117–123, 1996, Academic Press Inc., Orlando.)

et al., 1989; Bock and Koops, 1992; Ehrich et al., 1995; Sorokin et al., 1998). Three new species belonging to the genera *Nitrobacter*, *Nitrospina*, and *Nitrospira* are known but remain to be described in the literature. Differential traits of the four genera of nitrite-oxidizing bacteria are given in Table BXII.α.153, further properties in Table BXII.α.154, and fatty acid profiles in Table BXII.α.155. Motility and carboxysomes have been observed in *Nitrobacter* and *Nitrococcus* (Tables BXII.α.153 and BXII.α.154).

Nitrobacter can be separated from the other genera by means of monoclonal antibodies that recognize the key enzyme (Table BXII.α.153). Aamand et al. (1996) developed three monoclonal antibodies (MAbs) that recognize the α- and the β-subunit of the NOR of *Nitrobacter*. The key enzyme NOR is ubiquitous in *Nitrobacter* species; homologous reactions of the MAbs with the NOR of *Nitrobacter winogradskyi* and the NORs of *Nitrobacter vulgaris*, *Nitrobacter hamburgensis*, and *Nitrobacter alkalicus* were demonstrated. *Nitrobacter* sp. BS 5/6, identified as a new species by DNA–

DNA hybridization (Koops, personal communication), was not recognized by the otherwise genus-specific MAb Hyb 153-2. This antibody targets the α-NOR of *Nitrobacter*, whereas MAb Hyb 153-1 recognizes the β-NOSs of both *Nitrobacter* and *Nitrococcus*. The MAb Hyb 153-3 recognizes the β-NOSs of *Nitrobacter*, *Nitrococcus*, *Nitrospina* and *Nitrospira* (Bartosch et al., 1999). Thus, the suite of MAbs permits detection of all known nitrite oxidizers and allows discrimination of *Nitrobacter* and *Nitrococcus* from *Nitrospina* and *Nitrospira*.

TAXONOMIC COMMENTS

16S rDNA sequence analysis showed that all characterized strains of *Nitrobacter* form a closely related assemblage within the *Alphaproteobacteria* (Woese et al., 1984b) and that *Nitrobacter* strains cluster with species of the genera *Rhodopseudomonas*, *Bradyrhizobium*, *Blastobacter* and *Afipia* (Teske et al., 1994) in the family *Bradyrhizobiaceae*. (See Fig. 1 [p. 138] of the introductory chapter "Ni-

TABLE BXII.α.152. Three different media for lithoautotrophic (medium A for terrestrial strains; medium B for marine strains), mixotrophic (medium C), and heterotrophic (medium C without NaNO₂) growth of nitrite oxidizers

Ingredient	Culture medium		
	A[a,]	B[b]	C[c, d]
Distilled water (ml)	1000	300	1000
Seawater (ml)		700	
NaNO₂ (mg)	200–2000	69	200–2000
MgSO₄·7H2O (mg)	50	100	50
CaCl₂·2H₂O (mg)		6	
CaCO₃ (mg)	3		3
KH₂PO₄ (mg)	150	1.7	150
FeSO₄·7H₂O (mg)	0.15		0.15
Chelated iron (13%, Geigy) (mg)		1	
Na₂MoO₄·2H₂O (μg)		30	
(NH₄)₂Mo₇O₂₄·4H₂O (μg)	50		50
MnCl₂·6H₂O (μg)		66	
CoCl₂·6H₂O (μg)		0.6	
CuSO₄·5H₂O (μg)		6	
ZnSO₄·7H₂O (μg)		30	
NaCl (mg)	500		500
Sodium pyruvate (mg)			550
Yeast extract (Difco) (mg)			1,500
Peptone (Difco) (mg)			1,500
pH adjusted to[e]	8.6	6	7.4

[a]For terrestrial strains from Bock et al. (1983).

[b]For marine strains modified from Watson and Waterbury (1971).

[c]For terrestrial strains from Bock et al. (1983).

[d]For heterotrophic growth medium C without NaNO₂ is used.

[e]After sterilization pH should be 7.4–7.8.

TABLE BXII.α.153. Differentiation of the four genera of nitrite-oxidizing bacteria

Characteristic	Nitrobacter	Nitrococcus	Nitrospina	Nitrospira
Phylogenetic position	*Alphaproteobacteria*	*Gammaproteobacteria*	*Deltaproteobacteria* (preliminary)	Phylum *Nitrospira*
Morphology	Pleomorphic short rods	Coccoid cells	Straight rods	Curved rods to spirals
Intracytoplasmic membranes	Polar cap	Tubular	Lacking	Lacking
Size (μm)	0.5–0.9 × 1.0–2.0	1.5–1.8	0.3–0.5 × 1.7–6.6	0.2–0.4 × 0.9–2.2
Motility	+	+	−	−
Reproduction:	Budding or binary fission	Binary fission	Binary fission	Binary fission
Main cytochrome types[a]	*a, c*	*a, c*	*c*	*b, c*
Location of the nitrite oxidizing system on membranes	Cytoplasmic	Cytoplasmic	Periplasmic	Periplasmic
MAb-labeled subunits (kDa)[b]	130 and 65	65	48	46
Crystalline structure of membrane-bound particles	Rows of particle dimers	Particles in rows	Hexagonal pattern	Hexagonal pattern

[a]Lithoautotrophic growth.

[b]MAbs, monoclonal antibodies.

TABLE BXII.α.154. Properties of the nitrite-oxidizing bacteria

Characteristic	Nitrobacter winogradskyi	Nitrobacter alkalicus	Nitrobacter hamburgensis	Nitrobacter vulgaris	Nitrococcus mobilis	Nitrospina gracilis	Nitrospira marina	Nitrospira moscoviensis
Mol% G + C of the DNA	61.7	62	61.6	59.4	61.2	57.7	50	56.9
Carboxysomes	+	−	+	+	+	−	−	
Habitat:								
Fresh water	+			+				
Waste water	+			+				
Brackish water				+				
Oceans	+				+	+	+	
Soda lakes		+						
Soil	+		+	+				
Soda soil		+						
Stones	+			+				
Heating systems								+

TABLE BXII.α.155. Primary fatty acids of the described species of nitrite-oxidizing bacteria[a,b]

Fatty acid	Nitrobacter winogradskyi Engel	Nitrobacter alkalicus AN4	Nitrobacter hamburgensis X14	Nitrobacter vulgaris Z	Nitrococcus mobilis 231	Nitrospina gracilis 3	Nitrospira marina 295	Nitrospira moscoviensis M1
$C_{14:1\ \omega5c}$						+		
$C_{14:0}$	+		+		+	+ + +	+	+
$C_{16:1\ \omega9c}$							+ + +	+ +
$C_{16:1\ \omega7c}$	+	+	+	+	+ + +	+ + +		
$C_{16:1\ \omega5c}$							+ + +	+ + +
$C_{16:0\ 3OH}$						+		+
$C_{16:0}$	+ +	+ +	+ +	+ +	+ + +	+ +	+ + +	+ + +
$C_{16:0\ 11CH_3}$							+	+ + +
$C_{18:1\ \omega9c}$	+		+			+	+	
$C_{18:1\ \omega7c}$	+ + + +	+ + + +	+ + + +	+ + + +	+ + +	+	+	+
$C_{18:0}$	+	+	+		+	+	+ +	+
$C_{19:0\ cyclo\ \omega7c}$	+	+	+	+	+			

[a]Symbols: +, <5%; + +, 6–15%; + + +, 16–60%; + + + +, >60%.

[b]Stirred cultures were grown autotrophically at 28°C (*Nitrospira moscoviensis* at 37°C) and collected at the end of exponential growth. Modified from Lipski et al., (2001).

trifying Bacteria", Volume 2, Part A.) Common traits of *Nitrobacter*, *Rhodopseudomonas*, and *Blastobacter* include the ability to carry out denitrification and cell division by budding. The oligonucleotide data of Seewaldt et al. (1982) demonstrate a close relationship between the 16S rRNAs of *Nitrobacter winogradskyi*, *Nitrobacter hamburgensis* strain X14, and *Rhodopseudomonas palustris*. *Nitrobacter hamburgensis* strain X14 contains an unusual lipid A with 2,3 diamino-2,3-dideoxyglucose similar to that of *Rhodopseudomonas palustris*; this is significant because lipid A structure is considered to be conserved phylogenetically (Mayer et al., 1983). Furthermore, the soluble cytochrome *c* of *Nitrobacter* resembles the cytochrome c_2 of *Rhodopseudomonas viridis* (Yamanaka and Fukumori, 1988). The fact that both *Nitrobacter* spp. and *Rhodopseudomonas palustris* possess complex internal membrane systems suggests a common evolutionary origin. Similar membranes are also present in *Methylobacterium*, the closest methylotrophic relative of *Nitrobacter*.

ACKNOWLEDGMENTS

We thank Wolfgang Ludwig (Technical University in Munich, Germany) for phylogenetic trees and Dimitry Sorokin (Institute of Microbiology in Moscow, Russia) for providing cultures of *Nitrobacter alkalicus*.

FURTHER READING

Bartosch, S., I. Wolgast, E. Spieck and E. Bock. 1999. Identification of nitrite-oxidizing bacteria with monoclonal antibodies recognizing the nitrite oxidoreductase. Appl. Environ. Microbiol. 65: 4126–4133.

Bock, E., H.P. Koops, U.C. Möller and M. Rudert. 1990. A new facultatively nitrite oxidizing bacterium, *Nitrobacter vulgaris*, sp. nov. Arch. Microbiol. 153: 105–110.

Bock, E., H. Sundermeyer-Klinger and E. Stackebrandt. 1983. New facultative lithoautotrophic nitrite-oxidizing bacteria. Arch. Microbiol. 136: 281–284.

Kirstein, K. and E. Bock. 1993. Close genetic relationship between *Nitrobacter hamburgensis* nitrite oxidoreductase and *Escherichia coli* nitrate reductases. Arch. Microbiol. 160: 447–453.

Sorokin, D.Y., G. Muyzer, T. Brinkhoff, J.G. Kuenen and M.S.M. Jetten. 1998. Isolation and characterization of a novel facultatively alkaliphilic *Nitrobacter* species, *N. alkalicus* sp. nov. Arch. Microbiol. 170: 345–352.

Sundermeyer-Klinger, H., W. Meyer, B. Warninghoff and E. Bock. 1984. Membrane-bound nitrite oxidoreductase of *Nitrobacter*: evidence for a nitrate reductase system. Arch. Microbiol. 140: 153–158.

Watson, S.W., E. Bock, H. Harms, H.P. Koops and A.B. Hooper. 1989. Nitrifying bacteria. *In* Staley, Bryant, Pfennig and Holt (Editors), Bergey's Manual of Systematic Bacteriology, 1st Ed., Vol. 3, The Williams & Wilkins Co., Baltimore. pp. 1808–1833.

DIFFERENTIATION OF THE SPECIES OF THE GENUS *NITROBACTER*

Cells of the genus *Nitrobacter* have a comparable shape, size and ultrastructure. Since most strains are phenotypically similar, species differentiation is based on the mol% G + C content of the DNA, DNA–DNA hybridization, serological properties, patterns of membrane-bound heme proteins, and slight differences in growth characteristics. Differential characteristics are given in Table BXII.α.154 and fatty acid profiles in Table BXII.α.155. *Nitrobacter hamburgensis* and *Nitrobacter vulgaris* differ from *N. winogradskyi* in that mixotrophic growth of *N. hamburgensis* and *N. vulgaris* is more rapid than lithoautotrophic growth; *Nitrobacter alkalicus* can be separated from the other species by its ability to oxidize nitrite under alkaline conditions.

The DNA–DNA hybridization values between the pairs of species range from 20–49% (Bock et al., 1990). The mol% G + C of the DNA of *Nitrobacter* species ranges from 59–62%. Navarro et al. (1992) investigated the genomic relationships within this genus by examination of rRNA gene restriction patterns and by DNA–DNA hybridization. These authors differentiated three genomic species, two of which corresponded to the described species *N. winogradskyi* and *N. hamburgensis*. (No strains of *N. vulgaris* were examined in that study.) High-resolution phylogenetic investigations of *Nitrobacter* have also been carried out using 16S–23S rRNA intergenic spacer region sequences and partial sequences of the 23S rRNA gene (Grundmann et al., 2000). Analysis of 16S rDNA sequences showed that species of *Nitrobacter* form a coherent cluster with sequence similarities of 99.2–99.6% (Orso et al., 1994). Seewaldt et al. (1982) obtained an S_{AB} value of 0.82 for *N. winogradskyi* and *N. hamburgensis* X14. The 16S rRNA similarity between the newly described species *Nitrobacter alkalicus* and both *N. winogradskyi* and *N. hamburgensis* was 98.6–99.9% (Sorokin et al., 1998). The closest relative of the alkaliphilic species is *N. winogradskyi*. DNA–DNA hybridization studies of strains of *N. vulgaris* indicated that this species might contain subspecies. The hybridization values obtained in this study ranged between 55–97%; the distribution of hybridization values among strains agreed with the distribution of variable physiological traits; both sorts of variation may reflect adaptation to special environments (Bock et al., 1990).

List of species of the genus *Nitrobacter*

1. **Nitrobacter winogradskyi** Winslow, Broadhurst, Buchanan, Krumwiede, Rogers and Smith 1917, 552[AL] emended mut. char. Watson 1971b, 264.

   wi.no.grad'sky.i. M.L. gen. n. *winogradskyi* of Winogradsky; named after Winogradsky, the microbiologist who first isolated these bacteria.

   Short rods, often pear-shaped, sometimes coccoid with a size of 0.6–0.8 × 1.0–2.0 μm. Carboxysomes occur as cytoplasmic inclusions. Cells may be motile by a single subpolar to lateral flagellum. Many strains are facultative lithoautotrophs. During mixotrophic growth, nitrite is oxidized first, followed by the oxidation of organic material. Cells can also grow heterotrophically with pyruvate, acetate, and glycerol as energy sources and with yeast extract, casamino acids, peptone, ammonia, and nitrate as nitrogen sources. Under lithoautotrophic and mixotrophic conditions, the generation time varies from 8–14 h and under heterotrophic conditions from 70–100 h. The organism is able to grow anaerobically with nitrate as the electron acceptor (Bock et al., 1988) and forms nitrite, nitric oxide (NO), and nitrous oxide (N_2O). The cells are slightly sensitive to oxygen when undergoing a shift from anoxic to oxic conditions. Nitric oxide may be an alternative substrate for the aerobic oxidation to nitrate. The type strain was isolated from soils. Other habitats are fresh water, oceans, sewage disposal systems and compost piles.

   The mol% G + C of the DNA is: 61.7 (Bd).
   Type strain: ATCC 25391.

2. **Nitrobacter alkalicus** Sorokin, Muyzer, Brinkhoff, Kuenen and Jetten 1999, 1325[VP] (Effective publication: Sorokin, Muyzer, Brinkhoff, Kuenen and Jetten 1998, 352.)

   al.ka'li.cus. M.L. adj. *alkalicus* alkaline.

   Pear-shaped cells 0.6–0.9 × 1.2–1.8 μm. Cell wall contains an additional external layer with a regular subunit arrangement. In contrast to the other species of this genus, no carboxysomes have been detected. Lithoautotrophic growth with nitrite occurs in a broad pH range from 6.5–10.2 with an optimum at 9.5. Heterotrophic growth not observed. Obligately dependent on the presence of carbonate ions. Strains were isolated from the sediments of soda lakes and from soda soil in Siberia and Kenya. The 16S rRNA gene sequences of strains AN 1, AN 2, and AN 4 are deposited in the GenBank database.

   The mol% G + C of the DNA is: 62.0 (T_m).

   Type strain: AN 1, LMD 97.163.
   GenBank accession number (16S rRNA): AF 069956.
   The type strain is stored in the culture collection of the Department of Microbiologie, University of Technology, Delft, The Netherlands.

3. **Nitrobacter hamburgensis** Bock, Sundermeyer-Klinger and Stackebrandt 2001, 1[VP] (Effective publication: Bock, Sundermeyer-Klinger and Stackebrandt 1983, 283.)

   ham'bur.gen.sis. M.L. adj. *hamburgensis* pertaining to the city of Hamburg, Federal Republic of Germany, where the organism was first isolated.

   Morphology and ultrastructure are as described for the genus. Carboxysomes are present. Motile by a subpolar to lateral flagellum. Grows poorly lithoautotrophically but well mixotrophically. Optimal growth rates are obtained in a mixotrophic medium containing nitrite, pyruvate, yeast extract and peptone. One strain could grow only in the presence of organic material. Both nitrite and organic compounds are metabolized during mixotrophic growth. Growth by dissimilatory nitrate reduction is possible, but the cells are sensitive to oxygen when changing from anoxic to oxic conditions. Isolated from soils in Hamburg (Germany), Yucatan (Mexico) and Corse (France).

   The mol% G + C of the DNA is: 61.6 (T_m).
   Type strain: X 14.
   GenBank accession number (16S rRNA): L11663.
   The type strain is stored in the culture collection of the Institute of General Botany, Department of Microbiology, University of Hamburg, FRG.

4. **Nitrobacter vulgaris** Bock, Koops, Möller and Rudert 2001, 1[VP] (Effective publication: Bock, Koops, Möller and Rudert 1990, 109.)

   vul.ga'ris. L. adj. *vulgaris* common.

   The morphology and ultrastructure are as described for the genus. Carboxysomes are present. Most strains show diphasic growth, oxidizing nitrite first and then organic material. Lithoautotrophic growth is often slower than chemoorganotrophic growth. Cells of strain 'Z' double in number in 140 h during lithoautotrophic growth and in 25 h during mixotrophic growth. Either nitrate or oxygen can serve as electron acceptor during heterotrophic growth with acetate or pyruvate as electron donor. Cells produce extracellular polymers as biofilm and form microcolonies. In contrast to the other strains of this genus, cells of *N.*

vulgaris are insensitive to oxygen stress. Only one membrane-bound *c*-type cytochrome of 32 kDa is present, in contrast to *N. winogradskyi* and *N. hamburgensis*, which also possess a 30 kDa *c*-type cytochrome. Strains have been isolated from soils, groundwater, fresh and brackish water, sewage and a termite heap. *N. vulgaris* is the most abundant species of *Nitrobacter* in concrete and natural building stone.

The mol% G + C of the DNA is: 59.4 (T_m).

Type strain: Z.

Type strain is stored in the culture collection of the Institute of General Botany, Department of Microbiology, University of Hamburg, FRG.

Genus VII. **Oligotropha** Meyer, Stackebrandt and Auling 1994, 182[VP] (Effective publication: Meyer, Stackebrandt and Auling 1993, 391)

ORTWIN O. MEYER

O.li.go.tro'pha. Gr. n. *oligos* little, scanty; *tropha* nourishing, living on few substrates; M.L. fem. n. *Oligotropha* utilizer of few substrates.

Rod-shaped cells, slightly curved, 0.4–0.7 × 1.0–3.0 μm. **Motility variable**; when present, it is by means of **one subpolar flagellum** (Meyer and Schlegel, 1978). Star-shaped aggregates (rosettes)—held together by polar pili and excreted slime—are formed to a certain extent, especially in the post exponential and stationary growth phases. Gram negative. Colonies are white or cream colored, and no carotenoid pigment is produced. Colony formation on agar plates is slow (one to four weeks, depending on the medium). Aerobic, having a respiratory type of metabolism, with oxygen as the terminal electron acceptor; however, some strains can denitrify. **Facultatively chemolithoautotrophic.** Chemoorganoheterotrophic substrates are usually restricted to the salts of some organic acids such as pyruvate, although some strains can also utilize sugars and amino acids. Dinitrogen is not fixed. Not phototrophic. Not methanotrophic. Isolated from wastewater of sewage treatment settling ponds and from soil.

The mol% G + C of the DNA is: 62.5–63.1.

Type species: **Oligotropha carboxidovorans** (Meyer and Schlegel 1978) Meyer, Stackebrandt and Auling 1994, 182 (Effective publication: Meyer, Stackebrandt and Auling 1993, 391) (*"Pseudomonas carboxydovorans"* Meyer and Schlegel 1978, 42; *"Hydrogenomonas carboxydovorans"* Kistner 1954, 186.)

FURTHER DESCRIPTIVE INFORMATION

Cells of *Oligotropha* lack intracytoplasmic membranes or chlorosomes, and thus electron micrographs of the cytoplasm are in accord with the absence of phototrophic or methanotrophic metabolic capabilities (Meyer and Rohde, 1984; Rohde et al., 1984). For electron micrographs of the flagella, see Meyer and Schlegel (1978), Meyer and Rohde (1984), and Rohde et al. (1984). The flagellar fine structure (Aragno et al., 1977) and the cell wall (Walther-Mauruschat et al., 1977) are of type I.

Under heterotrophic conditions (e.g., on the mineral medium of Meyer and Schlegel (1983) supplemented with 0.1% nutrient broth and 0.1% pyruvate), colonies first become visible after about 1 week of incubation. Under chemolithoautotrophic conditions (e.g., the mineral medium of Meyer and Schlegel (1983) in a gas atmosphere of 50% CO, 5% CO_2 and 45% air, or of 50% H_2, 10% CO_2 and 40% air), colonies first become visible after a month or more of incubation. Colony sizes may vary, depending on whether a colony has been developed from a single bacterium or a number of aggregated cells.

Submerged growth in shaken liquid cultures is usually much faster than on solid media. Growth of strain OM5 with CO in submerged culture (e.g., in 30-l fermentors at 30°C) in the mineral medium of Meyer and Schlegel (1983) supplied with the indicated trace elements TS2 and a gas mixture of 20% CO and 80% air at a flow rate of 2 l/min occurs at generation times of 20 h. Under these conditions, exponential growth proceeds to an OD436 of 7 and a yield of 1 g wet weight/l. Cultivation of the bacteria with H_2 requires essentially the same conditions as with CO, except that a gas mixture of 5% CO_2, 20% H_2, and 75% air is employed. Under these conditions, the generation time is 15 h, and exponential growth proceeds to an OD_{436} of 7 and a yield of 1 g wet weight/l. Because *O. carboxidovorans* OM5 employs a membrane-bound heterodimeric NiFe-hydrogenase for the chemolithoautotrophic utilization of H_2 (Santiago and Meyer, 1997), Ni must be supplied for chemolithoautotrophic growth with H_2.

The metabolism is generally aerobic and facultatively chemolithoautrophic (Meyer and Schlegel, 1983). The respiratory metabolism is in accord with the presence of cytochromes (Cypionka and Meyer, 1983b). One strain can reduce nitrate to nitrite (Frunzke and Meyer, 1990). Obligate chemolithoautotrophic strains are not known. So far, only mesophilic strains have been characterized. Chemoorganoheterotrophic substrates utilized under aerobic conditions include the salts of pyruvate, formate, glyoxylate, lactate, ascorbate, acetate, malate, fumarate, oxoglutarate, and oxalate (Meyer et al., 1993). Vitamins are not required. Sugars and amino acids are not used by the strains OM3, OM4, and OM5 of *O. carboxidovorans*, whereas strain OM2 utilizes most sugars and amino acids. Suitable nitrogen sources of all strains are urea, ammonia, nitrate, and nitrite. Nitrate is not assimilated by strain OM5 when CO is the substrate.

The aerobic chemolithoautotrophic utilization of CO by *O. carboxidovorans* OM5 follows this equation: O_2 + 2.19 CO→1.83 CO_2 + 0.36 cell carbon (Meyer and Schlegel, 1978, 1983). The bacteria oxidize CO for the generation of energy (5 CO + 2.5 O_2→5 CO_2) and of reducing equivalents (2 CO + H_2O→2 CO_2 + 2 × 2 [H]). CO_2 derives from the oxidation of CO and is assimilated in the Calvin-Benson-Bassham-cycle (CO_2 + 2 × 2 [H]→[CH_2O] + H_2O). The sum of these equations (7 CO + 2.5 O_2 + H_2O→6 CO_2 + [CH_2O]) is in accordance with the actually observed stoichiometry of CO oxidation and indicates that 86% of the CO is oxidized for energy generation and 14% of the CO carbon is assimilated.

Carbon monoxide oxidation in *O. carboxidovorans* OM5 is catalyzed by the heterohexameric metalloenzyme CO dehydrogenase (CO + H_2O→CO_2 + 2 H^+ + 2 e^-) that is associated with the inner aspect of the cytoplasmic membrane (Meyer and Rohde, 1984; Rohde et al., 1984; Meyer et al., 2000). CO dehydrogenases are very much conserved in all aerobic CO oxidizing bacteria (Schübel et al., 1995; Santiago et al., 1999), and the CO dehydrogenases from the strains OM5, OM4, OM3, and OM2 of

O. carboxidovorans are very similar as indicated by a close immunological relationship, same molecular masses, same subunit structure, same types and number of cofactors, indistinguishable electron acceptor specificity, and co-migration upon non-denaturing or SDS-PAGE (Cypionka et al., 1980; Meyer and Rohde, 1984; Hugendieck and Meyer, 1992). The CO dehydrogenase from *O. carboxidovorans* is a 277-kDa Mo- and Cu-containing iron-sulfur flavoprotein (Meyer et al., 2000); therefore, growth with CO has a special requirement for Mo and Cu. The enzyme contains the cofactor molybdopterin cytosine dinucleotide. CO dehydrogenase has been crystallized in different states and structurally characterized (Gremer et al., 2000; Meyer et al., 2000; Dobbek, et al., 2002). The bimetallic [CuSMoO$_2$] active site of the enzyme has also been studied by x-ray spectroscopy (Gnida et al., 2003). The subunit structure of CO dehydrogenase is (LMS)$_2$, and it consists of a dimer of LMS heterotrimers. Each heterotrimer is composed of a 17.8-kDa iron-sulfur protein (S, 166 amino acid residues), which carries two types of [2Fe-2S] clusters, a 30.2-kDa flavoprotein (M, 288 amino acid residues), which contains a noncovalently bound FAD cofactor, and a 88.7-kDa molybdoprotein (L, amino acid 809 residues), which harbors a [CuSMoO$_2$] cluster in the active site of the enzyme.

The CO dehydrogenase structural genes *coxMSL* are an integral part of an elaborate CO oxidizing system (Santiago et al., 1999), which itself is part of an extended chemolithoautrophy module covering 39% of the entire sequence of the 133.056-kb plasmid pHCG3 of *O. carboxidovorans* OM5 Fuhrmann et al., 2003. The respiratory chain of *O. carboxidovorans*, particularly the terminal cytochrome oxidase, is insensitive to CO (Cypionka and Meyer, 1983a).

H$_2$ oxidation by *O. carboxidovorans* strain OM5 is catalyzed by a membrane-bound hydrogenase [H$_2$→2 H$^+$ + 2 e$^-$], which has been solubilized, isolated, and characterized (Santiago and Meyer, 1997). It belongs to the class I of hydrogenases and is a heterodimeric 101,692-Da NiFe-protein composed of the polypeptides HoxL and HoxS. The hydrogenase structural genes *hoxLS* are part of the chemolithoautotrophy module on pHCG3. HoxL comprises 604 amino acid residues and has a molecular mass of 67,163 Da. Pre-HoxS comprises 360 amino acid residues and is synthesized as a precursor protein that is cleaved after alanine at position 45, thus producing a mature HoxS of 33,767 Da. The leader sequence corresponds to the signal peptide of small subunits of hydrogenases. The hydropathy plots of HoxL and HoxS indicate the absence of transmembrane helices. Total DNA of species of *Oligotropha* did not hybridize with a *nifH* probe, indicating the absence of genes coding for the conventional, molybdenum-containing nitrogenase (Auling et al., 1988). In addition, growth with N$_2$ as a sole nitrogen source could not be demonstrated.

Megaplasmids have been identified in the strains OM5 (pHCG3: 133 kb) OM4 (pHCG5-a: 158 kb; pHCG5-b: 128 kb), and OM2 (pHCG4: 128 kb) of *O. carboxidovorans* (Kraut and Meyer, 1988). *O. carboxidovorans* strain OM3 does not carry a plasmid. *O. carboxidovorans* OM5 harbors the low copy number 133,058 bp circular DNA plasmid pHCG3, which has been sequenced and is required for the chemolithoautotrophic utilization of CO, H$_2$, and CO$_2$ by the bacterium. The calculated G + C content of the plasmid is 60.55 mol%. Loss of the plasmid is associated with loss of the ability to grow with CO or H$_2$ or to assimilate CO$_2$. The complete nucleotide sequence of the plasmid pHCG3 was obtained. Sequence analysis indicated 128 open reading frames. Of these, 95 are putative structural genes. The

most striking feature is the occurrence of the four gene clusters *cox*, *cbb*, *hox*, and *tra/trb*, which have functions in the utilization of CO, CO$_2$ or H$_2$, and the conjugal transfer of the plasmid, respectively. The clusters *cox*, *cbb*, and *hox* form a 51.2-kb chemolithoautotrophy module, containing 12 *cox* genes (14.54 kb), 13 *cbb* genes (13.33 kb), and 20 *hox* genes (23.35 kb). The 24.58-kb *tra/trb* cluster is separated from the chemolithoautotrophy module by two regions of 25.4- and 29.6-kb with miscellaneous or so far unknown functions. The *tra/trb* cluster carries 10 *tra* genes and 10 *trb* genes, in an arrangement very similar to the Ti-plasmid conjugal transfer system of *Agrobacterium tumefaciens*. The 25.4- and 29.6-kb regions with other functions carry a number of single genes coding for the replication and stabilization of the megaplasmid. Among these are an *oriV* coding for the replication proteins RepA, B, and C, two ORFs coding for transposase A, and a single ORF encoding transposase B. The plasmid pHCG3 also contains the insertion sequence SC1190 of *Sulfolobus solfataricus*, a gene function involved in the biosynthesis of acylated homoserine lactones, which participate in quorum sensing, and *rspD* and *rspE*, which function in rhizobiocine secretion. Most interestingly, the megaplasmid pHCG3 of *O. carboxidovorans* carries numerous genes with highest homologies to *Bradyrhizobium japonicum*, *Nitrobacter vulgaris*, *Agrobacterium tumefaciens*, *Rhizobium leguminosarum*, *Rhizobium rhizogenes*, *Rhodobacter capsulatus*, and *Rhodobacter sphaeroides*, which might be taken as an indication of horizontal gene transfer among the members of the α2 and the α3 subclasses of the *Alphaproteobacteria* (Fig. BXII.α.185).

As are all carboxidotrophic bacteria, the members of *Oligotropha* are cosmopolites inhabiting all sorts of environments, including sewage, water, and soil (Meyer and Schlegel, 1978, 1983). Natural enrichment can be found in the covering soil of smoldering charcoal piles (Meyer et al., 1991; Meyer, 1997).

ENRICHMENT AND ISOLATION PROCEDURES

Enrichment and isolation of pure cultures of *Oligotropha* and other aerobic carbon monoxide oxidizing microorganisms from natural habitats apply conditions selective for the ability to proliferate in a mineral medium with CO as a sole energy source and CO$_2$ as a sole carbon source under respiratory conditions with O$_2$ as the terminal electron acceptor. The CO present is counterselective for most contaminating microorganisms that cannot utilize CO. Alternatively, enrichments may be carried out under anoxic conditions in the presence of a suitable electron acceptor. Typically, enrichments are done in liquid batch cultures employing a mineral medium (e.g., that of Meyer and Schlegel, 1983, plus the indicated trace elements) incubated in desiccators supplied with a gas atmosphere composed of 5% CO$_2$, 10% O$_2$, and 85% CO (alternatively 5% CO$_2$, 45% CO, and 50% air). The concentrations of CO and O$_2$ can be varied to modify the selective conditions. About 1 g of environmental sample is suspended in 15 ml of sterile mineral medium contained in 100 ml Erlenmeyer flasks. The flasks are placed in desiccators and the appropriate gas atmosphere is established by evacuation and refilling. After about 1 month with or without shaking at 30°C in the dark, subcultures are made (10% inoculum). Further subcultures are prepared about every 2 weeks (5% inoculum). Aliquots of the suspensions are streaked on agar plates made of mineral medium supplied with nutrient broth and pyruvate (0.1% of each) and solidified with 1.4% agar. The plates are incubated in desiccators at 30°C under the same CO-containing atmosphere used for enrichment. The very fast growing colonies are generally CO-tolerant contaminants. The slow growing colonies appearing after

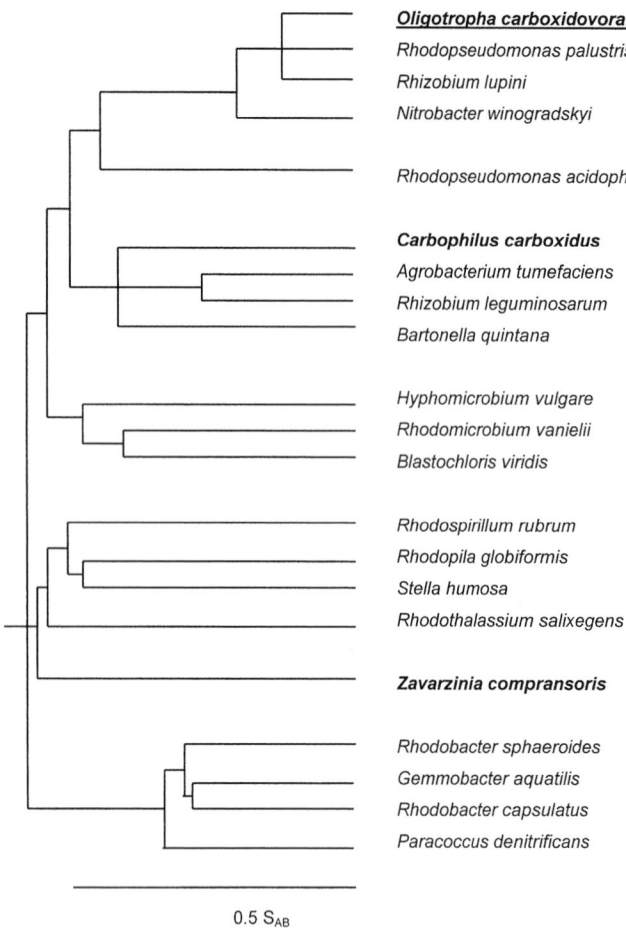

Oligotropha carboxidovorans
Rhodopseudomonas palustris
Rhizobium lupini
Nitrobacter winogradskyi

Rhodopseudomonas acidophila

Carbophilus carboxidus
Agrobacterium tumefaciens
Rhizobium leguminosarum
Bartonella quintana

Hyphomicrobium vulgare
Rhodomicrobium vanielii
Blastochloris viridis

Rhodospirillum rubrum
Rhodopila globiformis
Stella humosa
Rhodothalassium salixegens

Zavarzinia compransoris

Rhodobacter sphaeroides
Gemmobacter aquatilis
Rhodobacter capsulatus
Paracoccus denitrificans

0.5 S$_{AB}$

FIGURE BXII.α.185. Position of *Oligotropha carboxidovorans* strain OM5^T (DSM 1227, ATCC 49405) relative to other carboxidotrophic bacteria (*Carbophilus carboxidus* DSM 1086^T, *Zavarzinia compransoris* DSM 1231^T) in the phylogenetic tree of the *Alphaproteobacteria*. For a list of similarity coefficients refer to Table 6 of Auling et al. (1988). Modified from Auling et al. (1988). Refer to Meyer et al. (1993) for the transfer and amended descriptions of "*Pseudomonas carboxydovorans*" OM5^T to *Oligotropha*, gen. nov., as *Oligotropha carboxidovorans*, comb. nov., of "*Alcaligenes carboxydus*" DSM 1086^T to *Carbophilus*, gen. nov., as *Carbophilus carboxidus*, comb. nov., and of "*Pseudomonas compransoris*" DSM 1231^T to *Zavarzinia*, gen. nov., as *Zavarzinia compransoris*, comb. nov. Note that *O. carboxidovorans*, *Bradyrhizobium japonicum* strain 110spc4 (Lorite et al., 2000), *Carbophilus carboxidus*, and *Zavarzinia compransoris* are carboxidotrophic, i.e., they contain the molybdo- and copper-enzyme CO dehydrogenase and are capable of chemolithoautotrophic growth with CO as a sole energy and CO$_2$ as a carbon source under aerobic conditions.

1 week are purified by streaking. Because *Oligotropha* can form stellate aggregations, a pure culture will yield colonies of different diameter and dissimilar appearance. However, the microscopic observation of a single morphological type of cell in small and large colonies along with the formation of aggregates can be taken as evidence that a culture is axenic. Purity of a culture can be further confirmed by determining the viable cell numbers on different media (e.g., plates made up with the salts of pyruvate, lactate, acetate, succinate, or with nutrient broth, with sugars, or with mineral medium in the presence of 5% CO$_2$, 10% O$_2$, and 85% CO or 5% CO$_2$, 10% O$_2$, and 85% H$_2$). Pure cultures should reveal either identical colony numbers or no colonies at all. Because of the slow growth on CO or H$_2$ on solid media, the use of low agar concentrations (1.2–1.4%) is advisable.

DIFFERENTIATION OF THE GENUS *OLIGOTROPHA* FROM OTHER GENERA

Within the *Alphaproteobacteria*, the ability to form CO dehydrogenase and to utilize CO under aerobic chemolithoautotrophic conditions is known from individual strains of the genera *Oligotropha*, *Bradyrhizobium*, *Carbophilus*, and *Zavarzinia*. The genus *Oligotropha* is differentiated from *Bradyrhizobium* by the inability to nodulate leguminous plants, the absence of nitrogenase, and the inability to fix N$_2$. In addition, the variety of carbon sources utilized by the *Bradyrhizobium* species includes a much wider range than those utilized by *Oligotropha*. The genus *Carbophilus* is placed by 16S rRNA cataloging in an individual subline of descent within the α2 subclass of the *Alphaproteobacteria* (Fig. BXII.α.185). In contrast to *Oligotropha carboxidovorans*, *Carbophilus carboxidus* carries up to five peritrichous flagella, is unable to grow at the expense of H$_2$ plus CO$_2$ under aerobic chemolithoautotrophic conditions, and utilizes a broad spectrum of organic compounds, particularly sugars, as source of carbon and energy. In addition, *Carbophilus* and *Oligotropha* can be differentiated based on their sensitivity to different kinds of antibiotics and profiles of fatty acids and polyamines (see Table 1 of Meyer et al., 1993). The genus *Zavarzinia* has an isolated phylogenetic position intermediate between the α1 and α2 subclasses of the *Alphaproteobacteria* (Fig. BXII.α.185). *Zavarzinia compransoris* can be differentiated from *Oligotropha* by a very long polar flagellum, the requirement of thiamine as a growth factor, a higher G + C content of the DNA, different sensitivity to antibiotics, and different profiles of fatty acids and polyamines (see Table 1 of Meyer et al., 1993).

TAXONOMIC COMMENTS

On the basis of 16S rRNA cataloging, the presence of signature oligonucleotides in their 16S rRNA catalogues [see Table 5 of Auling et al. (1988)], and the presence of ubiquinone Q-10 as the major quinone (see Table 3 of Auling et al., 1988), *sym*-homospermidine as the major polyamine (see Table 4 of Auling et al., 1988), and *cis*-vaccinic acid as the major fatty acid (see Table 3 of Auling et al., 1988), *O. carboxidovorans* has been allocated to the class *Alphaproteobacteria* in the phylum *Proteobacteria* (Fig. BXII.α.185). For a dendrogram of relationship, derived from the S$_{AB}$ values by average linkage clustering, refer to Auling et al. (1988). The phylogenetic position of *O. carboxidovorans* indicates a high relationship to certain members of the *Alphaproteobacteria*, such as *Rhodopseudomonas palustris*, *Bradyrhizobium japonicum*, *Rhizobium lupini*, *Nitrobacter winogradskyi*, and *N. hamburgensis*.

The type species was formerly known as "*Pseudomonas carboxydovorans*" (Meyer and Schlegel, 1978) and was a re-isolate of "*Hydrogenomonas carboxydovorans*" (Kistner, 1953, 1954), which was lost or lost its properties (see discussion in Meyer and Schlegel, 1978).

Although they have been placed in a separate genus, the strains of *O. carboxidovorans* are phylogenetically related to the dinitrogen fixers *Bradyrhizobium japonicum* and *Rhodopseudomonas palustris* (Fig. BXII.α.185). Indeed, *B. japonicum* strain 110spc4 has been demonstrated to grow with CO chemolithoautotrophically and to contain a CO dehydrogenase that is very closely related to the corresponding enzyme from *O. carboxidovorans* strain OM5 (Lorite et al., 2000). However, *O. carboxidovorans* does not fix N$_2$ and no hybridization with a *nifH* gene probe points to the absence of a conventional nitrogenase.

DNA base ratios of total DNA from *O. carboxidovorans* strains OM5 (62.8 mol% G + C), OM4 (62.6 mol% G + C), OM3 (62.6 mol% G + C), and OM2 (63.1) based on T_m and analysis of the base content by reverse-phase HPLC are narrow (Auling et al., 1988).

So far, only a single species has been identified. The isolates designated OM2, OM3, OM4, and OM5 have all grouped together as strains of the species *Oligotropha carboxidovorans*.

List of species of the genus Oligotropha

1. **Oligotropha carboxidovorans** (Meyer and Schlegel 1978) Meyer, Stackebrandt and Auling 1994, 182[VP] (Effective publication: Meyer, Stackebrandt and Auling 1993, 391) (*"Pseudomonas carboxydovorans"* Meyer and Schlegel 1978, 42; *"Hydrogenomonas carboxydovorans"* Kistner 1954, 186.)

car.box.i.do'.vo.rans. L. n. *carbo* charcoal, carbon; Gr. adj. *oxys* sour, acid; L. v. *voro* devour; M.L. part. adj. *carboxidovorans* carbon-acid devouring; named for its ability to use CO as a sole carbon and energy source.

The description is as given for the genus. The strains OM4 and OM5 of *O. carboxidovorans* were isolated from wastewater of the settling pond of the municipal sewage treatment plant of the city of Göttingen (Germany). The strains OM2 and OM3 were isolated from soil.

The mol% G + C of the DNA is: 62.8 (T_m).

Type strain: OM5, ATCC 49405, DSM 1227.

Genus VIII. **Rhodoblastus** Imhoff 2001, 1865[VP]

JOHANNES F. IMHOFF

Rho.do.blas' tus. Gr. n. *rhodon* the rose; Gr. n. *blastos* bud shoot; M.L. masc. n. *Rhodoblastus* the budding rose.

Cells are rod shaped, motile by means of flagella; show polar growth, budding, and asymmetric cell division. Gram negative and belong to the *Alphaproteobacteria.* **Internal photosynthetic membranes appear as lamellae underlying and parallel to the cytoplasmic membrane. Photosynthetic pigments are bacteriochlorophyll** *a* **and carotenoids.** Straight-chain monounsaturated $C_{18:1}$, $C_{16:1}$, and saturated $C_{16:0}$ are the major cellular fatty acids. Contain ubiquinones, rhodoquinones, and menaquinones with 10 isoprene units (Q-10, MK-10, and RQ-10).

Preferred mode of growth is photoheterotrophically under anoxic conditions in the light. Photoautotrophic growth may be possible under anoxic conditions with hydrogen as electron donor. Chemotrophic growth occurs under microoxic to oxic conditions, but with some substrates also anaerobically by fermentation. Growth factors are not required by the type species. **Mesophilic freshwater bacteria with preference for acidic pH.**

Habitat: slightly acidic freshwater ponds.

The mol% G + C of the DNA is: 62.2–66.8.

Type species: **Rhodoblastus acidophilus** (*Rhodopseudomonas acidophila* Pfennig 1969a, 601) Imhoff 2001, 1865.

FURTHER DESCRIPTIVE INFORMATION

During growth on methanol, *Rhodoblastus acidophilus* assimilates its cell carbon via the ribulosebisphosphate cycle and carboxylation reactions of C_3 fatty acids. There was no evidence of the operation of a reduced C_1 fixation sequence (Sahm et al., 1976). A slow fermentative metabolism may be a property of most *Rhodopseudomonas* species. Gürgün et al. (1976) quantitatively demonstrated the formation from pyruvate of CO_2, formate, acetate, diacetyl, acetoin, and butandiol by *Rhodoblastus acidophilus*. Photoautotrophic growth with hydrogen has been found in *Rhodopseudomonas palustris* in *Rhodoblastus acidophilus* (Pfennig, 1969a). *Rhodoblastus acidophilus* also grows chemoautotrophically with hydrogen and utilizes methanol and formate under these conditions if the oxygen tension is kept at a low level (Siefert and Pfennig, 1979).

In contrast to *Rhodopseudomonas palustris*, sulfate is assimilated by *Rhodoblastus acidophilus* via the adenosine-5'-phosphosulfate

pathway (Imhoff, 1982). Nitrogenase has been found in all species investigated (Madigan et al., 1984). Ammonia assimilation proceeds via glutamine synthetase and glutamate synthase in *Rhodoblastus acidophilus* (NADH-linked) (Brown and Herbert, 1977; Herbert et al., 1978); glutamate dehydrogenase is lacking.

ENRICHMENT AND ISOLATION PROCEDURES

Standard procedures are used for the isolation of purple nonsulfur bacteria (PNSB). A mineral medium that is suitable for most PNSB is also applicable for the cultivation of *Rhodoblastus* species (Imhoff, 1988; Imhoff and Trüper, 1992; for this medium, see the footnote in the chapter describing the genus *Rhodospirillum*). For selective enrichments, the preference for low pH values is used for cultures of *Rhodoblastus acidophilus*. A succinate-mineral medium without growth factors with an initial pH of 5.2 is highly selective for *Rhodoblastus acidophilus* (and also *Rhodomicrobium vannielii*) (Pfennig, 1969a).

MAINTENANCE PROCEDURES

Cultures of *Rhodoblastus* can be preserved in liquid nitrogen, by lyophilization, or at −80°C in a mechanical freezer.

DIFFERENTIATION OF THE GENUS *RHODOBLASTUS* FROM OTHER GENERA

Differential characteristics of the genera and species of *Rhodoblastus*, *Rhodopseudomonas*, *Rhodobium*, *Rhodoplanes*, *Blastochloris*, and *Rhodomicrobium* and are shown in Tables 3 (pp. 125–126) and 4 (p. 127) of the introductory chapter "Anoxygenic Phototrophic Purple Bacteria", Volume 2, Part A. Their phylogenetic relationships are shown in Fig. 2 (p. 128) of that chapter. The properties of *Rhodoblastus* are compared to those of *Rhodopseudomonas* species in Tables BXII.α.156 and BXII.α.157.

TAXONOMIC COMMENTS

Rhodoblastus acidophilus contains glucosamine (not 2,3-diamino-2,3-dideoxyhexose) in the lipid A (Tegtmeyer et al., 1985). This is in contrast to *Rhodopseudomonas palustris* and to *Blastochloris* species (also formerly assigned to *Rhodopseudomonas*).

TABLE BXII.α.156. Differential characteristics of the anoxygenic phototrophic purple bacteria belonging to the family *Bradyrhizobiaceae* of the order *Rhizobiales*: genera *Rhodopseudomonas* and *Rhodoblastus*.[a]

Characteristic	*Rhodopseudomonas palustris*	*"Rhodopseudomonas cryptolactis"*	*Rhodopseudomonas julia*	*Rhodopseudomonas rhenobacensis*	*Rhodoblastus acidophilus*
Cell diameter (μm)	0.6–0.9	1.0	1.0–1.5	0.4–0.6	1.0–1.3
Type of budding	Tube	Sessile	Sessile	Sessile	Sessile
Rosette formation	+	+	+	+	−
Internal membrane system	Lamellae	Lamellae	Lamellae	Lamellae	Lamellae
Motility	+	+	+	+	+
Color of cultures	Brown-red to red	Red	Pink	Red	Red to orange-red
Bacteriochlorophyll	*a*	*a*	*a*	*a*	*a*
Salt requirement	None	None	None	None	None
Optimal pH	6.9	6.8–7.2	6.0–6.5	5.5	5.5–6.0
Optimal temperature	30–37	38–40	25–35	(28)	25–30
Sulfate assimilation	+ (PAPS)	nd	−	+	+ (APS)
Aerobic dark growth	+	+	+	+	+
Denitrification	+ / −	nd	nd	Reduction of nitrate	−
Fermentation of fructose	−	nd	nd	nd	−
Photoautotrophic growth with	H$_2$, thiosulfate, sulfide	−	Sulfide, sulfur	nd	H$_2$
Growth factors	*p*-Aminobenzoic acid, (biotin)	B$_{12}$, niacin, *p*-aminobenzoic acid	None	*p*-Aminobenzoic acid	None
Utilization of:					
Benzoate	+	−	−	−	−
Citrate	+	−	−	−	−
Formate	−	−	+	+	−
Glucose	−	−	+	−	−
Tartrate	−	−	−	+	−
Sulfide	+	−	+	nd	−
Thiosulfate	+	−	nd	−	−
Mol % G + C of the DNA	64.8–66.3 (Bd)	68.8	63.5	65.4 (HPLC)	62.2–66.8 (Bd)
Cytochrome *c*$_2$ size	Large	nd	nd	nd	Small
Major quinones	Q-10	nd	nd	Q-10	Q-10, MK-10, RQ-10
Major fatty acids					
C$_{14:0}$	trace	nd	nd	nd	0.8
C$_{16:0}$	5.2	nd	nd	11.7	14.8
C$_{16:1}$	3.1	nd	nd	9.5	37.2
C$_{18:0}$	7.3	nd	nd	7.8	0.8
C$_{18:1}$	79.7	nd	nd	66.1	46.0

[a]Symbols: +, positive in most strains; −, negative in most strains; + / −, variable in different strains; nd, not determined; (+), weak growth or microaerobic growth only; (APS), via adenosine-5′-phosphosulfate; (PAPS), via 3′-phosphoadenosine-5′-phosphosulfate; (biotin) biotin is required by some strains; Q-10, ubiquinone 10; MK-10, menaquinone 10; RQ-10, rhodoquinone 10. Bd, buoyant density.

List of species of the genus Rhodoblastus

1. **Rhodoblastus acidophilus** (*Rhodopseudomonas acidophila* Pfennig 1969a, 601[AL]) Imhoff 2001, 1865[VP]

   a.ci.do′phi.lus. L. adj. *acidus* sour; M.L. neut. n. *acidum* an acid; Gr. adj. *philos* loving; M.L. masc. adj. *acidophilus* acid-loving.

   Cells are rod shaped to elongate-ovoid, slightly curved, 1.0–1.3 × 2.0–5.0 μm, motile by polar flagella, and Gram negative. Daughter cells originate by polar growth as sessile buds at the pole opposite that bearing the flagella; there is no tube or filament between mother and daughter cells (Fig. BXII.α.186). When the daughter cell reaches the size of the mother cell, cell division is completed by constriction. In the next cycle, both cells form buds at the poles of the former cell division. Under certain conditions, rosettes and clusters are formed that are similar to those of the type species. In media lacking calcium ions, cells are immotile. Internal photosynthetic membranes are present as lamellae underlying and parallel to the cytoplasmic membrane. Color of anaerobic liquid cultures is purple-red to orange-brown. Cells grown under oxic conditions are colorless to light pink or orange. Absorption spectra of living cells show maxima at 375, 460, 490, 525, 590, 805, 855, and 890 nm. Photosynthetic pigments are bacteriochlorophyll *a* (esteri-

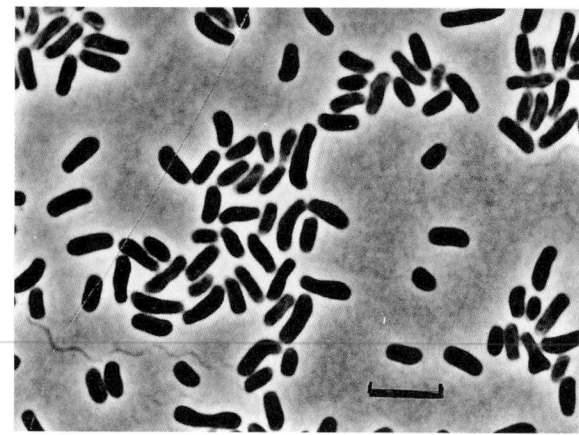

FIGURE BXII.α.186. *Rhodoblastus acidophilus* strain 2751. The shape of some of the cells is indicative of the polar type of cell growth. Tufts of detached flagella can be seen. Phase-contrast micrograph. Bar = 5 μm. (Courtesy of N. Pfennig.)

TABLE BXII.α.157. Growth substrates of the anoxygenic phototrophic purple bacteria belonging to the family *Bradyrhizobiaceae* of the order *Rhizobiales*: genera *Rhodopseudomonas* and *Rhodoblastus*[a]

Source/donor	*Rhodopseudomonas palustris*	*"Rhodopseudomonas cryptolactis"*	*Rhodopseudomonas julia*	*Rhodopseudomonas rhenobacensis*	*Rhodoblastus acidophilus*
Carbon source					
Acetate	+	+	+	+	+
Aspartate	+/−	−	+	nd	−
Benzoate	+	−	−	−	−
Butyrate	+	nd	+	+	+/−
Caproate	+	nd	−	nd	+/−
Caprylate	+	nd	−	nd	−
Citrate	+/−	nd	−	−	+/−
Ethanol	+/−	nd	−	+	+
Formate	+	nd	+	+	+/−
D-Fructose	+/−	−	+	−	−
Fumarate	+	nd	+	+	+
D-Glucose	+/−	−	+	−	+/−
Glutamate	+	−	−	−	−
Glycerol	+	nd	+	nd	+/−
Glycolate	+	nd	nd	nd	+/−
Lactate	+	+	nd	+	+
Malate	+	+	+	+	+
Malonate	+	nd	−	nd	+/−
Mannitol	+/−	nd	+	nd	−
Methanol	+/−	nd	−	−	+/−
Propanol	+	nd	−	nd	nd
Propionate	+	nd	+	−	+
Pyruvate	+	+	+	+	+
Sorbitol	+	nd	+	+	+
Succinate	+	+	+	+	+
Tartrate	−	nd	−	+	+/−
Valerate	+	nd	+	nd	+
Electron donor					
Sulfide	+	nd	+	nd	−
Thiosulfate	+	nd	nd	nd	−

[a]Symbols: +, positive in most strains; −, negative in most strains; +/− variable in different strains; nd, not determined.

fied with phytol) and carotenoids of the spirilloxanthin series with glucosides of rhodopin and rhodopinal. The latter are characteristic of this species.

Photoheterotrophic growth with a number of organic carbon sources is the preferred growth mode. Photoautotrophic growth is possible with hydrogen as electron donor; sulfide and thiosulfate cannot be used. Cells grow under microoxic to oxic conditions in the dark, with hydrogen as electron donor autotrophically. The organic carbon sources used are acetate, propionate, butyrate, lactate, pyruvate, fumarate, malate, succinate, valerate, formate, methanol, and ethanol. Not used are caprylate, pelargonate, glycerol, ben-

zoate, sugars, sugar alcohols, glutamate, and other amino acids. Sulfate can be used as sole sulfur source. Ammonia, dinitrogen, and some amino acids are used as nitrogen source. Growth factors are not required; yeast extract, or other complex nutrients do not increase the growth rate.

Mesophilic freshwater bacterium with optimal growth at 25–30°C and pH 5.5–6.0.

Ubiquinones and menaquinones with 10 isoprene units (Q-10 and MK-10) are present.

The mol% G + C of the DNA is: 62.2–66.8 (Bd).

Deposited strain: Pfennig 7050, ATCC 25092, DSM 137.

GenBank accession number (16S rRNA): M34128.

Genus IX. **Rhodopseudomonas** Czurda and Maresch 1937, 119[AL] emend. Imhoff, Trüper and Pfennig 1984, 341

JOHANNES F. IMHOFF

Rho.do.pseu.do.mo'nas. Gr. n. *rhodon* the rose; M.L. fem. n. *Pseudomonas* a bacterial genus; M.L. fem. n. *Rhodopseudomonas* the red *Pseudomonas*.

Cells are rod-shaped, motile by means of polar or subpolar flagella; show polar growth, budding and asymmetric cell division. Gram-negative and belong to the *Alphaproteobacteria*. Internal photosynthetic membranes appear as lamellae underlying and parallel to the cytoplasmic membrane. Photosynthetic pigments are **bacteriochlorophyll** *a* and carotenoids of the spirilloxanthin series. Straight-chain monounsaturated C$_{18:1}$ is the major component of cellular fatty acids. Contains ubiquinones with 10 isoprene units (Q-10).

Preferred mode of growth is photoheterotrophic under anoxic conditions in the light. **Photoautotrophic growth may be possible under anoxic conditions with hydrogen, thiosulfate, or sulfide** as electron donor. **Chemotrophic growth under microoxic to oxic conditions** is also possible. Growth factors may be required.

Mesophilic freshwater bacteria with preference for neutral pH.

The mol% G + C of the DNA is: 64.8–66.3.

Type species: **Rhodopseudomonas palustris** (Molisch 1907) van Niel 1944, 89 (*"Rhodobacillus palustris"* Molisch 1907, 14.)

FURTHER DESCRIPTIVE INFORMATION

Rhodopseudomonas species are characterized by asymmetric cells during cell division. Most characteristic is the formation of prosthecae and rosette-like cell aggregates.

A reductive and oxygen-inhibited degradation of benzoate has been demonstrated in *R. palustris* (Dutton and Evans, 1969, 1978; Gibson and Harwood, 1995). Citrate may be used by some strains of *Rhodopseudomonas palustris*. During fermentation, pyruvate is quantitatively transformed into CO_2, formate, acetate, lactate, butyrate, and acetoin by *R. palustris* (Gürgün et al., 1976). Photoautotrophic growth with hydrogen occurs in *R. palustris* (Klemme, 1968).

Sulfate can serve as sole sulfur source and is assimilated via the 3'-phosphoadenosine-5'-phosphosulfate (Imhoff, 1982). *R. palustris* is able to use reduced sulfur compounds as photosynthetic electron donors. Sulfide is oxidized by *R. palustris* to elemental sulfur (Hansen, 1974), and thiosulfate to sulfate (Rolls and Lindstrom, 1967).

In *R. palustris* (Zumft and Castillo, 1978) nitrogenase shows a "switch-off" effect by ammonia and some organic nitrogen compounds. This effect is reversible. Ammonia assimilation proceeds via glutamine synthetase and glutamate synthase in *Rhodopseudomonas palustris* (NADH-linked) (Brown and Herbert, 1977; Herbert et al., 1978). A glutamate dehydrogenase is present. Besides ammonia, *R. palustris* uses a great number of nitrogen sources: nitrate, dimethylamine, trimethylamine, azaguanine, L-histidine, L-glutamate, L-aspartate, L-arginine, L-cysteine, L-methionine, DL-lysine, DL-alanine, DL-leucine, casein hydrolysate (Malofeeva and Laush, 1976), guanine, uric acid, xanthine (Aretz et al., 1978), cytidine, and cytosine (Kaspari, 1979). Most strains of *R. palustris* are not able to use nitrate as a nitrogen source (Klemme, 1979). Three isolates, however, are capable of dissimilatory nitrate reduction and growth under anoxic dark conditions with nitrate as terminal electron acceptor (Klemme et al., 1980).

Rhodopseudomonas palustris and *Nitrobacter winogradskyi* show a high degree of 16S rDNA sequence similarity (Seewaldt et al., 1982), and the lipid A structure of the lipopolysaccharide of *Nitrobacter* species contains the unusual 2,3-diamino-2,3-deoxy-D-glucose that is characteristic for *Rhodopseudomonas palustris* (Mayer et al., 1983, 1984). This sugar has also been found in *Blastochloris viridis* (see Weckesser et al., 1979), *Blastochloris sulfoviridis* (Ahamed et al., 1982), *Pseudomonas diminuta*, and *Pseudomonas vesicularis* (Mayer et al., 1983).

ENRICHMENT AND ISOLATION PROCEDURES

Rhodopseudomonas palustris is a very common species of purple nonsulfur bacteria (PNSB) in nature, and many enrichments select for *R. palustris*. Benzoate is a carbon source particularly for enrichment of this species. Standard procedures for other PNSB can be used for isolation of *Rhodopseudomonas* species. A mineral medium that is suitable for most PNSB can be applied for isolation and cultivation of *Rhodopseudomonas* species (Imhoff, 1988; Imhoff and Trüper, 1992; see chapter Genus *Rhodospirillum* for this medium).

MAINTENANCE PROCEDURES

Cultures of *Rhodopseudomonas* species are well preserved in liquid nitrogen, by lyophilization or at $-80°C$ in a mechanical freezer.

DIFFERENTIATION OF THE GENUS *RHODOPSEUDOMONAS* FROM OTHER GENERA

Differential characteristics of the genera and species of *Rhodopseudomonas*, *Rhodobium*, *Rhodoplanes*, *Blastochloris*, *Rhodoblastus*, and *Rhodomicrobium* are shown in Tables 3 and 4 of the introductory chapter "Anoxygenic Phototrophic Purple Bacteria." (pp. 125–127, Volume 2, Part A). Their phylogenetic relationships are shown in Fig. 2 (p. 128) of that same chapter.

TAXONOMIC COMMENTS

After the removal of PNSB that contained vesicular internal photosynthetic membranes and those that were *Betaproteobacteria* from the genus *Rhodopseudomonas* (Imhoff et al., 1984), those that remained had lamellar internal membrane structures and a budding mode of growth and reproduction. Even after the removal of bacteria now recognized as species of *Rhodopila*, *Rhodobacter*, *Rhodovulum*, and *Rubrivivax*, the genus *Rhodopseudomonas* still represented a heterogeneous assemblage of species (Imhoff et al., 1984). Based on 16S rDNA sequence data (Kawasaki et al., 1993b), isolation and description of new species, and additional data, *Rhodopseudomonas rosea* was transferred to *Rhodoplanes roseus* (Hiraishi and Ueda, 1994b), *Rhodopseudomonas marina* to the new genus *Rhodobium* as *Rhodobium marinum*, and *R. viridis* and *R. sulfoviridis* to the genus *Blastochloris* as *Blastochloris viridis* and *Blastochloris sulfoviridis* (Hiraishi, 1997). *Rhodopseudomonas acidophila* was transferred to *Rhodoblastus acidophilus* (Imhoff, 2001).

Rhodopseudomonas blastica has many chemotaxonomic properties in common with bacteria in the informal group "alpha-3 proteobacteria", *Rhodobacter*, *Rhodovulum*, and *Paracoccus denitrificans*. Its 16S rDNA sequence is most similar to and clusters with the *Rhodobacter* species. Therefore, it was transferred to this genus and is now known as *Rhodobacter blasticus* (Kawasaki et al., 1993b).

Hiraishi et al. (1992b) considered *Rhodopseudomonas rutila* as identical to *R. palustris* and the species name as a later subjective synonym of *R. palustris*, although it has been described as a separate species (Akiba et al., 1983). In the original species description *R. palustris* was not included as a reference organism and properties of the new isolates were reinvestigated later (Hiraishi et al., 1992b). These authors established identity between the type strains of *R. palustris* and *R. rutila*. In particular, in regard to the utilization of acetate, propionate, benzoate, fructose, sulfide, and thiosulfate, Hiraishi et al. (1992b) disagreed with Akiba et al. (1983) and found conformity with the properties of six *R. palustris* strains. The former authors also found that *Rhodopseudomonas rutila* requires *p*-aminobenzoic acid as a growth factor. Both bacteria were identical in quinone, lipid and fatty acid composition and 16S rDNA sequence (accession number of the 16S rDNA sequence of strain ATCC 33872, the former type of *R. rutila* at EMBL is D14435). DNA–DNA similarity between *R. rutila* and *R. palustris* strains was between 80–90%.

In addition to *R. palustris*, three species of *Rhodopseudomonas* are known. *Rhodopseudomonas rhenobacensis* is a recently described bacterium notably reducing nitrate (Hougardy et al., 2000). *Rhodopseudomonas julia* is an acidic sulfur spring isolate with the distinct ability to grow photoautotrophically (Kompantseva, 1989a). *"Rhodopseudomonas cryptolactis"* was also isolated from a hot spring and is characterized by a lack of autotrophic growth and a quite restricted substrate spectrum (Stadtwalddemchick et al., 1990). This bacterium has not been validly published.

DIFFERENTIATION OF THE SPECIES OF THE GENUS *RHODOPSEUDOMONAS*

Characters that differentiate the species of the genus *Rhodopseudomonas* from each other and from *Rhodoblastus acidophilus* are given in Tables BXII.α.156 and BXII.α.157 of the chapter on *Rhodoblastus*.

List of species of the genus Rhodopseudomonas

1. **Rhodopseudomonas palustris** (Molisch 1907) van Niel 1944, 89[AL] (*"Rhodobacillus palustris"* Molisch 1907, 14.)
pa.lus' tris. L. fem. adj. *palustris* marshy, swampy.

Individual cells are rod-shaped to ovoid, occasionally slightly curved, $0.6–0.9 \times 1.2–2.0$ µm, motile by means of subpolar flagella, and reproduce by budding. The mother cell produces a slender prostheca 1.5–2.0 times the length of the original cell at the pole opposite to that bearing the flagella. The end of the prostheca swells, and the daughter cell grows, producing a dumbbell-shaped organism (Fig. BXII.α.187). Asymmetric division then takes place. Young individual cells are highly motile. The formation of rosettes and clusters in which the individual cells are attached to each other at their flagellated poles are characteristic in older cultures. In certain complex media, individual cells become up to 10 µm long and irregular in shape. Internal photosynthetic membranes appear as lamellae underlying and parallel to the cytoplasmic membrane; no lamellae are present in the prostheca. Color of cell suspensions is red to brownish-red. Living cells show absorption maxima at 375, 468, 493, 520–545, 589, 802, and 860–875 nm. Photosynthetic pigments are bacteriochlorophyll *a* (esterified with phytol) and carotenoids of the normal spirilloxanthin series.

Photoautotrophic growth occurs with hydrogen, sulfide, and thiosulfate as electron donors in the presence of small amounts of yeast extract. Photoheterotrophic growth is possible with various organic substrates. Chemotrophic growth occurs in the dark under microoxic to oxic conditions. With some substrates fermentation takes place under anoxic conditions. Sulfate can be used as sole sulfur source. The nitrogen source may be ammonia, dinitrogen, some amino acids, or for a few strains, nitrate.

Growth factors required are *p*-aminobenzoate and, for some strains, biotin; yeast extract stimulates growth considerably.

Ubiquinones with 10 isoprene units (Q-10) are present. Mesophilic freshwater bacteria with optimal growth at 30–37°C and pH 6.9 (pH range: 5.5–8.5).

The mol% G + C of the DNA is: 64.8–66.3 (Bd).

Type strain: ATH 2.1.6, ATCC 17001, DSM 123.

GenBank accession number (16S rRNA): D12700, D25312, L11664.

2. **Rhodopseudomonas julia** Kompantseva 1993, 188[VP] (Effective publication: Kompantseva 1989a, 258.)
ju.li.a. L. fem. adj. *julia* discovered and described in July.

Cells are straight or slightly curved rods, $1–1.5 \times 2.5$ µm. Formation of rosettes is characteristic. Propagation by budding; the bud is sessile. Motile by means of polar flagella. Photosynthetic membrane systems appear as concentric lamellae. Color of cell suspensions is pink to purple. The absorption spectra of living cell suspensions may have maxima at 380, 590, 803, and 850 nm (indicative of bacteriochlorophyll *a*) and at 490, 516, and 550 nm (indicating the presence of carotenoids of the normal spirilloxanthin series).

Facultative photoorganoheterotrophs. Can also grow chemoorganotrophically under microaerobic conditions and photolithoautotrophically, oxidizing H_2S or S^0 to sulfate. The intermediate product of H_2S oxidation is elemental sulfur, which is deposited both outside and inside the cells. Ammonia and Casamino acids, but not nitrate, are utilized as nitrogen sources; H_2S, S^0, cysteine, and glutathione, but not sulfate, serve as sulfur sources. No vitamins are required. Hydrogen donors and carbon sources: fatty acids to valerate, hydroxy acids of the tricarboxylic and acid cycle, formate, pyruvate, aspartate, glycerol, glucose, gluconate, mannitol, sorbitol, fructose, yeast extract, and Casamino acids. Substrates not utilized: arginine, benzoate, glutamate, malonate, tartrate, citrate, and alcohols. Storage material: poly-β-hydroxybutyrate. Catalase activity is present.

Optimal growth conditions: pH 6.0–6.5, 25–35°C, light intensity more than 2000 lux.

Habitat: slightly and moderately acid sulfide springs having a high content of elemental sulfur.

The mol% G + C of the DNA is: 63.5 ± 1 (method not available in the literature).

Type strain: KR-11-67, ATCC 52215, DSM 11549.

GenBank accession number (16S rRNA): AB087720.

3. **Rhodopseudomonas rhenobacensis** Hougardy, Tindall, Klemme 2000, 991[VP]
rhe.no.ba.cen' sis. M.L. adj. *rhenobacensis* pertaining to Rheinbach, a small German town.

Cells are rods, $0.4–0.6 \times 1.5–2$ µm. Multiply by budding and form characteristic rosette-like aggregates. Motile by means of polar flagella. Photosynthetic membrane system appears as lamellae. Color of cell suspensions is red. The absorption spectra of living cell suspensions show maxima at 376, 591, 805, and 878 nm (bacteriochlorophyll *a*) and at 471, 503, and 540 nm (carotenoids).

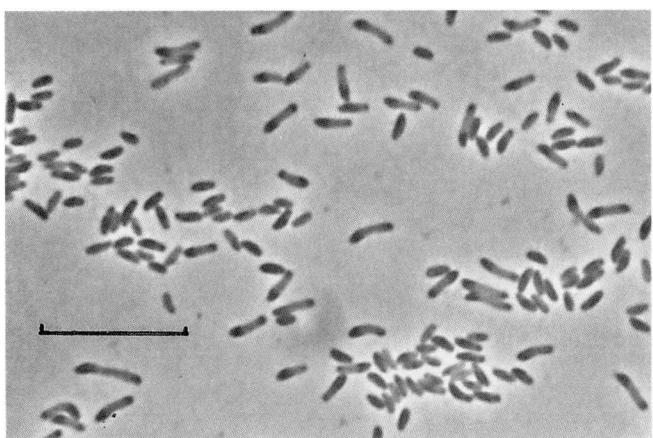

FIGURE BXII.α.187. *Rhodopseudomonas palustris* strain 1850 grown in succinate-yeast extract medium. The budding type of reproduction can be recognized in a number of cells. Phase-contrast micrograph. Bar = 10 µm. (Courtesy of N. Pfennig.)

Facultative photoorganoheterotrophs. Also grow chemoorganotrophically under aerobic conditions. Nitrate is reduced to nitrite, which accumulates in the medium. Hydrogen donors and carbon sources: formate, acetate, pyruvate, lactate, malate, succinate, fumarate, tartrate, gluconate, ethanol, and butyrate (plus carbonate). Substrates not utilized: citrate, benzoate, glucose, fructose, methanol, propionate, arginine, and glutamate. Gelatin is not hydrolyzed. Ammonia, glutamate dinitrogen, and yeast extract serve as nitrogen sources. *p*-Aminobenzoic acid required as growth factor.

Optimal pH 5.5 (range: 5.0–8.0). NaCl is not required and completely inhibits growth at 1%.

Habitat: freshwater lake sediment.

The mol% G + C of the DNA is: 65.4 (HPLC).

Type strain: Rb, DSM 12706.

GenBank accession number (16S rRNA): AJ132402.

4. **"Rhodopseudomonas cryptolactis"**

cryp.to′ lac.tis. Gr. adj. *cryptos* hidden; L. fem. n. *lactes* milk; M.L. fem. adj. *cryptolactis* hidden of milk.

Cells are rod-shaped, approximately 1.0×4 μm, motile by means of flagella, and reproduce by budding. The formation of rosettes is observed. Internal photosynthetic membranes appear as lamellae underlying the cytoplasmic membrane. Color of cell suspensions is red. Absorption maxima of living cells at 590, 800, and 857 nm is indicative of the presence of bacteriochlorophyll *a*.

Photoheterotrophic growth occurs with pyruvate, acetate, malate, succinate, and lactate (only in the presence of bicarbonate). Hydrogen and sulfide are not used as electron donors. Chemoheterotrophic growth occurs in the dark under oxic conditions. Ammonia, dinitrogen, urea, and L-glutamine are used as nitrogen source. Growth factors required are vitamin B_{12}, niacin, and *p*-aminobenzoate.

Mesophilic freshwater bacterium with optimal growth at 40°C (range: 35–46°C), and pH 6.8–7.2 (pH range: 6.4–8.5).

Habitat: thermopolis Hot Spring, Wyoming, USA.

The mol% G + C of the DNA is: 68.8.

Deposited strain: ATCC 49414.

Family VIII. **Hyphomicrobiaceae** Babudieri 1950, 589

GEORGE M. GARRITY, JULIA A. BELL AND TIMOTHY LILBURN

Hy.pho.mi.cro.bi.a′ ce.ae. M.L. neut. n. *Hyphomicrobium* type genus of the family; *-aceae* ending to denote family; M.L. fem. pl. n. *Hyphomicrobiaceae* the *Hyphomicrobium* family.

The family *Hyphomicrobiaceae* was circumscribed for this volume on the basis of phylogenetic analysis of 16S rRNA sequences; the family contains the genera *Hyphomicrobium* (type genus), *Ancalomicrobium*, *Ancylobacter*, *Angulomicrobium*, *Aquabacter*, *Azorhizobium*, *Blastochloris*, *Devosia*, *Dichotomicrobium*, *Filomicrobium*, *Gemmiger*, *Labrys*, *Methylorhabdus*, *Pedomicrobium*, *Prosthecomicrobium*, *Rhodomicrobium*, *Rhodoplanes*, *Seliberia*, *Starkeya*, and *Xanthobacter*.

The family is morphologically, metabolically, and ecologically diverse. Many members form hyphae or prosthecae; a number reproduce by budding. Includes organisms that are photosynthetic, facultatively methylotrophic, facultatively chemolithoautotrophic, and chemoheterotrophic.

Type genus: **Hyphomicrobium** Stutzer and Hartleb 1898, 76.

Genus I. **Hyphomicrobium** *Stutzer and Hartleb 1898, 76*[AL]

CHRISTIAN GLIESCHE, ANDREAS FESEFELDT AND PETER HIRSCH

Hy.pho.mi.cro′ bi.um. Gr. *hyphe* thread; Gr. adj. *micros* small; Gr. masc. n. *bios* life; M.L. neut. n. *Hyphomicrobium* thread-producing microbe.

Cells 0.3–1.2 × 1–3 μm; rod-shaped with pointed ends, or **oval, egg-, or bean-shaped**; produce **monopolar or bipolar filamentous outgrowths** (hyphae or prosthecae) of varying length and 0.2–0.3 μm in diameter. Hyphae are not septate, but hyphal cytoplasmic membranes show conspicuous constrictions. **Hyphae may be truly branched**; secondary branches are rare. Cells stain with carbol fuchsin, but stain weakly with aqueous aniline dyes. Gram negative and non-acid-fast. Do not form spores.

Multiplication: **daughter cell formation by a budding process at one hyphal tip at a time** (Figs. BXII.α.188 and BXII.α.189); mature buds become **motile swarmers** that break off and may attach to surfaces or other cells to form clumps or **rosettes**. Motility is lost soon after swarmer cell liberation and/or attachment. Older cultures may lack motile swarmer cells. **Poly-β-hydroxybutyrate is stored by most cells, usually at a distinct cell pole**.

Colonies on solid media are small, even after prolonged incubation; they are **brownish** in transmitted light and **bright beige or colorless** in reflected light. Colony surface is shiny or granular, folded or smooth. Older colonies often display concentric rings and change color to darker brown or bright yellow-orange.

Chemoorganotrophic, aerobic. Carbon dioxide is required for growth. **Oligocarbophilic**, i.e., growth can occur on mineral salts media without carbon sources added to the medium, possibly instead resulting from the presence of volatile carbon and energy sources. Growth may be stimulated by soil extract if the pH remains near neutral. **Good growth with 0.1–0.2% (w/v) of one-carbon compounds, such as methanol, methylamine, or even chloromethane**. NH_4^+ is a good nitrogen source, but organic nitrogen compounds (some amino acids) may also be utilized. No nitrification. Widely distributed in soils and aquatic habitats. Mesophilic. Optimal pH: above 7.0, except for one species with a lower optimal pH.

The mol% G + C of the DNA is: 59–65 (T_m).

Type species: **Hyphomicrobium vulgare** Stutzer and Hartleb 1898, 76.

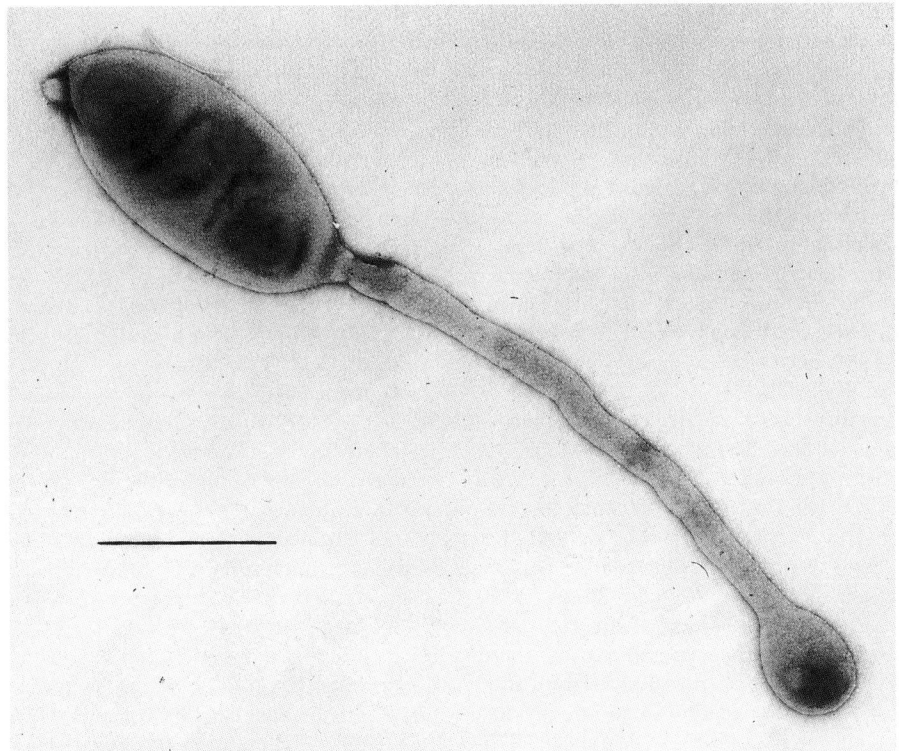

FIGURE BXII.α.188. *Hyphomicrobium facile* IFAM B-522. Mother cell with hyphae and young bud. Bar = 1.0 μm.

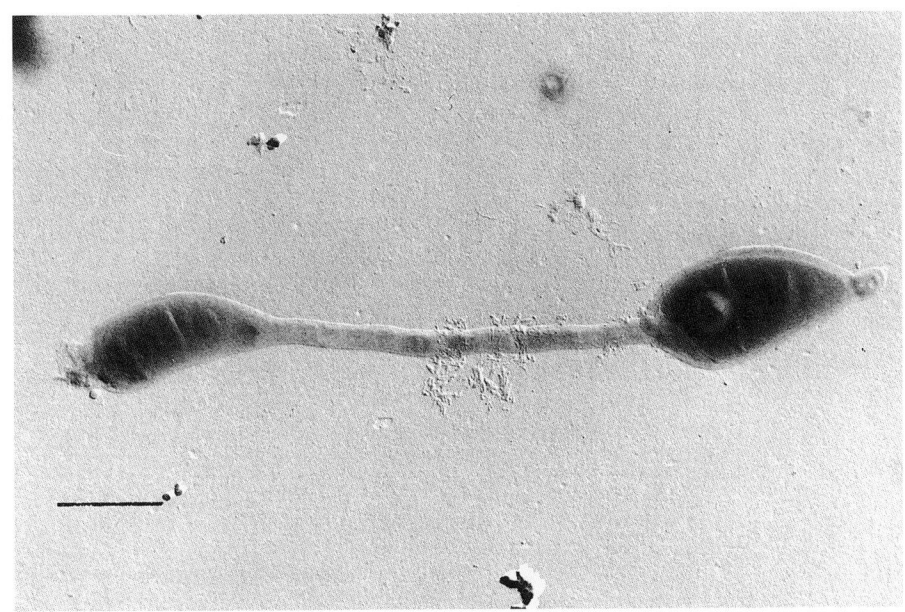

FIGURE BXII.α.189. *Hyphomicrobium facile* subsp. *ureaphilum* IFAM CO-582. Mother cell, hyphae, and mature bud. The mother cell contains a storage granule (poly-β-hydroxybutyrate). Shadow-cast. Bar = 1 μm.

FURTHER DESCRIPTIVE INFORMATION

Phylogeny and probes   Analysis of 5S rDNA (Stackebrandt et al., 1988a; Boulygina et al., 1993), 16S rRNA (Stackebrandt et al., 1988a), and 16S rDNA (Stackebrandt et al., 1988a; Tsuji et al., 1990; Tuhela et al., 1997; Rainey et al., 1998; Borodina et al., 2000; De Marco et al., 2000; Layton et al., 2000; McDonald et al., 2001), determination of 16S rRNA cistron similarities (Moore 1977; Roggentin and Hirsch 1989), membrane fatty acids and quinones (Urakami and Komagata 1979, 1986b, 1987a, b; Sittig and Hirsch, 1992) have confirmed the membership of hyphomicrobia in the Class *Alphaproteobacteria* (Order *Rhizobiales*). The coherency of the genus is supported by phage typing (Gliesche et al., 1988), low-molecular-mass RNA patterns (Höfle, 1990), and enzyme electrophoresis patterns (Urakami and Komagata,

1981). However, 16S rRNA cataloguing (Stackebrandt et al., 1988a) and 16S rDNA sequencing (Rainey et al., 1998; Borodina et al., 2000; Layton et al., 2000) show the genus *Filomicrobium* to branch within the radiation of *Hyphomicrobium*. Furthermore, 16S rRNA cistron similarities (Roggentin and Hirsch, 1989) and 16S rDNA sequences (Cox and Sly, 1997; Layton et al., 2000) confirm a close relationship of hyphomicrobia to the genus *Pedomicrobium*.

Based on data on morphological and physiological characteristics, 5S rRNA sequences, fatty acid and lipid composition of their membranes, and sensitivity to antibiotics, an evolutionary pathway from *Caulobacter* to *Hyphomicrobium* has been proposed by Nikitin et al. (1990). However, this phylogenetic view is not supported by 16S rDNA sequence analysis (Rainey et al., 1998; Sly et al., 1999; Layton et al., 2000).

The intrageneric structure based on 16S rDNA sequences shows two clusters (Rainey et al., 1998). Cluster I contains the neotype species of *H. vulgare*, as well as *H. aestuarii*, *H. hollandicum*, and *H. zavarzinii*. Cluster II comprises *H. facile*, *H. denitrificans*, *H. methylovorum*, and *H. chloromethanicum* (McDonald et al., 2001). 16S rDNA signature nucleotides that define clusters I and II of *Hyphomicrobium* species are given by Rainey et al. (1998). A detailed phylogenetic investigation of the 16S rDNA of hyphomicrobia isolated from a sewage treatment plant and its receiving lake (Holm et al., 1996) has revealed a significantly greater diversity and resulted in the creation of additional clusters, III and IV. Of these, cluster III comprises 16 strains (B 376, P 251, P 262, P 139, P 165, P 482, B 47, P 425, P 148, P 37, A 676, IFAM 1460, P 645, B 455, A 739, and A 679). Cluster IV consists of *Hyphomicrobium* sp. 502 and *Filomicrobium fusiforme* (Rainey and Stackebrandt, unpublished data). A dendrogram showing the phylogenetic position of the genera *Hyphomicrobium* and *Filomicrobium* among the closest relatives in the *Alphaproteobacteria* is given in Fig. BXII.α.190. This intrageneric structure is supported by a phylogenetic analysis of a fragment of the gene coding for the α-subunit of methanol dehydrogenase (*mxaF*). Fig. BXII.α.191 shows a dendrogram of hyphomicrobia based on published sequences of this gene fragment. A comparative sequence analysis of this *mxaF* gene fragment, including 150 strains, indicates a coherent evolution of this essential metabolic gene and the 16S rRNA gene in hyphomicrobia. The dendrogram based on the *mxaF* nucleic acid sequence has a finer resolution and

results in the five clusters A to E, which correspond exactly to the 16S rDNA-based subdivision (*mxaF* clusters B and C are identical to the 16S cluster II) (Fesefeldt, 1998). Furthermore, genomic DNA of *Filomicrobium fusiforme* DSM 5304[T] gives a strong hybridization signal with a fragment (position 1009–1553, according to *Paracoccus denitrificans* PD1207) of the *mxaF* gene of *Hyphomicrobium* sp. P 502 (*mxaF* cluster E) (Fesefeldt, 1998). This observation supports the close relationship between both genera and indicates that *Filomicrobium* retains parts of the methanol dehydrogenase gene cluster.

The distinctiveness of *Hyphomicrobium* species has been demonstrated by DNA–DNA hybridizations (Moore and Hirsch, 1972; Urakami et al., 1985; Gebers et al., 1986; Doronina et al., 1996b; McDonald et al., 2001); low-molecular-mass RNA patterns (Höfle, 1990); 16S rRNA cistron similarities (Roggentin and Hirsch, 1989); membrane fatty acids, phospholipids, and quinones (Sittig and Hirsch, 1992); and differences of more than 1.2% in 16S rDNA sequences for the closely related species (Rainey et al., 1998). Unfortunately, molecular analyses have not yet been conducted for undescribed strains for which species status has been discussed based on phenotypic analysis (Vedenina et al., 1991; Holm et al., 1996).

Probes for hyphomicrobia have been developed specifically for different taxonomic levels: (1) species-specific (Hfa-1 for *H. facile* [Fesefeldt et al., 1997] and Hvu-1 for *H. vulgare* [Gliesche et al., 1997]); (2) cluster-specific (S-S-HyphoCI-648-a-A-20 for *Hyphomicrobium* Cluster I and S-S-HyphoCII-654-a-A-18 for *Hyphomicrobium* Cluster II [Layton et al., 2000]); and (3) genus-specific (S-G-Hypho-1241-a-A-19 for *Hyphomicrobium* spp. [Layton et al., 2000]). Probe Hvu1034 (Neef et al., 1996) has exhibited an uncertain specificity (Neef and Fesefeldt, unpublished data).

Cell morphology The cell morphology may vary with growth conditions. In media with low nutrient concentrations, hyphae may elongate up to 300 μm. The degree of branching depends on the type and concentration of carbon source present. Some strains have helically twisted hyphae. Stirring laboratory cultures may result in increased intercalary bud formation in the hyphae. The number and attachment sites of flagella vary among the nine described species. Most hyphomicrobia that utilize C_1 compounds have 1–3 subpolar flagella, which are easily shed.

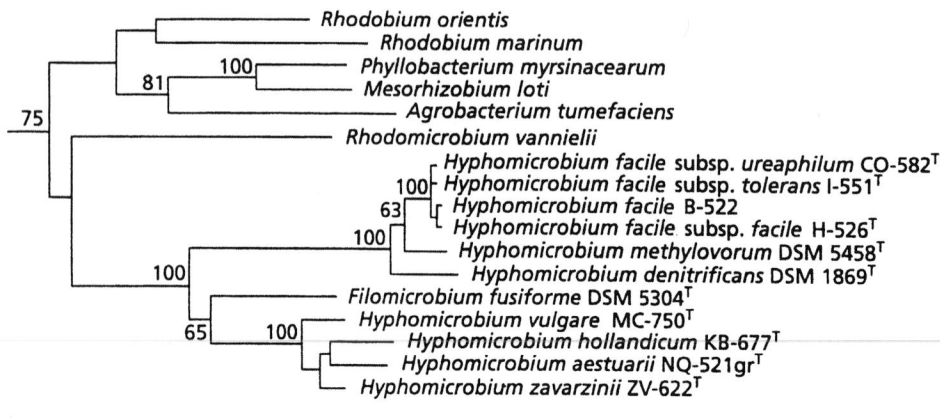

5 %

FIGURE BXII.α.190. Dendrogram based on 16S rDNA sequences, showing the phylogenetic position of the genera *Hyphomicrobium* and *Filomicrobium* among the closest relatives in the *Alphaproteobacteria* after Rainey et al. (1998).

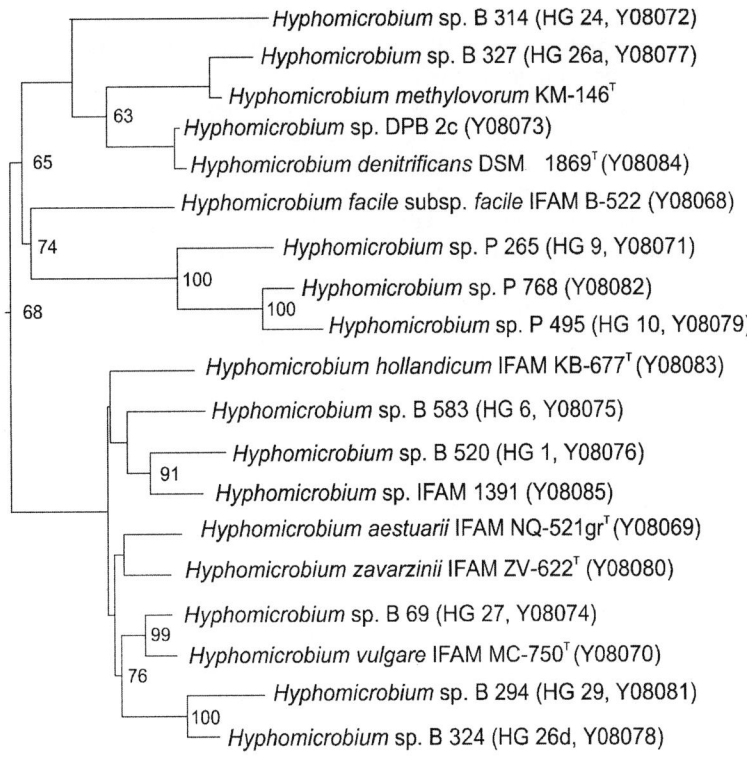

Hyphomicrobium sp. B 314 (HG 24, Y08072)

Hyphomicrobium sp. B 327 (HG 26a, Y08077)

Hyphomicrobium methylovorum KM-146^T

Hyphomicrobium sp. DPB 2c (Y08073)

Hyphomicrobium denitrificans DSM 1869^T (Y08084)

Hyphomicrobium facile subsp. *facile* IFAM B-522 (Y08068)

Hyphomicrobium sp. P 265 (HG 9, Y08071)

Hyphomicrobium sp. P 768 (Y08082)

Hyphomicrobium sp. P 495 (HG 10, Y08079)

Hyphomicrobium hollandicum IFAM KB-677^T (Y08083)

Hyphomicrobium sp. B 583 (HG 6, Y08075)

Hyphomicrobium sp. B 520 (HG 1, Y08076)

Hyphomicrobium sp. IFAM 1391 (Y08085)

Hyphomicrobium aestuarii IFAM NQ-521gr^T (Y08069)

Hyphomicrobium zavarzinii IFAM ZV-622^T (Y08080)

Hyphomicrobium sp. B 69 (HG 27, Y08074)

Hyphomicrobium vulgare IFAM MC-750^T (Y08070)

Hyphomicrobium sp. B 294 (HG 29, Y08081)

Hyphomicrobium sp. B 324 (HG 26d, Y08078)

10 %

FIGURE BXII.α.191. Phylogenetic relationship constructed for partial *mxaF* genes of selected *Hyphomicrobium* strains (based on the nucleotide sequence position 1009–1553 of *Paracoccus denitrificans* PD 1207). The dendrogram was generated by phylogenetic-distance analysis with a neighbor-joining algorithm. The bootstrap values indicate the percentage of 100 replicate trees supporting the branching order. Values below 50 were omitted. DNA–DNA-hybridization group and accession numbers are given in brackets. Bar = 10 mutations per 100 sequence positions.

Cell wall and membrane compositions Cell walls of one strain (IFAM B-522) have been analyzed and found to contain α,ε-diaminopimelic acid, as well as the other normal components of most Gram-negative walls (Jones and Hirsch, 1968). Mother cells have less D-alanine and, therefore, a less highly cross-linked murein than do hyphae or swarmer cells (Roggentin, 1980; Roggentin and Hirsch, 1982). Fine structural studies of a few isolates have shown intracytoplasmic membrane systems, which develop in older cells under certain growth conditions (Conti and Hirsch, 1965). The major cellular fatty acid in hyphomicrobia is straight-chain unsaturated $C_{18:1}$ (Urakami and Komagata, 1987b). Unusually long-chain hydroxy fatty acids with 24 and 26 carbon atoms have been found in some *Hyphomicrobium* strains (Sittig and Hirsch, 1992). All strains investigated so far contain ubiquinone Q-9 and a small amount of squalene, whereas Hopan-22(29)-ene and Hopan-22(29)-ol are present in only some strains (Urakami and Komagata, 1986b). 2-*o*-methyl-D-mannose has been reported to occur as a component of the extracellular acidic polysaccharide hyphomicran in *Hyphomicrobium* sp. JTS-811. Hyphomicran consists of D-glucose, D-mannose, 2-*o*-methyl-D-mannose, and pyruvic acid residues in relative proportions of 2:1:1:1 (Kanamaru et al., 1982a, b).

Colonial or cultural characteristics Liquid media may either become turbid or remain clear upon growth of hyphomicrobia, depending on the strain. In the latter case, growth occurs as a surface pellicle or ring on the glass walls near the medium sur-

face. Attachment to glass walls may be inhibited by light in some strains (Hirsch and Conti, 1964b). Surface pellicles, when shaken, fall to the bottom of the vessel, especially in older cultures. Colonies on solid media remain quite small, even after long incubation. Some pellicle-forming strains have colonies of variable sizes.

The life cycle of *Hyphomicrobium* species is complicated. It has been studied with special emphasis on the budding process (Kingma-Boltjes, 1936; Mevius, 1953) and on bud nucleation and the possibility of obtaining synchronous swarmer cells (Moore and Hirsch, 1973a; Matzen and Hirsch, 1982a). Up to 12 consecutive buds are formed on one hyphal tip, and the size of swarmer cells produced increases with mother-cell age. Synchronous swarmer cells of nearly identical size have been produced by Matzen and Hirsch (1982a) from chemostat cultures transferred to a glass wool-packed column and washed with fresh medium for several hours. Silver ions (5 ng/ml) increase the time required for growth initiation (Giangiordano and Klein, 1994). Motile swarmer cells of *Hyphomicrobium* sp. W1-1B display a positive chemotactic response toward methylamine, dimethylamine, and trimethylamine, but not toward methanol or arginine (Tuhela et al., 1998).

Metabolism and metabolic pathways All species described so far are methylotrophs; the highest growth yields are obtained with methanol or methylamine as the carbon and energy source. Most strains are also stimulated by a variety of other carbon sources; however, growth rates and yields are usually substantially lower than with the C_1 compounds (Matzen and Hirsch, 1982b). The uptake of methylamine occurs via a single inducible transport system, irrespective of whether methylamine is used as the carbon and energy source or only as the nitrogen source. Cells grown on other carbon and nitrogen sources do not possess this transport system, but it can be induced by methylamine within 45–50 min. Methylamine uptake is inhibited by azide, cyanide, N-ethylmaleimide, and carbonyl cyanide-m-chlorophenyl-hydrazone (Brooke and Attwood, 1984). Specialized *Hyphomicrobium* strains can use a variety of other specific C_1-compounds (Table BXII.α.158).

Hyphomicrobia do not possess an active pyruvate dehydrogenase complex. This makes it impossible to convert pyruvate into acetyl-CoA and to generate energy from carbon compounds containing three or more carbon atoms (Harder et al., 1975).

Some strains are able to grow with C_2-compounds, such as ethanol and acetate, or with the C_4-compound β-hydroxybutyrate. In these strains, the activities of the following enzymes have been measured in cell-free extracts: ethanol dehydrogenase, acetaldehyde dehydrogenase, acetothiokinase, 3-hydroxybutyrate dehydrogenase, and β-ketothiolase. It has been suggested that when cells are grown on ethanol, acetate, or 3-hydroxybutyrate, acetyl-CoA and energy are formed and reducing power is generated mainly by the tricarboxylic acid cycle, while carbon was assimilated via the glyoxylate cycle, with phosphoenolpyruvate carboxykinase functioning as the main gluconeogenetic enzyme (Attwood and Harder, 1974). A mutant lacking methanol dehydrogenase activity has been grown in a medium containing both methylamine and ethanol and does not show diauxic growth. When the concentration of methylamine in the medium falls to 9 ± 0.8 mM, ethanol is utilized and the two substrates are then metabolized simultaneously until the supply of methylamine is exhausted (Brooke and Attwood, 1983).

Methanol is oxidized by a periplasmic methanol dehydrogenase, which contains pyrroloquinoline-quinone (PQQ) as the redox cofactor (Duine and Frank, 1980a) and transfers the electrons to cytochrome c_L (Dijkstra et al., 1989). Methylamine is metabolized via N-methyl-glutamate using an N-methyl-glutamate dehydrogenase (Loginova et al., 1976; Meiberg and Harder, 1978). Di- and trimethylamine dehydrogenases are located in the cytoplasm (Kasprzak and Steenkamp, 1983, 1984). Hyphomicrobia contain tetrahydrofolate (H_4F)-linked, as well as tetrahydromethanopterin (H_4MPT)-linked, C_1-metabolites. Some interconverting enzymes have been determined (Marison and Attwood, 1982; Vorholt et al., 1999). Molecular analysis has indicated that the chloromethane utilization pathway in *H. chloromethanicum* is similar to the corrinoid-dependent methyl transfer system in *Methylobacterium chloromethanicum* (McAnulla et al., 2001).

The roles of some other NADP-dependent (plus unknown factor) or dye-linked (form)-aldehyde dehydrogenases, some of them with high activity and specificity, remain to be elucidated (Marison and Attwood, 1980; Köhler and Schwartz, 1982; Köhler et al., 1985; Poels and Duine, 1989; Klein et al., 1994; Kesseler and Schwartz, 1995).

The dichloromethane-utilizing *Hyphomicrobium* sp. DM2 possesses a type A dichloromethane dehalogenase (Schmid-Appert et al., 1997).

Hyphomicrobia use the serine pathway for carbon assimilation from reduced C_1-compounds. However, there has been controversy concerning the occurrence of isocitrate lyase in the genus *Hyphomicrobium*. Specific strains possess the icl$^+$ variant (Bellion

and Spain, 1976; Uebayasi et al., 1985; Yoshida et al., 1995a; Doronina et al., 1996b; Tanaka et al., 1997a). On the other hand, Attwood and Harder (1977), Doronina (1985), and Doronina et al. (1996b) have had evidence for the icl$^-$ variant of the serine pathway in certain strains. Recent observations have confirmed the distribution of unstable isocitrate lyase activities in the genus *Hyphomicrobium*, and the general operation of the icl$^+$ serine pathway has been suggested (Izumi et al., 1996).

Enzymes for the assimilatory pathways of either C_1- or C_2-compounds are regulated coordinately, but separately from the dissimilatory pathway enzymes associated with these compounds (Brooke and Attwood, 1985).

When the nitrogen source is ammonium sulfate or methylamine and the supply is in excess, NADPH-dependent glutamate dehydrogenase is used for the assimilation of nitrogen. In contrast, with a limited nitrogen supply, the cells express high levels of glutamine synthetase and NADH-dependent glutamine:2-oxoglutamate aminotransferase activity, while the activity of glutamate dehydrogenase is lower. When nitrate is the nitrogen source, the glutamine synthetase/glutamine oxoglutamate aminotransferase pathway is used, irrespective of the nitrogen concentration (Brooke et al., 1987). Some strains are able to use allantoin as a nitrogen source (van der Drift et al., 1981). The activity of NH_4^+-assimilating enzymes is regulated by the C/N ratio in the growth medium (Gräzer-Lampart et al., 1986; Duchars and Attwood, 1989).

Nitrate is reduced anaerobically by some strains; for these organisms, a special enrichment technique using KNO_3 and methanol has been described by Attwood and Harder (1972) and Sperl and Hoare (1971). A fundamental study of the *Hyphomicrobium* denitrifying capacity shows (Timmermans and van Haute, 1983) that these bacteria grow well anaerobically and with identical rates in the presence of methanol and either NO_3^- or NO_2^-. However, denitrification enzymes are present only to a limited extent (Sperl and Hoare, 1971; Vedenina et al., 1991; Kloos et al., 1995; Fesefeldt et al., 1998a, b) or are not expressed (Lebedinskii and Vedenina, 1981). Some strains have the ability to grow with methanol and nitrous oxide as the terminal electron acceptor (Lebedinskii, 1981). *Hyphomicrobium* sp. X can grow anaerobically on di- or trimethylamine in the presence of nitrate (Meiberg and Harder, 1978; Meiberg et al., 1980). *Hyphomicrobium* sp. DM2 has recently been shown to grow with dichloromethane in the absence of oxygen, using nitrate as a terminal electron acceptor (Kohler-Staub et al., 1995). White et al., (1987) have described a *Hyphomicrobium* strain that grows anaerobically on methylsulfate as a carbon and energy source at the expense of nitrate. Electron transport from methanol to nitrate has been described by Lebedinskii and Vedenina (1987).

Cells of *Hyphomicrobium* spp. VS and EG oxidize sulfide to thiosulfate very actively (Suylen et al., 1986; Pol et al., 1994). Yield data and ATP synthesis indicate that further oxidation of thiosulfate is possible (Suylen et al., 1986; Vedenina and Sorokin, 1992). *Hyphomicrobium* sp. 53-49 shows autotrophic growth with $H_2/CO_2/O_2$ and NH_4^+ (Uebayasi et al., 1981, 1984, 1985). Metabolism of monomethyl sulfate remains uncertain.^{13}C-NMR data are consistent with the hydroxylation of monomethyl sulfate via a monooxygenation mechanism and subsequent spontaneous hydrolysis of the methanediol monosulfate intermediate (Higgins et al., 1996).

Vitamin B_{12} stimulates growth of some species, especially their swarmer cells (Matzen and Hirsch, 1982b), but there is no absolute vitamin requirement in batch cultures. Most strains grow

TABLE BXII.α.158. Growth substrates used by specific undescribed *Hyphomicrobium* strains as sole carbon and energy source

Substrate	*Hyphomicrobium* species or strain
Acetate, 0.2% (w/w)[a]	MS 72, MS 75; MS 219, MS 223, MS 246, TMPO 1/8, TMPO 1/9, TMPO 1/10, DMMP 1/6, DMMP 1/9, DMMP 1/10, DMMP 1/11, DMMP 1/15, TMPO 2/3, TMPO 2/6, TMPO 2/9, TMPO 2/10, TMPO 2/13, DMPO/HW1-5
Ethanol, 0.5% (w/v)[a]	MS 72, MS 75, MS 219, MS 223, MS 246, TMPO 1/8, TMPO 1/9, TMPO 1/10, DMMP 1/6, DMMP 1/9, DMMP 1/10, DMMP 1/11, DMMP 1/15, TMPO 2/3, TMPO 2/6, TMPO 2/9, TMPO 2/10, TMPO 2/13, DMPO/HW1-5
Ethylamine, 0.5% (w/v)[b]	TMPO 1/8, TMPO 1/9, TMPO 1/10, DMMP 1/6, DMMP 1/9, DMMP 1/10, DMMP 1/11, DMMP 1/15, TMPO 2/3, TMPO 2/6, TMPO 2/9, TMPO 2/10, TMPO 2/13, DMPO/HW1-5
Ethylmethylamine, 0.5% (w/v)[c]	MS 72, MS 75, MS 219, MS 223, MS 246
N-formylglycineethylester, 0.3% (w/v)[d]	MS 72, MS 75, MS 219, MS 223, MS 246
N,N-dimethylformamide, 2,5% (w/v)[e]	TNBP221
Chloromethane, 0,1 mM[f, g]	CM1, CM2, CM9, CM29, CM35, MCl
Dichloromethane, 10 mM[h]	GJ21, DM2
Sodium monomethylsulfate, 0.5% (w/v)[i]	MS 72, MS 75; MS 219, MS 223, MS 246; *Hyphomicrobium* sp.
Methanesulfonate, 10 mM[j]	P2
Methanethiol[k,l]	I55, VS
Dimethylsulfone, 10 mM[m]	S1
Dimethylsulfoxide, 2–4 mM[n]	S1, I55, S, VS, EG
Dimethylsulfide, 0.5–1 mM[n]	S1, I55, VS, EG
Dimethyldisulfide, 1 mM[o]	VS, MS3
Dimethyltrisulfide, 1 mM[p]	VS
Diethylsulfone, 10 mM[m]	S1
Propanesulfonate, 10 mM[m]	S1
Butanesulfonate, 10 mM[m]	S1
Hexanesulfonate, 10 mM[m]	S1
Dimethylphosphate, 0.5% (w/v)[b]	TMPO 1/8, TMPO 1/9, TMPO 1/10, DMMP 1/6, DMMP 1/9, DMMP 1/10, DMMP 1/11, DMMP 1/15, TMPO 2/3, TMPO 2/6, TMPO 2/9, TMPO 2/10, TMPO 2/13, DMPO/HW1-5
Monomethylphosphate, 0.3% (w/v)[b]	TMPO 1/8, TMPO 1/9, TMPO 1/10, DMMP 1/6, DMMP 1/9, DMMP 1/10, DMMP 1/11, DMMP 1/15, TMPO 2/3, TMPO 2/6, TMPO 2/9, TMPO 2/10, TMPO 2/13, DMPO/HW1-5
Dimethylphosphonate, 0.3% (w/v)[b,q]	TMPO 1/8, TMPO 1/9, TMPO 1/10, DMMP 1/6, DMMP 1/9, DMMP 1/10, DMMP 1/11, DMMP 1/15, TMPO 2/3, TMPO 2/6, TMPO 2/9, TMPO 2/10, TMPO 2/13, DMPO/HW1-5
Diethylphosphate, 0.5% (w/v)[b]	TMPO 1/8, TMPO 1/9, TMPO 1/10, DMMP 1/6, DMMP 1/9, DMMP 1/10, DMMP 1/11, DMMP 1/15, TMPO 2/3, TMPO 2/6, TMPO 2/9, TMPO 2/10, TMPO 2/13, DMPO/HW1-5
Trimethylphosphate, 0.3% (w/v)[b]	TMPO 1/8, TMPO 1/9, TMPO 1/10, DMMP 1/6, DMMP 1/9, DMMP 1/10, DMMP 1/11, DMMP 1/15, TMPO 2/3, TMPO 2/6, TMPO 2/9, TMPO 2/10, TMPO 2/13, DMPO/HW1-5
Dimethylphosphite, 0.05% (w/v)[c]	MS 72, MS 75, MS 219, MS 223, MS 246
Trimethylphosphite, 0.05% (w/v)[c]	MS 72, MS 75, MS 219, MS 223, MS 246

[a]Ghisalba and Kuenzi (1983); Ghisalba et al. (1987).

[b]Ghisalba et al. (1987).

[c]Ghisalba and Kuenzi (1983).

[d]Ghisalba et al. (1985).

[e]Shuttleworth (1996).

[f]Doronina et al. (1996b).

[g]Hartmans et al. (1986).

[h]Diks et al. (1994a, b); Stucki et al. (1981).

[i]Ghisalba and Kuenzi (1983); White et al. (1987).

[j]De Marco et al. (2000).

[k]Concentration data not available.

[l]Cho et al. (1992); Pol et al. (1994).

[m]Borodina et al. (2000).

[n]Borodina et al. (2000); Cho et al. (1992); De Bont et al. (1981b); Pol et al. (1994); Suylen and Kuenen (1986).

[o]Pol et al. (1994); Smet et al. (1996).

[p]Pol et al. (1994).

[q]Used with very low growth rates.

in the presence of 2.5% NaCl, but they also develop at low salt concentrations, approaching that of distilled water. Milk is coagulated by one strain. H$_2$S evolution and gelatin liquefaction have also been observed in this organism. Some isolates of *Hyphomicrobium* grow at 4–6°C; others can multiply at 45°C (Hirsch, unpublished observations).

Genetics and bacteriophages Genome sizes of three species ranged from 2.13–2.62 × 10^9 Da (Moore and Hirsch, 1973b; Kölbel-Boelke et al., 1985).

In addition to amino acid auxotrophs, mutants have been found that are specifically defective in methanol oxidation, resistant against antibiotics, or nonmotile and/or morphologically altered (Wieczorek and Hirsch, 1979; Marison and Attwood, 1982; Gliesche and Eckhardt, 1991; Gliesche and Hirsch, 1992). Transposon mutagenesis has been established for *H. facile* IFAM B-522 using transposon Tn5 and its derivatives, Tn7, Tn10, and Tn501 (Gliesche and Hirsch, 1992).

Broad host range IncP-1 plasmids (RP1, RP4, RP4::Mu*cts*, R68, R68.45, pMO60, pLUB21, pLUB113) have been successfully transferred by interspecific matings to *H. facile* IFAM B-522 and *H. denitrificans* DSM 1869[T] (Dijkhuizen et al., 1984; Gliesche and Hirsch, 1992). Mermod et al. (1986) have reported the transfer of plasmid pNM185 (a pKT231 derivative) into *H. denitrificans* NCIB11706.

Chromosome mobilization has been demonstrated with the conjugative IncP-1 plasmids RP1, R68.45, and pMO60 into *H. facile* (Gliesche and Hirsch, 1992). Using the IncP helper pRK2013 plasmid, pLA2917 has been introduced into *Hyphomicrobium* species by triparental matings. However, pLA2917 is always found integrated into the chromosome (Gliesche, 1997). Genes *aceE*, *aceF*, and *lpd* from *Escherichia coli* K12 JC6310, which code for the pyruvate dehydrogenase complex, have been expressed in *H. denitrificans* DSM 1869[T], with RP4'*pdh*1 resulting in growth and denitrification with pyruvate as the sole carbon source (Dijkhuizen et al., 1984).

An efficient system for electroporation of *H. facile* IFAM H-526[T] and *H. denitrificans* DSM 1869[T] has been developed with vectors based on the broad-host-range plasmid pBBR1 (Gliesche, 1997).

The occurrence of genes coding for denitrification and nitrogen fixation enzymes has been shown by Southern or dot blot hybridization with gene probes specific for nitrate reductase (*narG*), cytochrome *c,d*$_1$-containing nitrite reductase (*nirS*), Cu-containing nitrite reductase (*nirK*), nitrous oxide reductase (*nosZ*), and nitrogenase reductase (*nifH*). In hyphomicrobia, the Cu-containing nitrite reductase appears to be more common than the cytochrome *c,d*$_1$-containing nitrite reductase (Suzuki et al., 1993; Kloos et al., 1995; Tuhela et al., 1997; Fesefeldt et al., 1998a, b). Southern blot hybridizations have indicated the presence of genes coding for methanesulfonic acid monooxygenase (*msn*) in a methanesulfonate-degrading *Hyphomicrobium* strain (De Marco et al., 2000).

DNA sequence data exist for genes coding for the 5S rDNA (Stackebrandt et al., 1988a; Boulygina et al., 1993), 16S rDNA (Stackebrandt et al., 1988a; Tsuji et al., 1990; Tuhela et al., 1997; Rainey et al., 1998; Borodina et al., 2000; De Marco et al., 2000; Layton et al., 2000; McDonald et al., 2001; Stein et al., 2001), methanol dehydrogenase (*mxaF*; Fesefeldt and Gliesche, 1997; McDonald and Murrell, 1997a; Tanaka et al., 1997b), dimethylamine dehydrogenase (*dmd*; Yang et al., 1995a), methyltransferase gene cluster (*cmu*; McAnulla et al., 2001), Cu-containing nitrite reductase (*nirK*; Braker et al., 1998), isocitrate lyase (Tanaka et al., 1997a), serine–glyoxylate aminotransferase (Hagishita et al., 1996a), serine hydroxymethyltransferase (Miyata et al., 1993), hydroxypyruvate reductase (Yoshida et al., 1994), dichloromethane dehalogenase (*dcmA*; Vuilleumier et al., 1997), methenyl tetrahydromethanopterin cyclohydrolase (*mch*; Vorholt et al., 1999), and unknown gene fragments involved in methanol oxidation (Fesefeldt et al., 1997; Gliesche et al., 1997). Specific PCR primer systems have been described for the amplification of these genes from pure cultures or environmental samples (Miyata et al., 1993; Yoshida et al., 1994; Yang et al., 1995a; Hagishita et al., 1996a; Fesefeldt and Gliesche, 1997; McDonald and Murrell, 1997a Tanaka et al., 1997a; Vuilleumier et al., 1997; Braker et al., 1998; Vorholt et al., 1999; McAnulla et al., 2001). An identification system for environmental *Hyphomicrobium* isolates, based on denaturing gradient gel electrophoresis of a fragment of the *mxaF* gene, has been described by Fesefeldt and Gliesche (1997).

Partial amino acid sequences exist for the enzymes dimethylamine dehydrogenase and trimethylamine dehydrogenase (Kas-przak et al., 1983). In *Hyphomicrobium* sp. DM2 and GJ21, the *dcm* region (dichloromethane utilization genes) is associated with the insertion sequences IS1354 (only in DM2), IS1355, and IS1357. Furthermore, multiple copies of these insertion sequences have been found outside the *dcm* region. The high degree of sequence conservation observed within the genomic region responsible for dichloromethane utilization in other aerobic methylotrophic bacteria and the occurrence of clusters of insertion sequences in the vicinity of the *dcm* genes suggest that a transposon is involved in the horizontal transfer of these genes among methylotrophic bacteria (Schmid-Appert et al., 1997).

Ribosomal RNA cistron homologies among *Hyphomicrobium* strains have been investigated by Moore (1977). Low molecular weight RNA profiles have been used for genotypic identification of several *Hyphomicrobium* species (Höfle, 1990). The 16S–23S rRNA internal transcribed spacer region has been investigated by Scheinert et al. (1996). Poly(A) sequences at the 3'-terminus have been observed in the RNA of *H. facile* (Schultz et al., 1978).

Lytic bacteriophages have been isolated for several strains of *Hyphomicrobium* (Voelz et al., 1971; Kaplan et al., 1976; Yelton et al., 1979; Gliesche et al., 1988; Preissner et al., 1988). The presence of temperate bacteriophages and bacteriocins in *Hyphomicrobium* strains has also been demonstrated (Gliesche et al., 1988; Holm, 1991). *Hyphomicrobium facile* IFAM B-522 (RP4) is not sensitive toward the donor-specific (IncP) phages PRD1 and GU5 (Gliesche and Hirsch, 1992).

Serological relationships among some hyphomicrobial isolates have been studied by Powell et al. (1980).

Pathogenicity Twelve strains have been tested for pathogenicity against mice or guinea pigs; all are avirulent (Famureva et al., 1983).

Ecology and use in biotechnology *Hyphomicrobium* species could be isolated from all soil samples tested so far; they are present in nearly all water samples as well (Hirsch and Conti, 1964a; Hirsch and Rheinheimer, 1968). In freshwater habitats, they are especially prevalent in the neuston layer, on submerged surfaces, and in the upper sediment layer, even under anaerobic conditions. Hyphomicrobia also come from temporary puddles, sewage treatment plants, and the surface of indoor flower-pots (Hirsch, 1974a).

An association of hyphomicrobia with methanotrophs is capable of denitrifying with methanol; growth is possible under high CH_4 and low O_2 conditions (Wilkinson et al., 1974; Amaral et al., 1995). On the other hand, co-immobilized mixtures of *Hyphomicrobium* sp. and methanogenic bacteria are very efficient in simultaneous denitrification and methanogenesis (Lin and Chen, 1995; Zellner et al., 1995).

Using the most-probable-number technique with methanol as the sole carbon source, hyphomicrobia have been enumerated as 0.2% ($= 10^6$ g^{-1} dry weight hyphomicrobia) of the total bacteria determined by acridine-orange direct counts in a clay loam soil (Aa and Olsen, 1996). With the same method, Fesefeldt et al. (1997) have found 2×10^4 g^{-1} dry weight hyphomicrobia in a garden soil (0.7% of colony-forming units of methylotrophic bacteria). The *H. facile* population amounted to 30% of total hyphomicrobia in this soil.

The existence of denitrifying hyphomicrobia is of special interest because of the necessity to remove nitrate at drinking water treatment plants (Liessens et al., 1993) and sewage treatment plants (Schmider and Ottow, 1986; Nyberg et al., 1992; Lee and Welander, 1996; Lemmer et al., 1997). Coenoses consisting of a

Hyphomicrobium sp. and a *Paracoccus* sp. are quite efficient in the removal of both methanol and nitrate (Claus and Kutzner, 1985; Vedenina and Govorukhina, 1988; Neef et al., 1996). Investigation of the community structure in sewage treatment plants has revealed a very high abundance and diversity of hyphomicrobia (Holm et al., 1996). A domestic wastewater treatment plant has been found to contain up to 2×10^4 hyphomicrobia ml^{-1} in the influent, 9×10^4–6×10^5 in activated sludge, 1–4×10^3 in the effluent, and 2–12 hyphomicrobia ml^{-1} in the receiving lake. DNA–DNA hybridizations have classified the isolates into 30 groups (Holm et al., 1996), which can be assigned to Clusters I to IV based on 16S rDNA sequences (Rainey and Stackebrandt, unpublished data). The population of denitrifying *Hyphomicrobium* DNA–DNA-hybridization group HG 27 amounts to approximately 30% of the total facultatively anaerobic hyphomicrobia found in this activated sludge (Gliesche and Fesefeldt, 1998). When Layton et al. (2000) extracted DNA from an industrial wastewater from a treatment plant, they found 16S rDNA sequences from representatives of *Hyphomicrobium* clusters I and II. *Hyphomicrobium* 16S rRNA comprised approximately 5% of the 16S rRNA in the activated sludge of this treatment plant.

16S rDNA genes related to *Hyphomicrobium* spp. have been amplified from total community DNA extracted from a water sample of the northern portion of Green Bay (40 m depth) (Stein et al., 2001).

Hyphomicrobia have also been used for the removal of odorous, volatile sulfur compounds, such as hydrogen sulfide, methanethiol, dimethylsulfide (Zhang et al., 1991; Cho et al., 1992; Pol et al., 1994; Smet et al., 1996), and dichloromethane (Diks et al., 1994a, b) from air and gases. Specific *Hyphomicrobium* strains and enzymes have been applied in biosensors for the detection of methanol (Argall and Smith, 1993), methylsulfates (Schär and Ghisalba, 1985; Ghisalba et al., 1986), trimethylamine (Large and McDougal, 1975; Wong and Gill, 1987), dihalomethanes (Gälli and Leisinger 1985; Henrysson and Mattiasson, 1993), L-serine, and glyoxylate (Yoshida et al., 1993). A coenosis consisting of *Hyphomicrobium* sp. DM2 and *Rhodococcus rhodochrous* OFS has efficiently degraded dichloromethane, 2-propanol, and methanol in a gas lift loop bioreactor (Vanderberg-Twary et al., 1997).

Hyphomicrobium methylovorum has been used for the production of L-serine (Yamada et al., 1986; Izumi et al., 1993).

Other studies *Hyphomicrobium* species have been investigated for fatty acids (Auran and Schmidt, 1972; Ikemoto et al., 1978a; Eckhardt et al., 1979; Urakami and Komagata, 1979, 1987a, b; Vedenina et al., 1991; Sittig and Hirsch, 1992), phospholipids (Goldfine and Hagen, 1968; Guckert et al., 1991; Sittig and Hirsch, 1992; Batrakov and Nikitin, 1996), hopanoids (Rohmer et al., 1984), ubiquinones (Köhler and Schwartz, 1981; Urakami and Komagata, 1981, 1986b; Sittig and Hirsch, 1992), cytochromes (Large et al., 1979; Köhler and Schwartz, 1983; Dijkstra et al., 1988a, b, 1989; Frank and Duine, 1990a), pyrroloquinoline quinone (Duine et al., 1980, 1981, 1990; De Beer et al., 1983; Houck et al., 1989), and poly-β-hydroxybutyrate (Jacobsen, 1975). The following enzymes have been purified and characterized in more detail: methanol dehydrogenase (Duine et al., 1978; Duine and Frank, 1979, 1980a, b; Schär et al., 1985; Miyazaki et al., 1987a; Frank et al., 1988; Frank and Duine, 1990b; Geerlof et al., 1994b), dimethylamine dehydrogenase (Kasprzak and Steenkamp, 1984; Meiberg and Harder, 1979; Steenkamp 1979; Steenkamp and Beinert, 1982a, b), trimethylamine dehydrogenase (Steenkamp 1979; Steenkamp and Beinert, 1982a, b;

Kasprzak et al., 1983), dye-linked aldehyde dehydrogenase (Marison and Attwood, 1980; Köhler and Schwartz, 1982), dye-linked formaldehyde dehydrogenase (Klein et al., 1994), NAD-linked, GSH- and factor-independent aldehyde dehydrogenase (Poels and Duine, 1989; Duine, 1990), NADP$^+$-dependent glutamate dehydrogenase (Duchars and Attwood, 1987), glutamine synthetase (Duchars and Attwood, 1991), hydroxypyruvate reductase (Goldberg et al., 1992, 1994; Yoshida et al., 1994; Hagishita et al., 1996b; Izumi et al., 1996), serine–glyoxylate aminotransferase (Hagishita et al., 1996a, b; Izumi et al., 1996), glycerate kinase (Hill and Attwood, 1974; Yoshida et al., 1992; Izumi et al., 1996), phosphoenolpyruvate carboxylase (Yoshida et al.,1995b; Izumi et al., 1996), methyl mercaptan oxidase (Suylen et al., 1987), dichloromethane dehalogenase (Kohler-Staub and Leisinger, 1985; Kohler-Staub et al., 1986; Leisinger and Kohler-Staub, 1990), dimethyl sulfoxide reductase (Hatton et al., 1994), phosphoglycerate mutase (Hill and Attwood, 1976a, b), serine hydroxymethyltransferase (Miyazaki et al., 1986, 1987b, c, d), and copper-containing nitrite reductase (Suzuki et al., 1993).

ENRICHMENT AND ISOLATION PROCEDURES

A variety of enrichment techniques has been proposed. In all cases, growth of hyphomicrobia is slow; they may be overgrown in the presence of other heterotrophic bacteria. Most enrichment cultures for nitrifying bacteria contain hyphomicrobia in large numbers (Stutzer and Hartleb, 1898; Hirsch and Rheinheimer, 1968; Hirsch, 1970). Under oligotrophic conditions and after prolonged incubation, hyphomicrobia usually outcompete other bacteria. A slow, but successful method consists of keeping a natural water sample at room temperature in the dark for several weeks or months. Eventually, hyphomicrobia become part of the dominant microflora, even in the presence of amoebae. Another method prescribes the addition to a natural water sample or soil suspension of methylamine hydrochloride at 3.38 g/l and/or incubation in an atmosphere of methanol (Hirsch, 1970). Such an enrichment should be monitored frequently in order to determine the optimal time for subculturing. Inoculation of a natural sample into mineral salts medium 337, containing 0.1–0.2% (w/v) of a C$_1$-compound, and dark incubation at 20–25°C for a few weeks usually yields hyphomicrobia in large numbers (Hirsch and Conti, 1965). Improvements of medium 337[1] that result in faster growth and higher yields have been reported by Matzen and Hirsch (1982b). Several modifications of medium 377-B (Matzen and Hirsch, 1982b) have been developed for specific genetic and physiological applications and for the isolation of a greater *Hyphomicrobium* diversity from natural habitats (Table BXII.α.159).

Isolation of rosette-forming and denitrifying hyphomicrobia from aquatic sediments has been achieved by inoculating sediment samples into a mineral salts medium containing 5 g/l KNO$_3$ and up to 0.5% (v/v) of methanol; incubation is anaerobic at room temperature (Attwood and Harder, 1972). It must be stressed, however, that all of these methods will yield different *Hyphomicrobium* species. The application of an oligotrophic medium containing low concentrations of peptone, yeast extract, and glucose ("PYG"; Staley, 1968) has often yielded morphologically *Hyphomicrobium*-like bacteria. Such isolates usually do not

---

1. Composition of medium 337: KH$_2$PO$_4$, 1.36 g; Na$_2$HPO$_4$, 2.13 g; MgSO$_4$·7H$_2$O, 0.3 g; (NH$_4$)$_2$SO$_4$, 0.5 g; CaCl$_2$·2H$_2$O, 1.99 mg; FeSO$_4$·7H$_2$O, 1.0 mg; MnSO$_4$·H$_2$O, 0.35 mg; Na$_2$MoO$_4$·2H$_2$O, 0.5 mg; vitamin B$_{12}$ (if needed), 2.5 µg; distilled water, to 1000 ml; pH, 7.2.

TABLE BXII.α.159. Growth media for specific applications with hyphomicrobia

Medium	Characteristics and composition	Application
337[a]	Mineral salts medium (carbon source: methylamine or methanol)	Enrichment, isolation, and cultivation of hyphomicrobia
337a[b]	Modification of medium 337 with 3.38 g/l methylamine and reduction of the trace elements	Enrichment, isolation, and cultivation of hyphomicrobia
337-MA[c]	337a with 3.38 g/l methylamine	Enrichment, isolation, and cultivation of hyphomicrobia
337-M[c]	337a with 0.5% (v/v) methanol	Enrichment, isolation, and cultivation of hyphomicrobia
337-B[c]	337a with 2.5 g/l vitamin B_{12}, 0.3 g/l L-lysine and 5.0 g/l Na-gluconate (C-source: 3.38 g/l methylamine–HCl or 0.5% (v/v) methanol)	Optimized medium for *Hyphomicrobium facile*
337-B1[d]	337-B with 3.38/l methylamine–HCl, but without vitamin B_{12}, L-lysine, or gluconate	Standard minimal medium for the cultivation of all hyphomicrobia growing on methylamine
337-B2[d]	337-B1 with 0.2% (w/v) Bacto nutrient broth and 0.2% (w/v) Casamino acids	Isolation of auxotrophic mutants
337-B3[d]	337-B1 and 0.075% (w/v) Bacto nutrient broth and 0.075% (w/v) Casamino acids	Interspecific matings
337-B4[e]	337-B1 and 2.5 g/l vitamin B_{12}	Cultivation of vitamin B_{12}-dependent hyphomicrobia
337-B5[e]	337-B4 and 0.25 g/l peptone, 20 ml/l HBM and 10 ml/l vitamin solution No. 6[f]	Enrichment, isolation, and cultivation of specific hyphomicrobia from sewage
337-B6[g]	337-B5 and 0.075% (w/v) Bacto nutrient broth and 0.075% (w/v) Casamino acids	Interspecific matings with hyphomicrobia with optimal growth on medium 337-B5
337-B7[h]	337-B5 with 5 mM $NaNO_3$ and 0.5% (v/v) methanol instead of methylamine	Determination of the denitrification activity
337-B8[i]	337-B1 with 0.5% (v/v) methanol instead of methylamine	Standard minimal medium for the cultivation of all hyphomicrobia growing on methanol
337-B9[j]	337-B8 with 0.5% (w/v) KNO_3	Enrichment, isolation, and cultivation of denitrifying hyphomicrobia
337-B10[k]	337-B5 with 0.5% (v/v) methanol instead of methylamine–HCl	Cultivation of specific hyphomicrobia from sewage
337-B11[k]	337-B5 with 5 mM KNO_3 and 0.5% (v/v) methanol instead of methylamine–HCl	Cultivation of denitrifying hyphomicrobia with optimal growth on medium 337-B5
DST[l]	Mineral salts, trace elements, 1% (w/v) Difco agar, and an atmosphere of 1% (v/v) CH_3Cl	Enrichment and cultivation of hyphomicrobia that utilize chloromethane

[a]Data from Hirsch and Conti (1964b).

[b]Data from Moore and Hirsch (1973a).

[c]Data from Matzen and Hirsch (1982b).

[d]Data from Gliesche and Hirsch (1992).

[e]Data from Holm et al. (1996).

[f]HBM and Vitamin Solution No. 6 according to Schlesner (1994).

[g]Data from Gliesche et al. (1997).

[h]Data from Fesefeldt et al. (1998b).

[i]Data from Gliesche and Fesefeldt (1998).

[j]Data from Gliesche and Fesefeldt (1998).

[k]Data from Fesefeldt (1998).

[l]Data from Doronina et al. (1996b).

grow on C_1 compounds and are often pigmented; they comprise a different group of generic rank (see *Hirschia* spp.; Schlesner et al., 1990).

Hyphomicrobia growing on chloromethane can be enriched for with the addition of 1% (w/v) CH_3Cl to a mineral salts medium. Subsequently, this can be plated on the same medium solidified with 1% Difco agar and containing 0.01% (w/v) bromothymol blue to indicate the release of HCl (Doronina et al., 1996b).

Once hyphomicrobia occur or predominate in liquid enrichments, they can be isolated and purified by repeatedly streaking enrichments on mineral salts medium "337" and incubating the plates in the dark at 20–25°C. To avoid excessive drying of the plates, they should be placed in plastic bags, or thicker layers of agar should be used. Concerning the purification procedure, it should be remembered that pellicle-forming strains (e.g., NQ-521Gr) usually produce colonies of quite different size. Subculturing small or large colonies results again in growth of both types.

Hyphomicrobium colonies are often tough and coherent; before spreading on plates, they should be ground up properly. Typical colonies of hyphomicrobia appear dark brown in transmitted light, often with folds and concentric rings. Under reflected light, the colonies of many strains are shiny and bright beige or even colorless.

MAINTENANCE PROCEDURES

Most hyphomicrobia can be kept well at 4–5°C when growing on slants. Subculturing every 5–6 months is sufficient. Lyophilization in skim milk is the optimal method for maintenance of most cultures. Suspension in phosphate buffer and sterile glycerol, followed by immediate vortexing and cooling down to −25°C, is another technique; such preparations may be kept for several years in the freezing compartment of a refrigerator. For subculturing, the glycerol suspension is streaked directly onto agar plates. Warming up of the glycerol suspension should be avoided, since rapid death of the cells results. Doronina and Trotsenko (1992) have successfully stored hyphomicrobia that grew with chloromethane by deep-freezing these on Whatman paper without cryoprotectants at −40°C to −80°C.

PROCEDURES FOR TESTING SPECIAL CHARACTERS

Cell shape and morphogenesis An agar slide culture is prepared with medium "337" and 1.8% Bacto Noble agar (Hirsch, unpublished). The sterile agar medium is spread thinly over sterile glass slides and allowed to solidify. Using sterile coverslips, agar is then cut off to leave two agar squares side-by-side and separated only by a small ditch. One of the squares is inoculated with a thin *Hyphomicrobium* suspension; the other square receives a small droplet of 0.5% methanol. Both squares are covered together with one large coverslip, and the edges can be sealed with a Vaseline–paraffin mixture. Spreading of methanol to the inoculated square should be avoided, since direct contact at this concentration is toxic for the bacteria. Growth and morphogenesis can then be followed over a period of up to 48 h or more.

Cell size measurements Only living cells, preferentially those in agar slide cultures, are used for size measurements. Phase-contrast light micrographs are prepared and enlarged 10 times; sizes are measured on these enlargements from at least 50–100 cells, since considerable size variations exist in asynchronous *Hyphomicrobium* cultures.

Growth on carbon sources Medium 337 is used as a base, and sterile-filtered carbon sources are added at 0.1–0.2% (w/v). Most hyphomicrobia are oligocarbophilic, i.e., they grow (although slowly) at the expense of contaminants in the laboratory air. It is mandatory, therefore, to have control plates inoculated, which do not contain the carbon source to be tested; growth on these controls has to be considered. Growth on plates can be scored after 1, 2, and 4 weeks. Furthermore, oligotrophic growth makes it necessary to subculture at least two additional times with the same carbon source to ensure that the growth observed is due to the substrate offered and not to substrate carried over with the inoculum. Liquid carbon utilization tests can be scored by measuring optical density at 650 nm (OD_{650}) if cells grow turbidly. However, dramatic changes of cell size and morphology can result from some carbon sources, and light microscopy is required to ascertain that this is not the case. Protein determinations are applied widely as a better method for growth estimation (Matzen and Hirsch, 1982b). In all such experiments, it is crucial to have very homogeneous suspensions in the initial inoculation procedure.

Growth stimulation by vitamins Growth of some hyphomicrobia is markedly stimulated by vitamin B_{12}, especially in chemostat cultures where the population consists mainly of very young cells. Obviously, nutrient requirements can change with cell age. Static asynchronous cultures contain large numbers of older mother cells and do not require B_{12} addition for growth. It has been found, however, that application of a vitamin mixture (Vitamin Solution no. 6; van Ert and Staley, 1971) leads to less stimulation than does the application of B_{12} alone (Matzen and Hirsch, 1982b).

Growth inhibition by visible light Hyphomicrobia that form surface pellicles rather than turbidity are markedly inhibited by light (Hirsch and Conti, 1964b). Experiments to determine the influence of light on *Hyphomicrobium* growth can be carried out with sunlight illumination of agar plates or liquid cultures. Care must be taken to ensure temperature constancy and to avoid water condensation and/or drying of the plates. In the case of liquid cultures, the light inhibition can be detected by the failure of swarmer cells to attach to glass walls on the illuminated side.

DNA extraction procedures Lysis of hyphomicrobia is often difficult; a variety of techniques has been described for DNA extraction (Gebers et al., 1985). For the "cell mill A" method, 1–2 g of bacterial wet weight are suspended in 20 ml of saline–EDTA supplemented with 1 mg proteinase K. Then, 50 g glass beads (0.1 mm diameter) are added, and the precooled mixture is shaken for 5–10 s in an MSK cell homogenizer (Braun, Melsungen, Germany). Cell lysis is completed by adding 20 mg sodium dodecyl sulfate ml^{-1} to the suspension. For DNA extraction, $NaClO_4$ and chloroform–isoamyl alcohol are added, and the suspension is shaken for 15 min at 100 rpm. Centrifugation at 1350 × g for 20 min results in separation of the emulsion into two layers. The nucleic acids can be precipitated from the aqueous phase. This is followed by a 45 min ribonuclease treatment and by a 2 h treatment with proteinase K (200 μg/ml) at 37°C. Then, 1 volume of phenol is added (saturated with 1 × SSC [0.15 M NaCl + 0.015 M Na$_3$-citrate], pH 7) and 0.1 volume of chloroform–isoamyl alcohol is added and the preparation agitated for 10 min at 100 rpm. Centrifugation at 27,000 × g for 20 min results in the separation of the emulsion into three layers, the upper one of which is used for the precipitation of DNA by ethanol. In some cases, a similar method described by Gebers et al. (1981b) and called the "enzyme A" method yields better results.

Extraction of genomic DNA for PCR

Procedure I of Gliesche et al. (1997) Late-exponential-phase cultures (500 ml; 400–500 mg wet weight) are harvested by centrifugation (8000 × g, 45 min, 4°C), resuspended in 1.0 ml double-distilled water with 4.0 ml acetone (4°C), and incubated for 30 min at 0°C. The pellet (4000 × g, 10 min) is dried and resuspended in 4.0 ml buffer 1 (50 mM EDTA, 30 mM Tris, and 5 mM NaCl). Then, 1.0 ml of a lysozyme solution (20 mg lysozyme ml^{-1} TE [10 mM Tris-HCl, 1 mM EDTA, pH 8.0]) is added and the mixture incubated for 45 min at 37°C. Cells are lysed by the addition of 1.0 ml SDS (10% w/v) and 130 μl of proteinase K solution (20 mg/ml phosphate buffer [0.1 M K_2HPO_4, 0.1 M KH_2PO_4, pH 6.25]), followed by incubation at 56°C for 3 h. After lysis, 275 μl of 5 M NaCl (final concentration 1%, w/v) are added and the suspension gently mixed and incubated for 1 h at 0°C. The cell debris is then pelleted for 30 min at 24,000 × g. High molecular weight genomic DNA is precipitated from the aqueous phase with 0.7–1 volume of isopropanol (20 min, 24,000 × g, 4°C) and air-dried (37°C, 20 min). The resulting pellet is resuspended in 4.0 ml TE. After gently shaking in a water bath (30°C, overnight), 60 μl of RNase A (10 mg/ml in TE) are added, and the suspension incubated for 45 min at 37°C. The lysate is extracted two to three times (6500 × g, 15 min, 4°C) with phenol (saturated with SSC [0.15 M NaCl + 0.015 M trisodium citrate, pH 7.5]), phenol/chloroform (until the interface is clear), and chloroform/isoamyl alcohol (24:1). DNA is precipitated from the aqueous phase with 1/10 volume of 3 M sodium acetate (pH 5.2; final concentration of 0.3 M), and three volumes of 96% ethanol (0°C, 30 min–4 h). The pellet (6000 × g, 30 min, 4°C) is washed once with 10 ml ethanol (70%), dried briefly in air at room temperature, and resuspended in 300–500 μl TE by gentle shaking.

Procedure II of Fesefeldt and Gliesche (1997) Cells from 200 ml late exponential phase *Hyphomicrobium* cultures are harvested by centrifugation (8000 × g, 45 min, 4°C). 100 mg cells (wet weight) are resuspended in 300 μl of 1 × SSC. This cell suspension, together with 200 mg glass beads (0.10–0.11 mm diameter; Braun, Melsungen) and 60 µl of a 10% (v/v) SDS are filled into 2.0 ml microtubes with screw cap closures (Sarstedt, Germany) and shaken at 2000 rpm on a rotary shaker 3300 (Eppendorf, Hamburg) for 5 (IFAM P-645) or 15 min. Then, 1.0 ml phenol (65°C; saturated with SSC) is added. After 2 min of gentle shaking by hand, the cell debris and glass beads are pelleted for 30 min at 14,000 rpm and 4°C (Centrifuge 5415C, Eppendorf). The aqueous phase (300–400 µl) is transferred to a fresh microtube. Purification and precipitation of nucleic acids are performed as described in procedure I. The nucleic acids are resuspended in 200 µl TE by gentle shaking. After adding 10 µl RNase A (50,000 U/ml in TE), the suspension is incubated for 30 min at 37°C. From this, DNA is purified and precipitated as in procedure I and resuspended in 100 µl TE by gentle shaking.

Isolation of plasmid DNA (Gliesche, 1997) Late exponential phase cultures (200 ml; 100–200 mg wet weight) are harvested by centrifugation (8000 × g, 45 min, 4°C), resuspended in 0.3 ml double distilled water with 1.5 ml acetone (4°C), and incubated for 15 min at 0°C. The pellet (10,000 × g, 10 min) is dried with air (5 min), resuspended in 300 µl buffer A (50 mM EDTA, 30 mM Tris, and 5 mM NaCl) and 45 µl of a lysozyme solution (20 mg/ml TE), and incubated for 45 min at 37°C. Then 60 µl of Proteinase K solution (20 mg/ml phosphate buffer [0.1 M K_2HPO_4, 0.1 M KH_2PO_4, pH 6.25]) are added, followed by incubation at 55°C for 2 h. Cells are harvested (10,000 × g, 10 min, 4°C) and resuspended in buffer 1 (QIAGEN, Germany). Plasmid DNA is isolated and purified with the QIAGEN Plasmid Maxi Kit, according to the manufacturer's instructions.

DIFFERENTIATION OF THE GENUS *HYPHOMICROBIUM* FROM OTHER GENERA

All *Hyphomicrobium* species described so far grow with C_1 compounds, especially methanol, although methanol was not mentioned in the initial description of the type species, *H. vulgare*, which grows well on formate (Stutzer and Hartleb, 1898). C_1 compounds do not support growth of any of the other morphologically similar genera listed in Table BXII.α.160. Cell shape, arrangement of the bud on the hyphal tip, and the tapering of prosthecae are the best differential characters for this group of budding bacteria. It should be pointed out, however, that the cell shape can vary with the carbon source, growth temperature, nutrient concentration, etc.

Natural samples may sometimes contain organisms that superficially resemble *Hyphomicrobium* species. There are bacteria, such as *Achromatium oxaliferum* from acid bog lakes, which undergo constrictive division, and the "umbilical cord" which holds the two daughter cells of this species together may be mistaken for a thin hyphae (Hirsch, 1974a). Fell (1966) has described *Sterigmatomyces* species, which are yeasts that have a surprising resemblance to hyphomicrobia, except that the cells are much larger and usually carry true vacuoles, as eucaryotes do.

TAXONOMIC COMMENTS

The genus *Hyphomicrobium*—considered monospecific until recently—has been defined mainly by morphological characters and by its conspicuous life cycle. Consequently, many authors have named their isolates *H. vulgare* because they saw mother cells, hyphae, and motile swarmer cells (Kingma-Boltjes, 1936; Mevius, 1953; Zavarzin, 1960; Hirsch and Conti, 1964a; Shishkina and Trotsenko, 1974; Lebedinskii, 1981). Often, the authors have

noted substantial morphological differences among their isolates (Hirsch and Conti, 1964a; Lebedinskii, 1981), and thus it has become increasingly clear that *H. vulgare* is apparently composed of several different species. A detailed study has been conducted on the taxonomy of some 80 C_1-utilizing hyphomicrobia (Hirsch, unpublished data). A computer analysis of these strains carried out by R. Colwell has revealed eight different groups, separated by similarity values of 50–65%. This must be taken as an indication of the need to separate these isolates on a higher level than that of species. A confirmation of these results comes from DNA–DNA hybridization data obtained with 19 *Hyphomicrobium* strains (Table BXII.α.161; Gebers et al., 1986). Those strains that fall within the eight computer-selected groups show high homology (86–100%), but the homology between members of different groups is, in most cases, below 10%. The differences among the *Hyphomicrobium* species shown in Tables BXII.α.162, BXII.α.163, and BXII.α.164 concern morphological, physiological, and ecological characteristics.

A large number of *Hyphomicrobium*-like bacteria that form hyphae and buds from mother cells and cannot usually grow with C_1 compounds can be found in culture collections. These require at least low concentrations of peptone, yeast extract, etc. In some cases, such organisms have been called "*Hyphomonas*", but information is still lacking on the properties of such strains. Mol% G + C values of these "organic hyphomicrobia" may be lower than those of the C_1 hyphomicrobia (Gebers et al., 1985).

Hyphomicrobium neptunium Leifson 1964 has been isolated from seawater and does not resemble C_1-utilizing hyphomicrobia, other than with respect to morphology and life cycle. Recently, this organism has been compared with *Hyphomonas* species (Moore and Hirsch, 1972; Hirsch, 1974a, b; Weiner et al., 1985), and Moore et al. (1984) has subsequently transferred *H. neptunium* to the genus *Hyphomonas*.

A bacterium isolated from seawater by Weisrock and Johnson (1966) and supposedly resembling hyphomicrobia has been described as *Hyphomicrobium indicum*. It has already been pointed out by Hirsch (1974a, b) that there exist substantial differences between this facultatively anaerobic organism and the type species, *Hyphomicrobium vulgare*. *H. indicum* forms acid from sugars, is indole, nitrate, and H_2S positive, and deaminates phenylalanine. It is psychrotrophic, requires 50–100% seawater, and lacks true budding and hyphal branching. The mol% G + C of the DNA from *Hyphomicrobium indicum* is 40, in contrast to all other, true hyphomicrobia, which have a range of 59–65. This bacterium should no longer be included in the genus *Hyphomicrobium*.

A budding bacterium labeled "*Hyphomicrobium variabile*" is available from culture collections; it is a patent strain, and further information on its properties is lacking.

The definition of the genus *Hyphomicrobium* has been hampered seriously by the lack of a type culture; *H. vulgare*, as described by Stutzer and Hartleb (1898), no longer exists. A search for a neotype culture has been difficult, since not all of the tests described in the 1898 publication can be used only partly or checked at the present time. Among the 80 C_1-utilizing hyphomicrobia mentioned above, strain IFAM MC-750 has been found to have properties essentially identical to those of *H. vulgare*. This strain also shows a high similarity to the hypothetical median organism of the 80 hyphomicrobia (Colwell and Hirsch, unpublished data). It has been proposed, therefore, to accept this strain as the neotype culture for the genus *Hyphomicrobium* (Hirsch, 1989). DNA–DNA hybridization studies of 19 hyphomicrobia indicate at least weak relationships between IFAM MC-750 (Institut

TABLE BXII.α.160. Differentiation of the genus *Hyphomicrobium* from other closely related or morphologically similar genera[a]

Characteristic	*Hyphomicrobium*	*Ancalomicrobium*	"Bacterium T"[b]	*Dichotomicrobium*[c]	*Filomicrobium*[d]	*Hirschia*[e]	*Hyphomonas*	*Pedomicrobium*	*Prosthecomicrobium*
Cells ovoid, pear-, or bean-shaped	+	−	+	−	−	+	+	+	+
Cells nearly spherical	−	−	−	−	−	−	+	−	−
Cells nearly tetrahedral	−	+	v	+	−	−	−	−	−
Cells fusiform	−	−	−	−	+	−	−	−	−
Hyphae formed regularly	+	−	+	+	+	+	+	+	−
Prosthecae normally tapering	−	+	−	−	−	−	−	−	+
Bud elongates with long axis of hypha	+	−	+	+	+	+	+	−	−
Bud elongates perpendicular to hyphal long axis	−	−	−	−	−	−	−	+	−
C$_1$ compounds support growth	+	−	−	−	−	−	−	−	−
Moderately thermophilic	−	−	−	+	−	−	−	−	−
Moderately halophilic or halotolerant	−	−	−	+	+	+	+	−	−
Fe and/or Mn are oxidized	v	−	−	−	−	−	−	+	−
May possess gas vesicles	−	+	−	−	−	−	−	−	+

[a]Symbols: +, 90% or more of strains are positive; −, 90% or more of strains are negative; v, variable, depending on growth conditions.

[b]Data from Eckhardt et al. (1979).

[c]Data from Hirsch and Hoffmann (1989a).

[d]Data from Schlesner (1987).

[e]Data from Schlesner et al. (1990).

für Allgemeine Mikrobiologie, Kiel, Germany) and *H. aestuarii* (18–24% homology), *H. zavarzinii* (14%), *H. hollandicum* (11–12%), and *H. facile* (4–11%) (Gebers et al., 1986; Table BXII.α.161).

Euzéby (1998b) has recommended that *Hyphomicrobium facilis*, *Hyphomicrobium facilis* subsp. *facilis*, *Hyphomicrobium facilis* subsp. *tolerans*, and *Hyphomicrobium facilis* subsp. *ureaphilum* (Hirsch, 1989) be changed to *Hyphomicrobium facile*, *Hyphomicrobium facile* subsp. *facile*, *Hyphomicrobium facile* subsp. *tolerans*, and *Hyphomicrobium facile* subsp. *ureaphilum*, respectively.

ACKNOWLEDGMENTS

Donation of cultures by M. Feil, C. Gliesche, W. Harder, N. Holm, T.Y. Kingma-Boltjes, E. Leifson, M. Macpherson-Kraviec, G.T. Sperl, B. Speralski, Y. Trotsenko, R. Weiner, G.A. Zavarzin, and many others is gratefully acknowledged. Critical discussions with R. Colwell, R. Gebers, W. Harder, T. Roggentin, A. Schwartz, and R. Weiner helped in organizing this material. This work would not have been possible without the able technical assistance of G. Maisch, A. Graeter, B. Hoffmann, M. Beese, M. Kusche, S. Liedtke, and K. Lutter-Mohr.

FURTHER READING

Harder, W. and M.M. Attwood. 1978. Biology, physiology and biochemistry of hyphomicrobia. Adv. Microb. Physiol. *17*: 303–359.

Moore, R.L. 1981. The biology of *Hyphomicrobium* and other prosthecate, budding bacteria. Annu. Rev. Microbiol. *35*: 567–594.

Moore, R.L. 1981. The genera *Hyphomicrobium*, *Pedomicrobium* and *Hyphomonas*. *In* Starr, Stolp, Trüper, Balows and Schlegel (Editors), The Prokaryotes. A Handbook on Habitats, Isolation, and Identification of Bacteria, 1st Ed., vol. I, Springer-Verlag, Berlin. pp. 480–487.

Poindexter, J.S. 1992. Dimorphic prosthecate bacteria: the genera *Caulobacter*, *Asticcacaulis*, *Hyphomicrobium*, *Pedomicrobium*, *Hyphomonas* and *Thiodendron*. *In* Balows, Trüper, Dworkin, Harder and Schleifer (Editors), The Prokaryotes. A Handbook on the Biology of Bacteria: Ecophysiology, Isolation, Identification, Applications, 2nd Ed., Springer-Verlag, New York. pp. 2176–2196.

DIFFERENTIATION OF THE SPECIES OF THE GENUS *HYPHOMICROBIUM*

The differential characteristics of the species of *Hyphomicrobium* are given in Table BXII.α.162, morphological characteristics of the species are listed in Table BXII.α.163, and physiological properties are listed in Table BXII.α.164.

List of species of the genus Hyphomicrobium

1. **Hyphomicrobium vulgare** Stutzer and Hartleb 1898, 76[AL]

   vul.ga′re. L. neut. adj. *vulgare* common.

   The following description is based on characteristics given by Stutzer and Hartleb (1898) and those of the neotype strain, IFAM MC-750. Mother cells are oval, pear-, or drop-shaped, 0.5–1.2 × 1–3 μm, with prosthecae (hyphae) of varying lengths and rather constant diameter (0.2–0.3 μm; up to 0.4 μm when stained). In liquids, growth occurs as turbidity or rarely as pellicle and turbidity; cells do not form rosettes. Swarmer cells have one to three subpolar flagella. Colonies remain small, even after long in-cubation; they are colorless or beige, and brownish in transmitted light. Colony surface shiny but granular, the edge is wavy.

   Chemoorganotrophic, aerobic, oligocarbophilic. Grow well with methanol, methylamine·HCl, formate, acetate, *n*-butyrate, isovalerate, propionate, lactate (except for IFAM MC-750), isobutanol, glycerol, L-arabinose, D-mannose, D-melibiose, raffinose, dextrin, amygdalin, esculin, D-glucosamine, *N*-acetylglucosamine, dilute human urine, succinate, or asparagine. Growth is slow, but stimulated by pyruvate, α-oxoglutarate, β-hydroxybutyrate, oxalate, galacturonate, chitin, lactose, or D-maltose. Most amino acids are inhibi-

TABLE BXII.α.161. Levels of DNA–DNA homologies of *Hyphomicrobium* species (Gebers et al., 1986)

Characteristic	*H. vulgare*[T]	*H. aestuarii*[T]	*H. aestuarii*	*H. aestuarii*	*H. aestuarii*	*H. denitrificans*[T]	*H. facile* subsp. *facile*[T]	*H. facile* subsp. *tolerans*[T]	*H. facile* subsp. *tolerans*	*H. facile* subsp. *ureaphilum*[T]	*H. facile*	*H. facile*	*H. hollandicum*[T]	*H. hollandicum*	*H. zavarzinii*[T]	*H. zavarzinii*
Collection number:																
IFAM	MC-750	NQ-521Gr	EA-617	MEV-533Gr	WH-563	TK 0415T	H-526	I-551	CO-558	CO-582	B-522	F-550	KB-677	MC-651	ZV-622	ZV-580
ATCC	27500	27483		27488			27485	27489	27491	27492	27484		27498	27497		
Mol% G + C[a]	61.38	64.11	63.46	64.68	63.09	60.0–61.0[b]	59.53	59.4	59.78	60.54	59.34	59.91	62.41	62.91	64.8[c]	61.77
% Homology with labeled DNA from:																
MC-750	100	18	20	24	19	10	5	5	7	11	4	5	12	11	8	14
NQ-521Gr	4	100	110	103	92	13		1		1	0		3			11
EA-617[d]	2	101	100	70	102		3				3		10			11
B-522	2	5	2	4	4		88	86	91	87	100	106	4			5
ZV-622	3	9						1		2	2		4		100	91
TK 0415[d]	10	13				100							13			

[a]Data are from T_m determinations (Gebers et al., 1985).

[b]Data are from Urakami et al. (1995b).

[c]Data are from Bd determinations (Mandel et al., 1972).

[d]Data are from Moore and Hirsch (1972).

TABLE BXII.α.162. Characteristics differentiating the species of the genus *Hyphomicrobium*[a, b]

Characteristics	*H. vulgare* Stutzer and Hartleb 1898	*H. vulgare* IFAM MC-750	*H. aestuarii*	*H. chloromethanicum*	*H. coagulans*	*H. denitrificans*	*H. facile* subsp. *facile*	*H. facile* subsp. *tolerans*	*H. facile* subsp. *ureaphilum*	*H. hollandicum*	*H. methylovorum*	*H. zavarzinii*
Mother cells bean-shaped	−	−	+	+	−	−	−	−	−	−	−	−
Mother cells oval or pear-shaped	+	+	−	+	+	+	+	+	+	+	+	+
Rosette formation	−	−	−	nd	−	+	−	−	−	−	−	+
Pellicle formed on liquids	v	−	+	nd	+	−	−	−	−	v	−	v
Maximum pH >7.5	−	−	−	−	+	−	−	−	−	−	−	−
Growth with CH$_3$Cl	nd	−	−	+	nd	−	−	−	−	−	nd	−
Growth with peptones	−	−	+	nd	+	−	−	+	−	−	nd	+
Growth with acetate	+	+	+	nd	−	(+)	(+)	−	−	−	−	(+)
Gelatin liquefied, milk coagulated	−	−	−	nd	+	nd	−	−	−	−	−	−
Isolated from soil	+	+	−	+	+	+	+	+	+	−	+	+
DNA mol% G + C (T_m)	nd	61	64	60	nd	60–61	60	59	61	62	61	65

[a]Symbols: −, 90% or more of strains are negative; +, 90% or more of strains are positive; v, variable, depending on growth conditions; nd, not determined; (+), growth stimulation weak but significant.

[b]Strains: *H. vulgare* Stutzer and Hartleb 1898; *H. vulgare* IFAM MC-750; *H. aestuarii* IFAM NQ-521Gr; *H. chloromethanicum* CM2; *H. coagulans* 10-2; *H. denitrificans* TK 0415; *H. facile* subsp. *facile* IFAM H-526; *H. facile* subsp. *tolerans* IFAM I-551; *H. facile* subsp. *ureaphilum* IFAM CO-582; *H. hollandicum* IFAM KB-677; *H. methylovorum* KM-146; *H. zavarzinii* IFAM ZV-622.

TABLE BXII.α.163. Morphological characteristics of the species of the genus *Hyphomicrobium*[a,b]

Characteristics	*H. vulgare* Stutzer and Hartleb 1989	*H. vulgare* IFAM MC-750	*H. aestuarii*	*H. chloromethanicum*	*H. coagulans*	*H. denitrificans*	*H. facile* subsp. *facile*	*H. facile* subsp. *tolerans*	*H. facile* subsp. *ureaphilum*	*H. hollandicum*	*H. methylovorum*	*H. zavarzinii*
Mother cells oval, pear-, or drop-shaped	+	+	−	+	+	+	+	+	+	+	+	+
Mother cells bean-shaped	−	−	+	+	−	−	−	−	−	−	−	−
Mother cell 0.5–1.2 μm wide	+	+	+	+	+	−	+	+	+	+	−	+
Mother cell 0.3–0.65 μm wide	−	−	−	−	−	+	−	−	−	−	+	−
Mother cell average length, μm	1–3[c]	2	1.64	1.3–1.8	1.2–2.0	1.0–3.0	2	1.93	1.9	1.66	0.5–1.2	1.78
Polar flagella	nd	−	−	nd	+	nd	−	−	−	−	−	−
Lateral flagella	nd	−	−	nd	−	nd	−	−	−	−	+	−
Subpolar flagella (1-3)	nd[d]	+	+	nd[d]	−	nd	+	+	+	+	−	+
Mother cells with polar holdfast	−	−	−	nd	−	nd	−	−	−	−	−	+
Rosette formation (R) or cell clumping (C)	C	C	C	−	C	C	−	−	−	C	−	R
Growth as turbidity (T) or pellicle (P)	T	T	P	T	P	T	T	T	T	v	T	v
Older cells yellow	−	−	−	−	+	−	−	−	−	−	+	−
Isolated from soil (S), water (W), or sewage (SW)	S	S	W	S	S	S	S	S	S	SW	S	S

[a]Symbols: −, 90% or more of strains are negative; +, 90% or more of strains are positive; v, variable, depending on growth conditions; nd, not determined.

[b]Strains: *H. vulgare* Stutzer and Hartleb 1898; *H. vulgare* IFAM MC-750; *H. aestuarii* IFAM NQ-521Gr; *H. chloromethanicum* CM2; *H. coagulans* 10-2; *H. denitrificans* TK 0415; *H. facile* subsp. *facile* IFAM H-526; *H. facile* subsp. *tolerans* IFAM I-551; *H. facile* subsp. *ureaphilum* IFAM CO-582; *H. hollandicum* IFAM KB-677; *H. methylovorum* KM-146; *H. zavarzinii* IFAM ZV-622.

[c]From stained preparations.

[d]Motility was observed.

TABLE BXII.α.164. Physiological properties of species of the genus *Hyphomicrobium*[a]

Characteristics	*H. vulgare*[b]	*H. vulgare*	*H. aestuarii*	*H. chloromethanicum*	*H. coagulans*	*H. denitrificans*	*H. facile* subsp. *facile*	*H. facile* subsp. *tolerans*	*H. facile* subsp. *ureaphilum*	*H. hollandicum*	*H. methylovorum*	*H. zavarzinii*
C-source:												
Acetate	+	+	+	nd	−	+	(+)	−	−	−	−	(+)
n-Butyrate	nd	+	(+)	nd	nd	−	−	+	+	−	nd	+
Lactate	+	−	+	nd	−	+	+	−	−	+	−	−
Succinate	+	+	+	nd	nd	−	−	−	−	+	−	−
Chloromethane	nd	−	−	+	−	nd	−	−	−	−	nd	−
Ethanol	nd	−	+	+	+	(+)	(+)	−	(+)	−	−	+
Glycerol	+	+	+	nd	nd	−	+	+	+	+	−	+
Amygdalin	nd	+	−	nd	nd	nd	(+)	−	−	+	nd	(+)
Peptones	−	−	+	nd	+	−	−	+	−	−	nd	+
N-Acetylglucosamine	nd	+	(+)	nd	nd	nd	+	+	(+)	+	nd	−
Formamide	nd	−	+	nd	nd	nd	−	−	−	−	+	−
Aspartate	nd	−	−	nd	+	nd	−	−	−	+	+	+
Asparagine	+	(+)	(+)	nd	nd	nd	−	−	−	−	nd	−
Oligocarbophilic	+	+	+	+	nd	nd	+	+	+	+	−	+
Nitrate reduction	+[c]	+	+	−	−	+	−	−	−	−	−	+
Growth at:												
5°C	nd	−	−	nd	nd	nd	+	−	−	−	−	−
15°C	+	+	+	nd	nd	nd	+	+	+	−	+	+
37°C	nd	+	+	nd	nd	−	+	+	+	+	−	+
45°C	nd	−	+	nd	nd	−	−	+	+	−	−	−
Maximum pH	nd	>7.0	>7.0	7.5	~6.0	>7.0	>7.0	>7.0	>7.0	>7.0	>7.0	>7.0
Inhibition by visible light	nd	−	+	nd	nd	nd	−	−	−	+	nd	+
Growth stimulated by B_{12}	nd	−	−	−	nd	nd	+	−	−	−	+	−
Gelatin liquefaction	−	−	−	nd	+	−	−	−	−	−	−	−
Milk coagulation	nd	−	−	nd	+	nd	−	−	−	−	nd	−
H_2 evolution	nd	−	−	nd	+	nd	−	−	−	−	−	−
α-Hemolysis of sheep blood	nd	+	−	nd	nd	nd	−	−	+	−	nd	+
Genome size: mol. wt. ($\times 10^9$)[d]	nd	2.13	2.62	nd	nd	nd	nd	nd	nd	2.43	nd	nd
Mol% G + C of DNA (T_m)	nd	61	64	60	nd	60–61	59	59	60	62	61	65

[a]Symbols: −, 90% or more of strains are negative; +, 90% or more of strains are positive; nd, not determined; (+), growth stimulation weak but significant.

[b]Strains: *H. vulgare* Stutzer and Hartleb 1898; *H. vulgare* IFAM MC-750; *H. aestuarii* IFAM NQ-521Gr; *H. chloromethanicum* CM2; *H. coagulans* 10-2; *H. denitrificans* TK 0415; *H. facile* subsp. *facile* IFAM H-526; *H. facile* subsp. *tolerans* IFAM I-551; *H. facile* subsp. *ureaphilum* IFAM CO-582; *H. hollandicum* IFAM KB-677; *H. methylovorum* KM-146; *H. zavarzinii* IFAM ZV-622.

[c]Anaerobic growth with KNO_3, but nitrate reduction was considered to be negative.

[d]Data from Kölbel-Boelke et al. (1985).

tory. The type strain grew well with propionate, isobutyrate, valerate, and mannitol. It did not grow with fructose or sucrose.

Nitrogen sources utilized (in order of growth stimulation) are: NH_4^+, NO_3^-, NO_2^-, and ureate. Do not use urea or fix N_2, although slow oligonitrophilic growth has been observed. Do not nitrify. Anaerobic growth occurs in the presence of NO_3^-, but NO_2^- has not been detected. Slow growth on sheep blood agar.

Strain IFAM MC-750 is inhibited by 30 μg disks of kanamycin, neomycin, and novobiocin and by 10 μg of streptomycin. It tolerates 5.5% NaCl, but growth is retarded at this concentration. The pH optimum is between 6.5 and 7.5; the temperature range for growth is 15–37°C. IFAM MC-750 is catalase and cytochrome oxidase positive, sheep's blood is hemolyzed (α-hemolysis), and most cells form poly-β-hydroxybutyrate as a storage product. MC-750 is not pathogenic for mice and guinea pigs.

The genome size is 2.13×10^9 Da (Kölbel-Boelke et al., 1985).

Habitat: soil. The neotype strain came from construction soil.

The mol% G + C of the DNA is: 61.4 (T_m) (Gebers et al., 1986) or 61.1 (HPLC) (Urakami et al., 1995b).

GenBank accession number (16S rRNA): Y14302.

Additional Remarks: IFAM MC-750 (ATCC 27500) is recommended as the neotype strain; the original type strain no longer exists.

2. **Hyphomicrobium aestuarii** Hirsch 1989b, 495[VP] (Effective publication: Hirsch 1989, 1901)

ae.stu.a′ri.i. L. n. *aestuarium* estuary; M.L. gen. n. *aestuarii* of the estuary.

Mother cells bean-shaped, often with short hyphae; the terminal bud on a hypha is also bean shaped, but is turned at a 90° angle from the mother cell. Mother cells 0.6 ×

1.6 μm (range: 0.5–1.0 × 0.6–5.0 μm). Older hyphae branched. Swarmers with 1–3 subpolar flagella. Cells do not form rosettes, but clump easily and grow in liquids as surface pellicles; shaking precipitates the pellicle. Growth also occurs on the glass surface near the top of the vessel. Attachment of swarmer cells to the glass walls is inhibited by light. Colonies on solid media are brownish, strongly and irregularly folded; surface often has concentric rings. Colony cells very cohesive.

Chemoorganotrophic, aerobic, oligocarbophilic. Grow well with methanol, methylamine·HCl, formate, formamide, dilute human urine, acetate, pyruvate, malate, ethanol, and acetamide. Growth is stimulated, but slow, with n-butyrate, isovalerate, lactate, α-oxoglutarate, succinate, crotonate, β-hydroxybutyrate, oxalate, glucuronate, ethanol, n-propanol, isobutanol, formaldehyde, glycerol, chitin, and Bacto peptone. Also stimulatory (as C sources) are: D,L-aspartate, L-asparagine, and N-acetylglucosamine.

Nitrogen sources utilized for growth: NH_4^+, NO_3^-, and urea; the organisms are oligonitrophilic. Slow growth on Bacto peptone with methylamine·HCl added and on sheep blood agar, but there is no hemolysis.

Inhibited by 30 μg each of kanamycin, neomycin, tetracycline, and erythromycin and by 10 μg of streptomycin (all administered on disks). Grow well in the presence of 3.5% NaCl and faintly with 5.5% NaCl. Temperature range: 5–45°C. Optimal pH: 6.5–7.5. Inhibited by visible light. With methanol and KNO_3, there is anaerobic growth and gas formation; grow anaerobically with methylamine in the presence of thioglycolate. Cytochrome oxidase and catalase positive, gelatin is not liquefied. Not pathogenic for mice or guinea pigs.

Habitat: isolated from brackish water of the Elbe River estuary near Cuxhaven, Germany (Mevius, 1953) and from harbor water of Woods Hole, Massachusetts (Hirsch and Rheinheimer, 1968). Originally present in enrichments of nitrifiers.

The genome size is $2.62 × 10^9$ Da (Kölbel-Boelke et al., 1985).

The mol% G + C of the DNA is: 64.1 (T_m) (Gebers et al., 1986) or 65.5–66.1 (HPLC) (Urakami et al., 1995b; Urakami and Komagata 1987b).

Type strain: ATCC 27483, IFAM NQ-521Gr.

Additional Remarks: Additional strains include IFAM MEV-533, IFAM MEV-533Gr (ATCC 27488), IFAM WH-563, IFAM EA-617, IFAM EN-616, IFAM NQ-521, and IFAM NQ-528. With the exception of strain IFAM WH-563, all of these strains are descendants of isolate "B", which was originally obtained by Mevius (1953).

3. **Hyphomicrobium chloromethanicum** McDonald, Doronina, Trotsenko, McAnulla and Murrell 2001, 121[VP]

chlo.ro.me.tha′ ni.cum. Gr. adj. *chloros* green, referring to the chlorine radical; M.L. adj. *methanicus* methane (utilizing); N.L. neut. n. *chloromethanicum* chloromethane (utilizing).

Monoprosthecate rods, budding on the prosthecal (hyphal) tips. Cells 0.5–0.6 × 1.3–1.8 μm, nonpigmented, may store poly-β-hydroxybutyrate. Daughter cells are motile. Growth is aerobic. Cells are restricted facultative methylotrophs, capable of growth on one-carbon compounds, such as chloromethane, methanol, or methylamine as sole sources of carbon and energy. The chloromethane-degrad-

ing enzymes are inducible. Growth occurs also on ethanol. The type strain, CM2[T], also oxidizes, but does not grow on, bromomethane or iodomethane. Do not grow autotrophically, but growth occurs on mineral salts medium or mineral agar (pH 7.2) with 1% (v/v) CH_3Cl as gas; pyruvate–dehydrogenase activity is negative. C_1 compound assimilation is via the serine pathway (icl[+]); NH_4^+ assimilation is via the glutamate cycle. Nitrate is not reduced; oxidase and catalase activity is positive. Vitamins are not required. Storage of viable cells on Whatman paper in sealed flasks or ampules kept at −40 to −80°C (Doronina and Trotsenko, 1992).

The temperature optimum for growth is 28–30°C. The pH optimum for growth is 6.5–7.5. The major fatty acid is $C_{18:1}$ and the major quinone is Q-9.

H. chloromethanicum shares high 16S rRNA gene sequence similarity with *H. facile* subsp. *facile* (97.1%), *H. denitrificans* (96.3%), and *H. methylovorum* (96.0%). The DNA–DNA similarity with *H. zavarzinii* ZV-622[T] has been found to be 29.4%.

Habitat: strain CM2 was isolated from polluted soil at the Nizhekamsk petrochemical factory, Tatarstan, Russia. Strain CM29 came from soil of the Alushta dendropark, Crimea, Ukraine.

The mol% G + C of the DNA is: 60 (T_m).

Type strain: CM2, NCIMB 13687, VKMB-2176.

GenBank accession number (16S rRNA): AF198623.

Additional Remarks: CM29 is identical to CM2.

4. **Hyphomicrobium coagulans** (ex Takada 1975) Hirsch 1989b, 495[VP] (Effective publication: Hirsch 1989, 1903.)

co.a′ gu.lans. L. part. adj. *coagulans* curdling, coagulating.

Mother cells ovoid or pear-shaped, 0.6–1.2 × 1.2–2.0 μm. Hyphae polar or bipolar, branched in older cultures. Swarmer cells motile with 1–3 polar flagella. In liquid media, growth occurs as turbidity or pellicle. Growth on solid media may become yellow.

Chemoorganotrophic, aerobic. Good growth on methanol, methylamine, and ethanol. Other carbon sources utilized are: n-propanol, n-butanol, isobutanol, benzyl alcohol, furfuryl alcohol, trimethylene glycol, formate, and citrate. Carbon sources not utilized: acetate, lactate, formaldehyde, and glycerate. Growth is stimulated by glucose, sucrose, and lactose. Methane is utilized as a carbon source.

Acid is formed (without gas) from glucose, lactose, glycerol, methyl-α-glucoside, sucrose, maltose, salicin, arabinose, galactose, mannitol, inulin, xylose, and soluble starch. Gelatin is liquefied, peptone is utilized, and milk is slowly coagulated, but not peptonized. Casein is not hydrolyzed, indole is not produced. H_2S-positive; methyl red test and Voges–Proskauer tests are positive. Nitrate reduction negative.

Nitrogen sources utilized are: Casamino acids, polypeptone, aspartate, KNO_3, and NH_4^+.

Optimal pH: ~6.0; range: 5.7–8.0. Temperature optimum: 35°C.

Habitat: rice field soil of Hirakata, Osaka, Japan.

The mol% G + C of the DNA is: not reported.

Type strain: Takada 10-2.

5. **Hyphomicrobium denitrificans** Urakami, Sasaki, Suzuki and Komagata 1995b, 531[VP]

de.ni.tri′ fi.cans. M.L. part. adj. *denitrificans* denitrifying.

Mother cells are rod-shaped, oval, egg-shaped or bean-shaped, often with pointed ends, 0.3–0.6 μm × 1.0–3.0 μm, with mono- or bipolar hyphae of varying lengths that are 0.3–0.4 μm in diameter. Swarmer cells motile, often attaching to surfaces or to other cells and forming clumps. Granules of poly-β-hydroxybutyrate accumulate in the cells.

Colonies on methanol-containing agar are shiny, smooth, raised, entire, white, and 1–2 mm after 3–6 d at 30°C. Chemoorganotrophic, aerobic, not fermentative. Good growth on methanol, mono-, di-, and trimethylamine, pectin, and acetate. Methanol is utilized by the serine pathway via activated formaldehyde incorporation.

Some strains utilize formate and ethanol. The following compounds are not utilized: L-arabinose, D-xylose, D-glucose, D-mannose, galactose, maltose, sucrose, lactose, trehalose, D-sorbitol, D-mannitol, inositol, glycerol, soluble starch, propionic acid, isobutyric acid, *n*-valeric acid, lactic acid, succinic acid, oxalic acid, and methane. Vitamins and amino acids are not required for growth. No growth on nutrient broth or peptone broth.

Nitrogen sources utilized: ammonia, nitrate, and urea. Nitrate is reduced to nitrite; nitrate reduction is strong. Methyl red, Voges–Proskauer, gelatin, and starch hydrolysis are negative. Indole, H_2S and ammonia are not produced. Oxidase, urease, and catalase positive. Acid is not produced from sugars oxidatively or fermentatively.

Optimal pH range for good growth: pH 6.0–8.0. Good growth at 30°C. No growth in nutrient broth, peptone broth, or in media with 3% sodium chloride.

The cellular fatty acids include large amounts of straight-chain, unsaturated $C_{18:1}$ acid. The hydroxy fatty acids include large amounts of $C_{14:0\ 3OH}$ and $C_{16:0\ 3OH}$ hydroxy acids. The major ubiquinone is Q-9.

Habitat: soil.

The mol% G + C of the DNA is: 60.4 (HPLC) (Urakami et al., 1995b) or 60.5 (HPLC) (Urakami and Komagata, 1987b).

Type strain: Attwood and Harder strain X, DSM 1869, NCIB 11706, TK 0145.

GenBank accession number (16S rRNA): Y14308.

6. **Hyphomicrobium facile** Hirsch 1989b, 495[VP] (Effective publication: Hirsch 1989, 1902.)
fa' ci.le. L. adj. *facile* ready, quick.

The three subspecies of *Hyphomicrobium facile* are described below.

The mol% G + C of the DNA is: 59.5–61.2

Type strain: see subspecies list below.

GenBank accession number (16S rRNA): see subspecies list below.

a. **Hyphomicrobium facile** *subsp.* **facile** Hirsch 1989b, 495[VP] (Effective publication: Hirsch 1989, 1902.)

    Mother cells pear- or drop-shaped, 0.95 × 2.0 μm (range: 0.6–1.5 × 0.9–6.0 μm); on some media, hyphae are richly branched and often formed bipolarly. Swarmer cells have 1–3 subpolar flagella, which are easily shed. Grow in liquid media as turbidity; colonies on solid media are light brownish to beige, with a smooth and shiny surface and entire edges. Vigorous stirring of liquid cultures results in intercalary bud formation in hy-

phae. Chemoorganotrophic, aerobic. Grow well with methanol, methylamine·HCl, *N*-acetylglucosamine, and gelatin (strain IFAM B-522). Growth is significantly stimulated by formate, acetate, lactate, α-oxoglutarate, succinate, β-hydroxybutyrate, oxalate, glucuronate, ethanol, glycerol, amygdalin, chitin, and poly-β-hydroxybutyrate. Oligocarbophilic. Growth is also stimulated significantly by DL-leucine, DL-lysine, and DL-phenylalanine. There is no growth in dilute human urine.

Nitrogen sources utilized: NH_4^+, NO_3^-, urea, and (slow) Bacto peptone. Slow growth oligonitrophilically. There is faint growth, but no hemolysis, on sheep blood agar. Inhibited by 30 μg each of kanamycin, neomycin, tetracycline, and erythromycin and by 10 μg streptomycin. Slow growth in the presence of 3.5% NaCl. Temperature range: 5–37°C. Optimal pH: 6.5–7.0. Light does not inhibit growth. Grow anaerobically with methylamine·HCl and thioglycolate; denitrification-negative; catalase and cytochrome oxidase positive. Gelatin is not liquefied, and poly-β-hydroxybutyrate is formed as a storage product. Lysine decarboxylase negative.

Not pathogenic for mice or guinea pigs.

The genome size is 2.35×10^9 Da (strain IFAM B-522).

Habitat: soil.

The mol% G + C of the DNA is: 59.5–61.2 (T_m, HPLC) (Gebers et al., 1986; Urakami et al., 1985, 1995b).

Type strain: ATCC 27485, IFAM H-526.

GenBank accession number (16S rRNA): Y14309.

Additional Remarks: Additional strains include IFAM D-524, E-525, G-527, K-529, L-530, and B-522 (ATCC 27484).

b. **Hyphomicrobium facile** *subsp.* **tolerans** Hirsch 1989b, 495[VP] (Effective publication: Hirsch 1989, 1902.)
to' le.rans. L. part. adj. *tolerans* tolerating; pertaining to the tolerance of high CO concentrations.

Morphology and cell sizes as in *H. facile* subsp. *facile*. Colonies on solid media are light brown to beige, smooth and shiny, with entire edges. Grow in liquids as turbidity.

Chemoorganotrophic, aerobic, oligocarbophilic. Good growth on methanol, methylamine·HCl, formate, *n*-butyrate, isovalerate, glycerol, and *N*-acetylglucosamine. Slow growth, but significant stimulation, by pyruvate, crotonate, β-hydroxybutyrate, isobutanol, formaldehyde, Bacto peptone, and ureate. Nitrogen source utilized: NH_4^+. Slow growth, but not hemolysis, on sheep blood agar. Inhibited by 30 μg each of kanamycin, neomycin, tetracycline, and erythromycin and by 10 μg of streptomycin. Growth in the presence of 2.5% NaCl has been observed; growth is faint with 5.5% NaCl. Temperature range: 15–45°C. Optimal pH: 6.5–7.5. Not inhibited by visible light. Do not denitrify, but growth occurs with methylamine·HCl anaerobically with thioglycolate. Catalase and cytochrome oxidase positive. Tolerate up to 90% (v/v) of CO in the atmosphere. Gelatin is not liquefied; poly-β-hydroxybutyrate is formed. Not pathogenic for mice or guinea pigs.

Habitat: soil.

The mol% G + C of the DNA is: is 59.4–60.5 (T_m, HPLC); (Urakami et al., 1985; Gebers et al., 1986; Urakami and Komagata, 1987b).

Type strain: ATCC 27489, IFAM I-551.

GenBank accession number (16S rRNA): Y14311.

Additional Remarks: Additional strains include IFAM O-545, P-546, Q-547, R-549, M-552, CO-553, CO-557, and CO-558 (ATCC 27491).

c. **Hyphomicrobium facile** *subsp.* **ureaphilum** Hirsch 1989b, 495[VP] (Effective publication: Hirsch 1989, 1902.)

u.re.a.phi' lum. Gr. n. *urum* urine; Gr. adj. *philus* loving; M.L. adj. *ureaphilum* loving urea.

Cell sizes and morphology are very similar to IFAM I-551 (Fig. BXII.α.189). Mother cell width 1.1 μm (range: 1.0–1.25 μm). Grow as turbidity in liquid media. Colonies on solid media brownish to beige, smooth, shiny, with entire edges. Physiological properties as for *H. facile* subsp. *tolerans* IFAM I-551, except for the following: growth is stimulated significantly by α-oxoglutarate, but not by crotonate; growth is stimulated by ethanol, but not by Bacto peptone, urea, and ureate as carbon sources. Urea is utilized as a nitrogen source in the presence of methanol. Temperature range: 25–37°C; there is weak growth at 15°C and 45°C. Grow weakly on sheep blood agar with α-hemolysis. Poly-β-hydroxybutyrate may be formed.

Habitat: soil.

The mol% G + C of the DNA is: 60.5 (T_m, HPLC) (Urakami et al., 1995b; Gebers et al., 1986).

Type strain: ATCC 27492, IFAM CO-582.

GenBank accession number (16S rRNA): Y14310.

Additional Remarks: Additional strains include IFAM CO-573, CO-574, CO-587, CO-611, CO-613, CO-614, CO-610, and CO-645.

7. **Hyphomicrobium hollandicum** Hirsch 1989b, 495[VP] (Effective publication: Hirsch 1989, 1902.)

hol.lan' di.cum. M.L. neut. adj. *hollandicum* pertaining to the isolation from the Netherlands.

Mother cells oval or pear-shaped, 0.70 × 1.66 μm (range: 0.6–0.8 × 1.5-3.5 μm). Older hyphae branched. Swarmer cells subpolarly flagellated. Growth in liquid media as pellicle or turbidity; cells may tend to clump. Colonies on solid media are brownish in transparent light and beige in reflected light, with entire edges and shiny surface. Chemoorganotrophic, aerobic, oligocarbophilic. Grow well with methanol, methylamine·HCl, formate, glycerol, cellobiose, raffinose, dextrin, amygdalin, esculin, *N*-acetylglucosamine, and ureate. Growth is stimulated with isovalerate, pyruvate, lactate, succinate, oxalate, isobutanol, formaldehyde, L-sorbose, D-maltose, chitin, DL-aspartate, D-glucosamine, and dilute urine.

Nitrogen sources utilized: NH_4^+, NO_2^-, NO_3^-. The bacteria grow slowly without an added nitrogen source, i.e., oligonitrophilically. Strain IFAM KB-677 is inhibited by 30 μg each of kanamycin, neomycin, and tetracycline. There is no growth in the presence of 2.5% NaCl. Temperature range: 25–37°C. Optimal pH: >7.5. Strain KB-677 is slightly inhibited by visible light. Catalase and cytochrome oxidase positive, gelatin liquefaction and hemolysis negative. Poly-β-hydroxybutyric acid is formed as a storage product. Not pathogenic for mice or guinea pigs.

The genome size is 2.43 × 10⁹ Da (Kölbel-Boelke et al., 1985).

Habitat: strain IFAM KB-677 was obtained from T.Y. Kingma-Boltjes, who isolated it from sewage.

The mol% G + C of the DNA is: 62.4 (T_m) (Gebers et al., 1986) or 62.9 (HPLC) (Urakami et al., 1995b).

Type strain: ATCC 27498, IFAM KB-677.

GenBank accession number (16S rRNA): Y14303.

Additional Remarks: Strain IFAM MC-651 (ATCC 27 497) is related and was isolated from soil by M. Macpherson-Kraviec.

8. **Hyphomicrobium methylovorum** Izumi, Takizawa, Tani and Yamada 1983, 439[VP] (Effective publication: Izumi, Takizawa, Tani and Yamada 1982, 373.)

me.thy.lo.vo' rum. M.L. n. *methyl* the methyl radical; L. v. *voro* devour; M.L. n. *methylovorum* methyl devourer.

Mother cells oval, 0.3–0.65 × 0.5–2.0 μm, with monopolar or bipolar hyphae. Swarmer cells motile with a single lateral flagellum. On solid media, colonies white to faintly yellow, circular, convex, smooth, and glistening.

Chemoorganotrophic, aerobic, aminopeptidase positive. Carbon sources utilized for growth: methanol, methylamine, dimethylamine, and trimethylamine. Poor growth on formate and formamide. Compounds not utilized are: methane, formaldehyde, ethanol, *n*-propanol, glycerol, D-glucose, D-fructose, D-lactose, D-arabinose, sucrose, trehalose, melibiose, cellobiose, mannitol, inositol, dextrin, starch, acetate, pyruvate, lactate, succinate, oxalate, glycolate, glyoxylate, citrate, fumarate, malate, tartrate, glycine, and serine.

Nitrogen sources utilized: NH_4^+ and L-glutamine; poor growth on peptone, Casamino acids, L-cysteine, and L-aspartate. Nitrogen compounds not utilized: nitrate, urea, yeast extract, meat extract, glycine, and L-serine. Nitrate reduction, Voges–Proskauer, indole, H_2S, starch hydrolysis, citrate, and urease are all negative. Catalase positive. Temperature range for growth: 14–33°C. Optimal temperature: around 28°C. Optimal pH: around 7.0. Vitamins are not required.

Habitat: soil.

The mol% G + C of the DNA is: 60.6 (T_m) (Urakami et al., 1985) or 58.5 (HPLC) (Urakami et al., 1995b).

Type strain: DSM 5458, IFO 1480, KM-146.

GenBank accession number (16S rRNA): Y14307.

9. **Hyphomicrobium zavarzinii** Hirsch 1989b, 495[VP] (Effective publication: Hirsch 1989, 1903.)

za.var.zin' i.i. M.L. gen. n. *zavarzinii* of Zavarzin, named for G.A. Zavarzin, the Russian microbiologist who isolated these bacteria.

Mother cells drop- or pear-shaped, somewhat slender, with hyphae that rarely branch. Mother cells 0.63 × 1.8 μm (range: 0.5–0.9 × 0.7–2.5 μm). Swarmer cells with 1–3 subpolar flagella. In liquid media under most growth conditions, rosettes are formed, since mother cells produce a polar holdfast. Growth in liquids initially as turbidity and later as a pellicle, with precipitation on the bottom. Colonies on solid media are colorless to light brownish or beige, smooth and shiny, with entire edges.

Chemoorganotrophic, aerobic, oligocarbophilic. Good growth with the following carbon sources: methanol, methylamine·HCl, formate, *n*-butyrate, isovalerate, crotonate, β-hydroxybutyrate, ethanol, *n*-propanol, isobutanol, and glycerol. Growth is stimulated significantly by acetate, *n*-valerate, α-oxoglutarate, galacturonate, formaldehyde, D-glucose, D-mannose, D-melibiose, amygdalin, esculin, chitin, Bacto peptone, DL-lysine, DL-aspartate, and dilute human urine. Nitrogen sources utilized are: NH_4^+, NO_2^-, NO_3^-, and (poorly) Bacto peptone. There is slow growth in the absence of added nitrogen sources (oligonitrophily). Poor growth on sheep blood agar with α-hemolysis. The following antibiotics inhibit growth at 30 μg (per disc): kanamycin, neomycin, and tetracycline. Streptomycin at 10 μg is also inhibitory. There is growth in the presence of 3.5% NaCl. Temperature range: 15–37°C. Optimal pH: 6.5–7.5. Visible light inhibits growth slightly.

Grow anaerobically with nitrate and gas formation (with methanol as the carbon source). With methylamine·HCl and thioglycolate, there is little growth. Catalase and cytochrome oxidase are positive; gelatin liquefaction is negative. Poly-β-hydroxybutyrate is a storage product.

Not pathogenic for mice or guinea pigs.

Genome size: 2.73×10^9 Da (strain ZV-580; Kölbel-Boelke et al., 1985).

Habitat: peaty and moist soil near Moscow, Russia.

The mol% G + C of the DNA is: 61.8–64.8 (Bd, T_m, HPLC) (Mandel et al., 1972; Gebers et al., 1986; Urakami and Komagata, 1987b; Urakami et al., 1995b).

Type strain: ATCC 27496, IFAM ZV-622.

GenBank accession number (16S rRNA): Y14305.

Additional Remarks: Additional strains include IFAM ZV-580, ZV-620, MY-619, MC-625, MC-629, MC-630, and MC-627.

Genus II. **Ancalomicrobium** Staley 1968, 1940[AL]

CHERYL JENKINS, PATRICIA M. STANLEY AND JAMES T. STALEY

An.ca' lo.mi.cro' bi.um. Gr. masc. n. *ancalos* arm; Gr. adj. *micros* small; Gr. masc. n. *bios* life; M.L. neut. n. *Ancalomicrobium* arm (-producing) microorganism.

Unicellular bacterium with conical cells ~1.0 μm in diameter. **Two to eight or more prosthecae** produced per cell. Prosthecae are cylindrical without crossbands and taper gradually from cell to a distal diameter of ~0.2 μm and a length of 2–5 μm when fully differentiated. Prosthecae may be bifurcated. **Budding bacterium**. Buds are formed directly from mother cell, never from prosthecae. Cells occur singly or in pairs prior to division; rarely form aggregates.

Gram negative. Flagella and holdfasts not produced. **Gas vacuoles** are formed by all strains investigated.

Facultatively anaerobic. Chemoheterotrophic. Use sugars anaerobically and aerobically. Ferments sugars by **mixed acid fermentation**. Some organic acids are used aerobically, but not fermented. Ammonium can be used as sole nitrogen source. Vitamins are required. Oxidase and catalase positive. Temperature range for type strain: 9–39°C. Optimal pH: 7.0, pH range 6.3–7.5. Found in freshwaters and pulp mill oxidation lagoons.

Type genus: **Ancalomicrobium adetum** Staley 1968, 1940.

FURTHER DESCRIPTIVE INFORMATION

A. adetum lives in freshwater habitats, including lakes and eutrophic habitats, such as ponds, rivers and pulp mill oxidation lagoons (Staley, 1971; Stanley et al., 1979; Staley et al., 1980).

The most distinctive feature of the genus is its cellular morphology. The size, shape, and location of the prominent prosthecae, combined with a lack of cell motility, are unique features of this genus. Under normal growth conditions, in which a low concentration of nutrients is supplied, the prosthecae attain a length of 2–5 μm (Fig. BXII.α.192). Shorter prosthecae, as formed by members of the genus *Prosthecomicrobium*, are not produced by *Ancalomicrobium*; however, aberrant cell morphology can be observed under *in vitro* cultivation conditions in the presence of high concentrations of nutrients.

Buds begin as small protuberances on a non-appendaged area of the "mother cell". The buds enlarge and differentiate, and new prosthecae are formed so that the daughter cell is a pseudo-mirror image of the mother cell at the time of cell separation. The mother cell retains its original prostheca, and each new bud is produced at the same location on the mother cell. The daughter buds are produced essentially *de novo* during the budding process (Staley, 1973b; Staley et al., 1981) as shown in Fig. BXII.α.193.

Gas vacuoles are produced by all strains studied. Gas vesicles may be produced in some cells under aerobic conditions with sugars as a carbon source; however, they are only observed during the stationary growth phase. In contrast, all strains studied produce gas vacuoles prodigiously during anaerobic growth at low temperatures (<18°C) (A. Van Neerven and J.T. Staley, unpublished observations). Gas vacuoles are also produced under microaerophilic conditions on organic acids, including acetic, pyruvic and succinic acids (R.L. Irgens and J.T. Staley, unpublished observations).

A variety of sugars, sugar alcohols, and organic acids can be used as carbon sources for aerobic growth. These include D-arabinose, L-arabinose, D-ribose, D-glucose, D-fructose, D-mannose, D-galactose, D-xylose, L-fucose, L-rhamnose, D-melibiose, maltose, lactose, cellobiose, trehalose, *N*-acetylglucosamine, acetate, pyruvate malate, glycerol, inositol, and mannitol. Agar is digested slowly.

Some carbon sources, including some sugars and sugar alcohols, are used anaerobically by fermentation. Sugars are fermented by a mixed acid fermentation (Van Neerven and Staley, 1988). Glucose is fermented to acetic acid, ethanol, lactic acid, formic acid, succinic acid, hydrogen gas, and carbon dioxide. These are the same products that *Escherichia coli* produces under the same conditions for growth.

Ammonium salts are used as a sole source of nitrogen. Pantothenic acid is required for growth, and biotin, thiamine, folic

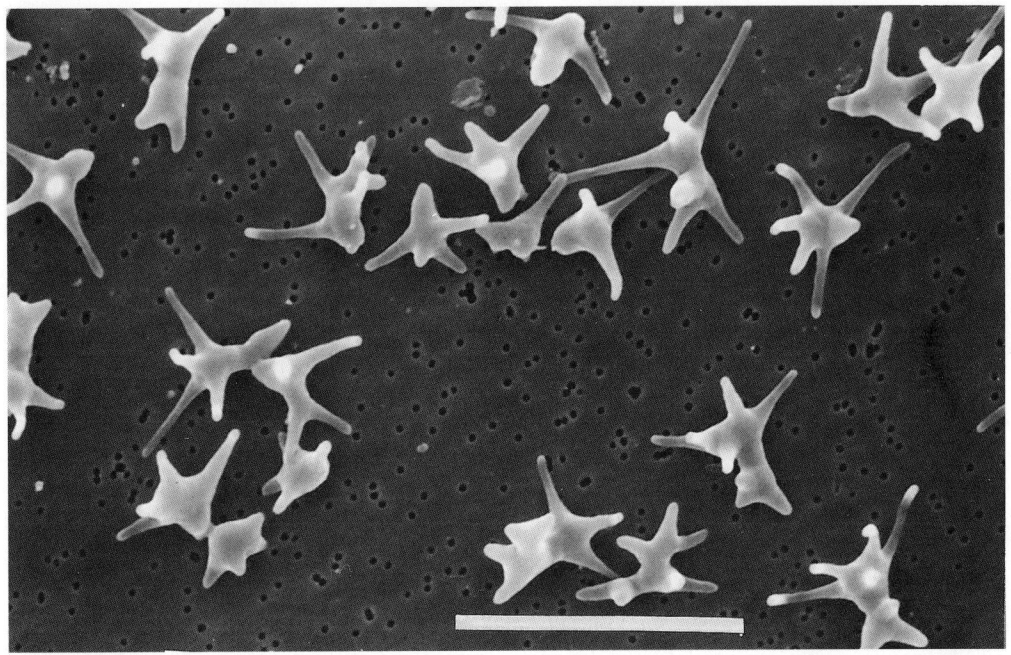

FIGURE BXII.α.192. *Ancalomicrobium adetum.* Scanning electron micrograph. Bar = 5.0 μm. (Courtesy of A.R.W. van Neerven.)

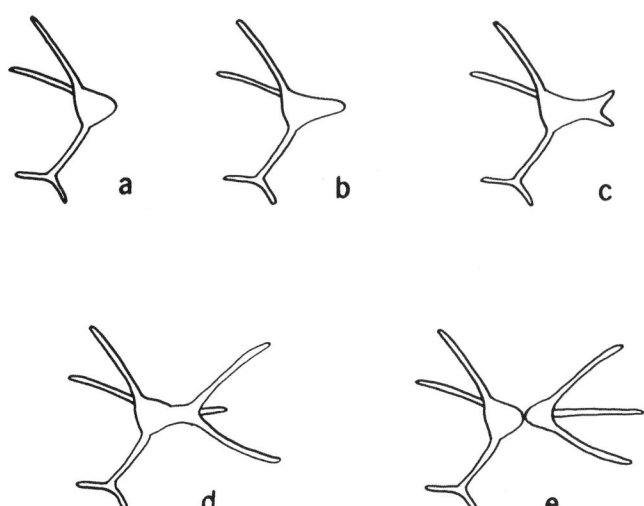

FIGURE BXII.α.193. A diagram of *Ancalomicrobium adetum*, illustrating the life cycle of this budding bacterium. The mother cell has three appendages, one of which is bifurcated (*a*). Bud formation occurs at the apex of the conical cell (*b*). The bud begins to differentiate (*c* and *d*) and ultimately separates from the mother cell (*e*). Following division, the mother cell, whose appearance after reproduction remains essentially identical to its initial appearance in *a*, will produce another daughter cell from the same location. This process will be repeated again and again, as long as conditions are favorable for growth. (Reproduced with permission from J.T. Staley, Journal of Bacteriology *95*: 1921–1942, 1968, ©American Society for Microbiology.)

acid, and nicotinamide are stimulatory for growth of the type strain.

The type strain grows on a minimal, defined medium containing ammonium salts, phosphate, a modified Hutner's salts solution, and vitamins (see *Bergey's Manual of Systematic Bacteriology*, first edition). Better growth is obtained in complex media, such as modified medium B (MMB; Staley, 1968), which contains yeast extract, in addition to the constituents provided in the defined medium. The highest yields have been obtained in a richer complex medium (R.L. Irgens, personal communication; Staley, 1989).

The minimal temperature for growth of the type strain is between 6 and 9°C, whereas the maximum lies between 39 and 43°C. The optimal pH for growth is between 6.9 and 7.3; however, growth has been observed at all pH values tested, ranging from 6.3 to 7.5.

ENRICHMENT AND ISOLATION PROCEDURES

Selective isolation procedures have not yet been developed. However, the addition of peptone to a final concentration of 0.01% may be used to enrich for these bacteria.

Enrichment cultures of 100 ml are normally used. Graduated 150-ml beakers containing 10 mg of peptone are covered with aluminum foil and autoclaved. Natural water samples are collected aseptically, and 100 ml aliquots are added to the beaker. Enrichment cultures are incubated at room temperature without shaking. Cultures should be examined periodically for the appearance of multiply appendaged prosthecate bacteria. Wet mounts should be prepared for phase microscopic observation, from the time the cultures become turbid (usually 4–7 days), until 2–3 weeks have elapsed. Whereas *Caulobacter* species are frequently observed attached to debris in the surface film of the enrichment, these planktonic multiply-appendaged forms are rarely encountered there. They are usually found moving with the currents in the wet mount and are most often seen in areas where free-floating forms have accumulated near debris or air bubbles. When multiply-appendaged forms comprise a significant portion of the enrichment culture (~1%), attempts should be made to isolate them by streaking onto MMB plates (see Staley,

1989). Alternatively, 10^{-3}–10^{-5} dilutions of the enrichment culture can be spread-plated onto MMB agar. Plates should be incubated at room temperature for 1–2 weeks before examination. Individual colonies are selected for examination by wet mount. When positive colonies are found, they are restreaked for purification.

MAINTENANCE PROCEDURES

A. adetum can be maintained for at least 1 month on MMB slants that have been refrigerated. Long-term preservation is effected by lyophilization of cultures grown on complex media.

DIFFERENTIATION OF THE GENUS *ANCALOMICROBIUM* FROM OTHER GENERA

The striking morphology of members of the genus *Ancalomicrobium* allows them to be readily distinguished from nonappendaged bacteria and from most of the other prosthecate bacteria. However, some of the multiply-appendaged bacteria may appear similar. The genus *Stella* can be differentiated from *Ancalomicrobium* because all of its appendages lie in one plane. The most morphologically similar genus is *Prosthecomicrobium*. Some cells of *P. hirschii*, in particular, appear identical to *A. adetum*, although short-appendaged cells are also produced by *P. hirschii*. In addition, all *P. hirschii* strains are motile, while *A. adetum* is nonflagellated. Furthermore, *Ancalomicrobium* strains produce gas vacuoles, while *P. hirschii* does not. More significantly, these two groups are physiologically distinct. All species of *Prosthecomicrobium* are obligately aerobic, whereas *Ancalomicrobium* strains are fermentative, facultative anaerobes. Table BXII.α.165 summarizes the important differential characteristics that distinguish these two genera.

TAXONOMIC COMMENTS

DNA–DNA hybridization studies have shown that *Ancalomicrobium* strains and *Prosthecomicrobium* strains form separate taxonomic groups (Moore and Staley, 1976). This finding is consistent with the current taxonomy of these two genera. According to 16S rRNA analyses, both *Ancalomicrobium* and *Prosthecomicrobium* species are members of the *Alphaproteobacteria*; however, based on sequence data from at least three strains, *Ancalomicrobium* has been shown to form a distinct genus (See Fig. BXII.α.194).

Although further taxonomic work on this genus is desirable, the difficulty encountered in obtaining isolates has hindered taxonomic investigations of the genus. However, since the initial isolation of the type strain, a number of strains have been isolated from pulp mill oxidation lagoons (A. Van Neerven and J.T. Staley, unpublished studies).

It is noteworthy that *A. adetum* is currently the only fermentative, heterotrophic, budding, and prosthecate bacterium in pure culture. Of special note in this regard is the discovery that *A. adetum* is a mixed acid fermenter whose fermentation products are identical to those of *E. coli* (Van Neerven and Staley, 1988). Furthermore, the phosphoenolpyruvate–sugar phosphotransferase system of this organism has been compared to that of certain enteric bacteria (Saier and Staley, 1977). The most interesting discovery from this investigation is that enzymatic cross-reactivity has been detected between the membrane-associated enzyme II complexes of *A. adetum* and the soluble enzyme I components of the enteric bacterium *Salmonella typhimurium*. It is noteworthy that *A. adetum* is a member of the *Alphaproteobacteria*, whereas the enteric bacteria are members of the *Gammaproteobacteria*.

FURTHER READING

Staley, J.T. 1989. Genus *Ancalomicrobium*. *In* Staley, Bryant, Pfennig and Holt (Editors), Bergey's Manual of Systematic Bacteriology, 1st Ed., Vol. 3, The Williams & Wilkins Co., Baltimore. pp. 1914–1916.

Staley, J.T. 1992. The genera: *Prosthecomicrobium, Ancalomicrobium* and *Prosthecobacter*. *In* Balows, Trüper, Dworkin, Harder and Schleifer (Editors), The Prokaryotes-A Handbook on the Biology of Bacteria: Ecophysiology, Isolation, Identification, Applications., 2nd Ed., Vol. III, Springer-Verlag, New York. pp. 2162–2166.

TABLE BXII.α.165. Differentiation of *Ancalomicrobium* from *Prosthecomicrobium*[a]

Characteristic	*Ancalomicrobium*	*Prosthecomicrobium*
Short prosthecae (i.e., <2.0 μm when fully developed)	−	+
Long prosthecae (i.e., 2–5 μm when fully developed)	+	D[b]
Gas vesicles	+	D
Glucose fermentation	+	−

[a]Symbols: −, no strains/species positive; +, all strains/species positive; D, varies depending on species.

[b]Some species of *Prosthecomicrobium*, such as *P. hirschii* and *P. pneumaticum*, produce longer prosthecae as well as shorter ones.

List of species of the genus Ancalomicrobium

1. **Ancalomicrobium adetum** Staley 1968, 1940[VP]

   a'de.tum. M.L. adj. *adetum* arm or appendage.

   The characteristics are as described for the genus. Occur in freshwater.

   The mol% G + C of the DNA is: 70.4 (Bd)

   Type strain: ATCC 23632, DSM 4722.

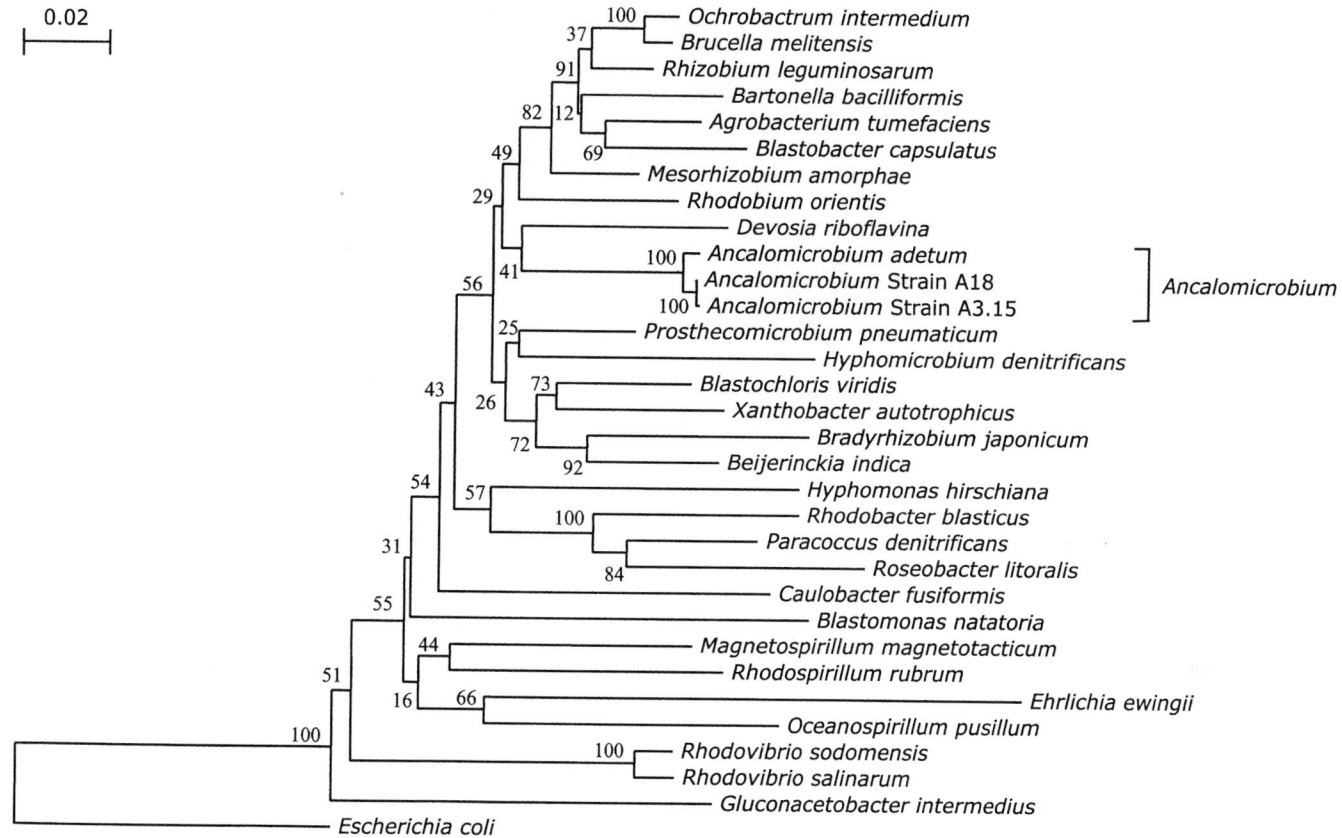

FIGURE BXII.α.194. Phylogenetic tree based on 16S rRNA sequences showing the position of *Ancalomicrobium* within the *Alphaproteobacteria*. The type strain of *Ancalomicrobium*, *A. adetum*, and two additional organisms isolated from pulp mill oxidation lagoons, strains A18 and A3.15, form a distinct genus. These strains form a separate lineage from other prosthecate bacteria, such as *Prosthecomicrobium* and *Hyphomicrobium*. The tree was constructed using Kimura distances and the neighbor-joining algorithm with 100 bootstrap replicates within the program Treecon. The distance scale indicates 0.02 substitutions per site.

Genus III. **Ancylobacter** *Raj 1983, 397*[VP]

JAMES T. STALEY, CHERYL JENKINS AND ALLAN E. KONOPKA

An'cy.lo.bac'ter. Gr. adj. *ankylos* sharply curved; Gr. n. *bakterion* rod; M.L. masc. n. *Ancylobacter* a curved rod.

Curved rods, 0.3–1.0 × 1.0–3.0 μm. **Rings** (0.9–3.0 μm outer diameter) occasionally formed prior to cell separation. Coiled, helical, and filamentous forms are not produced. Cells are encapsulated. Resting stages not known. **Some strains produce gas vacuoles.** Gram negative. **Generally nonmotile**, but motility occurs in one gas vacuolate strain by means of a single polar flagellum. **Obligately aerobic**, possessing a strictly respiratory type of metabolism, with oxygen as the terminal electron acceptor. Optimal temperature, 22–37°C. Colonies are translucent to opaque and white to cream colored. Pellicles are produced in liquid media. **Oxidase positive.** Catalase positive. Chemoorganotrophic, using a variety of sugars or salts of organic acids as carbon sources. Chemolithotrophic growth has been reported on molecular hydrogen. Strains that have been tested can use methanol and formate (**facultatively methylotrophic**). Occur in soil and freshwater environments.

The mol% G + C of the DNA is: 66–69 (Bd).

Type species: **Ancylobacter aquaticus** (Ørskov 1928) Raj 1983, 297 (*Microcyclus aquaticus* Ørskov 1928, 128.)

FURTHER DESCRIPTIVE INFORMATION

Cells of *Ancylobacter aquaticus* typically appear as curved rods (Figs. BXII.α.195 and BXII.α.196). Ring-like forms occur when the ends of a curved cell overlap prior to cell separation; this occurs infrequently under normal growth conditions. Pleomorphic forms have been reported under certain cultural conditions. For example, older cultures of the type strain contain swollen cells and other involution forms (Raj, 1970, 1977).

A number of gas-vacuolated strains of *Ancylobacter aquaticus* have been isolated (Nikitin, 1971; Van Ert and Staley, 1971; Konopka et al., 1977). These appear to be very similar to the type strain, so no new species have been proposed to accommodate them. We consider the type strain of *A. aquaticus* to be avacuolate, despite a claim to the contrary (Raj, 1977). In addition to a lack of convincing microscopic evidence for vacuoles in this strain in our laboratory, we have found that antiserum prepared against gas vacuoles is incapable of causing precipitation in lysates of the type strain, whereas precipitation does occur with the lysates of all gas-vacuolated strains.

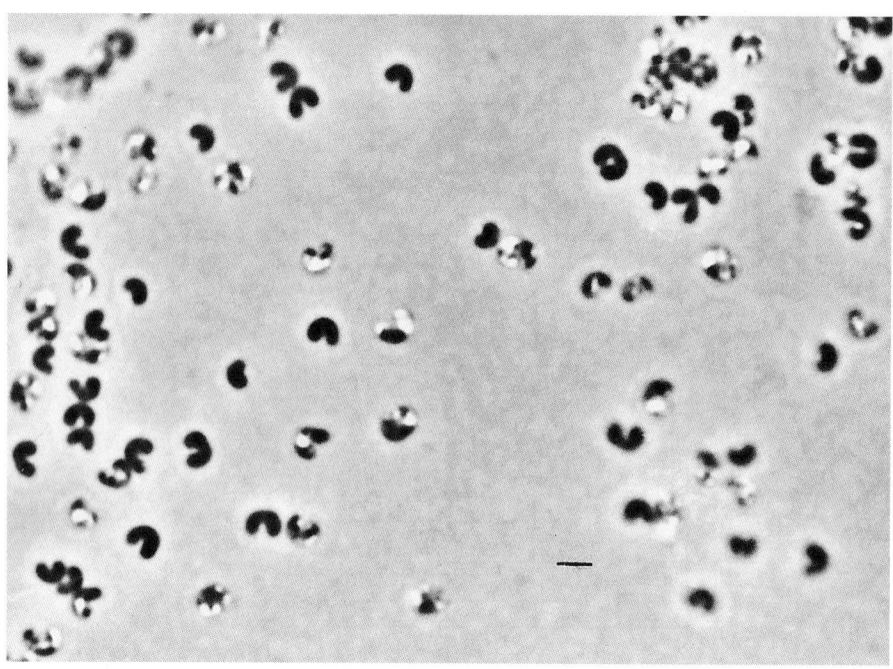

FIGURE BXII.α.195. A phase-contrast photomicrograph showing numerous cells of *Ancylobacter aquaticus*. The refractile intracellular areas of some cells are gas vacuoles. Bar = 2.0 μm.

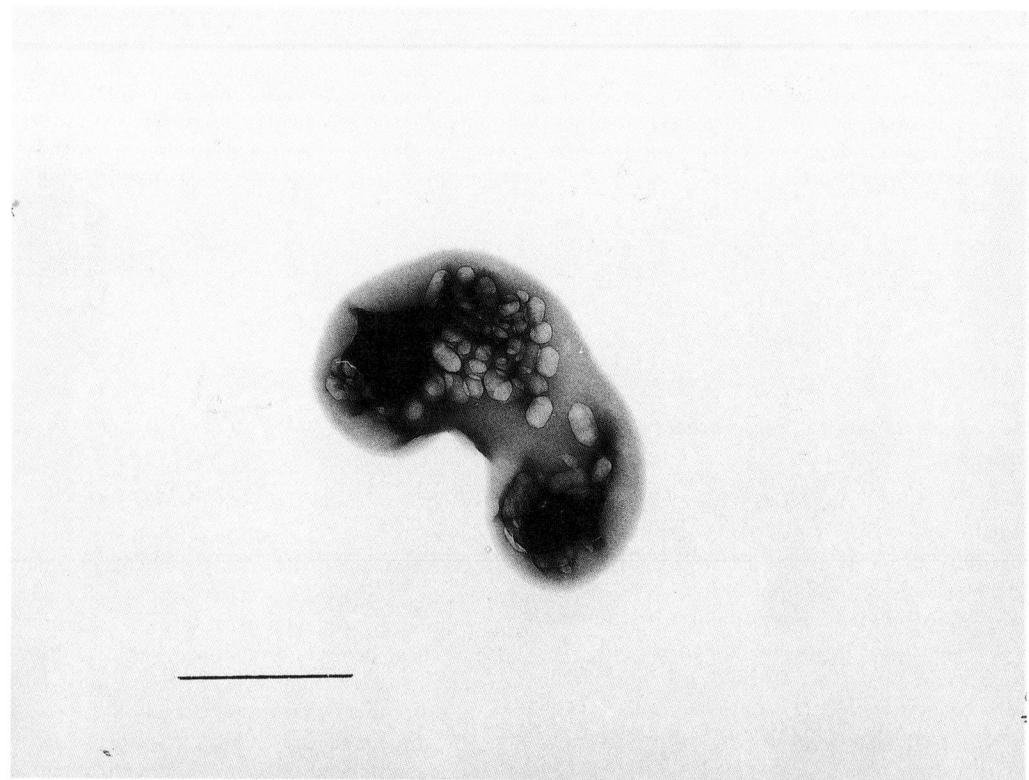

FIGURE BXII.α.196. An electron micrograph of a cell of a gas-vacuolated strain of *Ancylobacter aquaticus*. Individual gas vesicles can be seen. Bar = 1.0 μm. (Courtesy of Dr. J.C. Lara.)

Cells are encapsulated.

Motility has not been detected in *Ancylobacter*, except in the case of one of the gas-vacuolated strains of *A. aquaticus* (Konopka et al., 1976). In this strain (strain M6), the cells possess single polar flagellum. The occurrence of flagella is best seen in cells cultivated at higher growth temperatures, i.e., above 20°C, a condition that precludes extensive gas vacuole formation. The strain forms numerous gas vacuoles when cultivated at 20°C, but flagella are rarely formed at this temperature.

The maximum temperature at which growth occurs is usually

37°C, although one strain can grow at 43°C. The minimum temperature for growth is 5°C.

Colonies of *Ancylobacter aquaticus* are nonpigmented to cream colored, circular, and convex, with an entire margin. In gas-vacuolated strains, the colonies appear in gradations from translucent (if few vacuoles are formed) to opaque, chalky white (if vacuoles are abundant).

Acid is produced from a variety of carbon sources when the medium of Hugh and Leifson (1953) is incubated aerobically (Raj, 1970; Van Ert and Staley, 1971; Larkin et al., 1977). All strains of *Ancylobacter aquaticus* form acid from arabinose, xylose, glucose, fructose, galactose, mannose, melibiose, sucrose, mannitol, sorbitol, and inulin. Gelatin is not liquefied. Starch is not hydrolyzed. Additionally, some of the strains, including the type strain, have been shown to produce acid oxidatively from glycerol and ribose, to reduce nitrate to nitrite and ammonia, to curdle and peptonize litmus milk by rendering it alkaline, and to possess catalase, β-galactosidase, lipase, ornithine decarboxylase, oxidase, and urease. However, they have been shown not to hydrolyze agar, chitin, casein, cellulose, or esculin, nor to produce indole, acetyl methyl carbinol, or H_2S (Raj, 1970; Larkin et al., 1977).

Strains of *Ancylobacter aquaticus* may be grown in a defined medium[1] containing an ammonium salt as a sole nitrogen source; individual carbon sources may be tested in this medium as sole carbon sources for growth. Lactate and pyruvate are used by all strains of *A. aquaticus* under these conditions. In addition, the type strain and some others utilize acetate, citrate, formate, gluconate, malonate, oxaloacetate, and succinate as sole carbon sources (Raj, 1977). Some strains of *A. aquaticus* are able to degrade chlorinated compounds, such as 1,2-dichloroethane (DCE), 2-chloroethanol, chloroacetate, and 2-chloropropionate, and use them as sole carbon and energy sources (van den Wijngaard et al., 1992). All strains of *A. aquaticus* tested so far have been found to be facultative methylotrophs capable of utilizing C_1 compounds, such as methanol and formate (Van Ert and Staley, 1971; Namsaraev, 1973; Larkin et al., 1977). The one strain (Z-238) that has been analyzed for its enzyme content has been found to have ribulose bisphosphate carboxylase activity when grown on methanol (Loginova et al., 1978). Due to induced dehydrogenases, methanol is oxidized sequentially to formaldehyde, formate, and finally carbon. Carbon dioxide is assimilated by ribulose bisphosphate carboxylase and used in the Calvin cycle. Autotrophic growth of several strains has been achieved with hydrogen (Namsaraev and Nozhevnikova, 1978; Málek and Schlegel, 1981). A recent report indicates that some strains can grow as facultative chemolithoautotrophs using thiosulfate as an energy source (Stubner et al., 1998).

The pathways of glucose dissimilation have been determined by the use of position-radiolabeled glucose and gluconate (Kottel and Raj, 1973; Raj, 1977). In *Ancylobacter aquaticus* the Entner–Doudoroff pathway is the primary pathway and the pentose phosphate and Embden–Meyerhof–Parnas pathways are of lesser importance. In addition, a tricarboxylic acid cycle is present as a secondary pathway, as indicated by studies using specifically labeled acetate, glutamate, and pyruvate. The glyoxylic acid cycle does not appear to be operative in this species. Activities of key enzymes from each of these pathways have been demonstrated by assays of cell extracts (Kottel and Raj, 1973).

Strains that have been tested grow well on nitrogen-free media (Nikitin, 1971; Málek and Schlegel, 1981). Attempts to demonstrate nitrogenase activity by the acetylene reduction method or by $^{15}N_2$ incorporation, however, have been unsuccessful (Málek and Schlegel, 1981).

The type strain of *Ancylobacter aquaticus* is sensitive to the following antibiotics (concentrations, μg/disk): ampicillin, 10; cephalothin, 30; chlortetracycline, 30; demeclocycline, 30; dihydrostreptomycin, 10; doxycycline, 30; erythromycin, 15; kanamycin, 30; neomycin, 30; nalidixic acid, 30; nitrofurantoin, 300; novobiocin, 30; oxytetracycline, 30; streptomycin, 5; tetracycline, 5 (Raj, 1970, 1977; Larkin et al., 1977).

Ancylobacter is not known to be pathogenic to humans.

Ancylobacter aquaticus occurs in freshwater habitats, including ponds, creeks, and lakes (Van Ert and Staley, 1971; Konopka et al., 1976), and in rice paddies and soil environments (Stubner et al., 1998). A number of strains have also been isolated from pulp mill oxidation lagoons (Konopka et al., 1976; Fulthorpe et al., 1993; Fulthorpe and Allen, 1995).

ENRICHMENT AND ISOLATION PROCEDURES

The original strain of *Ancylobacter aquaticus* was isolated on a medium consisting of agar (2%) and water (Ørskov, 1928). Gasvacuolated strains can be isolated from enrichment cultures prepared with 100 ml of a freshwater source, such as creek water, added to a sterile aluminum foil-covered beaker containing 10 mg Bacto peptone (Difco). The enrichments are incubated at room temperature for 2 weeks, and dilutions are plated onto a hydrolysate medium[2] containing glucose (Van Ert and Staley 1971). These plates are incubated at 30°C for 1 week and examined for the development of chalky white colonies, which are indicative of gas vacuolate strains. Cells from such colonies should be observed by phase-contrast and electron microscopy to confirm that curved rods with refractile areas typical of gas vacuoles are in fact present.

MAINTENANCE PROCEDURES

Strains of *Ancylobacter aquaticus* are normally cultured on glucose–Casamino acids medium or TGEY medium.[3] After incubation at 20–30°C to allow abundant growth, the cultures may be maintained in a refrigerator (5°C) for at least 3 weeks. They may also be preserved indefinitely by lyophilization.

DIFFERENTIATION OF THE GENUS ANCYLOBACTER FROM OTHER GENERA

Table BXII.α.166 provides the primary characteristics that can be used to differentiate this genus from morphologically similar, aerobic, nonmotile, nonphotosynthetic bacteria of the genera *Flectobacillus*, *Spirosoma*, and *Runella*. Two other genera of gas-vacuolated curved rods that may be confused with *Ancylobacter* are "*Brachyarcus*" and *Meniscus*. "*Brachyarcus*", which has never been isolated, differs from *Ancylobacter* in having cells arranged in groups (coenobia) consisting of two, four, or more rings (Skuja, 1964). *Meniscus* is differentiated from *Ancylobacter* by be-

---

1. Defined medium has the following composition (per liter of distilled water): $(NH_4)_2SO_4$, 0.25 g; glucose, 0.25 g (or molar equivalent of other carbon source); Na_2HPO_4, 0.071 g; modified Hutner's salt solution, 20 ml; and vitamin solution, 10 ml. The salt solution is as described by Van Ert and Staley (1971), except that the amount of sodium molybdate is 12.67 mg. The vitamin solution is as described by Staley (1968).

2. Glucose–Casamino acids medium (per liter of distilled water): glucose, 1.0 g; Bacto Casamino acids (Difco), 1.0 g; modified Hutner's salt solution (see defined medium), 20 ml; vitamin solution (see defined medium), 10 ml; agar, 15.0 g.

3. TGEY medium is Bacto tryptone glucose extract agar (Difco) supplemented with 0.1% yeast extract (Raj, 1970).

TABLE BXII.α.166. Differentiation of *Ancylobacter* from morphologically similar heterotrophic genera

Property	*Ancylobacter*	*Spirosoma*	*Flectobacillus*	*Runella*
Shape	vibrios, rings	sinuous helices	large vibrios	curved rods
Colony pigmentation	none	yellow	pink	pink
Mol% G + C	66–69	51–53	39–41	49–50
Phylum:				
Proteobacteria	+			
Bacteroidetes		+	+	+

ing an aerotolerant anaerobe, rather than an aerobe, and by having a mol% G + C of 45 (Irgens, 1977). Furthermore, it is noteworthy that certain phototrophic bacteria, such as those of the purple nonsulfur genus *Rhodocyclus*, are morphological counterparts of *A. aquaticus*; however, *Rhodocyclus* species belong to a separate class of the *Proteobacteria*.

TAXONOMIC COMMENTS

Significant changes have occurred in the taxonomy of the genus *Ancylobacter*. In the previous edition of *Bergey's Manual of Systematic Bacteriology*, the former genus name *Microcyclus*, which is on the Approved List of names, was still in use. Based on morphological features, *Ancylobacter* was placed in Section 3 of the *Manual*, en-

titled "Nonmotile (or Rarely Motile), Gram-Negative Curved Bacteria", along with, but not included in, a family of bacteria called the *Spirosomaceae* (Larkin and Borrall, 1978). Genera placed in the family *Spirosomaceae* were *Spirosoma*, *Runella*, and *Flectobacillus*. Subsequent phylogenetic analyses based on 16S rRNA sequences have shown that all genera of the *Spirosomaceae* are members of the *Flavobacterium–Cytophaga–Bacteroides* group, whereas *Ancylobacter* are members of the *Alphaproteobacteria* (Woese et al., 1990). The genus *Ancylobacter* forms part of the *Blastochloris viridis* subgroup of the *Alphaproteobacteria*, with *Starkeya novella* (formerly *Thiobacillus novella*) as its closest relative (see Fig. BXII.α.197). Bacteria related to *Ancylobacter*, including members of the genera *Starkeya*, *Xanthobacter*, and *Blastochloris* (formerly *Rhodopseudo-*

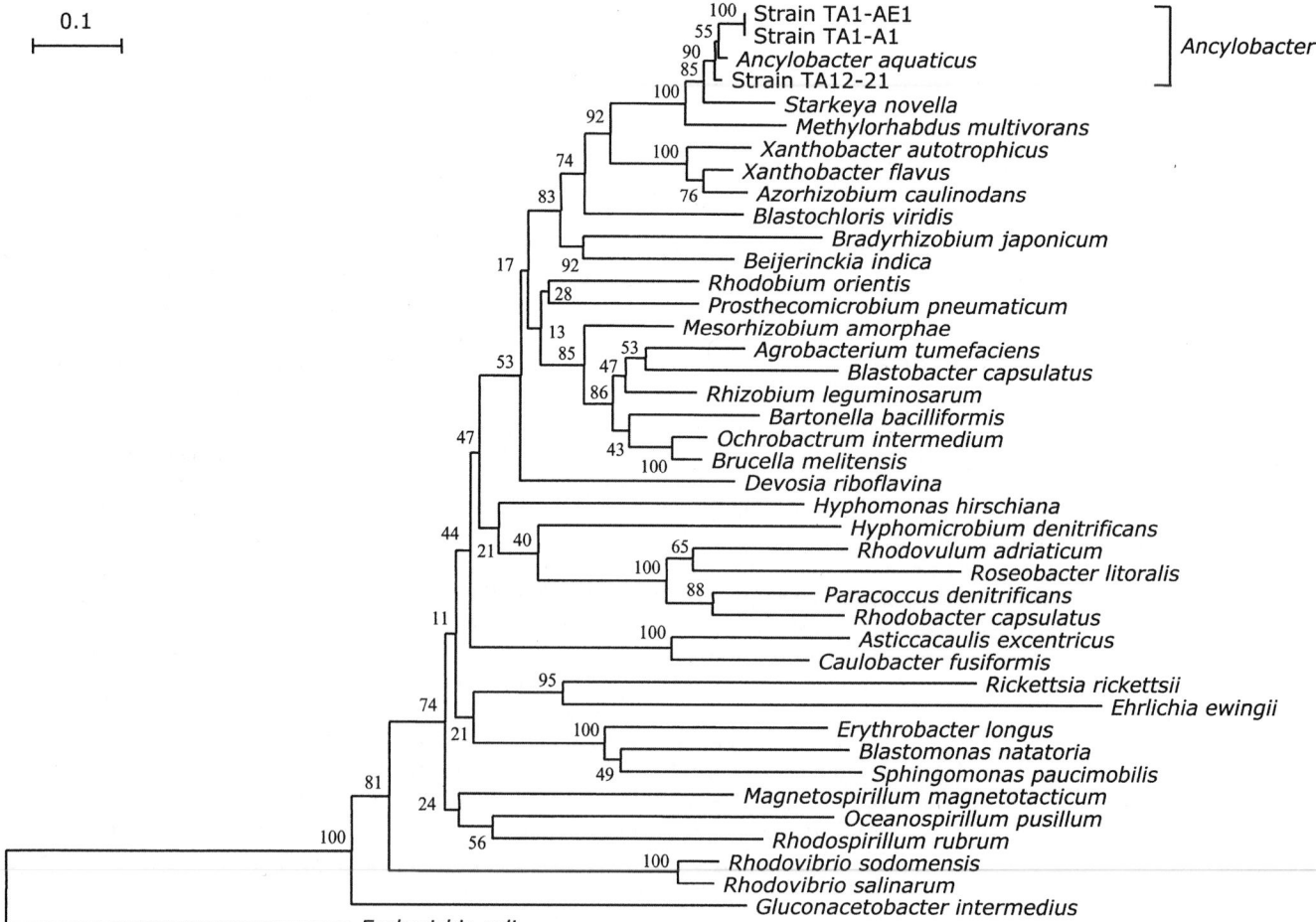

FIGURE BXII.α.197. Phylogenetic tree, based on 16S rRNA sequences, showing the position of *Ancylobacter* within the *Alphaproteobacteria*. Displayed in the tree are the type strain of *Ancylobacter aquaticus* and three additional strains, TA1-AE1, TA1-A1, and TA12-21, which were isolated from rice paddy soil. *Ancylobacter* forms part of the *Blastochloris viridis* subgroup of the *Alphaproteobacteria*, with the genera *Starkeya*, *Methylorhabdus*, and *Xanthobacter* as the closest relatives. The phylogenetic tree was constructed using Kimura distances, the neighbor-joining algorithm, and 100 bootstrap replicates within the program Treecon. The distance scale indicates 0.1 substitutions per site.

monas), are also physiologically similar to *Ancylobacter* in that they can grow autotrophically. In addition, *Xanthobacter* species and another close relative, *Methylorhabdus multivorans*, exhibit methylotrophy, another physiological feature shared with *Ancylobacter* strains.

DNA–DNA similarity studies of *A. aquaticus* strains B, H, M, M6, and W with the type strain show a much lower genetic relatedness (28–45%) than expected of strains of the same species, even though their mol% G + C values are very close (Konopka et al., 1976). These data indicate that the genus contains at least one other, as yet unnamed, species.

The finding that many strains of *Ancylobacter aquaticus* are facultative methylotrophs with a Calvin cycle and the report that several strains can grow as hydrogen autotrophs indicate these bacteria share some properties with methylotrophs and chemoautotrophic bacteria.

"*Renobacter vacuolatum*" Nikitin 1971, a new genus and species proposed for nonmotile, gas-vacuolated bacteria, is now regarded as a member of the genus *Ancylobacter* (Namsaraev, 1973; J. Larkin, personal communication).

The original strain described by Ørskov as the type strain of *A. aquaticus* has been lost (Ørskov, 1953). Subsequently, new strains have been isolated by Ørskov (1953), and Larkin and Borrall (1979) have proposed that one of these strains (ATCC 25396) be the neotype strain.

List of species of the genus Ancylobacter

1. **Ancylobacter aquaticus** (Ørskov 1928) Raj 1983, 297[VP] (*Microcyclus aquaticus* Ørskov 1928, 128.)
   a.qua' ti.cus. L. adj. *aquaticus* living in water.

   The characteristics are as described for the genus. Occur in soil and freshwater.

The mol% G + C of the DNA is: 66–69 (Bd).
Type strain: Ørskov, ATCC 25396.
GenBank accession number (16S rRNA): M62790.

Genus IV. **Angulomicrobium** Vasilyeva, Lafitskaya and Namsaraev 1986, 354[VP] (Effective publication: Vasilyeva, Lafitskaya and Namsaraev 1979, 1037)

LEANA V. VASILYEVA

An' gu.lo.mi.cro' bi.um. M.L. fem. n. *angularis* angular; Gr. adj. *micros* small; Gr. masc. n. *bios* life; M.L.neut. n. *Angulomicrobium* angular microbe.

Unicellular bacterium, having polygonal cells with radial symmetry, ranging in dimensions from 1.1–1.5 µm. The shape of the cells is tetrahedral or mushroom-like. Flat, triangular bacteria are provisionally included within the genus.

Cells divide by budding. Buds are produced on the tetrahedron directly on the conical point of elongation of the mother cell, with a shot tube connecting two cells. **Gram negative. Nonmotile.** Cells lack prosthecae, lamellated membranous structures, and gas vacuoles.

Obligately aerobic, nonfermentative, **chemoorganotrophic**. A variety of organic acids, monosaccharides, and amino acids are utilized. Methanol and formate can serve as energy sources only in the presence of yeast extract. Methane and hydrogen are not utilized. **Catalase and oxidase positive.** Monotypic.

The mol% G + C of the DNA is: 64.3–68.

Type species: **Angulomicrobium tetraedrale** Vasilyeva, Lafitskaya and Namsaraev 1986, 354 (Effective publication: Vasilyeva, Lafitskaya and Namsaraev 1979, 1037.)

FURTHER DESCRIPTIVE INFORMATION

Budding bacteria that resemble a "mushroom" during some stages of the growth cycle have been isolated from aquatic environments (Whittenbury and Nicoll, 1971; Namsaraev and Zavarzin, 1972; Lafitskaya and Vasilyeva, 1976; Stanley et al., 1976). Slide cultures indicate that the bacteria reproduce by a budding process. Newly divided organisms are rounded on one side, while the area where cell separation occurs appears conical. This conical section elongates to form a tube, and at this stage, the cell outline resembles a mushroom. The growing tube then enlarges. Just before cell division, the mother and daughter cells are of equal size. The region between the mother and daughter cells constricts, and the two cells separate. Both mother and daughter

cells synchronously produce lateral buds at the point of separation (see Fig. BXII.α.198).

The same cycle is observed for the flat, triangular bacteria. After separation, growth begins by elongation of the apex of the triangle at the point of division, and new buds are formed synchronously by the mother and daughter cells (Vasilyeva et al., 1979).

The cell wall structure is typical of Gram-negative bacteria, but the flat triangular form, in contrast to the form of the type strains, has no visible rigid layer in its cell wall. No membranous structures are observed, except for the small loops of membranes close to the cell wall (Fig. BXII.α.199).

All strains are capable of growth in the mineral medium with glucose or a variety of other carbohydrates as sole sources of energy and organic carbon. None of the strains grow anaerobically, with or without nitrate or sulfate. Sugars are used via the hexose–monophosphate and Entner–Doudoroff pathways.

Strain Z-2821 can grow with methanol or formate as a sole energy source, but in this case a limited amount of yeast extract should be added (Namsaraev and Zavarzin, 1974). Acetate enhances methylotrophic growth. Hydrolytic activity is absent (Lafitskaya and Vasilyeva, 1976).

DNA hybridization reassociation shows no relation of the mushroom-shaped bacteria to *Hyphomicrobium*, *Rhodopseudomonas*, *Ancalomicrobium*, or *Prosthecomicrobium* (Stanley et al., 1976).

The flat, triangular strain 1109 differs from the type strains and might be representative of another taxon. In particular, the shape of its flat cells suggests it may be closer to the prosthecobacteria *Stella* and *Labrys*. However, no further work has been done, and it is premature to establish a new species until additional isolates have been characterized.

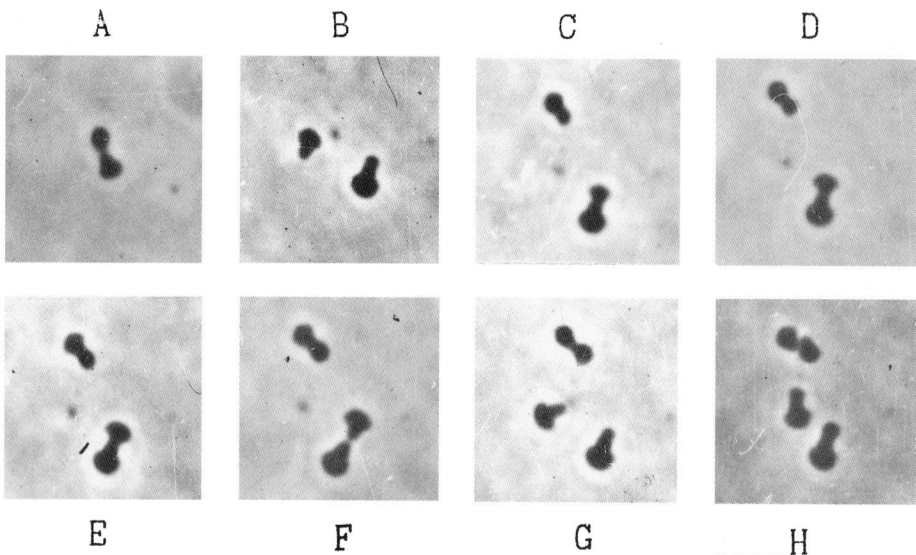

FIGURE BXII.α.198. *A. tetraedrale.* Sequential series of phase photomicrographs illustrating the growth cycle of the type strain. Time in hours: *A*, 0; *B*, 0.5; *C*, 1; *D*, 2; *E*, 2.5; *F*, 3.5; *G*, 4; and *H*, 4.5. Bar = 5 μm. (Reproduced with permission from L.V. Vasilyeva et al., Microbiologiya *48*: 1033–1039, 1979.)

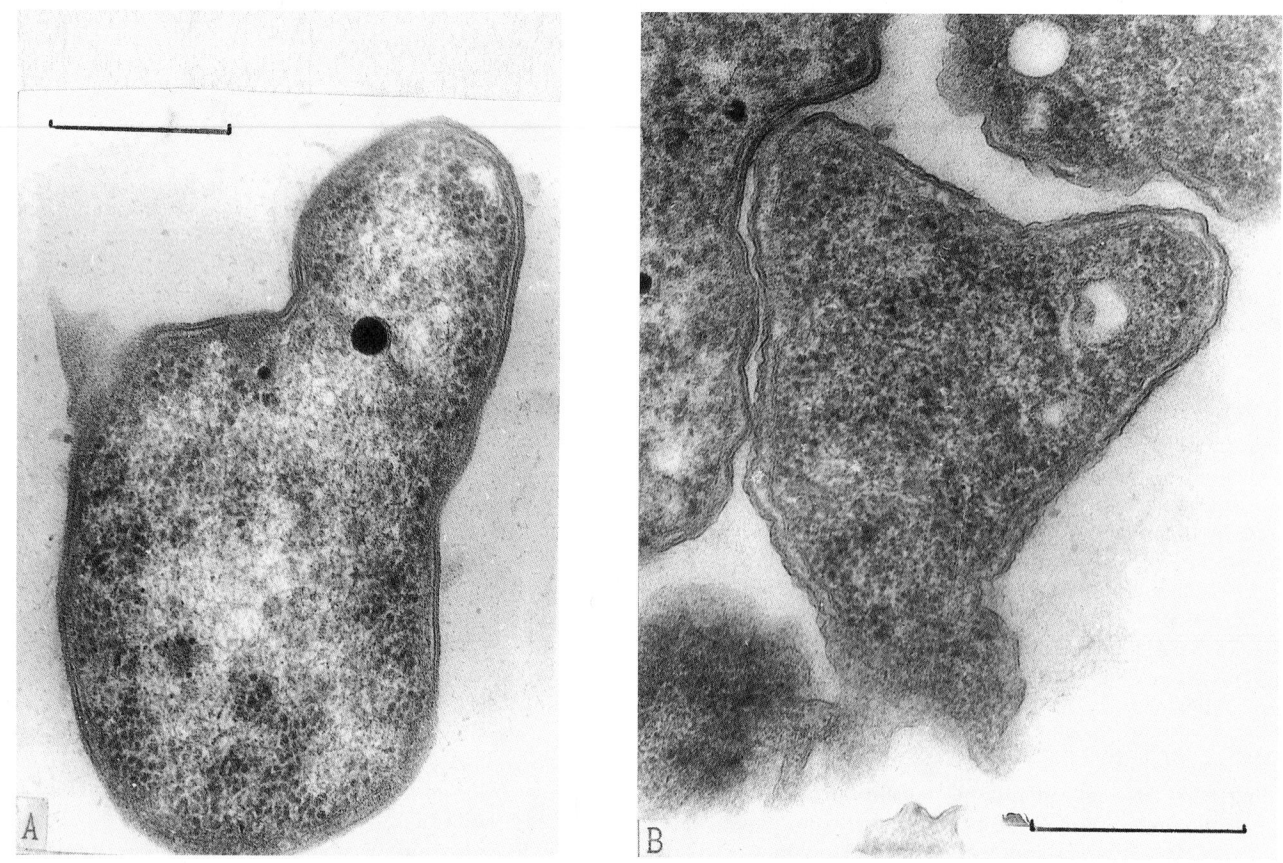

FIGURE BXII.α.199. Thin section of *Angulomicrobium*. *A*, type strain AUCM B-1335. Bar = 0.5 μm. *B*, flat triangular AUCM B-1336. Bar = 0.5 μm.

ENRICHMENT AND ISOLATION PROCEDURES

No selective procedure for enrichment is known. Strains have been isolated as colonies on a mineral base medium supplemented by low concentrations of organic substances, e.g., glucose. Numerous colonies are examined in wet mounts by phase-contrast microscopy, and those containing cells with unusual morphology isolated. The type strain was isolated as a satellite in a methane-oxidizing community from a swamp (Namsaraev and Zavarzin, 1972). Another strain has been isolated from freshwater (Whittenbury and Nicoll, 1971). A pulp mill aeration lagoon has been found to contain up to 10⁶ cells/ml (Stanley et al., 1976). The flat triangle was isolated from a peat bog in the Moscow River valley (Lafitskaya and Vasilyeva, 1976).

MAINTENANCE PROCEDURES

Strains can be maintained on slants in the refrigerator for at least 6 months. Potato infusion agar and mineral base medium (e.g., MMB agar, see *Ancalomicrobium*), with glucose or other sugars, are favorable growth media. Lyophilization can be used for long term preservation.

DIFFERENTIATION OF THE GENUS *ANGULOMICROBIUM* FROM OTHER GENERA

Angulomicrobium is differentiated from other genera of budding bacteria by its characteristic morphology. It differs from *Labrys* by the absence of prosthecae, and it differs from *Nitrobacter* and similar budding bacteria by the absence of lamellated membranous structures. The flat, triangular strain 1109 might be compared with the flat, dividing bacteria with radial symmetry of the genus *Stella*.

TAXONOMIC COMMENTS

Angulomicrobium represents a distinct morphological type in the oligocarbophilic "microflora of dispersal" (Zavarzin, 1970), or among the dissipotrophic microorganisms, as it is called now (Vasilyeva and Zavarzin, 1995). It utilizes either carbohydrates in low concentrations or methanol produced by incomplete oxidation of methane in methane-utilizing communities. Substrate utilization is similar for all strains, except the flat, triangular strain. It is not related to other genera of budding bacteria by DNA–DNA similarity studies. On the basis of DNA base composition (mol% G + C of the DNA = 68) the strains of *Angulomicrobium* cannot be closely related to other budding bacteria, such as *Gemmiger formicilis* (mol% G + C of the DNA = 59), *Nitrobacter* (mol% G + C of the DNA = 61–62), or *Hyphomicrobium* (mol% G + C of the DNA = 59–61); they are close to *Rhodopseudomonas palustris* (mol% G + C of the DNA = 65–66), but do not possess intracytoplasmic membranous structures (Stanley et al., 1976). DNA–DNA homology studies indicate a relatedness among all four strains of *Angulomicrobium* in the range of 40–85% (Stanley et al., 1976).

Sequencing 16S rRNA studies indicate that *Angulomicrobium tetraedrale* (strain WAL-4, Stanley et al., 1976) belongs to the *Alphaproteobacteria* and is most closely related to *Rhodopseudomonas acidophila* (now *Rhodoblastus acidophilus* and *R. palustris* (Vasilyeva, 1989).

List of species of the genus Angulomicrobium

1. **Angulomicrobium tetraedrale** Vasilyeva, Lafitskaya and Namsaraev 1986, 354[VP] (Effective publication: Vasilyeva, Lafitskaya and Namsaraev 1979, 1037.)

   tet' ra.ed' ra.le. Gr. pref. *tetra-* four; Gr. n. *edra* seat, face; Gr. adj. *tetraedralis* tetrahedral.

   Unicellular, budding bacterium. The cells have radial symmetry; cells of the type strain have a tetrahedral form, 1.1–1.5 µm (Fig. BXII.α.200); cells of other strains (1109) have a flat form, 0.95–1.2 µm (Fig. BXII.α.201). Single or in pairs. Gram negative. Nonmotile.

   Multiplication is by budding from the top of the tetrahedron (Fig.BXII.α.198) or from the top of the triangle. Mother and daughter cells are similar in form and size. Ultrastructure typical of Gram-negative bacteria, with no complex membranous system (Fig. BXII.α.199).

   Colonies are white, round, and mucous with a pearly shine.

   Chemoorganotroph. Carbohydrates, amino acids, organic acids, and alcohols are utilized as sole sources of carbon and energy. All strains utilize arabinose, glucose, L-histidine, L-proline, acetate, and mannitol. Some strains utilize mannose, ribose, xylose, propionate, citrate, ethanol, methanol, and methylamine; no strains utilize lactose, raffinose, pectin, glycogen, gelatin, L-tryptophan, L-arginine, L-leucine, DL-methionine, or Tween-80.

   Only the flat strain 1109 utilizes sucrose, cellobiose, and melibiose, but this strain does not utilize methanol, organic acids, or glycerol. Organic acids are not produced from carbohydrates.

   Sugars are used via the hexose–monophosphate and Entner–Doudoroff pathways. Type strain has enzymes of the Embden–Meyerhof pathway (Lafitskaya and Vasilyeva, 1976).

   Aerobic. Catalase and oxidase positive.

   Organic growth factors are not required, but they stimulate growth. Strain 1109 requires yeast extract.

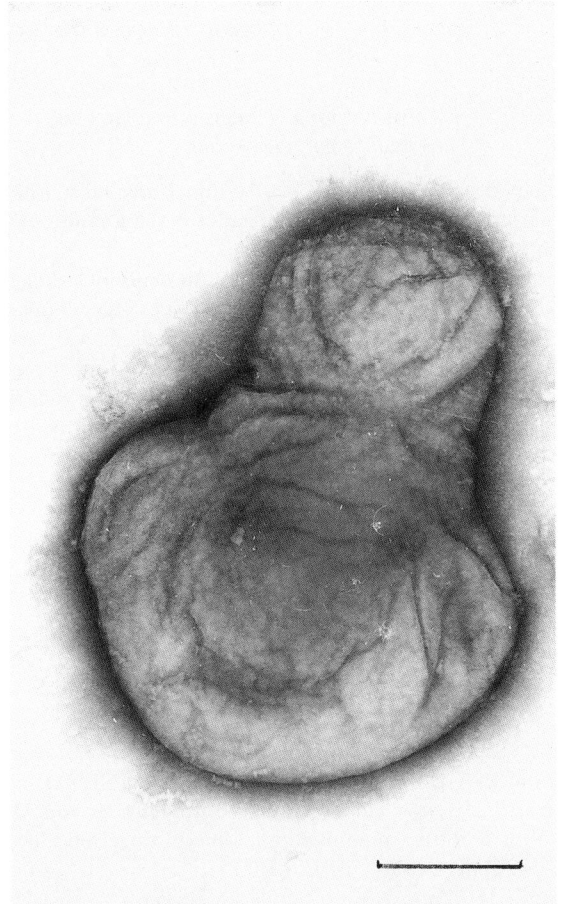

FIGURE BXII.α.200. Cell of type strain AUCM B-1335 with phosphotungstic acid. Bar = 0.5 µm. (Reproduced with permission from E.N. Michustin, Izvestiya Akademii Nauk S.S.S.R., Seriya Biologicheskaya *5*: 719–737, 1980.)

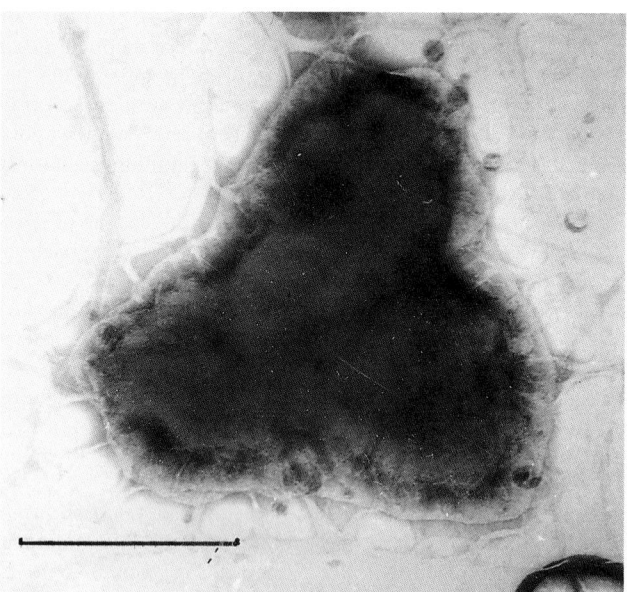

FIGURE BXII.α.201. Cell strain AUCM B-1336 with phosphotungstic acid. Bar = 0.5 μm. (Reproduced with permission from E.N. Michustin, Izvestiya Akademii Nauk S.S.S.R., Seriya Biologicheskaya *5:* 719–737, 1980.)

Temperature range: 15–35°C; optimal temperature: 28–30°C. Optimal pH for growth: near neutrality (6.8–7.0).

Sensitivity towards antibiotics (μg/ml) for type strain and 1109: penicillin G, 30 and 100, respectively; chloramphenicol, 100 and 0.5; erythromycin, 10 and 30; actinomycin D, 5.0 and 8.0; and polymyxin M, 50 and 10.

Habitat: type strain isolated from a swamp near Moscow. Other related strains isolated from a freshwater and pulp mill aeration lagoon.

The mol% G + C of the DNA is: 64.5 (T_m, type strain) (68.2 according to Stanley et al., 1976); 64.3 (strain 1109).

Type strain: Strain Z-2821, AUCM B-1335, DSM 5895.

GenBank accession number (16S rRNA): AS535708.

Genus V. **Aquabacter** *Irgens, Kersters, Segers, Gillis and Staley 1993, 864ᵛᴾ (Effective publication: Irgens, Kersters, Segers, Gillis and Staley 1991, 141)*

JAMES T. STALEY

Aq′ua.bac.ter. L. n. *aqua* water; Gr. neut. n. *bacter* masc. form of Gr. neut. n. *bacterion* a rod; N.L. masc. n. *Aquabacter* aquatic rod.

Rod-shaped cells, 0.5–1.0 × 1.5–3.0 μm. **Unicellular.** Encapsulated. **Motile by flagella, but only under certain conditions.** No resting stages known. **Gram negative.**

Aerobic. Chemoheterotrophic. Organic acids including some amino acids used aerobically as carbon source. Sugars not known to be used. Growth occurs on mineral medium with acetate, succinate, or pyruvate as sole carbon source and inorganic ammonium compounds as sole nitrogen source, along with B-vitamins. Oxidase and catalase positive. Type strain, which is gas vacuolate, was isolated from Spirit Lake, Washington following the eruption of Mt. St. Helens in 1980.

The mol% G + C of the DNA is: 67.

Type species: **Aquabacter spiritensis** Irgens, Kersters, Segers, Gillis and Staley 1993, 864 (Effective publication: Irgens, Kersters, Segers, Gillis and Staley 1991, 141.)

FURTHER DESCRIPTIVE INFORMATION

The 16S rDNA sequence has not yet been determined for the type species.

This monospecific genus belongs to the *Alphaproteobacteria* based on DNA–rRNA hybridization (Irgens et al., 1991). Within that group, it is most closely related to *Azorhizobium* and *Xanthobacter,* but it differs phenotypically and phylogenetically from these two genera (Table BXII.α.167).

Aquabacter spiritensis is not motile when grown in typical broth media. However, flagellation occurs when cells are grown on a semisolid medium as described by Lara and Konopka (1987).

Gas vacuoles are produced by the type strain when grown on typical broth or solid media.

The type strain uses a limited number of carbon sources. Some organic acids such as lactate, pyruvate, acetate, malate, succinate, propionate, and butyrate, as well as L-glutamate and L-alanine, are used as carbon sources in complex media. However, none of the sugars or alcohols that have been tested (D-glucose, D-fructose, DL-arabinose, rhamnose, sucrose, maltose, lactose, mannitol, sorbitol, glycerol, ethanol) are used as carbon sources for growth.

ENRICHMENT AND ISOLATION PROCEDURES

The sole strain of this genus was isolated from Spirit Lake, Washington, following the eruption of Mt. St. Helens volcano in 1980. A sample collected at 5.0 m depth in the lake on 10 April 1981, a little more than a year after the eruption, was diluted to 10^{-4} and plated out on dilute peptone agar containing (per liter): Bacto Peptone, 100 mg; Hutner's salts solution (see *Ancalomicrobium,* this volume), 20 ml; agar, 15 g. After prolonged incubation at room temperature, only one colony had the chalky white appearance typical of gas vacuolate bacteria.

MAINTENANCE PROCEDURES

The organism can be maintained by lyophilization.

DIFFERENTIATION OF THE GENUS *AQUABACTER* FROM OTHER GENERA

Table BXII.α.167 shows the differences between *Aquabacter, Xanthobacter,* and *Azorhizobium. Aquabacter spiritensis* does not use the typical carbon sources used by *Azorhizobium* and *Xanthobacter* spp. In addition, although all three genera resemble one another in morphology, the type strain and species of *Aquabacter* is gas vacuolate.

TABLE BXII.α.167. Differentiation of the genera *Aquabacter*, *Azorhizobium*, and *Xanthobacter*

Characteristics	*Aquabacter*	*Azorhizobium*	*Xanthobacter*
Motility	+[a]	+	D
Colony color	chalky white	cream	yellow
Gas vacuole formation	+[b]	−	−
Utilization of:			
Glucose	−	+	3/16[c]
DL-Proline	−	19/20[c]	3/16[c]
Methanol, ethanol	−	−	15/16[c]
Gluconate	−	+	12/16[c]
L-Lysine	−	+	−

[a]Motility is not ordinarily observed (see text).

[b]Only one strain has been described, and it is gas vacuolate.

[c]Fractions indicate the number of positive strains of the total number of strains tested.

List of species of the genus Aquabacter

1. **Aquabacter spiritensis** Irgens, Kersters, Segers, Gillis and Staley 1993, 864[VP] (Effective publication: Irgens, Kersters, Segers, Gillis and Staley 1991, 141.)

spi.ri.ten' sis. N.L. adj. *spiritensis* named after Spirit Lake, Washington, USA, from which the strain was isolated.

Gas vacuolate, nonflagellated rods. Typically nonmotile, however, flagella are produced under special conditions.

Obligately aerobic. Growth occurs at 25–37°C, but not at 20 or 40°C. Nitrate is reduced to nitrite. Nitrogen is not fixed. B-vitamins are required for growth on defined media. The following carbon sources are utilized: acetate, succi-

nate, pyruvate, lactate, propionate, butyrate, L-glutamate, and L-alanine.

Colonies are circular, slightly convex in elevation, with an entire margin and smooth, glistening surface. Colonies may appear translucent or chalky white if cells are gas vacuolate. Growth on older slant cultures has a rubbery texture.

The type strain was isolated from Spirit Lake, WA USA.

The mol% G + C of the DNA is: 67 (T_m).

Type strain: SPL-1, ATCC 43981, DSM 9035, LMG 8611.

Genus VI. *Azorhizobium* Dreyfus, Garcia and Gillis 1988, 89[VP]

L. DAVID KUYKENDALL

A.zo.rhi.zo' bi.um. Fr. n. *azote* nitrogen; M.L. neut. n. *Rhizobium* a bacterial generic name; M.L. neut. n. *Azorhizobium* a nitrogen (using) rhizobium.

Rods 0.5–0.6 × 1.5–2.5 μm. Motile by one polar or subpolar flagellum. **Nonsporeforming. Gram negative. Peritrichous flagella are formed on solid media, and a single lateral flagellum is formed in broth.** Growth temperature, 12–43°C. Broad optimum pH, 5.5–7.8. Colonies are circular and creamy. Urease negative. **Oxidase and catalase positive. Do not denitrify. Of sugars, only glucose is oxidized.** The organic acids lactic acid and succinic acid are preferred substrates for growth, both under and not under nitrogen-fixing conditions. Growth on malonate and DL-proline. Starch is not hydrolyzed. Nitrogen-fixing root- and stem-nodulating microsymbiont of *Sesbania rostrata*. Readily **fixes nitrogen *ex planta* under microaerobic conditions and with nicotinic acid provided.** Can grow on nitrogen-free medium, unlike all other legume microsymbionts, of the genera *Bradyrhizobium*, *Rhizobium*, and *Mesorhizobium*, which lie in distinct families.

The mol% G + C of the DNA is: 66–68.

Type species: **Azorhizobium caulinodans** Dreyfus, Garcia, and Gillis 1988, 89.

FURTHER DESCRIPTIVE INFORMATION

Nitrogen-fixing stem nodules are formed on *Sesbania rostrata* by *Azorhizobium caulinodans* ORS 571[T] (Fig. BXII.α.202). *Azorhizobium* Nod factors, responsible for the induction of nodule organogenesis, have been characterized by Mergaert et al., 1993. The nodulation genes, *nod*, have been studied by Goethals et al., (1989, 1990, 1992a, b), and by Geelen et al., (1993). A small rod-shaped cell of *Azorhizobium caulinodans* ORS 571[T] with a single lateral flagellum is depicted in Fig. BXII.α.203.

ENRICHMENT AND ISOLATION PROCEDURES

Azorhizobia are readily isolated from stem nodules on *Sesbania rostrata*, using the same procedures described for the isolation of *Bradyrhizobium* from soybean.

MAINTENANCE PROCEDURES

Azorhizobium will grow on a chemically defined minimal medium without added nitrogen. LO medium was defined by Dreyfus et al., (1983). Storage recommendations are the same as given for *Rhizobium*.

PROCEDURES FOR TESTING SPECIAL CHARACTERS

For verification of *Sesbania* nodulation, plant seeds are both scarified and surface-sterilized by immersion in concentrated H_2SO_4 for 15–30 min, then washed several times with sterile water and germinated on semisolid agar. Seedlings are placed in 25 × 150-mm plugged culture tubes on agar slopes of Jensen seedling agar (Vincent, 1970).

DIFFERENTIATION OF THE GENUS *AZORHIZOBIUM* FROM OTHER GENERA

Growth on a nitrogen-free medium (Dreyfus et al., 1988) clearly distinguishes *Azorhizobium* from the other, not very closely related, legume-nodulating bacteria *Bradyrhizobium*, *Mesorhizobium*, and *Rhizobium*. *Azorhizobium* is distinguished from close relatives like *Xanthobacter*, *Aquabacter*, and other members of the *Hyphomicrobiaceae* by the inability of these species to form nitrogen-fixing nodules on *Sesbania*.

FIGURE BXII.α.202. Stem nodules formed by *Azorhizobium caulinodans* ORS 1[T] on *Sesbania rostrata* (Reproduced with permission from B. Dreyfus et al., International Journal of Systematic Bacteriology *38*: 89–98, 1988, ©International Union of Microbiological Societies.)

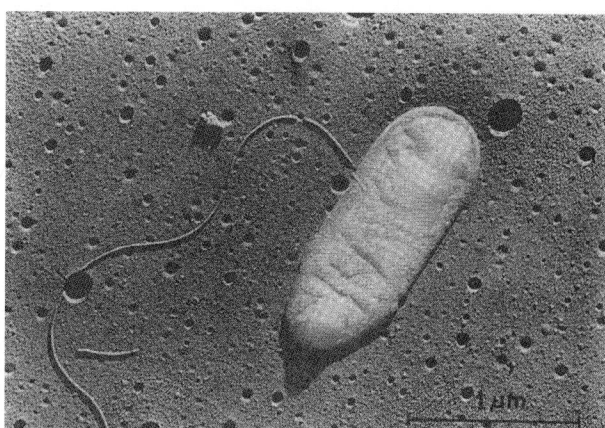

FIGURE BXII.α.203. A single lateral flagellum attached to a cell of *Azorhizobium caulinodans* ORS 1[T], as shown by negative staining and electron micrography of a diluted broth culture (Reproduced with permission from B. Dreyfus et al., International Journal of Systematic Bacteriology *38*: 89–98, 1988, ©International Union of Microbiological Societies.)

TAXONOMIC COMMENTS

At the time of the first edition of *Bergey's Manual of Systematic Bacteriology*, the bacteria that form stem nodules on *Sesbania rostrata* had not yet been named, *Bradyrhizobium* was still within the family *Rhizobiaceae*, and the only three *Rhizobium* species were *R. leguminosarum*, *R. meliloti*, and *R. loti*. *R. loti* is now in the separate genus *Mesorhizobium*. Since the microsymbionts of *Sesbania rostrata* (Dreyfus et al., 1983) are distinctly different from *Bradyrhizobium* and *Rhizobium*, both genotypically and phenotypically, Dreyfus et al. (1988) placed them in a new genus. FAME and 16S rRNA analysis have been used to confirm the clear separation of *Azo-*

rhizobium from both *Bradyrhizobium* and *Rhizobium* (So et al., 1994). Van Rossum et al. (1995) have also reconfirmed their distinct lineages. Because of their significant phylogenetic dissimilarity with the *Rhizobiaceae* and with the *Bradyrhizobiaceae*, it is appropriate that these organisms be in a distinct family, *Hyphomicrobiaceae*, based on 16S rRNA similarity.

Only one species of *Azorhizobium* is presently recognized. *Azorhizobium* has 98% 16S rDNA similarity with, and is thus very closely related to, *Xanthobacter* (Rainey and Wiegel, 1996) and these genera will probably be amalgamated, as might *Aquabacter*, since the latter shares approximately 97% 16S rDNA similarity with both *Azorhizobium* and *Xanthobacter*.

ACKNOWLEDGMENTS

The author is indebted to Dr. B. Dreyfus for permission to reprint the photographs.

List of species of the genus Azorhizobium

1. **Azorhizobium caulinodans** Dreyfus, Garcia, and Gillis 1988, 89.

    cau.li' no.dans. M.L. n. *caulis* stem; M.L. v. *nodare* to nodulate; M.L. part. adj. *caulinodans* stem-nodulating.

    The characteristics are as described for the genus. Strains effectively nodulate the roots and stems of *Sesbania rostrata*. Grow with 8% potassium nitrate present. Possess arginine and lysine decarboxylase. Do not utilize D-mannitol. Grow on azelate, maleate, adipate, pimelate, suberate, gluconate, mucate, crotonate, nicotinate, 2-ketogluconate, propionate,

butyrate, isobutyrate, valerate, isovalerate, caproate, laurate, 2-ketoglutarate, fumarate, glutarate, sebacate, DL-malate, citrate, pyruvate, aconitate, citraconitate, D-glucuronate, α-D-galacturonae, *m*-hydroxybenzoate, L-aspartate, quinate, L-α-alanine, L-lysine, L-asparagine, betaine, and sarcosine. Nicotinic acid must be supplied as a vitamin supplement for growth under nitrogen-limiting conditions.

    The mol% G + C of the DNA is: 66–68 (T_m).

    Type strain: ORS 571, LMG 6465.

    GenBank accession number (16S rRNA): D13948, D11342, X67221, X94200.

Genus VII. **Blastochloris** Hiraishi 1997, 218[VP]

JOHANNES F. IMHOFF

Blas.to.chlo' ris. Gr. n. *blastos* bud shoot; Gr. adj. *chloros* green; M.L. fem. n. *Blastochloris* green bud shoot.

Cells are rod shaped to ovoid and motile by means of subpolar flagella. They exhibit polar growth, budding, and asymmetric cell division and form rosette-like cell aggregates. They are **Gram negative and belong to the *Alphaproteobacteria*.** Internal photosynthetic membranes are present as lamellae underlying and parallel to the cytoplasmic membrane. Photosynthetic pigments are **bacteriochlorophyll *b* and carotenoids. Straight-chain monounsatu-**

rated $C_{18:1}$ **is the predominant component of cellular fatty acids. Ubiquinones and menaquinones are present,** and the lipopolysaccharides are characterized by a 2,3-diamino-2,3-deoxy-D-glucose (DAG)-containing, phosphate-free lipid A with amide-bound $C_{14:0\ 3OH}$.

Preferred growth mode is photoheterotrophically under anoxic conditions in the light. **Photoautotrophic growth may be**

possible under anoxic conditions with thiosulfate or sulfide as electron donor. **Chemotrophic growth is possible** under microoxic conditions in the dark. Growth factors may be required.

Mesophilic freshwater bacteria with a preference for neutral pH.

The mol% G + C of the DNA is: 66.3–71.4.

Type species: **Blastochloris viridis** (Drews and Giesbrecht 1966) Hiraishi 1997, 218 (*Rhodopseudomonas viridis* Drews and Giesbrecht 1966, 261.)

FURTHER DESCRIPTIVE INFORMATION

The cell morphology of the *Blastochloris* species is characterized by asymmetric growth and cell division. Most characteristic is the formation of prosthecae and rosette-like cell aggregates as in *Rhodopseudomonas* species.

Blastochloris sulfoviridis is unable to assimilate sulfate and depends on reduced sulfur compounds (Neutzling and Trüper, 1982), whereas *Blastochloris viridis* can grow with sulfate as the sole sulfur source and assimilates sulfate via the 3'-phospho-adenosine-5'-phosphosulfate pathway (Imhoff, 1982). Reduced sulfur compounds, sulfide, and thiosulfate are used as photosynthetic electron donors by *B. sulfoviridis*, but not *B. viridis*, and oxidized to sulfate (Keppen and Gorlenko, 1975). The lipid A structure of the lipopolysaccharide of *Blastochloris viridis* and *Blastochloris sulfoviridis* (Ahamed et al., 1982; Weckesser et al., 1995) contains the unusual 2,3-diamino-2,3-deoxy-D-glucose instead of glucosamine. This sugar has also been found in other species of the *Alphaproteobacteria* such as *Rhodopseudomonas palustris* and *Nitrobacter winogradskyi*, but not *Rhodomicrobium vannielii* and *Rhodoblastus acidophilus* (Weckesser et al., 1995).

ENRICHMENT AND ISOLATION PROCEDURES

Standard procedures for the isolation of purple nonsulfur bacteria (PNSB) can be applied. A mineral salts medium, which is suitable for most PNSB, is also applicable for the cultivation of *Blastochloris* species (Imhoff, 1988; Imhoff and Trüper, 1992; see chapter Genus *Rhodospirillum* for this medium recipe). *B. viridis* and *B. sulfoviridis* can be selectively enriched with appropriate light filters, that allow only long wavelength radiation to penetrate, since both species, owing to their content of bacterio-

chlorophyll *b*, show absorption maxima above 1000 nm. The dependence on reduced sulfur compounds demands the addition of low concentrations of sulfide or cysteine into media for *B. sulfoviridis*.

MAINTENANCE PROCEDURES

Cultures of *Blastochloris* species can be preserved in liquid nitrogen or at −80°C in a mechanical freezer.

DIFFERENTIATION OF THE GENUS *BLASTOCHLORIS* FROM OTHER GENERA

16S rDNA sequence analysis places the genus in the *Alphaproteobacteria*. *Rhodoplanes* species are the closest relatives among phototrophic bacteria. *Blastochloris* species may well be differentiated from other PNSB based on morphological and cultural characteristics. Their green to olive-green coloration, together with the characteristic long wavelength absorption maximum above 1000 nm, which is due to bacteriochlorophyll *b*, clearly separates *Blastochloris* species from other phototrophic *Alphaproteobacteria*. The rod-shaped cells, their asymmetric mode of growth and cell division, the formation of rosette-like aggregates, the lamellar structure of internal photosynthetic membranes lying parallel to and underlying the cytoplasmic membrane, as well as a number of chemotaxonomic characteristics and the sequences of 16S rDNA, differentiate *Blastochloris* species from other genera of the purple nonsulfur bacteria (see Tables 3 and 4 [pp. 125–127] as well as Fig. 2 [p. 128] of the introductory chapter "Anoxygenic Phototrophic Purple Bacteria" in Volume 2, Part A).

TAXONOMIC COMMENTS

Both species of this genus have been previously assigned to the genus *Rhodopseudomonas*, because of significant similarities in cell morphology and cell division to *Rhodopseudomonas palustris*. Significant differences in 16S rDNA sequences with respect to other PNSB have given reason to reconsider the taxonomic position of these bacteria and reevaluate phenotypic differences from the type species of *Rhodopseudomonas*, leading to their reassignment to the new genus *Blastochloris* as *Blastochloris viridis* and *Blastochloris sulfoviridis* (Kawasaki et al., 1993b; Hiraishi, 1997).

DIFFERENTIATION OF THE SPECIES OF THE GENUS *BLASTOCHLORIS*

Major differentiating properties between *Blastochloris* species and between *Blastochloris*, *Rhodomicrobium vannielii*, and *Rhodoplanes* species are shown in Tables BXII.α.168 and BXII.α.169.

List of species of the genus Blastochloris

1. **Blastochloris viridis** (Drews and Giesbrecht 1966) Hiraishi 1997, 218[VP] (*Rhodopseudomonas viridis* Drews and Giesbrecht 1966, 261.)

vi'ri.dis. L. adj. *viridis* green.

Cells are rod-shaped to ovoid, 0.6–0.9 × 1.2–2.0 μm. The mother cell produces a slender prostheca, 1.5–2.0 times the length of the original cell, at the pole opposite that bearing the flagella. The end of the prostheca swells, and the daughter cell grows, producing a dumbbell-shaped organism, just like *Rhodopseudomonas palustris* (see Fig. BXII.α.187 in the chapter describing the species *Rhodopseudomonas palustris*). Asymmetric division then takes place. In young cultures, swarmer cells motile by means of subpolar flagella are frequent. The formation of rosettes and clusters in which the individual cells are attached to each other at their flagellated poles are characteristic in older

cultures. Internal photosynthetic membranes are present as lamellae underlying and parallel to the cytoplasmic membrane. The color of photosynthetic cultures is first yellowish green, then green to olive green. Aerobic cultures are colorless to light yellowish green. *In vivo* absorption spectra show characteristic maxima at 400, 420, 451, 483, 604, 835, and 1020 nm. Photosynthetic pigments are bacteriochlorophyll *b* (esterified with phytol), with 1,2-dihydroneurosporene and 1,2-dihydrolycopene as major carotenoids.

Cells grow photoheterotrophically under anoxic conditions with organic carbon sources. Photoautotrophic growth has not been demonstrated. Chemotrophic growth is possible under microoxic conditions in the dark. Carbon sources used are acetate, pyruvate, malate, and succinate. Poor growth occurs in the presence of ethanol, glutamate, peptone, glucose, and xylose. Longer chain fatty acids in-

TABLE BXII.α.168. Differential characteristics of the anoxygenic phototrophic purple bacteria belonging to the family *Hyphomicrobiaceae* of the order *Rhizobiales*: genera *Blastochloris*, *Rhodomicrobium*, and *Rhodoplanes*[a]

Characteristic	Blastochloris viridis	Blastochloris sulfoviridis	Rhodomicrobium vannielii	Rhodoplanes roseus	Rhodoplanes elegans
Cell diameter (μm)	0.6–0.9	0.5–0.9	1.0–1.2	1.0	0.8–1.0
Type of budding	Tube	Sessile	Tube	Sessile	Tube
Rosette formation	+		Complex aggregates	−	+
Internal membrane system	Lamellae	Lamellae	Lamellae	Lamellae	Lamellae
Motility	+	+	+	+	+
Color of cultures	Green to olive-green	Olive-green	Orange-brown to red	Pink	Pink
Bacteriochlorophyll	*b*	*b*	*a*	*a*	*a*
Salt requirement	None	None	None	None	None
Optimal pH	6.5–7.0	7.0	6.0	7.0–7.5	7.0
Optimal temperature	25–30	28–30	30	30	30–35
Sulfate assimilation	+ (PAPS)	−	+ (APS)	nd	nd
Aerobic dark growth	(+)	(+)	+	+	+
Denitrification	−	−	nd	+	+
Fermentation of fructose	nd	nd	nd	−	−
Photoautotrophic growth with	−	Thiosulfate, sulfide	H₂, sulfide	Thiosulfate	Thiosulfate
Growth factors	Biotin, *p*-amino-benzoic acid	Biotin, *p*-amino-benzoic acid, pyridoxin	None	Niacin	Thiamine, *p*-amino-benzoic acid
Utilization of:					
Benzoate	−	−	−	−	−
Citrate	−	−	−	+	+
Formate	−	−	+/−	−	−
Glucose	(+)	+	−	−	−
Tartrate	−	−	−	+	+
Sulfide	−	+	+	−	−
Thiosulfate	−	+		+	+
Mol % G + C of the DNA	66.3–71.4 (Bd)	67.8–68.4 (CA)	61.8–63.8 (Bd)	66.8 (HPLC)	69.6–69.7 (HPLC)
Cytochrome *c*₂ size	Small	nd	Small	nd	nd
Major quinones	Q-9, MK-9	Q-8/10, MK-7/8	Q-10, RQ-10	Q-10, RQ-10	Q-10, RQ-10
Major fatty acids					
C₁₄:₀	0.5	2.5	2.4		
C₁₆:₀	8.4	8.6	3.7		
C₁₆:₁	5.5	9.2	0.6		
C₁₈:₀	2.2	1.7	3.6		
C₁₈:₁	74.6	76.5	85.6		

[a]+, positive in most strains; −, negative in most strains; +/−, variable in different strains; nd, not determined; (+), weak growth or microaerobic growth only; (APS), via adenosine-5′-phosphosulfate; (PAPS), via 3′-phosphoadenosine-5′-phosphosulfate; (biotin) biotin is required by some strains; Q-10, ubiquinone 10; MK-10, menaquinone 10; RQ-10, rhodoquinone 10. Bd, buoyant density; CA, chemical analysis.

hibit growth. Sulfide, thiosulfate, and hydrogen cannot be used as electron donors. Sulfate can be used as the sole sulfur source. Ammonia and dinitrogen are used as nitrogen sources; nitrate is not used. Most strains require biotin and *p*-aminobenzoate as growth factors; some strains require more growth factors and others none.

Mesophilic freshwater bacterium with optimal growth at 25–30°C and pH 6.5–7.0.

Ubiquinones and menaquinones with 9 isoprene (Q-9 and MK-9) units are present.

The mol% G + C of the DNA is: 66.3–71.4 (Bd).

Type strain: F, ATCC 19567, DSM 133.

GenBank accession number (16S rRNA): D25314.

2. **Blastochloris sulfoviridis** (Keppen and Gorlenko 1975) Hiraishi 1997, 218[VP] (*Rhodopseudomonas sulfoviridis* Keppen and Gorlenko 1975, 258.)

sul.fo.vi′ri.dis. L. n. *sulfur* sulfur; L. adj. *viridis* green; M.L. adj. *sulfoviridis* green and with sulfur.

Cells are rod-shaped to ovoid and form sessile buds and rosettes. Swarmer cells are motile by means of subpolar flagella. If cells age, they become immotile and encapsulated by slime. Internal photosynthetic membranes are pres-

ent as lamellae underlying and parallel to the cytoplasmic membrane. Color of cell suspensions is olive-green, sometimes with a brownish tinge. Photosynthetic pigments are bacteriochlorophyll *b* and carotenoids.

Growth is possible under anoxic conditions in the light or under microoxic conditions in the dark. Organic carbon sources, thiosulfate, and sulfide are used as photosynthetic electron donors. Sulfide and thiosulfate are oxidized to sulfate. Best growth occurs with glucose, fructose, maltose, sucrose, glycerol, propanol, fumarate, and malate as organic carbon sources. Cystine, cysteine, sulfide, and thiosulfate are used as sulfur sources, but sulfate is not used. Ammonia and casein hydrolysate are used as nitrogen sources, while arginine, glycine, alanine, asparagine, and urea inhibit growth. Yeast extract or biotin, pyridoxine, and *p*-aminobenzoate are required as growth factors.

Mesophilic freshwater bacterium with optimal growth at 28–30°C and pH 7.0.

The mol% G + C of the DNA is: 67.8–68.4 (chemical analysis).

Type strain: Pl, DSM 729.

GenBank accession number (16S rRNA): D86514.

TABLE BXII.α.169. Growth substrates of the anoxygenic phototrophic purple bacteria belonging to the family *Hyphomicrobiaceae* of the order *Rhizobiales*, *Blastochloris*, *Rhodomicrobium*, and *Rhodoplanes*[a]

Source/donor	Blastochloris viridis	Blastochloris sulfoviridis	Rhodomicrobium vannielii	Rhodoplanes roseus	Rhodoplanes elegans
Carbon source					
Acetate	+	+	+	+	+
Aspartate	nd	nd	−	−	−
Benzoate	−	−	−	−	−
Butyrate	−	+	+	+	+
Caproate	−	nd	+	−	+/−
Caprylate	nd	nd	+		
Citrate	−	−	−	+	+
Ethanol	+	+	+	−	−
Formate	−	−	+/−	−	−
D-Fructose	−	+	−	−	−
Fumarate	+	+	+	+	+
D-Glucose	+/−	+	−	−	−
Glutamate	+	nd	−	−	−
Glycerol	−	+	+/−	−	−
Glycolate	nd	nd	−	−	−
Lactate	+/−	+	+	+	+
Malate	+	+	+	+	+
Malonate	−	−	+	−	−
Mannitol	+/−	−	−	−	−
Methanol	−	−	+/−	−	−
Propanol	+/−	+	+	−	−
Propionate	−	−	+	+	+
Pyruvate	+	−	+	+	+
Sorbitol	+	+	−	−	−
Succinate	+	+	+	+	+
Tartrate	+/−	−	−	+	+
Valerate	−	−	+	+	+
Electron donor:					
Sulfide	−	+	+	−	−
Thiosulfate	−	+		+	+

[a]Symbols: +, positive in most strains; −, negative in most strains; +/− variable in different strains; nd, not determined.

Genus VIII. **Devosia** Nakagawa, Sakane and Yokota 1996, 20[VP]

YASUYOSHI NAKAGAWA, TAKESHI SAKANE AND AKIRA YOKOTA

De.vos' i.a. M.L. dim. ending -ia M.L. fem. n. *Devosia* honoring Paul De Vos, a Belgian microbiologist, for his contributions to the taxonomy of pseudomonads.

Rods, 0.4–0.8 × 2.0–8.0 μm. **Motile by means of several polar flagella.** Endospores are not formed. Gram negative. **Aerobic. Oxidase and catalase positive. The major respiratory quinone is ubiquinone-10. The major cellular hydroxy fatty acids are 3-hydroxytetracosenoic acid and 3-hydroxyhexacosenoic acid. 3-Hydroxy fatty acids that are shorter than 3-hydroxyoctadecanoic acid are absent.** Phylogenetically related to members of the *Alphaproteobacteria*.

The mol% G + C of the DNA is: 61.4.

Type species: **Devosia riboflavina** Nakagawa, Sakane and Yokota 1996, 20.

FURTHER DESCRIPTIVE INFORMATION

The following description pertains to the type strain of the type species. Cells grow vigorously on nutrient agar at 30°C. The quinone type is ubiquinone-10 (Q-10), which suggests that the organism belongs to *Alphaproteobacteria*. 2-Hydroxy (2-OH) fatty acids are not detected by thin-layer chromatography analysis. The cells contain 3-hydroxytetracosenoic acid ($C_{24:1\ 3OH}$) and 3-hydroxyhexacosenoic acid ($C_{26:1\ 3OH}$) as the major 3-OH fatty acids, and 3-hydroxyoctadecanoic acid ($C_{18:0\ 3OH}$), 3-hydroxylcosanoic acid ($C_{20:0\ 3OH}$), 3-hydroxylcosenoic acid ($C_{20:1\ 3OH}$), 3-hydroxy-docosanoic acid ($C_{22:0\ 3OH}$), and 3-hydroxydocosenoic acid ($C_{22:1\ 3OH}$) as minor components. The long-chain acids $C_{24:1\ 3OH}$, $C_{26:1\ 3OH}$, and $C_{22:1\ 3OH}$, as well as octadecanoic acid ($C_{18:0}$), octadecenoic acid ($C_{18:1}$), and hexadecanoic acid ($C_{16:0}$) as fatty acid components and glucose, galactose, mannose, and glucosamine as sugar components, exist in the lipopolysaccharides (LPS) purified from the cells by extraction with phenol–chloroform–petroleum ether. Thus, the long-chain 3-OH fatty acids are components of the LPS molecule, which is an important outer membrane component. No other bacteria that contain $C_{24:1\ 3OH}$ and $C_{26:1\ 3OH}$ as major components have been described.

ENRICHMENT AND ISOLATION PROCEDURES

Devosia riboflavina was originally isolated from riboflavin-rich soil (Foster, 1944). The procedures for this enrichment and isolation of the strain have also been described by Foster (1944). Soil is added to a solution of 0.1% riboflavin with small amounts of K_2HPO_4 and $MgSO_4·7H_2O$, then incubated at room temperature. After the riboflavin has disappeared, as indicated by the loss of the orange color, the culture is streaked onto solid medium that contains same components.

MAINTENANCE PROCEDURES

Cultures can be preserved by liquid drying (L-drying). For L-drying, cells grown on nutrient agar for 48 h at 30°C are suspended in the protective medium SM1 (Sakane et al., 1996), which contains 30 g of sodium L(+)-glutamate monohydrate, 15 g of ribitol, 5 g of L-cysteine hydrochloride monohydrate in 1 liter of 0.1 M potassium phosphate buffer (pH 7.0), and dispensed in ampules. The suspension is then vacuum-dried from the liquid state without freezing and stored at 4°C. The results of an accelerated storage test suggest that cells should survive more than 20 years (Sakane and Kuroshima, 1997). Cultures may be also preserved by freezing at −80°C or −196°C. For freezing, cells are suspended in nutrient broth (Difco) containing 10% glycerol or 7% DMSO as a cryoprotective agent.

DIFFERENTIATION OF THE GENUS *DEVOSIA* FROM OTHER GENERA

Table BXII.α.170 lists properties that are useful for distinguishing the genus *Devosia*. The genus *Devosia* can be differentiated from morphologically similar genera belonging to the *Alphaproteobacteria* by certain phenotypic and chemotaxonomic characteristics and 16S rRNA sequence signatures (Table BXII.α.171). The absence of 3-OH fatty acids that are shorter than $C_{18:0\ 3OH}$ and the presence of $C_{24:1\ 3OH}$ and $C_{26:1\ 3OH}$ as the major 3-OH fatty acids in the cells are the key differential characteristics.

TAXONOMIC COMMENTS

Phylogenetic analysis derived from 16S rRNA sequences shows that the genus *Devosia* occupies a distinct position in the *Alphaproteobacteria* (Fig. BXII.α.204). The phylogenetic independence

TABLE BXII.α.170. Some characteristics differentiating the genus *Devosia* from other morphologically similar genera of the *Alphaproteobacteria* and the genus *Pseudomonas*[a]

Characteristic	*Devosia*	*Acetobacter*	*Acidiphilium*	*Acidomonas*	*Agrobacterium*	*Bradyrhizobium*	*Brevundimonas*	*Gluconobacter*	*Methylobacterium*	*Mycoplana*	*Rhizobium*	*Rhizomonas*	*Pseudomonas*	*Sphingomonas*
Flagella:														
Lateral		+	+	+	+				+	+	+	+		
Polar	+		+			+	+	+	+		+	+	+	+
Oxidase	+	−	D	+	D	nd	+	−	D	+	nd	+	D	nd
Catalase	+	+	nd	+	+	nd	+	nd	+	nd	nd	+	+	+
Major hydroxy fatty acids:[b]														
2-OH:														
$C_{12:0}$													+	
$C_{14:0}$					+			+				+		+
$C_{16:0}$			+		+									
$(C_{16:0})$			+											
3-OH:														
$C_{10:0}$													+	
$C_{12:0}$						+	+			+			+	
$C_{14:0}$			+	+	+	+			+		+			
$(C_{14:0})$										+				
$C_{16:0}$		+			+			+						
$(C_{18:0})$			+											
$C_{24:1}$	+													
$C_{26:1}$	+													
$(C_{19:0\ cyclo})$								+						
$(C_{13:0\ iso})$												+		
Quinone[c]	Q-10	Q-10	Q-10	Q-10	Q-10	Q-10	Q-10	Q-10	Q-10	Q-10	Q-10	Q-10	Q-9	Q-10
Mol% G + C of DNA	61.4	51–65	62–70	63–65	55–63	61–65	65–68	56–64	60–70	63–68	59–64	58–65	59–68	59–68

[a]Symbols: +, positive; −, negative; D, different among species and/or strains; nd, not determined.

[b]Parentheses indicate major hydroxy fatty acids in some species.

[c]Q-9, ubiquinone 9; Q-10, ubiquinone 10.

TABLE BXII.α.171. 16S rRNA sequence signatures that distinguish the genus *Devosia* from other genera of *Alphaproteobacteria*

Position of base or base pair[a]			Base(s) in:						
	Devosia	*Rhizobiaceae*	*Erythrobacter/ Sphingobacterium* group	*Azospirillum/ Magnetospirillum* group	*Bradyrhizobium/ Methylobacterium* group	*Brevundimonas/ Caulobacter* group	*Paracoccus/ Rhodobacter/ Roseobacter* group	*Ehrlichia/ Rickettsia* group	*Acetobacter/ Gluconobacter* group
155:166	G:C	C:G	G/C:C/G	U:G	C:G	C:G	C/U:G/A	U/A:A/U	C/U:G
240:286	A:U	U:A	U:A	C:G	C:G	C:G	U:A	C/U:G/A	U:A
445:489	A:U	G:C	G:C	G:C	A/G:U/C	G:C	G:C	G:C/U	G:C
681:709	U:A	G:C	G:C	G/C:C/G	G:C	G:C	G/A:C/U	U:A	C:G
694	G	A	A	A	A	A	A	A	A
1419:1481	A:U	G:C	G:C	G:C	G:C	G:C	G/A:C/U	G:C/U	G:C

[a]*Escherichia coli* numbering system.

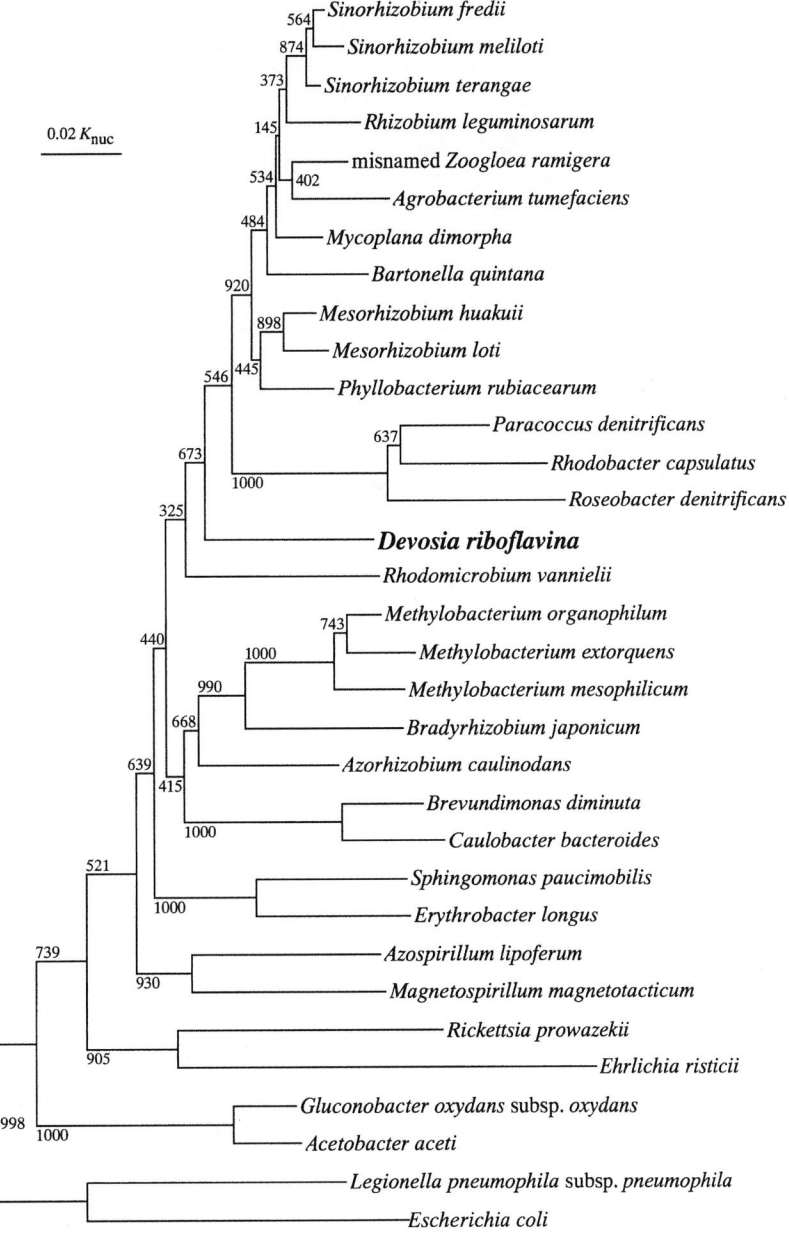

FIGURE BXII.α.204. Phylogenetic tree derived from 16S rRNA sequences of the *Alphaproteobacteria.* Clustal W ver. 1.7 (Thompson et al., 1994) was used to generate K_{nuc} values (Kimura, 1980) and to reconstruct a phylogenetic tree by the neighbor-joining method (Saitou and Nei, 1987). Scale bar = K_{nuc} values. The lengths of the vertical lines are not significant. The numbers of the branches refer to the confidence limits estimated by bootstrap analysis (Felsenstein, 1985) with 1000 replicates. The total number of nucleotides compared with 1143 after elimination of the positions at which secondary structures varied in the strains (Nakagawa et al., 1996) and all sites which were not determined in any sequences. *Escherichia coli* and *Legionella pneumophila* subsp. *pneumophila* were used as the outgroup. The accession numbers for nucleotide sequences are as follows: *Acetobacter aceti*, X74066; *Agrobacterium tumefaciens*, M11223; *Azorhizobium caulinodans*, D11342; *Azospirillum lipoferum*, M59601; *Bartonella quintana*, M11927; *Bradyrhizobium japonicum*, D11345; *Brevundimonas diminuta*, M59064; *Caulobacter bacteroides* (This species has recently been transferred to the genus *Brevundimonas* as *B. bacteroides*)., M83796; *Devosia riboflavina*, D49423; *Ehrlichia risticii*, M21290; *Erythrobacter longus*, M96744; *Escherichia coli*, J01695; *Gluconobacter oxydans* subsp. *oxydans*, X73820; *Legionella pneumophila* subsp. *pneumophila*, M59157; *Magnetospirillum magnetotacticum*, M58171; *Mesorhizobium huakuii*, D13431; *Mesorhizobium loti*, D14514; *Methylobacterium extorquens*, D32224; *Methylobacterium mesophilicum*, D32225; *Methylobacterium organophilum*, D32226; *Mycoplana dimorpha*, D12786; *Paracoccus denitrificans*, X69159; *Phyllobacterium rubiacearum*, D12790; *Rickettsia prowazekii*, M21789; *Rhizobium leguminosarum*, D12782; *Rhodobacter capsulatus*, D16428; *Rhodomicrobium vannielii*, M34127; *Roseobacter denitrificans*, M59063; *Sinorhizobium fredii*, D14516; *Sinorhizobium meliloti*, D01265; *Sinorhizobium terangae*, X68387; *Sphingomonas paucimobilis*, D16144; misnamed *Zoogloea ramigera*, X74915.

of the genus *Devosia* is also reflected by the 16S rRNA signature sequences (see above).

16S rRNA cataloging (Woese et al., 1984a) and DNA–rRNA hybridization (De Vos and De Ley, 1983; De Vos et al., 1985a, 1989) have revealed that members of the genus *Pseudomonas* belong to the *Alphaproteobacteria*, *Betaproteobacteria*, and *Gammaproteobacteria*. The level of heterogeneity in the genus *Pseudomonas* has been reduced by transferring various pseudomonads belonging to Palleroni's rRNA groups (Palleroni, 1984)—other than group I—to other existing new genera (Tamaoka et al., 1987; Willems et al., 1989, 1990; Yabuuchi et al., 1990a, 1992; Segers et al., 1994).

The organism presently named *Devosia riboflavina* is one of the misnamed pseudomonads. It was originally named *"Pseudomonas riboflavina"* Foster 1944, 30 and described as a soil bacterium that oxidized riboflavin to lumichrome. In the eighth edition of the *Determinative Manual*, *"P. riboflavina"* was treated as a *species incertae sedis* because it is not motile, even though the original description indicated that it was motile (Doudoroff and Palleroni, 1974). DNA–rRNA hybridization studies (De Vos et al.,

1989; Segers et al., 1994) have shown that *"P. riboflavina"* belongs to rRNA superfamily IV (*Alphaproteobacteria*), but its precise position in the *Alphaproteobacteria* remained unknown until its 16S rRNA sequence was determined. The results of a phylogenetic analysis based on 16S rRNA sequence data (Nakagawa et al., 1996) have placed *"P. riboflavina"* IFO 13584[T] in an independent position in the *Alphaproteobacteria*. The similarity rank analysis included in the Ribosomal Database Project (Maidak et al., 1997) suggests that the phylogenetic neighbors of the genus *Devosia* are members of the *Rhizobiaceae*, with which it exhibits 16S rRNA similarity values ranging from 91.9 to 93.6%.

Unfortunately, only one strain of the single species is available. Other strains belonging to the genus should be isolated and investigated in the future.

FURTHER READING

Foster, J.W. 1944. Microbiological aspects of riboflavin. I. Introduction. II. Bacterial oxidation of riboflavin to lumichrome. J. Bacteriol. *47*: 27–41.

List of species of the genus Devosia

1. **Devosia riboflavina** Nakagawa, Sakane and Yokota 1996, 20[VP]

ri.bo.fla′vi.na. M.L. fem. adj. *riboflavina* referring to the ability of the organism to oxidize riboflavin.

The characteristics are as described for the genus and listed in Tables BXII.α.170, BXII.α.171, and BXII.α.172. In addition, colonies are circular with entire or slightly undulate margins and cream colored. Urease positive. Organic nitrogenous substances, such as amino acids, are required for growth. Glycine, urea, and ammonium chloride cannot be substituted for organic material as nitrogen sources in media containing riboflavin as the sole energy source. Acid is produced from D-arabinose. Neither acid nor gas are produced from D-galactose, D-glucose, inositol, lactose, D-fructose, maltose, mannitol, sucrose, D-xylose, and ethanol. Esculin is hydrolyzed. Gelatin and starch are not hydrolyzed. Nitrate is slightly reduced to nitrite. Indole is not produced. Vigorous growth occurs on nutrient agar at 28°C. As determined by API 20NE tests, the following substrates are assimilated by the type strain: D-arabinose, glucose, maltose, D-mannitol, D-mannose, and *N*-acetyl-D-glucosamine. The following substrates are not assimilated by the type strain, as determined by API 20NE tests: adipic acid, *n*-capric acid, DL-malic acid, phenyl acetate, potassium gluconate, and sodium citrate. β-Galactosidase and β-glucosidase activities are present in the type strain, as determined by API 20NE tests. Arginine dihydrolase activity is not present in the type strain, as determined by API 20NE tests.

The mol% G + C of the DNA is: 61.4 (HPLC).

Type strain: 4R3337, ATCC 9526, DSM 7230, IFO 13584.

GenBank accession number (16S rRNA): D49423.

TABLE BXII.α.172. Phenotypic characteristics of *Devosia riboflavina* IFO 13584[a]

Test	Result
Color of colonies	Cream
Morphology of cells	Rods
Gram stain reaction	−
Motility	+
Spore formation	−
Oxidase	+
Catalase	+
Urease	+
Nitrate reduction	w
Hydrolysis of:	
Gelatin	−
Starch	−
Acid production from:	
D-Arabinose	+
D-Galactose	−
D-Glucose	−
Inositol	−
Lactose	−
D-Fructose	−
Maltose	−
Mannitol	−
Sucrose	−
D-Xylose	−
Decarboxylation of:	
L-Alanine	−
L-Lysine	−
L-Ornithine	−
Oxidation of riboflavin	+

[a]Symbols: +, positive; −, negative; w, weakly positive.

Genus IX. **Dichotomicrobium** *Hirsch and Hoffman 1989b, 495^VP (Effective publication: Hirsch and Hoffman 1989a, 300)*

PETER HIRSCH

Di.cho' to.mi.cro.bi.um. Gr. adj. *dichotomus* divided, forked; Gr. adj. *micrus* small; Gr. n. *bius* life; M.L. neut. n. *Dichotomicrobium* a forked microbe.

Tetrahedral to spherical cells, 0.8–1.8 × 0.8–2.0 μm, with up to four prosthecae (hyphae) of 0.2–0.3 μm width and varying length. Cells and hyphae may be covered with short, rigid, bent pili. **Hyphae and mother cells may branch (fork) dichotomously.** Multiplication by terminal bud formation on hyphae or by intercalary budding. Propagation cells initially spherical or pear-shaped, later tetrahedral and never motile. Gram negative, do not form spores. **Poly-β-hydroxybutyrate (PHBA) granules may be produced, even within the hyphae.** Aerobic, moderately thermophilic and halophilic, require yeast extract for growth. Carbon sources are acetate, malate, succinate, and 2-oxoglutarate. Utilize organic nitrogen sources, such as amino acids or yeast extract; ammonium, nitrite, nitrate, molecular nitrogen, and urea do not serve as nitrogen sources. **Oxidase, catalase, and peroxidase positive.** Do not grow anaerobically. Occur in saline ponds and lakes with temperatures above 20°C.

The mol% G + C of the DNA is: 62–64.

Type species: **Dichotomicrobium thermohalophilum** Hirsch and Hoffmann 1989b, 495 (Effective publication: Hirsch and Hoffmann 1989a, 300.)

FURTHER DESCRIPTIVE INFORMATION

Some information on *Dichotomicrobium* isolates was initially published with the strain designation "genus D" (Gebers et al., 1985; Kölbel-Boelke et al., 1985; Roggentin and Hirsch, 1989; Sittig and Hirsch, 1992) or as *"Dichotomicrobium"* sp. (Stackebrandt et al., 1988a). All of these studies were concerned with at least isolate IFAM 954^T, the type strain for *D. thermohalophilum*.

A phylogenetic survey of budding and prosthecate bacteria by 16S rRNA cataloguing (Fig. BXII.α.205) has shown that *Dichotomicrobium* strain IFAM 954^T belongs to the *Alphaproteobacteria*, close to *Hyphomicrobium*, *Filomicrobium*, and *Pedomicrobium* species (Stackebrandt et al., 1988a).

The general morphology and life cycle of 13 strains that have been investigated is very similar (Hirsch and Hoffmann, 1989a). Mature cells are tetrahedral, triangular, or even cubical, they may be dichotomously branched. Up to four prosthecae (hyphae) grow out from the corners (Fig. BXII.α.206). The average cell size is 0.9 × 1.1 μm, with a rather constant hyphal diameter of 0.25 μm. The hyphae may branch; nearly spherical buds are produced from the hyphal tips, and only from one at a time. The buds are separated by a cross wall upon maturation and can be released, but motility of these daughter cells has never been observed. Rarely, intercalary buds are produced by a local enlargement of a hyphal portion. Sometimes, terminal buds remain attached to the hyphae and begin to grow new hyphae with a terminal bud, thus resulting in the formation of a cell chain or even a network of cells after hyphal branching. The hyphal length is dependent on nutrient concentrations: lack of nutrients (especially yeast extract) results in longer and lesser-branched hyphae. In some strains, hyphae are covered with short, rigid, but bent pili to give the appearance of a fur (Fig. BXII.α.207).

The fine structure resembles that of other typical Gram-negative bacteria. Internal membranes are absent, except for an occasional mesosome. DNA nucleoids are found in sections of mother cells and mature buds, but cannot be demonstrated in sections of hyphae. Mother cells and often hyphae store PHBA, and most cells contain polyphosphate granules. Hyphal cross sections reveal an exceptionally dense and structured periplasm. Colonies of *Dichotomicrobium* strains grow slowly; they are rather flat, reddish brown, and have fuzzy edges. Growth in liquid media is pinkish orange; the pigment has been identified as canthaxanthin, a carotenoid (A. Pudleiner, personal communication).

All *Dichotomicrobium* isolates tested by Hirsch and Hoffmann (1989a) require yeast extract for growth, the optimal concentration for five strains ranges from 0.25 to 5.0 g/l. Two of these five strains require Vitamin Solution No. 6 (Van Ert and Staley, 1971). Mass cultures of *Dichotomicrobium* IFAM 954^T have been grown at 43°C with aeration; the yield after 10 d amounted to 7 g dry weight per 7.5 liters. Most strains show growth in media with a 0.2–5.5 fold concentration of artificial seawater (ASW; Lyman and Fleming, 1940), which corresponds to a total salinity of 8–222‰. The optimum ranges from 2.0 to 3.5 × ASW (salinity of 80–142‰). The temperature optimum of nine strains ranges from 37 to 50°C.

Several organic acids are utilized for growth in the presence of 0.025% (w/v) yeast extract; malate is especially favorable, but succinate, 2-oxoglutarate, and acetate also support good growth. Utilization of amino acids, sugars, and sugar alcohols varies with different strains (Table BXII.α.173). Inorganic nitrogen sources do not serve for growth, but yeast extract and some amino acids can be utilized. None of the strains grow with peptone as a nitrogen source. Enzymatic activities are limited. Neither acid nor gas is produced from 0.1% glucose or fructose. Gelatin, starch, casein, and DNA are not hydrolyzed, nitrate is not reduced, and NH_3, H_2S, indole, acetoin, and extracellular phosphatase are not formed. All strains are methyl-red negative, Fe(II) and Mn(II) are not oxidized. The strains are oxidase, catalase, and peroxidase positive.

Chemotaxonomic characteristics of *Dichotomicrobium* strains have been studied by Sittig and Hirsch (1992). Of nine isolates, strains IFAM 951, 954^T, 956, 958, 1185, and 1186 came from the Solar Lake (Sinai) and strains IFAM 1422, 1423, and 1424 from a pond in Brazil. All these isolates have ubiquinone Q-10 and the following phospholipids: phosphatidylglycerol, phosphatidylcholine, bisphosphatidylglycerol, and phosphatidyldimethylethanolamine. The three strains from Brazil also contain phosphatidylethanolamine. The fatty acid distribution is also typical (and alike) for the Solar Lake strains, but differs in the Brazilian isolates (Tables BXII.α.174 and BXII.α.175). The fatty acids $C_{20:1\ 3OH}$, $C_{22:0}$, and $C_{22:1\ \omega7}$ are especially characteristic of the genus *Dichotomicrobium*. A relatively high percentage of $C_{19:0\ \omega7c}$ has been found in strains IFAM 954^T and 1423 (35.2% and 13.5%, respectively); this fatty acid has also been found in other budding, hyphal bacteria, namely *Filomicrobium fusiforme* and *Pedomicrobium* sp. E 1129.

The DNA base compositions of strains IFAM 954^T, 958, and 1185 have been determined to range from 62.7 to 63.6 mol% G + C (T_m; Gebers et al., 1985). The intrageneric heterogeneity (nucleotide distribution) of three *Dichotomicrobium* strains and

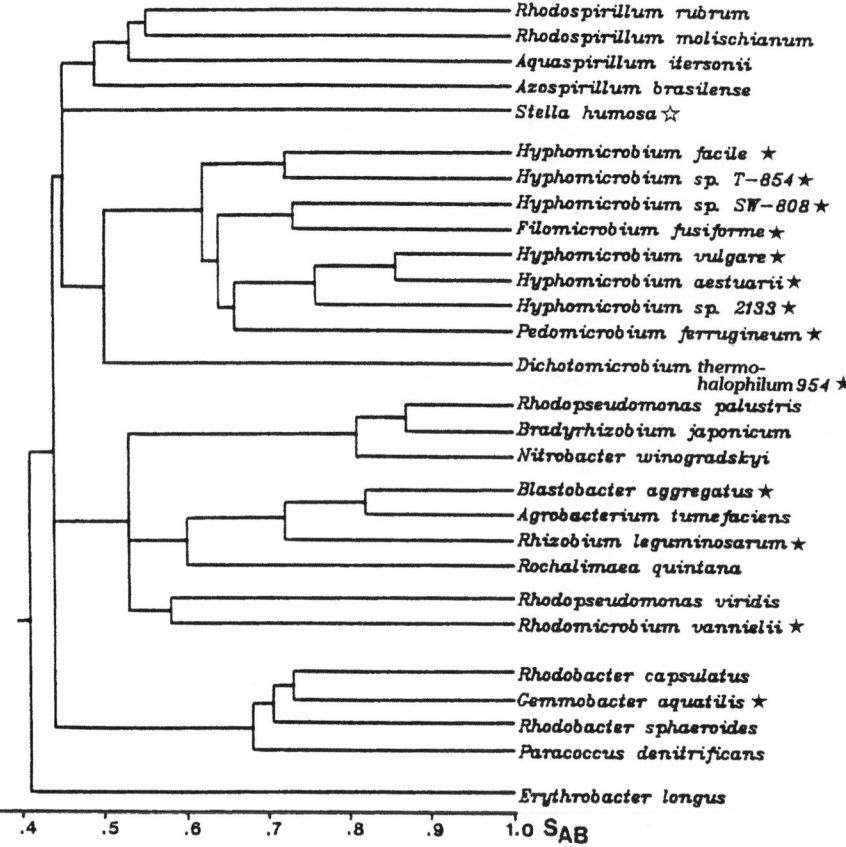

FIGURE BXII.α.205. 16S rRNA dendrogram showing the phylogenetic positions of *Dichotomicrobium thermohalophilum* IFAM 954[T] and other budding and/or prosthecate members of the *Alphaproteobacteria*. *open star*, budding and prosthecate; *closed star*, prosthecate (Reproduced with permission from E. Stackebrandt et al., Archives of Microbiology *149*: 547–556, 1988, ©Springer-Verlag, Berlin.)

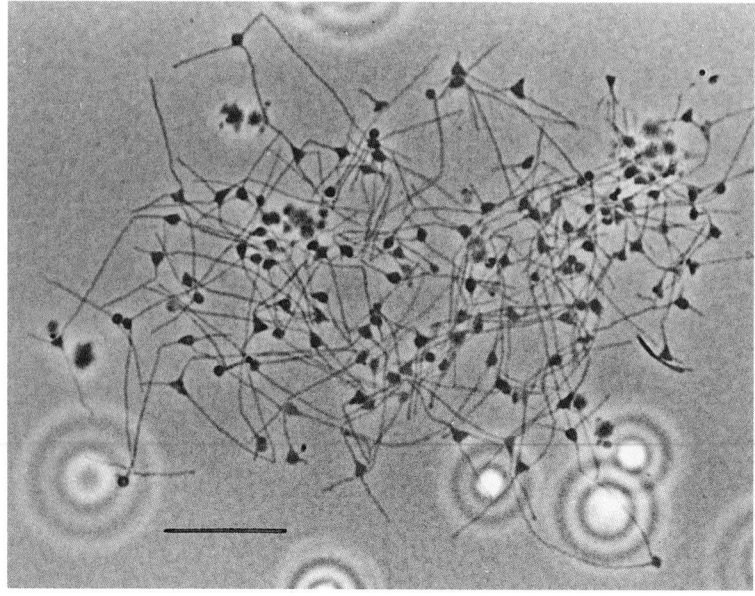

FIGURE BXII.α.206. *Dichotomicrobium thermohalophilum* IFAM 954[T] grown in PYGV + 2.5 × ASW at 35°C for 19 d. Phase-contrast light micrograph. Bar = 10 μm.

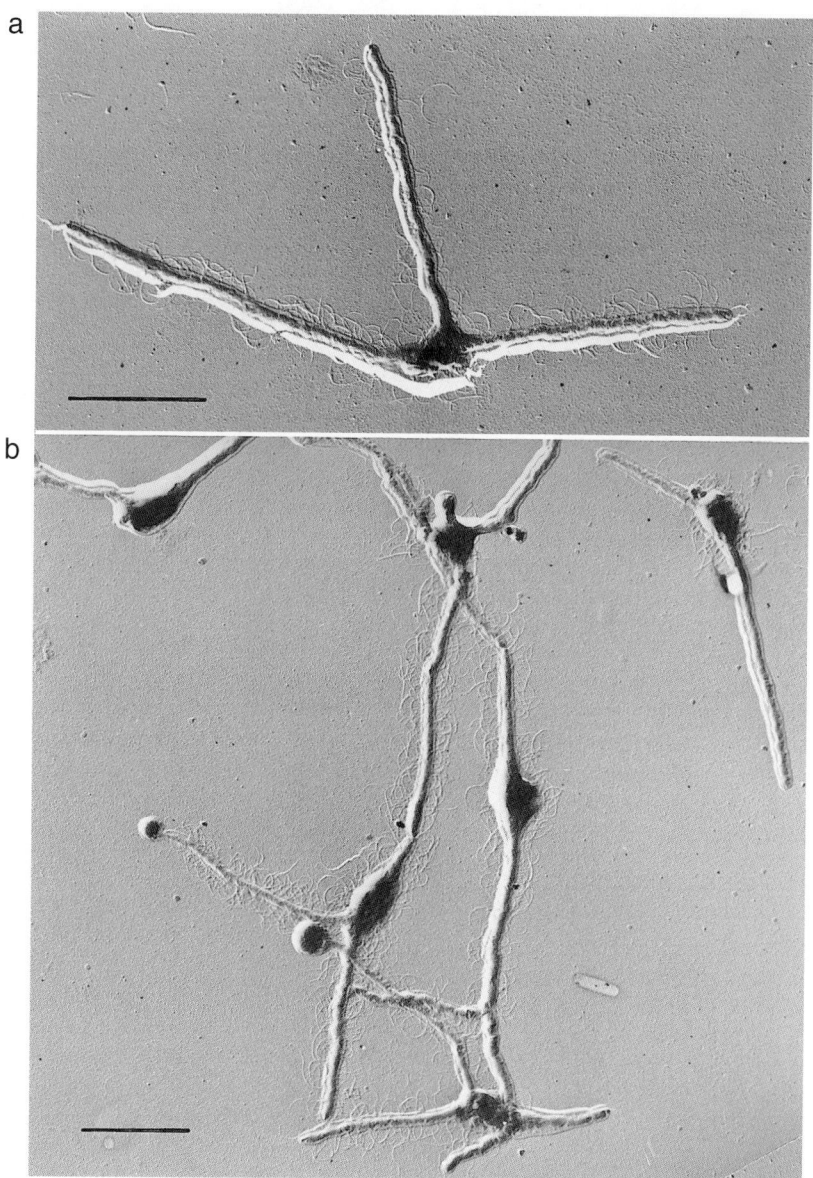

FIGURE BXII.α.207. *Dichotomicrobium thermohalophilum* IFAM 954[T] grown 9 d in PYGV + 2.5 × ASW at 43°C. Shadow-cast electron microscopic preparation. *(a)*, young cell with short stubby pili on hyphae; Bar = 2 μm. *(b)*, cells with hyphal branches and terminal buds; Bar = 2 μm (Reproduced with permission from P. Hirsch, Distribution and pure culture studies of morphologically distinct Solar Lake microorganisms. *In* Nissenbaum (Editor), Hypersaline Brine and Evaporitic Environments, ©Elsevier Science B.V, pp 41–60.)

differences among these and closely related genera (*Pedomicrobium*, "genus T") have also been studied by plotting DNA base compositions against the widths of the melting transitions. The three *Dichotomicrobium* strains share a high level of similarity with respect to their DNA properties. The genome size of *Dichotomicrobium* strain IFAM 954[T] has been determined by DNA renaturation kinetics and found to be 1.73×10^9 bp. When plotted against the mol% G + C ratio, strain 954[T] was found to be widely separated from *Hyphomonas*, *Hyphomicrobium*, and *Filomicrobium* species (Kölbel-Boelke et al., 1985).

Antibiotic sensitivity of *Dichotomicrobium* for four antibiotics has been tested by Hirsch and Hoffmann (1989a), with liquid cultures containing 1–100 μg antibiotic per ml. Significant inhibition occurs with 1–10 μg/ml penicillin G or streptomycin sulfate or with 10–100 μg/ml ampicillin or cephalothin.

Bacteria with a *Dichotomicrobium*-like morphology have so far been isolated only from the hypersaline Solar Lake (Sinai) and the Lagoa Vermelha pond in Brazil, which have salinities ranging from 68.7 to 183.3‰ (Table BXII.α.176) and temperatures of 30°C and higher. However, Güde et al., (1985) have observed similar bacterial morphotypes in Lake Constance between 10 and 150 m depth and at a temperature of only 4.5°C. Unfortunately, all attempts to cultivate these forms have failed.

Conditions for Solar Lake dichotomicrobia during sampling time varied considerably within the profile: from 43°C and 144.4‰ salinity near the aerobic surface, through 53°C and 156‰ at 2.5 m depth, to the anaerobic sediment (rich in sulfide) with 42°C and 183.3‰ salinity (Hirsch, 1980; Table BXII.α.176). The Solar Lake dichotomicrobia are evidently well adapted to tolerate this range of conditions.

TABLE BXII.α.173. Some physiological properties of selected *Dichotomicrobium* isolates[a,b]

Characteristic	IFAM 951	IFAM 954[T]	IFAM 955	IFAM 958	IFAM 1185	IFAM 1186
C-sources (0.1%, w/v):						
Malate	+	+	+	+	+	+
Glucose	(+)	+	−	nd	nd	nd
C-sources (0.01%, w/v):						
Serine	−	−	+	−	−	−
L-Aspartate	+	+	+	−	+	+
L-Glutamate	+	+	(+)	+	+	−
Peptone	−	−	nd	nd	nd	nd
Yeast extract	+	+	nd	nd	nd	nd
C-sources (0.2%, w/v) (+ 0.25 g/l yeast extract):						
Glucose	+	+	−	−	−	+
Fructose	+	+	−	+	+	+
Mannose	+	+	−	+	−	+
Galactose	+	+	−	+	−	−
Ribose	+	+	−	+	+	−
Arabinose	−	−	−	+	−	−
Mannitol	−	−	+	+	−	−
N-sources (0.1%, w/v) (+ 0.1%, w/v, malate):						
$(NH_4)_2SO_4$, urea	−	−	−	−	−	−
KNO_2, KNO_3	−	−	−	−	−	−
L-Glutamate/L-aspartate	+	+	−	−	+	−
L-Serine	−	(+)	−	−	−	−
Methionine	+	+	−	−	+	+
Arginine, cysteine, leucine	+	+	−	−	+	−
Yeast extract	+	+	+	+	+	+
Peptone	−	−	−	−	−	−

[a]Symbols: +, substrate utilized; −, substrate not utilized; (+), growth only weak and slow; nd, not determined.
[b]Data from Hirsch and Hoffmann (1989a).

TABLE BXII.α.174. Non-hydroxy fatty acids (%) of representative *Dichotomicrobium* strains[a]

Fatty acid	*D. thermohalophilum* IFAM 954[T]	*Dichotomicrobium* sp. IFAM 1423
$C_{14:1\ \omega12}$	0.5	0.1
$C_{14:1\ \omega11}$	0.2	0.1
$C_{14:0}$	0.1	0.1
$C_{16:1\ \omega7}$	0.2	0.2
$C_{16:0}$	2.2	2.1
$C_{17:0}$	0.1	0.1
$C_{18:1\ \omega7}$	39.3	57.8
$C_{18:0}$	15.2	18.7
$C_{18:1\ \omega6\ 11CH_3}$	1.4	1.4
$C_{19:0\ cyclo\ \omega7c}$	35.2	13.5
$C_{20:1\ \omega7}$	4.7	4.9
$C_{20:0}$	0.2	0.3
$C_{21:0}$	−	0.2
$C_{22:1\ \omega7}$	0.1	0.1
$C_{22:0}$	0.1	−

[a]Data from Sittig and Hirsch (1992).

TABLE BXII.α.175. Hydroxy fatty acids (%) of representative *Dichotomicrobium* strains[a]

Hydroxy fatty acid	*Dichotomicrobium* IFAM 954[T]	*Dichotomicrobium* IFAM 1423
$C_{14:0\ 3OH}$	10.8	3.1
$C_{16:0\ 3OH}$	2.9	1.8
$C_{18:1\ 3OH}$	3.5	5.7
$C_{18:0\ 3OH}$	72.1	73
$C_{20:1\ 3OH}$	9.9	15.2
$C_{20:0\ 3OH}$	0.7	1.2

[a]Data from Sittig and Hirsch (1992).

Three strains have been isolated from enrichments kept under N_2/CO_2 (95:5), but when tested, these isolates could not grow anaerobically. Since viable dichotomicrobia have been obtained from the anaerobic enrichments and from the anaerobic sediment, a lack of oxygen must be tolerated by these bacteria at least for some time. Survival and limited growth in the anaerobic enrichments could have been made possible by the presence of oxygen-producing cyanobacteria and dim light.

ENRICHMENT AND ISOLATION PROCEDURES

Two methods have successfully yielded *Dichotomicrobium* cultures. (1) Clean, sterile glass slides are exposed to the Solar Lake for

TABLE BXII.α.176. Origin and enrichment conditions of 13 strains of *Dichotomicrobium* sp.[a]

Characteristic	IFAM 951, 952	IFAM 953, 954[T]	IFAM 955, 956	IFAM 957, 958	IFAM 1185	IFAM 1186	IFAM 1422	IFAM 1423	IFAM 1424
Sample origin:									
Solar Lake (Sinai)	+	+	+	+	+	+			
Lagoa Vermelha (Brazil)							+	+	+
Depth (m)	2.5	3.5	1.5	1.5	4.3	2	0.2	0.2	0.2
Salinity (‰)	156.2	169.1	144.4	144.4	183.3	151.2	86.5	68.7	86.5
Enrichment conditions:									
Temperature (°C)	43	43	43	43	43	43	30	30	30
Atmosphere	O_2	N_2	O_2	O_2	N_2	N_2	O_2	O_2	O_2
Additions:									
Malate							+	+	+
Peptone			+						
Yeast extract and Vitamin solution No. 6	+								
Medium pH	7.2	7.2	7.2	7.2	7.2	7.2	8.5	7.2	8.5
Optimal conditions for growth:									
Salinity (‰)	121	142	111	111	121	142	80	80	80
Temperature (°C)	49	50	50	50	44	48	45	37	38
pH	nd[b]	8.3	nd	nd	nd	nd	8.5	7.1	8.3

[a]Data from Hirsch and Hoffmann (1989a).

[b]nd, not determined.

8–10 d to allow dichotomicrobia to attach. The slides are then transferred to a sample of Solar Lake water. In the laboratory, samples and slides are incubated either directly or with the addition of 0.025% (w/v) of Bacto yeast extract (sterilized in the same water). Alternatively, Bacto peptone is added to other samples with exposed slides. The samples (enrichments) are then incubated up to four weeks at 30–45°C (Hirsch, 1980; Hirsch and Hoffmann, 1989a). (2) The other successful procedure consists of directly streaking water samples containing dichotomicrobia onto agar plates with medium 398, of the following composition (g/l): Bacto yeast extract, 1; DL-malate, 1; Hutner's basal salts (Cohen-Bazire et al., 1957), 20 ml; 3× ASW, 980 ml; final pH 7.2. Some strains grow better on medium 399 with the same ingredients, but only 2.5× concentrated ASW and a final pH of 8.5. Identification and isolation of dichotomicrobia on the same media is facilitated by the reddish-brown color and fuzzy edges of the individual colonies.

MAINTENANCE PROCEDURES

Pure cultures can be stored in liquid medium at 20°C for at least one year, but storage at 4°C is much less effective. Lyophilization in growth medium (398, 399) (see Enrichment and Isolation Procedures above) is better than in skim milk. Slants kept at 20°C must be subcultured every 4–6 months, but liquid cultures survive better, as long as medium evaporation is controlled.

PROCEDURES FOR TESTING SPECIAL CHARACTERS

Mass cultures for the isolation of DNA contain (g/l): Bacto peptone, 1; Bacto yeast extract, 1; glucose, 1; vitamin solution no. 6, 10 ml; Hutner's basal salts, 20 ml; ASW (2.5–3×), 970 ml; final pH 7.5 (Gebers et al., 1985). Cells are harvested at 16,000× g for 20 min, washed twice with 0.85% (w/v) NaCl, and the pellet stored at −20°C for further use.

Lysis of dichotomicrobia is achieved by suspending 1 g (wet weight) of cells in 20 ml of 1 M NaCl/0.1 M EDTA containing 1 g proteinase K and 1% (w/v) N-acetyl-N,N,N-trimethylammonium bromide (E. Merck), as well as 50 g of glass beads (Gebers

et al., 1985). This mixture is treated for 5 sec in the cell homogenizer, followed by the addition of 1 volume chloroform–isoamyl alcohol and shaking at 100 rpm for 15 min. The resulting emulsion is centrifuged at 1350× g for 20 min; the upper aqueous phase is then removed and mixed with 1 volume double distilled water. Then, 0.6 volume of isopropanol is added, drop by drop, while the solution is stirred with a glass rod. Centrifugation at 1350× g for 10 min yields a nucleic acid pellet, which can be further purified as described by Marmur (1961).

DIFFERENTIATION OF THE GENUS *DICHOTOMICROBIUM* FROM OTHER GENERA

Differentiation of the genus *Dichotomicrobium* from other hyphal and budding bacteria is shown on Table BXII.α.177.

TAXONOMIC COMMENTS

Ten *Dichotomicrobium* strains came from the Solar Lake, with IFAM 954[T] as a representative isolate. Three other strains from the shallower Lagoa Vermelha (Brazil) are physiologically slightly different; a representative strain is IFAM 1423. These isolates have lower salinity and temperature optima (Table BXII.α.176), and they have different percentages of non-hydroxy (Table BXII.α.174) and hydroxy fatty acids (Table BXII.α.175). Additionally, the average cell size of these Brazilian strains was lower. For these reasons, only the ten nearly identical strains IFAM 951–1186 are described as *D. thermohalophilum* (Hirsch and Hoffmann, 1989a) and the Lagoa Vermelha isolates have been recommended for further studies as possibly different species.

ACKNOWLEDGMENTS

Skillful technical assistance rendered by B. Hoffmann and M. Beese is gratefully acknowledged. A. Höhn (Geesthacht) provided the Lagoa Vermelha water samples and physical data on the habitat. H. Güde (Langenargen) and M. Simon (Konstanz) contributed information on *Dichotomicrobium*-like morphotypes in Lake Constance. The Solar Lake studies were initiated by the late Prof. M. Shilo (Jerusalem) and supported by the Deutsche Forschungsgemeinschaft.

TABLE BXII.α.177. Differentiation of *Dichotomicrobium* from other related genera of budding and hyphal bacteria

Characteristics	*Dichotomicrobium*[a]	*Pedomicrobium*[b]	*Filomicrobium*[c]	*Hirschia*[d]	*Hyphomicrobium*[e]	*Hyphomonas*[f]	*Rhodomicrobium*[g]
Cell shape:							
Tetrahedral	+	V[h]	−	−	−	−	V
Ovoid, pear-, bean-shaped	V	+	−	+	+	+	+
Fusiform	−	−	+	−	−	−	−
Intracytoplasmic membranes	−	V	−	−	V	−	+
PHBA in mother cells and/or hyphae	+ / +	− / −	− / −	− / −	+ / −	−	+ / −
Bud perpendicular on hyphal tip	−	+			−	−	−
Daughter cells motile	−	+	−	+	+	+	+
Colony pigmentation[i]	Rb	Db	Lr	Y	Bb	C	R
Salinity range for growth (‰ ASW)	80–142	1 to <10	4.3–34.5	4.3–86.3	0–25[j]	5–150[j]	nd[k]
Anaerobic growth	−	−	−	−	d	−	V
Fe^{2+} or Mn^{2+} oxidized	−	+	−	−	V	−	−
Main respiratory quinone component	Q-10	Q-10	Q-9	Q-10	Q-9	Q-11/Q-10	Q-10
Mol% G + C (T_m)	62–64	62–67	62	46–47	59–65	57–62	62–64

[a]Data from Hirsch and Hoffmann (1989a).

[b]Data from Gebers (1981).

[c]Data from Schlesner (1987).

[d]Data from Schlesner et al., (1990).

[e]Data from Hirsch (1989).

[f]Data from Weiner et al., (1985).

[g]Data from Duchow and Douglas (1949).

[h]V denotes occasional mesosomal structures.

[i]Rb, reddish brown; Db, dark brown; Lr, light red; Y, yellow; Bb, brownish to beige; C, colorless; R, red.

[j]NaCl salinity.

[k]nd, not determined.

List of species of the genus Dichotomicrobium

1. **Dichotomicrobium thermohalophilum** Hirsch and Hoffmann 1989b, 495[VP] (Effective publication: Hirsch and Hoffmann 1989a, 300.)

   ther.mo.ha.lo'phi.lum. Gr. adj. *thermus* hot; Gr. n. *halus* salt; Gr. adj. *philus* loving; M.L. neut. adj. *thermohalophilum* heat and salt loving.

   Morphological description is the same as for the genus. Temperature optimum for growth 44–50°C, temperature minimum 20–30°C, temperature maximum 61–65°C. Temperature survival range 13–65°C. Optimal pH 8.0–8.5, pH range for growth 5.8–9.5. Optimal salinity for growth 80–142‰ ASW, minimal salinity 8 to >40‰, maximal salinity 182 to >222‰. Optimal concentration of yeast extract 0.25–5.0 g/l (w/v). Other physiological and biochemical characteristics as for the genus. Habitat: hypersaline, meromictic, and heliothermal Solar Lake (Sinai), throughout the whole profile (0–4.5 m).

   The mol% G + C of the DNA is: 62–64 (T_m).

   Type strain: IFAM 954, ATCC 49408, DSM 5002.

   Additional Remarks: Reference strains include IFAM 951, 952 (DSM 5001), 953, 955, 956, 957, 958, 1185, and 1186 (DSM 5006).

Genus X. Filomicrobium *Schlesner 1988, 220*[VP] *(Effective publication: Schlesner 1987, 65)*

HEINZ SCHLESNER

Fi.lo.mi.cro'bi.um. L. masc. n. *filum* thread; Gr. adj. *micros* small; Gr. n. *bios* life; M.L. neut. n. *Filomicrobium* threadlike microbe.

Fusiform cells, 0.5–0.7 × 1.5–4.0 μm, with two or three polar prosthecae, which are about 0.2 μm in diameter and have a length of up to 40 μm or more. Buds are formed at the tips of the prosthecae (Fig. BXII.α.208). **Nonmotile. Gram negative. Aerobic and chemoorganotrophic.** Utilize some organic acids. **C₁-compounds are not used as carbon sources.** The ubiquinone system is a **Q-9** system. Hydroxy fatty acids are of the **3-OH** type.

The mol% G + C of the DNA is: 62.

Type species: **Filomicrobium fusiforme** Schlesner 1988, 220 (Effective publication: Schlesner 1987, 65.)

FURTHER DESCRIPTIVE INFORMATION

Growth on solid media is poor. After 3 weeks of incubation at 25°C, microcolonies with diameters of about 0.1 mm can be ob-

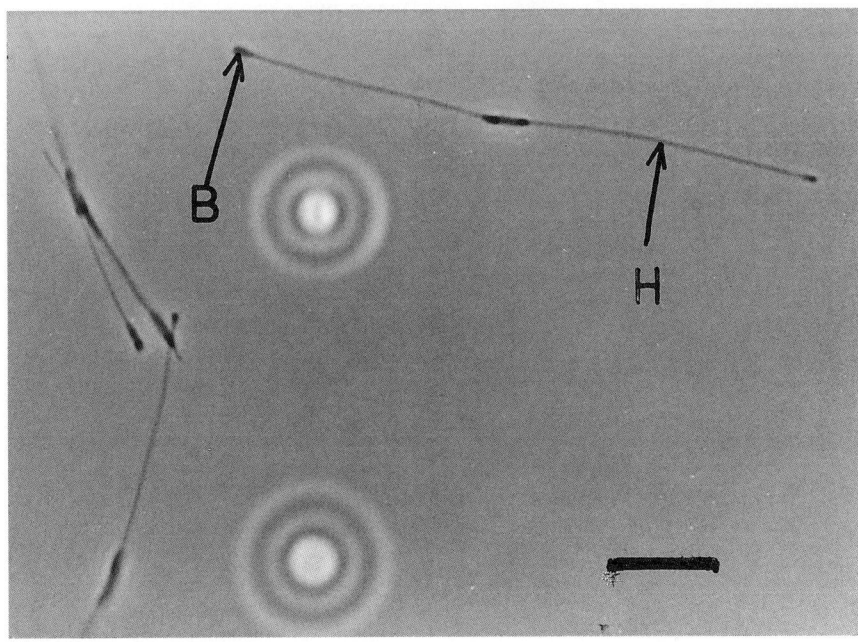

FIGURE BXII.α.208. *Filomicrobium fusiforme* IFAM 1315[T]. Cells with polarly inserted hyphae (*H*) at the tips of which buds (*B*) are being formed. Bar = 10 μm.

served. They are light red and have a rough surface. The red pigment is composed of canthaxanthin and echinone and can be extracted with ethanol and methanol but not with chloroform or petroleum ether. The genome size is 2.03×10^9 Da (Kölbel-Boelke et al., 1985).

Analysis of the 16S rDNA indicates that the genus *Filomicrobium* is a member of the *Alphaproteobacteria* and is related to two clusters of species of the genus *Hyphomicrobium* (Rainey et al., 1998).

ENRICHMENT AND ISOLATION PROCEDURES

The type strain was isolated from a sample taken from the Kiel Bight (part of the Baltic Sea, Germany) at a depth of 2 m. For enrichment, the sample was incubated at 20–23°C in daylight for 12 months and then streaked on agar plates of medium 17 (M17): sodium acetate, 1.0 g; KNO_3, 1.0 g; Hunter's basal salts (Cohen-Bazire et al., 1957), 20 ml; vitamin solution (Staley, 1968), 10 ml; artificial sea water (Lyman and Fleming, 1940), 500 ml; distilled water to 1000 ml (Schlesner, 1987).

MAINTENANCE PROCEDURES

Storage on solid media is not recommended. The cells are easily revived from lyophilized cultures.

PROCEDURES FOR TESTING SPECIAL CHARACTERS

Cells of *Filomicrobium* are easily recognized by microscopy.

DIFFERENTIATION OF THE GENUS *FILOMICROBIUM* FROM OTHER GENERA

Table BXII.α.55 in the entry describing the genus *Hirschia* lists the major features that differentiate *Filomicrobium* from other genera of budding, prosthecate bacteria.

FURTHER READING

Schlesner, H. 1987. *Filomicrobium fusiforme* gen. nov., sp. nov., a slender budding, hyphal bacterium from brackish water. Syst. Appl. Microbiol. *10*: 63–67.

List of species of the genus Filomicrobium

1. **Filomicrobium fusiforme** Schlesner 1988, 220[VP] (Effective publication: Schlesner 1987, 65.)
   fu.si.for′ me. L. n. *fusus* spindle; L. n. *forma* shape, form; M.L. adj. *fusiformis* spindle-shaped.

   Poor growth on solid media. Microcolonies 0.1–0.2 mm in diameter, with a rough surface, occur after 3 weeks incubation at 25°C. Growth is optimal between 20°C and 28°C; maximum growth temperature is 33°C. Vitamin B_{12} is required. Require eighth to full strength seawater for growth; optimal growth is with half-concentrated seawater. Carbon sources utilized for growth are acetate, propionate, fumarate, malate, succinate, and glutamic acid. Glucose is not fermented; nitrate is not reduced. Alginate, cellulose, chitin, and starch are not hydrolyzed. Ammonia, formamide, nitrate, and urea can serve as nitrogen sources. See Table BXII.α.178 for additional characteristics. Red pigment in type strain. Habitat is brackish water.

   The mol% G + C of the DNA is: 62 (T_m).

   Type strain: SH 128, ATCC 35158, DSM 5304, IFAM 1315.

   GenBank accession number (16S rRNA): Y14313.

TABLE BXII.α.178. characteristics of *Filomicrobium fusiforme*

Characteristic	Reaction/Result
Carbon source utilization:	
D-Arabinose	−
L-Arabinose	−
Cellobiose	−
Fructose	−
Galactose	−
Glucose	−
Lactose	−
Lyxose	−
Maltose	−
Mannose	−
Melibiose	−
Raffinose	−
Rhamnose	−
Ribose	−
Sorbose	−
Sucrose	−
Trehalose	−
Xylose	−
Methanol	−
Ethanol	−
Glycerol	−
Erythritol	−
Adonitol	−
Arabitol	−
Dulcitol	−
Inositol	−
Mannitol	−
Sorbitol	−
Acetate	+
Adipate	−
Benzoate	−
Butyrate	−
Caproate	−
Citrate	−
Formate	−
Fumarate	+
Glutarate	−
Lactate	−
Malate	+
Oxalate	−
2-Oxoglutarate	−
Phthalate	−
Propionate	+
Pyruvate	−
Salicylate	−
Succinate	+
Tartrate	−
Valerate	−
L-Alanine	−
L-Arginine	−
L-Asparagine	−
L-Aspartic acid	−
Glutamic acid	+
L-Glutamine	−
Glycine	−
L-Histidine	−

(continued)

TABLE BXII.α.178. *(cont.)*

Characteristic	Reaction/Result
L-Isoleucine	−
L-Leucine	−
L-Lysine	−
L-Methionine	−
D,L-Norleucine	−
L-Ornithine	−
L-Phenylalanine	−
L-Proline	−
L-Serine	−
L-Threonine	−
L-Tryptophan	−
L-Tyrosine	−
L-Valine	−
N-Acetylglucosamine	−
Amygdalin	−
Dextrin	−
Gluconate	−
Glucosamine	−
Glucuronate	−
Glycogen	−
Inulin	−
Pectin	−
Salicin	−
Utilization of nitrogen sources:	
N-Acetylglucosamine	−
Ammonia	+
Formamide	+
Nicotinic acid	−
Nitrate	+
Urea	+
Hydrolysis of:	
Alginate	−
Cellulose	−
Chitin	−
Starch	−
Fermentation of glucose	−
Hemolysis of horse blood	−
Ammonia produced from peptone	−
H$_2$S produced from thiosulfide	−
Catalase	+
Cytochrome oxidase	+
Urease	+
Sensitive to 1 µg/ml of:	
Chloramphenicol	+
Penicillin G	+
Streptomycin	+
Tetracycline	+
Fatty acids:	
C$_{18:1\ \omega7}$	+
C$_{18:2\ \omega7,\ 13}$	+
C$_{19:0\ cyclo\ \omega7c}$	+
Phospholipids:	
Phosphatidylglycerol (PG)	+
Phosphatidylethanolamine (PE)	−
Phosphatidylmonomethylethanolamine (MPE)	−
Phosphatidyldimethylethanolamine (DPE)	+
Phosphatidylcholine (PC)	−
Bisphosphatidylglycerol (BPG)	+

Genus XI. **Gemmiger** *Gossling and Moore 1975, 206*[AL]

JENNIFER GOSSLING

Gem'mi.ger. L. n. *gemma* a bud; L. v. *gero* to bear; M.L. masc. n. *Gemmiger* bud bearer.

Ovoid to hourglass-shaped bacteria, 0.9–2.5 × 1.0 µm, that apparently divide at a constriction, giving the **appearance of budding**. Rapidly growing organisms may form chains. Gram variable to Gram negative. Do not form spores. **Nonmotile** and with no external structures.

Obligately **anaerobic chemoorganotrophs using carbohydrate** as the only or major energy source. Growth on glucose or other sugars produces **butyrate**, usually lactate and formate, and sometimes small amounts of other compounds. Catalase-negative.

The mol% G + C of the DNA is: 59.

Type species: **Gemmiger formicilis** Gossling and Moore 1975, 206.

FURTHER DESCRIPTIVE INFORMATION

The cells most commonly have a bowling-pin shape and are 0.9–2.5 μm long; the diameter of the larger end is 0.5–1.0 μm, that of the smaller end is 0.2–0.8 μm, and that of the constriction is 0.2–0.5 μm (Figs. BXII.α.209 and BXII.α.210). The constriction may not be resolved under the light microscope, and the organism is seen as a pair of cocci, usually with one smaller than the other, resembling a budding yeast. Where the constriction is relatively long, secondary constrictions may be seen by electron microscopy. The bacteria are often seen in pairs, especially in logarithmic phase cultures; the two bacteria are attached to each other by the smaller ends (Figs. BXII.α.210 and BXII.α.211). Long chains are also formed, including pairs with the smaller ends together. At low concentrations of penicillin, sufficient to inhibit division but not growth, long filaments are formed (Salanitro et al., 1976).

There is a multilayered cell wall (Fig. BXII.α.211) (Gossling and Moore, 1975; Salanitro et al., 1976). Intracellular granules of a glycogen-like substance may be found (Salanitro et al., 1976).

Colonies are 1–2 mm in diameter after 48 h of growth. They are usually circular, entire, smooth, and low convex. They may be clear, translucent or opaque cream or white, depending on the medium. Broth cultures may become turbid initially, but the growth usually settles to form a ropy sediment.

Strict anaerobes, they have been cultured by using the roll tube, glove box, and steel wool (Mitsuoka, 1980) techniques but cannot be cultured on agar medium that has been exposed to air and is incubated in a conventional anaerobic jar (Gossling and Moore, 1975; Croucher and Barnes, 1983). Carbon dioxide is normally added to the gas phase for incubation and appears necessary for good growth.

Carbohydrates or related compounds are required for growth. Most strains use glucose; fresh isolates—from humans—that do not use glucose will use maltose (Holdeman et al., 1976; Mitsuoka, 1980). For strains from chickens, NH_4^+ has been shown to be the preferred nitrogen source (Salanitro et al., 1976; Croucher and Barnes, 1983). Various unidentified factors present in ruminal, fecal, liver or yeast extracts are necessary for growth. Growth of strains from chickens is stimulated by thiamine, riboflavin, pantothenate, and straight chain volatile fatty acids (Salanitro et al., 1976). Growth of some strains from humans is stimulated by Tween 80 (0.01–0.10%), but this is inhibitory to other strains (Gossling and Moore, 1975; Salanitro et al., 1976).

Glucose is fermented with the production of butyrate and usually formate and lactate; other products may include acetate, pyruvate, succinate, and malonate. All strains use and ferment a range of the following substances: amygdalin, arabinose, arabinoxylan, cellobiose, dextrin, esculin, fructose, galactose, glucose, glycogen, inulin, lactose, maltose, mannose, melibiose, raffinose, salicin, starch, sucrose, trehalose, and xylose. Lactate and nitrogenous compounds are not fermented.

A chicken isolate that has been tested for antibiotic susceptibility in broth was resistant to tetracycline (10 μg/ml), doxycycline (10 μg/ml), vancomycin (2.5 μg/ml), erythromycin

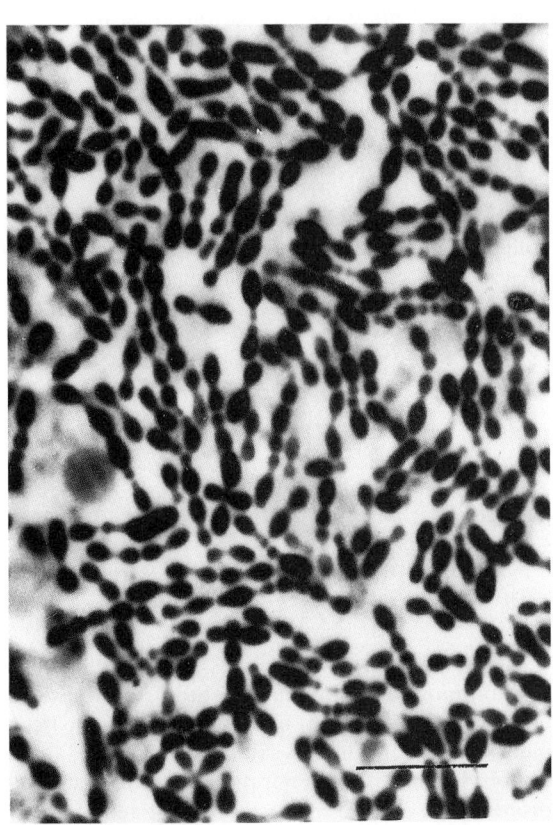

FIGURE BXII.α.209. *G. formicilis* strain SC3/5. Phase-contrast photomicrograph of broth culture. Bar = 5 μm. (Reproduced with permission from S.C. Croucher and E.M. Barnes, Revue de l'Institut Pasteur de Lyon 14: 95–102, 1981, ©Institut Pasteur de Lyon.)

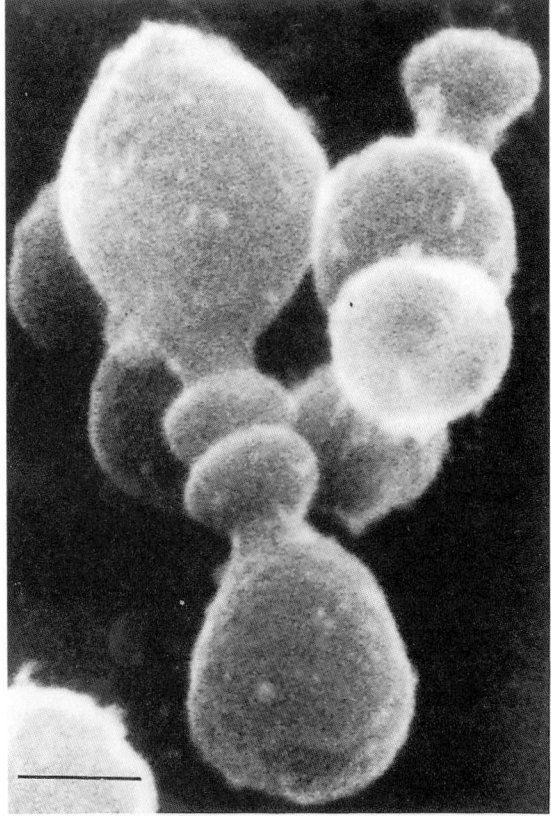

FIGURE BXII.α.210. *G. formicilis* strain L61. Scanning electron micrograph. Bar = 0.5 μm. (Reproduced with permission from J. Gossling and W.E.C. Moore, International Journal of Systematic Bacteriology 25: 202–207, 1975, ©International Union of Microbiological Societies.)

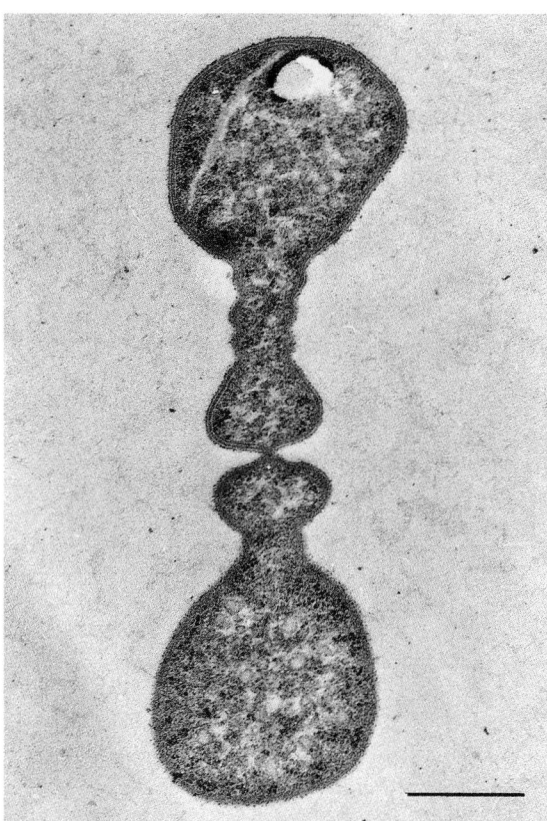

FIGURE BXII.α.211. *G. formicilis* strain L61. Transmission electron micrograph of a thin section. Bar = 0.5 μm. (Reproduced with permission from J. Gossling and W.E.C. Moore, International Journal of Systematic Bacteriology *25:* 202–207, 1975, ©International Union of Microbiological Societies.)

ENRICHMENT AND ISOLATION PROCEDURES

Via the roll tube technique, *Gemmiger* strains are isolated on media enriched with ruminal fluid, fecal extract, or liver extract and containing carbohydrates (glucose, maltose and/or cellobiose, soluble starch). No successful selection or enrichment techniques have been reported. Samples are diluted under strictly anaerobic conditions and streaked out on agar media for isolated colonies. After incubation for 3–7 days, large samples of colonies are picked for identification. *Gemmiger* can be distinguished microscopically by using phase contrast (Fig. BXII.α.209) or Gram stain (Mitsuoka, 1980).

MAINTENANCE PROCEDURES

Gemmiger strains are maintained by lyophilization, via methods suitable for strict anaerobes.

DIFFERENTIATION OF THE GENUS *GEMMIGER* FROM OTHER GENERA

The key feature that distinguishes *Gemmiger* from aberrant minicell-producing bacilli, budding cocci, or streptococci is the appearance of two small forms between two large ones (Fig. BXII.α.209). Metabolically, *Gemmiger* can be distinguished from most other intestinal anaerobes by their requirement for carbohydrates together with the production of butyrate and no gas.

TAXONOMIC COMMENTS

Gemmiger does not fit into any described higher taxon. Metabolically, these bacteria have been compared with other anaerobic intestinal bacilli, such as *Eubacterium* and the *Bacteroidaceae* (Moore and Holdeman, 1974; Salanitro et al., 1976), but *Gemmiger* has a higher mol% G + C of the DNA than do any of these. The mode of division may be similar to that of the facultative, prosthecate, freshwater bacterium *Ancalomicrobium* (Staley, 1968), but this bacterium has a higher mol% G + C of the DNA than does *Gemmiger*. Other bacteria that may have a similar mode of division are even more different metabolically (Whittenbury and Nicoll, 1971; Hirsch, 1974a).

Gemmiger have been grouped with the anaerobic cocci (Moore and Holdeman, 1974; Mitsuoka, 1980), but the formation of filaments in sublethal concentrations of penicillin (Salanitro et al., 1976) does not indicate a typical coccal morphology.

FURTHER READING

Croucher, S.C. and E.M. Barnes. 1983. The occurrence and properties of *Gemmiger formicilis* and related anaerobic budding bacteria in the avium caecum. J. Appl. Bacteriol. *54:* 7–22.

Gossling, J. and W.E.C. Moore. 1975. *Gemmiger formicilis*, n. gen., n. sp., an anaerobic budding bacterium from intestines. Int. J. Syst. Bacteriol. *25:* 202–207.

Salanitro, J.P., P.A. Muirhead and J.R. Goodman. 1976. Morphological and physiological characteristics of *Gemmiger formicilis* isolated from chicken ceca. Appl. Environ. Microbiol. *32:* 623–632.

(5 μg/ml), kanamycin (100 μg/ml), neomycin (100 μg/ml), and nalidixic acid (100 μg/ml); it was inhibited by tetracycline (100 μg/ml), doxycycline (100 μg/ml), vancomycin (5 μg/ml), erythromycin (10 μg/ml), and clindamycin (1 μg/ml) (Croucher and Barnes, 1983).

Gemmiger isolates have been obtained from chickens and humans. In chickens up to 6 weeks of age, these bacteria live in the lumen of the large intestine and the ceca, forming a significant proportion—often >10%—of the flora (Salanitro et al., 1976; Croucher and Barnes, 1983). In adult humans they comprise about 2% of the fecal bacteria, giving counts of about $10^{10}/g$ dry wt (Moore and Holdeman, 1974; Gossling and Moore, 1975; Holdeman et al., 1976). They presumably live on residual carbohydrates that are not digested or absorbed in the small intestine.

List of species of the genus Gemmiger

1. **Gemmiger formicilis** Gossling and Moore 1975, 206[AL]
   for.mi′ ci.lis. M.L. adj. *formicilis* pertaining to formic acid.

   Strictly anaerobic, mesophilic bacteria conforming to the generic description. Require glucose, fructose, maltose, or other carbohydrate-like substances (see generic description) for growth. Produce butyrate and usually formate and lactate and no gas from glucose and other sugars. The predominant product varies with the strain and the sugar used. The terminal pH in weakly buffered media is 4.8–6.0. Er-

ythritol, inositol, mannitol, melezitose, rhamnose, and ribose are not fermented. Indole, lecithinase, lipase, acetylmethylcarbinol, hydrogen sulfide, catalase, and urease are not produced. Nitrate is not reduced. Hippurate is not hydrolyzed. Casein and meat are not digested; there may be a weak action on gelatin (Gossling and Moore, 1975).

Study of chicken isolates indicates that most of them belong to one of two groups on the basis of fermentation products and the carbon sources used (Table BXII.α.179)

(Salanitro et al., 1976; Croucher and Barnes, 1983). Human isolates may be divided on the same characters, but the groups obtained do not correspond with the chicken groups.

All chicken isolates use raffinose and salicin; most human strains do not. However, the type strain is a human strain that uses salicin (Gossling and Moore, 1975).

It has been suggested that the human strains, which grow poorly in culture, do not use glucose on initial isolation, and do not produce formate, belong to a separate species (Holdeman et al., 1976); however, strains that normally produce formate may not do so when not growing well (Gossling and Moore, 1975).

The mol% G + C of the DNA is: 59 (T_m).

Type strain: ATCC 27749.

TABLE BXII.α.179. Differential characteristics of the groups of *Gemmiger formicilis* from poultry[a]

Characteristic	Group 1	Group 2
Major fermentation product from glucose:		
Butyrate	−	+
Lactate	+	−
Growth on:		
Cellobiose	+	−
Mannose	+	−
Sucrose	+	−
Trehalose	+	−

[a]Symbols:, −, 90% or more of strains are negative; and +, 90% or more of strains are positive.

Genus XII. **Labrys** *Vasilyeva and Semenov 1985, 375*[VP] *(Effective publication: Vasilyeva and Semenov 1984, 92)*

LEANA V. VASILYEVA

Lab' rys. Gr. n. *Labrys* double-headed ax, an organism resembling a Minoan ax, by the shape of the cell.

Unicellular flat bacterium; cells possess **triangular radial symmetry**. Dimensions are 1.1–1.3 × 1.3–1.5 µm. **Two to three tapering, short prosthecae** (<0.6 µm) protrude from two corners of the triangle; the third remains free and is associated with multiplication.

Cells divide by budding. Buds are produced directly from the mother cell at the tip of the triangle that lacks prosthecae. In this stage, the cell resembles a double-headed ax or labrys.

Gram negative, nonmotile, and do not possess fimbriae.

Obligately aerobic, nonfermentative, **chemoorganotrophic**. Utilize carbohydrates and some organic acids as sole carbon and energy sources. The type strain requires B vitamins for growth. Oxidase and catalase positive. Found in freshwater lakes.

The mol% G + C of the DNA is: 67.9.

Type species: **Labrys monachus** Vasilyeva and Semenov 1985, 375 (Effective publication: Vasilyeva and Semenov 1984, 92.)

ENRICHMENT AND ISOLATION PROCEDURES

Strain was isolated from the silt samples from Lake Mustjarv (Estonia). The medium is horse manure extract, obtained by heating dry manure 1% (w/v) in distilled water. The sediment is left to settle, and a liquid medium is prepared from the supernatant; 2% agar is added for solidification and colony isolation. Incubation is at 28°C for 10–14 d.

Minute colonies are isolated and examined under a phase-contrast microscope for the presence of bacteria with unusual morphology. Repeated subculturing of individual colonies produces a pure culture.

MAINTENANCE PROCEDURES

Potato agar provides a favorable growth medium. Growth occurs on MMB agar (see *Ancalomicrobium*) with glucose, but only if supplemented by up to 0.25% yeast extract (Vasilyeva and Semenov, 1984). The type strain can be maintained on slants in the refrigerator for at least 3 months. Lyophilization can be used for long-term preservation.

DIFFERENTIATION OF THE GENUS *LABRYS* FROM OTHER GENERA

Labrys combines morphological features typical of budding bacteria with prosthecae, such as *Prosthecomicrobium* and *Ancalomicrobium*, and of flat bacteria with radial symmetry, such as *Angulomicrobium* and *Stella* (Vasilyeva, 1980).

Physiological differences among these organisms are minimal.

Labrys, unlike *Angulomicrobium*, possesses prosthecae, which makes differentiation easy, in spite of the resemblance in cell shape and budding. Unlike *Ancalomicrobium*, *Labrys* has flat cells.

TAXONOMIC COMMENTS

Labrys combines features of a number of genera, and inclusion of these organisms into other genera would necessarily change their descriptions. Only a single strain has so far been isolated. DNA–DNA hybridization studies have revealed no homology with *Stella* (Lysenko et al., 1984). Further work is needed to reveal a phylogenetic interrelation among budding prosthecobacteria.

List of species of the genus Labrys

1. **Labrys monachus** Vasilyeva and Semenov 1985, 375[VP] (Effective publication: Vasilyeva and Semenov 1984, 92.)

   mon.ach' us. M.L., from ancient Gr. adj. *monachos* the only, unique, single.

   The description is as for the genus.

   Unicellular, flat, triangular, budding prosthecobacterium. Dimensions 1.1–1.5 µm. Prosthecae are short, <0.6 µm, tapering and protruding from two corners of triangle (Fig. BXII.α.212). Cells nonmotile. Gas vacuoles are not produced.

   Gram negative. On the outer cell surface is an irregular, visible capsular microlayer (Fig. BXII.α.213). No laminated membranous structures in the cells.

   Division is by budding on the corner that is free of prosthecae. The daughter cell separates when it approaches mother-cell size and shape (Fig. BXII.α.214).

   Utilizes the following carbohydrates as sole sources of carbon and energy: D-erythrose, D-ribose, L-arabinose, D-xylose, D-lyxose, D-glucose, L-sorbose, L-rhamnose, D-fruc-

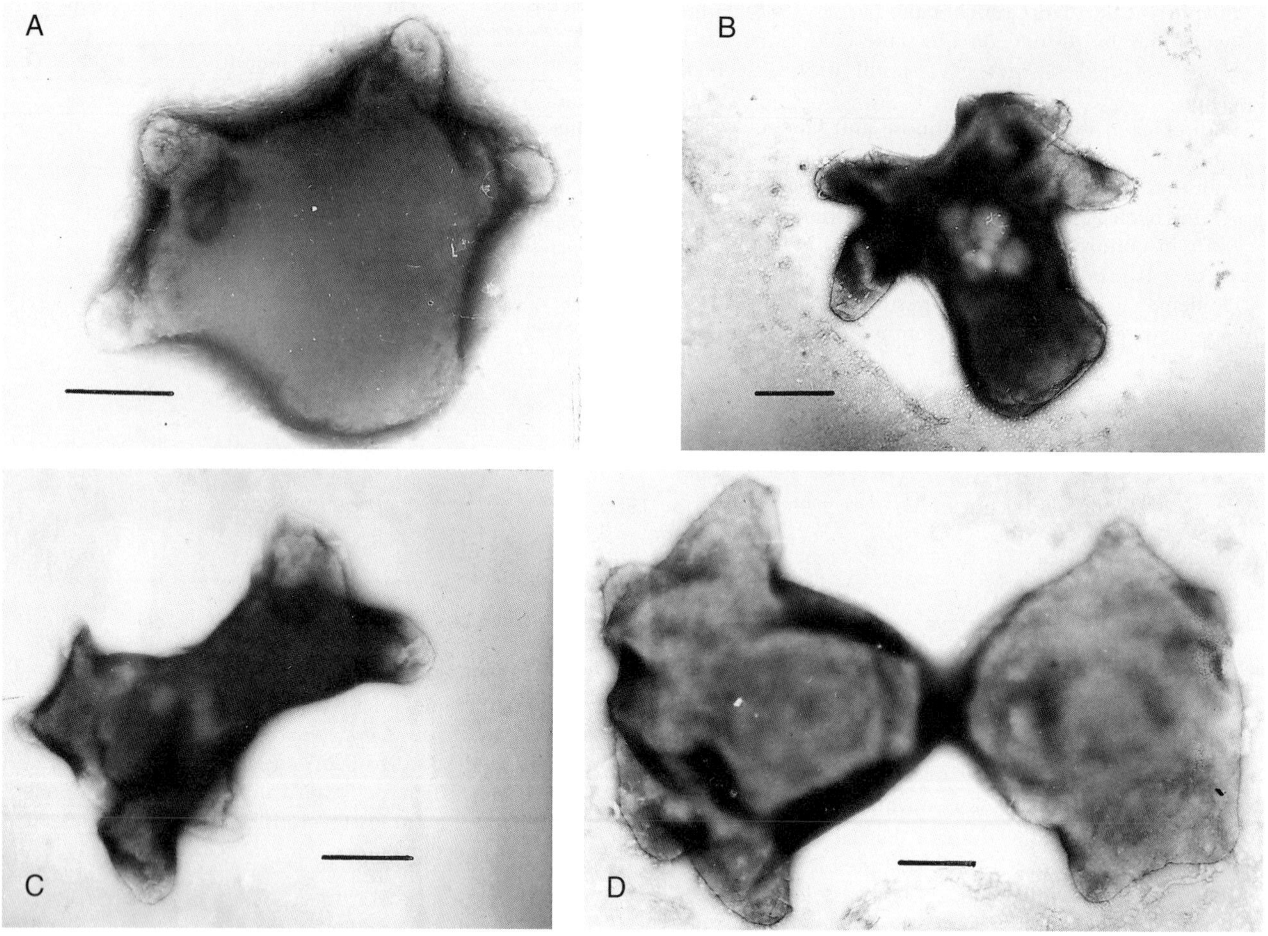

A

B

C

D

FIGURE BXII.α.212. *L. monachus. A–D,* electron microscopy showing cells in sequential stages of a multiplication. Note prosthecae protruding from two angles of the cell end, the "double ax" shape of the figure formed by the mother and daughter cells (*D*). Uranylacetate-negative stain. Bar = 0.5 μm.

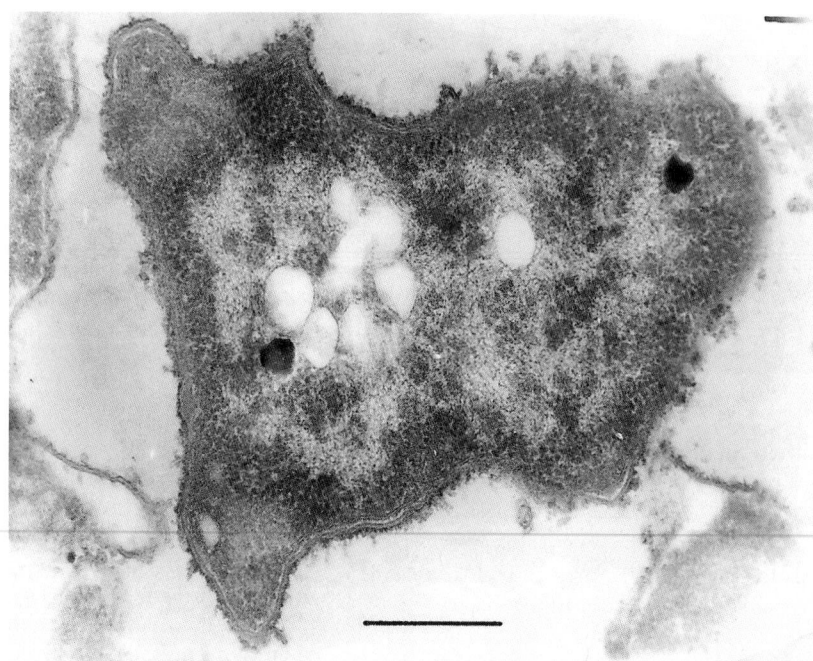

FIGURE BXII.α.213. *L. monachus.* Thin section. Note cytoplasm in prosthecae, granules of poly-β-hydroxybutyrate, and an external microcapsular layer on the outer side of the typical Gram-negative cell wall. Bar = 0.5 μm.

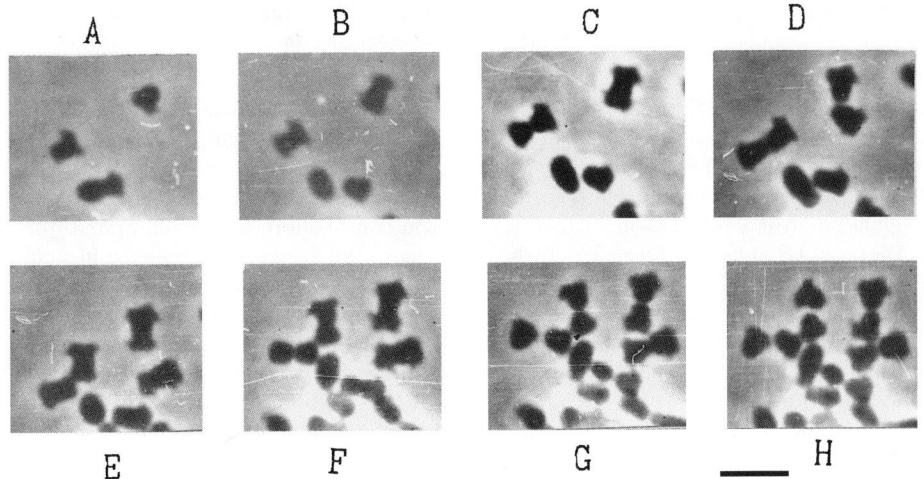

FIGURE BXII.α.214. *L. monachus.* Phase-contrast micrographs. Slide microculture. Hours: *A*, 0; *B*, 4; *C*, 7.5; *D*, 8.5; *E*, 11.5; *F*, 12.5; *G*, 13; and *H*, 15. Bar = 5.0 μm.

tose, D-talose, L-fucose, D-mannose, D-galactose, D-tagatose, trehalose, glycerol, mannitol, L-arabitol, adonitol, L-dulcitol, and D-sorbitol. Does not utilize D-lactose, D-cellobiose, D-melibiose, sucrose, D-raffinose, melezitose, starch, inulin, pectin, glycogen, dextrin, or inositol. The medium is acidified during growth on carbohydrates in all instances. Morphology is altered when luxurious growth occurs. Utilization of amino acids, organic acids, and alcohols is limited. Yeast extract above 0.25% inhibits growth; B vitamins are required for growth.

A considerable quantity of poly-β-hydroxybutyrate accumulates in cells immediately after inoculation on all utilized substrates.

Colonies circular, about 2 mm in diameter, colorless or gray, slightly convex, flat, opaque, glistening, smooth, and viscous.

Optimal temperature: 28°C; generation time: 8 h. Isolated from silt of Lake Mustjarv (Estonia).

The mol% G + C of the DNA is: 67.9 (T_m).

Type strain: 42, AUCM B-1479, ATCC 43932, DSM 5896.

Genus XIII. **Methylorhabdus** Doronina, Braus-Stromeyer, Leisinger and Trotsenko 1996a, 362[VP] (Effective publication: Doronina, Braus-Stromeyer, Leisinger and Trotsenko 1995, 97)

NINA V. DORONINA AND YURI A. TROTSENKO

Me.thy.lo.rhab' dus. Fr. *méthyle* the methyl radical; G. n. *rhabdos* rod; M.L. masc. n. *Methylorhabdus* methyl rod.

Rods 0.4–0.6 × 1.2–2.5 μm. No endospores. Gram negative. Nonmotile. Multiplication is by fission, with constriction. Aerobic, having a strictly respiratory type of metabolism with oxygen as the terminal electron acceptor. Nitrate is reduced to nitrite. Nonpigmented. Growth occurs on nutrient agar and peptone–yeast extract–glucose (PYG) agar. Optimal temperature, 28–34°C; optimal pH, 6.8–7.4. Methyl red and Voges–Proskauer negative. Catalase and urease positive. Oxidase negative. Indole is formed from L-tryptophan in the mineral medium with methanol as a sole carbon and energy source and with KNO_3 as the sole nitrogen source. Ammonium ions inhibit deamination of the tryptophan. Chemoorganotrophic. **Facultatively methylotrophic;** assimilate C_1 compounds (dichloromethane, methanol, and methylamine) by the isocitrate lyase-negative variant of the serine pathway. The major ubiquinone is Q-10. The cellular fatty acid profile is characterized by the presence of *cis*-vaccenic, cyclopropane, and palmitic acids ($C_{18:1 \omega 7}$, $C_{19:0 cyclo}$, and $C_{16:0}$, respectively). The dominant phospholipids are phosphatidylethanolamine, phosphatidylglycerol, phosphatidylcholine, and cardiolipin. Belongs to the *Alphaproteobacteria*.

The mol% G + C of the DNA is: 66.2 (T_m).

Type species: **Methylorhabdus multivorans** Doronina, Braus-Stromeyer, Leisinger and Trotsenko 1996a, 36 (Effective publication: Doronina, Braus-Stromeyer, Leisinger and Trotsenko 1995, 97.)

FURTHER DESCRIPTIVE INFORMATION

After division, cells remain connected by a constriction apparently formed by the outer membrane.

Facultative methylotroph. Dichloromethane-grown cells contain an inducible, glutathione-dependent dichloromethane dehalogenase, whereas methanol- or methylamine-grown cells possess the appropriate dehydrogenases. Formaldehyde is further oxidized to formate by glutathione-dependent formaldehyde dehydrogenase. The latter is oxidized to CO_2 by formate dehydrogenases. *Methylorhabdus multivorans* uses the isocitrate lyase-negative (icl^-) variant of the serine pathway and expresses hydroxypyruvate reductase, serine–glyoxylate aminotransferase, and malate lyase. Like other serine pathway methylobacteria, it has an NADP-dependent isocitrate dehydrogenase.

The primary assimilation of ammonia occurs by reductive amination of α-ketoglutarate.

ENRICHMENT AND ISOLATION PROCEDURES

Methylorhabdus multivorans was isolated on dichloromethane agar from an enrichment culture that had been inoculated with a groundwater sample (Doronina et al., 1995).

MAINTENANCE PROCEDURES

The bacteria can be stored in liquid mineral medium with an appropriate C_1 substrate at 4°C for 1 month. For longer-term preservation, freeze-drying can be performed with a protectant (skim milk).

DIFFERENTIATION OF THE GENUS *METHYLORHABDUS* FROM OTHER GENERA

Methylorhabdus differs from some serine-pathway methylobacteria by its formation of constrictions as morphological features (Doronina et al., 1995). *Methylorhabdus multivorans* is distinguished from the genus *Methylobacterium* by its absence of pigmentation and by the presence of cyclopropane carboxylic acid in its cellular fatty acid composition. It is distinguished from nonpigmented members of the genus *Aminobacter* by its ability to use methanol, its expression of the isocitrate lyase-negative variant of the serine pathway, and the presence of amine dehydrogenase. It also differs from members of the genus *Hyphomicrobium* by its inability to form hyphae, the presence of the Q-10 quinone system, and the icl^- variant of the serine pathway, and its oxidation of methylamine by amine dehydrogenase. It differs from *Methylopila* by its oxidase-negative reaction and the presence of catalase. The major characteristics differentiating the genus *Methylorhabdus* from other related genera are summarized in Table BXII.α.180.

TAXONOMIC COMMENTS

In the phylogenetic tree derived from 16S rDNA sequences, the genus *Methylorhabdus* forms a distinct branch within the *Alphaproteobacteria*. The highest degree of relationship is found with *Xanthobacter agilis* (92.6% similarity), *Thiobacillus novellus* (95.7% similarity), and *Ancylobacter aquaticus* (96.2% similarity).

There is less than 10% DNA–DNA similarity with members of *Methylopila*, *Methylobacterium*, *Aminobacter*, *Methylarcula*, and *Hyphomicrobium*.

TABLE BXII.α.180. Major differentiating characteristics of facultative serine pathway methylobacteria belonging to various genera[a]

Characteristics	*Methylarcula*	*Aminobacter*	*Hyphomicrobium*	"*Marinosulfonomonas*"	*Methylobacterium*	*Methylopila*	*Methylorhabdus*	"*Methylosulfonomonas*"
Morphology (flagella)	−	+	+	+	+	+	−	−
Reproduction by budding	−	−	+	+	−	−	−	−
Reproduction by division	+	+	−	−	+	+	+	+
Hyphae formation	−	−	−	+	−	−	−	−
Oxidase	−	+	+	−	+	−	+	+
Catalase	+	+	+	+	−	+	+	±
Carotenoids	−	+	−	−	−	−	−	−
Reduction of NO_3^- to NO_2^-	+	+	+	+	+	nd	nd	−
Methylamine metabolism:								
Amine dehydrogenase	+	+	−	−	+	nd	nd	−
N-Methylglutamate derivatives	−	+	+	+	−	nd	nd	+
γ-Glutamylmethylamide lyase	−	−	−	−	−	nd	nd	+
Isocitrate lyase	−	−	+	+	−	nd	nd	−
Cyclopropane acid, $C_{19:0\ cyclo}$	+	trace	+	+	+	+	+	+
Major ubiquinone	Q-10	Q-10	Q-10	Q-9	Q-10	nd	nd	Q-10
Tolerance to NaCl (%)	2	2.5	2.5	3	2	0.5	3.5	12
Growth at pH 10.0	−	−	−	−	−	−	−	+
Utilization of methanol	+	+	−	+	+	+	+	−
Mol% G + C of DNA	66.2	60–70	62-64	61–65	66–70	61	57	57–61

[a]Symbols: +, present; −, absent; nd, not determined.

List of species of the genus Methylorhabdus

1. **Methylorhabdus multivorans** Doronina, Braus-Stromeyer, Leisinger and Trotsenko 1996a, 36[VP] (Effective publication: Doronina, Braus-Stromeyer, Leisinger and Trotsenko 1995, 97.)

mul.ti'vo.rans. L. adj. *multus* much; L. part. adj. *vorans* devouring, digesting; M.L. part. adj. *multivorans* digesting many compounds.

The characteristics are as described for the genus, with the following additional features. Fig. BXII.α.215 illustrates the morphological features. Colonies on peptone agar medium are white, round with uneven edges, of uniform mucous consistency, and 2–3 mm in diameter. Growth occurs at 10–45°C and at pH 6.0–8.0. No growth occurs in the presence of 3% NaCl. Starch and gelatin are hydrolyzed slowly. Acidification of the medium occurs with glucose, but no gas is formed. Utilizable carbon sources are dichloromethane, methanol, methylamine, L-arabinose, D-xylose, D-glucose, D-mannitol, inositol, glycerol, dulcitol, adonitol, ethanol, isopropanol, acetate, α-ketoglutarate, fumarate, pyruvate, succinate, oxaloacetate, *cis*-aconitate, propionate, L-glutamate, L-alanine, sarcosine, and *N,N*-dimethylglycine. Neither methane nor $H_2/O_2/CO_2$ supports growth. Methylamine is oxidized to formaldehyde by amine dehydrogenase. The type (and only) strain was isolated from groundwater in Switzerland.

The mol% G + C of the DNA is: 66.2 (T_m).

Type strain: DM13, ATCC 51890, VKM B-2030.

GenBank accession number (16S rRNA): AF004845.

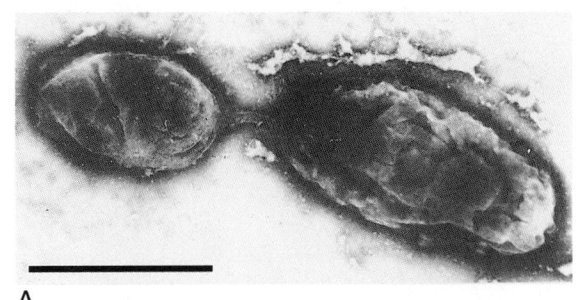

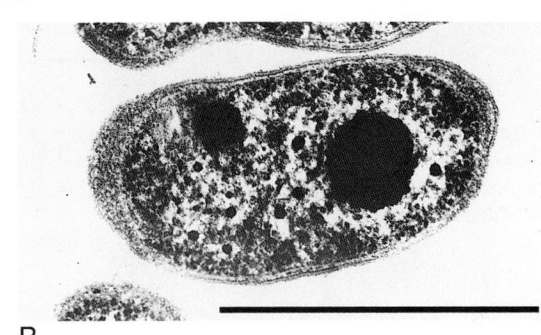

FIGURE BXII.α.215. *Methylorhabdus multivorans.* (*A*) Negatively stained cell. (*B*) Ultrathin section showing cell-wall structure and polyphosphate granules. Bars = 1 μm.

Genus XIV. **Pedomicrobium** Aristovskaya 1961, 957[AL] emend. Gebers and Beese 1988, 305

PETER HIRSCH AND RAINER GEBERS

Pe.do.mi.cro'bi.um. Gr. n. *pedon* soil; Gr. adj. *micros* small; Gr. masc. n. *bios* life; M.L. neut. n. *Pedomicrobium* soil microbe.

Cells oval, spherical, or rod-shaped, 0.4–2.0 × 0.4–2.5 μm. **Up to 5 hyphae** (i.e., cellular outgrowths or prosthecae of constant diameter with reproductive function) are formed per cell body. Hyphae are 0.15–0.2 μm in diameter. **At least one hypha originates laterally**; others may appear at the cell poles. Multiplication is by **budding at the hyphal tips, with the long axis of the bud arranged perpendicular to the hyphal axis** (Figs. BXII.α.216 and BXII.α.217). Mature buds either separate from the hyphae as uniflagellated swarmers or remain attached. Extracellular polymers can be stained with ruthenium red and sometimes are visible in India ink mounts as thick capsules around mother cells. **Oxidized iron or manganese compounds are deposited on mother cells and later on hyphae.** Resting stages are not known. **Gram negative**, older cells variable. Swarmer cells are motile by a single subpolar or polar flagellum (Fig. BXII.α.218). Other stages of the cell cycle (Fig. BXII.α.219) are nonmotile. Colonies are yellowish or reddish brown to dark brown, due to accumulated iron or manganese oxides. Aerobic. Catalase positive (test is in absence of MnO_2 and at neutral pH). **Chemoorganotrophic.** Acetate is utilized as a carbon source; most strains also grow on caproate or pyruvate. Protein digests, such as yeast extract, peptone, Casamino acids, and soytone, serve as carbon and/or ni-

trogen sources. Organic nitrogen sources utilized by most strains are glutamate, aspartate, glycine, serine, threonine, and valine. Inorganic nitrogen compounds allow only poor growth of some isolates; nitrate is reduced by most strains. **Slow and poor growth occurs on agar media with 0.1–1% fulvic acid iron sesquioxide complexes as sole carbon and nitrogen sources.** Vitamin mixtures stimulate growth; lack of vitamins results in pleomorphic cells, which produce large granules of poly-β-hydroxybutyric acid. Growth is inhibited by polymyxin B and neomycin.

A partial oligonucleotide catalogue of the type species, *P. ferrugineum* (IFAM S-1290) has been published (Stackebrandt et al., 1988a).

The mol% G + C of the DNA is: 63–66.8.

Type species: **Pedomicrobium ferrugineum** Aristovskaya 1961, 957, emend. Gebers and Beese 1988, 308.

FURTHER DESCRIPTIVE INFORMATION

Phylogenetic treatment The phylogenetic position of *Pedomicrobium ferrugineum* IFAM S-1290[T] and various other budding and prosthecate bacteria has been investigated by 16S rRNA cataloguing (Stackebrandt et al., 1988a), revealing an affiliation with *Alphaproteobacteria*. A coherent cluster has been found that em-

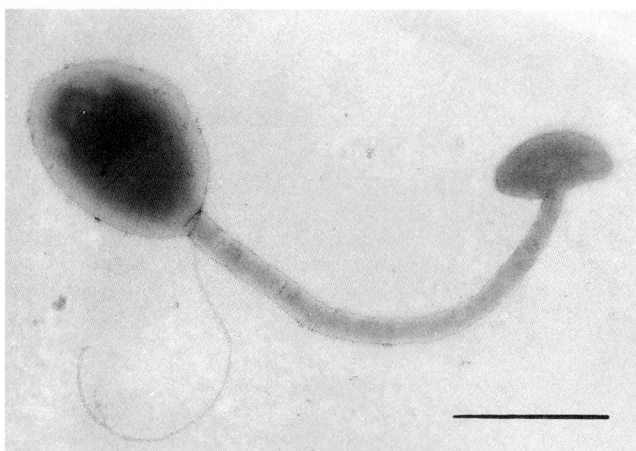

FIGURE BXII.α.216. *Pedomicrobium ferrugineum* IFAM S-1290[T]. Transmission electron micrograph showing a mother cell with bud and flagellum. Note the orientation of the bud perpendicular to the hyphal tip. Uranylacetate negatively stained. Bar = 1 μm.

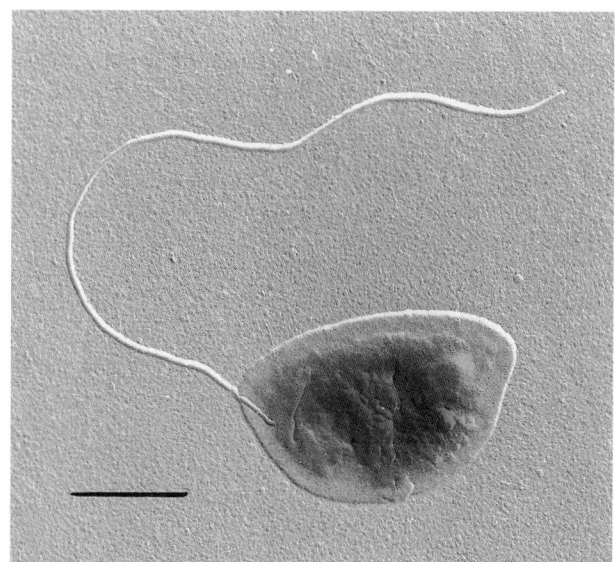

FIGURE BXII.α.218. *P. ferrugineum* IFAM S-1290[T]. Electron micrograph showing swarmer cell with subpolar flagellum. Pt/C-shadowed. Bar = 0.5μm.

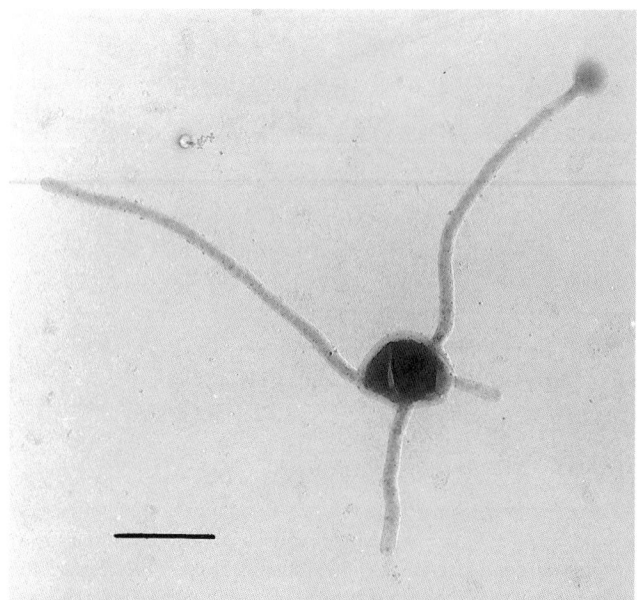

FIGURE BXII.α.217. *Pedomicrobium manganicum* IFAM E-1129[T]. Electron micrograph of mother cell with hyphae and young bud. Platinum/carbon-shadowed. Bar = 1 μm.

braces strains of the genera *Hyphomicrobium, Filomicrobium, Pedomicrobium,* and *Dichotomicrobium; Pedomicrobium ferrugineum* and *Filomicrobium fusiforme* are more closely related to some *Hyphomicrobium* strains than they are to each other (Fig. BXII.α.205; see chapter on *Dichotomicrobium*). All of these bacteria share morphological features, such as the formation of prosthecae (hyphae) and reproduction by a budding process. However, the validity of *Pedomicrobium* and *Filomicrobium* as genera separate from *Hyphomicrobium* has been questioned based on morphological traits, which are considered less important for taxonomic separation (Stackebrandt et al., 1988a).

Roggentin and Hirsch (1989), have made an attempt to determine the relationships among 19 *Hyphomicrobium* strains by DNA–rRNA hybrid thermal stability. This study also included *Pe-*

domicrobium ferrugineum IFAM S-1290[T], *Hyphomonas* (three strains), *Rhodomicrobium* (one strain), and *Dichotomicrobium* (one strain). The dendrogram derived from average linkage clustering depicts the distant relationships of the *Pedomicrobium* and *Dichotomicrobium* strains to the hyphomicrobia (Fig. BXII.α.220).

Cell morphology Cell shapes of some strains may vary; they can be rod- or spindle-shaped, tetrahedral, pear-, or bean-shaped. Hyphae vary in length according to cultural conditions; true branching occurs. Buds may arise in an intercalary fashion by localized hyphal swelling. Aristovskaya (1961) has occasionally observed direct budding and even cross division of single mother cells, but such observations have not been confirmed by other authors. Deposition of iron or manganese oxides has been observed among all strains studied so far (Aristovskaya, 1961; Khakmun, 1967; Gebers and Hirsch, 1978; Ghiorse and Hirsch, 1979). Many other bacteria with hyphae and buds may (depending on growth conditions) deposit iron or manganese oxides on their cell surfaces (Hirsch, 1968), but the perpendicular positioning of buds at the hyphal tips is a unique characteristic of pedomicrobia.

Chemotaxonomy The quinones, phospholipids, and fatty acids of ten *Pedomicrobium* strains have been studied (Sittig and Hirsch, 1992). All isolates have Q-10 as the respiratory quinone. The phospholipids present are phosphatidylglycerol, phosphatidylethanolamine, phosphatidylcholine, and bisphosphatidylglycerol. The fatty acid percentages are as follows: 8–22% normal, 64–89% unsaturated, 1–2% unsaturated/methylbranched, and 0–22% with a cyclopropane ring. Hydroxy fatty acids found in pedomicrobia were $C_{14:0\ 3OH}$ (82–84%), $C_{16:0\ 3OH}$ (5–8%), $C_{18:1\ 3OH}$ (4–5%), and $C_{18:0\ 3OH}$ (6–8%). A comparative study of pedomicrobia fatty acids (Eckhardt et al., 1979) indicates a strong influence of growth media and culture time on the percentage of octadecenoic acid.

The peptidoglycan of *P. ferrugineum* S-1290[T] has been studied following alternating treatments with Pronase P and Na-dodecylsulfate solution with mercaptoethanol (Roggentin and Hirsch, 1982). It has been found to have so little D-alanine that only

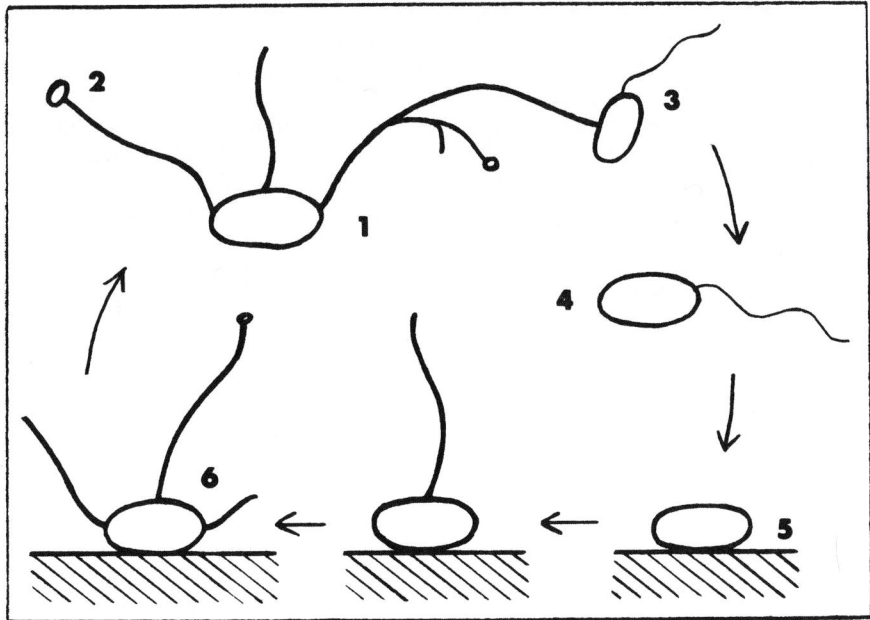

FIGURE BXII.α.219. Life cycle of *Pedomicrobium* species: *1*, mother cell with hyphae and buds; *2*, young bud; *3*, mature bud with flagellum; *4*, swarmer cell; *5*, young mother cell attached to solid surface; *6*, mature mother cell with hyphae and beginning bud formation.

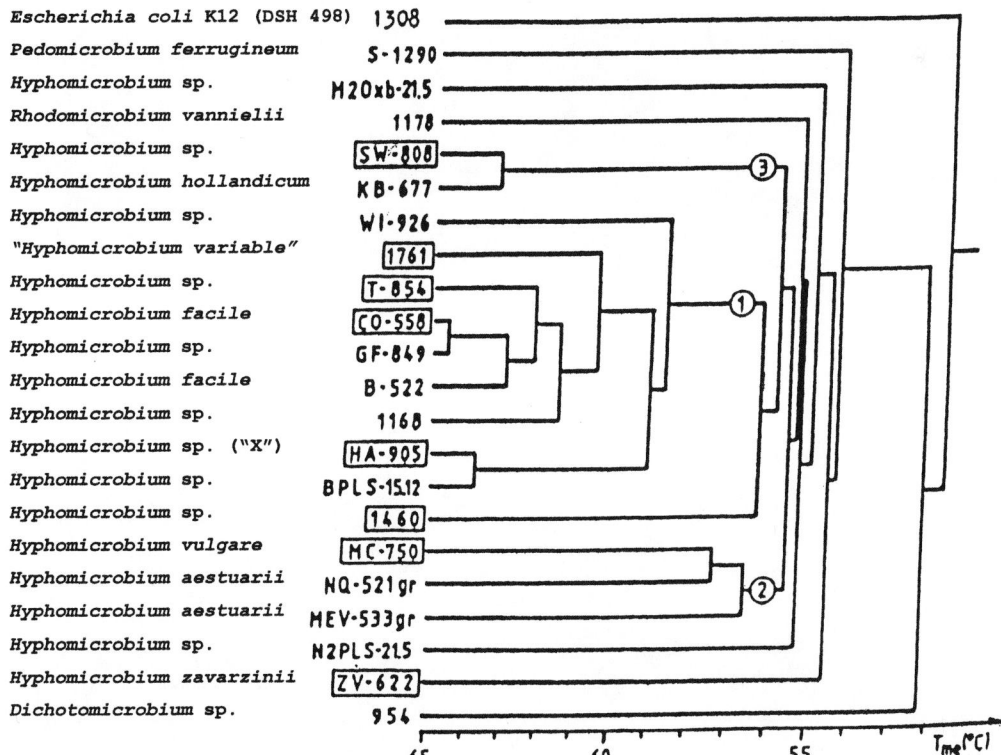

FIGURE BXII.α.220. Dendrogram derived from average linkage clustering showing relationships of *Pedomicrobium ferrugineum* IFAM S-1290[T], *Dichotomicrobium thermohalophilum* IFAM 954[T], *Rhodomicrobium vannielii* IFAM 1178[T] (ATCC 17100[T]), and various *Hyphomicrobium* strains. The data came from $T_{m(e)}$ values of DNA–rRNA duplexes. Strain numbers in boxes indicate strains that were used as ^{125}I labeled references (Roggentin and Hirsch 1989).

every eleventh side chain is cross-linked, a possible explanation as to why the cells can be pleomorphic on certain media.

DNA base compositions of pedomicrobia have been determined by Gebers et al. (1981a and 1985). They range from 64.8 to 65.7 mol% G + C (T_m) for *P. ferrugineum* isolates IFAM S-1290[T], P-1196, Q-1197, R-1198, and T-1130. *P. americanum* strains IFAM G-1381[T], BA-868, and BA-869 have 64.0, 64.7, and 64.4 mol% G + C, respectively (Gebers and Beese, 1988). For *P. manganicum* IFAM E-1129[T], the base ratio is 65.0 mol% G + C, and an unidentified strain (IFAM F-1225) has 64.9 mol% G + C. *Pedomicrobium australicum* strains IFAM ST-1306[T] and WD-1255 have 65.0 and 62.8 mol% G + C, respectively.

Fine structure Electron microscopy of ultrathin sections of *Pedomicrobium* isolates IFAM Ba-868 and IFAM Ba-869 reveals a structured surface layer external to the outer membrane layer of the Gram-negative wall. This layer is composed of subunits with approximately 18.5 nm center-to-center spacing (Ghiorse and Hirsch, 1979). Additionally, an electron-dense polymer material is discovered on the mother cell surface of both strains, especially when ruthenium red is included in the fixative (Fig. BXII.α.221). The iron or manganese oxide deposits are present on or within this surface polymer layer. Up to three granules of poly-β-hydroxybutyric acid are stored per cell. Small, dense granules stainable with Loeffler methylene blue are suggestive of polyphosphate storage.

Cultural characteristics All pedomicrobia are aerobic to microaerophilic and grow in the range of 1–21% pO$_2$ (Gebers and Beese, 1988). On solid media, two types of colonies may develop. Type 1 colonies are round and convex and have even or radially frayed edges; sometimes they exhibit concentric rings. These colonies have a soft consistency. Type 2 colonies are round, flat and even; they may be crateriform (due to growth into the agar), and the edges are even or radially frayed. The cells in the center of these colonies are often lysed, giving colonies a granular appearance. Type 2 colonies have a cartilaginous consistency and may be removed from the agar intact. Upon spreading on solid media, both colony types give rise to type 1 and 2 colonies. Cultivation with frequent transfers to fresh medium favors large colonies over small ones.

Nutrition and growth conditions Generally, organic acids appear to be the most appropriate carbon sources for *Pedomicrobium* species. Alcohols and carbohydrates are utilized only in some cases (Tables BXII.α.181 and BXII.α.182). Several amino acids serve as nitrogen sources. All strains grow at 20–30°C and in the pH range of 7–9.

Genetics, molecular data Levels of genetic relatedness of nine *Pedomicrobium* isolates with the labeled type strain of *Hyphomicrobium vulgare* (IFAM MC-750[T]) have been determined by DNA–DNA hybridization under optimal conditions; Table BXII.α.183 shows homologies of these strains with *H. vulgare* to be only 2–7% (Gebers et al., 1986). The *Pedomicrobium* hybridization data clearly indicate that the four type strains are distinct species.

Genome sizes have been determined for five *Pedomicrobium* strains by DNA renaturation kinetics. The M_r of these strains ranges from 2.81 to 3.0 × 10^9 Da (Kölbel-Boelke et al., 1985). Plotted against the G + C base ratios, pedomicrobia occupy an area quite distinct from those of other hyphal and budding gen-

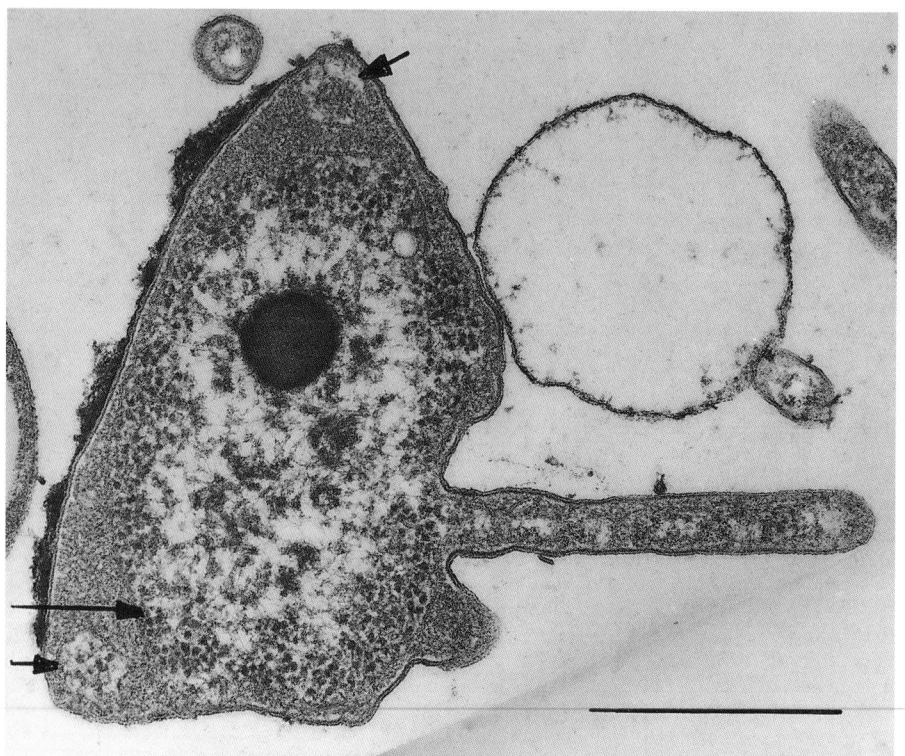

FIGURE BXII.α.221. Thin section of *Pedomicrobium australicum* IFAM ST-1306[T]. Note intracytoplasmic membranes (*arrows*) and dense polyphosphate granule. Deposition of heavy metal oxides occurred primarily on polymer located on the cell surface opposite to the hyphal outgrowth. Bar = 0.5 μm (Reproduced with permission from R. Gebers and M. Beese, International Journal of Systematic Bacteriology, *38*: 305–315, 1988, ©International Union of Microbiological Societies.)

TABLE BXII.α.181. Characteristics of species of the genus *Pedomicrobium*[a]

Characteristic	*Pedomicrobium ferrugineum*	*Pedomicrobium americanum*	*Pedomicrobium australicum*	*Pedomicrobium manganicum*
Cell size (μm)[b]	0.6–2.0 × 0.6–2.5	1.3 × 1.8	1.2 × 1.8	0.4–0.9 × 0.4–1.5
Flagella:				
Number	1	1	1	1
Attachment	polar or subpolar	subpolar	subpolar	subpolar
Intercalary buds formed occasionally	+	+	+	+
Fe oxide deposited with Fe powder	+	NT	NT	v[c]
Fe oxide deposited with FeS or iron paper clips	+	NT	NT	w
Mn oxide deposited with MnSO$_4$·H$_2$O	−	+	+	+
Utilization of acetate as carbon source	+	+	+	+
Utilization of ethanol	−[d]	−[d]	−[d]	NT
Growth with phenol	v[b]	−[d]	NT	NT
Growth with propanol	v[b]	−[d]	−[d]	NT
Growth with tartrate	w	NT	NT	NT
Growth with cholesterol or paraffin	w	NT	NT	NT
Utilization of amino acids for growth:[b]				
L-Arginine or L-lysine	+	+	w	NT
L-Isoleucine or L-leucine	+	+	w	NT
L-Aspartate or glycine	+	+	+	NT
L-Glutamate or L-histidine	+	+	+	NT
L-Phenylalanine	+	+	+	NT
L-Proline	+	−	w	NT
L-Serine, L-threonine, or D-valine	+	+	+	NT
L-Tyrosine	+	+	NT	NT
Growth with ammonia as N-source	+	+	+	NT
Growth with 0.05% (w/v) of yeast extract or peptone	+	+	+	w
Growth with urea as N-source	w	NT	NT	NT
Generation time (h) in PSM at 30°C	10	10	11	NT
Storage of PHBA and polyphosphate	+	+	+	+
Vitamin Solution No. 6	Required	Required	NT	Stimulation
Catalase activity	+	D	+	+
Nitrate reduction	+	+	+	NT
Pathogenic for guinea pigs	−	NT	NT	−

[a]Symbols: +, 90% or more strains positive; −, 90% or more strains negative; v, variable with growth conditions; D, differs among strains; w, growth weak and slow; NT, not tested or information lacking.

[b]Data from Gebers (1981) or Gebers and Beese (1988).

[c]W.C. Ghiorse, personal communication.

[d]Inhibition of growth.

era, such as *Hyphomicrobium*, *Hyphomonas*, *Dichotomicrobium* ("genus D"), *Filomicrobium* ("genus F"), or "genus T".

Bacteriophages Samples from a freshwater lake (Fuhlensee, Kiel, Germany), inoculated with cultures of *Pedomicrobium americanum* IFAM G-1381[T] and shaken at 30°C, have resulted in the isolation of phage Pe-60 (C. Gliesche, personal communication). This DNA phage adsorbs to *Pedomicrobium* mother cells as well as to the hyphae (Figs. BXII.α.222 and BXII.α.223). The following IFAM pedomicrobia are lysed by phage Pe-60: G-1381[T], WD-1355, BA-868, BA-869, ST-1306[T], and S-1290[T]; however, two *Hyphomicrobium* isolates (B-522 and WI-926) are not susceptible (C. Gliesche, unpublished). Pe-60 has been partially characterized by Majewski (1986). It has a base composition of 57.2 mol% G + C, the molecular weight is 66.86 MDa, and it can be stored under chloroform or lyophilized. Phage Pe-60 belongs to the *Styloviridae*.

Antibiotic sensitivities The strains of *P. ferrugineum*, *P. americanum*, and *P. australicum* all show high sensitivity to eight antibiotics, but low inhibition by sulfanilamide (Table BXII.α.184). *P. manganicum* is not inhibited by most of these antibiotics (Gebers, 1981; Gebers and Beese, 1988).

Pathogenicity *Pedomicrobium ferrugineum* IFAM S-1290[T] and *P. manganicum* IFAM E-1129[T] are not pathogenic for mice or guinea pigs within 7 days after intraperitoneal or intravenous injection of up to 10^8 cells. They are not β-hemolytic and do not utilize mannitol, but all strains cause antibody formation, with agglutination dilution titers reaching 1:2 (Famureva et al., 1983).

TABLE BXII.α.182. Growth of *Pedomicrobium* IFAM strains with selected carbon sources[a]

| Carbon source | *Pedomicrobium ferrugineum* | | | | | *Pedomicrobium americanum*[b] | | | *Pedomicrobium australicum*[b] | | *P. manganicum* |
	S-1290[Tb]	P-1196[c]	Q-1197[c]	R-1198[c]	T-1130[c]	G-1381[T]	BA-868	BA-869	ST-1306[T]	WD-1355	E-1129[Tc]
Formate	−	+	+	−	−	w	+	+	+	w	w
Acetate	+	+	+	+	+	+	+	+	+	+	+
Caproate	+	+	+	+	+	−	−	−	−	w	+
Pyruvate	+	+	+	+	+	+	+	−	−	+	+
D-Lactate	−	w	w	−	−	−	−	−	−	−	+
Citrate	−	NT	NT	NT	NT	−	−	+	−	−	NT
D,L-Malate	w	NT	NT	NT	NT	w	+	+	+	+	NT
Succinate	+	NT	NT	NT	NT	+	w	+	NT	w	NT
D-Glucose	−	+	+	−	−	−	−	−	+	−	w
Lactose	−	NT	NT	NT	NT	−	w	−	+	+	NT
D-Ribose	−	NT	NT	NT	NT	−	−	−	−	−	+
Ethanol	−	NT	NT	NT	NT	−	−	−	−	w	NT
Gluconate	−	NT	NT	NT	NT	w	w	+	+	+	NT
Glycerol	−	NT	NT	NT	NT	w	+	+	+	−	NT
D-Mannitol	w	NT	NT	NT	NT	w	+	−	+	−	NT
Methanol	−	NT	NT	NT	NT	−	−	−	−	w	NT

[a]Symbols: +, positive growth; −, no growth; w, growth weak and slow; NT, not tested or information lacking.

[b]Data from Gebers and Beese (1988).

[c]Data from Gebers (1981).

TABLE BXII.α.183. Levels of DNA–DNA similarity among *Pedomicrobium* and selected *Hyphomicrobium* strains[a]

| Taxon | Source of unlabeled DNA | | % Similarity with labeled DNA from strain | | | | |
	IFAM[b] strain no.	ATCC[c] strain no.	IFAM S-1290[T]	IFAM G-1381[T]	IFAM BA-869	IFAM ST-1306[T]	IFAM E-1129[T]
Pedomicrobium ferrugineum	S-1290[T]	33119[T]	100	16	14	11	14
	Q-1197	33117	99[d]	17	13	5	12
	T-1130	33120	97[d]	18	11	10	9
Pedomicrobium manganicum	E-1129[T]	33121[T]	13[d]	10	7	3	100
Pedomicrobium americanum	G-1381[T]	43612[T]	NT	100	100	28	8
	BA-868	43613	18	91	94	NT	8
	BA-869	43615	22	82	100	20	6
Pedomicrobium australicum	ST-1306[T]	43611[T]	18	25	26	100	8
	WD-1355	43614	17	25	29	87	8
Hyphomicrobium vulgare	MC-750[T]	27500[T]	7[d]/(4)[e]	NT	2/(4)	(5)	2/(5)
Hyphomicrobium sp.	T-854 (T37)		5	NT	2	NT	1

[a]Data from Gebers and Beese (1988). Similarity data of at least two reactions, corrected for background values obtained with self-reassociation controls.

[b]IFAM, Institut für Allgemeine Mikrobiologie, Universitt Kiel, Germany.

[c]ATCC, American Type Culture Collection, Rockville, Md., USA.

[d]Data from Gebers et al. (1981b).

[e]Data in parentheses are values for reciprocal reactions (Gebers et al., 1986).

Ecology *Pedomicrobium* species are widely distributed in podzolic and other soils, in freshwater lakes, ponds, and brooks, in iron springs, and in seawater; they are ubiquitous (Table BXII.α.185).

ENRICHMENT AND ISOLATION PROCEDURES

Isolation of *P. ferrugineum* and *P. manganicum* from soil, especially from podzolic soil, can be achieved by Aristovskaya's procedure (Aristovskaya, 1961) as described by Gebers and Hirsch (1978, 1979). Podzolic soil samples are suspended by repeatedly shaking in 0.85% (w/v) saline solution and are streaked onto humic gel agar[1]. Since fulvic acids serve as sole carbon and nitrogen sources, this medium is rather selective, but allows only slow growth of *Pedomicrobium* species. After 3–12 weeks incubation at 20 or 30°C

in the dark, *Pedomicrobium*-containing colonies of the agar plate "enrichment" may be recognized by their yellowish brown to dark brown color, due to accumulation of iron and/or manganese oxides. Identification of iron-depositing colonies by the Prussian blue reaction is difficult because humic gel usually contains Fe (III) and thus stains intensely. Screening for manganese-depositing colonies, however, is facilitated by flooding the enrichment plate with leuko-berbelin blue I or leuko-crystal violet. Since *Pedomicrobium*-containing colonies are strongly coherent, spreading onto agar plates is enhanced by sterile grinding of the inoculum in a drop of saline solution. Pedomicrobia obtained with this agar plate enrichment procedure grow, albeit slowly, on humic gel agar without any other additions.

For final purification streaks, humic gel agar is supplemented with a vitamin solution[2]. This is necessary because *Pedomicrobium*

1. Humic gel agar: Humic gel (fulvic acids complexed with metal sesquioxides) is prepared by hydrochloric acid extraction from (podzolic) humus soil (Ponomareva, 1964; Gebers and Hirsch, 1978). 5 g (wet weight) humic gel and 18 g agar (Difco) are suspended in 1000 ml of distilled water and autoclaved; the mixture is poured as thick layers into Petri plates; the final pH is 5–6.

2. Vitamin solution (Van Ert and Staley, 1971) consists of: 2 mg biotin, 2 mg folic acid, 5 mg thiamin·HCl, 5 mg calcium pantothenate, 0.1 mg cyanocobalamin, 5 mg riboflavin, 5 mg nicotinamide, 5 mg *p*-aminobenzoic acid, 10 mg pyridoxin·HCl, and 1000 ml distilled water.

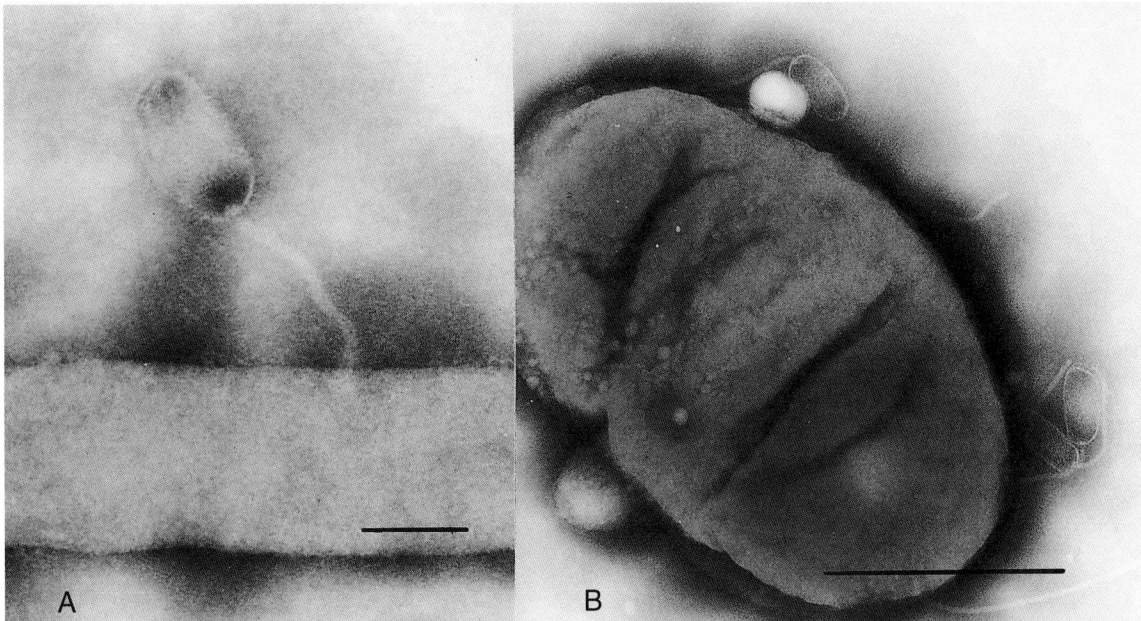

FIGURE BXII.α.222. Adsorption of phage Pe-60 of *Pedomicrobium americanum* IFAM G-1381[T] to a hypha (*A*) or mother cell (*B*). Negatively stained with 1% uranylacetate. Bars = 0.1 μm (*A*) and 0.5 μm (*B*). (Courtesy of C. Gliesche and D.M. Majewski, unpublished.)

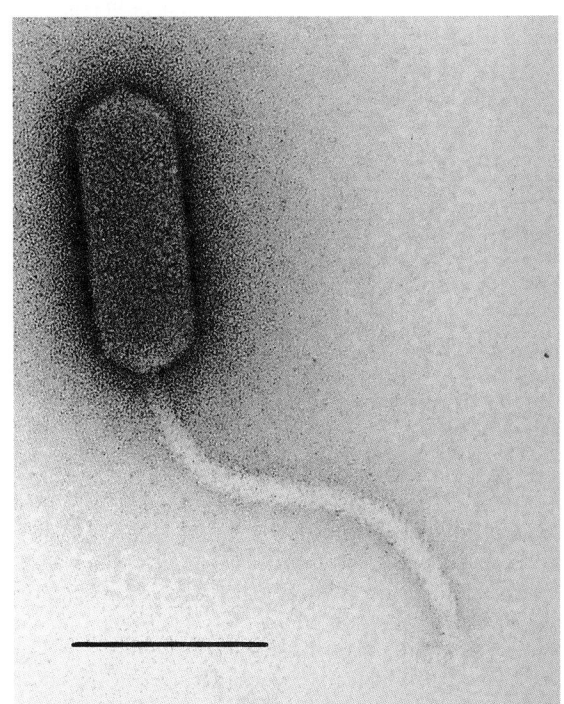

FIGURE BXII.α.223. Phage Pe-60 (*Styloviridae*) from *Pedomicrobium americanum* IFAM G-1381[T]. Electron micrograph on carbon-coated mica negatively stained with 1% uranylacetate. Bar = 0.1 μm. (Courtesy of C. Gliesche and D.M. Majewski, unpublished.)

cells have an irregular, pleomorphic appearance when grown without vitamins. Pedomicrobia also grow at pH 7.0 in synthetic media, which contain (besides some nutrients) either 400 mg/ 50 ml iron powder, an iron paper clip (400 mg), or 630 mg FeS (Gebers and Hirsch, 1979). *Pedomicrobium* standard medium

(PSM)[3] supplemented with 0.015% (w/v) Actidione (Roth, Germany) and 1.8% agar may serve as an alternative medium for enrichment and isolation of pedomicrobia. Actidione inhibits many fungi, but this medium also allows growth of contaminating bacteria; therefore, the selectivity for *Pedomicrobium* species is low.

Aquatic strains of *Pedomicrobium* spp. have been enriched by Staley et al. (1980) as follows: water samples are diluted in a series of test tubes with peptone medium DPM[4] up to the 10^{-9} dilution. Inoculated tubes are incubated at 20–30°C until surface pellicles develop in the highest dilution tubes. For isolation of pedomicrobia, the pellicles are streaked onto DPM solidified with 1.8% agar. A combination of growth on indicator medium PC and micromanipulation has enabled Sly and Arunpairojana (1987) to isolate manganese-oxidizing pedomicrobia from freshwater. The PC medium employed here contains (w/v): 0.005% Difco yeast extract, 0.002% $MnSO_4 \cdot 4H_2O$, 2% Bacto agar, and 50 μg/ml of Actidione (Upjohn, USA).

MAINTENANCE PROCEDURES

Pedomicrobium species grown on PSM agar slants survive for more than 2 years in tightly sealed screw-capped tubes at 20–30°C. For longer preservation, 1 part of an exponentially growing liquid

3. PSM consists of 10 mM sodium acetate, 0.5 g yeast extract (Difco), 1 ml "metals 44" (see below), 10 ml vitamin solution, and distilled water to 1000 ml. Adjust the pH to 9.0; final pH after autoclaving: 7.0. "Metals 44" (Cohen-Bazire et al., 1957) consists of: 125 mg EDTA, 547.5 mg $ZnSO_4 \cdot 7H_2O$, 250 mg $FeSO_4 \cdot 7H_2O$, 77 mg $MnSO_4 \cdot H_2O$, 19.6 mg $CuSO_4 \cdot 5H_2O$, 12.4 mg $Co(NO_3)_2 \cdot 6H_2O$, 8.85 mg $Na_2B_4O_4 \cdot 10H_2O$, and 50 ml distilled water. Adjust pH to 6.8.

4. DPM consists of 0.1 g peptone (Difco), 20 ml Hutner's modified salts solution (see below), 10 ml vitamin solution, and distilled water up to 1000 ml. Hutner's modified salts solution (Cohen-Bazire et al., 1957) contains (per liter, w/v): 10 g nitrilotriacetic acid, 29.7 g $MgSO_4 \cdot 7H_2O$, 3.3 g $CaCl_2 \cdot 2H_2O$, 12.7 mg $NaMoO_4 \cdot 2H_2O$, 99.0 mg $FeSO_4 \cdot H_2O$, and 50 ml "metals 44". Distilled water is added after pH adjustment. The nitrilotriacetic acid is first neutralized with potassium hydroxide. Then, the remaining ingredients are added before the pH is adjusted to 7.2 with KOH and H_2SO_4. Finally, distilled water is added to make 1 liter of solution.

TABLE BXII.α.184.　Growth inhibition of IFAM *Pedomicrobium* strains by antibiotics (100 μg/ml)

Antibiotic	*Pedomicrobium ferrugineum* S-1290[T]	*Pedomicrobium americanum*			*Pedomicrobium australicum*		*Pedomicrobium manganicum* E-1129[T]
		G-1381[T]	BA-868	BA-869	ST-1306[T]	WD-1355	
Liquid cultures:[a]							
Ampicillin	82	95	93	91	91	91	−
Penicillin G	87	93	97	90	97	95	NT
Cephalothin	+[b]	77	68	+[b]	+[b]	84	−
Cycloserine	82	95	96	92	94	95	+[c]
Polymyxin B	+[b]	+[b]	+[b]	+[b]	+[b]	+[b]	+[c]
Neomycin	84	84	82	84	83	95	+ +[c]
Chloramphenicol	79	92	98	88	98	82	−
Streptomycin	+[b]	+[b]	+[b]	+[b]	+[b]	+[b]	NT
Sulfanilamide	16	3	22	47	8	33	−
Plate diffusion tests:[c]							
Nalidixic acid	−	NT	NT	NT	NT	NT	−
Rifampin	+ +[c]	NT	NT	NT	NT	NT	−
Tetracycline	+ +[c]	NT	NT	NT	NT	NT	−
Gentamicin	−	NT	NT	NT	NT	NT	+ +[c]
Nitrofurazon	−	NT	NT	NT	NT	NT	−

[a]Data from Gebers and Beese (1988). Percent inhibition of protein formation as compared to untreated cultures. Symbols: +, strong inhibition (~60–100%); + +, inhibition zone >10 mm; +, inhibition zone 1–10 mm; −, no inhibition; NT, not tested.

[b]Inhibition estimated by visual comparison of culture turbidities.

[c]Data from Gebers 1981, determined by plate diffusion tests on PSM agar.

culture is freeze-dried at −55°C; addition of milk can be omitted. Viability of these lyophils needs to be tested frequently, and counting of the percentage of survivors is necessary. Medium PSM, either autoclaved or sterile-filtered, is recommended for the revival of such lyophilized cultures and for subculturing. Freeze-dried cultures are stored at 4°C and should not be kept for more than 2–3 years. Storage in liquid nitrogen may be required.

PROCEDURES FOR TESTING SPECIAL CHARACTERS

Testing for iron and manganese oxide accumulation from fulvic acid sesquioxide complexes is impractical because these humic substances are available solely from podzolic soils, which occur only in certain regions on earth. Therefore, it is recommended that instead of fulvic acid complexes embedded in agar media, elemental iron, FeS, or $MnSO_4$, be employed. The specificity of the test is reduced, however, since autoxidation of elemental iron and pyrite occurs at physiological pH, and various bacteria are capable of accumulating the oxidized products. The presence of ferric iron is indicated by a blue color around cells or colonies when a solution of 2% $K_4[Fe(CN)_6]$, acidified with HCl, is added (Prussian blue reaction). The presence of manganese (IV) is demonstrated by a blue color reaction with a 0.4% (w/v) solution of leuko-berbelin blue I (Krumbein and Altmann, 1973) or with Feigl's reagent (1% [w/v] benzidiniumhydrochloride in a 7% [v/v] acetic acid solution). In contrast to Feigl's reagent, leuko-berbelin blue I does not inhibit viability of the cells.

Methods used for cell wall disintegration, DNA extraction, and purification have been described by Gebers et al. (1985). Shearing of DNA and radioactive labeling have been described by Gebers et al. (1981b, 1986). DNA reassociation procedures, S1 nuclease treatment, and specific conditions for hybridization reactions are given by Gebers and Beese (1988).

DIFFERENTIATION OF THE GENUS *PEDOMICROBIUM* FROM OTHER GENERA

Table BXII.α.186 provides primary characteristics that can be used to differentiate the genus *Pedomicrobium* from morphologically similar taxa. The most important property of pedomicrobia

is the bud elongation perpendicular to the hyphal long axis; deposition of oxidized iron or manganese on the cell surface is not restricted to *Pedomicrobium* species, but either is often dependent on the production of polyanionic surface polymers (Ghiorse and Hirsch, 1978a).

TAXONOMIC COMMENTS

Originally, five species of *Pedomicrobium* were described: *P. ferrugineum*, *P. manganicum* (Aristovskaya, 1961), "*P. podsolicum*" (Aristovskaya, 1963), *P. americanum*, and *P. australicum* (Gebers and Beese, 1988). None of several *Pedomicrobium*-like isolates known fit the description of "*P. podsolicum*", and cultures of this species are not available. In addition, it is not mentioned in the Approved Lists of Bacterial Names.

Some manganese oxide-depositing bacteria discussed in the literature have been considered to be pedomicrobia, either because of their growth on organomineral complexes of fulvic acids and/or because of the presence of a large number of hyphae on the mother cells. Khakmun (1967) has observed such bacteria in soils of Sakhalin and named them *Pedomicrobium manganicum* biovar sachalinicum because of their more oval cell shape. Neither a pure culture nor a formal description of these is available. A micrograph published shows *Hyphomicrobium*-like, possibly budding bacteria, but lacks any indication of a perpendicular bud location on the hyphal tips. Epicellular iron-depositing aquatic bacteria studied by Hirsch (1968) or Hirsch and Rheinheimer (1968) were originally considered to represent pedomicrobia, but these bacteria also resemble hyphomicrobia and lack the perpendicular bud orientation on the hyphal tips. A bacterial isolate (strain T37) with manganese oxide deposition on the cell surface has been obtained from manganese oxide sludge on the inside of hydroelectric pipelines (Tyler and Marshall, 1967). This organism resembles hyphomicrobia morphologically, grows with methanol, and does not show the characteristic positioning of the buds. After subculturing, it eventually loses the ability to oxidize and deposit manganese, but instead oxidizes iron (Ghiorse and Hirsch, 1978b).

Real pedomicrobia have been observed by Kutuzova et al. (1972) after exposing slit peloscopes in ooze-containing vessels.

TABLE BXII.α.185. Differential characteristics of *Pedomicrobium* species[a]

Characteristic	*Pedomicrobium ferrugineum* IFAM S-1290	*Pedomicrobium americanum* IFAM G-1381	*Pedomicrobium australicum* IFAM ST-1306	*Pedomicrobium manganicum* IFAM E-1129
Cell shape:				
Coccoid	+			+
Oval	+	+	+	+
Rods	+	+	+	+
Spindles		+	+	
Tetrahedral	+	+	+	
Cell size (μm)	0.6–2.0 × 0.6–2.5[b]	1.3 × 1.8[b]	1.2 × 1.8[b]	0.4–0.9 × 0.4–1.5[c]
Hyphal origin:				
Laterally	+	+	+	+
Polarly	+	+	+	+
Subpolarly	+			
Number of hyphae per mother cell	1–3	1–3	1–3	1–5
Deposit Fe^{3+} (growth with fulvic acid sesquioxide complexes)	+	+	+	−
Deposit Fe^{3+} (growth in presence of Fe powder)	+	NT	NT	+[d]
Deposit Fe^{3+} (growth on PSM + $FeSO_4 \cdot 7H_2O$)	NT	+	+	NT
Deposit Mn oxide (growth on PSM + $MnSO_4 \cdot H_2O$)	−	+	+	+
Intracytoplasmic membranes may be formed	−	v	v	v
Growth with formate	−	w[b]	+	w
Growth with pyruvate	+	+	−	+
Growth with D,L-malate or D-mannitol	w[b]	w[b]	+	NT
Growth with lactose	−	−	+	NT
Growth with D-glucose	−	−	+	w
Growth with L-proline	+	−	w	NT
Utilization of $NaNO_3$ as N-source for growth	w	−	−	NT
Temperature range for growth (°C)	10[c]–43	15–41	15–36	NT
Temperature optimum for growth (°C)	29–30	32–38	29–32	30
Dependence on Vitamin Solution No.6 (Van Ert and Staley, 1971):				
Required	+	+		
Stimulated				+
Growth inhibition by ampicillin, cephalothin, or chloramphenicol	+	+	+	−
Habitat (source):				
Podsolic soil, Germany	+			
Freshwater pond and puddle, USA		+		
Fresh water, N. S. W., Australia			+	
Quartzite rock pool, France				+

[a]Symbols: +, positive; −, negative; w, weak or slow reaction; v, variable with growth conditions; NT, not tested or information lacking.

[b]Data from Gebers and Beese (1988).

[c]Data from Gebers (1981).

[d]W.C. Ghiorse (personal communication).

One of her electron micrographs labeled "*Pedomicrobium*-type organism" (Fig.1e) shows a tetrahedral mother cell, but with a straight bud attachment to the hypha, somewhat similar to that of *Dichotomicrobium* spp. However, Fig. 2g in that reference shows a bacterium with the morphology of a *Pedomicrobium* sp.; pure cultures have not been isolated.

Recently, Sly et al. (1988) have isolated eight strains of manganese-oxidizing pedomicrobia from "dirty drinking water" and biofilm samplings of southeast Queensland, Australia. Pure cultures have been obtained by micromanipulation and inoculation of PC agar, which contains (w/v): 0.005% Difco yeast extract, 0.002% $MnSO_4 \cdot 4H_2O$, and 2% Bacto agar (Sly and Arunpairojana, 1987). To control fungal growth, 50 μg/ml of Actidione (Pfizer, USA) were added. The morphology of these new isolates is thought to correspond closely to that of the *P. manganicum* type strain (IFAM E-1129[T]), although cells and mature swarmer

TABLE BXII.α.186. Differential characteristics of the genus *Pedomicrobium* and other morphologically similar taxa[a]

Characteristic	*Pedomicrobium*	*Hyphomicrobium*	*Filomicrobium*[b]	*Hyphomonas*	*Hirschia*[c]	*Dichotomicrobium*	*Rhodomicrobium*
Cell shape:							
Bean-shaped	+	+		+	+		+
Coccoid	+			+	+	+	+
Cubical						+	
Fusiform			+				+
Ovoid	+	+		+	+		+
Pear-shaped		+		+	+		+
Rod-shaped	+	+		+	+		+
Spindles		+					+
Tetrahedral	+					+	+
Origin of hyphae:							
Lateral	+					+	
Polar	+	+	+	+	+	+	+
Number of hyphae per mother cell	1–5	1–4	2–3	1–2	1–2	1–4	1–2
Hyphae with septa	−	−	−	−	−	−	+
Bud: long axis perpendicular to hypha	+	−	−	−	−	−	−
Accumulate Fe or Mn oxides on surface	+	v[d]	−	−	−	−	−
Utilization of:							
C$_1$ compounds	D	+	−	−	−	NT	v
Formate	D	D	−	−	−	NT	D
Acetate	+	D	+	−	+	+	+
Major quinone component:							
Q-9		+	+				
Q-10	+			+	+	+	+
Q-11				+			
Photosynthetic pigments	−	−	−	−	−	−	+
Nitrogen source:							
NaNO$_2$	−	D	NT	−[e]	NT	−	NT
NaNO$_3$	D	+	+	−	+	−	D
NH$_4$	+[f]	+	+	−	+	−	+
Grow in presence of 3.5% (w/v) NaCl	−[g]	D	+	+	+	+	NT
Genome size M_r ($\times 10^9$)[h]	2.81–3.43	2.13–2.73	2.03	1.67–2.00	NT	1.73	NT
Mol% G + C of DNA (T_m)	63–66[i,j]	59–65	62	57–62	45–47	62–64	62–64(Bd)

[a]Symbols: +, 90% or more strains positive; −, 90% or more strains negative; v, variable with growth conditions; D, differs among strains; NT, not tested or information lacking.

[b]Data from Schlesner (1987).

[c]Data from Schlesner et al. (1990).

[d]Data from Hirsch (1968).

[e]Data from Gebers et al. (1984).

[f]Data from Gebers and Beese (1988).

[g]Survival but no growth (Gebers, 1989).

[h]Data from Kölbel-Boelke et al. (1985).

[i]Data from Gebers et al. (1985).

[j]Data from Gebers et al. (1981a).

cells of the new isolates are slightly smaller: 0.4–0.6 × 0.8–1.2 μm and 0.4 × 0.6 μm, respectively, as compared to IFAM E-1129^T, whose cells and swarmer cells are 0.4–0.9 × 0.4–1.5 μm and 0.9 × 1.5 μm, respectively. The buds develop at hyphal tips in a lateral (perpendicular) as well as a polar orientation; mother cells have no more than four hyphae. There are also some physiological differences from the type strain. The eight new isolates fail to grow with acetate or glutamate as the sole carbon source; Vitamin Solution no. 6 does not stimulate growth, and besides manganese they oxidize iron, albeit much less rapidly. Their DNA base ratio range is 65.6–66.6 mol% G + C.

Sly et al. (1988) considered these eight isolates to be new strains of *P. manganicum*, which may have appeared justifiable in view of the fact that this species was based on only one isolate and thus its variability was not yet known. The observation of lateral, as well as polar, positions of buds on hyphal tips raises questions about an important differentiating characteristic of the genus. This aspect needs further investigation.

The phylogenetic relationships of six *Pedomicrobium* strains and some other hyphal and budding *Alphaproteobacteria* have been recently investigated by 16S rRNA sequence analysis (Cox and Sly, 1997, Fig. 1). Unfortunately, *Pedomicrobium australicum* strains IFAM ST-1306^T and IFAM WD-1355 could not be revived from lyophilized cultures (P. Hirsch, personal communication) and thus, DNA of these strains had to be extracted from the lyophils. The other pedomicrobia studied came as viable cultures from the Australian Culture Collection (ACM). The whole 16S rRNA genes were sequenced in both the forward and reverse directions and the sequences aligned manually with sequences of *Hyphomicrobium vulgare*, *Hyphomonas jannaschiana*, *Hirschia baltica*, *Rhodomicrobium vannielii*, *Escherichia coli*, *Neisseria gonorrhoeae*, *Desulfovibrio desulfuricans*, and *Campylobacter jejuni*. The pedomicrobia cluster coherently and separately from the other genera of budding *Alphaproteobacteria* and have sequence similarities of 96.2–99.9%. Within this cluster, *P. manganicum* IFAM E-1129^T is the most distantly related strain to the other pedomicrobia. Strain

ACM 3067, one of the eight manganese-oxidizing isolates from "dirty drinking water" (Sly et al., 1988), is highly related to *P. americanum* IFAM G-1381[T], and the two *P. australicum* strains (IFAM ST-1306[T] and IFAM WD-1355) are almost identical, with a sequence similarity of 99.9%. Also highly related, with sequence similarities greater than 99%, are *P. americanum*, the two *P. australicum* strains, and the isolate ACM 3067. Because of their data and some phenotypic differences, Cox and Sly (1997) have questioned the taxonomic validity of *Pedomicrobium australicum*, but suggested retaining *P. americanum*. Differences between these two species in the utilization of carbon sources (Table BXII.α.182) and temperature optima (Table BXII.α.185) may well be an expression of their variability, as similar differences also exist among the three isolates of *P. americanum*. However, since the low DNA–DNA similarity data (Table BXII.α.183) clearly indicate differences on the species level, more phenotypic information and

further studies of viable cultures (as well as a neotype culture) of *P. australicum* are needed to solve this problem.

Acknowledgments

We gratefully acknowledge provision of *Pedomicrobium*-like strains by J.A. Babinchak, E. Dale, W.C. Ghiorse, and J.T. Staley. Skillful technical assistance was rendered by M. Beese, B. Hoffmann, J. Kock, K. Lutter-Mohr, and U. Wehmeyer. We are also grateful for help and discussions with T.V. Aristovskaya (St. Petersburg).

Further Reading

Hirsch, P. 1974. Budding bacteria. Annu. Rev. Microbiol. *28*: 391–444.
Moore, R.L. 1981. The biology of *Hyphomicrobium* and other prosthecate, budding bacteria. Ann. Rev. Microbiol. *35*: 567–594.
Moore, R.L. 1981. The genera *Hyphomicrobium*, *Pedomicrobium* and *Hyphomonas*. *In* Starr, Stolp, Trüper, Balows and Schlegel (Editors), The Prokaryotes. A Handbook on Habitats, Isolation, and Identification of Bacteria, 1st Ed., vol. I, Springer-Verlag, Berlin. pp. 480–487.

Differentiation of the species of the genus *Pedomicrobium*

The differential characteristics of the species of *Pedomicrobium* are listed in Table BXII.α.185. Other descriptive characteristics are summarized in Tables BXII.α.181, BXII.α.182, BXII.α.183, and BXII.α.184.

List of species of the genus Pedomicrobium

1. **Pedomicrobium ferrugineum** Aristovskaya 1961, 957[AL] emend. Gebers and Beese 1988, 308.
   fer.ru.gi′ne.um. M.L. neut. adj. *ferrugineum* of iron color.

   Cells oval, spherical, rod-shaped, tetrahedral, or bean-shaped, 0.6–2.0 × 0.6-2.5 μm. Oxidized iron, but not manganese, is deposited on mother cells and occasionally later on hyphae. Swarmer cells with one polar or subpolar flagellum. Morphology of colonies as given for the genus. Further morphological and physiological characteristics are listed in Tables BXII.α.181, BXII.α.182, BXII.α.183, BXII.α.184, BXII.α.185, and BXII.α.186.

   Heterotrophic, aerobic to microaerophilic. The temperature for optimal growth is 29–30°C, and the temperature range is 10–40°C. Optimal pH for growth: 9.0; the pH range is 3.5–10.0. Growth occurs in the presence of up to 0.1% (w/v) NaCl; no growth, but survival, has been observed in the presence of 1–5% NaCl. Higher concentrations of NaCl are bacteriocidal. Requires cyanocobalamine (1 μg/l) for growth.

   Isolated from podzolic soils in northern Germany.
   The mol% G + C of the DNA is: 64.5–66.8 (T_m, Bd).
   Type strain: IFAM S-1290, ATCC 33119, DSM 1540.
   GenBank accession number (16S rRNA): X97690.
   Additional Remarks: Other strains include IFAM P-1196 (ATCC 33116), IFAM Q-1197 (ATCC 33117), IFAM R-1198 (ATCC 33118), and IFAM T-1130 (ATCC 33120).

2. **Pedomicrobium americanum** Gebers and Beese 1988, 310[VP]
   a.me.ri.ca′num. M.L. neut. adj. *americanum* obtained from America.

   Cells oval, tetrahedral, short rods, bean-, or spindle-shaped, up to 1.3 × 1.8 μm (Fig. BXII.α.224). Mother cells with 1–3 hyphae originating laterally or polarly. Bud attachment perpendicular to the hyphal tip; intercalary budding occurs. Swarmer cells with one subpolar flagellum. Extracellular polyanionic polymer located primarily on cell surfaces opposite lateral hyphae, with incrustations of iron or manganese oxides (depending on growth conditions). Intracytoplasmic membranes have been observed.

The temperature optimum for growth is 32–38°C; the range is 15–41°C. The pH optimum for growth in HEPES-buffered media is 7.6–8.4. Growth on carbon sources varies with the strains; all isolates grow on acetate, but none utilizes caproate, D-lactate, D-glucose, D-ribose, methanol, or ethanol. Strain IFAM BA-869 also grows on citrate, formate, gluconate, glycerol, D,L-malate, propanol, and succinate. Good growth occurs with (w/v) 0.05% yeast extract or 0.025% yeast extract plus 0.025% peptone. Nitrate does not

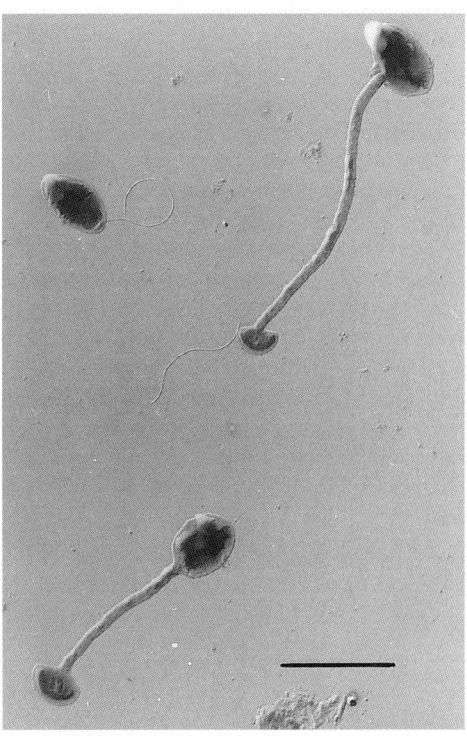

FIGURE BXII.α.224. *Pedomicrobium americanum* IFAM G-1381[T], Pt/C-shadowed. Note orientation of buds. Bar = 2 μm. (Electron micrograph courtesy of H. Völker, Kiel.)

serve as a sole nitrogen source, but ammonia and several amino acids support growth (Table BXII.α.181). Cyanocobalamin (1 μg/l) is required for growth of the type strain, IFAM G-1381[T]. Growth of strain IFAM BA-868 is stimulated by 0.1% NaCl. Catalase is produced, except by strain IFAM BA-869. Nitrate is reduced by all strains. The three isolates are inhibited by several antibiotics (Table BXII.α.184), but the type strain is resistant to sulfanilamide.

Habitat: bogwater and freshwater puddles in North America.

The mol% G + C of the DNA is: 64–65 (T_m).

Type strain: ACM 3090, ATCC 43612, IFAM G-1381.

GenBank accession number (16S rRNA): X97692.

Additional Remarks: Other strains include IFAM BA-868 (ATCC 43613) and IFAM BA-869 (ATCC 43615).

3. **Pedomicrobium australicum** Gebers and Beese 1988, 313[VP]

au.stra′ li.cum. M.L. neut. adj. *australicum* isolated from Australia.

Cells oval, tetrahedral, short rods, bean- or spindle-shaped, 1.2 × 1.8 μm with up to 3 hyphae originating laterally or polarly. Cells may produce up to 3 buds at the same time; intercalary budding has been observed. Motile swarmers have one subpolar flagellum. Extracellular polyanionic polymer is produced primarily on the cell surface opposite the lateral hypha (Fig. BXII.α.221). Iron oxides are deposited on the surface when cells are grown on fulvic acid iron sesquioxide complexes or in the presence of 5 mg/l of FeSO$_4$·7H$_2$O. Deposition of manganese oxides occurs on PSM agar in the presence of 1.54 mg/l MnSO$_4$·H$_2$O. Ultrathin sections reveal intracytoplasmic membranes, which form compartments, possibly due to unfavorable growth conditions.

The temperature optimum for growth is 29–32°C; the range is 15–36°C. The optimal pH for growth in buffered media is 7.3–7.6. Utilization of carbon sources varies between the two strains (Table BXII.α.182): both grow with acetate, D,L-malate, lactose, and gluconate, but not with D-lactate, citrate, or D-ribose. Good growth occurs with (w/v)

0.05% yeast extract or with 0.025% yeast extract plus 0.025% peptone. Organic nitrogen sources include several amino acids (Table BXII.α.181); ammonia may serve as a nitrogen source, but nitrate does not. Strain WD-1355 is stimulated by cyanocobalamin (1 μg/l) and by the presence of 0.1% NaCl. Both strains are positive for catalase and for nitrate reduction. The strains are inhibited by a variety of antibiotics (Table BXII.α.184), but not by sulfanilamide.

DNA–DNA similarity studies (Table BXII.α.183) clearly demonstrate species identity between the two isolates.

Habitat: a freshwater reservoir in New South Wales, Australia.

The mol% G + C of the DNA is: 65 (ST-1306[T]) and 63 (WD-1355) (T_m).

Type strain: ATCC 43611, IFAM ST-1306.

GenBank accession number (16S rRNA): X97693.

Additional Remarks: Other strains include IFAM WD-1355 (ATCC43614).

4. **Pedomicrobium manganicum** Aristovskaya 1961, 957[AL] emend. Gebers 1981, 315.

man.ga′ ni.cum. M.L. neut. adj. *manganicum* of manganese.

Cells primarily spherical, oval, or short rods, 0.4–0.9 × 0.4–1.5 μm, with 1–5 hyphae per cell. Colonial morphology as for the genus. Cells have polyanionic polymer and manganese oxide deposits, but only rarely deposits of iron oxide. Buds may be formed intercalary in hyphae. Intracytoplasmic membranes occur. Cells grow on acetate, caproate, formate, D-glucose, D-lactate, pyruvate, and D-ribose. Optimal growth temperature 30°C. Growth is stimulated by vitamin mixtures, but inhibited by gentamicin and bacitracin. Further physiological characteristics are listed in Tables BXII.α.181, BXII.α.182, BXII.α.183, BXII.α.184, BXII.α.185, and BXII.α.186.

Isolated from a quartzite rock pool in France.

The mol% G + C of the DNA is: 65 (T_m).

Type strain: ACM 3038, ATCC 33121, DSM 1545, IFAM E-1129.

GenBank accession number (16S rRNA): X97691.

Genus XV. *Prosthecomicrobium* Staley 1968, 1940[AL] emend. Staley 1984, 304

CHERYL JENKINS , FRED A. RAINEY, NAOMI L. WARD AND JAMES T. STALEY

Pros.the′ co.mi.cro.bi.um. Gr. fem. n. *prosthece* appendage; Gr. adj. *micros* small; Gr. masc. n. *bios* life;
M.L. neut. n. *Prosthecomicrobium* appendage (-bearing) microbe.

Unicellular bacterium with coccobacillary to rod-shaped cells ranging in diameter from 0.8 to 1.2 μm and containing **numerous prosthecae** extending from all locations on the cell surface. Prosthecae, which may number from 10 to more than 30/cell, are typically short (i.e., <1.0 μm in length); some species, however, also produce longer prosthecae (>2.0 μm). **Cells divide by budding**. Buds are produced directly from the mother cell, never from tips of prosthecae. **Gram negative**. Motile and nonmotile species exist. Motile organisms produce single polar to subpolar flagella; one species forms gas vacuoles but not flagella. **Obligately aerobic, nonfermentative, heterotrophic**. A variety of sugars and organic acids are used as energy sources for growth. All strains tested require one or more B vitamins for growth. Oxidase and catalase positive. Found in soils and fresh and marine waters.

The mol% G + C of the DNA is: 64–70.

Type species: **Prosthecomicrobium pneumaticum** Staley 1968, 1940.

FURTHER DESCRIPTIVE INFORMATION

The genus *Prosthecomicrobium* comprises a diverse collection of multiply-appendaged prosthecate bacteria. All produce short prosthecae, i.e., prosthecae that are typically <1.0 μm in length. The shortest prosthecae appear as small bumps on the surface of the cell and cannot be readily discerned by light microscopic examination of cells. The "corn cob" organism (now named "*P. polyspheroidum*") was so named because of the numerous short stubby prosthecae that are regularly arranged on the surface of the cell. Somewhat longer prosthecae are conical in shape and are typical for many species. These range in length from <0.25 to >1.0 μm, depending on the species and the growth medium (longer prosthecae may be produced when phosphate is limiting growth). Even longer prosthecae are produced by some strains. For example, *P. pneumaticum* occasionally produces prosthecae >2.0 μm long. This, however, occurs rarely (Staley, 1968). Some strains of the marine species *P. litoralum* also produce longer appendages. One species, *P. hirschii*, produces both short appendaged as well as long-appendaged cells. The long-appendaged cells closely resemble those of *Ancalomicrobium adetum* and could be confused with them if morphology is used as the sole criterion for identification of organisms from natural samples.

Buds are produced as outgrowths at or near one pole of the cell (Vasilyeva, 1972a; Staley, 1984). These typically appear as a forked protuberance from the dividing pole. The bud enlarges and differentiates to produce a mirror image of the mother cell. As in *Ancalomicrobium*, buds appear to be synthesized *de novo* (Staley, 1973b; Staley et al., 1981).

Some species of *Prosthecomicrobium* produce flagella. Flagella are always single and may be found in a polar or subpolar position on the cell. One species, *P. pneumaticum*, does not produce flagella but produces gas vacuoles. Some species are nonmotile and do not produce flagella or gas vacuoles.

All species are obligately aerobic chemoorganotrophs that use a variety of sugars, organic acids, and sugar alcohols for growth. All species can grow on a simple, defined medium (Staley, 1981a). All species require one or more water-soluble vitamins for growth.

Prosthecomicrobium species are not known to be pathogenic to humans.

Prosthecomicrobium species are found in soils, freshwater habitats of all trophic states (Staley et al., 1980), as well as in the marine habitat (Bauld et al., 1983). They have been reported as important components of biofilms in sewage treatment processes and in pulp mill aeration lagoons (Stanley et al., 1979). Bacteria that resemble these forms have also been reported in the intes-

tinal tracts of insects and other animals (Cruden and Markovetz, 1981); however, it is uncertain whether they have been concentrated from the foodstuff or are truly indigenous to this environment.

ENRICHMENT AND ISOLATION PROCEDURES

The same procedures described for *Ancalomicrobium* isolation can be used for isolation of *Prosthecomicrobium*. The dilute peptone enrichment procedure has been used successfully for the isolation of most strains. Recognition of short-appendaged species in the enrichment culture is difficult because they are not as noticeable as the longer appendaged species. Nonetheless, even with the phase-contrast microscope, they can be detected because of their slightly irregular surface (Fig. BXII.α.225). However, it may be desirable to observe preparations from wet mounts with the transmission electron microscope to confirm their presence in the enrichments. When their numbers have reached significant proportions of the total numbers of bacteria (i.e., 5%), then attempts at isolation should be made. Dilute peptone agar plates should be streaked and incubated at room temperature for 2 weeks or more. Colonies are typically small. Wet mounts of each colony type should be prepared and examined by phase-contrast microscopy for identification. The morphology of the resulting pure culture should be confirmed by examination of whole cells with the transmission electron microscope. Additional information on isolation has been published elsewhere (Staley, 1981a).

MAINTENANCE PROCEDURES

Strains can be maintained on slants in the refrigerator for at least 1 month. MMB agar (Staley, 1981a) with glucose as a carbon source provides a favorable growth medium. Lyophilization can be used for long-term preservation.

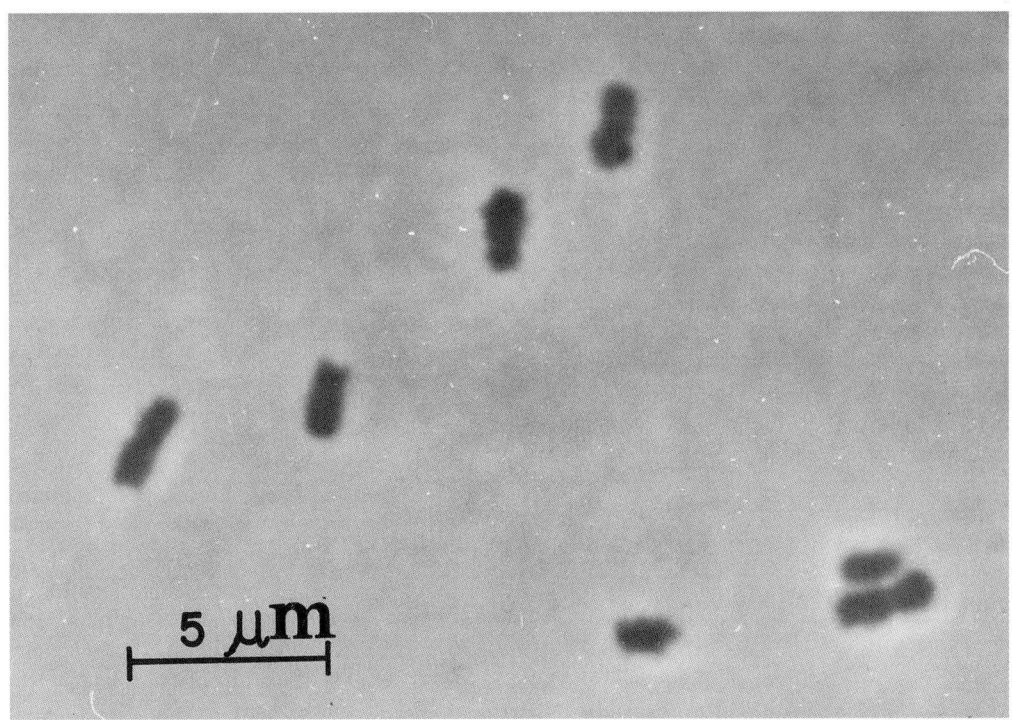

FIGURE BXII.α.225. Phase contrast photomicrograph of *P. enhydrum* showing several cells. Note the irregular surface of the cells due to the presence of numerous short prosthecae.

DIFFERENTIATION OF THE GENUS *PROSTHECOMICROBIUM* FROM OTHER GENERA

Only two genera pose major problems for differentiation. These are the genera *Ancalomicrobium* and *Verrucomicrobium*. Differentiation from the former genus is discussed in detail in the section on *Ancalomicrobium*. Morphologically, *Ancalomicrobium* can be distinguished from *Prosthecomicrobium* because of its conical cells and lack of short-appendaged cells. Furthermore, *Ancalomicrobium* is a genus of facultative anaerobes that ferments selected sugars. Thus, the Hugh–Leifson test can be used to distinguish between these two groups (cf. Table BXII.α.165 in *Ancalomicrobium*). The major phenotypic feature differentiating cells of the genus *Verrucomicrobium* from those of *Prosthecomicrobium* is the presence of fimbriae, which extend from the tips of the prosthecae. These are absent in cells of the genus *Prosthecomicrobium*.

TAXONOMIC COMMENTS

Morphologically, *Prosthecomicrobium* comprises a diverse collection of bacteria. There are differences in appendage length, motility, gas vacuole formation, habitat, and carbon source utilization among the species so far described.

The mol% G + C of the DNA within the genus ranges from 64 to 70 (Staley and Mandel, 1973). Results of DNA–DNA hybridization studies (Moore and Staley, 1976) suggest that additional species exist that have not yet been fully described. Several unnamed strains have been deposited in the ATCC (ATCC 27825, 27826, and 27833). *Ancalomicrobium adetum* shows only low levels of homology with *Prosthecomicrobium* species that have been tested.

Phylogenetic analysis of the 16S ribosomal RNA gene sequences of *P. pneumaticum*, *P. enhydrum*, and several uncharacterized *Prosthecomicrobium*-like strains demonstrates that these organisms belong to the *Alphaproteobacteria*. Morphologically similar organisms of the genus *Ancalomicrobium* also belong to the *Alphaproteobacteria* but form a separate and distinct genus based on 16S rRNA analyses. These findings are consistent with earlier 16S rRNA cataloging studies on these genera (Schlesner et al., 1989). Members of the genus *Verrucomicrobium* are not phylogenetically affiliated with *Prosthecomicrobium* strains, despite being morphologically similar, and fall within a separate bacterial division.

Analyses of the 16S rRNA genes of *Prosthecomicrobium* strains also demonstrate that the genus *Prosthecomicrobium* is polyphyletic (Fig. BXII.α.226). Of the strains that have been sequenced, only strain P4.10, an isolate from a pulp mill oxidation lagoon, clusters with the type species, *P. pneumaticum*. Four additional strains, including *P. enhydrum*, cluster with *Devosia riboflavina* (formerly "*Pseudomonas riboflavina*"), a non-prosthecate organism. One additional strain SCH71 does not cluster with either of these groups. Thus, the taxonomy of members of the genus *Prosthecomicrobium* should be reexamined.

16S rRNA sequencing of the remaining characterized species *P. litoralum* and *P. hirschii* has not yet been completed and is required to assess the extent of diversity within the *Prosthecomicrobium* genus. Such studies would also help to clarify the taxonomic relationships of the strains "*P. polyspheroidum*", "*P. consociatum*", "*P. mishustinii*", which are at present only unofficially named (Vasilyeva and Lafitskaya, 1976; Vasilyeva et al., 1991).

Reclassification of *P. enhydrum* and related strains is also required as they do not cluster with the type species of *Prosthecomicrobium*, *P. pneumaticum*. These strains should therefore be reassigned to a different genus.

DIFFERENTIATION OF THE SPECIES OF THE GENUS *PROSTHECOMICROBIUM*

The characteristics differentiating among the species of the genus *Prosthecomicrobium* are listed in Table BXII.α.187.

List of species of the genus Prosthecomicrobium

1. **Prosthecomicrobium pneumaticum** Staley 1968, 1940[AL]
   pneu.ma′ti.cum. N.L. adj. *pneumaticum* inflated, containing gas vacuoles.

   Most prosthecae are short (i.e., <1.0 μm in length); however, long prosthecae are occasionally produced (Fig. BXII.α.227). Gas vacuoles are formed by the type strain. Flagella are not produced. Ammonium, but not nitrate, can be used as a sole nitrogen source for growth. A variety of sugars, both hexoses and pentoses, monosaccharides and disaccharides, can be used as carbon sources for growth (Table BXII.α.187). Sugar alcohols and methyl sugars fucose and rhamnose can also be used. Biotin, thiamine, and vitamin B₁₂ are required for growth. The temperature range for the type strain is 9–42°C. Optimal pH: 6.0–6.5, although good growth also occurs at pH 7.0. Colonies are translucent to opaque white (the chalky white color is due to the formation of gas vacuoles). Colony size is quite variable.
   The mol% G + C of the DNA is: 69–70 (Bd).
   Type strain: ATCC 23633, VKM B-1389.

2. **Prosthecomicrobium enhydrum** Staley 1968, 1940[AL]
   en.hy′drum. N.L. adj. *enhydrum* living in water, aquatic.

   Prosthecae are always short, i.e., <0.5 μm, giving the cell an irregular surface when observed by phase-contrast microscopy (Fig. BXII.α.225). Cells are motile by a polar to subpolar flagellum (Fig. BXII.α.228). Gas vacuoles are not produced. Ammonium but not nitrate can be used as a sole source of nitrogen in media containing an appropriate carbon source and vitamins. Pentoses and hexoses are commonly used by the type strain. Some disaccharides and organic acids are also used as carbon sources (Table BXII.α.187). Thiamine is required for growth. The temperature range for growth of the type strain is 9–37°C. The pH optimum for growth is 7.0. Colonies may be white (type strain) or pigmented yellow or red.
   The mol% G + C of the DNA is: 65.8 (Bd).
   Type strain: ATCC 23634, VKM B-1376.

3. **Prosthecomicrobium hirschii** Staley 1984, 304[VP]
   hirsch′i.i. N.L. gen. *hirschii* of Hirsch; named in honor of P. Hirsch, an authority on budding bacteria.

   Prosthecae may be short (i.e.,< 1.0 μm in length) or long (>2.0 μm in length), depending on the cell. Both short- and long-appendaged cells occur simultaneously in culture (Fig. BXII.α.229). Cells may be motile by single polar or subpolar flagella. Gas vacuoles are not formed. Sugars and organic acids are commonly used as carbon sources. Long chain organic acids such as valerate and caproate are used. Methanol and ethanol can also be used as

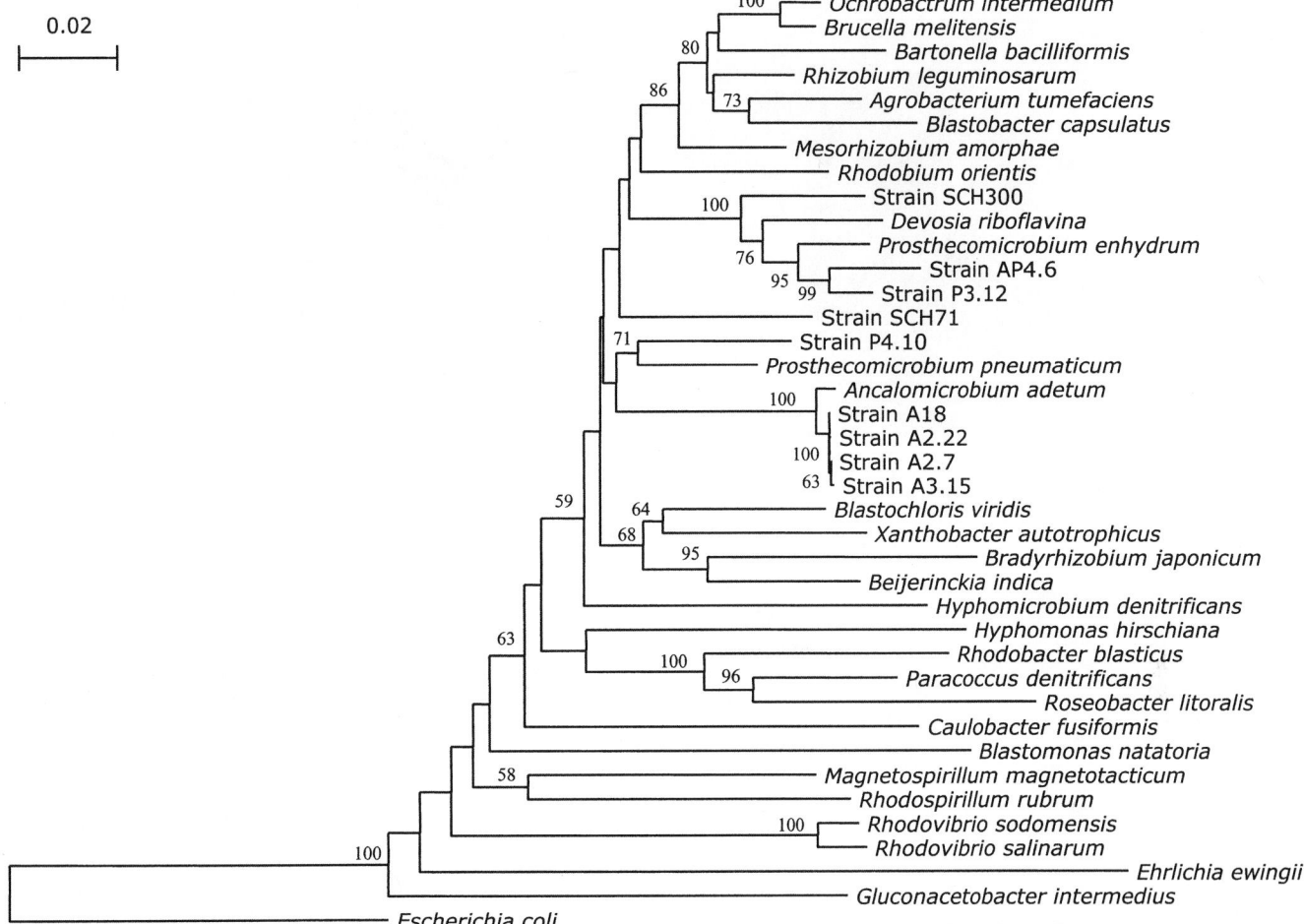

FIGURE BXII.α.226. Phylogenetic tree of 16S rRNA sequences showing the positions of various *Prosthecomicrobium* strains within the *Alphaproteobacteria*. The genus is polyphyletic, with the type strain *P. pneumaticum* and the pulp mill aeration lagoon isolate P4.10 forming one clade, while *P. enhydrum* and the *Prosthecomicrobium*-like strains AP4.6, P3.12, and SCH300 form a separate clade that includes the nonprosthecate organism *Devosia riboflavina*. Strain SCH71, which is also *Prosthecomicrobium*-like in morphology, forms a third group with no clear phylogenetic affiliation. The tree was constructed using Kimura-2-parameter distances and the neighbor joining algorithm with 100 bootstrap replications within the program TreeCon. Bootstrap values above 50% are shown. Bar indicates 0.02 substitutions per site.

TABLE BXII.α.187. Characteristics differentiating species of the genus *Prosthecomicrobium*[a]

Characteristic	*P. enhydrum*	*P. hirschii*	*P. litoralum*	*P. pneumaticum*	"*P. polyspheroidum*"	"*P. consociatum*"	"*P. mishustinii*"
Short prosthecae (<1 μm) on MMB	+	+	+	+	+	+	+
Long prosthecae (>2 μm) on MMB	−	−	−[b]	−	−	−	−
Lateral buds	−	−	−	−	+[c]	−	−
Flagella	+	+	−	−	+	−	−
Gas vacuoles	−	−	−	+	−	−	−
Sodium ion requirement; optimal salinity of 25‰	−	−	+	−	−	−	−
Carbon source utilization:							
Maltose, cellobiose, lactose	+	+	+	+	+	−	+
Melibiose	−	−	+	+	nd	−	−
Rhamnose	+	−	+	+	−	±	±
Sorbitol	−	−	+	+	±	−	−
Pyruvate	+	+	+	−	−	−	±
Propionate	−	+	−	−	±	−	−
Agar digestion	−	nd	+	−	−	−	−
Mol% G + C of DNA	66	68–70	66–67	69–70	64–67	66–68	64–65

[a]Symbols: +, 90% or more strains are positive; −, 90% or more strains are negative; nd, not determined; ±, indefinite.

[b]Strains of both species rarely produce long appendages.

[c]These are formed rarely.

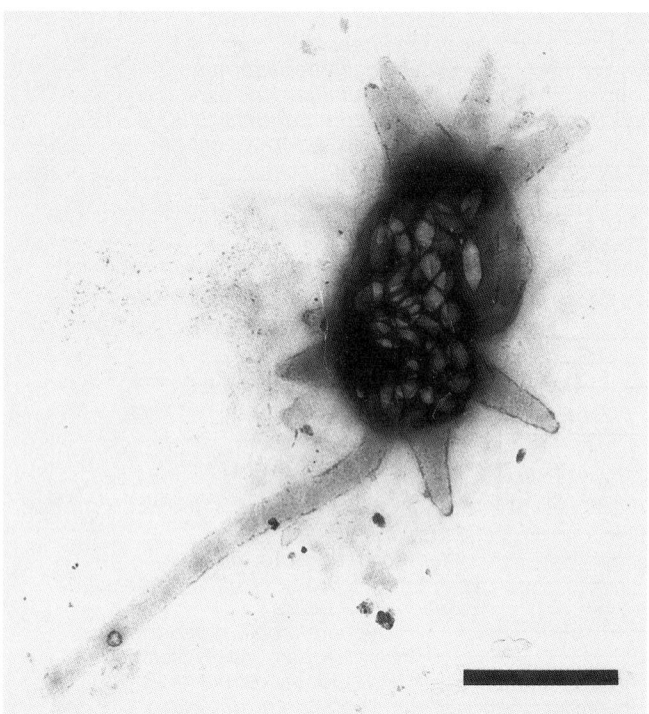

FIGURE BXII.α.227. *P. pneumaticum* as observed by electron microscopy. Note the gas vesicles within the cell and the single long prostheca. Bar = 1.0 μm.

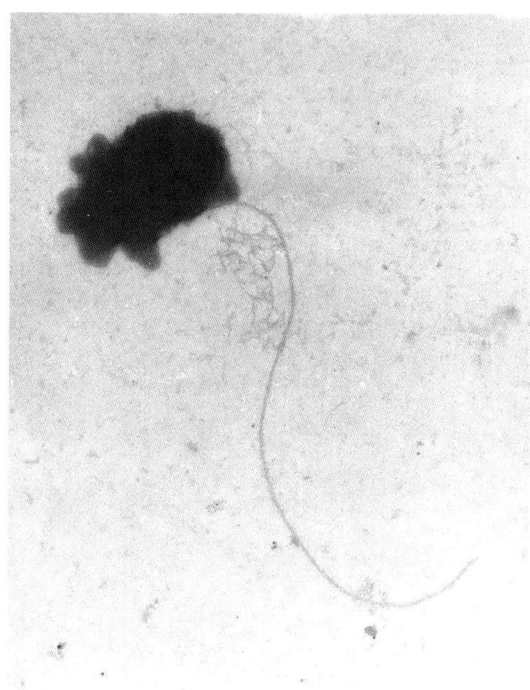

FIGURE BXII.α.228. Electron micrograph of *P. enhydrum.* Note the sub-polar location of the single flagellum.

carbon sources. Biotin, nicotinic acid, pantothenate, and thiamine are required for growth. Colonies are circular in form and umbonate in elevation. They have an entire margin and are pink in color.

The mol% G + C of the DNA is: 67.9–69.9 (Bd).

Type strain: ATCC 27832.

4. **Prosthecomicrobium litoralum** Bauld, Bigford and Staley 1983, 613[VP]

li.to.ra′ lum. L. adj. *litoralis* of the seashore.

Prosthecae are mostly short (i.e., <0.5 μm in length), although occasionally long prosthecae are formed. Non-motile. Neither flagella nor gas vacuoles are produced. Ammonium can be used as a sole source of nitrogen. A variety of pentoses and hexoses can be used for growth as well as some sugar alcohols and organic acids (Table BXII.α.187). Agar digestion occurs on prolonged incubation. Sodium ions are required for growth. The minimum salinity at which growth occurs is 5‰ the optimum, 25‰ (Table BXII.α.187). These bacteria are psychrotrophic. Their minimum temperature for growth is 1–5°C, depending on the strain, and their optimum is between 27° and 34°C. Colonies are white and raised initially and, after prolonged incubation, become cream-colored and umbonate. Agar digestion is noticeable around colonies on plates incubated at least 30 days at room temperature.

The mol% G + C of the DNA is: 66–67 (Bd).

Type strain: 524-16, ATCC 35022.

5. **"Prosthecomicrobium polyspheroidum"** (Nikitin and Vasilyeva 1968) Vasilyeva and Lafitskaya 1976, 768 (**"Agrobac-**

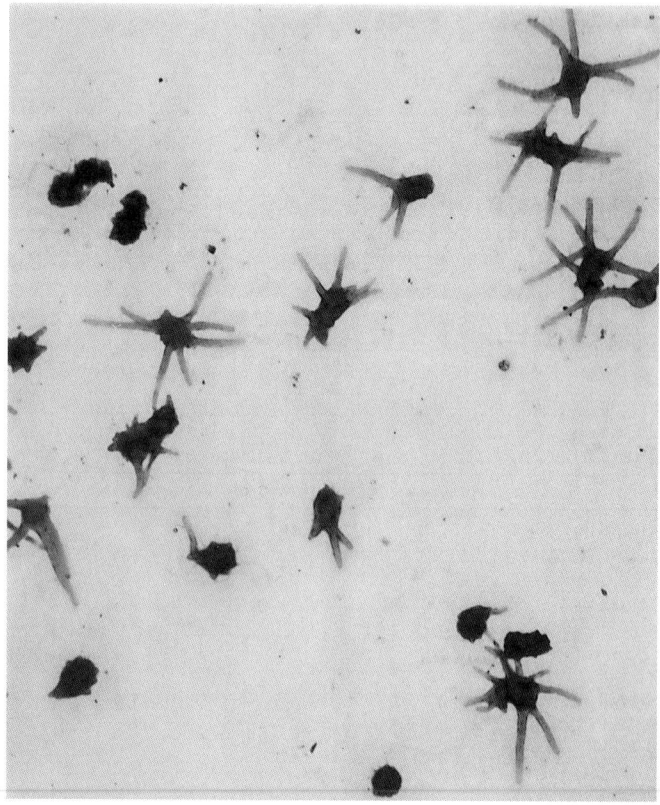

FIGURE BXII.α.229. Electron micrograph of *P. hirschii* showing short-appendaged cells, long-appendaged cells, and flagella.

terium polyspheroidum" Nikitin and Vasilyeva 1968, 444.)

po.ly.spher.oi′ dum. N.L. adj. *polyspheroidum* many spheroids, i.e., having numerous bumps.

Cells have many (up to 230/cell) short prosthecae (i.e., <0.1 µm in length). They are distributed in slightly spiral rows along the length of the rod-shaped cells, giving the appearance of a corn cob (Fig. BXII.α.230). Cells, which measure about 0.5 µm in diameter and up to 5.1 µm in length, are motile by a polar flagellum. Lateral buds may be formed under some conditions of growth. Prosthecae may not be evident in cultures containing more than 0.1% organic nutrients. Monosaccharides and disaccharides are metabolized by pentose phosphate and the Entner–Doudoroff pathway, with the formation of alcohols, organic acids, and aromatics (cf. Lafitskaya et al., 1976). Thiamine and riboflavin are required for growth. Nitrate reduced to nitrite. Voges–Proskauer and methyl red tests are negative. Colonies are punctiform after 2–3 days of growth and attain a diameter of about 1 mm after 10 days. They are raised with entire margins, opaque, cream colored, and slimy. Optimal temperature: 28–30°C; optimal pH: 6.5–7.1.

The mol% G + C of the DNA is: 64 to 67 (Bd).

Deposited strain: AUCM B-1313.

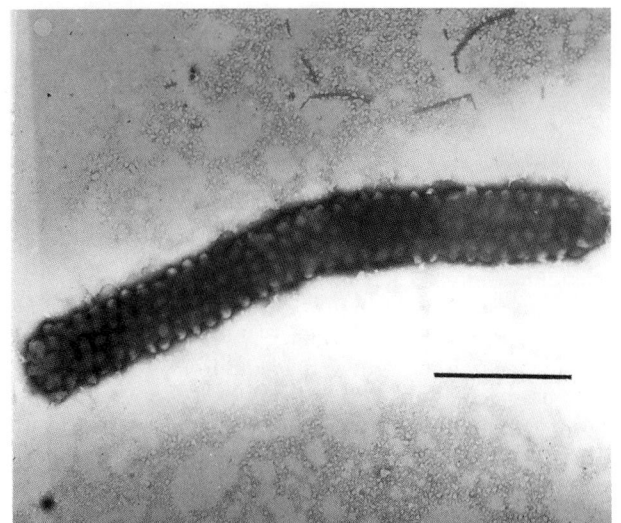

FIGURE BXII.α.230. Electron micrograph of *"P. polyspheroidum"*. Note the arrangement of the numerous short prosthecae in rows. Bar = 2.0 µm. (Photograph courtesy of L.V. Vasilyeva.)

Genus XVI. **Rhodomicrobium** *Duchow and Douglas 1949, 415^AL emend. Imhoff, Trüper and Pfennig 1984, 341*

JOHANNES F. IMHOFF

Rho.do.mi.cro′bi.um. Gr. n. *rhodon* the rose; Gr. adj. *micros* small; Gr. n. *bios* life; M.L. neut. n. *Rhodomicroibum* red microbe.

Ovoid to elongate-ovoid bacteria showing polar growth and performing a characteristic vegetative growth cycle. This cycle includes the formation of peritrichously flagellated swarmer cells and nonmotile "mother cells", which form prosthecae from one to several times the length of the mother cell. Daughter cells originate as spherical buds at the end of the prosthecae and may undergo differentiation in various ways. **Gram negative, belonging to the *Alphaproteobacteria*. Internal photosynthetic membranes are of the lamellar type. Photosynthetic pigments are bacteriochlorophyll *a* and carotenoids of the spirilloxanthin series.** The **predominant cellular fatty acid is C$_{18:1}$**, which comprises more than 80% of the membrane-bound fatty acids. **Ubiquinone and rhodoquinone with 10 isoprene units** are present, and the lipopolysaccharides are characterized by a glucosamine-containing, phosphate-free lipid A with amide-bound C$_{16:0\ 3OH}$.

Cells grow preferably photoheterotrophically under anoxic conditions in the light. Photoautotrophic growth may be possible with hydrogen and sulfide as electron sources. Various **organic substrates, molecular hydrogen, ferrous iron, and sulfide at low concentrations may be used as photosynthetic electron donors.** Cells are also **able to grow under microoxic to oxic conditions in the dark.**

Mesophilic freshwater bacteria with a preference for acidic pH, between 5.2 and 6.5.

The mol% G + C of the DNA is: 61.8–63.8.

Type species: **Rhodomicrobium vannielii** Duchow and Douglas 1949, 415.

FURTHER DESCRIPTIVE INFORMATION

Rhodomicrobium vannielii is the sole species of this genus. Its morphology and growth cycle resemble those of the nonphototrophic *Hyphomicrobium* species (Hirsch, 1974a). The characteristic growth cycle has been the object of intensive studies (Whittenbury and Dow, 1977) and is outlined in the species description. *Rhodomicrobium vannielii* grows best under anoxic conditions in the light with a number of carbon sources at low light intensities and acidic pH (5.5–6.5). Dinitrogen and ammonia are the best nitrogen sources. Ammonia is assimilated via glutamine synthetase/glutamate synthase (NADH-dependent); low activities of a glutamate dehydrogenase (NADPH-dependent) are also present (Brown and Herbert, 1977). Sulfate can be used as the sole sulfur source and is assimilated via adenosine-5′-phosphosulfate (Imhoff, 1982). Sulfide is tolerated at 2–3 mM and oxidized exclusively to tetrathionate in sulfide-limited chemostat culture. In batch culture, sulfide may react with the tetrathionate formed, and under these conditions, thiosulfate and elemental sulfur are the major end products (Hansen, 1974).

Particular similarities have been found in the amino acid sequences of the "small type" cytochromes c_2 of *Rhodomicrobium vannielii*, *Blastochloris viridis*, and *Rhodoblastus acidophilus* (Ambler et al., 1979). As in a few other purple nonsulfur bacteria (*Rhodopila globiformis*, *Rubrivivax gelatinosus*, and *Rhodocyclus tenuis*), a high potential iron–sulfur protein (HiPIP) is present (T.E. Meyer, personal communication). High proportions of unusual aminolipids have been found in *Rhodomicrobium vannielii* (Park and

Berger, 1967; Imhoff et al., 1982). The fatty acid composition is characterized by an extremely high proportion of $C_{18:1}$, comprising more than 85% of the membrane-bound fatty acids, as in *Hyphomicrobium* species, *Pedomicrobium* species, "*Nitrobacter agilis*", and *Nitrobacter winogradskyi* (Auran and Schmidt, 1972; Eckhardt et al., 1979). Ubiquinone and rhodoquinone with 10 isoprene units are present (Hiraishi and Hoshino, 1984). As is the case with *Rhodoblastus acidophilus*, but not with most of the phototrophic *Alphaproteobacteria*, the lipid A of *R. vannielii* contains glucosamine (not 2,3-diamino-2,3-dideoxyglucose) as the sole amino sugar and amide-bound $C_{16:0\ 3OH}$ (Weckesser et al., 1995).

A new isolate, which has been tentatively identified as *R. vannielii* and grown in the light with ferrous iron as the sole energy source, has been found to use ferrous iron as a photosynthetic electron donor, as has the type strain (Widdel et al., 1993; Heising and Schink, 1998). It has been noted that growth is not supported over prolonged time periods when ferrous iron is the exclusive electron source, and that acetate and succinate stimulate growth with ferrous iron (Heising and Schink, 1998). Ferric iron is not reduced in the dark, and manganese salts are neither oxidized nor reduced.

Enrichment and Isolation Procedures

Rhodomicrobium is commonly found in mud and water of ponds and lakes, as well as in wastewater. About 50% of enrichment cultures with freshwater sediments prove to contain *Rhodomicrobium* species (Whittenbury and Dow, 1977). *Rhodomicrobium* has also been isolated from brackish-water and seawater habitats (Hirsch and Rheinheimer, 1968).

Enrichment, isolation, and growth occur under conditions suitable for most of the other species of the purple nonsulfur bacteria (see chapter Genus *Rhodospirillum*) with organic acids or other organic substrates as carbon and electron sources (Imhoff, 1988; Imhoff and Trüper, 1992). For selective enrichments, a succinate–mineral medium with an initial pH of 5.2–5.5 should be used, and growth factors should be omitted (Pfennig, 1969a).

Maintenance Procedures

Cultures of *Rhodomicrobium vannielii* are well preserved in liquid nitrogen, by lyophilization, or at −80°C in a mechanical freezer. Late-exponential-phase cell suspensions should be mixed with glycerol or DMSO to yield the final cryoprotectant concentrations of 10% and 5%, respectively, kept at 0°C for 15 min, and then frozen immediately.

Differentiation of the Genus *Rhodomicrobium* from Other Genera

The lamellar structure of the internal photosynthetic membranes, the polar growth mode, and the budding type of multiplication are similar to those of other budding purple nonsulfur bacteria. Prostheca formation and the characteristic growth cycle are the most obvious distinguishing properties. In addition, according to rRNA–DNA hybridization studies, *Rhodomicrobium* is clearly distinguished from other purple nonsulfur bacteria (Gillis et al., 1982).

Among the closest phylogenetic relatives of *Rhodomicrobium*, based on 16S rDNA sequence analysis, are *Hyphomicrobium vulgare*, *Blastochloris viridis*, and *Rhodopseudomonas palustris* (Kawasaki et al., 1993b). Other outstanding characteristics are the compositions of lipid A, polar lipids, and fatty acids. Major differentiating properties between *Rhodomicrobium* and other phototrophic *Alphaproteobacteria* are shown in Tables BXII.α.168 and BXII.α.169 of the chapter on the genus *Blastochloris* as well as in Tables 3 (pp. 125–126) and 4 (p.127) in the introductory chapter "Anoxygenic Phototrophic Purple Bacteria", Volume 2, Part A. The phylogenetic relationships of these bacteria, based on 16S rDNA sequences, are shown in Fig. 2 (p. 128) of that chapter.

Further Reading

Gorlenko, V.M., N.N. Egorova and A.N. Puchkov. 1974. Fine structure of exospores of nonsulfur purple bacterium *Rhodomicrobium vannielii*. Mikrobiologia. *43*: 913–915.

Hirsch, P. 1974. Budding bacteria. Ann. Rev. Microbiol. *28*: 391–444.

Whittenbury, R. and C.S. Dow. 1977. Morphogenesis and differentiation in *Rhodomicrobium vannielii* and other budding and prosthecate bacteria. Bacteriol. Rev. *41*: 754–808.

List of species of the genus Rhodomicrobium

1. **Rhodomicrobium vannielii** Duchow and Douglas 1949, 415[AL]

   van.niel'i.i. M.L. gen. n. *vannielii* of van Niel; named for C.B. van Niel, an American microbiologist.

   Mature cells are ovoid to lemon-shaped, 1.0–1.2 × 2.0–2.8 μm, and multiply by polar growth and budding (Fig. BXII.α.231). Cells perform a characteristic vegetative growth cycle; the motile, peritrichously flagellated swarmer cells, which lose their flagella later in the growth cycle, form prosthecae of about 0.3 μm in diameter, and a daughter cell arises as a spherical bud at the end of each prostheca. This daughter cell may separate from the prostheca and start a new cycle as a swarmer cell. After the swarmer cell is released, the pole of the prostheca is free for the formation of another bud. Alternatively, the daughter cell may remain attached to the prostheca and form another prostheca at the opposite pole. Branching of prosthecae may occur by lateral outgrowth of new prosthecae from the primary prostheca, upon which the first daughter cell formed. A branched prostheca can be formed only on the most

   recently synthesized prostheca, and plug formation occurs in the prostheca of each daughter cell before that prostheca branches to form the next daughter cell. Only one daughter cell is formed at a time. Because cells tend to remain attached to the prosthecae, aggregates containing large numbers of cells are usually formed. In addition, smaller cells, called exospores, may eventually be produced. These polyhedral cells are 1.0–1.5 μm in diameter. One to four such cells are formed sequentially as buds at a common branching point at the end of a prostheca (Fig. BXII.α.231). The exospores are more resistant to dryness and heat than are normal vegetative cells.

   Internal photosynthetic membranes are present as lamellae underlying and parallel to the cytoplasmic membrane. No lamellae are present in the prosthecae. The color of cell suspensions is from salmon pink to deep orange-brown to red. Absorption spectra of living cells have maxima at 378, 461, 488–490, 522–525, 800–807, and 869–872 nm. Aerobically grown cells are colorless to pale orange-brown. Photosynthetic pigments are bacteriochlorophyll *a* (esteri-

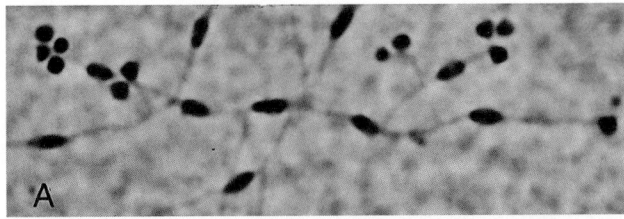

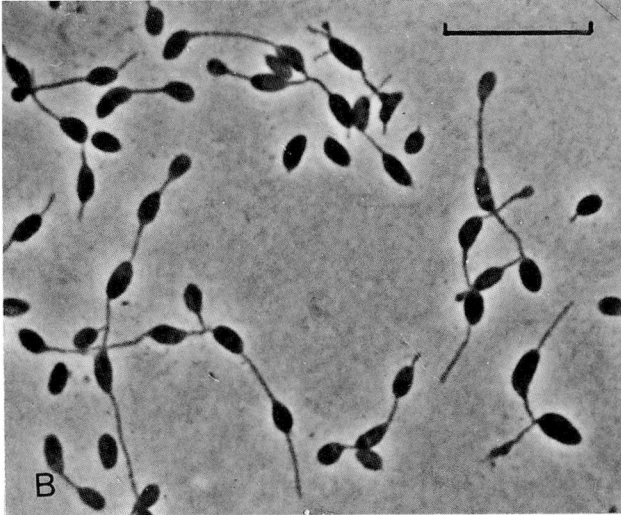

FIGURE BXII.α.231. *Rhodomicrobium vannielii* strain DSM 163. *A*, polyhedral exospores as buds at the ends of short filaments with common branching points. *B*, cells with filaments and buds of various sizes. Phase-contrast micrographs. Bar = 10 μm. (Courtesy of N. Pfennig.)

fied with phytol) and carotenoids of the spirilloxanthin series, with rhodopin as the major component and with small amounts of beta-carotene.

Cells grow preferably photoheterotrophically under anoxic conditions in the light, but photoautotrophic growth and chemotrophic growth under microoxic to oxic conditions is possible. Substrates used are acetate, propionate, butyrate, valerate, caproate, caprylate, ethanol, propanol, butanol, lactate, pyruvate, malate, fumarate, succinate, and malonate. Some strains also use methanol, formate, oxalacetate, OH-butyrate, and glycerol. Not used are citrate, tartrate, fructose, glucose, mannose, mannitol, sorbitol, glycolate, oxalate, pelargonate, benzoate, aspartate, arginine, and glutamate. Addition of HCO_3^- is essential only when CO_2 cannot be generated from the organic carbon source.

Molecular hydrogen and sulfide may serve as electron donors for photoautotrophic growth. The latter is oxidized to tetrathionate in sulfide-limited continuous culture, but thiosulfate and elemental sulfur are formed as major oxidation products in batch culture. In addition, ferrous iron is oxidized in the light, but may not support growth over prolonged times if supplied as the exclusive electron donor. Growth with ferrous iron is stimulated by acetate and succinate. Ammonia and dinitrogen, as well as Casamino acids and yeast extract, are used as nitrogen sources. Some strains grow poorly with nitrate and urea. Sulfate can serve as the sole sulfur source. No organic growth factors are required.

Mesophilic freshwater bacteria with a preference for slightly acidic pH showing optimal growth at 30°C and pH 6.0 (pH range: 5.2–7.5).

The mol% G + C of the DNA is: 61.8–63.8 (Bd).

Type strain: ATCC 17100, DSM 162.

Genus XVII. **Rhodoplanes** *Hiraishi and Ueda 1994b, 671*[VP]

AKIRA HIRAISHI AND JOHANNES F. IMHOFF

Rho.do.pla′ nes. Gr. n. *rhodon* rose; Gr. n. *planos* a wanderer; M.L. masc. n. *Rhodoplanes* a red wanderer.

Cells are rod-shaped, motile by means of polar, subpolar, or lateral flagella, and **multiply by budding and asymmetric cell division**. They are **Gram-negative and belong to the *Alphaproteobacteria*. Internal photosynthetic membranes are present as lamellar stacks** parallel to the cytoplasmic membrane (Fig. BXII.α.232). **Photosynthetic pigments are bacteriochlorophyll *a* and carotenoids of the spirilloxanthin series. Straight-chain, monounsaturated $C_{18:1}$ is the predominant component of the cellular fatty acids and $C_{16:0}$ is a second major component. Ubiquinones and rhodoquinones with 10 isoprene units (Q-10 and RQ-10) are present.**

Growth is preferably photoheterotrophic under anoxic conditions in the light, with simple organic substrates as carbon and electron sources. Photoautotrophic growth with sulfide as the electron donor does not occur. **Chemotrophic growth is possible under oxic conditions in the dark** at full atmospheric oxygen tension and by denitrification under anoxic conditions in the presence of nitrate. Growth factors may be required.

Mesophilic freshwater bacteria with preference for neutral pH. Some representatives may be thermotolerant.

Habitats: freshwater and wastewater environments.

The mol% G + C of the DNA is: 66.8–69.7.

Type species: **Rhodoplanes roseus** (Janssen and Harfoot 1991) Hiraishi and Ueda 1994b, 671 (*Rhodopseudomonas rosea* Janssen and Harfoot 1991, 27.)

FURTHER DESCRIPTIVE INFORMATION

Based on the 16S rDNA sequence information, *Rhodoplanes* species belong to the *Alphaproteobacteria* (Hiraishi and Ueda, 1994b). The nearest phylogenetic neighbors are *Blastochloris* species, which show approximately 94% identity of 16S rDNA sequences to those of *Rhodoplanes* species.

Although all *Rhodoplanes* species reproduce by budding and asymmetric cell division, the mode of budding differs somewhat with the species (Fig. BXII.α.233). *Rhodoplanes roseus* exhibits a sessile budding mode without formation of prosthecae, like *Rhodobium marinum* and *Rhodobacter blasticus*. *Rhodoplanes elegans*, however, does form prosthecae and rosette-like cell aggregates similar to those of *Blastochloris* species and *Rhodopseudomonas palustris*.

Cells of *Rhodoplanes* species are motile in young cultures, but motile cells become extremely rare at the late-exponential phase of growth. Electron microscopy of negatively stained cells shows that cells with single polar or subpolar flagella are dominant (Fig. BXII.α.234). In addition, cells having two lateral or subpolar

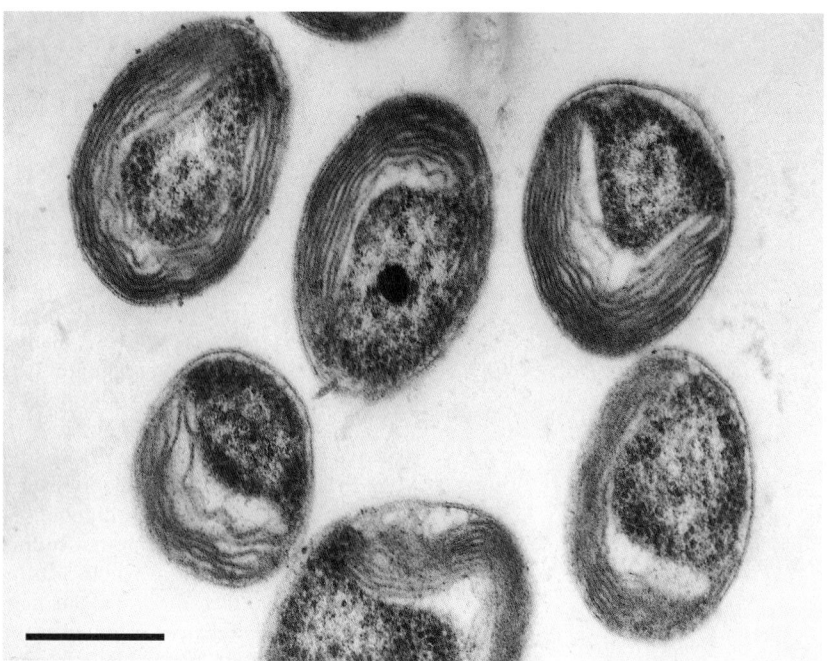

FIGURE BXII.α.232. Thin-section electron micrograph showing the internal photosynthetic membrane system of *Rhodoplanes roseus* DSM 5909. The lamellar type of photosynthetic membranes are observed. Bar = 0.5 μm.

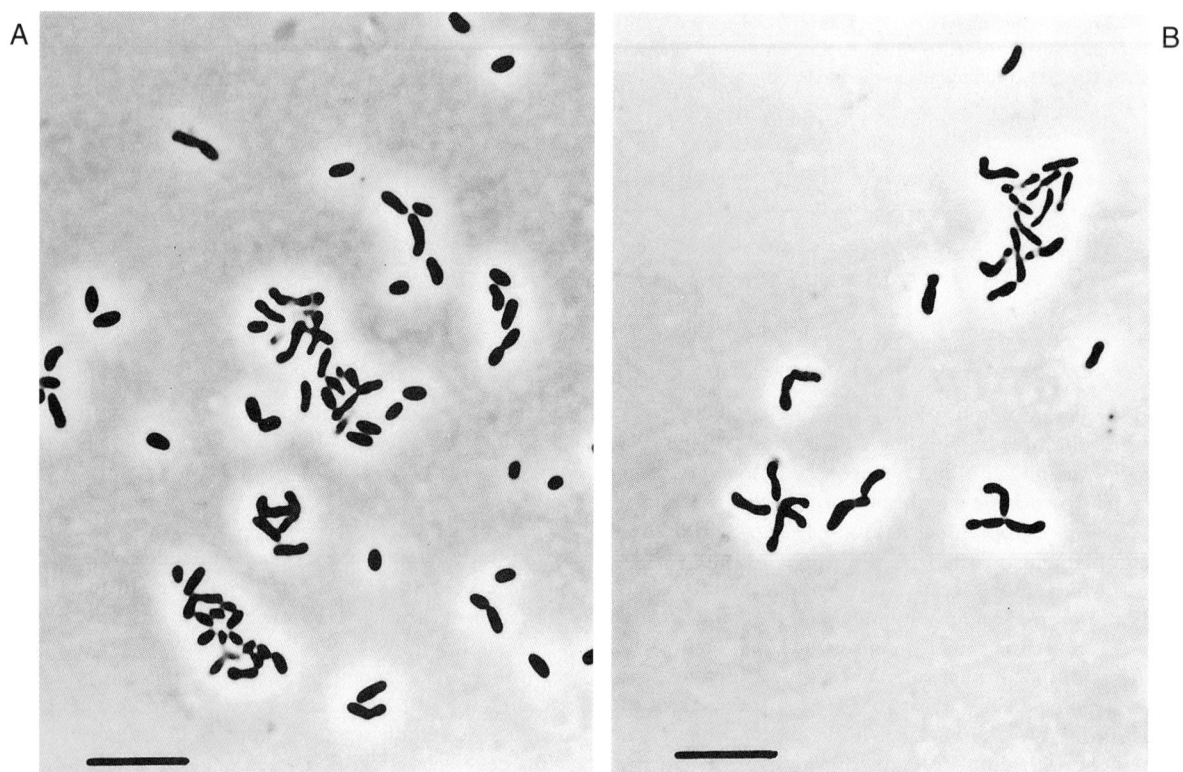

FIGURE BXII.α.233. Phase-contrast photomicrographs showing cell morphology of *Rhodoplanes*. A, *Rhodoplanes roseus* DSM 5909; B, *Rhodoplanes elegans* AS130. Bars = 8 μm.

flagella are frequently observed in *R. elegans* (Fig. BXII.α.235). Randomly distributed flagella are rarely found.

Photoheterotrophy, with simple organic compounds as electron donors and carbon sources, is the preferred mode of growth.

Pyruvate is the best substrate. Other simple organic acids, including acetate, lactate, and intermediates of the tricarboxylic acid cycle serve as good carbon sources. However, no growth occurs on sugars or sugar alcohols. While other species of pho-

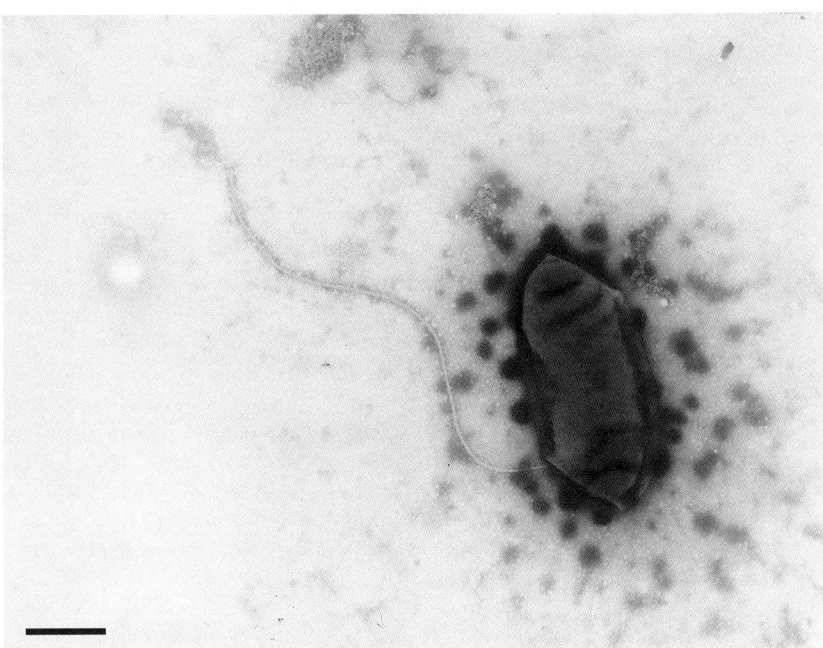

FIGURE BXII.α.234. Electron micrograph of a negatively stained cell of *Rhodoplanes roseus* DSM 5909 having a polar flagellum. Bar = 1 μm.

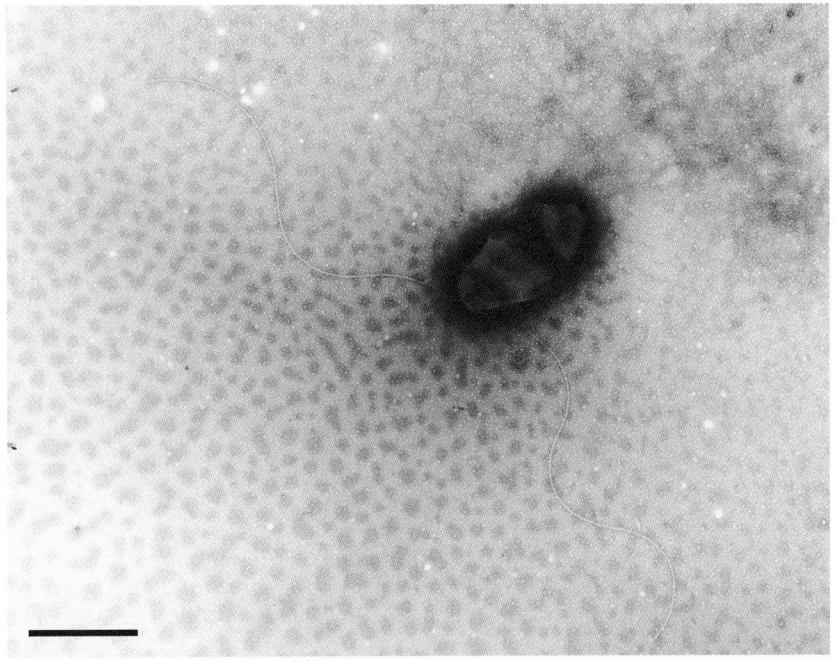

FIGURE BXII.α.235. Electron micrograph of a negatively stained cell of *Rhodoplanes elegans* AS130 having two subpolar flagella. Bar = 1 μm.

totrophic purple nonsulfur bacteria utilize straight-chain, saturated fatty acids as carbon sources, those acids ($>C_5$) have inhibitory effects on growth of *Rhodoplanes* (Janssen and Harfoot, 1987; 1991; Hiraishi and Ueda, 1994b).

Denitrification is a characteristic property of *Rhodoplanes* species. Growth under denitrifying conditions is very slow (doubling time >20 h) and is accompanied by little production of nitrogen gas. The growth yield under denitrifying conditions is 10–50% of the yield under the optimal phototrophic growth conditions.

Activities of denitrifying enzymes in *Rhodoplanes elegans* are much lower than those of other denitrifying phototrophic bacteria, such as *Rhodobacter sphaeroides* IL106 and *Rhodobacter azotoformans* (Hiraishi et al., 1995a).

The capacity for nitrogen fixation has not yet been studied intensively. However, this ability appears to be absent, as suggested by the observation that nitrogenase-dependent H_2 gas production does not occur in an ammonium-deficient medium (Hiraishi and Ueda, 1994b). Hydrolytic activities against starch, gelatin, casein, and Tween 80 are absent. In *R. elegans*, cardiolipin, phosphatidylglycerol, phosphatidylethanolamine, and phosphatidylcholine are present as the major phospholipids (A. Hiraishi, unpublished data). In all species, straight-chain, monounsaturated $C_{18:1}$ is the predominant component of cellular fatty acids, constituting 60–70% of the total content. Considerable amounts of $C_{16:0}$ (10–20%) are also present.

Since only a few strains of *Rhodoplanes* have hitherto been isolated and studied, their natural habitats and ecological significance are not yet fully understood. The type strain of *R. roseus* was isolated after enrichment in a Winogradsky column packed with mud from a duck pond in New Zealand (Janssen and Harfoot, 1991), whereas strains of *R. elegans* have been isolated from activated sludge (Hiraishi and Ueda, 1994b). As suggested by their sources and physiological characteristics, the natural habitats of *Rhodoplanes* species are freshwater environments highly polluted with organic matter, although sulfide-rich water bodies may not provide favorable conditions for their growth. Strains of *Rhodoplanes*, as well as those of *Rhodopseudomonas palustris*, *Rhodobacter blasticus*, and *Rhodobacter sphaeroides*, are frequently detected on agar plates inoculated with relatively high dilutions (10^{-5} to 10^{-6}) of activated sludge samples (A. Hiraishi, unpublished data). This suggests that *Rhodoplanes* may play an active role in the purification process of polluted water.

Phylogenetic and chemotaxonomic analyses of phototrophic purple nonsulfur bacteria that have been isolated recently from hot springs of New Zealand and are able to grow at elevated temperatures up to 46°C (P. Charlton, personal communication) have demonstrated that they belong to the genus *Rhodoplanes*.

ENRICHMENT AND ISOLATION PROCEDURES

Rhodoplanes species can be cultivated in a chemically defined medium containing simple organic compounds, such as carbon sources and yeast extract, as growth factor supplements, as is commonly employed for other members of the phototrophic purple nonsulfur bacteria (Imhoff, 1988; Imhoff and Trüper, 1992). A suitable medium for growth of *Rhodoplanes* consists of (per liter of distilled water): $(NH_4)_2SO_4$, 1 g; KH_2PO_4, 1 g; $MgSO_4 \cdot 7H_2O$, 0.2 g; $CaCl_2 \cdot 2H_2O$, 0.05 g; trace element solution SL12 (Pfennig and Trüper, 1992), 1 ml; yeast extract (Difco Laboratories), 1 g; 20 mM pyruvate (final concentration; filter-sterilized; pH 6.8).

Alternatively, a mineral medium supplemented with 0.2% tartrate and 0.1% succinate as carbon sources and niacin, thiamine, and *p*-aminobenzoic acid as growth factors is most effective for the enrichment of *Rhodoplanes*. The bacteria are enriched by anaerobic incubation in screw-capped test tubes filled with the medium under incandescent illumination. Direct plating or

membrane-filter plating techniques using anaerobic jars can also be used for isolation of *Rhodoplanes* species.

MAINTENANCE PROCEDURES

Cultures are well preserved in liquid nitrogen, by lyophilization, or at $-80°C$ in a mechanical freezer.

DIFFERENTIATION OF THE GENUS *RHODOPLANES* FROM OTHER GENERA

The genus *Rhodoplanes* is phenotypically similar to other genera of the budding phototrophic bacteria, especially the genus *Rhodopseudomonas*, but is separated from the latter genus by its phylogenetic position and outstanding physiological and chemotaxonomic characteristics. Characteristic features of the genus *Rhodoplanes* are denitrification, inability to grow on sugars, long straight-chain fatty acids ($>C_5$), and sulfide, as well as production of both Q-10 and RQ-10. In addition, *Rhodoplanes* species are differentiated from the phenotypically similar *Rhodopseudomonas palustris* by cell size, vitamin requirement, and carbon nutrition. It is quite easy to distinguish *Rhodoplanes* from its nearest phylogenetic neighbors, the *Blastochloris* species, because the latter produce bacteriochlorophyll *b* and have either Q-9 and MK-9 or Q-8 (Q-10) and MK-7 (MK-8) as the major quinones. Major differentiating properties between *Rhodoplanes* species and other phototrophic *Alphaproteobacteria* are shown in Tables BXII.α.168 and BXII.α.169 of the chapter on the genus *Blastochloris* as well as in Tables 3 (pp. 125–126) and 4 (p. 127) in the introductory chapter "Anoxygenic Phototrophic Purple Bacteria", Volume 2, Part A. The phylogenetic relationships of these bacteria, based on 16S rDNA sequences, are shown in Fig. 2 (p. 128) of that chapter.

TAXONOMIC COMMENTS

Rhodoplanes roseus was first described as *Rhodopseudomonas rosea* (Janssen and Harfoot, 1991). Later, this species was shown by 16S rDNA sequence analysis and chemotaxonomic information to be different at the generic level from any species of the genus *Rhodopseudomonas*. As a result, *Rhodopseudomonas rosea* was transferred to the new genus *Rhodoplanes* and designated as the type species of this genus, *Rhodoplanes roseus* (Hiraishi and Ueda, 1994b). At the same time, *Rhodoplanes elegans* was described as the second species of this genus. Furthermore, another isolate (strain IL245), classified as *Rhodopseudomonas* sp. (Kawasaki et al., 1993b), is closely related to the two *Rhodoplanes* species at a level of 97% sequence similarity of the 16S rDNA and may represent another species of *Rhodoplanes*. Phylogenetically, *Rhodoplanes* forms a major cluster in the *Alphaproteobacteria* together with the genus *Blastochloris* and the nonphototrophic genus *Ancylobacter*.

DIFFERENTIATION OF THE SPECIES OF THE GENUS *RHODOPLANES*

Two species, *Rhodoplanes roseus* and *Rhodoplanes elegans*, currently comprise this genus. They are phenotypically quite similar to each other, but major differences are noted in budding type, rosette formation, vitamin requirement, and growth at pH 6.0. The average level of genomic DNA–DNA hybridization between

R. roseus and *R. elegans* is 37% (Hiraishi and Ueda, 1994b). The two species show approximately 98% sequence similarity of their 16S rDNA. General and differential characteristics of the two species are shown in Tables BXII.α.168 and BXII.α.169 of the chapter on the genus *Blastochloris*.

List of species of the genus Rhodoplanes

1. **Rhodoplanes roseus** (Janssen and Harfoot 1991) Hiraishi and Ueda 1994b, 671[VP] (*Rhodopseudomonas rosea* Janssen and Harfoot 1991, 27.)

   ro' se.us. L. adj. *roseus* rose-colored, pink.

Cells are rod-shaped to elongate-ovoid, 1.0 × 1.8–2.5 μm, motile by means of single polar flagellum. They reproduce by budding without prosthecae and do not form rosettes and clusters. If cells age, they become nonmotile and encapsulated by slime. Internal photosynthetic membranes are present as lamellar stacks parallel to the cytoplasmic membrane. Phototrophically grown cultures are pink, while aerobic cultures are colorless. Absorption maxima of living cells are at 373–375, 468, 493, 527–530, 591–593, 799–801, and 850–854 nm. Photosynthetic pigments are bacteriochlorophyll *a* and carotenoids of the spirilloxanthin series.

Growth is preferably photoheterotrophic under anoxic conditions in the light. Photoautotrophic growth with thiosulfate, but not sulfide, occurs in the presence of 0.01% yeast extract. Chemotrophic growth is possible under oxic conditions in the dark at full atmospheric oxygen tension and by denitrification under anoxic conditions in the presence of nitrate.

Electron donors and carbon sources used for phototrophic growth are acetate, propionate, butyrate, valerate, lactate, pyruvate, succinate, fumarate, malate, citrate, tartrate, Casamino acids, yeast extract, and peptone. The following organic compounds are not utilized: formate, long-chain fatty acids with more than five carbon atoms, glycolate, malonate, benzoate, arabinose, xylose, rhamnose, fructose, glucose, mannose, galactose, dulcitol, mannitol, sorbitol, glycerol, methanol, ethanol, propanol, alanine, aspartate, asparagine, glutamate, and leucine. Sulfide is growth-inhibitory at 0.2 mM concentration and is not used as a photosynthetic electron donor. Ammonium salts are used as nitrogen sources. Sulfate and thiosulfate are assimilated as sulfur sources. Growth is also stimulated considerably by addition of 0.01% yeast extract. Niacin is required as growth factor.

Mesophilic freshwater bacterium with optimal growth at 30°C, pH 7.0–7.5 (pH range: 6.5–8.0). No growth occurs at pH 6.0 and 8.5 and at 1% NaCl.

Habitat: polluted freshwater lake sediments and wastewater environments.

Major quinone components are Q-10 and RQ-10.

The mol% G + C of the DNA is: 66.8 (HPLC) and 66.0 (T_m).

Type strain: 941, DSM 5909.

GenBank accession number (16S rRNA): D14429, D25313.

2. **Rhodoplanes elegans** Hiraishi and Ueda 1994b, 672[VP]

e' le.gans. L. adj. *elegans* choice, elegant.

Cells are rod-shaped, occasionally slightly curved, 0.8–1.0 × 2.0–3.5 μm, and motile by means of polar or subpolar flagella. They multiply by budding and asymmetric cell division, with a slender prostheca occurring between the mother and daughter cells. If cells age, they form rosette-like clusters. Internal photosynthetic membranes are present as lamellar stacks parallel to the cytoplasmic membrane. Phototrophically grown cultures are pink, while aerobic cultures are colorless. Absorption maxima of living cells are at 373–375, 466, 491, 527–530, 591–593, 799–801, and 854–856 nm. Photosynthetic pigments are bacteriochlorophyll *a* and carotenoids of the spirilloxanthin series.

Growth is preferably photoheterotrophic under anoxic conditions in the light. Photoautotrophic growth occurs with thiosulfate, but not sulfide, in the presence of 0.01% yeast extract. Chemotrophic growth is possible under oxic conditions in the dark at full atmospheric oxygen tension and by denitrification under anoxic conditions in the presence of nitrate.

The following organic compounds are photoassimilated: acetate, propionate, butyrate, valerate, caproate, lactate, pyruvate, succinate, fumarate, malate, citrate, tartrate, Casamino acids, and yeast extract. Not utilized are formate, long-chain fatty acids with more than six carbon atoms, glycolate, malonate, benzoate, arabinose, xylose, rhamnose, fructose, glucose, mannose, galactose, dulcitol, mannitol, sorbitol, glycerol, methanol, ethanol, propanol, alanine, aspartate, asparagine, glutamate, and leucine. Sulfide is growth-inhibitory at 0.2 mM concentration and is not used as an photosynthetic electron donor. Ammonium salts are used as nitrogen sources. Sulfate and thiosulfate are assimilated as sulfur sources. Thiamine and *p*-aminobenzoic acid are required as growth factors.

Mesophilic freshwater bacterium with optimal growth at 30–35°C and pH 7.0 (pH range: 6.0–8.5). No growth occurs at pH 5.5 or 8.5 or at 1% NaCl.

Habitat: activated sludge.

Major quinone components are Q-10 and RQ-10.

The mol% G + C of the DNA is: 69.6–69.7 (HPLC).

Type strain: AS130, ATCC 51906, JCM 9224.

GenBank accession number (16S rRNA): D25311.

Genus XVIII. **Seliberia** *Aristovskaya and Parinkina 1963, 56*[AL]

JEAN M. SCHMIDT AND SUZANNE V. KELLY

Se.li.be' ri.a. M.L. fem. n. *Seliberia* of Seliber, named for the Russian microbiologist, Professor G.L. Seliber.

Rods, 0.5–0.8 μm in diameter and 1–12 μm in length, with a **helically sculptured** or **furrowed** topography. The ends of the cell may be either blunt or rounded. **Stellate aggregates (rosettes) of sessile rods** joined at one pole, and individual, shorter **motile rods (swarmers)** occur in the same culture. An adhesive **holdfast**, secreted at one pole, mediates attachment into rosettes. Growth on appropriate soil extract media may permit formation of round to ovoid **"generative cells"**. Capsules are not produced, although a thin glycocalyx is produced later in the growth cycle. Resting stages are not known. Gram negative. Following **unidirectional polar cell growth**, a shorter motile cell (a **swarmer**) and a longer sessile cell are produced by **asymmetric transverse fission**. A sin-

gle **subpolar ensheathed flagellum** is characteristically present on the swarmer; **several lateral flagella, not ensheathed,** may also be present. **Strictly aerobic.** Optimal temperature, 25–30°C; maximum ca. 37°C, minimum 15–20°C. Chemoorganotrophic, having an **oxidative type of metabolism**. Catalase and oxidase positive. These organisms live in soil and fresh water environments as autochthonous microflora, often where oligotrophic conditions prevail.

The mol% G + C of the DNA is: 63–66.

Type species: **Seliberia stellata** Aristovskaya and Parinkina 1963, 55.

FURTHER DESCRIPTIVE INFORMATION

Phylogenetic Treatment   Limited phylogenetic information on the seliberias is available, due in large part to difficulties encountered in cell lysis for nucleic acid extractions. The analysis of 5S rRNA sequences has placed the genus *Seliberia* in the *Alphaproteobacteria* (Bulygina et al., 1990; Stackebrandt, 1992).

Cell morphology   The typical cells are helically structured rods of varying lengths (Figs. BXII.α.236, BXII.α.239, and BXII.α.240). In a growing culture, the shorter cells are motile swarmers; the longer cells are sessile and include the predivisional cells. The rods become very long (10 μm or more in length) when grown on fulvic acid medium (Aristovskaya, 1974); in aquatic environments with very dilute organic nutrient conditions, seliberias also occur as very long helically sculptured rods. In media containing ulmic acid complexes (Aristovskaya, 1974; Schmidt and Swafford, 1979) or when a plentiful supply of organic nutrients is available (Schmidt and Swafford, 1979), the cells are shorter, and the longest cells seldom exceed 5 μm in length. Growth of some strains of *Seliberia stellata* on soil extract media (Aristovskaya and Parinkina, 1963; Aristovskaya, 1964, 1974) may give rise to round or ovoid "generative cells"; production of the round to ovoid generative cells with the type strain of *S. stellata* (ICPB 4130) has not been verified (Schmidt and Starr, 1984).

Cell wall composition   The cell wall of *Seliberia stellata* is a bipartite structure with an outer layer of tightly bound protein, but lacking the appearance of an S layer. The outer protein layer is removed from the underlying layer of peptidoglycan with 0.5% KOH, but not with a variety of milder enzymatic and/or detergent treatments. No phospholipid or lipopolysaccharide (LPS) has been found in *Seliberia* cell wall preparations.

Fine structure   Ultrastructural features of *Seliberia stellata* are shown in Fig. BXII.α.237.

Colonial and cultural characteristics   On dilute peptone-yeast extract agar, colonies are slow-growing, requiring a week or more of incubation. The colonies are minute (0.5–1.0 mm diameter) and white in color. The colonies may adhere strongly to the agar. On ulmic acid or fulvic acid media, the colonies are also slow-growing and may show some browning, due to presence of iron oxide. In dilute liquid peptone-yeast extract medium, the bacteria adhere tightly to the surface of the growth vessel, and adherence is mediated by the polar holdfast.

Life cycle   Cell growth occurs unidirectionally, at the pole of the rod opposite the pole with the secreted holdfast. The end of the predivisional rod that consists of the newly made surface components eventually becomes a motile swarmer as a result of an asymmetric division. The remaining (longer) portion of the predivisional cell retains the surface components of the parent without redistributing them into the zone of new growth. The

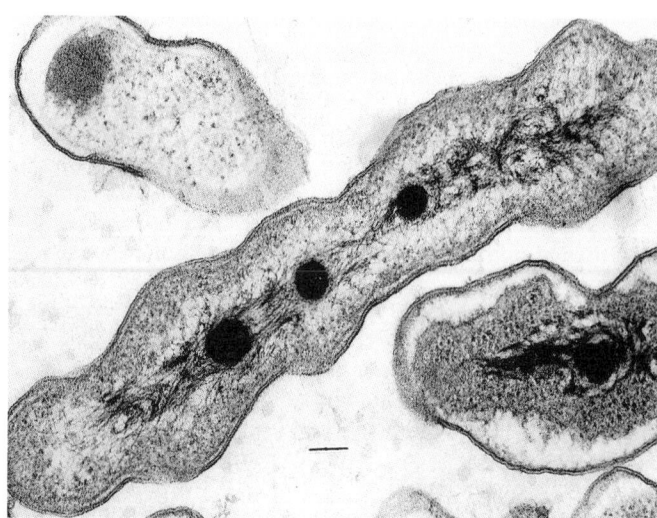

FIGURE BXII.α.237.   Thin section of *S. stellata* strain Z/A, prepared with a modification of the Ryter–Kellenberger fixation method (Kellenberger et al., 1958), 1.0% osmic acid in 0.1× buffer for 2 h, and embedded in Spurr resin. The thin cell envelope and wavy outline are characteristic of seliberias. Bar = 0.1μm (Courtesy of J.R. Swafford).

FIGURE BXII.α.236.   Scanning electron micrograph of *S. stellata* strain Z/A fixed with 2.0% glutaraldehyde, critical-point dried, and gold-coated, showing the helically sculptured topography of the bacteria. Bar = 1 μm (Courtesy of J.R. Swafford).

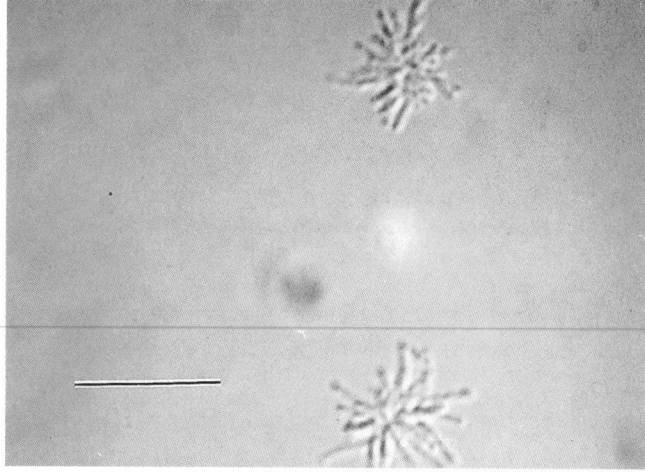

FIGURE BXII.α.238.   Nomarski interference contrast light micrograph of *S. stellata* rosettes. Bar = 5 μm (Courtesy of J.R. Swafford).

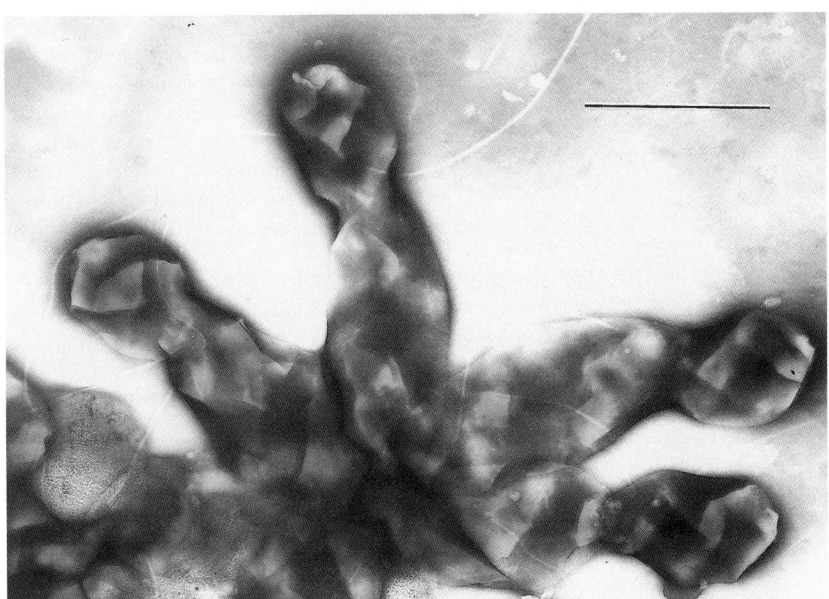

FIGURE BXII.α.239. Phosphotungstate (0.5%) negative stain of an aquatic *Seliberia*-like strain, ICPB 4141, illustrating the asymmetric plane of division. Bar = 1.0 μm.

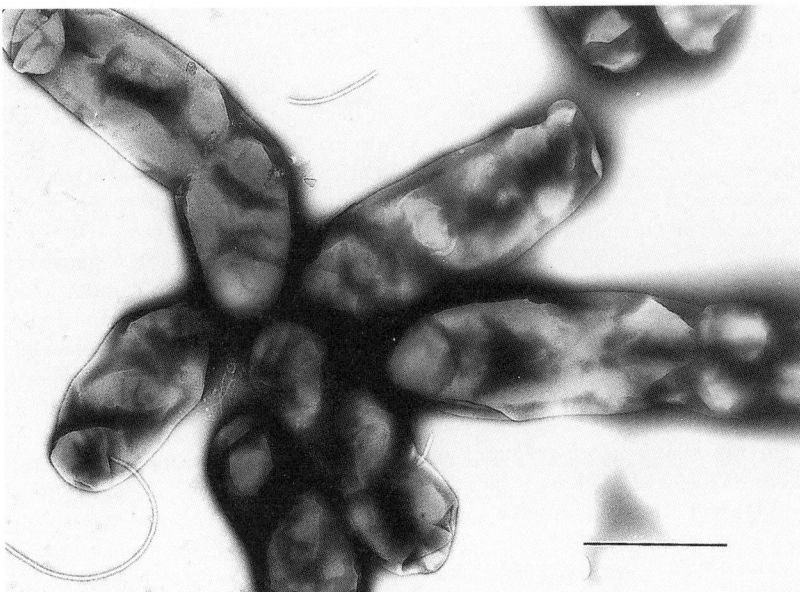

FIGURE BXII.α.240. Phosphotungstate (0.5%) negative stain of a rosette of *Seliberia*-like aquatic strain ICPB 4133. A subpolar flagellum is present. Bar = 1.0 μm (Courtesy of J.R. Swafford).

subpolar ensheathed flagellum of the new swarmer is found at the apical end of the predivisional cell, shortly before its division. The asymmetry perpendicular to the division plane results in unequal division products (Fig. BXII.α.239; Schmidt and Starr, 1984). Swarmer cell production by unidirectional polar growth in the genus *Seliberia* meets two of the major criteria (*de novo* synthesis of the bud surface and transverse asymmetry of division) of accepted definitions of budding (Staley, 1973a; Staley et al., 1981); however, a third feature typical of most budding bacteria, an increasing diameter of the bud during its development (Hirsch, 1974a), is not found in *Seliberia*; the developing daughter cell (swarmer) is (i) not initially narrower than the parent, and

(ii) does not increase in diameter (width) during growth (Schmidt and Starr, 1984).

The proportion of cells in stellate aggregates (rosettes) varies in *Seliberia* cultures, but rosettes are readily detectable microscopically under most cultural conditions (Figs. BXII.α.238, BXII.α.240). The stellate aggregates are formed by association of swarmers, which have produced an adhesive holdfast at one pole, with sessile cells–often with those stationary rods located in the immediate vicinity of the predivisional cell that produced the swarmer. In addition, at high cell densities in liquid cultures, polar aggregation of swarmer cells occurs (Schmidt and Starr, 1984), akin to the manner of formation of *Caulobacter* rosettes

(Poindexter, 1964, 1981). The *S. stellata* holdfast consists of a polysaccharide, containing glucose and/or mannose, galactose, *N*-acetylglucosamine, and *N*-acetylgalactosamine (Hood and Schmidt, 1996).

Nutrition and growth conditions Growth of *Seliberia* occurs on dilute organic media, such as yeast extract plus peptone or Casamino acids (Difco), soil extract agar, or pond water agar. Media containing organomineral complexes of either fulvic or ulmic acid with sesquioxides have been used (Aristovskaya and Parinkina, 1963; Aristovskaya, 1974; Schmidt and Swafford, 1981). With these media, aerobic incubation at 25°C gives satisfactory growth.

Metabolism and metabolic pathways Although the genus *Seliberia* has been previously described as facultatively anaerobic (Aristovskaya, 1974), the available type strain (Z/A = ICPB 4130) and all of the several aquatic strains are strictly oxidative; all lack the capacity to carry out anaerobic fermentations. *S. stellata* Z/A possesses glucose dehydrogenase and 6-phosphogluconate dehydrase activities, and can grow aerobically on D-glucuronic acid, suggesting the presence of the Entner–Doudoroff pathway. Strain Z/A also possesses strong NADH oxidase activity, isocitrate dehydrogenase, succinate dehydrogenase, and malate dehydrogenase, consistent with oxidative metabolism. *S. stellata* strain Z/A is a strong denitrifier and can grow anaerobically in the presence of 0.1% $NaNO_3$, indicating a nitrate/nitrite respiration. Most of the aquatic *Seliberia*-like strains are not able to denitrify anaerobically (Schmidt and Swafford, 1979, 1981).

Genetics and molecular data The seliberias have not received much attention, due to difficulties in obtaining DNA preparations, which are related to the problems encountered in obtaining cell lysis. No bacteriophages or plasmids have been reported for the seliberias.

Antigenic structure The outer layer of the cell envelope of the seliberias induces an antibody response in rabbits (Schmidt and Starr, 1984). Indirect fluorescent antibody and agglutination tests done with polyvalent antisera to strain Z/A and to two aquatic *Seliberia*-like strains indicated significant antigenic relatedness among the strains. The two major proteins making up the outer layer of strain Z/A (which lacks LPS) have respective molecular weights of 54–56 kDa and 82–84 kDa. Some information on flagellum-associated proteins is available: strain Z/A, which produces both ensheathed subpolar and nonsheathed lateral flagella, has flagella-associated proteins with molecular weights of 68, 36, and 35 kDa; several aquatic *Seliberia*-like strains which produced only the subpolar ensheathed flagellum gave major flagellar protein values of 56–60 kDa.

Antibiotic Sensitivity Seliberias are quite resistant to low concentrations (10 μg) of most of the common antibiotics; most aquatic *Seliberia*-like strains are inhibited by 100 μg concentrations of ampicillin, penicillin G, vancomycin, chlortetracycline, streptomycin, and rifampin. *S. stellata* strain Z/A is resistant to 100 μg concentrations of ampicillin, penicillin G, and vancomycin.

Pathogenicity No reports of seliberias causing overt or opportunistic infections in humans, animals, or insects have been reported. Their optimal growth temperature range of 25–30°C, poor growth at 37°C, and lack of LPS in their cell wall makes the seliberias unlikely candidates as pathogens of humans or warm-blooded animals.

Ecology The habitats of *Seliberia* include soil and freshwater environments. They have been reported as epiphytes of the cyanobacterium *Nodularia* (Smarda, 1985), and have been found in an activated sludge reactor (Baker et al., 1983). Although seliberias appear to be widely distributed, they have not received much attention from bacteriologists. Seliberias appear to be well suited to conditions of limited nutritional resources, although they can also occur in eutrophic environments.

ENRICHMENT AND ISOLATION PROCEDURES

Specific enrichment procedures for podzol-inhabiting seliberias have not been described, although it has been noted that these bacteria occur frequently in pedoscopes (Aristovskaya and Parinkina, 1963). Seliberias of both aquatic and soil origin can be enriched, nonspecifically, using oligotrophic conditions (Schmidt and Swafford, 1979, 1981). Glass beakers, to which water samples (for example, 200–800 ml of pond, lake, or even tap water) are added, with or without the addition of 0.001–0.005% peptone (Difco), are covered with plastic film to prevent evaporation and incubated for several days to several weeks or months at 24–26°C. To adapt this aquatic enrichment to soil inocula, suspensions of 1 g (or less) of soil are made in 500 ml of filtered pond or tap water to which (i) no nutrients or (ii) very low concentrations of peptone are added (as above, for aquatic sample enrichment). The seliberias usually occur at the air–water interface of the enrichment (in the surface pellicle, if heavy growth has occurred). They may be attached to other microbes or detritus. Seliberias will be easier to isolate subsequently if heavy growth of other, more rapidly growing bacteria has not occurred in the enrichment beaker. Hence, the use of several enrichment beakers with varying concentrations of added nutrients may facilitate the enrichment procedure. Observation of the bacteria at the surfaces of the enrichment, using transmission electron microscopy and negative stains (Schmidt and Swafford, 1992) is useful for determining whether significant numbers of seliberias are present. Use of genus-specific fluorescent dye-labeled oligonucleotide probes in glass slide hybridization reactions for these bacteria has not been reported yet, but it should be feasible once sequences are available.

To obtain pure cultures of seliberias, several types of solid culture media have been used. To isolate seliberias from soil samples, agar media containing dilute organomineral complexes (ulmic or humic acids, or fulvic acid) obtained from soil humus (Aristovskaya and Parinkina, 1961, 1963) have been used successfully. Soil samples thought to contain seliberias are streaked on the ulmic or fulvic acid agar media; after 4–7 d, the slow-growing colonies that show some ferric oxide–caused browning are examined for characteristic morphologic features. The accumulation of ferric hydroxide by seliberias of soil origin is characteristic of their behavior in mixed culture; in pure culture, an atmosphere of about 1% CO_2 stimulates deposition of iron oxide (Aristovskaya and Parinkina, 1963).

Seliberia-like bacteria of aquatic or soil origin may be isolated (from enrichment cultures) by streaking from the liquid enrichment surface onto a dilute organic medium; recipes for two useful media follow. PYE medium: 0.2% peptone (Difco), 0.1% yeast extract, 1.0% Hutner's vitamin-free mineral base (Cohen-Bazire et al., 1957), 1.2% or 1.5% Bacto-agar (Difco), and distilled water. Dilute medium I: 0.02% peptone, 0.01% yeast extract, 1.0% Hutner's mineral base, 1.2 or 1.5% Bacto-agar (Difco), and distilled water; addition of 0.1% glucose (filter-sterilized) is optional (Schmidt and Swafford, 1981). After two weeks or longer of incubation at 25°C, the plates are examined with a dissecting microscope for minute 0.5–1.0 mm diameter white colonies that may adhere to the agar. The selected colonies are transferred to patch plates, reincubated for several days, and then observed with phase-contrast light microscopy for rosette-forming, asymmetrically dividing, rod-shaped bacteria that have a motile stage. Transmission electron microscopy, using negative stains, can be used to check for the helical sculptured topography of the rods (Schmidt and Swafford, 1981). Specific oligonucleotide probes with fluorescent labels may be useful in identification of *Seliberia* colonies or patches.

MAINTENANCE PROCEDURES

Stock cultures grown aerobically on slants of PYE medium in screw-cap test tubes remain viable for several weeks at room temperature (24–26°C). The slant cultures survive several weeks longer at 4°C. Suspensions of cells scraped (with some difficulty) from agar surfaces and suspended in sterile distilled water also survive well for several weeks at room temperature. For longer-term preservation (several years), lyophilization is a satisfactory method.

PROCEDURES FOR TESTING SPECIAL CHARACTERS

The determination of cellular morphology at a level of resolution attainable by scanning electron microscopy or transmission electron microscopy (negative stains) is a key feature in the identification of seliberias. The helical sculpturing of the bacterial cell surface and the presence of a subpolar sheathed flagellum can be determined with electron microscopy.

DIFFERENTIATION OF THE GENUS *SELIBERIA* FROM OTHER GENERA

At the resolution achievable with light microscopes, seliberias can be confused with various kinds of caulobacters (members of the genera *Caulobacter* or *Asticcacaulis*), because of their occurrence in the same types of natural habitats and preference for aerobic environments, similarities in cell size and apparent rod-like shape, presence of shorter motile and longer sessile cells, and occurrence of rosette arrangements (star-like aggregates) of several to many cells. However, seliberias are not prosthecate, whereas predivisional and mature *Caulobacter* and *Asticcacaulis* cells have at least one prostheca or cellular stalk (Poindexter, 1964, 1981). Prosthecae can be observed with phase contrast high-resolution light microscopy but not without some difficulty

since prosthecae are usually 0.2 μm or less in diameter. Other rosette-forming bacteria, which also demonstrate motility, but which typically belong to the genera *Pseudomonas* or *Agrobacterium*, also can be difficult to distinguish from seliberias if only light microscopy is used.

Confirmation that an isolate belongs to the genus *Seliberia* can be established by phenetic characterization, although use of morphological and ultrastructural characteristics as a sole guide to generic assignment is not a desirable situation. Notably, there is still some lack of phylogenetic data for the seliberias, but obtaining phenetic information on the physiology, relationship to oxygen, mol% G + C, and mode of growth and division is useful in defining members of the genus *Seliberia*, distinguishing them from *Caulobacter*, pseudomonads, *Agrobacterium* spp., and other rosette-forming Gram-negative bacteria (Table BXII.α.188). Since the seliberias characteristically exhibit unidirectional polar growth, and their swarmers are generated at the end of the dividing cell (Schmidt and Starr, 1984), this relatively unusual mode of cell division is a useful distinguishing trait.

TAXONOMIC COMMENTS

The genus *Seliberia*, its single species *Seliberia stellata*, and the closely related but not quite identical aquatic *Seliberia*-like strains have not been examined for 16S rDNA sequences; 5S rRNA analysis has placed this genus in the *Alphaproteobacteria* (Bulygina et al., 1990; Stackebrandt, 1992). Difficulties in achieving cell lysis and DNA extractions have played a role in the delay in acquiring 16S rDNA phylogenetic data and DNA–DNA hybridization data among the phenetically quite similar *Seliberia*-like strains. The DNA–DNA hybridization data among *Seliberia*-like strains are important in determining species-level relationships. Seliberias also should be compared at the molecular level with other *Alphaproteobacteria*, including various species of *Caulobacter*, some pseudomonads, such as *Pseudomonas carboxydohydrogena* (Meyer et al., 1980) due to its morphology, and *Nitrobacter*, a budding bacterium with an unusual cell envelope similar in profile to that of *Seliberia stellata*, to establish the uniqueness of this genus.

ACKNOWLEDGMENTS

We thank G.A. Zavarzin, Institute of Microbiology, Academy of Sciences, Moscow, for the culture of *Seliberia stellata* strain Z/A (ICPB 4130). Micrographs were provided by J.R. Swafford. Data on cell wall composition and metabolic pathways were obtained in collaboration with Steven Baker and Michelle Santos.

TABLE BXII.α.188. Differential morphological and developmental characteristics of *Seliberia*[a] and other biochemically and morphologically similar taxa[b]

Characteristic	Seliberia	Agrobacterium	Caulobacter	Pseudomonas
Unidirectional polar cell growth	+	−	−	−
Asymmetric cell division products	+	−	+	−
Ensheathed polar flagellum	+	−	−	D
Prostheca production	−	−	+	−
Cell surface helically sculptured	+	−	−	−[c]
Stellate aggregate formation	+	D	+	D
Mol% G + C of DNA	63–66	58–62	62–67[d]	58–70[e]

[a]*S. stellata* and the aquatic and soil *Seliberia*-like strains.

[b]Symbols: +, 90% of strains positive; -, 90% of strains negative; D, different reactions in different species.

[c]Data from Meyer et al., 1980.

[d]Data from Poindexter, 1981.

[e]Data from Bergan, 1981.

FURTHER READING

Aristovskaya, T.V. 1974. *Seliberia. In* Buchanan and Gibbons (Editors), Bergey's Manual of Determinative Bacteriology, The Williams & Wilkins Co., Baltimore. p. 160.

Aristovskkaya, T.V. and V.V. Parinkina. 1963. New soil microorganism *Seliberia stellata* nov. gen., n. sp. Izv. Akad. Nauk S.S.S. R. Ser. Biol. *28*: 49–56.

Schmidt, J.M. and J.R. Swafford. 1979. Isolation and morphology of helically sculptured, rosette-forming, freshwater bacteria resembling *Seliberia.* Curr. Microbiol. *3*: 65–70.

Schmidt, J.M. and J.R. Swafford. 1992. The genus *Seliberia. In* Balows, Trüper, Dworkin, Harder and Schleifer (Editors), The Prokaryotes. A Handbook on the Biology of Bacteria: Ecophysiology, Isolation, Identification, Applications, 2nd Ed., Springer-Verlag, New York. pp. 2490–2494.

List of species of the genus Seliberia

1. **Seliberia stellata** Aristovskaya and Parinkina 1963, 55[AL]

stel.la' ta. L. adj. *stellata* starred.

Colonies on peptone-yeast extract (PYE medium, see Enrichment and Isolation Procedures, above) are smooth, convex, and white. The diameter is 0.5–1.5 mm, with a regular border. On ulmic acid agar, a soil extract medium (Aristovskaya and Parinkina, 1961, 1963), growth is sparse, colonies are less than 1.0 mm in diameter, and light brown. *Seliberia stellata* strain Z/A grows in a chemically defined medium containing an appropriate carbon source, 0.05% ammonium chloride, 1.0% Hutner's vitamin-free mineral base (Cohen-Bazire et al., 1957), and 0.001% phosphate. It grows over a wide pH range, from 4.5 to at least 9.0. No added vitamins are required. *S. stellata* is positive for urease production and starch hydrolysis; it can produce H_2S from cystine; it can reduce nitrate to nitrite under either aerobic or anaerobic conditions; it is unable to liquefy gelatin, utilize citrate, hydrolyze casein, or produce indole, and it gives negative methyl red and Voges–Proskauer reactions. *S. stellata* strain Z/A, the reference strain, can produce both a subpolar, ensheathed, typically single flagellum, and several lateral, ordinary (not ensheathed) flagella. This flagellation is characteristic of its swarmer cells grown on agar-solidified PYE medium at 25°C.

The mol% G + C of the DNA is: 66 (Bd) (strain Z/A).

Type strain: E-37, VKM B-1340.

Other Organisms

Several *Seliberia*-like strains from aquatic (freshwater) sources and a few from soil have been enriched and isolated in pure culture, and these are quite similar to *S. stellata*. Ten strains have been characterized in some detail. A typical aquatic *Seliberia*-like strain (ICPB 4133) is shown in Fig. BXII.α.240. Their description is the same as that for the genus, with the following additional characteristics: the aquatic *Seliberia*-like strains are unable to use any of 60 tested compounds as their sole carbon source; they appear to have complex growth factor requirements. They can be grown on the minimal medium described for *Seliberia stellata* supplemented with 0.001% Vitamin Assay Casamino Acids (Difco) and 0.1% (v/v) Staley's vitamin solution (Staley, 1968). They grow dependably but sparsely on PYE medium (see Enrichment and Isolation Procedures, above; Schmidt and Swafford, 1981).

Growth of aquatic *Seliberia*-like strains is very sparse on all complex media tested. Grown on PYE agar, colonies are white, minute (0.5–1.0 mm diameter), and very adherent to the agar medium. Growth in liquid PYE medium is granular, definitely not well dispersed, with a strong tendency for attachment to the submerged surfaces of glass culture vessels, so long as a shallow (about 1 cm deep) level of liquid medium is maintained. (The growth of *S. stellata* under similar circumstances is well dispersed, with some rosettes visible microscopically, but without the presence of macroscopic granules). The aquatic *Seliberia*-like strains demonstrate an obligate requirement for oxygen and they reduce nitrate under aerobic conditions, but not anaerobically. They do not produce H_2S from cystine. Most of the *Seliberia*-like strains can hydrolyze urea, and all can hydrolyze starch.

All of the aquatic strains produce swarmers with a subpolar ensheathed flagellum, but most have not been found to produce the lateral, ordinary (i.e., not ensheathed) flagella. Two of the aquatic strains, which have been examined with indirect immunoferritin surface labeling during growth, were found to exhibit unidirectional polar cell surface growth, with production of swarmers having *de novo* synthesized surface antigens. *Seliberia stellata* exhibited the same kind of unusual growth and division pattern in similar experiments (Schmidt and Starr, 1984). Nine of 10 aquatic *Seliberia*-like strains cross-react moderately to strongly with *S. stellata* in serological tests (tube agglutination and indirect immunofluorescence tests). The aquatic seliberias are less resistant to common antibiotics than is *S. stellata*.

The mol% G + C of the DNA is: 63–65 (Bd) (aquatic *Seliberia*-like strains).

Genus XIX. **Starkeya** *(ex Starkey 1934) Kelly, McDonald and Wood 2000, 1800*[VP]

DONOVAN P. KELLY AND ANN P. WOOD

Star.ke' ya. M.L. n. *Starkeya* of Starkey; named after Robert L. Starkey, who made important contributions to the study of soil microbiology and sulfur biochemistry.

Short rods, coccoidal or ellipsoidal cells 0.4–0.8 × 0.8–2.0 μm, occurring singly and, occasionally, in pairs. Nonmotile. Colonies grown on thiosulfate agar (with biotin) are small, smooth, circular, round, and opalescent, becoming white with sulfur. Thiosulfate liquid medium (lacking biotin) becomes turbid and sulfur precipitates during static incubation; thiosulfate is incompletely used, and the pH falls from 7.8 to 5.8. This poor development is due to the requirement for biotin exhibited by the type strain. The organism is **facultatively chemolithoautotrophic**, but optimal autotrophic development requires biotin, and optimal heterotrophic growth requires yeast extract, biotin, or other additions, such as pantothenate, depending on the organic substrate. Autotrophic growth is also observed with formate when high levels of ribulose 1,5-bisphosphate carboxylase are expressed. Some strains may degrade methylated sulfides. This organism is **strictly aerobic, both autotrophically and heterotrophically**, and is incapable of denitrification. **Oxidizes and grows on thiosulfate and tetrathionate but not on sulfur or thiocyanate.** Ammonium salts, nitrates, urea, and glutamate are used as nitrogen sources. Optimal temperature 25–30°C; growth range 10–37°C (no growth at 5°C or 42°C). Optimal pH 7.0; growth range pH 5.7–9.0. Contains ubiquinone Q-10. Major cellular fatty acids are octadecenoic acid and cyclopropane acid of C_{19}; lacks a major hydroxy fatty acid. The lipopolysaccharide lacks heptoses and has only 2,3-diamino-2,3-dideoxyglucose as the backbone sugar. Member of the *Alphaproteobacteria*. Isolated from soil. Presumably widely distributed.

The mol% G + C of the DNA is: 67.3–68.4.

Type species: **Starkeya novella** (Starkey 1934) Kelly, McDonald and Wood 2000, 1800 (*Thiobacillus novellus* Starkey 1934, 365.)

FURTHER DESCRIPTIVE INFORMATION

This newly created genus currently contains only one of the former *Thiobacillus* species, *T. novellus*, a facultatively heterotrophic and chemolithoautotrophic, rod-shaped organism in the *Alphaproteobacteria*. The properties, taxonomy, and differentiation of the genera of sulfur-oxidizing, chemolithoautotrophic, Gram-negative, rod-shaped bacteria are summarized in Tables BXII.β.70 and BXII.β.71 and Fig. BXII.β.60 in the chapter on the genus *Thiobacillus* in this *Manual*. *Starkeya* has been shown by 16S rRNA gene sequence analysis and comparison of physiological properties to not be a species of *Paracoccus* (which contains the physiologically comparable species *P. versutus* and *P. pantotrophus*), but rather to be significantly related only to *Ancylobacter* (Kelly et al., 2000). The description given here is based primarily on studies with the type strain, isolated from soil in the 1930s (Starkey, 1934, 1935a, b; Kelly et al., 2000). Another strain, isolated from sewage, has been shown to oxidize methanethiol, dimethylsulfide, and dimethyldisulfide (Cha et al., 1999), properties not yet reported for the type strain. Detailed studies of inorganic sulfur-compound oxidation have been undertaken with *S. novella*, including studies of its unique sulfite:cytochrome *c* oxidoreductase (Southerland and Toghrul, 1983; Toghrol and Southerland, 1983; Kappler et al., 2000) and its thiosulfate-oxidizing system. The latter was initially thought to be a multienzyme complex located exclusively in the cytoplasmic membrane system (Oh and Suzuki 1977), but Kappler et al. (2001) have shown that a periplasmic thiosulfate-oxidizing system not involving a membrane-anchored multienzyme system may also function (C. Dahl, personal communication).

List of species of the genus Starkeya

1. **Starkeya novella** (Starkey 1934) Kelly, McDonald and Wood 2000, 1800[VP] (*Thiobacillus novellus* Starkey 1934, 365.) *no.vel' la.* L. dim. adj. *novella* new.

   The species description is the same as the genus description, with the exception that oxidation of methylated sulfides has not been tested with the type strain. Biotin is required by the type strain for good growth on most sub-strates; yeast extract may be substituted for biotin and, in some cases, the biotin requirement may be replaced by lipoic acid or coenzyme A; good growth on methanol requires pantothenate or yeast extract rather than biotin.

   The mol% G + C of the DNA is: 67.3–68.4 (Bd, T_m).

   Type strain: ATCC 8093, DSM 506, IAM 12100, IFO 12443, NCIMB 9113.

   GenBank accession number (16S rRNA): D32247.

Genus XX. **Xanthobacter** *Wiegel, Wilke, Baumgarten, Opitz and Schlegel 1978, 573*[AL]

JÜRGEN K.W. WIEGEL

Xan.tho.bac' ter. Gr. adj. *xanthos* yellow; M.L. masc. n. *bacter* the equivalent of Gr. neut. n. *bacterion* rod, staff; M.L. masc. n. *Xanthobacter* yellow rod.

Cells are rod shaped, sometimes twisted, 0.4–1.0 × 0.8–6.0 μm. Pleomorphic cells are produced on media containing tricarboxylic cycle-intermediates (especially succinate), **whereas coccoid cells as well as cells up to 10 μm long are produced on media containing an alcohol** as the sole carbon source. Refractile (polyphosphate) and lipid (poly-β-hydroxybutyrate) bodies are evenly distributed in the cells. Resting stages are unknown. Key sporulation genes are absent. Depending on the species and growth conditions, the cells are nonmotile or motile (by peritrichous flagella). **The Gram reaction frequently appears falsely to be positive or variable due to polyphosphate granules; however, ultrastructurally and biochemically, the cell wall is of the negative Gram-type.**[1] **Aerobic**, having a strictly respiratory type of metabolism with oxygen as the terminal electron acceptor. Optimal temperature, 25–30°C. Optimal pH, 5.8–9.0. Colonies are opaque and slimy (although "slimeless" strains exist) and are **yellow** due to a water-insoluble carotenoid pigment (**zeaxanthin dirhamnoside**). The color intensity depends on the amount of slime produced by individual strains. **Catalase positive.** All strains can grow **chemolithoautotrophically** in mineral media under an atmosphere of H_2, O_2, and CO_2 (7:2:1, v/v) as well as **chemoorganoheterotrophically** on methanol, ethanol, n-propanol, n-butanol, and various organic acids as sole carbon sources. The carbohydrate utilization spectrum is limited, and neither volatile/nonvolatile fatty acids nor gas is produced from carbohydrates such as fructose, glucose, or mannose. Some strains require vitamins. Some strains can utilize substituted thiophenes as sole carbon, energy, and sulfur sources. **When degrading aliphatic epoxides, tested strains contain coenzyme M, which otherwise is a typical coenzyme of the obligate anaerobic methanogenic archaea. N_2 is fixed** in nitrogen-deficient media under heterotrophic or chemolithoautotrophic growth conditions, but by most strains

---

1. The terms "Gram-type negative" and "Gram-type positive" were proposed by Wiegel (1981) to describe bacteria according to ultrastructural and biochemical Gram characteristics (cell wall structure, presence or absence of indicator compounds such as lipopolysaccharide; Wiegel and Quandt, 1982). The terms are distinct from those referring to the results of the Gram-staining reaction: "Gram-reaction positive", "Gram-reaction negative", or "Gram-reaction variable".

only under a decreased O_2 pressure. The bacteria occur free-living in freshwater (mainly *X. agilis*), wet soil containing decaying organic material (*X. autotrophicus*, *X. flavus*), marine sediments (*X. flavus*), compost of root balls of *Tagetes* (thiophen-utilizing *X. tagetidis*), and associated with the roots of plants including wetland rice (*X. flavus*). *Xanthobacter* can be regarded an associative N_2-fixing bacterium (rice, tagetis, coconut palm). The induction of root or stem nodules has not been observed. 16S rDNA sequence analysis places the members into the class *Alphaproteobacteria*; however, the presently recognized species of *Xanthobacter* are intermixed with the single-species genera *Aquabacter* and *Azorhizobium*, and together they form a distinct cluster.

The mol% G + C of the DNA is: 65–70 (T_m) and 66–68 (Bd).

Type species: **Xanthobacter autotrophicus** (Baumgarten, Reh and Schlegel 1974) Wiegel, Wilke, Baumgarten, Opitz and Schlegel 1978, 580 ("*Corynebacterium autotrophicum*" Baumgarten, Reh and Schlegel 1974, 214.)

FURTHER DESCRIPTIVE INFORMATION

Colonial characteristics All species of *Xanthobacter* form colonies that are smooth, convex, circular, filiform, opaque, and either yellow, off-white, or—when abundant slime is produced—nearly colorless. Due to the slime formation, especially on nutrient broth agar plates, colonies develop into characteristic colony forms called "fried egg". The main component of the yellow pigment of all strains (the "white" strains contain just minor amounts) has been identified as zeaxanthin dirhamnoside (Hertzberg et al., 1976; J. Wiegel and K. Schmidt, unpublished data). It is water insoluble and is located in the cell wall (Eberhardt, 1971). In *Aquabacter* and *Azorhizobium*, this pigment has not been identified so far; their colonies are creamy white.

Cell morphology The cell shape of *Xanthobacter* depends strongly on the growth conditions, i.e., especially the carbon source, whereas the nitrogen source has a minor effect (Fig. BXII.α.241). Cells under N_2-fixing conditions are only slightly longer than in the presence of ammonia or other sources of combined nitrogen. Cells of both *X. autotrophicus* and *X. flavus* are coccoid when growing on *n*-propanol (Fig. BXII.α.241) and are long, filamentous rods when growing on *n*-butanol. One of the most significant morphological features is the formation of irregular, twisted (especially *X. tagetidis*) (Padden et al., 1997), and branched cells (especially *X. autotrophicus* and *X. flavus*) during growth on tricarboxylic acid cycle intermediates; the intermediates include succinate, which is a good growth substrate and promotes the development of the most irregular cell forms during the late exponential growth phase (Fig. BXII.α.241) (Wiegel and Schlegel, 1976). In contrast to strains of *X. autotrophicus*, *X. flavus*, and *X. tagetidis*, the strains of *X. agilis* (motile under nearly all growth conditions) are only slightly pleomorphic and exhibit branched cell formation only after prolonged incubation (3–7 d) on succinate-containing agar media (Aragno et al., 1977; J. Wiegel, unpublished data). Electron micrographs of ultrathin sections show that the branching cells do not contain septa at the branching points (Fig. BXII.α.242) (Wiegel et al., 1978). Scanning electron micrographs from *X. tagetidis* reveal twisted, nearly fusiliform (i.e., with tapered ends) spirals (Padden et al., 1997). The occurrence of a snapping type of cell division, as was assumed from light-microscopic inspection of *X. autotrophicus*, has not been confirmed by electron-microscopic-studies. The illusion of snapping cell division, as well as of "star formation" (Fig. BXII.α.241), is presumably due to cell aggregation by copious

amounts of slime. Surface patterns on the outer cell wall are not observed in electron micrographs. The cells of the exponential growth phase and those grown under heterotrophic conditions are usually larger than those of the early stationary growth phase and those grown under chemolithoautotrophic conditions; the latter are mainly straight or only slightly curved rods (Fig. BXII.α.241).

Cell wall type and Gram-reaction Many strains identified later as strains of *Xanthobacter* were originally described to give a positive or variable Gram reaction (Baumgarten et al., 1974). However, *Xanthobacter* belongs to the negative Gram-type bacterial group, since it contains lipopolysaccharides (Fig. BXII.α.243) (Wiegel and Mayer, 1978; Wiegel, 1981). The impression of a positive Gram-staining reaction is feigned by the content of refractile bodies, which were identified as polyphosphate granules by electron microscopy, as well as by volutin staining (Figs. BXII.α.242 and BXII.α.243). The polyphosphate content of the cells is high (15 mg of phosphate/g dry weight) and even at a low phosphate concentration in the growth medium (0.01 mM) volutin granules do not disappear (Wiegel et al., 1978). Using phosphate limitation or starvation, attempts to obtain cells of *X. autotrophicus* without polyphosphate granules have failed. Large dark bodies, frequently visible in the light microscope, are located at one or both (or in branched cells on several) ends of the cells. Electron microscopy of thin sections of cells (Walther-Mauruschat et al., 1977) indicated that these larger granules are poly-β-hydroxybutyrate granules (Fig. BXII.α.242).

The cell envelope of *X. autotrophicus* has a thin multilayered structure, resembling that of Gram-type negative cells (Fig. BXII.α.242) (Walther-Mauruschat et al., 1977). The peptidoglycan content of the cell wall of the *X. autotrophicus* strains (15–25%) is intermediate compared to that of typical Gram-type positive (30–70%) and Gram-type negative (10%) bacteria. Teichoic acid and teichuronic acid are absent (Schleifer and Kandler, 1972; O. Kandler and F. Fiedler, personal communication), but elevated concentrations of glycine can be present (see below for details).

Compared to other Gram-type negative bacteria the lipopolysaccharide content is relatively low for the strains tested. The lipopolysaccharide was identified both chemically after extraction from autotrophic cells and by the lipopolysaccharide polymyxin B interaction technique using electron microscopy (Wiegel and Mayer, 1978; Wiegel, 1981) (Fig. BXII.α.243). The citrate synthase of *Xanthobacter* species resembles that of Gram-type negative bacteria by having a molecular weight above 250,000 and by being inhibited by NADH (Weitzman and Jones, 1975; Berndt et al., 1976; Weitzman, 1987). All these properties are in agreement with the 16S rDNA sequence analysis placing *Xanthobacter* in the class *Alphaproteobacteria* (Rainey and Wiegel, 1996).

Slime formation The majority of the strains produce copious amounts of slime (Fig. BXII.α.244), which consists mainly of glucose, galactose, mannose, and uronic acid (unidentified; 30% or more of the monomers contain a carboxyl group (Andreesen and Schlegel, 1974; Opitz, 1977). At high C:N ratios, the slime formed can cause gelatinization and solidification of the medium without a significant increase in cell mass. The amount of slime varies with the strain as well as growth conditions. Most strains produce copious amounts of slime during growth on carbohydrates or lactate, but minor amounts are formed during growth on O_2/H_2 or on alcohols. No correlation between slime formation and oxygen tolerance during nitrogen fixation is ob-

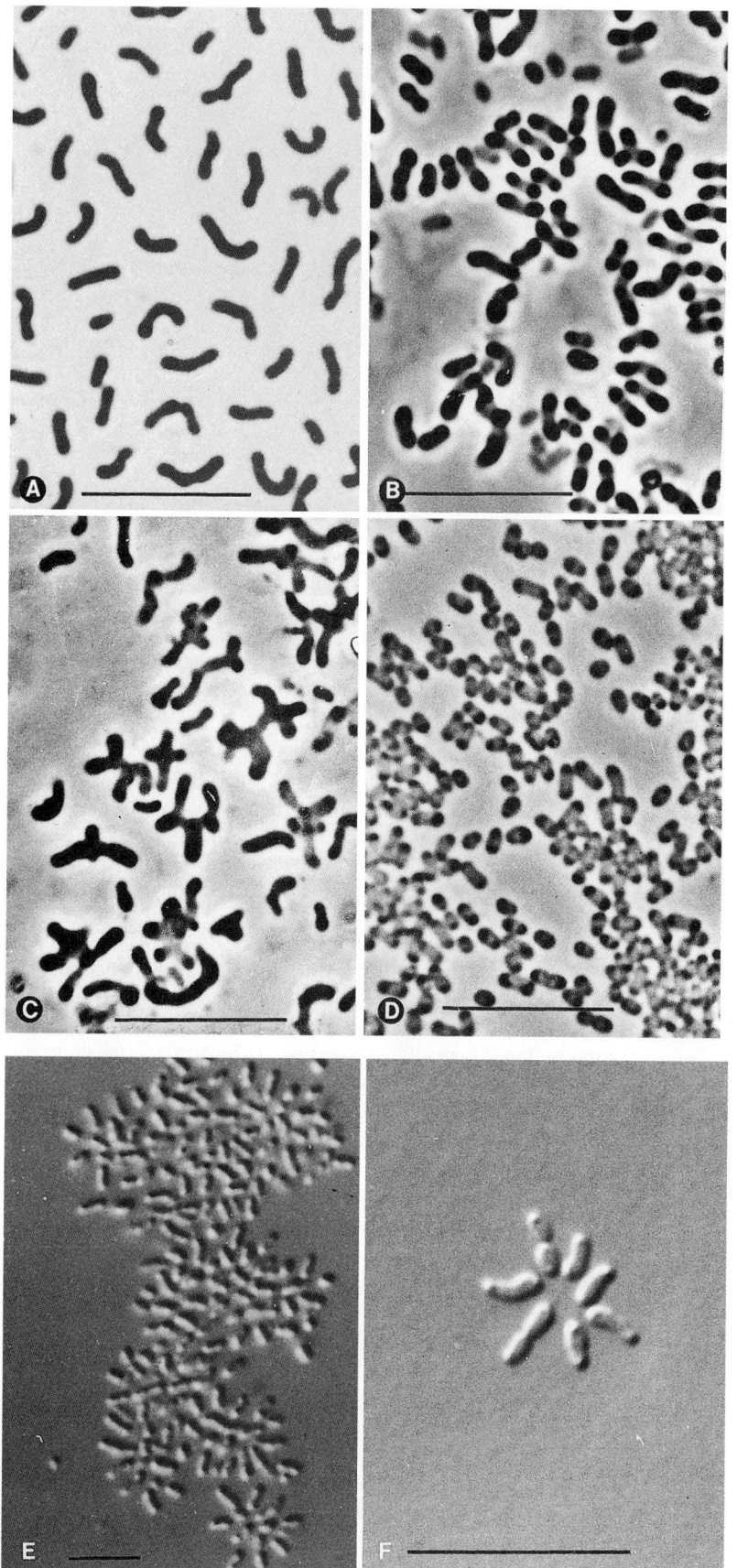

FIGURE BXII.α.241. Cells of *Xanthobacter autotrophicus*. *A*, strain 14g grown on malate in nonagitated liquid culture, showing irregular vegetative cells. *B–D*, strain JW 33 (reference strain). *B*, grown lithoautotrophically under conditions of nitrogen fixation, showing rod-shaped cells and palisade-like formations. *C*, grown in nutrient broth (0.7%) containing 0.1% succinate, showing the typical branched cell formation. *D*, grown on *n*-propanol in the presence of ammonium, showing coccoid cell formation. *E–F*, strain 14g, showing cell aggregation ("star-formation") on nutrient broth medium. This kind of aggregation is only observed after 2–5 h upon transfer into liquid nutrient broth medium under nonagitated growth conditions. All bars = 10 μm. (Parts *A* and *E* reproduced with permission from K. Schneider et al., Archives of Microbiology *93*: 179–193, 1973, ©Springer-Verlag, Heidelberg. Parts *B* and *C* reproduced with permission from J. Wiegel and H.G. Schlegel, Archives of Microbiology *107*: 139–142, 1976, ©Springer-Verlag, Heidelberg. Part *F* reproduced by permission from M. Reh.)

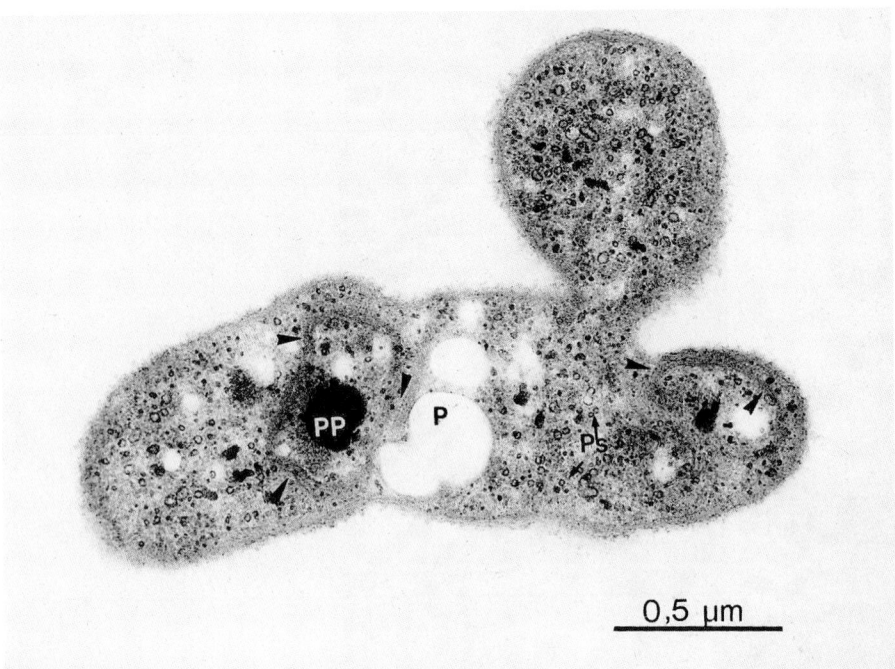

FIGURE BXII.α.242. Ultrathin section of *Xanthobacter autotrophicus* strain 14g grown on a succinate-containing medium. The cells were prepared by glutaraldehyde-osmium tetraoxide fixation, uranyl acetate block staining, and lead citrate poststaining; preparation by A. Walther-Mauruschat. *P*, poly-β-hydroxybutyrate; *PP*, polyphosphate; *Ps*, small-type polyphosphate. (Reproduced with permission from J. Wiegel et al., International Journal of Systematic Bacteriology *28:* 573–581, 1978, ©International Association of Microbiological Societies.)

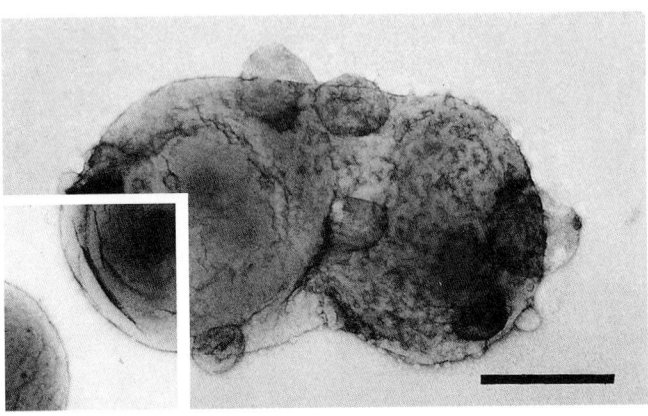

FIGURE BXII.α.243. Demonstration of the presence of lipopolysaccharides in *Xanthobacter autotrophicus*. Negatively stained with 3% uranyl acetate, pH 5. Polymyxin B-treated cell of reference strain JW 33 showing lipopolysaccharide-polymyxin B interactions (bleb formation). The inset shows a control without polymyxin B treatment. Bar = 0.5 μm. (Reproduced with permission from L. Quandt.)

served; however, many of the slime producers grow better than the nonproducers at a high (2–3%) oxygen concentration in the gas atmosphere. Moreover, the majority of the strains produce more slime under N$_2$-fixing conditions than in the presence of ammonium ions. Beside the motile strains and a very few strains of *X. autotrophicus* that produce traces of slime (e.g., strain JW50), slimeless mutants have been isolated (e.g., mutants of *X. autotrophicus* type strain 7c; Andreesen and Schlegel, 1974), mainly by applying carbon limitation in continuous culture (unpublished results) (Fig. BXII.α.244). The slime becomes soluble at

pH values below 4.5 and above 10.5. With alkaline treatment, the slimy cell aggregates can be separated into single cells as needed for isolation purposes (Wiegel and Schlegel, 1976). The slime is somewhat recalcitrant to microbial degradation under aerobic and anaerobic conditions (Andreesen and Schlegel, 1974), but it exhibits useful drag-reducing properties (e.g., reducing frictions in fluids moving through pipes) (Schubert et al., 1986 and personal communication) and could be used as a viscosifier in oil recovery (Kern, 1985; Wan et al., 1988). A different but interesting property is that *Xanthobacter* slime enhanced the efficiency of gene recombination experiments with *Agrobacterium tumefaciens* (Kawai, 1995).

Polyglutamine capsule All of the 20 strains of *X. autotrophicus* and *X. flavus* tested (other species have not been tested) produce an α-polyglutamine capsule that is located between the cell wall and the slime. This polymer is not separated from the cells during an alkaline (pH 11) treatment to solubilize the slime. The glutamine polymer is not found on cells grown under N$_2$-fixing or chemolithoautotrophic conditions. In contrast to the slime, this polymer is reutilized in the late stationary growth phase and under N-limitation (J. Wiegel, unpublished results). *Flexithrix* is the only other genus known that produces a similar α-polyglutamine capsule (Kandler et al., 1983).

Storage material *Xanthobacter* species deposit poly-β-hydroxybutyrate (PHB) as a reserve material under heterotrophic as well as under chemolithoautotrophic conditions (Fig. BXII.α.242). PHB formation occurred in all tested strains and was analyzed in more details in strains of *X. autotrophicus*, *X. flavus*, and *X. agilis*. In the stationary growth phase, heterotrophically grown cells contain between 5 and 600 mg PHB/g dry weight of cells, depending on the strain and on the carbon source. Growth on

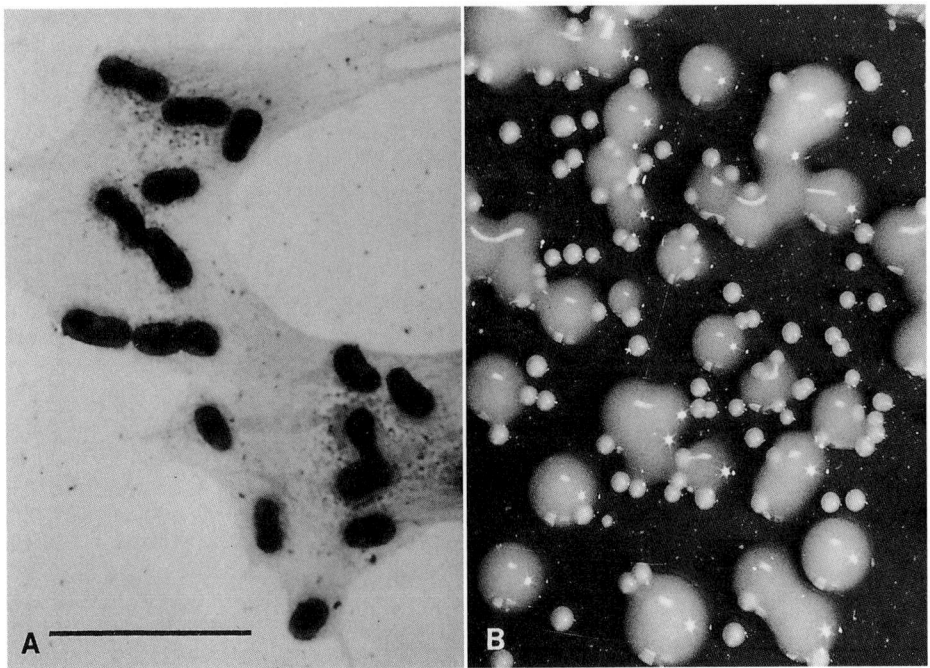

FIGURE BXII.α.244. *A,* Slime production by *Xanthobacter autotrophicus.* Cells and slime of the type strain (DSM 432) stained negatively by uranyl acetate. Bar = 5 μm. (Reproduced with permission from N. Tunail and H.G. Schlegel, Archives of Microbiology *100:* 341–350, 1974, ©Springer-Verlag, Heidelberg.) *B,* Colonies from the wild type (large, slimy colonies) and from the slimeless mutant (small, dark yellow colonies) of DSM 432. (Reproduced with permission from M. Andreesen and H.G. Schlegel, Archives of Microbiology *100:* 351–361, 1974, ©Springer-Verlag, Heidelberg.)

fructose usually results in a high PHB content, whereas growth on succinate results in very low values (less than 15 mg PHB/g dry weight). For example, reference strain *X. autotrophicus* JW33 has 668, 250, and 8 mg PHB/g dry weight when grown on fructose, sucrose, and succinate, respectively (J. Wiegel, unpublished data), based on quantification by spectroscopy after extraction (Jüttner et al., 1975).

Nitrogen fixation and nitrogenase The *Xanthobacter* species described so far are microaerophilic, heterotrophic, and autotrophic (requiring decreased H_2 and O_2 concentrations) N_2-fixing bacteria, as described in the original reports of nitrogen fixation in *X. flavus* (as *"Mycobacterium flavum"* 301) (Federov and Kalininskaya, 1961) and in *X. autotrophicus* (as *"Corynebacterium autotrophicum"*) (Gogotov and Schlegel, 1974; Wiegel and Schlegel, 1976). The nitrogen-fixing system and its relationship to oxygen have been studied mainly with *X. autotrophicus* strain GZ29 (Berndt et al., 1976, 1978) and *X. flavus* strain 301 (Biggins and Postgate, 1969, 1971). The composition of the nitrogenase components of heterotrophically grown cells of strain GZ29 and strain 301 is similar to that of other nitrogenases in regard to metal and sulfur content and amino acid composition. As in other aerobic diazotrophs, the nitrogenase system seems to be loosely associated with membranes. The protection of the nitrogen-fixing system in *Xanthobacter* might be due to some conformational protection for the nitrogenase, the respiration of associated aerobic microorganisms in the natural habitat, or the respiratory activity of *Xanthobacter* itself. The latter possibility seems unlikely, since the respiratory rate of *Xanthobacter* is about one to two orders of magnitude lower than that in *Azotobacter* (Biggins and Postgate, 1971). A possible contribution of slime production to protection against oxygen damage to the nitro-

genase could not be demonstrated for *Xanthobacter autotrophicus* (Wiegel, unpublished results). Additional information concerning N_2-fixation in two more well-characterized strains can be summarized as follows:

In *X. autotrophicus* strain GZ29 (Berndt et al., 1976, 1978), the growth rates with ammonium as the nitrogen source and sucrose as the carbon source are highest at an oxygen partial pressure of 0.15 atm, whereas with N_2 as the sole nitrogen source the maximum growth rates occur at 0.014 atm. In the whole cell assay for acetylene reduction, the optimal oxygen partial pressure is 0.0036 atm. Even in the absence of any detectable oxygen, acetylene is reduced linearly for more than 1 h. The acetylene reduction activity has a very narrow pH optimum at 6.8. The nitrogenase is not cold labile. The overall efficiency of nitrogen fixation is 22 mg nitrogen fixed/g of sucrose consumed, but in the early exponential growth phase, values up to 65 mg/g have been found.

In *X. flavus* strain 301 (Biggins and Postgate, 1969, 1971) heterotrophic growth in the presence of ammonium is not dependent on the O_2 partial pressure in the range of 0.01–0.20 atm, but with N_2 as the sole nitrogen source, growth and nitrogen fixation have an optimum at 0.1 atm N_2. The optimal pO_2 for acetylene reduction by whole cells is 0.05 atm. The nitrogenase activity of cell-free extracts stirred under 0.2 atm O_2 decays exponentially, with a half-life of about 5 min. ATP enhances the sensitivity to O_2, presumably because of the formation of an active enzyme conformation. As with nitrogenases from other sources, the enzyme reduces H^+ to H_2, KCN to CH_4, and CH_3NC to CH_4, C_2H_4 and C_2H_6. The optimal concentration of ATP in the cell-free extract for acetylene reduction is 4–8 mM. As in other aerobic diazotrophs, but in contrast to the systems of *Clostridium pasteurianum*, pyruvate fails to promote acetylene reduction by cell-free extracts.

Ammonium transport Wiegel and Kleiner (1982) postulated that *Xanthobacter* possesses an active ammonium (methyl ammonium) transport system similar to that of other diazotrophs. Methyl ammonium can serve as a sole carbon and nitrogen source (unpublished data). S. Kustu (personal communication) disputes this interpretation and maintains that there is no active ammonium transport system in any bacterium.

Hydrogenase The strains of *X. autotrophicus* studied (e.g., strains 14g, 7c, GZ29) contain a membrane-bound uptake hydrogenase. Like the majority of membrane-bound hydrogenases, this enzyme is not able to reduce NAD directly; instead, it channels electrons into the electron transport chain. A soluble, NAD-reducing hydrogenase is not present (Schneider and Schlegel, 1977). In strain 7c the enzyme is constitutive, and cells grown on various organic substrates and ammonium contain 10–100% of the hydrogenase activity compared to chemolithoautotrophically grown cells (Tunail and Schlegel, 1974). In strain 14g the hydrogenase is a strictly inducible enzyme (Schneider et al., 1973). In strain GZ29 the hydrogenase is apparently a constitutive enzyme; its specific activity (up to 31 mol H_2/h/mg protein in N_2-fixing cells) depends on the oxygen concentration during growth (Berndt and Wölfle, 1979; Pinkwart et al., 1979). Although the enzyme has been solubilized from the membranes and partially purified (Schink and Schlegel, 1980), a detailed description of it is still lacking. Immunological comparison of the membrane-bound hydrogenase of *Alcaligenes eutrophus* strain H16 reveals no relationship to that of *X. autotrophicus* GZ29 (Schink and Schlegel, 1980). Furthermore, an antiserum against the hydrogenase of *X. autotrophicus* strain GZ29 did not react with the membrane extracts of strains 7c, 14g, and 12/60/x. This indicates major differences between the strains. The hydrogenase of the motile strain MA2 is apparently a loosely membrane-bound enzyme that tends to form aggregates and exhibits an unusual high specific activity (Pinkwart et al., 1979, 1982).

Growth conditions The spectrum of carbohydrates utilized by about 80 tested strains of *X. autotrophicus* is normally limited to fructose and/or sucrose and/or—in the presence of traces of yeast extract—mannose. *X. tagetidis* can also grow on galactose and lactose as the sole carbon and energy source. Some strains (*X. autotrophicus* strain 14g and *X. agilis* strains MA2 and SA35 (Jenni and Aragno, 1987)) do not utilize carbohydrates at all, whereas Malik and Claus (1979) reported that strain *X. autotrophicus* DSM 685 could use a wider than normal range of carbohydrates. However, no 16S rRNA sequence or DNA–DNA hybridization data exist on these strains for a definite placement in the proposed species. Some strains of *X. flavus* can utilize a few more sugars, including glucose. Normally *X. autotrophicus*, *X. agilis*, and *X. tagetidis* cannot use glucose, although the pertinent catabolic enzymes are induced during growth on fructose (Opitz, 1977; Padden et al., 1997). However, glucose-utilizing *X. autotrophicus* and *X. flavus* (others not tested) cells appear spontaneously when cultures are incubated for a prolonged time (2–3 weeks) in the presence of glucose. These adapted strains (transport mutants) grow very well on glucose; however, they lose this ability again after growing on other carbohydrates for one or two generations (J. Wiegel, unpublished data). Using the API 20E test, *X. tagetidis* was positive for acetoin production from glucose.

Catabolite repression is exerted by H_2 on the utilization of organic substrates; however, the extent of this effect varies with the organic substrate and the strains studied. Some strains need CO_2/CO_3^{2-} for induction of the enzymes for heterotrophic growth after transition from lithotrophic to organotrophic conditions (Schneider et al., 1973).

The doubling times reported for various strains vary between 1.5 and 5.0 h (heterotrophic conditions) and 3.0 to over 12.0 h (chemolithoautotrophic conditions). Under N_2-fixing conditions, the doubling time is slightly longer (Wiegel and Schlegel, 1976; Berndt et al., 1978).

Traces of yeast extract shorten the lag phase after transition from heterotrophic to lithotrophic conditions. Many strains of *X. flavus* and three strains of *X. autotrophicus* are reported to require biotin (Aragno, 1975). High cell yields during mass culture under autotrophic conditions are obtained when oxygen levels are raised to match the increasing growth. For example, for autotrophic growth with ammonium as the nitrogen source, an O_2 level of 5–8% (v/v) is usually optimum; at optical densities higher than 1.0 (at 600 nm), 10–15% O_2 is sufficient, and at optical densities higher than 2.5 (at 600 nm), 20% O_2 in the gas atmosphere is required for good growth. The oxygen-resistant strain, *X. autotrophicus* Y38, requires nickel for optimal growth (Nakamura et al., 1985).

For strain *X. autotrophicus* GZ29, intact cells are able to reduce acetylene under anaerobic assay conditions, indicating the presence of a fermentative energy regenerating system. So far, however, growth has not been observed under anaerobic conditions and no strict anaerobic metabolic activity has been ascertained (Berndt et al., 1976; Arzumanyan et al., 1997). The finding of a typical methanogenic cofactor, coenzyme M, in *Xanthobacter*, suggests that this issue needs to be revisited, although the proposed role of the cofactor is in the carboxylation of aliphatic epoxide in the aerobic alkene utilization pathway (Allen et al., 1999; Krum and Ensign, 2000; Sauer and Thauer, 2000).

The typical pH for growth is between 6.5 and 8.0, with the optimum around 7.5 for most strains. Although usually no acids are formed from carbohydrates or alcohols, the formation of acidic slime and the consumption of ammonium ions cause the pH of the medium to decrease. This decrease has to be counterbalanced by adding alkali to maintain optimal growth. During growth on organic acids, the pH shift is small, but depending on the slime production of the strain used, it can be in either the acidic or alkaline direction.

Antibiotics Tested strains of *Xanthobacter autotrophicus* and *X. flavus* are sensitive to the following antibiotics at 100 µg/ml of growth medium: penicillin (MIC ~1 µg/ml), novobiocin, and polymyxin B. They are resistant to erythromycin and bacitracin at 200 µg. Violet red-bile medium (Oxoid), deoxycholate medium (Oxoid), and tellurite agar (selective for coryneform bacteria) support growth of the majority of the isolated strains. Both *Xanthobacter* species grow on mineral medium supplemented with an appropriate carbon source and 10^{-5} M crystal violet. However, the colonies formed are red instead of the blue color that is characteristic of other Gram-negative bacteria (Wiegel et al., 1978).

In contrast, *X. tagetidis* is resistant to 1 U penicillin and 5 µg/ml tetracycline but sensitive to 25 µg/ml of ampicillin and 30 µg/ml of chloramphenicol (Padden et al., 1997).

Metabolic pathways Carbohydrates and gluconate are degraded via the Entner–Doudoroff pathway, as has been shown for *X. autotrophicus* strains by determining enzyme activities after growth on various substrates (Tunail and Schlegel, 1972, 1974;

Schneider et al., 1973). In addition, radiorespirometric studies indicate that the pentose phosphate pathway is concomitantly used to a significant degree (Opitz and Schlegel, 1978). *X. autotrophicus* Py2 utilizes an ATP-utilizing acetone carboxylase when growing on acetone (Sluis et al., 1996).

C$_1$-carbon utilization CO$_2$ is mainly fixed via the ribulose bisphosphate pathway (Bowien and Schlegel, 1981; Lehmicke and Lidstrom, 1985; Meijer et al., 1991; Meijer 1994; Shively et al., 1998). The key enzyme, ribulose bisphosphate carboxylase, is inducible. In addition, phosphoenolpyruvate carboxylase activity occurs. Radiorespirometric experiments with ^{14}CO$_2$ or ^{14}CH$_3$OH as substrates yielded radioactive malate. Other experiments suggest that methanol is presumably utilized via CO$_2$ and the ribulose bisphosphate cycle (Opitz, 1977). Methanol is oxidized to CO$_2$ via a methanol dehydrogenase, which is of the normal type (mol. wt. of about 120,000) stimulated by NH$_4^+$ and contains 2 mol of the coenzyme pyrroloquinoline quinone (PQQ) (J. A. Duine, personal communication). The glucose-6-phosphate dehydrogenase is allosterically inhibited by phosphoenolpyruvate (PEP) (Tunail and Schlegel, 1972; Opitz and Schlegel, 1978). The triose phosphate isomerase is the same enzyme that is formed during chemolithoautotrophic and heterotrophic growth (Meijer et al., 1997).

Carbon monoxide is not oxidized to CO$_2$ by *X. autotrophicus* or *X. flavus* strains (others not tested) and cannot serve as an energy or carbon source for lithotrophic growth (O. Meyer, personal communication). Thiosulfate can substitute for H$_2$ as a substrate for lithotrophic growth and provide energy for CO$_2$-fixation (Friedrich and Mitrenga, 1981; J. Wiegel, unpublished data, six strains of *X. autotrophicus* tested).

Alkene utilization and epoxide formation The use of various alkenes such as ethene, propene, butene, and 1,3-butandien through oxidation by monooxygenases has been studied in detail on the physiological and biochemical level mainly in *X. autotrophicus* strain Py2. The oxidation leads to the formation of epoxides, which then are further metabolized by epoxide hydrolases, and carboxylases (Swaving and deBont, 1998, and older literature cited therein). The use of *Xanthobacter* strains for the commercial removal of propene from waste gas has been proposed (Reij and Hartmans, 1996).

Chloroalkenes and haloaromatic compounds Much work has been done on the haloalkene and haloaromatic compounds degradation by *Xanthobacter* strains, mainly with *X. autotrophicus* GJ10 and on chlorobenzenes with *X. flavus* 14p1 (Spiess et al., 1995; Spiess and Gorisch, 1996; Sommer and Gorisch, 1997). A novel dehalogenase has been found and purified and the crystal structure elucidated. Strain GJ10 contains two dehalogenating enzymes one for haloalkenes and the other for halogenated carboxylic acids (Prince, 1994). Strain *X. autotrophicus* Py2 has a novel pathway for degradation of epichlorohydrin. This strain also degrades trichloroethylene during growth on propene (Reij et al., 1995) as does *X. autotrophicus* GJ10 (Inguva and Shreve, 1999).

Genetics Intraspecific gene transfer has been detected in crosses involving the strains GZ29, GZ27, and JW50 of *X. autotrophicus*, which produce only traces of slime. Strain GZ29 was used to study the transfer more closely. The involvement of a defective generalized transducing bacteriophage as well as conjugational gene transfer have been described (Wilke and Schlegel, 1979; Wilke, 1980). A plasmid has also been identified in strain GJ10 (Bergeron et al., 1998). The bacteriophage CA3 was

detectable only by its transducing activity and by electron microscopy; it did not form plaques. The genetic markers (resistance, auxotrophy, pigmentation) were transducible at frequencies of about 10% per marker and per phage particle. No cotransduction of markers was detected. Mating experiments on agar plates have revealed a recombination system requiring direct cell contact. It allowed the transfer of large chromosomal segments at low frequency. All partners used functioned as donors as well as recipients. Two groups of closely linked markers have been found (Wilke, 1980). In addition, an electrotransformation system that is also applicable to other strains has been developed for *X. autotrophicus* GJ10. Transformation rates up to 2×10^6 cells transformed per μg DNA were obtained (Swaving et al., 1996; Swaving and deBont, 1998). A triparental mating system was used to express the dehalogenase in other *Xanthobacter* strains including type strain 7c (Wilke, 1980; Janssen et al., 1989). For *X. autotrophicus* strain Py2, mutants defective in epoxyalkane degradation were complemented with DNA fragments (Swaving et al., 1995).

Cytochromes, ubiquinones, and fatty acids *X. autotrophicus* (only strain 14g has been investigated) contains cytochromes *a*, *b* (two different ones), *c*, and *o*, irrespective of the growth conditions. The amounts of cytochromes *a*, *b*, and *c* type were 0.03, 0.4–0.52, and 0.36 μmol/g of particle protein, respectively (Bernard et al., 1974). Ubiquinones Q-10 (major), Q-9, and Q-8 are present in *X. autotrophicus* and *X. flavus* (Collins and Jones, 1981; Urakami et al., 1995a, and unpublished data). The cellular fatty acids are high in C$_{18:1}$ and include C$_{18:0}$, C$_{16:0 \, 3OH}$, and, in *X. agilis*, cyclo C$_{19:0 \, cyclo \, \omega 7c}$ (Urakami et al., 1995a; Wiegel, unpublished results) (Table BXII.α.189). Menaquinones have not been found in these species (Wiegel et al., 1978). *X. autotrophicus* and *X. flavus* contain the coenzyme pyrroloquinoline quinone after growth on methanol (J.A. Duine, personal communication).

Ferredoxin *Xanthobacter flavus* (previously called "*Mycobacterium flavum*" 301) contains a [4Fe–4S]$_2$- and a [4Fe–4S]-ferredoxin. It is likely that the latter is a [3Fe–3S]-ferredoxin. *X. autotrophicus* GZ29 contains two different [4Fe–4S]$_2$-ferredoxins (Bothe and Yates, 1976; Berndt et al., 1978; Yates et al., 1978; M.G. Yates, personal communication). The ferredoxin of *X. autotrophicus* GZ29 exhibits EPR features in the oxidized as well as in the reduced state, which is in contrast to the ferredoxins of *Azotobacter vinelandii* and *X. flavus*. Thus, it is possible that *X. autotrophicus*, at least strain GZ29, contains ferredoxins unique among N$_2$-fixing bacteria. Additional strains need to be analyzed before conclusions regarding generic diversity can be drawn. There is no evidence for the presence of constitutive flavodoxins in *X. flavus* 301 (Bothe and Yates, 1976) or in *X. autotrophicus* GZ29. In the latter strain, the ferredoxins probably serve as direct electron donors for the nitrogenase (Schrautemeier, 1981).

Habitats The type strain of *X. autotrophicus*, strain 7c, was isolated from black mud of a pond near Göttingen during enrichments for propane-oxidizing bacteria (Siebert, 1969). However, strain 7c and all other strains tested do not utilize propane. Typical strains of *X. autotrophicus* and *X. flavus* can be specifically isolated from wet soil and mud containing organic material including wetland rice fields (Oyaizu-Masuchi and Komagata, 1988; Reding et al., 1991). A highly specific isolation procedure is based on the ability to fix nitrogen under chemolithoautotrophic growth conditions and specific colony morphology and color on succinate-containing nutrient broth plates after an alkaline treatment (Wiegel and Schlegel, 1976). Yellow colonies containing

TABLE BXII.α.189. Fatty acid analysis of *Xanthobacter* strains[a]

Fatty acid	X. autotrophicus 7c	X. autotrophicus 7cSF	X. agilis SA35	X. flavus 301	X. flavus 4-14H	X. flavus JW-KR1	X. flavus W-KR2
$C_{16:0}$	4.83	3.48	10.32	2.67	3.84	2.68	1.19
$C_{16:0\ 3OH}$	1.32	0.98	0.74	0.73	1.45	1.69	1.45
$C_{16:1\ \omega 7c}$	5.04	6.24	3.45	0	1.86	0.52	0
$C_{17:0}$	0	0	0	0	0	0	0.76
$C_{17:0\ cyclo}$	2.45	1.23	0	0	0	0	0
$C_{18:0}$	1.98	1.53	3.25	2.65	1.19	1.79	1.99
$C_{18:1}$[b]	76.81	81.64	57.44	91.9	91.66	93.31	93.01
$C_{18:2\ \omega 6,9c}$ and $C_{18:0}$	0	0	1.13	0	0	0	0
$C_{19:0\ cyclo\ \omega 7c}$	5.99	3.02	19.68	1.84	0	0	0.86
$C_{19:1\ \omega 12t}$	0	0	1.4	0	0	0	0
$C_{20:1\ \omega 9t}$	1.55	1.9	1.35	0	0	0	0.74

[a]Data from Reding (1991). Samples were prepared and analyzed by Microbial ID Inc. (Newark, DE).

[b]Value given is the sum of $C_{18:1\ \omega 7c}$, $C_{18:1\ \omega 9t}$, and $C_{18:1\ \omega 12t}$.

branched cells (Fig. BXII.α.241) are probably strains of *Xanthobacter*. Thus, it appears that *Xanthobacter* is ubiquitous in sediment and wet soil samples—including from wet rice fields and marine mud samples—containing organic material.

Strains from *X. autotrophicus* and *X. flavus* can be isolated as associative nitrogen-fixing microaerophiles from rice roots (Reding et al., 1991) and grass roots (Wiegel, unpublished data). *X. tagetidis* was isolated from soil around the roots and roots of the marigold plant (*Tagetes*), suggesting that it is also an associative nitrogen-fixing species (Padden et al., 1997). *X. agilis* has been isolated from the water column of a Swiss lake (Aragno, 1975). Roots of coconut palms (*Cocos nucifera*) contained 10^3 cells per g dry weight (R. Prabhu, personal communication). One *X. flavus* strain originated from a marine sediment sample (Lidstrom-O'Connor et al., 1983; Meijer et al., 1990). *X. autotrophicus* and *X. flavus* isolates have also been found as members of a wood degrading community in wooden pipelines (2-m diameter) from hydroelectric power plants (Line, 1997). Another unusual environment in which *Xanthobacter* has been identified by 16S rDNA assay is in the accessory nidamental glands (ANG) of female cuttlefish and in the myopsid squids, where it occurs as a symbiont (Grigioni et al., 2000).

Pathogenicity No clinical reports of infections of humans or animals by *Xanthobacter* species have been published. In various laboratory settings, however, pathological effects have been observed. This includes the killing of embryos in normally pregnant mice by intraperitoneal injection of 2×10^8 cells of *X. autotrophicus* per g body weight (Manna and Sadhukhan, 1991). In addition, *X. flavus* has exhibited cytological effects and chromosomal aberrations in mouse and human cell cultures (Manna and Sadhukhan, 1992, 1993). No plant pathogenic effects have been reported.

MAINTENANCE PROCEDURES

All strains of *Xanthobacter* can be readily maintained autotrophically on minimal media containing vitamins and 0.02% yeast extract. Cultures grown chemolithoautotrophically can be kept at 2–5°C in liquid medium for about 10 months and in sealed agar slants for more than 15 months; in the presence of 60% (v/v) glycerol, they can be stored for at least 3 years at −20°C. Cultures in liquid nutrient broth or on nutrient agar plates tend to lose viability within 30 d at 4–25°C. For long-term preservation, the cultures are lyophilized in the presence of skim milk with 10% (wt/wt) honey; they have been kept in this form for more than 20 years without significant loss of viability (Malik, 1975, personal communication).

DIFFERENTIATION OF THE GENUS *XANTHOBACTER* FROM OTHER GENERA

Originally two main phenotypic characters were considered to be definitive for *Xanthobacter*: (a) nitrogen fixation under heterotrophic conditions as well as under chemolithoautotrophic conditions (with H_2 serving as the electron donor for respiration), and (b) the presence of the egg-yolk yellow color due to zeaxanthin dirhamnoside. However, nitrogen fixation with H_2 utilization also occurs in several other diazotrophic bacteria (e.g., *Rhizobium*, *Azotobacter*, and *Azospirillum*). Thus, besides determining the relatedness of new isolates to *Xanthobacter* based on the comparison of 16S rDNA sequences, the following special characteristics should be tested before strains are assigned to the genus *Xanthobacter*:

Chemolithoautotrophic growth on H_2 + CO_2 + O_2.

Microaerophilic N_2-fixation.

Pleomorphic cells when grown in nutrient broth containing TCA cycle intermediates.

The presence of zeaxanthin dirhamnoside (tested as described by Wiegel (1986)).

Since the Gram reaction (even with the modified procedures) is often doubtful with these species, other methods such as the polymyxin B-lipopolysaccharide interaction (Wiegel and Mayer, 1978; Wiegel and Quandt, 1982), Fig. BXII.α.243 should be used to determine the Gram-type (Wiegel, 1981).

The mol% G + C of the DNA should be in the range 65–70. See phylogenetic comments below for a differentiation from the closely related (based on 16S rDNA sequence analysis) *Aquabacter* and *Azospirillum*.

Xanthobacter strains were originally described as *Mycobacterium* and as *Corynebacterium* (coryneform bacteria) because of their pleomorphic cell shape, their tendency to exhibit a positive Gram staining reaction, their high mol% G + C value, and the impression of a "snapping-type" or "palisade" cell formation. However, in contrast to *Arthrobacter* and coryneform bacteria, snapping occurs because of the presence of adhesive slime and not because of the rupture of connective cell walls (Figs. BXII.α.241 and BXII.α.244). Moreover, *Xanthobacter* is a Gram-type negative organism (Wiegel, 1981). The main carotenoid, zeaxanthin dirhamnoside, separates recognized *Xanthobacter* from other yellow-pigmented chemolithotrophic or N_2-fixing bacteria of other genera. Zeaxanthin has also been found in some strains of *Flavobacterium* that have a high mol% G + C value; however, members of this group neither fix N_2 nor grow chemolithoautotrophically, and do not show pleomorphism similar to that of *Xanthobacter*

(Oyaizu-Masuchi and Komagata, 1988). The high mol% G + C of *Xanthobacter*, its pleomorphism (on succinate–nutrient broth media), and the ability to utilize short aliphatic alcohols are further properties that separate this genus from many other hydrogen-oxidizing bacteria.

TAXONOMIC COMMENTS

Xanthobacter belongs to the *Alphaproteobacteria* (Rainey and Wiegel, 1996), in agreement with its previous assignment to the fourth RNA superfamily of the Gram-type negative bacteria (De Smedt et al., 1980). Before 16S rDNA sequence analysis was completed, the genus *Xanthobacter* was regarded as a well-defined genus with easily recognizable traits (see above). However, 16S rDNA sequence analysis places two other organisms—*Azorhizobium caulinodans* (Dreyfus et al., 1988) and *Aquabacter spiritensis* (Irgens et al., 1991)—among the species of *Xanthobacter*, thereby converting the well-defined genus to a cluster—the *Xanthobacter-Azorhizobium-Aquabacter* cluster. The placement of *Azorhizobium* is in agreement with a previously published rRNA cistron similarity dendrogram (De Ley, 1992). *Azorhizobium caulinodans* differs from all *Xanthobacter* species by its absence of the characteristic yellow color, failure to grow on carbohydrates, methanol, or methylamine, and by its inability to fix CO_2; moreover, unlike *Xanthobacter*, it shows nodulation of plant stems. *Aquabacter spiritensis* (Irgens et al., 1991) differs from *Xanthobacter* by containing gas vacuoles, showing no sugar or methanol utilization, failing to grow under autotrophic conditions, and by an absence of pleomorphic cell shapes. For both *Aquabacter* and *Azorhizobium*, zeaxanthin dirhamnoside or a precursor thereof has not been demonstrated, although the organisms probably have not been care-

fully analyzed for its presence. Since *Aquabacter* and *Azorhizobium* apparently do not share several of the key properties used to define the genus *Xanthobacter*, it is presently not quite clear how to solve the problem of their taxonomy. To make a better decision on what defines this phylogenetic group beside its relatedness on the 16S rDNA sequence level, the author suggests waiting until more novel and unusual strains and 16S rDNA sequences have been described. One possibility is that, based on the 16S rDNA sequence data, all isolates including the strains presently named *Azorhizobium* or *Aquabacter* and similar strains might be combined as species in an emended genus *Xanthobacter*, since this genus name would have seniority. Another possibility is that *X. agilis*, *X. flavus*, and *X. tagetidis* might be transferred to new genera. At present, an assignment of novel strains to the genus *Xanthobacter*, *Aquabacter*, and *Azorhizobium* should definitely include 16S rDNA sequence analysis and if possible DNA–DNA hybridization data to aid in solving this question.

Two not validly published *Blastobacter* species were reassigned to two novel *Xanthobacter* species *X. viscosus* 7d and *X. aminoxidans* 14d (Doronina and Trotsenko, 2003). Both strains were originally isolated from activated sludge and exhibited yellow branching cells and other typical characteristics of *Xanthobacter* species. However, cells reproduced by budding, thus requiring an emendation of the *Xanthobacter* genus description to include this feature. Closest neighbors in the 16S rDNA based phylogenetic tree are *X. autotrophicus* and *X. flavus*, respectively. The creation of *X. aminoxidans* based on DNA–DNA hybridization data requires a re-evaluation of the *X. flavus* strains including H4-14 with respect to the type strain *X. flavus* 301, since the differences in the 16S rDNA values of the strains are small.

DIFFERENTIATION OF THE SPECIES OF THE GENUS *XANTHOBACTER*

Characteristics useful for the differentiation of the four species currently recognized in the genus *Xanthobacter* are listed in Table BXII.α.190. Because the designated type strain of the type species *X. autotrophicus* strain 7c is a relatively atypical strain, the reference strain *X. autotrophicus* JW33 (DSM 1618) should be used for

comparative studies (Wiegel et al., 1978). However, alkane oxidation and dehalogenation capabilities and thiophene carboxylic acid utilization have not been studied for either of these two strains.

TABLE BXII.α.190. Differential characteristics of the species of the genus *Xanthobacter*[a]

Characteristic	X. autotrophicus	X. agilis	X. flavus	X. tagetidis
Cell morphology: pointed ends, twisted	−	−	−	+
Highly pleomorphic on nutrient broth agar + succinate	+	−	+	+
Motility under autotrophic growth conditions	−	+	−	+
Vitamins required for growth (biotin, riboflavin)	d	−	+	−
Sensitivity to chloramphenicol	−	+	−	+
Autotrophic growth at 37°C	+	−	+	+
Growth on nutrient broth	+	−	+	+
Growth on glutamine as carbon source	+	−	−	nd
Growth on citrate	+	−	d	+

[a]For symbols, see standard definitions; nd, not determined.

List of species of the genus Xanthobacter

1. **Xanthobacter autotrophicus** (Baumgarten, Reh and Schlegel 1974) Wiegel, Wilke, Baumgarten, Opitz and Schlegel 1978, 580^{AL} ("*Corynebacterium autotrophicum*" Baumgarten, Reh and Schlegel 1974, 214.)

au.to.tro′ phi.cus. Gr. pref. *auto* self; Gr. n. *trophos* one who feeds; M.L. masc. adj. *autotrophicus* self feeding, referring to the ability of the organism to use CO_2 as a sole carbon source.

The characteristics are as described for the genus and as listed in Tables BXII.α.190 and BXII.α.191. The morphology is depicted in Fig. BXII.α.241.

Habitat: soil, mud, and water. Widely distributed in nature.

The mol% G + C of the DNA is: 69–70 (T_m) (Wiegel et al., 1978); 65–68 (T_m) (De Smedt et al., 1980); 66–68 (Bd) (M. Aragno, unpublished data).

Type strain: 7c, ATCC 35674, CIP 105432, DSM 432.

GenBank accession number (16S rRNA): X94201.

Additional Remarks: The type strain is atypical of the species, and reference strain DSM 1618 (strain JW33; Wiegel and Schlegel, 1976; Wiegel et al., 1978) should be used for comparative purposes. Reference strains T101 and T102 (ATCC7000551/2) degrade toluene and were isolated from a drainage ditch (Wilmington MA, USA). Strain GJ10 (DSM 3874; ATCC 43050) degrades haloalkanes including dichloromethane (Janssen et al., 1985). Other well-characterized strains can be obtained from the authors who described them.

2. **Xanthobacter agilis** Jenni and Aragno 1988, 136^{VP} (Effective publication: Jenni and Aragno 1987, 257.)

a′ gi.lis. L. adj. *agilis* quick, agile, indicating motility.

The general characteristics are as described for the genus and as listed in Tables BXII.α.190 and BXII.α.191. The morphology of *X. agilis* differs somewhat from that of *X. autotrophicus*, *X. flavus*, and *X. tagetidis* in that the branched-cell morphology on succinate-NB medium is much less pronounced and can be observed only after prolonged incubation.

At the time of its description, *Xanthobacter* species were described as nonmotile; this description was changed later (Reding et al., 1992; Reding and Wiegel, 1993) However, *X. agilis* strains are motile under nearly all growth conditions and exhibit a faster movement than that of other species.

Several strains were isolated from a freshwater lake in Switzerland (Jenni and Aragno, 1987). The type strain was isolated from the water column of a small lake in Neuchatel (Switzerland).

The mol% G + C of the DNA is: 68–69 (T_m).

Type strain: SA35, ATCC 43847, DSM 338, LMG 16336.

GenBank accession number (16S rRNA): X94198.

3. **Xanthobacter flavus** Malik and Claus 1979, 286^{AL}

fla′ vus. L. adj. *flavus* yellow.

The characteristics are as described for the genus and as listed in Tables BXII.α.190 and BXII.α.191. The properties of *X. flavus* are somewhat similar to those of *X. autotrophicus*, except the nitrogenase system of *X. flavus* 301 differs considerably from *X. autotrophicus* strain GZ29 (Berndt et al., 1978) and all *X. flavus* used to require the

addition of biotin to the culture medium. However, non-biotin-requiring strains have been isolated more recently and identified as *X. flavus* by 16S rDNA sequence analysis. The DNA–DNA hybridization values (T_m) between the type strains of *X. flavus* and *X. autotrophicus* is about 25%. Furthermore, in comparison with *X. autotrophicus*, the type strain 301 exhibits a higher sensitivity to oxygen under autotrophic conditions. The range of carbohydrates that can be used is no longer regarded as a valuable property for differentiating these two species as previously proposed (Malik and Claus, 1979).

Isolated from turf podozol soil in the U.S.S.R, rice patty sediments, marine sediments, sewage in the Netherlands, and wetland rice roots.

The mol% G + C of the DNA is: 68–69 (T_m).

Type strain: 301, ATCC 35867, DSM 3770; IFO 14759, NCIB 10071.

GenBank accession number (16S rRNA): X94199.

Additional Remarks: Reference strains include rice root isolate JW-KR2 (ATCC 51492); strain 14p1 (DSM 10330), which degrades 1,4-dichlorobenzene and was isolated from contaminated soil by Sommer and Gorisch (1997); and strain H4-14 of Meijer et al. (1990). The type strain, 301, was misclassified by Federov and Kalininskaya (1961) as "*Mycobacterium flavum*" strain 301.

4. **Xanthobacter tagetidis** Padden, Rainey, Kelly and Wood 1997, 400^{VP}

ta.ge.ti′ dis. M.L. n. *tagetidis* of *Tagetes* the marigold genus of flowering plants.

The main characteristics are as described for the genus and as listed in Tables BXII.α.190 and BXII.α.191. The main properties of *X. tagetidis* are similar to those of the other species in respect to the characteristic yellow color of zeaxanthin dirhamnoside, pleomorphism on succinate-NB medium, production of copious amounts of slime while growing on carbohydrates, and restricted utilization of carbohydrates. Although they grow poorly, the wild type strains are able to grow more readily on glucose than most other strains tested belonging to the other three species. It can be specifically isolated from soil around the roots of marigold plants (*Tagetes patula* and *T. erecta*). *X. tagetidis* is so far the only *Xanthobacter* species that can utilize substituted thiophenes. Whether or not the ability to grow on substituted thiophenes is restricted only to *X. tagetidis* (Padden et al., 1997, 1998) or also can be found among strains belonging to the other species is presently unknown. Similar to *X. agilis*, *X. tagetidis* is motile under most growth conditions.

To date, two strains were described that were isolated from sludge (strain DSM 11602) and from soil around marigold plants (strain TagT2C), respectively, in England. The type strain was isolated from composted root balls of *T. patula*.

The mol% G + C of the DNA is: 68–69 (T_m).

Type strain: TagT2C, ATCC 700314, DSM 11105.

GenBank accession number (16S rRNA): X99469.

Additional Remarks: Several strains with properties other than described above have been isolated and are under investigation. However, several of them, although biochemically well studied, have not been deposited in a culture collection or assigned a species name at this time.

TABLE BXII.α.191. Other characteristics of the species of the genus *Xanthobacter*[a]

Characteristic	*X. autotrophicus*	*X. agilis*	*X. flavus*	*X. tagetidis*
Cell diameter, μm	0.4–0.8	0.7	0.5–0.7	0.5
Cell length, μm	0.8–4.0[b]	1.1–3.6	1.0–2.5[c]	1
Motility	+[d] (on propanol)	+	+[d] (on propanol)	+
Water-insoluble zeaxanthin pigment produced[d]	+	+ (low conc.)	+	+
Slime produced[e]	+	Traces	+	+
α-Polyglutamine capsule-like material produced	+	nd	+	nd
Polyphosphate granules formed	+	+	+	+
Intracellular poly-β-hydroxybutyrate formed	+	+	+	nd
Growth at:				
15°C	Weak	+	Weak	+
28–32°C	+	+	+	+
37°C	Weak	−	Weak	+
45°C	d and weak	−	−	−
pH range for growth	5.0–9.0	nd	nd	6.8–8.7 (opt. 7.6–7.8)
Growth in the presence of:				
25% (w/v) NaCl	d	nd	+	nd
5.0% (w/v) NaCl	d	−	−	nd
7.5% (w/v) NaCl	−	−	−	nd
1.0% (w/v) dodecylsulfate	−	nd	−	nd
10 uM crystal violet	d	nd	+	nd
Growth on tellurite agar	d	nd	+	nd
Growth on H_2 (with O_2) as energy source	+	+	+	+
Growth on thiosulfate as energy source	+	+	+	+
Utilization of CO_2 as sole carbon source	+	+	+	+
Growth in submerged liquid culture	+	+	+	+
Strictly respiratory type of metabolism	+	+	+	+
Nitrate as terminal electron acceptor	−	−	−	−
Heterotrophic growth under air atmosphere	+	+	+	+
Autotrophic growth occurs:				
Up to 20% O_2 atmosphere	+	+	− (d)	+
Only below 5% O_2 atmosphere	(<10% air)	−	+	−
Nitrogen fixation:				
Microaerobic conditions	+	+	+	+
Air (100%)	−	−	−	+
Growth requirement: Biotin	−	−	(+)	−
Membrane-bound hydrogenase (uptake) that does not reduce NAD	+	+	+	nd
Hydrogenase activity:				
Inducible only	−	nd	+	nd
Inducible or constitutive	+	nd	−	nd
Oxygen-labile	+	nd	+	nd
Carbohydrates catabolized via the Entner–Doudoroff and pentose phosphate pathways	+	−	+	nd
Sole carbon source:				
Formate, acetate, propionate, butyrate, pyruvate, succinate, fumarate	+	+	+	+
Methanol, ethanol, *n*-propanol	+	+	+	+
α-Ketoglutarate	+	−	+	+
L(+)-Ascorbate	+	−	+	nd
Methylamines	+	+	+	nd
Carbonyl sulfide, carbon disulfide, dimethylsulfide, methanesulfide, (under anoxic conditions producing H_2S)	nd	nd	nd	+
Thiophene-2 carboxylate				
N-Caproate	+	d	−	nd
Citrate	+	−	D	+
Lactate	+	−	+	nd
Malonate	d	−	+	nd
Lactose, sorbose, raffinose, rhamnose	−	−	−	−
Fructose	+	−	d	+
Glucose (including weak or after adaptation)	+	−	d	+
Mannose	d	−	d	+
Sucrose	D	−	−	+
Alanine	−	+	+	+
Arginine, L-aspartate, L-phenylalanine, isoleucine, threonine, histidine	−	−	−	−
D-Glutamine	+	−	−	nd

(*continued*)

TABLE BXII.α.191. *(cont.)*

Characteristic	X. autotrophicus	X. agilis	X. flavus	X. tagetidis
Resistant to:				
Ampicillin, chloramphenicol	−	D	−	+
Penicillin, tetracycline	−	D	−	−
Erythromycin	−	+	−	nd
Nitrate reduced to nitrite	+	−	+	Weak + (− on API 20)
Tetrathionate to thiosulfate	+	+	+	+
Tetrazolium salts reduced	+	nd		nd
Lecithinase, deoxyribonuclease activities	−	nd	−	nd
Catalase, phosphate oxidase	+	+	+	nd
Urease activity	d	nd	−	nd
Tyrosinase activity	+	nd		nd
Indole production	−	nd		−
Litmus milk (alkaline reaction only change)	+	nd	+	nd
Voges–Proskauer test	−	−	−	nd
H₂S production	−	−	−	− (+ thiosulfate thiophene)
Gelatin liquefaction	−	nd	−	nd
Hydrolysis of Tween, starch, casein, or cellulose	−	−	−	−
Main component (60–90%) of fatty acid is 11-octadecenoic acid (cis)	+	+	+	nd
Salicin, β-hydroxybutyrate tartrate	−	nd	−	nd
Mol% G + C of DNA	69–70 (T_m); (65–68)[f]	66 ± 1(Bd)	69 (T_m); (68)f	67 ± 1 (Sp); 71 ± 3 (T_m)

[a]Symbols: see standard definitions; nd, not determined.

[b]Occasionally as long as 8.0 μm.

[c]Occasionally as long as 5.0 μm.

[d]Some white/whitish strains produce only minor amount.

[e]Amounts range from copious to small; slime contains acid sugars.

[f]Values in parenthesis are from De Smedt et al. (1980); T_m, obtained by buoyant density method; Bd, value obtained by thermal denaturation; Sp, value obtained by spectrometric assay.

Other Organisms

1. *"Xanthobacter methylooxidans"*

The species name *"X. methylooxidans"* has been proposed for the facultatively methylotrophic bacterial strain 32P (Doronina et al., 1996c). It is characterized by a high content of *cis*-vaccenic (55–75%) and lactobacillic (13–17%) acids. It does not require vitamins, forms indole, ammonia, and hydrogen sulfide, liquefies gelatin, and utilizes formate, methylamine, dimethylformamide, dimethylacetamide, and dimethyl sulfoxide as carbon and energy sources. The oxidation of methanol proceeds with the participation of the corresponding dehydrogenases via formaldehyde and formate to CO_2, which is assimilated via the ribulose bisphosphate pathway. Strain 32P oxidizes H_2 and fix N_2. The DNA–DNA hybridization values of strain 32P with type strains belonging to the genus *Xanthobacter* (*X. agilis* DSM 3770, *X. flavus* DSM 338, and *X. autotrophicus* DSM 432) did not exceed 25%; this, together with its phenotypic peculiarities, allowed strain 32P to be classified as a "new species", however, without being validly described and named).

The mol% G + C of the DNA is: 69.5 (T_m).
Deposited strain: 32P.

2. Several isolates have been described either as strains without attributing them to a genus, but that resemble *Xanthobacter* strongly in their properties, or were just described as strains of *Xanthobacter* but without assignment to a species. Many autotrophically isolated strains of *X. autotrophicus* from various habitats can be obtained from J. Wiegel. The others strains below should be available from the corresponding authors; unfortunately, the strains have not been deposited in public culture collections. Some of the early isolated and interesting strains are:

a. Sixteen strains isolated on alkenes (De Bont and Leijten, 1976)

b. *"Mycobacterium butanitrificans"* (Coty, 1967)

c. Group IV, 10 strains (Kouno and Ozaki, 1975)

d. Strains N61 and N63 (Jensen and Holm, 1975)

e. Strains PY 2 (Ensign, personal communication: *X. autotrophicus*; one of the best studied strains in respect to carbon metabolism; containing coenzyme M) PY10, By2 (van Ginkel and DeBont, 1986; van Ginkel et al., 1986) utilizing alkanes under epoxide formation

f. Strain 124X (ATCC 49450), degrading 4-hydroxyphenylacetate (Van den Tweel et al., 1986) and styrene (Hartmans et al., 1989)

g. Strain MAB2 (ATCC 48876) (Villarreal et al., 1991) and patent strain NW11 (ATCC 53272), (Robinson and Stipanovic, 1989) polysaccharide producing strains for use in oil recovery

h. Strain 25a (Meijer et al., 1990)

i. Strain Cimw 99 (Smith et al., 1991) degrading 3-chloro-2-methylproprionic acid

j. Strain CP (Ditzelmüller et al., 1989) degrading chlorophenols and 2,4-dichlorophenylacetic acid

Family IX. **Methylobacteriaceae** fam. nov.

GEORGE M. GARRITY, JULIA A. BELL AND TIMOTHY LILBURN

Meth.yl.o.bac.te.ri.a' ce.ae. M.L. neut. n. *Methylobacterium* type genus of the family; *-aceae* ending to denote family; M.L. fem. pl. n. *Methylobacteriaceae* the *Methylobacterium* family.

The family *Methylobacteriaceae* was circumscribed for this volume on the basis of phylogenetic analysis of 16S rRNA sequences; the family contains the genera *Methylobacterium* (type genus) and *Microvirga*. *Microvirga* was proposed after the cut-off date for inclusion in this volume (June 30, 2001) and is not described here (see Kanso and Patel (2003).

Chemoorganotrophic; *Methylobacterium* is facultatively methylotrophic. Form pink colonies.

Type genus: **Methylobacterium** Patt, Cole and Hanson 1976, 228 emend. Green and Bousfield 1983, 876.

Genus I. **Methylobacterium** *Patt, Cole and Hanson 1976, 228*[AL] *emend. Green and Bousfield 1983, 876*

PETER N. GREEN

Meth.yl.o.bac.te' ri.um. M.L. n. *methyl* the methyl radical; Gr. n. *bacterion* a small rod; M.L. neut. n. *Methylobacterium* methyl bacterium.

Rods 0.8–1.2 × 1.0–8.0 μm that occasionally branch and/or exhibit polar growth. Form pink to orange-red colonies on methanol salts agar. Stain Gram negative to Gram variable; Gram-negative cell wall type. Motile by means of a single polar, subpolar, or lateral flagellum. **Aerobic**, having a strictly respiratory type of metabolism with oxygen as the terminal electron acceptor. **Mesophilic. Chemoorganotrophic and facultatively methylotrophic**, able to grow with formaldehyde (often at micromolar concentrations), formate, and methanol; some strains grow on methylated amines. Widely distributed in nature.

The mol% G + C of the DNA is: 68.0–72.4.

Type species: **Methylobacterium organophilum** Patt, Cole and Hanson 1976, 228.

FURTHER DESCRIPTIVE INFORMATION

Morphology and cell structure All *Methylobacterium* strains are rods (0.8–1.2 × 1.0–8.0 μm) that occur singly or occasionally in rosettes (Heumann, 1962; Patt et al., 1974). They are often branched or pleomorphic, especially in older stationary-phase cultures (Fig. BXII.α.245). There is some evidence to suggest that they exhibit polar growth or a budding morphology (L.B. Perry, unpublished observations). All strains are motile by a single polar, subpolar or lateral flagellum, although some strains are not vigorously motile. Cells often contain large sudanophilic inclusions (poly-β-hydroxybutyrate) and sometimes volutin granules. They are Gram negative, although many strains stain as Gram variable. Representative strains have a multilayered cell wall structure and the type of citrate synthase (Green and Bousfield, 1982) characteristic of Gram-negative bacteria.

Cultural features Most strains grow slowly but a few grow poorly or not at all on nutrient agar. After 7 d incubation at 30°C, colonies on GP agar are 1–3 mm in diameter and pale pink to bright orange-red, whereas colonies on MMS agar are a more uniform pale pink. The pigment is nondiffusible, nonfluorescent, and is a carotenoid (Ito and Iizuka, 1971; Downs and Harrison, 1974, Urakami et al., 1993). In static liquid media, most strains form a pink surface ring and/or pellicle.

The optimal growth temperature for all *Methylobacterium* strains is in the range 25–30°C. Some strains will grow at 15°C or less, and some will grow at or above 37°C. Although growth is optimal around neutrality, some strains can grow at pH 4 and some at pH 10.

Pigments All strains contain carotenoid pigments that have absorption maxima at 465, 495, and 525 nm (Urakami et al., 1993). Sato (1978), Sato and Shimizu (1979), and Nishimura et al. (1981) have shown that strains of *Methylobacterium* can form bacteriochlorophyll *a* under specific cultural conditions, suggesting a common link with the phototrophs in their ancestry.

Metabolism and methylotrophy All strains are strict aerobes and are catalase and oxidase (often weakly) positive. They are chemoorganotrophs and facultative methylotrophs, capable of growth on a variety of C_1 compounds. All grow on formaldehyde (often at micromolar concentrations), formate, and methanol; some grow on methylated amines. Only one species (*M. organophilum*) is reported to have utilized methane as sole carbon and energy source, but the ability has since been lost by the type (and only) strain. Methane assimilability in this organism was thought to be plasmid borne and easily lost if cultures were not maintained on an inorganic medium in a methane atmosphere (R.S. Hanson, personal communication). This led Green and Bousfield (1983) to emend the genus description of *Methylobacterium* to exclude methane assimilation as a key taxonomic criterion. Representative *Methylobacterium* strains have been reported to assimilate C_1 compounds via the icl(-) serine pathway (the type of serine C_1-assimilatory pathway used by these organisms) (Quayle, 1972; Bellion and Spain, 1976) and to have a complete tricarboxylic acid cycle when grown on complex organic substrates.

Growth factors are not required by any strain although growth of some strains is stimulated by calcium pantothenate. Most strains do not degrade or hydrolyze casein, starch, gelatin, cellulose, lecithin, or DNA. Urease is produced by all strains, and some strains have weak lipolytic activity. The enzymes β-galactosidase, L-ornithine decarboxylase, L-lysine decarboxylase, and L-arginine dihydrolase are not produced. Indole (except for *M. thiocyanatum*) and H_2S are also not produced. The methyl red and Voges–Proskauer tests are negative, although some strains reduce nitrate to nitrite.

Carbon and nitrogen sources The following compounds were used as carbon sources by most (≥95%) strains of *Methylobacterium*: glycerol, malonate, succinate, fumarate, α-ketoglutarate, DL-lactate, DL-malate, acetate, pyruvate, propylene glycol, ethanol, methanol, and formate. Some strains (see Green and

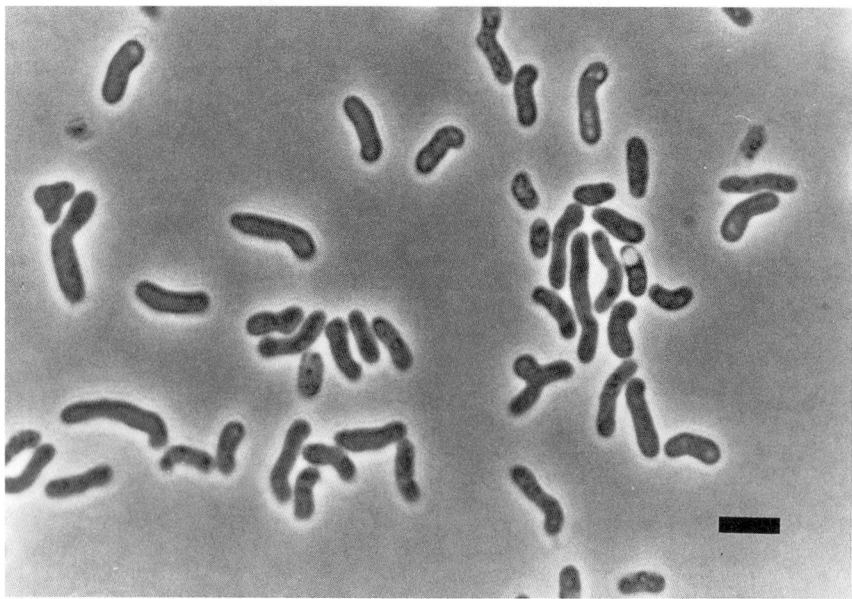

FIGURE BXII.α.245. Phase-contrast micrograph of *Methylobacterium* spp. Bar = 2 μm.

Bousfield, 1982; Urakami et al., 1993, and Table BXII.α.192) can also utilize L-arabinose, D-xylose, D-fucose, D-glucose, D-galactose, D-fructose, L-aspartate, L-glutamate, adipate, pimelate, sebacate, azelate, suberate, D-tartrate, citrate, citraconate, saccharate, monomethylamine, dimethylamine, trimethylamine, trimethylamine-*N*-oxide, ethanolamine, butylamine, formamide, *N*-methylformamide, dimethylglycine, betaine, tetramethylammonium hydroxide (TMAH), *N,N*-dimethylformamide (DMF), chloromethane, and dichloromethane. None of the strains appear to use any of the disaccharides or sugar alcohols examined (except for glycerol) (Green and Bousfield, 1982) or any of the following as sole carbon and energy source: propionate, DL-arginine, L-valine, glycine, geraniol, tryptamine, histamine, putrescine, *m*-hydroxybenzoate, testosterone, sarcosine, phenol, thiourea, tetramethyl urea, hexane, or benzene. Oxidative production of acid from sugars was variable among strains Urakami et al. (1993).

Ammonia, nitrate, and urea can serve as nitrogen sources (Urakami and Komagata, 1984). *M. thiocyanatum* can utilize thiocyanate or cyanate as sole source of nitrogen for growth (Wood et al., 1998).

Fatty acid composition The fatty acid composition of *Methylobacterium* strains is comprised largely (around 70–90%) of $C_{18:1}$ acid with small amounts of $C_{16:1}$, $C_{19:0\ cyclo}$, and $C_{14:0\ 3OH}$ hydroxy acid. Some strains contain hopanoids and small amounts of squalene. Phospholipids of all strains included large amounts of cardiolipin (diphosphatidylglycerol), phosphatidylethanolomine, phosphatidylcholine, and a small amount of phosphatidylglycerol (Urakami et al., 1993). The major isoprenoid quinone components are ubiquinones with 10 isoprene units (Urakami and Komagata, 1979). DNA base composition is 68.0– 72.4 mol% G + C (Hood et al., 1987).

Antibiotic sensitivity Most strains are sensitive to the chemotherapeutic agents kanamycin, gentamicin, albamycin T, streptomycin, framycetin, and especially the tetracyclines, whereas most are resistant to cephalothin, nalidixic acid, penicillin, bacitracin, carbenicillin, colistin sulfate, polymyxin B, and nitrofurantoin.

Habitats and sources Being ubiquitous in nature and a common airborne organism, strains of *Methylobacterium* spp. are found in a wide variety of environmental, industrial, and even occasionally clinical environments, mainly as part of a transient flora or as chance contaminants. Their ability to scavenge trace amounts of nitrogen and to resist a certain degree of desiccation contributes to their survival in hostile environments. Some strains have been shown to exhibit resistance to gamma irradiation 10–40 times higher than that exhibited by many other Gram-negative bacteria (Ito and Iizuka, 1971). *M. thiocyanatum* (Wood et al., 1998) has been shown to tolerate high (≥50 mM) thiocyanate or cyanate. Other *Methylobacterium* strains have not been tested for this feature.

ENRICHMENT AND ISOLATION PROCEDURES

Because of the ability of *Methylobacterium* spp. to grow on methanol as sole carbon and energy source, and because of their characteristic pigmentation, these organisms are relatively easy to isolate. Methanol mineral salts (MMS)[1] medium is a suitable selective medium for *Methylobacterium*.

Although many *Methylobacterium* strains can grow between 5°C and 37°C, all grow well at 30°C, and thus 25–30°C can be used for all isolation and subsequent growth experiments. These organisms are fairly slow growers, often taking 2–3 d at 30°C to produce clearly visible colonies or confluent growth, and often taking more than 7 d for colonies to reach their maxim size of 1–3 mm in diameter. Growth is sometimes more luxuriant, with

---

1. Methanol mineral salts medium consists of (per liter water): K_2HPO_4, 1.20 g; KH_2PO_4, 0.62 g; $CaCl_2·6H_2O$, 0.05 g; $MgSO_4·7H_2O$, 0.20 g; NaCl, 0.10 g; $FeCl_3·6H_2O$, 1.0 mg; $(NH_4)_2SO_4$, 0.5 mg; $CuSO_4·5H_2O$, 5.0 mg; $MnSO_4·5H_2O$, 10.0 mg; $Na_2MoO_4·2H_2O$, 10.0 mg; H_3BO_3, 10.0 mg; $ZnSO_4·7H_2O$, 70.0 mg; $CoCl_2·6H_2O$, 5.0 mg. The medium is sterilized by autoclaving at 121°C for 20 min and cooled to 50°C. A filter-sterilized vitamin solution (Colby and Zatman, 1973) is added if required, along with 0.1–0.2% v/v sterile methanol. The pH of the medium is adjusted to pH 7.0. A solidified medium (MMS agar) is prepared by the addition of 1.5–2% Oxoid purified agar before autoclaving. No *Methylobacterium* strain isolated to date has been shown to have an obligate requirement for vitamins or other added growth factors.

TABLE BXII.α.192. Substrates utilized as sole carbon source to differentiate strains of *Methylobacterium*[a,b]

Characteristic	M. organophilum	M. aminovorans	M. chloromethanicum[c]	M. dichloromethanicum[d]	M. extorquens	M. fujisawaense	M. mesophilicum	M. radiotolerans	M. rhodesianum	M. rhodinum	M. thiocyanatum[e]	M. zatmanii
D-Glucose	+	−	−	−		V	V	+	−	V	+	−
D-Fucose	−	−	nd	−	−	+	+	+	−	−	nd	−
D-Xylose	−	−	−	−	−	+	+	+	−	−	nd	−
L-Arabinose	−	−	−	−	−	+	+	+	−	−	V	−
Fructose	+	+	−	+	−	V	−	−	+	+	+	+
L-Aspartate/L-Glutamate	−	+	−	+	V	+	+	+	V	+	+[f]	−
Citrate	−	−	−	−	−	+	+	+	−	+	+	−
Sebacate[g]	−	−	nd	−	−	+	V	+	−	−	nd	−
Acetate	+	+	nd	+	+	V	−	+	+	+	+	+
Betaine	−	+	−	+	+	−	−	−	+	+	nd	−
Methylamine	+	+	+	+	+	−	−	−	+	+	+	+
Trimethylamine	+	+	−	−	−	−	−	−	−	−	−	V
Methane	V	−	−	−	−	−	−	−	−	−	nd	−
Tetramethylammonium hydroxide	−	+	nd	nd	−	−	−	−	−	−	nd	−
N,N-dimethylformamide	−	+	−	−	−	−	−	−	−	−	nd	−
Chloromethane	nd	nd	+	nd	nd	nd	nd	nd	nd	nd	nd	nd
Dichloromethane	nd	nd	nd	+	nd	nd	nd	nd	nd	nd	nd	nd
Growth on peptone-rich nutrient agar[h]	+	+	−	+	V	+	−	+	+	+	+	+

[a]Symbols: +, utilized as substrate; −, not utilized; V, variable result; nd, no data.

[b]Owing to the slow growth of some strains on certain substrates, carbon utilization tests were read after 14 days incubation.

[c]Data from McDonald et al. (2001).

[d]Data from Doronina et al. (2000b).

[e]Data from Wood et al. (1998).

[f]Tested for glutamate only.

[g]Most strains which utilize sebacate can also utilize pimelate, suberate, azelate, and adipate.

[h]Nutrient agar, e.g., Oxoid cm55.

a deeper pigmentation, on Glycerol–Peptone (GP) agar[2]. Although this medium is useful for subculturing stocks of pure *Methylobacterium* spp., it is less suitable for enrichment than the MMS medium, as other rapidly growing heterotrophs present in the sample can overgrow the pink-pigmented facultative methylotrophic (PPFM) bacteria. The use of certain antibiotics in selective media (see Further Descriptive Information, this chapter) can also be considered, as can the use of individual carbon sources, for the isolation of specific groups or species of *Methylobacterium*.

If MMS agar is used as a selective medium, the vast majority of the pink colonies which reach diameters of more than 1 mm will be strains of PPFM organisms. Pink methylotrophic yeasts are not uncommon, but bacterial pink methylotrophs other than *Methylobacterium* species are rare. Growth of PPFM organisms in liquid media is almost always characterized by a surface ring and/or thin pellicle, indicative of their aerobic nature.

When attempting to isolate PPFM strains from leaf surfaces, a leaf impression technique, using one of the above media, is recommended. Homogenization of whole leaves or embedding of leaves in molten agar are alternatives, although they are not as successful as the impression technique.

If fungal contamination is a problem with samples from particular habitats when attempting to isolate strains of PPFM, 20 µg/ml of cycloheximide can be added to the medium.

MAINTENANCE PROCEDURES

Strains of *Methylobacterium* can be maintained short term (2–4 weeks) on GP or methanol salts agar. For long-term preservation organisms can be stored at −70°C or in/over liquid nitrogen, using either of the above liquid growth media supplemented with 10% v/v glycerol; alternatively, the organisms can easily be lyophilized. Strains isolated on methane and dichloromethane must be maintained on a salts medium containing these compounds as sole carbon source or their ability to utilize these substrates may be lost.

PROCEDURES FOR TESTING SPECIAL CHARACTERS

Because the nine species of *Methylobacterium* are differentiated mainly by the pattern of compounds they utilize as sole carbon and energy source, care should be taken to standardize such tests since they are notoriously difficult to duplicate between laboratories. All carbon utilization tests shown in Table BXII.α.192 were carried out as described by Green and Bousfield (1982), who used faintly turbid suspensions of cells that had been thrice washed in sterile saline, to inoculate media. All tests were read only after 14 days incubation at 30°C, and growth was compared to a negative control with no added carbon source. This long incubation time was necessary to allow for slow growth on certain compounds. Tests for biochemical, physiological, and morpho-

---

2. Glycerol–Peptone Agar consists of (g/l): agar, 15.0; glycerol, 10.0; peptone (Difco), 10.0. The pH is adjusted to pH 7.0.

logical features as well as calculations of DNA base ratios are described by Green and Bousfield (1982) and Green (1992).

TAXONOMIC COMMENTS

Many of the strains which comprise the genus *Methylobacterium* have a checkered taxonomic history (Green and Bousfield, 1981) and based on their Gram variability, occasional cellular pleomorphism, and flagellar arrangements were previously included in the following genera: *Bacillus, Vibrio, Pseudomonas, Flavobacterium, "Protaminobacter", Mycoplana, Protomonas,* and *Methylobacterium*. Taxonomic studies (Wolfrum et al., 1986; Hood et al., 1987, 1988; Urakami et al., 1993) have shown the genus to be heterogeneous, with 12 validly published species to date (Green, 1992; Urakami et al., 1993; Wood et al., 1998; Doronina et al., 2000b; McDonald et al., 2001), as well as several unassigned strains representing centers of genetic variation which may equate to new taxa. Recent DNA–rRNA homology studies by Dreyfus et al. (1988) have placed *Methylobacterium* in the rRNA superfamily IV of De Ley (1978), along with other members of the *Agrobacterium–Rhizobium* complex.

DIFFERENTIATION OF THE SPECIES OF THE GENUS *METHYLOBACTERIUM*

Species within the genus *Methylobacterium* are differentiated mainly by the pattern of compounds they utilize as sole carbon and energy source (see Table BXII.α.192).

List of species of the genus Methylobacterium

1. **Methylobacterium organophilum** Patt, Cole and Hanson 1976, 228[AL]

   or.ga.no′phi.lum. Gr. n. *organo* organ, living bodies; Gr. adj. *philos* loving; Gr. adj. *organophilus* intended to mean preferring complex carbon sources.

   Characteristics are those given for the genus and for the carbon utilization spectra shown in Table BXII.α.192. Originally reported as being able to utilize methane as sole source of carbon and energy (Patt et al., 1974), but this feature has subsequently been lost (Green and Bousfield, 1982). The type strain was isolated from lake sediment.

   The mol% G + C of the DNA is: 70.5 (T_m); 71.1 (Bd).
   Type strain: XX, ATCC 27886, DSM 760, NCIMB 11278.
   GenBank accession number (16S rRNA): D32226.

2. **Methylobacterium aminovorans** Urakami, Araki, Suzuki and Komagata 1993, 510[VP]

   a.mi.no′vo.rans. N.L. n. *aminus* amine; L. part. adj. *vorans* devouring, digesting; N.L. part. adj. *aminovorans* amine digesting.

   Characteristics are those given for the genus and for the carbon utilization spectra shown in Table BXII.α.192. The type strain was isolated from soil.

   The mol% G + C of the DNA is: 68.0 (HPLC).
   Type strain: TH15, JCM 8240, NCIMB 13343.

3. **Methylobacterium chloromethanicum** McDonald, Doronina, Trotsenko, McAnulla and Marrell 2001, 121[VP]

   chlo.ro.me.tha′ni.cum. N.L. n. *chloromethanicum* chloromethane utilizing.

   Characteristics are those given for the genus and/or the carbon utilization spectra shown in Table BXII.α.192. The type strain was isolated from soil at the Nizhekamsk petrochemical factory, Tatarstan, Russia.

   The mol% G + C of the DNA is: 64.4 (T_m).
   Type strain: CM4, NCIMB 13688.
   GenBank accession number (16S rRNA): AF198624.

4. **Methylobacterium dichloromethanicum** Doronina, Trotsenko, Tourova, Kuznetsov and Leisinger 2000c, 1953[VP] (Effective publication: Doronina, Trotsenko, Tourova, Kuznetsov and Leisinger 2000b, 216.)

   di.chlo.ro.meth.an′ic.um. M.L. n. *methanum* methane; L. pref. *di* two; M.L. adj. *dichloromethanicum* dichloromethane utilizing.

   Characteristics are those given for the genus and/or the carbon utilization spectra shown in Table BXII.α.192. The type strain was isolated from activated sludge.

   The mol% G + C of the DNA is: 67.1 (T_m).
   Type strain: DM4, DSM 6343.
   GenBank accession number (16S rRNA): AF227128.

5. **Methylobacterium extorquens** (Bassalik 1913) Bousfield and Green 1985, 209[VP] (*Protomonas extorquens* (Bassalik 1913) Urakami and Komagata 1984, 198; *Bacillus extorquens* Bassalik 1913, 258.)

   ex.tor′quens. L. part. adj. *extorquens* twisting out.

   Characteristics are those given for the genus and for the carbon utilization spectra shown in Table BXII.α.192. The type strain was isolated from earthworm excreta in 1949 by L. Janota-Bassalik. Bassalik's original strain is no longer extant.

   The mol% G + C of the DNA is: 69.4 (Bd).
   Type strain: ATCC 43645, DSM 1337, NCIMB 9399.
   GenBank accession number (16S rRNA): L20847, D32224.

6. **Methylobacterium fujisawaense** Green, Bousfield and Hood 1988, 124[VP]

   fu.ji.sa′wa.en.se. N.L. neut. adj. *fujisawaense* coming from the Fujisawa region of Japan.

   Characteristics are those given for the genus and for the carbon utilization spectra shown in Table BXII.α.192.

   The mol% G + C of the DNA is: 70.8–71.8 (Bd).
   Type strain: 0-31 (Kouno and Ozaki 1975), ATCC 43884, DSM 5686, NCIMB 12417.

7. **Methylobacterium mesophilicum** (Austin and Goodfellow 1979) Green and Bousfield 1983, 876[VP] (*Pseudomonas mesophilica* Austin and Goodfellow 1979, 377.)

   me.so.phi′li.cum. Gr. n. *meson* the middle; Gr. adj. *philikos* friendly; N.L. neut. adj. *mesophilicum* friendly to the middle, because of its preference for moderate temperatures.

   Characteristics are those given for the genus and for the carbon utilization spectra shown in Table BXII.α.192. The type strain was isolated from *Lolium perenne* leaves.

   The mol% G + C of the DNA is: 69.9 (Bd).
   Type strain: A47, ATCC 29983, DSM 1708, NCIMB 11561.
   GenBank accession number (16S rRNA): D32225.

8. **Methylobacterium radiotolerans** (Ito and Iizuka 1971) Green and Bousfield 1983, 876[VP] (*Pseudomonas radiora* Ito and Iizuka 1971, 1568.)

   ra.di.o.to′le.rans. Eng. pref. *radio* pertaining to radiation; L. part. adj. *tolerans* tolerating; N.L. part. adj. *radiotolerans* tolerating radiation.

Characteristics are those given for the genus and for the carbon utilization spectra shown in Table BXII.α.192. The type strain was isolated from rice grains.

The mol% G + C of the DNA is: is 72.3 (Bd).

Type strain: 0-1, ATCC 27329, DSM 1819, NCIMB 10815.

GenBank accession number (16S rRNA): D32227.

9. **Methylobacterium rhodesianum** Green, Bousfield and Hood 1988, 124[VP]

rho′ de.si.an.um. M.L. neut. adj. *rhodesianum* named after the British taxonomist Muriel Rhodes–Roberts.

Characteristics are those given for the genus and for the carbon utilization spectra shown in Table BXII.α.192. The type strain was isolated from a fermentor operating with formaldehyde as sole carbon source.

The mol% G + C of the DNA is: 69.8–71.2 (Bd).

Type strain: *Pseudomonas* strain 1, ATCC 43882, DSM 5687, NCIMB 12249.

GenBank accession number (16S rRNA): L20850.

10. **Methylobacterium rhodinum** (Heumann 1962) Green and Bousfield 1983, 876[VP] (*Pseudomonas rhodos* Heumann 1962, 342.)

ro.di′ num. N.L. neut. adj. *rhodinum* rosy, because of its pink color.

Characteristics are those given for the genus and for the carbon utilization spectra shown in Table BXII.α.192. The type strain was isolated by Heumann (1962) from *Alnus* rhizosphere.

The mol% G + C of the DNA is: 71.8 (Bd).

Type strain: ATCC 14821, DSM 2163, NCIMB 1942.

GenBank accession number (16S rRNA): L20849, D32229.

11. **Methylobacterium thiocyanatum** Wood, Kelly, McDonald, Jordan, Morgan, Khan, Murrell and Borodina 1999, 341[VP] (Effective publication: Wood, Kelly, McDonald, Jordan, Morgan, Khan, Murrell and Borodina 1998, 157.)

thio.cy.an.a′ tum. M.L. n. *thiocyanatum* using thiocyanate.

Characteristics are those given for the genus and for the carbon utilization spectra shown in Table BXII.α.192. Able to tolerate high (≥50 mM) thiocyanate and/or cyanate and utilize these compounds as sole nitrogen source for growth. The type strain was isolated from soil in the rhizosphere of *Allium aflatunense.*

The mol% G + C of the DNA is: 69.8–71.0 (T_m).

Type strain: ALL/SCN-P, ATCC 700647, DSM 11490.

GenBank accession number (16S rRNA): U58018.

12. **Methylobacterium zatmanii** Green, Bousfield and Hood 1988, 124[VP]

zat.ma′ ni.i. M.L. gen. n. *zatmanii* named after the British biochemist L.J. Zatman.

Characteristics are those given for the genus and for the carbon utilization spectra shown in Table BXII.α.192. The type strain was isolated from a fermentor operating with formaldehyde as sole carbon source.

The mol% G + C of the DNA is: 69.4–70.3 (Bd).

Type strain: 125, ATCC 43883, DSM 5688, NCIMB 12243.

GenBank accession number (16S rRNA): L20804.

Family X. **Rhodobiaceae** *fam. nov.* *

GEORGE M. GARRITY, JULIA A. BELL AND TIMOTHY LILBURN

Rho.do.bi.a′ ce.ae. M.L. n. *Rhodobium* type genus of the family; *-aceae* ending to denote family; M.L. fem. pl. n. *Rhodobiaceae* the *Rhodobium* family.

The family *Rhodobiaceae* was circumscribed for this volume on the basis of phylogenetic analysis of 16S rRNA sequences; the family contains *Rhodobium* (type genus).

Grow phototrophically in the light and chemotrophically in the dark. Photosynthetic membranes present as lamellar stacks. Require NaCl for growth.

Type genus: **Rhodobium** Hiraishi, Urata and Satoh 1995d, 230.

Genus I. **Rhodobium** Hiraishi, Urata and Satoh 1995d, 230[VP]

JOHANNES F. IMHOFF AND AKIRA HIRAISHI

Rho.do′ bi.um. Gr. n. *rhodon* the rose; Gr. n. *bios* life; M.L. n. *Rhodobium* red life.

Cells are ovoid to rod shaped, motile by means of polar, subpolar or randomly distributed flagella; and cells multiply by budding and asymmetric cell division. Rosette formation is rare. **Gram negative and belong to the *Alphaproteobacteria*.** Phototrophically grown cells contain **internal photosynthetic membranes as lamellar stacks parallel to the cytoplasmic membrane.** Photosynthetic pigments are bacteriochlorophyll *a* and carotenoids of the spirilloxanthin series. **Straight-chain saturated and monounsaturated $C_{18:1}$ and $C_{18:0}$ are the major components of cellular fatty acids, with the former being predominant. Ubiquinones and menaquinones with 10 isoprene units (Q-10 and MK-10) are present as major quinones.**

Preferred growth is photoheterotrophically under anoxic conditions in the light with simple organic substrates as carbon and electron sources. **Chemotrophic growth is possible under oxic conditions in the dark.** Dark fermentative growth may also be possible. Growth factors are required. **Mesophilic, marine bacteria that require NaCl for growth** and grow at seawater salinity and also at slightly higher salt concentrations.

The mol% G + C of the DNA is: 61.5–65.7.

Type species: **Rhodobium orientis** Hiraishi, Urata and Satoh 1995d, 230.

FURTHER DESCRIPTIVE INFORMATION

Cells of *Rhodobium* multiply by budding or asymmetric cell division, and show polar growth (Fig. BXII.α.246), as do other species of phototrophic bacteria in the alpha-2 *Proteobacteria.* They

---

Editorial Note: As this subvolume went to press, *Roseospirillum* was moved from the Family *Rhodobiaceae* to the Family *Rhodospirillaceae.* Placement in either family is tentative.

form sessile buds, but prosthecae between mother and daughter cells, which are characteristic for other phototrophic budding bacteria, are only found in *R. orientis*. Rosette-like cell aggregates are occasionally found in *R. orientis* (Fig. BXII.α.246) but not in *R. marinum*. In particular, *R. marinum* contains very active and agile motile cells.

A quantitative analysis of the carotenoids of *R. marinum* revealed that spirilloxanthin and rhodovibrin constituted 65% and 24%, respectively, of the total content, whereas anhydrorhodovibrin, monodemethylated spirilloxanthin, and lycopene were minor components (Imhoff, 1983). Absorption spectra of living cells or membrane preparations of *R. orientis* show major absorption maxima at around 800 and 870 nm, suggesting that the core light-harvesting complex (LH I), together with the photosynthetic reaction center and a peripheral light-harvesting complex (LH II), are present. Cells and membrane preparations of *R. marinum*, however, exhibit an absorption maximum at 883 nm and have very low absorption at 800 nm, similar to *Rhodospirillum rubrum*. This is due to the presence of LH I as the only antenna complex. The core antenna complex of *R. marinum* consists of 24 polypeptides with 24 bacteriochlorophyll *a* molecules, and the B880 complex exhibits a six-fold symmetry of the ring-like structure (Meckenstock et al., 1994).

Major polar lipid components of *R. marinum* are phosphatidylglycerol, phosphatidylcholine, phosphatidylethanolamine, cardiolipin, and an ornithine lipid. Also unusually high proportions of a sulfolipid, most probably sulfoquinovosyl diglyceride, and an unidentified aminolipid are produced (Imhoff, 1983).

Based on 16S rDNA sequence information, *Rhodobium* species represent a deeply branching line of descent in the group of bacteria known as the alpha-2 *Proteobacteria*, where they are most closely related to representatives of *Rhizobium* and *Sinorhizobium*. 16S rDNA sequence analysis indicates that *R. orientis* and *R. marinum* share 93% similarity, a relatively low value for two species of the same genus. In addition, *R. marinum* has a shorter length

of 16S rDNA than *R. orientis*. This is due to a 21-base deletion in the conserved region of the 16S rRNA between positions 1256 and 1281 (according to the *Escherichia coli* numbering system), like the deletion found in the bacteria known as the alpha-2 group of the *Proteobacteria*. This sequence deletion is a good marker for differentiation of *R. marinum* from *R. orientis* and other species of the group of bacteria known as the alpha-2 *Proteobacteria*.

Rhodobium species occur in marine and hypersaline environments as their natural habitats. Strains of *R. orientis* have been isolated from tidal pools and coastal seashore of Japan. *R. marinum* appears to be one of the most common species of phototrophic purple bacteria in marine environments. Strains of this species have been isolated from diverse sites in the world, including a splash water pond at the Adriatic shore near Split (Yugoslavia), Solar Lake (Sinai), an intertidal flat near Inverary in Scotland (Imhoff, 1983), a microbial mat at Laguna Guerrero Negro in Mexico (Mangels et al., 1986), and surface seawater at the coast of Japan (A. Hiraishi, unpublished).

As with other members of the purple nonsulfur bacteria (PNSB), growth with limited concentrations of ammonium salts or amino acids as nitrogen sources leads to extensive H₂ photoevolution due to nitrogenase activity.

ENRICHMENT AND ISOLATION PROCEDURES

The methods and media used routinely for the cultivation of phototrophic PNSB (Imhoff, 1988; Imhoff and Trüper, 1992) can be employed for isolation and enrichment of *Rhodobium* species, if NaCl is added at appropriate concentrations. Media with 3–5% NaCl and pH 6.8–7.0 are suitable for *Rhodobium* species (Imhoff and Trüper, 1976; Imhoff, 1983; Hiraishi et al., 1995d).

MAINTENANCE PROCEDURES

Cultures are well preserved in liquid nitrogen, by lyophilization, or at −80°C in a mechanical freezer.

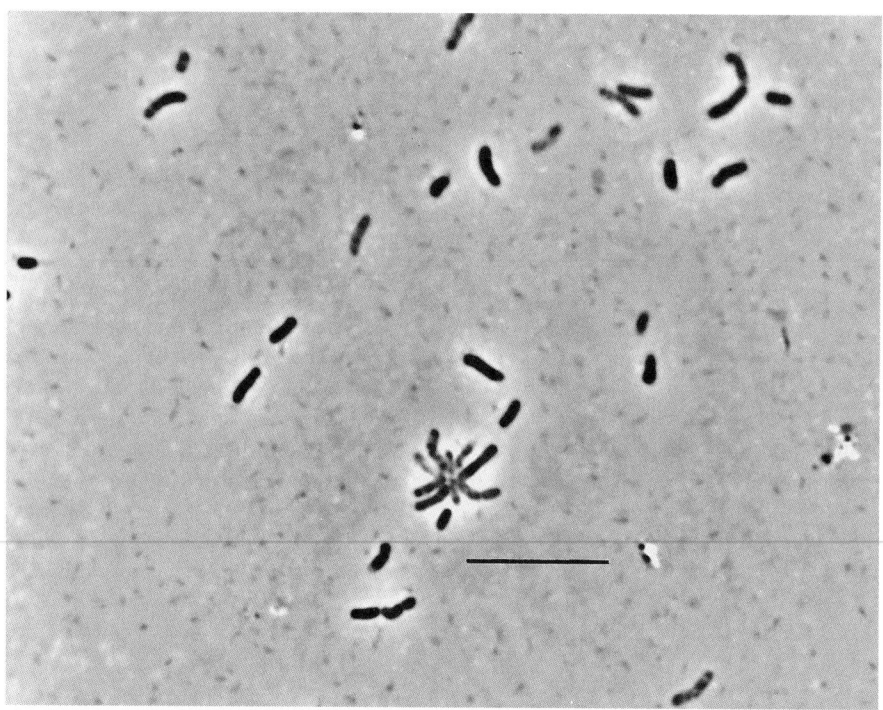

FIGURE BXII.α.246. Phase-contrast photomicrograph showing general cell morphology of *Rhodobium orientis* (strain MB 312). Bar = 10 μm.

DIFFERENTIATION OF THE GENUS *RHODOBIUM* FROM OTHER GENERA

The genus *Rhodobium* consists of species of the PNSB that are truly marine or slightly halophilic bacteria, multiply by budding or asymmetric cell division, and contain internal photosynthetic membrane systems of the lamellar type. In these respects, *Rhodobium* species are similar to other bacteria in the group of phototrophic bacteria belonging to the alpha-2 *Proteobacteria*, such as *Rhodoplanes*, *Blastochloris*, and *Rhodopseudomonas*. The genus *Rhodobium* is differentiated from the latter genera by its phylogenetic position, natural habitats, and chemotaxonomic characteristics. Major differentiating properties between *Rhodobium* and other phototrophic alpha-2 *Proteobacteria* are shown in Tables 3 (pp. 125–126) and 4 (p. 127) of the introductory chapter "Anoxygenic Phototrophic Purple Bacteria", Volume 2, Part A. The phylogenetic relationships of these bacteria based on 16S rDNA sequences are shown in Fig. 2 (p. 128) of that same chapter.

TAXONOMIC COMMENTS

At present, the genus *Rhodobium* consists of two species, *R. orientis* and *R. marinum*. After tentative assignment to *Rhodopseudomonas palustris* (Imhoff and Trüper, 1976), primarily based on morphological and physiological characteristics, the latter species was described as *Rhodopseudomonas marina* (Imhoff, 1983). Later, phylogenetic analysis based on 16S rDNA sequences revealed that this species was far distant from *Rhodopseudomonas palustris*, the type species of the genus *Rhodopseudomonas*, but closely related to new isolates from marine environments (Hiraishi et al., 1995d). These new isolates were assigned to the new genus *Rhodobium* and described as the type species of this genus, *Rhodobium orientis*, and *Rhodopseudomonas marina* was transferred to this genus as *Rhodobium marinum* (Hiraishi et al., 1995d).

Mangels et al. (1986) isolated a N$_2$-fixing marine strain of PNSB and designated it "*Rhodopseudomonas marina* var. *agilis*". Upon reinvestigation, no differences in physiological properties and 16S rDNA sequence between the type strain of *Rhodobium marinum* and "*Rhodopseudomonas marina* var. *agilis*" were found (Hiraishi et al., 1995d), confirming the phenotypic and genotypic coherency of the two organisms at the species level. Therefore, "var. *agilis*" isolates should be considered as strains of *Rhodobium marinum*.

DIFFERENTIATION OF THE SPECIES OF THE GENUS *RHODOBIUM*

Major differentiating properties between *Rhodobium* species are shown in Tables BXII.α.193 and BXII.α.194.

List of species of the genus Rhodobium

1. **Rhodobium orientis** Hiraishi, Urata and Satoh 1995d, 230^{VP}
   o.ri.en' tis. L. part. adj. *orientis* of the orient.

   Cells are ovoid to rod-shaped, 0.7–0.9 × 1.5–3.2 µm, motile by means of polar, subpolar, and randomly distributed flagella; cells multiply by budding and asymmetric cell division (Fig. BXII.α.246). Internal photosynthetic membranes are present as lamellae underlying and parallel to the cytoplasmic membrane (Fig. BXII.α.247). Phototrophically grown liquid cultures are pink-to-red, whereas aerobically grown or denitrifying cultures are faintly pink or colorless. Absorption maxima of living cells are at 377, 468, 500, 530, 591, 802, and 870 nm. Photosynthetic pigments are bacteriochlorophyll *a* (esterified with phytol) and carotenoids of the spirilloxanthin series.

   Preferred mode of growth is photoheterotrophically in the light. Photoautotrophic growth with thiosulfate, but not sulfide, as the electron donor occurs in the presence of 0.01% yeast extract. Chemotrophic growth occurs under oxic conditions in the dark at the full oxygen tension of air, and by denitrification under anoxic dark conditions with nitrate as a terminal electron acceptor. Good growth is possible with acetate, lactate, pyruvate, succinate, fumarate, malate, fructose, glucose, peptone, and yeast extract as carbon sources. Moderate growth occurs with butyrate, valerate, caproate, xylose, mannitol, and sorbitol. Formate, propionate, caprylate, citrate, tartrate, benzoate, arabinose, galactose, mannose, glycerol, methanol, ethanol, and Casamino acids are not utilized. Ammonium salts are used as nitrogen sources. Sulfate and thiosulfate are assimilated. Hydrolytic activities against starch, gelatin, casein, and Tween 80 are absent. Poly-β-hydroxybutyrate granules are formed as a storage material. Growth is stimulated considerably by addition of 0.05% yeast extract. Biotin and *p*-aminobenzoic acid are required as growth factors. Mesophilic, marine bacterium with optimal growth at 30–35°C, pH 7.0–7.5 (pH range: 6.0–8.5), and in the presence of 4–5% NaCl (salinity range: 2–8% NaCl). No growth in the absence of NaCl.

   Habitat: tidal seawater pools and similar marine environments.

   Major quinone components are Q-10 and MK-10.
   The mol% G + C of the DNA is: 65.2– 65.7 (HPLC).
   Type strain: MB312, ATCC 51972, DSM 11290, JCM 9337.
   GenBank accession number (16S rRNA): D30792.

2. **Rhodobium marinum** (Imhoff 1984b) Hiraishi, Urata and Satoh 1995d, 320^{VP} (*Rhodopseudomonas marina* Imhoff 1984b, 270.)
   ma.ri' num. L. adj. *marinum* of the sea, marine.

   Cells are ovoid to rod-shaped, 0.7–0.9 × 1.0–2.5 µm, multiply by budding cell division without prostheca, and are highly motile by means of randomly distributed flagella. Rosette formation is not observed. Internal photosynthetic membranes are present as lamellae underlying and parallel to the cytoplasmic membrane. Phototrophically or fermentatively grown liquid cultures are pink-to-red, whereas aerobically grown cultures are faintly pink or colorless. Absorption maxima of living cells are at 375, 483, 516, 533, 590, 803, and 883 nm, with only a low absorption maximum around 800 nm. Photosynthetic pigments are bacteriochlorophyll *a* (esterified with phytol) and carotenoids of the spirilloxanthin series with spirilloxanthin as the dominant component.

   Cells preferably grow photoheterotrophically under anoxic conditions in the light. Growth under microoxic dark conditions is also possible. Photoautotrophic growth with sulfide as a photosynthetic electron donor is poor, but the addition of 0.01% yeast extract enhances growth and sulfide tolerance. No photoautotrophic growth occurs with thio-

TABLE BXII.α.193. Differential characteristics of the species of the genus *Rhodobium*[a]

Characteristic	*Rhodobium orientis*	*Rhodobium marinum*
Cell diameter (μm)	0.7–0.9	0.7–0.9
Type of budding	Sessile	Sessile
Rosette formation	− / +	−
Internal membrane system	Lamellae	Lamellae
Motility	+	+
Color of cultures	Pink to red	Pink to red
Bacteriochlorophyll	*a*	*a*
Salt requirement	4–5%	1–5%
Optimal pH	7.0–7.5	6.9–7.1
Optimal temperature	30–35	25–30
Sulfate assimilation	nd	nd
Aerobic dark growth	+	(+)
Denitrification	+	−
Fermentation of fructose	−	+
Photoautotrophic growth with	Thiosulfate	Sulfide
Growth factors	Biotin, *p*-aminobenzoic acid	nd
Utilization of:		
Benzoate	−	−
Citrate	−	+ / −
Formate	−	(+)
Glucose	+	+
Tartrate	−	−
Sulfide	−	(+)
Thiosulfate	+	−
Mol % G + C of the DNA	65.2–65.7 (HPLC)	62.4–64.1 (HPLC)
Cytochrome c_2 size	nd	nd
Major quinones	Q-10, MK-10	Q-10, MK-10
Major fatty acids		
$C_{14:0}$		0.4
$C_{16:0}$		1.9
$C_{16:1}$		0.5
$C_{18:0}$		14.1
$C_{18:1}$		69.0

[a] +, positive in most strains; −, negative in most strains; + / −, variable in different strains; nd, not determined; (+), weak growth or microaerobic growth only; (APS), via adenosine-5′-phosphosulfate; (PAPS), via 3′-phosphoadenosine-5′-phosphosulfate; (biotin) biotin is required by some strains; Q-10, ubiquinone 10; MK-10, menaquinone 10; RQ-10, rhodoquinone 10. Bd, buoyant density.

sulfate as the electron donor. Fermentative growth in the dark occurs with fructose as substrate. Denitrification is not possible. All or most strains use the following compounds as electron donors and carbon sources: acetate, propionate, butyrate, valerate, caproate, lactate, pyruvate, succinate, fumarate, malate, ascorbate, fructose, glucose, sucrose, mannitol, sorbitol, ethanol, propanol, and Casamino acids. Weak growth occurs with formate, galactose, and glycerol. Benzoate, tartrate, and methanol are not utilized. Sulfate is assimilated and, in addition, thiosulfate, tetrathionate, cysteine, methionine, glutathione, sulfite, and sulfide can serve as assimilatory sulfur sources. Ammonia, glutamate, aspartate, alanine, and urea serve as nitrogen source. Hydrolytic activities against starch, gelatin, casein, and Tween 80 are absent. Growth is stimulated considerably by adding 0.01% yeast extract.

Mesophilic marine bacterium with optimal growth at 25–30°C, pH 6.9–7.1, and 1–5% NaCl. No growth in the absence of NaCl.

Habitats: marine and moderately hypersaline waters and coastal sediments.

Major quinone components are Q-10 and MK-10.

The mol% G + C of the DNA is: 61.5–64.1 (T_m), 62.4–64.1 (HPLC).

Type strain: BN 126, ATCC 35675, DSM 2698.
GenBank accession number (16S rRNA): D30790.

TABLE BXII.α.194. Growth substrates of the anoxygenic phototrophic purple bacteria belonging to the genus *Rhodobium*[a]

Source/donor	*Rhodobium orientis*	*Rhodobium marinum*
Carbon source		
Acetate	+	+
Aspartate	nd	nd
Benzoate	−	−
Butyrate	+	+
Caproate	+	+
Caprylate	−	+ / −
Citrate	−	−
Ethanol	−	+ / −
Formate	−	+
D-Fructose	+	+
Fumarate	+	+
D-Glucose	+	+
Glutamate	+ / −	nd
Glycerol	−	+ / −
Glycolate	nd	nd
Lactate	+	+ / −
Malate	+	+
Malonate	nd	nd
Mannitol	+ / −	+
Methanol	−	−
Propanol	−	+ / −
Propionate	−	+ / −
Pyruvate	+	+
Sorbitol	+	+
Succinate	+	+
Tartrate	−	−
Valerate	+	+
Electron donor		
Sulfide	−	+ / −
Thiosulfate	+	−

[a] Symbols: +, positive in most strains; −, negative in most strains; + / − variable in different strains; nd, not determined.

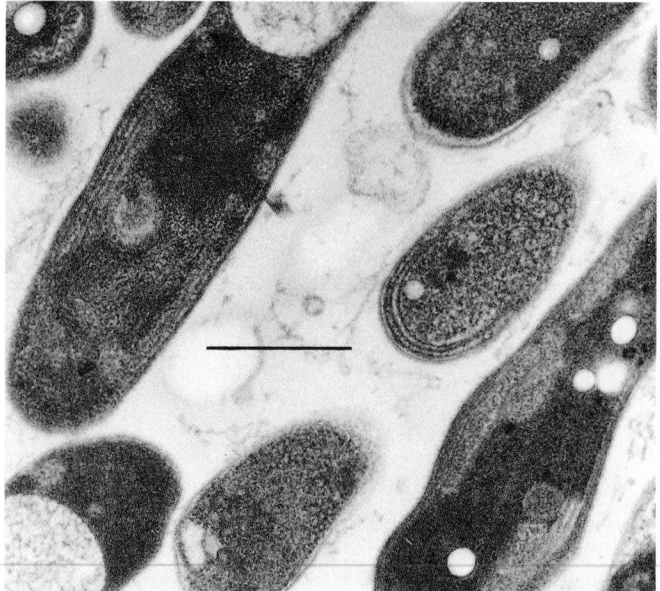

FIGURE BXII.α.247. Thin-section electron micrograph of *Rhodobium orientis* (strain BM 312) showing internal photosynthetic membranes of lamellar type (indicated by *arrow*). Bar = 0.5 μm. (Reprinted with permission from A Hiraishi et al., International Journal of Systematic Bacteriology *45:* 226–234, 1995, ©International Union of Microbiological Societies.)

Class II. **Betaproteobacteria** *class. nov.*

GEORGE M. GARRITY, JULIA A. BELL AND TIMOTHY LILBURN

Be.ta.pro.te.o.bac.te'ri.a. Gr. n. *beta* name of second letter of Greek alphabet; Gr. n. *Proteus* ocean god able to change shape; Gr. n. *bakterion* a small rod; M.L. fem. pl. n. *Betaproteobacteria* class of bacteria having 16S rRNA gene sequences related to those of the members of the order *Spirillales.*

The class *Betaproteobacteria* was circumscribed for this volume on the basis of phylogenetic analysis of 16S rRNA sequences; the class contains the orders *Burkholderiales, Hydrogenophilales, Meth-ylophilales, Neisseriales, Nitrosomonadales, "Procabacteriales"*, and *Rhodocyclales.*

Type order: **Burkholderiales** *ord. nov.*

Order I. **Burkholderiales** *ord. nov.*

GEORGE M. GARRITY, JULIA A. BELL AND TIMOTHY LILBURN

Burk.hol.de.ri.a'les. M.L. fem. n. *Burkholderia* type genus of the order; *-ales* ending to denote order; M.L. fem. n. *Burkholderiales* the *Burkholderia* order.

The order *Burkholderiales* was circumscribed for this volume on the basis of phylogenetic analysis of 16S rRNA sequences; the order contains the families *Burkholderiaceae, Oxalobacteraceae, Alcaligenaceae,* and *Comamonadaceae.*

Order is phenotypically, metabolically, and ecologically diverse. Includes strictly aerobic and facultatively anaerobic chemoorganotrophs; obligate and facultative chemolithotrophs; nitrogen-fixing organisms; and plant, animal, and human pathogens.

Type genus: **Burkholderia** Yabuuchi, Kosako, Oyaizu, Yano, Hotta, Hashimoto, Ezaki and Arakawa 1993, 398 (Effective publication: Yabuuchi, Kosako, Oyaizu, Yano, Hotta, Hashimoto, Ezaki and Arakawa 1992, 1268) emend. Gillis, Van, Bardin, Goor, Hebbar, Willems, Segers, Kersters, Heulin and Fernandez 1995, 286.

Family I. **Burkholderiaceae** *fam. nov.*

GEORGE M. GARRITY, JULIA A. BELL AND TIMOTHY LILBURN

Burk.hol.de.ri.a'ce.ae. M.L. fem. n. *Burkholderia* type genus of the family; *-aceae* ending to denote family; M.L. fem. pl. n. *Burkholderiaceae* the *Burkholderia* family.

The family *Burkholderiaceae* was circumscribed for this volume on the basis of phylogenetic analysis of 16S rRNA sequences; the family contains the genera *Burkholderia* (type genus), *Cupriavidus, Lautropia, Limnobacter, Pandoraea, Paucimonas, Polynucleobacter, Ralstonia,* and *Thermothrix. Limnobacter* was proposed after the cutoff date for inclusion in this volume (June 30, 2001) and is not described here (see Spring et al. (2001).

Family is phenotypically, metabolically, and ecologically diverse. Includes both strictly aerobic and facultatively anaerobic chemoorganotrophs and obligate and facultative chemolithotrophs.

Type genus: **Burkholderia** Yabuuchi, Kosako, Oyaizu, Yano, Hotta, Hashimoto, Ezaki and Arakawa 1993, 398 (Effective publication: Yabuuchi, Kosako, Oyaizu, Yano, Hotta, Hashimoto, Ezaki and Arakawa 1992, 1268) emend. Gillis, Van, Bardin, Goor, Hebbar, Willems, Segers, Kersters, Heulin and Fernandez 1995, 286.

Genus I. **Burkholderia** *Yabuuchi, Kosako, Oyaizu, Yano, Hotta, Hashimoto, Ezaki and Arakawa 1993, 398[VP] (Effective publication: Yabuuchi, Kosako, Oyaizu, Yano, Hotta, Hashimoto, Ezaki and Arakawa 1992, 1268) emend. Gillis, Van, Bardin, Goor, Hebbar, Willems, Segers, Kersters, Heulin and Fernandez 1995, 286**

NORBERTO J. PALLERONI

Burk.hol.de'ri.a. M.L. fem. n. *Burkholderia* named after W.H. Burkholder, American bacteriologist who discovered the etiological agent of onion rot.

---

*Editorial Note: The literature search for the chapter on *Burkholderia* was completed in January, 2000. During the course of unavoidable publication delays, a number of new species were described or reclassified after the chapter was completed. It was not possible to include these species in the text or to include their characteristics in the comparative tables. The reader is encouraged to consult the studies listed in the Further Reading section.

Cells single or in pairs, **straight or curved rods**, but not helical. Dimensions, generally 0.5–1 × 1.5–4 μm. Motile by means of one or, more commonly, **several polar flagella.** One species (*Burkholderia mallei*) lacks flagella and is nonmotile. Do not produce sheaths or prosthecae. No resting stages are known. **Gram negative.** Most species **accumulate poly-β-hydroxybutyrate (PHB) as**

carbon reserve material. Chemoorganotrophs. Have a strictly **respiratory type of metabolism with oxygen as the terminal electron acceptor. Some species can exhibit anaerobic respiration with nitrate.** Strains of some of the species (*B. cepacia, B. vietnamiensis*) are able to fix N_2. Catalase positive. A wide variety of organic compounds can be used as sources of carbon and energy for growth. Although hydroxylated fatty acids are present in the lipids of members of other genera of aerobic pseudomonads, species of *Burkholderia* are **characterized by the presence of hydroxy fatty acids of 14, 16, and 18 carbon atoms ($C_{14:0\ 3OH}$ and $C_{16:0}$, and $C_{16:0\ 2OH}$, $C_{16:1}$, and $C_{18:1}$). The most characteristic of these acids is the $C_{16:0\ 3OH}$.** Two different ornithine lipids are present in strains of some of the species. Over one-half of the species are **pathogenic for plants or animals (including humans)**. The genus belongs to the ribosomal RNA similarity group II, which can be differentiated from other groups of aerobic pseudomonads by rRNA/DNA hybridization experiments or by rDNA sequencing.

The mol% G + C of the DNA is: 59–69.6.

Type species: **Burkholderia cepacia** (Palleroni and Holmes 1981) Yabuuchi, Kosako, Oyaizu, Yano, Hotta, Hashimoto, Ezaki and Arakawa 1993, 398 (Effective publication: Yabuuchi, Kosako, Oyaizu, Yano, Hotta, Hashimoto, Ezaki and Arakawa 1992, 1271) (*Pseudomonas cepacia* Palleroni and Holmes 1981, 479.)

FURTHER DESCRIPTIVE INFORMATION

Cell morphology The cells of the genus *Burkholderia* correspond in their general characteristics to those of other aerobic pseudomonads: Gram-negative rods, straight or slightly curved, with rounded ends, usually motile when suspended in liquid. Motility is due to several polar flagella, but a single flagellum per cell has been reported for *B. andropogonis, B. glathei,* and *B. norimbergensis.** The single flagellum of *B. andropogonis* is sheathed (Fuerst and Hayward, 1969a). One species, *B. mallei,* is nonmotile and lacks flagella (Redfearn et al., 1966).

Intracellular granules The cells of species of the genus accumulate granules of carbon reserve material (poly-β-hydroxybutyrate, PHB, which may be part of a copolymer with poly-β-hydroxyvalerate, PHA). Proteins ("phasins") have been found to be associated with the granules (Wieczorek et al., 1996). Only *B. pseudomallei* and some strains of *B. mallei* use extracellular PHB for growth (Ramsay et al., 1990). All species able to accumulate PHB can use the intracellular polymer when needed; however, in the description of some species, the statement "can use PHB" has been included, without indicating whether the PHB was endo- or extracellular.

The colonies of most species of the genus are smooth, but those of the human pathogen *B. pseudomallei* often have a rough surface.

Lipids The first detailed analysis of the fatty acids of aerobic pseudomonads, using both saprophytic and phytopathogenic strains, demonstrated the possibility of establishing a correlation with the phylogenetic subdivision of the genus *Pseudomonas*, as classically defined, on the basis of rRNA–DNA hybridization (Oyaizu and Komagata, 1983). Years later, the results of this survey were confirmed and extended (Stead, 1992). The results of these investigations firmly established the fact that members of

rRNA similarity group II of Palleroni et al. (1973), which includes the genus *Burkholderia*, have a fatty acid composition containing $C_{14:0\ 3OH}$ and $C_{16:0}$, and $C_{18:1\ 2OH}$. Most strains also contained $C_{16:0\ 2OH}$ and $C_{16:1}$. Even though hydroxylated fatty acids are present in the lipids of members of other groups, group II including *Burkholderia* is the only one having hydroxylated fatty acids of 14, 16, and 18 carbon atoms (Table BXII.β.1). A recent evaluation of the taxonomic significance of fatty acid composition emphasizes the diagnostic value of the above results (Vancanneyt et al., 1996), confirming earlier findings on this approach for the characterization of major phylogenetic groups within the pseudomonads (Wollenweber and Rietschel, 1990).

Ornithine-containing lipids in *Pseudomonas aeruginosa, P. putida,* and *B. cepacia* represent from 2–15% of the total of extractable lipids. The amino acid was not found in the phospholipids that amount to more than 80% of all the extractable lipids (Kawai et al., 1988). An analysis of the polar lipids and fatty acids of *B. cepacia* has shown that the only significant phospholipids in this species are phosphatidyl-ethanolamine and bis(phosphatidyl)-glycerol. These characteristics, taken together with the unusual lipid profiles of *B. cepacia*, can be used as markers of chemotaxonomic importance (Cox and Wilkinson, 1989a).

A striking feature of the cellular composition of *B. cepacia* is the range of polar lipids, which include two forms (with and without 2-OH fatty acids) of phosphatidyl-ethanolamine and ornithine amide lipids. Variations in the lipid composition, as well as in pigmentation and flagellation, were observed as the consequence of changes in growth temperature and limiting oxygen, carbon, phosphorus, and magnesium supplies in the medium. Phosphorus limitation appears to be the only nutritional factor that results in a composition with polar lipids represented only by ornithine amide lipids (Taylor et al., 1998).

Interestingly, the 3-hydroxylated fatty acid of 10 carbon atoms is a component of the lipids of *B. gladioli* but not of those of *B. cepacia*, as indicated by lipid analysis performed on *B. gladioli* strains isolated from respiratory tract infections in cystic fibrosis patients (Christenson et al., 1989).

Differentiation of the plant pathogenic species of *Burkholderia* can be done by a direct colony thin-layer chromatographic

Editorial Note: Since submission of this manuscript, *Burkholderia norimbergensis* was reclassified as *Pandoraea norimbergensis* by Coenye et al. (2000). Readers are advised to review the chapter in that genus for additional details.

TABLE BXII.β.1. Fatty acid and ubiquinone composition of the genus *Burkholderia* (rRNA group II) and of aerobic pseudomonads of other rRNA groups[a,b]

Fatty acids	Ribosomal RNA groups				
	I	II	III	IV	V
3-OH					
$C_{10:0}$	+		+		+
$C_{11:0}$					+
$C_{11:0\ iso}$					+
$C_{12:0}$	+			+	+
$C_{12:0\ iso}$					+
$C_{13:0\ iso}$					+
$C_{14:0}$		+		+	
$C_{16:1}$		+			
2-OH					
$C_{12:0}$	(+)				
$C_{16:0}$		(+)			
$C_{16:1}$		(+)			
$C_{18:1}$		+			
Ubiquinones	Q-9	Q-8	Q-8	Q-10	Q-8

[a]Symbols: +, present; (+), not present in all strains of the group. A blank space means that the compound is not present in any strain of the group.

[b]Data taken from Oyaizu and Komagata (1983) and Stead (1992).

method. Only minor uncertainties have been noticed with respect to the composition of aminolipids (Matsuyama, 1995).

The general qualitative profile of hydroxylated fatty acids indicated above is constant for a given species, although at least in one case (*B. glumae*) a subdivision of strains into two types is possible based on differences in composition. One of the subgroups, which included the type strain, had a composition that was similar to the bulk of rRNA similarity group II. The other was represented by strains that had significant amounts of the $C_{10:0\ 3OH}$ fatty acid, and was unique in rRNA similarity group II in having the $C_{12:0\ 3OH}$ fatty acid (Stead, 1992). For some of the components of the fatty acid profile, significant quantitative variations can be observed among the strains of different species of *Burkholderia*.

In contrast to the hydroxylated fatty acids, which have their origin in lipid A, the more abundant, nonhydroxylated fatty acids are mainly located in the cytoplasmic membrane. Their value as taxonomic markers is significant at the species level and less at the higher level of the RNA similarity groups. All strains of group II have $C_{16:0}$, $C_{16:1\ cis}$, and $C_{18:1\ cis}$ nonhydroxylated fatty acids (Stead, 1992). The investigations of Komagata and his collaborators have established that Q-8 is the quinone characteristic of group II (Oyaizu and Komagata, 1983).

Hopanes have been detected in the composition of cells of *B. cepacia* (Rohmer et al., 1979), but the value of these compounds as chemotaxonomic markers is not known because later studies apparently did not include strains of other species of the group. This point perhaps warrants further attention.

Flagella Motility can be observed in young cultures of strains of all species of *Burkholderia*, with the exception of *B. mallei*, which lacks flagella. Cells of the latter species do not even show twitching motility on the surface of solid media (Henrichsen, 1975a). Motility in liquid is due to one or, more commonly, several polar flagella. A single flagellum per cell has been reported for *B. andropogonis*, *B. glathei*, and *B. norimbergensis* (see descriptions in the list of species at the end of this chapter). The best-known example is that of *B. andropogonis*, whose single flagellum is sheathed (Fuerst and Hayward, 1969a) (Fig. BXII.β.1).

SDS-polyacrylamide gel electrophoresis has been used for the characterization of the flagellins of different species of aerobic pseudomonads. Based on flagellin composition, *B. cepacia* strains have been divided into two groups. Group I has flagellin of molecular weight 31,000, whereas group II flagellin ranges from 44,000–46,000. Type I was serologically uniform, while group II was heterologous. The flagellin types of *B. cepacia* appear to be analogous to the two major flagellin types of *P. aeruginosa*, and they could be used as molecular epidemiological tools (Montie and Stover, 1983).

The methodology for the isolation of *B. pseudomallei* flagellin and its characterization has been described. Electrophoretic analysis under denaturing conditions results in monomer protein bands with an estimated M_r of 43,000 (Brett et al., 1994). O-polysaccharide-flagellin conjugates in this species have been described with respect to their structural and immunological characteristics (Brett and Woods, 1996).

Pili (fimbriae) Peritrichous pili have been identified years ago in *B. cepacia* (Fuerst and Hayward, 1969b). They are thought to facilitate adherence to mucosal epithelial surfaces (Kuehn et al., 1992; Sajjan and Forstner, 1993). Twitching motility is correlated with the presence of polar fimbriae, a correlation that is supported by the fact that this type of motility is absent from *B.*

cepacia, which has no polar fimbriae (Henrichsen, 1975a). No fimbriae have been observed in *B. andropogonis* (Fuerst and Hayward, 1969b).

One or more of five morphologically distinct classes of pili can be present in the cells of *B. cepacia*. Some of the types have been identified in cells of epidemically transmitted strains, and others in environmental isolates (Goldstein et al., 1995; Sajjan et al., 1995). The role of a 22-kDa pilin in binding *B. cepacia* to the mouth epithelial cells has been described (Sajjan and Forstner, 1993). Further details on fimbriae will be given in the section on pathogenesis for humans and animals.

Composition of cell envelope The earliest data on the cell wall composition of *B. cepacia* were obtained in S. Wilkinson's laboratory, where it was found that the major components, myristic, 3-hydroxymyristic, and 3-hydroxypalmitic acids, were indications that members of this group had a different composition than other species of aerobic pseudomonads. These results are in agreement with those of fatty composition of whole cells (Samuels et al., 1973). The core polysaccharide contains glucose, rhamnose, and heptoses, but at most a very low phosphorus content. The lipopolysaccharide (LPS) has a low content of 3-deoxy-D-manno-2-octulosonic acid (KDO), and the side chain is basically a mannan. One added peculiar feature is the presence of an acid-labile amino sugar phosphate presumably associated with the lipid A.

The results have been confirmed, at least in part, by Manniello et al. (1979). Analyses of *B. cepacia* performed by these workers revealed the presence of rhamnose, glucose, heptose, and hexosamine, but no KDO, and the phosphorus content was found to be about one-third of that of *P. aeruginosa*. The presence of KDO in the LPS of *B. cepacia*, however, was later confirmed (Straus et al., 1990).

A method of extraction of LPS from *B. pseudomallei* has been described. As in the case with *B. cepacia*, the link between the inner core and lipid A is stable to acid hydrolysis (Kawahara et al., 1992). The LPS of clinical strains of this last species was composed of two polymers made of different repeating units, but both polymers contained D-rhamnose and D-galactose residues (Cerantola and Montrozier, 1997).

Five major outer membrane proteins have been isolated from several strains of *B. pseudomallei*, with M_r values ranging from 17,000–70,000. One of the proteins associated with the peptidoglycan acts as a porin through which small saccharides may diffuse (Gotoh et al., 1994b).

Outer membrane proteins of *B. cepacia* and *B. pseudomallei* that are inducible by phosphate starvation appear to be similar to the *E. coli* PhoE porin protein. The latter does not have the binding sites for anions and phosphate that are present in the analogous proteins from *Pseudomonas* species (Poole and Hancock, 1986).

One porin of *B. cepacia* is an oligomer composed of two proteins. The purified 81-kDa whole protein (OpcPO) upon heating gives a major 36-kDa protein (OpcP1) and a minor one (OpcP2) of 27 kDa. The association of these two components is noncovalent (Gotoh et al., 1994a). Recently, the major porin protein, (OpcP1), was partially sequenced, and the information was used for cloning the gene. Its sequence showed an open reading frame of a length in agreement with that of OpcP1 (Tsujimoto et al., 1997).

Additional information about porins and other components of the cell envelopes may be found in the sections on antibiotic susceptibility and antigenic structure.

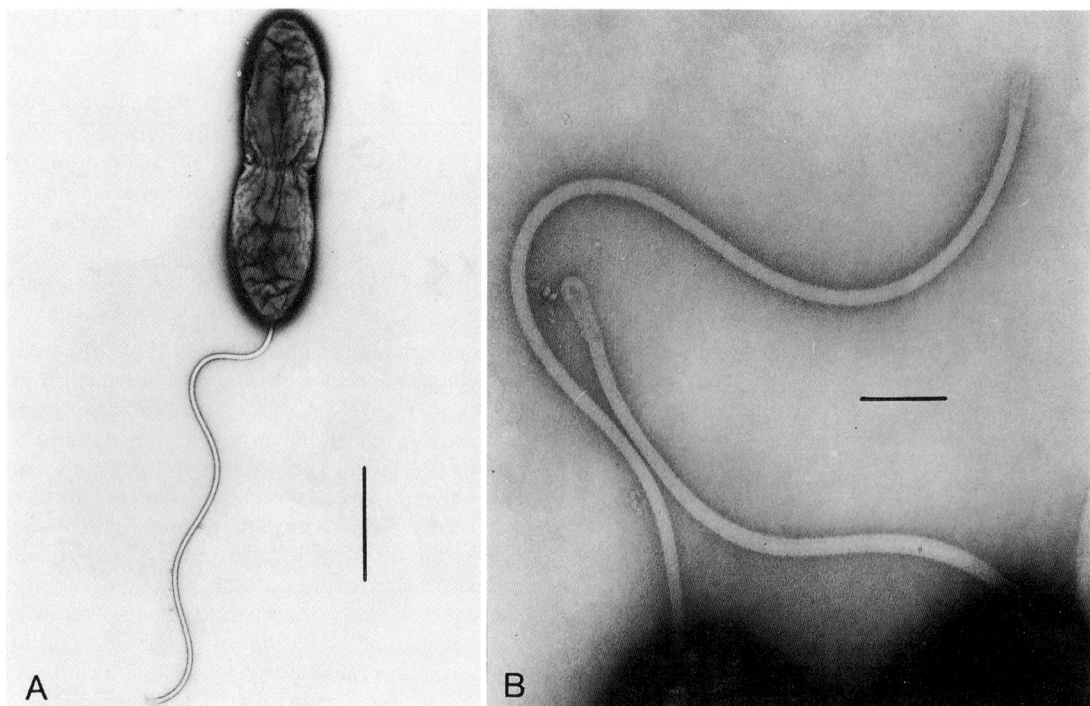

FIGURE BXII.β.1. *A, B. andropogonis* (syn. *"Pseudomonas stizolobii"*), with single-sheathed polar flagellum. Negative staining, 1% uranyl acetate, 0.4% sucrose. Bar = 1 μm. *B*, sheathed flagella of *B. andropogonis*. Same treatment as in *A*. Bar = 0.2 μm. (Courtesy of Dr. J.A. Fuerst.)

Pigmentation   Pigmentation is by no means a universal character of *Burkholderia*. Some *B. cepacia* strains are not pigmented, whereas others produce phenazine pigments of a bewildering variety of colors when grown on solid chemically defined media containing different carbon sources. Pigmented strains of the species can be subdivided into two types on the basis of their pigmentation: some are yellow on glucose yeast extract peptone agar and others are various shades of brown, red, violet, and purple (Morris and Roberts, 1959). Morris and Roberts isolated the pigment from a purple-pigmented strain and demonstrated that its basic structure was that of a phenazine. In fact, two phenazine pigments—one yellow and the other purple—may be synthesized, both of which are water-soluble under neutral or alkaline conditions. A single strain can produce both types, only one type, or none. Because the pigments are soluble in water, both the colonies and the medium appear pigmented. Growth on King A medium[1] often enhances pigment production.

In the author's experience, pigment production in *Burkholderia* is, as in many other cases, a striking but not very reliable taxonomic character, because pigment biosynthesis often requires conditions that cannot be precisely controlled.

Nutrition and growth conditions   The strains of *Burkholderia* species grow in media of minimal composition, without the addition of organic growth factors. Occasionally, strains are isolated from nature that grow extremely slowly but are stimulated by addition of complex organic mixtures such as yeast extract. Some of the fluorescent plant pathogens of the genus *Pseudomonas* fall in this category, and the phenomenon has also been observed for some *B. caryophylli* strains (Palleroni, 1984).

The ability to grow in media of very simple and chemically defined composition stimulated research on the nutritional versatility of the aerobic pseudomonads—among them, species later to be assigned to the genus *Burkholderia*. These studies have revealed a remarkable variety of organic compounds that can serve individually as carbon and energy sources for strains of some species (*B. cepacia, B. pseudomallei*). The types of compounds used for growth are the basis of the vast phenotypic information now available for many strains (Redfearn et al., 1966; Stanier et al., 1966; Ballard et al., 1970). One of the simplest chemically defined media used for nutritional studies is the one recommended for the hydrogen pseudomonads (Palleroni and Doudoroff, 1972).[2]

The nutritional investigations revealed that some strains of *B. cepacia* could utilize any of a list of 100 organic compounds (two-thirds of the list of tested substrates) (Stanier et al., 1966). Later work performed in various laboratories has enlarged the list considerably. This remarkable metabolic versatility of the species was unknown to plant pathologists, and its discovery rapidly converted *B. cepacia* into a fascinating subject for biochemical research. A sample of the nutritional properties of *Burkholderia* species is presented in Table BXII.β.2, and some of these properties, together with general characteristics of taxonomic importance, are summarized in Table BXII.β.3. Comparisons of nutritional properties of some related species are also to be found in later sections (Tables BXII.β.5 and BXII.β.7). To the remarkable metabolic versatility of *B. cepacia* we have to add the ability to fix N₂ that is exhibited by some of the strains (Bevivino et al., 1994;

---

1. Medium of King et al. (1954) (g/l distilled water): Bacto-peptone (Difco), 20.0; Bacto Agar (Difco), 15.0; glycerol, 10.0; K_2SO_4, 10.0; $MgCl_2$, 1.4; pH 7.2.

2. Medium of Palleroni and Doudoroff (1972) (g/l 0.33 M Na-K phosphate buffer, pH 6.8): NH_4Cl, 1.0; $MgSO_4 \cdot 7H_2O$, 0.5; ferric ammonium citrate, 0.05; $CaCl_2$, 0.005. The first two ingredients are added to the buffer and sterilized by autoclaving. The ferric ammonium citrate and $CaCl_2$ are added aseptically from a single stock solution that has been sterilized by filtration.

TABLE BXII.β.2. Utilization of carbon compounds by some *Burkholderia* species[a,b]

Compound[c]	*B. cepacia*	*B. andropogonis*	*B. caryophylli*	*B. cocovenenans*	*B. gladioli*	*B. glathei*	*B. glumae*	*B. graminis*	*B. mallei*	*B. phenazinium*	*B. plantarii*	*B. pseudomallei*	*B. pyrrocinia*	*B. vandii*	*B. vietnamiensis*
Carbohydrates/glycosides:															
N-Acetylglucosamine	+	−	+		+	+	+	+	+	+	+	+	+		+
Amygdalin	+	−	−		−	−	d^-	−		−	+		+		d^+
D-Arabinose	+	+	+		+	+	+	+	+	+	+	+	+		+
L-Arabinose	+	+	+	+	+	+	+	+	d	+	+	−	+		+
Arbutin	+	−			−	−	d^-	−		−	+		−		d^+
Cellobiose	d^+	−	d^-	+	d^+	−	d^+	d	+	−	+	+	+		+
L-Fucose	+	−	+	+	+	+	+	+		+	+		+		+
D-Fucose	+	d^-	+	+	+	+	d^+	d	+	−	+	+	−		+
Gentiobiose	d^+	−	d^-		−	−	d^-	d		−	+		+		+
Glucosamine	+	−	+		+		+				+				+
2-Ketogluconate	+	d^-	+		+	+	+	+	d	+	+	+	+		+
5-Ketogluconate	+	−	+		+	+	d	+		+	+		+		+
Lactose	−	+	−		−	+	−	+	−	−	−		−	+	d^-
D-Lyxose	+	+	+		+	+	+	+			+		+		+
Maltose	d^-	−	−	−	−	−	−	−	d	−	−	+	−		−
Melibiose	d^-	−	−	+	−	−	+	−		−	−		−		−
Raffinose	d^+	−	+	−	−	−	d^+	+		−	−		−		d^+
L-Rhamnose	d	d^+	+		−	+	−	+	−	+	±	−	−	−	d^-
D-Ribose	+	+	+	+	+	−	+	d	−	+	+	+	+		d^-
Salicin	+	−	−	+	−	−	d^-	−	d	−	+	+	−		d^+
Sucrose	d^+	−	+	−	−	+	−	+	+	+	±	+	+	−	+
Tagatose	+	−	−		+	+	d^+	−		−	−		+		+
Trehalose	+	d^-	+	+	+	−	+	+	+	−	−	+	+	+	+
D-Xylose	d^+	−	+	+	+	+	+	+	+	+	+	−	+	+	+
Polyalcohols:															
Adonitol	+	+	+	d^-	+	+	+	+		+	−	d	+	d^+	−
D-Arabitol	+	+	+		+	+	+	+		+	+		+		+
L-Arabitol	+	−	+		−	−	−	+		+	−		+		−
Dulcitol	+	−	−		+	+	+	−		−	+		+		+
Erythritol	−	−	−	−	−	−	−	−		−	−	+	−		−
Xylitol	+	−	+	+	d^+	+	−	+		+	−		+		−
Acids (anions):															
Aconitate	+		+	+	+		+		−		d^+	+			+
Adipate	+	−	−	+	+	+	−		+		−	+			+
Azelate	+	−	±	+	+	+	+		d		±	d	d^+		+
Butyrate	+	−	+	+	+	−	+		d		+	+			+
Caprate	+	−	−		+		+		−		±	+		+	+
Caproate	+	−	−		+		−		−		±	+		d^+	+
Caprylate	+	−	−	+	+		d^-		d		±	+	+	+	+
Citraconate	+	−	−	+	d^-		−		−		±	−		d^+	−
Citrate	+	+	+	+	+	+	+		d		±	+		±	+
Glutarate	+	−	−	−	d^+		−		d		−	d		−	+
Glycolate	+	−	d^-	−	d^-	+	−		−		−	−		−	−
Heptanoate	+	−	−		+		+		−		+	+		d^+	+
Isobutyrate	+	−	+	−	+	−	+		−		+	+			+
Isovalerate	d^+	−	d^-				−				−				+
Itaconate	−	−	−	−	−	+	−		−		−	−		−	+
α-Ketoglutarate	+	+	+	+	+		+		+		−	+			d^+
Levulinate	+	−	−	−	+		−		−		−	+		d	+
Malonate	+	−	−	+	+	+	d^-		d		d^+	−			d^-
Mesaconate	−	−	−	d^+	+		−		−		±	−		+	−
Oxalate	−	−	+	−	−		d^-		−		+	+			+
Pelargonate	+	−	−	+	+	−	d^+		−		+	+			+
Pimelate	+	−	−	d^+			−		d		−	d		d	−
Propionate	+	−	+		+		+		+		+	+			+
Sebacate	+	−	+	+	+	+	+		−		+	+		+	+
Suberate	+	−	d^+		d^+		d^+		d		−	+		±	d^-
Valerate	+	−	d^-		+	−	−		−		−	+			+
D(−)-Tartrate	−	−	−	+	+	+	−		−		−	−		±	−
L(+)-Tartrate	d^+	−	−	+	+	+	−		−		±	−		±	−
m-Tartrate	d^+	−	+	+	+	+	−		−		−	−			d^-

(continued)

TABLE BXII.β.2. *(cont.)*

Compound[c]	B. cepacia	B. andropogonis	B. caryophylli	B. cocovenenans	B. gladioli	B. glathei	B. glumae	B. graminis	B. mallei	B. phenazinium	B. plantarii	B. pseudomallei	B. pyrrocinia	B. vandii	B. vietnamiensis
Amino acids and related compounds:															
β-Alanine	+	−	+	+	+	+	d$^+$		+		−	+			+
DL-2-Aminobutyrate	+	−	−		+		−		d		d$_+$	−		d$^+$	d$^+$
DL-3-Aminobutyrate	+	−	+		+		−				−				d$^+$
DL-Aminovalerate	+	−	−		+		−				−				+
L-Arginine	+	−	+	+	+	+	+		+		d$^-$	+			+
L-Citrulline	+	−	−	d$^+$	+	+	−		−		−	−			d$^+$
L-Cysteine	+	+	+		+	+	+				+				+
Glycine	d$^-$	−	−	−	−	−	−		+		−	−		−	−
L-Histidine	+	−	+	+	+	+	+		+		+	+			+
L-Isoleucine	+	−	+	−	+	−	+		−		d$^-$	+		+	+
DL-Kynurenine	+	d$^-$	d$^-$		+		−		−		−	+			+
L-Leucine	+	−	+		+	+	d$^+$		−		+	−			+
L-Lysine	+	−	d$^-$	+	+	+	−		−		−	+			+
L-Norleucine	+	−	d$^-$	−	+		−		−		±	−		d$^-$	+
DL-Norvaline	+	−	−		+		−				d$^+$				d$^-$
L-Ornithine	+	−	+	−	+	+	−		−		−	+		+	+
L-Phenylalanine	+	−	+	d$^-$	+	+	+		d		+	+			+
L-Threonine	+	−	+	+	+	+	+		+	+	−	+		+	+
L-Tryptophan	+	−	+	−	+	+	+		+		−	+			+
L-Tyrosine	+	−	+	+	+	+	+		+		+	+			+
L-Valine	+	−	+	d$^-$	+	+	+		d		±	+		+	+
Amines:															
α-Amylamine	+	−	+		−		−		−		−	+		−	d$^+$
Benzylamine	d$^+$	−	d$^+$	−	d$^-$		−		−		−	−		d$^-$	+
Betaine	+	−	+		+	+	+		+		+	+			+
Butylamine	+	−	−		−		−				−				−
Diaminobutane	+	−	−		−		−				−				+
Ethanolamine	+	d$^+$	−	+	+		+		−		+	+		−	+
Histamine	d$^+$	−	−	−	−		−		−		−	−			+
Sarcosine	+	−	+	+	+	+	d$^+$		d		−	d		+	+
Spermine	d$^-$	−	−		−	+	−				−				d$^-$
Tryptamine	+	−	−		−		+				−			−	
Aromatic compounds:															
2-Aminobenzoate	+	d$^-$	−		+		+		−		−				+
4-Aminobenzoate	−	−	−	+	−		−		−		−	−	+	+	−
Benzoate	d$^+$	−	−	d$^-$	+	+	−		+		−	+	−	d$^+$	+
m-Hydroxybenzoate	+	−	−	d$^-$	d$^+$	−	−		−		−	−	−	−	−
o-Hydroxybenzoate	d$^+$	−	−		−		+				−	−			−
Phenylacetate	+	−	−		+		−		+		−	+			+
Terephthalate	−	−	−		+		−				−				−

[a]For symbols see standard definitions; ±, slow. The + or − superscripts of the d symbol refer to the result obtained with the type strain, when known.

[b]Data from Gillis et al. (1995), Viallard et al. (1998), and Palleroni (1984).

[c]Carbon compounds used by all strains (with few exceptions; see Gillis et al., 1995) are fructose, fumarate, galactose, gluconate, glucose, mannose, glycerol, *m*-inositol, mannitol, sorbitol, acetate, DL-glycerate, DL-lactate, L-malate, pyruvate, succinate, D-α-alanine, L-α-alanine, DL-aminobutyrate, L-aspartate, L-glutamate, L-serine, L-proline, *p*-hydroxybenzoate. Carbon compounds not used by any strain: glycogen, methyl-mannoside, D-melezitose, inulin, starch, D-turanose, aesculin, creatine, 3-aminobenzoate, D-mandelate, phthalate, isophthalate.

Tabacchioni et al., 1995). This ability is also present in a closely related taxon, *B. vietnamiensis* (Gillis et al., 1995).

Temperature for growth Usually a temperature of 30°C is used for growth of all strains of the genus. Many of them, however, can grow well at 37°C and even at 40°C. Additional information on temperature relationships will be given in the section dealing with the description of individual species.

Oxygen relationships All strains grow well under aerobic conditions. Some species (*B. mallei*, *B. pseudomallei*, *B. caryophylli*, *B. plantarii*, *B. vandii*) can also use nitrate as the terminal electron acceptor under anaerobic conditions. The original description of *B. vietnamiensis* (Gillis et al., 1995) indicates that the strains are able to reduce nitrate to nitrite, suggesting that this is the

final stage of the reduction process. However, *B. vietnamiensis* is described elsewhere as a denitrifier. In view of this, the species was omitted from Table 4, which presents the general properties of denitrifying aerobic pseudomonads. In spite of this, denitrification was included among the general properties useful as differential characteristics for *Burkholderia* species.

Metabolism and metabolic pathways Knowledge of many areas of the general metabolism of species of the genus *Burkholderia* is scanty and fragmentary. The following refers to some aspects of the general metabolism of the genus, with the exclusion of the catabolism of aromatic compounds, which will be treated in a second part.

Extensive nutritional information is available on a large num-

TABLE BXII.β.3. Characteristics useful for the differentiation of some *Burkholderia* species [a]

Characteristic	B. cepacia	B. andropogonis	B. caryophylli	B. cocovenenans	B. gladioli	B. glathei	B. glumae	B. mallei	B. plantarii	B. pseudomallei	B. vandii	B. vietnamiensis
Flagellar number	>1	1	>1	>1	>1	1	>1	0	>1	>1	>1	>1
Diffusible pigments[b]	+	−	+	+	+	−	+	−	+	−	−	−
Arginine dihydrolase	−	−	+	−	−	−	+	+	−	+	−	−
Denitrification	−	−	+	+	−	−	+	+	+	+	+	+
Growth at 40°C	+	−	−	+	+	+	+	+	+	+	−	+
Gelatin hydrolysis	d	−	−	+	+	−	+	+	+	+	+	+
Starch hydrolysis	−	w	−	−	−	−	d	d	−	+	−	−
Extracellular PHB hydrolysis	−	−	−		−	−	d		−		−	
Oxidase reaction	d	−	−	−	d	+	d	+	+	+	+	+
Growth on:												
Adonitol	+	+	+	d−	+	d	+	−	−	d	d+	d+
α-Amylamine	+	−	−	−	−		−	−	−	+	−	d+
Citraconate	+	−	−	d+			+	−	w	−	d+	−
Erythritol	−	−	−	−	d−	−	−	−	−	+	−	−
m-Hydroxybenzoate	+	−	−	d−	d−		−	−	−	−	d−	+
Levulinate	+	−	−	d+	+	+	−	−	−	+	+	+
Mesaconate	−	−	+	−	−	+	−	−	w	−	−	−
L-Rhamnose	d	d+	+	d+	−	+	−	−	w	−	−	−
D-Ribose	+	+	+	+	+	+	+	−	+	+	−	−
D(−)-Tartrate	−	−	−	+	+	+	−	−	−	−	w	d−
meso-Tartrate	d+	−	+	+	−	+	+	+	−	−	−	d−
Tryptamine	+	−	−	−	−		+		−	−	+	
D-Xylose	d+	−	+	+	+	+	+	+	+	−	+	+

[a] For the symbols and abbreviations see Table BXII.β.1. Data for *B. graminis*, *B. multivorans*, *B. norimbergensis*, *B. phenazinium*, *B. pyrrocinia*, and *B. thailandensis* have not been included since, aside from the general phenotypic features, for most of the characters in the table there is no information in the original sources.

[b] Strains of *B. cepacia* may produce nonfluorescent pigments of various colors; strains of *B. gladioli* and of *B. caryophylli* may excrete yellow-green nonfluorescent pigments; strains of *B. cocovenenans* produce greenish-yellow diffusible pigments.

TABLE BXII.β.4. Characteristics useful for differentiation of denitrifying aerobic pseudomonads[a]

Characteristics	Burkholderia caryophylli	B. mallei	B. plantarii	B. plantarii	B. vandii	Hydrogenophaga pseudoflava	Pseudomonas aeruginosa	P. alcaligenes	P. balearica	P. fluorescens and P. chlororaphis	P. mendocina	P. pseudoalcaligenes	P. stutzeri	Ralstonia solanacearum	R. pickettii
RNA group	II	II	II	II	II	III	I	I	I	I	I	I	I	II	II
Mol% G + C of DNA	65.3	69	64.8	69.5	68.5	66.5–68	67.2	64–68	64.1–64.4	59.4	62.8–64.3	62–64	60.7–66.3	66.5–68	64
Number of flagella	>1	0	>1	>1	>1	1	1	1	1	1	1	1	1	>1	1
PHB accumulation	+	+	+	+	+	+	−	−	−	−	−	d	−	+	+
H₂ autotrophy	−	−	−	−	−	−	−	−	−	−	−	−	−	−	−
Growth at 40°C	+	+	−	+	−	+	+	+	+	−	+	+	+	−	+
Gelatin liquefaction	−	+	+	+	+	−	+	d	−	+	−	d	−	−	−
Fluorescent pigment	−	−	−	−	−	−	+	−	−	+	−	−	−	−	−
Pyocyanin production	−	−	−	−	−	−	+	−	−	−	−	−	−	−	−
Yellow cellular pigment	−	−	−	d	−	+	−	d	−	−	+	−	−	−	−
Arginine dihydrolase	+	+	−	+			+	+	−	+	+	d	−	−	−
Starch hydrolysis	−	d	−	+	−		−	−	+	−	−	−	+	−	
Extracellular PHB hydrolysis	−	d	−	+											
Growth on:															
ʟ-Arginine	+	+	d-	+			+	+	d	+	+	+	−	−	−
Azelate	−	d	w	d	d		+	−		d	+	−	−	d	+
Betaine	+	+	+	+			−	−	−	+	+	+	−	−	−
2,3-Butylene glycol	+			−		−	+	−		d	−	−	d		d
Ethylene glycol	−	−	−	−		−	−	−		−	+	d	+	−	−
Geraniol	−		−	−			+	−		−	+	−	−	−	−
Glycolate	+	−	−	−	−	d	−	−	−	−	+	−	+	d	+
ʟ-Histidine	+	+	+	+		+	+	d	−	+	+	d	−	+	d
Levulinate	−	−	−	+	d		+	−	−	d	+	−	−	d	+
Maltose	−	d	−	+		+	−	−	+	−	−	−	+	−	−
Mannitol	+	+	+	+	+	+	+	−	+	−	+	−	d	d	−
Saccharate	+	−	−		d		−	−	d	+	−	d	+	+	+
Sarcosine	−	d	−	d	+	−	+	−		+	+	d	−	d	−
ʟ-Serine	+	d	+	+			d	−	−	d	+	d	d	d	+
ᴅ-Xylose	+	+	+	−	+	+	−	−	+	d	−	−	−	−	+

[a]For symbols see standard definitions. The denitrifying pseudomonads *P. azotoformans*, *P. mucidolens*, and *P. nitroreducens* have not been included in this table.

ber of carbon substrates used individually as sole sources of carbon and energy for *B. cepacia*, *B. gladioli*, *B. caryophylli*, *B. pseudomallei*, and *B. mallei* (Redfearn et al., 1966; Stanier et al., 1966; Ballard et al., 1970; Palleroni and Holmes, 1981; Palleroni, 1984). In some of these reports, the information refers to names that are synonyms of some of the above (*"Pseudomonas marginata"* for *B. gladioli*, and *"Pseudomonas multivorans"* for some of the strains of *B. cepacia*). The information has been summarized in several tables in this chapter.

For *B. vietnamiensis* and *B. andropogonis*, a source of information on nutritional spectra useful for a comparison with other species of the genus, is given by Gillis et al. (1995). Less extensive surveys have been performed with *B. glumae*, *B. vandii*, and *B. cocovenenans* (Azegami et al., 1987; Urakami et al., 1994; Zhao et al., 1995).

As mentioned in the section on nutrition, some of the *Burkholderia* species are extremely versatile from the metabolic standpoint. This is particularly true of *B. cepacia*, the most versatile of all known aerobic pseudomonads, but *B. pseudomallei* and *B. gladioli* are similarly remarkable for their capacity of living at the expense of any of a long list of organic compounds as sole source of carbon and energy. Unfortunately, these species also have an infamous reputation as direct or opportunistic human and animal pathogens, and investigations on the saprophytic activities of some of the species are sparse because of the danger of handling the organisms in the laboratory.

All species of the genus *Burkholderia* that have been examined are able to accumulate PHB as a carbon reserve material, which they can degrade when nutrients in the medium become exhausted. However, use of exogenous PHB is an uncommon property among the aerobic pseudomonads. In *Burkholderia*, only *B. pseudomallei* and some strains of *B. mallei* are capable of degrading extracellular PHB (Redfearn et al., 1966; Stanier et al., 1966).

Table BXII.β.5 summarizes data on arginine utilization and the occurrence of arginine deiminase in some representative members of the various rRNA similarity groups of aerobic pseudomonads. Interestingly, *B. cepacia*—a member of RNA group II that is notorious for its nutritional versatility and for the diversity of its catabolic pathways—can degrade arginine only through the use of the succinyl transferase pathway, although it can use 2-ketoarginine and agmatine, the products of arginine oxidase and arginine decarboxylase, respectively (Stalon and Mercenier, 1984; Vander Wauven and Stalon, 1985).

In a survey of lysine catabolic pathways in the pseudomonads,

TABLE BXII.β.5. Arginine utilization and deiminase system in some aerobic pseudomonads

Organisms	RNA group	Arginine utilization	Arginine deiminase
Fluorescent saprophytic *Pseudomonas* species	I	+	+
Burkholderia cepacia, B. gladioli	II	+	−
B. mallei, B. pseudomallei	II	+	+
Ralstonia solanacearum, R. pickettii	II	−	−
Comamonas, Hydrogenophaga, Acidovorax	III	−	−
Brevundimonas	IV	−	−
Stenotrophomonas, Xanthomonas	V	−	−
Stenotrophomonas, Xanthomonas	V	−	−

it has been reported that *B. cepacia* and *P. aeruginosa* use the pipecolate pathway and not the so-called oxygenase pathway. *P. aeruginosa* can also use the cadaverine pathway, but this is not operative in *B. cepacia* (Fothergill and Guest, 1977; Palleroni, 1984). These facts are summarized in Table 1 (BXII.γ.108, p. 334) of the genus *Pseudomonas* in Volume 2, Part B.

N_2 fixation has been detected in strains of *B. cepacia* isolated from plant rhizospheres (Bevivino et al., 1994; Tabacchioni et al., 1995). A different set of nitrogen fixing strains studied in another laboratory was found to be closely related to this species, and was assigned the new species name *B. vietnamiensis* (Gillis et al., 1995).

Some miscellaneous activities of interest of the lesser-known species of the genus include the following. An α-terpineol dehydratase—capable of converting the citrus compound limonene to α-terpineol—was partially solubilized from a particulate fraction obtained from *B. gladioli* cells (Cadwallader et al., 1992). A lipase from *B. glumae* has been purified and some of its properties have been described (Deveer et al., 1991; Cleasby et al., 1992). Cloning and sequencing of a lipase gene of *B. cepacia* (*lipA*) have been performed, and its expression is dependent on a second gene (*limA*) (Jorgensen et al., 1991).

In cells of *B. caryophylli*, a D-threo-aldolase dehydrogenase that catalyzes the oxidation of L-fucose and the "unnatural" sugars L-glucose, L-xylose, and D-arabinose, has been purified and characterized. It is inhibited by D-glucose and other natural aldoses (Sasajima and Sinskey, 1979). Some enzymatic activities of species other than *B. cepacia* may be of environmental importance. Thus, *B. pseudomallei* is capable of breaking the C-P bond in the utilization of the herbicide *N*-(phosphonomethyl)-glycine, which is known by the empirical name glyphosate. The genes controlling this activity have been cloned and sequenced (Peñaloza-Vazquez et al., 1995). Similarly, the phytopathogen and human opportunistic pathogen *B. gladioli* was found to be able to form and cleave C-P bonds (Nakashita et al., 1991; Nakashita and Seto, 1991). In a more general way, it has been suggested that *B. gladioli* probably participates in the degradation of some xenobiotic compounds in the environment (Cadwallader et al., 1992).

Several siderophores are produced by species of *Burkholderia*. In iron-deficient media, *Pseudomonas aeruginosa, P. fluorescens*, and *B. cepacia* synthesize salicylic acid, a compound of particular interest because of its siderophore capacity (Visca et al., 1993) and the fact that it is a precursor of another siderophore, pyochelin. Interestingly, salicylate is used as a source of carbon and energy by many strains of *Burkholderia* and of many other aerobic pseudomonads. An iron-binding compound produced by the great majority of *B. cepacia* strains isolated from the respiratory tract

was named azurechelin and is capable of releasing Fe from transferrins (Sokol et al., 1992). Later, however, azurechelin was found to be salicylic acid (Visca et al., 1993).

Pyochelin production was found in half of 43 strains of *B. cepacia* isolated from cystic fibrosis patients. The siderophore has in its structure one molecule of salicylic acid and two molecules of cysteine (Cox et al., 1981). A siderophore of a linear hydroxamate/hydroxycarboxylate type, was discovered in *B. cepacia* and also found in *B. vietnamiensis*. It was named ornibactin and functions as a specific iron-transport system equivalent to that of the pyoverdin system of fluorescent pseudomonads. In the composition of ornibactins, there is a peptide, an amine, and acyl groups of different lengths (Stephan et al., 1993a, b).

In nature, these siderophores are often successful in competing for iron with siderophores of various other sources (Yang et al., 1993). Thus, pyochelin allows *B. cepacia* to grow in the presence of transferrin (Sokol, 1986).

In low Fe-content medium, *B. cepacia* excretes both pyochelin and a low molecular weight compound (1-hydroxy-5-methoxy-6-methyl-2(1H)-pyridinone), which received the name of cepabactin. The structure resembles that of a cyclic hydroxamate, and it can also be considered a heterocyclic analogue of catechol (Meyer et al., 1989). The compound is related to synthetic hydroxy-pyridinones. Cepabactin had already been described as chelator (Winkler et al., 1986) and was known to have antibiotic properties (Itoh et al., 1979).

At least three different siderophores—ornibactins, pyochelin, and cepabactin—can be produced by a single *B. cepacia* strain. Other strains produce either two siderophores—ornibactins plus pyochelin or ornibactins plus cepabactin—and still other strains produce only pyochelin. In a survey of strains from a collection, 88% of the strains produced ornibactin, 50% pyochelin, and 14% cepabactin (J.M. Meyer, personal communication).

A recently identified member of the siderophore group produced by a strain of *B. cepacia* is cepaciachelin, a catecholate compound. The producing strain appears to be a unique example of a pseudomonad able to synthesize both hydroxamate and catecholate siderophores (Barelmann et al., 1996).

In summary, the variety of siderophores that *B. cepacia* is able to synthesize demonstrates once again the remarkable biochemical versatility of strains of this species.

All 84 strains of *B. pseudomallei* included in a study were found to produce a siderophore of approximately 1000 molecular weight. The compound, called malleobactin (Yang et al., 1991a), permitted cell growth in the presence of EDTA and of transferrin, and can trap iron from transferrin (at all pH values tested) and from lactoferrin. *B. cepacia* can use malleobactin as an iron-scavenging compound. However, pyochelin and azurechelin (salicylic acid) are more effective in trapping cell-derived iron as well as protein-bound iron (Yang et al., 1993).

Media with limiting phosphate concentrations enhance the production of a particular OM protein (OprP) of *Pseudomonas aeruginosa*. This protein is believed to be involved in phosphate transport and is only produced by species of *Pseudomonas*. Under similar conditions, both *B. cepacia* and *B. pseudomallei* produce proteins of a size similar to the PhoE protein of *Escherichia coli* and other enteric bacteria (Poole and Hancock, 1986).

Metabolism of aromatic and halogenated compounds For many years this subject has attracted the attention of biochemists because of the striking ability of prokaryotes to degrade aromatic compounds that are resistant to chemical attack and are not readily metabolized by other organisms. In recent years, this interest has increased due to concern for chemically polluted environments and the obvious convenience of favoring these degradative activities as part of bioremediation strategies.

The first edition of this *Manual* included information on basic catabolic activities of the aerobic pseudomonads on simple aromatic compounds because of their obvious taxonomic implications (Palleroni, 1984). Much research has since been done on the catabolism of many aromatic compounds that include benzoate and derivatives, polycyclic aromatic hydrocarbons, biphenyl (particularly the halogenated derivatives), metabolic intermediates such as the halogenated catechols, and some important herbicides (24D and 245T). This field of research now contains a multitude of references, of which only a selected minority will be discussed here.

The investigations of Ornston and his collaborators on the biochemistry of degradation of aromatic compounds are of particular interest because the corresponding pathways were also analyzed for their phylogenetic implications. The β-ketoadipate pathway has taken center stage in these investigations. Natural selection has adopted many permutations in the distribution of its components, their regulation, and the genetic makeup, all of which have been highlighted in an excellent review (Harwood and Parales, 1996). Immunological cross-reaction was observed between the enzymes of one of the species (*Pseudomonas putida*) and the corresponding enzymes of other fluorescent *Pseudomonas* species. Cross-reaction, however, was also detected between the γ-carboxymuconolactone decarboxylases of *P. putida* and *B. cepacia* (Patel and Ornston, 1976). Although the results suggested that the interspecific transfer of the structural gene for the enzyme was not common among pseudomonads, it nevertheless seemed to have occurred between these two distantly related species. The cross-reaction did not extend to other enzymes of the two species, including the salicylate hydroxylases, which are structurally different, and for which an explanation based on convergent evolution has been proposed (Kim and Tu, 1989).

As mentioned in a description of the metabolism of *para*-hydroxybenzoate (POB), *B. cepacia* is able to convert *meta*-hydroxybenzoate (MOB) to gentisate by the action of a 6-hydroxylase (Yu et al., 1987). Interestingly, this finding relates to an earlier report on the formation of gentisate from MOB by *Comamonas acidovorans*, which is a member of the *Betaproteobacteria* in which *Burkholderia* is located (Wheelis et al., 1967). A description of the induction of the hydroxylase from *B. cepacia* has been published (Wang et al., 1987).

The genetic organization and sequence of genes of the α and β subunits of protocatechuate 3,4-dioxygenase of *B. cepacia* has been investigated, and there is extensive similarity to genes of other *Pseudomonas* species, although this similarity does not extend to the promoter sequences (Zylstra et al., 1989b). The pattern of induction of this and other enzymes of the POB metabolism has been examined (Zylstra et al., 1989a).

As a member of a microbial community, *B. cepacia* was the most competitive member among other aerobic pseudomonads in the degradation of toluene (Duetz et al., 1994). A novel pathway of toluene catabolism was described for this organism, with the participation of toluene monooxygenase, which is able to hydroxylate toluene, phenol, and cresol, and also to catalyze the degradation of trichloroethylene (TCE) (Shields et al., 1991). Cometabolism of TCE and toluene has been described (Landa et al., 1994). A constitutive strain was selected for the degradation of TCE (Shields and Reagin, 1992). Degradation of TCE by *B. cepacia* using the toluene monooxygenase system is expressed at

higher capacity than the toluene dioxygenase systems present in other organisms (for instance, *P. putida*) (Leahy et al., 1996).

The toluene 2-monooxygenase from *B. cepacia* is a three-component system capable of oxidizing toluene to *o*-cresol and this to 3-methylcatechol. The catabolic features of this system resemble those of soluble methane monooxygenase from methanotrophic bacteria (Newman and Wackett, 1995). For the activity of toluene 2-monooxygenase on TCE, all its protein components and NADH are required. All protein components were modified during TCE oxidation, but reducing compounds such as cysteine protected the enzyme (Newman and Wackett, 1997).

Phthalate oxygenase is an enzyme specific for phthalate and closely related compounds. The system of *B. cepacia*, which requires the contribution of phthalate oxygenase reductase for efficient catalytic activity, is similar to other bacterial oxygenase systems. The *B. cepacia* enzyme can be isolated in large quantities and its stability is higher than that from other sources (Batie et al., 1987).

B. cepacia strains grow on fluorene and degrade this compound by a mechanism analogous to naphthalene catabolism. The system has wide specificity, and the range of substrates includes many other polycyclic aromatic compounds (Grifoll et al., 1995). But in spite of the remarkable catabolic versatility of the species, there are some limitations to the range of susceptible substrates. Thus, strains isolated on phenanthrene from polyaromatic hydrocarbon (PAH)-contaminated soils had limited capacity to use higher PAHs (Mueller et al., 1997).

For more than a decade, the degradation of the halogenated herbicides 2,4-dichlorophenoxyacetic acid and 2,4,5-trichlorophenoxyacetic acid (245T) by *B. cepacia* has been the object of attention by researchers, among them, A. Chakrabarty and his group (Haugland et al., 1990). In fact, *B. cepacia* grows luxuriantly on 245T. Although initially this species is not able to use phenoxyacetate, it can acquire this property by long selective pressure. Gene activation in the mutants seems to be due to translocation of insertion elements (Ghadi and Sangodkar, 1994). Polychlorinated phenols are produced in the degradation, and they are further metabolized by *B. cepacia* to the corresponding hydroquinones (Tomasi et al., 1995). The cloning, mapping, and expression of genes controlling the degradation by strains of this species have been under investigation (Sangodkar et al., 1988).

The study of spontaneous *B. cepacia* mutants unable to degrade 245T has permitted the identification of insertion sequences that facilitate or are required for growth at the expense of 245T (Haugland et al., 1991). The rapid evolution of the degradative pathway for this herbicide probably has been possible by insertion elements (such as IS1490) that play a central role in the transcription of 245T genes in *B. cepacia* (Hubner and Hendrickson, 1997). A 1477 bp sequence was repeated several times in a *B. cepacia* strain chromosome, and by the location it was concluded that genes involved in 245T degradation were actually recruited from foreign sources (Tomasek et al., 1989). Originally, foreign gene recruitment such as the mechanism postulated by Lessie and Gaffney (1986), may have been responsible for the acquisition of 245T degradation capacity. Better knowledge of the evolution of these pathways may facilitate the job of developing strains with enhanced degradative capacity (Daubaras et al., 1996a).

B. cepacia competes effectively in the microflora for the degradation of the herbicide 2,4-dichlorophenoxyacetic acid (24D) (Ka et al., 1994a). The enzyme that cleaves 3,5-dichlorocatechol in this pathway has been purified (Bhat et al., 1993). To monitor a strain in the degradation of 24D, a reporter gene system containing *luxAB* and *lacZY* was integrated into the chromosome of the strain, which thus could be readily identified in the community (Masson et al., 1993).

Oxygenase TftAB, capable of converting 245T to 2,4,5-chlorophenol, is an enzyme of wide specificity, since it can give phenolic derivatives of several other related compounds (Danganan et al., 1995). Two proteins, TftA1 and TftA2, characterized in the pathway, were sequenced and found to have similarity to BenA and BenB of the benzoate 1,2-dioxygenase system of *Acinetobacter calcoaceticus*, and to XylX and XylY from the equivalent system in *Pseudomonas putida* (Danganan et al., 1994). The *B. cepacia* enzyme that degrades 2,4,5-trichlorophenol, the intermediate in 245T catabolism, has been characterized and found to consist of two components (Xun, 1996).

1,2,4-trihydroxybenzene, another intermediate in 245T degradation, is a substrate for the enzyme hydroxyquinol 1,2-dioxygenase. The enzyme is specific, and it is a dimeric protein of 68 kDa (Daubaras et al., 1996b).

Environmental strains of *B. cepacia* are active in the degradation of polychlorinated biphenyls. As in other examples of aerobic metabolism of haloaromatic compounds, dehalogenation often occurs after ring cleavage (Arensdorf and Focht, 1995). *In vitro* constructed hybrids of *B. cepacia* could carry out total degradation of 2-Cl, 3-Cl, 2,4-dichloro-, and 3,5-dichloro-biphenyl (Havel and Reineke, 1991).

Chlorocatechols are key intermediates in the metabolism of haloaromatic compounds, and they are metabolized through reactions resembling those for catechol. A 2-halobenzoate 1,2-dioxygenase, a two-component system of *B. cepacia* strain 2CBS has a very broad specificity (Fetzner et al., 1992). However, in *B. cepacia* isolated from enrichment on 2-chlorobenzoate and in mutants blocked in different steps of the pathway, the 2,3-dioxygenase of the *meta* pathway predominated over the 1,2-dioxygenase (or *ortho*) system (Fetzner et al., 1989).

The dioxygenases and the cycloisomerases involved in the modified *ortho* pathways resemble the enzymes of pathways for nonhalogenated compounds, while the diene-lactone hydrolases needed in a later step are quite different (Schlomann et al., 1993). The results of work done on systems isolated from *Alcaligenes* and *B. cepacia* suggest that the hydrolases of the modified pathway may have been recruited from a different preexisting pathway, which was already operating before the start of industrial synthesis of halogenated compounds (Schlomann, 1994). When plasmid TOL is introduced into *B. cepacia* strains, these strains are able to grow with 3,4- and 3,5-dichlorotoluene, thus bypassing dead-end routes of chlorocatechols degradation (Brinkmann and Reineke, 1992).

The results of the experiments that have been briefly discussed in the preceding paragraphs often point to the high similarity among some components of peripheral metabolic pathways among organisms that are rather distantly related, suggesting horizontal transfer of the corresponding genetic determinants. One additional case is the high similarity between the genes coding for the *cis*-biphenyl dihydrodiol dehydrogenases of *P. putida* (gene *bphB*) and that of *B. cepacia* (Khan et al., 1997), and also between the *bphC* genes of both species, which code for the 2,3-dihydroxybiphenyl 1,2-dioxygenase (Khan et al., 1996b). On the other hand, in spite of having almost identical amino acid sequences, the biphenyl dioxygenase of *B. cepacia* (gene *bphA1*) and that of *Pseudomonas pseudoalcaligenes* have markedly different substrate ranges, the one for *B. cepacia* being wider (Kimura et al., 1997).

Additional details on the metabolism of aromatic compounds may be found in the section on plasmids.

Genetic characteristics The genome of a strain of *B. cepacia* (ATCC 17616) consists of three replicons whose respective sizes are 3.4, 2.5, and 0.9 Mb, which, together with the 170-kb cryptic plasmid present in the strain, gives an overall estimate for genome size of approximately 7 Mb (Cheng and Lessie, 1994). The three large replicons had ribosomal RNA genes, as well as the insertion elements previously described by Lessie and collaborators. Studies on mutants and the associated reductions in size of the replicons provide a convenient framework for genetic analysis of this strain.

The genome of another strain of *B. cepacia* (ATCC 25416) was analyzed and found to contain four circular replicons of sizes 3.65, 3.17, 1.07, and 0.2 Mb (Rodley et al., 1995). The values total 8.1 Mb. Again the interpretation is that the genome is made of three chromosomes and a large plasmid, because of the presence of ribosomal RNA genes only in the three large replicons. An interesting additional observation is the fact that multiple chromosomes are not confined to *B. cepacia* but also are found in other members of rRNA similarity group II (*B. glumae, Ralstonia pickettii*, and *R. solanacearum*) (Rodley et al., 1995). A very useful review that highlights the remarkable genomic complexity and plasticity of *B. cepacia* has been published (Lessie et al., 1996).

A *recA* gene has been identified in *B. cepacia* that is able to complement a *recA* mutation in *E. coli*, also restoring UV and methylmethane sulfonate resistance and proficiency in recombination (Nakazawa et al., 1990). Additionally, an SOS box related to LexA-regulated promoters and −10 and −35 consensus sequences have been detected in *B. cepacia*. The predicted RecA protein sequence shows 72% similarity with that of *P. aeruginosa*.

Transposable elements that activate gene expression in *B. cepacia* have been identified by Lessie and his collaborators (Scordilis et al., 1987). This opened a new horizon on insertional activation and on the significance of a high frequency of genomic rearrangements that presumably are related to the remarkable versatility of the species (Gaffney and Lessie, 1987; Lessie et al., 1996).

In spite of all these developments, at present there is incomplete knowledge of gene expression and of regulatory mechanisms in species of *Burkholderia* because of the lack of an appropriate gene exchange system. In the absence of a conventional bacterial genetic system, some alternatives have been developed. One of the proposed methods of genetic analysis is based on transposon mutagenesis and complementation of mutations by means of cloned genes. In a particular application, a shuttle plasmid was constructed that could be used for cloning genes of *B. cepacia* involved in protease production (Abe et al., 1996).

Studies on population genetics have been carried out on a population of *B. cepacia* isolated from a southwestern stream in the United States, to examine the allelic variation in a group of loci using multilocus enzyme electrophoresis. The studies showed a low degree of association between the loci, or extensive genetic mixing. This evidence of frequent recombination (and consequent low levels of linkage disequilibrium) indicates that the structure of the population was not clonal (Wise et al., 1995).

The topology of a 23S rRNA phylogenetic gene tree agrees with the 16S rRNA tree (Höpfl et al., 1989).

Plasmids A cryptic 170-kb plasmid that was discovered in *B. cepacia* ATCC 17616 has been the subject of much research. Derivatives of this strain carried versions of the plasmid containing various insertion sequences in different combinations. These elements, inserted in the broad host range plasmid RP1, served as probes to examine the extent of the reiteration of the various components in the genome of the organism. The results indicated a high frequency of genomic rearrangements, mainly the result of replicon fusions promoted by the insertion elements, that could help explain the remarkable biochemical versatility of the species (Barsomian and Lessie, 1986; Gaffney and Lessie, 1987).

Derivatives of a nonconjugative *Pseudomonas* plasmid (pVS1), carrying genes for mercury and sulfonamide resistance as well as segments required for stability and for mobilization by plasmid RP1, have been established in *B. cepacia* (Itoh et al., 1984). Some further constructions based on pVS1 could be used as cloning vectors (Itoh and Haas, 1985).

In many instances, the degradation of toxic compounds and environmental pollutants is controlled by genes located in plasmids. Some of them already have been mentioned, and only brief reference to some additional instances will be made here. A 50-kb plasmid is responsible for the degradation of *para*-nitrophenol by *B. cepacia*, following an oxidative route with the production of hydroquinone and nitrite. The plasmid can be conjugationally transferred (Prakash et al., 1996). A 70-kb plasmid in *B. cepacia* strain 2CBS carries a gene cluster with the determinant of an enzyme able to catalyze double hydroxylation of 2-halobenzoates with release of halogenide and CO_2 and producing catechol (Haak et al., 1995). A catabolic plasmid involved in the degradation of 4-methyl-*o*-phthalate was described and named MOP (Saint and Ribbons, 1990). Finally, a fragment of a plasmid involved in the catabolism of 4-methylphthalate in *B. cepacia* was sequenced and two open-reading frames were discovered, one of which encoded a permease that belongs to a group of symport proteins found in both pro- and eucaryotes. Information on this system could be used to improve the degradative capability for bioremediation purposes (Saint and Romas, 1996).

The novel pathway of toluene degradation mentioned in the section on metabolism of aromatic compounds is inducible and the corresponding genes are located in a plasmid. A strain of *B. cepacia* was found to carry two plasmids, one of 108 kb (named TOM) containing the genes for a toluene monooxygenase pathway that was expressed constitutively. The same strain also contained a small plasmid of less than 70 kb (Shields et al., 1995).

A new plasmid (pMAB1) with genes controlling the degradation of 24D in *B. cepacia* was characterized, and spontaneous negative mutants were isolated under nonselective conditions. Instead of the original 90-kb plasmid, these mutants had a smaller one (70-kb) or had lost it altogether. The activity could be regained by reintroducing the larger plasmid by electroporation. The 70-kb plasmid lacked a region that included the gene *tfdC* encoding the 3,5-dichlorocatechol 1,2-dioxygenase, whose sequence was identical with that of a well-characterized 24D degradative plasmid (pJP4) of *Alcaligenes eutrophus*. The similarity did not extend to the rest of the plasmids (Bhat et al., 1994). Another *B. cepacia* plasmid (pBS1502) was found to be able to control the early dehalogenation of 2,4-dichlorobenzoate (Zaitsev et al., 1991). There is also a report of the presence of a catabolic plasmid named MOP carrying genes involved in the catabolism of phthalate derivatives (Saint and Ribbons, 1990), and a small (2-

kb) plasmid was implicated in the degradation of phenylcarbamate herbicides (Gaubier et al., 1992).

Plasmid analyses in combination with other typing techniques have been proposed for epidemiological studies on nosocomial infections by *B. cepacia* (Yamagishi et al., 1993). Early studies based on agarose gel electrophoresis of *B. cepacia* extracts demonstrated the presence of one or more plasmids in several strains of *B. cepacia* of plant and clinical origin. The plasmid composition, together with bacteriocin production and sensitivity, and pectolytic activity, could have applications in epidemiological studies (González and Vidaver, 1979).

Strains of *B. cepacia* from clinical and pharmaceutical origin carried large plasmids (146–222 kb) containing antibiotic resistance genes (Lennon and DeCicco, 1991). The nonconjugative *B. cepacia* plasmid pVS1 contained Hg and sulfonamide resistance genes, and a segment required for mobilization by RP1 (Itoh et al., 1984).

Bacteriophages Most cultures of *B. pseudomallei* from collections have shown spontaneous phage production (Denisov and Kapliev, 1991). From the strains from a collection, 14 pure lines of bacteriophages belonging to two morphological types were isolated. The specificity of these phages was studied, and it was found that some strains of the host undergo poly-lysogeny, which was inferred from the fact that phages of different morphological types could be isolated from single strains (Denisov and Kapliev, 1995).

A generalized transducing phage was isolated from a lysogenic strain of *B. cepacia*, and half of more than 100 strains of the species were sensitive to it (Matsumoto et al., 1986).

Strain Berkeley 249 (ATCC 17616) of *B. cepacia*, which has been studied very intensively in Lessie's laboratory, carries an organic solvent-sensitive phage (Cihlar et al., 1978). Its sensitivity is attributed to alteration of a tail component provoked by the solvent. Results obtained by using other *B. cepacia* strains as hosts imply the occurrence of host restriction and modification systems. The phage has a head of 55 nm in diameter, a broad contractile tail of 15×145 nm, and double-stranded DNA of a molecular weight of about 3×10^7.

Many years ago a bacteriophage lytic for a wide range of aerobic pseudomonads was isolated and tested against strains of different species, among which was *"Pseudomonas multivorans"* (later identified as a synonym of *B. cepacia*) (Kelln and Warren, 1971). *B. cepacia* was insensitive; the phage was lytic only for species of rRNA similarity group I (*Pseudomonas sensu stricto*) and not for members of other rRNA groups.

Bacteriocins Early work performed on *B. cepacia* strains isolated from plants and from clinical specimens showed that the two groups could be differentiated by bacteriocin production patterns, onion maceration tests, and hydrolysis of pectate at low pH, thus suggesting the usefulness of these characteristics in epidemiological studies (González and Vidaver, 1979). Additional differences between strains from the two sources have been recorded (Bevivino et al., 1994) and will be discussed below (see Ecology, Habitats, and Niches).

A number of *B. cepacia* bacteriocins ("cepaciacins") were defined in work done on a collection of 34 strains isolated from plant rhizospheres and human patients (Dodatko et al., 1989a). One of the cepaciacins consisted of protein and carbohydrate in a 3:1 molecular ratio. The bacteriocin was thermolabile, stable within a narrow range of pH values, and it was destroyed by proteolytic action. UV irradiation or mitomycin C stimulated its biosynthesis (Dodatko et al., 1989b).

A typing scheme has been described based on bacteriocin susceptibility and production by *B. cepacia* strains using six producer strains and a set of eight indicator strains (Govan and Harris, 1985). The majority of strains of a large collection were typed into a total of 44 combinations, and the typing scheme was found to be useful for the possibility of its application to epidemiological studies.

Antigenic structure   The LPS structure of different *B. pseudomallei* strains is quite homogeneous. Antibodies prepared with material from one strain react with all others (Pitt et al., 1992). In agreement with the degree of their phylogenetic relationships, cross-reactions are observed with *B. mallei* and, to a lesser extent, with *B. cepacia*.

Many features of the biological activity of LPS isolated from *B. pseudomallei* cells have been described, and the strong mitogenic activity that it has toward murine splenocytes has been attributed to unusual chemical structures in the inner core attached to lipid A (Matsuura et al., 1996). In addition, information is available about the identification, isolation, and purification of an exopolysaccharide of this species. As in the instance described above, the compound having a molecular mass of >150 kDa did not show cross-reactivity with any of the species of all the *Pseudomonas* rRNA similarity groups, with the exception of the closely related species *B. mallei* (Steinmetz et al., 1995).

Purification to homogeneity of flagellin from several *B. pseudomallei* strains gave monomer flagellin bands of M_r 43,400 Da. Passive immunization studies showed that a specific antiserum could protect animals from challenge by a *B. pseudomallei* strain of different origin (Brett et al., 1994).

Two monoclonal antibodies were found to be highly specific for *B. pseudomallei* when tested by indirect enzyme-linked immunosorbent assay and immunoblotting against whole-cell extracts of other *Burkholderia* species, fluorescent pseudomonads, and *E. coli*. One of the antibodies could agglutinate all 42 *B. pseudomallei* strains included in the study, thus providing a tool for rapid identification of the species using primary bacterial cultures from clinical specimens (Pongsunk et al., 1996).

Of the serological typing schemes devised for *B. cepacia*, the one most widely used is that proposed by Werneburg and Monteil (1989). Originally, the scheme described procedures for the preparation, adsorption, and titration of O and H rabbit sera, and it could define seven O (O1 to O7) and five H antigens (H1, H3, H5, H6, and H7) for the slide agglutination test and the agglutination and immobilization test, respectively (Heidt et al., 1983). The scheme later was supplemented with new serotypes, using strains of a different geographical origin, to make a total of 9 O and 7 H antigens (Werneburg and Monteil, 1989). Other immunological typing schemes have been proposed (Nakamura et al., 1986).

Wilkinson and his collaborators have studied the composition of the O-specific polymers from the LPS of *B. cepacia* strains belonging to groups O1 (Cox and Wilkinson, 1990b), O3, O5 (Cox and Wilkinson, 1989b), O7 (Cox and Wilkinson, 1990a), and O9 (Taylor et al., 1994a). The O9 group has repeating units that are also present in *Serratia marcescens*. In the same laboratory it has been discovered that the O antigen of the LPS of *B. cepacia* serotype E (O2) is composed of two different trisaccharide repeating units in a 2:1 ratio (Beynon et al., 1995), and that the same O specific polymer is found in the two related species *B. cepacia* and *B. vietnamiensis*.

Interestingly, the lipopolysaccharides extracted from *B. cepacia* and *B. gladioli* have a higher endotoxic activity and provoke a higher cytokine response than that from *Pseudomonas aeruginosa* (Shaw et al., 1995). This adds to the importance to the human pathogenic propensies of *P. gladioli*, an example of a plant pathogen of medical importance similar to that of *B. cepacia*. Plant-associated *B. gladioli* can be differentiated from other pathogenic and symbiotic bacterial species by a rapid slide agglutination test using polyclonal antisera conjugated to protein-rich *Staphylococcus aureus* whole cells (Lyons and Taylor, 1990).

Cross-reactivity of *P. aeruginosa* antipilin monoclonal antibodies with heterogeneous strains of *P. aeruginosa* and *B. cepacia* has been reported (Saiman et al., 1989).

Antibiotic susceptibility   For obvious reasons, most information on antibiotic susceptibility of species of the genus *Burkholderia* refers to only a few species of medical importance (*B. pseudomallei*, *B. cepacia*, *B. gladioli*).

All strains of *B. cepacia* that have been tested are sensitive to sulfonamides and novobiocin. Most are also sensitive to trimethoprim plus sulfamethoxazole, and to minocycline and chloramphenicol (Santos Ferreira et al., 1985). Both *B. cepacia* and *B. gladioli* are resistant to a wide variety of antibiotics (ticarcillin by itself or mixed with clavulanic acid; cefsulodin, imipenem, aminoglycosides, colistin, and fosfomycin) (Baxter et al., 1997). A catechol-containing monobactam (BMS-180680) was quite active (MIC90, 1 µg/ml; MIC90 is a minimum concentration which inhibits 90% of the strains tested) (Fung-Tomc et al., 1997).

Eighty percent of a collection of strains of *B. cepacia* was tested for inhibition by ceftazidime (Tabe and Igari, 1994). A number of quinolone analogs and derivatives (trovafloxacin, ciprofloxacin, ofloxacin, levofloxacin, sparfloxacin, clinafloxacin, ceftazidime) were active on several Gram-negative, nonfermentative species, including *B. cepacia*. In comparison, the MIC for imipenem on *B. cepacia* and on *Stenotrophomonas maltophilia* was very high (Visalli et al., 1997).

In a comparison of the activity of many antibiotics against *B. pseudomallei*, a quinolone (tosufloxacine) and a tetracycline derivative (minocycline) appeared to be the most active (Yamamoto et al., 1990).

For years, the oral maintenance treatment of melioidosis has depended on the combined action of amoxicillin and the β-lactamase inhibitor clavulanic acid (Suputtamongkol et al., 1991). Biapenem, one of several carbapenem antibiotics tested against *B. pseudomallei*, was the most active against strains that showed a diminished susceptibility to third-generation cephalosporins (Smith et al., 1996).

Several resistance mechanisms operate in different species of *Burkholderia*. Resistance to cationic antibiotics in *B. cepacia* has been attributed to their ineffective binding to the outer membrane as a consequence of the low number of phosphate and carboxylate groups in the LPS, and the presence of protonated aminodeoxypentose (Cox and Wilkinson, 1991). The involvement of the outer membrane of *B. cepacia* in the resistance to polymyxin and aminoglycosides may also be related to a particular arrangement in the structure of the outer membrane, in which cation-binding sites on LPS are protected from polycations (Moore and Hancock, 1986).

An important factor in the antibiotic resistance in these organisms is porin permeability (Parr et al., 1987; Burns et al., 1996). However, a different interpretation is that resistance may not be a direct consequence of permeability of the porins, but instead may be related to the low number of porins per cell. A β-lactam-resistant mutant of *B. cepacia* and resistant strains of this species isolated from cystic fibrosis cases owe their resistance to a low porin content. These strains had reduced amounts of a 36-kDa outer membrane protein and did not express a 27-kDa outer membrane protein that can be a major porin or a major component of the porin complex of the cells (Aronoff, 1988). According to this view, the most common resistance mechanism in *B. cepacia* is the low porin-mediated outer membrane permeability, combined with multiple drug resistance due to an efflux pump system (Burns et al., 1996).

It may be of interest to mention here that growth in the presence of salicylate or other weak acids can induce resistance to antibiotics in *B. cepacia*, because these compounds have been found to be inhibitors of porin formation (Burns and Clark, 1992).

A number of β-lactams are susceptible to hydrolysis by a β-lactamase of *B. cepacia*, a metalloenzyme of type I that is induced by imipenem (Baxter and Lambert, 1994). A significant degree of similarity was found between the chromosomal β-lactamase of *B. cepacia* strain 249 and the enzymes of *Pseudomonas aeruginosa* and *E. coli*. Interestingly, in spite of differences in the mol% G + C content of the DNA of these organisms, the codon usage in *B. cepacia* resembled that of *E. coli* (Proenca et al., 1993).

The multiple resistance gene *oprM* of *P. aeruginosa* is part of a highly conserved efflux system. A gene homologous to *oprM* was identified in *B. cepacia* (Burns et al., 1996). Moreover, an intragenic probe hybridized the genomic DNA of several fluorescent pseudomonads and, in addition, the DNA of *B. pseudomallei* (Bianco et al., 1997). A lucid review is available on the participation of multidrug efflux pumps in the resistance of Gram-negative organisms to antibiotics (Nikaido, 1996).

The fusaric acid resistance gene of *B. cepacia* has been cloned and sequenced (Utsumi et al., 1991).

Antibiotic production Antifungal antibiotics have been identified in strains of *B. cepacia*. Cepacidine A, composed of two related forms, cepacidine A1 and A2, is a cyclic peptide and xylose connected to a 5,7-dihydroxy-3,9-diaminooctadecanoic acid (Lee et al., 1994a; Lim et al., 1994). Two antibiotics previously discovered and described under the names cepacin A and B are not related to the cepacidines. The cepacins showed good antistaphylococcal activities (MICs of 0.2 and 0.05 µg/ml for cepacin A and B, respectively), but no significant activity against Gram-negative bacteria (Parker et al., 1984).

Another antifungal antibiotic is pyrrolnitrin (Jayaswal et al., 1993). The producing organism was named *Pseudomonas pyrrocinia* (Imanaka et al., 1965), now *Burkholderia pyrrocinia*. The antibiotic is also produced by *B. cepacia* and by some strains of *Pseudomonas chlororaphis* (Elander et al., 1968). A study by Burkhead et al. (1994) has examined the conditions of pyrrolnitrin by *B. cepacia* in culture and in the wounded areas of potatoes colonized by the organism.

Some compounds related to pyrrolnitrin (amino-pyrrolnitrin, and monochloroamino-pyrrolnitrine) also have antifungal properties (McLoughlin et al., 1992). The antibiotic activity of *B. cepacia* has been reported to antagonize the pathogenic activity of the sunflower wilt fungus (McLoughlin et al., 1992) and to be a suppressor of maize soil-borne disease (Hebbar et al., 1992). The influence of some environmental factors on the antagonism of *B. cepacia* toward *Trichoderma viride* has been analyzed (Upadhyay et al., 1991), as well as some morphological alterations and inhibition of conidiation of plant pathogenic fungi (Upadhyay

and Jayaswal, 1992). A compound having both hemolytic activity and antifungal action was characterized and given the name cepalycin (Abe and Nakazawa, 1994).

Transposon mutagenesis could eliminate pyrrolnitrin production ability. However, the mutation failed to be complemented by the cloned gene because of difficulties encountered in mobilizing the carrier cosmids from *E. coli* to *B. cepacia* mutants (Jayaswal et al., 1992).

A group of eight cyclic peptides of antifungal activity, the xylocandins, has been isolated from *B. cepacia* and characterized (Bisacchi et al., 1987). A mixture of two of the forms (A1 and A2) showed potent anticandidal and antidermatophytic activities *in vitro* (Meyers et al., 1987).

The antagonistic activity of *B. cepacia* is not limited to antifungal activity. Of practical importance is the finding that metabolites produced by strains of this species can antagonize plant pathogenic agents such as *Ralstonia solanacearum* (Aoki et al., 1991).

Toxoflavin, an azapteridine antibiotic produced by *B. cocovenenans*, was identified more than 30 years ago (Lauquin et al., 1976). The production of toxoflavin is inhibited by the combined action of 2% NaCl and acidity (pH 4.5) in the medium (Buckle and Kartadarma, 1990). A monobactam antibiotic (MM 42842) is also produced by *B. cocovenenans*. It is related to a previously described antibiotic named sulfazecin. In addition, the *B. cocovenenans* strain synthesizes bulgecin, an antibiotic also produced by other aerobic pseudomonads (Box et al., 1988; Gwynn et al., 1988).

Tropolone, which is known to have antibacterial and antifungal activities (Lindberg, 1981), is produced by *B. plantarii* cultures (Azegami et al., 1987).

Plant pathogenicity Many species of the genus *Burkholderia* are pathogenic for animals or plants. Phytopathogenic pseudomonads are located in three of the five RNA similarity groups. One of them is rRNA group II, in which *Burkholderia* and *Ralstonia* are allocated. The various symptoms produced in plants by the phytopathogenic pseudomonads are tumorous outgrowth, rot, blight or chlorosis, and necrosis, which are caused by alteration of the normal metabolism of plant cells by pathogenicity factors excreted by the bacteria. These factors include enzymes capable of degrading components of plant tissues, toxins, and plant hormones.

The phytopathogenic species of the genus *Burkholderia* mainly produce rots, due to active pectinolytic enzymes and cellulases (Gehring, 1962; Hildebrand, personal communication). Further details on symptoms and on the list of hosts attacked by each species will be given in the section of species descriptions at the end of this chapter.

Aside from acting as agents of diseases, members of the genus also participate in producing beneficial effects by antagonizing other phytopathogenic organisms, mainly fungi. This effect has been reported in the literature for *B. cepacia* in a number of instances (Kawamoto and Lorbeer, 1976; Fantino and Bazzi, 1982; Janisiewicz and Roitman, 1988; Homma et al., 1989; Jayaswal et al., 1990; Parke et al., 1991).

Pathogenicity for humans and animals The most serious animal and human pathogenic species of the genus *Burkholderia* are *B. mallei* and *B. pseudomallei*, the agents of glanders or farcy of equids, and of melioidosis in humans, respectively. Detailed descriptions of these organisms and the diseases that they cause are available (Redfearn et al., 1966; Redfearn and Palleroni,

1975). As a free-living species present in the warm regions of the planet, *B. pseudomallei* has not spread far from the equator, although in China it has reached a northern latitude of at least 25.5 degrees (Yang et al., 1995b).

A useful updated treatment of the bacteriology of glanders and melioidosis is available (Pitt, 1998). Many years ago, melioidosis was recognized as a glanders-like disease (Whitmore, 1913), and subsequently the organism was assigned to several different genera. In fact, this has also been true of *B. mallei*, which has spent much of its scientific career in search of a proper generic allocation.

Two different biotypes have been recognized among various strains of *B. pseudomallei* isolated from patients and from soil in Thailand. However, all of the strains were recognized by using a specific polyclonal antibody. One of the biotypes may have low virulence or it may represent a different species altogether, based on the distribution of these phenotypes and the respective incidence of melioidosis in different areas (Wuthiekanun et al., 1996).

Strains of *B. pseudomallei* isolated in Australia were examined for their genomic relationships using random amplification of polymorphic DNA and multilocus enzyme electrophoresis. The strains could be divided into two groups that correlated with the clinical presentation and not with the geographic origin; in other words, there was a correlation between the clinical manifestation of the disease and the molecular characteristics of the pathogen (Norton et al., 1998).

Some of the human pathogens are "opportunistic pathogens" that create major medical problems in patients with reduced levels of natural resistance. This situation emerges because of an increasing use of instrumentation or drugs (including antibiotics) that reduce or bypass the level of natural resistance and/or the specific immune mechanisms (Spaulding, 1974).

In reference to *B. cepacia*, it seems appropriate here to highlight the main points of the abstract of an excellent article published more than a decade ago (Goldmann and Klinger, 1986). From its original habitat as a plant pathogen, *B. cepacia* has invaded the hospital environment as an important pathogen of compromised human hosts. Many nosocomial infections have their source in contaminated medicaments and even disinfectants and antiseptics. Various conditions in hospital patients can be complicated by *B. cepacia* infections, but the properties that define the virulence of strains of this species are still poorly defined. The last addition to the long list of pathological conditions that are further deteriorated by infections of this pathogen is cystic fibrosis. As in other conditions, some patients may be simply colonized, while other patients' initial conditions can be very seriously complicated, to which the remarkable resistance of *B. cepacia* to a wide range of antibiotics contributes very effectively. This bleak panorama has become even worse in the intervening years, particularly with respect to pulmonary exacerbations in cystic fibrosis patients, in many cases ending in rapid and fatal deterioration of lung function.

Characteristics that occur more frequently among strains isolated from infections in cystic fibrosis patients include catalase, ornithine decarboxylase, valine amino-peptidase, lipase, alginase, trypsin, ability to reduce nitrate, hydrolysis of xanthine and urea, complete hemolysis of bovine red blood cells, cold-sensitive hemolysis of human red blood cells, and greening of horse and rabbit red blood cells. Although some of these properties are associated with pathogenicity in other bacteria, their relationships to cystic fibrosis-associated pulmonary disease are far from

clear (Gessner and Mortensen, 1990). Acid phosphatase in *Burkholderia* species is a factor that may be related to pathogenicity. It is present in *B. cepacia* and *B. pseudomallei*, and in fact the activity in the latter species is remarkably high (Dejsirilert Butraporn et al., 1989).

In addition to these factors, as mentioned before in the section on antigenic structure, LPS preparations from clinical and environmental isolates of *B. cepacia* and from the closely related species *B. gladioli* exhibit a higher endotoxic activity and more pronounced cytokine response *in vitro* when compared to preparations of *P. aeruginosa* LPS. The latter species is also involved in infections in cystic fibrosis patients (Shaw et al., 1995).

B. gladioli, a species originally described as phytopathogenic, has been implicated in infections complicating cases of cystic fibrosis (Mortensen et al., 1988). In one study, the organism was isolated from respiratory tract specimens obtained from 11 cystic fibrosis patients and was identified by its biochemical properties, DNA hybridization, and fatty acid analysis. The authors recommend the inclusion of some of these criteria for the differentiation between *B. gladioli* and *B. cepacia*, two closely related species, and point out that most *B. gladioli* strains have $C_{10:0\ 3OH}$ fatty acid, which is absent from *B. cepacia* lipids (Christenson et al., 1989). Also based on the results of fatty acid analysis, a pseudomonad that was isolated from pleural fluid and pulmonary decortication tissue with granulomatous disease more closely resembled *B. gladioli* than *B. cepacia* (Trotter et al., 1990). Further information on the activity of *B. gladioli* as a human pathogen has been reported by Ross et al. (1995). A strain of this species was involved in empyema and bloodstream infection occurring after lung transplantation in a cystic fibrosis case (Khan et al., 1996c).

The pili of *B. cepacia* mediate adherence to mucous glycoproteins (Kuehn et al., 1992; Sajjan and Forstner, 1993) and epithelial cells in cystic fibrosis patients. Structural variant classes of pilus fibers have been identified in *B. cepacia* strains (Goldstein et al., 1995). One or more of five morphologically different types can be coexpressed. It has been noticed that, when present in the infective population, *Pseudomonas aeruginosa* cells enhance the adhesion of *B. cepacia* to epithelial cells (Saiman et al., 1990).

The major pilin subunit of *B. cepacia* corresponds to peritrichous fimbriae, which, based on their appearance as giant intertwined fibers, have been called "cable" (Cbl) pili. The *cblA* gene (the first pilin subunit gene to be identified in *B. cepacia*) has been detected in a DNA library (Sajjan et al., 1995).

One very infectious cystic fibrosis strain isolated as the agent of epidemics in England and Canada was found to have the cable pilin subunit gene. A conserved DNA marker in a 1.4-Kb fragment was present in epidemic strains, absent from the nonepidemic ones, and rare among the environmental strains. The presumed ORF was designated "epidemic marker regulator" (*esmR*) (Mahenthiralingam et al., 1997). Strains were recovered in cystic fibrosis centers in France. There was cross-colonization in 7 of 13 centers. The most chronically colonized patients harbored a single *B. cepacia* strain, which suggests a geographically clustered distribution of *B. cepacia*, with the exception of one genotype. This genotype was detected in four regions and proved to be different from the British-Canadian highly transmissible strain and was able to spread among cystic fibrosis units (Segonds et al., 1997).

Biochemical and genomic properties have been used for the typing of *B. cepacia* strains of nosocomial origin. In a first step, six enzymes and pigment production subdivided a collection of strains of this species into 12 groups, and the strains from one-third of the collection were further characterized by DNA fingerprinting, ribotyping, and plasmid analysis. By testing the typing scheme on strains isolated years later, the results showed the usefulness and consistency of the genomic typing, and the marked diversity of phenotypes among the strains of the species (Ouchi et al., 1995).

The amplified products of the internal transcribed spacers (ITS) separating the 16S and 23S rRNA genes in the DNA have been effective as tools for identification of reference strains of *Pseudomonas aeruginosa* and *B. pseudomallei*, but the primer pairs tested for *B. cepacia* have not provided much help in strain identification (Tyler et al., 1995). In contrast, PCR-ribotyping targeted on the 16S-23S intergenic spacer to determine the length heterogeneity of this region is a rapid and accurate method for *B. cepacia* strain typing (Dasen et al., 1994).

The internal diversity of *B. cepacia* strains is also manifested in differences in whole-cell protein profiles (Li and Hayward, 1994) and in the fatty acid composition of populations from various cystic fibrosis centers (Mukwaya and Welch, 1989).

A novel marker has been found to be associated with epidemic *B. cepacia* strains causing infections in cystic fibrosis patients. A highly infectious strain had both the cable pilin subunit gene (*cblA*) and a unique combination of insertion sequences. Although no specific marker was identified in common with other epidemic strains, a conserved DNA fragment among epidemic strains was detected. This fragment (called the "*B. cepacia* epidemic strain marker") was absent from other strains infecting individual cystic fibrosis patients and rarely found in environmental strains (Mahenthiralingam et al., 1997).

In a related study on 97 clinical and 2 environmental strains of *B. cepacia*, a search for possible correlations was made with respect to parameters such as certain insertion sequence (IS) elements, the *cblA* gene for a pilin subunit, the electrophoretic type (ET), and the ribotype (RT) (Tyler et al., 1996). No linkage was found between the presence of each of five different IS elements and ET or RT. All strains of a given ET also possessed *cblA*. One IS element different from the five initially identified was present in 72% of all isolates, and in half of them the new IS element was inserted in one of the original five. This hybrid IS element only was found in epidemic strains from Ontario, Canada, and the UK.

Methods were developed to detect the presence of *B. cepacia*, *Pseudomonas aeruginosa*, and *Stenotrophomonas maltophilia* in sputum, using primers based on 16S rRNA sequences. The results are in agreement with those of direct isolation of cultures from the samples (Campbell et al., 1995; Karpati and Jonasson, 1996).

Bongkreik acid (also named flavotoxin A), a product of fermentation by *B. cocovenenans*, is the cause of serious food poisoning outbreaks in China and other Far East countries (Hu et al., 1989). An isomer, isobongkreik acid, was later identified, and, like bongkreik acid, it acts as an uncompetitive inhibitor of ADP transport in mitochondria, although it is less active than bongkreik acid (Lauquin et al., 1976).

Ecology, habitats, and niches The habitats from which *Burkholderia* species have been isolated are quite diverse: soils and rhizospheres (*B. cepacia*, *B. vietnamiensis*, *B. pseudomallei*, *B. pyrrocinia*, *B. phenazinium*, *B. multivorans*, *B. thailandensis*, *B. glathei*), water (*B. norimbergensis*), plants (*B. cepacia*, *B. gladioli*, *B. caryophylli*, *B. glumae*, *B. plantarii*, *B. andropogonis*, *B. vandii*), foods (*B. cocovenenans*), animals (*B. mallei*, *B. pseudomallei*), and clinical specimens (*B. cepacia*, *B. vietnamiensis*, *B. gladioli*, *B. multivorans*). The occurrence of *B. pseudomallei* mainly in soils of tropical regions has been discussed elsewhere (Redfearn et al., 1966).

As mentioned in the section on metabolic properties, some species of the genus are metabolically very versatile, and there is little doubt that those normally found in soils and water must contribute substantially to the mineralization of carbon compounds in nature.

B. cepacia is commonly found in natural materials, particularly soil, and methods have been developed for its detection. One such method of detection in the environment was based on amplification of sequences in genomic DNA using primers specific for repetitive extragenic palindromic segments, followed by cloning of the amplified fragments. Probes were constructed based on the strain-specific sequences (Matheson et al., 1997). A striking example of the power of molecular approaches was the report of the detection of a single *B. cepacia* cell in a soil sample containing a population of 10^{10} procaryotic cells (Steffan and Atlas, 1988). Estimates of the relative abundance of *B. cepacia* in stream bacterioplankton can be performed after collecting the cells by various procedures, of which the use of filters made of inorganic materials was found to give the highest recoveries (Lemke et al., 1997).

Strains of *B. cepacia* can be isolated from rhizosphere environments (Tabacchioni et al., 1995). This is also the habitat from which *B. graminis* was originally obtained (Viallard et al., 1998). Rhizosphere strains of *B. cepacia* differed in phenotypic characteristics from those isolated from clinical materials. Among the differences were nitrogen fixation, indole-acetic acid production, a wide temperature range, antibiosis vs. phytopathogenic fungi, and growth promotion of *Cucumis sativus*—all positive for rhizosphere strains. These properties were absent from clinical strains, which instead possessed characteristics such as adhesion to human cells, protease production, and synthesis of siderophores different from those found in the nonclinical strains (Bevivino et al., 1994). The character of N_2 fixation is not clear-cut, since DNA preparations from clinical strains hybridized with the *nifA* gene probe from *Klebsiella pneumoniae*, and the DNA of one of the rhizosphere strains hybridized with the *nifHDK* from *Azospirillum brasilense* (Tabacchioni et al., 1995).

As far as animal pathogenicity is concerned, observations made in other laboratories seem to support somewhat different conclusions. Thus, strains isolated from plant or clinical materials did not differ in their lethality to mice, i.e., they have similar LD_{50} values (González and Vidaver, 1979). Indeed, strains of species known to be causal agents of human infections may be isolated from rhizosphere environments (Tabacchioni et al., 1995).

Although *B. cepacia* helps in the degradation of toxic compounds and has an inhibitory action on soil-borne plant pathogens, it can also behave as a serious opportunistic human pathogen. The key question of interest for its use in biotechnological projects is whether the two types of populations can interact to the point of transmission of the characteristics required for the pathogenic condition. Yohalem and Lorbeer (1994) state that "although there are (*B. cepacia*) strains with significant potential for the remediation of environmental toxins . . . and others with potential as biological controls of plant disease . . . environmental release of any strain may be prohibited because some strains of the nomenspecies have been implicated in nosocomial infections." The same considerations apply to other versatile species of the genus that manifest pathogenic propensities.

ENRICHMENT AND ISOLATION PROCEDURES

For the specific isolation of *B. cepacia* from environmental water samples or various aqueous solutions, a medium containing 9-chloro-9-(4-diethylaminophenyl)-10-phenylacridan (C-390) and polymyxin B sulfate (PBS) has been proposed.[3] The two drugs inhibited all the organisms tested, with the exception of *B. cepacia* and *Serratia marcescens* (Wu and Thompson, 1984).

A medium proposed for the isolation of *B. cepacia* contains tryptamine and azelaic acid as sole sources of nitrogen and carbon, respectively, in addition to the antifungal compound chlorothalonil (Diamond Shamrock Corp., Cleveland, Ohio) (Burbage and Sasser, 1982). Later on it was found that the effectiveness of this medium was rather low, and a different one, medium (TB-T),[4] based on the combined action of trypan blue (TB) and tetracycline (T) at a relatively low pH (5.5), was proposed. Using this medium, the efficiency of recovery was very variable among strains, and for some of them reached 76–86% (Hagedorn et al., 1987).

A medium for *B. cepacia* has been described in a patent (Lumsden and Sasser, 1986).

Three selective enrichment liquid media and four solid media were evaluated at two temperatures (35°C and room temperature) for their capacity of supporting growth of *B. pseudomallei* strains. Enrichment with trypticase soy broth with addition of 5 mg of crystal violet and 20 mg of colistin per liter and subculture in Ashdown medium[5] gave the highest recovery rates and the greatest suppression of other members of the soil microflora. The results were comparable at the two temperatures (room temperatures ranged between 20°C and 32°C).

DIFFERENTIATION OF THE GENUS *BURKHOLDERIA* FROM OTHER GENERA

Similarities between *Burkholderia* and other Gram-negative bacteria A sharp differentiation of members of *Burkholderia* and *Pseudomonas* is difficult because of many similarities between these organisms. In the course of biochemical studies carried out at Berkeley before the rRNA–DNA hybridization experiments showed that "*P. multivorans*" (*B. cepacia*) was not closely related to the fluorescent group, it was noticed that key steps of the metabolism of aromatic compounds were remarkably similar between the two groups. In the section on the metabolism of aromatic compounds, some comments have been made on the similarity of components of some of the enzymes involved in the degradation by distantly related organisms. In fact, many details of the metabolic constitution of the fluorescent pseudomonads are far closer to those found in some species of RNA similarity group II—in particular, *B. cepacia*—than to those characteristic

of RNA group V (*Stenotrophomonas* and *Xanthomonas*), which is closer to RNA group I. Such is the case of the enzymes codified by genes *bphB* and *bphC* in *P. putida* and *B. cepacia*, and the same is true for the gamma-carboxymuconolactone decarboxylases of these species. Fluorescent organisms and *B. cepacia* are able to synthesize salicylic acid, which acts both as a siderophore and also as the precursor of another siderophore found in both groups of organisms, pyochelin. Members of the fluorescent group and *B. pyrrocinia*, a species related to *B. cepacia*, produce the antifungal antibiotic pyrrolnitrin, and similarities also have been observed in the pilin antigenic structures of *P. aeruginosa* and *B. cepacia* (Saiman et al., 1989).

Similarities have also been observed between *B. cepacia* and bacteria of groups other than the aerobic pseudomonads. With respect to the production of chaperonins (so named for their relationship with eucaryotic chaperones), an evolutionary homolog of the protein involved in plants in the assembly of ribulose-bisphosphate carboxylase-oxygenase (the key enzyme in CO_2 fixation) was identified in *E. coli*. This protein, GroEL, is one of the chaperonins, a group widely present in procaryotes and organelles of procaryotic origin (chloroplasts and mitochondria) (Hemmingsen et al., 1988). Aside from *E. coli* and pseudomonads, chaperonins have been identified in species of *Legionella, Bacillus, Borrelia, Treponema, Mycobacterium*, and *Coxiella* (Kaijser, 1975; Houston et al., 1990). In work performed on the cloning and nucleotide sequencing of the *groE* operons of *P. aeruginosa* and *B. cepacia*, a high degree of similarity was found, which extended to the *E. coli groEL*. The level of similarity with the human protein was lower, but still considerable. Comparable results were obtained in the study of a second chaperonin, GroES (Jensen et al., 1995).

The above-cited findings on chaperonins indicate that the similarities between groups I and II (*Pseudomonas* and *Burkholderia*) in some cases go beyond the dispensable catabolic systems or the biosynthesis of secondary products.

Differentiation of *Burkholderia* from related genera The examples of similarities mentioned in the previous section show that a sharp differentiation of members of *Burkholderia* and *Pseudomonas* is difficult. Studies on DNA–DNA hybridization (Ballard et al., 1970) and on rRNA–DNA hybridization (Palleroni et al., 1973) demonstrated that species of the genus were members of RNA similarity group II. They were eventually allocated to a newly proposed genus under the name *Burkholderia* (Yabuuchi et al., 1992), but two of the species (*B. pickettii* and *B. solanacearum*) were later transferred to the new genus *Ralstonia*, and the criteria for the differentiation of this genus from *Burkholderia* were defined (Yabuuchi et al., 1995). Unfortunately, as in the case of the proposal of the genus *Burkholderia*, the published differential characteristics were limited to those of a single strain of each species, which makes it impossible to estimate the intraspecies diversity.

In a polyphasic study of *Burkholderia* species, Gillis et al. (1995) redefined this genus. Among the characteristics of differential value, these authors mention the presence of $C_{16:0\ 3OH}$ in the cellular fatty acid composition. Most of the properties given in the definition of the genus *Burkholderia* apply equally well to other genera of aerobic pseudomonads, and, as in the case of the genus *Pseudomonas*, the fatty acid composition and the 16S rRNA characteristics remain among the few useful differential criteria.

---

3. Plate Count Agar (PCA, Difco Laboratories) is rehydrated according to the manufacturer's instructions. Aqueous stock solutions containing 0.1% C-390 or 7.5% PBS are prepared. One milliliter of each of these solutions is added to each liter of rehydrated PCA to reach a final concentration of 1 µg/ml of C-390 and 75 µg/ml PBS. The medium is autoclaved at 121°C for 15 min.

4. TB-T agar medium (g/l): glucose, 2.0; L-asparagine, 1; NaHCO₃, 1.0; KH₂PO₄, 0.5; MgSO₄·7H₂O, 0.1; trypan blue, 0.05; and agar, 20.0. The pH is adjusted to 5.5 with phosphoric acid solution (4 ml/l of a 10% solution). After autoclaving, tetracycline (20 mg/l) is added from a filter-sterilized stock solution. When fungi are abundant in the sample, crystal violet (5 mg/l) and filter-sterilized nystatin (50 mg/l) are added.

5. Ashdown medium (Ashdown, 1979a, b) is trypticase soy agar (BBL) with the following additions (per liter): glycerol, 40 g; crystal violet, 5 mg; neutral red, 50 mg; and gentamicin, 4 mg.

TAXONOMIC COMMENTS

In the 1960s, a project that focused on the construction of a rational system of classification of *Pseudomonas* species was organized at the Department of Bacteriology of the University of California at Berkeley. This project had its main justification in the highly unsatisfactory situation of the taxonomy of this genus to which were assigned several hundreds of species names, many of which could not be identified based on published descriptions, and their type strains had been lost.

A thorough phenotypic characterization of a large collection of strains resulted in a subdivision of the genus into species and species groups (Stanier et al., 1966). With time, species not included in the original project were subjected to the same analysis and located in the classification scheme. Eventually, the phenotypic characterization received confirmation from the results of DNA–DNA hybridization experiments. More significantly, the results of these experiments revealed a very wide range of DNA similarity values among the species assigned to the genus, suggesting a considerable degree of genomic heterogeneity among the members of the genus. This hypothesis was corroborated by rRNA–DNA hybridization experiments. They clearly showed that *Pseudomonas*, as classically described, could be subdivided into five RNA similarity groups representing at least five different genera (Palleroni et al., 1973).

This demonstration that the genus *Pseudomonas* was in fact a complex entity of suprageneric hierarchy was taken as the basis for proposals from various other laboratories to give different generic names to designate members of the five rRNA similarity groups. Of the five groups, rRNA group I retained the name *Pseudomonas* and the type species *P. aeruginosa* was originally proposed for the genus. For rRNA group II, the new genus name *Burkholderia* was introduced (Yabuuchi et al., 1992), comprising seven new combinations: *B. cepacia* (Palleroni and Holmes, 1981), *B. mallei* (Zopf, 1885), *B. pseudomallei* (Whitmore, 1913), *B. caryophylli* (Burkholder, 1942), *B. gladioli* (Severini, 1913), *B. pickettii* (Ralston et al., 1973), and *B. solanacearum* (Smith, 1896). Since the heterogeneity of the genus even extended to the rRNA similarity group level, some of the groups could still be further subdivided to include more than one bacterial genus. Thus, the last two of the above-mentioned species, were assigned to a newly created genus, *Ralstonia* (Yabuuchi et al., 1995). The overall similarity of *Ralstonia solanacearum* and *R. pickettii* already had been noticed in studies on their phenotypic characteristics and on DNA homologies (Ralston et al., 1973).

The present treatment refers to those species of rRNA group II assigned to the genus *Burkholderia*.

In their studies on the aerobic pseudomonads, Stanier et al. (1966) described a group of metabolically versatile organisms under the new species name *"Pseudomonas multivorans"*. In the plant pathology department of the University of California, David Sands applied the same methodology to the study of phytopathogenic pseudomonads. Soon his studies resulted in the unexpected finding that the properties of the above strains were virtually identical to those of a species that had been known to phytopathologists for almost two decades under the name of *P. cepacia*. The remarkable versatility of strains of this species had been totally overlooked by the early workers.

DNA–DNA hybridization experiments confirmed the synonymy soon thereafter (Ballard et al., 1970). The collection of strains of *B. cepacia* examined at the time could be differentiated from other species on the basis of the capacity to grow at the expense of D-arabinose, D-fucose, cellobiose, saccharate, mucate, 2,3-butylene glycol, sebacate, *meso*-tartrate, citraconate, *o*-hydroxybenzoate, *m*-hydroxybenzoate, L-threonine, DL-ornithine, and tryptamine. This set of characteristics sharply separated *B. cepacia* from all other members of the various phenotypic groups, as can be seen in Table 52 of the original reference (Stanier et al., 1966).

The early observations that helped to uncover the remarkable metabolic versatility of *B. cepacia* and other species of the genus *Burkholderia* also contributed substantial lists of phenotypic characteristics for use in the identification of newly isolated strains. Some of the results are summarized in the tables included in this chapter. Substantial collections of strains of the species are now available, and studies on intraspecific diversity have been published. The interesting metabolic properties of members of this group of procaryotes stimulated the isolation of many strains, which has contributed to the recent proliferation of names for new species and for groups (biovars, genomovars) at the intraspecific level. On the one hand, the nutritional versatility of *Burkholderia* species—which includes the ability to degrade toxic environmental contaminants—suggests an important function in the carbon cycle in nature. However, research directed to the use of these characteristics in bioremediation projects has to take into account the fact that some of the species are serious pathogenic agents for plants and animals, and proper precautions should be taken to counteract these activities.

Early studies on *B. cepacia* and its synonym *"Pseudomonas multivorans"* (Stanier et al., 1966; Ballard et al., 1970) clearly indicated considerable intraspecific heterogeneity in phenotypic characteristics such as pigmentation and nutritional properties, and in the DNA similarity results of DNA–DNA hybridization experiments. One hundred strains of *B. cepacia* were examined, and a proposal for a subdivision of the species into biovars was suggested (Richard et al., 1981). In recent times, other comparative studies showed marked differences between *B. cepacia* strains isolated from the clinical environment and those found in plant rhizospheres (Bevivino et al., 1994; Tabacchioni et al., 1995). Excellent reviews are available on the intraspecific diversity of this species (Yohalem and Lorbeer, 1994) and on its genomic complexity and versatility (Lessie et al., 1996). All these reports have had a taxonomic impact, resulting in the proposal of converting some of the intraspecific groups into independent taxa at the species level, as discussed below.

The newly proposed species names include *B. vietnamiensis* and *B. multivorans*, related to *B. cepacia*, *B. thailandensis*, a relative of *B. pseudomallei*, and *B. graminis*, which is close to the group *B. caryophylli–B. glathei–B. phenazinium*. Some of the descriptions fail to create a precise circumscription of the proposed taxa, but it is to be expected that the interest in these organisms will eventually contribute to generate characterizations that are more comprehensive.

A genotypic analysis of 128 strains of *Burkholderia*, *Ralstonia*, and *Pseudomonas* was taken as the basis for a definition of the taxonomic structure of *B. cepacia* (Vandamme et al., 1997b). The strains isolated from cases of cystic fibrosis could be grouped into five so-called genomic species, very similar from the phenotypic standpoint. One of the genomic species of this "*B. cepacia* complex" corresponded to the previously described *B. vietnamiensis*, and a second one received the formal name *B. multivorans*, a revival of a synonym of *B. cepacia* used many years before (Stanier et al., 1966). The rest of the genomovars (I, III, and IV) have remained unnamed.

A description of *B. multivorans* is given in this chapter (see

List of Species of the genus *Burkholderia*), and a more complete report of its properties can be found in the original description (Vandamme et al., 1997b). Four of the strains of genomovar II—now *B. multivorans*—exhibited a high level of relatedness by DNA–DNA hybridization. The organisms are phenotypically similar to other biovars of *B. cepacia* and cannot be readily differentiated from them.

The name *B. thailandensis* was recently proposed for a group of strains closely related to *B. pseudomallei* based on a high 16S rRNA sequence similarity (Brett et al., 1998). The organisms differ phenotypically from *B. pseudomallei* by relatively few characteristics, including an ability to use L-arabinose, 5-ketogluconate, adonitol, erythritol, and dulcitol. Differences in biochemical profile and virulence also occur. DNA–DNA hybridization studies of the two species have not been reported.

The position occupied by *B. graminis* in the phylogenetic tree of Fig. BXII.β.2 indicates that this species is located in a cluster that includes *B. glathei* and *B. phenazinium*. This is further supported by a set of phenotypic properties of differential value and clear-cut differences in the DNA reassociation values, as can be seen in Tables 2 and 3 of the original publication (Viallard et al., 1998).

Finally, a new species with rather unique properties, *B. norimbergensis* (*Pandoraea norimbergensis*), has been described for a strain isolated from an oxic water layer above a sulfide-containing sediment (Wittke et al., 1997). This organism oxidizes several inorganic sulfur compounds including sulfur and produces sulfate. The reported characteristics of the new species do not fit the organization of the tables in this chapter. The description is given in the list of species for the genus *Pandoraea*, together with some comments on its phylogenetic relationships. The 16S rRNA sequence of a new species of *Burkholderia* that has been named *B. caribensis* has been deposited in GenBank under the accession numbers Y17009, Y17010, and Y17011. The sequence places this species near *B. graminis* (V. Viallard and J. Balandreau, personal communication).

Following the recommendation formulated by Vandamme et al. (1997b), buttressed by the evidence reported by Viallard et al. (1998), three species previously assigned to *Pseudomonas* are now being included in the list of *Burkholderia* species as *B. glathei*, *B. pyrrocinia*, and *B. phenazinium*.

FURTHER READING

Coenye, T., S. Laevens, A. Willems, M. Ohlen, W. Hannant, J.R.W. Govan, M. Gillis, E. Falsen and P. Vandamme. 2001. *Burkholderia fungorum* sp. nov., and *Burkholderia caledonica* sp. nov., two new species isolated from the environment, animals and human clinical samples. Int. J. Syst. Evol. Microbiol. *51*: 1099–1107.

Coenye, T., J.J. LiPuma, D. Henry, B. Hoste, K. Vandemeulebroecke, M. Gillis, D.P. Speert and P. Vandamme. 2001. *Burkholderia cepacia* genomovar VI, a new member of the *Burkholderia cepacia* complex isolated from cystic fibrosis patients. Int. J. Syst. Evol. Microbiol. *51*: 271–279.

Coenye, T., E. Mahenthiralingam, D. Henry, J.J. LiPuma, S. Laevens, M. Gillis, D.P. Speert and P. Vandamme. 2001. *Burkholderia ambifaria* sp. nov., a novel member of the *Burkholderia cepacia* complex including

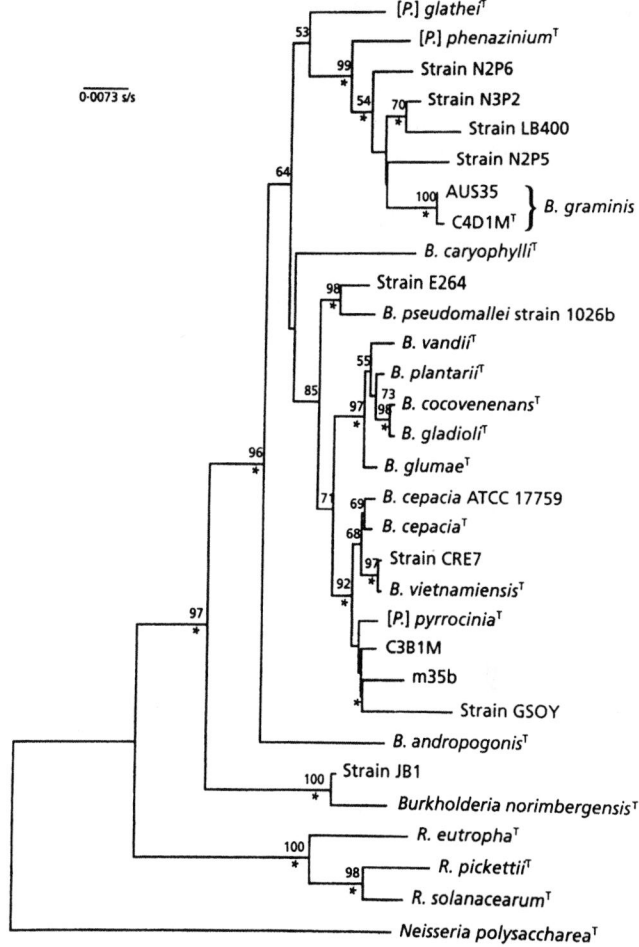

FIGURE BXII.β.2. Neighbor-joining tree obtained with the 16S rDNA sequences of *Burkholderia*, [*Pseudomonas*], and *Ralstonia* species. Bootstrap values (1000 resamplings) greater than 50% are indicated at the nodes, and asterisks indicate values higher than 50% found in the parsimony analysis. Bar = 0.0073 fixed mutations per nucleotide position. Three additional strains of *B. vietnamiensis* had the same sequence as the type strain represented in the tree. The tree shows the position of several unnamed strains whose source is indicated in the original paper (Viallard et al., 1998). Of these, strain E264 has been named recently *B. thailandensis* (Brett et al., 1997). [*P.*] *glathei*, [*P.*] *phenazinium*, and [*P.*] *pyrrocinia* are described in this chapter as *Burkholderia* species. (Reproduced with permission from V. Viallard et al., International Journal of Systematic Bacteriology *48*: 549–563, 1998, ©International Union of Microbiological Societies.)

biocontrol and cystic fibrosis-related isolates. Int. J. Syst. Evol. Microbiol. *51*: 1481–1490.

Zhang, H., S. Hanada, T. Shigematsu, K. Shibuya, Y. Kamagata, T. Kanagawa and R. Kurane. 2000. *Burkholderia kururiensis* sp. nov., a trichloroethylene (TCE)-degrading bacterium isolated from an aquifer polluted with TCE. Int. J. Syst. Evol. Microbiol. *50*: 743–749.

List of species of the genus Burkholderia

1. **Burkholderia cepacia** (Palleroni and Holmes 1981) Yabuuchi, Kosako, Oyaizu, Yano, Hotta, Hashimoto, Ezaki and Arakawa 1993, 398[VP] (Effective publication: Yabuuchi, Kosako, Oyaizu, Yano, Hotta, Hashimoto, Ezaki and Arakawa 1992, 1271) (*Pseudomonas cepacia* Palleroni and Holmes 1981, 479.)

ce.pa′ci.a. L. fem. n. *caepa* or *cepa* onion; M.L. fem. adj. *cepacia* of or like an onion.

Properties useful for differentiation from other species of the genus are given in Table BXII.β.2. Characteristics of the species are presented in Table BXII.β.3. As suggested in the description of "*Pseudomonas multivorans*" (Stanier et al., 1966), strains of *B. cepacia* can be differentiated from other species on the basis of their ability to grow at the expense of D-arabinose, D-fucose, cellobiose, saccharate, mucate, 2,3-butylene glycol, sebacate, *m*-tartrate, citraconate, *o*-hydroxybenzoate, *m*-hydroxybenzoate, L-threonine, DL-ornithine, and tryptamine. These characteristics, although not present in all strains, are useful for the identification of newly isolated strains. For further descriptive information see Ballard et al. (1970) and Palleroni and Holmes (1981). Optimal growth temperature, ~30–35°C.

Many strains have been isolated from rotten onions, soils, various natural materials, and clinical specimens. The species is considered a serious opportunistic human pathogen, and it has been found associated with infections of nosocomial origin. A study by Vandamme et al. (1997b) on strains isolated from cystic fibrosis cases has resulted in the identification of multiple genomovars of *B. cepacia*, among which some (genomovar II) are being described under the new species name *Burkholderia multivorans* (see comments in the description of this species below).

The mol% G + C of the DNA is: 67.4 (Bd).

Type strain: ATCC 25416, Ballard 717, DSM 7288, ICPB 25, NCTC 10743.

GenBank accession number (16S rRNA): M22518, U96927.

2. **Burkholderia andropogonis** (Smith 1911) Gillis, Van, Bardin, Goor, Hebbar, Willems, Segers, Kersters, Heulin and Fernandez 1995, 287[VP] (*Pseudomonas andropogonis* (Smith 1911) Stapp 1928, 27; *Pseudomonas andropogoni* (sic) Smith 1911, 63.)

an.dro.po′go.nis. M.L. n. *Andropogon* genus of widely distributed grasses; M.L. gen. n. *andropogonis* of the genus *Andropogon*.

The following description is slightly modified from the one given by Palleroni (1984) in the first edition of this *Manual*.

Slender rods with rounded ends, 0.5–0.7 × 1–2 μm, with one or rarely two polar flagella. Colonies change from butyrous to viscid with age. No fluorescent pigment is produced. Most strains are oxidase negative. Gelatin liquefaction, nitrate reduction, lipolysis, and arginine dihydrolase reactions negative. Production of sheathed flagella (flagella surrounded by a membrane that is a continuation of the cell wall) was reported by Fuerst and Hayward (1969a).

Although the species had been tentatively assigned to the rRNA similarity group III based on its enzymatic properties, it is now considered a member of the genus *Burkholderia* (Gillis et al., 1995). Further details of its phenotypic properties are given by Palleroni (1984) and in Tables BXII.β.2 and BXII.β.3. A comprehensive description is given in the Ph.D. thesis of Xiang Li (1993).

Recommendations to consider *P. stizolobii* and *P. woodsii* as synonyms of *B. andropogonis* were formulated by several workers (Goto and Starr, 1971; Hayward, 1972; Nishiyama et al., 1979; Shanks and Hale, 1984), and this opinion was reinforced by the demonstration of similar protein profiles (Vidaver and Carlson, 1978) and polyamine composition (Auling et al., 1991).

Pathogenic for sorghum, corn, clover, and velvet bean (*Stizolobium deeringianum*). To the host list should be added highbush blueberry, in which the pathogen causes leaf spot lesions in hardwood cuttings (Kobayashi et al., 1995). Per-

haps the host specialization may justify creating two pathovars—pathovar *andropogonis* for the causal agent of the bacterial stripe of sorghum, and pathovar *stizolobii* for that of the bacterial leaf spot of velvet bean (Palleroni, 1984). As mentioned in a previous section, strains of *B. andropogonis* are able to synthesize rhizobitoxine, a phytotoxin capable of causing foliar chlorosis.

The mol% G + C of the DNA is: 59–61.3 (T_m).

Type strain: ATCC 23061, DSM 9511, LMG 2129.

GenBank accession number (16S rRNA): X67037.

3. **Burkholderia caribensis** Achouak, Christen, Barakat, Martel and Heulin 1999, 792[VP]

ca.ri.ben′sis. M.L. adj. *caribensis* pertaining to the Caribbean Islands, where the strains were isolated.

The description is taken from the original paper.

Short rods, 1–2 × 0.5 μm. Motile and pleomorphic in actively growing cultures in LB medium. In sugar-rich media (with 2% glucose, xylose, fructose, sorbitol, arabinose, mannitol, or sorbitol) it produces abundant exopolysaccharide.

Oxidase, catalase, urease, arginine dihydrolase, and β-galactosidase are produced. A list of organic compounds that have been tested as substrates of oxidative activities is given in the paper. Isolated from a vertisol fraction in the island of Martinique.

The mol% G + C of the DNA is: 63.1 (T_m).

Type strain: LMG 18531, MWAP 64.

GenBank accession number (16S rRNA): Y17009.

4. **Burkholderia caryophylli** (Burkholder 1942) Yabuuchi, Kosako, Oyaizu, Yano, Hotta, Hashimoto, Ezaki and Arakawa 1993, 398[VP] (Effective publication: Yabuuchi, Kosako, Oyaizu, Yano, Hotta, Hashimoto, Ezaki and Arakawa 1992, 1273) (*Pseudomonas caryophylli* (Burkholder 1942) Starr and Burkholder 1942, 601; *Phytomonas caryophylli* Burkholder 1942, 143.)

ca.ry.o′phyl.li. M.L. masc. n. *caryophyllus* specific epithet of *Dianthus caryophyllus*, carnation; M.L. gen. n. *caryophylli* of the carnation.

The general characteristics of the species are presented in Tables BXII.β.2, BXII.β.3, and BXII.β.6. The main properties for differentiation from several other *Burkholderia* species and other denitrifying pseudomonads, are summarized in Table BXII.β.4. Optimal growth temperature ~30–33°C. For further descriptive information, see Ballard et al. (1970). Isolated from diseased carnations.

The mol% G + C of the DNA is: 65.3 (Bd).

Type strain: ATCC 25418, DSM 50341, ICPB PC113, NCPPB 2151, PDDCC 512.

GenBank accession number (16S rRNA): X67039.

5. **Burkholderia cocovenenans** (van Damme, Johannes, Cox and Berends 1960) Zhao, Qu, Wang and Chen 1995, 601[VP] (*Pseudomonas cocovenenans* van Damme, Johannes, Cox and Berends 1960, 255.)

co.co.ve′ne.nans. M.L. n. *Cocos* genus of coconut; L. v. *veneno* to poison; M.L. part. adj. *cocovenenans* coconut poisoning.

The following description is taken from the paper by Zhao et al. (1995). Rods 0.3–0.5 × 1.6–2.0 μm, occurring singly or in pairs. Motile by means of one to five polar flagella. No lipid soluble or fluorescent pigment is produced; however, a greenish-yellow diffusible pigment (tox-

TABLE BXII.β.6. Nutritional characteristics of *Burkholderia cepacia*, *Burkholderia gladioli*, and *Burkholderia caryophylli*[a]

Characteristics	B. cepacia	B. gladioli	B. caryophylli
Utilization of:			
D-Ribose, D-arabinose, L-arabinose, D-fucose, D-glucose, *N*-acetylglucosamine[b], D-mannose, D-galactose, D-fructose, sucrose, cellobiose, gluconate, 2-ketogluconate, sacchare, mucate, salicin, acetate, propionate, butyrate, isobutyrate, valerate, malonate, succinate, fumarate, D-malate, L-malate, *m*-tartrate, β-hydroxybutyrate, lactate, glycerate, hydroxymethyl-glutarate, citrate, α-ketoglutarate, pyruvate, aconitate, mannitol, sorbitol, *m*-inositol, adonitol[b], glycerol, *p*-hydroxybenzoate, phenylacetate, quinate, L-alanine, β-alanine, L-serine, L-cysteine[b], L-aspartate, L-glutamate, L-arginine, γ-aminobutyrate, L-histidine, L-proline, L-tyrosine, L-phenylalanine, L-tryptophan, betaine, hippurate	+	+	+
D-Xylose, *n*-propanol	d	+	+
L-Rhamnose, glycolate	d	−	+
Trehalose, L-threonine	+	+	d
Isovalerate, glutarate, citrulline, anthranilate, sarcosine	+	d	−
Heptanoate, caproate, caprylate, pelargonate, caprate, adipate, azelate, sebacate, citraconate, adonitol, dulcitol[b], benzoate, D-alanine, ornithine, kynurenate, ethanolamine, 2-aminobenzoate	+	+	−
Lyxose[b], tagatose[b], 5-ketogluconate[b], melibiose[b], gentiobiose[b], raffinose[b], amygdalin[b], arbutin[b], aesculin[b], salicin[b], L-arabitol[b], pimelate, suberate, levulinate, *m*-hydroxybenzoate, δ-aminovalerate, putrescine, spermine, butylamine, tryptamine, α-amylamine, diaminobutane[b]	+	−	
D(−)-Tartrate, mesaconate	−	+	−
L(+)-Tartrate, ethanol, L-isoleucine, nicotinate, trigonelline	d	+	−
Itaconate, propylene glycol, glycine, norleucine, α-aminobutyrate, α-aminovalerate	−	d	−
L-Fucose[b], 2,3-butylene glycol, D-arabitol[b], xylitol[b]	+	−	+
n-Butanol	d	d	+
Isobutanol	d	d	−
L-Mandelate, benzoylformate	d	−	d
o-Hydroxybenzoate, testosterone, benzylamine, histamine, acetamide	d	−	−
L-Leucine, L-valine	d	+	d
Dodecane, hexadecane	d		

[a]For symbols see standard definitions.

[b]Data for the type strains, taken from Yabuuchi et al. (1992).

oflavin) is formed after 1–2 d of incubation on potato dextrose agar or other media.

Growth temperature range goes from 6–41°C. Metabolism is respiratory and denitrification is negative, but nitrate is reduced to nitrite. Oxidase reaction is negative. Gelatinase, Tween 80 hydrolysis, and lecithinase reactions are all positive. Many organic compounds can be used as sole carbon and energy sources. Detailed information can be found in the original report (Zhao et al., 1995), in Gillis et al. (1995), and in Tables BXII.β.2 and BXII.β.3.

Strains have been isolated from fermented coconut food (bongkrek) in Indonesia, fermented cornmeal in China, deteriorated white fungus (*Tremella fuciformis*), and soil. DNA–DNA hybridization experiments have shown that this species is close to *B. cepacia* and *B. gladioli* (Zhao et al., 1995). The organism is not infectious to humans, but it is responsible for serious cases of food poisoning due to the production of a yellow poisonous compound, toxoflavin.

Recent studies by Coenye et al. (1999b) based on whole protein electrophoretic profiles, DNA–DNA hybridization, and comparison of many biochemical properties indicate that *B. cocovenenans* should be considered a junior synonym of *B. gladioli*.

The mol% G + C of the DNA is: 69 ± 0.5 (T_m).

Type strain: ATCC 33664, DSM 11318, LMG 11626, NCIB 9450.

GenBank accession number (16S rRNA): U96934.

6. **Burkholderia gladioli** (Severini 1913) Yabuuchi, Kosako, Oyaizu, Yano, Hotta, Hashimoto, Ezaki and Arakawa 1993, 398[VP] (Effective publication: Yabuuchi, Kosako, Oyaizu, Yano, Hotta, Hashimoto, Ezaki and Arakawa 1992, 1273) (*Pseudomonas gladioli* Severini 1913, 420.)

gla.di.o' li. L. n. *gladiolus* a small sword lily; M.L. masc. n. *gladioli* of gladiolus.

The characteristics of the species are summarized in Tables BXII.β.2, BXII.β.3, and BXII.β.6. Optimal growth temperature, ~30–35°C. Further descriptive information may be found in Ballard et al. (1970), where the species appears under the name *"Pseudomonas marginata"* (see also Hildebrand et al., 1973).

Isolated from decayed onions, *Gladiolus* spp, and *Iris* spp., for which the species is believed to be pathogenic. Two pathovar names have been proposed for the phytopathogenic strains, pathovar *gladioli* and pathovar *aliicola* (Young et al., 1978). To these, the new pathovar name pathovar *agaricicola* has been created to distinguish the *B. gladioli* strains producing rapid soft rot of cultivated mushrooms (*Agaricus bitorquis*) (Lincoln et al., 1991). *B. gladioli* was identified as the agent of leaf spot and blight of the bird's nest fern *Asplenium nidus* (Chase et al., 1984). As indicated in the section on pathogenicity for humans, strains of the species have been found to be serious opportunistic pathogens.

As a phytopathogen, *B. gladioli* pathovar *alliicola* behaves in a manner similar to that of *B. cepacia*, causing soft rot of onions (61-gl; 21-gl) (Tesoriero et al., 1982; Wright et al., 1993). It has also been identified as a probable agent of early blight in cherries (26-gl) (Li and Scholberg, 1992), and to cause rapid soft rot disease of the edible mushroom *Agaricus bitorquis* (38-gl; 29-gl) (Lincoln et al., 1991; Atkey et al., 1992). *B. gladioli* also contributes to postharvest dis-

TABLE BXII.β.7. Nutritional properties of *Burkholderia mallei* and *Burkholderia pseudomallei* [a]

Characteristics[b]	B. mallei	B. pseudomallei
Utilization of:		
Acetate, *N*-acetylglucosamine[c], adipate, D-alanine, L-alanine, β-alanine, γ-aminobutyrate, D-arabinose, L-arginine, L-aspartate, benzoate, betaine, cellobiose, D-fucose, fumarate, galactose, glucose, L-glutamate, glycerate, glycerol, hippurate, L-histamine, β-hydroxybutyrate, *p*-hydroxybenzoate, *m*-inositol, α-ketoglutarate, lactate, L-malate, mannitol, mannose, phenylacetate, poly-β-hydroxybutyrate, propionate, pyruvate, quinate, sorbitol, succinate, sucrose, L-threonine, trehalose, tryptamine, L-tyrosine	+	+
Aconitate, 2-aminobenzoate[c] α-amylamine, D-arabitol[c], benzoylformate, butylamine, caprate, caproate, L-cysteine[c], dulcitol[c], erythritol, ethanol, ethanolamine, L-fucose[c], glycogen[c], heptanoate, isobutyrate, L-isoleucine, isovalerate, kynurenate, kynurenine, levulinate, L-lysine, pelargonate, putrescine, ribose, sebacate, valerate	−	+
Glycine, D-xylose	+	−
α-Aminobutyrate, L-arabinose, malonate	d	−
δ-Aminovalerate, anthranilate, butyrate, caprylate, citrate, fructose, gluconate, 2-ketogluconate, maltose, L-phenylalanine, L-proline, putrescine, salicin, L-serine, suberate, starch, L-valine	d	+
Azelate, glutarate, D-malate, pimelate, sarcosine, trigonelline	d	d
Adonitol, ethanol, hexadecane, L-mandelate, spermine	−	d

[a]For symbols see standard definitions.

[b]The following compounds are not used by either species: acetamide, aesculin[c], *m*-aminobenzoate, *p*-aminobenzoate, α-amino-valerate, amygdalin[c], arbutin[c], benzylamine, *n*-butanol, 2,3-butylene glycol, citraconate, L-citrulline, creatine, dodecane, ethylene glycol, gentiobiose[c], geraniol, glycolate, histamine, *m*-hydroxybenzoate, *o*-hydroxybenzoate, hydroxymethylglutarate, inulin, iso-butanol, isophthalate, isopropanol, 5-ketogluconate, lactose, L-leucine, lyxose[c], maleate, D-mandelate, melibiose[c], mesaconate, methanol, methylamine[c], mucate, nicotinate, norleucine, L-ornithine, oxalate[c], pantothenate, phenol, phenylethanediol, phthalate, *n*-propanol, propylene glycol, raffinose[c], L-rhamnose, saccharate, salicin[c], tagatose[c], D(−)-tartrate, L(+)-tartrate, *meso*-tartrate, terephthalate, testosterone, tryptamine, D-tryptophan, xylitol[c].

[c]Data for the type strains, taken from Yabuuchi et al. (1992).

eases of fruits and vegetables, an activity in which the species has been identified by analyses of fatty acid composition (24-gl) (Wells et al., 1993).

The mol% G + C of the DNA is: 68.5 (Bd).

Type strain: ATCC 10248, DSM 4285, NCPPB 1891, PDDCC 3950.

GenBank accession number (16S rRNA): X67038.

Additional Remarks: This is also the reference strain for *B. gladioli* pathovar *gladioli*. The reference strain for *B. gladioli* pathovar *alliicola* is ATCC 19302 (PDDCC 2804; NCPPB 947). The reference strain for *B. gladioli* pathovar *agaricicola* is NCPPB 3580.

7. **Burkholderia glathei** (Zolg and Ottow 1975) Vandamme, Holmes, Vancanneyt, Coenye, Hoste, Coopman, Revets, Lauwers, Gillis, Kersters and Govan 1997b, 1199[VP] (*Pseudomonas glathei* Zolg and Ottow 1975, 296.)

gla'the.i. M.L. gen. *glathei* of Glathe, named after H. Glathe of Giessen, Germany.

The description is taken from Zolg and Ottow (1975). Rods to oval cocci, 0.5–0.7 × 1.5 μm, motile by a polar flagellum. Optimal growth temperature 30–37°C. Oxidase reaction positive. Negative for hydrolysis of starch, gelatin, lecithin (egg yolk reaction), esculin, and polypectate. Tributyrin, urea, and hippurate are hydrolyzed. Nitrate is reduced to nitrite. Denitrification and H₂S production are negative. No growth factor requirement has been found. The organism is capable of growth in nitrogen-deficient media, but acetylene reduction by cells grown under those conditions is negative. Acid tolerant (pH 4.5).

From the extensive phenotypic characterization of *B. glathei* following the methodology described by Stanier et al. (1966), it has been found that at least 68 organic com-pounds can be utilized as sole carbon and energy sources for growth. These include aldoses, ketoses, deoxysugars, sugar-alcohols, and sugar-acids. Except for lactose, melibiose, and melezitose, no di-, tri-, and polysaccharides are utilized. The only amino acids that are not utilized are glycine, L-serine, L-isoleucine, L-methionine, and β-alanine. Of 28 aliphatic organic acids, 23 are utilized. The list of utilizable acids includes oxalate. Some properties are summarized in Table BXII.β.3.

A 1,2-oxygenase responsible for the *ortho* cleavage of aromatic compounds is produced constitutively. This species differs in several important characteristics from other species of the genus. The main differences include the capacity for growing in nitrogen-deficient media, the acid tolerance, and the utilization of oxalate for growth. A study by Vandamme et al. (1997b), which includes fatty acid analysis, indicates that this species should be allocated to the genus *Burkholderia*, an opinion that has found confirmation by rRNA sequencing studies (Viallard et al., 1998).

The mol% G + C of the DNA is: 64.8 (T_m).

Type strain: N15, ATCC 29195, DSM 50014, LMG 14190.

GenBank accession number (16S rRNA): U96935, Y17052.

8. **Burkholderia glumae** (Kurita and Tabei 1967) Urakami, Ito-Yoshida, Araki, Kijima, Suzuki and Komagata 1994, 242[VP] (*Pseudomonas glumae* Kurita and Tabei 1967, 111.)

glu'mae. L. n. *gluma* hull; L. gen. n. *glumae* of a husk.

The following description is taken from the original paper by Kurita and Tabei (1967). Rods, 0.5–0.7 × 1.5–2.5 μm, motile by means of two to four flagella. A fluorescent pigment is produced in potato agar. Nitrate reduction, starch hydrolysis, and H₂S production are negative. Gelatin liquefaction was not reported in the original paper, but Urakami et al. (1994) found that the test was positive. Temperature limits for growth: 11–40°C; optimum: 30–35°C. Acid is produced from arabinose, glucose, fructose, galactose, mannose, xylose, glycerol, mannitol, and inositol. No acid is produced from rhamnose, sucrose, maltose, lactose, raffinose, dextrin, starch, inulin, or salicin. Milk is coagulated and peptonized. Pathogenic for the rice plant (*Oryza sativa*, fam. Gramineae).

To the above description, Urakami et al. (1994) have added a substantial amount of information on the nutritional spectrum of the type strain, and part of it has been summarized in Tables BXII.β.2 and BXII.β.3. A peculiar property of this species is the production of fluorescent pigment, but no further details are available on its relationship to the pigment characteristic of the fluorescent species of the genus *Pseudomonas*. This is a point of interest for further investigation. As mentioned before, two subgroups of *B. glumae* strains can be distinguished based on their fatty acid composition (Stead, 1992).

The mol% G + C of the DNA is: 68.2 (reversed HPLC) (Urakami et al., 1994).

Type strain: ATCC 33617, DSM 7169, LMG 2196, NCPPB 2981, NIAES 1169.

GenBank accession number (16S rRNA): U96931.

9. **Burkholderia graminis** Viallard, Poirier, Cournoyer, Haurat, Wiebkin, Ophel-Keller and Balandreau 1998, 560[VP]

gra' mi.nis. M.L. adj. *graminis* referring to its isolation from the rhizosphere of grasses.

The description is taken from the original paper (Viallard et al., 1998). Rods, 0.3–0.8 × 1.0–1.5 μm; motile. The number of flagella per cell is not reported. Colonies on LB agar are thin, brownish-yellow, and translucent. On agar prepared with the special medium PCAT (Burbage and Sasser, 1982) after 3 d at 28°C, the colonies are white, somewhat opaque and creamy, with entire margin. Oxidase, catalase, urease, and arginine deiminase reactions are positive. Nitrates are reduced, but there is no denitrification. Characteristics in common with other *Burkholderia* species are the assimilation of glycerol, D- and L-arabinose, ribose, galactose, glucose, fructose, mannose, inositol, mannitol, sorbitol, D-arabitol, gluconate, and 2-ketogluconate, and the inability to use L-sorbose, methyl-α-D-xyloside, methyl-α-D-mannoside, methyl-α-D-glucoside, inulin, melezitose, starch, glycogen, or D-turanose. Properties not found in some of the other species include the incapacity of acid formation from glucose, to hydrolyze esculin or to produce gelatinase. In addition, the strains grow on L-xylose, lactose, rhamnose, trehalose, D-lyxose, L-arabitol, xylitol, and raffinose, but not on dulcitol or D-tagatose. Isolated from the rhizosphere of wheat, corn, and pasture grasses.

The mol% G + C of the DNA is: 62.5–63.0 (HPLC).

Type strain: C4D1M, ATCC 700544.

GenBank accession number (16S rRNA): U96939.

10. **Burkholderia kururiensis** Zhang, Hanada, Shigematsu, Shibuya, Kamagata, Kanagawa and Kurane 2000a, 747[VP]

ku.ru.ri.en' sis. M.L. adj. *kururiensis* referring to Kururi, Chiba Prefecture, Japan, where the strain was isolated.

The following description is taken from the original paper. The cells are Gram negative and ovoid or rod shaped. (1 × 1.2–1.5 μm), occurring singly or in pairs. Growth

occurs between 15–42°C, with an optimum at 37°C. The optimal growth pH is 7.2. Under optimal conditions, the doubling time is about 1 h. Oxidase and catalase positive. Starch and gelatin are not hydrolyzed, but glycogen and Tween 80 are. The following organic compounds are degraded aerobically: arabinose, fructose, fucose, galactose, glucose, lactulose, maltose, mannose, psicose, rhamnose, adonitol, arabitol, glycerol, inositol, mannitol, sorbitol, xylitol, N-acetylgalactosamine, acetate, citrate, formate, galacturonate, gluconate, lactate, propionate, alanine, asparagine, aspartate, glutamate, glycine, histidine, leucine, phenylalanine, proline, serine, threonine, inosine, 2,3-butanediol, benzene, *p*-cresol, fluorobenzene, and phenol. The following compounds are not oxidized: cellobiose, lactose, melibiose, raffinose, sucrose, trehalose, dextrin, malonate, uridine, thymidine, glucose-1-phosphate, and glucose-6-phosphate. UQ-8 is the dominant respiratory quinone. Main cellular fatty acids are $C_{16:0}$, cyclopropanic acids $C_{17:0}$, cyclopropanic acid $C_{19:0}$, $C_{16:1}$, and $C_{18:1}$. $C_{13:1}$ and $C_{17:1}$ are also present. The organism was isolated from an aquifer polluted with trichloroethylene (TCE) in Kururi, Chiba Prefecture, Japan, and shows degradation activity for this contaminant when the cells are grown in the presence of phenol.

The mol% G + C of the DNA is: 64.8 (HPLC).

Type strain: KP23, ATCC 700977, CIP 106643, DSM 13646, LMG 19447.

GenBank accession number (16S rRNA): AB024310.

This species has some properties that are not common in the genus *Burkholderia*. The cells are nonmotile, a characteristic shared only with *B. mallei*. The strain uses maltose, which is rarely used by strains of other species, and fails to use disaccharides used by the latter. No DNA–DNA hybridization data have been reported.

11. **Burkholderia mallei** (Zopf 1885) Yabuuchi, Kosako, Oyaizu, Yano, Hotta, Hashimoto, Ezaki and Arakawa 1993, 398[VP] (Effective publication: Yabuuchi, Kosako, Oyaizu, Yano, Hotta, Hashimoto, Ezaki and Arakawa 1992, 1271) (*Pseudomonas mallei* (Zopf 1885) Redfearn, Palleroni and Stanier 1966, 305; *Bacillus mallei* Zopf 1885, 89.)

mal' le.i. L. n. *malleus* the disease of glanders; L. gen. n. *mallei* of glanders.

Characteristics of the species are listed in Tables BXII.β.2, BXII.β.3, and BXII.β.7, and those that are useful for the differentiation of the species from various other denitrifying pseudomonads are summarized in Table BXII.β.4. Optimal growth temperature ~37°C. For further descriptive information, see Redfearn et al. (1966) and Redfearn and Palleroni (1975). Parasitic on horses and donkeys, in which it causes glanders and farcy. The infection is transmissible to humans and to other animal species.

The mol% G + C of the DNA is: 69 (Bd).

Type strain: NBL 7, ATCC 23344.

GenBank accession number (16S rRNA): S55000.

12. **Burkholderia multivorans** Vandamme, Holmes, Vancanneyt, Coenye, Hoste, Coopman, Revets, Lauwers, Gillis, Kersters and Govan 1997b, 1198[VP]

mul.ti' vo.rans. L. adj. *multus* much; L. part. adj. *vorans* devouring, digesting; M.L. part. adj. *multivorans* digesting many compounds.

Rods, 0.6–0.9 × 1.0–2.0 µm. Motile at room temperature and at 37°C. Number of flagella per cell not reported. Able to grow at 42°C but not at 5°C. According to the original description, some strains grow at room temperature. No pigments are produced. Growth in some media used for enteric bacteria is reported. No information is given on utilization of carbon compounds for growth, but, instead, production of acids from some compounds is reported. Tolerance to cyanide is strain dependent. No fluorescence occurs in King's B medium. Tween 20 and 80 are hydrolyzed. Urease, catalase, oxidase, and lecithinase activities are all positive. Nitrate is reduced, but not nitrite. Gelatin liquefaction is negative. No hydrolysis of casein, starch, or esculin occurs. Arginine deiminase is negative. PHB is accumulated and, according to the description, is utilized. The report does not specifically state whether this refers to extracellular PHB.

The strains assigned to this species belonged to one of the genomovars (genomovar II) into which was subdivided a collection of *B. cepacia* isolated from cases of cystic fibrosis. Four strains of this genomovar shared high DNA sequence similarity as determined by hybridization methods, and the melting temperatures of rRNA–DNA hybrid molecules using *B. cepacia* as reference were lower than the homologous *B. cepacia* reassociated molecules. The phenotypic differences between *B. multivorans* and *B. cepacia* are limited to casein digestion, growth at 42°C, and acid production from sucrose and raffinose. In fact, these last two properties may be redundant, since the two saccharides may be hydrolyzed by the same enzyme, for instance, invertase. Acid may be produced from the resulting monosaccharides (glucose and fructose). Since the physiological properties used for the description of this species differ from those used for other species of the genus, it is not possible to use them in a more extensive comparison, and therefore the species has been excluded from the comparative Table BXII.β.3.

The mol% G + C of the DNA is: 68–69 (method unknown).

Type strain: LMG 13010, ATCC BAA-247, CCUG 34080, CIP 105495, DSM 13243, NCTC 13007.

GenBank accession number (16S rRNA): AF14855.

13. **Burkholderia phenazinium** (Bell and Turner 1973) Viallard, Poirier, Cournoyer, Haurat, Wiebkin, Ophel-Keller and Balandreau 1998, 5618[VP] (*Pseudomonas phenazinium* Bell and Turner 1973, 753.)

phe.na.zi′ni.um. Orthography and etymology uncertain; possibly refers to iodinin, which is a phenazine pigment.

The original bacteriological information on this species is fragmentary. It is included in a paper describing the production of iodinin under various conditions (Bell and Turner, 1973). The strain was isolated from soil after an enrichment using L-threonine as the sole carbon source. This property is mentioned by Viallard et al. (1998) in their short description. It has no value whatsoever as a diagnostic character, since many aerobic pseudomonads use this amino acid and the property is usually shared by all the strains of a given species.

An unusual feature mentioned in the original description of the species is its acidophilic character. The strain was acidophilic, the optimal pH for growth being 5.0. It failed to grow at pH 7.0. This feature apparently has not been mentioned by later researchers who have worked with this strain. Colonies on agar media produced iodine crystals, and the presence of L-threonine or glycine in the medium seemed to enhance pigment synthesis. In experiments designed to test iodinin production, growth was observed with threonine, glycine, fumarate, glycerol, sucrose, malate, succinate, DL-lactate, citrate, glucose, glutamate, pyruvate, and serine. Growth was poor with aminopropanol, and no growth was observed with propionate. Additional information, taken from Viallard et al. (1998), is summarized in Table BXII.β.2. The description does not mention specifically use of extracellular PHB as a growth substrate. Acid production from various sugars is given in the original paper.

The mol% G + C of the DNA is: (HPLC).

Type strain: ATCC 33666, DSM 10684, LMG 2247, NCIB 11027.

GenBank accession number (16S rRNA): U96936.

As a general comment, the phenotypic properties of the description for *B. phenazinium* do not clearly differentiate this species from *B. cepacia*, and no data on DNA similarity expressed by percent DNA hybridization are reported. Only the $T_{m(e)}$ values of the DNA–rRNA hybrids obtained with 23S rRNA seem to differentiate this species from *B. cepacia*. For the moment, it is advisable to consider this taxon as a genomovar of *B. cepacia* until more evidence is available to decide on its independent species rank.

14. **Burkholderia plantarii** (Azegami, Nishiyama, Watanabe, Kadota, Ochuchi and Fukazawa 1987) Urakami, Ito-Yoshida, Araki, Kijima, Suzuki and Komagata 1994, 242[VP] (*Pseudomonas plantarii* Azegami, Nishiyama, Watanabe, Kadota, Ochuchi and Fukazawa 1987, 151.)

plan′ tar.i.i. L. n. *plantarium* seedbed; L. gen. n. *plantarii* of seedbed.

The following description is summarized from that in the original paper by Azegami et al. (1987). Nonencapsulated straight rods (0.7–1.0 × 1.4–1.9 µm), motile with one to three polar flagella. They occur singly, in pairs, or in short chains. Colonies have a slightly yellow tint, and they produce a water-soluble reddish brown pigment depending on the conditions and the medium. Oxidase positive. Do not produce fluorescent pigment. Gelatinase, lecithinase, hydrolysis of Tween 80, and denitrification are all positive. Arginine dihydrolase negative. Acid production on a number of carbon compounds and nutritional properties are described in the original paper. Further details are summarized in Tables BXII.β.2, BXII.β.3, and BXII.β.4. The species causes rice seedling blight, and strains have been isolated from rice seedlings and from bed soil in nursery boxes in Japan.

The mol% G + C of the DNA is: 64.8 (Bd).

Type strain: ATCC 43733, AZ 6201, DSM 9509, JCM 5492, LMG 9035, NIAES 1723.

GenBank accession number (16S rRNA): U96933.

15. **Burkholderia pseudomallei** (Whitmore 1913) Yabuuchi, Kosako, Oyaizu, Yano, Hotta, Hashimoto, Ezaki and Arakawa 1993, 398[VP] (Effective publication: Yabuuchi, Kosako, Oyaizu, Yano, Hotta, Hashimoto, Ezaki and Arakawa 1992, 1273) (*Pseudomonas pseudomallei* (Whitmore 1913) Haynes and Burkholder 1957, 100; *Bacillus pseudomallei* Whitmore 1913, 9.)

pseu.do.mal′ le.i. Gr. adj. *pseudes* false; L. n. *malleus* the disease glanders; M.L. gen. n. *pseudomallei* of false glanders.

The general characteristics of the species and properties useful for differentiation from the related species *B. mallei* are given in Tables BXII.β.2, BXII.β.3, and BXII.β.7. Colonies can range in structure from extremely rough to mucoid, and in color from cream to bright orange. Optimal growth temperature, ~37°C. For further descriptive information see Redfearn et al. (1966) and Redfearn and Palleroni (1975). Isolated from human and animal cases of melioidosis and from soil and water in tropical regions, particularly Southeast Asia. Probably a soil organism and accidental pathogen, causing melioidosis.

The mol% G + C of the DNA is: 69.5 (Bd).

Type strain: ATCC 23343, NCTC 12939, WRAIR 286.

16. **Burkholderia pyrrocinia** (Imanaka, Kousaka, Tamura and Arima 1965) Vandamme, Holmes, Vancanneyt, Coenye, Hoste, Coopman, Revets, Lauwers, Gillis, Kersters and Govan 1997b, 1199[VP] (*Pseudomonas pyrrocinia* Imanaka, Kousaka, Tamura and Arima 1965, 205.)

pyr.ro.ci'ni.a. Etymology uncertain, possibly M.L. adj. *"pyrrocin"* referring to the antibiotic properties of pyrrolnitrin, which is produced by strains of this species.

The following description is taken from the original paper. Rods, 0.5–0.8 × 1.2–2.0 µm, occurring singly. Motile by means of polar flagella. No pigment is produced. Oxidase reaction, nitrate reduction, denitrification, and starch hydrolysis are all negative. H₂S is produced. Optimal temperature for growth, 26–30°C. Growth is scanty at 37°C and negative at 42°C. Acid produced from glucose, galactose, lactose, sucrose, and glycerol, but not from maltose, trehalose, mannose, raffinose, starch, or inulin. 2-Ketogluconate is produced from gluconate.

Growth on glucose, gluconate, 2-ketogluconate, and *p*-hydroxybenzoate as sole carbon sources. 5-Ketogluconate, citrate, ethanol, phenol, succinate, benzoate, salicylate, *m*-hydroxybenzoate, protocatechuate, gentisate, anthranilate, and *p*-aminobenzoate are not utilized for growth. Some of the properties are summarized in Table BXII.β.2. The strains produce the antibiotic pyrrolnitrin. This compound is also produced by some strains of *Pseudomonas chlororaphis*, *P. aureofaciens*, and *Burkholderia cepacia* ("*Pseudomonas multivorans*") (Elander et al., 1968), indicating that the synthesis occurs in organisms of different branches of the *Proteobacteria*.

The first suggestion of allocation of *B. pyrrocinia* to the RNA similarity group II was made by Byng et al. (1980). In their proposal for the transfer of *Pseudomonas pyrrocinia* to the genus *Burkholderia*, Vandamme et al. (1997b) add to the above description, the characteristic fatty acid composition. Allocation to the genus *Burkholderia* also has been recommended by Viallard et al. (1998) based on nucleic acid sequencing studies.

The mol% G + C of the DNA is: 65.

Type strain: ATCC 15958, DSM 10685, LMG 14191.

GenBank accession number (16S rRNA): U96930.

17. **Burkholderia thailandensis** Brett, Deshazer and Woods 1998, 318[VP]

thai.lan.den'sis. M.L. adj. *thailandensis* pertaining to Thailand, where the organism was originally isolated.

Rods, motile due to the presence of two to four polar flagella. Colonies are smooth and glossy with a pink pigmentation on modified Ashdown's selective medium (see Enrichment and Isolation Procedures), while *B. pseudomallei* colonies are rough and wrinkled, with a dark purple pigmentation. The API 20NE and API 50CH biochemical profiles are similar to those of *B. pseudomallei*. Exceptions are the capacity to use L-arabinose, 5-ketogluconate, and adonitol, and the inability to utilize erythritol and dulcitol. Strains of the species are avirulent for Syrian golden hamsters. Growth occurs at temperatures between 25°C and 42°C. Production of siderophore, lipase, lecithinase, and protease are positive. The strains are resistant to aminoglycosides but sensitive to tetracycline, ceftazidime, and trimethoprim. Type strain was isolated from a rice field soil sample in central Thailand.

The mol% G + C of the DNA is: unknown.

Type strain: E264, ATCC 700388.

GenBank accession number (16S rRNA): U91838.

A comparison of sequences of 16S rRNA genes indicates a close relationship of *B. thailandensis* to *B. pseudomallei*. No DNA–DNA hybridization experiments between strains of the two species have been reported. In view of the low resolving power of 16S rRNA sequence similarity at the species level and the limited number of phenotypic differences, it is hoped that future studies may provide evidence supporting the taxonomic position of this group of organisms as an independent species of *Burkholderia*.

18. **Burkholderia vandii** Urakami, Ito-Yoshida, Araki, Kijima, Suzuki and Komagata 1994, 242[VP]

van'di.i. M.L. gen. n. *vandii* of Vanda, a genus of orchids.

Description taken from Urakami et al. (1994). The cells are 0.5–1.0 × 1.5–3.0 µm and have rounded ends. They occur singly, rarely in pairs, and are motile by one or several polar flagella. Abundant growth occurs in nutrient broth and peptone water. The colonies are white to light yellow. No diffusible fluorescent pigment is produced.

The methyl red and the Voges–Proskauer reactions are negative. Indole and hydrogen sulfide are not produced. Starch is not hydrolyzed. Denitrification and hydrolysis of gelatin are positive. Acids are weakly produced from inositol and glycerol oxidatively, but not from L-arabinose, D-xylose, D-glucose, D-mannose, D-fructose, D-galactose, maltose, sucrose, lactose, trehalose, D-sorbitol, D-mannitol, or soluble starch. No fermentation of sugars occurs. Nitrate is not used as a nitrogen source. Results of nutritional studies at the expense of many organic compounds, as well as extensive physiological information, are given in the original paper (Urakami et al., 1994) and in Gillis et al. (1995). The single known strain was isolated from orchids of the genus *Vanda* as an antibiotic-producing bacterium active against the plant pathogenic organisms *Clavibacter michiganensis* and *Fusarium oxysporium*.

According to Coenye et al. (1999b), *B. vandii* is a junior synonym of *B. plantarii*.

The mol% G + C of the DNA is: 68.5 (HPLC).

Type strain: ATCC 51545, DSM 9510, JCM 7957, LMG 16020, VA-1316.

GenBank accession number (16S rRNA): U96932.

Studies by Coenye et al. (1999b) based on SDS-PAGE of whole cell proteins, DNA–DNA hybridization, and extensive biochemical characterization indicate that *B. vandii* should be considered a junior synonym of *B. plantarii*.

19. **Burkholderia vietnamiensis** Gillis, Van, Bardin, Goor, Hebbar, Willems, Segers, Kersters, Heulin and Fernandez 1995, 287[VP]

vi.et′ na.mi.en.sis. M.L. adj. *vietnamiensis* referring to Vietnam, where the rice strains were isolated.

Motile cells are 0.8–2 × 0.3–0.8 μm. No details on the number of flagella per cell are given. Colonies on nutrient agar are not pigmented and do not produce fluorescent pigment on King B medium, a property that is uniformly negative for species of *Burkholderia*. Growth occurs on nutrient agar between 20°C and 41°C. Strains are oxidase, catalase, β-galactosidase, and gelatinase positive. All strains fix N₂ and produce ornibactin siderophores, but not pyochelin or cepabactin. Nutritional properties and characteristics useful for differentiation from other *Burkholderia* species have been summarized in Tables BXII.β.2 and BXII.β.3.

Further details are to be found in Gillis et al. (1995). In one of the tables in that article, denitrification is recorded as positive, although the property is not specifically mentioned in the text, where *B. vietnamiensis* is described as capable of reducing nitrate to nitrite. Further reduction to gases (N₂O or N₂) and the ability to grow under anaerobic conditions in the presence of nitrate are not mentioned. It may be inferred that strains of this species, as those of *B. cepacia*, are unable to denitrify. Differences between *B. vietnamiensis* and *B. cepacia* are in the utilization of L-arabitol,

adonitol, butylamine, tryptamine, citraconate, and 5-ketogluconate by *B. cepacia*, and the ability of *B. vietnamiensis* to grow on itaconate. Strains of the latter do not synthesize the siderophores cepabactin and pyochelin (Gillis et al., 1995). *B. vietnamiensis* is capable of N₂ fixation, a property recorded earlier for rhizosphere strains of *B. cepacia* (Bevivino et al., 1994). However, extracts from clinical strains also gave a single hybridization signal with a *nifA* probe from *Klebsiella pneumoniae*, although they did not fix nitrogen (Tabacchioni et al., 1995). Interestingly, the clinical strains produced pyochelin and its precursor, salicylic acid, a property absent from the rhizosphere strains (Bevivino et al., 1994). Therefore, it is very likely that the latter may have corresponded to the taxon to be described later as *B. vietnamiensis* (Gillis et al., 1995).

There is little doubt, however, that *B. cepacia* and *B. vietnamiensis* are closely related, as suggested by their relative position in the tree of Fig. BXII.β.2. Aside from many phenotypic characteristics, their similarity extends to the production of the siderophore ornibactin (Meyer et al., 1995), and to the structure of the putative O-specific polymer isolated from the LPS of *B. vietnamiensis*, which resembles the O-antigen of *B. cepacia* serogroup J (Gaur and Wilkinson, 1996). The strains of *B. vietnamiensis* have been isolated from rice field soils and from clinical specimens. They are not pathogenic for onions.

The mol% G + C of the DNA is: 67.9 (*T_m*).

Type strain: DSM 11319, LMG 10929, TVV75.

GenBank accession number (16S rRNA): U96928, U96929.

Genus II. **Cupriavidus** Makkar and Casida 1987a, 325[VP]

DAVID L. BALKWILL

Cup.ri.a.vi′ dus. L. n. *cuprum* copper; L. adj. *avidus* eager for, loving; M.L. neut. n. *Cupriavidus* lover of copper.

Coccoid rods, 0.7–0.9 × 0.9–1.3 μm. Gram negative. **Motile by two to ten peritrichous flagella. Chemoheterotrophic.** An organic nitrogen source is not required. Glucose not utilized. **Strictly respiratory metabolism** with oxygen as the terminal electron acceptor. Oxidase positive. Catalase positive. Nitrate reduced. Gelatin, starch, and urea not hydrolyzed. Indole and H₂S not produced. **Can use any of several amino acids—but not L-lysine or L-methionine—as the sole source of carbon and nitrogen.** Optimal temperature, 27°C. Optimal pH, 7.0–8.0. NaCl at 3% inhibits growth. **Resistant to copper** at concentrations up to at least 800 μM. Growth initiation is stimulated by copper. Colonies on nutrient agar after 2 d at 27°C are off-white, glistening, mucoid, smooth, and convex, with an entire edge; 2–4 mm in diameter. Isolated from soil. **Nonobligate predator causing lysis of various Gram-positive and Gram-negative bacteria in soil.** Can lyse certain other nonobligate bacterial predators. Growth does not require presence of prey species.

The mol% G + C of the DNA is: 57 ± 1 (*T_m*).

Type species: **Cupriavidus necator** Makkar and Casida 1987a, 325.

FURTHER DESCRIPTIVE INFORMATION

C. necator cells are Gram-negative short rods measuring 0.7–0.9 × 0.9–1.3 μm, based on electron micrographs of negatively

stained cells. The cells decrease somewhat in size and become more rounded as cultures age, or when the organism is placed in contact with soil during soil column studies (Byrd et al., 1985). In soil, the rounded forms of *C. necator* appear to be dormant (Byrd et al., 1985). *C. necator* reproduces by binary fission and is motile by 2–10 peritrichous flagella.

Colonies of *C. necator* on nutrient agar after 2 d of incubation at 27°C are 2–4 mm in diameter, off-white, glistening, mucoid, smooth, and convex, with an entire edge. Nonmucoid variants of *C. necator* that form slightly smaller colonies appear after many transfers on laboratory media. The colonies of these variants are flat and do not glisten. A phage that lyses the mucoid form of the organism does not lyse the nonmucoid form, but the two variants are similar in all other respects.

C. necator grows aerobically on nutrient agar or synthetic medium with L-glutamic acid, but only scant growth occurs on these media under anaerobic conditions. In thioglycolate broth, the organism grows only in the oxidized zone. Good growth occurs in acetate or fructose broth media with NH₄Cl as the sole source of nitrogen. Excellent growth is obtained in N-1 synthetic medium (Makkar and Casida, 1987a) broth with L-glutamic acid as the nitrogen and carbon source. The optimal growth temperature for *C. necator* is 27°C, but it can grow well at 37°C. Growth occurs after a delay at 15°C, but there is no growth at 55°C. *C.*

necator grows over a pH range of 5.5–9.2, and its optimal growth pH is 7.0–8.0. It grows well in media that contain 1% NaCl, but growth is poor at 2% NaCl and inhibited fully at 3% NaCl.

C. necator does not grow in a modified Burk's N-free medium (see Makkar and Casida, 1987a) with glucose or sucrose as the sole source of carbon. Limited growth takes place with fructose but is not sustained during repeated transfers in N-free medium. *C. necator* can grow in a synthetic medium (Makkar and Casida, 1987a) without added magnesium. It is quite resistant to copper and grows well in the presence of 1200 μM $CaCl_2 \cdot 2H_2O$ (Casida, 1987). Copper also stimulates growth initiation of *C. necator* but has no significant effect on its subsequent growth rate at concentrations up to at least 800 μM (Casida, 1987; Makkar and Casida, 1987a).

The metabolism of *C. necator* is oxidative, as determined by the Hugh-Leifson test (Hugh and Leifson, 1953) performed in a synthetic medium with fructose as the carbon source. It grows and produces acid without gas in the top portion of aerobic tubes, but does not grow or produce acid in anaerobic tubes (Makkar and Casida, 1987a). *C. necator* is catalase positive and oxidase positive. It does not show hemolysis on blood agar. Hippurate, tributyrin, Tween 20, and Tween 80 are hydrolyzed, but esculin, gelatin, starch, and urea are not hydrolyzed. DNase activity is negative. *C. necator* degrades tyrosine. It does not produce indole or H_2S. In litmus milk, the reaction is basic with reduction of the litmus. Acetate, L-aspartate, citrate, fructose, fumarate, gluconate, L-glutamate, glutarate, β-hydroxybutyrate, lactate, L-leucine, oxalacetate, and succinate are utilized as sources of carbon for growth. Use of sucrose as a carbon source is equivocal. Utilization of acetate, fructose, L-glutamate, and L-leucine occurs after a brief delay (13–15 h). Somewhat longer delays are seen with L-alanine, and L-valine (61 and 92 h, respectively; Makkar and Casida, 1987a). *C. necator* does not utilize arabinose, adonitol, benzoate, glucose, glycerol, lactose, L-lysine, mannitol, mannose, melibiose, L-methionine, rhamnose, or xylose as carbon sources for growth.

The mol% G + C content of *C. necator* chromosomal DNA is 57 ± 1, as determined by the thermal melting point technique (Makkar and Casida, 1987a). The 16S rRNA gene sequence for the type strain of *C. necator* (ATCC 43291) has been determined (Balkwill, unpublished data; GenBank accession no. AF191737). Phylogenetic analyses of this sequence with distance matrix (see Fig. BXII.β.3), parsimony, and maximum likelihood methods indicated that *C. necator* should be assigned to the *Burkholderiaceae*. Moreover, all of the analytical methods placed *C. necator* within a cluster of strains representing the genus *Ralstonia* (Yabuuchi et al., 1995). *C. necator* was most closely related to two strains (including the type strain) of *Ralstonia eutropha* that were included in the analysis. Its 16S sequence was virtually identical to those of the two *Ralstonia* strains, differing from them by only a single base over 1338 bases compared (sequence similarity 99.9%). Technically, this high sequence similarity alone does not demonstrate that *C. necator* and *R. eutropha* are the same species. Stackebrandt and Goebel (1994) have suggested that DNA–DNA reassociation values should be used to determine whether strains of bacteria are members of the same species when their 16S rDNA sequence similarities are greater than 97.5%. Nevertheless, the phylogenetic analyses do provide strong evidence that *C. necator* is a member of the genus *Ralstonia*.

The habitat for *Cupriavidus* is soil. *C. necator* is a nonobligate predator that can cause lysis of various types of Gram-positive and Gram-negative bacteria in soils, including several other non-obligate predator species. However, growth of *C. necator* in soil or laboratory media does not require the presence of prey species. *C. necator* produces the heat-labile, magnesium-related growth initiation factor (GIF) that was first described by Casida (1984) and that is produced by various types of bacteria in soils. This GIF can initiate the growth of *Agromyces ramosus* and other species in soil (Casida, 1984; Byrd et al., 1985), thereby increasing the supply of suitable prey cells for nonobligate bacterial predators like *C. necator*. The actual mechanism by which *C. necator* kills its prey cells is not yet known, although Casida (1987) has suggested that a copper-related peptide used to scavenge copper from the environment (see below) might also be used to deliver excess (and toxic) amounts of copper to prey cells.

Zeph and Casida (1986) used an indirect phage analysis method involving soil percolator columns and plaque assays (see Byrd et al., 1985) to assess the ability of *C. necator* and several other nonobligate bacterial predators to attack various types of potential prey species in soil. Of the 11 predators examined, *C. necator* strain N-1 (the type strain) and an uncharacterized predator strain (designated L-2) were capable of attacking the widest variety of prey species. Nonpredatory bacteria attacked by *C. necator* included *Arthrobacter globiformis*, *Azotobacter vinelandii*, *Bacillus subtilis*, *Bacillus thuringiensis*, *Escherichia coli*, *Micrococcus luteus*, and *Staphylococcus aureus*. Other nonobligate bacterial predators attacked by *C. necator* included *Agromyces ramosus*, *Ensifer adhaerens*, strain C2 (an uncharacterized predator), and a streptomycete-like isolate (strain 3). *C. necator* did not attack *Agrobacterium tumefaciens*, *Nocardia salmonicolor*, *Salmonella typhi* (all of which are nonpredatory), or four uncharacterized predatory strains, including L-2. *C. necator* itself was attacked by *A. ramosus* and one uncharacterized predatory strain, but was not attacked by *E. adhaerens* or seven other uncharacterized predators. Comparison of prey ranges for different predators indicated that Gram-negative predators like *C. necator* are likely to be more important than Gram-positive predators, all of which had comparatively narrow prey ranges. It is also possible that nonobligatory predators such as *C. necator* control the populations of Gram-positive predators (like *A. ramosus*), and thus might be at the top of the hierarchy of bacterial predators in soils.

When *C. necator* interacts with *Agromyces ramosus* in a soil, a predator-on-predator attack and counterattack phenomenon occurs (Byrd et al., 1985). *A. ramosus* is a Gram-positive species that occurs in high numbers in soils and functions as a nonobligatory predator that destroys cells of several other bacterial species. It typically occurs in soils as coccoid rods but can produce a mycelium that eventually fragments. Within 2 d after *A. ramosus* and *C. necator* are added to a soil, *A. ramosus* forms a mycelium wherever the *C. necator* cells are situated, and roughly one-third of the *C. necator* cells become ghosted. *C. necator* apparently produces the magnesium-related GIF cited above, which stimulates the growth of *A. ramosus* and subsequent development of its mycelium. While this occurs, however, *A. ramosus* attacks and lyses some of the *C. necator* cells. The mycelium appears to deliberately seek out the cells of *C. necator*, possibly to make contact with them. Lysis of the *A. ramosus* mycelium in contact with cells of *C. necator* is seen within 4–5 d after adding the two bacteria to a soil. Proliferation of *C. necator* wherever the *A. ramosus* mycelium grew is also seen by this time, indicating that *C. necator* has counterattacked *A. ramosus*. *C. necator* only attacks the mycelial form of *A. ramosus*, so portions of the mycelium that fragment to the rod form before *C. necator* can attack them survive the attack-

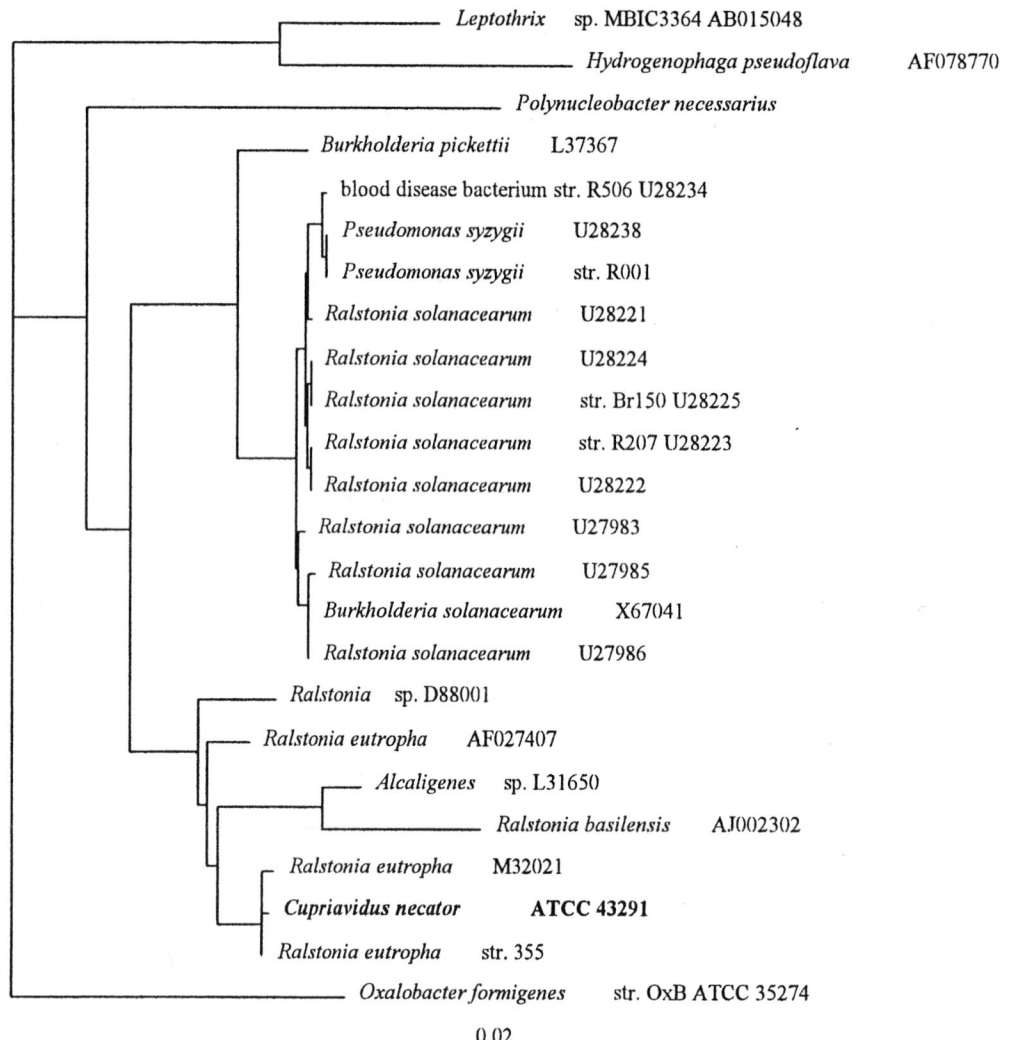

FIGURE BXII.β.3. Phylogenetic tree for *Cupriavidus necator* ATCC 43291 and selected strains of eubacteria, based on a distance matrix analysis. Bar = 2 substitutions per 100 bases. *Oxalobacter formigenes* was used as the outgroup. Parsimony and maximum likelihood analyses yielded virtually identical trees with respect to the positions of *C. necator* and *Ralstonia eutropha* and the branching patterns within the cluster containing strains of these two organisms. The numbers in parentheses are the GenBank/ EMBL accession numbers for the sequences used in the analysis. The accession number for the *C. necator* 16S rDNA sequence is AF191737.

counterattack interaction. This might explain why *A. ramosus* maintains high numbers in soils despite being a prey organism.

Casida (1989a) investigated the possibility that *Arthrobacter globiformis*, which is usually present in relatively high numbers in soils, might serve as a major reservoir of prey cells for *C. necator* and other copper-resistant Gram-negative bacterial predators. This study found that a large increase in the culturable numbers of an uncharacterized bacterium that hydrolyzes GELRITE occurs within 2.5 h. Because of the short time period involved, it was felt that exposure to *A. globiformis* caused the hydrolyzers to break dormancy (and thus become more readily culturable), rather than to multiply extensively. The numbers of *C. necator* and other Gram-negative predator cells increased in response to addition of *A. globiformis* cells as well. However, as the numbers of predator cells increased, the levels of hydrolyzers in the soil decreased extensively while the numbers of *A. globiformis* cells did not. It appeared, then, that the predators were responding directly to the hydrolyzers that were activated after *A. globiformis*

was added to the soil, rather than to *A. globiformis* itself. If so, *A. globiformis* most likely does not serve as the major reservoir of prey cells for *C. necator* and other copper-resistant Gram-negative predators, but it may play a role in enabling them to access prey cells such as the hydrolyzers.

Casida (1989b) also investigated the possibility that protozoa might be major predators of *C. necator* and other bacterial predators in soil. Protozoa that can attack *C. necator* were detected in soil, but no evidence was found that would indicate these protozoa and *C. necator* actually interact with each other in soils.

As noted above, *C. necator* is resistant to copper, and copper stimulates initiation of its growth. Casida (1987) found that growth initiation is delayed 13–120 h when *C. necator* is transferred to a synthetic medium with various carbon sources, the actual time depending on which carbon source is supplied in the medium. During this delay period, *C. necator* produces a GIF that is different from the heat-labile magnesium-associated GIF that it uses to stimulate the growth of *A. ramosus* and other bacteria in soil (Byrd et al., 1985). The newly discovered GIF was shown to specifically stimulate growth initiation of *C. necator* and to be a copper-related, heat-stable peptide with a molecular weight greater than 10,000 and a relatively low cysteine content. It is thought that *C. necator* uses this heat-stabile peptide GIF to marshal copper from the surrounding environment, as elevated levels of copper are apparently needed to initiate growth. After growth has been initiated, however, copper has no significant effect on the growth rate and production of the copper-associated peptide GIF stops. Casida (1987) also found that the copper-associated GIF strongly inhibits the growth of *A. ramosus*, perhaps by delivering lethal levels of copper to its cells. If so, *C. necator* may use a twofold strategy when it interacts with *A. ramosus* in soil. First, it could produce the heat-labile magnesium-associated GIF to stimulate growth of *A. ramosus* and development of the mycelium. Then, it could use the heat-stable copper-associated GIF to kill the mycelial cells.

ENRICHMENT AND ISOLATION PROCEDURES

C. necator can be enriched and isolated from soils with a baiting technique developed by Sillman and Casida (1986). This method appears to utilize chemotaxis, although that has not been confirmed. Air-dried soil is passed through a fine sieve and placed to a depth of 3–4 mm in a sterile Petri plate. This soil is tamped lightly to produce a uniformly flat surface, and about 2 ml of sterile distilled water is added along the walls of the plate. The soil should be moist but not flooded. A 0.65-µm pore size membrane filter is next placed on a circle (8.5 mm diameter) of sterile Whatman no. 2 filter paper in a separate sterile Petri plate. A concentrated washed suspension of early stationary-phase host cells (of *Micrococcus luteus*) is spread over a 28 mm diameter area at the center of the membrane filter, and excess moisture is absorbed by the underlying filter paper. The membrane filter is then placed on the surface of the soil with the side containing the host cells facing upward. After gently pressing the edges of the filter against the soil, the Petri plate is sealed with tape and incubated (right side up) at 27°C. Several filters are prepared in this way, so that one can check for the appearance of predators daily as incubation continues.

The following procedure is used to check for the appearance of predators during incubation. The host cells are removed from a filter by gentle scraping with an alcohol-flamed spatula and suspended in 5 ml of sterile tap water. This suspension is then mixed and diluted through a 1:2 dilution series. Aliquots (0.1 ml) of each dilution and the undiluted suspension are placed on the surfaces of agar plates, along with 0.1 ml of a fresh, concentrated host-cell suspension, and aseptically spread over the surfaces of the plates. Either 1.5% Noble agar (in distilled water) or a very dilute nutrient medium (such as 0.01 strength heart infusion broth) can be used in the plates for this part of the isolation procedure. The plates are then incubated for at least 3 d and checked for the development of plaques, colonies, or colonies within plaques. These are picked with a needle and spread through freshly prepared lawns of host cells on new plates of the same medium. After a sufficient period of incubation at 27°C, isolates growing on the plates are restreaked for purification on 0.01 strength heart infusion agar without added host cells. This approach establishes that the isolates are not obligatory predators.

C. necator typically appears on the initial isolation filters within 3 d of incubation at 27°C. Isolates can be tested for their ability to attach to host cells as follows. Microscope slides are positioned vertically along the inside walls of 250-ml wide-mouth Erlenmeyer flasks. The slides are then held in place by adding sterile glass beads to a depth of 4 cm. Sufficient sterile tap water is added to nearly submerge the glass beads, and host cells of *M. luteus* are added to produce a turbid suspension. One loopful of potential predator cells is also added, and the flasks are incubated on a rotary shaker at 27°C. After incubation, the slides are removed from the flask, stained with crystal violet, and examined for predator attachment to host. This should be done both at the splash line and on the part of the slide that was submerged.

The above procedure has been used to isolate a number of Gram-negative, nonobligate bacterial predators from soil, including several copper-resistant strains. If one in interested in isolating *C. necator* specifically, then it will be necessary to screen isolates for the key descriptive characteristics of this species.

MAINTENANCE PROCEDURES

C. necator can be maintained on 0.1-strength heart infusion agar. It can be lyophilized by common procedures used for aerobes. It can also be preserved by freezing in cryoprotectant and subsequent storage at −80°C or under liquid nitrogen.

DIFFERENTIATION OF THE GENUS *CUPRIAVIDUS* FROM OTHER GENERA

When the genus *Cupriavidus* was proposed by Makkar and Casida in 1987a, it was most similar to *Alcaligenes*. As a result it was compared directly to that genus, especially to *Alcaligenes faecalis*. Among the characteristics that both organisms had in common were cell morphology, the mol% G + C content of their DNA, and lack of the ability to utilize glucose as a sole carbon source. However, *C. necator* could be differentiated from *A. faecalis* on the basis of five traits: lack of a requirement for added magnesium in synthetic media, its predatory activity, ability to utilize fructose as a carbon source, copper resistance, and stimulation of growth initiation by elevated levels of copper. In all likelihood, the combination of copper resistance and predatory behavior also differentiates *Cupriavidus* from other Gram-negative aerobic rods.

Analyses of 16S rRNA gene sequences indicate that *C. necator* is a member of the *Burkholderiaceae* and is phylogenetically situated within the genus *Ralstonia* (Yabuuchi et al., 1995). Among comparable phenotypic traits published to date, *C. necator* differs from all currently recognized species of *Ralstonia* only in that it cannot utilize glycerol as a carbon source (see Makkar and Casida, 1987a; Yabuuchi et al., 1992, 1995; Urakami et al., 1994; Gillis et al., 1995; Coenye et al., 1999a; Vandamme et al., 1999). However, *C. necator* differs from the phylogenetically most closely related species of *Ralstonia* (*R. eutropha*) in two other ways: it does not utilize benzoate as a carbon source and it does not cause hemolysis on blood agar. *C. necator* also differs from all recognized *Ralstonia* species in that it has a considerably lower mol% G + C content (57 ± 1 versus 64–68.3).

TAXONOMIC COMMENTS

Phylogenetic analysis data and the apparent paucity of phenotypic differences between *C. necator* and *Ralstonia* spp. raise serious doubts that *Cupriavidus* and *Ralstonia* are distinct genera. The predatory behavior of *Cupriavidus* might distinguish it from *Ralstonia*, but no *Ralstonia* species have been tested for this characteristic. The reported mol% G + C content for *C. necator* (57 ± 1) is outside the range of values reported for *Ralstonia* spp. (64–68.3), and all determinations were done by the same method. A thorough direct comparison of *C. necator* with all recognized *Ralstonia* spp. might detect additional differences. Based on information available at this time, however, it appears that *C. necator* should be transferred to the genus *Ralstonia*.

Phylogenetic analyses indicated that *C. necator* and *Ralstonia eutropha* have virtually identical 16S rRNA gene sequences. At noted earlier, this very high sequence similarity does not prove that *C. necator* and *R. eutropha* are the same species; DNA–DNA reassociation values must be determined to resolve that question (see Stackebrandt and Goebel, 1994). This has not been done for *C. necator* and *R. eutropha*, but their differing mol% G + C contents (see above) imply that the reassociation values resulting from such an experiment might be quite low. Moreover, several phenotypic differences between these two strains have been reported in the literature (see above). Copper resistance and predatory behavior could be additional differences, but *R. eutropha* has not been examined for these traits. In addition, the copper resistance in *Cupriavidus* could be plasmid-coded (as metal resistance often is) and, if so, it should not be considered a characteristic that defines the species. At this point, there is insufficient information to know whether *C. necator* and *R. eutropha* are distinct species. DNA–DNA hybridization studies would be most helpful in clarifying the situation.

ACKNOWLEDGMENTS

For the most part, this chapter is a minor revision of the original description of *Cupriavidus* published by Makkar and Casida (1987a). The principal changes include the addition of information from analysis of 16S rDNA sequences and the addition of data from several more recent publications. The author thanks L.E. Casida, Jr. for providing reprints and other information related to his research on *Cupriavidus necator*. The strain of *C. necator* for which the 16S rDNA sequence was determined was obtained from the American Type Culture Collection.

List of species of the genus Cupriavidus

1. **Cupriavidus necator** Makkar and Casida 1987a, 325[VP]

ne.ca′ tor. L. n. *necator* slayer.

The characteristics are as described for the genus. During growth under laboratory conditions, *C. necator* may produce small numbers of a nonmucoid variant. The variant resembles the mucoid form except that its colonies are smaller and display a drier, flatter appearance. However, a bacteriophage isolated from soil that lyses the mucoid form does not lyse the nonmucoid variant.

The habitat is soil.

The mol% G + C of the DNA is: 57 ± 1 (T_m).

Type strain: ATCC 43291.

GenBank accession number (16S rRNA): AF191737.

Genus III. **Lautropia** *Gerner-Smidt, Keiser-Nielsen, Dorsch, Stackebrandt, Ursing, Blom, Christensen, Christensen, Frederiksen, Hoffmann, Holten-Andersen and Ying 1995, 418[VP] (Effective publication: Gerner-Smidt, Keiser-Nielsen, Dorsch, Stackebrandt, Ursing, Blom, Christensen, Christensen, Frederiksen, Hoffmann, Holten-Andersen and Ying 1994, 1795)*

PETER GERNER-SMIDT

Lau.tro′ pi.a. M.L. fem. n. *Lautropia* of Lautrop, named after Hans Lautrop.

Cocci may occur in at least **three forms: (1) encapsulated**, nonmotile 1–2 μm in diameter, forming aggregates of 10 to >100 cells; **(2) unencapsulated**, nonaggregated 1–2 μm in diameter, **motile** by a tuft of three to nine flagella; and **(3)** large (>5 μm in diameter), **spherical nonaggregated, nonmotile**. Do not form endospores. **Gram negative**. Nonpigmented. **Grow only on enriched media. Facultative**, but grow best under aerobic conditions. CO_2 is not required. Growth occurs between 30 and 44°C. **Oxidase, catalase, and urease positive**. A polysaccharide is produced on sucrose agar. Various carbohydrates are fermented. Belongs phylogenetically to a separate branch of the class *Betaproteobacteria*. Habitat: the oral cavity and upper respiratory tract of humans.

The mol% G + C of the DNA is: 65.

Type species: **Lautropia mirabilis** Gerner-Smidt, Keiser-Nielsen, Dorsch, Stackebrandt, Ursing, Blom, Christensen, Christensen, Frederiksen, Hoffmann, Holten-Andersen and Ying 1995, 418[VP] (Effective publication: Gerner-Smidt, Keiser-Nielsen, Dorsch, Stackebrandt, Ursing, Blom, Christensen, Christensen, Frederiksen, Hoffmann, Holten-Andersen and Ying 1994, 1795.)

FURTHER DESCRIPTIVE INFORMATION

Rough to smooth colonies grow on many plating media and are composed of extremely pleomorphic cocci with round cells with diameters from 1 to >10 μm. The smallest cells are often motile with circular movements. Glucose, fructose, sucrose, and mannitol are fermented.

An aggregate-forming coccus identified as *L. mirabilis* was the predominant microorganism in sputa from a cystic fibrosis patient on consecutive days (Ben Dekhil et al., 1997c). *L. mirabilis* has also been isolated from the oral cavities of 32 of 60 children infected with human immunodeficiency virus (HIV) and 3 of 25 HIV-uninfected controls (Rossmann et al., 1998); however, the bacterium was not associated with clinical disease in these children.

ENRICHMENT AND ISOLATION PROCEDURES

The organism may be isolated on most nonselective enriched media after aerobic incubation for 1–2 d at 35°C. Microscopy of wet mounts of different colony types is essential to identify the organism in mixed cultures. No selective enrichment procedures have been described for this organism.

MAINTENANCE PROCEDURES

Lautropia cells can either be stored in 10% glycerol broth at −80°C or lyophilized.

DIFFERENTIATION OF THE GENUS *LAUTROPIA* FROM OTHER GENERA

Organisms of the genus *Lautropia* are differentiated from other organisms by their requirement for enriched media, their characteristic colony morphology, their microscopic appearance, and their positive oxidase, urease, and nitrate reduction reactions.

Taxonomic Comments

In 1994, Gerner-Smidt et al. (1994), isolated six strains of an organism that seemed to be identical to an organism that had been described by Ørskov (1930). 16S rRNA sequencing revealed that the organisms belonged to class *Betaproteobacteria* of the phylum *Proteobacteria*, separate from all other described genera, but most closely related to *Burkholderia* (Gerner-Smidt et al., 1994). The organisms were assigned to a new genus, *Lautropia*, as the species *L. mirabilis*.

List of species of the genus Lautropia

1. **Lautropia mirabilis** Gerner-Smidt, Keiser-Nielsen, Dorsch, Stackebrandt, Ursing, Blom, Christensen, Christensen, Frederiksen, Hoffmann, Holten-Andersen and Ying 1995, 418^{VP} (Effective publication: Gerner-Smidt, Keiser-Nielsen, Dorsch, Stackebrandt, Ursing, Blom, Christensen, Christensen, Frederiksen, Hoffmann, Holten-Andersen and Ying, 1994, 1795.)

mi.ra' bi.lis. L. n. *mirabilis* wonderful.

The characteristics are as described for the genus and as follows. Grows on most enriched media, especially on chocolate, Levinthal, tryptose glucose yeast (TGY) extract, and Tween 80 agar. At least three colony morphologies are seen: (1) flat, dry, and circular; (2) larger, wrinkled, crisp, and crateriform; and (3) smooth, glistering, raised, round, and mucoid. Colonies adhere to the substrate; diameter of the colonies between pinpoint size and >5 mm. Growth in broth granular with a coarse sediment and granules adherent to the side of the tube. Acid is produced from D-glucose, D-fructose, maltose, sucrose, and mannitol; acid is not produced from lactose, trehalose, raffinose, inulin, salicin, adonitol, dulcitol, sorbitol, inositol, D-xylose, L-rhamnose, and L-arabinose. Negative for lysine decarboxylase, ornithine decarboxylase, arginine decarboxylase/dihydrolase, phenylalanine deaminase, Voges–Proskauer test, gelatinase, starch hydrolysis, and H$_2$S. Hippurate and Tween 80 not hydrolyzed. Most strains hydrolyze esculin. Some strains produce β-xylosidase. β-Galactosidase and β-glucuronidase are not produced.

Rossmann et al. (1998) found variable reactions for arginine dihydrolase, lysine decarboxylase, Voges–Proskauer test, gelatinase at 35°C, and sorbitol fermentation using a commercial identification system (API20E).

Unlike the strains isolated by Gerner-Smidt et al. (1994), the strain isolated by Ben Dekhil et al. (1997c) was hemolytic on blood agar. Rossmann et al. (1998) reported that, with their strains, the catalase test was weakly positive with 10% hydrogen peroxide, and catalase was undetectable with 3% H$_2$O$_2$.

Sensitive to penicillin G, ampicillin, piperacillin, cefuroxime, gentamicin, and erythromycin.

The mol% G + C of the DNA is: 64.6–65.4 (T_m).

Type strain: AB2188, ATCC 51599, CCUG 34794, NCTC 12852.

GenBank accession number (16S rRNA): X73223.

Additional Remarks: The type strain was isolated from human dental plaque.

Genus IV. **Pandoraea** Coenye, Falsen, Hoste, Ohlen, Goris, Govan, Gillis and Vandamme 2000, 895^{VP}

The Editorial Board

Pan.do.rae' a. N.L. fem. n. *Pandoraea* referring to Pandora's box in Greek mythology, the origin of diseases of mankind.

Gram-negative, motile, nonsporeforming rods (0.5–0.7 × 1.5–4.0 μm). **One polar flagellum.** Positive reactions: catalase, alkaline phosphatase, leucine arylamidase; assimilation of D-gluconate, L-malate, and phenylacetate; **growth on Drigalsky agar** and in 0.5 and 1.5% NaCl. Major fatty acids include C$_{12:0}$, C$_{12:0\ 2OH}$, C$_{16:0}$, C$_{16:0\ 2OH}$, C$_{16:0\ 3OH}$, C$_{17:0\ cyclo}$, C$_{18:1\ 2OH}$, and C$_{19:0\ cyclo\ \omega8c}$.

The mol% G + C of the DNA is: 61.2–64.3.

Type species: **Pandoraea apista** Coenye, Falsen, Hoste, Ohlen, Goris, Govan, Gillis and Vandamme 2000, 896.

Further descriptive information

Negative reactions include nitrite reduction and denitrification; liquefaction of gelatin; hydrolysis of esculin, poly-β-hydroxybutyrate, and Tween 80; production of acid or sulfide in TSI agar; production of indole; production of N-acetyl-β-glucosaminidase, C$_{14}$-lipase, chymotrypsin, DNase, α-fucosidase, α-galactosidase, β-galactosidase, α-glucosidase, β-glucuronidase, lysine decarboxylase, α-mannosidase, ornithine decarboxylase, trypsin, tryptophanase, and valine arylamidase; growth in 10% lactose and 6% NaCl; and growth in O/F medium containing adonitol, fructose, or xylose.

All strains grow on *Burkholderia cepacia* selective medium. Most strains of *Pandoraea* spp. have been isolated from clinical samples, primarily from respiratory tracts of patients suffering from cystic fibrosis but also from blood and from nonclinical sources such as water, sludge, soil, and dried milk (Coenye et al., 2000; Moore et al., 2001).

Coenye et al. (2000) and Daneshvar et al. (2001) provide tables of phenotypic traits that distinguish the five species and four described genomospecies of the genus *Pandoraea*. Coenye et al. (2001b) have described PCR primers that can be used to distinguish among the species of the genus *Pandoraea*. Coenye and LiPuma (2002) have described a PCR-RFLP scheme based on the *gyrB* locus that distinguishes four of the five *Pandoraea* species from each other and from members of three of the four genomospecies described by Daneshvar et al. (2001).

Differentiation of the genus *Pandoraea* from other genera

Because of their phenotypic similarity to *Burkholderia cepacia* and the importance of accurate identification for the treatment of cystic fibrosis patients, considerable effort has been devoted to the development of methods to distinguish *Pandoraea* spp. from *B. cepacia* and other organisms found in the respiratory tracts of cystic fibrosis patients. Phenotypic tests that distinguish *Pandoraea* species from *B. cepacia* genomovars I, III, and IV; *B. multivorans*; *B. vietnamiensis*; *Ralstonia paucula*; and *R. pickettii* are given by Coenye et al. (2000). Phenotypic tests that distinguish *Pandoraea* species from *B. cepacia* genomovars I–VII, *B. gladioli*, and *R. pickettii* are given by Henry et al. (2001). Coenye et al. (2001b) have

described PCR primers that can be used to separate isolates belonging to the genus *Pandoraea* from selected members of the genera *Alcaligenes*, *Burkholderia*, *Pseudomonas*, *Ralstonia*, and *Stentrophomonas* as well as distinguishing among the species of the genus *Pandoraea*.

List of species of the genus Pandoraea

1. **Pandoraea apista** Coenye, Falsen, Hoste, Ohlen, Goris, Govan, Gillis and Vandamme 2000, 896[VP]

   a.pis′ ta. Gr. adj. *apistos* disloyal, unfaithful, treacherous.

   As described for the genus with the following additional characteristics. Does not reduce nitrate. Grows on cetrimide agar; grows at 42°C; does not grow in O/F medium with glucose. Does not grow on acetamide. Assimilates caprate, citrate, and DL-lactate; does not assimilate maltose or sucrose. Produces amylase, arginine dihydrolase, C8-ester lipase, phosphoamidase, and urease.

   The mol% G + C of the DNA is: 61.8 (HPLC).

   Type strain: ATCC BAA-61, CCUG 38412, CIP 106627, LMG 16407.

   GenBank accession number (16S rRNA): AF139173.

2. **Pandoraea norimbergensis** (Wittke, Ludwig, Peiffer and Kleiner 1998) Coenye, Falsen, Hoste, Ohlen, Goris, Govan, Gillis and Vandamme 2000, 896[VP] (*Burkholderia norimbergensis* Wittke, Ludwig, Peiffer and Kleiner 1998, 631)

   no.rim.ber.gen′ sis. M.L. *Norimberga* Nurnberg (Bavaria, Germany); M.L. fem. adj. coming from Nurnberg, referring to its place of isolation.

   As described for the genus with the following additional characteristics. No reduction of nitrate. Grows on cetrimide agar. Does not grow on acetamide; does not grow at 42°C. Assimilates citrate; does not assimilate adipate, maltose, or sucrose. Produces oxidase and phosphoamidase. Does not produce cysteine arylamidase; some strains produce urease.

   The mol% G + C of the DNA is: 63.2 (HPLC).

   Type strain: ATCC BAA-65, CCUG 39188, CIP 105463, LGM 18379.

   GenBank accession number (16S rRNA): Y09879.

3. **Pandoraea pnomenusa** Coenye, Falsen, Hoste, Ohlen, Goris, Govan, Gillis and Vandamme 2000, 896[VP]

   pno.me.nu′ sa. Gr. n. *pnoe* breath, breathing; Gr. v. *meno* to reside, stay, live; (fem. part. pres. *menusa* N.L. part. adj. *pnomenusa* referring to the lung as the niche of these bacteria.

   As for the genus with the following additional charac-

   teristics. Most strains reduce nitrate. Grows at 42°C; does not grow in O/F medium with glucose. Assimilates caprate, citrate, and DL-lactate; does not assimilate adipate, D-glucose, maltose, or sucrose. Produces C8-ester lipase and urease; most strains produce C4-ester lipase and phosphoamidase.

   The mol% G + C of the DNA is: 64.3 (HPLC).

   Type strain: ATCC BAA-63, CIP 106626, CCUG 38742, LMG 18087.

   GenBank accession number (16S rRNA): AF139174.

4. **Pandoraea pulmonicola** Coenye, Falsen, Hoste, Ohlen, Goris, Govan, Gillis and Vandamme 2000, 896[VP]

   pul.mo.ni′ co.la. L. n. *pulmo* lung; L. suff. *cola* dwelling, occurring in; N.L. n. *pulmonicola* occurring in lungs.

   As for the genus with the following additional characteristics. No reduction of nitrate. Grows at 42°C; grows on cetrimide agar and in O/F medium with glucose. Does not grow on acetamide. Assimilates caprate, citrate, D-glucose, and DL-lactate; does not assimilate adipate, maltose, or sucrose. Produces oxidase and phosphoamidase; does not produce amylase, arginine dihydrolase, cysteine arylamidase, C4-esterase, C8-ester lipase, or urease.

   The mol% G + C of the DNA is: 61.8 (HPLC).

   Type strain: ATCC BAA-62, CIP 106625, CCUG 38759, LMG 18106.

   GenBank accession number (16S rRNA): AF139175.

5. **Pandoraea sputorum** Coenye, Falsen, Hoste, Ohlen, Goris, Govan, Gillis and Vandamme 2000, 896[VP]

   spu.to′ rum. L. n. *sputum* spit, sputum; L. gen. pl. n. *sputorum* of sputa.

   As for the genus with the following additional characteristics. Does not grow on acetamide or in O/F medium with glucose. Does not assimilate sucrose. Produces C8-ester lipase and phosphoamidase.

   The mol% G + C of the DNA is: 61.9.

   Type strain: CCUG 39682, LMG 18819.

   GenBank accession number (16S rRNA): AF139176.

Other Organisms

Note: this description, like those of the other organisms in this section, is based on a single strain. 16S rDNA sequence analyses place all of these organisms in the genus *Pandoraea*; other data show that they are distinct from the five described species (Coenye et al., 2000; Daneshvar et al., 2001).

1. *Pandoraea* sp. (genomospecies 1). Described by Coenye et al. (2000); originally isolated by Parsons et al. (1988). This organism was studied because of its ability to degrade chlorinated aromatic compounds. Reduces nitrate. Does not grow at 42°C, in O/F medium with glucose, or on acetamide or cetrimide agar. Assimilates DL-lactate; does not assimilate

adipate, caprate, or mannose. Produces C4-ester lipase, C8-ester lipase, oxidase, phosphoamidase, and urease.

   The mol% G + C of the DNA is: unknown.

   Deposited strain: JB1, R-5199, CCUG 39680.

   GenBank accession number (16S rRNA): X92188.

2. *Pandoraea* genomospecies 2. Described by Daneshvar et al. (2001). Does not reduce nitrate. Grows at 42°C. Does not grow on cetrimide agar. Produces oxidase; produces urease after 7 d on Christensen's agar.

   The mol% G + C of the DNA is: 65.2 (T_m).

Deposited strain: CDC G5084, ATCC BAA-108.
GenBank accession number (16S rRNA): AF247693.

3. *Pandoraea* genomospecies 3. Described by Daneshvar et al. (2001). Does not reduce nitrate. Grows at 42°C. Does not grow on cetrimide agar. Produces urease after 7 d on Christensen's agar; does not produce oxidase.
   The mol% G + C of the DNA is: 67.1 (T_m).
   Deposited strain: CDC G9805, ATCC BAA-109.

GenBank accession number (16S rRNA): AF247697.

4. *Pandoraea* genomospecies 4. Described by Daneshvar et al. (2001). Does not reduce nitrate. Grows at 42°C. Does not grow on cetrimide agar. Produces urease after 7 d on Christensen's agar; produces oxidase.
   The mol% G + C of the DNA is: 68.6 (T_m).
   Deposited strain: CDC H652, ATCC BAA-110.
   GenBank accession number (16S rRNA): AF247698.

Genus V. **Paucimonas** *Jendrossek 2001, 906*[VP]

THE EDITORIAL BOARD

Pau.ci.mo′ nas. L. adj. *paucus* little, few; Gr. fem. n. *monas* unit, cell; M.L. n. *Paucimonas* bacterium with restricted (few) metabolic capacities.

Gram-negative, strictly respiratory chemoorganotroph. Catalase positive; oxidase positive. **Preferred substrates are organic acids**; does not utilize alcohols or polyalcohols, amino acids or polypeptides, sugars or polysaccharides. **Accumulates poly-β-hydroxybutyrate.**

The mol% G + C of the DNA is: 59 ± 2.

Type species: **Paucimonas lemoignei** (Delafield, Doudoroff, Palleroni, Lusty and Contopoulos 1965) Jendrossek 2001, 906 (*Pseudomonas lemoignei* Delafield, Doudoroff, Palleroni, Lusty and Contopoulos 1965, 1460.)

FURTHER DESCRIPTIVE INFORMATION

Analysis of 16S rDNA sequences placed *Paucimonas lemoignei* in the *Betaproteobacteria* with *Herbaspirillum* spp. as its closest relatives (Jendrossek, 2001).

ENRICHMENT AND ISOLATION PROCEDURES

Organisms were isolated from soil. Soil samples were emended with poly-(3-hydroxyvalerate) and incubated; the soil was then diluted into a liquid enrichment medium containing poly-(3-hydroxyvalerate). Procedures and liquid and solid media are described by Mergaert et al. (1996).

List of species of the genus Paucimonas

1. **Paucimonas lemoignei** (Delafield, Doudoroff, Palleroni, Lusty and Contopoulos 1965) Jendrossek 2001, 906[VP] (*Pseudomonas lemoignei* Delafield, Doudoroff, Palleroni, Lusty and Contopoulos 1965, 1460.)

   le.moig′ ne.i. M.L. gen. n. *lemoignei* of Lemoigne; named after M.H. Lemoigne, a French bacteriologist.

   Gram-negative motile rods (0.6–0.8 × 1.5–3.0 μm). Nonsporeforming; do not produce fluorescent pigments. Poly-

hydroxyalkanoate depolymerases produced. Grow on acetate, butyrate, 3-hydroxybutyrate, 3-hydroxyvalerate, pyruvate, valerate, and succinate. Major fatty acids are $C_{12:0}$, $C_{16:0}$, $C_{16:1 \omega7c}$, $C_{10:0 3OH}$, $C_{12:0 3OH}$, $C_{14:0 2OH}$.

   The mol% G + C of the DNA is: 59 ± 2 (T_m).

   Type strain: ATCC 17989, CCUG 2114, CIP 103794, DSM 7445, LMG 2207.

   GenBank accession number (16S rRNA): X92555.

Genus VI. **Polynucleobacter** *Heckmann and Schmidt 1987, 456*[VP]

HANS-DIETER GÖRTZ AND HELMUT J. SCHMIDT

Pol.y.nuc′ le.o.bac.ter. Gr. adj. *polys* numerous; L. masc. n. *nucleus* nut, kernel; masc. *bacter* the equivalent of Gr. neut. n. *bactrum* a rod; *polynucleobacter* the bacterium with many nucleoids.

Bacterial endosymbiont formerly called omikron. Multiple nucleoids. Inhabits the cytoplasm of the following closely related *Euplotes* spp. (Ciliophora, Protozoa): *E. aediculatus, E. eurystomus, E. patella, E. plumipes, E. woodruffi, E. daidaleos,* and *E. octocarinatus.* Essential for their host species. Nonmotile.

The mol% G + C of the DNA is: 47.7 or 44.9.

Type species: **Polynucleobacter necessarius** Heckmann and Schmidt 1987, 456.

FURTHER DESCRIPTIVE INFORMATION

Only one species, *P. necessarius,* has been described. It has a number of features in common with other endosymbionts found in related *Euplotes* species (Heckmann et al., 1983). Approximately 900–1000 cells of *P. necessarius* inhabit the cytoplasm of *E. aediculatus* (Heckmann, 1975). If *P. necessarius* is stained with DNA-specific dyes, several intensely stained and regularly spaced dots become visible (Fig. BXII.β.4). They are considered to be nu-

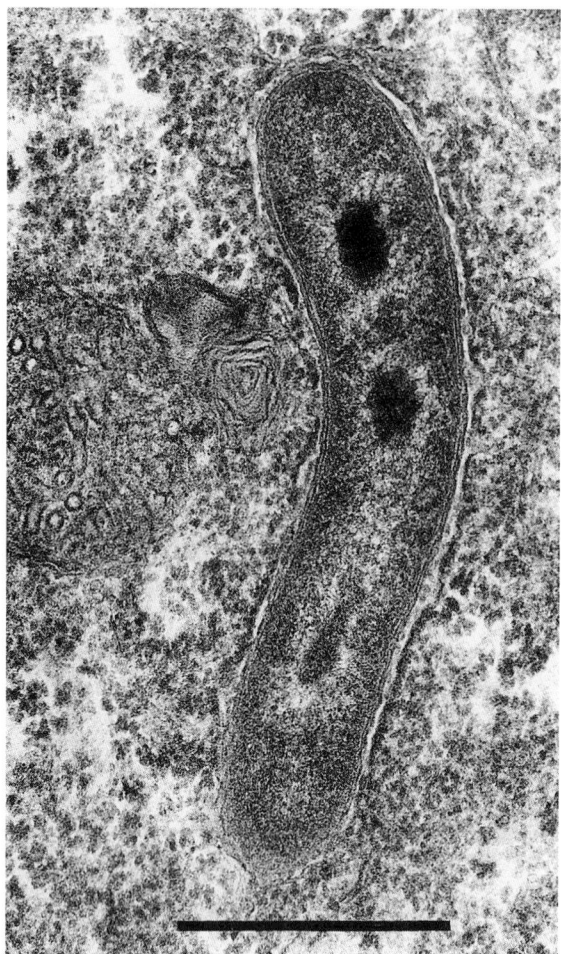

FIGURE BXII.β.4. *Polynucleobacter necessarius* (omikron), endosymbiont of *Euplotes aediculatus*. Longitudinal section showing nucleoids. Bar = 0.5 μm. (Reproduced with permission from K. Heckmann, Journal of Protozoology 22: 97–104, 1975, ©Society of Protozoologists.)

cleoids, and differ from those of most free-living bacteria by having an electron-dense central core. The symbionts reproduce by transverse binary fission. The fission products are often found to differ from each other in size. It is possible to isolate *P. necessarius* from *E. aediculatus* homogenates and to obtain pure preparations of the symbionts in quantities large enough for an ex-

traction and characterization of their DNA (Schmidt, 1982). The data are given below under the formal description of the species.

It is possible to remove the symbionts from *E. aediculatus* by treating a rapidly growing culture with penicillin (Heckmann, 1975). Aposymbiotic hosts eventually die. Heckmann concluded that *P. necessarius* is essential for the life of its host. Fauré-Fremiet (1952) made the same observation for *E. patella* and *E. eurystomus* and reached the same conclusions.

TAXONOMIC COMMENTS

Only the best-studied endosymbiont of *Euplotes*, *P. necessarius*, has been assigned a binomial name. This endosymbiont, previously known as omikron, lives obligately in *Euplotes aediculatus* and was named *Polynucleobacter necessarius* by Heckmann and Schmidt (1987). Similar, still unnamed endosymbionts inhabit related species of *Euplotes*, and are called omikron-like endosymbionts.

P. necessarius belongs to the *Betaproteobacteria* and is rather closely related to *Ralstonia eutrophus*, *R. solanacearum*, and *R. pickettii* (Fig. BXII.β.5). *In situ* hybridization with a specific oligonucleotide probe corroborated the assignment of the retrieved species to *P. necessarius* (Springer et al., 1996).

Omikron-like symbionts occur in several closely related species of freshwater *Euplotes* (*E. aediculatus*, *E. eurystomus*, *E. patella*, *E. plumipes*, *E. woodruffi*, *E. daidaleos*, and *E. octocarinatus*), but not in other unrelated freshwater species. In many stocks of these species, penicillin treatment has been found to remove the symbionts, which affected their hosts as described above, eventually resulting in their death (Heckmann et al., 1983; Heckmann, 1983). In this connection, it is of interest that Foissner (1977) described a species (*E. moebiusi f. quadricirratus* Kahl, 1930) in which he found omikron-like symbionts. This species is not closely related to the above *Euplotes* species. Foissner has not tested, however, whether or not the ciliate depends upon the symbionts. Fujishima and Heckmann (1984) tried to transfer symbionts between *Euplotes* species. They produced aposymbiotic cells by treating them with penicillin, and observed restoration of growth and the ability to divide when symbionts of *E. woodruffi*, of a stock collected in Japan, were introduced into a stock of *E. aediculatus*, which had been collected in France. The reverse combination, however, with *E. woodruffi* as the recipient and *E. aecliculatus* as the donor, did not work.

FURTHER READING

Heckmann, K. 1983. Endosymbionts of *Euplotes*. Int. Rev. Cytol. Suppl. *14*: 111–114.

List of species of the genus Polynucleobacter

1. **Polynucleobacter necessarius** Heckmann and Schmidt 1987, 456[VP]

   nec.es.sar′i.us. L. adj. *necessarius* indispensable, necessary.

   Description as given for the genus. Three to twelve nucleoids. Cells are 0.3 × 2.5–7.5 μm. Divides by binary fission.

   DNA molecular weight 3.5 × 10^9 daltons (analytic complexity) or 0.57 × 10^9 daltons (kinetic complexity).
   The mol% G + C of the DNA is: 47.7 (T_m) and 44.9 (Bd).
   Type strain: ATCC 30859.
   GenBank accession number (16S rRNA): X93019.
   Additional Remarks: The type strain is isolated from stock 15 of *E. aediculatus* (ATCC 30859).

Other Organisms

Other bacterial endosymbionts found in *Euplotes* have not been given binomial names, nor have they been validly described. Several endosymbionts of *Euplotes* produce the killer phenotype, and strains of killers with associated bacteria in the cytoplasm have been described for *E. minuta* and *E. crassus*. They were

named epsilon and eta, respectively, by Heckmann et al. (1967) and Rosati et al. (1976).

Certain stocks of *E. patella*, collected in Japan, were reported to be mate killers by Katashima (1965). Mate killers kill sensitive ciliates in the act of conjugation, when toxins produced by en-

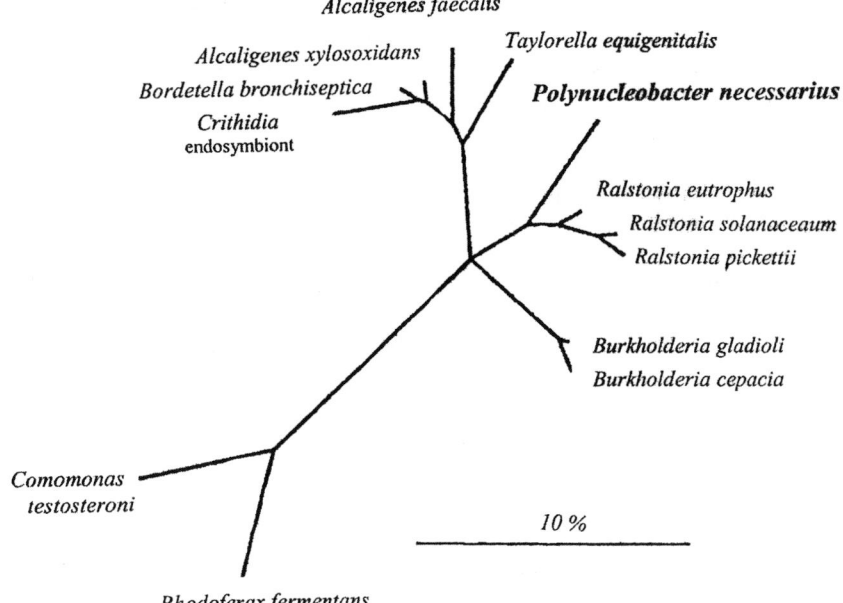

FIGURE BXII.β.5. Phylogenetic tree reflecting the relationship of *Polynucleobacter necessarius* and selected members of the *Betaproteobacteria*. The tree is based on a maximum likelihood analysis of a data set comprising a selection of 75 16S rRNA gene sequences containing no more than 140 ambiguities (for details see Springer et al., 1996) (Reproduced with permission from N. Springer et al., FEMS Microbiology Letters *135*: 333–336, 1996 ©Elsevier B.)

dosymbionts of mate killers are thought to be transferred to the sensitive mating partner not bearing endosymbionts. Mate killing was also observed for *E. minuta* by Heckmann et al. (1967) and for *E. crassus* by Rosati and Verni (1977). All these mate killers contained unnamed endosymbionts in their cytoplasm. *E. crassus* sometimes contains other endosymbionts with no known effects on their hosts. Rosati et al. (1976) observed them mostly in the cytoplasm. However, Rosati and Verni (1975) did find one endosymbiont in the macronucleus.

Genus VII. **Ralstonia** *Yabuuchi, Kosako, Yano, Hotta and Nishiuchi 1996, 625^VP (Effective publication: Yabuuchi, Kosako, Yano, Hotta and Nishiuchi 1995, 902)*

EIKO YABUUCHI, YOSHIAKI KAWAMURA AND TAKAYUKI EZAKI

Ral.sto′n.ia. M.L. dim. *-ia* ending; M.L. fem. n. *Ralstonia* named after E. Ralston, the American bacteriologist who first described *Pseudomonas pickettii.*

Gram-negative asporogenous rods. Motile or nonmotile; motile species have either a single polar flagellum or peritrichous flagella. Aerobic, having a strictly respiratory type of metabolism with oxygen as the terminal electron acceptor. Able to grow on ordinary peptone media. One species, *R. eutropha,* is a facultatively chemolithoautotrophic organism (a knallgas bacterium) and can oxidize H_2 as an electron donor. Furthermore, under anaerobic conditions, *R. eutropha* is able to utilize nitrate as the terminal electron acceptor. Colony color is beige in most species. Strains of *R. eutropha* are regarded as having outstanding biotechnological potential. **Oxidase and catalase positive.** Lysine and ornithine decarboxylase negative. Ubiquinone Q-8 is the major respiratory quinone. **None of the 26 carbohydrates tested are oxidized by the type strains of *R. campinensis* and *R. taiwanensis.*** Among 95 other organic compounds, 10—mainly salts of organic acids—were assimilated and 39 other compounds were not assimilated by all of the 11 type strains. **Cellular lipids of this** genus contain two kinds of phosphatidylethanolamine, PE-1 and PE-2. The latter possesses 2-hydroxy fatty acid at *sn*-2 position of the glycerol moiety. Major components of cellular fatty acids are $C_{16:0}$, a mixture of $C_{18:1\ \omega 9t}$ and $C_{18:1\ \omega 7c}$, and $C_{14:0\ 3OH}$. At present the genus is composed of 11 validated species. The sequence similarity of 11 type strains to that of *Ralstonia pickettii*—the type species—ranges from 95.0% to 98.1%. The genus contains **plant pathogens, human pathogens, knallgas bacteria, and metal-resistant bacteria.**

The mol% G + C of the DNA is: 64.0–68.0.

Type species: **Ralstonia pickettii** (Ralston, Palleroni and Doudoroff 1973) Yabuuchi, Kosako, Yano, Hotta and Nishiuchi 1996, 625 (Effective publication: Yabuuchi, Kosako, Yano, Hotta and Nishiuchi 1995, 903) (*Burkholderia pickettii* Yabuuchi, Kosako, Oyaizu, Yano, Hotta, Hashimoto, Ezaki and Arakawa 1993, 398; *Pseudomonas pickettii* Ralston, Palleroni and Doudoroff 1973, 18.)

FURTHER DESCRIPTIVE INFORMATION

Morphology The cell dimensions are 0.5–0.8 × 1.2–3.0 μm, except for *R. oxalatica*, whose cells are as small as 0.3–0.4 × 0.9–1.5 μm.

Poly-β-hydroxybutyrate (PHB) granules have been reported in both Gram-negative and Gram-positive bacteria as intracellular reserve material (Doudoroff and Stanier, 1959). Morphological detection of PHB granules by optical or fluorescent microscopy has been reported (Stanier et al., 1966; Ostle and Holt, 1982). Rapid detection of polyhydroxyalkanoate-accumulating bacteria isolated from the environment by colony PCR has been reported by Sheu et al. (2000). *R. eutropha* synthesizes short chain length (SCL; 3–5 carbon atoms) poly-β-hydroxyalkanoates (PHAs), and the resulting polymers are accumulated as intracellular granules. The synthesis of the PHAs is catalyzed by PHA synthase (Song et al., 2000c; Zhang et al., 2000c). PHAs are now of commercial value as thermoplastics (Holmes, 1985). Accumulated PHB is degraded and used for growth and survival when an exogenous carbon source is not available (Handrick et al., 2000). Compared with PHB homopolymers, the PHA copolymer (poly-β-hydroxybutyrate-co-β-hydroxyvalerate) [P(HB)-co-HV)] has more useful thermomechanical properties. *R. eutropha* accumulates larger amount of [P(HB)-co-HV] copolymer when provided with glucose and pentanoic acid rather than propionic acid under nitrogen-limited conditions (Ramsay et al., 1990).

Cloning of the gene encoding PHB polymerase and its expression by recombinant *E. coli* has been reported (Schubert et al., 1988). The PHB production by the recombinant *E. coli* was analyzed metabolically and kinetically (van Wegen et al., 2001). An intracellular PHB depolymerase gene from *R. eutropha* H16 was cloned and gene product was characterized (Saegusa et al., 2001).

Colonial morphology and pigmentation Colonies on tryptic soy agar are less than 1 mm in diameter after 48 h incubation at 28–30°C. When the colonies are fully-grown, their diameter is >1 mm. In six species, the colonies are beige-colored, domed, smooth, and glistening with entire margin. The colonies of two species (*R. campinensis* and *R. metallidurans*) sometimes have scalloped margins (Goris et al., 2001). *R. solanacearum* EY 2181 produces a water-soluble slightly brown pigment, but the color appears different from that of alcaptone.

Cells of *R. solanacearum* strain 1609 convert to a viable-but-nonculturable (VBNC) form in water microcosms kept at 4°C, but not in those at 20°C (van Elsas et al., 2001). The viability of a fraction of these VBNC forms was evidenced by the direct viable count staining with the redox dye 5-cyano-2,3-ditolyl tetrazolium chloride (van Elsas et al., 2001). Recently, it was reported that cells of *R. solanacearum* enter into a VBNC state under certain conditions, as in response to cupric sulfate when in a saline solution, and when placed in autoclaved soil (Grey and Steck, 2001). It was also verified that *R. solanacearum* cells in the VBNC state were able to infect and multiply in the tissue. Thus the VBNC form of plant pathogens might explain the persistence of infection in nature. The occurrence of VBNC cells of *R. solanacearum* in natural water poses new problems for the detection (van Elsas et al., 2001) and persistence of infection (Grey and Steck, 2001) of *R. solanacearum*.

Growth conditions and nutrition Growth occurs aerobically on ordinary peptone media at 25–41°C. The type strains of *R. basilensis*, *R. metallidurans*, and *R. solanacearum* do not grow at 41°C, but the type strains of the other species can do so. The organisms do not require any growth factors, including sodium chloride. *Ralstonia eutropha* is a H2-oxidizing (knallgas) chemolithoautotroph (Aragno and Schlegel, 1992); [NiFe] hydrogenases play a role as H2 sensors (Kleihues et al., 2000; Buhrke et al., 2001). Cells of *R. eutropha* also grow well aerobically on nutrient-rich ordinary media.

Physiology and metabolism *Ralstonia eutropha* strain H16 mediates the reduction of nitric oxide (NO) to nitrous oxide (N2O) by a single-component nitric oxide reductase (Cramm et al., 1999; Pohlmann et al., 2000). Resistance to nickel (Schmidt et al., 1991) and to both cobalt and nickel (Tibazarwa et al., 2000) has been reported in *R. eutropha*. Lead resistance was reported in *R. metallidurans* (Borremans et al., 2001). Genetically engineered *R. eutropha* strain MTB has an enhanced ability to immobilize external Cd^{2+} ions, and inoculation of this strain into Cd^{2+}-polluted soil reduces the toxic effects of heavy metal on the growth of tobacco plants (*Nicotiana bentamiana*) (Valls et al., 2000).

Recently, many reports concerning metabolic activities of *R. eutropha* have appeared in the literature (Lütke-Eversloh and Steinbuchel, 1999; Grzeszik et al., 2000; Happe et al., 2000; Padilla et al., 2000; Bernhard et al., 2001; Bramer and Steinbuchel, 2001; Drewlo et al., 2001; Schräder et al., 2001; Taguchi et al., 2001; York et al., 2001; Zarnt et al., 2001). The depolymerase gene for intracellular PHB of *R. eutropha* strain H16 was successfully cloned and its product was characterized (Saegusa et al., 2001). Under anaerobic conditions, *R. eutropha* undergoes nitrate respiration by utilizing nitrate as a terminal electron acceptor (Schwartz and Friedrich, 2001). Genetic determinants essential for these metabolic processes are linked to the megaplasmid pHG1.

Cellular lipids and fatty acids composition All of the two-dimensional TLCs (thin-layer chromatograms) of the type strains of eight *Ralstonia* species and of *Burkholderia cepacia* (Fig. BXII.β.6) reveal two spots of phosphatidylethanolamine (PE-1 and PE-2) and one spot of phosphatidylglycerol (PG). PE-2 possesses 2-hydroxy fatty acid, at the sn-2 position of the glycerol moiety. In addition, two spots of ornithine lipid (OL-1 and OL-2) have been visualized in *B. cepacia*. The major components of the nonpolar acids are $C_{16:0}$ followed by $C_{16:1\ \omega7c}$ and mixture of $C_{18:1\ \omega9t}$ and $C_{18:1\ \omega7c}$. The major polar acid is $C_{14:0\ 3OH}$. Percentage of minor components of fatty acids against the total amount of each of them cannot be a key of differentiation of species, because it could vary by differences in cultural condition and/or analyzing method.

Genetics and plasmids The complete genome sequence and its analysis of *R. solanacearum* strain GMI100 have been reported (Salanoubat et al., 2002). Part of their summary is as follows: "The 5.8-megabase (Mb) genome is organized into two replicons, one 3.7-Mb chromosome and another 2.1-Mb megaplasmid. Both replicons have a mosaic structure providing evidence for the acquisition of genes through horizontal gene transfer. Regions containing genetically mobile elements associated with the percentage of G + C bias may have an important function in genome evolution. The genome encodes many proteins potentially associated with a role in pathogenicity. In particular, many putative attachment factors were identified. The complete repertoire of type III secreted effector proteins can be studied. Over 40 candidates were identified. Comparison with other genomes

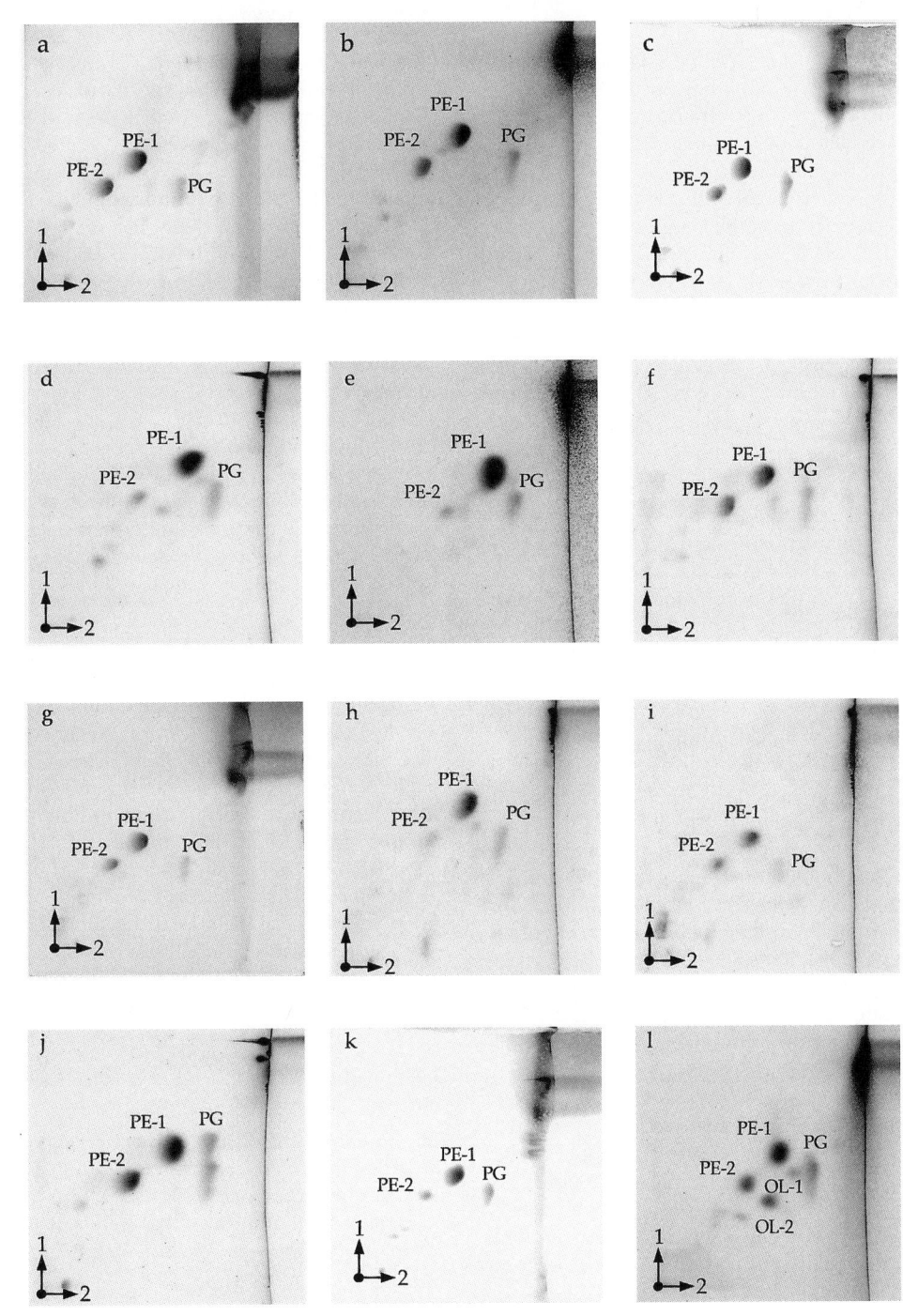

FIGURE BXII.β.6. Two-dimensional thin-layer chromatograms of cellular lipids of type strains of eight *Ralstonia* species and of *Burkholderia cepacia* (N. Fujiwara and T. Naka, unpublished data). The solvent system for the first direction was chloroform:methanol:water (65:25:4, v/v); for the second direction it was chloroform:methanol:acetic acid (65:25:10, v/v). (*a*) *R. pickettii* EY 4382; (*b*) *R. basilensis* EY4358; (*c*) *R. campinensis* EY 4379 (*d*) *R. eutropha* EY 3798; (*e*) *R. gilardii* EY 4363; (*f*) *R. mannitolilytica* EY 4364; (*g*) *R. metallidurans* EY 4380; (*h*) *R. oxalatica* EY 4365; (*i*) *R. paucula* EY 4366; (*j*) *R. solanacearum* EY 2181; (*k*) *R. taiwanensis* EY 4381; (*l*) *B. cepacia* EY 645. Abbreviations: PE, phosphatidylethanolamine; PG, phosphatidylglycerol; OL, ornithine lipid.

suggests that bacterial plant pathogens and animal pathogens harbor distinct arrays of specialized type III-dependent effectors."

The genome sequence of *R. solanacearum* is a first step toward an exhaustive functional analysis of pathogenicity determinants in this plant pathogen. Prior to this report, much genetic research on the virulence factors of *R. solanacearum* was reported (Rosenberg et al., 1982; Boucher et al., 1986; Allen et al., 1997; Huang and Allen, 1997; Lasserre et al., 1997; Huang et al., 1998; Aldon et al., 2000; Bertolla et al., 2000; Garg et al., 2000; Vasse et al., 2000; Wang et al., 2000; Belbahri et al., 2001; Schwartz and Friedrich, 2001; Staskawicz et al., 2001; van Overbeek et al., 2002).

The genome of *R. eutropha* strain H16 is composed of two circular chromosomes measuring 4.1 and 2.9 Mb. A physical map of the megaplasmid pHG1 has been given by Schwartz and Friedrich (2001).

A bacteriophage P4284 encoding a *R. solanacearum* bacteriolytic protein has been described (Ozawa et al., 2001).

Pathogenicity   The plant pathogens produce a pilus structure. Morphologically similar structures are associated with type III protein secretion systems found in animal pathogens, and are essential for the delivery of virulence proteins directly into host cells.

Ralstonia solanacearum causes a lethal vascular wilt disease in more than 200 plant species in 50 botanical families in the tropics, subtropics, and warm temperate regions of the world. Its agronomically important hosts include tomato, potato, tobacco, peanut, and banana. This pathogenic organism was named and described as early as 1896 by Smith (1896). The instability of virulence and correlation with colony morphology of this organism led to elucidation of the existence of a regulatory network modulating transcription of multiple virulence genes (Brumbley et al., 1993). A type IV pilus system is responsible for a twitching type of motility and seems to contribute significantly to the plant pathogeneses (Liu et al., 2001; Tans-Kersten et al., 2001). After *R. solanacearum* cells penetrate the xylem vessels of tomato plants, they rapidly travel to the upper part of the plant and become readily detectable throughout the stem. Viable counts of the organism can reach $>10^{10}$ cells per cm of stem by day 8. Infected plants rapidly collapse and die, and virulent organisms released from the host plant roots or collapsed stems in contact with the ground results in a return of the bacteria to a saprophytic life in the soil, where they await a new host (Schell, 2000). The primary virulence factor of wild type *R. solanacearum* is an exopolysaccharide I (EPS I)—a large, nitrogen-rich, acidic exopolysaccharide that occludes vascular tissues and inhibits water flow (Schell, 2000). A type II secretion system exports all the cell wall-degrading exoenzymes such as pectinolytic enzymes and cellulolytic enzymes (Saile et al., 1997). A type III secretion system seems to deliver toxic proteins directly into the plant cell cytoplasm. Production of several virulence determinants is governed by growth in the presence of a volatile extracellular factor (VEF) produced by the wild-type strain of the organism (Clough et al., 1994). The VEF is active at ≤ 1 nM for stimulating biosynthesis of extracellular polysaccharide (*eps*) (Flavier et al., 1997a). The virulence of *R. solanacearum* is regulated by a complex network, the core of which is the Phc (phenotype conversion) system (Schell, 2000). The PhcA system is composed of PhcA (Brumbley and Denny, 1990; Brumbley et al., 1993) and the products of the *phcBSRQ* operon (Clough et al., 1994, 1997). Transcriptional activity of PhcA is controlled by 3-OH PAME (Clough et al., 1997; Flavier et al. 1997b).

Three *Ralstonia* species infect predisposed humans: *R. pickettii*, *R. mannitolilytica*, and *R. gilardii*. *R. pickettii* has been associated with acute meningitis (Fass and Barnishan, 1976), osteomyelitis and intervertebral discitis (Wertheim and Markovitz, 1992), nosocomial infection (Phillips et al.; 1972; Kahan et al., 1983; Raveh et al., 1993; Maroye et al., 2000), bacteremia (Fujita et al., 1981; Roberts et al., 1990), pseudobacteremia (Verschraegen et al., 1985), and colonization (McNeil et al., 1985; Centers for Disease Control and Prevention, 1998; Labarca et al., 1999; Yoneyama et al., 2000). One case each of recurrent meningitis and hemoperitoneum infection due to *R. mannitolilytica* (Vaneechoutte et al., 2001) has been reported. In addition, a case of catheter sepsis associated with *R. gilardii* (Wauters et al., 2001) has been reported.

It is remarkable that a certain strain of *Pseudomonas aeruginosa* is not only pathogenic to predisposed humans but also capable of causing disease in plants. Such "interkingdom" pathogens are able to colonize the surface of both plant and animal cells and to escape from diverse host defense mechanisms. Such features that allow them to be pathogenic on both procaryotic and eucaryotic hosts should be elucidated in the future (Staskawicz et al., 2001).

Ecology   Organisms belonging to genus *Ralstonia* appear to be free-living in nature. However, because of their metabolic or pathogenic specialties, the type strains of certain species have been isolated from human clinical specimens (*R. pickettii*, *R. mannitolilytica*, and *R. paucula*), from a fixed-bed reactor with 2,6-dichlorophenol as the sole carbon and energy source (*R. basilensis*), from a zinc-decertified area in Lommel, Belgium (*R. campinensis*), from water (*R. eutropha* and *R. metallidurans*), from plants (*R. solanacearum* and *R. taiwanensis*), and from earthworms (*R. oxalatica*). Thus, *Ralstonia* organisms might be living in a variety of ecological niches. Furthermore, the long survival of *R. solanacearum* organisms in soil and rhizospheres might be possible by an asymptomatic infection of the roots of nonhost plants, such as bean, peas, soybean, corn, and rice (Granada and Sequeira, 1983).

DIFFERENTIATION OF THE GENUS *RALSTONIA* FROM OTHER GENERA

Differential characteristics of the type species of five genera in the order *Burkholderiales* (class *Betaproteobacteria*) with one genus each in the order *Sphingomonadales* and *Caulobacterales* (class *Alphaproteobacteria*) and the type genus of the order *Pseudomonadales* (class *Gammaproteobacteria*) are summarized in Table BXII.β.8.

TAXONOMIC COMMENTS

The genus *Ralstonia* was named after Ericka Ralston, the American bacteriologist who first named and described *Pseudomonas pickettii* and suggested a taxonomic relationship to *Pseudomonas solanacearum* (Ralston et al., 1973). Based on polyphasic taxonomic research, *Burkholderia pickettii*, *Burkholderia solanacearum*, and *Alcaligenes eutrophus* were transferred to a new genus, *Ralstonia*, by Yabuuchi et al. (1995, 1996). *R. pickettii* was designated as the type species for the genus, because neither plant nor animal quarantine law restricts its exchange. This is important for taxonomic research to proceed smoothly. The genus *Ralstonia*, when proposed, contained *R. pickettii*, a species pathogenic to predisposed humans and pathogenic to plants in the botanical

TABLE BXII.β.8. Differential characteristics of the type species of *Ralstonia* and the type species of seven other genera of *Proteobacteria*[a]

Characteristic	Ralstonia pickettii	Burkholderia cepacia	Oxalobacter formigenes	Alcaligenes faecalis[b]	Comamonas terrigena[c]	Sphingomonas paucimobilis	Brevundimonas diminuta[d]	Pseudomonas aeruginosa[e]
Classification in:								
Betaproteobacteria	+	+	+	+	+	−	−	−
Alphaproteobacteria	−	−	−	−	−	+	+	−
Gammaproteobacteria	−	−	−	−	−	−	−	+
Aerobic	+	+	−	+	+	+	+	+
Anaerobic	−	−	+	−	−	−	+	+
Flagellation:								
Single polar	+	−	−	−	−	+	+	+
Polar tuft	−	+	−	−	+	−	−	−
Peritrichous	−	−	−	+	−	−	−	−
Pigmentation:								
of colony	Beige	Yellow	−	Iridescent	−	Yellow	−	−
of medium	−	−	Clearance of oxalate crystals	−	−		−	Blue/green
Growth factor required	−	−	Rumen fluid or yeast extract	−	−		+	−
Denitrification	+	−		−	−	−		+
Esculin hydrolysis	−	+		−	−	+	−	−
Lysine decarboxylase	−	+		−	−	−	−	−
Urease	+	−		−	−	−		−
Oxidative formation of acid from glucose	+	+	−	−	−	+	−	+
Assimilation of:								
Glucose	+	+			−	+	−	+
Citrate	+	+		+		+	−	+
Sphingolipid	−	−		−		+		−
Characteristic hydroxy fatty acid:								
$C_{10:0\ 3OH}$					+			
$C_{12:0\ 2OH}$								+
$C_{12:0\ 3OH}$							+	+
$C_{14:0\ 2OH}$						+		
$C_{14:0\ 3OH}$	+	+		+	+			
$C_{16:0\ 3OH}$		+						
Isoprenoid quinone:								
Q-8	+	+		+	+	+		
Q-9								+
Q-10							+	
Mol% G + C of DNA	64	66.6	48–51	55.9–59.4[d]	64–66[c]	62.1–63.9	79.3–81.4	67.2[g]
Pathogenic for:								
Humans	+					+		+
Mammals		+						+
Plants								+
Regulates homeostasis of oxalic acid in gut			+[c]					

[a]Symbols: see standard definitions. Blank space, not determined or not applicable.

[b]Data from Kersters and De Ley (1984b).

[c]Data from De Vos et al. (1985b).

[d]Data from Segers et al. (1994).

[e]Data from Palleroni (1984).

family Soranaceae. The genus also contained *R. solanacearum* and the hydrogen-oxidizing knallgas bacterium, *R. eutropha*. Eight new species were also added.

A phylogenetic dendrogram of the 16S rDNA similarities of the type strains of 11 *Ralstonia* species and of *Burkholderia cepacia* is shown in Fig. BXII.β.7. The percent similarity values among the type strains of 10 *Ralstonia* species and *B. cepacia* versus the type strain of *R. pickettii* are summarized in Table BXII.β.9. The values range from 95.0% to 98.1% among the *Ralstonia* species and 91.3% between *R. pickettii* and *B. cepacia*. In view of the latter value, inclusion of the genus *Ralstonia* in the family *Burkholderiaceae* is supported.

DIFFERENTIATION OF THE SPECIES OF THE GENUS *RALSTONIA*

Tables BXII.β.10 and BXII.β.11 list the differential characteristics of the species of *Ralstonia* and of *Burkholderia cepacia*. Additional features are listed in Tables BXII.β.12, BXII.β.13, and BXII.β.14.

List of species of the genus Ralstonia

1. **Ralstonia pickettii** (Ralston, Palleroni and Doudoroff 1973) Yabuuchi, Kosako, Yano, Hotta and Nishiuchi 1996, 625[VP]

(Effective publication: Yabuuchi, Kosako, Yano, Hotta and Nishiuchi 1995, 903) (*Burkholderia pickettii* Yabuuchi, Ko-

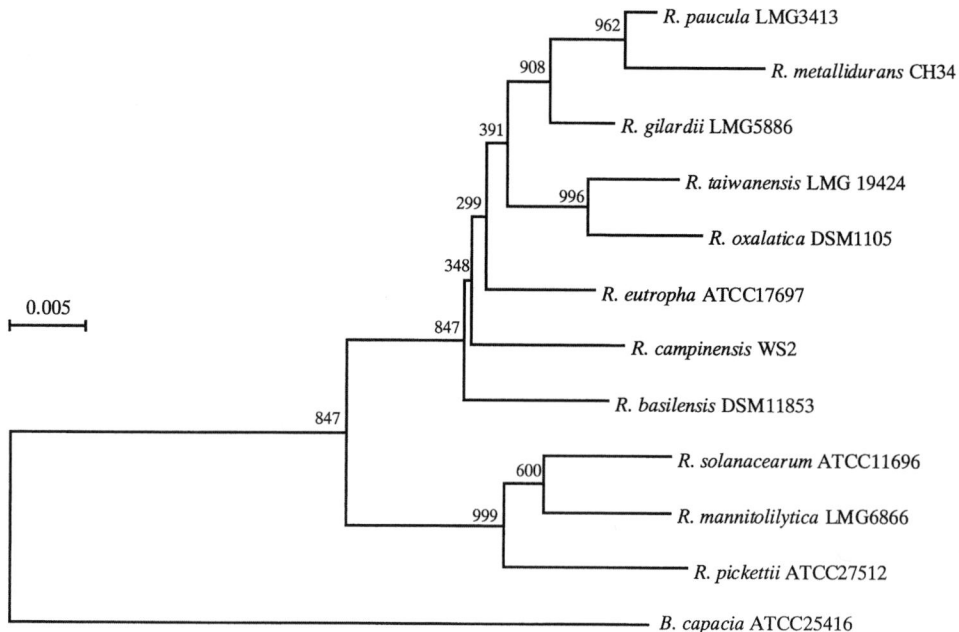

FIGURE BXII.β.7. Phylogenetic tree derived from 16S rDNA sequence analysis showing the relationships among the 11 *Ralstonia* species. Distance of the type species of the genus *Burkholderia*, *B. cepacia*, from the genus *Ralstonia* is also shown. Bar = K_{nuc} value.

TABLE BXII.β.9. 16S rDNA nucleotide sequence similarity of the type strains of 10 *Ralstonia* species and of *Burkholderia cepacia* vs. that of *Ralstonia pickettii*

Species	Type stain	Nucleotide sequence accession No.	% Sequence similarity to that of ATCC 27511
R. pickettii	ATCC 27511	S55004	100
R. mannitolilytica	LMG 6866	AJ270258	98.1
R. solanacearum	ATCC 10696	X67036	97.3
R. campinensis	WS2	AF312020	96.4
R. eutropha	ATCC 17697	M32021	96.4
R. gilardii	LMG 5886	AF076645	95.9
R. taiwanensis	LMG 19425	AF300325	95.7
R. paucula	LMG 3413	AF085226	95.6
R. basilensis	DSM 11853	AF312022	95.4
R. metallidurans	CH34	Y10824	95.3
R. oxalatica	Ox1, DSM 1105	AF155567	95
Burkholderia cepacia	ATCC 25416	M22518	91.3

sako, Oyaizu, Yano, Hotta, Hashimoto, Ezaki and Arakawa 1993, 398; *Pseudomonas pickettii* Ralston, Palleroni and Doudoroff 1973, 18.)

pick.et′ ti.i. M.L. gen. n. *pickettii* Pickett patronymic, of Pickett; named after M.J. Pickett.

The characteristics are as described for the genus and in Tables BXII.β.10, BXII.β.11, BXII.β.12, BXII.β.13, and BXII.β.14. The cell size is 0.5–0.6 × 1.5–3.0 μm. Accumulate poly-β-hydroxy-butyrate granules as a reserve carbon source (Ralston et al., 1973). Growth occurs at 41°C but not at 5°C. Nitrate respiration is positive. Cellular lipids of the type strain contain two kinds of ornithine lipids (Fig. BXII.β.6). The former CDC group Va-2 as well as Va-1 corresponds to *R. pickettii*, because the DNA–DNA hybridization value between type strain of *R. pickettii* and Va-2 strain was 84% (Pickett and Greenwood, 1980). The type strain was isolated from a patient who had undergone tracheostomy.

The mol% G + C of the DNA is: 64.0 (Bd).

Type strain: Ralston K-288, ATCC 27511, CCUG 3318, DSM 6297, EY 4382, GTC 1882, JCM 5969, LMG 5942.

GenBank accession number (16S rRNA): S55004, X67042.

2. **Ralstonia basilensis** Steinle, Stucki, Stettler and Hanselmann 1999, 1325[VP] (Effective publication: Steinle, Stucki, Stettler and Hanselmann 1998, 2569) emend. Goris, De Vos, Coenye, Host, Janssens, Brim, Diels, Mergeay, Kersters and Vandamme 2001, 1781.

ba.si.len′ sis. M.L. adj. *basilea, basilensis* pertaining to Basel, Switzerland, where the strain was isolated.

The characteristics are as described for the genus and listed in Tables BXII.β.10, BXII.β.11, BXII.β.12, BXII.β.13, and BXII.β.14. The cell size is 0.8 × 1.2–2.2 μm (Goris et al., 2001). Able to grow utilizing 2,6-dichlorophenol as the sole source of carbon and energy. The type strain was isolated from a fixed-bed reactor with 2,6-dichlorophenol as the sole carbon and energy source.

The mol% G + C of the DNA is: 65.0 (HPLC).

TABLE BXII.β.10. Differential physiological and biochemical characteristics of the type strains of 11 *Ralstonia* species and the type strain of *Burkholderia cepacia*[a]

Substrate or test	R. pickettii EY 4382	R. basilensis EY 4368	R. campinensis EY 4379	R. eutropha EY 3798	R. gilardii EY4363	R. mannitolilytica EY 4364	R. metallidurans EY 4380	R. oxalatica EY 4365	R. paucula EY 4366	R. solanacearum EY 2181	R. taiwanensis EY 4381	Burkholderia cepacia EY 645
Motility	+	−	+	+	+	+	+	+	+	−	+	+
Flagellation:												
Single polar	+			−	+	+		+				−
Polar tuft	−			+	−			+				+
Peritrichous			+	+					+			+
Growth at 41°C	+	−	+	+	+	+	−	−	+		+	+
Colony pigmentation	Beige	Beige			Beige	Beige	Beige	Beige	Beige		Beige	Yellow
Catalase	+	+	+	+	+	+	+	+	+	+	+	−
Oxidase	+	+	+	+	+	+	+	+	+	+	+	+
Citrate (Simmons)	+	+	+	+	+	+	+	+	+	−	+	+
Nitrate to gas	−	−	+	+	−	+	−	+	−	−	−	−
Nitrite test	+	−	+	+	−	−	−	−	+	+	+	+
Zn test on negative nitrite test		+						+	+			
Hydrolysis of:												
Esculin	−	−	−	+	−	−	−	−	−	−	−	+
Gelatin	+	+	−	+	−	−	−	−	−	+	−	+
Starch	+	−	−	+	+	−	−	−	−	−	−	−
Tween 80	+	+	+	+	−	+	+	+	+	+	+	+
Lysine decarboxylase	−	−	−	−	−	−	−	−	−	−	−	−
Phenylalanine deaminase	+	−	+	+	+	−	−	+	−	+	−	+
Urease	+	+	+	+	+	+	+	+	+	−	+	+
Acylamidase	+	−	−	+	−	−	−	−	−	−	−	+
Oxidative acid formed in OF base medium containing:												
Adonitol	−	−	−	−	−	−	−	−	−	−	−	+
Salicin	−	+	−	+	−	−	−	−	−	+	−	+
D-Ribose	+	−	−	+	+	−	−	−	−	−	−	+
Inositol	−	−	−	−	−	−	−	−	−	−	−	+
Mannitol	−	−	−	−	−	+	−	−	−	−	−	+
Sorbitol	−	+	+	+	−	−	−	+	−	+	−	+
L-Arabinose	+	−	−	+	+	+	+	+	+	+	+	+
Mannose	+	−	−	+	−	−	−	+	−	−	−	+
Dulcitol, D-arabinose	−	−	−	−	−	−	−	−	−	−	−	−
Raffinose, rhamnose	−	−	−	−	−	−	−	−	−	−	−	−
Melezitose	−	−	−	−	−	−	+	+	−	−	−	+
Melibiose	−	−	−	+	−	+	−	+	−	−	−	+
Inulin	−	−	−	−	−	−	−	−	−	−	−	−
Sucrose, trehalose	+	−	−	−	−	+	−	−	−	+	−	+
Maltose	+	−	−	+	+	+	−	+	−	+	−	+
Cellobiose	−	−	−	−	+	+	−	−	−	−	−	+
Glucose	+	+	+	+	+	+	+	+	+	+	+	+
Galactose	+	+	+	+	+	+	+	+	+	+	+	+
Lactose	+	−	−	+	+	+	−	+	+	+	+	+
Ethanol (3%)	−	+	+	+	+	+	+	+	+	+	+	+
Glycerol, xylose	+	−	−	+	−	+	−	+	−	+	−	+
Fructose	+	+	−	+	+	+	+	+	+	+	+	+
Alkaline reaction in OF base medium	+	+	+	+	+	+	+	+	+	+	+	+
Major respiratory quinone	Q-8	Q-8	Q-8	Q-8	Q-8	Q-8	Q-8	Q-8	Q-8	Q-8	Q-8	Q-8
Mol% G + C of DNA	64	65.4	66.6	66.5	67.2	66.2	63.7	68	63.7	66.5	67.3	66.6

[a]Symbols: +, positive; −, negative; blank space, not determined or not applicable.

TABLE BXII.β.11. Results of API 20 NE of type strains of 11 *Ralstonia* species and of *Burkholderia cepacia*[a]

Substrate or test	*Ralstonia pickettii* EY 4382^T	*Ralstonia basilensis* EY 4368^T	*Ralstonia campinensis* EY 4379^T	*Ralstonia eutropha* EY 3798^T	*Ralstonia gilardi* EY 4363^T	*Ralstonia mannitolilytica* EY 4364^T	*Ralstonia metallidurans* EY 4380^T	*Ralstonia oxalatica* EY 4365^T	*Ralstonia paucula* EY 4366^T	*Ralstonia solanacearum* EY 2181^T	*Ralstonia taiwanensis* EY 4381^Tb	*Burkholderia cepacia* EY 645^T
Urea	–	+	–	–	–	–	–	–	–	–	–	–
Esculin	–	–	–	–	–	–	–	–	+	+	+	+
p-Nitro-β-D-galactopyranoside	–	–	–	–	–	–	–	–	–	–	–	+
Assimilation of:												
Glucose	+	–	–	–	–	+	–	–	–	+	–	+
D-Mannitol	–	–	–	–	–	+	–	+	–	–	–	+
Maltose	+	+	–	+	–	–	+	+	–	–	–	–
Potassium gluconate	–	+	+	+	+	+	+	+	+	–	+	+
n-Caprate	–	+	+	+	+	+	+	+	+	–	+	+
Adipate	+	+	+	+	–	–	+	+	+	+	–	+
Sodium citrate	–	+	–	+	–	+	+	+	–	–	+	+
Phenylacetate	–	–	+	+	–	–	+	+	+	+	+	+
DL-Malate	+	+	–	+	+	+	+	+	+	+	–	+
L-Arabinose, D-Mannose, N-acetyl-D-glucosamine	–	–	–	–	–	–	–	–	–	–	–	+

[a]Symbols: +, positive; –, negative.

Type strain: Steinle RK1, DSM 11853, EY 4368, GTC 1873.
GenBank accession number (16S rRNA): AF312022, AJ002302.

3. **Ralstonia campinensis** Goris, De Vos, Coenye, Host, Janssens, Brim, Diels, Mergeay, Kersters and Vandamme 2001, 1780^VP

cam.pin.en'sis. L. adj. *campinensis* pertaining to the ikempen or Campine, the geographical region of northeast Belgium, where the strains were isolated.

The characteristics are as described for the genus and listed in Tables BXII.β.10, BXII.β.11, BXII.β.12, BXII.β.13, and BXII.β.14. The cell size is 0.8 × 1.2–1.8 μm. Colonies are round (sometimes with a slightly scalloped margin), smooth, convex, and transparent. The colony diameter is less than 0.5 mm in diameter on TSA agar plate after 24 h incubation at 30°C. Able to grow at 41°C, but not at 4°C. For enzyme activities determined with API Zyme, see Goris et al. (2001). The type strain was isolated from a zinc-decertified area in Lommel, Belgium.

The mol% G + C of the DNA is: 66.6–66.8 (HPLC).
Type strain: WS2, CCUG 44526, EY 4379, GTC 1881, LMG 19282.
GenBank accession number (16S rRNA): AF312020.

4. **Ralstonia eutropha** (Davis, Doudoroff, Stanier and Mandel 1969) Yabuuchi, Kosako, Yano, Hotta and Nishiuchi 1996, 625^VP (Effective publication: Yabuuchi, Kosako, Yano, Hotta and Nishiuchi 1995, 903) (*Alcaligenes eutrophus* Davis *in* Davis, Doudoroff, Stanier and Mansel 1969, 386.)

eu.troph'a. Gr. prep. *eu* good, beneficial; Gr. n. *trophus* one who feeds; M.L. n. *eutropha* good nutrition, well nourished.

The characteristics are as described for the genus and listed in Tables BXII.β.10, BXII.β.11, BXII.β.12, BXII.β.13, and BXII.β.14. Cells are 0.7 × 1.8–2.6 μm and have peritrichous flagella. Able to grow at 41°C. Resistant to nickel (Schmidt et al., 1991) and to a combination of cobalt and nickel (Tibazarwa et al., 2000). A genetically engineered strain has an enhanced ability to immobilize external Cd^{2+} ions and reduced the toxic effects of heavy metals on the growth of tobacco plants (Valls et al., 2000). Synthesizes short chain length (SCL) (3–5 carbon atoms) PHAs by means of PHA synthase (Song et al., 2000c; Zhang et al., 2000c) and accumulates them as intracellular granules. PHAs are now of commercial value as thermoplastics (Holmes, 1985), and have been extensively studied (Lütke-Eversloh and Steinbuchel, 1999; Grzeszik et al., 2000; Happe et al., 2000; Padilla et al., 2000; Bernhard et al., 2001; Bramer and Steinbuchel, 2001; Drewlo et al., 2001; Schräder et al., 2001; Taguchi et al., 2001; Zarnt et al., 2001). Accumulated PHB is degraded and used for growth and survival when an exogenous carbon source is not available (Handrick et al., 2000). Genetic determinants essential for synthesis of these polymers are linked to the megaplasmid pHG1. Genome of *R. eutropha* strain H16 is composed of two circular chromosomes measuring 4.1 and 2.9 Mb. A physical map of the megaplasmid pHG1 has been given (Schwartz and Fiedrich, 2001). The type strain was isolated from soil with hydrogen gas.

The mol% G + C of the DNA is: 65.5 (HPLC).
Type strain: ATCC 17697, DSM 531, EY 3798, GTC 1874, JCM 11282, LMG 1199.
GenBank accession number (16S rRNA): M32021.

TABLE BXII.β.12. Susceptibilities of type strains of 11 *Ralstonia* species and of the type strain of *Burkholderia cepacia*[a]

Compound (mg/disc)	*R. picketii* EY 4382	*R. basilensis* EY 4368	*R. campinensis* EY 4379	*R. eutropha* EY 3798	*R. gilardii* EY 4363	*R. mannitolilytica* EY 4364	*R. metallidurans* EY 4380	*R. oxalatica* EY 4365	*R. paucula* EY 4366	*R. solanacearum* EY 2181	*R. taiwanensis* EY 4381	*Burkholderia cepacia* EY 645
Minocycline (30)	S	S	S	S	S	S	S	S	S	S	S	S
Doxycycline (10)	S	S	S	S	S	S	S	S	IM	S	S	S
Tetracycline (30)	S	S	IM	S	S	S	S	S	IM	S	S	Rc
Levofloxacin (5)	S	S	S	S	S	S	S	S	S	S	S	R
Cefotaxime (30)	S	S	S	S	IM	S	S	S	S	S	S	R
Tosufloxacin (5)	S	S	S	S	S	S	S	S	R	S	S	R
Ciprofloxacin (5)	S	S	S	S	S	S	R	S	R	S	R	R
Ceftazidime (30)	S	S	S	S	R	IM	R	S	S	S	R	R
Norfloxacin (10)	S	S	IM	S	IM	R	R	S	R	S	R	R
Ofloxacin (5), sparfloxacin (5)	S	S	S	S	S	S	S	S	IM	S	S	S
Sulfamethoxazole-trimethoprim (23.75/1.25)	S	S	R	IM	S	R	R	S	R	S	S	R
Imipenem (10)	S	S	R	S	R	R	R	S	R	S	S	R
Carumonam (30)	S	S	R	R	R	R	R	R	R	S	S	R
Cefoperazon (75)	S	R	IM	S	R	IM	S	IM	R	S	S	R
Piperacillin (100)	S	R	R	R	S	R	S	S	R	S	S	R
Cefmetazole (30)	S	R	IM	S	R	R	R	S	IM	S	S	R
Trimethoprim (5)	IM	R	R	R	R	R	R	S	R	S	R	S
Flomoxef (30)	S	R	R	R	R	R	R	R	R	S	S	R
Cefaclor (30)	S	R	S	R	R	R	R	R	R	S	S	R
Cefazolin (30)	S	R	R	R	R	R	R	R	R	S	R	R
Clarithromycin (15)	R	IM	R	IM	R	R	R	S	R	R	R	R
Polymyxin B (300)	R	R	S	S	S	R	IM	S	S	R	S	R
Moxalactam (30)	R	IM	R	IM	R	R	R	S	R	S	R	R
Meropenem (10)	R	S	R	R	R	R	R	S	R	S	S	R
Roxithromycin (15)	R	R	R	R	R	R	R	IM	R	S	R	R
Erythromycin (15)	R	R	R	R	R	R	R	R	IM	S	IM	R
Penicillin (10), (25)	R	R	R	R	R	R	R	R	R	S	R	R
Ampicillin (10), amoxicillin (25)	R	R	R	R	R	R	R	S	R	S	S	S
AMPCd/CVCe (20/10), panipenem (10), aztreonam (30)	R	R	R	R	R	R	R	S	R	S	R	R
Gentamicin (10), amikacin (30), dibekacin (30)	R	R	S	R	R	R	R	S	R	S	R	R

[a]Symbols: S, susceptible; IM, intermediate; R, resistant.

TABLE BXII.β.13. Assimilation reactions of eleven type strains of *Ralstonia* species and the type strain of *Burkholderia cepacia* by the Biolog GN2 MicroPlate system[a]

Substrate	*Burkholderia cepacia* EY 645	*Ralstonia taiwanensis* EY 4381	*Ralstonia solanacearum* EY 2181	*Ralstonia paucula* EY 4366	*Ralstonia oxalatica* EY 4365	*Ralstonia metallidurans* EY 4380	*Ralstonia mannitolilytica* EY 4364	*Ralstonia gilardii* EY 4363	*Ralstonia eutropha* EY 3798	*Ralstonia campinensis* EY 4379	*Ralstonia basilensis* EY 4368	*Ralstonia pickettii* EY 4382
cis-Aconitic acid	+	+	+	+	+	+	+	−	+	−	+	+
Alaninamide	+	−	+	·	+	−	−	+	+	+	−	+
D-Alanine	+	+	+	−	+	+	−	+	+	+	+	+
L-Alanine	+	+	+	−	+	+	+	+	+	−	+	+
L-Alanyl-glycine	+	+	+	−	+	+	+	+	−	−	+	+
γ-Aminobutyric acid, D-glucuronic acid, α-D-glucose	+	−	+	−	−	+	+	−	+	−	+	+
L-Arabinose, glucuronamide	+	−	+	−	−	−	−	−	−	−	−	+
Bromosuccinic acid	+	+	+	+	+	+	+	−	+	+	+	+
Citric acid	+	+	+	+	+	+	+	+	+	+	+	+
Formic acid	+	+	+	+	+	+	+	−	+	+	+	+
D-Fructose	+	−	+	−	−	+	+	−	+	−	+	+
D-Galactonic acid lactone, D-galactose	−	−	−	−	−	−	+	−	−	−	+	−
D-Galacturonic acid	+	−	−	−	−	−	−	−	−	−	+	+
D-Gluconic acid	+	+	+	+	+	+	+	+	+	+	+	+
Glycerol	+	−	+	+	+	+	+	>	+	+	v	−
Glycogen	+	−	+	+	+	+	+	>	+	+	+	−
Glycyl-L-glutamic acid	−	+	+	+	+	+	−	+	+	+	+	+
L-Histidine, urocanic acid	+	+	+	+	+	+	+	+	+	+	+	+
α-Hydroxybutyric acid	+	−	+	+	+	+	−	−	+	−	+	−
γ-Hydroxybutyric acid	+	−	−	+	+	+	+	−	−	−	+	+
p-Hydroxyphenylacetic acid	−	−	+	+	+	+	−	−	+	−	+	−
Itaconic acid	+	−	+	+	+	+	+	−	+	−	+	−
α-Ketobutyric acid, L-leucine	−	+	+	+	+	+	−	+	+	+	+	−
α-Ketoglutaric acid	+	+	+	+	+	+	+	+	+	+	+	+
α-Ketovaleric acid	−	−	+	−	−	−	−	−	+	−	+	−
Malonic acid	+	+	+	+	+	+	+	−	+	+	+	+
Monomethylsuccinate, propionic acid,	+	+	+	+	+	+	>	+	+	+	+	>
L-Phenylalanine, succinamic acid,	+	+	+	+	+	+	>	>	+	>	+	−
Phenylethylamine	+	>	+	+	+	>	>	>	+	>	+	−
D-Psicose	−	−	−	−	−	−	−	−	−	−	−	−
Quinic acid	+	+	+	+	+	+	+	+	+	+	+	+
D-Saccharic acid	+	−	+	−	+	−	−	−	−	−	+	+
Sebacic acid	+	+	+	+	+	+	+	+	+	+	+	+
D-Serine	+	−	−	−	−	−	−	−	−	−	−	−
L-Serine	+	+	+	+	+	+	+	+	+	+	+	+
L-Threonine	+	−	+	+	+	+	−	−	+	+	+	−
Tween 40, Tween 80	+	+	+	+	+	+	+	+	+	+	+	+
Acetic acid, L-asparagine, L-aspartic acid, L-glutamic acid, β-hydroxybutyric acid, DL-lactic acid, methylpyruvate, L-proline, L-pyroglutamic acid, succinic acid	+	+	+	+	+	+	+	+	+	+	+	+
N-Acetyl-D-galactosamine, cellobiose, α-cyclodextrin, *i*-erythritol, glycyl-L-aspartic acid, inosine, D-raffinose, uridine, α-D-lactose, lactulose, maltose, D-melibiose, β-methyl-D-glucoside, L-rhamnose, D-trehalose, turanose, xylitol	−	−	−	−	−	−	−	−	−	−	−	−
N-Acetyl-D-glucosamine, adonitol, 2-aminoethanol, D-arabitol, 2,3-butanediol, DL-carnitine, dextrin, L-fucose, gentiobiose, glucose-1-phosphate, glucose-6-phosphate, D-glucosaminic acid, DL-α-glycerol phosphate, hydroxy-L-proline, *m*-inositol, D-mannitol, D-mannose, L-ornithine, putrescine, D-sorbitol, sucrose, thymidine	+	−	−	−	−	−	−	−	−	−	−	−

[a] +, positive; −, negative.

TABLE BXII.β.14. Cellular fatty acid composition (% of total) of the type strains of 11 *Ralstonia* species and of *Burkholderia cepacia* [a, b]

Fatty acids, % of total	*R. pickettii* strain EY 4382	*R. basilensis* strain EY 4368	*R. campinensis* strain EY 3797	*R. eutropha* strain EY 3798	*R. gilardii* strain EY 4363	*R. mannitolilytica* strain EY 4364	*R. metallidurans* strain EY 4380	*R. oxalatica* strain EY 4365	*R. paucula* strain EY 4366	*R. solanacearum* strain EY 2181	*R. taiwanensis* strain EY 4381	*Burkholderia cepacia* strain EY 645
$C_{14:0}$	4	3	5	4	3	3	4	2	4	5	2	5
$C_{15:0}$	tr[c]		tr	tr				tr	tr	1		tr
$C_{16:1\ \omega7c}$	24	14	11	26	12	8	18	12	5	13	23	5
$C_{16:0}$	23	18	30	32	19	20	24	19	18	33	27	36
$C_{17:0}$	1			tr	2		tr	tr		1	tr	tr
$C_{18:1\ \omega9v}$, $C_{18:1\ \omega7c}$	25	13	28	19	18	11	29	14	13	24	25	23
$C_{18:0}$	tr	2	tr	tr	4	3	1	2	3	2	2	5
$C_{20:0}$												tr
$C_{14:0\ 2OH}$	tr	3	2	4			tr	10	2		3	
$C_{16:1\ 2OH}$	4	10	3				2				tr	
$C_{16:0\ 2OH}$			tr		5	9	4		10	1		2
$C_{18:1\ 2OH}$	5	6	4		4		3			8	2	5
$C_{14:0\ 3OH}$	9	31	10	13	31	46	10	39	45	12	11	9
$C_{16:0\ 3OH}$												8
Other	3		6		2		4	tr			4	

[a]N. Fujiwara and T. Naka, unpublished data.

[b]By hydrolysis with HCl-methanol (1:5, v/v) at 100°C for 3 h.

[c]tr, <1%.

5. **Ralstonia gilardii** Coenye, Falsen, Vancanneyt, Hoste, Govan, Kersters and Vandamme 1999a, 412[VP]

gi.lar' di.i. M.L. gen. n. *gilardii* named after G.L. Gilardi, an American microbiologist who contributed much to our knowledge of *Alcaligenes* species.

The characteristics are as described for the genus and listed in Tables BXII.β.10, BXII.β.11, BXII.β.12, BXII.β.13, and BXII.β.14. The cell size is 0.7×1.8–2.6 µm. Motile by means of a single polar flagellum. Able to grow at 41°C. Growth occurs in the presence of 5.0% NaCl.

The mol% G + C of the DNA is: 68–69 (T_m).

Type strain: ATCC 700815, EY 4363, GTC 1875, JCM 11283, LMG 5886.

GenBank accession number (16S rRNA): AF076645.

6. **Ralstonia mannitolilytica** De Baere, Steyaert, Wauters, De Vos, Goris, Coenye, Suyama, Verschraegen and Vaneechoutte 2001, 556[VP]

man.ni.to.li.ly' ti.ca. N.L. adj. *mannitolilytica* cleaving mannitol.

The species accommodates strains previously known as the *Ralstonia pickettii* biovar 3/'*thomasii*' strains and at least some of the strains known as "*Pseudomonas thomasii*" Phillips et al. 1972. The characteristics are as described for the genus and listed in Tables BXII.β.10, BXII.β.11, BXII.β.12, BXII.β.13, and BXII.β.14. Motile by means of single polar flagellum, with the only exception being the type strain. Able to grow at 41°C. For enzymatic activity, see the results by means of API ZYM (De Baere et al., 2001). The type strain was isolated from blood of patient at St. Thomas' Hospital, London, UK, in 1971 (Phillips et al., 1972).

The mol% G + C of the DNA is: 66.2 (HPLC).

Type strain: LMG 6866, EY 4364, GTC 1876, JCM 11284, NCIB 10805.

GenBank accession number (16S rRNA): AJ270258.

7. **Ralstonia metallidurans** Goris, De Vos, Coenye, Host, Janssens, Brim, Diels, Mergeay, Kersters and Vandamme 2001, 1780[VP]

me.tal.li.du' rans. L. n. *metallum* metal; L. pres. part. *durans* enduring; N.L. part. adj. *metallidurans* enduring metal, to indicate that these strains are able to survive high heavy-metal concentrations.

The characteristics of the type strain are as described for the genus and listed in Tables BXII.β.10, BXII.β.11, BXII.β.12, BXII.β.13, and BXII.β.14. Cells are short motile rods 0.8×1.2–2.2 µm. Colonies are round (sometimes with a slightly scalloped margin), smooth, slightly convex, and beige colored when fully grown. Unable to grow either at 4° or 41°C. For enzyme activities, see Goris et al. (2001). The type strain was isolated from wastewater from a zinc factory at Liege, Belgium.

The mol% G + C of the DNA is: 63.7 (HPLC).

Type strain: DSM 2389, EY 4380, GTC 1882, LMG 1195.

GenBank accession number (16S rRNA): Y10824.

8. **Ralstonia oxalatica** (ex Khambata and Bhat 1953) Sahin, Isik, Tamer and Goodfellow 2000b, 1953[VP] (Effective publication: Sahin, Isik, Tamer and Goodfellow 2000a, 207.)

o.xa.la' ti.ca. M.L. fem. adj. *oxalatica* pertaining to oxalate.

The characteristics of the type strain are as described for the genus and listed in Tables BXII.β.10, BXII.β.11, BXII.β.12, BXII.β.13, and BXII.β.14. Motile by means of one polar or subpolar flagellum. Cell size is 0.3–0.4×0.9–1.5 µm. Able to grow on ordinary peptone media without supplementing with oxalate. Unable to grow autotrophically with hydrogen. The type strain was isolated from the alimentary tract of an Indian earthworm.

The mol% G + C of the DNA is: 68 (T_m) or 67 (Bd).

Type strain: ATCC 11883, DSM 1105, EY 4365, JCM 11285, NCIB 8642, LMG 2235.

GenBank accession number (16S rRNA): AF155567.

9. **Ralstonia paucula** Vandamme, Goris, Coenye, Hoste, Jenssens, Kersters, De Vos and Falsen 1999, 668[VP]

pau' cu.la. L. adj. *pauculus* rare, very few, to indicate that these strains only sporadically cause human infections.

The characteristics of the type strain are as described for the genus and listed in Tables BXII.β.10, BXII.β.11, BXII.β.12, BXII.β.13, and BXII.β.14. Cell size is 0.8 × 1.2–2.0 μm. Motile by means of peritrichous flagella. Does not denitrify. These organisms were previously placed in CDC group Ivc-2. Isolated from a variety of human clinical sources (Vandamme et al., 1999) as well as environmental sources such as pool water, groundwater, and bottled mineral water. Differentiated from other species by peritrichous flagellation, ability to form acid oxidatively from L-arabinose, and absence of cystine arylamidase, phosphoamidase, and lipase C$_{14}$ activity. The type strain was isolated from a human respiratory tract in USA.

The mol% G + C of the DNA is: 65–67 (T_m).

Type strain: CCUG 12507, CDC E6793, EY 4366, GTC 1878, JCM 11286, LMG 3244.

GenBank accession number (16S rRNA): AF085226.

10. **Ralstonia solanacearum** (Smith 1896) Yabuuchi, Kosako, Yano, Hotta and Nishiuchi 1996, 625^{VP} (Effective publication: Yabuuchi, Kosako, Yano, Hotta and Nishiuchi 1995, 903.) (*Pseudomonas solanacearum* (Smith 1896) Smith 1914, 178; *Bacillus solanacearum* Smith 1896, 10.)

so.la.na.ce.a′rum. M.L. fem. pl. n. *Solanaceae* the nightshade family; M.L. fem. pl. gen. n. *solanacearum* of the *Solanaceae*.

The characteristics of the type strain are as described for the genus and listed in Tables BXII.β.10, BXII.β.11, BXII.β.12, BXII.β.13, and BXII.β.14. Nonmotile. Pathogenic for tomato, potato, tobacco, banana, and peanut by the production of a copious amount of extracellular polysaccharide, which occludes vessels in the stem, inhibits water flow, and causes death of plant by wilt disease. Recently the nucleotide full sequence of the chromosome and megaplasmid were determined in order to elucidate and manipulate genes related to the virulence of the organism. The type strain was isolated from tomato.

The mol% G + C of the DNA is: 66.6 (HPLC).

Type strain: ATCC 11696, DSM 9544, EY 2181, GTC 1879, JCM 10489, LMG 2299, NCPPB 325.

GenBank accession number (16S rRNA): X67036.

11. **Ralstonia taiwanensis** Chen, Laevens, Lee, Coenye, De Vos, Mergeay and Vandamme 2001, 1734^{VP}

tai.wan.en′sis. N.L. fem. adj. *taiwanensis* pertaining to Taiwan, where the root nodule strains were isolated.

The characteristics of the type strain are as described for the genus and listed in Tables BXII.β.10, BXII.β.11, BXII.β.12, BXII.β.13, and BXII.β.14. Motile; type of flagellation unknown. Cell size, 0.5–0.7 × 0.8–2.0 μm. Growth is observed at 28°, 30°, and 37°C. Nitrate is reduced. Esculin is hydrolyzed.; gelatin and Tween 80 are not. Susceptible to tetracyclines and quinolines; resistant to β-lactams and aminoglycosides. Urease, β-galactosidase, and DNase are negative. Indole is not produced. Autotrophic growth does not occur. The type strain was isolated from root nodules of *Mimosa pudica*.

The mol% G + C of the DNA is: 67.3 (HPLC).

Type strain: R1, CCUG 44338, EY 4381, GTC 1883, LMG 19424.

GenBank accession number (16S rRNA): AF300324.

Genus VIII. **Thermothrix** *Caldwell, Caldwell and Laycock 1981, 217^{VP} (Effective publication: Caldwell, Caldwell and Laycock 1976, 1515)*

ANNA-LOUISE REYSENBACH, PAULA AGUIAR AND DOUGLAS E. CALDWELL

Ther′mo.thrix. Gr. adj. *thermos* hot; Gr. n. *thrix* hair; N.L. fem. n. *Thermothrix* hot hair.

Rod-shaped cells, usually 0.5–1.0 × 3–5 μm. **Filamentous cells are produced** under unfavorable growth conditions such as when grown at temperatures near the maximum temperature, at extreme pH values, or when oxygen limits growth. **Motile by means of a single polar flagellum.** Gram negative. No spores are produced. **Growth occurs between 63–86°C and pH 6–8.5. Facultatively or obligately chemolithoautotrophic aerobic.** Oxygen or nitrate may be used as the electron acceptor. Electron donors can be organic compounds or inorganic sulfur compounds such as thiosulfate, sulfur, hydrogen sulfide, and tetrathionate.

The mol% G + C of the DNA is: 39.7.

Type species: **Thermothrix thiopara** Caldwell, Caldwell and Laycock 1981, 217 (Effective publication: Caldwell, Caldwell and Laycock 1976, 1515.)

FURTHER DESCRIPTIVE INFORMATION

Elemental sulfur is often deposited extracellularly as spherical granules (Figs. BXII.β.8 and BXII.β.9), although *T. azorensis* accumulates sulfur intracellularly when the oxidation of thiosulfate is incomplete and the pH does not drop below pH 7.0. Respired substrates may include glucose, sulfide, and thiosulfate. When thiosulfate is oxidized to sulfate, elemental sulfur, sulfite, and polythionate accumulate as intermediates in batch culture (Brannan and Caldwell, 1980, 1983).

This genus was found within the sulfide-oxygen interfaces of neutral pH geothermal springs, including Mammoth Hot Springs (Yellowstone National Park, Wyoming, USA), Jemez Hot Springs (Jemez Springs, New Mexico, USA), and Furnas (Sao Miguel Island, Azores).

The isolates of *Thermothrix* were obtained from thermal springs characterized by "sulfur turf" mats. These mats generally consist of a mixture of filamentous and rod-shaped cells, intertwined to form "streamers" up to 10 cm in length. Although many *in situ* studies have been conducted on these mats, recent molecular phylogenetic assessments of the Mammoth Hot Springs, the Jemez Spring, and the Furnas "streamers" revealed the dominance of *Aquificales*-like 16S rRNA sequences and no beta-proteobacterial sequences (A.L. Reysenbach, P. Aguiar, and T. Kieft, unpublished data). Additionally, fluorescent *in situ* 16S rRNA probes, specific for the order *Aquificales*—in the class *Aquificae*, phylum *Aquificae*—confirmed that this group of organisms forms the primary matrix of these mats (Fig. BXII.β.10). Based on this new molecular evidence, it is therefore unclear how significant the role of *Thermothrix*—in the class *Betaproteobacteria*, phylum *Proteobacteria*—is in the formation of the mats.

ENRICHMENT AND ISOLATION PROCEDURES

Representatives of the genus are found within the sulfide-oxygen interface of terrestrial hot springs (40–80°C). Under these con-

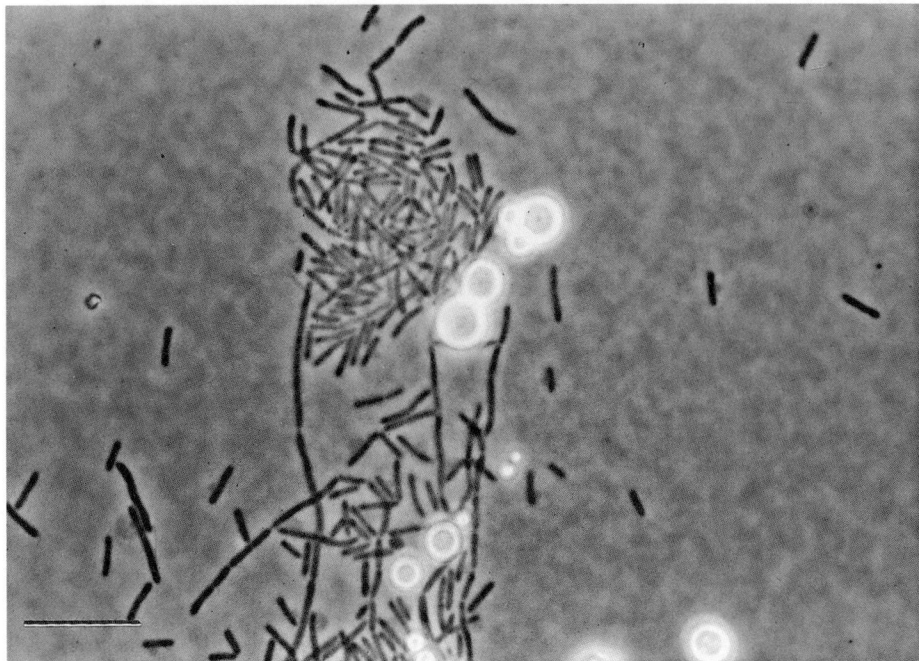

FIGURE BXII.β.8. *T. thiopara* during oxidation of thiosulfate, with extracellular deposition of elemental sulfur (spherical granules) shown. The culture was highly aerated, thus producing rod-shaped cells and cell chains but no cell filaments. Bar = 10 μm.

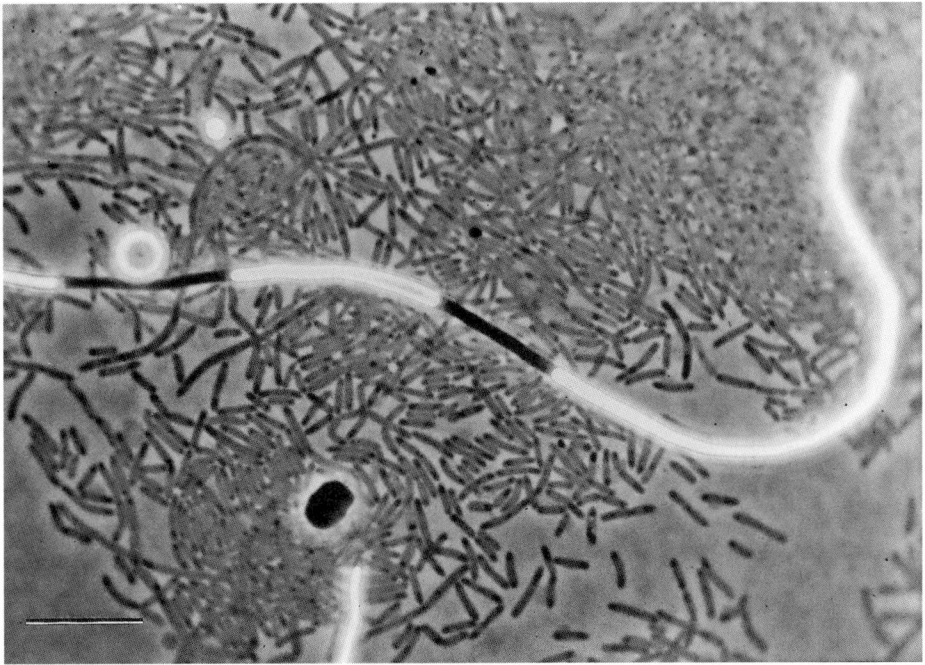

FIGURE BXII.β.9. Deposition of elemental sulfur in a smooth refractile layer partially surrounding a filament of *T. thiopara* (*center*) during the transition from nitrate broth to a synthetic medium with thiosulfate as sole energy source. Phase-contrast micrograph. Bar = 10 μm.

ditions, sulfide serves as the electron donor and oxygen as the electron acceptor. Consequently, a synthetic medium with either sulfide or thiosulfate as the sole energy source can be used for initial cultivation. The following medium has been used (g/l): $Na_2S_2O_3 \cdot 5H_2O$, 3.0; $NaHCO_3$, 2.0; NH_4Cl, 1.0; KNO_3, 2.0; $MgSO_4 \cdot 7H_2O$, 0.5; KH_2PO_4, 2.0; and $FeSO_4 \cdot 7H_2O$-EDTA, 0.02;

plus trace elements (pH 6.8) (Brannan and Caldwell, 1980). Care should be taken to avoid the loss of bicarbonate as carbon dioxide if the medium is autoclaved and when the medium is stored and incubated. The enrichments are incubated at 73°C and should be transferred during logarithmic phase (reached after about 24 h), or stored in log phase of growth at 4°C, as cell death may

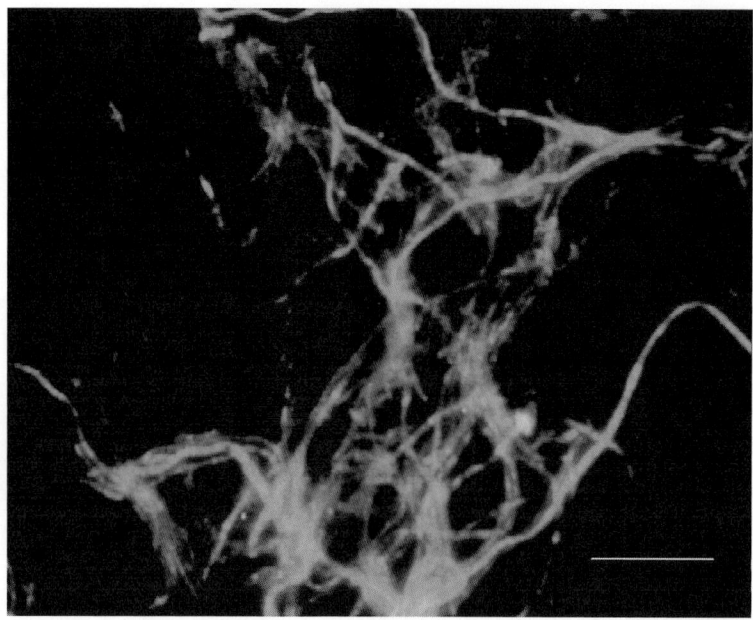

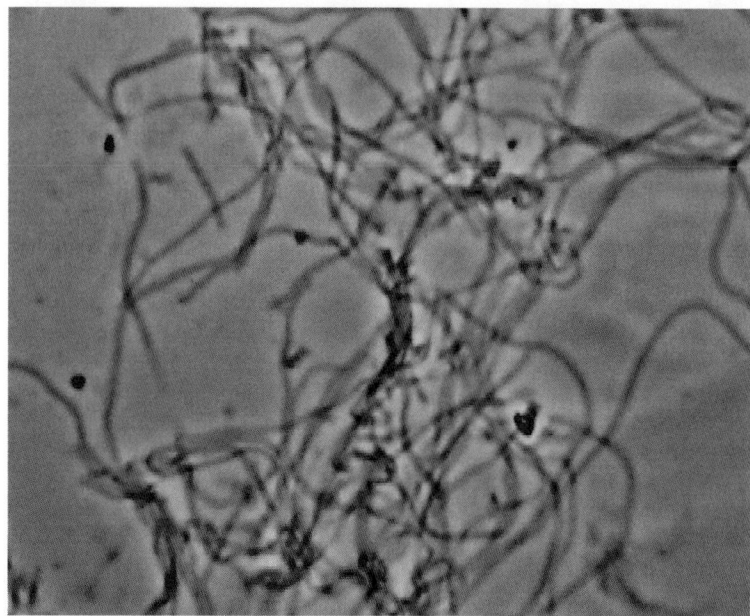

FIGURE BXII.β.10. *Upper panel*, streamers collected from Jemez Springs, New Mexico, probed with a fluorescein-labeled oligonucleotide 16S rDNA probe specific for the 16S rRNA from the *Aquificales* (Harmsen et al., 1997). *Bottom panel*, phase contrast micrograph of same frame. Bar = 10 μm.

occur within 4 h of reaching the stationary growth phase. Because of the low solubility of the oxygen at 73°C (approximately 3 mg/l), the organisms should be cultivated in shake flasks with a large headspace to provide sufficient aeration. Isolated colonies may be obtained on agar or Gelrite; however, recovery is poor. After initial isolation, some species of *Thermothrix* may be adapted, with difficulty, to grow on an organic medium such as a nutrient (aerobic) or nitrate (anaerobic) broth.

MAINTENANCE PROCEDURES

Cells may be frozen in a 15% glycerol (v/v) and stored at −80°C. Cultures of *T. thiopara* adapted to grow on nitrate broth (Difco)

were more easily maintained than were autotrophically grown cells.

DIFFERENTIATION OF THE GENUS *THERMOTHRIX* FROM OTHER GENERA

The primary basis for discrimination of *Thermothrix* species from thermophilic *Thiobacillus* species is the production of filaments when the concentration of the electron acceptor limits the rate of growth. Cells obtained during the initial isolation are usually motile and rod shaped. Induction of the filamentous growth form requires continuous cultivation under oxygen-limited conditions. The oxygen tension should be maintained at 1 mg/l or less.

FURTHER READING

Brannan, D.K. and D.E. Caldwell. 1980. *Thermothrix thiopara*: growth and metabolism of a newly isolated thermophile capable of oxidizing sul-

fur and sulfur compounds. Appl. Environ. Microbiol. *40*: 211–216.

Odintsova, E.V., H.W. Jannasch, J.A. Mamone and T.A. Langworthy. 1996. *Thermothrix azorensis* sp. nov., an obligately chemolithoautotrophic, sulfur-oxidizing, thermophilic bacterium. Int. J. Syst. Bacteriol. *46*: 422–428.

DIFFERENTIATION OF THE SPECIES OF THE GENUS *THERMOTHRIX*

Only two species are described: *Thermothrix thiopara* and *Thermothrix azorensis*. These have been distinguished by their 16S rRNA structure, temperature growth optima, and whether they are obligately or facultatively chemolithoautotrophs and obligately or facultatively aerobic.

List of species of the genus Thermothrix

1. **Thermothrix thiopara** Caldwell, Caldwell and Laycock 1981, 217[VP] (Effective publication: Caldwell, Caldwell and Laycock 1976, 1515.)

   thi.o'par.a. Gr. n. *thios* sulfur; L. v. *paro* produce; M.L. adj. *thiopara* sulfur-depositing.

   See the description of the genus for most features of the species. The cardinal temperatures are 60°, 73°, and 80°C (minimum, optimum, and maximum) for cells grown by using either inorganic or organic media (Caldwell et al., 1976; Brannan and Caldwell, 1982). The generation time at 73°C is approximately 2 h on either medium. When reduced sulfur compounds are used as the electron donor, elemental sulfur is frequently deposited as extracellular granules but is not deposited intracellularly (Fig. BXII.β.8). In batch culture, elemental sulfur, sulfite, and polythionates accumulate as transient intermediates but are subsequently oxidized to sulfuric acid. In continuous culture, elemental sulfur is not deposited during steady-state growth but is deposited during the transition from low to high dilution rates. During transfer from nitrate broth to autotrophic media containing thiosulfate, cell filaments often form and become encased in a coating of elemental sulfur (Fig. BXII.β.9).

   Denitrification occurs only heterotrophically (Brannan and Caldwell, 1982). During denitrification, N_2 is the primary end product. Nitrite accumulates as a transient intermediate.

   The mol% G + C of the DNA is: 39.7 (T_m).
   Type strain: ATCC 29244.
   GenBank accession number (16S rRNA): U61284.

   Additional Remarks: The original strain is no longer available from the ATCC.

2. **Thermothrix azorensis** Odintsova, Jannasch, Mamone and Langworthy 1996, 426[VP]

   a.zo'ren.sis. L. fem. adj. *azorensis* from the Azores.

   Long, thin rods ($2–5 \times 0.3–0.5$ μm) often in pairs. Filaments up to 70 μm long are produced under unfavorable growth conditions. Gram-negative nonsporulating cells, some contain sulfur inclusions. Strictly aerobic and obligately chemolithoautotrophic, using reduced sulfur compounds (hydrogen sulfide, tetrathionate, elemental sulfur, and thiosulfate) as electron donors and carbon dioxide as the carbon source. During the lag phase of growth, elemental sulfur oxidation is preceded by sulfide production. Grows best between 76° and 78°C, pH 7.0–7.5. Temperature and pH ranges for growth are 60–87°C and pH 6.0–8.5, respectively. Lipids include C_{14} to C_{22} fatty acids (95% C_{16} and C_{18}). No ribulose-1,5-bisphosphate carboxylase/oxygenase activity is present.

   Isolated from a hot spring located at Furnas, Sao Miguel, the Azores, Portugal.

   The mol% G + C of the DNA is: 39.7 (T_m).
   Type strain: TM, ATCC 51754.
   GenBank accession number (16S rRNA): U59127.

Family II. **Oxalobacteraceae** *fam. nov.*

GEORGE M. GARRITY, JULIA A. BELL AND TIMOTHY LILBURN

Ox.al.o.bac.ter.a′ce.ae. M.L. masc. n. *Oxalobacter* type genus of the family; *-aceae* ending to denote family; M.L. fem. pl. n. *Oxalobacteraceae* the *Oxalobacter* family.

The family *Oxalobacteraceae* was circumscribed for this volume on the basis of phylogenetic analysis of 16S rRNA sequences; the family contains the genera *Oxalobacter* (type genus), *Duganella*, *Herbaspirillum*, *Janthinobacterium*, *Massilia*, *Oxalicibacterium*, and *Telluria*. *Oxalicibacterium* was proposed after the cut-off date for inclusion in this volume (June 30, 2001) and is not described here (see Tamer et al. (2002)).

Family is metabolically diverse and includes strict anaerobes, strict aerobes, and nitrogen-fixing organisms.

Type genus: **Oxalobacter** Allison, Dawson, Mayberry and Foss 1985b, 375[VP] (Effective publication: Allison, Dawson, Mayberry and Foss 1985a, 6.)

Genus I. *Oxalobacter* Allison, Dawson, Mayberry and Foss 1985b, 375[VP] (Effective publication: Allison, Dawson, Mayberry and Foss 1985a, 6)

MILTON J. ALLISON, BARBARA J. MACGREGOR, RICHARD SHARP AND DAVID A. STAHL

Ox.al.o.bac'ter. Gr. n. *oxal* pertaining to oxalate; M.L. n. *bacter* the masculine form of the Gr. neut. n. *bactrum* a rod; M.L. masc. n. *Oxalobacter* an oxalate rod.

Straight or curved to vibrioid, Gram-negative, nonsporeforming rods 0.4–0.6 × 1.0–2.5 µm in length. Flagella may be present or absent. **Strictly anaerobic.** Chemoorganotroph. **Oxalate is used as the major carbon and energy source. Oxamate may also be used, but neither carbohydrates nor any of a wide variety of other compounds will replace oxalate as the growth substrate.** Acetate is assimilated for cell synthesis and is required by some, and perhaps by all, strains. Oxalate utilization is accompanied by alkalization of the medium, and formate is produced in approximately equimolar proportions to the amount of oxalate metabolized. Strains have been isolated from the rumens of cattle and sheep, from cecal and fecal samples from humans, guinea pigs, swine, domestic and wild rats, and from freshwater lake and marine sediments. It is probable that these bacteria colonize many other anaerobic habitats. *Oxalobacter* is currently classified in the class *Betaproteobacteria*, the order *Burkholderiales*, and the family *Oxalobacteraceae*.

The mol% G + C of the DNA is: 48–52.

Type species: **Oxalobacter formigenes** Allison, Dawson, Mayberry and Foss 1985b, 375 (Effective publication: Allison, Dawson, Mayberry and Foss 1985a, 6.)

FURTHER DESCRIPTIVE INFORMATION

Cultural and growth characteristics In anaerobic roll-tube cultures, colonies on the agar surface are colorless, transparent, and droplet-like, while subsurface colonies are colorless or white and are lens-shaped or globular. Colonies of anaerobic oxalate-degraders can be detected based on production of clear zones around colonies in media containing appropriate amounts of calcium oxalate.

Growth occurs under anaerobic conditions, under a gas phase of CO_2, in a defined carbonate–bicarbonate buffered medium that contains minerals, oxalate, and acetate, but growth is better with yeast extract in the medium. Specific growth rates (µ) may be as high as 0.3 h^{-1}. Isolations have been obtained on media containing 20–50 mM oxalate. Some freshly isolated strains are inhibited by 30 mM oxalate; laboratory adapted strains, however, grow well in media containing oxalate at concentrations of 100 mM or more. Acetate is required for growth of some, and perhaps all, intestinal strains, as well as by strains isolated from sediments. The temperature range for growth is 14–45°C, with gastrointestinal strains exhibiting a higher optimal temperature (about 37°C) than lake sediment strains.

Metabolism and metabolic pathways Oxalate serves as both the energy yielding substrate and the major source of carbon for growth. The products from oxalate metabolism are CO_2 and formate, with approximately 1 mol of each produced per mole of oxalate degraded. Approximately 1–1.7 g (dry wt.) of cells are produced per mole of oxalate degraded. Energy generation is centered around the development of a proton motive force through the electrogenic exchange of oxalate (in) and formate (out) across the cell membrane and the consumption of a proton inside the cell when the CoA-ester of oxalate is decarboxylated by oxalyl CoA-decarboxylase (Anantharam et al., 1989; Kuhner et al., 1996). An antiporter protein (OxlT) facilitates the oxalate–

formate exchange (Ruan et al., 1992). Oxalate serves as a major source of cell carbon, but acetate and carbon dioxide are also used for biosynthesis (Cornick and Allison, 1996b). In tests with benzyl viologen as the electron acceptor, formate dehydrogenase activity is detected in both *O. formigenes* and *O. vibrioformis* (Dehning and Schink, 1989b; Cornick and Allison, 1996a). However, since neither NAD nor NADP are reduced in such tests and since formate accumulation is roughly equivalent to the amount of oxalate degraded, no role for formate dehydrogenase in the energy metabolism of *Oxalobacter* is expected. Neither nitrate nor sulfate are reduced, indole is not formed, and cytochromes are not present.

Pathogenicity There is no evidence that these bacteria, which are normal inhabitants of the gastrointestinal tract, are pathogenic. Large quantities of cells (10^{10}), representing strains isolated from sheep rumen, pig cecum, human feces, guinea pig cecum, and wild rats, have been fed to laboratory rats without ill effects (Daniel et al., 1987a).

Ecology Oxalate, a product of both plant and animal metabolism, is widely distributed in many ecosystems. Metabolism of oxalate by *O. formigenes* in freshwater sediments and gut habitats occurs under anaerobic conditions and at near neutral pH. Metabolism is accompanied by alkalization, since 1 mol of protons is consumed per mole of oxalate metabolized. Formate produced from oxalate has been used for methanogenesis in enrichment cultures of rumen microbes (Dawson et al., 1980a). Normal diets of humans and animals usually contain enough oxalate to maintain intestinal populations of *O. formigenes*. Increased concentrations of these bacteria in the intestine are selected for when diets with elevated concentrations of oxalate are fed to animals. This selection leads to increased rates of degradation of otherwise potentially toxic amounts of oxalate, a phenomenon that has survival value for the host animal (Shirley and Schmidt-Nielson, 1967; Allison and Cook, 1981). The absence of *O. formigenes* in the intestinal tracts of some humans appears to be correlated with increased absorption of dietary oxalate, leading to increased risk for kidney stones and other oxalate-related disease (Sidhu et al., 1998).

Antibiotic sensitivity Strains HC1 and OxK are both sensitive to chloramphenicol, colistin, doxycycline, erythromycin, polymyxin B, rifampin, and tetracycline. Strain OxK is also sensitive to cefuroxime. Both strains are resistant to 22 other antibiotics that have been tested. Similar patterns have been observed with *O. formigenes* strains from rats (Daniel et al., 1987b) and with the type strain, OxB (Dawson, 1979).

ENRICHMENT AND ISOLATION PROCEDURES

Anaerobic media with oxalate as the main, or only, energy source are used for enrichment cultures from either rumen samples (Dawson et al., 1980a, b) or freshwater lake sediments (Smith et al., 1985; Dehning and Schink, 1989b). Carbonate-buffered media, heated to remove dissolved oxygen and reduced with cysteine and/or sulfide, are used for enrichment cultures. Media for enrichment from sediments also contain acetate and either yeast extract or a mixture of vitamins, while media for enrichment from the rumen contain rumen fluid.

Colonies of oxalate-degrading bacteria are identified based on the formation of clear zones around the colonies in media that are somewhat opaque due to the presence of calcium oxalate. Daniel et al. (1989) have compared variations in levels of calcium, yeast extract, and rumen fluid on culture counts of oxalate-degrading rumen bacteria, as detected by clear zones around colonies in roll-tube cultures. This method has been used for isolations of *O. formigenes* strains without prior enrichment steps and for enumeration of these bacteria from gastrointestinal samples from man and animals (Allison et al., 1985a).

MAINTENANCE PROCEDURES

Viable cultures may be recovered from cells stored at $-80°C$ for several years or from freeze-dried cells stored at 0–4°C.

PROCEDURES FOR TESTING SPECIAL CHARACTERS

Successful detection of *O. formigenes* in fecal samples can be accomplished using PCR amplification of selected sequences of the *oxc* gene (Sidhu et al., 1997a, 1999). Tests can be successfully conducted with fecal samples that have not been cultured or with enrichment cultures in a medium containing 10–30 mM oxalate. The loss of oxalate from such media is usually nearly complete, and a simple test for this loss can be made based on precipitate formation when a calcium chloride solution is added (Sidhu et al., 1997b).

DIFFERENTIATION OF THE GENUS *OXALOBACTER* FROM OTHER GENERA

Organisms of the genus *Oxalobacter* are the only known Gram-negative, anaerobic, oxalate degraders that are extreme specialists, to the extent that oxalate and/or the closely related compound oxamate are obligate requirements for growth. Oxalate serves as the major carbon and energy source, but a small amount of acetate may also be required.

Selected regions of the *oxc* gene (encoding oxalyl-CoA decarboxylase) and the *frc* gene (encoding formyl-CoA transferase) have been used to construct oligonucleotide probes able to distinguish *O. formigenes* DNA from DNA of any of a known group of strains of gastrointestinal bacteria. Further evidence for the specificity of these probes was obtained when it was found that they fail to hybridize or to amplify PCR products from whole fecal DNA isolated from fresh stool samples from an individual known not to be colonized with oxalate-degrading bacteria (Sidhu et al., 1997a).

TAXONOMIC COMMENTS

Similarity values based on 16S rRNA sequences (1440 nucleotides) that allow *Oxalobacter* (represented by strain OxCR of *O. formigenes*) to be compared with other selected species are given in Table BXII.β.15. The corresponding phylogenetic tree (Fig. BXII.β.11) illustrates that strains of *Oxalobacter* within the class *Betaproteobacteria* share a specific relationship with the other genera currently in the *Oxalobacteraceae* family, *Telluria*, *Janthinobacterium*, and *Duganella*.

List of species of the genus Oxalobacter

1. **Oxalobacter formigenes** Allison, Dawson, Mayberry and Foss 1985b, 375[VP] (Effective publication: Allison, Dawson, Mayberry and Foss 1985a, 6.)

form.i.ge′ nes. M.L. n. *acidum formicum* formic acid; Gr. v. *gennaio* produce; M.L. adj. *formigenes* formic acid producing.

The characteristics are as described for the genus with

TABLE BXII.β.15. 16S rRNA sequence similarities

	Oxalobacter formigenes OxCR	*Alcaligenes eutrophus* [a]	*Burkholderia gladioli* pathovar *gladioli* [a]	*Burkholderia solanacearum* [a]	*Escherichia coli*	*Iodobacter fluviatilis*	*Neisseria elongata*	*Pseudomonas pickettii* [a]	*Pseudomonas solanacearum* [a]	*Zoogloea ramigera* 4	*Zoogloea ramigera* 5
Oxalobacter formigenes OxCR	—										
Alcaligenes eutrophus	88.8	—									
Burkholderia gladioli pathovar *gladioli*	81.9	88.6	—								
Burkholderia solanacearum	88.4	96.1	91.4	—							
Escherichia coli	78.4	81.2	81.9	81.1	—						
Iodobacter fluviatilis	86.1	88.6	90.4	89.3	82.9	—					
Neisseria elongata	85.2	88.6	88.6	87.2	81.9	89.4	—				
Pseudomonas pickettii	91.2	96.1	89.2	97.8	81.2	89.2	89.2	—			
Pseudomonas solanacearum	88.2	95.8	89.6	99.2	81.1	89.6	89.6	97.8	—		
Zoogloea ramigera 4	88.4	88.5	91.8	90.1	81.1	88.6	88.6	90.4	91.2	—	
Zoogloea ramigera 5	89.3	90.3	90.5	90.1	80.2	88.1	86.4	89.4	90	98.2	—

[a]*Alcaligenes eutrophus* = *Ralstonia eutropha*; *Burkholderia solanacearum* = *Pseudomonas solanacearum* = *Ralstonia solanacearum*; *Pseudomonas pickettii* = *Ralstonia pickettii*.

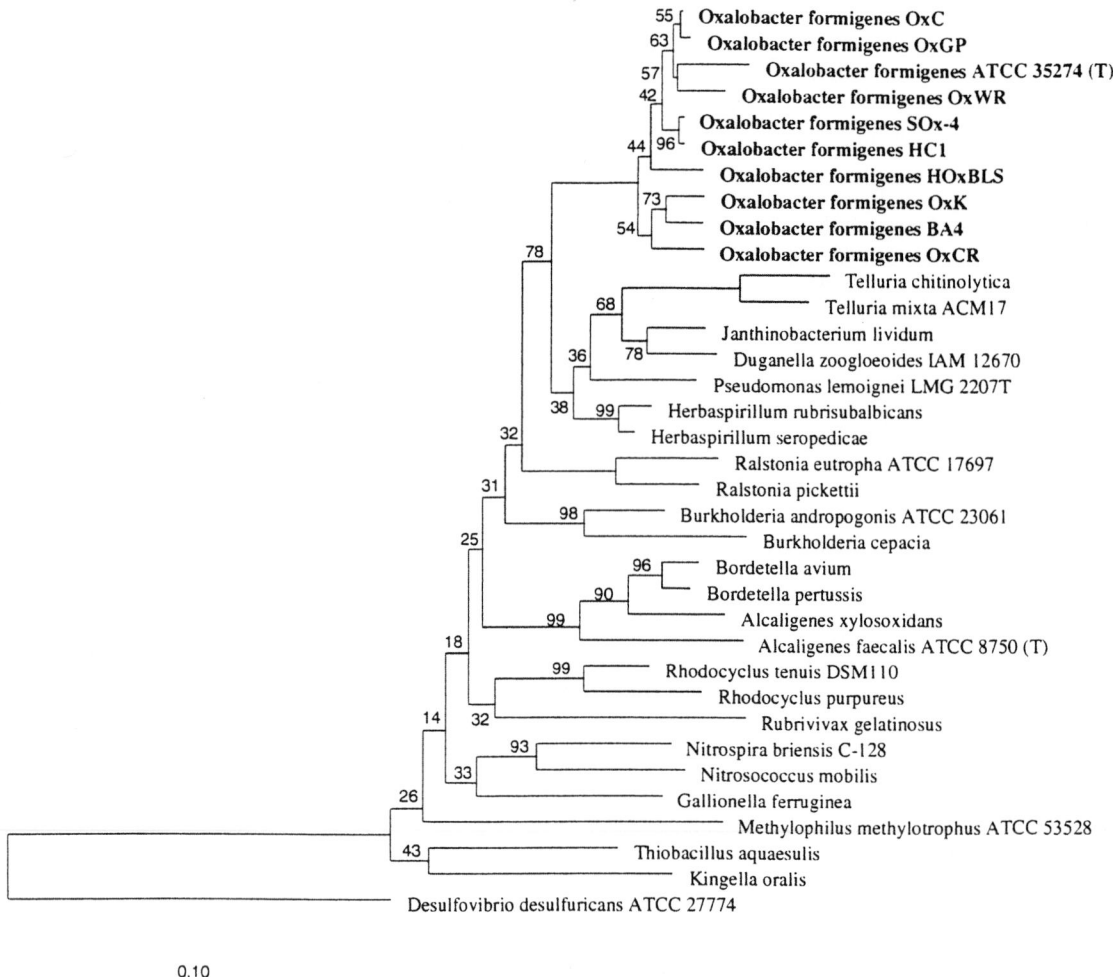

FIGURE BXII.β.11. Phylogenetic tree of *Oxalobacter formigenes* and other representatives of the *Betaproteobacteria* based on 16S rRNA sequence comparisons. The tree was constructed using the neighbor-joining method of Saitou and Nei (1987). Confidence limits were calculated by the bootstrap method using 100 samplings of the data set. Bar = 0.10 estimated substitutions per nucleotide position. *Desulfovibrio desulfuricans*, a member of the *Deltaproteobacteria*, was used to root the tree.

the following additional features. Rods 0.4–0.6 × 1.2–2.5 μm, with rounded ends, often curved, occurring singly, in pairs, or occasionally in chains, with curved cells leading to a spiral appearance. Flagella are not present.

High levels of 16S rRNA sequence similarity (96.6–99.8%) have been found for a group of strains of *O. formigenes* that includes both gastrointestinal and lake sediment strains (Table BXII.β.16).

The major total cellular fatty acids of representative

strains of *O. formigenes* are given in Table BXII.β.17. When strains are arranged based on descending values for quantities of 19-carbon cyclopropane ($C_{19:0 \text{ cyclo}}$) fatty acids, a separation into two groups is evident. When these fatty acid profiles are compared using a cluster analysis program (Sasser, 1990b), the same grouping is supported (Fig. BXII.β.12). Assignment of strains into Groups I and II based on cellular fatty acid profiles generally agrees with grouping based on 16S rRNA sequences (Fig. BXII.β.11), except for

TABLE BXII.β.16. 16S rRNA sequence similarity matrix

Oxalobacter formigenes strains	OxCR	HOxBLS	BA4	Ox-K	OxWR	Ox-B	OxGP	OxC	HC1	SOx4
OxCR	—	—	—	—	—	—	—	—	—	—
HOxBLS	96.9	—	—	—	—	—	—	—	—	—
BA4	97.7	97.7	—	—	—	—	—	—	—	—
Ox-K	96.9	97.1	98.2	—	—	—	—	—	—	—
OxWR	96.6	97.1	96.8	96.4	—	—	—	—	—	—
Ox-B	96.9	97.4	97.3	97	98.2	—	—	—	—	—
OxGP	97.4	97.7	97.5	97	98.4	98.7	—	—	—	—
OxC	97.4	97.7	97.7	97.1	98.6	99	99.5	—	—	—
HC1	97.1	97.9	98.2	96.9	97.9	98.2	98.4	98.7	—	—
SOx4	97.4	98.2	98.3	97	98.1	98.3	98.6	98.9	99.8	—

TABLE BXII.β.17. Major cellular fatty acids in different strains of *Oxalobacter formigenes*[a]

Fatty acid	$C_{14:0}$	$C_{16:0}$	$C_{17:0 \, cyclo}$	$C_{18:1}$	$C_{18:0}$	$C_{19:0 \, cyclo}$
Oxalobacter formigenes strains:						
GROUP I						
PoxC	1.41	39.03	31.14	1.97	1.20	12.54
Sox-6	1.05	40.19	29.74	2.43	0.78	14.88
OxDB12	0.87	35.61	28.76	5.76	0.93	16.05
OxWR	0.87	40.04	29.63	2.80	0.52	16.07
HC3	0.99	36.89	31.28	2.51	0.53	16.35
SOx4	0.98	40.35	29.72	1.35	0.68	16.45
HC-1	1.02	36.45	31.57	2.29	0.40	16.67
OxCC13	1.00	36.72	29.59	3.16	nr[b]	16.85
OxB	0.77	40.70	27.38	1.06	1.18	19.10
OxHM18	1.16	36.08	30.45	0.67	nr	20.06
GROUP II						
HOxBLS	3.50	35.01	12.90	8.45	1.11	31.63
HOxRW	3.56	35.95	12.53	7.84	1.58	31.80
BA1	2.97	32.46	5.69	11.03	2.71	37.61
BA6	3.85	34.04	6.16	7.05	1.64	39.24
OxK	3.21	28.54	2.91	12.63	2.86	42.83
OxGP	2.82	23.95	1.31	14.27	1.47	49.56
89-112	2.56	16.87	0.55	16.96	2.77	53.42

[a]Fatty acids separated as methyl esters by gas chromatography and identified as described by Sasser (1990a).

[b]nr, not reported.

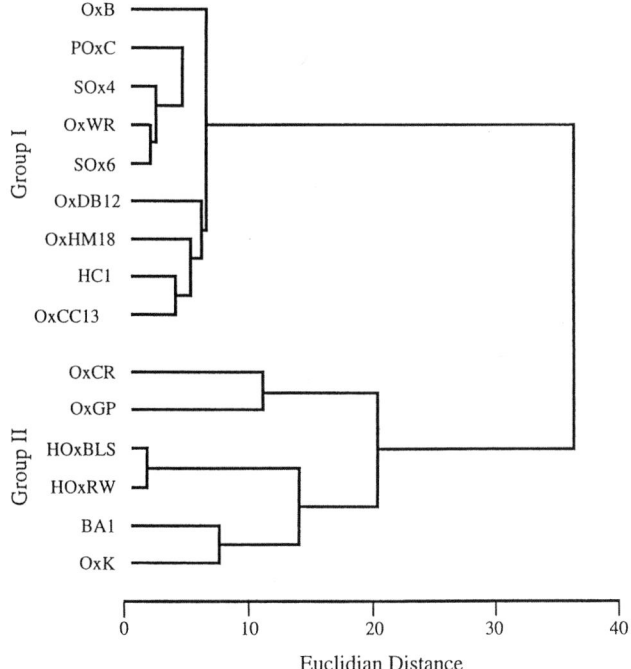

FIGURE BXII.β.12. Computer-generated dendogram (Sasser, 1990b) based on fatty acid profiles of strains of *Oxalobacter formigenes*.

probes and primers also support the concept that *O. formigenes* strains can be separated into two main groups that fit with the above separations. Further separation of Group II strains into three subgroups has also been suggested (Sidhu et al., 1997a).

The mol% G + C of the DNA is: 48–51 (T_m).

Type strain: OxB, ATCC 35274.

GenBank accession number (16S rRNA): U49757.

Additional Remarks: GenBank accession numbers (16S rRNA) for other strains are U49754 (strain BA4), U49750 (strain HOxBLS), U49751 (strain SOx4), U49752 (strain OxWR), U49753 (strain OxGP), U49754 (strain OxCR), U49755 (strain OxC), U49756 (strain OxK), and U49758 (strain HC1).

2. **Oxalobacter vibrioformis** Dehning and Schink 1990, 320[VP] (Effective publication: Dehning and Schink 1989b, 82.)

vi.bri.o.for' mis. L. v. *vibrare* to vibrate; M.L. masc. n. *vibrio* that which vibrates; L. fem. n. *forma* shape, e.g., a curved cell; M.L. masc. adj. *vibrioformis* shaped like a curved cell.

The characteristics are as described for the genus, with the following additional features. Vibrioid rods 0.4 × 1.8–2.4 μm; motile by 1–2 polar flagella. Cells occur singly, in pairs, or in spiral chains. Growth occurs in freshwater or brackish water mineral medium. Temperature range: 18–35°C; optimum, 30–32°C. pH range, 5.6–8.3; optimum, 6.8–7.0. Catalase negative. No reduction of nitrate, sulfate, sulfite, thiosulfate, or sulfur. Indole is not formed from tryptophan. Does not hydrolyze urea, gelatin, or esculin. Acetate (1–2 mM) is required for growth. Formate is produced from oxalate. Cell yields are 1.2–1.3 g cell (dry wt) per mol of oxalate metabolized in growth medium that does not contain yeast extract and 1.6–1.7 g/mol in the presence of 0.05% yeast extract. The type strain was isolated from freshwater sediments, but similar strains have been isolated from marine sediments (Dehning and Schink, 1989b).

The mol% G + C of the DNA is: 51.6 ± 0.6 (T_m).

Type strain: WoOx3, DSM 5502.

strain OxGP, which fits with Group I based on rRNA sequences but is placed in Group II based on fatty acid profiles. Representative strains from both Group I and Group II contain both 12 and 14-carbon hydroxy fatty acids (Allison et al., 1985a). Comparisons of profiles of cellular proteins separated by polyacrylamide gel electrophoresis also support the distinctions between strains and groupings found by other methods (Jensen and Allison, 1994).

Results of both RFLP and PCR tests with oligonucleotide

Genus II. **Duganella** Hiraishi, Shin and Sugiyama 1997b, 1251[VP]

THE EDITORIAL BOARD

Du.ga.nel′ la. M.L. dim. ending *-ella;* M.L. fem. n. *Duganella* named after P.R. Dugan, the American microbiologist who isolated the organism.

Straight or slightly curved rods, 0.6–0.8 × 1.8–3.0 μm. **Motile by single polar flagellum.** Gram negative. Nonsporeforming. **Aerobic,** having a strictly respiratory metabolism with O_2 as the terminal electron acceptor. Nitrate is not used as a terminal electron acceptor. No chemolithotrophic growth occurs with H_2. **In liquid media, either of two types of growth may occur: amorphous flocs (occasionally fingerlike), or dispersed growth with little or no formation of flocs. Colonies on nutrient agar are pale yellow to straw-colored.** Mesophilic. Neutrophilic. Chemoorganotrophic. Growth occurs on organic media or on mineral media supplemented with simple organic compounds. **Catalase and oxidase positive. Acid is produced oxidatively from glucose** and other carbohydrates. **Starch, gelatin, casein, and urea are hydrolyzed.** Major cellular fatty acids are $C_{16:0}$, $C_{16:1}$; $C_{10:0\ 3OH}$ is the major hydroxy fatty acid. Ubiquinone-8 is the sole respiratory quinone. Putrescine and hydroxyputrescine are intracellular polyamines. The genus is classified in the class *Betaproteobacteria.* Isolated from sewage and polluted water.

The mol% G + C of the DNA is: 63–64.

Type species: **Duganella zoogloeoides** Hiraishi, Shin and Sugiyama 1997b, 1251.

DIFFERENTIATION OF THE GENUS *DUGANELLA* FROM OTHER GENERA

Characteristics differentiating the genus *Duganella* from other related genera are listed in Table BXII.β.18.

TAXONOMIC COMMENTS

The type strain of *D. zoogloeoides* (Dugan strain 115) was originally classified in the genus *Zoogloea* as a strain of *Zoogloea ramigera* (Friedman and Dugan 1968). This placement was questioned by Unz (1984), who found extensive phenotypic dissimilarities between strain 115 and the type strain of *Zoogloea ramigera.* Phylogenetic studies by Hiraishi et al. (1997b) based on 16S rDNA sequencing have shown that strain 115 is not closely related to the type strain of *Z. ramigera,* but is instead located in a different cluster, with its closest relatives being the genus *Telluria* and the species *Pseudomonas lemoignei* (for which a new genus, *Paucimonas,* has been proposed by Jendrossek (2001). In this 2nd edition of the *Manual,* the genus *Duganella* is classified in the class *Betaproteobacteria,* order *Burkholderiales,* and the family *Oxalobacteraceae.* Other genera belonging to this family include *Oxalobacter, Herbaspirillum, Janthinobacterium, Massilia,* and *Telluria.*

List of species of the genus Duganella

1. **Duganella zoogloeoides** Hiraishi, Shin and Sugiyama 1997b, 1251[VP]

   zo.o.gloe.o′ i.des. M.L. n. *Zoogloea* bacterial genus name; Gr. suff. *-oides* similar to; M.L. adj. *zoogloeoides* similar to *Zoogloea.*

   The characteristics are as described for the genus, together with the following additional information. Growth occurs on nutrient media or mineral media supplemented with simple organic compounds. Colonies on nutrient agar are glistening, convex with entire margins, viscous, and pale yellow to straw colored. Colonies may also appear tough, leathery, dry, and wrinkled. Denitrification does not occur. No growth factors are required, but yeast extract stimulates

TABLE BXII.β.18. Characteristics that differentiate members of the genus *Duganella* from the genera *Paucimonas, Herbaspirillum, Oxalobacter, Janthinobacterium, Telluria, Zoogloea,* and *Massilia*[a]

Characteristic	*Duganella*	*Herbaspirillum*[b]	*Janthinobacterium*[c]	*Massilia*[d]	*Oxalobacter*[e]	*Paucimonas*[f]	*Telluria*[g]	*Zoogloea*[h]
Anaerobic	−	−	−	−	+	−	−	−
Vibrioid to helical cells	−	+	−	−	−	−	−	−
Cell diameter >1.0 μm	−	−	+	−	−	−	−	+
Flagellation in liquid media:								
Single polar flagellum	+	−		+	−	+	+	+
1–3 flagella at one or both poles	−	+		−	−	−	−	−
Flocculent growth	v	−	−	+		−	−	+
Oxidase	+	+	+	−			W	+
Purple pigment (violacein) produced	−	−	d	−		−	−	−
Growth occurs on nutrient agar	+		+		−	−	−	W
Plant root-associated N_2 fixer	−	+	−	−		−	−	−
Arginine dihydrolase	−			+			−	−
Hydrolysis of starch	+	−	−	+		−	+	+
Hydrolysis of gelatin	+	−	+	+		−	+	+
Urease	+			−			+	+
Mol% G + C of the DNA	63–64	60–65	61–67	64		57–61	67–72	65

[a]Symbols: see standard definitions. W, weak; blank space, data either not reported or not applicable.

[b]Data from Baldani et al. (1986a), Baldani et al. (1996), and Lincoln et al. (1999).

[c]Data from Lincoln et al. (1999).

[d]Data from La Scola et al. (1998a).

[e]Data from Allison et al. (1985a), Dehning and Schink (1989b), and Lincoln et al. (1999).

[f]Data from Delafield et al. (1965), Mergaert et al. (1996), and Jendrossek (2001).

[g]Data from Bowman et al. (1993b).

[h]Data from Unz (1984).

growth. Forms an ester-like, sweet odor when grown with organic acids. The following carbohydrates are used as carbon sources, and acid is formed from them oxidatively: L-arabinose, D-xylose, D-glucose, D-fructose, D-galactose, D-mannose, maltose, sucrose, cellobiose, lactose, and glycogen. No growth or acid production occurs on D-ribose, L-rhamnose, glycerol, D-mannitol, D-sorbitol, or inositol. The following noncarbohydrates are used: pyruvate, citrate, succinate, fumarate, malate, malonate, tartrate, ethanol, alanine, aspartate, asparagine, glutamate, and proline. No or little growth with formate, acetate, propionate, butyrate, caproate, caprylate, methanol, propanol, benzoate, *p*-hydroxybenzoate, glycine, histidine, arginine, lysine, ornithine, or tryptophan. Putrescine and hydroxyputrescine are intracellular polyamines.

The mol% G + C of the DNA is: 63-64 (method not reported).

Type strain: Dugan 115, ATCC 25935, IAM 12670.
GenBank accession number (16S rRNA): D14256.

Additional Remarks: The type strain shows two types of colonies, designated A and B. Colonies of the A type are glistening, convex, viscous, and cream to straw colored. The organisms show dispersed growth rather than floc formation when cultured in complex peptone-containing media with shaking, but they do produce flocs when cultured in a chemically defined medium supplemented with organic acids. In contrast, colonies of the B type are tough, leathery, dry, wrinkled, and pale yellow. Flocs are produced in both complex and chemically defined media and are amorphous, although occasionally flocs with fingerlike projections occur. The A and B types give 93–102% DNA–DNA hybridization to each other.

Genus III. **Herbaspirillum** Baldani, Baldani, Seldin and Döbereiner 1986a, 90[VP] emend. Baldani, Pot, Kirchhof, Falsen, Baldani, Olivares, Hoste, Kersters, Hartmann, Gillis and Döbereiner 1996, 808

JOSÉ IVO BALDANI, VERA LÚCIA DIVAN BALDANI AND JOHANNA DÖBEREINER

Her.ba.spi.ril' lum. L. fem. n. *herba* herbaceous, seed-bearing plant that does not produce persistent woody tissue; M.L. dim. neut. n. *spirillum* small spiral; *Herbaspirillum* small, spiral-shaped bacteria from herbaceous, seed-bearing plants.

Cells are generally **vibrioid**, sometimes spirilloid, approximately **0.6–0.7 μm in diameter**. The cell length varies from 1.5–5.0 μm depending on the culture medium. **Gram negative.** Motile, having 1–3 flagella at one or both poles. The organisms have a strictly **respiratory type of metabolism**, and sugars may be oxidized but not fermented. **The three named species presently included in the genus have the ability to fix atmospheric N₂ under microaerobic conditions** and grow well with N_2 as the sole nitrogen source, even in the presence of 10% sucrose. **Oxidase and urease positive.** Catalase variable or weak. Salts of organic acids such as malate, pyruvate, succinate and fumarate are the favored carbon source for NH_4^+ or N_2-dependent growth. Other carbon sources, such as glycerol, mannitol, D-glucose, and sorbitol, but not sucrose, are also catabolized. Optimal temperature, 30–34°C. **Optimal pH, 5.3–8.0.** Colonies on JNFb agar plates containing three times the usual concentration of bromothymol blue are smooth and white with blue or green centers after one week. Occur mainly in association with **graminaceous plants endophytically colonizing roots, stems and leaves.**

The mol% G + C of the DNA is: 60–65.

Type species: **Herbaspirillum seropedicae** Baldani, Baldani, Seldin and Döbereiner 1986a, 90 emend. Baldani, Pot, Kirchoof, Falsen, Baldani, Olivares, Hoste, Kersters, Hartmann, Gillis and Döbereiner 1996, 808.

FURTHER DESCRIPTIVE INFORMATION

Morphological features The flagellar arrangement is shown in Figs. BXII.β.13 and BXII.β.14. No lateral flagella have been observed for the species. Cells become very motile when close to air bubbles. Cells become elongated when grown in a semisolid medium containing glucose, galactose, arabinose, mannitol, or glycerol as the sole carbon source in the presence of 20 mM NH₄Cl (Figs. BXII.β.15 and BXII.β.16) (Baldani et al., 1992). With monosaccharides, *H. rubrisubalbicans* cells are more elongated and thinner, whereas *H. seropedicae* cells are wider. With organic acids such as malate, succinate, α-ketoglutarate, and fumarate, these characteristics have not been observed. The size of *H. seropedicae* and *H. rubrisubalbicans* cells are also different when they endophytically colonize sugarcane plant tissues. *H. seropedicae* has a cell size of 1.7 × 0.52 μm, whereas *H. rubrisubalbicans* cells are 2.1 × 0.45 μm (Olivares, 1997).

Cultural characteristics In semisolid JNFb medium[1], *H. seropedicae* and *H. rubrisubalbicans* form a fine white pellicle below the surface of the medium after incubation for 48 h at 30–34°C. The cells remain motile and vibrioid even after 1 week. Very pronounced swarming has been observed for *H. seropedicae* grown on nutrient agar (0.8% agar) plates at 35°C. *H. seropedicae* and *H. rubrisubalbicans* can grow and fix nitrogen in semisolid NFb medium, the medium routinely used to isolate *Azospirillum lipoferum* and *A. brasilense*. On this medium, the pellicles are similar to those of *Azospirillum*, except that the larger cell size of the latter differentiates *Herbaspirillum* cells under the microscope. No nitrogen fixation has been observed for either diazotrophic species grown at temperatures above 38°C, although growth can be observed at 41°C with inorganic nitrogen sources (Baldani et al., 1992).

The colonies are small, smooth, and whitish with blue or green centers when grown on JNFb agar plates containing three times

---

1. JNFb medium consists of L-malic acid, 5.0; K₂HPO₄, 0.6; KH₂PO₄, 1.8; MgSO₄·7H₂O, 0.2; NaCl, 0.1; CaCl₂·2H₂O, 0.02; trace element solution (Na₂MoO₄·2H₂O, 0.2 g; MnSO₄·H₂O, 0.235 g; H₃BO₃, 0.28 g; CuSO₄·5H₂O, 0.008 g; ZnSO₄·7H₂O, 0.024 g; distilled water, 1000 ml), 2.0 ml; bromthymol blue (0.5% aqueous solution (dissolve in 0.2 N KOH)), 2.0 ml; FeEDTA (1.64% solution), 4.0 ml; vitamin solution (biotin, 0.01 g; pyridoxine, 0.02 g; distilled water, 1000 ml), 1.0 ml; KOH, 4.5; pH adjusted to 5.8 with KOH. Bring the final volume to 1000 ml with distilled water. For a semisolid medium, add 1.75 to 1.9 g agar/l (agar should be dissolved before distribution into vials); for a solid medium, add 15.0 g agar/l.

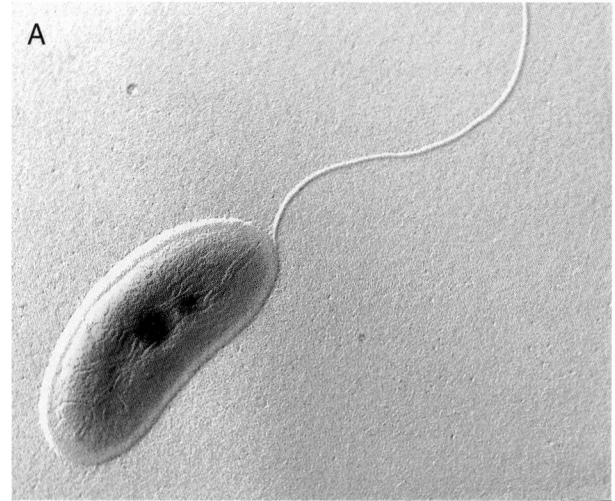

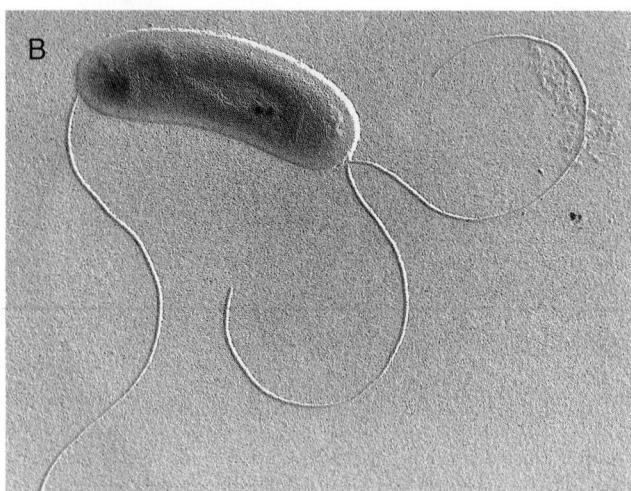

FIGURE BXII.β.13. *Herbaspirillum seropedicae* ATCC 35892 cells grown in nutrient broth and showing single (*A*) and double (*B*) polar flagella. (× 11,000) (Courtesy of N.R. Krieg).

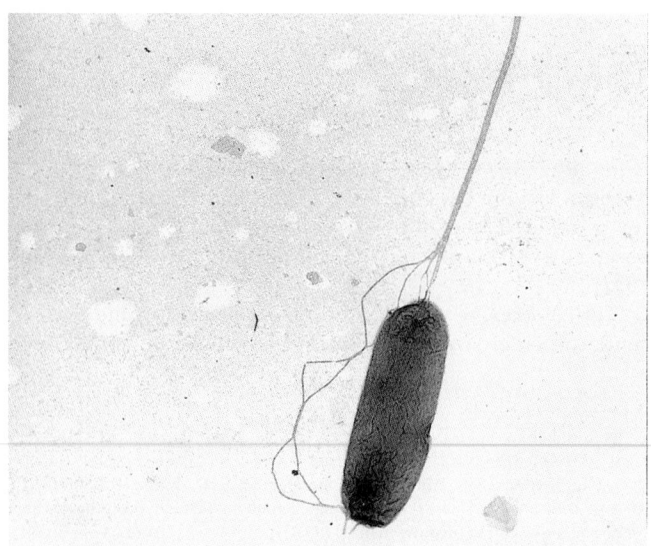

FIGURE BXII.β.14. *Herbaspirillum rubrisubalbicans* ATCC 19308 cells grown in nutrient broth and showing several polar flagella (× 12,000) (Courtesy of F.O. Olivares).

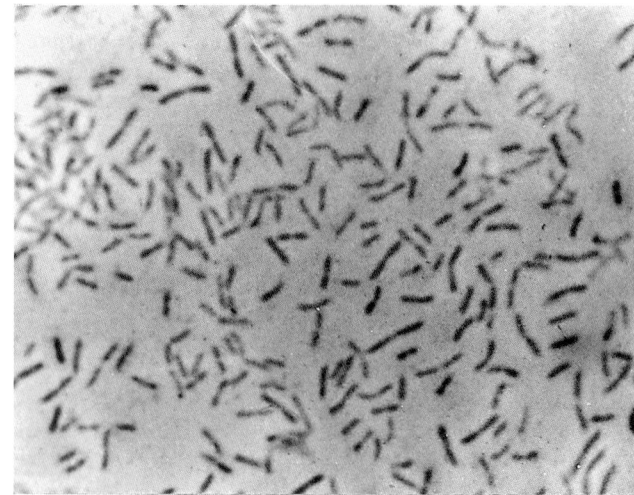

FIGURE BXII.β.15. *Herbaspirillum seropedicae* ATCC 35892 cultured in semisolid medium containing glucose as the sole carbon source and 20 mM NH_4Cl.

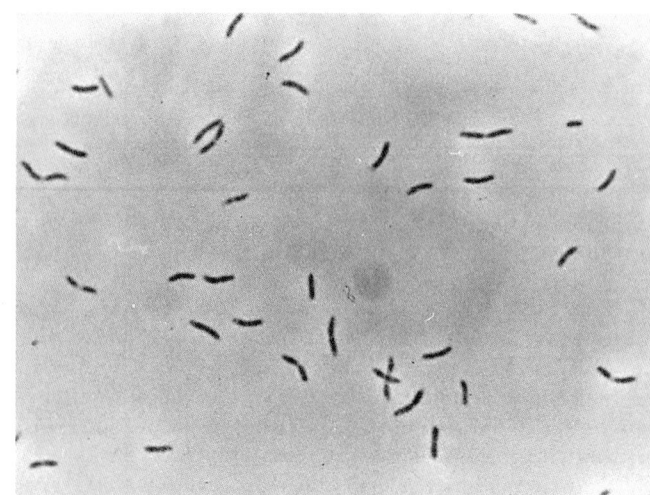

FIGURE BXII.β.16. *Herbaspirillum rubrisubalbicans* ATCC 19308 cultured in semisolid medium containing glucose as the carbon source and 20 mM NH_4Cl.

the usual amount of the bromothymol blue indicator. On BMS agar plates, colonies are moist and small with brown centers. (The recipe for BMS agar is given in the chapter on the genus *Azospirillum* located in the family *Rhodospirillaceae*.)

There is no growth or nitrogen fixation activity in liquid nitrogen-free medium under air (20.9% O_2). However, nitrogenase activity can be detected under an air when grown in liquid JNFb or NFbHP medium supplemented with L-glutamate and L-glutamine, but not with L-serine, L-alanine, or ammonium chloride. (The recipe for NFbHP medium is given in the chapter on the genus *Azospirillum* in the family *Rhodospirillaceae*.) Nitrogen fixation occurs only when the nitrogen source is exhausted from the culture medium. Other nitrogen sources, such as L-histidine, L-lysine, L-arginine, and the amines methylammonium chloride, tetramethylammonium chloride, and ethylenediamine chloride, do not support growth or nitrogen fixation by *H. seropedicae* (Klassen et al., 1997). Strains from this species also assimilate or dissimilate nitrate to nitrite under oxygen limitation, but no NO_3^--

dependent anaerobic growth or visible gas production from nitrate is observed either in solid or semisolid medium. Small amounts of N_2O are detected when 10% acetylene is added. Most strains from *H. rubrisubalbicans* reduce nitrate to nitrite, but denitrification has not been observed.

Compounds that can serve as sole carbon and energy sources for N_2-dependent growth of *H. seropedicae* and *H. rubrisubalbicans* strains include malate, succinate, citrate, α-ketoglutarate, fumarate, pyruvate, *trans*-aconitate, mannitol, glycerol, sorbitol, glucose, galactose, and L-arabinose. *N*-acetylglucosamine is also used as a sole carbon source for N_2-dependent growth by *H. seropedicae* strains. On the other hand, *meso*-erythritol is used by strains of *H. rubrisubalbicans* when ammonium chloride is added to the medium.

Biochemical tests (API 50CH, 50 A, and 50AA kits) reveal that catabolizabled carbon sources can be used to differentiate these species. All tested strains use the following substrates: glycerol, D- and L-arabinose, D-ribose, D-xylose, D-glucose, D-fructose, D-mannose, mannitol, sorbitol, xylitol, D-lyxose, DL-lactate, DL-glycerate, DL-3-hydroxybutyrate, pyruvate, citrate, succinate, fumarate, L-leucine, L-threonine, L-aspartate, L-glutamate, α-ketoglutarate, adonitol, D-galactose, L-fucose, D-and L-arabitol, gluconate, 2-ketogluconate, propionate, isobutyrate, *n*-valerate, isovalerate, D- and L-malate, *meso*-tartrate, butyrate, DL-4-aminobutyrate, aconitate, mesoaconitate, *p*-hydroxybenzoate, L-proline, D-arabinose, 5-ketogluconate, acetate, caprate, amylamine, D-fucose, and tryptamine (Baldani et al., 1996). Whether all of these carbon sources can support N_2-dependent growth has yet to be determined. On the other hand, all tested strains are unable to use the following carbon sources: arbutin, salicin, D-cellobiose, maltose, sucrose, trehalose, inulin, D-tartrate, D-raffinose, starch, L-phenylalanine, DL-5-aminovalerate, betaine, glutarate, L-histidine, L-sorbose, dulcitol, esculin, glycogen, phthalate, oxalate, and maleate.

Strains from this genus have a strictly respiratory metabolism, and sugars are oxidized but not fermented. Acid is produced from L-arabinose under N_2-fixation conditions, although it has also been observed from other carbon sources, such as glucose, fructose, galactose, mannitol, lactose, glycerol, and sorbitol, when using biochemical tests. The efficiency of N_2-fixation for *H. seropedicae* strains as evaluated in semisolid NFb medium is 12–15 mg N_2 per g DL-malate or 13 mg N_2 per g mannitol. Vitamins or other growth substances are not required. The plant-growth substance indoleacetic acid has been detected in strains of *H. seropedicae* and *H. rubrisubalbicans* when tryptophan is added.

Genetic features Many of the *nif* genes (*nif* A, B, HDK), as well as *glnA* and *B*, and *ntrBC* genes have already been identified in the chromosome of *H. seropedicae* species (Pedrosa et al., 1997). Plasmids have also been detected for some *H. seropedicae* strains, but no function has been ascribed to them (Pedrosa, personal communication).

Plasmids of the IncP1 incompatibility group are stable in *Herbaspirillum seropedicae* and can be transferred to this species by conjugation, electroporation, and transformation (Vande Broek et al., 1996; Pedrosa et al., 1997). Plasmids from this group have also been transferred by conjugation to a strain of *H. rubrisubalbicans* (Gitahy et al., 1997).

Antigenic features Polyclonal antibodies against whole cells of the type species of *H. seropedicae* (strain Z67) and against *H. rubrisubalbicans* strain M4 have been very useful for discriminating the plant tissue colonization by *H. seropedicae* and *H. rubrisubalbicans* when applying immunogold labeling (Olivares et al.,

1997). Purified antisera can discriminate among strains of both species by ELISA, and the immunotrapping technique using antibody-specific for *H. seropedicae* has permitted isolation of strains from sugarcane roots grown in the field (Reis et al., 1998).

Antibiotic susceptibility and resistance Strains of *H. seropedicae* and *H. rubrisubalbicans* are resistant to nalidixic acid, penicillin, and low concentrations of novobiocin and rifampicin. They are sensitive to several antibiotics, including streptomycin, kanamycin, spectinomycin, tetracycline, and chloramphenicol (Ureta et al., 1995; Baldani et al., 1986a).

Ecology The majority of the isolates of the genus *Herbaspirillum* has been isolated from graminaceous plants and show the ability to fix nitrogen. On the other hand, it has been shown that strains of *H. rubrisubalbicans* are mild pathogenic agents to some susceptible sugarcane varieties, causing "mottled stripe disease," mainly in crops highly fertilized with nitrogen. Experiments carried out in Brazil have shown that all commercial varieties are resistant to this disease, and neither *H. seropedicae* nor *H. rubrisubalbicans* strains cause the characteristic symptoms when artificially inoculated into leaves by injection (Olivares et al., 1997). In addition, strains of *H. seropedicae* and *H. rubrisubalbicans* cause "red stripe disease" in *Pennisetum purpureum* as well as in sorghum bicolor, although symptoms are very mild in sorghum leaves inoculated artificially with these bacteria (Pimentel et al., 1991; Olivares et al., 1997). However, symptoms are not observed in maize plants inoculated artificially with *H. seropedicae* and *H. rubrisubalbicans* strains (Olivares, 1997).

The ecological distribution of *Herbaspirillum seropedicae* and *H. rubrisubalbicans* has so far been restricted to tropical areas, where they have been found in numbers varying from 10^2–10^7 cells/g of fresh plant tissues. Strains of *H. seropedicae* were first isolated from washed, surface-sterilized roots of maize, sorghum, and rice grown in two different soils in Rio de Janeiro State, as well as from maize plants grown in a Cerrado soil in Brasilia, DF, Brazil (Baldani et al., 1986a). A few isolates have been obtained from rhizosphere soil (Baldani et al., 1986a). Since it is now known that *H. seropedicae* does not survive well in soil and could not be found in 10 different soil samples free of roots collected around Seropédica (Olivares et al., 1996), we suspect that small root pieces were present in the rhizosphere soil utilized by Baldani et al. (1986a). In addition, *H. seropedicae* has been isolated from roots and stems of *Saccharum* spp. and roots of *Echinola crusgalli*, *Pennisetum purpureum*, *Panicum maximum*, *Digitaria decumbens*, *Brachiaria decumbens*, and *Melinis minutiflora* (Olivares et al., 1996). In contrast, *H. rubrisubalbicans* has been isolated from roots, stems, and leaves of sugarcane plants as well as from roots of *Digitaria insularis*, a weed plant grown in the sugarcane field. No isolates of *H. seropedicae* and *H. rubrisubalbicans* could be found in many samples from seven different plant families: Compositae, Molluginaceae, Sterculaceae, Cyperaceae, Portulacaceae, Leguminosae, and Cucurbitaceae (Olivares et al., 1996). One isolate resembling *H. rubrisubalbicans* has been identified in banana plant material (Weber et al., 1997). One isolate of *H. seropedicae* and one resembling *H. rubrisubalbicans* have been identified in banana plant material (Weber et al., 1999).

Electron-microscopic analysis shows that *H. rubrisubalbicans* colonizes the xylem, intercellular, and substomatal cavities of a mottled-stripe-susceptible sugarcane variety, in which the bacteria are restricted to microcolonies encapsulated within membranes of plant cell origin (Olivares et al., 1997; Fig. BXII.β.17). Colonizations in intercellular spaces and xylem have also been found

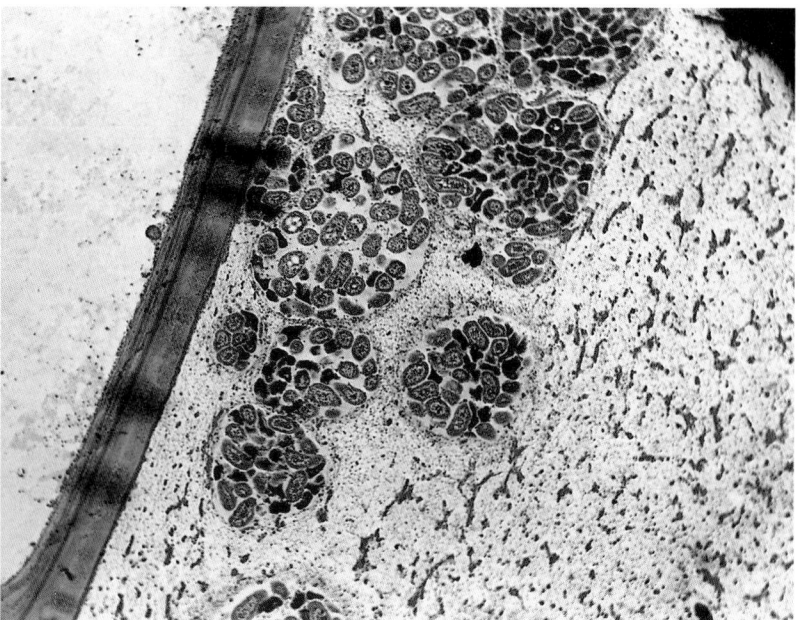

FIGURE BXII.β.17. Transmission electron micrograph (TEM) of xylem vessel from the leaf of a sorghum plant inoculated with *Herbaspirillum rubrisubalbicans* ATCC 19308 showing the microcolonies close to the walls of the vessels and that are surrounded by host-derived gums. (Reprinted with permission from E.K. James and F.L. Olivares, Critical Reviews in Plant Sciences *17*: 77–119, 1997, ©CRC Press, Boca Raton.)

for *Herbaspirillum* spp. in sugar cane roots (Olivares et al., 1997). *H. seropedicae* has also been localized within intercellular cavities of rice seeds and roots (Baldani et al., 1994).

Field inoculation experiments have shown yield increases of maize inoculated with *H. seropedicae* strains (Pereira et al., 1988). Increase of dry weight and grain yield has been observed in rice plants inoculated with selected strains of *H. seropedicae* (Döbereiner and Baldani, 1998).

ENRICHMENT AND ISOLATION PROCEDURES

Serial dilutions (0.1 ml) of root, stem, or leaf samples are inoculated into 10-ml cotton-plugged serum vials containing 5 ml of semisolid JNFb medium and incubated for 1 week at 32°C. Smashed pieces (5–8 mm long) of plant tissues can also be used, but the incubation time should be only 40–48 h. For vials exhibiting a fine white pellicle, cells are examined under the microscope for the presence of small, curved rods (0.6–0.7 × 4–6 μm) that move faster close to air bubbles. After a second transfer to JNFb semisolid medium and incubation for 24–48 h, cultures are streaked out on a solid JNFb medium containing 20 mg/l yeast extract and three times the bromothymol blue concentration of the JNFb medium. On these plates, *H. seropedicae* and *H. rubrisubalbicans* form small, moist colonies with a green or dark blue center (different from the white colonies of *Azospirillum lipoferum* and *A. brasilense*). For final purification, single colonies are again transferred to the JNFb semisolid medium, and cells from the typical pellicle are streaked onto BMS agar plates, on which moist, smooth and small brownish colonies develop. Colonies are selected and stored for further identification (Döbereiner et al., 1995).

MAINTENANCE PROCEDURES

Stock cultures can be maintained on JNFb or BMS medium under a layer of sterilized mineral oil in tubes tightly sealed with rubber caps. Under these conditions, *H. seropedicae* remains viable at room temperature for more than 12 years.

For routine use in the laboratory, it is recommended to maintain duplicate vials and subculture strains from time to time (e.g., every 6 months, with maintenance at room temperature). Strains can also be maintained in glycerol at −20°C by mixing equal volumes of sterilized glycerol and washed, resuspended cells from a 48-h-old culture grown in liquid JNFb medium (containing 20 mg of yeast extract and 5 mM ammonium chloride or sodium glutamate).

Strains can also be kept lyophilized for many years. Cells grown on slant JNFb medium, with malic acid replaced by D-glucose, for 48–72 h at 30°C are suspended in 2 ml of a 10% sucrose solution and 5% peptone in 100 ml water. Aliquots (0.2 ml) are then distributed into lyophilization ampoules and lyophilized.

PROCEDURES FOR TESTING SPECIAL CHARACTERS

***Meso*-erythritol and *N*-acetylglucosamine as carbon sources for growth that is dependent on nitrogen fixation or inorganic nitrogen** A loopful of culture grown in semisolid JNFb medium is inoculated into a vial of semisolid JNFb medium (in which malate is replaced by 0.5% *meso*-erythritol plus 1 g of NH₄Cl or 0.5% *N*-acetylglucosamine). *H. seropedicae* forms a fine pellicle on the surface of a semisolid medium containing *N*-acetylglucosamine after incubation at 32°C for 48 h, but no growth is observed in the presence of *meso*-erythritol. On the other hand, *H. rubrisubalbicans* forms a typical pellicle in the presence of *meso*-erythritol plus NH₄Cl, but not with *N*-acetylglucosamine as a carbon source. API galleries (API 50CH, API 50AO, and API 50AA; bioMérieux, Montalieu-Vercieu, France) can also be used to distinguish *H. seropedicae* from *H. rubrisubalbicans*.

Oligonucleotide probes Probe HS (5′-GTC CCG GTT TTT GCA TCG A-3′) and probe HR (5′-TAG TCG GTT TTT GCA

TCG A-3′) are species-specific for a highly variable stretch of helix 55–59 of the 23S rRNA of *H. seropedicae* and *H. rubrisubalbicans*. Bulk nucleic acids are isolated from strains, cultivated overnight in DYGS liquid medium, and transferred to a positively charged nylon membrane via spot blotting. Hybridization with radioactive or non-radioactive DIG-labeled probes is performed for 2–12 h (radioactive) or 16 h (nonradioactive) at 52°C according to the method of Kirchhof et al. (1997b). Hybridization signals are detected by autoradiography.

DIFFERENTIATION OF THE GENUS *HERBASPIRILLUM* FROM OTHER GENERA

The morphological, physiological and genetic characteristics that differentiate *Herbaspirillum* from other diazotrophic species within the class *Betaproteobacteria* and from *Azospirillum brasilense* are shown in Table BXII.β.19.

TAXONOMIC COMMENTS

This genus *Herbaspirillum* constitutes a separate rRNA cluster within the class *Betaproteobacteria*. Comparison of many bacteria within this group based on the $T_{m(e)}$ values of the DNA–rRNA hybridization experiments has shown that *Herbaspirillum* has a very close relationship to the genus *Janthinobacterium*. However, *Janthinobacterium* has the lowest $T_{m(e)}$ value within this rRNA cluster and because of its phenotypic distinction from the other members of this rRNA branch, including its inability to fix nitrogen, it has been maintained as a separate genus. The DNA–DNA hybridization experiments also confirm the results, since they show a very low degree of similarity with the *Herbaspirillum* species. More recently, a comparison of the 16S rRNA full sequence from *H. seropedicae* with those of several other non-nitro-

gen-fixing bacteria from the *Betaproteobacteria* has shown that this species, along with *Oxalobacter formigenes*, forms a separate lineage in the *Betaproteobacteria* (Sievers et al., 1998).

Other nitrogen-fixing bacteria are also found within the *Betaproteobacteria*, including the species *Burkholderia vietnamiensis*, *Azoarcus indigens*, and *Alcaligenes faecalis*. Although they share the ability to fix nitrogen, they have several other distinguishable phenotypes and genetic aspects that maintain them as separate species from the genus *Herbaspirillum*. *B. vietnamiensis* is able to use glucose, caprate, and itaconate as carbon sources, as does *Herbaspirillum*. However, in contrast to the latter genus, it is urease negative, the cells are not curved, the optimal temperature for growth is much lower, and it has been isolated mainly from rhizosphere rice roots. The mol% G + C is 66.9–68.1, which is higher than the 60–65 mol% G + C value established for *Herbaspirillum* species. The genus *Herbaspirillum* has some characteristics similar to those of *Azoarcus indigens*, such as cell shape, vibrioid movement, and an ability to use malic acid. However, *A. indigens* has a much higher oxygen tolerance for N_2 fixation (6.5%), in contrast to the 2.5% observed for the *Herbaspirillum*. In addition, it has a much higher optimal temperature for growth (40°C) and has been found associated only with Kallar grass plants. Although the $T_{m(e)}$ values show that they belong to the same rRNA superfamily, they are quite separate based on the DNA–rRNA hybridization dendrogram and therefore should not be included in the same genus. The other nitrogen-fixing species within the *Betaproteobacteria* has been identified as *Alcaligenes faecalis*. It has also been isolated from rice plants (You et al., 1991) and has an oxygen tolerance for N_2 fixation, size, and mol% G + C very close to those of *Herbaspirillum*.

Due to the similar shape, vibrioid movement, and mol% G

TABLE BXII.β.19. Differential morphological and physiological characteristics of the genus *Herbaspirillum*, selected diazotrophic species within the class *Betaproteobacteria*, and *Azospirillum brasilense* (class *Alphaproteobacteria*)[a]

Characteristic	*Herbaspirillum*	*Azoarcus indigens*[b]	*Azospirillum brasilense*[c]	*Burkholderia vietnamiensis*[d]
Arrangement of flagella:				
Location	Polar	Polar	Polar[e]	Polar or bipolar
Number	1–3	1	1[e]	1 or tuft
Vibrioid cell shape	+	+	+	−
Cell width, μm	0.5–0.7	0.5–0.7	1.0–1.2	0.3–0.8
Urease	+	+	+	−
Denitrification	−	−	v	+
Optimal growth temperature, °C	30–34	40	37	28
Carbon sources:				
D-Glucose	+	+	+	−
Sucrose	+	−	−	−
meso-Tartrate	+	d	nd	d
Caprate	+	−	nd	+
Itaconate	+	+	nd	+
Maximum O_2 tension (%) for N_2-fixation [f]	2.5	6.5	2.0	3.0
Habitats:				
Plant tissues	+	+	+	−
Soil	−	−	+	+
Clinical materials	−	−	−	−[g]
Mol% G + C of the DNA	60–65	66.6	70–71	66.9–68.1

[a]For symbols, see standard definitions, nd, not determined.

[b]Data from Reinhold-Hurek et al. (1993b).

[c]Date from Döbereiner (1991).

[d]Data from Gillis et al. (1995).

[e]In liquid medium, the cells possess only single polar flagella, but on agar media, several lateral flagella of shorter wavelength also occur.

[f]According to the experiments carried out by Vande Broek et al. (1996).

[g]A few strains are of clinical origin.

+ C content of DNA, *Herbaspirillum seropedicae* was provisionally placed into the genus *Azospirillum* (Baldani et al., 1984). However, results from RNA–RNA hybridization experiments comparing several strains of *H. seropedicae* with the three known species of *Azospirillum* and other possibly related N₂-fixing bacteria yielded very low RNA similarity values (<20%) (Falk et al., 1986). These differences were also confirmed by analysis of the SDS-PAGE membrane-protein pattern of strains of both genera: a characteristic major band at 37.5 kDa occurs only in *H. seropedicae* strains (Dianese et al., 1989). rRNA sequence analysis has shown that the genus *Herbaspirillum* belongs to the *Betaproteobacteria*, whereas the *Azospirillum* species were grouped into the *Alphaproteobacteria*. Even though *H. seropedicae* species show some physiological and morphological characteristics similar to those of *Azospirillum brasilense*, the genetic data show that it should not be included in the genus *Azospirillum*.

The genus *Herbaspirillum* was expanded when biochemical, DNA–rDNA, and DNA–DNA hybridization studies showed that *Pseudomonas rubrisubalbicans*, the causative agent of the "mottled strip" disease in sugarcane, has a high degree of rRNA similarity to *Herbaspirillum* and should therefore be included in this genus (Gillis et al., 1991). These data were confirmed by additional features of *P. rubrisubalbicans*, including its ability to fix nitrogen, culminating with the reclassification of the species as *Herbaspirillum rubrisubalbicans* (Baldani et al., 1996).

DNA–DNA hybridization studies have shown that there are three distinct groups, i.e., species, of bacteria within the *Herbaspirillum* genus. High values of DNA–DNA hybridization (71–100%) have been observed for *H. seropedicae* and *H. rubrisubal-*

bicans, and the full sequences of the 16S rRNAs of the type strains from both species show a 99.2% similarity (Kirchhof et al., 2001).

On the basis of physiological properties, phylogenetic analysis comparing 16S rDNA sequences and DNA–DNA hybridization experiments with chromosomal DNA, Kirchhof et al. (2001) created a new species, *Herbaspirillum frisingense*, for nitrogen-fixing bacteria isolated from the C4-fibre plants, *Spartina pectinata*, *Miscanthus sinensis*, *Miscanthus sacchariflorus*, and *Pennisetum purpureum*. Nitrogen-fixing capability was examined by PCR amplification of the *nifD* gene and an acetylene reduction assay and was found with all isolates tested. The 16S rDNA sequence similarity to the other two *Herbaspirillum* species is 98.5–99.1%.

ACKNOWLEDGMENTS

This chapter is dedicated to the memory of the late Dr. Johanna Döbereiner, the Brazilian soil microbiologist, who, during a long and distinguished career, identified and characterized many novel nitrogen-fixing bacteria associated with graminaceous plants.

FURTHER READING

Baldani, V.L.D., J.I. Baldani, F. Olivares and J. Döbereiner. 1992. Identification and ecology of *H. seropedicae* and the closely related *Pseudomonas rubrisubalbicans*. Symbiosis *13*: 65–73.

Döbereiner, J. 1991. The genera *Azospirillum* and *Herbaspirillum*. *In* Balows, Trüper, Dworkin, Harder and Schleifer (Editors), The Prokaryotes. A Handbook on the Biology of Bacteria: Ecophysiology, Isolation, Identification, Applications, 2nd Ed., Vol. 3, Springer-Verlag, New York. pp. 2236–2253.

Döbereiner, J 1992. History and new perspective of diazotrophs in association with non-leguminous plants. Symbiosis *13*: 1–13.

DIFFERENTIATION OF THE SPECIES OF THE GENUS *HERBASPIRILLUM*

The differential characteristics of the species *Herbaspirillum* are indicated in Table BXII.β.20. Other characteristics of the species are presented in Table BXII.β.21.

Although the 16S rDNA sequence analysis of *H. seropedicae* and *H. rubrisubalbicans* shows a high degree of similarity between

the two (99.2% Kirchhof et al., 2001), analysis of the hypervariable region of the 23S rDNA shows a higher degree of variation, allowing for the design of probes to differentiate between *H. seropedicae* and *H. rubrisubalbicans* (Kirchhof et al., 1997b).

List of species of the genus Herbaspirillum

1. **Herbaspirillum seropedicae** Baldani, Baldani, Seldin and Döbereiner 1986a, 90^VP emend. Baldani, Pot, Kirchoof, Falsen, Baldani, Olivares, Hoste, Kersters, Hartmann, Gillis and Döbereiner 1996, 808.

   se.ro.ped'i.cae. L. gen. n. *seropedicae* of Seropédica, Rio de Janeiro, Brazil, where the species was first isolated.

   The characteristics are as described for the genus and in Tables BXII.β.20 and BXII.β.21.

   Cells are vibrioid and sometimes spirilloid and become motile when close to O₂ sources. Two polar flagella are generally found at one or both poles, and pronounced swarming is observed on soft nutrient agar at 35°C. Oxidase, urease, and catalase positive. All strains except LMG 2284 fix nitrogen under microaerobic conditions and grow well with N₂ as a sole nitrogen source. Optimal pH and temperature for N₂-dependent growth in semisolid JNFb medium are 5.8 and 34°C, respectively. N₂-fixation can also occur in liquid medium when L-glutamate or L-glutamine

used as the sole nitrogen source is completely exhausted. Colonies on JNFb agar plates containing three times the usual concentration of bromothymol blue are small and moist with a green or dark blue center.

Salts of organic acids such as malate, fumarate, pyruvate, succinate, and *trans*-aconitate are favored carbon sources for both N₂- and NH₄-dependent growth, as are glucose, galactose, and L-arabinose. Fructose is not used for N₂-dependent growth. Growth and N₂ fixation occur in the presence of 10% sucrose. Vitamins and growth substances are not required. Sensitive to chloramphenicol, tetracycline, gentamicin, kanamycin, and streptomycin and resistant to penicillin, nalidixic acids, and low concentrations of novobiocin and rifampicin.

Habitats are roots, stems, and leaves of many members of the family Gramineae, but so far strains have not been isolated from soil, unless sorghum plants are used as trapping hosts. When artificially inoculated into leaves of sor-

TABLE BXII.β.20. Characteristics differentiating the species of the genus *Herbaspirillum*[a]

Characteristic	*H. seropedicae*	*H. frisingense*	*H. rubrisubalbicans*
Usual polar flagella arrangement	2 at one or both poles	Mostly two, occasionally 1 or 3; unipolar	Several polar
Carbon sources used for growth:			
meso-Erythritol	−	−	+
N-Acetylglucosamine, L-rhamnose, meso-inositol	+	+[b]	−
Adipate, pimelate, azelate, suberate, L-tartrate	−	−[c]	−
Hybridization with probes:			
HS	+	−	−
HR	−	−	+
Optimal temperature for growth, °C	34	30–37	30
Causes mottled stripe disease symptoms in susceptible sugar cane variety B-4362	−	−	+

[a]For symbols, see standard definitions.

[b]Positive only for N-acetylglucosamine.

[c]Only tested for adipate.

TABLE BXII.β.21. Additional characteristics of the species of the genus *Herbaspirillum*[a]

Characteristics	*H. seropedicae*	*H. frisingense*	*H. rubrisubalbicans*
Nitrogenase activity	+	+	+
Oxidase and urease	+	+	+
Cell length, μm	1.5–5.0	1.4–1.8	1.5–2.1
Hydrolysis of starch and gelatin	−	−	−
NO_3^- to NO_2^-	+	+	+
NO_2^- to N_2O	∓[b]	−	−
Optimal pH for growth	5.3–8.0	6.0–7.0	5.3–8.0
Maximum growth temperature (°C) in presence of organic nitrogen sources	38	37	41
$^{15}N_2$ incorporation into cells grown in semisolid media	+	nd	+
Growth in presence of 2% NaCl	−	nd	nd
Growth in presence of 10% sucrose	+	nd	+
Carbon sources used for growth (API tests):			
Sucrose, maltose, D-raffinose	−	−	−
Galactose, gluconate, adonitol, D-glucose,	+	+	+
D-Fructose,[c] mannitol, L-malate, pyruvate, citrate, succinate, 5-ketogluconate, acetate	+	+	+
Survival in soil	Poor	nd	Poor
Habitat	Tissues of several grasses	Tissue of C-4 fiber plants	Only leaves of sugar cane
Mol% G + C of DNA	64–65	61–65	62–63

[a]For symbols, see standard definitions; nd, not determined; ∓, several strains positive but the majority of strains negative.

[b]Several strains are able to reduce nitrite to N_2O, but the majority of strains are negative.

[c]All species grow poorly in D-fructose when dependent on nitrogen fixation.

ghum, *H. seropedicae* causes a mild symptom of red stripe disease.

The mol% G + C of the DNA is: 64–65 (T_m).

Type strain: Z67, ATCC35892, BR 11175, DSM 6445.

GenBank accession number (16S rRNA): Y10146.

Additional Remarks: Reference strains: ATCC 35893 (BR11177, Z78); ATCC 35894 (BR 11178, Z152).

2. **Herbaspirillum frisingense** Kirchhof, Eckert, Stoffels, Baldani, Reis and Hartmann 2001, 166[VP]

fri.sin.gen' se. frisingense L. gen. n. of Frisinga, now known as Freising, Germany, town where the species was first isolated.

The characteristics are as described for the genus.

Cells are vibrioid rods, but smaller and have a single polar flagellum. This group forms a fine pellicle on the surface of the semisolid NFb or JNFb medium quite similar to that of the other species. Colonies on JNFb agar plates containing three times the concentration of bromthymol blue are small and moist with a green or dark blue center. Oxidase, urease, and catalase positive. Gelatin is not hydrolyzed. All strains catabolize N-acetylglucosamine, arabinose, caprate, citrate, glucose, gluconate, malate, mannose, mannitol, and phenylacetate, but not maltose or adipate, as determined by biochemical analyses. All isolates reduce nitrate to nitrite but not to N_2. Possess a cytochrome oxidase. N_2-dependent growth (subsurface pellicle) was also observed with arabinose, N-acetylglucosamine, malate, mannitol, and glucose. In contrast to *H. seropedicae*, these isolates do not use L-rhamnose or inositol as sole carbon sources. They also differ from *H. rubrisubalbicans* by the ability to use *N*-acetylglucosamine but not *meso*-erythritol in the presence of NH₄Cl. Based on the sequence of a highly variable 23S rDNA stretch within domain III, an oligonucleotide probe named beta 20 (5′-GAT ACA AGA ACC GGG AC-3′) is used to distinguish *H. frisingense* from the other species of the genus.

Habitats are roots, stems, and leaves of the C-4 graminaceous plants *Spartina pectinata*, *Miscanthus sinensis*, and *Miscanthus sacchariflorus* grown in Germany and *Pennisetum purpureum* grown in Brazil.

The mol% G + C of the DNA is: 61–65 (T_m).

Type strain: DSM 13128, GSF30.

GenBank accession number (16S rRNA): AJ238358.

Additional remarks: Reference strains: DSM 13130 (strain Mb11) and ATCC 35894 (strain 75B).

3. **Herbaspirillum rubrisubalbicans** (Christopher and Edgerton 1930) Baldani, Pot, Kirchhof, Falsen, Baldani, Olivares, Hoste, Kersters, Hartmann, Gillis and Döbereiner 1996, 809[VP] (*Pseudomonas rubrisubalbicans* (Christopher and Edgerton 1930) Krasil'nikov 1949, 379; "*Phytomonas rubrisubalbicans*" Christopher and Edgerton 1930, 266.)

ru.bri.sub.al′ bi.cans. L. adj. *ruber* red; adj. *subalbicans* whitish; M.L. adj. *rubrisubalbicans* red-whitish, referring to the symptoms of mottled stripe disease.

Cells are slightly curved rods, motile by means of several polar flagella. Poly-β-hydroxybutyrate is accumulated. Colonies on 2% glucose–peptone agar are mucoid and similar to *H. seropedicae* when grown on JNFb agar plates. Oxidase positive. No hydrolysis of gelatin, starch, or Tween 80. Denitrification is negative, although most strains reduce nitrate to nitrite. Optimal pH and temperature for N₂-dependent growth in semisolid JNFb medium are 5.8 and 30°C, respectively, although growth can occur at temperatures as high as 40°C. N₂ is fixed as efficiently as by *H. seropedicae*. The favored carbon sources are those described for *H. seropedicae*. Growth and N₂ fixation occur in the presence of 10% sucrose, although this carbon source is not used (osmotolerance effect). The strains also do not use maltose, cellobiose, raffinose, salicin, and *meso*-inositol. The ability to tolerate high concentrations of these carbon sources was not tested. Sensitive to kanamycin, spectinomycin, and streptomycin and resistant to nalidixic acids and low concentrations of novobiocin and rifampicin.

Occurrence is apparently limited to sugar cane. Causes mottled stripe disease, mainly on sugar cane genotypes from regions in which high-nitrogen fertilizer applications are used. In addition, the organism can cause red stripe disease when artificially inoculated by injection into leaves of *Sorghum vulgare* and *Pennisetum purpureum*.

The mol% G + C of the DNA is: 62–63 (T_m).

Type strain: ATCC 19308, BR11192, DSM 9440, LMG 2286, NCPPB 1027.

GenBank accession number (16S rRNA): AB021424.

Other Organisms

A group, also called EF group 1, comprising two subgroups, 1a (8 strains) and 1b (14 isolates), and showing many physiological and genetics characteristics resembling *Herbaspirillum* was identified during the description of the *H. rubrisubalbicans* species. Subgroup 1a is differentiated from subgroup 1b by its ability to use *meso*-erythritol and benzoate as sole carbon sources. $T_{m(e)}$ (°C) values from the DNA–rRNA hybrids formed by the DNA hybridization of strains from this EF1 group with the 23S rRNA from *H. rubrisubalbicans* strain LMG 2286 vary from 77.0–79.3. These values are very close to those observed when strains of *H. seropedicae* were used (77.1–78.3). Because of the high level of DNA relatedness to the nitrogen-fixing species, this group has been tentatively included in the genus *Herbaspirillum* and has been provisionally called species 3. However, the correct taxonomic position of this group will be published elsewhere (Gillis, personal communication). This species has also been separated from other species based on its antigenic properties (Falsen, 1996). All of the strains are of clinical origin (i.e., wound, urine, otitis, gastric juice, respiratory tract, eye secretion, etc.) except for isolates LMG 2285, LMG 6421, and LMG 6416, which were isolated from sugarcane, sorghum, and maize plants, respectively, and previously described as *Pseudomonas rubrisubalbicans*. None of the strains fix nitrogen. The mol% G + C of the DNA of strain LMG 5523 is 61, as determined by the thermal denaturation method (De Ley and van Muylen, 1963) and calculated by the equation of Marmur and Doty (1962), modified by De Ley et al. (1970a).

Genus IV. **Janthinobacterium** De Ley, Segers and Gillis 1978, 164,[AL] emend. Lincoln, Fermor and Tindall 1999, 1586

MONIQUE GILLIS AND NIALL A. LOGAN

Jan.thin.o.bac.te′ ri.um. L. adj. *janthinus* violet-colored; Gr. n. *bakterion* a small rod; M.L. neut. n. *Janthinobacterium* a small, violet-colored rod.

Round-ended, straight rods, sometimes slightly curved, 0.8–1.5 × 1.8–6.0 μm, occurring singly and sometimes in pairs or short chains. Definite capsules are not evident, but intercellular slime may be produced. No resting stages are known. Gram negative, occasionally showing bipolar or barred staining and lipid inclusions. **Motile. Strict aerobes. Growth occurs from 4°C to about 30°C, with optimum growth at around 25°C.** Optimal pH 7–8, with no growth below pH 5. **Many strains produce the violet pigment violacein,** but strains producing partly pigmented or nonpigmented colonies are often encountered. Colonies on routine solid media are low convex and round. **Aerobic, having a strictly respiratory metabolism with oxygen as the terminal electron acceptor. Oxidase and catalase positive.** Chemoorganotrophs. Growth occurs on ordinary peptone media. Citrate and ammonium ions can be used as sole carbon and nitrogen sources, respectively. Small amounts of acid, but no gas, are produced from glucose and some other carbohydrates. **The genus is characterized by the following major phospholipids**: phosphatidyl ethanolamine, phosphatidylglycerol, and diphosphatidylglycerol as polar lipids. **The major fatty acids** are $C_{16:0}$, $C_{16:1\,\omega7c}$, and $C_{17:0\,cyclo}$. Only $C_{10:0\,3OH}$ and $C_{12:0\,2OH}$ are synthesized; the $C_{10:0\,3OH}$ are ester- and (presumably) amide-linked, and the $C_{12:0\,2OH}$ fatty acids are (presumptively) ester-linked. Q-8 is the major respiratory lipoquinone. Occur in soil and water and are common in temperate climates and cold climates, but can also be isolated from diseased mushroom tissue.

The mol% G + C of the DNA is: 61–67.

Type species: **Janthinobacterium lividum** (Eisenberg 1891) De Ley, Segers and Gillis 1978, 164 (*Chromobacterium lividum* Bergey, Harrison, Breed, Hammer and Huntoon 1923a, 119; *Bacillus lividus* Eisenberg 1891, 81.)

FURTHER DESCRIPTIVE INFORMATION

Pigmentation and colony characteristics The genus *Janthinobacterium* was created to contain the former oxidative, psychrotrophic, violacein-producing species *Chromobacterium lividum* (Sneath, 1984b) and contained solely that species for several years. A second species—*Janthinobacterium agaricidamnosum*—was proposed by Lincoln et al. (1999) based on the results of a polyphasic study on a group of isolates that provoked soft rot on mushrooms. The new species does not form a violet pigment, although in older cultures, a buff, nonfluorescent pigment is seen; the colonies are also slightly mucoid. The texture is more viscous than for the species *J. lividum*. Consequently, the description of the genus has been emended.

The pigment violacein is produced by *J. lividum* on or in media containing tryptophan. It is soluble in ethanol, but not in water or chloroform, and is readily identified by spectrophotometry and by testing with routine reagents (see "Procedures for Testing for Special Characters" in the chapter *Chromobacterium*). Subcultures of pigmented strains may contain partially or completely nonpigmented colonies. A nonpigmented strain is most readily recognized by a colony morphology similar to that of a pigmented strain isolated at the same time under the same conditions. Once it has been confirmed as an oxidase- and catalase-positive, Gram-negative rod, such an isolate should be stained to determine the type of flagellation. If it possesses the characteristic polar and lateral flagella, it should then be subjected to tests that differentiate the four genera *Chromobacterium*, *Iodobacter*, *Janthinobacterium*, and *Vogesella*—all of which belong to the class *Betaproteobacteria* and can produce a violet pigment (Table BXII.β.22). Although young cultures of *Aeromonas*, *Pseudomonas*, and *Vibrio* spp. grown on solid media sometimes produce lateral flagella in addition to their usual polar flagella, they give patterns of results that differ from those for the violet-pigmented organisms in the differential tests.

The characteristic flagellar arrangement in *J. lividum* is best seen with cells from young cultures on solid media. The single polar flagellum is inserted at the tip of the cell, shows long, shallow waves, and often stains faintly. The lateral flagella are usually long and 1–4 in number, although up to eight may occur. They may be inserted subpolarly or laterally, usually show deep, short waves, and stain readily (see the chapter on *Iodobacter*, Fig. BXII.β.71).

Atypical strains of *J. lividum*, characterized by gelatinous, tough colonies and differing from typical *J. lividum* in several biochemical test reactions (Tables BXII.β.22 and BXII.β.24), represented 16% of the *J. lividum* isolates in the studies of Moss and Ryall (1981) and Logan (1989). Atypical strains may prove difficult to subculture or maintain in the laboratory, and until further strains are isolated and studied, *J. lividum* must be regarded as a heterogeneous species. However, Sneath (1984b) has found that the strains with gelatinous colonies (atypical strains) cross-reacted serologically with other *J. lividum* strains and do not notably differ from them in other ways. The similarity between *J. agaricidamnosum* and the atypical *J. lividum* strains has not been studied.

The second species, *J. agaricidamnosum*, causes soft rot disease on *Agaricus bisporus*, the cultivated mushroom. The cells are motile, but the type of flagellation has not been determined. *J. lividum* strains either cause only a slight necrosis on mushroom tissue blocks or have no effect. *J. agaricidamnosum* and representative strains of *J. lividum* have been studied intensively by Lincoln et al. (1999), and chemotaxonomic and phenotypic characteristics have been found to distinguish both species. Unfortunately, the atypical *J. lividum* strains were not included in this study, and thus it is not possible to decide if some of the atypical strains belong in *J. agaricidamnosum*.

Chemotaxonomic features Whole-cell fatty acid analyses (Moss et al., 1980) have been performed on both species and on their phylogenetic neighbors. Special methods were used, which allowed the selective hydrolysis of ester- and amide-linked fatty acids (B. Tindall, personal communication). The respiratory lipoquinones and polar lipids have also been studied (Lincoln et al., 1999).

Sensitivity to antibiotics Members of *J. lividum* are sensitive to a number of antibiotics (Table BXII.β.24). *J. agaricidamnosum* strains are sensitive to erythromycin (15 μg/disc), streptomycin (10 μg/disc), tetracycline (30 μg/disc), and nalidixic acid (30 μg/disc).

ENRICHMENT AND ISOLATION PROCEDURES

Isolation of *J. lividum* from soil may be accomplished by the method of Corpe (1951). A few crumbs of soil are placed in a Petri dish and 10–25 ml of sterile water is added. About 5–6 grains of heat-sterilized, polished rice are added, and the plates are closed and incubated at 20°C for several weeks. Incubation at 4-10°C may be advantageous. Strains can be isolated by plating from violet patches on the rice onto nutrient agar and incubating at 20°C. Instead of nutrient agar, 0.025% yeast extract in water-agar or a medium consisting of 1% mashed and strained boiled rice in 1.5% water-agar supplemented with L-tryptophan (25 mg/l) may be used to improve pigmentation. *J. lividum* can also be enriched and isolated on the same media as given in detail for *Chromobacterium* (see chapter on *Chromobacterium*). According to Moss and Ryall (1981), *Chromobacterium* spp. grow rapidly and can be selected on a citrate ammonium salt agar medium[1]. Koburger and May (1982) have recovered *J. lividum* from a variety of food samples only on Bennett's agar[2] at 25°C. They found higher counts in water and soil samples on this medium than on the medium of Ryall and Moss (1975) or on *Aeromonas* agar (Rippey and Cabelli, 1979). Maximal numbers of

---

1. Citrate ammonium salt agar is composed of (g/l): NaCl, 1.0; $MgSO_4 \cdot 7H_2O$, 0.2; $NH_4H_2PO_4$, 1.0; K_2HPO_4, 1.0; Ionoagar no. 2 (Oxoid), 15.

2. Bennett's agar consists of (g/l): tryptose, 5; trehalose, 5; yeast extract, 2; NaCl, 3; KCl, 2; $MgSO_4 \cdot 7H_2O$, 0.2; $FeCl_3 \cdot 6H_2O$, 0.1; bromothymol blue, 0.04.

TABLE BXII.β.22. Patterns of results in tests useful for differentiating typical and atypical strains of *Janthinobacterium lividum* from *Chromobacterium*, *Iodobacter*, and *Vogesella* species[a,b]

Characteristic	*J. lividum* (typical)	*J. lividum* (atypical)	*Iodobacter fluviatilis*	*Vogesella*	*C. violaceum*
Number of strains tested	68	14	53	17	9
Pigment produced:					
Indigoidine				+	
Violacein	+	+	+		+
Fluorescence[c]	−[d]			+ (100)	−[d]
Colonies on 1/4 nutrient agar:					
Spreading	− (0)	− (0)	d (83)	nr	− (0)
Gelatinous	d (36)	+ (100)	− (4)	nr	− (0)
Tough	− (7)	d (71)	− (0)	nr	− (0)
Growth at:					
4°C	+ (94)	d (87)	+ (100)	+ (100)	− (0)
37°C	− (0)	− (0)	− (0)	+ (100)	+ (100)
Anaerobic growth	− (3)	− (0)	+ (100)	nr	+ (100)
Nitrate reduction	d (84)	nr	+ (98)	+ (100)	d (87)
Production of indole	− (0)	− (0)	− (0)	+ (100)	− (0)
Hydrolysis of Tween 80	+[d]	nr	nr	− (0)	+[d]
Growth on citrate	+ (95)	+ (94)	d (85)	− (0)	+ (100)
Glucose:					
Fermented	− (0)	− (0)[e]	+ (100)	− (0)	− (0)[f]
Oxidized	+ (97)	d (57)[e]	− (0)	− (0)	− (0)[f]
Acid from:					
L-Arabinose [g]	+ (100)	+ (87)	− (0)	− (0)	− (0)
D-Maltose	+ (98)	+ (94)	+ (100)	nr	− (0)
myo-Inositol [h]	+ (100)	− (0)	− (0)	d (35)	− (0)
Trehalose	− (1)	d (87)	+ (100)	nr	+ (100)
Gelatinase	+ (100)	d (69)	+ (100)	− (0)	+ (100)
Lactate utilization	+ (100)	d (75)	− (0)	nr	+ (100)
Esculin hydrolysis	+ (100)	− (0)	− (0)	nr	− (0)
Arginine hydrolysis	− (0)	− (0)	− (0)	nr	+ (100)
Chitin digestion	− (5)	nr	nr	− (0)	+ (100)[i]
Caseinase	− (5)	− (0)	+ (90)	− (0)	+ (100)
HCN production	− (0)	− (0)	− (0)	nr	+ (100)

[a]Symbols: see standard definitions; nr, data not reported. Numbers in parentheses indicate the % of strains giving a positive reaction.

[b]Data from Logan (1989) with later amendments.

[c]Fluorescence on chalk agar (Starr et al., 1960) is determined with short wavelength illumination. All *Vogesella* strains exhibit very weak fluorescence.

[d]Only tested for the type strain.

[e]Some strains give no reaction in this test.

[f]Occasional strains are oxidative.

[g]Similar patterns of results are obtained with D-cellobiose and D-galactose.

[h]Similar patterns of results are obtained with D-sorbitol.

[i]Negative for the type strain (Grimes et al., 1997).

J. lividum have been isolated at 25°C from turbot gills (Muddaris and Austin, 1988) on a casein–tryptone medium.[3]

J. agaricidamnosum can be isolated by transferring samples of diseased mushroom tissue to sterile 0.25× Ringer's solution and inoculating 10-fold dilution series onto *Pseudomonas* agar F (PAF, Merck) and nutrient agar, with incubation at 25°C for 24 h. Single colonies are isolated from the dilution plates and cultured at 25°C on the appropriate media. Each isolate is first "white line tested" for identity as *P. tolaasii* (Wong and Preece, 1979), a common pathogen for mushrooms, and then tested on mushroom blocks in an initial pathogenicity test (Gandy, 1968). The isolates are plated on *Pseudomonas* isolation agar (PIA, Difco). The cultures are reported to die if left more than 10 d without subculturing.

J. lividum strains have been isolated and identified in microbial mats in Antarctica (Brambilla et al., 2001) and in wet silk thread that became violet in color (Shirata et al., 2000). The question

has arisen as to whether violacein can be used to dye natural and synthetic fibers.

J. lividum JAC1 has also been described as producing metallo-β-lactamase upon exposure to β-lactams. The β-lactamase determinant encodes for a new member of the highly divergent subclass B3 lineage of metallo-β-lactamases (Rossolini et al., 2001).

MAINTENANCE PROCEDURES

Janthinobacterium lividum strains may survive for several years in dilute peptone water (0.1% peptone) at room temperature. They can also be preserved indefinitely by lyophilization or by freezing in nutrient broth containing 15% glycerol. Strains of *J. agaricidamnosum* can be stored freeze-dried or in liquid nitrogen. Working cultures must be subcultured weekly.

PROCEDURES FOR TESTING SPECIAL CHARACTERS

Methods for demonstrating special features in *J. lividum* (production of violacein, flagellar arrangement, growth temperature, oxidase reaction, oxidative attack of glucose, aerobic growth, acid production from carbohydrates, esculin hydrolysis, and lactate

---

3. Casein–tryptone medium (g/l): CaCl$_2$, 1.0; yeast extract, 1.0; beef extract, 5.0; casein, 6.0; tryptone, 2.0; agar no. 1, 15.0; aged sea water, 750 ml; pH 7.2.

TABLE BXII.β.23. Differential characteristics of three typical *J. lividum* strains and all strains of *J. agaricidamnosum* based on respiratory activity using Biolog GN plates[a]

Biolog GN test[b,c]	*J. lividum*	*J. agaricidamnosum*
Gentiobiose	−	d
Lactulose	−	d
D-Trehalose	−	+
Turanose	−	d
D-Glucuronic acid	−	d
m-Inositol	−	d
Urocanic acid	−	d
Dextrin	+	−
N-Acetyl-D-galactosamine	+	−
D-Fucose	+	d
Maltose	+	−
D-Mannose	+	−
D-Sorbitol	+	d
Xylitol	+	−
DL-Lactic acid	+	−
Propionic acid	+	d
Succinic acid	+	−
Succinamic acid	+	−
D-Alanine	+	−
L-Alanine	+	−
L-Asparagine	+	−
L-Aspartic acid	+	−
L-Glutamic acid	+	−
L-Proline	+	−
Inosine	+	−
Tween 40	d	+
Tween 80	d	+
L-Arabinose	d	−
L-Fucose	d	−
D-Galactose	d	−
D-Mannitol	d	−
D-Psicose	d	−
D-Raffinose	d	−
L-Rhamnose	d	−
Acetic acid	d	−
p-Hydroxyphenylacetic acid	d	d
α-Ketobutyric acid	d	−
α-Ketoglutaric acid	d	+
α-Ketovaleric acid	d	−
Malonic acid	d	+
D-Serine	d	−
L-Serine	d	−
L-Threonine	d	−
Glycerol	d	+

[a]Symbols: see standard definitions.

[b]The following tests are negative for all strains: α-xyclodextrin, glycogen, *N*-acetyl-D-glucosamine, adonitol, D-arabitol, cellobiose, *i*-erythritol, α-D-lactose, D-melibiose, methyl-β-D-glucoside, monomethylsuccinate, *cis*-aconitic acid, citric acid, D-galactonic acid lactone, D-galacturonic acid, D-gluconic acid, D-glucosaminic acid, α-hydroxybutyric acid, γ-hydroxybutyric acid, *p*-hydroxyphenylacetic acid, itaconic acid, quinic acid, D-saccharic acid, sebacic acid, glucuronamide, L-histidine, hydroxyl-L-proline, L-leucine, L-ornithine, L-phenylalanine, L-pyroglutamic acid, DL-carnitine, γ-aminobutyric acid, uridine, thymidine, phenylethylamine, putrescine, 2-amino-ethanol, and 2,3-butanediol.

[c]The following tests are positive for all strains: α-D-glucose, sucrose, methylpyruvate, formic acid, β-hydroxybutyric acid, bromosuccinic acid, alaninamide, L-alanylglycine, glycyl-L-aspartic acid, glycyl-L-glutamic acid, DL-glycerol phosphate, glucose-1-phosphate, and glucose-6-phosphate.

utilization) are described in the chapter on the genus *Chromobacterium*.

Janthinobacterium agaricidamnosum can be tested for pathogenicity on mushroom tissue blocks or on growing mushrooms. In the latter experiments, varying percentages of relative humidity (85–95%) have been applied, leading to the conclusion that there is a general trend in which no disease symptoms appear at below 71.5% relative humidity (Lincoln et al., 1999).

A special technique has been used by Lincoln et al. (1999) to study the release of ester-linked or amide-linked fatty acids.

DIFFERENTIATION OF THE GENUS *JANTHINOBACTERIUM* FROM OTHER GENERA

Table BXII.β.22 presents characteristics differentiating typical and atypical strains of *Janthinobacterium* from the other violet-pigment-producing genera of the class *Betaproteobacteria*—i.e., *Chromobacterium*, *Iodobacter*, and *Vogesella*. The percentage fatty acid composition and some phenotypic characteristics that differentiate the genus *Janthinobacterium* from *Duganella*, *Herbaspirillum*, *Oxalobacter*, *Massilia*, and *Paucimonas* are listed in Table BXII.β.25.

Janthinobacterium is phylogenetically more closely related to the following organisms: *Duganella*, *Herbaspirillum*, *Telluria*, *Oxalobacter*, *Massilia*, and to *Pseudomonas lemoignei*, for which a new genus, *Paucimonas*, has recently been proposed (Jendrossek, 2001). *Oxalobacter* can be differentiated from the other genera by the lower mol% G + C of its DNA (48–51), its obligate anaerobic growth requirement, and its strict use of only oxalate and oxamate as sole sources of carbon for growth. *Telluria* can be differentiated from *Janthinobacterium*, *Duganella*, *Paucimonas*, and *Herbaspirillum* by its relatively high mol% G + C (67–72) and its ability to grow on starch and glycogen. Chemotaxonomically, the genera *Janthinobacterium*, *Paucimonas*, *Duganella*, *Massilia*, and *Herbaspirillum* can be differentiated from each other by their quantitative fatty acid compositions (See Table BXII.β.25). Moreover, *Duganella* is the only member of this group to exhibit flocculent growth; *Massilia* has a tendency to form flocs.

TAXONOMIC COMMENTS

Sneath (1984b) described the genus *Chromobacterium* as containing two species: *C. violaceum* and *C. lividum*. The former species was fermentative and mesophilic, growing at 37°C, but not at 4°C, whereas the latter species was nonfermentative and psychrotrophic, with growth occurring at 4°C, but not at 37°C. Both species were described based on the numerical analysis of their phenotypic features. Sneath stressed that both groups were phenotypically very different from each other, but kept them in one genus because, at that time, no genomic results were available. Later genomic studies indicated that the two species differ too much to be contained in one genus, and the new genus *Janthinobacterium* was proposed for the former *Chromobacterium lividum* (De Ley et al., 1978; Sneath, 1984b). The phylogenetic affiliation was determined first by rRNA–DNA hybridization (De Ley et al., 1978) and later by 16S rDNA sequence analysis (Lincoln et al., 1999). *Janthinobacterium* is a member of the class *Betaproteobacteria*, in which it constitutes a cluster together with the genera *Herbaspirillum*, (Baldani et al., 1986a, 1996; Kirchhof et al., 2001) *Duganella* (Hiraishi et al., 1997b), *Oxalobacter* (Allison et al., 1985a; Dehning and Schink, 1989b), *Telluria* (Bowman et al., 1993b), *Massilia* (La Scola et al., 1998a), and *Paucimonas* (Jendrossek, 2001). The highest 16S rDNA sequence similarity is found among *Janthinobacterium*, *Herbaspirillum*, and *Duganella* (95–96.4%), raising the question as to whether the latter genera might be merged. Baldani et al. (1996) considered the single violacein-producing species *J. lividum* to be phenotypically different enough from *Herbaspirillum* to remain in a separate genus. With the inclusion of a second species in *Janthinobacterium*, Lincoln et al. (1999) raised this question again and performed a broad chemotaxonomic study within this rRNA cluster. The results obtained indicated clearly that the genera *Herbaspirillum*,

TABLE BXII.β.24. Characteristics of typical and atypical strains of *Janthinobacterium lividum* and of *Janthinobacterium agaricidamnosum*[a]

Characteristic	*J. lividum* (typical)	*J. lividum* (atypical)	*J. agaricidamnosum*
Formation of cell chains	− (9)	d (81)	nr
Colonies on 1/4 × nutrient agar:			
Spreading	−	−	nr
Gelatinous	d (36)	d (61)	nr
Pigmented	d (78)	d (50)	nr
Zoning of pigment	d (58)	d (61)	nr
Growth on:			
Minimal medium	+ (100)	− (7)	nr
1/4 × NA	d (47)	d (62)	nr
Growth at:			
4°C	+ (100)	d (87)	nr
30°C	d (87)	d (81)	+ (100)
37°C	− (0)	− (0)	+ (100)
Growth in the presence of:			
1% NaCl	+ (98)	d (19)	+ (100)
2% NaCl	d (28)	− (6)	+ (100)
4% NaCl	− (0)	− (0)	− (0)
Growth at:			
pH 4	d (22)	− (6)	+ (100)
pH 5	+ (98)	d (87)	+ (100)
Acid from:			
L-Arabinose	+ (100)	d (87)	nr
D-Cellobiose	+ (100)	+ (94)	nr
D-Fructose	+ (100)	d (12)	nr
D-Galactose	+ (100)	d (87)	nr
Gluconate	− (2)	− (0)	nr
D-Glucose	+ (100)	d (75)	+ (100)
Glycerol	+ (98)	− (0)	nr
Glycogen	− (2)	d (81)	nr
myo-Inositol	+ (100)	− (0)	nr
Inulin	d (52)	d (37)	nr
Lactose	d (55)	d (81)	nr
D-Maltose	+ (98)	+ (94)	nr
D-Mannitol	+ (100)	d (62)	nr
D-Mannose	+ (97)	d (87)	nr
Melezitose	− (0)	− (0)	nr
N-Acetylglucosamine	− (2)	− (0)	nr
D-Raffinose	− or w (30)	− (0)	nr
D-Sorbitol	+ (95)	− (0)	nr
Starch	− (0)	− or w (6)	nr
Sucrose	+ (92)	d (87)	nr
Trehalose	− (2)	d (87)	nr
D-Xylose	+ (100)	d (69)	nr
Catalase, oxidase	+ (100)	+ (100)	+ (100)
Nitrate reduction	d (84)	d (81)	− (0)
Nitrite reduction	d (50)	d (25)	− (0)
Production of HCN	− (0)	− (0)	nr
Egg yolk reaction	− (0)	− (0)	nr
Hemolysis	+ (90)	d (19)	nr
Carbon sources:			
Acetate	+ (100)	d (69)	nr
Citrate	+ (95)	+ (94)	d
Fumarate	+ (100)	+ (94)	nr
Glycerate	d (74)	+ (100)	nr
Lactate	+ (100)	d (75)	nr
Malate	+ (100)	+ (94)	nr
Propionate	+ (100)	d (12)	nr
Pyruvate	+ (100)	+ (100)	nr
Succinate	+ (100)	+ (94)	nr
Tartrate	− (0)	− (0)	nr
Hydrolysis of:			
Arginine	− (0)	− (0)	− (0)
Casein	− (5)	− (0)	nr
Esculin	+ (100)	− (0)	− (0)
Gelatin	+ (100)	d (69)	− (0)
Starch	− (0)	− (0)	nr
Resistant to:			
Ampicillin, 25 μg/disk	d (82)	d (80)	nr
Cephaloridine, 25 μg/disk	d (76)	d (75)	nr
Colistin sulfate, 10 μg/disk	+ (100)	+ (100)	nr

(*continued*)

TABLE BXII.β.24. *(cont.)*

Characteristic	*J. lividum* (typical)	*J. lividum* (atypical)	*J. agaricidamnosum*
Chloramphenicol, 10 μg/disk	d (57)	d (50)	nr
Chlortetracycline, 10 μg/disk	− (6)	− (0)	nr
Furazolidone, 50 μg/disk	d (27)	d (31)	nr
Kanamycin, 30 μg/disk	− (0)	− (0)	nr
Neomycin, 10 μg/disk	− (0)	d (12)	nr
Nalidixic acid, 30 μg/disk	d (12)	d (12)	− (0)
Nitrofurantoin, 200 μg/disk	+ (100)	+ (100)	nr
Oxytetracycline, 10 μg/disk	− (0)	d (18)	nr
Penicillin G, 1.5 IU	+ (100)	+ (94)	nr
Streptomycin, 10 μg/disk	− (0)	− (0)	− (0)
Sulfafurazole, 100 μg/disk	d (60)	− (0)	nr
Sulfafurazole, 500 μg/disk	− (0)	− (0)	nr
Vibriostatic agent O/129, 50 μg/disk	+ (100)	+ (100)	nr

[a]Symbols: see standard definitions; nr, data not reported. Numbers in parentheses indicate the % of strains giving a positive reaction.

Duganella, Oxalobacter, and *Paucimonas* share several chemotaxonomic features, but that the genus *Janthinobacterium* has peculiar features that distinguish it from its neighbors and thus deserve to be maintained in a separate genus. Moreover, Lincoln et al. (1999) have shown that isolates from diseased mushrooms share 99% 16S rDNA sequence similarity with *Janthinobacterium lividum,* thereby leading to the inclusion of this non-violacein-producing group in the genus *Janthinobacterium.* The phenotypic (auxanographic) results, DNA–DNA hybridizations (only 35% DNA–DNA hybridization has been reported between the type strains of both species), and chemotaxonomic results (cyclo-fatty acids and polar lipids) have shown that these mushroom isolates belong in a separate species, for which the name *J. agaricidamnosum* was proposed. Consequently, the genus *Janthinobacterium* has been emended to contain these mushroom isolates as a new species.

The 16S rDNA similarity between *Janthinobacterium, Duganella, Herbaspirillum, Paucimonas, Telluria, Massilia,* and *Oxalobacter* ranges from 92.7–96.4%.

DIFFERENTIATION OF THE SPECIES OF THE GENUS *JANTHINOBACTERIUM*

The phenotypic and chemotaxonomic features that differentiate *J. lividum* and *J. agaricidamnosum* are given in Table BXII.β.24.

Differential tests using the commercial Biolog GN and the API 50CH systems are listed in Tables BXII.β.23 and BXII.β.26.

List of species of the genus *Janthinobacterium*

1. **Janthinobacterium lividum** (Eisenberg 1891) De Ley, Segers and Gillis 1978, 164[AL] (*Chromobacterium lividum* Bergey, Harrison, Breed, Hammer and Huntoon 1923a, 119; *Bacillus lividus* Eisenberg 1891, 81.)

li'vi.dum. L. adj. *lividum* leaden-colored, dark blue.

The description of *J. lividum* is based on the studies of Leifson (1956), Sneath (1956, 1960), De Ley et al. (1978), and Logan (1989). The characteristics are as described for the genus and in Tables BXII.β.22, BXII.β.23, BXII.β.24, BXII.β.25, and BXII.β.26, with the following additional features. Pigmentation may be produced in concentric rings or in sectors within colonies, and subcultures may contain unpigmented colonies. Colonies may be butyrous, gelatinous, or rubbery or may comprise a gelatinous outer layer with butyrous or mucoid growth beneath. Growth in nutrient broth shows a violet ring at the junction of the liquid surface and the wall of the vessel; strains forming gelatinous or rubbery colonies may form a tough, violet pellicle in broth cultures.

Characteristics differentiating typical strains from atypical strains are listed in Tables BXII.β.22 and BXII.β.24.

Motile by means of a single polar flagellum and usually one or more subpolar or lateral flagella. Indole negative, Voges–Proskauer negative. Nitrate and nitrite are reduced, sometimes with visible gas production. Phosphatase positive. Arylsulfatase negative. Other characteristics are shown in Tables BXII.β.23, BXII.β.24, and BXII.β.25. Resistant to benzylpenicillin (10 μg/ml) and O/129 (2,4-diamino-6,7-diisopropyl pteridine) by disc diffusion method, (50 μg/disc).

Janthinobacterium lividum occurs in soil and water and is common in temperate and cold climates. It occasionally causes food spoilage, and these strains may produce active metalloproteinases (Dainty et al., 1978).

The mol% G + C of the DNA is: 61–67 (T_m).

Type strain: H-24, ATCC 12473, CCM 160, DSM 1522, NCIB 9130, NCTC 9796.

GenBank accession number (16S rRNA): Y08846.

2. **Janthinobacterium agaricidamnosum** Lincoln, Fermor and Tindall 1999, 1577[VP]

a.ga'ri.ci.dam.no.sum. L. masc. n. *agaricus* mushroom; L. adj. *damnosum* damaging; *agaricidamnosum* damaging mushroom.

The description of *J. agaricidamnosum* is based on the studies of Lincoln et al. (1999). The characteristics are as described for the genus and in Tables BXII.β.23, BXII.β.24, BXII.β.25, and BXII.β.26, with the following additional features. Low, convex, round, beige colonies are formed on solid media. Maximum growth temperature, 30°C; minimum, 2°C. No growth occurs below pH 5 or in media containing 2.9% NaCl. Acid is produced from glucose. Indole and Voges–Proskauer negative. Nitrate and nitrite are not reduced. No violet pigment is produced. Resistant to penicillin G (10 μg/disc) and vancomycin (30 μg/disc). The chemical composition is as described for the genus.

The type strain was isolated as the cause of soft rot disease of the cultivated mushroom *A. bisporus.*

The mol% G + C of the DNA is: 64.2 (T_m).

Type strain: W1r3, DSM 9628, NCPPB 3945.

GenBank accession number (16S rRNA): Y08845.

TABLE BXII.β.25. Percentage fatty acid composition and some phenotypic characteristics to differentiate members of the genera *Janthinobacterium*, *Duganella*, *Herbaspirillum*, *Oxalobacter*, *Massilia*, and *Paucimonas*[a]

Characteristic	Janthinobacterium[b]	Duganella[c]	Paucimonas[d]	Herbaspirillum[e]	Oxalobacter[f]	Telluria[g]	Massilia[h]
Fatty acid, % of total:							
$C_{10:0}$	0	2	0	0.0–0.9	0	nr	0.68
$C_{10:0\ 3OH}$	3.1–4.0	+ (89)[i]	3.6	3.1–4.0	0	nr	5.66
$C_{12:0}$	3.6–5.2	4	5.3	0.0–4.1	0-1	nr	6.49
$C_{12:0\ 2OH}$	TR-0.94	0	0	0.0–1.1	0	nr	2.77
$C_{12:0\ 3OH}$	0	0	5.7	2.6–3.7	4-6	nr	
$C_{14:0}$	1.0–1.4	3	1.3	0.0–3.5	1-2	nr	3.3
$C_{14:0\ 2OH}$	0	0	4.7	1.0–2.8	8-10	nr	nr
$C_{16:1\ \omega7c}$	22.4–40.7	41[j]	39.3	32.6–39.7	nr	nr	48.3
$C_{16:1}$		0	0	1.9–3.8	Tr-1[j]	nr	nr
$C_{16:0}$	35.3–40.9	38	12.8	19.8–23.7	28-33	nr	21.3
$C_{17:0\ cyclo}$	9.8–26.7	nr	1.9	2.9–5.9	3-34	nr	nr
$C_{18:1\ \omega7c}$	1.38–3.45	12[j]	25.5	19.4–23.2	nr	nr	nr
$C_{18:1}$	nr	nr	nr	0.8-2.7	Tr-13[j]	nr	8.14
$C_{18:0}$	0	nr	0	nr	Tr-5	nr	nr
$C_{19:0\ cyclo}$	0	nr	0	nr	14–38	nr	nr
Growth on:							
D-Ribose	+	−	−	+	−	−	nr
Glycerol	+	−	−	nr	−	v	nr
D-Mannitol	+	−	−	+	−	v	nr
Lactose	−	+	−	nr	−	v	+
Glucose	+	+	−	+	−[e]	+	−
D-Mannose	+	+	−	+	−	+	+
D-Fructose	+	+	−	+	−[e]	+	−
Sucrose	+	+	−	−	nr	−	−
Malate	+	+	−	+	−	nr	+
Starch	−	−	nr	−	−	+	nr
Glycogen	−	−	nr	−	nr	+	nr
Flocculent growth	−	+[k]	−	−	−	−	+[k]
Cell shape:							
Straight rods, sometimes slightly curved	+		+		+	−	nr
Straight rods only	−	+	−	−	+	+	+
Vibrioid to helical	−	−	−	+	−	−	nr
Hydrolysis of starch	−	+	−	nr	nr	+	+
Hydrolysis of gelatin	+	+	−	nr	−[l]	+	+
Mol% G + C of the DNA	61–67	63–64	57–61	60–65	48–51	67–72	64.6

[a]Symbols: see standard definitions; Tr, trace; nr, data not reported,

[b]Data from Lincoln et al. (1999) and from Table BXII.β.22.

[c]Data from Hiraishi et al. (1997b).

[d]Data from Delafield et al. (1965), Mergaert et al. (1996), and Jendrossek (2001).

[e]Data from Baldani et al. (1986a, 1996), and Lincoln et al. (1999).

[f]Data from Allison et al.(1985a), Dawson et al. (1980b), Dehning and Schink (1989b), and Lincoln et al. (1999).

[g]Data from Bowman et al. (1993b).

[h]Data from La Scola et al. (1998a).

[i]Reported as % of total 3-OH component by Hiraishi et al. (1997b).

[j]Hiraishi et al. (1997b) and Allison et al. (1985a) do not distinguish which isomers of 16:0 and 18:1 are present.

[k]Tendency to form flocs.

[l]Available only for *O. formigenes.*

TABLE BXII.β.26. Differential characteristics of three typical *J. lividum* strains and all *J. agaricidamnosum* strains based on carbohydrate utilization in API 50CH[a]

API 50CH test[b,c]	J. lividum	J. agaricidamnosum
Trehalose	−	+
β-Gentiobiose	−	+
D-Arabinose	+	−
L-Arabinose	+	−
D-Xylose	+	−
Galactose	+	−
Sorbitol	+	−
Arbutin	+	−
Salicin	+	−
Cellobiose	+	−
Maltose	+	−
Xylitol	+	−
D-Lyxose	+	−
L-Fucose	+	−

(continued)

TABLE BXII.β.26. *(cont.)*

API 50CH test[b,c]	J. lividum	J. agaricidamnosum
2-Ketogluconate	+	−
Rhamnose	d	−
N-Acetylglucosamine	d	−
Inulin	d	−
D-Raffinose	d	−
Citrate	+	d
Phenylacetate	+	−
Caprate	d	−
Adipate	d	−

[a]Symbols: see standard definitions.

[b]The following tests are negative for all strains: assimilation of gluconate, erythritol, L-xylose, adonitol, methyl-β-xyloside, L-sorbose, dulcitol, methyl-α-D-mannoside, methyl-α-D-glucoside, amygdalin, esculin, lactose, melibiose, melezitose, amidon, glycogen, D-turanose, D-tagatose, D-fucose, and L-arabitol.

[c]The following tests are positive for all strains: assimilation of glucose, malate, mannose, mannitol, glycerol, ribose, D-fructose, inositol, saccharose, and D-arabitol.

Genus V. **Massilia** *La Scola, Birtles, Mallet and Raoult 2000, 423^{VP} (Effective publication: La Scola, Birtles, Mallet and Raoult 1998a, 2852)*

THE EDITORIAL BOARD*

Mas.sil' i.a. L. n. *Massilia* Latin name of Marseille, France.

Gram-negative nonsporeforming motile rods (1.0 × 3.0 μm). **Strictly aerobic**; catalase positive and oxidase negative. **Grow in 3% NaCl.** No denitrification. Produce arginine dihydrolase and urease. Hydrolyze esculin, gelatin, and starch. **No acid from carbohydrates.** Utilize L-alanine, L-arabinose, D-cellobiose, fumarate, D-galacturonate, gentisate, D-glucuronate, α-ketoglutarate, α-lactose, D-malate, L-malate, malonate, maltose, D-mannose, protochachuate, α-L-rhamnose, succinate, D-tagatose, and D-xylose. Does not utilize D-alanine, L-aspartate, citrate, D-fructose, D-galactose, D-glucose, L-glutamate, DL-glycerate, β-hydroxybutyrate, DL-lactate, D-melezitose, mucate, D-saccharate, L-serine, sucrose, L-tartrate, or D-trehalose. Major fatty acids are $C_{16:1\,\omega7c}$, $C_{16:0}$, $C_{18:1}$, $C_{12:0}$, $C_{10:0\,3OH}$, $C_{14:0}$, and $C_{12:0\,2OH}$.

The mol% G + C of the DNA is: 64 ± 1.8 (HPLC).

Type species: **Massilia timonae** La Scola, Birtles, Mallet and Raoult 2000, 423 (Effective publication: La Scola, Birtles, Mallet and Raoult 1998a, 2852).

FURTHER DESCRIPTIVE INFORMATION

Colonies are straw-colored; flocs or films form in liquid medium; older cultures at 37°C may contain filamentous cells. Growth occurs on MacConkey agar, Columbia agar containing 5% sheep blood, chocolate agar, and trypticase soy agar.

The organism was isolated from the blood of an immunocompromised patient suffering from meningoencephalitis. The infection was viewed as opportunistic because the 16S rDNA sequence obtained from the organism was most similar to those of the soil microorganisms *Telluria mixta, Telluria chitinolytica*, and *Duganella zoogloeoides*, which are not known to be pathogenic to humans. Other characteristics, including biochemical test profiles and mol% G + C of the DNA, were consistent with this relationship.

ENRICHMENT AND ISOLATION PROCEDURES

The organism was isolated from blood using an automated blood culture system (BACTEC NR-860).

Editorial Note: The description in this chapter is based on a single strain—the type strain. Lindquist, et al., 2003, J. Clin. Microbiol. *41:* 192–196 reported the isolation of additional strains and proposed an emended species description to include them.

List of species of the genus Massilia

1. **Massilia timonae** La Scola, Birtles, Mallet and Raoult 2000, 423^{VP} (Effective publication: La Scola, Birtles, Mallet and Raoult 1998a, 2852)
ti.mon.ae. L. gen. n. belonging to Timone, because the organism was isolated from a patient at L'Hôpital de la Timone.

The species description and the genus description are identical.
The mol% G + C of the DNA is: 64 ± 1.8 (HPLC).
Type strain: UR/MT95, CIP 105350.
GenBank accession number (16S rRNA): U54470.

Genus VI. **Telluria** *Bowman, Sly, Hayward, Spiegel and Stackebrandt 1993b, 123^{VP}*

LINDSAY I. SLY AND MARK FEGAN

Tel.lu' ri.a. L. fem. n. *Tellus* a Roman goddess of the earth, also the ground or earth; M.L. fem. n. *Telluria* from the earth.

Cells are **rod-shaped**, 2–3 × 0.5–1.0 μm. **Gram negative**. Filamentous cells up to 30 μm long are formed occasionally in older cultures. **Cells occur singly, in pairs, or in short chains. Motile.** Exhibit **mixed flagellation**; cells in liquid media possess a single polar flagellum, while on solid media additional lateral flagella of shorter wavelength occur. **Accumulate poly-β-hydroxybutyrate. Strictly aerobic. Oxidative** metabolism. **Catalase and oxidase positive.** Surface pellicle formed in static liquid cultures. **Chemoheterotrophic.** Unable to grow chemolithotrophically with hydrogen. **Denitrification does not occur. Arginine dihydrolase is absent.** Good growth occurs on media containing carbohydrates and an inorganic or organic combined nitrogen source. Poor growth occurs on media lacking carbohydrates. **NaCl sensitive**; completely inhibited by NaCl concentrations greater than 1.5%, and only poor growth occurs at an NaCl concentration of 0.5%. **Utilize complex polysaccharides**, including starch and xylan. Cellulose is not hydrolyzed. Hydrolyze gelatin, casein, DNA, esculin, and Tween 40, Tween 60, and Tween 80. Produce phosphatase and arylsulfatase. Grow well at temperatures between 20° and 45°C and optimally at 30–35°C; optimal growth occurs at pH 7.0. The **major quinone is ubiquinone 8**. Belongs to the *Betaproteobacteria*. The only known habitat is soil, particularly the rhizosphere.

The mol% G + C of the DNA is: 67–72.

Type species: **Telluria mixta** (Bowman, Sly and Hayward 1989) Bowman, Sly, Hayward, Spiegel and Stackebrandt 1993b, 124 (*Pseudomonas mixta* Bowman, Sly and Hayward 1989, 205.)

FURTHER DESCRIPTIVE INFORMATION

The species of *Telluria* are characterized by their ability to degrade a range of complex carbohydrates. *T. mixta* strains are able to degrade dextran, starch, inulin, pectate, and xylan, but are unable to degrade cellulose or chitin. A few strains attack alginate and xanthan gum. The first isolate of *T. mixta* (*Pseudomonas* sp. ACM 733, UQM 733, ATCC 49107) was isolated from a sugar cane rhizosphere in Queensland, Australia (Richards and Streamer, 1972). A subsequent taxonomic study of similar dextranolytic strains led to the description of *Pseudomonas mixta* (Bowman et al., 1988, 1989), later transferred to the new genus *Telluria* based on phylogenetic evidence (Bowman et al., 1993b). The dextranases and other carbohydrate-degrading enzymes produced by strain UQM 733 have been studied extensively (Richards and Streamer, 1972; Covacevich and Richards, 1978, 1979). The only known strain of *T. chitinolytica* was isolated from soil in Israel and demonstrated to have bionematicidal activity and potential for the biocontrol of the root-knot nematode *Meloidogyne javinica* (Spiegel et al., 1991).

A distinctive morphological characteristic of the genus is mixed flagellation exhibited when cultures are grown on solid agar media (Bowman et al., 1988). Cells possess a single polar flagellum of long wavelength (2.6 μm) and at least two lateral

flagella of shorter wavelength (1.2 μm) (Fig. BXII.β.18). Cells usually possess uniform morphology, but occasionally may be distended by PHB inclusions or form longer cells (Fig. BXII.β.19).

Telluria strains have a predilection for carbohydrates and tricarboxylic cycle intermediates as sources of carbon and energy, although *T. chitinolytica* is slightly less biochemically versatile than *T. mixta*. The spectrum of polysaccharide degradation by *T. chitinolytica* is quite different from that of *T. mixta*, the former able to hydrolyze chitin, but not dextran or pectate (Table BXII.β.27).

Both species utilize D-xylose, L-rhamnose, D-fructose, D-galactose, D-glucose, D-mannose, D-melezitose, maltose, sucrose, cellobiose, trehalose, inulin, starch, xylan, gluconate, 2-ketogluconate, glucuronate, galacturonate, mucate, saccharate, β-hydroxybutyrate, malonate, succinate, DL-lactate, fumarate, DL-malate, pyruvate, L-(+)-tartrate, citrate, D-alanine, L-alanine, L-aspartate, and L-glutamate as sole carbon and energy sources. Unable to utilize D-ribose, acetate, propionate, isobutyrate, valerate, isovalerate, caproate, pelargonate, caprate, oxalate, glutarate, adipate, pimelate, suberate, azelate, sebacate, maleate, D-(−)-tartrate, meso-tartrate, citraconate, itaconate, mesaconate, levulinate, ethanol, propanol, butanol, 1,2-ethanediol, 1,3-propanediol, 2,3-butanediol, adonitol, *meso*-erythritol, *meso*-inositol, sorbitol, anthranilate, *p*-aminobenzoate, benzylamine, sarcosine, betaine, β-ala-

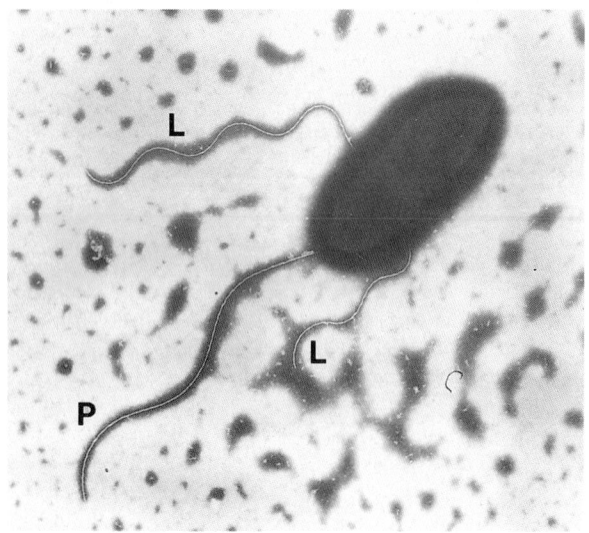

FIGURE BXII.β.18. Electron micrograph of *T. mixta* ACM 1762 showing mixed flagellation. *P*, polar; *L*, lateral flagellum. (Reprinted with permission from J.P. Bowman et al., Systematic and Applied Microbiology *11*: 53–59, 1988, ©Urban & Fischer Verlag GmbH & Co, KG, Jena, Germany.)

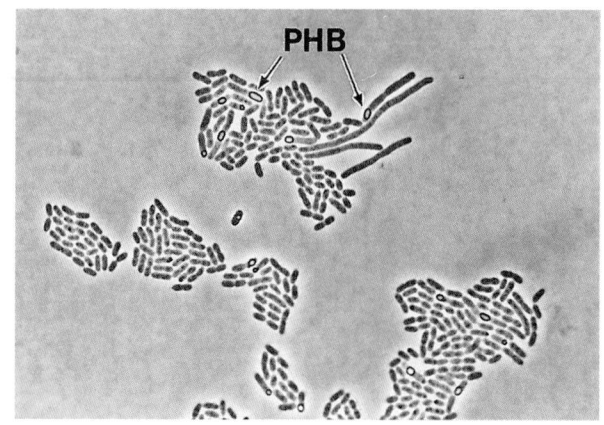

FIGURE BXII.β.19. Microcolony of *T. mixta* ACM 1762. PHB, poly-β-hydroxybutyrate granules. (Reprinted with permission from J. P. Bowman et al., Systematic and Applied Microbiology *11*: 53-59, 1988, ©Urban & Fischer Verlag GmbH & Co, KG, Jena, Germany.)

TABLE BXII.β.27. Differential and descriptive characteristics of the species of *Telluria*[a]

Characteristic	T. mixta	T. chitinolytica
Yellow pigmentation	−	+
Growth at 45°C	−	+
Nitrate reduced to nitrite	d	−
Urease and β-galactosidase activities	+	−
Lecithinase activity (egg yolk)	−	+
Hydrolysis of:		
Chitin and tributyrin	−	+
Dextran, pectate (pH 5 to 8.3), and Tween 20	+	−
Utilization of:		
DL-Arabinose, lactose, dextran, butyrate, DL-glycerate, glycolate, benzoate, *p*-hydroxybenzoate, DL-serine, DL-threonine, L-arginine, L-citrulline, L-ornithine, and L-phenylalanine	+	−
Glycerol and acetamide	−	+
L-Tyrosine, poly-β-hydroxybutyrate, heptanoate, caprylate, mannitol, phenol, *m*-hydroxybenzoate, benzoyl-formate, DL-mandelate, and quinate	d	−
L-Histidine	d	+

[a]Data from Bowman et al. (1988).

nine, DL-2-aminobutyrate, DL-4-aminobutyrate, DL-2-aminovalerate, DL-5-aminovalerate, L-valine, L-proline, L-hydroxyproline, L-glutamine, L-lysine, L-tryptophan, L-methionine, L-cystein, putrescine, spermine, ethanolamine, histamine, α-amylamine, tryptamine, and pantothenate (Bowman et al., 1993b).

Other biochemical differences include a lack of urease and β-galactosidase activities in *T. chitinolytica*, while the lipolytic activity (lecithinase and lipase) of *T. chitinolytica* is more extensive than *T. mixta*. The carbon source utilization pattern of *T. mixta* (57 of 118 compounds utilized) is more extensive that *T. chitinolytica* (36 of 118 compounds utilized). Aromatic compounds are not utilized by *T. chitinolytica* (Bowman et al., 1993b).

Although no further isolations of *Telluria* have been published, a partial 16S rRNA gene sequence of *T. mixta* from a bacterial endophyte in potato was submitted to GenBank (GenBank accession number AF297697) by van Elsas, van Overbeek and Garbeva (MIBU, PRI, Binenhaven 5, Wageningen 6700, The Netherlands) in 2000 supporting the association of the species with the plant rhizosphere.

ENRICHMENT AND ISOLATION PROCEDURES

Cultures of *T. mixta* are best enriched from soil or rhizosphere in a medium containing dextran minerals salts (Blackall et al., 1985; Bowman et al., 1988). Isolation of pure cultures is readily achieved by plating from the enrichment on dextran agar and observing for colonies surrounded by a clear zone after flooding with ethanol indicative of dextran hydrolysis (Bowman et al., 1988). Colonies are circular, low convex, with a rough or smooth surface, white to tan in color, and may have a highly elastic and cartilaginous consistency. *T. chitinolytica* was first isolated from potted soil emended with powdered crustacean shells kept in a glasshouse at 27–29°C for up to 45 d. After enrichment, isolation was made by plating on agar plates containing 0.2% (w/v) colloidal chitin as the sole carbon source and selecting yellow-pigmented colonies producing a halo of chitin degradation (Spiegel et al., 1991). Colonies of *Telluria* on isolation are highly elastic and cartilaginous.

MAINTENANCE PROCEDURES

Media containing carbohydrates are required for good growth and maintenance. Sucrose peptone agar (Hayward, 1964) is a good general medium for maintenance, and good growth occurs on glucose nitrate medium. Dextran agar and chitin agar are also suitable for *T. mixta* and *T. chitinolytica*, respectively, and for checking culture purity (Bowman et al., 1988, 1993b). Cultures may be preserved by cryogenic storage in liquid nitrogen when suspended in sucrose peptone broth containing 10% glycerol, and by freeze drying in glucose peptone broth containing horse serum.

DIFFERENTIATION OF THE GENUS *TELLURIA* FROM OTHER GENERA

The primary characters for differentiation of *Telluria* from other high mol% G + C, strictly oxidative aerobes in the *Betaproteobacteria* are possession of mixed flagellation on solid agar and the inability to grow well on nutrient agar without the addition of carbohydrates. Additional characters useful for differentiation from *Ralstonia*, *Burkholderia*, *Comamonas*, *Delftia*, *Acidovorax*, *Hydrogenophaga*, *Variovorax*, and *Xylophilus* are given in Table BXII.β.28.

TAXONOMIC COMMENTS

Phylogenetically, the genus *Telluria* belongs to a distinct, well-supported branch in the *Betaproteobacteria* comprising the two species of *Telluria* with 97% rDNA sequence similarity, and the species *Duganella zoogloeoides*, *Janthinobacterium lividum*, *Herbaspirillum* species, and *Paucimonas lemoignei* (Fig. BXII.β.20). The nearest relatives of *Telluria* species are *Duganella zoogloeoides* and *Janthinobacterium lividum*, which are only moderately related at 93% sequence similarity to *Telluria*. These phylogenetic relationships confirm studies by Bowman et al. (1993b), Anzai et al. (2000), and Jendrossek (2001).

DIFFERENTIATION OF THE SPECIES OF THE GENUS *TELLURIA*

General characteristics of the species are given in the generic description. Differential characteristics of the species of *Telluria* are indicated in Table BXII.β.27, and additional characteristics are provided in the section on further descriptive information.

List of species of the genus Telluria

1. **Telluria mixta** (Bowman, Sly and Hayward 1989) Bowman, Sly, Hayward, Spiegel and Stackebrandt 1993b, 124[VP] (*Pseudomonas mixta* Bowman, Sly and Hayward 1989, 205.)
   mix′ ta. L. adj. *mixtus* mixed; M.L. fem. adj. *mixta* mixed, referring to mixed flagellation.

   Colonies are circular, low convex, with a rough or smooth surface, white to tan in color, and may have a highly elastic and cartilaginous consistency. Unable to grow at 45°C. Nitrate is reduced to nitrite. Urease and β-galactosidase positive. Lecithinase not produced. Dextran, pectate, and Tween 20 are hydrolyzed, but not chitin or tributyrin. DL-Arabinose, lactose, dextran, butyrate, DL-glycerate, glycolate, benzoate, *p*-hydroxybenzoate, DL-serine, DL-threonine, L-arginine, L-citrulline, L-ornithine, and L-phenylalanine are utilized as carbon sources, but not glycerol or acetamide. L-Tyrosine, poly-β-hydroxybutyrate, heptanoate,

   caprylate, mannitol, phenol, *m*-hydroxybenzoate, benzoylformate, DL-mandelate, and quinate are utilized by most strains. Isolated from various soils in Queensland, Australia.

   The mol% G + C of the DNA is: 69 (T_i).
   Type strain: ACM 1762, ATCC 49108, UQM 1762.
   GenBank accession number (16S rRNA): X65589.

2. **Telluria chitinolytica** Bowman, Sly, Hayward, Spiegel and Stackebrandt 1993b, 124[VP]
   chi.tin.o.lyt′ i.ca. chitin clinical term for a polysaccharide; Gr. adj. *lytos* soluble; M.L. adj. *lytos* soluble; M.L. fem. adj. *chitinolytica* dissolving chitin.

   Colonies are yellow-pigmented and on isolation are highly elastic and cartilaginous. Able to grow at 45°C. Nitrate is not reduced to nitrite. Urease and β-galactosidase negative. Lecithinase is produced. Chitin and tributyrin are

TABLE BXII.β.28. Differentiation of the genus *Telluria* from other high mol% G + C, strictly oxidative aerobes in the *Betaproteobacteria*[a,b]

Characteristic	Telluria	Ralstonia	Burkholderia	Comamonas	Delftia	Acidovorax	Hydrogenophaga	Variovorax	Xylophilus
Flagellation:									
Mixed (1 polar and >1 lateral)	+	−	−	−	−	−	−	−	−
Polar (≥1)	+	D	D	+	+	+	+	−	+
Peritrichous	−	D	−	−	−	−	−	+	−
Bipolar tufts	−	−	−	+	+	−	−	−	−
Pigments	D	D	D	−	−	−	+	+	+
Hydrogen autotrophy	−	D	−	−	−	D	+	d	−
Oxidase	+	D	D	+	+	+	+	+	−
Hydrolysis of starch	+	D	D	−	−	−	−	−	d
Growth on nutrient agar	−	+	+	+	+	+	+	+	
Tolerance to 1.5% NaCl	−	+		+	+	+	+	+	−
Utilization of:									
Glucose	+	D	+	−	−	D	+	+	+
Fructose	+	D	+	−	−	+	+	D	−
Occurrence[c]	S	S, FW, CS, IP	S, FW, CS, IP	S, FW, CS	S, FW, CS	S, FW, CS	S, FW	S, FW	IP
Mol% G + C	67–72	64–69	59–70	63–66	67–69	62–66	65–69	66–68	68–69

[a]Symbols: +, present in all species; −, absent in all species; (+), weak reaction; d,11–89% of strains are positive; D, variable reaction in different species.

[b]Data from Bradbury (1984), Kersters and De Ley (1984b), Palleroni (1984), De Vos et al. (1985b), Tamaoka et al. (1987), Willems et al. (1987, 1989, 1990, 1991a, c, 1992a), Bowman et al. (1988, 1993b), Gillis et al. (1995), Urakami et al. (1995a), Yabuuchi et al. (1995), Zhang et al. (2000a), Brämer et al. (2001), Chen et al. (2001), Coenye et al. (2001a, c), and Goris et al. (2001).

[c]S, soil; FW, fresh water; CS, clinical sample; IP, infected plants.

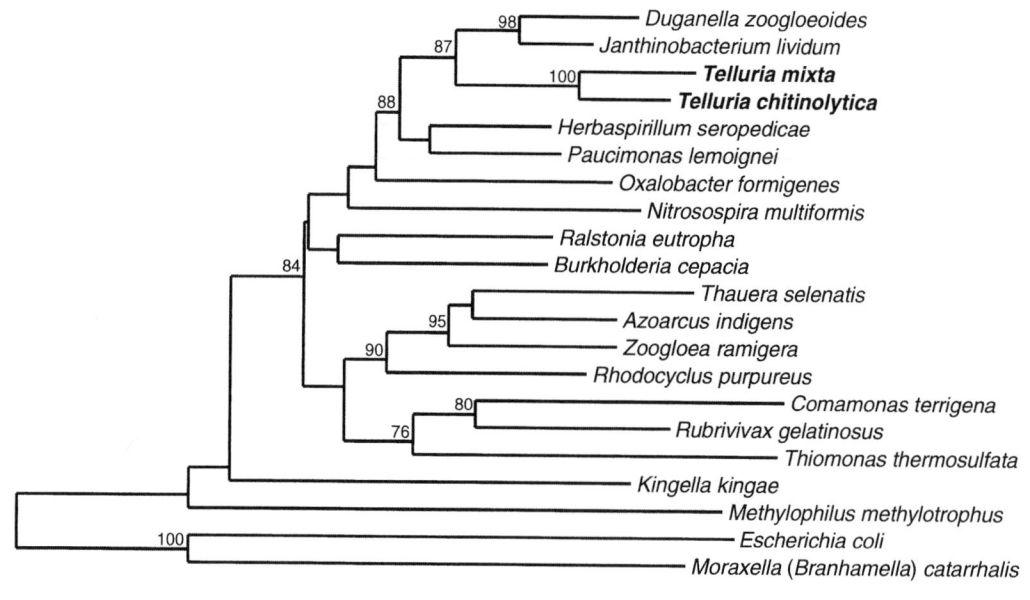

FIGURE BXII.β.20. Neighbor-joining tree reconstructed from 16S rRNA gene sequences using the PHYLIP programs (Felsenstein, 1993) as implemented in the ARB program (Software available from O. Strunk and W. Ludwig: ARB: a software environment for sequence data). Bootstrap values of 100 resamplings are shown at the branch points. Scale bar represents 1 nucleotide substitution per 100 nucleotide positions. The tree was constructed using the following sequences: *Duganella zoogloeoides* IAM 12670[T], D14256; *Janthinobacterium lividum* DSM 1522[T], Y08846; *Telluria mixta* ACM 1762[T], X65589; *Telluria chitinolytica* ACM 3522[T], X65590; *Herbaspirillum seropedicae* ATCC 35892[T] (DSM 6445[T]), Y10146; *Paucimonas lemoignei* ATCC 17989[T], AB021375; *Oxalobacter formigenes* OxB[T], U49757; *Nitrosospira multiformis* ATCC 25196[T], L35509; *Ralstonia eutropha* ATCC 17697[T], 335 (Stanier), M32021; *Burkholderia cepacia* ATCC 25416[T], M22518; *Thauera selenatis* ATCC 55363[T], X68491; *Azoarcus indigens* VB 32[T], L15531; *Zoogloea ramigera* ATCC 19544[T] (NCIMB 10706[T]) X74913; *Rhodocyclus purpureus* DSM 168[T], M34132; *Comamonas terrigena* IMI 359870[T], AF078772; *Rubrivivax gelatinosus* ATCC 17011[T] (DSM 1709[T]) D16213; *Thiomonas thermosulfata* ATCC 51520[T], U27839; *Kingella kingae* ATCC 23330[T], M22467; *Methylophilus methylotrophus* AS1[T], M29021; *Escherichia coli* ATCC 11775[T], X80725; *Moraxella catarrhalis* ATCC 25238[T], U10876.

hydrolyzed, but not dextran, pectate, or Tween 20. Glycerol and acetamide are utilized as carbon sources, but not DL-arabinose, lactose, dextran, butyrate, DL-glycerate, glycolate, benzoate, *p*-hydroxybenzoate, DL-serine, DL-threonine, L-arginine, L-citrulline, L-ornithine, and L-phenylalanine. L-tyrosine, poly-β-hydroxybutyrate, heptanoate, caprylate, mannitol, phenol, *m*-hydroxybenzoate, benzoyl-formate, DL-mandelate, and quinate are not utilized. The type strain was isolated from a loamy soil from Bet Dagan, Israel.

The mol% G + C of the DNA is: 72 (T_i).

Type strain: ACM 3522, CNCM I-804.

GenBank accession number (16S rRNA): X65590.

Family III. **Alcaligenaceae** De Ley, Segers, Kersters, Mannheim and Lievens 1986, 412^VP

HANS-JÜRGEN BUSSE AND GEORG AULING

Al.cal.li.ge.na′ce.ae. M.L. masc. n. *Alcaligenes* type genus of the family; M.L. masc. pl. n. *Alcaligenaceae* the *Alcaligenes* family.

Rods or coccobacilli 0.2–1.0 × 0.5–2.6 μm, occurring singly, in pairs, or rarely in chains. **Gram negative**. No resting stages are known. Motile by peritrichous flagella or nonmotile (*Bordetella parapertussis* and *Bordetella pertussis*). Aerobic, having a **strictly respiratory** type of metabolism with oxygen as the terminal electron acceptor. Some strains can use nitrate or nitrite as an alternative electron acceptor, allowing growth to occur anaerobically. Optimal temperature, 30–37°C. Colonies are usually nonpigmented. Most species are oxidase and catalase positive. Alkaline reaction in litmus milk. Gelatin not hydrolyzed. **Chemoorganotrophic**. Most species utilize a variety of organic acids and amino acids as carbon sources. Carbohydrates are usually not utilized. *Achromobacter xylosoxidans* subsp. *xylosoxidans* produces acid from D-glucose and D-xylose and utilizes both of these carbohydrates as carbon sources. Some species require nicotinamide, organic sulfur (e.g., cysteine), and organic nitrogen (amino acids).

As demonstrated by DNA–rRNA hybridization experiments (De Ley et al., 1986) and 16S rDNA gene sequence analyses (Vandamme et al., 1996a), the family *Alcaligenaceae* belongs to the *Betaproteobacteria*. In general, species of the family *Alcaligenaceae* are characterized by **ubiquinone-8** (Q-8) as the major isoprenoid quinone (Fletcher et al., 1987; Oyaizu-Masuchi and Komagata, 1988). Species analyzed for their polyamine content contain **putrescine** as the major compound and the β-subclass-specific polyamine **2-hydroxyputrescine** (HPUT) (Busse and Auling, 1988; Hamana and Takeuchi, 1998). All species contain in their fatty acid profile the major compounds **hexadecanoate** ($C_{16:0}$) and **cycloheptadecanoate** ($C_{17:0\ cyclo}$), but *Bordetella pertussis* lacks the acid $C_{17:0\ cyclo}$ (Oyaizu-Masuchi and Komagata, 1988; Vancanneyt et al., 1995; Weyant et al., 1995a; Vandamme et al., 1996a).

Species of the genera *Alcaligenes* and *Achromobacter* occur in soil, water, and the hospital environment and have been isolated from human clinical material. Species of the genus *Bordetella* have been isolated from humans and warm-blooded animals but not all are pathogenic for warm-blooded animals (*Bordetella trematum*). The single species of the genus *Pigmentiphaga* has been isolated from soil.

The mol% G + C of the DNA is: 56–70.

Type genus: **Alcaligenes** Castellani and Chalmers 1919, 936.

FURTHER DESCRIPTIVE INFORMATION

The family *Alcaligenaceae* was proposed by De Ley et al. (1986) to encompass the genera *Alcaligenes* and *Bordetella*, which were shown by DNA–rRNA hybridization to group in one cluster within the *Betaproteobacteria*. Phylogenetic analyses based on 16S rDNA sequence data confirm this clustering (Weyant et al., 1995a). Recently, the species *Alcaligenes denitrificans*, *Alcaligenes xylosoxidans*, and *Alcaligenes piechaudii* have been reclassified in the genus *Achromobacter* (Yabuuchi et al., 1998a). Sequence alignment to 120 primary structures of 23S rDNAs from other members of the domain *Bacteria* representing all known phyla confirmed a closer relationship between *Bordetella* species and the type species of the genus, *Alcaligenes faecalis*, in a stable subtree comprising the *Betaproteobacteria* (Ludwig et al., 1995). In this context it is of interest that species of both *Achromobacter* and *Bordetella* produce the macrocyclic siderophore alcaligin, which was originally isolated from *Alcaligenes xylosoxidans* (Nishio et al., 1988; Nishio and Ishida, 1990), and the alcaligin biosynthesis genes have been characterized in *Bordetella pertussis* (Kang et al., 1996).

On the other hand, 16S rDNA sequence data also demonstrate the distance of certain species to the family *Alcaligenaceae*, previously transferred from the genus *Alcaligenes* to other genera outside the family (Auling et al., 1988; Willems et al., 1991a; Dobson et al., 1993; Meyer et al., 1993; Yabuuchi et al., 1995; Dobson and Franzmann, 1996), such as *Ralstonia eutropha* (basonym *Alcaligenes eutrophus*), *Variovorax paradoxus* (basonym *Alcaligenes paradoxus*), *Halomonas aquamarina* (basonym *Alcaligenes aquamarinus*; basonym *Deleya aquamarina*), *Halomonas cupida* (basonym *Alcaligenes cupidus*; basonym *Deleya cupida*), *Halomonas pacifica* (basonym *Alcaligenes pacificus*; basonym *Deleya pacifica*), *Halomonas venusta* (basonym *Alcaligenes venustus*; basonym *Deleya venusta*), and *Carbophilus carboxidus* (formerly "*Alcaligenes carboxidus*"). The species reclassified as members of the genera *Ralstonia*, *Halomonas*, and *Carbophilus* can be easily distinguished from the genus *Alcaligenes* based on their fatty acid profiles, quinones, and/or polyamine patterns (Table BXII.β.29). *V. paradoxus* might be distinguished from members of the family *Alcaligenaceae* by its yellow pigmentation, utilization of the carbon sources D-galactose, L-arabinose, D-mannitol, and *p*-hydroxybenzoate, and the presence of urease (Kersters and De Ley, 1984b). Results from DNA–rRNA hybridization studies have demonstrated that *Alcaligenes latus* is most likely misnamed and a member of an as yet undescribed genus (Willems et al., 1991b). Differentiation of *A. latus* from the species of the genus *Alcaligenes* may be achieved by its ability to utilize maltose, sucrose, 2-ketogluconate, *p*-hydroxybenzoate, butylamine, betaine, sarcosine, and creatine as a carbon source, by chemolithotrophic growth with molecular hydrogen, and by hydrolysis of gelatin, starch, and Tween 80 (Kersters and De Ley, 1984b).

The genus *Bordetella* consists of seven species: *B. pertussis* (Moreno-López, 1952), *B. parapertussis* (Eldering and Kendrick, 1938), *B. bronchiseptica* (Moreno-López, 1952), *B. avium* (Kersters et al., 1984), *B. hinzii* (Vandamme et al., 1995c), *B. holmesii* (Weyant et al., 1995a), and *B. trematum* (Vandamme et al., 1996a). DNA–DNA hybridization (Kloos et al., 1981), multilocus enzyme electrophoresis (Musser et al., 1986), ERIC-PCR, and ARDRA (Vandamme et al., 1997a) studies have shown that the species *B. pertussis*, *B. parapertussis*, and *B. bronchiseptica* are closely related and may be considered different subspecies of a single species. In contrast, macrorestriction digests resolved by pulsed-field gel electrophoresis (Khattak and Matthews, 1993), whole-cell protein electrophoresis, and fatty acid profiles (Vancanneyt et al., 1995) generated species-specific profiles for these three species, supporting their present status as different species. Interestingly, it has been independently shown by two groups (Vancanneyt et al., 1995; Weyant et al., 1995a) that *B. pertussis* lacks the fatty acid $C_{17:0\ cyclo}$, which is a predominant compound in the fatty acid profile of the majority of members of the family. Thus, the important human pathogen *B. pertussis* can be easily identified by fatty acid analysis. The taxonomic status of the remaining species *B. avium*, *B. hinzii*, *B. holmesii*, and *B. trematum* is not a matter of discussion. These four species are phenotypically and genotypically sufficiently distinct from each other and from the other three species to justify their status as single species (Vandamme et al., 1997a).

Until the recent reclassification of the species *Alcaligenes denitrificans*, *Alcaligenes xylosoxidans*, and *Alcaligenes piechaudii* to the genus *Achromobacter* as *Achromobacter xylosoxidans* subsp. *denitrificans*, *Achromobacter xylosoxidans* subsp. *xylosoxidans*, and *Achromobacter piechaudii* (Yabuuchi et al., 1998a), taxonomic confusion arose from the phylogenetic clustering of species within the family. Based on 16S rDNA sequence comparisons, the species *Achromobacter xylosoxidans* subsp. *denitrificans*, *Achromobacter xylosoxidans* subsp. *xylosoxidans*, and *Achromobacter piechaudii* are more closely related to the type species of the genus *Bordetella* than to the type species of the genus *Alcaligenes*, *Alcaligenes faecalis* (Fig. BXII.β.21). Likewise, the newly described species *Alcaligenes defragrans* (Foss et al., 1998a) is phylogenetically clearly separated from all genera within the family *Alcaligenaceae* (Blümel et al., 2001b). Thus, it is most likely that *A. defragrans* will be reclassified as a member of a new genus within the family. The DNA base ratios reflect the separate position of *A. faecalis*. While *A. xylosoxidans* subsp. *denitrificans*, *A. xylosoxidans* subsp. *xylosoxidans*, *A. ruhlandii*, *A. piechaudii*, and *A. defragrans* have a mol% G + C content (64–70%) in the range of *Bordetella* species (62–70%), a lower DNA base ratio of 56–60% has been reported for *A. faecalis* (De Ley et al., 1986). Thus, the generic homogeneity within the family has been enhanced by removal of *A. denitrificans*, *A. xylosoxidans*, and *A. piechaudii* from the genus *Alcaligenes*. From the phylogenetic point of view, it would be acceptable to transfer the species *A. xylosoxidans* subsp. *denitrificans*, *A. xylosoxidans* subsp. *xylosoxidans*, *A. piechaudii*, and *A. defragrans* into the genus *Bordetella*. One major obstacle to this reclassification is the definition of the genus *Bordetella*, which describes its members as mammalian parasites and pathogens living among epithelial cilia of the respiratory tract (Pittman, 1984b). However, no association with the respiratory tract of any of the *Alcaligenes* and *Achromobacter* species in question has been described so far. Only *Achromobacter xylosoxidans* subsp. *xylosoxidans* is an important opportunistic pathogen as concluded from numerous reports (Reverdy et al., 1984; Chandrasekar et al., 1986; Reina et al., 1988;

Schoch and Cunha, 1988; Legrand and Anaissie, 1992; Cheron et al., 1994; Dunne and Maisch, 1995; Knippschild et al., 1996). The species *A. ruhlandii* and *A. defragrans* have not been isolated from clinical specimens (Foss et al., 1998a). In the last decade, new *Bordetella* species were isolated from the respiratory tract of birds, human blood cultures, ear infections, and wounds of humans (*B. avium* [Kersters et al., 1984], *B. hinzii* [Vandamme et al., 1995c], *B. holmesii* [Weyant et al., 1995a], and *B. trematum* [Vandamme et al., 1996a]). However, no pathogenic significance was demonstrated for *B. trematum*, and that of *B. holmesii* might be limited to opportunistic infections. *B. avium* is a pathogen only for animals. An eventual emendation of *Bordetella* to a more broadly defined genus that is not restricted to obligate pathogens that thrive between epithelial cilia of the respiratory tract would be an alternative to solve discrepancies between phylogeny and taxonomy. This would allow inclusion of the species *A. xylosoxidans* subsp. *denitrificans*, *A. xylosoxidans* subsp. *xylosoxidans*, and *A. piechaudii*, often isolated from clinical specimens, and even *A. ruhlandii*, into a redefined genus *Bordetella*.

Very recently, the genus *Pigmentiphaga* with the single species *Pigmentiphaga kullae* has been proposed as a new genus of the family *Alcaligenaceae* (Blümel et al., 2001b). The single strain of *P. kullae* was isolated after an aerobic enrichment with the azo compound 1-(4′-carboxyphenylazo)-4-naphthol as sole source of carbon and energy (Kulla et al., 1984). Analysis of the 16S rDNA sequence (95–96% similarity with members of the family *Alcaligenaceae*), designation of signature nucleotides, and chemotaxonomic characteristics (presence of ubiquinone Q-8, the polyamines putrescine and 2-hydroxyputrescine, and the fatty acids $C_{16:0}$ and $C_{17:0\ cyclo}$) placed *P. kullae* unambiguously within the family *Alcaligenaceae*.

Some taxonomic comments have to be addressed regarding *Taylorella equigenitalis*, *Taylorella asinigenitalis*, and *Pelistega europaea*. *T. equigenitalis* was originally described as *Haemophilus equigenitalis* (Taylor et al., 1978). A reinvestigation of the taxonomic position of *H. equigenitalis* based on phenotypic data, DNA base composition, and DNA–DNA hybridization data led to its transfer to the new genus *Taylorella*, as its type species *T. equigenitalis* (Sugimoto et al., 1983). The second species of this genus, *T. asinigenitalis*, shares 97.6% 16S rDNA sequence similarity with strains of *Taylorella equigenitalis* (Jang et al., 2001). 16S rDNA sequence analyses (Bleumink-Pluym et al., 1993; Jang et al., 2001) demonstrated that the genera *Pelistega*, *Alcaligenes*, *Achromobacter*, and *Bordetella* are the closest phylogenetic relatives of the genus *Taylorella*. As deduced from phylogenetic analyses based on 16S rDNA sequences, *T. equigenitalis* and *A. faecalis* share approximately the same distance to certain *Bordetella* species and *A. xylosoxidans* subsp. *xylosoxidans* (Bleumink-Pluym et al., 1993; Weyant et al., 1995a). These data indicate that the genus *Taylorella* may represent another genus within the family *Alcaligenaceae*. However, its inclusion in the family is not clearly supported by other data. *Taylorella* species have been shown to have a DNA base composition of ~38 mol% G + C. This value is substantially lower than the mol% G + C content of *A. faecalis*, which is known to have the lowest mol% G + C content (57.4% as determined for the type strain) within the family. Likewise, the fatty acid profile of *Taylorella* species with the major compounds $C_{18:1}$, $C_{16:0}$, $C_{18:0}$, and $C_{14:0\ 3OH}$ (Rossau et al., 1987; Jang et al., 2001) does not reflect the characteristics of the family, i.e., the compound $C_{17:0\ cyclo}$ (Table BXII.β.29) is missing and the predominant compound, $C_{18:1}$, is not a major fatty acid in the profile of the family members. However, the presence of $C_{16:0}$ and $C_{14:0\ 3OH}$ fatty acids

TABLE BXII.β.29. Differentiating characteristics of *Alcaligenes* species and species that previously have been transferred from the genus *Alcaligenes* to other genera, including *Ralstonia eutropha* (formerly *Alcaligenes eutrophus*), *Variovorax paradoxus* (formerly *Alcaligenes paradoxus*), *Halomonas aesta* (formerly *Alcaligenes aestus, Deleya aesta*), *Halomonas aquamarina* (formerly *Alcaligenes aquamarinus, Deleya aquamarina*), *Halomonas cupida* (formerly *Alcaligenes cupidus, Deleya cupida*), *Halomonas pacifica* (formerly *Alcaligenes pacificus, Deleya pacifica*), *Halomonas venusta* (formerly *Alcaligenes venustus, Deleya venusta*), and *Carbophilus carboxidus* (formerly "*Alcaligenes carboxidus*")[n]

Characteristic	*Alcaligenes faecalis*[a]	*Alcaligenes defragrans*[b]	*Alcaligenes latus*[c]	*Alcaligenes xylosoxidans* subsp. *denitrificans*[a]	*Alcaligenes xylosoxidans* subsp. *xylosoxidans*[d]	*Achromobacter ruhlandii*[e]	*Achromobacter piechaudii*[f]	*Ralstonia eutropha*[g]	*Variovorax paradoxus*[h]	*Halomonas aquamarina*[h]	*Halomonas cupida*[i]	*Halomonas pacifica*[i]	*Halomonas venusta*[i]	*Carbophilus carboxidus*[j]
Major fatty acids	$C_{17:0}$ cyclo, $C_{16:0}$	C $C_{16:1}$, $C_{16:0}$, $C_{18:1}$, $C_{17:0}$ cyclo[k]	$C_{16:1}$, $C_{18:1}$, $C_{16:0}$	$C_{16:0}$, $C_{17:0}$ cyclo	$C_{16:0}$, $C_{17:0}$ cyclo		$C_{17:0}$ cyclo, $C_{16:0}$	$C_{16:1}$, $C_{16:0}$	$C_{16:0}$, $C_{17:0}$ cyclo ($C_{15:0}$, $C_{16:1}$, $C_{16:0}$, $C_{18:1}$)	$C_{18:1}$, $C_{16:0}$, $C_{19:0}$ cyclo	$C_{18:1}$, $C_{16:0}$, $C_{19:0}$ cyclo	$C_{18:1}$, $C_{16:0}$, $C_{19:0}$ cyclo	$C_{18:1}$, $C_{16:0}$	$C_{18:1}$, $C_{16:0}$
Hydroxy fatty acids	$C_{14:0}$ 3OH, $C_{12:0}$ 2OH	$C_{14:0}$ 3OH	$C_{10:0}$ 3OH	$C_{14:0}$ 3OH, $C_{12:0}$ 2OH, $C_{16:0}$ 2OH	$C_{14:0}$ 3OH, $C_{12:0}$ 2OH, $C_{16:0}$ 2OH		$C_{14:0}$ 3OH, $C_{12:0}$ 2OH, $C_{16:0}$ 2OH	$C_{14:0}$ 3OH, $C_{14:0}$ 2OH, $C_{12:0}$ 3OH	$C_{12:0}$ 3OH, $C_{10:0}$ 3OH, $C_{16:0}$ 3OH ($C_{10:0}$ 3OH, $C_{14:0}$ 3OH, $C_{8:0}$ 3OH)[m]	$C_{12:0}$ 3OH	$C_{12:0}$ 3OH	$C_{12:0}$ 3OH	$C_{12:0}$ 3OH	$C_{12:0}$ 3OH
Respiratory quinone	Q-8		Q-8	Q-8	Q-8			Q-8	Q-8	Q-9	Q-9	Q-9	Q-9	Q-10
Characteristic polyamines[n]	PUT, HPUT			PUT, HPUT	PUT, HPUT	PUT, HPUT		PUT, HPUT	PUT, HPUT	SPD	SPD	SPD	SPD	HSPD, PUT
Yellow pigment	–	–	–	–	–	–	–	–	+	–	–	–	–	–
Chemolithotrophic growth on H_2	–	+	+	–	–	+	–	+	d	Not analyzed	–	–	–	–
Urease	–		–	–	–	–	–	–	+					
Hydrolysis of:														
Gelatin	–	–	+	–	–	–	–	–	d	d	–	–	–	–
Starch	–	–	+	–	–	–	–	+	–		–	–	–	–
Tween 80	–	+	+				–	+	+					

(*continued*)

TABLE BXII.β.29. (cont.)

Characteristic	Alcaligenes faecalis[a]	Alcaligenes defragrans[b]	Alcaligenes latus[c]	Alcaligenes xylosoxidans subsp. denitrificans[a]	Alcaligenes xylosoxidans subsp. xylosoxidans[d]	Achromobacter ruhlandii[e]	Achromobacter piechaudii[f]	Ralstonia eutropha[g]	Variovorax paradoxus[h]	Halomonas aquamarina[h]	Halomonas cupida[i]	Halomonas pacifica[i]	Halomonas venusta[i]	Carbophilus carboxidus[j]
Carbon source for growth:														
D-Galactose	−		−	−	+	+	Not analyzed	−	+	d	+	−	−	+
L-Arabinose	−		−	−	−	−	−	−	+	−	+	−	[+]	+
D-Mannitol	−		+	−	−	−	−	−	+	+	d	−	d	+
Maltose	−	−	+	−	−	−	Not analyzed	−	+	d	d	−	d	+
Sucrose			+	−	−		Not analyzed	−	−	−	d	−	d	+
2-Keto-gluconate			+	−	−		Not analyzed	+	d	d	d		d	
p-Hydroxy-benzoate	−		+	−				+	+	−	+	d	d	
Butylamine	−		+	−				−	d	−	−	+	d	
Betaine	−		+	−				−	−	−	+	+	+	+
Sarcosine	−		+	−				−	−	−	+	+	+	
Creatine	−		+	−				−	−				d	+
Mol% G + C of DNA	55.9–59.4	66.9	69.1–71.1	63.9–68.9	66.0–69.8	67.7	64.0–65.0	66.3–67.5	66.8–69.4	56.9–57.9	59.9–62.1	67.3–68.3	52.9–54.5	62.8

[a] Data from Gilardi (1978a); Pichinoty et al. (1978); Rarick et al. (1978); Yamasato et al. (1982); Kersters and De Ley (1984b); Busse and Auling (1988); Lipski et al. (1992); Vandamme et al. (1995c); Hamana and Takeuchi (1998).

[b] Data from Foss et al. (1998a).

[c] Data from Palleroni and Palleroni (1978); Oyaizu-Masuchi and Komagata (1988).

[d] Data from Yabuuchi et al. (1974); Gilardi (1978a); Pichinoty et al. (1978); Rarick et al. (1978); Rubin et al. (1980); Kiredjian et al. (1981); Yamasato et al. (1982); Kersters and De Ley (1984b); Busse and Auling (1988); Oyaizu-Masuchi and Komagata (1988); Vandamme et al. (1995c); Hamana and Takeuchi (1998).

[e] Data from Yamasato et al. (1982); Busse and Auling (1988); Yabuuchi et al. (1988); Vandamme et al. (1995c); Hamana and Takeuchi (1998).

[f] Data from Kiredjian et al. (1986); Vandamme et al. (1995c); Hamana and Takeuchi (1998).

[g] Data from Davis et al. (1969, 1970); Kersters and De Ley (1984b); Busse and Auling (1988); Oyaizu-Masuchi and Komagata (1988); Lipski et al. (1992).

[h] Data from ZoBell and Upham (1944); Baumann et al. (1972); Holding and Shewan (1974); Kersters and De Ley (1984b); Franzmann and Tindall (1990); Auling et al. (1991); Akagawa-Matsushita et al. (1992); Hamana (1997). The data include the characteristics of *D. aesta* recently reclassified as *H. aquamarina* (Akagawa and Yamasato, 1989).

[i] Data from Baumann et al. (1972); Franzmann and Tindall (1990); Akagawa-Matsushita et al. (1992); Hamana (1997).

[j] Data from Auling et al. (1988); Meyer et al. (1993).

[k] The fatty acid composition was analyzed from biomass that was grown under denitrifying conditions with acetate as the sole carbon source.

[l] In the original paper (Vandamme et al., 1995c) the authors pointed out that the microbial identification system (Microbial ID, Inc., Newark, Delaware) used for identification of fatty acids cannot distinguish between the two acids $C_{16:1\ iso}$ and $C_{14:0\ 3OH}$. Since branched-chain fatty acids have not been reported in *Alcaligenes* or *Bordetella* strains or are present only in trace amounts, the peak designated summed in feature 3 most probably corresponds to $C_{14:0\ 3OH}$.

[m] The values in brackets indicate the differing results as determined by Willems et al. (1989).

[n] PUT, putrescine; HPUT, 2-hydroxyputrescine; SPD, spermidine; HSPD, *sym*-homospermidine.

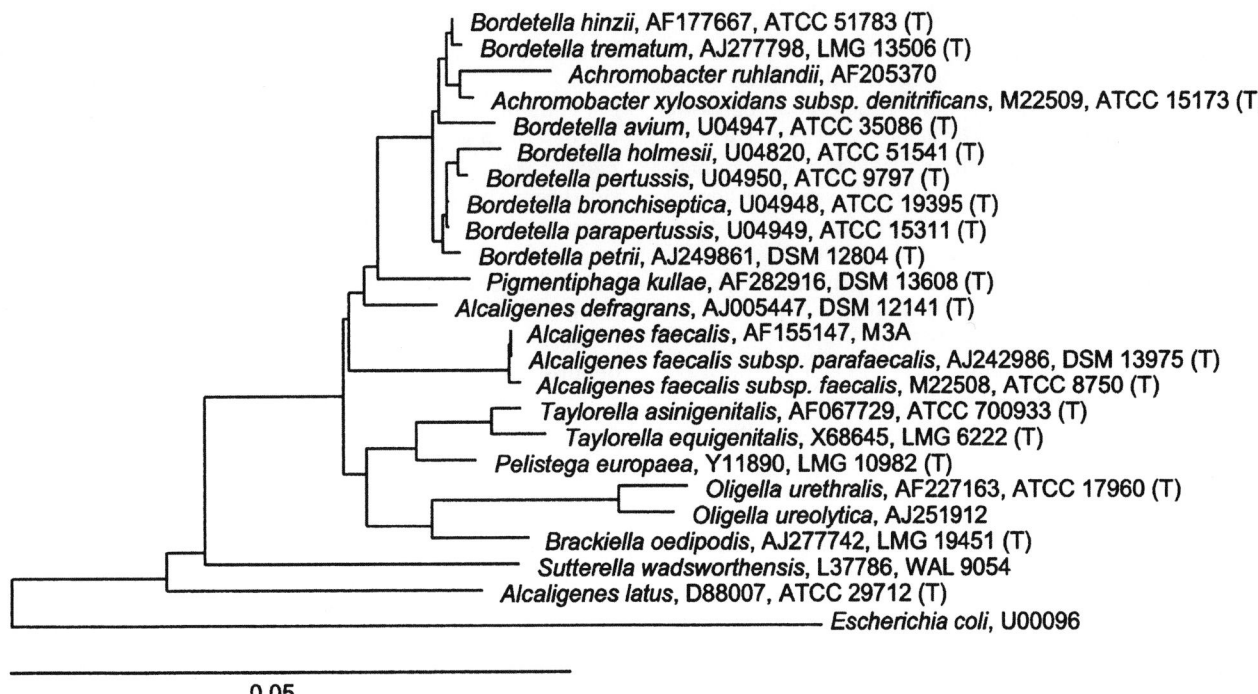

Bordetella hinzii, AF177667, ATCC 51783 (T)
Bordetella trematum, AJ277798, LMG 13506 (T)
Achromobacter ruhlandii, AF205370
Achromobacter xylosoxidans subsp. denitrificans, M22509, ATCC 15173 (T)
Bordetella avium, U04947, ATCC 35086 (T)
Bordetella holmesii, U04820, ATCC 51541 (T)
Bordetella pertussis, U04950, ATCC 9797 (T)
Bordetella bronchiseptica, U04948, ATCC 19395 (T)
Bordetella parapertussis, U04949, ATCC 15311 (T)
Bordetella petrii, AJ249861, DSM 12804 (T)
Pigmentiphaga kullae, AF282916, DSM 13608 (T)
Alcaligenes defragrans, AJ005447, DSM 12141 (T)
Alcaligenes faecalis, AF155147, M3A
Alcaligenes faecalis subsp. parafaecalis, AJ242986, DSM 13975 (T)
Alcaligenes faecalis subsp. faecalis, M22508, ATCC 8750 (T)
Taylorella asinigenitalis, AF067729, ATCC 700933 (T)
Taylorella equigenitalis, X68645, LMG 6222 (T)
Pelistega europaea, Y11890, LMG 10982 (T)
Oligella urethralis, AF227163, ATCC 17960 (T)
Oligella ureolytica, AJ251912
Brackiella oedipodis, AJ277742, LMG 19451 (T)
Sutterella wadsworthensis, L37786, WAL 9054
Alcaligenes latus, D88007, ATCC 29712 (T)
Escherichia coli, U00096

0.05

FIGURE BXII.β.21. Phylogenetic relationships of the family *Alcaligenaceae*. The distances in the tree were calculated using 1101 positions (the least-squares method, Jukes-Cantor model). (Courtesy T. Lilburn of the Ribosomal Database Project.)

as major compounds in *Taylorella* species is characteristic of all validly described species of the family *Alcaligenaceae*.

The closest phylogenetic relative of *Taylorella* is *Pelistega europaea* (Vandamme et al., 1998). Members of this species were reported to have a low mol% G + C content of genomic DNA (42–43%), and a fatty acid profile with the predominant acids $C_{18:1}$, $C_{16:1}$, $C_{16:0}$, summed feature 3 ($C_{14:0\ 3OH}$ or $C_{16:1\ iso\ I}$ or both), and $C_{14:0}$. *P. europaea* can be distinguished from members of the genus *Taylorella* based on 16S rDNA sequences dissimilarities (5.2%), and by qualitative and quantitative differences in the fatty acid profile and biochemical traits. To substantiate the allocation of *Taylorella* and *P. europaea* to an enlarged family *Alcaligenaceae*, the definition of signature nucleotides in the 16S rDNA sequence might be the most promising approach.

The genus *Sutterella* might be another member of the family *Alcaligenaceae* as deduced from 16S rDNA sequences of its single species *S. wadsworthensis* (Wexler et al., 1996a). This microaerophilic/anaerobic species has $C_{16:0}$ and $C_{16:1\ \omega7c}$ acids and summed feature 10 ($C_{18:1\ \omega7c,\ 9t,\ 12c}$ and/or an unknown peak). This profile and the relatively low mol% G + C content (36.5%) remind us of traits in members of *Taylorella* and *Pelistega*. Surprisingly, neither $C_{14:0\ 3OH}$ fatty acid present in all other members of the family *Alcaligenaceae* nor any other hydroxy fatty acid commonly associated to LPS of Gram-negative bacteria has been reported. On the other hand, inclusion of *Sutterella* into an emended family *Alcaligenaceae* appears questionable due to its phylogenetic equidistance (16S rDNA sequences similarities of 88–90%) to members of the family including *Taylorella* and *Pelistega* and to other members of the *Betaproteobacteria* such as *Ralstonia*, *Herbaspirillum*, and *Roseateles*.

Pathogenicity Strains of *A. faecalis*, *A. xylosoxidans* subsp. *denitrificans*, and *A. piechaudii* are frequently isolated from clinical specimens, and so they can be regarded as opportunistic path-

ogens (Kiredjian et al., 1981). Only *A. xylosoxidans* subsp. *xylosoxidans* is considered to be of clinical relevance. This species has been recognized as a causative agent of various human infections, mostly in immunocompromised hosts, i.e., patients with underlying diseases including endocarditis, meningitis, and ventriculitis after neurosurgery, pneumonia, bacteremia osteomyelitis, arthritis, and peritonitis (Shigeta et al., 1978; Reverdy et al., 1984; San Miguel et al., 1991; Legrand and Anaissie, 1992; Walsh et al., 1993). Several outbreaks of hospital-acquired infections have been observed in intensive care units or postsurgical recovery areas due to contamination with *A. xylosoxidans* subsp. *xylosoxidans* of antiseptic solutions (Shigeta et al., 1978), nonbacteriostatic solutions used in diagnostic tracer procedures (McGuckin et al., 1982), dialysis fluids (Reverdy et al., 1984), intravascular pressure transducers (Gahrn-Hansen et al., 1988), and a diagnostic contrast solution (Reina et al., 1988). Strains of *A. xylosoxidans* subsp. *xylosoxidans* have been reported to be susceptible to amoxicillin/clavulanic acid, piperacillin, piperacillin/tazobactam, azlocillin, ceftazidime, and imipenem, and resistant to cefazolin, cefuroxime, gentamicin, tobramycin, netilmicin, amikacin, tetracycline, ofloxacin, and ciprofloxacin (Knippschild and Ansorg, 1998). The increasing recovery of this species from clinical specimens might be related to the extensive use of third-generation antibiotics such as cephalosporins and fluoroquinolones not effective against *A. xylosoxidans* subsp. *xylosoxidans* (Glupczynski et al., 1988; Mensah et al., 1989). No pathogenic potential has been reported for *A. ruhlandii*, *A. defragrans*, or *P. kullae*.

The genus *Bordetella* consists of several pathogenic bacteria. These bacteria produce many virulence related factors, among which there are several toxins such as the pertussis toxin (PTX) that is produced only by *B. pertussis*, the adenylate cyclase toxin (CYA), and several adhesins such as the filamentous hemagglutinin (FHA), the pertactin (PRN), and the fimbriae (Weiss, 1992). *B. pertussis* is the etiological agent of whooping cough

(Weiss and Hewlett, 1986). *B. parapertussis*, which causes a milder form of whooping cough, has been assumed as a strictly human pathogen (Hewlett, 1990), but it has been isolated from both healthy sheep and sheep affected with chronic nonprogressive pneumonia (Cullinane et al., 1987; Chen et al., 1988a; Porter et al., 1994). *B. bronchiseptica* is found in a wide range of animals, and it is associated with atrophic rhinitis in swine and kennel cough in dogs (Goodnow, 1980; Harkness and Wagner, 1995; Keil and Fenwick, 1998; Speakman et al., 1999). *B. avium* is a pathogen of birds and causes rhinotracheitis (turkey coryza) in turkey poults (Simmons and Gray, 1979; Simmons et al., 1979; Saif et al., 1980; Arp and Cheville, 1984; Kersters et al., 1984). Only recently, *B. hinzii* (Vandamme et al., 1995c) has been reported as the causative agent of fatal septicemia (Kattar et al., 2000). *B. holmesii* has been associated most often with septicemia in patients with underlying conditions (Lindquist et al., 1995; Weyant et al., 1995a; Morris and Meyers, 1998; Tang et al., 1998). It also has been isolated from sputum from one patient with respiratory symptoms. *B. trematum* has been isolated from wounds and ear infections, but its pathogenic significance is unknown (Vandamme et al., 1996a).

T. equigenitalis is considered a commensal or opportunistic pathogen. It colonizes the urogenital membranes of the clitoral sinuses and fossa, urethra, and cervix of mares, and the urethra, urethral fosses, and penile sheath of stallions. So far, only infections of mares have been reported where *T. equigenitalis* causes vaginal discharge, infertility, or early abortion (Wada et al., 1983). *T. equigenitalis* is moderately resistant to a limited number of antibiotics including clindamycin, lincomycin, trimethoprim, and sulfamethoxazole but sensitive to penicillin G, ampicillin, carbenicillin, cephaloridine, cephalothin, erythromycin, tetracycline, kanamycin, gentamicin, chloramphenicol, and polymyxin B (Sugimoto et al., 1983). *T. asinigenitalis* is not reported to cause disease in jacks or mares (Jang et al., 2001).

Pelistega europaea strains have been mainly isolated from lungs, air sac exudate, and trachea mucosa and less frequently from other organs such as liver and spleen of pigeons.

Sutterella wadsworthensis has been isolated from infections of gastrointestinal origin (Finegold and Jousimies-Somer, 1997; Jousimies-Somer, 1997). The majority of *Sutterella* strains are susceptible to amoxicillin/clavulanate, ticarcillin/clavulanate, cefoxitin, ceftriaxone, clindamycin, piperacillin/tazobactam, ceftizoxime, ciprofloxacin, trovafloxacin, azithromycin, clarithromycin, erythromycin, and roxithromycin.

TAXONOMIC COMMENTS

The numerous reclassifications between the two genera *Alcaligenes* and *Achromobacter* suggest a common discussion of their history.

The type species of the genus *Alcaligenes* is *A. faecalis* Castellani and Chalmers 1919. This species also encompasses strains originally described as "*Pseudomonas odorans*" (Málek and Kazdová-Košiskova, 1946) Málek et al., 1963, "*A. odorans* var. viridans" Mitchell and Clarke, 1965 and "*Achromobacter arsenoxydans-tres*" Turner, 1954.

Based on the phylogenetic relationships within the family *Alcaligenaceae*, *A. faecalis* can be considered as the only representative of the genus *Alcaligenes sensu stricto*. Recently, a strain accumulating poly-β-hydroxybutyrate from acetone-butanol bioprocess residues was shown to be closely related with *A. faecalis* DSM 30030[T] (Schroll et al., 2001a). This strain shared 98.7% 16S rDNA sequence similarity, 56% DNA relatedness, and an almost iden-

tical protein pattern with the type strain of *A. faecalis*. Since it could only be distinguished from *A. faecalis* DSM 30030[T] based on few physiological and biochemical traits and the extremely low content of the diagnostic polyamine 2-hydroxyputrescine, it was described as a subspecies of *A. faecalis*, *A. faecalis* subsp. *parafaecalis*. As a result of this study, the species *A. faecalis* is subdivided into the two subspecies *A. faecalis* subsp. *faecalis* and *A. faecalis* subsp. *parafaecalis*.

A. defragrans has been described as a new species (Foss et al., 1998a), encompassing four strains isolated on the alkenoic monoterpenes (+)-menthene, α-pinene, 2-carene, and α-phellandrene, respectively, as the sole source of carbon and energy under denitrifying conditions. 16S rDNA sequence data indicate that *A. defragrans* occupies a separate position within the *Alcaligenaceae*. Thus, future reclassification of this species in a new genus is most likely. In agreement with a proposal of a new genus for *A. defragrans* is its unusual fatty acid profile. In contrast to the fatty acid profiles of *Alcaligenes*, *Achromobacter*, and *Bordetella* species, *A. defragrans* was reported to contain significant amounts of dodecanal and to lack $C_{12:0\ 2OH}$ acid. However, the fatty acid profiles of strains of this species show the characteristics of the members of the family, the predominant acids $C_{16:0}$ and $C_{17:0\ cyclo}$, and the presence of the hydroxy acid $C_{14:0\ 3OH}$. Unfortunately, the fatty acid profile of *A. defragrans* was not obtained from cells grown on the commonly used trypticase soy agar. Thus, it cannot be compared directly with other profiles of species within the family *Alcaligenaceae* available from the literature and from the MIDI fatty acid identification system (Microbial ID, Newark, NJ, USA).

The type species of the genus *Achromobacter* is *A. xylosoxidans* (ex Yabuuchi and Ohyama 1971; Yabuuchi and Yano 1981) Yabuuchi et al. 1998a. The history of the two subspecies *Achromobacter xylosoxidans* subsp. *denitrificans* (formerly *Alcaligenes denitrificans* subsp. *denitrificans*) and *Achromobacter xylosoxidans* subsp. *xylosoxidans* (formerly *Alcaligenes denitrificans* subsp. *xylosoxidans*) has been reviewed in detail by Kersters and De Ley (1984b) and will be only summarized here. The name *Alcaligenes denitrificans* was proposed by Leifson and Hugh (1954a) to accommodate two strains that reduce nitrate to nitrite and gas. Based on biochemical and nutritional characteristics, Hendrie et al. (1974) proposed that *A. denitrificans* was a subjective synonym of *A. faecalis* Castellani and Chalmers 1919. The name was not included in the Approved Lists of Bacterial Names (Skerman et al., 1980) and thus did not have standing in bacterial nomenclature. The name *A. denitrificans* was revived by Rüger and Tan (1983) to separate the original type strain of *A. denitrificans* (Leifson and Hugh, 1954a) from the species *A. faecalis*, based on DNA base composition, DNA reassociation, and nitrate reduction. In the first edition of *Bergey's Manual of Systematic Bacteriology*, *A. denitrificans* was named *A. denitrificans* subsp. *denitrificans*, *Achromobacter xylosoxidans* as *A. denitrificans* subsp. *xylosoxidans*, and *A. ruhlandii* was included in this second subspecies (Kersters and De Ley, 1984b). *Achromobacter xylosoxidans* was originally described by Yabuuchi and Ohyama (1971). Since this name was omitted from the Approved Lists of Bacterial names (Skerman et al., 1980), it had no standing in bacterial nomenclature. Thus, the name was revived by Yabuuchi and Yano (1981). Due to its high degree of similarity in the rRNA cistron to *A. denitrificans* subsp. *denitrificans*, Kersters and De Ley (1984b) proposed including as a second subspecies in *A. denitrificans*, namely *A. denitrificans* subsp. *xylosoxidans*. Additionally, the hydrogen-oxidizing species *A. ruhlandii* (Packer and Vishniac, 1955; Aragno and Schlegel,

1977), which appears in the Approved Lists, was transferred to *A. denitrificans* subsp. *xylosoxidans*, based on its similarity in genotypic and phenotypic features, and its indistinguishability from this subspecies in protein electrophoregrams.

Since the species epithet *xylosoxidans* had priority, Kiredjian et al. (1986) proposed the new combinations *A. xylosoxidans* subsp. *denitrificans* and *A. xylosoxidans* subsp. *xylosoxidans*. Vandamme et al. (1996a) demonstrated that strains of *A. xylosoxidans* subsp. *denitrificans* and *A. xylosoxidans* subsp. *xylosoxidans* can be readily distinguished by their whole-cell protein patterns, fatty acid components, and other phenotypic characteristics. These authors proposed that *A. xylosoxidans* subsp. *denitrificans* and *A. xylosoxidans* subsp. *xylosoxidans* should be elevated to species rank as *A. denitrificans* and *A. xylosoxidans*, respectively. Yabuuchi et al. (1998a) proposed combining *A. denitrificans* and *A. xylosoxidans* in a single species, *Achromobacter xylosoxidans*, with the two subspecies *Achromobacter xylosoxidans* subsp. *denitrificans* and *Achromobacter xylosoxidans* subsp. *xylosoxidans*, based on 63.4% DNA relatedness and 98.7% 16S rDNA sequence similarity between the type strains of the two subspecies. This proposal ignores the distinguishing characteristics reported by Vandamme et al. (1998) and thus creates a heterogeneous species.

The names *Alcaligenes denitrificans*, *Alcaligenes xylosoxidans* subsp. *denitrificans*, *Alcaligenes denitrificans* subsp. *denitrificans* and *Achromobacter xylosoxidans*, *Alcaligenes xylosoxidans*, *Alcaligenes xylosoxidans* subsp. *xylosoxidans*, and *Alcaligenes denitrificans* subsp. *xylosoxidans* are senior synonyms of the validated names *Achromobacter xylosoxidans* subsp. *denitrificans* and *Achromobacter xylosoxidans* subsp. *xylosoxidans*, respectively, as proposed by Yabuuchi et al. (1998a). Based on the low degree of DNA relatedness with other members of the genus *Achromobacter*, Yabuuchi et al. (1998a) also proposed the new species *Achromobacter ruhlandii* to accommodate the hydrogen-oxidizing strain of *Achromobacter xylosoxidans* originally described as "*Hydrogenomonas ruhlandii*" (Packer and Vishniac, 1955; Aragno and Schlegel, 1977).

The third species, *A. piechaudii*, encompasses isolates from human clinical material from different geographical regions (Kiredjian et al., 1986), as well as strains originally named *A. faecalis* CIB 60.75 and "*Achromobacter iophagus*" (Kiredjian et al., 1981; Holmes and Dawson, 1983), which were shown to be phenotypically and genotypically distinct from the other species transferred to the genus *Achromobacter*.

The marine *Alcaligenes* species *A. aestus*, *A. pacificus*, *A. cupidus*, and *A. venustus* have been reclassified and described as species of the new genus *Deleya* as *D. aesta*, *D. pacifica*, *D. cupida*, and *D. venusta* (Baumann et al., 1983). Akagawa and Yamasato (1989) showed that *A. aquamarinus* is another member of the genus *Deleya*. They also demonstrated that the type strains of *A. aquamarinus*, *A. faecalis* subsp. *homari* (Austin et al., 1981), and *D. aesta* are members of the same species, and due to priority of the species epithet *aquamarina* they were reclassified in the species *Deleya aquamarina*. Meanwhile, members of the genus *Deleya* have been transferred to the genus *Halomonas*, and the new combinations *H. aquamarina*, *H. pacifica*, *H. cupida*, and *H. venusta* have been proposed for the marine *Alcaligenes* species (Dobson and Franzmann, 1996).

FURTHER READING

Busse, H.J. and G. Auling. 1992. The genera *Alcaligenes* and "*Achromobacter*". *In* Balows, Trüper, Dworkin, Harder and Schleifer (Editors), The Prokaryotes. A Handbook on the Biology of Bacteria: Ecophysiology, Isolation, Identification, Applications., 2nd Ed., Vol. 3, Springer-Verlag, New York. pp. 2544–2555.

De Ley, J. 1992. Introduction to the Prokaryotes. *In* Balows, Trüper, Dworkin, Harder and Schleifer (Editors), The Prokaryotes: A Handbook on the Biology of Bacteria: Ecophysiology, Isolation, Identification, Application, 2nd Ed., Springer-Verlag, New York. pp. 2111–2140.

Weiss, A.A. 1992. The genus *Bordetella*. *In* Balows, Truper, Dworkin, Harder and Schleifer (Editors), The Prokaryotes: a Handbook on the Biology of Bacteria: Ecophysiology, Isolation, Identification, Applications., 2nd ed., Vol. 3, Springer-Verlag, Berlin, Germany. pp. 2530–2543.

Genus I. *Alcaligenes* Castellani and Chalmers 1919, 936[AL]

HANS-JÜRGEN BUSSE AND GEORG AULING

Al.ca.li'ge.nes. Arabic *al* the; Arabic n. *galiy* the ash of saltwort; French n. *alcali* alkali; Gr. v. *gennaio* to produce; M.L. masc. n. *Alcaligenes* alkali-producing (bacteria).

Rods or coccobacilli, 0.5–1.2 × 1.0–3.0 μm, usually occurring singly. Resting stages not known. **Gram negative. Motile** with one to nine **peritrichous** flagella. **Obligately aerobic**, possessing a strictly respiratory type of metabolism with oxygen as the terminal electron acceptor. Some strains are capable of anaerobic respiration in the presence of nitrate or nitrite. Optimal growth temperature: 20–37°C. Colonies on nutrient agar are **nonpigmented. Oxidase positive. Catalase positive.** Indole not produced. Cellulose, esculin, gelatin, and DNA usually not hydrolyzed. **Chemoorganotrophic, using a variety of organic acids and amino acids as carbon sources. Alkali produced** from several organic salts and amides. Carbohydrates usually not utilized. **Characteristic fatty acids are** $C_{17:0\ cyclo}$, $C_{16:0}$, $C_{14:0\ 3OH}$, $C_{16:1}$, and $C_{12:0\ 2OH}$. **Ubiquinone Q-8. Polyamine patterns with the predominant compound putrescine and the unusual diamine 2-hydroxyputrescine.** Isolated from water, soil, and clinical specimens such as blood, spinal fluid, pleural fluid, peritoneal fluid, pus, urine, stools, and swabs of eyes, ears, and pharynxes. Frequently found in unster-

ilized distilled water and in chlorhexidine solutions in hospitals. Occasionally causing opportunistic infections in humans.

The mol% G + C of the DNA is: 56–60.

Type species: **Alcaligenes faecalis** Castellani and Chalmers 1919, 936[AL] (*Achromobacter arsenoxydans-tres* Turner 1954, 475; "*Pseudomonas odorans*" (Málek and Kazdová-Košiskova 1946; Málek, Radochová and Lysenko 1963, 353; "*Alcaligenes odorans* var. viridans" Mitchell and Clarke 1965, 347.

FURTHER DESCRIPTIVE INFORMATION

For a broad and an extensive view the reader is referred to the excellent treatment of the genus by Kersters and De Ley (1984b) except for those characteristics that became available since the first edition of *Bergey's Manual of Systematic Bacteriology*. However, since the Kersters and De Ley publication, many of the xenobiotic-degrading soil bacteria, especially some of the strains well known for plasmid-encoded pathways that were originally allocated to *Alcaligenes* species, have been transferred to other genera

(Busse and Auling, 1992, cf. also the family *Alcaligenaceae* in this volume).

Dimethyl disulfide-producing bacteria have been isolated from activated sludge and some of them were phenotypically allocated to the genus *Alcaligenes* (Tomita et al., 1987). In the same study pure cultures from the authentic genus *Alcaligenes* were also shown to produce dimethyl disulfide from DL-methionine and S-methyl-L-cysteine. More recently, an *Alcaligenes*-like dimethyl sulfide-producing marine isolate, phenotypically identified (Vitek), was shown to contain a dimethylsulfoniopropion-ate lyase (de Souza and Yoch, 1995). This enzyme was purified and the authors argued that the K_m value and other properties observed may reflect the greater potential of facultative anaerobes over aerobes to metabolize lower levels of dimethylsulfoniopropionate in anoxic zones of seawater and salt marshes.

FURTHER READING

De Ley, J., P. Segers, K. Kersters, W. Mannheim and A. Lievens. 1986. Intrageneric and intergeneric similarities of the *Bordetella* ribosomal ribonucleic acid cistrons: proposal for a new family, *Alcaligenaceae*. Int. J. Syst. Bacteriol. *36*: 405–414.

List of species of the genus Alcaligenes

1. **Alcaligenes faecalis** Castellani and Chalmers 1919, 936[AL] (*Achromobacter arsenoxydans-tres* Turner 1954, 475; "*Pseudomonas odorans*" (Málek and Kazdová-Košiskova 1946; Málek, Radochová and Lysenko 1963, 353; "*Alcaligenes odorans* var. viridans" Mitchell and Clarke 1965, 347.)
   fae.ca' lis. L. n. *faex, faecis* dregs; M.L. adj. *faecalis* fecal.

   The species *Alcaligenes faecalis* is subdivided into the two subspecies, *A. faecalis* subsp. *faecalis* and *A. faecalis* subsp. *parafaecalis*.
   The mol% G + C of the DNA is: 56–60 (T_m, Bd).

   a. **Alcaligenes faecalis** subsp. **faecalis** Castellani and Chalmers 1919, 936[AL]

   The description of *Alcaligenes faecalis* subsp. *faecalis* is based on that given by Kersters and De Ley (1984b) for *A. faecalis*. This subspecies is the type subspecies of *Alcaligenes faecalis* and contains the type strain of the species. The morphological characteristics are as described for the genus. Colonies on nutrient agar are nonpigmented to grayish white, translucent to opaque, flat to low convex, margin usually entire, usually smooth, sometimes dull or rough. Most strains form colonies with a thin, spreading irregular edge. Some strains previously named "*A. odorans*" (Málek and Kazdová-Koziškova, 1946; Málek et al., 1963), produce a characteristic aromatic fruity odor and/or a green discoloration on blood agar. Those strains producing the latter characteristic were previously named "*A. odorans* var. viridans" (Mitchell and Clarke, 1965).

   Physiological and nutritional characteristics of *A. faecalis* are presented in Table BXII.β.29 of the chapter *Alcaligenaceae* and Tables BXII.β.30 and BXII.β.31 of this chapter. Carbohydrates are not utilized as sole carbon sources. Good growth is obtained on several organic acids and amino acids. Instead of using a monoamine oxidase, *Alcaligenes faecalis* initiates catabolism of aromatic amines alternatively by an aromatic amine dehydrogenase that is structurally similar to methylamine dehydrogenase and possesses the same tryptophan tryptophylquinone prosthetic group as the latter (Govindaraj et al., 1994). Chemolithotrophic growth using hydrogen gas has not been demonstrated.

   Nitrite, but not nitrate, is reduced. Anaerobic respiration with nitrite, but not with nitrate, as a sole electron acceptor is possible for most strains. Some strains oxidize arsenite. *Alcaligenes faecalis* is among the bacterial taxa that are able to combine aerobic denitrification and heterotrophic nitrification (van Niel et al., 1992; Anderson et al., 1993) and that may contribute to biogenic emissions of NO and N_2O into the atmosphere even when growing logarithmically (Papen et al., 1989; Otte et al., 1996). Since *Alcaligenes faecalis* is commonly found in soil, water, and waste water treatment plants, bacteria of this species, enriched by the denitrification process of activated sludge, may be responsible for significant trace gas emissions of suboptimally functioning systems of waste water treatment. Among a large collection of isolates from biofilters for off-gas treatment of animal-rendering plant emissions, 21 bacteria were allocated by a polyphasic taxonomic approach to either *Alcaligenes faecalis* or a new taxon (cluster J) within the family *Alcaligenaceae* (Ahrens et al., 1997).

   Degradation of microbial reserve polymers such as poly-(3-hydroxybutyrate) (PHB) is known to occur in marine environments and a PHB depolymerase was purified and characterized from a marine bacterium, *Alcaligenes faecalis* strain AE122 (Kita et al., 1995). From activated sludge an *Alcaligenes faecalis* strain, T1, was isolated that can hydrolyze not only water-insoluble PHB but also water-soluble D(−)-3-hydroxybutyrate oligomeric esters. The PHB depolymerase of this strain was cloned, and the structure and function of its three domains (catalytic, substrate-binding, and fibronectin type-like) were studied (Saito et al., 1989; Nojiri and Saito, 1997). Resting cells of *Alcaligenes faecalis* ATCC 8750 have been described as producing R-(−)-mandelic acid from mandelonitrile for synthesis of semisynthetic cephalosporins (Yamamoto et al., 1991). D-Aminoacylase with high stereospecificity has been purified from *Alcaligenes faecalis* strain DA1 (Yang et al., 1991b).

   A. faecalis subsp. *faecalis* has a fatty acid profile consisting of $C_{17:0\ cyclo}$, $C_{16:0}$, $C_{14:0\ 3OH}$, $C_{18:1}$, $C_{12:0\ 2OH}$, $C_{16:1\ \omega7c}$, $C_{12:0}$, $C_{19:0\ cyclo}$, $C_{14:0}$, and an unknown fatty acid (Vandamme et al., 1995c); phosphatidylethanolamine, phosphatidylglycerol, and an ornithine lipid (Yabuuchi et al., 1995), and an ubiquinone Q-8 (Lipski et al., 1992) are present; the polyamine pattern contains the predominant compound putrescine and the unusual 2-hydroxyputrescine (Busse and Auling, 1988).

   Isolated from soil, water, feces, urine, blood, sputum, wounds, pleural fluid, nematodes, and insects.
   The mol% G + C of the DNA is: 55.9–59.4 (T_m, Bd) (De Ley et al., 1970b; Pichinoty et al., 1978).
   Type strain: ATCC 8750, CCM 1052, CCUG 1325, CIP 60.80, DSM 30030, IAM 12586, IFO 13111, IMET 10443, JCM 1472, LMG 1229, NCDO 868, NCIB 8156.
   GenBank accession number (16S rRNA): M22508.

TABLE BXII.β.30. Differentiating characteristics of *Alcaligenes* and *Achromobacter*[a, b]

Characteristic	*Alcaligenes faecalis*	*Alcaligenes defragrans*[c]	*Alcaligenes latus*	*Achromobacter xylosoxidans* subsp. *denitrificans*	*Achromobacter xylosoxidans* subsp. *xylosoxidans*	*Achromobacter piechaudii*	*Achromobacter ruhlandii*
Gram reaction	−	−	−	−	−	−	−
Peritrichous flagella	+	+	+	+	+	+	+
Oxidase, catalase	+		+	+	+	+	+
Nitrate reduced to nitrite	−	+	+	[+]	+	+	+
Nitrite reduction	+	+	−	[+]	+	−	−
Nitrate respiration	−	+	−	[+]	+	−	
Nitrite respiration	+	+	−	[+]	+	−	
Chemolithotrophic growth with H$_2$	−		+	−	−	−	+
Hydrolysis of:							
Gelatin, starch, Tween 80	−		+	−	+	−	−
Urease	−			−	−	−	−
Acid from:							
D-Glucose	−			−	+	−	+
D-Xylose	−			−	+	−	+
Carbon source for growth:[d]							
D-Glucose	−	−	+	−	+		+
D-Xylose	−	−	−	−	+		+
D-Fructose	−	−	+	−	d		
D-Arabinose	−	−		−	d		+
D-Mannose	−		−	−	d		−
Adipate, pimelate	−	−	−	+	+	+	+
D-Gluconate	−	d[e]	+	[+]	+	+[f]	+
Sucrose	−	−[g]	+	−	−		
Trehalose	−	−[g]	d	−	−		
D-Arabitol	−		d	−	−		
Sebacate	−	−	d	+	+		+
Suberate	−	−	+	+	+		+
meso-Tartrate	−		−	+	+		+[f]
D-Tartrate	−		+	d	d		−[f]
D-Malate	−	+[e]	−	+	+		+[f]
Acetate	+	+	−	+	+	+	
Valerate, isovalerate	d		−	d	+	+	
Itaconate	−	−	+	+	+	+	+
Mesaconate	−		−	d	+	+	+
Glycerol	−		+	d	d		
β-Alanine	−	+[e]	+	d	d		+
Propionate	+	+	d	+	+		
Citrate	+		d	+	+		+
N-Acetylglucosamine	−			−	−		−
Pimelate	−	−	−	+	+	+	
Pimelate	−	−	−	+	+	+	
5-Ketogluconate, esculin	−			−	−	−	
D-Lyxose, L-xylose, inulin, D-tagatose, melezitose, L-fucose, arbutin, gentiobiose, turanose, L-arabitol, adonitol, glycogen, melibiose, amygdalin, spermine, histamine, ethanolamine, benzylamine, pentylamine	−			−	−		
Maltose, 2-ketogluconate, butylamine, betaine, creatine, sarcosine, *p*-hydroxybenzoate	−		+	−	−		
Butyrate, succinate, fumarate	+	+	+	+	+		
D- and L-α-alanine, L-glutamate	+	+[e]	+	+	+		
L-Malate	+	+[e]	+	+	+	+[f]	+[f]

(*continued*)

TABLE BXII.β.30. *(cont.)*

Characteristic	*Alcaligenes faecalis*	*Alcaligenes defragrans*[c]	*Alcaligenes latus*	*Achromobacter xylosoxidans* subsp. *denitrificans*	*Achromobacter xylosoxidans* subsp. *xylosoxidans*	*Achromobacter piechaudii*	*Achromobacter ruhlandii*
DL-Lactate, DL-β-hydroxybutyrate, L-proline, L-aspartate	+		+	+	+		
Isobutyrate	d		+	d	+		
Heptanoate	d	+	-	d	d		
Caproate, caprylate, pelargonate, maleate	d			d	d		-
Glycine	d		-	d	d		+
Benzoate	d	-	-	d	d		
Caprate	+		-	d	+[c]		+[a]
Malonate	+		+	d	-		-
Glutarate	d	+	-	+	+		
α-Ketoglutarate	d		-	+	+		
Azelate	-			+	+		+
Glycolate	+		d	d	-		
DL-Glycerate, L-ornithine	d		+	d	d		
L-Tartrate	-	-	d	d	-		
Pyruvate	d	+	-	+	d		
Aconitate	d		+	+	+		+[j]
Citraconate	d		-	d	+		
m-Hydroxybenzoate	-		+	d	-		+
L-Mandelate	d		+	d	-		
Phenylacetate	+		+	d	+	+	+
L-Leucine	+		+	d	+		-
Quinate			+	-			-
L-Serine	-		+	d	+		+
L-Threonine	d		+	+	+		-
L-Isoleucine	+		d	d	+		-
L-Valine	d	+[e]		d	+		
L-Lysine	d		-	d	d		+
L-Arginine	-	d[e]	-	-			
L-Citrulline	-		+	d	d		+[f]
γ-Aminobutyrate	-		+	d	d		+
DL-Norleucine	+		-	d	d		-[e]
L-Tryptophan	+		-	d	d		
δ-Aminovalerate	-		-	d	d		
L-Histidine	d		-	d	+		+
L-Tyrosine	d		d	d	d		
L-Phenylalanine	+	-[e]	-	+	+		
L-Cysteine, tryptamine, DL-kynurenine	d			d	d		
L-Methionine	d			-	-		
Kynurenate, putrescine			-				
Acetamide	d		-	d	-		

[a]Symbols: +, 90% or more of the strains are positive; [+], 80% or more of the strains are positive; d, 11–79% of the strains are positive; −, <11% of strains positive.

[b]Data from Yamasato et al. (1982), Kersters and DeLey (1984b), Kiredjian et al. (1986), Vandamme et al. (1996a), Foss et al. (1998a), Yabuuchi et al. (1998a).

[c]Utilization of carbon sources was measured under denitrifying conditions (Yamasato et al., 1982).

[d]The species *A. faecalis*, *A. xylosoxidans* subsp. *denitrificans*, *A. xylosoxidans* subsp. *xylosoxidans*, *A. defragrans*, and *A. latus* do not utilize any of the following carbon sources: D-ribose or *meso*-inositol (no data available for *A. piechaudii* and *A. ruhlandii*). The species *A. faecalis*, *A. xylosoxidans* subsp. *denitrificans*, *A. xylosoxidans* subsp. *xylosoxidans*, *A. ruhlandii*, and *A. latus* do not utilize any of the following carbon sources: L-arabinose or D-mannitol (no data available for *A. defragrans* and *A. piechaudii*). The species *A. faecalis*, *A. xylosoxidans* subsp. *denitrificans*, *A. xylosoxidans* subsp. *xylosoxidans*, and *A. latus* do not utilize any of the following carbon sources: L-arabinose, D-mannitol, D-galactose, raffinose, dulcitol, D-fucose, L-rhamnose, cellobiose, lactose, salicin, *meso*-erythritol, sorbitol, oxalate, levulinate, *o*-hydroxybenzoate, D-mandelate, phthalate, L-arginine, *m*-aminobenzoate, or *p*-aminobenzoate (no data available for *A. defragrans*, *A. ruhlandii*, and *A. piechaudii*).

[e]The mechanism of compound use not specified (Foss et al., 1998a).

[f]D-(−)-Tartrate, DL-malic acid, *trans*-aconitate, and DL-citrulline were subjected to assimilation tests (Yamasato et al., 1982).

[g]The D-isomer of the respective carbon source was not used (Foss et al., 1998a).

Additional Remarks: Other representative culture collection strains belonging to this taxon: ATCC 15554, CCEB 554, NCTC 10416, DSM 30033; LMG 1230; CIP 71.8 (formerly "*A. odorans*"); ATCC 19209, NCTC 10388, Burchill 1 (formerly "*A. odorans* var. viridans"); NCIB 8687; LMG 3368; LMG 3394 (formerly "*Achromobacter arsenoxidans*").

b. **Alcaligenes faecalis** *subsp.* **parafaecalis** Schroll, Busse, Parrer, Rölleke, Lubitz and Denner 2001b, 1619[VP] (Effective publication: Schroll, Busse, Parrer, Rölleke, Lu-

TABLE BXII.β.31. Physiological and biochemical traits distinguishing *Alcaligenes faecalis* subsp. *faecalis* and *Alcaligenes faecalis* subsp. *parafaecalis*[a,b]

Characteristic	A. faecalis subsp. faecalis (n = 2)	A. faecalis subsp. parafaecalis (n = 1)
Nitrite reduction	+	−
Gelatin liquefaction (37°C)	−	+
Assimilation of:[c]		
L-Histidine	+	−
L-Tryptophan	+	−
Benzoate	+	−
Gentisate	−	+
Biolog reactions:[d]		
L-Serine	+	−
L-Ornithine	+	−
L-Histidine	+	−
Aerobic growth at 42°C	+	−
Growth in the presence of 7% NaCl	+	−
Diagnostic polyamine content:		
2-hydroxyputrescine[e]	1.7–8.9	tr

[a]Symbols: +, positive reaction; −, no reaction.

[b]Data from Busse and Auling (1988), Ahrens et al. (1997), and Schroll et al. (2001a).

[c]Estimated by API Biotype 100 test.

[d]Estimated by Biolog GN system.

[e]mol/g dry weight; tr = trace amounts (<0.1 mol/g).

bitz and Denner 2001a, 41.)

para.fae.ca' lis. Gr. prep. *para* along side of, resembling; M.L. adj. *faecalis* specific epithet; M.L. *parafaecalis* intended to mean alongside of the species *A. faecalis.*

The description of this subspecies is taken from Schroll et al. (2001a). Cells are Gram negative, rod shaped, 0.75–1.0 × 1.5–3.0 μm, nonsporeforming, and motile. Colonies on standard bacteriological media are circular, entire, low convex, and smooth with a thin spreading irregular edge; diameter is up to 2.0 mm after 1 d incubation at 37°C. The following tests were positive: oxidase activity, catalase activity, alkalization, growth on MacConkey agar, citrate utilization, acetoin production, gelatin liquefaction at 37°C, assimilation of gentisate; enzymatic activities as follows: L-alanine aminopeptidase, alkaline phosphatase, esterase (C₄), leucine arylamidase, valine arylamidase, acid phosphatase, weak reactions for esterase lipase (C₈), lipase (C₁₄), naphthol-AS-BI-phosphohydrolase. No growth at 42°C or on MacConkey agar. Acid is not produced in Hugh/Leifson's glucose O-F medium, and L-histidine, L-tryptophan, and benzoate are not assimilated. Lysine decarboxylase, ornithine decarboxylase, tryptophan deaminase, indole production, H₂S production, nitrate reduction, nitrite reduction, urease activity, acetoin production, gelatin liquefaction at 28°C, hydrolysis of Tween 80 and casein, DNase activity, cystine arylamidase, trypsin, chymotrypsin, α-galactosidase, β-galactosidase, β-glucuronidase, α-glucosidase, β-glucosidase, N-acetyl-β-glucosamidase, α-mannosidase, and α-fucosidase are negative. The following compounds are assimilated (as tested by growth in M9 medium): acetate, propionate, butyrate, valerate, heptanoate, γ-hydroxybutyric acid, γ-butyrolactone, L-lactate, citrate, ethanol. By using the Biolog GN system the following reactions are positive: Tween 40, Tween 80, meth-

ylpyruvate, mono-methyl pyruvate, acetic acid, *cis*-aconitic acid, citric acid, formic acid; α-, β-, and γ-hydroxybutyric acid, *p*-hydroxyphenyl acetic acid, α-ketobutyric acid, D,L-lactic acid, malonic acid, propionic acid, succinic acid, bromosuccinic acid, succinamic acid, alaninamide, D-alanine, L-alanine, L-glutamic acid, L-leucine, L-phenylalanine, L-proline, urocanic acid, and phenylethylamine; weak reactions are observed for glycogen, L-alanylglycine, L-asparagine, L-aspartic acid, glycyl-L-glutamic acid, L-pyroglutamic acid, D-serine, L-threonine, and inosine. The following compounds are not assimilated (as tested by growth in M9 medium): 1,4 butandiol, 2,3 butandiol, methanol, 1-butanol, 2-butanol, 1-hexanol, glycerol, acetone, and tartric acid. Using the Biolog GN system, the following reactions are negative: α-cyclodextrin, dextrin, N-acetylglucosamine, adonitol, L-arabinose, D-arabitol, cellobiose, *i*-erythritol, D-fructose, L-fucose, D-galactose, gentiobiose, α-D-glucose, *meso*-inositol, α-D-lactose, lactulose, maltose, D-mannitol, D-mannose, D-melibiose, β-methyl-D-glucoside, D-psicose, D-raffinose, L-rhamnose, D-sorbitol, sucrose, D-trehalose, turanose, xylitol, D-galactonic acid lactone, D-galacturonic acid, D-gluconic acid, D-glucosaminic acid, D-glucuronic acid, itaconic acid, α-ketoglutaric acid, quinic acid, D-saccharic acid, sebacic acid, glucuronamide, glycyl-L-aspartic acid, L-histidine, hydroxy-L-proline, L-ornithine, L-pyroglutamic acid, L-serine, D,L-carnitine, γ-amino butyric acid, uridine, thymidine, putrescine, 2-amino ethanol, 2,3 butanediol, glycerol, D,L-α-glycerol phosphate, glucose-1-phosphate, and glucose-6-phosphate.

A. faecalis subsp. *parafaecalis* has a fatty acid profile consisting of $C_{16:0}$, $C_{16:1}$, $C_{17:0\ cyclo}$, $C_{18:1\ \omega7c/9t}$, $C_{14:0\ 3OH}$, $C_{14:0}$, $C_{12:0\ 2OH}$, $C_{12:0}$, $C_{18:0}$, $C_{15:0}$, and $C_{17:0}$; a ubiquinone Q-8; and a polyamine pattern with putrescine as the predominant compound and only trace amounts of the *Betaproteobacteria*-specific polyamine 2-hydroxyputrescine.

Isolated from water (garden pond, Lower Austria; Austria).

The mol% G + C of the DNA is: 56 (T_m).

Type strain: G, CIP 106866, DSM 13975.

GenBank accession number (16S rRNA): AJ242986.

2. **Alcaligenes defragrans** Foss, Heyen and Harder 1998b, 1083[VP] (Effective publication: Foss, Heyen and Harder 1998a, 243.)

de.fra.grans. L. prep. *de* away, from; L. adj. *fragrans* sweet scented; M.L. defragrans to annihilate fragrance, referring to the capacity to degrade monoterpenes.

The morphological characteristics are as described for the genus. The facultatively anaerobic chemoorganotrophic metabolism is strictly oxidative. Oxygen, nitrate, nitrite, or dinitrogen oxide can serve as electron acceptor. Under denitrifying conditions, acetate, propionate, butyrate, hexanoate, heptanoate, octanoate, pyruvate, malate, succinate, fumarate, 3-methylbutyrate, glutarate, glutamate, alanine, valine, ethanol, and the monoterpenes (+)-ρ-menth-1-ene, (+)-limonene, (−)-α-phellandrene, α-terpinene, γ-terpinene, (+)-sabinene, (+)-2-carene, (+)-3-carene, (−)-α-pinene, (−)-β-pinene, terpinolene, (+)-α-terpineol, and

(+)-terpinen-4-ol are used as carbon and energy sources. The majority of strains grow on myrcene, arginine, and gluconate. Under denitrifying conditions no growth occurs on D-glucose, D-fructose, D-sorbitol, *meso*-inositol, D-ribose, D-arabinose, D-xylose, D-saccharose, D-trehalose, D-cellobiose, ascorbate, phenylalanine, formate, L-tartrate, adipate, pimelate, suberate, sebacate, itaconate, methanol, cyclohexanol, cyclohexane-1,2-diol, cyclohexane-1,4-diol, decane, hexadecane, heptamethylnonane, cyclohexane, benzoate, toluene, 2,6-dimethyloctane, 3,7-dimethyl-1-octene, (−)-β-citronellene, 3,7-dimethyloctanol-1, (−)-β-citronellol, geraniol, nerol, linalol (+)-*trans*-isolimonene, ρ-cymene, (+)-perilla alcohol, (−)-carveol, (+)-dihydrocarveol, menthol, menthone, (+)-isopulegol, (+)-isomenthol, (+)-pulegone, (−)-carvone, (+)-dihydrocarvone, (−)-*trans*-pinane, eucalyptol, (−)-*cis*-myrtanol, (+)-*trans*-myrtanol, (−)-myrtenol, (+)-isopinocampheol, (−)-borneol, (+)-fenchol, (α+β)-thujone, (+)-fenchone, (−)-verbenone, (+)-camphor, and pine needle oil.

When grown on monoterpene and nitrogen, no vitamins are needed, and growth occurs at 15–40°C in the range of pH 5.9–8.4; pH-optimum is at 6.3–7.8.

The whole cell fatty acid profile is strongly dependent on the culture conditions. When grown under denitrifying conditions on monoterpene, the fatty acid profile is characterized by the predominant acids $C_{16:0}$ and $C_{17:0\ cyclo}$ and varying amounts of dodecanal, $C_{12:0}$, $C_{14:0}$, $C_{14:0\ 3OH}$, $C_{15:0}$, $C_{16:1}$, $C_{18:1}$, and $C_{19:0\ cyclo}$. When acetate is the only carbon and energy source $C_{16:1}$, $C_{16:0}$, $C_{17:0\ cyclo}$, and $C_{18:1}$ are the predominant fatty acids.

The species *Alcaligenes defragrans* was proposed by Foss et al. (1998a) to encompass four monoterpenes degrading strains. 16S rDNA sequence comparisons indicate that *A. defragrans* is more closely related to *Bordetella* and *Achromobacter* than *A. faecalis*. This relatedness is also reflected by its DNA base ratio, which is in the range of *Achromobacter* and *Bordetella*.

Isolated from activated sludge and a forest ditch.

The mol% G + C of the DNA is: 66.9 (HPLC).

Type strain: 54Pin, DSM12141.

GenBank accession number (16S rRNA): AJ005447.

Additional Remarks: Other representative culture collection strains belonging to this taxon: DSM 12142 (62Car), DSM 12143 (65Phen), DSM 12144 (51Men) (*Alcaligenes defragrans*).

Species Incertae Sedis

1. **Alcaligenes latus** Palleroni and Palleroni 1978, 423[AL]

la′ tus. L. adj. *latus* broad.

The cells are short to coccoid rods, 1.2–1.4 × 1.6–2.4 μm, occurring singly, in pairs, or in short chains. Gram negative. Motile by means of 5–10 peritrichous flagella. The cells are frequently heavily granulated. Under autotrophic growth conditions on a solid mineral medium colonies are round, grayish pink, and opaque (Palleroni and Palleroni, 1978). Colonies are wrinkled in fresh isolates but can become smooth upon subcultivation.

Facultatively chemolithotrophic in an atmosphere containing hydrogen, oxygen, and carbon dioxide. Can grow with dinitrogen as sole nitrogen source (H.G. Schlegel, personal communication). Physiological and nutritional characteristics are given in Table BXII.β.29 of *Alcaligenaceae* and Table BXII.β.30 of this chapter. Optimal growth temperature is about 35°C. A membrane-bound hydrogenase, but no soluble, NAD-reducing hydrogenase has been found in three strains of *A. latus* (Palleroni and Palleroni, 1978). *meta*-Hydroxybenzoate is metabolized via the gentisate pathway and protocatechuate is degraded by *meta*-cleavage when grown on *p*-hydroxybenzoate or quinate.

Three strains were isolated from soil (Palleroni and Palleroni, 1978). The fatty acid profile is characterized by the following compounds: $C_{16:1}$, $C_{18:1}$, $C_{16:0}$, $C_{18:0}$, $C_{14:0}$, $C_{12:0}$, $C_{17:0\ cyclo}$, $C_{17:0}$, $C_{15:0}$, $C_{10:0\ 3OH}$, and $C_{14:1}$ and the major respiratory quinone is ubiquinone Q-8 (Oyaizu-Masuchi and Komagata, 1988).

The mol% G + C of the DNA is: 69.1–71.1 (Bd, T_m) (Palleroni and Palleroni, 1978; Kersters and De Ley, 1984b).

Type strain: Palleroni H-4, ATCC 29712, CIP 10345, DSM 1122.

GenBank accession number (16S rRNA): D88007.

The species *Alcaligenes latus* was proposed by Palleroni and Palleroni (1978) to encompass hydrogen oxidizing, peritrichously flagellated bacteria. DNA–rRNA hybridization studies revealed that this species is most closely related to *Rubrivivax gelatinosus* and *Leptothrix discophora* (Willems et al., 1991b), and distant from the family *Alcaligenaceae*. Thus, *A. latus* cannot be considered as a member of the genus or even of the family. It is most likely that *A. latus* is a member of a genus that has not yet been described. The species description of *A. latus* is a revised version of that given by Kersters and De Ley (1984b).

Genus II. **Achromobacter** *Yabuuchi and Yano 1981, 477*[VP] *emend. Yabuuchi, Kawamura, Kosako and Ezaki 1998a, 1083*

HANS-JÜRGEN BUSSE AND GEORG AULING

A.chro.mo.bac′ ter. Gr. adj. *achromus* colorless; M.L. n. *bacter* the masc. equivalent of Gr. neut. n. *bactrum* a rod or staff; M.L. masc. n. *Achromobacter* colorless rodlet.

Straight rods, 0.8–1.2 × 2.5–3.0 μm with rounded ends. **Gram negative**. Nonsporeforming. **Motile with 1–20 sheathed flagella arranged peritrichously. Obligately aerobic** and nonfermentative. Some strains are capable of anaerobic respiration with nitrate as the electron acceptor. They perform nitrate respiration combined with nitrite and nitrous oxide respiration. Some strains are facultative lithoautotrophic hydrogen-oxidizers. **Oxidase positive. Catalase positive.** Urease, DNase, phenylalanine deaminase,

lysine and ornithine decarboxylase, arginine dihydrolase, and gelatinase negative. Nonhalophilic, nonhemolytic, nonpigmented. **Chemoorganotrophic, using a variety of organic acids and amino acids as carbon sources.** Carbohydrates usually not utilized. *Achromobacter xylosoxidans* subsp. *xylosoxidans* and *Achromobacter ruhlandii* utilize D-glucose as sole carbon source and produce acid from D-glucose, D-arabinose, and D-xylose. **Those strains analyzed contain the characteristic fatty acids $C_{17:0\ cyclo}$,**

$C_{16:0}$, $C_{14:0\ 3OH}$, $C_{16:1}$, $C_{12:0\ 2OH}$, and **ubiquinone Q-8**. The **polyamine patterns** display **putrescine** as the major compound and contain the unusual diamine **2-hydroxyputrescine**.

Isolated from water, soil; also from hospital environment and human clinical specimens with pathological significance or as contaminants.

The mol% G + C of the DNA is: 65–68.

Type species: **Achromobacter xylosoxidans** Yabuuchi and Yano 1981, 477.

FURTHER READING

De Ley, J., P. Segers, K. Kersters, W. Mannheim and A. Lievens. 1986. Intrageneric and intergeneric similarities of the *Bordetella* ribosomal ribonucleic acid cistrons: proposal for a new family, *Alcaligenaceae.* Int. J. Syst. Bacteriol. *36*: 405–414.

List of species of the genus Achromobacter

1. **Achromobacter xylosoxidans** (ex Yabuuchi and Ohyama 1971) Yabuuchi and Yano 1981, 477[VP] emend. Yabuuchi, Kawamura, Kosako and Ezaki 1998a, 436.

   xy.los.ox'.i.dans. Gr. n. *xylon* wood, xylose, wood sugar; Gr. adj. *oxys* sharp, acid; L. part. *dans* giving; M.L. pres part. *oxydans* acid-giving, oxidizing; M.L. *xylosoxydans* oxidizing xylose.

   The description is taken from Yabuuchi et al. (1998a). The morphological characteristics are as described for the genus. Colonies on nutrient agar are circular, nonpigmented to grayish white, translucent to opaque, flat to convex, usually smooth, sometimes dull or rough, margin usually entire. On heart infusion agar colonies are 1 mm in diameter, low convex with entire margin, moist, and with a glistening surface. Grows anaerobically in the presence of nitrate or nitrite by denitrification. In the API 20NE test citrate, adipic acid, DL-malic acid, and phenylacetate are assimilated. Assimilation of gluconate, capric acid, and glucose is variable. L-arabinose, D-mannitol, D-mannose, maltose, and *N*-acetyl-D-glucosamine are not assimilated.

   The mol% G + C of the DNA is: 63.9–69.8 (Bd, T_m).

   a. **Achromobacter xylosoxidans** *subsp.* **xylosoxidans** (ex Yabuuchi and Ohyama 1971) Yabuuchi and Yano 1981, 477[VP] (*Alcaligenes denitrificans* subsp. *xylosoxidans* (Yabuuchi and Yano 1981) Kersters and De Ley 1984b, 367; *Alcaligenes xylosoxidans* subsp. *xylosoxidans* (Yabuuchi and Yano 1981) Kiredjian, Holmes, Kersters, Guilvout and De Ley 1986, 285.)

   If not stated otherwise, the description is taken from Kersters and De Ley (1984b) and Yabuuchi et al. (1998a). Nearly all strains utilize D-glucose, D-xylose, D-gluconate, adipate, and pimelate as sole carbon sources and characteristically form acid from D-xylose in the O/F medium of Hugh and Leifson (1953). Further characteristics are shown in Table BXII.β.30.

   A plasmid-encoded resistance by two distinct nickel resistance loci (high- and low-level) against nickel, cobalt, and cadmium has been described for strain *Achromobacter xylosoxidans* 31A (Schmidt et al., 1991; Schmidt and Schlegel, 1994), originally isolated from a copper galvanization tank. Maintenance of the biodegradation capacities of *A. xylosoxidans* and other Gram-negative aerobic bacteria containing large plasmids during long-term preservation was studied and protective agents have been described (Lang and Malik, 1996).

   Mostly isolated from clinical specimens such as blood, sputum, wounds, purulent ear discharge, spinal fluid, cerebral tissue, urine, feces, and, in a few cases, also from disinfectant solutions.

   The fatty acid profile is characterized by the following compounds: $C_{16:0}$, $C_{17:0\ cyclo}$, $C_{14:0\ 3OH}$, $C_{16:1}$, $C_{14:0\ 2OH}$, $C_{12:0\ 2OH}$, $C_{16:0\ 2OH}$, $C_{18:0}$, $C_{14:0}$, and $C_{19:0\ cyclo}$ (Vandamme et al., 1995c). Phosphatidylethanolamine, phosphatidylglycerol, and diphosphadidylglycerol are the major compounds in the polar lipid profile (Yabuuchi et al., 1974). The polyamine pattern contains two major compounds: putrescine and the unusual 2-hydroxyputrescine (Busse and Auling, 1988).

   The mol% G + C of the DNA is: 66.9–69.8 (T_m) (De Ley et al., 1970b; Yabuuchi et al., 1974; Holmes et al., 1977b).

   Type strain: Hugh 2838, ATCC 27061, CIP 71.32, Yabuuchi KM 543, NCTC 10807.

   GenBank accession number (16S rRNA): X59163, D88005.

   Additional Remarks: The type strain was proposed as the type strain for *Achromobacter xylosoxidans* by Yabuuchi and Ohyama (1971). Other representative culture collection strains belonging to this taxon: CIP 58.72 and CIP 61.20 (formerly *A. denitrificans*). In the near future Coyne et al. will propose the reclassification of *Alcaligenes denitrificans* as *Achromobacter denitrificans* (Int. J. Syst. Evol. Microbiol., in press, 2003).

   b. **Achromobacter xylosoxidans** *subsp.* **denitrificans** (Rüger and Tan 1983) Yabuuchi, Kawamura, Kosako and Ezaki 1998b, 1083[VP] (Effective publication: Yabuuchi, Kawamura, Kosako and Ezaki 1998a, 436) (*Alcaligenes denitrificans* Rüger and Tan 1983, 88; *Alcaligenes denitrificans* subsp. *denitrificans* Rüger and Tan 1983; *Alcaligenes xylosoxidans* subsp. *denitrificans* (Rüger and Tan 1983) Kiredjian, Holmes, Kersters, Guilvout and De Ley 1986, 285.)

   de.ni.tri'fi.cans. L. prep. *de* away, from; L. n. *nitrum* soda; M.L. n. *nitrum* nitrate; M.L. v. *denitrificans* to denitrify; M.L. pres. part. *denitrificans* denitrifying.

   If not stated otherwise the description is taken from Kersters and De Ley (1984b) and Yabuuchi et al. (1998a). Physiological and nutritional characteristics of the species are presented in Table BXII.β.29 of *Alcaligenaceae* and Table BXII.β.30 of this chapter. Most strains characteristically use *meso*-tartrate, itaconate, adipate, pimelate, and other dicarboxylic acids as sole carbon sources. Some strains are auxotrophic and require organic nitrogenous compounds for growth. Nitrates and nitrites are usually reduced. Most strains carry out an anaerobic respiration in the presence of nitrate and nitrite.

   The fatty acid profile is characterized by the following compounds: $C_{16:0}$, $C_{17:0\ cyclo}$, $C_{14:0\ 3OH}$, $C_{14:0}$, $C_{16:1}$, $C_{12:0\ 2OH}$, $C_{18:1}$, $C_{12:0}$, $C_{18:0}$, $C_{16:0\ 2OH}$, and $C_{19:0\ cyclo}$ (Vandamme et al., 1995c). Ubiquinone Q-8 is the predominant respiratory quinone (Lipski et al., 1992); the polyamine pattern contains two major compounds: putrescine and the unusual 2-hydroxyputrescine (Busse and Auling, 1988).

Isolated from soil, and a variety of clinical specimens such as feces, urine, blood, pleural fluid, purulent ear discharges, prostatic secretions, and throat swabs.

Strains of *A. xylosoxidans* subsp. *denitrificans* or *A. xylosoxidans* subsp. *xylosoxidans* have been isolated from soil and examined for presence of D-aminoacylase in order to produce D-amino acids from *N*-acetyl-DL-amino acids enzymatically. The presence of D-aminoacylase appeared not to be a general characteristic of members of the genus, although inducible D-aminoacylases with high stereospecificity and higher specific activity than those described from *Pseudomonas* and *Streptomyces* were found (Moriguchi and Ideta, 1988; Tsai et al., 1988). A novel cyanide-hydrolyzing enzyme from *A. xylosoxidans* subsp. *denitrificans* has been described (Ingvorsen et al., 1991). Recently a gene from *A. xylosoxidans* subsp. *denitrificans* has been analyzed that confers albicidin resistance by reversible antibiotic binding. This gene appears to be a useful candidate for transfer to plants to protect plastid DNA replication from inhibition by albicidin phytotoxins involved in sugarcane leaf scald disease (Basnayake and Birch, 1995). Uptake of chloro-substituted benzoic acids that are formed as dead-end metabolites by polychlorinated biphenyl-degrading microorganisms has been studied in *A. xylosoxidans* subsp. *denitrificans* (Miguez et al., 1995). *A. xylosoxidans* subsp. *denitrificans* was shown to mineralize the structurally related chlorinated phenoxy alkanoics, including 2,4-dichlorophenoxyacetic acid (2,4-D), 2-methyl-4-chlorophenoxyacetic acid (MCPA), and 2-(2-methyl-4-chlorophenoxy) propionic acid (mecoprop), which are among the most widely used herbicides to control broad-leafed weeds in cereal crops throughout the world. The strain described uses the *ortho*-pathway for biodegradation (Tett et al., 1994, 1997). Reductive dechlorination of 2,4-dichlorobenzoate and hydrolytic dehalogenation of 4-halo-substituted aromatics has been studied with *A. xylosoxidans* subsp. *denitrificans* NTB-1 (van der Tweel et al., 1987). Gene transfer of plasmid-encoded 2,2-dichloropropionate halidohydrolase from *A. xylosoxidans* subsp. *denitrificans* to other soil bacteria has been demonstrated using a soil microcosm (Brokamp and Schmidt, 1991). Recently the plasmid (pFL40)-encoded D,L-haloalkanoic acid dehydrogenase gene (*dhlIV*) was sequenced and expressed in members of the *Gammaproteobacteria* (Brokamp et al., 1997). Gas-phase methyl ethylketone biodegradation in a novel type of bioreactor was demonstrated with a bacterial consortium dominated by *A. xylosoxidans* subsp. *denitrificans*, run under non-axenic conditions (Agathos et al., 1997).

The mol% G + C of the DNA is: 63.9–68.9 (T_m, Bd) (De Ley et al., 1970b; Pichinoty et al., 1978).

Type strain: Hugh 12, ATCC 15173, CIP 77.15, DSM 30026, NCTC 8582.

GenBank accession number (16S rRNA): M22509.

Additional Remarks: Other representative culture collection strains belonging to this taxon are ATCC 13138, CIP 60.81 (*A. xylosoxidans* subsp. *denitrificans*).

2. **Achromobacter piechaudii** (Kiredjian, Holmes, Kersters, Guilvout and De Ley 1986) Yabuuchi, Kawamura, Kosako and Ezaki 1998b, 1083^{VP} (Effective publication: Yabuuchi, Kawamura, Kosako and Ezaki 1998a, 436) (*Alcaligenes piechaudii* Kiredjian, Holmes, Kersters, Guilvout and De Ley 1986, 285.)

pie.chau' di.i. M.L. gen. n. *piechaudii* from Piechaud, to honor M. Piechaud, a bacteriologist at the Institut Pasteur, Paris, France.

The description is taken from the original paper. Gramnegative, obligately aerobic, straight rods, 0.5–1.0 × 1.0–1.5 μm, which have parallel sides and rounded ends. Cells occur singly and are nonsporulating and noncapsulated. Cells are motile by means of two to eight peritrichous flagella, which have one or two wavelengths. At the optimal temperature (28–30°C) visible growth within 24 h. Liquid cultures (in nutrient broth medium) are uniformly turbid; cultures on trypticase soy agar produce small smooth colonies (diameter, 0.2 mm). Colonies are 1 mm in diameter after 48 h of incubation; no pigments or odor produced. Circular, smooth, entire colonies on nutrient agar after 24 h of growth. Colonies are nonhemolytic on 5% (v/v) horse blood agar. The organisms are nonfluorescent on King medium B and do not produce a brown diffusible melanin-like pigment on tyrosine agar. Growth without NaCl in peptone–water medium and at a maximum NaCl concentration of 7% (w/v). Alkaline reaction in glucose oxidation–fermentation medium. Catalase positive. Oxidase positive. Hydrolyzes tributyrin and tyrosine but not gelatin, Tween 20, Tween 80, esculin, or starch. Casein is not digested. Extracellular deoxyribonuclease or opalescence on lecithovitellin agar is not produced. Does not produce indole, urease, arginine desimidase, arginine dihydrolase, lysine decarboxylase, ornithine decarboxylase, β-D-galactosidase (as determined by the *o*-nitrophenyl-β-D-galactopyranoside test) or phosphatase. Reduces nitrate but not nitrite; no anaerobic respiration in the presence of nitrate or nitrite. Produces hydrogen sulfide, as determined by the lead acetate paper method, but not as determined by the triple-sugar-iron agar method. Does not reduce thiosulfate. Grows on β-hydroxybutyrate (with production of lipid inclusion granules), on cetrimide, and on MacConkey agar. Produces alkali on Christensen citrate (as determined by the NCTC method; Kiredjian et al., 1986). Utilizes citrate (Simmons medium) as determined by the IP (Kiredjian et al., 1986) method but not as determined by the NCTC method; does not utilize malonate (as determined by the NCTC method). Oxidizes gluconate as determined by the IP method but not as determined by the NCTC method. Tolerates KCN at a concentration of 0.0075%. Does not produce 3-ketolactose. Does not reduce selenite and does not deaminate histidine, leucine, phenylalanine, or tryptophan. In minimal synthetic medium *n*-valerate, isovalerate, D-gluconate, mesaconate, and itaconate are assimilated. Does not assimilate carbohydrates in minimal synthetic medium. Produces acid in ammonium salt medium under aerobic conditions from ethanol. Does not produce acid in ammonium salt medium under aerobic conditions from D-glucose, adonitol, L-arabinose, D-cellobiose, dulcitol, D-fructose, glycerol, *m*-inositol, lactose, maltose, mannitol, raffinose, L-rhamnose, salicin, D-sorbitol, sucrose, trehalose, or D-xylose. Does not produce acid from 10% (w/v) D-glucose or 10% (w/v) lactose. Does not produce acid or gas from D-glucose in peptone–water medium.

Hydrolyzes the following substrates (determined by using API ZYM galleries): L-leucyl-2-naphthylamide, glycyl-β-naphthylamide, L-aspartyl-β-naphthylamide, arginyl-β-naphthylamide, L-alanyl-β-naphthylamide, and L-leucyl-glycyl-β-naphthylamide. Does not hydrolyze the following substrates (determined by using API ZYM galleries): 2-naphthyl phosphate (at pH 8.5), 2-naphthyl caprylate, 2-naphthyl myristate, L-valyl-2-naphthylamide, L-cystyl-2-naphthylamide, N-benzoyl-DL-arginine-2-naphthylamide, N-glutaryl-phenylalanine-2-naphthylamide, naphthol-AS-B1-phosphodiamide, 6-bromo-2-naphthyl-α-D-galactopyranoside, 2-naphthyl-β-D-galactopyranoside, naphthol-AS-BI-β-D-glucuronic acid, 2-naphthyl-α-D-glucopyranoside, 6-bromo-2-naphthyl-β-D-glucopyranoside, 1-naphthyl-N-acetyl-β-D-glucosaminide, 6-bromo-2-naphthyl-α-D-mannopyranoside, 2-naphthyl-α-L-fucopyranoside, α-D-galactopyranoside, β-D-galactopyranoside, β-D-galactopyranoside-6-phosphate, α-L-arabinofuranoside, α-D-glucopyranoside, β-D-glucopyranoside, β-D-galacturonide, β-D-glucuronide, α-maltoside, β-maltoside, N-acetyl-α-D-glucosaminide, N-acetyl-β-D-glucosaminide, α-L-fucopyranoside, β-D-fucopyranoside, β-D-lactoside, α-D-mannopyranoside, β-D-mannopyranoside, α-D-xylopyranoside, β-D-xylopyranoside, L-tyrosyl-β-naphthylamide, L-phenylalanyl-β-naphthylamide, L-hydroxyprolyl-β-naphthylamide, L-histidyl-β-naphthylamide, N-benzoyl-L-leucyl-β-naphthylamide, S-benzoyl-L-cystyl-β-naphthylamide, glycyl-L-prolyl-β-naphthylamide, N-carbobenzoxy-L-arginine-4-methoxy-β-naphthylamide hydrochloride, L-isoleucyl-β-naphthylamide, L-prolyl-β-naphthylamide hydrochloride, L-threonyl-β-naphthylamide, L-tryptophyl-β-naphthylamide and N-carbobenzoxy-glycyl-glycyl-L-arginine-β-naphthylamide.

Growth occurs at room temperature and 37°C but not at 5° or 42°C (optimal growth temperature 28–30°C). Although strains of *A. piechaudii* have been recovered mainly from human clinical material, few clinical details are available, so the clinical significance of the species is not yet determined.

The fatty acid profile is characterized by the following compounds: $C_{17:0\ cyclo}$, $C_{16:0}$, $C_{14:0\ 3OH}$, $C_{14:0}$, $C_{18:1}$, $C_{16:1}$, $C_{12:0\ 2OH}$, $C_{18:0}$, $C_{16:0\ 2OH}$, and $C_{19:0\ cyclo}$ (Vandamme et al., 1995c); the polyamine patterns contains two major compounds: putrescine and the unusual 2-hydroxyputrescine (Hamana and Takeuchi, 1998).

The mol% G + C of the DNA is: 64–65 (T_m).

Type strain: Hugh 366-5, CIP 60.75ATCC 43552, IAM 12591, LMG 1873.

GenBank accession number (16S rRNA): AB010841.

Additional Remarks: Other representative culture collection strains belonging to this taxon are LMG 6100 (CIP 55774), LMG 2828 (AB118), LMG 1861 (NCMB 1051), LMG 6102 (CL544/75), LMG 6101 (CL807/79), LMG 6103 (CL237/83).

3. **Achromobacter ruhlandii** (Aragno and Schlegel 1977) Yabuuchi, Kawamura, Kosako and Ezaki 1998b, 1083^VP (Effective publication: Yabuuchi, Kawamura, Kosako and Ezaki 1998a, 436) (*Alcaligenes ruhlandii* (Packer and Vishniac 1955) Aragno and Schlegel 1977, 280.)

ruh.lan' di.i. M.L. gen. n. *ruhlandii* of Ruhland, named for the German microbiologist W. Ruhland, who studied the physiology of the hydrogen bacteria.

Cells are Gram-negative, nonsporeforming rods. Nonfermentative, nonhemolytic, and nonhalophilic. Growth occurs on MacConkey agar, in the presence of 5% NaCl, and at 41°C. Catalase and oxidase positive. Simmons' citrate medium is alkalized within 2 d and malonate broth within 3 d of incubation. Nitrate is reduced to nitrite. Gas is not produced in nitrate broth. Oxidatively acid is produced in OF medium from D-arabinose within 2 d and from L-arabinose, glucose, D-ribose, and D-xylose within 3 d of incubation. No acid is produced from fructose, galactose, cellobiose, lactose, maltose, melibiose, sucrose, trehalose, raffinose, melezitose, glycerol, adonitol, dulcitol, inositol, mannitol, sorbitol, salicin, and inulin. Acylamidase test is positive within 3 d of incubation. Urease, phenylalanine deaminase, and DNase are negative. Esculin, gelatin, starch, and Tween 80 are not hydrolyzed. Lysine and ornithine decarboxylase and arginine dihydrolase are negative. No pigments are produced. Glucose, adipic acid, capric acid, malic acid, citrate, gluconate, and phenylacetate are assimilated.

Isolated from soil. The polyamine pattern contains two major compounds: putrescine and the unusual 2-hydroxyputrescine (Busse and Auling, 1988).

The only available strain, *A. ruhlandii* ATCC 15749 (DSM 653), can grow autotrophically with hydrogen. Hydrogen-oxidizing strains similar to the only available strain of the species *A. ruhlandii* or other autohydrogenotrophic denitrifying members of the genus *Achromobacter* could not be isolated from autohydrogenotrophic pilot-reactors for denitrification of drinking water (Vanbrabant, et al., 1993; Auling and Luo, unpublished results).

The mol% G + C of the DNA is: 68.1 (T_m) (De Ley et al., 1986) and 67.7 (HPLC) (Yabuuchi et al., 1998a).

Type strain: ATCC 15749, DSM 653, IAM 12600.

GenBank accession number (16S rRNA): AB010840.

Other Organisms

For *Achromobacter* species (formerly assigned to the genus *Alcaligenes*) a significant environmental role on biodegradation of aromatic or halogenated compounds, even in the deep subsurface (Boivin-Jahns et al., 1995), is proposed by reports on isolation of strains, tentatively assigned to *A. xylosoxidans* (Ewers et al., 1990) and *A. denitrificans* (Weissenfels et al., 1990). The descriptive information is limited to the three species comprising the genus *Achromobacter*—*A. xylosoxidans*, *A. piechaudii*, and *A. ruhlandii*—and close relatives. Nevertheless, strains preliminarily identified as members of the genus *Alcaligenes* are still awaiting final taxonomic characterization, although published and named. Thus, care has to be taken when extrapolating the properties of such interesting strains to the whole species or even the genus.

The main phylogenetic aspects have been treated in the description of the family. DNA–DNA hybridizations indicated the emergence of an additional species in the neighborhood of *A. denitrificans* that may be centered around a strain O-1 able to degrade sulfonated aromatic compounds (Jahnke et al., 1990). The G + C content (66.1 mol%) of *Alcaligenes* sp. strain O-1 corresponds to that of the species. The DNA relatedness values obtained (51% to the type strain of *A. xylosoxidans* subsp. *xylosoxidans* and 40% to the type strain of *A. xylosoxidans* subsp. *denitrificans*) would justify a status as a separate species. Polyclonal antibodies raised against strain O-1 displayed the highest cross-reactivity with *A. xylosoxidans*. Thus, *Alcaligenes* O-1 can be considered as another species of the genus *Achromobacter*. *Alcaligenes* sp. strain O-1 contains a 117-MDa conjugative plasmid (pSAH) encoding mineralization of sulfonated aromatic compounds. The conjugative transfer of the toluene transposon Tn 4651 from *Pseudomonas putida* harboring plasmid pWW0, the archetypal TOL plasmid (Assinder and Williams, 1990), into plasmid pSAH extends the expression range of the TOL catabolic genes (degradation of 3-methylbenzoic acid) to the genus *Achromobacter* (Jahnke et al., 1993). A chlorobenzoate transposon (Tn5271) from the indigenous plasmid pBRC60 in *Alcaligenes* sp. strain BR60, isolated from runoff waters adjacent to a chlorobenzoate contaminated landfill, has been intensively studied genetically, although the taxonomic characterization of the host strain is still lacking (Windham et al., 1994).

The expression of xenobiotic-degrading genes using naphthalene as substrate has been investigated in *Alcaligenes* sp. (strain NP-Alk) versus *Pseudomonas putida* ATCC 17484, using these microorganisms as models for bioremediation of contaminated environments by either indigenous bacteria or strains released into an ecosystem (Guerin and Boyd, 1995).

The two biodegradative isolates, strain A3-C, able to degrade naphthalene sulfonic acid (Brilon et al., 1981), and strain B1, able to mineralize toluene sulfonic acid (Thurnheer et al., 1986), were identified as members of *A. denitrificans*, or classified within the genus *Alcaligenes* close to *A. denitrificans*, respectively (Busse et al., 1992).

Another strain (L6), from a collection of aerobic isolates proficient of 3-chlorobenzoate degradation under reduced oxygen partial pressures by metabolization via gentisate or the protocatechuate pathway, was allocated to the genus *Alcaligenes* by 16S rDNA sequencing (Krooneman et al., 1996). The authors demonstrated an increased oxygen affinity of strain L6 in studies with continuous culture, as is expected for organisms that are involved in metabolism of aromatic compounds and play an important role in determining the fate of haloaromatics at oxic-anoxic interfaces. Given their previous allocation to *A. denitrificans* or *A. xylosoxidans*, the strains A3-C, B1, and L6 are considered members of the genus *Achromobacter*.

Genus III. Bordetella Moreno-López 1952, 178[AL]

GARY N. SANDEN AND ROBBIN S. WEYANT

Bor.de.tel' la. M.L. dim ending *-ella;* M.L. fem. n. *Bordetella* named after Jules Bordet, who with O. Gengou first isolated the organism causing pertussis.

Minute coccobacillus, 0.2–0.5 μm in diameter and 0.5–2.0 μm in length, often bipolar stained, and arranged singly or in pairs, more rarely in chains. Gram negative. **Nonmotile** or **motile** by peritrichous flagella. **Strictly aerobic. Optimal temperature, 35–37°C.** Colonies on Bordet–Gengou medium are smooth, convex, pearly, glistening, nearly transparent, and surrounded by a zone of hemolysis without definite periphery. **Respiratory metabolism.** Chemoorganotrophic. **Require nicotinamide, organic sulfur** (e.g., cysteine), and **organic nitrogen** (amino acids). **Utilize oxidatively** glutamic acid, proline, alanine, aspartic acid, and serine, with production of ammonia and CO_2. Litmus milk is made alkaline. Mammalian and avian parasite and pathogen. Most species localize and multiply among the epithelial cilia of the respiratory tract.

The mol% G + C of the DNA is: 66–70.

Type species: **Bordetella pertussis** (Bergey, Harrison, Breed, Hammer and Huntoon 1923a) Moreno-López 1952, 178 (Microbe de coqueluche Bordet and Gengou 1906, 731; *Haemophilus pertussis* Bergey, Harrison, Breed, Hammer and Huntoon 1923a, 269.)

FURTHER DESCRIPTIVE INFORMATION

Phylogeny and taxonomy The genus *Bordetella*, along with the closely related genus *Alcaligenes*, constitute the family *Alcaligenaceae* of the *Betaproteobacteria*. The close association of these genera has been demonstrated by phenotypic analysis, which identified *Alcaligenes* as the nearest neighbor to *Bordetella* and most closely associated strains of this genus with *B. bronchiseptica* (Johnson and Sneath, 1973). DNA–rRNA hybridization and G + C content confirm this association, together with their mutual divergence from other taxa (De Ley et al., 1986).

Historically, it has been difficult to differentiate between *Bordetella* and *Alcaligenes* based on phenotype because of the low reactivity and common responses in many tests. Formerly, *Bordetella* spp. were regarded as pathogens of the upper respiratory tract of humans and animals and *Alcaligenes* spp. as aquatic and terrestrial saprophytes. This distinction, based on ecological niche, has been confounded by recent revelations in phylogeny, reclassification of some *Alcaligenes* spp., and the designation of novel *Bordetella* spp. that are not etiologic agents of respiratory diseases.

The genus *Bordetella* contains seven species: *B. pertussis, B. parapertussis, B. bronchiseptica, B. avium, B. hinzii, B. holmesii,* and *B. trematum.* Various methods have been used to derive phylogeny within the genus including analyses of phenotypic characteristics, DNA base composition, nucleic acid hybridization, multilocus enzyme electrophoresis (MEE), gene-sequence analyses, and the distribution and copy numbers of insertion sequences (Johnson and Sneath, 1973; Kloos et al., 1979; Kiredjian et al., 1981; Kersters et al., 1984; De Ley et al., 1986; van der Zee et al., 1997). In 1973, Johnson and Sneath described a numerical taxonomic study of *B. pertussis, B. parapertussis,* and *B. bronchiseptica* based on 134 characteristics. They showed that the test strains aligned into three highly similar clusters corresponding to the respective species. Differentiation based on whole-cell protein and fatty acid analysis also supported the species status of these taxa (Vancanneyt et al., 1995). DNA hybridization results subsequently yielded similar clustering, but the tested isolates were sufficiently related as to constitute a single genomospecies (Kloos et al., 1979; Weyant et al., 1995a). Evaluations of the DNA base composition, DNA–rRNA hybridization results (De Ley et al., 1986), sequence similarity among the 23S rRNA genes (Muller and Hildebrandt, 1993), and genetic relatedness measured by multilocus enzyme electrophoresis (Musser et al., 1986) also supported the single-species concept. Conversely, lipopolysaccharide expression is nomenspecies-specific (van den Akker, 1998), and macro-restriction profiling of genomic DNA with a rarely cutting endonuclease demonstrated heterogeneity among the three species *B. pertussis*, *B. parapertussis*, and *B. bronchiseptica* (Khattak and Mathews, 1993). Although strains were more highly related within a given species, sufficient variability to discriminate between epidemiologically unrelated *B. pertussis* isolates has been documented.

Gene sequence analysis (Marchitto et al., 1987; Muller and Hildebrandt, 1993) and the data supporting the single-species concept suggest a relatively recent evolutionary origin and a common ancestor for *B. pertussis*, *B. parapertussis*, and *B. bronchiseptica* (Mooi et al., 1987; van der Zee, 1997). However, further elucidating the relationships within the genus was contentious until recently. *B. pertussis* and *B. parapertussis* were most closely related by the phenotypic characteristics evaluated (Johnson and Sneath, 1973). Sequence variations among pertussis toxin operons provided evidence for the evolution of *B. parapertussis* and then *B. bronchiseptica* from a common ancestor through *B. pertussis* (Arico and Rappuoli, 1987). DNA–rRNA hybridization studies separated *B. pertussis* and *B. bronchiseptica* with *B. parapertussis* between and overlapping them (De Ley et al., 1986). DNA hybridization (Kloos et al., 1979), multilocus enzyme electrophoresis (MEE) (Musser et al., 1986), and susceptibility to lytic bacteriophage (Rauch and Pickett, 1961) suggested that *B. parapertussis* is more highly related to *B. bronchiseptica* than to *B. pertussis*. The restricted diversity exhibited by MEE evaluations of *B. pertussis* and *B. parapertussis* isolates implies a shared ancestor (Musser et al., 1986). The comparatively greater heterogeneity found in *B. bronchiseptica* genetic structure, despite its general clonal nature, is consistent with an older lineage from which *B. pertussis* and *B. parapertussis* derived by host adaptation (Musser et al., 1986, 1987). Maximum parsimony evaluation of data from MEE and pertussis toxin operon sequences (Altschul, 1989) grouped all *B. pertussis* strains together with a unique common ancestor and designated *B. parapertussis* as a subtype of *B. bronchiseptica*.

MEE and DNA polymorphisms mediated by insertion sequence elements have recently yielded additional insights into phylogenetic relationships among *Bordetella* spp. (van der Zee et al., 1996, 1997). The distinction between sheep and human isolates of *B. parapertussis*, first based on differences in phenotype and electrophoretic types (Porter et al., 1994), and later reinforced by LPS expression (van den Akker, 1998), was confirmed by typing with insertion elements (van der Zee et al., 1996). In addition, this latter technique showed that transmission between the human and ovine populations was unlikely, and that the human strains were less divergent, suggesting that they evolved from *B. bronchiseptica* independently and more recently than sheep isolates.

Evaluation of 188 *Bordetella* isolates by MEE and the distribution of three insertion elements clarified the evolution and host adaptation in this genus (van der Zee et al., 1997). In general, host adaptation has restricted *B. pertussis* to humans and *B. parapertussis* to humans and sheep. Although some *B. bronchiseptica* clones tended to be more host- and geographically specific than others, it infects a much greater diversity of hosts, conforming with the greater genotypic diversity calculated for this species. Computer analyses of MEE along with distributions and sequences of insertion elements verified that *B. pertussis* and *B. parapertussis* descended from alternative clones of *B. bronchiseptica*, with *B. pertussis* infecting humans first. Indeed, *B. parapertussis* seems to have adapted to humans relatively recently from a *B. bronchiseptica* clade exclusive to pigs. In contrast, ovine strains of *B. parapertussis* most likely adapted to their host earlier.

Organisms belonging to four additional taxa are highly related to the historically recognized *Bordetella* spp. Among these were isolates classified as *Bordetella* or *Alcaligenes* causing an acute respiratory disease in birds that is symptomatically and pathologically analogous to pertussis in humans (Hinz et al., 1978; Simmons et al., 1981; Rimler and Semmons, 1983). The initial confusion over classifying such isolates was most likely due to their inactivity in conventional biochemical assays. Subsequently, several phenotypic approaches as well as DNA base composition and DNA–rRNA hybridization were used to assign such strains isolated from birds and causing coryza (rhinotracheitis) in turkey poults to a genotypically and phenotypically homogeneous fourth species, *B. avium* (Kersters et al., 1984). The DNA base composition of these strains was lower than those reported for the other *Bordetella* species and for the *Alcaligenes denitrificans–Achromobacter xylosoxidans* group (Kersters et al., 1984; De Ley et al., 1986). However, the 6.7 mol% G + C difference was reportedly insufficient to preclude *B. avium* from the genus, especially given the level of relatedness indicated by phenotypic properties, including isoprenoid quinone content (Fletcher et al., 1987), distinctive cellular fatty acid profiles (Moore et al., 1987), and DNA–rRNA hybridization (Kersters et al., 1984). Numerical analysis of phenotypic features related *B. avium* most closely to *B. parapertussis* (De Ley et al., 1986).

Bordetella holmesii was characterized in 1995 (Weyant et al., 1995a). Fifteen isolates representing this species were cultured from the blood of mostly young adults with underlying debilitations between 1983 and 1992. Initially referred to as CDC nonoxidizer group 2 (NO-2) based on common phenotypic characteristics, their cellular fatty acid profiles, 16S rRNA sequences, mol% G + C, and ubiquinone 8 content suggested close relatedness to the genus *Bordetella*. This was confirmed with DNA relatedness experiments that showed that these strains were sufficiently related as to constitute a single species (Wayne et al., 1987) and supported their placement within *Bordetella*. More recently, *B. holmesii*-like organisms were cultured from the sputum of a patient with acute pulmonary edema and bronchitis (Tang et al., 1998) and from 33 nasopharyngeal specimens collected from the nasopharynx of patients with cough illness as part of an active pertussis surveillance program (Yih et al., 1999). Additional studies are needed to define the role of *B. holmesii* in respiratory disease.

A sixth species is composed of isolates sharing similar features and resembling *B. avium* on biochemical assays that were cultured from the respiratory tracts of poultry and two human clinical specimens (Vandamme et al., 1995c). Although initially isolated in diverse geographical areas from turkeys and chickens with respiratory illness, there is insufficient evidence to attribute an etiologic role for these strains (Berkhoff and Riddle, 1984). Conversely, the human isolates from blood and sputum specimens were associated with bacteremia in an immunocompromised patient (Cookson et al., 1994; Vandamme et al., 1995c) and pulmonary complications in a cystic fibrosis patient (Funke et al., 1996), respectively. A third human strain was cultured from human sputum, but no clinical significance could be assigned due to lack of clinical data (Kiredjian et al., 1981). Twelve isolates from poultry and two from humans were evaluated by nucleic acid composition and hybridizations, and their phenotypic characteristics, including protein and fatty acid analysis. These results revealed that these strains are a new species, *Bordetella hinzii* (Vandamme et al., 1995c).

Previous efforts to characterize atypical members of the *Bordetella–Alcaligenes* group led to the evaluation of a collection of 10 isolates recovered from wounds and chronic otitis media (Dorittke et al., 1995; Vandamme et al., 1996a). Comparison of protein electrophoretic patterns of these isolates disclosed highly similar patterns that were distinguishable from the protein profiles of representative strains of the *Alcaligenaceae*. DNA base composition and hybridization analyses confirmed the homogeneity of this group. DNA–rRNA hybridization results placed these strains within the *Alcaligenaceae* and aligned a representative strain more closely to the *Bordetella* than the *Alcaligenes* type species. Consequently, the species has been named *Bordetella trematum* (Vandamme et al., 1996a).

Cellular morphology *Bordetella* cells usually appear as minute coccobacilli 0.2–0.5 × 0.5–2.0 μm. Cells of degraded strains may appear slightly larger. *B. trematum* cells may be slightly larger (0.5–0.6 × 1.0–2.4 μm) than those of the other species. Capsules have been observed with *B. pertussis*, *B. bronchiseptica*, and *B. trematum* (Lawson, 1940; Nakase, 1957; Vandamme et al., 1996a). Peritrichous flagella are observed on cells of the motile species, *B. bronchiseptica*, *B. avium*, *B. trematum*, and *B. hinzii*. Labow and Mosley have measured the periodic structure of *B. bronchiseptica* flagella as 19.0 by 13.9 nm (Labaw and Mosley, 1955). Pili-like filaments have been observed on cells of *B. pertussis*, *B. parapertussis*, *B. bronchiseptica*, and *B. avium* (Morse and Morse, 1970; Jackwood and Saif, 1987). Similar investigations have not as yet been reported for the more recently described species.

Nutrition and growth conditions Bordetellae are obligate aerobes with an optimal growth temperature range of 30–37°C. The metabolism of these organisms is based on the oxidation of amino acids. Carbohydrates are generally not utilized. Nicotinamide, organic nitrogen, and organic sulfur (cystine, cysteine, or glutathione) are required for growth (Weiss, 1992). With the exception of *B. pertussis*, growth can be obtained on a wide variety of culture media, including trypticase soy agar with 5% sheep blood, heart infusion agar with or without 5% rabbit blood, MacConkey agar, and peptone agar. Cultivation of *B. pertussis* is more difficult because of its susceptibility to various compounds (unsaturated fatty acids, colloidal sulfur, or sulfides) found in most growth media. Charcoal horse blood agar and Bordet–Gengou agar contain adsorbants for these inhibitors and allow for the growth of *B. pertussis* (Hoppe, 1988; Weyant et al., 1996). Charcoal horse blood agar has a longer shelf-life than Bordet–Gengou agar and is superior in its ability to support *B. pertussis* growth (Hoppe, 1988). Two chemically defined media, Stainer-Scholte broth and cyclodextrin solid medium (Aoyama et al., 1986), have also been developed to support the growth of *B. pertussis*.

Colonial morphology The time of development and the size of colonies differ among the species. *B. pertussis* is the most fastidious, with 3–6 d required for the development of pinpoint colonies on Bordet–Gengou medium. Colonies have a characteristic narrow zone of hemolysis and are usually convex and glistening, with an entire edge. Colony diameter rarely exceeds

3 mm. *B. parapertussis* and *B. holmesii* produce slightly larger convex, semi-opaque colonies at 2–3 d of incubation. Diffuse zones of browning may be observed around *B. parapertussis* and *B. holmesii* colonies, especially in media supplemented with tyrosine. Strains of the other *Bordetella* species usually produce colonies within 2 d of incubation. *B. bronchiseptica* colonies on MacConkey agar are reddish and surrounded by a small red zone with amber discoloration of the underlying medium. *B. trematum* strains produce convex, circular, and grayish cream white colonies with entire edges on blood agar (Vandamme et al., 1996a). Multiple colony types each have been described for *B. avium* and *B. hinzii*. For *B. avium*, type I colonies are small pearl-like with entire edges and glistening surfaces and <1 mm in diameter after 24 h of incubation. Type II colonies are larger, circular, and convex with entire edges and smooth surfaces. Colony types are usually strain-specific and stable (Kersters et al., 1984). A third colony type, characterized by a crenated edge and a flat surface with a wrinkled or ground-glass appearance, has also been described (Jackwood et al., 1991). Some strains of *B. hinzii* produce round, raised, glistening, grayish colonies about 2 mm in diameter at 48 h of incubation. Other strains produce flat, dry, crinkled colonies with diameters up to 5 mm at 48 h of incubation (Vandamme et al., 1995c).

Pathogenicity *Bordetella* species are responsible for respiratory tract infections in humans and various other warm-blooded animals. These organisms have a significant public health and economic impact. *B. pertussis* causes pertussis (whooping cough) in humans. This disease is characterized by violent paroxysms of coughing followed by a loud inspiratory "whoop" (von Lichtenberg, 1984). The incubation period is between 7 and 10 d. The disease is highly communicable, with person-to-person transmission occurring via aerosolized droplet nuclei. Pertussis infections result in over 350,000 deaths annually (Cherry, 1996; Hoppe, 1999).

Current pertussis-containing vaccines are effective in preventing or attenuating severe pertussis after three or more doses. Because vaccine-induced immunity wanes with time and vaccination does not protect from *B. pertussis* infection, vaccinees of any age may contribute to the reservoir of endemic pertussis. Booster immunizations may be licensed for adolescents and adults in the near future and have the potential to decrease pertussis morbidity in these populations. This strategy may also reduce transmission to young infants, reducing morbidity and mortality among this highly susceptible group (Guris et al., 1999; Centers for Disease Control and Prevention, 2001).

B. parapertussis causes a milder pertussis-like disease in humans and also causes chronic progressive pneumonia in sheep (Martin, 1996; Hoppe, 1999). Studies in mice and specific pathogen-free lambs indicate that prior inoculation with *B. parapertussis* predisposes these animals to pneumonia caused by *Mannheimia* species (previously *Pasteurella haemolytica* complex) (Porter et al., 1995a, c). Pulsed-field gel electrophoresis of cellular DNA, cellular fatty acid profiles, and LPS characterization with monoclonal antibodies have been used to differentiate human from ovine isolates of *B. parapertussis* (Porter et al., 1995b, 1996).

Bordetella avium and *Bordetella bronchiseptica* are predominately animal pathogens. *B. avium* causes coryza or rhinotracheitis in birds, particularly turkeys (Arp and Cheville, 1984; Skeeles and Arp, 1997). This disease is of significant economic concern to worldwide agriculture (Blackall and Doheny, 1987; Skeeles and Arp, 1997). *B. bronchiseptica* causes respiratory tract infections in various animals, including swine, dogs, guinea pigs, rabbits, cats, and horses (Roop, 1984). In swine, two forms of infection are recognized: atrophic rhinitis and bronchopneumonia in piglets. Atrophic rhinitis is a disease characterized by an initial episode of acute rhinitis, which is followed by chronic atrophy of the turbinate bones and deformity of the face. More severe cases of this disease are associated with dual infection by *B. bronchiseptica* and *Pasteurella multocida* (Rutter, 1985). Dogs (kennel cough) and guinea pigs in group housing are susceptible to outbreaks of respiratory infections caused by *B. bronchiseptica* (Bemis et al., 1977; Goodnow, 1980). *B. bronchiseptica* has also been isolated as a commensal organism in the upper respiratory tract of humans and has been implicated in infections of severely immunocompromised individuals (Woolfrey and Moody, 1991).

An extensive body of literature has been written to describe pathogenic mechanisms associated with *B. pertussis*, *B. parapertussis*, *B. bronchiseptica*, and *B. avium* (Wardlaw and Parton, 1988; Weiss, 1992; Parton, 1996). These species show a tropism for ciliated respiratory epithelial cells. Once attachment is achieved, various other virulence factors are produced that mediate damage and loss of ciliated cells and, in advanced cases, systemic pathology (Table BXII.β.32). Pili or fimbriae play a significant role in the early stages of pathogenesis. The major fimbrial subunits, agglutinogen 2/fimbria 2 and fimbria 3, mediate binding to host tissues (Miller et al., 1943; Sako, 1947; Preston, 1963; Preston and Stanbridge, 1972; Stanbridge and Preston, 1974; Preston et al., 1982, 1990; Mooi et al., 1992; Willems et al., 1992b; Hewlett, 1997; Preston and Matthews, 1998). Pertactin is a nonfimbrial surface protein associated with attachment (Charles et al., 1989, 1994; Hewlett, 1997). Nonsurface factors involved in the attachment process include tracheal colonization factor, serum resistance factor, and filamentous hemagglutinin (Van't Wout et al., 1992; Fernandez and Weiss, 1994; Finn and Stevens, 1995; Hewlett, 1997; Liu et al., 1997; Cotter et al., 1998).

Pertussis toxin is a complex protein that has attachment properties as well as other virulence functions (Katada and Ui, 1982; Pittman, 1984a; Weiss and Hewlett, 1986; Relman et al., 1990b; Sato and Sato, 1990; Granstrom et al., 1991; Van't Wout et al., 1992; Sandros and Tuomanen, 1993; Trollfors et al., 1995; Hewlett, 1997; Ui, 1998; Williamson and Matthews, 1999). The toxin contains two functional subunits: the B subunit facilitates binding to host cells and introduction of the enzymatically active A subunit into the host cell cytoplasm. The A subunit then acts as an ADP-ribosyltransferase that modifies and inactivates host cell G proteins. Other postattachment virulence factors include dermonecrotic toxin, which is a potent vasoconstrictor associated with ischaemia and skin necrosis in experimental animals (Livey and Wardlaw, 1984); tracheal cytotoxin, which inhibits DNA synthesis in ciliated epithelial cells (Goldman et al., 1982; Luker et al., 1993); adenylate cyclase toxin/hemolysin, which produces unregulated adenylate cyclase activity in host cells (Wolff et al., 1980; Hanski and Farfel, 1985; Friedman et al., 1987; Glaser et al., 1988; Gueirard et al., 1998; Harvill et al., 1999); and lipopolysaccharide, which produces endotoxin-like effects (Allen et al., 1998).

The expression of virulence in *B. pertussis*, *B. parapertussis*, *B. bronchiseptica*, and *B. avium* is an elegantly regulated phenomenon. In the premolecular era it was observed that *B. pertussis* isolates underwent a phase variation from a virulent (phase I) to avirulent (phase IV) phenotype in guinea pigs with passage *in vitro* (Leslie and Gardner, 1931; Pittman, 1984a). This reduction in virulence was accompanied by changes in colonial morphology and serologic activity. Phase variation occurs with a frequency of 1 per 10^3–10^6 organisms and is generally irreversible *in vitro*. The molecular basis for phase variation was described by Weiss et al., who demonstrated by transposon insertion experiments that pertussis toxin, adenylate cyclase toxin/hemolysis, filamentous hemagglutinin, and dermatonecrotic toxin were all regulated by a single regulatory locus, designated *bvg* (Weiss et al., 1983; Weiss and Falkow, 1984). Mutations or deletions affecting the *bvg* locus result in phase variation. The expression of virulence factors is also affected by growth conditions, including temperature and salt concentration (Lacy, 1960). Unlike phase variation, this phenomenon, designated antigenic modulation, is freely reversible.

The pathogenic mechanisms and clinical significance of the more recently described *Bordetella* species have not yet been worked out. *B. hinzii* has been isolated from human and avian sources. Studies by Blackall and Doheny (1987) and Jackwood et al. (1985) suggest that this species is not a significant avian pathogen. In humans, *B. hinzii* has been isolated from the blood of a human immunodeficiency virus-infected patient and the sputa of two patients, one of whom had cystic fibrosis (Cookson et al., 1994; Vandamme et al., 1995c; Funke et al., 1996). *Bordetella holmesii* has been isolated from human blood cultures and from nasopharyngeal specimens of patients suspected of having pertussis (Lindquist et al., 1995; Weyant et al., 1995a; Mazengia et al., 2000). Some of the patients with positive blood cultures had been previously splenectomized (Lindquist et al., 1995; Weyant et al., 1995a). *Bordetella trematum* has been isolated from human wound and otitis media specimens (Vandamme et al., 1996a).

Ecology *Bordetella* species are isolated from warm-blooded animals, including humans. Until recently, the habitat of this genus was thought to be limited to the respiratory tract of the host. However, *B. holmesii* and *B. hinzii* have been isolated from human blood, and *B. trematum* has been isolated from various human wound cultures (Vandamme et al., 1995c, 1996a; Weyant et al., 1995a). The phenotypic similarity of *B. bronchiseptica*, *B. trematum*, and *B. hinzii* to the free-living *Alcaligenes* species suggests the potential for their isolation from the natural environment, although this has yet to be documented.

Isolation and direct detection procedures Bordetellae are commonly isolated from human specimens, and suspected cases of pertussis should have a nasopharyngeal aspirate or swab obtained from the posterior nasopharynx for culture for *B. pertussis* and *B. parapertussis*. Nasopharyngeal aspirates have similar or higher rates of recovery for *B. pertussis* than nasopharyngeal swabs, but recovery from throat and anterior nasal swabs is unacceptably low (Hoppe and Weiss, 1987; Halperin et al.,1989; Hallander et al., 1993; Bejuk et al., 1995). An enriched agar medium optimizes recovery of *B. pertussis* on primary culture. Modified B-G medium (Kendrick and Eldering, 1969) and charcoal agar with horse blood (Regan and Lowe, 1977) are most commonly used for primary isolation: the former allows for detecting hemolytic activity of *Bordetella* spp., the latter is reportedly more sensitive (Muller et al., 1997a). Supplemental antibiotics (penicillin at 0.25–0.5 U/ml for B–G and cephalexin at 40 µg/ml for charcoal-horse blood agar) are effective in inhibiting the growth of some normal nasopharyngeal flora. Plates incubated at 36° ± 1°C under ambient air and high humidity are examined daily until positive or for 7 d. Other primary isolation media (Lautrop, 1960; Imaizumi et al., 1983) have been proposed, but

TABLE BXII.β.32. Virulence factors of *Bordetella pertussis* and other *Bordetella* species

Virulence factors[a]	Molecular mass (kDA)	Mechanism of action	Stage of infection	Protective immunity	Location and other features	*bvg* regulated	Expression in other *Bordetella* sp.[b]
Agglutinogen2/fimbria 2	22	Fim2 binds heparin; fimD binds heparin and integrin VLA-5	A	+	Located on fimbriae (major subunit); confers protective immunity against serotypes 1 and 2	−	Bp, Bb, Baa
Agglutinin 3			A	+	Either a somatic or fimbrial antigen; confers protective immunity against serotypes 1 and 3	−	Bp, Bb, Baa
Fimbria 3	21.5	FimD binds to heparin and integrin VLA-5	A	+	Major fimbrial subunit	+	Bp, Bb, Baa
P.69 Pertactin	69	RGD motif, probably binds CR3	A	+	Somatic antigen	+	Bp, Bb
Pertussis Toxin	Subunits of 26, 22, 22, 12, 12, 11	Binds ciliated epithelium and macrophages; mimics selectins and upregulates macrophage CR3 for FHA binding; ADP ribosylation of cellular G proteins	A,D,L,S	+	Synergistic adhesin with FHA	+	−
Filamentous hemagglutinin	220	Binds both the bacterium and macrophage CR3 to facilitate phagocytosis	A	−	Secreted; synergistic adhesin with PT	+	Bp, Bb
Adenylate cyclase/hemolysin	45	ATP hydrolysis with raised intracellular cAMP in macrophages and lymphocytes; induces apoptosis	D,L,S	−	Secreted by type I pathway requiring CyaB, D and E proteins; activated by eucaryotic calmodulin	+	Bp, Bb
Tracheal cytotoxin	921	DNA inhibition in ciliated epithelium	D,L	−	A muramyl peptide, derived from bacterial peptoglycan	−	Bp, Bb, ba
Dermonecrotic toxin	102 with subunits of 30, 24	Inhibition of Na^+-K^+ ATPase; vasoconstriction	L,S		Localized to the bacterial cytoplasm; part of molecule probably exposed at cell surface		Bp, Bb, Ba
Lipopolysaccharide		Endotoxin-like effects; pyrogenic, sensitization to histamine	L.S		Two lipids, A and X; two different oligosaccharides, I and II		Bp, Bb, Ba
Tracheal colonization factor	64	Tcf binding, probably mediated by RGD motif	A		Secreted by type IV pathway	+	−
Serum resistance factor	BrkA: 103; BRKB: 32		D		Secreted by type IV pathway		−

[a]Fim2, fimbria 2; Fim3, fimbria 3; FHA, filamentous hemagglutination; PT, pertussis toxin; tcf, tracheal colonization factor; ATP, adenosine triphosphate; ATPase, adenosine triphosphatase; VLA-5, very-late antigen 5; CR3, complement receptor 3; ADP, adenosine diphosphate; cAMP, cyclic adenosine monophosphate; brk, *Bordetella* resistance to killing; RGD, Arg-Gly-Asp sequence; A, attachment; D, evasion of host defense; L, local effects; S, systemic effects.

[b]Bp, *B. parapertussis*; Bb, *B. bronchiseptica*; Ba, *B. avium*; fimbriae are produced by *B. parapertussis*, *B. bronchiseptica*, and *B. avium*, but are antigenically distant from *B. pertussis* fimbriae.

are not widely used. If direct inoculation of a selective medium is not possible, swab specimens can be placed in Regan–Lowe transport medium (one-half strength charcoal agar supplemented with horse blood and cephalexin) (Regan and Lowe, 1977). Some investigators recommend incubating the specimen (swab in the transport medium) before transport to the laboratory because incubating the specimen before transport reportedly allows for growth of *B. pertussis* (Strebel et al., 1993). Others suggest that this practice can decrease the yield of *Bordetella* spp. due to overgrowth by normal flora and recommend storage and transport at 4°C (Morrill et al., 1988). Culture of *B. pertussis* and *B. parapertussis* is optimal at 35–36°C, aerobically

(without CO_2), and with sufficient humidity to avoid desiccation. Cultures can be incubated for up to 7 d; *B. pertussis* rarely is visible before 3–4 d, but *B. parapertussis* is commonly discernible at least 24 h earlier.

Numerous studies have demonstrated the potential for polymerase chain reaction (PCR) assays to detect *Bordetella* cells with greater sensitivity and more rapidly than by culture. However, no specific technique for PCR is universally accepted or validated among laboratories. In addition, there is no quality assurance program for PCR. The use of PCR without culture negatively impacts the monitoring of cases of disease, recruitment of isolates for epidemiologic studies, and surveillance for antibiotic resistance. For these reasons, PCR currently is best used as a presumptive assay in conjunction with culture under the conditions described in Meade and Bollen (1994).

Commercially available direct fluorescence assay (DFA) tests have been developed to detect *B. pertussis* in clinical specimens (Preston, 1970; Gilligan and Fisher, 1984; Ewanowich et al., 1993). Although these tests have been widely used to screen patients for *B. pertussis* infection, problems with specificity have been noted. Cross-reactions with normal nasopharyngeal flora accounted for false-positive results in up to 85% of tests (Ewanowich et al., 1993) and led to substantial unnecessary public health intervention. For the best results, DFA tests require care in all technical aspects and experienced personnel for their interpretation. If used, they should always be accompanied by a specimen for culture. Health care and public health workers can weigh the benefit of a presumptive diagnosis with the disadvantage of a high percentage of false-positive results. A mouse monoclonal antibody (BL-5)-based DFA test recently become available (Accu-MabTM, Biotex Laboratories Inc., Edmonton, Canada). Initial evaluation demonstrated 65% sensitivity and 99% specificity when compared with culture (McNicol et al., 1995). Further testing is necessary to confirm these results. The technical requirements and the careful standardization required for previous commercially available DFA tests also apply to this new monoclonal DFA; patients should also be cultured for isolation of *B. pertussis*.

Detection of antibodies to *B. pertussis* in humans   Several strategies have been used to measure the immune responses to *B. pertussis* in patients and in vaccinees, including serum antibodies that agglutinate *B. pertussis* cells and a cytopathogenic assay for antibodies that neutralize pertussis toxin (PT). The predictive values of these assays for infection or protection are unacceptably low. Recent reports show promise in defining the role of serum bactericidal antibodies in pertussis immunity, but additional studies are needed. Enzyme-linked immunosorbent assays (ELISA) have provided the most sensitive and specific results, but the materials and methods required for them are not uniform. These tests also fail to relate a specific antibody to a specific protective function. None of the techniques has been validated between laboratories or been approved for diagnostic use in the United States. Currently, the most generally accepted serologic criterion for diagnosis of pertussis is the use of ELISA to demonstrate a significant increase in serum antibody concentrations against PT between acute and convalescent specimens. Anti-PT concentrations in convalescent sera may have diagnostic relevance in populations with known age-related prevalence of these antibodies, or in investigations where the prevalence among controls matched to cases by age and vaccination status can be determined. For individual patients, serology generally is regarded as an adjunct to culture rather than a method for primary diagnosis.

Antibiotic susceptibility   Erythromycin is the antibiotic treatment of choice for *B. pertussis* infections (Hoppe et al., 1992). However, in recent years, erythromycin-resistant strains have been recovered (Lewis et al., 1995; Korgenski and Daly, 1997). The prevalence of these strains is not known but available evidence suggests that it is low. One of the erythromycin-resistant isolates was susceptible to trimethoprim-sulfamethoxazole (TMP-SMX), but cross-resistance to clarithromycin and clindamycin was demonstrated in the other. Susceptibility testing to a macrolide and to TMP-SMX can be considered for strains isolated from patients receiving at least 7 d of prophylaxis, or for purposes of

surveillance, but is not necessary for all patients. Disk-diffusion or E-Test (Korgenski and Daly, 1997) on charcoal agar containing horse blood have typically been used to screen for antibiotic susceptibilities of *B. pertussis*. There is less experience determining the minimum inhibitory concentrations (MIC) by agar dilution. Hoppe and Paulus suggest using Mueller-Hinton broth and agar with defibrinated horse blood and high inoculums of the test bacteria (Hoppe and Paulus, 1998). Erythromycin, TMP-SMX, and fluoroquinolones have high *in vitro* activities against *B. parapertussis* (Kurzynski et al., 1988; Hoppe and Simon, 1990; Hoppe and Tschirner, 1997).

In vitro antibiotic susceptibility studies with *B. bronchiseptica* indicate a resistance pattern similar to that observed with other nonfermentative Gram-negative bacilli (Kurzynski et al., 1988; Mortensen et al., 1989; Woolfrey and Moody, 1991). Strains of this species are usually sensitive to aminoglycosides, antipseudomonal penicillins, such as azlocillin, mezlocillin, and piperacillin, and broad-spectrum cephalothins, such as cefoperazone. High levels of resistance are usually seen with erythromycin and first- and second-generation penicillins and cephalosporins. Resistance to TMP-SMX and tetracyclines is variable. *In vitro* studies with *B. avium* show an overall lower level of antimicrobial resistance than with *B. bronchiseptica* (Mortensen et al., 1989). Strains of this species are usually sensitive to aminoglycosides, ampicillin, antipseudomonal penicillins, and tetracycline. Sensitivity to first- and second-generation cephalosporins is variable, and most strains are resistant to erythromycin and TMP-SMX (Mortensen et al., 1989; Blackall et al., 1995). *B. hinzii* has been reported to be susceptible to aminoglycosides and TMP-SMX, but resistant to ampicillin, erythromycin, and tetracycline (Blackall et al., 1995; Funke et al., 1996). Susceptibility studies with *B. holmesii* and *B. trematum* suffer from a lack of confirmed strains. Lindquist et al. (1995) described a single *B. holmesii* strain with disk diffusion results suggesting susceptibility to a wide range of agents, including aminoglycosides, TMP-SMX, tetracycline, ciprofloxacin, antipseudomonal penicillins, and ceftazidime.

MAINTENANCE PROCEDURES

For temporary preservation, *B. bronchiseptica*, *B. avium*, *B. hinzii*, *B. holmesii*, and *B. trematum* may be stored as stabs in motility medium containing 0.4% agar (Weyant et al., 1996). Screw-capped 15 × 125-mm tubes containing 8 ml of medium are inoculated with fresh agar-grown cultures by stabbing with a sterile needle or loop to approximately 2 cm below the agar surface. The tubes are then incubated at 35–37°C overnight with the caps loosened to allow for air exchange and then stored at room temperature with the caps tightened. *B. pertussis* and *B. parapertussis* may be grown and stored on B-G slants in closed containers at 4–6°C. Stored under these conditions, most strains will remain viable for 2–3 weeks. For permanent storage, fresh agar-grown cultures may be suspended in sterile skim milk of 10% powdered milk and freeze-dried (Novotny and Brookes, 1975). Alternatively, strains may be preserved by suspending fresh agar-grown cultures in defibrinated rabbit blood or 0.05 M Tris-HCl buffer (pH 7.6) with 20% glycerol and freezing in liquid nitrogen (Novotny and Brookes, 1975; Weyant et al., 1996).

DIFFERENTIATION OF THE GENUS *BORDETELLA* FROM OTHER GENERA

Phenotypic characteristics useful in the differentiation of *Bordetella* species are given in Table BXII.β.33. Although *Bordetella* species are highly related at the phylogenetic level, they are phenotypically heterogeneous. *B. pertussis* fails to grow on most commonly used growth media due to inhibition by various medium constituents including unsaturated fatty acids and metal ions (Rowatt, 1957). This characteristic is not observed with the other *Bordetella* species, although *B. holmesii* and *B. parapertussis* do not grow as rapidly on most media as do *B. bronchiseptica*, *B. avium*, *B. hinzii*, and *B. trematum*. *B. holmesii* and *B. parapertussis* also produce a characteristic brown diffusible pigment in heart infusion broth with tyrosine and are oxidase-negative. Urease activity is useful in differentiating *B. holmesii* from *B. parapertussis*. Of the four more rapidly growing species, *B. trematum* is unique in producing a negative oxidase reaction. The citrate, nitrate, and urease tests are useful in differentiating between *B. bronchiseptica*, *B. avium*, and *B. hinzii*. A more complete phenotypic characterization of the species of *Bordetella* is given in Table BXII.β.34.

Tests useful in differentiating *Bordetella* from other phenotypically similar genera are also given in Table BXII.β.33. From a phylogenetic prospective, *Bordetella*, *Achromobacter*, and *Alcaligenes* are closely related. This relatedness is reflected in the high level of phenotypic similarity observed between *Alcaligenes* and *Achromobacter* species and the motile and more rapidly growing species of *Bordetella* (*B. bronchiseptica*, *B. avium*, *B. hinzii*, and *B. trematum*). Historically, the ability of *Bordetella* to colonize mammalian and avian respiratory tracts was useful in differentiating these genera; however, the more recently described species—*B. holmesii*, *B. hinzii*, and *B. trematum*—have been isolated from nonrespiratory

sources. *Brucella* species, *Haemophilus* species, and asaccharolytic *Acinetobacter* species share phenotypic similarities with *B. pertussis*, *B. parapertussis*, and *B. holmesii*, and also may be isolated from clinical specimens. The requirement for X and/or V factors differentiates *Haemophilus* from *Bordetella*. Growth on blood agar, brown diffusible pigment production, and acidification of carbohydrates are useful in differentiating *Brucella* from nonmotile *Bordetella* species. Cellular fatty acid profiles have also been used successfully to differentiate among these organisms (Moore et al., 1987; Cookson et al., 1994; Vandamme et al., 1995c, 1996a; Weyant et al., 1995a; Funke et al., 1996). The predominant cellular fatty acids associated with each of these taxa are given in Table BXII.β.33.

TAXONOMIC COMMENTS

Tang et al. (1998) have described a series of *Bordetella holmesii*-like strains associated with septicemia, endocarditis, and respiratory failure in humans. These isolates are phenotypically consistent with *B. holmesii* and have 99.8% 16S rRNA gene sequence similarity. Additional studies will be required to determine the taxonomic status of these strains.

ACKNOWLEDGMENTS

This chapter is based in part on the genus *Bordetella* chapter of the 1st edition of *Bergey's Manual of Systematic Bacteriology*, authored by Dr. Margaret Pittman.

TABLE BXII.β.33. Differentiation of *Bordetella* species and other morphologically and phenotypically similar taxa[a]

Characteristic	B. pertussis	B. avium	B. bronchiseptica	B. hinzii	B. holmesii	B. parapertussis	B. trematum	Achromobacter sp.	Alcaligenes faecalis	Asaccharolytic Acinetobacter sp.	Brucella sp.	Haemophilus sp.
Motility	−	+	+	+	−	−	+	+	+	−	−	−
Growth on:												
Blood agar	−	+	+	+	+	+	+	+	+	+	+	V
MacConkey agar	−	+	+	+	+	+	+	+	+	V	V	V
Simmons citrate agar	−	−	+	+	−	−	−	+	+	V	−	−
Oxidase activity (Kovacs reagent)	+	+	+	+	−	−	−	+	+	−	+[b]	+
Acid production from D-glucose	−	−	−	−	−	−	−	V	−	−	V	+
Nitrate reduction	−	−	v	−	−	−	v	+	−	−	+	nd
Gas from nitrate	−	−	−	−	−	−	v	v	−	−	V	nd
Urease activity	−	−	+	−	−	+	−	−	−	V	+	V
Diffusible brown pigment[c]	−	−	−	−	+	+	−	−	−	V	−	−
Major cellular fatty acids:[d]												
$C_{14:0}$		+				+		+	+			+
$C_{16:0}$	+	+	+	+	+	+	+	+	+	+	+	+
$C_{16:1\ \omega7c}$	+		+					+	+	+		+
$C_{17:0\ cyclo}$		+	+	+	+	+	+	+	+			
$C_{18:0}$	+	+		+	+	+	+	+				
$C_{18:1\ \omega7c}$											+	
$C_{18:1\ \omega9c}$										+		
$C_{19:0\ cyclo\ \omega7c}$											+	
$C_{14:0\ 3OH}$	+											

[a]Symbols: +, >89% strains positive; V, 11–89% strains positive; −, fewer than 11% strains positive; nd, not determined. Data from Vandamme et al. (1996a) and Weyant et al. (1996).

[b]A small number of oxidase-negative *Brucella abortus* and *Brucella suis* strains have been reported (Weyant et al., 1996).

[c]Demonstrated on heart infusion agar with tyrosine (Weyant et al., 1996).

[d]Present at greater than 5% of total cellular fatty acids; $C_{16:1\ \omega7c}$, palmitoleic; $C_{16:0}$, palmitic; $C_{14:0\ 3OH}$, β-hydroxymyristic; $C_{18:0}$, stearic; $C_{17:0\ cyclo}$, $C_{17:0\ cyclo\ \omega7c}$; $C_{14:0}$, myristic; $C_{18:1\ \omega9c}$, oleic; $C_{19:0\ cyclo\ \omega7c}$, lactobacillic; $C_{18:1\ \omega7c}$, *cis*-vaccenic. Data from Weyant et al. (1996) and Vandamme et al. (1996a).

TABLE BXII.β.34. Characteristics of *Bordetella* species[a]

Characteristic	B. pertussis	B. avium	B. bronchiseptica	B. hinzii	B. holmesii	B. parapertussis	B. trematum
Aerobic growth at 25°C	−	+	+	+	−	−	+
Aerobic growth at 42°C	−	+	+	+	−	−	+
Growth on:							
MacConkey agar	−	+	+	+	−	+	+
MacConkey agar containing 320mg/l tellurite	nd	−	−	−	−	+	−
Simmons citrate agar	−	−	+	+	−	−	−
Salmonella-Shigella agar	nd	+	+	+	−	−	+
Cetrimide agar	nd	−	−	−	−	−	−
Growth in nutrient broth	nd	+	+	+	V	+	+
Growth in nutrient broth with 6% NaCl	nd	V	V	+	−	−	+
Motility	−	+	+	+	−	−	+
Diffusible brown pigment	nd	−	−	−	+	+	−
Pigment produced from tyrosine	nd	−	−	−	+	+	−
Oxidase activity:							
Kovacs reagent	+	+	+	+	−	−	−
Gaby-Hadley reagent	+	+	+	+	−	−	−
Nitrate reduction (classical method):							
Formation of nitrite	−	−	V	−	−	−	V
Denitrification	−	−	V	−	−	−	V
Nitrate reduction (API 20NE test):							
Formation of nitrite	−	−	+	−	−	−	V
Denitrification	−	−	−	−	−	−	−
Urease activity	−	−	+	−	−	+	−
Tetrazolium reduction	nd	+	+	+	−	+	+
Alkalinization of litmus milk	nd	−	+	−	−	−	−
Alkali production from:							
Acetamide	nd	+	−	+	−	−	V
Adipate	nd	+	+	+	−	−	+
Glycine	nd	−	+	+	−	−	−
Propionamide	nd	+	+	+	−	−	+
Valerate	nd	−	+	+	−	−	V
Malonamide	nd	−	+	+	−	−	+
Malonate	nd	−	+	+	−	−	+
Malonate (API ID 32E test)	−	−	−	−	−	−	−
Glucose fermentation	−	−	−	−	−	−	−
Assimilation of:							
D-Glucose	−	−	−	−	−	−	−
D-Xylose	−	−	−	−	−	−	−
D-Gluconate	−	−	−	−	−	−	−
D-Mannitol	−	−	−	−	−	−	−
Lactose	−	−	−	−	−	−	−
Sucrose	−	−	−	−	−	−	−
Maltose	−	−	−	−	−	−	−
Caprate	−	−	V	+	−	−	−
Adipate	−	+	+	+	−	−	+
L-Malate	−	+	V	+	−	−	+
Phenylacetate	−	+	+	+	−	−	+
Esculin	−	−	−	+	−	−	+
5-Keto-gluconate	−	−	−	−	−	−	V
Browning on 2-keto-gluconate	−	−	−	+	+	−	−
Alkaline phosphatase activity	−	V	V	+	+	−	+
Lysine decarboxylase activity	−	−	−	−	−	−	−
Ester C$_8$ lipase activity	+	V	V	+	−	+	+
Lipase C$_{14}$ activity	−	−	−	−	−	−	V
Trypsin activity	−	−	−	−	−	−	−
Chymotrypsin activity	+	V	−	−	+	+	V
Naphthol-AS-B1-phosphohydrolase activity	+	−	−	+	+	−	+
Arylamidase activity with:							
Valine	−	V	−	+	−	−	+
Cystine	−	−	−	−	−	−	−
Arginine	+	V	+	+	+	−	V
Proline	nd	−	−	−	+	−	−
Leucyl glycine	+	+	+	+	+	−	+
Phenylalanine	+	+	−	−	−	−	V
Pyroglutamic acid	−	−	−	−	−	−	−
Tyrosine	+	+	+	+	−	−	V
Glutamyl glutamic acid	+	−	−	−	−	−	V
Serine	+	+	+	+	−	−	+
L-Aspartic acid	+	+	+	+	−	+	+

[a]Symbols: +, >89% strains positive; V, 11–89% strains positive; −, <11% strains positive; nd, not determined. Data from Vandamme et al. (1996a) and Weyant et al. (1996).

FURTHER READING

Hoppe, J.E. 1999. *Bordetella. In* Murray, Baron, Phaller, Tenover and Yolken (Editors), Manual of Clinical Microbiology, American Society for Microbiology, Washington DC. pp. 614–624.

Parton, R. 1996. New perspectives on *Bordetella* pathogenicity. J. Med. Microbiol. **44**: 233–235.

Weiss, A.A. 1992. The genus *Bordetella. In* Balows, Trüper, Dworkin, Harder and Schleifer (Editors), The Prokaryotes: a Handbook on the Biology of Bacteria: Exophysiology, Isolation, Identification, Applications., 2nd Ed., Vol. 3, Springer-Verlag, Berlin, Germany. pp. 2530–2543.

List of species of the genus Bordetella

1. **Bordetella pertussis** (Bergey, Harrison, Breed, Hammer and Huntoon 1923a) Moreno-López 1952, 178[AL] (Microbe de coqueluche Bordet and Gengou 1906, 731; *Haemophilus pertussis* Bergey, Harrison, Breed, Hammer and Huntoon 1923a, 269.)

per.tus'sis. L. pref. *per* very, severe; L. n. *tussis* cough; M.L. gen. n. *pertussis* of a severe cough, of whooping cough.

Minute coccobacillus, 0.2–0.5 × 0.5–1.0 µm in length, encapsulated or surrounded by a slime sheath composed of extruded filaments or secreted blebs. Potato–glycerol–blood agar (Bordet–Gengou) has been preferred for primary isolation. Colonies are minute in size and are surrounded by a zone of hemolysis on media containing about 20% blood. Uniqueness in slow rate of growth and susceptibility to growth inhibitors are associated with the production of pertussis toxin. A defined broth medium may be used for mass growth. No pellicle is formed. Physiological and nutritional characteristics are presented in Tables BXII.β.33 and BXII.β.34. Parasitic, pathogenic, found only in the respiratory tract of humans, where it is the cause of pertussis (whooping cough).

The mol% G + C of the DNA is: 66–70 (T_m).

Type strain: 18-232 , ATCC 9797, DSM 5571.

GenBank accession number (16S rRNA): U04950.

2. **Bordetella avium** Kersters, Hinz, Hertle, Segers, Lievens, Siegmann and De Ley 1984, 65[VP]

a'vi.um. L. n. *avis* bird; L. gen. pl. n. *avium* of birds.

The morphology is as described for the genus. Colonies appear earlier on Bordet–Gengou medium and are larger than those of *B. pertussis* or *B. parapertussis*. Grows on MacConkey agar. The physiological and nutritional characteristics are presented in Tables BXII.β.33 and BXII.β.34. Found in avian species and pathogenic for turkey poults.

The mol% G + C of the DNA is: 61.6–62.6 (T_m).

Type strain: Hinz 591-77, ATCC 35086, DSM 11332, LMG 1852.

GenBank accession number (16S rRNA): U04947.

3. **Bordetella bronchiseptica** (Ferry 1912) Moreno-López 1952, 178[AL] (*Bacillus bronchicanis* Ferry 1911, 404; *Bacillus bronchisepticus* Ferry 1912, 377.)

bron.chi.sep'ti.ca. Gr. n. *bronchus* the trachea; Gr. adj. *septicus* putrefactive, septic; M.L. fem. adj. *bronchiseptica* intended to mean with an infected bronchus.

The morphology is as described for the genus. Colonies appear earlier on Bordet–Gengou medium and are larger than those of *B. pertussis* or *B. parapertussis*. Grows on MacConkey agar. A pellicle forms on liquid media. The physiological and nutritional characteristics are presented in Tables BXII.β.33 and BXII.β.34. Parasitic, pathogenic, found in the respiratory tract of domestic and wild mammalian animals (dogs, swine, guinea pigs, rabbits, raccoons, etc.), causes kennel cough in dogs.

The mol% G + C of the DNA is: 67–69 (T_m).

Type strain: 71, ATCC 19395, NCTC 452.

GenBank accession number (16S rRNA): U04948.

4. **Bordetella hinzii** Vandamme, Hommez, Vancanneyt, Monsieurs, Hoste, Cookson, Wirsing von König, Kersters and Blackall 1995c, 43[VP]

hin'zi.i. N.L. gen. n. *hinzii* of Hinz, named in honor of K.-H. Hinz.

The cellular morphology is as described for the genus. Two distinct colony types occur. Some strains produce round, raised, glistening, grayish colonies about 2 mm in diameter following 48 h of incubation at 37°C in air containing 5% CO_2. Under the same conditions other strains produce flat, dry, crinkled colonies that are up to 5 mm in diameter. The physiological and nutritional characteristics are presented in Tables BXII.β.33 and BXII.β.34. Strains have been isolated from respiratory tracts of turkeys and chickens. Pathogenic in humans, with strains isolated in blood and sputum specimens. The type strain was isolated from a chicken trachea in Australia.

The mol% G + C of the DNA is: 65–67 (T_m).

Type strain: TC58, LMG 13501, ATCC 51783, CIP 104527, DSM 11333, LMG 13501.

GenBank accession number (16S rRNA): AF177667.

5. **Bordetella holmesii** Weyant, Hollis, Weaver, Amin, Steigerwalt, O'Connor, Whitney, Daneshvar, Moss and Brenner 1995b, 619[VP] (Effective publication: Weyant, Hollis, Weaver, Amin, Steigerwalt, O'Connor, Whitney, Daneshvar, Moss and Brenner 1995a, 6.)

holmes'i.i. N.L. gen. n. *holmesii* named in honor of B. Holmes.

Cellular morphology is as described for the genus. Growth characteristics on blood agar are similar to those of *B. parapertussis*. Growth occurs on MacConkey agar at between 3 and 7 days of incubation at 35°C. The physiological and nutritional characteristics are presented in Tables BXII.β.33 and BXII.β.34. Strains have been isolated from blood and upper respiratory tract cultures of humans. The type strain was isolated from human blood.

The mol% G + C of the DNA is: 61.5–62.3 (T_m).

Type strain: ATCC 51541, CDC F5101.

GenBank accession number (16S rRNA): U04820.

6. **Bordetella parapertussis** (Eldering and Kendrick 1938) Moreno-López 1952, 178[AL] (*Bacillus parapertussis* Eldering and Kendrick 1938, 571.)

pa.ra.per.tus'sis. Gr. prep. *para* resembling; M.L. n. *pertussis* the specific epithet of *Bordetella pertussis*; M.L. adj. *parapertussis* resembling *Bordetella pertussis*.

The morphology is similar to that of *B. pertussis*. Colonies are slightly larger and appear earlier on Bordet–Gengou medium than *B. pertussis*. Growth on peptone agar is accompanied by a brown discoloration of the medium that is attributed to the action of tyrosinase on the tyrosine in the medium (Ensminger, 1953). A pellicle forms on liquid media. Growth is more rapid than with *B. pertussis*. The physiological and nutritional characteristics are presented in Tables BXII.β.33 and BXII.β.34. Parasitic, pathogenic, found in the respiratory tract of humans and sheep. Causes a milder pertussis-like syndrome in humans and chronic progressive pneumonia in ovines.

The mol% G + C of the DNA is: 66–70 (T_m).

Type strain: 522, ATCC 15311, NCTC 5952.

GenBank accession number (16S rRNA): U04949.

7. **Bordetella trematum** Vandamme, Heyndrickx, Vancanneyt, Hoste, De Vos, Falsen, Kersters and Hinz 1996a, 857[VP]

tre.ma' tum. Gr. neut. n. *trema* referring to something pierced or penetrated, an aperture, or a gap; N.L. gen. pl. n. *trematum* pertaining to penetrated or open things, referring to the presence of these bacteria in wounds and other exposed parts of the human body.

Cellular morphology is as described for the genus. Strains do not require special growth factors and grow on conventional media. Colonies may be observed at 16–24 h of incubation at 37°C under aerobic conditions. The physiological and nutritional characteristics are presented in Tables BXII.β.33 and BXII.β.34. Strains have been isolated from wound and ear infections of humans.

The mol% G + C of the DNA is: 64–65 (T_m).

Type strain: CCUG 32381, DSM 11334, LMG 13506.

GenBank accession number (16S rRNA): AJ277798.

Genus IV. **Derxia** *Jensen, Petersen, De and Bhattacharya 1960, 193*[AL]

CHRISTINA KENNEDY

Derx' i.a. M.L. fem. n. *Derxia* named after the Dutch microbiologist H.G. Derx (1894–1953).

Cells are rod-shaped with rounded ends, 1.0–1.2 μm × 3.0–6.0 μm, occurring singly or in short chains. **Cells are rather pleomorphic**, depending on age and the medium. In aging cultures cells often remain together forming long filaments of sometimes locally swollen or distorted cells. Some cells may assume very large lengths (up to 30 μm). **Young cells have a homogeneous cytoplasm; older cells show typical large refractile bodies—probably poly-β-hydroxybutyrate—throughout the whole cell**. Resting stages are not known. Gram negative. **Motile by a short polar flagellum**; motile cells are numerous in liquid glucose media containing combined nitrogen, but rare on nitrogen-deficient solid media. **Aerobic**, having a strictly respiratory type of metabolism with oxygen as the terminal electron acceptor. **N₂ is fixed under aerobic or microaerobic conditions**. Optimal temperature, 25–35°C; growth is slow at 15°C, feeble at 40°C; no growth at 50°C. **Growth occurs between pH 5.5 and ~9.0**; no growth at pH 4.4. **Broth cultures turn into a gelatinous mass**, but growth near the surface is more luxuriant and forms a thick, tough pellicle. **Colonies on agar media are at first slimy and semitransparent, later massive and opaque, highly raised with a wrinkled surface. Older colonies develop a dark mahogany-brown color. Catalase negative**. A wide range of sugars, alcohols, and organic acids are oxidized mostly to CO_2 and a small amount of acid, probably acetic, when growing in an alkaline medium. **Can grow as a facultative hydrogen autotroph. Found in tropical soils** (Asia, Africa, South America).

The mol% G + C of the DNA is: 69.2–72.6.

Type species: **Derxia gummosa** Jensen, Petersen, De and Bhattacharya 1960, 193.

FURTHER DESCRIPTIVE INFORMATION

The description, above and below, of the genus, as given by J.-H. Becking in the 1st edition of the *Manual* remains largely unchanged; information has been updated.*

Cell morphology The appearance of cells from young cultures is depicted in Fig. BXII.β.22. Older cells on sugar-rich media contain large refractile bodies. On glucose–peptone agar especially, very elongated cells are produced containing many refractile and misshapen bodies (Becking, 1984b). The refractile material is probably poly-β-hydroxybutyrate since it stains with Sudan III and Sudan black, but some vacuoles do not stain (Becking, 1984b). On nitrogen-free glucose agar older cells undergo shrinkage and are finally enclosed by a slime envelope (Fig. BXII.β.23).

Motile cells may become numerous in liquid glucose medium containing ammonia or glutamate as the nitrogen source. The cells usually have a single polar flagellum, but some may have a flagellum at each pole. According to Thompson and Skerman (1979), the single polar flagellum is less than a full wave and is less than 3 μm in length.

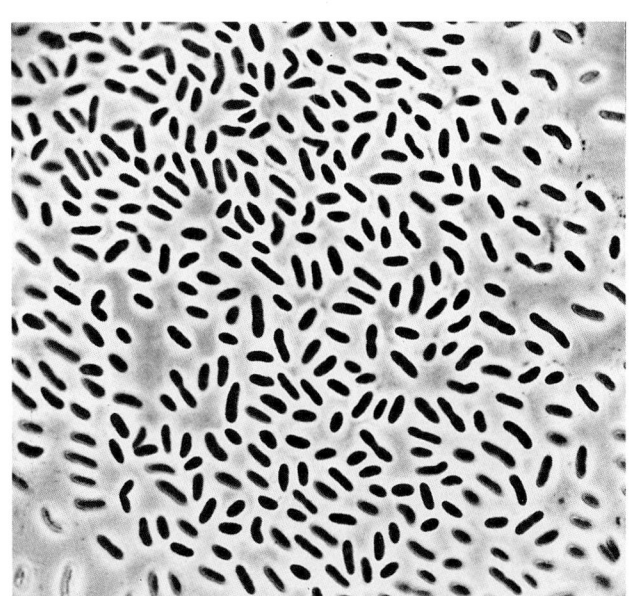

FIGURE BXII.β.22. Seven-day-old cells of *D. gummosa* on nitrogen-free agar containing 2% glucose. Phase-contrast microscopy (× 950).

---

*Editorial Note: *Derxia* was placed in the *Beijerinckiaceae* in the first edition of the *Manual*, because the 16S rDNA sequence of *Derxia gummosa* had not been published. However, now because of the 16S rDNA placement, the genus *Derxia* has been moved to the family *Alcaligenaceae* and is no longer listed in *Beijerinckiaceae*.

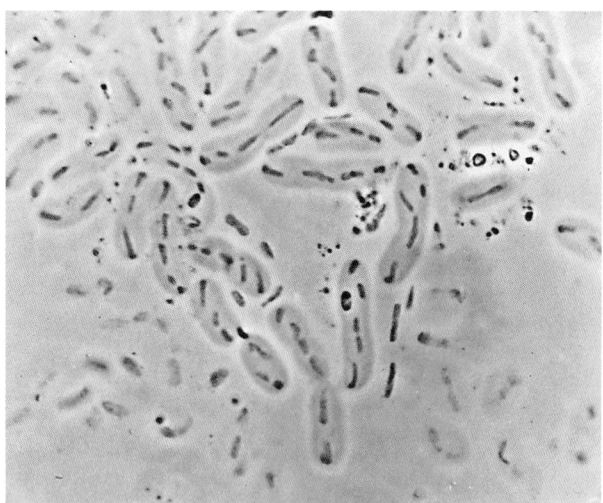

FIGURE BXII.β.23. Three-month-old cells of *D. gummosa* on nitrogen-free glucose agar. The cells show shrinkage and are enclosed by a thick slime envelope. Phase contrast microscopy (× 950).

Colonial and cultural characteristics Growth in liquid media usually starts as a ring at the glass-liquid interface and develops into a thick, wrinkled, tough pellicle. Shallow layers of medium change into a firm gelatinous mass after 2 weeks. The color gradually becomes a dark red-brown.

Growth on nitrogen-deficient agar media begins as thin, whitish or semitransparent scattered colonies. Later, more massive, highly raised or dome-shaped colonies emerge that rapidly assume a diameter of 1 cm or more (giant colonies) (Fig. BXII.β.24). These colonies are very like those of *Beijerinckia* spe-

cies. For further comparisons of the differential characteristics and carbon compounds of *Beijerinckia* and *Derxia*, see Tables BXII.α.141 and BXII.α.142 in *Beijerinckia*. Colonies are at first whitish or dull yellow with a smooth surface, but the surface soon becomes coarse and wrinkled and the color deepens to a dark mahogany-brown. The slime of these colonies is very tenacious and gumlike, but in the other developmental stages it is more soft and smeary.

As noted by Jensen et al. (1960) in the original description of *D. gummosa*, aerobic growth on nitrogen-free, mineral glucose agar[1] consists of a mixture of a few "massive" colonies among many "thin," whitish colonies. It is the former that are associated with N$_2$ fixation; their slime affords some protection to the oxygen-sensitive nitrogenase system by decreasing the penetration of oxygen to the cells (Hill and Postgate, 1969). The occurrence of two kinds of colonies under aerobic conditions is probably due to the occurrence of local areas of decreased oxygen concentration on the plates. It is only in these areas that N$_2$ fixation and extensive cell multiplication can occur that leads to the formation of the "massive" colonies. Under microaerobic conditions (pO$_2$ <0.2 atm), all of the colonies on a plate develop into the "massive" type (Hill, 1971).

Metabolism and metabolic pathways As for *Beijerinckia* species, *Derxia* can utilize many sugars, organic acids, and alcohols for growth (see Table BXII.α.142 in Genus *Beijerinckia*). Growth on methane or methanol as sole carbon source has also been reported for *D. gummosa* (Sampaio et al., 1981).

The efficiency of N$_2$ fixation by *D. gummosa* varies between 9 and 25 mg N/g of glucose consumed, but in most strains it is distinctly lower than in *Azotobacter* or *Beijerinckia*. There is no requirement for amino acids, vitamins, or growth factors, but trace elements, particularly molybdenum, are required. Vanadium cannot replace the molybdenum for N$_2$ fixation, suggesting that the alternative V nitrogenase (nitrogenase-2) is not present (for further explanation see the genus *Azotobacter*). Nitrite but not nitrate inhibits nitrogenase activity as do ammonia and glutamine (Wang and Nicholas, 1986c).

Growth with combined nitrogen sources is much faster than with N$_2$ and is completely uniform, in contrast to the uneven growth on nitrogen-free agar. Colonies change from a pale yellow through rust-brown to almost black (darkest if nitrate is present) and sometimes a light brown, water-soluble pigment is produced. Growth with glutamic acid, ammonium acetate, alanine, sodium nitrate, and urea decreases from abundant to good in approximately the same sequence (Becking, 1984b). Aspartic acid, asparagine, and peptone give a much slower growth that is uneven and mostly confined to scattered colonies. Glycine seems to be toxic.

Although Thompson and Skerman (1979) reported four of six strains tested as able to utilize nitrate as N source, these strains gave negative results in tests of nitrite production from nitrate (see also Becking, 1984b). The reason for this discrepancy is not clear, but it is certain that some strains have nitrate reductase. The enzyme was purified from *D. gummosa* (unnamed strain supplied by Y.T. Tchan) and contained two dissimilar subunits, 80 and 88 kDa in size (Wang and Nicholas, 1986a). Induction of

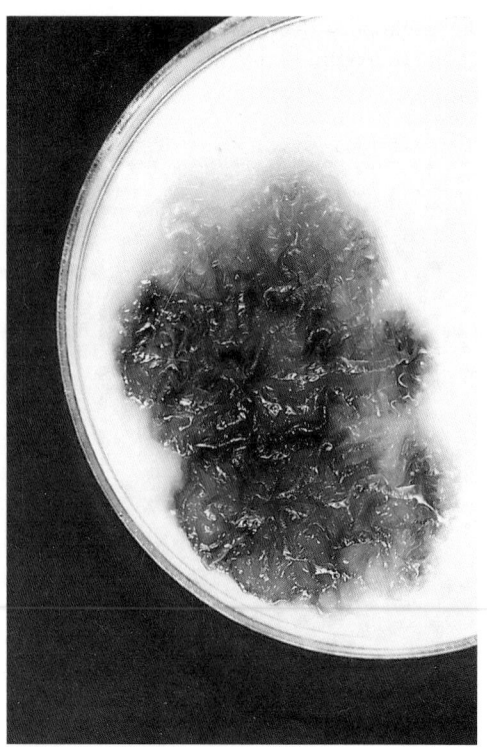

FIGURE BXII.β.24. Colony type of *D. gummosa* on nitrogen-free glucose agar with calcium carbonate (× 1.0).

---

1. Nitrogen-free, glucose mineral medium (g/l distilled water): K$_2$HPO$_4$, 0.5; MgSO$_4$·7H$_2$O, 0.25 (or 0.2); NaCl, 0.25; FeSO$_4$·7H$_2$O, 0.1; CaCl$_2$ or CaCO$_3$, 0.1; Na$_2$MoO$_4$·2H$_2$O, 0.005; and glucose, 10.0 (or 20.0); pH, 6.9. For a solid medium, add 15.0 g agar per liter.

nitrate reductase formation by nitrate was inhibited by ammonium or glutamine. Nitrite reductase was also purified (this enzyme also reduces hydroxylamine) (Wang and Nicholas, 1986d). Thus, at least some isolates of *D. gummosa* have both enzymes needed for conversion of nitrate to ammonia (assimilation).

As with most other *Proteobacteria*, *Derxia gummosa* assimilates ammonia via glutamine synthetase/glutamate synthase. Both enzymes were purified (Wang and Nicholas, 1985) and found to have physical and biochemical properties similar to these enzymes in other microorganisms, including adenylation of glutamine synthetase in response to ammonium.

D. gummosa grows autotrophically in an atmosphere containing $H_2/CO_2/O_2$, with either N_2 or NH_4^+ as the nitrogen source (Pedrosa et al., 1980). Indeed, it appears to grow nearly as well autotrophically as it does heterotrophically. Ribulose-1,5-bisphosphate activity, which mediates the CO_2 fixation, occurs in autotrophically grown cells but not in cells grown heterotrophically. Shankar et al. (1986) found that rapid O_2 depletion in a closed system caused variable, sometimes extremely poor, autotrophic growth; these authors developed a flow-through culturing technique that gave consistent results. Hydrogen recycling also takes place via an electron transport chain consisting of flavoprotein, ubiquinone, cytochrome *b*, and cytochrome *a* (Wang and Nicholas, 1986b).

Indole may or may not be produced from tryptophan. Starch is not hydrolyzed. *D. gummosa* antagonizes the growth of Gram-positive bacteria (Thompson and Skerman, 1979). Growth occurs on 1% peptone.

Ecology *Derxia* occurs mainly in tropical soils. It was isolated originally by Jensen et al. (1960) from a West Bengal soil having a pH of 6.5. It is widely distributed in Brazil (Campêlo and Döbereiner, 1970), Indonesia, China, and southern Africa (Becking, 1981). In Brazil, Campêlo and Döbereiner (1970) found it to be most frequent in flooded soils, less frequent in humid soils, and least frequent in dry soils. In comparing isolations made on a nitrogen-free, mineral starch agar, they found that *Derxia* could be more often isolated from root pieces of plants (mostly Gramineae) than from soil samples obtained from the same locality. In soil, the presence of the organisms seemed to be favored when the soil pH was between 5.1 and 5.5, although isolations were made from soils over a pH range of 4.5–6.5.

ENRICHMENT AND ISOLATION PROCEDURES

The sieved soil method may be used for isolation. Soil particles are regularly distributed over the surface of nitrogen-free mineral media with glucose[2] (Becking, 1984b). Mannitol and starch were previously suggested to be suitable C sources for isolation of *Derxia* (Jensen et al., 1960; Campêlo and Döbereiner, 1970), but Thompson and Skerman (1979) found all six strains tested unable to hydrolyze starch, and three of them were unable to utilize mannitol. Consequently, the use of mannitol or starch for isolation/enrichment of *Derxia* is not recommended. After growth on glucose N-free medium, one looks for the development of yellowish colonies around the soil particles. These colonies eventually become much larger and acquire a rust-brown color. Isolates should be further purified by repeated streaking.

Derxia strains are acid-tolerant and have been isolated from cultures designed to enrich for *Beijerinckia* strains by use of a liquid acidic nitrogen-free glucose medium[3] (Becking, 1984b). This medium is dispensed into Petri dishes so as to form a thin layer 2–3 mm deep.

MAINTENANCE PROCEDURES

The maintenance procedures are the same as those described for the genus *Beijerinckia*.

PROCEDURES FOR TESTING SPECIAL CHARACTERS

The ability to grow autotrophically can be tested on a solid or liquid medium[4] in the presence or absence of $(NH_4)_2SO_4$ (1.0 g/l). Cultures are incubated at 30°C in sealed vessels under the following atmospheres: (a) O_2/N_2, 1:99, v/v; (b) $O_2/CO_2/N_2$, 1:5:94; (c) $O_2/H_2/N_2$, 1:5:94; and (d) $O_2/H_2/CO_2/N_2$, 1:5:5:89. (Under atmospheres containing CO_2 the medium may also need to contain $NaHCO_3$ (1.0 g/l) to prevent acidification.) Significant growth should occur only when both H_2 and CO_2 are provided, even when N_2 is the sole nitrogen source. Under conditions of H_2/CO_2 dependent autotrophic growth under N_2-fixing or non-N_2-fixing conditions, the average doubling time of *D. gummosa* is 34.1 h (Pedrosa et al., 1980). A flow-through system for culturing *D. gummosa* autotrophically was also described (Shankar et al., 1986).

DIFFERENTIATION OF THE GENUS *DERXIA* FROM OTHER GENERA

Derxia can be distinguished from other genera of N_2-fixing bacteria by its very slimy (gummy) growth, both on agar plates and in liquid media, combined with appearance of the cells (Figs. BXII.β.22 and BXII.β.23) depending on age and type of medium (cells on peptone medium are very pleomorphic). *Azotobacter* and *Azomonas* species never produce such slimy colonies. Confusion with *Beijerinckia* is unlikely since *Beijerinckia* cells usually show very characteristic cells with two polar bodies containing PHB, whereas *Derxia* cells contain numerous PHB-containing bodies. Moreover, in contrast to *Beijerinckia* colonies, the colonies of *Derxia* acquire on aging a typical dark mahogany-brown color. In addition, *Derxia* is catalase negative, whereas *Beijerinckia* is catalase positive.

Distinction of *Derxia* and *Beijerinckia* species is also demonstrated by the differential ability to utilize compounds as carbon sources (see Table BXII.α.142 in the genus *Beijerinckia*). Most useful are possibly malonate, aspartate, glutamate, and ethylamine, carbon sources on which *D. gummosa* can grow but not species of *Beijerinckia* (Thompson and Skerman, 1979). In addition, resistance/sensitivity to antibiotics distinguish these genera. *D. gummosa* is more resistant to chlortetracycline and sulfanilamide than are *Beijerinckia* species (25 versus 1 μg/ml and

---

2. Nitrogen-free, mineral glucose agar for sieved-soil method (g/l distilled water): glucose, 20.0; K_2HPO_4, 0.8; KH_2PO_4, 0.2; $MgSO_4 \cdot 7H_2O$, 0.5; $FeCl_3 \cdot 6H_2O$, 0.025 (or 0.05); $Na_2MoO_4 \cdot 2H_2O$, 0.005; $CaCl_2$, 0.05; and agar, 15.0. The pH is adjusted to 6.9.

3. Liquid acidic nitrogen-free glucose medium (g/l distilled water): glucose, 20.0; KH_2PO_4, 1.0; and $MgSO_4 \cdot 7H_2O$, 0.5. The pH is adjusted to 5.0.

4. Medium for testing autotrophy (per liter of distilled water): KH_2PO_4, 1.2 g; K_2HPO_4, 0.8 g; $MgSO_4 \cdot 7H_2O$, 0.2 g; NaCl, 0.2 g; $CaCl_2 \cdot 2H_2O$, 0.02 g; $FeSO_4 \cdot 7H_2O$, 0.002 g; trace element solution (see below), 2.0 ml; biotin, 10 μg; and agar, 15.0 g. Trace element solution (g/l): $Na_2MoO_4 \cdot 2H_2O$, 1.0; $MnSO_4 \cdot H_2O$, 1.75; H_3BO_3, 1.4; $CuSO_4 \cdot 5H_2O$, 0.04; and $ZnSO_4 \cdot 7H_2O$, 0.12. This medium was employed by Pedrosa et al. (1980) in their studies of *Derxia*. In those studies, a very low concentration of potassium malate (0.1 g/l) was sometimes added to the medium. It should be noted that most strains of *Derxia* (including the type strain) give only scant growth with malate. Thompson and Skerman (1979) reported no growth on DL-malate in all of the six strains that they tested.

125 versus 1–5 µg/ml, respectively). *Beijerinckia* species are more resistant to chloramphenicol and penicillin than is *D. gummosa* (125 versus 5 µg/ml and 25 versus 1 µg/ml, respectively) (Thompson and Skerman, 1979). *D. gummosa* can utilize glutamate as a nitrogen source, whereas species of *Beijerinckia* cannot. Unlike most other aerobic diazotrophic species, *Derxia* can grow as a facultative hydrogen autotroph.

TAXONOMIC COMMENTS

Derxia, like *Beijerinckia*, was previously classified in the family *Beijerinckiaceae* in the order *Rhizobiales*, class *Alphaproteobacteria*. This placement was based on phenotypic characterization, not 16S rDNA sequence. De Smedt et al. (1980), based on rRNA hybridization, placed *D. gummosa* in the third rRNA superfamily, which also contains *Pseudomonas solanacearum*, *Chromobacterium violaceum*, *Janthinobacterium lividum*, and *Alcaligenes faecalis*. Phenotypically, both *Derxia* and *Beijerinckia* species produce a viscous, tenacious slime and form colonies that are similar in several respects. They utilize similar carbon sources, and they have a molybdenum requirement for N₂ fixation that cannot be replaced by vanadium (Thompson and Skerman, 1979; Becking, 1984a, b) (see also Tables BXII.α.141 and BXII.α.142 in the genus *Beijerinckia*). In addition, both are isolated from similar habitats. It is likely that other species of *Derxia* besides *D. gummosa* exist, but none have been formally named. Roy and Sen (1962) described a new species of *Derxia* from a sample of jute from Uttar Pradesh, India. Wu and Chen (1991) proposed a new species, "*Derxia peritricha*", for an isolate that shared many cultural characteristics with *D. gummosa*, and DNA hybridization studies also indicated significant similarity.

FURTHER READING

Becking, J.H. 1992. The Genus *Derxia*. *In* Balows, Trüper, Dworkin and Schleifer (Editors), The Prokaryotes. A Handbook on the Biology of Bacteria: Ecophysiology, Isolation, Identification, Applications, Springer-Verlag, New York. pp. 2605–2611.

List of species of the genus Derxia

1. **Derxia gummosa** Jensen, Petersen, De and Bhattacharya 1960, 193[AL]

gum.mo'sa. L. fem. adj. *gummosa* slime (gum) producing.

The characteristics are as described for the genus. Originally isolated from a soil of West Bengal of pH 6.5, but later also from slightly acidic or neutral soils of South America (Brazil, Surinam), South Africa, and Java (Becking, 1984b).

The mol% G + C of the DNA is: 70.4–71.7 (T_m) (De Ley and Park, 1966); 69.2–72.6 (T_m) (De Smedt et al., 1980).

Type strain: ATCC 15994, DSM 723, NCIB 9064.

Genus V. Oligella Rossau, Kersters, Falsen, Jantzen, Segers, Union, Nehls and De Ley 1987, 205[VP]

KAREL KERSTERS AND MARC VANCANNEYT

O.li.gel'la. Gr. adj. *oligos* little, scanty; M.L. dim. ending -*ella*; M.L. fem. n. *Oligella* referring to small bacterium with limited nutritional properties.

Small rods or coccobacilli, usually not exceeding 1 µm in length and often occurring in pairs; cell width is seldom larger than 0.6 µm. Gram negative, noncapsulated, and nonsporeforming. **Motility, if present, is by means of peritrichous flagella. Aerobic**, having a strictly respiratory type of metabolism with oxygen as the terminal electron acceptor. Moderately fastidious chemoorganotrophs growing at 30°C and 37°C. Grow on nutrient agar, but growth is enhanced by the addition of yeast autolysate, serum, or blood. No pigments and no odor are produced. Nonhemolytic. **Biochemically they are rather inert**; only a few organic acids and amino acids are oxidized or utilized as the sole carbon source. **Carbohydrates are neither fermented nor oxidized. Oxidase positive and usually catalase positive.** Indole and H₂S are not formed. Gelatin is not hydrolyzed. The urease test is useful to differentiate the two species. The major cellular fatty acids are *cis*-vaccenic acid (C₁₈:₁ ω7c) and palmitic acid (C₁₆:₀); two 3-hydroxylated acids (C₁₄:₀ ₃OH and C₁₆:₀ ₃OH) are also present, whereas branched-chain acids are absent. Mainly isolated from the genitourinary tract of humans. Pathogenicity is unknown but likely low. Belongs to the *Betaproteobacteria*; the genera *Brackiella*, *Pelistega*, and *Taylorella* are phylogenetic neighbors.

The mol% G + C of the DNA is: 46–48.

Type species: **Oligella urethralis** (Lautrop, Bøvre and Frederiksen 1970) Rossau, Kersters, Falsen, Jantzen, Segers, Union, Nehls and De Ley 1987, 206[VP] (*Moraxella urethralis* Lautrop, Bøvre and Frederiksen 1970, 255.)

FURTHER DESCRIPTIVE INFORMATION

Oligella urethralis is nonmotile, whereas most strains of *Oligella ureolytica* are motile by peritrichous flagella. However, motility of *Oligella ureolytica* may be difficult to demonstrate. Colonies on blood agar develop rather slowly and are more overtly white than those of all recognized species of *Moraxella*.

The following descriptive information is mainly based on data from Clark et al. (1984), Schreckenberger and von Graevenitz (1999), Schreckenberger et al., (1999), and, in particular, on the properties of 15 *O. urethralis* strains and 11 *O. ureolytica* strains as reported by Rossau et al. (1987). *Oligella* strains can grow in the presence of up to 3% NaCl. *O. ureolytica* and some strains of *O. urethralis* can also grow in the presence of 4.5% NaCl. Nitrite is reduced. The following physiological features are negative: acid production in oxidative-fermentative (O/F) media containing D-glucose, D-fructose, D-xylose, maltose, and *meso*-ribitol; formation of fluorescent pigment on King B medium; growth on cetrimide; hydrolysis of acetamide, DNA, esculin, gelatin, starch, and Tween 80; lysine and ornithine decarboxylase; and arginine dihydrolase. The following organic compounds are utilized as the sole carbon source (as tested with API auxanographic galleries): acetate, D-alanine, benzoate, fumarate, L-glutamate, glutarate, DL-3-hydroxybutyrate, itaconate, 2-ketoglutarate, DL-lactate, L-malate, pyruvate, succinate, and *n*-valerate. The following compounds are not utilized as carbon source (as tested with API auxanographic galleries): acetamide, *N*-acetylglucosamine, aconitate, adipate, β-alanine, 2-aminobenzoate, 3-aminobenzoate, 4-aminobenzoate, DL-2-aminobutyrate, DL-3-aminobutyrate, DL-4-aminobutyrate, amygdalin, amylamine, D-arabinose, L-arabinose, D-arabitol, L-arabitol, arbutin, L-arginine, L-aspartate, azelate, benzylamine, betaine, butylamine, caprate, caprylate, cellobiose, citraconate, L-citrulline, creatine, L-cysteine, diaminobutane, dulcitol, *meso*-erythritol, esculin, ethanolamine, ethylamine, D-fructose, D-fucose, L-fucose, D-galactose, gentiobiose, D-gluconate, D-glucosamine, D-glucose, DL-glycerate, glycerol, glycine, glycogen, glycolate, heptanoate, histamine, L-histidine, *o*-hydroxybenzoate, *meso*-inositol, inulin, 2-ketogluconate, 5-ketogluconate, DL-kynurenine, lactose, levulinate, L-lysine, D-lyxose, maleate, malonate, maltose, D-mandelate, L-mandelate, D-mannitol, D-mannose, melezitose, melibiose, L-methionine, methyl-α-D-glucoside, methyl-α-D-mannoside, methyl-β-D-xyloside, L-ornithine, oxalate, pelargonate, L-phenylalanine, phthalate, *iso*-phthalate, pimelate, raffinose, L-rhamnose, *meso*-ribitol (adonitol), D-ribose, salicin, sarcosine, sebacate, L-serine, sorbitol, L-sorbose, spermine, starch, suberate, sucrose, D-tagatose, D-tartrate, *meso*-tartrate, terephthalate, L-threonine, trehalose, trigonelline, tryptamine, D-tryptophan, L-tryptophan, turanose, L-tyrosine, urea, *meso*-xylitol, D-xylose, and L-xylose. The following substrates are hydrolyzed (as tested with API-ZYM galleries): L-leucyl-2-naphthylamide, 2-naphthyl butyrate, and 2-naphthyl caprylate. The following substrates are not hydrolyzed (using API-ZYM galleries): *N*-benzoyl-DL-arginine-2-naphthylamide, 6-bromo-2-naphthyl-α-D-galactopyranoside, 6-bromo-2-naphthyl-α-D-mannopyranoside, 6-bromo-2-naphthyl-β-D-glucopyranoside, L-cystyl-2-naphthylamide, *N*-glutarylphenylalanine-2-naphthylamide, naphthol-AS-BI-β-D-glucuronate, 1-naphthyl-*N*-acetyl-β-D-glucosaminide, 2-naphthyl-α-L-fucopyranoside, 2-naphthyl-β-D-galactopyranoside, 2-naphthyl-α-D-glucopyranoside, 2-naphthylmyristate, 2-naphthylphosphate (at pH 5.4), and L-valyl-2-naphthylamide.

According to version 3.5 of the Biolog database, the following substrates of the Biolog GN Microplate yield positive results with *Oligella* strains: acetic acid, alaninamide, D-alanine, L-alanine, bromosuccinic acid, formic acid, L-glutamic acid, 2-hydroxybutyric acid, 3-hydroxybutyric acid, itaconic acid, 2-ketobutyric acid, 2-ketoglutaric acid, 2-ketovaleric acid, DL-lactic acid, L-leucine, methylpyruvate, monomethylsuccinate, L-proline, propionic acid, succinamic acid, and succinic acid.

The major cellular fatty acids of *Oligella* are *cis*-vaccenic acid ($C_{18:1\ \omega7c}$) and palmitic acid ($C_{16:0}$) (comprising respectively approximately 35–45% and 20–25% of the total fatty acid content). Two 3-hydroxylated acids $C_{14:0\ 3OH}$ and $C_{16:0\ 3OH}$ are present (3–5%), whereas branched chain acids are reported to be absent (Rossau et al., 1987; Moss et al., 1988; Schreckenberger and von Graevenitz, 1999). Only *O. ureolytica* strains characteristically possess in addition lactobacillic acid ($C_{19:0\ cyclo\ \omega8c}$) (Rossau et al., 1987). However, when cultivated on tryptic soy medium, no $C_{19:0\ cyclo\ \omega8c}$ was detected for the type strain of *O. ureolytica* (M. Vancanneyt, unpublished data).

O. urethralis contains ubiquinone with eight isoprene units as the major isoprenolog (Moss et al., 1988).

O. urethralis produces 2-hydroxyputrescine and putrescine as predominant polyamine compounds (respectively 22 and 10 μmol/dry weight), and smaller amounts of spermidine (4 μmol/g dry weight) (H.-J. Busse, personal communication).

Oligella urethralis strains are generally susceptible to most antibiotics, including penicillin, whereas *O. ureolytica* is susceptible to a limited number of antibiotics (Welch et al., 1983). Quinolone resistance has been reported in *O. urethralis* (Riley et al., 1996).

The size of *Oligella* genomes is $1.2–1.4 \times 10^9$ Da (Rossau et al., 1987).

Oligella strains have been mainly isolated from the human genitourinary tract, and both species have been reported to cause urosepsis (Rockhill and Lutwick, 1978; Pugliese et al., 1993). Some strains of *O. urethralis* were also isolated from the ear, blood, and a foot wound. Cases of septic arthritis and chronic ambulatory peritoneal dialysis peritonitis caused by *O. urethralis* have been reported (Mesnard et al., 1992; Riley et al., 1996). *O. urethralis* has also been isolated from the conjunctiva of rabbits (Marini et al., 1996).

Enrichment and Isolation Procedures

The nutritional requirements of *Oligella* have not been studied in detail. Strains grow on ordinary blood media, heart infusion agar, tryptic soy agar, and Drigalski agar. Incubation in the presence of 5% CO_2 will usually enhance growth. Growth on MacConkey agar is variable. The majority of the *O. urethralis* strains will also grow on nutrient agar. No procedures for the selective isolation of *Oligella* have been worked out.

Maintenance Procedures

Oligella strains can be maintained by classical preservation techniques such as lyophilization, storage in liquid nitrogen, and storage in a freezer at $-80°C$.

Differentiation of the genus *Oligella* from other genera

Table BXII.β.35 summarizes the differential characteristics between *Oligella* and the phylogenetically closely related genera *Pelistega* and *Taylorella*, as well as the phenotypically similar species *Bordetella bronchiseptica*, *Moraxella osloensis*, and *Ralstonia paucula*, a new species encompassing bacteria of the CDC group IVc-2 (Vandamme et al., 1999).

Rossau et al. (1987), Holt et al. (1994), Weyant et al. (1996), Schreckenberger et al. (1999), and Schreckenberger and von Graevenitz (1999) have compiled other useful tables for differentiating *Oligella* from morphologically or biochemically similar bacterial taxa such as *Moraxella*. Gas chromatographic analysis of the cellular fatty acids yields usually a good discrimination between these taxa; e.g., *Oligella* cells possess $C_{18:1\ \omega7c}$ as the major acid, whereas authentic *Moraxella* species lack this compound, and typically contain oleic acid ($C_{18:1\ \omega9c}$) (Moss et al., 1988) (Table BXII.β.35). The biochemical inertness of the *Oligella* strains renders the identification of *Oligella* and phenotypically similar species difficult. Due to its fast and strong urease reaction, *O. ureolytica* can be confused with *B. bronchiseptica* or *Ralstonia paucula* (CDC group IVc-2) (Table BXII.β.35).

Taxonomic Comments

The genus *Oligella* currently contains two species: *Oligella urethralis*, which was previously classified in the genus *Moraxella*, and *Oligella ureolytica*, which was formerly referred to as CDC group IVe bacteria. Based on comparative 16S rDNA-analysis *Oligella* belongs to the *Betaproteobacteria*, its closest phylogenetic neighbors are the genera *Brackiella*, *Pelistega*, and *Taylorella* (see Fig. BXII.β.25 in this chapter and Fig. BXII.β.21 in the chapter on the family *Alcaligenaceae*).

Oligella urethralis was first described as *Moraxella urethralis* by Lautrop et al. (1970), who stated that the allocation of these bacteria in the genus *Moraxella* was only a matter of convenience. Hence in the previous edition of *Bergey's Manual*, *M. urethralis* was considered as a *species incertae sedis* of the genus *Moraxella*

TABLE BXII.β.35. Differential characteristics of the genus *Oligella* and phylogenetically related or phenotypically similar bacteria[a,b]

Feature	*Oligella*[c]	*Pelistega europaea*[d]	*Taylorella equigenitalis*[e]	*Ralstonia paucula* (CDC group IVc-2)[f]	*Bordetella bronchiseptica*[g]	*Moraxella osloensis*[h]
Cell diameter, μm	0.6	0.2–0.4	0.7	0.8	0.2–0.5	1.0–1.5
Cell length, μm	<1.0	1.0–2.0	0.7–1.8	1.2–2.0	0.5–2.0	1.5–2.5
Motility (peritrichous flagella)	D	–	–	+	+	–
Growth at 42°C	D	+	+	+	+	–
Growth on MacConkey agar	d	–	–	+		d
Growth in presence of 3% NaCl	+	–	–	–		–
Urease	D	d	–	+	+	–
Nitrate reduction	D	–	–	–	+	d
Nitrite reduction	+	–	–	–	–	–
Denitrification	d	–	–	–	–	–
Carbon sources for growth:						
Adipate	–	–		+	+	–
D-Gluconate	–	–		+	–	–
Benzoate	+			d	–	–
p-Hydroxybenzoate	D			–	–	–
L-Malate	+	+		+	+	–
Characteristic cellular fatty acids[i]:						
$C_{10:0}$						+
$C_{12:0}$		+				
$C_{14:0}$	+	+		+	+	
$C_{16:0}$	+	+	+	+	+	+
$C_{16:1}$	+		+	+		
$C_{18:0}$			+		+	+
$C_{16:1\,\omega7c}$		+			+	+
$C_{17:0\,cyclo}$				+	+	
$C_{18:1\,\omega7c}$	+	+	+	+	+	
$C_{18:1\,\omega9c}$						+
$C_{12:0\,2OH}$					+	
$C_{12:0\,3OH}$						+
$C_{14:0\,3OH}$	+		+			+
$C_{16:0\,3OH}$	+	+	+	+		
$C_{16:1\,2OH}$				+		
$C_{18:1\,2OH}$				+		
SF 3[j]		+		+	+	
Mol% G + C of DNA	46–48	42–43	36–38	65–67	68–69.5	43–46

[a]Symbols: see standard definitions.

[b]Data from Clark et al. (1984), Rossau et al. (1987), Moss et al. (1988), Vandamme et al. (1998, 1999), and Schreckenberger and von Graevenitz (1999).

[c]Isolated from human genitourinary tract.

[d]Isolated from respiratory tracts of pigeons.

[e]Isolated from genital tract of mares; caused endometritis and cervicitis.

[f]Isolated from various human clinical and environmental sources.

[g]Isolated from respiratory tract of animals and humans.

[h]Isolated from clinical sources.

[i]The references for the fatty acid data are as follows: *Oligella* (Rossau et al., 1987), *Pelistega* (Vandamme et al., 1998), *Taylorella* (Rossau et al., 1987), *Ralstonia paucula* (P. Vandamme, personal communication), *Bordetella bronchiseptica* (Vancanneyt et al., 1995), and *Moraxella osloensis* (Moss et al., 1988; Vandamme et al., 1993b).

[j]SF 3 (summed feature 3) comprises $C_{14:0\,3OH}$ or $C_{16:1\,iso\,I}$, or any combination of these fatty acids.

(Bøvre, 1984b), which is phylogenetically quite distinct from *Oligella* (Fig. BXII.β.25); the authentic *Moraxella* species belong to the *Gammaproteobacteria* (Rossau et al., 1991; Pettersson et al., 1998b). Although *O. urethralis* displays several phenotypic features in common with *Moraxella* species (particularly *M. osloensis*), genetic transformation studies with streptomycin resistance as a marker indicated that the species *Oligella urethralis* (previously named *Moraxella urethralis*) displayed no distinct genetic affinities to the authentic *Moraxella* species (Bøvre, 1980). Consequently, Bøvre (1984b) suggested that the species might belong to a separate genus. Moreover, *Oligella* cells are also smaller and lack the typical plumpness of moraxellae, and their cellular fatty acid composition differs significantly (Moss et al., 1988; Table BXII.β.35). *O. urethralis* was formerly also referred to as "CDC group M-4", and includes some of the strains previously called "*Mima polymorpha* biovar oxidans".

O. ureolytica was previously known as CDC group IVe, displaying some phenotypic similarities (e.g., rapid and strong urease test) with *Bordetella bronchiseptica* and *Ralstonia paucula* (formerly CDC group IVc-2) (Clark et al., 1984; Vandamme et al., 1999).

The grouping of *Oligella* in the *Betaproteobacteria* was initially established by Rossau et al. (1987) based on rRNA–DNA hybridization studies. Recently the knowledge of the phylogenetic relationships of the genus has been refined by an analysis of the nearly complete 16S rDNA sequences of the type strains of the species *O. urethralis* and *O. ureolytica* (M. Vancanneyt and B. Hoste, unpublished data) (Fig. BXII.β.25). The type strains of the two *Oligella* species display a significant 16S rDNA similarity (97%). The nearest phylogenetic neighbors of *Oligella* are *Pelistega europaea* (92.1%) and *Taylorella equigenitalis* (91.8%) (Fig. BXII.β.25). At present *Pelistega* encompasses one species, *P. europaea* (Vandamme et al., 1998), which is involved in the path-

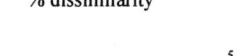

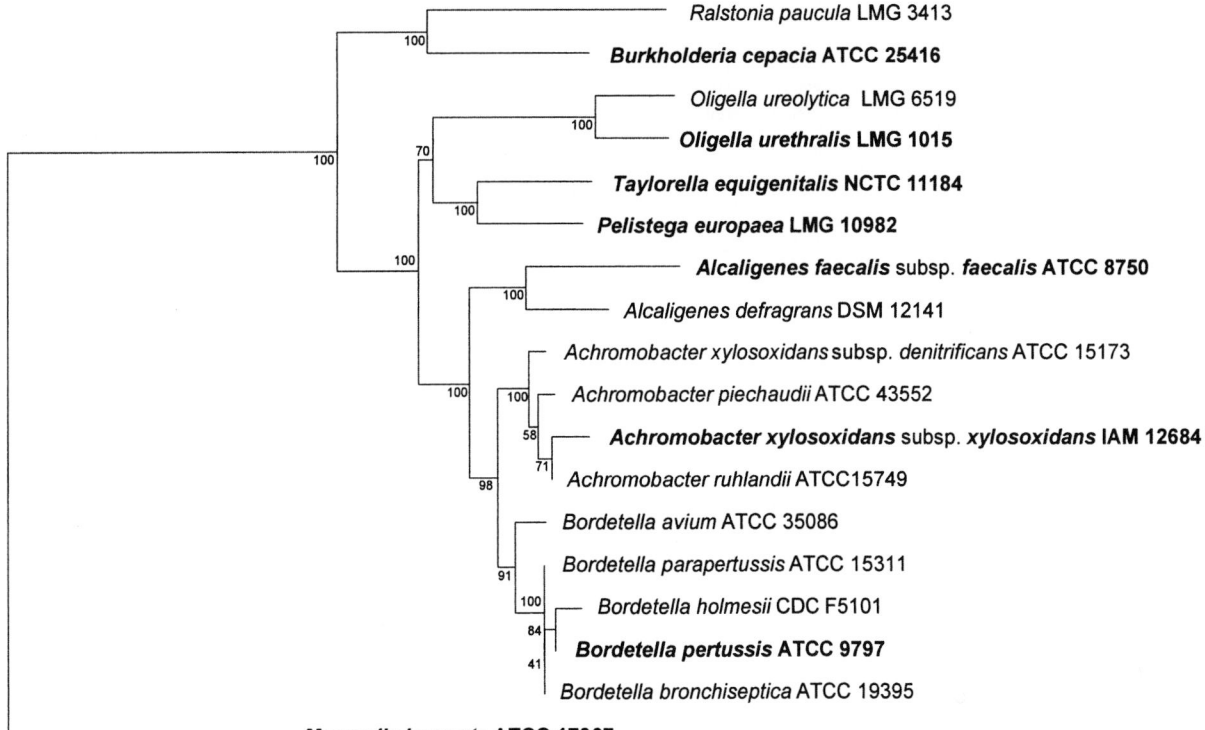

FIGURE BXII.β.25. Phylogenetic tree of members of the genus *Oligella* and related bacteria. The analysis was performed using the software package GeneCompar (Applied Maths, Kortrijk, Belgium) after including the consensus sequence in a multiple alignment of small ribosomal subunit sequences collected from the international nucleotide sequence library EMBL. A resulting tree was constructed using the neighbor-joining method. Unknown bases and gaps were not considered in the numerical analysis. Bootstrap probability values are indicated at the branch-points (100 trees resampled). All strains are type strains and the type species are indicated in bold.

ogenesis of respiratory diseases in pigeons. *Taylorella equigenitalis*, the sole species of the genus *Taylorella* (Sugimoto et al., 1983) is the causative agent of contagious equine metritis, a sexually transmitted bacterial disease of horses. The phylogenetic grouping of *T. equigenitalis* within the *Betaproteobacteria* was reported by Bleumink-Pluym et al. (1993) and Rossau et al. (1987). The 16S rDNA similarity values of *Oligella* versus the investigated type strains of the various species of the family *Alcaligenaceae* vary between 91.3% and 92.6%, whereas similarity values with the type species of the genus *Moraxella* are only 82% (Fig. BXII.β.25). It is surprising that the phylogenetically related taxa *Achromobacter*, *Alcaligenes*, *Bordetella*, *Oligella*, *Pelistega*, and *Taylorella* (Fig. BXII.β.25) cover in their genomic DNA a very broad mol% G + C range (36–70) (Table BXII.β.35).

The majority of the above-mentioned taxa, including *Oligella* but excluding some of the *Bordetella* species, can be circumscribed as oxidase positive, indole-negative, asaccharolytic nonfermenting Gram-negative bacteria. Strains of the genera *Alcaligenes*, *Achromobacter* (Yabuuchi et al., 1998a), and *Bordetella* can be implicated in human and animal diseases.

ACKNOWLEDGMENTS

We thank B. Hoste for help in the determination of the sequence of the rRNA gene of *O. urethralis* and *O. ureolytica*.

FURTHER READING

Rossau, R., K. Kersters, E. Falsen, E. Jantzen, P. Segers, A. Union, L. Nehls and J. De Ley. 1987. *Oligella*, a new genus including *Oligella urethralis* comb. nov. (formerly *Moraxella urethralis*) and *Oligella ureolytica* sp. nov. (formerly CDC group IVe): relationship to *Taylorella equigenitalis* and related taxa. Int. J. Syst. Bacteriol. *37*: 198–210.

Schreckenberger, P.C. and A. von Graevenitz. 1999. *Acinetobacter, Achromobacter, Alcaligenes, Moraxella, Methylobacterium,* and other nonfermentative Gram-negative rods. *In* Murray, Baron, Pfaller, Tenover and Yolken (Editors), Manual of Clinical Microbiology, 7th Ed., ASM Press, Washington D.C. pp. 539–560.

DIFFERENTIATION OF THE SPECIES OF THE GENUS *OLIGELLA*

Characteristics useful in differentiating the species of *Oligella* are given in Table BXII.β.36.

List of species of the genus Oligella

1. **Oligella urethralis** (Lautrop, Bøvre and Frederiksen 1970) Rossau, Kersters, Falsen, Jantzen, Segers, Union, Nehls and De Ley 1987, 206[VP] (*Moraxella urethralis* Lautrop, Bøvre and Frederiksen 1970, 255.)

    u.re.thra′ lis. Gr. *ourethra* urethra; M.L. gen. n. *urethralis* of the urethra.

    The description is as given for the genus and as listed in Tables BXII.β.35 and BXII.β.36. Cells are nonmotile. Inclusions of intracellular poly-β-hydroxybutyrate have been reported. Grows at 42°C. Urease negative (Christensen's).

No growth on *p*-hydroxybenzoate. Nitrate is not reduced, but denitrification is usually positive. Usually highly sensitive to penicillin. Lactobacillic acid ($C_{19:0 \text{ cyclo } \omega 8c}$) is not detectable among the cellular fatty acids.

Has been isolated from human urine, the human genitourinary tract, and the ear.

The mol% G + C of the DNA is: 46–47.5 (T_m).

Type strain: ATCC 17960, CCUG 13463, CDC 7603, DSM 7531, LMG 5303.

GenBank accession number (16S rRNA): AF133538, AF227163.

2. **Oligella ureolytica** Rossau, Kersters, Falsen, Jantzen, Segers, Union, Nehls and De Ley 1987, 209[VP]

    ur′ e.o.ly.ti.ca. M.L. n. *urea* urea; Gr. adj. *lyticus* dissolving; M.L. adj. *ureolytica* dissolving (hydrolyzing) urea.

    The description is as given for the genus and as listed in Tables BXII.β.35 and BXII.β.36. The species accommodates the CDC group IVe strains.

    Some strains are motile by means of long peritrichous flagella. No growth at 42°C. Grows on *p*-hydroxybenzoate as carbon source. The urease test (Christensen's) is usually strongly positive after 4 h. Most strains reduce nitrate and are resistant to penicillin. Among the cellular fatty acids lactobacillic acid ($C_{19:0 \text{ cyclo } \omega 8c}$) is present in moderate amounts.

    Isolated from human urine, particularly from males.

    The mol% G + C of the DNA is: 46–47 (T_m).

    Type strain: ATCC 43534, CCUG 1465, CDC C379, LMG 6519.

    GenBank accession number (16S rRNA): AJ251912.

TABLE BXII.β.36. Differential features of the species of the genus *Oligella*[a,b]

Characteristic	O. urethralis	O. ureolytica
Motility	−	d
Growth at 42°C	+	−
Growth in the presence of 4.5% NaCl	d	+
Urease	−	+[c]
Nitrate reduction	−	d
Denitrification	+[d]	d[d]
Growth on:		
p-Hydroxybenzoate	−	+
DL-5-Aminovalerate	+	−
Presence of $C_{19:0 \text{ cyclo}}$ fatty acid (>1%)	−	+

[a]Symbols: see standard definitions.

[b]Data from Clark et al. (1984), Rossau et al. (1987), and Schreckenberger and von Graevenitz (1999).

[c]The urease reaction is strongly positive.

[d]Demonstration of gas may be difficult.

Genus VI. **Pelistega** *Vandamme, Segers, Ryll, Hommez, Vancanneyt, Coopman, De Baere, Van De Peer, Kersters, De Wachter and Hinz 1998, 437[VP]*

PETER VANDAMME AND KARL-HEINZ HINZ

Pe.li′ ste.ga. Gr. n. *peleia* pigeon; Gr. fem. n. *stege* house, stay, residence; N.L. fem. n. *Pelistega* refers to the bacteria living in pigeons.

Cells in 16–24-h-old cultures on blood agar are 0.2–0.4 × 1–2 μm, with **variable morphological forms**. Gram negative. **Non-sporeforming.** Capsulated. **Nonmotile. Optimal growth under microaerobic conditions.** Growth is weak under aerobic conditions and absent under anaerobic conditions. Growth at 37°C and 42°C, not at 24°C. Does not require growth factors on conventional media. No growth on MacConkey agar. **Glucose is oxidized with the production of alkali,** no fermentation of, or acid production from, glucose. **Catalase and oxidase activity is present.** No reduction of nitrate, no denitrification, no production of acetylmethylcarbinol, indole, or methyl red, no esculin hydrolysis, β-galactosidase, DNase, chondroitin sulfatase, hyaluronidase, lysine decarboxylase, ornithine decarboxylase, lecithinase, or phenylalanine deaminase activity. **Alkaline and acid phosphatase activity is present.** Thus far only isolated from pigeon samples.

The mol% G + C of the DNA is: 42–43.

Type species: **Pelistega europaea** Vandamme, Segers, Ryll, Hommez, Vancanneyt, Coopman, De Baere, Van De Peer, Kersters, De Wachter and Hinz 1998, 437.

FURTHER DESCRIPTIVE INFORMATION

The genus *Pelistega* belongs to the *Betaproteobacteria* with the genus *Taylorella* as its closest relative (Fig. BXII.β.26). Other genera belonging to the same phylogenetic neighborhood are *Bordetella*, *Alcaligenes*, and *Burkholderia*.

Strains of the genus *Pelistega* have been isolated from samples of lungs, air sac exudate, and trachea mucosa of pigeons; less frequently from other organs such as the liver and spleen. The pathogenicity is unknown.

ENRICHMENT AND ISOLATION PROCEDURES

At present, blood agar is the only medium by which *Pelistega* strains have been successful isolated. MacConkey agar and litmus

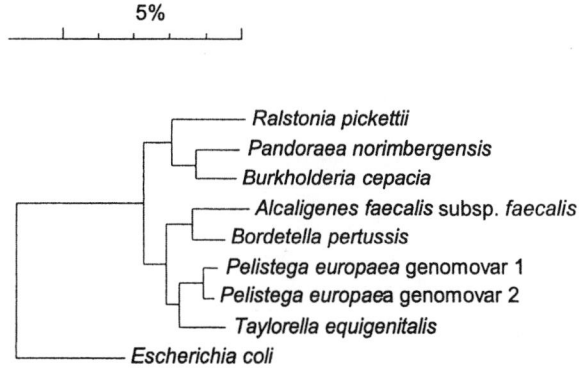

5%

FIGURE BXII.β.26. Phylogeny of *Pelistega europaea*. Bar = 5% difference in 16S rRNA nucleotide sequences. The neighbor-joining method was used for tree construction. Accession numbers for the sequences used are: *Alcaligenes faecalis* subsp. *faecalis* LMG 1229[T], D88008; *Bordetella pertussis* LMG 14455[T], U04950; *Burkholderia cepacia* LMG 1222[T], M22518; *Pandoraea norimbergensis* LMG 18379[T], Y09879; *Pelistega europaea* genomovar 1 LMG 10982[T], Y11890 and *Pelistega europaea* genomovar 2 LMG 15725, AF190911; *Ralstonia pickettii* LMG 5942[T], X67042; *Taylorella equigenitalis* LMG 6222[T], X68645; *Escherichia coli* (J01695) was chosen as an outgroup.

lactose agar are not suitable for primary isolation. Selective procedures have not been developed. Clinical specimens are plated on blood agar plates and incubated at 37°C for 24–48 h in a microaerobic atmosphere with a high humidity (e.g., candle jar conditions). *Pelistega* strains can be identified especially by the characters given in Table BXII.β.37. The numerical analysis of the whole-cell fatty acid patterns by using the MIDI system also allows a rapid and reliable identification.

MAINTENANCE PROCEDURES

Cultures may be stored for many years by lyophilization, freezing at −80°C, or in liquid nitrogen. Cryoprotective agents such as 10% glycerol or DMSO should be added to cultures before freezing. Strains also remain viable for over 10 years when stored as dense suspensions in skimmed milk in a frozen state at −70°C. Twenty-four-hour blood agar cultures will survive refrigeration (4°C) in ambient air for at least 1 week.

DIFFERENTIATION OF THE GENUS *PELISTEGA* FROM OTHER GENERA

Comparative 16S rDNA sequence analysis revealed that *Pelistega europaea* belongs to the *Betaproteobacteria* with *Taylorella equigenitalis* as its nearest neighbor (Fig. BXII.β.26). Like *Pelistega europaea*, *Taylorella equigenitalis* (the type species of the genus *Taylorella*) is a microaerobic, Gram-negative, nonmotile, rod-shaped organism that does not acidify carbohydrates and exhibits catalase, oxidase, and acid and alkaline phosphatase activity. *Taylorella equigenitalis* differs from *Pelistega europaea* by the absence of urease and arginine dihydrolase activity, and by alkali production from glucose. *Taylorella equigenitalis* strains can also be differentiated from *Pelistega europaea* by means of whole-cell protein and fatty acid analysis and by its mol% G + C content (about 36–37% for the former and 42–43% for the latter). The most salient differences between *Pelistega europaea* and *Taylorella equigenitalis*, however, are the fastidious growth requirements of the latter, while *Pelistega europaea* grows abundantly when incubated under microaerobic conditions. Vandamme et al. (1998) reported a remarkable stim-

ulation of growth of *Taylorella equigenitalis* strains by the addition of growth factors, which was not recorded for *Pelistega europaea* strains. Earlier studies by Sugimoto et al. (1983), however, did not reveal this growth stimulation.

Table BXII.β.37 lists differential diagnostic characteristics that can be used to distinguish *Pelistega europaea* from other species that may be encountered in specimens from pigeons: *Riemerella anatipestifer*, *Ornithobacterium rhinotracheale*, *Chryseobacterium meningosepticum*, and *Pasteurella* species.

TAXONOMIC COMMENTS

Vandamme et al. (1998) described a polyphasic taxonomic study of a collection of 24 pigeon isolates that remained unidentified after preliminary examination by conventional biochemical tests. Additional extensive biochemical characterization by using a considerable number of conventional tests and several API microtest systems further substantiated the high phenotypic similarity among all of the strains examined. Unexpectedly, whole-cell fatty acid analysis demonstrated a marked subdivision of the pigeon isolates into two main clusters. Strains of both clusters had the same major fatty acid components but there were marked quantitative differences. However, the same strains were again further subdivided into four subgroups and several strains with unique patterns by means of one-dimensional whole-organism protein electrophoresis. The DNA base ratio of representative strains of these subgroups was determined, and all values were between 42–43 mol% G + C. Results obtained by whole-cell protein electrophoresis were a good indicator of the level of DNA–DNA hybridization, as only within the protein electrophoretic subgroups high binding levels were detected (>82%); all other values were low or insignificant.

In taxonomic practice, the DNA–DNA hybridization level within a bacterial species is mostly above 50–70%. This level is primarily determined by the phenotypic consistency of the taxon studied, as species should be identifiable by phenotypic characteristics (Wayne et al., 1987; Ursing et al., 1995). The term "genomovar" refers to closely related genomic species that cannot be distinguished by phenotypic characteristics (Ursing et al., 1995). The group of pigeon isolates obviously comprised several subgroups that were genotypically distinct enough to warrant their classification as different species. However, in the absence of differential phenotypic characteristics between these genomovars, all isolates were included in a single nomenspecies, *Pelistega europaea*.

Strain LMG 10982 (representing genomovar 1) was chosen as the type strain for the nomenspecies. Strains LMG 15725, LMG 11609, and LMG 12985 were selected as reference strains for genomovars 2, 3, and 4, respectively. In addition, several strains with unique whole-organism protein patterns were not formally classified into different genomovars until the DNA–DNA hybridization levels between these strains and the four designated genomovars were fully determined.

It therefore should be clear that *Pelistega europaea* presently encompasses multiple genomic species. It is not unlikely that this name may be restricted in the future to only one or a few of these genomovars, upon discovery of differential diagnostic features.

ACKNOWLEDGMENTS

P. Vandamme is indebted to the Fund for Scientific Research–Vlaanderen (Belgium) for a position as a postdoctoral research fellow.

TABLE BXII.β.37. Differential characteristics between *Pelistega europaea* and other bacteria from pigeons[a]

Characteristic	*Pelistega europaea*	*Chryseobacterium meningosepticum*	*Ornithobacterium rhinotracheale*	*Pasteurella* species	*Riemerella columbina*
Pigmentation	−	V[b]	−	−	−
Growth on MacConkey agar	−	+	−	−	−
Aerobic growth on blood agar at 37°C	+ or W	+	W	+	+
Catalase activity	+	+	−	+	+
Urease activity	V	−	+	−	V
Nitrate reduction	−	−	−	+	−
Indole production	−	V[b]	−	V	−
Gelatinase activity	−	+	−	−	+
Chondroitin sulfatase activity	−	ND	+	V	−
Hyaluronidase activity	−	ND	+	V	−
Esculin hydrolysis[c]	−	+	−	−	+
β-Galactosidase activity	−	+	+	V[b]	−
Acid production from glucose	−	V[b]	+	+	+
Utilization of carbon sources for growth[c]	+	+	−	−	−
DNA base ratio	42–43	36–37	37–39	40–45	36–37
Host spectrum	Pigeons	Humans, birds	Turkeys, chickens, wild birds	Mammals, birds	Pigeons

[a]Symbols: +, characteristic present; −, characteristic absent; V, strain-dependent reaction; W, weak reaction; ND; not determined.

[b]Over 80% of the strains contains this feature.

[c]As determined using the API 20 NE microtest system.

FURTHER READING

Sugimoto, C., Y. Isayama, R. Sakazaki and S. Kuramochi. 1983. Transfer of *Haemophilus equigenitalis* Taylor et al., 1978 to the genus *Taylorella*, gen. nov. as *Taylorella equigenitalis*, comb. nov. Curr. Microbiol. *9*: 155–162.

Vandamme, P., P. Segers, M. Ryll, J. Hommez, M. Vancanneyt, R. Coopman, R. de Baere, Y. van de Peer, K. Kersters, R. De Wachter and K.H. Hinz. 1998. *Pelistega europaea* gen. nov., sp. nov., a bacterium associated with respiratory disease in pigeons: taxonomic structure and phylogenetic allocation. Int. J. Syst. Bacteriol. *48*: 431–440.

List of species of the genus Pelistega

1. **Pelistega europaea** Vandamme, Segers, Ryll, Hommez, Vancanneyt, Coopman, De Baere, Van De Peer, Kersters, De Wachter and Hinz 1998, 437[VP]

   eu.ro.pae′ a. L. adj. *europaea* of Europe, because the first collection of strains was isolated in different European countries.

   Cells in 16–24-h-old cultures on blood agar are 0.2–0.4 × 1–2 μm, with variable morphological forms. Gram negative. Nonsporeforming. Capsulated. Nonmotile.

   Strains produce convex, circular, and grayish-white to yellowish colonies with entire edge and smooth surface on blood agar.

   Optimal growth under microaerobic conditions. Growth is weak under aerobic conditions, and absent under anaerobic conditions. Growth at 37°C and 42°C, not at 24°C. Does not require growth factors on conventional media. In standard II nutrient broth, growth appears as moderate turbidity and pellicle near the surface of the broth after 24–48 h of incubation. In most cases, a small, button-like sediment could be seen. No growth on litmus lactose agar at the primary isolation from the infected tissues. However, after some passages on artificial media some strains showed a poor growth. No growth on MacConkey agar.

   All strains produce catalase and oxidase activity; urease and arginine dihydrolase activity is strain dependent. No reduction of nitrate, no denitrification, no production of acetylmethylcarbinol, indole, or methyl red, no esculin hydrolysis, gelatinase (except strain LMG 11609), β-galactosidase, DNase, chondroitin sulfatase, hyaluronidase, lysine decarboxylase, ornithine decarboxylase, lecithinase, or phenylalanine deaminase activity. Glucose is oxidized with the production of alkali, no fermentation of, or acid production from, glucose. Malonate is not used as a carbon source. Strain-dependent utilization of citrate on Simmons' citrate agar.

   No acid production from D-glucose, D-fructose, sucrose, lactose, maltose, D-galactose, galacturonate, D-mannose, rhamnose, cellobiose, palatinose, dextrin, N-acetyl-D-glucosamine, lactulose, L-sorbose, adonitol, D-mannitol, L-arabinose, D- and L-arabitol, salicin, D-sorbitol, trehalose, D-xylose, dulcitol, inositol, or *myo*-inositol.

   There is no assimilation of D-glucose, maltose, D-mannose, N-acetyl-D-glucosamine, D-mannitol, L-arabinose, D-gluconate, caprate, adipate, or phenyl acetate; most strains (18 out of 19 examined) assimilate L-malate.

   Alkaline and acid phosphatase, esterase C4, ester lipase C8, and leucine, arginine, alanine, glycine, and L-aspartic acid arylamidase activity is present. Activity of naphthol-AS-BI-phosphohydrolase and lipase (as present in the API ID 32E system) is strain dependent. Lipase C14, valine, leucyl glycine, phenylalanine, histidine, glutamyl glutamic acid, serine, and cystine arylamidase, trypsin, chymotrypsin, α-galactosidase, α-maltosidase, β-glucuronidase, α- and β-glucosidase, N-acetyl-β-glucosaminidase, α-mannosidase, and α-fucosidase activity has not been detected. Proline, pyroglutamic acid, and tyrosine arylamidase activity is strain dependent.

   Two major clusters of strains are delineated by means of cellular fatty acid analysis. The major fatty acid components of all strains examined were $C_{12:0}$, $C_{14:0}$, $C_{16:0}$, $C_{16:1\ \omega7c}$, $C_{16:1\ \omega5c}$, $C_{16:0\ 3OH}$, and summed features 3 and 7. However, the percentages of these fatty acids vary strongly among the different genomovars (Table BXII.β.38).

   Strains have been isolated on common nonselective blood agar media under microaerobic (candle jar) conditions mainly from samples of lungs, air sac exudate, and

TABLE BXII.β.38. Fatty acid composition of *Pelistega europaea* strains[a]

Fatty acid	Genomovars 1, 2, and 3	Genomovar 4
$C_{12:0}$	4.4 ± 0.7	4.4 ± 1.0
$C_{14:0}$	9.7 ± 1.5	10.8 ± 2.5
$C_{16:0}$	15.7 ± 2.9	25.2 ± 2.8
$C_{18:0}$	Tr	Tr
$C_{16:1\ \omega7c}$	21.7 ± 3.5	41.4 ± 2.7
$C_{16:1\ \omega5c}$	5.2 ± 2.6	Tr
$C_{16:0\ 3OH}$	1.2 ± 0.3	1.2 ± 0.2
$C_{19:0\ 10CH_3}$	Tr	nd
Summed feature 1[b]	Tr	2.7 ± 1.6
Summed feature 3[c]	12.7 ± 2.1	12.4 ± 1.1
Summed feature 7[d]	27.8 ± 9.4	1.5 ± 0.9

[a]Those fatty acids for which the average amount for all taxa was less than 1% are not given. Therefore, the percentages for each group do not total 100%. Mean percentages and standard deviations are given; Tr, trace amount (less than 1%); nd, not detected.

[b]Summed feature 1 comprises $C_{14:1\ \omega5c}$, $C_{14:1\ \omega5t}$, or both.

[c]Summed feature 3 comprises $C_{14:0\ 3OH}$, $C_{16:1\ iso\ I}$, an unidentified fatty acid with equivalent chain length value of 10.928, or 12:0 ALDE, or any combination of these fatty acids.

[d]Summed feature 7 comprises $C_{18:1\ \omega7c}$, $C_{18:1\ \omega9t}$, or $C_{18:1\ \omega12t}$, or any combination of these fatty acids.

trachea mucosa, and less frequently from other organs such as the liver and spleen. They also can be isolated from swabs taken from the palatine cleft or trachea of living acutely diseased pigeons. At present the knowledge with regard to pathogenicity is incomplete. The present clinical observations suggest that they are pathogenic and involved especially in pathogenesis of respiratory diseases in pigeons. However, the role of cofactors and the interaction with other agents are uncertain. Bacteria associated with similar clinical signs in pigeons were described by Andreasen and Sandhu (1993). All morphological and biochemical characteristics reported by the latter authors corresponded with our findings.

The type strain was isolated from a pigeon in Belgium. *The mol% G + C of the DNA is*: 42–43 (43 for the type strain) (T_m).

Type strain: Strain Hommez N57, LMG 10982.

GenBank accession number (16S rRNA): Y11890.

Additional Remarks: GenBank accession number (16S rRNA) for LMG 15725 (reference strain for genomovar 2) is AF190911.

Genus VII. **Pigmentiphaga** *Blümel, Mark, Busse, Kämpfer and Stolz 2001b, 1870*[VP]

HANS-JÜRGEN BUSSE

Pig.men.ti.pha.ga. L. n. *pigmentum* dye; Gr. n. *phagos* eater; M.L. fem. adj. *Pigmentiphaga* eating dyes.

Cells are Gram-negative, motile, rod-shaped, nonsporeforming bacteria 1.3–4 × 0.7–1.2 µm. The colonies are opaque, circular, convex with an entire margin. Growth occurs at 30°, 37°, and 42°C, no growth is found at 4°C.

Oxidase and catalase positive. L-Alanine-*p*-nitroanilide is hydrolyzed. The following compounds are not hydrolyzed: *p*-nitrophenyl-β-D-galactopyranoside, *p*-nitrophenyl-β-D-glucuronide, *p*-nitrophenyl-α-D-glucopyranoside, *p*-nitrophenyl-β-D-glucopyranoside, *p*-nitrophenyl-β-D-xylopyranoside, bis-*p*-nitrophenyl-phosphate, bis-*p*-nitrophenyl-phenyl-phosphonate, bis-*p*-nitrophenyl-phosphoryl-choline, L-aniline-*p*-nitroanilide, γ-L-glutamate-*p*-nitroanilide, and L-proline-*p*-nitroanilide. The following compounds are assimilated: acetate, propionate, *cis*-aconitate, *trans*-aconitate, adipate, azelate, citrate, fumarate, glutarate, DL-3-hydroxybutyrate, itaconate, DL-lactate, mesaconate, 2-oxoglutarate, pyruvate, suberate, L-aspartate, 3-hydroxybenzoate, and 4-hydroxybenzoate. The following compounds are not assimilated: *N*-acetylgalactosamine, *N*-acetylglucosamine, L-arabinose, L-arbutin, D-cellobiose, D-fructose, D-galactose, gluconate, D-glucose, D-maltose, D-mannose, α-D-melibiose, L-rhamnose, D-ribose, D-sucrose, salicin, D-trehalose, D-xylose, adonitol, *i*-inositol, maltitol, D-mannitol, D-sorbitol, putrescine, 4-aminobutyrate, L-alanine, L-histidine, L-leucine, L-ornithine, L-phenylalanine, L-serine, L-tryptophan, and phenylacetate. No acids are produced from glucose, lactose, sucrose, D-mannitol, dulcitol, salicin, adonitol, inositol, sorbitol, L-arabinose, raffinose, rhamnose, maltose, D-xylose, trehalose, cellobiose, methyl-D-glucoside, erythritol, melibiose, D-arabitol, and D-mannose.

The **major fatty acids** are $C_{16:0}$, **summed feature 3** ($C_{16:1\ iso\ I}$ and $C_{14:0\ 3OH}$), **summed feature 7** ($C_{18:1\ \omega7c}$, $C_{18:1\ \omega9t}$, and/or $C_{18:1\ \omega12t}$), **and** $C_{10:0\ 3OH}$, $C_{14:0\ 2OH}$, $C_{16:0\ 2OH}$, and $C_{17:0\ cyclo}$ **and** $C_{19:0\ cyclo\ \omega8c}$; the major isoprenoid quinone is **ubiquinone Q-8**. The **polyamine pattern** contains two major compounds: **putrescine** and the unusual **2-hydroxyputrescine**. **Phosphatidylethanolamine** is the major **polar lipid** and **phosphatidylglycerol** and **diphosphatidylglycerol** are minor components.

The mol% G + C of the DNA is: 68.5 ± 0.3.

Type species: **Pigmentiphaga kullae** Blümel, Mark, Busse, Kämpfer and Stolz 2001b, 1870.

List of species of the genus Pigmentiphaga

1. **Pigmentiphaga kullae** Blümel, Mark, Busse, Kämpfer and Stolz 2001b, 1870[VP]

   kul.lae. in honor of Hans G. Kulla, who initiated the work about the aerobic degradation of azo dyes by bacteria, which resulted in the isolation of strain K24.

   The characteristics of the species is identical to that of the genus. Strain K24[T] has been isolated from soil after continuous enrichment with 1-(4'-carboxyphenylazo)-4-naphthol.

   The mol% G + C of the DNA is: 68.5 ± 0.3 (HPLC).

   Type strain: K24, DSM 13608, NCIMB 13708.

   GenBank accession number (16S rRNA): AF282916.

Genus VIII. **Sutterella** Wexler, Reeves, Summanen, Molitoris, McTeague, Duncan, Wilson and Finegold 1996a, 257^{VP}

HANNAH M. WEXLER

Sut.ter.el' la. M.L. dim. fem. n. *Sutterella* named in memory of Vera Sutter, respected colleague and director of the Wadsworth Anaerobe Laboratory for twenty years.

Straight rod, 0.5–1 × 1–3 µm. Gram negative. Grows in a microaerophilic atmosphere (2% and 6% oxygen) or under anaerobic conditions. Isolated mainly from the intestinal tract and from infections of gastrointestinal origin. Urease negative. Oxidase negative. Indoxyl acetate negative. Reduces nitrate. **Resistant to 20% bile disks.** Asaccharolytic. Cannot reduce tetrazolium tetrachloride under aerobic conditions if formate and fumarate are added to the medium. **May be resistant to metronidazole.**

Type species: **Sutterella wadsworthensis** Wexler, Reeves, Summanen, Molitoris, McTeague, Duncan, Wilson and Finegold 1996a, 257.

FURTHER DESCRIPTIVE INFORMATION

Biochemical characteristics and oxygen tolerance Like *B. ureolyticus, C. gracilis,* and other *Campylobacter* species, members of *Sutterella* are asaccharolytic and reduce nitrate to nitrite. Both *Campylobacter* and *Sutterella* were able to grow on solid media at 2% and 6% oxygen. *C. gracilis* grew only in 6% oxygen on plates containing *Brucella* agar with formate/fumarate but no blood added; if blood was added, *C. gracilis* did not grow. *Sutterella* is resistant to both human bile and oxgall (Table BXII.β.39); this may be important in their ability to survive in the biliary tract and bowel.

Dehydrogenase profiles Three general patterns of dehydrogenase activity were seen among the typical *C. gracilis, Sutterella wadsworthensis,* and *Campylobacter* strains studied. The group including the type strain of *C. gracilis* (as well as *C. gracilis* type 2 WAL 10733/FDC 20A1) showed no bands or very faintly reactive malate-specific bands (which appeared only when the gels were very heavily loaded), whereas strains of *Sutterella* had strongly reactive single or multiple bands of a nonspecific dehydrogenase enzyme (with anaerobic incubation only). *Campylobacter* strains (including WAL 10732/FDC 286) had different MDH-specific bands of variable migration distances, both under aerobic and anaerobic incubation conditions.

Cellular fatty acid analysis The fatty acid methyl esters (FAME) found in the different groups of organisms are listed in Table BXII.β.40. Cluster analysis of the cellular fatty acid data indicated three major groups: (1) typical *C. gracilis* (including the strain ATCC 33236^T), (2) the *Sutterella* group, and (3) *Campylobacter* species. Among other differences, the absence of $C_{12:0}$ fatty-acid-methyl-ester (FAME) distinguished *Sutterella* from typical *C. gracilis* and other *Campylobacter* species. The typical *C. gracilis, Campylobacter,* and *B. ureolyticus* also cluster quite independently of each other. The strain chosen as the type strain of *Sutterella* (WAL 9799) was selected because it resided in the center of the *Sutterella* cluster. *C. gracilis* ATCC strain 33236^T clustered at the edge of the cellular fatty acid cluster analysis of *C. gracilis.*

16S rRNA sequencing Phylogenetic relationships of *Sutterella* are shown in the chapter describing the family *Alcaligenaceae* (Fig. BXII.β.21).

Sequencing of the 16S rRNA gene was performed for two *Sutterella* strains (WAL 9054 [GenBank L37786] and WAL 7877 [GenBank L37785], *Campylobacter gracilis* strain ATCC 33236^T (GenBank Collection L37787), and *C. rectus* WAL 7943. The 16S rRNA sequences of the two *Sutterella* strains (WAL 9054 and WAL

7877 [1454 and 1419 base pairs sequenced, respectively]) were identical to each other but did not cluster with either the *C. gracilis* or *Campylobacter* strains analyzed or those contained in the databases GenBank or RDP. *C. gracilis* strain ATCC 33236^T branches at the same point as *Campylobacter rectus.*

DNA hybridization The *Sutterella wadsworthensis* type strain did not hybridize with DNA from either *C. gracilis* (10 strains) or other *Campylobacter* species (9 strains), but hybridized strongly (>96%) with DNA from 68% of the *Sutterella* (19 strains) and exhibited hybridization values of >67% with all of the strains tested, indicating that these strains are of the same species or very closely related species that cannot be differentiated phenotypically. Ten *Sutterella* strains did not hybridize at all with the *C. gracilis* type strain, and the DNA of three *Sutterella* strains hybridized very slightly (3%, 5%, and 12%); seven strains of *Campylobacter* did not hybridize at all with *C. gracilis* ATCC 33236^T, and two strains of *C. rectus* hybridized slightly (15%). *C. gracilis* strain ATCC 33236^T hybridized moderately with other typical *C. gracilis* strains (2/9 strains hybridized at 70–80%, 3/9 at 40–50%, 3/9 at 30–50%, and 1/9 at 22%). *C. gracilis* strain WAL 8030 hybridized more strongly with other *C. gracilis* strains, did not hybridize at all with *Sutterella,* and hybridized slightly (23%) with the one *C. rectus* strain tested.

Clinical source The strains belonging to the *Sutterella* group were isolated mainly from infections of gastrointestinal origin (Finegold and Jousimies-Somer, 1997; Jousimies-Somer, 1997). Strains belonging to *C. gracilis* were found virtually only in infections above the diaphragm (e.g., brain, pleural fluid, etc.); only one strain of *C. gracilis* was found from an appendiceal specimen. The resistance of *Sutterella* to both human bile (unpublished data from our laboratory) and oxgall is notable; this may account for their ability to survive in the biliary tract and bowel.

Antimicrobial susceptibility Over 95% of *Sutterella* strains are susceptible to amoxicillin/clavulanate, ticarcillin/clavulanate, cefoxitin, ceftriaxone, and clindamycin. 85–95% of strains are susceptible to piperacillin, piperacillin/tazobactam, ceftizoxime, ciprofloxacin, trovafloxacin, azithromycin, clarithromycin, erythromycin, and roxithromycin.

ENRICHMENT AND ISOLATION PROCEDURES

Sutterella wadsworthensis is isolated on *Brucella* blood agar with 5% lysed sheep blood, 1 µg/ml Vitamin K$_1$, 1 µg/ml hemin, 1% formate/fumarate (used as a growth supplement). Colonies are seen after 48 h at 37°C in an anaerobic chamber. Colonies are circular, entire, convex, yellow to brown, translucent to opaque, 1–1.5 mm in diameter.

MAINTENANCE PROCEDURES

Strains are kept in skim milk (20%) at −70°C.

DIFFERENTIATION OF THE GENUS *SUTTERELLA* FROM OTHER GENERA

Sutterella can be differentiated from *Campylobacter gracilis* and from other *Campylobacter* species as shown in Table BXII.β.39.

Dehydrogenase patterns are distinctive from those of *C. gracilis* and other campylobacters when the gels are incubated anaerobically. Cellular fatty acid analysis indicates that the cluster of *Sutterella* is distinct from the (separate) clusters formed by *C. gracilis*, *Bacteroides ureolyticus*, or other *Campylobacter* species. 16S rRNA sequence and DNA hybridization data indicate that *Sutterella* is distinct from *C. gracilis* and other *Campylobacter* species at the genus and species levels.

TAXONOMIC COMMENTS

Historical notes During biochemical characterization and susceptibility testing of isolates from clinical specimens in the Wadsworth Anaerobe laboratory, we found heterogeneity among the strains previously identified as *Campylobacter gracilis*. These organisms differed with regard to their bile resistance and ability to reduce triphenyltetrazolium chloride (TTC). Isolates that were bile resistant and unable to reduce TTC aerobically were compared to other strains of *C. gracilis* (including the ATCC type strain) and five species of *Campylobacter* using biochemical characteristics, cellular fatty acid profiles, DNA relatedness, and 16S rRNA sequence homology.

Campylobacter gracilis was originally known as *Bacteroides gracilis*, one of the members of the *B. ureolyticus* group. These are asaccharolytic, nitrate-positive organisms that require formate or hydrogen as an electron donor and may pit the agar. *B. ureolyticus* and *B. gracilis* clearly do not belong in *Bacteroides sensu stricto* (currently limited to the *B. fragilis* group; Shah and Collins, 1989), and 16S rRNA sequence analysis and DNA–rRNA hybridization studies support description of a tight homology group that includes *B. ureolyticus*, *B. gracilis*, and *Campylobacter* (including some previous *Wolinella* species) (Paster and Dewhirst, 1988; Whiley and Beighton, 1991). Both *B. ureolyticus* and *B. gracilis* are microaerophiles and not anaerobes; it has been proposed that both be included in the group of "true campylobacters" (Han et al., 1991). *B. gracilis* was reclassified as *Campylobacter gracilis*; *B. ureolyticus* was not renamed along with *C. gracilis*; its recommended status was a *species incertae sedis* pending further investigation (Vandamme et al., 1995a).

Strains belonging to *C. gracilis* were (with the exception of one strain) isolated from infections above the diaphragm, while the *Sutterella* strains were mostly isolated from infections below the diaphragm; the metronidazole-resistant strains all belonged to *Sutterella* (Molitoris et al., 1997).

List of species of the genus Sutterella

1. **Sutterella wadsworthensis** Wexler, Reeves, Summanen, Molitoris, McTeague, Duncan, Wilson and Finegold 1996a, 257[VP]

   wads.worth' en.sis. M.L. adj. *wadsworthensis* from Wadsworth,

   referring to the Wadsworth Anaerobe Laboratories, VAMC, West Los Angeles, where the strains were identified.

   The characteristics are as described for the genus and as shown in Tables BXII.β.39 and BXII.β.40.

TABLE BXII.β.39. Differentiation of *Sutterella wadsworthensis* from related taxa

Characteristic	S. wadsworthensis	B. ureolyticus	Campylobacter	Campylobacter gracilis
Oxidase	−	+	+	−
Urease	−	+	−	−
Indoxyl acetate	−	NT[a]	Var[b]	+
Growth in 2% O_2	+	+	+	+
Growth in 6% O_2	+(−)	+ or w	+	−[c]
Resistance to 20% bile	R	S	Var[d]	S
TTC reduction[e]	−	NT	+	+

[a]NT, not tested.

[b]Variable, even within the same species.

[c]*C. gracilis* only grew in 6% oxygen on plates containing *Brucella* agar with formate/fumarate but no blood added; if blood was added, *C. gracilis* did not grow.

[d]Strains tested included *C. curvus* (Res), 2; *C. rectus* (Sens), 4; *C. concisus* (Sens), 1; *Campylobacter* sp. 1 (Strain WAL 4864), Intermediate.

[e]Ability to reduce TTC in an aerobic atmosphere when formate/fumarate is added to the media.

TABLE BXII.β.40. Cellular fatty acid methyl esters of *Sutterella* and related taxa

Characteristic	S. wadsworthensis (30 strains)	Campylobacter gracilis (formerly B. gracilis) (10 strains)	Campylobacter (16 strains)[a]
$C_{12:0}$ FAME[b]	<1	10–20	4–13
$C_{14:0}$ FAME	3–8.5	8–13	8.5–13
$C_{16:1 \, \omega7c}$ FAME	17–24	<5	9–18
$C_{16:0}$ FAME	26–44	6–12	22–34
$C_{16:0}$ DMA[c]	0.2–6	15–23	2–5[d]
$C_{18:1 \, \omega9c}$ DMA[c]	0–1	10–12	0
Summed feature 10[e]	11–30	10–20	16–30

[a] *C. rectus*, 8 strains; *C. curvus*, 3 strains; *C. concisus* 3 strains; *Campylobacter* sp., 2 strains.

[b]Fatty acid methyl ester.

[c]DMA, dimethyl acetal.

[d] *C. curvus*, 2–5%; other *Campylobacter* strains, 0.

[e]18:1 C11/t9/t6 FAME and/or unknown peak at 17.834.

The mol% G + C of the DNA is: unknown.

Type strain: Strain WAL 9799, ATCC 51579.

Additional Remarks: The type strain was selected because

it resided in the center of the *Sutterella* cluster in the cellular fatty acid analysis.

Genus IX. **Taylorella** *Sugimoto, Isayama, Sakazaki and Kuramochi 1984, 503^(VP) (Effective publication: Sugimoto, Isayama, Sakazaki and Kuramochi 1983, 155)*

NANCY M.C. BLEUMINK-PLUYM AND BERNARD A.M. VAN DER ZEIJST

Tay'lor.el.la. M.L. dim. ending *-ella* M.L. fem. n. *Taylorella* named after C.E.D. Taylor, who first studied the organism.

Generic definition: **Short rods**, 0.7×0.7–1.8 μm. **Gram negative**. Nonmotile. Requires an atmosphere enriched with **5–10% CO_2**. Growth on rich peptone based chocolate agar medium at 35–37°C. Colonies are circular, smooth, grayish-white and 0.5–3.0 mm in diameter after an incubation period of 3–6 d. **Oxidase positive.** Catalase, phosphatase, and phosphoamidase are produced. Acid is not produced from carbohydrates. The phylogenetic position based on 16S rDNA analysis is in the *Betaproteobacteria*. The 16S rRNA contains the signature sequences of the *Betaproteobacteria* in 63 of 66 positions.

The mol% G + C of the DNA is: 36.5.

Type species: **Taylorella equigenitalis** (Taylor, Rosenthal, Brown, Lapage, Hill and Legros 1978) Sugimoto, Isayama, Sakazaki and Kuramochi 1984, 503 (Effective publication: Sugimoto, Isayama, Sakazaki and Kuramochi 1983, 155) (*Haemophilus equigenitalis* Taylor, Rosenthal, Brown, Lapage, Hill and Legros 1978, 136.)

FURTHER DESCRIPTIVE INFORMATION

The initial name of this bacterium was *Haemophilus equigenitalis* (Taylor et al., 1978). It was classified based on its mol% G + C content in the family *Pasteurellaceae* as a *species incertae sedis* belonging to the genus *Haemophilus*. In 1983, based on DNA hybridization data, Sugimoto proposed the new genus *Taylorella* with *Taylorella equigenitalis* as the only species (Sugimoto et al., 1983). This was validated in the IJSB in 1984. The 16S rDNA sequence of *Taylorella equigenitalis* was determined in 1993 (Bleumink-Pluym et al., 1993) and a phylogenetic analysis of this sequence revealed a position in the *Betaproteobacteria* apart from the position of *Haemophilus influenzae*, which belongs to the *Gammaproteobacteria*.

Cells of *Taylorella equigenitalis* are short rods or usually coccobacilli. The outer membrane can be covered by a layer of capsular-like material (Hitchcock et al., 1985; Bleumink-Pluym, 1995). Fimbriae have only been demonstrated *in vivo* (Kanemaru et al., 1992). The most abundant outer membrane protein has a molecular mass of 39–41 kDa (Sugimoto et al., 1988). The N-terminus of this immunodominant OMP is similar to that of porins from *Bordetella pertussis, Neisseria gonorrhoeae*, and *Neisseria meningitidis* (Bleumink-Pluym, 1995).

T. equigenitalis grows well on agar medium containing a rich peptone base with 5% chocolated sheep blood. Horse blood is not recommended because of the possible presence of antibodies directed toward *T. equigenitalis*. The medium should be supplemented with sodium sulfite (200 mg/l) and L-cysteine hydrochloride (100 mg/l). The plates should be incubated in an atmosphere enriched with 5–10% CO_2. Growth is observed over a temperature range of 30–41°C; optimal temperature is 37°C.

Visible colony formation requires at least 48 h. Colonies are circular, smooth, raised, and shiny. They vary in size from 0.5–3.0 mm, are grayish-white, and can subsequently turn yellowish-brown. Sometimes the colonies can be moved freely over the surface of the agar.

Acid is not produced from carbohydrates. Oxidase, catalase, phosphatase, and phosphoamidase are produced. Nitrates and nitrites are not reduced. Lysine and ornithine decarboxylase, arginine dihydrolase, gelatinase, lipase, urease, and deoxyribonuclease are not produced. Indole and H_2S not produced (Sugimoto et al., 1983).

The 16S rDNA sequence of *T. equigenitalis* was compared with entries present in the EMBL and GenBank database. Close relationships with 16S rDNA sequences of *Alcaligenes xylosoxidans* (94.2%) and *Bordetella bronchiseptica* (93.5%) were detected. Two variable regions that distinguish the three closest relatives most effectively are located at the 5′ end of the 16S rDNA at positions 65–97 and 412–478 (*E. coli* numbering).

Hybridizations between DNAs from two different *Taylorella* strains and DNAs from strains of *Bordetella, Moraxella, Kingella, Legionella, Haemophilus*, and *Brucella* species revealed no significant relatedness. Relatedness of DNAs from five *Taylorella* strains showed reassociation values of about 68% and a thermal stability difference of 2.5% (Sugimoto et al., 1983), essentially fulfilling the genetic definition of a species.

The genome size of *T. equigenitalis* is 1.5×10^6 bp. DNA of different isolates can be cleaved in a few large fragments by the restriction endonuclease *ApaI*, which recognizes the infrequently occurring GGGCCC sequence. Different genomic restriction patterns can be observed by using field inversion gel electrophoresis (FIGE) (Bleumink-Pluym et al., 1990; Engval et al., 1991; Matsuda et al., 1993). *T. equigenitalis* strains from different outbreaks generally have different genomic restriction patterns.

By immunoblotting, Sugimoto et al. demonstrated that mares experimentally infected with *T. equigenitalis* developed antibodies to a 41-kDa major outer membrane protein (Sugimoto et al., 1988). The N-terminus of this immunodominant OMP is similar to that of porins from *Bordetella pertussis, Neisseria gonorrhoeae*, and *Neisseria meningitidis* (Bleumink-Pluym, 1995).

T. equigenitalis is moderately resistant to clindamycin, lincomycin, trimethoprim, and sulfamethoxazole (MIC 32 μg/ml); sensitive to penicillin G, ampicillin, carbenicillin, cephaloridine, cephalotin, erythromycin, tetracycline, kanamycin, gentamicin, neomycin, chloramphenicol, and polymyxin B (Sugimoto et al., 1983). The majority of *T. equigenitalis* strains isolated internationally are resistant to streptomycin (MIC >512), but streptomycin-susceptible strains exist.

Clinical signs of *T. equigenitalis* infection in horses have been reported only in the mare and can be vaginal discharge, infertility, or early abortion. In the acute state of infection a widespread endometritis with local erosion of surface epithelium can be observed. The effect of the disease on fertility can be attributed to acute endometrial inflammation. The principal lesions are in the uterus and are characterized by destruction of endometrial epithelial cells and infiltration of neutrophils into the epithelium and lamina propria of the uterus. Histopathological studies have demonstrated that *T. equigenitalis* adheres to the cilia of the epithelial cells and proliferates on the endometrium (Wada et al., 1983). *In vitro, T. equigenitalis* adheres to and invades equine dermal cells. These abilities differ between strains and do not correlate with their genomic restriction patterns (Bleumink-Pluym et al., 1996).

The principal sites of colonization in mares are the urogenital membranes of the clitoral sinuses and fossa, urethra, and cervix. In stallions, the principal sites of colonization are the urogenital membranes of the urethra, urethral fossa, and penile sheath. *Taylorella* can be regarded as a commensal or opportunist as the bacterium is mostly isolated from asymptomatic carriers. Likewise, the prevalence of *Taylorella* in the horse population is high as demonstrated by PCR assay (Bleumink-Pluym et al., 1994).

ENRICHMENT AND ISOLATION PROCEDURES

Taylorella requires rather specific growth conditions and is rapidly overgrown by other bacteria resident in the genital tract of horses. Genital swabs are inoculated on Colombia blood agar base (Oxoid Ltd)-chocolated agar supplemented with sodium sulfite (200 mg/l) and L-cysteine hydrochloride (100 mg/l) (Atherton, 1983) with and without the addition of 5 µg of clindamycin, 1 µg trimethoprim, and 5 µg of amphotericin-B per ml. Plates are incubated with 7% CO_2 in air at 37°C for 6 d. Colonies are circular, smooth, grayish-white, and vary in size from 0.5–3.0 mm. Colonies suspected of being *T. equigenitalis* are Gram stained and tested for oxidase. Slide agglutination, indirect immunofluorescence with hyperimmune serum, or PCR detection based on 16S rDNA sequences can be used for confirmation.

MAINTENANCE PROCEDURES

Plate cultures will survive for 3 weeks without subcultivation but should be kept in an atmosphere enriched with 5–10% CO_2. All strains can be maintained for up to 4 yr at −70°C in brain heart infusion broth containing 15% glycerol. Lyophilization, e.g., in skim milk, can also be used for conserving cultures.

DIFFERENTIATION OF THE GENUS *TAYLORELLA* FROM OTHER GENERA

The differential characteristics of the genus *Taylorella* are listed in Table BXII.β.41.

TAXONOMIC COMMENTS

The phylogenetic position of *Taylorella equigenitalis* in the *Betaproteobacteria* differs from that of *Haemophilus influenzae*, which belongs to the *Gammaproteobacteria*, and supports its exclusion from the family *Pasteurellaceae*. *Taylorella* does not seem to fit in one of the known bacterial families. Based on 16S rDNA analysis *Taylorella* is most closely related to *Bordetella* and *Alcaligenes*. Although these species share substantial 16S rDNA sequences (93.5% and 94.2%, respectively), it is not likely that they belong to the same family. No relatedness was detected in a DNA–DNA hybridization between *Taylorella* and *Bordetella*, which is in accord with the variation in the mol% G + C contents of their genomes. *Bordetella* and *Alcaligenes* have a mol% G + C content of ~69, while *Taylorella* has a mol% G + C content of 36.5. Members of the genera *Bordetella*, *Alcaligenes*, and *Taylorella* occur in different ecological niches. The principal habitat of *Taylorella* is the mucosal membrane surface of the genital tract of horses. Certain phenotypic characteristics are shared with other bacteria harboring mucous membranes of the genital tract, like *Neisseria gonorrhoeae*. Both species are transmitted sexually, cause similar clinical signs of infection (*Taylorella* in horses; *N. gonorrhoeae* in humans), have similar culture conditions, are oxidase positive, have variable colony morphology, and have a similarity in the N-terminal amino acid sequences of the 41-kDa and 17-kDa outer membrane protein.

List of species of the genus Taylorella

1. **Taylorella equigenitalis** (Taylor, Rosenthal, Brown, Lapage, Hill and Legros 1978) Sugimoto, Isayama, Sakazaki and Kuramochi 1984, 503[VP] (Effective publication: Sugimoto, Isayama, Sakazaki and Kuramochi 1983, 155) (*Haemophilus equigenitalis* Taylor, Rosenthal, Brown, Lapage, Hill and Legros 1978, 136.)

   e.qui.ge.ni.ta' lis. L. gen. n. *equi* of the horse; L. adj. *genitalis* genital; L. adj. *equigenitalis* genital of the horse.

All characteristics are as described for the genus *Taylorella.*

This strain was isolated in 1977 from a cervical swab of a thoroughbred mare with contagious metritis.

The mol% G + C of the DNA is: 36.5 (T_m).

Type strain: Strain 61717/77, CIP 7909, DSM 10668, NCTC 11184.

GenBank accession number (16S rRNA): X68645.

TABLE BXII.β.41. Differentiation of the genus *Taylorella* from other closely related taxa[a]

Characteristic	*T. equigenitalis*	*Bordetella avium*	*Bordetella bronchiseptica*	*Bordetella holmesii*	*Bordetella parapertussis*	*Bordetella pertussis*	*Alcaligenes faecalis*	*Alcaligenes xylosoxidans*
Strictly aerobic	−	+	+	+	+	+	+	+
Urease	−	−	+	−	+	−	−	−
Oxidase	+	+	+	−	−	+	+	+
Motility	−	+	+	−	−	−	+	+
Mol% G + C	36.5	61.6	68.9	61.9	66.7	67.7	56	69

[a]For symbols see standard definitions.

Family IV. **Comamonadaceae** Willems, De Ley, Gillis and Kersters 1991a, 447[VP]

ANNE WILLEMS AND MONIQUE GILLIS

Co.ma.mo.na.da' ce.ae. M.L. fem. n. *Comamonas* type genus of the family; suff. *-aceae* to denote a family; M.L. fem. pl. n. *Comamonadaceae* the *Comamonas* family.

Cells are **straight** or **slightly curved rods or spirilla**. Gram negative. Most genera are **motile by a single polar flagellum or by bipolar tufts of 1–5 flagella**; one genus (*Variovorax*) is motile by sparse peritrichous flagella, and another (*Brachymonas*) lacks flagella. No endospores are formed. Coccoid bodies are rarely formed, but do occur in some (*Aquaspirillum*) species. (The use of square brackets indicates these species are not true aquaspirilla, which are related to type species *Aquaspirillum serpens*.) **Chemoorganotrophic or facultatively chemolithotrophic with H₂ or CO oxidation.** Possess a strictly respiratory type of metabolism, with oxygen as the terminal electron acceptor. Some species can also use nitrates. **Oxidase positive**, except for the genus *Xylophilus*. **Optimal growth temperature ranges from 24 to 35°C**, except for *Polaromonas* and [*Aquaspirillum*] *psychrophilum*, which prefer 4 and 18°C, respectively. Yellow, insoluble pigments are produced by the genera *Hydrogenophaga*, *Variovorax*, and *Xylophilus*. Nitrogen fixation has been reported for some *Hydrogenophaga* species. A wide variety of organic acids, including amino acids, is used, but sugars are rarely attacked. **Major fatty acids are palmitoleic acid ($C_{16:1}$), palmitic acid ($C_{16:0}$) and *cis*-vaccenic acid ($C_{18:1\ \omega7c}$),** and the major quinone is **ubiquinone Q-8.**

Members of the *Comamonadaceae* have been **isolated from soil, mud, and water, in natural and industrial environments**. Strains from the genera *Comamonas* and *Acidovorax* have also been isolated from clinical samples, but are not regarded as pathogenic. The genus *Xylophilus* is pathogenic to grapevines and some members of the genus *Acidovorax* are pathogenic to grasses, orchids, and members of the cucumber family.

The *Comamonadaceae* belong to the *Betaproteobacteria*. At present the following genera belong to this family, on phylogenetic grounds: *Comamonas*, *Acidovorax*, *Alicycliphilus*, *Brachymonas*, *Caldimonas*, *Delftia*, *Diaphorobacter*, *Hydrogenophaga*, *Lampropedia*, *Macromonas*, *Polaromonas*, *Ramlibacter*, *Rhodoferax*, *Variovorax*, *Xenophilus*, and *Xylophilus*. In addition, a number of misnamed *Aquaspirillum* species belong to the *Comamonadaceae* on phylogenetic grounds. The genera *Aquabacterium*, *Ideonella*, *Leptothrix*, *Roseateles*, *Rubrivivax*, *Schlegelella*, *Sphaerotilus*, *Tepidimonas*, *Thiomonas*, and *Xylophilus* have been placed in the *Comamonadaceae* as *genera incertae sedis*. The genera *Alicycliphilus*, *Caldimonas*, *Diaphorobacter*, *Ramlibacter*, and *Schlegelella* were described after the cut-off date for inclusion in this volume. Differentiating features for a number of these taxa are presented in Table BXII.β.42.

The mol% G + C of the DNA is: 52–70.

Type genus: **Comamonas** De Vos, Kersters, Falsen, Pot, Gillis, Segers and De Ley 1985b, 450 emend. Tamaoka, Ha and Komagata 1987, 57; emend. Willems, Pot, Falsen, Vandamme, Gillis, Kersters and De Ley 1991c, 438.

TAXONOMIC COMMENTS

The family *Comamonadaceae* has been proposed as a formal taxon for the so-called acidovorans rRNA complex (Willems et al., 1991a). This was described as one of the rRNA groups within the large genus *Pseudomonas* in the previous edition of this *Manual* (Palleroni, 1984). Previously, it had been apparent from extensive phenotypic studies that this genus contained several species groups (Stanier et al., 1966). The genotypic basis for these groupings was revealed when the genus *Pseudomonas* was shown by DNA–rRNA hybridizations to consist of at least five main rRNA groups (Palleroni et al., 1973), and further DNA–rRNA hybridizations have shown these groups to be only remotely related to each other (De Vos and De Ley, 1983). The same groupings also have been supported by serological data (Baumann and Baumann, 1978), regulatory patterns in the biosynthesis of aromatic amino acids (Byng et al., 1983), coliphage QB host factor activity and antigenicity (Dubow and Ryan, 1977), fatty acid composition (Oyaizu and Komagata, 1983), and 16S rRNA oligonucleotide cataloging (Woese et al., 1984a).

In addition to at least 10 different *Pseudomonas* species, the *Pseudomonas acidovorans* rRNA group also contained *Comamonas terrigena*, *Xanthomonas ampelina*, *Alcaligenes paradoxus*, and several *Aquaspirillum* species (De Vos and De Ley, 1983; De Vos et al., 1985a, b; Willems et al., 1991a). It thus included common water and soil inhabitants, clinical isolates, plant pathogens, and hydrogen and CO oxidizers. Except for *Comamonas terrigena*, all of these species were phylogenetically unrelated to the type species of their respective genera, and in a series of polyphasic studies, they were therefore gradually transferred to other new or existing genera. An overview of these changes is presented in Table BXII.β.43. Since the original proposal of the family *Comamonadaceae*, four new genera (*Brachymonas*, *Delftia*, *Polaromonas*, and *Rhodoferax*) have been proposed within this group and these are listed in Table BXII.β.43. Not included is *Xenophilus*, which was described after the completion of this chapter (Blümel et al., 2001a).

This subdivision of the *Comamonadaceae* into different genera has relied heavily on DNA–rRNA hybridization data and has now been confirmed and extended by rRNA gene sequence analysis (Wen et al., 1999). Most members of the *Comamonadaceae* share at least 94–95% rDNA sequence similarity. A dendrogram showing the phylogenetic relationships of the *Comamonadaceae* and their nearest neighbors is shown in Fig. BXII.β.27. It was calculated using 16S rRNA gene sequences available from the EMBL database. From these data, it is clear that the genera *Acidovorax*, *Hydrogenophaga*, *Delftia*, *Polaromonas*, *Brachymonas*, and *Rhodoferax* and the species [*Aquaspirillum*] *gracile* and [*Aquaspirillum*] *sinuosum* are well separated. It is also apparent that the relationships between *Xylophilus* and *Variovorax*, and between [*Aquaspirillum*] *metamorphum* and [*Aquaspirillum*] *psychrophilum* need further study. The bootstrap value for the grouping of *Comamonas terrigena* and *Comamonas testosteroni* is relatively low (70; Fig. BXII.β.27), indicating that this is not a very robust grouping. It may be rearranged as more sequences are included.

It seems quite likely that additional genera will be included in this family, not only because of reclassification of some of the above listed species and genera, but also because of newly isolated groups. Many of the members of the *Comamonadaceae* are common inhabitants of water, soil, and polluted environments, and are regularly isolated in ecological studies. Some *Comamonadaceae*

TABLE BXII.β.42. Differentiating features for the members of the family *Comamonadaceae*[a]

Characteristic	*Comamonas*	*Acidovorax*	*Brachymonas*	*Delftia*	*Hydrogenophaga*	*Polaromonas*	*Rhodoferax*	*Variovorax*	*Xylophilus*	*[Aquaspirillum]*[b]
Cell morphology:										
Rods		+		+	+	+		+	+	
Curved rods							+			
Rods to spirilla	+									
Cocci to short rods			+							
Spirilla										+
Gas vacuoles	−	−	−	−	−	+	−	−	−	−
Flagella:										
Number	1–5	1		1–5	1–2	1	1		1	
Location:										
Polar	+	+			+	+	+		+	+[c]
Polar tufts	+			+						
Bipolar tufts				+						
Subpolar					+					+
Peritrichous								+		
Isolated from:										
Soil	+	+	−	+	+			+	−	−
Fresh water	+	+	−	+	+	−	−	+	−	+
Seawater	−	−	−	−	−	+	−	−	−	−
Activated sludge	+	+	+	+	−	−	+	−	−	−
Plant tissue	−	+							+	
Clinical samples	+	+	−	+	−		−	−	+	−
Optimal growth temperature, °C	30	30–35	30–35	30	30–35	4	25–30	30	24	30–32[d]
Growth at 37°C	+	d	+	d	d	−	−	nd	−	D
Growth on nutrient agar	+	+	+	+	+	+	+	+	−	
Yellow pigment	−	−	−	−	+	−	−	+	+	−
Growth with 3% NaCl	d	d	+	d	nd	+	−	nd	−	−
Growth on glucose	−	D[e]	−	−	+	d	−	+	−	D[f]
Growth factors[g]	M[h] N	nd	−	−	−	aa	B, T	nd	G	−
Autotrophic growth with H_2	−	d	−	−	+	−	nd	d	−	−
Phototrophic growth	−	−	−	−	−	−	+	−	−	−
Anaerobic growth with nitrate as terminal electron acceptor	−	d	+	−	D	−	nd	−	−	D[i]
Denitrification	−	d	+	−	D	−	nd	−	−	D[i]
Major quinones:[j]										
Q-8	+	+	+	+	+			+		
RQ-8			+				+			
Major 3-OH fatty acids:										
$C_{8:0}$				+	+			+		
$C_{10:0}$	+	+	+	+				+		
Mol% G + C of the DNA	60–69	62–70	63–65	67–90	65–69	52–57	60	67–69	68–69	57–59

[a]For symbols see standard definitions; nd, not determined.

[b]In this table the *[Aquaspirillum]* species are listed grouped in a single column, but they are likely to represent several additional genera (Wen et al., 1999).

[c] *[Aquaspirillum] delicatum* has 1 or 2 flagella at one pole (Krieg, 1984b).

[d]The optimal temperature for *[Aquaspirillum] psychrophilum* is 20°C (Krieg, 1984b).

[e]Only *A. konjaci* does not use glucose (Willems et al., 1992a).

[f]Glucose is used only by *[Aquaspirillum] gracile* (Krieg, 1984b).

[g]M, methionine; N, nictotinamide; aa, amino acids; B, biotin; T, thiamine; G, glutamic acid.

[h]Methionine and nicotinamide are required only by *C. terrigena* and the former *[Aquaspirillum] aquaticum* requires niacin (Pot et al., 1992b).

[i]Denitrification and anaerobic growth with nitrate is positive for *[Aquaspirillum] psychrophilum* (Krieg, 1984b).

[j]R, ubiquinone; RQ, rhodoquinone.

have the capacity to degrade complex organic compounds and are of potential use in degradation applications and bioremediation, topics which are the focus of intense research and which result in many new isolates. Some of these new strains may belong to new groups within the *Comamonadaceae*, as has occurred with *Rhodoferax* (Hiraishi, 1994), *Brachymonas* (Hiraishi et al., 1995b) and *Polaromonas* (Irgens et al., 1996).

As mentioned above, the family *Comamonadaceae* encompasses a remarkable diversity of phenotypic traits. In the past, different properties have often been studied in members of this family belonging to different phenotypic groups. This makes phenotypic comparison between members difficult and often incomplete. The accumulation of more 16S rDNA and other sequence data will become increasingly important for the classification within this group. As was shown previously (Amann et al., 1996), it may be possible to derive oligonucleotide probes for the identification of the various genera. The restriction analysis of amplified rRNA genes of a limited number of *Comamonadaceae* has

TABLE BXII.β.43. Members of the family *Comamonadaceae*

Current name[a]	Subjective synonym(s)	Reference
Acidovorax delafieldii	*Pseudomonas delafieldii*, EF group 13	Willems et al., 1990
Acidovorax facilis	*Pseudomonas facilis*	Willems et al., 1990
Acidovorax temperans	EF group 16	Willems et al., 1990
Acidovorax avenae subsp. *avenae*	*Pseudomonas avenae*, *Pseudomonas rubrilineans*, "*Pseudomonas setariae*"	Willems et al., 1992a
Acidovorax avenae subsp. *citrulli*	*Pseudomonas pseudoalcaligenes* subsp. *citrulli*	Willems et al., 1992a
Acidovorax konjaci	*Pseudomonas pseudoalcaligenes* subsp. *konjaci*	Willems et al., 1992a
Brachymonas denitrificans	new	Hiraishi et al., 1995b
Comamonas terrigena	"*Comamonas terrigena*", *Aquaspirillum aquaticum*, EF group 10	De Vos et al., 1985b, Willems et al., 1991c
Comamonas testosteroni	*Pseudomonas testosteroni*	Tamaoka et al.,1987
Delftia acidovorans	*Comamonas acidovorans*, *Pseudomonas acidovorans*	Wen et al., 1999
Hydrogenophaga flava	*Pseudomonas flava*	Willems et al., 1989
Hydrogenophaga palleronii	*Pseudomonas palleronii*	Willems et al., 1989
Hydrogenophaga pseudoflava	*Pseudomonas pseudoflava*	Willems et al., 1989
Hydrogenophaga taeniospiralis	*Pseudomonas taeniospiralis*	Willems et al., 1989
Polaromonas vacuolata	new	Irgens et al., 1996
Rhodoferax fermentans	new	Hiraishi et al., 1991a
Variovorax paradoxus	*Alcaligenes paradoxus*	Willems et al., 1991a
Xylophilus azovorans	new	Blümel et al., 2001a
Xylophilus ampelinus	*Xanthomonas ampelina*	Willems et al., 1987
[Aquaspirillum] anulus		Willems et al., 1991c
[Aquaspirillum] delicatum		Willems et al., 1991c
[Aquaspirillum] giesbergeri		Willems et al., 1991c
[Aquaspirillum] gracile		Willems et al., 1991c
[Aquaspirillum] metamorphum		Willems et al., 1991c
[Aquaspirillum] psychrophilum		Willems et al., 1991c
[Aquaspirillum] sinuosum		Willems et al., 1991c

[a]Square brackets are used to indicate that a genus allocation is phylogenetically unsound because the species is phylogenetically unrelated to its type species.

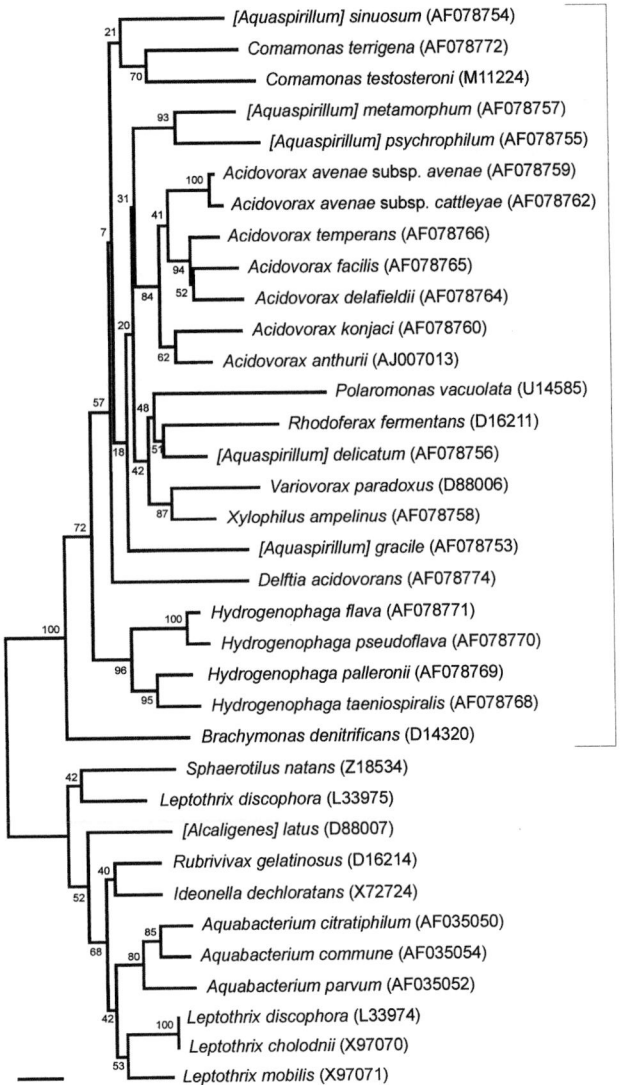

FIGURE BXII.β.27. Dendrogram showing the phylogenetic relationships of the *Comamonadaceae* and their closest relatives. The dendrogram is based on 16S rRNA gene sequences available from the European Molecular Biology Laboratory data library (accession numbers are given in brackets) and was constructed after multiple alignment of data, calculation of distances (corrected according to the Kimura-2 model, using a stretch of 1360 aligned positions) and clustering with the neighbor-joining method. Bootstrap values based on 500 replications are listed at the branching points. Programs from the GCG and PHYLIP packages (Felsenstein, 1982; Devereux et al., 1984) were used on the BEN computer facility operated at the Brussels Free University Computing Centre. Bar = 1 estimated substitution per 100 bases.

been shown to allow differentiation of species, but not always genera (Vaneechoutte et al., 1992).

The nearest phylogenetic neighbors of the *Comamonadaceae* are the photosynthetic genus *Rubrivivax*, the sheath-forming *Leptothrix* and *Sphaerotilus*, the hydrogen oxidizers *[Alcaligenes] latus* and *[Pseudomonas] saccharophila* (Willems et al., 1991b), *Ideonella* (Malmqvist et al., 1994a), and *Aquabacterium* (Kalmbach et al., 1999).

ACKNOWLEDGMENTS

We are grateful to Dr. L.I. Sly for allowing consultation and use of results in press. Anne Willems is indebted to the Fund for Scientific Research–Flanders for her position as a postdoctoral research fellow.

Genus I. **Comamonas** De Vos, Kersters, Falsen, Pot, Gillis, Segers and De Ley 1985b, 450[VP] emend. Tamaoka, Ha and Komagata 1987, 57, emend. Willems, Pot, Falsen, Vandamme, Gillis, Kersters and De Ley 1991c, 438

ANNE WILLEMS AND MONIQUE GILLIS

Co.ma.mo'nas. L. n. *coma* lock of hair; Gr. n. *monas* a unit, monad; M.L. fem. n. *Comamonas* cell with a polar tuft of flagella.

Straight or slightly curved rods or spirilla, 0.3–0.8 × 1.1–4.4 µm; occasionally longer (5–7 µm), irregularly curved cells or spirilla may occur. Cells occur separately or in pairs and **are motile by means of polar or bipolar tufts of 1–5 flagella** except for *C. koreensis*, which is nonmotile. **Gram negative.** No diffusible pigments are produced on nutrient agar. **Oxidase and catalase positive. Aerobic. Chemoorganotrophic**, oxidative carbohydrate metabolism with oxygen as the terminal electron acceptor. *C. nitrativorans* is also capable of denitrification. Good growth on media containing organic acids, amino acids, or peptone; few carbohydrates are used. Major fatty acids are hexadecanoic acid ($C_{16:0}$), hexadecenoic acid ($C_{16:1}$) and octadecenoic acid ($C_{18:1}$); 3-hydroxydecanoic acid ($C_{10:0\ 3OH}$) is always present. **The major quinone is ubiquinone Q-8.**

The mol% G + C of the DNA is: 60–69.

Type species: **Comamonas terrigena** (ex Hugh 1962) De Vos, Kersters, Falsen, Pot, Gillis, Segers and De Ley 1985b, 450, emend. Willems, Pot, Falsen, Vandamme, Gillis, Kersters and De Ley 1991c, 439.

FURTHER DESCRIPTIVE INFORMATION

Cell morphology Cells are rods that are straight or slightly curved in one plane or spirilla. Polar pili have been reported in *Comamonas testosteroni* (Fuerst and Hayward, 1969b).

Cell wall composition Cell walls have a typical Gram-negative structure consisting of an outer membrane and a cytoplasmic membrane, separated by a dense peptidoglycan layer. Analysis of the lipid A component of the lipopolysaccharides in the outer membrane of *C. testosteroni* has revealed the lipid A backbone to consist of 6-*O*-(2-deoxy-2-amino-β-D-glucopyranosyl)-2-deoxy-2-amino-α-D-glucose, which is phosphorylated in positions 1 and 4′. The hydroxyl groups at positions 4 and 6′ are unsubstituted, and position 6′ of the non-reducing terminal residue is the attachment side for the polysaccharide component. The 2, 2′, 3, and 3′ positions of the sugar backbone are *N*-acylated or *O*-acylated by 3-hydroxydecanoic acid, and the hydroxyl groups of the amide-linked residues at positions 2 and 2′ of the backbone are *O*-acylated by tetradecanoic and dodecanoic acids, respectively (Iida et al., 1996).

Colonial characteristics On nutrient agar, colonies are round and convex with a smooth to wavy margin and a smooth to granular surface. Most strains are unpigmented, but some strains may produce a brown diffusible pigment on nutrient agar. Colony diameters can reach 0.4–3 mm after 3 d at 28°C. Occasionally, two colony types, one mucoid and the other non-mucoid, can be isolated from one strain. Comparison of such types by whole-cell protein electrophoresis shows identical protein profiles, and they are therefore considered morphological variants of the same strain (Willems et al., 1991c). For a *C. testosteroni* strain isolated from activated sludge on the basis of its ability to coaggregate with yeast cells, two colony types have been reported to occur together on solid agar media, resulting in composite colonies with sectors of different morphologies (Bossier and Verstraete,

1996). In liquid medium, in the absence of agitation, and in contact with glass, cultures shift towards and are dominated by mucoid-colony-forming cells. In such media, the addition of stress factors such as hydrogen peroxide, sodium dodecyl sulfate, or starvation results in the rapid formation of non-mucoid cells. Non-mucoid cell types rapidly coaggregate and settle in flocs with yeast cells, whereas mucoid cells do not coaggregate with yeast cells (Bossier and Verstraete, 1996).

Nutrition and growth conditions *Comamonas* strains grow well on most organic acids and amino acids but use very few sugars (De Vos et al., 1985b). They are capable of degrading a wide variety of complex aromatic compounds, steroids, and many man-made complex organic molecules. According to Tamaoka et al. (1987), *C. terrigena* requires methionine and nicotinamide as growth factors in a mineral medium with L-glutamate as the carbon source. Optimal growth temperature is 30°C.

Metabolism *Comamonas* strains have a respiratory metabolism using oxygen as the terminal electron acceptor. Although most *Comamonas* strains are capable of nitrate reduction, most cannot reduce nitrites. Only 3 of 42 *C. terrigena* strains have tested positive for nitrite reduction (Willems et al., 1991c). In a study of the denitrifying biofilms at a water treatment plant, *Comamonas* strains were identified, but they were thought to have washed in from previous stages and to not be true members of the denitrifying biofilms (Lemmer et al., 1997). An isolate named *Comamonas* strain SGLY2, capable of aerobic denitrification and nitrogen production without nitrite build-up, has been studied for application in mixed-culture aerated reactors (Patureau et al., 1997).

C. testosteroni strains capable of degrading polycyclic aromatic hydrocarbons, such as phenanthrene, naphthalene, and anthracene, have been described (Goyal and Zylstra, 1996). Their genes for naphthalene and phenanthrene degradation have been cloned and characterized and found to differ from the classical naphthalene degradation genes (*nah*) of *Pseudomonas putida*. Among different *C. testosteroni* strains, at least two different sets of genes for phenanthrene degradation may be present. In addition, in one of the strains, the arrangement of these genes differs from that in *Pseudomonas* species (Goyal and Zylstra, 1996).

The gene *phtD* for 4,5-dihydroxyphthalate decarboxylase, an enzyme involved in phthalate metabolism, has been identified and sequenced. The deduced amino acid sequence shows more than 77% similarity to the *pht5* gene from *Pseudomonas putida*. The sequence upstream of *phtD* and its deduced amino acid sequence show high similarity to *pht1* and its deduced amino acid sequence, which is thought to be the positive regulator for the other *pht* genes of *Pseudomonas putida*. These data suggest a common origin for the genes involved in phthalate metabolism (Lee et al., 1994c).

C. testosteroni uses the meta cleavage pathway for the degradation of 4-chlorophenol and related compounds, as has been demonstrated by the isolation and identification of several of the key enzymes and intermediary metabolites of this pathway (Hollender et al., 1997). The *bphD* gene from strain B-356, encoding for 2-hydroxy-6-oxo(phenyl/chlorophenyl) hexa-2,4-dienoic acid hydrolase has been sequenced, and the gene product has a mechanism of action similar to that of classical lipases and serine hydrolases (Ahmad et al., 1995).

The first step in the degradation of biphenyl and chlorobiphenyl compounds by *C. testosteroni* is their conversion to 2,3-dihydro-2,3-dihydroxybiphenyl by biphenyl/chlorobiphenyl dioxygenase. This enzyme system consists of three components, the genes of which have been sequenced: a terminal oxygenase, which is an Fe–S protein consisting of 2 subunits (encoded by *bphA* and *bphE*); a ferredoxin, encoded by *bphF*; and a ferredoxin reductase encoded by the *bphG* gene, which is not located near the *bphAEF* genes in strain B-356 (Sylvestre et al., 1996b). Most of these enzymes or recombinant forms of them have been characterized in some detail (Hurtubise et al., 1995, 1996). Further degradation steps involve 2,3-dihydro-2,3-dihydroxybiphenyl dehydrogenase (*bphB*), a recombinant form of which has been characterized (Sylvestre et al., 1996a), and 2,3-dihydroxybiphenyl-1,2-dioxygenase (*bphC*), which catalyzes *meta*-1,2 fission of the aromatic ring (Bergeron et al., 1994).

In *Comamonas* sp. strain JS765, nitrobenzene is broken down by nitrobenzene 1,2-dioxygenase to nitrohydrodiol, which spontaneously decomposes into catechol and nitrite. Catechol is then broken down via the meta cleavage pathway. The catechol 2,3-dioxygenase gene *cdoE*, a ferredoxin gene *cdoT*, and a regulatory gene *cdoR* from strain JS765 have been sequenced and their amino acid sequences deduced (Parales et al., 1997).

Experimental evidence indicates that the mechanism of uptake of 4-toluene sulfonate by *C. testosteroni* strain T-2 involves an inducible secondary proton symport system (Locher et al., 1993). Strain T-2 degrades 4-toluene sulfonate and 4-toluene carboxylate by oxygenation of the side chains to 4-sulfobenzoate and terephthalate, respectively, before further oxidation to protocatechuate. The first three enzymes involved in this pathway are a 4-toluenesulfonate methyl-monooxygenase system, which consists of a reductase B and an oxygenase M (*tsaMB* genes), a 4-sulfobenzyl alcohol dehydrogenase (*tsaC*), and a 4-sulfobenzaldehyde (*tsaD*). Their genes are coexpressed under the regulation of *tsaR*. The components of this system have been purified and partially characterized. The *tsaMB* oxygenase system is a class IA mononuclear iron oxidase, with *tsaM* having a Rieske [2Fe–2S] center. *TsaC* is a short-chain zinc-dependent dehydrogenase (Junker et al., 1997). The conversion of 4-sulfobenzoate to protocatechuate requires a 4-sulfobenzoate-3,4-dioxygenase system encoded by the *psbAC* genes and consisting of a reductase C and an oxygenase A, which again contains a Rieske [2Fe–2S] center (Junker et al., 1996). Both sets of genes (*tsaMBCD*, *tsaR*, and *psbAC*) are located on conjugative plasmids in some *C. testosteroni* strains (Junker and Cook, 1997). The degradation of 4-toluene sulfonate by two different strains is comparable in liquid cultures but varies in biofilms (Khlebnikov et al., 1997). Terephthalate is broken down by a terephthalate dioxygenase system to (1*R*,2 *S*)-dihydroxy-3,5- cyclohexadiene-1,4-dicarboxylic acid, which is further degraded by (1*R*,2 *S*)-dihydroxy-1,4-dicarboxy-3,5-cyclohexadiene dehydrogenase to protocatechuate (Oppenberg et al., 1995). The dioxygenase system is a Rieske [2Fe–2S] protein consisting of two subunits, α and β, which are thought to form an $\alpha_2 \beta_2$ structure (Schläfli et al., 1994). Protocatechuate is further degraded via the meta cleavage pathway. Enzymes involved in these processes are thought to be located on the chromosome (Junker and Cook, 1997).

Metabolism of steroids   Möebus et al. (1997) have observed that testosterone, used as a sole carbon source by *Comamonas testosteroni*, induces expression of steroid- and aromatic hydrocarbon-catabolizing enzymes and represses at least one amino acid degrading enzyme. It has been suggested that steroids may play a regulative role in catabolic enzyme synthesis in adaptive growth. Several enzymes involved in the steroid metabolism of *C. testosteroni* have been studied. A 3-α-hydroxysteroid dehydrogenase has been isolated and characterized (Oppermann and Maser, 1996), and its gene has been sequenced, cloned and expressed (Abalain et al., 1995). Similarity to ribosomal proteins L10 and L7/12 has led to the suggestion that this enzyme may be formed by fusion of two ribosomal proteins (Baker, 1996). The gene for 3-ketosteroid-δ^4-5-α-dehydrogenase has been sequenced and is located downstream of a gene for 3-ketosteroid-δ^1-dehydrogenase. The gene product is a flavoprotein. Although both of these dehydrogenases are functionally similar, their genes are probably not derived from a common ancestor (Florin et al., 1996). The 3β/17β hydroxysteroid dehydrogenase of *Comamonas testosteroni* has been studied by mutagenic replacement within the active site to identify important residues (Oppermann et al., 1997). The structure and function of δ^5-3-ketosteroid isomerase and 3-oxo-δ^5-steroid isomerase, which catalyze the conversion of δ^5- to δ^4-3-ketosteroids by an intramolecular proton transfer, have been studied extensively (Brothers et al., 1995; Zhao et al., 1996, 1997). A bile acid 3-α-sulfate sulfohydrolase from *Comamonas testosteroni* has been purified and characterized (Tazuke et al., 1994).

Metabolism of other organic compounds   *Comamonas* strains capable of degrading terpenes, such as abietane and pimarane-type resin acids, have been isolated by enrichment from wastewater from an aerated stabilization basin of a bleached kraft pulp mill (Morgan and Wyndham, 1996).

Extracellular poly-β-hydroxybutyrate depolymerase has been purified from a *Comamonas* sp. strain. In addition to poly-β-hydroxybutyrate, it can also hydrolyze poly(β-hydroxybutyrate-co-β-hydroxyvalerate) and poly-β-hydroxyvalerate. It differs from other poly-β-hydroxybutyrate depolymerases in that it is insensitive to phenylmethylsulfonyl fluoride and that it hydrolyzes poly-β-hydroxybutyrate to β-hydroxybutyrate monomers (Jendrossek et al., 1993). Part of the encoding gene has been cloned and sequenced, and the deduced protein structure has been comparatively analyzed (Jendrossek et al., 1995). Shinomiya et al. (1997) have since cloned, sequenced, and analyzed a gene for the same enzyme from *C. testosteroni* YM1004 and obtained a very similar sequence.

An aliphatic nitrilase, active on adiponitrile and cyanovaleric acid, has been isolated and purified from *Comamonas testosteroni*, and its gene has been sequenced and overexpressed. The Cys163 residue is thought to play an essential role in the active site (Levy-Schil et al., 1995).

A quinohemoprotein ethanol dehydrogenase, which catalyzes the NAD-independent oxidation of a broad range of alcohols to the corresponding aldehydes and on to the carboxylic acids, has been isolated and purified from *Comamonas testosteroni* cells grown on ethanol. The active holoenzyme has been found to contain pyrroquinoline–quinone, calcium ions, and heme c (De Jong et al., 1995). The encoding gene (*qhedh*) has been cloned, sequenced, and expressed in *E. coli* (Stoorvogel et al., 1996). The enzyme has been investigated for use in the production of (*S*)-solketal (2,2-dimethyl-1,3-dioxolane-4-methanol) by enantioselective oxidation of racemic solketal (Geerlof et al., 1994a) and has been applied in electrodes through immobilization in a redox polymer network (Stigter et al., 1997). Quinoline 2-oxidoreductase and 2-oxo-1,2-dihydroquinoline 5,6-dioxygenase, the enzymes involved in the first two steps of the degradation of quinoline and 3-methyl quinoline, have been isolated from *Comamonas testosteroni* strain 63 and their structure and cofactors studied (Schach et al., 1995). Quinoline 2-oxidoreductase is a molybdo-iron/sulfur flavoprotein, and 2-oxo-1,2-dihydroquinoline 5,6-dioxygenase is a single component protein (Schach et al., 1995).

The iron-containing nitrile hydratase from *C. testosteroni* strain Nil, which catalyzes the hydration of 5-cyanovaleric acid to adipamic acid, is inactivated by stoichiometric amounts of NO, and is reactivated by photoirradiation. In these reactions, it is similar to a *Rhodococcus* R312 nitrile hydratase, which has a quite different amino acid sequence (Bonnet et al., 1997).

5-Aminolevulinic acid, a precursor to tetrapyrroles, is formed from aminoacylated tRNA–Glu in a two-step pathway by *C. testosteroni*. This pathway involves glutamyl–tRNA reductase, encoded by *hemA*, and glutamine-1-semialdehyde-2,1-aminomutase, encoded by *hemL* (Hungerer et al., 1995). In a study on the synthesis of new agricultural chemicals, *Comamonas testosteroni* strain CMI 2848 possessed the highest activity for hydroxylating 3-cyanopyridine to 3-cyano-6-hydroxypyridine among 4600 isolates screened (Yasuda et al., 1995).

C. testosteroni is capable of selectively degrading L-lysine from racemized lysine crystals, a characteristic which has been applied in the large-scale production of D-lysine (Takahashi et al., 1997).

The various pathways for the biosynthesis of aromatic amino acids and their regulation have been investigated extensively among pseudomonad bacteria. *C. testosteroni* possesses prephenate dehydrogenase and arogenate dehydrogenase reactive with either NAD or NADP. Arogenate dehydratase activity is absent; prephenate dehydratase is present, and its activity is increased 6–12-fold by L-tyrosine (Byng et al., 1983).

Chemotaxonomic data *Comamonas* strains contain ubiquinone Q-8 as the major quinone (>90%), with smaller amounts of Q-7 and Q-9 (Tamaoka et al., 1987). Major fatty acids (comprising 10–40% of the total amount) are palmitic acid ($C_{16:0}$), palmitoleic acid ($C_{16:1}$), and *cis*-vaccenic acid ($C_{18:1\ \omega7c}$). Small amounts (2–7%) of lauric acid ($C_{12:0}$) are present. Myristic acid ($C_{14:0}$) is present as 3–4% of the total in *C. terrigena* and less than 1% of the total in *C. testosteroni*. 2-Hydroxyhexadecanoic acid ($C_{16:0\ 2OH}$) is present as less then 1% of the total in *C. terrigena* and 2–7% in *C. testosteroni* (Tamaoka et al., 1987; Willems et al., 1989). For some fatty acids, the amounts reported by Tamaoka et al. (1987) and Willems et al. (1989) are quite different: 3-hydroxydecanoic acid ($C_{10:0\ 3OH}$), 3–5% and 7–17%, respectively; *n*-heptadecanoic acid ($C_{17:0}$), <2% and traces to 7%; cyclopropane-substituted methylene-hexadecanoic fatty acid ($C_{17:0\ cyclo}$), <2% and 1–15%. The polyamine content of *C. testosteroni* strain DSM 1622 has been determined. The major polyamines (>50%) are 2-hydroxyputrescine and putrescine; spermidine, spermine, and 1,3-diaminopropane are present in much smaller quantities (<5%). Cadaverine has not been detected (Busse et al., 1992).

Plasmids Plasmids have been isolated from and successfully transferred to *Comamonas* strains and may contribute to the catabolic versatility and flexibility of these organisms. *C. testosteroni* strains T-2, PSB-4, and the type strain have been reported to contain two plasmids (pTSA and pT2T), one plasmid (pPSB), and no plasmids, respectively (Junker and Cook, 1997). Plasmid pTSA (85 kb) is a conjugative plasmid belonging to the IncP1 group. It carries the genes *tsaMBCD*, *tsaR*, and *psbAC*, involved in the degradation of 4-toluene sulfonate to protocatechuate, as well as two copies of insertion element IS 1071. Plasmid pPSB (85 kb) is also a conjugative plasmid and carries the *psbAC* genes, as well as two copies of IS 1071. Results of conjugation experiments with the type strain suggest that the *psb* genes are located in a composite transposon (Junker and Cook, 1997).

Comamonas strains have been used in various experiments as recipients of plasmids carrying catabolic genes. Plasmid RP4::Tn4371, carrying genes for biphenyl and 4-chlorobiphenyl degradation, has been transferred from *Enterobacter agglomerans* to indigenous soil bacteria, including *Comamonas* sp., where these catabolic genes are expressed (De Rore et al., 1994). Plasmid PR4::Mu3A, carrying chromosomal DNA fragments encoding the ability to use biphenyl as the sole carbon source, has been transferred to *C. testosteroni* and the catabolic genes effectively expressed (Springael et al., 1996). The plasmids pACK5 and pACT72, which each include a replicon of the cryptic plasmid pAC1 from *Acetobacter pasteurianus*, have been transferred to and successfully expressed in *Comamonas terrigena*, as well as in several other Gram-negative and Gram-positive taxa (Grones and Turna, 1995).

Antibiotic sensitivity *C. testosteroni* possesses steroid-inducible hydroxysteroid dehydrogenases/carbonyl reductases. This contributes to increased resistance of cells grown on a steroid substrate to the fungal steroid fusidic acid and to faster uptake and alternative metabolism of the anti-insect agent NKI 42255 (2-(1-imidazolyl)-1-(4-methoxyphenyl)-2-methyl-1-propane). These steroid-inducible pathways provide protection against natural and synthetic toxic compounds present in the soil and in the intestinal tracts of mammals (Oppermann et al., 1996).

Ecology *Comamonas* strains have been isolated from sites heavily contaminated with various complex organic compounds and heavy metals. Resistance of *Comamonas* strains to a number of heavy metals has been reported. A cadmium-resistant strain of *C. testosteroni* has been isolated from soil contaminated with heavy metals (Kanazawa and Mori, 1996), and nickel-resistant *Comamonas* strains have been isolated from naturally nickel-percolated soils from New Caledonia (Stoppel and Schlegel, 1995). Among other chromate-reducing bacteria isolated from the cooling water of an electricity generating plant, a strain tentatively identified as *Comamonas testosteroni* has been found to be one of the strongest chromate-reducers and has therefore been implicated in blockage of pipes due to precipitation of chromium (III) oxide (Cooke et al., 1995).

Because of their ability to degrade a wide variety of complex organic compounds, *Comamonas* strains are of potential interest in bioremediation. *C. testosteroni* has been used to degrade 4-toluenesulfonic acid in a continuously operated fixed-bed biofilm reactor (Khlebnikov and Peringer, 1996). *Comamonas* strains are among the rare isolates of the *Betaproteobacteria* that have been identified in activated sludge (Kämpfer et al., 1996). Cells of *C. terrigena* strain N3H are capable of degrading the anion active surfactant dihexyl-sulfosuccinate and have been applied for this purpose immobilized in alginate gel (Huska et al., 1996). Cells starved of a carbon source for 16 h have the highest biotransformation rate of dihexyl-sulfosuccinate (Toth et al., 1996).

Among isolates from a PCB-polluted site, *C. testosteroni* strains comprise the majority and are capable of degrading biphenyl and various polychlorinated biphenyls (Joshi and Walia, 1995).

Comamonas strains that degrade poly-β-hydroxybutyrate (PHB) have frequently been isolated from various environments (Mergaert and Swings, 1996), and the extracellular PHB depolymerase of one strain has been purified and characterized (Jendrossek et al., 1993).

Because of the broad substrate specificity of its 3-α-hydroxysteroid dehydrogenases of *C. testosteroni* and the occurrence of this and related bacterial species in the intestinal tract of vertebrates, it has been suggested that these enzymes may contribute to bioactivation or inactivation of hormones, bile acids, and xenobiotics (Oppermann and Maser, 1996).

ENRICHMENT AND ISOLATION PROCEDURES

Comamonas strains are common inhabitants of soil, mud, and water, in both natural and polluted environments. They have also been isolated from various clinical samples, the hospital environment, and horse and rabbit blood (Willems et al., 1991c). *Comamonas* strains can generally be isolated from water by plating on nutrient agar, but specific selective isolation procedures yielding only *Comamonas* isolates have not been described. The original *C. terrigena* type strain was isolated on a medium of hay infusion filtrate (Mudd and Warren, 1923). Often, selective enrichment can be obtained by targeting the capacity of *Comamonas* strains to degrade particular aromatic compounds, hydrocarbons, and higher dicarboxylic acids by using these compounds as the sole carbon source in a mineral medium. Tamaoka et al. (1987) reported that *C. terrigena* requires methionine and nicotinamide as growth factors and, consequently, these should be added to isolation media.

The following compounds have been used to isolate or enrich for *Comamonas* strains: phenol and *m*-cresol (Gray and Thornton, 1928), abietic and dehydroabietic acid (Morgan and Wyndham, 1996), and poly-β-hydroxybutyrate (Jendrossek et al., 1993). The isolation of *C. testosteroni* strains has been reported with the following compounds: testosterone (Talalay et al., 1952), imidazolylpropionate and imidazolyl-lactate (Coote and Hassal, 1973), *p*-cresol (Dagley and Patel, 1957), fumarate, bromosuccinate, anthranilate, kynurenate, and poly-3-hydroxybutyrate (Stanier et al., 1966), naphthalene (García-Valdés et al., 1988), phenanthrene (Goyal and Zylstra, 1996), chlorophenol and methylphenol (Hollender et al., 1997), and polychlorinated biphenyls (Joshi and Walia, 1995). *C. testosteroni* strains have also been isolated by selecting for cadmium resistance (Kanazawa and Mori, 1996).

Comamonas strains have been isolated increasingly from the clinical environment (Gilardi, 1971, 1985; Ben-Tovim et al., 1974; De Vos et al., 1985b; Willems et al., 1991c) and are regarded as rare opportunistic pathogens (Gilardi, 1985). Sources include blood, pus, urine, pharyngeal mucosae, kidneys, feces, burst appendix, intravenous tubing, and urinary catheters. *Comamonas* strains are isolated from such samples using methods for the isolation of Gram-negative glucose-nonfermenters. This includes the use of a blood agar medium, such as tryptic soy agar plus defibrinated blood, and a selective enteric medium, such as MacConkey agar. The commonly used incubation regime for primary isolation media of 24 h at 35°C should be extended with 24 h at 30°C to permit growth of glucose-nonfermenters that grow slowly at 35°C and may be masked by other bacteria (Rubin et al., 1985).

MAINTENANCE PROCEDURES

Comamonas strains can be maintained on nutrient agar at 4°C for up to 2 months. For long-term preservation, strains can be lyophilized using standard procedures. This can be done routinely by freeze-drying the cells in horse serum (70%) supplemented with glucose (7%) and nutrient broth (0.6%).

DIFFERENTIATION OF THE GENUS *COMAMONAS* FROM OTHER GENERA

Table BXII.β.42 lists features differentiating the genus *Comamonas* from other genera in the family.

Restriction analysis of a 2.4 kb amplified ribosomal DNA fragment containing the 16S rRNA gene, the spacer region between 16S and 23S genes, and part of the 23S rRNA gene with the enzyme *Hin*fI allows differentiation of *Comamonas* species from members of *Acidovorax*, *Delftia* (formerly *Comamonas acidovorans*), *Hydrogenophaga*, and *Variovorax* (Vaneechoutte et al., 1992).

An oligonucleotide probe targeting 16S rRNA has been used to identify members of the genus *Comamonas*, among other members of the *Comamonadaceae* (Amann et al., 1996). Its sequence is 5'-ACCTACTTCTGGCGAGA-3', which is homologous to positions 1424–1440 on the rRNA.

TAXONOMIC COMMENTS

The nomenclatural history of the genus *Comamonas* is complex and was initially quite obscure because of incomplete descriptions and lack of original cultures. *Comamonas* was therefore not included in the Approved Lists of Bacterial Names (Skerman et al., 1980), nor was it listed in the previous edition of this *Manual*, although in the 8th edition of *Bergey's Manual of Determinative Bacteriology*, *Comamonas terrigena*, "*Vibrio cyclosites*", and "*Vibrio neocistes*" were included as *species incertae sedis* in addendum IV of the genus *Pseudomonas*. The genus became valid only in 1985, when the name was revived (De Vos et al., 1985b).

The name *Comamonas* was originally proposed by Davis and Park (1962) to replace the invalid name *Lophomonas*. This latter genus, with one species (*Lophomonas alcaligenes*), had been proposed for a group of Gram-negative rod-shaped bacteria attacking few carbohydrates and possessing 2–4 lophotrichous flagella (Galarneault and Leifson, 1956). *L. alcaligenes* was considered to be a subjective synonym of the peritrichously flagellated *Vibrio alcaligenes*, which had itself been created (Lehmann and Neumann, 1927) to replace *Bacillus faecalis alcaligenes*, a species which contained Gram-negative, rod-shaped, human fecal isolates that did not attack carbohydrates (Petruschky, 1896). When the genus *Comamonas* was created to replace *Lophomonas*, the former type species (*Lophomonas alcaligenes*) was not maintained because of differences between the original descriptions of *Vibrio alcaligenes* and the new genus *Comamonas*. Instead, *Comamonas percolans*, formerly *Vibrio percolans* (Mudd and Warren, 1923), was proposed as the type species. *Vibrio alcaligenes*, *Vibrio neocistes*, and *Vibrio cyclosites* were also assigned to the new genus *Comamonas*, but were not allocated to a particular species (Davis and Park, 1962). *Comamonas percolans* was eventually shown to be a later subjective synonym of *Vibrio terrigenus*, an organism isolated from soil and motile by means of bipolar tufts of flagella (Günther, 1894), and the organism was therefore renamed *Comamonas terrigena* (Hugh, 1962). Later still, Hugh (1965) reported that the type strains of *Comamonas terrigena* and *Pseudomonas testosteroni* were very similar and suggested both species be united.

The genus *Comamonas* was revived in 1985 (De Vos et al., 1985b) after a polyphasic study that included some of the old preserved isolates. Initially, only four strains, each motile by a polar tuft of flagella, were included in the revived species *Comamonas terrigena*: NCIB 8193^T (formerly *Vibrio percolans*), NCIB

2581 (formerly *Vibrio cyclosites*), NCIB 2582 (formerly *Vibrio neocistes*) and CCUG 12940. Later, *Pseudomonas acidovorans* and *Pseudomonas testosteroni* were transferred to an emended genus *Comamonas* as *Comamonas acidovorans* and *Comamonas testosteroni* (Tamaoka et al., 1987). A polyphasic study that included many unnamed clinical isolates, as well as strains representing related genera, resulted in the inclusion of many more strains in *C. terrigena* and the recognition of three subgroups in this species. These subgroups can be distinguished by DNA–rRNA and DNA–DNA hybridizations, whole cell protein electrophoresis, and immunotyping, but not by morphological, auxanographic, or biochemical characterization; consequently, they have been retained in a single species (Willems et al., 1991c). The first subgroup contains the four strains assigned to *C. terrigena* by De Vos et al. (1985b) and one additional clinical strain. The second subgroup comprises the type strain of *Aquaspirillum aquaticum* and 13 clinical isolates previously designated EF group 10. The third subgroup comprises 13 clinical strains of EF group 10 as well as three misnamed *Pseudomonas* strains from blood. Several new *Comamonas terrigena* strains cannot be assigned to any of the three groups, indicating that this species may contain even more diversity than was previously thought (Willems et al., 1991c). The name *Aquaspirillum aquaticum* was abandoned since it was a later synonym of *Comamonas terrigena*. Since this organism has bipolar tufts of flagella, its inclusion in *Comamonas* brought the genus description, at least for this aspect of morphology, back in line with the original description of *Vibrio terrigenus* (Günther, 1894).

From DNA–rRNA hybridizations it was apparent that the three *C. terrigena* subgroups and *C. testosteroni* were probably more closely related to each other than to *C. acidovorans* (Willems et al., 1991c). This has recently been confirmed by 16S rRNA gene sequence analysis and, consequently, it has been proposed that *C. acidovorans* be transferred to a new genus *Delftia* as *Delftia acidovorans* (Wen et al., 1999). By 16S rDNA analysis, *C. terrigena* and *C. testosteroni* cluster together, but their grouping is not supported by a high bootstrap value. This indicates that the grouping of these branches is not very stable, and their position in the dendrogram may change as sequence data for additional strains are included. It is therefore at present not clear whether both species form a phylogenetically coherent genus, comparable to, for example, *Hydrogenophaga* or *Acidovorax*.

Comamonas spp. are regularly reported in ecological or applied studies involving isolates from soil and water. Although some of these isolates may escape species identification because of inadequacy of the identification methods used, other isolates may represent additional species within the genus. Wen et al. (1999) have described a *Comamonas* strain with an taxonomic position that is intermediate between *C. terrigena* genomic subgroup 1 (*sensu* Willems et al., 1991c) and *C. testosteroni*, but no representatives of the other two genomic subgroups were included in their analysis. It is possible, therefore, that this strain may belong to one of these two *C. terrigena* genomic subgroups.

"*Comamonas compransoris*", which has been described as a facultatively lithotrophic carbon monoxide or hydrogen oxidizer (Nozhevnikova and Zavarzin, 1974), has been shown to belong to the *Alphaproteobacteria* and has been assigned to the new genus *Zavarzinia* as *Zavarzinia compransoris* (Meyer et al., 1993).

ACKNOWLEDGMENTS

Anne Willems is indebted to the Fund for Scientific Research–Flanders for a position as a postdoctoral research fellow.

DIFFERENTIATION OF THE SPECIES OF THE GENUS *COMAMONAS*

Table BXII.β.44 lists characteristics differentiating *Comamonas terrigena* from *Comamonas testosteroni*.

It is possible to distinguish the three *C. terrigena* groups, as well as *C. testosteroni*, by restriction analysis with the enzyme *Hinf*1 of a 2.4 kb amplified ribosomal DNA fragment containing the 16S rRNA gene, the spacer region between 16S and 23S genes, and part of the 23S rRNA gene (Vaneechoutte et al., 1992).

TABLE BXII.β.44. Characteristics differentiating *Comamonas* species[a,b]

Characteristic	C. terrigena	C. testosteroni
Occurrence of 2-hydroxy fatty acids	−	+
Growth on:		
L-Histidine	−	+
2-Aminobenzoate (anthranilate)	−	+
Citrate	−	+
Glycolate	−	+
Benzoate	−	+[c]
Testosterone	−	+

[a]For symbols see standard definitions.

[b]Data taken from Palleroni (1984), De Vos et al. (1985b), Tamaoka et al. (1987), and Willems et al. (1991c).

[c]Negative according to Tamaoka et al. (1987); 11–89% of strains are positive (d score) according to Palleroni (1984).

List of species of the genus Comamonas

1. **Comamonas terrigena** (ex Hugh 1962) De Vos, Kersters, Falsen, Pot, Gillis, Segers and De Ley 1985b, 450[VP] emend. Willems, Pot, Falsen, Vandamme, Gillis, Kersters and De Ley 1991c, 439.

   ter.ri.ge′ na. L. n. *terra* soil; L. v. *gignere* to bear; L. n. *terrigena* borne by the earth (soil), child of the earth.

   The characteristics are as described for the genus and as indicated in Tables BXII.β.44 and BXII.β.45.

   The type strain was isolated from a hay infusion filtrate.

   The mol% G + C of the DNA is: 59.7–63.3 (species); 64 (type strain) (T_m).

   Type strain: ATCC 8461, IAM 12409, LMG 1253, NCIB 8193.

   GenBank accession number (16S rRNA): AB021418.

2. **Comamonas denitrificans** Gumaelius, Magnusson, Petterson and Dalhammar 2001, 1005[VP]

   de.ni.tri′fi.cans. L. prep. *de* away from; L. n. *nitrum* soda; N.L. n. *nitrum* nitrate; M.L. v. *denitrifico* to denitrify; N.L. part.adj. *denitrificans* denitrifying.

   The characteristics are the same as those of the genus. This species is not included in Tables BXII.β.44 and BXII.β.45 because it was published after the completion of

TABLE BXII.β.45. Additional characteristics of *Comamonas* species[a,b]

Characteristic[c]	C. terrigena	C. testosteroni
Growth at 4°C	nd	−
Growth at 42°C	d	−
Growth with 3% NaCl	d	d
Growth with 4.5% NaCl	d	−
Susceptibility to penicillin (10 µg/disk)	d	−
Growth on cetrimide	−	d
Hydrolysis of Tween 80	−	d[d]
Christensen urease	−	d
Phosphoamidase	+	d
Nitrate reduction	+	d
Denitrification	nd	−
Chemolithoautotrophic growth with H₂ as electron donor	nd	−
Levan from sucrose	nd	−
Poly-β-hydroxybutyrate accumulation	nd	+
Extracellular poly-β-hydroxybutyrate hydrolysis	nd	d
Growth on:		
Ethanol	nd	−
Glycerol	−	d[e,f]
Gluconate	d	+
Ethylene glycol, propylene glycol, 2,3-butylene glycol, *n*-hexadecane	nd	−
DL-Kynurenine	−	d
L-Threonine, L-tryptophan	−	d[e,f]
Glycine	−	d[d,g]
D-α-Alanine	d	d[d]
L-Alanine, L-valine, DL-norvaline, DL-norleucine	d	d
DL-4-aminobutyrate	−	−[h]
L-Isoleucine, L-norleucine, L-tyrosine	d	+
L-Phenylalanine	d	+[i]
L-Asparagine, glucono-δ-lactone	nd	+
DL-Methionine, putrescine	nd	−
Acetate, succinate	+	+[i]
Isobutyrate	+	+[h]
Heptanoate, caprate	d	d[f]
Pyruvate, aconitate	d	d[d]
Mesaconate	d	d
Caprylate	d	d[e,f]
L-Tartrate, L-mandelate	−	d[f]
m-Tartrate, isophthalate, terephthalate	−	d
Phthalate	−	d[e]
Pelargonate	−	d[e,f]
D-Tartrate, trigonelline	−	−[h]
DL-Glycerate, citraconate, itaconate, *m*-hydroxybenzoate, *p*-hydroxybenzoate	d	+
α-Ketoglutarate	d	+[e]
Saccharate, mucate, hydroxymethylglutarate, butanol, L-kynurenine, kynurenate	nd	+
DL-Malate, hippurate, *n*-propanol, geraniol, poly-β-hydroxybutyrate, benzoylformate, 2-aminovalerate, nicotinate	nd	d
Phenol	nd	d[f]
DL-Tartrate, anthranilate, isobutanol, phenylethanediol, naphthalene, pantothenate, dodecane, hexadecane, quinate	nd	−
Hydrolysis of:		
2-Naphthylphosphate (pH 5.4)	d	d
2-Naphthylcaprylate	d	+
Naphthol-AS-BI-phosphodiamide	+	d

[a]For symbols see standard definitions; nd, not determined.

[b]Data taken from Palleroni (1984), De Vos et al. (1985b), Tamaoka et al. (1987), and Willems et al. (1991c).

[c]The following characteristics are present in both species: growth at 30 and 37°C, growth on Drigalski-Conradi agar; growth in the presence of 0.5 or 1.5% NaCl; growth on L-proline, L-leucine, L-aspartate, L-glutamate, propionate, butyrate, *n*-valerate, levulinate, isovalerate, *n*-caproate, D-malate, fumarate, glutarate, adipate, pimelate, suberate, azelate, sebacate, DL-lactate, DL-β-hydroxybutyrate, and L-malate; and hydrolysis of 2-naphthylbutyrate, L-leucyl-2-naphthylamide. The following characteristics are absent in both species: growth in the presence of 6.5% NaCl; acid production in 10% lactose, in triple sugar iron medium, and in OF medium with D-glucose, D-fructose, D-xylose, maltose, or adonitol; α-hemolysis; hydrolysis of esculin, gelatin, starch, acetamide, and DNA; lysine and ornithine decarboxylases; arginine dihydrolase; indole production; nitrite reduction; and β-galactosidase. Both species fail to grow on D-arabinose, L-arabinose, D-lyxose, D-ribose, D-xylose, L-xylose, D-fructose, L-fucose, D-fucose, D-tagatose, adonitol, D-arabitol, L-arabitol, *m*-xylitol, *m*-erythritol, *m*-inositol, D-mannitol, D-galactose, D-glucose, D-mannose, L-rhamnose, L-sorbose, dulcitol, sorbitol, D-cellobiose, D-gentiobiose, lactose, maltose, D-melibiose, sucrose, trehalose, D-turanose, D-melezitose, D-raffinose, methyl-β-D-xyloside, methyl-α-D-glucoside, methyl-α-D-mannoside, 2-ketogluconate, 5-ketogluconate, N-acetylglucosamine, amygdalin, arbutin, salicin, esculin, glycogen, inulin, starch, L-cysteine, L-ornithine, L-lysine, L-citrulline, β-alanine, L-serine, L-methionine, D-tryptophan, L-arginine, DL-2-aminobutyrate, DL-3-aminobutyrate, DL-5-aminovalerate, 3-aminobenzoate, 4-aminobenzoate, ethanolamine, ethylamine, butylamine, amylamine, benzylamine, acetamide, diaminobutane, urea, betaine, creatine, spermine, sarcosine, histamine, tryptamine, glucosamine, oxalate, maleate, malonate, phenylacetate, *o*-hydroxybenzoate, or D-mandelate. Neither species hydrolyzes 2-naphthylmyristate, 2-naphthylphosphate (pH 8.5), L-cystyl-2-naphthylamide, L-valyl-2-naphthylamide, N-benzoyl-DL-arginine-2-naphthylamide, N-glutaryl-phenylalanine-2-naphthylamide, 6-bromo-2-naphthyl-β-D-galactopyranoside, 2-naphthyl-β-D-galactopyranoside, naphthol-AS-BI-β-D-glucuronate, 2-naphthyl-α-D-glucopyranoside, 6-bromo-2-naphthyl-β-D-glucopyranoside, 1-naphthyl-N-acetyl-β-D-glucosaminide, 6-bromo-2-naphthyl-α-D-mannopyranoside, and 2-naphthyl-α-L-fucopyranoside.

[d]Positive according to Palleroni (1984).

[e]Negative according to Tamaoka et al. (1987).

[f]Negative according to Palleroni (1984).

[g]Positive according to Tamaoka et al. (1987).

[h]11–89% of strains are positive (d score) according to Palleroni (1984).

[i]11–89% of strains are positive (d score) according to Tamaoka et al. (1987).

the chapter. The description below is taken from the original description (Gumaelius et al., 2001).

Cells are straight to slightly curved rods, 1–2 × 2–6 μm. On nutrient agar plates, yellow-white colonies are formed, and strain P17 produces a brownish pigment. On this medium cells occur singly or as filaments, motile by means of polar flagella. Growth at 20, 30, and 37°C, but not at 4°C. Reduces nitrate to nitrogen gas and is the only *Comamonas* species to do so. Contains cd_1-type nitrate reductase. All strains utilize fumarate, L-malonate, pyruvate, glycolate, D-β-hydroxybutyrate, α-ketovalerate, L-lactate, L-glutamate, L-lysine, D-saccharate, salicin, L-tartrate, D-glucuronate, succinate, L-alanine, and L-arginine. The following substrates are used by some of the four strains only: citrate, urea, *m*-hydroxybenzoate, *trans*-aconitate, maleinate, D-tartrate, gentisate, *p*-coumarate, hippurate, DL-2-γ-aminobutyrate, L-serine, and esculin. None of the strains use mannoic acid γ-lactone, L-arabinose, D-xylose, D-galactose, maltose, D-cellobiose, D-trehalose, palatinose, sucrose, D-lactose, melibiose, lactulose, β-gentiobiose, D-melezitose, L-raffinose, inosine, adonitol, *meso*-inositol, D-arabitol, glycerol, maltitol, D-sorbitol, dulcitol, L-sorbose, 2-deoxy-D-ribose, L-rhamnose, D-fucose, L-fucose, D-tagatose, D-amygdalin, arbutin, methyl-β-D-galactopyranoside, 5-keto-D-gluconate, D-gluconate, 6-O-α-D-galactopyranosyl-D-gluconic acid, D-galactonic acid γ-lactone, D-ribose, L-xylose, D-glucose, D-mannose, L-arabitol, *meso*-erythritol, D-mannitol, xylitol, D-fructose, 6-deoxy-D-galactose, 2'-deoxyinosine, inulin, methyl-α-D-mannopyranoside, methyl-α-D-xylopyranoside, methyl-α-D-galactopyranoside, starch, D-galacturonate, D-arabinose, D-turanose, D-glucuronolactone, glycogen, D-lyxose, N-acetyl-D-glucosamine, maltose, D-gluconate, caprate, adipate, maleate, phenylacetate, gelatin, p-nitrophenyl-β-D-galactopyranoside, L-tryptophan, L-histidine, L-ornithine, and arabic acid. Able to grow in 2% saline solution and to survive in 5% NaCl, but not in 9% NaCl. All strains are sensitive to chloramphenicol (30 μg), erythromycin (15 μg), streptomycin (30 μg), tetracycline (30 μg), and ampicillin (10 μg). Sensitivity to rifampicin (5 μg), sulfisoxazole (250 μg), and penicillin G (10 μg) varies among strains.

Isolated from activated sludge with biological nitrogen removal properties.

The mol% G + C of the DNA is: 60.4–60.8 (HPLC).

Type strain: 123, ATCC 700936, CCUG 44425.

GenBank accession number (16S rRNA): AF233877.

3. **Comamonas koreensis** Chang, Han, Chun, Lee, Rhee, Kim and Bae 2002, 380^{VP}

ko.re.en' sis. N.L. fem. adj. *koreensis* of Korea, the geographical origin of isolation.

The characteristics are the same as those of the genus. The species is not included in Tables BXII.β.44 and BXII.β.45 because it was published after the completion of the chapter. The description below is taken from the original description (Chang et al., 2002).

Grows at 10, 25, and 37°C, but not at 5 or 42°C. Best growth at pH 7 and 30°C. Grows on nutrient agar with 3% (w/v) NaCl, but not 4.5 or 6.5% NaCl. Aerobic, but also able to grow anaerobically in an atmosphere of $H_2/CO_2/N_2$ (7:5:88) on nutrient agar. Nonmotile, and no flagella are produced. Nitrate is reduced to nitrite; no denitrification. Arginine dihydrolase, lysine decarboxylase, ornithine de-

carboxylase, and urease are not produced. The following substrates are utilized: cellobiose, malate, maltose, D-psicose, D-raffinose, L-rhamnose, D-glucose, aconitate, D-tagatose, *m*-inositol, and D-mannitol. The following substrates are not used: citrate, caprate, N-acetyl-D-galactosamine, N-acetylglucosamine, adonitol, esculin, amygdalin, DL-arabinose, DL-arabitol, arbutin, dulcitol, erythritol, DL-fucose, galactose, α-gentiobiose, glycogen, inulin, 2-ketogluconate, 5-ketogluconate, lactose, lactulose, D-lyxose, D-mannose, melibiose, melezitose, methyl-β-D-glucoside, methyl-α-D-glucoside, methyl-α-D-mannoside, methyl-β-D-xyloside, ribose, salicin, sorbitol, sorbose, sucrose, trehalose, D-turanose, DL-xylose, xylitol, adipate, D-gluconate, glycerol, propionate, sebacate, L-threonine, D-alanine, L-alanine, L-aspartate, L-leucine, L-histidine, L-phenylalanine, phenylacetate, D-fructose or starch. Hydrolyzes Tween 80, acetic acid, D-galactonic acid lactone, D-gluconic acid, itaconic acid, methyl pyruvate, monomethyl succinate, quinic acid, D-saccharic acid, succinamic acid, succinic acid, bromosuccinic acid, L-leucyl-2-naphthylamide, 2-naphthylcaprylate, 2-naphthyl phosphate (pH 5.4 and 8.5), and napthol-AS-BI phosphate. Does not hydrolyze gelatin, 2-naphthylbutyrate, 2-naphthyl phosphate (pH 8.5), alaninamide, L-alanylglycine, γ-aminobutyric acid, 2-aminoethanol, arginine, L-asparagine, DL-carnitine, glucuronamide, L-glutamic acid, glycyl-L-aspartic acid, glycyl-L-glutamic acid, hydroxy-L-proline, phenylethylamine, L-proline, putrescine, L-pyroglutamic acid, DL-serine, citric acid, α-hydroxybutyric acid, β-hydroxybutyric acid, γ-hydroxybutyric acid, *p*-hydroxyphenylacetic acid, formic acid, D-galacturonic acid, D-glucosaminic acid, D-glucuronic acid, inosine, α-ketobutyric acid, α-ketoglutaric acid, α-ketovaleric acid, DL-lactic acid, malonic acid, thymidine, Tween 40, urocanic acid, uridine, 2,3-butanediol, α-cyclodextrin, dextrin, gelatin, DL-α-glycerol phosphate, glucose-1-phosphate, glucose-6-phosphate, N-benzoyl-DL-arginine-2-naphthylamide, 6-bromo-2-naphthyl-α-D-galactopyranoside, 6-bromo-2-naphthyl-β-D-glucopyranoside, 6-bromo-2-naphthyl-α-D-mannospyranoside, L-cystyl-2-naphthylamide, N-glutaryl-phenylalanine-2-naphthylamide, 2-naphthylmyristate, 2-naphthyl-β-D-galactopyranoside, naphthol-AS-BI β-D-glucuronate, 2-naphthyl-α-D-glucopyranoside, 1-napthyl-N-acetyl β-D-glucosaminide, 2-naphthyl α-L-fucopyranoside, and L-valyl-2-naphthylamide. Acetoin, H_2S, and indole are not produced. No β-galactosidase and tryptophan deaminase.

Major fatty acids are hexadecanoic acid ($C_{16:0}$), *cis*-9-hexadecanoic acid ($C_{16:1\,\omega7c}$), methylene-hexadecanoic acid ($C_{17:0\,cyclo}$), and octadecanoic acid ($C_{18:1}$); 2-hydroxy fatty acids ($C_{15:0\,2OH}$ and $C_{16:0\,2OH}$) and 3-hydroxy fatty acids ($C_{10:0\,3OH}$) are present in smaller amounts (<4 %).

Isolated from a wetland sample in Woopo, Republic of Korea.

The mol% G + C of the DNA is: 66 (T_m).

Type strain: YH12, KCTC 12005, IMSNU 11158.

GenBank accession number (16S rRNA): AF275377.

4. **Comamonas nitrativorans** Etchebehere, Errazquin, Dabert, Moletta and Muxí 2001b, 982^{VP}

ni.tra.ti.vo' rans. N.L. n. *nitras* nitrate; L. adj. part. *vorans* devouring, digesting; N.L. adj. *nitrativorans* nitrate-consuming.

The characteristics are the same as those of the genus. The species is not included in Tables BXII.β.44 and

BXII.β.45 because it was published after the completion of the chapter. The description below is taken from the original description (Etchebehere et al., 2001b).

Cells are motile with two tufts of polar flagella. On TSA colonies are cream-colored, circular, and 1–2 mm in diameter after 24 h. Growth on acetate, butyrate, *n*-caproate, *i*-butyrate, *i*-valerate, propionate, *n*-valerate, lactate, alanine, benzoate, L-phenylalanine, and ethanol. Weak growth on maleate. No growth on glucose, arabinose, fructose, galactose, xylose, mannitol, malonate, tartrate, *p*-aminobenzoate, gluconate, pyruvate or citrate. Anoxic reduction of nitrate, nitrite, and nitrous oxide to nitrogen. Under anaerobic conditions, the same substrates (except benzoate and phenylalanine) can be used as under aerobic conditions. Optimal pH is 7, and optimal temperature for growth is 30°C.

Isolated from a denitrifying reactor from a landfill leachate treatment system in Montevideo, Uruguay.

The mol% G + C of the DNA is: unknown.
Type strain: 23310, CIP 107121, DSM 13191, NCCB 100007, CCT 7062.
GenBank accession number (16S rRNA): AJ251577.

5. **Comamonas testosteroni** (Marcus and Talalay 1956) Tamaoka, Ha, and Komagata 1987, 58^VP emend. Willems, Pot, Falsen, Vandamme, Gillis, Kersters and De Ley 1991c, 440 (*Pseudomonas testosteroni* Marcus and Talalay 1956, 661.)

tes.tos.te.ro′ni. M.L. gen. n. *testosteroni* of testosterone, a chemical compound.

The characteristics are as described for the genus and as indicated in Tables BXII.β.44 and BXII.β.45.
Isolated from soil.
The mol% G + C of the DNA is: 62.5–64.5 (species); 62.5 (type strain) (T_m).
Type strain: ATCC 11996, LMG 1786, NCTC 10698.
GenBank accession number (16S rRNA): M11224.

Genus II. **Acidovorax** *Willems, Falsen, Pot, Jantzen, Hoste, Vandamme, Gillis, Kersters and De Ley 1990, 394^VP emend. Willems, Goor, Thielemans, Gillis, Kersters and De Ley 1992a, 115*

ANNE WILLEMS AND MONIQUE GILLIS

A.ci.do.vo′rax. L. neut. n. *acidum* acid; L. adj. *vorax* voracious; M.L. masc. n. *Acidovorax* acid-devouring (bacteria).

Straight to slightly curved rods, 0.2–1.2 × 0.8–5.0 μm, occurring singly, in pairs, or in short chains. **Gram negative. Motile by means of one or rarely two or three polar flagella. Aerobic, having a strictly oxidative type of metabolism with O₂ as the terminal electron acceptor**; some strains of two species (*Acidovorax delafieldii* and *Acidovorax temperans*) are capable of **heterotrophic denitrification** of nitrate. Most strains do not produce pigments on nutrient agar, but some phytopathogenic strains may produce a yellow to slightly brown diffusible pigment. **Oxidase positive; urease activity varies among strains. Chemoorganotrophic**, although strains of two species (*A. facilis* and *A. delafieldii*) can grow **lithoautotrophically**, using the oxidation of H₂ as an energy source. Good growth occurs on organic acids, amino acids, and peptone, but organisms show less versatile growth on carbohydrates. Fatty acids present always include 3-hydroxyoctanoic acid ($C_{8:0\ 3OH}$) and 3-hydroxydecanoic acid ($C_{10:0\ 3OH}$); 2-hydroxy-substituted fatty acids are absent. *Acidovorax* strains can be isolated from soil, water, clinical samples, activated sludge, and infected plants.

The mol% G + C of the DNA is: 62–70.

Type species: **Acidovorax facilis** (Schatz and Bovell 1952) Willems, Falsen, Pot, Jantzen, Hoste, Vandamme, Gillis, Kersters and De Ley 1990, 394 (*Pseudomonas facilis* (Schatz and Bovell 1952) Davis, Doudoroff, Stanier and Mandel 1969, 385; "*Hydrogenomonas facilis*" Schatz and Bovell 1952, 88.)

FURTHER DESCRIPTIVE INFORMATION

Flagellation and pili *Acidovorax* cells have one or, rarely, two to three polar flagella (Willems et al., 1992a). In a study of the cell morphology and flagellation of Gram-negative hydrogen bacteria, Aragno et al. (1977) defined three types of flagella on the basis of fine structural details as revealed by electron microscopy. They have reported *A. facilis* to possess a polar monotrichous

flagellation, with flagella of 19–20 nm in diameter, a wavelength of 1.4–1.7 mm, and a fine structure of type I. In *A. facilis*, pili may be spread over the total cell surface, but they are observed only rarely (Aragno et al., 1977).

Cell wall composition In a study of the cell envelope of Gram-negative hydrogen bacteria, Walther-Mauruschat et al. (1977) distinguished three types of cell walls, differing mainly in the visibility and location of the peptidoglycan layer. *Acidovorax* possesses type I cell walls, typical of most Gram-negative bacteria and characterized by a multilayered structure, consisting of an outer membrane and a cytoplasmic membrane of similar dimensions and appearance separated by a dense layer of peptidoglycan (Walther-Mauruschat et al., 1977). *A. delafieldii* exhibits a crystalline S layer in close contact with the outer membrane and covering the whole cell surface (Lapchine, 1979; Chalcroft et al., 1986). A second major type of outer membrane protein, Omp34, has been purified and characterized as a typical porin, forming anion-selective channels. The channel conductance depends on ion concentration (Brunen et al., 1991). Functional properties of Omp34 have been studied and are largely determined by positively charged amino acid residues (Brunen and Engelhardt, 1995). The finding of anion-selective porins is compatible with the preference of *Acidovorax* strains for acidic substrates.

Fine structure Intracellular mesosome-like membrane systems with a spiral appearance, often located in the area of cell division or at the cell poles, have been reported in *A. facilis* (Walther-Mauruschat et al., 1977). However, there is a growing consensus that such structures are mostly artifacts resulting from the preparation of cells for electron microscopy. Intracellular inclusions of poly-β-hydroxybutyrate and polyphosphate and translucent glycogen-like inclusions have been detected in *A. facilis* (Walther-Mauruschat et al., 1977). *A. avenae* has been reported to contain inclusion bodies consisting of a proteinaceous ribbon, which is rolled up to form a hollow cylinder (Wells and Horne, 1983). The function of these so-called R-bodies remains uncertain. Similar structures have been described in some kappa-particles, the endosymbionts in killer paramecia, where they are thought to be involved in toxin release (Lalucat et al., 1982).

Colony morphology On nutrient agar, colonies are round with smooth to slightly scalloped or spreading margins. Occasionally, two different but unstable colony margin types may be observed in a culture. A translucent marginal zone may be present. Colonies are convex, smooth to slightly granular, and beige to faintly yellow. At 30°C, colonies can attain diameters of 0.5–3 mm in 3 d and 4 mm in 7 d. Most species do not produce pigments on nutrient agar, but some phytopathogenic strains may produce a yellow to slightly brown diffusible pigment (Willems et al., 1992a; Gardan et al., 2000). Optimal growth temperature is 30–35°C.

Nutrition and metabolism A variety of organic compounds can be used as sole carbon sources (Tables BXII.β.46 and BXII.β.47). *A. facilis* and some strains of *A. delafieldii* are able to grow lithoautotrophically, using the oxidation of hydrogen as an energy source. The hydrogenase is membrane-bound and does not reduce NAD (Schneider and Schlegel, 1977).

Chemotaxonomic characteristics *A. facilis*, *A. delafieldii*, and *A. avenae* have been reported to contain putrescine and 2-hydroxyputrescine as major polyamines and spermidine and spermine in smaller quantities. Some strains contain traces to small amounts of cadaverine and 1,3-diaminopropane (Busse and Auling, 1988; Auling et al., 1991). The fatty acid contents of the four nonphytopathogenic species *A. delafieldii*, *A. facilis*, *A. defluvii* and *A. temperans* have been determined (Willems et al., 1992a; Schulze et al., 1999a). For the phytopathogenic species *A. avenae* and *A. konjaci*, data are limited to five strains of *A. avenae* (Willems et al., 1990) and two strains of *A. anthurii* (Gardan et al., 2000). For all these species major fatty acids (representing at least 11% of total fatty acids) are palmitic acid ($C_{16:0}$), palmitoleic acid ($C_{16:1}$), and *cis*-vaccenic acid ($C_{18:1 \, \omega 7c}$); small amounts (less than 6%) of lauric acid ($C_{12:0}$), myristic acid ($C_{14:0}$), and 3-hydroxydecanoic acid ($C_{10:0 \, 3OH}$), and smaller amounts still (less than 1%) of *n*-pentadecanoic acid ($C_{15:0}$), *n*-heptadecanoic acid ($C_{17:0}$), oleic acid ($C_{18:1 \, \omega 7c}$), stearic acid ($C_{18:0}$), and cyclopropane-substituted methylene-hexadecanoic acid ($C_{17:0 \, cyclo}$) are present. In addition, the nonphytopathogenic species contain 1–2.5% 3-hydroxyoctanoic acid ($C_{8:0 \, 3OH}$). No 2-hydroxy substituted fatty acids are present (Willems et al., 1990).

Plasmids Mitomycin treatment of *A. delafieldii* cells results in high numbers of mutants unable to oxidize hydrogen, suggesting that in this species, hydrogenase may be plasmid encoded (Pootjes, 1977); the species has been reported to contain a plasmid (Gerstenberg et al., 1982). However, mutants of *A. facilis* defective in lithoautotrophy, obtained by Tn5 mutagenesis, mitomycin C treatment, and incubation at sublethal temperature, all show wild-type plasmid patterns with the same restriction map. Preliminary hybridization experiments have revealed a transposon location in the chromosomal DNA and not in the plasmid (Warrelmann and Friedrich, 1986).

In a study of the influence of earthworm activity on the transfer of plasmid pJP4 from an inoculated *Pseudomonas fluorescens* strain to indigenous soil bacteria, *Acidovorax* species have been identified among transconjugants (Daane et al., 1996).

A highly efficient transformation system using pBR322-derived plasmid vectors with *A. avenae* as a host bacterium has been developed, following observations of efficient transformation of plasmid pBR322 and derived plasmids in *A. avenae* strain K1 and its proline-auxotrophic mutant Pr47 (Fukumoto et al., 1997).

Phages Phages have been reported in *Acidovorax* only in the original description of "*Hydrogenomonas facilis*", where the occurrence of plaques in autotrophic cultures was observed approximately two months after isolation (Schatz and Bovell, 1952). Repeated streaking on mineral agar eliminated the plaques, but they reappeared six months later and again a year after purification. Plaques were 3–4 mm in diameter and contained numerous small resistant colonies. It was suggested that the abundance and rapid growth of these resistant strains explained why liquid cultures never cleared. Phages were never observed in heterotrophic cultures (Schatz and Bovell, 1952).

Ecology *A. delafieldii* and *A. facilis* constitute the dominant microorganisms from soils, sludge, old compost, and freshwater sources that are able to degrade poly-3-hydroxybutyrate (PHB) and poly-3-hydroxybutyrate-co-3-hydroxyvalerate (PHBV) *in vitro* (Mergaert and Swings, 1996). *A. avenae* subsp. *avenae* is able to depolymerize synthetic thermoplastic polymers, such as poly-butylene succinate-co-butylene adipate and poly-caprolactone and poly-caprolactone-starch composites, and has been successfully applied in an *in vitro* biodegradation test system (Scandola et al., 1998). Elevated copper concentrations inhibit the degradation of powdered PHB by *A. delafieldii*, as demonstrated using an overlay agar technique (Birch and Brandl, 1996). *A. delafieldii* biofilms have been shown to contribute to copper solvency in laboratory reactors, suggesting that they may contribute to copper dissolution and increased bulk phase copper levels in domestic water systems (Davidson et al., 1996).

Antibacterial and antifungal activities may give pathogenic bacteria an advantage over competitors and, therefore, such activities of phytopathogenic *Acidovorax* species have been investigated. Effects against various Gram-negative and Gram-positive bacteria have been tested, but only *A. avenae* subsp. *cattleyae* shows antibacterial activity, and this is limited to just one of the bacteria tested, *Listeria innocua*. When antifungal activity was tested against *Rhodotorula mucilaginosa*, most strains of *A. avenae* subsp. *avenae* and *A. avenae* subsp. *cattleyae* showed clear antifungal activity, *A. avenae* subsp. *citrulli* strains showed weak or variable antifungal activity, and *A. konjaci* showed no antifungal activity (Hu and Young, 1998).

Pathogenicity Three *Acidovorax* species are the causal agents of diseases on various plants. *A. avenae* subsp. *avenae* causes symptoms of leaf blight in many members of the Poaceae family, including corn (*Zea mays*), sugar cane (*Saccharum officinarum*), and rice (*Oryza sativa*) (Claflin et al., 1989). Small round to oval-shaped, grayish white spots, with a red margin occur and may merge to form red longitudinal stripes. Sometimes, water-soaked lesions occur, which may develop into large necrotic zones. Stalk rot and bud rot have also been reported (Manns, 1909; Rosen, 1922; Lee et al., 1925; Okabe, 1934; Hayward, 1962; Baraoidan, 1981). *A. avenae* subsp. *cattleyae* causes brown spot disease on *Cattleya* and *Phalaenopsis* orchids. Large water-soaked lesions occur on leaves and quickly turn brown, with the surrounding tissue showing yellow-green halos. Large necrotic spots result, and the plant may die if the growth tip is affected (Ark and Thomas, 1946). *A. avenae* subsp. *citrulli* is pathogenic for watermelon (*Ci-*

TABLE BXII.β.46. Characteristics differentiating the species of the genus *Acidovorax*[a]

Characteristics	A. facilis	A. avenae subsp. avenae	A. avenae subsp. cattleyae	A. avenae subsp. citrulli	A. delafieldii	A. konjaci	A. temperans
Growth in the presence of 3% NaCl	−				−	+	−
Peptonization of milk		−	−	d		+	
Soluble starch utilization	−	d	+	d[b]	−	−	−
Hydrolysis of Tween 80	+	+		+	d	+	−
Hydrolysis of gelatin	+	−	−	d	d	−	−
Nitrite reduction	−	d		d	d	+	+
Autotrophic growth with H₂	+				d		−
Production of H₂S	−		−		−	+	
Growth on:							
D-Glucose	+	+	+	+[b]	+	−	+
D-Mannose	+	−	−	−	+	−	−
L-Arabinose	+	+	+	+[b]	+	−	−
D-Ribose	+	d	+	+	+	+	−
D-Xylose	−	+	−	d[c,d]	−	−	−
D-Galactose	+	+	+	+	+		
D-Mannitol	+	+	+	−	+	+	d
D-Arabitol	+	+	+	−	+	+	d
Sorbitol	+	+	+	−[e]	+	−	d
D-Fucose	−	+[f]	+	d	+[c]	−	−
Dextrin		+		−		−	
2-Ketogluconate	−	−	−	d	+	−	d
L-Threonine, L-histidine	+	+	+	−	+	+	d
L-Tryptophan	+[c]	+	+	d	d	+	−
Ethanolamine	−	d	+	+[e]	−	−[e]	−
DL-2-Aminobutyrate	−	d	+	−	−	−	d
DL-3-Aminobutyrate	−	d	+	d	−	+	−
2-Aminopentanoate		+		+		−	
Acetamide	−	d	+	−		d	
Isobutyrate	−	+	+	d	d	+	d
Isovalerate	+[c]	+	+	−	+	+	+
n-Caproate	−	d	+	−	d	−	d
Adipate	−	+	+	+	+	+	+
D-Tartrate	−	+	+	+[g]	d	+[g]	−
m-Tartrate	−	d	−	+[e,h]	+	+	d
2-Ketoglutarate	−	+	+	+	+[c]	+	d
Citraconate	−	+	+	+	+	+	d
Aconitate	−	d	−	−	d	+	d
Citrate	−	d[h]	+	d[h,i]	d	d[c,j]	−
Butane-2,3-diol		+		d		−	
p-Hydroxybenzoate	d	−	−	−	+	−	d
Anthranilate		+		d		−	
Acid from:							
D-Glucose	−	+	+		−		−
D-Fructose	−		+		−	+	−
D-Xylose	−		+		−		−
Mol% G + C of DNA	64–65	68–70	69	67–68	65–66	68	62–66

[a]for symbols see standard definitions. Data taken from Haynes and Burkholder (1957) (for *A. avenae* subsp. *cattleyae*), Schaad et al. (1978), Goto (1983), Ramundo and Claflin (1990), Willems et al. (1990, 1992a), and Hu et al. (1991).

[b]Negative according to Schaad et al. (1978).

[c]Late positive reaction (5 days).

[d]Positive according to Hu et al. (1991).

[e]Eleven to 89% of strains are positive (d score) according to Hu et al. (1991).

[f]The type strain is negative.

[g]Negative according to Hu et al. (1991).

[h]Weak growth.

[i]Positive according to Schaad et al. (1978).

[j]Positive according to Goto (1983).

trullus lanatus) and other members of the family Cucurbitaceae, in which it causes bacterial fruit blotch. The disease appears first on the upper side of the fruit, where dark olive green stains rapidly increase in size. The rind ruptures, allowing access to organisms causing secondary rotting and ultimately leading to the collapse of the fruit. Leaf lesions are inconspicuous, small, dark brown spots with margins that, when viewed from the bottom of the leaf, appear water-soaked (Latin and Hopkins, 1995).

A. konjaci causes leaf blight on *Amorphophallus rivieri* cv. Konjac, a food crop in Japan. It is characterized by small water-soaked spots, which soon extend to become larger lesions with yellowish halos. The whole leaf may ultimately rot away and drop, leaving just bare stalks (Goto, 1983). *A. anthurii* causes leaf spot on *Anthurium*; first symptoms consist of necrotic lesions close to veins and leaf margins, which blacken and turn gray (Gardan et al., 2000). From these lesions the bacteria enter the leaf and spathe

TABLE BXII.β.47. Other characteristics of the species of the genus *Acidovorax*[a]

Characteristics	*A. facilis*	a. *A. avenae* subsp. *avenae*	b. *A. avenae* subsp. *cattleyae*	c. *A. avenae* subsp. *citrulli*	*A. delafieldii*	*A. konjaci*	*A. temperans*
Growth at:							
4°C		−	−	−		−	
30°C	+		+		+		+
37°C	d				+		+
41°C		d		d[b]	+	−[c]	
42°C	−				d		d
Growth in the presence of:							
0.5% NaCl	+				+		+
1.5% NaCl	d				d		d
4.5 and 6.5 % NaCl	−				−		−
Growth on Drigalski–Conradi agar					d		d
Urease	d	+		d	d		d
Catalase	+				d	+	d
ONPG (β-Galactosidase)	−				d	+	d
Lysine and ornithine decarboxylase	−				−		
Arginine dihydrolase	−	−		−	−	d[d]	−
Phenylalanine deaminase, arginine decarboxylase						+	
Hemolysis (α and β)	−				−		−
Alkaline reaction in milk		−	−	−[b]		−	
Acid reaction in milk		−	−	−		−	
Potato starch utilization		d		d		d	
Hydrolysis of DNA, esculin, acetamide	−				−		
Pectate lyase		d		d		−	
Polygalacturonase		−		−		−	
Nitrate reduction	+	d	+	+[e]	+	+	+
Denitrification	−				d	−	d
Acid production in 10% lactose and in triple sugar iron medium	−				−		−
Production of 2-ketogluconate				+			
Levan production		−		−		−	
Fluorescence	−			−			
Growth on cetrimide	−				−		−
Indole production	−		−		−		−
Resistance to penicillin (10 µg/disk)	d		−		d		d
Growth on:[f]							
Xylan, glucitol, *meso*-erythritol, *myo*-inositol, ribitol, galactitol, carboxymethyl cellulose						−	
D-Melibiose	−	−[g]	−	−[h]	−	−[h]	−
L-Rhamnose, lactose, D-cellobiose, trehalose, salicin, inuline, erythritol	−	−[g]	−	−	−	−	−
Maltose	−	−	−	−[g]		−	
Sucrose	−	−[g]	−	−[g]	−	−[g]	
D-Raffinose, D-melizitose	−	−	−	−	−	−[g]	−
L-Fucose	−	d	−	−	−	−	−
Gluconate	+	+	d	+	+	+[d]	d
D-Lactate	d	+		+	+	+	+
D-Lactate plus methionine	d				+		+
D-α-Alanine, L-norleucine	d	+	+	+	d	+	d
L-α-Alanine, L-leucine, L-isoleucine, L-serine, L-phenylalanine, L-tyrosine, L-aspartate, β-alanine	+	+	+	+	+	+	d
L-Cysteine	−	−	−	d	−	−	−
DL-Norleucine	−				d		d
L-Valine	+	+	+	d	d	+	d
DL-Norvaline	−	d	d	−	−	−	d
D-Serine				+			
L-Arginine	−	−	−		−	d	−
L-Methionine	d[i]	−	−	−	−	−	d
L-Ornithine	d[i]	d	+	d	d	d	d
L-Citrulline	−	−	−	−	d	−	d
DL-Kynurenine, DL-5-aminovalerate	−	d	−	−	−	−	−
DL-4-Aminobutyrate	+	+	+	+	+	+	d
2-Aminopentanoate		+		+			
Formate, hippurate						−	
Acetate	d	+	+	+	d	d[c,i]	+
Propionate	+[i]	+	+	d[i]	d	+[g]	d
Butyrate, sebacate	+	+	+	+	+	+[d]	+
n-Valerate	+	+	+	d	+	+	+
Caprate	−	d			−	−[c]	
Oxalate	−	−[g]	−	d[b,i]	−	d[d,i,j]	−
Succinate	+	+[g]	+	+	+	+[g]	+

(continued)

TABLE BXII.β.47. *(cont.)*

Characteristics	A. facilis	A. avenae subsp. avenae	A. avenae subsp. cattleyae	A. avenae subsp. citrulli	A. delafieldii	A. konjaci	A. temperans
Malonate	−	−[g]	−	−	−	+[g]	d
Heptanoate	−	−	−	−	−	−	d
Maleate	−	−	−	d[i]	d	−[c]	d
Glutarate	d	+	+	+	+	+	d
Pimelate	d	+	+	+	+	+	+
DL-Glycerate	+	+	+	+	+	+	d
L-Tartrate	−	−[g]	−	−	−	d[h]	−
Pyruvate	+	+	+	+	d	+	d
Mesaconate	−	−	−	−	−	d[i]	−
Levulinate	d	+	+	+	d	+[g]	d
Ethanol, propanol					+	+	
m-Hydroxybenzoate, phthalate	−	−	−	−	d	−	−
Acid from:							
Maltose, adonitol	−				−		−
D-Ribose, ethanol						+	
Glycerol		+	+			+	
D-Mannitol			+			+	
Galactose, arabinose, lactose, sucrose, dulcitol			+				
Raffinose		−	−				
myo-Inositol, rhamnose		−					
Hydrolysis of:							
2-Naphthylbutyrate, 2-naphthylcaprylate, L-leucyl-2-napthylamide	+					+	+
2-Naphthylphosphate (pH 8.5)	d					d	d
2-Naphthylphosphate (pH 5.4)	d					−	−
2-Naphthylmyristate	d					d	d
L-Cystyl-2-naphthylamide	−					d	d

[a]For symbols see standard definitions. Data taken from Haynes and Burkholder (1957) (for *A. avenae* subsp. *cattleyae*), Schaad et al. (1978), Goto (1983), Ramundo and Claflin (1990), Willems et al. (1990, 1992a), and Hu et al. (1991).

[b]Positive according to Schaad et al. (1978).

[c]Positive according to Goto (1983).

[d]Negative according to Goto (1983).

[e]Negative according to Schaad et al. (1978).

[f]Growth on the following substrates was positive for all *Acidovorax* species: D-fructose, glycerol, D-malate, L-malate, fumarate, suberate, azelate, DL-lactate, DL-3-hydroxybutyrate, L-glutamate, and L-proline. Growth on the following substrates was negative for all *Acidovorax* species: D-arabinose, L-xylose, adonitol, D-lyxose, L-arabitol, methyl-xyloside, L-sorbose, D-tagatose, dulcitol, inositol, methyl-D-mannoside, methyl-D-glucoside, *N*-acetylglucosamine, amygdalin, arbutin, esculin, salicin, β-gentiobiose, D-turanose, starch, glycogen, xylitol, 5-ketogluconate, L-mandelate, caprylate, pelargonate, glycolate, itaconate, phenylacetate, benzoate, *o*-hydroxybenzoate, D-mandelate, isophthalate, terephthalate, glycine, D-tryptophan, trigonelline, L-lysine, betaine, creatine, 3-aminobenzoate, 4-aminobenzoate, urea, sarcosine, ethylamine, butylamine, amylamine, benzylamine, diaminobutane, spermine, histamine, tryptamine, and glucosamine. In addition, *A. facilis*, *A. delafieldii*, and *A. temperans* do not hydrolyze the following substrates: L-valyl-2-naphthylamide, *N*-benzoyl-DL-arginine-2-naphthylamide, *N*-glutaryl-phenylalanine-2-naphthylamide, 6-bromo-2-phosphodiamide-3-naphthoic acid-2-methoxyanilide (= napthol-AS-BI-phosphodiamide), 6-bromo-2-naphthyl-α-D-galactopyranoside, 2-naphthyl-β-D-galactopyranoside, 6-bromo-2-hydroxy-3-naphthoic acid-2-methoxyanilide-β-D-glucuronate (=naphthol-AS-BI-β-D-glucuronate), 2-naphthyl-α-D-glucopyranoside, 6-bromo-2-naphthyl-β-D-glucopyranoside, 1-naphthyl-*N*-acetyl-β-d-glucosaminide, 6-bromo-2-naphthyl-α-D-mannopyranoside, and 2-naphthyl-α-L-fucopyranoside.

[g]11–89% of strains are positive (d score) according to Hu et al. (1991).

[h]Positive according to Hu et al. (1991).

[i]Late positive reaction (5 days).

[j]Negative according to Hu et al. (1991).

parenchymas and become systemic, resulting in tissue discoloration and plant death.

Pathogenicity tests with *A. avenae* subsp. *avenae* on watermelon, cantaloupe, squash, cucumber, tomato, and cowpea are negative (Schaad et al., 1978). Similar tests with *A. avenae* subsp. *citrulli* on tomato, cowpea, corn, and Konjac are also negative (Schaad et al., 1978; Goto, 1983). Tests with *A. konjaci* on cantaloupe, squash, and cucumber are negative and on watermelon results in a hypersensitive reaction (Goto, 1983). Phytopathogenicity of the other *Acidovorax* species has not been reported.

ENRICHMENT AND ISOLATION PROCEDURES

For most *Acidovorax* species, no special isolation procedures have been described, and most strains have been isolated using standard techniques from soil, water, sludge, or infected plant material. These normally result in the isolation of strains of various taxa, except for phytopathogenic species, which are isolated from infected tissue that often contains only the causal bacterium. *A. facilis* was originally isolated from lawn soil in a mineral medium[1] supplemented with 0.05% NaHCO$_3$ and incubated at 25°C under an atmosphere of CO$_2$/air/H$_2$ (10:30:60) (Schatz and Bovell, 1952). *A. delafieldii* was originally isolated from soil, using poly-3-hydroxybutyrate as the sole carbon source in a mineral liquid enrichment culture (Delafield et al., 1965). This procedure yielded strains belonging to various species of Gram-negative, polarly flagellated, straight to slightly curved, nonsporeforming rods, one of which was later named *Pseudomonas delafieldii* (Davis et al., 1970). A phenanthrene-degrading *A. delafieldii* strain has been isolated from creosote-contaminated soil by enrichment cultures on phenanthrene (Shuttleworth and Cerniglia, 1996).

1. The mineral medium contains (g/l distilled water) KH$_2$PO$_4$, 1.0; NH$_4$NO$_3$, 1.0; MgSO$_4$·7H$_2$O, 0.2; FeSO$_4$·7H$_2$O, 0.01; and CaCl$_2$·2H$_2$O, 0.01; pH 6.8–7.2 (Schatz and Bovell, 1952).

Aznar et al. (1992) have isolated *A. delafieldii* from a hypertrophic freshwater lagoon on casein peptone starch agar and have studied *A. delafieldii* strains isolated from healthy and diseased eels from eel farms. Some isolates of *A. delafieldii* and *A. temperans* have originated from various clinical sources, where they were isolated using routine clinical techniques (Willems et al., 1990). *A. defluvii* was isolated from activated sludge form a wastewater treatment plant in Munich (Schulze et al., 1999a).

The phytopathogenic species can be isolated from infected plant material. Small pieces (about 1 cm²) of diseased tissue are cut from lesions in leaf blades, surface-sterilized (for example, with 70% ethanol (Goto, 1983), 0.5% sodium hypochlorate (Schaad et al., 1978), or 70% ethanol followed by 0.001% mercuric chloride and again 70% ethanol (Hayward, 1962) and rinsed with sterile distilled water. The tissue may be further mashed mechanically. The suspension can then be plated out immediately or left to diffuse at room temperature for several hours before plating out on a suitable medium, such as yeast extract–peptone agar. Plates are incubated at 28–30°C, and colonies usually appear within 2 days (Hayward, 1962; Schaad et al., 1978; Goto, 1983).

Maintenance Procedures

Acidovorax strains can be maintained on nutrient agar at 4°C for up to 2 months. Unlike some autotrophic species, *A. facilis* does not readily lose its H₂-oxidizing capacity when grown heterotrophically over long periods of time or in the presence of 100% O₂ (Schatz and Bovell, 1952).

For long-term preservation, strains can be lyophilized using standard procedures. For example, at the BCCM-LMG (Ghent, Belgium) culture collection this is routinely done in horse serum.

Differentiation of the genus *Acidovorax* from other genera

See Table BXII.β.42 for the family *Comamonadaceae* for features differentiating *Acidovorax* from the other genera of this family.

An oligonucleotide probe for the genus *Acidovorax*, which hybridizes to variable region II of the 16S rRNA gene, has been designed to allow differentiation from species of *Comamonas* and *Hydrogenophaga*, but hybridization conditions are crucial. In the presence of 0–10% formamide, *Acidovorax* strains, and *Variovorax paradoxus*, hybridize with the probe. In the presence of 20% formamide, *Variovorax paradoxus* no longer hybridizes with the probe, but *A. delafieldii* is also negative (Amann et al., 1996).

Taxonomic Comments

Pseudomonas facilis was originally described as "*Hydrogenomonas facilis*", a hydrogen oxidizing isolate from lawn soil (Schatz and Bovell, 1952). When the genus "*Hydrogenomonas*" was later abandoned, "*Hydrogenomonas facilis*" was transferred to the genus *Pseudomonas* (Davis et al., 1969). *Pseudomonas delafieldii* was proposed for a number of strains phenotypically similar to *Pseudomonas facilis* but unable to oxidize hydrogen (Davis et al., 1970). By DNA–rRNA hybridization, it was later shown that the genus *Pseudomonas* was polyphyletic and consisted of five distinct and remotely related groups that could not be maintained in a single genus (De Vos and De Ley, 1983). *Pseudomonas facilis* and *Pseudomonas delafieldii* belong to the acidovorans rRNA complex in the *Betaproteobacteria*. In a polyphasic study of the relationships within this group, it was shown that both species, together with a number of clinical isolates that had been provisionally grouped in the so-called EF groups 13 and 16, form a separate rRNA

subbranch, for which the genus *Acidovorax* was proposed. Initially, three species were proposed: *Acidovorax facilis* for *Pseudomonas facilis*, *Acidovorax delafieldii* for *Pseudomonas delafieldii* and most of the strains of EF group 13, and *Acidovorax temperans* for most of the strains from EF group 16 and several misnamed *Pseudomonas* and *Alcaligenes* strains (Willems et al., 1990).

In an extension of this polyphasic study, several phytopathogenic *Pseudomonas* species were also transferred to the genus *Acidovorax* (Willems et al., 1992a). Several species causing disease on Poaceae have been united in *Acidovorax avenae* subsp. *avenae*. The causal agent of leaf blight on oats in Ohio was first described by Manns (1909) as *Pseudomonas avenae*. A similar disease of foxtail in Arkansas was attributed to a new species "*Pseudomonas alboprecipitans*" (Rosen, 1922). Later, isolates from diseased corn from Georgia and Florida were compared with the descriptions of both species, and Schaad et al. (1975) concluded that "*Pseudomonas alboprecipitans*" was a later synonym of *Pseudomonas avenae*. *Pseudomonas rubrilineans*, the causal agent of red stripe disease on sugarcane, was originally described as *Phytomonas rubrilineans* (Lee et al., 1925), but was later transferred to *Xanthomonas* (Starr and Burkholder, 1942) and then to *Pseudomonas* (Hayward, 1962). Later Ramundo and Claflin (1990) demonstrated the synonymy between *Pseudomonas rubrilineans* and *Pseudomonas avenae* by comparing physiological, biochemical, and serological data. "*Pseudomonas setariae*", the causal agent of brown stripe disease on *Setaria italica*, was first described as "*Bacterium setariae*" (Okabe, 1934), but was later transferred to *Pseudomonas* (Săvulescu, 1947). It was not included on the Approved Lists (Skerman et al., 1980) because the only extant strain was considered a member of *Pseudomonas avenae*.

A second subspecies, *Acidovorax avenae* subsp. *cattleyae*, has been proposed for a pathogen of *Cattleya* orchids (Willems et al., 1992a). This organism was first described as "*Bacterium cattleyae*" by Pavarino (1911); later, brown spot disease on *Cattleya* and *Phalaenopsis* species was attributed to *Phytomonas cattleyae* (Ark and Thomas, 1946). This species was later transferred to *Pseudomonas* (Săvulescu, 1947). Based on a phenotypic analysis and DNA–DNA hybridizations, Hu et al. (1991) considered *Pseudomonas cattleyae* to be a synonym of *Pseudomonas avenae*.

A third subspecies, *Acidovorax avenae* subsp. *citrulli*, has been proposed as the causal agent of leaf blight on watermelon. It had been described as *Pseudomonas pseudoalcaligenes* subsp. *citrulli*, mainly because of its phenotypic similarity to the *Pseudomonas alcaligenes* group (Schaad et al., 1978). Hu et al. (1991) transferred this subspecies to *Pseudomonas avenae* as *Pseudomonas avenae* subsp. *citrulli*.

Acidovorax konjaci has been proposed as a separate species (Willems et al., 1992a) comprising *Pseudomonas pseudoalcaligenes* subsp. *konjaci*, a pathogen isolated from the leaves of the Konjac plant (*Amorphophallus konjac*) (Goto, 1983). In addition, this subspecies was first transferred as a separate subspecies to *Pseudomonas avenae* by Hu et al. (1991).

Thus, the genus *Acidovorax* now contains seven species, which can be separated in two groups: the soil and water inhabitants *A. facilis*, *A. delafieldii*, *A. defluvii*, and *A. temperans*, which also include some opportunistic clinical isolates, and the phytopathogenic species *A. avenae* with its three subspecies, *A. konjaci* (Willems et al., 1992a), and *A. anthurii* (Gardan et al., 2000). This separation into two groups is reflected in the 16S rDNA phylogeny of these organisms (Willems et al., 1992a), but they have been maintained in a single genus because of phenotypic similarities (Willems et al., 1992a). More recently, the 16S rRNA gene sequence analysis of the type strains of the all *Acidovorax* species has confirmed that *Acidovorax* does form a separate phylogenetic cluster within the *Comamonadaceae*, with all species showing at least 95.5% sequence similarity, thus supporting the grouping of the previously described rRNA subbranches into a single genus. DNA–DNA hybridizations do not show any significant DNA-binding among any of the *Acidovorax* species. They demonstrate that *A. avenae* consists of at least three subgroups for which separate subspecies have been proposed (Willems et al., 1992a). DNA-binding values of 54–67% have been obtained among the subspecies, whereas within the subspecies, values of at least 75% are obtained. These three subspecies can also be distinguished by 16S rDNA sequence comparison (Wen et al., 1999).

Acknowledgments

Anne Willems is indebted to the Fund for Scientific Research, Flanders for a position as a postdoctoral research fellow.

List of species of the genus Acidovorax

1. **Acidovorax facilis** (Schatz and Bovell 1952) Willems, Falsen, Pot, Jantzen, Hoste, Vandamme, Gillis, Kersters and De Ley 1990, 394[VP] (*Pseudomonas facilis* (Schatz and Bovell 1952) Davis, Doudoroff, Stanier and Mandel 1969, 385; "*Hydrogenomonas facilis*" Schatz and Bovell 1952, 88.)

   fa'ci.lis. L. adj. *facilis* ready, quick.

   Wilde and Schlegel (1982) have reported strains to be tolerant to 60% O_2 when grown autotrophically and 100% O_2 when grown heterotrophically. Characteristics differentiating *A. facilis* from the other *Acidovorax* species are presented in Table BXII.β.46. Further descriptive information is provided in Table BXII.β.47. Isolated from lawn soil in the United States.

   The mol% G + C of the DNA is: 64–65 (T_m); type strain, 65 (T_m).

   Type strain: ATCC 11228, CCUG 2113, DSM 649, LMG 2193.

   GenBank accession number (16S rRNA): AF078765.

2. **Acidovorax anthurii** Gardan, Dauga, Prior, Gillis and Sadler 2000, 245[VP]

   an.thu'ri.i. L. fem. n. *Anthurium* anthurium; L. gen. n. *anthurii* of anthurium, referring to the plant from which the phytopathogenic bacterium was first isolated.

   The characteristics are the same as those of the genus. This species is not included in Tables BXII.β.46 and BXII.β.47 because it was published after the completion of the chapter. The description below is taken from the original description (Gardan et al., 2000).

   On YBGA, colonies are circular, raised with an entire margin. A white-creamy, and a brown diffusible pigment is produced.

   Catalase and urease are positive. Poly-β-hydroxybutyrate is accumulated in the cell; arginine is used as the sole source of carbon. H_2S is produced from cysteine and cellulose is hydrolysed. Indole, levane, and acetoin are not produced. Casein and esculin are not hydrolyzed. Acid is produced from galactose, arabinose, and glycerol. Acetate, formate, glycerol, DL-5-aminobutyrate, D(−)tartrate, and azelate are utilized. Trehalose, caprylate, D-ribose, D-glucose, N-acetyl-glucosamine, L-arginine, saccharose, inositol, sarcosine, itaconate, D-xylose, L-tryptophan, esculin, and mannitol are not utilized. The sole hydroxylated fatty acid present is 3-hydroxy-decanoic acid.

   All strains elicit a hypersensitive reaction on tobacco leaves (HR) and are pathogenic on anthurium, producing typical leaf-spot symptoms.

   The mol% G + C of the DNA is: 63.5 (T_m).

   Type strain: CFBP 3232, CIP 107058.

   GenBank accession number (16S rRNA): AJ007013.

3. **Acidovorax avenae** (Manns 1909) Willems, Goor, Thielemans, Gillis, Kersters and De Ley 1992a, 115[VP] (*Pseudomonas avenae* Manns 1909, 1933; "*Pseudomonas alboprecipitans*" Rosen 1922, 383.)

   a've.nae. M.L. n. *avena* genus of plants; M.L. gen. n. *avenae* of *Avena*.

   This species contains only phytopathogenic strains, which are divided into three subspecies.

   The mol% G + C of the DNA is: 67–70 (T_m).

   a. **Acidovorax avenae** subsp. **avenae** (Manns 1909) Willems, Goor, Thielemans, Gillis, Kersters and De Ley 1992a, 117[VP] (*Pseudomonas avenae* Manns 1909, 1933; *Pseudomonas avenae* subsp. *avenae* Manns 1909, 1933; "*Pseudomonas alboprecipitans*" Rosen 1922, 383.)

   Characteristics differentiating *A. avenae* subsp. *avenae* from the other *Acidovorax* species are presented in Table BXII.β.46. Further descriptive information is provided in Table BXII.β.47. This subspecies is pathogenic for various species of Poaceae, including oats, corn, wheat, barley, sorghum, rye, sugarcane, rice seedlings, Italian millet, pearl millet, barnyard millet, foxtail millet, finger millet, Proso millet, and Vasey grass. There are phytopathological indications that strains pathogenic to rice are different from strains pathogenic to other plants, supported also by the detection of a protein specific to the rice-pathogenic strains (Kadota et al., 1996).

   The mol% G + C of the DNA is: 68–70 (T_m); type strain, 70 (T_m).

   Type strain: ATCC 19860, LMG 2117, NCPPB 1011.

   GenBank accession number (16S rRNA): AF078759.

   Additional Remarks: The type strain was isolated from *Zea mays* in the United States in 1958.

   b. **Acidovorax avenae** subsp. **cattleyae** (Pavarino 1911) Willems, Goor, Thielemans, Gillis, Kersters and De Ley 1992a, 118[VP] (*Pseudomonas avenae* subsp. *cattleyae* (Pavarino 1911) Hu, Young and Triggs 1991, 524; *Pseudomonas cattleyae* (Pavarino 1911) Săvulescu 1947, 11; "*Bacterium cattleyae*" Pavarino 1911, 234.)

   catt'ley.ae. M.L. fem. n. *Cattleya* a genus of orchids; M.L. gen. n. *cattleyae* of *Cattleya*.

   Characteristics differentiating *A. avenae* subsp. *cattleyae* from the other *Acidovorax* species are presented in Table BXII.β.46. Further descriptive information is provided in Table BXII.β.47. Causes leaf spot and bud rot on *Cattleya* and *Phalaenopsis* orchids.

   The mol% G + C of the DNA is: 69 (T_m).

Type strain: ATCC 33619, LMG 2364, NCPPB 961.
GenBank accession number (16S rRNA): AF078762.

c. **Acidovorax avenae** *subsp.* **citrulli** (Schaad, Sowell, Goth, Colwell and Webb 1978) Willems, Goor, Thielemans, Gillis, Kersters and De Ley 1992a, 118^{VP} (*Pseudomonas avenae* subsp. *citrulli* (Schaad, Sowell, Goth, Colwell and Webb 1978) Hu, Young and Triggs 1991, 524; *Pseudomonas pseudoalcaligenes* subsp. *citrulli* Schaad, Sowell, Goth, Colwell and Webb 1978, 123.)

ci' trul.li. M.L. masc. n. *Citrullus* a genus in the cucumber family; M.L. gen. n. *citrulli* of *Citrullus.*

Characteristics differentiating *A. avenae* subsp. *citrulli* from the other *Acidovorax* species are presented in Table BXII.β.46. Further descriptive information is provided in Table BXII.β.47. Pathogenic for several species of Cucurbitaceae, including watermelon, cantaloupe, cucumber, and squash.

The mol% G + C of the DNA is: 67–68 (T_m); type strain, 68 (T_m).

Type strain: ATCC 29625, ICMP 7500, LMG 5376.

GenBank accession number (16S rRNA): AF078761.

Additional Remarks: The type strain was isolated from a water-soaked lesion on a cotyledon of *Citrullus lanatus* in the United States in 1977.

4. **Acidovorax delafieldii** (Davis *in* Davis, Stanier, Doudoroff and Mandel 1970) Willems, Falsen, Pot, Jantzen, Hoste, Vandamme, Gillis, Kersters and De Ley 1990, 396^{VP} (*Pseudomonas delafieldii* Davis *in* Davis, Stanier, Doudoroff and Mandel 1970, 12.)

de.la.fiel' di.i. M.L. gen. n. *delafieldii* of Delafield, named after F.P. Delafield, who first isolated this organism.

Characteristics differentiating *A. delafieldii* from the other *Acidovorax* species are presented in Table BXII.β.46. Further descriptive information is provided in Table BXII.β.47. Isolated from soil, water, and various samples from clinical origin.

The mol% G + C of the DNA is: 65–66 (T_m); type strain, 66 (T_m).

Type strain: ATCC 17505, DSM 64, LMG 5943.

GenBank accession number (16S rRNA): AF078764.

Additional Remarks: This strain was isolated from soil with PHB as the sole carbon source. Since it is not the most representative of the redefined species, strain CCUG 23830B (LMG 8909) has been proposed as an alternative reference strain for this species (Willems et al., 1990).

5. **Acidovorax defluvii** Schulze, Spring, Amann, Huber, Ludwig, Schleifer and Kämpfer 1999b, 1325^{VP} (Effective publication: Schulze, Spring, Amann, Huber, Ludwig, Schleifer and Kämpfer 1999a, 205.)

de.flu' vi.i. L. n. *defluvium* sewage; L. gen. n. *defluvii* of sewage.

The characteristics are the same as those of the genus. This species is not included in Tables BXII.β.46 and BXII.β.47 because it was published after the completion of the chapter. The description below is taken from the original description (Schulze et al., 1999a).

The following characteristics are positive for all strains: growth at 28°C, storage of polyhydroxybutyrate, and hydrolysis of L-alanine-pNA. The following characteristics are negative for all strains: growth at 37°C or in the presence of 1.5% NaCl. Utilization of the following substrates for growth: N-acetyl-glucosamine, L-arabinose, p-arbutin, D-cellulose, D-fructose, D-galactose, gluconate, D-glucose, D-mannose, D-maltose, α-D-melibiose, L-rhamnose, D-ribose, sucrose, salicin, D-trehalose, D-xylose, adonitol, i-inositol, D-mannitol, D-sorbitol, propionate, azelate, citrate, itaconate, mesaconate, β-alanine, L-aspartate, L-histidine, L-leucine, L-ornithine, L-phenylalanine, L-serine, L-tryptophan, m-hydroxybenzoate, p-hydroxybenzoate, and phenylacetate. Hydrolysis of esculin, pNP-β-D-galactopyranoside, pNP-α-D-glucopyranoside, and pNP-β-glucopyranoside.

The hydroxylated fatty acids 3-hydroxyoctanoic ($C_{8:0\ 3OH}$) and 3-hydroxydecanoic acid ($C_{10:0\ 3OH}$) are present. Isolated from activated sludge of a municipal wastewater treatment plant. The type strain is nonmotile under laboratory conditions, is catalase positive, and reduces nitrate.

The mol% G + C of the DNA is: 62–64 (T_m); value for the type strain, 62.

Type strain: BSB411, DSM12644, CIP 106824.

GenBank accession number (16S rRNA): Y18616.

6. **Acidovorax konjaci** (Goto 1983) Willems, Goor, Thielemans, Gillis, Kersters and De Ley 1992a, 118^{VP} *Pseudomonas avenae* subsp. *konjaci* (Goto 1983) Hu, Young and Triggs 1991, 524; *Pseudomonas pseudoalcaligenes* subsp. *konjaci* Goto 1983, 544.)

kon.ja' ci. M.L. gen. n. *konjaci* of the konjac plant, *Amorphophallus rivieri* cv. Konjac.

Characteristics differentiating *A. konjaci* from the other *Acidovorax* species are presented in Table BXII.β.46. Further descriptive information is provided in Table BXII.β.47. Causal agent of leaf blight of *Amorphophallus rivieri* cv. Konjac, a Japanese food crop.

The mol% G + C of the DNA is: 68 (T_m).

Type strain: ATCC 33996, ICMP 7733, LMG 5691.

GenBank accession number (16S rRNA): AF078760, AF137507.

Additional Remarks: The type strain was isolated from *Amorphophallus rivieri* cv. Konjac.

7. **Acidovorax temperans** Willems, Falsen, Pot, Jantzen, Hoste, Vandamme, Gillis, Kersters, and De Ley 1990, 396^{VP}

tem' pe.rans. L. v. *temperare* to moderate; M.L. pres. part. *temperans* moderate, referring to the moderate metabolic versatility of the species.

Characteristics differentiating *A. delafieldii* from the other *Acidovorax* species are presented in Table BXII.β.46. Further descriptive information is provided in Table BXII.β.47. Isolated from various samples from clinical environments; one strain was isolated from active sludge from a waste water purification plant in Sweden.

The mol% G + C of the DNA is: 62–66 (T_m); type strain 62 (T_m).

Type strain: CCUG 11779, DSM 7270, LMG 7169.

GenBank accession number (16S rRNA): AF078766.

Additional Remarks: The type strain was isolated from a urine sample of a 68-year-old male patient in Göteborg, Sweden.

Genus III. **Brachymonas** *Hiraishi, Shin and Sugiyama 1995b, 879*[VP] *(Effective publication: Hiraishi, Shin and Sugiyama 1995c, 110)*

AKIRA HIRAISHI

Bra.chy.mo′ nas. Gr. adj. *brachy* short; Gr. n. *monas* a unit; M.L. fem. n. *Brachymonas* a small short unit.

Cells are coccobacilli or short rods, 0.6–1.0 × 0.8–1.5 μm, with rounded ends, occurring singly, in pairs, and in chains. Nonsporeforming, nonencapsulated, and nonmotile. Gram negative.

Aerobic chemoorganotrophs having a strictly respiratory type of metabolism, with oxygen as the terminal electron acceptor. **Growth occurs anaerobically with nitrate** as the terminal electron acceptor. Denitrification positive. Liquid cultures and colonies are cream to pale yellow. Grow well in ordinary nutrient media containing peptone or in mineral media supplemented with simple organic compounds as electron donors and carbon sources. **Do not utilize sugars as carbon or energy sources.** Mesophilic, neutrophilic, and nonhalophilic. Catalase and oxidase positive. No hydrolysis of polysaccharides, protein, or lipids.

Straight-chain hexadecanoic acids ($C_{16:0}$ and $C_{16:1}$) are the major components of cellular fatty acids. 3-Hydroxydecanoic acid ($C_{10:0\ 3OH}$) is present. Ubiquinones and rhodoquinones with eight isoprene units (Q-8 and RQ-8) are the major respiratory quinones. Isolated from activated sludge systems, such as soybean-curd wastewater sludge and municipal sewage sludge.

The mol% G + C of the DNA is: 63–65.

Type species: **Brachymonas denitrificans** Hiraishi, Shin and Sugiyama 1995b, 879 (Effective publication: Hiraishi, Shin and Sugiyama 1995c, 110.)

FURTHER DESCRIPTIVE INFORMATION

The description of the genus *Brachymonas* is based on only one species, *B. denitrificans*. Morphological features are as shown in Fig. BXII.β.28. Electron microscopy of negatively stained cells shows that the cells have no flagella. Colonies on agar media are of intermediate size (3–5 mm after 3 d), smooth, circular, convex, and cream to pale yellow. Some strains produce mucoid colonies upon subculture, and it is hard to harvest cells from such colonies by ordinary centrifugation. In liquid cultures with shaking, growth occurs as a uniform cell suspension. No flocculent growth is found at any growth stage.

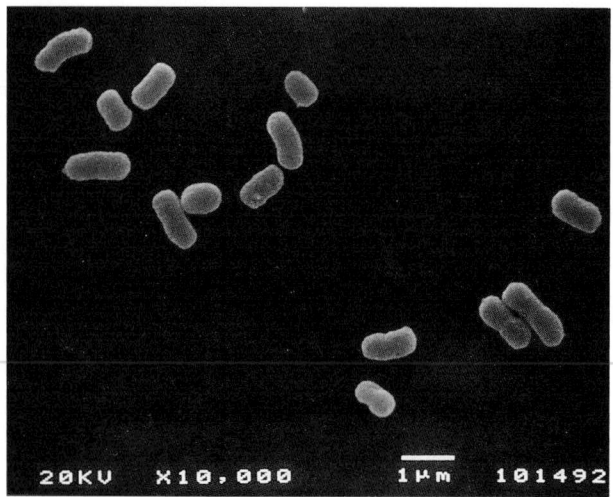

FIGURE BXII.β.28. Scanning electron micrograph showing general cell morphology of *Brachymonas denitrificans* (strain AS-P1[T]).

Cells exhibit a doubling time of 1.5–2.0 h under optimal aerobic growth conditions (in peptone–yeast extract medium or lactate–mineral salt medium). Anaerobic growth with nitrate is somewhat slower (doubling time 2–3 h) and is accompanied with nitrogen gas production. Some strains are also capable of growing anaerobically with trimethylamine-*N*-oxide (TMAO) as the terminal electron acceptor, but growth is much slower (doubling time 9–10 h). No strains grow anaerobically with fumarate as the terminal oxidant, but cell-free extracts from these strains exhibit fumarate-reducing activity with reduced FMN as an electron donor. It is unknown whether this activity is related to the occurrence of the low-potential quinone rhodoquinone or if it has some biological significance. A wide variety of simple organic compounds are used as electron donors and carbon sources for both aerobic and denitrifying growth. Good growth occurs with the salts of organic acids (e.g., acetate, lactate, pyruvate, fumarate, and succinate) and amino acids (e.g., alanine, glutamate, leucine, phenylalanine, and proline). Carbohydrates are never utilized.

The optimal temperature for growth is 30–35°C and the optimal pH is 7.0–7.5. NaCl is not required for optimal growth, but multiplication occurs at NaCl concentrations of up to 3%. Growth factors are not required. Ammonium salts are used as nitrogen sources. Sulfate is assimilated as the sulfur source. The following characteristics are negative: indole production; Voges–Proskauer reaction; urease; phenylalanine deaminase; decarboxylation of arginine, lysine, and ornithine; and hydrolysis of aliginic acid, chitin, starch, gelatin, casein, tributyrin, and Tween 80.

Palmitoleic acid ($C_{16:1}$) and palmitic acid ($C_{16:0}$) are the major components of whole-cell fatty acids, constituting 33–43% and 21–28% of the total content, respectively. In addition, a considerable amount (10–17%) of $C_{18:1}$ acid is present. 3-Hydroxydecanoic acid ($C_{10:0\ 3OH}$) is the main fatty acid component of the outer membranes. The major respiratory quinones are Q-8 and RQ-8. The RQ/Q molar ratio is higher in cells growing in complex media (0.09–0.10) than in those in a chemically defined medium with lactate as the sole carbon source (<0.05). In addition, the RQ/Q ratio is higher in aerobically grown cells than in anaerobically grown cells using nitrate or TMAO as the terminal electron acceptor.

Strains of *B. denitrificans* have hitherto been isolated from activated sludge systems, such as soybean-curd wastewater sludge and municipal sewage sludge (Hiraishi and Komagata, 1989b). In addition, rhodoquinone-producing denitrifying strains that are phenotypically similar to *Brachymonas* have been isolated from photosynthetic wastewater treatment sludge (Hiraishi et al., 1991b). As suggested by their capacity for denitrification, *Brachymonas* strains may play important roles in nitrogen removal and organic matter removal in biological wastewater treatment systems.

ENRICHMENT AND ISOLATION PROCEDURES

Selective isolation and enrichment of *Brachymonas* from the environment are difficult, but the following mineral medium supplemented with 20 mM acetate and 20 mM potassium nitrate is recommended (per liter distilled water): $(NH_4)_2SO_4$, 1 g; KH_2PO_4, 1 g; $MgSO_4 \cdot 7H_2O$, 0.2 g; $CaCl_2 \cdot 2H_2O$, 0.05 g; trace element solution SL8, 1 ml (Biebl and Pfennig, 1978) (pH 6.8). The medium is incubated anaerobically (under denitrifying conditions).

Brachymonas strains grow well in ordinary complex nutrient media, such as PBY medium (0.5% peptone, 0.2% beef extract, and 0.1% yeast extract) (Hiraishi and Komagata, 1989a). Good growth also occurs in the mineral medium described previously with 20 mM lactate or acetate as the sole carbon source.

MAINTENANCE PROCEDURES

Cultures are well preserved in liquid nitrogen or by lyophilization. Preservation can also be done in a mechanical freezer at $-80°C$.

PROCEDURES FOR TESTING SPECIAL CHARACTERS

The presence of both respiratory quinones Q-8 and RQ-8 is a characteristic feature of *Brachymonas* and can be determined by TLC and reverse-phase HPLC (Hiraishi and Hoshino, 1984). In silica-gel TLC, rhodoquinones migrate just behind ubiquinones and are detectable as a pink to purple band under visible light. The type of rhodoquinone homologs, as well as ubiquinones, can be determined by reverse-phase HPLC with an ODS column and methanol–isopropyl ether (3:1, v/v) as the mobile phase.

DIFFERENTIATION OF THE GENUS *BRACHYMONAS* FROM OTHER GENERA

The genus *Brachymonas* can be differentiated from phenotypically and phylogenetically related genera by the features listed in Table BXII.β.48.

TAXONOMIC COMMENTS

Brachymonas strains were isolated from activated sludge and first designated unidentified rhodoquinone-containing chemoorganotrophic bacteria (Hiraishi and Komagata, 1989b). This was the first demonstration of the existence of rhodoquinone-containing chemoorganotrophic bacteria, as, before this study, rhodoquinones had been known to be distributed only in phototrophic purple bacteria among the procaryotes (Hiraishi and Hoshino, 1984; Hiraishi, 1988b). A phylogenetic analysis based on 16S rDNA sequences has shown that *B. denitrificans* strains form a major cluster in the *Betaproteobacteria*, together with members of the genera *Comamonas*, *Variovorax*, and *Rhodoferax* (Hiraishi et al., 1995b). The nearest phylogenetic neighbor is *Comamonas testosteroni* (95% similarity). Also genomic DNA–DNA hybridization studies have shown that *B. denitrificans* forms a genetically coherent group, in that the levels of hybridization among strains of the species are 74–100%. Based on this phylogenetic evidence, together with the chemotaxonomic information, creation of a new genus and species with the name *Brachymonas denitrificans* was proposed for these strains (Hiraishi et al., 1995b). *Brachymonas* may belong to the family *Comamonadaceae*.

DIFFERENTIATION OF THE SPECIES OF THE GENUS *BRACHYMONAS*

The genus *Brachymonas* includes only one species, *B. denitrificans*, at this time. The general characteristics of *B. denitrificans* are indicated in Table BXII.β.49.

List of species of the genus Brachymonas

1. **Brachymonas denitrificans** Hiraishi, Shin and Sugiyama 1995b, 879[VP] (Effective publication: Hiraishi, Shin and Sugiyama 1995c, 111.)

   de.ni.tri′fi.cans. M.L. adj. *denitrificans* denitrifying.

   The characteristics are as described for the genus and indicated in Tables BXII.β.48 and BXII.β.49. The habitat is activated sludge.

   The mol% G + C of the DNA is: 63–65 (HPLC).

   Type strain: AS-P1, JCM 9216.

   GenBank accession number (16S rRNA): D14320.

TABLE BXII.β.48. Differential characteristics of *Brachymonas* and phenotypically and phylogenetically related genera[a]

Characteristic	*Brachymonas*	*Acidovorax*	*Comamonas*	*Hydrogenophaga*	*Variovorax*
Cell shape	Short rods, coccobacilli	Rods	Rods	Rods	Rods
Motility by flagella	−	+	+	+	+
Yellow pigment	d	−	−	+	−
Denitrification	+	−	D	−	−
Growth on glucose	−	D	−	+	+
Major quinone(s)	Q-8, RQ-8	Q-8	Q-8	Q-8	Q-8
Mol% G + C of DNA	63–65	62–66	60–69	65–69	67–69

[a]For symbols see standard definitions; Q-8, ubiquinone-8; RQ-8, rhodoquinone-8.

TABLE BXII.β.49. Other characteristics of *Brachymonas denitrificans*[a]

Characteristic	Result or reaction
Cell size (µm)	0.6–1.0 × 0.8–1.5
Yellow pigment	d
Temperature range for growth (°C)	10–40
pH Range for growth	5–9
Growth with 3% NaCl	+
Growth with 5% NaCl	−
Anaerobic growth with:	
Fumarate, dimethylsulfoxide, HCO$_3^-$/pyruvate	−
Nitrate	+
TMAO	d
N$_2$ gas production from nitrate	+
Growth factor requirement	−
Catalase, oxidase	+
Hydrolysis of starch, alginate, chitin, casein, gelatin, tributyrin, Tween 80	−
Electron donors and carbon sources:	
Ethanol, propylene glycol, acetate, butyrate, lactate, pyruvate, succinate, fumarate, malate, glutarate, alanine, asparagine, glutamate, leucine, phenylalanine, proline	+
Propionate, benzoate	d
L-Arabinose, D-xylose, D-ribose, L-rhamnose, D-glucose, D-mannose, D-fructose, cellobiose, lactose, sucrose, adonitol, mannitol, sorbitol, glycerol, methanol, formate, caproate, caprylate, pelargonate, citrate, malonate, tartrate, glycolate, gluconate, aminobutyrate, arginine, glycine, histidine, lysine, ornithine, tryptophan	−
Major fatty acids	C$_{16:0}$, C$_{16:1}$
3-OH fatty acid	C$_{10:0}$
Major quinones	Q-8, RQ-8
Mol% G + C of DNA	63–65

[a]For symbols see standard definitions; Q-8, ubiquinone-8; RQ-8, rhodoquinone-8.

Genus IV. **Delftia** Wen, Fegan, Hayward, Chakraborty and Sly 1999, 573[VP]

LINDSAY I. SLY, AIMIN WEN AND MARK FEGAN

Delf'tia. M.L. fem. n. *Delftia* referring to the city of Delft, the site of isolation of the type species, and in recognition of the role of Delft research groups in the development of bacteriology.

Cells are **straight to slightly curved rods**, 0.4–0.8 × 2.5–4.1 µm (occasionally up to 7 µm), occurring **singly or in pairs. Motile by means of polar or bipolar tufts of one to five flagella.** Do not produce endospores. Gram negative. **Oxidase and catalase positive.** No fluorescent pigment produced. Poly-β-hydroxybutyrate is accumulated. **Aerobic**, having a strictly respiratory type of metabolism with oxygen as the terminal electron acceptor. Nonfermentative. Reduce nitrate to nitrite. Do not denitrify. **Chemoorganotrophic.** Good growth occurs on media containing organic acids, amino acids, peptone, and carbohydrates. Utilize mannitol and fructose as the sole carbon source, but not glucose. No autotrophic growth with H$_2$. Hydrolyze acetamide. No levan formation from sucrose. Arginine dihydrolase absent. *Meta* cleavage of protocatechuate. Lipase (Tween 80 hydrolysis) positive. **Main polyamines are putrescine and 2-hydroxyputrescine. The major quinone is ubiquinone Q-8**; minor quinones are Q-7 and Q-9. Major fatty acids are hexadecanoic acid (C$_{16:0}$), hexadecenoic acid (C$_{16:1}$), and octadecenoic acid (C$_{18:1}$). 3-hydroxy fatty acids (C$_{10:0\ 3OH}$ and C$_{8:0\ 3OH}$) but not 2-hydroxy fatty acids are present. DNA–rRNA hybridization and 16S rRNA sequence analysis places the genus in the family *Comamonadaceae*. Isolated from soil, sediment, activated sludge, crude oil, oil brine, water, and various clinical samples.

The mol% G + C of the DNA is: 67–69.

Type species: **Delftia acidovorans** (den Dooren de Jong 1926) Wen, Fegan, Hayward, Chakraborty and Sly 1999, 573 (*Pseudomonas acidovorans* den Dooren de Jong 1926, 106; *Comamonas acidovorans* (den Dooren de Jong 1926) Tamaoka, Ha and Komagata 1987, 58.)

FURTHER DESCRIPTIVE INFORMATION

The description of *Delftia* is based on data from a number of studies (Ikemoto et al., 1978b; Palleroni, 1984; De Vos et al., 1985b; Tamaoka et al., 1987; Busse and Auling, 1988; Willems et al., 1989, 1991c, 1992a; Wen et al., 1999).

Growth occurs at 30°C, but not at 4°C and 41°C. Growth occurs in the presence of 0.5 or 1.5% NaCl. No pigments are produced on nutrient agar.

The following organic compounds can be utilized as carbon and energy sources: acetate, acetamide (reported as a variable reaction by Tamaoka et al., 1987 and Willems et al., 1991c), aconitate, adipate, D-alanine (reported as a variable reaction by Willems et al., 1991c), L-alanine (reported as a variable reaction by Willems et al., 1991c), 2-aminobutyrate (reported as a variable reaction by Willems et al., 1991c), δ-aminovalerate, L-aspartate, azelate, butanol, 2,3-butylene glycol, butyrate, caproate (reported as a variable reaction by Willems et al., 1991c), citraconate, citrate (reported as a variable reaction by Willems et al., 1991c), ethanol, D-fructose, fumarate, gluconate, L-glutamate, glutarate, glycerate, glycine (reported as a variable reaction by Willems et al., 1991c), glycolate, hippurate, L-histidine, *m*-hydroxybenzoate, *p*-hydroxybenzoate, β-hydroxybutyrate, hydroxymethylglutarate, isobutyrate, L-isoleucine (reported as a variable reaction by Willems et al., 1991c), isovalerate, itaconate, α-ketoglutarate, kynurenate, L-kynurenine, lactate, levulinate, L-leucine, D-malate (reported as a variable reaction by Willems et al., 1991c), L-malate, maleate (reported as a variable reaction by Willems et al., 1991c), malonate (reported as a variable reaction by Willems et al., 1991c), mannitol, mesaconate, mucate, nicotinate, DL-norleucine (reported as a variable reaction by Willems et al., 1991c), L-norleucine (reported as a variable reaction by Willems et al., 1991c), DL-norvaline, phenylacetate (reported as a variable reaction by Willems et al., 1991c), L-phenylalanine (reported as a variable reaction by Willems et al., 1991c), pimelate, L-proline, n-propanol, propionate, pyruvate, quinate, saccharate, sebacate, suberate, succinate, L(+)-tartrate (reported as a variable reaction by Willems et al., 1991c), m-tartrate, trigonelline (reported as a negative reaction by Willems et al., 1991c), L-tryptophan (reported as a variable reaction by Willems et al., 1991c and Tamaoka et al., 1987), D-tryptophan (reported as a variable reaction by Willems et al., 1991c), L-tyrosine, and valerate.

The following compounds are not utilized: N-acetylglucosamine, adonitol, 3-aminobenzoate, amylamine, anthranilate, D-arabinose, L-arabinose, D-arabitol, L-arabitol, arbutin, L-arginine, benzoate (reported as a variable reaction by Willems et al., 1991c), benzoylformate, benzylamine, betaine, butylamine, caprylate, cellobiose, L-citrulline, creatine, L-cysteine, diaminobutane, dodecane, dulcitol, erythritol, esculin, ethanolamine, ethylamine, ethylene glycol (reported as a positive reaction by Tamaoka et al., 1987), D-fucose (reported as a variable reaction by Willems et al., 1991c), D-galactose, β-gentiobiose, geraniol, glucosamine, D-glucose, methyl-α-D-glucoside, glycogen, heptanoate, hexadecane, histamine, o-hydroxybenzoate, poly-β-hydroxybutyrate, DL-β-hydroxybutyrate, inulin, isophthalate, 2-ketogluconate, 5-ketogluconate, lactose, L-lysine, D-lyxose, maltose, D-mandelate, L-mandelate (reported as a variable reaction by Willems et al., 1991c), D-mannose, methyl-α-D-mannoside, D-melezitose, D-melibiose, naphthalene, L-ornithine, oxalate, pantothenate, pelargonate (reported as a variable reaction by Willems et al., 1991c), phenol, phenylethanediol, phthalate (reported as a variable reaction by Willems et al., 1991c), propylene glycol (reported as a positive reaction by Tamaoka et al., 1987), putrescine, D-raffinose, L-rhamnose, D-ribose, salicin, sarcosine (reported as a variable reaction by Willems et al., 1991c), L-serine, sorbitol, L-sorbose, spermine, sucrose, starch, D(−)-tartrate, terephthalate, testosterone, L-threonine (reported as a variable reaction by Willems et al., 1991c and a positive reaction by Tamaoka et al., 1987), trehalose, tryptamine, D-turanose, urea, L-valine (reported as a variable reaction by Willems et al., 1991c), m-xylitol, D-xylose, L-xylose, and methyl-β-D-xyloside.

Variable utilization among different strains occurs with β-alanine, 2-aminobenzoate, 4-aminobenzoate, 3-aminobutyrate, 4-aminobutyrate, 5-aminobutyrate, α-aminovalerate, amygdalin, 2,3-butylene glycol, caprate, m-erythritol, glycerol, heptanoate, m-inositol, isobutanol, DL-kynurenine, L-methionine, and tagatose.

The following characteristics are absent: growth in the presence of 6.5% NaCl, acid production in 10% lactose, in triple sugar iron medium, and in oxidative-fermentative medium containing D-glucose, D-fructose, D-xylose, maltose, or adonitol; production of H₂S in triple sugar iron medium; hydrolysis of esculin, gelatin, and DNA; indole production, and β-galactosidase activity; hydrolysis of 2-naphthylmyristate, L-valyl-2-naphthylamide, N-benzoyl-DL-arginine-2-naphthylamide, N-glutaryl-phenylalanine-2-naphthylamide, 6-bromo-2-naphthyl-α-D-galactopyranoside, 2-naphthyl-β-D-galactopyranoside, naphthol-AS-BI-β-D-glucuronate, 2-naphthyl-α-D-glucopyranoside, 6-bromo-2-naphthyl-β-D-glucopyranoside, 1-naphthyl-N-acetyl-β-D-glucosaminide, 6-bromo-2-naphthyl-α-D-mannopyranoside, and 2-naphthyl-α-L-fucopyranoside.

ENRICHMENT AND ISOLATION PROCEDURES

The type strain was isolated from soil enriched with acetamide in Delft in the Netherlands in 1926. Since then the bacterium has been shown to be ubiquitous. Strains have been isolated from soil, soil enriched with indole, soil enriched with p-hydroxybenzoate, sediment, activated sludge, sludge enriched with testosterone, crude oil, oil brine, water, and various clinical samples such as urine, pus, and pharyngeal swabs (Willems et al., 1991c). There is increasing evidence of the diverse metabolic capability of Delftia in the environment, particularly with respect to the degradation and mineralization of xenobiotic pollutants such as aniline, 2-chloroaniline, and 3-chloroaniline (Ferschl et al., 1991, Hinteregger et al., 1994; Boon et al., 2001), and the enantiomers of 2-(4-sulfophenyl)butyrate (Schulz et al., 2000).

MAINTENANCE PROCEDURES

Cultures are easily maintained on nutrient agar in the laboratory. Cryogenic storage in liquid nitrogen in nutrient broth containing 10% glycerol, and by freeze drying in glucose peptone broth containing horse serum are successful for long-term preservation of Delftia cultures.

DIFFERENTIATION OF THE GENUS DELFTIA FROM OTHER GENERA

The characteristics for differentiating Delftia from other members of the Comamonadaceae are given in Table BXII.β.50.

TAXONOMIC COMMENTS

Phylogenetically, Delftia belongs to the family Comamonadaceae in the class Betaproteobacteria according to DNA–rRNA hybridization (Willems et al., 1991a) and rDNA sequence analysis (Wen et al., 1999) (Fig. BXII.β.29). The type species Delftia acidovorans belongs to a deep distinct branch. Although the position of the Delftia acidovorans branch is not well supported by bootstrap analysis, D. acidovorans is consistently the only member of the branch that does not cluster with species of Comamonas—where it had been assigned until its transfer to Delftia by Wen et al. (1999). The separation of D. acidovorans from Comamonas is supported by results from DNA–rRNA hybridization (Willems et al., 1991c), 16S rRNA cataloging (Woese et al., 1984a, c), 16S rDNA sequences (Wen et al., 1999), and conventional and chemotaxonomic methods (Tamaoka et al., 1987).

List of species of the genus Delftia

1. **Delftia acidovorans** (den Dooren de Jong 1926) Wen, Fegan, Hayward, Chakraborty and Sly 1999, 573[VP] (*Pseudomonas acidovorans* den Dooren de Jong 1926, 106; *Comamonas acidovorans* (den Dooren de Jong 1926) Tamaoka, Ha and Komagata 1987, 58.)

a.ci.do'vo.rans. L. neut. n. *acidum* acid; L. v. *voro* to devour; M.L. part. adj. *acidovorans* acid devouring.

The characteristics are the same as for the genus and as given in Table BXII.β.50. Strains have been isolated from soil, sediment, activated sludge, crude oil, oil brine, water, and various clinical samples.

The mol% G + C of the DNA is: 67 (T_m).

Type strain: Stanier 14, ACM 489, ATCC 15668, DSM 39, IAM 12409, IMET 10620, JCM 5833, KS 0057, LMG 1226, NCIB 9681.

GenBank accession number (16S rRNA): AF078774, AB021417.

TABLE BXII.β.50. Differential characteristics of the genus *Delftia* and other genera in the family *Comamonadaceae*[a]

Characteristic	*Delftia*	*Comamonas*[b]	*Acidovorax*[c]	*Hydrogenophaga*[d]	*Variovorax*[e]	*Xylophilus*[f]	*Rhodoferax*[g]	*Brachymonas*[h]	*Polaromonas*[i]	*Aquaspirillum*[j]
Cell shape	Rods	Rods or spirilla	Rods	Rods	Rods	Rods	Curved rods	Cocco-bacilli or short rods	Rods	Spirilla or curved rods
Flagella	Polar or bipolar tufts	Polar or bipolar tufts	1–2 polar	1 polar	Peritrichous	1 polar	1 polar	−	1 polar	Bipolar tufts or 1–2 flagella at only 1 pole
Pigments	−	−	−	+	+	+	−	(+)	−	−
Occurrence:										
Soil	+	+	+	+	+	−	−	−	−	−
Fresh water	+	+	+	+	+	−	−	−	−	+
Seawater	−	−	−	−	−	−	−	−	+	−
Infected plants	−	−	+	−	−	+	−	−	−	−
Clinical samples	+	+	+	−	−	−	−	−	−	−
Activated sludge	+	+	+	−			+	+	−	−
Phototrophy	−	−	−	−	−	−	+	−	−	−
Oxidase	+	+	+	+	+	−	+	+	+	+
Chemolithotrophic growth with H_2	−	−	D	+	D	−			−	
Psychrophilic growth	−	−	−	−	−	−	−	−	+	−
Growth factors	−	D	−	−	−	L-Glutamate	Biotin and thiamine	−	−	D
Denitrification	−	−	D	D	−	−	−	+	−	D
Carbon sources:										
Acetamide	d	−	D	−	−					
β-Alanine	d	−	D	−	d			−		
2-Aminobutyrate	d	−	D	D				−		
3-Aminobutyrate	d	D	D	D				−		
D-Fructose	+	−	+	D		−	+	−	−	−
D-Glucose	−	−	D	+	+	+	+	−	+	D
Glycerol	d	D	+	+	+	+	−	−	+	D
Malonate	d	−	D	−	d	−		−	−	D
D-Mannitol	+	−	D	D		−	+	−		
Maleate	d	−	D	−	d	−				
Phenylacetate	d	−	−	−	d					
L-(+)-Tartrate	d	−	−	D	d			−		D
D-Tryptophan	d	D	−	−				−		−
L-Tryptophan	d	−	D	D	d			−		−
Major quinone system:										
Q-8	+	+	+	+	+	+				+
Q-8 + RQ-8							+	+		
Major cellular fatty acid(s):										
$C_{16:0}$	+	+	+	+	+		+	+	+	+
$C_{16:1}$	+	+	+	+	+		+	+		+
$C_{16:1 \, \omega 7c}$									+	
$C_{18:1}$	+	+	+		+					
$C_{18:1 \, \omega 7c}$									+	
$C_{18:1 \, \omega 9t}$									+	
$C_{18:1 \, \omega 12t}$									+	
Major 3-OH acids:										
$C_{8:0}$	+		+	+			+			
$C_{10:0}$	+	+	+	D	+			+		+
Mol% G + C of DNA	67–69	63–66	67–70	65–69	66–68	68–69	59–61 (HPLC)	63–65 (HPLC)	52–57	56–62

[a]Symbols: +, present in all species; −, absent in all species; (+), weak reaction; d, 11–89% of strains are positive; D, variable reaction in different species.

[b]Data from Palleroni (1984), De Vos et al. (1985b), Tamaoka et al. (1987), and Willems et al. (1991c).

[c]Data from Palleroni (1984) and Willems et al. (1990, 1992a).

[d]Data from Palleroni (1984) and Willems et al. (1989).

[e]Data from Kersters and De Ley (1984b), Willems et al. (1991a), and Urakami et al. (1995a).

[f]Data from Bradbury (1984) and Willems et al. (1987, 1991a).

[g]Data from Hiraishi et al. (1991a).

[h]Data from Hiraishi et al. (1995b).

[i]Data from Irgens et al. (1996).

[j]Data from Krieg (1984a), Pot et al. (1992a, b), Hamana et al. (1994), and Sakane and Yokota (1994).

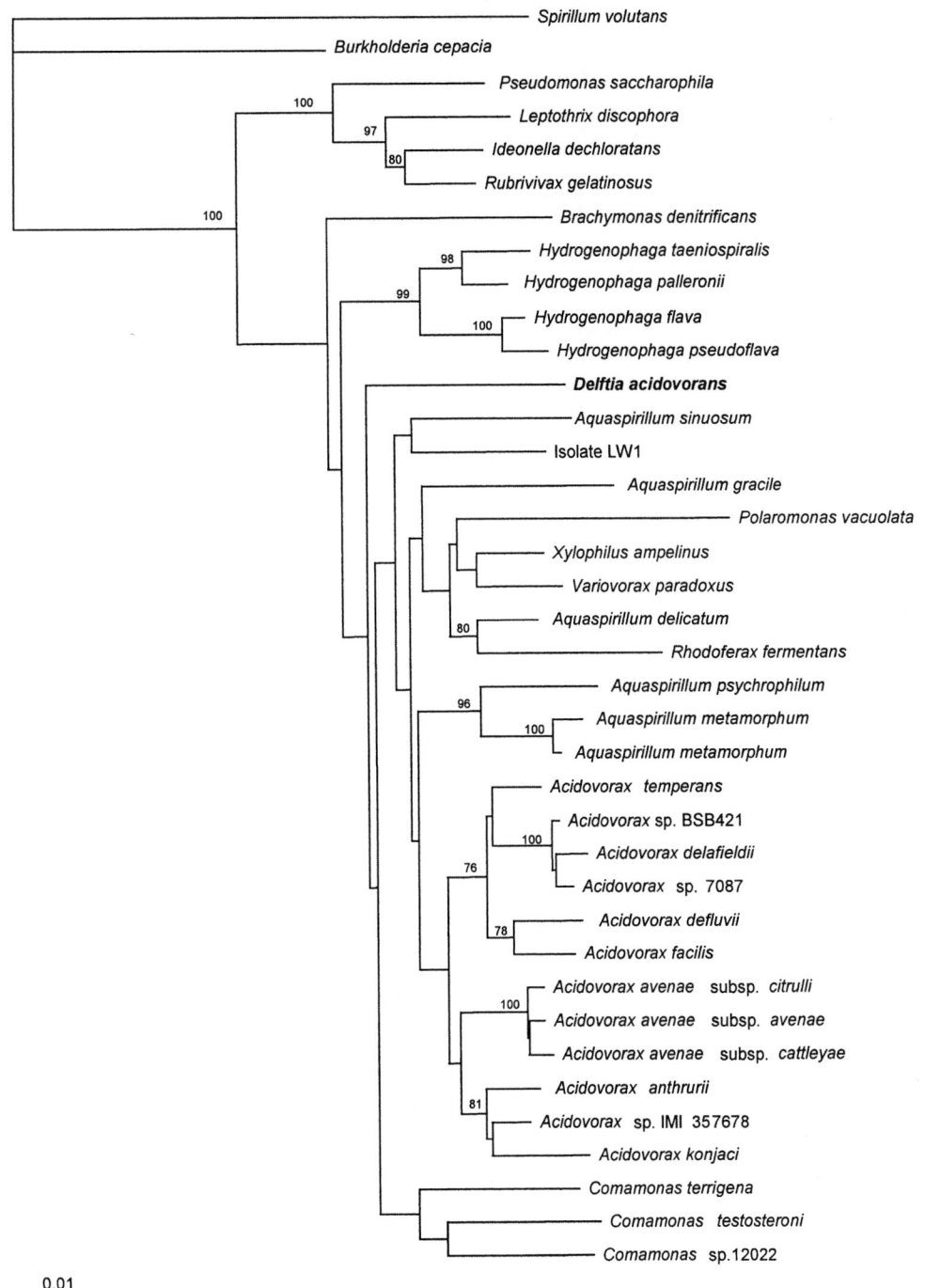

0.01

FIGURE BXII.β.29. Neighbor-joining tree showing the phylogenetic relationships of *Delftia acidovorans* with members of the family *Comamonadaceae* and related species based on 1290 nucleotide positions of their 16S rDNA sequences. *Burkholderia cepacia* was used as the outgroup. Scale bar represents one nucleotide substitution per 100 nucleotides. Bootstrap values of 100 resamplings are shown at the branch points. Only bootstrap values >75% are shown. The following sequences were used to generate this tree: *Acidovorax avenae* subsp. *avenae* ATCC 19860ᵀ, AF078759; *A. avenae* subsp. *citrulli* ATCC 29625ᵀ, AF078761; *A. avenae* subsp. *cattleyae* NCPPB 961ᵀ, AF078762; *A. konjaci* ATCC 33996ᵀ, AF078760; *A. facilis* CCUG 2113ᵀ, AF078765; *A. temperans* CCUG 11779ᵀ, AF078766; *A. delafieldii* ATCC 17505ᵀ, AF078764; *Acidovorax* sp. IMI 357678, AF078763; *Acidovorax* sp. 7087, AF078767; *Acidovorax anthurii* CFBP 3232ᵀ, AJ007013; *Acidovorax defluvii* BSB411ᵀ, Y18616; *Acidovorax* sp. BSB421, Y18617; *Acidovorax* sp. LW1, AJ130765; *Aquaspirillum gracile* ATCC 19624ᵀ, AF078753; *Aquaspirillum metamorphum* LMG 4339ᵀ, AF078757; *Aquaspirillum metamorphum* DSM 1837ᵀ, Y18618; *Aquaspirillum delicatum* LMG 4328ᵀ, AF078756; *Aquaspirillum psychrophilum* LMG 5408ᵀ, AF078755; *Aquaspirillum sinuosum* LMG 4393ᵀ, AF078754; *Brachymonas denitrificans* JCM 9216ᵀ, D14320; *Burkholderia cepacia* ATCC 25416ᵀ, M22518; *Comamonas testosteroni* ATCC 11996ᵀ, M11224; *Comamonas terrigena* IMI 359870ᵀ, AF078772; *Delftia acidovorans* ACM 489ᵀ, AF078774; *Comamonas* sp. 12022, AF078773; *Hydrogenophaga flava* CCUG 1658ᵀ, AF078771; *H. pseudoflava* ATCC 33668ᵀ, AF078770; *H. palleronii* CCUG 20334ᵀ, AF078769; *H. taeniospiralis* ATCC 49743ᵀ, AF078768; *Ideonella dechloratans* ATCC 15173ᵀ, X72724; *Leptothrix discophora* ATCC 43182, Z18533; *Polaromonas vacuolata* ATCC 51984ᵀ, U14585; *Rhodoferax fermentans* JCM 7819ᵀ, D16211; *Rubrivivax gelatinosus* ATCC 17011ᵀ, D16213; *Spirillum volutans* ATCC 19554, M34131; *Variovorax paradoxus* IAM 12373ᵀ, D30793; *Xylophilus ampelinus* ATCC 33914ᵀ, AF078758; *Pseudomonas saccharophila* ATCC 15946ᵀ, unpublished.

Genus V. **Hydrogenophaga** *Willems, Busse, Goor, Pot, Falsen, Jantzen, Hoste, Gillis, Kersters, Auling and De Ley 1989, 329*[VP]

ANNE WILLEMS AND MONIQUE GILLIS

Hy.dro.ge.no'pha.ga. Gr. n. *hydro* water; Gr. n. *gennao* to create; M.L. *hydrogenum* hydrogen, that which produces water; Gr. v. *phagein* to eat; M.L. fem. n. *Hydrogenophaga* eater of hydrogen.

Straight to slightly curved rods, 0.3–0.6 × 0.6–5.5 µm, occurring singly or in pairs. **Gram negative. Motile by means of one or, rarely, two polar to subpolar flagella.** Colonies are yellow due to the presence of **carotenoid pigments. Aerobic. Oxidase positive** and **catalase positive**, except for *H. pseudoflava*, which is catalase variable. **Chemoorganotrophic or chemolithoautotrophic,** using the oxidation of H_2 as an energy source and CO_2 as a carbon source. Oxidative carbohydrate metabolism occurs, with oxygen as the terminal electron acceptor; alternatively, two species (*H. pseudoflava* and *H. taeniospiralis*) are capable of **heterotrophic denitrification** of nitrate. Good growth occurs on media containing organic acids, amino acids, or peptone; there is less versatility in the use of carbohydrates. Cellular lipids contain 3-hydroxyoctanoic acid ($C_{8:0\ 3OH}$) either alone or together with 3-hydroxydecanoic acid ($C_{10:0\ 3OH}$); 2-hydroxy-substituted fatty acids are absent. **Ubiquinone Q-8** is the main quinone. Putrescine and 2-hydroxyputrescine are present in approximately equimolar concentrations, either exclusively or as the dominant polyamines.

The mol% G + C of the DNA is: 65–69.

Type species: **Hydrogenophaga flava** (Niklewski 1910) Willems, Busse, Goor, Pot, Falsen, Jantzen, Hoste, Gillis, Kersters, Auling, and De Ley 1989, 329 (*Pseudomonas flava* (Niklewski 1910) Davis, Doudoroff, Stanier, and Mandel 1969, 385; *Hydrogenomonas flava* Niklewski 1910, 123.)

FURTHER DESCRIPTIVE INFORMATION

Flagellation *Hydrogenophaga* cells have one or, rarely, two flagella with polar or subpolar insertion. Aragno et al. (1977) have defined three types of flagella in Gram-negative bacteria based on fine structural details as revealed by electron microscopy. They have reported the flagellar fine structure of *Hydrogenophaga flava* and *Hydrogenophaga palleronii* to be of type I, with flagella having diameters of 13.5–14.0 nm and 14–16 nm, respectively, and wavelengths of 1.2–1.3 µm and 1.2–2.0 µm, respectively. For *Hydrogenophaga pseudoflava*, the flagellar diameter is 13.5–14.0 nm and the wavelength is 1.4–1.9 µm, but the interpretation of flagellar fine structure is uncertain. Pili occur over the total cell surface in *H. flava* and *H. pseudoflava* and are inserted at the polar caps in *H. palleronii* (Aragno et al., 1977) and *Hydrogenophaga taeniospiralis* (Lalucat et al., 1982).

Cell wall composition Walther-Mauruschat et al. (1977) have distinguished three types of cell walls in Gram-negative bacteria, differing mainly in the visibility and location of the peptidoglycan layer. *Hydrogenophaga* possesses type I cell walls, typical of most Gram-negative bacteria, in which the outer membrane and the cytoplasmic membrane are of similar dimensions and appearance and are separated by a dense peptidoglycan layer (Walther-Mauruschat et al., 1977). Cytochromes of the *a*, *b*, and *c* types are present in the membrane fraction, and cytochrome *c* is found in the soluble fraction of autotrophically or heterotrophically grown cells of *H. pseudoflava*, autotrophically grown cells of *H. taeniospiralis*, and heterotrophically grown cells of *H. palleronii* and *H. pseudoflava* (Auling et al., 1978; Lalucat et al., 1982). In addition, *Hydrogenophaga taeniospiralis* contains cytochrome *a* in the soluble fraction (Lalucat et al., 1982), and *H. pseudoflava*

contains small amounts of cytochrome *d* when grown autotrophically (Auling et al., 1978).

Fine structure Intracellular mesosome-like membrane systems with a spiral appearance, often located in the area of cell division or at the cell poles, have been reported in *H. flava*, *H. pseudoflava*, and *H. palleronii* (Walther-Mauruschat et al., 1977). The significance of these structures is unclear, and there are indications they may be artifacts resulting from preparation of cells for electron microscopy. Intracellular inclusions of poly-β-hydroxybutyrate and polyphosphate are present, and translucent glycogen-like inclusions may also be detected (Walther-Mauruschat et al., 1977; Lalucat et al., 1982). *H. taeniospiralis* has been reported to contain R-bodies, which are inclusion bodies consisting of a proteinaceous ribbon that is rolled up to form a hollow cylinder. The function of these structures remains uncertain. Similar structures have been described in some kappaparticles, the endosymbionts in killer paramecia, where they are thought to be involved in toxin release (Lalucat et al., 1982).

Pigments Colonies on nutrient agar are yellow due to the presence of carotenoid pigments. Their absorption maxima in acetone have been reported to be approximately 405, 425, and 446 nm for *H. pseudoflava* (Auling et al., 1978) and *H. palleronii* (Davis et al., 1970; Auling et al., 1978) and approximately 440 and 465 nm for *H. flava* (Davis et al., 1970). More recently, Urakami et al. (1995a) have reported an absorption maximum of 450 nm for the carotenoid pigments of all *Hydrogenophaga* species except *H. taeniospiralis*, for which no absorption spectrum has been determined.

Nutrition and metabolism A variety of organic compounds can be used as sole carbon sources (Tables BXII.β.51 and BXII.β.52). In general, *H. palleronii* is less versatile in its carbohydrate usage than are the other *Hydrogenophaga* species (Willems et al., 1989). The carbohydrate metabolism of *H. pseudoflava* has been studied, and doubling times in various growth conditions have been determined. Glucose and fructose are degraded via the Entner–Doudoroff pathway and 6-phosphogluconate dehydrogenases are not detected in this species (Lee and Schlegel, 1981). Protocatechuate is cleaved via the *meta* mechanism by *H. palleronii* and *H. flava* (Davis et al., 1970); no data are available for the other two species. No organic growth factors are required. All *Hydrogenophaga* species are facultatively chemolithoautotrophic, using the oxidation of hydrogen as an energy source. Hydrogenase is membrane-bound and does not reduce NAD (Aragno and Schlegel, 1992). Some strains of *H. pseudoflava* are capable of nitrogen fixation and carry genes homologous to *Klebsiella pneumoniae nifHDK* genes. It is therefore assumed that the *nif* genes are located on the chromosome (Jenni et al., 1989). Such nitrogen-fixing strains are isolated under heterotrophic selection conditions with combined nitrogen (Jenni et al., 1989). The strains of *H. pseudoflava* that were previously named "*Pseudomonas carboxydoflava*" are capable of autotrophic growth using CO or hydrogen oxidation as an energy source. These organisms can also grow mixotrophically, using the oxidation of CO or hydrogen as an additional energy source while growing hetero-

TABLE BXII.β.51. Characteristics differentiating *Hydrogenophaga* species[a,b]

Characteristic	H. flava	H. palleronii	H. pseudoflava	H. taeniospiralis
Growth on L-arabinose, sucrose, D-galactose, D-fructose, D-mannose, mannitol, D-arabitol, sorbitol, cellobiose, ethanolamine	+	−	+	+
Growth on maltose	+	−	+	−
Growth on L-mandelate, azelate	−	+	+	+
Growth on D-xylose, 4-aminobutyrate, DL-5-aminovalerate, glutarate, L-ornithine, spermine	−	−	+	+
Growth on lactose	−	−	+	−
Growth on 4-hydroxybenzoate	−	+	+	−
Heterotrophic denitrification	−	−	+	+
Reduction of nitrate	+	−	+	+
Reduction of nitrite	−	−	+	+
Hydrolysis of Tween 80	−	+	+	+
Hydrolysis of gelatin	−	−	−	+
Urease	+	−	−	−
Presence of 3-hydroxy-decanoic acid	+	−	+	−

[a]For symbols see standard definitions.

[b]Data from Lalucat et al. (1982), Palleroni (1984), Jenni et al. (1989), and Willems et al. (1989).

trophically (Kiessling and Meyer, 1982). The CO dehydrogenase from *H. pseudoflava* is a molybdenum-containing iron–sulfur flavoprotein with molybdopterin cytosine dinucleotide as the molybdenum cofactor (Tachil and Meyer, 1997). Urothine, which is excreted by *H. pseudoflava*, is produced through the degradation of this cofactor (Volk et al., 1994). *H. pseudoflava* produces poly(3-hydroxybutyrate) when grown on glucose, xylose, or arabinose as the sole carbon source. With propionic acid as a cosubstrate, co-polymers of 3-hydroxybutyric and 3-hydroxyvaleric acids are produced (Bertrand et al., 1990). When grown on combined substrates of glucose and lactones, *H. pseudoflava* strains also accumulate large amounts of copolyesters. The copolyester produced depends on the lactone cosubstrate provided: γ-butyrolactone yields poly(3-hydroxybutyrate-co-4-hydroxybutyrate), while γ-valerolactone yields poly(3-hydroxybutyrate-co-3-hydroxyvalerate) (Choi et al., 1995b). *H. palleronii* strain S1 has been reported to break down 4-aminobenzenesulfonate (sulfanilate) in coculture with *Agrobacterium radiobacter* strain S2. Strain S1 fails to grow as an axenic culture in a mineral medium with the same carbon source, but does grow on this substrate after addition of either a sterile supernatant from strain S2 or a combination of 4-aminobenzoate, biotin, and vitamin B$_{12}$ (Dangmann et al., 1996).

Other biochemical characteristics *Hydrogenophaga* species possess a Q-8 ubiquinone system (Willems et al., 1989; Urakami et al., 1995a). Putrescine and hydroxyputrescine are the main polyamines and are present in approximately equimolar amounts; in addition, *H. pseudoflava* contains small amounts of spermidine and spermine (Willems et al., 1989). Major fatty acids (representing at least 7% of total fatty acids) are palmitic acid (C$_{16:0}$), palmitoleic acid (C$_{16:1}$), and *cis*-vaccenic acid (C$_{18:1\ \omega7c}$). No 2-hydroxy substituted fatty acids are present, but all species contain small amounts (1–5%) of 3-hydroxyoctanoic acid (C$_{8:0\ 3OH}$), and *H. flava* and *H. pseudoflava* contain small amounts (1–6%) of 3-hydroxydecanoic acid (C$_{10:0\ 3OH}$) (Willems et al., 1989; Urakami et al., 1995a). In addition, *H. pseudoflava*, *H. taeniospiralis*, and *H. palleronii* contain about 2%, 5%, and 10–20 %, respectively, of a cyclopropane substituted fatty acid (C$_{17:0\ cyclo}$) (Willems et al., 1989).

Plasmids Two plasmids have been found in *"Pseudomonas carboxydoflava"* (Gerstenberg et al., 1982; Kraut and Meyer, 1988), now incorporated into *H. pseudoflava*. Plasmids have not been detected in *H. pseudoflava* strains previously classified as *Pseudomonas pseudoflava* (Gerstenberg et al., 1982; Jenni et al., 1989) or in *H. palleronii* (Gerstenberg et al., 1982).

Phages Auling et al. (1978) have tested susceptibility of *H. pseudoflava*, *H. palleronii*, and *H. flava* strains to a number of temperate phages that were isolated originally from autotrophically grown *Pseudomonas pseudoflava* strains. The bacteriophages have a very restricted host range, showing no effects on *H. flava*, *H. palleronii*, *Variovorax paradoxus*, or *Ralstonia eutropha*. Lalucat et al. (1982) have reported that old heterotrophically-grown cultures of *H. taeniospiralis*, when treated with mitomycin C, occasionally contain phage tails and empty phage-head-like particles, and even complete phages with contracted tails in one instance. Suspensions of these structures do not have a lytic effect on any of the four *Hydrogenophaga* species, nor on *V. paradoxus*. The use of ultraviolet light or a temperature shift does not induce the formation of similar phage-like structures. The presence of defective phages in *H. taeniospiralis* may be seen as an argument in favor of the hypothesis that the characteristic R-bodies present in this species may be coded by defective phages, in analogy to a hypothesis put forward to explain the presence of R-bodies in *Caedibacter*, obligate endosymbionts of certain species of paramecia (Lalucat et al., 1982).

Antibiotic sensitivity Most *Hydrogenophaga* species are susceptible to a wide range of antibiotics and bacteriostatic agents (Auling et al., 1978). Table BXII.β.52 contains a detailed listing of antibiotic sensitivities for all species, except *Hydrogenophaga taeniospiralis*, for which few such data are available.

Ecology *Hydrogenophaga* species occur in water, mud, and soil, from which they are usually isolated using enrichment procedures for autotrophic hydrogen oxidizers. All strains isolated are initially capable of autotrophic growth, but some species, such as the type species *Hydrogenophaga flava*, may lose this ability. Nitrogen fixation has been reported in some strains of *H. pseudoflava*, but not in the type strain. *Hydrogenophaga* strains are among the bacteria detected in biofilms in a sand filter used for denitrification of treated waste water and occur in both aerobic and anoxic conditions (Lemmer et al., 1997). *Hydrogenophaga* strains are present in the culturable microbial flora isolated from a waste-oil contaminated site during the course of a 2-year bio-

TABLE BXII.β.52. Additional characteristics of *Hydrogenophaga* species[a, b]

Characteristic	H. flava	H. palleronii	H. pseudoflava	H. taeniospiralis
Tolerance to 20% oxygen	−	+	+	nd
Aerobic autotrophic growth with CO	nd	nd	d	−
Growth at:				
37°C	nd	nd	+	+[c]
40°C	nd	nd	+	+
41°C	−	−	+	nd
42°C	nd	nd	−	−
Use of ammonia as sole nitrogen source	+	+	+	nd
Nitrogen fixation	−	−	d	−
Hydrolysis of acetamide, casein, DNA	nd	nd	−	−
Hydrolysis of esculin	+	nd	−[d]	−[d]
Lysine and ornithine decarboxylase	nd	nd	−	
Production of pyoverdins or phenazine pigments	−	−	−	nd
β-Galactosidase	nd	nd	+	+
Hemolysis	nd	nd	−	−
Growth on:				
Trehalose	+	−	+	+[c]
D-Turanose	+	−	+[c]	−
L-Rhamnose	+	−	v	+
L-Fucose	+	−	−[e]	+
D-Ribose, D-fucose	−	−	−[e]	−
D-Arabinose	−[d,f]	−	−[e]	−
D-Lyxose	nd	nd	d	−
D-Melibiose, D-raffinose, D-tagatose, L-sorbose, L-xylose, melezitose, β-gentiobiose, dulcitol, L-arabitol, xylitol, glycogen, α-methyl-D-glucoside, α-methyl-D-mannoside, α-methylxyloside	nd	nd	−	−
Salicin	−[d,f]	−	+	−
Amygdalin, arbutin, esculin	nd	nd	−[e]	−
Ethanol	+	+	v[f]	nd
n-Propanol	+[g]	+	v[e]	nd
n-Butanol	−	+	v[c]	nd
Isobutanol	−	+	nd	nd
Methanol, phenylethanediol	−	−	nd	nd
Ethylene glycol, 2,3-butylene glycol	−	−	−	nd
Propylene glycol	+	+	nd	nd
Phenol	−	+	+	nd
Testosterone	−	−	−	nd
Acetate	+	+[c,g]	+[e]	−
Propionate	−	−[e,h]	−[f]	−
Butyrate	−[f]	+	v[d,f]	−[d,f]
Isobutyrate	−	−[f]	−	−
Formate	−	−[f]	−	+
Valerate	−	−[e,h]	v[f]	−
Isovalerate	−	−[d,f]	v[c]	−
2-Ketoglutarate	+[c,g]	+[c,g]	v[e]	+
DL-Glycerate	+	+	v[d,f]	+
D(−)-Tartrate	+	+	v[c]	−
L(+)-Tartrate	+[c,g]	+[c,g]	v	−
meso-Tartrate	−	+	−[e]	−
3-Hydroxybutyrate	+	+	+	nd
Glycerate	+	+	nd	nd
Glycolate	−[d]	+	v[f]	−
Citrate	−	+[c,g]	+[e]	−
Aconitate	−	+[c,g]	+[c,g]	−
Levulinate	−[d,f]	−	+	+
m-Hydroxybenzoate	+[g]	+	d	−
Suberate	−	+	v[d,f]	+
Saccharate	−	+	v	nd
Adipate	−	d	+[c]	−
Caproate, oxalate, phthalate, 2-ketogluconate, citraconate, benzoate	−	−	−[e]	−
5-Ketogluconate	nd	nd	−[e]	−
Quinate	−	−	+	nd
Kynurenate, mucate	−	+	nd	nd
Benzoylformate	−	d	nd	nd
Terephthalate	−	−[h]	−	−
Itaconate		d	−	−
Pimelate	−	−[e,h]	v[c,f]	−
Sebacate	−	−[d,f]	+	+
Anthranilate, hippurate, hydroxymethylglutarate, nicotinate, pantothenate, poly-β-hydroxybutyrate	−	−	−	nd

(continued)

TABLE BXII.β.52. *(cont.)*

Characteristic	*H. flava*	*H. palleronii*	*H. pseudoflava*	*H. taeniospiralis*
Eicosanedioate	−	−	nd	nd
L-Phenylalanine	+	+[e,h]	+	+
L-Tyrosine	+[e]	+[e,h]	+	+
L-Alanine	+[c,g]	+[e,h]	+[e]	−
D-Alanine	−	+[e,h]	+	−
L-Leucine, L-tryptophan	−	+[e,h]	+	+
L-Isoleucine	−[d,f]	d	+	+
L-Histidine	−[d,f]	−	+	−
L-Threonine, citrulline	−	−	−[e]	−
L-Lysine	−	−	+	+[c,g]
L-Aspargine	−	+	+	nd
Glycine	−	−	v[c,g]	−
L-Arginine	−	−	v	+
L-Valine	−	−	v	−
β-Alanine	−	−	d[c]	−
DL-2-Aminobutyrate	−	−	d	−
DL-3-Aminobutyrate	−	−	−[e]	+
L-Serine, benzylamine	−	−	d	−
DL-Arginine, DL-citrulline, DL-ornithine, methylamine, DL-2-aminovalerate	−	−	nd	nd
L-Cysteine, L-methionine, DL-kynurenine, ethylamine, glucosamine, trigonelin, urea	nd	nd	−	−
L-Norleucine	nd	d	d	−
DL-Norvaline	nd	nd	d	−
Butylamine	+	−	d[d]	+
L-Kynurenine	−	−	−	nd
3-Aminobenzoate, 4-aminobenzoate	−	−	−	−
2-Aminobenzoate, phenylacetate	nd	nd	−	−
N-Acetyl glucosamine	nd	nd	−[e]	−
Amylamine	nd	nd	+	−
Diaminobutane	nd	nd	+	+
Cetrimide	nd	nd	−	−
Growth in the presence of:				
NaCl, 0.5%	nd	nd	+	−
NaCl (1.5%, 3%, 4.5%, 6.5%)	nd	nd	−	−
Aerobic acid production from:				
D-Glucose, maltose	+	−	+[c]	−
Adonitol	nd	nd	−	−
Saccharose, trehalose, cellobiose	+	−	+	nd
D-Fructose	+	−	+[c]	−
Arabinose	+	−	+	−
Mannose, galactose, rhamnose	+	−	d	nd
Xylose	−	−	d	−
Fucose	−	−	−	nd
Growth inhibited by:				
SDS, 100 μg/ml	+	+	+	nd
SDS, 50 μg/ml	−	+	−	nd
5% NaCl, 3% Glycine	+	+	+	nd
2% Glycine	−	+	−	nd
3% Tween 20	+	+	+	nd
3% Tween 40, 3% Tween 80	+	+	−	nd
Sodium azide 100 mg/ml	+	+	+	nd
2% Methionine	+	+	−	nd
Susceptibility to:[i]				
Chloramphenicol (30 μg), erythromycin (15 μg), neomycin (30 μg), novobiocin (30 μg), streptomycin (10 μg), tetracycline (30 μg)	+	+	+	nd
Ampicillin (10 μg), kanamycin (30 μg), methicillin (5 μg), polymyxin B (300 U)	+	d	+	nd
Penicillin (10 μg)	nd	nd	+	+

[a]For symbols see standard definitions; nd, not determined.

[b]Data taken from Auling et al. (1978), Aragno and Schlegel (1992), Davis et al. (1969, 1970), Palleroni (1984), and Willems et al. (1989). All strains can grow at 30°C and use nitrate as sole nitrogen source and grow on the following substrates: D-malate, DL-β-hydroxybutyrate, fumarate, pyruvate, succinate, gluconate, lactate, glucose, glycerol, m-inositol, L-aspartate, L-glutamate, L-malate, and L-proline. None of the strains hydrolyzes starch or poly-β-hydroxybutyrate, possesses arginine dihydrolase or produces acid from ribose. All strains fail to grow on caprate, caprylate, heptanoate, maleate, malonate, mesaconate, o-hydroxybenzoate, pelargonate, D-mandelate, isophthalate, adonitol, erythritol, inulin, starch, D-tryptophan, creatine, betaine, histamine, acetamide, sarcosine, and tryptamine.

[c]Negative according to Willems et al. (1989).

[d]Positive according to Willems et al. (1989).

[e]Mixed reaction reported by Aragno and Schlegel (1992).

[f]Positive according to Aragno and Schlegel (1992).

[g]Negative according to Aragno and Schlegel (1992).

[h]Mixed reaction reported by Willems et al. (1989).

[i] +, inhibition zone >5 mm; −, inhibition zone <2 mm; d, inhibition zone 2–5 mm.

remediation program (Kämpfer et al., 1993). The mineralization and colonization of palmitic acid by *H. pseudoflava* has been studied to investigate the role of bacterial colonization in the degradation of water-insoluble organic compounds (Thomas and Alexander, 1987). In a study on the influence of temperature on the growth rate of and competition among psychrotolerant Antarctic bacteria, *H. pseudoflava* was used as a model organism in a coculture with a *Brevibacterium* strain (Nedwell and Rutter, 1994). Studies of the biphenyl and polychlorinated biphenyl (PCB)-degrading bacteria isolated from a PCB-contaminated site have identified *H. pseudoflava* as a minor component of the microflora dominated by *Comamonas testosteroni* (Joshi and Walia, 1995). A root-growth promoting effect is attributed to *H. pseudoflava* strains occurring in the rhizosphere of hybrid spruce seedlings (Chanway and Holl., 1993).

ENRICHMENT AND ISOLATION PROCEDURES

Facultatively chemolithotrophic H_2-oxidizing strains of *Hydrogenophaga* can be isolated from soil, mud, or water by enrichment in a mineral medium, when incubated under an atmosphere composed of H_2, O_2, and CO_2. Many of the earlier isolates were obtained using the liquid basal mineral medium[1] described by Palleroni and Doudoroff (1972). After inoculation, cultures are incubated at 30°C under an atmosphere of $H_2/O_2/CO_2/N_2$ (50:4–20:5:25–41) (Davis et al., 1970). A varying proportion of *H. pseudoflava* strains isolated under these conditions may be capable of N_2 fixation, despite the presence of organic nitrogen in the medium (Jenni et al., 1989). Nitrogen-fixing *H. pseudoflava* strains have also been isolated from nitrogen-limited soils that were flushed with a gas mixture composed of $H_2/O_2/CO_2/N_2$ (10:5:55:30) (Aragno and Schlegel, 1992). Heterotrophic strains can be isolated from soil by enrichment with pantothenate (Davis et al., 1970) or poly-β-hydroxybutyrate (Delafield et al., 1965). Other strains have been isolated from soil and water samples using various complex chemicals for enrichment (Komagata et al., 1997; Suyama et al., 1998a; Thomas et al., 1998).

MAINTENANCE PROCEDURES

Hydrogenophaga strains can be maintained under autotrophic or heterotrophic conditions, but autotrophic properties may be lost under a prolonged heterotrophic regime, and it may therefore be advisable to maintain cultures under autotrophic conditions. In particular, *H. flava* has been reported to have lost its capacity for autotrophic hydrogen oxidation (Kluyver and Manten, 1942). In general, autotrophic cultures of hydrogen-oxidizing bacteria remain viable for up to 6 months at 4°C, the exception being *H. pseudoflava*, which requires transferring about every 2 months (Aragno and Schlegel, 1992). For heterotrophic growth, convenient media, such as nutrient agar, can be used; for the maintenance of specific metabolic properties, special selective media may be required.

Hydrogenophaga strains can be lyophilized in skim milk in the presence of a suitable cryoprotectant, such as 5% glutamate, 5% *meso*-inositol, or 10% honey (Aragno and Schlegel, 1992). For preservation of the autotrophic properties, autotrophically grown cells should be used for lyophilization. The use of activated charcoal (5%, w/v) in the suspension medium, together with skim milk (20%, w/v) and raffinose or *meso*-inositol (5% w/v),

is also reported to result in good recovery of *Hydrogenophaga* strains after lyophilization (Malik, 1990b).

DIFFERENTIATION OF THE GENUS *HYDROGENOPHAGA* FROM OTHER GENERA

See Table BXII.β.42 for the family *Comamonadaceae* for features differentiating *Hydrogenophaga* from other genera.

An oligonucleotide probe for the genus *Hydrogenophaga*, located in variable region II of the 16S rRNA gene, has been designed and shown to allow differentiation from the species of *Comamonas* and *Acidovorax* when hybridized in the presence of 30% formamide (Amann et al., 1996).

TAXONOMIC COMMENTS

The genus *Hydrogenophaga* consists of four species previously classified in the genus *Pseudomonas*. Earlier still, one of those species, *Pseudomonas flava*, was classified in *Hydrogenomonas*, a large genus containing Gram-negative, facultatively autotrophic, hydrogen bacteria. Davis et al. (1969) proposed that this genus should be abandoned and transferred the various species to other genera. The yellow-pigmented *Hydrogenomonas flava* was transferred to the genus *Pseudomonas*. Several other species of yellow-pigmented hydrogen oxidizers have subsequently been described in the genus *Pseudomonas*. Davis et al. (1970) created *Pseudomonas palleronii*, which differed from *Pseudomonas flava* in its carotenoid pigments and its inability to grow on a number of carbohydrates. Auling et al. (1978) described *Pseudomonas pseudoflava*, a species highly related to *Pseudomonas flava*, and Lalucat et al. (1982) described *Pseudomonas taeniospiralis*, a species containing characteristic R-bodies. By DNA–rRNA hybridizations, the genus *Pseudomonas* was later shown to consist of five distinct groups, which were only remotely related and could not be maintained in a single genus (Palleroni et al., 1973; De Vos and De Ley, 1983). This initiated a gradual removal of species from this genus as new data became available, leading to a classification reflecting phylogenetic relationships. The yellow hydrogen-oxidizing *Pseudomonas* species belonged to the acidovorans rRNA complex, together with a number of other taxa. Based on genotypic, phenotypic, and chemotaxonomic data, it was shown that they were more closely related to each other than to the other members of the acidovorans rRNA complex, and, therefore, a new genus *Hydrogenophaga* was proposed to accommodate them (Willems et al., 1989). More recently, analysis of the 16S rRNA genes of members of the *Comamonadaceae* has confirmed that the genus *Hydrogenophaga* forms a separate cluster in this family (Wen et al., 1999).

Several strains or groups of strains very similar to, but different from, *H. pseudoflava* have been described. They may represent additional separate species, but so far have not been formally named. For instance, Davis et al. (1969) have described *Pseudomonas* strain 450, isolated from soil in Berkeley, California, as being similar to, but phenotypically distinct from, *Pseudomonas flava*. The main differences are its ability to grow autotrophically in the presence of 20% O_2 and its use of a wider variety of carbohydrates for growth. These characteristics also distinguish *H. flava* from *H. pseudoflava* (Auling et al., 1978). A comparison of the metabolic properties of strain 450 and *H. pseudoflava* shows that they are quite similar, the main differences being the absence of denitrification in strain 450 and its inability to use *p*-hydroxybenzoate, phenol, quinate, D-alanine, or L-tryptophan. Auling et al. (1978) did not include this strain in their study when they described *Pseudomonas pseudoflava*, so the precise classification of strain 450 remains uncertain.

---

1. Basal mineral medium consists of 0.033 M Na-K phosphate buffer, pH 6.8; 0.1% NH_4Cl; 0.05% $MgSO_4 \cdot 7H_2O$; 0.005% ferric ammonium citrate; and 0.0005% $CaCl_2$ (Palleroni and Doudoroff, 1972).

DNA–DNA hybridizations have confirmed that *H. flava* and *H. pseudoflava* are highly related species, with levels of DNA-binding of 48–62%, whereas *H. palleronii* and *H. taeniospiralis* are clearly separate species, each showing no significant DNA-binding with the other *Hydrogenophaga* species (Willems et al., 1989). No significant DNA binding has been detected between representative strains of *Hydrogenophaga* species and *Variovorax paradoxus*, another yellow-pigmented, hydrogen-oxidizing genus of the *Comamonadaceae* (Willems et al., 1991a). In line with these data, comparison of the 16S rRNA gene sequences of the type strains of the *Hydrogenophaga* species demonstrates that *H. flava* and *H. pseudoflava* are highly related (sequence similarity 99.1%; Wen et al., 1999).

Several strains similar to *Pseudomonas pseudoflava* have been isolated from water of a freshwater eutrophic lake by membrane filtration and incubation on mineral medium containing NH_4Cl and pyruvate under an atmosphere of H_2 and O_2. Most of these strains are capable of N_2 fixation, and by phenotypic comparison and DNA–DNA hybridization, they are similar to, but distinct from, *Pseudomonas pseudoflava* (Jenni et al., 1989). Comparison of published phenotypic data for these strains (Aragno and Schlegel, 1992) with those for strain 450, *H. flava*, and *H. pseudoflava* shows this group to be distinguishable. More genotypic data are required to assess whether it represents a separate species.

ACKNOWLEDGMENTS

Anne Willems is indebted to the Fund for Scientific Research–Flanders for a position as a postdoctoral research fellow.

FURTHER READING

Aragno, M. and H.G. Schlegel. 1992. The mesophilic hydrogen-oxidizing (Knallgas) bacteria. *In* Balows, Trüper, Dworkin, Harder and Schleifer (Editors), The Prokaryotes. A Handbook on the Biology of Bacteria: Ecophysiology, Isolation, Identification, Applications., 2nd Ed., Vol. 1, Springer-Verlag, New York. pp. 344–384.

DIFFERENTIATION OF THE SPECIES OF THE GENUS *HYDROGENOPHAGA*

Characteristics useful in differentiating the various species of *Hydrogenophaga* are listed in Table BXII.β.52.

List of species of the genus Hydrogenophaga

1. **Hydrogenophaga flava** (Niklewski 1910) Willems, Busse, Goor, Falsen, Jantzen, Hoste, Gillis, Kersters, Auling and De Ley 1989, 329[VP] (*Pseudomonas flava* (Niklewski 1910) Davis, Doudoroff, Stanier and Mandel 1969, 385; *Hydrogenomonas flava* Niklewski 1910, 123.)

   fla'va. L. fem. adj. *flava* yellow.

   The characteristics are as described for the genus and as listed in Tables BXII.β.51 and BXII.β.52. Optimal temperature, 30°C.

   The only available strain was isolated from mud from a ditch.

   The mol% G + C of the DNA is: 66.7 (T_m).and 67.3 (Bd).

   Type strain: ATCC 33667, CCUG 1658, DSM 619, LMG 2185.

   GenBank accession number (16S rRNA): AF078771.

   Additional Remarks: This strain has been reported to have lost its ability to grow autotrophically soon after isolation (Kluyver and Manten, 1942) and to grow rather slowly and unreliably (Auling et al., 1978). Therefore, *Hydrogenophaga pseudoflava*, a genotypically and protein electrophoretically very similar species, has been suggested as an alternative reference species for the genus (Willems et al., 1989). However, in later experiments the *H. flava* LMG 2185[T] did grow well autotrophically using H_2 as its energy source (A. Willems, unpublished observation).

2. **Hydrogenophaga intermedia** Contzen, Moore, Blümel, Stolz and Kämpfer 2001, 793[VP] (Effective publication: Contzen, Moore, Blümel, Stolz and Kämpfer 2000, 492.)

   in.ter.me'di.a. L. neut. adj. *intermedium* intermediate, referring to the phylogenetic position of this organism within the genus *Hydrogenophaga*.

   The characteristics are the same as those of the genus. This species is not included in Tables BXII.β.51 and BXII.β.52 because it was published after the completion of the chapter. The description below is taken from the original description (Contzen et al., 2000).

   H. intermedia is not able to grow chemolithoautotrophically with H_2. The overall polar lipid pattern is characterized by nearly equal amounts of phosphatidylglycerol, phosphatidylethanolamine, and diphosphatidylglycerol. The fatty acid pattern is typical of the genus *Hydrogenophaga* and includes 3-hydroxyoctanoic acid ($C_{8:0\ 3OH}$). The presence of $C_{19:0\ cyclo\ \omega8-9}$ differentiates *H. intermedia* from other *Hydrogenophaga* species.

   Colonies are circular, entire, slightly convex, smooth, and pale yellow on YPG agar at 25°C. Growth at 40°C, but not at 10°C. The following compounds can be used as sole carbon sources: gluconate, D-mannitol, adipate, lactate, 3-hydroxybutyrate, 2-oxoglutarate, suberate, L-leucine, phenylalanine, L-proline, 3-hydroxybenzoate, 4-hydroxybenzoate, and phenylacetate. Hydrolysis of L-alanine-pNA positive. The type strain does not utilize: *N*-acetyl-D-glucosamine, L-arabinose, arbutin, D-cellobiose, D-fructose, D-galactose, D-glucose, maltose, D-mannose, D-melibiose, L-rhamnose, ribose, sucrose, salicin, trehalose, D-xylose, adonitol, inositol, sorbitol, putrescine, acetate, propionate, *cis*-aconitate, *trans*-aconitate, 4-aminobutyrate, azelate, citrate, fumarate, glutarate, itaconate, D-malate, mesaconate, pyruvate, L-alanine, β-alanine, L-aspartate, L-histidine, L-leucine, L-ornithine, L-serine, and L-tryptophan as sole carbon sources. Hydrolysis of the following substrates is negative: pNP-α-D-glucopyranoside, pNP-β-D-glucopyranoside, pNP-β-D-galactopyranoside, pNP-β-D-glucuronide, pNP-phenylphosphonate, pNP-phosphorylcholine, 2-deoxythymidine-5′-pNP-phosphate, glutamate-γ-3-carboxy-pNP-ester, and L-proline-pNP.

   Features for the differentiation from other *Hydrogenophaga* species include the use of mannitol (negative in *H. palleroni*) and the inability to use the following substrates: L-arabinose, sucrose, D-galactose, D-fructose, D-mannose, sorbitol, and D-cellobiose (positive for *H. flava*, *H. pseudoflava* and *H. palleroni*), maltose and L-histidine (positive for *H. flava* and *H. pseudoflava*), azelate (positive for *H. pseudoflava*, *H. taeniospiralis* and *H. palleroni*) and D-xylose (positive for *H. pseudoflava* and *H. taeniospiralis*).

   The mol% G + C of the DNA is: 68.6 ± 0.1 (HPLC).

   Type strain: S1, DSM 5680.

   GenBank accession number (16S rRNA): AF019037.

3. **Hydrogenophaga palleronii** (Davis, Stanier, Doudoroff and Mandel 1970) Willems, Busse, Goor, Falsen, Jantzen, Hoste, Gillis, Kersters, Auling and De Ley 1989, 330^{VP} (*Pseudomonas palleronii* Davis, Stanier, Doudoroff and Mandel 1970, 11.) *pal.le.ro' ni.i.* M.L. gen. n. *palleronii* of Palleroni, named after N.J. Palleroni, who first isolated this organism.

The characteristics are as described for the genus and as listed in Tables BXII.β.51 and BXII.β.52. In addition, some strains have been reported to grow on DL-norleucine (Willems et al., 1989). Optimal temperature is 30°C.

Isolated from soil and water by enrichment for hydrogen autotrophs in minimal media with an atmosphere of $H_2/O_2/CO_2$.

The mol% G + C of the DNA is: 67.3–68.5 (T_m).

Type strain: Stanier 362, ATCC 17724, DSM 63, LMG 2366.

GenBank accession number (16S rRNA): AF019073.

Additional Remarks: This strain contains two stable colony types, which have identical gel electrophoretic protein profiles.

4. **Hydrogenophaga pseudoflava** (Auling, Reh, Lee and Schlegel 1978) Willems, Busse, Goor, Falsen, Jantzen, Hoste, Gillis, Kersters, Auling and De Ley 1989, 330^{VP} (*Pseudomonas pseudoflava* Auling, Reh, Lee and Schlegel 1978, 93.) *pseu.do.fla' va.* Gr. adj. *pseudes* false; L. fem. adj. *flava* yellow; M.L. fem. adj. *pseudoflava* not the true (*Hydrogenophaga*) *flava*, referring to the close relationship to *Hydrogenophaga flava*.

The characteristics are as described for the genus and as listed in Tables BXII.β.51 and BXII.β.52. In addition, none of the strains tested produce indole, H_2S in triple sugar iron medium, or acid from 10% lactose and triple sugar iron medium. All strains fail to grow on allantoin, DL-threonine, salicylate, 2-methyl-2-propanol, and uric acid, but do grow on L-glutamine and shikimic acid.

Some *H. pseudoflava* strains, but not the type strain, are capable of N_2 fixation (Jenni et al., 1989). The strains previously classified as "*Pseudomonas carboxydoflava*" are capable of mixotrophic growth on organic substrates and CO or H_2 and CO_2. For CO-autotrophic growth, the presence of molybdopterin cytosine dinucleotide is required as a cofactor (Meyer and Schlegel, 1983; Volk et al., 1994).

H. pseudoflava is highly related to *H. flava* (DNA homology values 48–62%; Willems et al., 1989), but can be distinguished by its ability to grow at higher temperatures (up to 41°C) and at oxygen concentrations of 20% and by its ability to use a wider variety of substrates for growth (Auling et al., 1978).

Optimal temperature is 35–38°C.

Isolated from soil, mud, or water by liquid enrichment for hydrogen bacteria in an atmosphere consisting of $O_2/CO_2/H_2$ (10:10:80).

The mol% G + C of the DNA is: 66.2–68.6 (T_m).

Type strain: GA3, ATCC 33668, CCUG 13799, DSM 1034, LMG 5945.

GenBank accession number (16S rRNA): AF078770.

5. **Hydrogenophaga taeniospiralis** (Lalucat, Parés and Schlegel 1982) Willems, Busse, Goor, Falsen, Jantzen, Hoste, Gillis, Kersters, Auling and De Ley 1989, 330^{VP} (*Pseudomonas taeniospiralis* Lalucat, Parés and Schlegel 1982, 337.) *tae.ni.o.spi.ra' lis.* Gr. n. *taenia* ribbon; L. adj. *spiralis* coiled; M.L. adj. *taeniospiralis* ribbon coiled, after *Caedibacter taeniospiralis*, an organism with which it shares characteristics.

The characteristics are as described for the genus and as listed in Tables BXII.β.51 and BXII.β.52. In addition, this species does not produce indole or acetoin, or H_2S from cysteine, and it does not grow on D-sorbose. No strains able to fix N_2 have been reported. A variable portion of cells in stationary growth phase may contain inclusions, referred to as R-bodies, which are coiled proteinaceous ribbons with an average diameter of 0.25 μm and an average height of 0.21 μm. Similar structures have been described in *Caedibacter taeniospiralis*, a bacterial endosymbiont of *Paramecium*. These R-bodies are thought to originate from defective prophages and extrachromosomal DNA elements and to be involved in conferring the killer trait upon the host paramecium (Lalucat et al., 1982). The R-bodies of *Caedibacter* species have been shown to be quite different from those of *H. taeniospiralis* (Kanabrocki et al., 1986), but the fact that a defective prophage has been detected in *H. taeniospiralis* indicates a possible similar origin in this species (Lalucat et al., 1982).

Optimal temperature is 37°C.

Isolated from soil in Spain.

The mol% G + C of the DNA is: 65.0 (T_m).

Type strain: 2K1, ATCC 49743, CCUG 15921, DSM 2082, LMG 7170.

GenBank accession number (16S rRNA): AF078768.

Genus VI. *Lampropedia* Schroeter 1886, 151^{VP}

R.G.E. MURRAY

Lam.pro.pe' di.a. Gr. adj. *lampros* bright, radiant; Gr. n. *pedia* a plain, flat country; M.L. fem. n. *Lampropedia* a shining flat sheet (of cells).

Sheets of rounded, almost cubical cells, arranged in square tablets of 16–64 cells, occasionally separated into pairs or tetrads. **Divide synchronously in a sheet and alternately in two planes.** The cells of a tablet are enclosed within a **complex, structured envelope.** Each cell is enclosed in a **Gram-negative type of cell wall.** Intracellular granules of **poly-β-hydroxybutyrate** are prominent. No flagella occur. **Twitching movements** of small groups of cells occur during active growth. **Obligately aerobic,** having a strictly respiratory type of metabolism with oxygen serving as the terminal electron acceptor. **Growth occurs as a thin, hydrophobic, extending pellicle** on the surface of both liquid and solid media. Nonpigmented. Optimal temperature, 30°C. Optimal pH, 7.0. Oxidase and catalase positive. Chemoorganotrophic. **Energy sources are limited to intermediates of the tricarboxylic acid cycle.** Carbohydrates, alcohols, glucosides, and fatty acids are not utilized. Ammonium salts or certain amino acids can serve as sole nitrogen sources. **Vitamins may be required for growth.** The ecological niche is unknown but observations and isolations indicate an environment rich in organic matter.

The mol% G + C of the DNA is: 61 (Bd).

Type species: **Lampropedia hyalina** (Ehrenberg 1832) Schroeter 1886, 151 (*Gonium hyalinum* Ehrenberg 1832, 63.)

FURTHER DESCRIPTIVE INFORMATION

This is a genus based mainly on morphological characteristics, and it consists of a single species, *L. hyalina*. The organism is cultivable and strains are available for study. Three other species have been named in the distant past (de Toni and Trevisan, 1889), including pigmented species, but they have not been re-isolated in the intervening years and have no validity today.

The characteristic appearance of growth is a hydrophobic pellicle consisting of square tablets of Gram-negative cells, apparently dividing synchronously in two planes (Fig. BXII.β.30), forming continuous but rumpled sheets on the surface of media (Kuhn and Starr, 1965; Puttlitz and Seeley, 1968; Seeley, 1974).

The distinctive morphology (Fig. BXII.β.30) allows recognition in its natural surroundings. In rich organic environments, and especially in laboratory media loaded with sodium acetate, the cells are full of poly-β-hydroxybutyrate (PHB) (see Fig. BXII.β.31) as several small granules or one large granule (Kuhn and Starr, 1965).

L. hyalina exhibits a peculiar form of motility involving sudden, irregular shifts in the position of small groups of cells. Pringsheim (1955) and Puttlitz and Seeley (1968) describe this phenomenon, and it has been compared (Pringsheim, 1966) to movements seen in the *Chroococcales* as an argument for relationship to cyanobacteria. However, such a "twitching" motility is also seen in other nonflagellated bacteria and most of them have fimbriae (Henrichsen, 1972). *Lampropedia* strains have not been shown to have either flagella or fimbriae. Twitching, generally (Henrichsen, 1972) and in this case (Puttlitz and Seeley, 1968), is a function of living, metabolizing cultures.

The cell wall is of the Gram-negative type with an outer membrane and a thin, underlying peptidoglycan layer. It has not been isolated and characterized in chemical terms. All layers of this wall intrude at once to separate the sister cells (Fig. BXII.β.32).

The structure of the cells is not remarkable, but the envelope that encloses each tablet of cells has unusual features (Chapman et al., 1963; Murray, 1963; Pangborn and Starr, 1966). The separate tablets of 16, 32, or 64 cells are surrounded by a hexagonal array of complex spindle-shaped units (spacing is 23–26 nm) on a thin continuous but perforated layer (7.5 nm holes; spacing is 13.5–14.5 nm). This highly structured integument bridges over the spaces between the cells in the tablet (Fig. BXII.β.31). The space between cells, between the walls and the structured envelope, and within the intruding septa (Figs. BXII.β.31 and BXII.β.32) contains fibrous materials and is called the intercalated layer. Two layers form the enveloping structure and each have p6 symmetry: an outer complex of three polypeptides joined

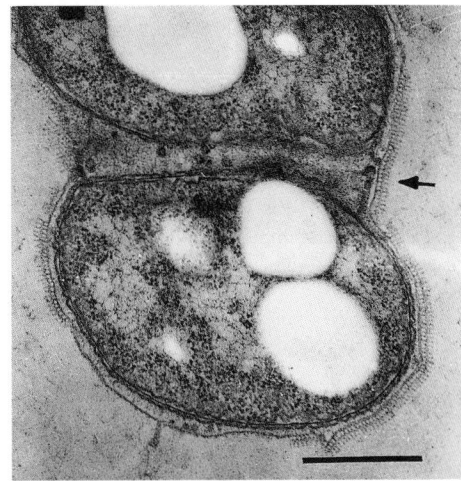

FIGURE BXII.β.31. Section transverse to the edge of a tablet of cells showing a bipartite external structured envelope. The inner layer is obvious in an area exposed by stripping. The envelope encloses the tablet and bridges over the matrix separating adjacent cells (*arrow*). The low density vesicles represent poly-β-hydroxybutyrate granules. Bar = 0.5 µm.

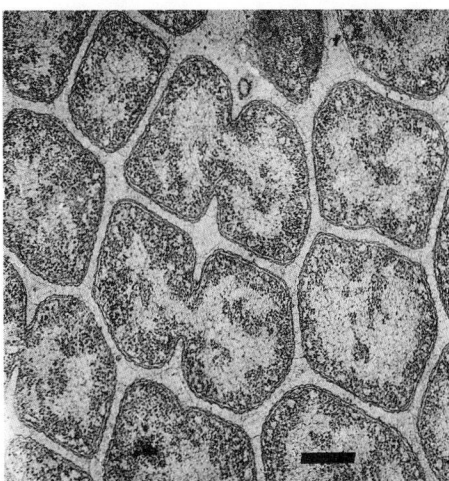

FIGURE BXII.β.32. Electron micrograph of a section in the plane of a tablet to show division and the relatively large volume of nucleoplasm. A matrix substance separates the cell walls (of usual Gram-negative character; see Fig. BXII.β.36) of adjacent cells. Divisions always show a "constrictive" form. Bar = 0.5 µm.

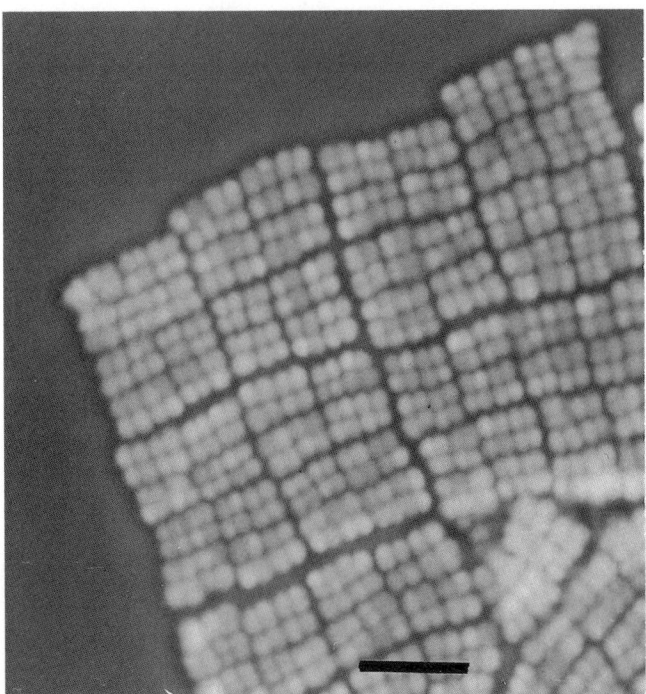

FIGURE BXII.β.30. Light micrograph of a nigrosin preparation of *L. hyalina* showing a corner of a sheet of actively growing tablets of cells. Adjacent tablets are almost synchronized in division. Bar = 5 µm.

to an inner layer formed by one polypeptide (Austin and Murray, 1987, 1990). The nature of the intercalated materials is unknown. The enveloping layers together accomplish the division and separation of tablets of cells.

Colonial characteristics are determined by the growth habit and the physical nature of the surface of the enveloping layer. The sheets of cells, one cell thick, extend as they grow over the surface of liquid or solid media and tend to wrinkle as they grow to a large size and meet some obstacle to spreading. Many microcolonies are square. An irregularly shaped piece transferred to a liquid medium floats on the surface and maintains that odd shape as it grows; old cultures provide a rain of cells and debris that sink to the bottom. The capacity to float and grow as a pellicle suggests a hydrophobic surface, which is believed to be mediated by fine fibrous material (possibly proteinaceous) external to the envelope array (Lanys, 1972). There is no life cycle.

The description of the genus and its growth characteristics assumes morphological stability. However, mutants are easily derived by isolation of natural variants or stimulated by a mutagen such as nitrosoguanidine (Lanys, 1972, and unpublished results) producing different smooth or rough colonies rather than sheets. Many of these have lost the ability to form tablets or even tetrads, and most have lost the envelope layers (Figs. BXII.β.33 and BXII.β.34). Rough variants usually have retained the envelope arrays (Fig. BXII.β.35), and these most commonly have been derived with the help of nitrosoguanidine. Therefore, it is possible—even probable—that clones exist in nature that are phylogenetically related to *Lampropedia* but show none of the unique morphological characteristics that we use to identify *L. hyalina* in the absence of clear metabolic distinctions. The cultural, physiological, and biochemical characteristics of the available strains have been studied by Puttlitz and Seeley (1968). It is evident that this respiratory, catalase-positive chemoorganotroph requires Krebs cycle intermediates as energy sources and utilizes ammonium salts and a limited palette of amino acids as nitrogen sources. It is slightly fastidious in that it requires biotin and thiamine for growth, but can be cultured in a simple defined medium[1] if these vitamins are supplied. It is a mesophile that grows best at neutral pH and does not tolerate either 0.5% bile or 1.5% salt. It does not produce exoenzymes that degrade proteins, lipids, or the major carbohydrate polymers. So that without its peculiar morphology it is not a distinguished aerobe and would be hard to recognize (see Figs. BXII.β.33, BXII.β.34, and BXII.β.35).

Nothing is known about the genetics, antigenic structure or antibiotic sensitivity of this organism. The mol% G + C of the DNA (M. Mandel, personal communication) ranges from 60.7–61.2 for all available strains.

The ecological niche is unknown. The microbe is sufficiently distinctive (Starr and Skerman, 1965) to be recognized in its tablet or sheeting form by simple microscopy of natural specimens, and undoubtedly it has been seen many times (Starr, 1981), even if there have been no more than three isolations (Pringsheim, 1955; Frank Kovács' strain "Mac 583", described by Schad et al., 1964; Julius Kirchner, noted by Hungate, 1966). The isolations and sightings (sometimes called "window-pane sarcinas") involved waters infused with quantities of organic material, probably well digested and populated with many other microbes. A partial listing includes waters polluted with sugar refinery wastes, swamp water (Schroeter, 1886), stagnant water including aquatic plants (de Toni and Trevisan, 1889), liquid manure from a barnyard (Pringsheim, 1955), rumen fluid (see Smiles and Dobson, 1956; Eadie, 1962; Hungate, 1966; Clarke, 1979), intestinal content of herbivorous reptiles and their nematodes (Schad et al., 1964), and sewage-polluted, muddy water (R. Kolkwitz, 1909, cited by Starr, 1981). Considering the need for organic acids (Krebs cycle intermediates) and for vitamins, as well as the lack of exoenzymes and growth as a pellicle at the air/water interface, it is not surprising that it thrives in many of these situations. Its real habitat remains to be discovered and could be many other places accessible to these sources, perhaps associated with plants or the soil around them. It is hard to believe that this strict aerobe can be even an irregular inhabitant of cattle or sheep rumen unless the lampropedias found in this site can use some hydrogen acceptor other than oxygen (R.E. Hungate, personal communication). However, there is no doubt of the occasional presence of enough window-pane sarcinas to be detected by microscopy.

ENRICHMENT AND ISOLATION PROCEDURES

L. hyalina will grow on many rich media (both liquid and solidified with agar), but isolation is made difficult by enormous numbers of other bacteria in the source material. Pringsheim (1955) used a capillary to transfer individual tablets of cells to dishes with water over a layer of soil in which starch particles or a grain of wheat was embedded; the lampropedias multiplied as a pellicle in the surface film, and, after 1–2 d at room temperature, could be streaked on an agar medium (0.1% sodium acetate, 0.2% yeast extract, and 0.1% peptone) to give a few isolated microcolonies. Microcolonies can be identified readily by low-power microscopy and subcultured. Kirschner (cited by Hungate, 1966) left rumen fluid standing for several days in an open flask and isolated *Lampropedia* by streaking the pellicle that formed onto rumen fluid

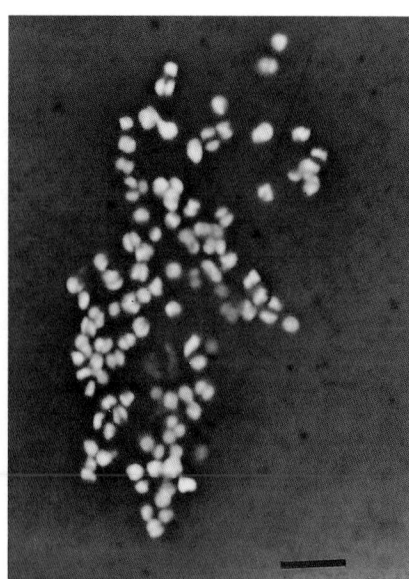

FIGURE BXII.β.33. Light micrograph of a nigrosin preparation of a nonsheeting strain derived from the original isolation of *L. hyalina* ATCC strain 11041 (Murray, 1963). Bar = 5.0 μm.

---

1. The chemically defined medium for growth of *L. hyalina* (Puttlitz and Seeley, 1968) consists of the following (per 1 distilled water): basal salts solution (CaCl₂·2H₂O, 0.1 g; FeSO₄·7H₂O, 0.2 g; MgSO₄·7H₂O, 0.2 g; MnCl₂·4H₂O, 0.25 g; distilled water, 1000 ml), 100 ml; sodium pyruvate, 3.0 g; NH₄Cl, 3.0 g; phosphate buffer solution (KH₂PO₄, 7.0 g; K₂HPO₄, 13.0 g; distilled water, 1000 ml), 50 ml; thiamine-HCl, 1.0 mg; biotin, 1.0 mg; and NaOH, 70 mg. The vitamins are added aseptically to the sterile basal medium from stock solutions sterilized by filtration.

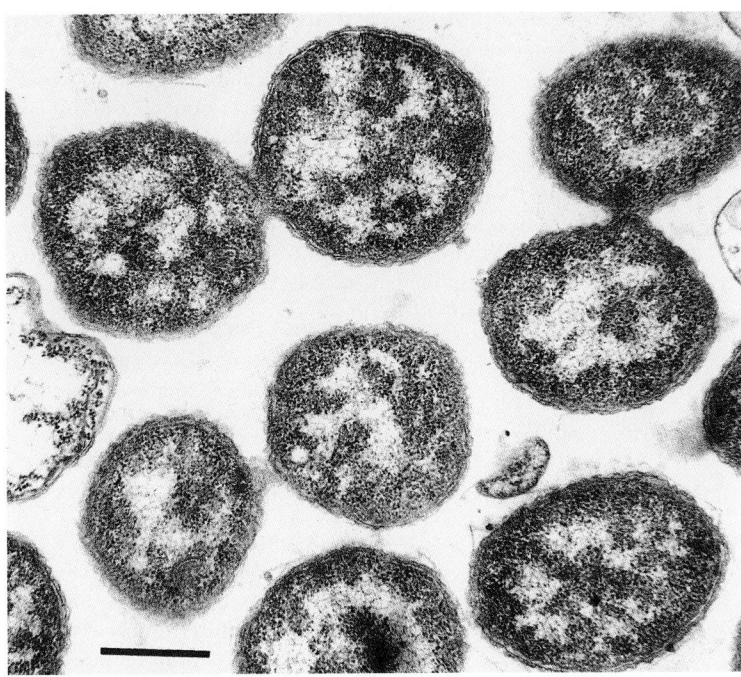

FIGURE BXII.β.34. A derivative of *L. hyalina* ATCC 11041 in section showing that it is almost, if not completely, devoid of the enveloping layers. Bar = 0.5 μm.

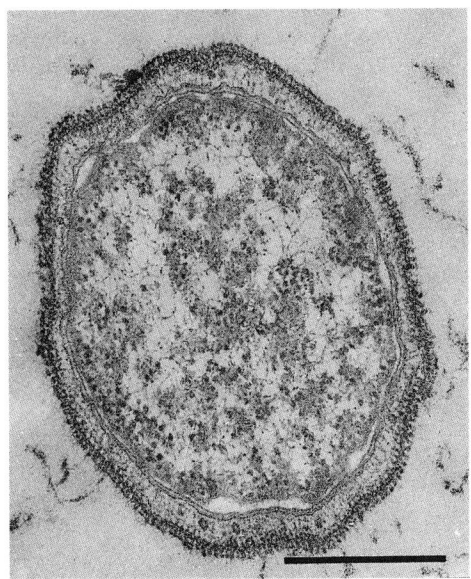

FIGURE BXII.β.35. Electron micrograph of a section of a nonsheeting derivative of *L. hyalina* ATCC 13691, which possesses both the structured and the intercalated enveloping layers outside the cell wall. Bar = 5 μm.

agar, which was then incubated aerobically at room temperature. There are a number of remarks on pellicle formation in the literature, so it seems likely that simple retention at room temperature is the preliminary approach to enrichment.

MAINTENANCE PROCEDURES

The common laboratory practice is maintenance by monthly subculture at room temperature on a neutral pH yeast extract-acetate-peptone agar (0.1–0.5% of each). The organisms survive lyophilization to a fair degree. Undoubtedly, constant subculture can lead to selection of mutants. Murray (1963) observed that ATCC strain 11041, derived from Pringsheim's isolation, had lost its ability to form sheets (Figs. BXII.β.33 and BXII.β.34). Nowadays, safe stocks can be kept in liquid N₂ suspended in the above fluid medium with 20% glycerol added.

DIFFERENTIATION OF THE GENUS *LAMPROPEDIA* FROM OTHER GENERA

The ability of *Lampropedia* to grow as sheets of cells on liquid or solid media distinguishes it from other genera of nonphotosynthetic bacteria. An organism known as *Thiopedia rosea* has a morphological resemblance to *Lampropedia*, but is photosynthetic (see Taxonomic Comments, below).

TAXONOMIC COMMENTS

Sequence analysis of rDNA shows that *L. hyalina* belongs among the *Betaproteobacteria* (T.M. Schmidt, personal communication) in a grouping that includes *Comamonas*, *Brachymonas*, and more distantly *Leptothrix* and *Variovorax*. However, some aspects of its unique tablet-forming morphology are repeated in *Thiopedia rosea*, a nonsulfur purple, photosynthetic bacterium (Hirsch, 1977b). A similar type of photosynthetic organism was studied in the electron microscope (Lauritis, 1967) and identified as *"Rhodothece"* (now *Amoebobacter*), which did not get beyond the diplococcal form. The extraordinary feature of this organism was the possession of a superficial wall array virtually identical in fine structure to the complex envelope array of *L. hyalina*. In fact, apart from the chromatophore membranes it was similar to some of the enveloped but nonsheeting variants of *L. hyalina* (Fig. BXII.β.35; S.G. Lanys and R.G.E. Murray, unpublished observations). Natural variants isolated from cultures often lose the enveloping structures and the ability to form tablets (Murray, 1963). Associations dependent upon such similarities are fraught with danger because we have no understanding of the genetics of such structures and whether or not the determinants are genetically transferable to any receptive organisms. It is probable that at least one of the old named species, "*L. violacea*" (see Seeley, 1974), may have been one of the *Chromatiaceae* (such as *Lamprocystis roseopersicina*) because the color of cell suspensions is purple to purple-violet (Pfennig and Trüper, 1974).

No less a problem is presented by Pringsheim's (1966) proposal that *L. hyalina* should be regarded as an apochlorotic species of *Merismopedia*, a large coccoid cyanobacterium that forms squared sheets of cells embedded in a glutinous matrix but without evidence of an enveloping structured array. Somewhat similar associations are evident in *Thiocapsa* and *Lamprocystis* among the *Chromatiaceae*.

Few, if any, of these taxonomic hypotheses can be given any credence without strong supportive evidence based, in all probability, on several accepted molecular markers for phylogenetic relationship. Furthermore, comparative studies of fine structure of a wider range of genera, as well as studies, if possible, of heterotrophically grown *Thiopedia* have yet to be undertaken. Uncertainty about taxonomic associations is exemplified by association with sulfur bacteria in the sixth edition of *Bergey's Manual of Determinative Bacteriology*, complete omission from the seventh edition, and relegation to a genus of uncertain affiliation among Gram-negative cocci in the eighth edition. Obviously, we are no further ahead now, but the tools to solve these problems may be available.

FURTHER READING

Starr, M.P. 1981. The Genus *Lampropedia*. *In* Starr, Stolp, Trüper, Balows and Schlegel (Editors), The Prokaryotes: a Handbook of Habitats, Isolation and Identification of Bacteria, Springer-Verlag, Berlin. pp. 1530–1536.

List of species of the genus Lampropedia

1. **Lampropedia hyalina** (Ehrenberg 1832) Schroeter 1886, 161[AL] (*Gonium hyalinum* Ehrenberg 1832, 63.)
   hya.li'na. Gr. adj. *hyalinos* glassy, shiny; M.L. fem. adj. *hyalina* hyaline.

   Cells are 1.0–1.5 × 1.0–2.5 μm. Morphological characteristics are as described for the genus and as depicted in Figs. BXII.β.30, BXII.β.31, BXII.β.32, and BXII.β.36.

   Physiological and histochemical characteristics are as described for the genus and as listed in Table BXII.β.53. Utilizes pyruvate, lactate, butyrate, fumarate, malate, succinate (and acetate in the presence of catalytic levels of pyruvate) as sole energy sources. Utilizes NH_2Cl, alanine, arginine, and tyrosine as sole nitrogen sources. Biotin and thiamine are required for growth.

   Temperature range for growth, 10–35°C; optimum, 30°C. pH range for growth, 6.0–8.6, optimum, 7.0.

   The mol% G + C of the DNA is: 60.7–61.2 (Bd; M. Mandel, personal communication).

   Type strain: ATCC 11041.

   Additional Remarks: The type strain is inappropriate because it has lost the characteristic of forming tablets and sheets of cells, and has lost the ability to cover its surface with a structured envelope (Murray, 1963). A possible candidate for a neotype is reference strain ATCC 13871.

FIGURE BXII.β.36. A high magnification of a section showing the complex structured envelope (*el*), the envelope matrix (*m*), the cell wall (*cw*), and plasma membrane (*pm*). Bar = 0.5 μm.

TABLE BXII.β.53. Physiological and biochemical characteristics of *Lampropedia hyalina*[a,b]

Characteristic	Reaction or result
Growth under anaerobic conditions	−
Intracellular poly-β-hydroxybutyrate formed	+
Oxidase test	+
Catalase test	+
Indole production	−
Acetyl methyl carbinol production (Voges–Proskauer test)	−
Litmus milk	No change
Benzidine test for heme groups	+
Hemolysis on blood agar	−
Arginine deaminase	+
Hydrolysis of gelatin, casein, fats and fatty acids, starch, hippurate, DNA and urea	−
Growth in the presence of:	
1.0% NaCl	+
1.5% NaCl	−
2.0% sucrose	+
4.0% sucrose	−
0.5% bile	−
Final pH in culture media	8.4–8.6

[a]For symbols see standard definitions.
[b]Data from Puttlitz and Seeley (1968).

Other Organisms

It is possible that the "window-pane sarcina" seen in rumen contents (which usually do not show poly-β-hydroxybutyrate granules; R.E. Hungate, personal communication) is separable from *L. hyalina*, but recognition would require cultivation and metabolic studies. An early sighting gave rise to the name *"Bacterium merismopedioides"* (Zopf, 1883), which might be an appropriate name if a rumen strain is ever differentiated from *L. hyalina*.

Many other synonyms probably have been created. Among them the generic name *"Pedioplana"* (Wolff 1907, 10) was considered to be synonymous with *Lampropedia* by Seeley (1974) but, as Starr (1981) points out, it was described as motile and as possessing flagella.

The *species incertae sedis* included by Seeley (1974) are no longer valid, but memory of them should not be erased because they may represent other physiological variations for rediscovery: *"Lampropedia reitenbachii"* (Caspary) de Toni and Trevisan 1889, 1048; *"Lampropedia violacea"* (Brébisson) de Toni and Trevisan 1889, 1048; and *"Lampropedia ochracea"* (Mattenheimer) de Toni and Trevisan 1889, 1049.

Genus VII. **Macromonas** Utermöhl and Koppe in Koppe 1924, 632[AL]

GALINA A. DUBININA, FRED A. RAINEY AND J. GIJS KUENEN

Ma'cro.mo'nas. Gr. adj. *macrus* large; Gr. n. *monas* a unit, monad; M.L. fem. n. *Macromonas* a large monad.

Large cells are colorless, cylindrical, bean-shaped, or slightly bent. Gram negative. Sluggish or rapidly motile by means of a polar tuft of flagella. Strictly aerobic or microaerophilic. A typical characteristic is the presence of several large inclusions of calcium oxalate, previously believed to be calcium carbonate. Sulfur globules may also be present. Multiplication by constriction followed by fission. No resting stages are known. The genus is composed of two species; the type species has not been grown in pure culture. Several strains of the second species have been studied in pure culture.

Phylogenetic analyses based on 16S DNA sequence comparisons place **Macromonas bipunctata** within the *Betaproteobacteria*. Closest related validly described taxa are the species of the genus *Hydrogenophaga*.

Type species: **Macromonas mobilis** (Lauterborn 1915) Utermöhl and Koppe *in* Koppe 1924, 632 (*"Achromatium mobile"* Lauterborn 1915)

FURTHER DESCRIPTIVE INFORMATION

Cells of the two species are morphologically similar, but differ markedly in size (Table BXII.β.54). *Macromonas bipunctata* cells that lack large inclusions move rapidly. However, as the number of inclusions increases, the motility rate slows down to 600–800 μm/min, which is probably a result of the increased specific gravity of the cells. Between one and four large, or numerous smaller, optically dense inclusions may almost fill the cell (Fig. BXII.β.37). The nature of the refractile intracellular inclusions can be determined by infrared spectroscopy. Absorption spectra of *M. bipunctata* cells grown on medium with succinate have maxima at 1600 and 1300 cm^{-1}, which are typical of the carboxyl groups of oxalate. The polar tuft of flagella ranges in length from 10–40 μm and can sometimes be seen under the light microscope.

Aerobic, aerotactic. Strains of *M. bipunctata* studied thus far are heterotrophs, but can partially oxidize sulfide and thiosulfate by means of hydrogen peroxide.

Both species are found in freshwater environments with low oxygen and hydrogen sulfide concentrations, including hypolimnia, chemoclines, the upper layers of the mud in lakes and ponds, and the surface of sediment in sewage treatment plants. *Macromonas mobilis* has also been reported in acid bog waters (Schultz and Hirsch, 1973).

ENRICHMENT AND ISOLATION PROCEDURES

No methods are known for the isolation and maintenance of cultures of the type species *M. mobilis*. *M. bipunctata* can be enriched and isolated by the following procedure (Dubinina and Grabovich, 1984). A suitable inoculum (for example, the white mat found on the surface of bottom sediment in a sewage aeration tank) is added to tubes containing 10 ml portions of the following semisolid medium (g/l distilled water): sodium acetate, 1; CaCl$_2$, 0.1; casein hydrolysate, 0.1; yeast extract, 0.1; agar, 1.0. After sterilization, vitamins and trace elements (Pfennig and Lippert, 1966) are added, along with 200 mg freshly precipitated FeS; pH 7.2–7.4. After 2–3 d at 28°C, a white surface film should appear. A suspension of this film is streaked on plates containing the above medium solidified with 10 g agar/l. After 2–3 d, flat colonies of *Macromonas* should appear, and these can be subcultured on solid medium.

MAINTENANCE PROCEDURES

Cultures of *M. bipunctata* can be maintained in semisolid (0.15% agar) medium without FeS but supplemented with thiosulfate (1 g/l), but they should be transferred at about one-month intervals. Lyophilization with cells suspended in skimmed milk has proven to be unsuccessful: the cells remain viable for no more than one or two months.

TAXONOMIC COMMENTS

Phylogenetic analyses based on 16S rDNA–RNA sequence comparisons demonstrate that *Macromonas bipunctata* belongs to the *Betaproteobacteria* and clusters within the radiation of the genera *Hydrogenophaga*, *Comamonas*, *Acidovorax*, *Variovorax*, and *Rhodoferax*, with its closest relatives being the species of the genus *Hydrogenophaga* (Fig. BXII.β.38).

The 16S rDNA sequence of *Macromonas bipunctata* strain VKM 1366^T has 92.8–96.2% similarity to the 16S rDNA sequences of the type strains of each of the species of the genera *Hydrogenophaga*, *Comamonas*, *Acidovorax*, *Variovorax*, and *Rhodoferax*. The highest 16S rDNA sequence similarities of the *Macromonas bipunctata* sequence are found to members of the genus *Hydrogenophaga*: 95.9% to *H. taeniospiralis*, 95.9% to *H. pseudoflava*, 96.1% to *H. flava*, and 96.2% to *H. palleronii*. The 16S rDNA sequence similarity values between the species of the genus *Hydrogenophaga*

TABLE BXII.β.54. Differentiation between *M. mobilis* and *M. bipunctata*

Characteristic	*M. mobilis*	*M. bipunctata*
Cell size	6–14 × 10–30 μm	2.2–4 × 3.3–6.5 μm

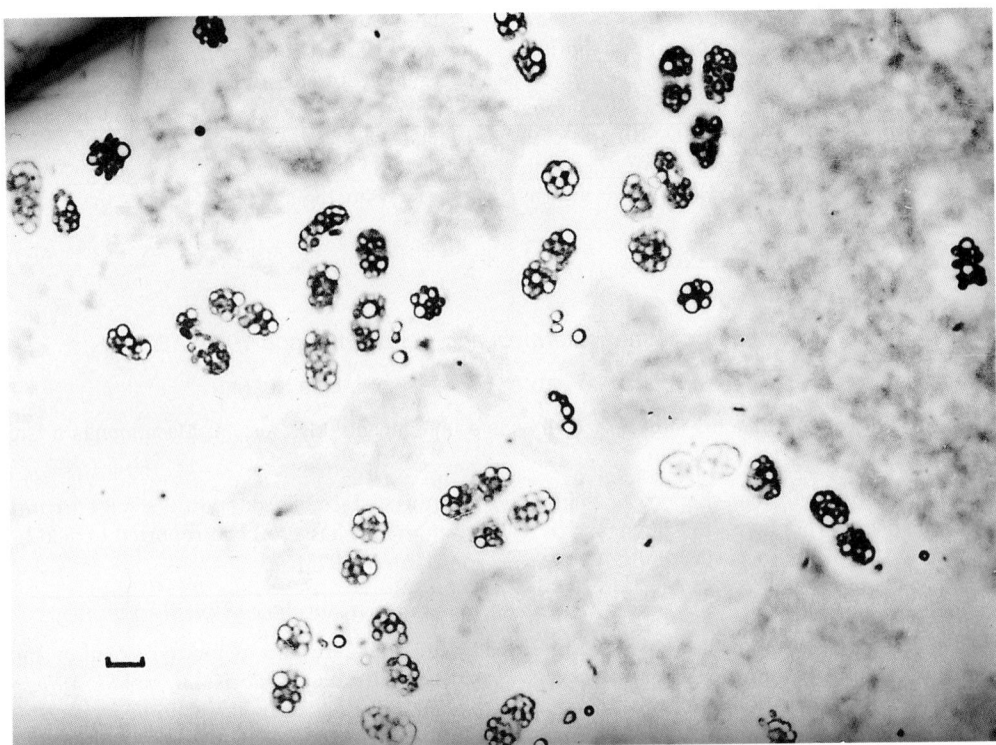

FIGURE BXII.β.37. Phase-contrast and electron micrographs of *M. bipunctata* VKM 1366[T] showing inclusions of calcium oxalates in cells grown on medium with succinate; Bar = 5 μm.

are in the range of 97.3–99.0%, and so, as shown in the phylogenetic dendrogram (Fig. BXII.β.38), *Macromonas bipunctata* is related to the *Hydrogenophaga* species cluster, but branches off at the deepest point of the cluster.

Although the phylogenetic position of the species *Macromonas bipunctata* is clearly demonstrated by the 16S rDNA analysis of the type strain VKM 1366[T] (which exists as a pure culture), the phylogenetic position of *Macromonas mobilis* (the as yet uncultured type species of the genus) is still unknown. Until the 16S rDNA sequence of *Macromonas mobilis* is determined and analyzed, the phylogenetic position of the genus is unclear. However, considering the high degree of similarity between *Macromonas bipunctata* and *Macromonas mobilis* in terms of morphology and ecology, the genus *Macromonas* is probably a member of the *Betaproteobacteria*, with a close phylogenetic relationship to species of the genus *Hydrogenophaga*. Future comparisons of the physiological characteristics of *Macromonas bipunctata* are required to further establish the relationship to the genus *Hydrogenophaga*.

FURTHER READING

la Rivière, J.W.M. and K. Schmidt. 1981. Morphologically conspicuous sulfur-oxidizing bacteria. *In* Starr, Stolp, Trüper, Balows and Schlegel (Editors), The Prokaryotes: a Handbook on Habits, Isolation and Identification of Bacteria, Springer-Verlag, Berlin. pp. 1037–1048.

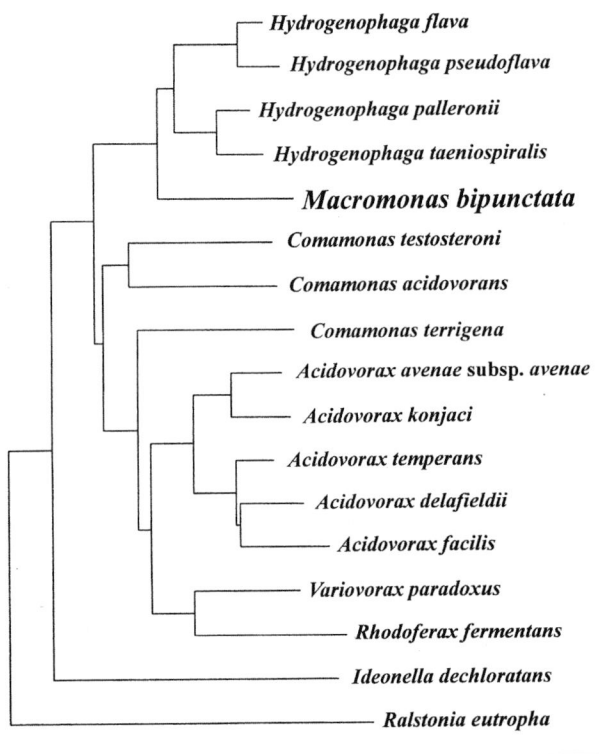

FIGURE BXII.β.38. Phylogenetic dendrogram, based on comparison of 16S rDNA sequences, indicating the position of *Macromonas bipunctata* strain VKM 1366[T] and its closest relatives within the *Betaproteobacteria*. The dendrogram was reconstructed from evolutionary distances (Jukes and Cantor, 1969) by the neighbor-joining method (Saitou and Nei, 1987). Bar = 2.5 inferred nucleotide changes per 100 nucleotides.

List of species of the genus Macromonas

1. **Macromonas mobilis** (Lauterborn 1915) Utermöhl and Koppe *in* Koppe 1924, 632[AL] (*"Achromatium mobile"* Lauterborn 1915)

mo' bi.lis L. fem. adj. *mobilis* motile.

The main feature by which this species is currently distinguished from *M. bipunctata* is its large cell size, which is usually 9 × 20 µm, and sometimes 6–14 × 10–30 µm. Its polar tuft of flagella is 20–40 µm long. For further description see that of genus.

The mol% G + C of the DNA is: unavailable.

Type strain: no culture isolated.

2. **Macromonas bipunctata** (ex Utermöhl and Koppe *in* Koppe 1924, 632) Dubinina and Grabovich 1989, 496[VP] (Effective publication: Dubinina and Grabovich 1984, 754.)

bi.punc.ta' ta. L. *bis* twice; L. part. adj. *punctatus* punctate, dotted; M.L. fem. adj. *bipunctata* twice punctate.

Cells single or in pairs, pear-shaped, cylindrical or slightly curved, 2.2–4 × 3.3–6.5 µm, motile by means of a polar tuft of flagella made up of 16–20 fine flagella and ranging in length from 20–30 µm.

Electron micrographs of ultrathin sections of *M. bipunctata* cells reveal the typical structure of Gram-negative bacteria. The cytoplasm is surrounded by a three-layered cell membrane that occasionally forms invaginations. The five-layered cell wall, with a thin layer of peptidoglycan, is separated from the protoplast by a periplasmic space. The cells are covered by a thin layer of fibrillar polysaccharide. In cells grown under intense aeration, this layer may increase in thickness up to 200–500 nm and form slime capsules (Fig. BXII.β.39).

At least four types of inclusion have been observed in *M. bipunctata* cells. These are calcium oxalate, calcium poly-

phosphate, S^0, and poly-β-hydroxybutyrate. When *M. bipunctata* is grown on media containing succinate or malate, refractile irregularly-shaped calcium oxalate inclusions (previously believed to be calcium carbonate) accumulate in the cells. In *M. bipunctata*, calcium oxalate is synthesized mainly via oxaloacetate conversion to acetate and oxalate, with the participation of oxaloacetate hydrolase in the TCA cycle (Grabovich et al., 1995). Depending on the concentration of organic compounds in the medium, oxalates may appear as ballast or serve as a pool of reserve organic matter that can maintain metabolism under unfavorable conditions. When used as ballast, the oxalates may be removed from the cells by oxidation to CO_2 using oxalate oxidase. The reaction is not associated with either biosynthetic or energy metabolism.

S^0 accumulates in cells grown in media containing FeS or CaS, but not in cells grown with thiosulfate. Sulfur globules also appear in cells that have been grown in the absence of sulfides and then incubated for 20 minutes in a 0.005–0.01% polysulfide solution. Electron micrographs of ultrathin sections show that the sulfur globules are located in the periplasm and within the invaginations formed by the cell membrane.

Colonies on solid media are non-pigmented, of a fine-grained structure, slightly opalescent, and flat, measuring 0.5–4 mm. Growth in static liquid occurs in the upper 3–5 mm layer.

M. bipunctata strains thus far isolated have been chemoorganotrophic, but they do not tolerate or utilize high concentrations of organic compounds. Low concentrations of growth substrates (5–10 mM) are apparently optimal. High concentrations induce rapid autolysis of cells due to hydrogen peroxide accumulation in the periplasm. The addition of sulfides, thiosulfate, or catalase to the growth medium increases the biomass yield by 20–40% and extends the survival of stored cultures from several days to a month. Acetate, succinate, malate, fumarate, oxalacetate, benzoate, and pyruvate are good substrates, but sugars, alcohols (with the exception of ethanol), and amino acids are not used. Ammonium salts and organic nitrogen compounds can serve as nitrogen sources. Vitamins are required for growth. Further physiological and biochemical characterization of *M. bipunctata* is shown in Table BXII.β.55.

Optimal temperature 28°C, optimal pH 7.5–8.2.

The complete TCA cycle functions in *M. bipunctata*, its only unusual feature being a low fumarate hydratase activity, leading to the accumulation of fumarate in the cells. The glyoxylate cycle also functions in *M. bipunctata*, regardless of the carbon source used. It has been suggested that the glyoxylate cycle compensates for the low fumarate hydratase activity, shunting fumarate to be converted to malate. Calcium oxalate formation and its intracellular accumulation have been shown to be another peculiarity of carbon metabolism in *M. bipunctata* (Grabovich et al., 1993).

M. bipunctata oxidizes sulfide to S^0, and thiosulfate to tetrathionate by means of hydrogen peroxide, a process that does not provide useful energy for the cells (Dubinina and Grabovich, 1984; Chekanova and Dubinina, 1990).

Isolated from the sediment in an aeration tank of a sewage treatment plant.

The mol% G + C of the DNA is: 67.6 (T_m).

Type strain: DSMZ 12705, VKM 1366.

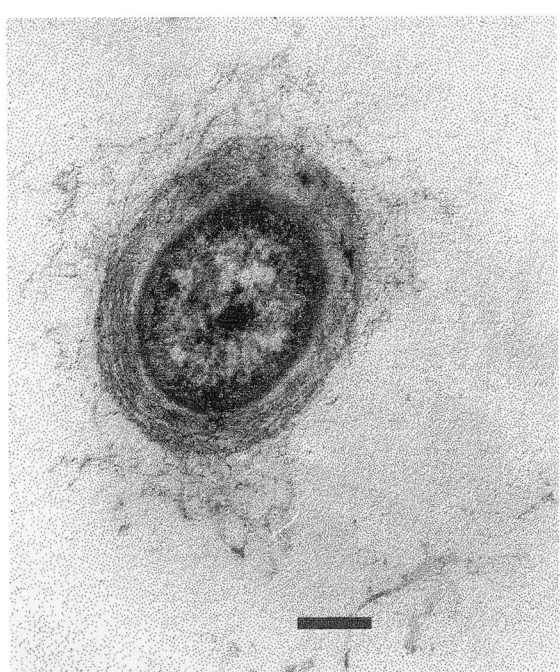

FIGURE BXII.β.39. Electron micrographs of ultrathin sections of ruthenium red-stained *M. bipunctata* VKM 1366[T]. The ruthenium red has stained the extracellular polysaccharides. Bar = 0.5 µm.

TABLE BXII.β.55. Characterization of *Macromonas bipunctata* strains[a]

Characteristics	Reaction
Use as a sole carbon source:	
Organic acids:	
Acetate, benzoate, fumarate, malate, oxaloacetate, succinate	+
Citrate, isocitrate, aconitate, malonate, glycolate	−
α-Ketoglutarate, glyoxylate, lactate, formate, pyruvate, propionate, oxalate	d
Alcohols:	
Ethanol	d
Methanol, butanol, isobutanol, glycerol	−
Amino acids:	
Serine, aspartate, lysine, cysteine, tryptophan, alanine, histidine, phenylalanine, asparagine, methionine, proline, tyrosine	−
Peptone	−
Casein hydrolysate, yeast extract	+
Use as a sole nitrogen source:	
Ammonium	+
Peptone, casein hydrolysate	+
Nitrate and nitrite	−
Glutamate	+
Alanine, aspartate, serine, cysteine, cystine, methionine	d
Reduction of NO_3^- to NO_2^-	−
N_2 fixation	−
H_2S production from cysteine or $S_2O_3^{2-}$	−
S^0 production from sulfide	+
Hydrolysis of gelatin, starch, casein	−
Indole production	−
Catalase activity	+[b]
Oxidase activity	+
Urease activity	+
Anaerobic growth with nitrate, sulfate, thiosulfate, or fumarate as acceptors of electrons	−
Mol% G + C of DNA (T_m)	66–68

[a]Symbols: +, 90% or more of the strains are positive; −, 90% or more of the strains are negative; d, 11–89% of the strains are positive.

[b]Activity of catalase is weak.

Genus VIII. Polaromonas *Irgens, Gosink and Staley 1996, 825*[VP]

JOHN J. GOSINK

Po'lar.o.mo'nas. M.L. adj. *polaris* pertaining to the geographic poles; Gr. fem. n. *monas* unit; M.L. fem. n. *Polaromonas* polar bacterium.

Cigar-shaped rods, $0.8 \times 2.0–3.0$ μm. Gram negative and encapsulated. Motile by a polar flagellum. Aerobic. Chemoorganotrophic, catalase and oxidase positive. Require amino acids, but not vitamins for growth. **Cellular fatty acids predominantly (74–79%) $C_{16:1\ \omega7c}$, and $C_{16:0}$ (14–17%), and some (7–9%) $C_{18:1\ \omega7c}$, $C_{18:1\ \omega9t}$, or $C_{18:1\ \omega12t}$. Organisms may contain gas vesicles.** Psychrophilic; growth temperature maximum of known strains is 15°C. 16S rDNA-based phylogeny shows this is a member of the *Comamonadaceae* in the *Betaproteobacteria*.

The mol% G + C of the DNA is: 52–57.

Type species: **Polaromonas vacuolata** Irgens, Gosink and Staley 1996, 825.

FURTHER DESCRIPTIVE INFORMATION

Five putative strains (34-P, 41-P, 54-P, JA, and JB) of *Polaromonas* have been isolated from Antarctic waters. Only one strain, 34-P, has been examined in depth by phenotypic, genotypic, and phylogenetic methods and hence officially recognized as a member of this genus.

Phylogenetic trees of *Polaromonas* and closely related taxa have been generated using distance, parsimony, and maximum likelihood methods. Analyses using various forms of parsimony (Swofford, 1991; Maddison and Maddison, 1992), with different character weighting masks and substitution matrix weights, give a number of different trees with similar structures (Fig.

BXII.β.40A). Distance (Fig. BXII.β.40B) and likelihood (Fig. BXII.β.40C) have also been performed on the data, using a transition to transversion ratio of 1.3 (Felsenstein, 1981, 1989; Olsen et al., 1994).

The results of these analyses show slightly different trees, depending on which assumptions for the model of evolution are used. All of these methods show that *Polaromonas* is a member of the *Betaproteobacteria* and shares a moderately close relationship with *Rhodoferax fermentans* (Hiraishi et al., 1991a), *Variovorax paradoxus,* (Davis et al., 1969; Willems et al., 1991a) and the environmental sequence "str. Stripa" (Ekendahl et al., 1994). Recent inspection of the Ribosomal Database Project (Maidak et al., 1999) small subunit database reveals additional partial "environmental population shotgun" sequences closely related to those of these organisms. More accurate or reproducible phylogenetic trees would be produced if full length sequences were available from the environmental clones. In addition, since these sequences have been obtained from uncultured bacteria, there are no phenotypic data available for the bacteria that produced these sequences, so comparison with these organisms can not proceed beyond the phylogenetic stage. The exact phylogenetic relationship among these taxa remains unclear pending future detailed examination.

Polaromonas strains appear as short, unicellular, Gram-negative rods ($0.8 \times 2–3$ μm) that typically produce gas vacuoles, which

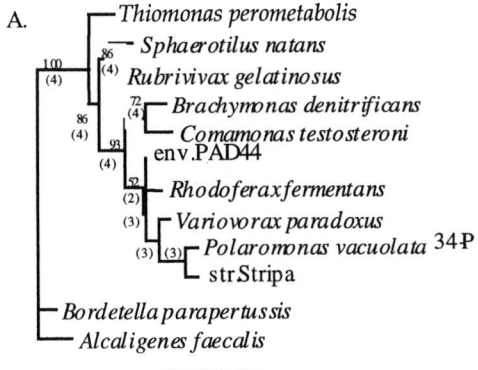

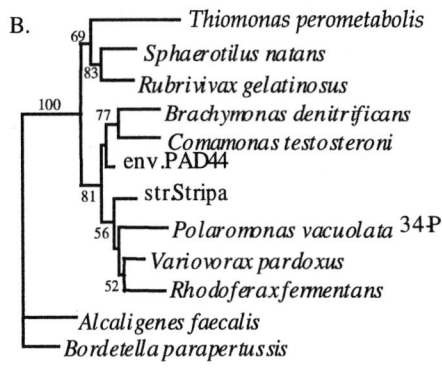

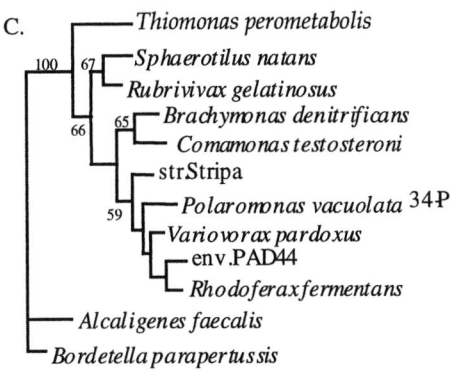

FIGURE BXII.β.40. The phylogenetic relatedness of *Polaromonas vacuolata* strain 34-P to the most closely related species. Numbers near the branches but not in parentheses indicate percent bootstrap support for that clade from 100 bootstrap resamplings. Only bootstrap values of 50% or greater are shown. Bars = 0.1 changes/average nucleotide position. A maximum parsimony tree (*A*), was determined using PAUP 3.0s by an exact (branch and bound) search, using a substitution matrix to correct for the various rates of nucleotide substitutions (Swofford, 1991; Maddison and Maddison, 1992). Numbers in parentheses near the branches indicate how many of the four equally most parsimonious trees share that branch structure. A neighbor joining tree (*B*) was made using PHYLIP 3.2 and a maximum likelihood tree (*C*) was made with fast DNAmL; both were made with a Kimura 2 parameter correction, using a transition to transversion ratio of 1.3 (Felsenstein 1989; Olsen et al., 1994).

appear as bright refractile areas within the cells. Gas vesicles are confirmed by transmission electron microscopy (Fig. BXII.β.41). Although cells are nonmotile under usual culture conditions, they produce polar flagella in addition to gas vacuoles (Fig. BXII.β.41). These bacteria produce circular, convex colonies with a smooth, glistening surface and an entire edge on agar plates. The colonies are chalky white in pigmentation. All strains produce large amounts (74–79%) of $C_{16:1 \, \omega7c}$ and smaller amounts of $C_{16:0}$ (14–17%) (Irgens et al., 1996). A third fatty, $C_{18:1 \, \omega7c}$, $C_{18:1 \, \omega9t}$, $C_{18:1 \, \omega12t}$, or possibly a combination of more than one of these, is present in lower amounts (7–9%). Such predominance of a single fatty acid is unusual in bacteria, and this is among the highest level of $C_{16:1 \, \omega7c}$ in any bacterial species. Additional features are shown in Table BXII.β.56.

Although many members of the *Alphaproteobacteria* and *Gammaproteobacteria* have been shown to produce gas vesicles, this is the first described gas vacuolate member of the *Betaproteobacteria*, though there are likely many more gas vacuolate members among this phylogenetic group. Although gas vacuolate heterotrophic bacteria are well known inhabitants of aquatic ecosystems, until recently, none had been observed or isolated from marine habitats. In 1989, several types from Antarctica were reported growing in association with the sea ice microbial community (Irgens, et al., 1989; Staley et al., 1989). The ecological role of *Polaromonas* and the function that gas vesicles serve in that role are unknown.

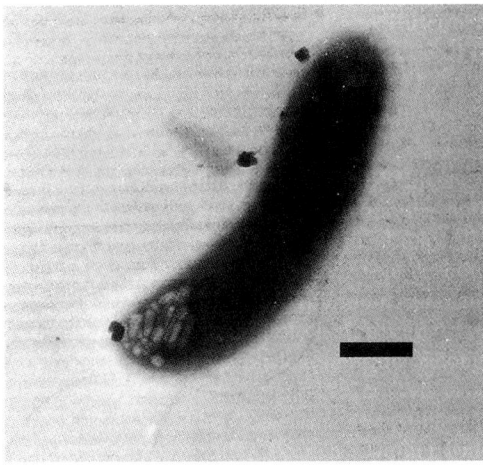

FIGURE BXII.β.41. An electron micrograph of *Polaromonas vacuolata* strain 34-P showing a cell containing several gas vesicles. Flagella are not attached. Bar = 0.5 μm. (Reprinted with permission from R.L. Irgens et al., International Journal of Systematic Bacteriology *46*: 822–26, 1996 ©International Union of Microbiological Societies.)

ENRICHMENT AND ISOLATION PROCEDURES

All strains were isolated from Antarctic waters near the U.S. Palmer Station, Anvers Island, Antarctica. These organisms are

TABLE BXII.β.56. Phenotypic and genotypic features of *Polaromonas* strains 34-P, 41-P, 54-P, JA, and JB [a,b]

Characteristic	*Polaromonas vacuolata* strain 34-P	*Polaromonas* sp. strain 41-P	*Polaromonas* sp. strain 54-P	*Polaromonas* sp. strain JA	*Polaromonas* sp. strain JB
Catalase	+	+	+	+	+
Oxidase	+	+	+	+	+
Urease	+	nd	nd	v	nd
Deaminase	+	−	−	−	−
Lipase	+	+	+	+	+
Agarase	−	nd	nd	nd	nd
Amylase, proteinase, tryptophanase, nitrate reductase, cysteine desulfurase	−	−	−	−	−
Lactate	+	+	+	+	+
Malate	+	+	−	+	−
Fumarate	+	−	−	−	−
Succinate	+	−	+	+	−
D-Glucose	+	−	−	−	+
Sucrose	−	−	−	−	+
Lactose	−	−	+	−	−
Ethanol	−	−	−	−	−
Acetate, pyruvate, propionate, citrate, oxaloacetate, butyrate, 2-oxoglutarate, glycerol, sorbitol	+	nd	nd	nd	nd
Formate, benzoate, malonate, maltose, D-fructose, D-xylose, D-ribose, cellobiose, D-mannose, L-fucose, melibiose, melezitose, L-rhamnose, sorbose, trehalose, methanol, erythritol, propanol	−	nd	nd	nd	nd
DL-Alanine	+	+	−	+	−
DL-Glutamate	+	+	+	+	−
DL-Aspartate	+	+	−	−	−
DL-Arginine	−	−	−	−	+
DL-Ornithine	−	−	−	−	−
DL-Proline, DL-asparagine	+	nd	nd	nd	nd
Glycine, DL-serine, DL-isoleucine, DL-lysine, DL-histidine, DL-methionine, DL-valine, DL-threonine, DL-tryptophan	−	nd	nd	nd	nd
Growth temperature range (°C)	≤ − 1.5–10	≤ − 1.5–7	≤ − 1.5–9	≤ − 1.5–9	≤ − 1.5–8
Growth salinity range, %	0–6	0.15–2.5	0.15–2.5	0.15–2.5	0.15–2.5
Vitamin requirement	none	nd	nd	nd	nd
Anaerobic growth	−	−	−	−	−
Mol% G + C	52	57	52	53	52

[a]For symbols see standard definitions; nd, not determined.

[b]Data from Irgens et al., 1989, 1996.

isolated by plating water samples onto SWCm (Staley et al., 1989) agar plates and incubating at 6°C for 6 weeks. Gas vacuolate colonies can be identified by their chalky appearance. *Polaromonas* strains are delineated from other gas vacuolate strains by their color, colony morphology, and fatty acid composition.

Maintenance Procedures

Stocks are maintained for general work on SWCm (Staley et al., 1989) at 4°C. Care must be taken when transferring or working with the strains to keep them and the growth media at 4°C. These cultures can be killed by leaving them out at room temperature for several hours or overnight. Long-term storage is best done by transferring cells to SWCm (Staley et al., 1989) broth with 25% glycerol or 10% DMSO and freezing at −80°C.

Differentiation of the Genus *Polaromonas* from Other Genera

Analyses of 16S rDNA sequences using various phylogenetic methods produce slightly different trees. For all the trees, how-

ever, it is clear that strain 34-P is most closely related to *Rhodoferax fermentans* (Hiraishi et al., 1991a), *Variovorax paradoxus* (Davis et al., 1969; Willems et al., 1991a), "str. Stripa", and "env. PAD44" (Ueda, unpublished results). It is, however, phenotypically and genotypically quite different from these organisms. Strain 34-P is not photosynthetic and does not grow as a nonsulfur purple bacterium under conditions used for the growth of *R. fermentans*. Strain 34-P differs by 5% and 7% in 16S rDNA base homology from *V. paradoxus* (Davis et al., 1969; Willems et al., 1991a) and *R. fermentans* (Hiraishi et al., 1991a), respectively. Furthermore, other genotypic and phenotypic differences indicate marked differences among *Polaromonas vacuolata*, *V. paradoxus*, and *R. fermentans* (Table BXII.β.57). For example, the mol% G + C values are 52–57 versus 67–69 and 60, respectively (Davis et al., 1969; Irgens et al., 1989; Hiraishi et al., 1991a; Willems et al., 1991a). In addition, *V. paradoxus* and *R. fermentans* are both pigmented, are not gas vacuolate, and differ from *P. vacuolata* in cell shape and motility (Table BXII.β.57).

TABLE BXII.β.57. Phenotypic comparison of *Polaromonas vacuolata* to the two phylogenetically most closely related species[a]

Characteristic	Polaromonas vacuolata	Rhodoferax fermentans	Variovorax paradoxus
Shape of cells	rods	curved rods	straight or curved rods
Gas vesicles	+	−	−
Photosynthetic	−	+	−
Flagella:			
Peritrichous			+
Polar	+	+	
O₂ requirement:			
Facultative aerobe		+	
Obligate aerobe	+		+
Temperature relations:			
Mesophilic		+	+
Psychrophilic	+		
Colony pigmentation[b]	W	Pb	Y
Mol% G + C	52–57	60	67–69

[a]Data from Davis et al., 1969; Irgens et al., 1989; Hiraishi et al., 1991a.

[b]W, white; Pb, peach-brown; Y, yellow.

TAXONOMIC COMMENTS

The complete region of 16S rDNA phylogeny delineated by this genus remains uncertain pending nucleotide sequence analysis of additional members. In addition, the large number of partial nucleotide sequences closely related to the 16S rDNA sequence of this genus may blur resolution of this region of the phylogenetic tree.

List of species of the genus Polaromonas

1. **Polaromonas vacuolata** Irgens, Gosink and Staley 1996, 825[VP]

va.cu.o.la′ ta. L. adj. *vacuus* empty; N.L. part. adj. *vacuolata* equipped with gas vacuoles.

Cells contain gas vesicles. Temperature for optimum growth is 4°C, with a range of 0–12°C. Colonies are snowy white, circular, and convex, with a smooth surface and an entire edge. The more gas vesicles within the cells, the whiter the colony. Grows on fumarate, glucose, alanine, glutamate, and aspartate. Does not grow on sucrose, lactose, or arginine. There is good growth in media with NaCl concentrations ranging from 0–6.0% but no growth at 7.0%. Fatty acid composition is 75% $C_{16:1\ \omega7c}$, 17% $C_{16:0}$, and 8% $C_{18:1\ \omega7c}$, $C_{18:1\ \omega9t}$, or $C_{18:1\ \omega12t}$.

The mol% G + C of the DNA is: 52.0 (T_m).

Type strain: 34-P, ATCC 51984.

GenBank accession number (16S rRNA): U14585.

Genus IX. **Rhodoferax** Hiraishi, Hoshino, and Satoh 1992a, 192[VP] (Effective publication: Hiraishi, Hoshino, and Satoh 1991a, 334)

AKIRA HIRAISHI AND JOHANNES F. IMHOFF

Rho.do.fe′ rax. Gr. n. *rhodon* rose; L. adj. *ferax* fertile; M.L. masc. n. *Rhodoferax* red and fertile.

Cells are curved or vibrioid rods, motile by means of polar flagella, that multiply by binary fission and do not form endospores and capsules. **Gram negative, belonging to the *Betaproteobacteria*.** Internal membrane systems are poorly developed or absent. **Photosynthetic pigments are bacteriochlorophyll *a* and carotenoids of the spheroidene series. Contain ubiquinones and rhodoquinones with eight isoprene units (Q-8 and RQ-8). Straight-chain $C_{16:1}$ and $C_{16:0}$ acids are the major components of cellular fatty acids. $C_{8:0\ 3OH}$ acid is present.** Membrane-bound fumarate reductase activity occurs with $FMNH_2$ as the electron donor.

Growth is possible by photosynthesis, aerobic respiration, or fermentation. Photoheterotrophy with various organic compounds as carbon sources is the preferred mode of growth. Cells grow well with simple organic compounds as electron donors and carbon sources and in complex media containing peptone, yeast extract, or Casamino acids. **Mesophilic and psychrophilic, neutrophilic fresh water bacteria.** Habitat: ditchwater, activated sludge, Antarctic microbial mats.

The mol% G + C of the DNA is: 59.8–61.5.

Type species: **Rhodoferax fermentans** Hiraishi, Hoshino and Satoh 1992a, 192 (Effective publication: Hiraishi, Hoshino and Satoh 1991a, 334.)

FURTHER DESCRIPTIVE INFORMATION

According to 16S rDNA sequences, the closest phylogenetic relative to the *Rhodoferax* species is the purely chemotrophic *Variovorax paradoxus* (sequence similarity of 96%). The nearest phototrophic phylogenetic neighbor is *Rubrivivax gelatinosus* (91% sequence similarity; see Fig. 4 [p.132] of the introductory chapter "Anoxygenic Phototrophic Purple Bacteria", Volume 2, Part A).

In addition, the aerobic, bacteriochlorophyll-containing bacterium *Roseateles depolymerans* is within the same phylogenetic group.

R. fermentans is a facultatively photoheterotrophic bacterium that grows anaerobically in the light and aerobically in darkness at full atmospheric oxygen tension and exhibits a doubling time of 3–4 h under optimal growth conditions. Colonies grown on agar media are of intermediate size (2–4 mm after one week of incubation), round, smooth, circular, convex, and peach brown. However, rigid colonies frequently occur upon subculture. Phototrophic liquid cultures become peach brown, whereas aerobic, chemotrophic cultures are colorless or faintly pink. Absorption spectra of phototrophically grown cells or membrane preparations show major absorption maxima at around 800 and 850 nm, indicating that the cells contain the core light-harvesting complex (LH I), together with the photosynthetic reaction center and the peripheral light-harvesting complex (LH II).

One of the most remarkable properties of *Rhodoferax fermentans* is its capability of anaerobic fermentation in darkness. In Hugh-Leifson's OF test, which is commonly used to characterize chemoorganotrophic bacteria, *R. fermentans* produces acid from glucose in both open and sealed tubes within a few days of incubation (Hiraishi and Kitamura, 1984). Such a rapid glucose fermentation has not been found in other species of phototrophic bacteria described so far. Fermentative growth in darkness occurs on pyruvate and sugars, among which fructose is the best substrate. The addition of bicarbonate significantly enhances anaerobic growth in the dark. The doubling time for fermentative growth on 20 mM fructose plus 30 mM bicarbonate is 4–5 h (Hiraishi, 1988b). The end products of fructose fermentation are acetate, formate, lactate, succinate, and ethanol. Hypophosphite, a potent inhibitor of pyruvate–formate lyase, completely suppresses the production of formate and increases the amount of succinate excreted. These observations suggest that the bicarbonate-dependent fermentative growth may be linked to CO_2 fixation via part of the reductive TCA cycle and the subsequent reduction of fumarate to succinate. GDP-dependent phosphoenolpyruvate carboxykinase has been suggested to function as a key enzyme for CO_2 fixation (Hiraishi, 1988b), and reduced FMN-linked fumarate reductase activity that is possibly associated with a low-potential rhodoquinone within the cells has been found in the membrane preparations (Hiraishi, 1988a).

While many species of phototrophic purple nonsulfur bacteria exhibit anaerobic growth in darkness coupled with reduction of trimethylamine-*N*-oxide or dimethylsulfoxide as the terminal electron acceptor, *R. fermentans* lacks these properties. Most strains of this species are also devoid of nitrate reductase activity, although strain DSM 10139 is highly active at nitrate reduction (Hougardy and Klemme, 1995).

Some information has been available on the components of the respiratory and photosynthetic electron transport systems of *R. fermentans*. Cells produce two quinone molecules, Q-8 and RQ-8, constitutively. The RQ/Q molar ratio is 0.5–0.7 under phototrophic growth conditions with malate as the carbon source (Hiraishi and Hoshino, 1984). This ratio is higher (0.9–1.7) in cells grown fermentatively or phototrophically and becomes much lower (0.1–0.2) in aerobically grown cells (Hiraishi et al., 1991a). It has been shown that the oxidative electron transport chain is branched at the ubiquinone level and does not involve rhodoquinone (Hochkoeppler et al., 1995b). Aerobically grown cells contain four *b*-type and three *c*-type membrane-bound cytochromes but lack soluble *c*-type cytochromes. A soluble, high-potential iron-sulfur protein (HiPIP) functions as an alternative

to the soluble cytochrome *c* in linking the bc_1 complex to the terminal oxidase in respiratory electron transfer (Hochkoeppler et al., 1995a). Phototrophic cells contain the reaction center complex with tetraheme *c* as direct electron donors to the bacteriochlorophyll dimer. HiPIP is also competent as an alternative to the soluble *c*-type cytochrome in photosynthetic electron transfer (Hochkoeppler et al., 1995a, 1996).

The genes (*puf*) coding for the L and M protein subunits of the photosynthetic reaction center have been analyzed by PCR cloning (Nagashima et al., 1997a). The deduced amino acid sequences of the puf proteins of *R. fermentans* are most similar to those of *Rubrivivax gelatinosus* (77% identity). A phylogenetic analysis based on the nucleotide sequences of DNA fragments coding for the L and M subunits has shown that the sequences of *R. fermentans* and of two other species of the *Betaproteobacteria*, *Rubrivivax gelatinosus* and *Rhodocyclus tenuis*, are positioned among those of *Alphaproteobacteria*. This contrasts with their phylogenetic relations, as based on 16S rDNA sequences (see Fig. 4 [p. 132] in the introductory chapter "Anoxygenic Phototrophic Purple Bacteria", Volume 2, Part A). The inconsistency is explained by possible horizontal transfer of the genes encoding the reaction center during evolution of photosynthesis.

As with other phototrophic *Betaproteobacteria*, the major phospholipid components of *Rhodoferax fermentans* are cardiolipin, phosphatidylethanolamine, and phosphatidylglycerol (A. Hiraishi, unpublished data). Straight-chain $C_{16:1}$ and $C_{16:0}$ acids are the main components of cellular fatty acids (see Table BXII.β.58).

Rhodoferax antarcticus is characterized by its preference for low temperatures and grows optimally at 12–18°C, but not above 25°C (Madigan et al., 2000b).

ENRICHMENT AND ISOLATION PROCEDURES

The natural habitats of *Rhodoferax* species are freshwater environments that are rich in organic matter. Sulfide-rich water bodies may not provide favorable conditions for *R. fermentans* because of its inability to use sulfide as the electron donor for growth and its weak tolerance toward sulfide. Strains of *Rhodoferax fermentans* have been isolated from pond water, sewage, and activated sludge. *R. antarcticus* was isolated from an Antarctic microbial mat.

Enrichment and isolation are possible under conditions also suitable for most of the species of phototrophic purple nonsulfur bacteria. Although it is not easy to perform selective enrichment of *Rhodoferax fermentans* from environmental samples, the addition of 0.5 mM EDTA to the enrichment medium may be effective for suppressing the overgrowth of possibly co-existing, fast growing phototrophic species, such as *Rhodobacter capsulatus* and *Rhodobacter sphaeroides* (Hougardy and Klemme, 1995). A suitable enrichment medium consists of basal mineral medium, one or more simple organic compounds (e.g., 0.1% acetate or 0.2% glucose), and a vitamin mixture or 0.01% yeast extract. Incubation is under anoxic conditions under incandescent illumination at 1000–2000 lux at 28°C for *R. fermentans* or at 12–18°C for *R. antarcticus*.

A simple medium for growth and purification of *Rhodoferax* is MYCA medium (Hiraishi et al., 1991a), which contains 0.1% DL-malate, 0.3% yeast extract, 0.2% Casamino acids, and 0.05% $(NH_4)_2SO_4$ and is adjusted to pH 6.6–6.8.

MAINTENANCE PROCEDURES

Cultures are well-preserved in liquid nitrogen or by lyophilization. Preservation is also possible in a mechanical freezer at −80°C.

TABLE BXII.β.58. Differential characteristics of anoxygenic phototrophic *Betaproteobacteria* of the order *Burkholderiales*: genera *Rhodoferax* and *Rubrivivax*[a]

Characteristic	Rhodoferax fermentans	Rhodoferax antarcticus	Rubrivivax gelatinosus
Cell diameter (μm)	0.6–0.9	0.7	0.4–0.7
Cell shape	Curved rods	Curved rods	Straight to curved rods
Motility	+	+	+
Slime production	−	−	+
Color	Peach brown	Peach brown	Brown
Major carotenoids	Spheroidene, OH-spheroidene, spirilloxanthin	Most likely spheroidene and OH-spheroidene	Spheroidene, OH-spheroidene, spirilloxanthin
Growth factors	thiamine, biotin	biotin	thiamine, biotin[b]
Gelatin liquefaction	+	nd	+
Fructose fermentation	+	−	−
Starch hydrolysis	−	nd	+
Tween 80 lysis	−	nd	+
Carbon sources:			
Benzoate	−	−	−
C_{10} to C_{18} fatty acids	nd	nd	+
Citrate	+	+	+
Mannitol	+	−	−
Sorbitol	+	nd	+
N_2-fixation	+	+	+
Fumarate reductase activity:			
With reduced methylviologen	Low	nd	High
With $FMNH_2$	High	nd	Low
Major fatty acids:			
$C_{16:0}$	33–39	nd	24–35
$C_{16:1}$	52–54	nd	35–45
$C_{18:0}$	<1	nd	1–3
$C_{18:1}$	5	nd	16–25
3-OH fatty acid	8:0	nd	10:0
Major quinones	Q-8 + RQ-8	nd	Q-8 + MK-8
Mol% G + C of DNA			
by HPLC	59.8–60.3	nd	71.2–72.1
by Bd	nd	nd	70.5–72.4
by T_m	nd	61.5	70.2–71.9

[a]Symbols: +, positive in most strains; −, negative in most strains; Q-8, ubiquinone - 8; RQ-8, rhodoquinone-8; MK-8, menaquinone-8.

[b]Some strains may also require pantothenate.

DIFFERENTIATION OF THE GENUS *RHODOFERAX* FROM OTHER GENERA

The genus *Rhodoferax* is differentiated from other genera of phototrophic *Proteobacteria* by its phylogenetic position and its unique physiological and chemotaxonomic properties. The characteristic features of *Rhodoferax* include fructose fermentation, Q-8 and RQ-8 production, and relatively low mol% G + C contents of the DNA. Differential characteristics of *Rhodoferax* species and other phototrophic *Betaproteobacteria* are given in Tables 6 (p. 130) and 7 (p. 131) of the introductory chapter "Anoxygenic Phototrophic Purple Bacteria", Volume 2, Part A. The phylogenetic relationships of the phototrophic *Betaproteobacteria* are shown in Fig. 4 (p. 132) of the introductory chapter "Anoxygenic Phototrophic Purple Bacteria", Volume 2, Part A. *Rhodoferax* is compared to *Rubrivivax gelatinosus* in Tables BXII.β.58 and BXII.β.59.

TAXONOMIC COMMENTS

Strains of *R. fermentans* were first recognized as the " *Rhodocyclus gelatinosus*-like (RGL)" group (Hiraishi and Hoshino, 1984), because they were phenotypically similar to *Rubrivivax gelatinosus* (at that time known as *Rhodocyclus gelatinosus*), in particular appearing to resemble *Rubrivivax gelatinosus* subgroup II (Weckesser et al., 1969). More detailed physiological, chemotaxonomic, and genetic studies revealed major differences between the RGL group and *Rubrivivax gelatinosus*, and these observations led to the proposal for the classification of the RGL group into the new genus *Rhodoferax* (Hiraishi et al., 1991a) as *R. fermentans*. Later, a phylogenetic study based on 16S rDNA sequences showed that the genera *Rhodoferax*, *Rubrivivax*, and *Rhodocyclus* are phylogenetically distinct groups within the *Betaproteobacteria* (Hiraishi, 1994).

List of species of the genus Rhodoferax

1. **Rhodoferax fermentans** Hiraishi, Hoshino and Satoh 1992a, 192[VP] (Effective publication: Hiraishi, Hoshino and Satoh 1991a, 334.)

fer.men' tans. M.L. part. adj. *fermentans* fermenting.

Cells are curved or vibrioid rods, occurring singly or in

TABLE BXII.β.59. Carbon sources and electron donors used by anoxygenic phototrophic *Betaproteobacteria* of the order *Burkholderiales*: genera *Rhodoferax* and *Rubrivivax*[a]

Source/donor	*Rhodoferax fermentans*	*Rhodoferax antarcticus*	*Rubrivivax gelatinosus*
Carbon source			
Acetate	+	+	+
Arginine	−	nd	nd
Aspartate	+	+	+
Benzoate	−	−	−
Butyrate	+	+	+/−
Caproate	−	−	nd
Caprylate	−	−	nd
Citrate	−	+	+
Ethanol	+/−	−	+
Formate	−	−	+/−
Fructose	+	+	+
Fumarate	+	+	+
Glucose	+	+	+
Glutamate	+	−	+
Glycerol	−	−	−
Glycolate	−	−	nd
Lactate	+/−	+	+
Malate	+	+	+
Malonate	−	nd	nd
Mannitol	+	−	−
Mannose	+	−	+
Methanol	−	−	+/−
Pelargonate	nd	nd	nd
Propionate	−	−	+/−
Pyruvate	+	+	+
Sorbitol	+	nd	−
Succinate	+	+	+
Tartrate	−	nd	+/−
Valerate	nd	−	+
Electron donor:			
Hydrogen	nd	+	+
Sulfide	−	−	−
Sulfur	nd	−	−
Thiosulfate	−	−	−

[a] Symbols: +, positive in most strains; −, negative in most strains; +/− variable in different strains; nd, not determined.

pairs (Fig. BXII.β.42), 0.6–0.9 × 2–5 μm under optimal growth conditions. Long chains and helical filaments may be produced, especially when cells are grown with sugars under phototrophic or fermentative growth conditions. Spheroplast-like coccoid cells occasionally occur as cultures age (see Fig. BXII.β.42). **Motile by single polar flagella** (Fig. BXII.β.43). Internal membrane systems are poorly developed or absent (Fig. BXII.β.44). The color of photosynthetic cultures and colonies is peach brown. Absorption spectra of living cells show maxima at 377–378, 457–458, 484–485, 589–590, 799–801, and 850–851 nm. **Photosynthetic pigments are bacteriochlorophyll *a* esterified with phytol and carotenoids of the spheroidene series** with spheroidene, spirilloxanthin, and OH-spheroidene are present as major components.

Photoheterotrophic growth under anoxic conditions in the light or chemotrophic growth under oxic conditions in the dark at full atmospheric oxygen tension is possible. Photoorganotrophy with sugars, peptone, yeast extract, and Casamino acids as electron donors and carbon sources is the preferred mode of growth. In addition, simple organic compounds, such as acetate and intermediates of the tricarboxylic acid cycle, also support growth as carbon sources. The carbon sources utilized are listed in Table BXII.β.59.

Also utilized are xylose, arabinose, galactose, gluconate, and asparagine. Bicarbonate-dependent dark fermentative growth is also possible. Sulfide, thiosulfate, and other reduced sulfur compounds do not serve as electron donors for photoautotrophic growth. Sulfide tolerance is weak; only 0.2 mM sulfide inhibits growth in the presence of 0.05% yeast extract. Ammonium salts and glutamate, but not nitrate, serve as nitrogen sources. The capacity for nitrogen fixation appears to be present, as suggested by the observation that nitrogenase-dependent H_2 production occurs in an ammonium-deficient medium. Sulfate is assimilated as a sulfur source. Catalase is negative or weakly positive. Oxidase reaction is positive. Proteolytic activities are present, but weak compared to those of the phylogenetic relative *Rubrivivax gelatinosus*. Gelatin is weakly hydrolyzed. Thiamine and biotin are required as growth factors.

Mesophilic and neutrophilic fresh water bacterium with optimal growth at 25–30°C (no growth at 37°C) and pH 6.5–7.0 (pH-range: pH 5–9). No growth occurs in the presence of 1.5% NaCl. Habitat: freshwater ponds, sewage ditches, and activated sludge. Major quinone components are Q-8 and RQ-8.

The mol% G + C of the DNA is: 59.8–60.3 (HPLC); type strain, 60.1 (HPLC).

Type strain: FR2, ATCC 49787, DSM 10138, JCM 7819.

GenBank accession number (16S rRNA): D16211.

2. **Rhodoferax antarcticus** Madigan, Jung, Woese, Achenbach 2000b, 275[VP]

ant.arc′ti.cus. M.L. adj. *antarcticus* belonging to, coming from Antarctica.

Cells are curved or vibrioid rods, 0.7 × 2–3 μm under optimal growth conditions. Highly motile by polar flagella. Internal membrane systems are poorly developed or absent. The color of photosynthetic cultures and colonies is peach brown. Absorption spectra of living cells show maxima at 427, 452, 487, 518, 582, 799, 819, and 866 nm. Photosynthetic pigments are bacteriochlorophyll *a* and carotenoids, most likely of the spheroidene series.

Photoheterotrophic growth under anoxic conditions in the light or chemotrophic growth under oxic conditions in the dark at full atmospheric oxygen tension is possible. Acetate pyruvate, lactate, succinate, malate, fumarate, glucose, fructose, sucrose, citrate, and aspartate support growth as carbon sources. Not utilized are methanol, ethanol, propanol, butanol, mannitol, glycerol, lactose, benzoate, formate, propionate, butyrate, valerate, caproate, heptanoate, glutamate, glutamine, and oxoglutarate. Bicarbonate-dependent dark fermentative growth with glucose and fructose is not possible. Sulfide at 2 mM inhibits growth. Photoautotrophic growth with hydrogen is slow. Ammonium salts, glutamate, glutamine, and aspartate serve as nitrogen sources. The capacity for dinitrogen fixation is present. Sulfate is assimilated as a sulfur source. Biotin is required as growth factor.

Psychrophilic and neutrophilic freshwater bacterium with optimal growth at 12–18°C (range: 0–25°C, no growth above 25°C). NaCl is not required and growth inhibitory above 2%.

Habitat: Antarctic microbial mat.

The mol% G + C of the DNA is: 61.5 (T_m).

Type strain: ATCC 700587.

GenBank accession number (16S rRNA): AF084947.

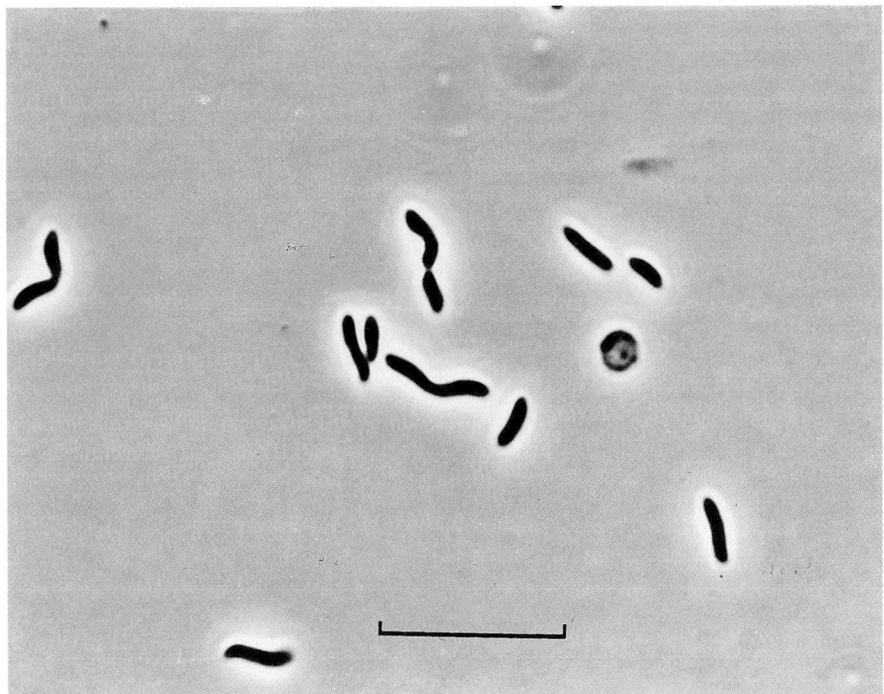

FIGURE BXII.β.42. Phase-contrast micrograph showing cell morphology of *Rhodoferax fermentans* (strain FR2). Curved rods occurring singly and in pairs and a spheroplast-like cell are observed. Bar = 10 μm.

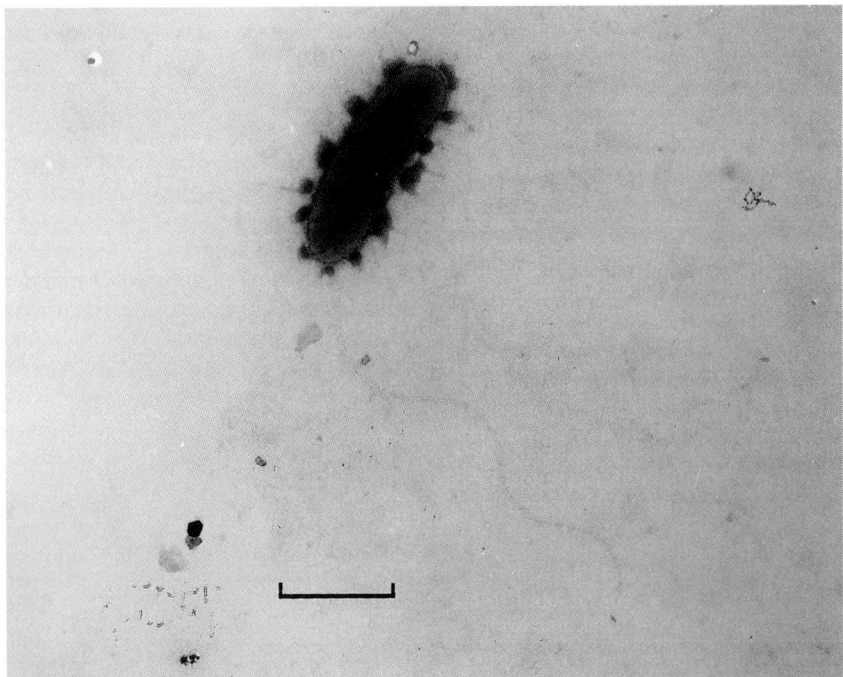

FIGURE BXII.β.43. Electron micrograph of a negatively stained cell of *Rhodoferax fermentans* (strain FR2) with a polar flagellum. Bar = 1 μm.

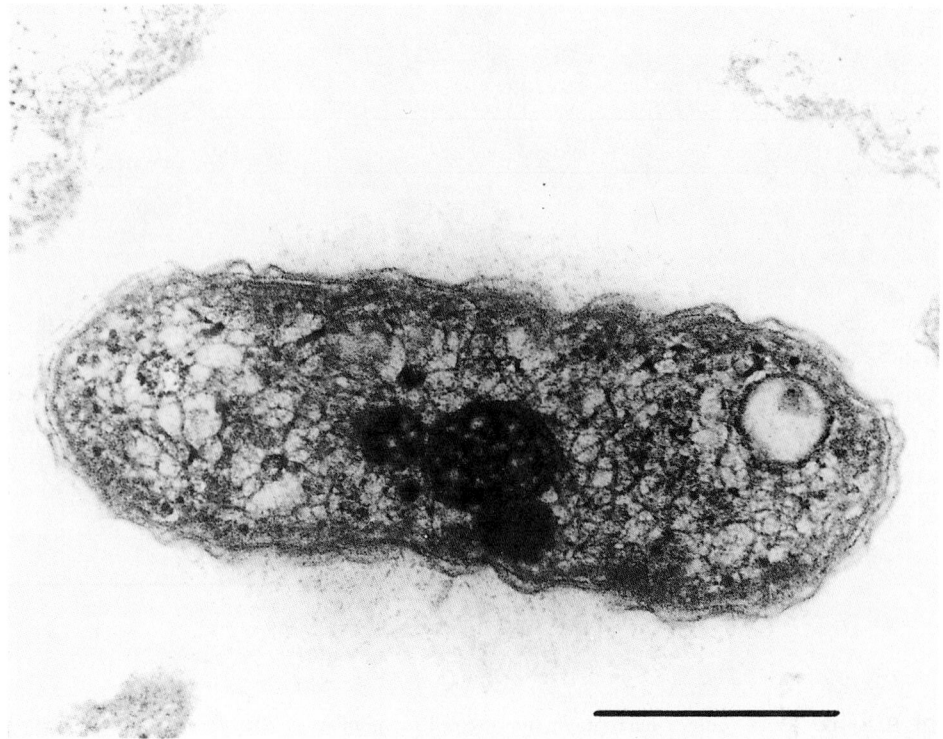

FIGURE BXII.β.44. Thin-section electron micrograph showing ultrastructure of *Rhodoferax fermentans* (strain FR2) grown phototrophically. Bar = 0.5 μm.

Genus X. *Variovorax* Willems, De Ley, Gillis and Kersters 1991a, 446[VP]

ANNE WILLEMS, JORIS MERGAERT AND JEAN SWINGS

Va.ri.o' vo.rax. L. adj. *varius* various; L. adj. *vorax* voracious; M.L. masc. n. *Variovorax* (bacteria) devouring a variety (of substrates).

Straight to slightly curved rods, 0.5–0.6 × 1.2–3.0 μm, occurring singly or in pairs. **Gram negative. Motile by means of sparse, peritrichous flagella.** Colonies are yellow due to the presence of carotenoid pigments. **Aerobic, having a strictly respiratory type of metabolism with O₂ as the terminal electron acceptor. Oxidase and catalase positive. Chemoorganotrophic. Some strains are capable of lithoautotrophic growth**, using H₂ as an energy source. These are referred to as biotype I, whereas nonlithoautotrophic strains are referred to as biotype II (Davis et al., 1969). Good growth occurs on media containing carbohydrates, organic acids, amino acids, or peptone.

The mol% G + C of the DNA is: 66.8–69.4.

Type species: **Variovorax paradoxus** (Davis 1969) Willems, De Ley, Gillis and Kersters 1991a, 447 (*Alcaligenes paradoxus* Davis *in* Davis, Doudoroff, Stanier and Mandel 1969, 387.)

FURTHER DESCRIPTIVE INFORMATION

Flagellation In a study of the cell morphology and flagellation of Gram-negative hydrogen bacteria, Aragno et al. (1977) defined three types of flagella based on electron microscopy. They reported that *V. paradoxus* cells have one to three laterally inserted type II flagella. This oligotrichous flagellation was originally described as "degenerately peritrichous" (Davis et al., 1969). The flagella have diameters of 15–17 nm. Flagellar wave-

lengths have been observed to be of two distinct types, one about half as long (1.2–1.3 μm) as the other (2.2–2.5 μm) occurring in the same culture (Aragno et al., 1977). Pili are inserted at the polar caps (Aragno et al., 1977).

Cell wall composition Walther-Mauruschat et al. (1977) have studied the cell envelopes of Gram-negative hydrogen bacteria and distinguished three types of cell walls, differing mainly in visibility and location of the peptidoglycan layer. *Variovorax* possesses type I cell walls, typical of most Gram-negative bacteria and characterized by a multilayered structure, consisting of an outer membrane and a cytoplasmic membrane of similar dimensions and appearance, which are separated by a dense layer of peptidoglycan.

Ultrastructure Intracellular mesosome-like membrane systems with a spiral appearance are present and often located in the area of cell division or at the cell poles (Walther-Mauruschat et al., 1977). Their significance is unclear, and they are now regarded mainly as artifacts resulting from the preparation of cells for electron microscopy. Intracellular inclusions of poly-3-hydroxybutyrate and polyphosphate are present, and translucent, glycogen-like inclusions may also be detected (Walther-Mauruschat et al., 1977). Cytochromes of the *a*, *b*, and *c* types are present, including a cytochrome *a2* with a peak at 625 nm in the reduced/oxidized spectrum (Davis et al., 1969).

Colonial characteristics Colonies on nutrient agar are glistening and shiny, yellow or greenish yellow. Occasionally, two types of colonies may occur in a single culture. Often, these types prove to be unstable and cannot be separated. Sometimes, nonmucoid variants may be separated from mucoid strains, but SDS–polyacrylamide electrophoregrams of whole-cell proteins of such types are always identical. The yellow pigments are carotenoids with absorption maxima in acetone of approximately 405 and/or 425 nm (Davis et al., 1969). More recently, Urakami et al. (1995a) have reported an absorption maximum of 420 nm for the carotenoid pigments of *V. paradoxus* DSM 30034[T].

Nutrition and metabolism A large variety of organic compounds may be used as sole carbon sources (Table BXII.β.60). Of a total of 143 organic compounds tested, 99 are used by *V. paradoxus* strains (Davis et al., 1969). No growth factors are required. Two biotypes are recognized. Strains of biotype I are able to grow autotrophically using the aerobic oxidation of hydrogen as an energy source. The hydrogenase is membrane-bound and does not reduce NAD (Schneider and Schlegel, 1977). It is similar to that of *Ralstonia eutropha* in that PCR tests with primers specific for the *R. eutropha* hydrogenase gene yield products of the correct size for *V. paradoxus* but no product for other hydrogen bacteria (Lechner and Conrad, 1997). No soluble hydrogenase is present (Schneider and Schlegel, 1977). Strains of biotype II cannot grow autotrophically or oxidize hydrogen (Davis et al., 1969).

Nitrate, but not atmospheric nitrogen, can be used by many strains as a nitrogen source. *Meta*- and *para*-hydroxybenzoate are metabolized via the *ortho* cleavage of protocatechuate. A new pathway for the degradation of homovanillic acid, distinct from that of *Pseudomonas putida*, has been reported for a *V. paradoxus* strain isolated from farmyard soil by enrichment on vanillylpyruvate. Homovanillic acid is degraded via ring hydroxylation to a dihydroxymonomethoxyphenylacetic acid, which is then cleaved by homogentisate 1,2-dioxygenase (Allison et al., 1995). Catechol 1,2-dioxygenase from *V. paradoxus* has been purified and characterized (Matevosyan et al., 1989).

Biosynthesis of aromatic amino acids *V. paradoxus* strains possess prephenate dehydrogenase and arogenate dehydrogenase that are reactive with either NAD or NADP. Arogenate dehydratase activity is absent; prephenate dehydratase is present, and its activity is increased 2–3-fold by 0.5 mM L-tyrosine (Byng et al., 1983). 3-Deoxy-D-arabinoheptulosonate 7-phosphate synthetase is inhibited by phenylalanine and chorismate (Whitaker et al., 1981). The various pathways for the biosynthesis of aromatic amino acids and their regulation have been shown to be useful indicators for the differentiation of different groups of pseudomonad bacteria (Byng et al., 1983).

Chemotaxonomic characteristics *Variovorax* possesses a Q-8 ubiquinone system (Urakami et al., 1995a). Major fatty acids (representing at least 5% of total fatty acids) are palmitic acid ($C_{16:0}$), a cyclopropane substituted fatty acid ($C_{17:0\ cyclo}$), *cis*-vaccenic acid ($C_{18:1}$), palmitoleic acid ($C_{16:1}$), 3-hydroxydecanoic acid ($C_{10:0\ 3OH}$), and 2-hydroxytetradecanoic acid ($C_{14:0\ 2OH}$). (Willems et al., 1989; Urakami et al., 1995a). Characteristic polyamines are putrescine and 2-hydroxyputrescine (Busse and Auling, 1988).

Plasmids Plasmid DNA is regarded as particularly important for the genetic adaptation of microorganisms in the natural environment, because as a mobile form of DNA, it can provide new phenotypes, allowing bacteria to survive and even thrive under

TABLE BXII.β.60. Characteristics of *Variovorax paradoxus*[a,b]

Characteristic	Reaction
Oxidase, catalase, urease	+
Denitrification	−
Nitrite reduction	−
Intracellular accumulation of poly-3-hydroxybutyrate	+
Extracellular hydrolysis of poly-3-hydroxybutyrate, Tween 80	+
Hydrolysis of starch, gelatin	−
Growth with D-glucose, D-fructose, D-galactose, L-arabinose, D-fucose, D-xylose, D-mannose, sorbitol, D-mannitol, D-arabitol, ethanol, glycerol, D,L-glycerate, citrate, mesaconate, succinate, fumarate, D,L-lactate, L-malate, D,L-3-hydroxybutyrate, α-ketoglutarate, D-gluconate, 5-ketogluconate, acetate, pyruvate, sebacate, mesaconate, pantothenate, adipate, pimelate, hydroxymethylglutarate, quinate, *m*-hydroxybenzoate, *p*-hydroxy-benzoate, L-leucine, L-histidine, L-α-alanine, L-glutamate, L-proline, L-aspartate, L-phenylalanine, L-ornithine, D,L-4-aminobutyrate[c]	+
Growth with L-xylose, L-sorbose, D-melibiose, sucrose, trehalose, melezitose, D-raffinose, maltose, lactose, D-turanose, D-tagatose, β-gentiobiose, α-methylxyloside, α-methyl-D-mannoside, α-methyl-D-glucoside, amygdalin, arbutin, esculin, salicin, inulin, starch, glycogen, dulcitol, *meso*-erythritol, heptanoate, caprylate, pelargonate, caprate, oxalate, anthranilate, nicotinate, testosterone, benzoate, *o*-hydroxybenzoate, benzylamine, 2-aminobenzoate, 3-aminobenzoate, 4-aminobenzoate, D-mandelate, isophthalate, terephthalate, L-valine, L-lysine, L-norleucine, L-cysteine, L-methionine, D-tryptophan, L-citrulline, L-arginine, D,L-kynurenine, D,L-2-amino-butyrate, δ-aminovalerate, trigonellin, urea, amylamine, ethylamine, diaminobutane, glucosamine, putrescine, spermine, histamine, tryptamine, betaine, sarcosine, creatine, acetamide, *N*-acetylglucosamine[c,d]	−
Utilization of D-fucose, L-fucose, cellobiose, D-ribose, L-rhamnose, D-arabinose, D-lyxose, adonitol, *meso*-inositol, xylitol, L-arabitol, 2-ketogluconate, saccharate, malonate, suberate, D-malate, *meso*-tartrate, D-tartrate, L-tartrate, mucate, glycolate, propionate, butyrate, isobutyrate, valerate, isovalerate, malonate, caproate, maleate, glutarate, azelate, levulinate, aconitate, citraconate, itaconate, L-mandelate, phthalate, phenylacetate, kynurenate, 2,3-butylene glycol, phenol, D-α-alanine, β-alanine, L-tryptophan, L-serine, L-threonine, L-isoleucine, L-tyrosine, L-tryptophan, D,L-norvaline, D,L-3-aminobutyrate, butylamine, pentylamine, glycine, ethanolamine[d,e]	d
Susceptible to:	
Chloramphenicol (30 μg/disk), tetracycline (30 μg/disk), kanamycin (30 μg/disk), neomycin (30 μg/disk), polymyxin B (300 U/disk)	+
Erythromycin (15 μg/disk), streptomycin (10 μg/disk)	d
Ampicillin (10 μg/disk), methicillin (5 μg/disk), novobiocin (30 μg/disk)	−

[a]For symbols see standard definitions.

[b]Data taken from Davis et al. (1969, 1970) (14 strains), Auling et al. (1978), Kersters and De Ley (1984b) (14 strains), and Aragno and Schlegel (1992).

[c]Aragno and Schlegel (1992) reported 11–89% of strains grew on citrate, acetate, *p*-hydroxybenzoate, L-leucine, L-histidine, L-α-alanine, L-glutamate, L-aspartate, L-phenylalanine, L-valine, and δ-aminovalerate in API auxanographic galleries.

[d]Aragno and Schlegel (1992) reported all strains grew on oxalate, 2-ketoglutarate, and glutarate in API auxanographic galleries.

[e]Aragno and Schlegel (1992) reported all strains failed to grow on cellobiose, *meso*-inositol, butylamine, and ethanolamine in API auxanographic galleries.

changing and often adverse environmental conditions (Leahy and Colwell, 1990). Pathways for the degradation of various complex organic molecules have been shown to be plasmid encoded in *Pseudomonas*-like bacteria (Chakrabarty, 1976). Since *V. para-*

doxus strains are soil and water bacteria, it is not surprising that various plasmids have been reported to occur. Strains isolated from soil samples contaminated with the herbicide 2,4-dichlorophenoxyacetic acid (2,4-D) contain plasmids pJP2, which carries genes for the degradation of 2,4-D, and pJP3, which carries genes conferring resistance to merbromin, phenyl-mercury acetate, and mercuric ions, and the ability to degrade *m*-chlorobenzoate (Don and Pemberton, 1981). Another 2,4-D-degradative plasmid has been isolated from a *V. paradoxus* strain from agricultural soil and is capable of integrating in the chromosome without loss of the 2,4-D$^+$ phenotype. In the same study, a very similar plasmid pKA4 was found in a 2,4-D-degrading *Ralstonia pickettii* strain, suggesting natural horizontal gene transfer among genera (Ka and Tiedje, 1994). This has been supported by use of *V. paradoxus* as the recipient organism for the large catabolic plasmid pJP4 of *Ralstonia eutropha* in a study of horizontal gene transfer in soil (Neilson et al., 1994).

Di Giovanni et al. (1996) have observed that several different plasmids may occur per strain in a study of 32 strains of 2,4-D-degrading *V. paradoxus* isolates from a sample of contaminated soil. Based on the size and number of plasmids, they have distinguished 6 groups among the strains, with each group containing identical and unique plasmids of diverse size. Curing of the plasmids results in the loss of the ability to break down 2,4-D. Comparison of restriction patterns of plasmids from 2,4-D$^+$ and 2,4-D$^-$ strains shows that plasmids involved in 2,4-D-degradation in different strains can have different origins. The data suggest that plasmids are frequently exchanged within a population, enhancing gene transfer and recombination events (Di Giovanni et al., 1996).

Hydrogen metabolism and autotrophic growth are thought to be, at least in part, plasmid-linked, because exposure of a *V. paradoxus* strain to the plasmid-curing agent mitomycin C results in the loss of the autotrophic phenotype (Lim et al., 1980).

Antibiotic sensitivity Auling et al. (1978) have studied the susceptibility to antibiotics of five *V. paradoxus* strains and reported all strains to be sensitive to five of the ten antibiotics tested (see Table BXII.β.60).

Ecology *Variovorax* is a common soil inhabitant, also occurring in contaminated soils, where it may harbor plasmids carrying genes involved in the breakdown of complex chemicals and resistance to toxic compounds. *Variovorax* strains dominate the 2,4-dichlorophenoxyacetic acid (2,4-D)-degrading population in agricultural soil amended with this herbicide (Dunbar et al., 1995). A strain of *V. paradoxus* capable of degrading 2,4-D has been isolated from pristine Hawaiian volcanic soil. It has been found to contain a *tfdA* gene (encoding the enzyme for the first step of 2,4-D mineralization) that is transmissible to *Ralstonia eutropha* JMP228 carrying a plasmid with a mutant *tfdA* gene (Kamagata et al., 1997). In an analysis of a bacterial community decomposing polyethylene glycol, *V. paradoxus* has been identified as the most active decomposer (Sedina and Ivanov, 1991). *V. paradoxus* is able to degrade poly(3-hydroxybutyrate) (PHB), poly(3-hydroxybutyrate-co-3-hydroxyvalerate) (PHBV), and polypropiolactone plastics *in vitro* (Mergaert and Swings, 1996; Kobayashi et al., 1999). *V. paradoxus* is among the most frequently isolated species of polyhydroxyalkanoate-degrading microorganisms from PHB and PHBV plastics buried in soils and sludge, and is sporadically isolated from PHB and PHBV plastics buried in household compost or immersed in fresh water (Mergaert and Swings, 1996). An extracellular PHB depolymerase from *V. paradoxus* S2 has been isolated and characterized (Shiraki et al., 1995; Kobayashi

et al., 1999). *Variovorax* strains from river water have been reported to degrade the aliphatic polycarbonates polyhexamethylene carbonate and polytetramethylene carbonate (Suyama et al., 1998a). *Variovorax* also has been detected in a mixed culture capable of degrading metal–EDTA complexes, isolated from water from the river Mersey (UK) mixed with sludge from an industrial effluent treatment plant (Thomas et al., 1998). In a study of the influence of different chemical treatments on the transport of bacteria through porous media, *V. paradoxus* has been used as a model organism. Retention of bacteria is reduced by treatments making cells more hydrophobic and less electrostatically charged. Such effects may be important in improving the penetration of bacteria used in bioremediation programs of subsurface pollutants (Gross and Logan, 1995). *V. paradoxus* has been found to be a major bacterial group in the rhizosphere of chicory (*Cichorium intybus* L. var. *foliosum* Hegi), mainly early in the growing season (Van Outryve et al., 1988).

ENRICHMENT AND ISOLATION PROCEDURES

Facultatively chemolithotrophic, hydrogen-oxidizing strains of *V. paradoxus* can be isolated from soil, mud, and water by enrichment in the liquid basal mineral medium[1] described by Palleroni and Doudoroff (1972). After inoculation, cultures are incubated at 30°C under an atmosphere of $H_2/O_2/CO_2/N_2$ (50:4–20:5:25–41) (Davis et al., 1970). Heterotrophic strains can be isolated from soil by enrichment with pantothenate (Davis et al., 1970) or poly-3-hydroxybutyrate (Delafield et al., 1965). Other strains have been isolated from soil and water samples using various complex chemicals for enrichment (Ka and Tiedje, 1994; Kamagata et al., 1997; Suyama et al., 1998a; Thomas et al., 1998).

MAINTENANCE PROCEDURES

For heterotrophic growth, convenient media, such as nutrient agar, can be used. Biotype I strains can be maintained under autotrophic or heterotrophic conditions. No reports are available regarding the preservation of autotrophic properties after prolonged periods of heterotrophic growth. However, Aragno and Schlegel (1992) have reported autotrophic cultures to remain viable at 4°C for longer periods (up to 6 months) than do heterotrophic ones.

Variovorax strains can be lyophilized in skim milk in the presence of a suitable cryoprotectant, such as 5% glutamate, 5% *meso*-inositol, or 10% honey (Aragno and Schlegel, 1992). For the preservation of autotrophic properties, the use of autotrophically grown cultures for lyophilization is preferred. Cells can also be stored at −80°C in nutrient broth (or another suitable medium) plus 10% glycerol.

DIFFERENTIATION OF THE GENUS *VARIOVORAX* FROM OTHER GENERA

See Table BXII.β.42 for the family *Comamonadaceae* for features differentiating *Variovorax* from the other genera of this family.

TAXONOMIC COMMENTS

In their phenotypic study of 65 hydrogen bacteria and related nonautotrophic bacteria, Davis et al. (1969) described a group of yellow pigmented, peritrichously flagellated organisms comprising both facultatively autotrophic and nonautotrophic

---

1. Basal mineral medium contains: Na-K phosphate buffer (pH 6.8) 0.033 M; NH$_4$Cl, 0.1%; MgSO$_4$·7H$_2$O, 0.05%; ferric ammonium citrate, 0.005%; CaCl$_2$, 0.0005% (Palleroni and Doudoroff, 1972).

strains. The authors proposed that the genus *Hydrogenomonas*, which until then had been used to group all Gram-negative facultatively autotrophic, hydrogen-oxidizing bacteria, be abandoned because its type species, *Hydrogenomonas pantropha*, was considered a *nomen dubium*, of which no strains were available or could be isolated. Most of the polarly flagellated *Hydrogenomonas* species could be accommodated in the genus *Pseudomonas*, but the choice was less evident for the peritrichously flagellated species because of the lack of a broadly defined genus with a good type species. By elimination, the only acceptable existing peritrichously flagellated genus with a well defined type species was *Alcaligenes*. Thus, *Alcaligenes paradoxus* was created to comprise the yellow-pigmented, peritrichously flagellated strains, even though these strains did not have a true peritrichous type of flagellation, but rather an oligotrichous flagellation with 1–3 laterally inserted flagella (Aragno et al., 1977). Within *A. paradoxus*, two biotypes were proposed to accommodate the facultatively autotrophic and heterotrophic strains, respectively (Davis et al., 1969). Based on immunological comparisons of glutamine synthetase (Baumann and Baumann, 1978) and DNA–rRNA hybridizations (De Vos and De Ley, 1983), it became clear that *A. paradoxus* was more closely related to the *Pseudomonas acidovorans* group of species than to *Alcaligenes faecalis* and *Alcaligenes eutrophus*. More extensive DNA–rRNA hybridizations and further polyphasic characterization of the acidovorans group finally led to a proposal to transfer *Alcaligenes paradoxus* to a new genus *Variovorax* as *Variovorax paradoxus* (Willems et al., 1991a).

Variovorax is a member of the so-called "acidovorans rRNA complex" (Kersters and De Ley, 1984b), now the family *Comamonadaceae* (Willems et al., 1991a), in the *Betaproteobacteria*. Within this group, it is equidistantly related to each of the other taxa by DNA–rRNA hybridizations. No significant DNA binding is detected between *Variovorax* and representative strains of the other yellow pigmented genera *Hydrogenophaga* and *Xylophilus* (Willems et al., 1991a). In a recent phylogenetic analysis of the 16S rRNA gene sequences of members of the family *Comamonadaceae*, *V. paradoxus* has been found to group with *Xylophilus ampelinus*, a yellow pigmented pathogen of grapevine. The level of 16S rDNA similarity of 97.9% suggests that both species may belong to the same genus (Wen et al., 1999). Until further evidence of genotypic similarities is provided, the phenotypic differences between the slow-growing plant pathogen *Xylophilus* and the more versatile hydrogen-oxidizing *Variovorax* justify maintaining both as separate genera.

ACKNOWLEDGMENTS

Anne Willems is indebted to the Fund for Scientific Research–Flanders for a position as a postdoctoral research fellow.

FURTHER READING

Aragno, M. and H.G. Schlegel. 1992. The mesophilic hydrogen-oxidizing (Knallgas) bacteria. *In* Balows, Trüper, Dworkin, Harder and Schleifer (Editors), The Prokaryotes. A Handbook on the Biology of Bacteria: Ecophysiology, Isolation, Identification, Applications., 2nd Ed., Vol. 1, Springer-Verlag, New York. pp. 344–384.

List of species of the genus Variovorax

1. **Variovorax paradoxus** (Davis 1969) Willems, De Ley, Gillis and Kersters 1991a, 447[VP] (*Alcaligenes paradoxus* Davis *in* Davis, Doudoroff, Stanier and Mandel 1969, 387.)
   pa.ra.dox′us. Gr. prep. *para* amis, contrary to; Gr. n. *doxus* an opinion; M.L. n. *paradoxus* contrary to expectation, in reference to the chemolithotrophic and/or organotrophic metabolism of the organism.

   The morphological and cellular characteristics are as described for the genus. Additional descriptive information is presented in Table BXII.β.60, which is based on data from Davis et al. (1969, 1970), Kersters and De Ley (1984b), and Aragno and Schlegel (1992).

   Biotype I comprises facultatively autotrophic strains that reduce nitrate to nitrite in organic media. Some strains isolated under low partial pressures of oxygen have been reported to show "oxygen sensitivity" and initially do not grow autotrophically with 20% O_2. These strains can produce mutants indistinguishable with respect to oxygen sensitivity from other strains initially isolated with 20% or 30% O_2 (Davis et al., 1969). Wilde and Schlegel (1982) have reported strains to be tolerant to 80% O_2 when grown autotrophically and 100% O_2 when grown heterotrophically. Autotrophic cultures produce a characteristic unpleasant odor similar to that of soapy water in a laundry (Davis et al., 1969). Typical biotype I strains are the type strain of the species, as well as strains ATCC 17712, ATCC 17715, ATCC 17716, ATCC 17722, and ATCC 17723.

   Biotype II strains are nonautotrophic and rarely reduce nitrate to nitrite. A typical biotype II strain is strain ATCC 17549; other strains are ATCC 17716, ATCC 17719, and ATCC 11720 (Davis et al., 1969). More strains are available in several culture collections, but it is not always known which biotype they represent.

   Isolated by enrichment from soil, mud and water.

   The mol% G + C of the DNA is: 66.8–69.4 (T_m).

   Type strain: ATCC 17713, DSM 30034, IFO 15149, LMG 1797.

Genus Incertae Sedis XI. **Aquabacterium** *Kalmbach, Manz, Wecke and Szewzyk 1999, 775*[VP]

WERNER MANZ, SIBYLLE KALMBACH AND ULRICH SZEWZYK

A.qua.bac′te.ri.um. L. n. *aqua* water; Gr. n. *bakterion* rod; *Aquabacterium* a rod-shaped bacterium isolated from drinking water biofilms.

Rod-shaped cells, 0.5×1–4 μm. **Motile** by means of **monotrichous polar flagella**. Gram negative. **Polyalkanoate** and **poly-** **phosphate** inclusion bodies have been frequently observed as storage materials. **Extracellular polymeric substances are formed,**

even under oligotrophic conditions. **Oxidase positive. Catalase negative. Microaerophilic**, having a strictly respiratory type of metabolism with oxygen as the terminal electron acceptor. Nitrate can be used as an alternate electron acceptor for anaerobic respiration, but nitrite, chlorate, sulfate, and iron (III) are not used as electron acceptors. **Growth by fermentation does not occur. Manganese is not oxidized. Carbohydrates are not used.** Tweens 20, 40, 60, and 80, acetate, butyrate, valerate, caproate, caprylate, succinate, adipate, pimelate, azelate, sebacate, fumarate, β-hydroxybutyrate, malate, and butanol can be used as carbon sources. Starch, esculin, gelatin, and DNA are not hydrolyzed. NaCl levels are tolerated up to 1.8%. Optimal pH, ~7; pH growth range, 5.5–10, depending on the species. Optimal temperature, ~20°C; temperature range, 6–36°C. Habitat: drinking water biofilm originating from various raw water sources (groundwater, surface water, artificially recharged groundwater) used for drinking water production.

The mol% G + C of the DNA is: 65–66.

Type species: **Aquabacterium commune** Kalmbach, Manz, Wecke and Szewzyk 1999, 776.

FURTHER DESCRIPTIVE INFORMATION

Aquabacterium is a slow-growing organism forming colonies of 1.5–3 mm in diameter on solid media after 10 d at 20°C. Colonies on modified R2A agar[1] are flat, transparent, white to cream-white, with a smooth margin and transparent edges. Cells are

---

1. Modified R2A agar (Kalmbach et al., 1999) has the following composition (g/l of distilled water): yeast extract, 0.5; Difco protease peptone no. 3, 0.5; Casamino acids, 0.5; glucose, 0.5; Tween 80, 0.1% (v/v); sodium pyruvate, 0.3; K$_2$HPO$_4$, 0.3; MgSO$_4$·7H$_2$O, 0.05; agar, 15. Adjust the medium at final pH 7.2 with crystalline K$_2$HPO$_4$ or KH$_2$PO$_4$ before adding agar, and sterilize for 15 min at 121°C.

notable for the presence of both polyphosphate and polyalkanoate inclusion bodies (Figs. BXII.β.45 and BXII.β.46).

In a survey of drinking water biofilms grown on different raw water sources in Europe for the occurrence of *Aquabacterium, in situ* probing with fluorescently labeled, rRNA-targeted oligonucleotide probes have shown *Aquabacterium commune* to be a widespread bacterial species (Kalmbach et al., 2000).

ENRICHMENT AND ISOLATION PROCEDURES

For the enrichment and isolation of *Aquabacterium*, bacteria have been detached from young drinking water biofilms that had been grown on glass and polyethylene slides in a modified Robbins device installed in a house installation system at the Technical University of Berlin (Kalmbach et al., 1997). The temperature in the Berlin drinking water varies from 9.4°C to 15.6°C and the pH values ranges from 7.2 to 7.7. Detached bacteria are pooled in a total volume of 2 ml sterile drinking water and vigorously mixed with a vortex mixer. Pure cultures are obtained by plating serial dilutions of the bacterial suspension on R2A agar (Reasoner and Geldreich, 1985) with subsequent incubation at 20°C for 10 d in the dark. Liquid cultures are agitated constantly at 100 rpm at 20°C. For routine cultivation of strains, R2A medium is modified by replacing starch with 0.1% (v/v) Tween 80.

MAINTENANCE PROCEDURES

Aquabacterium can be cultivated on either R2A agar or broth (Reasoner and Geldreich, 1985) that is modified by replacing starch with 0.1% (v/v) Tween 80 (Sigma). Strains grow well in liquid cultures under constant agitation at 100 rpm and 20°C. Storage is possible in the frozen state in culture medium containing 50% (v/v) glycerol. For long-term preservation of *Aquabacterium*, freeze-drying is recommended.

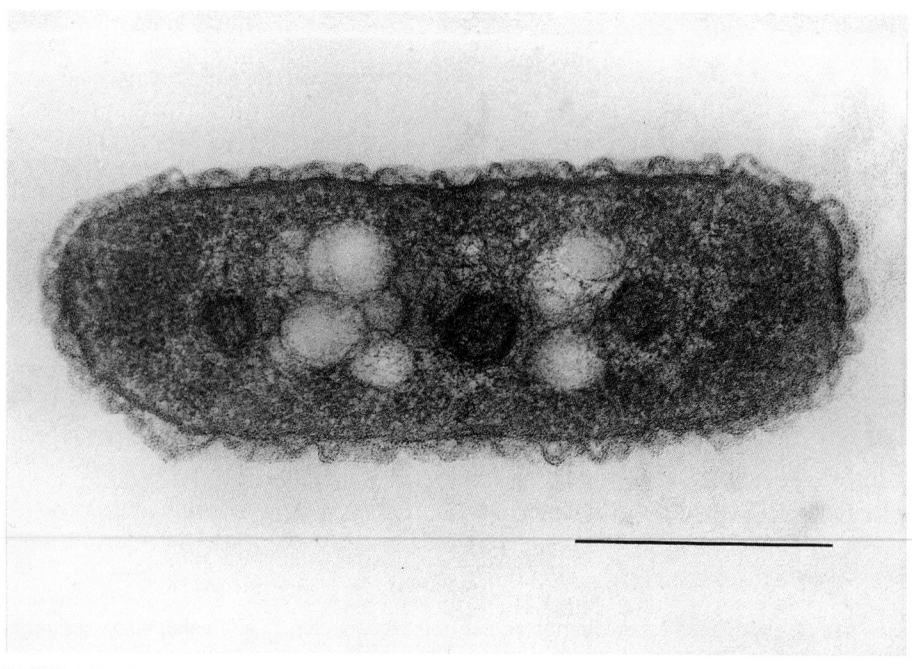

FIGURE BXII.β.45. Transmission electron micrograph of longitudinal thin section of an *A. commune* cell grown in modified R2A liquid medium, showing the typical cell morphology with polyphosphate (black) and polyalkanoate (white) inclusion bodies. Bar = 0.5 μm.

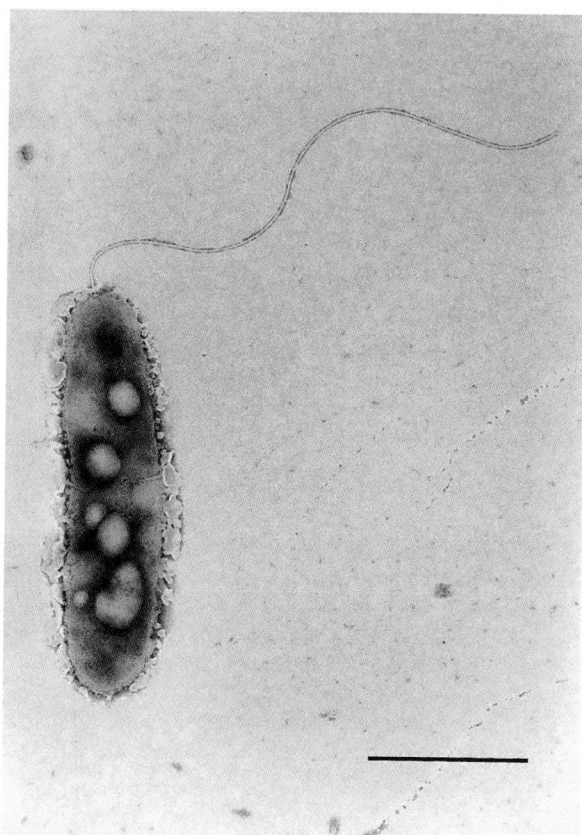

FIGURE BXII.β.46. Thin-section transmission electron micrograph of a negatively stained cell of *A. citratiphilum* showing the polar flagellum, polyphosphate (black), and large polyalkanoate inclusion bodies (white). Flagellation is similar for *A. parvum* and *A. commune*. Bar = 1.0 μm. (Reprinted with permission from S. Kalmbach et al., International Journal of Bacteriology, *49:* 769–777, 1999, ©International Union of Microbiological Societies.)

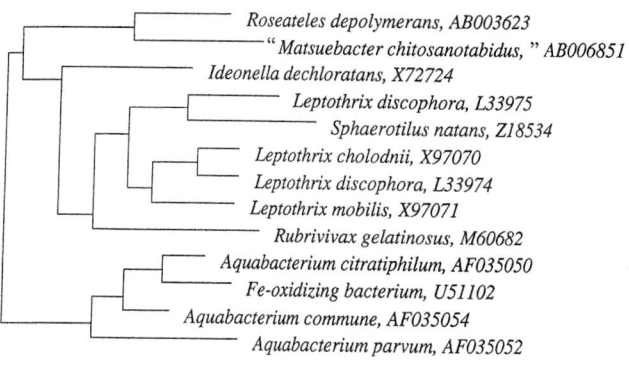

FIGURE BXII.β.47. Phylogenetic tree reflecting the affiliation of the three species of the genus *Aquabacterium* among the next closely related genera within the *Comamonadaceae*. The phylogenetic tree was reconstructed on the basis of full-length 16S rDNA sequences by the maximum parsimony algorithm, using the ARB software package (Strunk and Ludwig, 1998). Bar = 5% sequence divergence.

DIFFERENTIATION OF THE GENUS *AQUABACTERIUM* FROM OTHER GENERA

The genus *Aquabacterium* can be differentiated from closely related genera within the family *Comamonadaceae*—particularly the genera *Ideonella* and *Leptothrix*—by its inability to metabolize carbohydrates. In addition, in contrast to *Ideonella dechloratans*, *Aquabacterium* shows no catalase activity and does not reduce chlorate. Members of the genus *Leptothrix* are morphologically characterized by sheath formation, whereas *Aquabacterium* does not form sheaths. *Leptothrix* can be further differentiated from *Aquabacterium* by its ability to oxidize manganese.

TAXONOMIC COMMENTS

From a phylogenetic perspective, *Aquabacterium* is placed within the class *Betaproteobacteria*, the order *Burkholderiales*, and the family *Comamonadaceae*. At present, a total of 16 genera are included in this family. The genera most closely related to *Aquabacterium* are shown in Fig. BXII.β.47.

Spectrophotometrically determined reassociation rates of the genomic DNA (De Ley et al., 1970a) among the *Aquabacterium* species show the following levels of DNA–DNA relatedness: 44.9% between *A. citratiphilum* and *A. commune*, 45.4% between *A. citratiphilum* and *A. parvum*, and 51.3% between *A. parvum* and *A. commune*. 16S rDNA sequence similarity values are 98.2% between *A. citratiphilum* and *A. commune*, 97.2% between *A. parvum* and *A. commune*, and 96.5% between *A. citratiphilum* and *A. parvum*.

The genus *Aquabacterium* is characterized by an oligonucleotide signature (sequence 5′-CUUGCUGUUCAGUAACGAAGC-3′) within the 16S rRNA located at position 841 according to the *E. coli* numbering (Brosius et al., 1981).

ACKNOWLEDGMENTS

The preparation of ultrathin sections, staining, and electron microscopy were kindly performed by J. Wecke at the Robert Koch-Institute, Berlin, Germany.

DIFFERENTIATION OF THE SPECIES OF THE GENUS *AQUABACTERIUM*

The differential characteristics of *A. citratiphilum*, *A. parvum*, and *A. commune* are summarized in Table BXII.β.61. Other characteristics of the species are listed in Table BXII.β.62.

List of species of the genus Aquabacterium

1. **Aquabacterium commune** Kalmbach, Manz, Wecke and Szewzyk 1999, 776[VP]

   com′mu.ne. L. adj. *communis, -e* common, referring to the predominance of the species in drinking water biofilms of the Berlin distribution system.

   The characteristics are as given for the genus and as listed in Tables BXII.β.61 and BXII.β.62. Colonies are flat, transparent, have a smooth margin, and are 1.5–2 mm in diameter after 10 d at 20°C on modified R2A agar. *A. commune* is the only species of *Aquabacterium* that utilizes ben-

TABLE BXII.β.61. Differential features of the species of the genus *Aquabacterium*[a]

Property	A. commune	A. citratiphilum	A. parvum
Dimensions (µm)	0.5–2-4	0.5–2-4	0.5–1-2
Temperature range, °C	6–34	10–36	14–34
Tolerated NaCl concentration, %	0–0.4	0–1.8	0–0.8
Urea hydrolysis	−	+	+
Casein hydrolysis	+	−	−
pH range for growth	6.5-9.5	5.5-10.0	6.5-10.0
Carbon and energy sources:			
Bromosuccinate	+	+	−
Propionate	+	+	−
Pyruvate	+	+	−
Benzoate	+	−	−
Casamino acids	+	−	−
Glutamate	+	−	−
Citrate	−	+	−
Glycerol	−	+	−
γ-Hydroxybutyrate	−	+	−
Lactate	−	+	−

[a]Adapted from Kalmbach et al. (1999).

zoate, Casamino acids, and glutamate. Among *Aquabacterium* species, *A. commune* shows the lowest temperature growth limit.

The mol% G + C of the DNA is: 66 (T_m).

Type strain: B8, ATCC BAA-209, CIP 106984, DSM 11901.

GenBank accession number (16S rRNA): AF035054.

2. **Aquabacterium citratiphilum** Kalmbach, Manz, Wecke and Szewzyk 1999, 776[VP]

ci.tra.ti' phi.lum. L. n. *citrus* lemon tree; L. n. *acidum* acid; L. neut. adj. *acidum citri* citric acid; Gr. adj. *philos* loving; M.L. neut. adj. *citratiphilum* citrate-loving, referring to the preferred utilization of citrate as a carbon and energy source.

The characteristics are as given for the genus and as listed in Tables BXII.β.61 and BXII.β.62. Colonies are flat, cream-white, have a smooth margin, and are 2–3 mm in diameter after 10 d at 20°C on modified R2A agar. The pH range for growth is 5.5–10.0. Differs from other *Aquabacterium* species by its preferred growth on citrate and its utilization of lactate, γ-hydroxybutyrate, and glycerol.

The mol% G + C of the DNA is: 66 (T_m).

Type strain: B4, ATCC BAA-207, CIP 106985, DSM 11900.

GenBank accession number (16S rRNA): AF035050.

3. **Aquabacterium parvum** Kalmbach, Manz, Wecke and Szewzyk 1999, 776[VP]

TABLE BXII.β.62. Other characteristics of the species of the genus *Aquabacterium*[a,b]

Characteristic	A. commune	A. citratiphilum	A. parvum
Rod-shaped cells	+	+	+
Motile by a single polar flagellum	+	+	+
Oxidase test	+	+	+
Catalase test	−	−	−
Reduction of:			
NO³⁻	+	+	+
NO²⁻, Fe³⁺, SO₄²⁻, ClO₃⁻	−	−	−
Hydrolysis of esculin, DNA, starch, and gelatin	−	−	−
Carbon and energy sources:[c]			
Acetate	+	+	+
Adipate	+	+	+
Azelate	+	+	+
Butanol	+	+	+
Butyrate	+	+	+
Caproate	+	+	+
Caprylate	+	+	+
Fumarate	+	+	+
β-Hydroxybutyrate	+	+	+
Malate	+	+	+
Succinate	+	+	+
Pimelate	+	+	+
Sebacate	+	+	+
Tweens 20, 40, 60, and 80	+	+	+
Valerate	+	+	+
Habitat: drinking water biofilm	+	+	+

[a]Symbols: see standard definitions.

[b]Adapted from Kalmbach et al. (1999).

[c]Substrates which are not utilized by aquabacteria include: N-acetylglucosamine, L-arabinose, ascorbate, caprate, ethanol, formate, D-fructose, D-galactose, galacturonate, gluconate, D-glucose, glutarate, glyoxylate, D-lactose, malonate, D-maltose, D-mannose, D-mannitol, D-melibiose, methanol, oxalate, phthalate, L-rhamnose, D-ribose, D-ribulose, sucrose, tartrate, D-trehalose, and D-xylose.

par' vum. L. adj. *parvus* small.

The characteristics are as given for the genus and as listed in Tables BXII.β.61 and BXII.β.62. Colonies are flat with a smooth margin, white in the center with transparent edges, and 1.5–2 mm in diameter after 10 d at 20°C on modified R2A agar.

The mol% G + C of the DNA is: 65 (T_m).

Type strain: B6, ATCC BAA-208, CIP 106983, DSM 11968.

GenBank accession number (16S rRNA): AF035052.

Genus Incertae Sedis XII. **Ideonella** *Malmqvist, Welander, Moore, Ternström, Molin, Stenström 1994b, 595*[VP] *(Effective publication: Malmqvist, Welander, Moore, Ternström, Molin, Stenström 1994a, 63)*

Åsa Malmqvist, Edward R.B. Moore and Anders Ternström

I.de.o.nel' la. M.L. fem. n. derived from *Ideon* the research center where the bacterium was isolated and described; M.L. dim. ending *ella*.

Cells are straight or slightly curved rods, 0.7–1.0 × 2.5–5.0 µm, occurring singly, in pairs or in short filaments of four to five cells. When occurring in pairs or filaments, individual cells have a fusiform appearance with pointed ends. **Motile by two or more polar or subpolar flagella.** Gram negative. Prosthecae, sheaths, or endospores are not produced. No resting stages are known. Bacteriochlorophyll, carotenoids, or other pigments are not produced. **Aerobic,** having a strictly respiratory metabolism with oxygen as the terminal electron acceptor; however, **chlorate can serve as an alternate terminal electron acceptor under anaerobic conditions. Oxidase positive. Weakly catalase positive;** catalase is produced only in minute amounts and the conventional agar colony test may be interpreted as negative. Chemoorganotrophic, utilizing organic acids, amino acids, and carbohydrates as carbon sources. Mesophilic, growing in the temperature range of 12–42°C. Belongs to the class *Betaproteobacteria*.

The mol% G + C of the DNA is: 68.1.

Type species: **Ideonella dechloratans** Malmqvist, Welander, Moore, Ternström, Molin, Stenström 1994b, 595 (Effective publication: Malmqvist, Welander, Moore, Ternström, Molin, Stenström 1994a, 63.)

FURTHER DESCRIPTIVE INFORMATION

Growth of *Ideonella dechloratans* occurs in a simple mineral medium containing an appropriate carbon source, but the growth yield in an anaerobic environment, using chlorate as the electron acceptor, is increased by the addition of 50 mg/l beef extract.

Chlorate is reduced to chloride in an anaerobic environment. The highly oxidized chlorate ion is an excellent electron acceptor for *I. dechloratans*, being even more efficient, theoretically, than oxygen ($\Delta G° = -132$ kJ/mol and -110 kJ/mol of electrons exchanged, respectively), and chlorate-contaminated wastewater effluents have been detoxified by a bacteria-mediated, anaerobic reduction of chlorate to chloride (Malmqvist et al., 1991). In practice, however, *I. dechloratans* prefers oxygen, which is evidenced by the fact that chlorate is not reduced in aerobic conditions.

Nitrate can serve as an alternative electron acceptor, but this ability is frequently lost after repeated subcultures on chlorate. Nitrate is reduced to nitrite.

ENRICHMENT AND ISOLATION PROCEDURES

Ideonella dechloratans was originally enriched anaerobically from a chemostat of chlorate-containing sewage at 37°C, using a defined medium of 10 mM $NaClO_3$, 25 mM acetic acid, 3 mM NH_3, 0.25 mM KH_2PO_4, 0.1 mM $MgSO_4$, 0.1 mM $Ca(OH)_2$, 0.05 mM $FeSO_4 \cdot 7H_2O$, 0.05 mM $MnSO_4 \cdot H_2O$, 5 μM $NiCl \cdot 6H_2O$, 5 μM $CoCl_2 \cdot 6H_2O$, 5 μM $ZnSO_4 \cdot 7H_2O$, 0.1 μM H_3BO_3, 0.1 μM Na_2SeO_4, 0.1 μM Na_2WO_4, and 0.1 μM Na_2MoO_4. The pH was 7.0. Samples were taken from the reactor and streaked onto the same medium supplemented with 1.5% agar and incubated anaerobically in a GasPak jar (BBL), at 37°C. After 4 d, nonpigmented colonies were observed.

DIFFERENTIATION OF THE GENUS *IDEONELLA* FROM OTHER GENERA

Ideonella is recognized to be related genotypically (i.e., as determined by 16S rRNA gene sequence analysis) to the genera *Aquabacterium* (Kalmbach et al., 1999), *Leptothrix* (Siering and Ghiorse, 1996), and *Roseateles* (Suyama et al., 1999), within the *Betaproteobacteria*. Phenotypically, all of the species of the genera composing the *Ideonella–Aquabacterium–Leptothrix–Roseateles* 16S rDNA phylogenetic lineage produce Gram-negative, motile, rod-shaped cells. They are all chemoorganotrophic, collectively being able to utilize a range of organic acids, amino acids, and carbohydrates. They are all aerobic, possessing a strictly respiratory metabolism. *Ideonella dechloratans* may be distinguished from the species of *Aquabacterium, Leptothrix,* and *Roseateles* by its unique ability to reduce chlorate and by the production of catalase. It is differentiated from *Aquabacterium* species in its ability to metabolize

carbohydrates and its ability to grow at temperatures as high as 42°C. It is differentiated from *Leptothrix* species by the absence of sheaths, by possessing more than one polar or subpolar flagellum, and by failing to oxidize Mn^{2+}. It is differentiated from *Roseateles* species by failing to produce bacteriochlorophyll or carotenoids, but is able to reduce nitrate and grow anaerobically.

TAXONOMIC COMMENTS

Ideonella dechloratans was recognized initially as a typical "pseudomonad", i.e., a Gram-negative, respiring, cytochrome *c* oxidase positive, motile rod, capable of utilizing a wide spectrum of organic molecules. At that time, however, the genus *Pseudomonas* was being acknowledged as comprising a range of phylogenetically diverse species, which would be ultimately reclassified in different genera. Analysis of the 16S rDNA gene sequence and estimation of the phylogenetic position (Malmqvist et al., 1994a) demonstrated that *Ideonella* belonged within the β-subclass (Woese et al., 1984c) of the *Proteobacteria* (Stackebrandt et al., 1988b) and was most closely related to species of the *Comamonadaceae* or "acidovorans rRNA complex" (Willems et al., 1991a). However, the analysis shown in Fig. BXII.β.48 indicates that *I. dechloratans* belongs within a phylogenetic lineage related to, yet

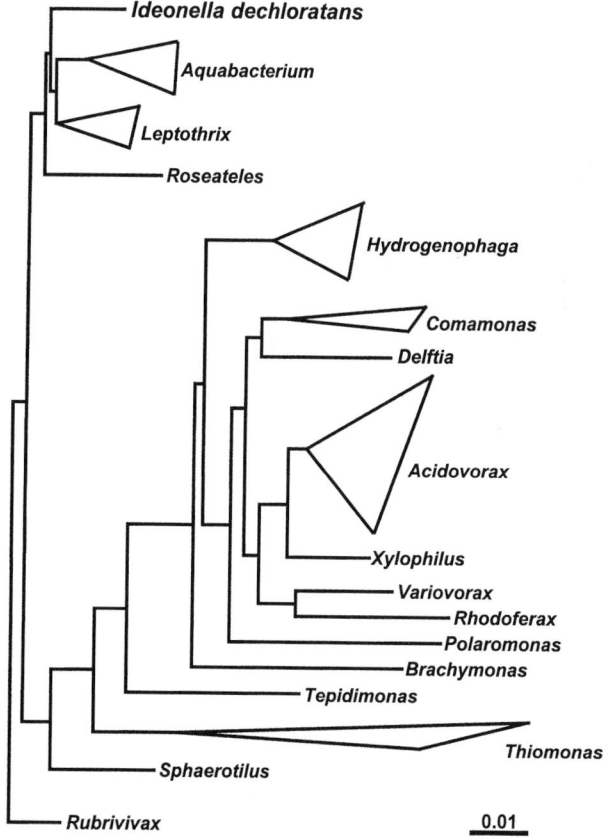

FIGURE BXII.β.48. The inferred phylogenetic position of *Ideonella dechloratans* among other genera of the *Betaproteobacteria*. Evolutionary distances were derived from pair-wise dissimilarities of 16S rRNA gene sequences. The unrooted dendrogram was generated using the neighbor-joining algorithm. The sequences used were obtained from the EMBL Nucleotide Sequence Database (Stoesser et al., 2001) and included those for all validly published species of each genus depicted in the dendrogram. The cluster length for each genus is depicted as the estimated length of the longest branch within each cluster.

distinct from, the *Comamonadaceae* and more closely related to species of the genera *Aquabacterium*, *Leptothrix*, and *Roseateles*.

The species of the genera composing the *Ideonella–Aquabacterium–Leptothrix–Roseateles* 16S rDNA phylogenetic lineage possess DNA composition values ranging from 65 to 71 mol% G + C (*Ideonella*, 68%; *Aquabacterium*, 65–68%; *Leptothrix*, 69–71%; *Roseateles*, 66%). However, *I. dechloratans* is distinguished from the species of *Aquabacterium*, *Leptothrix*, and *Roseateles*, possessing a monophyletic lineage with 16S rDNA gene sequence differences of 3–4%.

It is not clear whether the ability to utilize chlorate as an energy-yielding terminal electron acceptor is limited to a specialized group of bacteria or whether it is distributed throughout different bacterial taxa, as is the ability to grow with nitrate as terminal electron acceptor. Observations of enriched chlorate-reducing bacteria with cell morphologies different from that of

I. dechloratans (Malmqvist et al., 1991) suggest that other chlorate reducers exist, although it is not possible to draw conclusions as to their relatedness to *I. dechloratans*. Other studies have identified N_2-fixing bacteria associated with rice, which are closely related to *I. dechloratans*, as determined by 16S rDNA sequence similarities greater than 99% (unpublished, sequence data deposited in the EMBL nucleotide sequence database). The ability to fix N_2 has not been determined for *I. dechloratans*. Yet other studies have detected 16S rDNA sequence types (unpublished, sequence data deposited in the EMBL nucleotide sequence database) similar to that of *I. dechloratans* in samples of an industrial, nitrifying/denitrifying activated sludge and in petroleum-contaminated groundwater. Such analyses suggest that other bacteria related to *I. dechloratans* exist within a range of different ecosystems and may or may not have the capability of utilizing chlorate as a terminal electron acceptor.

List of species of the genus Ideonella

1. **Ideonella dechloratans** Malmqvist, Welander, Moore, Ternström, Molin, Stenström 1994b, 595[VP] (Effective publication: Malmqvist, Welander, Moore, Ternström, Molin, Stenström 1994a, 63.)

   de.chlor.at' ans. L. *de* from; M.L. adj. *chloratans* referring to chlorate; M.L. adj. *dechloratans* derived from chlorate; i.e., chlorate-utilizing bacterium.

   The description of the species is the same as that for the genus, with the following additional information. Cells may contain several inclusions of unknown composition. Colonies are circular, smooth, and nonpigmented. Chlorate is reduced to chloride in an anaerobic environment. Nitrate can serve as an alternative electron acceptor, being reduced to nitrite. Nitrite and sulfate do not serve as electron acceptors. Acetate, alanine, asparagine, butyrate, fructose, glucose, lactate, propionate, pyruvate, and succinate are utilized as sole sources of carbon in mineral medium, both

   aerobically and anaerobically, in the latter case with chlorate as the electron acceptor. Adipate and fumarate are also utilized, with chlorate, but not with oxygen. Aminobenzoate, phenol, and phenylalanine are not utilized. No acid is produced from cellobiose, glucose, or maltose in Hugh–Leifson medium. Proteolytic activity is produced against casein but not against egg yolk. Lipolytic activity is evident against egg yolk, Tween 20, Tween 40, Tween 60, and Tween 80, but not against Tween 85.

   Mesophilic, growing in the temperature range of 12–42°C; no growth occurs at 10–46°C. No growth occurs in media with 3% NaCl.

   Isolated from anaerobic enrichments of chlorate-containing sewage in a chemostat reactor.

   The mol% G + C of the DNA is: 68 (HPLC) (Mesbah et al., 1989).

   Type strain: CCUG 30898.

   GenBank accession number (16S rRNA): X72724.

Genus Incertae Sedis XIII. *Leptothrix* Kützing 1843, 184[AL]

STEFAN SPRING AND PETER KÄMPFER

Lep' to.thrix. Gr. adj. *leptus* fine, small; Gr. n. *thrix* hair; M.L. fem. n. *Leptothrix* fine hair.

Straight rods, 0.6–1.5 × 2.5–15 μm, occurring in chains within a sheath or free swimming as single cells, pairs, or short chains. Most strains are motile by **one polar flagellum.** One species is characterized by holdfasts[1] and a subpolar tuft of flagella. Poly-β-hydroxybutyrate is stored in globules as reserve material. Gram negative. **Sheaths can become encrusted by the deposition of iron and manganese oxides.**

Chemoorganotrophic. Strictly aerobic, respiratory metabolism. Good growth at low oxygen tensions. The optimal temperature for growth is around 25°C, and the optimal pH is 6.5– 7.5. A variety of sugars and organic acids can be used as carbon and energy sources by most species. The need for additional growth factors, including vitamin B₁₂, biotin, thiamine, adenine, and guanine, has been reported for several strains (Rouf and Stokes, 1964; Stokes and Johnson, 1965).

Widely distributed in a variety of freshwater habitats, ranging from unpolluted springs and slowly running water rich in soluble iron and manganese compounds to sediments and sewage.

Type species: **Leptothrix ochracea** (Roth 1797) Kützing 1843, 198 (*Conferva ochracea* Roth 1797.)

FURTHER DESCRIPTIVE INFORMATION

For several species of the genus *Leptothrix*, including *L. discophora*, *L. mobilis*, and *L. cholodnii*, 16S rRNA gene sequences have been determined. In phylogenetic trees, these species cluster together with a group of diverse bacteria including *Sphaerotilus natans*, *Ideonella dechloratans*, *Rubrivivax gelatinosus*, *Alcaligenes latus*, *Roseateles depolymerans*, and *Aquabacterium* species. This branch of bacteria is phylogenetically distinct from the *Comamonadaceae* as defined by Willems et al. (1991a) and is also referred to as the *Rubrivivax* subgroup of the *Betaproteobacteria* (Wen et al., 1999). No 16S rRNA gene sequences are currently available for *Leptothrix ochracea*, the type species of the genus, and *L. lopholea*.

Cells are straight rods, which may be as long as 15 μm. They

---

1. Evaginations or blebs formed at one cell pole by which organisms attach to walls of containers, submerged plants, stones, and other surfaces.

occur as filaments within characteristic sheaths or free swimming as single cells or short chains. Enlarged cells and coccoid bodies are formed in older cultures of some strains. A typical trait of several strains is false branching, which develops if a cell attaches to an existing filament and forms a new filament.

Sheaths formed by *Leptothrix* species are tube-like, extracellular structures that resemble those of *Sphaerotilus natans*. The sheath consists of a matrix of heteropolysaccharide and protein fibrils associated with the outer membrane of the Gram-negative wall (Emerson and Ghiorse, 1993b). The heteropolysaccharides contain a mixture of uronic acids and amino sugars and are negatively charged. The sheath proteins contain a high proportion of cysteine residues, which covalently cross-link the fibrils by disulfide bonds, resulting in a stable, mesh-like fabric. Numerous free sulfhydryl and carboxyl groups are distributed throughout the sheath and provide sites for binding of metal cations (Emerson and Ghiorse, 1993a).

During growth in media containing iron and manganese, sheaths become impregnated with metal oxides, which give them a yellow to dark-brown appearance. In this way, cells within the sheaths are protected from inhibitory concentrations of soluble iron or manganese compounds in the environment (Rogers and Anderson, 1976a, b). Further deposition of metal oxides can lead to massive incrustations, thereby increasing the diameter of the bare sheaths several fold (Fig. BXII.β.49). The deposition of iron and manganese oxides by *Leptothrix* species appear to be separate processes, which are catalyzed by proteins. The corresponding enzymes are excreted into the medium or bound to the extracellular sheath (Adams and Ghiorse, 1987; Corstjens et al., 1992). Several studies have shown that *Leptothrix* species do not gain energy from the oxidation of iron or manganese (Van Veen et al., 1978). Nevertheless, it has been observed that the addition of Mn^{2+} to the medium stimulates growth. It has been postulated that this beneficial effect may be mainly due to the oxidation product, MnO_2, which catalytically inactivates H_2O_2, which is excreted in toxic amounts by *Leptothrix* species during growth (Dubinina, 1978a, b).

Isoprenoid quinone Q-8 is the predominant quinone type of *L. discophora* and *L. cholodnii* (Kämpfer, 1995). The fatty acid composition, as determined by gas chromatographic analysis, of strains representing *L. discophora*, *L. cholodnii*, and *L. mobilis* differ only slightly (Spring et al., 1996). All strains contain the fatty acids dodecanoic acid, hexadecanoic acid, *cis*-9-hexadecenoic acid, *cis*-9,11-octadecenoic acid, and 3-hydroxydecanoic acid; this fatty acid profile is typical among members of the *Betaproteobacteria*.

The size and shape of colonies depends largely on the ability of the respective strain to form sheaths (Fig. BXII.β.50). Rough and filamentous colonies (Fig. BXII.β.50a) are characteristic of sheath-forming cells, whereas smooth and regular colonies (Fig. BXII.β.50b) consist mainly of sheathless cells. Due to the oxidation of iron and manganese, a black-brown precipitate is formed within the colonies and occasionally also as a halo around them. Cells tend to aggregate and form flocs in liquid culture.

ENRICHMENT AND ISOLATION PROCEDURES

Several methods for the enrichment of *Leptothrix* species from the environment are based on the tendency of these filamentous organisms to attach to surfaces. Mulder and van Veen (1963) have used continuous-flow devices in order to imitate the natural growth conditions of these organisms. A continuous flow of a dilute nutrient solution that has been enriched with soluble iron and manganese compounds is guided through an apparatus consisting of several vessels arranged one after the other. As in unpolluted, slowly running water, bacteria that are able to attach to solid surfaces have a selective advantage within this device, because they obtain more nutrients than do bacteria moving with the medium. An alternative method that is less laborious has been applied by Spring et al. (1996). Microscope slides are put into a sample of freshwater sediment, and after several weeks,

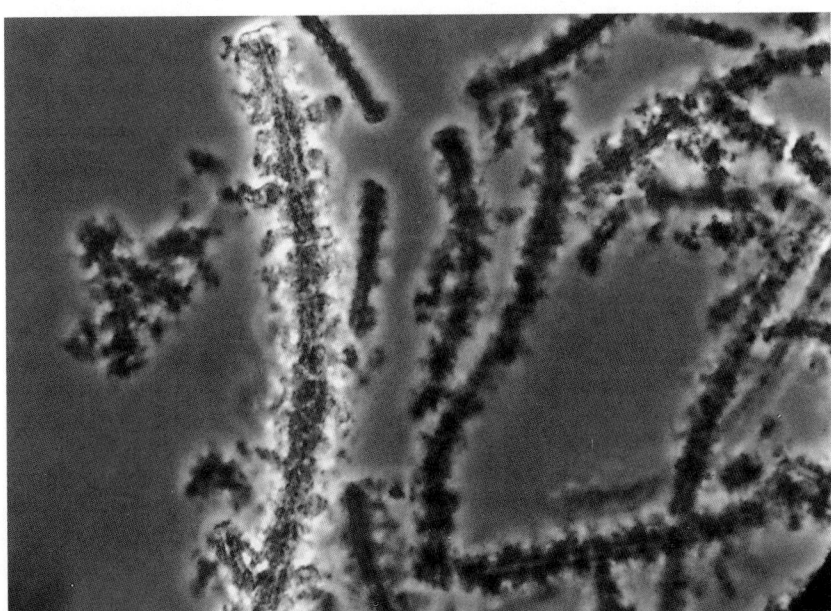

FIGURE BXII.β.49. Unknown strain, presumably belonging to *L. discophora*, grown under laboratory conditions in slowly running iron- and manganese-containing soil extract. Sheaths are covered with Fe(III) and Mn(IV) oxides. Light micrograph ($\times$ 1268).

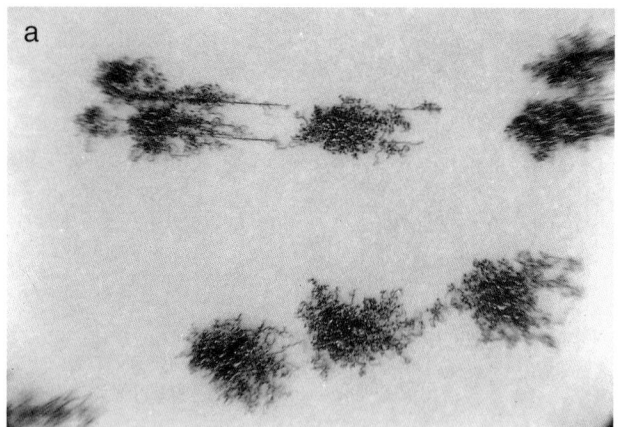

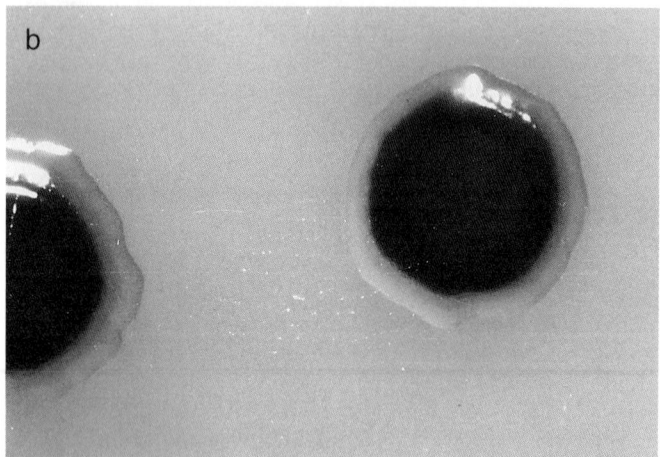

FIGURE BXII.β.50. Examples of various colony morphologies of *Lepto-thrix* strains grown on Mn(II)-containing agar media. (*a*) Filamentous colonies of an unknown strain presumably belonging to *L. discophora*. Light micrograph (× 18). (*b*) Smooth colonies of *L. mobilis* Feox-1 (DSM 10617). Light micrograph (× 29)

sheaths of filamentous bacteria encrusted with iron and manganese oxides covered parts of the slides. Subsequently, the slides may be removed from the sediment and used for isolation. Another simple enrichment method has been introduced by Rouf and Stokes (1964), who filled glass cylinders with water taken from the environment and added extracted alfalfa straw as a nutrient source, as well as $MnCO_3$ and freshly precipitated $Fe(OH)_3$. After several days, the flocculent growth of sheathed bacteria adhering to the walls of the cylinder indicates the enrichment of *Leptothrix* bacteria.

Isolation of pure cultures may be achieved by streaking material from enrichment cultures on previously dried agar plates containing low levels of nitrogen and carbon sources. The isolation medium used by Rouf and Stokes (1964) has the following composition (per liter tap water): peptone, 5.00 g; ferric ammonium citrate, 0.15 g; $MgSO_4 \cdot 7H_2O$, 0.20 g; $CaCl_2$, 0.05 g; $MnSO_4 \cdot H_2O$, 0.05 g; $FeCl_3 \cdot 6H_2O$, 0.01 g; agar, 12.00 g.

Colonies of *Leptothrix* strains can be easily distinguished from most contaminating bacteria on this medium by their dark-brown color.

Enrichment cultures may not be necessary if natural environments are studied where *Leptothrix* bacteria can be detected by their flocculent growth. Flocculent cell material from these sites may be washed several times with sterile tap water and streaked directly on solid media.

MAINTENANCE PROCEDURES

Stock cultures can be stored on agar slants of the medium of Rouf and Stokes (1964) at 4°C for about three months. Most *Leptothrix* strains do not survive lyophilization. For the long-term preservation of these strains, freezing in liquid nitrogen is recommended.

DIFFERENTIATION OF THE GENUS *LEPTOTHRIX* FROM OTHER GENERA

The genus that is most closely related to *Leptothrix* is *Sphaerotilus*. Traits that distinguish both genera are discussed in the corresponding section of the genus *Sphaerotilus* and listed in Table BXII.β.66 of that chapter.

TAXONOMIC COMMENTS

Some distinguishing characteristics of *Leptothrix* species are not very stable and are regularly lost upon cultivation. The ability to form sheaths has been irreversibly lost by most strains available from public culture collections, with the exception of *Leptothrix cholodnii* LMG 8142. The manganese-oxidizing activity is independent from the formation of sheaths and is more stable in most strains, but loss of this trait has also been reported.

A major problem in the taxonomy of the genus *Leptothrix* is the availability of reference strains. The type species of the genus *Leptothrix*, *L. ochracea*, has not been cultured, and the description is based only on morphological observations. The type strains of *Leptothrix lopholea*, LVMW 124, and *L. cholodnii*, LVMW 99, are not available from public culture collections and have apparently been lost. The strain LMG 7171 may be used as reference strain for *L. cholodnii* until a neotype is designated, but for *L. lopholea*, no other strains are available, preventing detailed taxonomic studies on this species.

Thus, species descriptions of *L. ochracea*, "*L. pseudo-ochracea*", and *L. lopholea* have not been emended and adopted from the first edition of *Bergey's Manual of Systematic Bacteriology* (Mulder, 1989a). Emended or new descriptions of *L. discophora*, *L. cholodnii*, and *L. mobilis* are based on the work of Spring et al. (1996).

FURTHER READING

Spring, S., P. Kämpfer, W. Ludwig and K.-H. Schleifer. 1996. Polyphasic characterization of the genus *Leptothrix*: new descriptions of *Leptothrix mobilis* sp. nov. and *Leptothrix discophora* sp. nov. nom. rev. and emended description of *Leptothrix cholodnii* emend. Syst. Appl. Microbiol. *19*: 634–643.

van Veen, W.L., E.G. Mulder and M.H. Deinema. 1978. The *Sphaerotilus–Leptothrix* group of bacteria. Microbiol. Rev. *42*: 329–356.

List of species of the genus Leptothrix

1. **Leptothrix ochracea** (Roth 1797) Kützing 1843, 198[AL] (*Conferva ochracea* Roth 1797.)
   o.chra′ce.a. Gr. n. *ochra* yellow-ochre; M.L. adj. *ochracea* like ochre.

L. ochracea is probably the most common iron-precipitating ensheathed bacterium all over the world, occurring in slowly running ferrous iron-containing waters poor in readily decomposable organic matter. Its contribution to the oxidation of Fe^{2+} is uncertain, since (a) the organism normally grows at a pH value of 6–7, at which Fe (II) is readily oxidized nonbiologically, and (b) pure cultures of this bacterium have never been obtained. For the latter reason, its Mn (II)-oxidizing capacity, which probably occurs under natural conditions, has never been confirmed. The pronounced development and activity of *L. ochracea* in iron- and manganese- containing waters give rise to the accumulation and deposition of large masses of ferric oxide and, probably, MnO_2, which are thought to be responsible for the formation of bog ore (see, for instance, Ghiorse and Chapnick, 1983).

Authors who have studied and described *L. ochracea* under natural conditions have been unable to obtain pure cultures (Cholodny, 1926; Charlet and Schwartz, 1954). Those who thought they had isolated *L. ochracea* had, in fact, described one of the other species of this genus (Winogradsky, 1888, 1922; Molisch, 1910; Lieske, 1919; Cataldi, 1939; Präve, 1957).

The most typical characteristic of *L. ochracea* is the formation of large numbers of almost empty sheaths within a relatively short time. The mechanism of this procedure can be followed under a phase-contrast microscope in a slide culture of the organism in an iron-containing soil extract medium. In this way, the behavior of *L. ochracea* in crude culture can be observed continuously. Chains of cells leave their sheaths at the rate of 1–2 µm/min, continuously producing new hyaline sheaths connected with the old envelope (Fig. BXII.β.51). Impregnation and covering of the sheaths with iron probably take place after the cells have left the envelopes. Aged, golden-brown sheaths are brittle and easily broken into relatively short fragments (Fig. BXII.β.52).

Isolation of *L. ochracea* has not been achieved, either from natural enrichments or in the laboratory from enrichment cultures of slowly running soil extract, so few details of the organism are available. In some instances, an organism resembling *L. ochracea* has been isolated, viz. "*L. pseudo-ochracea*".

The mol% G + C of the DNA is: not available.

Type strain: No culture available.

2. **Leptothrix cholodnii** Mulder and van Veen 1963, 137[AL]

cho.lod' ni.i. M.L. gen. n. *cholodnii* of Cholodny, named for N. Cholodny, a Russian bacteriologist.

Cells of freshly isolated strains are usually found in long chains inside the sheaths. Single, motile cells may be seen outside the sheaths. In the presence of Mn^{2+}, the cells become covered with granular MnO_2 (Fig. BXII.β.53). At some sheath locations, the MnO_2 deposits may even exceed 10 µm. *L. cholodnii*, in contrast to other *Leptothrix* species, responds to an increased supply of organic nutrients (Table BXII.β.63), resulting in relatively large colonies (up to 5 mm in diameter) on nutrient-rich agar media. On manganese (II)-containing agar, black-brown hairy colonies are formed, particularly when the organism is seeded densely. Most strains display a strong tendency to dissociate spontaneously and to produce smooth, rather than the typical rough, colonies. Such mutant strains are largely sheathless and oxidize manganese slightly or not at all (Mulder and van Veen, 1963; Rouf and Stokes, 1964; Stokes and Powers, 1965).

Growth takes place between 10 and 35°C; the pH range for growth is from pH 6.5 to 8.5. Oxidase positive. The following compounds support growth in GMBN medium (Kämpfer et al., 1995): DL-3-hydroxybutyrate, DL-lactate, L-glutamate, L-leucine, L-proline, phenylpyruvate, and quinate; no growth is obtained with D-fucose, fumarate, or glutarate. Poly-β-hydroxybutyrate granules are formed.

Major fatty acids (>5% each) present in all strains are

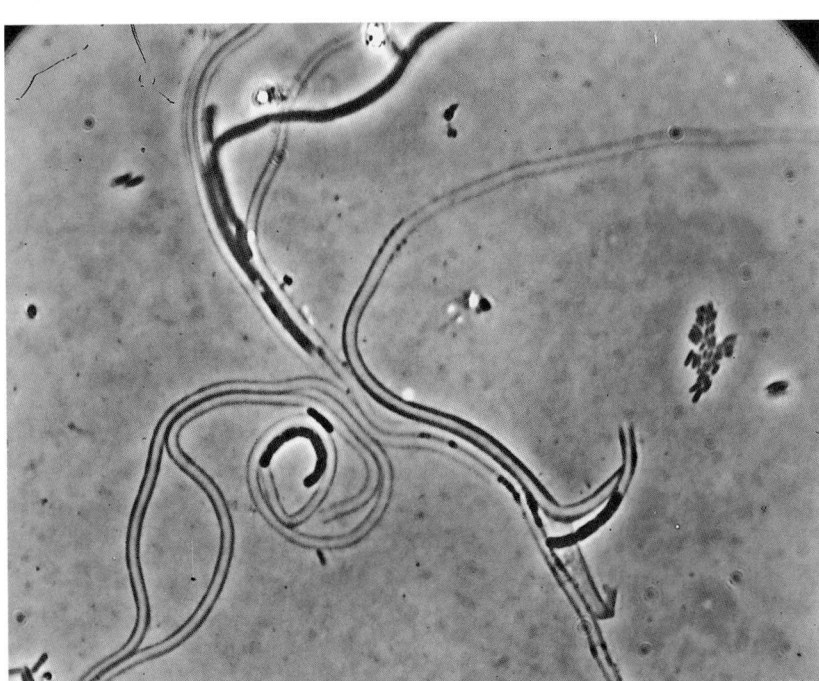

FIGURE BXII.β.51.   *L. ochracea* in iron- and manganese-containing soil extract. Cells are continuously leaving sheaths and forming new sheaths. Light micrograph (× 1268).

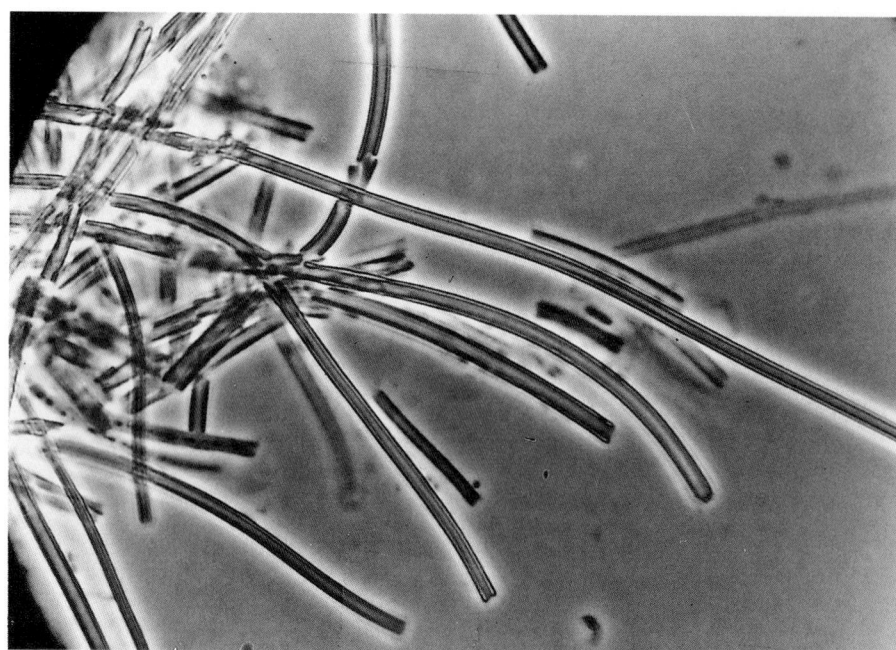

FIGURE BXII.β.52. Broken empty sheaths of *L. ochracea* in crude culture in slowly flowing iron- and manganese-containing soil extract. Light micrograph (× 1186). (Reproduced with permission from E.G. Mulder and W.L. van Veen, Antonie van Leeuwenhoek Journal of Microbiology and Serology *29:* 121–153, 1963, ©Kluwer Academic Publishers.)

TABLE BXII.β.63. Differential characteristics of *Leptothrix* species[a]

Characteristic	*L. ochracea*	*L. cholodnii*	*L. discophora*	*L. lopholea*	*L. mobilis*	*"L. pseudo-ochracea"*
Cell dimensions:						
Width (μm)	1	0.7–1.5	0.6–0.8	1.0–1.4	0.6–0.8	0.8–1.3
Length (μm)	2–4	2.5–15	2.5–12	3–7	1.5–12	5–12
Monotrichous polar flagella	+	+	+	−	+	+
Polytrichous subpolar flagella	−	−	−	+	−	−
Holdfasts	−	−	−	+	−	−
False branching	−	−	+	+	−	−
Growth at:						
35°C	nd	+	−	nd	+	nd
pH 8.5	nd	+	−	nd	+	nd
Growth on:[b]						
D-Fucose	nd	−	+	nd	−	nd
Fumarate	nd	−	−	nd	+	nd
DL-Lactate	nd	+	−	nd	−	nd

[a]For symbols see standard defininitions; nd, not determined.

[b]Growth was determined in GMBN medium supplemented with the respective carbon source (Kämpfer et al., 1995).

hexadecanoic acid, *cis*-9-hexadecenoic acid, dodecanoic acid, and *cis*-9,11-octadecenoic acid. Tetradecanoic acid is found in small amounts in all strains, whereas other minor fatty acids have not been detected.

Results of DNA–DNA hybridization experiments have indicated that, currently, three different strains belong to this species, viz. LMG 8142, LMG 8143, and LMG 7171, sharing DNA–DNA similarity values well above 70% (Spring et al., 1996). The strain LMG 7171 served as the reference strain in this study, because the original designated type strain has apparently been lost.

In agreement with its nutritional requirements, *L. cholodnii* is found in slowly running iron- and manganese-containing, unpolluted waters or in polluted waters, particularly in activated sludge.

The mol% G + C of the DNA is: 68–70 (T_m).

Type strain: LVMW 99 (not available).

3. **Leptothrix discophora** Spring, Kämpfer, Ludwig and Schleifer 1997, 601[VP] (Effective publication: Spring, Kämpfer, Ludwig and Schleifer 1996, 640.)

dis.co'phor.a. Gr. n. *discos* a disk; Gr. adj. *phoros* bearing; M.L. adj. *discophora* disk-bearing.

Cells are relatively small compared to those of the other *Leptothrix* species described (Table BXII.β.63). They may occur in narrow sheaths or be free swimming; free cells are motile by thin polar flagella at one or both poles. The manganese-oxidizing and ferric oxide-storing capacities of this organism are very pronounced. In enrichment cultures containing nutrient media with Mn (II) salts, the sheaths are heavily but irregularly encrusted with MnO_2, giving rise to sheaths of up to 10 μm thickness. In media with both

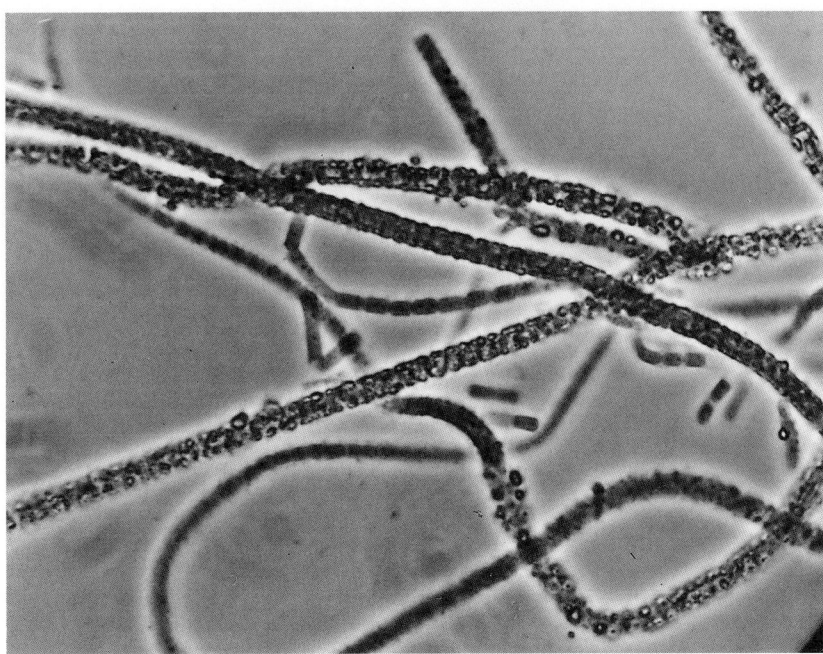

FIGURE BXII.β.53. *L. cholodnii* sheaths encrusted with granulated MnO_2 (× 1280). (Reproduced with permission from E.G. Mulder and W.L. van Veen, Antonie van Leeuwenhoek Journal of Microbiology and Serology *29*: 121–153, 1963, ©Kluwer Academic Publishers.)

Mn (II) and Fe (II), as in slowly flowing soil extract, the sheaths become covered with a thick, dark brown, fluffy layer of ferric oxide and MnO_2, which may increase the diameter of the trichomes to up to about 20–25 μm (Fig. BXII.β.49). Under these conditions, the sheaths may taper toward the growing tips. Following isolation, the ability to form sheaths is easily lost in this species (Adams and Ghiorse, 1986). False branching is regularly observed, even with sheathless strains. In older cultures, coccoid bodies and cell evaginations are formed.

Colonies on the solid medium of Rouf and Stokes are about 1 mm in diameter, irregular in shape, flat, and dark-brown. Under certain conditions, filamentous colonies may be formed. Increasing the supply of nutrients, such as glucose, peptone, methionine, purine bases, vitamin B_{12}, biotin, and thiamine, only increases growth slightly. Visible aggregates are formed when cells are grown in liquid media.

Growth occurs between 15 and 33°C; the pH range for growth is from pH 6.0 to 8.0. Oxidase positive. D-fucose, L-proline, and protocatechuate support growth in GMBN medium; no growth is obtained with fumarate, glutarate, DL-3-hydroxybutyrate, DL-lactate, L-glutamate, L-leucine, phenylpyruvate, and quinate. Poly-β-hydroxybutyrate granules are formed. Major fatty acids (>5%) are hexadecanoic acid, *cis*-9-hexadecenoic acid and dodecanoic acid.

Minor fatty acids are *cis*-9,11-octadecenoic acid, 3-hydroxydecanoic acid and octadecanoic acid.

The normal habitat is slowly running, unpolluted, iron- and manganese-containing water of ditches, rivers, and ponds.

The mol% G + C of the DNA is: 71 (T_m).

Type strain: SS-1, LMG 8141.

GenBank accession number (16S rRNA): L33975.

4. **Leptothrix lopholea** Dorff 1934, 33[AL]

loph.o.le′ a. Gr. n. *lophos* a crest; M.L. dim. fem. adj. *lopholea* somewhat crested or tufted.

L. lopholea resembles *S. natans* to a greater extent than do the other *Leptothrix* species. It produces polytrichous, subpolar flagellation and forms holdfasts and false branches. Strains may also grow in rich media. Cells usually develop short, sheathed filaments radiating from a cluster of holdfasts, giving rise to many tiny flocs when the cells are grown in liquid media (Fig. BXII.β.54). Deposition of iron and manganese oxides is more pronounced on hold-fasts than on filaments. On Mn (II)-containing agar media, encrustation of sheaths with MnO_2 is retarded, so that colonies are white at first and later become black-brown. Cell growth responds poorly to an increased supply of organic nutrients. Strains that do show a good response oxidize manganese more slowly.

This species may be isolated from slowly flowing, unpolluted or polluted freshwater and from activated sludge.

The mol% G + C of the DNA is: not available.

Type strain: LVMW 124 (not available).

5. **Leptothrix mobilis** Spring, Kämpfer, Ludwig and Schleifer 1997, 601[VP] (Effective publication: Spring, Kämpfer, Ludwig and Schleifer 1996, 640.)

mo′ bi.lis. L. adj. *mobilis* motile, movable.

Cells are similar in width to those of *L. discophora*, but are usually shorter. They are highly motile by single polar flagella. Sheaths are not formed under laboratory conditions. Colonies on the agar medium of Rouf and Stokes are about 1 mm in diameter, circular, flat, smooth, and dark-brown. Visible aggregates are formed when cells are grown in liquid media.

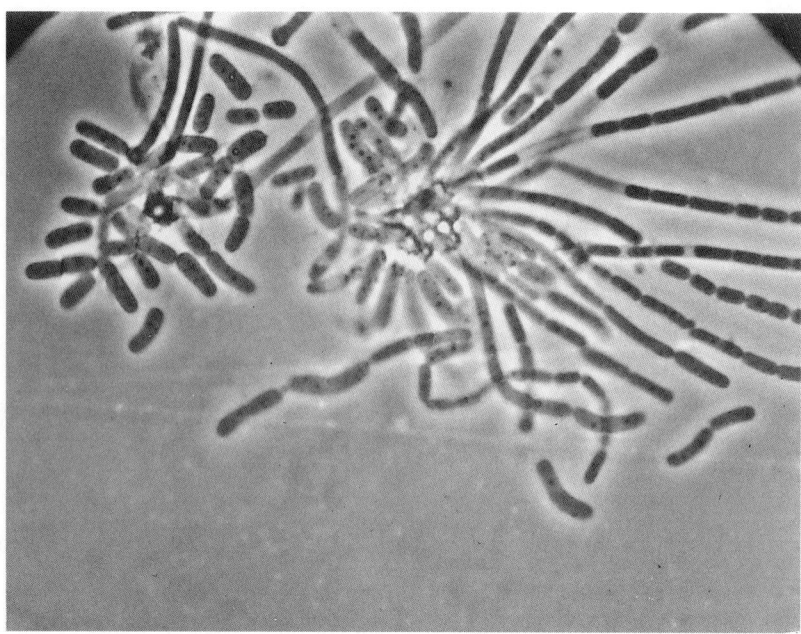

FIGURE BXII.β.54. *L. lopholea* in culture solution. Many trichomes can be seen radiating from common holdfasts. Light micrograph (× 1316). (Reproduced with permission from E.G. Mulder and W.L. van Veen, Antonie van Leeuwenhoek Journal of Microbiology and Serology *29:* 121–153, 1963, ©Kluwer Academic Publishers.)

Growth takes place between 10 and 37°C; the pH range for growth is from pH 6.5 to 8.5. Oxidase positive. Fumarate and glutarate support growth in GMBN medium; no growth is obtained with D-fucose, DL-3-hydroxybutyrate, DL-lactate, L-glutamate, L-leucine, L-proline, phenylpyruvate, protocatechuate, and quinate. Poly-β-hydroxybutyrate granules are formed.

Major fatty acids (>5%) are hexadecanoic acid, *cis*-9-hexadecenoic acid, and *cis*-9,11-octadecenoic acid. Minor fatty acids are 3-hydroxydecanoic acid, dodecanoic acid, octadecanoic acid, tetradecanoic acid, and *cis*-9,10-methyleneoctadecanoic acid.

Isolated from the sediment of a freshwater lake (Chiemsee) in Upper Bavaria, Germany.

The mol% G + C of the DNA is: 68 (T_m).

Type strain: Feox-1, LMG 17066, DSM 10617.

GenBank accession number (16S rRNA): X97071.

6. **"Leptothrix pseudo-ochracea"** Mulder and van Veen 1963, 135.

pseu.do.o.chra'ce.a. Gr. adj. *pseudes* false; M.L. adj. *ochracea* specific epithet; M.L. adj. *pseudo-ochracea* not the true *L. ochracea.*

Cells are more slender than those of the other *Leptothrix* species (Table BXII.β.63), and are very motile by single, thin, polar flagella. Even chains of 6–10 cells may show an undulatory locomotion after leaving their sheaths. This characteristic may account for the relatively large number of empty sheaths in culture, as compared with the number found in most *Leptothrix* species; however *L. ochracea* possesses even more empty sheaths. In slowly flowing ferrous iron-containing soil extract, the sheaths become impregnated with ferric oxide and appear yellow-brown. In this respect, the organism resembles *L. ochracea*. In media with added manganese compounds, the sheaths are covered with small granules of MnO_2. On manganese (II)-containing agar, the black-brown colonies are very filamentous and may exceed a width of 10 mm. On basal agar media containing 0.1% peptone and 0.1% glucose, the organism may grow in concentric rings.

The normal habitat of *"Leptothrix pseudo-ochracea"* is the slowly running, unpolluted, iron- and manganese-containing freshwater of ditches and brooklets. This species may also be found in slightly polluted water.

Deposited strain: none designated.

Genus Incertae Sedis XIV. **Roseateles** *Suyama, Shigematsu, Takaichi, Nodasaka, Fujikawa, Hosoya, Tokiwa, Kanagawa and Hanada 1999, 455*[VP]

AKIRA HIRAISHI AND JOHANNES F. IMHOFF

Ro.se.a.te'les. L. adj. *roseus* rose-colored, pink; Gr. adj. *ateles* defective, incomplete; M.L. masc. n. *Roseateles* the rose-colored incomplete (photosynthetic bacterium).

Cells are straight slender rods, 0.5–0.6 μm wide, motile by means of polar flagella, reproduce by binary fission. Gram negative. **Belong to the** *Betaproteobacteria.* Do not form any type of internal membranes. **Synthesize bacteriochlorophyll** *a* **and carotenoids as photosynthetic pigments.**

Strictly aerobic, chemoorganoheterotrophic bacteria. Do not grow and produce photosynthetic pigments under anoxic conditions in the light. Carbon sources and electron donors supporting growth are simple organic compounds, Casamino acids, peptone, skim milk, and yeast extract. Accumulate poly-β-hydroxybutyrate. Catalase negative, oxidase positive. **The major respiratory quinone is ubiquinone-8.**

Mesophilic and neutrophilic freshwater bacteria.

The mol% G + C of the DNA is: 66.2–66.3.

Type species: **Roseateles depolymerans** Suyama, Shigematsu, Takaichi, Nodasaka, Fujikawa, Hosoya, Tokiwa, Kanagawa and Hanada 1999, 455.

FURTHER DESCRIPTIVE INFORMATION

The genus *Roseateles* is at this time monotypic, with *R. depolymerans* as the type species (Suyama et al., 1999). Based on 16S rDNA sequence analysis, *R. depolymerans* belongs to the *Betaproteobacteria*, with *Rubrivivax*, *Ideonella*, *Leptothrix*, and *Sphaerotilus* species as phylogenetic neighbors. The 16S rDNA sequence of the phototrophic, purple, nonsulfur bacterium *Rubrivivax gelatinosus* is 96.3% similar to that of *Roseateles depolymerans*.

Colonies on agar media are circular, smooth, and soft. Rough and opaque colonies occur in some cases. The color of colonies is white at the early stage of cultivation but becomes pink in older cultures due to the production of carotenoids. Liquid cultures exhibit no or little pigmentation. The intensity of colony pigmentation depends upon the growth medium. The best medium for pigment production is an agar medium containing 0.02% Casamino acids. Thin-section electron microscopy indicated the absence of any type of internal membranes and the presence of poly-β-hydroxybutyrate granules (cells stained positively with Nile blue A).

Photosynthetic pigments are bacteriochlorophyll *a* esterified with phytol and carotenoids. The major carotenoid of *R. depolymerans* is spirilloxanthin (88–89 mol% of the total content). Biosynthetic precursors of spirilloxanthin, OH-spirilloxanthin, anhydrorhodovibrin, and 3,4-dehydrorhodopin are present as minor components. The molar ratios of total carotenoids/bacteriochlorophyll *a* were 0.58–0.65 (Suyama et al., 1999). Ultrasonically disrupted cells have absorption maxima at 482, 515, 550, 590, 800, and 870 nm. The spectral pattern in the near-infrared region indicates that the cells contain a core light-harvesting complex (B870, LH I), together with the photosynthetic reaction center, but lack a peripheral light-harvesting complex (LH II). Like other aerobic bacteriochlorophyll-containing bacteria, *R. depolymerans* produces BChl *a* only under aerobic growth conditions in darkness.

R. depolymerans is an obligately aerobic chemoorganoheterotrophic bacterium, which is unable to grow under anaerobic conditions, neither in the dark nor in the light. The doubling time exhibited under optimal growth conditions is approx. 2 h. One of the most characteristic features of *Roseateles depolymerans* is the ability to degrade aliphatic polycarbonates, including poly(tetramethylene carbonate), poly(hexamethylene carbonate), and poly(ε-caprolactone), which are known as biodegradable plastics. On polycarbonate agar plates, clear zones are produced around the colonies and the turbidity of opaque polycarbonate suspension in liquid culture is reduced within a few days of incubation. Gel permeation chromatography showed that poly(hexamethylene carbonate) was degraded via the production of diester, di(6-hydroxyhexyl) carbonate (Suyama et al., 1998a). The accumulation of diesters as intermediate products suggests that lipolytic enzymes may be the key enzymes in this degradation process.

Strains of *R. depolymerans* were isolated together with other polycarbonate-degrading bacteria from Hanamuro River in Ibaraki Prefecture, Japan (Suyama et al., 1998a). Polycarbonate-degrading pigmented strains that are phylogenetically very close to *R. depolymerans* have also been isolated from soil environments (Suyama et al., 1998b). In view of these results, *R. depolymerans* and related bacteria may inhabit a wide variety of freshwater and terrestrial environments.

ENRICHMENT AND ISOLATION PROCEDURES

Surface water of freshwater environments is a possible source for *Roseateles*. For selective enrichment and isolation, the use of mineral medium supplemented with poly(hexamethylene carbonate) is recommended (Suyama et al., 1998a). The polycarbonate liquid medium is inoculated with a suitable water sample and incubated with shaking at 35°C in darkness. Cultures positive for reducing the turbidity of polycarbonate are transferred to agar media of the same composition. After incubation for at least one week, pink colonies with clear zones may appear on the agar medium. Growth media and cultural conditions commonly used for the isolation of freshwater aerobic bacteria can be used for the isolation of *R. depolymerans*.

MAINTENANCE PROCEDURES

Cultures are well-preserved in liquid nitrogen or by lyophilization. Preservation in a mechanical freezer at −80°C is also possible.

DIFFERENTIATION OF THE GENUS *ROSEATELES* FROM OTHER GENERA

Differential characteristics of the genus *Roseateles* and related genera are shown in Table BXII.β.64.

TAXONOMIC COMMENTS

Strains of *Roseateles depolymerans* were first isolated as obligately aerobic chemoorganotrophic bacteria capable of degrading aliphatic polycarbonates (Suyama et al., 1998a). Detailed phenotypic studies showed that these polycarbonate-degrading strains produce bacteriochlorophyll *a* and carotenoids with spirilloxanthin as the main component. However, photosynthetic pigments were formed only under aerobic growth conditions, as found in other obligately aerobic bacteriochlorophyll-containing bacteria. 16S rDNA-based sequence analysis indicated that the polycarbonate-degrading, bacteriochlorophyll-producing strains were closely related to the phototrophic, purple, nonsulfur bacterium *Rubrivivax gelatinosus*. However, major phenotypic differences between the polycarbonate-degrading strains and *Rubrivivax gelatinosus* and other phylogenetically related bacteria led to the proposal of a new genus and species, *Roseateles depolymerans* (Suyama et al., 1999). The genus *Roseateles* belongs to the *Betaproteobacteria* and is the first described genus of aerobic bacteriochlorophyll-containing bacteria that does not fall within the *Alphaproteobacteria*.

ACKNOWLEDGMENTS

The authors are indebted to T. Suyama for his helpful comments and discussions.

List of species of the genus Roseateles

1. **Roseateles depolymerans** Suyama, Shigematsu, Takaichi, Nodasaka, Fujikawa, Hosoya, Tokiwa, Kanagawa and Hanada 1999, 455[VP]

de.po.ly′ me.rans. M.L. v. *depolymerare* depolymerize; M.L. part. adj. *depolymerans* depolymerizing.

Cells are straight slender rods, 0.5–0.6 × 2–5 μm, motile

TABLE BXII.β.64. Differential characteristics of *Roseateles depolymerans* and phylogenetically related bacteria[a]

Characteristic	Roseateles depolymerans	Alcaligenes latus	Ideonella dechloratans	Leptothrix discophora	Rubrivivax gelatinosus	Sphaerotilus natans
Mode of flagellation:						
Lateral		+				
Polar tuft	+		+			+
Single polar				+	+	
Sheath formation	−	−	−	−	−	+
PHB deposition	+	+	nd	+	nd	+
Bacteriochlorophyll *a*	+	−	−	−	+	−
Carotenoid pigment:						
Sprilloxanthin	+				+	
Spheroidene					+	
OH-spheroidene					+	
Anaerobic phototrophy	−	−	−	−	+	−
Autotrophy with H$_2$	−	+	−	−	+	−
Nitrogen fixation	−	+	−	−	+	−
Nitrate reduction	−	−	+	−	−	+
Catalase	+	+	+	nd	+	+
Cytochrome oxidase	+	+	+	nd	+	+
Growth factors required:						
B$_{12}$				+		+
Biotin					+	
Thiamine					+	
Major quinone:						
MK-8					+	
Q-8	+	+	+	+	+	+
Habitats:						
Freshwater	+		+	+		+
Mud					+	
Soil		+			+	
Mol% G + C of DNA	66.2–66.3	69.1–71.1	68.1	67.8–71.1	70.0–72.5	69.9
Carbon source utilized:						
Acetate	d	−	+	−	+	+
Pyruvate	+	−	+	+	+	+
Lactate	+	+	+	−	+	+
Citrate	+	d	nd	−	+	+
Malate	+	+	nd	+	+	+
Succinate	+	+	+	+	+	+
D-Ribose	−	−	nd	+	nd	nd
D-Glucose	+	+	+	+	+	+
D-Fructose	+	+	+	−	+	+
D-Galactose	+	−	nd	nd	nd	+
Sucrose	−	+	nd	+	nd	+
Glycerol	−	+	nd	+	−	+
Mannitol	+	−	nd	nd	−	+
Sorbitol	−	−	nd	nd	−	+

[a]Symbols and abbreviations: +, 90% or more of strains positive; −, 90% or more of strains negative; d, 11–89% of strains positive; nd, not determined; PHB, poly-β-hydroxybutyrate.

by means of a single flagellum or several polar flagella. Motility is observed only in the early exponential phase of growth. Spores and sheaths are not formed. Cell suspensions and colonies are white to pink. Pigmentation varies depending upon growth media and culture age. Absorption maxima of living cells are at 482, 515, 550, 590, 800, and 870 nm. Bacteriochlorophyll *a* and carotenoids with spirilloxanthin as the major component are synthesized as photosynthetic pigments only under aerobic conditions in the dark.

Obligately aerobic chemoorganoheterotrophic bacteria producing photosynthetic pigments. Do not grow phototrophically under anoxic conditions in the light. Do not grow by anaerobic respiration with nitrate, dimethyl sulfoxide, or trimethylamine *N*-oxide as the terminal electron acceptors. Autotrophic growth with hydrogen is absent. Good carbon sources for growth are D-glucose, D-fructose, D-galactose, mannitol, pyruvate, lactate, citrate, succinate, and L-malate. Good growth also occurs with Casamino acids, peptone, skim milk, and yeast extract. Most characteristic is the ability to co-metabolize and degrade aliphatic polycarbonates. Not utilized are D-ribose, sucrose, glycerol, and sorbitol. Hydrolytic activities against starch and gelatin are present. Acid production from glucose in Hugh–Leifson's OF medium is weak or negative. Oxidase and β-galactosidase (ONPG), but not catalase, are produced. Nitrate is not reduced to nitrite. Nitrogen fixation is negative. No growth factors are required.

Optimal growth occurs at 35°C (range: 5–43°C) and pH 6.5 (range: pH 5–8).

Habitat: wide variety of freshwater environments.

The mol% G + C of the DNA is: 66.2–66.3 (HPLC).

Type strain: Suyama 61A , DSM 11813.

GenBank accession number (16S rRNA): AB003623.

Genus Incertae Sedis XV. **Rubrivivax** Willems, Gillis and De Ley 1991b, 70[VP]

JOHANNES F. IMHOFF

Rub.ri.vi'vax. L. neut. adj. *rubrum* red; L. masc. adj. *vivax* long living; M.L. masc. n. *Rubrivivax* the red and long living (bacterium).

Cells are straight or curved rods, are motile by means of polar flagella, and multiply by binary fission. **Gram negative and belonging to the *Betaproteobacteria.*** Internal membrane systems are poorly developed or absent. **Photosynthetic pigments are bacteriochlorophyll *a* and carotenoids of the spheroidene series. Ubiquinones and menaquinones with eight isoprene units (Q-8 and MK-8) are present. Straight-chain C$_{16:1}$ and C$_{16:0}$ are the major components of cellular fatty acids. C$_{10:0\ 3OH}$ is present.**

Photoautotrophic and photoheterotrophic growth may occur with hydrogen and a variety of carbon compounds as electron donors. **Chemotrophic growth is possible** by respiration under microoxic to oxic conditions in the dark or by fermentation. Cells grow well with simple organic compounds as electron donors and carbon sources, as well as in complex media containing peptone, yeast extract or Casamino acids. Growth factors are required. **Mesophilic and neutrophilic freshwater bacteria.** Habitat: freshwater ponds, sewage ditches, activated sludge.

The mol% G + C of the DNA is: 70.5–72.4 (Bd), 71.2–72.1 (HPLC).

Type species: **Rubrivivax gelatinosus** (Molisch 1907) Willems, Gillis and De Ley 1991b, 71 (*Rhodocyclus gelatinosus* (Molisch 1907) Imhoff, Trüper and Pfennig 1984, 341; *Rhodopseudomonas gelatinosa* (Molisch 1907) van Niel 1944, 98; *Rhodocystis gelatinosa* Molisch 1907, 22.)

FURTHER DESCRIPTIVE INFORMATION

Characteristic of *Rubrivivax gelatinosus* is the liquefaction of gelatin, which is catalyzed by an extracellular protease (Klemme and Pfleiderer, 1977). *R. gelatinosus* grows well with citrate as the carbon source and thereby excretes large amounts of acetate, which serves as the carbon source after citrate is exhausted (Schaab et al., 1972). Citrate lyase, the key enzyme for growth on citrate, has been characterized in this species (Giffhorn et al., 1972; Beuscher et al., 1974). *R. gelatinosus* can also be adapted to grow with CO as the sole energy and carbon source under anoxic conditions in the dark (Uffen, 1976). Under these conditions, the activities of ribulosebisphosphate carboxylase and enzymes of the serine pathway are enhanced (Uffen, 1983).

As with other phototrophic *Betaproteobacteria*, the major phospholipid components of *Rubrivivax gelatinosus* are cardiolipin, phosphatidylethanolamine, and phosphatidylglycerol (Imhoff and Bias-Imhoff, 1995). Straight-chain C$_{16:1}$ and C$_{16:0}$ acids are the main components of cellular fatty acids (see Table BXII.β.58 of the chapter on the genus *Rhodoferax*).

ENRICHMENT AND ISOLATION PROCEDURES

R. gelatinosus is widely distributed in freshwater habitats where purple nonsulfur *Bacteria* occur. Media for enrichment, isolation, and growth generally employed for freshwater purple nonsulfur *Bacteria* are also suitable for *Rubrivivax* species. Though citrate is not a commonly utilized substrate among purple nonsulfur

Bacteria, it can be used by several species. Nevertheless, it has been proven to be useful for the selective enrichment of *R. gelatinosus*. Standard techniques for the isolation of anaerobic bacteria in agar dilution series and on agar plates also can be applied for *Rubrivivax* (see Biebl and Pfennig, 1981; Trüper and Imhoff, 1992).

MAINTENANCE PROCEDURES

Rubrivivax species are readily maintained by standard procedures in liquid nitrogen. Preservation is also possible by lyophilization or storage at −80°C in a mechanical freezer.

DIFFERENTIATION OF THE GENUS *RUBRIVIVAX* FROM OTHER GENERA

Rubrivivax is clearly distinguished from other phototrophic purple bacteria by its separate phylogenetic line within the *Betaproteobacteria*. The high mol% G + C content of >70 is quite characteristic. Properties to differentiate *Rubrivivax* species from other phototrophic *Betaproteobacteria* are given in Tables 6 (p. 130) and 7 (p. 131) of the introductory chapter "Anoxygenic Phototrophic Purple Bacteria", Volume 2, Part A. The phylogenetic relationships of the phototrophic *Betaproteobacteria* are shown in Fig. 4 (p. 132) of that same chapter.

TAXONOMIC COMMENTS

Currently, only one species is known. *Rubrivivax gelatinosus* has been reported to occur in two distinct morphological forms (Biebl and Drews, 1969). Form I cells are clearly curved, 0.4–0.7 μm in diameter, and produce little slime during active growth. Form II cells are more or less straight rods, although they are sometimes also bent, and produce more slime during active growth, causing sedimentation of the cells in a gelatinous layer. Exemplified by two strains from each of the morphological groups, form I cells utilize a greater variety of carbon sources and have a much shorter doubling time than do form II cells (Weckesser et al., 1969). Another strain of this species, described by Klemme (1968), is intermediate in carbon substrate utilization with respect to the two groups established by Weckesser et al. (1969). Furthermore, the lipopolysaccharides of *R. gelatinosus* show two different serotypes, which do not cross-react with each other (Weckesser et al., 1975). It is not known whether correlations exist between serotypes, morphological types, and nutritional types. Detailed and careful studies are required to clarify the taxonomic status of the strains assigned to this species.

Further information is required on the specificity of gelatin liquefaction by *Rubrivivax gelatinosus*. Siefert et al. (1978) have identified about half of the strains with this property as belonging to *Rhodobacter capsulatus*, while other strains, with the morphology typical of *Rubrivivax gelatinosus*, do not liquefy gelatin. Weak hydrolytic activity of gelatin is also present in *Rhodoferax fermentans*.

List of species of the genus Rubrivivax

1. **Rubrivivax gelatinosus** (Molisch 1907) Willems, Gillis and De Ley 1991b, 71[VP] (*Rhodocyclus gelatinosus* (Molisch 1907) Imhoff, Trüper and Pfennig 1984, 341; *Rhodopseudomonas*

gelatinosa (Molisch 1907) van Niel 1944, 98; *Rhodocystis gelatinosa* Molisch 1907, 22.)

ge.la.ti.no'sus. L. part. adj. *gelatus* frozen, stiffened; M.L. n.

gelatinum gelatin, that which stiffens; M.L. masc. adj. *gelatinosus* gelatinous.

Cells are rod shaped, straight, or slightly curved, 0.4–0.7 × 1–3 μm and in older cultures up to 15 μm long and irregularly curved. Most strains show abundant mucous production in all media, which causes the cells to clump together and appear immotile. In young cultures, cells are highly motile by means of polar flagella. Internal photosynthetic membranes appear as small fingerlike intrusions of the cytoplasmic membrane. Cultures grown anaerobically in the light are pale peach to dirty yellowish brown; aerobically grown cells appear colorless to light yellowish brown. Cells contain bacteriochlorophyll *a* (esterified with phytol) and carotenoids of the spheroidene series, with spheroidene, OH-spheroidene and spirilloxanthin as major components.

Preferably grow photoheterotrophically under anoxic conditions in the light, with a variety of organic compounds as electron and carbon sources. Photoautotrophic growth is also possible with hydrogen as electron source in the presence of growth factors. Chemotrophic growth is possible under microoxic to oxic conditions in the dark. Some strains can also adapt to grow anaerobically in the dark,

with CO as sole carbon and energy sources. Pyruvate is fermented anaerobically in the dark. The carbon sources utilized are listed in Table BXII.β.59 of the chapter on the genus *Rhodoferax*. In addition, a variety of amino acids, yeast extract, and peptone are used. Most characteristic is the liquefaction of gelatin. Fatty acids are utilized only at low concentrations. Suitable nitrogen sources are ammonia, dinitrogen, and a number of amino acids; some strains may also utilize uracil, thymine, guanine, and uric acid both anaerobically in the light and aerobically in the dark, but xanthine and adenine only under oxic conditions. Sulfate can be used as the sole sulfur source. Biotin and thiamine are required as growth factors; some strains also require pantothenate.

Mesophilic freshwater bacterium, with optimal growth at 30°C and pH 6.0–8.5. Habitat: freshwater ponds, sewage ditches, and activated sludge. Major quinone components are Q-8 and MK-8.

The mol% G + C of the DNA is: 70.5–72.4 (Bd), 70.2–72.0 (T_m), 71.2–72.1 (HPLC); type strain, 71.9 (T_m).

Type strain: ATCC 17011, DSM 1709, LMG 4311.

GenBank accession number (16S rRNA): D16213, M60682.

Genus Incertae Sedis XVI. Sphaerotilus *Kützing 1833, 386*[AL]

PETER KÄMPFER AND STEFAN SPRING

Sphae.ro′ti.lus. Gr. n. *sphaera* a sphere; Gr. n. *tilus* anything shredded, floc, down; M.L. masc. n. *Sphaerotilus* spherical flock.

Straight rods, 1.2–2.5 × 2–10 μm, usually arranged in single chains within sheaths of uniform width, which may be attached by means of holdfasts to walls of containers, submerged plants, stones, and other surfaces. True, rather than false, branching of the filaments does not occur. **Single or paired cells released from the sheaths are motile by means of a bundle of subpolar flagella**, sometimes so intertwined as to give the appearance of a single large "unit flagellum". Sheaths usually thin without encrustation by ferric and manganic oxides. They cannot always be easily recognized when completely filled with cells, but if parts of the sheaths are vacated by the cells, recognition of the organism cannot be misinterpreted. Resting stages are not known. Gram negative. Has a propensity for storing **poly-β-hydroxybutyrate** in granules. **Aerobic**, having a strictly respiratory type of metabolism with oxygen as the terminal electron acceptor. Can grow at very low concentrations of dissolved oxygen (below 0.1 mg/l). Temperature range: 10–40°C; optimum: between 20° and 30°C. pH range: 5.4–9; Optimal pH: between 6.5 and 7.5. **Chemoorganotrophic.** Alcohols, organic acids, and sugars are used as sources of carbon and energy. Ammonium salts and nitrates may serve as nitrogen source in the presence of vitamin B_{12} or methionine. Peptone, Casamino acids, and mixtures of aspartic and glutamic acids, and vitamin B_{12}, or methionine give better results.

The mol% G + C of the DNA is: 70.

Type species: **Sphaerotilus natans** Kützing 1833, 386.

FURTHER DESCRIPTIVE INFORMATION

Pure cultures, upon prolonged incubation, may sometimes show large, circular bodies resembling protoplasts. Their formation is probably due to the production of enzymes involved in the decomposition of the cell walls during the death phase. Incorpor-

ation of 0.4 g glycine per liter of nutrient medium favors this phenomenon (Phaup, 1968).

Fig. BXII.β.55 illustrates partly filled and empty sheaths. The surface of the sheaths has a smooth structure, which is in contrast with the rough structure of the sheath surface of *Leptothrix* species (Fig. BXII.β.56). The sheaths of *S. natans* are covered with a cohering slime layer of variable thickness. For the composition of the slime, see Gaudy and Wolfe (1962). The composition of the sheath was investigated by Romano and Peloquin (1963), Petitprez et al. (1969), and Hoeniger et al. (1973). Takeda et al. (1998) found carbohydrate (54.1%), protein (12.2%), and lipid (1–3%) in the sheaths by colorimetric reactions and solvent extraction. Gas-liquid chromatography showed glucose and galactosamine to be present in the molar ratio of 1:4. The most abundant amino acids in the sheath protein were glycine (49.2 mol%) and cysteine (24.6 mol%). The sheaths were resistant to agents that reduce disulfide bonds (dithiothreitol and 2-mercaptoethanol) and to protease treatment. Sheaths could be degraded completely by hydrazine, and a heteropolysaccharide composed of glucose and galactosamine (1:4) was released. A specific enzyme produced by a *Paenibacillus* sp. acting on the polysaccharide moiety was described in detail by Takeda et al. (2000).

The size and shape of colonies depends largely on the ability of the respective strain to form sheaths. Typical colonies are rough (Fig. BXII.β.57). Often the colonies are characterized by a smooth central region and a "short-haired" periphery (Pellegrin et al., 1999). In broth cultures, the organisms often develop a pellicle on the broth surface composed of long, unbranched, sheathed filaments.

The normal habitat of *S. natans* is slowly running freshwater that is heavily contaminated with sewage or wastewater from pa-

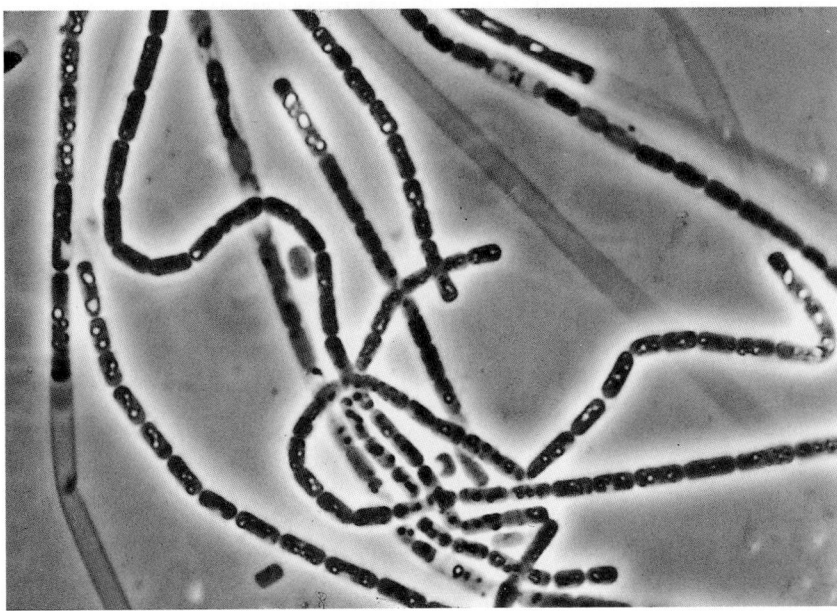

FIGURE BXII.β.55. *S. natans* filaments from a rough colony grown on agar medium containing glucose and peptone at 1 g/l each. Partly filled and empty sheaths can be seen. Many cells contain globules of poly-β-hydroxybutyrate. × 1006. (Reproduced with permission from E.G. Mulder and W.L. van Veen, Antonie van Leeuwenhoek Journal of Microbiology and Serology *29:* 121–153, 1963, ©Kluwer Academic Publishers, Dordrecht.)

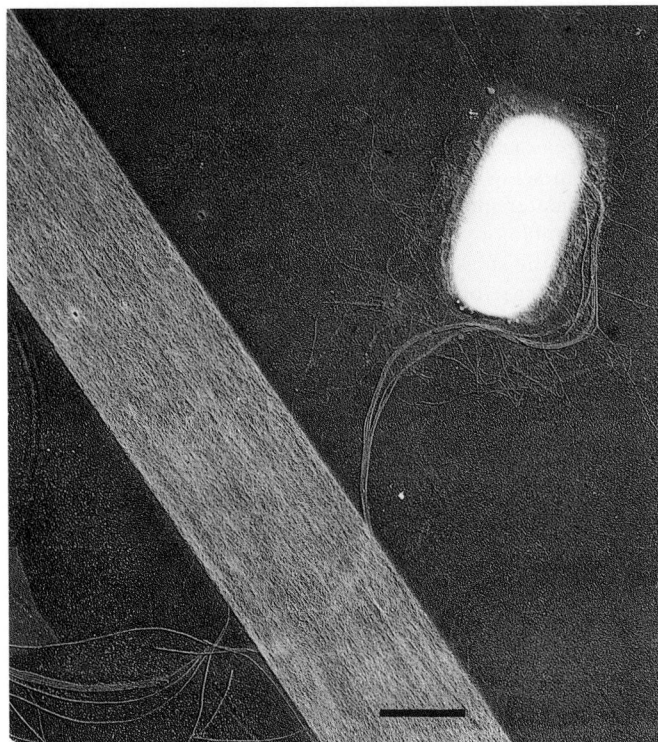

FIGURE BXII.β.56. Empty sheath of *S. natans* showing a smooth surface, and a single cell with a tuft of subpolar flagella. Electron micrograph. Bar = 1 μm. (Reproduced with permission from E.G. Mulder and M.H. Deinema. 1981. *In* Starr, Stolp, Trüper, Balows and Schlegel (Editors), The Prokaryotes. A Handbook on Habitats, Isolation, and Identification of Bacteria. Springer-Verlag, Berlin, pp. 425–440.)

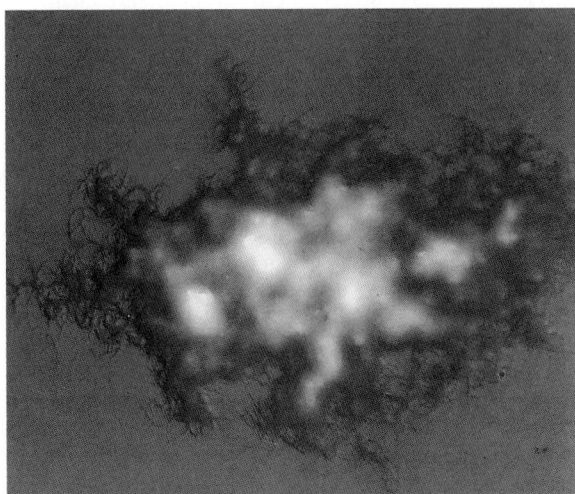

FIGURE BXII.β.57. Rough colony of *S. natans.* × 22. (Reproduced with permission from E.G. Mulder and W.L. van Veen, Antonie van Leeuwenhoek Journal of Microbiology and Serology. *29:* 121–153, 1963, ©Kluwer Academic Publishers. Dordrecht.)

per, potato, dairy, or other agricultural industries. The organism also occurs regularly in activated sludge, particularly when this material is settling poorly—so-called bulking. The relatively good growth under conditions of low oxygen concentrations and its capability to utilize a large number of organic carbon sources may be responsible for its relative dominance in biological deposits (Stokes, 1954; Dondero, 1975), including paper mill slimes (Väisänen et al., 1994; Pellegrin et al., 1999).

S. natans is one of several types of filamentous bacteria that may cause pipe clogging and bulking of activated sludge (Eikelboom, 1975). A ready settling of sludge flocs is one of the requisites for the successful operation of the activated-sludge process, which includes the aerobic biological purification of sewage and industrial wastewater. After absorption of the soluble wastes by the sludge organisms, the sludge flocs should readily settle, so that flocs and purified water can be separated. When the sludge flocs are densely populated by *S. natans* or some other filamentous organism, they are voluminous with many trichomes protruding into the surrounding water, thereby preventing a ready settling. This phenomenon is the cause of bulking sludge. Several different factors may cause the dominant growth of filamentous bacteria in activated sludges, among them: (a) low or high concentrations of available nutrients found in systems continuously fed with wastewater and (b) the low oxygen tensions that occur in such systems. With respect to the relatively high proportion of cell surface to cell contents of protruding filamentous bacteria compared with clumps of floc-forming bacteria, the former organisms occupy a more favorable position for nutrient uptake and growth. The ability of *S. natans* to thrive at very low pO_2 values is an additional factor favoring the competition with floc-forming bacteria. According to Wanner and Grau (1989), *Sphaerotilus* prefers environmental conditions with high sludge retention times and low concentrations of dissolved oxygen combined with high substrate concentrations. The organism does not often occur in anaerobic/aerobic selector plants. However, other factors may also lead to sludge bulking by other organisms. More details can be found in Eikelboom (1975), Jenkins et al. (1986), Wanner and Grau (1989), and Kämpfer (1997).

Although *S. natans* prefers a growth medium containing adequate amounts of easily assimilable organic nutrients, the organism is found sporadically in unpolluted water of brooklets, ditches, and ponds, where unknown compounds are the substrate. In the former habitat, the sheaths are thin and colorless; in the latter, particularly in the presence of soluble iron compounds, they may turn yellow-brown and sometimes become encrusted with ferric oxide. This characteristic can be clearly observed in a laboratory apparatus in which *S. natans* is grown in slowly running soil extract enriched with Fe (II). Under these conditions, the sheaths of *S. natans* resemble those of *Leptothrix ochracea*, and a number of authors (Pringsheim, 1949a, b; Stokes, 1954; and others) have assumed that both organisms are identical. However, 16S rRNA sequencing studies have shown the separate position of organisms belonging to both genera, although an unambiguous pattern of the respective branches of both genera is not possible on the basis of the available data set (Pellegrin et al., 1999), mainly because the 16S rRNA sequence of *L. ochracea*, which has not been cultivated in pure cultures until now, is not available.

Isoprenoid quinone Q-8 is the predominant quinone type for *S. natans* (Kämpfer et al., 1995, 1998). In a study of the fatty acid composition as determined by gas chromatographic analysis, the composition of 15 strains differed only slightly (Kämpfer, 1998). All strains contain the fatty acids $C_{12:0}$, $C_{16:0}$, $C_{16:1 \omega 7c}$, $C_{17:1}$, and $C_{18:1 \omega 7, 9c}$, and $C_{10:0 3OH}$; a fatty acid profile that is very similar to that of *Leptothrix* species and typical for members of the *Betaproteobacteria* (Table BXII.β.65).

ENRICHMENT AND ISOLATION PROCEDURES

Several isolation techniques for *Sphaerotilus* have been described. When slimy masses of the organism—attached to submerged sur-

TABLE BXII.β.65. Main differential characteristics of the genera *Sphaerotilus* and *Leptothrix* [a]

Characteristic	Sphaerotilus	Leptothrix
Mn^{2+} oxidation[b]	−	+
F_2O_3 accumulation on the sheaths[b]	+	+
Reserve material: [c]		
Polysaccharide	+	−
Carbon sources used for growth (in GMBN base):[d]		
L-Alanine	+	−
L-Asparagine	+	−
L-Aspartate	+	−
Butyrate		+
D-Fructose	+	D
D-Glucose	+	D
D-Gluconate	+	−
L-Ornithine	+	−

[a]Symbols: see standard definitions.

[b]Data from Mulder (1989b).

[c]Data from Willems et al. (1991b).

[d]Data from Spring et al. (1996) and Kämpfer (1998). For additional results of carbon substrate utilization tests, see the references by these authors and also by Willems et al. (1991b).

faces in polluted, slowly running water—are available, direct isolation of *S. natans* may succeed. Pellegrin et al. (1999) collected slime samples from paper machines that were rinsed fivefold in sterile distilled water. The deposits were placed on the surface of FIL agar containing 0.005 g/l cycloheximide and incubated 72 h at 30°C. For details see Pellegrin et al. (1999). Rough colonies (Fig. BXII.β.57) were selected and further investigated. A similar procedure can be applied to activated sludge containing many filaments of the organism (i.e., bulking sludge). When the sheathed bacteria occur in low numbers in activated sludge or in nonpolluted water samples, the use of enrichment cultures may be desirable (Mulder and Deinema, 1981). Mulder (1989b) described a technique, based on extracted alfalfa straw (Stokes, 1954) or extracted pea straw (Mulder and van Veen, 1963, 1965), which can serve as the nutrient material. Most of the soluble organic matter should be removed to prevent the accumulation of undesirable organisms. This can be achieved by boiling and extracting the straw after it has been cut into pieces of about 2 cm. In the case of alfalfa straw, a 1% suspension is extracted three or four times by boiling with large amounts of tap water. The extracted straw medium is distributed in 50-ml quantities in 125-ml Erlenmeyer flasks, which are then inoculated with about 10 ml of water from various sources. The preparation of the pea straw medium differs slightly from the preceding technique. The straw is extracted for 10 h at 100°C with tap water that is renewed every hour. One or 2 g of extracted pea straw (dry weight) in 25 ml of tap water is autoclaved twice (15 min at 110°C) and used as an enrichment medium. After inoculation with small amounts of river or ditch water or activated sludge and subsequent incubation for about 1 week at 22–25°C, tufts of filaments of *S. natans* may be seen after microscopic observation.

Isolation may be achieved by streaking the enrichment cultures on previously dried agar plates containing low levels of nitrogen and carbon sources. Activated sludge containing many filaments of *S. natans* is streaked directly on such plates. Slimy masses of sheathed bacteria grown in slowly running, polluted water are washed several times with sterile water. Homogenization of the washed flocs by blending for a very short time may be advisable.

The use of a nutritionally poor agar medium limits the size

of undesirable bacterial colonies, leaving large areas for the filamentous organisms. This medium has the following basal composition (per liter of glass-distilled water): KH_2PO_4, 27 mg; K_2HPO_4, 40 mg; $Na_2HPO_4 \cdot 2H_2O$, 40 mg; $CaCl_2$, 50 mg; $MgSO_4 \cdot 7H_2O$, 75 mg; $FeCl_3 \cdot 6H_2O$, 10 mg; $MnSO_4 \cdot H_2O$, 5 mg; $ZnSO_4 \cdot 7H_2O$, 0.1 mg; $CuSO_4 \cdot 5H_2O$, 0.1 mg; $Na_2MoO_4 \cdot 2H_2O$, 0.05 mg; cyanocobalamin, 0.005 mg. This medium is enriched with peptone, 1 g/l; glucose, 1 g/l; and agar (Davis), 7.5 g/l. To inhibit the rapid spreading of contaminating bacteria, the excess surface moisture of the sterile agar plates should be evaporated by overnight storage of these plates at a temperature of 37–45°C. Upon inoculation and incubation of these plates at 20–25°C, colonies of *S. natans* may be seen and tentatively identified within a few days by their characteristically flat, dull, cotton-like appearance. The edges of the colonies are irregular, owing to curly filamentous growth extending in all directions (Fig. BXII.β.57). Confirmation of the identification may be achieved by microscopic observation (Fig. BXII.β.55).

Sphaerotilus may also be isolated by spread plate techniques with or without centrifugation or vortex treatment (Williams and Unz, 1985; Ziegler et al., 1990). Growth has been reported on I and SCY media (Eikelboom, 1975), SS media (Williams and Unz, 1985), R2A agar (Seviour et al., 1994), and GMBN-Agar (Kämpfer et al., 1995).

MAINTENANCE PROCEDURES

Stock cultures of *S. natans* on agar slants of the previously described media can be stored for about 3 months at 4°C. Addition of 2–3 ml of sterile tap water to the agar slants may prolong the viability for another 3 months. Preservation for longer periods is accomplished by common lyophilization techniques; however, it must be stressed that some *Sphaerotilus* strains do not survive lyophilization. Freezing in liquid nitrogen can be used for the long-term preservation of these strains.

DIFFERENTIATION OF THE GENUS *SPHAEROTILUS* FROM OTHER GENERA

Differential characteristics of the genera belonging to the *Rubrivivax gelatinosus–Leptothrix discophora* rRNA cluster of the *Betaproteobacteria* group are listed in Table BXII.β.66, and the main characteristics of the genera *Sphaerotilus* and *Leptothrix* are given in Tables BXII.β.65 and BXII.β.67. It can be seen that *Sphaerotilus* is closely related to *Leptothrix*. This applies to the motility of separate cells when released from the sheaths, to the formation of poly-β-hydroxybutyrate as reserve material, to the accumulation of ferric oxide on the sheaths, the quinone type and the fatty acid patterns, and to the mol% G + C of the DNA. However, several other properties are clearly different. They include morphological as well as physiological characteristics, such as size of cells; flagellation; structure of sheath surface; ability of the *Leptothrix* species to oxidize Mn^{2+} to Mn^{4+} (MnO_2), which is absent

in *Sphaerotilus*; and the pronounced response of *S. natans* to organic nutrients as contrasted with no or poor response of most *Leptothrix* species to added nutrients; the latter factor is especially of considerable ecological significance. *S. natans* thrives in water heavily contaminated with organic nutrients (wastewater); *Leptothrix* species are never found in such environments, except *L. cholodnii* and *L. lopholea*, which respond more clearly to added nutrients than do the other *Leptothrix* species. Pellegrin et al. (1999) found that *Sphaerotilus* isolates from paper-mill slimes were adapted to the specific environmental conditions, regarding temperature tolerance and utilization of cellulose and starch.

TAXONOMIC COMMENTS

The only known species of the genus is *S. natans*. The 16S rRNA gene sequences have been determined for several reference strains and isolates of *Sphaerotilus* (Corstjens and Muyzer, 1993; Siering and Ghiorse, 1996, 1997; Pellegrin et al., 1999) and are highly similar to each other (>99.5%). In phylogenetic trees, the highest similarities are found with *Leptothrix* species. Together with *Ideonella dechloratans*, *Rubrivivax gelatinosus*, *Alcaligenes latus*, *Roseateles depolymerans*, and *Aquabacterium* species (sequence similarities between 16S rRNA genes within this group are above 93%), they form a branch of bacteria that is phylogenetically distinct from the *Comamonadaceae* as defined by Willems et al. (1991b), and that is also referred to as the *Rubrivivax* subgroup of the *Betaproteobacteria* (Wen et al., 1999). A phylogenetic tree of members of the *Rubrivivax gelatinosus–Leptothrix discophora* rRNA cluster of the *Betaproteobacteria* is presented in Fig. BXII.β.58.

As with the genus *Leptothrix*, some characteristics that distinguish *Sphaerotilus natans* from *Leptothrix* species are not very stable and regularly lost upon cultivation. The ability to form sheaths has been lost by some strains available from public culture collections. In addition, it has been pointed out by Pellegrin et al. (1999) that based on comparison of 16S rDNA sequences, no unambiguous branching of *Sphaerotilus* and *Leptothrix* have been obtained with different treeing methods.

However, in view of the pronounced differences between *Sphaerotilus natans* and *Leptothrix* species, the genus allocation seems to be firm.

FURTHER READING

Dondero, N.C. 1975. The *Sphaerotilus-Leptothrix* group. Annu. Rev. Microbiol. *29*: 407–428.

Kämpfer, P. 1998. Some chemotaxonomic and physiological properties of the genus *Sphaerotilus*. Syst. Appl. Microbiol. *21*: 156–162.

Pellegrin, Y., S. Juretschko,, M. Wagner, and G. Cottenceau,. 1999. Morphological and biochemical properties of *Sphaerotilus* sp. isolated from paper mill slimes. Appl. Environ. Microbiol. *65*: 156–162.

van Veen, W.L., E.G. Mulder and M.H. Deinema. 1978. The *Sphaerotilus-Leptothrix* group of bacteria. Microbiol. Rev. *42*: 329–356.

List of species of the genus Sphaerotilus

1. **Sphaerotilus natans** Kützing 1833, 386[AL]

   na'tans. L. part. adj. *natans* swimming.

   The cell morphology is as described for the genus. Loss of the sheath-forming capacity by mutation has been reported (Stokes, 1954). Colonies of sheathless cells are smooth (they have lost their filamentous edges). Discontinuation of sheath formation may also be due to nongenetic factors, particularly nutritional conditions (Gaudy and

TABLE BXII.β.66. Characteristics of the genera belonging to the *Rubrivivax gelatinosus–Leptothrix discophora* rRNA cluster[a,b]

Characteristic	Rubrivivax	Ideonella	Roseateles	Leptothrix	Sphaerotilus	Pseudomonas saccharophila	Aquabacterium	Alcaligenes latus
Number of species	1	1	1	5[c]	1		3	
Flagellation	1, polar	2 or more, polar	2 or more, polar	1, polar	Tuft, subpolar	1, polar	1, polar	Peritrichous
Formation of sheaths	−	−	−	+	+	−	−	−
Photoautotrophic growth	+	−	−	−	−	−	−	−
Autotrophic growth with H_2	+	−	−			+	−	+
Oxidation of:								
Carbon monoxide	+							
Manganese (Mn^{2+})				+	−		−	
Accumulation of Fe_2O_3 on the sheaths				+	+			
Anaerobic growth with chlorate		+					−	
Nitrogen fixation	+		−			+	−	+
Gelatinase	+				Slow	+	−	+
Accumulation of:								
Poly-β-hydroxybutyrate			+	+	+	+	+	+
Polysaccharides				−	+		−	
Carotenoid pigments	+	−	+	−	−	−	−	−
Growth factors required:								
Biotin, thiamine	+	−	−	−	−			
Vitamin B_{12}	−	−	−	+[d]	+			
Carbon sources used for growth:								
Acetate	+	+	D	D	+	+	+	−
Pyruvate	+	+	+	D	+	+	D	−
Butyrate	D				+	+	+	+
Lactate	+	+	+	D	+	+	D	+
L-Malate	+		+	D	+	+	+	+
Succinate	+	+	+		+	+	+	+
Fumarate	+	+	+	D	+	+	+	+
Citrate	+		+		+	+	D	D
D-Ribose			−	D		+	−	−
Glucose	+		+	D	+	+[e]	−	+
Fructose			+	D			−	
D-Galactose			+	D	+	+	−	−
Sucrose			−	D	+	+	−	+
Glycerol	−		−	D	+	−	D	+
Mannitol	−		+	D	+	−	−	−
Sorbitol	−		−		+	−		−
Isolation source:								
Mud	+					+		
Sewage		+						
Soil								+
Water		+	+	+	+		+	
Mol% G + C content of DNA[f]	70.0–72.5	68.1	66.2–66.3	67.8–71.1	69.1	69.7	65–66	69.1–71.1

[a]Symbols: see standard definitions. Blank space, not determined or not applicable.

[b]Data adapted from Willems et al. (1991b), Malmqvist et al. (1994a), Spring et al. (1996), Kämpfer (1998), Kalmbach et al. (1999), and Suyama et al. (1999).

[c]Data for *Leptothrix lopholea* and *Leptothrix ochracea* are not available.

[d]Thiamine and biotin may be required by some strains (Mulder, 1989b).

[e]Growth may require mutation.

[f]Most G + C values were determined by use from thermal denaturation curves; the exception was the value for *Leptothrix cholodnii*, which was taken from Van Veen et al. (1978).

Wolfe, 1961; Mulder and van Veen, 1963). When the organism is grown on a basal medium with glucose and peptone at 1 g/l each, normal hairy colonies are formed (Fig. BXII.β.57), as contrasted with the smooth, almost circular colonies formed when 5 g of these nutrients are supplied. High concentrations of peptone are more effective in producing this effect than are sugars. Smooth colonies consist of sheathless cells that have larger dimensions than cells in sheaths. Transfer of the former cells to a poor medium restores sheath formation. Growth is supported on several media (see genus description).

False branching of the filaments occurs in every strain of *S. natans*, but it occurs in some strains more than others. Its occurrence depends on cultural conditions (relatively poor media) rather than on strain specificity (Pringsheim, 1949a).

Utilization of fructose, glucose, maltose, sucrose, lactate, pyruvate, and succinate as sole sources of carbon has been reported by Stokes (1954), Höhnl (1955), Mulder and van Veen (1963), Kämpfer (1998), and Pellegrin et al. (1999). Numerous other carbon sources can be utilized (for details see Kämpfer, 1998 and Pellegrin et al., 1999). Strains of *S.*

TABLE BXII.β.67. Other characteristics of the genera *Sphaerotilus* and *Leptothrix*[a]

Characteristic	*Sphaerotilus*	*Leptothrix*
Phylogenetic affiliation:		
Rubrivivax subgroup of the *Betaproteobacteria*	+	+
Mol% G + C of DNA[b,c]	70	68–71
Major quinone type	Q-8	Q-8
Major fatty acids:		
$C_{16:1}$	+	+
$C_{16:0}$	+	+
$C_{18:1 \, \omega7, \, 9c}$	+	+
$C_{18:1}$	+	+
Hydroxyl fatty acid:		
$C_{10:0 \, 3OH}$	+	+
Cell dimensions:[b]		
Width (μm)	1.2–2.5	0.6–1.5
Length (μm)	1–10	1.5–14
Reserve material: [b]		
Poly-β-hydroxybutyrate	+	+

[a]Symbols: see standard definitions.

[b]Data from Mulder (1989b).

[c]Data from Willems et al. (1991b).

natans differ widely in their capacity to utilize other carbon compounds. In contrast to most *Leptothrix* strains, *S. natans* utilizes relatively high concentrations of assimilable substrates, from which it synthesizes considerable amounts of cellular material.

Cells may contain large amounts of poly-β-hydroxybutyrate either as numerous small globules or as a few large globules. Polysaccharides may also accumulate. The synthesis of both reserve compounds is stimulated by a high carbon/nitrogen ratio in the medium (Mulder and van Veen, 1963) and by oxygen deficiency.

The major quinone is ubiquinone Q-8. Major fatty acids (>5%) are hexadecanoic acid, *cis*-9-hexadecenoic acid, and *cis*-9,11-octadecenoic acid. Minor fatty acids are 3-hydroxydecanoic acid, dodecanoic acid, octadecanoic acid, tetradecanoic acid, and *cis*-9,10-methyleneoctadecanoic acid.

The mol% G + C of the DNA is: 70 (T_m).

Type strain: ATCC 13338, DSM 6575.

GenBank accession number (16S rRNA): L33980.

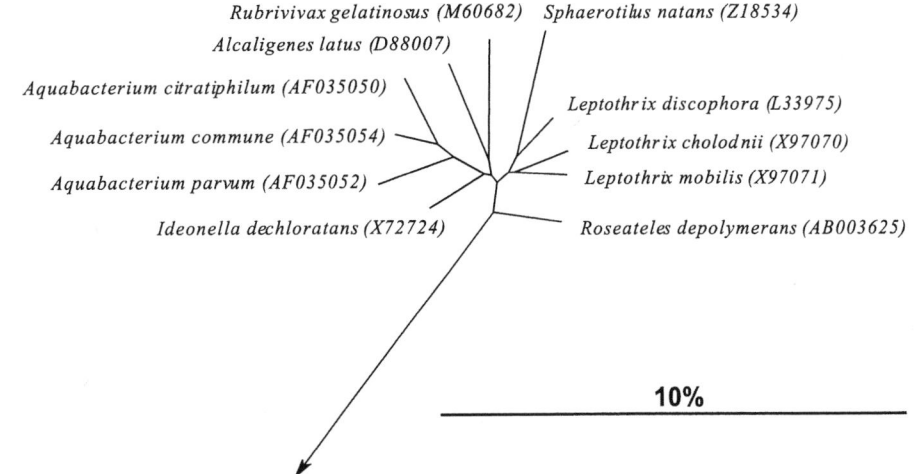

FIGURE BXII.β.58. Phylogenetic tree showing members of the *Rubrivivax gelatinosus–Leptothrix discophora* rRNA cluster of the *Betaproteobacteria*. This tree is based on a distance matrix analysis of almost complete sequences of 16S rRNA genes (at least 1300 nucleotides, GenBank accession numbers are given in parentheses). Phylogenetic distances were calculated as described by Felsenstein (1982). The sequence of *Escherichia coli* (L10328) was used as an outgroup (not shown). The scale bar represents 10% estimated sequence divergence.

Genus Incertae Sedis XVII. **Tepidimonas** *Moreira, Rainey, Nobre, da Silva and da Costa 2000, 741[VP]*

MILTON S. DA COSTA, M. FERNANDA NOBRE AND FRED A. RAINEY

Te.pi.di.mo' nas. L. adj. *tepidus* warm; Gr. n. *monas* unit; N.L. fem. n. *tepidimonas* warm monad.

Short rod-shaped cells 0.5–1.0 × 1.0–2.0 μm. Motile by means of one polar flagellum. Endospores are not formed. Cells stain Gram negative. Colonies are not pigmented. **Strictly aerobic.** Nitrate is not used as a terminal electron acceptor. Oxidase and catalase positive. **Optimal temperature for growth, between 50 and 55°C.** Optimal pH for growth, between 7.5 and 8.5. Fatty acids are of the straight-chain type. Major polar lipids are phosphatidylethanolamine and phosphatidylglycerol; ubiquinone 8 is the major respiratory quinone. **Chemolithoheterotrophic.** Reduced sulfur compounds are oxidized to sulfate in the presence

of a metabolizable organic carbon source. Organic acids and amino acids are used as carbon and energy sources, but sugars, polysaccharides and polyols are not assimilated.

The mol% G + C of the DNA is: 69.7.

Type species: **Tepidimonas ignava** Moreira, Rainey, Nobre, da Silva and da Costa 2000, 741.

FURTHER DESCRIPTIVE INFORMATION

The optimal growth temperature of the type strain of *Tepidimonas ignava* is between 50 and 55°C; the organism does not grow above 60°C or below 35°C. It is interesting to note that related strains ac-15 and DhA-73 (see Other Organisms, below) also have cardinal growth temperatures in this range and that the related environmental clone tmbr15-22 (see Other Organisms) was obtained from thermophilic aerobic treatment of synthetic wastewater. These results imply that the genus *Tepidimonas* contains several slightly thermophilic species with optimal growth rates in the neighborhood of 50°C.

The optimal pH for growth of *T. ignava* is between 7.5 and 8.5.

The polar lipid pattern, the respiratory quinones, and the fatty acids of this organism, are as expected for mesophilic or slightly thermophilic members of the *Betaproteobacteria*, although $C_{16:0}$, $C_{17:0}$, and $C_{18:0}$ are the predominant fatty acids.

Perhaps the most interesting characteristic of *T. ignava*—which is not restricted to this species—among the species of the *Betaproteobacteria* relates to its inability to assimilate any of the carbohydrates or polyols examined and to grow only on amino acids and organic acids. The inability to assimilate carbohydrates and polyols has also been reported for strain DhA-73, indicating that one or more steps of glycolysis are absent in both organisms. Strain DhA-73 is capable of degrading tricyclic diterpenes such as abietic acid, dehydroabietic acid, and palustric acid, found in pulp and paper mill effluent and derived from conifer resin (Yu and Mohn, 1999). The degradation of these acids was not examined in *T. ignava*. Strain ac-15 has never been characterized and comparisons between this organism and *T. ignava* cannot be made.[1]

Like several other members of the *Betaproteobacteria*, *T. ignava* oxidizes thiosulfate and tetrathionate to sulfate. However, this organism does not appear to be chemolithoautotrophic, since the oxidation of these reduced sulfur compounds occurs only in the presence of a metabolizable carbon source, indicating that it is chemolithoheterotrophic.

ENRICHMENT AND ISOLATION PROCEDURES

Only one strain of *T. ignava* was isolated from the hot spring at São Pedro do Sul in central Portugal, despite repeated attempts to isolate other strains of this organism. The isolation of this organism was achieved on a medium composed of one part of

---

1. Recently a second species of the genus *Tepidimonas*, designated *Tepidimonas aquatica*, was described by Freitas et al. (2003). The new species has characteristics that are very similar to those of *T. ignava*. The major differences between the two species are in their fatty acid compositions. The 16S rRNA gene sequences of the two species have 97% similarity. This species and *T. ignava*, unlike strain DhA-73, do not degrade resinic acids.

Kligler's iron agar (Difco) and four parts of *Thermus* agar. Samples of water were filtered through cellulose nitrate membrane filters (pore diameter, 0.45 μm), placed on the surface of the Kligler's iron/*Thermus* medium, and incubated for several days at 50°C. However, the isolation of *T. ignava* appears to have been completely fortuitous, since it did not show any enhanced growth on this medium. The organism grows well on several low-nutrient media containing yeast extract and tryptone, namely *Thermus* medium (Williams and da Costa, 1992) and Degryse medium 162 (Degryse et al., 1978). The addition of thiosulfate (1.0–5.0 g/l) enhances growth of the organism.

MAINTENANCE PROCEDURES

Stock cultures of *T. ignava* remain viable for years at −80°C in *Thermus* medium containing 15% glycerol. Cultures have also been maintained lyophilized for several years without the loss of viability.

DIFFERENTIATION OF THE GENUS *TEPIDIMONAS* FROM OTHER GENERA

The bacteria of the genus *Tepidimonas* can be easily distinguished from all members of the family *Comamonadaceae* because of their high growth temperature range. Other characteristics, namely the polar lipids and the fatty acids, are similar.

TAXONOMIC COMMENTS

Phylogenetic analysis of the 16S rRNA gene sequence (Fig. BXII.β.59) shows that *Tepidimonas ignava* clusters within the class *Betaproteobacteria*, but represents a distinct lineage along with other undescribed strains, namely strain ac-15, isolated some years ago from a hot spring microbial mat at Yellowstone National Park, USA (EMBL accession number U46749; Nold et al., 1996), and strain DhA-73 (EMBL accession number AF125877; Yu and Mohn, 1999), isolated from a bioreactor treating bleached kraft mill effluent. An environmental 16S rRNA gene clone, designated tmbr15-22 (EMBL accession number AF309815) also belongs to this lineage. *Tepidimonas ignava* shares 96.2% 16S rRNA gene sequence similarity with strain ac-15, 97.5% sequence similarity with strain DhA-73, and 97.4% sequence similarity with the environmental clone tmbr15-22. These results indicate that the undescribed organisms (strains ac-15 and strain DhA-73) and the organism detected as the environmental clone tmbr15-22 represent novel species within the genus *Tepidimonas*. Moreover, this group shares a low 16S rRNA gene sequence similarity (<94.0%) with any previously described species within the *Betaproteobacteria*, and can be considered to represent a distinct lineage of, at least, genus status within the family *Comamonadaceae*.

List of species of the genus Tepidimonas

1. **Tepidimonas ignava** Moreira, Rainey, Nobre, da Silva and da Costa 2000, 741[VP]

ig.na′va L. adj. *ignavus* lazy, pertaining to the organism's trait of not using sugars for growth.

The characteristics are as described for the genus, with the following additional information. Colonies on Degryse 162 medium are not pigmented and are 1–2 mm in diameter after 60 h of incubation. Growth occurs above 35°C and below 65°C. Growth does not occur at pH 6.0 or pH 10.0. The major fatty acid is $C_{16:0}$; unsaturated fatty acids are also present in large amounts. Ubiquinone 8 is the major respiratory quinone. Yeast extract, or growth factors,

are required for growth. Nitrate is not reduced to nitrite. Thiosulfate is oxidized to sulfate and serves as an energy source coupled to the assimilation of organic substrates. Chemolithoheterotrophic growth occurs on reduced sulfur compounds. Xylan, starch, casein, elastin, and fibrin are not degraded. Several amino acids and organic acids are utilized for growth, but hexoses, disaccharides, pentoses, and polyols are not used.

Isolated from the hot spring at São Pedro do Sul in central Portugal.

The mol% G + C of the DNA is: 69.7 (HPLC).

Type strain: SPS-1037, DSM 12034.

GenBank accession number (16S rRNA): AF177943.

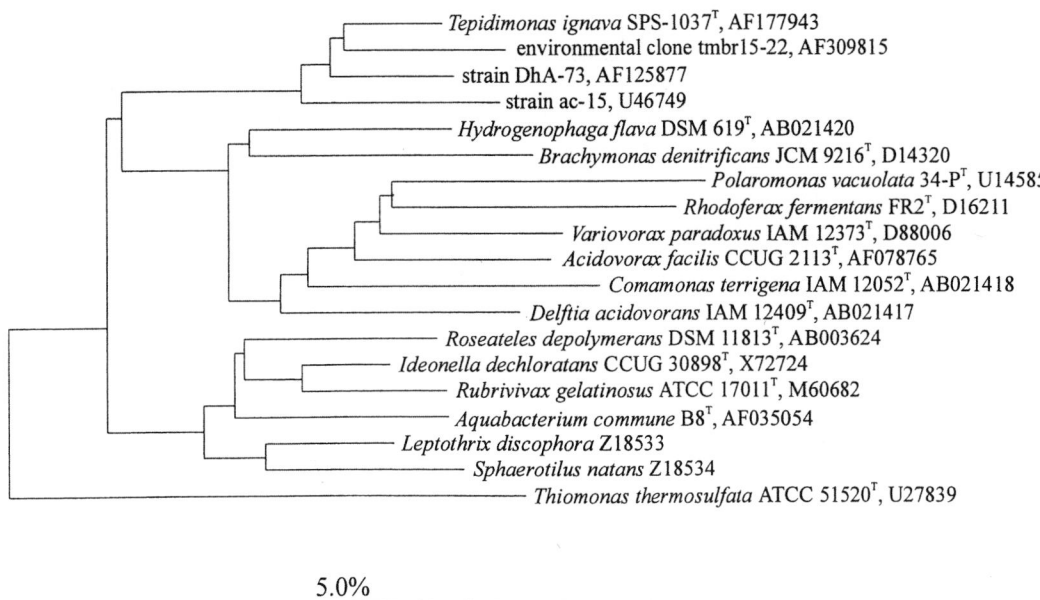

5.0%

FIGURE BXII.β.59. Phylogenetic dendrogram based on 16S rRNA gene sequence comparisons indicating the position of the genus *Tepidimonas* within the radiation of genera considered to represent the family *Comamonadaceae*. The scale bar represents 5 inferred nucleotide substitutions per 100 nucleotides.

Other Organisms

Strain ac-15 (EMBL accession number U46749) was isolated from a hot spring microbial mat at Yellowstone National Park, USA, by Nold et al. (1996). *Tepidimonas ignava* shares 96.2% 16S rRNA gene sequence similarity with strain ac-15.

Strain DhA-73 (EMBL accession number AF125877) was isolated from a bioreactor treating bleached kraft mill effluent by Yu and Mohn (1999). *Tepidimonas ignava* shares 97.5% sequence similarity with strain DhA-73.

The environmental 16S rRNA gene clone tmbr15-22 (EMBL accession number AF309815) was obtained from the thermophilic aerobic treatment of synthetic wastewater. *Tepidimonas ignava* shares 97.4% sequence similarity with strain tmbr15-22.-

---

Genus Incertae Sedis XVIII. **Thiomonas** *Moreira and Amils 1997, 527*[VP]

DONOVAN P. KELLY AND ANN P. WOOD

Thi.o.mo′ nas. Gr. n. *thios* sulfur; Gr. fem. n. *monas* a unit, monad; M.L. fem. n. *Thiomonas* sulfur monad.

The phenotypic description is that of the general traits for group II of *Thiobacillus* (Kelly and Harrison, 1989; Moreira and Amils, 1997). Gram-negative, nonsporeforming, short rods that are about $0.3–0.5 \times 1–3$ μm. Cells motile by means of single polar flagella. Obligate aerobes. Optimal temperature 30–36°C for mesophilic species and 50°C for moderately thermophilic species. Optimal pH between 3 and 6. Facultative chemolithoautotrophs; optimal growth occurs in mixotrophic media supplemented with reduced sulfur compounds and organic supplements (yeast extract, peptone, some sugars, and amino acids). Chemoorganotrophic growth on yeast extract, Casamino acids, peptone, and meat extract. Chemolithoautotrophic on thiosulfate, tetrathionate, S^0, and H_2S. Do not oxidize ferrous iron. Sensitive to ampicillin. Contain ubiquinone Q-8. Members of the *Betaproteobacteria*.

The mol% G + C of the DNA is: 61–69.

Type species: **Thiomonas intermedia** (London 1963) Moreira and Amils 1997, 527 (*Thiobacillus intermedius* London 1963, 335.)

FURTHER DESCRIPTIVE INFORMATION

This genus was created by Moreira and Amils (1997), who determined complete sequences of the 5S and 16S rRNA genes for *Thiobacillus cuprinus* and compared them with homologous sequences from *Thiobacillus intermedius*, *Thiobacillus perometabolis*, and *Thiobacillus thermosulfatus*. These four species were found to form a phylogenetic cluster within the *Betaproteobacteria*, but to be only remotely related to the type species of *Thiobacillus*, *Thiobacillus thioparus*. The properties of the species of *Thiomonas* are summarized in Table BXII.β.68. All the species are moderately acidophilic and facultatively chemolithotrophic, with *Thiomonas cuprina* also able to grow autotrophically on pyrite (Huber and Stetter, 1990; Moreira and Amils, 1997). The properties, taxonomy, and differentiation of the genera of sulfur-oxidizing, che-

TABLE BXII.β.68. Basic characteristics of species of the genus *Thiomonas*

Characteristic	*T. intermedia*	*T. cuprina*	*T. delicata*	*T. perometabolis*	*T. thermosulfata*
Mol% G + C	65–67	67–68	66–67	65–66	61
Cell size (μm)	0.6–0.8 × 1.0–1.4	0.3–0.5 × 1.0–4.0	0.4–0.6 × 0.7–1.6	0.4–0.5 × 1.1–1.7	0.9 × 1.3–2.3
Motility	+	+	−	+	+
Carboxysomes	+	nd	nd	−	nd
Facultatively chemolithoautotrophic[a]	+	+	+	+	+
Growth on complex media	+	+	+	+	+
Optimal pH	5.5–6.0	3.0–4.0	5.5–6.0	5.5–6.0	5.2–5.6
pH limits	5.0–7.5	2.0–6.5	5.0–7.0	5.0–7.0	4.3–7.8
Optimal temperature, °C	30–35	30–36	30–35	35–37	50–53
Nitrate reduction to N_2	−	−	−[b]	−	−
Growth on:					
Sulfur	+	+	+	+	+
Thiosulfate	+	−	+	+	+
Tetrathionate	+	−	+	+	+
Trithionate	nd	−	nd	nd	nd
Metal sulfide ores[c]	−	+	nd	−	nd
Thiocyanate	−	−	−	−	−

[a]Optimal growth is mixotrophic in media with reduced sulfur compounds plus yeast extract, peptone, some sugars and amino acids.

[b]Nitrate reduced to nitrite.

[c]Chalcopyrite, arsenopyrite, galena, sphalerite.

molithoautotrophic, Gram-negative, rod-shaped bacteria are summarized in Tables BXII.β.70 and BXII.β.71 and Fig. BXII.β.60 in the chapter on the genus *Thiobacillus*. *T. cuprina* (DSM 5495)

contains a circular chromosome (3.8 Mb) and appears to contain a linear 50 Mb megaplasmid that is inducible during chemolithotrophic growth (Marín et al., 1995; Amils et al., 1998).

List of species of the genus Thiomonas

1. **Thiomonas intermedia** (London 1963) Moreira and Amils 1997, 527[VP] (*Thiobacillus intermedius* London 1963, 335.)
   in.ter.me' di.a. L. prep. *inter* between, among; L. adj. *media* middle; M.L. adj. *intermedia* in between, intermediate.

   Thin, short rods, 0.6–0.8 × 1.0–1.4 μm. Motile by means of a polar flagellum. On thiosulfate agar, small colonies (up to 1 mm) with raised centers develop that are yellowish and opaque with precipitated sulfur and surrounded by veil-like fringes. This organism is facultatively mixotrophic. Capable of chemolithotrophic, autotrophic growth on sulfur, thiosulfate, or tetrathionate, but not on thiocyanate. Also oxidizes sulfide. Aerobic; unable to denitrify. Unable to grow in heterotrophic media such as nutrient broth or with single organic substrates in the absence of thiosulfate. Grows very poorly on yeast extract alone but produces substantial growth (after lags of 1–10 d) on yeast extract supplemented with glucose, fructose, sucrose, maltose, aspartate, or glutamate. It is possible that a reduced sulfur compound stimulates growth under these conditions. Best growth is with mixotrophic media containing thiosulfate and yeast extract, alanine, malate, succinate, citrate, 2-oxoglutarate, serine, lactate, or the supplements listed for heterotrophic growth. This organism can use ammonium salts, nitrate, urea, glutamate, or aspartate as nitrogen sources. Optimal temperature: 30–35°C; growth range: 15–37°C. Optimal pH: 5.5–6.0; growth range pH: 5.0–7.5, although for mixotrophic media, the pH may be lowered to about 2.8. Isolated from freshwater mud; presumably widely distributed. DNA hybridization with *Thiomonas perometabolis* has been reported as 31–35 (Katayama-Fujimura et al., 1983) and 56–78 (Harrison, 1983).

   The mol% G + C of the DNA is: 65–67 (T_m).

   Type strain: ATCC 15466.

2. **Thiomonas cuprina** Moreira and Amils 1997, 527[VP]
   cu.pri' na. L. adj. *cuprina* copper, describing its ability to extract copper ions from ores.

   Cells are Gram-negative rods, about 1–4 × 0.3–0.5 μm. Each is motile by one polar flagellum. Colonies on agar plates (yeast extract medium) have a brownish color. Optimal temperature 30–36°C, no growth at 15 or 50°C. Optimal pH 3–4, no growth at pH 1 or 7.5. Facultative chemolithotroph, aerobic. Chemoorganotrophic growth on yeast extract, Casamino acids, peptone, meat extract; some strains grow on pyruvate. Chemolithoautotrophic growth on chalcopyrite, sphalerite, arsenopyrite, galena, S^0, and H_2S, forming sulfuric acid. No oxidation of ferrous iron. Sensitive to ampicillin; *meso*-diaminopimelic acid present. Contain ubiquinone Q-8 but no rusticyanin. Lives in continental solfataric fields and mines. Insignificant DNA hybridization with *Thiobacillus ferrooxidans*, *Thiobacillus thiooxidans*, *Thiobacillus thioparus*, *Thiobacillus neapolitanus*, *Thiobacillus prosperus*.

   The mol% G + C of the DNA is: is 66–69 (T_m, HPLC).

   Type strain: DSM 5495.

   GenBank accession number (16S rRNA): U67162.

3. **Thiomonas perometabolis** (London and Rittenberg 1967) Moreira and Amils 1997, 527[VP] (*Thiobacillus perometabolis* London and Rittenberg 1967, 218.)
   pe.ro.me.ta' bo.lis. Gr. adj. *peros* maimed, crippled; Gr. v. *metabole* alter, change; M.L. part. adj. *perometabolis* with a maimed metabolism.

   Thin, short rods with rounded ends, 0.4–0.5 × 1.1–1.7 μm. Motile by means of a polar flagellum. Colonies grown on yeast extract–thiosulfate agar (1–3 mm after 1 week) are circular, entire, convex, smooth, creamy white, and opaque; the center of old colonies becomes pink-orange. Colonies grown on thiosulfate agar (0.5 mm after 10 d) are circular, entire, convex, smooth, creamy white, and opaque, developing a brown center with age. The original isolate of *T. perometabolis* did not exhibit chemolithotrophically autotrophic or heterotrophic growth on single-carbon substrates (London and Rittenberg, 1967), and its "maimed metabolism" was described as "obligately mixotrophic" (Vishniac, 1974). Further study led to an emended description (Katayama-Fujimura and Kuraishi, 1983; Katayama-Fujimura et al., 1984a) on which this entry is based. Facultative chemolithoautotroph. Chemolithotrophic autotrophic growth occurs on thiosulfate, tetrathionate, or sulfur, but not on thiocyanate. Little or no tetrathionate or trithionate accumulates during growth on thiosulfate. This organism exhibits diauxic growth on a mixture of thiosulfate and glutamate, with preferential use of the thiosulfate. Grows slowly after a lag time of about 2 weeks in heterotrophic media containing one of the following: alanine, glutamate, aspartate, malate, citrate, or succinate. Lags are shortened by the presence of thiosulfate. Best growth occurs in mixotrophic media with thiosulfate and organic supplements, such as yeast extract, casein hydrolysate, 2-oxoglutarate, some sugars, and some amino acids. Probably requires a reduced inorganic sulfur compound during heterotrophic growth. Obligate aerobe. Ammonium salts, nitrate, and urea are used as nitrogen sources, and glutamate and aspartate are used as both carbon and nitrogen sources. Optimal temperature: 35–37°C; growth range: 15–42°C. Optimal pH: 5.5–6.0; growth range pH: 5.0–7.0, although for mixotrophic media, the pH is lowered to about 2.8. Isolated from soil. Distribution unknown. DNA hybridization with *T. intermedia* has been reported as both 31–35% (Katayama-Fujimura et al., 1983) and 56–78% (Harrison, 1983).

The mol% G + C of the DNA is: 65–66 (T_m).

Type strain: ATCC 23370.

4. **Thiomonas thermosulfata** (Shooner, Bousquet and Tyagi 1996) Moreira and Amils 1997, 527[VP] (*Thiobacillus thermosulfatus* Shooner, Bousquet and Tyagi 1996, 414.)

ther.mo.sul.fa' ta. Gr. n. *thermus* heat; L. n. *sulfatus* sulfur; L. adj. *thermosulfata* organism that produces sulfate and grows at high temperatures.

Cells are Gram-negative rods, 0.9 × 1.3–2.3 μm, motile by means of single polar flagella. Typically, cells contain 2–3 polyphosphate inclusions, and polyhedral bodies. Colonies on thiosulfate agar are small (<1 mm), and round and either are translucent or have sulfur deposits in the center. Strictly aerobic; grows chemolithoautotrophically on thiosulfate, tetrathionate, and sulfur, and chemoorganotrophically on yeast extract, succinate, and glutamate. Tetrathionate, trithionate, and sulfur are produced during growth on thiosulfate. Autotrophic growth on thiosulfate occurs between pH 4.3–7.8 (optimal pH 5.2–5.6) and 34–65°C (optimal 50–52.5°C). pH should be decreased to 2.5 on thiosulfate and sulfur media. Does not grow on carbohydrates, pyruvate, acetate, or formate; does not denitrify. Cells adhere to S^0 by means of a glycocalyx. Contains ubiquinone Q-8.

The mol% G + C of the DNA is: 61 (UV ratios).

Type strain: ATCC 51520.

GenBank accession number (16S rRNA): U27839.

Other Organisms

1. *Thiomonas delicata comb. nov.* (*Thiobacillus delicatus* Katayama-Fujimura, Kawashima, Tsuzaki and Kuraishi 1984a, 142.)

del.i.cat' a. L. adj. *delicata* delicate.

Rods, usually single, rarely in pairs, 0.4–0.6 × 0.7–1.6 μm. Nonmotile. Colonies grown on yeast extract–thiosulfate agar (1 mm in diameter) are smooth and circular and change from transparent to whitish-yellow with sulfur. Facultative chemolithotroph and mixotroph. Grows autotrophically with sulfur, thiosulfate, or tetrathionate, but not with thiocyanate; accumulates tetrathionate and trithionate transiently during growth on thiosulfate. Incapable of heterotrophic growth on single carbon compounds. Grows mixotrophically in thiosulfate media supplemented with tricarboxylic acid cycle intermediates or amino acids. Optimum growth requires both organic substances and thiosulfate or sulfur. Facultative anaerobe; reduces nitrate and produces nitrite in mixotrophic and autotrophic media with thiosulfate or tetrathionate. Ammonium salts, nitrate, urea, glutamate, or aspartate can be used as nitrogen sources. Optimal temperature: 30–35°C; growth range: 15–42°C (no growth at 10°C or 45°C). Optimal pH: 5.5–6.0; growth range: pH 5.0–7.0. Isolated from mine water. Distribution unknown.

The mol% G + C of the DNA is: 66–67 (T_m, chemical analysis).

Deposited strain: THI 091, IAM 12624.

Additional Remarks: This species cannot yet be firmly assigned to the genus *Thiomonas*, but its G + C content, mixotrophy, and physiological similarities to *T. perometabolis* indicate that it should be reassigned to *Thiomonas* (Y. Katayama, personal communication). Determination of its 16S rRNA sequence is essential to confirm its phylogenetic relationships (Kelly and Harrison, 1989; Moreira and Amils, 1997).

Genus Incertae Sedis XIX. **Xylophilus** Willems, Gillis, Kersters, Van Den Broecke and De Ley 1987, 428[VP]

ANNE WILLEMS AND MONIQUE GILLIS

Xy.lo' phi.lus. Gr. n. *xylon* wood; Gr. n. *philos* friend; M.L. masc. *Xylophilus* friend of wood.

Straight to slightly curved rods, 0.4–0.8 × 0.6–3.3 μm, occurring singly, in pairs or short chains. **Long filamentous cells may occur in older cultures** (length may be 30 μm or more). **Gram negative.** **Motile by a single polar flagellum. Aerobic, having a strictly respiratory type of metabolism with oxygen as the only terminal electron acceptor. *In vitro* growth is generally very slow** and rather poor, even at the optimal growth temperature of 24°C. **Oxidase negative**, catalase positive. **Chemoorganotrophic** with oxidative carbohydrate metabolism. *Xylophilus* causes bacterial necrosis and canker of grapevine and can be isolated from infected plant material.

The mol% G + C of the DNA is: 68–69.

Type species: **Xylophilus ampelinus** (Panagopoulos 1969) Willems, Gillis, Kersters, Van Den Broecke and De Ley 1987, 428 (*Xanthomonas ampelina* Panagopoulos 1969, 75.)

FURTHER DESCRIPTIVE INFORMATION

As the genus contains only one species, all of the characteristics provided below describe the species *Xylophilus ampelinus*.

Colonial characteristics   On nutrient agar, colonies are circular, semitranslucent, slightly raised, glistening, and pale yellow with entire margins; colony diameters are 0.2–0.3 mm after 6 d and 0.6–0.8 mm after 15 d. Better growth is obtained on GYCA medium[1] and best growth occurs on YGC medium[2]. On the latter medium colonies are yellow and a brown diffusible pigment is produced. *Xylophilus* strains may produce two stable colony types with one type (t1) consisting of relatively large yellow colonies (diameter 0.8–2.0 mm after 15 d on GYCA) and the other (t2) consisting of smaller, paler and more slowly growing colonies (diameter 0.4–1.0 mm after 15 d). Both types were highly similar microscopically and when analyzed by whole-cell protein gel electrophoresis and by DNA–DNA hybridization (Willems et al., 1987).

Pigments   The yellow pigments produced by *Xylophilus* strains are different from xanthomonadins, the yellow water-soluble brominated aryl-polyene pigments produced by another plant-pathogenic genus, *Xanthomonas*. *Xylophilus* pigments appear sensitive to potassium isobutoxide but are generally very hard to purify and therefore little is known about them (Starr et al., 1977).

Growth conditions   Minimum and maximum growth temperature are 6°C and 30°C, respectively (Panagopoulos, 1969). Optimal growth is obtained at 24°C. In a medium consisting only of a salt solution, ammonium chloride, and glucose or galactose, addition of 0.1% glutamic acid is required for growth (Panagopoulos, 1969).

Nutrition and metabolism   *Xylophilus* strains have a strictly aerobic chemoorganotrophic metabolism. They use only a limited number of carbohydrates, organic acids, and amino acids for growth. In a study on the use of 60 substrates by nine French *Xylophilus* isolates, growth was observed on only D-glucose, D-galactose, L-glutamic acid, Na-succinate, Na-fumarate, K,Na-tartrate, Na-L-malate, Na₃-citrate, and Ca-gluconate (Van den Mooter and Swings, 1990). In general, *Xylophilus* strains show little variation from one another: strains from different geographic origin were shown to be highly similar by means of whole-cell protein gel electrophoresis, comparison of 106 enzymatic features and DNA–DNA hybridization (Willems et al., 1987). However, certain physiological characteristics do show considerable variation: the use of glucose and tartaric acid for growth and tyrosinase activity may vary between different populations (Ridé, 1996). Metabolic fingerprints using Biolog™ GN plates

were produced for a limited number of strains, but the recommended sucrose peptone agar for preparation of cell cultures was replaced by nutrient agar. Under these conditions the strains oxidized only acetic acid, propionic acid, L-aspartic acid, L-glutamic acid, and L-pyroglutamic acid (Serfontein et al., 1997).

Biosynthesis of aromatic amino acids   A study of the regulation mechanisms involved in the biosynthesis of aromatic amino acids has shown that 3-deoxy-D-arabinoheptulonate-7-phosphate synthetase is inhibited by tryptophan, chorismate, prephenate, phenylalanine, and tyrosine. Co^{2+} is needed for maximum activity. Prephenate-dehydrogenase is NAD^+-specific and is not inhibited by tyrosine (Whitaker et al., 1981; Byng et al., 1983).

Chemotaxonomic characteristics   *X. ampelinus* contains putrescine as the main polyamine and smaller amounts of 2-hydroxyputrescine, spermidine, and spermine (Auling et al., 1991).

Phage susceptibility   Various bacteriophages have been isolated from soil or water near infected vines and are differentiated by the morphology of their plaques on exponentially growing plates of *Xylophilus* strains. All strains of *Xylophilus* are sensitive to phage P *X.a.*15, which produces plaques that keep expanding over several days with the formation of a halo. Other phages are specific to certain groups of strains (Ridé, 1996).

Antibiotic sensitivity   Nine *Xylophilus* strains tested were susceptible to 16 of 20 antibiotics tested (see Table BXII.β.69) and all nine strains were resistant to methicillin. Mixed results were obtained for ampicillin, colistin sulfate, and polymyxin B (Van den Mooter and Swings, 1990).

Pathogenicity   *Xylophilus* strains are plant pathogens. They are responsible for bacterial necrosis and canker of grapevine (*Vitis vinifera*) in the Mediterranean region and South Africa (Panagopoulos, 1969; Erasmus et al., 1974). Although first described in France in 1895 as "Maladie d'Oléron" (Ravaz, 1895), the causal agent of bacterial necrosis of grapevine was not isolated until 1969 (Panagopoulos, 1969). The disease becomes apparent in early spring, when buds on affected shoots fail to open. Longitudinal cracks and cankers appear as they develop from hyperplasiae in the cambial tissue. Underlying vascular tissue shows a brown discoloration and will eventually die. Other parts of the plant that may be infected include petioles, flower stalks, and fruit stalks, resulting in death of leaves, flowers, or fruits. Leaves infected through hydathodes or stomata will show reddish-brown lesions. Roots can also be affected, resulting in retarded growth of shoots. Severity of symptoms may vary considerably for different varieties of grapevine (Panagopoulos, 1969; Grasso et al., 1979; Ridé, 1984, 1996; López et al., 1987). From being a rather rare disease at the beginning of the 20th century, it has now become more significant in the Mediterranean area through a combination of factors such as favorable environmental conditions, pruning procedures, and the increased mechanization of various viticultural practices (Ridé, 1996).

An *in vitro* test to assess cultivar susceptibility has been described by Peros et al. (1995). It involves inoculation of 2-month-old plantlets (having 8–12 internodes) by decapitation with scissors dipped in a bacterial suspension. After several weeks of incubation, the number of internodes with symptoms is recorded as an estimate for the progression of infection.

ENRICHMENT AND ISOLATION PROCEDURES

Isolation from affected vines is possible year round. The extremely slow and poor growth of *Xylophilus* strains often com-

---

1. GYCA medium (g/l distilled water): yeast extract, 5.0; glucose, 10.0; CaCO₃, 30.0; and agar, 20.0.

2. YGC medium (g/l distilled water): yeast extract, 10.0; galactose, 20.0; CaCO₃, 20.0; and agar, 20.0.

TABLE BXII.β.69. Characteristics of *Xylophilus ampelinus*[a,b]

Characteristic[c]	Reaction
Hydrolysis of Tween 80	+
Hydrolysis of starch, tributyrin	d
Polygalacturonase, tyrosinase	d
Urease, catalase	+
Alkaline reaction in litmus milk (18 days)	+
Acetoin	d
Growth at 28°C	+
Growth at pH 6.5	d
Tolerance to 1% and 0.5% NaCl	d
Tolerance to 0.01% actidione on GYCA medium	+
Tolerance to 2,3,5-triphenyltetrazolium chloride (TTC) at concentrations of 0.005%, 0.01%, and 0.02%	d
Production of H₂S from thiosulfate	+
Production of H₂S from L-cysteine, peptone	d
Acid production from L-arabinose	+
Acid production from D-galactose, glycerol, maltose	d
Growth on L-arabinose, acetate (0.2%), L-glutamic acid, sodium L-malate, trisodium citrate, potassium sodium tartrate	+
Growth on D-glucose, D-galactose, sodium succinate, sodium fumarate, calcium gluconate, calcium lactate	d
Growth on 0.5% yeast extract	+
Growth on a medium containing NH₄Cl, D-glucose, mineral salts, and vitamins	+
L-Glutamate (0.1%) required for growth	+
Stimulatory effect on growth in basal medium by:	
L-Asparagine, L-aspartic acid, L-glutamic acid, L-glutamine, L-serine, L-tryptophan	+
L-Alanine, L-arginine, glycine, L-histidine, L-hydroxyproline, L-isoleucine, L-leucine, L-methionine, L-phenylalanine, L-proline, L-threonine, L-tyrosine, L-lysine, L-valine, and L-ornithine	d
Glycine and L-lysine as sole nitrogen sources	d
L-Glutamic acid as sole nitrogen and carbon source	+
L-Glutamine as sole source of nitrogen and carbon	d
Hydrolysis of the following substrates in API enzymatic tests:[d]	
2-Naphthyl-butyrate, 2-naphthyl-caprylate, 2-naphthyl-valerate, 2-naphthyl-caproate, 2-naphthyl-phosphate (pH 5.4), L-leucine-naphthylamide, L-tyrosine-naphthylamide, L-serine-naphthylamide, L-lysine-naphthylamide, glycine-naphthylamide, L-phenylalanine-naphthylamide, L-aspartic acid-naphthylamide, L-arginine-naphthylamide, L-alanine-naphthylamide, DL-methionine-naphthylamide, L-tryptophan-naphthylamide, L-ornithine-naphthylamide, L-glutamine-naphthylamide hydrochloride, α-L-glutamic acid-naphthylamide, glycyl-glycine-naphthylamide hydrobromide, glycyl-L-phenylalanine-naphthylamide, L-seryl-L-tyrosine-naphthylamide, L-alanyl-L-arginine-naphthylamide, L-prolyl-L-arginine-naphthylamide, glycyl-L-alanine-naphthylamide, glycyl-L-arginine-naphthylamide, L-leucyl-L-alanine-naphthylamide, L-leucyl-glycine-naphthylamide, L-phenyl-alanyl-L-arginine-naphthylamide, L-seryl-L-methionine-naphthylamide, L-histidyl-L-phenylalanine-naphthylamide, 2-naphthyl-nonanoate	+
Naphthylphosphate (pH 8.5), 2-naphthyl-caprate, L-pyrrolidonyl-naphthylamide, S-benzyl-L-cysteine-naphthylamide, L-arginyl-L-arginine-naphthylamide, L-lysyl-L-alanine-naphthylamide, L-phenylalanyl-L-prolyl-L-alanine-naphthylamide, glycyl-L-tryptophan-naphthylamide	v
L-Phenylalanyl-L-proline-naphthylamide	d
Growth in the presence of the following metal compounds or dyes:	
0.01% KSCN, 0.01% (NH₄)₆Mo₇O₂₄, 0.001% antimony sodium tartrate, 0.001% NH₄CuCl₃, 0.001% CuI₂, 0.01% sodium salicylate	+
0.01% KIO₃, 0.01% K₂SO₄, 0.01% CuO, 0.01% BaCl₂, 0.01% Ba(NO₃)₂, 0.01% MnCl₂, 0.01% MnSO₄, 0.01% H₃BO₃, 0.05% neutral red, 0.0001% crystal violet, 0.01% Nile blue, 0.01% Bismarck brown, 0.01% Congo red	d

(continued)

TABLE BXII.β.69. *(cont.)*

Characteristic[c]	Reaction
Growth in the presence of:	
10 µg/disk methicillin	+
10 µg/disk ampicillin, 10 µg/disk colistin sulfate, 300 U/disk polymyxin B	d

[a]For symbols see standard definitions.

[b]Data taken from Bradbury (1973, 1984), Panagopoulos (1969), Willems et al. (1987), and Van den Mooter and Swings (1990).

[c]All strains are negative for the following characteristics: oxidase; nitrate reduction; indole production; ammonia production; Voges–Proskauer test; phenylalanine deaminase; arginine dihydrolase; lysine and ornithine decarboxylase; lecithinase; hydrolysis of esculin, gelatin, casein, arbutin, sodium hippurate, and DNA; methyl red test; production of a fluorescent pigment on King's medium B; Simmons citrate; potato soft rot test; the following reactions in litmus milk: acid formation, coagulation, peptonization and reduction; β-glucosidase, lecithinase. None of the strains grew at 37°C, at pH 4.5, and 5.5, and in the presence of 30%, 20%, or 10% glucose, 2–10% NaCl, 0.1% actidione on GYEA medium (0.5% D-glucose, 0.1% yeast extract, 2% agar, trace of FePO₄), 0.1%, 0.05%, 0.01%, and 0.005% TTC. None of the strains produced acid from cellobiose, D-fructose, D-mannose, D-ribose, sucrose, D-xylose, salicin, *meso*-erythritol, raffinose, melibiose, trehalose, L-rhamnose, D-lactose, D-glucose, sorbitol, dulcitol, amygdalin, glycogen, inulin, dextrin, adonitol, D-mannitol, L-sorbose, α-methylglucoside, arbutin, and *meso*-inositol. None of the strains grew on D-xylose, D-mannose, D-ribose, sucrose, sorbose, lactose, raffinose, trehalose, maltose, cellobiose, L-rhamnose, D-fructose, melibiose, D-mannitol, D-sorbitol, inositol, dulcitol, adonitol, sodium-2-ketogluconate, salicin, inulin, glycogen, dextrin, arbutin, α-methylglucoside, glycerol, D-saccharic acid, mucic acid, D-glyceric acid, D-alanine, L-alanine, norleucine, L-valine, L-isoleucine, L-threonine, L-serine, L-asparagine, ethanolamine, *n*-butyric acid, ethanol, *iso*-butanol, formate, acetate (0.5%), calcium lactate, sodium propionate, malonate, maleate, oxalate, benzoate. All strains failed to grow on 0.5% peptone, on a medium containing NH₄Cl and D-glucose, and on a medium containing NH₄Cl, D-glucose, and mineral salts. None of the strains required L-methionine for growth or showed a stimulated growth in basal medium in the presence of L-cysteine; none used the following amino acids as sole nitrogen source: L-alanine, L-arginine, L-asparagine, L-aspartic acid, L-cysteine, L-glutamic acid, L-glutamine, L-histidine, L-hydroxyproline, L-isoleucine, L-leucine, L-methionine, L-phenylalanine, L-proline, L-serine, L-threonine, L-tyrosine, L-tryptophan, L-valine, and L-ornithine; use the following amino acids as sole nitrogen and carbon source: L-alanine, L-hydroxyproline, L-proline, and L-asparagine. All strains failed to grow in the presence of the following metal compounds and dyes: 0.01% (NH₄)₂S₂O₈, 0.001% HgCl₂, 0.001% Hg(NO₃)₂, 0.001% CuCl, 0.001% CuCl₂, 0.001% CuSO₄, 0.001% Cd(CH₈COO)₂, 0.0001% Cd(CH₃COO)₂, 0.01% CoCl₂, 0.01% ZnCl₂, 0.01% Pb(CH₃COO)₂, 0.001% pyronine Y, 0.005% safranine T, 0.001% methylene blue, 0.001% malachite green, 0.005% methyl green, 0.07% methyl green, 0.005% crystal violet, 0.05% Nile blue, 0.005% diamond fuchsine, 0.01% basic fuchsin, 0.0001% brilliant green, and 0.01% thionine, and the following antibiotics (disks): 10 µg erythromycin, 10 U bacitracin, 10 µg streptomycin, 30 µg nalidixic acid, 30 µg kanamycin, 100 µg sulfafurazole, 10 U penicillin G, 25 µg cephaloridin, 10 µg gentamicin, 30 µg chloramphenicol, 10 µg fucidin, 30 µg tetracycline, 30 µg novobiocin, 30 µg neomycin, 200 µg nitrofurantoin. None of the strains hydrolyzed the following substrates in API enzymatic tests: 2-naphthyl-α-D-glucopyranoside, 2-naphthyl-β-D-galactopyranoside, 2-naphthyl-α-D-galactopyranoside, 6-Br-2-naphthyl-α-D-galactopyranoside, 6-Br-2-naphthyl-β-D-glucopyranoside, 6-Br-2-naphthyl-α-D-mannopyranoside, 1-naphthyl-N-acetyl-β-D-glucosaminide, naphthol-AS-BI-β-glucuronic acid (6-bromo-2-hydroxy-3-naphthoic acid-2-methoxyanilide-β-D-glucuronate), naphthol-AS-BI-phosphodiamide (6-bromo-2-phosphodiamide-3-naphthoic acid-2-methoxyanilide, 2-naphthyl-myristate, 2-naphthyl-laurate, 2-naphthyl-palmitate, 2-naphthyl-stearate, L-cysteine-naphthylamide, β-alanine-naphthylamide, L-histidine-naphthylamide, L-hydroxyproline-naphthylamide, L-isoleucine-naphthylamide, L-valine-naphthylamide, L-threonine-naphthylamide, γ-L-glutamic acid-naphthylamide, L-proline-naphthylamide hydrochloride, N-benzoyl-L-leucine-naphthylamide, N-benzoyl-DL-arginine-naphthylamide, N-benzoyl-L-alanine-4-methoxy-naphthylamide, N-carbobenzyloxy-L-arginine-4-methoxy-naphthylamide, N-glutaryl-DL-phenylalanine-naphthylamide, glycyl-L-proline-naphthylamide, L-lysyl-L-lysine-naphthylamide, α-L-aspartyl-L-alanine-naphthylamide, aspartyl-L-arginine-naphthylamide, α-L-glutamyl-α-L-glutamic acid-naphthylamide, α-L-glutamyl-L-histidine-naphthylamide, L-histidyl-L-serine-naphthylamide, N-acetyl-glycyl-L-lysine-naphthylamide, L-lysyl-L-serine-4-methoxy-naphtylamide, L-alanyl-L-phenylalanyl-L-proline-naphthylamide, L-histidyl-L-leucyl-L-histidine-naphthylamide, L-valyl-L-tyrosyl-L-serine-naphthylamide, N-carbobenzyloxy-glycyl-glycyl-L-arginine-naphthylamide, L-alanyl-L-phenylalanyl-L-prolyl-L-alanine-naphthylamide, L-leucyl-L-leucyl-L-valyl-L-tyrosyl-L-serine-naphthylamide, *p*-nitrophenyl-α-D-galactopyranoside, *p*-nitrophenyl-β-D-galactopyranoside, *p*-nitrophenyl-β-D-galactopyranoside-6-phosphate, *p*-nitrophenyl-α-L-arabinofuranoside, *p*-nitrophenyl-β-D-glucopyranoside, *p*-nitrophenyl-β-D-glucopyranoside, *p*-nitrophenyl-β-D-galacturonide, *p*-nitrophenyl-β-D-glucuronide, *p*-nitrophenyl-α-maltoside, *p*-nitrophenyl-β-maltoside, *p*-nitrophenyl-N-acetyl-α-D-glucosaminide, *p*-nitrophenyl-N-acetyl-β-D-glucosaminide, *p*-nitrophenyl-α-L-fucopyranoside, *p*-nitrophenyl-β-D-fucopyranoside, *p*-nitrophenyl-β-L-fucopyranoside, *p*-nitrophenyl-β-D-lactoside, *p*-nitrophenyl-α-D-mannopyranoside, *p*-nitrophenyl-β-D-mannopyranoside, *p*-nitrophenyl-α-D-xylopyranoside, *p*-nitrophenyl-β-D-xylopyranoside.

[d]The following API test strips were used: API-ZYM, Osidases, Esterases, AP1, AP2, and AP3 (Biomérieux, France).

plicates isolation because fast-growing saprophytes may rapidly overgrow isolation cultures. Serfontein et al. (1997) reported the isolation of bacteria from infected material collected in hot and dry periods to be difficult. Under cool and wet conditions the same plants did yield large numbers of the bacterium. The most frequently used isolation materials are small pieces of infected wood taken aseptically from diseased vines and soaked in sterile water for 20 min. The resulting bacterial suspension is streaked onto nutrient agar (Panagopoulos, 1969; Erasmus et al., 1974; Grasso et al., 1979). Small pale-yellow colonies will appear after 5–6 d of incubation; after 8–10 d diameters of 0.4–0.6 mm are attained (Panagopoulos, 1969). An enrichment technique to improve isolation yields was proposed by Serfontein et al. (1997). Cuttings from diseased shoots that include a node were incubated in closed plastic bags together with wet cotton plugs. After 3 d in the dark at 15°C, extracts from the shoots contained significantly more *Xylophilus ampelinus* bacteria than controls analyzed before incubation. This method also permits isolation of bacteria from latently infected shoots with no apparent signs of disease (Serfontein et al., 1997). The most widely used isolation medium is nutrient agar, but better growth is obtained by adding 5% glucose to nutrient agar. YGC and GYCA medium also result in better growth, but because the $CaCO_3$ renders these media opaque they are less suitable for the study of colony morphology. YEGAL medium[3] is more convenient and also yields good growth (Starr et al., 1977). Repeated sterilization or re-melting of solid media inhibits growth of *Xylophilus* (Panagopoulos, 1969).

MAINTENANCE PROCEDURES

Cultures grown on screw-capped slants at 24°C for 2–3 d can be stored tightly closed at 4°C and should be transferred at least every 2 months. For long-term preservation, strains can be lyophilized.

DIFFERENTIATION OF THE GENUS *XYLOPHILUS* FROM OTHER GENERA

The most typical feature of *Xylophilus* strains is their extremely slow and poor growth on most media at the optimal growth temperature of 24°C. They can be differentiated from most *Xanthomonas* species by the following characteristics: very low salt tolerance, inability to produce acid from numerous sugars, failure to hydrolyze gelatin and esculin, positive urease reaction, growth on K,Na-tartrate, L-glutamic acid, absence of growth on sucrose, and the use of L-glutamic acid as a sole carbon and nitrogen source (Van den Mooter and Swings, 1990). Other slow-growing organisms can also be isolated from grapevines but many of these are Gram positive. Gram reaction, catalase and oxidase tests, urease production, and lipolysis of Tween 80 have been proposed as preliminary confirmation tests to distinguish *Xylophilus ampelinus* from other slow growers (Serfontein et al., 1997).

Serological tests with specific antisera are used for rapid identification of the pathogen (Erasmus et al., 1974; Ridé, 1996).

TAXONOMIC COMMENTS

Xylophilus ampelinus was first classified in the genus *Xanthomonas* because it is a Gram-negative, aerobic, nonsporeforming plant pathogen with monotrichously flagellated rod-shaped cells, produces a yellow water-insoluble pigment and metabolizes sugars oxidatively (Panagopoulos, 1969). Later it became evident from additional data, such as pigment analyses (Starr et al., 1977), comparative studies of the biosynthesis of aromatic amino acids (Whitaker et al., 1981; Byng et al., 1983) and DNA–rRNA hybridization studies comparing *Xanthomonas ampelina* DNA with *Xanthomonas campestris* (De Vos and De Ley, 1983) that the vine pathogen was not a genuine *Xanthomonas*. As a result of a study of 34 strains from different geographic origins by means of SDS-polyacrylamide gel electrophoresis of whole-cell proteins, numerical analysis of 106 enzymatic features, and DNA–DNA and DNA–rRNA hybridization, we created a separate genus *Xylophilus* for these organisms (Willems et al., 1987). DNA–rRNA hybridization demonstrated that *Xylophilus* is a member of the family *Comamonadaceae* (previously called the "acidovorans rRNA complex"). Using this method none of the other diverse organisms of this family seemed especially closely related to the vine pathogen. Some phytopathogenic *Pseudomonas* species (now *Acidovorax*) and *Alcaligenes paradoxus* strains yielded slightly more stable DNA–rRNA hybrids when their DNA was hybridized with labeled rRNA of *X. ampelinus*, but when taking into account the experimental error, this was not considered significant (Willems et al., 1987). More recently the 16S rDNA sequences of *Xylophilus ampelinus* and various other members of the *Comamonadaceae* were determined, and these data revealed a closer relationship between *Xylophilus ampelinus* and *Variovorax paradoxus*. A sequence similarity of 97.9% was reported between both type strains (Wen et al., 1999). In other genera of the *Comamonadaceae* (*Acidovorax*, *Comamonas*, and *Hydrogenophaga*), this level of similarity is found between species within a genus. Together with the absence of DNA binding between the type strains of *Xylophilus* and *Variovorax* (Willems et al., 1991a), these would be arguments in support of a merger of these genera. Both are yellow-pigmented, although it is unknown whether these pigments have a similar structure. On the other hand, *Xylophilus* grows extremely slowly and is a phytopathogen with a rather restricted metabolic versatility, whereas *Variovorax* uses a wide variety of substrates and some of its strains are capable of chemolithotrophic growth using hydrogen oxidation as an energy source. In view of these differences and until further genotypic similarities are reported, the maintenance of separate genera may be justified for these organisms.

ACKNOWLEDGMENTS

Anne Willems is indebted to the Fund for Scientific Research–Flanders for a position as postdoctoral research fellow.

FURTHER READING

Ridé, M. 1996. La nécrose bactérienne de la vigne: données biologiques et épidémiologiques, base d'une stratégie de lutte. Comptes rendus de l'Académie d'Agriculture de France. *82*: 31–50.

---

3. YEGAL medium (g /l distilled water): yeast extract, 5.0; galactose, 10.0; K_2HPO_4, 4.01; NaH_2PO_4, 4.55; NH_4Cl, 1.0; $MgSO_4 \cdot 7H_2O$, 0.5; ferric ammonium citrate, 0.05; $CaCl_2$, 0.005. Yeast extract and galactose are each dissolved in 100 ml of water and autoclaved separately and added aseptically (Starr et al., 1977).

List of species of the genus Xylophilus

1. **Xylophilus ampelinus** (Panagopoulos 1969) Willems, Gillis, Kersters, Van Den Broecke, and De Ley 1987, 428[VP] (*Xanthomonas ampelina* Panagopoulos 1969, 75.)

am.pe.li′nus. Gr. n. *ampelos* grape vine; Gr. adj. *ampelinos* M.L. masc. adj. *ampelinus* of the vine.

The morphological and cellular characteristics are as

described for the genus. Additional descriptive information is presented in Table BXII.β.69, which is based on data from Panagopoulos (1969), Bradbury (1973, 1984), Willems et al. (1987), and Van den Mooter and Swings (1990).

Isolated from *Vitis vinifera*, where it causes bacterial necrosis and canker, in the Mediterranean area and South Africa. Similar symptoms on grape vine have been reported from Argentina, Austria, Bulgaria, the Canary Islands, and Switzerland (Bradbury, 1973).

The mol% G + C of the DNA is: 68–69 (T_m).

Type strain: ATCC 33914, DSM 7250, LMG 5856, NCPPB 2217.

GenBank accession number (16S rRNA): AF078758.

Order II. **Hydrogenophilales** *ord. nov.*

GEORGE M. GARRITY, JULIA A. BELL AND TIMOTHY LILBURN

Hy.dro.ge.no.phi.la' les. M.L. masc. n. *Hydrogenophilus* type genus of the order; *-ales* ending to denote order; M.L. fem. n. *Hydrogenophilales* the *Hydrogenophilus* order.

The order *Hydrogenophilales* was circumscribed for this volume on the basis of phylogenetic analysis of 16S rRNA gene sequences; the order contains the family *Hydrogenophilaceae*.

Description is the same as for the family *Hydrogenophilaceae*.

Type genus: **Hydrogenophilus** Hayashi, Ishida, Yokota, Kodama and Igarashi 1999, 785.

Family I. **Hydrogenophilaceae** *fam. nov.*

GEORGE M. GARRITY, JULIA A. BELL AND TIMOTHY LILBURN

Hy.dro.ge.no.phi.la' ce.ae. M.L. masc. n. *Hydrogenophilus* type genus of the family; *-aceae* ending to denote family; M.L. fem. pl. n. *Hydrogenophilaceae* the *Hydrogenophilus* family.

The family *Hydrogenophilaceae* was circumscribed for this volume on the basis of phylogenetic analysis of 16S rRNA gene sequences; the family contains the genera *Hydrogenophilus* (type genus) and *Thiobacillus*.

Obligate or facultative chemolithotrophs, using H_2 (*Hydro-genophilus*) or reduced sulfur compounds (*Thiobacillus*) as electron donors. Fix carbon by the Calvin-Benson cycle.

Type genus: **Hydrogenophilus** Hayashi, Ishida, Yokota, Kodama and Igarashi 1999, 785.

Genus I. **Hydrogenophilus** Hayashi, Ishida, Yokota, Kodama and Igarashi 1999, 785[VP]

THE EDITORIAL BOARD

Hy.dro.ge.no' phi.lus. Gr. n. *hydro* water; Gr. v. *genein* to produce; M.L. neut. n. *hydrogenum* hydrogen (that which produces water); Gr. adj. *philo* loving, friendly to; M.L. masc. n. *Hydrogenophilus* hydrogen lover.

Straight rods 0.5–0.8 × 1.0–3.0 μm during exponential growth. Occur singly. Motile or nonmotile. Gram negative. Nonsporulating. Aerobic or microaerophilic, having a strictly respiratory type of metabolism, with oxygen as the terminal electron acceptor. **Colonies are yellow. Thermophilic**; one species grows optimally at 50–52°C; and a second, at 60–65°C. **Facultatively chemolithoautotrophic**; can use H_2 as an electron donor and CO_2 as a carbon source. CO_2 is fixed via the Calvin-Benson cycle. Carbohydrates are not utilized. Acetate, pyruvate, DL-lactate, succinate, and DL-malate can be used as electron donors and carbon sources. Ammonium can be used as a nitrogen source. The major quinone system is ubiquinone 8. Isolated from hot springs and surrounding soil.

The mol% G + C of the DNA is: 61–65.

Type species: **Hydrogenophilus thermoluteolus** Hayashi, Ishida, Yokota, Kodama and Igarashi 1999, 785.

FURTHER DESCRIPTIVE INFORMATION

Hayashi et al. (1999) found the type strain (TH-1) of *H. thermoluteolus* to be nonmotile; however, Goto et al. (1978) reported the same strain to be motile. Reference strain TH-4 is nonmotile (Hayashi et al., 1999). The type strain of *H. hirschii* is motile by a single polar flagellum.

ENRICHMENT AND ISOLATION PROCEDURES

Isolation and culture conditions for *H. thermoluteolus* have been described by Goto et al. (1977). The isolation of *H. hirschii*—a microaerophilic organism—has been described by Stöhr et al. (2001); briefly, the autotrophic medium of Huber et al. (1992) was used lacking NaCl but containing 0.02% $CaCl_2 \cdot 2H_2O$. The medium was deoxygenated with N_2 and dispensed in 20-ml portions into 120 ml serum bottles under N_2. The gas phase in the bottles was changed to H_2/CO_2 (80:20, v/v), and the medium was sterilized at 100°C for 90 min and cooled. To each bottle was added 100 μl of a 10% sterile solution of $CaCO_3$ and 20 ml of filter-sterilized air. After several passages of serially diluted cultures, the cultures were plated onto the isolation medium solidified with 1.5% agar to obtain yellow colonies consisting of motile rods.

DIFFERENTIATION OF THE GENUS *HYDROGENOPHILUS* FROM OTHER GENERA

The genus *Hydrogenophilus* differs from the genera *Calderobacterium* and *Hydrogenobacter* in that it consists of facultative hydrogen autotrophs instead of obligate hydrogen autotrophs. Moreover, the optimal growth temperatures of *H. thermoluteolus* and *H. hirschii* (50–52°C and 63°C, respectively) are lower than that of *Calderobacterium* (74–76°C; Kryukov et al., 1983) and of *Hydrogenobacter* (70–75°C; Kawasumi et al., 1984). In addition, the cell diameter of *H. thermoluteolus* and *H. hirschii* (0.5–0.6 μm and 0.6–0.8 μm, respectively) is greater than that of *Calderobacterium* (0.25–0.5 μm) and *Hydrogenobacter* (0.3–0.5 μm) cells.

TAXONOMIC COMMENTS

Based on DNA–DNA hybridization experiments, Hayashi et al. (1999) found strain TH-1—the type strain of *H. thermoluteolus*—and reference strain TH-4 to have a hybridization value of 89%,

indicating that the two strains belong to the same species. Moreover, 16S rDNA analysis indicated that, although the two strains belonged to the class *Betaproteobacteria*, they were sufficiently unrelated to this class as to warrant their placement in a new genus, *Hydrogenophilus* (Hayashi, et al., 1999). Analysis of 16S rDNA sequences by Stöhr et al. (2001) indicated that strain TH-1—the type strain of *H. thermoluteolus*—has a phylogenetic distance of 0.0257 from the type strain of *H. hirschii* (Yel5a); thus the two type strains should be placed within the same genus.

In this edition of the *Manual*, the genus *Hydrogenophilus* is classified in the Order *Hydrogenophilales* and the Family *Hydrogenophilaceae* within the Class *Betaproteobacteria*. The only other member of the family is the genus *Thiobacillus*, which differs markedly in its phenotypic properties from *Hydrogenophilus*. Other genera of Gram-negative, thermophilic, hydrogen chemolithotrophs include *Calderobacterium* and *Hydrogenobacter*, which are classified in the Phylum *Aquificae*, Class *Aquificae* Order *Aquificales*, and Family *Aquificaceae*.

DIFFERENTIATION OF THE SPECIES OF THE GENUS *HYDROGENOPHILUS*

Hydrogenophilus thermoluteolus grows optimally at 50–52°C, whereas *H. hirschii* grows best at 63°C. In regard to their relationship to oxygen, *H. thermoluteolus* grows best at 22% O_2, whereas *H. hirschii*

is a microaerophile that grows best at 2.5% O_2 and fails to grow at O_2 levels higher than 5%.

List of species of the genus Hydrogenophilus

1. **Hydrogenophilus thermoluteolus** Hayashi, Ishida, Yokota, Kodama and Igarashi 1999, 785[VP]

ther.mo.lu.te′o.lus. Gr. adj. *thermos* hot; L. adj. n. *luteolus* light yellow; M.L. masc. adj. *thermoluteolus* hot and light yellow.

Exponentially growing cells are 0.5–0.6 × 2.0–3.0 μm. Optimal temperature, 50–52°C. Optimal pH, 7.0. Aerobic. Other characteristics are as given for the genus, with the following additional information. Carbon sources include propionate, butyrate, and α-ketoglutaric acid. No growth occurs on lactose, D-glucose, D-galactose, sucrose, citrate, ethanol, benzoate, and *m*-hydroxybenzoate. Ammonium, nitrate, and urea can be used as sole nitrogen sources; nitrite and N_2 are not used. Major fatty acids: $C_{16:0}$, $C_{18:0}$. The major 3–hydroxy cellular fatty acid is $C_{10:0\ 3OH}$. Cyclic fatty acids have not been described in *H. thermoluteolus*. Isolated from soil around a hot spring in Izu peninsula, Shizuoka Prefecture, Japan (Goto et al., 1977).

The mol% G + C of the DNA is: 63–65 (T_m).
Type strain: TH-1, IFO 14978.
GenBank accession number (16S rRNA): AB009828.

2. **Hydrogenophilus hirschii** Stöhr, Waberski, Liesack, Völker, Wehmeyer and Thomm 2001, 488[VP]

hir′schi.i. N.L. gen. n. *hirschii* in honor of Peter Hirsch, in recognition of his fundamental contributions to the taxonomy of unusual bacteria.

Cells are 0.6–0.8 × 1.0–2.0 μm. Motile by a single polar flagellum. Optimal growth temperature, 63°C; upper limit, 68°C. No growth at 45°C. Microaerophilic, growing best at 2.5% O_2 and failing to grow at O_2 levels higher than 5%. Other characteristics are as given for the genus, with the following additional information. Growth occurs anaerobically with nitrate. Growth occurs on yeast extract, peptone, meat peptone, and meat extract. Fumarate, glutamate, and gluconate can be used as carbon sources. No growth occurs on carbohydrates, aromatic compounds, L-alanine, L-proline, citric acid, methanol, or ethanol. Neither thiosulfate nor sulfur is used as an electron donor. Major fatty acids: $C_{16:0}$, $C_{17:0\ cyclo}$, and $C_{19:0\ cyclo}$. Isolated from a water sample from Angel Terrace in Yellowstone National Park, U.S.A.

The mol% G + C of the DNA is: 61 (HPLC).
Type strain: Yel5a, DSM 11420, JCM 10831.
GenBank accession number (16S rRNA): AJ131694.

Genus II. **Thiobacillus** Beijerinck 1904b, 597[AL]

DONOVAN P. KELLY, ANN P. WOOD AND ERKO STACKEBRANDT

Thi.o.ba.cil′lus. Gr. n. *thios* sulfur; L n *bacillus* a small rod; M. L. masc n. *Thiobacillus* sulfur rodlet.

Small, Gram-negative, rod-shaped cells (0.3–0.5 × 0.9–4.0 μm). Some species are motile by means of polar flagella. No resting stages known. Energy is derived by the oxidation of one or more reduced sulfur compounds, including sulfides, sulfur, thiosulfate, polythionates, and thiocyanate. Sulfate is the end product of sulfur-compound oxidation, but sulfur, sulfite, and polythionates may be accumulated by most species, sometimes transiently. All

species can fix carbon dioxide by means of the Benson–Calvin cycle and are capable of autotrophic growth; some species are obligately chemolithotrophic, while others are chemoorganotrophic. The genus currently includes obligate aerobes and facultative denitrifiers. Optimal pH of 2–8 with optimal temperature of 28–43°C. Distribution is seemingly ubiquitous in marine, freshwater, and soil environments, especially where oxidizable sulfur is abundant (e.g., sulfur springs, sulfide minerals, sulfur deposits, sewage treatment areas, and sources of sulfur gases, such as sediments or anaerobic soils releasing H_2S).

The mol% G + C of the DNA is: 62–67.

Type species: **Thiobacillus thioparus** Beijerinck 1904b, 597.

FURTHER DESCRIPTIVE INFORMATION

The chemolithotrophic, sulfur-compound-oxidizing, Gram-negative, rod-shaped members of the genera and species described (Table BXII.β.70) are usually 0.5×1.0–4.0 μm in size, occurring singly, in pairs or in short chains; some are motile by means of single polar flagella; some possess pili and other specialized surface features. All can obtain energy from the oxidation of reduced inorganic sulfur compounds, and, in most cases, elemental sulfur. Some species are obligately chemolithotrophic and autotrophic, others are facultatively heterotrophic, and, in some strains, optimal growth occurs mixotrophically. Both electron-transport-dependent and substrate-level phosphorylation occur during sulfur-compound oxidation, and reduction of NAD(P) requires an energy-dependent flow of electrons from cytochrome *c* (or *b*). Carbon dioxide fixation occurs mainly by means of the Benson–Calvin cycle, with some fixation occurring through pyruvate or phosphoenolpyruvate carboxylation. The obligately autotrophic species possess an incomplete tricarboxylic acid cycle (lacking 2-oxoglutarate dehydrogenase), which is used as a biosynthetic "horseshoe" pathway via oxaloacetate or succinate and 2-oxoglutarate, respectively, down each arm of the "horseshoe". Some species contain characteristic plasmids, and some have been shown to be susceptible to introduction of *Pseudomonas* plasmids. Little work on the genetics of *Thiobacillus* spp. is available, but it is known that autotrophic and drug-resistant mutants can be obtained and that mutations that decrease autotrophic efficiency can be induced. No species is known to be pathogenic. *Thiobacillus* spp. are ubiquitous, with the facultative species occurring in soil, freshwater, and marine environments as heterotrophs or mixotrophs. Sulfur-compound-oxidizing species have been isolated from Arctic, temperate, and tropical waters, soil, salt marshes, freshwater lakes, rivers, canals, hot springs, sulfur-rich mine or acid-mine wastewaters and other environments where oxidizable sulfur compounds occur naturally or anthropogenically.

ENRICHMENT AND ISOLATION PROCEDURES

Most of the species can be isolated from natural habitats by the use of mineral media containing elemental sulfur or thiosulfate as the energy-yielding substrate. Use of media of different pH will assist in differential selection of the neutrophilic and acidophilic species, use of acid ferrous sulfate medium will frequently select for *Acidithiobacillus ferrooxidans*, and use of anaerobic thiosulfate medium (pH 7) supplemented with nitrate will select for *T. denitrificans*. A procedure has been described for the enrichment of facultatively autotrophic, mixotrophic strains, using a continuous-flow chemostat with both organic and inorganic substrates (Gottschal and Kuenen, 1980). This provides a means of avoiding the predomination by heterotrophs in standard batch enrichment media containing supplements such as thiosulfate and glucose or thiosulfate and acetate. In the latter medium, a mixture of obligately chemolithotrophic thiobacilli and chemoorganotrophs normally develops. *Acidiphilium acidophilum* was originally isolated as a commensal of *A. ferrooxidans*, although "*T. organoparus*" (now considered to be a strain of *A. acidophilum*) was enriched directly from an acid mine water environment. Media suitable for the different species are summarized in Kelly and Wood (1998) and are available from the original literature describing the species. In addition, culture collection catalogues also provide guidance for media suitable for their cultures.

Most strains are able to produce colonial growth on appropriate media solidified with agar. Some strains, especially those of *A. ferrooxidans*, grow poorly on agar media. In some cases, this difficulty is due to the toxicity of agar hydrolysis products and has been overcome in a number of ways (e.g., Tuovinen and Kelly, 1973). Toxic effects are generally avoided by the use of a minimal concentration of agar, screening of suitable brands of purified agars, use of agarose, and, in the case of *A. ferrooxidans*, use of a combination of media with low agar concentrations, at

TABLE BXII.β.70. Differentiation of the genera of chemolithotrophic, sulfur-oxidizing, rod-shaped bacteria

Character	Thiobacillus	Acidiphilium	Acidithiobacillus	Halothiobacillus	Paracoccus	Starkeya	Thermithiobacillus	Thiomonas
Obligate chemolithoautotroph[a]	+	−	+	+	−	−	+	−
Heterotrophic growth on defined media	−	+[b]	−	−	+[c]	+	−	−[d]
Mol% G + C of DNA	62–67	63–68	52–64	56–67	63–71	67–68	66–67	61–67
Class:								
Alphaproteobacteria		+			+	+		
Betaproteobacteria	+							+
Gammaproteobacteria			+	+			+	
Respiratory ubiquinone	Q-8	Q-10	Q-8	Q-8	Q-10	Q-10	Q-8	Q-8
Facultative denitrification	+[e]	−	−	−	+[c]	−	−	−
Optimal temperature (°C)	28–43	25–37	30–45	28–30	25–37	25–30	43–45	30–50
Optimal pH	6.8–8.0[f]	3.0–3.5	2.0–3.5	6.5-8.0	6.5–9.0	7	6.8–7.5	5.2–6.0[g]
Halophilic or halotolerant	−	−	−	+	−	−	−	−
Contains photosynthetic reaction centers	−	+	−	−	−	−	−	−

[a]With inorganic sulfur compounds as sole energy substrates.

[b]Most species of the genus are not chemolithoautotrophic.

[c]Not all species exhibit all these features, and only *P. denitrificans*, *P. versutus*, and *P. pantotrophus* are facultative, sulfur-oxidizing chemolithoautotrophs.

[d]Most strains will grow on complex rich media, and some grow best mixotrophically with thiosulfate plus organic supplements.

[e]Denitrification to dinitrogen in one species only: *T. denitrificans*.

[f]*T. plumbophilus* grows only between pH 4.0–6.5.

[g]Optimum for *T. cuprinus* is pH 3.0–4.0.

pH 2.2–2.5, and with ferrous sulfate at only about 20 mM. The use of silica gel media as an alternative is normally unnecessary.

MAINTENANCE PROCEDURES

After cultivation on suitable media, most species survive storage at 5°C for periods of weeks to months, especially if the media for neutrophiles have not become too acidic before storage. *A. ferrooxidans* survives in culture on pyrite for very long periods when stored at 5 to 15°C, and "*T. plumbophilus*" survives for at least a year at room temperature on galena (Drobner et al., 1992). Many strains have been successfully lyophilized or have survived storage at −20°C, at liquid nitrogen temperature, or in glycerol suspension at −20°C.

DIFFERENTIATION OF THE GENUS *THIOBACILLUS* FROM OTHER GENERA

At the time of the previous edition of this *Systematics*, information on ubiquinone and fatty acid content, DNA base composition, and interspecific DNA–DNA hybridization had been obtained for the thiobacilli (Kelly and Harrison, 1989). This enabled division of the species into distinct groups on the basis of physiological and biochemical characteristics (Harrison, 1982, 1983; Katayama-Fujimura et al., 1982; Katayama-Fujimura and Kuraishi, 1983). It has also complicated consolidation of species, as DNA–DNA hybridization studies in some cases show considerable diversity among strains regarded as members of the same species, while also showing very high levels of similarity between putatively different species. The confirmation by 16S rRNA sequencing of the phylogenetic diversity of the formally approved species now forces major revision of this genus.

Primary separation of most of the former 21 species of *Thiobacillus* is relatively easy to achieve by virtue of differences in gross physiological characteristics, such as pH and temperature requirements for growth, ability or inability to grow heterotrophically as well as chemolithotrophically, and differences in the ability to grow anaerobically with denitrification or to use elemental sulfur. Most species designated as *Thiobacillus* may precipitate elemental sulfur into the medium during growth on sulfide, thiosulfate, and, in some cases, trithionate or tetrathionate. One of the exceptions is *T. versutus*. Sulfur precipitation is not a highly distinctive diagnostic characteristic, as it is a variable property influenced by growth conditions, such as oxygen availability or perturbation of steady state in chemostat culture. The precipitation of sulfur is superficially comparable to the extracellular precipitation observed with species of *Chlorobium* and contrasts with the intracellular accumulation of sulfur by members of the *Chromatiaceae* such as *Chromatium*. It is likely, however, that some species deposit sulfur internally, subsequently oxidizing it to sulfate. Thus, *T. albertensis* grown on thiosulfate is believed to contain a sulfur granule bounded by a membrane (Bryant et al., 1983), the appearance of which is correlated with the production of a large amount of extracellular elemental sulfur at the end of the growth phase. Morphologically, these inclusions resemble the sulfur found in *Beggiatoa* and *Thiothrix*. These granules do not appear during mid-log growth (Bryant et al., 1983), and had not been reported in earlier ultrastructural surveys of other thiobacilli (Mahoney and Edwards, 1966; Shively et al., 1970), but may have been observed in "*T. kabobis*" (Reynolds et al., 1981), which is now regarded as a synonym of *T. thiooxidans* (Kuenen et al., 1992). The possibility of both extracellular deposition and intracellular accumulation of elemental sulfur in some thiobacilli is somewhat of a physiological anomaly,

possibly indicative of differing mechanisms or locations of sulfur compound oxidation. Extracellular sulfur precipitation is best explained as a consequence of the conversion of sulfide or the sulfane groups of thiosulfate (or polythionates) to sulfur at the surface of the cell, presumably in the periplasmic space and catalyzed by enzyme systems located in or external to the bounding membrane of the cell. This seems to be more plausible than intracellular generation of sulfur and its excretion to the outside (as has indeed been suggested). The physicochemical nature of the sulfur at the moment of formation is uncertain, and, of course, the cell wall would likely be a barrier to the excretion of large granules from inside the cell or even to their generation in the periplasmic space. The production of intracellular sulfur from a soluble substrate (thiosulfate) implies intracellular oxidation of that substrate and transport of S^0 across the membrane. Although it is clear that this process would involve membrane-associated electron transport systems in the oxidation of sulfur compounds to sulfate, the mechanism remains to be elucidated. Clearly, the present stage of our knowledge of the mechanism of transport of sulfur and its compounds and of the biochemistry of their conversion to sulfate is inadequate. Therefore, characteristics such as intracellular sulfur accumulation cannot be employed as reliable taxonomic features.

TAXONOMIC COMMENTS

Since the description of *Thiobacillus* was published in the first edition of the *Systematics* (Kelly and Harrison, 1989), the genus has undergone a number of revisions and emendations. These changes are summarized in Table BXII.β.70 and Table BXII.β.71. Table BXII.β.71 also presents the DNA base composition and class within the *Proteobacteria* to which each species is now assigned, based on sequence analysis of either partial or complete 16S rRNA genes and recently proposed reassignments to new genera (Kelly and Wood, 2000a, b; Kelly et al., 2000). A phylogenetic tree illustrating the new and different genera to which most of the *Thiobacillus* species have been assigned is shown in Fig. BXII.β.60.

First, *T. versutus* and *T. acidophilus* (both members of the class *Alphaproteobacteria*) have been formally transferred from *Thiobacillus* into other genera (Katayama et al., 1995; Hiraishi et al., 1998; see the chapter by Hiraishi and Imhoff on the genus *Acidiphilium* in the family *Acetobacteraceae* in this volume), and each appears elsewhere within this edition. The remaining species group within the classes *Alphaproteobacteria*, *Betaproteobacteria*, and *Gammaproteobacteria*. The type species, *T. thioparus*, falls in the *Betaproteobacteria*, along with the closely related *T. denitrificans* and the moderate thermophile *T. aquaesulis* (McDonald et al., 1997; Kelly and Wood, 2000a, b). These three species are the only ones that can be justified based on molecular criteria to be members of the genus *Thiobacillus*. A fourth validly named species, *Thiobacillus delicatus*, is physiologically similar to members of the genus *Thiomonas* and may be reassigned to that genus when more data are available (Y. Katayama, personal communication). *Thiobacillus denitrificans* has been regarded as a close relative of *T. thioparus* since it was originally isolated by Beijerinck (1904a, b), and that view is now strongly supported by the confirmation that it is a member of the *Betaproteobacteria* (Lane et al., 1992; H.G. Trüper, personal communication, 1999; Kelly and Wood, 2000a). Of the remaining species, only *Thiobacillus novellus* has been shown to be a member of the *Alphaproteobacteria*. *T. novellus* is facultatively autotrophic (as are the *Alphaproteobacteria* species formerly classified as *T. versutus* (now *Paracoccus versutus*) and *T.*

TABLE BXII.β.71. Species of *Thiobacillus* recognized in the first edition of this *Manual* (Kelly and Harrison, 1989), or described subsequently, showing those species assigned, or proposed for assignment, to new or different genera

Basonym	Revised genus designations (published or proposed)
Thiobacillus thioparus [a]	*Thiobacillus thioparus*[T]
Thiobacillus denitrificans[b]	*Thiobacillus denitrificans*
Thiobacillus aquaesulis[c]	*Thiobacillus aquaesulis*
Thiobacillus plumbophilus[d]	Uncertain (currently *Thiobacillus*)
Thiobacillus intermedius[e]	*Thiomonas intermedia*[T]
Thiobacillus perometabolis[e]	*Thiomonas perometabolis*
Thiobacillus cuprinus[e]	*Thiomonas cuprina*
Thiobacillus thermosulfatus[e]	*Thiomonas thermosulfata*
Thiobacillus delicatus[f]	*Thiomonas delicata*
Thiobacillus neapolitanus[g]	*Halothiobacillus neapolitanus*[T]
Thiobacillus halophilus[g]	*Halothiobacillus halophilus*
Thiobacillus hydrothermalis[g]	*Halothiobacillus hydrothermalis*
Thiobacillus tepidarius[g]	*Thermithiobacillus tepidarius*[T]
Thiobacillus thiooxidans[g]	*Acidithiobacillus thiooxidans*[T]
Thiobacillus ferrooxidans[g]	*Acidithiobacillus ferrooxidans*
Thiobacillus caldus[g]	*Acidithiobacillus caldus*
Thiobacillus albertensis[d]	(*Acidithiobacillus*)
Thiobacillus prosperus[d]	Uncertain ("*Acidihalobacter*")
Thiobacillus novellus[h]	*Starkeya novella*
Thiobacillus acidophilus[i]	*Acidiphilium acidophilum*
Thiobacillus versutus[j]	*Paracoccus versutus*

[a]Kelly and Harrison, 1989.

[b]Kelly and Wood, 2000a; see text.

[c]McDonald et al., 1997.

[d]See text.

[e]Moreira and Amils, 1997.

[f]Requires investigation.

[g]Kelly and Wood, 2000b.

[h]Kelly et al., 2000.

[i]Hiraishi et al., 1998.

[j]Katayama et al., 1995.

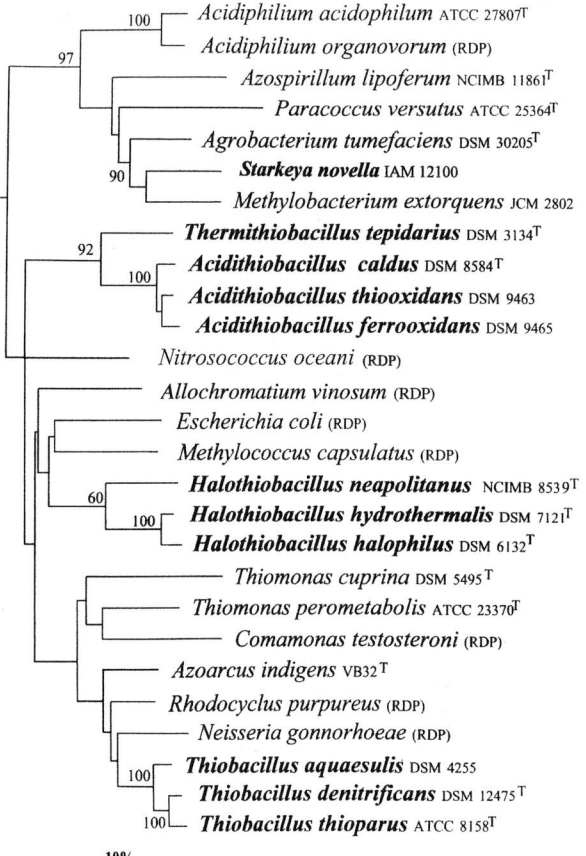

FIGURE BXII.β.60. Phylogenetic tree based on 16S rRNA gene sequences of representatives of the genera to which species formerly described as *Thiobacillus* are now assigned (*Thiobacillus*, *Acidiphilium*, *Starkeya*, *Thermithiobacillus*, *Acidithiobacillus*, *Halothiobacillus*, and *Thiomonas*). Reference species are taken from the Ribosomal Database Project (RDP) or from the sequence database libraries. *T* indicates the type strain of the species. Bar = 10 inferred nucleotide changes per 100 nucleotides. Numbers at some nodes indicate the number of times the species to the right occurred in 100 bootstrap replicates.

acidophilus (now *Acidiphilium acidophilum*), as well as *Thiosphaera pantotropha* (now *Paracoccus pantotrophus*) and has been proposed to be the type species of a new genus, *Starkeya* (Kelly et al., 2000). Of the other species in the *Betaproteobacteria*, four have been reassigned to the new genus *Thiomonas* (Moreira and Amils, 1997). A fifth acidophilic, lead-sulfide-leaching species reportedly also in the *Betaproteobacteria*, "*Thiobacillus plumbophilus*" (Drobner et al., 1992), has been too little studied to know if it should also be reassigned to *Thiomonas*, so for the present time it is retained in the genus *Thiobacillus*. Of the species within the *Gammaproteobacteria*, a broad division can be seen between two principal groups: one comprising the proposed new genus *Halothiobacillus* (McDonald et al., 1997; Kelly et al., 1998a; Kelly and Wood, 2000b), and the other (*Acidithiobacillus*) containing at least three of the acidophilic species, some of which also oxidize Fe^{2+} and sulfide minerals (Table BXII.β.71; Fig. BXII.β.60). The former *Thiobacillus tepidarius* also seems to be somewhat distantly related to this group (McDonald et al., 1997; Goebel et al., 2000), but we consider its moderate thermophily and lack of acidophily sufficient grounds to regard it as a distinct genus (*Thermithiobacillus*) at this time (Kelly and Wood, 2000b). *Thiobacillus albertis* (whose specific epithet is here corrected to *albertensis*) is also acidophilic and strictly chemolithotrophic, making it physiologically similar to *Acidithiobacillus thiooxidans*, but it differs from that species in possessing a glycocalyx and having a mol% G + C of 61.5, which is significantly higher than that of *A. thiooxidans* (50–

52 mol% G + C). Based on 16S rRNA sequences, we confirm the placement of *T. albertensis* in *Acidithiobacillus* (Table BXII.β.71), whose members have DNA mol% G + C values ranging between 50–64 (Table BXII.β.71). "*T. prosperus*" is more difficult to define taxonomically, as it appears to be completely unrelated to the *T. ferrooxidans*–*T. thiooxidans*–*T. caldus* cluster based on 16S rRNA sequence comparisons (Goebel et al., 2000) and shows negligible DNA–DNA hybridization with other thiobacilli. We have reported the acidophilic species *T. ferrooxidans*, *T. thiooxidans*, and *T. caldus*, as well as *T. tepidarius*, to fall in the *Gammaproteobacteria*, whereas some reports in the literature assign these to the *Betaproteobacteria*. It seems clear to us from recent studies that these species, while close to the beta–gamma separation, are rightly placed in the *Gammaproteobacteria* (McDonald et al., 1997; Goebel et al., 2000; Kelly and Wood, 2000b).

Descriptions of the former *Thiobacillus* species *T. versutus* and *T. acidophilus* are found in the chapters on *Paracoccus* and *Acidiphilium*, respectively. Some characteristics of the genus *Thiobacillus* of the newly created genera, and of *Paracoccus* and *Acidiphilium* are given in Table BXII.β.70.

DIFFERENTIATION OF THE SPECIES OF THE GENUS *THIOBACILLUS*

The characteristics of each of the six species are summarized in Table BXII.β.72.

List of species of the genus Thiobacillus

1. **Thiobacillus thioparus** Beijerinck 1904b, 597[AL]

   thi.o'par.us. G. n. *thios* sulfur; M.L. adj. *thioparus* sulfur producing.

   Rods, averaging 0.5 × 1.7 μm. Motile with a polar flagellum. Gram negative. Colonies grown on thiosulfate agar (1–2 mm in diameter) are circular and whitish-yellow due to precipitated sulfur. Turn pink, then brown on aging, especially in the center of old colonies. In static culture in liquid thiosulfate medium, sulfur is precipitated, and the medium becomes turbid with a pellicle of sulfur and cells. Sulfur granules and tetrathionate and/or trithionate may accumulate, accompanied by a pH drop to 4.5. In well-aerated or chemostat culture, sulfur oxidation of thiosulfate to sulfate may occur without precipitation. Some strains oxidize thiocyanate, thiosulfate, trithionate, tetrathionate, sulfur, and sulfide. Obligately chemolithotrophic and autotrophic. Ammonium salts and nitrates used as nitrogen sources. Aerobic. Optimal temperature: 28°C. Optimal pH: 6.6–7.2, with growth occurring between pH 4.5 and 7.8. Some strains are claimed to grow at pH 10.0. Found in mud, soil, canal water, and other freshwater sources. Presumably widely distributed.

   The mol% G + C of the DNA is: 62–63 (Bd, T_m).

   Type strain: ATCC 8158, DSM 505.

   GenBank accession number (16S rRNA): M79426.

2. **Thiobacillus aquaesulis** Wood and Kelly 1995, 418[VP] (Effective publication: Wood and Kelly 1988, 342.)

   a.quae.su'lis. L. n. *aquae* waters; L. n. *Sulis* pertaining to the Temple of Sulis Minerva (Minerva, the Roman goddess of wisdom); M.L. adj. *aquaesulis* from the waters of Sulis Minerva.

   Short rods, 0.3 × 0.9 μm, containing some polyphosphate inclusions. Motile. Gram negative. Non-sporeforming. Chemolithoautotrophic growth on thiosulfate, trithionate, or tetrathionate. Colonies on thiosulfate agar at 43°C are small (1–2 mm), circular, convex, and smooth, becoming white or yellow with precipitated sulfur. In liquid batch culture, sulfur precipitation and a drop in pH without tetrathionate accumulation occur. Initiates growth at pH 7–9 (30–55°C), dropping the pH to 6–7. No growth at pH 6.4 or 9.4 or at 26°C or 58°C. Chemostat cultures do not accumulate sulfur or other intermediates during growth on thiosulfate, trithionate, or tetrathionate at pH 7.6 and 43°C. Facultatively heterotrophic on complex media (yeast extract or nutrient broth) but unable to grow on common sugars, organic acids, formate, or methylamine as single substrates. Uses ammonium salts as nitrogen sources. Capable of anaerobic, autotrophic growth during thiosulfate-dependent denitrification in batch culture, producing nitrite and sulfur from thiosulfate and nitrate. Optimal temperature: 40–50°C. Optimal pH: 7.5–8.0. Contains ubiquinone-8. Isolated from the thermal springs at Bath, Avon, England.

   The mol% G + C of the DNA is: 65.7 (T_m).

   Type strain: ATCC 43788, DSM 4255.

   GenBank accession number (16S rRNA): U58019.

3. **Thiobacillus delicatus** (ex Mizoguchi, Sato and Okabe 1976)[VP] Katayama-Fujimura, Kawashima, Tsuzaki and Kuraishi 1984a, 142.

   del.i.cat'us. L. masc. adj. *delicatus* delicate.

   Rods, usually single, rarely in pairs, 0.4–0.6 × 0.7–1.6 μm. Nonmotile. Colonies grown on yeast extract–thiosulfate agar (1 mm. diameter) are smooth and circular and

TABLE BXII.β.72. Basic characteristics of species of the genus *Thiobacillus*

Characteristic	*T. thioparus*	*T. aquaesulis*	*T. denitrificans*	"*T. plumbophilus*"	*T. delicatus*	"*T. prosperus*"
Mol% G + C	62–63	66	63	66	66–67	64
Cell size (μm)	0.5 × 1.7	0.3 × 0.9	0.5 × 1.0–3.0	0.25 × 3.0	0.4–0.6 × 0.7–1.6	0.3 × 3–4
Motility	+	+	+	+	−	+
Carboxysomes	+	nd	−	nd	nd	+
Obligately chemolithoautotrophic	+	−	+	+	−	+
Optimal pH	6.6–7.2	7.5–8.0	6.8–7.4	4.0–6.5	5.5–6.0	nd
pH limits	4.5–7.8	6.5–9.0	4.5–8.0	4.0–6.5	5.0–7.0	1.0–4.5
Optimal temperature (°C)	28	40–50	28–32	21–34	30–35	37
Nitrate reduction:						
To nitrite	+	+	+	−	+	nd
To N₂	−	−	+	−	−	nd
Growth on:						
Hydrogen sulfide	+	nd	+	+	nd	+
Thiosulfate	+	+	+	−	+	+
Thiocyanate	+	−	+	−	−	+
PbS	nd	nd	nd	+	nd	+
Methylated sulfides	+	−	−	nd	nd	nd
Complex media	−	+[a]	−	−	nd	nd

[a]Unable to grow on common sugars, organic acids or one-carbon organic substrates.

change from transparent to whitish-yellow with sulfur accumulation. Facultatively chemolithotrophic and mixotrophic. Grow autotrophically with sulfur, thiosulfate, or tetrathionate, but not with thiocyanate. Accumulates tetrathionate and trithionate transiently during growth on thiosulfate. Incapable of heterotrophic growth on single-carbon compounds. Grows mixotrophically in thiosulfate media supplemented with tricarboxylic acid cycle intermediates or amino acids. Optimal growth requires both organic and inorganic substances and thiosulfate or sulfur. Facultatively anaerobic; reduces nitrate and produces nitrite in mixotrophic and autotrophic media with thiosulfate or tetrathionate. Ammonium salts, nitrate, urea, glutamate, or aspartate can be used as the nitrogen source.

Optimal temperature 30–35°C, range 15–42°C (no growth at 10 or 45°C). Optimal pH 5.5–6.0, range 5.0–7.0. Isolated from mine water. Distribution unknown.

The mol% G + C of the DNA is: 66–67 (T_m, chemical analysis).

Type strain: IAM 12624.

Additional Remarks: Until 16S rDNA sequence data become available for this species, it cannot be firmly reassigned to the genus *Thiomonas*. Physiologically, the species is more similar to *Thiomonas perometabolis* than to the type species *Thiobacillus thioparus*.

4. **Thiobacillus denitrificans** (ex Beijerinck 1904b) Kelly and Harrison 1989, 1855 emend. Kelly and Wood 2000a, 548[VP]
de.ni.tri′fi.cans. M.L. v. *denitrifico* denitrify; M.L. part. adj. *denitrificans* denitrifying.

Short rods, 0.5 × 1.0–3.0 μm. May be motile by means of a polar flagellum. Under anaerobic conditions, colonies are clear or weakly opalescent when grown on thiosulfate–nitrate agar, On aging, colonies may become white with sulfur. Vigorous nitrogen production when grown under anaerobic conditions, leading to splitting of agar when solid media are used. Facultatively anaerobic. Grows autotrophically and aerobically on thiosulfate or tetrathionate, on which it produces growth yields approximately double those of *T. thioparus* or *T. neapolitanus*. Grows anaerobically on thiosulfate, tetrathionate, or sulfide by using nitrate, nitrite, or nitrous oxide as the terminal electron acceptor. Oxidizes sulfur, sulfide, thiosulfate, tetrathionate, and probably sulfite, but not thiocyanate. Chemostat culture can be switched easily and repeatedly between aerobic and anaerobic growth modes, with adaptation involving derepression of nitrate and nitrite reductase synthesis. Ammonium salts and, in at least some strains, nitrate are used as nitrogen sources.

Obligately chemolithotrophic and autotrophic. Optimal temperature: 28–32°C. Optimal pH: 6.8–7.4. Found in soil, mud, and freshwater and marine sediments, especially under anoxic conditions. Probably very widely distributed.

The original isolation of Beijerinck (1904a) may not have been a pure culture (Vishniac and Santer, 1957), and a viable sample is not available in the Delft Culture Collection (L.A. Robertson, personal communication) or in any other culture collection. Later work has demonstrated unambiguously that Beijerinck's designation was of a legitimate species exhibiting stable physiological character-

istics (Lieske, 1912; Baalsrud and Baalsrud, 1954; Taylor et al., 1971; Justin and Kelly, 1978; Katayama-Fujimura et al., 1982). Earlier claims that the capacity of this organism to denitrify was lost on aerobic subculture and that it was facultatively heterotrophic were erroneous. The name has therefore been revived by Kelly and Harrison (1989, in Validation List No. 31), who suggested a neotype strain. It is now proposed that NCIB 9548 (the strain isolated by White and Hutchinson and used by Justin and Kelly, 1978; AB7, ATCC 23644, JCM 3870) be accepted as this reference strain (Kelly and Wood, 2000a).*

The mol% G + C of the DNA is: 63 (Bd, T_m).
Type strain: ATCC 23644, DSM 12475, NCIB 9548.
GenBank accession number (16S rRNA): AJ243144.

5. **"Thiobacillus plumbophilus"** Drobner, Huber, Rachel and Stetter 1992, 217.
plum.bo′philus. L. neut. n. *plumbum* lead; Gr. v. *philein* to love; M. L. adj. *plumbophilus* loving lead, referring to its ability to grow with lead sulfide as a sole energy source.

Cells are rod-shaped, Gram negative, 0.25 × 3 μm, and motile by one polar flagellum. Optimal growth between 21°C and 34°C, with growth occurring at up to 41°C. Growth between pH 4.0 and 6.5. Strictly chemolithoautotrophic and aerobic. Oxidation of galena (PbS), H_2S, and hydrogen. Sensitive to ampicillin and rifampicin. Possesses 96.5% ubiquinone Q-8. Isolated from a uranium mine in Germany. Insignificant DNA hybridization to *A. ferrooxidans* and "*T. cuprinus*".

The mol% G + C of the DNA is: 66 (T_m, HPLC).
Deposited strain: DSM 6690.

6. **"Thiobacillus prosperus"** Huber and Stetter 1989, 484.
pros′pe.rus. L. masc. adj. *prosperus* prosperous, referring to its ability to gain precious metals by ore "leaching".

Cells are Gram-negative rods, 0.3 × 3–4 μm. Motile by one polar flagellum. Optimal growth around 37°C and up to 41°C. Growth from 0–3.5% NaCl (some strains tolerate 6% NaCl) and at pH 1.0–4.5. Strictly chemolithoautotrophic and aerobic. Growth on sulfidic ores, like pyrite, sphalerite, chalcopyrite, arsenopyrite, and galena, and on H_2S. Poor growth on elemental sulfur and ferrous iron. Produces sulfuric acid from reduced sulfur compounds. Sensitive to ampicillin and vancomycin. Possesses *meso*-diaminopimelic acid and ubiquinone Q-8. Lives in marine sediments in hydrothermal areas. Insignificant DNA hybridization to *T. ferrooxidans*, *T. thiooxidans*, *T. neapolitanus*, and *T. thioparus*. Member of the class *Gammaproteobacteria*.

The mol% G + C of the DNA is: 64 (T_m, HPLC).
Deposited strain: DSM 5130.
GenBank accession number (16S rRNA): AY034139.

Additional Remarks: "*T. prosperus*" will in due course be removed from the genus, as recent work has shown its closest phylogenetic relatives to be "*Acidihalobacter aerolicus*" (DSM 14174) and "*Acidihalobacter ferrooxidans*" (DSM 14175), with both of which it shares 95% 16S rRNA sequence identity (P.R. Norris, K.B. Hallberg and B. Johnson, personal communication, 2001).

Editorial Note: In the original description, the authors erroneously designated the neotype strain as NCIMB 8327.

Order III. **Methylophilales** *ord. nov.*

GEORGE M. GARRITY, JULIA A. BELL AND TIMOTHY LILBURN

Me.thy.lo.phi.la' les. M.L. masc. n. *Methylophilus* type genus of the order; *-ales* ending to denote order; M.L. fem. n. *Methylophilales* the *Methylophilus* order.

The order *Methylophilales* was circumscribed for this volume on the basis of phylogenetic analysis of 16S rRNA sequences; the order contains the family *Methylophilaceae*.

Description is the same as for the family *Methylophilaceae*.
Type genus: **Methylophilus** Jenkins, Byrom and Jones 1987, 447.

Family I. **Methylophilaceae** *fam. nov.*

GEORGE M. GARRITY, JULIA A. BELL AND TIMOTHY LILBURN

Me.thy.lo.phi.la' ce.ae. M.L. masc. n. *Methylophilus* type genus of the family; *-aceae* ending to denote family; M.L. fem. pl. n. *Methylophilaceae* the *Methylophilus* family.

The family *Methylophilaceae* was circumscribed for this volume on the basis of phylogenetic analysis of 16S rRNA sequences; the family contains the genera *Methylophilus* (type genus), *Methylobacillus*, and *Methylovorus*.

Aerobic respiratory metabolism. Oxidize methanol but not methane. Some grow on a limited range of other compounds. Found in a variety of habitats.

Type genus: **Methylophilus** Jenkins, Byrom and Jones 1987, 447.

Genus I. **Methylophilus** *Jenkins, Byrom and Jones 1987, 447*[VP]

THE EDITORIAL BOARD

Me.thy.lo.phi' lus. Fr. n. *méthyle* the methyl radical; Gr. adj. *philos* loving; M.L. masc. n. *Methylophilus* methyl radical loving.

When grown on methanol–mineral salts agar or in methanol–mineral salts liquid medium, the cells are straight or slightly curved rods, usually 0.3–0.6 × 0.8–1.5 μm, occurring singly or in pairs. Gram negative, but the stain is often not taken up well. **Motile by polar flagella or nonmotile.** Endospores are absent. **No cellular inclusions.** No sheaths or prosthecae detected. No capsules formed, but **slime may be produced by some strains.** Colonies on methanol–mineral salts agar plates incubated for 2 days at 30°C or 37°C are **nonpigmented**, circular, 1–2 mm in diameter, with entire edge, convex, translucent to opaque. Pyocyanin and fluorescein are not produced. No, or extremely poor, growth on nutrient agar and in nutrient broth incubated at 30°C or 37°C for 2 days. No, or extremely poor, growth on blood agar; no hemolysis. Optimal temperature, 30–37°C; no growth occurs at 4°C or 45°C. Optimal pH, 6.5–7.2. **Aerobic. Metabolism respiratory**; very little or no acid is produced from glucose. **Methanol is oxidized as the sole carbon and energy source by all strains.** In addition, a limited range of other carbon compounds such as methylamines, formate, glucose, and fructose may be utilized as sole carbon and energy sources. Methane is not used. Nutritionally nonexacting; nitrate and ammonium salts serve as nitrogen sources. Catalase positive and **oxidase positive.** The fatty acid composition is primarily of the nonhydroxylated straight-chain saturated and monounsaturated types with $C_{16:0}$ and $C_{16:1}$ predominating. The major isoprenoid quinone components are ubiquinones with eight isoprene units (Q-8). Isolated from activated sludge, mud, and river and pond water.

The mol% G + C of the DNA is: 50 (T_m).

Type species: **Methylophilus methylotrophus** Jenkins, Byrom and Jones 1987, 447.

FURTHER DESCRIPTIVE INFORMATION

Hexulose phosphate synthetase and hexulose phosphate isomerase, key enzymes of the ribulose monophosphate pathway (RMP) for methanol assimilation, are present (Beardsmore et al., 1982; ElRayes et al., 1991). Methanol dehydrogenase, and NAD^+-linked formaldehyde and formate dehydrogenases also occur (Beardsmore et al., 1982). A variant of the ribulose monophosphate cycle of formaldehyde fixation occurs that involves cleavage of hexose phosphate by 2-keto-3-deoxy-6-phosphogluconate aldolase and a rearrangement sequence involving transketolase and transaldolase (Beardsmore et al., 1982).

Hydroxypyruvate reductase, isocitrate lyase, malyl-CoA-lyase, and glyoxylate aminotransferase, which are characteristic of the serine assimilation pathway, are absent (ElRayes et al., 1991).

A glucose-6-phosphate dehydrogenase is present that is active with both $NADP^+$ and NAD^+. Two separate 6-phosphogluconate dehydrogenases occur, one active with both $NADP^+$ and NAD^+ and the other active only with NAD^+ (Beardsmore et al., 1982).

Ammonia is assimilated via the glutamate dehydrogenase pathway (ElRayes et al., 1991). Acetamide and acrylamide are hydrolyzed by a cytoplasmic amidase (Silman et al., 1991).

The most positive redox potential ever recorded for a flavin adenine dinucleotide-containing protein has been measured for an electron-transfer flavoprotein (ETF) synthesized by *Methylophilus methylotrophus* (Byron et al., 1989).

Urea is hydrolyzed to ammonia by a cytoplasmic urease that is inducible by urea and short-chain amides and repressed by excess ammonia (Greenwood et al., 1998).

TAXONOMIC COMMENTS

Studies of the type strain by 16S rRNA analysis have placed the genus in the class *Betaproteobacteria* (Tsuji et al., 1990).

List of species of the genus Methylophilus

1. **Methylophilus methylotrophus** Jenkins, Byrom and Jones 1987, 447[VP]

me.thy.lo.tro′phus. M.L. n. *methyl* the methyl radical; Gr. adj. *tropho* pertaining to nutrition; M.L. adj. *methylotrophus* methyl radical-consuming.

The cells are motile by single flagella. Colonies on methanol–mineral-salts agar are grayish-white. In addition to growth on methanol as the sole carbon and energy source, good growth occurs on glucose and may or may not occur on methylamines as sole carbon and energy sources. Different strains give different results with fructose as the sole carbon and energy source. Poor growth, which varies between strains, may occur on lactose, sucrose, D-ribose, D-xylose, ethanol, propanol, butanol, acetate, and formate. Acid is not produced from glucose. Acetoin, tested by the Voges–Proskauer method, may or may not be produced. Tween 20, 40, and 60 are hydrolyzed. Tween 80 is not hydrolyzed. Urease is produced. Leucine arylamidase is produced. Phosphatase production is weak and differs between different strains. Sulfatase is not produced. H_2S is not produced. Gelatin is not liquefied. Extracellular DNase and RNase are not produced. 2,3,5-Triphenyltetrazolium chloride (0.01%) is reduced. No growth occurs in the presence of 0.01% potassium tellurite, or with 5% NaCl. Resistant to penicillin, oleandomycin; sensitive to nalidixic acid, streptomycin, and a number of other antibiotics.

The type strain produces significant amounts of a low-viscosity extracellular polysaccharide from methanol under conditions of nitrogen limitation in chemostat culture.

The mol% G + C of the DNA is: about 50 (T_m), 50.3 (T_m) (NCIB 10515[T]), 49.8 (T_m) (ATCC 31226[T]).

Type strain: AS1, ATCC 53528, DSMZ 46235, NCIB 10515.

GenBank accession number (16S rRNA): L15475.

2. **Methylophilus leisingeri** corrig. Doronina and Trotsenko 2001b, 1[VP] (*Methylophilus leisingerii* (sic)) (Effective publication: Doronina and Trotsenko 1994, 529.

lei.sing′e.ri. M. L. gen. n. *leisingeri* named after the Swiss microbiologist, Thomas Leisinger, who isolated the organism.

Gram-negative rods 0.4–0.6 × 1.1–1.7 μm. Nonmotile (unlike *M. methylotrophus*). A slender polysaccharide capsule is present. Colonies on methanol salt agar are circular, 1–4 mm in diameter, white or pale pink in color, opaque, convex, with even edges and a smooth surface. Temperature range, 10–37°C; optimum, 30–35°C. pH range, 6.5–7.8; optimum, 6.8–7.2. The generation time is 2 h on media with methanol or dichloromethane, 10 h with glucose. No vitamins or other growth factors are required. Cellulose is not digested; gelatin is not liquified; starch is weakly hydrolyzed. Acetoin, indole, hydrogen sulfide, and ammonia are not produced. Oxidase, catalase, and urease positive. Strictly aerobic, although able to reduce nitrates to nitrites. Carbon and energy sources include methanol, dichloro- and dibromomethane, glucose, and galactose; methylated amines are not used. Nitrogen sources include ammonium and nitrate. No growth occurs on media with 3% NaCl.

One-carbon units are assimilated via the 2-keto-3-deoxy-6-phosphogluconate (KDPG) variant of the ribulose monophosphate pathway. The Krebs cycle lacks an α-ketoglutarate dehydrogenase. Glyoxylate shunt enzymes are not present. Ammonium is assimilated via the glutamate cycle; glutamate dehydrogenase is lacking. Cells contain a constitutive methanol dehydrogenase (PQQ) and may produce an inducible dichloromethane dehalogenase and hexokinase. Palmitic ($C_{16:0}$) and palmitoleic ($C_{16:1}$) acids predominate among the cellular fatty acids. Cardiolipin is absent from cellular phospholipids. The major ubiquinone is Q-8. The type strain exhibits a DNA hybridization value of 26% with the type strain of *M. methylotrophus*. The type strain was isolated from wastewater in Switzerland.

The mol% G + C of the DNA is: 50.2 (T_m).

Type strain: VKM B-2013, DM11, DSM 6813.

Genus II. **Methylobacillus** Yordy and Weaver 1977, 254[AL], emend. Urakami and Komagata 1986a, 509

THE EDITORIAL BOARD

Meth.yl.o.ba.cil′lus. Fr. *méthyle* the methyl radical; L. dim. n. *bacillus* a small rod; M.L. masc. n. *Methylobacillus* methyl rodlet.

Rods 0.3–0.6 × 0.8–2.0 μm. **Motile by means of a single polar flagellum or nonmotile.** Gram negative. **Most strains are obligate methylotrophs that grow on one-carbon compounds other than methane;** however, some strains can also use fructose. **Aerobic,** having a strictly respiratory type of metabolism with oxygen as the terminal electron acceptor. The major cellular fatty acids are straight-chain saturated $C_{16:0}$ and unsaturated $C_{16:1}$. The major quinone is Q-8; Q-7 and Q-9 are minor components.

The mol% G + C of the DNA is: 50–56.

Type species: **Methylobacillus glycogenes** Yordy and Weaver 1977, 254; emend. Urakami and Komagata 1986a, 510.

MAINTENANCE PROCEDURES

Cultures of *M. glycogenes* have been maintained on Medium B[1] at 30°C.

---

1. Medium B (Urakami and Komagata, 1986a) contains (per liter of distilled water): $(NH_4)_2SO_4$, 3.0 g; KH_2PO_4, 1.4 g; Na_2HPO_4, 3.0 g; $MgSO_4 \cdot 7H_2O$, 0.2 g; ferric citrate, 0.03 g; $CaCl_2 \cdot 2H_2O$, 0.03 g; $MnCl_2 \cdot 4H_2O$, 0.005 g; $ZnSO_4 \cdot 7H_2O$, 0.005 g; $CuSO_4 \cdot 5H_2O$, 0.0005 g; yeast extract, 0.2 g; vitamin solution, 1.0 ml; methanol, 10.0 ml. The vitamin solution contains (per liter of distilled water): biotin, 2 mg; calcium pantothenate, 400 mg; pyridoxine hydrochloride, 400 mg; *p*-aminobenzoic acid, 200 mg; folic acid, 2 mg; inositol, 2 g; nicotinic acid, 400 mg; riboflavin, 200 mg.

DIFFERENTIATION OF THE GENUS *METHYLOBACILLUS* FROM OTHER GENERA

Table BXII.β.73 lists characteristics differentiating *Methylobacillus* from other nonmethane-utilizing methylotrophs.

TAXONOMIC COMMENTS

In this edition of the *Manual*, the genus *Methylobacillus* is placed in the class *Betaproteobacteria*, the order *Methylophilales*, and the family *Methylophilaceae*. This family includes three genera: *Methylophilus*, *Methylobacillus*, and *Methylovorus*.

List of species of the genus Methylobacillus

1. **Methylobacillus glycogenes** Yordy and Weaver 1977, 254[AL]; emend. Urakami and Komagata 1986a, 510.

   gly.co'gen.es. Gr. adj. *glykus* sweet; Gr. v. *gennairo* to produce; Gr. adj. *glycogenes* sweet-producing, intended to mean sugar-producing, glycogen-producing.

   Gram-negative rods with rounded ends, 0.3–0.5 × 0.8–2.0 μm, occurring singly or rarely in pairs. Motile by means of single polar flagella. Do not form spores. Do not form capsules. Do not accumulate poly-β-hydroxybutyrate. No growth occurs in nutrient or peptone broth. Colonies are shiny, smooth, raised, entire, white to light yellow, 1–3 mm in diameter after 3 days at 30°C on methanol-containing agar. Nitrate reduced to nitrite. Methyl red and Voges–Proskauer negative. Do not produce indole, hydrogen sulfide, or ammonia. Do not hydrolyze gelatin or starch. No denitrification. No acid from D-glucose or D-fructose. Utilize methanol but not methane as the sole carbon source. Obligate methylotroph. Do not utilize L-arabinose, D-xylose, D-glucose, D-mannose, galactose, maltose, sucrose, lactose, trehalose, D-sorbitol, D-mannitol, inositol, glycerol, soluble starch, succinic acid, citric acid, acetic acid, ethanol, or hydrogen. Variable utilization of D-fructose and methylamine. Utilize ammonia, urea, and nitrate as sole nitrogen sources. Urease and oxidase negative. Catalase positive. Growth at 30°C (most strains also grow at 37°C) and pH 6–8. Most strains do not grow in the presence of 3% NaCl.

   The mol% G + C of the DNA is: 50–56; type strain 54 (Bd).

   Type strain: T-11, ATCC 29475, DSM 5685, JCM 2850, NCIB 11375.

   GenBank accession number (16S rRNA): M95652.

2. **Methylobacillus flagellatus** Govorukhina, Kletsova, Tsygankov, Trotsenko and Netrusov 1998, 631[VP] (Effective publication: Govorukhina, Kletsova, Tsygankov, Trotsenko and Netrusov 1987, 676.)

TABLE BXII.β.73. Characteristics differentiating *Methylobacillus* species from each other and from other nonmethane-utilizing methylotrophs[a,b]

Characteristic	*Methylobacillus glycogenes*	*Methylobacillus flagellatus*	*Acidomonas methanolica*	*Methylobacterium*	*Methylophaga*	*Methylophilus methylotrophus*	*Methylovorus glucosotrophus*	*Xanthobacter*
Methane utilization	−	−	−	D	−	−	−	−
Oxidizes ethanol to acetic acid	−	−	+	−	−	−	−	−
Grows only below pH 5.5	−	−	+	−	−	−	−	−
Na+ or seawater required for growth	−	−	nd	−	+	−	−	−
Poly-β-hydroxybutyrate accumulation in cells	−	nd	nd	+	nd	−	nd	nd
Colony pigmentation	white to light yellow	milky	nd	pink to orange-red	pale pink	none	pink, creamy, or milky	yellow
Known to fix N₂	−	−	−	−	−	−	−	+
Urease	nd	−	nd	+	+	+	+	nd
Oxidase	−	+	nd	+ or W	+	+	+	+
Glucose can be used as a carbon and energy source	−	−	nd	D	−	+	+	W
Fructose can be used as a carbon and energy source	d	nd	nd	D	+	v and W	−	+
Starch hydrolysis	−	+	nd	nd	nd	nd	+	nd
Major 1-C pathway:								
Ribulose monophosphate	+	nd	nd	−	+	+	+	nd
Serine	−	nd	nd	+	−	−	−	nd
Ammonia assimilation:								
Glutamate cycle	−	nd	nd	nd	nd	+	+	nd
Glutamate dehydrogenase	+	nd	nd	nd	nd	−	−	nd
Presence of 6-phosphogluconate dehydrogenase (NADP-linked)	+	nd	nd	nd	nd	+	−	nd
Presence of branched C₁₇ fatty acids	−	nd	nd	nd	nd	+	−	nd
Presence of diphosphatidyl glycerol	nd	+	nd	nd	−	+	nd	nd

[a]Symbols: see standard definitions; nd, not determined; w, weak.

[b]Data from and Yordy and Weaver (1977), Urakami and Komagata (1986a), Govorukhina et al. (1987), Jenkins et al. (1987), Govorukhina and Trotsenko (1991), Holt et al. (1994), Padden et al. (1997).

fla.gel.la' tus. L. *flagellum;* L. *flagellatum* for attached flagella, flagellated.

Gram-negative rods, 0.5–0.6 × 1.4–1.8 μm. Motile. Reproduce by division. Do not form spores or capsules. Colonies on agar with methanol are round, translucent, with milky color, smooth, convex, with an even edge and viscous consistency. Obligate methylotroph. Methanol and methylamine are used as carbon and energy sources. Methylamine, ammonia salts, dimethylamine, nitrates, and peptone are used as nitrogen sources. Optimal temperature 42°C; range, 10–52°C. Optimal pH 7.2–7.3; range 5.8–8.4. Generation time is 2 h. Vitamins are not required. No acid or gas is produced from glucose. Urease, oxidase, and catalase positive. Nitrate reduced to nitrite. Hydrolyze starch but not cellulose, gelatin, or Tween 80. Major fatty acids are $C_{16:0}$ and $C_{16:1}$. The ratio of *cis*-vaccenic acid to palmitoleic acid is 0.32.

The mol% G + C of the DNA is: 55.5.

Type strain: KT, ATCC 51484, DSM 6875, VKM B-1610.

GenBank accession number (16S rRNA): M95651.

Genus III. **Methylovorus** Govorukhina and Trotsenko 1991, 161[VP]

THE EDITORIAL BOARD

Me.thyl.o' vo.rus. Fr. *méthyle* the methyl radical; L. v. *voro* to consume; M.L. adj. *methylovorus* methyl-consuming.

When grown on methanol–mineral salts medium, cells are straight or slightly curved rods, usually 0.5–0.6 × 1.0–1.3 μm, occurring singly or in pairs. Gram negative. **Motile** by a single polar flagellum. Endospores are absent. No complex intracellular membranes. No sheaths or prosthecae detected. No capsules formed, but slime may be produced by some strains. Colonies on methanol–mineral salts agar incubated for 2 days at 30°C are circular, 1–2 mm in diameter, with entire edge, convex, and translucent to opaque; may be pink, creamy, or milky in color. Pyocyanin and fluorescein not produced. Cells multiply by binary fission. No aggregation or pigmentation occurs in liquid medium. No growth or extremely poor growth occurs on nutrient agar and in nutrient broth at 30–37°C; no growth occurs under an atmosphere of CH_4/O_2 or $H_2/CO_2/O_2$. Optimal pH for growth, 7.0–7.2; temperature, 35–37°C. **Aerobic,** having a strictly respiratory type of metabolism with oxygen as the terminal electron acceptor. **Only methanol and glucose are used as carbon and energy sources.** In addition, some strains are able to grow slowly on methylated amines and betaine. Nitrates, ammonium salts, methylated amines, glutamate, and peptone serve as nitrogen sources. Acetoin, indole, H_2S, and NH_3 are not produced in test medium. Milk is not hydrolyzed. **Oxidase positive.** Urease and catalase positive. Peroxidase variable. Arginine dihydrolase negative. Strains may hydrolyze starch, but not cellulose, gelatin, or Tween 80. Acid, but not gas, is produced from glucose. Isolated from activated sludge, mud, soil, and pond water.

The mol% G + C of the DNA is: 56–57.

Type species: **Methylovorus glucosotrophus** Govorukhina and Trotsenko 1991, 162.

FURTHER DESCRIPTIVE INFORMATION

All strains assimilate methanol carbon through the ribulose monophosphate (RuMP) pathway and ammonia via the glutamate cycle (glutamate synthase and glutamine synthetase). Neither α-ketoglutarate dehydrogenase nor glyoxylate shunt enzymes are present. The fatty acid composition is primarily of the nonhydroxylated straight-chain saturated and monounsaturated types with $C_{16:0}$ and $C_{16:1\ \omega7}$. Branched C_{17} fatty acids are absent. The major phospholipids are phosphatidylethanolamine and phosphatidylglycerol. The strains also possess diphosphatidylglycerol.

DIFFERENTIATION OF THE GENUS *METHYLOVORUS* FROM OTHER GENERA

Table BXII.β.74 indicates some features that differentiate *Methylovorus* from *Methylophilus* and *Methylobacillus*.

TAXONOMIC COMMENTS

Results of DNA–DNA hybridization and 5S rRNA sequencing studies on some strains of obligately and restricted facultatively methylotrophic bacteria support the establishment of *Methylovorus* as a genus separate from other methylotrophs (Wolfrum and Stolp, 1987; Bulygina et al., 1993). The genus is closely related to *Methylobacillus* and *Methylophilus* and belongs to the class *Betaproteobacteria.*

TABLE BXII.β.74. Characteristics differentiating *Methylovorus* from related genera[a,b]

Characteristic	*Methylovorus*	*Methylophilus*	*Methylobacillus*
Presence of:			
6-Phosphogluconate dehydrogenase (NADP-linked)	−	+	+
Branched C_{17} fatty acids	−	+	−
Diphosphatidyl glycerol	+	−	+
Assimilation of NH_3 occurs by:			
Glutamate cycle (glutamate synthase and glutamine synthetase)	+	+	−
Glutamate dehydrogenase	−	−	+
Mol% G + C of the DNA	56–57	50–53	50–56

[a]For symbols see standard definitions.

[b]Data from Govorukhina and Trotsenko (1991).

List of species of the genus Methylovorus

1. **Methylovorus glucosotrophus** Govorukhina and Trotsenko 1991, 162[VP]

    glu.co.so.tro' phus. M.L. n. *glucosum* glucose; Gr. adj. *tropho* pertaining to nutrition; M.L. adj. *glucosotrophus* glucose-consuming.

    The characteristics are as described for the genus. In addition, generation times when grown on methanol and glucose are 2 and 17 h, respectively. Vitamins and co-factors are not required.

    The mol% G + C of the DNA is: 56–57 (T_m).

    Type strain: 6B1, DSM 6874, UCM B-1475.

Order IV. **Neisseriales** *ord. nov.*

TONE TØNJUM

Neis.se.ri.a' les. M.L. fem. n. *Neisseriaceae* type family of the order; *-ales* ending to denote an order; M.L. fem. pl. n. *Neisseriales* the *Neisseriaceae* family.

Organisms are coccal, coccoid, or distinctly rod-shaped, occurring singly, in pairs, in masses, or in short chains. Cells of *Aquaspirillum* are helical, while *Simonsiella* and *Alysiella* may exhibit a characteristic multicellular micromorphology. Endospores are not formed. Gram-negative, but there may be a tendency to resist decolorization. Flagella and swimming motility are absent except in the genus *Aquaspirillum*. The cells are nonmotile in liquid media but surface-bound motility ("twitching motility") is frequently observed. Fimbriae (pili) are often present.

All species grow aerobically. Strains of all recognized species usually have an optimal growth temperature of approximately 32–36°C. Capsules may be present. Colonies of most genera are not pigmented, with the exception of strains of *Vogesella* and *Chromobacterium*. Several species have complex growth factor requirements, while some species grow readily in simple defined media containing a single organic carbon and energy source.

The mol% G + C of the DNA is: 46–67.

Type genus: **Neisseria** Trevisan 1885, 105.

TAXONOMIC COMMENTS

The order *Neisseriales* comprises a major branch of the *Betaproteobacteria* with the single family *Neisseriaceae* as the type family. *Neisseriaceae* consists of the type genus *Neisseria* as well as the heterogeneous genera (in alphabetical order) *Alysiella, Aquaspirillum, Catenococcus, Chromobacterium, Eikenella, Formivibrio, Iodobacter, Kingella, Microvirgula, Prolinoborus, Simonsiella, Vitreoscilla,* and *Vogesella*. In order of decreasing relatedness to the type genus *Neisseria* the genera included in *Neisseriaceae* include *Kingella, Eikenella, Alysiella, Simonsiella* (Rossau et al., 1986, 1989; Dewhirst et al., 1989), *Microvirgula* (Patureau et al., 1998), *"Laribacter"* (Yuen et al., 2001), *Vogesella* (Grimes et al., 1997), *Vitreoscilla* (Joshi et al., 1998), *Chromobacterium* (Dewhirst et al., 1989), *Aquaspirillum* (Rossau et al., 1989; Wen et al., 1999), *Prolinoborus* (Pot et al., 1992b), *Formivibrio* (Hippe et al., 1999), and *Iodobacter* (Fig. BXII.β.61). Other entities that, based on 16S rRNA phylogenetic analysis, show the closest relatedness to *Neisseriales* are *Taylorella* spp. encompassing *Taylorella equigenitalis* and *Taylorella asinigenitalis* (Bleumink-Pluym et al., 1993; Jang et al., 2001) and *Pelistega europaea* (Vandamme et al., 1998), through their relationship to *Iodobacter fluviatilis*; these genera appear closer to the *Neisseriaceae* type genus *Neisseria* than does the genus *Catenococcus*. However, the genera *Taylorella* and *Pelistega* are currently placed in the new family *Alcaligenaceae* in the order *Burkholderiales*.

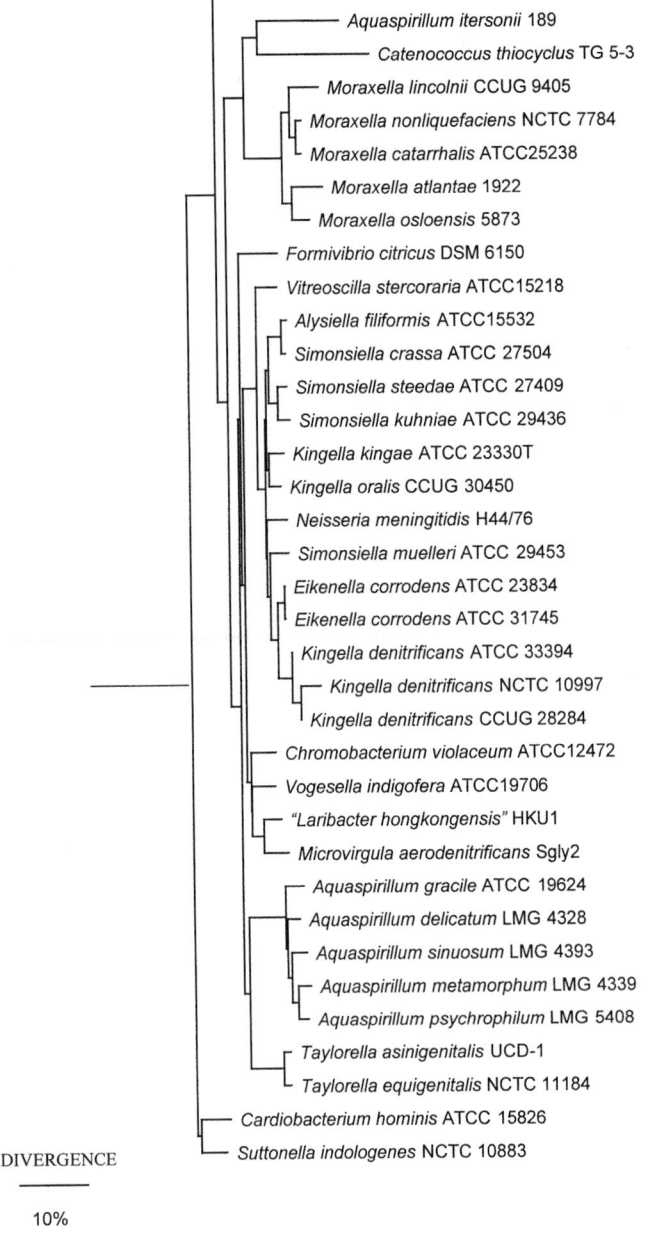

DIVERGENCE

10%

FIGURE BXII.β.61. Phylogenetic neighbor-joining tree of the family *Neisseriaceae* based on 1355 nucleotide positions of the 16S rRNA genes. *Oligella urethralis* CIP 103 116 was used as an outgroup (K.K. Garborg, S.A. Frye, B. Pettersson and T. Tønjum, unpublished studies).

Family I. **Neisseriaceae** Prèvot 1933, 119[AL] emend. Dewhirst, Paster and Bright 1989, 265

TONE TØNJUM

Neis.se.ri.a' ce.ae. M.L. fem. n. *Neisseria* type genus of the family; *-aceae* ending to denote a family; M.L. fem. pl. n. *Neisseriaceae* the *Neisseria* family.

Organisms are coccal, occurring singly, in pairs, or in masses, often with adjacent sides flattened (different planes of division), or distinctly rod-shaped or coccoid (one plane of division), frequently occurring in pairs or short chains. Cells of *Aquaspirillum* are helical, while *Simonsiella* and *Alysiella* may exhibit a characteristic multicellular micromorphology. Endospores are not formed. Gram negative, but there may be a tendency to resist decolorization. Flagella and swimming motility are absent. The cells are nonmotile in liquid media but surface-bound motility ("twitching motility") is frequently observed. Fimbriae (pili) are often present.

All species grow aerobically. Strains of all recognized species usually have an optimal growth temperature of approximately 32–36°C. Capsules may be present. Colonies are not pigmented. Several species have complex growth factor requirements, while some species grow readily in simple defined media containing a single organic carbon and energy source.

All strains except those of *Formivibrio* are oxidase positive. Usually catalase positive. Indole is not produced. True waxes are not present. Fatty acid profiles show the presence of mainly unbranched saturated and mono- or di-unsaturated fatty acids composed of 16 and 18 carbon atoms. Most strains reside indigenously on mucosal membranes of humans and animals, although environmental species were recently included in this family.

The mol% G + C of the DNA is: 46–67.

Type genus: **Neisseria** Trevisan 1885, 105.

FURTHER DESCRIPTIVE INFORMATION

Since the 1980s, mainly as a result of the application of new molecular techniques, major changes have occurred within the taxonomy and classification of the family *Neisseriaceae*. This family comprises a major branch of the *Betaproteobacteria*. Studies with DNA–DNA and DNA–rRNA hybridizations, and subsequently 16S rRNA sequence comparisons, have redefined the relationships among species belonging to the *Neisseriaceae* (Dewhirst et al., 1989; Rossau et al., 1989; Enright et al., 1994; Tønjum et al., 1995a; Harmsen et al., 2001). Findings from DNA-mediated transformation to streptomycin resistance, DNA base composition data, DNA–DNA and DNA–rRNA hybridization, and rRNA sequencing have revealed differences that mandate the separation of the genera *Moraxella* and *Acinetobacter* (belonging to the *Gammaproteobacteria*) from *Neisseriaceae* (Stackebrandt et al., 1988b; Dewhirst et al., 1989). In addition, using transformation analysis, the four coccal species of the "false neisseriae" (previously "*N. catarrhalis*", *N. caviae*, *N. ovis*, and *N. cuniculi*) have been transferred from *Neisseria* to the *Gammaproteobacteria* (Catlin, 1970; Bøvre, 1979). The moraxellae, including the "false neisseriae", psychrobacters, and acinetobacters, were grouped to form the new family *Moraxellaceae* (Catlin, 1991; Rossau et al., 1991; Pettersson et al., 1998b), leaving the genera *Neisseria*, *Kingella*, and *Simonsiella*, the genospecies *Eikenella corrodens* and *Alysiella filiformis*, and the CDC groups EF-4 and M-5 as members of *Neisseriaceae* (Fig. BXII.β.61; Dewhirst et al., 1989; Rossau et al., 1989). Cur-

rently, *Aquaspirillum*, *Catenococcus*, *Chromobacterium*, *Formivibrio*, *Iodobacter*, *Microvirgula*, *Prolinoborus*, *Vitreoscilla*, and *Vogesella* are also assigned to *Neisseriaceae*. Table BXII.β.75 lists some differential characteristics of the genera of the family.

Many species of *Neisseriaceae* contain strains that are highly competent for genetic transformation (see Table BXII.β.77 in the *Neisseria* chapter). In addition to representing the most important form of genetic exchange causing genomic heterogeneity (Spratt et al., 1992), transformation also provides a tool for definite identification of numerous, although not all, species of the *Neisseriaceae*, and also forms an important basis for allocation of species within the genera of this family (Bøvre, 1980; Bøvre and Hagen, 1981; Tønjum et al., 1995b).

Genetic taxonomy in *Neisseriaceae* was pioneered by Catlin and Cunningham in 1961 when they demonstrated close relationships between most *Neisseria* species (except the species then misnamed "*N. catarrhalis*") and by Bøvre and Henriksen in 1962. That a genus may contain both coccal and rod-shaped species was shown for genus *Neisseria* by the detection of the rod-shaped organism known as *Neisseria elongata* (Bøvre and Holten, 1970). More recently, another rod-shaped organism, *N. weaveri*, has been included in the genus *Neisseria* (Andersen et al., 1993; Holmes et al., 1993). The genus *Kingella* was proposed by Henriksen and Bøvre in 1976 for the organism known previously as *Moraxella kingae* (now *Kingella kingae*), and was subsequently enlarged by the proposed inclusion of two more species, *K. denitrificans* and *K. indologenes* (Snell and Lapage, 1976). The genus *Kingella* now consists of the species *K. kingae*, *K. denitrificans*, and *K. oralis* (Dewhirst et al., 1993), while *K. indologenes* was transferred to the family *Cardiobacteriaceae* as *Suttonella indologenes* (Dewhirst et al., 1990). A slight genetic affinity between *K. kingae* and *N. elongata* has been demonstrated by genetic transformation and DNA–DNA hybridization (Bøvre and Holten, 1962; Tønjum et al., 1989).

The cellular lipid composition of *Neisseriaceae* is characterized by the absence of branched fatty acids. The quantitative and qualitative distributions of straight-chain fatty acids, including hydroxy acids, generally distinguish the genera as genetically circumscribed and can also be important in differentiation among several species (Jantzen et al., 1974, 1975; Bøvre et al., 1976). True waxes, i.e., simple esters of fatty alcohols and fatty acids, are absent in the genera *Neisseria* and *Kingella*, while activities corresponding to the enzymes thymidine phosphorylase, nucleoside deoxyribosyltransferase, and thymidine kinase have been shown to be present in the genus *Kingella*, but absent in the genus *Neisseria* (Jyssum and Bøvre, 1974). Carbonic anhydrase has been detected in all *Neisseria* species (Berger and Issi, 1971).

TAXONOMIC COMMENTS

The taxonomic placement of *Neisseriaceae* corresponds to rRNA superfamily III according to DNA–rRNA hybridizations performed by Rossau et al. (1986, 1989). Based on 16S rDNA phylogeny, *Neisseriaceae* consists of the type genus *Neisseria*, as well as the genera (in alphabetical order) *Alysiella*, *Aquaspirillum*, *Caten-*

TABLE BXII.β.75. Differential characteristics of the genera of the family *Neisseriaceae*[a]

Characteristic	*Neisseria*	*Alysiella*	*Aquaspirillum*	*Chromobacterium*	*Eikenella*	*Formivibrio*	*Iodobacter*	*Kingella*	*"Laribacter"*	*Microvirgula*	*Simonsiella*	*Vitreoscilla*	*Vogesella*
Cell morphology:													
Cocci	+	−	−	−	−			−	−	−	−	−	−
Rods	+[b]	+	+	+	+	+	+	+	+	+	+	+	+
Oxidase test	+	+	+	+	+		+	+[c]	+	+	+	+	+
Catalase test	+	+		+	−	−		−		+	+	+	+
Acid from glucose	[+]	+	D	+	−		−	+	−	−	[+]		−
Nitrite reduction	+	−		+	+			+	+		[+]		+
Mol% G + C of DNA	46.5–58.0	44–48	56–62	65–68	56–58	59	50–52	47–55		68	40–52		65–69

[a]Symbols: +, positive for the majority of strains and some strains of each species; [+], positive for all strains of the majority of species (only one species uniformly negative); D, positive and negative species (or strains of *Acinetobacter*) about equally represented; −, all strains negative.

[b]Only two species, *N. elongata* and *N. weaveri*, consist of rods.

[c]*K. denitrificans* may be negative or weakly positive with the least sensitive test reagent, dimethyl-*p*-phenylenediamine, whereas it is distinctly positive when tetramethyl-*p*-phenylenediamine is used. For other organisms of this table (including *K. kingae*) the result is the same with either reagent.

ococcus, Chromobacterium, Eikenella, Formivibrio, Iodobacter, Microvirgula, Prolinoborus, Simonsiella, Vitreoscilla, and *Vogesella.* In order of decreasing relatedness to the type genus *Neisseria* the genera included in *Neisseriaceae* encompass *Kingella, Eikenella, Alysiella, Simonsiella* (Rossau et al., 1986, 1989; Dewhirst et al., 1989), *Microvirgula* (Patureau et al., 1998), *"Laribacter"* (Yuen et al., 2001), *Vogesella* (Grimes et al., 1997), *Vitreoscilla* (Joshi et al., 1998), *Chromobacterium* (Dewhirst et al., 1989; Martin and Brimacombe, 1992; Duran and Menck, 2001), *Aquaspirillum* (Rossau et al., 1989; Wen et al., 1999), *Prolinoborus* (Pot et al., 1992b), *Formivibrio* (Hippe et al., 1999), and *Iodobacter* (Fig. BXII.β.61). Other entities that, based on 16S rRNA phylogenetic analysis, show the closest relatedness to *Neisseriales* are *Taylorella* spp., encompassing *Taylorella equigenitalis* and *Taylorella asinigenitalis* (Bleumink-Pluym et al., 1993; Jang et al., 2001), and *Pelistega europaea* (Vandamme et al., 1998) through their relationship to *Iodobacter fluviatilis*. These genera appear closer to the genus *Neisseria* than does the genus *Catenococcus*. The phylogenetic maps obtained by 16S rDNA sequence analysis show that species within *Neisseriaceae* are polyphyletic. This is due to the high frequency of horizontal gene transfer by transformation combined with frequently occurring recombination events and a relatively high spontaneous mutation rate (Spratt et al., 1992). These factors contribute to the limitations of the use of 16S rDNA sequences as the basis for taxonomy in this bacterial family.

Microvirgula aerodenitrificans (Patureau et al., 1998) are denitrifying and very motile curved rods isolated from activated sludge. Although phenotypically similar to *Comamonas testosteroni*, phylogenetic analysis based on the 16S rRNA sequence shows that *Microvirgula aerodenitrificans* is most closely related to the genus *Vogesella*. Another recent addition to the genera in *Neisseriaceae* is the genus *"Laribacter"*, represented by *"Laribacter hongkongensis"* (Yuen et al., 2001). This bacterium was isolated from the blood and empyema of a cirrhotic patient. The cells are seagull shaped or spiral rods. Phylogeny based on 16S rDNA sequence analysis shows that *"L. hongkongensis"* is most closely related to *Microvirgula aerodenitrificans, Vogesella indigofera,* and *Chromobacterium* species, respectively.

Vogesella indigofera (Grimes et al., 1997) has blue-pigmented colonies with a metallic copper-colored sheen. These blue-pig-

mented bacteria were isolated from freshwater samples, and numerical analysis of morphological and biochemical characteristics has revealed 90.0% relatedness between *Vogesella* spp. and *Burkholderia cepacia* and *Janthinobacterium lividum*. A phylogenetic analysis in which both 5S rRNA and 16S rRNA were used also revealed that the *Vogesella* strains were closely related to *Chromobacterium violaceum*, but sufficiently distinct to warrant placement in a separate genus, *Vogesella*, named in honor of Otto Voges, who first isolated and described this blue-pigmented eubacterium in 1893.

Selected strains/biovars currently classified as *Aquaspirillum* spp., *Aquaspirillum gracile, Aquaspirillum delicatum, Aquaspirillum sinuosum, Aquaspirillum metamorphum,* and *Aquaspirillum psychrophilum* also belong to the *Neisseriaceae* (Wen et al., 1999), together with *Prolinoborus* (Pot et al., 1992b). *Iodobacter fluviatilis* (previously *Chromobacterium fluviatile*) (Wen et al., 1999) and *Catenococcus thiocycli* (partial 16S rDNA sequence submitted by Tourova, T.P. and Kuznetsov, B.B., Moscow, Russia) are the entities most distantly related to the type genus *Neisseria* (Fig. BXII.β.61).

ACKNOWLEDGMENTS

Stimulating discussions with Kjell Bøvre and Bertil Pettersson are deeply appreciated. The information provided in the first version of *Bergey's Manual of Systematic Bacteriology* (1984) by Kjell Bøvre is gratefully acknowledged.

FURTHER READING

Bøvre, K. 1980. Progress in classification and identification of *Neisseriaceae* based on genetic affinity. *In* Goodfellow and Board (Editors), Microbiological Classification and Identification, Academic Press Inc. (London), Ltd., London. pp. 55–72.

Bøvre, K. and N. Hagen. 1981. The family *Neisseriaceae*: rod-shaped species of the genera *Moraxella, Acinetobacter, Kingella,* and *Neisseria,* and the *Branhamella* group of cocci. *In* Starr, Stolp, Trüper, Balows and Schlegel (Editors), The Prokaryotes. A Handbook on Habitats, Isolation, and Identification of Bacteria, Vol. 2, Springer-Verlag, Berlin. pp. 1506–1529.

Grimes, D.J., C.R. Woese, M.T. MacDonell and R.R. Colwell. 1997. Systematic study of the genus *Vogesella* gen. nov. and its type species, *Vogesella indigofera* comb. nov. Int. J. Syst. Bacteriol. *47*: 19–27.

Tønjum, T., K. Bøvre and E. Juni. 1995. Fastidious Gram-negative bacteria: Meeting the diagnostic challenge with nucleic acid analysis. APMIS *103*: 609–627.

Genus I. **Neisseria** *Trevisan 1885, 105*[AL]

Tone Tønjum

Neis.se' ri.a. M.L. fem. n. *Neisseria* named after Albert Neisser, who discovered the etiological agent of gonorrhea in the pus of patients in 1889.

Cocci 0.6–1.9 µm in diameter, occurring singly but often in pairs with adjacent sides flattened; two species (*Neisseria elongata* and *N. weaveri*) are exceptions and consist of short rods 0.5 µm wide, often arranged as diplobacilli or in chains. Division of the coccal species is in two planes at right angles to each other, sometimes resulting in tetrads. Capsules and fimbriae (pili) may be present. Endospores are not present. Gram negative, but there is a tendency to resist decolorization. Swimming motility does not occur and flagella are absent. Aerobic. Some species produce a yellow carotenoid pigment. Some species are nutritionally fastidious and hemolytic. Optimal temperature, 35–37°C. Oxidase positive. Catalase positive except most strains of *N. elongata*. Carbonic anhydrase is produced by all species. All species reduce nitrite except *N. gonorrhoeae* and *N. canis. Neisseria* spp. produce acid from carbohydrates by oxidation, not fermentation. Chemoorganotrophic. Exotoxins are not produced. Some species are saccharolytic. Inhabitants of the mucous membranes of mammals. Some species are primary pathogens of humans. The genus *Neisseria* belongs to the family *Neisseriaceae* of the *Betaproteobacteria*.

The mol% G + C of the DNA is: 48–56.

Type species: **Neisseria gonorrhoeae** (Zopf 1885) Trevisan 1885, 106 (*Merismopedia gonorrhoeae* Zopf 1885, 54.)

Further descriptive information

Cells of most *Neisseria* spp. are cocci that divide in two planes at right angles to each other, often resulting in the formation of diplococci and tetrads. However, cells of the species *N. elongata* and *N. weaveri* are short, slender rods (Bøvre and Hagen, 1981; Andersen et al., 1993). *Neisseria* spp. produce oxidase and catalase, carbonic anhydrase, and nitrite reductase, but do not produce thymidine phosphorylase, nucleoside deoxyribosyl transferase, or thymidine kinase. In contrast to *Moraxella* spp., the cell walls of *Neisseria* spp. do not contain true waxes. Summaries of the occurrence and isolation (Bøvre and Hagen, 1981), phylogeny and taxonomy (Bøvre and Hagen, 1981), biochemistry, physiology, and identification (Morse and Knapp, 1989; Knapp and Koumanis, 1999), and clinical significance (Cartwright, 1995; Knapp and Koumanis, 1999) of neisseriae are available.

Phylogeny and classification The genus *Neisseria* contains 17 species and biovars that may be isolated from humans including six species that may be isolated from animals (Table BXII.β.76). The taxonomy of the species of human origin has undergone many changes; these and the names by which individual species have been known have been summarized (Rossau et al., 1989; Tønjum et al., 1995b; Knapp and Koumanis, 1999). *Neisseria* spp. may be identified by phenotypic characteristics including their patterns of acid production from carbohydrates, production of polysaccharide from sucrose, and reduction of nitrate (Table BXII.β.76), as well as 16S rDNA sequence analysis (Fig. BXII.β.62).

Based on conventional characterization and 16S rDNA sequence analysis, *Neisseria* spp. can be divided into three major groups. The first group includes the closely related species *N. gonorrhoeae, N. meningitidis, N. lactamica, N. cinerea,* and *N. polysaccharea.* Strains of these species grow as nonpigmented, translucent colonies. Members of the second group, which consists of *N. subflava* (including biovar flava, biovar perflava, and biovar

subflava), *N. sicca,* and *N. mucosa,* referred to as the saccharolytic species, usually grow as opaque, yellow-pigmented colonies on solid media. Some strains of *N. subflava* biovar perflava have translucent colonies. Although asaccharolytic, *N. flavescens* grows as yellow, opaque colonies and appears to be a member of this group (Kingsbury, 1967; Rossau et al., 1989). The third group contains *N. elongata* and the animal species *N. weaveri, N. denitrificans, N. animalis, N. canis, N. macacae, N. iguanae,* and *N. dentiae.* However, based on 16S rDNA sequence analysis, *N. denitrificans* is just as related to *Simonsiella muelleri* and *Kingella denitrificans* as to *Neisseria* spp., and *N. canis* and the other animal neisserial species are related to *K. kingae* and *K. oralis.* The genetic basis for interrelatedness of strains of *Neisseria* and other genera of the *Neisseriaceae* is addressed below.

Cell morphology Although the genus *Neisseria* has traditionally consisted only of cocci (Fig. BXII.β.63), two rod-shaped species, *N. elongata* and *N. weaveri,* are currently included in the genus.

Cell wall composition Neisserial cell surface variation serves as an adaptive mechanism that can modulate tissue tropism, immune evasion, and survival in the changing host environment.

Outer membrane The outer membrane of the neisseriae is typical for a Gram-negative bacterium and consists of phospholipids, several proteins, and LPS. The gonococcal outer membrane is more permeable to hydrophobic compounds such as fatty acids, detergents, and certain antibiotics than the membranes of other Gram-negative bacteria (Miller et al., 1977). This may be due in part to the presence of phospholipids in the outer leaflet of the outer membrane (Lysko and Morse, 1981).

Outer membrane proteins *N. gonorrhoeae* and *N. meningitidis* possess several outer membrane proteins that have been studied at the molecular level and found to be analogous to one another. The nomenclature of these proteins has been devised to reflect the genetic relatedness of these organisms (Hitchcock, 1989).

Porins The gonococcal (protein I, PI) and meningococcal (class 1, 2, and 3 proteins) outer membranes contain trimeric proteins that form hydrophilic pores (porins) termed Por (Newhall et al., 1980a; Douglas et al., 1981; Derrick et al., 1999; Müller et al., 1999). The porin function is necessary for the survival of the bacterium, allowing nutrients and other solutes to enter the cell. The neisserial porin proteins are important as serotyping antigens, putative vaccine components, and for their proposed role in the intracellular colonization of humans. Gonococci express one of two classes of porins, termed PIA and PIB or PorA and PorB (Sandstrøm et al., 1982), which are encoded by alleles of the same gene (Gotschlich et al., 1987b; Carbonetti et al., 1988). Antigenic heterogeneity in the PIA and PIB molecules provides the basis for gonococcal serotyping. In contrast to gonococci, many meningococcal strains express two porin molecules, one encoded by *porA* (designated the PorA or class I protein) and a second encoded by *porB* (designated as the PorB or class II/III). PorB is the meningococcal counterpart of gonococcal porin with class II corresponding to PIA and class III corresponding to PIB, while PorA/class I expression is unique to meningococci. Evidence of interspecies recombination of *porB*

TABLE BXII.β.76. Differential characteristics of the species of the genus *Neisseria*[a]

Characteristic	*N. gonorrhoeae*	*N. animalis*	*N. canis*	*N. cinerea*	*N. denitrificans*	*N. dentiae*	*N. elongata*	*N. flavescens*	*N. iguanae*	*N. lactamica*	*N. macacae*	*N. meningitidis*	*N. mucosa*	*N. polysaccharea*	*N. sicca*	*N. subflava*	*N. weaveri*
Shape of cells:																	
Cocci	+	+	+	+	+	+	−	+	+	+	+	+	+	+	+	+	−
Rods[b]	−	−	−	−	−	−	+	−	−	−	−	−	−	−	−	−	+
Arrangement of cells:																	
Pairs	+	+	+	+	+	+	+	+	+	+	+	+	+	+	+	+	+
Tetrads	−	+	−	−	−	−	−	−	+	−	−	−	−	−	−	−	−
Short chains	−	−	−	−	−	−	+	−	−	−	−	−	−	−	−	−	+
Yellowish pigment	−	−	+	D	D	(+) 71	weak	+	+	+	+	−	D	+	D	+	−
Hemolysis on blood agar:																	
Sheep	−	−	−	−	−	−	−	−	+	−	−	−	−	−	D	−	+
Horse	−	−	d	−	−	−	−	−	+	D	+	−	−	−	D	−	−
Rabbit	−	−	−	−	−	−	−	−	−	−	+	−	−	−	D	−	−
Human	−	−	−	−	−	−	−	−	−	−	−	−	−	−	D	−	−
Acid produced from:																	
Glucose	+	−	−	−	+	+	D	−	+w	+	+	+	+	+	+	+	−
Maltose	−	−	−	−	+	−	−	−	−	+	+	+	+	+	+	+	−
Fructose	−	−	−	−	+	+	−	−	−	−	+	−	+	−	+	D	−
Sucrose	−	−	−	−	+	+	−	−	−w	−	+	−	+	−	+	D	−
Mannose	−	−	−	−	+	−	−	−	−	−	+	−	+	−	−	−	−
Lactose	−	−	−	−	+	−	−	−	+	+	−	−	−	−	−	−	−
Nitrate reduction	−	−	+	−	+	−	+	−	+	−	−	−	+	−	−	−	−
Nitrite reduction	−[c]	−	−[c]	+	+	+	+	+	+	+	+	−[c]	+	+	+	+	−
Gas from nitrite	−	−	−	−	+	−	−	−	−	−	+	−	+	−	−	−	−
Synthesis of polysaccharide (iodine test)	−	−	−	−	+	+	+	+	−	−	+	−	+	+	+	+	+
Tributyrin hydrolysis	−	−	−	−	+	ND	−	−	−	−	−	−	−	−	−	D[d]	−
Mol% G + C of DNA	50–53	ND	50	49–51	56	ND	53	46–50	51–52	52	50–51	50–52	50–52	53–56	49–52	48–51	51–52

[a] Symbols: +, positive for the majority of strains of each species; (+), positive for all strains of the majority of species, only one strain known to be negative; D, positive and negative strains about equally represented; −, negative for the majority of strains for each species; −w, reported as both negative and weakly positive for most strains of each species; +w, reported as both positive and weakly positive for most strains of each species; ND, not determined. Data compiled from: Hollis et al. (1969), Reyn (1974), Riou (1977), Catlin (1978), Morello and Bonhoff (1980), Bøvre and Hagen (1981), Hoke and Vedros (1982a), and Andersen et al. (1993).

[b] Only two species, *N. elongata* and *N. weaveri*, consist of rods.

[c] Nitrite in low concentrations can be reduced by *N. gonorrhoeae* and by serogroups A, D, and Y of *N. meningitidis* (Berger, 1970).

[d] *N. subflava* biovar perflava is positive; the reaction differs among strains of *N. subflava* biovar subflava.

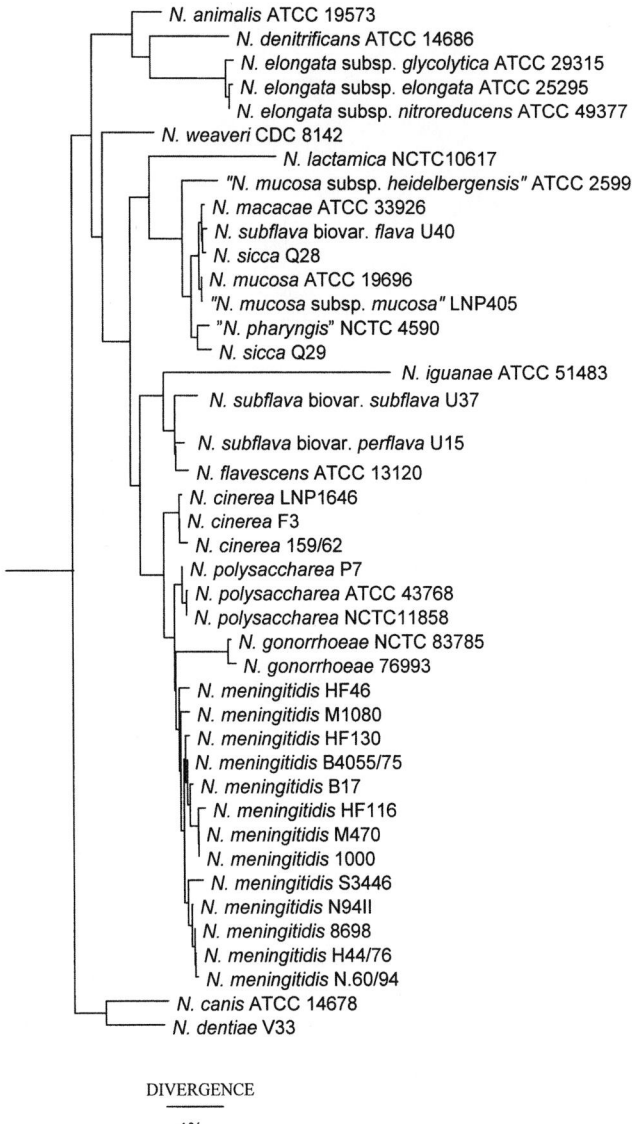

N. animalis ATCC 19573
N. denitrificans ATCC 14686
N. elongata subsp. glycolytica ATCC 29315
N. elongata subsp. elongata ATCC 25295
N. elongata subsp. nitroreducens ATCC 49377
N. weaveri CDC 8142
N. lactamica NCTC10617
"N. mucosa subsp. heidelbergensis" ATCC 25999
N. macacae ATCC 33926
N. subflava biovar. flava U40
N. sicca Q28
N. mucosa ATCC 19696
"N. mucosa subsp. mucosa" LNP405
"N. pharyngis" NCTC 4590
N. sicca Q29
N. iguanae ATCC 51483
N. subflava biovar. subflava U37
N. subflava biovar. perflava U15
N. flavescens ATCC 13120
N. cinerea LNP1646
N. cinerea F3
N. cinerea 159/62
N. polysaccharea P7
N. polysaccharea ATCC 43768
N. polysaccharea NCTC11858
N. gonorrhoeae NCTC 83785
N. gonorrhoeae 76993
N. meningitidis HF46
N. meningitidis M1080
N. meningitidis HF130
N. meningitidis B4055/75
N. meningitidis B17
N. meningitidis HF116
N. meningitidis M470
N. meningitidis 1000
N. meningitidis S3446
N. meningitidis N94II
N. meningitidis 8698
N. meningitidis H44/76
N. meningitidis N.60/94
N. canis ATCC 14678
N. dentiae V33

DIVERGENCE

1%

FIGURE BXII.β.62. Neighbor-joining tree of the genus *Neisseria* based on 1355 nucleotide positions of the 16S rRNA genes. *Moraxella nonliquefaciens* NCTC 7784 was used as an outgroup (Reproduced with permission from K.K. Garborg., S.A. Frye, B. Pettersson and T. Tønjum, unpublished studies).

has been documented in the isolation of a meningococcal stain whose *porB* allele was virtually identical to that of gonococcal allelle (Vazquez et al., 1995). Although gonococci fail to express PorA/class I protein, they carry a complete but defective copy of the *porA* gene (Feavers and Maiden, 1998). The PorB of *N. gonorrhoeae* is able to translocate from the outer bacterial membrane into host cell membranes where it modulates the infection process (Müller et al., 1999). Most evidence suggests that meningococcal PorB/class II/III proteins behave analogously in function and structure to the gonococcal counterpart, while PorA/class I is structurally distinct but with overlapping function. For example, meningococcal strains lacking PorA/class I protein can be constructed as long as they express PorB/class II/III protein and vice versa (Tommassen et al., 1990). In the case of both porins, immunization has been demonstrated to be capable of engendering bactericidal (van der Voort et al., 1996) and opsonic

antibodies (Lehmann et al., 1999). PorA/class I protein appears to be an important component for noncapsular meningococcal vaccines, but like PorB/class II/III is polymorphic due to horizontal gene exchanges (Bart et al., 1999). Moreover, the efficacy of vaccines composed solely of PorA/class I protein may be compromised because strains failing to express the PorA/class I molecule have been shown to arise due to 1) phase variation owing to slipped strand mispairing within its promoter region (van der Ende et al., 1995), 2) complete deletions of the gene (van der Ende et al., 1999), and 3) gene inactivation via insertion of IS1301 (Newcombe et al., 1998). Phase variation appears to account for the absence of PorA expression in most strains (Bart et al., 1999).

Several studies have demonstrated the association of particular PI serotypes with either disseminated or localized gonococcal infections (Cannon et al., 1983; Sandstrøm et al., 1984; Brunham et al., 1985). Strains of *N. gonorrhoeae* isolated from patients with disseminated disease (PIA strains), and strains of *N. meningitidis* were more efficient at porin insertion than strains isolated from a mucosal site (PIB strains). PIA has also been shown to be specifically associated with increased bacterial invasiveness for human cells, which may explain its association with invasive disease (van Putten et al., 1998). Treatment with PI also decreased the ability of human neutrophils to exocytose granules and affected other cellular functions (Haines et al., 1988). PI has been considered as a potential vaccine candidate against gonococcal disease. Antibodies against PI are effective in both bactericidal and opsonic assays (Virji et al., 1986, 1987), and infection with one PI serotype appears to provide protection against subsequent infection with the same serotype (Buchanan et al., 1980). However, even the use of a meningocccal hexavalent PorA vaccine candidate in a clinical trial did not result in adequate protection (Rouppe van der Voort et al., 2000).

OPACITY-ASSOCIATED (OPA) PROTEINS Both *N. gonorrhoeae* and *N. meningitidis* express a family of closely related but size-variable surface-exposed outer membrane proteins termed opacity-associated or Opa proteins. The gonococcal proteins have been named protein IIs while those from meningococci are referred to as class 5 proteins. A similar protein has been identified in *N. lactamica* and *N. cinerea* (Aho et al., 1987). Most, but not all, gonococcal opacity-associated proteins are associated with pronounced colony opacity because they increase the aggregation of gonococcal cells and thereby contribute to autoagglutination and adherence (Blake et al., 1995). There is no relationship between colony opacity and expression of the opacity-associated proteins in *N. meningitidis* (Hagman and Danielsson, 1989). With regard to function, gonococcal and meningococcal opacity-associated protein expression have been demonstrated to promote adherence and uptake into mammalian cells by virtue of their ability to engage cell surface receptors including CD66 (Chen et al., 1997a) and proteoglycan receptors (van Putten and Paul, 1995).

The opacity-associated proteins undergo phase variation, are immunogenic, and appear to have an important role in bacterial adherence and invasion. The *in vivo* importance of opacity-associated protein expression for neisserial infection is suggested by the findings that gonococci recovered after urogenital infection are typically Opa[+], as are bacteria recovered after the inoculation of human volunteers with transparent (Opa[−]) bacteria (Swanson et al., 1988). In *N. gonorrhoeae*, *N. meningitidis*, and *N. lactamica*, 11, 3–4, and 2 unlinked chromosomal alleles, respec-

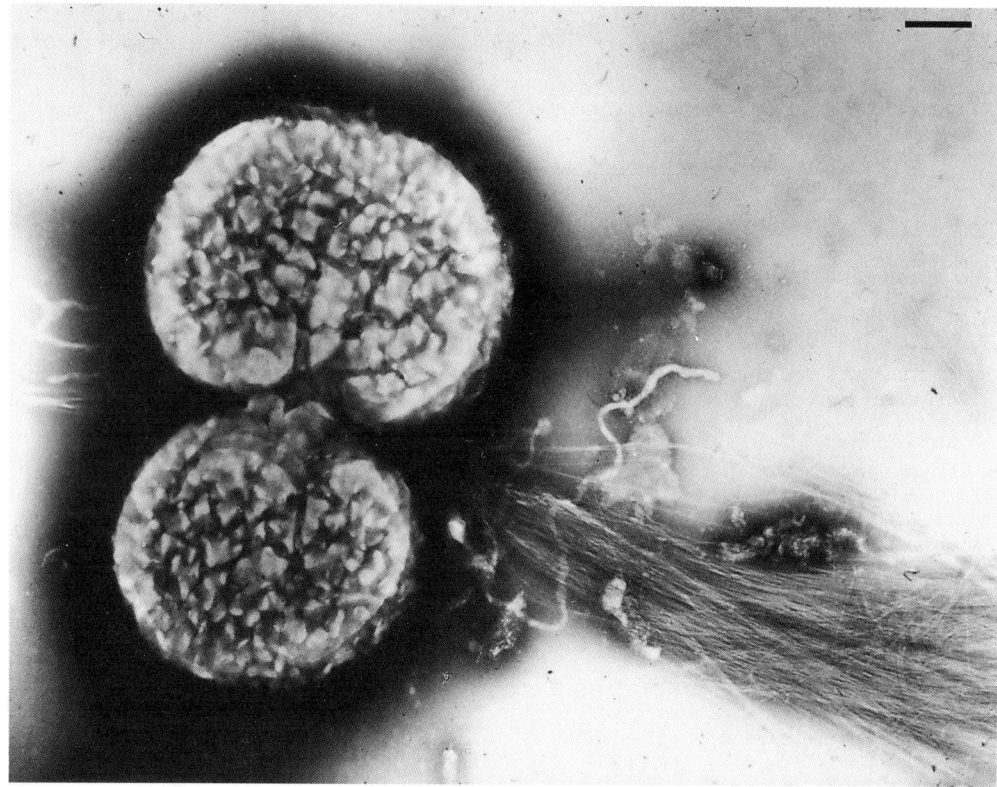

FIGURE BXII.β.63. Electron microscopic image of negatively stained *Neisseria meningitidis* strain M1080. Bar = 0.25 mm.

tively, have been identified that encode distinct opacity-associated protein variants (Meyer and van Putten, 1989). Each of these is a complete gene with a functional promoter and is transcribed (Stern and Meyer, 1987). Phase variable expression results from slipped strand mispairing DNA rearrangements, which occur at high frequencies and alter the number of pentanucleotide CTCTT coding repeat units in the N-terminal part of the open reading frames (Murphy et al., 1989). The opacity-associated proteins of *N. gonorrhoeae* and *N. meningitidis* can undergo phase variation during natural infection (Zak et al., 1984; Tinsley and Heckels, 1986).

Opa proteins also display antigenic variability and diversity within strains occurring by recombinational reassortment, which has been documented both in gonococci (Connell et al., 1988) and meningococci (Hobbs et al., 1998).

Opc   The meningococcal Opc outer membrane protein and its gene were first identified in a genetic cloning study using antibodies to the meningococcal Class 5C protein (Olyhoek et al., 1991). Further analysis revealed, however, that Opc is structurally and antigenically unrelated to the Class 5 proteins (Merker et al., 1997). Most meningococcal strains are capable of expressing Opc, and intrastrain phase variation occurs owing to changes in the number of contiguous cytosine residues that separate the *opc*-10 promoter and the start codon sequences (Sarkari et al., 1994). Although originally thought to be meningococcal-specific, gonococci carry an allele of *opc* in their genome and Opc is weakly expressed in gonococci (Zhu et al., 1999). Meningococcal Opc promotes binding to and uptake by mammalian cells by virtue of binding proteoglycan receptors (de Vries et al.,

1998). The potential benefits of a humoral immune response to Opc remain unclear (Thiesen et al., 1997).

H.8 PROTEIN   The DNA sequence predicts that the H.8 protein is a lipoprotein based on the presence of a lipoprotein signal-peptide processing site and an N-terminal cysteine residue. The predicted protein is 71 amino acids in length and is composed entirely of 13 repeats of the AAEAP consensus sequence that was also present in another neisserial protein, the lipid-modified azurin (Gotschlich et al., 1987a; Kawula et al., 1987). This protein is not essential for the growth and survival of gonococci in a complex medium, and the function of this protein remains unknown.

PILQ   PilQ was first identified as the outer membrane protein-macromolecular complex (OMP-MC), a major protein component of the outer membrane suggested to account for about 10% of its protein mass (Newhall et al., 1980b; Hansen and Wilde, 1984). This protein is required for type IV pilus biogenesis in the pathogenic neisseriae and was renamed PilQ based on its homology to the *Pseudomonas aeruginosa* PilQ protein (Drake and Kommey, 1995; Tønjum et al., 1998). Neisserial mutants expressing defective forms of the protein are devoid of pili and pilus-associated phenotypes. The PilQ complex has a molecular mass of 800–900 kDa and is composed of 12 identical subunits of 76–80 kDa (Tsai et al., 1989; Drake and Koomey, 1995; Tønjum et al., 1998). The C-terminus of PilQ shares identity with members of a large family of proteins associated with membrane translocation of macromolecular complexes, termed secretins. PilQ appears to function in pilus biogenesis in its complex form by serving as a pore through which pili are extruded. The PilQ monomer exhibits in its N-terminus a polymorphic region in which an octapeptide repeat occurs 4–7 times in *N. meningitidis* and only 2–3 times in *N. gonorrhoeae*. Antibodies against PilQ are produced during a natural infection (Tønjum and Koomey, personal communication), and PilQ antibodies are bactericidal for both homologous and heterologous gonococcal and meningococcal strains in a complement-dependent assay system (Corbett et al., 1986; Tønjum et al., 1998).

PILC A 110-kDa protein termed PilC was originally identified by virtue of its co-purification with the neisserial pilus fibers has also been identified in the neisserial outer membrane (Jönsson et al., 1991). Neisserial mutants failing to express PilC were reported to show a dramatic reduction in piliation (Jönsson et al., 1991; Nassif et al., 1994). More recently, PilC has been implicated as the molecule responsible for pilus-associated epithelial cell adherence (Rudel et al., 1995a, b).

OUTER MEMBRANE PROTEINS RELATED TO IRON METABOLISM Growth of gonococci and meningococci under iron-limiting conditions leads to the increased production of a variety of outer membrane or periplasmic proteins, collectively termed Irps (iron repressed proteins) (Norqvist et al., 1978). Consistent with the notion that levels of available iron are low in mammalian tissue and fluids, studies of seroconversion following disease indicate that Irps are expressed *in vivo* (Ferreiros et al., 1994). Irps are involved in iron uptake and their expression is coordinately regulated by the action of the iron-dependent regulatory protein Fur (Thomas and Sparling, 1996). Gonococci and meningococci express surface proteins engaged in specific uptake of iron from human transferrin, lactoferrin, and hemoglobin, and, in most cases, proteins found in these two species in each uptake system are structurally and functionally related (Schryvers and Stojiljkovic, 1999). Uptake from transferrin and lactoferrin involves expression of two large proteins, a lipoprotein (Tbp2 and LbpB) and a TonB dependent integral membrane protein (Tbp1 and LbpA), which act in concert to bind their respective ligand (Pettersson et al., 1994). In both systems, expression of the TonB dependent integral membrane proteins is required for function, while that of the lipoprotein is not (Anderson et al., 1994; Pettersson et al., 1998a).

Only 50% of gonococcal strains produce a functional lactoferrin receptor (Biswas and Sparling, 1995), and, as such, the ability to utilize lactoferrin as an iron source is not necessary for gonococcal infection. In contrast, transferrin receptor activity appears to be essential since mutants lacking Tbp1/2 are incapable of colonization in the human urethral challenge model (Cornelissen et al., 1998). Given their surface exposure and expression *in vivo*, meningococcal Irps have been examined for their abilities to engender bactericidal or opsonic antibodies (Lissolo et al., 1995; Ala'Aldeen and Borriello, 1996; Lehmann et al., 1999). Like other surface-exposed molecules, transferrin binding proteins display heterogeneity, which arises due to genetic exchange (Rokbi et al., 1997; Legrain et al., 1998), but such diversity may not automatically preclude their utility as protective immunogens since broadly cross-reactive antibodies against Tbps have also been reported to restrict growth under conditions of iron limitation by virtue of blocking the interaction with transferrin (Ala'Aldeen, 1996).

OTHER MEMBRANE PROTEINS Rmp is a highly conserved, reduction modifiable outer membrane protein that is found only in pathogenic neisseriae (Wolff and Stern, 1995). It is so named because its migration in SDS-PAGE is altered following treatment with reducing agents and was previously designated as Protein III (PIII) in gonococci and the class 4 protein in meningococci. Although mutants lacking Rmp have no discernible phenotypes, the protein does co-purify with the Por proteins (Wetzler et al., 1992b). In addition, antibodies directed against Rmp appear to block the bactericidal activity of antibodies directed toward other outer membrane proteins (Munkley et al., 1991), which may account for the observation that Rmp antibodies are associated with an increased susceptibility to gonococcal infection (Plummer et al., 1993).

Several other neisserial outer membrane proteins have been described. Some of these proteins appear to be both highly conserved and surface exposed. NspA is a conserved membrane protein that is reported to elicit protective antibody responses against *N. meningitidis* serogroups A, B, and C in mice (Martin et al., 1997). Although originally found in meningococci, it has also recently been shown to be expressed by *N. gonorrhoeae* strains (Plante et al., 1999).

Lipo-oligosaccharide Lipopolysaccharide (LPS) is a major component of the outer membrane of Gram-negative bacteria. The LPS of enteric bacteria such as *E. coli* consists of a lipid A moiety, a core oligosaccharide, and a repeating O-antigen polysaccharide. In contrast, the corresponding molecule of the pathogenic *Neisseria* spp. lacks the repeating O-antigen polysaccharides and is therefore termed lipo-oligosaccharide (LOS) (Mintz et al., 1984; Yamasaki et al., 1994). Differences in LOS composition appear to account for the difference in M_r. The oligosaccharide component consists of 1) a partially conserved and highly substituted basal oligosaccharide that branches at a heptose residue; 2) a linear segment consisting of $(hexose)_n$ residues that determines the length of the oligosaccharide; and 3) terminal sequences that are similar to those of glycosphingolipids (Griffiss et al., 1988; Mandrell et al., 1988). The resemblance of the lactoneoseries glycolipids in gonococcal (Yamasaki et al., 1994) and meningococcal (Tsai et al., 1998) LOS to human glycolipids appears to represent a form of host mimicry that may play a role in immune evasion (Moran et al., 1996). Gonococci expressing these LOS forms appear to be selected for *in vivo* and may have reduced infective doses (Schneider et al., 1991, 1995). Differences in the chemical composition of individual oligosaccharides relate to their ability to inhibit the killing of serum-sensitive strains of gonococci (Griffiss et al., 1987). *N. meningitidis* releases large amounts of the potent LOS through blebs or lysis. The oligosaccharide of neisserial LOS can be bound by host cell-expressed lectins and, therefore, might also contribute to bacterial adhesion (Apicella et al., 1986; Porat et al., 1995) and invasion (Song et al., 2000d).

Gonococci have the ability to produce a variable set of the oligosaccharide components of LOS at a high frequency. This variation occurs primarily due to phase variable expression of tandemly arrayed glycosyl transferase genes that ensue from slipped strand mispairing within poly-G tracts (Yang and Gotschlich, 1996). The lactoneoseries glycolipids in gonococcal LOS are also able to become modified *in vivo* by sialylating the oligosaccharide with host-derived CMP-*N*-acetyl neuraminic acid (CMP-NANA) (Nairn et al., 1988; Parsons et al., 1989). The enzyme responsible for this modification, sialyl transferase, is of gonococcal origin. The sialylation of the LPS results in the conversion of a serum-sensitive organism to one that is serum-resistant (Parsons et al., 1989) and is one of several mechanisms responsible for this phenomenon in gonococci. The sialylated

LPS appears to be rapidly lost after growth *in vitro* in the absence of CMP-NANA and may explain the type of resistance referred to as unstable serum resistance (Ward et al., 1970; Wetzler et al., 1992a). Curiously, sialylation *in vitro* prior to urethral challenge in men has been reported to reduce infectivity (Schneider et al., 1996). Meningococcal serogroup B and C strains are able to endogenously sialylate LOS since they have sialyl transferase and produce CMP-NANA as a substrate for capsule biosynthesis (Tsai et al., 1998). In contrast to other Gram-negative species, meningococcal mutants lacking LOS are viable (Steeghs et al., 1998). This remarkable finding may be attributable to compensatory changes in the phospholipid composition.

Immunoglobulin M (IgM) class antibodies directed against LPS epitopes are responsible for the complement-dependent bactericidal activity of normal human serum for serum-sensitive strains of *N. gonorrhoeae* (Apicella et al., 1986). The LPS from serum-resistant strains also contains bactericidal epitopes; however, normal human serum often lacks antibodies to these particular epitopes (Schneider et al., 1985). Nevertheless, antibodies to these epitopes are present in convalescent serum samples from patients with disseminated infections (Rice and Kasper, 1977).

Fimbriae (pili) Fimbriae (pili) are hairlike filamentous appendages emanating several micrometers from the bacterial cell surface. Fimbriae have been found on the surface of *N. gonorrhoeae* (Jephcott et al., 1971; Swanson et al., 1971), *N. meningitidis* (DeVoe and Gilchrist, 1975; McGee et al., 1977), and the nonpathogenic *Neisseria* spp., including *N. perflava* and *N. elongata* (Wistreich and Bakter, 1971; Frøholm et al., 1973; Bøvre et al., 1977). A PCR-based screen detected homology to the conserved N-terminal region of the pilus expression locus PilE in 12 out of 15 *Neisseria* species investigated, including all human commensal isolates (Aho et al., 2000).

Neisserial pili are ordered arrays of polymerized protein subunits termed pilin, with a molecular weight that varies between strains in the range of 16–22 kDa (Robertson et al., 1977; Brinton et al., 1978; Buchanan, 1978). Pili from *N. gonorrhoeae* and *N. meningitidis* are antigenically and structurally similar (Hermodson et al., 1978; Olafson et al., 1985; Potts and Saunders, 1988). However, meningococcal pili can be divided into two classes: those cross-reacting with the SM1 monoclonal antibody that recognizes a conserved epitope in gonococcal pili (class I), and those not showing this cross-reactivity (class II). The pili demonstrated on a number of commensal *Neisseria* spp. are homologous to the class II pili of *N. meningitidis* (Aho et al., 2000). The neisserial prepilins at short leader sequences and proximal 30 amino acids (Meyer et al., 1984; Potts and Saunders, 1988) show a high degree of homology with pilins of other Gram-negative human pathogens including *P. aeruginosa* (Strom and Lory, 1986), *Vibrio cholerae* (Faast et al., 1989), enterotoxigenic *E. coli* (ETEC) (Giron et al., 1994), and enteropathogenic *E. coli* (EPEC) (Giron et al., 1991), as well as opportunistic pathogens in the genera *Eikenella* (Tønjum et al., 1993) and *Moraxella* (Frøholm and Sletten, 1977; Tønjum et al., 1991). Collectively, this family of surface appendages has been termed type IV pili (Ottow, 1975) or Tfp (Wolfgang et al., 1998b).

These related pilus colonization factors appear to be essential for infectivity and disease. Pili mediate the attachment of *N. gonorrhoeae* to a wide range of different cell types, including tissue culture cells (Swanson, 1973; Heckels, 1989), vaginal epithelial cells (Mardh et al., 1975), fallopian tube epithelium (Ward et al., 1974), and buccal epithelial cells (Punsalang and Sawyer,

1973). Pili also mediate the attachment of *N. meningitidis* to nasopharyngeal cells (Stephens and McGee, 1981). Evidence for the critical roles of gonococcal and meningococcal pili can be found in the invariable recoveries of piliated organisms from primary cultures (Jyssum and Lie, 1965; Kellogg et al., 1986; Swanson et al., 1987) and the capacities of these structures to undergo antigenic variation (Tinsley and Heckels, 1986; Swanson et al., 1987). Pili appear to be multifunctional organelles. In addition to adherence, their expression is associated with other phenotypes that may be relevant to the host-parasite interaction. Pilus expression is correlated with high level (sequence-specific) competence for DNA transformation with frequencies of transformation being reduced 1000-fold in nonpiliated mutants (Sparling, 1966; Frøholm et al., 1973; Seifert et al., 1990; Zhang et al., 1992; Drake and Koomey, 1995; Freitag et al., 1995; Tønjum et al., 1995b; Drake et al., 1997). The consensus DNA uptake sequence recognized (5'-GCCGTCTGAA) is distributed throughout the gonococcal and meningococcal genomes (Goodman and Scocca, 1988).

Pilus expression is also associated with multicellular behavior. Bacterial autoagglutination (independent of that associated with Opa proteins) is a pilus-dependent phenomenon (Swanson et al., 1971). Although flagella and motility in liquid media are absent in *Neisseria* species, a form of bacterial surface translocation termed "twitching motility" is only displayed by piliated organisms (Henrichsen, 1975b; Swanson, 1978; Wolfgang et al., 1998a). In *P. aeruginosa*, twitching motility is proposed to be a consequence of pilus retraction (Bradley, 1980) and mutants that retain piliation but have lost twitching motility have been shown to carry mutations in a gene designated *pilT* (Whitchurch et al., 1991). Likewise, neisserial PilT mutants do not display twitching motility (Wolfgang et al., 1998b), and laser tweezer analysis of gonococci supports the notion that twitching motility is a consequence of pilus retraction (Merz et al., 2000).

Neisserial pilus expression is subject to both phase and antigenic variation. Gonococcal pilus variation results from homologous recombination between a single complete pilin gene or expression locus (*pilE*) and multiple partial pilin gene copies or silent alleles (*pilS*) (Haas and Meyer, 1986; Swanson et al., 1986; Koomey et al., 1987), and an analogous mechanism is thought to operate in meningococcal pilus variation (Aho and Cannon, 1988; Perry et al., 1988; Blake et al., 1989). Antigenic variation results from the unidirectional transfer of genetic information from the variant-encoding *pilS* genes to the active expression locus. The pilin amino acid changes reflect the nucleotide changes in *pilE*, which result from nonreciprocal RecA-dependent recombination events with numerous silent loci, *pilS*. The most convincing results show that productive pilin gene rearrangements in *N. gonorrhoeae* mainly arise by gene conversion and not by transformation (Zhang et al., 1992). The frequency of phase variation (on–off) from p$^+$ to p$^-$ (10^{-3}) is much higher than reversion from p$^-$ to p$^+$ (10^{-6}) (Swanson et al., 1987; Swanson and Koomey, 1989). The expression of pili is therefore often lost on repeated nonselective subculture. The high-frequency changes in the primary amino acid sequence of the pilin subunit alter pilus expression and aid in adjusting attachment abilities and the avoidance of the host immune response. The frequently occurring antigenic variation may result in phase variation, and loss of fimbriation may occur. Cultivating *N. gonorrhoeae*, *N. meningitidis*, and *N. elongata* in special growth conditions such as pellicle formation and selective subculture based on the observation of colony morphology will maintain pilus expression

(McGee et al., 1979; Swanson et al., 1987; Swanson and Koomey, 1989).

The fimbriae are produced and exported to the surface by the action of a multi-component machinery homologous to type II secretion systems (Tønjum and Koomey, 1997). A number of genes encoding products involved in neisserial pilus biogenesis have been identified (Tønjum and Koomey, 1997). By chromosomal mapping (Dempsey et al., 1995), it has become clear that these genes are distributed throughout neisserial genomes. The machinery dedicated to pilus biogenesis comprises, in addition to the pilin subunit itself, cytoplasmic and cytoplasmic membrane components as well as periplasmic and outer membrane components. A locus carrying the genes *pilF*, *pilD*, and *pilG* related to pilus biogenesis encodes cytoplasmic and cytoplasmic membrane components. Mutants failing to express functional PilD, PilF, or PilG were nonpiliated and showed over a 1000-fold decrease in frequencies of competence for natural transformation (Freitag et al., 1995; Tønjum et al., 1995b). Based on homology to *P. aeruginosa* PilD, the neisserial PilD is thought to function as a prepilin peptidase and methylase of the first amino acid in the mature protein (Strom et al., 1991; Freitag et al., 1995; Tønjum et al., 1995b). PilD is the only component engaged in pilus biogenesis that to date has a defined function (Nunn and Lory, 1991). The role of PilF is still obscure, but its identity with other related molecules (Pugsley, 1993; Sandkvist et al., 1995) and evidence from studies of its homologues in *P. aeruginosa* suggests that it may function as an ATPase or kinase (Turner et al., 1993). The deduced amino acid sequence of PilG reveals the absence of a signal sequence and three membrane spanning domains, which indicates it is a cytoplasmic membrane protein. The exact functional role of PilG remains to be established, but it has been proposed that the homologous PilC in *P. aeruginosa* may be needed for optimal localization or stabilization of PilD (Koga et al., 1993; Tønjum et al., 1995b). Gonococcal PilT, a protein homologous to PilF and belonging to a large family of molecules sharing a highly conserved nucleotide binding domain motif, has been shown to be dispensable for pilus biogenesis but essential for twitching motility and competence for genetic transformation (Wolfgang et al., 1998b).

Outer membrane components engaged in pilus biogenesis characterized to date are PilQ, PilP, and PilC. The PilQ protein, which is essential for type IV pilus assembly, is a member of the secretin family (see above) (Russel, 1998). The common functions served by the secretin homologues appear to involve the translocation of large, macromolecular complexes across the outer membrane. PilQ of the pathogenic neisseriae forms a multimer composed of 10–12 subunits that is resistant to SDS and heating (Newhall et al., 1980a), and the conserved C-terminal residues are necessary for expression of the multimerized form and function in pilus extrusion (Drake and Koomey, 1995; Tønjum et al., 1998). The meningococcal PilQ protein is constitutively expressed in abundant amounts. Analysis of the quaternary structure of meningococcal PilQ by transmission electron microscopy and self-rotation analysis suggest that PilQ is organized as a ring of 12 identical subunits (Collins et al., 2001, 2003). The *pilP* gene, immediately upstream of *pilQ*, encodes a lipoprotein that is predicted to be localized to the outer membrane, and PilP⁻ mutants are nonpiliated (Drake et al., 1997). It seems that PilP is required for stable expression of the multimerized form of PilQ. Gonococcal PilP⁻ as well as PilQ⁻ mutants release PilC (Drake et al., 1997). PilQ and PilC may interact during the terminal stages of pilus biogenesis (Drake et al., 1997; Wolfgang et al., 2000).

The presence or absence of fimbriae is of no taxonomic value in the genus, but is indirectly important because of the use of the associated property competence for natural transformation in qualitative and semiquantitative assays (see below) (Bøvre and Hagen, 1981; Tønjum et al., 1995b).

Secreted proteins *N. meningitidis* and *N. gonorrhoeae* produce IgA1-protease, which degrades and inactivates immunoglobulin of the IgA1 subtype found in mucosal secretions and serum (Koomey and Falkow, 1987).

Peptidoglycan The peptidoglycan composition of those species of *Neisseria* that have been analyzed is typical of other Gram-negative bacteria. The peptidoglycans of *N. gonorrhoeae* and *N. perflava* belong to chemotype 1 and are composed of muramic acid, glucosamine, alanine, diaminopimelic acid, and glutamic acid in an approximate molar ratio of 1:1:2:1:1, respectively (Martin et al., 1973; Hebeler and Young, 1976). The percentage of peptide cross-linking in gonococcal peptidoglycan was found to be approximately 41%, which is relatively high for a Gram-negative bacterium (Rosenthal et al., 1980). *N. gonorrhoeae* and *N. flava* peptidoglycans are extensively O-acetylated (Martin et al., 1973; Blundell and Perkins, 1981). O-acetylated peptidoglycan is more resistant to lysozyme and other human peptidoglycan hydrolases (Rosenthal et al., 1980; Blundell and Perkins, 1981), and this resistance to hydrolases might enable gonococci to persist *in vivo* or potentiate the biological effects in pathogenesis of peptidoglycan *in vivo*.

Polysaccharides Several *Neisseria* sp. are known to produce polysaccharide capsules and exopolysaccharides. The acidic and hydrophilic capsular polysaccharides provide the surface charge and humid environment that are critical for survival of the bacteria in aerosol droplets. Serogrouping of *N. meningitidis* is based on antigenic differences in capsular polysaccharides, and presently recognized serogroups include A, B, C, H, I, K, L, X, Y, Z, W135, and 29E (Frasch, 1989). The capsular polysaccharide may be a homopolymer or a heteropolymer composed of a monosaccharide-glycerol, disaccharide, or trisaccharide repeating unit. 1,2-Diacylglycerols have been identified as components of meningococcal capsular polysaccharides (Gotschlich et al., 1981). It has been postulated that the presence of these hydrophobic moieties may be responsible for the association of the capsular polysaccharide with the outer membrane of the meningococcus, giving rise to the structure recognized as a capsule. The disease-associated serogroup A capsule is composed of mannoseamine-phosphate, while the disease-associated serogroups B, C, Y, and W-135 all have sialic acid in their capsular polysaccharide. Sialic acid-containing polysaccharides confer resistance to host complement-mediated attack mechanisms (Jarvis and Vedros, 1987). *N. meningitidis* serogroup B strains are a predominant cause of meningococcal disease in the developed world. The serogroup B polysaccharide is a homopolymer of alpha-2,8 polyneuraminic acid and is poorly immunogenic, probably because of immunotolerance resulting from cross-reactivity between this polysaccharide and polysialic acid expressed on host neural cell adhesion molecules (NCAMs) (Finne et al., 1983). *N. meningitidis* serogroup B capsular polysaccharide is also chemically and immunologically identical to capsular polysaccharide in *E. coli* K-1 and surface polysaccharide found in a large proportion of *Moraxella nonliquefaciens* strains (Bøvre et al., 1983; Krambovitis et al., 1987).

The genes encoding all the enzymes necessary for capsular polysaccharide biosynthesis in *N. meningitidis* serogroup B are located on a 5-kb DNA fragment within the chromosomal capsule (*cps*) gene cluster (Frosch et al., 1989). It contains genes encoding enzymes required for capsular polysaccharide biosynthesis in the cytoplasm, for phospholipid substitution, and for the production of proteins whose function is the translocation of polysaccharide across the inner and outer bacterial cell membranes. The full-size capsular polysaccharide with a phospholipid anchor is synthesized intracellularly, and the lipid modification is a strong requirement for translocation of the polysaccharide to the cell surface (Frosch and Müller, 1993). Regulation of capsule expression is shown to be performed through slipped strand mispairing of a polyC stretch in the *siaD* gene (Hammerschmidt et al., 1996).

It seems evident that the capsule of *N. meningitidis* functions as a virulence factor, since the virulence of this organism depends, in part, on the antiphagocytic properties of its capsule (Masson and Holbein, 1983). In general, meningococcal capsular polysaccharides are immunogenic in humans; antibodies directed against the capsule are bactericidal. However, serogroup B capsular polysaccharide is poorly immunogenic in humans, possibly because of its similarity to sialic acid moieties in human tissues. Furthermore, the serogroup A and C capsules are poorly immunogenic in infants and young children (Lieberman, 1996). *N. gonorrhoeae* does not have a capsule that can be confirmed by electron microscopy, using India ink (Melly et al., 1979) or wheat germ agglutinin (Frasch, 1980). However, *N. gonorrhoeae* produces a high-molecular-weight extracellular polyphosphate (Noegel and Gotschlich, 1983), and a similar polyphosphate is also produced by *N. meningitidis*, *N. lactamica*, *N. sicca*, and *N. flava* (Noegel and Gotschlich, 1983). It has been postulated that the extracellular polyphosphate may serve the function of a capsule in gonococci.

N. perflava possesses a novel system for the synthesis of a glycogen-like polysaccharide. The mechanism involves the transfer of the glucosyl moiety of sucrose to a 1,4-α-D-glucan by the enzyme amylosucrase (Okada and Hehre, 1974). A similar D-glucan is produced by *N. polysaccharea* (Riou et al., 1986), *N. canis*, *N. cinerea*, *N. cuniculi*, *N. denitrificans*, *N. sicca*, and *N. subflava* (MacKenzie et al., 1977, 1978a, b, c) when grown on a medium containing sucrose. A galactosamine polymer has been isolated from the cell walls of *N. sicca* (Adams and Chaudhari, 1972; Wagner et al., 1973). The function of this polymer is presently unknown.

Some bacterial species produce a starch-like polysaccharide from sucrose that stains dark blue-purple to black with iodine. Among the *Neisseria* spp., *N. polysaccharea*, *N. subflava* biovar perflava, *N. mucosa*, *N. sicca*, and *N. flavescens* produce polysaccharide from sucrose. This test is invaluable for differentiating between strains of *N. meningitidis* (polysaccharide-negative) and *N. polysaccharea* (polysaccharide-positive); as many as 25% of organisms identified as nontypable *N. meningitidis* strains were found to be *N. polysaccharea* when tested for polysaccharide production.

Colony morphology and cultural characteristics   Colony morphology varies with the species and ranges from small smooth, transparent, butyrous colonies to wrinkled, dry, adherent colonies (Catlin, 1978; Swanson and Koomey, 1989). Colony morphology has been particularly useful in characterization of *N. gonorrhoeae* and *N. elongata*, since agglutinating and aggregating bacterial properties display various forms of colony consistency

and edge, as well as spreading and/or corroding growth (Kellogg et al., 1963; Reyn et al., 1971; Swanson and Koomey, 1989). The autoagglutination of bacterial cells into multicellular aggregates induces clumping in broth and variations in colony morphology that can be observed in oblique light in a dissecting stereoloupe/microscope (Juni and Heym, 1977). Important components affecting autoagglutination are pilus expression, opacity proteins, and LPS. Colonies of cells expressing pili and opacity proteins are wrinkled and well defined with a clear edge, while colonies from non-piliated cells have a less defined edge and are smoother and more glistening (Swanson and Koomey, 1989; Koomey et al., 1991). Opaque and transparent variants of the colony types have been described (Swanson, 1978). The autoagglutinating colonies are typically isolated from the urethra of infected males and the cervix of females (see review by James and Swanson, 1978; Swanson et al., 1987). The autoagglutination properties can be camouflaged in strains expressing a capsule.

Colonies of *N. meningitidis* are larger than those of gonococci (≥1.0 mm in diameter) and are smooth and moist with a glistening surface and entire edge (Morello and Bonhoff, 1980). The colony morphology of meningococci is difficult to distinguish, probably due to the capsule giving rise to mucoid colonies. Colonies of *N. lactamica* closely resemble those of meningococci, but may be less moist and smaller (Morello and Bonhoff, 1980). Some strains of *N. lactamica* produce a yellow pigment (Hollis et al., 1969). A yellowish green pigment is also produced by *N. subflava*, *N. flavescens*, and *N. sicca*. *N. elongata* may also have a slight yellow tinge due to pigment production. Pigmentation of neisseriae varies with growth conditions, and quantitative analysis of the extracted pigment indicates that it is of little taxonomic value (Berger, 1961a; Hoke and Vedros, 1982c).

Nutrition and growth conditions   *Neisseria* species other than *N. meningitidis* and *N. gonorrhoeae* will grow on plain nutrient agar at 35–37°C. *N. meningitidis* requires mineral salts, lactate, a few amino acids, and glutamic acid as a carbon source (Reyn, 1974; Catlin, 1978). Cystine is required by approximately 10% of the strains (Catlin, 1978), and some strains can be adapted to grow with ammonium salts as the sole nitrogen source (Jyssum, 1959). *N. gonorrhoeae* is more fastidious and requires glutamine for primary isolation of approximately 20% of strains and co-carboxylase for 1% of strains (Reyn, 1974). Several defined media have been developed for growth of *N. gonorrhoeae* (Catlin, 1973; LaScolea and Young, 1974; Wong et al., 1980b). Iron is an essential growth factor for *N. gonorrhoeae* (Kellogg et al., 1963) and its availability in culture media influences the virulence of both meningococci and gonococci (Payne and Finkelstein, 1975). The optimal temperature for growth of *Neisseria* spp. is 35–37°C with a range of 22–40°C, except for *N. meningitidis* and *N. gonorrhoeae*, which grow poorly or not at all below 30°C. A high relative humidity (50%) is beneficial to the growth of all species, and CO_2 (3–10%) is required for the growth of gonococci and enhances the growth of meningococci on solid media. After many passages, laboratory strains become less fastidious in their growth requirements than fresh isolates. See Catlin (1977) for a detailed review of gonococcal nutritional requirements.

N. meningitidis and *N. gonorrhoeae* do not survive long after the cessation of growth, and decreases in viability appear to be related to cellular lysis (autolysis) (Morse and Bartenstein, 1974). Several peptidoglycan-degrading enzymes have been described in *N. gonorrhoeae*, including D-alanine carboxypeptidase (or endopeptidase) (Davis and Salton, 1975; Chapman and Perkins, 1983),

N-acetylmuramyl-L-alanine amidase (Hebeler and Young, 1976), transglycosylase (Rosenthal, 1979; Sinha and Rosenthal, 1980), exo-*N*-acetyl-glucosaminidase (Chapman and Perkins, 1983), and endo-*N*-acetylglucosaminidase (Gubish et al., 1982; Chapman and Perkins, 1983). It is likely that some of these enzymes normally have a biosynthetic role in cell growth, but that under nongrowth conditions (i.e., no cell wall biosynthesis) enzyme activity results in peptidoglycan hydrolysis.

Metabolism and biochemistry *Neisseria* species produce oxidase and catalase, carbonic anhydrase, and nitrite reductase, but do not produce thymidine phosphorylase, nucleoside deoxyribosyl transferase, or thymidine kinase (Table BXII.β.76).

Carbohydrate metabolism *Neisseria* species do not catabolize many carbohydrates and some species (*N. cinerea*, *N. flavescens*, and *N. elongata*) are asaccharolytic. Glucose is the only carbohydrate that can be used as an energy source by *N. gonorrhoeae* and is apparently the only monosaccharide that is used by the other saccharolytic species.

Neisseria species produce acid from carbohydrates by oxidation, not fermentation. Because these species are oxidative and produce less acid from carbohydrates such as glucose, maltose, fructose, and sucrose than do fermentative organisms, and because they also produce ammonia from peptones (which may neutralize acid produced from carbohydrates), acid production must be determined in a medium with a low protein/carbohydrate ratio and a sensitive indicator such as phenol red (Knapp, 1988).

Phosphoglucose isomerase (Pgi) in *E. coli* is a dimeric enzyme that catalyzes the reversible isomerization of glucose-6-phosphate and fructose-6-phosphate in glycolysis (Froman et al., 1989; Fraenkel, 1992). In *E. coli* there is a single *pgi* gene, and Pgi appears to be dispensable because strains lacking it are also able to grow on glucose minimal media, apparently utilizing glucose primarily by the pentose phosphate shunt (Fraenkel, 1992). Expression analysis in starch gel electrophoresis and isoenzyme analysis revealed the presence of two distinct isoforms of Pgi in *N. meningitidis* and *N. gonorrhoeae*. Indeed, Southern blot analysis of *N. meningitidis* and *N. gonorrhoeae* chromosomal DNAs and the complete genome sequences confirmed that there are two complete *pgi* genes on different parts of the chromosomes in pathogenic *Neisseria* (Tønjum et al., 1994). This gene duplication may be significant since Pgi expression in other bacteria has been implicated in pathogenicity (Tung and Kuo, 1999).

Tricarboxylic acid cycle *Neisseria* species have a functional tricarboxylic acid cycle (Holten, 1975). All tricarboxylic acid cycle enzymes, except a soluble pyridine nucleotide-dependent malate dehydrogenase, have been detected in *N. gonorrhoeae* and *N. meningitidis* (Jyssum, 1960). Instead of the pyridine nucleotide-dependent malate dehydrogenase, *N. gonorrhoeae* and *N. meningitidis* possess a flavine adenine dinucleotide (FAD)-dependent malate oxidase (Holten, 1976b). A similar enzyme is present in the other *Neisseria* spp. (Holten, 1976b) in spite of the observation that these species also have a pyridine nucleotide-dependent malate dehydrogenase (Holten and Jyssum, 1974).

N. gonorrhoeae lacks a glyoxylate bypass (Holten, 1976b). With the exception of *N. elongata*, *Neisseria* spp. do not oxidize acetate in the absence of other substrates, providing additional evidence for the absence of a glyoxylate bypass (Holten, 1976b). Indirect evidence of acetate oxidation in the absence of other substrates suggests that *N. elongata* may have a glyoxylate bypass (Holten, 1976b, 1977).

Amino acid metabolism The biosynthesis of amino acids by *Neisseria* spp. appears to occur by pathways similar to those in other microorganisms. *Neisseria* species vary widely with respect to their amino acid requirements. In general, the nonpathogenic *Neisseria* species are able to grow in a defined medium containing one to five amino acids (McDonald and Johnson, 1975), while the amino acid requirements of the pathogenic species are more complex (Catlin, 1973). *N. gonorrhoeae* exhibits an absolute requirement for cysteine (or cystine) (Catlin, 1973). Strains of *N. gonorrhoeae* often require one or more amino acids in addition to cysteine for growth in a chemically defined medium (Janik et al., 1976). Amino acid requirements have been used to differentiate among isolates (auxotyping) for epidemiologic purposes (Carifo and Catlin, 1973) or for identification by genetic transformation by returning auxotrophic strains to phototrophy (Juni and Heym, 1980). Auxotyping data have also demonstrated that the requirement for certain amino acid biosynthesis in *N. gonorrhoeae* is associated with spontaneous mutations in the genes encoding enzymes involved in the biosynthesis of amino acids (Lerner et al., 1980; Shinners and Catlin, 1982).

Amino acids can be used as energy and carbon sources by many *Neisseria* spp. via their oxidation by the tricarboxylic acid cycle (Holten, 1976a). Glutamate, proline, and to a lesser extent aspartate are the preferred amino acids (McDonald and Johnson, 1975; Pillon et al., 1982). Glutamate dehydrogenase is a key enzyme in the catabolism of glutamate and proline, and all *Neisseria* species contain two forms of glutamate dehydrogenase (Holten, 1973; Holten and Jyssum, 1973). The NAD-dependent enzyme serves a catabolic function, and the NADP-dependent enzyme serves an anabolic function, depending on the needs of the cell (Holten, 1973).

Electron transport *Neisseria* spp. characteristically have high cytochrome *c* oxidase activity (Jurtshuk and Milligan, 1974). The oxidation of tetramethyl-*p*-phenylendiamine by this cytochrome provides the basis for the oxidase test used in the identification of members of the genus *Neisseria*.

Proteolytic enzymes The degradation of proteins or peptides requires specific enzymes. *Neisseria* species possess aminopeptidases that are capable of hydrolyzing L-amino-acid-β-naphthylamide derivatives of various amino acids (D'Amato et al., 1978). *N. meningitidis* can be distinguished from *N. gonorrhoeae* by the presence of *N*-γ-glutamyl aminopeptidase (D'Amato et al., 1978). However, some of the nonpathogenic species (*N. mucosa*, *N. sicca*, and all biovars of *N. subflava*) also produce *N*-γ-glutamyl aminopeptidase (Riou et al., 1982).

N. gonorrhoeae and *N. meningitidis* express a highly specific IgA protease that cleaves human IgA of the IgA1 subclass (Plaut et al., 1975; Koomey and Falkow, 1984). This enzyme is an important virulence factor in that its production promotes bacterial immune evasion. Each isolate of *N. gonorrhoeae* and *N. meningitidis* produces only one of two distinct types of IgA protease (Mulks et al., 1980). The two types of IgA protease can be distinguished by their specificity for two different peptide bonds in the hinge region of human IgA1 (Mulks and Knapp, 1987). The type I protease cleaves the proline-serine bond at position 237/238 in the IgA1 hinge region (Mulks et al., 1980), while the type 2 protease cleaves the proline-threonine peptide bond at position 235–236 (Plaut et al., 1975). The IgA protease is an extracellular enzyme that is secreted throughout exponential growth (Pohlner et al., 1987). The gene encoding IgA protease in *N. gonorrhoeae* was cloned, inactivated, and reintroduced into *N. gonorrhoeae* to obtain an IgA protease negative mutant (Koomey et al., 1982). Gonococcal IgA protease has also been suggested to play a role in adherence and invasion processes (Lorenzen et al., 1999).

Periplasmic proteins The gonococcal periplasm is very similar to that of *E. coli* (Judd and Porcella, 1993). Several proteinases that are disulfide oxidoreductases, termed DsbA, are located in the periplasmic space (Jose et al., 1996). These autotransporters are subjected to secretion across the inner membrane in a Sec-dependent manner followed by self-directed passage and release through the outer membrane (Maurer et al., 1997). The periplasmic meningococcal superoxide dismutase C catalyzes the conversion of the superoxide radical anion to hydrogen peroxide, preventing the production of toxic hydroxyl free radicals (Wilks et al., 1998).

Iron The free iron concentration in the human and other eucaryotic hosts is extremely low and cannot support sustained microbial growth. Therefore, *Neisseria* species rely on mechanisms for obtaining growth-essential iron from their host. The pathogenic neisseriae are capable of satisfying their iron requirements with human iron-binding proteins such as transferrin and lactoferrin (see iron-regulated outer membrane proteins).

Oxygen requirements *Neisseria* species are generally considered to consist of aerobic and facultatively anaerobic organisms. *N. gonorrhoeae*, however, has been isolated from body sites where anaerobes usually are found, suggesting that *N. gonorrhoeae* can grow under reduced oxygen tension. *N. gonorrhoeae* will also grow *in vitro* under oxygen tension suitable for anaerobes (Kellogg et al., 1983). Nitrite is toxic to *N. gonorrhoeae*, so that the reduction of nitrite is only observed at low concentrations (Knapp and Clark, 1984). The addition of subtoxic concentrations of nitrite to the medium enables gonococci to grow under anaerobic conditions to a level comparable to that observed aerobically (Knapp and Clark, 1984). Gonococcal nitrate reductase is a constitutive enzyme that is synthesized under aerobic conditions in the absence of nitrite (Knapp and Clark, 1984). Since nitrite is present in biological fluids, the ability to grow aerobically or anaerobically by anaerobic nitrite respiration may be one of the factors responsible for the diversity of body sites from which gonococci can be isolated. Nitrite is reduced by all *Neisseria* spp. isolated from humans, with the possible exception of some serogroups of *N. meningitidis*, and some strains of *N. lactamica*, *N. cinerea*, and *N. polysaccharea* (Morse and Knapp, 1987).

N. gonorrhoeae grown *in vitro* under anaerobic conditions expresses several new outer membrane proteins (Clark et al., 1987). The presence of antibodies in serum samples from patients with gonorrhea that react with one or more of the anaerobically induced proteins suggests that gonococci grow anaerobically *in vivo* (Clark et al., 1988).

Carbon dioxide requirements *N. gonorrhoeae* and *N. meningitidis* both require an increased CO_2 tension for isolation from clinical specimens (Tuttle and Scherp, 1952; Griffin and Racker, 1956). Gaseous CO_2 can be replaced by the addition of bicarbonate to the medium (Talley and Baugh, 1975). Little is known about the CO_2 requirements of the nonpathogenic *Neisseria* spp. The enzymes known to be involved in the assimilation of CO_2 by *Neisseria* spp. are carbonic anhydrase, which appears to be located in the cytoplasmic membrane (MacLeod and DeVoe, 1981), and phosphenolpyruvate carboxylase (Jyssum and Jyssum, 1962; Holten and Jyssum, 1974).

Genetics

Transformation Transformation is the binding, uptake, and chromosomal integration of naked DNA. Many neisserial species are constitutively competent throughout their entire life cycle (Table BXII.β.77). DNA uptake in *Neisseria* spp. is dependent on the presence of a 10-basepair specific DNA uptake sequence (DUS, 5′-GCCGTCTGAA) (Goodman and Scocca, 1988). Genome sequencing shows that the pathogenic *Neisseria* contains approximately 1900 copies of the DNA uptake sequence (Parkhill et al., 2000; Tettelin et al., 2000) and that this sequence is found very infrequently in the genomes of other organisms. Some reports have claimed that most DUSs are located as inverted repeats downstream of open reading frames throughout the genome, serving as putative transcriptional regulators (Smith et al., 1999a). However, it has recently been shown that many of the genes engaged in genome maintenance harbor DUSs within their open reading frames. Genome-wide analysis of DUS representation in the coding sequences of *N. meningitidis* has indeed demonstrated that the group of genes engaged in DNA repair, recombination, restriction–modification, and replication exhibited the highest rate of DUSs among all designated gene groups (Davidsen and Tønjum, personal communication). A genome-wide overrepresentation of the *H. influenzae*-specific DUS was also found in the genome maintenance genes of this bacterial species. The biased distribution of the respective DUS within sequences encoding genome maintenance genes of the phylogenetically divergent *N. meningitidis* and *H. influenzae* is evidence for fundamental evolutionary prioritization and conservation. However, the complete role of DUS in this context is yet to be unraveled.

The expression of natural competence may vary since it is associated with type IV pilus expression in most species (Bøvre and Frøholm, 1971, 1972; Swanson et al., 1971; Frøholm et al., 1973). Therefore, it is often critical to perform selective culture passages to maintain pilus expression; nonselective subcultures should be minimized to avoid loss of pilus expression and thus a loss of competence. The transformation yield, however, is higher in early phases of growth, before autolysis of bacteria occurs in the culture and causes inhibition of transformation by

TABLE BXII.β.77. List of species of *Neisseria* and *Neisseriaceae* known to have naturally competent isolates

Species/Subspecies	References[a]
Neisseria elongata subsp. *elongata*	Bøvre and Holten (1970)
Neisseria elongata subsp. *glycolytica*	Bøvre et al. (1977)
Neisseria flava	Catlin and Cunningham (1961)
Neisseria flavescens	Catlin and Cunningham (1961)
Neisseria gonorrhoeae	Sparling (1966)
Neisseria lactamica	Hoke and Vedros (1982c)
Neisseria meningitidis	Catlin (1960); Jyssum and Lie (1965)
Neisseria mucosa	Bøvre and Hagen, unpublished
Neisseria perflava	Catlin and Cunningham (1961)
Neisseria sicca	Catlin and Cunningham (1961)
Neisseria subflava	Catlin and Cunningham (1961)
Neisseria weaveri	Bøvre and Hagen, unpublished
Eikenella corrodens	Tønjum et al. (1985)
Kingella denitrificans	Weir and Marrs (1992)
Kingella kingae	Henriksen and Bøvre (1968b, 1976)

[a]Additional references that relate to the taxonomic application and consequences of genetic transformation with recipients of these species/subspecies: Henriksen and Bøvre (1968b), Bøvre and Frøholm (1972), Juni (1972, 1990), Henriksen (1976), Bøvre (1979, 1980); Bøvre and Hagen, 1981), Tønjum et al. (1995b).

exogenous DNA due to saturation of DNA binding sites with residual autologous DNA.

Many reports strongly suggest that transformation is the genetic mechanism that contributes the most to the large amount of horizontal gene transfer taking place in the *Neisseria* (Feil et al., 1996; Koomey, 1998). Transformation takes place most often between closely related species, but also between more distantly related ones, contributing to species diversity and fitness for survival. Furthermore, transformation represents an invaluable tool in the study of specific determinants by virtue of the ease with which strains carrying defined mutations in chromosomal genes can be constructed.

Competence for genetic transformation provides a unique tool for measuring genetic distances for classification and identification purposes, since the relative efficiency of transformation depends on the degree of DNA homology as well as on DNA uptake sequences (Albritton et al., 1986). The widespread occurrence of natural competence for genetic transformation among *Neisseria* spp. has enabled the use of this method for critical classification schemes and strain identification (see also chapter on the *Moraxellaceae*) (Bøvre, 1965, 1980; Bøvre et al., 1977). In the 1950s, Henriksen made observations causing subdivision of bacterial groups into species of *Neisseria*, *Moraxella*, and other genera (Henriksen, 1952). These early studies demonstrated that it was possible to show relatedness of bacterial strains because of the ability of their DNAs to undergo genetic recombination, provided that at least one of these strains was competent for genetic transformation (Henriksen, 1952; Leidy et al., 1956; Catlin and Cunningham, 1961; Bøvre, 1964). This procedure involved quantitative transformation of streptomycin-resistance markers and has later been applied in many neisserial species (Table BXII.β.77). Juni and co-workers developed another useful and easy-to-perform transformation assay for distinction of species (Juni and Janik, 1969; Juni, 1972, 1990). This technique involved transformation of auxotrophic bacterial strains to prototrophy, using crude lysates of bacterial cells as transforming DNA. The use of lysates as transforming DNA greatly facilitates the practical application of this method. To establish nutritional transformation assays it is necessary to isolate nutritionally defective mutants of the competent strain, and also to design a minimal or defined medium on which transformants of the nutritionally deficient mutants can grow and thus demonstrate that transformation has taken place. For example, *N. gonorrhoeae* strains grow on a simple sodium lactate–mineral medium with ammonium chloride as the nitrogen source, no growth factors being required (Juni, 1974). Metabolic markers in transformation can give an increased opportunity for discrimination between closely related strains. However, this approach may in some cases be of reduced utility due to a relatively higher frequency of divergence occurring in these genes.

Conjugation Several neisserial conjugative plasmids have been described (see Plasmids and Antibiotic susceptibility).

Recombination The plasticity of the neisserial genomes for large and small genome rearrangements demands highly facilitated DNA recombination and repair systems. High-frequency recombination events may be RecA-dependent, such as gene conversion and homologous recombination, or RecA-independent, such as replicative DNA slippage. Indeed, a number of recombination genes have been identified in *Neisseria* (Carrick et al., 1998; Salvatore et al., 2002; Skaar et al., 2002). However, an in-

ducible DNA repair and recombination response (SOS response) has not been identified in *Neisseria* (Black et al., 1998).

Consequences of natural transformation, recombination, and spontaneous mutation on phylogenetic sequence analyses of **Neisseria** *species* The pattern of nucleotide sequence variation within genes shows that species within the genus *Neisseria* can be assigned to five phylogenetic groups, but that sequence divergence within *N. meningitidis* and closely related species is inconsistent with a bifurcating tree-like phylogeny and is better represented by an interconnected network (Smith et al., 1999a). New data indicate that although the human commensal *Neisseria* species can be separated into discrete groups of related species, the relationships both within and among these groups, including those reconstructed using 16S rDNA sequences, have been distorted by interspecies recombination events (Smith et al., 1999a). The sequence data indicate that there has been a history of interspecies recombination within selected genes of the human *Neisseria* species, which has obscured the phylogenetic relationships between the species (Feil et al., 1996). The neisseriae are therefore polyphyletic species, most probably due to the frequent DNA exchange and recombination events, both intragenomic and subsequent to transformation, taking place in these species. One consequence of the high spontaneous mutation and recombination rates detected and horizontal gene transfer is that the bifurcating phylogeny that can be inferred for other genetic bacterial entities are not valid for genus *Neisseria*, except for very closely related strains reflecting short-term genetic distance and exchange.

Among the neisseriae, the limitations of 16S rDNA based phylogeny are demonstrated due to 1) the high level of horizontal gene transfer, 2) high frequency and variability of mutation rates, as well as 3) high frequencies of gene duplication (and extinction) occurring in bacteria with high numbers of repetitive genetic elements of various kinds, such as those demonstrated by the genome sequences of the pathogenic *Neisseria* (Parkhill et al., 2000; Tettelin et al., 2000). Any phylogeny created depends on the gene chosen for analysis, and no deep evolutionary tree can be determined for the neisseriae due to the continuously high spontaneous mutation rate and recombination including incorporation of DNA from other species. If an analysis is performed that does not force the data into a tree-like phylogeny (split decomposition), then a network phylogeny can be obtained with all genes examined to date, although the precise network will differ from gene to gene.

Genomics The complete genome sequences of meningococcal isolates of both *N. meningitidis* serogroup A Z2491 and *N. meningitidis* serogroup B strain MC58 have been published, and the genomes are 2.2 and 2.3 Mbp in size, respectively (Parkhill et al., 2000; Tettelin et al., 2000). These genomes have a little over 2000 open reading frames, with 83% of each genome being coding sequence, which is the lowest percentage coding sequence among the intact bacteria so far sequenced. The mol% G + C is variable and some low mol% G + C regions are associated with open reading frames predicted to code for surface structures. Several partial genes and pseudogenes have been identified that represent the C-terminal portion of upstream genes; these may allow variation by the use of alternate C-termini. Approximately 1900 copies of the *Neisseria* uptake sequence are present throughout the genome. A comparison of the two genome sequences reveal 91.2% similarity with one inversion of

955 kb being the major difference. However, two inversion events have occurred, both of which center around the origin of replication. Repeat arrays are present near the points of inversion. A related bacteriophage has been identified in each genome, although in different positions. Three major islands of horizontal DNA transfer have been identified in the genomes; most of these contain genes encoding proteins involved in pathogenicity. However, the most notable feature of these genomes is the presence of many hundreds of repetitive elements, ranging from short repeats, positioned either singly or in large multiple arrays, to insertion sequences and gene duplications of 1 kb or more. Many of these repeats appear to be involved in genome fluidity and antigenic variation. *N. meningitidis* contains more genes that undergo phase variation than any pathogen studied to date, a mechanism that controls their expression profiles and contributes to the evasion of the host immune system.

The genome sequence of *N. meningitidis* serogroup C strain FAM18 and *N. gonorrhoeae* strain F1090 are almost completed. The genome of *N. gonorrhoeae* is approximately 2.1 Mbp with a prediction of around 2000 open reading frames and 2000 uptake sequences. Approximately 30% of the gonococcal genome is repetitive DNA with 6000 repeats less than 2 kb in size. Comparisons show that it has approximately 95% conservation with the *N. meningitidis* isolate Z2491 with one large rearrangement. Of the two isolates sequenced to date, the serogroup A Z2491 isolate was more similar in gene order to the sequenced gonococcal isolate than it was to the MC58 variant sequenced.

Plasmids and transposons Plasmids in *Neisseria* spp. have attracted considerable attention because of their potential role in virulence and association with antibiotic resistance. Beta-lactamase plasmids have been identified in *N. gonorrhoeae* and *N. meningitidis* and have been associated with resistance to penicillin and other beta-lactam antibiotics (Roberts, 1989). Several beta-lactamase-conferring plasmids have been described; these nonconjugative plasmids range in size from 2.9 to 4.4 mDa and share considerable homology with each other (Dillon and Young, 1989; Pagotto et al., 2000; Pagotto and Dillon, 2001).

Two types of larger nonconjugative plasmids have been described in *Neisseria* spp. The 24.5-mDa plasmid carries no detectable markers for antibiotic resistance. However, it efficiently mobilizes itself and the beta-lactamase plasmids between strains of gonococci. A 25.2-mDa conjugative plasmid has been described in isolates of *N. gonorrhoeae* and *N. meningitidis* (Morse et al., 1986; Knapp et al., 1988). This plasmid carries tetracycline resistance and was formed by the transposition of the TetM determinant onto the 24.5 conjugative plasmid. The 25.2-mDa plasmid will also mobilize beta-lactamase plasmids and has an extended host range (Roberts and Knapp, 1988a, b). A group of plasmids that are genetically related to the enteric plasmid RSF1010 and range in size from 4.9 to 9.4 mDa has been described in *N. meningitidis*, *N. mucosa*, *N. subflava*, and *N. sicca* (Pintado et al., 1985; Rotger et al., 1986; Facinelli and Varaldo, 1987). Some of these plasmids confer resistance to sulfonamide alone, whereas others specify resistance to sulfonamide, streptomycin, and penicillin (Rådstrøm et al., 1992). Many *Neisseria* spp. contain cryptic plasmids, i.e., plasmids with no measurable phenotype. The 2.6-mDa plasmid from *N. gonorrhoeae* is present in auxotypes of *N. gonorrhoeae*, except those strains that require proline, citrulline, and uracil (Dillon and Pauze, 1981). Some strains of *N. meningitidis*, *N. lactamica*, *N. mucosa*, and *N. cinerea* carry plasmids that are homologous to the gonococcal 2.6-mDa

plasmids (Ison et al., 1986) and that are not homologous with this plasmid (Aalen and Gundersen, 1985; Prere et al., 1985).

Conjugal transfer of genetically modified plasmids to appropriate gonococcal recipients has been most useful for research purposes (Eisenstein et al., 1977; Kupsch et al., 1996).

Phages A nontransducing bacteriophage has been isolated for *N. subflava* biovar perflava (Stone et al., 1956; Phelps, 1967). Evidence for the existence of a prophage can be found in the analysis of the pathogenicity island of the *N. meningitidis* serogroup A strain Z2491 (Klee et al., 2000).

Bacteriocins *N. meningitidis* and *N. gonorrhoeae* may release bacteriocins during growth (Kingsbury, 1966; Flynn and McEntegart, 1972; Senff et al., 1976; Allunans and Bøvre, 1996; Allunans et al., 1998). On average, 0.8% of gonococcal strains were reported bacteriocin producers (Lawton et al., 1976). In systemic meningococcal strains and isolates from healthy carriers of Norwegian origin, the overall reported frequency varied from 1.5% to 2.8% (Andersen et al., 1987; Allunans and Bøvre, 1996). While a high frequency of bacteriocin producers (13.5%, systemic strains) coincided with the peak incidence of the ongoing epidemic in Norway in 1975, the involvement of bacteriocins in meningococcal epidemiology and disease remains unclear.

Habitats The principal habitats of those *Neisseria* spp. isolated from humans are the mucous membrane surfaces. Similarly, for domestic and experimental animals the mucosal surfaces of the oropharynx are the principal habitats. Only *N. meningitidis* and *N. gonorrhoeae* are considered to be primary pathogens of humans. Other *Neisseria* species isolated from humans have been reported to be responsible for disease, e.g., *N. mucosa* (Berger et al., 1974), *N. flavescens* (Branham, 1930), and *N. subflava* (Lewin and Hughes, 1966), but it is generally considered that these species are opportunists that rarely cause infection. The species isolated from animals other than humans have caused infections, possibly as primary pathogens (*N. iguanae*) or as opportunists, also in humans (*N. weaveri* and *N. canis*) (Hoke and Vedros, 1982b; Anderson et al., 1994). These species are generally part of the normal flora of their respective animal hosts, but may have a broader host range as indicated by the isolation of *N. mucosa* from marine mammals (Vedros et al., 1973).

N. meningitidis strains are carried as normal flora in the oro- and nasopharynx of adults and children. The prevalence of *N. meningitidis* carriage varies geographically (Holten et al., 1978; Knapp and Koumanis, 1999). It has been suggested that *N. meningitidis* may occur more frequently in adults with gonorrhea and in homosexual men. Fewer than 1% of children are colonized by *N. meningitidis* during the first 4 years of life; the carriage rate increases after this age.

N. lactamica colonizes the throats of children more frequently than adults (Holten et al., 1978). *N. lactamica* colonizes the throats of as many as 4% of infants to a peak of 21% in children 18–24 months; colonization by *N. lactamica* then declines to 2% by age 14–17 years (Holten et al., 1978; Blakebrough et al., 1982; Knapp and Koumanis, 1999). It is estimated that 59% of children have been colonized by *N. lactamica* at least once by the age of 4 years. This pattern of colonization may reflect the fact that young children may drink large volumes of milk. Among *Neisseria* species, strains of *N. lactamica* are unique in their ability to use lactose; this characteristic may enhance populations of *N. lactamica* in the throats of younger children. Unlike *N. gonorrhoeae* and *N. meningitidis*, *N. lactamica* has not been implicated as a primary pathogen, although it may be an opportunistic pathogen. The other commensal *Neisseria* species are normal inhabitants of the oro- or nasopharynx. These species are occasionally isolated from other sites but are not considered to be normal flora of sites other than the throat.

Because strains of the commensal *Neisseria* species rarely grow on selective media used to isolate *N. meningitidis* and *N. gonorrhoeae*, the prevalence of these species must be determined on a medium that does not contain colistin, the antibiotic incorporated in neisserial selective medium to inhibit the growth of the commensal species. It therefore was not possible to accurately determine the carriage rate of commensal *Neisseria* species in the early 1900s studies (Berger, 1961a). Most studies of the prevalence of *Neisseria* spp. were performed with nonselective media, e.g., blood agar, which neither inhibited the growth of other bacterial species nor permitted differentiation between colonies of *Neisseria* and related species. Thus, it is probable that the carriage rates of some species (*N. cinerea*, *N. polysaccharea*, and *N. lactamica*) were underestimated because these species occur in relatively small numbers and were probably overgrown by either non-neisserial species or the sucrose-positive *Neisseria* species (*N. subflava* biovar perflava, *N. sicca*, and *N. mucosa*). Furthermore, in these studies, strains were identified with acid detection tests that were not appropriate for the detection of the relatively small amounts of acid produced by *Neisseria* species, and additional differential tests now used to accurately identify commensal *Neisseria* species were unknown.

Some studies have been performed using selective differential media that inhibited the growth of non-neisserial species and selected for commensal *Neisseria* spp., by differentiating either between the asaccharolytic species or among several different groups of species.

The commensal *Neisseria* species have been determined to colonize the throats of adults as follows: *N. sicca*, 45%; *N. perflava*, 40%; and *N. subflava*–*N. flava*, 11% (Stechmann and Berger, 1964). It must be remembered that because nitrate reduction was not used as a differential test for the identification of *Neisseria* species at that time, strains of *N. mucosa* were not recognized by these authors. Although not verifiable, it is reasonable to assume that *N. mucosa* strains were present in the throats of these individuals. Thus, in this study, strains of *N. mucosa* were identified as either *N. sicca* or *N. perflava*, and the prevalences of these latter species were overestimated by the inclusion of strains of *N. mucosa* in their numbers (Berger and Miersch, 1970). Using a selective medium at a later time, Berger found that asaccharolytic strains, i.e., strains that do not produce detectable acid from glucose, maltose, sucrose, or fructose, accounted for 15% of all neisserial isolates, but, because nitrate reduction was not used to differentiate between *N. cinerea* and *M. catarrhalis* until a few years later, the relative colonization rates of these individual species were not given.

Pathogenicity Among the *Neisseria* and related species, only *N. gonorrhoeae* is always considered to be pathogenic and cause disease; *N. gonorrhoeae* is not considered to be normal flora under any circumstances. *N. gonorrhoeae* strains may infect the mucosal surfaces of urogenital sites (cervix, urethra, rectum) and the oro- and nasopharynx (throat), causing symptomatic or asymptomatic infections. Gonococcal infections of the urogenital sites are more frequently symptomatic than asymptomatic; however, asympto-

matic infections may occur. Gonococcal infections of the oro- and nasopharynx and the rectum may be asymptomatic more frequently than symptomatic. The prevalence of gonorrhea varies geographically (Lind, 1990; van der Heyden et al., 2000).

N. meningitidis causes epidemic meningitis in many parts of the world such as sub-Saharan Africa. Certain types of *N. meningitidis* are usually associated with meningitis. Of a total of 13 serogroups of *N. meningitidis*, strains belonging to the serogroups A, B, C, and W-135 have most frequently been associated with epidemics. Group A strains have been associated with most epidemics, whereas group B, C, W-135, and Y strains have caused sporadic epidemics. Strains of *N. meningitidis* may be carried as normal flora in the throat or nasopharynx. Between 3% and 30% of healthy persons in nonepidemic geographic areas may be asymptomatic carriers of *N. meningitidis*, i.e., meningococci have colonized their throats without causing disease. The carrier state may persist for many months. *N. meningitidis* may be considered as part of the oral normal flora particularly in crowded human settings, such as child care facilities and military camps. *N. meningitidis* is a major cause of bacterial meningitis and septicemia worldwide. From its commensal state in the pharyngeal mucous membrane, *N. meningitidis* may opportunistically disseminate to the bloodstream and, in the absence of bactericidal serum activity, cause sudden onset of disease. Systemic meningococcal disease affects primarily small children and adolescents, often leading to neurological sequelae or a fatal outcome. Despite the unique disease manifestations associated with *N. meningitidis* and *N. gonorrhoeae*, many of their basic strategies for successful colonization of their exclusive human hosts are highly conserved.

The ability of *N. meningitidis* to inflict damage on its host is correlated with adherence to mucosal epithelial cells in the nasopharynx and further invasion of subepithleial tissues and blood vessels (Nassif, 1999; Hardy et al., 2000). The most important virulence factors contributing to disseminated disease are pili, IgA1 protease, LOS, outer membrane proteins, and capsule polysaccharides (Poolman et al., 1995). *N. meningitidis* releases large amounts of the potent LOS through blebs or lysis. Meningococcal LOS exerts its impact on pathogenesis by facilitating close attachment to and inducing a potent cytokine response in host cells, mainly through TNF-α and interleukin-1 (Brandtzæg, 1995).

N. gonorrhoeae and *N. meningitidis* can cause conjunctivitis, and epidemics of gonococcal ophthalmia occur in developing countries (Koch, 1883; Myerhoff, 1911; Maxwell-Lyons and Amies, 1949). "Gonococcal" eye infection was common in Egypt even after the introduction of antimicrobial agents (Maxwell-Lyons and Amies, 1949). For example, in one study, conjunctival scrapings from children with trachoma and conjunctivitis contained intracellular diplococci. The Gram-negative bacterial strains isolated resembled *N. gonorrhoeae* in growth characteristics and sugar utilization patterns but could not be typed as gonococci. The colony morphology resembled that of meningococci more than that of gonococci. The DNA of the Egyptian isolates was cleaved by the restriction enzyme *Hae*III, like that of *N. meningitidis* but not like that of *N. gonorrhoeae*. The frequency of transformation of a temperature-sensitive mutant of *N. gonorrhoeae* to the ability to grow at the non-permissive temperature was 5–10 fold lower when DNA from two of the Egyptian isolates or from two *N. meningitidis* strains was used for transformation than when DNA from two *N. gonorrhoeae* strains was used (Mazloum et al., 1986). One of the Egyptian isolates exhibited 68–73% DNA relatedness to *N. gonorrhoeae* DNA and 57–63% DNA relatedness to *N. meningitidis* DNA. The Egyptian isolates were therefore thought to be a variant of *N. gonorrhoeae*, which Mazloum et al. (1986) termed "*N. gonorrhoeae* subsp. *kochii*". 16S rDNA analysis confirmed "*N. gonorrhoeae* subsp. *kochii*" as an intermediate between *N. gonorrhoeae* and *N. meningitidis*.

Other *Neisseria* species are considered to be commensals; they colonize the host without causing disease. Strains of these species are normal flora in the throat and appear to be opportunistic pathogens; they may cause infections although they are not routinely associated with specific types of infections or infections of specific sites. Most *Neisseria* species have been isolated occasionally from blood, cerebrospinal fluid, abscesses, etc., but no consistent association between any species and syndrome has been established that would warrant designating any of these species as pathogens. Some infections caused by commensal *Neisseria* species have occurred in persons who have deficient immune systems and who therefore may be predisposed to infections with organisms that would not normally cause disease.

Antibiotic susceptibility   Neisserial species are naturally susceptible to most antibiotics active against Gram-negative bacteria. The antibiotic susceptibilities are monitored through disk diffusion or agar dilution methods. However, antibiotic resistance is now widespread among the pathogenic neisseriae, more so in *N. gonorrhoeae* than in *N. meningitidis*, and occurs as both chromosomally mediated resistance to a variety of agents and plasmid-mediated resistance to penicillins (penicillinase/beta-lactamase producing strains) and to tetracycline (Dillon et al., 1983; Mendelman et al., 1988; Rice and Knapp, 1994). Penicillin resistance due to changes in the genes encoding PBPs is acquired through horizontal gene transfer by transformation (Spratt et al., 1992). Penicillin insensitivity in neisserial species is associated with changes in penicillin-binding protein PBP-2 occurring by horizontal gene transfer (Spratt, 1988; Spratt et al., 1989; Maggs et al., 1998). Resistance to fluoroquinolones in *N. gonorrhoeae* has already emerged and resistance to narrow-spectrum cephalosporins in *N. meningitidis* and *N. gonorrhoeae* is emerging (Fox and Knapp, 1999). Spectinomycin resistance in *N. gonorrhoeae*, due to point mutations in the gene encoding the ribosomal protein S12, and chloramphenicol-resistant *N. meningitidis* strains are increasing problems (Galimand, 1999, 2000). Sulfonamide resistance due to changes in the gene encoding dihydropteroate (*dhps*) was acquired through transformation (Rådstrøm et al., 1992). Comprehensive antibiotic susceptibility data are not available for commensal *Neisseria* spp. Generally, the commensal *Neisseria* spp. have been reported to be susceptible at least *in vitro* to penicillin, ampicillin, and tetracycline, although some strains may possess the TetM determinant. Strains of *N. cinerea*, however, appear to exhibit uniformly decreased susceptibility or resistance to erythromycin when tested by procedures for *N. gonorrhoeae* (Knapp and Koumanis, 1999).

Sulfonamide resistance is predominant in the *N. meningitidis* ET-5 complex, a distinctive group of genetically closely related clones that has been responsible for an epidemic of meningococcal disease in Norway since the mid-1970s. Clones of the *N. meningitidis* ET-5 complex have been identified as the causative agents of recent outbreaks and epidemics in many parts of the world. Analysis of sulfonamide susceptibility of isolates of the ET-5 complex from various geographic sources showed that there was no difference in resistance according to geographic source of the isolates, demonstrating that sulfonamide resistance is an essentially invariant property of clones of the ET-5 complex (Caugant et al., 1989).

Antigens and vaccines   Vaccines directed against the polysaccharide capsule are available for four of the five pathogenic meningococcal serogroups (A, C, Y, W), but these vaccines offer only short-duration protection, may induce tolerance after repeated immunization, and are ineffective in infants. New conjugate polysaccharide vaccines against serogroups A, C, W-135, and Y are now available. The meningococcal C glycoconjugate vaccine currently available is approximately 70% effective for 15–17 year olds. The challenge remains in finding an effective vaccine against meningococci that express the serogroup B polysaccharide, as the sialic acid parts of this capsule mimic carbohydrates found in human fetal nervous tissue and other tissues (Bøvre, 1980). This problem is exacerbated by the continual variation of meningococcal protein antigens generated by both mutation and recombination. It is likely that vaccines directed against these molecules will become less effective with the passage of time.

Antigenic variation of surface components such as fimbriae, Opa, and LPS precludes their being relevant vaccine candidates. Other noncapsular vaccine candidates are being sought, but many are variable and give weak protection. The descending order of preference for a protein vaccine candidate employed by some manufacturers is: 1) secreted, 2) outer membrane, 3) lipoprotein, 4) periplasmic, 5) integral membrane. The genome projects have opened up new approaches in the search for new vaccine candidates. The entire genome sequence of a virulent serogroup B strain (MC58) was used to identify vaccine candidates. A total of 350 candidate antigens were expressed in *E. coli*, purified, and used to immunize mice. The sera allowed the identification of proteins that are surface exposed, that are conserved in sequence across a range of strains, and that induce a bactericidal antibody response, a property known to correlate with vaccine efficacy in humans. Among the candidates identified were four lipoproteins, two outer membrane proteins, a membrane protein, murein lytic transglycolase, HtrA serine protease, an autotransporter, and members of the iron transport system (Pizza et al., 2000). Other multiple component approaches including exploiting class I outer membrane proteins, pilin, and LOS conjugates are being pursued. Targeting the semi-conserved section of the pilus had produced some degree of cross-reactivity and an LOS conjugate had induced antibody production. In some animal models, however, the production of bactericidal antibodies may not be necessary to give protection. On the other hand, the detection of bactericidal and opsonophagocytic antibodies in mice may not guarantee that the same will be induced in the human host.

ENRICHMENT AND ISOLATION PROCEDURES

Details of the isolation and processing of *Neisseria* species from humans have been provided by Morello and Bonhoff (1980) and Bartlett and Finegold (1978). In brief, specimens considered to yield pure cultures (blood, spinal fluid, urethral pus, joint fluid) are plated on prewarmed or room temperature chocolate agar (blood agar heated to 80–90°C to rupture the blood cells) and incubated at 36–37°C for a minimum of 48 h in 3–10% CO_2

atmosphere having high humidity. Specimens from body sites that may contain contaminants (cervix, oropharynx) are plated on modified Thayer–Martin medium (Martin et al., 1974) or New York City medium (Faur et al., 1973) and incubated as above. Selective media for *N. meningitidis* and *N. gonorrhoeae* contain four antimicrobial agents—vancomycin, 3–4 µg/ml; colistin, 7.5 µg/ml; trimethoprim, 5 µg/ml; and nystatin, 13.5 µg/ml—to inhibit Gram-positive bacteria, Gram-negative bacteria including commensal *Neisseria* spp., swarming *Proteus* species, and fungi, respectively. Nasopharyngeal samples may be inoculated onto chocolate and blood agar media. Generally, only morphology, Gram stain, oxidase reaction, and acidification of certain sugar media (particularly glucose, maltose, lactose, and sucrose) are used for routine identification. Immunofluorescence is often used for the gonococci and serological grouping by means of agglutination tests for the meningococci. Other *Neisseria* species from humans or animals can be isolated on Mueller–Hinton agar with or without 3% defibrinated sheep blood; once obtained in pure culture they can be identified by the characteristics listed in Table BXII.β.76.

Transport The best method for preserving viable organisms is the inoculation of specimens directly onto a nutrient medium and incubation at 35–37°C in a CO_2-enriched atmosphere immediately after collection. If specimens must be transported and it is not possible to incubate the inoculated media immediately before transport, it is more important to place the inoculated plates in a CO_2-enriched atmosphere than to incubate them at 35–37°C. Inoculated media may be held at room temperature in a CO_2-enriched atmosphere either in a candle extinction jar or commercial CO_2-generating zip-lock bags for up to 5 h without considerable loss of viability. If specimens must be transported through very high or low temperatures, the samples should be transported in a Styrofoam container. Transport of clinical samples is still often performed in semisolid non-nutritive, charcoal-containing transport media, but this method is inferior to the use of a nutrient transport system in a CO_2-enriched atmosphere. If specimens must be transported a long distance, the inoculated media should be incubated for 18–24 h before being transported, and the specimen should arrive within 48 h.

MAINTENANCE PROCEDURES

Isolates should be subcultured every 18–24 h to maintain maximum viability. A neisserial isolate will usually survive for no longer than 48 h in culture, although some isolates may survive for 72–96 h. *N. meningitidis* and *N. gonorrhoeae* are particularly sensitive to cold temperatures and autolysis (Morse and Bartenstein, 1974) and therefore need to be subcultured frequently. Frequent subculturing may lead to the loss of important factors such as fimbriae unless piliated colonies are picked, as observed in a stereomicroscope.

The best method of long-term preservation is by lyophilization in a rich broth (e.g., trypticase soy broth) containing 6% lactose, with storage of the vials at 4°C (Heckly, 1961). For long-term storage in −70°C or liquid nitrogen the bacteria are resuspended in broth containing 20% glycerol or 10–50% serum (Morello and Bonhoff, 1980). The loss of cells during freezing may be minimized by rapid freezing of the specimen in an ethanol bath containing dry ice.

Direct detection in clinical specimens Direct nonculture tests are available for detecting *N. meningitidis* and *N. gonorrhoeae*. Com-

mercial latex agglutination tests and coagglutination tests are used to detect meningococcal polysaccharide capsular antigens in body fluids such as spinal fluid. These kits contain polyvalent antibodies against the serogroups A, C, Y, and W135 and a separate reagent for serogroup B that also detects the cross-reacting *E. coli* K1 antigen. Antigen detection should always be performed along with Gram staining and culture. Direct detection of nucleic acids can be performed by PCR and other amplification techniques (Crotchfelt et al., 1997; Backman et al., 1999; Palmer et al., 2003). Relevant target DNAs for these assays are 16S rDNA, *porA* (Ni et al., 1992), IS1106, and the gene encoding CtrA, which is involved in the excretion of capsular polysaccharide (Porritt et al., 2000). Amplifying the gene encoding SiaD can provide species and serogroup identification in the same assay (Borrow et al., 1997). A direct nucleic acid detection of 16S ribosomal RNA from *N. gonorrhoeae* with a chemiluminescence-enhanced probe assay is useful for the direct detection of gonococci in clinical samples (Granato and Franz, 1990). More recently, a ligase chain reaction-based nucleic acid amplification has provided increased sensitivity in detecting *N. gonorrhoeae* in clinical samples (Hook et al., 1997).

PROCEDURES FOR TESTING SPECIAL CHARACTERS

Initial identification of neisseriae is based on observing Gram-negative diplococci (except for *N. elongata* and *N. weaveri* cells, which are short rods) taken from the colonies that are oxidase positive when tested with tetramethyl-*p*-phenylendiamine (Kovács, 1956). Another routine test used in initial identification is the production of acid from glucose, maltose, lactose, and sucrose, either by inoculation of cystine trypticase agar supplemented with 1% sugar or by measuring preformed enzymes in a rapid sugar fermentation tests (Table BXII.β.76) (Knapp, 1988). *N. lactamica* is the only neisserial species that possesses a typical β-galactosidase that is produced constitutively and will hydrolyze the chromogenic substrate *o*-nitro-phenyl-β-D-galacto-pyranoside (ONPG). Rapid methods, such as RapidNE or Quad-FERM (Bio-Merieux, Vitek, France), permit the detection of acid from neisseriae within 4 h. However, since many of the reactions in different neisserial species are identical, other phenotypic tests are needed for rapid, presumptive identification. All *Neisseria* species except *N. elongata* are catalase positive when tested with 3% hydrogen peroxide and observing the prompt evolution of bubbles of gas. The superoxol test is analogous to the catalase test, but is performed with a 30% hydrogen peroxide solution (Arko and Odugbemi, 1984). This reagent is most useful in differentiating between *N. gonorrhoeae* (strongly positive), *N. cinerea* (weakly positive), and *Kingella denitrificans* (superoxol negative). The biochemical reactions with the least variability among the species are the nitrate/nitrite reduction test (Cowan, 1974) and the synthesis of polysaccharide from 5% sucrose. The latter test employs bacteria grown on heart infusion agar containing 5% sucrose; after incubation for 2 days the colonies are tested with Lugol's iodine solution diluted 1:4 and the immediate development of a blue color indicates a positive reaction.

The presence of carbonic acid anhydrase is assayed as described by Berger and Issi (1971) and Berger and Piotrowski (1974). The test is performed by determining the minimum inhibitory concentration (MIC) of acetezolamide for cultures grown on heart infusion agar (Difco) containing 5% bovine serum. The cultures are incubated under an air atmosphere and also under air $+10\%$ CO_2. In general, if the MIC for an air-grown strain is approximately 32 mg/ml or lower, the strain produces carbonic anhydrase; this is confirmed by finding a much higher MIC for the strain when grown under 10% CO_2. The confirmation is especially important when the MIC is borderline (e.g., 16–62 mg/ml), as may occur, for example, with some strains of *N. elongata*.

Molecular strain typing The nature of gonococcal infection as a sexually transmitted disease, the rapid course of meningococcal disease, and the capacity of some serogroups to cause large-scale epidemics necessitate the use of sensitive, reliable, and rapid typing methods of characterizing strains. Because of the high plasticity of the neisserial genomes, the choice of typing methods is dependent on the epidemiological questions to be answered and on the population genetics of the organism under investigation. With highly clonal populations comprising independent non-recombining lineages such as gonococci and serogroup A meningococci, ribotyping, multilocus enzyme electrophoresis (MLEE), pulsed-field gel electrophoresis (PFGE), multilocus sequence typing (MLST), and PCR with short primers for random arbitrary polymorphism detection (RAPD) or with other gene-based primers each provides a constant measure of the relationship among strains (Yakubu et al., 1999). A more restricted portfolio of molecular methods, such as PFGE, MLEE, and MLST, is appropriate for the investigation of strains of the less clonal *N. meningitidis* serogroup B and C from localized outbreaks.

The worldwide spread of distinct ET types of *N. meningitidis* has been defined by MLEE (Caugant et al., 1989). However, MLST (Maiden et al., 1998) has been introduced to improve the resolution of bacterial isolates and contributed to the identification and tracking of the spread of virulent or drug-resistant pathogens. In MLST the nucleotide sequences of 450–500 bp of at least seven nonselected housekeeping genes are determined, revealing all of the variation at each locus. Sequences that differ at even a single nucleotide are assigned as different alleles. The large number of alleles at each of the different loci provides the ability to distinguish billions of different allelic profiles. It is extremely unlikely that two unrelated isolates would have the same allelic profile. The relatedness of isolates is displayed as a dendrogram constructed using the matrix of pair-wise differences between their allelic profiles; the pattern of nucleotide sequence variation within genes clearly resolves the major meningococcal lineages known to be responsible for invasive meningococcal disease around the world. MLST exploits the unambiguous nature and electronic portability of nucleotide sequence data for the characterization of organisms (Maiden et al., 1998). With this method the strain associations obtained are consistent with clonal groupings previously determined by MLEE. The advantage of MLST over other molecular typing methods is that sequence data are truly portable between laboratories, permitting expanding global databases for each species to be utilized on a World Wide Web site, thus enabling the exchange of molecular typing data for global epidemiology via the Internet (Enright and Spratt, 1999). The need for a method like MLST is clearly demonstrated by *N. meningitidis*, which is constitutively competent for transformation, is highly recombinogenic, and has an increased spontaneous mutation rate, representing an unusual challenge for DNA repair mechanisms to handle genome instability. MLST analysis of more than 100 isolates of *N. meningitidis* indicates that 1) identical alleles are disseminated among genetically diverse

isolates, with no evidence for linkage disequilibrium, 2) different loci give distinct and incongruent phylogenetic trees, and 3) allele sequences are incompatible with a bifurcating treelike phylogeny at all loci (Holmes et al., 1999). These observations are consistent with the hypothesis that meningococcal populations consist of organisms assembled from a common gene pool, with alleles and allele fragments spreading independently, together with the occasional importation of genetic material from other species (Feil et al., 1996, 2000, 2001). Further, they support the view that recombination is an important genetic mechanism in the generation of new meningococcal clones and alleles. Consequently, for anything other than the short-term evolution of *N. meningitidis*, some researchers state that a bifurcating treelike phylogeny is not an appropriate model (Holmes et al., 1999).

Recombinational exchanges in relatively diverse species such as *Neisseria* spp. are very likely to introduce multiple polymorphic nucleotide sites, whereas point mutation results in only a single polymorphism. MLST analysis of seven housekeeping genes in a large number of meningococcal isolates (n = 126) reveals that a single nucleotide site in a meningococcal housekeeping gene is at least 80-fold more likely to change as a result of recombination than as a result of mutation (Feil et al., 1996). This per-site recombination/mutation parameter value is estimated to be 10–50-fold for *E. coli* and approximately 50-fold for *Streptococcus pneumoniae*, another naturally competent species.

DIFFERENTIATION OF THE GENUS *NEISSERIA* FROM OTHER GENERA

The genus *Neisseria* contains mostly cocci. Distinguishing cocci from short rods by microscopic observation alone may sometimes be difficult, and a reliable test that can be applied in doubtful cases is to culture the organisms in the presence of subinhibitory levels of penicillin: rod-shaped organisms form long, stringy cells, whereas cocci retain their coccal morphology. It should be noted that *N. elongata* and *N. weaveri*, the rod-shaped species in the genus *Neisseria*, produce long cells by this method (Bøvre and Holten, 1970; Andersen et al., 1993).

Table BXII.β.75 of the chapter on the family *Neisseriaceae* lists other characteristics that differentiate the genus *Neisseria* from other members of the family. The rod-shaped species *N. elongata* and *N. weaveri* can be distinguished from acinetobacters by their oxidase positive reaction and from the rod-shaped moraxellae by their ability to reduce nitrite. They can be distinguished from kingellae by their generally larger and more opaque colonies and frequent inability to produce acid from glucose; more specifically, they differ from *K. kingae* by being nonhemolytic, from *S. indologenes* by being indole negative, and from *K. denitrificans* by failing to reduce nitrate (although some exceptional nitrate-positive *N. elongata* strains have been found).

Strains of the genus *Neisseria* can be differentiated from related genera such as *Kingella* and *Eikenella* as well as the more distantly related genera *Moraxella* and *Acinetobacter* by cellular morphology, catalase, production of acid from glucose, and nitrite reduction (Table BXII.β.78).

TAXONOMIC COMMENTS

Classification of the neisseriae into defined species is the subject of intense studies and has been in a state of continuous flux for the last four decades (see reviews by Henriksen, 1976; Bøvre and Hagen, 1981; Vedros, 1981; Tønjum et al., 1995b). The high frequency of horizontal gene transfer among the neisseriae obscures the phylogenetic relationships and species definitions in

TABLE BXII.β.78. Differentiation of the species of the genus *Neisseria* from other genera of the *Neisseriaceae* and two genera of *Moraxellaceae*[a]

Characteristic	*Neisseria*	*Kingella*	*Eikenella*	*Moraxella*	*Acinetobacter*
Cell morphology:					
Cocci	+	−	−	+	−
Rods	+[b]	+	+	+	+
Oxidase test	+	+	+	+	−
Catalase test	+	−	−	+	+
Acid from glucose	(+)	+	−	−	(+)
Nitrite reduction	+	+	+	(+)	−
True waxes present in cell wall	−	−	−	(+)	D
Mol% G + C of DNA	46–58	47–55	56–58	40–48	38–47

[a]Symbols: +, positive for the majority of strains of each species; (+), positive for all strains of the majority of species, only one strain known to be negative; D, positive and negative strains about equally represented. Data compiled from Hollis et al. (1969), Catlin (1978), Bøvre and Hagen (1981), Hoke and Vedros (1982a).

[b]Only two species, *N. elongata* and *N. weaveri*, consist of rods.

this genus. The degree of genetic relatedness between *N. gonorrhoeae* and *N. meningitidis* is extremely high. Kingsbury et al. (1969) found by thermal stability of hybrid DNA duplexes that the two species had at least 80% similarity in their nucleotide sequences, and the DNA–DNA hybridization studies by Hoke and Vedros (1982c) indicated a relatedness of 93%. More recently performed subtractive hybridization analysis and genome sequencing data indicate that the genome sequence similarity among the pathogenic neisseriae is higher than 96% (Tinsley and Naissif, 1996), which has been confirmed by the complete genome sequences (Parkhill et al., 2000; Tettelin et al., 2000). On purely genetic grounds, the two species therefore should be considered as subspecies of a single species. Yet, *N. meningitidis* and *N. gonorrhoeae* cause distinctly different kinds of clinical infections, and from a practical viewpoint it seems desirable to continue to consider these organisms as separate species.

Lactose-positive strains of organisms resembling *N. meningitidis* were recognized as early as 1934 (Jessen, 1934), but were largely ignored until the report by Hollis et al. (1969). *N. lactamica* showed a close relationship to *N. meningitidis* by transformation studies and to the other "true neisseriae" by DNA–DNA hybridization, cellular fatty acid composition, and mol% G + C values (Bøvre et al., 1972; Hoke and Vedros, 1982a, c).

The close relationships between all of the "true neisseriae" have been demonstrated consistently by a variety of techniques (see Bøvre, 1980; Bøvre and Hagen, 1981; Hoke and Vedros, 1982a, c). Of particular interest have been the similarities among *N. flava*, *N. perflava*, and *N. subflava*. It is difficult to differentiate these three species by cultural and biochemical reactions, and in the eighth edition of *Bergey's Manual* they were incorporated in the single species *N. subflava* (Reyn, 1974). This close similarity is supported by DNA–DNA hybridization and genetic transformation studies (Hoke and Vedros, 1982a), as well as by 16S rDNA sequence analysis (Fig. BXII.β.62). Although a high level of DNA relatedness between *N. sicca* and the *N. subflava* group has been found (Hoke and Vedros, 1982a), biochemical distinction between *N. perflava* and *N. sicca* has also been reported (Berger and Catlin, 1975). Further confusion in the classification of these chromogenic neisseriae has been added by the finding of identical lipopolysaccharides in *N. subflava* and *N. canis* (Johnson et al., 1976).

New species in genus *Neisseria* since the last edition of *Bergey's Manual of Systematic Bacteriology* (1984) are *N. polysaccharea* (Riou and Guibourdenche., 1987), *N. macacae, N. iguanae, N. weaveri,*

N. animalis, and *N. dentiae*. The last four of these all have mammalian hosts other than humans.

Although the species *N. elongata* consists of rod-shaped cells, genetic transformation studies have shown that *N. elongata* has very high genetic affinities to the coccal species of *Neisseria* (Bøvre and Holten, 1970; Bøvre et al., 1977) and has no affinities to other genera of the family *Neisseriaceae* except a low affinity to *Kingella kingae*. *N. elongata* possesses carbonic anhydrase, which is characteristic of the "true neisseriae", and also possesses a fatty acid composition similar to other neisserial species (Jantzen et al., 1974; Hoke and Vedros, 1982c). The evidence shows that the rod-shaped *N. weaveri* belongs in the genus *Neisseria*, most closely related to *N. elongata*, *N. flavescens*, and *N. animalis*.

The taxonomic status of the previous "false neisseriae", that is the animal species *N. caviae*, *N. ovis*, and *N. cuniculi*, as well as *M. catarrhalis*, has been reassigned to the *Moraxellaceae* (Rossau et al., 1989; Pettersson et al., 1998b). These species were previously considered as a group within the *Neisseria* spp., because they share several phenotypic characteristics (Henriksen, 1976; Bøvre, 1984a). They are listed and described in the article on the genus *Moraxella* in this *Manual*.

ACKNOWLEDGMENTS

Stimulating discussions with Kjell Bøvre, Michael Koomey, and Bertil Pettersson are deeply appreciated. We are grateful for the receipt of *N. dentiae* strains from P.H. Sneath. The information provided in the first version of *Bergey's Manual of Systematic Bacteriology* (1984) by N.A. Vedros is greatly acknowledged.

FURTHER READING

Bøvre, K. 1980. Progress in classification and identification of *Neisseriaceae* based on genetic affinity. *In* Goodfellow and Board (Editors), Microbiological Classification and Identification, Academic Press Inc. (London), Ltd., London. pp. 55–72.

Morse, S.A. and J.S. Knapp. 1989. The genus *Neisseria*. *In* Starr, Stolp, Trüper, Balows and Schlegel (Editors), The Prokaryotes. A Handbook on Habitats, Isolation, and Identification of Bacteria, Vol.2, Springer-Verlag, Berlin. pp. 2495–2529.

Andersen, B.M., A.G. Steigerwalt, S.P. O'Connor, D.G. Hollis, R.S. Weyant, R.E. Weaver and D.J. Brenner. 1995. *Neisseria weaveri* sp. nov., formerly CDC group M-5, a Gram-negative bacterium associated with dog bite wounds. J. Clin. Microbiol. *31*: 2456–2466.

Spratt, B.G., L.D. Bowler, Q.Y. Zhang, J. Zhou and J.M. Smith. 1992. Role of interspecies transfer of chromosomal genes in the evolution of penicillin resistance in pathogenic and commensal *Neisseria* species. J. Mol. Evol. *34*: 115–125.

Neisseria species are relatively inert biochemically, but those activities and characteristics that are of determinative value are shown in Table BXII.β.76.

List of species of the genus Neisseria

1. **Neisseria gonorrhoeae** (Zopf 1885) Trevisan 1885, 106[AL] (*Merismopedia gonorrhoeae* Zopf 1885, 54.)
   go.nor.rhoe′ ae. Gr. n. *gonorrhoeae* gonorrhoea; M.L. gen. n. *gonorrhoeae* of gonorrhoea.

   Common name: gonococcus.

   The biochemical characteristics are as described for the genus and as listed in Table BXII.β.76. Primary isolation is made on chocolate agar at temperatures of 35–36°C under an atmosphere containing 3–10% CO_2 and high relative humidity. Minimum growth temperature, 30°C. At 48 h, colonies are 0.6–1.0 mm in diameter, opaque, grayish white, raised, finely granular, glistening, and convex. They become mucoid with further incubation. The cocci are arranged in pairs. Primarily found in gonorrhea (purulent venereal discharges), oropharyngeal infection, anorectal infection, endometritis, conjunctivitis, and disseminated gonococcal infection. Also found in blood, the conjunctiva, petechiae, pharynx, and cerebrospinal fluid. Found only in humans.

   The mol% G + C of the DNA is: 49.5–53.3 (T_m, chromatography).

   Type strain: ATCC 19424, CCUG 26876, CIP 79.18, DSM 9188, NCTC 8375.

   GenBank accession number (16S rRNA): X07714.

   This species shows very high genetic relatedness to *N. meningitidis* (Tinsley and Nassif, 1996; Fig. BXII.β.62). "*N. kochii*" is considered to be a subspecies of *N. gonorrhoeae* and has no independent taxonomic status.

2. **Neisseria animalis** Berger 1960, 160[AL]
   an.i.mal′ is. L. n. *animal* animal; L. gen. n. *animalis* of an animal.

   Typical diplococci. Colonies are smooth, round, graywhite. Produces acid from saccharose, no acid from fructose and maltose (which is positive for *N. sicca* and *N. perflava*, the other saccharose-positive *Neisseria*) (Berger, 1962). Isolated from the throats of guinea pigs. Appears to be closely related to *N. denitrificans* (Fig. BXII.β.62).

   The mol% G + C of the DNA is: not determined.

   Type strain: ATCC 14678, CCUG 808, CIP 72.15, NCTC 10212.

   GenBank accession number (16S rRNA): AJ239288.

3. **Neisseria canis** Berger 1962, 455[AL]
   ca′ nis. L. gen. n. *canis* of the dog.

   The cells are typical diplococci, rarely tetrads. Colonies are smooth, butyrous, with a light yellowish tinge. Absorption peaks of the extracted pigment are similar to those of *N. lactamica* (Hoke and Vedros, 1982c). The cellular fatty acids are similar to those of the "true" neisseriae. Carbonic anhydrase is produced (Berger and Issi, 1971). Other characteristics are listed in Table BXII.β.76. Isolated from the throats of cats (Berger, 1962) and as an opportunist in a cat-bite wound of a human (Hoke and Vedros, 1982b).

The mol% G + C of the DNA is: 49.6 (T_m).

Type strain: ATCC 14687, CIP 103347, LMG 8383, NCTC 10296.

GenBank accession number (16S rRNA): L06170.

16S rDNA sequence analysis of the type strain suggests that this species is more closely related to genus *Kingella* than to *Neisseria* spp.

4. **Neisseria cinerea** (von Lingelsheim 1906) Murray 1939, 283[AL] (*Micrococcus cinereus* von Lingelsheim 1906, 396.)
   ci.ne′ re.a. L. fem. adj. *cinerea* gray.

   The cocci are plump and arranged in pairs or more often in scattered clusters. Colonies are small (1.0–1.5 mm in diameter), grayish white with entire edges, and slightly granular. The percentage of fatty acids having a chain length of over 16 carbon atoms is similar to that of the other neisseriae (Hoke and Vedros, 1982c). Carbonic anhydrase is produced (Berger and Issi, 1971). Other characteristics are listed in Table BXII.β.76. Found in the nasopharynx of humans. Opportunistic pathogen (e.g., newborn ocular infections) (Bourbeau et al., 1990).

   The mol% G + C of the DNA is: 49.0–50.9 (T_m, Bd).

   Type strain: ATCC 14685, CCUG 2156, CIP 73.16, DSM 4630, LMG 8380, NCTC 10294.

5. **Neisseria denitrificans** Berger 1962, 455[AL]
   de.ni.tri′ fi.cans. L. prep. *de* away from; L. n. *nitrum* soda; M.L. n. *nitrum* nitrate; M.L. v. *denitrifico* to denitrify; M.L. part. adj. *denitrificans* denitrifying.

   The percentage of fatty acids with chain length over 16 carbon atoms and the fatty acid profile are similar to those of the "true" neisseriae (Hoke and Vedros, 1982c). Carbonic anhydrase is produced (Berger and Issi, 1971). Other characteristics are as listed in Table BXII.β.76. Isolated from the throats of guinea pigs.

   The mol% G + C of the DNA is: 55.6 (T_m).

   Type strain: ATCC 14686, CCUG 2155, CIP 72.16, NCTC 10295.

   GenBank accession number (16S rRNA): L06173, M35020.

6. **Neisseria dentiae** Sneath and Barrett 1997, 915[VP] (Effective publication: Sneath and Barrett 1996, 357.)
   den′ ti.ae. M.L. fem. gen. sing. n. *dentiae* of (Dr.) Dent; named for Vija E. Dent (Mrs. Pratley) in recognition of her pioneering work on neisserias from dental plaque.

   Cells are small cocci about 1 μm in diameter, arranged mainly as diplococci with occasional tetrads, but without diplobacilli. Nonmotile. No endospores. Growth is aerobic with optimal temperature close to 35°C. No growth under anaerobic conditions. Colonies on horse blood agar are round transparent and domed, without obvious yellow pigment, about 1 mm in diameter after 24 h growth at 35°C. No hemolysis on horse blood; weak hemolysis on sheep blood is occasionally found. Oxidase positive. Catalase positive. Indole negative. Gelatin negative. Nitrate negative. Most strains reduce nitrite. Found in dental plaques of domestic cows. Resembles *N. animalis, N. canis,* and *N. iguanae* phenotypically, but is distinguished from the first two by being positive for acidification of gluconate, D-glucose, and usually D-fructose, and from the third by lacking predominant tetrad arrangement and distinct α-hemolysis and by growing on nutrient agar and usually acidifying D-fructose. May have impact on dental microbiology because members of the genus *Neisseria* rapidly utilize oxygen and this may contribute to the anaerobic microenvironment found in dental plaques.

The mol% G + C of the DNA is: not determined.

Type strain: V33, SHI/3848, ATCC 700276, CIP 106968.

GenBank accession number (16S rRNA): AF487709.

16S rDNA sequence analysis shows highest homology to *N. canis* (as for the other neisseriae from animal habitats other than humans).

7. **Neisseria elongata** Bøvre and Holten 1970, 73[AL]

e.lon'ga.ta. L. fem. part. adj. *elongata* elongated, stretched out.

Rods, short and slender, ~0.5 μm in diameter, often arranged as diplobacilli or in short chains. A marked elongation effect of sublethal concentrations of penicillin occurs during growth, with formation of very long filaments. Often fimbriated. Colonies on blood agar are shiny and low convex with an entire edge, ~2–3 mm in diameter after 48 h of incubation. The colonies are semi-opaque to grayish white with a yellowish tinge due to pigment production. Older colonies may attain a diameter of 4–5 mm and often show granular spreading zones around the periphery, or the colonies become irregular in outline with spreading projections. The colony texture is usually clay-like and coherent, and the growth mass when collected is lumpy and difficult or impossible to disperse. No hemolysis occurs. Agar corrosion with a peripheral groove and a central pit is often observed under the colony. The spreading and corrosion are related to the presence of fimbriae on the cells; nonfimbriated or less fimbriated variants give rise to colonies that are often smaller, nonspreading, and noncorroding. Optimal growth temperature 33–37°C. Capable of growing weakly on blood agar under anaerobic conditions. Grows on simple peptone media. The specific growth requirements are not known.

Capsules are not formed. Nonmotile. Oxidase positive. Usually acid is not formed from glucose, but some strains may form small amounts of acid. Nitrite reducing, but usually not nitrate reducing (see below). Catalase activity is usually not detectable but may be positive or weakly positive in some strains. No liquefaction of coagulated serum or gelatin occurs. Urease negative. Phenylalanine is not produced except for weak reactions observed with some strains. Carbonic anhydrase is produced (Berger and Issi, 1971). Highly sensitive to penicillin. The species lacks true cellular waxes (Jantzen et al., 1976) and contains heptose (Jantzen et al., 1976). Other characteristics are listed in Table BXII.β.76. Isolated from the pharynx of healthy individuals and from cases of pharyngitis. Also isolated from bronchial aspirates, pus from perimandibular abscesses, blood cultures during endocarditis, and from the urinary tract. So far recorded only from human sources. Considered as a largely harmless parasite.

The mol% G + C of the DNA is: 53.0–53.5 (T_m).

Type strain: ATCC 25295, CCUG 2043, CCUG 2130 A, CIP 72.27, LMG 5124, NCTC 10660.

GenBank accession number (16S rRNA): L06171.

The cells are frequently competent in genetic transformation and can be identified by transformation. The competence is apparently associated with the fimbriated state (Bøvre and Holten, 1970; Bøvre et al., 1977). There is a distinct genetic affinity by transformation between *N. elongata* and the coccal species of the "true" neisseriae. The species also fits in with the "true" neisseriae with respect to cellular lipid and carbohydrate composition and the characteristics of glycolytic enzymes (see reviews by Bøvre, 1980; Bøvre and Hagen, 1981).

The species is genetically somewhat heterogeneous (Bøvre et al., 1972) and it is now proposed to consist of three subspecies: *Neisseria elongata* subsp. *elongata, Neisseria elongata* subsp. *glycolytica,* and *Neisseria elongata* subsp. *nitroreducens.*

Another subspecies, "*N. elongata* subsp. *intermedia*", was proposed by Berger and Falsen (1976). This subspecies presently has no standing in nomenclature. It is catalase positive and immunologically distinct from the type strain of the elongata. It may possibly be identical to the subsp. *glycolytica.*

a. **Neisseria elongata** *subsp.* **elongata** Bøvre and Holten 1970, 73[AL]

Oxidase positive, aerobic, nonmotile, rod-shaped bacterium. No acid production from glucose or other carbohydrates. Catalase negative. Urease, indole, and motility negative. Nitrite reducing. Differs from subsp. *glycolytica* and subsp. *nitroreducens* by showing no acidification of glucose media and being catalase negative.

The mol% G + C of the DNA is: 56 (T_m).

Type strain: ATCC 25295, CCUG 2043, CCUG 2130 A, CIP 72.27, LMG 5124, NCTC 10660.

GenBank accession number (16S rRNA): L06171.

b. **Neisseria elongata** *subsp.* **glycolytica** Henriksen and Holten 1976, 480[AL]

gly.co.ly'ti.ca. Gr. glyko- from Gr. glykys sweet; Gr. adj. lyticus dissolving; M.L. fem. adj. *glycolytica* meant to indicate an ability to attack glucose.

Oxidase positive, aerobic, nonmotile, rod-shaped bacterium. Nitrite reducing. Catalase positive. Urease, indole, and motility negative. No acid production from carbohydrates. Differs from elongata by being catalase positive, by causing a weak acidification of glucose media, and by forming colonies with a smooth texture. Quantitative genetic transformation data show identity reactions between subsp. *elongata* and subsp. *glycolytica,* and they are also indistinguishable in terms of fatty acid composition (Bøvre et al., 1977).

The mol% G + C of the DNA is: 56 (T_m).

Type strain: ATCC 29315, CCUG 6508, CIP 82.85, NCTC 11050.

Although subspeciation of *N. elongata* subsp. *elongata* and *N. elongata* subsp. *glycolytica* has been proposed based on differences in glycolytic properties (Henriksen and Holten, 1976), genetic and other evidence for subspeciation are rather weak. In fact, Bøvre et al. (1977) has reported general species identification of the glycolytic strain 6171/75 ATCC 23915 with the type strain M2 = ATCC 25295 by genetic transformation. They found DNA homology of up to 86% between the glycolytic strain (6171/75) and the type strain (M2) by quantitative streptomycin resistance transformation.

c. **Neisseria elongata** *subsp.* **nitroreducens** Grant, Brenner, Steigerwalt, Hollis and Weaver 1990, 2596^VP

nit.ro.re′ du.cens. Gr. *nitroreducens* the agent that reduces nitrate.

Oxidase positive, aerobic, nonmotile, rod-shaped bacterium. Reduction of nitrate and nitrite with no gas formation. Negative reactions for catalase, urease, indole, and motility; and no acid production from carbohydrates. Differences from subsp. *elongata*: reduces nitrate and sometimes causes a weak acidification of glucose media. Differences from *N. elongata* subsp. *glycolytica*: catalase negative, reduces nitrate. Found in throat or sputum and from blood, with many of the systemic isolates being associated with endocarditis. Rarely occurring, but can cause serious human infections (Hofstad et al., 1998). The fact that it is found in association with endocarditis and other systemic diseases differentiates it from the other *N. elongata* subspecies.

The mol% G + C of the DNA is: 55–58 (*T_m*).

Type strain: B109, ATCC 49377, CCUG 30802, CIP 103511, NCTC 12736.

N. elongata biovar nitroreducens was previously termed CDC group M-6. By the hydroxyapatite method, DNAs from the M-6 strains showed an average of 78% relatedness to M-6 reference strain B1019 in reactions at 60°C and 73% relatedness in reactions at 75°C (Grant et al., 1990). This organism is biochemically similar to *Kingella denitrificans* and displays a cellular fatty acid profile consistent with CDC groups M-5 and EF-4 as well as with *N. elongata*. The relatively high mol% G + C content of DNA supported the establishment of this subspecies. 16S rDNA sequence analysis show high similarities among the three subspecies of *N. elongata* (Fig. BXII.β.62).

8. **Neisseria flavescens** Branham 1930, 849^AL

fla.ves′ cens. L. v. *flavus* to become golden yellow; L. part. adj. *flavescens* becoming golden yellow.

The characteristics are as described for the genus and as listed in Table BXII.β.76. Cocci in pairs and tetrads. Colonies are smooth and opaque with golden yellow pigment. Nitrate and nitrite reducing. Synthesis of polysaccharide positive (iodine test). Found in cerebrospinal fluid of patients with meningitis and in cases of septicaemia. Rare (Branham, 1930; Wertlake and Williams, 1968). Found in the pharynx of humans (rare).

The mol% G + C of the DNA is: 46.5–50.1 (*T_m*, Bd, chromatography).

Type strain: ATCC 13120, CCUG 345, CCUG 17913, CIP 73.15, LMG 5297, NCTC 8263.

GenBank accession number (16S rRNA): L06168.

9. **Neisseria iguanae** Barrett, Schlater, Montali and Sneath 1994b, 852^VP (Effective publication: Barrett, Schlater, Montali and Sneath 1994a, 201.)

i.gu′ an.ae. M.L. fem. gen. sing. n. *iguanae* of the iguana lizard; from Sp. fem. n. iguana.

Cells are small cocci about 0.8 μm in diameter, largely arranged as diplococci but show numerous tetrads in culture. Nonmotile. Surface colonies grown on 5% sheep or horse blood agar are nonpigmented, round, transparent, domed, about 1 mm in diameter after 72 h at 35°C with marked zones of α-hemolysis. Oxidase positive, catalase positive, urease and deoxyribonuclease negative. Weak acidity usually from glucose, sucrose, and gluconate. Strongly positive for alkaline phosphatase. Gluconate positive. Strongly positive for alkaline phosphatase 2. No endospores. Aerobic growth, no growth under anaerobic conditions. Mesophilic. Grows between 25°C and 37°C. Chemoorganotrophic. May be isolated from the oral cavity of healthy lizards. Associated with septicemia and abscesses in iguanid lizards (*Iguana iguana* and *Cyclura cornuta*).

The mol% G + C of the DNA is: 50.8–52.0 (*T_m*).

Type strain: ATCC 51483, NVSL 85737.

10. **Neisseria lactamica** Hollis, Wiggins and Weaver 1969, 72^AL

lac.ta.mi′ ca. L. n. *lac* milk, from *lactose* milk sugar; L. adj. *amicus* fond of; M.L. fem. adj. *lactamica* fond of lactose.

The characteristics are as described for the genus and as listed in Table BXII.β.76. Cocci arranged in pairs. Colonies are small, smooth, translucent, slightly butyrous, and often have a yellowish tinge. Primary isolation is made on chocolate agar at temperatures of 35–36°C under an atmosphere containing 3–10% CO_2 and high relative humidity. Minimum growth temperature 30°C. At 48 h, colonies are 0.6–1.0 mm in diameter, opaque, grayish white, raised, finely granular, glistening, and convex. They become mucoid with further incubation. The only *Neisseria* species that produces acid from lactose. Primarily found in the nasopharynx, most commonly found in children and young teenagers. Rarely pathogenic, but has been identified as a possible cause of meningitis.

The mol% G + C of the DNA is: 49.5–53.3 (*T_m*, chromatography).

Type strain: ATCC 23970, CCUG 5853, CIP 72.17, DSM 4691, NCTC 10617.

GenBank accession number (16S rRNA): AJ239286.

Genetic transformation, DNA–DNA hybridization, and 16S rDNA sequence analysis have indicated a close relationship between *N. lactamica*, *N. meningitidis*, and *N. gonorrhoeae*, but not as close as that between *N. meningitidis* and *N. gonorrhoeae* (Siddiqui and Goldberg, 1975; Hoke and Vedros, 1982a; Garborg and Tønjum, personal communication).

11. **Neisseria macacae** Vedros, Hoke, and Chun 1983, 519^VP

ma′ ca.cae. Port. n. *macaco* female monkey; N.L. fem. gen. *macacae* of a monkey, referring to the source of the isolates.

Cocci occurring in pairs with adjacent sides flattened or singly; cells are nonmotile and 0.6–1.0 μm in diameter. Colonies on Mueller–Hinton agar or nutrient agar are slightly raised, yellowish green, and glistening, have entire edges, and are about 1–1.5 mm in diameter after 17 h of incubation at 37°C in a humid atmosphere. Aerobic. Optimal growth at 35–37°C; less growth at 30°C. Growth enhanced by 5–8% CO_2 at 30°, 35°, or 37°C. Slight or no growth at 22° and 42°C. Moderate hemolysis of horse blood agar and rabbit blood agar. Oxidase positive. Catalase positive. Nitrate is not reduced, nitrite is reduced with the production of gas. Polysaccharides are produced from sucrose, DNA is hydrolyzed, tributyrin is not hydrolyzed. Isolated from the oropharynges of rhesus monkeys.

The mol% G + C of the DNA is: 50–51 (T_m).

Type strain: M-740, ATCC 33926.

GenBank accession number (16S rRNA): L06169.

12. **Neisseria meningitidis** (Albrecht and Gohn 1901) Murray 1929, 8[AL] (*Micrococcus meningitidis* Albrecht and Gohn 1901, 498.)

me.nin.gi′ti.dis. Gr. n. *meninx, meningis* the membrane enclosing the brain; M.L. fem. n. *meningitis, meningitidis* inflammation of the meninges.

Common name: meningococcus.

Cocci arranged in pairs. Cellular division occurs in two planes with the second division at a right angle to the first. Transient tetrads can therefore be observed in wet mounts of young, growing cultures (less than 8-h-old). Primary isolation is made on blood agar, chocolate agar, or Mueller–Hinton agar. An atmosphere containing 3–10% CO_2 and high relative humidity enhances growth. Optimal temperatures 36–37°C. Colonies vary in size depending on the medium, extent of crowding, and length of incubation; in 18–24 h they are approximately 1.0 mm in diameter. Colonies are small, round, smooth, glistening, sometimes mucoid and translucent on Mueller–Hinton agar, and are often iridescent. Due to autolysis with age, colonies become more butyrous and rubbery to the touch of an inoculating needle. Other characteristics are listed in Table BXII.β.76. Some strains are erratic in producing acid from maltose and glucose (Jyssum and Jyssum, 1968; Bøvre, 1969). Found in cerebrospinal fluid as the causative agent of cerebrospinal meningitis and in blood as the cause of septicemia (including Waterhouse–Friderichsen syndrome), lower genital tract infections (rare), and pneumonia (rare). Can be cultivated from petechiae, joints, nasopharynx, and conjunctiva, occasionally found in venereal discharges. Frequently found in a commensal state in the oro- or nasopharynx of asymptomatic carriers.

The mol% G + C of the DNA is: 50–52 (T_m, chromatography).

Type strain: M1027, ATCC 13077, CCUG 3269, CIP 73.10, DSM 10036, NCTC 10025.

N. meningitidis shows a very high genetic relatedness to *N. gonorrhoeae* by DNA–DNA hybridization (>80%) (Kingsbury et al., 1969) and by subtractive hybridization (96% similarity) (Tinsley and Nassif, 1996).

13. **Neisseria mucosa** (von Lingelsheim 1906) Véron, Thibault and Second 1959, 508[AL] (*Diplococcus mucosus* von Lingelsheim 1906, 395.)

mu.co′sa. L. fem. adj. *mucosa* slimy.

The characteristics are as described for the genus and as listed in Table BXII.β.76. Cocci arranged in pairs. Colonies are large, mucoid, and often adherent. Most strains are nonpigmented (Berger and Miersch, 1970) or grayish to buff yellow colonies, but one strain has been described

as slightly yellow (Véron et al., 1959). Found in the nasopharynx of humans and, in one report, as part of the normal flora of the respiratory tissues in dolphins (Vedros et al., 1973). Occasionally pathogenic for humans, causing pneumonia in children (rare). Pathogenic for mice.

The mol% G + C of the DNA is: 50.5–52.0 (T_m, Bd).

Type strain: ATCC 19696, CCUG 26877, CIP 59.51, DSM 4631, NCTC 12978.

An organism called "*N. mucosa* biovar heidelbergensis" was described by Berger (1971). It differed from "*N. mucosa* biovar mucosa" in that some strains were pigmented, possessed deoxyribonuclease, and produced more gas from nitrite. "*N. mucosa* mucosa" has white and larger colonies of more dry consistence than "*N. mucosa* biovar heidelbergensis". Growth in broth results in pellet forming. "*Neisseria mucosa* biovar heidelbergensis" has smaller colonies of more butyrous consistence than "*N. mucosa* biovar mucosa". Yellow pigment on blood agar. Diffuse growth in broth. 16S rDNA sequence analysis shows significant differences between "*N. mucosa* biovar mucosa" and "*N. mucosa* biovar heidelbergensis". By 16S rDNA sequence analysis *N. mucosa* strains are most closely related to *N. sicca* and *N. subflava* biovar flava.

14. **Neisseria polysaccharea** Riou and Guibourdenche 1987, 163[VP]

pol.y.sac.cha.re.a. Gr. adj. *poly* many; Gr. n. *Saccharum* sugar; M.L. fem. adj. *polysaccharea* with many saccharides.

Unencapsulated cocci arranged in pairs or tetrads. Oxidative respiratory metabolism. Oxidase and catalase produced. All strains grow on selective medium producing small, grayish yellow colonies that are translucent and raised of about 2 mm in diameter after 24 h at 37°C. No growth occurs at 22°C. Large amounts of polysaccharide are produced on solid or in liquid medium containing 1% or 5% sucrose. Requires cystine–cysteine for growth on *Neisseria* defined medium (Catlin, 1973); in addition, some strains require arginine. Acid is produced from glucose and maltose, rarely from sucrose, but never from fructose, D-mannitol, or lactose. Tributyrin and *o*-nitrophenyl-β-D-galactopyranoside are not hydrolyzed. Nitrites are reduced, nitrates are not. Deoxyribonuclease, gelatinase, and γ-glutamyltransferase are not produced. No hemolysis is observed on horse blood agar. Produces extracellular polysaccharide and exhibits γ-glutamyltransferase activity. It is found in the nasopharynx of infants and children. No strain of this species has been recognized as a cause of human disease.

The mol% G + C of the DNA is: 52.5–54. (T_m, Bd).

Type strain: ATCC 43678, CCUG 18030, CIP 100113, NCTC 11858.

GenBank accession number (16S rRNA): L06167, AJ239289.

15. **Neisseria sicca** (von Lingelsheim 1908) Bergey, Harrison, Breed, Hammer and Huntoon 1923a, 43[AL] (*Diplococcus siccus* von Lingelsheim 1908, 476.)

sic′ca. L. fem. adj. *sicca* dry.

The characteristics are as described for the genus and as listed in Table BXII.β.76. Cocci arranged in pairs and tetrads. Colonies are usually large (up to 3 mm), grayish white, opaque, dry, wrinkled, and adherent. Forms opaque, dry, wrinkled, adherent, colonies but this may vary with some strains. Some strains may produce a xanthophyll pigment (Berger, 1961b); when extracted, this pigment shows absorption peaks similar to the pattern exhibited by the pigments of *N. perflava*, *N. flava*, and *N. mucosa* (Hoke and Vedros, 1982c). Spontaneous agglutination occurs in saline. The strain used in the report on envelope proteins is questionable (Russell et al., 1975). *N. sicca* may be serologically distinct from *N. subflava* (*N. flava*, *N. perflava*) and *N. flavescens* (Berger and Brunhoeber, 1961; Berger and Wulff, 1961). The species is serologically related to *N. mucosa* (Véron et al., 1959). Found in the nasopharynx, saliva, and sputum of humans (very common); opportunistic pathogen.

The mol% G + C of the DNA is: 49.0–51.5 (T_m, Bd, chromatography).

Type strain: ATCC 29256, CCUG 23929, CCUG 24959, CIP 103345, LMG 5290, NRL 30,016.

Genetic transformation and DNA–DNA hybridization indicate that *N. sicca* should be considered for inclusion with *N. perflava* and *N. flava* into the single species *N. subflava* (Catlin and Cunningham, 1961; Hoke and Vedros, 1982a). However, 16S rDNA sequence analysis shows closer relationship to *N. mucosa* and *N. macacae* (Pettersson and Tønjum)

16. **Neisseria subflava** (Flügge 1886) Trevisan 1889, 32^{AL} (*Micrococcus subflavus* Flügge 1886, 159; *Neisseria flava* Bergey, Harrison, Breed, Hammer and Huntoon 1923a, 43.)

sub.fla′va. L. pref. *sub* less than; L. adj. *flavus* yellow; L. fem. adj. *subflava* yellowish.

The characteristics are as described for the genus and as listed in Table BXII.β.76. Cocci arranged in pairs and tetrads. Colonies are smooth, transparent or opaque, often adherent. Grows on blood agar at 22°C. Some strains produce a yellowish pigment. Often agglutinates spontaneously in saline (Reyn, 1974). Found in secretions from the human nasopharynx and rarely in cerebrospinal fluid in cases of meningitis (Noguchi et al., 1963; Baraldes et al., 2000).

The mol% G + C of the DNA is: 48.0–51.0 (T_m, Bd, chromatography).

Type strain: ATCC 49275, CCUG 23930, CCUG 33675, CIP 103343, LMG 5313, NRL 30,017.

Neisseria subflava contains the previous species *Neisseria flava*, *Neisseria perflava*, and *Neisseria subflava*. *N. subflava* biovar flava and *N. subflava* biovar perflava produce acid from fructose, whereas *N. subflava* biovar subflava does not. *N. subflava* biovar perflava produces acid and polysaccharide from sucrose which *N. subflava* biovar flava and *N. subflava* biovar subflava do not. 16S rDNA sequence analysis shows that *N. subflava* biovar subflava and biovar perflava are considerably more closely related to each other than they are to *N. subflava* biovar flava. Inclusion of *N. sicca* into this species could be considered based on its relatedness to *N. subflava* biovar flava.

17. **Neisseria weaveri** Holmes, Costas, On, Vandamme, Falsen and Kersters 1993, 691^{VP}

wea′ver.i. M.L. gen. n. *weaveri* was named in honor of Robert E. Weaver.

The cells are broad, plump, medium-to-large, straight rods of varying length when grown on slants and plates, with a tendency to grow in chains or longer rods in broth cultures. They are nonmotile, aerobic, and non-salt requiring. Grow well between 25° and 35°C; most strains grow at 42°C. Colonies are gray-white with an entire border, flat, somewhat glistening, and smooth and variable in size. They are 1–2 mm in diameter after 24 h of incubation at 35°C and 2–4 mm after 48 h of incubation. A zone of α-hemolysis is produced on sheep blood agar plates in areas of heavy growth. The oxidase and catalase reactions are strongly positive. The bacterium does not utilize carbohydrates; it uses nitrite but not nitrate and has a weakly positive phenylalanine deaminase reaction from culture grown on sheep blood agar plates. It is found as normal oral flora in dogs and is associated with human wound infections resulting from dog bites.

The mol% G + C of the DNA is: 50.8–52.0 (T_m).

Type strain: ATCC 51410, CCUG 4007, CCUG 33675, CIP 103940, ISL775/91, LMG 5135, NCTC 12742.

Genus II. *Alysiella* Langeron 1923, 116^{AL}

TONE TØNJUM

A.ly.si.el′ la. Gr. fem. n. *alysion* small chain; M.L. *-ella* dim. ending; M.L. fem. n. *Alysiella* small chain.

Organisms that exist in characteristic **flat, ribbon-like multicellular filaments**. **The long axis of the individual cells is perpendicular to the long axis of the filament**. The **cells within the filament are paired**, and in axenic culture the filament often breaks up into groups of two or four cells. The width of an individual cell (the width of a filament) is about 2.0–3.0 μm, and the length of a cell is about 0.6 μm. The length of the filament is quite variable. The filament does not show either a dorsal-ventral differentiation or a convex-concave curvature in transverse cross-section. The **ends of the individual filaments are square**. Gram negative. **Motile by gliding** of the entire filament in the direction of the long axis. Aerobic. Chemoorganotrophic. Some may produce acid aerobically from carbohydrates. Optimal temperature: 37°C. Found in the oral cavity of warm-blooded vertebrates

The mol% G + C of the DNA is: 44–48.

Type species: **Alysiella filiformis** (Schmid *in* Simons 1922) Langeron 1923, 118, *Simonsiella filiformis* Schmid *in* Simons 1922, 509.)

FURTHER DESCRIPTIVE INFORMATION

The filaments of *Alysiella* are distinctive, and members of this genus can be recognized by their morphology alone (Fig. BXII.β.64). The filament is flat and ribbon-shaped rather than cylindrical and consists of continuous pairs of cells (Figs. BXII.β.65 and BXII.β.66). The individual cells are oblong, disk-shaped, and several times greater in width than in length. Fibrils are produced from only one side of the filament (Fig. BXII.β.66), and the fibrils appear to be involved in anchoring the filament to epithelial cells of the oral cavity (Kaiser and Starzyk, 1973; McCowan et al., 1979). Cells anchored in this way give rise to a typical palisade arrangement (Fig. BXII.β.64).

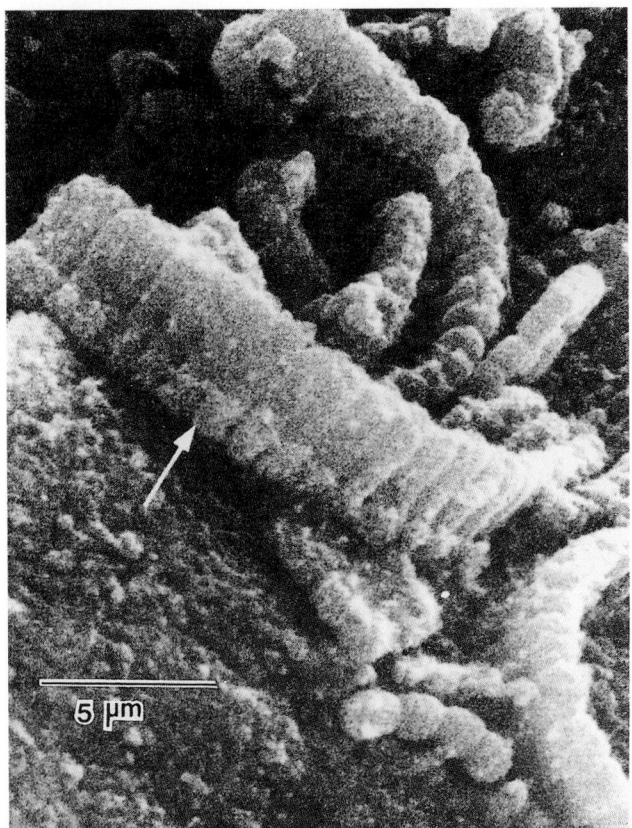

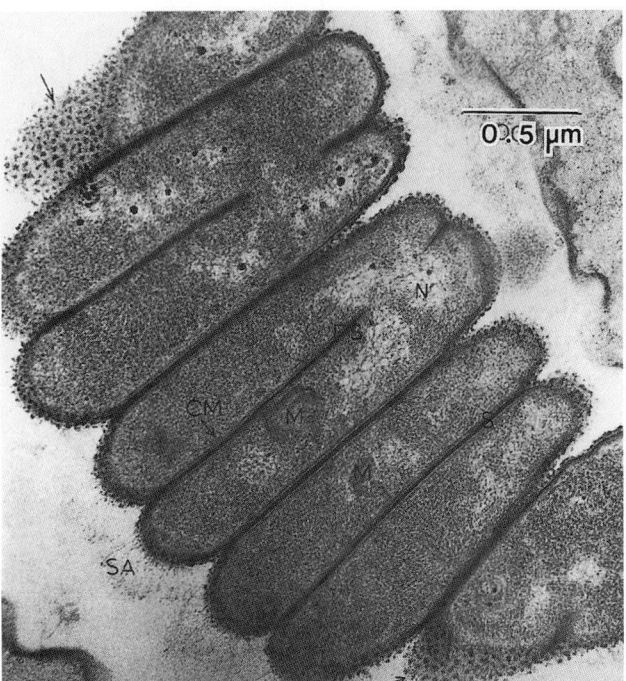

FIGURE BXII.β.65. Longitudinal electron micrograph of *Alysiella* from the oral cavity of a rabbit. The paired nature of the cells within the filament is typical of the genus. (Reproduced with permission from G.E. Kaiser and M.J. Starzyk, Canadian Journal of Microbiology *19:* 325–327, 1973, ©National Research Council of Canada.)

FIGURE BXII.β.64. Scanning electron micrograph of an *Alysiella* filament attached to the epithelium of the bovine tongue. The arrow indicates the fringe of fibers that attach the bacterial filament by its side to the substrate. The palisade organization of the filament is characteristic of *Alysiella.* (Reproduced with permission from R.P. McCowen et al., Applied and Environmental Microbiology *37:* 1224–1229, 1979, ©American Society for Microbiology.)

Gliding motility occurs when the flat surface of the filament is in contact with an agar surface. No organs of locomotion have been detected. Colonies on Oxoid nutrient agar containing 10% horse or ox serum are nonpigmented and about 1.0–1.5 mm in diameter (Steed, 1962). On BSTSY agar (see *Simonsiella* for ingredients), the colonies are about 1.0–2.0 mm in diameter and exhibit a pale yellow pigmentation (Kuhn, 1981). No resting stage has been detected. Generally, the cellular carbohydrate pattern exhibits some similarities with that of *S. muelleri* (Heiske and Mutters, 1994). However, although *A. filiformis* has a cellular carbohydrate pattern different from all other taxa described, the results have to be regarded as preliminary because only the type strain has so far been analyzed.

All reports of *Alysiella* indicate that it is restricted to the oral cavity of warm-blooded vertebrates, where it apparently is nonpathogenic.

ENRICHMENT AND ISOLATION PROCEDURES

At the present time there is no enrichment procedure for *Alysiella.* Direct isolation from the oral cavity can be achieved; however, Steed (1962) isolated *Alysiella* from sheep and rabbits with oral swabs that were plated directly onto Oxoid nutrient agar containing 10% horse or ox serum. After 6 h the microcolonies were transferred to new media using a micromanipulator.

McCowan et al. (1979) isolated *Alysiella* from cows by im-

pressing a portion of tongue on the agar surface and then spreading the organisms or by washing a portion of tongue in phosphate-buffered saline, homogenizing the tongue with a Waring blender, and plating the resulting suspension on agar. They used both Tryptose-blood agar (Difco Laboratories) with 10% sheep blood or an agar consisting of the following ingredients: nutrient agar (Difco), 2.5%; sodium acetate, 0.01%; and yeast extract (Difco), 0.5%.

Kuhn (1981) suggested that *Alysiella* can be isolated on BSTSY agar by using the procedures that she used for isolating *Simonsiella* (see the description of the genus *Simonsiella*).

MAINTENANCE PROCEDURES

Alysiella should be grown at 37°C, and freshly isolated cultures should be transferred at intervals of 2–3 days. Older cultures must be transferred at weekly intervals.

Refrigeration is not recommended for preservation of cultures, but they can be preserved by freezing in liquid nitrogen using glycerol as a cryoprotectant or by lyophilization.

PROCEDURES FOR TESTING SPECIAL CHARACTERS

The procedures that should be used for characterizing isolates of *Alysiella* are identical to those used for *Simonsiella*, and they are presented in the description of that genus.

DIFFERENTIATION OF THE GENUS *ALYSIELLA* FROM OTHER GENERA

The unusual morphology of the *Alysiella* filaments serves to differentiate the genus from all other procaryotic organisms. *Alysiella* filaments differ from *Simonsiella* in consisting of continuous pairs of cells instead of being segmented into units of eight cells,

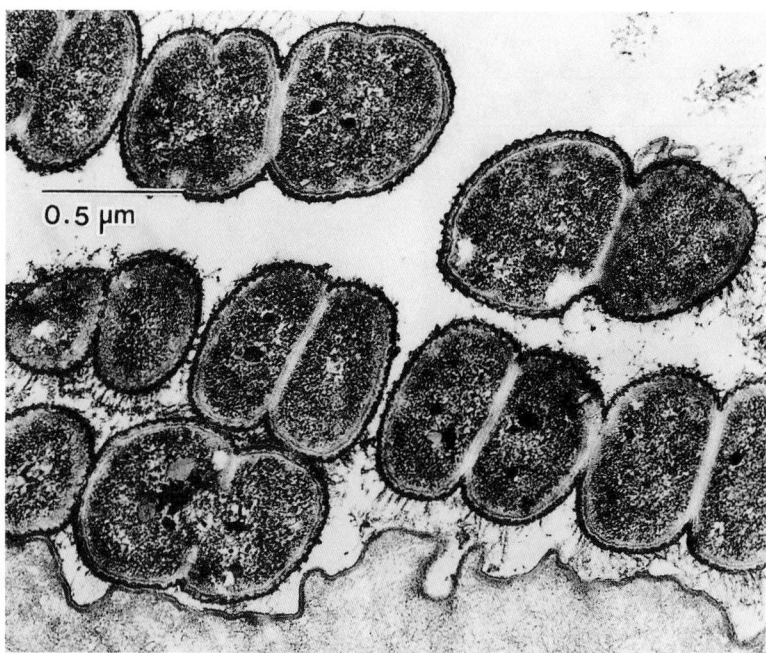

FIGURE BXII.β.66.   Section of *Alysiella* from the bovine tongue, showing ruthenium red-stained fibrils emanating from one side of the filament only. (Photo courtesy of J.W. Costerton.)

and the terminal cells are not rounded, but square. In addition, *Alysiella* has a fringe of fibers on the side of the filament rather than on the bottom as in *Simonsiella*. Furthermore, *Alysiella* does not show the dorsal–ventral, convex–concave curvature of *Simonsiella*.

Taxonomic Comments

Alysiella has been reported from the oral cavity of many animals including chickens, sheep, horses, cows, goats, pigs, rabbits, and guinea pigs. Isolates have only been obtained from guinea pigs (Berger, 1963), sheep (Steed, 1962), and cows (McCowan et al., 1979); only a few strains from sheep have been described in detail. Strains from other animals have resisted isolation. It is probable that as strains from other animals are isolated, there will prove to be additional species, just as with *Simonsiella*. The

analysis of misnamed *Alysiella* strains has confused the taxonomic work on this entity (Rossau et al., 1989). However, the combined and repeated findings on *Alysiella* morphology (Kaiser and Starzyk, 1973), cellular carbohydrate pattern (Heiske and Mutters, 1994), ribosomal RNA hybridization (Rossau et al., 1989), and 16S rDNA sequence analysis on the verified type strain clearly show that *Alysiella* belongs to the family *Neisseriaceae*. The complete data also indicate that *Alysiella* is a separate genus within this family, closely affiliated with the genus *Simonsiella* (see Fig. BXII.β.76 under chapter on the genus *Simonsiella* and Fig. BXII.β.61 under chapter on the order *Neisseriales*).

Acknowledgments

The information provided in the first version of *Bergey's Manual of Systematic Bacteriology* (1984) by J.M. Larkin is greatly acknowledged.

Differentiation of the species of the genus *Alysiella*

Only a single species of *Alysiella* is currently recognized. The multicellular spore formers found in the rumen of domestic animals and in the cecum of guinea pigs that Grassè (1924) termed

Alysiella filiformis are probably members of the genus *Arthromitus* (Kuhn, 1981). Some strains in the literature have been misnamed as *A. filiformis* (Rossau et al., 1989).

List of species of the genus Alysiella

1. **Alysiella filiformis** (Schmid *in* Simons 1922) Langeron 1923, 118[AL] *Simonsiella filiformis* Schmid *in* Simons 1922, 509.)

   fi.li.for′mis. L. n. *filum* thread; L. n. *forma* shape; M.L. adj. *filiformis* filiform.

   See the generic description for additional features. *Alysiella* is aerobic, possesses cytochrome oxidase, and produces catalase. Good growth occurs between 33° and 40°C, with an optimal temperature at 37°C. Growth also occurs at 43°C but not at 27°C. This organism grows in the presence of 1% but not 1.5% NaCl. Grows at pH 7.3 and 9.0 but not 6.0. Acid is produced from D-fructose, α-D-glucose, maltose,

   sucrose, and trehalose. No acid is produced from the following: L-arabinose, cellobiose, dulcitol, erythritol, D-galactose, glycerol, inositol, α-lactose, mannose, melibiose, melizitose, raffinose, rhamnose, salicin, sorbitol, sorbose, or xylose. Variable results occur on inulin and ribose. A rich medium containing 10% serum is best for growth. A slight hydrolysis of gelatin may occur, but agar, casein, starch, esculin, and hippurate are not hydrolyzed. No change occurs in litmus milk. Indole, MR, VP, and reduction of nitrates are negative. Urease is not produced. H_2S production is variable and inconsistent. Hemolytic on rabbit or horse blood agar. The type strain was isolated from sheep saliva.

The mol% G + C of the DNA is: 44–48 (T_m, Bd).

Type strain: ATCC 15532, ATCC 29469, CIP 103342, ICPB 3653, HIM 928-7,NCTC 10282.

GenBank accession number (16S rRNA): AF487710.

Genus III. **Aquaspirillum** Hylemon, Wells, Krieg and Jannasch 1973b, 36[AL]*

BRUNO POT AND MONIQUE GILLIS

Aq.ua.spi.ril' lum. L. *aqua* water; Gr. n. *spira* a spiral; N.L. dim. neut. n. *spirillum* a small spiral; *Aquaspirillum* a small water spiral.

Rigid, helical cells, 0.2–1.4 µm in diameter; however, one species is vibrioid and one species contains straight-to-curved rods. A polar membrane underlies the cytoplasmic membrane at the cell poles in all species so far examined for this characteristic by electron microscopy. **Intracellular poly-β-hydroxybutyrate** is formed, except in two species. Some species form thin-walled coccoid bodies, which predominate in cultures of three to four weeks. **Gram negative. Motile by polar flagella, generally bipolar tufts**; one species is monotrichous, others have a single flagellum at one or at each pole. **Aerobic to microaerophilic**, having a respiratory type of metabolism with **oxygen as the terminal electron acceptor**; a few species can grow anaerobically with nitrate. The optimal growth temperature for most species is 30–32°C. **Chemoorganotrophic**; however, one species is a facultative hydrogen autotroph. **Oxidase positive. Usually catalase and phosphatase positive. Indole and sulfatase negative**. Casein, starch, and hippurate are not hydrolyzed. No growth occurs in the presence of 3% NaCl. A few species can denitrify. Nitrogenase activity occurs in some species, but only under microaerobic conditions. Carbohydrates are not usually metabolized, but a few species can attack a limited variety. Amino acids or the salts of organic acids serve as carbon sources. Vitamins are not usually required. Usually occur in stagnant, freshwater environments.

The mol% G + C of the DNA is: 49–66.

Type species: **Aquaspirillum serpens** (Müller 1786) Hylemon, Wells, Krieg and Jannasch 1973b, 366 (*"Vibrio serpens"* Müller 1786, 48.)

FURTHER DESCRIPTIVE INFORMATION

Cellular morphology Most species of *Aquaspirillum* have helical cells; however, *A. delicatum* is vibrioid (has less than one complete turn or twist). Variants that are nearly straight rods have been obtained from helical species after prolonged transfer (Terasaki, 1972). For helical aquaspirillae, the cells within a given species have a constant type of helix clockwise (right-handed) or counter-clockwise (left-handed) (Terasaki, 1972). The effect of the beta-lactam antibiotic cephalexin on the spiral conformation has been examined by scanning electron microscopy (Konishi and Yoshii, 1986). *A. itersonii* and *A. peregrinum*, which have a left-handed spiral shape, maintained this shape in elon-

gated cells in medium containing cephalexin. The spiral conformation of the elongated cells is therefore considered to represent the natural condition (Konishi and Yoshii, 1986). Photographs showing the comparative size and shape of various aquaspirillae are presented in Fig. BXII.β.67.

Although aquaspirillae are more rigid than spirochetes, they do have a certain degree of flexibility. For example, during rapid swimming the helical cells tend to become straighter. Also, cells embedded in glycerol gelatin can be stretched to three times their original length (Isaac and Ware, 1974).

As unusual elaboration of the plasma membrane, the "polar membrane," occurs in all of the species examined so far (I.J. Beveridge and R.G.E. Murray, unpublished results). It is attached to the inside of the plasma membrane by bar-like links and is most commonly located in the region surrounding the polar flagella (Murray and Birch-Andersen, 1963; Fig. BXII.β.68). Such a membrane has been found mainly in genera of helical bacteria, such as *Spirillum, Oceanospirillum, Campylobacter, Ectothiorhodospira,* and *Rhodospirillum*.

Intracellular poly-β-hydroxybutyrate occurs in all species except *A. gracile* and *A. psychrophilum*. The granules of this polymer stain with metachromatic dyes such as toluidine blue (Martinez, 1963), as well as with lipid-soluble dyes such as Sudan black.

In certain species, the cells develop into thin-walled coccoid bodies (sometimes termed "microcysts") within several days to several weeks. All species may show a few such forms in old cultures, but in *A. itersonii, A. peregrinum* subsp. *peregrinum*, and *A. polymorphum* they predominate. Such coccoid bodies are also formed by members of the genus *Oceanospirillum* and the genus *Campylobacter*. In *A. itersonii*, the development of the helical cells into coccoid bodies can be greatly accelerated by treatment with mitomycin or ultraviolet light; this effect has been correlated with the induction of a defective bacteriophage (Clark-Walker, 1969). Whether the coccoid bodies of aquaspirillae are resistant to desiccation, or whether they are viable, is not known.

Most species of *Aquaspirillum* are motile by means of bipolar tufts or fascicles of flagella; however, *A. delicatum* usually has 1–2 flagella at a single pole, and *A. polymorphum* has bipolar single flagellum. Aquaspirillae generally have flagella that are crescent shaped or that have a long wavelength (over 3 µm), with less than one complete wave. Such flagella are especially likely to occur with the larger aquaspirillae such as *A. serpens, A. metamorphum,* and *A. putridiconchylium*, and the motility of such spirilla is similar to that described for *Spirillum volutans* (i.e., the flagellar fascicles form cones of revolution). However, some aquaspirillae, especially those that are small or medium in cell diameter such as *A. dispar* or *A. delicatum*, have more conventional, helical flagella (see Hylemon et al., 1973b).

A. serpens has been studied extensively with regard to its flagella cell-wall association (Coulton and Murray, 1978), cell-wall ultrastructure, and cell-wall chemical composition (Murray et al.

---

*Editorial Note: Phylogenetically, the genus *Aquaspirillum* is heterogeneous; however, it is difficult to definitively delineate new genera for potentially misclassified species. Therefore these organisms are retained in the genus *Aquaspirillum*. The following species are not true aquaspirillae (i.e., not phylogenetically related to the type species, *Aquaspirillum serpens*, at the genus level) but have not been transferred to other genera: *A. anulus, A. autotrophicum, A. delicatum, A. dispar, A. giesbergeri, A. gracile, A. metamorphum, A. polymorphum, A. psychrophilum, A. putridiconchylium,* and *A. sinuosum*. Aquaspirillae that have been generically renamed since the first edition of the *Bergey's Manual of Systematic Bacteriology* (Krieg, 1984a), or for which a new genus name is proposed here, include: *Aquaspirillum aquaticum, A. bengal, A. fasciculus, A. itersonii, A. magnetotacticum,* and *A. peregrinum.*

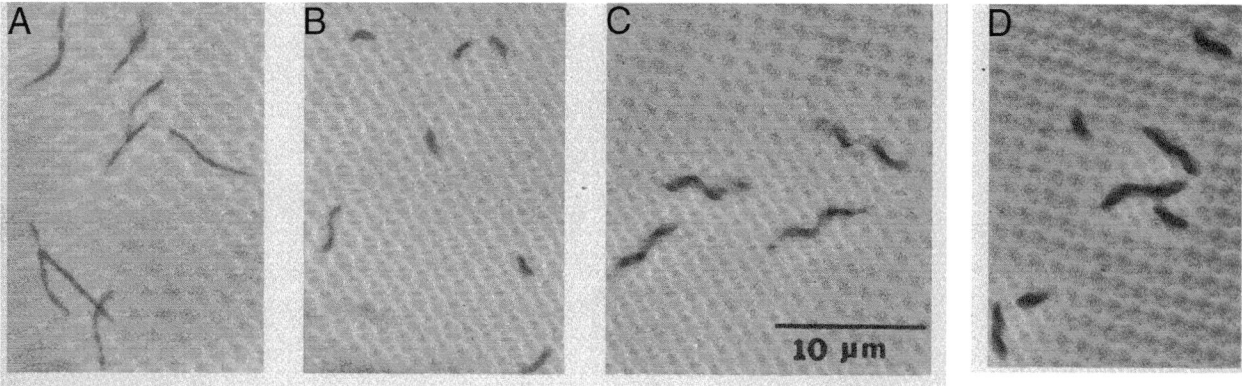

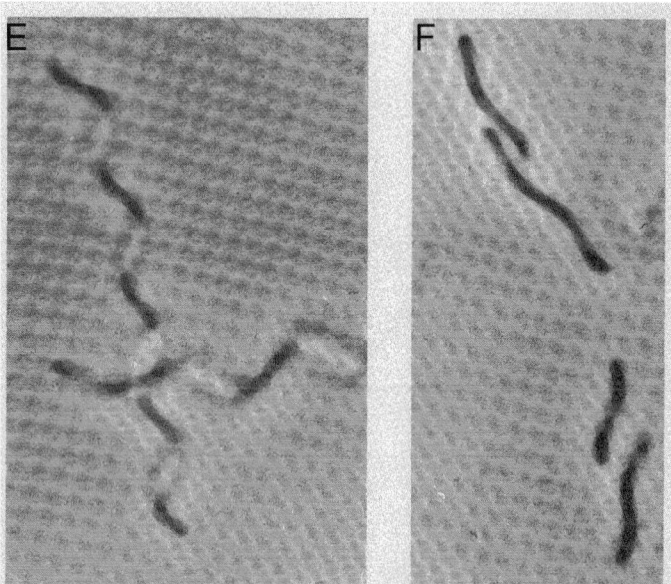

FIGURE BXII.β.67. Phase-contrast photomicrographs of several species of the genus *Aquaspirillum.* The spirillae were cultured in MPSS broth for 24–48 h at 30°C; however, *A. serpens* subsp. *bengal* was incubated at 37°C and *A. delicatum* was cultured in nutrient broth, since its morphology and mobility are more characteristic in this medium. All photographs were taken at the same magnification. A, *A. gracile* ATCC 19624. B, *A. delicatum* ATCC 14667. C, *A. polymorphum* NCIB 9072. D, *A. dispar* ATCC 27510. E, *A. sinuosum* ATCC 9786. F, *A. putridiconchylium* ATCC 15279. G, *A. serpens* subsp. *serpens* strain VH. H, *A. serpens* subsp. *serpens* ATCC 12638. I, *A. serpens* subsp. *bengal* ATCC 27641. J, *A. metamorphum* ATCC 15280. K, *A. anulus* NCIB 9012. L, *A. giesbergeri* NCIB 8320. (Reproduced with permission from N.R. Krieg, Bacteriological Reviews *40:* 55–115, 1976, ©American Society for Microbiology.)

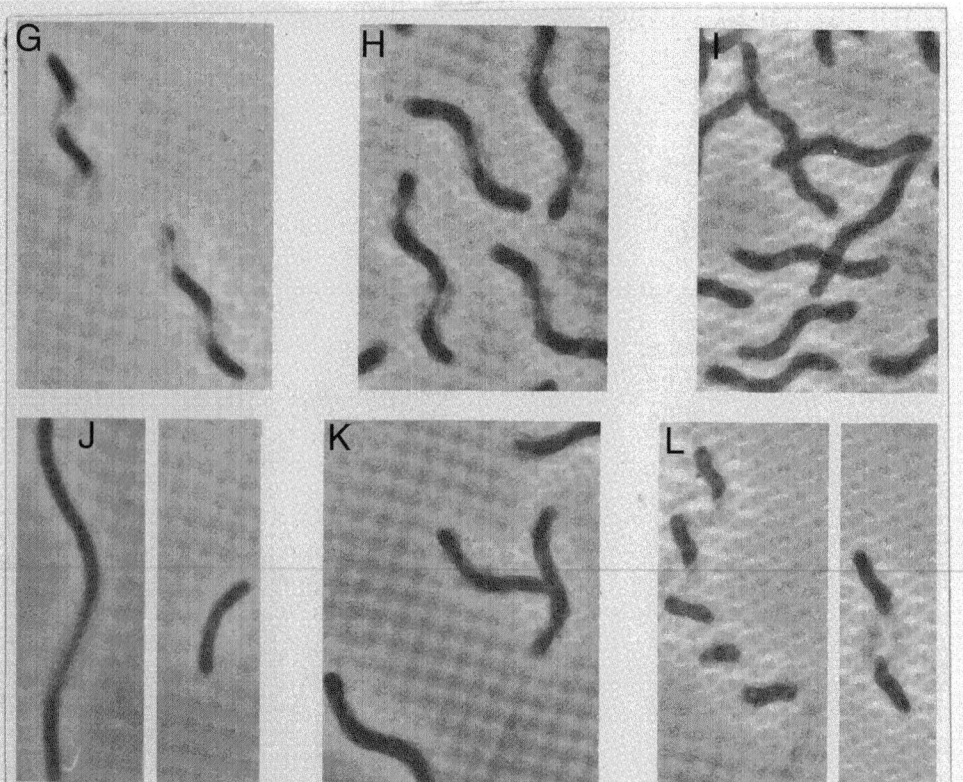

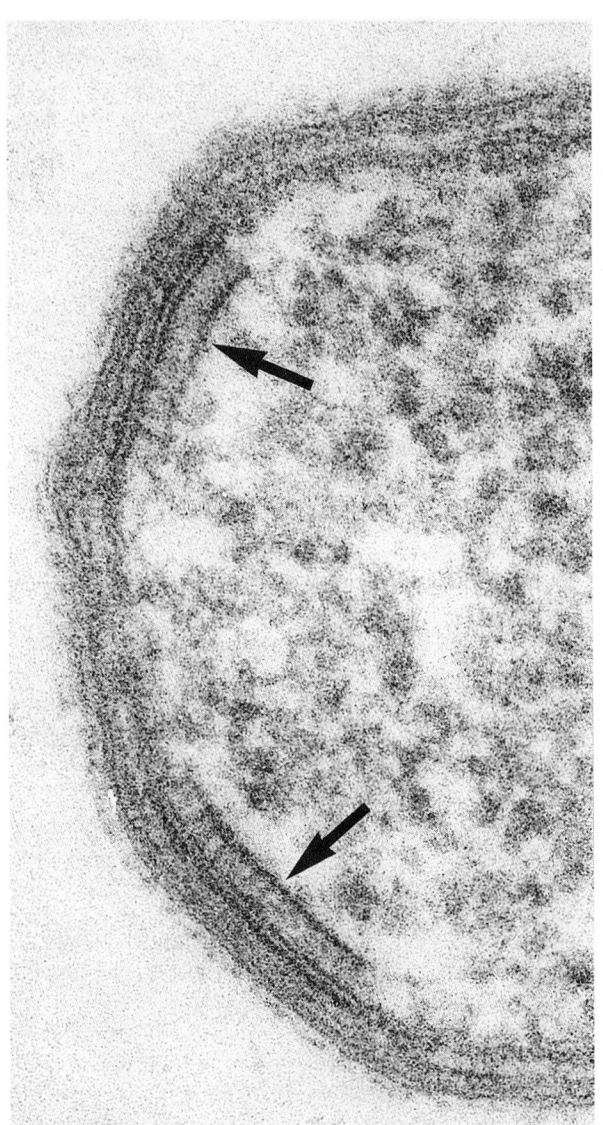

FIGURE BXII.β.68. Thin-section through the polar region of a cell of *Aquaspirillum serpens* strain VHA, showing the polar membrane (*arrows*). A protein layer can also been seen external to the outer wall membrane. × 262,000. (Reproduced with permission from R.G.E. Murray, University of Western Ontario, London, Canada.)

1965; Chester and Murray, 1975, 1978). A protein layer, consisting of a regular array or mosaic of subunits, surrounds the cell walls of certain species of aquaspirillae (Buckmire and Murray, 1970; Beveridge and Murray, 1976; Stewart et al., 1980). Such protein layers can be dissociated by agents such as sodium dodecyl sulfate or guanidine, and can subsequently be reassembled onto templates *in vitro* in the presence of Ca^{2+}. One function of the protein layer of *A. serpens* is to protect against attack by bdellovibrios (Buckmire, 1971). The cell wall lipopolysaccharide of *A. serpens* differs from that of the majority of other Gram-negative bacteria in that it lacks 2-keto-3-deoxyoctonic acid (Chester and Murray, 1975); this compound is also present in the lipopolysaccharide of *A. itersonii* and *A. peregrinum*. In *A. serpens*, lipid A of the lipopolysaccharide differs from that found in members of *Enterobacteriaceae* in that 3-hydroxydodecanoic acid, rather than 3-hydroxytetradecanoic acid, is the *N*-acylating acid.

Physiology All species of *Aquaspirillum* are aerobic. Although *A. itersonii* and *A. peregrinum* can acidify fructose-containing media sealed with a layer of oil or petrolatum, significant turbid growth does not occur, and these species should be considered to have an essentially oxidative type of metabolism. *A. itersonii*, *A. dispar*, and *A. psychrophilum* can grow anaerobically using nitrate and possess a dissimilatory nitrate reductase. *A. itersonii*, *A. dispar*, and *A. psychrophilum* can reduce nitrate beyond nitrite, but only *A. psychrophilum* appears to form visible amounts of gas from nitrate (Terasaki, 1972, 1979).

The respiratory chain of *A. itersonii* has been studied in detail and appears to be an unbranched, membrane-bound electron transport chain from NADH and succinate to oxygen (Dailey, 1976). Cytochromes of both the *b* and *c* type, but not of the *a* type, are present, and their biosynthesis and properties have been investigated (Clark-Walker et al., 1967; Clark-Walker and Lascelles, 1970; Ho and Lascelles, 1971; Dailey and Lascelles, 1974). *A. itersonii* synthesizes higher levels of cytochromes *b* and *c* under semi-anaerobic conditions with nitrate in the medium than it does anaerobically without nitrate (Clark-Walker et al., 1967), and much of the cytochrome *c* is present in a soluble form in the periplasmic space (Gauthier et al., 1970). This soluble cytochrome *c*, and other periplasmic proteins, can be selectively liberated from the cells by the use of a mixture of Tris buffer and EDTA (Garrard, 1971). Biosynthesis of the soluble cytochrome *c* has been investigated by Garrard (1972). The complete amino acid sequence of the cytochrome c_{550} from *A. itersonii* has been elucidated (Woolley, 1987). The sequence is a single polypeptide chain of 111 residues and shows a high degree of sequence homology with the cytochrome c_2 from the photosynthetic bacterium *Rhodospirillum rubrum*. This homology is in agreement with other phylogenetic data, as shown below.

Aquaspirillum species cannot grow in the presence of 3% NaCl, and many species cannot tolerate even 1% NaCl. This lack of salt tolerance distinguishes *Aquaspirillum* from *Oceanospirillum*, because the latter genus requires seawater or Na^+ for growth.

Commonly used culture media for aquaspirillae are PSS broth, MPSS broth, and nutrient broth[1]. Aquaspirillae generally produce moderate-to-abundant turbid growth in 2–3 d in PSS broth (Hylemon et al., 1973b). In nutrient broth, membranous masses are often formed at the surface and can be dispersed with shaking to yield turbid cultures (Terasaki, 1972).

Colonies of aquaspirillae generally develop within 2–3 d on PSS agar and are usually white, circular, and convex, ranging from pinpoint to 1.5 mm in diameter (Hylemon et al., 1973b). Colonies on nutrient agar are generally pinpoint in size at 48 h but become larger (up to 2.0 mm in diameter) at 7 d; they are usually convex or umbonate, glistening, opaque, pale yellow, and butyrous (Terasaki, 1972). S–R variation has been found in several species (Terasaki, 1972). Some species produce a water-soluble, yellow-green, fluorescent pigment on PSS agar.

Most species grow best at 30–32°C, except *A. psychrophilum*, which grows best at 20°C and cannot grow above 26°C, and *A. bengal*, which grows best at 41°C. The optimal pH for most species

---

1. PSS broth (g/l): Bacto Peptone (Difco), 10.0; succinic acid (free acid), 1.0; $(NH_4)_2SO_4$, 1.0; $MgSO_4\cdot7H_2O$, 1.0; $FeCl_3\cdot6H_2O$, 0.002; and $MnSO_4\cdot H_2O$, 0.002. The pH is adjusted to 6.8 with KOH before autoclaving the medium. For PSS agar, 16.0 g of agar is added/l; for PSS semisolid medium, 1.5 g of agar is added/l. For MPSS media, use 5.0 g of peptone rather than 10.0 g. Nutrient broth (g/l): peptone, 5.0; meat extract, 3.0 g; the pH is adjusted to 7.0–7.2 with KOH before autoclaving. For nutrient agar, 15.0 g/l of agar is added.

is 6.5–7.5, but many species can grow at pH values as high as 8.5 or 9.0 (Terasaki, 1972).

The nutrition of aquaspirillae is generally simple. Most species grow in simple defined media with amino acids or the salts of organic acids as carbon sources and ammonium salts as the nitrogen source. Only *A. gracile* is known to require biotin. Few species can catabolize sugars, although a limited set of sugars can be used by *A. gracile*, *A. itersonii*, and *A. peregrinum*. Acidification of sugar media by these species occurs only when the peptone level is kept low (0.2% or less). A listing of the carbon sources for aquaspirillae is given in Table BXII.β.79. In this table, some contradictions exist between the results obtained in different laboratories, although the intralaboratory results are reproducible. These contradictions are likely due to the differences in methodology, to the definition of what constitutes a positive growth response, and, in some cases, to the use of different strains.

It is clear that the genus *Aquaspirillum* is quite heterogeneous (see below). Some of the phenotypic and nutritional differences confirm the genotypic differentiation. The subdivision of the species in the Tables BXII.β.79, BXII.β.80, BXII.β.81, BXII.β.82, and BXII.β.83 corresponds with the genotypic groupings obtained from 16S rRNA sequencing and rRNA–DNA hybridization data, as discussed below.

A. peregrinum exhibits nitrogenase activity, but only under microaerobic conditions (Strength et al., 1976). In this respect, it is similar to members of the genus *Azospirillum*. Atmospheric nitrogen fixation has also been reported in certain strains of *Aquaspirillum itersonii* (Ketkar, 1967; Ketkar and Dhala, 1978). Phylogenetic data also suggest that both species are phylogenetically more closely related to members of the genus *Azospirillum* than to other species of the genus *Aquaspirillum*.

The intermediary metabolism of sugars has been studied in *A. itersonii* and *A. gracile*. *A. itersonii* can acidify glucose media under semi-anaerobic conditions, but not under aerobic conditions (Terasaki, 1972, 1979); this observation has not yet been explained. Under aerobic conditions, *A. itersonii* is impermeable to glucose despite the occurrence of high levels of glucokinase activity (Hylemon et al., 1974). Fructose is transported and phosphorylated by means of a fructose-specific phosphoenolpyruvate phosphotransferase system (P.V. Phibbs, unpublished results). The Embden–Meyerhof–Parnas and Entner–Doudoroff pathways occur in *A. itersonii* and *A. gracile*, but the hexose monophosphate pathway is absent (Hylemon et al., 1974; Laughon and Krieg, 1974). *A. gracile* acidifies sugar media by formation of sugar acids, such as gluconic acid; other organic acids are not formed (Laughon and Krieg, 1974).

The tricarboxylic acid cycle has been demonstrated in *A. serpens* and *A. itersonii* (Cole and Rittenberg, 1971). Whether the glyoxylate shunt occurs is not known.

Serology McElroy and Krieg (1972) reported serological differentiation of most species of *Aquaspirillum*. Antisera are prepared against whole cells and adsorbed with heated cells, leaving only antibodies against thermolabile cell components. The use of such antisera in agglutination tests with a limited number of strains suggests that most species can be distinguished from one another and from organisms of other genera.

Genetics Electroporation methods and conjugal mating have been used to transfer several plasmid vectors (including the incompatibility P class plasmid RP4 by conjugation from *Escherichia coli* HB101) to *A. dispar* and *A. itersonii*. The transconjugants

were able to donate the plasmid to plasmid-free *E. coli* and *A. dispar* strains by conjugal mating (Eden and Blakemore, 1991). *A. dispar* and *A. itersonii* were transformed at efficiencies as high as 3×10^4 transformants/µg plasmid DNA by high-voltage electrotransformation. RP4 DNA from *Spirillum* hosts, but not RP4 from *E. coli*, was successfully transferred to *A. dispar* and to *A. itersonii*, indicating the possible presence of a restriction/modification system in these *Aquaspirillum* species. The restriction endonuclease *Ase*I has been isolated from *A. serpens*. This enzyme recognizes the sequence 5′-AT–TAAT-3′ (Polisson and Morgan, 1988).

There has been only one report of bacteriophages for *Aquaspirillum*. An icosahedral, double-stranded DNA phage specific for a particular strain of *A. itersonii* was isolated from raw sewage in Australia by Clark-Walker and Primrose (1971). It produced plaques on plate cultures, but was unable to lyse broth cultures. Oddly, the host strain of *A. itersonii* was originally isolated from Lake Erie, U.S.A., rather than from Australia.

Ecology Aquaspirillae are considered to be nonpathogenic for humans and animals. An organism known as *"Spirillum minus"* (see *Species Incertae Sedis*) is the cause of one of the two forms of rat-bite fever in man, and another organism, *"Spirillum pulli"* (see *Species Incertae Sedis*), is apparently the cause of a diphtheritic stomatitis in chickens. Neither of these species belong to the genus *Spirillum* or to the genus *Aquaspirillum*, and their affiliation with other established genera is uncertain. A few cases of human infection have been reported to be caused by, or associated with, organisms resembling aquaspirillae (e.g., see Edwards and Kraus, 1960; Kowal, 1961). The identity of these strains is uncertain.

A total of seven strains were isolated from blood cultures of six patients and from the cerebrospinal fluid of one patient (five cases of pneumonia and two cases of pneumonia associated with meningitis). These unusual Gram-negative rods had highest protein profile similarity with *A. serpens* and *Chromobacterium violaceum* (Casalta et al., 1989), as measured by sodium dodecyl sulfate polyacrylamide gel electrophoresis (SDS-PAGE). However, no serologic cross-reactions were observed (Western blot and immunofluorescence), and considerable differences in morphological and biochemical characteristics were seen. It was concluded that additional studies, including DNA–DNA hybridization, were required to classify this bacterium.

Spirillum-like organisms have also been detected in diseased mosquito larvae (Fulton et al., 1974), but whether they are aquaspirillae is uncertain. Giant spore-forming spirillae have been described in the intestinal contents of tadpoles by Delaporte (1964) (see *Genus Incertae Sedis "Sporospirillum"*); little is known of these organisms, but they do not appear to be *Aquaspirillum*. Tiunov et al. (1997) also reported that bacteria of the genera *Aquaspirillum* and *Cytophaga* invariably predominated in the saprotrophic bacterial community at 2 mm distance from the burrow wall of the earthworm *Lumbricus terrestris* (drilosphere). The earthworm burrow walls differed from the surrounding soil by the high density of bacteria (10 times more, or $15–20 \times 10^6$ CFU/g, at 7–10 cm depth) and by the composition. Spirillae predominated in the 2 mm zone around the burrow walls over the entire experimentation period. At distances of 5 and 10 mm from the burrow, the taxonomic structure of the bacterial community was highly diverse and varied significantly depending on the season. Identification, however, was based only on microscopic examination of isolated colonies; therefore possible con-

TABLE BXII.β.79. Carbon sources used by existing and former species of *Aquaspirillum*

	Aquaspirillum spp. from the *Betaproteobacteria*													
	Aquaspirillum (sensu stricto)			*Aquaspirillum* spp. (phylogenetically not correctly named)				*Aquaspirillum* spp. belonging to the *Comamonadaceae*						
	A. serpens subsp. *serpens*		*A. serpens* subsp. *bengal*[c]	*A. autotrophicum*[d]	*A. dispar*[a]	*A. putridiconchylium*		*A. anulus*		*A. delicatum*		*A. gracile*[e]	*A. giesbergeri*	
Carbon source	A[a]	B[b]				A[a]	B[b]	A[a]	B[b]	A[a]	B[b]		A[a]	B[b]
Citrate	−	−	−	+	+	−	−	−	−	−	−	−	−	−
Aconitate	−			+	+	+		−		+		−		
Isocitrate	−			+	+	+		−		+		−		
α-Ketoglutarate	d			+	+	+		−		+		+	−	
Succinate	−	+	+	+	+	+	d	−	w	+	w	+	−	d
Fumarate	−	+	+	+	+	+	+	−	w	+	w	−	d	d
Malate	−	d	+	+	+	+	+	−	d	+	w	−	−	d
Oxaloacetate	d			+	+	+		−		+		−		
Pyruvate	d	+	+	+	+	+	+	d	d	+	w	+	+	d
Lactate	−	d	+	+	+	+	d	−	w	+	w	+	−	d
Malonate	−	−		+	+	+	−	w	−	+	w			d
Tartrate	−			+		−		−		−			−	−
Acetate	−	d	+	+	+	+	w	−	w	+	w	+	−	d
Propionate	−	d		+	+	−	d	−	−	−	−	−	−	−
Butyrate		+		+			+		d		w			d
Caproate	−			−	+	−		−		−			−	
β-Hydroxybutyrate	−			+	+	+		−		−			−	
p-Hydroxybenzoate	−			+	−	−		−		−			−	
Ethanol	−	−		−	d	−	w	−	−	−	w		−	−
n-Propanol	−	−	−	−	−	−	w	−	−	−	−		−	−
n-Butanol	−	−	−	−	−	−	w	−	−	−	w		−	−
Glycerol	−	−	−	−	−	−	−	−	−	−	w	+	−	−
D-Fructose	−	−	−	−	−	−	−	−	−	−	−	−	−	−
D-Glucose	−	−	−	−	−	−	−	−	−	−	−	+	−	−
D-Xylose	−	−	−	−	−	−	−	−	−	−	−	+	−	−
L-Arabinose	−	−	−	−	−	−	−	−	−	−	−	+	−	−
L-Histidine	−		−	−	+			−		−			−	
L-Tyrosine	−		−	+	−	−		−		−			−	
L-Phenylalanine	−			+	−	−		−		−			−	
L-Alanine	d	+	+	+	+	+		−		−			−	
L-Glutamate	+	+	+	+	d	+		+		−		+	+	
L-Aspartate	+	+	+	+	+	+		−		−		+	+	
L-Glutamine	d	+	+	+	−	+		−		−			−	
Asparagine	d	+	+	+	−	+		−		−			−	
L-Proline	d	+	+	+	+	+		−		−			−	
L-Hydroxyproline	−			−	−			−		−			−	
L-Ornithine	−			−	+	−		−		−			−	
L-Citrulline	−			−	+	−		−		−			−	
L-Arginine	−	−		−	−			−		−			−	
L-Lysine	−	−		−	−			−		−			−	
Putrescine	−			−	−			−		+			−	
L-Methionine	−		−	−	−			−		−			−	
L-Serine	d		−	−	−			−		−			−	
L-Cysteine	−		−	−	−			−		−			−	
Glycine	−			+	−			−		−			−	
L-Leucine	−		−	−	−			−		−			−	
L-Isoleucine	−		−	+	−			−		−			−	
L-Valine	−		−	−	−			−		−			−	
L-Tryptophan	−			+	−	−		−		−			−	

[a]As determined by the method of Hylemon et al. (1973b). A turbidimetrically standardized cell suspension in physiological saline was inoculated into a defined, vitamin-free medium containing the carbon sources (0.1%) and ammonium sulfate as the nitrogen source. Growth responses were measured turbidimetrically after one 72-h serial transfer from the initial cultures, using a Klett colorimeter with the blue (420 nm) filter and 16-mm cuvettes. Symbols: +, 10 or more Klett units of turbidity for all strains tested; −, less than 10 Klett units of turbidity; d, differs among strains; blank space, not determined.

[b]As determined by the method of Terasaki (1972, 1979). A cell suspension washed in basal, defined, vitamin-free medium (Williams and Rittenberg, 1957) lacking carbon sources was inoculated into similar media containing the test compounds (0.05%) and ammonium chloride as the nitrogen source. After 7 days, growth was estimated turbidimetrically. Symbols: +, a turbidity of 0.025 absorbance units or greater for all strains tested; w, a turbidity of less than 0.025; −, no growth (turbidity equals the same as the appearance of controls without a carbon source); d, differs among strains; blank space, not determined.

[c]As determined by Kumar et al. (1974), using a modification of the method of Hylemon et al. (1973b). Symbols: +, a turbidity of 0.03 absorbance units or more, using a green filter and 16-mm cuvettes; −, turbidity less than 0.03; blank space, not determined.

[d]The utilization of compounds as sole carbon sources was tested on agar plates as described by Stanier et al. (1966) using a velvet-disk replicator. The medium was the basal mineral agar described under Procedures for Testing Special Characters, supplemented with 0.2% carbohydrates or 0.1% of other compounds. Symbols: +. growth greater than on control plate with no carbon source; −, growth no greater than on control plate; blank space, not determined.

[e]As determined by the method of Canale-Parola et al. (1966). A complex growth-limiting medium containing the carbon sources at 0.5% was used. Growth in the presence of the test compounds was compared turbidimetrically or by microscopic count to that occurring in the absence of the compounds. Symbols: +, >10% increase in the growth of all strains in the presence of the test compound; −, 10% or less increase in growth; blank space, not determined.

(*continued*)

TABLE BXII.β.79. *(cont.)*

| | Aquaspirillum spp. from the *Betaproteobacteria* / Aquaspirillum spp. belonging to the Comamonadaceae | | | | | Aquaspirillum spp. from the Alphaproteobacteria | | Former *Aquaspirillum* spp. | | | | | |
| | *A. metamorphum* | | *A. psychrophilum*[b,f] | *A. sinuosum* | | *A. polymorphum* | | *Levispirillum itersonii* | | *Levispirillum peregrinum* | | *Comamonas terrigena* biovar aquaticum[i] | *Prolinoborus fasciculus*[j] |
Carbon source	A[a]	B[b]		A[a]	B[b]	A[a]	B[b]	A[a,g]	B[b]	A[a,h]	B[b]		
Citrate	−	w	−	−	−	−	−	−	d	−	+	−	
Aconitate	−			−		−		+		+			
Isocitrate	−			−		−		−		−			
α-Ketoglutarate	+			−		−		+		+		+	−
Succinate	+	w	−	−	w	−	+	+	d	+	+	+	+
Fumarate	+	d	−	−	w	−	+	+	d	+	+	+	+
Malate	+	w	−	+	w	−	+	+	d	+	+	+	+
Oxaloacetate	+			+		−			d	+			+
Pyruvate	+	w	−	+	w	−	+	d	d	+	+	+	+
Lactate	+	w	−	−	w	−	+	d	d	+	+	+	+
Malonate	+	−	−	−	−	+	w	d	−	+	+	−	−
Tartrate		−	−	−			+	+		−		−	
Acetate	−	w	−	−	w	+	w	−	+	+	+	+	−
Propionate	−	d	−	−	−	−	w	d	d	−	+	+	−
Butyrate		w	−		d	−			d		+	−	
Caproate	−			−		−		−		−			
β-Hydroxybutyrate	−			−		−		+		+		+	+
p-Hydroxybenzoate	−			−		−		−		−			
Ethanol	−	−	−	−	−	−	−	+	d	−	+	−	−
n-Propanol	−	−	−	−	−	−	−	+	d	−	+	−	−
n-Butanol	−	−	−	−	−	−	−	d	d	−	+	−	−
Glycerol	−	−	−	−	−	−	−	d	+	−	d	−	−
D-Fructose	−	−	−	−	−	−	−	+	+	+	+	−	−
D-Glucose	−	−	−	−	−	−	−	−	−	−	−	−	−
D-Xylose	−	−	−	−	−	−	−	−		−		−	
L-Arabinose	−	−	−	−	−	−	−	−		−			
L-Histidine	−			−		−		+		−			
L-Tyrosine	−			−		−		−		−			
L-Phenylalanine	−			−		−		+		−			
L-Alanine	+			−		−		d		+		−	+
L-Glutamate	+			+		+		+		+		+	+
L-Aspartate	+			+		+		+		+		+	+
L-Glutamine	+			+		+		+		+		+	+
Asparagine	+			−		−		+		+		+	+
L-Proline	−			−		+		+		+		+	+
L-Hydroxyproline	−			−		−		d		+		−	−
L-Ornithine	−			−		−		d		−		−	
L-Citrulline	−			−		−		d		−			
L-Arginine	−			−		−		d		−		−	+
L-Lysine	−			−		−		d		−		−	
Putrescine	−			−		−		d		−			
L-Methionine	−			−		−		−		−			−
L-Serine	−			−		−		−		−			
L-Cysteine	−			−		−		d		−			
Glycine	−			−		−		−		−			
L-Leucine	−			−		−		d		−			
L-Isoleucine	−			−		−		−		−			
L-Valine	−			−		−		d		−			−
L-Tryptophan	−			−		−		−		−			

[f]The nutritional requirements of *A. psychrophilum* have not yet been determined.

[g]Strains of subsp. *nipponicum* were not tested.

[h]Strains of subsp. *integrum* were not tested.

[i]As determined by the method of Kropinski (1975). Samples (0.1 ml) of a washed cell suspension were spread on plates of a minimal agar medium containing ammonium sulfate and niacin. Approximately 8 mg of the test compounds were placed in small areas on the plates. After incubation for 48 h, the growth response was estimated. Symbols: + growth in the area around the test compound; −, no growth; blank space, not determined.

[j]As determined by the method of Strength et al. (1976). A washed, turbidimetrically standardized suspension was inoculated into a defined, vitamin-free, semisolid medium containing the test compounds (on an equal carbon basis relative to 0.2% fumaric acid) and ammonium sulfate as the nitrogen source. Growth responses were measured turbidimetrically at 36 h after gently inverting the semisolid cultures several times to obtain an even distribution of cells. A Klett colorimeter was used (blue filter, 420 nm) with 16 mm cuvettes. Symbols: +, production of at least 10 Klett units of turbidity; −, less than 10 Klett units; blank space, not determined.

TABLE BXII.β.80. Cellular quinone systems in *Aquaspirillum*, *Magnetospirillum*, and former *Aquaspirillum* species[a,b]

Species	Strain number	Group number	Q-6	Q-7	Q-8	Q-9	Q-10
Aquaspirillum species from the *Betaproteobacteria*							
Aquaspirillum (sensu stricto)							
Aquaspirillum serpens subsp. *serpens*	14924[T]	IV	2	2	96		
Aquaspirillum serpens subsp. *serpens*	14923	IV	16	11	73		
Aquaspirillum serpens subsp. *serpens*	15465	IV	3	2	95		
Aquaspirillum spp. (phylogenetically not correctly named)							
Aquaspirillum autotrophicum	15327[T]	Vf		3	97		
Aquaspirillum dispar	15328[T]	Ve		4	94	2	
Aquaspirillum putridiconchylium	13962[T]	IV	2	2	96		
Aquaspirillum spp. belonging to the *Comamonadaceae*							
Aquaspirillum anulus	14917[T]	Va			98	2	
Aquaspirillum delicatum	14919[T]	Vd	1	1	98		
Aquaspirillum giesbergeri	14959[T]	Va	2	1	96	1	
Aquaspirillum gracile	14920[T]	Vc		1	98	1	
Aquaspirillum metamorphum	13960[T]	Va	1	1	97	1	
Aquaspirillum psychrophilum	13611[T]	Vb	3	1	95	2	
Aquaspirillum sinuosum	14925[T]	Va		1	97	1	1
Aquaspirillum spp. from the *Alphaproteobacteria*							
Aquaspirillum polymorphum	13961[T]	VII			2	3	95
Former *Aquaspirillum* spp.							
Comamonas terrigena biovar *aquaticum*	14918[T]	Va	1	3	85	11	
Levispirillum peregrinum subsp. *peregrinum*	14922[T]	VI		3	3	93	2
Levispirillum peregrinum subsp. *integrum*	13617[T]	VI		2	4	92	2
Levispirillum itersonii subsp. *itersonii*	14921	VII			1	3	96
Levispirillum itersonii subsp. *nipponicum*[c]	13615[T]	VII			1	6	93
Magnetospirillum							
Magnetospirillum gryphiswaldense	15271[T]	VII			10	4	86
Magnetospirillum magnetotacticum	15272[T]	VII				3	97

[a]Data from Sakane and Yokota (1994).

[b]Numbers refer to the percentage of a quinone system relative to total cellular quinone systems.

[c]Strain ATCC 33333[T] has been shown to be different from strain IFO 13615[T] (Pot, 1996).

fusion with azospirillae cannot be excluded. These "spirillae" are common in soil communities. A similar remark can probably be made regarding the description of some nitrogen-fixing *Aquaspirillum* spp. from the endorhizosphere of rice (Garcia et al., 1983; Mishustin, et al., 1984), sorghum, winter rye, annual rye grass, meadow fescue, timothy grass, meadow soft grass, hogweed cow parsnip, and common colewort growing on different soils (Berestetsky et al., 1985). Genotypic identification is needed to confirm the identity of the *Aquaspirillum* species reported.

Aquaspirillae have been isolated from a wide variety of freshwater sources, especially stagnant ones or those containing organic matter: ditch water, canal water, stagnant ponds, primary oxidation ponds, and eutrophic lakes. They have also been isolated from storage tanks of distilled water in laboratories, where the organisms and the nutrients to support their growth or survival apparently come from the surrounding air. Aquaspirillae have also been isolated from hay infusions made with pond water (the water is probably the source of the organisms) and from putrid infusions of freshwater mussels (where the mud adherent to the shellfish is probably the source).

Aquaspirillae, including strains of *A. itersonii*, have been isolated from soils polluted with chloroanilines (Surovtseva et al., 1996). Under aerobic conditions, these strains were able to grow and degrade 3- and 4-chloroaniline and 3,4-dichloroaniline as sole sources of carbon and nitrogen. Enzymes with different substrate specificity were synthesized during cultivation on different substrates, indicating a potential use for these organisms as bioremediators in soil polluted with chloroanilines (Vasilyeva et al., 1996). Identification of these microorganisms, however, has so far been based only on morphological and physiological characteristics.

Although widely distributed in nature, aquaspirillae comprise only a very small proportion of the total flora of natural habitats. Helical bacteria in general represent only 0.1–0.6% of the flora of pond mud, surface water, slime on stones, or trickling filter effluents, and less than 0.01% in most other habitats (Scully and Dondero, 1973). Consequently, an enrichment procedure is usually necessary before spirillae can be isolated from these habitats.

Chemotaxonomic characteristics In a comparison of the chemotaxonomic characteristics of 34 spirillae, Sakane and Yokota (1994) found two groups with different nonpolar fatty acid profiles. The first group consisted of aquaspirillae (and oceanospirillae) with the Q-8 quinone system and containing hexadecanoic acid ($C_{16:0}$) and hexadecenoic acid ($C_{16:1}$) as major fatty acids. This group consisted of all the aquaspirillae and oceanospirillae that belonged to the *Betaproteobacteria* and *Gammaproteobacteria*, respectively. A second group, consisting of aquaspirillae, oceanospirillae, and some magnetospirillae, had Q-9 and Q-10 quinones and contained octadecenoic acid ($C_{18:1}$) as a major fatty acid. Phylogenetically all these spirillae belong to the *Alphaproteobacteria*. With regard to the 3-hydroxy fatty acids, more than eight different profiles were found among the 34 spirillae in-

TABLE BXII.β.81. Cellular concentrations of nonpolar and hydroxy fatty acids in *Aquaspirillum*, *Magnetospirillum*, and former *Aquaspirillum* species[a]

Species	Strain[d]	Nonpolar fatty acid[b]													3-hydroxy fatty acids[c]							2-hydroxy fatty acid[e]
		$C_{10:1}$	$C_{12:0}$	$C_{12:1}$	$C_{14:0}$	$C_{14:1}$	$C_{15:0}$	$C_{16:0}$	$C_{16:1}$	$C_{17:0}$	$C_{17:1}$	$C_{18:0}$	$C_{18:1}$	$C_{19:0}$	$C_{8:0}$	$C_{10:0}$	$C_{12:0}$	$C_{14:0}$	$C_{14:1}$	$C_{16:0}$	$C_{18:0}$	
Aquaspirillum spp. from the *Betaproteobacteria*																						
Aquaspirillum (*sensu stricto*)																						
Aquaspirillum serpens subsp. *serpens*	14924[T]		3			1		8	51			3	33				100					—
Aquaspirillum serpens subsp. *serpens*	14923		5			1		9	45	2	1	7	29				100					—
Aquaspirillum serpens subsp. *serpens*	15465		3		1	2		13	42		1	1	37				100					—
Aquaspirillum spp. (phylogenetically not correctly named)																						
Aquaspirillum autotrophicum	15327[T]		3					33	33	3	3	1	19	6		19	81					+ ($C_{11:0}$, $C_{18:1}$)
Aquaspirillum dispar	15328[T]		5	2	2		4	29	63	3			13				100					—
Aquaspirillum putridiconchylium	13962[T]		1	1	1			11	73	2		2	9				100					—
Aquaspirillum spp. belonging to the *Comamonadaceae*																						
Aquaspirillum anulus	14917[T]		2		3		2	33	52	3		3	4			100						—
Aquaspirillum delicatum	14919[T]		1					32	35	2		1	23	8	100							—
Aquaspirillum giesbergeri	14959[T]		4		4			24	60			1	8			100						—
Aquaspirillum gracile	14920[T]	5	4		4			35	50	3		9				58	42					—
Aquaspirillum metamorphum	13960[T]		2		3			25	55	1		1	13			100						—
Aquaspirillum psychrophilum	13611[T]		2		3			19	63	1		2	10		10	90						—
Aquaspirillum sinuosum	14925[T]		2		3			31	58				6			100						—
Aquaspirillum spp. from the *Alphaproteobacteria*																						
Aquaspirillum polymorphum	13961[T]		2	2	4			20	13			1	58					66	6		28	+ ($C_{18:1}$, $C_{19:0 \ iso}$)
Former *Aquaspirillum* spp.																						
Comamonas terrigena biovar aquaticum	14918[T]		2		4		7	28	40	3			16			100						—
Levispirillum peregrinum subsp. *peregrinum*	14992[T]		3					16	19	1	2	2	57					39	45		15	+ ($C_{18:1}$)
Levispirillum peregrinum subsp. *integrum*	13617[T]		2					12	12			2	73					42	34		21	+ ($C_{18:1}$)
Levispirillum itersonii subsp. *itersonii*	14921		4			1		15	13			2	63					40	30		30	+ ($C_{18:1}$)
Levispirillum itersonii subsp. *nipponicum*[f]	13615[T]		3			1		14	20			2	59					42	31		27	+ ($C_{18:1}$)
Magnetospirillum																						
Magnetospirillum gryphiswaldense	15271[T]				3	2		7	25		1	5	56					66	18		16	+ ($C_{18:1}$, $C_{19:0 \ iso}$)
Magnetospirillum magnetotacticum	15272[T]		2		1			25	30			1	41					58	39		3	+ ($C_{19:0 \ iso}$)

[a]Data from Sakane and Yokota (1994).

[b]The numbers refer to the percentage of an acid relative to the total nonpolar acids.

[c]The numbers refer to the percentage of an acid relative to the total 3-hydroxy acids.

[d]All strain numbers are from the IFO (NBRC) culture collection.

[e]—, absent; +, present

[f]Strain ATCC 33333[T] has been shown to be different from strain IFO 13615[T] (Pot, 1996).

TABLE BXII.β.82. Cellular concentrations of polyamines in *Aquaspirillum*, former *Aquaspirillum* species, *Magnetospirillum*, and *Spirillum*[a,b]

Organism	Strain number (IFO)	Medium	Dap	H-Put	Put	Cad	Spd	HSpd
Aquaspirillum species from the *Betaproteobacteria*								
Aquaspirillum (*sensu stricto*)								
Aquaspirillum serpens subsp. *serpens*	IFO 14923 (ATCC 11335)	199	−	2.00	1.12	−	−	−
		PY	−	0.46	0.94	−	−	−
	IFO 14924[T] (ATCC 12638[T])	PY	−	1.25	1.55	−	−	−
Aquaspirillum serpens subsp. *bengal*	IFO 15485[T] (ATCC 27641[T])	199	−	0.55	0.80	−	−	−
Aquaspirillum spp. (phylogenetically not correctly named)								
Aquaspirillum autotrophicum	IFO 15327[T] (ATCC 29984[T])	199	−	0.65	1.45	−	−	−
Aquaspirillum dispar	IFO 15328[T] (ATCC 27510[T])	199	0.20	0.95	0.01	1.30	−	−
		199[c]	0.09	0.35	0.02	0.51	−	−
Aquaspirillum putridiconchylium	IFO 13962[T] (ATCC 15279[T])	199	−	0.41	0.06	−	−	−
		PY	−	0.40	1.06	−	−	−
Aquaspirillum spp. belonging to the *Comamonadaceae*								
Aquaspirillum anulus	IFO 14917[T] (ATCC 11879[T])	PY	−	1.15	2.00	−	−	−
Aquaspirillum delicatum	IFO 14919[T] (ATCC 14667[T])	PY	−	0.02	0.58	−	−	−
Aquaspirillum giesbergeri	IFO 13959[T] (ATCC 11334[T])	199	−	0.50	0.70	−	−	−
		199[c]	−	0.29	0.76	−	−	−
		PY	−	0.25	0.50	−	−	−
Aquaspirillum gracile	IFO 14920[T] (ATCC 19624[T])	PY	−	1.40	1.10	−	−	−
Aquaspirillum metamorphum	IFO 12012	199	−	1.25	3.00	−	−	−
	IFO 13960[T] (ATCC 15280[T])	199	−	0.77	0.49	−	−	−
Aquaspirillum psychrophilum	IFO 13611[T]	199	−	0.15	0.93	−	0.15	−
		PY	−	0.25	1.54	−	0.08	−
Aquaspirillum sinuosum	IFO 14925[T] (ATCC 9786[T])	PY	−	0.10	0.64	−	−	−
Aquaspirillum spp. from the *Alphaproteobacteria*								
Aquaspirillum polymorphum	IFO 13961[T] (ATCC 11332[T])	199	−	0.30	0.32	−	0.88	0.14
		PY	−	1.75	0.95	−	0.74	0.50
Former *Aquaspirillum* spp.								
Comamonas terrigena biovar *aquaticum*	IFO 14918[T] (ATCC 11330[T])	199	−	0.81	1.14	−	0.06	−
		199[c]	−	0.08	0.62	−	0.02	−
Levispirillum peregrinum subsp. *peregrinum*	IFO 14922[T] (ATCC 11332[T])	199	−	−	0.47	−	0.91	−
Levispirillum peregrinum subsp. *integrum*	IFO 13617[T]	199	−	−	0.85	−	0.65	−
		PY	−	−	1.50	−	0.25	−
Levispirillum itersonii subsp. *itersonii*	IFO 14921[T] (ATCC 11331[T])	199	−	−	0.24	−	1.40	0.34
		199[c]	−	−	0.21	−	0.48	0.11
Levispirillum itersonii subsp. *nipponicum*[d]	IFO 13615[T]	199	−	−	0.41	−	0.64	0.26
		199[c]	−	−	1.25	−	0.42	0.14
Magnetospirillum								
Magnetospirillum magnetotacticum	IFO 15272[T] (DSM 3856[T])	ATY	−	−	0.71	0.21	0.21	−
Magnetospirillum gryphiswaldense	IFO 15271[T] (DSM 6361[T])	ATY	−	−	1.30	−	0.33	−
Spirillum								
Spirillum volutans	ATCC 19554[T]	199	−	−	0.20	0.43	0.60	−

[a]Data from Hamana et al. (1994).

[b]Abbreviations: Dap, diaminopropane; H-Put, 2-hydroxyputrescine; Put, putrescine; Cad, cadaverine; Spd, spermidine; HSpd, homospermidine; IFO, Institute for Fermentation, Osaka, Japan; IAM, Institute of Applied Microbiology, The University of Tokyo, Tokyo, Japan; ATCC, American Type Culture Collection, Rockville, Maryland, U.S.A.; DSM, German Collection of Microorganisms and Cell Cultures, Braunschweig, Germany; T, type strain; −, not detectable (<0.005).; ATY, growth medium containing 0.1% sodium acetate, 0.05% sodium thioglycolate, 0.01% yeast extract, 0.01% NH_4Cl, 0.01% $MgSO_4 \cdot 7H_2O$, 0.05% K_2HPO_4, 20 µM ferric citrate, pH 6.9; PY, growth medium containing 1% peptone, 0.2% yeast extract, 0.1% $MgSO_4$, pH 7.0; 199, polyamine-free growth medium from Flow Lab., Irvine, U.K., pH 7.0.

[c]Harvested at logarithmic growth phase (others were harvested at stationary growth phase).

[d]Strain ATCC 3333[T] has been shown to be different from strain IFO 13615[T] (Pot, 1996).

vestigated. In regard to the 2-hydroxy fatty acids, all group 2 strains contained 2-hydroxy-octadecenoic acid ($C_{18:1\ 2OH}$) and/or 2-hydroxy-*iso*-nanodecenoic acid ($C_{19:0\ iso\ 2OH}$). Only one group 1 species with Q-8, namely *A. autotrophicum*, contained a single 2-hydroxy-fatty acid, namely undecanoic acid ($C_{11:0\ 2OH}$). *A. autotrophicum* is the only member of the genus known to be a facultative hydrogen autotroph, i.e., capable of growing with CO_2 as a sole carbon source under an atmosphere containing $H_2/O_2/CO_2$ (Aragno and Schlegel, 1978). Although hydrogen autotrophy may not have been tested in other species, the separate genotypic characteristics of this species (Gillis et al., unpublished results) support these aberrant chemotaxonomic and phenotypic characteristics and its separate taxonomic position among the aquaspirillae (see below sub taxonomic comments). Chemotaxonomic data on polyamines separated the *Aquaspirillum* species into six groups (Hamana et al., 1994). The species

A. metamorphum, *A. giesbergeri*, *A. anulus*, *A. delicatum*, *A. gracile*, *A. sinuosum*, *A. autotrophicum*, *A. putridiconchylium*, and *A. serpens* belong to the 2-hydroxyputrescine–putrescine type. *A. aquaticum* and *A. psychrophilum* belong to the 2-hydroxyputrescine–putrescine–spermidine type and differ from the above seven species by the presence of spermidine in addition to putrescine and 2-hydroxyputrescine, suggesting a corresponding phylogenetic difference. *A. dispar* belongs to the diaminopropane–2-hydroxyputrescine–putrescine–spermidine–cadaverine type. *A. peregrinum* belongs to the putrescine–spermidine type, *A. itersonii* to the putrescine–spermidine–homospermidine type, and *A. polymorphum* to the 2-hydroxyputrescine–putrescine–spermidine–homospermidine type. These last three species differ from the species mentioned above in their dominant polyamines; it has been shown (see below) that these species occupy a separate phylogenetic position.

TABLE BXII.β.83. Differential characteristics for species of the genus *Aquaspirillum* and comparison to related and former species of the genus[a]

Differential characteristics	*Spirillum volutans*	*Propioniborus fasciculus*	*Comamonas terrigena* biovar *aquaticum*	*Aquaspirillum delicatum*	*Aquaspirillum gracile*	*Aquaspirillum metamorphum*	*Aquaspirillum psychrophilum*	*Aquaspirillum sinuosum*	*Aquaspirillum giesbergeri*	*Aquaspirillum anulus*	*Aquaspirillum itersonii* subsp. *nipponicum* (ATCC 33333)	*Aquaspirillum serpens* subsp. *serpens*	*Aquaspirillum serpens* subsp. *bengal*	*Aquaspirillum arcticum*[b]	*Aquaspirillum autotrophicum*
			Betaproteobacteria												
Cell diameter, μm[c]	1.4–1.7	0.7–0.9	0.5–0.6	0.3–0.4	0.2–0.3	0.7–1.3	0.7–0.9	0.6–0.9	0.7–1.4	0.8–1.4	0.5–0.8	0.6–1.1	0.9–1.2	1.0	0.6–0.8
Cell morphology	H	SR	H	V	H	H	H	H	H	H	H	H	H	SR-V	H
Type of helix[d]			C	–	C	C	C	C	C	H	CC	C	C	–	C
Wavelength of helix, μm[e]	16–28		2.0–5.0	–	2.8–3.5	7.5–12.0	5.5–6.5	8.6–10.5	4.5–8.4	5.0–13.0	2.5–6.0	3.5–12.0	4.6–8.1	–	3.0–4.0
Helix diameter, μm[e]	5.0–8.0		0.8–1.0	0.4–0.7	0.5–2.1	2.2–3.5	1.0–1.4	1.4–3.5	1.2–5.0	1.7–4.5	1.2–2.0	1.2–4.2	1.7–2.3	–	
Length of helix, μm[e]	14–60	3.6–43.0	2.5–13.0	3.0–5.0	3.5–14.0	3.5–11.0	1.5–14.0	5.0–42.0	4.0–40.0	4.0–52.0	2.0–10.0	3.5–42.0	5.2–22.0	3.0–7.0	2.0–5.0
Polar membrane present	+	+	+	+	+	+	+	+	+	+	+	+	+	+	+
Poly-β-hydroxybutyrate formed	+	+	+	+	–	+	–	+	+	+	+	+	+	+	+
Type of flagellation	BT	BT	BT	U(1–2)	BT	BT	BT	BT	BT	BT	BT	BT	BT	BS	BT
Coccoid bodies dominant after 3–4 weeks	–	+	–	–	–	–	–	–	–	–	+	–	–	+	+
Acid produced from sugars[f]	–	–	–	+	+	–	+	–	–	–	–	–	–	+	–
Temperature range for growth in °C			12–42	09–40	10–42	03–38	02–26	09–37	09–36	3–36	12–42	12–44	15–42	0–20	10–35
Optimal temperature for growth is 20°C and no growth at >26°C	–	–	–	–	–	–	+	–	–	–	–	–	–	–	–
Optimal growth temperature is 5°C	–	–	–	–	–	–	–	–	–	–	–	–	–	+	–
Optimal growth temperature is 41°C	–	–	–	–	–	–	–	–	–	–	–	–	+	–	–
Obligately microaerophilic	+	–	–	–	–	–	–	–	–	–	–	–	–	–	–
Oxidase	+	+	+	+	+	+	+	+	+	+	+	+	+	+	+
Catalase	–	+	+	+	+	+	+	+	+	+	+	+	+	+	+
Phosphatase	+	+	+	+	+	+	+	+	+	+	+	+	+	+	+
Urease[g]	–	+	+	–	+[g]	–	–	+	+	–	W	d	–	–	+
Indole test	–	–	–	–	–	–	–	–	–	–	–	d	–	–	–
Nitrate only reduced to nitrite	–	+	+	+	+	–	–	–	–	–	–	–	–	–	–
Anaerobic growth with nitrate	–	+	–	–	–	–	+	–	–	–	+	–	–	+	–
Denitrification	–	–	–	–	–	–	+	–	–	–	+	–	–	–	–
Hydrolysis of esculin	–	–	–	–	–	–	+	–	–	–	+	–	–	–	–
Hydrolysis of casein and starch	–	–	–	–	–	–	–	–	–	–	+	–	–	–	–
Hydrolysis of hippurate	–	–	–	–	–	–	–	–	–	–	–	d	–	–	–
Hydrolysis of gelatin at 30°C after 4 d	–	–	–	–	–	+	–	–	–	–	–	d	–	–	–

(continued)

Table (continued). Symbol key given in footnote a.

Characteristic	1	2	3	4	5	6	7	8	9	10	11	12	13	14	15
Liquefaction of gelatin at 20°C after:															
7 d	−	−	+	+	+	+	+	+	+	+	d	−	−		−
28 d	−	−	+	+	+	+	+	+	+	+	d	−	−		−
Growth factors required[h]	+	−	−	+	−	−	−	−	+	+	+	+	+		+
Glutamate used as sole carbon source	+	−	−	+	+	+	−	−	+	+	+	+	+		−
Histidine used as sole carbon source	−	−	−	−	−	−	+	+	−	−	−	−	−		−
Tryptophan and glycine used as sole carbon source	−	−	−	−	−	−	−	−	−	−	−	−	−		+
Nitrogenase activity[g]	+	+	−	−	−	−	−	−	+	−	d	−	−		−
Hydrogen autotrophy[g]	−	−	+	+	+	+	+	+	+	+	−	−	+		+
pH range for growth	5.5–8.5	5.5–9.0	5.5–8.5	6.0–9.0	5.5–9.0	6.0–9.0	6.0–9.0	6.0–8.5	5.5–9.0	6.0–9.0	6.0–9.0	6.0–8.4	5.5–7.6		5.0–8.0
Growth in the presence of:															
1% Oxgall	+	+	+	+	+	+	+	+	+	+	+	+	−		+
1% Glycine	−	+	−	−	−	−	−	−	+	−	−	−	−		−
3% NaCl	−	−	−	−	−	−	−	−	−	−	−	−	−		−
Water-soluble brown pigment formed in the presence of:															
0.1% tyrosine	−	−	−	−	−	−	−	−	−	−	−	+	−		−
0.1% tryptophan	−	−	−	−	−	−	−	−	−	−	−	+	−		−
Water-soluble yellow-green fluorescent pigment formed	W	+	d	+	+	+	+	−	−	−	d	−	−		−
Alkaline reaction in litmus milk	−	−	−	−	−	−	+	+	−	−	+	−	−		−
Growth on:															
Eosine methylene blue agar	+	+	+	+	+	+	−	−	−	−	+	+	d		−
MacConkey agar	+	+	−	−	+	+	−	−	−	−	+	+	d		+
Triple-Sugar-Iron agar	−	+	+	−	+	+	−	−	+	−	+	+	d		+
Sellers agar	+	+	+	+	+	+	+	+	+	−	+	+	d		+
Methyl red-Voges-Proskauer broth	+	+	+	+	−	−	+	+	−	+	+	−	−		+
Reduction of 0.3% H₂SeO₃	+	+	+	+	+	−	−	−	−	−	+	+	+		−
H₂S from 0.2% cysteine in PSS broth after 7 d	− or W	+	+	+	+	+	+	+	−	+	+	+	+		−
H₂S from 0.01% cystine in NA after 7 d		+	+	−	+	+	+	−	W	W	+	+	+		+
Deoxyribonuclease	−	+	+	−	+	+	+	+	+	−	+	+	−	d	d
Ribonuclease	−	+	+	+	+	+	−	−	+	+	+	−	−	+[j]	−
Mol% G + C of the DNA	38	62–65	64–65	63	64–65	63	57–59	57–59	57–58	58–59	66	49–51	52	NA	60–62

[a]Symbols: +, positive for all strains; −, negative for all strains; W, weak reaction; blank space, not determined; H, helical (one or more complete turns or twists); V, vibrioid (less than one complete turn or twist); SR, straight rod; C, clockwise helix; CC, counterclockwise helix; BT, bipolar tufts; U (1–2), 1 or 2 flagella at each pole; US, single flagellum at one pole; L, lophotrichous flagellation; d, differs among strains.

[b]The phylogenetic position of *A. arcticum* has not been investigated.

[c]By phase-contrast microscopy of 24- to 48-h-old broth cultures.

[d]Determined by focusing on the bottom of the cells. The pattern //// indicates a clockwise (right-handed) helix, whereas the pattern \\\\ indicates a counterclockwise (left-handed) helix.

[e]*Prolinoborus fasciculus* is a straight rod, the length of the helix refers to the length of the rod; *A. delicatum* is vibrioid rather than helical; length of helix can not be determined; helix diameter refers to the width of the vibrio; length of helix refers to the length of the vibrio.

[f]For *A. gracile*, acid from D-glucose, D-galactose, and L-arabinose (aerobically). For *Lewispirillum itersonii*, acid from glycerol (aerobically), fructose (aerobically and anaerobically), and glucose (anaerobically only). For *Lewispirillum peregrinum*, acid from fructose (aerobically and anaerobically). Peptone concentrations must be kept low (0.2% or less) in order to detect changes in pH indicator. Although *Lewispirillum itersonii* and *Lewispirillum peregrinum* acidify sugar media anaerobically, turbid growth does not occur and the organisms should be considered to have mainly a respiratory rather than fermentative type of metabolism. *A. arcticum* uses fructose, glucose, and ribose, although the use of fructose is listed positive as well as negative in the original publication.

[g]See Procedures for Testing Special Characters.

[h]*A. gracile* requires biotin and *Comamonas terrigena* biovar aquaticum requires niacin.

[i]Acid formation from D-glucose and D-fructose belong to the differentiating features for the discrimination of *Azospirillum* species.

[j]Positive at 37°C but not at 41°C.

TABLE BXII.β.83. *(cont.)*

Betaproteobacteria: *Aquaspirillum putridiconchylium*, *Aquaspirillum dispar*, *Aquaspirillum polymorphum*. Alphaproteobacteria: all remaining taxa.

Differential characteristics	*A. putridiconchylium*	*A. dispar*	*A. polymorphum*	*"L. itersonii* subsp. *itersonii"*	*"L. peregrinum* subsp. *peregrinum"*	*"L. peregrinum* subsp. *integrum"*	*M. magnetotacticum*	*O. pusillum*	*A. brasilense*	*A. lipoferum*	*A. amazonense*	*A. halopraeferens*	*A. doebereinerae*	*A. irakense*	*A. largimobile*
Cell diameter, μm[c]	0.7–1.2	0.5–0.7	0.3–0.5	0.4–0.6	0.5–0.7	0.5–0.7	0.2–0.4	0.3–0.5	0.3–0.4	0.2–0.3	0.7–1.3	0.7–1.4	1.0–1.5	0.6–0.9	0.7–1.5
Cell morphology	H	C	H	H	H	H	H	H	V	H	H	H	H	H	H
Type of helix[d]	C	C	CC	CC	CC	CC	C	CC	C	C	C	C			
Wavelength of helix, μm[e]	4.5–7.0	2.0–3.5	4.0–5.0	2.5–6.0	3.0–4.5	3.0–4.5	1.0–2.0	1.7–2.0	−	2.8–3.5	7.5–12.0	5.5–6.5			
Helix diameter, μm[e]	1.2–2.0	1.0–2.1	1.0–1.5	1.0–2.2	1.4–2.0	1.4–2.2		1.0–1.2	0.4–0.7	0.5–2.1	2.2–3.5	1.0–1.4			
Length of helix, μm[e]	4.0–23.0	2.1–6.5	3.5–8.4	2.0–10.0	1.5–22.0	1.5–22.0	4.0–6.0	1.2–4.0	3.0–5.0	3.5–14.0	3.5–11.0	1.5–14.0			
Polar membrane present	+	+	+	+	+	+	+	+	+	+	+	+	+	+	
Poly-β-hydroxybutyrate formed	+	+	+	+	+	+	+	−	+	+	+	−			
Type of flagellation	BT	BT	BS	BT	BT	BT	BS	BS	U(1–2)	BT	BT	BT	US	US + L	US + L
Coccoid bodies dominant after 3–4 weeks	−	−	+	+	+	+	+	−	−	−	+	−			
Acid produced from sugars[f]	−	−	−	+	+	+	−	−	−	+	−	−	d[i]	−	+
Temperature range for growth in °C	8–40	10–44	14–36	12–42	11–40	11–40		11–39	09–40		03–38	02–26	25–37	30–43	14–43.5
Optimal temperature for growth is 20°C and no growth at >26°C	−	−	−	−	−	−	−	−	−	−	−	+	−	−	−
Optimal growth temperature is 5°C	−	−	−	−	−	−	−	−	−	−	−	−	−	−	−
Optimal growth temperature is 41°C	−	−	−	−	−	−	−	−	−	+	−	−	−	−	−
Obligately microaerophilic	−	−	−	−	−	−	+	−	−	−	−	−	−	−	−
Oxidase	+	+	+	+	+	+	+	W or −	+	+	+	+	+	−	W
Catalase	+	+	+	+	+	+	−	W	+	+	+	+	+	−	W
Phosphatase	−	−	−	+	+	+	+	−	−	+	−	+	+	−	−
Urease[g]	−	−	−	−	+	+	−	+	+	−	−	+	−	−	−
Indole test	−	−	−	−	−	−	−	−	−	+	−	−	−	−	−
Nitrate only reduced to nitrite	−	−	+	−	−	−	−	−	+	+	−	−	−	d	−
Anaerobic growth with nitrate	−	+	+	+	−	+	−	−	+	+	−	+	−	−	−
Denitrification	−	+	+	+	−	−	+	−	+	+	−	+	+	−	−
Hydrolysis of esculin	−	+	+	+	−	−	−	−	−	−	−	+	−	−	−
Hydrolysis of casein and starch	−	−	−	−	−	+	−	−	−	−	−	+	−	−	−
Hydrolysis of hippurate	−	−	−	−	−	−	−	−	−	−	−	−	−	−	−
Hydrolysis of gelatin at 30°C after 4 d	−	−	−	−	−	−	−	−	−	−	−	+	−	−	−

(continued)

Liquefaction of gelatin at 20°C after:														
7 d	−	+	−	−	−	−	−	−	−	−	−	−	−	−	−
28 d	+	+	+	+	+	+	+	+	+	+	+	+	+	+	+
Growth factors required[h]	−	d	−	−	−	+	+	−	−	−	+	−	−	−	
Glutamate used as sole carbon source	+	+	+	+	+	+	+	−	+	+	+	−	−	+	
Histidine used as sole carbon source	+	−	−	+	+	−	−	−	−	−	−	−	−	+	
Tryptophan and glycine used as sole carbon source	−	−	−	−	−	−	−	−	−	−	−	−	−	−	
Nitrogenase activity[g]	−	−	d	+	+	+	+	−	+	−	−	−	+	+	
Hydrogen autotrophy[g]	−	+	−	−	−	−	−	−	+	−	−	−	−	−	
pH range for growth	5.5–8.5	6.0–8.5	5.5–9.0	5.5–9.0	5.5–9.0				5.5–8.5	6.0–9.0	6.8–8			5.5–8.5	
Growth in the presence of:															
1% Oxgall	+	+	+	+	+	+	+	−	+	+	+	−	−	+	
1% Glycine	−	+	−	−	−	−	+	−	−	−	−	−	−	+	
3% NaCl	−	−	−	−	−	−	−	−	−	−	+	−	−	+	
Water-soluble brown pigment formed in the presence of:															
0.1% tyrosine	−	−	−	+	+	−	−	−	−	−	−	−	−	−	
0.1% tryptophan	−	−	+	+	+	−	−	−	+	d	−	−	−	−	
Water-soluble yellow-green fluorescent pigment formed	−	+	+	+	+	+	−	−	−	d	−	−	−	−	
Alkaline reaction in litmus milk	−	−	−	−	−	+	−	−	−	−	−	−	−	−	
Growth on:															
Eosine methylene blue agar	+	+	+	+	+	+	+		+	+	+			+	
MacConkey agar	−	+	+	+	+	+	+		+	−	−			+	
Triple–Sugar–Iron agar	+	+	−	+	+	+	+		+	+	+			+	
Sellers agar	−	+	−	+	−	+	−		−	−	+			+	
Methyl red–Voges–Proskauer broth	−	−	+	+	+	+	+		+	+	+			+	
Reduction of 0.3% H_2SeO_3	−	−	+	d	+	−	−		−	−	−			−	
H_2S from 0.2% cysteine in PSS broth after 7 d	+	+	+	+	+	+	+		+	+	+			+	
H_2S from 0.01% cystine in NA after 7 d	+	−	−	+	+	+	−		+	+	+			+	
Deoxyribonuclease	−	−	−	+	−	−	−		−	+	+			+	
Ribonuclease	+	+	+	+	+	+	+		+	+	+			+	
Mol% G + C of the DNA	52	63–65	61–62	60–64	60–62	64	65	51	63	64–65	63	69–70	69.6	64–67	70

[a] Symbols: +, positive for all strains; −, negative for all strains; W, weak reaction; blank space, not determined; H, helical (one or more complete turns or twists); V, vibrioid (less than one complete turn or twist); SR, straight rod; C, clockwise helix; CC, counterclockwise helix; BT, bipolar tufts; U (1–2), 1 or 2 flagella at only one pole; US, single flagellum at one pole; L, lophotrichous flagellation; d, differs among strains.

[b] The phylogenetic position of A. arcticum has not been investigated.

[c] By phase-contrast microscopy of 24- to 48-h-old broth cultures.

[d] Determined by focusing on the bottom of the cells. The pattern //// indicates a clockwise (right-handed) helix, whereas the pattern \\\\ indicates a counterclockwise (left-handed) helix.

[e] Prolinoborus fasciculus is a straight rod, the length of the helix refers to the length of the rod; A. delicatum is vibrioid rather than helical; length of helix can not be determined; helix diameter refers to the width of the vibrio; length of helix refers to the length of the vibrio.

[f] For A. gracile, acid from D-glucose, D-galactose, and L-arabinose (aerobically). For Levispirillum itersonii, acid from glycerol (aerobically), and glucose (aerobically and anaerobically). For Levispirillum peregrinum, acid from fructose (aerobically and anaerobically). Peptone concentrations must be kept low (0.2% or less) in order to detect changes in pH indicator. Although Levispirillum itersonii and Levispirillum peregrinum acidify sugar media anaerobically, turbid growth does not occur and the organisms should be considered to have mainly a respiratory rather than fermentative type of metabolism. A. arcticum uses fructose, glucose, and ribose, although the use of fructose is listed positive as well as negative in the original publication.

[g] See Procedures for Testing Special Characters.

[h] A. gracile requires biotin and Comamonas terrigena biovar aquaticum requires niacin.

[i] Acid formation from D-glucose and D-fructose belong to the differentiating features for the discrimination of Azospirillum species.

[j] Positive at 37°C but not at 41°C.

ENRICHMENT AND ISOLATION PROCEDURES

A number of enrichment methods have been used, usually taking advantage of the ability of aquaspirillae to grow with levels of nutrients low enough to discourage active growth of many other organisms. Two methods employed by Williams and Rittenberg (1957) have yielded excellent results:

a. Peptone or yeast autolysate (1%) is added to a sample of the source water. The samples are incubated at room temperature for ~1 week or until the spirillae become numerous. A portion of this culture is then added to an equal quantity of the source water and the mixture is sterilized by autoclaving. It is then inoculated from the unsterilized portion of the initial culture. After 1–3 transfers through successively nutrient-exhausted medium, the spirillae predominate.

b. A second method is to enrich the initial sample of source water with 1% calcium malate or lactate and incubate for ~1 week. A serial transfer is then made into more source water similarly supplemented with malate or lactate. Spirillae predominate after 3 or 4 such transfers.

For isolation, the enrichments are diluted 1:100 to 1:100,000 with sterile tap water. The dilution bottles are shaken vigorously and allowed to stand at room temperature for 20 min to allow migration of spirillae to the surface of the diluent. Isolation is then accomplished by streaking the surface water onto a suitable medium such as PSS agar or nutrient agar.

For enrichment by the use of putrid infusions of mussels or mud and sand samples, see Terasaki (1963, 1970, 1980); Jannasch (1965) summarized other general methods. Special methods have been used for the following organisms: *A. gracile*, see Canale-Parola et al. (1966); *A. autotrophicum*, see Aragno and Schlegel (1978); *A. bengal*, see Kumar et al. (1974).

MAINTENANCE PROCEDURES

Aquaspirillae may be maintained in semisolid PSS medium at 30°C (except for *A. psychrophilum*, which is maintained at 15°C) with weekly transfer (Hylemon et al., 1973b). Cultures may also be maintained as nutrient agar stabs at room temperature (except for *A. psychrophilum* which is maintained in a refrigerator) with monthly transfer (Terasaki, 1972).

Preservation is most easily accomplished by adding a dense suspension of cells to nutrient broth containing 10% (v/v) dimethylsulfoxide, with subsequent freezing in liquid nitrogen. Terasaki (1975) reported a method for freeze-drying spirillae.

PROCEDURES FOR TESTING SPECIAL CHARACTERS

Characterization methods for aquaspirillae have been described in detail by Terasaki (1972, 1979) and by Hylemon et al. (1973b). The following comments refer to certain aspects of these procedures. Cell dimensions are best measured in wet mounts of broth cultures by phase-contrast microscopy, rather than by dark-field microscopy or by light microscopy of stained smears. To determine whether the cells have a clockwise or counter-clockwise type of helix, see footnote d of Table BXII.β.83. The presence of intracellular poly-β-hydroxybutyrate is best determined by chemical analysis; for example, *A. delicatum* has no visible granules but does make the polymer. The type and number of flagella is best determined by electron microscopy rather than by flagella staining (Williams, 1960). With regard to coccoid bodies, all strains have a few such forms in old cultures; however, it is only in certain species that coccoid bodies become predomi-

nant in old cultures and have taxonomic significance. For testing the acidification of sugar media it is important to use a low concentration of peptone (0.2% or less). For the urease test, cells should be cultured in PSS broth for 24 h and centrifuged and suspended in sterile water to a dense concentration. An aliquot of 0.5 ml of this suspension is then added to 2.0 ml of a medium consisting of 0.1% BES buffer (*N,N*-bis (2-hydroxyethyl)-2-aminoethane sulfonic acid), 2% urea, and 0.001% phenol red; pH 7.0. This medium must be sterilized by filtration because of the thermolability of the urea. A red or magenta color after incubation at 30°C for 24 h indicates a positive reaction, provided that controls in similar media lacking urea remain colorless. For detection of a water-soluble fluorescent pigment, cultures are streaked in a line across plates of PSS agar and incubated for 48–72 h; the covers of the plates are removed and the plates examined with an ultraviolet lamp of the type used for mineralogical specimens (254-nm wavelength). The occurrence of a distinct, yellow-green, fluorescent zone in the agar medium surrounding the growth constitutes a positive test. Cultures to be tested for nitrogenase activity should be cultured in nitrogen-deficient semisolid malate medium (see the genus *Azospirillum*) supplemented with 0.005% yeast extract. Cultures are incubated for 3 d at 30°C and then sealed with rubber vaccine bottle stoppers. Acetylene is injected to a final concentration of 10% (v/v) and the cultures are tested for ethylene production by gas chromatography after 1 h of further incubation. Controls using liquid rather than semisolid medium, and semisolid medium containing 0.1% $(NH_4)_2SO_4$, should be negative for ethylene production. For testing hydrogen autotrophy, the mineral medium of Aragno and Schlegel (1978) is used[2]. Cultures are incubated under an atmosphere of $O_2/CO_2/H_2$ (5:10:85). A requirement for both H_2 and CO_2 should be demonstrated. For testing sole carbon sources, the procedures of Terasaki (1972, 1979) or Hylemon et al. (1973b) should be followed for most species (see Tables BXII.β.79 and BXII.β.83 for additional methods). It is recommended that the type strain of the suspected species be subjected to the same battery of characterization tests as used for the new isolate in order to confirm an identification. Chemotaxonomic characteristics of the aquaspirillae and *Magnetospirillum* can be tested as described by Sakane and Yokota (1994) and Hamana et al. (1994).

DIFFERENTIATION OF THE GENUS *AQUASPIRILLUM* FROM OTHER GENERA

Table BXII.β.84 indicates the characteristics of *Aquaspirillum* that distinguish it from other genera with similar morphological or physiological features. The phylogenetic diversity observed is used to delineate the different subgroups considered within the genus *Aquaspirillum*. The renamed *Aquaspirillum* species (see below) have been represented by their new generic names.

TAXONOMIC COMMENTS

In the eighth edition of *Bergey's Manual of Determinative Bacteriology* (Krieg, 1974), a single genus, *Spirillum*, contained all of the

---

2. Mineral medium (g/l): $Na_2HPO_4 \cdot 12H_2O$, 9.0; KH_2PO_4, 1.5; $MgSO_4 \cdot 7H_2O$, 0.2; NH_4Cl, 1.0; ferric ammonium citrate, 0.005; $CaCl_2 \cdot 2H_2O$, 0.01; R trace element solution (see below), 3.0 ml; pH 7.1. For a solid medium, 17.0 g of agar is added. For both liquid and solid media, 0.05% $NaHCO_3$ should be incorporated aseptically into the sterilized medium to buffer against changes in pH caused by the CO_2 of the gas atmosphere. Trace element solution (mg/l): $ZnSO_4 \cdot 7H_2O$, 10.0; $MnCl_2 \cdot 4H_2O$, 3.0; H_3BO_3, 30.0; $CoCl_2 \cdot 6H_2O$, 20.0; $CuCl_2 \cdot 6H_2O$, 0.79; $NiCl_2 \cdot 6H_2O$, 2.0; $Na_2MoO_4 \cdot 2H_2 0$, 3.0.

TABLE BXII.β.84. Differential characteristics of the genus *Aquaspirillum* and other genera of oxidase-positive, motile, curved, vibrioid, or helical Gram-negative rods[a]

Differential characteristics	Aquaspirillum	Spirillum	Oceanospirillum	Campylobacter	Azospirillum	Bdellovibrio	Pseudomonas	Vibrio	Alteromonas	Protoborus	Comamonas	Leisispirillum	Herbaspirillum	Marinospirillum
Predominant cell shape:														
Helical	+[b]	+	+	-[c]	-	-	-	-	-	-	+	+	+	+
Vibrioid or curved	-[b]	-	-	+[c]	+[d]	+	-[e]	d	d	+	+	-	+	-
Straight rod	-[b]	-	-	-	-[d]	-	+[e]	d	d	+	+	+	-	-
Cell diameter, μm[b]	0.2–1.4	1.4–1.7	0.3–1.4	0.2–0.5	1.0	0.2–0.5	0.5–1.0	0.5–0.8	0.7–1.5	0.7–0.9	0.3–0.8	0.4–0.8	0.6–0.7	0.3–1.2
Polar membrane present	+	+	+	+	+	-	-	-	-	+	d	+	-	+
Unusual arrangement of polar flagella:														
Bipolar tufts	+[f]	+	+[g]	-	-	-	-	-[h]	-	+	+	+	+	+
Tuft at one pole	-[f]	-	-	-	-	-	d	+[h]	+	-	+	-	+	-
Single flagellum at one or both poles	-[f]	-	-[g]	+	+	+	d	d	-	-	-	-	+	-
Lateral flagella also formed under certain conditions	-	-	-	-	+	-	d	d	-	-	-	-	-	+
Intracellular poly-β-hydroxybutyrate formed	+[i]	+	+	-	+	-	d	-	-	+	+	+	+	+
Relation to oxygen:														
Aerobic	+	-	+	-	+	+	+	-	+	+	+	+[m]	+	d
Facultative	-	-	-	-	-	-	-	+	-	-	-	-[m]	+	d
Microaerophilic	-	+	-	+	-	-	-	-	-	-	-	-[m]	+	d
CO$_2$ required for growth	-	-	-	+	-	-	-	-	-	-	-	-	-	-
Seawater or Na$^+$ required for growth	-	-	+	-	-	d	d	+[j]	+	-	-	-	-	+
Tolerant to 3% NaCl	-[k]	-	+	d	d	d	d	+	+	-	-	-	-	+[n]
Indole test	-	-	-	-	-	-	-	d	+	-	d[l]	d[l]	+	-[n]
Carbohydrates are catabolized	-	-	-	-	+	-	d	+	+	-	-	-	+	-
Habitat:														
Freshwater	+	+	-	-	-	d	d	d	-	+	+	+	+	-
Marine	-	-	+	-	-	d	d	d	+	-	+	-	-	+[o]
Soil	-	-	-	+	+	d	d	-	-	-	+	-	+	-
Humans or mammals	-	-	-	+	-	-	-	d	-	-	-	-	-	-
Capable of multiplying in the periplasmic space of other bacteria	-	-	-	-	-	+	-	-	-	-	-	-	-	-
Nitrogenase activity	-	-	-	-	+	-	-	-	-	-	-	d[j]	d[j]	-
Mol% G + C	49–66	38	42–51	30–38	69–71	33–37, 42–51	57–70	38–51	38–50	62–65	60–69	60–66	60–65	42–45

[a]Symbols: +, all species positive except where noted; −, all species negative except where noted; d, differs among species.

[b]*A. delicatum* is mainly vibrioid.

[c]Cells in chains resemble spirilla.

[d]In pure cultures, a proportion of the cells may be straight rods.

[e]The genus *Pseudomonas* contains straight rods and rods that are curved in one plane, but not helically curved rods.

[f]*A. delicatum* mainly has a single flagellum at one pole; *A. polymorphum* mainly has a single flagellum at each pole.

[g]*Oceanospirillum pusillum* mainly has a single polar flagellum, but two species (*V. fischeri* and *V. logei*) have a tuft of polar flagella.

[h]Most species of *Vibrio* have a single polar flagellum, but two species (*V. fischeri* and *V. lugei*) have a tuft of polar flagella.

[i]*A. gracile* and *A. psychrophilum* lack this polymer.

[j]Growth of all species is stimulated by NaCl, and most species have an absolute requirement for Na$^+$.

[k]*A. gracile*, *Levispirillum itersonii*, and *Levispirillum peregrinum* can catabolize a very restricted number of sugars. All other species are incapable of catabolizing any carbohydrates.

[l]Only a few carbohydrates are used.

[m]Some species fix nitrogen in microaerophilic conditions, but with a source of fixed nitrogen they grow as aerobes.

[n]Only known for *M. minutulum*.

[o]*M. magaterium* has been isolated from kusaya gravy (used to produce traditional dried fish).

various aerobic and microaerophilic spirillae. However, the DNA base composition for this genus ranged from 38 to 65 mol% G + C, a range much greater than expected for a well-defined bacterial genus. Within the genus, three groups became evident: (a) the aerobic freshwater spirillae that could not tolerate 3% NaCl (mol% G + C is 49–66); (b) the aerobic marine spirillae that required seawater for growth (mol% G + C is 42–51); and (c) the large microaerophilic spirillae that belonged to the species *Spirillum volutans* (mol% G + C is 38). Accordingly, Hylemon et al. (1973a, b) divided the genus into three genera *Aquaspirillum*, *Oceanospirillum*, and *Spirillum*, respectively.

In the first edition of the *Bergey's Manual of Systematic Bacteriology* (Krieg, 1984a), for practical purposes, the single genus *Aquaspirillum* was maintained for a considerably heterogeneous group of microorganisms, as indicated by the wide range of mol% G + C (49–66). Phenotypically, however, the genus *Aquaspirillum* was based largely on a pattern of core properties considered typical for the genus. These included, besides the lack of tolerance to 3% NaCl: a helical shape; bipolar tufts of flagella; poly-β-hydroxybutyrate formation; a strictly respiratory type of metabolism; positive oxidase, catalase, and phosphatase reactions; an inability to attack sugars; hydrolysis of starch and casein; a negative indole test; an optimal growth temperature of 30°C; and a simple chemoheterotrophic nutrition with amino acids or the salts of organic acids serving as carbon sources. Species were assigned to the genus based on a similarity between their characteristics and this pattern of core features, with the recognition that exceptional characteristics may occur. Such exceptional characteristics are: a vibrioid shape or a straight rod shape, nitrogen-fixing ability, hydrogen autotrophy, lack of poly-β-hydroxybutyrate, high or low temperature optima for growth, a single flagellum at one or both poles, catabolism of a limited variety of sugars, and vitamin requirements.

Other than the wide mol% G + C range of the DNA, genotypic evidence to support the lack of phylogenetic relationship at the genus level of the various species of *Aquaspirillum* was scarce, and mainly based on 16S rRNA oligonucleotide catalogs of three *Aquaspirillum* species (Woese et al., 1982). *A. serpens* was shown to belong to group II of the phototrophic bacteria as defined by Gibson et al. (1979), or to the β-subdivision of the purple bacteria (*Proteobacteria*) (Woese et al., 1984c; Vandenberghe et al., 1985). It was found to be related to *Rhodospirillum tenue* and, to a lesser extent, to *Spirillum volutans*. *A. gracile* was also shown to be a member of group II, but was more closely related to *Rhodopseudomonas gelatinosa* than to *A. serpens*, *S. volutans*, or *R. tenue*. Importantly, *A. itersonii* was found to be a member of group I or the α-subdivision of the purple bacteria (*Proteobacteria*) (Woese et al., 1984b) and to be related to *Rhodospirillum rubrum* and *Azospirillum brasilense*. The relationship of *A. itersonii* and *A. polymorphum* to the genus *Azospirillum* was also demonstrated by rRNA–DNA hybridization studies (De Smedt et al., 1980).

Subsequent rRNA–DNA hybridizations (Willems et al., 1991a, c; Pot et al., 1992b; Pot, 1996), DNA–DNA hybridizations (Boivin et al., 1985; Pot, 1996), and 16S rRNA sequencing results (Kawasaki et al., 1997; Wen et al., 1999) confirmed this extensive genotypic heterogeneity and revealed some unexpected relationships. It is striking that most of these genotypic variations could not be linked to clear phenotypic descriptions of well-defined separate taxa. Still, some formal taxonomic changes have recently been introduced.

As suggested by Krieg (1984a), DNA–DNA hybridization showed that indeed some species did not deserve separate species status; e.g., it was shown that *A. serpens* and *A. bengal* belonged to a single species (Boivin et al., 1985; Pot, 1996), with DNA–DNA hybridization values between 59 and 73%. Consequently, *A. bengal* was included in *A. serpens* as *A. serpens* biovar bengal. *A. serpens* biovar bengal showed an average of 58% DNA–DNA relatedness to three reference strains of *A. serpens*, while the average DNA–DNA relatedness between these *A. serpens* strains (ATCC 12638^T, 15278, and 27050) was 72%. Considering these values and the clear phenotypic differences (Tables BXII.β.79 and BXII.β.83), *A. serpens* biovar bengal could be regarded as a subspecies of *A. serpens* or even as a separate species. The description of the two subspecies, *A. serpens* subsp. *bengal* and *A. serpens* subsp. *serpens*, was first mentioned by Pot (1996) and will be proposed here.

The position of *A. serpens* biovar azotum is, at present, not very clear. Pot (1996) showed by SDS-PAGE of whole-cell proteins that strain ATCC 11335 belonged to *A. serpens* (comparison with strains Murray VHA, Murray St. Rhodes, Murray VHL, Murray VH, ATCC 27050, and ATCC 12638^T), confirming some of the serological relationships observed by McElroy and Krieg (1972) and the phenotypic similarity found by Hylemon et al. (1973b) and Carney et al. (1975). However, strain ATCC 11335 was shown by SDS-PAGE of whole-cell proteins to be different from its supposed homologous subculture NCIB 9011 (Pot, 1996). The latter strain was found to be as far removed from *A. serpens* as the genera *Neisseria* and *Chromobacterium* and had no significant DNA–DNA relatedness with the other *A. serpens* strains (Pot, 1996). Possibly this subculture of NCIB 9011 can be regarded as a contaminant.

A polyphasic study resulted in the inclusion of *A. aquaticum* and 13 clinical isolates previously designated EF group 10 (Falsen, 1996) in *Comamonas terrigena* (Willems et al., 1991c; and the chapter on *Comamonas* in this *Manual*). Consequently *A. aquaticum* should be considered as a junior synonym of *Comamonas terrigena*. The synonymy between *C. terrigena* and *A. aquaticum* was previously suggested by 16S rRNA cataloging data (Woese et al., 1984c).

Seven other *Aquaspirillum* species were classified in the *Comamonadaceae* based on rRNA–DNA hybridizations and 16S rDNA sequencing (Willems et al., 1991c; Wen et al., 1999; the chapter on the *Comamonadaceae* in this *Manual*): *A. anulus*, *A. delicatum*, *A. giesbergeri*, *A. gracile*, *A. metamorphum*, *A. sinuosum*, and *A. psychrophilum*. These species are phylogenetically too distantly removed from the type species, *A. serpens*, to be considered members of the genus *Aquaspirillum*. *A. anulus*, *A. giesbergeri*, and *A. sinuosum* share a relatively low mol% G + C content (57–58). SDS-PAGE analysis of whole-cell proteins showed that the type strains of *A. giesbergeri* and *A. sinuosum* are very similar (a Pearson correlation of 0.9), with only small differences in the zone with molecular weights between 40,000 and 70,000, and therefore possibly belong to a single species. Further DNA–DNA hybridization studies are necessary to confirm this similarity. *A. giesbergeri* and *A. sinuosum* together constitute a separate rRNA branch in the *Comamonadaceae*. The position of *A. anulus* as a separate rRNA branch has been solely determined by rRNA–DNA hybridizations (Pot et al., 1992b).

A. delicatum, *A. gracile*, *A. metamorphum*, and *A. psychrophilum* have mol% G + C values of 62–66. *A. psychrophilum* and *A. metamorphum* have 97.2% 16S rDNA sequence similarity and constitute a separate rRNA branch in the *Comamonadaceae*. *A. delicatum* belongs to the same rRNA branch as *Rhodoferax* (96.8% 16S rDNA sequence similarity). More data are needed to determine whether *A. delicatum* should be considered as a third species in *Rhodoferax* (Hiraishi et al., 1991a; Hiraishi, 1994; Madigan et al., 2000b), which currently contains two species of curved rods to spirillae, able to grow phototrophically and having a mean mol% G + C of ~61. A possible relationship between *A. delicatum* and *Comamonas* had already been suggested (Krieg, 1984a) because *A. delicatum* is vibrioid rather than helical, and has one or two flagella at only one pole rather than bipolar flagellar tufts. Although Leifson (1962) placed the organism in the genus *Spirillum* (*Aquaspirillum*), he recognized that it differed morphologically from typical spirillae and suggested the possibility of creating a new genus for such non-carbohydrate-utilizing vibrios.

A. gracile belongs to a separate group deeply branching within the *Comamonadaceae*. The detailed chemotaxonomic data on polyamines, quinones, 2- and 3-hydroxy fatty acids, and main nonpolar fatty acids (Hamana et al., 1994; Sakane and Yokota, 1994) are displayed in Tables BXII.β.80, BXII.β.81, and BXII.β.82, but do not completely support the division on the basis of 16S rDNA sequence analysis. For example, *A. psychrophilum* differs considerably from *A. metamorphum* in its polyamine content and in its 3-hydroxy fatty acid content.

The considerable phenotypic inertness of the aquaspirillae, the lack of uniform phenotypic testing procedures, and the large phenotypic differences observed with the other genera of the *Comamonadaceae* make it difficult to phenotypically describe *Aquaspirillum* species within the family *Comamonadaceae*. Until more phenotypic characters are available, or until genotypic parameters can be used exclusively for the description of taxonomic entities, the species will be maintained in the genus *Aquaspirillum*.

Aquaspirillum fasciculus received special attention in the first edition of *Bergey's Manual of Systematic Bacteriology* (Krieg, 1984a). This species showed many characteristics atypical for the aquaspirillae (straight-rod shape, forming of viscous flocs, failing to swim except in a viscous medium, and nitrogenase activity). However, *A. fasciculus* did possess bipolar tufts of flagella, formed poly-β-hydroxybutyrate, had a strictly respiratory type of metabolism, did not attack sugars, could not grow with 3% NaCl (or even 1% NaCl), had a simple nutrition, grew best at 30°C, was indole-negative, did not hydrolyze starch or casein, possessed oxidase, catalase, and phosphatase activity, and had a mol% G + C content of the DNA of 62–65. Moreover, *A. fasciculus* formed coccoid bodies in abundance, which is a characteristic of certain aquaspirillae and other helical or vibrioid bacteria (*Oceanospirillum, Campylobacter, Desulfovibrio*, and *Vibrio*; see Williams and Rittenberg, 1957; Ogg, 1962; Levin and Vaughn, 1968; Felter et al., 1969; Baker and Park, 1975). Furthermore, the presence of a polar membrane is also a characteristic mainly associated with helical bacteria (Murray and Birch-Andersen, 1963; Hickman and Frenkel, 1965a, b; Keeler et al., 1966; Ritchie et al., 1966), although it has also been found in the rod-shaped organism *Chromatium* (Murray and Birch-Andersen, 1963). Nitrogenase activity, typical for *A. fasciculus*, but unusual for aquaspirillae, did occur in two helical species. The morphology of some aquaspirillae is variable: Strength et al. (1976) reported the isolation of a strain with rods very similar to *A. fasciculus*, but after prolonged transfer many cells were found to be curved or even S-shaped. Moreover, some typical helical species develop variants

that are nearly straight rods upon prolonged transfer (Williams, 1959a, b; Terasaki, 1973). Genotypic comparison of *A. fasciculus* to other *Aquaspirillum* species by rRNA–DNA hybridization showed that the species has to be regarded as belonging to a separate genus, for which the name *Prolinoborus* has been proposed (see the genus *Prolinoborus* in this family).

Another species with a separate phylogenetic position is *A. autotrophicum*, which can grow autotrophically under an atmosphere of oxygen, carbon dioxide, and hydrogen. Other members of *Aquaspirillum* are not facultative hydrogen autotrophs or have not been tested for this character. The other characteristics of *A. autotrophicum* are completely consistent with the description of the genus *Aquaspirillum*. However, rRNA–DNA hybridization data showed that this species is phylogenetically more closely related to *Janthinobacterium lividum* and *Herbaspirillum* than to *A. serpens*, the type species of the genus *Aquaspirillum* (Pot, 1996). 16S rDNA sequence analysis confirmed the relationship between *Herbaspirillum* and *Janthinobacterium* (Kirchhof et al., 2001). Whether *A. autotrophicum* can be considered as a separate species of *Herbaspirillum*, or deserves a separate genus status, requires further study.

The taxonomic position of *A. dispar* has been partially clarified. The two available strains of *A. dispar* (ATCC 27510^T and ATCC 27650) showed 88–100% DNA–DNA binding, and have nearly identical electrophoretic protein profiles (Aragno and Schlegel, 1978; Pot, 1996). Both strains shared the highest rRNA cistron similarity with *Chromobacterium violaceum* NCTC 9757^T, although according to 16S rRNA cataloging, *A. dispar* appeared to be more closely related to *A. serpens* than to *Chromobacterium violaceum*. Because of a $T_{m(e)}$ difference of 6°C (Pot, 1996) or more, and some considerable phenotypic differences, *A. dispar* cannot be assigned to the genus *Chromobacterium*. More data are required to determine its exact relationship to a new genus, *Vogesella* (see the genus *Vogesella* in this family), which is phylogenetically the closest neighbor of *Chromobacterium*.

A. putridiconchylium also belongs in the *Betaproteobacteria*, where it constitutes a separate branch (Pot, 1996); a close taxonomic relationship with *A. dispar* is excluded. Because neither of these species can be discriminated phenotypically from the *Aquaspirillum* species, we continue to consider them as separate species of the genus *Aquaspirillum*.

Three species of the genus *Aquaspirillum* belong to the *Alphaproteobacteria* based on rRNA–DNA hybridization and 16S rRNA gene sequencing. *A. itersonii* and *A. peregrinum* were more closely related to *Rhodospirillum rubrum* and *Rhodospirillum photometricum* (Kawasaki et al., 1997), and formed a separate branch in the *Alphaproteobacteria*. *A. polymorphum* was shown to be most closely related to *Magnetospirillum gryphiswaldense*. Schleifer et al. (1991) proposed this generic name for the former species *Aquaspirillum magnetotacticum* (Blakemore et al., 1979, 1989) and the new isolate *M. gryphiswaldense*. Also, it was shown by rRNA–DNA hybridization that *A. itersonii* and *A. peregrinum* form a separate rRNA branch at the node of the *Azospirillum* and *Rhodospirillum rubrum* branch (Pot, 1996). The different subspecies of *A. peregrinum* and *A. itersonii* were shown by 16S rRNA gene sequence comparison to be very highly related within each species (Kawasaki et al., 1997). *A. peregrinum* subsp. *peregrinum* and *A. peregrinum* subsp. *integrum* showed 91% DNA–DNA relatedness and clearly belong to a single species (Pot, 1996). In our hands, however, the two subspecies of *A. itersonii* were not closely related. *A. itersonii* subsp.

itersonii was situated on the *A. peregrinum* rRNA branch, but *A. itersonii* subsp. *nipponicum* was found to belong to the *Betaproteobacteria* (A. Willems unpublished results). In these studies, different subcultures of the type strain were used (*Aquaspirillum itersonii* subsp. *nipponicum* ATCC 33333[T] by Pot (1996) and *Aquaspirillum itersonii* subsp. *nipponicum* IFO 13615[T] by Kawasaki et al., 1997).

In contrast to the aquaspirillae of the *Betaproteobacteria*, the aquaspirillae of the *Alphaproteobacteria* all contain a counter-clockwise type of helix and can hydrolyze esculin. Using these characteristics, they can easily be discriminated from the authentic aquaspirillae and from *Prolinoborus*. Given these genotypic and phenotypic differences, it is possible to allocate two of these species to a new genus. The name *Levispirillum*, with *L. itersonii* and *L. peregrinum*, has been proposed (Pot, 1996) and has been effectively published but not yet validly published. The position of *A. polymorphum* is not clear: phenotypically it could belong to *Levispirillum*, but genotypically it definitely belongs to the genus *Magnetospirillum* (Kawasaki et al., 1997). It certainly does not belong phylogenetically to *Aquaspirillum*. A formal transfer, however, has not been proposed, probably because of the extensive phenotypic differences between *A. polymorphum* and *Magnetospirillum*.

Although the genus *Oceanospirillum* belongs to the *Gammaproteobacteria*, *O. pusillum* has been shown to belong to the *Alphaproteobacteria* (Woese et al., 1985; Kawasaki et al., 1997) by 16S rDNA sequence analysis. This position was confirmed by rRNA–DNA hybridization results (Pot, 1996). Using either method, the type strain occupies a separate branch in the *Alphaproteobacteria*. Although this species also has a counter-clockwise type of helix, is not capable of esculin hydrolysis, requires 3% NaCl for growth, and occupies a phylogenetically distinct position, its inclusion in the genus *Levispirillum* is questionable. Kawasaki et al. (1997) suggested creation of a new genus for this species.

Butler et al. (1989) isolated and described a new psychrophilic, straight-to-curved rod-shaped bacterium, with an optimal growth temperature of 5°C, from an Arctic sediment. It was assigned to the genus *Aquaspirillum* as *Aquaspirillum arcticum* based on its morphological and physiological characteristics. It was distinguished from *A. psychrophilum* (Terasaki, 1973, 1979) by a different optimal temperature for growth (5°C rather than 20°C), and by its ability to grow at 0°C. Unlike *A. psychrophilum*, *A. arcticum* will also grow in vitamin-free, defined, basal salts–glucose medium. Other biochemical characteristics are listed in Table BXII.β.83. *A. arcticum* was found to acquire thermotolerance when heat shocked or treated with nalidixic acid, two conditions responsible for the introduction of heat shock- or stress proteins (McCallum and Inniss, 1990; Gounot, 1991). This indicates that growth temperature is not necessarily a reliable parameter for delineation of species. Genotypic or phylogenetic information is not presently available for *A. arcticum*.

Taxonomic conclusions   In the past, species descriptions have been based mainly on differences in morphology, nutrition, and DNA base composition. In some cases, such species are obviously different; for example, a small spirillum such as *A. polymorphum* would not belong to the same species as a large spirillum (such as *A. serpens*), and a spirillum having a mol% G + C of 50 would not be placed in the same species as one with a mol% G + C of 60. However, species distinctions within a particular morphological group of strains having a similar DNA base composition are less firm. Another difficulty is that many species are represented by only one or two strains, and the limits of variation within such a species may be broader than is presently assumed. Therefore, it was suggested by Krieg (1984a) that some species may not deserve separate species status while other species may represent more than a single species. Genotypic analysis of most aquaspirillae confirmed these observations. The true aquaspirillae should be restricted to the type species *Aquaspirillum serpens*, including the former species *Aquaspirillum bengal*. All other species are phylogenetically too distantly related to be regarded as "true" or authentic aquaspirillae. However, with the exception of a few species, the majority cannot be described reliably from the phenotypic point of view. Therefore they are presently retained in the genus *Aquaspirillum*. The genus definition needed only minor adaptations so as to compensate for the formal removal of *Aquaspirillum fasciculus*, *Aquaspirillum aquaticum*, *Aquaspirillum peregrinum*, and *Aquaspirillum itersonii*.

The two *species incertae sedis*, "*Spirillum minus*" and "*Spirillum pulli*", do not belong to the genera *Aquaspirillum*, *Oceanospirillum*, or *Spirillum*, and their placement is uncertain. Studies of these species have been hampered by the lack of reproducible *in vitro* cultivation methods. They do not appear on the Approved Lists of Bacterial Names because no type or reference strains are available and the organisms are not well characterized. The disease syndromes caused by these species are distinct and recognizable, however. If possible, neotype strains should be designated and either maintained by animal passage or preserved in a recognized culture collection.

ACKNOWLEDGMENTS

We acknowledge Dr. N.R. Krieg for the template text, figures, and tables on which this chapter has been based.

FURTHER READING

Krieg, N.R. 1976. Biology of the chemoheterotrophic spirilla. Bacteriol. Rev. *40*: 55–115.

Krieg, N.R. and P.B. Hylemon. 1976. The taxonomy of the chemoheterotrophic spirilla. Annu. Rev. Microbiol. *30*: 303–325.

Pot, B., M. Gillis and J. De Ley. 1992. The genus *Aquaspirillum*. *In* Balows, Trüper, Dworkin, Harder and Schleifer (Editors), The Prokaryotes. A Handbook on the Biology of Bacteria: Ecophysiology, Isolation, Identification, Applications, 2nd Ed., Vol. 4, Springer-Verlag, New York. pp. 2569–2582.

List of species of the genus Aquaspirillum

1. **Aquaspirillum serpens** (Müller 1786) Hylemon, Wells, Krieg and Jannasch 1973b, 366[AL] (*"Vibrio serpens"* Müller 1786, 48.)

   ser'pens. L. v. *serpo* to crawl or creep; L. part. adj. *serpens* creeping.

The morphological characters are depicted in Fig. BXII.β.67 and described in Table BXII.β.83. Optimal growth temperature 35°C. Chemotaxonomic and physiological characters are described in Tables BXII.β.80, BXII.β.81, BXII.β.82, and BXII.β.83. Sole carbon sources are listed in Table BXII.β.79; the best growth occurs with glutamate. Nitrate is not used as a sole nitrogen source; results with ammonium salts have been conflicting (Terasaki, 1972; Hylemon et al., 1973b). A defined medium suitable for batch and continuous cultures has been described by Whitby and Murray (1980).

Habitat: pond waters.

Aquaspirillum serpens belongs phylogenetically to the *Betaproteobacteria* (Pot, 1996). The former species *Aquaspirillum bengal* (Kumar et al., 1974, 457) has been assigned to *A. serpens* as *A. serpens* biovar bengal (Boivin et al., 1985). Based on sufficient phenotypic differences and a low level of DNA relatedness, *A. serpens* biovar bengal could be regarded as a subspecies of *A. serpens* or even as a separate species.

The mol% G + C of the DNA is: 49–51 (T_m).

Type strain: ATCC 12638, DSM 68.

a. **Aquaspirillum serpens** subsp. **serpens** subsp. nov. (*Aquaspirillum serpens* (Müller 1786) Hylemon, Wells, Krieg and Jannasch 1973b, 366; "*Vibrio serpens*" Müller 1786, 48.)

Cell diameter 0.6–1.1 μm. Description as for the species. No brown pigments in media with aromatic components; some strains do produce water-soluble fluorescent pigments. Growth at temperatures from 12 to 44°C with 35°C as optimum. Can grow at pH values up to 9. Some strains produce gelatinase and ribonuclease. Can grow on EMB agar. KNO_3 can sometimes be reduced to KNO_2 (*Aquaspirillum serpens* subsp. *serpens* var. azotum ATCC 11335).

The mol% G + C of the DNA is: 49–51 (T_m).

Type strain: ATCC 12368, DSM 68.

b. **Aquaspirillum serpens** subsp. **bengal** subsp. nov. (*Aquaspirillum bengal* Kumar, Banerjee, Bowdre, McElroy and Krieg 1974, 457.)

ben' gal. M.L. n. *bengal* Bengal.

The description is as for the species, but the cells are usually shorter. They have a diameter of 0.9–2.2 μm, the wavelength of the helix is 4.6–8.1 μm, and the helix length ranges from 5.2 to 22 μm. Brown pigments can be formed on media containing 0.1% tyrosine or tryptophan. Water-soluble fluorescent pigments are not produced. Can grow at temperatures from 12 to 42°C; 41°C is the optimal growth temperature. Can grow at pH values up to 8.4. No growth on EMB agar, no gelatinase and ribonuclease activity. Prefer to grow in microaerophilic conditions. KNO_3 can not be reduced to KNO_2.

The mol% G + C of the DNA is: 52 (T_m).

Type strain: ATCC 27641.

2. **Aquaspirillum anulus** (Williams and Rittenberg 1957) Hylemon, Wells, Krieg and Jannasch 1973b, 368[AL] (*Spirillum anulus* Williams and Rittenberg 1957, 86.)

a' nu.lus. L. masc. n. *anulus* a ring.

The morphological characters are shown in Fig. BXII.β.67 and described in Table BXII.β.83. Optimal growth temperature 27–30°C. Chemotaxonomic and physiological characters are described in Tables BXII.β.80, BXII.β.81, BXII.β.82, and BXII.β.83. Sole carbon sources are listed in Table BXII.β.79. Ammonium salts can be used as sole nitrogen sources; nitrate is not used.

Isolated from pond water and from putrid infusions of fresh water mussels.

Aquaspirillum anulus belongs phylogenetically to the *Comamonadaceae* (Willems et al., 1991c; Wen et al. 1999; Willems and Gillis, The *Comamonadaceae*, this volume) in which it constitutes a separate branch.

The mol% G + C of the DNA is: 58–59 (T_m).

Type strain: ATCC 35958, NCIB 9012.

3. **Aquaspirillum arcticum** Butler, McCallum, Inniss 1990, 320[VP] (Effective publication: Butler, McCallum, Inniss 1989, 265.)

arc' ti.cum. L. adj. *arcticum* of the Arctic, where the species was first isolated.

The morphological characters are described in Table BXII.β.83. The species is atypical in that it possesses only a single polar flagellum rather than bipolar tufts. The species is considered psychrophilic (defined as organisms having an optimal temperature for growth of ~15°C or lower, a maximum temperature for growth of ~20°C, and a minimum growth temperature of 0°C or below; (Morita, 1975). Optimal growth temperature is 5°C. Differential characteristics are described in Table BXII.β.83. The species is atypical in that it produces acid from carbohydrates (fructose, glucose, and ribose).

Isolated from Arctic sediment; other habitats unknown.

There are no phylogenetic data available for *Aquaspirillum arcticum*.

The mol% G + C of the DNA is: not reported.

Type strain: Res-10, ATCC 49402, DSM 6444.

4. **Aquaspirillum autotrophicum** Aragno and Schlegel 1978, 116[AL]

au.to.tro' phi.cum. Gr. n. *autos* self; Gr. adj. *trophikos* nursing, tending or feeding; M.L. neut. adj. *autotrophicum* self-nursing or self-feeding.

The morphological characters are described in Table BXII.β.83. Optimal growth temperature 28°C. Chemotaxonomic and physiological characters are described in Tables BXII.β.80, BXII.β.81, BXII.β.82, and BXII.β.83. Sole carbon sources are listed in Table BXII.β.79. Ammonium salts and nitrate can be used as sole nitrogen sources.

Aquaspirillum autotrophicum is phylogenetically not closely related to the type species of *Aquaspirillum* (Pot, 1996), but belongs to the *Herbaspirillum–Janthinobacterium* rDNA branch (Gillis et al., 1991; Baldani et al., 1996; Willems et al., unpublished results). Further studies are required to determine whether it belongs in *Herbaspirillum* and/or if it deserves a separate genus status.

Isolated from an eutrophic lake in Switzerland.

The mol% G + C of the DNA is: 60–62 (T_m).

Type strain: SA 32, ATCC 29984, DSM 732.

5. **Aquaspirillum delicatum** (Leifson 1962) Hylemon, Wells, Krieg and Jannasch 1973b, 371[AL] (*Spirillum delicatum* Leifson 1962, 164.)

de.li.ca' tum. L. neut. adj. *delicatum* delicate.

Vibrioid. Chains of cells may resemble spirillae. The morphological characters are shown in Fig. BXII.β.67 and described in Table BXII.β.83, and are most characteristic for cultures grown in nutrient broth rather than PSS broth. Intracellular poly-β-hydroxybutyrate granules are not evident, but chemical tests indicate the presence of this polymer. Optimal growth temperature 30–32°C. Chemotaxonomic and physiological characters are described in Tables BXII.β.80, BXII.β.81, BXII.β.82, and BXII.β.83. Sole carbon sources are listed in Table BXII.β.79. Growth is poor in defined media; the best growth occurs with malate as the carbon source and glutamine as the nitrogen source. Ammonium salts or potassium nitrate are used poorly as nitrogen sources.

Aquaspirillum delicatum belongs phylogenetically to the *Comamonadaceae* (Willems et al., 1991c; Wen et al. 1999; Willems and Gillis, The *Comamonadaceae*, this volume).

Isolated from stored distilled water.

The mol% G + C of the DNA is: 63 (T_m).

Type strain: 146, ATCC 14667, CCUG 15846, DSM 11558, LMG 4328, NCIB 9419.

GenBank accession number (16S rRNA): AF078756.

6. **Aquaspirillum dispar** Hylemon, Wells, Krieg and Jannasch 1973b, 372[AL]

dis'par. L. neut. adj. *dispar* unlike.

The morphological characters are shown in Fig. BXII.β.67 and described in Table BXII.β.83. Optimal growth temperature 30°C. Moderate growth at 25 and 37°C; no growth at 10 or 45°C. Chemotaxonomic and physiological characters are described in Tables BXII.β.80, BXII.β.81, BXII.β.82, and BXII.β.83. Although originally described as being unable to grow anaerobically with nitrate (Hylemon et al., 1973b), the species does grow well under these conditions (Aragno and Schlegel, 1978). Sole carbon sources are listed in Table BXII.β.79. Ammonium salts are used well as sole nitrogen sources; nitrate is not used.

Isolated from fresh water.

Aquaspirillum dispar is phylogenetically not closely related to the type species of *Aquaspirillum* (Pot, 1996); because more than 84% DNA–DNA hybridization was found with the type strain of *Microvirgula aerodenitrificans* (Cleenwerck et al., 2003), both names can be considered as subjective synonyms.

The mol% G + C of the DNA is: 63–65 (T_m).

Type strain: ATCC 27510, DSM 736.

7. **Aquaspirillum giesbergeri** (Williams and Rittenberg 1957) Hylemon, Wells, Krieg and Jannasch 1973b, 368[AL] (*Spirillum giesbergeri* Williams and Rittenberg 1957, 88.)

gies'ber.ger.i. M.L. n. *giesbergeri* of Giesberger, the first investigator to study certain physiological characteristics of spirillae.

The morphological characters are depicted in Fig. BXII.β.67 and described in Table BXII.β.83. Optimal growth temperature 30°C. Chemotaxonomic and physiological characters are as described in Tables BXII.β.80, BXII.β.81, BXII.β.82, and BXII.β.83. Sole carbon sources are listed in Table BXII.β.79. Ammonium salts can be used as sole nitrogen sources; nitrate is not used.

This species includes organisms previously assigned to the two species "*Spirillum giesbergeri*" and "*Spirillum graniferum*" by Williams and Rittenberg (1957). The two species were combined into a single species by Hylemon et al.

(1973b) based on a high degree of similarity in phenotypic characters and in DNA base composition. *Aquaspirillum giesbergeri* belongs phylogenetically to the *Comamonadaceae* (Willems et al., 1991c; Wen et al., 1999; Willems and Gillis, The *Comamonadaceae*, this volume) where it constitutes a separate rRNA branch together with *A. sinuosum*. By SDS-PAGE of whole-cell proteins, it was shown that the type strains of *A. giesbergeri* and *A. sinuosum* are very similar (a correlation of 0.9) and therefore possibly belong to a single species. Further DNA–DNA hybridization data are necessary to confirm this homology.

Isolated from pond water.

The mol% G + C of the DNA is: 57–58 (T_m).

Type strain: ATCC 11334, DSM 9157, NCIB 9073, NRRL B-2060.

8. **Aquaspirillum gracile** (Canale-Parola, Rosenthal and Kupfer 1966) Hylemon, Wells, Krieg and Jannasch 1973b, 369[AL] (*Spirillum gracile* Canale-Parola, Rosenthal and Kupfer 1966, 124.)

gra.ci'le. L. neut. adj. *gracile* slender or thin.

The smallest of the aquaspirillae. The morphological characters are shown in Fig. BXII.β.67 and described in Table BXII.β.83. When originally isolated, all strains formed subsurface, spreading, semitransparent colonies on a medium containing 1.0% agar, the spreading occurring within the medium (Canale-Parola et al., 1966). After prolonged subculturing, some of the spirillae in each strain lost the ability to diffuse through 1.0% agar and formed small, nonspreading colonies. *A. gracile* was originally described as being microaerophilic, based on the growth of the organisms a few millimeters below the surface of semisolid media (Canale-Parola et al., 1966). However, later analysis of the type strain indicated that in liquid media a maximum growth response occurred under an air atmosphere (Laughon, 1973). Optimal growth temperature 30°C. Scanty growth in PSS broth at 25°C; no growth at 10 or 42°C. Chemotaxonomic and physiological characters are described in Tables BXII.β.80, BXII.β.81, BXII.β.82, and BXII.β.83. Carbon sources are listed in Table BXII.β.79. Ammonium chloride can be used as a nitrogen source but potassium nitrate cannot (Canale-Parola et al., 1966). A chemically defined medium has been devised by Laughon (1973; see also Krieg, 1976). Biotin is required for growth. Isolation is accomplished by allowing the spirilla to pass through a membrane filter disk (0.45 μm pore size) to an underlying agar medium (Canale-Parola et al., 1966).

Aquaspirillum gracile phylogenetically belongs to the *Comamonadaceae* (Willems et al., 1991c; Wen et al., 1999; see also Willems and Gillis in the chapter in this book on *Comamonas*).

Isolated from pond or stream water.

The mol% G + C of the DNA is: 64–65 (T_m).

Type strain: D4, ATCC 19624, DSM 9158.

GenBank accession number (16S rRNA): AF078753.

Additional Remarks: Reference strains are ATCC 19625 and 19626.

9. **Aquaspirillum metamorphum** (Terasaki 1961b) Hylemon, Wells, Krieg and Jannasch 1973b, 366[AL] (*Spirillum metamorphum* Terasaki 1961b, 220.)

me.ta.mor'phum. Gr. neut. adj. *metamorphum* changing.

The morphological characters are shown in Fig. BXII.β.67 and described in Table BXII.β.83. Optimal growth temperature 30°C. Chemotaxonomic and physiological characters are described in Tables BXII.β.80, BXII.β.81, BXII.β.82, and BXII.β.83. Sole carbon sources are listed in Table BXII.β.79. Ammonium salts can be used as a sole nitrogen source; nitrate is not used.

Isolated from the putrid infusion of a freshwater mussel.

Aquaspirillum metamorphum belongs phylogenetically to the *Comamonadaceae* (Willems et al., 1991c; Wen et al., 1999; Willems and Gillis, The *Comamonadaceae*). With *A. psychrophilum*, it constitutes a separate rRNA branch within this family.

The mol% G + C of the DNA is: 63 (T_m).

Type strain: ATCC 15280, CCUG 13794, DSM 1837, IFO 12012, LMG 4339, NCIB 9509.

GenBank accession number (16S rRNA): AF078757.

10. **Aquaspirillum polymorphum** (Williams and Rittenberg 1957) Hylemon, Wells, Krieg and Jannasch 1973b, 371[AL] (*Spirillum polymorphum* Williams and Rittenberg 1957, 85.)
po.ly.mor'phum. Gr. adj. *poly* many; Gr. n. *morphus* form, shape; M.L. neut. adj. *polymorphum* of many shapes.

Although originally reported to have bipolar tufts of flagella (Williams and Rittenberg, 1957), this species appears to have only a single flagellum at each pole (Terasaki, 1972; Hylemon et al., 1973b). The morphological characters are shown in Fig. BXII.β.67 and described in Table BXII.β.83. Optimal growth temperature 30°C. Chemotaxonomic and physiological characters are described in Tables BXII.β.80, BXII.β.81, BXII.β.82, and BXII.β.83. Although originally reported to grow anaerobically with nitrate (Williams and Rittenberg, 1957), more recent studies have not confirmed this result (Terasaki, 1972; Hylemon et al., 1973b). Sole carbon sources are listed in Table BXII.β.79. The best sole carbon sources are glutamate and aspartate; they are also the best sole nitrogen sources. Ammonium salts can serve as nitrogen sources; there are conflicting results concerning the utilization of nitrate as a sole nitrogen source (Terasaki, 1972; Hylemon et al., 1973b). Growth in PSS broth abundant, cloudy. Colonies on PSS agar are circular, convex, translucent, pinpoint.

Aquaspirillum polymorphum is a member of the *Alphaproteobacteria;* it is possibly most closely related to the genus *Magnetospirillum* (Pot, 1996; Kawasaki et al., 1997).

Isolated from pond water.

The mol% G + C of the DNA is: 61–62 (T_m).

Type strain: ATCC 11332, DSM 9160, IAM 14441, IFO 13961, NCIB 9072, NRRL B-2066.

11. **Aquaspirillum psychrophilum** (Terasaki 1973) Terasaki 1979, 138[AL] (*Spirillum psychrophilum* Terasaki 1973, 57.)
psy.chro'phi.lum. Gr. adj. *psychros* cold; Gr. adj. *philus* liking, preferring; M.L. neut. adj. *psychrophilum* preferring cold.

The morphological characters are described in Table BXII.β.83. Grows in nutrient broth or PSS broth. Optimal growth temperature 20°C. No growth above 26°C. The organism may be best described as psychrotolerant or psychrotrophic rather than psychrophilic (Morita, 1975). Che-

motaxonomic and physiological characters are described in Tables BXII.β.80, BXII.β.81, BXII.β.82, and BXII.β.83. This species has not been cultured in vitamin-free defined media and most likely has a growth factor requirement.

Isolated from Antarctic mosses.

Aquaspirillum psychrophilum belongs phylogenetically to the *Comamonadaceae* (Willems et al., 1991c, Wen et al., 1999; Willems and Gillis, The *Comamonadaceae*, this volume). With *A. metamorphum*, it constitutes a separate rRNA branch within this family.

The mol% G + C of the DNA is: 65 (T_m).

Type strain: CA 1, ATCC 33335, DSM 11588, IFO 13611, LMG 5408.

GenBank accession number (16S rRNA): AF078755.

12. **Aquaspirillum putridiconchylium** (Terasaki 1961a) Hylemon, Wells, Krieg and Jannasch 1973b, 367[AL] (*Spirillum putridiconchylium* Terasaki 1961a, 80.)
pu'tri.di.con.chy.li.um. L. adj. *putridus* putrid, decayed; L. n. *conchylium* a shellfish; L. n. *putridiconchylium* decayed shellfish.

The morphological characters are shown in Fig. BXII.β.67 and described in Table BXII.β.83. Optimal growth temperature 32°C. Chemotaxonomic and physiological characters are described in Tables BXII.β.80, BXII.β.81, BXII.β.82, and BXII.β.83. Sole carbon sources are listed in Table BXII.β.79. Ammonium salts can be used as a sole nitrogen source; nitrate is not used.

Isolated from the putrid infusion of a freshwater mussel.

Phylogenetically, *Aquaspirillum putridiconchylium* is not closely related to the type species of *Aquaspirillum* (Pot, 1996). It constitutes a separate branch in the *Betaproteobacteria*.

The mol% G + C of the DNA is: 52 (T_m).

Type strain: ATCC 15279, IFO 12013, NCIB 9508.

13. **Aquaspirillum sinuosum** (Williams and Rittenberg 1957) Hylemon, Wells, Krieg and Jannasch 1973b, 368[AL] (*Spirillum sinuosum* Williams and Rittenberg 1957, 94.)
si.nu.o'sum. L. neut adj. *sinuosum* full of curves.

The morphological characters are shown in Fig. BXII.β.67 and described in Table BXII.β.83. Optimal growth temperature 30°C. Chemotaxonomic and physiological characters are described in Tables BXII.β.80, BXII.β.81, BXII.β.82, and BXII.β.83. Sole carbon sources are listed in Table BXII.β.79. Ammonium salts can be used as a sole nitrogen source; nitrate is not used.

Aquaspirillum sinuosum phylogenetically belongs to the *Comamonadaceae* (Willems et al., 1991c; Wen et al., 1999; Willems and Gillis, The *Comamonadaceae*, this volume); it constitutes a separate rRNA branch together with *A. giesbergeri*. By SDS-PAGE of whole-cell proteins, it was shown that the type strains of *A. sinuosum* and *A. giesbergeri* are very similar (a correlation of 0.9) and therefore possibly belong to a single species. Further DNA–DNA hybridization studies are necessary for confirmation.

Isolated from freshwater.

The mol% G + C of the DNA is: 57–59 (T_m).

Type strain: ATCC 9786, CCUG 13728, LMG 4393, NCIB 9010, NCMB 59, NRRL B-2065, VPI 18.

GenBank accession number (16S rRNA): AF078754.

Other Organisms

1. "*Aquaspirillum bipunctata*" Dubinina, Grabovich, Lysenko, Chernykh and Churikova 1993.

   bi.punc.ta' ta. L. *bis* twice; L. part. adj. *punctatus* punctate, dotted; M.L. fem. adj. *bipunctata* twice punctate.

   Dubinina et al. (1993) discussed the diagnostic criteria of the genus *Thiospira* and the taxonomic validity of this genus. Six strains of gigantic sulfur spirillae, whose characteristics corresponded to the species diagnosis of *Thiospira winogradski* and "*Thiospira bipunctata*", were shown to belong to the genera *Spirillum* and *Aquaspirillum* according to phenotypic properties, nucleotide composition, and DNA relatedness studies. Strains D-405, D-409, D-410, D-411, and D-423, which were morphologically similar to "*Thiospira bipunctata*" formed a tight cluster phenotypically and genotypically and were assigned to a new species within the genus *Aquaspirillum* as "*Aquaspirillum bipunctata*" comb. nov. They are isolated from sediments of sewage from aerotanks and a water pond containing up to 7 mg/l H_2S. Strain D-427, which corresponded to the description of *Thiospira winogradski*, was proposed as a species of the genus *Spirillum*. The species name has not been validly published.

   The mol% G + C of the DNA is: 48–51 (T_m).
   Deposited strain: D-411.

2. "*Aquaspirillum denitrificans*" Dubinina, Churikova, Grabovich, Chernykh, Raichenstein and Petukhova 1989.

   de.ni.tri' fi.cans. M.L. adj. *denitrificans* pertaining to the property of denitrification.

   Spiral-shaped cells, from 1 to 1.4 µm wide, 2–4.1 to 25–30 µm long. Motile with bipolar flagella. Gram negative. Poly-β-hydroxybutyrate, volutin, and elemental sulfur (in the presence of sulfides) are accumulated in the cells. Colonies are flat, nonpigmented, 2–4 mm in diameter. Facultatively anaerobic growth with nitrates, fumarate, or molecular oxygen as electron acceptors. Under anaerobic conditions, nitrates are reduced to molecular nitrogen. Growth within a pH range of 6.0–8.5; optimal pH is 7.0–8.2. Optimal temperature for growth is 28°C. Chemoorganotrophs, using a wide variety of organic acids for growth: acetate, succinate, malate, fumarate, benzoate, α-ketoglutarate, oxaloacetate, pyruvate, lactate, and glyoxylate. Use several amino acids as carbon source: tyrosine, proline, asparagine, and cysteine. Growth on yeast extract and casein hydrolysate; vitamin solution is essential for growth. Does not use carbohydrates and alcohols. Nitrates, ammonium salts, casein hydrolysate, yeast extract, peptones, cysteine, and serine are used as nitrogen source. Catalase, oxidase, and urease positive. Casein and starch not hydrolyzed. Nongelatinolytic. Does not produce indole. Nitrate reduced to nitrites. Hydrogen sulfide produced from cysteine. Does not grow with 3% NaCl. Forms brown products on benzoate medium. Cannot fix nitrogen and does not contain proteolytic enzymes. Two very similar strains have been isolated from the sewage sludge of a putrefying aero tank. A limited phenotypic comparison with *A. metamorphum*, *A. giesbergeri*, *A. itersonii*, and *A. anulus* has been published. The species name has not been validly published.

   The mol% G + C of the DNA is: 60 (T_m).
   Deposited strain: D-415.

3. "*Aquaspirillum elegans*" Grabovich, Churikova, Chernykh, Leshcheva, Pushkina, Shipilova and Panteleyeva 1990.

   e' le.gans. L. adj. *elegans* elegant.

   Helical cells with pointed ends. Cell diameter varies from 0.6 to 1.0 µm. The length of the helix is 6.5–11 µm. Motile with bipolar flagella. Poly-β-hydroxybutyrate, volutin, and elemental sulfur (in the presence of sulfides) are stored in the cells. Colonies are flat, nonpigmented, 2–4 mm in diameter. Facultatively anaerobic growth with a strictly respiratory type of metabolism. Under anaerobic conditions, nitrates are reduced to nitrites. Grows in pH range 6.5–8.5; optimal pH is 7.5. Optimal temperature for growth is 28°C. Chemoorganotrophs, using a wide variety of organic compounds for growth: organic acids (acetate, succinate, malate, fumarate, citrate, isocitrate, oxaloacetate, aconitate, α-ketoglutarate, pyruvate, lactate), amino acids (lysine, tyrosine, proline, histidine, alanine, aspartate, and ornithine), alcohols (ethanol, glycerol, mannitol), and carbohydrates (fructose and sorbose), as well as yeast extract and casein hydrolysate. Vitamins and trace elements are essential for growth. Of the nitrogen sources investigated, growth occurred on nitrates, nitrites, ammonium salts, casein hydrolysate, yeast extract, peptone, aspartate, glutamate, cysteine, and serine. Nitrate reduced to nitrite. Casein, gelatin, and starch not hydrolyzed. Catalase, oxidase, and urease positive. Does not produce indole. Hydrogen sulfide produced, from cysteine. Does not grow with 3% NaCl. Isolated from sewage of purification installations. A limited phenotypic comparison with *A. peregrinum*, *A. dispar*, *A. itersonii*, *A. autotrophicum*, *A. sinuosum*, and *A. putridiconchylium* has been published. The species name has not been validly published.

   The mol% G + C of the DNA is: 61.6 (T_m).
   Deposited strain: D-425.

4. "*Spirillum pleomorphum*" Inoue and Komagata 1976, 170.

   ple.o.mor' phum. L. adj. *pleomorphum* pleomorphic.

   Helical cells, curved rods, crescent-shaped cells, U-form cells, and nearly ring-like forms occur on peptone–yeast extract–glucose (PYG) agar. Cell size: 0.7–1.0 × 2.0–4.5 µm. Motile by a single polar flagellum. Growth in PYG broth is turbid with sediment. Colonies on PYG agar are circular, smooth, convex, entire, opaque, and pale brown. Optimal growth temperature, 9°C; maximum, 20°C; minimum below 0°C. Aerobic. Oxidase and catalase positive. Indole, Voges–Proskauer, and methyl red tests are negative. No growth with 5% NaCl. Acid but no gas from xylose (aerobically). No acid or gas from glucose, lactose, sucrose, maltose, arabinose, or glycerol aerobically or anaerobically. No change in litmus milk. H_2S is not produced. Starch and cellulose are not degraded. Nitrate is reduced to nitrite. No growth occurs anaerobically with nitrate. Succinate, formate, acetate, fumarate, and propionate are assimilated. Citrate, lactate, protocatechuate, p-hydroxybenzoate, and hippurate are not assimilated. Isolated from Antarctic soil. Differences from *A. arcticum* include its high number of pleomorphic cells, its ability to reduce nitrate, its production of acid from xylose but not from glucose, its assimilation of formate but not lactate, and its higher temperature optimal for growth (9 versus 5°C).

The mol% G + C of the DNA is: 63 (T_m).
Deposited strain: 22-o-d, IAM 12028.

5. *"Aquaspirillum voronezhense"* Grabovich, Churikova, Chernykh, Kononykhina and Popravko 1987.

vo.ro.ne.zhen.se. M.L. adj. *voronezhense* pertaining to the city of Voronezh, former USSR.

Spiral-shaped cells, from 1.5–2.1 to 3 μm wide, with 1–3 turns of the spiral, 2.9–6.8 μm in diameter. Motile with bipolar flagella, in tufts of up to 50. Poly-β-hydroxybutyrate, volutin, and elemental sulfur (in the presence of sulfides) are stored in the cells. Colonies are flat, nonpigmented, 1–4 mm in diameter. Aerobic growth within a pH range of 6.0–9.0; optimal pH is 7.2–8.5. Optimal temperature for growth is 28°C. Chemoorganotrophs, using a wide variety of organic acids for growth including acetate, succinate, malate, fumarate, isocitrate, formate, α-ketoglutarate, oxaloacetate, pyruvate, salicylate, lactate, and glyoxylate. Strain D-420 differs from D-419 by its ability to use oxalate and aconitate and by the inability to grow on isocitrate and formate. Strain D-419 uses the amino acids proline, glycine, aspartate, glutamate, and valine; strain D-420 can use aspartate and tyrosine as carbon sources. A vitamin solution supplement is essential for growth. Ammonium salts, casein hydrolysate, yeast extract, peptone, aspartate, glutamate, cysteine, and serine (strain D-419) are used as nitrogen sources. Nitrate not reduced to nitrites. Casein and starch not hydrolyzed. Nitrates, sulfate, thiosulfate, and fumarate not used as electron acceptor. Catalase, oxidase, and urease positive. Hydrogen sulfide produced from cysteine. Do not produce indole. Form colored products on benzoate medium. The strains cannot fix nitrogen and do not contain proteolytic enzymes. Two very similar strains have been isolated from the sewage sludge of a putrifying aero tank. A limited number of DNA–DNA hybridization experiments showed that *"A. voronezhense"* is not related to *A. metamorphum, Levispirillum peregrinum,* and *"A. winogradskyi"*. The species name has not been validly published. Isolated from sewage sludge.

The mol% G + C of the DNA is: 58.5–60 (T_m).
Deposited strain: D-419.

Species Incertae Sedis

1. **"Spirillum minus"** Carter 1888, 47 (*Spirillum minor* (sic) Carter 1888, 47)*

mi' nus. L. neut. adj. *minus* less, smaller and refers to the small cell size.

Rigid cells; usually described as spiral with two or three turns, although the waves have been reported to be planar (McDermott, 1928). The ends of the cell may be blunt or pointed. Cell diameter, ~0.2 μm; cell length, 3–5 μm; wavelength, 0.8–1.0 μm. Actively motile by one or more flagella at each pole.

Causes one of the two forms of rat-bite fever in man. The disease caused by *"S. minus"* is often termed "Sodoku", it occurs worldwide but has its greatest frequency in the Far East. The organisms are usually transmitted to humans through the bite of an infected rat, although mice, squirrels, and rodent-ingesting animals such as cats, dogs, ferrets, and weasels have also been implicated. *"S. minus"* appears to be a natural parasite of rats, which act as carriers; the infection is usually not lethal in rats. The natural infection frequency for rats varies from country to country but may be as high as 25% (see Babudieri, 1973, for pertinent literature).

The clinical aspects of rat-bite fever and the distinctions between the form caused by *"S. minus"* and that caused by *Streptobacillus moniliformis* have been summarized by Joklik et al. (1980) and by Rogosa (1980). Experimental infections of humans and animals by *"S. minus"* have been described by Babudieri (1973).

"S. minus" is best observed in blood or exudates from patients by dark-field or phase-contrast microscopy of wet mounts; staining with Giemsa or Wright's stain or by silver impregnation is also useful.

"S. minus" is cultured *in vivo* by intraperitoneal inoculation of patients' blood or exudates from lesions, or blood from naturally infected rats injected into spirillum-free mice or guinea pigs (Rogosa, 1980); mice are the animals most susceptible to *"S. minus"* infection (Babudieri, 1973). It is questionable whether the organism has ever been cultured successfully in artificial media. Numerous attempts have failed, and various claims of successful cultivation have not been confirmed. One report that may indicate successful cultivation is that of Hitzig and Liebesman (1944), who inoculated blood from a patient into 2% dextrose–veal infusion broth and into 10% tomato extract–veal infusion broth. The addition of citrated human or rabbit blood was required for successful subculturing; also, the organisms initially required incubation in a candle jar but eventually were able to grow aerobically after five months of serial transfer. Confirmation of this report is needed. Considering the morphology, pathogenicity, and sources of *"S. minus"*, serious attention should be given to the possibility that the organism might belong to, or be related to, the genus *Campylobacter*, and the microaerophilic techniques employed for campylobacters might also prove useful for *"S. minus"*.

The species name has not been validly published. There are no type or reference strains.

The mol% G + C of the DNA is: not determined.

2. **"Spirillum pulli"** Mathey 1956, 745.

pul' li. L. gen. n. *pulli* of a young chicken.

Rigid spiral cells. By dark-field microscopy, the cell diameter is ~1 μm and the cell length is from 5 to 12 μm. Actively motile by means of a single flagellum at each end of the cell. Believed to be the cause of a diphtheroid stomatitis in the mouths of adult chickens. The lesions are yellowish white, rather firm, and adherent to the underlying tissue; often they are symmetrically ovoid, one at each side of the lower jaw. Lesions also occur on the palate, the lower surface of the tongue, the floor of the mouth, between the larynx and the transverse row of papillae on the tongue, around the larynx, and on the walls of the pharynx. The lesions vary in size from ~2 to 20 mm.

Attempts to culture *"S. pulli"* in artificial media have been unsuccessful. Experimental passage of the disease in chickens has been accomplished by contact and by experimental inoculation.

There are no type or reference strains. The species name has not been validly published.

The mol% G + C of the DNA is: not determined.

Editorial Note: The specific epithet *minor* is grammatically incorrect as noted by Robertson (1924) and the correct form is *minus*.

Genus IV. **Chromobacterium** *Bergonzini 1881, 153*[AL]

Monique Gillis and Niall A. Logan

Chro.mo.bac.te' ri.um. Gr. n. *chroma* color; Gr. n. *bakterion* a small rod; M.L. neut. n. *Chromobacterium* a small, colored rod.

Cells straight, round-ended, often coccoid, but usually occuring as rods, 0.6–0.9 × 1.5–3.0 μm, occurring singly, sometimes in pairs or short chains. No definite capsules are evident. No resting stages known. Gram negative. Usually contain poly-β-hydroxy-butyrate crystals (80% of strains positive), but rarely contain metachromatic granules. **Motile by means of both a single polar flagellum and, usually, one or more subpolar or lateral flagella. Facultative anaerobes. Minimum temperature for growth 10–15°C; maximum about 40°C; optimal growth at 30–35°C.** Optimal pH 7–8; no growth occurs below pH 5. Grow on ordinary peptone media; no distinctive organic growth factor requirements. Colonies are smooth, but rough variants may occur; colonies are of butyrous consistency and are easily emulsified in water. **Most strains produce the violet pigment violacein,** but strains producing nonpigmented colonies are sometimes encountered, and subcultures of pigmented strains often contain partially or completely unpigmented colonies. Growth in nutrient broth produces a violet ring at the surface with a fragile pellicle. Chemoorganotrophs; **most strains (80%) attack carbohydrates fermentatively, some (20%) oxidatively, producing small amounts of acid** but usually no gas. Lactate is oxidized to CO_2. Usually oxidase positive by the method of Sivendra and Lo (1975). No growth in media containing 6% or more NaCl. Catalase positive, but very sensitive to hydrogen peroxide. Indole negative, Voges–Proskauer negative. Nitrate and nitrite reduced. Ammonia produced from peptone. Arginine hydrolyzed. **Hydrogen cyanide produced. Resistant to benzylpenicillin,** 10 mg/ml. **Sensitive to tetracycline,** 30 mg/ml. **Resistant to O/129 (2,4-diamino-6,7-diisopropyl pteridine)** by disc diffusion method, 50 mg/disc. Soil and water organisms, occasionally causing infections of humans and other mammals. 16S rRNA gene sequence data place *Chromobacterium* in the class *Betaproteobacteria*.

The mol% G + C of the DNA is: 65–68 (T_m).

Type species: **Chromobacterium violaceum** Bergonzini 1881, 153.

Further descriptive information

The pigment violacein is produced on or in media containing tryptophan. It is soluble in ethanol, but not in water or chloroform, and is readily identified by spectrophotometry and by testing with routine reagents (see Procedures for Testing for Special Characters). Subcultures of pigmented strains may contain partially or completely unpigmented colonies. An unpigmented strain is most readily recognized by having a colony morphology similar to that of a pigmented strain isolated at the same time under the same conditions. Once it has been confirmed as an oxidase- and catalase-positive, Gram-negative rod, such an isolate should be flagella-stained. If it possesses the characteristic polar and lateral flagella, it should then be subjected to the differential tests for the three genera *Chromobacterium*, *Iodobacter*, and *Janthinobacterium*. Although young cultures of *Aeromonas*, *Pseudomonas*, and *Vibrio* spp. grown on solid media sometimes produce lateral flagella in addition to their usual polar flagella, they give patterns of results different from those of the violet-pigmented organisms in the differential tests.

The characteristic flagellar arrangement is best seen in young cultures on solid media. The single polar flagellum is inserted at the tip of the cell, shows long, shallow waves, and often stains faintly. The lateral flagella are usually long. Usually 1–4 such flagella are produced, although up to eight may occur. They may be inserted subpolarly or laterally, usually show deep, short waves, and stain readily (see *Iodobacter fluviatilis* Fig. BXII.β.71).

Chromobacterium violaceum occurs in soil and water and is common in tropical climates. It occasionally causes serious pyogenic or septicemic infections of mammals, including humans (reviewed by Sneath, 1960; Gillis and De Ley,1992; with subsequent reports by Hassan et al., 1993; Ti et al., 1993; Roberts et al., 1997; Midani and Rathore, 1998; Desjardins et al., 1999). The disease can be cured by tetracyclines if treated at an early stage (Moss and Ryall, 1981) as well as by ciprofloxacin, norfloxacin, perfloxacin (Aldridge et al., 1988), gentamicin, imipenem, and trimethoprim–sulfamethazole (Midani and Rathore, 1998). In some cases, a sepsis mimicking melioidosis has been described (Chong and Lam, 1997), and potentially misidentification as *Burkholderia pseudomallei* might occur (Inglis et al., 1998). Pathogenic bacteria of carp have been characterized as closely resembling *Chromobacterium violaceum* (Rahmatullah and Beveridge, 1993; Kozinska and Antychowicz, 1996).

Enrichment and Isolation Procedures

Selective and enrichment media have not been developed for *C. violaceum*, but colonies on routine growth media are readily recognized by their violet pigmentation, and isolates may be screened for the ability to grow at 37°C and for other characters that distinguish this organism from the species of *Iodobacter* and *Janthinobacterium* (Table BXII.β.85; see Procedures for Testing for Special Characters).

Maintenance Procedures

The organisms may survive for several years in dilute peptone water (0.1% peptone) at room temperature. They can also be preserved indefinitely by lyophilization or by freezing in nutrient broth containing 15% glycerol.

Procedures for Testing Special Characters

To identify the pigment violacein, pigmented growth is shaken from a plate culture into a small volume of 96% (v/v) ethanol, and cells are then removed by centrifugation, followed by filtration through a 0.22 μm membrane; the solution shows an absorption maximum of 579 nm and a minimum of 431 nm if violacein is present. In 10% (v/v) H_2SO_4 in ethanol (96% v/v), the pigment gives a green solution with an absorption maximum at 699 nm and a minimum at 502 nm. Color reactions may be observed by mixing loopfuls of pigmented growth in a few drops of 96% (v/v) ethanol on a white tile; stirring in a drop of 25% (v/v) H_2SO_4 gives a green color, and stirring a drop of 10% (w/v) NaOH into a fresh growth/ethanol mixture gives first a green, then a reddish brown color. The biosynthesis of violacein has been studied intensively, and different metabolites have been characterized e.g., pseudodeoxyviolacein (Hoshino et al., 1994), deoxyviolacein, and chromopyrrolic acid (Hoshino et al., 1993a).

TABLE BXII.β.85. Patterns of results in tests useful for distinguishing *Chromobacterium* from *Iodobacter*, *Janthinobacterium*, and *Vogesella*

Characteristic	Percentage positive in species or group[a]				
	C. violaceum	*Iodobacter fluviatilis*	*J. lividum* (typical)	*J. lividum* (atypical)	*Vogesella*
Pigment:					
Indigoidin					+
Violacein	+	+	+	+	
Fluorescence[b]	− (T)[c]		− (T)[c]		100
Colonies on 1/4 NA:					
Spreading	0	83	0	0	
Gelatinous	0	4	36	100	
Tough	0	0	7	71	
Growth at:					
4°C	0	100	94	87	100
37°C	100	0	0	0	100
Anaerobic growth	100	100	3	0	
Nitrate reduction	87	98	84		100
Production of indole	0	0	0	0	100
Hydrolysis of Tween 80	+ (T)[c]		+ (T)[c]		0
Growth on:					
Citrate	100	85	95	94	0
Glucose:					
Fermented	10[d]	100	0	0[e]	0
Oxidized	0[d]	0	97	57[e]	0
Acid from:					
L-Arabinose [f]	0	0	100	87	0
D-Maltose	0	100	98	94	
myo-Inositol [g]	0	0	100	0	35
Trehalose	100	100	1	87	
Gelatinase	100	100	100	69	0
Lactate utilization	100	0	100	75	
Esculin hydrolysis	0	0	100	0	
Arginine hydrolysis	100	0	0	0	
Chitin digestion	100[h]		5		0
Caseinase	100	90	5	0	0
HCN production	100	0	0	0	

[a]From Logan (1989) with later amendments. Number of strains studied for each species: *C. violaceum*, 9; *Iodobacter fluviatilis*, 53; *J. lividum* (typical), 68; *J. lividum* (atypical), 14; *Vogesella*, 17.

[b]Fluorescence on chalk agar (Starr et al., 1960) with short wavelength illumination. All *Vogesella* strains exhibited very weak fluorescence.

[c]Only tested for the type strain; − (T), type strain negative; + (T), type strain positive.

[d]Occasional strains are oxidative.

[e]Some strains give no reaction in this test.

[f]Similar patterns of results are obtained with D-cellobiose and D-galactose.

[g]Similar patterns of results are obtained with D-sorbitol.

[h]Grimes et al. (1997) found that the type strain was negative under conditions different from those used by the authors.

L-Tryptophan is the carbon source for violacein, and isatin and indole 3-acetic acid are important metabolic intermediates in violacein production and can also serve as substrates. Violacein has trypanocide activity (Duran et al., 1994).

Because the production of violacein by *Chromobacterium violaceum* is dependent on *N*-acetylhomoserine lactones (AHLs), it provides a simple assay for the detection of these molecules, which are signal molecules in quorum sensing that interact with a transcriptional activator protein to couple gene expression to cell population density (McClean et al., 1997; Bachofen and Schenk, 1998).

The procedure of Mayfield and Inniss (1977) is recommended for studies of flagellar arrangement, and the even simpler technique of Heimbrook et al. (1989) appears, on limited experience, to be equally dependable. For the latter method, cultures are grown overnight on slopes of nutrient agar prepared at a quarter of the usual nutrient strength (1/4 NA), which have drops of condensate at their bases. A drop of the condensate is removed with a small loop and touched to a small drop of water on a clean slide.

The oxidase reactions of pigmented strains are commonly obscured by the pigment, but may be revealed by the method of Sivendra and Lo (1975). Smears of pigmented growth and control organisms are placed along the edge of a strip of filter paper that has been soaked in the reagent (1% [w/v] tetramethyl-*p*-phenylenediamine HCl) and supported on a glass slide. When the positive control has oxidized the reagent to a violet color, further drops of reagent are placed above the smears and the slide tilted so as to flush reacted reagent onto a white background.

Unless indicated otherwise, all media are autoclaved at 121°C for 15 min. Media are inoculated with one drop or loopful of quarter strength nutrient broth (1/4 NB) culture grown overnight at 25°C and, with the exception of those used for growth temperature tests, incubated at 25°C. This temperature permits strains of *Chromobacterium*, *Iodobacter*, and *Janthinobacterium* to grow and be compared in the differential tests. For assessment of growth temperatures, tubes of 1/4 NB are incubated in waterbaths at 37°C and 4°C and examined for growth at 2 and 7 d, respectively. To assess anaerobic growth, plates of nutrient agar (full strength) are streak-inoculated and incubated in an anaerobic jar (GasPak, BBL) for 7 d at 25°C.

To investigate oxidative and fermentative attack of glucose, the medium of Ward et al. (1986) is used without its H₂S indicator system. This modified medium contains (g/l): glucose, 10; casitone, 2; yeast extract, 1; phenol red, 0.04; bromothymol blue, 0.02; agar, 15. After heating to dissolve components and adjustment to pH 7.1, the medium is dispensed into tubes in 10-ml amounts, autoclaved, and left to set as deep-butted slopes, which are then inoculated by streaking the slope and stabbing the butt and incubated for up to 7 d. Fermentative organisms readily produce a yellow color throughout the medium, whereas a yellow or light green color restricted to the slope indicates oxidative attack. *Chromobacterium violaceum* usually attacks carbohydrates fermentatively without gas production, but aerogenic isolates have been reported (Sivendra, 1976), and as many as 20% of strains may show oxidative attack (Sneath, 1984a). *Iodobacter fluviatilis* is always fermentative and anaerogenic. Typical strains of *J. lividum* are oxidative, but atypical strains may give no reaction in this test.

Acid production from carbohydrates may be determined using a medium containing (g/l): (NH₄)₂HPO₄, 1; KCl, 0.2; MgSO₄·7H₂O, 0.2; yeast extract, 0.2; phenol red, 0.06; and agar, 12. After adjustment to pH 7.4 (orange), autoclaving, and addition of a filter-sterilized solution of L-arabinose, *myo*-inositol, D-maltose, D-mannitol, or trehalose to give a final concentration of 0.5%, the medium is dispensed into tubes and left to set as deep-butted slopes. Tubes are examined at 2, 4, and 7 d for acid (yellow) reactions in the slopes, butts, or both.

Utilization of lactate is assessed using a medium containing (g/l): lactic acid (sodium salt), 2; NaCl, 1; (NH₄)₂HPO₄, 1; KH₂PO₄, 0.5; MgSO₄·7H₂O, 0.2; phenol red, 0.06; agar, 12. After adjustment to pH 6.8 and autoclaving (115°C for 20 min), the medium is distributed into tubes and left to set as slopes. The medium is streak-inoculated and examined for alkaline (red) reactions at 2, 4, and 7 d.

For esculin hydrolysis, the medium of Sneath (1979) is used, which contains (g/l): peptone, 10; esculin, 1; sodium citrate, 1; ferric citrate, 0.05. After adjustment to pH 7.0, 5-ml quantities are dispensed in tubes and autoclaved. Tubes are examined at 3 and 4 d for browning.

Arginine-hydrolysis medium contains (g/l): L-arginine HCl, 10; NaCl, 5; peptone, 1; K₂HPO₄, 0.3; phenol red, 0.06; and agar, 3. It is adjusted to pH 7.2 (orange), dispensed in tubes to a depth of 2 cm, and autoclaved. Following stab-inoculation and sealing with sterile paraffin oil, tubes are incubated for 4 d and examined for alkaline (red) reactions.

Hydrogen cyanide indicator papers are prepared by dipping filter paper strips in saturated aqueous picric acid, drying (potentially explosive, care and doing under chemical hood recommended), and then dipping in 10% (w/v) aqueous sodium carbonate and drying again. A bottle of half-strength (semi-solid) nutrient agar is inoculated by stabbing and an indicator paper is trapped between the cap and rim; HCN production is indicated by the paper strip changing color from orange-yellow to brick-red within 3 d.

DIFFERENTIATION OF THE GENUS *CHROMOBACTERIUM* FROM OTHER GENERA

Table BXII.β.85 presents characteristics differentiating the genus *Chromobacterium* from the genera *Janthinobacterium*, *Iodobacter*, and *Vogesella*. *Microvirgula aerodenitrificans* is not included because comparable test results are not available.

TAXONOMIC COMMENTS

The present description of the genus *Chromobacterium* is based primarily on the description given by Sneath (1984a). The phylogenetic affiliation of *Chromobacterium* was studied first by DNA–rRNA hybridization (De Ley et al., 1978), and later by 16S rDNA sequence analysis. It is a member of the class *Betaproteobacteria*, in which it constitutes a separate lineage in the *Neisseriaceae*. Its closest neighbor is the recently proposed genus *Vogesella* (see the genus *Vogesella* in this family). *Vogesella* was created to contain the former *Pseudomonas indigofera* and new isolates of blue-pigmented bacteria isolated from fresh water bacterial communities. A single species, *Vogesella indigofera*, has been proposed because no DNA–DNA hybridizations between this species and the group of new isolates are available. *Vogesella* members do not contain violacein, but rather indigoidine. Also belonging to the *Neisseriaceae*, but more distantly related, is the genus *Iodobacter* (see Genus *Iodobacter*, this *Manual*). Earlier results from DNA–rRNA hybridization (Pot et al., 1992a) have shown that *Aquaspirillum dispar* and *Aquaspirillum putridiconchylium* also belong to this phylogenetic lineage. A new genus, *Microvirgula*, with the single species *Microvirgula aerodenitrificans* recently has been created for a denitrifying bacterium occupying a separate phylogenetic position between the *Vogesella*–*Chromobacterium violaceum* lineage and *Iodobacter fluviatilis* (Patureau et al., 1998). It is an aerobic as well as anoxenic heterotroph that has an atypical respiratory type of metabolism, in which oxygen and nitrogen oxides can be used simultaneously as terminal electron acceptors.

In addition to species of *Chromobacterium*, *Janthinobacterium*, and *Iodobacter*, which are found in soil and fresh water, violacein-producing organisms have been isolated from marine waters. Hamilton and Austin (1967) have named their strain "*C. marinum*", but Gauthier (1976) has considered his isolates to be excluded from *Chromobacterium* by their low mol% G + C range, and he assigned them to the genus *Alteromonas* as the new species *A. luteoviolaceus* (now *A. luteoviolacea*).

Species producing pigments that are not violacein, or even violet-colored in some cases, have at various times been assigned to *Chromobacterium*, and still occasionally appear in the literature as such. Of these, "*C. iodinum*" is established as a species of *Brevibacterium* (Collins et al., 1980), and "*C. marismortui*" has been placed in the new genus *Chromohalobacter* (Ventosa et al., 1989). The production of lipases by "*Chromobacterium viscosum*" has been reported. This bacterium does not belong in *Chromobacterium*; it is a Gram-positive bacterium whose exact taxonomic affiliation is unknown and which has been cited as *Corynebacterium* sp., *Micrococcus* sp., or *Arthrobacter* sp., (De Ley et al., 1978). Two strains isolated from *Psychrotia nairobensis* and *Ardisia crispa* and formerly named "*C. lividum*" or "*C. folium*" have been reclassified as *Sphingomonas yanoikuyae* (Takeuchi and Hatano, 1998). "*Chromobacterium chocolatum*" has been reclassified as *Microbacterium chocolatum* (Takeuchi et al., 1995), and "*C. typhiflavum*" strains (Pedersen et al., 1970) have been reclassified as *Chryseomonas luteola* and *Flavimonas oryzihabitans* (Holmes et al., 1987).

The genus *Chromobacterium* is of interest because it produces substances of importance to agriculture, medicine, and industry. Several of these have been reported in Gillis and De Ley (1992). *Chromobacterium* spp. have been reported to produce chitinase and to reduce the hatch of the nematode *Globodera rostochiensis* (Cronin et al., 1997); the production of a set of chitinolytic enzymes is regulated by an endogenous AHL (Chernin et al., 1998).

Depsipeptide, a cyclic peptide produced by *Chromobacterium violaceum* WB968, has potent antitumor activity against human tumor cells (Ueda et al., 1994; Chan et al., 1997). *Chromobacterium violaceum* strain DSMZ 30191 accumulate a homopolyester of 3-hydroxyvaleric acid up to high percentages of the cellular dry matter when grown on valeric acid (Steinbüchel et al., 1993; Marchessault et al., 1995; Steinbüchel and Schmack, 1995). Several enzymes and metabolites from *Chromobacterium* members have been intensively studied, including phenylalanine hydroxylase (Carr and Benkovic, 1993; Carr et al., 1995), L-tryptophan 2′,3′-oxidase (Hoshino et al., 1993b; Genet et al., 1995; Hammadi et al., 1997), indoloxygenase (Cheah et al., 1998), cytosine deaminase (Kim and Yu, 1998), and serine hydroxylmethyltransferase (Shirazi-Beechey and Knowles, 1984).

List of species of the genus Chromobacterium

1. **Chromobacterium violaceum** Bergonzini 1881, 153[AL]

vi.o.la′ ce.um. L. adj. *violaceum* violet colored.

The description is the same as that given for the genus. Other characters are as shown in Table BXII.β.86.

The mol% G + C of the DNA is: 65–68 (T_m).
Type strain: MK, ATCC 12472, NCIMB 9131, NCTC 9757.
GenBank accession number (16S rRNA): M22510.

TABLE BXII.β.86. Characteristics of *Chromobacterium violaceum*[a]

Characteristics	*Chromobacterium violaceum*
Colonies on 1/4 nutrient agar:	
Spreading	−
Gelatinous	−
Pigmented	[+] (87)
Zoning of pigment	[−] (25)
Growth on:	
Minimal medium	+
4X Strength nutrient agar	+
Growth at:	
4°C	−
30°C	+
37°C	+
Growth in:	
1% NaCl	+
2% NaCl	+
4% NaCl	−
Growth at:	
pH 4	+
pH 3	−
Formation of cell chains	[−] (12)
Formation of filaments	[−] (15)
Acid from:	
L-Arabinose	−
D-Cellobiose	−
D-Fructose	+
D-Galactose	−
Gluconate	+
D-Glucose	+
Glycerol	+
Glycogen	+
myo-Inositol	−
Inulin	[−] (12)
Lactose	−
D-Maltose	−
D-Mannitol	−
D-Mannose	+
Melezitose	−
N-acetylglucosamine	+
D-Raffinose	−
D-Sorbitol	−
Starch	[+] (87)
Sucrose	[−] (25)
Trehalose	+
D-Xylose	−

(continued)

TABLE BXII.β.86. *(cont.)*

Characteristics	*Chromobacterium violaceum*
Catalase	+
Oxidase	+
Nitrate reduction	[+] (87)
Nitrite reduction	[−] (25)
Production of HCN	+
Egg yolk reaction	+
Hemolysis	+
Carbon sources:	
Acetate	[+] (87)
Citrate	+
Fumarate	+
Glycerate	−
Lactate	+
Malate	+
Propionate	[+] (87)
Pyruvate	+
Succinate	+
Tartrate	−
Hydrolysis of:	
Arginine	+
Casein	+
Esculin	−
Gelatin	+
Starch	−
Antimicrobial agents (per disk):	
Ampicillin, 25 μg	R
Cephaloridine, 25 μg	R
Colistin sulfate, 10 μg	R
Chloramphenicol, 10 μg	s
Chlortetracycline, 10 μg	S
Furazolidone, 50 μg	S
Kanamycin, 30 μg	S
Neomycin, 10 μg	s
Nalidixic acid, 30 μg	S
Nitrofurantoin, 200 μg	S
Oxytetracycline, 10 μg	S
Penicillin G, 1.5 U	R
Streptomycin, 10 μg	s
Sulfafurazole, 100 μg	R
Sulfafurazole, 500 μg	s
Vibriostatic agent O/129, 50 μg	R

[a]Symbols: +, all strains positive; [+], positive for 80% or more strains; d, positive for 31–79% strains; [−], positive for 30% or fewer strains; −, negative for all strains; R, resistant; S, sensitive; s, slightly sensitive. Numbers in parentheses indicate the % of strains giving positive reactions.

Genus V. **Eikenella** *Jackson and Goodman 1972, 74^{AL}*

EDWARD J. BOTTONE, FRANCIS L. JACKSON AND YVONNE E. GOODMAN

Ei.ke.nel' la. M.L. dim. ending *-ella*; M.L. fem. n. *Eikenella* named after M. Eiken, who first named the type species of the genus.

Straight rods, 0.3–0.4 × 1.5–04.0 µm, unbranched, with rounded ends and a regular morphology. Short filaments are occasionally formed. Nonsporeforming. Gram negative. **Nonmotile**, possessing no flagella; however, a "twitching motility" may occur on agar surfaces. **Facultatively anaerobic.** Optimal growth temperature, 35–37°C. **Colonies may appear to corrode the surface of the agar**; noncorroding strains may also occur. Nonhemolytic; a slight greening of blood media around colonies may occur. **Oxidase positive** (Kovacs method). **Negative for catalase, urease, arginine dihydrolase, and indole. Lysine-decarboxylase positive.** Nitrates are reduced to nitrites. **No acid is formed from glucose or other carbohydrates. Hemin is usually required for growth under aerobic conditions.** Occur in the human mouth and intestine; can be opportunistic pathogens.

The mol% G + C of the DNA is: 56–58.

Type species: **Eikenella corrodens** (Eiken 1958) Jackson and Goodman 1972, 75 (*Bacteroides corrodens* Eiken 1958, 415.)

FURTHER DESCRIPTIVE INFORMATION

Electron micrographs of *E. corrodens*, the only species, negatively stained with phosphotungstate show a finely convoluted (cerebral) cell surface. Sections stained with ruthenium red-OsO$_4$ show a cytoplasmic membrane and outer membrane characteristic of Gram-negative bacteria. A slime layer, loosely organized and fibrous, was present on strains examined by Progulske and Holt (1980). The slime layer is associated with the outer surface of the outer membrane, and it is possible that the pilus-like structures demonstrable by negative staining represent components of this layer modified by preparation techniques (Jackson et al., 1971; Progulske and Holt, 1980). Scanning electron microscopy of 7-d-old cultures shows fibrillar material connecting cells (Progulske and Holt, 1980). Adherence of *E. corrodens* to cell surfaces of host tissues is mediated by a cell-associated *N*-acetyl-D-galactosamine-specific, lectin-like substance (Ebisu and Okada, 1983). The lectin is also involved in coaggregation with other bacteria, forming dental plaque and causing proliferation of B-cells in peridontal lesions (Ebisu et al., 1988). The gene encoding a component protein of the lectin-like adhesin complex has been cloned (Yumoto et al., 1996).

Lipopolysaccharide from the organism has been reported to have endotoxin activity, whereas the slime layer has little endotoxin activity, but may be immunosuppressive (Behling et al., 1979). A component with endotoxin activity and containing 0.5% ketodeoxyoctonate, probably a lipopolysaccharide, has been obtained from cells by phenol–water extraction followed by differential centrifugation and gel filtration. It may represent a group antigen, and is distinct from type-specific outer membrane protein antigens, which may also be present (Maliszewski et al., 1983). This major outer membrane protein has a molecular weight of 42,000 Da. *E. corrodens* lipopolysaccharide also demonstrates mitogenic activity different from that elicited by conventional lipopolysaccharide (Progulske et al., 1984).

The organisms appear to be nonmotile according to conventional tests. Corroding colonies (see below) may show spreading edges, and microscopic observation of corroding strains growing on agar surfaces has shown that a form of surface translocation termed "twitching motility" occurs (Henrichsen, 1975a; Henrichsen and Blom, 1975; Schröter, 1975). This motility is correlated with the presence of asymmetrically arranged pili, each with a molecular mass of about 14.8 kDa and an N-terminal amino acid sequence, suggestive of type IV pilins (Hood and Hirschberg, 1995). "Twitching motility" involves small, intermittent jerks leading to displacement over only short distances, not regularly related to the long axis of the cell, at speeds of 1–2 to 2–5 mm/min. The use of cover slips over growth on agar surfaces may prevent this movement, as organisms adhere to the glass, and the phenomenon is dependent on the presence of a thin film of water at the surface of the medium, as is found in cultures incubated in a humidified atmosphere.

Two types of colonies may occur on agar media. Typical "corroding" strains are so named because the colonies appear to corrode the surface of the agar. The organisms penetrate into the surface of the medium and, under humid conditions, tend to spread by twitching motility. The colonies appear as if they are in small depressions in the agar surface, probably because of a combination of spreading surface growth, localized physical alterations of the medium, and optical properties of the colony. The appearance of pitting is not produced on inspissated serum medium. Problems of interpretation of these features have been discussed by Jackson et al. (1971) and by Khairat (1967). Colonies of corroding strains growing on Columbia or trypticase soy agar with 5% sheep's blood at 36°C under 10% CO$_2$ are dry, flat, and radially-spreading, with irregular peripheries (Bottone et al., 1973). With more prolonged incubation, colonies display a characteristic morphology consisting of a clear, moist, glistening center devoid of growth, encircled by a highly refractile, speckled, pearl-like zone of growth that is in turn surrounded by an outer nonrefractile perimeter of spreading growth (Fig. BXII.β.69). Colonies are small (0.2–0.5 mm at 24 h; 0.5–1.0 mm at 48 h). The spreading edge may give colonies a final diameter of ~3.0 mm (Jackson et al., 1971). Colony sizes are similar on cystine–hemin agar. Noncorroding strains, forming colonies 0.5–1.0 mm in diameter that are translucent and dome-shaped, may be isolated by selection from corroding strains and may also be encountered upon primary isolation. These variants do not produce colonies with spreading edges, do not exhibit twitching motility and lack pilus-like surface appendages (Henrichsen, 1975a).

Plate cultures have an odor described as "bleach-like" or as resembling that of *Haemophilus* and *Pasteurella* species.

Growth in fluid media is usually described as poor, but may be improved by addition of cholesterol (10 g/ml; Henriksen, 1969b) or 3% blood serum. The addition of 0.2% agar improves growth (Jackson et al., 1971). In thioglycollate broth, granular growth develops after 3 days in a band 1 cm below the surface. In glucose broth, growth may produce uniform turbidity or small granules adherent to the sides of the tube (Bottone et al., 1973). Granule formation may be related to the presence of pili.

The optimal temperature for growth is 35–37°C. At 25°C, minute colonies are visible in 5–7 days. Growth is good at 40°C but poor at 42°C. No growth occurs at 44°C. The optimal pH for growth is 7.3.

Hemin (5–25 g/ml) is required for aerobic growth of freshly

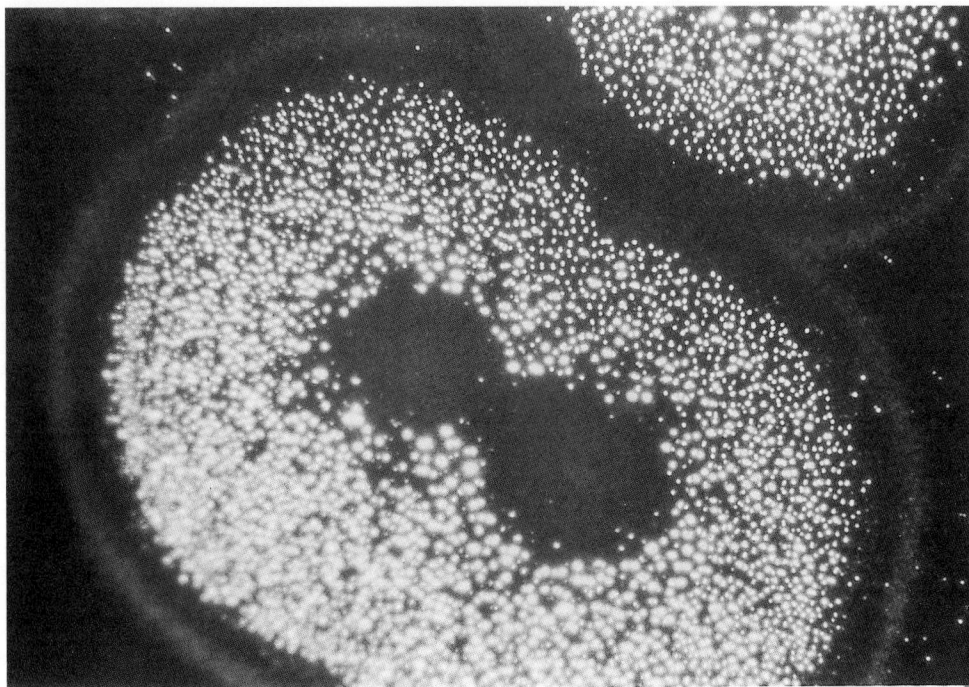

FIGURE BXII.β.69. Colony of *E. corrodens* viewed by oblique lighting, showing a clear, moist center circled by a highly refractive, speckled, pearl-like zone that is surrounded by a perimeter of spreading growth. (Reproduced with permission from E.J. Bottone et al., American Journal of Clinical Pathology *59:* 560–566, 1973 ©American Society of Clinical Pathologists.)

isolated strains, but not for anaerobic growth. The hemin requirement may be influenced by other constituents of the medium, as indicated by the demonstration that addition of cystine (0.005%) raises the minimum hemin requirement to 20–25 g/ml. Under these conditions, the colony size is increased. Cystine at 0.1% is strongly inhibitory to growth at all hemin concentrations that support growth in the absence of added cystine (Jackson et al., 1971).

Under aerobic or anaerobic conditions, growth of freshly isolated strains is enhanced by 5–10% CO_2. Repeated subculture in air leads to a loss of this response (Hill et al., 1970), but it is retained if strains are repeatedly transferred in a CO_2-enriched atmosphere (Jackson et al., 1971). Strains that have been tested for bile tolerance are inhibited under aerobic conditions by 5 or 10% bile; under anaerobic conditions, growth has been reported to occur in the presence of 10% bile, and there is evidence that this resistance may be higher under microaerobic conditions in the presence of nitrate (Jackson et al., 1971).

Organisms grown anaerobically or aerobically, with or without added CO_2, are strongly oxidase-positive by the method of Kovács (1956), using either dimethyl-*p*-phenylenediamine or the tetramethyl reagent. The standardized tube test (Jackson and Goodman, 1972) can also be used, and it shows that the reaction is azide-sensitive.

Spectroscopic examination has shown the presence of cytochromes of the *b* and *c* groups. Ubiquinone, but not demethylmenaquinone, is present (Holländer and Mannheim, 1975). Growth yields of oxygen-limited cultures are not increased by fumarate. Energy metabolism in *E. corrodens* proceeds via oxidative deamination of key amino acids (proline, glutamate, serine, and glutamine) linked to a respiratory chain, with nitrate,

the ultimate electron acceptor, reduced to nitrite in the process (Gully and Rogers, 1996).

An examination of six strains of *E. corrodens* by gas–liquid chromatography/mass spectrophotometry has shown that the organisms contain hexadecanoic and octadecenoic acids as major fatty acid constituents (Prefontaine and Jackson, 1972). This is in contrast to *Bacteroides ureolyticus*, which has a high content of octadecenoic acid but a low content of hexadecanoic acid.

Antigenic differences between strains have been demonstrated by agglutination reaction and also by immunodiffusion studies, in which up to four antigens have been detected (although some strains lack one or two components) (Jackson et al., 1971). Badger and Tanner (1981) have divided 46 strains into four groups by microagglutination tests. Maliszewski and Badger (1980) have isolated a group antigen common to three serotypes. It appears to be a lipopolysaccharide containing 0.5% ketodeoxyoctonate and showing endotoxin activity by the Limulus amoebocyte lysate test. A type-specific antigen (protein, sensitive to trypsin and heat) has been prepared from one strain. According to Johnson et al. (1978), endotoxin prepared from a corroding strain of *E. corrodens*, in contrast to typical endotoxins, which tend to increase microviscosity of cell membranes, has the unusual effect of causing a decrease in microviscosity of cell membranes, a finding these authors attributed to unique characteristics of the *E. corrodens* endotoxin. Indeed, Progulske et al. (1984) have shown that *E. corrodens* endotoxin is nonclassical, as it contains little 2-keto-3-deoxyoctulosonic acid and heptose and no detectable 9-hydroxy fatty acids.

Some anaerobic Gram-negative rods may show a degree of cross-antigenicity with *E. corrodens* (Robinson and James, 1973), but as pointed out by Jackson and Goodman (1978) the difference in mol% G + C content of the DNA between *E. corrodens* and these anaerobes precludes a close genetic relationship, and the designation by Robinson and James of strain NCL-20 (now known to be a strain of *Bacteroides ureolyticus*) as a "link" strain between the facultative and anaerobic species is not justifiable.

For determination of antibiotic susceptibility, best results are obtained by standardized-inoculum plate-dilution methods, but other techniques have been used with variable results (Hill et al., 1970; Jackson et al., 1971; Zinner et al., 1973; Brooks et al., 1974; Robinson and James, 1974; Labbé et al., 1977; Goldstein et al., 1978; Slee and Tanzer, 1978). The reports presently available indicate that most strains are susceptible to penicillin G, ampicillin, and cefoxitin, but resistant to penicillinase-resistant penicillins and moderately susceptible or resistant to cephalothin, cephapirin, and cephaloridine. Resistance to aminoglycosides is variable, but usually sufficient to preclude clinical effectiveness. Strains are often susceptible to chloramphenicol, tetracycline, rifampicin, and colistin. Resistance to lincomycin, clindamycin and metronidazole is a constant feature. Two β-lactamase-producing strains of *E. corrodens* have been isolated from an intra-abdominal abscess and an abdominal wound abscess of a 10-year-old and a 2-year-old boy, respectively, hospitalized in San Sebastian, Spain (Trallero et al., 1986).

There is good evidence that *E. corrodens* alone may be pathogenic in humans. When isolated from lesions, it is usually present in mixed culture with other facultative bacteria or with anaerobic bacteria; however, in 10–15% of positive specimens, it is present in pure culture, and it may cause serious diseases. In some cases, it may be the sole survivor in antibiotic-treated mixed infections. Holm (1950) has noted the presence of "corroding bacilli" in actinomycotic lesions, and Reinhold (1966) has studied strains from human sources. Marsden and Hyde (1971) and Kaplan et al. (1973) have reported infections in children. Khairat (1967) has recovered corroding bacilli from 16% of blood cultures drawn 1 min after dental extraction, but the true identity of some of the strains remains uncertain. King (1964) included *E. corrodens* (termed "HB-I") in a discussion of unusual pathogenic Gram-negative bacteria. Brooks et al. (1974) have produced abscesses by injecting mixtures of *E. corrodens* and streptococci, with further potentiation by the addition of methylphenidate, thus simulating lesions found in drug addicts. Experimental endocarditis, rarely fatal and seldom bacteremic, has been produced in catheterized rabbits (Badger et al., 1979); the organism can be shown in vegetation by fluorescent-antibody staining. *E. corrodens* has been associated with endocarditis, osteomyelitis, and septicemia in intravenous drug abusers, and may be related to the practice of "licking needles or skin" prior to injection (Angus et al., 1994; Olopoenia et al., 1994; Swisher et al., 1994).

The possible role of *E. corrodens* in the production of periodontal disease with bone destruction has been discussed and investigated (Socransky, 1977; Johnson et al., 1978; Listgarten et al., 1978; Behling et al., 1979; Progulske and Holt, 1980). *E. corrodens* is frequently found in subgingival plaque in patients with advanced periodonitis (Müller et al., 1997b). Coaggregation with other bacterial species (*Capnocytophaga*, *Actinobacillus*, streptococci) in dental plaque and the ability to elaborate a toxin in gingival sulcus (Levine and Miller, 1996) that may damage surrounding cells both contribute to periodontal disease. Other potential virulence factors operative in periodontal disease include a thiol-dependent hemolysin and other hydrolytic enzymes (Allaker et al., 1994). Infection of the mouth, upper and lower respiratory tract, sinuses, lips, and face is sometimes reported

(Schröter and Stawru, 1970; Jackson et al., 1971; Marsden and Hyde, 1971; Bottone et al., 1973; Carruthers and Sommers, 1973; Kaplan et al., 1973; Brooks et al., 1974; Rubenstein et al., 1976; Goodman, 1977; Piéron and Mafart, 1977; Dudley et al., 1978; DeMello and Leonard, 1979; Jones and Romig, 1979; Colloc et al., 1980; Mégraud et al., 1981; Knudsen and Simko, 1995). Serious infections, including brain abscesses, endocarditis, pneumonia, osteomyelitis, and septic arthritis (Johnson and Pankey, 1976), have been encountered. Transfer from the mouth through human bites and "clenched fist" injuries has resulted in more than 60 cases (Bilos et al., 1978; Goldstein et al., 1978).

The ability of *E. corrodens* to survive in the intestine leads to its presence in abdominal infections, including wound infection, abscesses, and peritonitis of the liver and other viscera (Quinlivan et al., 1996).

E. corrodens is regarded by several investigators as an opportunistic pathogen, particularly likely to produce infection in compromised hosts. Infection is often polymicrobic and is particularly associated with a *Streptococcus* species (Jackson et al., 1971; Bottone et al., 1973; Dorff et al., 1974; Shinhar et al., 1980; Flesher and Bottone, 1989; Quinlivan et al., 1996). Young et al. (1996) have shown that both coaggregation and growth stimulation occur between *E. corrodens* and viridans streptococci of the "*Streptococcus milleri*" group, which may enhance the establishment of mixed infections with these species.

ENRICHMENT AND ISOLATION PROCEDURES

Specimens should be plated on blood agar medium (5% sheep or horse blood, Columbia, or Oxoid No. 2 base, or similar media) and incubated anaerobically and aerobically (in each case with 5–15% CO_2) at 35–37°C. About half of the strains isolated have been recovered from the anaerobic plates but will grow aerobically on first subculture. This may be partly accounted for by suppression of associated organisms and by faster initial growth anaerobically upon transfer from relatively anaerobic body sites to artificial media. In addition, the typical corroding appearance of colonies may develop better under the humid conditions that are usual in anaerobic cultures, leading to easier recognition of the organism. For further information on media and atmospheres of incubation, see Goldstein et al. (1981).

Isolation from mixed infections is facilitated by the use of the medium described by Slee and Tanzer (1978), which contains clindamycin (5 µg/ml), KNO_3 (2.0 mg/ml), and hemin (5.0 g/ml) in agar-solidified Todd–Hewitt medium.

Aerobic plates should be maintained in a humid atmosphere during incubation to encourage production of typical colonies, e.g., in properly humidified CO_2 incubators, or in plastic bags gassed with a CO_2/air mixture and containing wet absorbent paper. Plate cultures should be examined daily for up to 5 days.

MAINTENANCE PROCEDURES

Cultures on blood agar plates should be subcultured at weekly intervals to avoid loss of viability. They may be stored in plastic bags at 4°C after 3 d of incubation.

For preservation in liquid nitrogen, a portion of the growth from a plate is transferred to 1 ml of horse serum, placed in a sterile screw-capped plastic vial and stored in a liquid-nitrogen freezer. Survival for at least 2 years has been reported (Labbé et al., 1977).

For freeze-drying, growth from a single plate is washed off with a suspending medium (either bovine serum containing 7.5% glucose or double-strength reconstituted dried skim milk), placed in 0.2-ml volumes in freeze-drying tubes, and lyophilized. Freeze-dried cultures in vacuum-sealed glass ampules should be stored in a refrigerator (4°C). Some strains may be serum-sensitive, and, for these, the milk medium may be preferable. The milk should be free from antibiotics. Samples of cultures preserved by any method should be tested for viability before storage of batches.

DIFFERENTIATION OF THE GENUS *EIKENELLA* FROM OTHER GENERA

Differentiation of *Eikenella* from morphologically similar microorganisms (Chadwick et al., 1995), especially *H. paraphrophilus*, is shown in Table BXII.β.87. Additional features that distinguish *E. corrodens* from *Bacteroides ureolyticus* are presented in Table BXII.β.88.

TAXONOMIC COMMENTS

The genus has been defined to include facultative organisms formerly grouped under the name *"Bacteroides corrodens"* (Eiken, 1958). Examination of strains to which this name had been applied revealed that they are heterogeneous. Most are not strict anaerobes, but rather grow in the presence of oxygen on hemin-containing media (Henriksen, 1969a; Hill et al., 1970; Jackson et al., 1971). Certain less oxygen-tolerant strains that are urease-positive have been assigned to the species *Bacteroides ureolyticus* (Jackson and Goodman, 1978). It has been found that some other anaerobic "corroding bacteria" are poorly characterized flagellated organisms (Henrichsen, 1975a; Smibert and Holdeman, 1976; Jackson and Goodman, 1978; Tanner et al., 1981). The statement made by Smibert and Holdeman (1976) that the anaerobic, urease-positive strain VPI 7814 is flagellated is incorrect, as is the report by Brooks et al. (1974) that this strain has a mol% G + C DNA content of 55.0. The true value is 28.4%, and this strain is *B. ureolyticus* and not an atypical *Eikenella* (Jackson and Goodman, 1978).

The organisms termed HB-1 by King (1964) are the same as *E. corrodens* (Jackson and Goodman, 1972; Riley et al., 1973). Similar organisms were included among those referred to as "corroding bacilli" by Holm (1950). The Gram-negative anaerobes forming spreading colonies described by Henriksen (1948) were heterogeneous.

Coykendall and Kaczmarek (1980) have found that 22 *E. corrodens* strains examined have an overall mean mol% G + C DNA content of 56.3, and show at least 70% DNA–DNA relatedness. Variations in relatedness values (70–100%) do not correlate with antigenic differences. It has been concluded that *E. corrodens* can be regarded as a "molecularly homogeneous species," taking into account DNA relatedness, reported mol% G + C ratios, and fatty acid profiles.

FURTHER READING

Bottone, E.J., J. Kittick and S.S. Schneierson. 1973. Isolation of *Bacillus* HB-1 from human clinical sources. Am. J. Clin. Pathol. *59*: 560–566.

Coykendall, A.L. and K.S. Kaczmarek. 1980. DNA homologies among *Eikenella corrodens* strains. J. Periodontal Res. *15*: 615–620.

Dorff, G.F., L.J. Jackson and M.W. Rytel. 1974. Infections with *Eikenella corrodens*, a newly recognized human pathogen. Ann. Intern. Med. *80*: 305–309.

Ebisu, S., H. Nakae and H. Okada. 1988. Coaggregation of *Eikenella corrodens* with oral bacteria mediated by bacterial lectin-like substance. Adv. Dent. Res. *2*: 323–327.

Eiken, M. 1958. Studies on an anaerobic rod-shaped Gram-negative microorganism: *"Bacteroides corrodens"* N. sp. Acta Pathol. Microbiol. Scand. *43*: 404–416.

Goldstein, E.J.C., E.O. Agyare and R. Silletti. 1981. Comparative growth of *Eikenella corrodens* on fifteen media in three atmospheres of incubation. J. Clin. Microbiol. *13*: 951–953.

Jackson, F.L. and Y.E. Goodman. 1972. Transfer of the facultatively anaerobic organism *"Bacteroides corrodens"* Eiken to a new genus, *Eikenella*. Int. J. Syst. Bacteriol. *22*: 73–77.

Jackson, F.L., Y.E. Goodman, F.R. Bel, P.C. Wong and R.L.S. Whitehouse. 1971. Taxonomic status of facultative and strictly anaerobic "corroding bacilli" that have been classified as *"Bacteroides corrodens"*. J. Med. Microbiol. *4*: 171–184.

Levine, M. and F.C. Miller. 1996. An *Eikenella corrodens* toxin detected by plaque toxin-neutralizing monoclonal antibodies. Infect. Immun. *64*: 1672–1678.

Maliszewski, C.R., C.W. Shuster and S.J. Badger. 1970. A type-specific

TABLE BXII.β.87. Differentiation of *Eikenella* from similar Gram-negative bacilli that may also produce "corroding" colonies[a]

Characteristic	Eikenella corrodens	Cardiobacterium hominis	Actinobacillus actinomycetemcomitans	Haemophilus aphrophilus	Haemophilus paraphrophilus	Kingella kingae	Kingella denitrificans	Moraxella atlantae	Pasteurella multocida	Bacteroides ureolyticus
Oxidase[b]	+	+	+	V	V	+	+	+	W[c]	W
Catalase	−	−	[+][d]	−	−	−	−	+	+	−
Nitrate reduction	+	−	+	+	+	−	+	−	+	+
Nitrite reduction	+	−	−	−	−	−	V	−	−	−
Indole	−	+	−	−	−	−	−	−	+	−
Urease	−	−	−	−	−	−	−	−	−	+
Acid from glucose	−	+	+	+	+	+	+	−	+	−
Lysine decarboxylase	V	−	−	−	−	−	−	−	−	−
Ornithine decarboxylase	+	−	−	−	−	−	−	−	+	−
Gelatin hydrolysis	+	−	−	−	−	−	−	−	−	−
Acid from lactose	−	−	−	+	+	−	−	−	−	−
Anaerobic growth only	−	−	−	−	−	−	−	−	−	+
V-factor dependent	−	−	−	−	+	−	−	−	−	−

[a]Adapted from Weyant et al. (1996); and Chadwick et al. (1995).

[b]Using dimethyl-*p*-phenylenediamine.

[c]W, weakly positive.

[d][+], positive delayed.

antigen of *Eikenella corrodens* is the major outer membrane protein. Infect. Immun. *42*: 208–213.

Progulske, A., R. Mishell, C. Trummel and S.C. Holt. 1970. Biological activities of *Eikenella corrodens* outer membrane and lipopolysaccharide. Infect. Immun. *43*: 178–182.

Young, K.A., R.P. Allaker, J.M. Hardie and R.A. Whiley. 1996. Interactions between *Eikenella corrodens* and "*Streptococcus milleri* group" organisms: possible mechanisms of pathogenicity in mixed infections. Antonie Leeuwenhoek *69*: 371–373.

Yumoto, H., H. Azakami, H. Nakae, T. Matsuo and S. Ebisu. 1996. Cloning, sequencing and expression of an *Eikenella corrodens* gene encoding a component protein of the lectin-like adhesin complex. Gene *183*: 115–121.

TABLE BXII.β.88. Differentiation of *Eikenella corrodens* from *Bacteroides ureolyticus*[a]

Characteristics	E. corrodens	B. ureolyticus
Aerobic growth, with hemin	+	−
Urease	−	+
Oxidase test (dimethyl-*p*-phenylenediamine)	+	v
Lysine decarboxylase[b]	+	−
Ornithine decarboxylase[b]	d	−
Gelatin hydrolysis[c]	−	+
Anaerobic growth enhanced by formate–fumarate; succinate is major end product	−	+
Susceptible to clindamycin (5 μg/ml)	−	+
Susceptible to metronidazole (5 μg/ml)	−	+
Anaerobic growth in the presence of 10% bile	+	−
Growth with 0.02% sodium azide	−	+
Odor of plate cultures is "bleach-like"	+	
Mol% G + C of DNA	56–58	27–29

[a]For symbols see standard definitions.

[b]Møller's method (Møller, 1955).

[c]Frazier (1926) method preferred. 30% trichloroacetic acid may be substituted for mercuric chloride. If conventional 12% gelatin is used, formate–fumarate supplementation is necessary to ensure growth of *B. ureolyticus* (Smibert and Holdeman, 1976). The use of different methods for any test may affect the results. For suitable methods, see Lapage et al. (1968); Hill et al. (1970); Midgley et al. (1970); Jackson et al. (1971); Brooks et al. (1974); and Cowan (1974, appendix C).

List of species of the genus Eikenella

1. **Eikenella corrodens** (Eiken 1958) Jackson and Goodman 1972, 75[AL] ("*Bacteroides corrodens*" Eiken 1958, 415.) *cor.ro′ dens*. M.L. part. adj. *corrodens* gnawing.

The characteristics are as described for the genus and as listed in Tables BXII.β.87, BXII.β.88, and BXII.β.89. Probably normal inhabitants of the human mouth and intestine. Can be opportunistic.

The mol% G + C of the DNA is: 56–58 (T_m).

Type strain: 33/54-55, ATCC 23834, DSM 8340, NCTC 10596.

GenBank accession number (16S rRNA): M22512.

TABLE BXII.β.89. Other characteristics of *Eikenella corrodens*[a]

Test	Reaction/Result
Anaerobic growth can occur	+
Catalase	−
Oxidase test (di- or tetramethyl-*p*-phenylenediamine)	+
Cytochromes of *b* and *c* groups present	+
Nitrates reduced to nitrites[b]	+
Anaerobic growth enhanced by nitrate	+
Arginine dihydrolase[c]	−
Indole	−
Acid from glucose and other carbohydrates	−
Flagella present	−
Corroding and noncorroding colonies occur	+

[a]For symbols, see standard definitions.

[b]The method of Cook (1950) is convenient for this test (Jackson et al., 1971), but strains giving a negative reaction by this method have been encountered (Hill et al., 1970).

[c]Method of Møller (1955).

Genus VI. **Formivibrio** Tanaka, Nakamura and Mikami 1991c, 580[VP] (Effective publication: Tanaka, Nakamura and Mikami 1991a, 494)

THE EDITORIAL BOARD

For.mi.vib′ ri.o. M.L. n. *acidum formicum* formic acid; M.L. masc. n. *Vibrio* that which vibrates, a generic name; M.L. masc. n. *Formivibrio* the formic acid forming vibrio.

Genus description is the same as for the description of *Formivibrio citricus*.

Type species: **Formivibrio citricus** Tanaka, Nakamura and Mikami 1991c, 580 (Effective publication: Tanaka, Nakamura and Mikami 1991a, 494.)

List of species of the genus Formivibrio

1. **Formivibrio citricus** Tanaka, Nakamura and Mikami 1991c, 580[VP] (Effective publication: Tanaka, Nakamura and Mikami 1991a, 494.)

ci′ tri.cus. M.L. adj. *citricus* pertaining to citric acid.

Gram-negative, curved rods with tapered or rounded ends, 0.5–0.6 × 1.1–2.5 μm, occurring singly or in pairs. Motile in an active tumbling manner by means of a single polar flagellum. Colonies are colorless, translucent, circular, convex, with an entire margin, up to 0.2 mm in diameter. Anaerobic, having a fermentative type of metabolism. Catalase negative. Do not reduce nitrate, sulfate, sulfite, thiosulfate, or S^0. Do not hydrolyze esculin, urea, or gelatin. Do not produce indole from tryptophan. S-Citramalate, citrate, mesaconate, and pyruvate are fermented to acetate and formate and serve as carbon and energy sources. Do not utilize acetoin, acrylate, citraconate, R-citramalate, crotonate, fumarate, L-glutamate, glycerol, H_2/CO_2/acetate, lactate, DL-threo-β-methylaspartate, L-malate, maleate, oxalate/acetate, peptones, yeast extract, or carbohydrates.

Cells contain β-hydroxymyristic acid ($C_{14:0\ 3OH}$). Cells wall contains meso-diaminopimelic acid. Temperature range 30–35°C; no growth at 25°C or 40°C. Optimal pH 7.6 (range 7.1–7.9); no growth at pH 6.7 or 8.2. Specific growth rate on S-citramalate under optimal conditions is 0.12/h. Habitat: anoxic freshwater mud.

The mol% G + C of the DNA is: 59–61 (HPLC).

Type strain: CreCit1, ATCC 49791, DSM 6150.

GenBank accession number (16S rRNA): Y17602.

Genus VII. Iodobacter Logan 1989, 455^{VP}

NIALL A. LOGAN

I.o.do.bac' ter. Gr. adj. ioeides violet-colored; M.L. masc. n. bacter the equivalent of Gr. neut. n. bakterion a small rod; M.L. masc. n. Iodobacter a violet-colored, small rod.

Straight, **round-ended rods, 0.7 × 3.0–3.5 µm, occurring singly, in pairs**, sometimes in chains, and occasionally as long filaments. Definite capsules are not evident. No resting stages known. Gram negative. **Motile by means of both a single polar flagellum and usually one or more subpolar or lateral flagella. Facultative anaerobes.** Grow on ordinary peptone media. **On low-nutrient media, such as nutrient agar containing a quarter of the normal concentration of nutrients, colonies are usually very thin, with rough surfaces and irregular edges, and may spread to 1 cm or more in diameter**; they are of butyrous consistency and are easily emulsified in water. Most strains produce the violet pigment violacein, and this shows most intensely in the centers of spreading colonies, but strains producing nonpigmented colonies (which are easily overlooked) are sometimes encountered. Occasional strains may spread only poorly and may even produce slightly gelatinous colonies, whose appearances are reminiscent of those produced by Chromobacterium and Janthinobacterium species. On full-strength nutrient agar and other routine, richer media, colonies may show less tendency to spread. Growth in quarter-strength nutrient broth is moderate after 24 h at 25°C, with uniform turbidity and a violet ring at the surface, but no pellicle; strains that produce long filaments may appear to gel in broths. Chemoorganotrophs; **attack carbohydrates fermentatively, producing small amounts of acid, but no gas. Grow at 4–30°C, with optimal growth at ~25°C.**

The mol% G + C of the DNA is: 50–52.

Type species: **Iodobacter fluviatilis** (Moss, Ryall and Logan 1978) Logan 1989, 455 (Chromobacterium fluviatile Moss, Ryall and Logan 1981, 216 (Effective publication: Moss, Ryall and Logan 1978, 18.))

FURTHER DESCRIPTIVE INFORMATION

The pigment violacein is produced on or in media containing tryptophan. It is soluble in ethanol but not in water or chloroform, and is readily identified by spectrophotometry and by testing with routine reagents (see section on Chromobacterium violaceum).

Colonies of Iodobacter on low-nutrient media are characteristically irregular, thin, spreading, and butyrous, with uneven surfaces like beaten copper. The centers show denser growth and thus more intense pigmentation than do the diffuse peripheries (Fig. BXII.β.70). Occasional strains produce slightly gelatinous, non-spreading colonies. Nonpigmented strains are, without doubt, frequently overlooked when isolated. Subcultures of pigmented strains may contain partially or completely unpigmented colonies. An unpigmented strain is most readily recognized by

a colony morphology similar to that of a pigmented strain isolated at the same time under the same conditions. Once it has been confirmed as an oxidase- and catalase-positive, Gram-negative rod, such an isolate should be flagella-stained. If it possesses the characteristic polar and lateral flagella, it should then be subjected to the differential tests for the three genera Chromobacterium, Iodobacter, and Janthinobacterium. Although young cultures of Aeromonas, Pseudomonas, and Vibrio spp. grown on solid media sometimes produce lateral flagella in addition to their usual polar flagella, they give different patterns of results in the differential tests than do the violet-pigmented organisms.

The characteristic flagellar arrangement is best seen in young cultures on solid media. The single polar flagellum is inserted at the tip of the cell, exhibits long, shallow waves and often stains faintly. The lateral flagella are usually long, and 1–4 such flagella, but as many as 8, may occur. They may be inserted subpolarly or laterally, usually exhibit deep, short waves, and stain readily (Fig. BXII.β.71).

Iodobacter fluviatilis strains have been isolated from running fresh water in England and Scotland, and have also been reported from Antarctic lakes and their sediments (Wynn-Williams, 1983).

ENRICHMENT AND ISOLATION PROCEDURES

Iodobacter fluviatilis appears to be a minor member of fresh water floras, and competition with other organisms often results in poor growth and pigmentation and infrequent isolations. For general purposes, the selective medium developed by Ryall and Moss (1975) is recommended: nutrient broth (Oxoid No. 1) at one-quarter strength (¼× NB) is solidified with 1.2% agar (Oxoid No. 3) (to make ¼× Nutrient Agar; ¼× NA) and supplemented, immediately prior to pouring, with filter-sterilized solutions of colistin sulfate, cycloheximide, and sodium deoxycholate (final concentrations, 15, 30, and 300 µg/ml, respectively); the cycloheximide may be omitted if fungal contamination is negligible. Unsupplemented ¼× NA should also be prepared and used both alongside the supplemented medium for isolation and later for subcultures. Spread plates are prepared from undiluted water and from low dilutions and are incubated at 25°C for 5–7 d. Limited trials of the medium described by Keeble and Cross (1977) for isolating Janthinobacterium suggest that it is unsuitable for the isolation of Iodobacter (Logan, 1989).

MAINTENANCE PROCEDURES

The organisms survive for weeks to months in ¼× NB and for several years in dilute peptone water (0.1% peptone) at 4°C. They

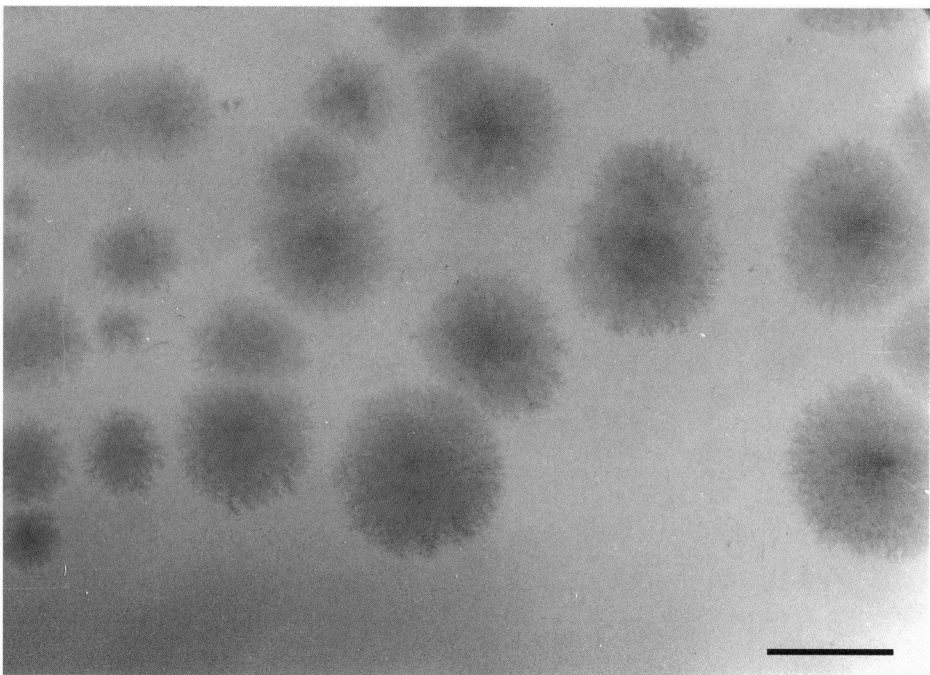

FIGURE BXII.β.70. Colonies of *Iodobacter fluviatilis* ATCC 33051 on quarter strength nutrient agar after 48–72 h at 25°C. Bar = 5 mm.

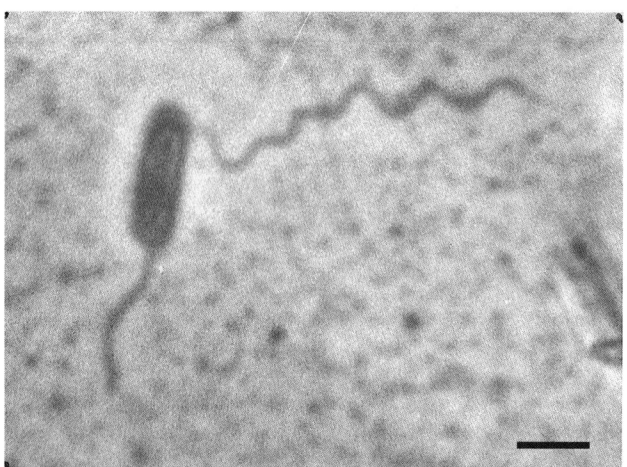

FIGURE BXII.β.71. *Iodobacter fluviatilis* ATCC 33051 stained for flagella by the method of Mayfield and Innis (1977) and showing a long, lateral flagellum with deep waves of short wavelength and a polar flagellum with shallow waves of long wavelength. Bar = 1 μm.

can also be preserved indefinitely by lyophilization or by freezing in nutrient broth containing 15% glycerol.

PROCEDURES FOR TESTING SPECIAL CHARACTERS

Methods for demonstrating the production of violacein, for the study of flagellar arrangement and growth temperature, and for demonstrating oxidase reaction, fermentative attack of glucose, anaerobic growth, and acid production from carbohydrates are described in the section on *Chromobacterium violaceum*.

DIFFERENTIATION OF THE GENUS *IODOBACTER* FROM OTHER GENERA

Table BXII.β.90 presents characteristics differentiating the genus *Iodobacter* from the genera *Chromobacterium* and *Janthinobacterium*.

Poorly growing and weakly reacting strains are not unusual (see Logan, 1989, for example), so an identification should be based upon a pattern of reactions rather than on the result of a single test.

TAXONOMIC COMMENTS

Violacein production is a distinctive and important diagnostic feature of *Iodobacter*, but it cannot be emphasized as a taxonomic character, since unpigmented strains are not uncommon (Logan, 1989) and species producing the pigment are found in the genera *Alteromonas*, *Chromobacterium*, and *Janthinobacterium* (Logan and Moss, 1992). Although its mol% G + C range is much lower than that of the other two species in the genus, Moss et al. (1978) allocated their new species of violet chromogen to the genus *Chromobacterium* as *C. fluviatile*, as it is an oxidase- and catalase-positive Gram-negative rod producing violacein and showing the characteristic flagellar arrangement, and because its taxonomic affiliation was unknown. Similarly, *C. violaceum* and "*C. lividum*" were long retained in the same genus on account of these features, even though they show many differences in other respects. In the same year that *C. fluviatile* was described, De Ley et al. (1978) reclassified "*C. lividum*" in the new genus *Janthinobacterium*. Based on nucleic acid hybridization and phenotypic properties and following further nucleic acid hybridization studies (Moss and Bryant, 1982) and numerical analysis of phenotypic characters, *C. fluviatile* was transferred to the new genus *Iodobacter* (Logan, 1989). Phylogenetic relationships inferred from rRNA–DNA hybridization and 16S rRNA sequencing studies support the recognition of these three genera, which represent separate deeply branching lineages of the *Betaproteobacteria*, with *Chromobacterium* and *Iodobacter* being outlying members of the family *Neisseriaceae* (Dewhirst et al., 1989). The adjectival epithet of the only species of the genus, *Iodobacter fluviatile*, has been revised to *Iodobacter fluviatilis* in order to agree in gender with the masculine genus name (Euzéby, 1997).

TABLE BXII.β.90. Patterns of results in tests useful for distinguishing *Iodobacter* from *Chromobacterium* and *Janthinobacterium*[a, b]

Characteristic	*Iodobacter fluviatilis*	*Chromobacterium violaceum*	*Janthinobacterium lividum* (typical)	*Janthinobacterium lividum* (atypical)
Colonies on ¼× NA:				
Spreading	89	0	0	0
Gelatinous	4	0	38	100
Tough	0	0	7	71
Growth at:				
4°C	100	0	97	93
37°C	0	100	0	0
Anaerobic growth	100	100	3	0
D-Glucose:				
Fermented	100	100[c]	0	0[d]
Oxidized	0	0[c]	97	57[d]
Acid from:				
L-Arabinose[e]	0	0	100	100
D-Maltose	100	0	98	93
D-Mannitol[f]	0	0	97	0
Trehalose	100	100	1	100
Lactate utilization	100	0	100	100
Esculin hydrolysis	0	0	100	0
Arginine hydrolysis	0	100	0	0
HCN production	0	100	0	0

[a]Values represent the percentage of strains positive for each characteristic. Number of strains studied: *Iodobacter fluviatilis*, 53; *Chromobacterium violaceum*, 9; *Janthinobacterium lividum* (typical), 68; *Janthinobacterium lividum* (atypical), 14.

[b]From Logan (1989).

[c]Occasional strains are oxidative.

[d]Some strains give no reaction in this test.

[e]Similar patterns of results are obtained with D-cellobiose and D-galactose.

[f]Similar patterns of results are obtained with *myo*-inositol and D-sorbitol.

List of species of the genus Iodobacter

1. **Iodobacter fluviatilis** (Moss, Ryall and Logan 1978) Logan 1989, 455[VP] (*Chromobacterium fluviatile* Moss, Ryall and Logan 1981, 216 (Effective publication: Moss, Ryall and Logan 1978, 18.))

flu.vi.a.ti′lis. L. masc. adj. *fluviatilis* of rivers.

Rods, 0.7 × 3.0–3.5 μm, occurring singly or in short chains, with occasional elongated forms.

Colonies on low-nutrient media are flat, very thin, irregular in outline, spreading and pale violet, with a copper-beaten, slightly rough surface, and they are not gelatinous. A uniform turbidity is produced in nutrient broth with a violet ring at the surface, but usually no pellicle.

Grows on ordinary peptone media. Grows on Mac-Conkey agar, giving violet colonies.

Facultatively anaerobic. Optimal temperature, 25°C; minimum, 4°C; maximum, ~30°C.

Other characteristics are given in Tables BXII.β.90 and BXII.β.91.

Isolated from running fresh water in England and Scotland, and from Antarctic lakes and their sediments.

The mol% G + C of the DNA is: 50–52 (T_m).

Type strain: Moss 165/Sp7, ATCC 33051, DSM 3764, NCIMB 2053, NCTC 11159.

GenBank accession number (16S rRNA): M22511.

TABLE BXII.β.91. Characteristics of *Iodobacter fluviatilis*[a]

Characteristic	*Iodobacter fluviatilis*
Colonies on ¼× NA:	
Spreading	[+] (83)
Gelatinous	[−] (4)
Pigmented	[+] (83)
Zoning of pigment	[−] (6)
Growth on:	
Minimal medium	[+] (96)
4NA	d (45)
Growth at:	
30°C	[+] (83)
37°C	−
Growth in:	
1% NaCl	+
2% NaCl	[−] (9)
4% NaCl	−
Growth at:	
pH 9	[+] (98)
pH 4	[−] (19)
pH 3	−
Formation of cell chains	d (34)
Formation of filaments	[−] (15)
Acid from:	
L-Arabinose	−
Cellobiose	−
Fructose	+
D-Galactose	−
Gluconate	[+] (96)
D-Glucose	+
Glycerol	[−] (13)
Glycogen	−
myo-Inositol	−
Inulin	−
Lactose	−
Maltose	+

(*continued*)

TABLE BXII.β.91. *(cont.)*

Characteristic	*Iodobacter fluviatilis*
D-Mannitol	−
D-Mannose	+
Melezitose	−
N-Acetylglucosamine	+
Raffinose	−
D-Sorbitol	−
Starch	−
Sucrose	−
Trehalose	+
D-Xylose	−
Catalase	+
Oxidase	+
Nitrate reduction	[+] (98)
Nitrite reduction	−
Production of HCN	−
Egg yolk reaction	[−] (13)
Hemolysis	[+] (92)
Carbon sources:	
Acetate	[+] (89)
Citrate	[+] (85)
Fumarate	[+] (89)
Glycerate	d (60)
Lactate	−
Malate	+
Propionate	−
Pyruvate	d (79)
Succinate	[+] (94)

(continued)

TABLE BXII.β.91. *(cont.)*

Characteristic	*Iodobacter fluviatilis*
Tartrate	−
Hydrolysis of:	
Arginine	−
Casein	[+] (90)
Esculin	−
Gelatin	+
Starch	−
Antimicrobial agents (per disk):	
Ampicillin, 25 µg	R
Cephaloridine, 25 µg	R
Colistin sulfate, 10 µg	R
Chloramphenicol, 10 µg	S
Chlortetracycline, 10 µg	S
Furazolidone, 50 µg	S
Kanamycin 30 µg	S
Neomycin 10 µg	s
Nalidixic acid 30 µg	S
Nitrofurantoin, 200 µg	S
Oxytetracycline, 10 µg	S
Penicillin G, 1.5 IU	R
Streptomycin, 10 µg	s
Sulfafurazole, 100 µg	s
Sulfafurazole, 500 µg	s
Vibriostatic agent O/129, 50 µg	R

[a]Symbols: +, all strains positive; [+], positive for 80% or more strains; d, positive for 31–79% strains; [−], positive for 30% or fewer strains; −, negative for all strains; R, resistant; S, sensitive; s, slightly sensitive. Numbers in parentheses indicate the % of strains giving positive reactions.

Genus VIII. **Kingella** *Henriksen and Bøvre 1976, 449*[AL] *emend. Dewhirst, Chen, Paster and Zambon 1993, 497*

Robbin S. Weyant

King.el′ la. M.L. *-ella* dim. ending; M.L. fem. n. *Kingella* named after Elizabeth O. King, an American bacteriologist.

Straight rods, 0.6–1.0 × 1.0 to 3.0 µm, with rounded or square ends. Occur in pairs and sometimes in short chains. Endospores are not formed. **Gram negative, but there is a tendency to resist Gram-decolorization.** Nonmotile by normal tests, but may be fimbriated (piliated) and show "twitching motility." **Aerobic or facultatively anaerobic**; grow best aerobically, but can grow weakly under anaerobic conditions on blood agar. Optimal growth temperature, 33–37°C. Two types of colonies occur on blood agar: (a) a spreading, corroding type associated with "twitching motility," fimbriation, and transformation competence, and (b) a smooth, convex type not showing "twitching motility", fimbriation, or competence. **Oxidase positive** (when tested with tetramethyl-*p*-phenylene diamine; the dimethyl reagent may give weak or negative reactions). **Catalase negative.** Coagulated serum is not liquefied. Urease- and indole-negative. Phenylalanine deaminase activity is negative or weak. Chemoorganotrophic. **D-Glucose and a limited number of other carbohydrates are fermented with production of acid, but no gas.** Occur as normal flora in human mucous membranes of the upper respiratory tract. *K. kingae* is an emerging human pathogen, and *K. denitrificans* has been isolated from human infections.

The mol% G + C of the DNA is: 47–58.

Type species: **Kingella kingae** (Henriksen and Bøvre 1968a) Henriksen and Bøvre 1976, 449 (*Moraxella kingae* (Henriksen and Bøvre 1968a) Bøvre, Henriksen and Jonsson 1974, 307; *Moraxella kingii* (sic) Henriksen and Bøvre 1968a, 383.)

FURTHER DESCRIPTIVE INFORMATION

Phylogenetic treatment The genus *Kingella* is in the family *Neisseriaceae*, class *Betaproteobacteria*, and contains three species, *K. kingae*, *K. denitrificans*, and *K. oralis* (Snell, 1984; Dewhirst, et al., 1993).

Cellular morphology and colonial characteristics Cells of *Kingella* strains are characteristically plump, Gram-negative rods or coccobacilli, occurring in pairs or chains. Films from 18-h-old cultures show a tendency to retain the crystal violet of the Gram stain. *Kingella* cells show some pleomorphism, with swollen, irregularly-stained variants. Cells of *Kingella* may show "twitching motility" (Henrichsen et al., 1972; Dewhirst et al., 1993). Twitching motility and competence in genetic transformation are associated with the presence of pili on the cell surface (Frøholm and Bøvre, 1972; Henrichsen, 1972; Weir et al., 1996). *K. kingae* and *K. denitrificans* express type 4 pili that are characteristically long and thin (5–6 nm in diameter) (Weir and Marrs, 1992). Colonies formed by piliated cells are spreading and corroding, in contrast to the smooth, entire, convex colonies formed by nonpiliated cells. Freshly isolated strains of *K. oralis* and *K. denitrificans* produce spreading, corroding colonies (Snell, 1984; Dewhirst et al., 1993). Colonies of *K. kingae* are easily recognized on blood agar in mixed culture by a distinct zone of β-hemolysis surrounding the colonies.

Nutritional and growth requirements All species in the genus are nutritionally fastidious; little or no growth occurs on unsupplemented peptone media. Growth on nutrient agar is only marginally improved by addition of blood or serum, and the colonies remain small, typically 0.5–1.0 mm in diameter after incubation for 48 h. There is no requirement for X or V factors. Although there is no strict requirement for a CO_2-enriched atmosphere, growth of some strains may be enhanced by incubation in 5% CO_2. Growth occurs at 30°C and 37°C but not at 5°C or 45°C. Strains differ in their ability to grow at 22°C. There is some disagreement about of *Kingella* species to grow anaerobically. Henriksen and Bøvre (1976) have described *K. kingae* as aerobic, but Snell and Lapage (1976) have found growth of *K. kingae* and *K. denitrificans* on blood agar in an atmosphere of H_2/CO_2, (95:5), and Dewhirst et al. (1993) have demonstrated growth of *K. oralis* in anaerobic conditions.

Cellular fatty acid composition Cells of *K. kingae* and *K. denitrificans* contain the following fatty acids: *n*-dodecanoic, 3-hydroxydodecanoic, *n*-tetradecanoic, 3-hydroxytetradecanoic, *cis*-9-hexadecenoic, *n*-hexadecanoic, *cis*-9,12-octadecadienoic, *cis*-9-octadecenoic, and *n*-octadecanoic (Weyant, et al. 1996). The cells do not contain *n*-pentadecanoic, 3-hydroxyhexadecanoic, or heptadecenoic acids (Jantzen et al., 1974). Waxes have not been found in *K. kingae* (Bryn et al., 1977).

Antibiotic sensitivity Although *Kingella* strains are normally sensitive to β-lactam antibiotics, sulfonamides, erythromycin, tetracycline, chloramphenicol, ciprofloxacin, and streptomycin, rare cases of antibiotic resistance have been reported. Sordillo et al. (1993) and Minamoto and Sordillo (1992) have reported human infections caused by β-lactamase-positive *K. kingae* and *K. denitrificans* strains. Knapp et al. (1988) have described *K. denitrificans* strains with plasmid-mediated, high-level tetracycline resistance, and Jensen et al. (1994) have identified trimethoprim-resistant *K. kingae* strains. *Kingella* strains are usually resistant to vancomycin.

Pathogenicity *Kingella* species have been associated with infections in humans. Since the 1990s, a significant increase of *K. kingae* septicemia, bone, and joint infections in children under the age of 3 years has been reported (Goutzmanis et al., 1991; Birgisson et al., 1997; Yagupsky and Dagan, 1997; La Scola et al., 1998b; Lundy and Kehl, 1998). *K. kingae* has also been isolated from human corneal ulcer specimens (Mollee et al., 1992). *K. denitrificans* and *K. kingae* have been isolated from blood cultures of individuals with endocarditis (Kerlikowske and Chambers, 1989; Hassan and Hayak, 1993). *K. oralis* has not been associated with invasive infections in humans, although Chen (1996) has reported a significantly higher concentration of this species in dental plaque of individuals with juvenile peridontitis. There have been no published reports of *Kingella* infections in animals or plants.

Natural habitat The natural habitat of *Kingella* is the upper respiratory tract and oral cavity of humans and possibly other primates, where the organisms are present on the mucous membranes as part of the normal flora. Although most *Kingella* isolates have been from human respiratory tract and oral sources, a few isolates of *K. denitrificans* have been obtained from chimpanzee throat and human urogenital specimens (Weyant et al., 1996).

ENRICHMENT AND ISOLATION PROCEDURES

Recovery of *Kingella* from normally sterile clinical specimens, such as blood and synovial fluid, is greatly enhanced by inoculating the specimens into blood culture media (Yagupsky et al., 1992; Yagupsky and Press, 1997; Lejbkowicz et al., 1999). *Kingella denitrificans* has been isolated from specimens with mixed flora by culturing on Thayer–Martin agar (Hollis et al., 1972). *K. kingae* has been selectively isolated on sheep blood agar with 2 µg/ml vancomycin (BAV medium; Yagupsky et al., 1995), and a complex medium containing 5% sheep blood, 5 µg/ml hemin, 0.5 µg/ml menadione, and 1 µg/ml clindamycin has been described by Chen (1996) for the selective recovery of *K. oralis*. Incubation for 48 h may be required for development of colonies of reasonable size on any selective medium. Once isolated, *Kingella* strains may be subcultured on a general, enriched laboratory media, such as blood agar.

MAINTENANCE PROCEDURES

Kingella strains are difficult to maintain by serial transfer, because blood agar cultures become sterile after 6–12 d at room temperature. Preservation is best achieved by freeze drying. Horse serum containing 5% (w/v) *i*-inositol is a suitable suspending medium for freeze drying (Redway and Lapage, 1974). Alternatively, strains may be preserved by suspending fresh cultures in defibrinated rabbit blood and freezing in liquid nitrogen (Weyant et al., 1996).

PROCEDURES FOR TESTING SPECIAL CHARACTERS

The main difficulty in characterizing members of the genus is the poor growth obtained in nonenriched media. Growth is better on solid or semisolid media than in liquid media. Serum promotes a slight, although not a dramatic, improvement of growth, and addition of 5% horse or rabbit serum to test media is worthwhile. For testing the acidification of carbohydrates, rabbit serum is preferable because horse serum contains endogenous maltase activity (Hollis et al., 1983). Due to the slow growth of the organisms, longer than usual incubation periods (3–7 days) may be required to confirm negative results.

Methods suitable for characterization of the genus have been described by Snell et al. (1972), Snell and Lapage (1976), and Weyant et al. (1996).

DIFFERENTIATION OF THE GENUS *KINGELLA* FROM OTHER GENERA

Characteristics by which *Kingella* may be differentiated from phenotypically similar genera and species within the family *Neisseriaceae* are shown in Table BXII.β.92.

TAXONOMIC COMMENTS

These bacteria were historically associated with the genus *Moraxella*. *K. kingae* was originally named *Moraxella kingii* by Henriksen and Bøvre (1968a). The masculine epithet *kingii* was later corrected to the feminine *kingae* by Bøvre et al. (1974). The species was transferred to a newly created genus *Kingella* by Henriksen and Bøvre (1976). In 1972, Hollis et al. described TM-1, a fastidious Gram-negative rod isolated on Thayer–Martin agar from human pharyngeal cultures (Hollis et al., 1972). TM-1 was later classified as *Kingella denitrificans* by Snell and Lapage, who also described *K. indologenes* in the same paper (Snell and LaPage, 1976). In 1990, Dewhirst et al., using 16S rRNA sequence-based methods, found that *K. indologenes* was sufficiently distant from the other *Kingella* species to justify its inclusion in the new genus *Suttonella* as *S. indologenes* (Dewhirst et al., 1990). With the description of *K. oralis* in 1993 (Dewhirst et al., 1993), the genus was again expanded to the 3 currently recognized species.

TABLE BXII.β.92. Differentiation of the genus *Kingella* from phenotypically similar genera and species[a,b]

Characteristics	*Kingella*	*Actinobacillus actinomycetemcomitans*	*Cardiobacterium hominis*	*Eikenella corrodens*	*Haemophilus aphrophilus*	*Moraxella*	*Neisseria*	*Suttonella indologenes*
Cell shape:								
Cocci	−	−	−	−	−	D[c]	D[d]	−
Rods	+	+	+	+	+	D[c]	D[d]	+
Catalase	−	+	−	−	−	+	D[d]	−
Acid from D-glucose	+	+	+	−	+	−	D[d]	+
Acid from D-sorbitol	−	−	+	−	−	−	−	−
Ornithine decarboxylase	−	−	−	+	−	−	−	−
Nitrate reduction	D	+	−	+	+	D	D	−
Indole	−	−	+	−	−	−	−	+

[a]For symbols see standard definitions.

[b]Data from Snell (1984) and Weyant et al. (1996).

[c]*Moraxella catarrhalis* is coccoid. Other *Moraxella* species are bacillary.

[d]*Neisseria elongata* is bacillary. *N. elongata* subsp. *elongata* and *N. elongata* subsp. *nitroreducens* are catalase negative. *N. flavescens*, *N. cinerea*, *N. elongata* subsp. *elongata*, and *N. elongata* subsp. *nitroreducens* usually do not form acid from D-glucose.

ACKNOWLEDGMENTS

This chapter is based in part on the Genus *Kingella* chapter of the 1st edition of *Bergey's Manual of Systematic Bacteriology*, authored by Dr. J.J.S. Snell. The author thanks Dr. Snell for his generous contributions.

FURTHER READING

Dewhirst, F.E., C.K.C. Chen, B.J. Paster and J.J. Zambon. 1993. Phylogeny of species in the family *Neisseriaceae* isolated from human dental plaque and description of *Kingella orale*, sp. nov. Int. J. Syst. Bacteriol. *43*: 490–499.

Henriksen, S.D. and K. Bøvre. 1976. Transfer of *Moraxella kingae* Henriksen and Bøvre to the genus *Kingella* gen. nov. in the family *Neisseriaceae*. Int. J. Syst. Bacteriol. *26*: 447–450.

Yagupsky, P. and R. Dagan. 1997. *Kingella kingae*: an emerging cause of invasive infections in young children. Clin. Infect. Dis. *24*: 860–866.

DIFFERENTIATION OF THE SPECIES OF THE GENUS *KINGELLA*

Characteristics differentiating the species of *Kingella* are listed in Table BXII.β.93. Other characteristics of the species are listed in Table BXII.β.94.

TABLE BXII.β.93. Characteristics useful in the differentiation of *Kingella* species[a,b]

Characteristic	*K. kingae*	*K. denitrificans*	*K. oralis*
β-hemolytic	+	−	−
Nitrate reduction	−	+	−
Nitrite reduction	−	+	v[c]
Gas produced from nitrite	−	+	−
Phosphatase activity	+	−	+
Casein digestion	+	−	nd

[a]For symbols see standard definitions; nd, not determined.

[b]Data from Snell (1984) and Weyant et al. (1996).

[c]The type strain of *K. oralis* (CCUG 30450[T]) fails to reduce 0.1% nitrite, but reduces 0.01% nitrite with no gas (data from author's laboratory).

List of species of the genus Kingella

1. **Kingella kingae** (Henriksen and Bøvre 1968a) Henriksen and Bøvre 1976, 449[AL] (*Moraxella kingae* (Henriksen and Bøvre 1968a) Bøvre, Henriksen and Jonsson 1974, 307; *Moraxella kingii* (sic) Henriksen and Bøvre 1968a, 383.) king'ae. M.L. gen. n. *kingae* of King, named after Elizabeth O. King, an American bacteriologist.

   The characteristics are as described for the genus and as listed in Tables BXII.β.93 and BXII.β.94.

   When grown on blood agar, *K. kingae* colonies produce a characteristic zone of β-hemolysis. This characteristic, along with the tendency of this species to resist Gram-stain decolorization and its failure to produce catalase, may cause some confusion between *K. kingae* and β-hemolytic streptococci in the analysis of specimens from the human throat and oral cavity. The oxidase test, which is generally positive for *K. kingae* and negative for *Streptococcus* species, is a simple and effective differential method.

   In recent years, there has been an increased recognition of *Kingella kingae* as an etiologic agent of invasive infections of humans, especially children less than 2 years of age. Since 1985, numerous case reviews describing *Kingella kingae* septic arthritis, osteomyelitis, septicemia, or endocarditis in pediatric patients have been reported. These cases were from various geographic locations, including Sweden (9 cases)

TABLE BXII.β.94. Characteristics of the species of the genus *Kingella*[a,b]

Characteristic	*K. kingae*	*K. denitrificans*	*K. oralis*
Catalase test	−	−	−
Oxidase test	+	+	+
Motility (swimming)	−	−	−
Anaerobic growth (blood agar)	w	w	w
Temperature tolerance, growth at:			
5°C	−	−	nd
22°C	d	d	+
30°C	+	+	+
37°C	+	+	+
45°C	−	−	−
Growth in the presence of 6% NaCl	−	−	−
Bile tolerance, growth in:			
10% bile	−	−	nd
40% bile	−	−	nd
Growth stimulation by bile	−	−	nd
Esculin hydrolysis	−	−	−
Growth on MacConkey agar	−	−	−
Liquefaction of gelatin and coagulated serum	−	−	−
Citrate utilization	−	−	−
Growth in mineral medium with β-hydroxybutyrate	−	−	nd
Formation of intracellular poly-β-hydroxybutyrate (nutrient medium)	−	−	nd
Urease activity	−	−	−
Lipase activity: hydrolysis of:			
Tween 20	−	d	nd
Tween 80	−	−	nd
Lecithinase activity	−	−	nd
Arginine dihydrolase	−	−	−
Ornithine decarboxylase	−	−	−
Lysine decarboxylase	−	−	−
Deoxyribonuclease activity	−	d	nd
H_2S production (detection by lead acetate strips)	d	−	+
Susceptible to penicillin (1.0 U/ml)	+	+	nd
β-galactosidase activity (ONPG test)	−	−	nd
Starch hydrolysis	−	−	nd
Oxidation/fermentation test	F	F	F
Acid produced from:			
D-Glucose	+	+	+
Sucrose	−	−	−
Maltose	+	−	−
D-Xylose	−	−	−
Lactose	−	−	−
D-Mannitol	−	−	−
Fructose	−	−	nd
Dextrin	−	d	nd
Adonitol	−	−	nd
L-Arabinose	−	−	nd
Cellobiose	−	−	nd
Dulcitol	−	−	nd
Ethanol	−	−	nd
D-Galactose	−	−	nd
Glycerol	−	−	nd
i-Inositol	−	−	nd
D-Mannose	−	−	nd
Raffinose	−	−	nd
L-Rhamnose	−	−	nd
Salicin	−	−	nd
D-Sorbitol	−	−	nd
Trehalose	−	−	nd

[a]For symbols see standard definitions; w, weak growth; nd, not determined.

[b]Data from Snell (1984), Dewhirst et al. (1993), and Weyant et al. (1996).

(Claesson et al., 1985), Iceland (5 cases) (Birgisson et al., 1997), Australia (10 cases) (Goutzmanis et al., 1991), France (5 cases) (La Scola et al., 1998b), Israel (42 cases) (Yagupsky and Dagan, 1997), and the United States (10 cases) (Lundy and Kehl, 1998). In many cases, *K. kingae* infections occur as sequelae to viral infections, with the viral agents presumably compromising mucosal immunity to allow the seeding of the bloodstream with *K. kingae* from the oral cavity (Waghorn and Cheetham, 1997; Amir and Yagupsky, 1998). Lundy and Kehl (1998) have observed that the apparent increase in *K. kingae* infections corresponds with the vaccine-mediated decrease in *Haemophilus influenzae* infections in young children, suggesting that *K. kingae* is replacing *H. influenzae* as a significant etiologic agent in this population.

The mol% G + C of the DNA is: 47.3–47.4 (T_m).

Type strain: ATCC 23330.

GenBank accession number (16S rRNA): M22517.

2. **Kingella denitrificans** Snell and Lapage 1976, 456[AL]

de.ni.tri′fi.cans. L. prep. *de* away from; L. n. *nitrum* soda; M.L. n. *nitrum* nitrate; M.L. v. *denitrifico* to denitrify; M.L. part. adj. *denitrificans* denitrifying.

The characteristics are as described for the genus and as listed in Tables BXII.β.93 and BXII.β.94.

Kingella denitrificans was first described by Hollis et al. (1972), who encountered isolates of this species on Thayer–Martin selective agar while performing a study to determine pharyngeal carrier rates of *Neisseria meningitidis* and *N. lactamica* in healthy adults. The isolates were originally designated "TM-1" in recognition of their ability to grow on Thayer–Martin agar.

Kingella denitrificans is commonly found as part of the normal flora of the human pharynx and is generally considered to be of low pathogenicity. Of 60 strains studied at the U.S. Centers for Disease Control and Prevention, only 2 have been from sources, i.e., blood and mandibular abscess, suggestive of invasive infection (Weyant et al., 1996). Rare cases of endocarditis (Hassan and Hayek, 1993), and a case of granulomatous disease in an AIDS patient (Minamoto and Sordillo, 1992) have been reported.

The mol% G + C of the DNA is: 54.1–54.8 (T_m).

Type strain: ATCC 33394, DSM 10202, NCTC 10995.

GenBank accession number (16S rRNA): M22516.

3. **Kingella oralis** Dewhirst, Chen, Paster, and Zambon 1993, 498[VP] (*Kingella orale* (sic) Dewhirst, Chen, Paster, and Zambon 1993, 498.)

o.ra′lis. L. adj. *oralis* oral, pertaining to the mouth.

The characteristics are as described for the genus and as listed in Tables BXII.β.93 and BXII.β.94.

Kingella oralis was originally described by Chen et al. (1990a) as *Eikenella corrodens*-like isolates obtained from human oral cavity specimens. In the original 1993 classification paper, this species was named *K. orale*, but the spelling of the species name was changed to "*oralis*" in 1994 (Dewhirst et al., 1993, 1994). *K. oralis* is found in the human oral cavity and produces corroding colonies similar to those of *E. corrodens*. Biochemical tests useful in differentiating *K. oralis* from *E. corrodens* include acidification of D-glucose and ornithine decarboxylase.

The mol% G + C of the DNA is: 56–58 (T_m).

Type strain: ATCC 51147, CCUG 30450.

GenBank accession number (16S rRNA): L06164.

Genus IX. **Microvirgula** *Patureau, Godon, Dabert, Bouchez, Bernet, Delgenes and Moletta 1998, 781*[VP]

DOMINIQUE PATUREAU

Mi.cro.vir'gu.la. Gr. adj. *micros* small; L. fem. n. *virgula* twig or rod; M.L. n. *Microvirgula* small twig or rod.

Curved rods, occurring singly, in pairs, or in clusters. **Motile. Gram negative. Oxidase and catalase positive. Aerobic,** having a strictly respiratory type of metabolism with oxygen as the terminal electron acceptor; however, anaerobic respiration occurs in the presence of nitrate, nitrite, and nitrous oxide, with the formation of N_2. **Denitrification is not repressed under aerobic conditions.** Chemoorganotrophic. **Growth occurs on a variety of carbon sources including acetate, succinate, propionate, ethanol, methanol, and glycerol, but not on sugars.** Belongs to the *Betaproteobacteria.* Isolated from an up-flow anoxic filter inoculated with activated sludges.

The mol% G + C of the DNA is: 65.

Type species: **Microvirgula aerodenitrificans** Patureau, Godon, Dabert, Bouchez, Bernet, Delgenes and Moletta 1998, 781.

DIFFERENTIATION OF THE GENUS *MICROVIRGULA* FROM OTHER GENERA

Phenotypically, the genus most closely resembles *Comamonas testosteroni, Alcaligenes faecalis,* and *Thauera mechernichensis.* Table BXII.β.95 lists characteristics that differentiate the genus from these species and from the phylogenetically related organisms *Chromobacterium violaceum* and *Vogesella indigofera.*

TAXONOMIC COMMENTS

In this edition of the *Manual,* the genus *Microvirgula* is placed in the class *Betaproteobacteria,* the order *Neisseriales,* and the family *Neisseriaceae.* Based on studies of the 16S rDNA sequence by Patureau et al. (1998), the genus is most closely related to *Vogesella indigofera* (87.3%) and *Chromobacterium violaceum* (87.9%) (Fig. BXII.β.72).

List of species of the genus Microvirgula

1. **Microvirgula aerodenitrificans** Patureau, Godon, Dabert, Bouchez, Bernet, Delgenes and Moletta 1998, 781[VP]

ae.ro.de.ni.tri'fi.cans. Gr. n. *aer* air; M.L. v. *denitrificare* to denitrify; M.L. part. *aerodenitrificans* denitrifying with or in air.

The characteristics are as described for the genus, with the following additional information. Oxidizable carbon and energy sources include methylpyruvate, acetate, formate, β-hydroxybutyrate, propionate, sebacic acid, succinate, L-asparagine, L-aspartic acid, L-glutamic acid, L-leucine, L-proline, methanol, ethanol, and glycerol.

The following compounds are not utilized: dextrin, glycogen, xylitol, citrate, α-ketoglutarate, α-ketovalerate, malonate, D-alanine, L-phenylalanine, D-serine, L-threonine, D,L-carnitine, inosine, uridine, 2,3-butanediol, and glucose-1-phosphate, D-galacturonic acid, and D-gluconic acid.

Optimal pH, 7. Optimal temperature, 35°C; range, 15–45°C. Maximum growth rates (μmax) are 0.37 h^{-1} under aerobic conditions, 0.23 h^{-1} under mixed oxic–nitrate conditions, and 0.11 h^{-1} under strict anaerobic conditions.

The type strain can co-respire oxygen and nitrogen oxides under O_2-saturated conditions with activity and synthesis of the four denitrifying enzymes, regardless of the aeration conditions. A similar behavior is observed with *Paracoccus pantotrophus, Alcaligenes faecalis* strain TUD, and *Thauera mechernichensis.*

Lipase activity occurs.

The mol% G + C of the DNA is: 65.

Type strain: SGLY2, LMG 18919.

GenBank accession number (16S rRNA): U89333.

TABLE BXII.β.95. Characteristics differentiating the genus *Microvirgula* from other bacteria[a,b]

Characteristic	*Microvirgula*	*Comamonas testosteroni*	*Alcaligenes faecalis*	*Chromobacterium violaceum*	*Vogesella indigofera*	*Thauera mechernichensis*
Nitrate reduced to nitrite	+	+	−	+	+	+
Nitrite reduced to N_2	+	+	+	+	+	+
Capable of aerobic denitrification	+	−	+	nd	nd	+
Colonies are blue or violet	−	−	−	+	+	−
Growth at 45°C	+	−	nd	−	nd	nd
Oxidative metabolism only; cannot ferment	+	+	+	−	+	+
Substrates utilized:						
Methanol	+	nd	nd	nd	nd	−
Ethanol	+	−	nd	nd	+	+
Glucose	−	−	−	+	−	nd
Citrate	−	+	+	+	+	nd
Acetate	+	−	+	d	nd	+

[a]Symbols: see standard definitions; nd, not determined.

[b]Data from Grimes et al. (1997), Patureau et al. (1998), and Scholten et al. (1999).

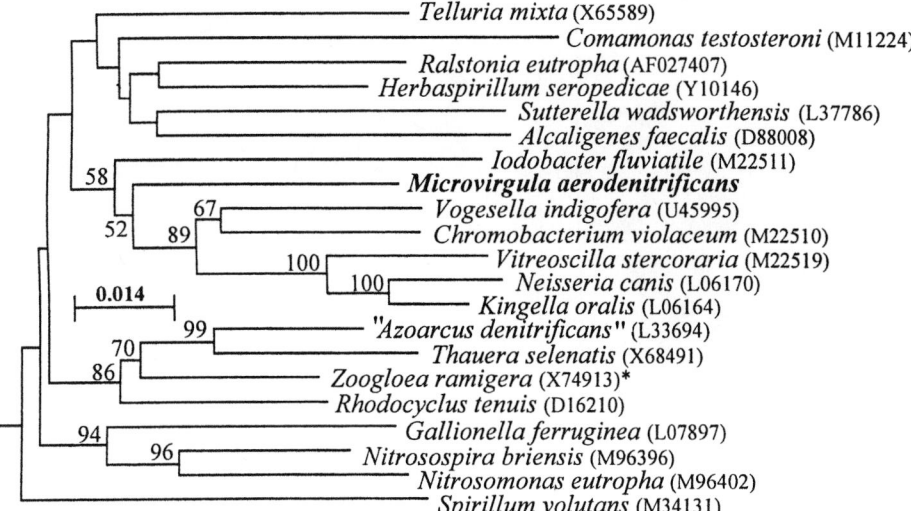

FIGURE BXII.β.72. Phylogenetic tree showing the relationship of *Microvirgula aerodenitrificans* to other members of the class *Betaproteobacteria*. This tree was inferred from 16S rRNA gene sequence data by the neighbor-joining method (Saitou and Nei, 1987). A total of 1120 nucleotide positions were included in the analysis. The tree was rooted with the sequence of *Paracoccus denitrificans*. The scale bar indicates the percentage difference per nucleotide position using the Jukes and Cantor correction (Jukes and Cantor, 1969). Numbers refer to bootstrap values up to 50% for each node out of a total of 500 replicate samplings (Felsenstein, 1985). *, accession number corresponds to strain ATCC 19544. Names are given as cited in the GenBank database. (Reproduced with permission from D. Patureau et al., International Journal of Systematic Bacteriology *48:* 775–782, 1998, ©International Union of Microbiological Societies.)

Genus X. **Prolinoborus** *Pot, Willems, Gillis and De Ley 1992b, 52*[VP]

NOEL R. KRIEG

Pro.li.no′bo.rus. L. n. *prolina* the amino acid proline; Gr. adj. *boros* voracious M.L. masc. n. *Prolinoborus* (bacteria) that readily consume proline.

Straight rods, 0.7–0.9 μm in width.. Cells are approximately 5–10 μm in length, but old cultures may contain cells up to 42 μm long. Curved or S-shaped variants have been reported to occur in one strain after prolonged serial transfer. A polar membrane underlies the cytoplasmic membrane. **Intracellular poly-β-hydroxybutyrate granules are present** in the rods. Extensive conversion of the rod-shaped cells to round forms (**"coccoid bodies"**) occurs in older cultures. **Upon initial isolation, the cells form highly viscous flocs, within which the cells swim steadily in straight lines.** The floc-forming ability is gradually lost during subsequent serial transfers, and the growth eventually becomes turbid. Cells have **bipolar flagellar fascicles composed of up to 11 flagella.** The fascicles can be seen clearly by dark-field microscopy and show **unusual and distinctive behavior** when the cells are suspended in ordinary, nonviscous media: helical wave propagation with waves progressing from base to tip, an ability to coil up like springs, and basal bending accompanied by a change in wavelength. **In such nonviscous media the cells do not swim; instead, they exhibit an ineffectual "floundering about" movement;** however, when they are suspended in a medium of high viscosity (10–200 centipoise, obtained by the use of agents such as DNA or methylcellulose "400 centipoise"), they swim steadily in straight lines. Optimal temperature, 30°C; no growth at 20°C or 40°C. **Catalase and oxidase positive. Carbohydrates are not catabolized.** Pyruvate and proline are the most effective sole carbon sources. **Proline can be used a sole source of nitrogen**

and carbon. Nitrogenase activity (ability to reduce acetylene) **occurs under microaerobic conditions.** Habitat: pond water.

The mol% G + C of the DNA is: 62–65.

Type species: **Prolinoborus fasciculus** (Strength, Isani, Linn, Williams, Vandermolen, Laughon and Krieg 1976) Pot, Willems, Gillis and De Ley 1992b, 53 (*Aquaspirillum fasciculus* Strength, Isani, Linn, Williams, Vandermolen, Laughon and Krieg 1976, 266.)

FURTHER DESCRIPTIVE INFORMATION

The cell morphology and behavior of the flagellar fascicles is illustrated in Fig. BXII.β.73 and BXII.β.74.

Upon initial isolation, the cells form highly viscous flocs which, when homogenized in a small quantity of water, contain free-swimming cells that move in straight lines within the viscous matrix. The tailing flagellar fascicle is extended behind each cell, while the leading fascicle is either coiled into a polar loop or coiled around the cell. Free cells or cells at the periphery of the flocs are not motile, or their motility is irregular over short distances. After several serial transfers, the floc-forming ability is gradually lost, and the cells no longer swim, although the flagellar fascicles exhibit helical wave propagation, basal bending, and the ability to coil up like springs. The ability to swim only in viscous media may represent an adaptation to the viscous conditions that occur within cell flocs.

The mass conversion of the rod form to the coccoid form is

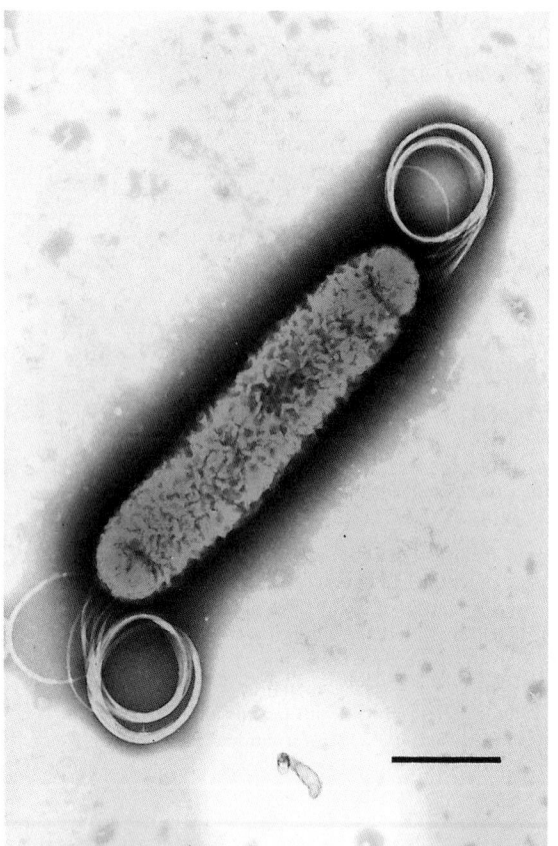

FIGURE BXII.β.73. Formalin-fixed cells of *Prolinoborus fasciculus* showing the bipolar fascicles of flagella coiled into loops. When the fascicles are extended, they have a helical configuration with several waves. Bar = 1.0 μm. (Reproduced with permission from W.J. Strength and N.R. Krieg, Canadian Journal of Microbiology *17*: 1133–1137, 1971, ©National Research Council of Canada.)

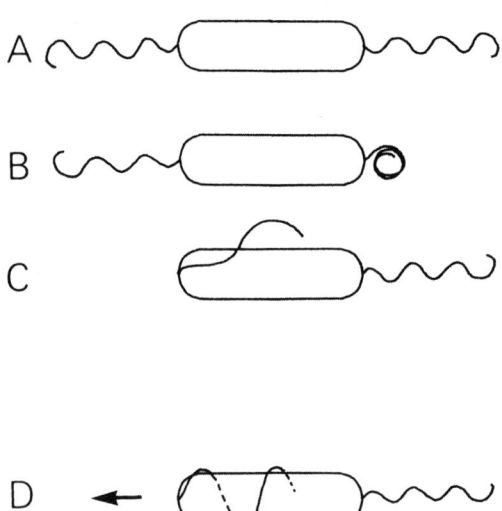

FIGURE BXII.β.74. Flagellar behavior in *P. fasciculus*. *A–C*, flagellar orientation observed in nonviscous media: *A*, flagellar fascicles extended; *B*, coiling into a polar loop; and *C*, basal bending accompanied by a change in wavelength. *D*, orientation of the fascicles in motile cells suspended in a viscous medium; the arrow indicates the direction of swimming.

shown in Fig. BXII.β.75. Koechlein and Krieg (1998) have found that chloramphenicol does not prevent the conversion of the rods to the coccoid form. Attempts to obtain variants that do not convert to the coccoid form have been unsuccessful. Although the coccoid form fluoresces with acridine orange, extensive rRNA degradation has occurred, as indicated by agarose gel electrophoresis. Poly-β-hydroxybutyrate, abundant in the vegetative rods, is not detectable in the coccoid cells. The results suggest that the coccoid form of *P. fasciculus* may be a degenerative form, rather than part of a life cycle.

ENRICHMENT AND ISOLATION PROCEDURES

In studies by Strength et al. (1976), organisms with the characteristic morphological features were found to reach maximal numbers in the surface scum of pond water–hay infusions in 3 d at 30°C. Some enrichment can be achieved by subsequent cultivation of the scum in Pringsheim soil medium[1]. A loopful of

surface pellicle from a 4-d-old enrichment culture is transferred to 10 ml of sterile water. After vigorous shaking to disperse the cell flocs, the suspension is serially diluted in sterile water, and each dilution is used to seed plates of melted, cooled PR medium[2] containing 0.75% agar. After 36 h at 30°C, numerous small (0.1–0.4 mm diameter), white, irregularly shaped colonies (as well as other types of colonies) develop. Each small colony is removed with sterile capillary tubes under a dissecting microscope and the agar plug containing the isolated colony is blown out into a tube containing semisolid (0.15% agar) PR medium. The agar plugs are crushed with a glass rod after the transfer.

After growth occurs, the organisms can be transferred to PFS[3] broth. On initial isolation, the organism forms highly viscous flocs. This floc-forming ability is gradually lost during subsequent transfers, and eventually the strains exhibit homogeneous, turbid growth.

MAINTENANCE PROCEDURES

Strains can be maintained by serial transfer at 48-h intervals in semisolid PFS medium. Incubation beyond 48 h usually results in lack of viability, probably because of extensive formation of the coccoid form. For long-term preservation, centrifuged cells are suspended to high density in PFS broth containing 15% glycerol and stored in liquid nitrogen.

---

1. Pringsheim soil medium (Rittenberg and Rittenberg, 1962): place one wheat or barley grain in a large test tube, and cover it with 3–4 cm of garden soil. Fill the tube almost to the top with tap water. Sterilize the medium at 121°C for 30 min.

---

2. Proline-salts (PR) medium of Strength et al. (1976) (g/l): L-proline, 0.5; K_2HPO_4, 0.45; $MgSO_4 \cdot 7H_2O$, 0.25; $MnSO_4 \cdot H_2O$, 0.001; and $FeCl_3 \cdot 6H_2O$, 0.001. Adjust to pH 7.0 with KOH.

3. Peptone-fumarate-salts (PFS) broth of Strength et al. (1976) (g/l): Bacto peptone (Difco), 10.0; fumaric acid, 2.0; $(NH_4)_2SO_4$, 1.0; $MgSO_4 \cdot 7H_2O$, 0.5; $FeCl_3 \cdot 6H_2O$, 0.002; $MnSO_4 \cdot H_2O$, 0.002; Adjust to pH 7.0 with KOH. The medium can be used in liquid, semisolid (0.15% agar), or solid (1.5% agar) form.

PROCEDURES FOR TESTING SPECIAL CHARACTERS

Cultures to be tested for nitrogenase activity should be cultured in nitrogen-free, semisolid malate medium (see the genus *Azospirillum*) supplemented with 0.005% yeast extract. Cultures are incubated for 3 d at 30°C and then sealed with rubber vaccine-bottle stoppers. Acetylene is injected to a final concentration of 10% (v/v), and the cultures are tested for ethylene production by gas chromatography after 1 h of further incubation. Controls using liquid rather than semisolid medium and semisolid medium containing 0.1% $(NH_4)_2SO_4$ should be negative for ethylene production.

DIFFERENTIATION OF THE GENUS *PROLINOBORUS* FROM OTHER GENERA

In most of its physiological properties, *Prolinoborus* is similar to aerobic, heterotrophic, freshwater spirilla such as *Aquaspirillum serpens* (its closest relative). However, the rod shape of the cells, floc formation, and especially the unusual behavior of the flagellar fascicles clearly differentiate *Prolinoborus* from these spirilla. The unusual flagellar behavior, formation of viscous flocs, ability to use proline as a sole carbon and nitrogen source, inability to use carbohydrates, and nitrogenase activity readily differentiate *Prolinoborus* from other aerobic Gram-negative freshwater rods.

TAXONOMIC COMMENTS

Prolinoborus fasciculus was initially classified in the genus *Aquaspirillum* by Strength et al. (1976). However, DNA–rRNA hybridization studies by Pot et al. (1992b) indicated that the genus *Prolinoborus* belongs to rRNA superfamily III (now the class *Betaproteobacteria*), with a difference in $T_{m(e)}$ of more than 6°C with its closest relative, *Aquaspirillum serpens*. Pot et al. regarded this $T_{m(e)}$ difference as support for the classification of *A. fasciculus* into a new genus, *Prolinoborus*. *Prolinoborus fasciculus* is currently the only species in this genus.

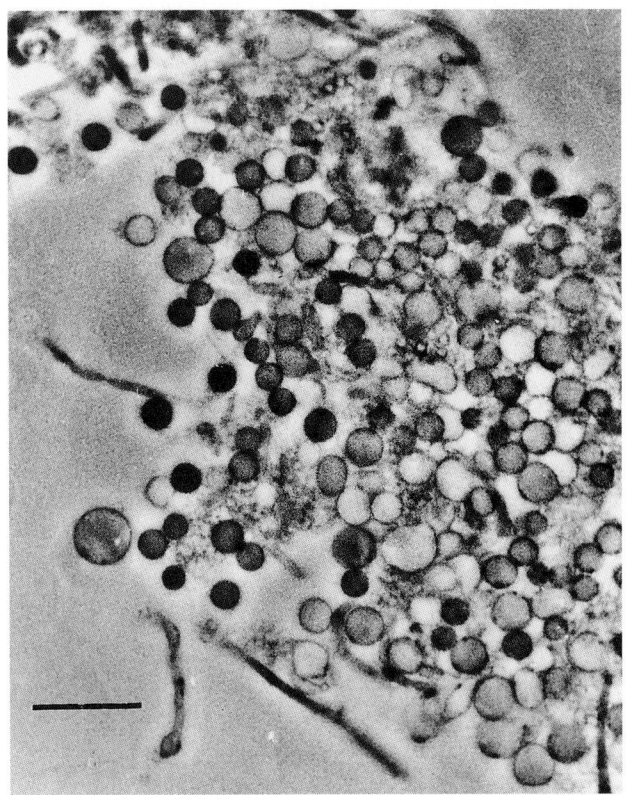

FIGURE BXII.β.75. Coccoid bodies of *P. fasciculus* in peptone–fumarate–salts broth cultures incubated on a shaking machine at 30°C for 48 h. Phase contrast microscopy. Bar = 10 μm.

List of species of the genus Prolinoborus

1. **Prolinoborus fasciculus** (Strength, Isani, Linn, Williams, Vandermolen, Laughon and Krieg 1976) Pot, Willems, Gillis and De Ley 1992b, 53^VP (*Aquaspirillum fasciculus* Strength, Isani, Linn, Williams, Vandermolen, Laughon and Krieg 1976, 266.)

   fas.ci' cu.lus. L. masc. dim. n. *fasciculus* a small bundle.

   The characteristics are as described for the genus, with the following additional information. No water-soluble brown pigment is formed in the presence of 0.1% tyrosine or tryptophan. Growth occurs in the presence of 1% oxgall. No growth occurs in the presence of 1% glycine and 3% NaCl. No hydrolysis of casein, starch, esculin, or hippurate. Gelatin is hydrolyzed in 4 d at 30°C. Anaerobic growth occurs with nitrate. KNO_3 is reduced only to KNO_2. Phosphatase and urease positive. Indole negative.

   Sole carbon sources, as determined by the method of Strength et al. (1976) include succinate, fumarate, malate, oxaloacetate, pyruvate, lactate, β-hydroxybutyrate, L-alanine, L-glutamate, L-aspartate, L-glutamine, L-asparagine, L-proline, and L-arginine. The following carbon sources are not used: citrate, α-ketoglutarate, malonate, acetate, propionate, butyrate, ethanol, n-propanol, n-butanol, glycerol, D-fructose, D-glucose, D-xylose, L-arabinose, L-histidine, L-tyrosine, L-phenylalanine, L-hydroxyproline, L-ornithine, L-lysine, L-methionine, L-serine, L-cysteine, L-leucine, L-isoleucine, L-valine, and L-tryptophan.

   The mol% G + C of the DNA is: 62–65 (T_m).

   Type strain: ATCC 27740, LMG 6233.

Genus XI. **Simonsiella** Schmid in Simons 1922, 504[AL]

BRIAN P. HEDLUND AND TONE TØNJUM

Si.mon.si.el' la. M.L. dim. *-ella* ending; M.L. fem. n. *Simonsiella* (organism of) Simons; named for H. Simons, who studied the species of this genus.

Organisms that exist in characteristic **multicellular filaments** that are **flat rather than cylindrical** and often segmented into groups of eight cells. The **width of an individual cell is greater than its length. The long axis of an individual cell is perpendicular to the long axis of the filament.** The diameter of the filaments (the width of the individual cells) may vary from about 2.0 to 8.0 μm, and the length of filaments may vary from about 10.0 to over 50.0 μm. Individual cells within the filaments may be from about 0.5 to 1.3 μm long. In thin sections cut perpendicular to the long axis of the filament, the **cells are flattened and curved to yield a crescent-shaped, convex–concave (dorsal–ventral) asymmetry. The ends of the individual filaments are rounded.** Gram negative. **Gliding motility** of the entire filament in the direction of the long axis when the flat side of the filament is in contact with a surface. Chemoorganotrophs. Aerobic. Some may produce acid aerobically from carbohydrates. Optimal temperature: 37°C. **Found in the oral cavity of warm-blooded vertebrates.**

The mol% G + C of the DNA is: 41–55.

Type species: **Simonsiella muelleri** Schmid *in* Simons 1922, 504.

FURTHER DESCRIPTIVE INFORMATION

The genus *Simonsiella* is characterized by a unique multicellular morphology. The filaments of *Simonsiella* are distinctive and members of this genus can be recognized by their morphology alone (Figs. BXII.β.76, BXII.β.77, and BXII.β.78). The dorsal-ventral flattening of the filaments is quite striking, as is the fact that the individual cells of the filament are wider than they are long if the long axis of the filament is considered to represent the length of the cells. The cells toward either end of the filament decrease in width, and the terminal cells may be rounded, giving the filaments a tapered appearance with rounded ends. In isolated colonies, some of the filaments may be turned on their sides, which shows the flattening quite clearly. Starr and Skerman (1965) suggested that this type of structure should be called a

trichome—chain of closely apposed bacterial cells—found, for example, in certain genera of the cyanobacteria, the genus *Toxothrix* and the genus *Caryophanon*. From the convex (bottom) side of the filaments there are numerous fine fibrils (Fig. BXII.β.78), which appear to be involved in the adhesion of the filaments to a surface and perhaps in locomotion (Pangborn et al., 1977). In thin sections, the cells appear to have a typical Gram-negative cell wall structure.

Individual filaments glide over the agar surface only when the broad, ventral surface is in contact with the agar. Movement of the filaments over the agar leaves depressed tracks in the agar surface (Fig. BXII.β.77), perhaps indicating a change in the structure of the agar (Pangborn et al., 1977). No flagella have been demonstrated, but the presence of pili (fimbriae) should be assessed. The speed of gliding varies from about 5 to 24 μm/min (Buchanan and Kuhn, 1978). Colonies on BSTSY agar (see Enrichment and Isolation Procedures) may have a pale yellow pigmentation. Most, but not all, strains produce a zone of hemolysis on agar containing horse, sheep, or rabbit blood. No resting stage has been detected.

All isolates of *Simonsiella* are chemoorganotrophic, and the nutrition of the isolates (as well as their classification) parallels their source of origin. The isolates from sheep (*S. crassa*) are generally the most physiologically active, being both proteolytic and saccharolytic (Tables BXII.β.96, BXII.β.97, BXII.β.98, BXII.β.99, and BXII.β.100). In contrast, the isolates from dogs (*S. steedae*) are neither proteolytic nor saccharolytic. Kuhn et al. (1978) have described over 50 isolates of *Simonsiella*. All strains are aerobic, possess cytochrome oxidase, and produce catalase. Good growth occurs between 3° and 40°C, is optimal at 37°C, and does not occur at 45°C. Growth occurs in 1% NaCl but not

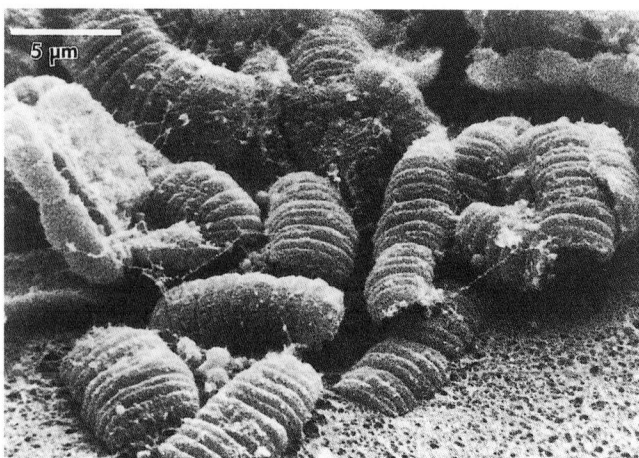

FIGURE BXII.β.76. Scanning electron micrograph of the edge of a colony of *Simonsiella* obtained from a cat. (Micrograph taken by J. Pangborn.) (Reproduced with permission from J. Pangborn et al., Archives of Microbiology *113:* 197–204, 1977, ©Springer-Verlag.)

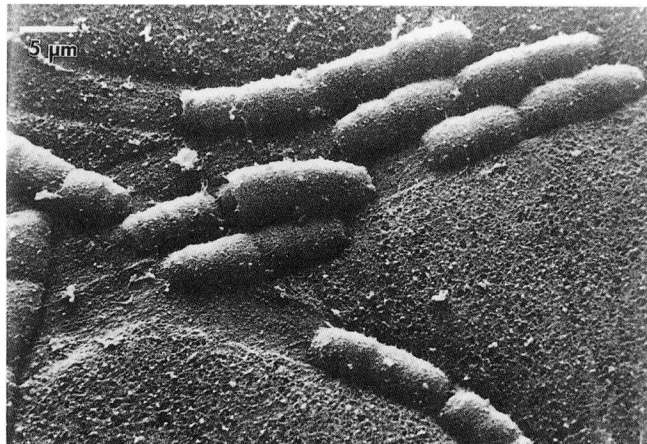

FIGURE BXII.β.77. Scanning electron micrograph of a culture of *S. steedae* ATCC 27411 growing on BSTSY agar (see Enrichment and Isolation Procedures). The depressed tracks where the organism has glided over the surface of the agar are clearly shown. The dorsal surface of the filaments are covered with a capsular material that obscures the individual cells. (Micrograph taken by J. Pangborn.) (Reproduced with permission from J. Pangborn et al., Archives of Microbiology *113:* 197–204, 1977, ©Springer-Verlag.)

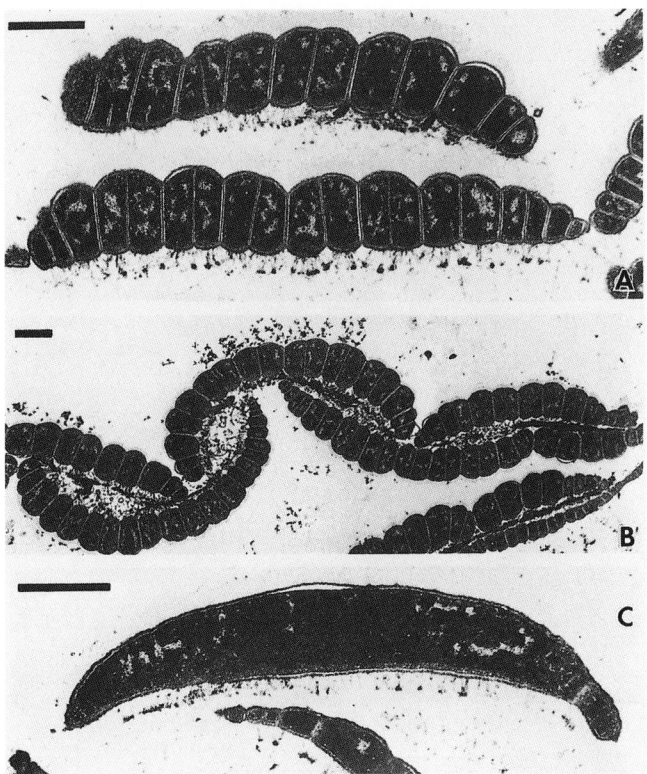

FIGURE BXII.β.78. Transmission electron micrograph of thin sections of *S. steedae*. The multicellular nature of the filaments and the dorsal-ventral differentiation are shown. (*A*) Strain ATCC 27411 from an agar surface. (*B*) Strain ATCC 27396 showing a curvature of the filaments that often causes them to stand on their edge in a colony. (*C*) Section of strain ATCC 27411 cut perpendicular to the long axis of the filament. The dorsal-ventral, convex–concave curvature that results in a crescent-shaped transverse section is shown. Notice the fine fibrillar structures on the ventral surface of all of the filaments. Bars = 1 μm. (Micrograph taken by J. Pangborn.) (Reproduced with permission from J. Pangborn et al., Archives of Microbiology *113*: 197–204, 1977, ©Springer-Verlag.)

TABLE BXII.β.96. Characteristics differentiating the species of the genus *Simonsiella*[a]

Characteristic	*S. muelleri*	*S. crassa*	*S. steedae*
Acid from:			
Glucose	+	+	−
Maltose	+	+	−
Trehalose	−	+	−
Ribose	−	+	−
Fructose	−	+	−
Sucrose	−	+	−
Mannitol	−	+	−
Growth at:			
27°C	+	+	−
43°C	−	+	−
pH 6.0	+	+	−
pH 8.0	−	+	−
Source	Humans	Sheep	Dogs

[a]Symbols: +, 90% or more of strains are positive; −, 90% or more of strains are negative

in 2% NaCl. No strains hydrolyze starch or agar or produce urease or indole. No strains produce acid from cellobiose, dulcitol, erythritol, galactose, glycerol, inositol, lactose, mannose, melibiose, melizitose, raffinose, rhamnose, salicin, sorbitol, sorbose, or xylose. Some species may produce acid from other carbohydrates. A rich medium is best for growth, and the addition of serum to the medium has been found to be necessary for some strains and to be advantageous for all other strains (Kuhn et al., 1978).

The cell-bound fatty acid profiles of 48 strains of *Simonsiella* matched the general pattern of Gram-negative bacteria, in which high percentages of even-numbered saturated and monounsaturated fatty acids occur (Jenkins et al., 1977). Tetradecanoic acid was the predominant saturated fatty acid followed by hexadecanoic acid. 9-Hexadecanoic and 9-octadecanoic acids were the predominant monounsaturated fatty acids. Results of stepwise discriminant analysis of the mean relative percentages of tetradecanoic, hexadecanoic, and 9-octadecanoic acids demonstrated that 85% of the isolates (with two cat strains and one dog strain being the exceptions) were correctly identified in their source-of-origin groups.

Habitat In healthy human populations, the incidence of *Simonsiella* is in the range of 30–40% (Simons, 1922; Kuhn et al.,

1974). Children possibly have a higher incidence than adults (Fellinger, 1924; Richardson et al., 1966). In dogs and cats *Simonsiella* are common and abundant, the incidence approaching 100% in specimens from the palate or from the buccal cavity (Nyby et al., 1977). In specimens from oral cavities and gingival margins, simonsiellas are less likely to be found (~20% incidence; Saphir and Carter, 1976; Bailie et al., 1978). The incidence varies not only with the site in the mouth from which the specimens are obtained, but also with the method of observation. Direct microscopy yields a higher incidence than cultivation, and microscopic observation of oral specimens growing on plates gives a higher yield than the detection of macroscopic colonies after prolonged incubation (Richardson et al., 1966; Saphir and Carter, 1976; Bailie et al., 1978). Even though these bacteria have been considered as members of the normal flora (Kuhn et al., 1974), they are often underrecognized or not even mentioned as part of the oral microflora. Pathological changes of disease have not been associated with the presence of *Simonsiella* in the mouth. In fact, the incidence of *Simonsiella* in human patients with obvious oral pathology has been considerably lower than in groups of healthy people (Simons, 1922; Bruckner and Fahey, 1969). Despite their low pathogenicity, *Simonsiella* spp. have been isolated from erosive lesions of the human oral cavity (Carandina et al., 1984) and from the gastric aspirate of a neonate (Whitehouse et al., 1987).

Phylogeny and classification Steed designated the family *Simonsiellaceae* to include *Simonsiella* and the morphologically similar genus *Alysiella* (Steed, 1962). Rossau and co-workers demonstrated close relationships between *Simonsiella* strains and strains of *Neisseria*, *Kingella*, *Eikenella*, and *Alysiella* by DNA–rRNA hybridization and suggested that *Simonsiella* should be included in the emended family *Neisseriaceae* (Rossau et al., 1989). Dewhirst et al. (1989) found that the type strain of *S. muelleri* by 16S rRNA sequence analysis clustered within the *Neisseriaceae* in the *Betaproteobacteria*. Appropriately, these authors also supported the emended family *Neisseriaceae* to include genus *Simonsiella*.

Kuhn et al. (1977, 1978) have isolated nearly 50 strains of *Simonsiella* from dogs, cats, sheep, and humans. Based on morphology, physiology, and the mol% G + C of the DNA, the strains can be separated into distinct groups that correlated with their source of origin. Three of these groups (dogs, sheep, and humans) have been designated as separate species, while the isolates from cats and other vertebrates need additional characterization

TABLE BXII.β.97. Selected phenotypic features of *Simonsiella* and *Alysiella* species[a,b]

Characteristic	*A. filiformis* ATCC 15532[T]	*S. crassa* ATCC 15533[T]	*S. muelleri* ATCC 29453[T]	*S. steedae* ATCC 27409[T]
Number of strains investigated	1	2	2	3
Corroding colonies	−	−	−	−
β-Hemolysis	−	+	W	+
Fructose	−	+	+	W
Galactose	−	−	−	−
Glucose	+	+	+	−
Maltose	−	−	D	−
Malonate	+	+	+	−
Mannose	−	−	−	−
Sucrose	+	+	−	−
Trehalose	−	+	−	−
OF test (glucose)	O	O	O	O,W
H_2S production	−	−	−	−
Catalase	+	+	+	+
Nitrate reduction	D	+	+	+
Nitrite reduction	−	+	+	−
Phosphatase (alkaline)	−	−	−	−
Gelatine liquefaction	+	+	−	−
Indole production	−	−	−	−
Arginine dihydrolase	−	−	−	−
Lysine decarboxylase	−	−	−	−
Ornithine decarboxylase	−	−	−	−
Glutamyltransferase	−	−	−	−

[a]Symbols: +, positive result; −, negative result; W, weak result; D, different results with different strains; O, oxidative reaction; F, fermentative reaction.

[b]All strains were positive for TMPD-oxidase. All strains were negative for arabinose, dextrose, esculin, lactose, raffinose, rhamnose, salicin, sorbose, starch hydrolysis, xylose, citrate, urease, ONPG test, adonitol, dulcitrol, inositol, and mannitol.

TABLE BXII.β.98. Semiquantitative determination of enzymatic activities in *Simonsiella* and *Alysiella* species using the API-ZYM system (selected from Heiske and Mutters, 1994)[a,b]

Characteristic	*A. filiformis* ATCC 15532[T]	*S. crassa* ATCC 15533[T]	*S. muelleri* ATCC 29453[T]	*S. steedae* ATCC 27409[T]	*Simonsiella* sp. 1 (ATCC 29466)	*Simonsiella* sp. 2 ("*S. kuhniae*" ATCC 27381)	*Simonsiella* sp. 3 (HIM 942-7)
Number of strains investigated	1	2	2	3	1	1	1
Phosphatase, alkaline	1	1	1	1	1	1	1
Esterase (C_4)	3	3	3	3	2	3	3
Esterase lipase (C_8)	2	2	2	2	2	2	2
Leucine arylamidase	2	2	2	3	2	5	2
Valine arylamidase	1	1	1	1	1	2	1
Cysteine arylamidase	1	1	1	1	1	1	1
Phosphatase (acidic)	1	1	1	1	1	1	1

[a]All strains were positive: naphthol-AS-BI-phosphohydrolase. Tests negative for all strains: lipase (C14), trypsin, α-chymotrypsin, α-galactosidase, β-galactosidase, β-glucuronidase, N-acetyl-β-glucosaminidase, α-mannosidase, α-fucosidase.

[b]0, negative reaction; 1, very weak reaction; 2, weak reaction; 3, strong reaction; 5, very strong reaction.

before any decisions about species designations can be made (Kuhn et al., 1978; Kuhn and Gregory, 1978). Numerical taxonomy studies showed that most of the *Simonsiella* strains grouped according to the mammalian host from which they were isolated (Kuhn et al., 1978; Heiske and Mutters, 1994) (Tables BXII.β.96, BXII.β.97, BXII.β.98, BXII.β.99, and BXII.β.100). *Simonsiella* isolates from humans and sheep were each monophyletic in a dendrogram derived from the numerical taxonomy data. Isolates from cats and dogs tended to cluster together with strains from the same host; however, neither group was strictly monophyletic based on the phenotypic data. These observations led to the proposal that each of these mammals hosts a unique type of *Simonsiella* (Kuhn et al., 1978). Kuhn suggested that these host groups represented ecospecies: species of bacteria that each occupied a niche in a unique ecosystem, in this case the mouths of different animals. Accordingly, three of the *Simonsiella* groups were assigned to separate species. *S. muelleri*, *S. crassa*, and *S. steedae* were proposed for *Simonsiella* strains native to humans,

sheep, and dogs, respectively. Hedlund and Staley assessed the relationships between *Simonsiella* strains and other members of the *Neisseriaceae* further by 16S rDNA sequence analysis of 16 *Simonsiella* strains (Hedlund and Staley, 2002) (Fig. BXII.β.79). Most *Simonsiella* strains grouped according to established *Simonsiella* species designations and the mammalian hosts from which they were isolated. The host groups corresponded to the three existing *Simonsiella* species— *S. muelleri*, *S. steedae*, and *S. crassa* —which are commensals of humans, dogs, and sheep, respectively. The fourth group consisted of *Simonsiella* isolates from domestic cats, and the phenotypic results previously obtained by Kuhn were common to the *Simonsiella* cat isolates (Kuhn et al., 1978). Although this entity was phylogenetically more diverse than the other groups and phenotypically similar to *S. steedae*, the distinct phylogeny and ecological habitat of this group supported the establishment of a new species. Thus, the phylogeny together with phenotypic data supported the suggestion of "*S. kuhniae*" to encompass *Simonsiella* strains isolated from domestic

TABLE BXII.β.99. Results of the API 20NE test and selected reaction of the API rapid ID 32A test activities in *Simonsiella* and *Alysiella* species (from Heiske and Mutters, 1994)[a,b]

Characteristic	*Alysiella filiformis* ATCC 15532[T]	*S. crassa* ATCC 15533[T]	*S. muelleri* ATCC 29453[T]	*S. steedae* ATCC 27409[T]
Number of strains	1	2	2	3
Nitrate reduction	W	+	+	+
Indole production	−	−	−	−
Glucose acidification	+	+	+	+
Arginine dihydrolase	−	−	−	−
Gelatin hydrolysis	−	+	−	−
Assimilation of:				
Glucose	+	+	+	−
Mannose	−	−	−	−
N-Acetylglucosamine	−	−	−	−
Maltose	−	−	−	−
Gluconate	−	−	−	−
Malate	−	−	+	−
Arginine arylamidase	+	+	+	+
Proline arylamidase	−	−	−	−
Leucylglycine arylamidase	+	+	+	+
Phenylglycine arylamidase	−	+	−	−
Leucine arylamidase	+	+	+	+
Tyrosine arylamidase	+	+	W	W
Alanine arylamidase	+	+	+	+
Glycine arylamidase	+	+	+	+
Glutamic acid decarboxylase	+	−	+	−
Glutamylglutamate arylamidase	+	−	+	−
Serine arylamidase	+	+	+	+

[a]Symbols: +, positive result; −, negative result; W, weak result.

[b]Test negative with all strains: assimilation of adipate, arabinose, caprate, citrate, mannitol, phenyl acetate; hydrolysis of esculin; indole production; urease.

cats, as proposed by Hedlund and Staley (2002). Hedlund and Staley (2002) have provided evidence that the different *Simonsiella* strains may have co-evolved with their respective hosts. The bacteria seem to be evolving more rapidly than their hosts, most likely because they are growing more quickly (Interview with J.T. Staley in Stencel, C. Microbial diversity: Eyeing the big picture. ASM News 66:146, 2000). These considerations indicate that one needs to look beyond 16S rDNA for phylogenetic analysis of these species.

Members of the genus *Simonsiella* possess a morphology that is striking and unique among bacteria. For this reason it could be expected that they would make up a coherent phylogenetic group that excluded other bacteria. However, the *Simonsiella* are polyphyletic in 16S rDNA based phylogenetic analysis (Fig. BXII.β.61 in Chapter *Neisseriales*; Hedlund and Staley, 2002). Such instabilities may be attributable to horizontal gene transfer between members of the *Neisseriaceae*, including *Simonsiella*. Most members of the family *Neisseriaceae* are naturally competent for transformation (Bøvre, 1980). The discrepancies between 16S rDNA phylogenies of the genus *Neisseria* and those derived from analyses of other loci (Smith et al., 1999b) or from chemotaxonomic data (Barrett and Sneath, 1994) have suggested that the irregularities in the phylogenetic trees were due to interspecies gene exchange and that certain *Neisseria* 16S rDNA sequences are hybrids (Zhou et al., 1997; also see chapter on Genus *Neisseria*). Thus, members of the genus *Neisseria* could have acquired and recombined with *Simonsiella* 16S rDNA sequences. It is not known whether *Simonsiella* strains themselves are competent; if they are, the horizontal gene transfer from *Neisseria* or other oral flora to *Simonsiella* could have added to the phylogenetic complexity.

ENRICHMENT AND ISOLATION PROCEDURES

At the present time, there are no enrichment procedures for *Simonsiella*. Isolation directly from the oral cavity has been achieved by several people, however. The easiest method for isolation of *Simonsiella* was described by Kuhn et al. (1978). A sterile cotton swab is rubbed over the palate, tongue, or inner surface of the cheeks of the animal and immediately rolled over the surface of a thin layer of BSTSY agar (tryptic soy broth without dextrose (Difco), 27.5 g; yeast extract (Difco), 4 g; agar, 15.0 g; water, 900 ml; and sterile bovine serum, 100 ml, which is added after autoclaving and cooling the other ingredients to 45°C) in a plastic Petri dish. Without delay, the dish is then placed into a 37°C incubator for about 6–10 h. During this short incubation period the filaments glide away from the oral epithelial cells and multiply. The Petri dish is then scanned via a microscope with a magnification of up to ×125 or a dissecting microscope. Isolated filaments or microcolonies are picked from the agar surface and transferred to fresh plates of BSTSY agar. Suitable instruments for transfer could include a dissecting needle, inoculating needle, toothpick, dental probe, light bulb filament, or other finetipped instrument. Macroscopically visible colonies generally appear within 16–24 h after transfer. Although this method was used to isolate nearly 50 strains of *Simonsiella*, some strains did not grow on this medium and could not be isolated.

Steed (1962) isolated *Simonsiella* from sheep with a medium consisting of Oxoid nutrient agar plus 10% horse or ox serum. This medium was not suitable for the isolation of strains from humans (Kuhn et al.,1978). Berger (1963) used blood agar to detect and isolate *Simonsiella*, but this medium suffers from the opaqueness of the blood, which prevents easy observation of the colonies with a microscope.

TABLE BXII.β.100. Cellular carbohydrate patterns in *Simonsiella* and *Alysiella* species (selected from Heiske and Mutters, 1994)

RT (min)	Carbohydrate	*A. filiformis* ATCC 15532[T] 1[a]	*S. crassa* ATCC 15533[T] 2	*S. muelleri* ATCC 29453[T] 2	*S. steedae* ATCC 27409[T] 3	*Simonsiella* sp. 1 (ATCC 29466) 1	*Simonsiella* sp. 2 ("*S. kuhniae*") (ATCC 27381) 1	*Simonsiella* sp. 3 (HIM 942-7) 1
11.69	*Meso*-crythrol	0[b]	3	0	0	0	3	0
12.02	*Meso*-crythrol	4	3	0	0	0	0	0
12.1	Ribose/lyxose (A)[c]	5	5	5	4	5	5	5
12.14	Rhamnose (A)	0	0	0	0	0	0	0
12.5	Fucose (A)	0	0	0	0	0	0	0
13.08	Fucose (O)	0	0	0	0	0	0	0
13.31	Ribose/lyxose (O)	5	5	5	5	4	5	5
13.36	Rhamnose (O)	0	0	0	0	0	0	0
13.64	Ribose/lyxose (O)	7	8	8	8	7	7	7
13.68	Rhamnose (O)	0	0	4	2	2	2	3
13.72	Fucose (O)	0	0	0	0	0	0	0
13.84	Threose	0	0	0	0	0	0	0
13.92	Arabinose (O)	2	1	0	1	2	0	2
15.39	C-6	0	1	2	3	2	3	2
15.46	C-6	2	0	0	3	3	3	2
15.5	C-6	0	2	3	3	2	0	0
15.54	Glucose (A)	3	3	3	4	3	3	3
15.82	Mannose	0	2	0	0	0	0	0
15.86	Galactose (A)	3	3	3	3	2	2	3
14.34	Galactose (O)	3	3	4	3	3	3	3
16.62	Glucose (O)	3	3	3	3	3	2	3
16.75	Mannose (O)	0	2	0	1	1	0	2
16.8	Glucose (O)	5	5	4	5	6	4	5
16.9	Galactose (O)	5	4	6	5	4	4	6
17.04	Sorbose/tagatose	2	0	0	1	2	1	0
17.19	Glucosamine (A)	4	5	5	5	4	4	4
17.3	Inositol	1	0	0	0	1	3	2
17.75	Lyxose (phenyl) (A)	2	2	2	1	0	2	3
18.08	Galactosamine (A)	1	2	3	3	2	0	0
18.18	Heptose (A)	0	1	0	2	0	0	0
18.24	Muramic acid	2	3	3	3	4	4	3
18.4	Heptulose (A)	0	0	0	0	0	0	0
18.42	Glucosamine (O)	3	3	3	3	3	2	2
18.58	Glucosamine (O)	3	5	4	4	5	4	4
18.77	Galactosamine (O)	3	3	4	4	3	2	3
18.93	Heptose (O)	0	1	0	2	0	0	0
19.18	Glucoheprose (O)	1	3	2	3	1	0	0
19.48	Glucoheprose (O)	3	2	2	3	0	2	1
19.57	Muarmic acid	1	2	2	2	2	2	0
20.14	Talose (phenyl) (A)	0	0	1	0	0	0	0

[a]Number of strains tested.

[b]% of total carbohydrate amounts: 0, absent; 1, 0.1–0.5%; 2, 0.51–1.0%; 3, 1.1–3.0%; 4, 3.1–6.0%; 5, 6.1–10.0%; 6, 10.1–20.0%; 7, 20.1–30.0%; 8, 30.1–50.0% (RT value of *n*-ocradecane, 14.11).

[c](A), peracetylated aldononitrile; (O), peracetylated O-methyloxime; C-6, unknown carbohydrate with six C atoms; phenyl, prenylated carbohydrate; no specification of reaction type, only acetylation could be observed (in case of sugar alcohols, amino sugars, and similar).

MAINTENANCE PROCEDURES

Freshly isolated cultures should be grown at 37°C and transferred about every 2–3 d to the same medium on which they were isolated. Older cultures must be transferred at intervals of about 1 week.

Refrigeration of cultures is not recommended for preservation, but they can be preserved by freezing in liquid nitrogen using glycerol as a cryoprotectant or by lyophilization.

PROCEDURES FOR TESTING SPECIAL CHARACTERS

The procedures for characterizing *Simonsiella* were first described by Steed (1962) and later modified by Kuhn et al. (1978). The procedures generally parallel those that are commonly used for other taxa, except that for *Simonsiella* most media contain 10% serum. All incubations are carried out at 37°C. The procedures below are from Kuhn et al. (1978).

Casein hydrolysis BSTSY agar containing 2% skim milk is inoculated, incubated for 3 d, and examined for evidence of hydrolysis.

Starch hydrolysis TSY agar (BSTSY agar without serum) containing 5% soluble starch is inoculated, incubated for 4 d, and then flooded with Lugol's iodine.

Gelatin hydrolysis TSY broth containing 10% serum and 12% gelatin is inoculated and tested for liquefaction after 5 d of incubation by chilling the tube to 4°C.

Peptonization of litmus milk Litmus milk (Difco) containing 10% serum is inoculated and examined for peptonization over a 14-d incubation period.

Action on carbohydrates A medium is poured into Petri dishes, inoculated, and examined after 48 h for the production of acid (yellow color). It consists of the following ingredients:

2%

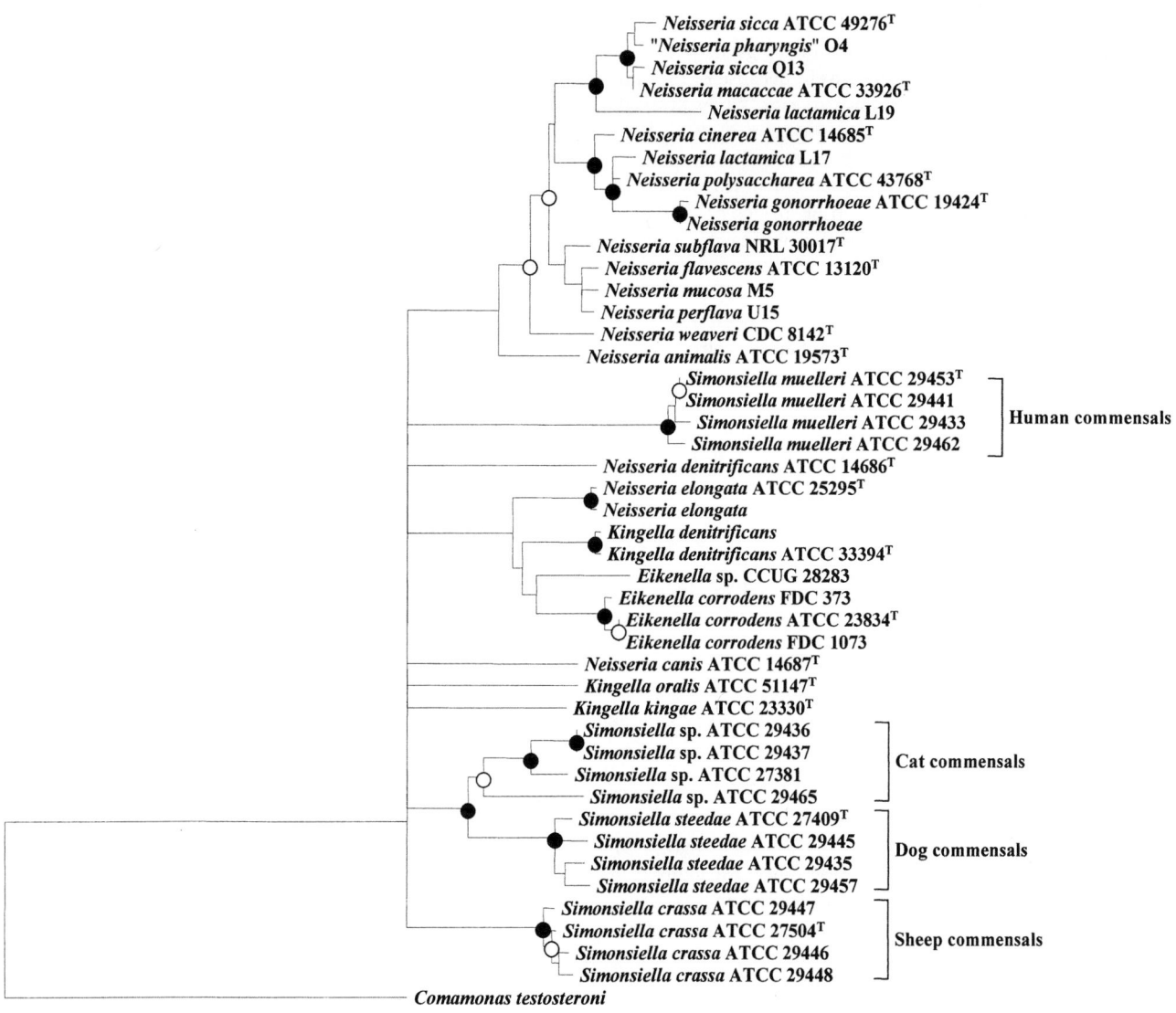

FIGURE BXII.β.79. Phylogenetic neighbor-joining tree of the genus *Simonsiella* of the family *Neisseriaceae* based on nucleotide sequence of the 16S rRNA genes. *Comamonas testosteroni* was used as an outgroup. ●, Nodes with >90% bootstrap support for all analyses; ○, with >75% bootstrap support. Values below 50% are not shown. Bar = 2% nucleotide divergence. (Reproduced with permission from B.P. Hedlund and J.T. Staley, Journal of Systematic and Evolutionary Microbiology *52:* 1377–1382, 2002, ©International Union of Microbiological Societies.)

phenol red broth (Difco), 1.6%; yeast extract (Difco), 0.2%; agar (Difco), 1.5%; phenol red (Hartman Leddon Co., Philadelphia, Pennsylvania), 0.007%; serum, 10%; and carbohydrate, 1%.

Indole production BSTSY broth containing 1% tryptone is inoculated, incubated for 3 d, and examined for evidence of indole production with Kovac's reagent.

Nitrate reduction TSY broth containing 0.2% KNO₃ in a Durham tube assembly is inoculated and examined for 5 d by the dimethyl-α-naphtholamine/sulfanilic acid method (Miller and Neville, 1976).

DIFFERENTIATION OF THE GENUS *SIMONSIELLA* FROM OTHER GENERA

The unusual morphology of the *Simonsiella* filament serves to differentiate the genus from all other procaryotic organisms. *Aly-*

siella resembles *Simonsiella*, but filaments of *Alysiella* do not show the unusual dorsal-ventral differentiation and do not have rounded ends, and the individual cells are paired within the filament.

Caryophanon filaments may be mistaken for *Simonsiella* in stained preparations. However, in live preparations the cylindrical form of *Caryophanon* and the flattened shape of *Simonsiella* are easily distinguished. Also, *Caryophanon* are motile by flagella; they do not glide.

TAXONOMIC COMMENTS

The citation "Schmid *in* Simons" following the generic name in the title is the result of Simons (1922) crediting the name *Simonsiella* to G. Schmid. However, no record can be found that Schmid published the name *Simonsiella*.

When Simons (1922) first described the genus *Simonsiella*

from the human oral cavity, he named two species *S. muelleri* and *S. crassa*, which he differentiated based on the width of the filaments. Later, Steed (1962), examining strains from sheep, and Berger (1963), examining strains from guinea pigs, named their isolates *S. crassa* and *S. muelleri*, respectively, because of the similarity in filament width to those known species. When Kuhn et al. (1978) carried out an extensive investigation of 49 isolates from humans, dogs, cats, and sheep, they determined that each animal had its own species and that filament width was not a suitable indicator for the differentiation of species.

Simonsiella has been reported from the oral cavity of many warm-blooded vertebrates including cats, chickens, cows, dogs, goats, guinea pigs, humans, horses, pigs, rabbits, and sheep. Only those from cats, dogs, humans, and sheep have been described in detail (Steed, 1962; Kuhn et al., 1978). When those from other animals are sufficiently characterized, it may be necessary to revise the number of species within this genus and the means of differentiation among the species. Moreover, some *Simonsiella* have been refractory to isolation, possibly indicating that they were a different strain or species (Kuhn et al., 1978) from those that grew and were isolated on BSTSY agar.

Kuhn et al. (1978) characterized several *Simonsiella* strains from cats. 16S rDNA sequence analysis of *Simonsiella* strains indicated that they are members of the family *Neisseriaceae* (Hedlund and Staley, 2002).

ACKNOWLEDGMENTS

The information provided in the first version of *Bergey's Manual of Systematic Bacteriology* (1984) by J.M. Larkin is greatly acknowledged. The contribution of J.T. Staley to the phylogenetic analysis of *Simonsiella* is deeply appreciated.

FURTHER READING

Heiske, A. and R. Mutters. 1994. Differentiation of selected members of the family *Neisseriaceae* (*Alysiella*, *Eikenella*, *Kingella*, *Simonsiella* and CDC groups EF-4 and M-5) by carbohydrate fingerprints and selected phenotypic features. Zentbl. Bakteriol. *281*: 67–79.

Kuhn, D.A. 1981. The genera *Simonsiella* and *Alysiella*. *In* Starr, Stolp, Trüper, Balows and Schlegel (Editors), The Prokaryotes. A Handbook on Habitats, Isolation and Identification of Bacteria, Springer-Verlag, New York. pp. 390–399.

McCowen, R.P., K.J. Cheng and J.W. Costerton. 1979. Colonization of a portion of the bovine tongue by unusual filamentous bacteria. Appl. Environ. Microbiol. *37*: 1224–1229.

DIFFERENTIATION OF THE SPECIES OF THE GENUS *SIMONSIELLA*

Some differential features of the four recognized species of *Simonsiella* are shown in Tables BXII.β.96, BXII.β.97, BXII.β.98, BXII.β.99, and BXII.β.100.

List of species of the genus Simonsiella

1. **Simonsiella muelleri** Schmid *in* Simons 1922, 504^AL

   muel' le.ri. M.L. gen. n. *muelleri* Müller, named for R. Müller, who first described these organisms.

   See Tables BXII.β.96, BXII.β.97, BXII.β.98, BXII.β.99, and BXII.β.100 and the generic description for many features. The long axis of the cells (width of filament) varies from 2.1 to 3.5 μm, with an average of about 2.5–3.2 μm; the short axis (length of individual cell) varies from about 0.5 to 0.9 μm, with an average of about 0.8 μm. Casein, gelatin, esculin, and hippurate are not hydrolyzed. No change occurs in litmus milk. The ability to reduce nitrates varies among strains, with about half of them being positive; when reduction occurs, nitrite is produced, and a few strains produce N_2 gas. H_2S production is variable and inconsistent. Found in the oral cavity of humans.

   The mol% G + C of the DNA is: 40–42 (Bd) for 16 of 18 strains; two other strains had a G + C mol% 44 and 50 (Bd).

   Type strain: ATCC 29453, CCUG 30554, CIP 103436, DSM 2579, ICPB 3636, LMG 7828.

   GenBank accession number (16S rRNA): M59071, AF328147.

2. **Simonsiella crassa** Schmid *in* Simons 1922, 504^AL

   cras' sa. L. fem. adj. *crassa* thick.

   See Tables BXII.β.96, BXII.β.97, BXII.β.98, BXII.β.99, and BXII.β.100 and the generic description for many features. The long axis of the cells (width of filament) varies from 1.9 to 3.6 μm, with an average of about 2.9–3.5 μm; the short axis (length of individual cell) varies from 0.7 to 0.9 μm, with an average of about 0.8 μm. Steed (1962) reported that her isolates produced acid from inulin and

that four of six isolates produced acid from arabinose. Kuhn et al. (1978) reported negative results with both carbohydrates for her isolates, as well as for one of Steed's strains. The discrepancy may be due to differences in the method of testing. Gelatin and casein are hydrolyzed. Litmus milk is peptonized. Nitrate is reduced to nitrogen gas. MR-negative. VP-negative. Steed reported that all of her isolates produced H_2S, but Kuhn et al. (1978) stated that H_2S production is variable and inconsistent. Found in the oral cavity of sheep.

   The mol% G + C of the DNA is: 44–45 (Bd).

   Type strain: ATCC 27504, CCUG 25927, CIP 103341, DSM 2578, ICPB 3651, LMG 782, NCTC 10283.

   GenBank accession number (16S rRNA): AF328141.

3. **Simonsiella steedae** Kuhn and Gregory 1978, 13^AL

   stee' dae. M.L. gen. n. *steedae* of Steed; named for P. Steed (Glaister) who first isolated axenic cultures of *Simonsiella* and erected the family *Simonsiellaceae*.

   See Tables BXII.β.96, BXII.β.97, BXII.β.98, BXII.β.99, and BXII.β.100 and the generic description for many features. The long axis of the cells (width of filament) varies from 2.5 to 7.1 μm, with an average of about 3.1–3.8 μm; the short axis (length of individual cell) varies from 0.7 to 1.3 μm, with an average of about 1.1 μm. Casein, gelatin, esculin, and hippurate are not hydrolyzed. No change in litmus milk. Most strains reduce nitrate to nitrite without gas production. H_2S production is variable and inconsistent. Found in the oral cavity of dogs.

   The mol% G + C of the DNA is: 48–52 (Bd).

   Type strain: ATCC 27409, CCUG 30555, CIP 103435, DSM 2580, ICPB 3604, LMG 7830.

   GenBank accession number (16S rRNA): AF328153.

4. "Simonsiella kuhniae"

kuhn.i.ae. F.L. gen. n. kuhniae of Kuhn, named after Daisy A. Kuhn, in honor of her work on the isolation, characterization, and autecology of *Simonsiella* in domestic animals and humans.

Multicellular filaments, segmented into groups of 8–12 cells; cells possess dorsal-ventral asymmetry, gliding on ventral surface. Cell width 2.7–4.4 μm; cell length 0.5–1.1 μm; cell thickness 0.5–1.1 μm. Terminal segments often decreased in size. Colonies are entire, low convex <1–2 mm diameter, smooth and butyrous, opaque with pale, nondiffusible yellow pigmentation. Strict aerobe. Catalase and oxidase positive. Chemoorganotrophic. Acid not produced from D-glucose, maltose, D-ribose, D-fructose, sucrose, mannitol, salicin. Weakly proteolytic on casein. No activity on gelatin, inspissated serum, or litmus milk. Growth at 37°C, pH 7.2, and pH 8.0; no growth at 27°C, 43°C, or pH 6.0. Nitrate may or may not be reduced. Found in the oral cavity of domestic cats, *Felis domesticus*.

Deposited strain: ATCC 29436, ICPB 3618.

GenBank accession number (16S rRNA): AF328149.

Additional Remarks: Strains ATCC 29437 (ICPB 3619), ATCC 27381 (ICPB 3601), and ATCC 29465 (ICPB 3648) also belong to this species.

Genus XII. Vitreoscilla Pringsheim 1949c, 70[AL]

WILLIAM R. STROHL

Vit.re.os.cil′ la. L. adj. *vitreus* glassy, clear; L. n. *oscillum* a swing; M.L. fem. n. *Vitreoscilla* transparent oscillator.

Cylindrical cells (two species) or sausage-shaped cells (one species) ranging from 1.0 to 3.0 μm in diameter. The organisms occur as **colorless filaments** with lengths up to several hundred μm. Two species have cells that are not normally visible within filaments; one species has cells that are clearly visible within the filaments. **Gram negative. Gliding motility.** No locomotor organelles known. **Resting stages are not known. Sheaths and holdfasts are not produced. Nonpigmented. Aerobic to microaerophilic**, having a respiratory type of metabolism with oxygen as the terminal electron acceptor. No growth anaerobically. **Sulfur inclusions are not formed from hydrogen sulfide or thiosulfate. Chemoorganotrophic.** Two species are oligotrophic and grow best at low nutrient concentrations; one species can grow luxuriantly on rich media, e.g., 0.5% peptone broth. Various organic acids and amino acids are used as carbon and energy sources. Found in dung, soil, water with decaying plant material, sediments, and in association with oscillatorian mats.

The mol% G + C of the DNA is: 42–63.

Type species: **Vitreoscilla beggiatoides** Pringsheim 1949c, 70, emend. Strohl, Schmidt, Lawry, Mezzino and Larkin 1986a, 311.

FURTHER DESCRIPTIVE INFORMATION

Morphological characteristics The genus *Vitreoscilla* consists of two morphological types. One type is characterized by *V. beggiatoides* (the type species) and *V. filiformis*. The cells are ultrastructurally similar to those of *Beggiatoa* in having (i) continuous outer layers of the filaments (Figs. BXII.β.80 and BXII.β.81), (ii) extra cell wall layers outside of the "lipopolysaccharide-like" layer (Strohl et al., 1986a), (iii) large accumulations of poly-β-hydroxybutyrate (PHB; sometimes >50% of dry cell mass), and (iv) membrane invaginations in the cytoplasm. If these filaments are mechanically broken up into individual cells, the cells die—an indication that these are truly multicellular bacteria like the beggiatoas. The cells within a filament of these species divide via septation, with only the cytoplasmic membrane and peptidoglycan layers invaginating, like the closure of an iris diaphragm, as in *Beggiatoa* (Strohl and Larkin, 1978a). In thin sections observed by transmission electron microscopy, cells of *V. beggiatoides* and *V. filiformis* appear similar to cells of *Beggiatoa* (Strohl et al., 1986a). Furthermore, the sacrificial-cell-death life cycle described for beggiatoas (Strohl and Larkin, 1978a) appears to apply to *V. beggiatoides* as well, and may also occur with *V. filiformis* strains.

The second morphological type of *Vitreoscilla* is exemplified by *V. stercoraria*. Cells of this species have discontinuous outer layers (Figs. BXII.β.82 and BXII.β.83), and the filaments appear similar to chains of bacilli. The cells are held together by a ruthenium-red-staining material, but they can be mechanically broken apart without accompanying massive cell death. Multicellular filaments are formed seemingly only because the cells do not completely detach after division. Reports have indicated that certain growth conditions promote growth of *V. stercoraria* as single

FIGURE BXII.β.80. Thin-section electron micrograph of *V. beggiatoides* strain B23SS showing the continuous cell wall and extra cell layers. Also note the large depositions of PHB. Bar = 0.5 μm. (Reproduced with permission from W.R. Strohl et al., International Journal of Systematic Bacteriology *36*: 302–316, 1986, ©International Union of Microbiological Societies.)

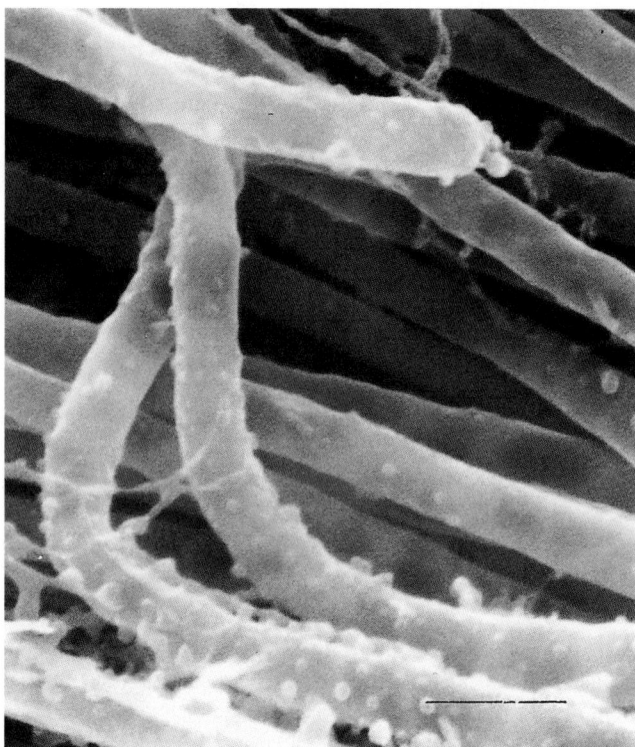

FIGURE BXII.β.81. Scanning electron micrograph of *V. beggiatoides* strain B23SS, with continuous cell envelope and slime-like matrix shown. Bar = 5 μm.

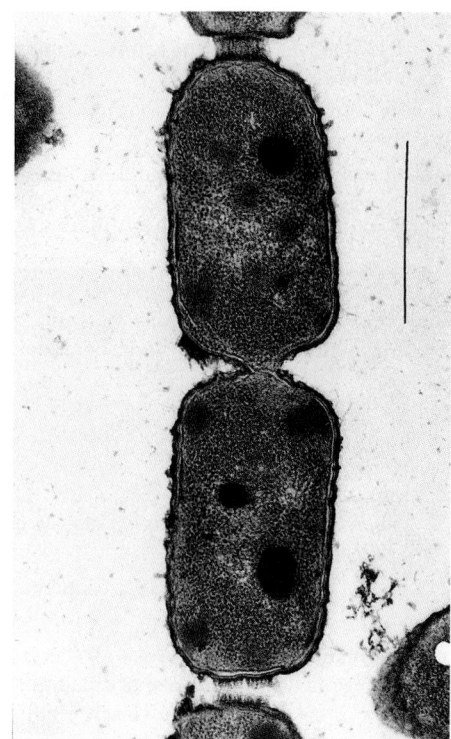

FIGURE BXII.β.82. Thin-section micrograph of ruthenium-red-stained *V. stercoraria* VT-1, with chains of cells held together by connecting material, and the Gram-negative cell envelope shown. Bar = 1 μm.

cells (Brzin, 1966a, b). Ultrastructurally, *V. stercoraria* cells have a typical Gram-negative cell envelope, with the addition of an external, ruthenium-red-staining surface layer (Fig. BXII.β.82; Strohl et al., 1986a). Perhaps the most notable feature of *V. stercoraria* is the presence in the cells of a bacterial form of hemoglobin (Wakabayashi et al., 1986; Khosla and Bailey, 1988b). The *V. stercoraria* hemoglobin gene has been cloned, sequenced, and characterized in detail, and has been used in recombinant form in several attempts to improve oxygen transfer to industrially significant organisms (Khosla and Bailey, 1988a; Brunker et al., 1998).

Cultural conditions For all *Vitreoscilla* species, the growth conditions determine the shape and characteristics of the colonies. On agar media containing low amounts of nutrients, wavy, curly, or spiral colonies are produced, from which trichomes glide radially outward from the central colony area. On solid media containing high nutrient concentrations, the trichomes spread very little and can form drop-like colonies resembling those of most eubacteria.

Nutritional characteristics *Vitreoscilla* species were formerly differentiated based on trichome size alone, although critical nutritional (Strohl et al., 1986a) and phylogenetic (Stahl et al., 1987; Dewhirst et al., 1989) differences among them are now known. *V. beggiatoides* and *V. filiformis* strains grow best on media with very low nutrient concentrations (e.g., 0.05–0.1% peptone, acetate, etc.), but differ in the nutrients utilized. *V. beggiatoides* can grow on a simple defined medium containing acetate, ammonium, and basal salts and is fairly restrictive in the nutrients it utilizes, i.e., certain C_2–C_5 organic acids and amino acids. *V.*

filiformis can grow on the minimal medium described above, but it also can utilize glucose and citrate. Neither species, however, tolerates high concentrations of nutrients or nutritionally rich media well. Similarly, neither species grows to great cell masses, with 50–80 mg/l of dry cell mass being the approximate maximum. These organisms can withstand moderate concentrations of sulfide, even though they do not form sulfur depositions from it. *V. beggiatoides* B23SS and three *V. filiformis* strains can fix molecular nitrogen, as determined by acetylene reduction (Polman and Larkin, 1990). In all four cases, the presence of a fixed ammonium salt source in the medium abolishes the ability of the organisms to reduce acetylene (Polman and Larkin, 1990).

Vitreoscilla stercoraria differs from the other two species in that it grows luxuriantly on rich media (0.5% peptone, w/v). *V. stercoraria* is an obligate amino-acid-utilizing organism with fairly complex nutritional requirements (Mayfield and Kester, 1972, 1975), although it does not utilize glucose. *V. stercoraria* grows very well in nutritionally complex and rich media, and it achieves much higher cell masses during growth than do the other vitreoscillas.

Metabolic characteristics Cultures of *V. beggiatoides* and *V. filiformis* are able to oxidize both the methyl and carboxyl carbons of acetate to CO_2 (Strohl et al., 1986b). The rate of acetate oxidation is very high, as it is with the beggiatoas. These organisms may use the glyoxylate bypass cycle when growing on acetate, but this has been investigated very little. *V. stercoraria* contains all of the enzymes of the tricarboxylic and glyoxylate bypass cycles (W.R. Strohl and G.W. Luli, unpublished data). The activities of the enzymes isocitrate lyase and malate synthase are stimulated significantly by the addition of 1% sodium acetate to the *V. ster-*

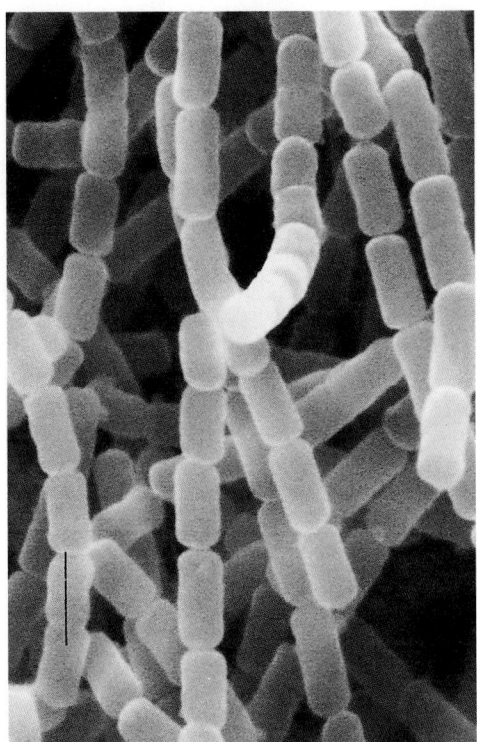

FIGURE BXII.β.83. Scanning electron micrograph of *V. stercoraria* strain VT-1, with the discontinuous cell walls and the connecting material between the cells shown. Bar = 1 μm.

coraria growth medium (W.R. Strohl and G.W. Luli, unpublished data). This is in apparent contradiction to Pringsheim's original statement that acetate does not stimulate growth of *V. stercoraria* (Pringsheim, 1951).

Plasmids *V. beggiatoides* and *V. filiformis* strains are apparently devoid of small plasmids such as the 12.3 and 12.8 × 10⁶ M_r (18.9 and 19.7 kb, respectively) plasmids observed in *Beggiatoa alba* (Minges et al., 1983). Strains of *V. stercoraria* contain small plasmids with an apparent 1.4 × 10⁶ M_r (~2.2 kb; Minges et al., 1983). No functions have yet been ascribed to these plasmids. DNA transformation procedures have been described for *V. stercoraria* (Navani et al., 1996), but not for the trichome-forming vitreoscillas.

Habitat Vitreoscillas of the *V. stercoraria* type are usually found in dung, soil, water with decaying plant material, and in association with oscillatorian mats (Pringsheim, 1949c, 1951). *V. beggiatoides* strain B23SS (group B strains, as described in Strohl and Larkin, 1978b) was isolated from a sandy, lightly sulfide-emanating sediment that also included sulfur-containing beggiatoas.

ENRICHMENT AND ISOLATION PROCEDURES

Enrichment and isolation procedures for *V. beggiatoides*, *V. filiformis*, and other trichome-forming strains are basically the same as those described in the chapter on the genus *Beggiatoa* (this volume; see also Pringsheim, 1964 and Strohl and Larkin, 1978b). The medium used, however, should reflect the type of *Vitreoscilla* being sought (i.e., lack of sulfide in the media and the type and amount of organic supplements). A medium that is excellent for the isolation and maintenance of *V. beggiatoides* is "MY medium",

described in the article on *Beggiatoa* (this volume), but without the added reduced sulfur source. For *V. filiformis* and similar strains, a modification of MY medium in which glucose or citrate is substituted for acetate might be more appropriate.

For isolation of *V. stercoraria*, small samples of dung can be placed directly onto plates containing a nutritionally rich medium. After several days, trichomes gliding away from the inoculum can be observed with a dissecting microscope. Those filaments can be retrieved with a flame-sterilized 26-gauge needle, as described for *Beggiatoa*. A good medium for this purpose is 0.1-CAYTS medium, which contains 0.1% each of casitone, tryptone, sodium acetate, and yeast extract.

MAINTENANCE PROCEDURES

Vitreoscillas can survive for about three weeks on plates of standard media employed for their growth. In general, the lower the concentration of nutrients in the media, the longer the time of survival. Care should be taken to transfer from the edges of the spreading colonies, so as to transfer the youngest and most active filaments. *V. stercoraria* can be lyophilized. *V. beggiatoides* and *V. filiformis* cannot be lyophilized, but they can be stored at −70°C or −196°C in the presence of 20–30% glycerol.

PROCEDURES FOR TESTING SPECIAL CHARACTERS

Because the larger species of *Vitreoscilla* are distinguished from beggiatoas by their inability to deposit sulfur, it is critical that the tests for sulfide oxidation and sulfur deposition be sensitive and reproducible. Investigators frequently have viewed the filaments with a phase microscope and assumed that refractile objects are sulfur. In fact, the strains now included in *V. filiformis* were once considered to be *Beggiatoa* strains because the phase-bright bodies (presumably polyphosphate) in these organisms were interpreted as sulfur (Pringsheim, 1964; Kowallik and Pringsheim, 1966).

One method for determining if inclusions contain sulfur is a chemical analysis of ethanol-extracted (Jørgensen and Fenchel, 1974) or carbon-disulfide-extracted (Nelson and Castenholz, 1981) filaments. A much more sensitive, but more difficult and expensive, method for analyzing sulfur deposition is through the oxidation of ³⁵S-labeled Na₂S at a physiological pH to internal ³⁵S⁰ (Strohl and Schmidt, 1984; Strohl et al., 1986a; Schmidt et al., 1987b). Controls should be included for the adsorption of Na₂³⁵S by autoclaved cells and for Na₂³⁵S assimilation by known sulfide-oxidizing organisms (e.g., *Beggiatoa* species) and non-sulfide-oxidizing organisms (e.g., *V. stercoraria*). Once it is demonstrated that labeled sulfide is assimilated, it should be at least 50–70% extractable by washing with warm ethanol or, alternatively, ~90% extractable by warm ethanol treatment followed by a wash with ethanol/diethyl ether (1:1) (W.R. Strohl, unpublished data).

DIFFERENTIATION OF THE GENUS *VITREOSCILLA* FROM OTHER GENERA

Although the vitreoscillas make up a heterogeneous group, they still can be differentiated from members of other genera (Table BXII.β.101). The primary distinguishing characteristics include their ability to glide but not deposit sulfur, their inability to produce sheaths or holdfasts, and their absence of pigmentation.

TAXONOMIC COMMENTS

Briefly, the three species that currently make up the genus *Vitreoscilla* are not phylogenetically coherent, and it is proposed that each of the three species within the genus *Vitreoscilla* should be placed into a separate genus. While these organisms share some general features, they are vastly different in morphological, physiological, and phylogenetic characteristics.

Phylogenetic analyses have clarified the taxonomy of these organisms to some degree. Whereas the resemblance of *Vitreoscilla* filaments to those of cyanobacteria caused Pringsheim to label them as colorless forms of cyanobacteria (Pringsheim, 1949c), none of the species is phylogenetically related to any cyanobacterium (Reichenbach et al., 1986; Stahl et al., 1987). Moreover, phylogenetic analysis of the vitreoscillas, beggiatoas, and similar organisms has demonstrated that gliding motility is not a particularly useful taxonomic or phylogenetic marker, as has been pointed out by Stahl et al. (1987). In another example, members of the gliding cytophagas are now grouped together with non-gliding flavobacteria and bacteroides, and, in some instances, former *Cytophaga* species have been reclassified as *Flavobacterium* species (Nakagawa and Yamasato, 1993).

Sequence analysis of 5S rRNA has placed *V. beggiatoides* strain B23SS (the neotype strain) in the γ3-subgroup (Woese, 1987) of the *Gammaproteobacteria* (Stahl et al., 1987) (Fig. BXII.β.84A). The authors state that *V. beggiatoides* is "specifically, but distantly, related to *B. alba*" strain B18LD (type strain), and point out that the evolutionary distance between the two organisms is greater than that separating *E. coli* and *Proteus vulgaris* (Stahl et al., 1987; see also Fig. BXII.β.84A).

Similar 5S rRNA analysis indicates that *V. stercoraria* VT1 and two strains of *V. filiformis* (ATCC 15551 and L-1401-7) are members of the *Betaproteobacteria* (Stahl et al., 1987) (Fig. BXII.β.84A). The 5S rRNA analyses carried out by Stahl et al. (1987) indicate that *V. filiformis* is related to *Rhodocyclus gelatinosus* in the *Chromobacterium* branch of the β-subdivision. Our revised analysis of the 5S rRNA sequence of *V. filiformis* places it in the *Rubrivivax–Leptothrix* sub-branch (Kalmbach et al., 1999) of the β1 subgroup of the *Proteobacteria*, most closely related to *Thiomonas cuprina* and *Leptothrix discophora* (Fig. BXII.β.84A).

Based on early 5S rRNA sequence data, *V. stercoraria* was originally placed at the base of the β1-β2 subgroups (Stahl et al., 1987) or within the β2 subgroup (Woese, 1987) of the β-*Proteobacteria*, and was considered to be somewhat related to *Neisseria* (Stahl et al., 1987). Later 16S rRNA analyses by Dewhirst et al. (1989) have confirmed the phylogenetic relatedness of *V. stercoraria* and *Neisseria* species (see Fig. BXII.β.84B), although those authors argued against including *V. stercoraria* in the *Neisseriaceae*. They pointed out that although the 16S rRNA of *V. stercoraria* has an average sequence identity of 93.7% with members of *Neisseria* (they average 95.5% within the genus), it should be separated from them because it lacks a key unique "*Neisseria* signature" sequence, it is a free-living organism, and it is oxidase negative (Dewhirst et al., 1989). It should be noted, however, that the current "Taxonomy" browser of the National Center for Biotechnology Information places the genus *Vitreoscilla* within the family *Neisseriaceae* (based on the 16S rRNA sequences of *V. stercoraria*).

Certain unclassified *Vitreoscilla* strains, morphologically similar to *V. stercoraria* (Costerton et al., 1961; Nichols et al., 1986; Strohl et al., 1986a), should now be considered as strains of *V. stercoraria*. Strain VT-1 is physiologically identical with *V. stercoraria* ATCC 15218, the type strain of the species (Strohl et al., 1986a). *Vitreoscilla* strains 389 and 390 (Costerton et al., 1961) have also been determined to be nearly 100% related by DNA hybridization to *Vitreoscilla stercoraria* strain ATCC 15218 (Nichols et al., 1986).

On the other hand, *Filibacter limicola*, an organism morphologically similar to *V. stercoraria*, has been shown to be completely unrelated to *V. stercoraria* (Clausen et al., 1985; Nichols et al., 1986).

Additional 16S rRNA sequences analyses and strain–strain nucleic acid hybridizations are clearly needed to place these organisms into their proper taxonomic locations with respect to related organisms. In particular, a serious taxonomic problem, which needs to be rectified, exists with the genus *Vitreoscilla. V. beggiatoides* was designated as the type species of the genus, based on the description of vitreoscillas as *Beggiatoa*-like organisms that do not deposit sulfur from hydrogen sulfide (Pringsheim, 1949c, 1951). Until the mid-1980s, however, the only well-characterized species of *Vitreoscilla* was *V. stercoraria*, since no other species from Pringsheim's collection had survived (Koch, 1964). Thus, strains did not exist for the type species of the genus until the description (Strohl et al., 1986a) of strains that fit the original description of *V. beggiatoides*. Now that the neotype strain of *V. beggiatoides* has been reasonably well characterized, and has been phylogenetically placed near, yet far enough to be distinct from, *Beggiatoa alba* in the class *Gammaproteobacteria*, the genus *Vitreoscilla* should be modified to contain only this single species. *V. stercoraria* is morphologically, ultrastructurally, nutritionally, physiologically, and ecologically very different from the other species (Strohl et al., 1986a) (Table BXII.β.102; Figs. BXII.β.80, BXII.β.81, BXII.β.82, BXII.β.83, and BXII.β.84). *V. stercoraria* and *V. filiformis*, both members of the class *Betaproteobacteria*, are not related to *V. beggiatoides*. Interestingly, *V. filiformis*, which more closely resembles *V. beggiatoides* morphologically, physiologically, and ecologically, is actually more closely related to *V. stercoraria*, although these two organisms are reasonably separated within the *Betaproteobacteria*. Because *V. beggiatoides* is the type species and the intent of the genus was to contain non-sulfur-depositing, beggiatoa-like organisms, it is proposed that *V. filiformis* and *V. stercoraria* be removed from the genus. *V. filiformis* and *V. stercoraria* should be taxonomically reclassified into separate genera more in line with their phylogenetic lineages. A fourth species, "*Vitreoscilla proteolytica*", was proposed in 1970 (Perschmann and Gräf, 1970) but the strain described possesses carotenoids (a characteristic never ascribed to *Vitreoscilla*), was never accepted as a new species of *Vitreoscilla*, and has not been further characterized phylogenetically; therefore it will not be considered here.

ACKNOWLEDGMENTS

I sincerely thank Annaliesa Anderson for her considerable assistance with the phylogenetic analysis of the vitreoscillas. I also thank Bo B. Jørgensen for sharing a preliminary version of the *Thioploca* chapter, which helped to stimulate thoughts about the taxonomy of these organisms.

FURTHER READING

Dewhirst, F.E., B.J. Paster and P.L. Bright. 1989. *Chromobacterium, Eikenella, Kingella, Neisseria, Simonsiella,* and *Vitreoscilla* species comprise a major branch of the beta group *Proteobacteria* by 16S ribosomal ribonucleic acid sequence comparison: transfer of *Eikenella* and *Simonsiella* to the family *Neisseriaceae* (emend). Int. J. Syst. Bacteriol. *39*: 258–266.

Stahl, D.A., D.J. Lane, G.J. Olsen, D.J. Heller, T.M. Schmidt and N.R. Pace. 1987. Phylogenetic analysis of certain sulfide-oxidizing and related morphologically conspicuous bacteria by 5S ribosomal ribonucleic acid sequences. Int. J. Syst. Bacteriol. *37*: 116–122.

Strohl, W.R., T.M. Schmidt, N.H. Lawry, M.J. Mezzino and J.M. Larkin. 1986. Characterization of *Vitreoscilla beggiatoides* and *Vitreoscilla filiformis* sp. nov., nom. rev., and comparison with *Vitreoscilla stercoraria* and *Beggiatoa alba*. Int. J. Syst. Bacteriol. *36*: 302–313.

A

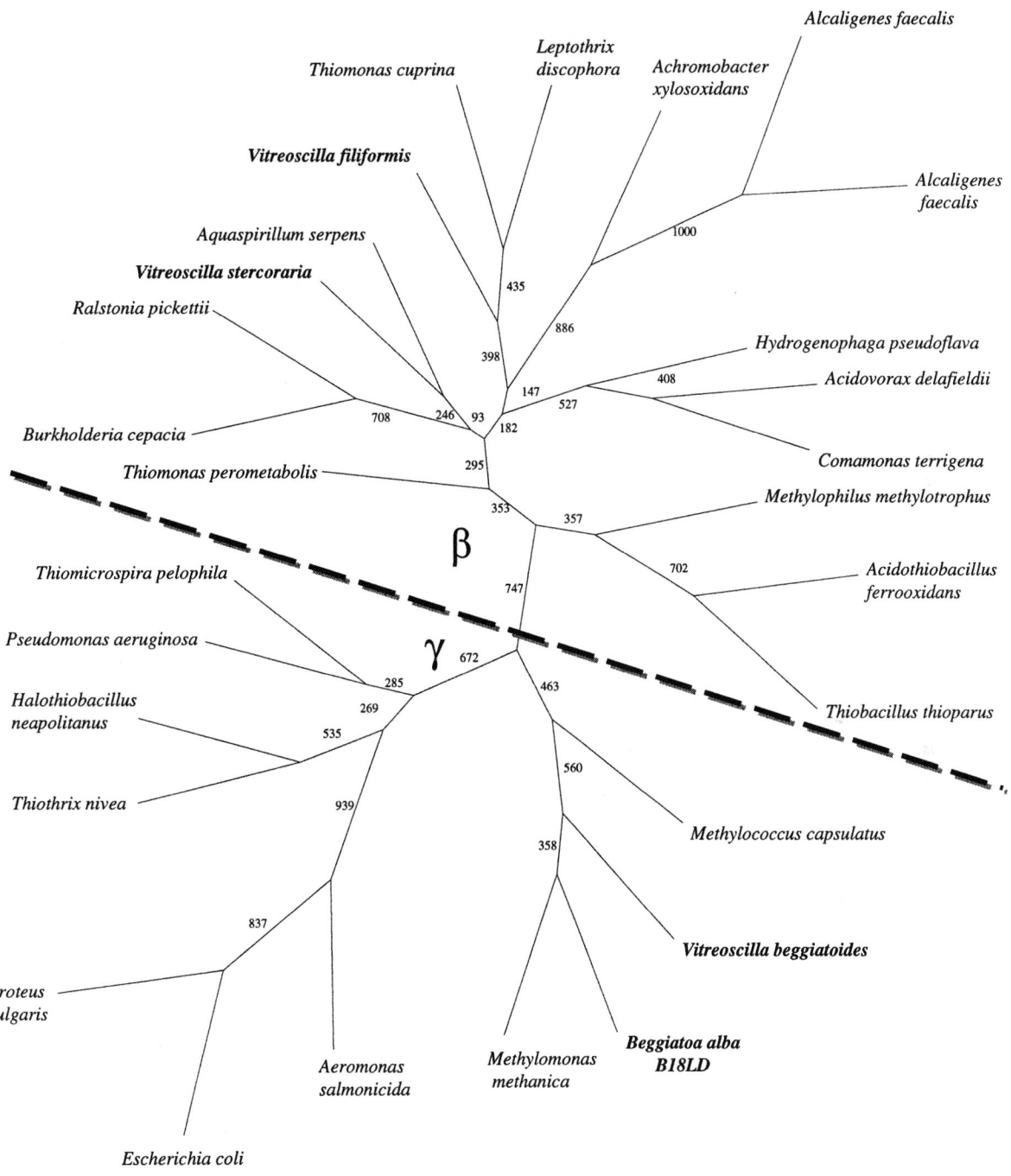

_100

FIGURE BXII.β.84. Unrooted distance trees of *Vitreoscilla* species, along with representative members of the classes *Betaproteobacteria* and *Gammaproteobacteria*. The sequences were aligned using the PILEUP program of the GCG Package, and the aligned sequences were compared using the DNADIST and NEIGHBOR programs of the PHYLIP (Felsenstein, 1993) software package. Bootstrap analyses were obtained from 1000 iterations. *A*, Phylogenetic analyses based on 5S rRNA sequences (in which all three species of *Vitreoscilla* are represented). *B*, Phylogenetic analyses based on 16S rRNA sequences, for which only *V. stercoraria* is the only *Vitreoscilla* species represented. The approximate positions of *V. filiformis* and *V. beggiatoides* are noted based on their nearest relatives with both 5S and 16S rRNA sequence alignments.

(continued)

B

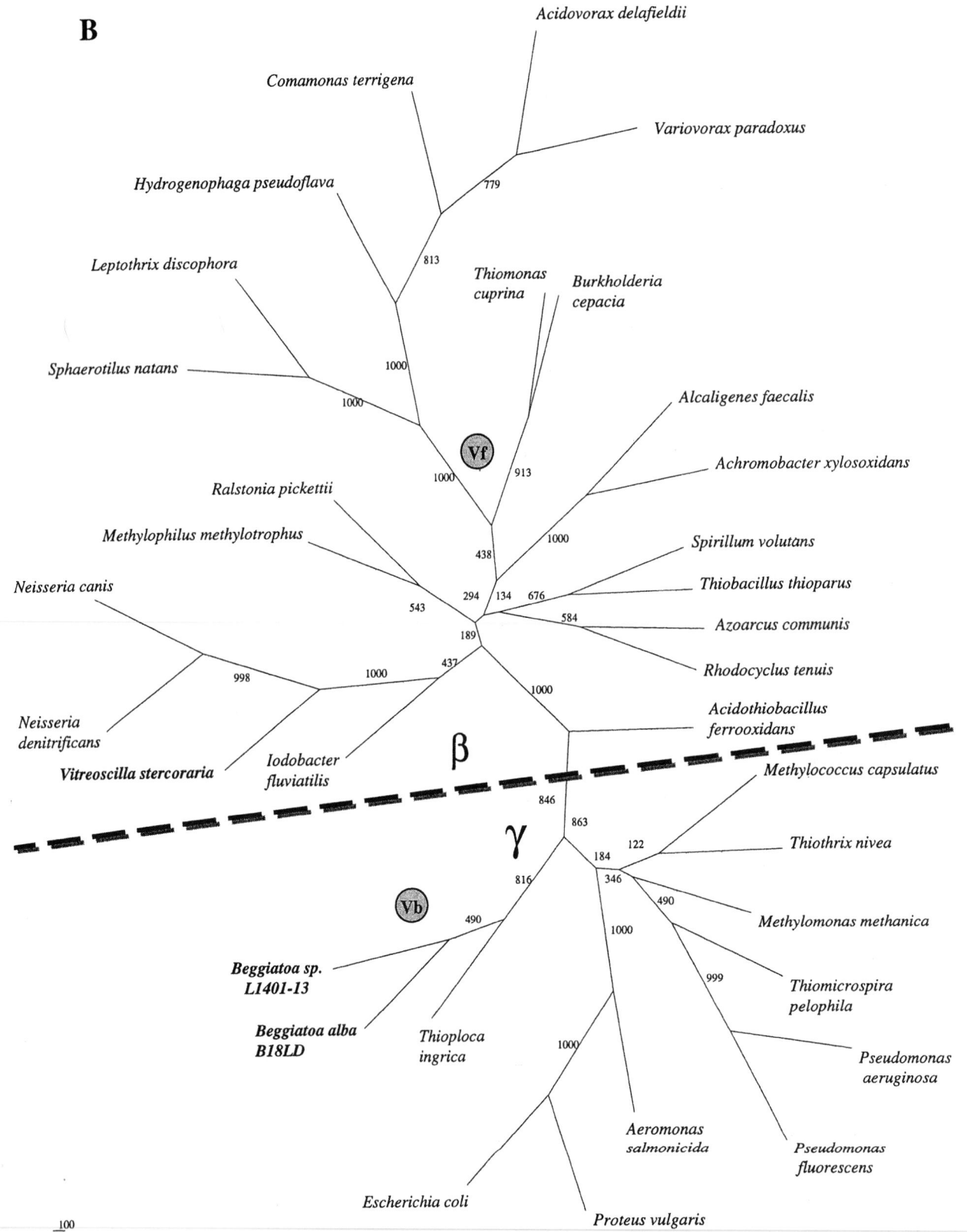

List of species of the genus Vitreoscilla

1. **Vitreoscilla beggiatoides** Pringsheim 1949c, 70,[AL] emend. Strohl, Schmidt, Lawry, Mezzino and Larkin 1986a, 311.

beg.gi.a.toi' des. M.L. fem. n. *Beggiatoa* a generic name; Gr. n. *idos* shape, form; M.L. adj. *beggiatoides Beggiatoa*-like.

TABLE BXII.β.101. Differentiation of the genus *Vitreoscilla* from other genera[a]

Characteristic	*Vitreoscilla*	*Beggiatoa*	*Leptothrix*	*Neisseria*	*Thioploca*
Colorless	+	+	+	±	+
Strains deposit sulfur	−	+	−	−	+
Ensheathed	−	−	+	−	+
Holdfast	−	−	±	−	−
Gliding motility	+	+	−	−	+
Flagellated	−	−	+	−	−
Filament formation	+	+	+	−	+
Individual cells observable within filaments	±	−	+	na	−
Cell division in two planes	−	−	−	+	−
Coccoid	−	−	−	+	−
Class:					
Betaproteobacteria	+		+	+	
Gammaproteobacteria	+	+			+

[a]Symbols: +, positive in >90% of strains; −, negative in >90% of strains; ±, positive in 10–90% of strains; na, not applicable.

The characteristics are as described for the genus and as listed in Table BXII.β.102, with the following additional features. Cells are cylindrical, 2.5–3.0 × 3–6 μm, and occur in filaments. The cells are not normally visible within the filaments. The filaments have a length of up to several hundred μm. Filaments lack visible septation, and the ends of the filaments are rounded. Necridia and hormogonia are produced. Motile by a gliding motility with speeds of ~3 μm/sec. Produce colonies of the circuitans type when grown on low-nutrient media and of the linguiformis type when grown in high-nutrient media. Trichomes often glide singly, away from other trichomes. Sulfur is not deposited when cells are grown on sulfide; however, sulfide at a concentration of ~1 mM is apparently not inhibitory. PHB is always present when cells are cultured on acetate-containing media and often comprises >50% of the cell dry weight. Polyphosphate bodies are often present.

Grows best at low nutrient concentrations. Acetate, ethanol, and some C_4 organic acids are utilized as sole carbon and energy sources. Ammonium and nitrate are used as nitrogen sources. Microaerophilic to aerobic. Gelatin, starch, and casein are not hydrolyzed. Catalase negative. Cytochrome oxidase positive. Growth occurs in 0.5% but not 1.0% NaCl. Sensitive to the antibiotics polymyxin B, neomycin, and furadantin/macrodantin; resistant to the antibiotics bacitracin and streptomycin. Found in sandy, but lightly sulfide-emanating, sediments of freshwater streams. Phylogenetic analysis places this species within the class *Gammaproteobacteria*, relatively close to *Beggiatoa alba*.

The mol% G + C of the DNA is: 42 (T_m, Bd).

Type strain: B23SS, ATCC 43189.

2. **Vitreoscilla filiformis** Strohl, Schmidt, Lawry, Mezzino and Larkin 1986a, 310[VP]

fi.li.for′ mis. L. n. *filum* a thread; L. n. *forma* shape; M.L. adj. *filiformis* thread-shaped.

The characteristics are as described for the genus and as listed in Table BXII.β.102, with the following additional features. Cells occur in flexible filaments with diameters of about 1.0–1.5 μm and, when cells are grown in liquid media, lengths of up to several hundred μm. Filament width may vary slightly. The cells are not clearly visible within the filaments, although slight indentations of the trichomes may sometimes be observed at septal regions between cells. The filament walls are continuous, and the ends of the filaments are rounded. Hormogonia are sometimes produced. Motile by gliding motility; gliding is relatively slow, at ~0.1–

0.5 μm/sec. Produce linguiformis-type colonies. PHB and condensed phosphate deposits may be present.

Growth is best at low nutrient concentrations. Can use many 2-, 3-, 4-, and 6-carbon organic acids, several amino acids, ethanol, and glucose as sole carbon and energy sources. Organic and inorganic forms of nitrogen can be used as nitrogen sources. Nitrate is reduced to nitrite. Aerobic to microaerophilic. Gelatin and starch are weakly hydrolyzed. Casein is not hydrolyzed. Catalase negative, cytochrome oxidase positive. No growth occurs in the presence of 0.5% NaCl. Sensitive to the antibiotics bacitracin, streptomycin, furadantin/macrodantin, and polymyxin B. Found in freshwater sediments, usually in association with decaying matter. Phylogenetic analysis of *V. filiformis* places it into the *Betaproteobacteria*, with the closest relatives being *Thiomonas cuprina* and *Leptothrix*.

The mol% G + C of the DNA is: 59–63 (T_m).

Type strain: L1401-2, ATCC 43190.

3. **Vitreoscilla stercoraria** Pringsheim 1951, 136[AL] emend. Strohl, Schmidt, Lawry, Mezzino and Larkin 1986a, 312.

ster.co.ra′ ri.a. L. fem. adj. *stercoraria* pertaining to dung.

Cells usually occur as flexible chains with diameters of about 1.0 μm and, when grown in liquid media, lengths of up to about 100 μm. Deep constrictions separate the individual cells, yielding discontinuous filaments. Cells can occur singly, especially if grown at temperatures of ~22°C. The cells are sausage-shaped, 1.0 × 1.5–12.0 μm, and are connected in filaments by extracellular material. Division is by binary fission. PHB and condensed phosphate deposits may be present.

Grows luxuriantly on rich media such as 0.5% peptone broth. Growth occurs on Casamino acids plus acetate. Requires amino acids; arginine, tyrosine, tryptophan, and glutamine are required for good growth. Combinations of amino acids from the glutamate and aspartate families plus arginine are required minimally for growth. Cytochromes *a*, *c*, and the non-CO-binding *b* are not present. Contains unique bacterial hemoglobin. Aerobic. No growth microaerophilically or anaerobically. Catalase positive. Cytochrome oxidase negative. Gelatin, casein, and starch not hydrolyzed. Usually isolated from cow dung. Phylogenetic analysis of *V. stercoraria* places it into the class *Betaproteobacteria*, with the closest relatives being members of the genus, *Neisseria*.

The mol% G + C of the DNA is: 50–51 (Bd).

Type strain: SAG 1488-6, ATCC 15218, DSM 513.

GenBank accession number (16S rRNA): L06174.

TABLE BXII.β.102. Differential characteristics of the species of the genus *Vitreoscilla*[a]

Characteristic	V. beggiatoides	V. filiformis	V. stercoraria
Colony type	L, C	L	L
Filament type:[b]			
Continuous cell wall	+	+	−
Discontinuous cell wall	−	−	+
Filament diameter (µm)	2.5–3.0	1.0–1.5	1
Habitat:			
Freshwater sediments	+	+	−
Cow dung	−	−	+
Obligatory requirement for amino acid mixtures	−	−	+
Growth on:			
Nutrient agar	−	−	+
Acetate plus salts	+	+	−
Use as sole carbon and energy source:			
Glucose, citrate, lactate, glutamate	−	+	−
Succinate, acetate	+	+	−
Utilization of nitrate as sole nitrogen source	+	+	−
Nitrogen fixation	+	+	nd
Cytochromes:			
CO-binding types	+	+	+
c-type	+	+	−
Bacterial hemoglobin	nd	nd	+
Mol% G + C of DNA	42	59–63	50–51

[a]For symbols see standard definitions; L, linguiformis type colonies; C, circuitans type colonies (see Pringsheim, 1964 and Strohl and Larkin, 1978b for a description of these colony types); nd, not determined.

[b]Cells within filaments having continuous cell walls share the outer cell wall layers, with at most only shallow constrictions between cells noticeable (cf. Figs. BXII.β.80 and BXII.β.81). Strains having discontinuous filament cell walls consist of chains of individual cells held together by connective material (c.f. Figs. BXII.β.82 and BXII.β.83).

Genus XIII. **Vogesella** Grimes, Woese, Macdonell and Colwell 1997, 25[VP]

NOEL R. KRIEG

Vo.ges.el'la. Ger. *Voges* proper name; M.L. *-ella* dim. ending; M.L. fem. n. *Vogesella* named after Otto Voges to honor his original isolation of *Bacillus indigoferus* on gelatin plates inoculated with tap water from the central water supply system in Kiel, Germany, in 1893.

Straight rods, 0.5 × 3.5 µm, mainly occurring singly but also in pairs and short chains. Occasional vibrioid rods may occur. Most strains form filamentous rods and long chains when grown on nitrogen-limiting agar. **Motile by means of a single polar flagellum.** Poly-β-hydroxybutyrate granules occur, especially under nitrogen limitation. **Colonies are a deep royal blue with a metallic copper-colored sheen at 36 to 48 h, due to production of indigoidine** ($C_{10}H_8N_4O_4$ or 5,5′-diamino-4,4′-dihydroxy-3,3′-diazadiphenoquinone-[2,2′]). **Oxidase and catalase positive. Aerobic**, having a strictly respiratory type of metabolism with oxygen as the terminal electron acceptor. All strains can carry out denitrification with production of gas. **Nonfermentative.** Citrate, several amino acids and peptides, and some monosaccharides are utilized as carbon sources. Starch, pectin, and sugars associated with pectin (i.e., arabinose and galactose) are not metabolized. Nonlipolytic. **Casein, gelatin, and esculin are not hydrolyzed. Indole positive.** Belong to the *Betaproteobacteria* and are closely related to the genus *Chromobacterium*. Occur in freshwater habitats and oxidation pond sediment.

The mol% G + C of the DNA is: 65.4–68.8.

Type species: **Vogesella indigofera** (Voges 1893) Grimes, Woese, Macdonell and Colwell 1997, 25 (*Pseudomonas indigofera* (Voges 1893) Migula 1900, 950; *Bacillus indigoferus* Voges 1893, 307.)

FURTHER DESCRIPTIVE INFORMATION

Colonies begin as translucent slightly yellowish colonies at early stages of growth (i.e., 16 to 20 h), then assume a faint bluish hue (24 h), and finally become a deep royal blue with a metallic copper-colored sheen (36 to 48 h). The pigment is localized within the cytoplasmic membrane (Kolar, 1974). Slightly acidic conditions (ca. pH 6.5) favor pigment production on plate count agar (PCA, Difco Laboratories, Detroit, MI). Excess pigment may occur in the colonies in the form of bluish crystals. When a colony is removed from an agar plate, a blue imprint or stain is left in the agar. Abundant blue pigmentation occurs on PCA, Thayer-Martin agar, chocolate agar, blood agar, *Pseudomonas* F agar, and especially chalk agar (Starr et al., 1960). Indigoidine production appears to be a stable trait of *Vogesella* strains.

Violacein (a violet pigment characteristic of *Chromobacterium* and *Janthinobacterium* species) and pyoverdin are not formed. All *Vogesella* strains exhibit a very weak fluorescence on chalk agar with short-wavelength UV illumination.

Growth occurs at 4°C and 37°C but not at 2°C and 40°C. Growth occurs at pH 6.2 and most strains grow at pH 8.6, but not at pH 4.0.

DNase is produced. The following tests are negative: methyl red; Voges–Proskauer; malonate; H_2S production; arginine, lysine, ornithine, and tryptophan decarboxylases; phenylalanine deaminase; urease; hydrolysis of esculin, *o*-nitrophenyl-β-D-galactopyranoside (ONPG), Tween 80, and corn oil.

All or most strains can use D-fructose, D-glucose, glycerol, ethanol, D-alanine, L-alanine, L-arginine, L-histidine, L-ornithine, L-proline, γ-aminobutyrate, fumarate, and putrescine as sole carbon sources. Propionate is not used.

Blue-pigmented bacteria appear to be a consistent feature of freshwater bacterial communities in terms of temporal and spatial distribution (Grimes et al., 1997).

ENRICHMENT AND ISOLATION PROCEDURES

Isolates can be selected from PCA plates that have been spread with water or sediment samples and incubated at 20 to 25°C for 48 h. Colonies are selected that have a blue pigmentation and a metallic copper-colored sheen.

MAINTENANCE PROCEDURES

Pure cultures can be maintained in deeps of $0.5 \times$ plate count broth (Difco) containing 0.3% agar (Difco) and stored at room temperature (ca. 20°C). Cultures maintained in this manner can remain viable for more than 10 years. Cultures can also be stored indefinitely under liquid nitrogen in tryptic soy broth (Difco) containing 5% glycerol.

DIFFERENTIATION OF THE GENUS VOGESELLA FROM OTHER GENERA

Arthrobacter spp., *Corynebacterium insidiosum*, and *Erwinia chrysanthemi* also form indigoidine. *Arthrobacter* and *Corynebacterium* are Gram positive, whereas *Vogesella* is Gram negative. Unlike *Vogesella*, *Erwinia* has peritrichous flagella, is oxidase negative, and can ferment sugars.

Chromobacterium species form violet-pigmented colonies that might be confused with those of *Vogesella*, although the pigment is violacein, not indigoidine. Unlike *Vogesella*, *Chromobacterium* is facultatively anaerobic, has a mainly fermentative attack on carbohydrates, is indole negative, digests casein, and does not denitrify.

Janthinobacterium species, which have an oxidative type of metabolism, produce violet-pigmented colonies due to formation of violacein. Unlike *Vogesella* strains, *Janthinobacterium* is indole negative, hydrolyzes esculin, does not grow at 37°C, and does not denitrify.

TAXONOMIC COMMENTS

A phylogenetic study of 13 new isolates by Grimes et al. (1997), in which both 5S rRNA and 16S rRNA analyses were employed, indicated that the isolates were closely related to each other and to strains *Pseudomonas indigofera* ATCC 19706[T] and *P. indigofera* ATCC 14036. In addition, both 5S rRNA and 16S rRNA analyses demonstrated that the new isolates and strains ATCC 19706[T] and ATCC 14036 were members of the *Betaproteobacteria*. Moreover, although these strains were closely related to *Chromobacterium violaceum* ATCC 12742[T], they were sufficiently distinct to warrant placement in a new genus, *Vogesella*.

List of species of the genus Vogesella

1. **Vogesella indigofera** (Voges 1893) Grimes, Woese, Macdonell and Colwell 1997, 25[VP] (*Pseudomonas indigofera* (Voges 1893) Migula 1900, 950; *Bacillus indigoferus* Voges 1893, 307.)

   in.di.go'fe.ra. Fr. n. *indigo* the dye indigo [from India]; L. suff. *fer* from L. v. *fero* to bear; M.L. adj. *indigofera* bearing indigo.

The characteristics are as described for the genus.

The mol% G + C of the DNA is: 65.4–68.8 (T_m).

Type strain: ATCC 19706, CCUG 2873, CCUG 32860, CIP 103306, DSM 3303, LMG 6867.

GenBank accession number (16S rRNA): U45995.

Genus Incertae Sedis XIV. **Catenococcus** Sorokin 1994, 852[VP] (Effective publication: Sorokin 1992, 2291)

DIMITRY Y. SOROKIN

Caten.o.coc'cus. L. n. *catena* chain; Gr. n. *coccus* berry; *Catenococcus* a chain of berries.

Cells **coccoid**, occurring mostly in **chains. Gram negative. Nonmotile.** Metabolism **respiratory** and **fermentative; facultative anaerobe.** Neutrophilic and mesophilic. **Obligately heterotrophic;** able to utilize some sugars, sugar alcohols, and organic acids as carbon and energy sources. **Oxidize thiosulfate, sulfide, and S^0 incompletely to tetrathionate** during heterotrophic growth. Able to **reduce S^0** and **ferric iron** in the presence of fermentable substrate under microaerobic and anaerobic conditions. Oxidase and catalase positive.

Isolated from sulfidic marine water.

The mol% G + C of the DNA is: 49.8 ± 0.5 (T_m).

Type species: **Catenococcus thiocycli** Sorokin 1994, 852 (*Catenococcus thiocyclus* (sic)) (Effective publication: Sorokin 1992, 2291.)

FURTHER DESCRIPTIVE INFORMATION

Cells grown in batch culture occur in long chains (Fig. BXII.β.85A and B), connected by slime discs. Cells grown in continuous cultures usually occur in pairs or in short chains. The cell wall has a structure that is typical of Gram-negative bacteria

(Fig. BXII.β.85C). Growth is NaCl-dependent and occurs over a range of 0.5–8% (w/v) NaCl (optimal, 2.5–3.5%). Cell wall stability demands high Mg^{2+} concentrations. The pH range for growth is 5.6–7.8 (optimal, 6.7–6.9). Growth above pH 7.5 and with low Mg^{2+} concentrations results in sphaeroplast formation and rapid cell lysis. The temperature range for growth is 10–35°C (optimal, 25–28°C).

Catenococcus grows best aerobically with starch and sugars (D-glucose, D-maltose, sucrose, D-fructose, L-arabinose, D-trehalose, and D-cellobiose) as carbon and energy sources, presumably because of slight acidification of the medium. All of these sugars also support anaerobic growth, except L-arabinose. The main products of glucose fermentation are formate and acetate. Aerobic growth also occurs with acetate, succinate, malate, fumarate, pyruvate, gluconate, citrate, D-lactate, malonate, 2-oxoglutarate, propionate, butyrate, ethanol, propanol, glycerol, D-mannitol, L-aspartate, L-glutamate, L-proline, L-cysteine, and L-histidine. Growth is vitamin independent. Tween-80 and gelatin are not hydrolyzed. Ammonium salts, nitrate, and nitrite serve as nitrogen sources. Nitrate is reduced to nitrite.

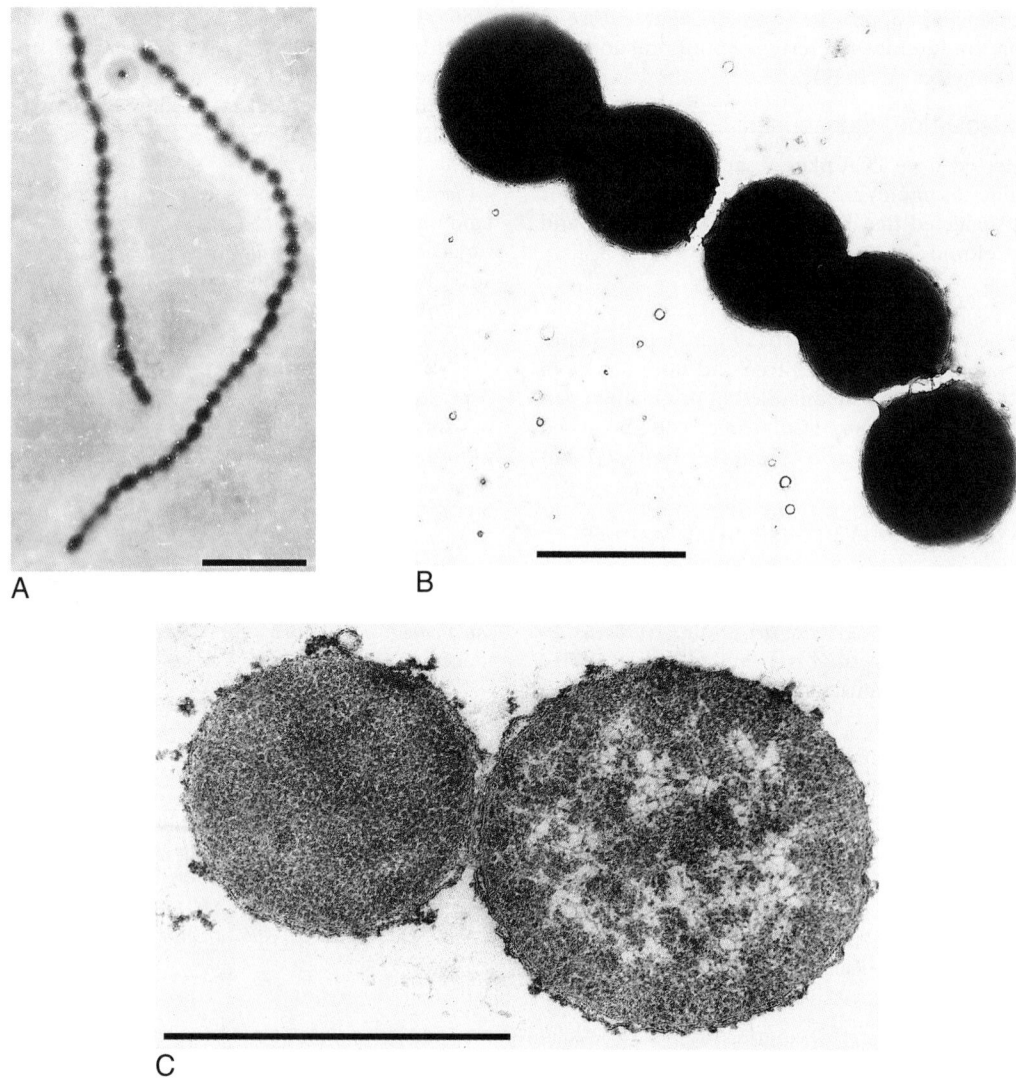

FIGURE BXII.β.85. Morphology of *Catenococcus thiocycli*. *(A)* phase-contrast photograph (Bar = 10 μm); *(B)* electron micrograph, total preparation stained with ammonium molybdate (Bar = 1 μm); *(C)* electron micrograph, thin section, stained with uranyl acetate + lead citrate (Bar = 1 μm).

Cultures oxidize thiosulfate to tetrathionate with high activity and sulfide to S^0 with much less activity. They are unable to grow autotrophically with reduced sulfur compounds, but can utilize metabolic energy released during thiosulfate oxidation to tetrathionate, as has been revealed in chemostat experiments (thiosulfate-dependent yield increase), and with washed cells (thiosulfate-dependent ATP synthesis). These data can be taken to imply the ability of *C. thiocycli* to grow chemolithoheterotrophically. Moreover, the bacterium is also capable of the reduction of S^0 to sulfide and of ferric iron to ferrous iron in the presence of glucose under either microaerobic or anaerobic conditions. However, these reactions do not support anaerobic respiration.

C. thiocycli is sensitive to chloramphenicol, streptomycin, tetracycline, erythromycin, ampicillin, kanamycin, colistin sulfate, nitrofurantoin, sulfafurazole, and penicillin G.

C. thiocycli has been isolated as a dominant tetrathionate-forming heterotroph from a near-shore volcanic thermal region (Papua, New Guinea) rich in sulfide and ferric iron. Its versatile metabolic potential suggests that such heterotrophs may play an important role in inorganic, as well as organic, cycling.

ENRICHMENT AND ISOLATION PROCEDURES

Catenococcus can be enriched, isolated, and successfully grown on a medium with the following composition (g/l): NaCl, 25; NH_4Cl, 0.5; potassium phosphate buffer pH 7, 2–5; $MgSO_4 \cdot 7H_2O$, 0.5–1; $CaCl_2 \cdot 2H_2O$, 0.1; yeast extract, 0.05; acetate or D-glucose, 10–20 mM; sodium thiosulfate, 20 mM; trace elements (Pfennig and Lippert, 1966), 1 ml/l.

MAINTENANCE PROCEDURES

Active cultures can be maintained in liquid culture or on agar slants at 4°C for 6 months without significant loss of viability. They survive lyophilization and may also be stored in liquid nitrogen.

DIFFERENTIATION OF THE GENUS *CATENOCOCCUS* FROM OTHER GENERA

C. thiocycli differs from the morphologically similar *Paracoccus* by its low mol% G + C and inability to grow autotrophically, from *Neisseria* by its fermentative potential, and from fermentative bacteria by its morphology. The main differences between *Catenococcus* and *Neisseria* are given in Table BXII.β.103.

TAXONOMIC COMMENTS

At the time this genus was described, phylogenetic analysis was not widespread. Therefore, until recently, the true relatives of this organism and its taxonomy remained unclear. Based on phenotypic characteristics, *Catenococcus* was placed in the family *Neisseriaceae* of the *Betaproteobacteria*, which includes mostly nonfermentative, often pathogenic, coccoid bacteria. Recently, partial

16S rDNA sequencing of *C. thiocycli* was performed which demonstrated that this bacterium is within the radiation of the *Vibrionaceae* of the *Gammaproteobacteria* and is closely related to *Listonella pelagia* (96% sequence identity). The fermentative capability of *C. thiocycli* is in agreement with this positioning, but strong morphological differences exist between *Catenococcus* and all other members of the *Vibrionaceae*.

FURTHER READING

Sorokin, D.Y., L.A. Robertson and J.G. Kuenen. 1996. Sulfur cycling in *Catenococcus thiocyclus*. FEMS Microbiol. Ecol. *19*: 117–125.

TABLE BXII.β.103. Comparison of *Catenococcus thiocycli* with the genus *Neisseria*

Characteristic	*C. thiocycli*	*Neisseria*
Morphology	Cocci in chains	Diplococci or cocci in short chains; one species rod-shaped
Division	One plane	Two planes
Acid production from sugars	+	D
Anaerobic growth with sugars	+	−
Anaerobic growth with nitrite	−	D
Mol% G + C	49.8	46.5–55.6
Habitat	Sulfidic seawater	Mammalian material

List of species of the genus Catenococcus

1. **Catenococcus thiocycli** Sorokin 1994, 852[VP] (*Catenococcus thiocyclus* (sic)) (Effective publication: Sorokin 1992, 2291.) thi.o.cy′cli. Gr. n. *thios* sulfur; Gr. n. *cyclos* circle.

Characteristics are those of the genus. The type strain was isolated from sulfidic seawater in Papua New Guinea.

The mol% G + C of the DNA is: 49.8 ± 0.5 (T_m).

Type strain: TG 5-3, ATCC 51228, DSM 9165, LMD 92.12.

Genus Incertae Sedis XV. **Morococcus** Long, Sly, Pham and Davis 1981, 300[VP]

LINDSAY I. SLY

Mo.ro.coc′cus. L. neut. n. *morum* mulberry; Gr. n. *coccus* a grain or berry; M.L. masc. n. *Morococcus* the mulberry coccus.

Colorless, **spherical organisms**, <1 μm in diameter, bound firmly together in tightly packed, **mulberry-like aggregates** of 10–20 cells, with **adjacent sides often flattened. Gram negative. Non-motile.** Endospores not formed. **Aerobic.** Nitrate is reduced. Catalase and cytochrome oxidase are produced. H_2S is produced. Acid is produced from carbohydrates. **Hemolytic.** Complex growth factors not required. Growth occurs between 23° and 42°C and between pH 5.5 and 9.0. Ecological niche unknown, but originally isolated from a human brain abscess.

The mol% G + C of the DNA is: 52.

Type species: **Morococcus cerebrosus** Long, Sly, Pham and Davis 1981, 300.

FURTHER DESCRIPTIVE INFORMATION

The description of the genus *Morococcus* is based on the characteristics of a single isolate designated as the type strain of the type species *Morococcus cerebrosus*. Thus, the diversity of the species characteristics is unknown. The type strain was isolated from a cerebellar abscess in a 54-year-old woman in 1971 in Australia. The following comments are based on the original description by Long et al. (1981). The type strain grows well on a variety of nutrient media including peptone yeast extract agar (PYEA) and chocolate agar. The addition of glucose enhances growth. Growth occurs on agar containing only peptone or yeast extract, but better growth occurs in media containing both components.

Growth is poor on vitamin-free Casamino acids medium, but improved when yeast extract is added.

On PYEA after 24 h, colonies are 1 mm in diameter and have a frosted-glass appearance under reflected light. The colonies remain intact when moved and are difficult to emulsify. Colonies grown on sucrose peptone agar stain black with Burke iodine stain and the cells are surrounded by a slime layer when stained with nigrosin.

Colonies appear to be constructed of masses of subunits 3–5 μm in diameter. Each subunit contains 10–20 individual cells approximately 0.8 μm in diameter (Fig. BXII.β.86). Individual cells, pairs, or tetrads are rarely observed. Morphological examination by light microscopy is difficult due to the cellular aggregates (Fig. BXII.β.87). Gram stains may appear Gram positive unless the ethanol decolorization step is extended for 20–40 s, and irregular staining is often observed.

Despite obvious morphological differences, the type strains of *Morococcus cerebrosus* and *Neisseria mucosa* show considerable biochemical similarity. However, the two type strains can be differentiated by several characteristics. *Neisseria mucosa* strongly reduces litmus milk and grows on MacConkey agar, but fails to produce acid reactions in Falkow decarboxylase tests. *Morococcus cerebrosus*, on the other hand, reduces litmus milk weakly, fails to grow on MacConkey agar, and gives a strong decarboxylase reaction for ornithine.

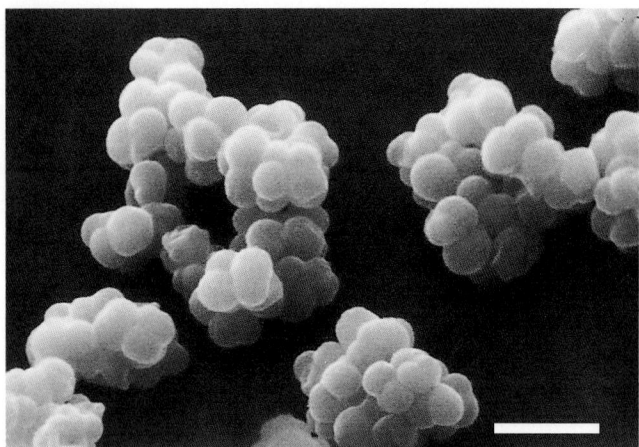

FIGURE BXII.β.86. Scanning electron micrograph of *Morococcus cerebrosus* ACM 858[T] showing mulberry-like cellular aggregates (Bar = 4 μm).

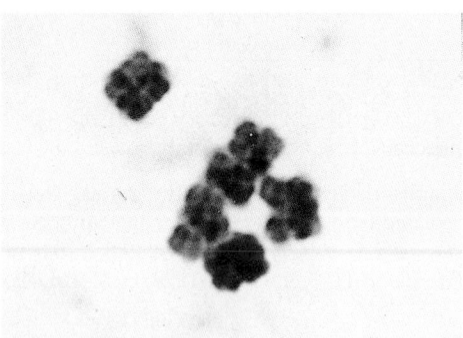

FIGURE BXII.β.87. Photomicrograph of methylene blue-stained cellular aggregates of *Morococcus cerebrosus* ACM 858[T] (× 3000 magnification).

Mice inoculated intraperitoneally with 0.5×10^6 colony-forming units of *Morococcus cerebrosus* became ruffled within 24 h but completely recovered by 48 h. Mice inoculated with 0.5×10^9 colony-forming units died within 24 h. Aggregates were observed microscopically in homogenized spleen samples from dead animals. Mice inoculated subcutaneously developed pyogenic lesions of varying sizes at the sites of inoculation and showed lymph node involvement that persisted for more than 7 d. The larger lesions contained free pus, in which the bacteria were observed by Gram staining and recovered by culture.

Antiserum raised against *Morococcus cerebrosus* in rabbits showed no affinity to *Neisseria mucosa* and when this antiserum was adsorbed with *Neisseria mucosa* it remained reactive to *Morococcus cerebrosus*, indicating the serological difference between these two type strains. These serological differences were confirmed by gel immunodiffusion and gel immunoelectrophoresis.

MAINTENANCE PROCEDURES

Morococcus cerebrosus may be grown on a variety of nutrient media and maintained routinely on peptone yeast extract agar (PYEA). The organism may be preserved by cryogenic storage in liquid nitrogen when suspended in PYE broth containing 10% glycerol, and by freeze drying in glucose peptone broth containing horse serum.

DIFFERENTIATION OF THE GENUS *MOROCOCCUS* FROM OTHER GENERA

The description of *Morococcus* supports its membership of the family *Neisseriaceae*. The closest phenotypic relatives are *Neisseria mucosa* and *Neisseria macacae* (Long et al. 1981; Barrett and Sneath 1994; Ben Dekhill, Stackebrandt and Sly, unpublished data). The cellular aggregates formed by *Morococcus* most clearly differentiate the genus from other members of the *Neisseriaceae*. Barrett and Sneath (1994) confirmed the unique mulberry-like aggregates of *Morococcus cerebrosus* and concluded that numerical analysis clustering supported the inclusion of the genus *Morococcus* within the *Neisseriaceae*. However, unpublished phylogenetic evidence (see Taxonomic Comments below) supports the inclusion of *Morococcus cerebrosus* in the genus *Neisseria*, a taxonomic change that would require emendation of the genus *Neisseria* to accommodate an aggregate forming species. The characteristics that differentiate the genus *Morococcus* from related Gram-negative genera with coccoid morphology are shown in Table BXII.β.104.

TAXONOMIC COMMENTS

In the original description of *Morococcus cerebrosus*, Long et al. (1981) made considerable efforts to distinguish *Morococcus cerebrosus* from *Neisseria mucosa* and concluded that *Morococcus cerebrosus* could be sufficiently differentiated from *Neisseria mucosa* based on physiological, cultural, serological, and morphological characteristics to warrant its description as a separate species in a new genus. Barrett and Sneath (1994) made an extensive numerical phenotypic study of *Neisseria* including *Morococcus cerebrosus*. *M. cerebrosus* was placed alone on a separate branch as a satellite of phenon 24 that contained strains from dental plaque

TABLE BXII.β.104. Differentiation of the genus *Morococcus* from related Gram-negative genera with coccoid morphology[a]

Characteristic[b]	*Morococcus cerebrosus*	*Neisseria*	*Moraxella (Branhamella)*
Cellular morphology:			
Cocci	+	+	+
Rods	−	+[c]	−
Cell aggregates	+	−	−
Acid from glucose	+	D	−
Nitrate reduced	+	D	D
Hemolysis of blood cells	+	D	D
Mol% G + C of DNA	52	46–56	40–47

[a]Symbols: +, present in all species; −, absent in all species; (+), weak reaction; d, 11–89% of strains are positive; D, variable reaction in different species.

[b]Data from Bøvre (1984a, b), Vedros (1984).

[c]*Neisseria elongata* only.

from gorilla and cows. The neighboring cluster (phenon 25) contained one of the replicates of the type strain of *Neisseria mucosa*, although the authors regarded this relationship as unusual. The other replicate was placed in phenon 9 together with one replicate of the type strain of *Neisseria macacae*. The other replicate of *N. macacae* was placed as a satellite of phenons 9–12.

Phylogenetically, *Morococcus cerebrosus* forms a clade with *Neisseria mucosa* and *Neisseria macacae* based on 16S rRNA gene sequence similarity (Ben Dekhill, Stackebrandt, and Sly, unpublished data). *Morococcus cerebrosus* should therefore be considered a species of the genus *Neisseria* but its relationship with *Neisseria mucosa* and *Neisseria macacae* requires further investigation. In a recent study of the phylogeny of the genus *Neisseria*, Harmsen et al. (2001) showed that *Neisseria mucosa* and *Neisseria macacae* had identical 16S rDNA (*E. coli* positions 54–510) and 23S rDNA (*E. coli* position 1400–1600) sequences. Taxonomic change at this time is considered premature until further strains of *Morococcus cerebrosus* are obtained, if possible, and their relationship to *N. mucosa* and *N. macacae* is determined. However, it appears that in the 30 years since the first isolation of *M. cerebrosus* no further isolates have been reported. Aggregate formation as observed in *M. cerebrosus* is inconsistent with the description of the genus *Neisseria* and the description would require emendation to include *M. cerebrosus*.

List of species of the genus Morococcus

1. **Morococcus cerebrosus** Long, Sly, Pham and Davis 1981, 300[VP]

ce.re.bro' sus. L. adj. *cerebrosus* pertaining to the brain, the original source of isolation of this organism.

Colorless, spherical organisms, less than 1 μm in diameter, bound firmly together in tightly packed, mulberry-like aggregates of 10–20 cells, with adjacent sides often flattened. Gram negative. Nonmotile. Endospores not formed. Poly-β-hydroxybutyrate not produced. Capsules are produced in sucrose-containing media.

Growth occurs on vitamin-free Casamino acids, peptone, yeast extract, and chocolate agars. No growth occurs on Sabouraud dextrose agar, Czapek Dox agar, or peptone yeast extract agar (PYEA) containing 40% bile, 0.1% potassium tellurite, or 10% NaCl, or in glucose ammonium sulfate broth. After 24 h colonies on PYEA are convex or umbonate, entire, matt, buff colored, 1 mm in diameter, and difficult to emulsify. Colonies on sucrose agar give a black reaction with iodine. Granular growth occurs in peptone yeast extract broth. Growth occurs between 23° and 42°C, and between pH 5.5 and 9.0.

Aerobic, but may give weak acid reactions in both tubes of the Hugh and Leifson test. Nitrate is reduced to nitrite and gas but the reaction may be variable. Litmus milk is reduced. Hemolyses horse erythrocytes. H$_2$S is produced from cysteine. Deoxyribonuclease, ornithine decarboxylase, catalase, cytochrome oxidase, and methyl red positive. Acid is produced from glucose, fructose, sucrose, and maltose, but not from arabinose, ribose, xylose, rhamnose, galactose, mannose, sorbose, salicin, cellobiose, lactose, melibiose, trehalose, melezitose, raffinose, dextrin, inulin, starch, ethanol, glycerol, erythritol, adonitol, arabinol, dulcitol, mannitol, sorbitol, or inositol. Citrate and malonate are not oxidized. Does not hydrolyze starch, esculin, gelatin, casein, or Tween 80. Lecithinase, phosphatase, phenylalanine deaminase, and urease are not produced. Arginine, glutamic acid, and lysine decarboxylases are not produced. Indole is not produced. Serum liquefaction is negative.

Susceptible to the following antibiotic disks: kanamycin (30 μg), gentamicin (10 μg), streptomycin (10 μg), penicillin G (10 U), ampicillin (2 μg), erythromycin (15 μg), nalidixic acid (30 μg), chloramphenicol (10 μg), and tetracycline (10 μg). Resistant to cloxacillin (5 μg) and methicillin (5 μg).

Antigenically unrelated to *Neisseria mucosa* ACM 1903[T] (ATCC 19696[T]). Pathogenic to mice. Ecological niche unknown, but originally isolated from a human brain abscess in Australia.

The mol% G + C of the DNA is: 52 (T_m).

Type strain: ACM 858, ATCC 33486, NCTC 11393, UQM 858.

Order V. **Nitrosomonadales** *ord. nov.*

GEORGE M. GARRITY, JULIA A. BELL AND TIMOTHY LILBURN

Ni.tro.so.mo.na.da' les. M.L. fem. n. *Nitrosomonas* type genus of the order; *-ales* ending to denote order; M.L. fem. n. *Nitrosomonadales* the *Nitrosomonas* order.

The order *Nitrosomonadales* was circumscribed for this volume on the basis of phylogenetic analysis of 16S rRNA sequences; the order contains the families *Nitrosomonadaceae*, *Gallionellaceae*, and *Spirillaceae*.

Order is morphologically, metabolically, and ecologically diverse. Includes organisms having spiral-shaped, stalked, coccal, rod-shaped, and pleomorphic cells. Includes organisms that can grow chemolithotrophically, mixotrophically, and chemoorganotrophically. Found in a variety of habitats.

Type genus: **Nitrosomonas** Winogradsky 1892, 127 (Nom. Cons. Opin. 23 Jud. Comm. 1958, 169.)

Family I. **Nitrosomonadaceae** *fam. nov.*

GEORGE M. GARRITY, JULIA A. BELL AND TIMOTHY LILBURN

Ni.tro.so.mo.na.da'ce.ae. M.L. fem. n. *Nitrosomonas* type genus of the family; *-aceae* ending
to denote family; M.L. fem. pl. n. *Nitrosomonadaceae* the *Nitrosomonas* family.

The family *Nitrosomonadaceae* was circumscribed for this volume
on the basis of phylogenetic analysis of 16S rRNA sequences; the
family contains the genera *Nitrosomonas* (type genus), *Nitrosolobus,*
and *Nitrosospira.*

Lithoautotrophic bacteria that oxidize ammonia.
   Type genus: **Nitrosomonas** Winogradsky 1892, 127 (Nom. Cons.
Opin. 23 Jud. Comm. 1958, 169.)

Genus I. **Nitrosomonas** *Winogradsky 1892, 127*[AL] *(Nom. Cons. Opin. 23 Jud. Comm. 1958, 169)*

HANS-PETER KOOPS AND ANDREAS POMMERENING-RÖSER

Ni.tro.so.mo'nas. M.L. adj. *nitrosus* nitrous; Gr. fem. n. *monas* a unit, monad; M.L. fem. n. *Nitrosomonas*
nitrite monad, i.e., the monad producing nitrite.

Spherical, ellipsoidal, or rod-shaped cells (Figs. BXII.β.88a,
BXII.β.89a, BXII.β.90a, and BXII.β.91a). Strains belonging to the
same species are generally very similar in the shape and size of
their cells. Cells occur singly or occasionally in short chains. Depending on the growth conditions, cells are **free-living or embedded in slimy aggregates**. Gram-negative cell wall, but **some**

marine representatives show an additional outer cell-wall layer
composed of subunits arranged in a macromolecular array (Watson and Remsen, 1969). Motile cells possess **polar flagella. Intracytoplasmic membranes are arranged as flattened vesicles, primarily in the peripheral cytoplasm** (Figs. BXII.β.88b, BXII.β.89b,
BXII.β.90b, and BXII.β.91b). **Carboxysomes** (Fig. BXII.β.92) are

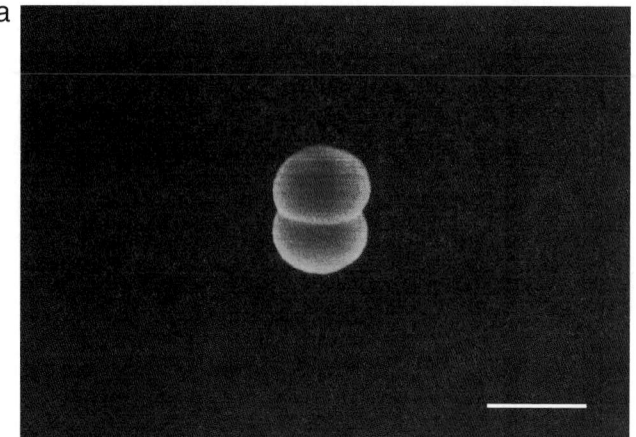
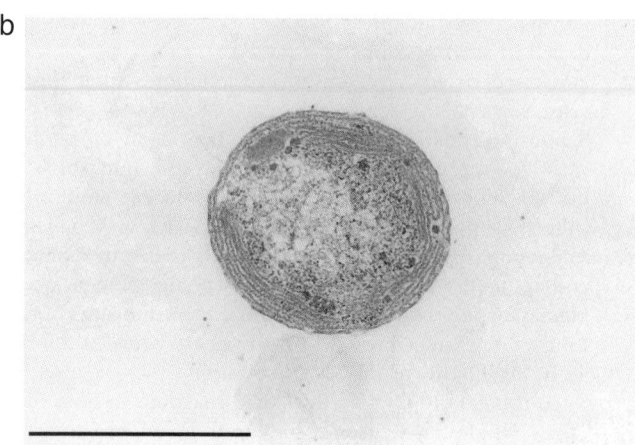

FIGURE BXII.β.88.   (*a*) Scanning electron micrograph and (*b*) electron micrograph of a thin section of cells of *Nitrosomonas mobilis* (*"Nitrosococcus
mobilis"*). Bars = 1 µm.

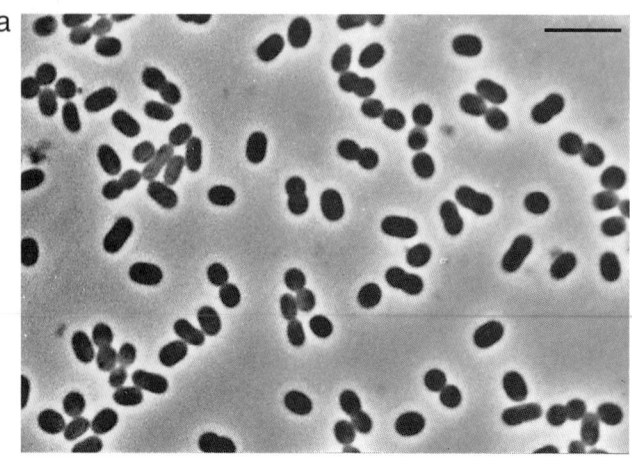
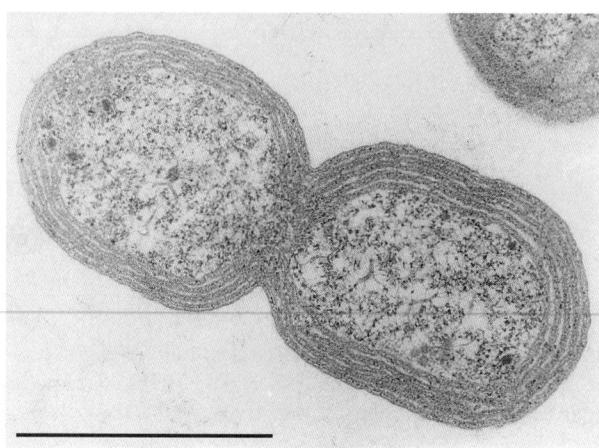

FIGURE BXII.β.89.   (*a*) Phase contrast photomicrograph and (*b*) electron micrograph of a thin section of cells of *Nitrosomonas communis.* Bars = 5 µm
(*a*) and 1 µm (*b*).

present in some species. **Urea** can be used as ammonia source by many, but not all, species. The K_s values for oxidation of NH_3 range between 0.6 and 158 μM. Most, if not all, species are predominantly or exclusively present in special environments.

The mol% G + C of the DNA is: 45–54 (T_m).

Type species: **Nitrosomonas europaea** Winogradsky 1892, 127, emend. mut. char. Watson 1971b, 266.

TAXONOMIC COMMENTS

Eleven species of the genus *Nitrosomonas* are described. Together with other cultured, but undescribed, species of the genus and the reclassified *"Nitrosococcus mobilis"* (proposed here as *Nitrosomonas mobilis* comb. nov.), they form one of the two clusters of ammonia oxidizers within the *Betaproteobacteria*. Seven distinct phylogenetic lineages can be distinguished within this cluster, reflecting ecophysiologically distinct groupings (Table BXII.β.105).

DIFFERENTIATION OF THE SPECIES OF THE GENUS *NITROSOMONAS*

Table BXII.β.105 lists characteristic features useful for phenotypic differentiation of the cultured species of *Nitrosomonas*.

List of species of the genus Nitrosomonas

1. **Nitrosomonas europaea** Winogradsky 1892, 127[AL] emend. mut. char. Watson 1971b, 266.*

   eu.ro.pae′ a. Gr. adj. *europaeus* of Europe, European.

Editorial Note: Readers are advised that although the species of *Nitrosomonas* listed here appeared on Validation List No. 83 (IJSEM 2001, *51:* 1945) type material was not deposited in two public service collections as required by Rule 27 and 30 of the Bacteriological Code as amended in 1999. As such, these names may be illegitimate and invalid, and their usage may be called into question until this matter is satisfactorily addressed.

Cells are rod shaped, with rounded or pointed ends, 0.8–1.1 × 1.0–1.7 μm, generally occurring as single cells. Gram-negative cell envelope. Motility is not observed. Carboxysomes are not present. Urease negative. At pH ~7.8, cultures tolerate ammonium salt concentrations of up to 400 mM. The K_s values of NH_3 oxidation range between 30 and 56 μM. No obligate salt requirement, but cultures exhibit salt tolerance of up to 500 mM NaCl. Most common in sewage disposal plants and in eutrophic waters, but occasionally found in soils.

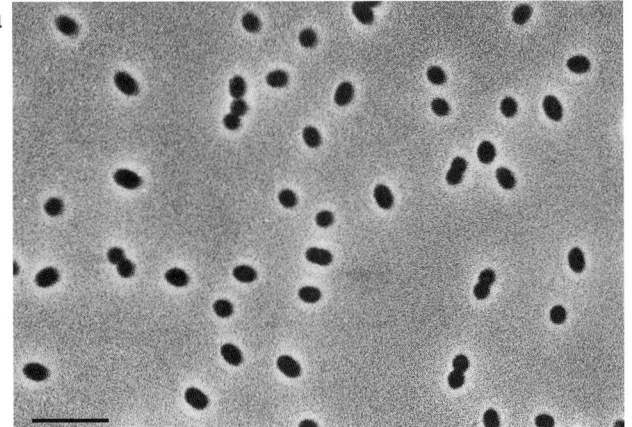
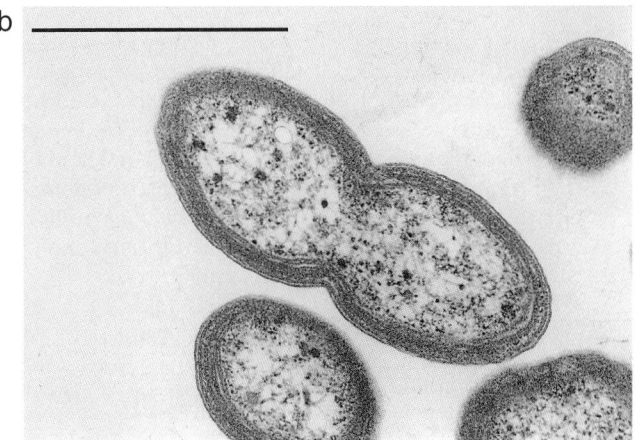

FIGURE BXII.β.90. (*a*) Phase contrast photomicrograph and (*b*) electron micrograph of a thin section of cells of *Nitrosomonas europaea*. Bars = 5 μm (*a*) and 1 μm (*b*).

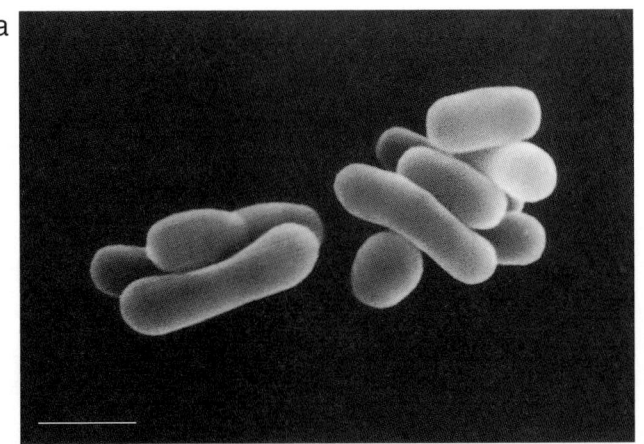
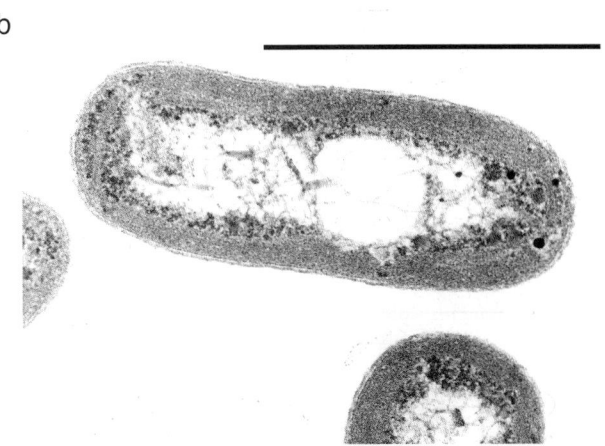

FIGURE BXII.β.91. (*a*) Scanning electron micrograph and (*b*) electron micrograph of a thin section of cells of *Nitrosomonas marina*. Bars = 1 μm.

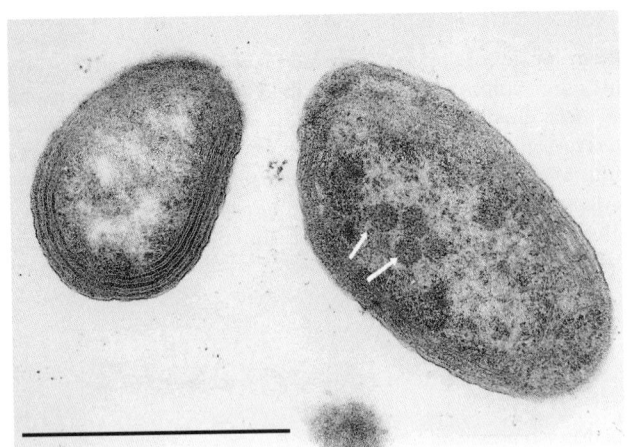

FIGURE BXII.β.92. Electron micrograph of a thin section of a cell of *Nitrosomonas eutropha* showing carboxysomes (*arrows*). Bar = 1 μm.

The mol% G + C of the DNA is: 50.6–51.4 (T_m).

Type strain: ATTC 25978.

GenBank accession number (16S rRNA): M96399.

2. **Nitrosomonas aestuarii** Koops, Böttcher, Möller, Pommerening-Röser and Stehr 1991, 1697.*

ae.stu.a′ ri.i. L. n. *aestus* tides; M.L. gen. n. *aestuarii* of the estuary.

Cells are rod shaped with rounded ends, 1.0–1.3 × 1.4–2.0 μm. Gram-negative cell wall. Additional layers, as found in *N. marina*, are not observed. Nonmotile. No carboxysomes. Cells can use urea as an ammonia source. The maximal ammonium salt tolerance is about 300 mM at pH 7.8. Cultures have an obligate salt requirement, with optimum growth at NaCl concentrations around 300 mM; 700 mM NaCl is tolerated. Isolated from marine environments, chiefly coastal waters.

Editorial Note: Readers are advised that although the species of *Nitrosomonas* listed here appeared on Validation List No. 83 (IJSEM 2001, *51:* 1945) type material was not deposited in two public service collections as required by Rule 27 and 30 of the Bacteriological Code as amended in 1999. As such, these names may be illegitimate and invalid, and their usage may be called into question until this matter is satisfactorily addressed.

The mol% G + C of the DNA is: 45.7–46.3 (T_m).

Type strain: Nm 36.

GenBank accession number (16S rRNA): AJ298734.

3. **Nitrosomonas communis** Koops, Böttcher, Möller, Pommerening-Röser and Stehr 1991, 1697.

com.mu′ nis. L. adj. *communis* common.

Cells are short rods or ellipsoidal with rounded ends, 1.0–1.4 × 1.7–2.2 μm. Gram-negative cell wall. Motility is not observed. No carboxysomes. Urea not used as an ammonia source. Ammonium salts tolerance is up to 200 mM at pH 7.8. The K_s value of NH_3 oxidation is 18–19 μM. No salt requirement; NaCl is tolerated up to 300 mM. Distributed mainly in agriculturally treated neutral soils.

The mol% G + C of the DNA is: 45.6–46.0 (T_m).

Deposited strain: Nm 2.

GenBank accession number (16S rRNA): AJ298732, Z46981.

Additional Remarks: At least two species, *Nitrosomonas* spp. I and II (Table BXII.β.105), exist that are physiologically not distinguishable from *N. communis* and are closely related to this species.

4. **Nitrosomonas eutropha** Koops, Böttcher, Möller, Pommerening-Röser and Stehr 1991, 1697.

eu.tro′ pha. Gr. pref. *eu-* good; *trophos* one who feeds; M.L. fem. n. *eutropha* good nutrition.

Pleomorphic cells, rod to pear shaped, with one or both ends pointed, 1.0–1.3 × 1.6–2.3 μm, occasionally in short chains. Gram-negative cell wall. Most strains are motile. Carboxysomes are present. Cells are urease negative. At pH ~7.8, cultures tolerate concentrations of ammonium salts of up to 600 mM. The K_s value of ammonia oxidation is 35–36 μM. No salt requirement, but up to 500 mM NaCl is tolerated. Commonly abundant in sewage disposal plants and in eutrophic waters. Occasionally found in soils.

The mol% G + C of the DNA is: 47.9–48.5 (T_m).

Type strain: C-91, Nm 57.

GenBank accession number (16S rRNA): AJ298739, AY123795, M96402.

5. **Nitrosomonas halophila** Koops, Böttcher, Möller, Pommerening-Röser and Stehr 1991, 1698.*

hal.o′ phi.la. Gr. n. *halos* salt; Gr. adj. *halophila* salt loving.

TABLE BXII.β.105. Differentiation of the species of the genus *Nitrosomonas*

Characteristic	*N. europaea*	*N. aestuarii*	*N. communis*	*N. eutropha*	*N. halophila*	*N. marina*	*N. mobilis*	*N. nitrosa*	*N. oligotropha*	*N. ureae*	"*N. cryotolerans*"	*Nitrosomonas* sp. I	*Nitrosomonas* sp. II	*Nitrosomonas* sp. III
Phylogenetic linage	1	5	4a	1	1	5	2	4b	3	3	6	4a	4a	5
Mol% G + C content of DNA	51.0	45.8	45.8	48.2	53.8	47.7	49.3	47.9	49.5	45.8	45.8	45.8	45.8	47.7
Salt requirement	−	+	−	−	+	+	+	−	−	−	+	−	−	+
Limit of NaCl tolerans (mM)	500	700	300	500	1000	800	600	200	200	300	600	300	300	800
K_s-value of ammonia oxidation (μM)	36		19.2	36				46	3.6	2.4		14.4	43	
Limit of NH₄Cl tolerance at pH 7.8 (mM)	400	300	200	500	400	200	300	100	50	100	400	200	200	200
Use of urea	−	+	−	−	−	+	−	+	+	+	+	−	−	+
Carboxysomes	−	−	−	+	+	−	−	+	−	−	−	−	−	+

Cells 1.1–1.5 × 1.5–2.2 µm, with a Gram-negative cell wall. Motility has not been observed. Carboxysomes are present. Urease negative. At pH ~7.8, cultures tolerate ammonium salts of up to 400 mM. Moderately halophilic; optimal NaCl concentration around 250 mM; tolerate up to 1000 mM NaCl. Strains were isolated from a brackish water environment (North Sea) and from soda lakes.

The mol% G + C of the DNA is: 53.8 (T_m).

Type strain: Nm 1.

GenBank accession number (16S rRNA): AJ298731, Z46987.

6. **Nitrosomonas marina** Koops, Böttcher, Möller, Pommerening-Röser and Stehr 1991, 1697.*

ma.ri′na. L. fem. adj. *marina* of the sea, marine.

Cells are generally slender rods with rounded ends, 0.7–0.9 × 1.4–2.3 µm. Gram-negative cell wall, but with an additional layer composed of subunits arranged in a macromolecular array. Motility is not observed. No carboxysomes. Cultures have an obligate salt requirement; the optimal NaCl concentration is around 400 mM; 800 mM NaCl is tolerated. Ammonium salts are tolerated at up to 200 mM at pH 7.8. Urease positive. Commonly distributed in marine environments.

The mol% G + C of the DNA is: 47.4–48.0 (T_m).

Type strain: Nm 22

GenBank accession number (16S rRNA): Z46990.

Additional Remarks: At least one further species, *Nitrosomonas* sp. III, exists (Table BXII.β.105) that is physiologically indistinguishable from *N. marina*, but cells are short rods and possess carboxysomes.

7. **Nitrosomonas mobilis** *comb. nov.* ("*Nitrosococcus mobilis*" Koops, Harms and Wehrmann 1976, 281).*

mo′bi.lis. L. adj. *mobilis* movable.

Cells generally spherical, 1.5–1.7 µm in diameter. However, some strains are rod shaped, 1.5–1.7 × 1.7–2.5 µm. Cells occur singly, in pairs, and occasionally as short chains. Gram-negative cell wall. Motile cells have a tuft of 1–22 flagella about 12 nm wide and 3–5 µm long. No carboxysomes. Cells are moderately halophilic, with an optimum NaCl concentration of about 100 mM. NaCl is tolerated at concentrations up to 600 mM. At pH 7.8, ammonium compounds are tolerated up to 300 mM. Urease negative. Strains have been isolated from brackish water environments and sewage disposal plants.

The mol% G + C of the DNA is: 49.3 (T_m).

Type strain: Nc 2.

GenBank accession number (16S rRNA): AF287297, AJ298728, M96403.

8. **Nitrosomonas nitrosa** Koops, Böttcher, Möller, Pommerening-Röser and Stehr 1991, 1697.*

ni.tro′sa. M.L. fem adj. *nitrosa* nitrous.

Spheres or short rods with rounded ends, 1.3–1.5 × 1.4–2.2 µm. Gram-negative cell wall. Motility not observed. Carboxysomes are present. Urea can serve as an ammonia source. Maximum tolerance of ammonium salts at pH 7.8 is 100 mM. The K_s value of NH_3 oxidation is 45–46 µM. No salt requirement; 200 mM NaCl is tolerated. Isolates originate from industrial sewage disposal plants, ponds, and occasionally rivers and soils.

The mol% G + C of the DNA is: 47.9 (T_m).

Type strain: Nm 90.

GenBank accession number (16S rRNA): AJ298740.

9. **Nitrosomonas oligotropha** Koops, Böttcher, Möller, Pommerening-Röser and Stehr 1991, 1698.*

o.li.go.tro′pha. Gr. adj. *oligos* little; Gr. n. *trophos* one who feeds; M.L. fem. n. *oligotropha* little nutrition.

Cells are rod shaped or spherical, 0.8–1.2 × 1.1–2.4 µm in size, with rounded ends. Gram-negative cell wall. Motility not observed. Slimy cell aggregates generally occur in cultures after exponential growth has ceased. No carboxysomes. Urease positive. Sensitive to increasing concentrations (>50 mM at pH 7.8) of ammonium salts. The K_s value of NH_3 oxidation is 2.4–4.2 µM. No salt requirement. NaCl is tolerated at up to 200 mM. Most strains have been isolated from oligotrophic freshwater environments.

The mol% G + C of the DNA is: 49.5–50 (T_m).

Type strain: Nm 45.

GenBank accession number (16S rRNA): AJ298736.

10. **Nitrosomonas ureae** Koops, Böttcher, Möller, Pommerening-Röser and Stehr 1991, 1698.*

u′re.ae. M.L. n. *urea* urea; M.L. gen. n. *ureae* of urea.

Cells are rod shaped, 0.9–1.1 × 1.5–2.5 µm, with rounded ends. Gram-negative cell wall. Motility is not observed. No carboxysomes. Urease positive. Ammonium salts are tolerated at up to 100 mM at pH 7.8. The K_s value of NH_3 oxidation is 1.9–3.6 µM. No salt requirement. NaCl is tolerated at up to 300 mM. Mainly found in oligotrophic fresh water environments and occasionally in soils.

The mol% G + C of the DNA is: 45.6–46.0 (T_m).

Type strain: Nm 10.

GenBank accession number (16S rRNA): AJ298730.

11. **"Nitrosomonas cryotolerans"** Jones, Morita, Koops and Watson 1988, 1122.

cry.o.to.le′rans. Gr. n. *cryos* cold, frost; L. pres. part. *tolerans* tolerating; M.L. part. adj. *cryotolerans* tolerating cold.

Cells are rod shaped with rounded ends, 1.2–2.2 × 2.0–4.0 µm. Gram-negative cell wall, but with an additional outer, electron-dense layer. Besides the flattened peripheral intracytoplasmic membranes that are typical of *Nitrosomonas*, membranes may intrude deep into the cytoplasm, forming what appear to be membrane-bound vesicles. Motility not observed. No carboxysomes. Urease positive. Cultures tolerate ammonium salts at up to 400 mM at pH 7.8. Cells have an obligate salt requirement, with optimal growth around 300 mM NaCl; 600 mM NaCl is tolerated. Cultures can grow at temperatures down to −5°C. The only existing strain originates from the Gulf of Alaska.

The mol% G + C of the DNA is: 45.8 (T_m)

Deposited strain: NW430, Nm 55.

Editorial Note: Readers are advised that although the species of *Nitrosomonas* listed here appeared on Validation List No. 83 (IJSEM 2001, *51*: 1945) type material was not deposited in two public service collections as required by Rule 27 and 30 of the Bacteriological Code as amended in 1999. As such, these names may be illegitimate and invalid, and their usage may be called into question until this matter is satisfactorily addressed.

Genus II. **Nitrosolobus** *Watson, Graham, Remsen and Valois 1971a, 200*[AL]

HANS-PETER KOOPS AND ANDREAS POMMERENING-RÖSER

Ni.tro.so.lob' us. M.L. *nitrosus* nitrous; M.L. n. *lobus* a lobe; M.L. fem. *nitrosolobus* nitrous lobe, a lobe producing nitrite.

Pleomorphic, lobate cells (Fig. BXII.β.93a) **that divide by constriction.** Gram-negative cell wall. Motile cells possess 1–20 **peritrichous flagella**, each 15 nm wide and 2.5–5.0 μm long. **Cells are partially compartmentalized by intracytoplasmic membranes** (Fig. BXII.β.93b). Carboxysomes are not observed. Most, but not all, strains are **urease positive. Glycogen inclusions and polyphosphates** are located primarily in the peripheral compartments of the cell. Commonly distributed in agricultural amended soils. Occasionally observed in freshwater environments.

The mol% G + C of the DNA is: 53.2–56.5.

Type species: **Nitrosolobus multiformis** Watson, Graham, Remsen and Valois 1971a, 200.

DIFFERENTIATION OF THE SPECIES OF THE GENUS *NITROSOLOBUS*

Besides the type species, *N. multiformis*, at least one other species exists. It is distinguishable from the type species by a mol% G + C content of DNA of 56.5 and by its significantly smaller cells (0.8–1.2 × 1.0–1.5 μm). In contrast to *N. multiformis*, this species is occasionally found in sewage disposal plants. This second species of the genus *Nitrosolobus* has not yet been named.

List of species of the genus Nitrosolobus

1. **Nitrosolobus multiformis** Watson, Graham, Remsen and Valois 1971a, 200.

   mul.ti.for' mis. L. adj. *multus* many; L. n. *forma* shape; M.L. adj. *multiformis* many shapes.

   The general characteristics are the same as those described for the genus. Cells are 1.0–1.2 × 1.0–2.5 μm.

   The mol% G + C of the DNA is: 53.2–54.6 (Bd, T_m).

   Type strain: C-71, ATCC 25196.

   GenBank accession number (16S rRNA): L35509.

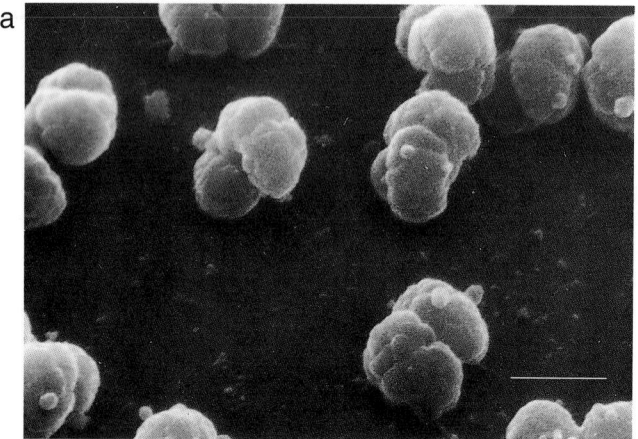
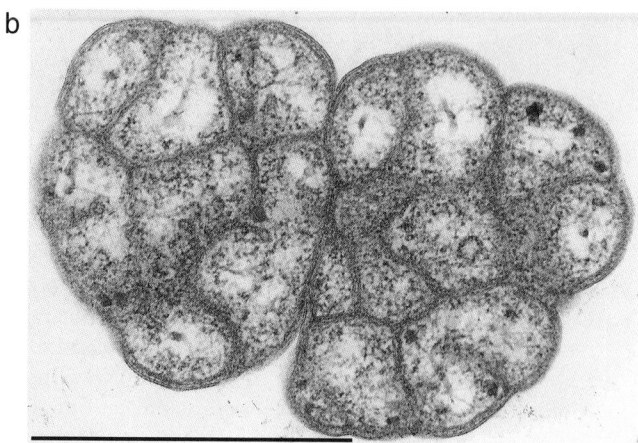

FIGURE BXII.β.93. (*a*) Scanning electron micrograph and (*b*) electron micrograph of a thin section of cells of *Nitrosolobus multiformis* and *Nitrosolobus* sp. N1 5, respectively. Bars = 1 μm.

Genus III. **Nitrosospira** *Winogradsky and Winogradsky 1933, 406*[AL]

HANS-PETER KOOPS AND ANDREAS POMMERENING-RÖSER

Ni.tro.so.spi' ra. M.L. adj. *nitrosus* nitrous; Gr. n. *spira* a coil, spiral; M.L. fem. n. *Nitrosospira* nitrous spiral.

Cells are tightly coiled spirals (Fig. BXII.β.94a), 0.3–0.4 μm in width, with 3–20 turns. Occasionally, vibrioid cells occur in cultures. By phase-microscopy, **spherical forms** 0.8–1.2 μm in diameter may be observed in cultures. Gram-negative cell wall. **Intracytoplasmic membranes are rare** (Fig. BXII.β.94b), but tubular invaginations are observed. Motile strains possess **peritrichous flagella**. 1–6 flagella are observed, each 3–5 μm in length. Urease-positive strains, as well as urease-negative strains, exist in all species. Common in grasslands, heath, forest soils, and mountainous environments. Some strains have been isolated from building stones and from acid soils (pH 4.0–4.5). Occasionally observed in freshwater environments. Using molecular ecological methods, the existence of *Nitrosospira*-like groups in marine environment has repeatedly been indicated (Stephen et al., 1996; Phillips et al., 1999; Bano and Hollibaugh, 2000). However this has not yet been proven via isolation of strains.

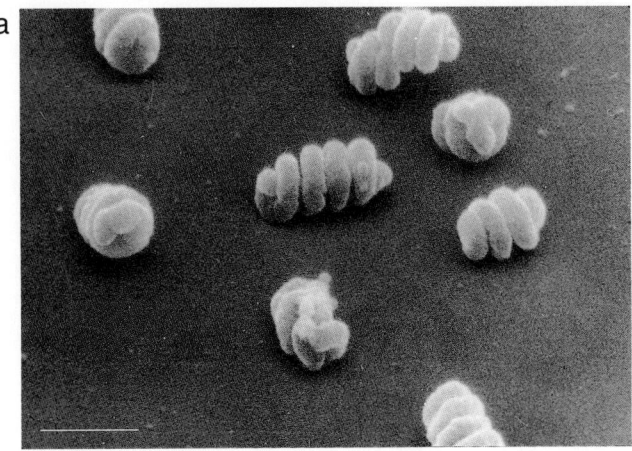
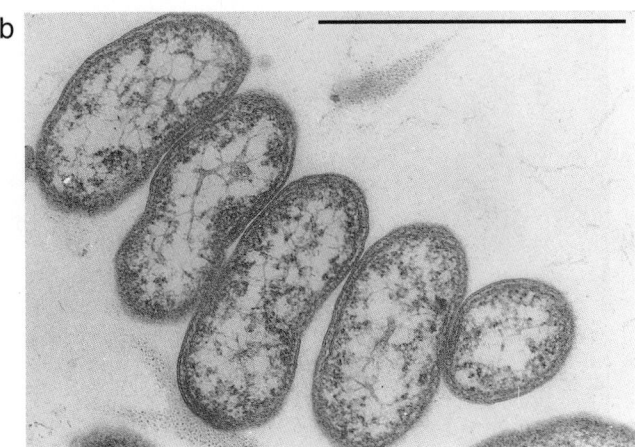

FIGURE BXII.β.94. (*a*) Scanning electron micrograph and (*b*) electron micrograph of a thin section of cells of *Nitrosospira briensis* and *Nitrosospira* sp. Nsp 17, respectively. Bars = 1 μm.

The mol% G + C of the DNA is: 53.2–55.4.

Type species: **Nitrosospira briensis** Winogradsky and Winogradsky 1933, 407.

TAXONOMIC COMMENTS

Originally, two species, *N. briensis* and *"N. antarctica"*, were described by Winogradsky and Winogradsky (1933). Since significant differential characteristics were missing, *"N. antarctica"* was stated by Watson (1971a) to be a subjective synonym of *N. briensis*. However, in DNA reassociation experiments, Koops and Harms (1985) have demonstrated that at least five species of the genus *Nitrosospira* exist. Two groups, containing three and two species, respectively, could be distinguished with 52–53 and 55–56 mol% G + C of their DNA. However, new species have not been described because further phenotypic differential characteristics have not yet been determined.

List of species of the genus Nitrosospira

1. **Nitrosospira briensis** Winogradsky and Winogradsky 1933, 407.

    bri.en' sis. N.L. adj. *briensis* of Brie, a French place name.

    General characteristics are those described for the genus. The original type strain has not been preserved as laboratory culture.

    The mol% G + C of the DNA is: 53.8–54.1 (*T_m*, Bd).

    Type strain: C-76.

Genus IV. "Nitrosovibrio" Harms, Koops and Wehrmann 1976, 110

HANS-PETER KOOPS AND ANDREAS POMMERENING-RÖSER

Ni.tro.so.vib' ri.o. M.L. adj. *nitrosus* nitrous; L. v. *vibrio* to move rapidly to and from, to vibrate; M.L. masc. n. *Nitrosovibrio* a vibrio producing nitrite.

Slender, curved rods (Fig. BXII.β.95a), 0.3–0.4 × 1.1–3.0 μm. **Spherical forms**, 1.0–1.2 μm in diameter, may occur in cultures. Gram-negative cell wall. **Intracytoplasmic membranes are rare** (Fig. BXII.β.95b), but when present, appear as tubular invaginations. Motile cells possess 1–4 **subpolar to lateral flagella** (Fig. BXII.β.96), about 18 nm wide and 4.2–7.5 μm long. Carboxysomes observed in one strain. Most, but not all, strains are **urease positive**. Commonly distributed in oligotrophic soils, such as grasslands, heath, and forest soils, as well as in mountainous environments. Some isolates originate from acid tea soils and from building stones.

The mol% G + C of the DNA is: 53.9.

Type species: **"Nitrosovibrio tenuis"** Harms, Koops and Wehrmann 1976, 110.

TAXONOMIC COMMENTS

Besides the type species, at least one further species exists, but has not yet been distinguished by phenotypic characters or named.

a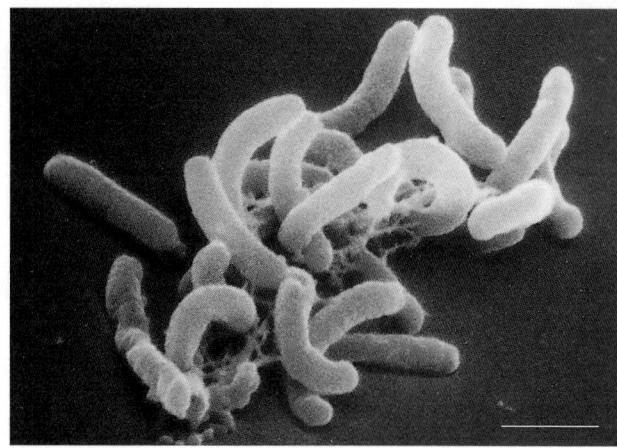

b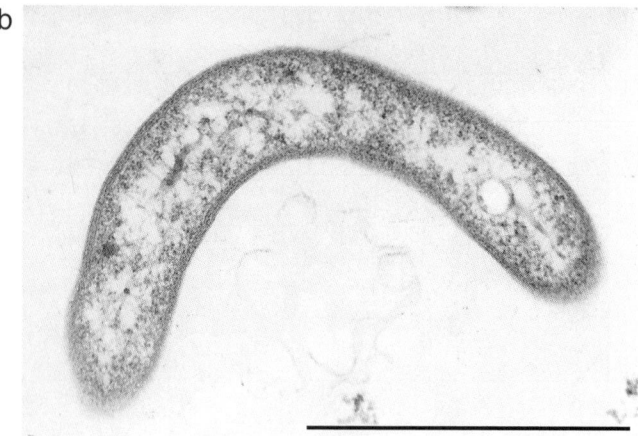

FIGURE BXII.β.95. (*a*) Scanning electron micrograph and (*b*) electron micrograph of a thin section of cells of *"Nitrosovibrio tenuis"* and *"Nitrosovibrio"* sp. Nv 12, respectively. Bars = 1 μm.

List of species of the genus "Nitrosovibrio"

1. **"Nitrosovibrio tenuis"** Harms, Koops and Wehrmann 1976, 110.

   te′ nu.is. L. adj. *tenuis* slender.

   Characteristics are the same as those described for the genus.

   Deposited strain: Nv 1.

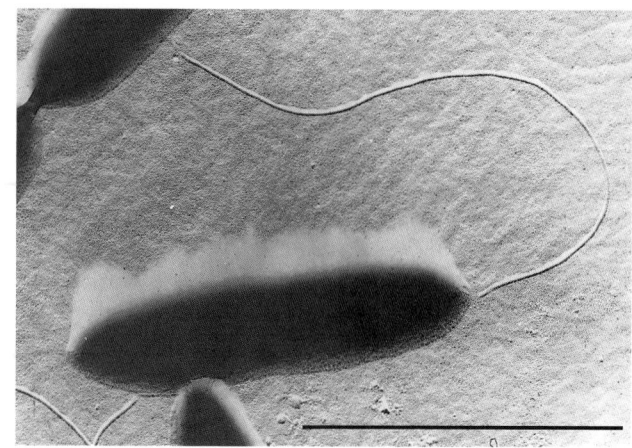

FIGURE BXII.β.96. Electron micrograph of a *"Nitrosovibrio tenuis"* cell shadowed with chromium and showing a single subpolar flagellum. Bar = 1 μm.

Family II. **Spirillaceae** Migula 1894, 237[AL]

GEORGE M. GARRITY, JULIA A. BELL AND TIMOTHY LILBURN

Spi.ril.la′ ce.ae. M.L. dim. neut. n. *Spirillum* type genus of the family; *-aceae* ending to denote family; M.L. fem. pl. n. *Spirillaceae* the *Spirillum* family.

The family *Spirillaceae* was circumscribed for this volume on the basis of phylogenetic analysis of 16S rRNA sequences; the family contains the genus *Spirillum* (type genus).

Description is the same as for the genus *Spirillum.*
Type genus: **Spirillum** Ehrenberg 1832, 38.

Genus I. **Spirillum** Ehrenberg 1832, 38[AL]

NOEL R. KRIEG

Spi.ril′ lum. Gr. n. *spira* a spiral; M.L. dim. neut. n. *Spirillum* a small spiral.

Rigid, helical cells, 1.4–1.7 × 14–60 μm. A **polar membrane** underlies the cytoplasmic membrane at the cell poles and is visible in ultrathin sections. **Intracellular poly-β-hydroxybutyrate granules are formed.** Coccoid bodies are not formed. Gram neg-

ative. **Motile by large bipolar tufts of flagella** having a long wavelength and about one helical turn; these are composed of approximately 75 flagella and are easily visible by dark-field or phase-contrast microscopy. **Microaerophilic** in ordinary liquid media, but can grow aerobically in special media or with certain supplements. Colonies on solid media can be obtained only under special conditions. Have a **strictly respiratory type of metabolism** with oxygen as the terminal electron acceptor. Growth does not occur anaerobically with nitrate. Optimal temperature, 30°C. **Oxidase and phosphatase positive. Catalase negative.** Indole and sulfatase negative. Casein, starch, esculin, gelatin, DNA, and RNA are not hydrolyzed. Inhibited by extremely low levels of hydrogen peroxide in the culture medium. NaCl levels above 0.02% are inhibitory. Phosphate levels greater than 0.01 M are inhibitory. **Carbohydrates are not catabolized.** The salts of certain organic acids are used as carbon sources; succinate is used especially well. Vitamins are not required. Occur in **stagnant, freshwater environments**.

The mol% G + C of the DNA is: 38 or 36.

Type species: **Spirillum volutans** Ehrenberg 1832, 38.

FURTHER DESCRIPTIVE INFORMATION

Structural features An unusual elaboration of the plasma membrane, the "polar membrane," occurs in *S. volutans*. It is attached to the inside of the plasma membrane by barlike links and is located in the region surrounding the polar flagella. Such a membrane has been found mainly in genera of helical bacteria, such as *Aquaspirillum*, *Campylobacter*, *Ectothiorhodospira*, and *Rhodospirillum*. In *Campylobacter*, it is an assembly of ATPase molecules at the poles of the cell (Brock and Murray, 1988). Intracellular poly-β-hydroxybutyrate occurs in the form of prominent granules which are refractile by phase-contrast microscopy and which stain with metachromatic stains such as Ponder's stain (Wells and Krieg, 1965) or with the fluorescent dye Nile blue A (Ostle and Holt, 1982).

Motility The cells are actively motile and swim in straight lines with frequent reversal of direction. The bipolar flagellar fascicles are exceptionally large and consist of many individual flagella (Fig. BXII.β.97). As noted by Metzner (1920), during its rotation the fore fascicle appears to describe a wide bell which is opened toward the rear of the cell; the aft fascicle extends behind the cell and appears to describe a wide goblet (Fig. BXII.β.98). There is no true anterior cell pole: when the fascicles change their orientation the cell reverses its direction of swimming (Fig. BXII.β.98). Because of the wide zones of rotation of the fascicles, Metzner believed that the mechanical effect of the flagella was mainly indirect, i.e., to cause an opposite rotation of the cell body, which, because of its helical shape, would then screw through the medium. Winet and Keller (1976) provided evidence that the aft fascicle of helical cells beats in a helical fashion just as other bacterial flagella do, and Padgett et al. (1983) reported that straight mutant cells could swim at nearly the same speed as helical cells. Thus, it is likely that the flagella operate in a manner similar to that of other bacterial flagella and that the bell and goblet conformations are merely due to the long wavelength of the flagella. Ramia and Swan (1994) have used high-speed cinemicrography to record the swimming of unipolarly flagellated cells. The geometry of these cells was numerically modeled with curved isoparametric boundary elements (from the measured geometrical parameters), and an existing boundary element method (BEM) program was applied to predict the mean swimming linear and angular speeds. Inhibition of the motility of *S. volutans* by low levels of various heavy metals and other toxicants led to development of a method for the monitoring of pollutants in industrial effluents (Bowdre and Krieg, 1974). (See Dutka et al. (1983), Goatcher et al. (1984), Moore (1984), Cortes et al. (1996), Ghosh et al. (1996) and Lacava and Ortolono (1997) for applications, modifications, and evaluations.)

Cultivation *S. volutans* is a microaerophile but can be cultivated easily in a semisolid medium, such as MPSS broth or CHSS

FIGURE BXII.β.97. Flagellar fascicle of *Spirillum volutans* (× 13,000).

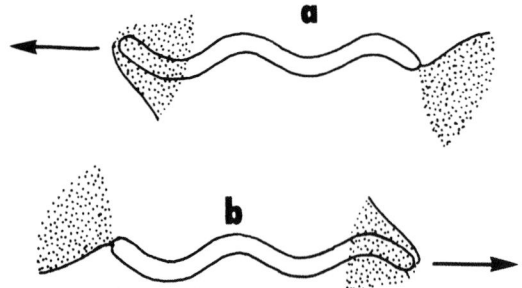

FIGURE BXII.β.98. Diagram of *Spirillum volutans* showing the orientation of the bipolar flagellar fascicles. (*a*) During swimming the fascicles form oriented cones of revolution. (*b*) During reversal of swimming direction, both fascicles reorient simultaneously. *Arrows* in (*a*) and (*b*) indicate the direction of swimming. (Reproduced with permission from J.H. Bowdre and N.R. Krieg, Virginia Polytechnic Institute and State University Water Resources Research Center Bulletin No. 69, 1974, ©Virginia Water Resources Research Center.)

broth[1] prepared with 0.15% agar, incubated under an air atmosphere. Growth is initiated as a thin band or disc several millimeters or centimeters below the surface of the medium, where the respiratory rate of the cells matches the rate of diffusion of oxygen to the cells. As the cell numbers increase, the band becomes denser and migrates closer to the surface. Dense growth just beneath the surface occurs in 48 h.

In MPSS broth, *S. volutans* grows only when incubated under atmospheres of 1–12% oxygen, despite the occurrence in the cells of a superoxide dismutase (SOD) of the iron type (Cover, 1978; Padgett, 1981). Growth does not occur under anaerobic conditions or under an air atmosphere. Addition of catalase (0.8 U/ml) or bovine erythrocyte SOD (4 U/ml) to MPSS broth allows growth to occur under an air atmosphere with static incubation; the two enzymes are most effective when used in combination (0.08 U/ml each) and exert a synergistic effect (Padgett et al., 1982). Potassium metabisulfite (0.005%), acting in conjunction with the $FeCl_3$ component of the MPSS broth, also permits aerobic growth to occur (Padgett et al., 1982).

Aerobic growth can also be obtained by the use of CHSS broth; turbid cultures are obtained in 24 h in 20×125-mm loosely screw-capped tubes incubated in a slanted position under an air atmosphere; daily transfer is required. The pH of the medium rises to 8.0 or higher due to oxidation of the succinate component of the medium. A chemically defined medium for *S. volutans* devised by Bowdre et al. (1976) supports growth under a 6% oxygen atmosphere. When supplemented with norepinephrine (10^{-5} to 10^{-6} M), this medium also supports growth under an air atmosphere.

Growth of *S. volutans* on the surface of solid media (MPSS or CHSS broth solidified with 15.0 g/l agar) is difficult to obtain and depends on several factors: (a) protection of the plates from exposure to illumination during preparation and incubation; (b) addition of potassium bisulfite (0.002%), catalase (130 U/ml),

or SOD (30 U/ml) to the medium (the enzymes must be added aseptically to the molten medium just before dispensing into Petri dishes); and (c) incubation of the plates in an atmosphere of high humidity for 24 h prior to inoculation and for 5 d after inoculation (Padgett et al., 1982). Even with these precautions, colonies have developed only under oxygen atmospheres of 12% or less, and only 22–72% of the cells spread onto the surface of the plates develop into colonies. Colony counts approaching direct microscopic counts have been obtained by inoculating culture dilutions into a semisolid Colony Count Medium (CCM)[2] and using this to overlay a thicker layer of sterile medium, and by supplementing the medium with pyruvate, which destroys hydrogen peroxide. Both the pyruvate and the overlay are necessary for optimal results (Alban and Krieg, 1996). Colonies are pinpoint in size.

Respiration   Respiration rates of *S. volutans* suspended in 0.05 M phosphate buffer are very low (Cole and Rittenberg, 1971), but are higher and, in fact, comparable to those for aerobes when a less inhibitory buffer is used, such as BES buffer (*N,N*-bis(2-hydroxyethyl)-2-aminoethanesulfonic acid) (Caraway and Krieg, 1974). Succinate supports the highest rate of oxygen uptake; fumarate, malate, oxaloacetate, and pyruvate are also readily oxidized. Lactate, butyrate, and β-hydroxybutyrate are oxidized to a lesser extent, and citrate, aconitate, isocitrate, α-ketoglutarate, aspartate, glutamate, casein hydrolysate, and carbohydrates are oxidized only very slightly or not at all. When placed in a nonnutritive medium, *S. volutans* continues to retain motility for up to 24 h, with intracellular poly-β-hydroxybutyrate serving as an endogenous energy source (Caraway and Krieg, 1974). Cytochromes of the *b*, *c*, and *o* types have been detected in *S. volutans*, as well as cytochrome oxidase, NADH oxidase, and various tricarboxylic acid cycle enzymes (Cole and Rittenberg, 1971). Aerotactic responses of *S. volutans* to self-created oxygen gradients have been described by Wells and Krieg (1965) and Caraway and Krieg (1974).

Sensitivity to hydrogen peroxide   *S. volutans* lacks catalase and is extraordinarily sensitive to H_2O_2. Inhibition of growth occurs by addition of as little as 0.29 µM H_2O_2 to culture media (Padgett et al., 1982), and the organism is killed rapidly by levels greater than 10 µM (Alban and Krieg, 1998). Exposure of MPSS broth to moderate or strong illumination causes the generation of H_2O_2 at levels sufficient to inhibit growth, and culture media should be protected from illumination. Potassium bisulfite, and especially a combination of SOD and catalase, can help to prevent the inhibitory effects caused by illumination. A hydrogen peroxide-resistant mutant obtained by single step mutagenesis with diethyl sulfate was reported to survive and grow after exposure to 40 µM H_2O_2 and had high NADH peroxidase activity, whereas the wild type had no detectable activity; however, the mutant was no more oxygen tolerant than the wild type. The mutant con-

---

1. MPSS broth (g/l): Bacto peptone (Difco), 5.0; succinic acid (free acid), 1.0; $(NH_4)_2SO_4$, 1.0; $MgSO_4 \cdot 7H_2O$, 1.0; $FeCl_3 \cdot 6H_2O$, 0.002; and $MnSO_4 \cdot H_2O$, 0.002. The pH is adjusted to 6.8 with 2 N KOH before autoclaving. For CHSS broth, 2.5 g of vitamin-free, salt-free acid-hydrolyzed casein (ICN Biochemicals Inc., Aurora, Ohio) is substituted for the peptone component, and 0.1 g NaCl is added (*S. volutans* requires a low level of Na^+). Different lots of casein hydrolysate may vary in their ability to support good growth. For semisolid media, 1.5 g of agar is added.

2. CCM medium (g/l): vitamin-free, salt-free CH (ICN Biochemicals Inc., Aurora, Ohio), 2.6 g ; succinic acid (free acid), 1.0; $(NH_4)_2SO_4$, 1.0 ; $MgSO_4 \cdot 7H_2O$, 1.0; KH_2PO_4, 0.12; NaCl, 0.04; sodium pyruvate, 0.3; $FeCl_3 \cdot 6H_2O$, 0.002; and $MnSO_4 \cdot H_2O$, 0.002. The pH is adjusted to 7.3 with KOH, 0.05 g potassium metabisulfite is added, and the pH is readjusted to 7.3. Agar (7 g/l) is added and dissolved by boiling, and the medium is sterilized by autoclaving. The pH of the cooled medium after autoclaving should be 6.8. Fifteen-ml volumes of sterile semisolid medium (CCM plus 0.7% agar) are dispensed into Petri dishes and allowed to gel for 30 min. A 0.1-ml volume of dilution of the cell suspension is inoculated into 10 ml of semisolid CCM at 45°C and is poured onto the plates as an overlay. After this has gelled, the plates are incubated at 30°C in an atmosphere of O_2/N_2 (6:94) and incubated for 3–4 days.

stitutively expressed a 21.5-kDa protein that showed high relatedness to rubrerythrin and nigerythrin based on amino acid sequence comparison and was undetectable and noninducible in the wild type cells (Alban et al., 1998).

Nutrition The nutritional requirements of *S. volutans* are not well understood, since the defined medium of Bowdre et al. (1976) may not necessarily be a minimal medium. If succinate is omitted from this medium, however, no growth occurs, and omission of any of the amino acids present (threonine, methionine, histidine, isoleucine, and cystine) gives a decreased growth response. A very low level of NaCl is required. The phosphate concentration must be no higher than 10^{-4} M for aerobic growth to occur, although growth under microaerobic conditions will occur if the level is increased to 10^{-2} M. Because of the toxic effects of low levels of heavy metals, glassware used for cultivation of *S. volutans* must be cleaned with acid and washed extensively in tap and distilled water; even growth in CHSS broth is dependent upon the use of exceptionally clean glassware.

Polyamines Using a polyamine-free medium, Hamana et al. (1994) reported that *S. volutans* cells contain putrescine, cadaverine, and spermidine, but not diaminopropane, 2-hydroxyputrescine, or homospermidine.

Habitat *S. volutans* has been isolated from stagnant pond water in Virginia and from the cooling water of a sugar beet refinery in England (Rittenberg and Rittenberg, 1962; Wells and Krieg, 1965); however, the organism is widely distributed in stagnant freshwater sources and can be demonstrated in nearly any hay infusion prepared from such sources. In hay infusions, the organism occurs in greatest numbers just beneath the surface scum (composed of aerobic organisms), presumably at a location where the dissolved oxygen level is most suitable or where hydrogen peroxide is being destroyed by other bacteria.

ENRICHMENT AND ISOLATION PROCEDURES

Initial enrichment from hay infusion is accomplished by inoculation of Pringsheim's soil medium (Rittenberg and Rittenberg, 1962). This medium is prepared by placing one wheat or barley grain in a large test tube, covering the grain with 3–4 cm of garden soil, filling the tube nearly to the top with tap water, and sterilizing in an autoclave. Even with this enrichment, *S. volutans* is vastly outnumbered by other bacteria. At present, the only successful method for isolation is a mechanical method based on the ability of *S. volutans* to out-swim other bacteria. This method, first devised by Giesberger (1936) for isolation of other spirilla, has been successfully used for *S. volutans* by Rittenberg and Rittenberg (1962) and Wells and Krieg (1965).

A capillary is prepared by heating sterile 5-mm glass tubing in a flame, pinching the softened portion with square-ended forceps until almost closed, then reheating the flattened portion and drawing it out. The resulting oval capillary should be 15–30 cm long and 0.1–0.3 mm in diameter. After cooling, the capillary is broken with sterile forceps at the tip and 10–20 cm of sterile medium (the supernatant fluid from Pringsheim's soil medium) are drawn into it, followed by 2–4 cm of enrichment culture. There should be no air space between the sterile medium and the culture. The tip is then sealed in a flame, leaving a small air space at the tip of the capillary. The capillary is mounted horizontally on the stage of a microscope and observed at 100X. Because of its rapid motility, *S. volutans* will often be able to outdistance the contaminants and be the first to arrive in the distal portion of the sterile medium. (*S. volutans* will frequently

form a band of cells that migrates along the capillary in response to a self-created oxygen gradient; however, the migration rate of such a band is relatively slow, and it is more fruitful to watch for faster-swimming cells well in advance of such a band.) As soon as some spirilla have entered the distal regions of the sterile medium, the capillary is broken behind them and sealed in a flame. After the outside of the capillary has been sterilized with strong hypochlorite solution, and the latter removed with sterile thiosulfate solution, the tip is broken and the spirilla are expelled into a suitable medium. The medium used by Rittenberg and Rittenberg (1962) and Wells and Krieg (1965) was sterile Pringsheim's soil medium contained in a dialysis sac which was suspended in a mixed culture of other bacteria; however, it seems likely that simply expelling the spirilla into a tube of semisolid MPSS or CHSS medium might be a satisfactory alternative. Purity of the cultures is verified by microscopic examination.

MAINTENANCE PROCEDURES

S. volutans may be maintained in semisolid MPSS or CHSS medium at 30°C with transfer every 4–5 days. Cultures may also be maintained in CHSS broth (incubated statically under an air atmosphere) at 30°C with daily transfer.

Preservation by lyophilization has not yet been possible, but cultures may be preserved indefinitely in liquid nitrogen (Pauley and Krieg, 1974). Cells from a broth culture are harvested at 3500 × g, washed once with sterile nutrient broth, and suspended to a dense concentration in nutrient broth containing 10% (v/v) DMSO. After incubation for 30 min to allow penetration of the cells by the cryoprotective agent, the suspension is dispensed into vials. After sealing, the vials are frozen in a mixture of dry ice and alcohol and stored by submersion in liquid nitrogen.

PROCEDURES FOR TESTING SPECIAL CHARACTERS

The characteristics listed in Table BXII.β.106 were determined by methods described by Hylemon et al. (1973a). These authors employed PSS broth (containing 1.0% peptone) rather than MPSS broth, but the latter medium is more satisfactory. Cultures in liquid media are incubated under an atmosphere containing 6% oxygen; this oxygen level is most easily obtained by exhausting the air from the culture vessel and refilling the vessel several times with a mixture of O_2/N_2 (6:94). An alternative procedure is to exhaust the air in the culture vessel until the pressure becomes 0.29 atm, then fill the vessel with N_2 to 1 atm. For certain biochemical tests such as phosphatase activity, the surface of plates of PSS (or MPSS) medium containing 0.7% agar is inoculated heavily (to give confluent growth) and the plates are incubated in a humid atmosphere under 6% O_2.

DIFFERENTIATION OF THE GENUS *SPIRILLUM* FROM OTHER GENERA

Table BXII.β.107 lists the features that distinguish the genus *Spirillum* from other genera of chemoheterotrophic, microaerophilic, motile, curved bacteria.

The large size of the cells, the ease with which the flagellar fascicles can be seen by phase microscopy, and the microaerophilic nature of the organism, serve to differentiate the genus from chemotrophic freshwater spirilla in the genus *Aquaspirillum*. Inhibition of growth by NaCl differentiates the genus from members of *Oceanospirillum*.

The genus *Campylobacter* exhibits certain similarities to *Spirillum*, in that both genera contain cells with polar flagella and are microaerophilic, with a strictly respiratory type of metabolism.

Neither genus can catabolize carbohydrates. The DNA base composition is similar (32–35 mol% G + C for *C. fetus*; 38 mol% G + C for *S. volutans*). Moreover, campylobacters, although nominally vibrioid in shape, can often exhibit a spirillum-like appearance, and, like *S. volutans*, *Campylobacter* species have a polar membrane. Despite these similarities, there are marked differences between the two species: *S. volutans* has a much greater cell diameter, large bipolar tufts of flagella rather than a single flagellum at one or both poles, does not form coccoid bodies, and is not associated with animals or humans.

Distinctive morphological similarities exist between *S. volutans* and members of the phototrophic genus *Thiospirillum*. *Thiospirillum jenense* has flagellar fascicles which are remarkably similar to those of *S. volutans*, and both species have cells of large diameter. However, *T. jenense* is obligately phototrophic and anaerobic, and has a higher mol% G + C for its DNA (45%).

"*Aquaspirillum voronezhense*" has been described by Grabovich et al. (1987) and has a cell diameter of 1.5×3.0 μm and large flagellar fascicles with up to 50 flagella, as well as a number of other features similar to those of *S. volutans*; however, it is aerobic, catalase- and urease-positive, forms colonies on PSS medium under aerobic conditions, and has a mol% G + C of the DNA of 58–60.

TAXONOMIC COMMENTS

At present, the genus *Spirillum* is represented by only a single species, *Spirillum volutans*, of which only two isolates are available. *S. volutans* is a member of the *Betaproteobacteria* and the family *Spirillaceae*, based on the *Bergey's Manual* revision of the RDP tree.

The two species "*Spirillum minus*" and "*Spirillum pulli*" do not belong to the genus *Spirillum* and are described under *Species Incertae Sedis*.

List of species of the genus Spirillum

1. **Spirillum volutans** Ehrenberg 1832, 38[AL]
   vo' lu.tans. L. v. *voluto* to tumble about; L. part. adj. *volutans* tumbling about.

   The morphological characteristics are as described for the genus, listed in Tables BXII.β.106 and BXII.β.107, and depicted in Figs. BXII.β.98 and BXII.β.99. Isolated from stagnant freshwater sources.

   The mol% G + C of the DNA is: 38 (T_m) or 36 (Bd).
   Type strain: ATCC 19554.
   GenBank accession number (16S rRNA): M34131.
   Additional Remarks: Reference strain ATCC 19553.

Other Organisms

1. "*Sporospirillum*" Delaporte 1964, 257.
   spo.ro.spi.ril' lum. Gr. n. *sporos* a seed (spore); Gr. n. *spira* a spiral; M.L. dim. n. *spirillum* a small spiral; M.L. neut. n. *Sporospirillum* a small spore (-forming) spiral.

   Rigid, helical bacteria of enormous size, $1.8–4.8 \times 40–100$ μm. Structures that morphologically resemble endospores occur within the cells, but their thermal resistance has not been determined. The sporelike structures have the ability to rotate and to migrate within the cytoplasm of the bacteria. They initially develop near the cell poles and later migrate to the center where they are released after the cell ruptures and disintegrates. The Gram reaction has not been reported. The cells are motile, but no organs of locomotion are evident. The relationship of the cells to oxygen is unknown. Occur in the intestinal contents of tadpoles. Have not been isolated. No type species designated.

   a. "*Sporospirillum praeclarum*" (Collin 1913) Delaporte 1964, 259 (*Spirillum praeclarum* Collin 1913, 62.)
      prae.cla' rum. L. adj. *praeclarum* distinguished, famous.

      Cells $3.0–4.0 \times 50–100$ μm. Diameter of helix, 5–10 μm. Wavelength, 17–23 μm. A single endospore is present, $3–4 \times 9–12$ μm.

   b. "*Sporospirillum gyrini*" Delaporte 1964, 259.
      gy.ri' ni. L. n. *gyrinus* a tadpole; L. gen. n. *gyrini* of a tadpole.

      Cells $1.8–2.6 \times 40–100$ μm. Diameter of helix 3–6 μm. Wavelength, 13–20 μm. A single endospore is present, $2 \times 5–7$ μm.

   c. "*Sporospirillum bisporum*" Delaporte 1964, 260.
      bi.spo' rum. L. adv. *bis* twice; G. n. *sporos* a seed; M.L. gen. pl. n. *bisporum* of two seeds (spores).

      Cells $3.5–4.8 \times 50–90$ μm. Diameter of helix, 11–15 μm. Wavelength, 27–35 μm. At each pole an endospore occurs, $2–4 \times 10–14$ μm.

Species Incertae Sedis

1. "**Aquaspirillum voronezhense**" Grabovich, Churikova, Chernykh, Kononyhina and Popravko 1987, 666.
   vo.ro.ne.zhen' se. M.L. adj. *voronezhense* pertaining to the town of Voronezh.

   This species has cells that resemble *Spirillum volutans* in size but the mol% G + C of the DNA differs markedly. Consequently, the placement of this species is uncertain.

   Helical cells 1.5–3.0 μm wide, with 1–3 turns per cell. Diameter of helix, 2.9–6.8 μm. Motile by means of bipolar flagellar fascicles, each fascicle being composed of up to 50 flagella. Gram negative. Poly-β-hydroxybutyrate and volutin are accumulated within the cells. S^0 is accumulated in the presence of sulfides. Colonies are flat and nonpigmented with a diameter of 1–4 mm. Aerobic. Optimal pH for growth, 7.2–8.5; pH range for growth, 6.0–9.0. No growth occurs in the presence of 3% NaCl. Optimal temperature for growth, 28°C. Chemoorganotrophic. Carbohydrates are not utilized. A wide range of organic acids, including acetate, succinate, malate, fumarate, benzoate, α-ketoglutarate, oxaloacetate, pyruvate, salicylate, lactate, and glyoxylate, can be utilized as carbon sources. Citrate, ethanol, glycerol, butanol, and mannitol are not utilized. Utilization of isocitrate, aconitate, oxalate, and formate differs among strains. Aspartate can be utilized as a carbon source; utilization of proline, tyrosine, histidine, glycine, glutamine, and valine differs among strains. Tryptophan, methionine, serine, lysine, phenylalanine, asparagine, cysteine, cystine, alanine, leucine, valine, arginine, and ornithine are not utilized. Vitamins are required. Sources of nitrogen used by the type strain include ammonium salts, casein hydrolysate, yeast extract, peptone, aspartate, glutamate, cysteine; utilization of serine differs among strains. Nitrate, nitrite, and methionine are not utilized. Starch, casein, and gelatin

TABLE BXII.β.106. Characteristics of *Spirillum volutans*

Characteristics	Reaction or Result
Cell diameter, μm[a]	1.4–1.7
Wavelength, μm[a]	16–28
Diameter of helix, μm[a]	5–8
Length of helix, μm[a]	14–60
Number of turns[a]	1–5
Intracellular poly-β-hydroxybutyrate granules present[a]	+
Coccoid bodies present in older cultures[a]	−
Bipolar tufts of flagella present[a]	+
Oxidase (moistened test disc inoculated with centrifuged cells)[a]	+
Catalase[b]	−[c]
Phosphatase (0.01% phenolphthalein diphosphate)[d]	+
Sulfatase (0.01% phenolphthalein disulfate)[d]	−
H$_2$S from 0.2% cysteine (detector strip)[a]	+
Liquefaction of 12% gelatin[a]	−
Hydrolysis of casein (single-strength milk)[d]	−
Hydrolysis of esculin[a]	−
Hydrolysis of 0.1% DNA or RNA (clear zone after acidification)[d]	−
Indole production from 0.1% tryptophan[a]	−
Hydrolysis of 10% soluble starch[d]	−
Aerobic reduction of 0.1% KNO$_3$[a]	−
Aerobic reduction of 0.1% H$_2$SeO$_2$ (by pink color)[a]	−
Visible growth with 1% bile or 1% glycine[a]	−
Anaerobic growth with 0.1% KNO$_3$ (sealed with petrolatum)[b]	−
Acid reaction from carbohydrates (38 compounds tested)[e]	−
Urease[f]	−
Mol% G + C of DNA (T_m)	38

[a]Basal medium is PSS broth.

[b]Basal medium is PSS broth + 0.15% agar.

[c]When a few drops of 3% H$_2$O$_2$ are added, a few bubbles of oxygen form after 30-min incubation. However, this reaction is probably attributable to the alkalinity of the cultures rather than to catalase activity.

[d]Basal medium is PSS broth + 0.7% agar.

[e]Basal medium is PSS broth lacking succinate and with peptone decreased to 0.2%; 0.0018% phenol red indicator added.

[f]Cells suspended in distilled water to a dense, milky concentration; 0.5 ml added to 2.0 ml of the following medium: BES buffer, 0.1065%; urea, 2.0%; phenol red, 0.001%; pH 7.0. Controls without urea are used. The test is incubated for 24 h.

are not hydrolyzed. Nitrates are not reduced to nitrites. Anaerobic growth does not occur with nitrate, sulfate, thiosulfate, or fumarate as terminal electron acceptors. Catalase, oxidase, and urease positive. Indole negative. H$_2$S is formed from cysteine. Colored products are formed in media containing benzoate. Source of isolates: sludge in a purification system air tank for the treatment of domestic sewage.

The mol% G + C of the DNA is: 58.5–60.0 (method unknown).

Deposited strain: D-419.

Additional Remarks: Reference strain D-420.

2. **"Spirillum minus"** Carter 1888, 47 (*Spirillum minor* (sic) Carter 1888, 47.)

mi' nus. L. neut. adj. *minus* less, smaller.

"*Spirillum minus*" and "*Spirillum pulli*" (below) do not belong to the genera *Aquaspirillum*, *Oceanospirillum*, or *Spirillum*, and their placement is uncertain. Studies of these species have been hampered by lack of reproducible *in vitro* cultivation methods. They do not appear on the *Approved*

Lists of Bacterial Names because no type or reference strains are available, and the organisms are not well characterized. The disease syndromes caused by these species are distinct and recognizable, however. If possible, neotype strains should be designated and either maintained by animal passage or preserved in a recognized culture collection.

Rigid cells; usually described as spiral with two or three turns, although the waves have been reported to be planar (McDermott, 1928). The ends of the cell may be blunt or pointed. Cell size ~0.2 × 3–5 μm; wavelength, 0.8–1.0 μm. Actively motile by one or more flagella at each pole.

Causes one of the two forms of rat-bite fever in man. The disease caused by "*S. minus*" is often termed "Sodoku"; it occurs worldwide but has its greatest frequency in the Far East. The organisms are usually transmitted to humans through the bite of an infected rat, although mice, squirrels and rodent-ingesting animals such as cats, dogs, ferrets, and weasels have also been implicated. "*S. minus*" appears to be a natural parasite of rats, which act as carriers; the infection is usually not lethal in rats. The natural infection frequency for rats varies from country to country but may be as high as 25% (see Babudieri, 1973, for pertinent literature).

The clinical aspects of rat-bite fever and the distinctions between the form caused by "*S. minus*" and that caused by *Streptobacillus moniliformis* have been summarized by Joklik et al. (1980) and by Rogosa (1980). Experimental infections of humans and animals by "*S. minus*" have been described by Babudieri (1973).

"*S. minus*" is best observed in blood or exudates from patients by dark-field or phase-contrast microscopy of wet mounts; staining with Giemsa or Wright's stain or by silver impregnation is also useful.

"*S. minus*" is cultured *in vivo* by intraperitoneal inoculation of patients' blood or exudates from lesions, or blood from naturally infected rats, into spirillum-free mice or guinea pigs (Rogosa, 1980); mice are the animals most susceptible to "*S. minus*" infection (Babudieri, 1973). It is questionable whether the organism has ever been cultured successfully in artificial media. Numerous attempts have failed, and various claims of successful cultivation have been unable to be confirmed. One report that may indicate successful cultivation is that of Hitzig and Liebesman (1944), who inoculated blood from a patient into 2% dextrose–veal infusion broth and into 10% tomato extract–veal infusion broth. The addition of citrated human or rabbit blood was required for successful subculturing; also, the organisms initially required incubation in a candle jar but eventually were able to grow aerobically after five months of serial transfer. Confirmation of this report is needed. Considering the morphology, pathogenicity, and sources of "*S. minus*", serious attention should be given to the possibility that the organism might belong to, or be related to, the genus *Campylobacter*, and the microaerophilic techniques employed for campylobacters might also prove useful for "*S. minus*".

Deposited strain: none.

3. **"Spirillum pulli"** Mathey 1956, 745.

pul' li. L. gen. n. *pulli* of a young chicken.

Rigid spiral cells. By dark-field microscopy, the cell size is ~1 × 5–12 μm. Actively motile by means of a single flagellum at each end of the cell.

Believed to be the cause of a diphtheroid stomatitis in the mouths of adult chickens. The lesions are yellowish

TABLE BXII.β.107. Differential features of the genus *Spirillum* and other chemoheterotrophic, motile, aerobic or microaerophilic, Gram-negative vibrioid or helical bacteria[a]

Characteristic	*Aquaspirillum*	"*Aquaspirillum voronezhense*"	*Arcobacter*	*Azospirillum*	*Bdellovibrio*	*Campylobacter*	*Cellvibrio*	*Halomonas variabilis*	*Helicobacter*	*Herbaspirillum*	*Marinomonas communis*
Predominant shape:											
Helical	+[b]	+	−	−	−	−	−	−	−	−	−
Curved in one plane	−	−	−	−	−	+[d]	−	−	+[e]	+[f]	+
Vibrioid	−	−	+	+[c]	+	+	+	+	+	+	+
Cell diameter, μm	0.2–1.4	1.5–3.0	0.2–0.9	0.9–1.5	0.2–0.5	0.2–0.9	0.2–0.5	0.5–0.8	0.5–1.0	0.6–0.7	0.7–1.5
Cultivable on inanimate laboratory media	+	+	+	+	−[g]	+	+	+	+	+	+
Require host or host cells for cultivation	−	−	−	−	+[g]	−	−	−	−	−	−
Predacious on other Gram-negative bacteria	−	−	−	−	+	−	−	−	−	−	−
Grow in periplasmic space of host cell	−	−	−	−	+	−	−	−	−	−	−
Exoparasitic growth	−	−	−	−	−	−	−	−	−	−	−
Predacious on eucaryotic algae	−	−	−	−	−	−	−	−	−	−	−
Structures morphologically resembling endospores are present	−	−	−	−	−	−	−	−	−	−	−
Usual arrangement of polar flagella:											
Monotrichous	−	−	+	+	+	−	+	+	−	−	+
Bipolar tufts	+[h]	+	−	−	−	−	−	−	−	−	−
1 at one or both poles	−	−	−	−	−	−	−	−	−	−	−
1–3 at one or both poles	−	−	−	−	−	−	−	−	−	+	−
Multiple at one or both poles	−	−	−	−	−	−	−	−	+[j]	−	−
1 at each pole	−	−	−	−	−	+[i]	−	−	−	−	−
1 or more at each pole	−	−	−	−	−	−	−	−	−	−	−
Sheathed flagella present	−	−	−	−	+	−	−	−	D[k]	−	−
Lateral flagella occur in addition to polar flagella	−	−	−	D[l]	−	−	−	−	D[m]	−	−
Optimal growth temperature 5–9°C; maximum temperature 20°C	−[n]	−	−	−	−	−	−	−	−	−	−
Pathogenic for humans or animals	−	−	D	−	−	D	−	−	+	−	−
Inhibited by 3.5% NaCl	+	−	D	D	D	D	−	−	+	+	−
Sea water or Na+ required for growth	−	−	−	−	D[o]	−	−	+	−	−	+
Require at least 7% NaCl for growth; can grow with 28% NaCl	−	−	−	−	−	−	−	+	−	−	−
Exhibit magnetotaxis; contains magnetosomes	D[p]	−	−	−	−	−	−	−	−	−	−
Nitrogenase activity under microaerobic conditions	−	−	−	+	−	−	−	−	−	+	−
Relation to oxygen under non-N2-fixing conditions:											
Aerobic	+	+	+[q]	+	+	−	+	+	−	+	+
Microaerophilic	−	−	−	−	−	+	−	−	+	−	−
Grows anaerobically by using H_2 or formate as electron donor and fumarate as electron acceptor	−	−	−	−	−	D	−	−	−	−	−
Some carbohydrates catabolized	−[r]	−	−	+	−	−	+	−	−	+	+

(continued)

TABLE BXII.β.107. (cont.)

Characteristic	Aquaspirillum	"Aquaspirillum voronezhense"	Arcobacter	Azospirillum	Bdellovibrio	Campylobacter	Cellvibrio	Halomonas variabilis	Helicobacter	Herbaspirillum	Marinomonas communis
Glucose catabolized	−[s]	−	−	D			+		−	+	+
Cellulose hydrolyzed	−	−	−	−			+		−	−	−
Urease		+				−[t]	−	+	+[u]	−	−
Habitat:											
Freshwater	+	+	−	−	D	−	−	−	−	−	−
Marine	−	−	−[v]	−	D	−	−	+	−	−	+
Soil	−[w]	−	D	+	D	−	+	−	−	+	−
Within plant roots	−	−	D	+	−	−	−	−	−	+	−
Humans and/or warm-blooded animals	−	−	−	−	−	+	−	−	+	−	−
Intestinal contents of tadpoles	−	−	−	−	−	−	−	−	−	−	−

[a] Symbols: +, all species positive except where noted; −, all species negative except where noted; D, differs among species.

[b] A. delicatum is mainly vibrioid.

[c] Some cells in Azospirillum cultures are straight rods.

[d] Chains of Campylobacter cells may have a helical appearance. Campylobacter rectus, Campylobacter showae, and Campylobacter gracilis are straight rods.

[e] H. trogontum and H. bilis are fusiform straight rods. H. bizzozeronii and H. felis cells are long helices. H. cholecystus cells are coccoid to short curved rods.

[f] Some cells in cultures of Herbaspirillum are helical.

[g] All wild-type strains of Bdellovibrio upon initial isolation are dependent on intraperiplasmic growth in susceptible bacterial prey. Mutants capable of axenic growth (prey-independent strains) have been derived from the predacious strains, and some strains are facultative, i.e., capable of growth in the presence or absence of prey cells.

[h] A. delicatum has mainly a single flagellum at one pole; A. polymorphum has mainly a single polar flagellum at each pole; and A. arcticum has a single polar flagellum.

[i] Campylobacter gracilis is nonmotile.

[j] H. cinaedi, H. fennelliae, H. hepaticus, H. pametensis, H. pullorum, and H. rodentium have a single flagellum at one of both poles.

[k] H. pullorum and H. rodentium do not have sheathed flagella.

[l] Especially (or in the case of azospirilla, only) when cells are grown on solid media.

[m] H. mustelae is reported to have multiple lateral flagella in addition to polar flagella.

[n] A. arcticum is positive.

[o] A. nitrofigilis requires Na+. Marine bdellovibrios require Na+.

[p] A. peregrinum and some strains of A. itersonii have nitrogenase activity. C. nitrofigilis have nitrogenase activity.

[q] Arcobacter species can grow aerobically but may be microaerophilic on primary isolation. A. nitrofigilis can grow aerobically on complex media such as Brucella agar.

[r] A. gracile, A. itersonii, A. peregrinum, and A. arcticum can catabolize a very restricted variety of sugars.

[s] A. gracile can produce acid from glucose aerobically. A. itersonii can produce acid from glucose anaerobically but not aerobically. A. arcticum can grow on glucose, fructose, and ribose but not on other carbohydrates.

[t] Some strains of C. lari are urease positive.

[u] H. cinaedi and H. fennelliae are urease negative.

[v] C. nitrofigilis occurs in the roots of salt-marsh grasses.

[w] A. arcticum occurs in arctic sediments.

[x] "Sporospirillum" spp. are motile but no organelles of locomotion have been observed.

[y] O. pusillum has mainly a single flagellum at each pole.

[z] Although Wolinella succinogenes has been regarded as anaerobic, it is in fact a microaerophile. It is capable of respiring with oxygen when it is provided at low concentrations and cannot grow under an air atmosphere. It can also grow anaerobically by using fumarate, polysulfide, or S^0 as a terminal electron acceptor for anaerobic respiration. For further information see Wolin et al. (1961) and Ringel et al. (1996).

[aa] Occurs in sewage waters.

(continued)

TABLE BXII.β.107. (cont.)

Characteristic	Magnetospirillum	Micavibrio	Oceanospirillum (helical species)	Pseudoalteromonas undina	Spirillum	Spirillum minus	"Spirillum pleomorphum"	"Spirillum pulli"	Sporospirillum	Vampirovibrio	Wolinella succinogenes
Predominant shape:											
Helical	+	−	+	−	+	+	−	+	+	−	−
Curved in one plane	−	−	−	−	−	−	+	−	−	−	−
Vibrioid	−	+	−	+	−	−	−	−	−	+	+
Cell diameter, μm	0.2–0.7	0.25–0.35	0.3–1.4	0.7–1.5	1.4–1.7	0.2	0.7–1.0	1.0	1.4–4.8	0.3	0.5–1.0
Cultivable on inanimate laboratory media	+	−	+	+	+	−	+	+	−	−	+
Require host or host cells for cultivation	−	+	−	−	−	+	−	−	+	+	−
Predacious on other Gram-negative bacteria	−	+	−	−	−	−	−	−	−	−	−
Grow in periplasmic space of host cell	−	−	−	−	−	−	−	−	−	−	−
Exoparasitic growth	−	+	−	−	−	−	−	−	−	+	−
Predacious on eucaryotic algae	−	−	−	−	−	−	−	−	−	+	−
Structures morphologically resembling endospores are present	−	−	−	−	−	−	−	−	+	−	−
Usual arrangement of polar flagella:											
Monotrichous		+		+	−		+		−x	+	+
Bipolar tufts			+y		+			+			
1 at one or both poles											
1–3 at one or both poles											
Multiple at one or both poles											
1 at each pole	+										
1 or more at each pole						+					
Sheathed flagella present	−										
Lateral flagella occur in addition to polar flagella	−										
Optimal growth temperature 5–9°C; maximum temperature 20°C	−						+				
Pathogenic for humans or animals	−					+		+			
Inhibited by 3.5% NaCl	+		−	−	+			+			
Sea water or Na+ required for growth	−		+	+	−						
Require at least 7% NaCl for growth; can grow with 28% NaCl	−		−	−	−						
Exhibit magnetotaxis; contains magnetosomes	+										
Nitrogenase activity under microaerobic conditions	D		−		−						
Relation to oxygen under non-N2-fixing conditions:											
Aerobic	−	+	+	+	−	−	+			+	−
Microaerophilic	+	−	−	−	+	+	−	+		−	+z
Grows anaerobically by using H2 or formate as electron donor and fumarate as electron acceptor	−		−	−	−						+

(continued)

TABLE BXII.β.107. *(cont.)*

Characteristic	Magnetospirillum	Micavibrio	Oceanospirillum (helical species)	Pseudalteromonas undina	Spirillum	Spirillum minus	"Spirillum pleomorphum"	"Spirillum pulli"	Sporospirillum	Vampirovibrio	Wolinella succinogenes
Some carbohydrates catabolized	−		−	+	−	−	−	−			−
Glucose catabolized				+							
Cellulose hydrolyzed											
Urease											−
Habitat:											
Freshwater	+	+[aa]	−	−	+	−	−	−		+	−
Marine	−		+	+	−	−	−	−	−		−
Soil	−		−	−	−	−	+	−	−		−
Within plant roots	−		−	−	−	−	−	−	−	−	−
Humans and/or warm-blooded animals	−		−	−	−	+	−	+	−	−	+
Intestinal contents of tadpoles	−		−	−	−	−	−	−	+	−	−

[a] Symbols: +, all species positive except where noted; −, all species negative except where noted; D, differs among species.

[b] *A. delicatum* is mainly vibrioid.

[c] Some cells in *Azospirillum* cultures are straight rods.

[d] Chains of *Campylobacter* cells may have a helical appearance. *Campylobacter rectus*, *Campylobacter showae*, and *Campylobacter gracilis* are straight rods.

[e] *H. trogontum* and *H. bilis* are fusiform straight rods. *H. bizzozeronii* and *H. felis* cells are long helices. *H. cholecystus* cells are coccoid to short curved rods.

[f] Some cells in cultures of *Herbaspirillum* are helical.

[g] All wild-type strains of *Bdellovibrio* upon initial isolation are dependent on intraperiplasmic growth in susceptible bacterial prey. Mutants capable of axenic growth (prey-independent strains) have been derived from the predacious strains, and some strains are facultative, i.e., capable of growth in the presence or absence of prey cells.

[h] *A. delicatum* has mainly a single flagellum at one pole; *A. polymorphum* has mainly a single flagellum at each pole; and *A. arcticum* has a single polar flagellum.

[i] *Campylobacter gracilis* is nonmotile.

[j] *H. cinaedi*, *H. fennelliae*, *H. hepaticus*, *H. pametensis*, *H. pullorum*, and *H. rodentium* have a single flagellum at one of both poles.

[k] *H. pullorum* and *H. rodentium* do not have sheathed flagella.

[l] Especially (or in the case of azospirilla, only) when cells are grown on solid media.

[m] *H. mustelae* is reported to have multiple lateral flagella in addition to polar flagella.

[n] *A. arcticum* is positive.

[o] *A. nitrofigilis* requires Na^+. Marine bdellovibrios require Na^+.

[p] *A. peregrinum* and some strains of *A. itersonii* have nitrogenase activity. *C. nitrofigilis* has nitrogenase activity.

[q] *Arcobacter* species can grow aerobically but may be microaerophilic on primary isolation. *A. nitrofigilis* can grow aerobically on complex media such as *Brucella* agar.

[r] *A. gracile*, *A. itersonii*, *A. peregrinum*, and *A. arcticum* can catabolize a very restricted variety of sugars.

[s] *A. gracile* can produce acid from glucose aerobically. *A. itersonii* can produce acid from glucose anaerobically but not aerobically. *A. arcticum* can grow on glucose, fructose, and ribose but not on other carbohydrates.

[t] Some strains of *C. lari* are urease positive.

[u] *H. cinaedi* and *H. fennelliae* are urease negative.

[v] *C. nitrofigilis* occurs in the roots of salt-marsh grasses.

[w] *A. arcticum* occurs in arctic sediments.

[x] "*Sporospirillum*" spp. are motile but no organelles of locomotion have been observed.

[y] *O. pusillum* has mainly a single flagellum at each pole.

[z] Although *Wolinella succinogenes* has been regarded as anaerobic, it is in fact a microaerophile. It is capable of respiring with oxygen when it is provided at low concentrations and cannot grow under an air atmosphere. It can also grow anaerobically by using fumarate, polysulfide, or S^0 as a terminal electron acceptor for anaerobic respiration. For further information see Wolin et al. (1961) and Ringel et al. (1996).

[aa] Occurs in sewage waters.

white, rather firm, and adherent to the underlying tissue; often they are symmetrical ovoids, one at each side of the lower jaw. Lesions also occur on the palate, the lower surface of the tongue, on the floor of the mouth, between the larynx and the transverse row of papillae on the tongue, around the larynx, and on the walls of the pharynx. The lesions vary in size from approximately 2 to 20 mm.

Attempts to culture *"S. pulli"* in artificial media have been unsuccessful. Experimental passage of the disease in chickens has been accomplished by contact and by experimental inoculation.

Deposited strain: none.

Family III. **Gallionellaceae** Henrici and Johnson 1935b, 4[AL]

GEORGE M. GARRITY, JULIA A. BELL AND TIMOTHY LILBURN

Gal.li.o.nel.la' ce.ae. M.L. fem. n. *Gallionella* type genus of the family; *-aceae* ending to denote family; M.L. fem. pl. n. *Gallionellaceae* the *Gallionella* family.

The family *Gallionellaceae* was circumscribed for this volume on the basis of phylogenetic analysis of 16S rRNA sequences; the family contains the genus *Gallionella* (type genus).

Description is the same as for the genus *Gallionella*.

Type genus: **Gallionella** Ehrenberg 1838, 166.

Genus I. **Gallionella** Ehrenberg 1838, 166[AL]

LOTTA E-L. HALLBECK AND KARSTEN PEDERSEN

Gal.li.o.nel' la. M.L. dim. ending *-ella* ; M.L. fem. n. *Gallionella* named for B. Gaillon, a customs agent and zoologist (1782–1839) in Dieppe, France.

Gram-negative, **bean-shaped cells**, usually 0.5–0.8 × 1.6–2.5 μm, that secrete an extracellular **twisted stalk** from the concave side, 0.3–0.5 μm in width and up to 400 μm or more in length. The stalk is composed of numerous 2 nm-wide fibers and is produced under microaerophilic conditions when cells are in late exponential or stationary growth phase. Motile by means of a polar flagellum. **Microaerophilic; chemolithotrophic growth** can be obtained *in vitro* using oxygen and ferrous iron concentration gradients in a salt medium with CO_2 **as sole carbon source** (Table BXII.β.108). **Mixotrophic metabolism** has been demonstrated with glucose, fructose, and sucrose. Can be found where anaerobic **groundwater with ferrous iron** reaches an oxygen-containing environment. Belongs to the **Betaproteobacteria**, family *Gallionellaceae*, with one known species, *G. ferruginea*. Most closely related species according to 16S rDNA sequence analysis is the chemolithotroph *Nitrosospira multiformis*, distantly related with a 16S rDNA sequence similarity of 90%.

The mol% G + C of the DNA is: 51–54.6 (Hanert, 1989).

Type species: **Gallionella ferruginea** Ehrenberg 1838, 166.

FURTHER DESCRIPTIVE INFORMATION

The 16S rRNA gene of *Gallionella ferruginea* (strain Johan) has been sequenced between base numbers 47 and 1405 (*E. coli* numbering) (Hallbeck et al., 1993). Phylogenetic analysis of these sequence data placed *Gallionella* among the *Betaproteobacteria*. *G. ferruginea* is distant from other species in the tree, with a 10% sequence difference compared to the closest species, the chemolithotroph *Nitrosospira multiformis*. The remote position of *G. ferruginea* in relation to other species and its utilization of iron as energy source and electron donor compared to ammonia for the closest gene cluster, *Nitrosomonadaceae*, indicate that a separate family, *Gallionellaceae*, is justified. There is only one named species at this time. The capacity for chemolithotrophic iron oxidation among bacteria has been suggested to be evolutionarily widespread (Lane et al., 1992). This is supported by, for instance, the phylogenetic distance (85% 16S rRNA gene sequence simi-

larity) between *Gallionella* and the iron-oxidizing genus *Thiobacillus*.

The size, shape, and ultrastructure of *Gallionella* are illustrated in Figs. BXII.β.99, BXII.β.100, BXII.β.101, and BXII.β.102. Cells are curved, bean-shaped, and may have a polar flagellum (Fig. BXII.β.99). A twisted extracellular stalk is secreted from the concave side of the cell (Figs. BXII.β.100 and BXII.β.101A). It consists of numerous 2 nm-wide fibers at the point of excretion (Vatter and Wolfe, 1956). The stalk becomes continuously encrusted with precipitated ferric iron oxide (Fig. BXII.β.101B), which may totally cover old stalks. The composition of the stalk has not been conclusively demonstrated, but inorganic as well as organic compositions have been suggested. Hanert (1989) proposed the stalk to consist of colloidal ferric hydroxide, based on its disappearance in 0.12% sodium thioglycolate. Hallbeck and Pedersen (1995) found a higher carbon-to-nitrogen ratio (C/N 6.8) in stalk-forming cultures compared to a non-stalk-forming culture (C/N = 4.3), and concluded the stalk to be composed of an extracellular carbon skeleton without a dominating protein component. The cell wall is Gram-negative (Fig. BXII.β.102D–E).

Two strains of *Gallionella* have been described (Table BXII.β.109). Strain BD from a drainpipe in Braunschweig was described by Hanert (1989) and strain Johan from a 60 m-deep drinking water well was described by Hallbeck and Pedersen (1990, 1991, 1995) and Hallbeck et al. (1993). Strain BD was reported to have intracytoplasmic membranes (Hanert, 1989) while strain Johan does not (Fig. BXII.β.102). The isolation procedure for *Gallionella* is by serial dilution and therefore the possibility of contaminants in the cultures cannot be conclusively excluded. Serial thin sectioning should show a stalk connected to the sectioned cell, as in Figure BXII.β.102A–C, to confirm it to be a cell of *Gallionella*. Inclusions that might be poly-β-hydroxybutyrate have been noted but apart from this, strain Johan does not show any specific intracellular fine structures (Fig. BXII.β.102D–E).

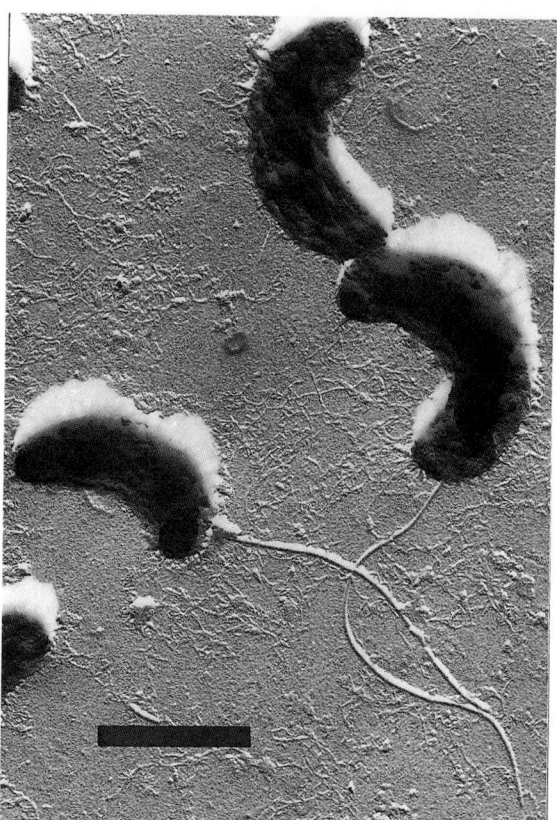

FIGURE BXII.β.99. *G. ferruginea* strain BD cell morphology and flagella arrangement in cell suspension after the stalks were dissolved in sodium thioglycolate. Bar = 1 μm.

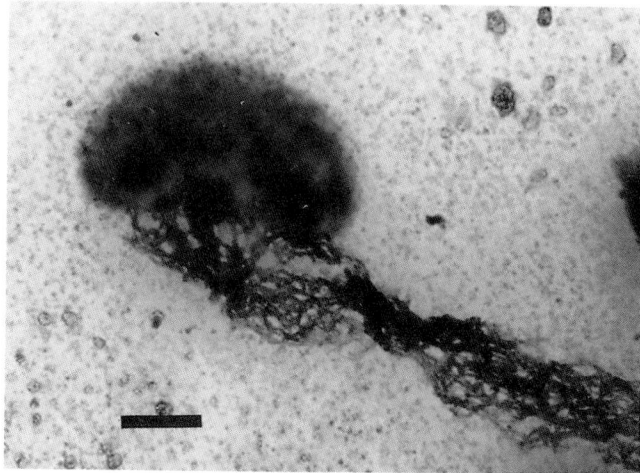

FIGURE BXII.β.100. Apical cell, region of stalk secretion, and ultrastructure of the stalk of attached *G. ferruginea* strain BD from the drain pipe from which it was isolated. Bar = 0.5 μm. (Reproduced with permission from H.H. Hanert, *Archives of Microbiology 60*: 348–376, 1968, ©Springer-Verlag, Berlin.)

Gallionella can be cultured *in vitro* in screw-capped test tubes using oxygen and ferrous iron concentration gradients in a salt medium with carbon dioxide as sole carbon source (Fig. BXII.β.103). Solid-phase ferrous iron is placed on the bottom of the tube in a fresh, autoclaved, and oxygen-free salt medium. With this procedure, the concentration of ferrous iron will decrease from the bottom to the top of the tube as it dissolves and diffuses away from the solid phase. Oxygen will decrease in concentration from the top of the tube downwards as it diffuses into the salt medium from the air above the medium. The optimal growth temperature is 17–20°C and temperatures above 25–30°C are lethal. Iron sulfide or iron carbonate can be used as a source of ferrous iron. The culture grows to a maximum of 5×10^6 cells/ml culture (Fig. BXII.β.101C), which makes many traditional techniques for strain characterization impossible. Strain BD develops small, circular, colonies attached to the wall of the tube, 3–5 d after inoculation. Strain Johan predominantly forms a ring at a specific level in the concentration gradient. The cells of strain Johan colonize as new rings from upper levels in the tube as the culture gets old, i.e., 10 days or more. Eventually, most of the tube will be filled by a brittle mass of stalks, iron oxides, and cells. Hallbeck and Pedersen (1990) have demonstrated that *Gallionella* strain Johan is free living *in vitro* in its exponential growth phase and does not produce stalks until late exponential and stationary phases. The generation time at optimal temperature for strain Johan is 8.3 h. Stalk production continues for many days in stationary phase. Some cell division may still occur but not at the growth rate observed during the first 4–5 d after inoculation. The maximum mean stalk length

per cell of strain Johan has been determined to be 60 μm in a 16-day-old culture. Individual cells may produce much longer stalks; 400 μm has been reported for a single cell of strain BD. With prolonged subculturing on iron carbonate as energy source, some of the *Gallionella* strain Johan cultures were found to have irreversibly lost their ability to form a stalk (Hallbeck et al., 1993). Their identity was confirmed by 16S rRNA gene sequencing. They still form a ring of oxidized iron as the stalk-forming variant does, but the ring is very thin.

Gallionella is chemolithotrophic with ferrous iron as energy source and electron donor and with carbon dioxide as sole carbon source. CO_2 fixation by strain Johan was revealed using hydrogen [^{14}C]-carbonate (Hallbeck and Pedersen, 1991), while *in vivo* activity of the Calvin cycle key enzyme, ribulose bisphosphate-carboxylase, has been reported for strain BD (Hanert, 1989). Mixotrophic metabolism has been shown on glucose, fructose, and sucrose. Growth did not occur on these sugars without ferrous iron and carbon dioxide. Ammonium and nitrate can be used as nitrogen sources; the capacity for nitrogen fixation is unknown.

The environment where stalk-forming *Gallionella* can be found, commonly attached to surfaces, is slowly flowing groundwater that is rich in ferrous iron but has a low organic carbon content. Typical places to search for *Gallionella* are in drainpipes, storage basins for groundwater from deep wells, in tunnels, and on rock walls with seeping groundwater. A common feature of these environments is that cold (below 20°C), reduced, anaerobic, and ferrous-iron-bearing groundwater reaches an oxygen-containing atmosphere. Such environments are suitable for chemolithotrophic growth with ferrous iron as energy source and electron donor, and oxygen as electron acceptor. A slow flow, supplying ferrous iron and possibly carbonate from the groundwater to the attached cells, seems to be an absolute prerequisite for growth of *Gallionella*. In this situation, the stalk may act as a holdfast and prevent the cells from being washed out to a more oxidized environment without ferrous iron.

Hallbeck and Pedersen (1995) have demonstrated an additional function for the stalk. The iron oxidation that occurs in a typical *Gallionella* environment can be differentiated into two

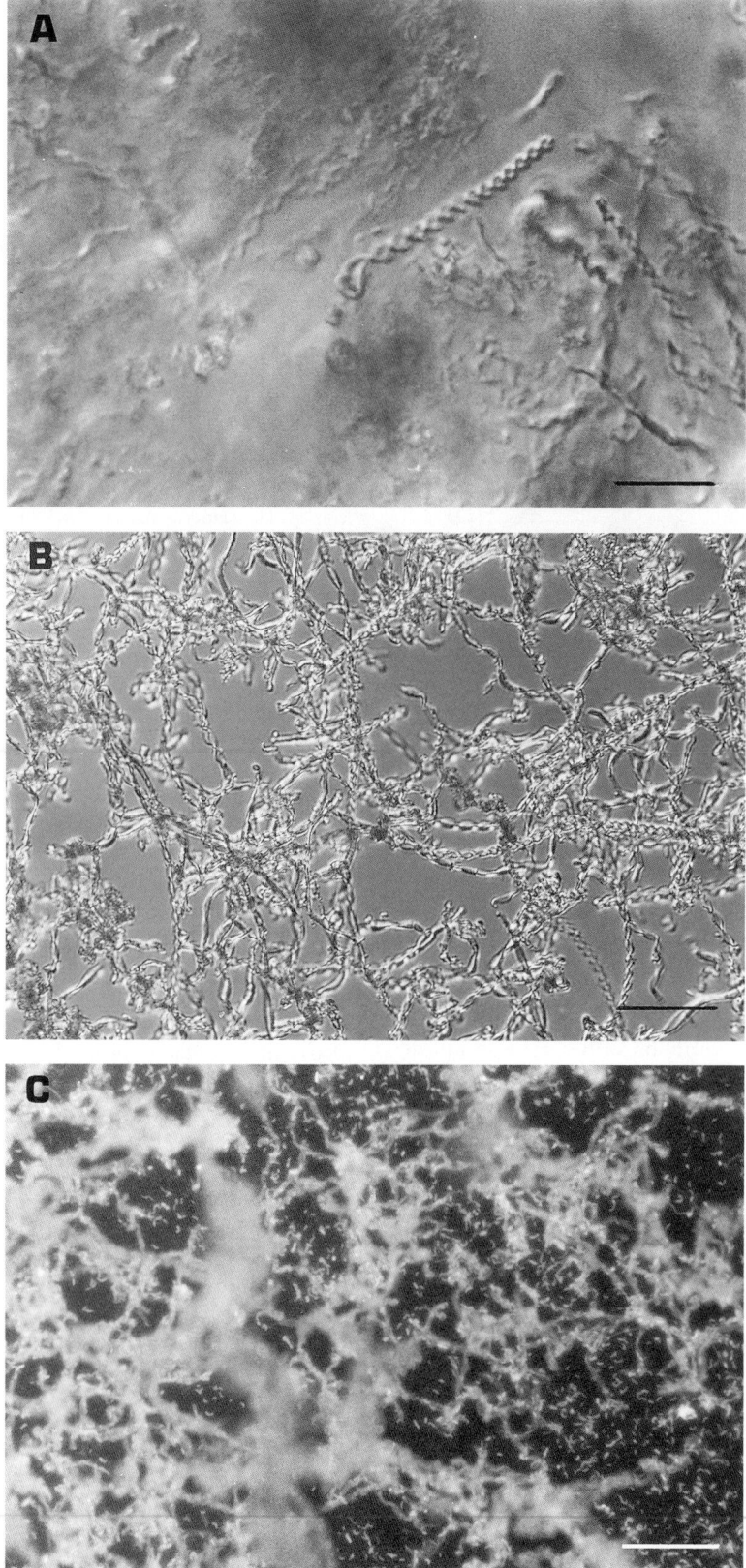

FIGURE BXII.β.101. Light microscopy images of cells and stalks of *G. ferruginea* strain Johan. *A*, Normarski microscope image of a stalked cell from an environmental sample. Bar = 10 μm. *B*, Normarski microscope image of stalks with precipitated iron from a culture. Bar = 50 μm. *C*, Fluorescent microscopy image of an acridine orange stained culture, filtered on a 0.2 μm membrane filter. Note the abundant cells among the stalks. Bar = 50 μm.

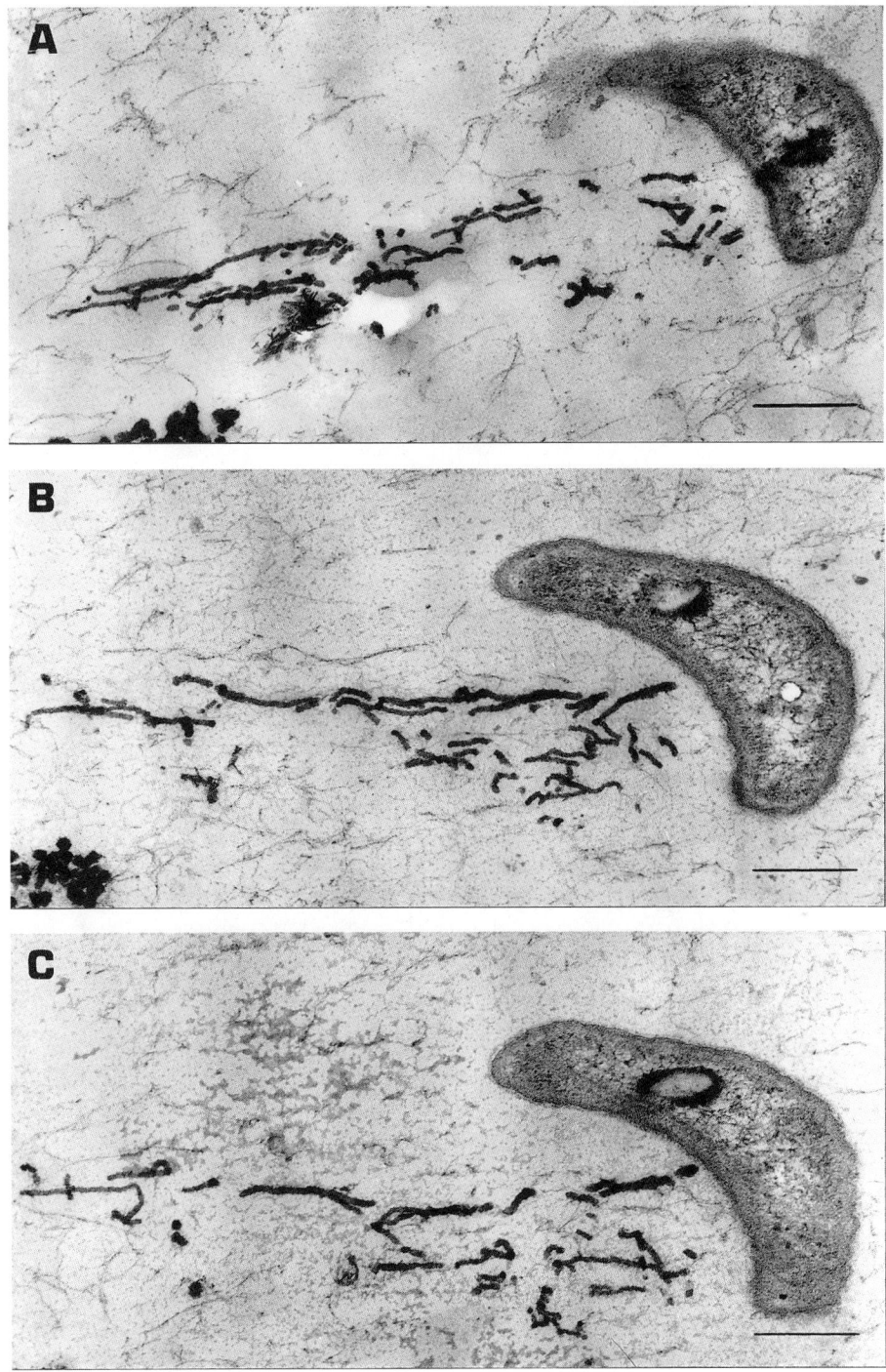

FIGURE BXII.β.102. Ultra thin section of *G. ferruginea* strain Johan in stationary phase. *A–C*, Serial thin sections of a cell with an adjacent stalk. Bars = 0.5 μm. *D*, Cross section of strain Johan with cell wall outer membrane (see next page). Bar = 0.2 μm. *E*, Longitudinal section of strain Johan. Bar = 0.25 μm (see next page).

(continued)

parts: a) The respiratory iron oxidation performed by the cells in their energy metabolism and b) the nonmetabolic iron oxidation induced by the increasing oxygen tension as the anoxic groundwater reaches the atmosphere. Ferrous iron reacting with oxygen participates in a chain of reactions yielding highly reactive oxygen such as perhydroxyl (HO_2), hydrogen peroxide (H_2O_2), and the hydroxyl radical (HO) (Stumm and Morgan, 1996). The survival of stalk-forming *Gallionella* strain Johan (Sta^+) in media with low and high potential for oxygen radical formation was compared with a variant of strain Johan that irreversibly lost the ability to form a stalk (Sta^-) (Hallbeck and Pedersen, 1995). It was found that *Gallionella* Sta^+ survived longer (9 weeks) than Sta^- (6 weeks) in cultures with a high potential for oxygen radical formation. It was therefore suggested that the stalk of *Gallionella* protects the cells against the toxic oxygen species discussed above, by directing the oxidation of iron to the stalk. This phe-

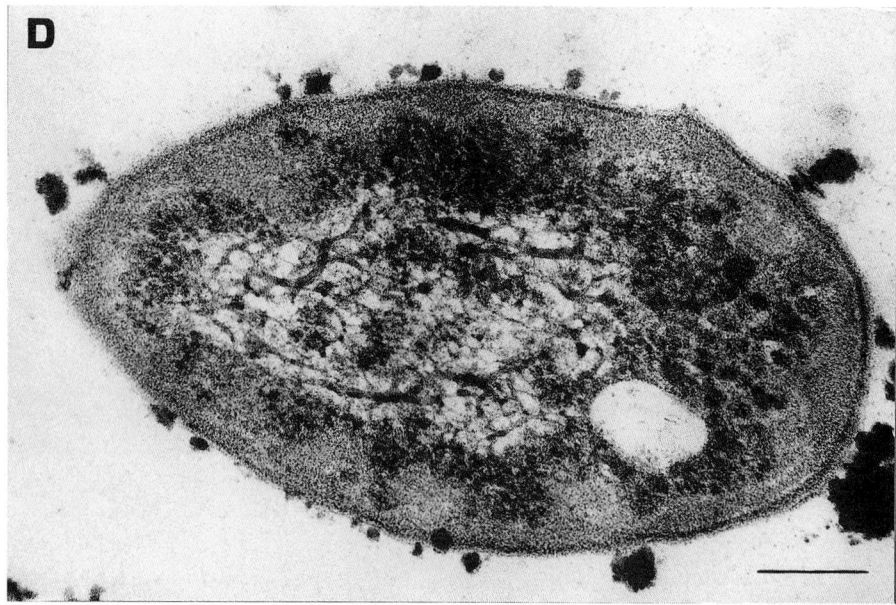

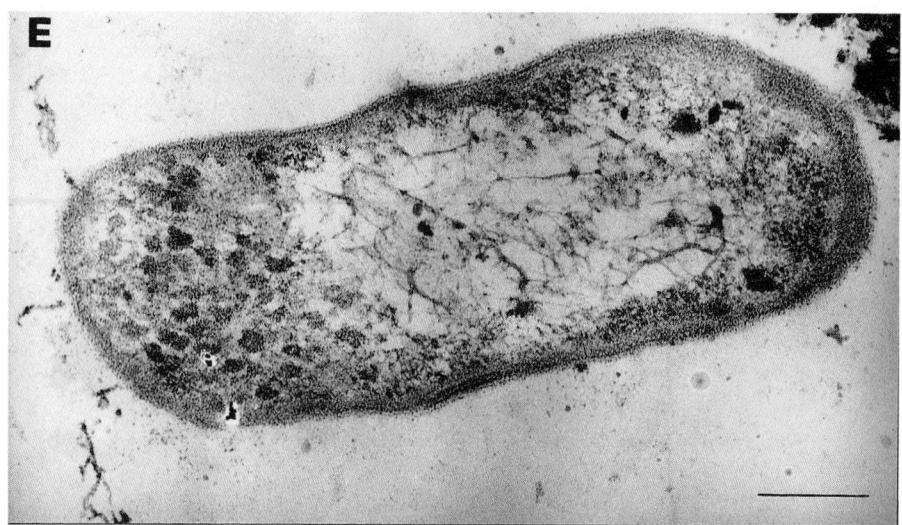

FIGURE BXII.β.102. *(continued)*

TABLE BXII.β.108. Characteristics for identification of the genus *Gallionella*

Characteristic	*Gallionella*
Cell morphology	Bean shaped
Cell dimension, μm	0.5–0.8 × 0.8–2.5
Produce an extracellular twisted stalk composed of numerous 2 nm-wide fibers	+
Grows with CO_2 as sole carbon source	+
Grows with ferrous iron as energy source	+
Temperature range (°C)	5–25
Optimal temperature (°C)	17–20
pH range	5.0–6.5
Motile	+

nomenon could be compared to the action of the protein ferritin, proposed to perform iron oxidation in both procaryotic and eucaryotic cells (Artymiuk et al., 1991). It is not known whether the iron oxidation on the stalk is enzymatic or the stalk

acts as a surface catalyst for the oxidation reaction. Thus, the stalk acts as a holdfast that allows *Gallionella* to colonize and survive in an ecological niche, with high ferrous iron content and some oxygen, that is unavailable for bacteria without a defense system against the oxygen radicals formed during inorganic oxidation of ferrous iron.

ENRICHMENT AND ISOLATION PROCEDURES

Various media for enrichment and cultivation of *Gallionella* have been tested. To create proper growth conditions, ferrous iron and CO_2 must be in the medium. Lieske (1911) designed a culture medium composed of carbonic water and metallic iron. The use of iron sulfide as a source of reduced iron was first suggested by Van Niel (Vatter and Wolfe, 1956). Kucera and Wolfe (1957) published a growth medium composed of a salt solution and iron sulfide, which made it possible to obtain pure cultures of *Gallionella*. The salt solution was initially prepared with tap water, because the medium lacked a crucial component. This component was later found to be calcium. Since then the ferrous iron medium has been widely used with some minor modifications.

TABLE BXII.β.109. Differential characteristics of *Gallionella ferruginea* strains Johan and BD

Characteristic	*Gallionella ferruginea* strain Johan	*Gallionella ferruginea* strain BD
Cells bean-shaped	+	+
Cell dimensions (μm)	0.5–0.8 × 1.6–2.5	0.5–0.7 × 0.8–1.8
Temperature range for growth (°C)	5–25	nd
Optimal temperature (°C)	20	17
pH range for exponential growth	5.0–6.5	nd
Generation time in exponential growth	8.3 h	nd
Motility without stalks	+	nd
Motility with stalks	−	+
Stalks not produced in exponential growth phases	+	−
Length of stalks (μm)	average 60/cell	nd
Maximum cell number in *in vitro* culture (cells/ml)	5 × 10⁶	up to 400/cell
Colony form *in vitro*	Ring on the tube wall	nd
Growth with CO₂ as sole carbon source	+	Circular colonies
Growth with ferrous iron as sole energy source	+	+
		+

ᵃFor symbols, see standard definitions; nd, not determined.

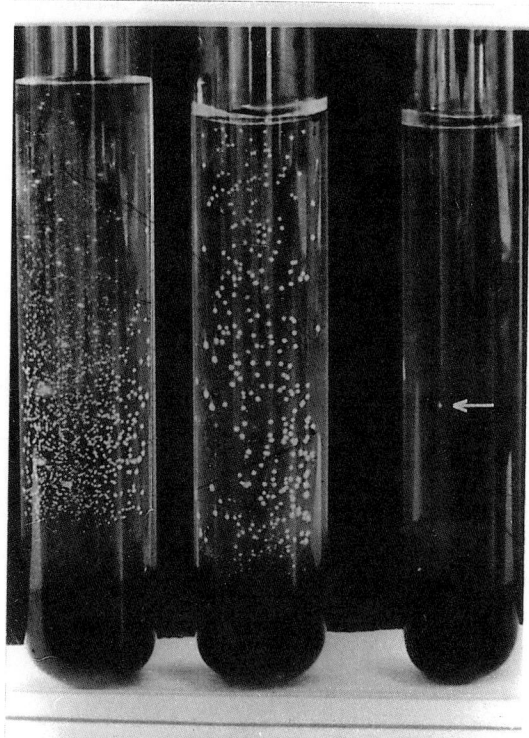

FIGURE BXII.β.103. Isolation procedure for *G. ferruginea* strain BD (*arrow*, one-colony culture). (Reproduced with permission from H.H. Hanert, Archives of Microbiology *60*: 348–376, 1968, ©Springer-Verlag, Berlin.)

Wolfe's modified medium is made as follows: Screw-capped tubes (180 × 16 mm) are filled with 10 ml salt medium consisting of 1.0 g NH₄Cl, 0.4 g MgSO₄·7H₂O, 0.1 g CaCl₂·2H₂O, 0.05 g K₂HPO₄ and 1 liter double-distilled water. The salt medium is autoclaved, chilled to 5°C and infused with sterile filtered CO₂ to pH 4.6–4.8. A ferrous sulfide or ferrous carbonate precipitate (0.5 ml) is added slowly to the bottom of the tubes with a Pasteur pipette and the tubes are left for four to six hours to allow gradient conditions to establish before inoculation. The ferrous sulfide and ferrous carbonate have to be prepared in the laboratory. Ferrous sulfide is prepared by dissolving 7.8 g FeSO₄(N₆H₄)₂SO₄ (Mohr's salt) and 4.8 g sodium sulfide separately, each in 200

ml boiling, distilled water and subsequently pouring the ferrous solution into the sulfide solution. Use two 500 ml beakers, mix with a glass rod. Fill the beaker up to the top and seal with a rubber stopper to avoid oxidation of the ferrous iron. Let the iron sulfide sediment for at least four hours. Decant and wash with boiling water five times. Centrifuge at high velocity, collect the iron sulfide in small bottles, fill up with water and close with an airtight lid. Sterilize at 121°C for 20 min. Store cool in airtight vials. Ferrous carbonate is prepared by dissolving 3.9 g FeSO₄(NH₄)₂SO₄ and 1.0 g anhydrous Na₂CO₃ separately, each in 100 ml boiling distilled water, and subsequently pouring the carbonate solution into the ferrous solution, preferably under a nitrogen atmosphere. The precipitated ferrous carbonate is then washed five times with boiling double-distilled water and sterilized in closed vials at 121°C for 20 min. Store cool in airtight vials. It is recommended that all solutions used for preparation of the medium and source of ferrous iron are filter sterilized (0.2 μm) to remove any cells or particles that may give a background during microscopic counts.

Isolation by serial dilution and serial transfer has been applied successfully to *Gallionella* strains BD and Johan (Fig. BXII.β.103). For strain BD, an enrichment test tube with colonies attached to the wall of the test tube is washed several times (Hanert, 1989). One colony is subsequently suspended in 10 ml fresh sterile medium and inoculated into new test tubes with Wolfe's modified medium at dilutions of maximally 10⁻⁶ (30–50 parallel cultures starting with a 10⁻⁴ serial dilution). Five to ten serial transfers, each starting from one colony, are necessary to achieve pure cultures in this manner. This procedure requires up to 10 weeks but is a very certain method for continually reducing the number of contaminants and obtaining a pure culture. Purity is checked microscopically and by use of a variety of heterotrophic and autotrophic media i.e., yeast extract bouillon, nutrient agar, *Nitrosomonas* medium, and *Thiobacillus ferrooxidans* medium. Stalk material of strain Johan can be collected, suspended in new medium, diluted and inoculated according to the strain BD procedure.

MAINTENANCE PROCEDURES

Maintaining pure stock cultures of *Gallionella* strain BD and Johan over the past years was performed using the described culture conditions with serial dilution transfers every four to eight weeks. Preservation of *Gallionella* culture material for at least 13 weeks by freezing at −80°C in 15% glycerol has been reported by Nunley and Krieg (1968). This procedure, however, did not result in survival of strain Johan.

TAXONOMIC COMMENTS

The genus *Gallionella* is characterized by its chemolithotrophic growth with ferrous iron, and its production of a twisted stalk consisting of a bundle of numerous fibers that makes *Gallionella* very easy to identify. *Gallionella* has been described under several different names such as *Spirophyllum ferrugineum* (Ehrenberg, 1836; Adler, 1904), *Didymohelix ferrugineum* (Griffith, 1853), *Gloeosphaera ferruginea* (Rabenhorst, 1854), and *Gallionella filamenta* (Balashova, 1967a, b). Most, if not all, of the attempts to categorize species of the genus *Gallionella* under various names arise from observed differences in the appearance of the stalk. The current phylogenetic position of the family *Gallionellaceae* is based on only a single 16S rRNA gene sequence of *Gallionella*.

More than 160 years ago, in 1836, Ehrenberg first discovered the stalks of *Gallionella* when he studied ochre masses. In this description (Ehrenberg, 1836), *Gallionella* was referred to as a fossil infusorian and he called it "die Eisenochertierchen", the small iron ochre animal. Haeckel (1866) presented a phylogenetic tree that for the first time included the kingdom *Monera* for unicellular organisms and Zopf (1879) first included *Gallionella* with the *Bacteria*. *Gallionella* has been observed and described by many more scientists since these early days, and there has been continuous discussion about its morphology and physiology, but most studies have focused on the stalks. Beger and Bringmann (1953) and van Iterson (1958) have summarized the first 100 years of discussion about the intriguing characteristics of *Gallionella*. Winogradsky (1888, 1922) proposed an autolithotrophic life for the so-called "iron bacteria", including *Gallionella*. He mentioned *Leptothrix*, *Cladothrix*, and *Gallionella* as examples of lithotrophic iron bacteria. Adler (1904) found only small amounts of *Gallionella* in fresh water from Karlspader, but when the water was left in bottles for several days they "ausserordentlich stark vermehrt" i.e., they had grown, or more correctly, the stalks had become elongated. Lieske (1911) succeeded in the cultivation of *Gallionella* in carbonic water with metallic iron as ferrous iron source. In 1924–1929, Cholodny made microscopic studies on cover slips that he had submerged in habitats of *Gallionella* (Cholodny, 1924, 1929). He sketched admirable pictures of cells attached to the end of stalks, and he showed that the stalk was excreted by the cell and was not a living part of it. Teichmann (1935) made cultures according to Lieske (1911) and found a great number of bean-shaped cells in the fluid. This observation influenced Pringsheim (1949b) to suggest

that "It is not impossible that motile cells are formed under certain conditions", a conclusion in accordance with current knowledge of free-living cells in exponential growth phase. Beger and Bringmann (1953) made comparisons between earlier drawings of the stalk of *Gallionella* and their own electron microscopy studies and proposed that the genus *Gallionella* consisted of five species. Vatter and Wolfe (1956) presented electron microscopy images of cells with stalks and in 1957, Kucera and Wolfe (1957) introduced an excellent growth medium containing iron sulfide as the source of ferrous iron. Wolfe could have succeeded in working with *Gallionella* but he concluded that "these organisms (iron bacteria) are too difficult to be profitable" (Wolfe, 1964). In 1958, a thesis on *Gallionella* was presented by van Iterson (1958). Excellent electron microscopy images of the organism were presented and it was suggested that the stalk was a living part of the organism with sporangia in the form of membrane sacs on the stalk. Balashova (1967a,b, 1968) and Balashova and Cherni (1970) made electron microscopy observations of the stalk and concluded that the stalk might have zoogloeal forms and budding cells on the stalks. Hanert (1989) made detailed studies on stalk elongation using single cells and measurements of iron oxidation in both natural samples and lab cultures. Intracytoplasmic membranes and evidence for autotrophic growth was presented. In 1990, Lütters-Czekalla (1990) reported growth of *Gallionella* BD with reduced sulfur compounds as electron donor instead of ferrous iron. This observation remains to be confirmed. In 1990–1995 Hallbeck and Pedersen (1990, 1991, 1995) and Hallbeck et al. (1993) reported the 16S rDNA sequence of *Gallionella* and showed that *Gallionella* has a free-living stage without a stalk in exponential growth phase. They demonstrated chemolithotrophic growth with CO_2 as the sole carbon source and mixotrophic metabolism. They also demonstrated an organic composition of the stalk and found that the stalk was important for survival in the environments where *Gallionella* is found. A thesis summarizing these results was published in 1993 (see further reading).

ACKNOWLEDGMENTS

We are grateful to Professor W.G. Ghiorse for many valuable discussions about *Gallionella* and its life style and to Professor Grant Ferris for valuable comments on the manuscript. Our work with *Gallionella* was supported by the Swedish Natural Science Research Council.

FURTHER READING

Hallbeck, L. 1993. On the biology of the iron-oxidizing and stalk-forming bacterium *Gallionella ferruginea*, Thesis, Göteborg University, Göteborg Sweden.

List of species of the genus Gallionella

1. **Gallionella ferruginea** Ehrenberg 1838, 166[AL]

   fer.ru.gi' ne.a. L. fem. adj. *ferruginea* rust-colored.

   Morphology and description the same as those of the genus. Organisms occur in groundwater habitats, especially in places where iron-bearing groundwater reaches an oxygen-containing environment. Microaerophilic, possibly fac-

ultatively anaerobic, chemolithotrophic, and mixotrophic. Gram negative. G + C buoyant density was calculated by using a micromethod with novel collimating optics (Hanert 1989). (*E. coli* B was used as a reference.)

The mol% G + C of the DNA is: 51–54.6. (Bd).

Type strain: no culture isolated.

GenBank accession number (16S rRNA): L07897.

Order VI. **Rhodocyclales** ord. nov.

GEORGE M. GARRITY, JULIA A. BELL AND TIMOTHY LILBURN

Rho.do.cy.cla' les. M.L. masc. n. *Rhodocyclus* type genus of the order; -*ales* suffix to denote order; M.L. fem. n. *Rhodocyclales* the *Rhodocyclus* order.

The order *Rhodocyclales* was circumscribed for this volume on the basis of phylogenetic analysis of 16S rRNA sequences; the order contains the family *Rhodocyclaceae*.

Description is the same as for the family *Rhodocyclaceae*.
Type genus: **Rhodocyclus** Pfennig 1978, 285.

Family I. **Rhodocyclaceae** fam. nov.

GEORGE M. GARRITY, JULIA A. BELL AND TIMOTHY LILBURN

Rho.do.cy.cla' ce.ae. M.L. masc. n. *Rhodocyclus* type genus of the family; -*aceae* ending to denote family; M.L. fem. pl. n. *Rhodocyclaceae* the *Rhodocyclus* family.

The family *Rhodocyclaceae* was circumscribed for this volume on the basis of phylogenetic analysis of 16S rRNA sequences; the family contains the genera *Rhodocyclus* (type genus), *Azoarcus*, *Azonexus*, *Azospira*, *Azovibrio*, *Dechloromonas*, *Dechlorosoma*, *Ferribacterium*, *Propionibacter*, *Propionivibrio*, *Quadricoccus*, *Sterolibacterium*, *Thauera*, and *Zoogloea*. *Quadricoccus* and *Sterolibacterium* were proposed after the cut-off date for inclusion in this volume (June 30, 2001) and are not described here (see Maszenan et al., 2002, and Tarlera and Denner, 2003).

Family is phenotypically, metabolically, and ecologically diverse. Includes photoheterotrophs; aerobes, anaerobes, and facultative anaerobes utilizing a number of electron acceptors; fermentative organisms; and nitrogen-fixing organisms.

Type genus: **Rhodocyclus** Pfennig 1978, 285.

Genus I. **Rhodocyclus** Pfennig 1978, 285[AL]

JOHANNES F. IMHOFF

Rho.do.cy' clus. Gr. n. *rhodon* the rose; Gr. n. *cyclos* a circle; M.L. masc. n. *Rhodocyclus* red circle.

Cells are slender, curved, or straight, thin rods. Motile by means of polar flagella or nonmotile. Multiply by binary fission. **Gram negative, belonging to class** ***Betaproteobacteria.*** Internal photosynthetic membranes may appear as small, single, finger-like intrusions of the cytoplasmic membrane or may be absent. **Photosynthetic pigments are bacteriochlorophyll** *a* **and carotenoids. Contain ubiquinones and menaquinones with eight isoprene units (Q-8 and MK-8). Straight-chain** $C_{16:1}$ **and** $C_{16:0}$ **acids are the major components of cellular fatty acids.** $C_{10:0\ 3OH}$ **is present.**

Preferably grow photoheterotrophically under anoxic conditions in the light with different organic substrates as carbon and electron sources. **Photoautotrophic growth with molecular hydrogen** may be possible if growth factors are supplied. **Chemotrophic growth is also possible under microoxic to oxic conditions in the dark.** Reduced sulfur compounds are not used as photosynthetic electron donors. Assimilatory sulfate reduction is possible. Growth factors may be required. **Mesophilic and neutrophilic freshwater bacteria.** Habitat: freshwater ponds, sewage ditches, swine waste lagoon.

The mol% G + C of the DNA is: 64.1–65.1.
Type species: **Rhodocyclus purpureus** Pfennig 1978, 285.

FURTHER DESCRIPTIVE INFORMATION

Only a few carbon compounds can be assimilated by *Rhodocyclus purpureus*. Benzoate and cyclohexane carboxylate are both used, which may indicate that *Rhodocyclus purpureus* uses the same pathway for anaerobic benzoate degradation as *Rhodopseudomonas palustris* (Dutton and Evans, 1969; Pfennig, 1978). Neither of these carbon substrates is used by other *Rhodocyclus* species, and they are used only rarely by other purple nonsulfur bacteria.

Nitrogen metabolism of *R. purpureus* and *R. tenuis* has been studied in some detail (Masters and Madigan, 1983). Alanine dehydrogenase is absent in both species. Glutamate dehydrogenase (NADPH-dependent) is found in *R. purpureus* at unusually high activity levels under all growth conditions, and the glutamine synthetase inhibitor methionine sulfoximine exerts no growth inhibition. This may indicate that the major route of nitrogen assimilation in *R. purpureus* is via glutamate dehydrogenase (unlike that in all other investigated purple nonsulfur bacteria). *R. tenuis* employs the glutamine synthetase/glutamate synthase (NADPH-dependent) pathway for the assimilation of ammonia (Masters and Madigan, 1983).

As with other phototrophic *Betaproteobacteria*, the major phospholipid components of *Rhodocyclus* species are cardiolipin, phosphatidylethanolamine and phosphatidylglycerol (Imhoff and Bias-Imhoff, 1995). Straight-chain $C_{16:1}$ and $C_{16:0}$ acids are the main components of cellular fatty acids (see Table BXII.β.110).

ENRICHMENT AND ISOLATION PROCEDURES

R. purpureus was isolated from a swine waste lagoon in Ames, Iowa (USA), where it was the dominant phototrophic bacterium. It has not been observed in other localities and is probably a rare species. Media for enrichment, isolation and growth of *Rhodocyclus* species are the same as those generally employed for other purple nonsulfur bacteria. While suitable conditions for selective enrichment of *R. tenuis* are not available, the vitamin B_{12} requirement and its unusual carbon nutrition are properties that can be exploited for the selective enrichment and isolation of *R. purpureus*. From a suitable habitat, it should be possible to selectively enrich this species with benzoic acid as the carbon source, in the presence of vitamin B_{12}, and in the absence of reduced sulfur compounds.

TABLE BXII.β.110. Differential characteristics of the species of the genus *Rhodocyclus*[a]

Characteristic	*Rhodocyclus purpureus*	*Rhodocyclus tenuis*
Cell diameter (μm)	0.6–0.7	0.3–0.5
Cell shape	Half-circle to circle	Curved rods
Motility	−	+
Slime production	−	+
Color	Purple-violet to violet	Brownish-red or purple-violet
Major carotenoids	Rhodopin, rhodopinal	Rhodopin, rhodopinal, lycopene[b]
Growth factors	B$_{12}$, *p*-aminobenzoic acid, biotin	None[c]
Gelatin liquefaction	−	−
Fructose fermentation	−	−
Starch hydrolysis	nd	nd
Tween 80 lysis	nd	nd
Carbon sources:		
Benzoate	+	−
C$_{10}$ to C$_{18}$ fatty acids	−	+
Citrate	−	−
Mannitol	−	−
Sorbitol	−	+
N$_2$-fixation	−	+
Fumarate reductase activity:		
With reduced methylviologen	High	High
With FMNH$_2$	Low	Low
Major fatty acids:		
C$_{16:0}$	33–35	33-3-6
C$_{16:1}$	40–45	43–50
C$_{18:0}$	<1	<1
C$_{18:1}$	18	15–18
3-OH fatty acid	C$_{10:0}$	C$_{10:0}$
Major quinones	Q-8 + MK-8	Q-8 + MK-8
Mol% G + C of DNA		
by HPLC	65.1	64.1–64.8
by Bd	65.3	64.8
by T$_m$	67.7	64.4–67.2

[a]Symbols: +, positive in most strains; −, negative in most strains; Q–8, ubiquinone-8; MK-8, menaquinone-8.

[b]Some strains may contain carotenoids of the spirilloxanthin series and lack rhodopinal (Schmidt, 1978).

[c]Some strains may require vitamin B$_{12}$ (Siefert and Koppenhagen, 1982).

MAINTENANCE PROCEDURES

Rhodocyclus species are easily maintained by standard procedures in liquid nitrogen or by storage at −80°C in a mechanical freezer.

DIFFERENTIATION OF THE GENUS *RHODOCYCLUS* FROM OTHER GENERA

Rhodocyclus species are distinguished from other phototrophic purple bacteria by their separate phylogenetic position within the class *Betaproteobacteria*. Physiological properties are, in general, not significantly different from those of other purple nonsulfur bacteria, although the utilization of benzoic acid by *R. purpureus* is not a common property among other purple nonsulfur bacteria. The requirement for vitamin B$_{12}$ is also unusual among the purple nonsulfur bacteria, being more common among the green and purple sulfur bacteria. Besides *Rhodocyclus purpureus*, only single strains of *Rhodocyclus tenuis* and *Rhodopseudomonas palustris* have been reported to require vitamin B$_{12}$ (Siefert and Koppenhagen, 1982). Properties that differentiate *Rhodocyclus* species from other phototrophic *Betaproteobacteria* are given in Tables 6 (p. 130) and 7 (p. 131) of the introductory chapter "Anoxygenic Phototrophic Purple Bacteria", Volume 2, Part A. The phylogenetic relationships among the phototrophic *Betaproteobacteria* are shown in Fig. 4 (p. 132) of that same chapter.

TAXONOMIC COMMENTS

At present, this genus comprises two species, *Rhodocyclus purpureus* and *Rhodocyclus tenuis* (formerly *Rhodospirillum tenue*). Only a single strain of *R. purpureus* is known. A dichotomy has been observed in *R. tenuis* strains based on carotenoid composition (Schmidt, 1978), color of cell suspensions, and absorption spectra (Biebl, 1973). Some strains of *R. tenuis* have carotenoids of the rhodopinal series and others have carotenoids of the spirilloxanthin series. The type strain (DSM 109) belongs to those strains that transform rhodopin further to spirilloxanthin, whereas another group of strains (including strain DSM 110) does not form anhydrorhodovibrin and spirilloxanthin, but rather accumulates major amounts of rhodopinal, rhodopinol, and lycopenal (Schmidt, 1978). Both of these strains (DSM 109 and 110) show a reasonably close genetic relationship to each other (see Fig. 4 [p.132] of the introductory chapter "Anoxygenic Phototrophic Purple Bacteria", Volume 2, Part A).

DIFFERENTIATION OF THE SPECIES OF THE GENUS *RHODOCYCLUS*

While cells of *Rhodocyclus purpureus* are nonmotile and form a half or full circle, those of *Rhodocyclus tenuis* are slender, slightly curved, and rapidly motile under optimal growth conditions. *R. purpureus* and *R. tenuis* also differ significantly in their nitrogen nutrition. Whereas the former species uses only ammonia and glutamine as nitrogen sources and is unable to fix dinitrogen (a property common to most species of the purple nonsulfur bacteria), the latter species utilizes a greater number of amino acids, urea, dinitrogen, yeast extract, peptone, and Casamino acids

(Masters and Madigan, 1983). In addition, differences in carbon nutrition and vitamin requirements clearly differentiate the two species. Diagnostic properties of *Rhodocyclus* species are listed in Table BXII.β.110. Their carbon substrates are shown in Table BXII.β.111.

List of species of the genus Rhodocyclus

1. **Rhodocyclus purpureus** Pfennig 1978, 285[AL]

pur.pu′ re.us. L. adj. *purpureus* purple or red-violet.

Cells are half-ring-shaped to ring-shaped before cell division and are 0.6–0.7 μm wide. The diameter of a circle is 2.0–3.0 μm. Half-circle-shaped cells are about 2.7 μm long (Fig. BXII.β.104). Open or compact spirals or coils of variable length may be formed. In sulfide-containing media, closely wound spirals are united in compact cell aggregates. Cells are nonmotile under all growth conditions. The color of phototrophically grown cultures is purple-violet to violet. Aerobically grown cells are colorless to pale violet. Living cells have absorption maxima at 379, 408, 510, 535, 597, 813, and 866 nm. Photosynthetic pigments are bacteriochlorophyll *a* (esterified with phytol) and carotenoids of the rhodopinal series, with rhodopinal as the major component.

Photoheterotrophic growth occurs under anoxic conditions in the light with a relatively small number of organic

substrates as carbon and electron sources. Cells grow photoautotrophically with hydrogen as the electron donor in the presence of growth factors. Chemotrophic growth is possible under microoxic to oxic conditions in the dark. Carbon sources utilized are listed in Table BXII.β.111. In addition, cyclohexane carboxylate is used, but propanol, yeast extract, and Casamino acids are not utilized as sole carbon sources. Photoheterotrophically grown cells use only ammonia and glutamine as nitrogen sources; dinitrogen is not assimilated. Sulfate can be used as the sole sulfur source. Vitamin B_{12}, *p*-aminobenzoic acid, and biotin are required as growth factors.

Mesophilic freshwater bacterium with optimal growth at 30°C and pH 7.2 (pH range with acetate: pH 6.5–7.5). Habitat: swine waste lagoon. Major quinone components are Q-8 and MK-8.

The mol% G + C of the DNA is: 65.3 (Bd), 65.1 (HPLC), 67.7 (T_m).

Type strain: Ames 6770, DSM 168.

GenBank accession number (16S rRNA): M34132.

2. **Rhodocyclus tenuis** (Pfennig 1969b) Imhoff, Trüper and Pfennig 1984, 341[VP] (*Rhodospirillum tenue* Pfennig 1969b, 619.)

te′ nu.is. L. masc. adj. *tenuis* slender, thin.

Cells are curved in spirals of one to two complete turns, 0.3–0.5 × 1.5–6.0 μm, sometimes even longer. One complete turn of a spiral is ~0.8–1.0 μm wide and 3 μm long (Fig. BXII.β.105). Photosynthetically grown cells are brownish-red or purple-violet, depending on the strain. Aerobically grown cells may be colorless or pigmented. Absorption maxima of brownish cells that have carotenoids of the spi-

TABLE BXII.β.111. Carbon sources and electron donors used by the species of the genera *Rhodocyclus*[a]

Source/donor	Rhodocyclus purpureus	Rhodocyclus tenuis
Carbon source		
Acetate	+	+
Arginine	−	−
Aspartate	−	−
Benzoate	+	−
Butyrate	+	+
Caproate	+	+
Caprylate	−	+ / −
Citrate	−	−
Ethanol	−	+ / −
Formate	−	−
Fructose	−	−
Fumarate	+	+
Glucose	−	−
Glutamate	−	−
Glycerol	−	−
Glycolate	−	−
Lactate	−	+
Malate	+	+
Malonate	−	−
Mannitol	−	−
Mannose	−-	−
Methanol	−	−
Pelargonate	−	+
Propionate	−	+ / −
Pyruvate	+	+
Sorbitol	−	nd
Succinate	−	+
Tartrate	−	−
Valerate	−	+
Electron donor:		
Hydrogen	+	+
Sulfide	−	−
Sulfur	−	−
Thiosulfate	−	−

[a] Symbols: +, positive in most strains; −, negative in most strains; + / − variable in different strains; nd, not determined.

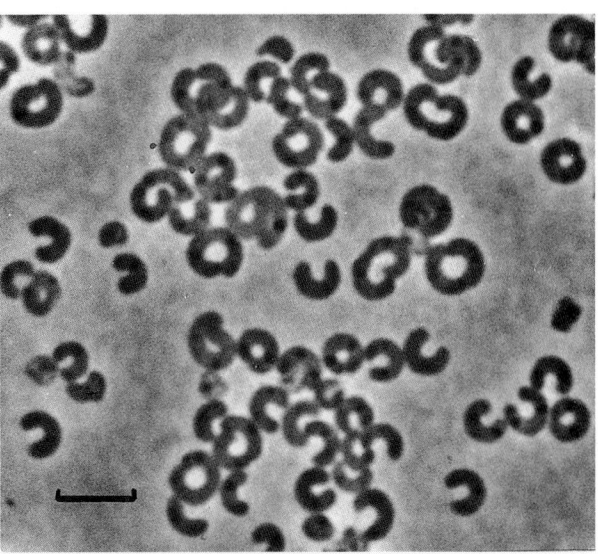

FIGURE BXII.β.104. *Rhodocyclus purpureus* DSM 168. Phase-contrast micrograph. Bar = 5 μm (Courtesy of N. Pfennig).

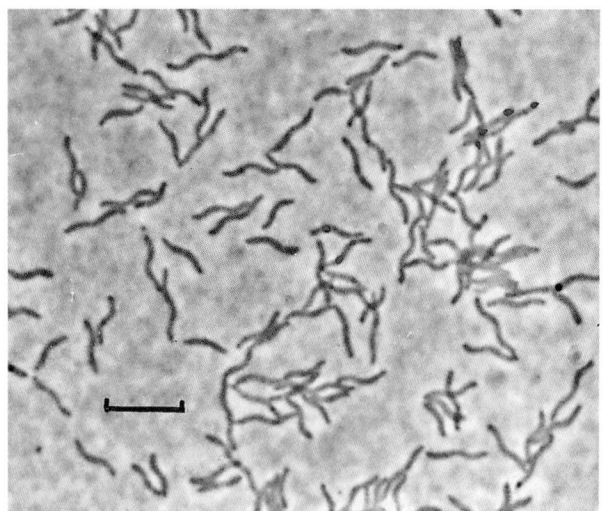

FIGURE BXII.β.105 *Rhodocyclus tenuis* DSM 109. Phase-contrast micrograph. Bar = 5 μm (Courtesy of N. Pfennig).

rilloxanthin series are at 378–380, 465, 492–495, 528, 592–594, 799–801, and 868–871 nm. Absorption maxima of purple cells that have carotenoids of the rhodopinal series are at 377–378, 469, 495–500, 529–533, 590–592, 798–801, and

856–858 nm. Both types of cells contain bacteriochlorophyll *a* esterified with phytol.

Growth occurs preferably under anoxic conditions in the light with numerous carbon substrates as carbon and electron sources. Photoautotrophic growth with molecular hydrogen is possible. Chemotrophic growth is possible under microoxic to oxic conditions in the dark. The organic substrates used are listed in Table BXII.β.111. In addition, Casamino acids and yeast extract are utilized, but cyclohexane carboxylate is not. Aspartate, glutamate, glutamine, ammonia, and dinitrogen are used as nitrogen sources; also utilized are Casamino acids, peptone, yeast extract, alanine, arginine, lysine, methionine, serine, threonine, and urea. Sulfate, sulfite, sulfide, thiosulfate, cysteine, and reduced glutathione can serve as assimilatory sulfur sources. Growth factors are not required. Growth is stimulated, however, in the presence of complex organic nutrients or yeast extract, and some strains may need vitamin B_{12}.

Mesophilic freshwater bacterium with optimal growth at 30°C and pH 6.6–7.4. Habitat: freshwater ponds, sewage ditches. Major quinone components are Q-8 and MK-8.

The mol% G + C of the DNA is: 64.1–64.8 (HPLC), 64.4–67.2 (T_m); type strain: 64.8 (Bd) and 66.1 (T_m).

Type strain: ATCC 25093, DSM 109.

GenBank accession number (16S rRNA): D16208.

Genus II. **Azoarcus** Reinhold-Hurek, Hurek, Gillis, Hoste, Vancanneyt, Kersters and De Ley 1993b, 582^{VP}

BARBARA REINHOLD-HUREK, ZHIYUAN TAN AND THOMAS HUREK

A.zo'ar.cus. Fr. n. *azote* nitrogen; L. masc. n. *arcus* arch, bow; M.L. masc. n. *Azoarcus* nitrogen (-fixing) bow.

Straight to slightly curved rods, 0.4–1.5 × 1.1–4.0 μm. Cell pairs often appear slightly S-shaped. In most species, some elongated cells (8–12 μm) occur in late-log or stationary-phase cultures on medium containing malic acid and N_2 or ammonium as nitrogen source. **Cells are highly motile by means of one polar flagellum. Accumulate poly-β-hydroxybutyrate granules.** Gram negative. **Some species are nitrogen fixers**; these require microaerobic conditions for growth on N_2. On semisolid nitrogen-free media, microaerophilic growth can be observed as veil-like pellicles developing several mm below the surface and moving to the medium surface during growth. In most species, colonies on VM agar[1] supplemented with ethanol develop **a nondiffusible yellowish pigment.** Optimal temperature for growth 30–40°C; no growth occurs at 45°C. **Chemoorganoheterotrophic. Bacteria have a strictly respiratory metabolism with O_2 as the terminal electron acceptor,** except one species. Alternatively, under anaerobic conditions, nitrate can be used for dissimilatory nitrate reduction. **Oxidase positive. Grow well on salts of organic acids** such as L-malate, succinate, fumarate, DL-lactate; also grow well on ethanol, on L-glutamate, but not on mono- or disaccharides except for species that are not plant-associated. These soil-borne species utilize a variety of **aromatic substrates as sole carbon sources under denitrifying conditions**. Nitrate can be used as a

nitrogen source (assimilatory nitrate reduction). Growth factor requirements vary: some strains depend on *p*-aminobenzoic acid or on cobalamine. All investigated species have $C_{16:1}$ cellular fatty acids; all species except one have $C_{16:1 \omega 7c}$ and $C_{18:1}$ as the major cellular fatty acids.

The mol% G + C of the DNA is: 62–68 (T_m).

Type species: **Azoarcus indigens** Reinhold-Hurek, Hurek, Gillis, Hoste, Vancanneyt, Kersters and De Ley 1993b, 583.

FURTHER DESCRIPTIVE INFORMATION

The members of *Azoarcus* comprise two biogroups: (1) the soil-borne species *A. tolulyticus, A. toluclasticus, A. toluvorans, A. evansii,* and *A. anaerobius,* and (2) the plant-associated species *A. indigens, A. communis,* and an unnamed *Azoarcus* strain, strain BH72 (Fig. BXII.β.106).

Morphology Cells are straight to slightly curved rods that are highly motile by one, rarely two, polar flagella (Fig. BXII.β.107). All species have a cell width of ≤1 μm, with the exception of *Azoarcus anaerobius,* which has a cell width of 1.5 μm (Springer et al., 1998). The cell length ranges from 1.5–4.0 μm. Elongated cells of 8–12 μm length are rarely found in the plant-associated species (*A. communis, A. indigens,* and *Azoarcus* strain BH72) in stationary phase cultures grown in semisolid N-free malate medium.

Growth The optimal growth temperature for most species is 30°C or 37–40°C depending on species, but is lower (28°C) for *A. anaerobius.* The optimal pH is 7.

---

1. VM agar contains (g per liter): K_2HPO_4, 0.6; KH_2PO_4, 0.4; NH_4Cl, 0.5; $MgSO_4 \cdot 7H_2O$, 0.2; NaCl, 1.1; $CaCl_2 \cdot 2H_2O$, 0.026; $MnSO_4 \cdot H_2O$, 0.01; $Na_2MoO_4 \cdot 2H_2O$, 0.002; Fe (III)-EDTA, 0.066; yeast extract, 1; Bacto-Tryptone, 3; pH 6.8; ethanol, 6 ml per l, sterilized by filtration.

Metabolism The optimal growth temperature for most species is 30°C or 37–40°C depending on species, but is lower (28°C) for *A. anaerobius*. The optimal pH is 7.

All species except *A. anaerobius* use oxygen as the terminal electron acceptor. In the original description of *Azoarcus*, standard procedures had failed to reveal an ability to denitrify (Reinhold-Hurek et al., 1993b). On closer examination, the ability to use nitrate as the terminal electron acceptor was demonstrated for the plant-associated strains (Hurek and Reinhold-Hurek, 1995), a feature that they share with all other species (Anders et al., 1995; Zhou et al., 1995; Springer et al., 1998; Song et al., 1999). Most strains belonging to *A. tolulyticus*, *A. evansii*, *A. toluclasticus*, *A. toluvorans*, and *A. anaerobius* were enriched or isolated anaerobically on nitrate, whereas the plant-associated species were enriched under nitrogen-fixing conditions.

In general, carbohydrates are not the preferred carbon sources of *Azoarcus* species. None of the plant-associated species is able to utilize any of the 50 mono- and disaccharides or sugar alcohols tested (Reinhold-Hurek et al., 1993b). Similarly, *A. anaerobius* shows no growth on common carbohydrates such as D(+)-glucose or D(−)-fructose (Springer et al., 1998). In contrast, all strains of soil-borne species tested so far are able to utilize at least some carbohydrates (see Table BXII.β.112). All strains grow well on organic acids and a few amino acids (Table BXII.β.112). The carbon sources listed in Table BXII.β.112 were tested with O_2 as the terminal electron acceptor (except in the case of *A. anaerobius*). For some strains, tests were also carried out under anaerobic conditions with nitrate as electron acceptor. For the majority of carbon sources, results were identical; however, for several carbon sources discrepancies were found between aerobic and anaerobic conditions as well as between strains (Song et al., 1999). Most soil-borne species grow on aromatic compounds such as toluene or phenol (Zhou et al., 1995; Song et al., 1999), benzoate (Anders et al., 1995), or resorcinol (Springer et al., 1998) under denitrifying conditions, in contrast to plant-associated species (Reinhold-Hurek et al., 1993b; Hurek and Reinhold-Hurek, 1995). Due to the anaerobic degradation of aromatic compounds, this bacterial group has received particular attention for its role in biodegradation and biotransformation. Whereas aerobic metabolism is characterized by the extensive use of molecular oxygen, which is essential for the hydroxylation and cleavage of the ring structures, anaerobic degradation uses other strategies that are currently being studied. Toluene, which can be decomposed anaerobically by three species, is activated by the addition of fumarate to form benzylsuccinate (Beller and Spormann, 1997). The reaction is catalyzed by benzylsuccinate synthase and involves a glycyl radical (Krieger et al., 2001). See the description of *A. evansii* for further details on degradation of aromatic compounds.

In the original description of the genus *Azoarcus*, the ability to fix nitrogen is listed as a genus character (Reinhold-Hurek et al., 1993b). Subsequently, nitrogen fixation or the occurrence of a nitrogenase gene *nifH* was also demonstrated for *A. tolulyticus* (Fries et al., 1994; Zhou et al., 1995; Hurek et al., 1997a). However, nitrogen fixation can no longer be considered a universal characteristic of the genus *Azoarcus*, due to the addition of new species that do not fix nitrogen. Nitrogen fixation was not detected in *A. anaerobius* (Springer et al., 1998). Although there are no physiological data available (Anders et al., 1995), *A. evansii* is unlikely to be diazotrophic since there is no evidence for occurrence of a PCR-amplifiable *nifH* fragment (Hurek and Reinhold-Hurek, unpublished observations). For most other isolates enriched under denitrifying conditions, this character was not tested at all.

In those species and strains that carry out nitrogen fixation (*A. indigens*, *A. communis*, *A. tolulyticus*, *Azoarcus* sp. strain BH72), nitrogenase activity occurs only under microaerobic conditions, probably due to a lack of efficient oxygen protection mechanisms. When cultured on N-free semisolid medium, the nitrogen-fixing strains develop a veil-like pellicle, which, due to aerotaxis, moves to the medium surface when the culture becomes denser. For strain BH72, nitrogen fixation is more tolerant to oxygen than in *Azospirillum* spp., reaching steady states in an oxygen-controlled chemostat up to 25 µM dissolved O_2 (Hurek et al., 1987). In accordance with these physiological data, the expression of nitrogenase genes is transcriptionally regulated in response to oxygen (fully repressed by 4% oxygen in the headspace) or combined nitrogen (repressed by 0.5 mM ammonium or nitrate) (Egener et al., 1999). In empirically optimized batch cultures, strain BH72 shows augmented rates and efficiency of nitrogen fixation, called hyperinduction, when shifting into extremely low oxygen concentrations (30 nM) (Hurek et al., 1994a). In the course of hyperinduction, novel intracytoplasmic membrane stacks are formed that might participate in efficient nitrogenase activity, since the iron protein of nitrogenase is mainly associated with these membranes (Hurek and Reinhold-Hurek, 1995). The formation of these so-called diazosomes is most abundant and reproducible in coculture of strain BH72 and an endophytic fungus isolated from Kallar grass (Hurek and Reinhold-Hurek, 1995) related to *Acremonium alternatum* (Hurek and Reinhold-Hurek, 1999). As in many *Proteobacteria*, the nitrogenase genes are organized in a *nifHDK* operon (Egener et al., 2001). Phylogenetically, the nitrogenase in the genus *Azoarcus* either follows the organismal phylogenetic tree or appears to have been acquired by lateral gene transfer, depending on the species. In *A. indigens*, *A. communis*, and *Azoarcus* spp. strain BH72, a fragment of the iron protein of nitrogenase encoded by the *nifH* gene is most closely related to nitrogenases that occur in diazotrophs of *Gammaproteobacteria*, whereas in *A. tolulyticus* it is located within a clade of nitrogenases that occur in species of *Alphaproteobacteria* (*Bradyrhizobium* and *Azospirillum*) (Hurek et al., 1997a).

Exoenzymes There are few studies of the exoenzymes of *Azoarcus* sp. In all three plant-associated species, an exoglucanase having β-glucosidase and cellobiohydrolase activity was detected, whereas an endoglucanase was found in strain BH72 and in *A. communis* only (Reinhold-Hurek et al., 1993a). Although these enzymes cannot be used for direct metabolic purposes because these strains do not grow on carbohydrates, they may be involved in plant infection.

Antigens Polyclonal antibodies have been raised against *Azoarcus* sp. strain BH72 and mainly bind to the cell surface. They cross-react weakly with *A. indigens* and *A. communis* cells. These antibodies have been used for histochemical detection of *Azoarcus* in grass roots (Hurek et al., 1994b).

Genetics DNA reassociation experiments with the three plant-associated species of *Azoarcus* have led to the estimation of a genome size from 4.5–5 Mb. Nothing is known about bacteriophages of the genus *Azoarcus*. Similarly, the plasmid content has not been well studied. In strain BH72, plasmids could not be detected in Eckhardt gels or by pulsed-field gel electrophoresis (Reinhold-Hurek and Hurek, unpublished observations). Several strains of *Azoarcus* species have been shown to be transformable.

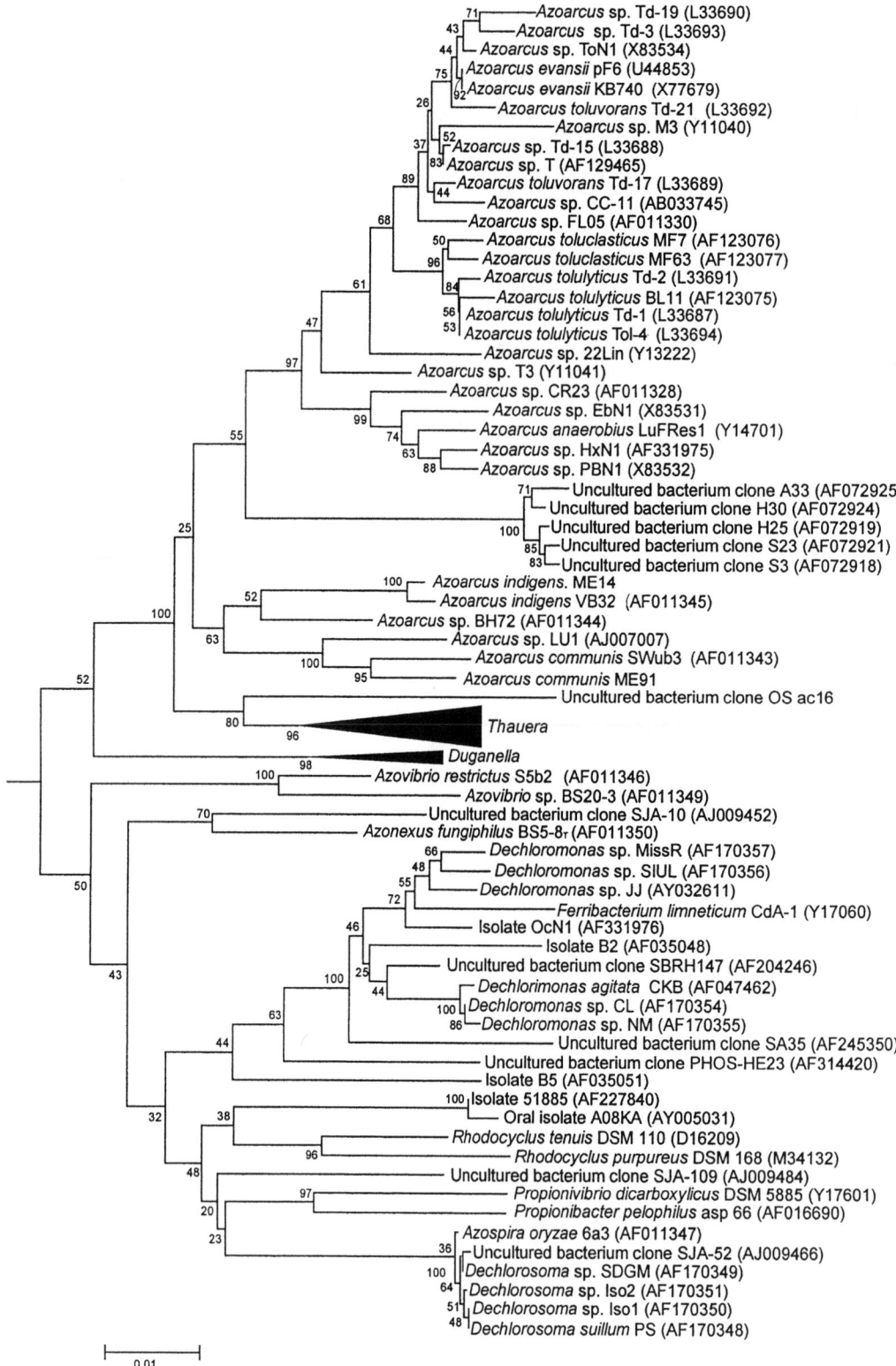

FIGURE BXII.β.106 Phylogenetic analysis of 16S rDNA sequences (1358 positions). Shown is a subtree of the *Rhodocyclus/Thauera/Azoarcus* group derived from an analysis of 158 sequences of the *Betaproteobacteria*. Tree inference was carried out using the neighbor-joining algorithm with a Jukes-Cantor correction with 125 bootstrap repetitions. Sequence accession numbers are given in parentheses.

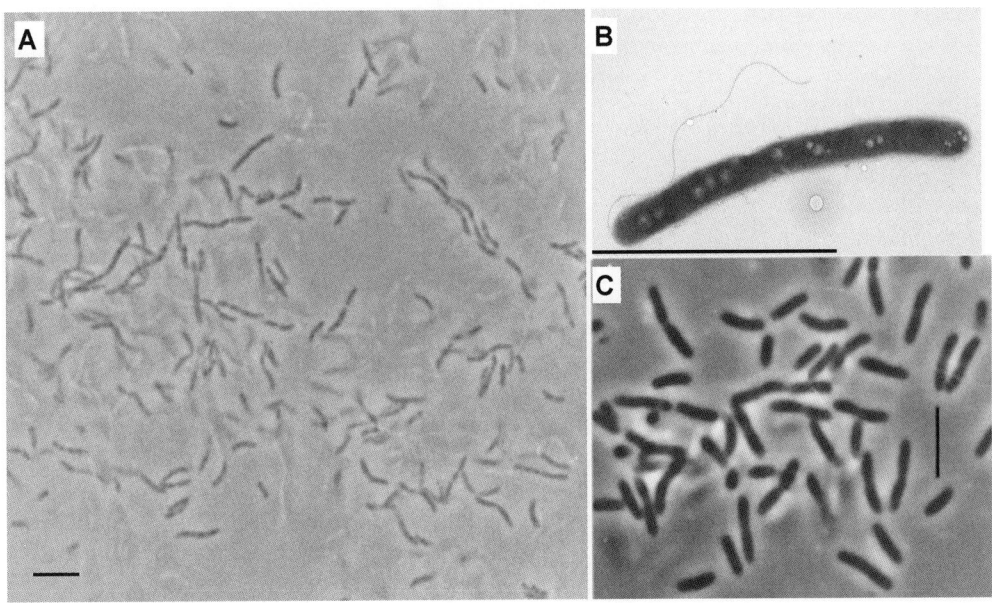

FIGURE BXII.β.107. Phase contrast (*A, C*) and transmission electron (*B*) microscopic images. (*A*) *Azoarcus indigens* VB32^T cultured on N$_2$ in semisolid SM medium for 24 h. (*B*) *Azoarcus* sp. strain BH72 cultured in liquid VM medium for 24 h. (*C*) *Azoarcus tolulyticus* BL2 grown on nutrient plates for 48 h. Bars, 5 μm. (Reproduced with permission from B. Reinhold-Hurek et al., International Journal of Systematic Bacteriology *43*:574–584, 1993, ©International Union of Microbiological Societies (*A, B*); (Reproduced with permission from B. Song, International Journal of Systematic Bacteriology *49*: 1129–1140, 1999, ©International Union of Microbiological Societies, (*C*).)

TABLE BXII.β.112. Differential characteristics of the genera *Azoarcus, Azovibrio, Azospira, Azonexus,* and morphologically similar diazotrophs or related bacteria of the *Proteobacteria*[a]

Characteristic	Azoarcus	Azonexus	Azospira	Azovibrio	Azospirillum	Burkholderia vietnamiensis	Herbaspirillum	Gluconacetobacter diazotrophicus	Thauera
Class									
Alphaproteobacteria	−	−	−	−	+	−	−	+	−
Betaproteobacteria	+	+	+	+	−	+	+	−	+
Cells curved	+	+	+	+	−	−	+		−
Cell width, μm	0.4–1.0[b]	0.6–0.8	0.4–0.6	0.6–0.8	0.8–1.4	0.3–0.8	0.6–0.7	0.7–0.9	0.7–1.0
Colony color	Yellow to beige	Ochreous	Pink translucent	Beige	Pink opaque	Cream	Cream	Brown	White-yellow[c]
Fermentative	−	−	−	−	D	−	−	+	−[d]
Growth on sugars	D[e]	−	−	−	+	+	+	+	−[f]
Requirement for cobalamine	−	+	−	−	−	−	−	−	−
Growth on:									
n-Butylamine	+[g]	−	−	−		−	D		
n-Caproate	D	−	+	−		+	−		D
Glutarate	+[h]	−	−	−		+	−		+[i]
4-Hydroxybenzoate	+[j]	−	−	−		+	+		+
D-Mannose	−	−	−	−	D	+	+		
Phenylacetate	+[k]	−	−	−	−	+	−		+[d]
L-Proline	D	+	−	−	+	+	+	+	+[d]
Propionate	D	−	+	+	+	+	+		+[d]
2-Oxoglutarate	+[l]	+	+	−	D	d	+		

[a]Symbols: see standard definitions. Blank space, not determined.

[b]Except for *A. anaerobius,* whose cells are 1.5 μm.

[c]Several strains do not grow on nutrient agar.

[d]Tested for *T. aromatica.*

[e]Positive for soil-borne species, negative for plant-associated species.

[f]Except for *T. selenatis.*

[g]ND for *A. toluvorans, A. toluclasticus, A. anaerobius.*

[h]Negative for *A. indigens,* ND for *A. toluvorans, A. toluclasticus, A. anaerobius.*

[i]D for *T. mechernichensis.*

[j]Negative for *A. evansii.*

[k]Negative for *A. toluclasticus,* d for *A. tolulyticus.*

[l]Negative for *A. tolulyticus,* ND for *A. toluvorans, A. toluclasticus, A. anaerobius.*

Broad host range plasmids based on RK2 such as pRK290 or pAFR3 can be transferred by triparental mating and are stably replicated in strain BH72 (Egener et al., 2001). Transformation can also be achieved by electroporation (Hurek et al., 1995), and mutagenesis by allelic exchange (Hurek et al., 1995) or transposon mutagenesis (Dörr et al., 1998) is possible.

Antibiotic resistance Resistance of *Azoarcus* species to antibiotics has not been extensively tested. *A. indigens*, *A. communis*, and strain BH72 are not resistant to ampicillin, kanamycin, streptomycin, spectinomycin, or tetracycline (Reinhold-Hurek and Hurek, unpublished observation).

Pathogenicity None of the strains has been reported to be pathogenic. Although the plant-associated species colonize the interior of grasses as endophytes, no symptoms of plant disease have been reported (Reinhold-Hurek and Hurek, 1998).

Ecology The two groups of *Azoarcus* species—plant-associated and soil-borne species—differ strongly in their ecology. Strains of *A. indigens*, *A. communis*, and *Azoarcus* sp. BH72 occur inside roots or on the root surface of Gramineae and have never been isolated from root-free soil (Reinhold-Hurek and Hurek, 1998), except for a strain of *A. communis* that originated not from plants but from French refinery oily sludge (Laguerre et al., 1987; Reinhold-Hurek et al., 1993b). In contrast, all isolates that belong to the clade of *A. tolulyticus*, *A. toluvorans*, *A. toluclasticus*, *A. evansii*, or *A. anaerobius* have not been isolated from living plants but instead from soil and sediments; therefore, their ecology will be treated separately. The plant-associated strains were originally detected in association with roots of Kallar grass (*Leptochloa fusca* (L.) Kunth), a flood-tolerant salt marsh grass grown as a pioneer species on salt-affected, flooded, low-fertility soils in the Punjab of Pakistan since the 1970s (Reinhold et al., 1986; Reinhold-Hurek et al., 1993b). These nitrogen-fixing *Azoarcus* strains were found in high numbers in surface-sterilized roots and only rarely on the root surface (*A. communis*). Isolates were later obtained from rice (*Oryza sativa*) from Nepal (Engelhard et al., 2000) or from resting stages (sclerotia) of a plant-associated basidiomycete found in rice field soil from Pakistan (Hurek et al., 1997b). More detailed studies on plant–microbe interactions of a strain originating from Kallar grass—strain BH72—showed that in gnotobiotic culture in the laboratory, these bacteria had a wider host range. They could also invade rice roots and stems (Hurek et al., 1994b), where they mainly colonize the cortex tissue intercellularly and, rarely, the stele including xylem cells (Hurek et al., 1994b). Despite a high density of colonization, the bacteria do not cause symptoms of plant disease and thus have an endophytic and not a pathogenic lifestyle. Unlike rhizobia, they do not form an endosymbiosis in living plant cells. Nevertheless, they show endophytic nitrogen fixation, expressing nitrogenase genes in the apoplast of aerenchymatic air spaces of flooded rice seedlings (Egener et al., 1999) or field-grown Kallar grass plants (Hurek et al., 1997a). They might even be distributed, based on molecular-ecological studies on root material or fungal spores. *Azoarcus* 16S rDNA genes (Hurek and Reinhold-Hurek, 1995) or *nifH* genes (Ueda et al., 1995; Engelhard et al., 2000) have been retrieved that did not correspond to the genes of cultivated strains or species. Because the corresponding bacteria could not be isolated from the same samples, they may occur in an as yet unculturable state and thus be overlooked by classical microbiological techniques. Interestingly, the sequences that have been retrieved so far from plant material have not clustered with genes from soil-borne species, confirming that they do not appear to be plant-associated.

The soil-borne species are very widespread. Strains belonging to the valid species have been isolated from uncontaminated soils (Song et al., 1999), from soils containing unknown contaminants in industrial areas (Song et al., 1999), and from soils containing known contaminants such as petroleum (Fries et al., 1994; Zhou et al., 1995). Many strains have also been cultured from sediments of uncontaminated or contaminated creeks (Anders et al., 1995), aquifers (Fries et al., 1994, 1997; Zhou et al., 1995), or activated sewage sludge (Springer et al., 1998). Soil-borne *Azoarcus* spp. are also widespread with respect to their geographical distribution. For example, they have been found in North America (USA and Canada), South America (Puerto Rico, Brazil), and Europe (Germany, Switzerland). Their occurrence in anoxic sediments or sewage sludge indicates that their lifestyle *in situ* might be anaerobic (using nitrate as terminal electron acceptor) rather than microaerobic (fixing nitrogen, and using O_2 as terminal electron acceptor) in contrast to the plant-associated strains. Further information on strains that do not belong to validly published species is given below (see Other Organisms).

ENRICHMENT AND ISOLATION PROCEDURES

Plant-associated species and soil-borne species of *Azoarcus* will be treated separately. Plant-associated strains are best enriched on media free of combined nitrogen. Washed roots or roots that have been surface-sterilized in 5% NaOCl for 2 min and then thoroughly washed are macerated aseptically in the enrichment medium free of carbon source. To avoid overgrowth by faster growing diazotrophs, enrichment cultures should be inoculated with serial dilutions of this material. For enrichment of nitrogen-fixing bacteria, N-free semisolid synthetic malate medium (SM-N) is used.[2] The medium is solidified in screw-cap tubes and is inoculated with 10 μl of macerate or root pieces below the medium surface. The tubes are incubated without shaking at 30°C (37°C for tropical or subtropical strains) and checked for development of a subsurface pellicle. Samples for streaking should be taken before the pellicle has moved to the medium surface and become very dense. Isolation of single colonies is carried out on SM-N agar supplemented with 20 mg of yeast extract per liter and 10 g of agar per liter. Enrichment cultures or isolates are analyzed for nitrogenase activity in semisolid medium by the acetylene reduction method. Purity and colony color are checked on a complex medium, VM ethanol, which is based on SM-N medium but supplemented with the following ingredients (per liter): Lab Lemco Powder (Oxoid) or Bacto Peptone (Difco), 3.0 g; yeast extract, 1.0 g; NaCl, 1.0 g; NH_4Cl, 0.5 g; and agar, 15.0 g. Alternatively, for salt-affected habitats, the SSM medium of Reinhold et al. (1986) can be used instead of SM medium for enrichment and isolation.

---

2. N-free semisolid synthetic malate medium (SM-N) consist of (per liter of distilled water): $MgSO_4 \cdot 2H_2O$, 0.2 g; NaCl, 0.1 g; $CaCl_2$, 0.02 g; $MnSO_4 \cdot H_2O$, 0.01 g; $Na_2MoO_4 \cdot 2H_2O$, 0.002 g; Fe(III)-EDTA, 0.066 g; phosphate buffer consisting of KH_2PO_4, 0.6 g, and K_2HPO_4, 0.4 g, which is adjusted to pH 6.8; DL-malate solution consisting of DL-malic acid, 5 g; and KOH, 4.5 g, adjusted to pH 6.8; agar, 2 g; and vitamin solution, 1 ml. The vitamin solution contains (mg per liter of distilled water) myoinositol, 10,000; niacinamide, 100; pyridoxine-HCl, 100; thiamine-HCl, 100; calcium pantothenate, 50; folic acid, 20; choline chloride, 50; riboflavin, 10; ascorbic acid, 100; *p*-aminobenzoic acid, 1.0; vitamin A, 0.5; vitamin D_3, 0.5; vitamin B_{12}, 0.5; and D-biotin, 0.5. Phosphate buffer and vitamin solution are sterilized separately and added to the autoclaved medium after autoclaving.

For soil-borne strains, enrichment is routinely done under strictly anaerobic conditions with nitrate as the terminal electron acceptor. A variety of aromatic carbon sources has been used, depending on the species or strain under study. The preparation of media and cultivation of bacteria are carried out under strictly anoxic conditions. A typical medium is that used for *A. evansii*.[3] Ascorbic acid (4 mM) can be used to reduce the medium. Potentially toxic carbon sources are added at low concentrations (toluene, 5 ppm; phenol, 1 mM; Na-benzoate, 5 mM), and cultures are spiked again with the carbon source after its degradation. Enrichment is carried out in anoxic sealed serum bottles. When rich sediments are used as an inoculum, they may have to be depleted of readily oxidizable carbon sources by repeated incubation in a medium free of these sources (Fries et al., 1994). The enriched samples are plated on agar media such as M-R2A medium[4] or used for an agar dilution series (Widdel and Bak, 1992). For cultivation on poorly water-soluble compounds such as alkylbenzenes, special procedures are required (see Rabus and Widdel, 1995).

MAINTENANCE PROCEDURES

Due to alkalinization of the medium in stationary phase, subculturing *Azoarcus* strains on the salts of malic acid should be avoided and VM ethanol medium should be used instead (soil-borne strains may also be subcultured on M-R2A medium).

Preservation can be done by lyophilization. Strains may also be stored in liquid nitrogen, with 5% DMSO as a cryoprotectant.

PROCEDURES FOR TESTING SPECIAL CHARACTERS

The genus *Azoarcus* can be specifically detected by 16S rDNA-targeted PCR. Primers TH3 (5′-GATTGGAGCGGCCGATGTC-3′) and TH5 (5′-CTGGTTCCCGAAGGCACCC-3′), which correspond to *E. coli* positions 222 to 240 and 1040 to 1022, respectively, yield a specific amplification product at an annealing temperature of 70°C for 30 cycles (Hurek et al., 1993).

DIFFERENTIATION OF THE GENUS *AZOARCUS* FROM OTHER GENERA

Table BXII.β.112 gives the characteristics of *Azoarcus* that differentiate it from other morphologically or physiologically similar genera. The phylogenetically closely related genus *Thauera*,

which is physiologically the most similar to the soil-borne species, is difficult to differentiate by classical tests.

TAXONOMIC COMMENTS

Azoarcus spp. belong to the order *Rhodocyclales* of the class *Betaproteobacteria* according to phylogenetic analysis of almost complete 16S rRNA sequences. Within the genus *Azoarcus*, the 16S rDNA phylogenetic distances are as high as 6%; thus, the genus represents a rather heterogeneous group. Especially within the phylogenetic branch containing soil-borne strains, the 16S rDNA sequence analysis does not always resolve species well from each other (see *A. toluvorans*, Fig. BXII.β.106). *Thauera* is the most closely related genus with a 16S rDNA phylogenetic distance of 6–7%.

Several members of the genus *Azoarcus sensu lato* (Reinhold-Hurek et al., 1993b) were recently reclassified as separate genera; these are *Azovibrio*, *Azonexus*, and *Azospira*, which are closely related to the *Azoarcus/Thauera* clade (Fig. BXII.β.106) (Reinhold-Hurek and Hurek, 2000). Members of the residual genus *Azoarcus* are phylogenetically most closely related to the genus *Thauera*, with which they form the *Azoarcus/Thauera* branch, which is reasonably supported by statistical analysis (Fig. BXII.β.106). *Azoarcus* species are located in two different clades, which in part reflects their physiology and ecology.

The genus *Azoarcus* consists of the following validly described species based on polyphasic taxonomic approaches including DNA–DNA hybridization, protein profiles, fatty acid analysis, and nutritional profiles: *A. indigens* (type species), *A. communis*, *A. tolulyticus*, *A. toluclasticus*, *A. toluvorans*, *A. evansii*, and *A. anaerobius*. The unnamed strain BH72 differs from these at the species level according to DNA–DNA homology studies (≤ 25% DNA binding); however, a species name has not been given because there is only a single strain (Reinhold-Hurek et al., 1993b). Numerous additional soil-borne strains are localized in the *Azoarcus* clade according to 16S rDNA sequence analysis (examples given in Fig. BXII.β.106). Most have been isolated under conditions of denitrification with aromatic hydrocarbons as carbon sources, but the lack of nutritional/physiological data and DNA–DNA homology values do not allow a species assignment. Interestingly, a deeply branching clade of as yet uncultured bacteria is also localized within the *Azoarcus* cluster.

The analysis of phylogenetic relationships in the *Azoarcus/Thauera* 16S rDNA cluster is rendered difficult because the branching pattern between *Thauera*, the soil-borne *Azoarcus* species, and the plant-associated *Azoarcus* species is unstable when tested with different tree-building methods. This was also observed previously (Hurek et al., 1997a; Reinhold-Hurek and Hurek, 2000). Similarly, in other taxa closely related to each other, such as the rhizobia, the resolution of phylogenetic analysis is sometimes limited. Nevertheless, all named species in the genus *Thauera* cluster in one clade with a significant level of support by bootstrap analysis (Fig. BXII.β.106), allowing a reliable assignment to this genus. Likewise, most soil-borne species and strains of *Azoarcus* fall into one clade significantly supported by bootstrap analysis, and none of the species containing plant-associated strains are located on this branch. The phylogenetic distances within these three clades are similar; the consequence of this finding—that the clades might deserve the rank of different genera—is discussed below. All of the valid species of *Azoarcus* are well resolved except for the strains of *A. toluvorans*, which are interspersed among several strains of uncertain affiliation (Fig. BXII.β.106). Former strains of *A. tolulyticus*—Td-3 and Td-

---

3. *Azoarcus evansii* medium consists of (g per liter of distilled water): KH₂PO₄, 0.816 g; K₂HPO₄, 5.920 g; NH₄Cl, 0.53 g; MgSO₄·7H₂O, 0.200 g; KNO₃, 0.5 g; and CaCl₂·2H₂O, 0.025 g. The phosphate is dissolved separately from the other ingredients. Both solutions are adjusted to pH 7.8 for *A. evansii* or to pH 7 for other strains. The two solutions are autoclaved and combined after cooling. To this solution are added 10 ml of sterile trace elements SL-1 and 5 ml of vitamin solution. SL-1 solution contains (per liter of distilled water): HCl (25%; 7.7 M), 10 ml; FeCl₂·4H₂O, 1.5 g; ZnCl₂, 70 mg; MnCl₂·4H₂O, 100 mg; H₃BO₃, 6 mg; CoCl₂·6H₂O, 190 mg; CuCl₂·2H₂O, 2 mg; NiCl₂·6H₂O, 24 mg; and Na₂MoO₄·2H₂O, 36 mg. The FeCl₂ is dissolved in the HCl and then diluted in water. The other salts are then added and dissolved. The vitamin solution contains (mg per liter of distilled water): vitamin B₁₂, 50 mg; pantothenic acid, 50 mg; riboflavin, 50 mg; pyridoxamine-HCl, 10 mg; biotin, 20 mg; folic acid, 20 mg; nicotinic acid, 25 mg; nicotinamide, 25 mg; α-lipoic acid, 50 mg; *p*-aminobenzoic acid, 50 mg; and thiamine-HCl·2H₂O, 50 mg.

4. M-R2A medium (Fries et al., 1994) contains (per liter of distilled water): yeast extract, 0.5 g; peptone, 0.5 g; Casamino acids, 0.5 g; dextrose, 0.5 g; soluble starch, 0.5 g; sodium pyruvate, 0.3 g; K₂HPO₄, 0.4 g; KH₂PO₄, 0.25 g; KNO₃, 0.505 g; CaCl₂·2H₂O, 0.015 g; MgCl₂·2H₂O, 0.02 g; FeSO₄·7H₂O, 0.007 g; Na₂SO₄, 0.005 g; NH₄Cl, 0.8 g; MnCl₂·4H₂O, 5 mg; H₃BO₃, 0.5 mg; ZnCl₂, 0.5 mg; CoCl₂·6H₂O, 0.5 mg; NiSO₄·6H₂O, 0.5 mg; CuCl₂·2H₂O, 0.3 mg; NaMoO₄·2H₂O, 0.01 mg; agar, 15 g. The pH is adjusted to 7.0 before autoclaving.

19—have recently been removed from this species due to low DNA–DNA hybridization values (<40%), which are also reflected in the 16S rDNA sequence analysis (Fig. BXII.β.106).

In the original description of the genus *Azoarcus*, two groups of bacteria that were distinct at the species level and located on the *Azoarcus* rRNA branch—albeit at low $T_{m(e)}$ values—were classified as *Azoarcus sensu lato* (Reinhold-Hurek et al., 1993b). Availability of additional strains and 16S rDNA sequences allowed the reassessment of the taxonomic structure of *Azoarcus sensu lato*. The unnamed groups C and D (Reinhold-Hurek et al., 1993b) were classified as the genera *Azovibrio* and *Azospira*, respectively (Reinhold-Hurek and Hurek, 2000), and *Azoarcus sensu lato* group E, which had been described later (Hurek et al., 1997b), was proposed as *Azonexus* (Reinhold-Hurek and Hurek, 2000). These bacteria are very similar in physiology and ecology to the plant-associated *Azoarcus* species.

The definition of the genus *Azoarcus* is complex. It includes strains with a variety of physiological and ecological attributes, which place them into at least two groups—the plant-associated species and the soil-borne species. In addition to their ecological differences, the species containing plant-associated strains differ from the soil-borne species in some nutritional features, e.g., the inability to use carbohydrates or, under conditions of denitrification, certain aromatic compounds, as sole carbon source. Moreover, the phylogenetic analysis of 16S rDNA sequences also points to a separation of these two groups. Since the phylogenetic distances within the three clades in the *Azoarcus/Thauera* group are similar, both subgroups of *Azoarcus* might also deserve the rank

of different genera in future. However, this will require a rigid polyphasic taxonomic analysis, which subjects the strains to the same tests. As can be seen from Tables BXII.β.113 and BXII.β.114, many features have not been tested for all species or strains. For instance, nitrogen fixation and aerobic nutritional profiles are not tested in all soil-borne strains, whereas nutritional profiles under conditions of denitrification are lacking for plant-associated species. These studies should also include the numerous soil-borne isolates that have not yet been assigned to any species. Because some strains are deeply branching in the phylogenetic analysis (e.g., strains 22Lin, T3, and M3; Fig. BXII.β.106) the description of new species might be expected.

FURTHER READING

Harwood, C.S., G. Burchhardt, H. Herrmann and G. Fuchs. 1998. Anaerobic metabolism of aromatic compounds via the benzoyl-CoA pathway. FEMS Microbiol. Rev. *22*: 439–458.

Reinhold-Hurek, B. and T. Hurek. 1998. Life in grasses: diazotrophic endophytes. Trends Microbiol. *6*: 139–144.

Reinhold-Hurek, B. and T. Hurek. 2000. Reassessment of the taxonomic structure of the diazotrophic genus *Azoarcus sensu lato* and description of three new genera and new species, *Azovibrio restrictus* gen. nov., sp. nov., *Azospira oryzae* gen. nov., sp. nov. and *Azonexus fungiphilus* gen. nov., sp. nov. Int. J. Syst. Evol. Microbiol. *50*: 649–659.

Song, B., M.M. Haggblom, J. Zhou, J.M. Tiedje and N.J. Palleroni. 1999. Taxonomic characterization of denitrifying bacteria that degrade aromatic compounds and description of *Azoarcus toluvorans* sp. nov. and *Azoarcus toluclasticus* sp. nov. Int. J. Syst. Bacteriol. *49 Pt 3*: 1129–1140.

DIFFERENTIATION OF THE SPECIES OF THE GENUS *AZOARCUS*

The differential characteristics of the species of *Azoarcus* are indicated in Table BXII.β.113. Other characteristics of the species are presented in Table BXII.β.114.

List of species of the genus Azoarcus

1. **Azoarcus indigens** Reinhold-Hurek, Hurek, Gillis, Hoste, Vancanneyt, Kersters and De Ley 1993b, 583[VP]

in' di.gens. L. v. *indigere* to be in need of; M.L. pres. part. *indigens* being in need of, referring to the vitamin requirements.

This species can be differentiated from the other species by its requirement for *p*-aminobenzoic acid, by its ability to grow on itaconate, and by a combination of characteristics given in Table BXII.β.113. Additional characteristics are given in Table BXII.β.114. Cells are thin (0.5–0.7 μm wide) and curved; cell pairs appear slightly S-shaped. Colonies are very compact and difficult to disperse. Growth in liquid media is clumpy; aggregation is very strong on peptone media. Optimal temperature, 40°C. Diazotrophic. The major fatty acids are $C_{10:0\ 3OH}$, $C_{12:0}$, $C_{16:1\ \omega7c}$, $C_{17:0\ cyclo}$, and $C_{18:1}$. Isolated from roots and stem bases of *Leptochloa fusca* (L.) Kunth from Punjab of Pakistan (Reinhold-Hurek et al., 1993b), black sclerotia of an *Ustilago*-related basidiomycete from rice soil in the Punjab of Pakistan (Hurek et al., 1997b), and from rice roots (*Oryza sativa*) from Nepal (Engelhard et al., 2000).

The mol% G + C of the DNA is: 62.4 (T_m).

Type strain: VB32, DSM 12121, LMG 9092.

GenBank accession number (16S rRNA): AF011345.

2. **Azoarcus anaerobius** Springer, Ludwig, Philipp and Schink 1998, 954[VP]

an.a.e.ro' bi.us. Gr. pref. *an* not; Gr. n. *aer* air; Gr. n. *bios* life; N.L. adj. *anaerobius* not living in air, anaerobic.

This species can be differentiated from the others by its strictly anaerobic lifestyle with nitrate as the only electron acceptor. Nitrate is quantitatively reduced to N_2 gas, nitrite is accumulated as an intermediate, and N_2O is not detected. Oxygen is not reduced, even at low partial pressures. Sulfate, thiosulfate, sulfite, sulfur, trimethylamine *N*-oxide, DMSO, $Fe(OH)_3$, $K_3[Fe(CN)_6]$, and fumarate are not reduced. Superoxide dismutase-positive. Additional characteristics are given in Tables BXII.β.113 and BXII.β.114. Optimal temperature, 28°C. pH range, 6.5–8.2; optimal pH, 7.2. Enhanced salt concentration impairs growth. Not diazotrophic. Sole carbon sources for growth include propanol, valerate, pyruvate, cyclohexanecarboxylate, phenol, resorcinol, and *p*-cresol. No growth occurs with L-malate, formate, 5-oxocaproate, pimelate, catechol, hydroquinone, 2-hydroxybenzoate, *o*-cresol, and *m*-cresol. No autotrophic growth occurs with hydrogen or thiosulfate. Isolated from sewage sludge.

The mol% G + C of the DNA is: 65.5 ± 0.5 (T_m).

Type strain: LuFRes1, DSM 12081.

GenBank accession number (16S rRNA): Y14701.

3. **Azoarcus communis** Reinhold-Hurek, Hurek, Gillis, Hoste, Vancanneyt, Kersters and De Ley 1993b, 583[VP]

com' mu.nis. L. masc. adj. *communis* common, referring to diverse habitats.

TABLE BXII.β.113. Characteristics differentiating species of the genus *Azoarcus*[a]

Characteristic	*A. indigens*	*A. anaerobius*	*A. communis*	*A. evansii*	*A. toluclasticus*	*A. tolulyticus*	*A. toluvorans*	Unnamed strain BH72
Cell width (μm)	0.5–0.7	1.5	0.8–1.0	0.4–0.8	0.6–0.8	0.8–1.0	0.8–1.0	0.6–0.8
O_2 terminal electron acceptor	+	−	+	+	+	+	+	+
Catalase	+	−	+	+	−	d	+	+
Nitrogen fixation	+	−	−	−		+		+
Requirement for *p*-aminobenzoic acid	+	−	−	−		−	−	
Sole carbon sources:								
Adipate, D-ribose	−		−	−	+	+	+	−
p-Aminobenzoate	+		−	+		−		−
L-Arabinose, D-xylose	−	−	−	−	+	+	+	−
n-Caproate	−		+	−	−	d	−	+
Citrate	−		+	−		−		
D-Fructose	−	−	−	+	d	+	−	−
D-Galactose, L-proline, sucrose	−		−	+	+	+	+	−
D-Glucose	−	−	−	d	d	+	−	−
3-Hydroxybenzoate	+	+	+	+	−	d	−	+
Isovalerate	−	+	+	+	−	+		+
Itaconate	+		−	−		−	−	−
Maltose	−		−	d	+	+	+	−
D-Mandelate	+		+	−		−		−
Phenylacetate	+	+	+	d	−	d	+	+
L-Phenylalanine	+	+	+	+	d	d	±	−
D-Tartrate	+	D		+	−	−	−	+

[a]Symbols: see standard definitions. Blank space, not determined. ±, weakly positive.

This species can be differentiated from the others by its cell width (0.8–1.0 μm), by its growth on D-mandelate and on citrate, and by a combination of characteristics given in Table BXII.β.113. Additional characteristics are given in Table BXII.β.114. Cells are plump and only slightly curved. Optimal temperature, 37°C. Diazotrophic. Some strains grow well at 2% NaCl. The major fatty acids are $C_{10:0\ 3OH}$, $C_{14:0}$, $C_{16:1\ \omega7c}$, $C_{16:0}$, and $C_{18:1}$. Isolated from roots of *L. fusca* (L.) Kunth from Punjab of Pakistan (SWub3) (Reinhold-Hurek et al., 1993b; Engelhard et al., 2000) and from refinery oily sludge in France (Laguerre et al., 1987).

The mol% G + C of the DNA is: 62.4 (T_m).

Type strain: SWub3.

GenBank accession number (16S rRNA): AF011343.

Additional Remarks: The species affiliation of strain LU1, which was isolated from a compost biofilter in Canada (Juteau et al., 1999), is not clear; it has a phylogenetic distance of 2.9% with the type strain.

4. **Azoarcus evansii** Anders, Kaetzke, Kämpfer, Ludwig and Fuchs 1995, 331[VP]

e'van.si.i. L. gen. n. *evansii* of Evans, in honor of the late W.C. Evans, a pioneer in studies of anaerobic aromatic metabolism.

This species can be differentiated from other species by a combination of characters given in Table BXII.β.113. Cells are rods with rounded ends, motile by means of a subpolar flagellum. Yeast extract (0.1%) inhibits growth. Optimal temperature 35–37°C. Optimal pH, 7.8. Does not fix nitrogen. During denitrifying growth on aromatic compounds, nitrite is an intermediate and is reduced mainly to N_2O (strain KB740). Characteristics in addition to those listed in Table BXII.β.114 are as follows. Under anaerobic conditions, the type strain KB740 uses benzoate, phenylacetate, phenylglyoxylate, 3-and 4-hydroxybenzoate, 2-aminobenzoate, 4-hydroxyphenylacetate, phenylalanine, *p*-cresol, 2-fluorobenzoate, benzaldehyde, benzyl alcohol, indolylacetate, *o*-phthalate, adipate, pimelate, cyclohexanecarboxylate, succinate, fumarate, L-malate, acetate, acetone, D-fructose, and D-maltose. Slow growth occurs on glutarate and D-glucose but not on toluene, phenol, 2-hydroxybenzoate, protocatechuate, *o*-and *m*-cresol, indole, ethanol, D-ribose, and D-lactose. Pyridine is used by reference strain pF6 under aerobic and anaerobic conditions. The aerobic metabolism of benzoate is unusual in the type strain KB740, since none of the known pathways—i.e., the conversion of benzoate to catechol (1,2-dihydroxybenzoene) or protocatechuate (3,4-dihydroxybenzoate)—appear to operate in this species (Mohamed et al., 2001). The first step is the activation of benzoate to benzoyl-CoA by a benzoate-CoA ligase, and the second step involves the hydroxylation of benzoyl-CoA by a novel benzoyl-CoA oxygenase (Mohamed et al., 2001). The first step of phenylacetate degradation is catalyzed by two different phenylacetate-CoA-ligases under aerobic and anaerobic conditions, respectively (Mohamed, 2000). Reference strain PF6 degrades pyridine aerobically and anaerobically when growing on nitrate (Rhee et al., 1997). Benzoyl-CoA is also a central intermediate in the anaerobic degradation of aromatic compounds (Harwood et al., 1998). In the type strain KB740, the nucleotide sequence analysis of the gene cluster—including a gene for benzoate-CoA ligase—indicates that the degradation of benzoate is probably similar to the benzoate-CoA pathway in *Thauera aromatica* (Harwood et al., 1998). The major cellular fatty acids are $C_{16:1\ \omega7c}$, $C_{16:0}$, $C_{12:0}$, $C_{10:0\ 3OH}$, and $C_{18:1\ \omega7,\ 9c}$. The major quinone is ubiquinone 8. The species was described with one strain isolated from creek sediment in the United States (Braun and Gibson, 1984). However, a second strain,

TABLE BXII.β.114. Further characteristics of species of the genus Azoarcus[a]

Characteristic	Azoarcus indigens	Azoarcus anaerobius	Azoarcus communis	Azoarcus evansii[b]	Azoarcus toluclasticus	Azoarcus tolulyticus[b]	Azoarcus toluvorans	Azoarcus sp. BH72	Azonexus fungiphilus	Azospira oryzae	Azovibrio restrictus	Azovibrio sp. BS20-3
Number of strains examined	5	1	3	2	5	9	2	1	5	12	5	1
Cell width (μm)	0.5–0.7	1.5	0.8–1.0	0.4–0.8	0.6–0.8	0.8–1[b]	0.8–1.0	0.6–0.8	0.6–0.8	0.4–0.6	0.6–0.8	0.6–0.8
Cell length (μm)	2.0–4.0	2.7–3.3	1.5–3.0	1.5–3.0	1.7–4.0	1.4–2.8	1.4–2.8	1.5–4.0	1.5–4.0	1.1–2.5	1.5–3.5	1.5–3.5
Elongated cells in stationary cultures	r	ND	R	–	ND	–	ND	r	+	r	R	ND
Colony diameter (mm, VM/CR)[c]	2–3	0	2–4	0.2–0.7[b]	ND	1–1.5[b]	ND	2–3/1	1–2/0.7	1.0–2.0	1.5–2/1	Negligible
Colony color (VM)[c]	Opaque yellowish Whitish pink; white margin	–	Translucent yellowish Whitish pink; pink center	Translucent beige[b]	ND	Opaque yellowish[b]	ND	Translucent yellowish	Opaque ocher	Translucent pink-salmon	Opaque beige	Negligible
Colony color (CR)[c]		–		Orange red[b]	ND	Orange red[b]	ND	Orange red	Dark red	Translucent orange	Orange red	Negligible
Colony surface	Rough	–	Smooth	Shining	ND	Shining	ND	Shining	Shining	Shining	Shining	ND
Growth at 40°C	+	–	+	–[b]	–	–[b]	–	+	+	+	d	–
Requirement for p-aminobenzoic acid	+	–	–	–	–	–	–	–	–	–	–	–
Requirement for cobalamin	–	–	D	–	–	–	–	–	+	d	–	+
Nitrogen fixation	+	–	+	–[b]	ND	+[b]	ND	+	+	+	+	+
Oxidation/fermentation of glucose	–	–	–	+[d]	ND	ND	ND	–	–	–	–	–
Catalase	+	–	+	+	+	D	+	+	+	+	+	ND
Sole carbon sources for growth:[e]												
L(+) Arabinose, D(+)-xylose	–	ND	–	–	–	+	+	–	–	–	–	–
Adipate	–	ND	–	–	+	+	+	–	–	–	–	–
D-Alanine, glycerol	–	ND	D	–	ND	–	ND	–	–	–	–	–
L-Alanine	d	ND	+	+[b]	ND	+[b]	ND	–	–	–	d	+
p-Aminobenzoate	+	ND	–	+	+	–[b]	ND	–	–	–	–	–
L-Arginine	–	ND	V	–	+	+[b]	+	–	–	–	–	–
L-Aspartate	+	ND	D	+[b]	+	+[b]	+	+	+	+	+	–
Benzoate	d	+	V	+	+	+[b]	+	–	–	–	–	–
Benzylamine	d	ND	–	+[b]	ND	–	ND	–	–	–	–	–
Betain, meso-tartrate[b]	d	ND	+	+[b]	ND	+[b]	ND	–	–	–	–	–
n-Butylamine	+	ND	+	+[b]	ND	d	ND	–	–	+	–	–
n-Caproate	–	ND	+	–[b]	–	–[b]	–	–	–	–	–	–
Citrate	–	ND	–	–	ND	–[b]	ND	–	–	–	–	–
m-Coumarate	d	ND	+	–[b]	ND	–[b]	ND	–	–	+	–	–
D(+)-Fructose	–	–	–	+[b]	d	+	–	–	–	–	–	–

(continued)

TABLE BXII.β.114. (*cont.*)

Characteristic	*Azoarcus indigens*	*Azoarcus anaerobius*	*Azoarcus communis*	*Azoarcus evansii* [b]	*Azoarcus toluclasticus*	*Azoarcus tolulyticus* [b]	*Azoarcus toluvorans*	*Azoarcus* sp. BH72	*Azonexus fungiphilus*	*Azospira oryzae*	*Azovibrio restrictus*	*Azovibrio* sp. BS20-3
Fumarate	+	+	+	+	d	+	+	+	+	+	+	+
D(+)-Galactose, sucrose	−	ND	−	+[b]	+	−[b]	+	+	−	−	−	−
Gentisate	+	ND	+	ND	ND	−[b]	ND	+	−	−	−	−
D(+)-Glucose	−	−	−	d	d	+	−	+	−	−	v	−
Glutarate	−	ND	+	+[b]	ND	+[b]	ND	+	−	−	−	−
DL-Glycerate	d	ND	−	+[b]	ND	+[b]	ND	+	−	−	−	−
3-Hydroxybenzoate	+	+	+	+[b]	−	d	−	+	−	−	−	−
4-Hydroxybenzoate	+	+[f]	+	−	+[f]	+[f]	+[f]	+	+	+	−	+
3-Hydroxybutyrate	+	ND	+	+[b]	ND	+[b]	ND	+	−	v	d	+
Isobutyrate	−	+	+	+[b]	−	+	±	+	+	v	−	−
Isovalerate	+	ND	+	−[b]	ND	−[b]	ND	+	−	+	−	+
Itaconate	−	ND	−	−	+	−[b]	ND	−	−	−	−	−
Lactose	−	ND	−	−[b]	ND	−[b]	d	−	−	−	−	−
L-Mandelate, protocatechuate,	+	ND	+	−[b]	ND	+[b]	ND	+	−	−	d	−
D(+) Malate	+	−	+	+	ND	+[b]	ND	+	−	+	+	+
L-Malate	+	−	+	+	ND	+[b]	ND	+	+	+	+	+
Malonate	−	ND	−	−	ND	−	ND	−	d	−	−	−
Maltose	−	ND	−	d	+	+	+	−	−	−	−	−
Maltotriose, palatinose, D(+)-melezitose	−	ND	−	+[b]	ND	+	ND	−	−	−	−	−
D-Mandelate	+	ND	+	−[b]	ND	−[b]	ND	−	−	−	−	−
2-Oxoglutarate	+	ND	+	+[b]	ND	−[b]	ND	+	+	+	−	−
Phenylacetate	+	+	+	d	−	d	−	+	−	−	−	−
L-Phenylalanine	+	+	+	+	d	d	±	+	−	−	−	−
L-Proline	−	ND	−	+	+	−[b]	+	−	+	−	−	−
Propionate	−	+	+	+	−	d	−	+	+	+	+	+
(−)-Quinate	−	ND	D	−	ND	−[b]	ND	−	−	−	−	−
D(−)-Ribose	−	ND	D	−[b]	+	+	+	−	−	−	−	−
D-Tartrate	+	ND	D	+	−	−	−	+	−	v	−	−
L(+)-Tartrate	−	ND	−	−[b]	d	−[b]	ND	+	−	+	+	+
Toluene (denitrifying)	+	ND	V	ND	ND	−	ND	−	−	−	−	−
Tryptamine	−	ND	−	−[b]	ND	−[b]	ND	+	−	−	−	−
L-Tyrosine	−	+	−									

[a]Data from Reinhold-Hurek et al. (1993b); Anders et al. (1995); Hurek and Reinhold-Hurek (1995); Zhou et al. (1995); Hurek et al. (1997a); Rhee et al. (1997); Springer et al. (1998); Song et al. (1999); Engelhard et al. (2000); Reinhold-Hurek and Hurek (2000). For all characteristics: ±, intermediate reaction; r, rare; ND, not determined. All strains have the following features: cells are straight to curved rods; oxidase positive; no growth in the presence of 5% NaCl and no growth-rate increase when NaCl is added to medium (ND for *A. toluvorans*, *A. toluclasticus*, *A. anaerobius*); denitrification (ND for *Azonexus* sp.); no spore formation; no starch hydrolysis (ND for *A. evansii* and *A. anaerobius*).

[b]Marked characters were determined for *A. tolulyticus* strain Td-1 or *A. evansii* strain KB740 (Reinhold-Hurek and Hurek, 2000).

[c]Growth on VM ethanol agar (VM) or Congo red agar (CR) at 37°C for 4 d.

[d]Not tested for strain pF6

[e]All strains grow on DL-lactate, succinate, acetate, L-glutamate, butyrate, ethanol (ND for *A. toluvorans*, *A. toluclasticus*, and *A. anaerobius*) (no substrate but L-malate tested for *A. evansii* pF6); no growth on D(+)-mannose, maltitol, N-acetyl-glucosamine, D-gluconate, caprate; no growth on[b] (ND for *A. toluclasticus*, *A. toluvorans*, *A. anaerobius*) D(+)-trehalose, L(+)-sorbose, α-D(+)-melibiose, D(+)-raffinose, lactulose, 1-O-methyl-β-galactopyranoside, 1-O-methyl-α-galactopyranoside, D(+) cellobiose, β-gentiobiose, 1-O-methyl-β-D-glucopyranoside, α-L-rhamnose, α-L-fucose, D(+)-arabitol, xylitol, dulcitol, D-tagatose, myo-inositol, D-sorbitol, adonitol, hydroxyquinoline-β-glucuronide, D-lyxose, erythritol, 1-O-methyl-α-D-glucopyranosid, 3-O-methyl-glucopyranose, D-saccharate, mucate, D-glucuronate, D-galacturonate, 2-keto-D-gluconate, 5-keto-D-gluconate, 3-phenylpropionate, 4-aminobutyrate, DL-α-amino-n-valerate, trigonelline, putrescine, ethanolamine, D-glucosamine.

[f]Under denitrifying conditions.

pF6, sharing 100% 16S rDNA sequence identity with the type strain, was isolated on pyridine from industrial wastewater in Korea (Rhee et al., 1997).

The mol% G + C of the DNA is: 67.5.

Type strain: DSM 6898, KB740.

GenBank accession number (16S rRNA): X77679.

5. **Azoarcus toluclasticus** Song, Häggblom, Zhou, Tiedje and Palleroni 1999, 1139[VP]

to.lu.clas' ti.cus. N.L. n. Fr. Sp. *tolu* balsam from Santiago de Tolu, toluene; Gr. adj. *clasticus* breaking; M.L. adj. *toluclasticus* toluene-breaking.

This species can be differentiated from the other species by the lack of catalase activity despite its aerobic growth, and also by a combination of characters given in Table BXII.β.113. Additional characteristics are given in Table BXII.β.114. Cells are short motile rods. Optimal temperature, 30°C. Under denitrifying conditions, growth occurs on acetate, benzoate, pyruvate, succinate, D-xylose, L-arabinose, D-ribose, D-galactose, sucrose, lactose, maltose, adipate, lactate, mannitol, aspartate, proline, and arginine. All strains except strain MF23 can use toluene as a growth substrate; strains MF58 and MF63T can also grow on phenol under denitrifying conditions. The strains grow on brainheart infusion, nutrient and trypticase soy agar, except for strain MF63T, which does not grow on nutrient agar. The predominant fatty acids are $C_{16:0}$ and $C_{16:1 \omega 7c}$. DNA–DNA hybridizations show intermediate similarities (47–55%) among genomovar I (strains MF7 and MF23) and genomovar II (strains MF58, MF63T, and MF441). Isolated from aquifer sediments in the United States.

The mol% G + C of the DNA is: 67.3 (HPLC).

Type strain: MF63, ATCC 700605.

GenBank accession number (16S rRNA): AF123077.

6. **Azoarcus tolulyticus** Zhou, Fries, Chee-Sanford and Tiedje 1995, 505[VP]

to.lu.ly' ti.cus. N.L. n. Fr. Sp. *tolu* balsam from Santiago de Tolu, toluene; Gr. adj. *lyticus* dissolving; N.L. masc. adj. *tolulyticus* toluene dissolving.

This species can be differentiated from the other species by a combination of characters given in Table BXII.β.113. Additional characteristics are given in Table BXII.β.114. Cells are short motile rods that are slightly elongated (to 2.8 μm) when grown on M-R2A agar. Diazotrophic. Optimal temperature, 30°C; growth can occur at 37°C. Under denitrifying conditions, growth occurs on acetate, adipate, arginine, L-arabinose, aspartate, benzoate, fumarate, D-galactose, D-glucose, lactate, lactose, maltose, mannitol, proline, pyruvate, D-ribose, succinate, sucrose, toluene, D-xylose, and 4-hydroxybenzoate. Growth occurs on brain-heart infusion agar but either not at all or only poorly on nutrient and trypticase soy agar. The predominant fatty acids are $C_{16:0}$ and $C_{16:1 \omega 7c}$. Isolated from aquifer sediments and petroleum-contaminated soils in the United States (Fries et al., 1994, 1997; Chee-Sanford et al., 1996).

The mol% G + C of the DNA is: 66.9 (HPLC).

Type strain: Tol-4, CC 51758.

GenBank accession number (16S rRNA): L33694.

Additional Remarks: Strains Td-19 and Td-3 (Zhou et al., 1995) have been removed from the species due to low DNA–DNA hybridization values with the type strain (Song

et al., 1999). Other strains of *A. tolulyticus* include strains MF66, 2a1, 3a1, 7a1, BL2, and BL11. The cells of strain MF66 are short motile rods with monopolar flagellation. This strain can grow under aerobic and denitrifying conditions with acetate, benzoate, butyrate, D-fructose, fumarate, D-glucose, phenol, pyruvate, succinate, toluene, D-xylose, L-arabinose, D-ribose, D-galactose, sucrose, lactose, maltose, adipate, lactate, mannitol, aspartate, proline, phenylalanine, or arginine. Under denitrifying conditions, 4-hydroxybenzoate is used as a growth substrate. The cells are oxidase-positive and catalase-negative and grow on brain-heart infusion plates, but do not grow on nutrient and trypticase soy agar plates. The predominant fatty acids are $C_{16:0}$ and $C_{16:1 \omega 7c}$, similar to those of other members of *A. tolulyticus*. It belongs to the species *A. tolulyticus* based on 16S rRNA gene sequence analysis, DNA–DNA hybridization, similar patterns of whole-cell proteins, and genomic DNA analysis. Strains 2a1, 3a1, 7a1, BL2, and BL11 were isolated from northern Michigan, USA, and are short motile rods with monopolar flagellation. They can grow on acetate, benzoate, butyrate, D-fructose, fumarate, D-glucose, pyruvate, succinate, D-xylose, L-arabinose, D-ribose, D-galactose, sucrose, lactose, maltose, adipate, lactate, mannitol, aspartate, proline, or arginine under aerobic and denitrifying conditions and use 4-hydroxy-benzoate and toluene for growth under denitrifying conditions. They grow on brain-heart infusion, nutrient, and trypticase soy agar plates, and give positive reactions in the catalase and oxidase tests. A different colony morphology of these strains on half-strength trypticase soy agar plus nitrate at 30°C for 48 h has been reported (Chee-Sanford et al., 1996). The predominant fatty acids are $C_{16:0}$ and $C_{16:1 \omega 7c}$. Whole cell protein and genomic DNA fragmentation analyses show identical patterns to the other members of *A. tolulyticus*, and 16S rRNA sequence analysis of one strain (BL11) shows a close relationship with the members of *A. tolulyticus*. Thus, they belong to the species *A. tolulyticus*. Because the DNA hybridization values between these isolates and the strains of *A. tolulyticus* are approximately 55%, they constitute a genomovar of this species.

7. **Azoarcus toluvorans** Song, Häggblom, Zhou, Tiedje and Palleroni 1999, 1139[VP]

to.lu.vo' rans. N.L. n. Fr. Sp. *tolu* balsam from Santiago de Tolu, toluene; L. part. adj. *vorans* devouring; M.L. part. adj. *toluvorans* toluene-devouring.

This species can be differentiated from the other species by a combination of characters given in Table BXII.β.113. Additional characteristics are given in Table BXII.β.114. Optimal temperature, 30°C. Growth occurs on brain-heart infusion, nutrient, and trypticase soy agar. Under denitrifying conditions, growth occurs on acetate, benzoate, butyrate, fumarate, phenylacetate, pyruvate, succinate, toluene, D-xylose, L-arabinose, D-ribose, D-galactose, sucrose, maltose, adipate, lactate, mannitol, aspartate, proline, phenylalanine, arginine, 4-hydroxybenzoate, and phenol, and under aerobic conditions on benzene or ethylbenzene. Isolated from soil from industrial area in Brazil (Td17) and uncontaminated organic soil in the United States (Td21) (Fries et al., 1994).

The mol% G + C of the DNA is: 67.8 (HPLC).

Type strain: Td21, ATCC 700604.

GenBank accession number (16S rRNA): L33692.

Other Organisms

1. *Azoarcus* sp. strain BH72 Reinhold-Hurek, Hurek, Gillis, Hoste, Vancanneyt, Kersters and De Ley 1993b, 582.

This unnamed strain differs at the species level by having a DNA–DNA hybridization value of ≤25% with other species. It can be differentiated from the other species by its lack of growth on L-phenylalanine and a combination of characteristics given in Table BXII.β.113. Additional characteristics are given in Table BXII.β.114. Cells are long, thin (0.6–0.8 μm wide) and slightly curved, elongated (8–12 μm) cells occurring in late log or in stationary phase culture on N-free or ammonium-supplemented SM medium. The optimal temperature for growth is 40°C. Diazotrophic. The major fatty acids are $C_{16:0}$, $C_{16:1\ \omega7c}$, $C_{18:1}$, $C_{18:1}$, and $C_{14:0}$. Isolated from the root interior of *Leptochloa fusca* (L.) Kunth from Punjab of Pakistan (Reinhold et al., 1986).

The mol% G + C of the DNA is: 67.6 (T_m).

GenBank accession number (16S rRNA): AF011344.

Additional Remarks: A wide range of strains have been isolated under conditions of denitrification, mostly on aromatic carbon sources, which cluster within the clade of soil-borne *Azoarcus* species according to phylogenetic analysis of 16S rDNA sequences (Fig. BXII.β.106); however, data are not sufficient for the assignment to given species. As described above, some are only distantly related and may thus deserve the rank of separate species in future. Examples of strains and habitats are: EbN1, PbN1, isolated on ethylbenzene from freshwater mud, Germany (Rabus and Widdel, 1995); Lin22, isolated on cyclohexane-1,2-diol from activated sludge, Germany (Harder, 1997); pCyN1, isolated on *p*-cymene from freshwater mud, Germany (Harms et al., 1999a); CC-11, isolated from a 3-year-enrichment culture on phenol, Japan (Shinoda et al., 2000); M3, isolated on toluene from a diesel-fuel contaminated aquifer, Switzerland (Hess et al., 1997); CR23, isolated on phenol from creek in Costa Rica (van Schie and Young, 1998).

Genus III. **Azonexus** *Reinhold-Hurek and Hurek 2000, 658*[VP]

THE EDITORIAL BOARD

A.zo'nex.us. Fr. n. *azote* nitrogen; L. masc. n. *nexus* coil; M.L. masc. n. *Azonexus* nitrogen-fixing coil.

Gram-negative somewhat curved motile rods (0.6–0.8 × 1.5–4.0 μm). **Cells elongated and coiled in stationary phase.** One polar flagellum. Chemoorganoheterotrophic and aerobic; microaerophilic when fixing nitrogen. Strictly respiratory; O_2 is the electron acceptor. Oxidase and catalase positive. **Fix N₂.** Require cobalamine. Grow on acetate, ethanol, fumarate, DL-lactate, L-malate, 2-oxoglutarate, L-proline, and succinate.

The mol% G + C of the DNA is: not known.

Type species: **Azonexus fungiphilus** Reinhold-Hurek and Hurek 2000, 658.

FURTHER DESCRIPTIVE INFORMATION

Azonexus fungiphilus is notable for its tendency to produce elongated (up to 50 μm) cells in early stationary phase cultures (Reinhold-Hurek and Hurek, 2000). These elongated cells appear to break up into curved segments later in stationary phase.

Azonexus fungiphilus strains were recovered from fungal resting stages obtained from the rhizosphere of rice plants in Pakistan (Hurek et al., 1997b).

Analysis of 16S rDNA sequences placed *Azonexus* in the *Betaproteobacteria*; in these analyses, *Azonexus* was most closely related to members of the genera *Thauera* and *Rhodocyclus*. *Azonexus* strains could be separated from *Azoarcus*, *Azospira*, and *Azovibrio* strains by comparisons of fatty acid composition, SDS-soluble pro-

tein electrophoretic patterns, and BOX-PCR genomic fingerprinting. DNA–DNA hybridization results supported these divisions (Hurek et al., 1997b; Reinhold-Hurek and Hurek, 2000).

ENRICHMENT AND ISOLATION PROCEDURES

Enrichment and isolation procedures are described in Reinhold et al. (1986).

MAINTENANCE PROCEDURES

Maintenance and long-term storage procedures are described in Reinhold-Hurek et al. (1993b).

DIFFERENTIATION OF THE GENUS *AZONEXUS* FROM OTHER GENERA

Reinhold-Hurek and Hurek (2000) provide tables of characteristics that differentiate the genera *Azoarcus*, *Azonexus*, *Azospira*, and *Azovibrio* from each other and from *Acetobacter diazotrophicus*, *Burkholderia vietnamensis*, the genus *Azospirillum*, and the genus *Herbaspirillum* (see also Table BXII.β.112 in the chapter on *Azoarcus*).

TAXONOMIC COMMENTS

This genus was first described as *Azoarcus* sp. group E (Hurek et al., 1997b; Reinhold-Hurek and Hurek, 2000).

List of species of the genus Azonexus

1. **Azonexus fungiphilus** Reinhold-Hurek and Hurek 2000, 658[VP]

fun.gi'phil.us. M.L. masc. n. *fungi* mushrooms, fungi; Gr. adj. *philos* loving; M.L. masc. adj. *fungiphilus* loves mushrooms or fungi, referring to its source of isolation.

Description as for the genus with the following additional characteristic: grows on L-aspartate, L-glutamate, and 3-hydroxybutyrate.

The mol% G + C of the DNA is: unknown.

Type strain: BS5-8.

GenBank accession number (16S rRNA): AF011350.

Genus IV. **Azospira** *Reinhold-Hurek and Hurek 2000, 658*^{VP}

THE EDITORIAL BOARD

A.zo'spi.ra. Fr. n. *azote* nitrogen; L. fem. n. *spira* winding, turn; M.L. fem. n. *Azospira* nitrogen-fixing spiral.

Gram-negative curved motile rods (0.4–0.6 × 1.1–2.5 μm). One polar flagellum. **Cells elongated in stationary cultures.** Chemoorganoheterotrophic and aerobic; microaerophilic when fixing N_2. Strictly respiratory; O_2 and nitrate are electron acceptors. Oxidase and catalase positive. **Fix N_2.** Grow on acetate, *n*-caproate, ethanol, fumarate, L-glutamate, DL-lactate, L-malate, 2-oxoglutarate, propionate, and succinate.

The mol% G + C of the DNA is: 65–66.

Type species: **Azospira oryzae** Reinhold-Hurek and Hurek 2000, 658.

FURTHER DESCRIPTIVE INFORMATION

Azospira oryzae cells become elongated (up to 8 μm) in stationary phase cultures and acquire at least one helical turn (Reinhold-Hurek and Hurek, 2000).

Azospira oryzae 6a3 was isolated from the roots of Kallar grass collected in Pakistan (Reinhold-Hurek et al., 1993b). Other strains have been isolated from roots of *Oryza officinalis*, *O. minuta*, and *O. sativa*.

Analysis of 16S rDNA sequences placed *Azospira* in the *Betaproteobacteria*; in these analyses, *Azospira* was most closely related to members of the genera *Thauera* and *Rhodocyclus*. *Azospira* strains could be separated from *Azoarcus*, *Azonexus*, and *Azovibrio* strains by comparisons of fatty acid composition, SDS-soluble protein electrophoretic patterns, and BOX-PCR genomic finger-printing. DNA–DNA hybridization results supported these divisions (Hurek et al., 1997b; Reinhold-Hurek and Hurek, 2000).

ENRICHMENT AND ISOLATION PROCEDURES

Enrichment and isolation procedures are described in Reinhold et al. (1986).

MAINTENANCE PROCEDURES

Maintenance and long-term storage procedures are described in Reinhold-Hurek et al. (1993b).

DIFFERENTIATION OF THE GENUS *AZOSPIRA* FROM OTHER GENERA

Reinhold-Hurek and Hurek (2000) provide tables of characteristics that differentiate the genera *Azoarcus*, *Azonexus*, *Azospira*, and *Azovibrio* from each other and from *Acetobacter diazotrophicus*, *Burkholderia vietnamensis*, the genus *Azospirillum*, and the genus *Herbaspirillum* (see also Table BXII.β.112 in the chapter on *Azoarcus*).

TAXONOMIC COMMENTS

This genus was first described as *Azoarcus* sp. group D (Reinhold-Hurek et al., 1993b; Hurek et al., 1997b; Reinhold-Hurek and Hurek, 2000).

List of species of the genus Azospira

1. **Azospira oryzae** Reinhold-Hurek and Hurek 2000, 658^{VP}

   o'ry.zae. L. fem. *Oryza* genus name of rice; L. gen. n. *oryzae* from rice, referring to its frequent occurrence in association with rice roots.

   Description as for the genus with the following additional characteristics. Optimal growth at 40°C. Utilizes L-aspartate, L-glutamate, 3-hydroxybutyrate, isovalerate, and D-tartrate.

   The mol% G + C of the DNA is: 65.2.

   Type strain: 6a3, LMG9096.

   GenBank accession number (16S rRNA): AF011347.

Genus V. **Azovibrio** *Reinhold-Hurek and Hurek 2000, 657*^{VP}

THE EDITORIAL BOARD

A.zo'vi.bri.o. Fr. n. *azote* nitrogen; L. v. *vibrare* move rapidly to and fro, vibrate; M.L. masc. n. *Azovibrio* nitrogen-fixing organism which vibrates.

Gram-negative somewhat curved motile rods (0.6–0.8 × 1.5–3.6 μm). One polar flagellum. Chemoorganoheterotrophic and microaerophilic. Strictly respiratory; O_2 and nitrate are electron acceptors. Oxidase positive. **Fix N_2.** Grow on acetate, ethanol, fumarate, L-glutamate, DL-lactate, L-malate, propionate, and succinate.

The mol% G + C of the DNA is: 64–65.

Type species: **Azovibrio restrictus** Reinhold-Hurek and Hurek 2000, 657.

FURTHER DESCRIPTIVE INFORMATION

Azovibrio restrictus does not form the elongated cells seen in stationary cultures of *Azospira oryzae* and *Azonexus fungiphilus* (Reinhold-Hurek and Hurek, 2000).

Azovibrio restrictus S5b2 was isolated from the roots of Kallar grass collected in Pakistan (Reinhold-Hurek et al.,1993b). Other strains were isolated from roots of *Oryza sativa* and from fungal sclerotia found in soil on which rice was grown.

Analysis of 16S rDNA sequences placed *Azovibrio* in the *Betaproteobacteria*; in these analyses, *Azovibrio* was most closely related to members of the genera *Thaurea* and *Rhodocyclus*. *Azovibrio* strains could be separated from *Azoarcus*, *Azospira*, and *Azonexus* strains by comparisons of fatty acid composition, SDS-soluble protein electrophoretic patterns, and BOX-PCR genomic finger-printing. DNA–DNA hybridization results supported these divisions (Hurek et al., 1997b; Reinhold-Hurek and Hurek, 2000).

ENRICHMENT AND ISOLATION PROCEDURES

Enrichment and isolation procedures are described in Reinhold et al. (1986).

MAINTENANCE PROCEDURES

Maintenance and long-term storage procedures are described in Reinhold-Hurek et al. (1993b).

DIFFERENTIATION OF THE GENUS AZOVIBRIO FROM OTHER GENERA

Reinhold-Hurek and Hurek (2000) provide tables of characteristics that differentiate the genera *Azoarcus*, *Azonexus*, *Azospira*, and *Azovibrio* from each other and from *Acetobacter diazotrophicus*, *Burkholderia vietnamensis*, the genus *Azospirillum*, and the genus *Herbaspirillum*.

TAXONOMIC COMMENTS

This genus was first described as *Azoarcus* sp. group C (Reinhold-Hurek et al., 1993b; Hurek et al., 1997b; Reinhold-Hurek and Hurek, 2000).

List of species of the genus Azovibrio

1. **Azovibrio restrictus** Reinhold-Hurek and Hurek 2000, 657[VP]
   re' stric.tus. L. adv. *restrictus* limited, restricted, referring to the restricted spectrum of carbon sources used for growth.

Description as for the genus with the following additional characteristic: grows on L-aspartate.
The mol% G + C of the DNA is: 64.8 (T_m).
Type strain: S5b2.
GenBank accession number (16S rRNA): AF011346.

Genus VI. **Dechloromonas** Achenbach, Michaelidou, Bruce, Fryman and Coates 2001, 531[VP]

THE EDITORIAL BOARD

De.chlo.ro.mo' nas. L. pref. *de* from; Gr. adj. *chloros* green (chlorine); Gr. fem. n. *monas* unit, monad; N.L. fem. n. *Dechloromonas* a dechlorinating monad.

Gram-negative, nonsporeforming, facultatively anaerobic, motile rods (0.5 × 2 µm). Strictly respiratory metabolism. **Chlorate and perchlorate reduced to chloride**; organic electron donors. O_2 also serves as an electron acceptor; some strains use nitrate as an electron acceptor.
The mol% G + C of the DNA is: 63.5.
Type species: **Dechloromonas agitata** Achenbach, Michaelidou, Bruce, Fryman and Coates 2001, 531.

FURTHER DESCRIPTIVE INFORMATION

Dechloromonas agitata was isolated from paper plant pulp sludge (Achenbach et al., 2001).

Dechloromonas agitata can use acetate, butyrate, fumarate, malate, propionate, succinate, yeast extract, Fe(II), sulfide, and reduced 2,6-anthrahydroquinone as electron donors; it does not use benzene, benzoate, citrate, Casamino acids, citrate, ethanol, formate, glucose, hexadecane, hydrogen, methanol, *N*-octane, palmitate, phenol, or toluene. Electron acceptors include chlorate, perchlorate, and oxygen (Bruce et al., 1999).

Analysis of 16S rDNA sequences of *Dechloromonas agitata* and ten other isolates placed them in the *Betaproteobacteria* and showed that they are allied to the genera *Rhodocyclus* and *Azoarcus* (Achenbach et al., 2001).

ENRICHMENT AND ISOLATION PROCEDURES

Enrichment cultures were prepared using environmental samples suspended and incubated in anaerobic liquid medium that contains 10 mM acetate as electron donor and 10 mM chlorate as electron acceptor (Bruce et al., 1999; Coates et al., 1999). Pure cultures can be achieved using the same medium and the agar-shake serial dilution method described by Pfennig and Biebl (1981).

DIFFERENTIATION OF THE GENUS DECHLOROMONAS FROM OTHER GENERA

Dechloromonas agitata can be distinguished from *Dechlorosoma suillum* by the ability of the former to use Fe(II), sulfide, and reduced 2,6-anthrahydroquinone sulfonate as electron donors for chlorate reduction (Achenbach et al., 2001).

List of species of the genus Dechloromonas

1. **Dechloromonas agitata** Achenbach, Michaelidou, Bruce, Fryman and Coates 2001, 531[VP]
   a.gi.ta' ta. L. fem. part. adj. *agitata* agitated, highly active.

   Description as for the genus with the following additional characteristics. Electron donors include acetate, butyrate, fumarate, malate, propionate, succinate, yeast ex-

tract, Fe(II), sulfide, and reduced 2,6-anthrahydroquinone. Electron acceptors include chlorate, perchlorate, and oxygen. Growth optima at 1% NaCl, pH 7.5, and 35°C.
The mol% G + C of the DNA is: 63.5 (HPLC).
Type strain: CKB, ATCC 700666, DSM 13637.
GenBank accession number (16S rRNA): AF 047462.

Genus VII. **Dechlorosoma** Achenbach, Michaelidou, Bruce, Fryman and Coates 2001, 531[VP]

THE EDITORIAL BOARD

De.chlo.ro.so' ma. L. pref. *de* from; Gr. adj. *chloros* green (chlorine); Gr. neut. n. *soma* body; N.L. neut. n. *Dechlorosoma* dechlorinating body.

Gram-negative, nonsporeforming motile rods (0.3 × 1.0 μm). **Nonfermentative facultative anaerobes. Chlorate and perchlorate reduced to chloride; organic electron donors.** O₂ and nitrate also serve as electron acceptors.

The mol% G + C of the DNA is: 65.8.

Type species: **Dechlorosoma suillum** Achenbach, Michaelidou, Bruce, Fryman and Coates 2001, 532.

FURTHER DESCRIPTIVE INFORMATION

Dechlorosoma suillum strains were isolated from samples from a swine waste treatment lagoon (Coates et al., 1999).

Dechlorosoma suillum PS is able to grow with acetate, butyrate, ethanol, fumarate, isobutyrate, lactate, malate, pyruvate, succinate, valerate, and Casamino acids as electron donors; it does not use benzoate, catechol, citrate, formate, glucose, glycerol, methanol, or hydrogen as electron donors. Chlorate, nitrate, and oxygen can serve as electron acceptors for growth on acetate; 2,6-anthraquinone disulfonate, fumarate, malate, selenate, sulfate, Fe(III), and Mn(IV) cannot (Coates et al., 1999).

Analysis of 16S rDNA sequences of *Dechlorosoma suillum* PS and three other isolates placed them in the *Betaproteobacteria* and showed that they are allied to the genera *Rhodocyclus* and *Azoarcus* (Achenbach et al., 2001).

ENRICHMENT AND ISOLATION PROCEDURES

Enrichment cultures can be prepared using environmental samples suspended and incubated in anaerobic liquid medium that was 10 mM in both acetate and chlorate as electron donor and acceptor, respectively (Bruce et al., 1999; Coates et al., 1999). Pure cultures can be achieved using the same medium and the agar-shake serial dilution method described by Pfennig and Biebl (1981).

DIFFERENTIATION OF THE GENUS *DECHLOROSOMA* FROM OTHER GENERA

Dechlorosoma suillum can be distinguished from *Dechloromonas agitata* by the ability of the latter to use 2,6-anthrahydroquinone sulfonate, Fe(II), and sulfide as electron donors for chlorate reduction (Achenbach et al., 2001).

List of species of the genus Dechlorosoma

1. **Dechlorosoma suillum** Achenbach, Michaelidou, Bruce, Fryman and Coates 2001, 532[VP]

   su.il' lum. L. neut. adj. *suillum* pertaining to swine.

   Description as for the genus with the following additional characteristics. Electron donors include acetate, butyrate, Casamino acids, ethanol, lactate, and propionate.

   Electron acceptors include chlorate, nitrate, oxygen, and percholorate. Reduce nitrate to N₂. Growth optima at 0% NaCl, pH 6.5, and 37°C.

   The mol% G + C of the DNA is: 75.8 (HPLC).

   Type strain: PS, ATCC BAA-33, DSM 13638.

   GenBank accession number (16S rRNA): AF 170348.

Genus VIII. **Ferribacterium** Cummings, Caccavo, Spring and Rosenzweig 2000, 1953[VP](Effective publication: Cummings, Caccavo, Spring and Rosenzweig 1999, 187)

THE EDITORIAL BOARD

Fer' ri.bac.te' ri.um. L. neut. n. *ferrum* iron; Gr. neut. n. *bacterion* a small rod; M.L. neut. n. *Ferribacterium* rod-shaped iron bacterium.

Strictly anaerobic straight or curved rods (0.3–0.5 × 1.4–2.0 μm); contain poly-β-hydroxybutyrate granules. Nonsporeforming. **Oxidizes acetate, benzoate, formate, and lactate using Fe(III) as the electron acceptor**; also oxidizes acetate using fumarate, or nitrate as electron acceptors.

The mol% G + C of the DNA is: unknown.

Type species: **Ferribacterium limneticum** Cummings, Caccavo, Spring and Rosenzweig 2000, 1953 (Effective publication: Cummings, Caccavo, Spring and Rosenzweig 1999, 187.)

FURTHER DESCRIPTIVE INFORMATION

Ferribacterium limneticum CdA-1 was isolated from a sediment core taken in Lake Coeur d'Alene, Idaho, USA (Cummings et al., 1999).

Ferribacterium limneticum CdA-1 is a strict anaerobe that grows only when Fe (III) is present in the medium; it does not grow in a complex organic medium. Electron donors are acetate, benzoate, formate, and lactate. Citrate, ethanol, glucose, isopropanol, methanol, propionate, and succinate are not used as electron donors. As (V), Mn (IV), Se (VI), nitrite, sulfate, sulfite, thiosulfate, and trimethylamine are not used as electron acceptors when acetate is the electron donor (Cummings et al., 1999).

An analysis of 16S rDNA sequences showed that *Ferribacterium limneticum* CdA-1 belongs to the class *Betaproteobacteria* (Cummings et al., 1999).

ENRICHMENT AND ISOLATION PROCEDURES

Enrichment in liquid medium was carried out under anaerobic conditions using acetate as the electron donor and amorphous ferric oxyhydroxide as the electron acceptor (Cummings et al., 1999). Incubation for four months resulted in blackening of the culture; the organism was recovered by streaking onto solid medium containing Fe (III) pyrophosphate as electron acceptor.

List of species of the genus Ferribacterium

1. **Ferribacterium limneticum** Cummings, Caccavo, Spring and Rosenzweig 2000, 1953^{VP} (Effective publication: Cummings, Caccavo, Spring and Rosenzweig 1999, 187.)
lim.ne' ti.cum. L. neut. n. *limne* lake; N.L. adj. *limneticum* from a lake.

The description is the same as that of the genus.
The mol% G + C of the DNA is: unknown.
Type strain: CdA-1.
GenBank accession number (16S rRNA): Y17060.

Genus IX. **Propionibacter** *Meijer, Nienhuis-Kuiper and Hansen 1999, 1042^{VP}*

THEO A. HANSEN

Pro.pioni.bac' ter. M.L. n. *acidum propionicum* propionic acid; M.L. masc. n. *bacter* equivalent of Gr. neut. n. *bakterion* small rod; M.L. masc. n. *Propionibacter* propionic acid rod.

Rods 0.9–1.1 × 0.5–0.6 μm. Motile by means of a single polar flagellum. Multiply by binary fission. Gram negative. Do not form spores. **Anaerobic.** Can grow in stationary cultures under an air atmosphere, but not in well-aerated cultures. In the absence of nitrate, **substrates are fermented to propionate and acetate.** Growth occurs on simple organic compounds (sugars, dicarboxylic acids, sugar alcohols, and aspartate). Nitrate is reduced to nitrite. **Utilize N_2 as nitrogen source.** Isolated from estuary mud.
The mol% G + C of the DNA is: 61.
Type species: **Propionibacter pelophilus** Meijer, Nienhuis-Kuiper and Hansen 1999, 1043.

ENRICHMENT AND ISOLATION PROCEDURES

Propionibacter has been obtained by direct anaerobic isolation from black surface mud of the Ems-Dollard estuary at the border between the Netherlands and Germany. Roll tubes with aspartate medium are used (Hansen et al., 1990). Strain asp 66, the type strain of *Propionibacter pelophilus*, is the most numerous aspartate-fermenting bacterium.

MAINTENANCE PROCEDURES

DSM Medium 503 (FWM medium; Deutsche Sammlung von Mikroorganismen und Zellkulturen GmbH, Braunschweig, Germany) supplemented with 0.2 g/l yeast extract and 20 mM aspartate or another suitable energy substrate can be used for routine cultivation. For long-term preservation, the organism can be stored in ampules in medium supplemented with 8% glycerol under liquid nitrogen or at −80°C.

DIFFERENTIATION OF THE GENUS *PROPIONIBACTER* FROM OTHER GENERA

Propionibacter is most closely related to *Propionivibrio* (Tanaka et al., 1990; Hippe et al., 1999) based on the 16S rDNA sequences (approximately 3% difference between the type species). Both genera have a (mainly) fermentative metabolism with propionate and acetate as characteristic products. *Propionibacter* cells are straight rods, whereas *Propionivibrio* consists of curved rods. *Propionibacter* uses a wider range of substrates than does *Propionivibrio*; the latter is restricted to C_4-dicarboxylic acids.

Characteristics that differentiate *Propionibacter* from other genera and species of bacteria that carry out propionic fermentation are listed in Table BXII.β.115.

TAXONOMIC COMMENTS

In this edition of the *Manual*, the genus *Propionibacter* is placed in the class *Betaproteobacteria*, the order *Rhodocyclales*, and the family *Rhodocyclaceae*. Other genera included in this family are *Rhodocyclus, Azoarcus, Azonexus, Azospira, Azovibrio, Ferribacterium, Propionivibrio, Thauera,* and *Zoogloea*.

The phylogenetic positions of *Propionibacter* and *Propionivibrio* have become clear only recently (Hippe et al., 1999; Meijer et al., 1999). The small difference in 16S rRNA gene sequence (approximately 3%) and the similarity in fermentative patterns would support a possible inclusion of *Propionibacter* in *Propionivibrio*; this, however, would require an emendation of the genus *Propionivibrio* as described by Tanaka et al. (1990).*

List of species of the genus Propionibacter

1. **Propionibacter pelophilus** Meijer, Nienhuis-Kuiper and Hansen 1999, 1043^{VP}
pe.lo' phi.lus. Gr. n. *pelos* mud; Gr. adj. *philus* loving; M.L. adj. *pelophilus* mud-loving.

Characteristics are as described for the genus, with the following additional features. Gram-negative cell envelope ultrastructure with an outer membrane. Colonies on anaerobic plates with aspartate are slimy white. Best growth occurs under anoxic conditions in sulfide-reduced media. Growth occurs in mineral media with a single fermentable substrate.

The following are utilized as energy sources: L-aspartate, fumarate, L-malate, pyruvate, oxaloacetate, citrate, fructose, glucose, xylose, gluconate, arabitol, xylitol, mannitol, arabinose, and mannose. The following substrates are not used:

L-alanine, L-serine, L-threonine, L-methionine, L-glutamate, L-histidine, L-arginine, L-lysine, lactate, α-ketoglutarate, glycolate, succinate, propionate, butyrate, glycerol, ethanol, propanol, 2,3-butanediol, acetoin, cellulose, cellobiose, lactose, sucrose, maltose, melibiose, starch, sorbose, rhamnose, galactose, sorbitol, pectin, peptone, xylan, and inulin.

Indole and urease negative. Gelatin is not liquefied. Growth occurs on glucose and aspartate at pH 6.5–8.5 (optimal pH 7.5–8.0). The final pH of glucose-fermenting cul-

---

Editorial Note: While this volume of the *Manual* was in preparation, Brune et al. (2002) published a valid proposal to transfer the type species of the genus *Propionibacter, Propionibacter pelophilus,* to the genus *Propionivibrio.* Since the genus *Propionibacter* was monospecific, bacteriologists adhering to the proposal recognize that the genus *Propionibacter* thereby ceased to exist.

TABLE BXII.β.115. Characteristics that differentiate the genus *Propionibacter* from other genera and species of anaerobic bacteria that carry out a propionate fermentation[a]

Characteristic	*Propionibacter*	*Pelobacter propionicus*	*Desulfobulbus propionicus*	*Propionibacterium*	*Propionivibrio*
Gram-staining reaction	−	−	−	+	−
Cell shape:					
Straight rods	+	+	−	+[b]	−
Curved rods	−	−	−	−	+
Ovoid, rod-shaped, or lemon-shaped	−	−	+	−	−
Sugars are fermented	+	−	−	+	−
N₂ can be fixed	+	+	+	−	
SO₄²⁻ can be used as a terminal electron acceptor for anaerobic respiration	−	−	+	−	−

^{a}For symbols, see standard definitions.

^{b}Pleomorphic.

tures is pH 5.5. Sulfate can be used as a sulfur source; growth in the absence of sulfide leads to cell clumps.

Optimal temperature 27–30°C; no growth at 35°C. Habitat: anoxic freshwater and estuarine sediments.

The mol% G + C of the DNA is: 60.8 (T_m).

Type strain: asp 66, DSM 12018.

GenBank accession number (16S rRNA): AF016690.

Genus X. **Propionivibrio** Tanaka, Nakamura and Mikami 1991b, 331^{VP} (Effective publication: Tanaka, Nakamura and Mikami 1990, 327) emend. Brune, Ludwig and Schink 2002, 443

THE EDITORIAL BOARD

Pro.pi.o.ni.vi'bri.o. M.L. n. *acidum propionicum* propionic acid; M.L. masc. n. *Vibrio* that which vibrates, a generic name; M.L. masc. n. *Propionivibrio* the propionic acid vibrio.

Gram-negative, curved or straight **rods. Motile by means of a single polar flagellum.** Do not form spores. **Strict anaerobe or aerotolerant.** Multiply by binary fission. Chemoorganotrophic. Propionate and acetate formed as end products of fermentation. Succinate decarboxylated to propionate. Mesophilic. Member of the class *Betaproteobacteria*.

The mol% G + C of the DNA is: 60.8–61.6.

Type species: **Propionivibrio dicarboxylicus** Tanaka, Nakamura and Mikami 1991b, 331 (Effective publication: Tanaka, Nakamura and Mikami 1990, 327.)

List of species of the genus Propionivibrio

1. **Propionivibrio dicarboxylicus** Tanaka, Nakamura and Mikami 1991b, 331^{VP} (Effective publication: Tanaka, Nakamura and Mikami 1990, 327.)

di.car.bo.xy'li.cus. M.L. adj. *dicarboxylicus* pertaining to dicarboxylic acid.

Gram-negative, curved to helical rods with tapered and rounded ends, 0.5–0.6 × 1.0–2.0 µm. Chains are helical. Motile in an active tumbling motion by means of a single polar flagellum. Catalase negative. Do not reduce nitrate, sulfate, sulfite, thiosulfate, or S^0. Utilize maleate, fumarate, and L-malate as carbon and energy sources, forming propionate and acetate as end products. Succinate decarboxylated to propionate. Do not utilize acetoin, acrylate, corn oil, crotonate, formate, glycerol, H₂/CO₂/acetate, lactate, peptone, pyruvate, yeast extract, monosaccharides, disaccharides, or trisaccharides. Optimal temperature 30°C (range 26–34°C); no growth at 20°C or 40°C. Optimal pH 6.7 (range 6.2–7.1). Habitat: anaerobic mud of freshwater sediments.

The mol% G + C of the DNA is: 61.

Type strain: CreMal1, DSM 5885, JCM 7784.

GenBank accession number (16S rRNA): Y17601.

2. **Propionivibrio limicola** Brune, Ludwig and Schink 2002, 443^{VP}

li.mi'co.la. L. n. *limus* mud; L. v. *colere* to inhabit; N.L. n. *limicola* living in mud.

Gram-negative; straight, slender rods, 0.6–0.7 × 1.5–2.5 µm. Motile by single polar flagella. Oxidase and catalase negative; superoxide dismutase positive. Do not form spores. Chemoorganotrophic. Fermentative metabolism, utilizing quinic acid and shikimic acid and producing acetate, propionate, and CO₂. No external electron acceptors utilized. Does not utilize sugars, alcohols, carboxylic acids, amino acids, or aromatic compounds. Aerotolerant; grows in unreduced media when incubated under air without agitation. Optimal temperature: 37°C; no growth at 45°C. pH range: 6.0–8.0; optimal pH: 7.0–7.5. Optimal growth in freshwater media. Growth inhibited in brackish media with 10 g/l NaCl and 1.0 g/l MgCl₂. Habitat: anoxic freshwater sediments.

The mol% G + C of the DNA is: 61.6 ± 2.

Type strain: GolChi1, ATCC BAA-290, DSM 6832.

GenBank accession number (16S rRNA): AJ307983.

3. **Propionivibrio pelophilus** (Meijer, Nienjuis-Kuiper and Hansen 1999) Brune, Ludwig and Schink 2002, 444[VP] (*Propionibacter pelophilus* Meijer, Nienhuis-Kuiper and Hansen 1999, 1043.)*

pe.lo' phi.lus. Gr. n. *pelos* mud; Gr. adj. *pelophilus* mud-loving.

Gram-negative cell envelope ultrastructure with an outer membrane. Colonies on anaerobic plates with aspartate are slimy white. Best growth occurs under anoxic conditions in sulfide-reduced media. Growth occurs in mineral media with a single fermentable substrate. The following are utilized as energy sources: L-aspartate, fumarate, L-malate, pyruvate, oxaloacetate, citrate, fructose, glucose, xylose, gluconate, arabitol, xylitol, mannitol, arabinose, and mannose.

Editorial Note: This description of *Propionivibrio pelophilus* is that of *Propionibacter pelophilus* and has been reproduced here from the preceding chapter because of a proposal to transfer *Propionibacter pelophilus* to the genus *Propionivibrio*.

The following substrates are not used: L-alanine, L-serine, L-threonine, L-methionine, L-glutamate, L-histidine, L-arginine, L-lysine, lactate, α-ketoglutarate, glycolate, succinate, propionate, butyrate, glycerol, ethanol, propanol, 2,3-butanediol, acetoin, cellulose, cellobiose, lactose, sucrose, maltose, melibiose, starch, sorbose, rhamnose, galactose, sorbitol, pectin, peptone, xylan, and inulin. Indole and urease negative. Gelatin is not liquefied. Growth occurs on glucose and aspartate at pH 6.5–8.5 (optimal pH 7.5–8.0). The final pH of glucose-fermenting cultures is pH 5.5. Sulfate can be used as a sulfur source; growth in the absence of sulfide leads to cell clumps. Nitrate reduced to nitrite. N_2 utilized as nitrogen source. Optimal temperature 27–30°C; no growth at 35°C. Habitat: anoxic freshwater and estuarine sediments.

The mol% G + C of the DNA is: 60.8.

Type strain: asp 66, CIP 106101, DSM 12018.

GenBank accession number (16S rRNA): AF016690.

Genus XI. **Thauera** Macy, Rech, Auling, Dorsch, Stackebrandt and Sly 1993, 139[VP] emend. Song, Young and Palleroni 1998, 893

JOHANN HEIDER AND GEORG FUCHS

Thau' e.ra. M.L. fem. n. *Thauera* named after R.K. Thauer, a German microbiologist.

Rods 0.5–1.4 × 1.4–3.7 μm, usually occurring singly. Some species exhibit great variability of cell form, from rods to coccoid forms of different sizes. Gram negative. **Most species are motile.** No resting stages known. **Oxidase positive.** Catalase positive. Colonies on minimal medium are nonpigmented. **Aerobic,** having a strictly respiratory type of metabolism; never fermentative. **Molecular oxygen, nitrate, nitrite, and nitrous oxide are used as terminal electron acceptors,** and the nitrogen oxides are reduced to N_2O or N_2; one species also uses selenate, which is reduced to elemental Se. Optimal temperature 28–40°C. All species grow at pH 7.5; pH optima of different species range from 7.2 to 8.4. **Chemoorganotrophic, using various organic acids, amino acids, and aromatic and aliphatic compounds** as sole substrates. Only a **few carbohydrates are utilized.** Ammonium and nitrate salts can be used as sole nitrogen sources. **No N_2 fixation** known. Not proteolytic on casein or gelatin. Starch, cellulose, chitin, and agar are not hydrolyzed. Occur in polluted freshwater and wet soil environments and wastewater treatment plants.

The mol% G + C of the DNA is: 64–69.

Type species: **Thauera selenatis** Macy, Rech, Auling, Dorsch, Stackebrandt and Sly 1993, 140.

FURTHER DESCRIPTIVE INFORMATION

Cell morphology and fine structure Cells of *Thauera* spp. are usually rods 0.5–1.0 μm in width and 1.4–3.4 μm in length. Variation in cell size is commonly observed in liquid cultures. Most strains have been found to accumulate poly-β-hydroxybutyrate granules in the late growth phase. Flagella are usually present; flagellar insertion in the described species is either polar monotrichous or peritrichous with up to eight flagella; cells of some strains have either no flagella or a single flagellum inserted at any point, which has been termed "degenerately peritrichous" (Song et al., 1998). No sheaths, capsules, or other extracellular features are known.

Colonial and cultural characteristics A red-brown pigmentation is observed in densely packed cells or in colonies on solid medium, which is due to normal cell constituents, e.g., ferredoxins or cytochromes. Therefore, *Thauera* spp. can be classified as nonpigmented. Exceptions are the production of large amounts of insoluble red elemental selenium by *T. selenatis* growing under selenate-reducing conditions and the production of an as yet uncharacterized red pigment (metabolic intermediate or side product) by *T. aromatica* strain AR-1 growing anaerobically on α-resorcylate in the presence of nitrate (Gallus et al., 1997).

Lipid composition Major fatty acids contained in membrane lipids are *cis*-9-hexadecenoate ($C_{16:1 \omega 7c}$), hexadecanoate ($C_{16:0}$), and *cis*-11-octadecenoate ($C_{18:1 \omega 7c}$). Relative amounts of these fatty acids in different species are 41–58%, 19–32%, and 10–16%, respectively. Less abundant fatty acids found in amounts of <8% in all species are dodecanoate ($C_{12:0}$) and 3-hydroxydecanoate ($C_{10:0 3OH}$). Small amounts (0.7–2.1%) of a cyclic fatty acid ($C_{17:0 cyclo}$) produced in the stationary growth phase have been reported as a characteristic property of *T. aromatica* and *T. chlorobenzoica* strains, whereas they are lacking in *T. selenatis*, *T. linaloolentis*, *T. terpenica*, and *Azoarcus* spp. (Song et al., 2000a, 2001).

Nutrition and growth conditions *Thauera* species grow in mineral medium with ammonium ions or nitrate as the nitrogen source and simple organic compounds as sole carbon and energy sources. Only a few species require organic growth factors. All species are chemoorganotrophic and obligately respiratory, requiring either oxygen or an alternative final electron acceptor, such as nitrate, nitrite, nitrous oxide, or selenate. The optimal growth temperature for most species is around 30°C, but one species has an optimal of 40°C.

Metabolism and metabolic pathways All species are obligately respiratory, but can shift between aerobic respiration and denitrification (dissimilatory nitrate reduction to N_2O or N_2). One species is capable of anaerobic respiration on selenate. The first step, reduction of selenate to selenite, is catalyzed by a unique periplasmic selenate reductase, an enzyme containing a molybdenum cofactor, iron–sulfur clusters, and a cytochrome *b*

(Schröder et al., 1997; Krafft et al., 2000). Three of the known species catabolize various aromatic compounds anaerobically under denitrifying conditions. Two other species catabolize terpenoids, but not aromatic compounds, in the absence of oxygen. Whereas not much is known about the biochemistry involved in terpenoid metabolism, the anaerobic metabolism of aromatic substrates has been well studied. All tested species can catabolize some aromatic compounds under aerobic conditions. The aromatic substrates utilized aerobically and under denitrifying conditions are not necessarily the same, and the presence of completely different pathways for the same substrates has been demonstrated for *T. aromatica* and the physiologically similar *Azoarcus evansii* (Ziegler et al., 1989; Altenschmidt et al., 1993; Heider et al., 1998; Mohamed, 2000). Many organic acids and several amino acids serve as sole substrates aerobically and anaerobically for all species, whereas most species are very limited in their capacities to catabolize sugars.

Metabolism of aromatic compounds Aromatic substrates that are utilized under anaerobic conditions by strains of *T. aromatica* include benzoate, 3- and 4-hydroxybenzoate, phenol, toluene, phenylacetate, 4-hydroxyphenylacetate, and halogenated benzoates. Most of these substrates are first converted via diverse peripheral metabolic pathways to a common intermediate, benzoyl-CoA (Heider and Fuchs, 1997a, b). Benzoyl-CoA is then further catabolized by a conserved metabolic pathway, as shown in Fig. BXII.β.108. The first step is a two-electron reduction to a non-aromatic derivative, cyclohexa-1,5-diene-1-carboxy-CoA, by the key enzyme benzoyl-CoA reductase. This enzyme catalyzes benzoyl-CoA reduction, coupling this reaction to the hydrolysis of two moles of ATP (Boll et al., 1997). Further metabolic steps proceed via a beta-oxidation-like pathway, which leads to the first open-chain intermediate 3-hydroxypimelyl-CoA (Fig. BXII.β.108), generated by hydrolytic ring opening (Laempe et al., 1998, 1999). The benzoyl-CoA reductase of several analyzed *T. aromatica* strains appears to be immunologically distinct from the enzyme of closely related *Azoarcus* strains (Mechichi et al., 2002). The intermediate 3-hydroxypimelyl-CoA is degraded to three acetyl-CoA units and one CO_2, and acetyl-CoA is finally oxidized to CO_2 via the tricarboxylic acid cycle. Anaerobic catabolism of some aromatic compounds, such as α-resorcylate, apparently does not occur via the benzoyl-CoA intermediate. The first steps of a completely different pathway have recently been described for *T. aromatica* strain AR-1 that leads to oxidation of α-resorcylate to the intermediate hydroxyhydroquinol (Gallus and Schink, 1998). Catabolism of other aromatic substrates by strain AR-1 appears to proceed via the benzoyl-CoA pathway (Philipp and Schink, 2000). All tested *Thauera* species catabolize aromatic compounds aerobically. Aromatic substrates used under aerobic conditions include benzoate, phenylacetate, and 2-aminobenzoate. The catabolic pathways used for aerobic catabolism of aromatic compounds differ from those used under denitrifying growth conditions and involve the participation of oxygenases in all known instances.

Habitats Members of the genus *Thauera* are common saprophytic inhabitants of polluted freshwater and wet soil environments. They occur in wastewater treatment plants and are apparently naturally enriched in groundwater aquifers, river, lake, and pond sediments that are contaminated with aromatic or aliphatic organic compounds or toxic inorganic compounds such as selenate. They seem to be involved in mineralization and detoxification of xenobiotic contaminants in these habitats.

FIGURE BXII.β.108. Anaerobic benzoyl-CoA catabolic pathway of *T. aromatica*. Benzoate is converted to benzoyl-CoA by a coenzyme A ligase (AMP forming). For other substrates converted to this intermediate, see Heider and Fuchs (1997a, b).

Pathogenicity No pathogenic or symbiotic associations are known with plants or animals.

ENRICHMENT AND ISOLATION PROCEDURES

There is no selective isolation procedure for *Thauera* strains. The known species have been obtained from enrichments designed for special physiological properties that are shared by strains of various other taxa. For example, anaerobic catabolism of aromatic compounds is also found in the related genus *Azoarcus* and even in some less related recent isolates of denitrifying *Alphaproteobacteria*, *Betaproteobacteria*, and *Gammaproteobacteria* (Spormann and Widdel, 2000), and the anaerobic catabolism of terpenoids has also been reported for an *Alcaligenes* species (Foss et al., 1998a). *Thauera* strains are commonly isolated from wastewater treatment plants or contaminated soil or freshwater habitats after enrichment under anaerobic respiratory conditions. The enrichments are performed under mesophilic conditions at pH values of 7–8 in a buffered mineral salt medium containing either nitrate or selenate as the electron acceptor. Organic acids, aromatic compounds, or terpenoids may be used as sole substrates. Growth on some phenolic compounds or on acetone requires CO_2 or bicarbonate in the medium. Pure cultures are obtained after serial dilution in agar shakes under denitrifying conditions. However, highly purified agar (e.g., Oxoid purified agar) should be used for growth on solid media, because several strains are known to be sensitive to impurities in the agar. Alternative solidifying agents such as Gelrite should be tested to improve growth of some strains.

MAINTENANCE PROCEDURES

Stock cultures of *Thauera* strains grown under denitrifying conditions can be maintained in liquid cultures at 4°C for up to 3 months. Cultures of *Thauera* strains can be preserved for years in liquid nitrogen if they are supplemented with 5% (v/v) dimethylsulfoxide.

PROCEDURES FOR TESTING SPECIAL CHARACTERS

No specific 16S rRNA targeted probes have yet been derived for members of the genus *Thauera*. However, such oligonucleotide probes have been designed for *in situ* detection of strains of the closely related genus *Azoarcus* (Hess et al., 1997). Since no cross-reaction with *T. aromatica* has been recorded with these probes, it should be possible to design analogous probes for *Thauera* species. Another potentially useful procedure for characterizing the *Thauera* species that are capable of anaerobic catabolism of aromatic compounds is the immunological detection of benzoyl-CoA reductase. Antibodies raised against purified benzoyl-CoA reductase from the type strain of *T. aromatica* specifically react with extracts of several different strains of *Thauera* sp., but do not react with any of several tested *Azoarcus* spp. grown anaerobically on benzoate (Mechichi et al., 2002).

DIFFERENTIATION OF THE GENUS *THAUERA* FROM OTHER GENERA

The most distinctive characteristics of *Thauera* are the ability to grow on unusual organic substrates and the limited ability to catabolize carbohydrates. These properties are shared with *Achromobacter*, *Acidovorax*, *Alcaligenes*, *Bordetella*, *Burkholderia*, *Comamonas*, *Pseudomonas*, *Ralstonia*, and *Zoogloea* species, as well as with the closely related genus *Azoarcus*. Useful phenotypic properties to discriminate these species are listed in Table BXII.β.116. The currently described species of the above-mentioned genera, except *Azoarcus*, are differentiated by their inability to degrade aromatic compounds such as benzoic acid in the absence of oxygen, but it can be expected that new isolates capable of anaerobic degradation of aromatic compounds will also be affiliated with some of these genera. *Azoarcus* strains are distinguished from *Thauera* sp. by their ability to fix dinitrogen and to use various sugars. Most above-mentioned genera belong to the class *Betaproteobacteria*, as does *Thauera*.

TAXONOMIC COMMENTS

The genus *Thauera* is defined mainly based on 16S rDNA phylogeny. The few criteria listed in Table BXII.β.116 to differentiate most species of the genus from physiologically similar genera (e.g., absence of N_2 fixation) are not tested for all known species, and it may be expected that *Thauera* species will be isolated that deviate from some of these criteria. The six validly described species have been obtained following very different enrichment procedures and vary considerably in their physiological properties.

Phylogeny within the genus is currently based on 16S rDNA and DNA–DNA reassociation data. As shown in Fig. BXII.β.109, a phylogenetic tree of all known 16S rRNA sequences of *Thauera* strains currently in the database shows that the genus is monophyletic and well separated from *Azoarcus* spp., the closest known relatives. The positions of the *T. terpenica* and *T. linaloolentis* strains justify their treatment as separate species. The 16S rDNA sequences of the other four described species, *T. selenatis*, *T. aromatica*, *T. chlorobenzoica*, and *T. mechernichensis* are highly similar to one another, but treatment as separate species is supported by mutual DNA–DNA reassociation values that are significantly lower than 70% (Song et al., 1998, 2000a, 2001; Scholten et al., 1999). These four species apparently form a subcluster of related species within the genus (Fig. BXII.β.109). An additional strain, ATCC 700265, is included in the species *T. aromatica*, as justified by the observed similarity (99.2% 16S rRNA identity and 77% DNA–DNA reassociation; Song et al., 1998). Recently, a three-dimensional representation of the DNA–DNA hybridization values among *Thauera* strains has indicated that *T. chlorobenzoica* ATCC 700723^T, which is capable of degrading halobenzoates under denitrifying conditions, clusters together with additional isolates that are similarly capable of degrading halobenzoates. As a result, these strains have been separated from *T. aromatica* and placed into a separate species, *T. chlorobenzoica* (Song et al., 2001).

A number of recently isolated bacterial strains also belong to the genus *Thauera* based on their 16S rDNA sequences, but have not yet been fully described. Several of these strains belong to the species *T. aromatica*. One of these, strain mXyN1, which was isolated under denitrifying conditions with *m*-xylene as the sole carbon and energy source (Rabus and Widdel, 1995), exhibits 99.7% 16S rRNA identity with the type strain and 90% relatedness by DNA–DNA reassociation (Scholten et al., 1999). Another strain, AR-1 (DSM 11528), is capable of growing anaerobically on α-resorcylate and shares 99.9% 16S rDNA identity with the type strain (Gallus et al., 1997). Several other strains with 16S rDNA sequence identities of 99 ± 0.1% with the *T. aromatica* type strain have been characterized in a survey of halobenzoate-degrading denitrifying bacteria; these strains are 3CB-2, 3CB-3, 3BB-1, 4FB-1, and 4FB-2 in Fig. BXII.β.109 (Song et al., 2000b). Strains 4FB-1, 4FB-2, and 3BB-1 were later reclassified as reference strains of *T. chlorobenzoica*, and strains 3CB-2 and 3CB-3 were reclassified as reference strains of *T. aromatica*, on the basis of DNA–DNA hybridization studies (Song et al., 2001). The 16S rDNA sequences of two other recently characterized strains cluster between those of *T. selenatis* and *T. chlorobenzoica* (Fig. BXII.β.109). One of these, strain mz1t, was recently isolated directly from a bacterial floc of activated sludge using a micromanipulator (Lajoie et al., 2000), the other, strain PN-1, was isolated 30 years ago from a denitrifying enrichment culture with 4-hydroxybenzoate (Taylor et al., 1970). Strain PN-1 had been classified as *Achromobacter xylosoxidans* by physiological properties (Blake and Hegeman, 1987) and has only recently been iden-

TABLE BXII.β.116. Characteristics differentiating the genus *Thauera* from other bacterial genera[a]

Characteristics	*Thauera*	*Azoarcus*	*Zoogloea*	*Acidovorax*	*Comamonas*	*Burkholderia*	*Alcaligenes*	*Ralstonia*	*Achromobacter*	*Bordetella*	*Pseudomonas*
Flagella	1–8	1	1	1	>1	>1	1–8	1–4	1–8	1–8	1 or >1
Animal pathogens in genus	−	−	−	−	−	−	+	−	+	+	+
Plant pathogens in genus	−	−	−	−	−	+	−	+	−	−	+
Plant symbionts in genus	−	+	−	−	−	−	−	−	−	−	−
Saprophytes in genus	+	+	+	+	+	+	+	+	+	−	+
Denitrification	+	+[b]	+	−	−	−[c]	+	+	+	D	D
Urease	−	−[d]	D		−		−			+	
Growth factor requirement	D	−	−	−	−	−		−	D	+	−
Autotrophic growth with H_2	−[e]	−	−	+	−	−	−	D	D	−	−
Aerobic growth on:											
Glucose/fructose	−	+[b]	D	+	−	+	−	D	D		+
Ribose	−	−	D	+	−	+	−	D	−		+
Glutarate	+	+		+	+	+	D	+			+
Adipate	D	D		+	+	+	−	+	+		D
Pimelate	−	D		D	+	+	−	+	+		−
Ethanol	D	−[b]	+	−	D	+					D
4-Hydroxybenzoate	+	D		−	+	+	−	+	−	−	+
Benzoate	+	+[b]	+	−	D	+					D
Phenylacetate	+	D		−	D	+	+	D	D		
Anaerobic growth with aromatic substrates	D	D	−	−	−	−	−	−	−	−	−
N_2 fixed	−	+[f]	−	−	−	−	−	−	−	−	−
Poly-β-hydroxybutyrate accumulation	+	+	+	D	+	+	D	+	+		−
Mol% G +C of DNA	64–69	62–67	65	62–69	63–69	67	56–60	64–68	64–70	69	58–70
Class:											
Betaproteobacteria	+	+	+	+	+	+	+	+	+	+	
Gammaproteobacteria											+

[a]Symbols: see standard definitions. Blank space, data not available or not applicable.

[b]Not for plant-symbiotic species.

[c]Nitrate reduced to nitrite.

[d]Urease present in *A. indigens*.

[e]Autotrophy has been described in *T. selenatis.*

[f]Not all strains tested.

tified as a *Thauera* sp. based on its 16S rRNA sequence (GenBank AF170281). Two other strains of *Thauera* sp., strains O and 1917, isolated from an anoxic leachate treatment reactor, are apparently less related to the species clustering with *T. aromatica*, as indicated by branching of their 16S rDNA sequences before branching of the *T. linaloolentis* sequence (Fig. BXII.β.109; Etchebehere et al., 2001a). Strain O has been reported to degrade benzoate under denitrifying conditions (Etchebehere et al., 2001a), but there is no more information available on the physiology of these strains. Finally, a number of bacterial strains isolated under denitrifying conditions on different aromatic substrates (strains S100, LG356, SP, B4P, and S2; Tschech and Fuchs, 1987; Seyfried et al., 1991) have also been found to be *Thauera* spp. on the basis of >98% 16S rDNA sequence identity with the *T. aromatica* type strain. New species have recently been established for two of these strains, *T. aminoaromatica* (strain S2) and *T. phenylacetica* (strain B4P), while the others were described as

further reference strains of *T. aromatica* (for details, see Mechichi et al., 2002).

Nomenclatural problems within the genus *Thauera* arise mainly from the fact that very high degrees of 16S rDNA identity are accompanied by wide physiological variability between the species and even between several strains of one species. Therefore, taxonomy cannot be based on 16S rDNA sequences alone, but rather needs to be complemented by DNA–DNA reassociation studies and the physiological properties of the strains. For example, the establishment of *T. aromatica* and *T. selenatis* as separate species was much more dependant on the profound physiological differences between the strains than on the highly similar 16S rRNA sequences, and this speciation is corroborated by low DNA–DNA reassociation values (Anders et al., 1995). From the speed of description of new *Thauera* isolates in the last few years, it may be expected that the number of strains and species in the genus will expand rapidly in the future.

DIFFERENTIATION OF THE SPECIES OF THE GENUS *THAUERA*

Characteristics useful for the differentiation of the species of the genus *Thauera* are listed in Tables BXII.β.117 and BXII.β.118.

List of species of the genus Thauera

1. **Thauera selenatis** Macy, Rech, Auling, Dorsch, Stackebrandt and Sly 1993, 140[VP]

se.le.na′ tis. M.L. gen. n. *selenatis* of selenate, according to the electron acceptor used for isolation.

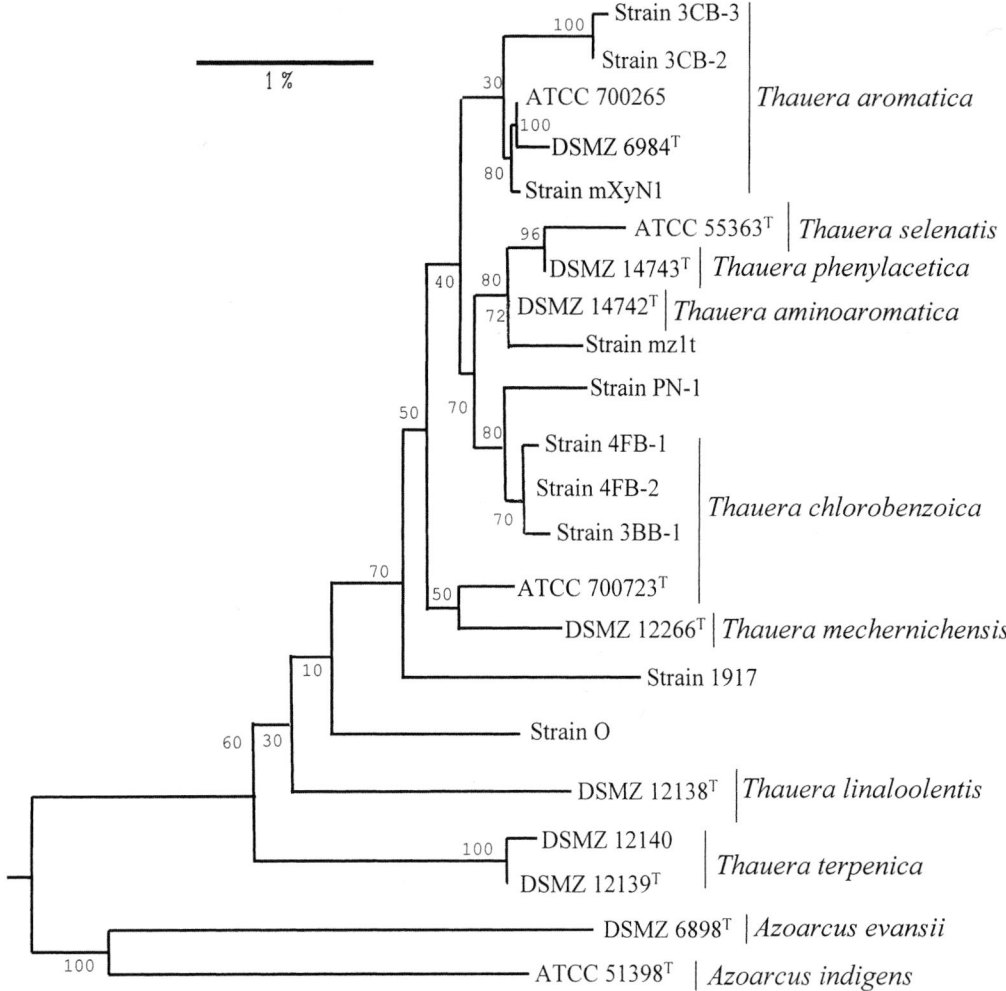

FIGURE BXII.β.109. Phylogenetic tree of the 16S rRNA sequences of bacterial strains affiliated with the genus *Thauera* and the type strains of *Azoarcus evansii* and *Azoarcus indigens*. All sequences larger than 1 kb currently in the database were included. The tree was derived from a multiple alignment of a segment of 1070 bases present in all sequences (bases 177–1244 in 16S rDNA of *T. aromatica* DSM 6984[T]) by the neighbor-joining method. Bootstrapping values of the branch points are indicated. Currently affiliated species names are indicated to the right of the strain designations. Strain ATCC 700723 has recently been reclassified as the type strain of the species *T. chlorobenzoica*. Similarly, strains 4FB-1, 4FB-2, and 3BB-1 have been reclassified as reference strains of *T. chlorobenzoica*, and strains 3CB-2 and 3CB-3 as reference strains of *T. aromatica* (Song et al., 2001). Accession numbers for 16S rDNA sequences of not validly described strains, which are not included in the list of species are: strain mXyN1, X83533; strain mz1t, AF110005; strain PN-1, AF170281; strain 1917, AJ277705; strain O, AJ277704. Accession numbers of *T. aminoaromatica* and *T. phenylacetica* are AJ315677 and AJ315678, respectively. Note that database entry Y17591 was used for the 16S rRNA of *T. selenatis*.

TABLE BXII.β.117. Characteristics differentiating the species of the genus *Thauera*[a]

Characteristic	*T. selenatis*	*T. aromatica*	*T. chlorobenzoica*	*T. linaloolentis*	*T. mechernichensis*	*T. terpenica*
Flagellation:						
Monotrichous	+	−	−	−	+	+
Peritrichous	−	+	+	−	−	
Optimal pH	8	7.2–7.8	7.5–8.0	7.0–7.3		7.9–8.8
Optimal temperature, °C	28	28–37	30–37	32	40	32
SeO$_4^{2-}$ can be used as a terminal electron acceptor	+	−	−	−	−	−
Mol% G + C of DNA	66	67	69	66	65	64

[a]Symbols: see standard definitions. Blank space, data not reported or not applicable.

TABLE BXII.β.118.   Growth of *Thauera* species on various substrates under aerobic (O_2) and denitrifying (NO_3^-) conditions[a]

Substrate	*T. selenatis*		*T. aromatica*		*T. chlorobenzoica*		*T. linaloolentis*		*T. mechernichensis*		*T. terpenica*	
	O_2	NO_3^-	O_2	NO_3^-	O_2	NO_3^-	O_2	NO_3^-	O_2	NO_3^-	O_2	NO_3^-
Autotrophic on H_2/CO_2	+	−										
Benzoate, 3-hydroxybenzoate, 4-hydroxybenzoate	+	−	+	+	+	+		−	+	+		−
3-Aminobenzoate			−	+	−	+						
Toluene			−	+	−	−			−			−
Phenol			−	D	−	−			−	+		
Phenylacetate, phenylalanine	+		+	+				−	+	+		−
4-Hydroxyphenylacetate			+	+	−	−						
Benzyl alcohol			−	−	−	+						
Vanillate			+	+	+	+						
2-Fluorobenzoate, 4-fluorobenzoate			D	D	+	+						
3-Chlorobenzoate, 3-bromobenzoate, 3-iodobenzoate			−	D	−	D						
Linalool, geraniol								+				−
Terpenes (e.g., menthol, eucalyptol)								−				+
Ethanol			D	+	−	D		+				−
Formate	+			−				−				−
Acetate, butyrate	+	+	+	+	+	+		+	+	+		+
Caproate			D	D	D	+		+				+
Propionate	+		+	+	−	−		+				+
Isobutyrate	+		+									
Glutarate			+	+				−				+
Adipate			D	D	D	D						
Pimelate			−	−								
Succinate, fumarate	+		+	+	+	+		+				+
Aspartate	+		−	−								
Glutamate	+		+	+	+	+		+	+	+		+
Proline	+		−	−								
Leucine	+		+	+					+	+		
Valine			−	−				+				+
Serine	+		D	D								
Alanine			D	D	−	−						
β-Alanine	+		D	−								
Glucose, fructose, ribose, lactose	+		−	−				−				−

[a]Symbols: see standard definitions. Blank space, data not reported or not applicable.

Characteristics are as described for the genus and as listed in Tables BXII.β.117 and BXII.β.118. Additional features are as follows. Cells possess a single polar flagellum. Selenate is reduced to selenite by a periplasmic selenate reductase, and the selenite is further reduced to elemental (red) selenium, which precipitates in the culture medium. No aromatic compounds are catabolized anaerobically. For further descriptive information see Macy et al. (1993). Optimal temperature, 28°C. Isolated from selenate-contaminated water by use of enrichment cultures under selenate-reducing conditions.

The mol% G + C of the DNA is: 66 (T_m).

Type strain: AX, ATCC 55363.*

GenBank accession number (16S rRNA): X68491, Y17591.

2. **Thauera aromatica** Anders, Kaetzke, Kämpfer, Ludwig and Fuchs 1995, 331[VP]

a.ro.ma' ti.ca. M.L. adj. *aromatic* referring to the nutritional preferences of this organism.

Characteristics are as described for the genus and as listed in Tables BXII.β.117 and BXII.β.118. Additional features are as follows. Cells may have peritrichous flagellation with up to eight flagella or may have lost all but one flagellum ("degenerately peritrichous"). Benzoate and various other aromatic compounds are catabolized anaerobically via the benzoyl-CoA pathway. Selenate is not reduced. Optimal temperature, 28°C. Isolated from (contaminated) water and soil, particularly from denitrification stages of sewage treatment plants. Enrichment is possible under denitrifying conditions with aromatic substrates. Different isolates vary in their catabolic capacities for aromatic compounds.

The mol% G + C of the DNA is: 66–67 (HPLC).

Type strain: K172, DSM 6984.

GenBank accession number (16S rRNA): X77118.

Additional Remarks: The accession numbers for reference strains ATCC 700265, CB-2, and CB-3 are U59176, AF229881, and AF229882, respectively.

3. **Thauera chlorobenzoica** Song, Palleroni, Kerkhof and Häggblom 2001, 600[VP]

*Editorial Note: The type strain of *Thauera selenatis* is currently unavailable as it was deposited in association with a patent application that was subsequently abandoned.

chlor.o.ben.zo'i.ca. M.L. adj. *chloro* pertaining to chlorine; from Gr. adj. *chloros* pale green; M.L. adj. *benzoicus* pertaining to benzoic acid; M.L. fem. adj. *chlorobenzoica* indicating the ability to utilize chlorobenzoic acid.

Characteristics are as described for the genus and as listed in Tables BXII.β.117 and BXII.β.118. Additional features are as follows. Cells are 0.85–0.97 × 1.2–2.7 μm and have peritrichous flagellation. In contrast to *T. aromatica,* all known strains of the species grow under denitrifying conditions on 2- and 4-fluorobenzoate; some strains also grow on 3-chlorobenzoate, 3-bromobenzoate, or 3-iodobenzoate. Isolated from river and estuarine sediments and from agricultural soil.

The mol% G + C of the DNA is: 69 (HPLC).

Type strain: 3CB-1, ATCC 700723.

GenBank accession number (16S rRNA): AF123264.

Additional Remarks: The accession numbers for reference strains 4FB-1, 4FB-2, and 3BB-1 are AF229867, AF229868, and AF229887, respectively.

4. **Thauera linaloolentis** Foss and Harder 1999, 2^VP (Effective publication: Foss and Harder 1998, 370.)

li.na.lo.o.len' tis. Sp. n. *linaloe* wood of American trees, from which an oil containing mainly linalool is extracted; L. *olere* smell; M.L. adj. *linaloolentis* linaloe-smelling, referring to the ability of the organism to catabolize linalool.

Characteristics are as described for the genus and as listed in Tables BXII.β.117 and BXII.β.118. Additional features are as follows. Catabolizes some open-chain terpenoids, such as linalool or geraniol, anaerobically. No anaerobic catabolism of benzoate or other aromatic compounds occurs. For further descriptive information, see Foss and Harder (1998). Isolated from activated sewage sludge after enrichment for denitrifying bacteria with linalool as the sole substrate. Optimal temperature, 32°C.

The mol% G + C of the DNA is: 66 (HPLC).

Type strain: 47Lol, DSM 12138.

GenBank accession number (16S rRNA): AJ005816.

5. **Thauera mechernichensis** Scholten, Lukow, Auling, Kroppenstedt, Rainey and Kiekmann 1999, 1049^VP

me.cher.ni.chen' sis. M.L. adj. *mechernichensis* pertaining to Mechernich, Germany, the site of isolation of the organism.

Characteristics are as described for the genus and as listed in Tables BXII.β.117 and BXII.β.118. Additional features are as follows. The cells have single polar flagella. Anaerobic catabolism of benzoate and other aromatic compounds occurs. Denitrification is not repressed under aerobic conditions. For further descriptive information, see Scholten et al. (1999). Isolated from the nitrification step of a leachate treatment plant after aerobic continuous culture with acetate as the carbon source. Optimal temperature, 40°C.

The mol% G + C of the DNA is: 65 (HPLC).

Type strain: TL1, DSM 12266.

GenBank accession number (16S rRNA): Y17590.

6. **Thauera terpenica** Foss and Harder 1999, 2^VP (Effective publication: Foss and Harder 1998, 372.)

ter.pe' ni.ca. M.L. adj. *terpenica* related to terpenes, referring to the nutritional preferences of these organisms.

Characteristics are as described for the genus and as listed in Tables BXII.β.117 and BXII.β.118. Additional features are as follows. Catabolize several monoterpenes anaerobically. The substrate range of catabolized terpenes varies for different isolates. For example, the terpenoid substrates used by strain DSM 12139^T include eucalyptol, sabinene, β-pinene, α-phellandrene, and other chemically similar compounds, whereas those used by strain DSM 12140 include menthol, isomenthol and some similar compounds. No anaerobic catabolism of benzoate or other aromatic compounds occurs. For further descriptive information, see Foss and Harder (1998). Isolated from forest ditches after enrichment culture for denitrifying bacteria with defined monoterpenes as sole substrates. Optimal temperature, 32°C.

The mol% G + C of the DNA is: 64 (HPLC).

Type strain: 58Eu, DSM 12139.

GenBank accession number (16S rRNA): AJ005817.

Additional Remarks: The accession number for the 16S sequence for strain DSM 12140 is AJ005818. The recently described *p*-cymene degrading bacterial strain pCyN2 (Harms et al., 1999a) may also belong to this species based on its very similar 16S rDNA sequence (accession number Y17285).

Other Organisms

While this volume was in press, two more *Thauera* species were validly published, *T. aminoaromatica* (DSM 14742) and *T. phenylacetica* (DSM 14743). For further details, see Mechichi et al. (2002). Furthermore, an aerobic butane-degrading bacterium previously desgnated as *"Pseudomonas butanovora"* was recently shown to be affiliated with *T. linaloolentis* on the basis of its 16S rDNA sequence, but has not yet been formally described as a new *Thauera* species (Anzai et al., 2000).

Genus XII. **Zoogloea** Itzigsohn 1868, 39^AL emend. Shin, Hiraishi and Sugiyama 1993, 830

RICHARD F. UNZ

Zo.o.gloe' a. Gr. adj. *zoos* living; Gr. n. *gloia* glue; M.L. fem. n. *Zoogloea* living glue.

The following description is representative of the genus based on well-characterized isolates recovered directly from naturally occurring, typical, fingered zoogloeae: **Straight to slightly curved, plump rods,** 1.0–1.3 × 2.1–3.6 μm, with rounded ends; sometimes tapered to a blunt point at one or both poles. Nonsporeforming and noncystforming. **Cells in older cultures are demonstrably encapsulated.** Gram negative. **Actively motile,** especially in young cultures, by means of a **single polar flagellum. Intracellular granules of poly-β-hydroxybutyrate are formed** on media containing the salts of organic acids. Cultures enter into formation of flocs and films in liquid media at late growth stages; **the cells become embedded in gelatinous matrices to form zoo-**

gloeae, which may be distinguished by a "tree-like" or "finger-like" morphology. Young colonies on solid media under a normal air atmosphere are translucent and punctiform, but may increase to 1 or 2 mm in diameter and exhibit opaque centers. Nonpigmented. **Aerobic, having a strictly respiratory type of metabolism with oxygen or nitrate as the terminal electron acceptor. Denitrification occurs with formation of N$_2$.** Optimal temperature for growth, 28–37°C. Optimal pH, 7.0–7.5. **Oxidase positive.** Weakly catalase positive. Chemoorganotrophic. **Acid is not formed from carbohydrates except for xylose, glycerol, and ethanol, which are attacked oxidatively by a few strains.** Proteolytic on gelatin. Most strains are urease positive. Litmus milk is unchanged. Hydrogen sulfide is usually not produced from cysteine. Major carbon sources include salts of several organic acids (e.g., lactate, pyruvate, and fumarate), dicarboxylic amino acids (e.g., aspartate, glutamate, and asparagine), alcohols, and salts of certain aromatic acids (e.g., benzoate and *m*-toluate). Benzene derivatives are attacked by *meta* cleavage of the ring structure. Organic nitrogen compounds (e.g., dicarboxylic amino acids) and ammonia serve as nitrogen sources; nitrate is unsuitable. Specific growth factor requirements, if any, are unknown. Major cellular fatty acids are palmitoleic (C$_{16:1}$) and 3-hydroxydecanoic (C$_{10:0\ 3OH}$) acids. Major respiratory quinones are Q-8 and RQ-8. The major polyamine is 2-hydroxyputrescine. **Occur free-living in organically polluted fresh waters and in wastewaters at all stages of treatment.**

The mol% G + C of the DNA is: 67.3–69.0.

Type species: **Zoogloea ramigera** Itzigsohn 1868, 30.

FURTHER DESCRIPTIVE INFORMATION

Zoogloea strains form flocculent masses of zoogloeae in both complex and defined media containing suitable carbon sources (Fig. BXII.β.110). Arrangement of the bacteria into sharply demarcated columns or "fingers" (Fig. BXII.β.111), which protrude from a cluster or aggregate of cells, constitutes the historically recognized growth form of *Z. ramigera* (Koch, 1877; Butterfield, 1935; Dugan and Lundgren, 1960; Unz and Farrah, 1976b). Cells are embedded in an extracellular matrix, which is discernible by treatment of wet mount specimens with a contrasting agent (Fig. BXII.β.112) or by scanning electron microscopy (Fig. BXII.β.113). However, taken by itself, *Zoogloea* morphology is an unreliable character upon which to base the identification of *Z. ramigera*, since (a) fragmented portions of flocs and pellicles and artifacts created by random coalescence of bacteria may be mistaken under microscopic observation for finger-like zoogloeae, and (b) *Z. ramigera* may form amorphous, rather than finger-like, zoogloeae (Fig. BXII.β.114). The extent of zoogloea production varies among strains and with the culture conditions and may diminish greatly or be lost, especially during frequent transfer of strains in rich culture media.

Cells of *Zoogloea* are plump rods that are motile by means of a single, monopolar flagellum (Fig. BXII.β.115). Chains of cells are rare. Cells in very old cultures may appear elongated.

Colonies produced on CY agar[1] are initially punctiform. After 3 or 4 days, the colonies reach 1 mm in diameter and appear circular, slightly raised, and translucent with opaque centers (Fig. BXII.β.116) or completely gray-white. Colony edges are entire or lobate. Mature colonies are distinctively tenacious and cohesive and may be lifted intact from the agar surface with a needle. Colonies develop poorly on ordinary nutrient agar.

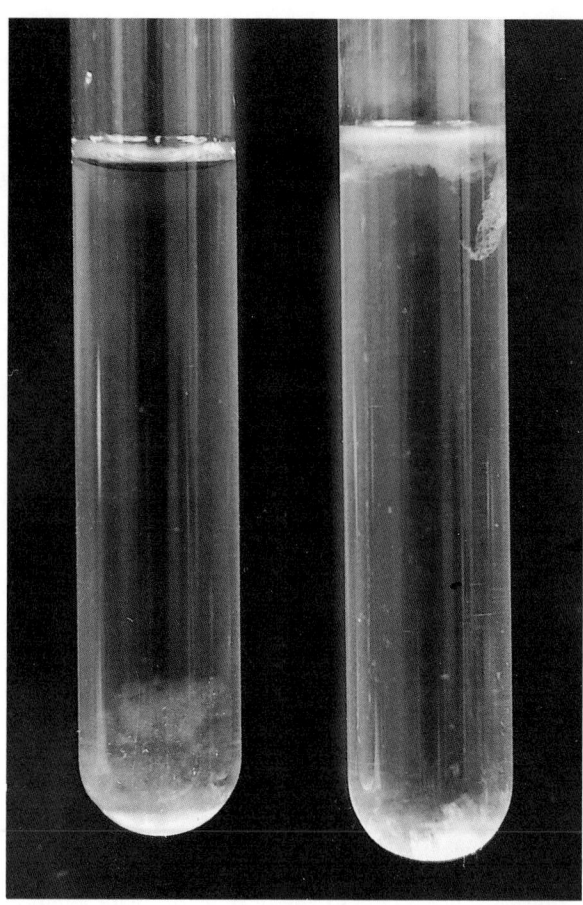

FIGURE BXII.β.110. Flocculent growth habit of *Zoogloea ramigera*. *Left*, ATCC strain 19544. *Right*, strain G4, freshly isolated and exhibiting the development of a thick, straggly, pellicle. Casitone–yeast autolysate medium, 28°C, 72 h.

Growth of *Zoogloea* strains is slow at 9°C and nonexistent at 45°C. The pH range for growth lies between pH 5.0 and 10.0, with poor development occurring at extreme pH values.

Strains survive, but do not grow, under strictly anaerobic conditions in the absence of nitrate. They exhibit a microaerophilic tendency in semisolid agar deeps, as evidenced by the appearance of culture bands 3–5 mm below the agar surface (Unz and Dondero, 1967b). Approximately 50% of strains tested reduce trimethylamine oxide to trimethylamine (Unz and Dondero, 1967b).

Acid was found to be produced oxidatively from xylose by 3 of 65 strains in one study (Unz and Dondero, 1967a), and by 5 of 37 strains in another study (Unz and Farrah, 1972). Acid was also found to be produced oxidatively from glycerol and/or ethanol by 10 of 65 strains (Unz and Dondero, 1967a).

No growth on citrate was found to occur in one study (Unz and Dondero, 1967a), but growth by 20 of 37 strains occurred on Koser's citrate in another study (Unz and Farrah, 1972).

Zoogloea strains are not fastidious nutritionally and may be cultured on a variety of organic carbon sources in a simple defined medium.[2] Carbon sources supporting growth of at least

---

1. CY agar contains (g/l of distilled water): Casitone (Difco), 5.0; yeast autolysate, 1.0; agar, 15.0.

2. Defined medium (g/l distilled water): carbon source, 0.25–0.5; (NH$_4$)$_2$SO$_4$, 0.375; MgSO$_4$·7H$_2$O, 0.2; CaCl$_2$, 0.2; K$_2$HPO$_4$, 0.1; and FeSO·7H$_2$O, 0.005. The pH is adjusted to 7.2 with 0.1 N NaOH. Yeast autolysate (0.01 g) and vitamin B$_{12}$ (1 × 10^{-6} g) may be included as sources of growth factors to decrease growth lag.

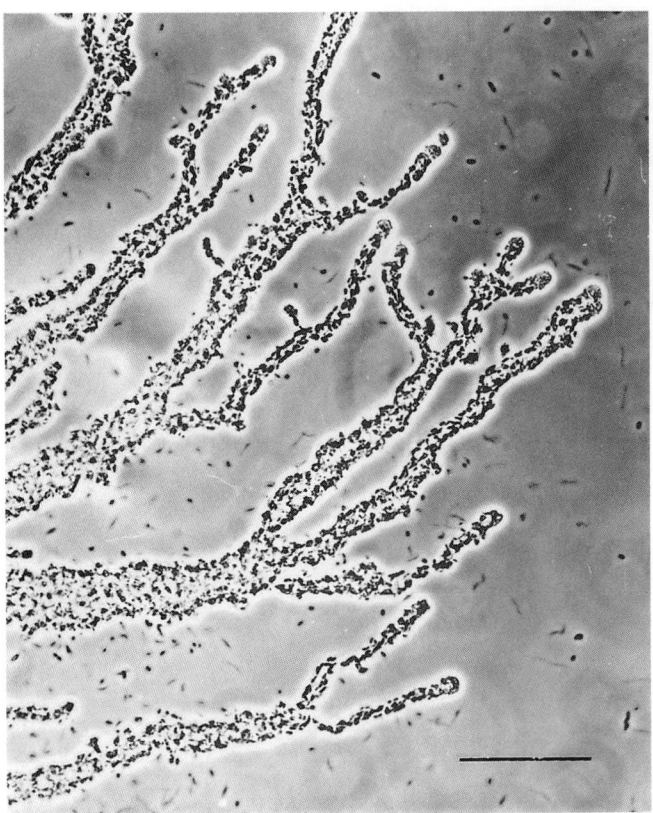

FIGURE BXII.β.111. Finger-like zoogloea. *Zoogloea ramigera* ATCC strain 19544 grown in lactate–mineral salts medium at 28°C for 60 h. Phase contrast. Bar = 50 μm.

90% of *Zoogloea* strains are lactate, pyruvate, α-hydroxybutyrate, fumarate, ethanol, butanol, benzoate, glutamate, aspartate, and asparagine (Unz and Dondero, 1967b).

Resistance to 2.5 U of penicillin G occurs in 67% of the strains tested (Unz and Dondero, 1967a).

Zoogloea strains are found principally in organically polluted fresh waters, wastewaters, and aerobic biological wastewater treatment systems, e.g., activated sludge and trickling filter units.

ENRICHMENT AND ISOLATION PROCEDURES

Finger-producing strains of *Zoogloea* may be cultivated best in enrichment media inoculated with activated sludge or with the films from aerobic wastewater treatment devices. Approximately 2 ml of an inoculum are added to 15 ml of mineral salts solution[3] overlaying 5 ml of a nutrient-enriched agar[4] contained at the bottom of a metal-capped test tube (20 mm OD × 150 mm). The enrichment culture is incubated at 28°C until a pellicle develops, usually within 2 or 3 d if activated sludge is the inoculum. A simpler method for enrichment of fingered zoogloeae involves storage of activated sludge in covered glass containers at room temperature until a surface film appears (Amin and Ganapati, 1967). The latter method is not always reliable, and success may depend on obtaining a proper ratio of the height of activated sludge solids to the total volume of liquid stored in the container.

A wet mount of enrichment-culture surface film is examined at 100× magnification by phase contrast microscopy to confirm the presence of fingered zoogloeae. Several loopfuls of film are

---

3. Mineral salts solution (g/l): $(NH_4)_2SO_4$, 0.3; NaCl, 5.85; $CaCl_2 \cdot 2H_2O$, 0.2; K_2HPO_4, 0.1; $MgSO_4 \cdot 7H_2O$, 0.14; $FeSO_4 \cdot 7H_2O$, 0.0003; $MnCl_2 \cdot 4H_2O$, 0.0063; $CoSO_4 \cdot 7H_2O$, 0.0001; H_3BO_4, 0.0006; $ZnCl_2$, 0.00022; and $CuSO_4 \cdot 5H_2O$, 0.00008. The medium components are prepared from stock solutions and the pH of the medium is adjusted to 8.5 ± 0.1 pH unit with 0.5 N NaOH.

4. Nutrient-enriched agar: to mineral salts solution containing agar (20 g/l), any of the following sole carbon sources are added in the amount shown per liter: starch, 2.4 g; *m*-toluic acid (neutralized), 1.35 g; *n*-butanol, 1.5 ml; lactic acid (85%), 1.35 g; ethanol (95%), 1.5 ml; or glucose, 2.4 g. After adding the carbon source to the liquid mineral salts–agar mixture, the pH is adjusted to 8.5 ± 0.1 with 0.5 N NaOH.

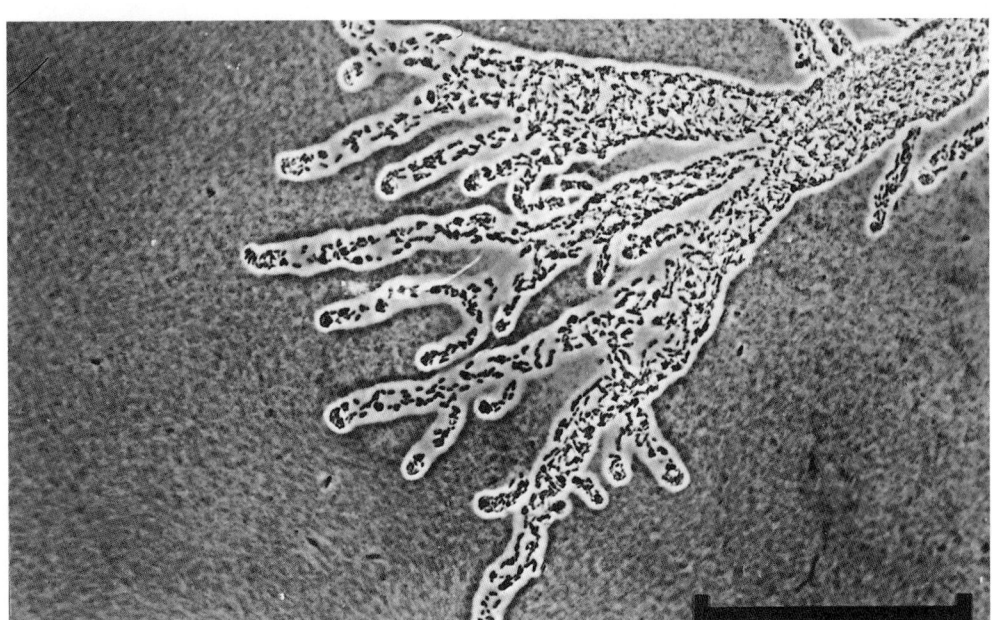

FIGURE BXII.β.112. Finger-like zoogloea treated in wet mount with skim milk to accentuate the exopolymer in which the cells are embedded. *Zoogloea ramigera* ATCC strain 19544. Lactate–mineral salts medium, 28°C, 60 h. Phase contrast. Bar = 100 μm. (Reproduced with permission of R.F. Unz, International Journal of Systematic Bacteriology *21:* 91–99, 1971 ©International Union of Microbiological Societies.)

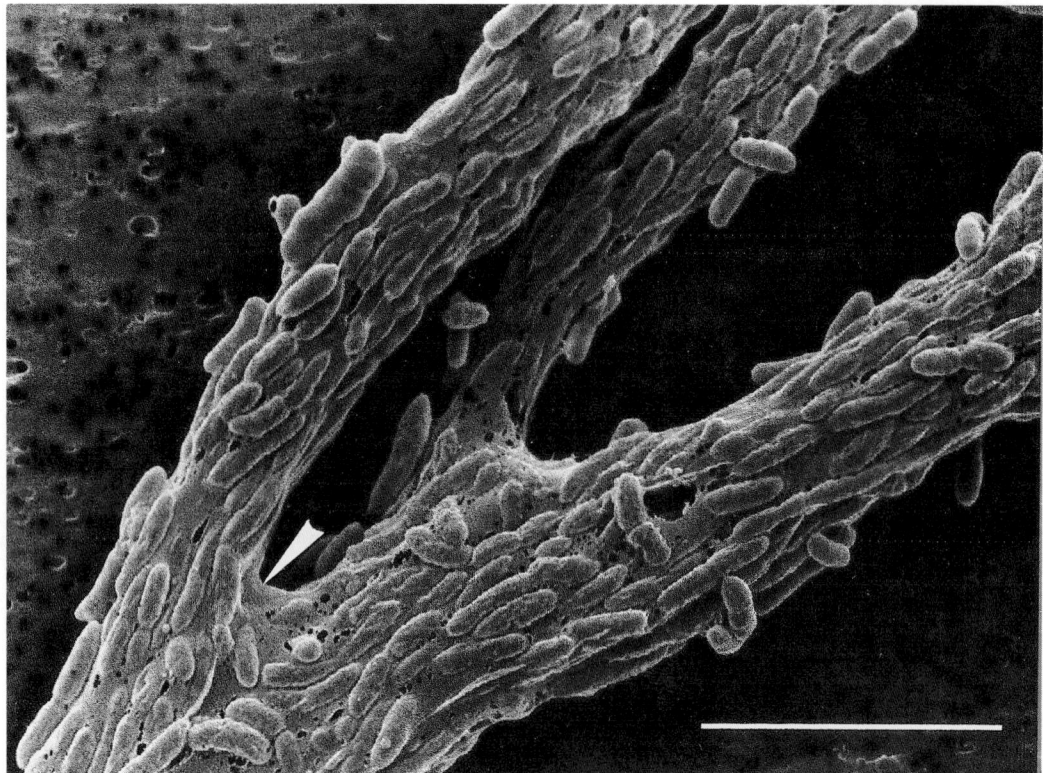

FIGURE BXII.β.113. Finger-like zoogloea mounted on a nucleopore filter, dehydrated, fixed by gradient ethanol dehydration, washed with hexa-methyldisilizane, dried, sputter-coated with gold, and observed by field emission scanning electron microscopy. Cells embedded in a gelatinous matrix (*arrow*). *Zoogloea ramigera* ATCC 19544 grown in yeast extract (2.5 g/l)–peptone (2.5 g/l), 28°C. Bar = 7.2 μm.

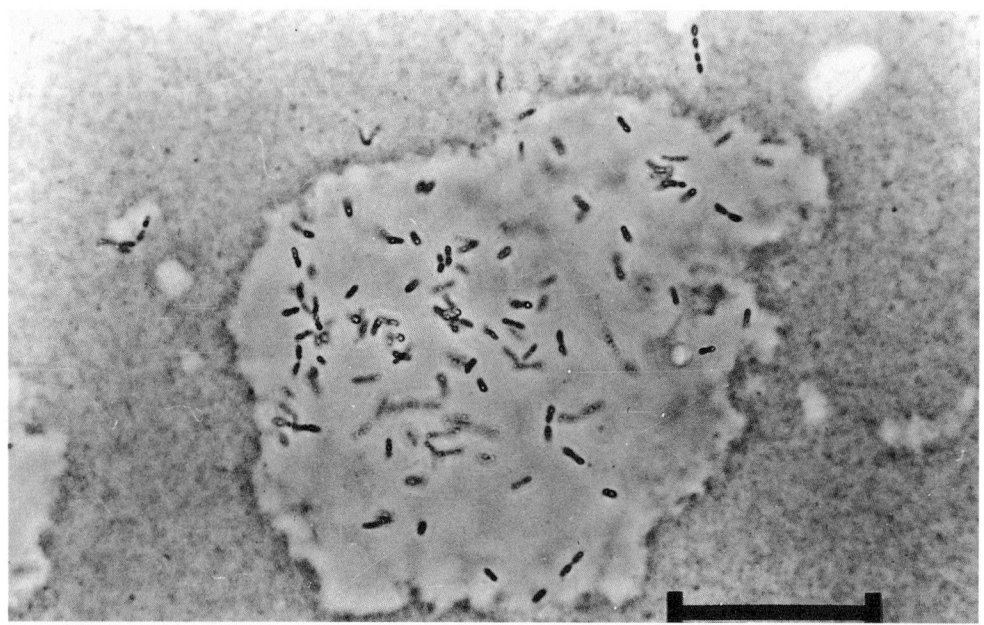

FIGURE BXII.β.114. Amorphous zoogloea treated in wet mount with skim milk to accentuate the exopolymer in which the cells are embedded. *Zoogloea ramigera* ATCC strain 19544. Lactate–mineral salts medium, 28°C, 60 h. Phase contrast. Bar = 100 μm. (Reproduced with permission of R.F. Unz, International Journal of Systematic Bacteriology *21:* 91–99, 1971 ©International Union of Microbiologial Societies.)

transferred to a 2-ml droplet of CY broth contained in a Petri plate. The fingered zoogloeae may be located easily in the droplet under 45-60× magnification. Approximately 10–12 of the fin-

gered zoogloeae are transferred individually and successively by a micropipette through each of four 0.7-ml droplets of CY medium in order to free loosely attached debris and microorganisms

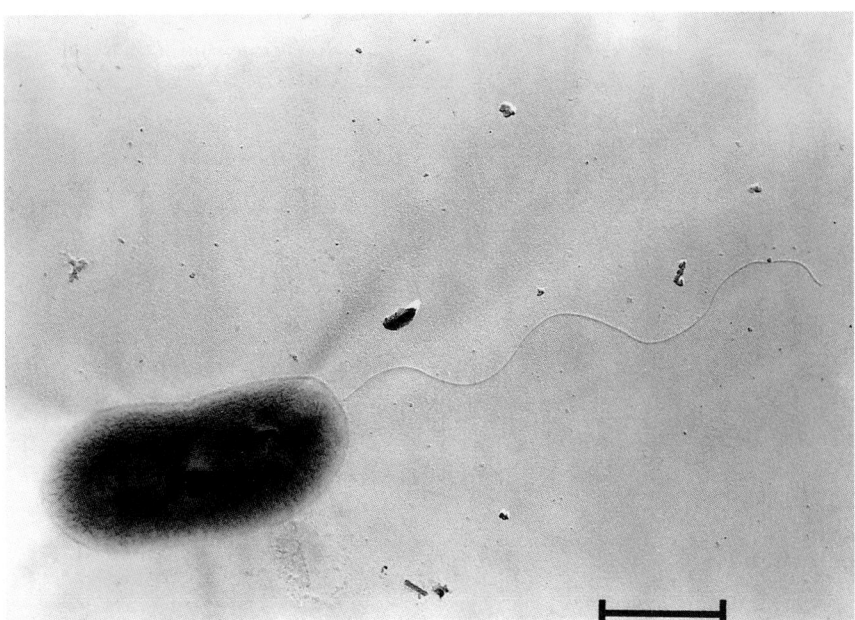

FIGURE BXII.β.115. *Zoogloea ramigera* ATCC strain 19544 showing the single, polar flagellum. Casitone–yeast autolysate medium, 28°C, 24 h. Platinum–carbon shadowed. Bar = 1 μm. (Reproduced with permission of R.F. Unz, International Journal of Systematic Bacteriology *21:* 91–99, 1971 ©International Union of Microbiological Societies.)

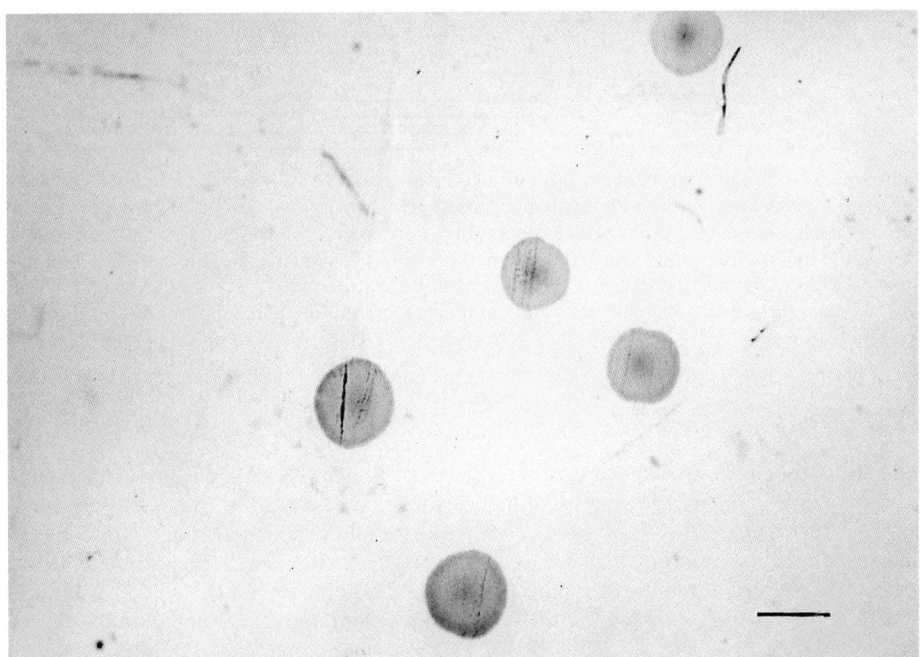

FIGURE BXII.β.116. Colonies of *Zoogloea ramigera* ATCC strain 19544. Casitone–yeast autolysate agar, 28°C, 72 h. Photographed by reflected light. Bar = 1 mm. (Reproduced with permission of R.F. Unz, International Journal of Systematic Bacteriology *21:* 91–99, 1971 ©International Union of Microbiological Societies.)

from the zoogloeae. Finally, the washed zoogloeae are collectively transferred to 3 ml of CY medium and subjected briefly to sonic oscillation (e.g., 30 s at 50 watts) to release the cells. A loopful of the sonicate is streaked onto solid CY medium and incubated at 28°C. Typically cohesive colonies of *Zoogloea* are large enough to be transferred intact to CY broth after 3–4 d. Inoculation of CY broth with a single colony usually results in the appearance of slight turbidity and a slippery, glistening pellicle after 3 d at 28°C. Pellicles may be composed entirely of fingered zoogloeae or amorphous zoogloeae, which, following detachment, settle and give rise to a flocculent sediment.

MAINTENANCE PROCEDURES

Zoogloea strains may be maintained in half-strength CY medium at 20°C for at least 2 months between subcultures. The formation of zoogloeal flocs is visibly reduced upon continuous transfer in

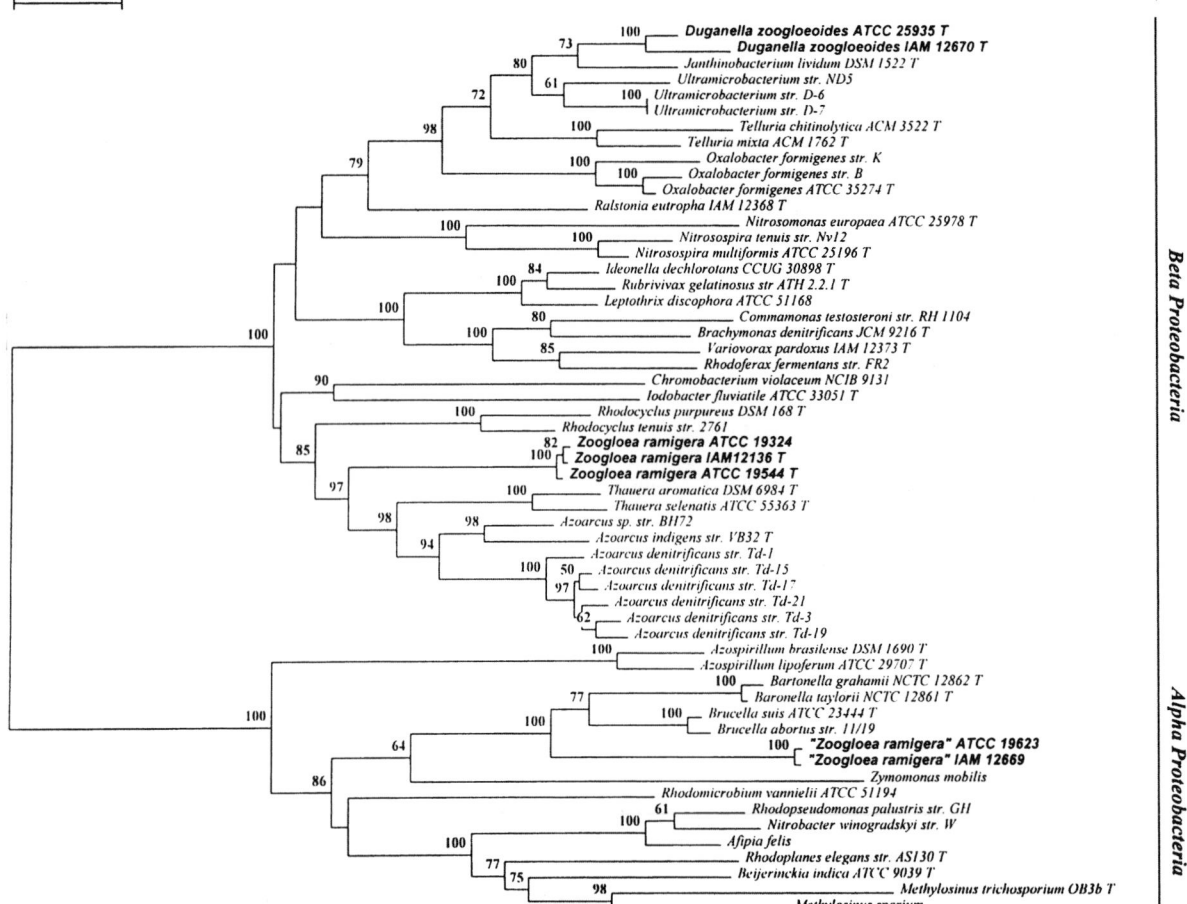

FIGURE BXII.β.117. Dendrogram revealing the phylogenetic position of *Zoogloea ramigera*. The phylogenetic distance tree was generated from 1309 bases of unambiguously aligned 16S rRNA sequences from 39 members of the class *Betaproteobacteria* and 17 members of the class *Alphaproteobacteria*. Positions corresponding to 33–68 plus 104–1376 (*Escherichia coli* numbering) were used. Phylogenetic distances were calculated with correction for multiple substitutions at a single position (Jukes and Cantor, 1969), and the tree was generated using the neighbor-joining method (Saitou and Nei, 1987), as implemented in the TREECON software package (Van de Peer and de Wachter, 1994). Bootstrap resampling (100 replicates) was used to indicate the confidence in the sequence clusters obtained. The values at nodes represent the percentage bootstrap support for the groupings to the right of the node. Bootstrap values below 50% are not shown. Bar = 2 nucleotide changes per 100 nucleotides in the 16S rRNA sequence (see text next page).

laboratory culture media. The zoogloeal growth habit may be restored by plating a broth culture and transferring a colony back to liquid medium.

Strains do not survive prolonged refrigeration; however, they may be preserved indefinitely by lyophilization or cryopreservation.

DIFFERENTIATION OF THE GENUS *ZOOGLOEA* FROM OTHER GENERA

The taxonomic status of the genus *Zoogloea* within the family *Pseudomonadaceae* has been called into question based on 16S rDNA sequence analyses and chemotaxonomic data (Shin et al., 1993). The most recent phylogenetic positioning of the type species of the genus *Zoogloea* lies within the *Betaproteobacteria* as shown in Fig. BXII.β.117. Inspection of the phylogenetic tree reveals that the three most closely related genera to the type strain of *Z. ramigera* (ATCC 19544) are *Thauera*, *Azoarcus*, and *Rhodocyclus*. Strains of authentic and misnamed *Z. ramigera* that have received molecular characterization are given in bold face. *Z. ramigera* ATCC 19324 was isolated by single-cell microdissection of a nat-

ural, finger-like zoogloea sampled from a trickling filter in a similar fashion as the type strain. Sequence data from independent studies of the type strain obtained from two different culture collections, namely *Z. ramigera* ATCC 19544 (Rosselló-Mora et al., 1993) and *Z. ramigera* IAM 12136 (Shin et al., 1993), are reflected in the dendrogram. Although mol% G + C of the DNA provides a basis for comparison, differentiation of the genus *Zoogloea* from the genera *Thauera*, *Azoarcus*, and *Rhodocyclus* is possible (Table BXII.β.119). Cell flocculation, which was historically held to be a distinctive feature of the genus *Zoogloea*, is widespread among eubacteria, including a species of the dinitrogen-fixing genus *Azoarcus* (Reinhold-Hurek et al., 1993b). The chemical composition of the exopolymers formed by organisms designated as *Zoogloea*, e.g., hexosamine, glucose, xylose, and arabinose (Crabtree et al., 1966); glucose, mannose, and galactose (Wallen and Davis, 1972); glucose and galactose (Parsons and Dugan, 1971); and glucosamine and fucosamine (Tezuka, 1973), is diverse and has not been useful in the taxonomic sense. The mucopolysaccharide (Farrah and Unz, 1976) produced by the neotype strain of *Z. ramigera* (ATCC 19544) appears to be chemically related to

TABLE BXII.β.119. Differential characteristics of the genus *Zoogloea* and phylogenetically related genera of the *Betaproteobacteria*[a]

Characteristics	*Zoogloea*	*Azoarcus*[b]	*Rhodocyclus*[c]	*Thauera*[d]
Cell diameter typically 1.0 μm or slightly larger	+	−	−	−
Pigment formed by colonies	−	+	+	+
Photoautotrophic	−	−	+	−
Obligately chemooganotrophic	+	+	−	−
Denitrification	+	−[e]	−	+
Urease	+	D	D	
Oxidase	+	+		+
Catalase	w +[f]	+		+
Gelatinase	+	−	D	−
N$_2$ fixation	−	+	D	−
Poly-β-hydroxybutyrate	+			+
Predominant polyamine: 2-OH putrescine	+[g]			−
Mol% G + C of DNA	67.3–69.0[h]	62–68	64.8–72.4	66
Major respiratory quinones	Q-8, RQ-8[i]		Q-8, MK-8	Q-8

[a]For symbols see standard definitions.

[b]Data from Reinhold-Hurek et al. (1993b).

[c]Data from Imhoff and Trüper (1989).

[d]Data from Macy et al. (1993).

[e]Denitrification demonstrated by some strains (Hurek and Reinhold-Hurek (1995).

[f]Weakly positive.

[g]Data from Hamana and Matsuzaki (1993).

[h]Data from Shin et al. (1993).

[i]Data from Hiraishi et al. (1992c).

the exopolymer described by Tezuka (1973); moreover, it exhibits a fine, strand-like mesh (Fig. BXII.β.118), rather than the coarse, fibrillar network of cellulose-like glycans observed for other zoogloeal and nonzoogloeal bacteria (Friedman et al., 1968, 1969) or the cellulose of *Acetobacter* species (Ohad et al., 1962). The exocellular homo- and heteropolysaccharide exopolymers that have been characterized for species of *Xanthomonas*, *Pseudomonas*, *Arthrobacter*, and *Alcaligenes* are variably water-soluble and may increase the consistency of the culture medium (Powell, 1979; Sutherland, 1979a). In contrast, the mucopolysaccharide-producing neotype strain of *Z. ramigera* produces a water-insoluble exopolymer at room temperature and the culture medium is not visibly thickened.

The genus *Comamonas* (Davis and Park, 1962) has been used to classify non-floc-forming bacteria to distinguish them from otherwise similar floc-forming bacteria considered to be *Zoogloea* strains (Dias and Bhat, 1964).

TAXONOMIC COMMENTS

Important developments concerning the genus *Zoogloea* have occurred since the last edition of the *Manual* (Unz, 1984). These pertain to solidifying the position of ATCC 19544 (strain 106; Unz, 1971) as the neotype strain of the species *Z. ramigera* (Judicial Commission, 1979) and resolving the issue of two strains of uncertain affiliation. The type strain and ATCC 25935 (strain 115; Friedman and Dugan, 1968), both members of the class *Betaproteobacteria*, differ on the basis of the presence of rhodoquinone-8 and palmitoleic acid (major cellular fatty acid of the nonpolar fraction) in the former (Hiraishi et al., 1992c), putrescine and 2-hydroxyputrescine content (Hamana and Matsuzaki, 1993), and their 16S rRNA sequences (Rosselló-Mora et al., 1993). ATCC 19623 (strain I-16-M; Crabtree and McCoy, 1967) is confined to the class *Alphaproteobacteria* based on spermidine content (Hamana and Matsuzaki, 1993), ubiquinone Q-10 content (Hiraishi et al., 1992c), and 16S rDNA (Shin et al., 1993) and 16S rRNA (Rosselló-Mora et al., 1993) sequence analyses. The use of molecular biological techniques has resolved the am-

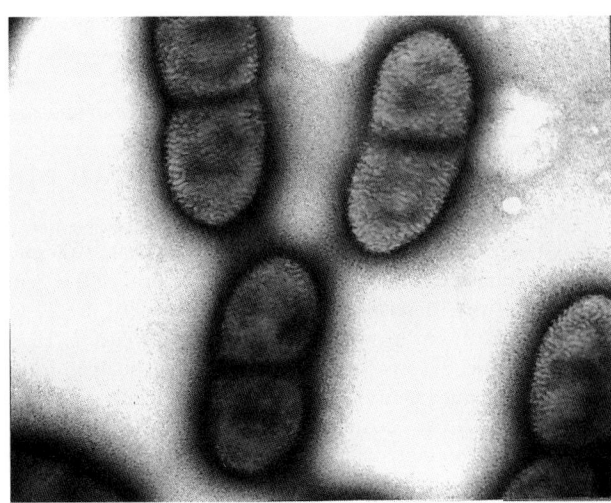

FIGURE BXII.β.118. Cells of *Zoogloea ramigera* ATCC strain 19544 embedded in a fine meshwork of exopolymer. Casitone–yeast autolysate medium, 28°C, 72 h. Preparation negatively stained with 2% phosphotungstic acid for 30 s × 27,000.

biguous taxonomic status of the genus *Zoogloea*, originally attributed to nomenclatural difficulties arising from (a) uncertainties about the type species, which was originally described on the basis of a growth form of an organism in mixed culture, (b) acceptance of the flocculent growth habit as the principal characteristic for the identification of sundry aggregative bacteria as strains of *Zoogloea* or of *Z. ramigera*, and (c) conflicting descriptions of bacteria stated to be *Z. ramigera*.

Z. ramigera ATCC 19544 is one of 147 *Zoogloea* strains of certain origin, in that the bacteria were isolated as single cells during micromanipulation of natural, finger-like zoogloeae under microscopic observation (Unz and Dondero, 1967a). *Z. ramigera* ATCC 19544 forms fingered zoogloeae in liquid culture, which are similar in appearance to those of the natural type from which it came and to the zoogloeae referred to as "tree-like" ramifications of spherical gelatinous masses in the original description of the species (Itzigsohn, 1868). Recent evidence of a relationship of the type strain to cells in natural, finger-like zoogloeae is provided in the exclusive hybridization between the bacteria of natural zoogloeae and an oligonucleotide probe complementary to a region of 16S rRNA of *Z. ramigera* ATCC 19544, with little or no reaction taking place with probes designed for *Z. ramigera* ATCC 19623 or *Z. ramigera* ATCC 25935 (Rosselló-Mora et al., 1995).

At least 90–100% of *Zoogloea* strains denitrify, demonstrate positive oxidase and catalase reactions, hydrolyze gelatin, carry out *meta* cleavage of aromatic ring structures, and fail to form acid from carbohydrates (Unz and Dondero, 1967a; Unz and Farrah, 1972). Approximately 82% of strains demonstrate urease activity. To define the salient characteristics of the genus is difficult, because minor attributes, such as acid formation from ethanol and glycerol, are often located in strains of a common natural provenance—i.e., progeny of a single zoogloea or of several zoogloea found in a specific wastewater treatment plant or polluted environment.

Floc formation (Finstein, 1967) and major exopolymer production (Unz and Farrah, 1976a) by *Zoogloea* strains take place during the decreasing and stationary growth phases. Consequently, bioflocculation and exopolymer formation are to be regarded as conditional characters of *Zoogloea* strains, albeit ones that are exhibited reliably by freshly-isolated organisms.

Degradation of the benzenoid ring structure by a *meta* cleavage mechanism is a distinctive character of *Zoogloea* strains and is considered taxonomically significant for the acidovorans group of *Pseudomonas* (Stanier et al., 1966). *Z. ramigera* differs from *P. acidovorans* and *P. testosteroni* by its ability to hydrolyze gelatin and to denitrify and by its inability to hydrolyze starch or to utilize glycolate, caproate, and several amino acids.

ACKNOWLEDGMENTS

The phylogram presented in Figure BXII.β.117 was kindly generated by Dr. Ian M. Head, University of Newcastle, United Kingdom. Figures BXII.β.110 and BXII.β.116 were prepared by Paul Wichlacz and Terry Williams, formerly of Pennsylvania State University. Figure BXII.β.113 was contributed by Fuha Lu, Jerzy Lukasik, and Samuel R. Farrah, University of Florida, Gainesville.

FURTHER READING

Zvirbulis, E. and H.D. Hatt.. 1967. Status of the generic name *Zoogloea* and its species. Int. J. Syst. Bacteriol. *17*: 11–21.

DIFFERENTIATION OF THE SPECIES OF THE GENUS *ZOOGLOEA*

Because of the taxonomic difficulties described previously, the characteristics of the neotype strain of *Z. ramigera* are presented for comparison with those of two other strains whose inclusion in the species is considered by this author to be unwarranted (Table BXII.β.120). See also the section on Other Organisms.

List of species of the genus Zoogloea

1. **Zoogloea ramigera** Itzigsohn 1868, 30[AL.]

   ra.mi′ge.ra. L. n. *ramus* branch; L. v. *gero* to bear; M.L. adj.*ramigera* branch-bearing.

   The morphological, cultural, physiological, and nutritional characters are given in the description of the genus and in the description of the neotype strain (Table BXII.β.120). No growth occurs on Koser's citrate. H₂S is not produced from cysteine or on Kligler iron agar. Indole is not produced. Ammonia is produced from asparagine. Lipolytic activity occurs.

   Growth on benzoate, *m*-toluate, and *o*-cresol occurs with *meta* cleavage of the aromatic ring.

   Methanol, formate, and formaldehyde are not utilized as sole carbon sources.

   This strain was cultivated from a single cell isolated by micromanipulation of a finger-like zoogloea present in a trickling filter sample collected in Freehold, New Jersey.

   The mol% G + C of the DNA is: 68.9 (HPLC).

   Type strain: 106 (neotype), ATCC 19544.

   GenBank accession number (16S rRNA): X74913.

2. **Zoogloea resiniphila** Mohn, Wilson, Bicho and Moore 1999b, 935[VP] (Effective publication: Mohn, Wilson, Bicho and Moore 1999a, 76.)

   re.si.ni′phi.la. L. fem. n. *resina* resin; Gr. adj. *philos* loving, friendly to; M.L. fem. n. *resiniphila* loving resins.

   Gram-negative, nonsporeforming rods. Catalase negative and oxidase positive. Aerobic, using nitrate, but not fermentative. Mesophilic, growing at 45°C but not at 50°C; however, also reported to grow at 30°C and 37°C, but not at 40°C (Mohn, 1995). Cells grown on dehydroabietic acid are 1.5–2.8 × 1 μm and motile, forming clumps bound by exopolymer; colonies are white, smooth, punctiform, and convex, tenaciously stick to agar.

   In mineral medium, use abietic, dehydroabietic, palmitic, benzoic, acetic, and pyruvic acids, as well as D-xylose, phenol, ethanol, and β-sitosterol. Use poorly 12- and 14-chlorodehydroabietic acids. Do not use pimaric, isopimaric, or linoleic acids.

   Main cellular fatty acids are $C_{16:1\ \omega7c}$, $C_{16:0}$, and $C_{18:1}$.

   The 16S rDNA sequences (EMBL AJ011506) groups within the class *Betaproteobacteria*.

   Isolated from a laboratory sequencing batch reactor in Vancouver, British Columbia, Canada.

   The mol% G + C of the DNA is: not determined.

   Type strain: DhA-35, ATCC 700687.

Other Organisms

In addition to the neotype strain of *Z. ramigera*, there exist two other strains that have attained some prominence through experimental use: strain 1-16-M (Crabtree and McCoy, 1967) and strain 115 (Friedman and Dugan, 1968). The three strains all share the property of floc formation, and each was obtained from polluted environments; however, phenotypic (Table BXII.β.120) and phylogenetic (Fig. BXII.β.117) dissimilarities among the strains indicate that strains 1-16-M and 115 should not be included in the genus *Zoogloea*. Additional evidence for this view is that cross-reactions are not observed between strains 1-16-M

TABLE BXII.β.120. Characteristics of *Zoogloea ramigera* ATCC 19544 and two misidentified *Zoogloea ramigera* strains of historical importance[a]

Characteristics	ATCC 19544	ATCC 19623	ATCC 25935[b]
Cell diameter is 1.0 μm or slightly larger	+	−	−
Flagellar arrangement:			
Monotrichous only	+	−	+
Monotrichous and polytrichous (lateral)	−	+	−
Zoogloeae are produced	+	−[c]	+
Straw-colored colonies	−	+	+
Growth on potato	−	+	
Hydrolysis of gelatin and casein	+	−	+
Tyrosine agar cleared	−	+	
Arginine dihydrolase	−	+	
Sensitivity to 0/129[d]	+	−[c]	
Growth in presence of 3% NaCl	−	+	
Litmus milk:			
Alkalinity produced	−	+	+
Reduction occurs	−	+	−
Denitrification (to N₂)	+	−	
Urease activity	+	−	+
Hydrolysis of starch	−	−	+
Acid formed oxidatively from:			
Arabinose, fructose, sucrose	−	+[c]	+
Galactose, maltose	−	+[e]	+
Cellobiose, glycogen, lactose, mannose	−	−[e]	+
Glucose	−	+[c]	+[f]
Rhamnose, mannitol		+[c]	−[f]
Ribose	−	+[c]	
Utilized as sole carbon source:			
Acetate	+	+	+
Citrate	+	−	+[g]
Malate, pyruvate, fumarate	+	+	
Butyrate, α-ketoglutarate, propionate	+		+[g]
n-Propanol, ethanol, *n*-butanol	+	+[e]	+
Acetaldehyde, succinate, oxalacetate, lactate, α-hydroxybutyrate, palmitate, myristate	+		
Benzoate, *m*-toluate, *p*-toluate, phenol, *o*-cresol, *m*-cresol, *p*-cresol	+[h]	−[h]	−[i]
Methanol	−	+[e]	−
Mol% G + C of DNA[j]	68.9	63.6	63.4
Ubiquinone-10[k]	−	+	−
Ubiquinone-8[k]	+	−	+
Rhodoquinone-8[k]	+	−	−
Spermadine[l]	−	+	
Putrescine[l]	+	−	+
2-Hydroxyputrescine[l]	+	−	+
3-Hydroxydecanoic acid[k]	+	−	+
3-Hydroxylauric acid[k]	+	−	−
3-Hydroxymyristic acid[k]	−	+	+
3-Hydroxypalmitic acid[k]	−	+	−
Palmitoleic acid[k]	+	trace	+

[a]Unless otherwise indicated, the sources of data for strains are as follows: ATCC 19544 (strain 106), Unz (1971); ATCC 19623 (strain I-16-M), Crabtree and McCoy (1967); and ATCC 25935 (strain 115), Friedman and Dugan (1968). For symbols see standard definitions.

[b]Proposed as *Duganella zoogloeoides* (Hiraishi et al., 1997b).

[c]Data from Unz (1971).

[d]Vibriostatic agent 0/129 is 2,4-amino-6,7-diisopropylpteridine.

[e]Data from Friedman and Dugan (1968).

[f]Data from Hiraishi et al. (1997b).

[g]Data from Joyce and Dugan (1970).

[h]Data from Unz and Farrah (1972).

[i]Only benzoate tested.

[j]Data from Shin et al. (1993).

[k]Data from Hiraishi et al. (1992c).

[l]Data from Hamana and Matsuzaki (1993).

or 115 and fluorescein-labeled, whole-cell antiserum against the neotype strain (Farrah and Unz, 1975). Weak, whole-cell antigen–antibody reactions do occur between strain 115 and strain 1-16-M, although strain 115 bears a greater antigenic relationship to *Gluconobacter oxydans* subsp. *suboxydans* (ATCC strain 621) (Chorpenning et al., 1978).

Recently, based on 16S rDNA sequence analysis, strain 115 (*Z. ramigera* ATCC 25935) has been proposed as the type strain for the renamed type species, *Duganella zoogloeoides*, sp. nov., of a proposed genus, *Duganella* gen. nov. (Hiraishi et al., 1997b).

Nonextant strains of *Z. ramigera*, which bear some resemblance to the neotype strain according to published descriptions, are strain Z-1 (Butterfield, 1935) and an early isolate of questionable purity (Bloch, 1918).

Class IV. **Deltaproteobacteria** *class nov.*

JAN KUEVER, FRED A. RAINEY AND FRIEDRICH WIDDEL

Del.ta.pro.te.o.bac.te′ri.a. Gr. n. *delta* name of fourth letter of Greek alphabet; Gr. n. *Proteus* ocean god able to change shape; Gr. n. *bakterion* a small rod; M.L. fem. pl. n. *Deltaproteobacteria* class of bacteria having 16S rRNA gene sequences related to those of the members of the order *Myxococcales*.

The class is defined solely based on sequence similarity of 16S rRNA. It comprises several bacterial groups that had been previously treated as separate systematic assemblages according to phenotypic characteristics. 16S rRNA sequence analyses are not only informative for the definition of orders and families within the *Deltaproteobacteria*. This method is usually also relevant for the establishment of genera and sometimes species; nevertheless, phenotypic features such as nutritional characteristics or chemotaxonomic properties may be equally important at the genus level, and, in combination with DNA–DNA hybridization, at the species level.

The class *Deltaproteobacteria* comprises morphologically diverse, Gram-negative, nonsporeforming bacteria that exhibit either anaerobic or aerobic growth. Members with facultatively anaerobic/aerobic growth are not known so far.

Most anaerobic members can use inorganic electron acceptors that allow energy conservation by anaerobic respiration. By performing such reactions, these bacteria play a major role in the global cycling of elements. Utilization of inorganic electron ac-

ceptors is an important physiological and taxonomic characteristic. However, in a number of isolates the reduction of some electron acceptors (e.g., sulfur, ferric iron) may not be associated with growth (as in the case of oxygen reduction). Therefore, the utilization of an electron acceptor (especially in sulfate-reducing bacteria) must be confirmed by definite growth in subcultures. This has not been proven in every case.

Some anaerobic members are fermentative and/or exhibit syntrophic growth by proton reduction and interspecies hydrogen transfer.

One striking feature of the aerobic representatives is the ability to digest other bacteria. Several of these members are important constituents of the microflora in soil and waters.

Orders *Desulfurellales, Desulfovibrionales, Desulfobacterales, Desulfarcales, Desulfuromonales, Syntrophobacterales.*

Orders *Myxococcales* and *Bdellovibrionales* are exclusively aerobic.

Type order: **Myxococcales** Tchan, Pochon and Prévot 1948, 398.

Order I. **Desulfurellales** *ord. nov.*

JAN KUEVER, FRED A. RAINEY AND FRIEDRICH WIDDEL

De.sul.fu.rel.la′ les. M.L. fem. n. *Desulfurella* type genus of the order; *-ales* ending to denote an order; M.L. fem. pl. n. *Desulfurellales* the order of *Desulfurellaceae*.

Cells are rod-shaped and usually motile. Strictly anaerobic chemoorganotrophs or chemolithoautotrophs with a respiratory type of metabolism; in addition, limited fermentative capacities may occur in some members. The common electron acceptor is elemental sulfur (or polysulfide), which is reduced to sulfide; thiosulfate may be used. Sulfate does not serve as electron acceptor. Simple organic compounds serve as electron donors and carbon sources that are completely oxidized to CO_2. So far as known, the mechanism of terminal oxidation is the citric acid cycle. All members are moderate thermophiles with temperature optima between 50 and 60°C. A striking biochemical feature in

species investigated so far is the absence of cytochromes, despite a respiratory metabolism. Menaquinones are present. Species have been isolated from geothermally heated, sulfidic freshwater and marine environments. The order currently contains only one family, *Desulfurellaceae*.

Type genus: **Desulfurella** Bonch-Osmolovskaya, Sokolova, Kostrikina, and Zavarzin 1993, 624 emend. Miroshnichenko, Rainey, Hippe, Chernyh, Kostrikina and Bonch-Osmolovskaya 1998, 478 (Effective publication: Bonch-Osmolovskaya, Sokolova, Kostrikina, and Zavarzin 1990, 155.)

Family I. **Desulfurellaceae** *fam. nov.*

JAN KUEVER, FRED A. RAINEY AND FRIEDRICH WIDDEL

De.sul.fu.rel.la' ce.ae. M.L. fem. n. *Desulfurella* type genus of the family; *-aceae,* ending to denote a family; M.L. fem. pl. n. *Desulfurellaceae* the family of *Desulfurella.*

The main properties have been indicated in the description of the order. The family currently comprises two genera, *Desulfurella* and *Hippea.*

Type genus: **Desulfurella** Bonch-Osmolovskaya, Sokolova, Kos-

trikina, and Zavarzin 1993, 624 emend. Miroshnichenko, Rainey, Hippe, Chernyh, Kostrikina and Bonch-Osmolovskaya 1998, 478 (Effective publication: Bonch-Osmolovskaya, Sokolova, Kostrikina, and Zavarzin 1990, 155.)

Genus I. **Desulfurella** *Bonch-Osmolovskaya, Sokolova, Kostrikina, and Zavarzin 1993, 624[VP] emend. Miroshnichenko, Rainey, Hippe, Chernyh, Kostrikina and Bonch-Osmolovskaya 1998, 478 (Effective publication: Bonch-Osmolovskaya, Sokolova, Kostrikina, and Zavarzin 1990, 155)*

FRED A. RAINEY AND BECKY HOLLEN

De.sul.fur.el' la. L. pref. *de* from; L. n. *sulfur* sulfur; M.L. dim. ending *-ella;* M.L. fem. n. *Desulfurella* a small sulfur reducer.

Short rods. Motile by means of a single polar flagellum or nonmotile. The cell wall is of the Gram-negative type and has an outer S-layer. Spore formation not observed. **Obligately anaerobic, having a mainly respiratory type of metabolism. Energy conservation occurs by S[0] or thiosulfate respiration or by pyruvate fermentation. Molecular hydrogen, organic acids and fatty acids are utilized. Lithotrophic growth can occur. CO_2 and H_2S are produced.** Menaquinones are present. Cytochromes are not detected. Growth occurs optimally at pH 6.8–7.0. **Moderately thermophilic** with a temperature range between 45 and 75–80°C; optimal, 55–60°C. Occur in warm sediments and in thermally heated cyanobacterial or bacterial communities that are rich in organic compounds and elemental sulfur and are at temperatures in the range 50–70°C. The detailed properties of described species of this genus are given in Table BXII.δ.1.

The mol% G + C of the DNA is: 31–34.

Type species: **Desulfurella acetivorans** Bonch-Osmolovskaya, Sokolova, Kostrikina and Zavarzin 1993, 624 (Effective publication: Bonch-Osmolovskaya, Sokolova, Kostrikina and Zavarzin 1990, 155.)

ENRICHMENT AND ISOLATION PROCEDURES

These thermophilic sulfur-reducing bacteria are enriched by using an anaerobically prepared medium containing (per liter): NH_4Cl, 0.33 g; KCl, 0.33 g; KH_2PO_4, 0.33 g; $CaCl_2 \cdot 2H_2O$, 0.33 g; $MgCl_2 \cdot 6H_2O$, 0.33 g; yeast extract, 0.1 g; $Na_2S \cdot 9H_2O$, 0.5 g; $NaHCO_3$, 1.5 g; resazurin, 0.002 g; trace elements solution (Pfennig, 1965), 1.0 ml, and vitamin solution (Pfennig and Lippert, 1966), 1.0 ml. An aqueous suspension of elemental sulfur is added to give a final concentration of 10 g/l. Various organic substrates and possible electron acceptors can be added at 0.5% (w/v) and 0.2% (w/v), respectively. Medium is dispensed in 15-ml Hungate tubes, and the headspace is replaced with a N_2/CO_2 or H_2/CO_2 mixture (80:20). Pure cultures can be obtained by either serial dilution or colony isolation in agar roll tubes. In agar roll tubes, polysulfide can be used to replace the elemental sulfur.

MAINTENANCE PROCEDURES

Desulfurella strains can be preserved indefinitely by suspending cells in an anaerobic medium (see above) containing 5% dimethylsulfoxide and storing in liquid nitrogen.

TABLE BXII.δ.1. Differential characteristics of the species of the genus *Desulfurella*[a]

Characteristics[b]	D. acetivorans	D. kamchatkensis	D. multipotens	D. propionica
Cell size, mm	0.5–0.7 × 1.0–2.0	1.5–2.0 × 0.5–0.8	0.5–0.7 × 1.5–1.8	1.5–2.0 × 0.5–0.8
Motility	+	+	+	−
Growth substrates:[c]				
Butyrate	−	−	+	−
Formate	−	−	−	−
Fumarate, lactate, malate	−	+	−	+
Propionate	−	−	−	+
Pyruvate	+	+	+	+ (with external electron acceptor)
Growth requirement(s)	−	elemental sulfur	−	−
Growth products	H_2S, CO_2	H_2S, CO_2	H_2S, CO_2	CO_2
Thiosulfate used as an electron acceptor	−	−	−	+
Temperature range, °C	44–70	40–70	42–77	33–63
Optimal temperature, °C	52–55	54	58–60	55
Optimal pH	6.4–6.8	6.9–7.2	6.4–6.8	6.9–7.2
Mol% G + C of the DNA	31.4	31.6	33.5	32.2

[a]For symbols, see standard definitions.

[b]All species consist of short rods that can live by anaerobic respiration with elemental sulfur.

[c]All species can use acetate, stearate, palmitate, and H_2 as growth substrates.

DIFFERENTIATION OF THE GENUS *DESULFURELLA* FROM OTHER GENERA

Desulfurella can be differentiated from *Hippea* based on its lack of a requirement for NaCl and yeast extract. Moreover, *Desulfurella* has an optimal temperature of 55–60°C, whereas *Hippea* has an optimal temperature of 52–54°C. Also, the mol% G + C content of the DNA of *Desulfurella* is 31–34 , whereas the value for *Hippea* is 40.

TAXONOMIC COMMENTS

A phylogenetic analysis based on 16S rDNA sequence comparisons showed *Desulfurella acetivorans* to be equidistant to members of the four then-recognized subclasses of the *Proteobacteria* (Rainey et al., 1993). The highest levels of 16S rDNA sequence similarity were found between *Desulfurella* and sulfate-reducing members of the *Deltaproteobacteria*, but the similarity levels were less than 84%. With the isolation and description of *Hippea maritima* (Miroshnichenko et al., 1999), an additional member was added to this distinct lineage within the *Proteobacteria*. The 16S rRNA gene sequence determined for *Desulfurella* shares 88% sequence similarity with that of *Hippea maritima*.

List of species of the genus Desulfurella

1. **Desulfurella acetivorans** Bonch-Osmolovskaya, Sokolova, Kostrikina and Zavarzin 1993, 624[VP] emend. Miroshnichenko, Gongadze, Lysenko and Bonch-Osmolovskaya 1994, 92 (Effective publication: Bonch-Osmolovskaya, Sokolova, Kostrikina and Zavarzin 1990, 155.)

   a.ce.ti.vo′rans. L. n. m. *acetat* acetate; L. adj. fem. *vorans* eating; L. adj. fem. *acetivorans* acetate-eating.

   The characteristics are as given for the genus and as listed in Table BXII.δ.1. Cells occur singly or in pairs. Motile by a single polar flagellum. White, irregular colonies 1 mm in diameter develop on agar containing polysulfide. Acetate, pyruvate, pentadecanate, palmitate, and stearate are used for growth. Growth is stimulated by yeast extract. Growth by-products include H_2S and CO_2. Cells are obligately sulfur-respiring. The temperature range for growth is 44–70°C; optimal, 52–55°C. The pH range for growth is 4.3–7.5; optimal, 6.4–6.8.

   The mol% G + C of the DNA is: 31.4 (T_m).
   Type strain: A-63, ATCC 51451, DSM 5264.
   GenBank accession number (16S rRNA): X72768.

2. **Desulfurella kamchatkensis** Miroshnichenko, Rainey, Hippe, Chernyh, Kostrikina and Bonch-Osmolovskaya 1998, 478[VP]

   kam.chat.ken′sis. M.L. adj. *kamchatkensis* pertaining to the Kamchatka Peninsula in the Russian Far East.

   The characteristics are as given for the genus and as listed in Table BXII.δ.1. Cells occur singly or in pairs. Obligate anaerobe. Motile by a single polar flagellum. H_2/CO_2, acetate, pyruvate, lactate, fumarate, malate, palmitate, and stearate are used for growth. Glucose, sucrose, starch, peptone, yeast extract, formate, propionate, benzoate, hexadecane, succinate, ethanol, and methanol are not used for growth. Elemental sulfur is required for growth. Yeast extract is not required for growth. Growth products include H_2S and CO_2. The temperature range for growth is 40–70°C; optimal, 54°C. The optimal pH range for growth is 6.9–7.2.

   The mol% G + C of the DNA is: 31.6 (T_m).
   Type strain: K-119, ATCC 700655, DSM 10409.
   GenBank accession number (16S rRNA): Y16941.

3. **Desulfurella multipotens** Miroshnichenko, Gongadze, Lysenko and Bonch-Osmolovskaya1996, 625[VP] (Effective publication: Miroshnichenko, Gongadze, Lysenko and Bonch-Osmolovskaya 1994, 92.)

   mul.ti.po′ tens, L. adj. *multum* many, L. part. *potens* able, *multipotens* having many abilities.

   The characteristics are as given for the genus and as listed in Table BXII.δ.1. Cells occur singly or in pairs. Motile by a single polar flagellum. Colonies on agar with polysulfide are white, irregular, and measure 1 mm in diameter. H_2/CO_2, acetate, pyruvate, butyrate, and saturated acids are used for growth. Formate, propionate, lactate, isobutyrate, caproate, alcohols, carbohydrates, and peptides are not used for growth. Cell growth produces H_2 and CO_2. Yeast extract is not required for growth. Cells are lithoautotrophic when grown on mineral medium containing H_2, CO_2, and S^0. Obligately sulfur-respiring. The temperature range for growth is 42–77°C; optimal, 58–60°C. The pH range for growth is 6.0–7.2; optimal, 6.4–6.8.

   The mol% G + C of the DNA is: 33.5 (T_m).
   Type strain: RH-8, DSM 8415.
   GenBank accession number (16S rRNA): Y16943.

4. **Desulfurella propionica** Miroshnichenko, Rainey, Hippe, Chernyh, Kostrikina and Bonch-Osmolovskaya 1998, 479[VP]

   pro.pi.o′ ni.ca. N.L. fem. adj. *propionica* propionate-utilizing.

   The characteristics are as given for the genus and as listed in Table BXII.δ.1. Cells occur singly or in pairs. Nonmotile. Cells have multiple pilus-like structures. H_2/CO_2, acetate, propionate, pyruvate, lactate, fumarate, malate, palmitate, and stearate are used for growth. Glucose, sucrose, starch, peptone, yeast extract, formate, benzoate, hexadecane, succinate, ethanol, and methanol are not used for growth. All growth substrates are oxidized to CO_2. Growth on pyruvate occurs in the absence of sulfur. Elemental sulfur and thiosulfate are reduced to H_2S. The temperature range for growth is 33–63°C; optimal, 55°C. The optimal pH range for growth is 6.9–7.2.

   The mol% G + C of the DNA is: 32.2 (T_m).
   Type strain: U-8, ATCC 700656, DSM 10410.
   GenBank accession number (16S rRNA): Y16942.

Genus II. **Hippea** *Miroshnichenko, Rainey, Rhode and Bonch-Osmolovskaya 1999, 1037*[VP]

THE EDITORIAL BOARD

Hip'pea. L. fem. n. *Hippea* named after Hans Hippe, a German microbiologist, in recognition of his significant contribution to the characterization of new, obligately anaerobic procaryotes and the understanding of their physiology.

Obligately anaerobic and **moderately thermophilic** Gram-negative rods. **Grow lithotrophically using sulfur and hydrogen gas.** Able to oxidize fatty acids and alcohols. Habitat is marine hot vents.

The mol% G + C of the DNA is: 40.

Type species: **Hippea maritima** Miroshnichenko, Rainey, Rhode and Bonch-Osmolovskaya 1999, 1037.

FURTHER DESCRIPTIVE INFORMATION

The three *Hippea maritima* strains studied were isolated from marine hot vent environments that have high levels of both elemental sulfur and organic matter. Like *Desulfurella* spp., which inhabit similar terrestrial environments, *Hippea maritima* is a moderate thermophile that can grow lithotrophically on H_2 and sulfur as well as coupling sulfur reduction to the oxidation of organic compounds.

The temperature range for growth was 40–65°C; the pH range was 5.7–6.5.

Yeast extract (0.02% w/v), NaCl (2.0–3.0% w/v), and elemental sulfur were required. The organisms could not use cystine, ferric iron, fumarate, sulfate, sulfite, or thiosulfate as electron acceptors. Substrates used by all three strains studied included acetate, hydrogen gas, pyruvate and saturated fatty acids; one strain could also use ethanol. Substrates that could not be used by any strain included glucose, butyrate, formate, hexadecane, lactate, pyruvate, succinate, methanol, starch, or peptone. All three strains produced only CO_2 and H_2S as growth products on all substrates (Miroshnichenko et al., 1999).

Analysis of 16S rDNA sequences showed that the three *Hippea maritima* strains were most closely related to the four described species of the genus *Desulfurella* (Miroshnichenko et al., 1999).

ENRICHMENT AND ISOLATION PROCEDURES

The three *Hippea maritima* strains were isolated from marine sands collected from hot vents in the tidal zone of Matupi Harbour, Papua New Guinea, and at 40 m depth in the Bay of Plenty, New Zealand. The medium contained 10 g/l elemental sulfur (as sulfur flowers), 25 g/l NaCl, 5 g sodium acetate, 0.1 g/l yeast extract, vitamins, and trace elements (Miroshnichenko et al., 1999). Samples were incubated with this medium in Hungate tubes under an atmosphere of N_2 and CO_2 (4:1 v/v) at 55°C. Isolated colonies were obtained by serial dilution of the liquid enrichment cultures in an agar medium modified from the enrichment medium by the addition of agar and the substitution of polysulfide for the sulfur flowers. One isolate was obtained from an enrichment in which 0.5% ethanol was used in place of the sodium acetate (Miroshnichenko et al., 1999).

DIFFERENTIATION OF THE GENUS *HIPPEA* FROM OTHER GENERA

Hippea maritima strains differ from *Desulfurella* strains in having slightly lower optimal temperature and pH ranges for growth as well as in requiring 2–3% NaCl and yeast extract for growth. The three *Hippea maritima* strains examined by Miroshnichenko et al. (1999), showed optimal growth at 52–54°C and pH 5.8–6.0; the corresponding values for *Desulfurella* spp. are 55–60°C and pH 6.8–7.0.

List of species of the genus Hippea

1. **Hippea maritima** Miroshnichenko, Rainey, Rhode and Bonch-Osmolovskaya 1999, 1037[VP]

   ma.ri' ti.ma. L. fem. adj. *maritima* inhabiting marine environments.

   Short, obligately anaerobic rods (0.4–0.8 × 1.0–3.0 µm). Single polar flagellum. Grow at 40–65°C (optimal 53–54°C) and pH 5.4–6.5 (optimal 6.0). Substrates for growth include acetate, ethanol, hydrogen gas, pyruvate and saturated fatty acids. Require elemental sulfur, 2.5–3.0% NaCl, and 0.02% yeast extract. Do not use cystine, ferric iron, fumarate, sulfate, sulfite, or thiosulfate as electron acceptors. Sole growth products CO_2 and H_2S.

   The mol% G + C of the DNA is: 40.

   Type strain: MH2, DSM 10411.

   GenBank accession number (16S rRNA): Y18292.

Order II. **Desulfovibrionales** *ord. nov.*

JAN KUEVER, FRED A. RAINEY AND FRIEDRICH WIDDEL

De.sul.fo.vi.bri.o.na' les. M.L. masc. n. *Desulfovibrio* type genus of the type family of the order; suff. *ales* denotes order; M.L. fem. pl. n. *Desulfovibrionales* the order that comprises *Desulfovibrionaceae* and other families.

Cells are curved or rod shaped and often motile. **Strictly anaerobic chemoorganotrophic or chemolithoheterotrophic** (H_2, acetate, sulfate) **growth by respiratory metabolism**; in addition, fermentative abilities are found in several members. **The common electron acceptor is sulfate, which is reduced to sulfide.** Sulfite or thiosulfate may be used in addition; some members may also reduce elemental sulfur (or polysulfide) to sulfide or nitrate to ammonia. **Members of the genera *Bilophila* and *Lawsonia* are unable to use sulfate as an electron acceptor.** Simple organic compounds serve as electron donors and carbon sources. Oxidation is usually incomplete, leading to formation of acetate; the capacity for complete oxidation has been observed so far only in *Desulfothermus naphthae*. Growth by disproportionation of sulfur, thiosulfate, and sulfite may occur. Most members are mesophilic; in addition, some moderate thermophiles with temperature optima between 50 and 60°C have been isolated. Cells contain various cytochromes, other redox proteins, and menaquinones. Mesophilic species have been isolated from almost every type of aquatic environment, including marine habitats, freshwater habitats, technical water systems, and digestive tracts; few

clinical isolates without obvious pathogenic properties have been reported. Thermophilic members have been isolated from geothermally heated marine environments. Whereas most members grow preferentially under neutral conditions, some alkaliphilic isolates are known.

The order comprises the families *Desulfovibrionaceae*, *Desulfomicrobiaceae*, *Desulfohalobiaceae*, and *Desulfonatronumaceae* (see Fig. BXII.δ.1).

Type genus: **Desulfovibrio** Kluyver and van Niel 1936, 397.

Family I. **Desulfovibrionaceae** *fam. nov.*

Jan Kuever, Fred A. Rainey and Friedrich Widdel

De.sul.fo.vi.bri.o.na.ce′ae. M.L. masc. n. *Desulfovibrio* type genus of the family; suffix *-ceae* denotes family; M.L. fem. pl. n. *Desulfovibrionaceae* the family that comprises the genus *Desulfovibrio*.

The usual electron acceptor for the genera *Desulfovibrio* and *Desulfonatronum* is sulfate, which is reduced to sulfide. Members of the genera *Bilophila* and *Lawsonia* are unable to use sulfate as electron acceptor. All known members are mesophilic bacteria. Organic substrates are incompletely oxidized to acetate.

The family comprises three described genera, *Desulfovibrio*, *Bilophila*, and *Lawsonia*. Detailed properties of described species of these genera are given in Table BXII.δ.2.

Type genus: **Desulfovibrio** Kluyver and van Niel 1936, 397.

Genus I. **Desulfovibrio** *Kluyver and van Niel 1936, 397*[AL]

Jan Kuever, Fred A. Rainey and Friedrich Widdel

De.sul.fo.vi′bri.o. L. pref. *de* from; L. n. *sulfur* sulfur; L. v. *vibrio* to vibrate; M.L. masc. n. *Vibrio* that which vibrates, a generic name; M.L. n. *Desulfovibrio* a vibrio that reduces sulfur compounds.

Curved or occasionally straight **rods**, sometimes **sigmoid or spirilloid, 0.5–1.5 × 2.5–10.0 μm**. The morphology is influenced by age and environment; descriptions refer to freshly grown cultures in anoxic sulfate media. **Spore formation is absent.** Gram negative. **Motile** by means of a single or lophotrichous polar flagella. Obligately anaerobic growth in pure cultures. **Possess mainly a respiratory type of metabolism with sulfate or other sulfur compounds as the terminal electron acceptors, being reduced to H₂S**; however, the metabolism is sometimes fermentative. Media containing a reducing agent are required for growth. In a few cases, a vitamin requirement has been reported. Some species and subspecies are moderately halophilic. Optimal growth temperature, usually 25–35°C; upper limit normally 44°C. No thermophilic species have been reported. Thermophilic *Desulfovibrio* species formerly described have been reclassified and currently belong to the genera *Thermodesulfovibrio* and *Thermodesulfobacterium*. Chemoorganotrophic. **Most species oxidize organic compounds such as lactate incompletely to acetate, which cannot be oxidized further.** Carbohydrates are utilized by few species. One species, *D. inopinatus*, can use hydroquinone as electron donor and carbon source for growth. Cells contain *c*-type cytochromes (such as *c₃*) and usually *b*-type cytochromes. All members of the genus *Desulfovibrio* contain desulfoviridin. Hydrogenase is usually present. Strains of some species may show chemolithoheterotrophic growth, using H₂ as electron donor and assimilating acetate and CO₂, or yeast extract, as carbon sources. Gelatin is not liquefied. Nitrate is sometimes reduced to ammonia. Some species can reduce oxygen or metal ions, but growth has never been observed with these electron acceptors in pure cultures. Molecular nitrogen is sometimes fixed. Species generally show some degree of antigenic cross reaction. Habitats:

anoxic mud of fresh and brackish water and marine environments; intestines of animals; manure and feces.

The mol% G + C of the DNA is: 46.1–61.2.

Type species: **Desulfovibrio desulfuricans** (Beijerinck 1895) Kluyver and van Niel 1936, 397 (*Spirillum desulfuricans* Beijerinck 1895, 113.)

FURTHER DESCRIPTIVE INFORMATION

Young (i.e., early stationary phase) cultures are morphologically rather homogenous, though morphology is somewhat influenced by the composition of the culture medium. Sigmoid or spirilloid forms are rare, but the species *D. africanus* has a tendency to an elegant sigmoid form and *D. salexigens* tends to be stubby. Old, stressed, or very sulfide-rich cultures may show aberrant cell forms. Rod-shaped strains have been reported but are a minority. Internal structure is rarely visible except in the large species *D. gigas*, which often possesses refractile zones.

Desulfovibrio species usually exhibit a random, rapid, progressive motility. Motility may be suppressed at high sulfide concentrations. Flagella are always polar and are lophotrichous in certain species.

All described *Desulfovibrio* strains contain desulfoviridin as sulfite reductase, and MK-6 or MK-6H₂ as the major menaquinone (Collins and Widdel, 1986). The major cellular fatty acids of *Desulfovibrio* strains are $C_{15:0 iso}$, $C_{17:0 iso}$ and $C_{17:1 iso \omega9c}$ (Vainshtein et al., 1992; Kohring et al., 1994).

Desulfovibrio species have been investigated in much detail with respect to enzymes (e.g., hydrogenases, enzymes of the sulfur metabolism), other redox proteins (e.g., cytochromes *c₃*), and genes encoding these proteins. In addition, hypothetical models of energy conservation in sulfate-reducing bacteria were mostly

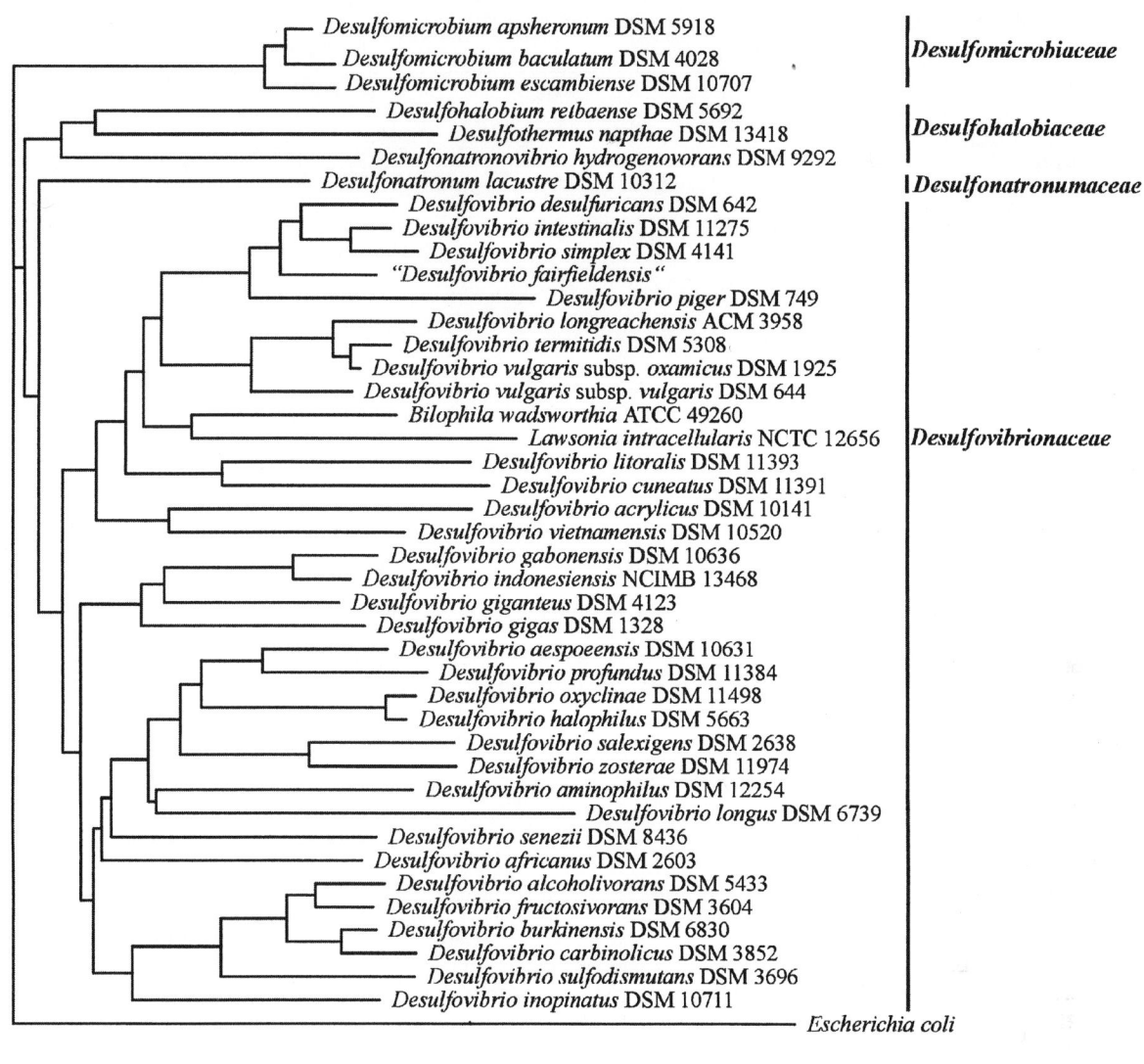

FIGURE BXII.δ.1. Phylogenetic relationships based on 16S rRNA gene sequence analyses of the taxa within the order *Desulfovibrionales* for which 16S rRNA gene sequences are available. The scale bar represents 10 inferred nucleotide substitutions per 100 nucleotides. The dendrogram was reconstructed from distance matrices using the neighbor-joining method.

developed from studies with *Desulfovibrio* species (for overview see Odom and Singleton, 1993; Voordouw, 1995; Rabus et al., 2000a).

Although *Desulfovibrio* spp. are not usually considered pathogenic, there are several reports concerning the isolation of *Desulfovibrio* spp. or closely related organisms from humans and animals. Some strains have been isolated from human patients showing clinical symptoms (Postgate, 1984b, a; Gebhart et al., 1993; McOrist et al., 1995a; Tee et al., 1996; Cooper et al., 1997; McDougall et al., 1997; La Scola and Raoult, 1999c; Loubinoux et al., 2000). An intracellular isolate from intestines of pigs with proliferative enteropathy disease was described as *Lawsonia intracellularis*; it shares 91% 16S rRNA gene sequence similarity with *D. desulfuricans* (ATCC 27774) (McOrist et al., 1995a). It therefore remains unclear whether certain close relatives of *Desulfovibrio* strains are potentially pathogenic to man, or whether they are opportunistic.

Desulfovibrio species are robust sulfate-reducing bacteria with relatively simple growth demands. They may grow rapidly (often

with doubling times of only 3–5 h) in various media if oxygen is excluded and sulfate and a suitable electron donor and carbon source are present. Either lactate or a mixture of H_2 and CO_2 in the presence of acetate (as additional carbon source for cell synthesis) is a highly suitable growth substrate. Two different media are commonly used. One is the traditional yeast-extract-containing high-iron medium and its modifications[1] developed

---

1. The most commonly used of Postgate's media (often referred to as medium B) has the following composition (g/l of tap water): KH_2PO_4, 0.5; NH_4Cl, 1; $CaSO_4$, 1; $MgSO_4 \cdot 7H_2O$, 2; sodium lactate, 3.5; yeast extract, 1; ascorbic acid, 0.1; thioglycollic acid (fresh), 0.1; $FeSO_4 \cdot 7H_2O$, 0.5. pH is adjusted between 7 and 7.5. For marine strains, addition of 25 g/l NaCl is recommended. Agar may be used up to 10 g/l for gelling. One modification is a low-iron medium (0.004 g/l $FeSO_4 \cdot 7H_2O$) with or without sodium citrate dihydrate (0.3 g/l). For details see Postgate (1984a). The reducing capacity of thioglycollic acid and ascorbic acid is best maintained if these are not directly dissolved and autoclaved in the medium, but rather dissolved separately in 10 ml water, filter sterilized, and added to the medium after autoclaving. Also, preparation of medium and solutions under nitrogen (instead of air) favors maintenance of anoxic conditions.

TABLE BXII.δ.2. Characteristics of the genera of the family *Desulfovibrionaceae*[a]

Genus	*Desulfovibrio*	*Bilophila*	*Lawsonia*
Morphology:			
Rod-shaped	+	+	−
Vibrio-shaped	+	−	+
Motility	+[b]	−	−
Mol% G + C content	47–60	39–40	nr
Sulfite reductase	DV	DV[c]	nr
Major menaquinone	MK-6 or MK-6(H$_2$)	nr	nr
Substrate oxidation	I	I	I
CO-DH[d]	−	nr	nr
Typical electron donors and carbon sources:			
H$_2$/acetate	+		
Alcohols	±	+	
Formate + acetate	±	+	−
Lactate	+	+	−
Pyruvate	+	+	−
Complex media	±	+	+
Unknown			+
Chemolithoautotrophic growth	−	−	−
Fermentative growth	±	+	nr
Growth by disproportionation of reduced sulfur compounds	±	nr	nr
Typical electron acceptors: [e]			
Cysteate	−	+	
Dimethylsulfoxide	±	+	
Laurine	−	+	
Nitrate	±	+	
Sulfate	+	−	
Sulfite	+	+	
Thiosulfate	+	+	
Optimal growth temperature (°C)	30–37	35	30–37
NaCl requirement	±	−	−
Habitat (anoxic)	Freshwater and marine	Freshwater or clinical isolates	Obligate intracellular parasite of intestinal cells of pigs

[a]Symbols: nr, not reported; +, present, observed in all species; −, not present, not observed; ±, variable; I, incomplete; C, complete; DV, desulfoviridin; DR, desulforubidin.

[b]Most species.

[c]Molecular evidence (Laue et al., 2001).

[d]Carbon monoxide dehydrogenase, a key enzyme indicative of the C$_1$-pathway for acetyl-CoA oxidation.

[e]*Lawsonia* exhibits no sulfate reduction.

by Postgate (1984a). The other medium is the defined bicarbonate-buffered medium used for *Desulfobacter* species and other sulfate-reducing bacteria, as described in later sections of this chapter (vitamins may be omitted for *Desulfovibrio* species). Reductants are usually included in the media, but fresh cells of *Desulfovibrio* may often grow upon transfer if oxygen is removed only by physical means (e.g., sparging with N$_2$), without any chemical reduction. In the case of brackish water or marine strains, concentrations of NaCl and MgCl$_2$ should be added to match roughly the salinity of the original habitat. The medium of Postgate usually contains high concentrations of ferrous iron, forming a whitish precipitate. It reacts with sulfide to produce intensely black ferrous sulfide, thus acting as an indicator of metabolic activity and growth.

ENRICHMENT AND ISOLATION PROCEDURES

Enrichment of *Desulfovibrio* species is usually straightforward, and highly enriched and even pure cultures can be obtained within 3–4 weeks. Sulfate media with 1–3.5 g sodium lactate, or alternatively 0.1 g sodium acetate trihydrate, per liter medium and

with 80–95% H$_2$ and 5–20% CO$_2$ in a head space above the medium (head space–medium ratio approx. 1:1) usually allow selective, rapid growth. In addition, ethanol (0.05%, v/v; presence of CO$_2$ or bicarbonate needed) or other substrates described for *Desulfovibrio* species (see species descriptions) may be used. In some instances, use of sodium lactate may lead to the development of fermentative bacteria that do not form sulfide. Yeast extract and ascorbate may be omitted since they favor development of various fermentative bacteria that may interfere with isolation procedures. If iron-rich medium (Postgate's medium B) is used, development of sulfate-reducing bacteria is recognized by blackening of the precipitate (formation of FeS). The salinity of the medium should be roughly adjusted to that of the material used as inoculum. For marine cultures, seawater may also be used for medium preparation.

Screw-capped bottles (50 or 100 ml) or culture tubes (20 ml) are satisfactory containers for enrichment cultures. Screw-capped bottles or tubes are completely filled to maintain anoxic conditions. Alternatively, tubes or bottles with anoxic head space (obligatory in the case of H$_2$ as electron donor) sealed with butyl rubber may be used. A detailed description of anoxic techniques is given by Widdel and Bak (1992). Enrichment may be achieved in batch cultures, with 1–5% (v/v) sample (mud, sediment) added to the medium, or in liquid dilution series (subsequent dilution steps of 1:10), as used for determination of the most probable number (so-called MPN method). Whereas the former method selects for the most rapidly growing strains, the latter method usually reveals the most abundant type of sulfate-reducing bacterium, which is not necessarily a *Desulfovibrio* species, in a sample.

Sediment-free enrichment cultures are obtained by subsequent transfer of inocula (2–10% of culture volume) to fresh media. Pure cultures are obtained by repeated application of the agar dilution method. A solution of agar (33 g/l) is prepared in distilled water; the agar should be washed several times in cold distilled water before preparing the solution. The agar solution is dispensed in 3-ml amounts into test tubes, which are stoppered with cotton or aluminum caps and autoclaved. Agar tubes may be stored for 2–4 months. The agar tubes are kept molten (or remelted at 100°C) and placed in a water bath at 55°C. Defined culture medium is prewarmed to 41°C and 6-ml amounts are added to the tubes of liquefied agar; exposure to air is minimized by dipping the tip of the pipette into the agar medium. To compensate for the diluting effect of the aqueous agar, it is recommended to increase the amounts of NaCl and $MgCl_2 \cdot 6H_2O$ to 30 and 4.5 g, respectively, per liter of medium (e.g., by addition of sterile salt concentrate) before the medium is added to the agar. Addition of salts to the agar before autoclaving is not recommended as an alternative, because this may promote hydrolysis of the agar. Starting with 0.1 to 0.5 ml of an enrichment culture as inoculum, serial dilutions of approximately 1:10 are made using 6–8 tubes. In general, for isolation of *Desulfovibrio* species addition of dithionite is not required. All agar tubes are then hardened in cold water and immediately sealed with butyl rubber stoppers under a mixture of N_2/CO_2 (90:10). The tubes are incubated in the dark for 2–4 weeks at 25–35°C. Well-separated colonies that occur in the higher dilutions can be removed with sterile Pasteur pipettes (thinned upon heating in a flame); the cells are used as inoculum for subsequent agar dilution cultures. The process is repeated until the cultures are pure, as indicated by homogenous colonies and homogenous cells.

Purity of cultures is checked by microscopy of growth in various media that may allow growth of non-sulfate-reducing bacteria. Contaminant aerobes are excluded by plating out on any nutrient agar containing glucose, yeast extract, and peptone, and incubation under air. No colonies should appear. Contaminant anaerobes are excluded by inoculation of parallel sets of media containing glucose (0.5 g/l) and sodium pyruvate (1 g/l) with or without yeast extract (1 g/l; routinely present in Postgate's medium). Glucose supports growth of only a few *Desulfovibrio* species, but growth of many non-sulfate-reducing anaerobes. Pyruvate is a common growth substrate for *Desulfovibrio*. Microscopy should always reveal the same cell type.

MAINTENANCE PROCEDURES

Pure cultures are best maintained in a medium that is rich in FeS precipitate. Cultures are refrigerated before, or at the beginning of, the stationary phase. Transfers are made every 4–6 months. Lyophilization in the presence of skim milk is used by culture collections for preservation of *Desulfovibrio* spp.

DIFFERENTIATION OF THE GENUS *DESULFOVIBRIO* FROM OTHER GENERA

A curved cell morphology and progressive motility often provide a provisional identification of *Desulfovibrio* species. Members of the genus do not have heat-resistant forms and sensitivity to heating for 5 min at 100°C distinguishes them unequivocally from *Desulfotomaculum*. All members of the genus *Desulfovibrio* give a positive reaction to the desulfoviridin test: a characteristic red fluorescence (due to the sirohydrochlorin chromophore of that pigment) appears when a cell suspension is inspected in light of wavelength 365 nm immediately after addition of a few drops of 2.0 N NaOH. This test and morphological characters such as motility and vibrioid form distinguish most strains from *Desulfobacter*, *Desulfococcus*, *Desulfosarcina*, and *Desulfobulbus*. However, a clear differentiation requires a comparative 16S rDNA sequence analysis.

Characteristics useful in differentiating the species of *Desulfovibrio* are listed in Table BXII.δ.3.

TAXONOMIC COMMENTS

The classification of *Desulfovibrio* species given in the present edition of the *Manual* is based on comparative analysis of the 16S rDNA sequences. The closest relatives to the genus *Desulfovibrio* are members of the genera *Desulfomicrobium*, *Bilophila*, and *Lawsonia*. A set of 16S rRNA oligonucleotide probes which allows the detection of most described *Desulfovibrio* strains was developed by Manz et al. (1998). The signature sequences for nearly all described *Desulfovibrio* species and other closely related sulfate-reducing bacteria are 5′-CCCGATCGTCTGGGCAGG-3′ (oligonucleotide probe DSD131), 5′-CCGAAGGCCTTCTTCCCT-3′ (oligonucleotide probe DSV407), 5′-GTTCCTCCAGATATCT-ACGG-3′ (oligonucleotide probe DSV698), and 5′-CAATCCGGA-CTGGGACGC-3′ (oligonucleotide probe DSV1292), which bind to the 16S rRNA corresponding to positions 5′-131–148-3′, 5′-407–424-3′, 5′-698–717-3′, and 5′-1292–1309-3′ of the *Escherichia coli* 16S rRNA sequence, respectively (Manz et al., 1998).

FURTHER READING

Postgate, J.R. 1984. The Sulfate-Reducing Bacteria, 2nd ed., Cambridge University Press, Cambridge; New York.

List of species of the genus Desulfovibrio

1. **Desulfovibrio desulfuricans** (Beijerinck 1895) Kluyver and van Niel 1936, 397^{AL} (*Spirillum desulfuricans* Beijerinck 1895, 113.)

   de.sul.fu′ri.cans. L. pref. *de* from; L. n. *sulfur* sulfur; M.L. part. adj. *desulfuricans* reducing sulfur compounds.

   The morphology is as described in Table BXII.δ.3 and as depicted in Fig. BXII.δ.2. Sigmoid forms may occur. Other characteristics are listed in Table BXII.δ.3. Sulfite or thiosulfate can be used as electron acceptors instead of sulfate. Organic substrates are lactate, pyruvate, formate, choline, or certain simple primary alcohols, including ethanol, 1-propanol, and 1-butanol. The latter two serve merely as hydrogen donors and are oxidized to propionate and butyrate, respectively (acetate is required as a carbon source). Nitrate can be used as an electron acceptor and is reduced to ammonia (Seitz and Cypionka, 1986). Oxygen,

TABLE BXII.δ.3. Morphological and physiological characteristics of described species of the genus *Desulfovibrio*[a]

Characteristics	*D. desulfuricans*[b]	*D. acrylicus*[c]	*D. aespoeensis*[d]	*D. africanus*[e]	*D. alcoholovorans*[f]	*D. aminophilus*[g]	*D. burkinensis*[h]	*D. carbinolicus*[i]	*D. cuneatus*[j]	*D. fructosovorans*[k]	*D. furfuralis*[l]	*D. gabonensis*[m]	*D. giganteus*[n]	*D. gigas*[o]
Cell morphology	Vibrio	Vibrio	Vibrio	Vibrio	Vibrio	Vibrio	Curved rod	Rod	Curved rod	Vibrio	Vibrio	Rod	Vibrio or rod	Large vibrio
Cell size (μm)	0.5–0.8 × 1.5–4.0	0.8 × 2–4	0.5 × 1.7–2.5	0.5–0.6 × 2–3	0.7–0.9 × 2.8–3.2	0.2 × 3.0–4.0	0.8–1.2 × 2.2–3.1	0.6–1.1 × 1.5–5	0.4–0.6 × 1.6–3.0	0.5–0.7 × 2–4	0.3–1.2 × 0.8–3	0.4 × 2–4	1 × 5–10	0.8–1 × 6–11
Motility	+ (sp)	+	+	+ (lo)	+ (sp)	+ (sp)	+ (sp)	–	+ (sp)	+ (sp)	+ (sp)	+ (sp)	+ (sp)	+ (lo)
G + C content (mol%)	59	45.1	61	65	64.5	66	67	65	52.7	64	61	59.5	56	65
Sulfite reductase	DV	DV	DV	DV	DV	DV	DV	DV	DV	DV	DV	DV	DV	DV
Major menaquinone	MK-6	nr	nr	MK-6(H₂)	nr	nr	nr	nr	nr	nr	nr	nr	nr	MK-6
Optimal growth temperature (°C)	30–36	30–37	25–30	30–36	35–37	35	37	37–38	28	35	38	30	35	30–36
Oxidation of substrate	I	I	I	I	I	I	I	I	I	I	I	I	I	I
Electron donors used:														
H₂	+	+	+	+	+	+	+	+	+	+	nr	+	+	–
Formate	+	+	–	+	+	+	+	+	+	+	nr	(+)	–	+
Acetate	–	–	–	–	–	–	–	+	nr	–	–	–	–	+
Fatty acids	+	–	–	–	–	–	–	–	–	–	–	–	–	–
Ethanol	+	+	–	+	+	+	+	+	–	(+)	+	+	+	(+)
Lactate	+	+	+	+	+	+	+	+	+	+	+	+	+	+
Pyruvate	+	+	+	+	+	+	+	+	+	+	+	+	+	+
Fumarate	+	+	+	–	+	–	+	+	+	+	–	+	–	+
Succinate	–	–	nr	–	+	nr	+	+	–	–	nr	+	–	+
Malate	+	+	–	+	+	–	+	+	–	+	–	–	–	–
Benzoate	–	nr	–	nr	nr	–	nr	–	nr	nr	nr	–	–	nr
Fermentative growth on:														
Fumarate	+	nr	–	–	+	–	+	+	+	+	nr	+	–	+
Malate	+	nr	–	–	+	–	+	+	+	+	–	+	–	+
Pyruvate	+	nr	+	–	+	+	+	+	+	+	+	+	+	+
Electron acceptors used:														
Sulfate	+	+	+	+	+	+	+	+	+	+	+	+	+	+
Sulfite	+	–	nr	nr	+	+	+	+	+	+	+	+	+	+
Sulfur	–	–	+	nr	+	–	+	+	+	+	nr	+	–	+
Thiosulfate	+	+	+	nr	+	+	+	+	+	+	+	+	+	+
Fumarate	+	nr	–	–	nr	–	+	+	+	–	+	+	–	–
Nitrate	+	–	–	–	–	–	–	–	–	–	+	–	–	–
Growth factor requirement	–	Yeast extract	nr	–	–	–	Vitamins or yeast extract	Yeast extract	–	–	nr	vitamins	bi	bi

(continued)

TABLE BXII.δ.3. *(cont.)*

Characteristics	D. desulfuricans[b]	D. acrylicus[c]	D. aespoeensis[d]	D. africanus[e]	D. alcoholovorans[f]	D. aminophilus[g]	D. burkinensis[h]	D. carbinolicus[i]	D. cuneatus[j]	D. fructosovorans[k]	D. furfuralis[l]	D. gabonensis[m]	D. giganteus[n]	D. gigas[o]
NaCl requirement (g/l)	–	nr	–	–		–	–	–	–	–	–	10	2	–

[a]Symbols: nr, not reported; +, present, observed in all species; –, not present, not observed; ±, variable; (+), poorly motile, poor growth; (sp), flagellation type: single, polar; (lo), flagellation type: lophotrichous; DV, desulfoviridin; DR, desulforubidin; I, incomplete; C, complete; bi, biotin; ni, nicotinate; pa, *p*-aminobenzoate; pt, pantothenate; th, thiamine.

[b]Data from Postgate, 1984a, b; Brauman et al., 1990; Devereux et al., 1990; Dilling and Cypionka, 1990; Lovley and Phillips, 1992; Lovley et al., 1993b.

[c]Data from van der Maarel et al., 1996.

[d]Data from Motamedi and Pedersen, 1998.

[e]Data from Postgate, 1984a, b.

[f]Data from Qatibi et al., 1991.

[g]Data from Baena et al. 1998.

[h]Data from Ouattara et al., 1999.

[i]Data from Nanninga and Gottschal, 1987.

[j]Data from Sass et al., 1998a.

[k]Data from Ollivier et al., 1988.

[l]Data from Folkerts et al., 1989a.

[m]Data from Tardy-Jacquenod et al., 1996.

[n]Data from Esnault et al., 1988a; Brauman et al., 1990.

[o]Data from Postgate, 1984a, b; Esnault et al., 1988a.

[p]Data from Caumette et al., 1991a.

[q]Data from Feio et al., 1998.

[r]Data from Reichenbecher and Schink, 1997.

[s]Data from Fröhlich et al., 1999a.

[t]Data from Sass et al., 1998a.

[u]Data from Redburn and Patel, 1994.

[v]Data from Magot et al., 1992.

[w]Data from Krekeler et al., 1997.

[x]Data from Moore et al., 1976; Biebl and Pfennig, 1977; Loubinoux et al., 2002.

[y]Data from Bale et al., 1997.

[z]Data from Postgate 1984a, b; Zellner et al., 1989; Bale et al., 1997.

[aa]Data from Tsu et al., 1998.

[bb]Data from Zellner et al., 1989.

[cc]Data from Bak and Pfennig, 1987.

[dd]Data from Trinkerl et al., 1990.

[ee]Data from Nga et al., 1996.

[ff]Data from Postgate 1984a, b; Lovley et al., 1993b; Lovley and Phillips, 1994b.

[gg]Data from Nielsen et al., 1999.

(continued)

TABLE BXII.δ.3. (cont.)

Characteristics	D. halophilus[p]	D. indonesiensis[q]	D. inopinatus[r]	D. intestinalis[s]	D. litoralis[t]	D. longreachensis[u]	D. longus[v]	D. oxyclinae[w]	D. piger[x]	D. profundus[y]	D. salexigens[z]	D. senezii[aa]	D. simplex[bb]	D. sulfodismutans[cc]	D. termitidis[dd]	D. vietnamensis[ee]	D. vulgaris subsp. vulgaris[ff]	D. zosterae[gg]
Cell morphology	Vibrio	Rod	Vibrio	Vibrio	Curved Rod	Vibrio	Rod	Curved rod	Rod	Vibrio	Vibrio	Vibrio	Vibrio	Vibrio	Vibrio	Vibrio	Vibrio	Curved rod
Cell size (μm)	0.6 × 2.5–5	0.5–1.2 × 3–5	1–1.5 × 4–12	0.4–0.5 × 1–1.4	0.4–0.6 × 1.8–2.7	0.5 × 2.0–4.0	0.4–0.5 × 5–10	0.5 × 2–3	0.8–1.3 × 1.2–5	0.5–1 × 1–2	0.5–0.8 × 1.3–2.5	0.3 × 1.0–1.3	0.5–1.0 × 1.5–3.0	0.5–1.0 × 3.0–5.0	0.4 × 3.0	0.8–1 × 2.5–3	0.5–0.8 × 1.5–4.0	0.5 × 3
Motility	+ (sp)	+ (sp)	+ (sp)	+ (mo)	+ (sp)	+ (sp)	+ (sp)	+	–	+	+ (sp)	+ (sp)	+ (sp)	+	+ (sp)	+ (sp)	+ (sp)	+ (sp)
G + C content (mol%)	60.7	Nr	49.7	55	36.7	69	62.3	nr	64	54	49	62	48	64	67.5	60.6	65	42.7
Sulfite reductase	DV	DV	DV	DV	DV	DV	DV	DV	DV	DV	DV	DV	DV	DV	DV	DV	DV	DV
Major menaquinone	nr	nr	nr	nr	nr	nr	nr	nr	MK-6	nr	MK-6(H$_2$)	nr	nr	nr	nr	nr	MK-6	nr
Optimal growth temperature (°C)	35	nr	30	37	28	37	35	nr	37	25	30–36	37	37	30–35	35	37	30–36	32–34
Oxidation of substrate	I	I	I	I	I	I	I	I	I	I	I	I	I	I	I	I	I	I
Electron donors used:																		
H$_2$	+	nr	+	+	+	+	+	+	+	+	+	+	+	(+)	–	nr	+	+
Formate	+	nr	+	+	+	nr	+	+	+	+	+	+	+	+	+	+	+	(+)
Acetate	–	–	–	–	–	–	–	–	–	–	–	–	–	–	–	–	–	–
Fatty acids	+	–	–	–	nr	nr	–	–	–	–	–	–	–	–	nr	nr	(+)	–
Ethanol	+	–	+	–	+	+	+	+	+	+	+	+	+	+	+	+	+	+
Lactate	+	+	+	+	+	+	+	+	+	+	+	+	+	+	+	+	+	+
Pyruvate	+	+	+	+	+	+	+	+	–	+	+	+	+	+	nr	+	+	+
Fumarate	–	nr	–	+	+	nr	+	–	nr	+	+	+	–	+	+	+	±	+
Succinate	–	nr	+	+	–	–	–	–	nr	nr	+	–	–	–	–	–	+	–
Malate	–	nr	–	+	–	–	–	+	nr	–	–	+	+	nr	nr	+	nr	+
Benzoate	–	nr	–	–	nr	+	–	nr	nr	–	–	–	nr	–	–	–	–	–
Fermentative growth on:																		
Fumarate	–	nr	–	+	+	+	+	–	–	–	nr	+	nr	nr	nr	–	±	+
Malate	–	nr	–	+	+	–	–	–	nr	–	–	+	nr	nr	–	–	+	+
Pyruvate	+	nr	+	+	+	+	+	+	–	+	+	+	nr	+	+	+	+	+
Electron acceptors used:																		
Sulfate	+	+	+	+	+	+	+	+	+	+	+	+	+	+	+	+	+	+
Sulfite	+	nr	–	+	+	+	+	+	nr	+	+	+	+	+	+	+	+	+
Sulfur	(+)	nr	+	–	+	nr	+	(+)	–	–	–	+	–	–	+	nr	–	+
Thiosulfate	+	nr	nr	nr	+	+	+	+	nr	+	+	+	nr	+	nr	+	+	+
Fumarate	–	nr	–	–	–	+	+	–	nr	(+)	(+)	+	+	+	+	+	nr	–
Nitrate	–	nr	–	–	–	nr	–	–	pa	+	–	–	–	–	nr	–	–	–
Growth factor requirement	–	nr	–	–	–	nr	bi, pa	—	pa	–	–	Unknown	Unknown	bi, pt	nr	Yeast extract	–	—

(continued)

TABLE BXII.δ.3. *(cont.)*

Characteristics	D. halophilus[P]	D. indonesiensis[q]	D. inopinatus[r]	D. intestinalis[s]	D. litoralis[t]	D. longreachensis[u]	D. longus[v]	D. oxyclinae[w]	D. piger[x]	D. profundus[y]	D. salexigens[z]	D. senezii[aa]	D. simplex[bb]	D. sulfodismutans[cc]	D. termitidis[dd]	D. vietnamensis[ee]	D. vulgaris subsp. vulgaris[ff]	D. zosterae[gg]
NaCl requirement (g/l)	0.9	–	10	–	–	–	–	2.5	–	0.2	20	–	–	–	–	–	–	12

[a]Symbols: nr, not reported; +, present, observed in all species; −, not present, not observed; ±, variable; (+), poorly motile, poor growth; (sp), flagellation type: single, polar; (lo), flagellation type: lophotrichous; DV, desulfoviridin; DR, desulforubidin; I, incomplete; C, complete; bi, biotin; ni, nicotinate; pa, p-aminobenzoate; pt, pantothenate; th, thiamine.

[b]Data from Postgate, 1984a, b; Brauman et al., 1990; Devereux et al., 1990; Dilling and Cypionka, 1990; Lovley and Phillips, 1992; Lovley et al., 1993b

[c]Data from van der Maarel et al., 1996.

[d]Data from Motamedi and Pedersen, 1998.

[e]Data from Postgate, 1984a, b.

[f]Data from Qatibi et al., 1991.

[g]Data from Baena et al. 1998.

[h]Data from Ouattara et al., 1999.

[i]Data from Nanninga and Gottschal, 1987.

[j]Data from Sass et al., 1998a.

[k]Data from Ollivier et al., 1988.

[l]Data from Folkerts et al., 1989a.

[m]Data from Tardy-Jacquenod et al., 1996.

[n]Data from Esnault et al., 1988a; Brauman et al., 1990.

[o]Data from Postgate, 1984a, b; Esnault et al., 1988a.

[p]Data from Caumette et al., 1991a.

[q]Data from Feio et al., 1998.

[r]Data from Reichenbecher and Schink, 1997.

[s]Data from Fröhlich et al., 1999a.

[t]Data from Sass et al., 1998a.

[u]Data from Redburn and Patel, 1994.

[v]Data from Magot et al., 1992.

[w]Data from Krekeler et al., 1997.

[x]Data from Moore et al., 1976; Biebl and Pfennig, 1977; Loubinoux et al., 2002.

[y]Data from Bale et al., 1997.

[z]Data from Postgate 1984a, b; Zellner et al., 1989; Bale et al., 1997.

[aa]Data from Tsu et al., 1998.

[bb]Data from Zellner et al., 1989.

[cc]Data from Bak and Pfennig, 1987.

[dd]Data from Trinkerl et al., 1990.

[ee]Data from Nga et al., 1996.

[ff]Data from Postgate 1984a, b; Lovley et al., 1993b; Lovley and Phillips, 1994b.

[gg]Data from Nielsen et al., 1999.

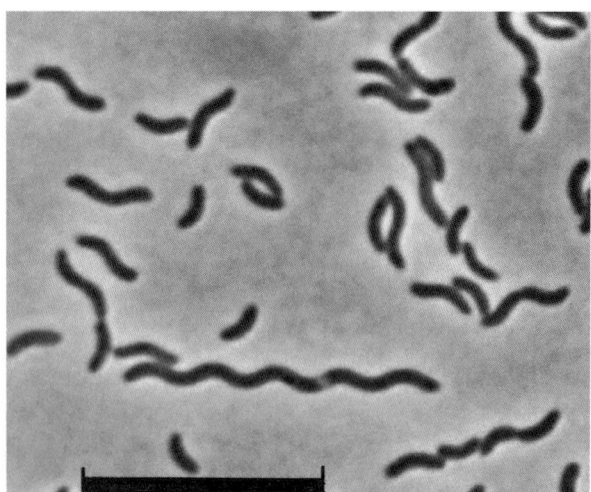

FIGURE BXII.δ.2. Phase-contrast micrograph of *Desulfovibrio desulfuricans* (viable cells). Bar = 10 μm.

chelated Fe(III), and U(VI) can be used as electron acceptors, but these processes are not connected to growth (Dilling and Cypionka, 1990; Lovley and Phillips, 1992; Lovley et al., 1993b). Choline and pyruvate support fermentative growth without sulfate.

Found in freshwater, in particular in polluted waters showing blackening and sulfide formation in the sediment; also found in anoxic or waterlogged soils rich in organic materials, and in marine or brackish waters.

The mol% G + C of the DNA is: 55 (Bd), 59 (T_m), (HPLC).
Type strain: Essex 6, ATCC 29577, DSM 642, NCIB 8307.
GenBank accession number (16S rRNA): M34113, AF192153.

Several subspecies of *D. desulfuricans* (like strain El Agheila Z, ATCC 27774, and ATCC 7757) and *D. vulgaris* have to be reclassified and are not closely related to the type strains (Devereux et al., 1990; Vainshtein et al., 1992). This is also true for strain MB (ATCC 27774), which is often referred to as *Desulfovibrio desulfuricans*; however, it does not resemble the type strain and has a different phylogenetic position within the genus *Desulfovibrio*. Newly published 16S rRNA gene sequences indicate that strain MB (ATCC 27774) is more closely related to *Desulfovibrio intestinalis* (sequence identity of 98.5%) than to the type strain of *D. desulfuricans* (ATCC 29577, strain Essex 6; sequence identity of 97.3%) (Loubinoux et al., 2000).

2. **Desulfovibrio acrylicus** Van der Maarel, Van Bergeijk, Van Werkhoven, Laverman, Meijer, Stam and Hansen 1997, 242^VP (Effective publication: Van der Maarel, Van Bergeijk, Van Werkhoven, Laverman, Meijer, Stam and Hansen 1996, 114.)

a.cry'li.cus. L. adj. *acrylicus* derived from *acidum acrylicum* acrylic acid.

Characteristics are summarized in Table BXII.δ.3. Isolated from anoxic marine sediments.

The mol% G + C of the DNA is: 65 (HPLC).
Type strain: W218, DSM 10141.
GenBank accession number (16S rRNA): U32578.

3. **Desulfovibrio aespoeensis** Motamedi and Pedersen 1998, 313^VP

ae.spoe.en'sis. M.L. adj. *aespoeensis* pertaining to Äspo, in southeastern Sweden.

Characteristics are summarized in Table BXII.δ.3. Isolated from deep, subterranean, granitic groundwater (600 m) of southeastern Sweden.

The mol% G + C of the DNA is: 61 (T_m).
Type strain: Aspo-2, DSM 10631.
GenBank accession number (16S rRNA): X95230.

4. **Desulfovibrio africanus** Campbell, Kasprzycki and Postgate 1966, 1127^AL

af.ri.ca'nus. L. adj. *africanus* pertaining to Africa.

Characteristics are similar to those of *D. vulgaris*; exceptions are noted in Table BXII.δ.3. Contain cytochrome *c*. Isolated from salt and fresh waters in Africa. Has a wide salt tolerance (Campbell et al., 1966; ; Skyring and Jones, 1972; Singleton et al., 1979).

The mol% G + C of the DNA is: 61 (Bd), 65 (T_m).
Type strain: Benghazi, DSM 2603, NCIB 8401.
GenBank accession number (16S rRNA): M37315, X99236.

5. **Desulfovibrio alcoholovorans** Qatibi, Nivière and Garcia 1995, 879^VP (Effective publication: Qatibi, Nivière and Garcia 1991, 147.)

al.co.ho.lo.vor.ans. E. M. alcohol; L. v. *voro* to devour; M.L. part. adj. *alcoholovorans* devouring alcohol.

Characteristics are summarized in Table BXII.δ.3. Isolated from a pilot fermenter containing alcohol distillery waste-water, France. Optimal growth at a NaCl concentration of 5–10 g/ l.

The mol% G + C of the DNA is: 64.5 (HPLC).
Type strain: SPSN, DSM 5433.
GenBank accession number (16S rRNA): AF053751.

6. **Desulfovibrio aminophilus** Baena, Fardeau, Labat, Ollivier, Garcia and Patel 1999, 341^VP (Effective publication: Baena, Fardeau, Labat, Ollivier, Garcia and Patel 1998, 503.)

a.mi.no.phi.lus. Gr. n. *aminus*; adj. *philos* loving; M.L. *aminophilum* amino acid loving.

Characteristics are summarized in Table BXII.δ.3. Isolated from anoxic sludge of a dairy wastewater treatment plant in Santa Fe de Bogota, Columbia, with alanine as electron donor and carbon source. Does not require NaCl. Can grow by fermentation of peptone, Casamino acids, serine, glycine, threonine, and cysteine. Thiosulfate and sulfite are disproportionated to sulfate and sulfide.

The mol% G + C of the DNA is: 66 (HPLC).
Type strain: Ala-3, DSM 12254.
GenBank accession number (16S rRNA): AF067964.

7. **Desulfovibrio burkinensis** Ouattara, Patel, Cayol, Cuzin, Traore and Garcia 1999, 642^VP

bur.ki.nen'sis. N.L. adj. *burkinensis* pertaining to Burkina Faso, West Africa.

Characteristics are summarized in Table BXII.δ.3. Isolated from an anoxic layer of a rice field in Burkina Faso, West Africa, using lactate.

The mol% G + C of the DNA is: 67 (HPLC).
Type strain: HDv, DSM 6830.
GenBank accession number (16S rRNA): AF053752.

8. **Desulfovibrio carbinolicus** Nanninga and Gottschal 1995, 879^VP (Effective publication: Nanninga and Gottschal 1987, 807.)

car.bi.no'li.cus. M.L. adj. *carbinolicus* referring to carbinols syn. alcohols; *carbinolicus* metabolizing alcohols.

Characteristics are summarized in Table BXII.δ.3. Isolated from an anaerobic purification plant, the Netherlands, using ethanol (Nanninga and Gottschal, 1986, 1987).

The mol% G + C of the DNA is: 65 (T_m).

Type strain: EDK82, DSM 3852.

9. **Desulfovibrio cuneatus** Sass, Berchtold, Branke, König, Cypionka and Babenzien 1998b, 1083[VP] (Effective publication: Sass, Berchtold, Branke, König, Cypionka and Babenzien 1998a, 218.)

cu.ne'a.tus. M.L. adj. *cuneatus* cuneiform, referring to the tapered ends of the cells.

Characteristics are summarized in Table BXII.δ.3. Isolated using lactate from an MPN series inoculated with material from the uppermost 10 mm of the littoral sediment of the oligotrophic Lake Stechlin, Germany. Typical cells are tapering off at one end; conical ends sometimes appear as a short stalk. Cells form aggregates and attach to the surface of the culture flasks. Oxygen can be used as an electron acceptor for the oxidation of H_2, formate, pyruvate, lactate, sulfite, and sulfide, but does not lead to growth. Catalase is present. Grows by disproportionation of sulfite in the presence of acetate.

The mol% G + C of the DNA is: 52.7 (HPLC).

Type strain: STL1, DSM 11391.

GenBank accession number (16S rRNA): X99501.

10. **Desulfovibrio fructosovorans** Ollivier, Cord-Ruwisch, Hatchikian and Garcia 1990, 105[VP] (Effective publication: Ollivier, Cord-Ruwisch, Hatchikian and Garcia 1988, 449.)

fruc.to.so.vo.rans. E.M. fructose; L. v. *voro* to devour; M.L. part. adj. *fructosovorans* devouring fructose.

Characteristics are summarized in Table BXII.δ.3. Isolated from a defined mixed culture on sucrose. Originally isolated from estuarine sediment, USA, with H_2 as electron donor (Jones et al., 1984). Methanol is used as electron donor after the cells have been grown on pyruvate. Fermentative growth occurs on fructose and other substrates (Cord-Ruwisch et al., 1986; Ollivier et al., 1988).

The mol% G + C of the DNA is: 64.1 (Bd).

Type strain: JJ, ATCC 49200, DSM 3604.

GenBank accession number (16S rRNA): AF050101.

11. **Desulfovibrio furfuralis** Folkerts, Ney, Kneifel, Stackebrandt, Witte, Forstel and Schoberth 1989b, 495[VP] (Effective publication: Folkerts, Ney, Kneifel, Stackebrandt, Witte, Forstel and Schoberth 1989a, 168.)

fur.fur.al.is. M.L. adj. *furfuralis* pertaining to furfural (2-furaldehyde).

Characteristics are summarized in Table BXII.δ.3. Isolated from anaerobic wastewater of a paper-producing plant using furfural.

The mol% G + C of the DNA is: 61 (T_m).

Type strain: F1, DSM 2590.

12. **Desulfovibrio gabonensis** Tardy-Jacquenod, Magot, Laigret, Kaghad, Patel, Guezennec, Matheron and Caumette 1996, 714[VP]

ga.bo'nen.sis. M.L. adj. *gabonensis* pertaining to Gabon, a country in West Africa.

Characteristics are summarized in Table BXII.δ.3. Isolated from an oil pipeline, Gabon, West Africa, using lactate. Moderately halophilic, requires at least 10 g NaCl/l for growth.

The mol% G + C of the DNA is: 59.5 (T_m).

Type strain: SEBR 2840, DSM 10636.

GenBank accession number (16S rRNA): U31080.

13. **Desulfovibrio giganteus** Esnault, Caumette and Garcia 1988b, 328[VP] (Effective publication: Esnault, Caumette and Garcia 1988a, 150.)

gi.gan'te.us. L. adj. *giganteus* giant, gigantic.

Characteristics are summarized in Table BXII.δ.3. Isolated from anoxic sediments of Berre Lagoon, Marseilles, France. Slightly halophilic, requires at least 2 g NaCl/l for growth. Can use cysteine and glycerol as electron donors.

The mol% G + C of the DNA is: 55.5 (T_m).

Type strain: 8601, DSM 4123.

14. **Desulfovibrio gigas** Le Gall 1963, 1120[AL]

gi'gas. L. n. *gigas* giant.

The morphology is as described in Table BXII.δ.3 and as depicted in Fig. BXII.δ.3. Characteristics are summarized in Table BXII.δ.3. The cells are often in chains, appearing as spirilla. Young organisms show areas of low contrast when examined by phase-contrast microscopy. Growth is slower than that of other species. Isolated from Etang de Berre, near Marseilles, France. Despite its saltwater origin, saline media are not needed for its cultivation. A slightly different strain was isolated from a sewage plant (Schoberth, 1973).

The mol% G + C of the DNA is: 60.2 (Bd); 65 (T_m).

Type strain: VKM B-1759, ATCC 19364, DSM 1382, NCIB 9332.

GenBank accession number (16S rRNA): M34400.

15. **Desulfovibrio halophilus** Caumette, Cohen and Matheron 1991b, 331[VP] (Effective publication: Caumette, Cohen and Matheron 1991a, 36.)

ha.lo'phi.lus. Gr. n. *halos* salt; Gr. v. *philein* love, like; M.L. adj. *halophilus* salt-loving.

Characteristics are summarized in Table BXII.δ.3. Isolated from an anoxic, benthic, microbial mat of Solar Lake, Sinai, Egypt, with lactate as electron donor. Requires at least 30 g NaCl/l for growth.

The mol% G + C of the DNA is: 60.7 (Bd).

Type strain: SL 8903, DSM 5663.

GenBank accession number (16S rRNA): U48243.

16. **Desulfovibrio indonesiensis** Feio, Beech, Carepo, Lopes, Cheung, Franco, Guezennec, Smith, Mitchell, Moura and Lino 2000, 1415[VP] (Effective publication: Feio, Beech, Carepo, Lopes, Cheung, Franco, Guezennec, Smith, Mitchell, Moura and Lino 1998, 128.)

in.do'ne.si.en.sis. L. adj. *indonesiensis* from Indonesia, referring to the country of isolation.

Characteristics are summarized in Table BXII.δ.3. Isolated from the corroded hull of an oil storage vessel off the Indonesian coast, Indonesia, with lactate as electron donor and carbon source. Does not require NaCl.

The mol% G + C of the DNA is: not reported.

Type strain: Ind 1, NCIMB 13468.

GenBank accession number (16S rRNA): Y09504.

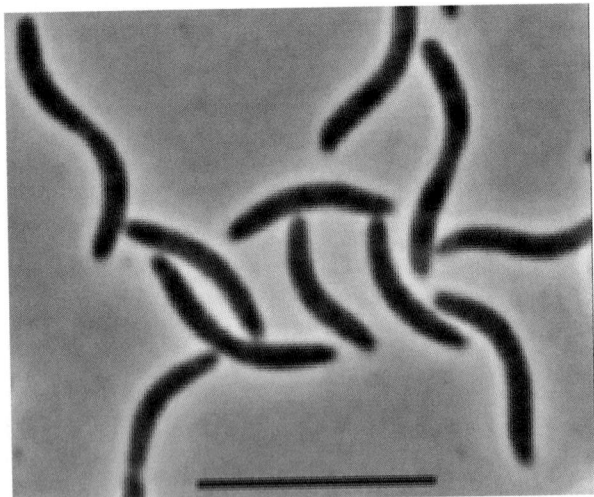

FIGURE BXII.δ.3. Phase-contrast micrograph of *Desulfovibrio gigas* (viable cells). Bar = 10 μm.

17. **Desulfovibrio inopinatus** Reichenbecher and Schink 1999, 1^VP (Effective publication: Reichenbecher and Schink 1997, 343.)

in.o.pi.na' tus. L. adj. *inopinatus* unexpected, referring to the utilization of an aromatic compound by a member of the genus *Desulfovibrio*.

Characteristics are summarized in Table BXII.δ.3. Isolated from anoxic mud (Venice, Italy), with hydroxyhydroquinone as electron donor and carbon source. Requires at least 10g NaCl/l and 1.5 g/l MgCl$_2$·6H$_2$O for growth. Can oxidize ribose, fructose, and hydroxyhydroquinone incompletely to acetate and CO$_2$ with sulfate as electron acceptor.

The mol% G + C of the DNA is: 49.7 (HPLC).
Type strain: HHQ 20, DSM 10711.
GenBank accession number (16S rRNA): AF177276.

18. **Desulfovibrio intestinalis** Fröhlich, Sass, Babenzien, Kuhnigk, Varma, Saxena, Nalepa, Pfeiffer and König 1999b, 1325^VP (Effective publication: Fröhlich, Sass, Babenzien, Kuhnigk, Varma, Saxena, Nalepa, Pfeiffer and König 1999a, 150.)

in.tes.ti' na.lis. M.L. adj. *intestinalis* the isolate lives in the intestine of termites.

Characteristics are summarized in Table BXII.δ.3. Isolated from the hindgut of the lower termite *Mastotermes darwiniensis* Frogatt, with lactate as electron donor and carbon source.

The mol% G + C of the DNA is: 55.0 (HPLC).
Type strain: KMS2, DSM 11275.
GenBank accession number (16S rRNA): Y12254.

19. **Desulfovibrio litoralis** Sass, Berchtold, Branke, König, Cypionka and Bebenzien 1998b, 1083^VP (Effective publication: Sass, Berchtold, Branke, König, Cypionka and Bebenzien 1998a, 218.)

li.to' ra.lis. M.L. adj. *litoralis* from the shore, pertaining to the habitat from where the organism was isolated.

Characteristics are summarized in Table BXII.δ.3. Isolated using lactate from a MPN series inoculated with material from a depth of 20–30 mm of littoral sediment of the oligotrophic Lake Stechlin, Germany. Typical cells are ta-

pering off at one end; conical ends sometimes appear as a short stalk. Cells form aggregates and attach to the surface of the culture flasks. Oxygen can be used as electron acceptor for the oxidation of H$_2$, formate, pyruvate, lactate, ethanol, sulfite, and sulfide, but does not lead to growth. Catalase is present.

The mol% G + C of the DNA is: 36.7 (HPLC).
Type strain: STL6, DSM 11393.
GenBank accession number (16S rRNA): X99504.

20. **Desulfovibrio longreachensis** Redburn and Patel 1995, 879^VP (Effective publication: Redburn and Patel 1994, 37.) *long.reach' en.sis.* M.L. gen. n. *longreachensis* of Longreach; named after the town in Australia.

Characteristics are summarized in Table BXII.δ.3. Isolated from the water of a free-flowing bore of the Great Artesian Basin in Australia, using lactate.

The mol% G + C of the DNA is: 62.3 (HPLC).
Type strain: AB16910a, ACM 3958.
GenBank accession number (16S rRNA): Z24450.

21. **Desulfovibrio longus** Magot, Caumette, Desperrier, Matheron, Dauga, Grimont and Carreau 1992, 402^VP *long' us.* L. adj. *longus* long.

Characteristics are summarized in Table BXII.δ.3. Isolated from production fluid from an oil-producing well in the Paris Basin, France. Does not require NaCl, but best growth occurs in the presence of 10–20 g/l NaCl.

The mol% G + C of the DNA is: 62.3 (HPLC).
Type strain: DSM 6739, SEBR 2582.
GenBank accession number (16S rRNA): X63623.

22. **Desulfovibrio oxyclinae** Krekeler, Sigalevich, Teske, Cypionka and Cohen 2000, 1699^VP (Effective publication: Krekeler, Sigalevich, Teske, Cypionka and Cohen 1997, 373.)

o.xy' cli.nae. Gr. adj. *oxys* acidic; Gr. v. *clinein* decline; *oxyclinae* referring to the oxycline as habitat.

Characteristics are summarized in Table BXII.δ.3. Isolated from the oxic zone of a hypersaline cyanobacterial mat of Solar Lake, Sinai, Egypt. Requires at least 22 g/l NaCl, with a growth optimal at 5–10 g NaCl. Oxygen can be used as electron acceptor for the oxidation of H$_2$, lactate, sulfite, and sulfide, but does not lead to growth. Catalase is present. Growth by disproportionation of sulfite and thiosulfate in the presence of acetate.

The mol% G + C of the DNA is: not reported.
Type strain: P1B, DSM 11498.
GenBank accession number (16S rRNA): U33316.

23. **Desulfovibrio piger** (Moore, Johnson and Holdeman 1976) Loubinoux, Valente, Pereira, Costa, Grimont and Le Faou 2002, 1307^VP (*Desulfomonas pigra* Moore, Johnson and Holdeman 1976, 238.)

pig' er. L. adj. *piger* lazy (referring to the limited number of substrates utilized by the species).

The morphology is as described in Table BXII.δ.3 and as depicted in Fig. BXII.δ.4. Characteristics are summarized in Table BXII.δ.3. Comparative 16S rDNA sequence analysis and testing of biochemical characteristics clearly indicate that *Desulfomonas pigra* should be reclassified as *Desulfovibrio piger*.

The mol% G + C of the DNA is: 64 (HPLC).
Type strain: ATCC 29098, DSM 749.

GenBank accession number (16S rRNA): M34404.

24. **Desulfovibrio profundus** Bale, Goodman, Rochelle, Marchesi, Fry, Weightman and Parkes 1997, 520[VP]

pro.fun' dus. L. adj. *profundus* deep.

Characteristics are summarized in Table BXII.δ.3. Isolated from deep marine sediment (500 m below sea floor), Japan Sea, Japan, with lactate as electron donor and carbon source. Barophilic. Requires at least 2 g/l NaCl for growth; the optimal NaCl concentration for growth is 6–8 g/l. Nitrate, ferric iron, dimethyl sulfoxide, and lignosulfonates are used as alternative electron acceptors.

The mol% G + C of the DNA is: 53 (HPLC).

Type strain: 500-1, DSM 11384.

GenBank accession number (16S rRNA): U90726.

25. **Desulfovibrio salexigens** Postgate and Campbell 1966, 735[AL]

sal.ex' i.gens. L. n. *sal* salt; L. v. *exigo* to demand; M.L. part. adj. *salexigens* salt demanding.

The morphology is as described in Table BXII.δ.3 and as depicted in Fig. BXII.δ.5. Characteristics are summarized in Table BXII.δ.3. Requires Cl⁻ ions for growth, which are supplied as NaCl (>6 g/l, usually 25–50 g/l). Found in seawater, marine and estuarine muds, and pickling brines.

The mol% G + C of the DNA is: 46 (Bd).

Type strain: ATCC 14822, DSM 2638, NCIB 8403.

GenBank accession number (16S rRNA): M34401.

26. **Desulfovibrio senezii** Tsu, Huang, Garcia, Patel, Cayol, Baresi and Mah 1999, 341[VP] (Effective publication: Tsu, Huang, Garcia, Patel, Cayol, Baresi and Mah 1998, 316.)

se.ne' zi.i. M.L. gen. *senezii* named to honor the French microbiologist Jacques C. Senez.

Characteristics are summarized in Table BXII.δ.3. Isolated from solar saltern of Chula Vista (San Diego Bay area, California, U.S.A.). Halotolerant.

The mol% G + C of the DNA is: 62 (HPLC).

Type strain: CVL, DSM 8436.

GenBank accession number (16S rRNA): AF050100.

27. **Desulfovibrio simplex** Zellner, Messner, Kneifel and Winter 1990, 470[VP] (Effective publication: Zellner, Messner, Kneifel and Winter 1989, 333.)

sim' plex. M.L. masc. *simplex* meaning ordinary, simple, inconspicuous.

Characteristics are summarized in Table BXII.δ.3. Isolated from an anaerobic digester for the biomethanation of sour whey (originally inoculated with sewage sludge), Regensburg, Germany. Requires at least 2 g/l NaCl for growth, optimal NaCl concentration for growth is 6–8 g/l. Nitrate is used as alternative electron acceptor for growth after adaptation. Can use aromatic aldehydes as electron donors for growth.

The mol% G + C of the DNA is: 47.5 (T_m).

Type strain: XVI, DSM 4141.

28. **Desulfovibrio sulfodismutans** Bak and Pfennig 1988, 136[VP] (Effective publication: Bak and Pfennig 1987, 189.)

sul.fo.dis' mu.tans. L. n. *sulfur* sulfur; L. pref. *dis* apart; L. part. adj. *mutans* that which changes; M.L. part. adj. *sulfodismutans* dismutating sulfur compounds.

Characteristics are summarized in Table BXII.δ.3. Isolated from anoxic freshwater sediments, Konstanz, Germany, with sulfite as electron donor and acetate and bicarbonate as carbon source. Requires at least 0.1% (w/v) NaCl and 0.04% (w/v) MgCl₂·6H₂O for growth. Thiosulfate, sulfite, and dithionite are disproportionated to sulfate and sulfide and allow growth in the presence of acetate and bicarbonate.

The mol% G + C of the DNA is: 64.1 (Bd).

Type strain: ThAc01, DSM 3696.

GenBank accession number (16S rRNA): Y17763.

29. **Desulfovibrio termitidis** Trinkerl, Breunig, Schauder and König 1991, 178[VP] (Effective publication: Trinkerl, Breunig, Schauder and König 1990, 376.)

ter.mi' ti.dis. M.L. nom. *termes* the termite; M.L. gen. nom. *termidites* of a termite.

Characteristics are summarized in Table BXII.δ.3. Isolated from the hindgut of the termite *Heterotermes indicola* (Wasmann).

The mol% G + C of the DNA is: 67 (T_m).

Type strain: HI1, ATCC 49858, DSM 5308.

GenBank accession number (16S rRNA): X87409.

30. **Desulfovibrio vietnamensis** Nga, Ha, Lai and Stan-Lotter

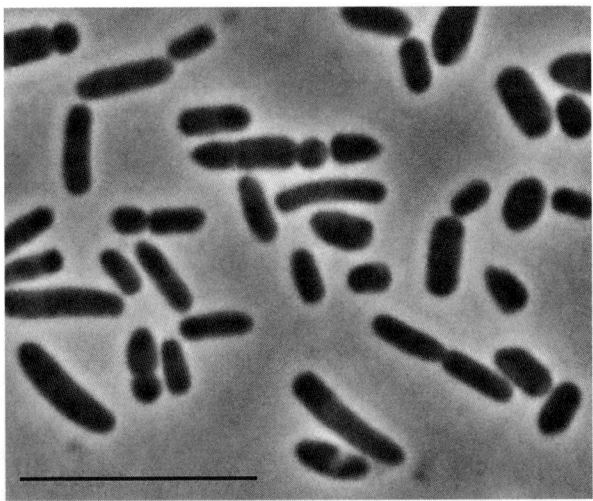

FIGURE BXII.δ.4. Phase-contrast micrograph of *Desulfovibrio piger* (basonym *Desulfomonas pigra*) (viable cells). Bar = 10 μm.

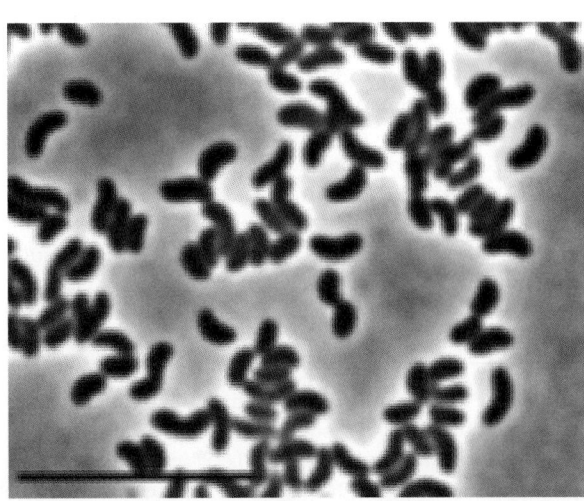

FIGURE BXII.δ.5. Phase-contrast micrograph of *Desulfovibrio salexigens* (viable cells). Bar = 10 μm.

2002, 1075[VP] (Effective publication: Nga, Ha, Lai and Stan-Lotter 1996, 391.)

viet.na'men.sis. L. adj. *vietnamensis* from Vietnam, referring to the country of isolation.

Characteristics are summarized in Table BXII.δ.3. Isolated from the production fluid of an offshore oil field, Vietnam. Does not require NaCl, although optimal growth occurs in the presence of 5% (w/v) NaCl.

The mol% G + C of the DNA is: 60.6 (HPLC).

Type strain: G3 100, DSM 10520.

GenBank accession number (16S rRNA): X93994.

31. **Desulfovibrio vulgaris** Postgate and Campbell 1966, 734[AL]

vul.ga'ris. L. adj. *vulgaris* common.

Characteristics are summarized in Table BXII.δ.3. Similar to *D. desulfuricans*, except for some characteristics indicated in Table BXII.δ.3; notably, it does not show sulfate-free growth on pyruvate and has a higher G + C content of the DNA (see below). The type strain can use Cr(VI) and chelated Fe(III) as electron acceptors, but the reduction is not connected to growth (Lovley et al., 1993b; Lovley and Phillips, 1994b).

The mol% G + C of the DNA is: 61 (Bd).

Type strain: DSM 644, NCIB 8303.

GenBank accession number (16S rRNA): M34399.

The subspecies *D. vulgaris* subsp. *oxamicus* of Postgate and Campbell (1966) (NCIB 9442, DSM 1925) and strain Marburg should be regarded as separate species and should be reclassified.

32. **Desulfovibrio zosterae** Nielsen, Liesack and Finster 1999, 864[VP]

zo.ste.ra'e. N.L. bot. n. *zostera;* N.L. gen. n. *zosterae* denoting that the bacterium was isolated from the plant *Zostera marina.*

Characteristics are summarized in Table BXII.δ.3. Isolated from roots of the marine macrophyte *Zostera marina* in Logstor Broad, Denmark, with lactate as electron donor and carbon source. Grows diazotrophically. Catalase present. Aerotolerant.

The mol% G + C of the DNA is: 42.7 (HPLC).

Type strain: lac, DSM 11974.

GenBank accession number (16S rRNA): Y18049.

Other Organisms

1. *"Desulfovibrio fairfieldensis"* Tee, Dyall-Smith, Woods and Eisen 1996.

fair.field'en.sis. L. adj. *fairfieldensis* from Fairfield, referring to the location of isolation.

The information about this proposed species is rather limited.

The mol% G + C of the DNA is: 62 (HPLC).

Deposited strain: None available.

GenBank accession number (16S rRNA): U42221.

Genus II. **Bilophila** *Baron, Summanen, Downes, Roberts, Wexler and Finegold 1990, 320[VP]*
(Effective publication: Baron, Summanen, Downes, Roberts, Wexler and Finegold 1989, 3410)

ELLEN JO BARON

Bi.lo'phi.la. L. n. *bilis* gall, bile; Gr. adj. *philus* loving; M.L. fem. n. *Bilophila* bile-loving organism.

Regularly shaped rods 0.7–1.0 × 1.0–2.0 μm when grown in peptone–yeast glucose broth with 1% pyruvate. Rods have rounded ends and occasionally form long filaments, but no internal structures are visible by light microscopy. Cells do not possess pili or flagella and are nonmotile. Nonsporeforming. Gram negative. **Anaerobic.** Nonfermentative; metabolism is by anaerobic respiration. Chemoorganotrophs. Growth is enhanced by 20% bile and 1% pyruvate. **Metabolize peptone but not carbohydrates. The major product from peptone is acetic acid with lesser to trace amounts of succinic acid.** Able to reduce nitrate to nitrite; some strains reduce nitrate to nitrogen gas. Hydrogen sulfide is produced from sulfur-containing amino acids, but the organisms do not reduce sulfur. In the presence of formate, *B. wadsworthia* utilizes taurine as an electron acceptor, oxidizing it to acetate with the reduction of sulfonate sulfur to sulfide. **Strongly catalase positive** (when tested with 15% H_2O_2). **Approximately 75–90% of strains produce urease, depending on the test system.** (Claros et al., 1999). In the presence of formate, *B. wadsworthia* utilizes taurine as an electron acceptor, with the production of acetate and the reduction of sulfonate sulfur to sulfide (Laue et al., 1997; Cook et al., 1999). The dissimilatory sulfite reductase (desulfoviridin) of *B. wadsworthia* is related to that of *Desulfovibrio* (Laue et al., 2001). Isolated from gastrointestinal, genital, and oral (periodontal) cavities of humans and other mammals (dogs).

The mol% G + C of the DNA is: 59.2 (HPLC).

Type species: **Bilophila wadsworthia** Baron, Summanen, Downes, Roberts, Wexler and Finegold 1990, 320 (Effective publication: Baron, Summanen, Downes, Roberts, Wexler and Finegold 1989, 3410.)

FURTHER DESCRIPTIVE INFORMATION

One described species, *Bilophila wadsworthia.* Cells are small to medium rods, 0.7–1.1 × 1.0–2.0 μm, that are regularly shaped when cultured on pyruvate-containing media. Cultures on less supportive media exhibit pleomorphic cell structures including filaments. Freeze-fracture electron micrographs show a smooth outer membrane with few structures (Summanen et al., 1989). There is no obvious peptidoglycan layer visible in transmission electron micrographs of the cell wall (Summanen et al., 1989, 1993).

Fatty acid methyl ester analysis reveals a unique series of whole cell fatty acids, which include $C_{15:0\ iso}$, $C_{16:0}$, $C_{17:0\ cyclo\ \omega7c}$, and $C_{19:0\ cyclo\ \omega9c}$ (Baron et al., 1989).

Under anaerobic conditions, small amounts of malate dehydrogenase and glutamate dehydrogenase can be detected in some species (Baron, 1997).

Bilophila strains require strictly anaerobic atmospheric conditions for growth. Freshly prepared or anaerobically stored media may be necessary. The organisms grow slowly, with colonies often visible only after 48–72 h. Supplementation of media, particularly broth media, with 1% pyruvate may be required for growth or metabolic reactivity of most strains. Growth at temperatures other than 35–37°C has not been attempted.

Strains have been recovered from numerous infected tissues in humans and dogs, usually in mixed culture, but occasionally in pure culture, such as from blood (Bennion et al., 1990a; Baron et al., 1992a, b; Baron, 1997; Gerardo et al., 1997; Schumacher and Bucheler, 1997). Although the organism comprises only 0.01% of the normal human fecal flora, it is the third most common isolate among anaerobes recovered from patients with perforated or gangrenous appendicitis (Bennion et al., 1990a,b; Baron et al., 1992a). Environmental strains (sewage) and urease-negative strains may have PCR profiles different from those of clinical strains when specific primers are used for amplification (Claros et al., 1999). Unpublished observations suggest that some strains can produce abscesses in the mouse peritoneal abscess model (A. Onderdonk, personal communication) without other associated enhancers, as do *Bacteroides fragilis* species. Some strains possess adherence mechanisms for laminin (Schumacher, 1997), but not for lipids (Schumacher et al., 2000). Cell walls contain endotoxin, as demonstrated by limulus amebocyte lysate assay, but endotoxin activity is not as great as that of other Gram-negative bacilli (Mosca et al., 1995).

ENRICHMENT AND ISOLATION PROCEDURES

Anaerobic blood-containing media supplemented with Vitamin K and hemin minimally support the growth of *Bilophila* species. *Brucella* agar base has been used primarily; comparisons with other agar bases have not been published. The organism grows slowly on *Bacteroides* bile esculin (BBE) agar; therefore, plates should be held for at least 7 days before they are discarded as negative. Addition of 20% bile (oxgall) and 1% pyruvate to all media enhances growth and promotes the formation of regular rod-shaped cell morphology. Pyruvate enrichment may be necessary for growth in some broths, for accurate growth-dependent

susceptibility testing, and for demonstration of some metabolic activities, such as β-lactamase production (Bennion et al., 1990a; Summanen et al., 1992; Mochida et al., 1998).

MAINTENANCE PROCEDURES

Fresh subcultures can be maintained for at least one month on anaerobic blood agar sealed in an oxygen-free atmosphere in a small, gas-impermeable foil bag or a plastic anaerobic pouch containing a strong oxygen-removing catalyst packet (Downes et al., 1990). Other methods for maintaining anaerobic bacteria, such as freezing a heavy suspension made from a fresh subculture in 10% glycerol, defibrinated sheep blood, or chopped meat broth, are also acceptable.

DIFFERENTIATION OF THE GENUS *BILOPHILA* FROM OTHER GENERA

Table BXII.δ.4 shows key characteristics of organisms that may appear similar to *Bilophila* in some respects, especially when recovered from anaerobic blood agar. If colonies are first noted on BBE agar, the typical small, translucent, black-centered colony should suggest the possibility that the colony represents *Bilophila*. The strong catalase reaction of *Bilophila* is unlike that of virtually any other anaerobic organism tested and can therefore rule out other anaerobes and the occasional *Enterococcus* that yields a similar colony on BBE.

TAXONOMIC COMMENTS

Based on 16S rRNA sequence analyses, *Bilophila* is located within the *Deltaproteobacteria*, along with a number of species of *Desulfovibrio* (Sapico et al., 1994) within the family *Desulfovibrionaceae*.

Crude whole chromosomal dot-blots have shown that, of twenty ATCC type strains of *Bacteroides, Porphyromonas, Prevotella, Desulfomonas, Desulfovibrio,* and *Mitsuokella* tested, only *Bacteroides vulgatus* exhibits some homology with *Bilophila* (at the 30% level) (Baron et al., 1989).

List of species of the genus Bilophila

1. **Bilophila wadsworthia** Baron, Summanen, Downes, Roberts, Wexler and Finegold 1990, 320[VP] (Effective publication: Baron, Summanen, Downes, Roberts, Wexler and Finegold 1989, 3410.)

   wads.wor' thi.a. M.L. fem. adj. *wadsworthia* originating from the Wadsworth Anaerobe Laboratories of the Wadsworth Veterans Administration Medical Center.

   Characteristics are as described for the genus and as listed in Table BXII.δ.5, with the following additional characteristics. After a minimum of 48 h incubation, surface

colonies on *Brucella* blood agar are 0.6–0.8 mm in diameter, transparent to translucent, raised with a circular, erose edge. After 72 h incubation, surface colonies on *Bacteroides* bile esculin agar are 1–2 mm in diameter, and either low convex, black, opaque, and irregular, or translucent, umbonate, circular with an erose edge, and often showing a black central area of precipitation of hydrogen sulfide. Subsurface colonies (created by stabbing the agar during inoculation) display heavy black precipitated H_2S. To maximize detection of these small colonies, BBE agar should be incubated for 7 d.

TABLE BXII.δ.4. Differential characteristics of species similar to *Bilophila*[a]

Characteristic[a]	Bilophila	Bacteroides putredinis	Bacteroides ureolyticus	Campylobacter gracilis	Desulfomonas	Desulfovibrio	Fusobacterium species	Sutterella wadsworthensis
Metabolic end products by GLC	A (s)	S, P, IV	A, S	S (a)	A	A	B (and others)	S (a)
Catalase	Strong +	+ (−)	− (+)	−	V weak	V weak	−	−
Nitrate reduction	+	−	+	+	− (+)	V	−	+
Urease	+ (−)	−	+ (−)	−	− (+)	− (+)	−	−
Motility	−	−	−	−	−	+	−	−
Growth enhanced by bile (oxgall)	+	+ (−)	−	−	V	V	V	+
Inhibited by 5 µg colistin by special potency disk diffusion	+	+	+	+	−	−	+	+

[a]Symbols: A, major acetic; (s), variable minor succinic; S, major succinic: P, major propionic; IV, major isovaleric; (a), variable minor acetic; B, major butyric; V, variable reactions; values in parentheses are those of a minority of strains.

TABLE BXII.δ.5. Characteristics of the genus *Bilophila*[a]

Characteristic	Result or reaction
Acid products formed in cultures grown in peptone–yeast extract glucose broth supplemented with 1% pyruvate	+[b]
H₂S produced	+
Starch and esculin hydrolyzed	–
Nitrate reduced to nitrite and occasionally to N₂	+
Indole produced	–
Motility	–
Urease	d (75%)
Growth in 20% bile	+
Catalase (tested with 15% H₂O₂)	+ (strong)
Gelatin digested	–
Lecithinase, lipase	–
Milk reaction	–
Acid from carbohydrates	–
Acid phosphatase (API ZYM)	+
Sulfate reduced	–

[a]For symbols see standard definitions.

[b]Mainly acetic acid plus smaller amounts of succinic acid.

Peptone broth cultures with 1% pyruvate show diffuse, cloudy growth. Many strains produce acid phosphatase and the cellular enzymes malate and glutamate dehydrogenase. Testing in an anaerobic atmosphere is often required for detection (Baron, 1997). Esculin is not hydrolyzed. Approximately 75% of strains hydrolyze urea, as demonstrated either by a rapid urease disk test, by overnight anaerobic incubation on Christensen's urea agar, or with PRAS urea broth. All strains tested so far have been susceptible to metronidazole (8.0 μg/ml) (Baron et al., 1993; Mochida et al., 1998). Strains appear to be inhibited by clindamycin (4 μg/ml), imipenem (32 μg/ml), ampicillin/sulbactam (16/8 μg/ml), and cefoxitin (32 μg/ml), but method and inoculum differences contribute to major differences in endpoints (Baron et al., 1993; Mochida et al., 1998). Although a hazy, trailing endpoint in agar dilution studies with imipenem has been associated with viable spheroplast formation; the clinical implications of this finding are not known (Summanen et al., 1993). Detection of β-lactamase, present in >85% of strains, requires growth on 1% pyruvate-containing medium (Summanen et al., 1992; Mochida et al., 1998). *B. wadsworthia* has been isolated from inflamed and gangrenous appendices, other intra-abdominal infections, brain abscesses, Bartholin cyst abscesses, liver abscesses, breast abscesses, empyema, gingivitis, Fournier's gangrene, osteomyelitis, and other common sites of mixed aerobic and anaerobic infections in humans (Baron et al., 1992b; Summanen et al., 1995; Baron, 1997), as well as otitis media, and blood cultures (Kasten et al., 1992; Bernard et al., 1994; Gerardo et al., 1997; Schumacher and Bucheler, 1997). *B. wadsworthia* is found in human feces, saliva, and vaginal discharge in the absence of infection (Baron et al., 1989, 1992b; Baron, 1997) and in periodontal pockets of dogs (Baron et al., 1992b).

The mol% G + C of the DNA is: 59.2 (HPLC, Laue et al., 2001; originally published as 39–40 (*T_m*).

Type strain: ATCC 49260 (WAL 7959).

Genus III. **Lawsonia** *McOrist, Gebhart, Boid and Barns 1995a, 824*[VP]

STEVEN MCORIST AND CONNIE J. GEBHART

Law.so'ni.a. L. dim. ending -ia M.L. masc. n. *Lawsonia* named after Gordon H.K. Lawson, the Scottish veterinarian who first recognized the organism causing porcine proliferative enteropathy.

Curved, vibrioid, or occasionally straight rods, 0.3–0.4 × 1.5–2.0 μm. The morphology is influenced by age and environment; descriptions refer to organisms within naturally infected intestinal epithelial cells or freshly grown co-cultures of intestinal epithelial cells. Gram negative. Single polar bacterial cell appendages have been observed in some preparations of co-cultured strains, with rapid darting motion apparent in active cultures. Endospores not formed. **Obligately intracellular, cell-dependent, and microaerobic.** Optimal temperature usually 35–37°C. Strains have demonstrated survival, but not growth, at lower temperatures. The organisms react with the IG4 monoclonal antibody shown to be specific for *Lawsonia* sp. Habitats: intestines of pigs and other animals; feces. The causal agent of proliferative enteropathy. Key aspects of the 16S rDNA sequence are given in McOrist et al., 1995a. A 16S rRNA signature sequence for the genus is 5′-GAUUAAGAGGUGCCUUUCGGGGAGCCUCAA-3′, corresponding to the nucleotide positions 1015–1042 of the *E. coli* 16S rRNA sequence.

The mol% G + C of the DNA is: 34; total genome length 1.713 Mbp.

Type species: **Lawsonia intracellularis** McOrist, Gebhart, Boid and Barns 1995a, 824.

FURTHER DESCRIPTIVE INFORMATION

The placement of the *Lawsonia* genus within the *Desulfovibrionaceae* is based on 16S rDNA sequence similarities and morphologic similarities (Gebhart et al., 1993; McOrist et al., 1995a). A full-length (1552 nucleotide) 16S rDNA sequence was obtained from a genomic clone of the type strain purified from porcine intestinal mucosa. Comparison of this sequence with sequences available in the Ribosomal RNA Database Project (RDP) indicated highest similarity with the 16S sequences of the *Desulfovibrionaceae*. Extensive phylogenetic analysis of this sequence and those of the *Desulfovibrionaceae*, using distance matrix, maximum parsimony, and maximum likelihood techniques, has suggested placement within this group.

Cells of *Lawsonia* are typical vibrioid-shaped, curved to straight rods (Rowland and Lawson, 1974). This morphology appears to be similar in organisms observed within host cells, associated with host cells, or present within artificial co-cultures (Rowland and Lawson, 1974; Frisk and Wagner, 1977; Lawson et al., 1993). The life cycle of *Lawsonia* within infected cells closely resembles that of another (unrelated) obligately intracellular bacterium *Orientia tsutsugamushi* (McOrist et al., 1995b). Following a brief (<3 h) association with the mammalian cell border, *Lawsonia* cells enter via induced endocytosis. Viability of the mammalian cells, but not of the bacteria, is required for uptake (Lawson et al., 1995). Following a further brief (<3 h) period within endocytic vacuoles, bacteria quickly escape into the cytosol (McOrist et al., 1995b). It is possible that *Lawsonia* possesses a phospholipase or some other enzymatic method for quick release from vacuoles, but this has not been definitely established. *Lawsonia* spp. have only been observed to grow and multiply within the cytosol, often in close proximity to cell mitochondria. It is possible that *Lawsonia* possesses a requirement for preformed triphosphates. The trilaminar cell wall and fine structure of *Lawsonia* are typical of Gram-negative bacteria (Rowland and Lawson, 1974); however, no S-layer was observed in other limited studies (McOrist et al., 1997).

The *Lawsonia* organism was originally described in various papers as a *Campylobacter*-like organism on the sole basis of a morphologic similarity. However, unlike both the *Desulfovibrio* and *Campylobacter* genera, *Lawsonia* is an obligately intracellular bacterium. A single polar flagellum has been observed on extracellular organisms in some preparations of co-cultured *Lawsonia* strains. This structure has not been observed on *Lawsonia* present inside cells or *in vivo*. There is no evidence of sulfate reductase gene sequences in *Lawsonia*. Presumably, active sulfate reduction would be an evolutionary disadvantage in its intracellular habitat. When the relationship of *Lawsonia* to *Desulfovibrio* became known, many workers, including the authors, attempted cell-free growth in conditions suitable for the latter genus. Many (over 80) media preparations and various atmospheres have been tested on isolates already adapted to laboratory co-culture, but the growth conditions of *Lawsonia* have remained resolutely obligately intracellular. The nutrition and growth requirements of *Lawsonia* are therefore the fastidious ones listed below for co-culture in cell lines. Growth of the organism in various chambers of chicken eggs has not been successful, despite several efforts by different groups.

The DNA sequences of the 16S ribosomal subunits of various strains of *Lawsonia* have been determined and are similar; see GenBank accession numbers L15739, U30147 (cultured porcine strains; Gebhart et al., 1993; McOrist et al., 1995a), U06423, U65995 (cultured hamster strains; Peace et al., 1994; Cooper et al., 1997), U65996, U65997, U65998, U07570 (uncultured strains purified from the intestines of deer, ostrich, horse, and ferret, respectively; see Fox et al., 1994a; Cooper et al., 1997). The partial DNA sequence of a chaperonin (GroEL) of one strain of *Lawsonia* has been determined; see GenBank accession number U45241 (uncultured strain purified from the intestines of a pig; Dale et al., 1998). No plasmids have been detected in *Lawsonia* spp. (McOrist et al., 1990).

The antigenic structure of *Lawsonia* has been identified in PAGE and immunoblotting studies of the outer membranes of several strains. The major structural protein is probably an observed 55-kDa protein, while highly immunoreactive components of the outer cell membrane, probably glycoproteins, are 27–29 kDa (McOrist et al., 1989a, 1995a).

Because *Lawsonia* is an animal pathogen, the antibiotic and drug sensitivity of several strains have been determined (McOrist et al., 1995c). The methods used were adapted from ones used successfully for the evaluation of the microaerobic, cell-dependent *Treponema pallidum*. *Lawsonia* appears to be resistant to the actions of aminoglycosides and various anti-Gram-positive compounds such as bacitracin and vancomycin, at standard concentrations. *Lawsonia* spp. appear to be relatively sensitive to the actions of compounds such as macrolides, lincosamides, and pleuromutilins. This is probably due to an ability of these compounds to concentrate in the cytosol of mammalian cells in concentrations that are likely to allow both bacterial uptake and inhibition of bacterial protein synthesis.

Lawsonia is an obligately intracellular bacterium that has only been identified in the intestinal epithelial cells of mammals (particularly pigs and hamsters) or ratite birds. It is one of the few bacterial genera consistently associated with the proliferation of a monotype of mammalian cells, namely the infected intestinal epithelial cells. Studies of *Lawsonia* pathogenesis indicate that once the epithelial cells of an intestinal crypt become infected, these infected cells fail to mature and instead continue as the undifferentiated, proliferative phenotype (Johnson and Jacoby, 1978; McOrist et al., 1989b). *Lawsonia* isolates have therefore been derived solely from the intestinal mucosa of animals demonstrated, by histologic assay of adjacent portions of mucosa, to have been suffering from proliferative enteropathy (ileitis). Oral inoculation of pigs or hamsters with cultured *Lawsonia* leads to characteristic intracellular infection and resulting intestinal lesions (Stills 1991; McOrist et al., 1993, 1994b). Staining of these sections by specific *in situ* immunoassay or by DNA probes for *Lawsonia* spp. confirms the diagnosis and presence of the causative agent at this site (McOrist et al., 1993; Gebhart et al., 1994). No such lesions or bacteria have ever been observed in pigs or hamsters not exhibiting proliferative enteropathy, or in humans, even those with enteric disease. The pathogenesis of proliferative enteropathy is discussed more fully in the referenced publications (see McOrist et al., 1989b, 1993, 1996). Once *Lawsonia* are cleared from intestinal cells, normal cell differentiation and apoptosis resume, with normal architecture returning within two weeks (McOrist et al., 1996). Presumably, the consequent proliferation of target cells following infection is a useful evolutionary adaptation for the transmission of *Lawsonia* among susceptible animals. Very few other bacterial genera are so closely associated with mammalian cell proliferation. *Bartonella bacilliformis* are consistently associated with proliferation of human endothelial cells and *Citrobacter rodentium* causes a colonic hyperplasia in mice, but neither organism is an intracellular bacterium *in vivo*. Presumably, some members of these three unrelated genera have the ability to produce substances capable of influencing either the mammalian genome or other factors active on it, but no definitive work addressing this trait has appeared with regard to any of the genera.

ENRICHMENT AND ISOLATION PROCEDURES

Initial purification and isolation of the intracellular bacteria from fresh intestinal mucosa samples can be accomplished by following established co-culture techniques (Lawson et al., 1993). Cell cultures used for co-culture should be demonstrated to be free of *Chlamydia* spp., *Mycoplasma* spp., and extracellular bacteria by both assays with commercially available kits and routine bacteriologic culture under aerobic, microaerophilic, and anaerobic atmospheres. Homogenization and trypsinization of the intestinal mucosa from infected intestines to release the bacteria, followed by differential centrifugation or filtration steps, gives a purified *Lawsonia* suspension suitable for inoculation into established cell lines of intestinal epithelial cells, such as IEC-18 or INT 407. Addition of antibiotics that retard contaminants from the initial bacterial suspension, but do not retard *Lawsonia* growth, is required until *Lawsonia* growth is advanced and the culture is purified by cell passage. Antibiotics suitable for this purpose include neomycin and vancomycin. *Lawsonia* spp. prefer rapidly dividing cells for intracellular growth. The intracellular growth curve typically reaches peak infection 1–5 d after initial infection in established cultures (peak infection is reached more slowly in initial passages). Examination of infected cell cultures by routine light microscopy during this active growth phase does not reveal any visible changes or evidence of infection. Older, mature co-cultures exhibit reduced bacterial growth, increased cell death, and bacterial release into the culture medium. Growth of *Lawsonia* can be confirmed by staining of the co-cultures with a modified Zeehl–Neilsen stain or immunostain with IG4 *Lawsonia*-specific monoclonal antibody. To date, co-culture of more than a small number of isolates (~14) has not been possible due to the tedious nature of intracellular bacterial cultivation from a contaminated environment such as the intestine (Knittel et al., 1996). A *Lawsonia* co-culture is illustrated in Fig. BXII.δ.6.

A general procedure for isolation and cultivation of *Lawsonia* is as follows: The experimental system is the laboratory culture of the bacterium *Lawsonia* in rat intestinal epithelial cells. An established cell line of rat intestinal epithelial cells, designated IEC-18 (ATCC CRL 1569, see Quaroni and May, 1980), is the cell line of choice. The cells are maintained in a standard cell culture medium, namely Dulbecco's modification of Eagle's medium (DMEM) supplemented with 5% bovine fetal serum, and 2 mM L-glutamine. Cells are transferred weekly into fresh tissue culture flasks by standard trypsin/EDTA methods. Following trypsinization of a seven-day culture of the cells, up to 2.5×10^4 cells/ml of fresh DMEM medium are inoculated into fresh flasks for establishment of cell growth. Cultures are incubated in an atmosphere of 5% CO_2 in air at 37°C for one day. The 1-d-old cell cultures are inoculated with 2×10^4 *Lawsonia*/ml suspended in medium and then incubated in an atmosphere of $O_2/CO_2/N_2$ (8%:8.8%:83.2%) at 37°C for 5–7 d.

Maintenance procedures and preparation of *Lawsonia* suspensions by lysis of infected cells are as described below. At intervals of 1 d, 2 d and 5 d following inoculation, the medium overlying the co-cultures is replaced with fresh medium and co-cultures re-incubated. Because flasks are difficult to stain, small cell culture vessels, each containing a glass coverslip, may be used for parallel sentinel co-cultures. During the period 1–7 d after inoculation, individual glass coverslips carrying each co-culture may be removed and stained for *Lawsonia*. For immunostaining, co-cultures are washed, fixed in acetone, and stained by an indirect immunoassay incorporating the mouse monoclonal antibody (IG4) specific to *L. intracellularis*. A count of the number of heavily infected cells (>30 bacteria/cell) may be used as a guide to infection. Upon serial passages over a period of weeks, *Lawsonia* infections of cells should attain 100%. Tumbling or darting motion of *Lawsonia* may be evident in heavily infected cells in these active infections. Co-cultures are considered optimized when at least 90% of the surface area of each culture flask and glass coverslip is covered by cells. The confluence of the cell

monolayer in each test system may be recorded as a percentage of the entire surface area. Careful monitoring of cell culture growth has not yet established any specific *in vitro* effect on cell proliferation (McOrist et al., 1997). Dedicated incubators and gas supplies, laminar flow hoods, and clean work areas are essential to optimize cultures.

MAINTENANCE PROCEDURES

Preparation of *Lawsonia* suspensions for inoculation of fresh cell cultures in order to maintain co-cultures requires the lysis of infected cells. Passage of whole infected cells, trypsinized cells, or bacteria from cells lysed by vigorous methods does not lead to successful passages. Cell lysis and bacterial release are best accomplished by gentle lysis of infected cells within flasks. The addition of 0.1% potassium chloride in water as an overlay to monolayers for 10 minutes at 20°C, followed by physical removal of the infected cell suspension and further disruption of cells by passage through a small gauge needle, is recommended. Subsequent differential centrifugation supplies a pelleted *Lawsonia*, ready for re-suspension in DMEM and maintenance inoculations.

Bacterial storage is best achieved in SPG buffer with 5% bovine fetal serum (Lawson et al., 1993) at −70°C. Freeze-drying has not been widely tested.

PROCEDURES FOR TESTING SPECIAL CHARACTERS

A specific DNA probe has been developed for detection of *Lawsonia* in clinical specimens by dot blot or *in situ* hybridization (Gebhart et al., 1991, 1994). Successful and specific PCR assays have also been developed using primer pairs derived from the DNA sequence of porcine *Lawsonia intracellularis* (Jones et al., 1993). These probes and PCRs have detected *Lawsonia* DNA in organisms identified in intestinal tissues (fresh or formalin-fixed) or feces of pigs, hamsters, horses, deer, ostriches, and rabbits (Gebhart et al., 1991; McOrist et al., 1994a; Cooper and Gebhart, 1998).

PCR primers were developed from *Lawsonia intracellularis* NCTC 12656[T]. Primers C and D were derived from a closed genomic segment (GenBank accession number L08049) and primers 878F and 1050R from a segment of the 16S rDNA sequence (L15739). The sequences of these primers are: C (antisense), 5′-TTACAGGTGAAGTTATTGGG-3′; D (sense), 5′-CTTTCTCATGTCCCATAAGC-3′; 878F (sense), 5′-TAACGCGTTAAGCA(C or T)C-3′; 1050R (antisense), 5′-GTCTTGAGGCTCCCCGAAAGGCACCTCTTAATC-3′.

PCR is performed in a 50 μl volume containing PCR buffer (10 mM Tris-HCl, pH 8.3, 50 mM KCl, 3 mM MgCl₂), 62 ng each C and D primers, 177 ng each 878F and 1050R primers, 0.2 mM each dATP, dGTP, dCTP, dTTP, and 1 μl extracted sample DNA (5–10 ng) or 1 μl dH₂O as a negative control. The mixture is heated to 95°C for 5 min and cooled to 4°C before adding 2.5 units of Taq polymerase. The sample is subjected to 35 cycles of PCR in a Perkin-Elmer thermal cycler with an amplification program of 93°C for 30 sec, 45°C for 1 min, and 72°C for 1 min. The expected sizes of the PCR products from these primers are 258 bp for the C and D primer set and 182 bp for the 878F and 1050R primer set.

The monoclonal antibody IG4 was prepared in 1986 and has since been used as a specific marker for immunostaining of *Lawsonia* (McOrist et al., 1987). No reactions have been detected with *Desulfovibrio* spp. or other related genera. IG4 was originally prepared as a fusion between splenic lymphocytes from a mouse immunized with whole *Lawsonia intracellularis* and NS0 cells. It

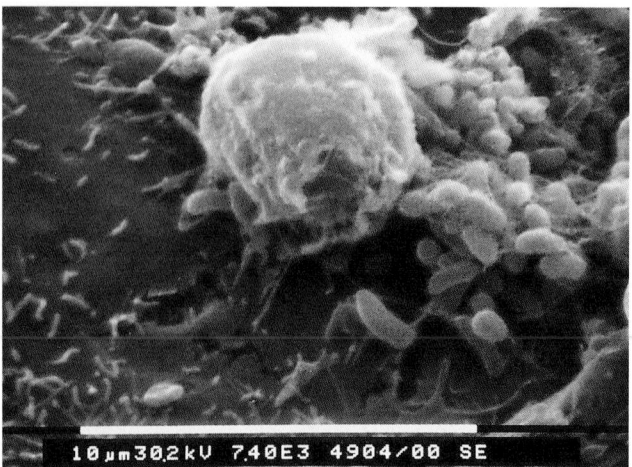

FIGURE BXII.δ.6. Scanning electron micrograph of *Lawsonia intracellularis* co-culture.

reacts with a 27–29-kDa glycoprotein on the outer surface of *Lawsonia* (McOrist et al., 1989a, b).

DIFFERENTIATION OF THE GENUS *LAWSONIA* FROM OTHER GENERA

Lawsonia may be differentiated from other vibrioid-shaped organisms in the intestine by its obligately intracellular habitat and its special characters, namely, the PCR reaction and monoclonal antibody staining. The specific 16S rDNA sequence of *Lawsonia* can be confirmed by PCR (see above), and its specific outer surface components can be detected by reaction with the IG4 specific monoclonal antibody. A 16S rRNA signature sequence for the genus is 5′-GAUUAAGAGGUGCCUUUCGGGGAGCCU-CAA-3′, corresponding to the nucleotide positions 1015–1042 of the *E. coli* 16S rRNA sequence.

TAXONOMIC COMMENTS

Lawsonia was placed within the *Desulfovibrionaceae* in 1995 based on 16S rDNA sequence data. According to this classification, its closest genetic relatives are *Bilophila wadsworthia*, with 92% sequence identity, and *Desulfovibrio desulfuricans*, with 91% sequence identity (Gebhart et al., 1993; McOrist et al., 1995a). The most recent analysis of all available sequences from *Lawsonia* isolates confirms that they form a closely related cluster within the family *Desulfovibrionaceae*, with *Bilophila wadsworthia* as the closest non-*Lawsonia* relative. *B. wadsworthia* is an anaerobic organism capable of extracellular growth. It has been isolated from the large intestine and other sites in humans (Sapico et al., 1994). Analysis of PCR-amplified rDNA from *Lawsonia* organisms derived from a variety of infected host animals (pig, hamster, deer, horse, ostrich, ferret) has shown very high levels of sequence similarity (98–100%). This indicates that isolates from a variety of hosts are all members of the *Lawsonia* genus and perhaps of the same species as well.

In one DNA study, a *Lawsonia* isolate from a ferret was erroneously placed in the *Desulfovibrio* genus purely on DNA sequence data (Fox et al., 1994a). Biologic, phenotypic, and epigenetic characteristics, such as the smaller size of *Lawsonia*, its novel habitat, pathogenicity for animals, lack of anaerobiosis, lack of black pigment, lack of extracellular growth, and different bacterial protein profiles, as compared to *D. desulfuricans* and other members of the *Desulfovibrio* genus, all clearly indicate that *Law-sonia* is a separate genus. A single polar flagellum has been observed on organisms in some extracellular preparations of co-cultured *Lawsonia* strains. This is consistent with the rapid, darting motility of *Lawsonia* observed when in the cytoplasm of co-culture cell monolayers. However, the flagellum structure has not been observed on *Lawsonia* present inside cells or *in vivo*. The type and exact role of the *Lawsonia* flagellum is not known. There is no evidence of sulfate reductase gene sequences present in *Lawsonia*. The relatively small size of the *Lawsonia* genome is consistent with the genome reduction noted in many other obligate intracellular bacteria.

Following the first identification of *Lawsonia* in diseased pig tissues in 1973 (Rowland and Lawson, 1974), it was another 20 years before successful culture of the organism occurred, with subsequent fulfillment of Koch's postulates (Lawson et al., 1993; McOrist et al., 1993). During this 20-year period there was considerable confusion about the exact identity of the bacterium, particularly as a range of *Campylobacter* spp. were cultured from diseased tissues. Prior to definitive 16S rDNA sequencing work, *Lawsonia* spp. were shown to be separate from these *Campylobacter* spp. by DNA restriction enzyme analysis and DNA probe analysis (McOrist et al., 1990; Gebhart et al., 1991). Comparison of the outer membranes of these genera by immunoblotting with monoclonal antibodies also indicated their separate status (McOrist et al., 1989b). This led to the nomenclatural form of *Campylobacter*-like organisms, then to the vernacular form, ileal symbiont intracellularis, both of which are considered synonyms of *Lawsonia* when referring to vibrioid organisms present within the intestinal epithelial cells of pigs, horses, or hamsters affected with proliferative enteropathy.

Further work needs to be done to address the relationship among the isolates from various host species and geographical sites. All DNA sequencing and protein analysis work to date indicates that *Lawsonia* has just one genetic monotype, with a unique life cycle in a particular cell type. *In vitro* work corroborates the observation *in vivo* that the organism can rapidly and readily enter intestinal epithelial cells (Johnson and Jacoby, 1978; McOrist et al., 1989b, 1995b). It is possible that there is limited, if any, variation in strains derived from various host species, merely one well-adapted organism, which has been spread widely by contact among different host species.

List of species of the genus Lawsonia

1. **Lawsonia intracellularis** McOrist, Gebhart, Boid and Barns 1995a, 824[VP]

   in.tra.cel.lu.la' ris. L. prep. *intra* within; L. n. *cella* small room; L. adj. *intracellularis* within cell.

   The morphology, phenotypic characters, and co-culture conditions are as described for the *Lawsonia* genus. Originally isolated from intestinal mucosa of pigs affected with proliferative enteropathy.

   Naturally occurring bacteria resembling the original pig isolates and containing 16S rDNA sequences identical to these isolates have been identified in or co-cultured from the following host species: hamster, ostrich, horse, deer, and ferret. In all cases, these bacteria were prepared from intestines affected with proliferative enteropathy, reacted with the IG4 antibody, and are considered to be *Lawsonia intra-cellularis*. Bacteria resembling the original pig isolates have also been identified in other host species in intestines affected with proliferative enteropathy, and these have shown positive reactions with IG4 antibody. These bacteria are considered highly likely to be *Lawsonia intracellularis*, although they have not been cultured nor had their DNA examined. These bacteria have been identified in rat, rhesus monkey, and emu. Other naturally occurring bacteria that resemble the original pig isolates but have not been further identified have been found in the host species dog, guinea pig, and fox.

   The mol% G + C of the DNA is: 34 (enumeration in complete genome sequence; the genome size is 1.713 Mbp).

   Type strain: 1482/89, NCTC 12656.

   GenBank accession number (16S rRNA): U30147.

Family II. **Desulfomicrobiaceae** *fam. nov.*

JAN KUEVER, FRED A. RAINEY AND FRIEDRICH WIDDEL

De.sul.fo.mi.cro.bi.a.ce'ae. M.L. masc. n. *Desulfomicrobium* type genus of the family; suffix-*ceae* denotes family; M.L. fem. pl. n. *Desulfomicrobiaceae* the family that comprises the genus *Desulfomicrobium*.

All known members are mesophilic sulfate-reducing bacteria. Organic substrates are incompletely oxidized to acetate. The family comprises one genus so far, *Desulfomicrobium*.

Type genus: **Desulfomicrobium** Rozanova, Nazina and Galushko 1994, 370 (Effective publication: Rozanova, Nazina and Galushko 1988, 518.)

Genus I. **Desulfomicrobium** Rozanova, Nazina and Galushko 1994, 370[VP] (Effective publication: Rozanova, Nazina and Galushko 1988, 518)

BARBARA R. SHARAK GENTHNER AND RICHARD DEVEREUX

De.sul.fo.mi.cro'bi.um. L. pref. *de* from; L. n. *sulfur* sulfur; G. adj. *micros* small; G. n. *bios* life; M.L. neut. n. *Desulfomicrobium* sulfate-reducing, small life.

Short, straight rod- or ellipsoidal-shaped cells, 0.5–0.9 × 1.3–2.9 μm, with rounded ends, either singly or in pairs (Sharak Genthner et al., 1994). **Gram-negative** stain reaction and cell-wall structure. Cells are **motile, usually by a single polar flagellum.** Endospores not formed. **Anaerobic.** Pre-reduced medium or reducing agent required in medium for growth. **Growth can occur by anaerobic respiration with sulfate or sulfoxyanions as terminal electron acceptor, producing H₂S.** Simple organic compounds that serve as electron donors during sulfate respiration include lactate, pyruvate, ethanol, formate, and hydrogen. Sulfate respiration with lactate as electron donor is incomplete, with the formation of acetate and CO_2. Hydrogenase present. **Cells contain *b*- and *c*-type cytochromes. Metabolism can also be fermentative** on simple organic substrates, including pyruvate, malate, or fumarate. **Carbohydrates are not fermented.** No specific vitamins required. **Nitrate not reduced.** No requirement for NaCl. **Desulfoviridin absent.** Optimal temperature 25–30°C. Found in freshwater to brackish to marine anaerobic sediments and mud, in anaerobic stratal or overlying water, and in saturated mineral or organic deposits.

The mol% G + C of the DNA is: 52.5–59.6.

Type species: **Desulfomicrobium baculatum** (Rozanova and Nazina 1976) Rozanova, Nazina and Galushko 1994, 370 (Effective publication: Rozanova, Nazina and Galushko 1988, 519) (*Desulfovibrio baculatus* Rozanova and Nazina 1976, 825; *Desulfomicrobium baculatus* (sic) Rozanova, Nazina and Galushko 1988, 519.)

ENRICHMENT AND ISOLATION PROCEDURES

Stringent anaerobic techniques (Breznak and Costilow, 1994) must be used for enrichment and cultivation of species belonging to the genus *Desulfomicrobium*. Strains can be enriched from various anaerobic habitats such as freshwater to brackish to marine anaerobic sediments and mud; anaerobic stratal or overlying water; and saturated mineral or organic deposits. Samples can be enriched in anaerobic lactate/sulfate mineral medium (Widdel and Bak, 1992; Sharak Genthner et al., 1994, 1997a; DSM medium 63) with the addition of yeast extract (0.02% w/v); however, this medium is nonselective and can result in growth of *Desulfovibrio* species or other lactate-utilizing, sulfate-reducing bacteria. Initial enrichment in lactate/sulfate or pyruvate/sulfate, followed by enrichment in a fermentation medium containing pyruvate, malate and/or fumarate as the energy source, may be useful for enrichment and isolation of *Desulfomicrobium* spp.

MAINTENANCE PROCEDURES

SRB medium described above and prepared by the serum bottle/tube method of Balch and Wolfe (1976) can be used for short-term (weeks) storage of cultures. Stab-cultures in anaerobic agar deeps can be employed for longer-term storage (months) at −70°C without breakage of tubes when heavy-walled anaerobic culture tubes (Bellco Glass, Inc.) are employed. Liquid nitrogen or freeze-drying can be employed for very long term storage of cultures, but care must be taken to use stringent anaerobic techniques when preparing cultures and storage medium (Ljungdahl and Wiegel, 1986; Malik, 1990a; Impey and Phillips, 1991; Gherna, 1994).

DIFFERENTIATION OF THE GENUS *DESULFOMICROBIUM* FROM OTHER GENERA

Desulfomicrobium contains dissimilatory sulfate-reducing species that are physiologically similar to *Desulfovibrio* species but have a nonvibrioid cellular morphology and lack desulfoviridin (Miller and Saleh, 1964), a pigment considered diagnostic for the genus *Desulfovibrio* (Postgate, 1959). The simple desulfoviridin test (Postgate, 1959) can be used to distinguish members of the genus *Desulfomicrobium* from members of the genus *Desulfovibrio*. The ability to ferment pyruvate, malate, or fumarate distinguishes *Desulfomicrobium* spp. from *Desulfovibrio* spp. other than *Desulfovibrio desulfuricans*, although growth is minimal with some of the *Desulfomicrobium* spp. *Desulfomicrobium* spp. are further distinguished from *Desulfovibrio* by the inability to ferment choline.

TAXONOMIC COMMENTS

Macrorestriction analysis using pulsed-field gel electrophoresis revealed unique fragmentation patterns among the four strains (Fig. BXII.δ.7), indicating genotypic variability between *D. apsheronum*, *D. escambiense*, *D. baculatum*, and *D. norvegicum*. *D. escambiense* strain ESC 1 (Lane B) and *D. escambiense* ATCC 51164 (Lane D), the ATCC-deposited culture of *D. escambiense* strain ESC 1, were compared as a control. These two cultures have identical fragmentation patterns, except for a few faint bands not visible in the photograph that were observed during visual inspection of the gel.

Comparisons of 16S rRNA gene sequences and DNA–DNA homologies show the genus to be composed of closely related species (Fig. BXII.δ.8). Similarities of 16S rRNA gene sequences range from 96.7% to 99.7% (Table BXII.δ.6). *D. apsheronum* and

D. baculatum are the most similar species phylogenetically but are physiologically distinct (Tables BXII.δ.7 and BXII.δ.8). Separa-tion of *Desulfomicrobium* from *Desulfovibrio* species is clearly indi-cated by the low (~87%) similarity in 16S rRNA gene sequence.

DIFFERENTIATION OF THE SPECIES OF THE GENUS *DESULFOMICROBIUM*

Tables BXII.δ.7 and BXII.δ.8 present characteristics that differ-entiate the species of *Desulfomicrobium*. Additional characteristics of the species are presented in Table BXII.δ.9. A requirement for yeast extract, or more defined nutritional supplements, may explain earlier discrepancies in reported electron donor usage and fermentation of substrates by the four *Desulfomicrobium* spe-cies.

List of species of the genus Desulfomicrobium

1. **Desulfomicrobium baculatum** (Rozanova and Nazina 1976) Rozanova, Nazina and Galushko 1994, 370[VP] (Effective pub-lication: Rozanova, Nazina and Galushko 1988, 519) (*De-sulfovibrio baculatus* Rozanova and Nazina 1976, 825; *Desul-fomicrobium baculatus* (sic) Rozanova, Nazina and Galushko 1988, 519.)

   ba.cu.la' tum. M.L. adj. *baculatum* shaped like a rod; isolated from water-saturated manganese carbonate ore.

   Characteristics are as described for the genus and as presented in Tables BXII.δ.7, BXII.δ.8, and BXII.δ.9. Tung-state or molybdate are required for ethanol utilization (Hensgens et al., 1994). Oxidation of lactate with sulfate as the electron acceptor is incomplete to acetate and CO_2. Preferentially reduces fumarate and malate to succinate over reduction of sulfate to sulfide when both are present in medium (Sharak Genthner et al., 1997a). Originally re-ported to not use S^0 (Rozanova et al., 1988) as an electron acceptor, but some sulfate-reducing bacterial strains thought to belong to the species *D. baculatum* have been found to use S^0 (Biebl and Pfennig, 1977). Catalyzes S^0 to sulfate with Mn(IV) as an electron acceptor, although the reaction does not support growth (Lovley and Phillips,

FIGURE BXII.δ.7. Macrorestriction fragment analysis of the *Desulfomi-crobium* spp.; *D. apsheronum* (*lane A*), *D. escambiense* ATCC 51164 (*lane B*), *D. baculatum* (*lane C*), *D. escambiense* strain ESC 1 (*lane D*), *D. norvegicum* (*lane E*), and a phage 8 ladder size standard (*lane F*) by pulsed-field gel electrophoresis (Reproduced with permission from B. Sharak Genthner et al., Int. J. Syst. Bacteriol. *47:* 889–892, 1997, ©International Union of Microbiological Societies.)

FIGURE BXII.δ.8. Phylogenetic relationship of *Desulfomicrobium* species to members of the family *Desulfovibrionaceae* and other sulfate-reducing bacteria. The root was determined using the 16S rRNA gene sequence of *Escherichia coli* as the outgroup. The scale bar is in nucleotide substi-tutions per sequence position (Knuc).

TABLE BXII.δ.6. DNA–DNA and 16S rRNA gene sequence similarity values for *Desulfomicrobium* and *Desulfovibrio* strains

Species/strain	% Similarity values[a]				
	D. apsheronum DSM 5918	*D. baculatum* DSM 4028	*D. escambiense* ATCC 51164	*D. norvegicum* NCIMB 8310	*Desulfovibrio desulfuricans* ATCC 27774
D. apsheronum		0.997	11.1	0.992	0.87
D. baculatum	19.8		12.3	0.993	0.877
D. escambiense	0.981	0.981		0.976	0.871
D. norvegicum	13.1	27.4	13		0.877
Desulfovibrio desulfuricans	0.14	1.8	1.5	2.1	

[a]DNA–DNA similarity values are at the lower left; 16S rRNA gene sequence similarity values are at the upper right.

TABLE BXII.δ.7. Differential characteristics of the species of the genus *Desulfomicrobium*[a]

Characteristic	D. baculatum	D. apsheronum	D. escambiense	D. norvegicum
Cell dimensions, μm	$0.5-0.7 \times 0.9-1.9$	$0.7-0.9 \times 1.4-2.9$	$0.5 \times 1.7-2.2$	$0.5-1.0 \times 3.0-5.0$
Single polar flagellum	+	+	nr	+[b]
Cytochromes *b* and *c* present	+	+	nr	+
Can grow autotrophically	No	Yes	No	No
Terminal electron acceptors:				
Sulfite	+	+	nr	+
Sulfur	−	−	nr	+
Nitrate	−	−	−	nr
Oxygen	−	−	−	−
Electron donors:[c]				
Fumarate, malate	−[d]	+	+	−[d]
Growth with fermentation substrates:[c]				
Pyruvate	+	W	+	W
Fumarate	+	+	W	W
Malate	W	+	W	+
Mol% G + C of DNA (T_m)	56.8	52.5	59.6	56.3

[a]For symbols, see standard definitions; nr, not reported; W, weak.

[b]Peritrichous flagellation reported by Biebl and Pfennig (1977).

[c]Yeast extract required at supplemental concentrations.

[d]Poor, or no, sulfate reduction in the presence of fumarate or malate.

TABLE BXII.δ.8. Differentiation of the species of the genus *Desulfomicrobium* by growth, sulfate reduction and/or fermentation, and by fermentation end products with fumarate, malate, or pyruvate in the absence of sulfate[a]

Electron donor and electron acceptor	D. baculatum	D. apsheronum	D. escambiense	D. norvegicum
Fumarate and sulfate:				
Growth response	+ + +	+ + +	+ + +	+ +
Major organic products	Ac, Su	Ac, Su	Ac, La, Su	Ac, La, Su
Sulfate reduction	−	+ +	+ +	−
Malate and sulfate:				
Growth response	+ +	+ + +	+ + +	+ +
Major organic products	Ac, Su	Ac	Ac, La	Ac, La, Su
Sulfate reduction	−	+ +	+ +	−
Pyruvate and sulfate:				
Growth response	+ +	+ +	+ +	+ +
Major organic products	Ac	Ac	Ac, La	Ac
Sulfate reduction	+	+	+	+
Fumarate only:				
Growth response	+ + +	+ + +	+	+
Major organic products	Ac, Su	Ac, Su	Ac, La, Su	Ac, La, Su
Malate only:				
Growth response	+	+	+	+
Major organic products	Ac, Su	Ac, La, Su	Ac, La, Su	Ac, La, Su
Pyruvate only:				
Growth response	±	+	+	±
Major organic products	Ac, Su	Ac, La, Su	Ac, La, Su	Ac, La, Su

[a]Symbols: + + +, excellent; + +, good; +, fair; ±, poor; −, none; Ac, acetate; Su, succinate; La, lactate.

1994a). Reported to fix nitrogen (Rozanova et al., 1988). Uses sulfite or thiosulfate as electron acceptor. Growth in the absence or presence of up to 6% NaCl with 1% NaCl optimal. Optimal growth temperature, 28–37°C; range, 2–41°C. Isolated from water-saturated manganese carbonate ore.

The mol% G + C of the DNA is: 56.8 (T_m).

Type strain: X, DSM 4028, VKM B-1378.

GenBank accession number (16S rRNA): AF030438.

2. **Desulfomicrobium apsheronum** Rozanova, Nazina and Galushko 1994, 370[VP] (Effective publication: Rozanova, Nazina and Galushko 1988, 519.)

ap.she.ron′um. M.L. adj. *Apsheronum* pertaining to waters of the Apsheronum peninsula in Russia; isolated from stratal waters of oil-bearing deposits in the Apsheronum peninsula in Russia.

Characteristics are as described for the genus and as presented in Tables BXII.δ.7, BXII.δ.8, and BXII.δ.9. Oxidation of lactate during sulfate respiration is incomplete: to acetate and CO_2. Autotrophic growth with formate or H_2 has been reported during sulfate respiration. Sulfite and thiosulfate, but not elemental sulfur, are used as electron acceptors (Rozanova et al., 1988). Growth occurs in the absence of NaCl or in the presence of up to 6% NaCl.

TABLE BXII.δ.9. Other characteristics of the species of the genus *Desulfomicrobium*[a]

Characteristic	D. baculatum	D. apsheronum	D. escambiense	D. norvegicum
Cell morphology and arrangement:				
Rods with rounded ends	+	+	+	+
Occur as singles or pairs	+	+	+	+
Endospores formed	−	−	−	−
Optimal temperature, 25–30°C	+	+	+	+
NaCl or vitamins required for growth	−	−	−	−
Desulfoviridin present	−	−	−	−
Desulforubidin present	nr	nr	nr	+
Terminal electron acceptors:				
Sulfur, thiosulfate	+	+	+	+
Sulfite	+	+	nr	+
Nitrate	−	−	−	nr
Oxygen	−	−	−	−
Electron donors:[b]				
Lactate, pyruvate, ethanol, formate, H₂	+	+	+	+
Succinate, acetate, propionate, butyrate, methanol, propanol, butanol, choline	−	−	−	−
Oxalate	−	−	nr	nr
Citrate	nr	−	nr	nr
Glucose, fructose, ribose	−	−	−	nr
Lactose	−	nr	−	−
Growth with fermentation substrates:[b]				
Choline, lactate, glucose	−	−	−	−[c]

[a]For symbols, see standard definitions; nr, not reported.

[b]Yeast extract required at supplemental concentrations.

[c]Other fermentation substrates not used include adonitol, amygdalin, arabinose, arginine, cellobiose, dulcitol, esculin, fructose, galactose, glycerol, glycogen, inositol, inulin, lactose, maltose, mannitol, mannose, melezitose, melibiose, raffinose, rhamnose, salicin, sorbitol, soluble starch, sucrose, threonine, trehalose, and xylose.

Isolated from stratal waters of oil-bearing deposits in the Apsheronum peninsula.

> The mol% G + C of the DNA is: 52.5 (T_m).
> Type strain: 1105, DSM 5918, VKM B-1804.
> GenBank accession number (16S rRNA): U64865.

3. **Desulfomicrobium escambiense** Sharak Genthner, Mundfrom and Devereux 1996, 1189[VP] (Effective publication: Sharak Genthner, Mundfrom and Devereux 1994, 218.)

es.cam'bi.ens.e. M.L. adj. *escambiensis* pertaining to the Escambia River in northwest Florida.

Characteristics are as described for the genus and as presented in Tables BXII.δ.7, BXII.δ.8, and BXII.δ.9. Sulfate and thiosulfate are reduced. Incomplete oxidation of lactate occurs during sulfate respiration with the formation of acetate and CO_2. Reduces benzoate and halogenated benzoate derivatives to corresponding benzyl alcohols during fermentative growth with pyruvate (Sharak Genthner et al., 1997b). Casein and esculin are not hydrolyzed. Gelatin is not liquefied. Isolated from a 3-chlorobenzoate-dechlorinating coculture originally inoculated with anaerobic freshwater sediment.

> The mol% G + C of the DNA is: 59.6 (T_m).

Type strain: ESC 1, ATTC 51164, DSM 10707.
GenBank accession number (16S rRNA): U02469.

4. **Desulfomicrobium norvegicum** Sharak Genthner, Friedman and Devereux 1997a, 891[VP]

nor.ve'gi.cum. M.L. adj. *norvegicum* Norwegian.

Characteristics are as described for the genus and as presented in Tables BXII.δ.7, BXII.δ.8, and BXII.δ.9. Tungstate or molybdate are required for ethanol utilization (Hensgens et al., 1994). Uses S^0 as a terminal electron acceptor (Biebl and Pfennig, 1977). Cytochrome c_3 is reported to be the sulfur reductase in this species (Fauque et al., 1979; Cammack et al., 1984). Reduces Fe(III) to Fe(II) with hydrogen and oxidizes elemental sulfur to sulfate with Mn(IV), but does not gain energy for growth from these processes (Lovley and Phillips, 1994a; *D. norvegicum* designated *D. baculatum* DSM 1743 in this publication). Fumarate has been reported as an additional product of malate fermentation (Miller et al., 1970). Preferentially reduces fumarate and malate to succinate over reduction of sulfate to sulfide when both are present in medium (Sharak Genthner et al., 1997a).

> The mol% G + C of the DNA is: 56.8 (Bd).
> Type strain: Norway 4, DSM 1741, NCIB 8310.

Other Organisms

Several sulfate-reducing strains that have characteristics in common with the members of the genus *Desulfomicrobium* have been isolated. Additional physiological and phylogenetic characterization will be necessary to confirm placement of these strains in the genus *Desulfomicrobium* as either new species or strains of accepted species.

Strain 4474, isolated from an alder swamp near Hamburg, Germany, and strain 5174, isolated from a forest pond near Braunschweig, Germany and designated DSM 1742, were obtained from an enrichment culture on S^0 (Biebl and Pfennig, 1977). Strain 9974 was isolated from a culture designated "*Chloropseudomonas ethylica* N2" (DSM 1743). All three strains consist of short, straight, nonsporeforming, sulfate-reducing rods that are motile by a single polar flagellum and that ferment pyruvate, and contain cytochrome c_3 but do not contain desulfoviridin (Biebl and Pfennig, 1977), thus resembling members of the genus *Desulfomicrobium*. All three strains reduce S^0 to sulfide using ethanol as the electron donor in a medium supplemented with yeast extract (0.1%), as does *D. norvegicum* NCIB 8310 (Biebl and Pfennig, 1977). Cytochrome c_3 is the sulfur reductase (Fauque et al., 1979) in *D. norvegicum* and strain 9974 (DSM 1743). Strain 9974 transports sulfate electroneutrally in symport with sodium ions (Kreke and Cypionka, 1995; designated *Desulfovibrio baculatus* DSM 1743 in this study).

Strain H.L21 is a rod-shaped, sulfate-reducing bacterial strain isolated from intertidal mud (Hoogwatum, Netherlands) in lactate/sulfate marine enrichment medium (Laanbroek and Pfennig, 1981; also designated DSM 2555); it can also grow in a freshwater lactate/sulfate medium. In addition to lactate, strain H.L21 uses H_2, formate, malate, and ethanol as an electron donor for sulfate reduction in the presence of acetate. It also ferments pyruvate and fumarate but not lactate or glucose. It was designated a strain of *Desulfovibrio baculatus* by Laanbroek and Geerligs (1983), i.e., *Desulfomicrobium baculatum*.

Desulfovibrio desulfuricans subsp. *desulfuricans* strain New Jersey (NCIMB 8313) was proposed to belong to the genus *Desulfomicrobium* and was designated *Desulfomicrobium baculatus* strain New Jersey by Pereira et al. (1996). Inclusion in *Desulfomicrobium baculatum* was based on a high degree of similarity between proteins (hydrogenase, cytochrome c_3, adenylyl sulfate reductase, sulfite reductase, i.e., desulforubidin) isolated from this strain and from *Desulfomicrobium baculatus* strain Norway 4, i.e., *Desulfomicrobium norvegicum*. This strain was isolated from a corroding cast iron heat exchanger and was implicated in corrosion of the steel.

The sulfate-reducing bacterial strain WHB (DSM 10293) was tentatively assigned to the genus *Desulfomicrobium* by Drzyzga et al. (1996) based on shared morphological and physiological characteristics. Strain WHB was isolated from anoxic, black, marine sediment from tidal shallows (near Wilhelmshaven, Germany) in lactate/sulfate marine (20 g/l NaCl) enrichment medium, and requires 10–20 g/l NaCl for growth. This strain consists of straight, Gram-negative, motile rods ($0.5–0.6 \times 1.6–2.4$ μm) that lack desulfoviridin, contain cytochrome c_3, and perform incomplete oxidation during sulfate reduction using lactate, H_2, formate, ethanol, and malate but not fumarate, acetone, acetate, butyrate, 2-methylbutyrate, or several aromatic compounds (phenol, benzoate, *para*-hydroxybenzoate, or phenyl acetate). The medium used was chemically defined. Strain WHB is apparently an autotroph because the authors did not indicate that addition of yeast extract or acetate was necessary to support growth on formate or H_2. Fermentative capabilities were not examined. Strain WHB reduced dinitrophenylamines to aminodiphenylamines during oxidation of lactate. The major differences between strain WHB and accepted species of *Desulfomicrobium* are its inability to grow in fumarate/sulfate medium, and the high mol% G + C content of the DNA (82.2% compared with 52–56% for *Desulfomicrobium* species).

A sulfate-reducing bacterium (strain CN-A) which was recently isolated from subsurface sandstone (New Mexico, USA) has been tentatively placed in the *Desulfomicrobium* genus as a new species, "*D. hypogeium*". (L. Krumholz, personal communication). This strain is a nonsporeforming, motile (single, polar flagellum), Gram-negative, short rod ($0.6–0.8 \times 1.4–1.9$ μm) that lacks desulfoviridin. It incompletely oxidizes lactate to acetate and uses ethanol, pyruvate, formate, and H_2 as electron donors during sulfate reduction, requiring either acetate or yeast extract for growth. Growth does not occur with sulfate when acetate, fumarate, fructose, methanol, malate, or choline are electron donors. Pyruvate, lactate, malate, fumarate, and choline do not support fermentative growth. Sulfite and thiosulfate are also used as electron acceptors, but fumarate and nitrate are not. Strain CN-A grows better at 30°C than at either 23°C or 37°C. Based on 16S rDNA sequence analysis, strain CN-A exhibits 98.4–99.0% similarity to accepted *Desulfomicrobium* species and is most closely related to *D. baculatum* and *D. norvegicum* (98.9% and 99.0% similarity, respectively). Failure to use fumarate as an energy source in any capacity and failure to ferment pyruvate, fumarate, or malate even poorly differentiates strain CN-A from the other *Desulfomicrobium* species.

Family III. **Desulfohalobiaceae** *fam. nov.*

JAN KUEVER, FRED A. RAINEY AND FRIEDRICH WIDDEL

De.sul.fo.ha.lo.bi.a.ce'ae. M.L. masc. n. *Desulfohalobium* type genus of the family; suffix *ceae* denotes family; M.L. fem. pl. n. *Desulfohalobiaceae* the family that comprises the genus *Desulfohalobium* and other genera.

Members are either mesophilic or moderately thermophilic sulfate-reducing bacteria. Known mesophilic members oxidize organic substrates incompletely to acetate, whereas the thermophilic member (*Desulfothermus*) performs complete oxidation of organic substrates. The family comprises three genera so far, *Desulfohalobium*, *Desulfonatronovibrio*, and *Desulfothermus*. Detailed properties of all described species of these genera are given in Table BXII.δ.10.

Type genus: **Desulfohalobium** Ollivier, Hatchikian, Prensier, Guezennec and Garcia 1991, 78.

TABLE BXII.δ.10. Differential characteristics of the genera of the family *Desulfohalobiaceae*[a]

Genus	*Desulfohalobium*	*Desulfonatronovibrio*	*Desulfothermus*
Cell morphology:			
Rod-shaped	+	−	+
Vibrio-shaped	−	+	−
Motility	+	+	(+)
Mol% G + C content	57	49	37
Sulfite reductase	DR	nr	nr
Major menaquinone	nr	nr	nr
Substrate oxidation	I	I[c]	C
CO-DH[b]	nr	+	+
Typical electron donors and carbon sources:			
Alkanes (C_6–C_{14})	nr	nr	+
Fatty acids (C_4–C_{18})	nr	nr	+
H_2/acetate	+	+	−
Ethanol	+	−	nr
Formate	+	+	−
Lactate	+	−	−
Pyruvate	+	−	−
Chemolithoautotrophic growth	−	−[c]	−
Fermentative growth	+	−	−
Growth by disproportionation of reduced sulfur compounds	nr	nr[d]	nr
Typical electron acceptors:			
Sulfate	+	+	+
Sulfite	+	+	nr
Sulfur	+	−	nr
Thiosulfate	+	−	+
Optimal pH	6.5–7.0	9.0–9.7	6.5–6.8
Optimal growth temperature (°C)	37–40	37	60–65
NaCl requirement	+	±	+
Habitat (anoxic)	Hypersaline lake	Alkaline soda lakes	Marine

[a]Symbols: nr, not reported; +, present, observed in all species; −, not present, not observed; ±, variable; (+), poorly motile, poor growth; DV, desulfoviridin; DR, desulforubidin; I, incomplete; C, complete.

[b]Carbon monoxide dehydrogenase, a key enzyme indicative of the C_1-pathway for acetyl-CoA.

[c]High CO-DH activity was reported.

[d]Disproportionation of thiosulfate to sulfide and sulfate might indicate growth.

Genus I. **Desulfohalobium** *Ollivier, Hatchikian, Prensier, Guezennec and Garcia 1991, 78*[VP]

BERNARD OLLIVIER, BHARAT K.C. PATEL AND JEAN-LOUIS GARCIA

De.sul.fo.hal.o'bium. L. pref. *de* from; L. n. *sulfur* sulfur; Gr. n. *hals* the sea, salt; Gr. n. *bios* life; M.L. adj. *halobius* living on salt; M.L. masc. n. *Desulfohalobium* a sulfate-reducing, salt-requiring, rod-shaped bacterium.

Rod-shaped cells, 0.7–0.9 × 1–3 μm, with rounded ends. Occur singly or in pairs when H_2 is the energy source. Longer cells, up to 20 μm in length, occur in lactate-containing medium. Motile by one or rarely two polar flagella. Spores are not observed. Gram negative. **Anaerobic, having a respiratory type of metabolism with sulfate or other oxidized sulfur compounds serving as terminal electron acceptors and being reduced to H_2S. The cells contain desulforubidin.** Desulfoviridin is absent. Chemoorganotrophic, using simple organic compounds as carbon sources and as electron donors for anaerobic respiration. **These compounds are incompletely oxidized to acetate and CO_2. Moderately halophilic** with NaCl and $MgCl_2$ required for growth. Vitamins, which can be replaced by yeast extract or bio-Trypcase and acetate, act as carbon sources when cells are grown on H_2. The optimal temperature range for growth is 37–40°C. Isolated from sediments of a hypersaline lake.

The mol% G + C of the DNA is: 57.

Type species: **Desulfohalobium retbaense** Ollivier, Hatchikian, Prensier, Guezennec and Garcia 1991, 78.

FURTHER DESCRIPTIVE INFORMATION

The cells are rods (Fig. BXII.δ.9) that form long chains at the end of growth. Electron micrographs of thin sections of the bacterium reveal a multilayered cell wall that is typical of Gram-negative bacteria. *Desulfohalobium retbaense* grows optimally in the presence of 10% NaCl. Its doubling time when cultivated on lactate medium[1] is about 5 h. It is mesophilic, with 45°C being the upper temperature limit for growth; however, the first subculture from an inoculum incubated at 37°C will grow at temperatures between 45°C and 52°C.

A soluble extract of *D. retbaense* exhibits the characteristic absorption bands of cytochrome c_3. Desulfoviridin, the dissimilatory sulfite reductase present in most species of the genus *Desulfovibrio* (Devereux et al., 1989), is absent. *Desulfohalobium retbaense* pos-

---

1. Lactate medium (g/l of the basal medium described in Enrichment and Isolation Procedures): bio-Trypcase (Bioměrieux, Craponne, France), 1.0; yeast extract, 1.0; and NaCl, 100.0; sodium lactate (instead of sodium acetate), 1.0.

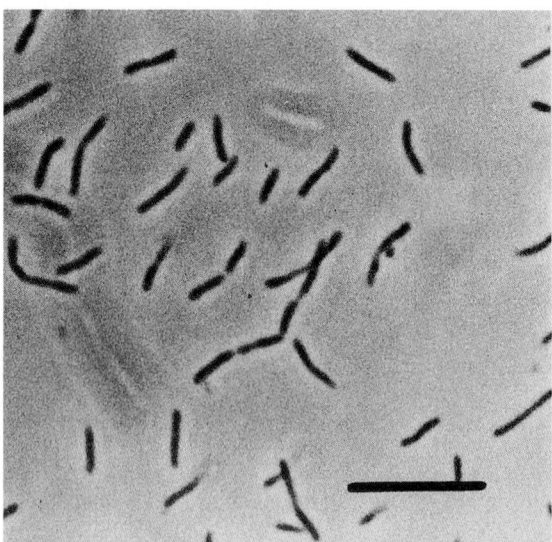

FIGURE BXII.δ.9. Phase contrast micrograph of *Desulfohalobium retbaense*. The isolate was cultivated in lactate medium containing 1 g/l yeast extract and 1 g/l bio-Trypcase. Bar = 10 μm. (Reproduced with permission from B. Ollivier et al., International Journal of Systematic Bacteriology *41:* 74–81, 1991, ©International Union of Microbiological Societies.)

sesses a desulforubidin bisulfite reductase similar to that of *Desulfomicrobium norvegicum* (Lee et al., 1973; Sharak Genthner et al., 1994, 1997a). However, the absorption spectrum of *Desulfohalobium retbaense* bisulfite reductase differs slightly from that of the homologous protein of *Desulfomicrobium norvegicum* in that the former has an additional weak absorption band at 700 nm and a more distinguishable shoulder in the 580-nm region.

The fatty acid distribution of the membrane phospholipids indicates that branched saturated fatty acids account for 30% of the total fatty acids. The following monounsaturated fatty acids are also present: $C_{16:1 \, \omega 9}$, $C_{16:1 \, \omega 7}$, $C_{18:1 \, \omega 9}$, $C_{18:1 \, \omega 7c}$, and $C_{17:1 \, \omega 8}$ fatty acids. The fatty acid profile also contains branched-chain components, such as iso-$C_{17:1 \, iso \, \omega 7c}$ (4%) and $C_{18:1 \, \omega 6}$.

Halophilic microorganisms are divided into three major groups and include (i) the slight halophiles, which exhibit optimal growth at salinities of 2–5%, (ii) the moderate halophiles, which grow optimally at 6–10% NaCl, and (iii) the extreme halophiles, which grow optimally with 20–25% NaCl (Ollivier et al., 1994). Currently, the only moderately halophilic sulfate-reducing bacteria known to exist in hypersaline environments are *Desulfohalobium retbaense*, *Desulfovibrio halophilus* (Caumette et al., 1991a), *Desulfovibrio gabonensis* (Tardy-Jacquenod et al., 1996), *Desulfovibrio oxyclinae* (Krekeler et al., 1997), *Desulfovibrio profundus* (Bale et al., 1997), *Desulfovibrio vietnamensis* (Nga et al., 1996), and *Desulfocella halophila* (Brandt et al., 1999). Extremely halophilic sulfate-reducing bacteria have so far not been isolated from these environments and, therefore, it has been suggested that the reduction of sulfate in hypersaline ecosystems may be due only to the activity of moderately halophilic microorganisms (Caumette, 1993). However, *D. retbaense*, while not an extreme halophile, can grow at NaCl concentrations as high as 24%. Interestingly, the salt content of the environment from which *D. retbaense* was isolated is 34%, which indicates either that this bacterium cannot be very active in its ecosystem (Lake Retba, Senegal, West Africa) and perhaps prefers to grow in niches with lower concentrations of NaCl (e.g., areas of freshwater river in-

flow where the salt concentration is decreased), or that the artificial conditions under which this bacterium is cultured in the laboratory differ from the mineral or organic content of the natural environment and thus might influence and alter the growth response of the bacterium.

ENRICHMENT AND ISOLATION PROCEDURES

Desulfohalobium retbaense can be enriched from sediments of hypersaline lakes using an anaerobic medium (Ollivier et al., 1991) that contains NaCl at a concentration of 100–240 g/l, H_2 or lactate as the energy source, and Na_2S or sodium dithionite as the reductive agent[2]. At least three subcultures under the same growth conditions with incubation at temperatures of up to 45°C are necessary for initiating enrichments. Subsequently, the enrichment cultures are serially diluted using the method of Hungate (1969) in anaerobic roll tubes containing the basal medium, H_2 as the energy source, sulfate as the electron acceptor, and purified agar at a concentration of 2%. Several colonies are picked, and the process of serial dilution in roll tubes is repeated. Culture purity is tested on (1) a complex sulfate-free medium containing 0.1% bio-Trypcase, 0.1% yeast extract, and 20 mM glucose and (2) a rich complex NaCl-free medium. There should be no growth on these media.

MAINTENANCE PROCEDURES

Stock cultures are maintained in the medium described by Ollivier et al. (1991) by monthly serial transfers. Liquid cultures retain viability after several weeks of storage at 4°C. Cultures can be maintained indefinitely by lyophilization or by storage at −80°C in the basal medium containing 20% glycerol (v/v). Viability is best maintained if the cells are obtained from mid-log phase cultures.

DIFFERENTIATION OF THE GENUS *DESULFOHALOBIUM* FROM OTHER GENERA

Differentiation of the genus from the genera *Desulfomicrobium* and *Desulfonatronovibrio*, with which it might be confused, is presented in Table BXII.δ.11. Major differences between *Desulfohalobium retbaense* and the six other moderately halophilic sulfate-reducing species, namely *Desulfovibrio halophilus*, *Desulfovibrio gabonensis*, *Desulfovibrio oxyclinae*, *Desulfovibrio profundus*, *Desulfovibrio vietnamensis*, and *Desulfocella halophila* are listed in Table BXII.δ.12.

TAXONOMIC COMMENTS

Desulfohalobium retbaense is currently a member of the family *Desulfohalobiaceae* in the *Deltaproteobacteria*, together with the genera *Desulfonatronovibrio* and *Desulfothermus* (Fig. BXII.δ.10). However, we believe that future revision of the family *Desulfohalobiaceae* should be considered in view of the following: (1) *Desulfohalobium retbaense* together with *Desulfonatronovibrio hydrogenovorans* and *De-*

---

2. Anaerobic medium of Ollivier et al. (1991) (per liter of distilled water): NH_4Cl, 1.0 g; K_2HPO_4, 0.3 g; KH_2PO_4, 0.3 g; Na_2SO_4, 3.0 g; NaCl, 100.0 g; sodium acetate, 1.0 g; resazurin, 0.001 g; $MgCl_2 \cdot 6H_2O$, 20.0 g; KCl, 4.0 g; trace element solution of Imhoff-Stuckle and Pfennig (1983), 1 ml; sodium selenite solution of Pfennig et al. (1981), 1 ml. After autoclaving, 0.2 ml of 2% (w/v) $Na_2S \cdot 9H_2O$ and 1 ml of 10% (w/v) $NaHCO_3$ (sterile, anaerobic solutions), as well as 0.2 ml of the vitamin solution of Pfennig et al. (1981) and 0.1 ml of a 0.2% (w/v) sodium dithionite solution (filter-sterilized solutions), are injected into the 60-ml serum bottles (20 ml of medium) before inoculation. In the case of roll tubes, 2% agar is added to the medium in Hungate tube (5 ml). A gas mixture containing H_2/CO_2 (80:20) is added at a pressure of 200 kPa after sterilization.

TABLE BXII.δ.11. Differentiation of *Desulfohalobium retbaense* from its closest phylogenetic relatives *Desulfomicrobium* spp. and *Desulfonatronovibrio hydrogenovorans*[a]

Characteristic	*Desulfohalobium retbaense*[b]	*Desulfomicrobium escambiense*[c]	*Desulfomicrobium norvegicum*[d]	*Desulfomicrobium baculatum*[e]	*Desulfomicrobium apsheronum*[f]	*Desulfonatronovibrio hydrogenovorans*[g]
Morphology:						
Cocco bacilli-rods		+				
Rods	+		+	+	+	
Vibrios						+
Cell dimensions, μm	0.7-0.9 × 1–20	0.5 × 1.7–2.2	0.5–1 × 3–5	0.6 × 1.3	0.7–0.9 × 1.4–2.9	0.5 × 1.5–2.0
Flagellar arrangement	1-2 polar	nd	nd	1 polar	1 polar	1 polar
Growth at 45°C	+	nd	nd	−	−	−
Upper NaCl concentration for growth (%)	24	nd	nd	5–6	8	12
Electron donors (with sulfate):						
Lactate	+	+	+	+	+	−
Ethanol	+	+	+	−	+	−
Fumarate	−	−	+	−	+	−
Malate	−	−	+	+	+	−
Pyruvate fermentation (without sulfate)	+	+	+	−	+	−
Mol% G + C of DNA	57	60	57	57	52	49

[a]For symbols, see standard definitions; nd, not determined.
[b]Data from Ollivier et al. (1991).
[c]Data from Sharak Genthner et al. (1994).
[d]Data from Sharak Genthner et al. (1997a).
[e]Data from Rozanova and Nazina (1976).
[f]Data from Rozanova et al. (1988).
[g]Data from Zhilina et al. (1997).

TABLE BXII.δ.12. Differentiation of *Desulfohalobium retbaense* from other moderately halophilic sulfate-reducing bacteria[a]

Characteristic	*Desulfohalobium retbaense*[b]	*Desulfovibrio halophilus*[c]	*Desulfovibrio gabonensis*[d]	*Desulfovibrio oxyclinae*[e]	*Desulfovibrio profundus*[f]	*Desulfovibrio vietnamensis*[g]	*Desulfocella halophila*[h]
Cell shape:							
Rods	+						
Curved rods		+	+	+	+	+	+
Dimensions, μm	0.7–0.9 × 1–20	0.6 × 2.5–5	0.4 × 2.4	0.5 × 2.3	0.5–1 × 1–2	0.8–1 × 2.5–3	0.5–0.7 × 2–4
Desulfoviridin present	−	+	+	+	+	+	−
Desulforubidin present	+	−	−	−	−	−	nd
Salinity:							
Optimal, %	10	6–7	5–6	6–10	0.6–8	5	4–5
Range, %	3–24	3–18	1–17	2.5–22.5	0.2–10	0–10	2–19
Mol% G + C of DNA	57	61	60	61	53	61	35
Electron donors (with sulfate):							
Hydrogen/acetate	+	+	+	+	+	−	−
Ethanol	+	+	+	+	−	+	−
Lactate	+	+	+	+	+	+	−
Fumarate	−	−	+	−	−	+	−
Malate	−	−	+	+	−	+	−
Succinate	−	nd	+	−	−	−	−
Fermentation of pyruvate (without sulfate)	+	−	+	+	+	−	−

[a]For symbols, see standard definitions; nd, not determined.
[b]Data from Ollivier et al. (1991).
[c]Data from Caumette et al. (1991a).
[d]Data from Tardy-Jacquenod et al. (1996).
[e]Data from Krekeler et al. (1997).
[f]Data from Bale et al. (1997).
[g]Data from Nga et al. (1996).
[h]Data from Brandt et al. (1999).

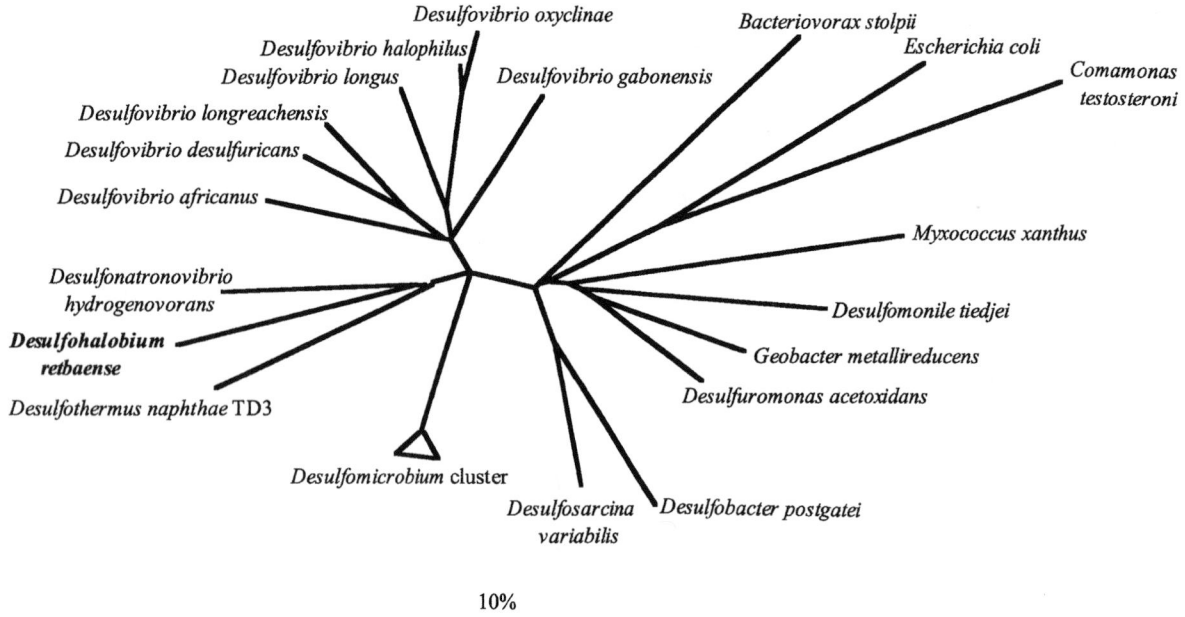

FIGURE BXII.δ.10. Phylogenetic relationship, based on 16S rRNA sequence analysis, of *Desulfohalobium retbaense*, a member of the family *Desulfohalobiaceae*, with other members of that family, including *Desulfonatronovibrio hydrogenovorans* and *Desulfothermus naphthae* TD3; and with species of the genera *Desulfovibrio* (family *Desulfovibrionaceae*) and *Desulfomicrobium* (family *Desulfomicrobiaceae*). All sequences were obtained from the Ribosomal Database Project (RDP) (Maidak et al., 1996), with the exception of the sequences for *Desulfonatronovibrio hydrogenovorans* and *Desulfovibrio oxyclinae*, which were obtained from EMBL (accession number X99234) and GenBank (accession number U33316), respectively. The methods for alignment and phylogenetic evaluation were performed as described previously (Andrews and Patel, 1996). Bar = 10 nucleotide changes per 100 nucleotides.

sulfothermus naphtae are closely related to members of the genus *Desulfomicrobium*; (2) desulfoviridin is absent in *Desulfohalobium retbaense*, in *Desulfonatronovibrio hydrogenovorans*, in *Desulfothermus naphtae*, and in members of the genus *Desulfomicrobium*, but is found in all members of the genus *Desulfovibrio* (Devereux et al., 1989); and (3) desulforubidin is present in *Desulfohalobium retbaense* (Ollivier et al., 1991), *Desulfomicrobium baculatum* (DerVartanian, 1994), and *Desulfomicrobium norvegicum* (Lee et al., 1973; Sharak Genthner et al., 1994, 1997a). However, the presence of desulforubidin has yet to be investigated in *Desulfonatronovibrio hydrogenovorans* (Zhilina et al., 1997), *Desulfothermus naphtae* (Rueter et al., 1994), and *Desulfomicrobium* spp. (Rozanova et al., 1988; Gogotova and Vainshtein, 1989; Sharak Genthner et

al., 1994; Krumholz et al., 1999; Langendijk et al., 2001). If it is detected in these species, then the establishment of a new family should be considered in order to accommodate members of the genera *Desulfohalobium*, *Desulfonatronovibrio*, *Desulfothermus*, and *Desulfomicrobium*.

ACKNOWLEDGMENTS

The authors thank P. Roger for helpful discussion.

FURTHER READING

Ollivier, B., C.E. Hatchikian, G. Prensier, J. Guezennec and J.-L. Garcia. 1991. *Desulfohalobium retbaense* gen. nov., sp. nov., a halophilic sulfate-reducing bacterium from sediments of a hypersaline lake in Senegal. Int. J. Syst. Bacteriol. *41*: 74–81.

List of species of the genus Desulfohalobium

1. **Desulfohalobium retbaense** Ollivier, Hatchikian, Prensier, Guezennec and Garcia 1991, 78[VP]
   ret.ba.en′ se. M.L. adj. *retbaense* pertaining to Retba Lake in Senegal.

   The characteristics are as described for the genus and as listed in Tables BXII.δ.11 and BXII.δ.12. Additional characteristics are as follows. Cells contain cytochrome c_3 and desulforubidin. Lactate, ethanol, and pyruvate are incompletely oxidized to acetate and CO_2. Pyruvate is fermented slowly. Glycerol, fructose, glucose, fumarate, malate, succi-

nate, choline, yeast extract, and bio-Trypcase are not used. NaCl and $MgCl_2$ are required for growth. Moderately halophilic, growing in media with up to 24% NaCl. The optimal NaCl concentration for growth is 10%.

Optimal temperature, 37–40°C. The pH range for growth is 5.5–8.0; optimal, 6.5-7.

Isolated from sediments of a hypersaline African lake, Retba Lake, near Dakar, Senegal.

The mol% G + C of the DNA is: 57 (HPLC).

Type strain: HR100, DSM 5692.

GenBank accession number (16S rRNA): U48244, X99235.

Genus II. **Desulfonatronovibrio** Zhilina, Zavarzin and Rainey in Zhilina, Zavarzin, Rainey, Pikuta, Osipov and Kostrikina 1997, 149[VP]

GEORGE A. ZAVARZIN AND TATJANA N. ZHILINA

De.sul.fo.nat.ro.no.vib' rio. L. pref. *de* from; L. n. *sulfur* sulfur; M.L. n. *natron* soda; M.L. masc. n. *vibrio* that which vibrates; L. n. *Desulfonatronovibrio* sulfate-reducing vibrio from soda environment.

Cells are minute vibrios, single or sigmoid in pairs, sometimes spirilloid, 0.5×1.5-2 μm. Gram negative. **Motile by a single polar flagellum.** Endospores not formed. **Anaerobic. Possess a respiratory type of metabolism, with sulfate or other oxidized sulfur compounds serving as terminal electron acceptors and being reduced to sulfide. Extremely alkaliphilic.** No growth at pH 7; optimal pH 9.0–9.7; maximum pH 10.2. **Obligately dependent on sodium and carbonates in the growth medium.** Optimal temperature 37–40°C. **Lithoheterotrophic.** Utilize H_2 or formate for reduction of sulfur compounds. Utilize acetate plus vitamins or yeast extract for anabolism. Desulfoviridin is absent. Cells contain *c*-type cytochromes. Habitat: soda lakes. The genus currently includes only a single species.

The mol% G + C of the DNA is: 47.9–48.5.

Type species: **Desulfonatronovibrio hydrogenovorans** Zhilina, Zavarzin and Rainey *in* Zhilina, Zavarzin, Rainey, Pikuta, Osipov and Kostrikina 1997, 149.

FURTHER DESCRIPTIVE INFORMATION

The cells are vibrioid with a single polar flagellum (Fig. BXII.δ.11) and occur singly or in S-shaped pairs. Short spirilla develop under suboptimal conditions (Fig. BXII.δ.12) The cell wall has a typical Gram-negative structure in ultrathin sections.

The cell membrane and cytoplasmic fraction contain *c*-type cytochromes with absorption maxima at 552, 521.8, and 417.8 nm.

Sulfate and sulfite serve as electron acceptors. Sulfur, fumarate, and nitrate are not reduced. Thiosulfate may be disproportioned: equimolar quantities of sulfate and sulfide have been observed to form from 10 mmol of thiosulfate. Sulfide is the only product of catabolism.

The fatty acid profile of type strain *D. hydrogenovorans* contains compounds that are typical for bacterial membranes. Saturated fatty acids account for 66.7% of the total fatty acids, with $C_{16:0}$ and $C_{18:0}$ acids predominating. Branched saturated fatty acids account for only 11% of the total fatty acids. The following monounsaturated fatty acids, which account for 12% of the total fatty acids, are also present: $C_{16:1}$, $C_{18:1 \omega9}$, $C_{18:1 \omega11}$, and $C_{19:1}$.

D. hydrogenovorans was first isolated from bottom deposits of the equatorial soda-depositing Lake Magadi. In central Asia, it is ubiquitous in alkaline lakes south of Baikal in the cryoarid steppe zone (Pikuta et al., 1997).

As a H_2-scavenging alkaliphile, *D. hydrogenovorans* may act as a hydrogen sink in an anaerobic, alkaliphilic community decomposing organic matter (Zavarzin et al., 1996). For instance, it develops in co-culture with alkaliphilic fermentative spirochetes growing on carbohydrates and produces an acetogenic shift in their metabolism (Pikuta et al., 1998b). As an extremely alkaliphilic sulfate-reducer, *Desulfonatronovibrio* participates in the alkalinization of soda lakes where sulfate reduction was for a long time considered to be an important process for the development of alkalinity. However, sulfate reducers capable of growing at pH values near 10 were previously unknown.

ENRICHMENT AND ISOLATION PROCEDURES

Strains of *Desulfonatronovibrio* can be enriched and isolated from the alkaline soda lakes under strictly anaerobic conditions in the selective alkaline mineral medium II[1] of Zhilina and Zavarzin (1994), which has a high sodium carbonate content. Incubation is under an atmosphere of H_2. N_2 can substitute for the H_2 if Na-formate (5 g/l) is added as the electron donor.

Isolation is by serial dilution in liquid medium II with H_2 as

---

1. Medium II has the following composition (g/l distilled water): $NaHCO_3$, 15.0; Na_2CO_3, 10.0; NaCl, 10.0; Na_2SO_4, 3.0; NH_4Cl, 0.5; KCl, 0.2; K_2HPO_4, 0.2; $Na_2S\cdot9H_2O$, 0.5; yeast extract, 0.5; and 1 ml trace element solution (Whitman et al., 1982).

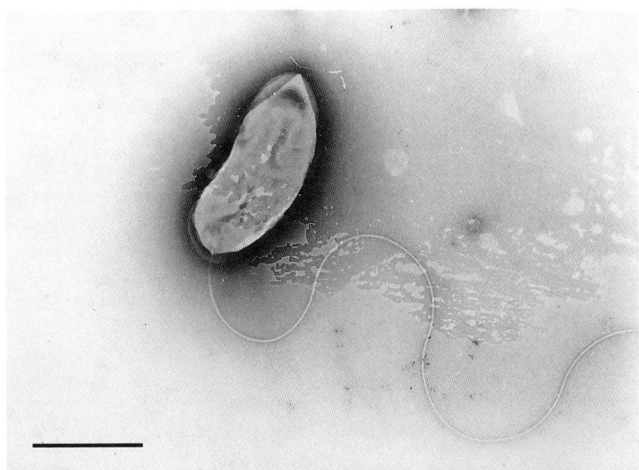

FIGURE BXII.δ.11. Negatively stained cell of *Desulfonatronovibrio hydrogenovorans* Z-7952 isolated from Lake Shara-Nur (Tuva) showing the monopolar flagellum. Bar = 1 μm.

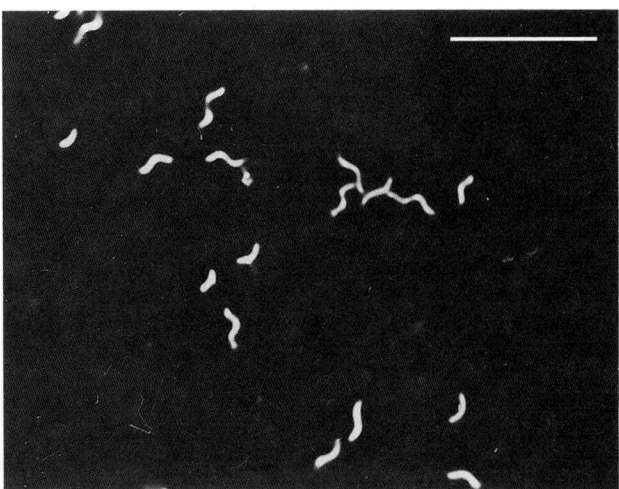

FIGURE BXII.δ.12. Cells of *Desulfonatronovibrio hydrogenovorans* Z-7935 as viewed with the anoptral microscope. Bar = 10 μm.

the substrate, followed by cloning by single colony isolation in roll tubes. For roll tube cultivation, 2% (w/v) agar (Difco) is added to 4-ml portions of carbonate-free medium II with sodium formate as the substrate. The carbonate solution is injected into the Hungate tubes along with melted agar after sterilization. Colonies are yellowish, translucent, lens shaped, and less than 0.2 mm in diameter.

MAINTENANCE PROCEDURES

Desulfonatronovibrio is maintained by regular subculturing in liquid medium and can be stored in a refrigerator at 6°C. Stock cultures are transferred every 2–3 months, but lysis of cells may occur. Strains of *Desulfonatronovibrio* may be stored indefinitely in liquid nitrogen.

DIFFERENTIATION OF THE GENUS *DESULFONATRONOVIBRIO* FROM OTHER GENERA

Genus *Desulfonatronovibrio* is differentiated from other genera of sulfate-reducing bacteria by its extreme alkaliphily and its obligate dependence on sodium and carbonate ions. The strains are morphologically and nutritionally similar to lithotrophic species

of *Desulfovibrio* type but lack desulfoviridin. The fatty acid profile of *Desulfonatronovibrio* contains compounds that are associated with sulfate-reducing species; however, based on its fatty acid profile, *Desulfonatronovibrio* resembles only *Desulfohalobium* and *Desulfomicrobium* (correlation coefficient, 0.23). It lacks $C_{17:1\ iso}$ fatty acid, which is a biomarker of *Desulfovibrio*. *Desulfonatronovibrio* differs from halophilic *Desulfohalobium* (Ollivier et al., 1991) by its lower salinity range, substrate specificity, morphology, 8.5% lower mol% G + C content, and 16S rDNA sequence phylogenetic position. *Desulfonatronovibrio* also has an unusually narrow range of electron donors. Another alkaliphilic sulfate reducer *Desulfonatronum* (Pikuta et al., 1998b) utilizes a number of organic compounds, such as ethanol and lactate, that are not used by *Desulfonatronovibrio*. It also differs from *Desulfonatronovibrio* by a 9% higher mol% G + C content, 16S rDNA sequence positioning, and cell size and shape.

TAXONOMIC COMMENTS

The phylogenetic position of *Desulfonatronovibrio* was determined by F. Rainey based on 16S rDNA sequence analysis (Fig. BXII.δ.13). *Desulfonatronovibrio* belongs to the cluster of sulfate-

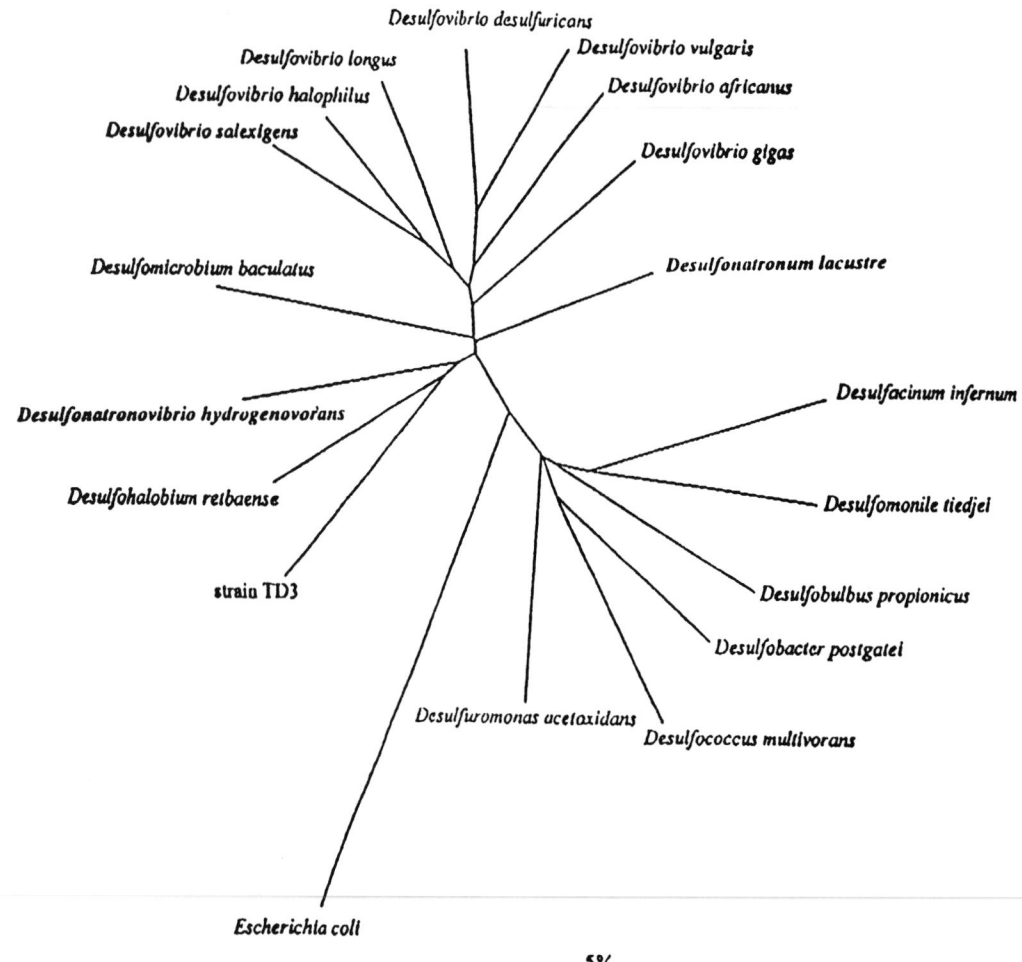

FIGURE BXII.δ.13. Phylogenetic dendrogram indicating the position of *Desulfonatronovibrio* and *Desulfonatronum* within the radiation of sulfate-reducing bacteria of the *Deltaproteobacteria*. The position of the root was determined by including *Escherichia coli* as an outgroup organism. Strain TD3 is proposed in this chapter as the type strain of *Desulfothermus naphthae* gen. nov., sp. nov. Bar = 5 inferred substitutions per 100 nucleotides.

reducing eubacteria within the *Deltaproteobacteria*. It represents a distinct lineage, clustering with extremely halophilic *Desulfohalobium retbaense* (Ollivier et al., 1991) and hydrocarbon-utilizing strain TD3 (Rueter et al., 1994), with 16S rDNA sequence similarities of 88.7% and 86.6%, respectively. Its status as a separate genus is based on its distinct branch point and its 16S rDNA sequence difference of >11% compared with its nearest relative, as well as its distinct phenotypic characteristics.

FURTHER READING

Pikuta, E.V., A.M. Lysenko and T.N. Zhilina. 1997. Distribution of *Desulfonatronovibrio hydrogenovorans* in soda lakes of Tuva. Mikrobiologiya *66*: 216–221.

Zhilina, T.N., G.A. Zavarzin, F.A. Rainey, E.N. Pikuta, G.A. Osipov and N.A. Kostrikina. 1997. *Desulfonatronovibrio hydrogenovorans* gen. nov., sp. nov., an alkaliphilic, sulfate-reducing bacterium. Int. J. Syst. Bacteriol. *47*: 144–149.

List of species of the genus Desulfonatronovibrio

1. **Desulfonatronovibrio hydrogenovorans** Zhilina, Zavarzin and Rainey *in* Zhilina, Zavarzin, Rainey, Pikuta, Osipov and Kostrikina 1997, 149[VP]

   hy.dro.ge.no.vo'rans. M.L. n. *hydrogen* hydrogen; M.L. part. adj. *vorans* devouring, utilizing; M.L. part. adj. *hydrogenovorans* hydrogen utilizing.

   The characteristics are as given for the genus and under Further Descriptive Information. The following are additional characteristics. The pH range for growth is 8.0–10.2, and the pH optima for different strains lie in the range 9.0– 9.7 (that of the type strain is 9.5–9.7). Sodium and carbonate ions are required for growth. The optimal NaCl concentration is 3% (w/v); growth of the type strain occurs within the range 1–12%. No growth is evident in the type strain without NaCl, whereas other strains need only sodium carbonates.

   Other strains of *D. hydrogenovorans* are Z-7952, Z-7953 (DSM 10321), and Z-7954, which were isolated from soda lakes in Tuva, Central Asia. These are similar to the type strain *D. hydrogenovorans*, with small differences in pH and temperature optima as well as mol% G + C content. Strains have a high degree of DNA–DNA hybridization similarity (90–100% with the type strain *D. hydrogenovorans* from Lake Magadi), despite their remote geographic habitats. All the Tuva strains differ from the type strain by their ability to grow in the absence of NaCl, instead requiring only sodium carbonates (Zavarzin et al., 1996 Pikuta et al., 1997).

   The optimal temperature for the type strain is 37°C; maximum, 43°C.

   The mol% G + C of the DNA is: 47.9–48.5 (T_m).

   Type strain: Z-7935, DSM 9292.

   GenBank accession number (16S rRNA): X99234.

Genus III. **Desulfothermus** *gen. nov.*

JAN KUEVER, FRED A. RAINEY AND FRIEDRICH WIDDEL

De.sul.fo.ther'mus. L. pref. *de* from; L. n. *sulfur* sulfur; Gr. adj. *thermus* hot; M.L. masc. n. *Desulfothermus* sulfate reducer living in hot places.

Rod-shaped to slightly curved cells, 0.8–1.0 × 2.0–3.5 µm. Occur singly or in pairs. May grow as individual cells or as compact clumps. Cells tend to stick to hydrophobic surfaces. Spore formation is not observed. Gram negative. Slow motility may occur. Strict anaerobe with a respiratory type of metabolism. **Chemoorganotroph**, using alkanes (C_6–C_{14}), and long-chain fatty acids (C_4–C_{18}) **as electron donors and as carbon sources**; these compounds are completely oxidized to CO_2 via the anaerobic C_1-pathway (carbon monoxide dehydrogenase [CO-DH] pathway, Wood pathway) as indicated by CO–DH activity. **Sulfate and thiosulfate serve as terminal electron acceptors and are reduced to H_2S.** The optimal pH range for growth is 6.5–6.8. The optimal temperature range for growth is 60–65°C. Mesophilic species have not been described. Anoxic media containing a reductant (usually sulfide), NaCl, and $MgCl_2$ are necessary for growth. Vitamins are not required. Occur in anoxic marine sediment rich in aliphatic hydrocarbons.

The mol% G + C of the DNA is: 37.4.

Type species: **Desulfothermus naphthae** *sp. nov.*

FURTHER DESCRIPTIVE INFORMATION

The type strain of *D. naphthae* has rod-shaped to slightly curved cells which may be slightly motile, and stains Gram-negative. The temperature range for growth is between 50 and 69°C; the optimal is 60–65°C. The pH range for growth is 6.1–7.1, with optimal growth between 6.5 and 6.8.

D. naphthae reduces sulfate and thiosulfate to H_2S. Nitrate cannot serve as an electron acceptor. Alkanes (C_6–C_{14}), and long-

chain fatty acids (C_4–C_{18}) are utilized as electron donors and organic carbon sources. Pentadecane and hexadecane are utilized very slowly. No growth on sugars, amino acids, dicarboxylic acids, aromatic compounds, and complex substrates such as yeast extract or peptone. 1-Alkenes, primary alcohols, and aldehydes, which are regarded as possible intermediates during growth on alkanes, are not used. All organic substrates are completely oxidized to CO_2 via the anaerobic C_1-pathway (carbon monoxide dehydrogenase [CO-DH] pathway, Wood pathway) as indicated by CO–DH activity (Ehrenreich, 1996). Ammonium ions are used as the nitrogen source. Vitamins are not required as growth factors. Desulfoviridin is not present (Rueter et al., 1994). Growth cannot be monitored by increase in turbidity, as even after production of 20 mM sulfide only a slight increase in turbidity has been observed.

Desulfothermus naphthae strains occur in anoxic marine sediments rich in aliphatic hydrocarbons.

D. naphthae may be cultured in a defined medium (e.g., as indicated for *Desulfobacter*) with sulfate as electron acceptor and alkanes as the electron donor and carbon source. The type strain is dependent on an elevated salt concentration; best growth is obtained in a medium containing 25 mM Mg^{2+}. The presence of 380 mM Cl^- ions is essential. This corresponds to 21 g/l NaCl and 5 g/l $MgCl_2 \cdot H_2O$.

ENRICHMENT AND ISOLATION PROCEDURES

For the selective enrichment of *Desulfothermus* spp., the defined marine sulfate medium described for the isolation of *Desulfobacter* spp. may be used. Instead of acetate, the medium is covered with *n*-decane (10% in 2,3,4,6,8,8-heptamethylnonane as inert carrier phase) or crude oil. Bottles of 100-ml capacity are inoculated with marine sediment slurry (approximately 5 ml) and kept anoxic by flushing with a mixture of N_2/CO_2 (90:10) and sealing with butyl rubber stoppers. Enrichment cultures are incubated at 60°C. Cultures should not be shaken, because the type strain shows better growth without shaking. After intense formation of H_2S has occurred (approximately three weeks), transfers to fresh medium are made. The enrichments should be mixed well before subculturing.

Pure cultures of *Desulfothermus* strains are obtained by repeated application of the agar dilution method, using a short-chain fatty acid as substrate. The type strain was isolated using caproate. The choice of substrate should depend on whether it was used by the enrichment culture. The method is similar to that described for the genus *Desulfobacter*, except that caproate medium is added to the molten concentrated agar solution.

MAINTENANCE PROCEDURES

Stock cultures are transferred every 2–3 months. Towards the end of growth or at the beginning of the stationary phase, cultures are refrigerated and stored at 2–5°C in the dark. *Desulfo-*

thermus strains may be preserved indefinitely by suspension of cells in anoxic medium containing 5% dimethylsulfoxide and storage in liquid nitrogen.

DIFFERENTIATION OF THE GENUS *DESULFOTHERMUS* FROM OTHER GENERA

A distinctive characteristic of *Desulfothermus* is the capacity for growth with *n*-alkanes as the only organic substrate at temperatures around 60°C. However, a definite differentiation requires comparative 16S rDNA sequence analysis and DNA–DNA hybridization.

TAXONOMIC COMMENTS

Based on 16S rRNA gene sequence data, *Desulfothermus naphthae* forms a distinct lineage within the *Deltaproteobacteria*, but clusters with members of the family *Desulfovibrionaceae*. The closest relatives are members of the genera *Desulfohalobium* and *Desulfonatronovibrio*.

FURTHER READING

Rueter, P., R. Rabus, H. Wilkes, F. Aeckersberg, F.A. Rainey, H.W. Jannasch and F. Widdel. 1994. Anaerobic oxidation of hydrocarbons in crude oil by new types of sulphate-reducing bacteria. Nature *372*: 455–458.

List of species of the genus Desulfothermus

1. **Desulfothermus naphthae** *sp. nov.*
   naph' thae. Gr. fem. *naphtha* naphtha, crude oil; M.L. n. *naphthae* meant to indicate the ability to oxidize crude oil.

   Characteristics are as described for the genus. Occur in anoxic marine sediments rich in aliphatic hydrocarbons. The type strain was enriched from an anoxic sediment sample from The Guaymas Basin (Gulf of California) with crude oil and isolated with caproate as sole electron donor and carbon source.

   The mol% G + C of the DNA is: 37.4 (HPLC).
   Type strain: TD3, DSM 13418, JCM 12298.
   GenBank accession number (16S rRNA): X80922.

Family IV. **Desulfonatronumaceae** *fam. nov.*

JAN KUEVER, FRED A. RAINEY AND FRIEDRICH WIDDEL

De.sul.fo.na.tro.nu.ma' ce.ae. M.L. masc. n. *Desulfonatronum* type genus of the family; *-aceae* ending to denote family; M.L. fem. pl. n. *Desulfonatronumaceae* the *Desulfonatronum* family.

Members are mesophilic and alkaliphilic sulfate-reducing bacteria. The known member oxidizes a limited range of organic substrates incompletely to acetate. The family comprises the genus *Desulfonatronum*.

Type genus: **Desulfonatronum** Pikuta, Zhilina and Rainey 1998a, 631[VP] (Effective publication: Pikuta, Zhilina and Rainey *in* Pikuta, Zhilina, Zavarzin, Kostrikina Osipov and Rainey 1998b, 112.)

Genus I. **Desulfonatronum** *Pikuta, Zhilina and Rainey 1998a, 631[VP] (Effective publication: Pikuta, Zhilina, and Rainey in Pikuta, Zhilina, Zavarzin, Kostrikina Osipov and Rainey 1998b, 112)*

TATJANA N. ZHILINA

De.sul.fo.na.tro' num. L. pref. *de* from; L. n. *sulfur* sulfur; M.L. n. *natron* soda; M.L. masc. n. *Desulfonatronum* a sulfate reducer inhabiting soda lakes.

Cells are vibrios 0.7–0.9 × 2.0–3.0 μm, occurring singly or sigmoid in pairs. Cells may be spirilloid under suboptimal conditions. Asporogenous. Gram negative. Motile by a single polar flagellum. **Anaerobic. Possess a strictly respiratory type of metabolism with sulfate and other oxidized sulfur compounds serving as terminal electron acceptors** and being reduced to H_2S. **Extremely alkaliphilic.** No growth at pH values below 8; pH optimum, 9.5; pH maximum, 10. **Obligately dependent on sodium**

and carbonates. **Lithoheterotrophic.** Utilizes H_2, formate, ethanol, or lactate for reduction of sulfur compounds. Utilizes acetate plus vitamin or yeast extract for anabolism. Desulfoviridin is not present. Cells contain *c*-type cytochromes. The habitat is bottom sediments of alkaline lakes. The genus presently includes only a single species.

The mol% G + C of the DNA is: 56–57.

Type species: **Desulfonatronum lacustre** Pikuta, Zhilina and

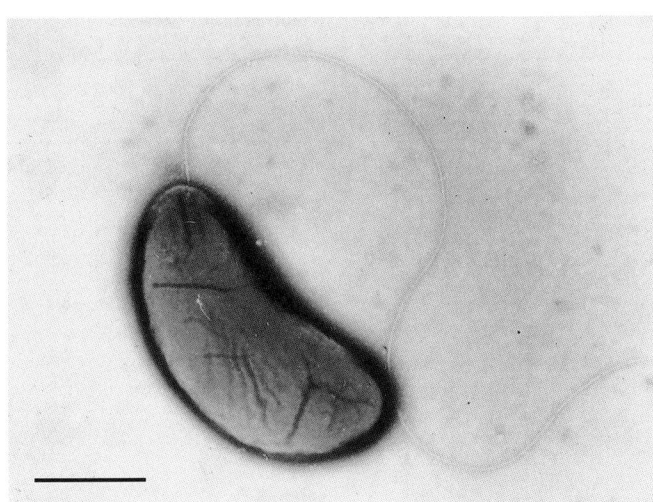

FIGURE BXII.δ.14. Negatively-stained cell of *Desulfonatronum lacustre* Z-7951 with monopolar flagellum. Bar = 1 μm.

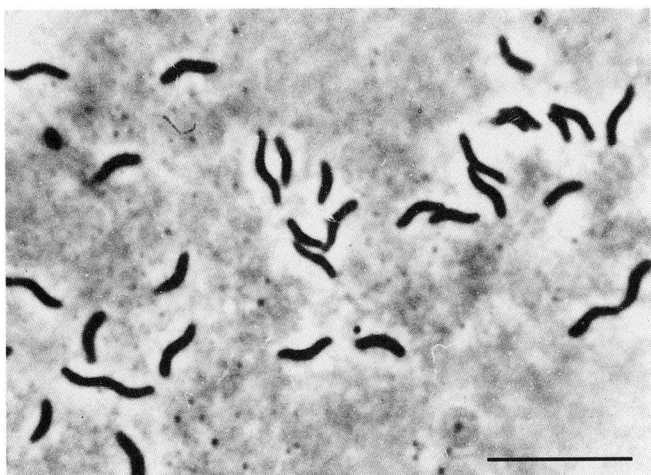

FIGURE BXII.δ.15. Cells of *Desulfonatronum lacustre* Z-7951 as viewed by phase-contrast microscopy. Bar = 10 μm.

Rainey 1998a, 631 (Effective publication: Pikuta, Zhilina and Rainey *in* Pikuta, Zhilina, Zavarzin, Kostrikina, Osipov and Rainey 1998b, 112.)

FURTHER DESCRIPTIVE INFORMATION

The cells are vibrioid with one polar flagellum (Fig. BXII.δ.14). Multiplication is by binary fission (Fig. BXII.δ.15). The cell wall has a typical Gram-negative structure in ultrathin section. The wall is resistant to enzymatic disintegration methods but not to ultrasound.

The cells contain *c*-type cytochromes with maxima at 552, 521.8, and 417.8 nm.

The type strain utilizes hydrogen, formate, and ethanol as electron donors. Some strains, such as Z-7955 from Lake Khylganta, Burjatia, also utilize lactate. Acetate is produced from ethanol or lactate. Sulfate, sulfite, and thiosulfate serve as electron acceptors. Fumarate, nitrate, DMSO, $Fe(OH)_3$, and S^0 are not reduced. Thiosulfate or sulfite may be disproportionated. Sulfide is the only product of catabolic reduction of sulfur compounds. Acetate or yeast extract is utilized for anabolism.

The prevailing fatty acids of *D. lacustre* are unsaturated fatty acids (about 60%), with major ones $C_{16:1}$ and $C_{18:1}$.

D. lacustre was first isolated from the sandy bottom deposits of low mineralized alkaline Lake Khadyn, Central Asia, south of Baikal in the cryoarid steppe zone. As an H_2-scavenging alkaliphile, *D. lacustre* might function as a hydrogen sink in an anaerobic, alkaliphilic community decomposing organic matter. For instance, it develops in coculture with alkaliphilic fermentative spirochetes growing on carbohydrates and produces an acetogenic shift in their metabolism (Pikuta et al., 1998b). As an extremely alkaliphilic sulfate-reducer, *Desulfonatronum* might participate in the alkalinization of soda lakes.

ENRICHMENT AND ISOLATION PROCEDURES

Strains of *Desulfonatronum* can be enriched from alkaline soda lakes under strictly anaerobic conditions in the selective alkaline mineral medium no. 1[1] (Zavarzin et al., 1996), which has a high sodium carbonate content. Incubation is under an atmosphere of H_2. N_2 can substitute for H_2 if Na-formate or ethanol is added as the electron donor. Isolation is by serial dilution in liquid medium no. 1 where yeast extract is replaced by 2 mM sodium acetate with subsequent cloning by single colony isolation in roll tubes. For roll tube cultivation, 2% (w/v) agar (Difco) is added to 4-ml portions of carbonate-free medium no. 1. The carbonate solution is injected into Hungate tubes after sterilization. Colonies are isolated by the capillary technique under a dissecting microscope. Colonies are lenticular, yellowish, translucent, with even edges.

MAINTENANCE PROCEDURES

Desulfonatronum is maintained by regular subculturing in liquid medium and storage in the refrigerator at 6°C. Stock cultures are transferred every 2–3 months but lysis of cells may occur. Strains of *Desulfonatronum* may be stored indefinitely in liquid nitrogen.

DIFFERENTIATION OF THE GENUS *DESULFONATRONUM* FROM OTHER GENERA

The genus *Desulfonatronum* is differentiated from other genera of sulfate-reducing bacteria by its extreme alkaliphily. It differs from the genus *Desulfonatronovibrio* (which also includes true alkaliphiles) by its larger cell size, substrate specificity, utilization of thiosulfate as electron acceptor, lipid profile, 10% higher G + C content of the DNA, and almost twofold larger genome size (Pikuta et al., 1997). Few sulfate reducers have mol% G + C values similar to *Desulfonatronum*; however, *Desulfohalobium* (Ollivier et al., 1991) and *Desulfomicrobium* (Rozanova et al., 1988) are among them. *Desulfonatronum* differs from halophilic *Desulfohalobium* species by its independence from NaCl for growth, substrate specificity, lipid profile, and 16S rDNA sequence-based phylogenetic position. It differs from *Desulfomicrobium* by its cell morphology, substrate specificity, lipid profile, and 16S rDNA sequence-based phylogenetic position (see Table BXII.δ.13). The $C_{17:1\ iso}$ typical of *Desulfovibrio* species is found in only minor amounts in *Desulfonatronum*. The predominance of unsaturated fatty acids in *Desulfonatronum* distinguishes it from *Desulfonatronovibrio*.

---

1. Medium no.1 has the following composition (per liter of distilled water): Na_2CO_3, 2.76 g; NaCl, 10.0 g; Na_2SO_4, 3.0 g; NH_4Cl, 0.5 g; KCl, 0.2 g; K_2HPO_4, 0.2 g; $Na_2S \cdot 9H_2O$, 0.5 g; yeast extract, 0.5 g; substrate (electron donor), 5.0 g; and 1 ml of trace element solution (Whitman et al., 1982).

TABLE BXII.δ.13. Differentiation of *Desulfonatronum lacustre* and *Desulfonatronovibrio hydrogenovorans* from other sulfate-reducing bacteria with similar mol% G + C values[a]

Characteristic	*Desulfonatronum lacustre* Z-7951[b]	*Desulfonatronovibrio hydrogenovorans* Z-7935[c]	*Desulfomicrobium baculatum*[d]	*Desulfohalobium retbaense* HR 100[e]	Strain TD3[f]
Cell shape	Vibrio	Vibrio	Rods	Curved rod	Curved rod
Motility (monopolar)	+	+	+	+[g]	NR
Cell size (μm)	0.7–0.9 × 2.5–3.0	0.5 × 1.5–2.0	0.6 × 1.3	0.7–0.9 × 1–3	0.8 × 2.0
pH range (optimal pH)	8.0–10.1 (9.3–9.5)	8.0–10.2 (9.5–9.7)	6.8–7.4 (7.2)	5.5–8.0 (6.5–7.0)	NR (6.8)
Optimal temperature (°C)	40	37	28–37	37–40	55–65
Requirement for NaCl	−	+	−	+	+
Range (optimal) NaCl%	<10	1–12 (3)	NA	1.0–24.0 (10.0)	NR (3.5)
Requirement for HCO_3^-	+	+	−	−	−
Yeast extract requirement	+	+	+	+	NR
mol% G + C content of DNA	57.3	48.6	57	57.1	NR
Utilization of electron donors (with SO_4^{2-}):					
H_2/CO_2	−	−	−	−	−
H_2/acetate	+	+	+	+	−
Formate/acetate	+	+	+	+	−
Acetate	−	−	−	−	NR
Ethanol	+	−	−	+	−
Lactate	−[h]	−	+	+	−
Pyruvate	−	−	−	+	NR
Malate	−	−	−	−	NR
Fumarate	−	−	−	−	NR
Utilization of electron acceptors:					
Sulfate	+	+	+	+	+
Sulfite	+	+	+	+	NR
Thiosulfate	+	−	+	+	NR
Sulfur	−	−	−	+	NR

[a]For symbols see standard definitions; NR, not reported; NA, not applicable.
[b]Data from Pikuta et al. (1998b).
[c]Data from Zhilina et al. (1997).
[d]Data from (Rozanova et al., 1988).
[e]Data from Ollivier et al. (1991).
[f]Data from Rueter et al. (1994).
[g]One or two flagella are present.
[h]Some strains utilize lactate.

TAXONOMIC COMMENTS

The phylogenetic position of *Desulfonatronum* was determined based on 16S rDNA sequence analysis. The genus belongs to the cluster of sulfate-reducing bacteria within the *Deltaproteobacteria* (see Fig. BXII.δ.13). It represents a distinct lineage, with 16S rDNA sequence similarity between *Desulfonatronum* and other taxa within this group in the range of 83.4–89.7%. Its status as a separate genus is based on phenotypic and phylogenetic analysis, including a distinct branching point, 16S rDNA sequence difference of >10% from the nearest relative, and lipid composition. When the fatty acid profile of *Desulfonatronum* was compared with the lipid profiles of *Desulfomicrobium baculatum*, *Desulfohalobium retbaense*, and *Desulfonatronovibrio hydrogenovorans*, computer-assisted analysis revealed no correlation with any of these species or with strains of other genera of sulfate-reducing bacteria in the database.

FURTHER READING

Pikuta, E.V., A.M. Lysenko and T.N. Zhilina. 1997. Distribution of *Desulfonatronovibrio hydrogenovorans* in soda lakes of Tuva. Mikrobiologiya *66*: 216–221.

Pikuta, E.V., T.N. Zhilina, G.A. Zavarzin, N.A. Kostrikina, G.A. Osipov and F.A. Rainey. 1998. *Desulfonatronum lacustre* gen. nov., sp. nov.: a new alkaliphilic sulfate-reducing bacterium utilizing ethanol. Mikrobiologiya *67*: 105–113.

List of species of the genus Desulfonatronum

1. **Desulfonatronum lacustre** Pikuta, Zhilina and Rainey 1998a, 631[VP] (Effective publication: Pikuta, Zhilina and Rainey in Pikuta, Zhilina, Zavarzin, Kostrikina, Osipov and Rainey 1998b, 112.)

   la.cus′tre. F. or Ital. *lacustre* lacustrine, of lakes; from L. *lacus* basin, lake; probably influenced by L. masc. *paluster, palustris* marshy; M.L. adj. *lacustre* inhabiting lakes.

   The characteristics are as given for the genus and under

   Further Descriptive Information. The following are additional characteristics. Optimal temperature, 37–40°C; range, 20–45°C. No growth below pH 8; maximum pH for growth, ~10; optimal pH, 9.5. Sodium and carbonate ions are required for growth. NaCl is not required but up to 10% (w/v) NaCl can be tolerated.

   The mol% G + C of the DNA is: 57.3 (T_m).

   Type strain: Strain Z-7951, DSM 10312.

   GenBank accession number (16S rRNA): Y14594.

Order III. **Desulfobacterales** *ord. nov.*

JAN KUEVER, FRED A. RAINEY AND FRIEDRICH WIDDEL

De.sul.fo.bac.ter.a'les. M.L. masc. n. *Desulfobacter* type genus of family; suffix *-ales* denotes order; M.L. fem. pl. n. *Desulfobacterales* the order that comprises *Desulfobacteraceae* and other families.

The order comprises **morphologically diverse sulfate-reducing bacteria** including cocci, oval, rod-shaped, and curved cells; some may form sarcina-like cell aggregates or multicellular, uniseriately arranged filaments. Many forms are motile by means of single polar flagella; multicellular filaments exhibit gliding motility. **Strictly anaerobic chemoorganotrophs, chemolithoheterotrophs, or chemolithoautotrophs with respiratory metabolism**; in addition, fermentative abilities are found in some members. With the exception of one isolate (*Desulfocapsa sulfexigens*), **the common electron acceptor is sulfate, which is reduced to sulfide.** Sulfite or thiosulfate may be used in addition; a few members also reduce elemental sulfur (or polysulfide) to sulfide, or nitrate to ammonia. Simple organic compounds such as long-chain fatty acids, alcohols, polar aromatic compounds, and in some cases even aliphatic or aromatic hydrocarbons, serve as electron donors and carbon sources. Most members of the family *Desulfobacteraceae*

perform a complete oxidation of organic substrates. Most members of the family *Desulfobulbaceae* are incomplete oxidizers that form acetate as an end product. Some members can grow chemolithoautotrophically with H_2 and sulfate as the source of energy and CO_2 as the source of carbon for cell synthesis. Growth by disproportionation of sulfur, thiosulfate, and sulfite may occur. Members are mesophilic or psychrophilic; thermophiles have not been isolated in this order. Cells contain various cytochromes, other redox proteins, and menaquinones. Mesophilic species have been isolated from various aquatic environments, including marine, brackish water, and freshwater habitats.

The order comprises the families *Desulfobacteraceae* and *Desulfobulbaceae* (see Fig. BXII.δ.16).

Type genus: **Desulfobacter** Widdel 1981, 382 (Effective publication: Widdel 1980, 376.)

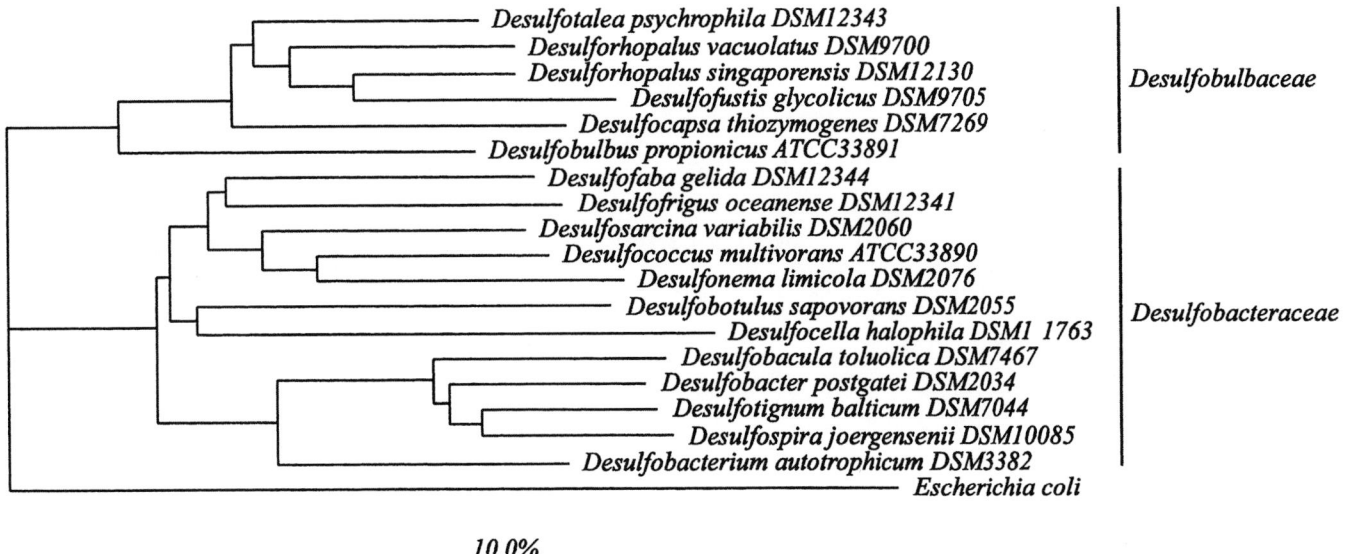

Desulfotalea psychrophila DSM12343
Desulforhopalus vacuolatus DSM9700
Desulforhopalus singaporensis DSM12130
Desulfofustis glycolicus DSM9705
Desulfocapsa thiozymogenes DSM7269
Desulfobulbus propionicus ATCC33891
} *Desulfobulbaceae*

Desulfofaba gelida DSM12344
Desulfofrigus oceanense DSM12341
Desulfosarcina variabilis DSM2060
Desulfococcus multivorans ATCC33890
Desulfonema limicola DSM2076
Desulfobotulus sapovorans DSM2055
Desulfocella halophila DSM1 1763
Desulfobacula toluolica DSM7467
Desulfobacter postgatei DSM2034
Desulfotignum balticum DSM7044
Desulfospira joergensenii DSM10085
Desulfobacterium autotrophicum DSM3382
Escherichia coli
} *Desulfobacteraceae*

10.0%

FIGURE BXII.δ.16. Phylogenetic relationships based on 16S rRNA gene sequence analyses of the genera within the order *Desulfobacterales*. The scale bar represents 10 inferred nucleotide substitutions per 100 nucleotides. The dendrogram was reconstructed from distance matrices using the neighbor-joining method.

Family I. **Desulfobacteraceae** *fam. nov.*

JAN KUEVER, FRED A. RAINEY AND FRIEDRICH WIDDEL

De.sul.fo.bac.ter.a.ce'ae. M.L. masc. n. *Desulfobacter* type genus of the family; suff. *ceae* denotes family; M.L. fem. pl. n. *Desulfobacteraceae* the family that comprises the genus *Desulfobacter* and other genera.

The family comprises morphologically diverse sulfate-reducing bacteria including cocci, oval, rod-shaped, and curved cells; some may form sarcina-like cell aggregates or multicellular, uniseriately

arranged filaments. Many forms are motile by means of a single polar flagellum; multicellular filaments exhibit gliding motility. Strictly anaerobic chemoorganotrophs or chemolithoautotrophs with respiratory metabolism; fermentative abilities are found in some members. The common electron acceptor is sulfate, which is reduced to sulfide. Sulfite or thiosulfate may also be used. Simple organic compounds such as long-chain fatty acids, alcohols, polar aromatic compounds, and in some cases even aliphatic or aromatic hydrocarbons, serve as electron donors and carbon sources. Most members of the family *Desulfobacteraceae* perform a complete oxidation of organic substrates. Some members can grow chemolithoautotrophically with H_2 and sulfate as the source of energy and CO_2 as the source of carbon for cell synthesis. Members are mesophilic or psychrophilic; thermophiles have not been isolated. Cells contain various cytochromes, other redox proteins, and menaquinones. Mesophilic species have been isolated from various aquatic environments, including marine, brackish water, and freshwater habitats.

The family comprises 14 described genera, *Desulfobacter, Desulfobacterium, Desulfobacula, Desulfobotulus* (proposed below), *Desulfocella, Desulfococcus, Desulfofaba, Desulfofrigus, Desulfomusa, Desulfonema, Desulforegula, Desulfosarcina, Desulfospira,* and *Desulfotignum.* Properties of genera are given in Table BXII.δ.14. Detailed properties of described species of each genus are given in the genus description. *Desulfomusa* and *Desulforegula* were validly published after the cut-off date for inclusion in this edition of the *Manual* (June 30, 2001).

Type genus: **Desulfobacter** Widdel 1981, 382 (Effective publication: Widdel 1980, 376.)

TABLE BXII.δ.14. Differential characteristics of the genera of the family *Desulfobacteraceae*[a]

Genus	*Desulfobacter*	*Desulfobacterium*	*Desulfobacula*	*Desulfobotulus*	*Desulfocella*	*Desulfococcus*	*Desulfofaba*	*Desulfofrigus*	*Desulfonema*	*Desulfosarcina*	*Desulfospira*	*Desulfotignum*
Morphology	Rod- or vibrio-shaped	Rod-shaped	Oval- to rod-shaped	Vibrio-shaped	Vibrio-shaped	Spherical or lemon-shaped	Rod-shaped	Rod-shaped	Multicellular filaments	Rod-shaped or sarcina-like aggregates	Vibrio-shaped	Rod-shaped
Motility	±	+	±	+	+	±	±	±	Gliding	±	−	+
Mol% G+C content	45–49	45–48	41–42	53	35	56–57	52–53	52–53	35–55	51–59	50	62
Sulfite reductase	Most species, DR	nr	nr	nr	nr	DV	nr	nr	DV[b]	DR[b]	nr	nr
Major menaquinone	MK-7 or MK-7(H_2)	MK-7 or MK-7(H_2)	MK-7(H_2)	MK-7	nr	MK-7	MK-8	MK-9	MK-7 or MK-9	MK-7	MK-7 and MK-7(H_2)	nr
Substrate oxidation	C	C	C	I	I	C	I	I or C	C	C	C	C
CO-DH[c]	−	+	+	−	−	+	nr	nr	+	+	nr	+
Typical electron donors and carbon sources:												
Acetate	+	+	±	−	−	+	−	±	+	+	−	+
Benzoate	−	−	+	−	−	+	−	−	±	+	−	+
Ethanol	±	+	+	−	−	+	+	+	±	+	−	+
Fatty acids: number of carbon atoms	−	4–18	4	4–18	4–18	3–16	3–4	4	3–10	3–10	−	4–10
Formate	−	+	±	−	−	+	+	+	+	+	+	+
Fumarate	−	+	+	−	−	−	+	±	+	+	+	+
Lactate	−	+	+	+	−	±	+	+	±	+	+	+
Pyruvate	+	+	+	+	+	+	+	+	±	+	+	+
H_2/CO_2	±	+	−	−	−	−	−	±	±	±	+	+
Chemolithoautotrophic growth	±[b]	+	−	−	−	±[b]	−	−	±	±	+	+.
Fermentative growth	−	±	nr	+	−	±	+	+	−	±	−	+
Growth by disproportionation of reduced sulfur compounds	nr	nr	nr	nr	nr	nr	−	−	nr	nr	nr	nr
Typical electron acceptors:												
Sulfur	−	−	−	−	−	−	−	−	−	−	+	−
Sulfate	+	+	+	+	+	+	+	+	+	+	+	+
Sulfite	+	±	+	+	−	+	+	±	±	+	+	+
Thiosulfate	+	+	+	+	−	+	+	+	±	+	+	+
NaCl requirement	+	+	+	−	+	±	+	+	+	±	+	+
Habitat (anoxic):												
Freshwater	−	−	−	+	−	+	−	−	±	+	−	−
Brackish water	+	+	+	−	−	+	−	+	+	+	−	+
Marine	+	+	+	−	+	+	+	+	+	+	+	+

[a]Symbols: nr, not reported; +, present, observed in all species; −, not present, not observed; ±, variable; I, incomplete; C, complete; DV, desulfoviridin; DR, desulforubidin.

[b]One species only.

[c]Carbon monoxide dehydrogenase, a key enzyme indicative of the C_1-pathway for acetyl-CoA oxidation.

Genus I. *Desulfobacter* Widdel 1981, 382^{VP} (Effective publication: Widdel 1980, 376)

JAN KUEVER, FRED A. RAINEY AND FRIEDRICH WIDDEL

De.sul.fo.bac′ ter. L. pref. *de* from L. n. *sulfur* sulfur; M.L. n. *bacter* masc. equivalent of Gr. fem. n. *bacteria* rod or staff; M.L. masc. n. *Desulfobacter* a rod-shaped sulfate-reducing bacterium.

Oval to rod-shaped or slightly curved to vibrio-shaped cells, 0.5– 2.5 × 1.5–8 µm. Occur singly or in pairs. Spore formation is not observed. Gram negative. Cells are motile by single polar flagella or nonmotile, but motility can be lost during cultivation. Strict anaerobe with a respiratory type of metabolism. **Chemoorgano- trophs or chemoautotrophs. Acetate is the preferred general elec- tron donor and carbon source,** and is completely oxidized to CO_2. Some species also use ethanol, and a few species use H_2/CO_2 or pyruvate. H_2-utilizing *Desulfobacter hydrogenophilus* grows autotrophically with CO_2 as the sole carbon source. **Sulfate, and usually sulfite and thiosulfate, serve as terminal electron acceptors and are reduced to H_2S. Sulfur and nitrate are not used as terminal electron acceptors.** Fermentative growth has not been observed. Desulfoviridin is absent, but desulforubidin is present in some species. The major menaquinone is MK-7. Growth occurs in simple defined media containing sulfide as a reductant. Most species require vitamins. Many members of this genus can fix N_2. The addition of ≥7 g NaCl and ≥1 g

$MgCl_2·6H_2O$ per liter medium is usually stimulatory or required for growth. The optimal temperature range for growth is 28– 34°C. Thermophilic species have not been described. The opti- mal pH range for growth is 6.5–7.4. Carbon monoxide dehydro- genase activity is absent. Acetyl-CoA oxidation and CO_2 fixation with H_2 as electron donor in *D. hydrogenophilus* is accomplished via a modified TCA cycle. *Desulfobacter* species are most common in anoxic brackish or marine sediments, but some types may be found in anoxic freshwater sediments or in activated sludge.

The mol% G + C of the DNA is: 45–49.

Type species: **Desulfobacter postgatei** Widdel 1981, 382 (Effec- tive publication: Widdel 1980, 373.)

FURTHER DESCRIPTIVE INFORMATION

Many *Desulfobacter* species have ellipsoidal to short, rod-shaped cells with rounded ends (Figs. BXII.δ.17, BXII.δ.18, BXII.δ.19, and BXII.δ.20); some strains form longer rods or curved cells;

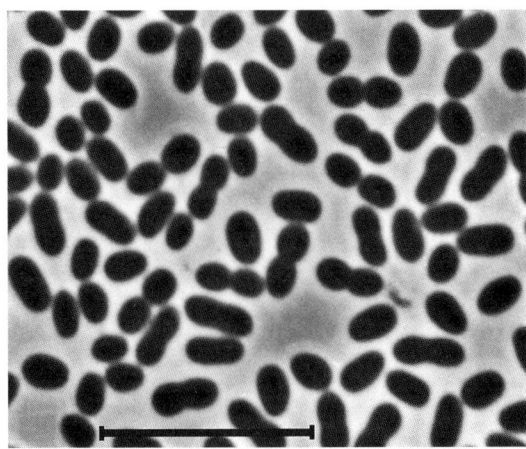

FIGURE BXII.δ.17. Phase-contrast micrograph of *Desulfobacter postgatei* (viable cells). Bar = 10 µm.

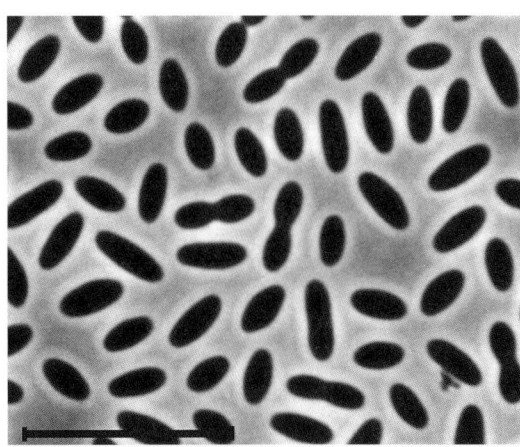

FIGURE BXII.δ.19. Phase-contrast micrograph of *Desulfobacter hydrogeno- philus* (viable cells). Bar = 10 µm.

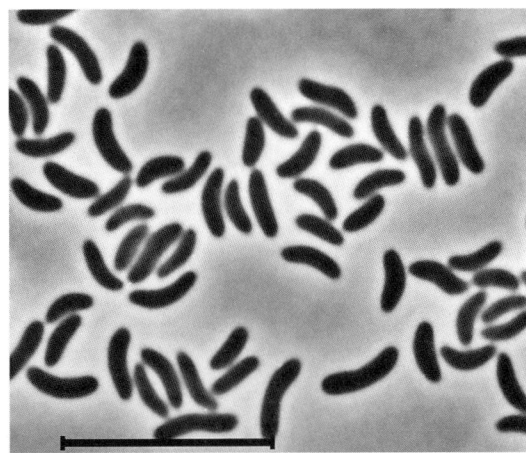

FIGURE BXII.δ.18. Phase-contrast micrograph of *Desulfobacter curvatus* (viable cells). Bar = 10 µm.

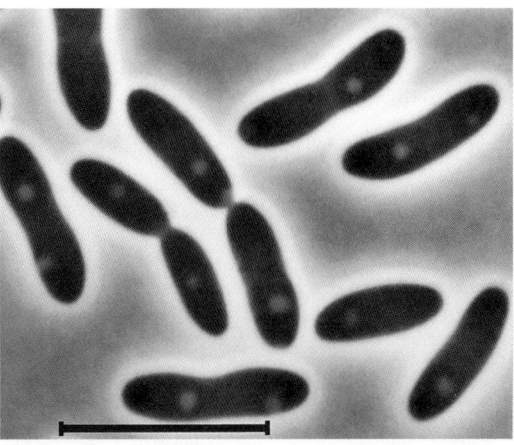

FIGURE BXII.δ.20. Phase-contrast micrograph of *Desulfobacter latus* (vi- able cells). Bar = 10 µm.

a few strains tend to form pleomorphic cells. The cells are often in pairs. Cells of marine strains or species have the tendency to stick together in clumps, which settle to the bottom of culturing vessels. In the first few enrichment passages on acetate and sulfate, motile cells can often be seen; after isolation most strains become nonmotile.

The optimal growth temperature is usually 28–32°C; the maximum is 37°C and the minimum usually 10°C; however, growth at 0°C has been observed in *D. hydrogenophilus*. The pH range for growth is 6.2–8.5, with an optimal at 7.3.

Growth occurs only in the presence of sulfate, sulfite, or thiosulfate as electron acceptors. Nitrate, fumarate, and malate are not reduced. Elemental sulfur inhibits growth. The type strain of *D. postgatei* uses only acetate as the electron donor and carbon source; however, other species may also use ethanol, pyruvate, or lactate (see Table BXII.δ.15). *D. hydrogenophilus* can grow autotrophically with H_2 as electron donor and CO_2 as carbon source. Sugar, pyruvate, and fumarate are not fermented. *Desulfobacter curvatus* can disproportionate sulfite and thiosulfate to

sulfate and sulfide (Bak and Pfennig, 1987). Carbon monoxide dehydrogenase activity is absent. Acetyl-CoA oxidation (and CO_2 fixation with H_2 as electron donor in *D. hydrogenophilus*) is accomplished via a modified TCA cycle, as indicated by 2-oxoglutarate dehydrogenase activity (Brandis-Heep et al., 1983; Gebhardt et al., 1983; Schauder et al., 1986; Lien and Beeder, 1997).

D. postgatei contains membrane-bound and soluble cytochromes of the *b* and *c* type. The sulfite reductase desulforubidin is present in the type strain and in *D. vibrioformis*; other species were not investigated. All members of the genus *Desulfobacter* investigated so far contain MK-7 or MK-7(H_2) as the major menaquinone (Collins and Widdel, 1986). Major cellular fatty acids of the *Desulfobacter* strains are $C_{14:0}$, $C_{16:0}$, $C_{16:0\ 10CH_3}$, $C_{16:1\ \omega7c}$ and $C_{17:0\ cyclo}$ (Vainshtein et al., 1992; Kohring et al., 1994).

Desulfobacter strains occur in black anoxic sediments from brackish water and marine habitats. Occurrence in anoxic freshwater habitats (ditches, activated sludge) has also been shown by use of enrichment cultures prepared with brackish water media.

TABLE BXII.δ.15. Morphological and physiological characteristics of described species of the genus *Desulfobacter*[a]

Species	D. postgatei	D. curvatus	D. halotolerans	D. hydrogenophilus	D. latus	D. vibrioformis
Morphology	Oval-shaped	Vibrio-shaped	Rod-shaped	Rod-shaped	Oval-shaped	Vibrio-shaped
Cell size (μm)	1–1.5 × 1.7–2.5	0.5–1 × 1.7–3.5	0.8–1.2 × 3–5	1–1.3 × 2–3	1.6–2.4 × 5–7	1.9–2.3 × 4.5–8
Motility	±	+	+ (sp)	−	±	+ (sp)
Mol% G + C content	46	46	49	45	44 (48)	47
Sulfite reductase	DR	nr	nr	nr	nr	DR
Major menaquinone	MK-7	MK-7(H₂)	nr	MK-7(H₂)	MK-7	nr
Optimal temperature (°C)	28–32	28–30	32–34	28–30	28–32	33
Oxidation of substrate	C	C	C	C	C	C
Compounds used as electron donors and carbon sources:						
H₂/CO₂	−	(+)	−	+[b]	−	−
Formate	−	−	nr	−	−	−
Acetate	+	+	+	+	+	+
Fatty acids	−	−	−	−	−	−
Isobutyrate	−	−	−	−	−	nr
2-Methylbutyrate	−	−	nr	−	−	nr
3-Methylbutyrate	−	−	nr	−	−	nr
Ethanol	−	+	+	−	−	−
Lactate	−	−	−	−	−	−
Pyruvate	−	+	(+)	+	−	−
Fumarate	−	−	−	−	−	−
Succinate	−	nr	−	nr	nr	−
Malate	−	−	−	−	−	−
Benzoate	−	−	−	−	−	−
Fermentative growth	−	−	−	−	−	−
Compounds used as electron acceptors:						
Sulfate	+	+	+	+	+	+
Sulfite	+	+	+	+	+	+
Sulfur	−	−	−	−	−	nr
Thiosulfate	+	+	+	+	−	+
Nitrate	−	nr	−	nr	nr	nr
Growth factor requirement	bi, pa	bi	bi, pa	bi, pa	bi, th	−
NaCl requirement (g/l)	7	10	5	20	20	10
Literature	Widdel and Pfennig, 1981	Widdel, 1987	Brandt and Ingvorsen, 1997	Widdel, 1987	Widdel, 1987	Lien and Beeder, 1997; Beeder pers. com.

[a]Symbols: nr, not reported; +, present, observed in all species; −, not present, not observed; ±, variable; (+), poorly motile, poor growth; (sp), flagellation type: single, polar; DR, desulforubidin; C, complete; bi, biotin; ni, nicotinate; pa, *p*-aminobenzoate; th, thiamine.

[b]Autotrophic growth.

Strains of *Desulfobacter* may be cultured in a defined medium[1] with sulfate as the electron acceptor and acetate as the electron donor and carbon source. Sulfide is added as reductant. Marine strains often require 20.0 g/l NaCl and 3.0 g/l MgCl$_2$·6H$_2$O; brackish water strains tolerate lower levels of these salts. *p*-Aminobenzoic acid and biotin are required as growth factors by some species. Initiation of growth is favored, especially when inoculation is done from old cultures, by the addition of 10–30 mg/l of fresh sodium dithionite as a further strong reductant.

ENRICHMENT AND ISOLATION PROCEDURES

For the selective enrichment of *Desulfobacter* strains, acetate-sulfate medium is used. Bottles or tubes (100-, 50-, or 20-ml capacity) can serve as culture vessels and are kept anoxic with tight screw caps (only if completely filled), or under a head space containing a mixture of N$_2$/CO$_2$ (90:10) and sealed with butyl rubber stoppers. The medium is inoculated with black anoxic mud from brackish water or marine habitats; 2–5% of the total volume should be added. *Desulfobacter* species may also be enriched and isolated via direct serial dilution of anoxic environmental samples in liquid medium (according to the most-probable-number method) or agar medium (see next paragraph) with propionate as organic substrate. The NaCl concentration of the medium should correspond approximately to the salinity of the natural source. The amount of MgCl$_2$·6H$_2$O added need not be greater than 3 g/l. Brackish water medium containing 7 g NaCl and 1.2 g MgCl$_2$·6H$_2$O per liter should be used for enrichment of *Desulfobacter* strains from anoxic freshwater habitats; typical freshwater media containing only 1 g NaCl and 0.5 g MgCl$_2$·6H$_2$O (or less) per liter are not suitable for such enrichment, but once strains grow well in brackish water medium they may subsequently be cultured at lower salinities. Addition of dithionite to enrichment cultures is recommended. Enrichment cultures are incubated at 25–30°C and are mixed by brief shaking every second day. After intense formation of H$_2$S has occurred (generally after 8–20 d), transfers to fresh medium are made. Because cells of *Desulfobacter* often attach to sediment particles or many form aggregates that sediment to the bottom of culture vessels, the cultures should be mixed well before subculturing.

Pure cultures are obtained by repeated application of the agar dilution method. A 3.3% (w/v) solution of agar is prepared in distilled water; the agar should be washed several times in cold distilled water before preparing the solution. The agar solution is dispensed in 3-ml amounts into test tubes, which are closed with aluminum caps and autoclaved. Agar tubes may be stored for 2–4 months. The agar tubes are kept molten (or remelted at 100°C) and placed in a water bath at 55°C. Defined culture medium is prewarmed to 41°C and 6-ml amounts are added to the tubes of liquefied agar; exposure to air is minimized by dipping the tip of the pipette into the agar medium. To compensate for the diluting effect of the aqueous agar, it is recommended to increase the amounts of NaCl and MgCl$_2$·6H$_2$O to 30 and 4.5 g, respectively, per liter of medium (e.g., by addition of sterile salt concentrate) before this is added to the agar. Addition of salts to the agar before autoclaving is not recommended as an alternative because this may promote hydrolysis of agar. Starting with 0.1–0.5 ml of an enrichment culture as inoculum, serial dilutions of approx. 1:10 are made using 6–8 tubes. Before the agar medium solidifies, dithionite solution is added aseptically to each tube from a 0.1-ml pipette which is at the same time used for gently mixing, starting at the highest dilution. All agar tubes are then hardened in cold water and immediately sealed with butyl rubber stoppers under a mixture of N$_2$/CO$_2$ (90:10) (v/v). The tubes are incubated in the dark for 2–4 weeks at 25–30°C. Well-separated colonies that occur in the higher dilutions can be removed with sterile Pasteur pipettes (thinned upon heating in a flame); the cells are used as inoculum for subsequent agar dilution cultures. The process is repeated until the cultures are pure, as indicated by homogenous colonies and homogenous cells.

MAINTENANCE PROCEDURES

Stock cultures are transferred every 3–5 months. Towards the end of growth or at the beginning of the stationary phase, cultures are refrigerated and stored at 2–5°C in the dark. *Desulfobacter* strains may be preserved indefinitely by suspension of cells in anoxic medium containing 5% dimethylsulfoxide and storage in liquid nitrogen.

DIFFERENTIATION OF THE GENUS *DESULFOBACTER* FROM OTHER GENERA

In addition to standard 16S rRNA gene sequence analysis (see next section) and DNA–DNA hybridization, methods testing physiological characteristics may also be included in the differentiation of *Desulfobacter* from other morphologically similar sulfate-reducing bacteria (Widdel and Pfennig, 1981). *Desulfobacter* strains grow well (with a 15–20 h doubling time) in seawater or brackish water media containing acetate and sulfate. Sulfide concentrations of 20 mM, or sometimes even higher, may be produced. In contrast to *Desulfotomaculum acetoxidans*, which also grows on acetate and sulfate, *Desulfobacter* cells have rounded ends or are vibrio-shaped and do not form spores; moreover, they have a lower optimal growth temperature and they utilize only a few organic substrates beside acetate. A unique feature of all *Desulfobacter* strains is the presence of 2-oxoglutarate dehydrogenase, indicating the operation of a complete tricarboxylic acid cycle for oxidation of acetyl-CoA. All other completely oxidizing sulfate-reducing bacteria investigated so far use the anaerobic C$_1$ pathway (carbon monoxide dehydrogenase pathway, Wood pathway) for oxidation of acetyl-CoA.

Some *Desulfobacter* species resemble cells of *Desulfococcus multivorans* and *Desulfosarcina variabilis*; however, strains of the latter two species develop only slowly on acetate. They are capable of

---

1. The defined medium has the following composition (g/l of distilled water): Na$_2$SO$_4$, 3.0; KH$_2$PO$_4$, 0.2; NH$_4$Cl, 0.3; NaCl, 7.0 (for brackish water strains) or 20.0 (for marine strains); MgCl$_2$·6H$_2$O, 1.2 (for brackish water strains) or 3.0 (for marine strains); KCl, 0.5; CaCl$_2$·2H$_2$O, 0.15. After the medium has been autoclaved and cooled under an atmosphere of N$_2$/CO$_2$ (90:10), the following components are added per liter of medium from sterile stock solutions, while access of air is prevented by continuous flushing with a mixture of N$_2$/CO$_2$ (90:10): acetate solution (CH$_3$COONa·3H$_2$O, 280 g/l), 10.0 ml; trace element solution (see below), 1.0 ml; bicarbonate solution (NaHCO$_3$, 84 g/l, saturated with CO$_2$ and autoclaved under a CO$_2$ atmosphere), 30.0 ml; sulfide solution (Na$_2$S·9H$_2$O, 120 g/l, autoclaved under an N$_2$ atmosphere), 3.0 ml; vitamin solution (*p*-aminobenzoic acid, 40 mg/l, and biotin, 10 mg/l), 1.0 ml. Trace element solution (without chelating agent) contains (per liter): HCl (25% solution), 10 ml; FeSO$_4$·7H$_2$O, 2.1 g; CoCl$_2$·6H$_2$O, 190 mg; ZnSO$_4$, 144 mg; MnCl$_2$·4H$_2$O, 100 mg; Na$_2$MoO$_4$·2H$_2$O, 36 mg; NiCl$_2$·6H$_2$O, 24 mg; H$_3$BO$_3$, 6 mg; CuCl$_2$·2H$_2$O, 2 mg. The FeSO$_4$ is initially dissolved in the HCl solution and distilled water is added, followed by the other components. The pH of the complete defined medium is adjusted to 7.0–7.3. After the medium is inoculated, sodium dithionite (10–30 mg/l) is added from a freshly prepared 5% solution sterilized by filtration under anaerobic conditions; alternatively, dry crystals can be added with a sterile spatula.

utilizing a larger variety of organic compounds, such as 1-propanol, 1-butanol, higher fatty acids, and benzoate.

TAXONOMIC COMMENTS

Based on 16S rRNA gene sequence data, *Desulfobacter* strains form one cluster within the *Deltaproteobacteria* and cluster with members of the family *Desulfobacteraceae*. The closest relatives are members of the genus *Desulfobacterium*, *Desulfobacula toluolica*, and *Desulfospira joergensenii*. A signature sequence for all described *Desulfobacter* species and other closely related sulfate-reducing bacteria is 5′-CACAGGATGTCAAACCCAG-3′ (oligonucleotide probe DSB985), which binds to the 16S rRNA at *E. coli* position 5′-985–1004-3′ (Devereux et al., 1992; Manz et al., 1998).

FURTHER READING

Widdel, F. and N. Pfennig. 1981. Studies on dissimilatory sulfate-reducing bacteria that decompose fatty acids. I. Isolation of new sulfate-reducing bacteria with acetate from saline environments. Description of *Desulfobacter postgatei* gen. nov., sp. nov. Arch. Microbiol. *129*: 395–400.

List of species of the genus Desulfobacter

1. **Desulfobacter postgatei** Widdel 1981, 382[VP] (Effective publication: Widdel 1980, 373.)

   post.ga′ te.i. M.L. gen. n. *postgatei* of Postgate; in honor of J.R. Postgate, an English microbiologist.

   The morphology is as described in Table BXII.δ.15 and as depicted in Fig. BXII.δ.17. Characteristics are as described for the genus and summarized in Table BXII.δ.15. Cells contain desulforubidin as sulfite reductase. Occur in anoxic parts of brackish water and marine habitats; may also be enriched from freshwater mud by use of saltwater media.

   The mol% G + C of the DNA is: 49 (HPLC).
   Type strain: 2ac9, DSM 2034.
   GenBank accession number (16S rRNA): M26633.

2. **Desulfobacter curvatus** Widdel 1988b, 328[VP] (Effective publication: Widdel 1987, 290.)

   cur.va′ tus. L. part. *curvatus* curved, bent.

   The morphology is as described in Table BXII.δ.15 and as depicted in Fig. BXII.δ.18. Characteristics are summarized in Table BXII.δ.15. Growth requires 0.7 g/l NaCl and 0.13 g/l MgCl$_2$·6H$_2$O. The type strain was isolated from marine mud from Venice (Italy) using acetate; however, vibrio-shaped, acetate-oxidizing, sulfate-reducing bacteria may also be found in freshwater sediments.

   The mol% G + C of the DNA is: 46 (T_m).
   Type strain: AcRM3, ATCC 43919, DSM 3379.
   GenBank accession number (16S rRNA): M34413.

3. **Desulfobacter halotolerans** Brandt and Ingvorsen 1998, 327[VP] (Effective publication: Brandt and Ingvorsen 1997, 371.)

   ha.lo′ to.le.rans. Gr. n. *halos* salt; L. v. *tolerare* tolerate; M.L. adj. *halotolerans* salt-tolerating.

   Characteristics are summarized in Table BXII.δ.15. An unusual characteristic of this species is the ability to grow even at an NaCl concentration of 130 g/l, all other *Desulfobacter* spp. can grow up to an NaCl concentration of only 60 g/l. The type strain was isolated from moderately hypersaline, thalassohaline sediment of the Great Salt Lake (Utah, USA).

   The mol% G + C of the DNA is: 49 (HPLC).

   Type strain: GSL-Ac1, DSM 11383.
   GenBank accession number (16S rRNA): Y14745.

4. **Desulfobacter hydrogenophilus** Widdel 1988b, 328[VP] (Effective publication: Widdel 1987, 290.)

   hy.dro.ge.no′ phi.lus. M.L. n. *hydrogenum* hydrogen, from Gr. n. *hydro* water, and Gr. v. *genein* form or constitute; Gr. v. *philein*-like, love; M.L. adj. *hydrogenophilus* that likes hydrogen.

   The morphology is as described in Table BXII.δ.15 and as depicted in Fig. BXII.δ.19. Characteristics are summarized in Table BXII.δ.15. This is the only described species of the genus *Desulfobacter* that can grow chemolithoautotrophically with H$_2$ and CO$_2$. The type strain was isolated from marine mud from Venice (Italy) using acetate.

   The mol% G + C of the DNA is: 45 (T_m).
   Type strain: AcRS1, ATCC 43915, DSM 3380.
   GenBank accession number (16S rRNA): M34412.

5. **Desulfobacter latus** Widdel 1988b, 328[VP] (Effective publication: Widdel 1987, 290.)

   la′ tus. L. adj. *latus* large, wide.

   The morphology is as described in Table BXII.δ.15 and as depicted in Fig. BXII.δ.20. Characteristics are summarized in Table BXII.δ.15. The type strain was isolated from marine mud from Venice (Italy) using acetate.

   The mol% G + C of the DNA is: 44 (HPLC).
   Type strain: AcRS2, ATCC 43918, DSM 3381.
   GenBank accession number (16S rRNA): M34414.

6. **Desulfobacter vibrioformis** Lien and Beeder 1997, 1127[VP]

   vi.bri.o.for′ mis. L. v. *vibrio* vibrate; M.L. n. *vibrio* that which vibrates, a generic name; L. adj. suff. *formis*-like, of the shape of; *vibrioformis* vibrio shaped.

   Characteristics are summarized in Table BXII.δ.15. Cells of *D. vibrioformis* contain desulforubidin as sulfite reductase. *D. vibrioformis* is able to grow without supplementation of vitamins. The type strain was isolated from a water-oil separation system on the deck of an oil field platform in the North Sea, with acetate as the only substrate.

   The mol% G + C of the DNA is: 47 (HPLC).
   Type strain: B54, DSM 8776.
   GenBank accession number (16S rRNA): U12254.

Genus II. **Desulfobacterium** Bak and Widdel 1988, 136[VP] (Effective publication: Bak and Widdel 1986, 175)

JAN KUEVER, FRED A. RAINEY AND FRIEDRICH WIDDEL

De.sul.fo.bac.te'ri.um. L. pref. *de* from L. n. *sulfur* sulfur; Gr. dim. n. *bakterion* a small rod; M.L. neut. n. *Desulfobacterium* a rod-shaped sulfate reducer.

Oval to rod-shaped, or spherical cells, 0.9–2 × 1.5–3 μm. Occur singly or in pairs; sometimes also in loose chains. Spore formation is not observed. Gram negative. Some species are motile, but motility may be lost during cultivation. Strict anaerobes with a respiratory and a fermentative type of metabolism. **Chemoorganotrophs or chemoautotrophs,** using formate, butyrate, higher fatty acids, other organic acids, alcohols, and $H_2 + CO_2$ **as electron donors and as carbon sources**; these compounds are completely oxidized to CO_2. One species can use nicotinate as sole electron donor and carbon source. Growth on acetate and propionate is usually very slow. **Sulfate and other oxidized sulfur compounds serve as terminal electron acceptors and are reduced to H_2S. Sulfur and nitrate are not used as terminal electron acceptors. Some species may grow slowly in the absence of an external electron acceptor by fermentation of lactate, pyruvate, malate, and fumarate.** Optimal growth temperature, 26–29°C. Anoxic media (with sulfide as a reductant) and vitamins are required for growth. *Desulfobacterium* species require brackish or marine concentrations of NaCl and MgCl₂. Cells contain cytochromes of the *b*- and *c*-type. Carbon monoxide dehydrogenase activity is commonly observed, indicating the operation of the anaerobic C₁ pathway (carbon monoxide dehydrogenase pathway, Wood pathway) for complete oxidation of acetyl-CoA, or for CO₂ fixation during autotrophic growth. *Desulfobacterium* species are widespread in brackish or marine sediments, but occur less frequently in freshwater habitats.

The mol% G + C of the DNA is: 45–48.

Type species: **Desulfobacterium autotrophicum*** Brysch, Schneider, Fuchs and Widdel 1988, 328 (Effective publication: Brysch, Schneider, Fuchs and Widdel 1987, 272.)

FURTHER DESCRIPTIVE INFORMATION

D. autotrophicum uses sulfate, thiosulfate, or fumarate as electron acceptor, whereas sulfite, sulfur, and nitrate are not used. Formate, butyrate, long chain fatty acids, other organic acids, and alcohols serve as electron donors and as carbon sources; these compounds are oxidized completely to CO₂ via the anaerobic C₁ pathway (carbon monoxide dehydrogenase pathway, Wood pathway) (Schauder et al., 1986). Despite the capacity for complete oxidation, high substrate concentrations can lead to an excretion of acetate, implying an incomplete oxidation; the acetate may be slowly utilized again.

D. niacini can grow with nicotinate as electron donor and carbon source. Phenylpropionate is oxidized to benzoate, which is not degraded. Chemolithoautotrophic growth occurs with H₂ as the electron donor and CO₂ as the carbon source. Growth on acetate and propionate is very slow. *D. vacuolatum* is able to grow on several amino acids as electron donor and the sole carbon and nitrogen source (Rees et al., 1998). In the absence of an external electron acceptor, some species grow slowly by fermentation of lactate, pyruvate, malate, and fumarate. Ammonium

ions are used as the nitrogen source. Some strains require vitamins (Table BXII.δ.16). All members of the genus *Desulfobacterium* contain MK-7(H₂) or MK-7 as the major menaquinone (Collins and Widdel, 1986). Desulfoviridin is not present. Major cellular fatty acids of the *Desulfobacterium* strains are $C_{16:0}$, $C_{16:1\,\omega7c}$, and $C_{17:1\,\omega6c}$ (Vainshtein et al., 1992; Kohring et al., 1994).

Desulfobacterium species are abundant in anoxic marine sediments, especially those rich in organic compounds.

All described strains require NaCl and MgCl₂ for growth; best growth is obtained with 15–20 g NaCl and 3 g MgCl₂·6H₂O per liter.

ENRICHMENT AND ISOLATION PROCEDURES

For the selective enrichment of *Desulfobacterium*, the same defined marine sulfate medium described for the isolation of *Desulfobacter* spp. may be used. *D. autotrophicum* may be enriched with H₂ and CO₂ or with formate as electron donors and carbon sources. To avoid the development of mixed cultures of acetate-producing *Acetobacterium* spp. and lithoheterotrophic *Desulfovibrio* spp., which will also become established using these enrichment substrates, marine sediment samples should be diluted in anoxic liquid or agar medium under H₂/CO₂ (90:10) (v/v) (Brysch et al., 1987). Autotrophic growth of *D. autotrophicum* with H₂ and CO₂ occurs with a doubling time of around 16.5 h. *D. vacuolatum* may be enriched with isobutyrate (5 mM), whereas *D. niacini* is usually enriched with nicotinate (5 mM).

Bottles or tubes (100-, 50-, or 20-ml capacity) can serve as culture vessels and are kept anoxic with tight screw caps (if completely filled), or by providing an anoxic head space of N₂/CO₂ (90:10) and sealing with butyl rubber stoppers. The medium is inoculated (2–5% of the total volume) with dark anoxic sediment or sludge from sewage digesters. Enrichment cultures are incubated at 28–30°C. Cultures should be briefly mixed every day. After intense formation of H₂S has occurred, transfers to fresh medium are made. Enrichment cultures should be mixed well before subculturing. Pure cultures of *Desulfobacterium* strains are obtained by repeated application of the agar dilution method (as described for *Desulfobacter* in this edition of the *Manual*). The same substrates used for the enrichment cultures may be used to obtain pure cultures.

MAINTENANCE PROCEDURES

Stock cultures are transferred every 2–3 months. Towards the end of growth or at the beginning of the stationary phase, cultures are refrigerated and stored at 2–5°C in the dark. *Desulfobacterium* strains may be preserved indefinitely by suspension of cells in anoxic medium containing 5% dimethylsulfoxide and storage in liquid nitrogen.

DIFFERENTIATION OF THE GENUS *DESULFOBACTERIUM* FROM OTHER GENERA

Strains of *Desulfobacterium* can be distinguished from other mesophilic sulfate-reducing bacteria by physiological and morphological properties. All *Desulfobacterium* strains known so far can grow lithoautotrophically with H₂/CO₂ (or with formate), or orga-

---

Editorial Note: The type species of the genus is *Desulfobacterium autotrophicum* and not *Desulfobacterium indolicum*, as indicated in List No. 24 in *International Journal of Systematic Bacteriology* (1988).

TABLE BXII.δ.16. Morphological and physiological characteristics of all described species of the genus *Desulfobacterium*[a]

Species	*Desulfobacterium autotrophicum*	*Desulfobacterium niacini*	*Desulfobacterium vacuolatum*
Morphology	Oval-shaped	Oval to irregular-shaped	Oval-shaped or spherical
Cell size (μm)	0.9–1.3 × 1.5–3	1.5–3	1.5–2 × 2–2.5
Motility	+ (sp)	+ (sp)	−
Mol% G + C content	48	46	45
Sulfite reductase	nr	nr	nr
Major menaquinone	MK-7(H$_2$)	MK-7	MK-7(H$_2$)
Optimal temperature (°C)	20–26	29	25–30
Oxidation of substrate	C	C	C
Compounds used as electron donors and carbon sources:			
H$_2$/CO$_2$	+[b]	+[b]	+[b]
Formate	+[b]	+[b]	+[b]
Acetate	(+)	(+)	(+)
Fatty acids: C atoms	(3)–16	(3)–16	(3)–16
Isobutyrate	+	−	+
2-Methylbutyrate	+	−	(+)
3-Methylbutyrate	−	−	(+)
Ethanol	+	(+)	(+)
Lactate	+	−	+
Pyruvate	+	+	+
Fumarate	+	+	+
Succinate	+	+	+
Malate	+	+	+
Benzoate	−	−	−
Alanine	nr	nr	+
Glutamate	+	+	+
Glutamine	nr	−	+
Glutarate	nr	+	+
Glycine	nr	nr	+
Isoleucine	nr	nr	+
Leucine	nr	nr	+
Nicotinate	nr	+	nr
Pimelate	nr	+	nr
Proline	nr	nr	+
Serine	nr	nr	+
Valine	nr	nr	+
Fermentative growth on:			
Fumarate	+	−	nr
Malate	+	−	nr
Pyruvate	+	−	nr
Compounds used as electron acceptors:			
Sulfate	+	+	+
Sulfite	−	+	nr
Sulfur	−	nr	nr
Thiosulfate	+	+	nr
Fumarate reduction	+	−	nr
Nitrate	−	−	nr
Other electron acceptors	nr	nr	nr
Growth factor requirement	bi, ni, th	bi, th	−
NaCl requirement (g/l)	20	15	20
Literature	Brysch et al., 1987	Imhoff-Stuckle and Pfennig, 1983	Widdel, 1988a; Rees et al., 1998

[a]Symbols: sp, single polar flagellum; nr, not reported; C, complete oxidation; +, good growth; (+), poor growth; −, no growth; bi, biotin; th, thiamin; ni, nicotinate.
[b]Autotrophic growth.

noheterotrophically with long chain fatty acids, but are not able to grow with homocyclic aromatic compounds. *Desulfosarcina* and *Desulfococcus* strains can also use long chain fatty acids, but differ morphologically from *Desulfobacterium* species. Unambiguous differentiation requires 16S rDNA sequence analysis and DNA–DNA hybridization.

TAXONOMIC COMMENTS

Based on 16S rRNA gene sequence data, all members of the genus *Desulfobacterium* form one cluster within the *Deltaproteobacteria* and cluster with members of the family *Desulfobacteraceae*. The closest relatives of *Desulfobacterium* species are members of the genera *Desulfobacter* and *Desulfobacula*. A signature sequence for all described *Desulfobacterium* species and other closely related sulfate-reducing bacteria is 5'-TGCGCGGACTCATCTTCAAA-3' (oligonucleotide probe 221), which binds to the 16S rRNA at *E. coli* position 5'-221–240-3' (Devereux et al., 1992; Manz et al., 1998).

The following organisms are no longer included in the genus, because 16S rDNA sequence analysis performed after the original description revealed the need for reclassification as members of other genera or new genera: *Desulfobacterium anilini* (new genus within the *Desulfobacteraceae*), *Desulfobacterium catecholicum* (new genus within the *Desulfobulbaceae*), *Desulfobacterium cetonicum* (see genus *Desulfosarcina*), *Desulfobacterium indolicum* (new genus within the *Desulfobacteraceae*), *Desulfobacterium macestii* (see genus *Desulfomicrobium*), *"Desulfobacterium oleovorans"* (new genus), and *Desulfobacterium phenolicum* (see genus *Desulfobacula*).

FURTHER READING

Brysch, K., C. Schneider, G. Fuchs and F. Widdel. 1987. Lithoautotrophic growth of sulfate-reducing bacteria, and description of *Desulfobacterium autotrophicum* gen. nov., sp. nov. Arch. Microbiol. *148*: 264–274.

List of species of the genus Desulfobacterium

1. **Desulfobacterium autotrophicum** Brysch, Schneider, Fuchs and Widdel 1988, 328[VP] (Effective publication: Brysch, Schneider, Fuchs and Widdel 1987, 272.)

   au.to.tro'phi.cum. Gr. pron. *autos* by itself; Gr. v. *trophein* nourish; *autotrophicum* that nourishes itself from inorganic compounds.

   The morphology is as described in Table BXII.δ.16 and as depicted in Fig. BXII.δ.21. Characteristics are summarized in Table BXII.δ.16. The type strain was isolated from marine mud (Venice, Italy) using H_2 and CO_2.

   The mol% G + C of the DNA is: 47.6 (T_m).
   Type strain: HRM2, ATCC 43914, DSM 3382.
   GenBank accession number (16S rRNA): M34409.

2. **Desulfobacterium niacini** *sp. nov.* (*"Desulfococcus niacini"* Imhoff-Stuckle and Pfennig 1983, 197.)

   ni.a.ci'ni. M.L. n. *niacinum* niacin or nicotinic acid; L. gen. n. *niacini* of nicotinic acid.

   The morphology is as described in Table BXII.δ.16 and

as depicted in Fig. BXII.δ.22. Characteristics are summarized in Table BXII.δ.16. The type strain was isolated from marine sediment (Venice, Italy) using nicotinate.

   The mol% G + C of the DNA is: 45.8 (T_m).
   Type strain: NAV-1, DSM 2650, JCM 12294.
   GenBank accession number (16S rRNA): M34406.

3. **Desulfobacterium vacuolatum** *sp. nov.*

   va.cu.o.la'tum. M.L. n. *vacuolatum* vacuolated due to the morphology of the cells.

   The morphology is as described in Table BXII.δ.16 and as depicted in Fig. BXII.δ.23. Characteristics are summarized in Table BXII.δ.16. The type strain was isolated from marine mud (Venice, Italy) using isobutyrate. It can utilize several amino acids as sole electron donor and carbon source (Rees et al., 1998).

   The mol% G + C of the DNA is: 45 (T_m).
   Type strain: IbRM, DSM 3385, JCM 12295.
   GenBank accession number (16S rRNA): M34408.

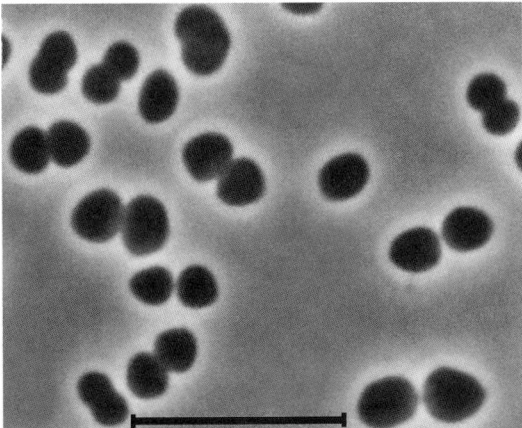

FIGURE BXII.δ.21. Phase-contrast micrograph of *Desulfobacterium autotrophicum* (viable cells). Bar = 10 μm.

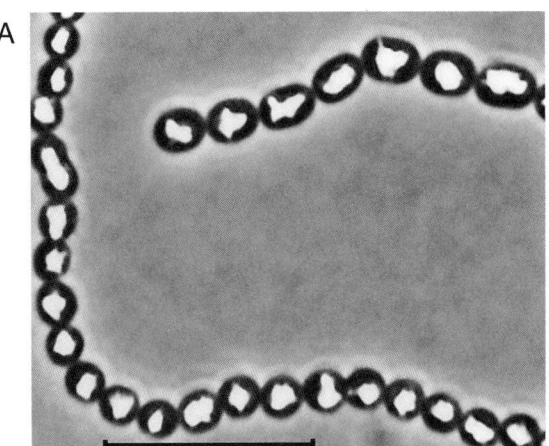

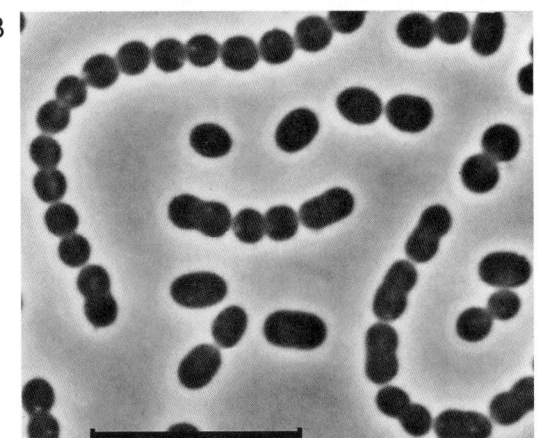

FIGURE BXII.δ.23. (*A*) Phase-contrast micrograph of *Desulfobacterium vacuolatum* (viable cells). Bar = 10 μm; and (*B*) Phase-contrast micrograph of *Desulfobacterium vacuolatum* (viable cells), after application of pressure which caused collapse of gas vesicles. Bar = 10 μm.

FIGURE BXII.δ.22. Phase-contrast micrograph of *Desulfobacterium niacini* (*"Desulfococcus niacini"*) (viable cells). Bar = 10 μm.

Genus III. **Desulfobacula** Rabus, Nordhaus, Ludwig and Widdel 2000b, 1415[VP] (Effective publication: Rabus, Nordhaus, Ludwig and Widdel 1993, 1450) emend. Kuever, Könneke, Galushko and Drzyzga 2001, 176

JAN KUEVER, FRED A. RAINEY AND FRIEDRICH WIDDEL

De.sul.fo.ba'cu.la. L. pref. *de* from; L. n. *sulfur* sulfur; L. n. dem. *bacula* of baca, berry, olive-like fruit; *Desulfobacula* sulfate-reducing small berry.

Oval to sometimes coccoid or slightly curved cells, 1.2–1.5 × 1.2–3.0 μm. Occur singly or in pairs. Spore formation is not observed. Gram negative. Motility may occur. Strict anaerobes with a respiratory type of metabolism. **Chemoorganotrophs,** using toluene, other aromatic compounds, and a number of low-molecular mass aliphatic acids and alcohols **as electron donors and also as carbon sources**; these compounds are completely oxidized to CO_2 via the anaerobic C_1 pathway (carbon monoxide dehydrogenase [CO-DH] pathway, Wood pathway) as indicated by CO-DH activity. **Sulfate and other oxidized sulfur compounds serve as terminal electron acceptors and are reduced to H_2S.** Optimal growth temperature, 28°C. Anoxic media (with sulfide as a reductant) containing NaCl, $MgCl_2$, and vitamins are necessary for growth of the type species. Colonies in anoxic agar media are ochre to yellowish brown. Occur in anoxic marine sediments that are rich in organic compounds.

The mol% G + C of the DNA is: 41–42.

Type species: **Desulfobacula toluolica** Rabus, Nordhaus, Ludwig and Widdel 2000b, 1415 (Effective publication: Rabus, Nordhaus, Ludwig and Widdel 1993, 1450.)

FURTHER DESCRIPTIVE INFORMATION

The type strain of *D. toluolica* has oval-to-coccoid cells, which may be motile and stain Gram negative. The cells occasionally adhere to glass surfaces and are surrounded by slime capsules. In contrast, cells of *D. phenolica* are more curved and have a subpolar flagellum.

The type species has an optimal growth temperature of 28°C. The pH optimal for growth is between 7.0 and 7.1.

Desulfobacula reduces sulfate; other electron acceptors have not been tested. *D. phenolica* can also reduce thiosulfate, whereas sulfite, elemental sulfur, and nitrate cannot serve as electron acceptors. Butyrate, pyruvate, dicarboxylic acids, alcohols, and aromatic compounds including toluene serve as electron donors and organic carbon sources. *D. phenolica* can grow on toluene (Rabus et al., 1993). All organic substrates are completely oxidized to CO_2 via the anaerobic C_1 pathway (carbon monoxide dehydrogenase pathway, Wood pathway) (Rabus et al., 1993). The major menaquinone in the type strain of *D. phenolica* is MK-7(H_2) (Collins and Widdel, 1986). Major cellular fatty acids of *Desulfobacula* strains are $C_{16:0}$, $C_{16:1\ \omega7c}$, and $C_{16:0\ 10CH_3}$ (Kuever et al., 2001). Desulfoviridin is not present.

Desulfobacula species occur in anoxic marine sediments rich in organic material.

Desulfobacula species may be cultured in a defined medium with sulfate as electron acceptor and toluene or benzoate as the electron donor and carbon source. Both described species are dependent on an increased salt concentration; best growth is obtained with 20 g NaCl and 3 g $MgCl_2 \cdot 6H_2O$ per liter. Ammonium ions are used as the nitrogen source. The type strain of *D. toluolica* requires vitamins.

ENRICHMENT AND ISOLATION PROCEDURES

For the selective enrichment of *Desulfobacula*, the same marine sulfate medium described for the isolation of *Desulfobacter* spp.

may be used. For *D. toluolica*, the anoxic medium is overlaid with mineral oil or 2,2,4,4,6,8,8-heptamethynonane as carrier phase containing 2–5% (v/v) toluene (higher concentrations might be toxic). For *D. phenolica*, the medium is supplemented with 2 mM phenol.

Bottles or tubes (100-, 50-, or 20-ml capacity) can serve as culture vessels and are kept anoxic with tight screw caps (if completely filled), or by providing an anoxic head space of N_2/CO_2 (90:10) and sealing with butyl rubber stoppers. The medium is inoculated (2–5% of the total volume) with dark anoxic sediment. Enrichment cultures are incubated at 28–30°C. Cultures should be briefly mixed once a week. After intense formation of H_2S has occurred (approximately 8 weeks), transfers to fresh media are made. Enrichment cultures should be mixed well before subculturing.

Pure cultures of *Desulfobacula* species are obtained by repeated application of the agar dilution method. The method is similar to that described for the genus *Desulfobacter*. The same organic substrates as in the enrichment cultures may be used in agar medium; in the case of *D. toluolica*, the agar is overlaid with toluene in the carrier phase. Cells developing next to the toluene-containing carrier phase are largest in size. Alternatively, benzoate (2–3 mM) may be added as a soluble, almost non-toxic substrate to the agar for both species.

MAINTENANCE PROCEDURES

Stock cultures are transferred every 3–5 months. Towards the end of growth or at the beginning of the stationary phase, cultures are refrigerated and stored at 2–5°C in the dark. *Desulfobacula* strains may be preserved indefinitely by suspension of cells in anoxic medium containing 5% dimethylsulfoxide and storage in liquid nitrogen.

DIFFERENTIATION OF THE GENUS *DESULFOBACULA* FROM OTHER GENERA

The ability to readily utilize toluene and phenol distinguishes *D. toluolica* and *D. phenolica* from most sulfate-reducing bacteria. However, a definite differentiation requires comparative 16S rDNA sequence analysis and DNA–DNA hybridization.

TAXONOMIC COMMENTS

Based on 16S rRNA gene sequence data, *Desulfobacula* species cluster with members of the family *Desulfobacteraceae*. The closest relatives of members of the genus *Desulfobacula* are *Desulfospira joergensenii*, *Desulfotignum balticum*, and members of the genus *Desulfobacter*. A signature sequence for all described *Desulfobacula* species, *Desulfobacter* species, and other closely related sulfate-reducing bacteria is 5′-CACAGGATGTCAAACCCAG-3′ (oligonucleotide probe DSB985), which binds to the 16S rRNA at *E. coli* position 5′-985–1004-3′ (Devereux et al., 1992; Manz et al., 1998).

FURTHER READING

Bak, F. and F. Widdel. 1986. Anaerobic degradation of indolic compounds by sulfate-reducing enrichment cultures, and description of *Desulfobacterium indolicum* gen. nov., sp. nov. Arch. Microbiol. *146*: 170–176.

Kuever, J., M. Konneke, A. Galushko and O. Drzyzga. 2001. Reclassification of *Desulfobacterium phenolicum* as *Desulfobacula phenolica* comb. nov. and description of strain SaxT as *Desulfotignum balticum* gen. nov., sp. nov. Int. J. Syst. Evol. Microbiol. *51*: 171–177.

Rabus, R., R. Nordhaus, W. Ludwig and F. Widdel. 1993. Complete oxidation of toluene under strictly anoxic conditions by a new sulfate-reducing bacterium. Appl. Environ. Microbiol. *59*: 1444–1451.

List of species of the genus Desulfobacula

1. **Desulfobacula toluolica** Rabus, Nordhaus, Ludwig and Widdel 2000b, 1415[VP] (Effective publication: Rabus, Nordhaus, Ludwig and Widdel 1993, 1450.)

tol.u.o' li.ca. L. adj. *toluolica* pertaining to toluene.

Characteristics are described in Table BXII.δ.17. Occurs in anoxic marine sediments that are rich in organic compounds. Isolated with toluene as the sole electron donor and carbon source.

The mol% G + C of the DNA is: 42 (T_m).

Type strain: Tol2, DSM 7467.

GenBank accession number (16S rRNA): X70953, AJ237606.

2. **Desulfobacula phenolica** (Bak and Widdel 1988) Kuever, Könneke, Galushko and Drzyzga 2001, 176[VP] (*Desulfobacterium phenolicum* Bak and Widdel 1988, 136.)

phe.no' li.ca. L. adj. *phenolica* pertaining to phenol.

Characteristics are described in Table BXII.δ.17. Occurs in anoxic marine sediments that are rich in organic compounds. The type strain was isolated from marine mud (Venice, Italy) using phenol. It shows rapid growth on benzoate and glutarate. Pleomorphism may be observed, depending on the growth substrate.

The mol% G + C of the DNA is: 40.6 (T_m).

Type strain: Ph01, ATCC 43956, DSM 3384.

GenBank accession number (16S rRNA): AJ237606.

TABLE BXII.δ.17. Morphological and physiological characteristics of described species of the genus *Desulfobacula*[a]

Species	Desulfobacula toluolica	Desulfobacula phenolica
Cell morphology	Oval	Oval to vibrio
Cell size (μm)	1.2–1.4 × 1.2–2.0	1–1.5 × 2–3
Motility	(+)	+ (sp)
Mol% G + C content	42	41
Sulfite reductase	nr	nr
Major menaquinone	nr	MK-7(H$_2$)
Optimal temperature (°C)	28	28
Oxidation of substrate	C	C
Compounds used as electron donors and carbon sources:		
H$_2$/CO$_2$	−	−
Formate	−	(+)
Acetate	−	(+)
Fatty acids: C atoms	4	(4)
Isobutyrate	−	−
2-Methylbutyrate	−	−
3-Methylbutyrate	−	−
Ethanol	+	(+)
Lactate	−	−
Pyruvate	+	+
Fumarate	+	(+)
Succinate	+	(+)
Malate	+	(+)
Benzoate	+	+
2-Aminobenzoate	−	+
1-Butanol	+	+
p-Cresol	+	+
Glutarate	nr	+
Phenol	−	+
1-Propanol	+	+
Toluene	+	+
Compounds used as electron acceptors:		
Sulfate	+	+
Sulfite	nr	−
Sulfur	nr	−
Thiosulfate	nr	+
Nitrate	nr	−
Growth factor requirement	vitamins	−
NaCl requirement (g/l)	20	20
Literature	Rabus et al., 1993	Bak and Widdel, 1986

[a]Symbols: sp, single polar flagellum; nr, not reported; C, complete oxidation; +, good growth; (+), poor growth; −, no growth.

Genus IV. *Desulfobotulus* gen. nov.

JAN KUEVER, FRED A. RAINEY AND FRIEDRICH WIDDEL

De.sul.fo.bo' tu.lus. L. pref. *de* from; L. n. *sulfur* sulfur; M.L. masc. n. *botulus* sausage; M.L. masc. n. *Desulfobotulus* a sausage-shaped sulfate reducer.

Vibrio-shaped cells, 1.5 × 3–5.5 μm. Occur singly or in pairs. Sigmoid or spirilloid cells sometimes occur. Spore formation is not observed. Gram negative. The type species is motile. Strict anaerobe with a respiratory type of metabolism. **Chemoorganotrophic**, using butyrate, 2-methylbutyrate, higher fatty acids up to 18 carbon atoms, lactate, and pyruvate as electron donors and as carbon sources; **carbon sources are incompletely oxidized to acetate as an end product. Sulfate and sulfite serve as terminal electron acceptors and are reduced to H₂S. Thiosulfate, sulfur, fumarate, and nitrate are not used as terminal electron acceptors. Growth in the absence of an external electron acceptor is only possible with pyruvate.** Optimal temperature, 34°C. Anoxic medium, with sulfide as a reductant, is necessary for growth. Vitamins are not required. Cells contain cytochromes of the *b*- and *c*-type. Storage granules of poly-β-hydroxybutyrate occur. *Desulfobotulus* species are widespread in freshwater mud, but might occur in brackish or marine sediments as well.

The mol% G + C of the DNA is: 53.

Type species: **Desulfobotulus sapovorans** *comb. nov.* (*Desulfovibrio sapovorans* Widdel 1981, 382.)

FURTHER DESCRIPTIVE INFORMATION

The type strain of *D. sapovorans* has large vibrio-shaped cells under all growth conditions (Fig. BXII.δ.24). The cells are motile and stain Gram negative. The flagellum is sheathed. Sigmoid or spirilloid cells sometimes occur. The temperature range for growth is 15–38°C. The type strain has an optimal growth temperature of 34°C. The pH range for growth is 6.5–9.3, with an optimal pH of 7.7.

Desulfobotulus sapovorans, the only species of the genus described so far, is one of the relatively few members of the *Desulfobacteraceae* that cannot perform a complete oxidation of organic compounds, but rather form acetate as an end product. Fatty acids with an odd number of C-atoms are oxidized to acetate and propionate, in accordance with a β-oxidation pathway. A striking, although not unique, feature is the utilization of long-chain fatty acids. Good growth occurs with fatty acids up to 16 C-atoms (palmitate), whereas stearate (18 C-atoms) is utilized very slowly. H₂, formate, propionate, isobutyrate, 3-methylbutyrate (isovalerate), cyclohexanecarboxylate, succinate, fumarate, monovalent alcohols, and sugars are not utilized. *D. sapovorans* reduces only sulfate and sulfite to H₂S. Thiosulfate, elemental sulfur, fumarate, nitrate, and oxygen cannot serve as electron acceptors.

The membrane fraction and cytoplasmic fraction contain *b*- and *c*-type cytochromes, the former being extractable with acetone plus HCl. Desulfoviridin is not present. The major menaquinone is MK-7 (Collins and Widdel, 1986). Major cellular fatty acids of the *Desulfobotulus* strains are C₁₅:₀ anteiso, C₁₆:₁ ω7c, C₁₈:₁ ω9c, and C₁₈:₁ ω7c (Vainshtein et al., 1992; Kohring et al., 1994).

Desulfobotulus strains occur in anoxic mud of freshwater environments. Enrichment can be accomplished with palmitate (see below). However enrichment is not highly specific.

Strains of *Desulfobotulus* may be cultured in a defined medium with sulfate as electron acceptor and fatty acids as the electron donors and carbon sources. Growth requires anoxic mineral media; sulfide is usually added as reductant. Vitamins are not required. Ammonium ions are used as the nitrogen source.

ENRICHMENT AND ISOLATION PROCEDURES

For the enrichment of *Desulfobotulus* strains, palmitate-sulfate medium[1] may be used. However, a highly selective medium for *D. sapovorans* has not been reported. Palmitate may also support growth of other sulfate-reducing bacteria, which may also have vibrio-shaped cells (but often smaller than *D. sapovorans*); oxidation of palmitate by such bacteria is usually also incomplete.

Bottles or tubes (100-, 50-, or 20-ml) can serve as culture vessels and are kept anoxic with tight screw caps (if completely filled), or by providing an anoxic head space of N₂/CO₂ (90:10) and sealing with butyl rubber stoppers. The medium is inoculated (2–5% of the total volume) with dark anoxic sediment or sludge from sewage digesters. Enrichment cultures are incubated at 34–37°C. Cultures should be briefly mixed every day. After intense formation of H₂S has occurred, transfers to fresh medium are made. Enrichment cultures should be mixed well before subculturing. Pure cultures of *Desulfobotulus* strains are obtained by repeated application of the agar dilution method (as described for *Desulfobacter* in this edition of the *Manual*). In the agar cultures, the insoluble fatty acid palmitate should be replaced by the butyrate–caproate–caprylate mixture (see footnote).

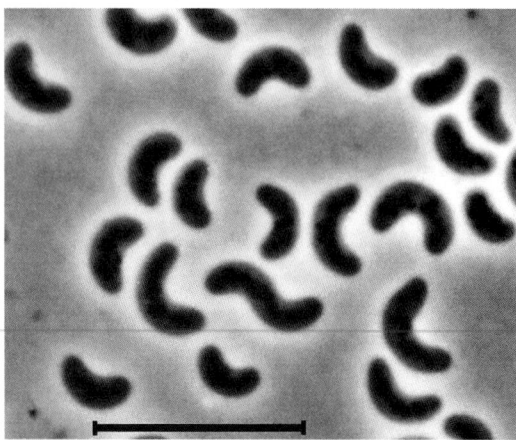

FIGURE BXII.δ.24. Phase-contrast micrograph of *Desulfobotulus sapovorans* with granules of poly-β-hydroxyalkanoic acid (viable cells). Bar = 10 μm.

---

1. For cultivation of *D. sapovorans*, the mineral medium described in the *Desulfobacter* section (this *Manual*) is first prepared without NaCl and MgCl₂. The sodium salt of palmitic acid is added from a stock solution (26 g/l of palmitic acid plus 4 g/l NaOH, heated in a boiling water bath until the solution is clear; sterilized by autoclaving. The solutions must be remelted by heating before use; prewarmed pipettes should be used to take aliquots). Added volumes per liter of medium: 10–20 ml palmitate solution, followed by 1.0 ml of MgCl₂ solution (400 g/l MgCl₂·6H₂O) and 4.0 ml of NaCl solution (300 g/l). Pure cultures can be alternatively grown on soluble monocarboxylic acids. For instance, 5–10 ml of a butyrate–caproate–caprylate mixture (containing per liter: 6.0, 2.5, and 1.0 g, respectively, of acids neutralized with NaOH) may be added per liter of medium.

MAINTENANCE PROCEDURES

Stock cultures are transferred every 3–5 months. Towards the end of growth or at the beginning of the stationary phase, cultures are refrigerated and stored at 2–5°C in the dark. *Desulfobotulus* strains may be preserved indefinitely by suspension of cells in anoxic medium containing 5% dimethylsulfoxide and storage in liquid nitrogen.

DIFFERENTIATION OF THE GENUS *DESULFOBOTULUS* FROM OTHER GENERA

Together with members of the genera *Desulfocella* and *Desulfarculus*, *Desulfobotulus* strains can be distinguished by their vibrio-shaped cells from other sulfate-reducing bacteria (except *Desulfocella*, *Desulfarculus*, and some unnamed types) that are able to grow on long-chain fatty acids. *Desulfarculus* strains perform complete substrate oxidation and autotrophic growth on formate, whereas members of the genera *Desulfocella* and *Desulfobotulus* oxidize substrates only incompletely to acetate and are not capable of autotrophic growth. The mol% G + C content of DNA from *Desulfocella halophila* is much lower than from *D. sapovorans* (35 vs. 53). However, a definite differentiation requires comparative 16S rDNA sequence analysis and DNA–DNA hybridization.

TAXONOMIC COMMENTS

Based on 16S rRNA gene sequence data, *D. sapovorans* clusters with members of the family *Desulfobacteraceae*. The closest relatives of *D. sapovorans* are *Desulfocella halophila* and *Desulfococcus* species. A signature sequence for *D. sapovorans* is 5′-GGGACGCGGACT-CATCCTC-3′ (oligonucleotide probe DSBO 224), which binds to the 16S rRNA at *E. coli* position 5′-224–240-3′ (Manz et al., 1998).

List of species of the genus Desulfobotulus

1. **Desulfobotulus sapovorans** *comb. nov.* (*Desulfovibrio sapovorans* Widdel 1981, 382[VP])

   sa.po′vo.rans. L. n. *sapo* soap; L. *voro* to devour; M.L. part. adj. *sapovorans* devouring soap (i.e., higher fatty acids).

The morphology is as depicted in Fig. BXII.δ.24. The characteristics are as described for the genus.

The mol% G + C of the DNA is: 53 (T_m).

Type strain: 1pa3, Lindhorst, ATCC 33892, DSM 2055.

GenBank accession number (16S rRNA): M34402.

Genus V. **Desulfocella** Brandt, Patel and Ingvorsen 1999, 198[VP]

KRISTIAN K. BRANDT, KJELD INGVORSEN AND BHARAT K.C. PATEL

De.sul.fo.cel′la. L. pref. *de* from; L. n. *sulfur* sulfur; L. n. *cella* small room, cell; M.L. fem. n. *Desulfocella* sulfate-reducing cell.

Vibrio-shaped cells, 0.5–0.7 × 2–4 μm. Motile by means of a single polar flagellum. Gram negative. Do not form spores. **Anaerobic, having a respiratory type of metabolism with sulfate as the terminal electron acceptor and producing H₂S. Chemoorganotrophic. A range of fatty acids (C₄–C₁₆) serve as electron donors and are incompletely oxidized. No fermentative growth.** Mesophilic. Neutrophilic. **NaCl is required for growth.** Vitamins are not required, but growth is stimulated by yeast extract. The genus belongs to the family *Desulfobacteraceae* within the *Deltaproteobacteria*. Isolated from hypersaline, thalassohaline sediments of the Great Salt Lake (Utah, USA).

The mol% G + C of the DNA is: 35.

Type species: **Desulfocella halophila** Brandt, Patel and Ingvorsen 1999, 199.

FURTHER DESCRIPTIVE INFORMATION

Desulfocella halophila strain GSL-But2 is capable of growth in a defined medium containing up to ~20% (w/v) NaCl and 620 mM Mg²⁺. It may thus be active in its natural habitat, which has an *in situ* salinity of approximately 13% NaCl and 185 mM Mg²⁺. However, like all other sulfate reducers described from hypersaline environments, it exhibits a salinity optimal for growth significantly below the *in situ* salinity of the environment from which it was isolated. Recently, direct evidence for the presence of slightly to moderately halophilic and extremely halotolerant, sulfate-reducing bacteria was obtained from a study of Great Salt Lake sediments (Brandt et al., 2001). *Desulfocella halophila* strain GSL-But2 so far represents the only described sulfate reducer isolated from a hypersaline environment that is able to grow by oxidation of butyrate and long-chain fatty acids. Long-chain fatty acids are abundant in cell membranes, and butyrate is an important fermentation product during anaerobic breakdown of organic matter. *Desulfocella halophila* and similar organisms might thus occupy important niches within the anaerobic degrader community in hypersaline sediments. In a recent study, addition of butyrate to Great Salt Lake sediments was found to significantly stimulate bacterial sulfate reduction activity (Brandt et al., 2001).

ENRICHMENT AND ISOLATION PROCEDURES

Desulfocella halophila strain GSL-But2 was isolated from hypersaline surface sediment collected from the deepest part of the southern arm of the Great Salt Lake (Station AS2, Utah Geological Survey station code). Enrichment was performed in anaerobic, bicarbonate-buffered, dithionite-reduced liquid medium containing 10% (w/v) NaCl, with a mixture of volatile fatty acids (acetate, propionate, and butyrate) as electron donors and sulfate as electron acceptor. After 4 weeks, the enrichment culture contained cells of different morphologies and showed high sulfide production. The enrichment culture was subsequently transferred (5% v/v) to a similar medium with butyrate as the sole electron donor. After 2–3 transfers in butyrate medium, strain GSL-But2 was isolated using the deep agar dilution technique (Widdel and Bak, 1992).

DIFFERENTIATION OF THE GENUS *DESULFOCELLA* FROM OTHER GENERA

Desulfocella halophila strain GSL-But2 is the only known truly halophilic and halotolerant sulfate reducer able to grow with butyrate and long-chain fatty acids (C₄–C₁₆) as electron donors. With the exception of *Desulfobotulus sapovorans* and *Desulforegula conservatrix* (Rees and Patel, 2001), *Desulfocella halophila* strain GSL-But2 can also be differentiated from all known mesophilic dissimilatory sulfate-reducing bacteria by its ability to incompletely oxidize long-chain fatty acids (C₄–C₁₆), with acetate as the major degradation product.

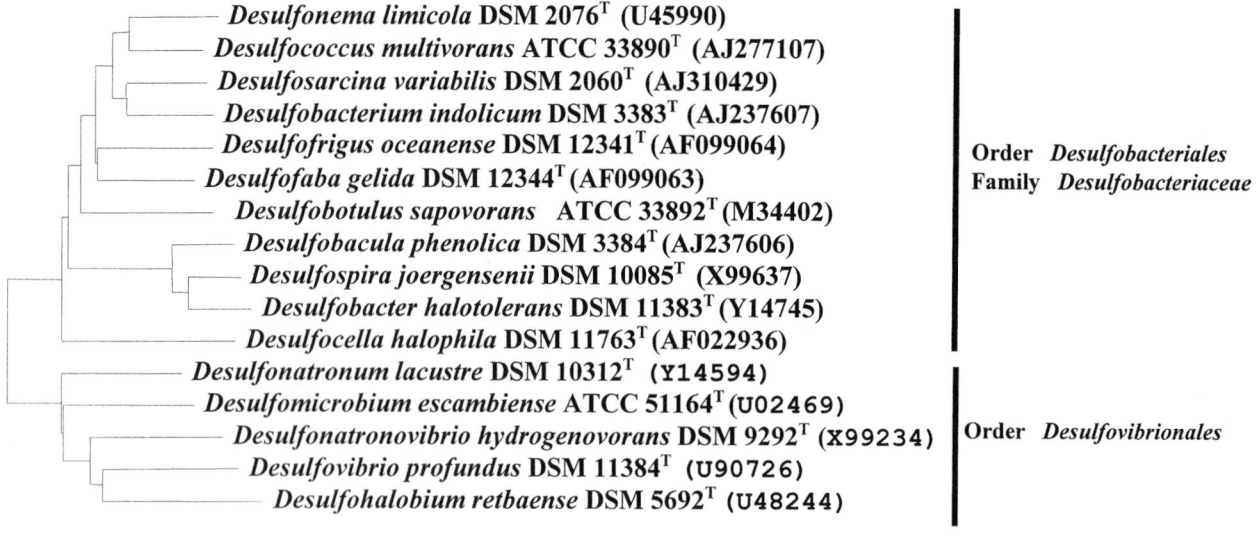

10%

FIGURE BXII.δ.25. Dendrogram showing the phylogenetic position of *Desulfocella halophila* strain GSL-But2[T] amongst members of the family *Desulfobacteraceae* in the order *Desulfobacterales* of the class *Deltaproteobacteria* of the phylum *Proteobacteria*. The strains used in the analysis, their culture collection numbers, and corresponding 16S rRNA gene sequences which were extracted from GenBank/EMBL (in parenthesis) are shown. Phylogenetic analysis was performed on 1130 unambiguous nucleotides using dnadist and neighbor-joining programs which form part of the PHYLIP suite of software. Bar indicates 10 nucleotide changes per 100 nucleotides. The following abbreviations have been used: T, type culture; ATCC, American Type Culture Collection; DSM, Deutsche Sammlung von Mikroorganismen und Zellkulturen GmbH.

TAXONOMIC COMMENTS

Genetically, *Desulfocella halophila* differs from all known described organisms by virtue of its unique 16S rDNA sequence (Fig. BXII.δ.25), showing 87.5% similarity to its closest known relatives, and by its low DNA mol% G + C content of 35.

List of species of the genus Desulfocella

1. **Desulfocella halophila** Brandt, Patel and Ingvorsen 1999, 199[VP]

   ha.lo′phi.la. Gr. n. *hals* salt; Gr. adj. *philos* friendly to; M.L. fem. adj. *halophila* salt-loving.

   The characteristics are as described for the genus, with the following additional features. Does not reduce thiosulfate, sulfite, S⁰, ferric iron, nitrate, or fumarate. The following compounds are used as electron donors: pyruvate, L-alanine, 2-methylbutyrate, and many straight-chain saturated fatty acids with 4–16 carbons. Lactate is not utilized.

   Does not grow by fermentation of pyruvate, L-alanine, or butyrate when sulfate is absent.

   Temperature range for growth, 14–37°C; optimal, ~34°C. Optimal salinity 4–5% NaCl; halotolerance 2–20% NaCl. Optimal growth occurs in the presence of 2–100 mM Mg^{2+}; growth occurs at up to 620 mM Mg^{2+}. pH Range for growth, 5.8–7.6; optimal, 6.5–7.3. Isolated from hypersaline, thalassohaline sediment of the Great Salt Lake (Utah, USA).

   The mol% G + C of the DNA is: 35 (HPLC).
   Type strain: GSL-But2, ATCC 700426, DSM 11763.
   GenBank accession number (16S rRNA): AF022936.

Genus VI. **Desulfococcus** *Widdel 1981, 382[VP] (Effective publication: Widdel 1980, 376)*

JAN KUEVER, FRED A. RAINEY AND FRIEDRICH WIDDEL

De.sul.fo.coc′cus. L. pref. *de* from L. n. *sulfur* sulfur; M.L. masc. n. *coccus* equivalent of Gr. masc. n. *kokkos* grain, berry; M.L. masc. n. *Desulfococcus* a berry shaped (spherical) sulfate-reducer.

Cells are spherical or lemon-shaped, 1.4–2.3 µm in diameter. Occur singly or in pairs. Spore formation is not observed. Cells stain Gram negative and often contain granules of poly-β-hydroxybutyrate. The cells are motile by a single polar flagellum, or are nonmotile. Slime capsules may occur. Strict anaerobes with a respiratory and fermentative type of metabolism. Nutritionally versatile. **Many species use formate and higher monocarboxylic acids up to C_{16}, lactate, pyruvate, and alcohols as electron donors and as carbon sources.** Acetone and phenyl-substituted organic acids may be utilized. These compounds are completely oxidized to CO_2 via the anaerobic C_1 pathway (carbon monoxide dehydrogenase pathway, Wood pathway). **Sulfate and other oxidized sulfur compounds serve as terminal electron acceptors and are reduced to H_2S.** In the absence of an external electron acceptor, slow growth occurs by fermentation of lactate or pyruvate to acetate and propionate. The optimal temperature range for growth is 28–35°C. The optimal pH range for growth is 6.7–7.6. Growth occurs in simple, defined media containing a reductant (usually sulfide) and vitamins. Colonies in anoxic agar media are whitish to yellowish (with sometimes grayish appearance) and tend to be slimy. Desulfoviridin is present in all species. Thermophilic species have not been described. Occur in anoxic mud of freshwater, brackish water, and marine habitats; also occur in sludge of anaerobic sewage digestors.

The mol% G + C of the DNA is: 56–57.

Type species: **Desulfococcus multivorans** Widdel 1981, 382 (Effective publication: Widdel 1980, 377.)

FURTHER DESCRIPTIVE INFORMATION

The type strain of *D. multivorans* has spherical cells under all growth conditions (Fig. BXII.δ.26). Cells of other strains may sometimes be irregular. The type strain stains Gram negative, but a morphologically and nutritionally similar strain that stains Gram positive has been described; nevertheless, cells of this strain exhibited an ultrastructure characteristic of Gram-negative bacteria.

The type strain of *D. multivorans* is nonmotile, but cells from other strains and from enrichment cultures may exhibit a slow motility.

The optimal growth temperature for the type strain of *D. multivorans* is 35°C. Slow growth has been observed at 15°C.

D. multivorans reduces sulfate, sulfite, or thiosulfate to H_2S. Elemental sulfur, fumarate, malate, and nitrate cannot serve as electron acceptors. The following compounds are utilized as electron donors and organic carbon sources: formate, acetate, propionate, butyrate, isobutyrate, valerate, 2-methylbutyrate, 3-methylbutyrate, higher fatty acids up to 16 carbon atoms, lactate, pyruvate, ethanol, 1-propanol, and 1-butanol. Growth also occurs with aromatic compounds such as benzoate, phenylacetate, 3-phenylpropionate, 2-hydroxybenzoate, and with cyclohexanecarboxylate. All organic substrates are completely oxidized to CO_2 via the anaerobic C_1 pathway (carbon monoxide dehydrogenase pathway, Wood pathway) (Schauder et al., 1986). Growth on formate requires no additional organic carbon source.

In the absence of an external electron acceptor, slow growth is possible with pyruvate or lactate, which are fermented to acetate and propionate. Sugars are not fermented.

Cytochromes of the *b* and *c* type were shown to be present mainly in the membrane fraction. The sulfite reductase desulfoviridin has been identified in the cytoplasm of *D. multivorans*. The major menaquinone is MK-7 (Collins and Widdel, 1986).

Major cellular fatty acids of the *Desulfococcus* strains are $C_{15:0\ anteiso}$, $C_{16:0}$, and $C_{17:1\ anteiso\ \omega 8c}$ (Kohring et al., 1994).

Strains of *Desulfococcus* may be cultured in a defined medium[1] with sulfate as electron acceptor and benzoate or another carbon compound as the electron donor and carbon source. The best growth is obtained with 5–20 g/l NaCl and 1–3 g/l $MgCl_2\cdot6H_2O$, although growth also occurs with lower concentrations of these salts. Ammonium ions are used as the nitrogen source. When *D. multivorans* is cultured on benzoate or phenyl-substituted fatty acids, selenite and molybdate are required as trace compounds. Growth on benzoate is stimulated by low levels of fatty acids, especially those having an odd number of carbon atoms; with such supplementation the doubling time on benzoate is around 24 h. *p*-Aminobenzoate, biotin, and thiamin are required as growth factors. Sulfide may serve as a reductant. When inoculation is done from old cultures, initiation of growth is favored by the addition of 10–30 mg of sodium dithionite per liter of medium as a further strong reductant.

ENRICHMENT AND ISOLATION PROCEDURES

For the selective enrichment of *Desulfococcus multivorans* strains, the sulfate medium described for *Desulfobacter* (see *Desulfobacter* section), but with the addition of benzoate (2–3 mM) as electron donor and carbon source, may be used. For *D. biacutus* acetone should be used instead of benzoate.

Bottles or tubes (100-, 50-, or 20-ml) can serve as culture vessels and are kept anoxic with tight screw caps (if completely filled), or by providing an anoxic head space of N_2/CO_2 (90:10) and sealing with butyl rubber stoppers. The medium is inoculated (2–5% of the total volume) with dark anoxic mud from ditches, brackish water, or marine habitats, or with sludge from anaerobic sewage digesters. Enrichment cultures are incubated at 30–35°C. Cultures should be briefly mixed every day. After intense formation of H_2S has occurred (generally after 2–4 weeks), transfers to fresh medium are made. Enrichment cultures should be mixed well before subculturing.

Pure cultures of *Desulfococcus* strains are obtained by repeated application of the agar dilution method (as described for *Desulfobacter* in this edition of the *Manual*).

MAINTENANCE PROCEDURES

Stock cultures are transferred every 3–5 months. Towards the end of growth or at the beginning of the stationary phase, cultures are refrigerated and stored at 2–5°C in the dark. *Desulfococcus* strains may be preserved indefinitely by suspending the cells in anaerobic medium containing 5% dimethylsulfoxide and storing in liquid nitrogen.

DIFFERENTIATION OF THE GENUS *DESULFOCOCCUS* FROM OTHER GENERA

Typical phenotypic characteristics of *Desulfococcus* species are spherical or lemon-shaped cells and the ability to grow well with

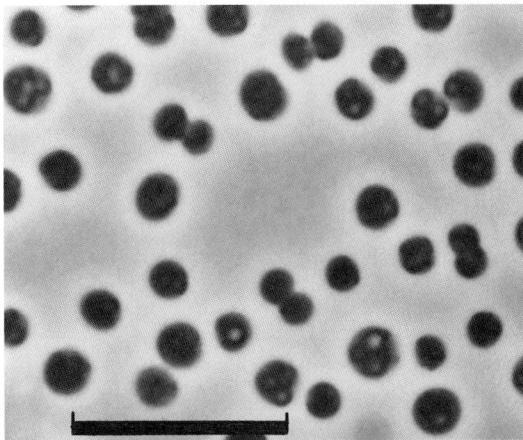

FIGURE BXII.δ.26. Phase-contrast micrograph of *Desulfococcus multivorans* viable cells, with granules of poly-β-hydroxyalkanoic acid. Bar = 10 μm.

---

1. The defined medium is prepared as described in the *Desulfobacter* section (this *Manual*), with the following additions (g/l): NaCl, 7.0; $MgCl_2\cdot6H_2O$, 1.2. After autoclaving, the following components are added aseptically from sterile stock solutions (per liter of medium): sodium benzoate solution (150 g/l), 5.0 ml; vitamin solution (*p*-aminobenzoate, 40 mg/l; biotin, 10 mg/l; thiamine hydrochloride, 100 mg/l; solution sterilized by filtration), 1.0 ml; sodium selenite solution ($Na_2SeO_3\cdot5H_2O$, 3 mg/l; NaOH, 0.5 g/l), 1.0 ml. The pH of the medium is adjusted to 7.2–7.4. For stimulation of growth on benzoate, 50–200 μM *n*-valerate (or a mixture of propionate, *n*-butyrate, isobutyrate, *n*-valerate, 2-methylbutyrate, 3-methylbutyrate and *n*-caproate; approximately 50 μM each) may be added from a sterile stock solution.

benzoate, other phenyl-substituted fatty acids, or acetone as electron donors for sulfate reduction. However, these properties are insufficient for unequivocal taxonomic assignment. Several oval-shaped sulfate-reducing bacteria may also occasionally form spherical cells, and the number of isolates with oval cells that grow on aromatic compounds has been increasing. *Desulfosarcina*, the sulfate-reducing bacterium with the highest phenotypic and phylogenetic similarity to *Desulfococcus* species, also utilizes benzoate and tends to form free-living spherical cells; however, unlike *Desulfococcus*, *Desulfosarcina* also forms cell aggregates, which are especially favored in colonies growing on agar media. A clear differentiation of *Desulfococcus* from other species is only possible by analysis of 16S rRNA genes and DNA–DNA hybridization.

TAXONOMIC COMMENTS

Based on 16S rRNA gene sequence data, members of the genus *Desulfococcus* cluster with other members of the family *Desulfobacteraceae*. The closest relatives are *Desulfosarcina variabilis*, *Desulfonema magnum*, *Desulfonema limicola*, and *Desulfonema ishimotonii*. A signature sequence for all described *Desulfococcus* species is 5′-CCCAAACGGTAGCTTCCT-3′ (oligonucleotide probe DCC209), which binds to the 16S rRNA corresponding to *E. coli* position 5′-209–226-3′ (Ravenschlag et al., 2000).

FURTHER READING

Platen, H. , A. Temmes and B. Schin. 1990. Anaerobic degradation of acetone by *Desulfococcus biacutus* spec. nov. Arch. Microbiol. *154*: 355–361.

List of species of the genus Desulfococcus

1. **Desulfococcus multivorans** Widdel 1981, 382[VP] (Effective publication: Widdel 1980, 377.)

   mul.ti.vo′rans. L. adj. *multus* many, numerous; L. v. *voro* to devour, swallow; M.L. part. adj. *multivorans* devouring numerous kinds of substrates.

   The morphology is as described in Table BXII.δ.18 and as depicted in Fig. BXII.δ.26. Characteristics are as described in Table BXII.δ.18. Isolated from an anaerobic sewage digestor with benzoate as electron donor and carbon source.

   The mol% G + C of the DNA is: 57.4 (T_m).
   Type strain: 1be1, ATCC 33890, DSM 2059.
   GenBank accession number (16S rRNA): M34405.

2. **Desulfococcus biacutus** Platen, Temmes and Schink 1991, 580[VP] (Effective publication: Platen, Temmes and Schink 1990, 360.)

   bi.a.cu′tus. L. adj. *biacutus* twice pointed, referring to atypical cell shape.

   Characteristics are as described in Table BXII.δ.18. Isolated from an anaerobic sewage digestor with acetone as electron donor and carbon source.

   The mol% G + C of the DNA is: 56.5 (T_m).
   Type strain: KMRactS, DSM 5651.
   GenBank accession number (16S rRNA): AJ277887.

TABLE BXII.δ.18. Morphological and physiological characteristics of described species of the genus *Desulfococcus*[a]

Species	Desulfococcus multivorans	Desulfococcus biacutus
Cell morphology	Spherical	Lemon-shaped
Cell size (μm)	1.5–2.2	1.4 × 2.3
Motility	−	−
Mol% G + C content	57	56.5
Sulfite reductase	DV	DV
Major menaquinone	MK-7	nr
Optimal temperature (°C)	35	28–30
Oxidation of substrate	C	C
Compounds used as electron donors and carbon sources:		
H$_2$/CO$_2$	−	nr
Formate	+[b]	nr
Acetate	(+)	(+)
Fatty acids: C atoms	3–16	3-nr
Isobutyrate	+	nr
2-Methylbutyrate	+	+
3-Methylbutyrate	+	+
Ethanol	+	+
Lactate	+	−
Pyruvate	+	+
Fumarate	−	−
Succinate	−	−
Malate	−	−
Benzoate	+	−
Acetone	+[c]	+
Butanone	nr	+
Phenlyacetate	+	nr
Fermentative growth on:	Pyruvate	−
Compounds used as electron acceptors:		
Sulfate	+	+
Sulfite	+	+
Sulfur	−	nr
Thiosulfate	+	nr
Nitrate	−	nr
Other electron acceptors	nr	nr
Growth factor requirement	bi, pa, th	unknown
NaCl requirement (g/l)	5	−
Literature	Widdel, 1980	Platen et al., 1990

[a]Symbols: sp, single polar flagellum; nr, not reported; C, complete oxidation; +, good growth; (+), poor growth; −, no growth; bi, biotin; th, thiamin; pa, *p*-aminobenzoate.

[b]Autotrophic growth.

[c]Some strains.

Genus VII. *Desulfofaba* Knoblauch, Sahm and Jørgensen 1999, 1641[VP]

JAN KUEVER, FRED A. RAINEY AND FRIEDRICH WIDDEL

De.sul.fo.fa' ba. L. pref. *de* from; L. n. *sulfur* sulfur; L. fem. n. *faba* a bean; M.L. fem. n. *Desulfofaba* a sulfate-reducing bean.

Cells are large rods, 3.1 × 5.4–6.2 μm, which form aggregates. Motile cells may occur in old cultures. Cells stain Gram negative and do not form spores. Strict anaerobe with a respiratory and fermentative type of metabolism. Chemoorganotroph, using formate, propionate, butyrate, lactate, pyruvate, other organic acids, alcohols, and alanine as electron donors and also as carbon sources; organic compounds are incompletely oxidized to acetate and CO_2. Sulfate, thiosulfate, and sulfite serve as terminal electron acceptors and are reduced to sulfide. Fermentative growth on pyruvate or other carbon substrates is possible. Obligately psychrophilic. The temperature optimal for growth is 7°C; growth still occurs at −1.8°C. The optimal pH for growth is between 7.1 and 7.6. Mesophilic and thermophilic species are not known. Sodium chloride and magnesium chloride are required, and optimal growth occurs at marine concentrations of these salts. Anoxic media containing a reductant (usually sulfide) are necessary for growth. Vitamins are not required. Desulfoviridin is absent. Occur in permanently cold marine sediments.

The mol% G + C of the DNA is: 52.5.

Type species: **Desulfofaba gelida** Knoblauch, Sahm and Jørgensen 1999, 1641.

FURTHER DESCRIPTIVE INFORMATION

Desulfofaba gelida, the only known species of the genus, is one of the first obligately psychrophilic sulfate-reducing bacteria that has been described (Knoblauch et al., 1999). *D. gelida* is nutritionally relatively versatile. Formate, propionate, butyrate, lactate, pyruvate, malate, succinate, fumarate, ethanol, 1-propanol, 1-butanol, glycerol, glycine, and alanine serve as carbon sources and electron donors. No growth on fructose and glucose. As electron acceptor *D. gelida* uses sulfate, thiosulfate, and sulfite, which are reduced to sulfide. Elemental sulfur, nitrate, nitrite, Fe(III)-oxyhydroxide, and Fe(III)-citrate are not reduced. Fermentative growth occurs on pyruvate and fumarate. Sulfur and thiosulfate are not disproportionated.

Cells contain phosphatidylethanolamine and phosphatidylglycerol as lipids; MK-8 is the sole menaquinone. Major cellular fatty acids are odd numbered and unbranched, $C_{15:0}$ and $C_{15:1\ \omega6c}$. The only described species contains no desulfoviridin.

D. gelida is cultivated in defined, synthetic, marine medium (see chapter on *Desulfobacter*) with propionate or another organic substrate and sulfide as reductant.

ENRICHMENT AND ISOLATION PROCEDURES

For specific enrichment, synthetic marine mineral medium reduced with sulfide should be used (see chapter on *Desulfobacter*). *D. gelida* may be especially enriched with propionate (5 mM) that is added from sterile stock solutions. Bottles or tubes (100-, 50-, or 20-ml) can serve as culture vessels and are kept anoxic with tight screw caps (if completely filled), or by providing an anoxic head space of N_2/CO_2 (90:10) and sealing with butyl rubber stoppers. The medium is inoculated (2–5% of the total volume)

with dark anoxic sediment. *Desulfofaba* species may be also enriched and isolated via direct serial dilution of anoxic environmental samples in liquid medium (according to most-probable-number method) or agar medium (see next paragraph) with propionate as organic substrate. Enrichment cultures are incubated at 4°C. After intense formation of H_2S has occurred (~8–10 weeks), transfers to fresh medium are made. Enrichment cultures should be mixed well before subculturing. Pure cultures of *Desulfofaba* strains are obtained by repeated application of the agar dilution method (as described for *Desulfobacter* in this edition of the *Manual*, but modified after Isaksen and Teske, 1996).

MAINTENANCE PROCEDURES

Stock cultures are transferred every 6 weeks. Towards the end of growth or at the beginning of the stationary phase, cultures are refrigerated and stored at 2–5°C in the dark. *Desulfofaba* strains may be preserved indefinitely by suspension of cells in anoxic medium containing 5% dimethylsulfoxide and storage in liquid nitrogen.

DIFFERENTIATION OF THE GENUS *DESULFOFABA* FROM OTHER GENERA

Desulfofaba gelida can be phenotypically distinguished from other sulfate-reducing bacteria by physiological and morphological properties. Distinctive features are optimal growth at low temperature, large cells, and utilization of propionate. Apart from low growth temperature and large cells, incomplete substrate oxidation and the inability to grow on long-chain fatty acids or aromatic compounds allows differentiation from completely oxidizing members of the genera *Desulfococcus* and *Desulfosarcina*. In contrast to other incomplete oxidizers able to grow with propionate, like *Desulfobulbus* and *Desulforhopalus* strains, *Desulfofaba gelida* is able to grow on butyrate. However, a clear identification can only be done by comparative analysis of 16S rDNA sequences and DNA–DNA hybridization.

TAXONOMIC COMMENTS

Based on 16S rRNA gene sequence data, *Desulfofaba gelida* clusters with other members of the family *Desulfobacteraceae*. The closest relatives of *D. gelida* are members of the genus *Desulfofrigus* and *Desulfosarcina variabilis*. A signature sequence for described *Desulfofaba* species and *Desulfofrigus* species is 5′-CCTCTACACA-CCTGGAATTCC-3′ (oligonucleotide probe DSf672), which binds to the 16S rRNA corresponding to *E. coli* position 5′-672–690-3′ (Ravenschlag et al., 2000).

FURTHER READING

Knoblauch, C., K. Sahm and B.B. Jørgensen. 1999. Psychrophilic sulfate-reducing bacteria isolated from permanently cold Arctic marine sediments: description of *Desulfofrigus oceanense* gen. nov., sp. nov., *Desulfofrigus fragile* sp. nov., *Desulfofaba gelida* gen. nov., sp. nov., *Desulfotalea psychrophila* gen. nov., sp. nov and *Desulfotalea arctica* sp. nov. Int. J. Syst. Bacteriol. *49*: 1631–1643.

List of species of the genus Desulfofaba

1. **Desulfofaba gelida** Knoblauch, Sahm and Jørgensen 1999, 1641[VP]

   ge′ li.da. L. adj. *gelide* ice-cold; referring to the low temperature optimal for growth.

   Characteristics are as indicated for the genus. Isolated from anoxic sediment from Hornsund (Svalbard) with propionate as electron donor and carbon source and sulfate as electron acceptor.

   The mol% G + C of the DNA is: 52.5 (HPLC).

   Type strain: PSv29, DSM 12344.

   GenBank accession number (16S rRNA): AF099063.

Genus VIII. **Desulfofrigus** *Knoblauch, Sahm and Jørgensen 1999, 1640*[VP]

JAN KUEVER, FRED A. RAINEY AND FRIEDRICH WIDDEL

De.sul.fo.fri′ gus. L. pref. *de* from; L. n. *sulfur* sulfur; L. neut. n. *frigus* cold; M.L. neut. n. *Desulfofrigus* sulfate reducer living in the cold.

Cells are large, straight or slightly curved rods, 0.8–2.1 × 3.2–6.1 μm. Motile cells may be observed in old cultures. Cells stain Gram negative and do not form spores. Strict anaerobe with a respiratory and fermentative type of metabolism. **Chemoorganotroph**, using fatty acids, other carboxylic acids including amino acids, and alcohols as electron donors and also as carbon sources. No growth on fructose and glucose. **Organic substrates completely oxidized to CO_2 or incompletely oxidized to acetate. Sulfate serves as terminal electron acceptor and is reduced to sulfide.** Ferric citrate may also serve as electron acceptor. The type strain can also use thiosulfate and sulfite as electron acceptor. Elemental sulfur, Fe (III)-oxohydroxide, nitrate, and nitrite are not used as electron acceptors. Fermentative growth on pyruvate, malate, or other carbon substrates is possible. Sulfur and thiosulfate are not disproportionated. Only the type species is obligately psychrophilic. The temperature optimal for growth is 10–18°C, but growth still occurs at −1.8°C. Thermophilic species are not known. The optimal pH for growth is between 7.0 and 7.5. *Desulfofrigus* species require sodium chloride and magnesium chloride, and optimal growth occurs at marine salt concentrations. Media containing a reductant are necessary for growth. Occur in permanently cold marine sediments.

The mol% G + C of the DNA is: 52–53.

Type species: **Desulfofrigus oceanense** Knoblauch, Sahm and Jørgensen 1999, 1640.

FURTHER DESCRIPTIVE INFORMATION

Desulfofrigus species were among sulfate-reducing bacteria isolated during a microbiological study of permanently cold, Arctic sediment. Special adaptation to conditions in such sediments are indicated by the ability to grow at −1.8°C, viz. near the freezing point of seawater. Whereas *D. oceanense* may be regarded as a true psychrophile, *D. fragile* is psychrotolerant and characterized by a relatively wide temperature range for growth, up to 27°C. In *D. oceanense*, the capacity for complete oxidation of organic substrates to CO_2 is evident from utilization of acetate as growth substrate. The mode of substrate oxidation (complete versus incomplete) in *D. fragile* has not been unequivocally proven. Acetate added to cultures is not utilized.

Cells contain the polar lipids phosphatidylethanolamine and phosphatidylglycerol, and MK-9 is the sole menaquinone. Major cellular fatty acids are $C_{16:1 \omega 7c}$ and $C_{18:1 \omega 7c}$. The only described species contains no desulfoviridin. Obligate or facultatively psychrophilic, although *Desulfofrigus fragile* can grow up to 27°C.

Desulfofrigus species are cultivated in defined synthetic marine media (see chapter on *Desulfobacter*) with lactate, acetate, or another organic substrate and sulfide as reductant.

ENRICHMENT AND ISOLATION PROCEDURES

For specific enrichment, synthetic marine mineral medium reduced with sulfide should be used (see chapter on *Desulfobacter*). *Desulfofrigus* species may be especially enriched with lactate (10 mM) or acetate (10 mM) that is added from sterile stock solutions. Bottles or tubes (100-, 50-, or 20-ml) can serve as culture vessels and are kept anoxic with tight screw caps (if completely filled), or by providing an anoxic head space of N_2/CO_2 (90:10) and sealing with butyl rubber stoppers. The medium is inoculated (2–5% of the total volume) with dark anoxic sediment. *Desulfofrigus* species may also be enriched and isolated via direct serial dilution of anoxic environmental samples in liquid medium (according to most-probable-number method) or agar medium with lactate or acetate as organic substrates. Enrichment cultures are incubated at 4°C. After intense formation of H_2S has occurred (~8–10 weeks), transfers to fresh medium are made. Enrichment cultures should be mixed well before subculturing. Pure cultures of *Desulfofrigus* strains are obtained by repeated application of the agar dilution method (as described for *Desulfobacter* in this edition of the *Manual*).

MAINTENANCE PROCEDURES

Stock cultures are transferred every 6 weeks. Towards the end of growth or at the beginning of the stationary phase, cultures are refrigerated and stored at 2–5°C in the dark. *Desulfofrigus* strains may be preserved indefinitely by suspension of cells in anoxic medium containing 5% dimethylsulfoxide and storage in liquid nitrogen.

DIFFERENTIATION OF THE GENUS *DESULFOFRIGUS* FROM OTHER GENERA

Desulfofrigus species can be phenotypically distinguished from other sulfate-reducing bacteria by physiological properties. A distinctive feature is optimal growth at low temperature. The inability to grow with propionate is one characteristic that allows differentiation from other psychrophilic or mesophilic sulfate reducers, such as *Desulfobulbus*, *Desulfofaba*, and *Desulforhopalus* species. However, a clear identification can only be done by comparative analysis of 16S rDNA sequences and DNA–DNA hybridization.

TAXONOMIC COMMENTS

Based on 16S rRNA gene sequence data, *Desulfofrigus* species cluster with other members of the family *Desulfobacteraceae*. The closest relatives of *Desulfofrigus* species are members of the genus *Desulfofaba* and *Desulfosarcina variabilis*. A signature sequence for described *Desulfofaba* species and *Desulfofrigus* species is 5'-CCTCTACACACCTGGAATTCC-3' (oligonucleotide probe DSf672), which binds to the 16S rRNA corresponding to *E. coli* position 5'-672–690-3' (Ravenschlag et al., 2000).

FURTHER READING

Knoblauch, C., K. Sahm and B.B. Jørgensen. 1999. Psychrophilic sulfate-reducing bacteria isolated from permanently cold Arctic marine sed-iments: description of *Desulfofrigus oceanense* gen. nov., sp. nov., *Desulfofrigus fragile* sp. nov., *Desulfofaba gelida* gen. nov., sp. nov., *Desulfotalea psychrophila* gen. nov., sp. nov and *Desulfotalea arctica* sp. nov. Int. J. Syst. Bacteriol. 49: 1631–1643.

List of species of the genus Desulfofrigus

1. **Desulfofrigus oceanense** Knoblauch, Sahm and Jørgensen 1999, 1640[VP]

   o.ce.a.nen' se. L. adj. *oceanensis* belonging to the ocean.

   Characteristics are listed in Table BXII.δ.19. Isolated from anoxic sediment from Hornsund (Svalbard) with acetate as electron donor and carbon source and sulfate as electron acceptor.

   The mol% G + C of the DNA is: 52.8 (HPLC).
   Type strain: ASv26, DSM 12341.
   GenBank accession number (16S rRNA): AF099064.

2. **Desulfofrigus fragile** Knoblauch, Sahm and Jørgensen 1999, 1641[VP]

   fra'gi.le. L. adj. *fragilis* referring to the rapid lysis of the type strain in the stationary phase.

   Characteristics are listed in Table BXII.δ.19. Isolated from anoxic sediment from Hornsund (Svalbard) with lactate as electron donor and carbon source and sulfate as electron acceptor.

   The mol% G + C of the DNA is: 52.1 (HPLC).
   Type strain: LSv21, DSM 12345.
   GenBank accession number (16S rRNA): AF099065.

TABLE BXII.δ.19. Morphological and physiological characteristics of described species of the genus *Desulfofrigus*[a]

Species	*Desulfofrigus oceanense*	*Desulfofrigus fragile*
Morphology	Rod	Rod
Width × length (μm)	2.1 × 4.2–6.1	0.8 × 3.2–4.2
Motility	±	±
Mol% G + C content	52.8	52.1
Sulfite reductase	nr	nr
Major menaquinone	MK-9	MK-9
Optimal temperature (°C)	10	18
Oxidation of substrate	C	I
Compounds used as electron donors and carbon sources:		
H₂/CO₂	(+)	−
Formate	+	+
Acetate	+	−
Fatty acids: C atoms	4–5	4,6,10,16
Ethanol	+	+
Lactate	+	+
Pyruvate	+	+
Fumarate	−	+
Succinate	−	−
Malate	+	+
Benzoate	−	−
Alanine	−	+

(continued)

TABLE BXII.δ.19. *(cont.)*

Species	*Desulfofrigus oceanense*	*Desulfofrigus fragile*
Glycerol	+	+
Glycine	+	−
1–Butanol	+	+
1–Propanol	+	+
Serine	+	+
Fermentative growth on:		
Lactate	+	−
Malate	+	+
Pyruvate	+	+
Compounds used as electron acceptors:		
Sulfate	+	+
Sulfite	+	−
Sulfur	-	−
Thiosulfate	+	−
Fumarate reduction	nr	nr
Nitrate	−	−
Fe(III)	Fe(III)-citrate	Fe(III)-citrate
Growth factor requirement	−	−
NaCl requirement (g/l)	15	10

[a]Symbols: sp, single polar flagellum; nr, not reported; C; complete; I, incomplete; +, good growth; (+), poor growth; −, no growth. Data from Knoblauch et al. (1999).

Genus IX. **Desulfonema** *Widdel 1981, 382[VP] (Effective publication: Widdel 1980, 378)*

JAN KUEVER, FRED A. RAINEY AND FRIEDRICH WIDDEL

De.sul.fo.ne' ma. L. pref. *de* from; L. n. *sulfur* sulfur; L. n. *nema* thread; M.L. neut. n. *Desulfonema* thread-forming sulfate reducer.

Cells arranged in uniseriately multicellular, flexible filaments with gliding motility. Filaments are 2.5–8 μm in diameter and sometimes about 2 mm in length; the nearly cylindrical cells are 2.5–13 μm long. Cross-walls are visible (Figs. BXII.δ.27 and BXII.δ.28C). Granules of poly-β-hydroxyalkanoic acid are commonly stored. Filaments are always attached to surfaces that serve as substrata for gliding motility. Cell walls characteristic of Gram-negative bacteria have been revealed by electron microscopy of ultrathin sections, even though Gram stain may be positive in some instances (unequally stained). The outer membrane has a wavy structure. Strictly anaerobic, having a respiratory type of metabolism. **Chemoorganotrophs or chemoautotrophs**, using acetate and higher monocarboxylic acids up to C₁₄, fumarate, and succinate **as electron donors and carbon sources**. Lactate or benzoate may also serve as organic substrates. Organic substrates are completely oxidized to CO₂ via the anaerobic C₁ pathway (carbon monoxide dehydrogenase [CO-DH], Wood pathway) as indicated by CO-DH activity. For chemoautotrophic growth of

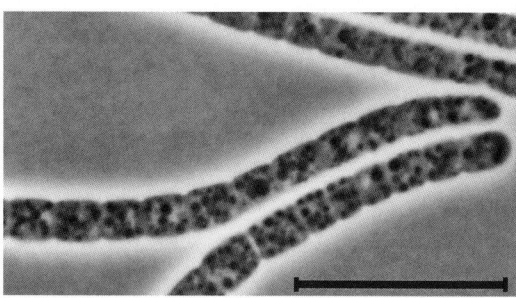

FIGURE BXII.δ.27. Phase-contrast micrograph of *Desulfonema limicola* with granules of poly-β-hydroxyalkanoic acid (viable cells). Bar = 10 µm.

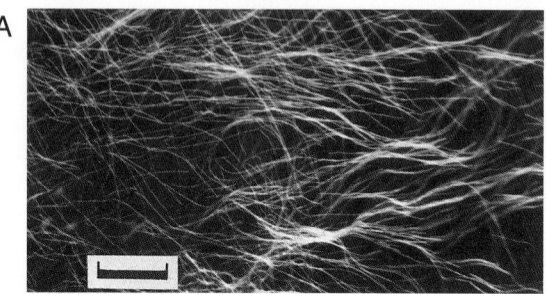

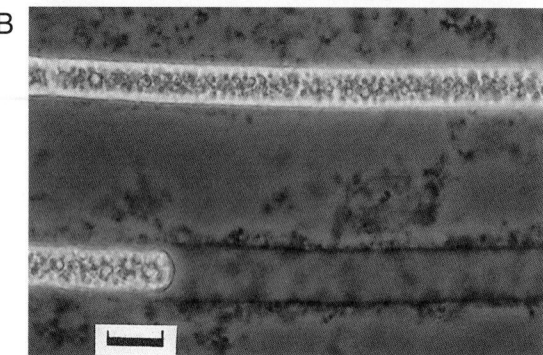

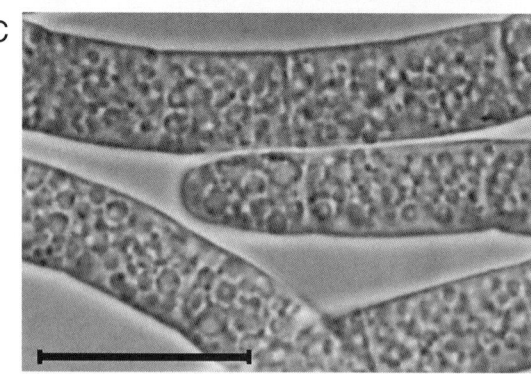

FIGURE BXII.δ.28. (*A*) *Desulfonema magnum,* layer of filaments at the wall of a culture bottle (viable cells). Bar = 1 mm. (*B*) Phase-contrast micrograph of *Desulfonema magnum* with granules of poly-β-hydroxyalkanoic acid; filaments glide in synthetic sediment of aluminum phosphate which makes slime trails visible (viable cells). Bar = 10 µm. (*C*) Phase-contrast micrograph of *Desulfonema magnum* with granules of poly-β-hydroxyalkanoic acid (viable cells). Bar = 10 µm.

some species, **H₂ can serve as the electron donor and CO₂ as the carbon source. Sulfate serves as terminal electron acceptor and is reduced to H₂S; sulfite and thiosulfate may or may not**

be used. Elemental sulfur does not serve as electron acceptor but is inhibitory. **Growth does not occur by fermentation in the absence of an external electron acceptor.**

All species are able to grow in simple defined media containing a reductant. Vitamins are required for growth. The optimal pH range for growth is 7.0–7.6. The optimal temperature range for growth is 28–32°C. Thermophilic species are not known. The addition of >15 g NaCl and >2 g $MgCl_2 \cdot 6H_2O$ is required for growth. Gliding movement and growth are promoted by addition of insoluble substrata such as agar or inorganic precipitates. Desulfoviridin is present or absent. Cytochromes are present.

Occur in organic-compound-rich, anoxic sediments. Only marine species have been isolated so far. *Desulfonema* species also occur in freshwater habitats, but have never been obtained in pure culture.

The mol% G + C of the DNA is: 35–55.

Type species: **Desulfonema limicola** Widdel 1981, 382 (Effective publication: Widdel 1980, 379.)

FURTHER DESCRIPTIVE INFORMATION

The multicellular filaments of *Desulfonema* species are not sheathed; however, remnants of lysed cells within a filament can simulate a sheath. Filaments of *D. limicola* and *D. ishimotonii* do not glide regularly; the most conspicuous motility characteristic of this species is a slow twitching or jerky swinging of the filaments when they creep out of sediment particles. The gliding movement of *D. magnum* is more regular, and a speed of about 4 µm/s is reached. Gliding filaments reverse the direction of movement if they meet obstacles; a reversion of the gliding direction may also occur without obvious reason. Gliding filaments often form trails from sediment particles, probably by slime excretion.

The ultrastructure of *D. limicola* and *D. magnum* has been examined (Widdel et al., 1983). Ultrathin sections of both species exhibited intracytoplasmic membranes and a wavy outer membrane. The wavy structure was more regular in *D. magnum* than in *D. limicola.*

Desulfonema species are nutritionally versatile sulfate-reducing bacteria that oxidize a range of fatty acids and dicarboxylic acids; hydrogen, lactate, benzoate, or higher phenyl-substituted organic acids may also be oxidized. Sulfate is used as an external electron acceptor; alternatively sulfite or thiosulfate may be reduced. All organic substrates are completely oxidized to CO_2 via the anaerobic C_1 pathway (carbon monoxide dehydrogenase pathway, Wood pathway) (Fukui et al., 1999). Nitrate, fumarate, or malate do not serve as electron acceptors. Sugars, pyruvate, or fumarate are not fermented. Elemental sulfur inhibits growth, probably by increasing the redox potential. Filaments are damaged by oxygen.

Desulfonema species occur in anoxic sediments, especially those rich in decomposing algal material. In samples taken from such habitats, the filaments may be observed under the microscope without a preceding enrichment in the laboratory. Only marine, salt-requiring *Desulfonema* species have been isolated so far in pure culture. However, observation of morphologically similar, sometimes thinner filaments grown in anaerobic enrichments from pond and ditch sediments and application of 16S rRNA-targeted fluorescent probes (Fukui et al., 1999) suggest that freshwater forms may also occur.

Even though some organic acids (e.g., isobutyrate, benzoate) may promote growth of *Desulfonema* species in natural samples, none of the electron donors known to be oxidized by *Desulfonema* species allows a reliable selective enrichment. Since the filamentous, gliding, sulfate reducers grow relatively slowly (doubling times around 14–30 h or longer), they are often outcompeted in enrichment cultures by unicellular sulfate reducers, after some transfers. A selective advantage of *Desulfonema* in nature is probably the capacity for gliding movement that allows filaments to spread in sediments and exploit favorable growth conditions. Moreover, the long filaments resist grazing by protozoa (Widdel, 1983).

Desulfonema species may be cultivated under strictly anoxic conditions in reduced synthetic media[1] with sulfate as the electron acceptor and acetate, propionate, higher fatty acids, succinate, lactate, or benzoate as electron donors. Good growth may be obtained using combinations of organic substrates. Growth on acetate as sole organic substrate is very slow but can be stimulated by low concentrations (~50 mg/l) of additional electron donors, e.g., formate, fatty acids, or succinate. Previously sterilized extracts from anaerobically digested complex organic matter (sludge extracts) have been used to promote growth on organic substrates (Widdel, 1983; Widdel et al., 1983). Ammonium ions are used as nitrogen source. *D. limicola* and *D. ishimotonii* are routinely grown with 20 g NaCl and 3 g $MgCl_2 \cdot 6H_2O$ per liter medium. The standard medium contains biotin or a vitamin mixture. For optimal growth of *D. magnum*, 20 g NaCl, 5 g $MgCl_2 \cdot 6H_2O$, and at least 0.8 g $CaCl_2 \cdot 6H_2O$ per liter medium are necessary; biotin, 4-aminobenzoic acid, and vitamin B_{12} are required as growth factors. The high requirement for calcium ions has not been reported for any other sulfate-reducing bacterium. Gliding movement and growth are supported by addition

of artificial sediments such as precipitated aluminum phosphate or sloppy agar (2 g/l).

ENRICHMENT AND ISOLATION PROCEDURES

Filamentous, gliding, sulfate-reducing bacteria can be enriched in raw cultures if anoxic sediment samples from sulfate-rich freshwater or marine habitats are mixed with algal material or cellulose powder. In case of low sulfate concentrations, $CaSO_4$ should be added. Enrichments are incubated under an anoxic gas phase in a sealed bottle in the dark at room temperature. After 1–2 weeks, cell filaments may form dense silky layers covering the sediment and the glass wall.

Of the defined substrates known for *D. ishimotonii* and *D. magnum*, isobutyrate (5 mmol/l) and benzoate (3 mmol/l), respectively, are most suitable for enrichment in synthetic media. Acetate and a fatty acid mixture (see footnote) may be used to enrich *D. limicola*, but the procedure has low selectivity and often leads to the growth of *Desulfobacter*-like sulfate-reducing bacteria. An inoculum of 2–5% (v/v) black marine mud should be used. No artificial substrata for gliding movement should be added to the enrichment subcultures, so that filaments are forced to grow at the glass wall. If possible, transfers should be made from the glass wall. Parts of the layers may be sucked into a Pasteur pipette. However, despite these precautions, unicellular sulfate reducers often become dominant. *Desulfonema* may be obtained in pure culture only if the number of these competitors is not significantly higher than the number of filaments. Before the enrichment medium becomes turbid due to spreading of unicellular competitors, tufts of filaments are taken from the glass wall into Pasteur pipettes and washed anaerobically to remove the bulk of other cells. For washing, a fine-mesh copper grid (as used in electron microscopy) fixed in a conically drawn glass tube by means of a thin resin layer (Widdel 1980, 1983) or a small cotton plug in such a glass tube (Fukui et al., 1999) may be used as a sieve (Widdel, 1980, 1983). Anoxic medium is kept above the grid or cotton plug by the level of an elevated, bent outlet tube. After addition of filament tufts, sterile medium passes through the grid or cotton plug; *Desulfonema* filaments are retained, whereas the bulk of unicellular competitors and commensals pass through. Growth of *D. ishimotonii* is affected if exclusive addition of the initial, selective substrate isobutyrate is continued (Fukui et al., 1999). Therefore, an acetate–isobutyrate mixture (see above) should be added to subcultures.

In the case of *D. magnum*, a pure culture may be finally obtained from the washed filaments by transferring one of the relatively thick, visible filaments through a series of small portions of sterile anoxic medium. *D. limicola* may be isolated from washed filaments in agar dilution series. A medium with agar at lower concentration (8 g/l) than is found in standard agar media is recommended. The agar is prepared as described for *Desulfobacter*. The same concentrations of NaCl, $MgCl_2$, and $CaCl_2$ as in the enrichment medium are added in concentrated form to the agar before mixing with medium. The dilution procedure is analogous to that for *Desulfobacter*.

Filamentous sulfate reducers may form fluffy colonies. If, however, the gliding filaments migrate through the agar (as in *D. magnum*), the described method is not useful for isolation, since separate colonies are not obtained. Colonies of filaments are picked with medium-containing Pasteur pipettes. These may previously be drawn in a flame to an appropriate diameter. If the filaments cannot be reached without touching colonies of other bacteria, a part of the soft agar is carefully sucked off with a Pasteur pipette connected to a vacuum pump; however, a min-

---

1. Defined medium has the following composition (in g/l distilled water): Na_2SO_4, 4.0; KH_2PO_4, 0.2; NH_4Cl, 0.2; NaCl, 20; KCl, 0.5; $MgCl_2 \cdot 6H_2O$, 3.0 for *D. limicola* and 5.0 for *D. magnum*; and $CaCl_2 \cdot 2H_2O$, 0.15 for *D. limicola* and 1.4 for *D. magnum*. The dissolved salts are autoclaved. For the preparation of soft agar medium that supports homogeneous growth and gliding movement, 2 g of repeatedly washed (with cold H_2O) separately autoclaved agar in 70 ml H_2O are added per liter (final volume) of the hot autoclaved medium and mixed immediately; agar is omitted if artificial sediment is added later (see below). After the medium is cooled under an anaerobic atmosphere, the following volumes from separately sterilized stock solutions are added per liter medium while access of air is prevented by flushing with a mixture of N_2/CO_2 (CO_2 content: 5–20%, depending on the desired pH): solution of 84 g/l $NaHCO_3$ (autoclaved under CO_2 atmosphere), 30 ml; trace element solution (see below), 1 ml; solution of 3 mg/l $Na_2SeO_3 \cdot 5H_2O$ + 0.5 g/l NaOH, 1 ml; solution of 120 g/l $Na_2S \cdot 9H_2O$ (autoclaved under N_2 atmosphere), 3 ml; filter-sterilized vitamin mixture (see below), 1 ml; filter-sterilized thiamin solution (100 mg/l of 25 mM NaH_2PO_4/H_3PO_4 buffer, pH 3.4), 1 ml; and filter-sterilized B_{12} solution (50 mg/l), 1 ml. The pH of the mixed medium is adjusted with sterile HCl or Na_2CO_3 solution to 7.6 for *D. limicola* and to 7.0 for *D. ishimotonii* and *D. magnum*. The medium may be dispensed into tubes (10ml) or culture bottles (50 or 100 ml); bottles are completely filled and tightly sealed with screw caps or are provided with an anaerobic gas phase and sealed with butyl rubber stoppers. The desired organic substrates are added from sterile stock solutions. For *D. limicola*, 5–10 ml from a solution of 280 g/l sodium acetate trihydrate are added per liter of medium. Growth on acetate is stimulated by the addition of 1 ml from a mixture of organic acids (see below) per liter of culture medium. For *D. ishimotonii*, 5–10 ml from a solution of 140 g/l sodium acetate trihydrate and 22 g/l sodium isobutyrate are added per liter of medium. The substrates for *D. limicola* are also suitable for *D. ishimotonii*, and vice versa. For *D. magnum*, 3–5 ml from a solution of 150 g/l sodium benzoate are added per liter of medium. A sediment of aluminum phosphate per liter of medium is precipitated before inoculation: 5 ml of an autoclaved solution of 48 g/l $AlCl_3 \cdot 6H_2O$ are added, and the pH is readjusted with 1.6 ml from a solution of 106 g/l Na_2CO_3.

Trace element solution (without complexing agent) as described for *Desulfobacter* should be used.

Vitamin mixture contains per liter 10 mM Na_2HPO_4/NaH_2PO_4 buffer (pH 7.1): 4-aminobenzoic acid, 40 mg; D(+)-biotin, 10 mg; nicotinic acid, 100 mg; calcium D(+)-pantothenate, 50 mg; pyridoxine hydrochloride, 150 mg.

Mixture of organic acids (modified after Bryant, 1973) contains (per l): isobutyric acid, 5 g; valeric acid, 5 g; isovaleric acid, 5 g; 2-methylbutyric acid, 5 g; caproic acid, 2 g; heptanoic acid, 2 g; octanoic acid, 2 g; and succinic acid, 45 g. The acids are neutralized with NaOH.

imum distance of 8 mm should be kept between filaments and agar surface. In order to prevent contamination by bacteria liberated from disrupted colonies, the agar surface is sterilized, e.g., by addition of bromine vapor to the head space of the agar tube. The slow penetration of bromine into the agar is visible by a yellow zone. After 1 min, the bromine is removed with a gas stream of N_2. The bromine dissolved in the upper agar zone is reduced by injection of 0.5 ml SO_2 gas (e.g., made from sodium bisulfite plus acid in a serum bottle) into the head space. Immediately after decolorization of the bromine zone, the SO_2 is removed with N_2.

A few isolated filaments often do not start to grow if transferred into large volumes (>20 ml) of liquid medium in common tubes or bottles. Such filaments should be transferred into media in special anaerobic glass tubes (15 ml) with pointed, conical bottoms. The filaments sink to this narrow part of the tube where they can establish a growth-favoring microenvironment and where growth can be observed under the dissecting microscope. Initiation of growth may be promoted by addition of sludge extract (supernatant from centrifuged sediment of the habitat; 10–20% [by volume] added to the medium and dithionite as strong reductant). In the beginning, tubes are incubated without shaking. If growth is visible as increasing numbers of filaments attached to the glass wall, the culture may be shaken twice a day.

Freshly isolated filamentous sulfate reducers usually spread on surfaces and form silky layers that may cover the whole inner wall of the culture vessel. After a number of subcultures, however, *Desulfonema* filaments tend to creep together and form irregular clumps; such cultures grow only poorly, and filaments in clumps may soon die off. Clump formation is avoided by providing an insoluble substratum (aluminum phosphate or sloppy agar; see above) for gliding movement and by shaking the culture once or twice a day.

MAINTENANCE PROCEDURES

Pure cultures are maintained in liquid medium with an added sediment in completely filled screw-capped bottles, or under an anaerobic gas phase in bottles sealed with butyl rubber stoppers. Filaments remain viable longest with substrates on which growth is relatively slow. The enrichment substrate isobutyrate usually affects *D. ishimotonii* after a number of transfers. Stock cultures of *D. limicola* may be kept on acetate plus low concentrations of other fatty acids, and those of *D. magnum* may be kept on benzoate. *D. limicola* is stored at 2–5°C. *D. magnum* is damaged at such low temperatures and, therefore, is stored at −20°C. Stock cultures are transferred every 2–4 months. Strains may also be preserved indefinitely by suspending the filaments in anaerobic medium containing 5% (v/v) dimethylsulfoxide and storing in liquid nitrogen.

DIFFERENTIATION OF THE GENUS *DESULFONEMA* FROM OTHER GENERA

Desulfonema species are most easily differentiated from all other genera of sulfate-reducing bacteria by their striking filamentous morphology. The filaments consist of regularly arranged, nearly cylindrical cells. Single cells are usually not observed but may be liberated by shearing forces.

Desulfonema is distinguished from *Beggiatoa* species that may be abundant at the oxic–anoxic interface in sediments by the absence of highly refractile sulfur globules. Granules of poly-β-hydroxyalkanoic acid that are often present in *Desulfonema* are less refractile than sulfur globules. In sediment samples kept under completely anoxic conditions, *Desulfonema* continues to move and to multiply, whereas *Beggiatoa* becomes nonmotile, loses the sulfur droplets, and dies off after some days. *Desulfonema* can be differentiated from filamentous gliding cyanobacteria and green bacteria by the absence of pigments. Layers of *Desulfonema* are whitish, and filaments appear colorless in the brightfield microscope.

In addition, 16S rRNA gene sequence analysis (see next section) and DNA–DNA hybridization, as well as physiological characteristics (Table BXII.δ.20), may allow differentiation from other genera.

TAXONOMIC COMMENTS

The closest relatives of *Desulfonema* species are members of the genera *Desulfosarcina* and *Desulfococcus*. A signature sequence for *Desulfonema magnum* and *Desulfonema limicola* species is 5′-TAUGGGAGAGGGAAGCGGAA-3′ (oligonucleotide probe DNM 657), which binds to the 16S rRNA corresponding to position 5′-657–676-3′ of the *Escherichia coli* 16S rRNA sequence (Fukui et al., 1999). Despite one mismatch in the signature sequence of *Desulfonema ishimotonii*, good hybridization signals can be obtained with this species. Members of the genera *Desulfococcus* and *Desulfosarcina* have also only one mismatch and might result in positive signals, but are morphologically differentiated from *Desulfonema* (see above).

FURTHER READING

Fukui, M., A. Teske, B. Assmus, G. Muyzer and F. Widdel. 1999. Physiology, phylogenetic relationships, and ecology of filamentous sulfate-reducing bacteria (genus *Desulfonema*). Arch. Microbiol. *172*: 193–203.

Widdel, F. 1983. Methods for enrichment and pure culture isolation of filamentous gliding sulfate-reducing bacteria. Arch. Microbiol. *134*: 282–285.

Widdel, F., G.-W. Kohring and F. Mayer. 1983. Studies on dissimilatory sulfate-reducing bacteria that decompose fatty acids III. Characterization of the filamentous gliding *Desulfonema limnicola* gen. nov. sp. nov., and *Desulfonema magnum* sp. nov. Arch. Microbiol. *134*: 286–294.

List of species of the genus Desulfonema

1. **Desulfonema limicola** Widdel 1981, 382[VP] (Effective publication: Widdel 1980, 379.)

   li.mi′ co.la. L. suff. verbal n. *cola* dweller; M.L. masc. n. *limicola* mud-dweller.

   The morphology is as depicted in Fig. BXII.δ.27. Characteristics are summarized in Table BXII.δ.20. The type strain requires a saline medium with at least 13 g/l NaCl and 2 g/l $MgCl_2 \cdot 6H_2O$ for optimal growth. Isolated from marine mud, Jadebusen area, Germany. Morphologically

   similar filaments have been also observed in freshwater sediments.

   The mol% G + C of the DNA is: 34.5 (T_m).
   Type strain: 5ac10, DSM 2076, ATCC 33961.
   GenBank accession number (16S rRNA): U45990.

2. **Desulfonema ishimotonii** Fukui, Teske, Assmus, Muyzer and Widdel 2000, 1415[VP] (Effective publication: Fukui, Teske, Assmus, Muyzer and Widdel 1999, 202.)

   i.shi.mo′ to.ni.i. M.L. gen. n. *ishimotonii* of Ishimoto, named

TABLE BXII.δ.20. Morphological and physiological characteristics of described species of the genus *Desulfonema*[a]

Species	*Desulfonema limicola*	*Desulfonema ishimotonii*	*Desulfonema magnum*
Morphology	Multicellular filament, 10–400 cells	Multicellular filament	Multicellular filament, 10–200 cells
Cell size (μm)	2.5–3 × 2.5–3	2.5–3 × 3–6;	6–8 × 9–13
Motility	Gliding	Gliding	Gliding
Mol% G + C content	35	55	42
Sulfite reductase	DV	DV	P582
Major menaquinone	MK-7	nr	MK-9
Optimal temperature (°C)	30	30	32
Oxidation of substrates	C	C	C
Compounds used as electron donors and carbon sources:			
H$_2$/CO$_2$	+[b]	+[b]	−
Formate	+[b]	+[b]	+
Acetate	(+)	+	(+)
Fatty acids: C atoms	3–14	3–14	3–10
Isobutyrate	+	+	+
2-Methylbutyrate	+	+	(+)
3-Methylbutyrate	+	+	+
Ethanol	−	+	−
Lactate	+	+	−
Pyruvate	+	+	−
Fumarate	+	+	+
Succinate	+	+	+
Malate	−	−	(+)
Benzoate	−	−	+
1–Butanol	nr	+	nr
1–Propanol	nr	+	nr
Compounds used as electron acceptors:			
Sulfate	+	+	+
Sulfite	+	nr	−
Sulfur	−	nr	−
Thiosulfate	+	nr	−
Fumarate reduction	−	nr	−
Nitrate	−	nr	−
Growth factor requirement	bi	nr	bi, pa, B$_{12}$
NaCl requirement (g/l)	15	20	20
Literature	Widdel et al., 1983	Fukui et al., 1999	Widdel et al., 1983

[a]Symbols: nr, not reported; C, complete oxidation; + , good growth; (+), poor growth; −, no growth; bi, biotin; th, thiamin; pa, *p*-aminobenzoate; DV, desulfoviridin.

[b]Autotrophic growth.

after Maoto Ishimoto, a Japanese biochemist and microbiologist.

Characteristics are summarized in Table BXII.δ.20. Optimal growth of the type strain requires 20 g/l NaCl, 3 g/l MgCl$_2$·6H$_2$O, and at least 0.2 g/l CaCl$_2$·2H$_2$O in the culture medium. *D. ishimotonii* is morphologically indistinguishable from *D. limicola*, but exhibits faster growth (14–18 h doubling time in comparison to ~30 h in case of the latter). Isolated from marine mud from Tokyo Bay, Japan.

The mol% G + C of the DNA is: 55.0 (HPLC).

Type strain: Tokyo 01, DSM 9680.

GenBank accession number (16S rRNA): U45992.

3. **Desulfonema magnum** Widdel 1981, 382[VP] (Effective publication: Widdel 1980, 381.)

mag′num. L. adj. *magnus* large, big.

The morphology is as depicted in Fig. BXII.δ.28. Characteristics are summarized in Table BXII.δ.20. Optimal growth of the type strain requires 20 g/l NaCl, 5 g/l MgCl$_2$·6H$_2$O, and at least 0.8 g/l CaCl$_2$·2H$_2$O in the culture medium. Isolated from marine mud, Montpellier, France.

The mol% G + C of the DNA is: 41.6 (T_m).

Type strain: 4beI3, DSM 2077, ATCC 35288.

GenBank accession number (16S rRNA): U45989.

Genus X. **Desulfosarcina** *Widdel 1981, 382[VP] (Effective publication: Widdel 1980, 382)*

Jan Kuever, Fred A. Rainey and Friedrich Widdel

De.sul.fo.sar′ ci.na. L. pref. *de* from; L. n. *sulfur* sulfur; *sarcina,* *Desulfosarcina* sarcina-shaped sulfate reducer.

Irregularly shaped cells occurring singly and in large, sarcina-like packets. Cells are 0.8–1.5 × 1.5–7.0 μm, occurring singly or in pairs. Spore formation is not observed. Granules of poly-β-hydroxyalkanoic acid frequently occur within the cells. Gram negative. Usually nonmotile, but cells motile by means of a single polar flagellum may occur. Strictly anaerobic, having both a respiratory and a fermentative type of metabolism. **Chemoorgano-trophs or chemoautotrophs**, using formate, acetate, propionate, butyrate, higher fatty acids, other organic acids, alcohols, and **benzoate or similar aromatic compounds as electron donors and also as carbon sources**; these compounds are completely oxidized to CO_2. Chemoautotrophic growth occurs with **H_2 as the electron donor and CO_2 as the carbon source. Sulfate and other oxidized sulfur compounds serve as terminal electron acceptors and are reduced to H_2S.** In the absence of an external electron acceptor, slow growth of the type species occurs by fermentation of lactate or pyruvate to acetate and propionate. Carbon monoxide dehydrogenase activity is commonly observed, indicating the operation of the anaerobic C_1 pathway (carbon monoxide dehydrogenase pathway, Wood pathway) for complete oxidation of acetyl-CoA, or for CO_2 fixation during autotrophic growth. Media containing a reductant and not less than 10 g/l NaCl and 2 g/l $MgCl_2 \cdot 6H_2O$ are necessary for growth. Optimal growth temperature, 28–33°C. Thermophilic species have not been described. The optimal pH range for growth is 7.2–7.6. Cells contain desulforubidin as sulfite reductase, and cytochromes of the b and c type. Colonies in anaerobic agar media are grayish to yellowish, compact, and irregular in shape. Occur in anoxic part of brackish water and marine habitats.

The mol% G + C of the DNA is: 51–59.

Type species: **Desulfosarcina variabilis** Widdel 1981, 382 (Effective publication: Widdel 1980, 383.)

FURTHER DESCRIPTIVE INFORMATION

D. variabilis forms large free-living cells (Fig. BXII.δ.29A) and sarcina-like packets with irregular arrangement of cells or distorted appearance (Fig. BXII.δ.29B). These packets form pellets in liquid media and layers at the glass wall. Cells liberated from the packets (e.g., by squeezing) are irregularly shaped. Repeated subculturing from the culture fluid favors the development of coccoidal or ellipsoidal cells which occur singly or in pairs (Fig. BXII.δ.29A). Single cells of the type strain are nonmotile, although some possess a single polar flagellum. Motile cells have been observed in enrichment cultures of *Desulfosarcina*.

The temperature range for growth is 15–38°C, with optimal growth occurring at 33°C. The pH range for growth is 6.7–9.0, with pH 7.4 being the optimal.

D. variabilis reduces sulfate, sulfite, or thiosulfate to H_2S. Elemental sulfur and nitrate cannot serve as electron acceptors. *D. cetonicum* can use sulfur as electron acceptor. Chemoautotrophic growth occurs with H_2 as the electron donor and CO_2 as the carbon source. The following compounds serve as both electron donors and organic carbon sources for chemoorganotrophic growth: formate, propionate, butyrate, valerate, 2-methylbutyrate, 3-methylbutyrate, higher fatty acids up to 14 carbon atoms, lactate, pyruvate, succinate, fumarate, ethanol, 1-propanol, 1-butanol, benzoate, and phenyl-substituted organic acids. Some species can additionally use acetone, toluene, *m*-cresol, or *o*-xylene. Growth on acetate alone is very slow. All organic substrates are completely oxidized to CO_2 via the anaerobic C_1 pathway (carbon monoxide dehydrogenase pathway, Wood pathway; Schauder et al., 1986; Janssen and Schink, 1995a, b; Harms et al. 1999b).

In the absence of an external electron acceptor, slow growth is possible with pyruvate or lactate, which are fermented to acetate and propionate. Fumarate can be slowly fermented to acetate, propionate, and succinate; it does not act as terminal electron acceptor for anaerobic respiration. Sugars are not fermented.

Membrane-bound and soluble cytochromes of the b and c type have been identified in *D. variabilis*. Cells contain desulforubidin as sulfite reductase (Arendsen et al., 1993) and menaquinone MK-7 as the predominant quinone (Collins and Widdel, 1986). Major cellular fatty acids of *Desulfococcus multivorans* strains are $C_{15:0\ anteiso}$ and $C_{16:0}$ (Kohring et al., 1994).

D. variabilis is sensitive toward light and has to be incubated in the dark. Diffuse daylight in the laboratory may inhibit growth completely.

Desulfosarcina occurs in anoxic black mud of brackish water and marine habitats.

Strains of *Desulfosarcina* may be cultured in a defined medium[1] with sulfate as electron acceptor and benzoate as the electron donor and carbon source. For *D. cetonicum*, brackish medium should be used. Ammonium ions are used as the nitrogen source. For growth on benzoate, addition of molybdate as a trace element is necessary. It is not yet known whether selenite is required, but

---

1. The defined medium is prepared as described in the footnote of the section on the genus *Desulfobacter*. The following additions are recommended (g/l): NaCl, 13.0 and $MgCl_2 \cdot 6H_2O$, 2.0. After autoclaving of the medium, the following additional components are added aseptically from sterile stock solutions (per liter of medium): sodium benzoate solution (150 g/l), 5.0 ml; sodium selenite solution ($Na_2SeO_3 \cdot 5H_2O$, 3 mg/l and NaOH, 0.5 g/l), 1.0 ml. The pH of the medium is adjusted to 7.3–7.5.

A

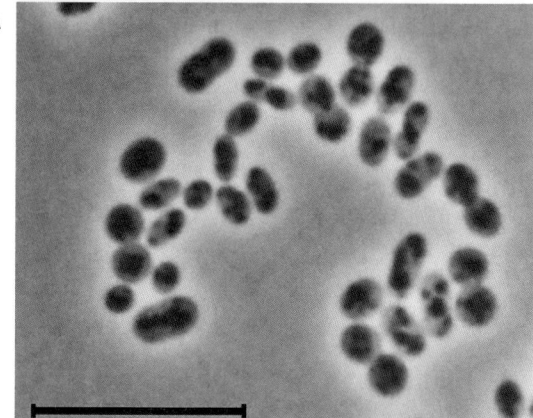

B

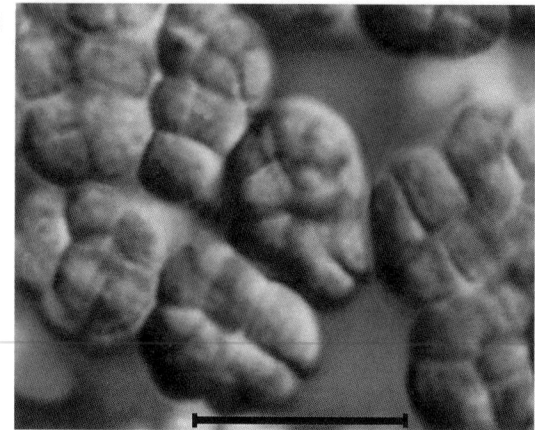

FIGURE BXII.δ.29. (*A*) Phase-contrast micrograph of *Desulfosarcina variabilis*, viable single cells from liquid medium. Bar = 10 μm; (*B*) phase-contrast micrograph of *Desulfosarcina variabilis*, cell packets from an agar colony. Bar = 10 μm.

selenite is added routinely to the medium. The type strains of *D. variabilis* and *D. cetonicum* do not require vitamins; other strains have not been studied in this regard. When inoculation is done from old cultures, initiation of growth is favored by the addition of 10–30 mg sodium dithionite per liter of medium as a further strong reductant, in addition to routinely added sodium sulfide.

ENRICHMENT AND ISOLATION PROCEDURES

For the selective enrichment of *Desulfosarcina* strains, benzoate–sulfate medium is used. Other aromatic compounds or acetone will favor growth of species other than the type species, *D. variabilis*. Bottles or tubes (100-, 50-, or 20-ml capacity) can serve as culture vessels and are kept anoxic with tight screw caps (if completely filled), or by providing an anoxic headspace of N_2/CO_2 (90:10) and sealing with butyl rubber stoppers. The medium is inoculated with black anoxic mud from brackish or seawater sediments; 2–5% of the total volume should be added. Enrichment cultures are incubated at 25–30°C and mixed by shaking every second day. After intense formation of H_2S has occurred (generally after 10–20 d), transfers to fresh medium are made. Enrichment cultures should be mixed well before subculturing.

Pure cultures of *Desulfosarcina* strains are obtained by repeated application of the agar dilution method. The method is similar to that described for the genus *Desulfobacter*, except that benzoate medium is added to the molten concentrated agar solution. Before inoculation of the agar medium, cell packets of *Desulfosarcina* should be broken carefully with a homogenizer in medium under anoxic conditions. In agar media, *Desulfosarcina* forms irregularly shaped, compact colonies, which can be removed with a sterile Pasteur pipette.

Desulfosarcina ovata strains may be enriched and isolated with *o*-xylene as electron donor and carbon source. Pure cultures are obtained by repeated application of the agar dilution method. The method is similar to that described for *Desulfobacula toluolica* (see genus *Desulfobacula*), except that the overlying carrier phase contains 2% (v/v) *o*-xylene (Harms et al., 1999b).

MAINTENANCE PROCEDURES

Pure cultures are maintained in liquid medium in screw-capped bottles or in anoxically gassed, butyl rubber-sealed bottles, with storage at 2–5°C. Stock cultures are transferred every 2–3 months. *Desulfosarcina* strains may be preserved indefinitely by suspending the cells in anaerobic medium containing 5% dimethylsulfoxide and storing in liquid nitrogen.

DIFFERENTIATION OF THE GENUS *DESULFOSARCINA* FROM OTHER GENERA

Desulfosarcina differs from many other anaerobic sulfate-reducing bacteria by its formation of dense sarcina-like cell packets. A characteristic physiological property of *Desulfosarcina* is its growth with benzoate and phenyl-substituted fatty acids. This property is also shared by members of the genera *Desulfococcus* and *Desulfobacterium*; however, species of the latter genera are coccoid or small rods under all cultural conditions and never form cell packets.

TAXONOMIC COMMENTS

The closest relatives of the genus *Desulfosarcina* are members of the genus *Desulfonema*, and *Desulfococcus multivorans*. A signature sequence for all described *Desulfosarcina* species except *D. cetonicum* is 5′-AGGCCACCCTTGATCCAA-3′ (oligonucleotide probe DSC193), which binds to the 16S rRNA corresponding to position 5′-193–210-3′ of the *Escherichia coli* 16S rRNA sequence (Ravenschlag et al., 2000).

FURTHER READING

Galushko, A.S. and E.P. Rosanova. 1991. *Desulfobacterium cetonicum* spec. nov., a sulfate-reducing bacterium oxidizing fatty acids and ketones. Mikrobiologiya *60*: 102–107.

List of species of the genus Desulfosarcina

1. **Desulfosarcina variabilis** Widdel 1981, 382^VP (Effective publication: Widdel 1980, 383.)

   va.ri.a' bi.lis. L. adj. *variabilis* changeable, variable.

   The morphology is as depicted in Fig. BXII.δ.29A and B. Characteristics are as described for the genus and listed in Table BXII.δ.21. Occurs in anoxic mud of brackish water and marine habitats.

   The mol% G + C of the DNA is: 51.2 (T_m).
   Type strain: 3be13, Montpellier, DSM 2060.
   GenBank accession number (16S rRNA): M34407, M26632.

2. **Desulfosarcina cetonicum** *comb. nov.* (*Desulfobacterium cetonicum* Galushko and Rozanova 1994, 370.)

   ce' to.ni.cum. L. adj. *cetonicum* pertaining to ketones; *Desulfosarcina cetonicum* a sulfate-reducing bacterium oxidizing ketones.

   Characteristics are as described in Table BXII.δ.21. The type strain was isolated from flooded oil deposits of the Apsheron peninsula (Azerbaijan) using butyrate.

   The mol% G + C of the DNA is: 59.0 (T_m).
   Type strain: 480, DSM 7267, JCM 12296, VKM B-1975.
   GenBank accession number (16S rRNA): AJ237603.

3. **Desulfosarcina ovata** *sp. nov.*

   o' va.ta. M.L. adj. *ovatus* from L. n. *ovum* egg; L. fem. adj. *ovata* referring to the oval (egg-like) cell shape.

   Characteristics are as described in Table BXII.δ.21. The type strain was isolated from a North Sea oil tank in Wilhelmshaven (Germany) with *o*-xylene as sole organic substrate.

   The mol% G + C of the DNA is: 51.0 (T_m).
   Type strain: oXyS1, DSM 13228, JCM 12297.
   GenBank accession number (16S rRNA): Y17286.

TABLE BXII.δ.21. Morphological and physiological characteristics of described species of the genus *Desulfosarcina*[a]

Species	Desulfosarcina variabilis	Desulfosarcina cetonicum[b]	Desulfosarcina ovata[b]
Morphology	Oval rod, packages	Oval	Rod
Cell size (μm)	1–1.5 × 1.5–2.5	0.8–1.2 × 1.8–2.7	0.8–1.0 × 2.5–4.0
Motility	± (sp)	−	nr
Mol% G + C content	51	59	51
Sulfite reductase	DR	nr	nr
Major menaquinone	MK-7	nr	nr
Optimal temperature (°C)	33	30	32
Oxidation of substrate	C	C	C
Electron donors used for sulfate reduction:			
H_2	+[c]	nr	nr
Formate	+[c]	+[c]	+
Acetate	(+)	+	+
Fatty acids: C atoms	3–14	3–16	4–nr
Isobutyrate	−	nr	nr
2-Methylbutyrate	(+)	nr	nr
3-Methylbutyrate	(+)	nr	nr
Ethanol	+	+	+
Lactate	+	+	+
Pyruvate	+	+	+
Fumarate	+	−	nr
Succinate	+	−	+
Malate	−	−	+
Benzoate	+	+	+
Acetone	nr	+	nr
1–Butanol	+	+	nr
m-Cresol	nr	+	−
Phenylacetate	+	nr	nr
1–Propanol	+	+	nr
Toluene	nr	+	+
o-Xylene	nr	nr	+
Fermentative growth on:			
Fumarate	+	nr	nr
Lactate	+	nr	nr
Pyruvate	+	nr	nr
Compounds used as elelectron acceptors:			
Sulfate	+	+	+
Sulfite	+	nr	nr
Sulfur	−	+	nr
Thiosulfate	+	+	nr
Nitrate	−	−	nr
Other electron acceptors	nr	nr	nr
Growth factor requirement	−		nr
NaCl requirement (g/l)	15	−	nr[d]
Literature	Widdel, 1980	Galushko and Rozanova, 1991; Mueller et al., 1999	Harms et al.,1999b

[a]Symbols: sp, single polar flagellum; nr, not reported; C, complete oxidation; + , good growth; (+), poor growth; −, no growth; DR, desulforubidin.

[b]So far not a validated species.

[c]Autotrophic growth.

[d]So far only grown in marine medium.

Genus XI. **Desulfospira** *Finster, Liesack and Tindall 1997e, 1274*[VP] *(Effective publication: Finster, Liesack and Tindall 1997c, 207)*

WERNER LIESACK AND KAI FINSTER

De.sul.fo.spi′ra. L. pref. *de* from; L. n. *sulfur* sulfur; Gr. n. *spira* a coil; M.L. fem. n. *Desulfospira* a sulfate-reducing coil.

Curved, often spirilloid cells 0.7–0.8 × 1–2 μm. Nonmotile. Gram negative. Endospores not formed. Anaerobic, having a respiratory type of metabolism. Fermentative growth does not occur. Catalase positive. **Chemoorganotrophic or chemoautotrophic**, using a wide range of organic compounds as electron donors and also as carbon sources, e.g., fatty acids, dicarboxylic acids, oxoacids, hydroxyacids, or compatible solutes; **these substrates are completely oxidized to CO₂. Maleate and proline are used as** electron donors. Aromatic compounds and sugars are not utilized. **Sulfate, sulfite, thiosulfate, and S⁰ serve as terminal electron acceptors** and are reduced to H_2S. Na^+ and Mg^{2+} are required for growth. Biotin is required for growth. Cells contain *c*-type cytochromes and menaquinones with seven isoprenoid units. Desulfoviridin is not present. Isolated from iron-rich, oxidized, marine surface sediment.

The mol% G + C of the DNA is: 50.

Type species: **Desulfospira joergensenii** Finster, Liesack and Tindall 1997e, 1274 (Effective publication: Finster, Liesack and Tindall 1997c, 207.)

FURTHER DESCRIPTIVE INFORMATION

Growth in the presence of Fe(III)-oxyhydroxide in nonreduced medium may be interpreted as an adaptation to oxidized conditions which prevail in the surface sediment from which *D. joergensenii* strain B331 was isolated. In the absence of Fe(III)-oxyhydroxide, substrate utilization ceased when sulfide concentrations reached 8–9 mM. This was probably due to the toxicity of hydrogen sulfide at higher concentrations. In the presence of Fe(III)-oxyhydroxide, larger amounts of substrate were utilized because hydrogen sulfide was continuously removed by chemical reoxidation to S^0 and/or precipitation as FeS.

ENRICHMENT AND ISOLATION PROCEDURES

Desulfospira strains are selectively enriched in a sulfate-free, anaerobic mineral medium (Widdel and Bak, 1992; Finster and Bak, 1993) with highly purified S^0 (200 mg per 50 ml medium) as electron acceptor in the presence of large amounts of amorphous Fe(III)-oxyhydroxide (1 ml 0.3 M Fe(III)-oxyhydroxide per 10 ml medium). Reducing agents are omitted. Sodium butyrate (5 mM) is used as the electron donor. S^0 is prepared and added as described by Pfennig and Biebl (1976). Amorphous Fe(III)-oxyhydroxide is prepared by titration of an acidic FeCl$_3$ solution with NaOH to pH 7. Oxidized marine surface sediment serves as inoculum. The formation of black spots around small sulfur particles indicates the reduction of S^0 to H_2S, which results in the precipitation of FeS. Black particles can be dissolved by addition of a few drops of 1 M HCl solution to a subsample of the enrichment culture. The release of H_2S is indicated by its characteristic odor. Cells are primarily present in the precipitates and not in the supernatant.

Pure cultures are obtained by repeated application of the agar shake dilution method (Pfennig, 1978). A 3% (w/v) solution of agar is prepared in distilled water. The agar should be washed several times in distilled water before preparing the solution. The agar shakes consist of two different phases, one containing amorphous Fe(III)-oxyhydroxide in 3% agar, and a second containing substrates and inoculum. For preparation of the iron-containing phase, 1 ml of a 0.3 M Fe(III)-oxyhydroxide suspension is added to test tubes containing 3 ml of molten sterile agar. After mixing, the suspension is allowed to solidify while placed in a horizontal position. The solidified iron-containing phase should cover one half of the glass wall of the test tube.

The second phase contains an agar dilution series supplemented with sodium butyrate and S^0. The agar medium is kept molten after inoculation and then rapidly transferred to test tubes containing the iron-agar phase. The tubes are immediately transferred to a cold water bath, sealed with butyl rubber stoppers, and gassed with a mixture of N_2/CO_2 (90:10). The formation of black FeS in the iron-agar layer is indicative of the reduction of S^0 to H_2S. Well-separated colonies that occur in the higher dilutions can be removed with sterile Pasteur pipettes and used as the inoculum for subsequent agar shake cultures or liquid cultures containing butyrate, S^0, and Fe(III)-oxyhydroxide. Alternatively, *Desulfospira* strains can be grown in anaerobic, sulfide-reduced medium with sulfate as electron acceptor instead of S^0. Addition of Fe(III)-oxyhydroxide is then omitted.

MAINTENANCE PROCEDURES

Pure cultures are maintained in liquid culture medium in screw-capped bottles or in anaerobically gassed bottles sealed with butyl rubber stoppers, with storage at 2–5°C. Stock cultures are transferred every 1–3 months.

DIFFERENTIATION OF THE GENUS *DESULFOSPIRA* FROM OTHER GENERA

The genus *Desulfospira*, currently represented only by its type species *D. joergensenii*, is characterized by the ability to use both oxidized sulfur compounds and S^0 as terminal electron acceptors for anaerobic respiration. This phenotypic trait has only been observed for a few species belonging to the genera *Desulfovibrio* and *Desulfomicrobium* (Widdel and Pfennig, 1992), and for *Desulfofustis glycolicus* (Friedrich and Schink, 1995; Friedrich et al., 1996). A unique nutritional characteristic of *D. joergensenii* among known sulfate reducers is the utilization of maleate and proline. *D. joergensenii* and *Desulfobacula toluolica* (Rabus et al., 1993) differ in the pattern of utilized substrates (Table BXII.δ.22). The presence of two types of menaquinones with seven isoprenoid units (MK-7 and MK-7 [VII-H2]) and the absence of an internally branched hexadecyl fatty acid ($C_{16:0 CH_3}$) distinguishes *D. joergensenii* from members of the genus *Desulfobacter* (Taylor and Parkes, 1983; Dowling et al., 1986; Vainshtein et al., 1992).

TAXONOMIC COMMENTS

Based on comparative sequence analysis of the 16S rRNA-encoding gene, the genus *Desulfospira* belongs to the family *Desulfobacteraceae* in the *Deltaproteobacteria*. Common characteristics of all described members of this family are the complete oxidation of organic compounds to CO_2 (Widdel and Hansen, 1992) and the presence of menaquinones with seven isoprenoid units. Within the family *Desulfobacteraceae*, *D. joergensenii* forms a coherent line of descent with *D. toluolica* and members of the genus *Desulfobacter* (Fig. BXII.δ.30). The overall dissimilarity value(s) of the 16S rRNA gene sequence from *D. joergensenii* to that of *D. toluolica* is 4.6%, and to those of members of the genus *Desulfobacter* between 5.2% and 7.3%.

List of species of the genus Desulfospira

1. **Desulfospira joergensenii** Finster, Liesack and Tindall 1997e, 1274[VP] (Effective publication: Finster, Liesack and Tindall 1997c, 207.)

   joer.gen.se' ni.i. M.L. gen. n. *joergensenii* of Joergensen; named after B.B. Jørgensen, a Danish microbiologist who has made important contributions to our current knowledge of the sulfur cycle.

   The description is based on the type strain B331 (Finster et al., 1997c). The morphology is as indicated in the generic description and as depicted in Fig. BXII.δ.31. The main characteristics are as listed in the generic description.

   In liquid media, chains consisting of two or more cells often occur. Colonies in anaerobic agar media are lens-shaped with a faint orange-red color, whether grown with elemental sulfur as electron acceptor in the presence of Fe(III)-oxyhydroxide, or with sulfate as electron acceptor in the absence of Fe(III)-oxyhydroxide.

   Formate and CO_2 can serve as the sole electron donor and carbon source, respectively, for chemoautotrophic

TABLE BXII.δ.22. Some characteristics of the species of the genera *Desulfospira* and *Desulfobacula*[a]

Characteristic	*Desulfospira joergensenii*	*Desulfobacula toluolica*
Morphology	curved	oval-coccoid
Width (μm)	0.7–0.8	1.2–1.4
Length (μm)	1.0–2.0	1.2–2.0
Motility	−	+
Electron donors used for sulfate reduction:		
Hydrogen	+	−
Formate	+	−
Lactate	+	−
Maleate	+	−
Proline	+	−
Long chain fatty acids	+	−
Aromatic compounds	−	+
Primary (short-chain) alcohols	−	+
Electron acceptors:		
Elemental sulfur	+	nd
Sulfite	+	nd
Thiosulfate	+	nd
Requirement for vitamins	+	+
Catalase present	+	−
Optimal pH	7.0–7.4	7.0–7.1
Optimal temperature, °C	26–30	28
Mol% G + C of DNA	50	42
16S rRNA gene sequence similarity, %	95.4	95.4

[a]Symbols: +, utilized; −, not utilized; nd, not determined.

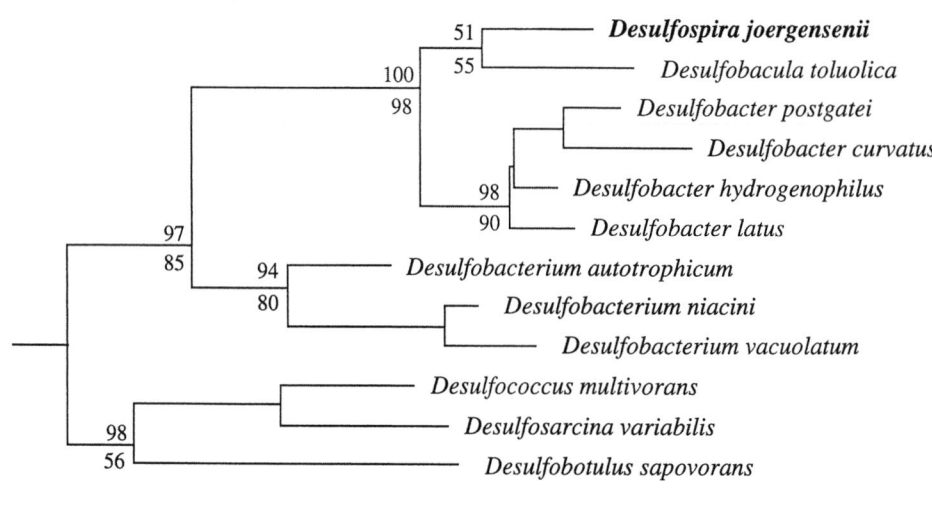

0.10

FIGURE BXII.δ.30. Dendrogram showing the phylogenetic position of *Desulfospira joergensenii* in relation to other members of the family *Desulfobacteraceae* in the *Deltaproteobacteria*. The rooting was inferred using the 16S rRNA gene sequence of *Escherichia coli* as an outgroup. The tree was constructed using a maximum likelihood method (FastDNAml), Maidak et al., 1997, and the ARB program package, Strunk and Ludwig, 1998. Bootstrap values were determined using maximum parsimony (PHYLIP), Felsenstein, 1993; 100 data resamplings) and neighbor-joining (ARB; 1000 data resamplings) methods. The values are given as percentage of outcome for the maximum parsimony test above, and for the neighbor-joining method below, the respective branches. The region of the 16S rRNA molecule between positions 28 and 1357 (International Union of Biochemistry of *Escherichia coli* 16S rRNA numbering) was used in the analysis. Bar = estimated number of base changes per nucleotide sequence position.

growth. The following substrates are utilized as electron donor for growth: hydrogen, formate, butyrate, caprylate, laurinate, myristinate, crotonate, α-ketoglutarate, glutarate, 3-oxoglutarate, fumarate, maleate, succinate, oxalacetate, pyruvate, lactate, glycolate, glycerol, choline chloride, betaine, DMSP, proline, and yeast extract.

The following substrates support sulfate reduction but not growth: valerate, palmitate, stearate, methanol, glycine, alanine, L-threonine, glutamate, acetone, ethylene-glycol, CE-broth, nicotinate, and benzoate.

The following substrates are not utilized: 1-propanol, 1-butanol, dimethylamine, trimethylamine, acetate, isobuty-

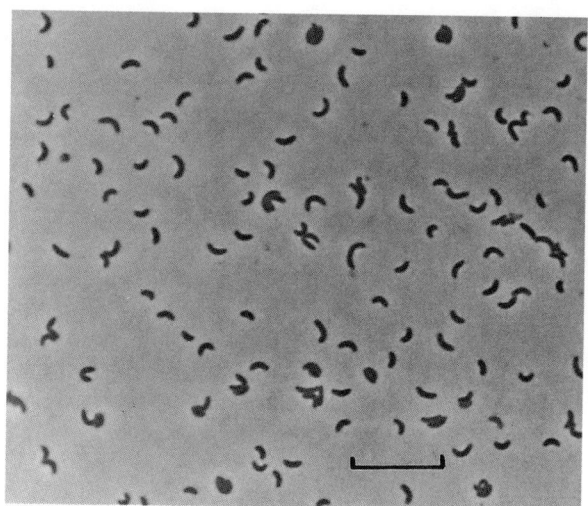

FIGURE BXII.δ.31. Phase-contrast photomicrograph showing cells of *Desulfospira joergensenii* strain B331 isolated from iron-rich, oxidized, marine surface sediment. Bar = 10 μm.

cose, fructose, mannitol, sorbitol, trehalose, methionine, tyrosine, L-arginine, tryptophan, phenylalanine, phenol, *p*-cresol, toluene, cyclohexanecarboxylate, vanillin, pyrimidine, pyridine, pyrogallol, and 3-phenylpropionate.

No fermentative type of metabolism occurs with lactate, pyruvate, fumarate, maleate, malate, glutamate, oxalacetate, and aspartate as substrates. Nitrate, nitrite, Fe(III)-oxyhydroxide, Fe(III)-citrate, trimethylamine oxide, dimethyl sulfoxide, and fumarate are not utilized as electron acceptors.

Optimal NaCl concentration, 1.2–2%; range for growth, 0.6–4%. Optimal MgCl$_2$·6H$_2$O concentration, 0.1–0.3%; minimum concentration for growth, 0.1%. Compared to growth obtained at the optimal concentration, only slight inhibition with 2.1% MgCl$_2$·6H$_2$O. Optimal temperature, 26–30°C; range for growth, 8–30°C. Optimal pH, 7.4–7.6; range for growth, pH 6.5–7.9. Catalase positive.

MK-7 and MK-7 (VII-H$_2$) menaquinones are present. The fatty acid composition comprises mainly unbranched fatty acids. The major polar lipids are phosphatidyl glycerol and phosphatidyl ethanolamine.

Isolated from oxidized, iron-rich, surface sediment of a mud flat overgrown with *Zostera noltii*.

The mol% G + C of the DNA is: 49.9 (HPLC).

Type strain: Strain B331, DSM 10085.

GenBank accession number (16S rRNA): X99637.

rate, 2-methylbutyrate, isovalerate, pelargonate, caprinate, oxamate, citrate, tartrate, malonate, malate, adipate, pimelate, aspartate, oxalate, glyoxylate, croton-aldehyde, glu-

---

Genus XII. **Desulfotignum** *Kuever, Könneke, Galushko and Drzyzga 2001, 176*^{VP}

JAN KUEVER, FRED A. RAINEY AND FRIEDRICH WIDDEL

De.sul.fo.tig′num. L. pref. *de* from; L. n. *sulfur* sulfur; M.L. n. *tignum* stick; M.L. neut. n. *Desulfotignum* sulfate-reducing stick.

Rod-shaped, sometimes slightly curved, cells, 0.5–0.7 × 1.5–3 μm. Occur singly or in pairs. Spore formation is not observed. Gram negative. Motile by a single polar flagellum. Strictly anaerobic, having both a respiratory and a fermentative type of metabolism. **Chemoorganotroph or chemoautotroph**, using formate, acetate, butyrate, higher fatty acids, other organic acids, alcohols, and **benzoate or similar aromatic compounds as carbon sources and also as electron donors for anaerobic respiration**; these compounds are completely oxidized to CO$_2$. Chemolithoautotrophic growth occurs with **H$_2$ as the electron donor and CO$_2$ as the carbon source. Sulfate, sulfite, and thiosulfate serve as terminal electron acceptors, and are reduced to H$_2$S.** In the absence of an external electron acceptor, slow growth occurs by fermentation of pyruvate. Carbon monoxide dehydrogenase activity is commonly observed, indicating the operation of the anaerobic C$_1$ pathway (carbon monoxide dehydrogenase pathway, Wood pathway) for CO$_2$ fixation and complete oxidation of acetyl-CoA. Media containing vitamins, a reductant, and not less than 10 g NaCl/l are necessary for growth. The optimal growth temperature is 28–32°C, the temperature range 10–42°C. Thermophilic species have not been described. The optimal pH for growth is 7.3, the pH range 6.5–8.2. Cells contain no desulfoviridin, cytochromes of the *c* type are present. Occur in anoxic part of marine habitats.

The mol% G + C of the DNA is: 62.

Type species: **Desulfotignum balticum** Kuever, Könneke, Galushko and Drzyzga 2001, 176.

FURTHER DESCRIPTIVE INFORMATION

Formate, acetate, crotonate, C$_4$–C$_{12}$ straight-chain fatty acids, pyruvate, lactate, fumarate, succinate, maleinate, malate, benzoate, 4-hydroxybenzoate, phenol, phenylalanine, and phenylacetate serve as electron donors and also as carbon sources; these compounds are completely oxidized to CO$_2$. *D. balticum* uses sulfate, sulfite, and thiosulfate as electron acceptors, whereas sulfur and nitrate are not used. Cells contain cytochrome *c* but no desulfoviridin. Major cellular fatty acids of the type strain are C$_{16:0}$, C$_{16:0\ 10CH_3}$, and C$_{17:0\ cyclo}$. Ammonium ions are used as the nitrogen source. *Desulfotignum* strains occur in anoxic marine sediments rich in organic material. Media containing a reductant such as sulfide and vitamins are necessary for growth.

ENRICHMENT AND ISOLATION PROCEDURES

For the selective enrichment of *Desulfotignum*, the same defined marine sulfate medium described for the isolation of *Desulfobacter* spp. may be used. Instead of acetate, benzoate (2 mM) is added as electron donor and carbon source. Bottles or tubes (100-, 50- or 20-ml capacity) can serve as culture vessels and are kept anoxic with tight screw caps (if completely filled), or by providing an anoxic head space of N$_2$/CO$_2$ (90:10) and sealing with butyl rubber stoppers. The medium is inoculated (2–5% of the total volume) with dark anoxic sediment. Enrichment cultures are incubated at 30°C. Cultures should be briefly shaken once a week. After intense formation of H$_2$S has occurred, transfers to fresh medium are made. Enrichment cultures should be mixed well before subculturing. Pure cultures of *Desulfotignum* strains are obtained by repeated application of the agar dilution method (as described for *Desulfobacter* in this edition of the *Manual*). The same substrates used for the enrichment cultures may be used for obtaining pure cultures.

MAINTENANCE PROCEDURES

Stock cultures are transferred every 3–5 months. Towards the end of growth or at the beginning of the stationary phase, cultures are refrigerated and stored at 2–5°C in the dark. *Desulfobacter* strains may be preserved indefinitely by suspension of cells in anoxic medium containing 5% dimethylsulfoxide and storage in liquid nitrogen.

DIFFERENTIATION OF THE GENUS *DESULFOTIGNUM* FROM OTHER GENERA

From other sulfate-reducing bacteria that utilize benzoate (*Desulfobacula*, *Desulfococcus*, and *Desulfosarcina*), *Desulfotignum balti-cum* is easily distinguished by its pronounced rod-shaped, elongate cells. Unambiguous differentiation from these other sulfate-reducing bacteria requires comparative 16S rDNA sequence analysis and DNA–DNA hybridization.

TAXONOMIC COMMENTS

The closest relatives of *Desulfotignum balticum* are members of the genera *Desulfobacula*, *Desulfobacter*, and *Desulfospira*. A signature sequence for *Desulfobacter*, *Desulfobacula*, *Desulfospira*, and *Desulfotignum* is 5′-CACAGGATGTCAAACCCAG-3′ (oligonucleotide probe DSB985), which binds to the 16S rRNA corresponding to *E. coli* position 5′-985–1003-3′ (Manz et al., 1998).

FURTHER READING

Kuever, J., M. Konneke, A. Galushko and O. Drzyzga. 2001. Reclassification of *Desulfobacterium phenolicum* as *Desulfobacula phenolica* comb. nov. and description of strain SaxT as *Desulfotignum balticum* gen. nov., sp. nov. Int. J. Syst. Evol. Microbiol. *51*: 171–177.

List of species of the genus Desulfotignum

1. **Desulfotignum balticum** Kuever, Könneke, Galushko and Drzyzga 2001, 176[VP]

   bal.ti′ cum. M.L. neut. adj. *balticum* from the Baltic Sea.

   Characteristics are as described for the genus. The type strain was isolated from anoxic, marine, organic-rich sediment (Saxild, Denmark) using benzoate.

   The mol% G + C of the DNA is: 52.7 (HPLC).
   Type strain: sax, ATCC BAA-19, DSM 7044.
   GenBank accession number (16S rRNA): AF233370.

Family II. **Desulfobulbaceae** fam. nov.

JAN KUEVER, FRED A. RAINEY AND FRIEDRICH WIDDEL

De.sul.fo.bul.ba.ce′ ae. M.L. masc. n. *Desulfobulbus* type genus of the family; suffix *ceae* denotes family; M.L. fem. pl. n. *Desulfobulbaceae* the family that comprises the genus *Desulfobulbus* and other genera.

The order comprises morphologically diverse sulfate-reducing bacteria including cocci, oval, rod-shaped, and curved cells. Many forms are motile by means of single polar flagella. Some members contain gas vesicles. **Strictly anaerobic chemoorganotrophs, chemolithoheterotrophs, or chemolithoautotrophs with respiratory metabolism**; in addition, fermentative abilities are found in some members. With the exception of one isolate (*Desulfocapsa sulfexigens*), **the common electron acceptor is sulfate,** which is reduced to sulfide. **Sulfite or thiosulfate may also be used; some members may also reduce elemental sulfur (or polysulfide) or nitrate**; nitrate is reduced to ammonia. **Simple organic compounds, such as long-chain fatty acids and alcohols, serve as electron donors and carbon sources**. Most members of the family are incomplete oxidizers that form acetate as an end product. Growth by disproportionation of sulfur, thiosulfate, and sulfite may occur. Members are mesophilic or psychrophilic; thermophiles have not been isolated. Cells contain various cytochromes, other redox proteins, and menaquinones. Desulfoviridin is not present; desulforubidin has been detected in most species. Mesophilic species have been isolated from various aquatic environments, including marine, brackish water, and freshwater habitats.

The family comprises the genera *Desulfobulbus*, *Desulfocapsa*, *Desulfofustis*, *Desulforhopalus*, and *Desulfotalea*. Comparative properties of all described genera are given in Table BXII.δ.23. Detailed properties of described species of each genus are given in the genus descriptions.

Type genus: **Desulfobulbus** Widdel 1981, 382 (Effective publication: Widdel 1980, 374.)

Genus I. **Desulfobulbus** Widdel 1981, 382[VP] (Effective publication: Widdel 1980, 374)

JAN KUEVER, FRED A. RAINEY AND FRIEDRICH WIDDEL

De.sul.fo.bul′ bus. L. pref. *de* from; L. n. *sulfur* sulfur; L. n. *bulbus* onion; M.L. masc. n. *Desulfobulbus* onion-shaped sulfate reducer.

Cells are ovoid to rod shaped, or lemon shaped with pointed ends, 0.6–1.3 × 1.5–3.5 μm. Occur singly, in pairs, or in chains. Spore formation is not observed. Gram negative. **Cells are motile by a single polar flagellum, or are nonmotile.** Strictly anaerobic, having both a respiratory and a fermentative type of metabolism. **Chemoorganotrophs**, using **propionate**, lactate, pyruvate, ethanol, or 1-propanol as electron donors and also as carbon sources; organic compounds are incompletely oxidized to acetate. H$_2$ can

TABLE BXII.δ.23. Differential characteristics of the genera of the family *Desulfobulbaceae*[a]

Characteristic	*Desulfobulbus*	*Desulfocapsa*	*Desulfofustis*	*Desulforhopalus*	*Desulfotalea*
Cell morphology	Rods	Rods	Rods	Rods	Rods
Motility	±	+	+	−	±
Mol% G + C content	47–60	47–50	56	48	42–47
Sulfite reductase	Most species DR	nr	DR	nr	nr
Major menaquinone	MK-5(H$_2$)	nr	MK-5(H$_2$)	nr	Mk-6 or Mk-6(H$_2$)
Substrate oxidation	I	I	I[b]	I	I
CO–DH[c]	−	nr	+	nr	nr
Typical electron donors:					
H$_2$ + acetate	+	−	+	+	+
Formate + acetate	+[d]	−	−	+	+
Fumarate	+[d]	−	+	+[d]	+[d]
Glycolate	nr	nr	+	nr	nr
Glyoxylate	nr	nr	+	+[d]	nr
Lactate	+	−	+	+	+
Malate	+[d]	−	+	+[d]	+[d]
Primary alcohols	+	+[d]	−	+	+
Propionate	+	−	−	+	−
Pyruvate	+	−	nr	+	+
Chemolithoautotrophic growth	−	+[d]	−	−	−
Fermentative growth	+	−	+[e]	+	+
Growth by disproportionation of reduced sulfur compounds	±	+	nr	nr	−
Typical electron acceptors:					
Sulfate	+	+[d]	+	+	+
Sulfite	+	+	+	+	+[d]
Sulfur	−	+	+	−	−
Thiosulfate	+	+	−	+	+[d]
Optimal growth temperature (°C)	29–39	30	28	18–19	10–18
NaCl requirement	±	±	+	+	+
Habitat (anoxic):					
Freshwater	+	+	−	−	−
Marine	+	+	+	+	+

[a]Symbols: nr, not reported; +, present, observed in all species; −, not present, not observed; ±, variable; I, incomplete; DR, desulforubidin.

[b]Presence of carbon monoxide dehydrogenase may indicate ability for complete oxidation.

[c]Presence of carbon monoxide dehydrogenase, a key enzyme indicative of the C$_1$-pathway for acetyl-CoA.

[d]One species only.

[e]Only on glycolate.

be used as electron donor in the presence of acetate as organic carbon source (chemolithoheterotrophic growth). **Sulfate and often sulfite or thiosulfate serve as terminal electron acceptors and are reduced to H$_2$S.** Sulfur is never reduced. **In the absence of an external electron acceptor, growth may occur by fermentation of lactate, pyruvate, and ethanol (+ CO$_2$), malate, or fumarate.** The optimum pH range for growth is 6.6–7.5. The optimum temperature range for growth is 25–40°C. Thermophilic species are not known. Media containing a reductant and *p*-aminobenzoate are required for growth. Colonies in anaerobic agar media are whitish to grayish and smooth. All described species contain desulforubidin as sulfite reductase. Occur in anoxic parts of freshwater, brackish water, and marine habitats; also isolated from rumen contents, animal dung, and sewage sludge.

The mol% G + C of the DNA is: 50–60.

Type species: **Desulfobulbus propionicus** Widdel 1981, 382 (Effective publication: Widdel 1980, 374.)

FURTHER DESCRIPTIVE INFORMATION

Cells of *D. propionicus* are generally lemon or onion shaped (Fig. BXII.δ.32), other species have a more rod-like cell shape and less pointed ends. Cells are often in pairs and chains. Many strains are motile by a single polar flagellum. However, the type strain of *D. propionicus* is nonmotile; electron microscopy of this strain has shown the presence of pili but not of flagella.

The temperature range for growth is 10–43°C. *D. propionicus*

has an optimum growth temperature of 39°C; other strains may have lower temperature optima of 28–31°C. The pH range for growth is 6.0–8.6, with an optimum pH of 7.2.

All members of the genus reduce sulfate, sulfite, and thiosulfate to H$_2$S. *D. propionicus* can also use nitrate and ferric iron

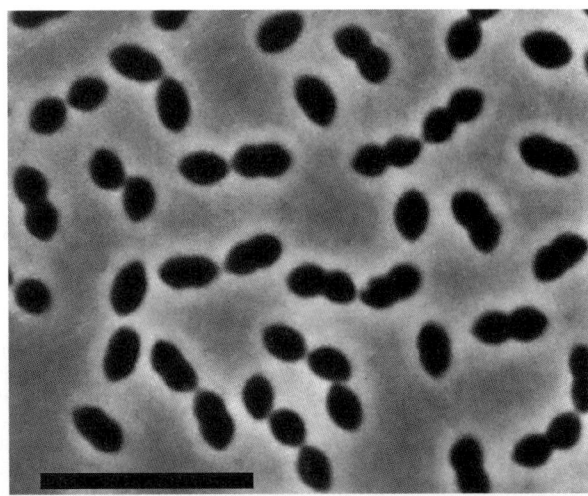

FIGURE BXII.δ.32. Phase-contrast micrograph of *Desulfobulbus propionicus* (viable cells). Bar = 10 μm.

as terminal electron acceptors, which are reduced to ammonia and ferrous iron, respectively (Widdel and Pfennig, 1982; Lovley et al., 1993b). Elemental sulfur, fumarate, and malate cannot serve as electron acceptors. Slow growth by disproportionation of sulfur was observed with *D. propionicus* if acetate was supplied as the carbon source (Lovley and Phillips, 1994a). This process was accelerated by the presence of Mn(IV). Propionate, lactate, pyruvate, ethanol, and 1-propanol are utilized as electron donors and organic carbon sources; *D. rhabdoformis* can also use malate, fumarate, and succinate. Oxidation of these substrates usually leads to the formation of acetate as an end product. *Desulfobulbus* spp. can grow with H_2 as an electron donor, but growth does not occur chemoautotrophically; acetate is necessary as a carbon source in the presence of bicarbonate (chemolithoheterotrophic growth; see also chapter *Desulfovibrionales*).

In the absence of a terminal electron acceptor, growth is possible with pyruvate or lactate, which are fermented to acetate and propionate. *D. propionicus* can also grow by fermentation of ethanol and CO_2 (Laanbroek and Pfennig, 1981; Tasaki et al., 1990), whereas *D. rhabdoformis* grows by fermentation of malate and fumarate (Lien et al., 1998). Sugars are not fermented.

Membrane-bound and soluble cytochromes of the *b* and *c* type have been identified in *D. propionicus* and other species. All described *Desulfobulbus* spp. contain desulforubidin as sulfite reductase and menaquinone-5(H_2) as the predominant quinone (Collins and Widdel, 1986). Major cellular fatty acids (PFLA) are $C_{17:1\,\omega6c}$ and $C_{18:1\,\omega7c}$ (Kohring et al., 1994).

Desulfobulbus occurs in the anoxic black mud of freshwater ditches and lakes, brackish water, and marine habitats. It also occurs in bovine rumen fluid, in the intestinal tract of animals, and in the anaerobic sludge of animal manure deposits and sewage plants (Tasaki et al., 1990).

Strains of *Desulfobulbus* may be cultured in a defined anoxic medium[1] with sulfate as electron acceptor and propionate, lactate, or 1-propanol as the electron donor and carbon source. For *D. rhabdoformis*, malate, fumarate, and succinate may be used as alternative electron donors and carbon sources. Ammonium ions serve as the nitrogen source. When inoculation is done from old cultures, initiation of growth is favored by the addition of 10–30 mg sodium dithionite per liter of medium as a further strong reductant, in addition to routinely added sodium sulfide.

ENRICHMENT AND ISOLATION PROCEDURES

For the selective enrichment of *Desulfobulbus* strains, propionate–sulfate medium is used. Bottles or tubes (100-, 50-, or 20-ml capacity) can serve as culture vessels and are kept anoxic with tight screw caps, or by using a headspace of N_2/CO_2 (90:10) and sealing with butyl rubber stoppers. The medium is inoculated with black anoxic mud from freshwater ditches or ponds, brackish water, or marine habitats, manure deposits, sludge from sewage digestors, or rumen fluid; 3–5% of the total culture volume

should be added. *Desulfobulbus* species may be also enriched and isolated via direct serial dilution of anoxic environmental samples in liquid medium (according to most-probable-number method) or agar medium (see next paragraph) with propionate as organic substrate. Furthermore, selective enrichment of *Desulfobulbus* species is possible under fermentative conditions without sulfate in anoxic bicarbonate-buffered medium in a chemostat, using ethanol as the growth-limiting substrate (Laanbroek et al., 1982); such cultures form propionate and acetate as fermentation products.

The NaCl concentration of the medium should correspond approximately to the salinity of the natural source, but the added amount of $MgCl_2 \cdot 6H_2O$ need not be higher than 3 g/l. Addition of dithionite to enrichment cultures is recommended. Enrichment cultures are incubated at 28–36°C and mixed by shaking every second day. When intense formation of H_2S has occurred (generally after 10 d), transfers to fresh medium are made. Enrichment cultures should be mixed well before subculturing.

Pure cultures of *Desulfobulbus* strains are obtained by repeated application of the agar dilution method. The method is similar to that described for the genus *Desulfobacter*, except that propionate medium is added to the molten concentrated agar solution.

In enrichment cultures with seawater or brackish water medium, cells of *Desulfobacter* spp. may develop on the acetate excreted by *Desulfobulbus* spp. In the highest dilutions of agar cultures, however, only *Desulfobulbus* develops into large colonies.

MAINTENANCE PROCEDURES

Stock cultures are transferred every 2–3 months. Towards the end of growth or at the beginning of the stationary phase, cultures are refrigerated and stored at 2–5°C in the dark. *Desulfobulbus* strains may be preserved indefinitely by suspension of cells in anoxic medium containing 5% dimethylsulfoxide and storage in liquid nitrogen.

DIFFERENTIATION OF THE GENUS *DESULFOBULBUS* FROM OTHER GENERA

Desulfobulbus spp. grow well (with doubling times of ~10 h) with sulfate and propionate, and the latter substrate is a characteristic organic substrate for the genus. Other sulfate-reducing bacteria, like *Desulfococcus multivorans*, *Desulfosarcina variabilis*, or *Desulfobacterium* spp., can also utilize propionate; however, these organisms carry out complete oxidation of the propionate and also grow more slowly than does *Desulfobulbus*.

TAXONOMIC COMMENTS

The closest relatives of *Desulfobulbus* species are member of the genera *Desulfocapsa* and *Desulforhopalus*. A signature sequence for all described *Desulfobulbus* species and other closely related sulfate-reducing bacteria is 5′-GAATTCCACTTTCCCCTCTG-3′ (oligonucleotide probe 660), which binds to the 16S rRNA corresponding to position 5′-660–679-3′ of the *Escherichia coli* 16S rRNA sequence (Devereux et al., 1992; Manz et al., 1998).

FURTHER READING

Widdel, F. and N. Pfennig. 1982. Studies on dissimilatory sulfate-reducing bacteria that decompose fatty-acids. II. Incomplete oxidation of propionate by *Desulfobulbus propionicus* gen. nov., sp. nov. Arch. Microbiol. *131*: 360–365.

---

1. The defined medium is prepared as described in the footnote on the genus *Desulfobacter*, with the following additions (g/l): NaCl, 1.0 (for freshwater strains) or 20.0 (for marine strains); $MgCl_2 \cdot 6H_2O$, 0.5 (for freshwater strains) or 3.0 (for marine strains). Sodium propionate (final concentration, 1.5 g/l) is added aseptically from a sterile stock solution to provide an organic substrate; however, the best growth occurs with sodium lactate (final concentration, 2.0–4.0 g/l) in the presence of sulfate. The only growth factor required by *Desulfobulbus* is *p*-aminobenzoic acid. The pH of the complete medium is adjusted to 7.3–7.4.

List of species of the genus Desulfobulbus

1. **Desulfobulbus propionicus** Widdel 1981, 382[VP] (Effective publication: Widdel 1980, 374.)

pro.pi.o′ ni.cus. M.L. n. *acidum propionicum* propionic acid; M.L. adj. *propionicus* pertaining to propionic acid.

The morphology is as described in Table BXII.δ.24 and as depicted in Fig. BXII.δ.32. Characteristics are summarized in Table BXII.δ.24. Small concentrations of sulfide and sulfate are formed by disproportionation of sulfur (Janssen et al., 1996). The type strain has been isolated from an anoxic freshwater ditch with propionate as selective substrate.

The mol% G + C of the DNA is: 59.9 (T_m).
Type strain: 1pr3, ATCC 33891, DSM 2032.
GenBank accession number (16S rRNA): M34410.

2. **Desulfobulbus elongatus** Samain, Dubourguier and Albagnac 1985, 223[VP] (Effective publication: Samain, Dubourguier and Albagnac 1984, 399.)

e.lon.ga′ tus. L. adj. *elongatus* elongated.

Characteristics are summarized in Table BXII.δ.24. Small concentrations of sulfide and sulfate are formed by disproportionation of sulfur (Janssen et al., 1996). The type strain has been isolated from an anaerobic digester with propionate as selective substrate.

The mol% G + C of the DNA is: 59 (T_m).
Type strain: FP, ATCC 43118, DSM 2908.
GenBank accession number (16S rRNA): X95180.

3. **Desulfobulbus marinus** *sp. nov.*

ma.ri′ nus. synonym of L. adj. *maritimus* pertaining to the sea.

The morphology is as described in Table BXII.δ.24 and as depicted in Fig. BXII.δ.33. Characteristics are summarized in Table BXII.δ.24. Type strain isolated from an anoxic mud flat of the Jadebusen (Germany) with propionate as selective substrate.

The mol% G + C of the DNA is: 47.3 (HPLC).

TABLE BXII.δ.24. Morphological and physiological characteristics of described species of the genus *Desulfobulbus*[a]

Species	*Desulfobulbus propionicus*	*Desulfobulbus elongatus*	*Desulfobulbus marinus*	*Desulfobulbus rhabdoformis*
Cell morphology	Oval- or lemon-shaped	Rod-shaped	Oval-shaped	Rod-shaped
Cell size (μm)	1.0–1.3 × 1.5–2.0	0.6–1.0 × 1.5–2.5	1.0–1.3 × 1.8–2.0	0.6–1.0 × 1.7–3.5
Motility	−	+ (sp)	+ (sp)	
Mol% G+C content	60	59	47.3	50.6
Sulfite reductase	DR	DR	nr	DR
Major menaquinone	MK-5(H_2)	MK-5(H_2)	MK-5(H_2)	MK-5(H_2)
Optimal temperature (°C)	39	31	29	31
Oxidation of substrate	I	I	I	I
Compounds used as electron donors and carbon sources:				
H_2[b]	+	+	+	+
Formate	−	−	+	−
Acetate	−	−	−	−
Fatty acid: C atoms	3	3	3	3
Isobutyrate	−	nr	nr	nr
2-Methylbutyrate	−	nr	nr	nr
3-Methylbutyrate	−	nr	nr	nr
Ethanol	+	+	+	+
Lactate	+	+	+	+
Pyruvate	+	+	+	+
Fumarate	−	−	−	+
Succinate	−	−	−	+
Malate	−	−	−	+
Benzoate	−	nr	−	nr
1-Propanol	+	+	+	+
Fermentative growth on:				
Fumarate	−	−	−	+
Malate	−	−	−	+
Pyruvate	+	+	+	+
Compounds used as electron acceptors:				
Sulfate	+	+	+	+
Sulfite	+	+	+	+
Sulfur	−	−	nr	nr
Thiosulfate	+	+	+	+
Fumarate	−	−	nr	+
Nitrate	+	−	nr	−
Other electron acceptors	Fe(III)	nr	nr	nr
Growth factor requirement	pa	pa	pa	−
NaCl requirement (g/l)	−	−	20	15
Literature	Widdel and Pfennig, 1982	Samain et al., 1984	Widdel and Pfennig, 1982; Janssen et al., 1996	Lien et al., 1998

[a]Symbols: nr, not reported; +, present, observed in all species; −, not present, not observed; ±, variable; (+), poorly motile, poor growth; (sp), flagellation type: single, polar; DR, desulforubidin; I, incomplete; pa, *p*-aminobenzoate.

[b]Used as electron donor only in the presence of acetate as carbon source.

Deposited strain: 3pr10, DSM 2058.
GenBank accession number (16S rRNA): M34411.

4. **Desulfobulbus rhabdoformis** Lien, Madsen, Steen and Gjerdevik 1998, 473[VP]

rhab.do.for' mis. Gr. fem. n. *rhabdos* rod; L. adj. suff. *formis*-like, of the shape; *rhabdoformis* rod-shaped.

Characteristics are summarized in Table BXII.δ.24. Isolated from a water-oil separation system (North Sea oil platform) with propionate as selective substrate.

The mol% G + C of the DNA is: 50.6 (HPLC).
Type strain: M16, DSM 8777.
GenBank accession number (16S rRNA): U12253.

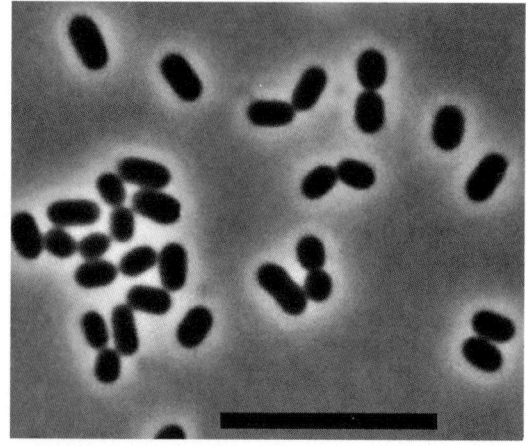

FIGURE BXII.δ.33. Phase-contrast micrograph of *Desulfobulbus marinus* (viable cells). Bar = 10 μm.

Genus II. **Desulfocapsa** Janssen, Schuhmann, Bak and Liesack 1997, 601[VP] (Effective publication: Janssen, Schuhmann, Bak and Liesack 1996, 190)

JAN KUEVER, FRED A. RAINEY AND FRIEDRICH WIDDEL

De.sul.fo.cap' sa. L. pref. *de* from; L. n. *sulfur* sulfur; L. n. *capsa* box; M.L. fem. n. *Desulfocapsa* a sulfate-reducing box.

Cells are round-ended rods, or elongated rods with pointed ends, 0.5–0.9 × 2.0–4.0 μm. Cells are motile by means of single subpolarly inserted flagella, but motility can be limited to only a few cells in the population of some species. No spores are formed. Metabolism is strictly anaerobic. Simple organic compounds are incompletely oxidized, with sulfate as electron donor that is reduced to sulfide. One species of the genus, *D. sulfexigens*, is not able to use sulfate as electron acceptor. **Inorganic sulfur compounds are disproportionated to sulfate and sulfide and support chemolithoautotrophic growth with CO₂ as carbon source.** Fermentative growth with organic compounds is not possible. The optimal pH range for growth is 7.3–7.5. The optimal temperature range for growth is 20–30°C. Thermophilic species are not known. The NaCl requirement varies within the genus; growth of the type strain is inhibited at an NaCl concentration higher than 15 g/l. The genus comprises marine and freshwater strains. All described species contain cytochromes and no desulfoviridin. Occur in sediments of freshwater or marine habitats close to the anoxic/oxic interface.

The mol% G + C of the DNA is: 47–51.

Type species: **Desulfocapsa thiozymogenes** Janssen, Schuhmann, Bak and Liesack 1997, 601 (Effective publication: Janssen, Schuhmann, Bak and Liesack 1996, 190.)

FURTHER DESCRIPTIVE INFORMATION

Ethanol, 1-propanol, and butanol are used as electron donors with sulfate as electron acceptor and are oxidized to the respective fatty acids. *D. sulfexigens* is not able to grow with sulfate as the electron acceptor; only small amounts of sulfate were reduced to sulfide in the presence of formate and hydrogen, without growth (Finster et al., 1998).

Disproportionation of thiosulfate and sulfite added to media in concentrations of a few mmol/l is thermodynamically favorable with these compounds serving as the only sources to sustain growth. The energetically less favorable disproportionation of elemental sulfur to sulfide and sulfate allows growth only if prod-

ucts are scavenged, for instance by the addition of ferric hydroxide as a sulfide trap. Cell synthesis occurs autotrophically with CO₂ as carbon source; the pathway for CO₂ fixation is unknown.

The optimal pH range for growth of *D. thiozymogenes* is 7.3–7.5, whereas *D. sulfexigens* shows optimal growth between pH 6.7 and 7.3. The optimal temperature range for growth is 20–30°C. *D. sulfexigens* requires an NaCl concentration of 15 g/l, whereas growth of *D. thiozymogenes* is inhibited at this concentration. The genus contains marine and freshwater strains.

ENRICHMENT AND ISOLATION PROCEDURES

For specific enrichment, a sulfate-free marine or freshwater mineral medium (e.g., as described for *Desulfobacter*) reduced with sulfide should be used. Thiosulfate at a concentration of 10 mM is added as the only substrate from sterile stock solution. Acetate may or may not be added at low concentration (1–2 mM) for enrichment of potential nonautotrophic (chemolithohetero-trophic) species, or autotrophic species that are stimulated by acetate. Bottles or tubes (100-, 50-, or 20-ml capacity) can serve as culture vessels and are kept anoxic with tight screw caps, or by using a headspace of N₂/CO₂ (90:10) and sealing with butyl rubber stoppers. The medium is inoculated with sediment from the anoxic/oxic interface; 3–5% of the total culture volume should be added. *Desulfocapsa* species may also be enriched and isolated via direct serial dilution of anoxic environmental samples in liquid medium (according to most-probable-number method) or agar medium with thiosulfate as the only substrate. Incubations should be done at 30°C. Pure cultures are obtained by repeated application of agar dilutions as described in the chapter for the genus *Desulfobacter*, using the combined substrates described above. Alternative substrates are sulfite (10mM) and sulfur. For specific enrichment of *D. sulfexigens*, sulfur in the presence of amorphous ferric hydroxide is used (Finster et al., 1998), followed by isolation of pure cultures with thiosulfate as electron donor. Ferric hydroxide is prepared as described by Lovley and

Phillips (1986). Stock solutions of thiosulfate and sulfite are filter sterilized, whereas amorphous ferric hydroxide is autoclaved. Sulfur is prepared as described in the chapter for the genus *Desulfuromonas*.

MAINTENANCE PROCEDURES

Maintenance procedures have not been described. Procedures described for other genera in this chapter may be attempted.

DIFFERENTIATION OF THE GENUS DESULFOCAPSA FROM OTHER GENERA

Members of the genus *Desulfocapsa* can be distinguished from other sulfate-reducing bacteria by the limited number of electron donors used for the reduction of sulfate or other sulfur compounds. A unique feature of this genus is the ability to grow autotrophically by disproportionation of thiosulfate, sulfite, or sulfur (in the presence of ferric iron); other sulfate-reducing bacteria may also be able to disproportionate these compounds, but have not been described to grow autotrophically. *Desulfovibrio*

sulfodismutans and *Desulfobulbus propionicus* require acetate as carbon source for growth by disproportionation of reduced sulfur compounds. However, a clear differentiation can only be obtained by comparative analysis of 16S rDNA sequences.

TAXONOMIC COMMENTS

D. thiozymogenes and *D. sulfexigens* fall into the *Desulfobulbaceae* of the *Deltaproteobacteria*. The closest relatives are *Desulfofustis glycolicus*, *Desulforhopalus vacuolatus*, and members of the genus *Desulfobulbus*.

FURTHER READING

Finster, K., W. Liesack and B. Thamdrup. 1998. Elemental sulfur and thiosulfate disproportionation by *Desulfocapsa sulfexigens* sp. nov., a new anaerobic bacterium isolated from marine surface sediment. Appl. Environ. Microbiol. *64*: 119–125.

Janssen, P.H., A. Schuhmann, F. Bak and W. Liesack. 1996. Disproportionation of inorganic sulfur compounds by the sulfate-reducing bacterium *Desulfocapsa thiozymogenes* gen. nov., sp. nov. Arch. Microbiol. *166*: 184–192.

List of species of the genus Desulfocapsa

1. **Desulfocapsa thiozymogenes** Janssen, Schuhmann, Bak and Liesack 1997, 601[VP] (Effective publication: Janssen, Schuhmann, Bak and Liesack 1996, 190.)

   thi.o.zy.mo'ge.nes. Gr. n. *thios* sulfur; Gr. n. *zyme* leaven, ferment; Gr. v. *gennaio* to produce; M.L. adj. *thiozymogenes* causing a fermentation of sulfur.

   Characteristics are summarized in Table BXII.δ.25. Ethanol, 1-propanol, and 1-butanol are used as electron donors with sulfate as electron acceptor and are oxidized to the respective fatty acids. Thiosulfate and sulfite are disproportionated to sulfate and sulfide with concomitant growth. Sulfur is disproportionated to sulfate and sulfide, but supports growth only in the presence of ferric iron. The type strain has been isolated from anoxic freshwater sediment from Lake Brabrand, Århus (Denmark) with thiosulfate as the sole energy source and acetate to stimulate carbon assimilation.

   The mol% G + C of the DNA is: 50.7 (HPLC).
   Type strain: Bra2, DSM 7269.

   GenBank accession number (16S rRNA): X95181.

2. **Desulfocapsa sulfexigens** Finster, Liesack and Thamdrup 2000, 1699[VP] (Effective publication: Finster, Liesack and Thamdrup 1998, 124.)

   sulf.ex' i.gens. L. n. *sulfur* sulfur; L. v. *exigo* to demand; M.L. part. adj. *sulfexigens*, demanding sulfur for growth.

   Characteristics are summarized in Table BXII.δ.25. Thiosulfate and sulfite are disproportionated to sulfate and sulfide with concomitant growth. Sulfur is disproportionated to sulfate and sulfide and supports growth only in the presence of ferric iron. The type strain was isolated from oxidized marine sediment of the basin of Arcachon (France) with sulfur as the electron donor and ferric iron as hydrogen sulfide scavenging agent. The isolate cannot grow with sulfate as the electron acceptor.

   The mol% G + C of the DNA is: 47.2 (HPLC).
   Type strain: SB164P1, DSM 10523.
   GenBank accession number (16S rRNA): Y13672.

TABLE BXII.δ.25. Morphological and physiological characteristics of described species of the genus *Desulfocapsa*[a]

Species	*Desulfocapsa thiozymogenes*	*Desulfocapsa sulfexigens*
Cell morphology	Rod-shaped	Rod-shaped
Cell size (μm)	0.8–0.9 × 2–3.5	0.5 × 2–4
Motility	+ (sp)	(+)
Mol% G + C content	50.7	47.2
Sulfite reductase	nr	nr
Major menaquinone	nr	nr
Optimal temperature (°C)	30	30
Oxidation of substrate	I	unknown
Compounds used electron donors and carbon sources:		
H$_2$	−	(+)[b]
Formate	−	(+)[b]
Acetate	−	−
Fatty acid: C atoms	−	−
Isobutyrate	−	nr
2-Methylbutyrate	−	nr
3-Methylbutyrate	−	nr
Ethanol	+	−
Lactate	−	−

(continued)

TABLE BXII.δ.25. *(cont.)*

Species	*Desulfocapsa thiozymogenes*	*Desulfocapsa sulfexigens*
Pyruvate	−	−
Fumarate	−	−
Succinate	−	−
Malate	−	−
Benzoate	−	−
Others	1-Propanol[c], 1-butanol[c]	nr
Fermentative growth on:	−	−
Compounds used as electron acceptors:		
Sulfate	+	(+)[b]
Sulfite	+	+
Sulfur	+	+
Thiosulfate	+	+
Fumarate reduction	−	nr
Nitrate	−	−
Ferric hydroxide	−	−
Growth by disproportionation of reduced sulfur compounds[d]	+	+
Growth factor requirement	−	−
NaCl requirement (g/l)	−	15
Literature	Janssen et al., 1996	Finster et al., 1998

[a]Symbols: nr, not reported; +, present, observed in all species; −, not present, not observed; ±, variable; (+), poorly motile, poor growth; (sp), flagellation type: single, polar; I, incomplete.

[b]Some utilization occurs without growth.

[c]Used as electron donor only in the presence of acetate as carbon source.

[d]Chemolithoautotrophic growth.

Genus III. Desulfofustis *Friedrich, Springer, Ludwig and Schink 1996, 1067*[VP]

MICHAEL FRIEDRICH AND BERNHARD SCHINK

De.sul.fo.fus′ tis. L. pref. *de* from; L. n. *sulfur* sulfur; L. n. *fustis* club; M.L. masc. n. *Desulfofustis* a club-shaped sulfate reducer.

Straight or slightly curved rods, spore formation not observed. Gram negative. Metabolism anaerobic, respiratory. **Organic substrates oxidized incompletely to acetate (and CO₂); glycolate or glyoxylate completely oxidized to CO₂. Sulfate or sulfite used as electron acceptors and reduced to H₂S.** Grouped with *Deltaproteobacteria* based on 16S rDNA gene sequence analysis.

Type species: **Desulfofustis glycolicus** Friedrich, Springer, Ludwig and Schink 1996, 1067.

FURTHER DESCRIPTIVE INFORMATION

Desulfofustis glycolicus clusters with the *Deltaproteobacteria*, close to *Desulforhopalus vacuolatus* and *Desulfobulbus propionicus* (90.6% and 86.5% 16S rDNA sequence similarity, respectively). These microorganisms represent a subgroup within the *Deltaproteobacteria*.

D. glycolicus reduces sulfate, sulfite, and elemental sulfur to H₂S. Other terminal electron acceptors such as thiosulfate, nitrate, and fumarate cannot be reduced. The following compounds serve as electron donor and carbon sources: glycolate, glyoxylate, L-lactate, L-malate, fumarate, succinate, and yeast extract (0.1%). *D. glycolicus* oxidizes only glycolate and glyoxylate completely to CO₂. All other electron donors are oxidized neither completely to CO₂ nor incompletely to 1 mol acetate per mol of electron donor, as expected for incomplete oxidation stoichiometries. Reduced acetate formation is counterbalanced by increased H₂S formation. This unusual substrate degradation balance seen in the type strain may result from the presence of carbon monoxide dehydrogenase, which may be responsible for

partial oxidation of acetate. However, the mechanism of acetate excretion and activation in *D. glycolicus* is still unknown, since phosphotransacetylase and acetate kinase are lacking. *D. glycolicus* grows with H₂ as electron donor but acetate is required for assimilatory metabolism.

Glycolate is oxidized by a membrane-associated methylene-blue-dependent glycolate dehydrogenase (E.C.1.1.99.14), but the further reaction sequence leading to complete oxidation is not known.

D. glycolicus strains may be cultured in defined medium[1] with sulfate as electron acceptor and glycolate, lactate, malate, fumarate, or succinate as electron donor and carbon source. Cells grown with glyoxylate as carbon source and electron donor exhibit long and unpredictable lag phases upon transfer on the same substrate. An inoculum size of 10% is recommended, especially with older cultures as inoculum source.

Cytochromes of *b*- and *c*-type are present in both the membranes and the cytoplasmic fraction, and a menaquinone-5 (H₂) as major quinone. *D. glycolicus* contains a sulfite reductase of

---

1. The defined medium is prepared as described for *Desulfobacter*, with the following additions (g/l): NaCl, 3.0; MgCl₂·2H₂O, 3.0; vitamin solution (*p*-aminobenzoic acid, 40 mg/l; biotin, 10 mg/l; nicotinic acid, 100 mg/l; Ca-D(+) pantothenate, 50 mg/l; pyridoxamine HCl, 150 mg/l; thiamine dichloride, 100 mg/l; folic acid, 30 mg/l; DL-α-liponic acid; solutions sterilized by filtration), and cyanocobalamin (100 mg/l), 1 ml/liter each; sodium selenite-tungsten solution (Na₂SeO₃·2H₂O, 3 mg/l; Na₂WO₄·2H₂O, 4 mg/l; KOH, 0.7 g/l), 1 ml/l. Sodium glycolate (10 mM final concentration) is added from a filter-sterilized stock solution as organic substrate.

desulforubidin type, as identified by its characteristic absorption maxima (279, 396, 545, and 580 nm wavelength). The purified protein is a heteropolymer consisting of three subunits with molecular masses of 42.5 (α), 38.5 (β), and 13 (γ) kDa.

The ecological role of *D. glycolicus* in its natural habitat, a marine sediment, may be glycolate degradation. However, in nature glycolate is formed mainly as a result of photorespiration of photoautotrophs under conditions of high light intensity and low CO_2 availability, and this type of metabolism is rarely encountered in marine sediments. Therefore, other ecologically relevant habitats for glycolate-oxidizing sulfate-reducing bacteria may be mats of photoautotrophs, such as a cyanobacterial mat in a hypersaline pond system. In the upper, hyperoxic layer of this system, Fründ and Cohen (1992) found evidence for sulfate-dependent glycolate oxidation.

ENRICHMENT AND ISOLATION PROCEDURES

D. glycolicus can be selectively enriched using glycolate-sulfate medium with NaCl concentrations corresponding to the salinity of the inoculum source. Anoxic black mud from marine habitats (2–5% of the culture volume) is used to inoculate medium in screw-capped bottles. Enrichments are incubated at 28°C with occasional mixing of the bottles. After strong H_2S formation (usually within 3 to 4 weeks) in glycolate enrichments as compared to controls without organic substrate added, enrichments can be transferred to fresh culture media. Subsequently, pure cultures can be obtained by the agar shake dilution method (for details see section on *Desulfobacter*). *D. glycolicus* forms disc-shaped brownish-to-black colonies in agar shake dilutions.

MAINTENANCE PROCEDURES

Pure cultures may be stored anaerobically in liquid medium at room temperature in the dark. Stock cultures should be transferred every 1–2 months.

For long-term preservation, a dense cell suspension of *D. glycolicus* may be suspended in anaerobic medium without bicarbonate containing DMSO (5% v/v), glycerol (10% v/v), and peptone (1% w/v), sealed in sterile glass capillaries, and stored in liquid nitrogen.

DIFFERENTIATION OF THE GENUS *DESULFOFUSTIS* FROM OTHER GENERA

The genus *Desulfofustis* can be differentiated from other genera mainly by 16S rDNA gene sequence comparison, and by its growth with glycolate as electron donor and carbon source. However, since only a single species has been described so far, it is not known whether glycolate is a unique electron donor and carbon source for other potential members of this genus.

TAXONOMIC COMMENTS

Based on 16S rDNA sequence analysis, *D. glycolicus* forms a subgroup within the *Deltaproteobacteria*, together with *Desulforhopalus vacuolatus* and *Desulfobulbus propionicus* (90.6% and 86.5% 16S rDNA sequence similarity, respectively). Several cloned partial 16S rDNA sequences (clones A30, A36, A33; RDP tree), and partial sequences of *Desulfobulbus* spp. (Dbb.sp1 and Dbb.sp.2; RDP tree) are more closely related to *D. glycolicus* than *Desulforhopalus vacuolatus* and *Desulfobulbus propionicus*; however, only partial sequences are available for phylogenetic comparison.

List of species of the genus Desulfofustis

1. **Desulfofustis glycolicus** Friedrich, Springer, Ludwig and Schink 1996, 1067[VP]

   gly.co' li.cus. M.L. n. *acidum glycolicum* glycolic acid; M.L. adj. *glycolicus* referring to glycolic acid as the key substrate of this species.

   Straight or slightly curved rods, 0.55 × 2.0–4.5 μm. Motile by means of one subterminally inserted flagellum. Anaerobic, respiratory metabolism. Chemoorganotrophic, using glycolate, glyoxylate, L-lactate, L-malate, fumarate, succinate, yeast extract, or hydrogen (in the presence of CO_2 and acetate) as electron donors and carbon sources for anaerobic respiration; only glycolate and glyoxylate are completely oxidized to CO_2. Sulfate, sulfite, and S^0 used as electron acceptors and are reduced to H_2S. Thiosulfate or nitrate not reduced. Temperature range for growth is 15–37°C, optimal at 28°C. pH range 6.7–8.3, optimal at pH 7.4. Occur in anoxic sediments from brackish and marine water sources.

   The mol% G + C of the DNA is: 56.2 ± 0.2 (HPLC).
   Type strain: PerGlyS, DSM 9705.
   GenBank accession number (16S rRNA): X99707.

Genus IV. **Desulforhopalus** *Isaksen and Teske 1999, 935*[VP] *(Effective publication: Isaksen and Teske 1996, 167)*

JAN KUEVER, FRED A. RAINEY AND FRIEDRICH WIDDEL

De.sul.fo.rho' pa.lus. L. pref. *de* from; L. n. *sulfur* sulfur; L. n. *rhopalus* cudgel; M.L. masc. n. *Desulforhopalus* cudgel-formed sulfate reducer.

Cells are oval to rod shaped with rounded ends, 1.5–1.8 × 3–5 μm. Longer cells occur in the stationary phase. Cells may contain gas vesicles. Occur singly, in pairs, or short chains. Spore formation is not observed. Gram negative and nonmotile. Strictly anaerobic, having both a respiratory and a fermentative type of metabolism. **Chemoorganotrophs**, using propionate, lactate, pyruvate, ethanol, and 1-propanol **as electron donors and carbon sources. Organic substrates are incompletely oxidized to acetate and CO_2. Chemolithoheterotrophic growth occurs with H_2 as the electron donor in the presence of acetate** as the organic carbon source. **Sulfate, sulfite, and thiosulfate serve as terminal electron acceptors and are reduced to H_2S.** Sulfur and oxygen are not reduced; nitrate is reduced to ammonia only by *D. singaporensis*. **In the absence of an external electron acceptor, growth occurs by fermentation of pyruvate and lactate.** *D. singaporensis* can grow by fermentation of 2-aminoethansulfonate (taurine) and by disproportionation of sulfite in the presence of acetate.

All species are able to grow in simple defined media containing a reductant. Some species require the vitamins pyridoxamine, nicotinate, and 4-aminobenzoate for growth. The optimal pH range for growth is 6.8–7.2. The optimal temperature range for growth is 20–30°C. Thermophilic species are not known. At least

5 g NaCl and 1 g $MgCl_2 \cdot 6H_2O$ per liter are required for growth. Desulfoviridin is absent. Occur in the upper part of anoxic marine or brackish sediment.

The mol% G + C of the DNA is: 48.4–50.6.

Type species: **Desulforhopalus vacuolatus** Isaksen and Teske 1999, 935 (Effective publication: Isaksen and Teske 1996, 167.)

FURTHER DESCRIPTIVE INFORMATION

The temperature range for growth of *D. vacuolatus* is 0–24°C, with an optimal growth temperature of 18–19°C. In contrast, *D. singaporensis* has a temperature range for growth between 20 and 35°C, with an optimal at 31°C.

Propionate, lactate, pyruvate, propanol, and ethanol are utilized as electron donors and organic carbon sources. Oxidation is incomplete and leads to formation of acetate as an organic end product. Disproportionation of sulfite allows growth only in the presence of acetate as carbon source.

Desulfoviridin is not present. Major cellular fatty acids of *D. vacuolatus* are $C_{15:1\ \omega6c}$ and $C_{17:1\ \omega6c}$ (Knoblauch et al., 1999). *Desulforhopalus* strains occur in the upper part of reduced marine or brackish sediments.

D. vacuolatus may be cultured in a defined medium (e.g., as described for *Desulfobacter*) with sulfate as electron acceptor and lactate or various other substrates as the electron donor and carbon source. The type strain requires at least 5 g NaCl and 1 g $MgCl_2 \cdot 6H_2O$ per liter for growth. Ammonium ions are used as the nitrogen source; *D. singaporensis* can also use taurine. The vitamins pyridoxamine, nicotinate, and 4-aminobenzoate are required as growth factors for *D. vacuolatus*.

ENRICHMENT AND ISOLATION PROCEDURES

For the selective enrichment of *D. vacuolatus* strains, a marine medium (e.g., as described for *Desulfobacter*) with lactate or propionate as electron donor and thiosulfate as electron acceptor may be used. Bottles or tubes (100-, 50-, or 20-ml capacity) can serve as culture vessels and are kept anoxic with tight screw caps, or by using a headspace of N_2/CO_2 (90:10) and sealing with butyl rubber stoppers. The medium is inoculated with black anaerobic mud from a marine habitat; 2–5% of the total culture volume should be added. *D. vacuolatus* strains may also be enriched and isolated via direct serial dilution of anoxic environmental samples in liquid medium (according to most-probable-number method) or agar medium (see next paragraph) with lactate or propionate added as organic substrate. Enrichment cultures are incubated at 10°C. Occasional mixing may favor growth. After intense formation of H_2S has occurred, transfers to fresh medium are made. Enrichment cultures should be mixed

before subculturing. Use of thiosulfate as electron acceptor under psychrophilic growth conditions (around 4–8°C) favors selection of *Desulforhopalus* strains over marine *Desulfobulbus* strains. For enrichment of *D. singaporensis*, a combination of malate and taurine (each 10 mM) in the absence of sulfate should be used. The NaCl concentration used should be the same as for sulfate-reducing bacteria from brackish water habitats (5–10 g/l).

Pure cultures of *Desulforhopalus* strains are obtained by repeated application of the agar dilution method. The method is similar to that described for the genus *Desulfobacter*, except that another electron donor (see above) is added to the molten concentrated agar solution.

MAINTENANCE PROCEDURES

Maintenance procedures have not been described. Procedures described for other genera in this chapter may be attempted.

DIFFERENTIATION OF THE GENUS DESULFORHOPALUS FROM OTHER GENERA

Physiologically, *Desulforhopalus* species share many features with members of the genus *Desulfobulbus*. Both oxidize propionate incompletely to acetate and CO_2 and can grow by fermentation. Clear differentiation requires comparative 16S rDNA sequences analysis. A more or less reliable detection of *Desulforhopalus* cells in naturally enriched populations may be possible using the oligonucleotide probe described below.

TAXONOMIC COMMENTS

The closest relatives of *Desulforhopalus* species are members of the genera *Desulfotalea*, *Desulfocapsa*, *Desulfobulbus*, and *Desulfofustis glycolicus*. A signature sequence for *Desulforhopalus* species, *Desulfofustis glycolicus*, *Desulfocapsa sulfexigens*, and other closely related sulfate-reducing bacteria is 5'-CCCCCTCCAGTACTCAAG-3' (oligonucleotide probe DSR651), which binds to the 16S rRNA corresponding to position 5'-651–668-3' of the *Escherichia coli* 16S rRNA sequence (Manz et al., 1998).

As indicated by its different physiological properties and phylogenetic distance, inclusion of *D. singaporensis* within the genus remains questionable. More detailed characterization in the future might justify the reclassification of *D. singaporensis* as a new genus within the *Desulfobulbaceae*.

FURTHER READING

Isaksen, M.F. and A. Teske. 1996. *Desulforhopalus vacuolatus* gen. nov., sp. nov., a new moderately psychrophilic sulfate-reducing bacterium with gas vacuoles isolated from a temperate estuary. Arch. Microbiol. *166*: 160–168.

List of species of the genus Desulforhopalus

1. **Desulforhopalus vacuolatus** Isaksen and Teske 1999, 935[VP] (Effective publication: Isaksen and Teske 1996, 167.)

   va.cu.o.la′tus. M.L. n. *vacuolatus* vacuolated due to the morphology of the cells.

   Characteristics are described in Table BXII.δ.26. The type strain was isolated from anoxic marine sediment, Kysing Fjord (Denmark), with lactate or propionate as only organic substrate and thiosulfate as electron acceptor.

   The mol% G + C of the DNA is: 48.4 (HPLC).

   Type strain: itk10, DSM 9700.

   GenBank accession number (16S rRNA): L42613.

2. **Desulforhopalus singaporensis** Lie, Clawson, Godchaux

and Leadbetter 2000, 1699[VP] (Effective publication: Lie, Clawson, Godchaux and Leadbetter 1999, 3333.)

   sin.ga.po′ren.sis. M.L. n. *singaporensis* referring to the place of isolation, Republic of Singapore.

   The characteristics are described in Table BXII.δ.26. Cells produce nonprosthecate structures called spinae. The type strain was isolated from anoxic sediment from a salt marsh (Singapore) using 10 mM malate and 10 mM taurine. Taurine allows growth in the absence of a nitrogen source.

   The mol% G + C of the DNA is: 50.6 (HPLC).

   Type strain: T1, DSM 12130.

   GenBank accession number (16S rRNA): AF118453.

TABLE BXII.δ.26. Morphological and physiological characteristics of described species of the genus *Desulforhopalus*[a]

Species	*Desulforhopalus vacuolatus*	*Desulforhopalus singaporensis*
Cell morphology	Rod-shaped	Rod-shaped
Cell size (μm)	1.5–1.8 × 3.9–5.0	0.9–1.2 × 1.7–2.3
Motility	−	−
Mol% G + C content	48.4	40.6
Sulfite reductase	nr	nr
Major menaquinone	nr	MK-5(H_2)
Optimal temperature (°C)	18–19	31
Oxidation of substrate	I	I
Compounds used as electron donors and carbon sources:		
H_2[b]	+	nr
Formate	−	+
Acetate	−	−
Fatty acid : C atoms	3	3–4
Isobutyrate	nr	+
Ethanol	+	+
Lactate	+	+
Pyruvate	+	+
Fumarate	−	+
Succinate	nr	+
Malate	nr	+
Benzoate	−	−
Alanine	nr	+
Butanol[b]	+	+
Propanol[b]	+	+
Casamino acids	nr	+
Fermentative growth on:		
Lactate	+	nr
Pyruvate	+	+
Taurine	nr	+
Compounds used as electron acceptors:		
Sulfate	+	+
Sulfite	+	+
Sulfur	−	nr
Thiosulfate	+	+
Fumarate	−	nr
Nitrate	−	+
Growth by disproportionation of reduced sulfur compounds	−	+
Growth factor requirement	Py, pa, ni	−
NaCl requirement (g/l)	At least 5	5
Literature	Isaksen and Teske, 1996	Lie et al., 1999

[a]Symbols: nr, not reported; +, present, observed in all species; −, not present, not observed; ±, variable; (+), poorly motile, poor growth; I, incomplete; py, pyrodoxine; ni, nicotinate; pa, *p*-aminobenzoate.

[b]Used as electron donor only in the presence of acetate as carbon source.

Genus V. **Desulfotalea** *Knoblauch, Sahm and Jørgensen 1999, 1641*[VP]

JAN KUEVER, FRED A. RAINEY AND FRIEDRICH WIDDEL

De.sul.fo.ta'le.a. L. pref. *de* from; L. n. *sulfur* sulfur; L. fem. n. *talea* a rod; M.L. fem. n. *Desulfotalea* a sulfate-reducing rod.

Cells are large rods of variable length, 0.6–0.7 × 1.6–7.4 μm. Cells may be motile, but motility may not be observed in cultures at the beginning of growth. Cells stain Gram negative. Spores are not formed. Strictly anaerobic, having a respiratory and fermentative type of metabolism. **Chemoorganotrophs**, using formate, carboxylic acids, amino acids, and alcohols **as electron donors and also as carbon sources. Oxidation is incomplete and leads to acetate. Sulfate and ferric citrate serve as terminal electron acceptors and are reduced to sulfide and ferrous citrate, respectively.** The type species can also use thiosulfate and sulfite as electron acceptors. Elemental sulfur, Fe(III)-oxohydroxide, nitrate, and nitrite are not used as electron acceptors. **Fermentative growth on pyruvate and fumarate is possible. Obligately or fac-** ultatively psychrophilic. The temperature optimal for growth is 10–18°C, and growth still occurs in saltwater at −1.8°C. Thermophilic species are not known. The optimal pH for growth is between 7.2 and 7.9. The species require sodium chloride and magnesium chloride, and optimal growth occurs at brackish or marine salt concentrations, depending on the strain. Media containing a reductant are necessary for growth. Vitamins are not required. Desulfoviridin is absent. Occur in permanently cold marine sediments.

The mol% G + C of the DNA is: 42–47.

Type species: **Desulfotalea psychrophila** Knoblauch, Sahm and Jørgensen 1999, 1641.

FURTHER DESCRIPTIVE INFORMATION

Formate, lactate, pyruvate, fumarate, amino acids, and alcohols serve as electron donors and also as carbon sources for *Desulfotalea* species. Growth on H_2 requires low concentrations (1–2 mM) of acetate as an organic carbon source in addition to CO_2. No growth on fructose and glucose. The ability to use thiosulfate and sulfite instead of sulfate varies from species to species. *D. arctica* reduces sulfur without growth. Sulfur and thiosulfate are not disproportionated.

The temperature range for growth of *D. psychrophila* is −1.8–19°C, with an optimal growth temperature of 10°C. *D. arctica* has a temperature range for growth between −1.8 and 26°C; with an optimal at 18°C. Mesophilic and thermophilic species are not known.

Optimal growth of the type species requires an NaCl concentration of 10 g/l, whereas *D. arctica* needs 19 g/l. Vitamins are not required for growth. *Desulfotalea* species contain phosphatidylethanolamine, diphospatidylglycerol, and phosphatidylglycerol as lipids; MK-6 or MK-6(H_2) are the sole menaquinones. Major cellular fatty acids are $C_{16:1\ \omega7c}$ and $C_{16:1\ \omega5c}$. Described species do not contain desulfoviridin. Their habitat is permanently cold marine sediments.

ENRICHMENT AND ISOLATION PROCEDURES

For specific enrichment, synthetic marine mineral medium (e.g., as described for *Desulfobacter*) reduced with sulfide should be used. As electron donor and carbon source, 10 mM lactate is added from sterile stock solutions. Bottles or tubes (100-, 50-, or 20-ml capacity) can serve as culture vessels and are kept anoxic with tight screw caps, or by using a headspace of N_2/CO_2 (90:10) and sealing with butyl rubber stoppers. The medium is inoculated with anoxic sediment from a marine habitat; 3–5% of the total culture volume should be added. *Desulfotalea* species may also be enriched and isolated via direct serial dilution of anoxic environmental samples in liquid medium (according to most-probable-number method) or agar medium (see next paragraph) with lactate as organic substrate. Incubations should be done at 4°C. Sulfide production after 8–10 weeks indicates development of sulfate-reducing bacteria. After some transfers, a stable enrichment is obtained. This can be used for isolation of pure cultures by the repeated application of agar dilutions, as described in the chapter for the genus *Desulfobacter* (modified after Isaksen and Teske, 1996), or by serial dilution in liquid media using lactate as electron donor.

MAINTENANCE PROCEDURES

Stock cultures are transferred every 4–6 weeks. *Desulfotalea* strains may be preserved indefinitely by suspension of cells in anoxic medium containing 5% dimethylsulfoxide and storage in liquid nitrogen. Transfers should be done every six weeks. Incubation at 4°C and an inoculum size of 10% are recommended.

DIFFERENTIATION OF THE GENUS *DESULFOTALEA* FROM OTHER GENERA

Strains of *Desulfotalea* can be distinguished from other sulfate-reducing bacteria by physiological properties. The most important characteristic of *Desulfotalea* species for differentiation from other species is the requirement for a lower temperature for optimal growth, compared to mesophilic sulfate-reducers. The inability to grow with propionate allows a separation from other psychrophilic or mesophilic sulfate-reducers, like *Desulfobulbus* spp., *Desulfofaba gelida*, and *Desulforhopalus* spp. Growth by disproportionation of reduced sulfur compounds as observed in psychrotolerant *Desulfocapsa* spp. does not occur within the genus *Desulfotalea*. However, a clear differentiation can only be done by comparative analysis of 16S rDNA sequences.

TAXONOMIC COMMENTS

The closest relatives of *Desulfotalea* species are *Desulforhopalus vacuolatus*, *Desulfofustis glycolicus*, and members of the genus *Desulfocapsa*. A signature sequence for *Desulfotalea* species, *Desulfofustis glycolicus*, and other closely related sulfate-reducing bacteria is 5′-CCATCTGACAGGATTTTAC-3′ (oligonucleotide probe Sval428), which binds to the 16S rRNA corresponding to position 5′-428–446-3′ of the *Escherichia coli* 16S rRNA sequence (Sahm et al., 1999).

FURTHER READING

Knoblauch, C., K. Sahm and B.B. Jørgensen. 1999. Psychrophilic sulfate-reducing bacteria isolated from permanently cold Arctic marine sediments: description of *Desulfofrigus oceanense* gen. nov., sp. nov., *Desulfofrigus fragile* sp. nov., *Desulfofaba gelida* gen. nov., sp. nov., *Desulfotalea psychrophila* gen. nov., sp. nov. and *Desulfotalea arctica* sp. nov. Int. J. Syst. Bacteriol. *49*: 1631–1643.

List of species of the genus Desulfotalea

1. **Desulfotalea psychrophila** Knoblauch, Sahm and Jørgensen 1999, 1641[VP]

   psy.chro′phi.la. Gr. adj. *psychros* cold; *philos* loving; M.L. adj. *psychrophilus, -a, -um*, cold loving.

   Characteristics are listed in Table BXII.δ.27. Isolated from enrichment cultures inoculated with anoxic sediment from Starfjord (Svalbard) with lactate as electron donor and carbon source and sulfate as electron acceptor.

   The mol% G + C of the DNA is: 46.8 (HPLC).
   Type strain: LSv54, DSM 12343.
   GenBank accession number (16S rRNA): AF099062.

2. **Desulfotalea arctica** Knoblauch, Sahm and Jørgensen 1999, 1642[VP]

   arc′ ti.ca. L. adj. *arcticus* from the Arctic, referring to the site where the type strain was isolated.

   Characteristics are listed in Table BXII.δ.27. Isolated from enrichment cultures inoculated with anoxic sediment from Starfjord (Svalbard) with lactate as electron donor and carbon source and sulfate as electron acceptor.

   The mol% G + C of the DNA is: 41.8 (HPLC).
   Type strain: LSv514, DSM 12342.
   GenBank accession number (16S rRNA): AF099061.

TABLE BXII.δ.27. Morphological and physiological characteristics of described species of the genus *Desulfotalea*[a,b]

Species	*Desulfotalea psychrophila*	*Desulfotalea arctica*
Cell morphology	Rods	Rods
Cell size (μm)	0.6 × 4.5–7.4	0.7 × 1.6–2.7
Motility	±	−
G + C content (mol%)	46.8	41.8
Sulfite reductase	nr	nr
Major menaquinone	MK-6H$_2$	MK-6
Optimal temperature (°C)	10	18
Oxidation of substrate	I	I
Compounds used as electron donors and carbon source:		
H$_2$[c]	+	+
Formate	+	+
Acetate	−	−
Fatty acid: C atoms	−	−
3-Methylbutyrate	−	−
Ethanol	+	+
Lactate	+	+
Pyruvate	+	+
Fumarate	+	−
Succinate	−	−
Malate	+	−
Benzoate	−	−
Alanine	+	−
Butanol	+	−
Glycerol	−	+
Glycine	+	−
Propanol	+	−
Serine	+	+
Fermentative growth on:		
Fumarate	+	−
Pyruvate	+	+
Compounds used as electron acceptors:		
Sulfate	+	+
Sulfite	+	−
Sulfur	−	(+)[d]
Thiosulfate	+	−
Nitrate	−	−
Other electron acceptors	Fe(III)-citrate	Fe(III)-citrate
Growth factor requirement	−	−
NaCl requirement (g/l)	10	19

[a]Symbols: nr, not reported; +, present, observed in all species; −, not present, not observed; ±, variable; (+), poorly motile, poor growth; (sp), flagellation type: single, polar; (lo), flagellation type: lophotrichous; DV, desulfoviridin; DR, desulforubidin; I, incomplete; C, complete; bi, biotin; ni, nicotinate; pa, *p*-aminobenzoate; pt, pantothenate; th, thiamine.

[b]Data from Knoblauch et al. (1999).

[c]Used as electron donor only in the presence of acetate as carbon source.

[d]Reduced to sulfide, but growth was not obtained.

Family III. **Nitrospinaceae** *fam. nov.*

GEORGE M. GARRITY, JULIA A. BELL AND TIMOTHY LILBURN

Ni.tro.spi.na′ce.ae. M.L. masc. n. *Nitrospina* type genus of the family; *-aceae* ending to denote family; M.L. fem. pl. n. *Nitrospinaceae* the *Nitrospina* family.

The family *Nitrospinaceae* was circumscribed for this volume on the basis of phylogenetic analysis of 16S rRNA sequences; the family contains the genus *Nitrospina* (type genus).

Description is the same as for the genus *Nitrospina*.

Type genus: **Nitrospina** Watson and Waterbury 1971, 225.

Genus I. **Nitrospina** *Watson and Waterbury 1971, 225*[AL]

EVA SPIECK AND EBERHARD BOCK

Ni.tro.spi′ na. L. n. *nitrum* nitrate; L. n. *spina* spine; M.L. masc. n. *Nitrospina* nitrate sphere.

Rod-shaped cells 0.3–0.5 × 1.7–6.6 μm. Spherical forms 1.4–1.5 μm in diameter are found in old cultures. Cells reproduce by binary fission. **An intracytoplasmic membrane system is miss**ing, but occasionally invaginations of the cytoplasmic membrane occur. Gram negative. Nonmotile. **Aerobic.** The major source of energy and reducing power is from the **oxidation of nitrite to**

nitrate. **Obligate lithoautotrophs**. Optimal growth requires 70–100% seawater. Cells occur in marine environments.

The mol% G + C of the DNA is: 57.7.

Type species: **Nitrospina gracilis** Watson and Waterbury 1971, 225.

FURTHER DESCRIPTIVE INFORMATION

The two strains of *Nitrospina* are peripherally related to the *Deltaproteobacteria* (family *Desulfarculaceae*) (Teske et al. 1994), the class in which the genus *Nitrospina* was first placed. Since no phototrophic organisms are known so far in this class, it is questionable whether *Nitrospina* is derived from photosynthetic ancestry. Additional details and a comparison of the biochemical properties of *Nitrospina* to those of other nitrite-oxidizing genera can be found in the introductory chapter "Lithoautotrophic Nitrite-Oxidizing Bacteria." Detailed treatments of the ecology of nitrite-oxidizing bacteria and of the phylogeny of these organisms can be found in the introductory chapter "Nitrifying Bacteria."

Cells are straight, slender rods without intracytoplasmic membranes (Fig. BXII.δ.34). Resting forms are shown in Fig. BXII.δ.35. Several modes of electron microscopy reveal the presence of particles associated with the cell membrane (Figs. BXII.δ.36, BXII.δ.37, and BXII.δ.38). In negatively stained preparations, irregular particles with a diameter of 13–17 nm are bound to the outside of the cytoplasmic membrane (Fig. BXII.δ.36). In ultrathin sections, electron-dense material is visible in the periplasmic space (Watson and Waterbury, 1971). The most abundant proteins of crude extracts of *Nitrospina* have molecular masses of 130 and 47 kDa (Ehrich, personal communication). The β-subunit of the nitrite-oxidizing system (β-NOS) of *Nitrospina* has been shown to be located in the periplasmic space by labeling with monoclonal antibodies (MAbs) against the β-subunit of the nitrite oxidoreductase (β-NOR) of *Nitrobacter* (Fig. BXII.δ.37). This protein has a molecular mass of 48 kDa, which is similar to the β-NOS of *Nitrospira*. Cytochromes of the *c*-type but not of the *a*-type were detected. Cell suspensions show characteristic absorption peaks at 425, 532 and 553 nm. Glycogen-like inclusion bodies may represent storage material. No carboxysomes have been found in this organism. The temperature range for growth is 20–30°C and the pH range is 7.0–8.0. The two strains of *Nitrospina gracilis* have been isolated from Atlantic and Pacific seawater. 16S rDNAs with high similarities to *Nitrospina* were found in the microbial community of marine sponges (Hentschel et al., 2002). No terrestrial strains are known, although 16S rDNA clones related to *Nitrospina* were obtained recently from uranium waste piles (Selenska-Pobell, 2002). A further species, represented by strain 347 from the Black Sea, was separated by 16S rRNA sequence analysis (Ludwig, personal communication).

ENRICHMENT AND ISOLATION PROCEDURES

Nitrite oxidizers can be isolated using a mineral medium containing nitrite; the compositions of media for lithotrophic, mixotrophic, and heterotrophic growth are given in Table BXII.δ.28. Serial dilutions of enrichment cultures must be incubated for one to several months in the dark. Since nitrite oxidizers are sensitive to high partial pressures of oxygen, cell growth on agar surfaces is limited. Nitrite oxidizers can be separated from heterotrophic contaminants by Percoll gradient centrifugation and subsequent serial dilution (Ehrich et al., 1995).

MAINTENANCE PROCEDURES

Nitrifying organisms can survive starvation for more than one year when kept at 17°C in liquid medium. Nevertheless, cells should be transferred to fresh media every four months. Three different growth media for nitrite oxidizers are listed in Table

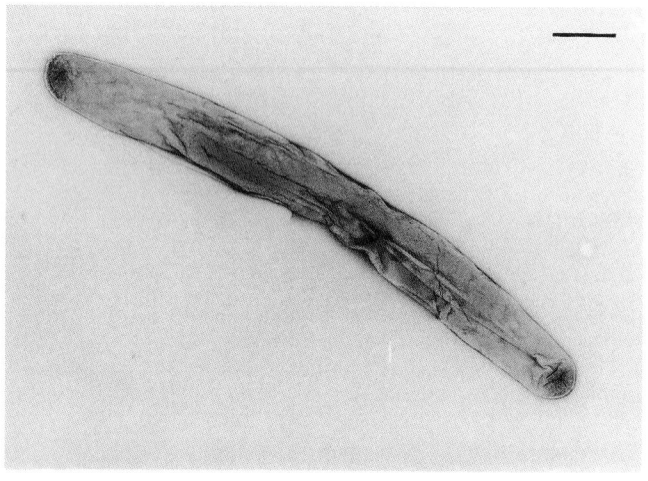

FIGURE BXII.δ.34. Negatively stained cell of *Nitrospina gracilis*. Bar = 500 nm.

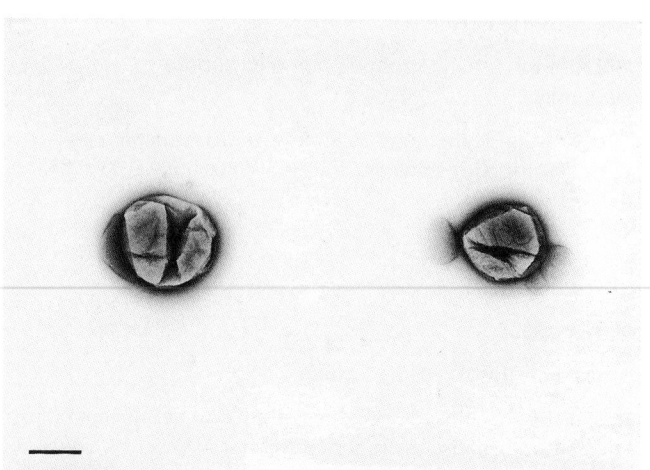

FIGURE BXII.δ.35. Resting forms of *Nitrospina gracilis*. Bar = 500 nm.

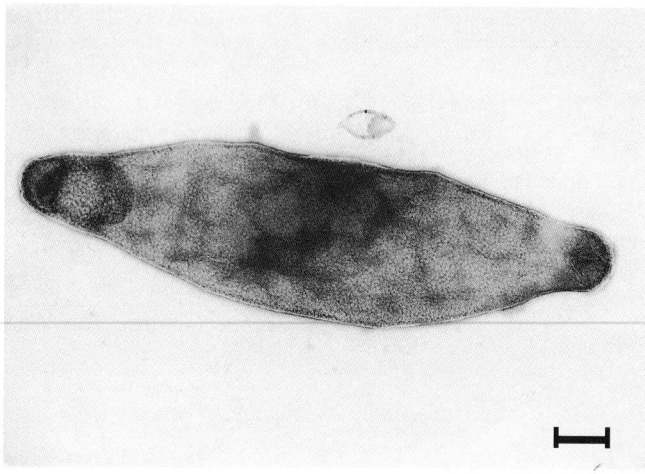

FIGURE BXII.δ.36. Electron micrograph of a partly destroyed cell of *Nitrospina gracilis* in negative contrast. Hexagonal particles are shining through the outer membrane. Bar = 250 nm.

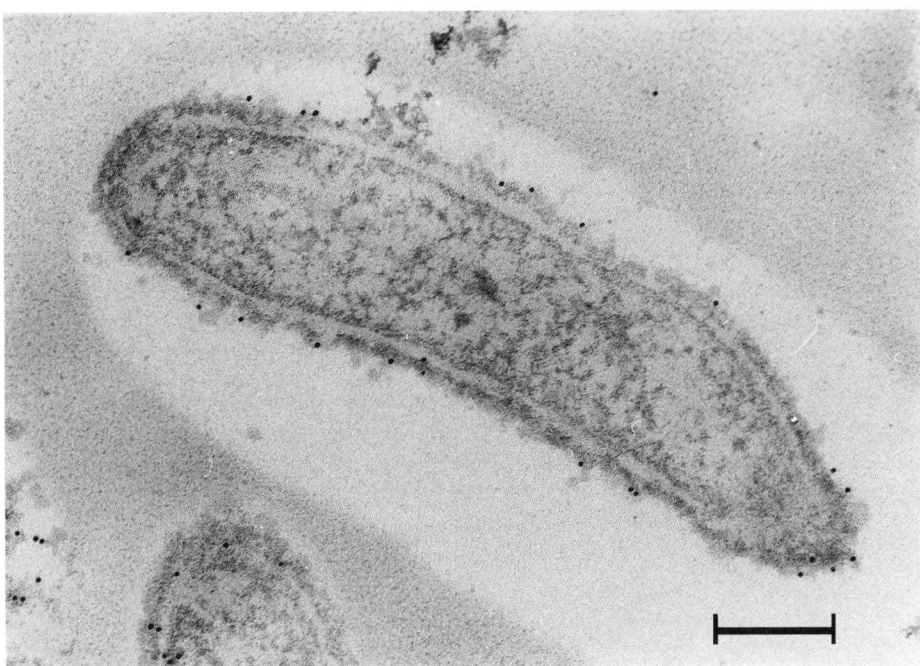

FIGURE BXII.δ.37. Ultrathin section of *Nitrospina gracilis*. The β-NOS was localized in the periplasmic space as shown by immunogold-labeling (MAbs 153-3). Bar = 250 nm.

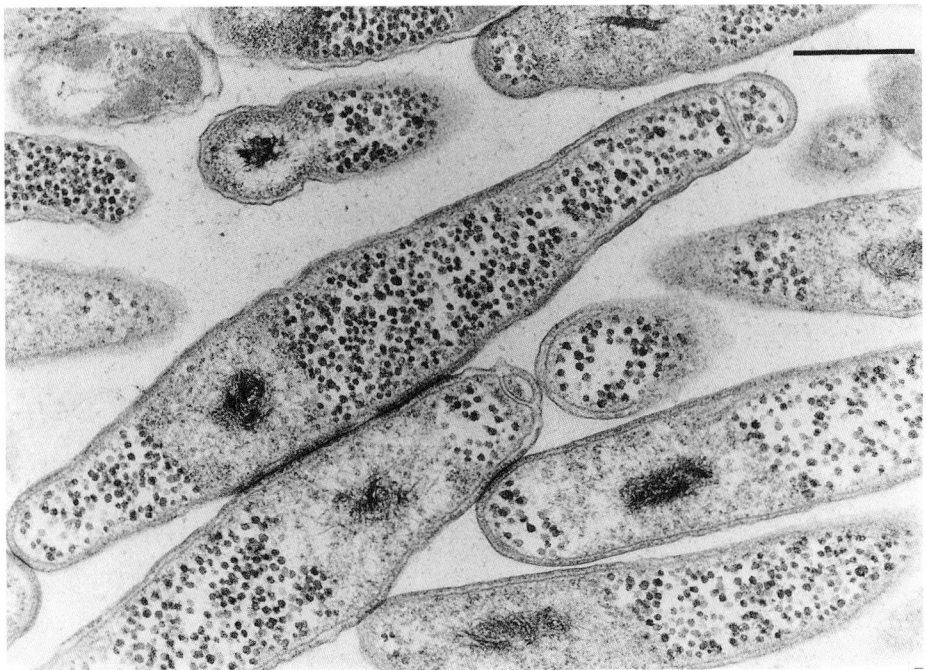

FIGURE BXII.δ.38. Ultrathin section of *Nitrospina* spec. 347. Regularly arranged particles are located in the periplasmic space. Bar = 500 nm.

BXII.δ.28. Freezing in liquid nitrogen is a suitable technique for maintenance of stock cultures suspended in a cryoprotective buffer containing sucrose and histidine.

DIFFERENTIATION OF THE GENUS *NITROSPINA* FROM OTHER GENERA

The nitrite oxidizers are a diverse group of long or short rods, cocci, and spirilla. *Nitrospina* is nonmotile and has no carboxysomes or internal membranes, whereas the genera *Nitrobacter* and

Nitrococcus possess a complex arrangement of intracytoplasmic membranes in the form of flattened vesicles or tubes. The taxonomic categorization is based on the work of Sergei and Helene Winogradsky (Winogradsky, 1892). Traditionally, the classification of genera is performed primarily on cell shape and arrangement of intracytoplasmic membranes. So far, four morphologically distinct genera (*Nitrobacter, Nitrococcus, Nitrospina,* and *Nitrospira*) have been described; the four genera contain a total of eight species (Watson et al., 1989; Bock and Koops, 1992; Ehrich

TABLE BXII.δ.28. Three different media for lithoautotrophic (medium A for terrestrial strains; medium B for marine strains), mixotrophic (medium C), and heterotrophic (medium C without NaNO₂) growth of nitrite oxidizers

Ingredient	Culture medium		
	A[a]	B[b]	C[c, d]
Distilled water (ml)	1000	300	1000
Seawater (ml)		700	
NaNO₂ (mg)	200–2000	69	200–2000
MgSO₄·7H₂O (mg)	50	100	50
CaCl₂·2H₂O (mg)		6	
CaCO₃ (mg)	3		3
KH₂PO₄ (mg)	150	1.7	150
FeSO₄·7H₂O (mg)	0.15		0.15
Chelated iron (13%, Geigy) (mg)		1	
Na₂MoO₄·2H₂O (µg)		30	
(NH₄)₂Mo₇O₂₄·4H₂O (µg)	50		50
MnCl₂·6H₂O (µg)		66	
CoCl₂·6H₂O (µg)		0.6	
CuSO₄·5H₂O (µg)		6	
ZnSO₄·7H₂O (µg)		30	
NaCl (mg)	500		500
Sodium pyruvate (mg)			550
Yeast extract (Difco) (mg)			1,500
Peptone (Difco) (mg)			1,500
pH adjusted to[e]	8.6	6	7.4

[a]For terrestrial strains from Bock et al. (1983).

[b]For marine strains modified from Watson and Waterbury (1971).

[c]For terrestrial strains from Bock et al. (1983).

[d]For heterotrophic growth medium C without NaNO₂ is used.

[e]After sterilization pH should be 7.4–7.8.

et al., 1995; Sorokin et al., 1998). Differential traits of the four genera of nitrite-oxidizing bacteria are given in Table BXII.δ.29. Further properties are given in Table BXII.δ.30, and fatty acid profiles are listed in Table BXII.δ.31.

Nitrospina can be separated from *Nitrobacter* and *Nitrococcus* by means of monoclonal antibodies that recognize the key enzyme (Table BXII.δ.29). Aamand et al. (1996) developed three monoclonal antibodies (MAbs) that recognize the α- and the β-subunit of the NOR of *Nitrobacter*. The key enzyme NOR is ubiquitous in nitrite-oxidizing bacteria. MAb Hyb 153-2 was specific for the α-NOR of *Nitrobacter*, whereas MAb Hyb 153-1 recognized the β-NOSs of both *Nitrobacter* and *Nitrococcus*. The MAb Hyb 153-3 recognized the β-NOSs of *Nitrobacter*, *Nitrococcus*, *Nitrospina* and *Nitrospira* (Bartosch et al., 1999). Thus, the suite of MAbs permits detection of all known nitrite oxidizers and allows discrimination of *Nitrobacter* and *Nitrococcus* from *Nitrospina* and *Nitrospira*.

Taxonomic Comments

The classification of *Nitrospina* as a distinct genus is based on the unique morphology and ultrastructure of this organism. 16S rRNA sequence analysis confirmed the phylogenetic position of this organism. However, the phylogenetic affiliation of *Nitrospina* with the *Deltaproteobacteria* remains provisional due to the very limited data set (Ludwig, personal communication). Euclidian distances derived from fatty acid profiles indicated a higher similarity between *Nitrospina* and *Nitrospira* than between either genus and the nitrite-oxidizing bacteria of the *Proteobacteria* (*Nitrobacter* and *Nitrococcus*) (Lipski et al., 2001).

Further Reading

Watson, S.W. and J.B. Waterbury. 1971. Characteristics of two marine nitrite oxidizing bacteria, *Nitrospira gracilis* nov. gen. nov. sp. and *Nitrococcus mobilis* nov. gen. nov. sp. Arch. Microbiol. *77*: 203–230.

Watson, S.W., E. Bock, H. Harms, H.P. Koops and A.B. Hooper. 1989. Nitrifying bacteria. *In* Staley, Bryant, Pfennig and Holt (Editors), Bergey's Manual of Systematic Bacteriology, 1st Ed., Vol. 3, The Williams & Wilkins Co., Baltimore. pp. 1808–1833.

List of species of the genus Nitrospina

1. **Nitrospina gracilis** Watson and Waterbury 1971, 225[AL]

   gra' ci.lis. L. adj. *gracilis* slender.

   The morphological, cultural and biochemical characteristics are as described for the genus. No growth occurs below 14°C or above 40°C. The minimum generation time is 24 h. High concentrations of nitrite are toxic. Growth is inhibited by many organic compounds but mixotrophic growth with low concentrations of pyruvate, yeast extract and peptone is possible. The type strain was isolated from a depth of 13 m in open Atlantic Ocean waters approximately 200 miles from the mouth of the Amazon River.

   The mol% G + C of the DNA is: 57.7 (Bd).

   Type strain: ATCC 25379.

TABLE BXII.δ.29. Differentiation of the four genera of nitrite-oxidizing bacteria

Characteristic	Nitrobacter	Nitrococcus	Nitrospina	Nitrospira
Phylogenetic position	Alphaproteobacteria	Gammaproteobacteria	Deltaproteobacteria (preliminary)	Phylum Nitrospirae
Morphology	Pleomorphic short rods	Coccoid cells	Straight rods	Curved rods to spirals
Intracytoplasmic membranes	Polar cap	Tubular	Lacking	Lacking
Size (µm)	0.5–0.9 × 1.0–2.0	1.5–1.8	0.3–0.5 × 1.7–6.6	0.2–0.4 × 0.9–2.2
Motility	−	−	−	−
Reproduction	Budding or binary fission	Binary fission	Binary fission	Binary fission
Main cytochrome types[a]	a, c	a, c	c	b, c
Location of the nitrite oxidizing system on membranes	Cytoplasmic	Cytoplasmic	Periplasmic	Periplasmic
MAb-labeled subunits (kDa)[b]	130 and 65	65	48	46
Crystalline structure of membrane-bound particles	Rows of particle dimers	Particles in rows	Hexagonal pattern	Hexagonal pattern

[a]Lithoautotrophic growth.

[b]MAbs, monoclonal antibodies.

TABLE BXII.δ.30. Properties of the nitrite-oxidizing bacteria

Characteristic	Nitrobacter winogradskyi	Nitrobacter alkalicus	Nitrobacter hamburgensis	Nitrobacter vulgaris	Nitrococcus mobilis	Nitrospina gracilis	Nitrospira marina	Nitrospira moscoviensis
Mol% G + C of the DNA	61.7	62	61.6	59.4	61.2	57.7	50	56.9
Carboxysomes	+	−	+	+	+	−	−	−
Habitat:								
Fresh water	+			+				
Waste water	+			+				
Brackish water				+				
Oceans	+				+	+	+	
Soda lakes		+						
Soil	+		+	+				
Soda soil		+						
Stones	+			+				
Heating system								+

TABLE BXII.δ.31. Primary fatty acids of the described species of nitrite-oxidizing bacteria[a,b]

Fatty acid	Nitrobacter winogradskyi Engel	Nitrobacter alkalicus AN4	Nitrobacter hamburgensis X14	Nitrobacter vulgaris Z	Nitrococcus mobilis 231	Nitrospina gracilis 3	Nitrospira marina 295	Nitrospira moscoviensis M1
$C_{14:1\ \omega5c}$						+		
$C_{14:0}$	+		+		+	+ + +	+	+
$C_{16:1\ \omega7c}$							+ + +	+ +
$C_{16:1\ \omega9c}$	+	+	+	+	+ + +	+ + +		
$C_{16:1\ \omega11c}$							+ + +	+ + +
$C_{16:0\ 3OH}$						+		+
$C_{16:0}$	+ +	+ +	+ +	+ +	+ + +	+ +	+ + +	+ + +
$C_{16:0\ 11CH_3}$							+	+ + +
$C_{18:1\ \omega9c}$	+		+			+	+	
$C_{18:1\ \omega11c}$	+ + + +	+ + + +	+ + + +	+ + + +	+ + +	+	+	+
$C_{18:0}$	+	+	+		+	+	+ +	+
$C_{19:0\ cyclo\ \omega7c}$	+	+	+	+	+			

[a]Symbols: +, <5%; + +, 6–15%; + + +, 16–60%; + + + +, >60%.

[b]Stirred cultures were grown autotrophically at 28°C (*Nitrospira moscoviensis* at 37°C) and collected at the end of exponential growth. Modified from Lipski et al., (2001).

Order IV. **Desulfarcales** *ord. nov.*

JAN KUEVER, FRED A. RAINEY AND FRIEDRICH WIDDEL

De.sulf.ar.ca' les. M.L. masc. n. *Desulfarculus* type genus of the type family of the order; suffix -*ales* denotes order; M.L. fem. pl. n. *Desulfarcales* the order that comprises the family *Desulfarculaceae.*

The order comprises one family so far, *Desulfarculaceae.* The properties of the order and the family resemble the description given for the genus and the only described species, *Desulfarculus baarsii.* Even though this species shares nutritional features (utilization of fatty acids) with several other sulfate-reducing bacteria, a separate order is established to account for the deep phylogenetic branching within the *Deltaproteobacteria.*

Type genus: **Desulfarculus** *gen. nov.*

Family I. **Desulfarculaceae** *fam. nov.*

JAN KUEVER, FRED A. RAINEY AND FRIEDRICH WIDDEL

De.sulf.ar.cu.la.ce' ae. M.L. masc. n. *Desulfarculus* type genus of the family; suffix -*aceae* denotes family; M.L. fem. pl. n. *Desulfarculaceae* the family that comprises the genus *Desulfarculus.*

The family comprises one genus so far, *Desulfarculus.* Detailed properties of the genus and the only described species are given in the genus description.

Type genus: **Desulfarculus** *gen. nov.*

Genus I. *Desulfarculus* gen. nov.

JAN KUEVER, FRED A. RAINEY AND FRIEDRICH WIDDEL

De.sulf.ar' cu.lus. L. pref. *de* from; L. n. *sulfur* sulfur; M.L. masc. n. *arculus* a small bow; M.L. masc. n. *Desulfarculus* a bow-shaped sulfate-reducer.

Vibrio-shaped cells, 0.5–0.7 × 1.5–4 μm. Occur singly or in pairs. Spore formation is not observed. Gram negative. The type species is motile. Strict anaerobe with respiratory type of metabolism. **Chemoorganotroph, using formate, butyrate, and higher fatty acids as electron donors and carbon sources**; these compounds are completely oxidized to CO_2. Growth on acetate and propionate is very slow. **Sulfate and other oxidized sulfur compounds serve as terminal electron acceptors and are reduced to H_2S. Sulfur, fumarate, and nitrate are not used as terminal electron acceptors. Growth does not usually occur in the absence of an external electron acceptor.** Optimal temperature for growth, 35–39°C. Anoxic media (with sulfide as a reductant) are necessary for growth. Vitamins are not required. Cells contain cytochromes of the *b*- and *c*-type. Carbon monoxide dehydrogenase activity is commonly observed, indicating the operation of the anaerobic C_1-pathway (carbon monoxide dehydrogenase pathway, Wood-pathway) for CO_2 fixation and complete oxidation of acetyl-CoA. *Desulfarculus* species are widespread in freshwater habitats, but might occur in brackish or marine sediments as well.

The mol% G + C of the DNA is: 66.

Type species: **Desulfarculus baarsii** *comb. nov.* (*Desulfovibrio baarsii* Widdel 1981, 382.)

FURTHER DESCRIPTIVE INFORMATION

The type strain of *D. baarsii* has vibrio-shaped cells under all growth conditions. Cells are motile and stain Gram negative. The temperature range for growth is 20–39°C. The type strain has an optimal growth temperature of 35–39°C. The pH range for growth is 6.5–8.2, with an optimal pH of 7.3.

D. baarsii reduces sulfate, sulfite, and thiosulfate to H_2S. Elemental sulfur, fumarate, and nitrate cannot serve as electron acceptors. The following compounds are utilized as electron donors and carbon sources: formate, acetate, propionate, butyrate, isobutyrate, 2-methylbutyrate, valerate, isovalerate, and higher fatty acids up to 18 carbon atoms. All organic substrates are completely oxidized to CO_2. Growth on formate does not require an additional organic carbon source. Growth does not usually occur in the absence of an external electron acceptor. However, very slow growth with presumably only a few cell divisions may occur after adaptation to sulfate-free medium with formate that is converted to acetate. Carbon monoxide dehydrogenase activity is commonly observed, indicating the operation of the anaerobic C_1-pathway (carbon monoxide dehydrogenase pathway, Wood-pathway) for formate and CO_2 fixation or complete oxidation of acetyl-CoA (Schauder et al., 1986).

Ammonium ions are used as a nitrogen source. Vitamins are not required. Desulfoviridin is not present.

Desulfarculus strains occur in anoxic zones of freshwater and brackish water habitats.

Strains of *Desulfarculus* may be cultured in a defined medium with sulfate as electron acceptor and fatty acids as the electron donor and carbon source. Even in the case of isolates from freshwater sediment, best growth is usually obtained with 7–20 g NaCl and 1.2–3 g $MgCl_2 \cdot 6H_2O$ per liter, but growth is nearly as rapid at lower concentrations.

ENRICHMENT AND ISOLATION PROCEDURES

The only known substrate for selective enrichment of *Desulfarculus* is stearate, which is added to defined medium with sulfate.[1] Bottles or tubes (100-, 50-, or 20-ml capacity) can serve as culture vessels and are kept anoxic by using tight screw caps (if completely filled), or by providing an anoxic head space of N_2/CO_2 (90:10) and sealing with butyl rubber stoppers. The medium is inoculated (2–5% of the total volume) with dark anoxic sediment or sludge from sewage digesters. Enrichment cultures are incubated at 37–39°C, because this temperature will favor growth of *Desulfarculus* spp. Cultures should be briefly mixed every day. After intense formation of H_2S has occurred, transfers to fresh medium are made. Enrichment cultures should be mixed well before subculturing. For isolation of *D. baarsii*, the agar dilution method (as described for *Desulfobacter* in this edition of the *Manual*) is recommended. The insoluble fatty acid stearate should be replaced by soluble fatty acids. Formate may also be used as the only organic substrate in agar. After inoculation, strains may be grown on any substrate including stearate.

MAINTENANCE PROCEDURES

Stock cultures are transferred every 2–3 months. Towards the end of growth or at the beginning of the stationary phase, cultures are refrigerated and stored at 2–5°C in the dark. *Desulfarculus* strains may be preserved indefinitely by suspension of cells in anoxic medium containing 5% dimethylsulfoxide and storage in liquid nitrogen.

DIFFERENTIATION OF THE GENUS *DESULFARCULUS* FROM OTHER GENERA

Desulfarculus baarsii can be distinguished from other sulfate-reducing bacteria capable of growing on fatty acids (except *Desulfocella* and *Desulfobotulus*) by its vibrio-shaped morphology. *Desulfarculus baarsii* differs from *Desulfocella* and *Desulfobotulus* in its capacity for complete substrate oxidation to CO_2 and growth on formate without addition of other organic compounds. The mol% G + C content of DNA from *Desulfocella halophila* is much lower than that of *D. baarsii* (32 vs. 66). This property, together with the ability of *D. baarsii* to grow in freshwater media, allows a clear differentiation of these species. However, a definite differentiation requires comparative 16S rDNA sequence analysis and DNA–DNA hybridization.

TAXONOMIC COMMENTS

On the basis of 16S rRNA gene sequence analysis, *D. baarsii* falls into the *Deltaproteobacteria* and forms a distinct lineage. The closest relatives are *Desulfosarcina variabilis*, *Desulfonema magnum*, *De-*

---

1. For cultivation of *D. baarsii*, the mineral medium described in the *Desulfobacter* section (this *Manual*) is first prepared without NaCl and $MgCl_2$. The sodium salt of stearic acid is added from a stock solution (28 g of stearic acid plus 4 g/l NaOH, heated in a boiling water bath until the solution is clear; sterilized by autoclaving. The solution must be remelted by heating before use; prewarmed pipettes should be used to take aliquots). Volumes added per liter of medium: 10 ml stearate solution, followed by 3.0 ml of $MgCl_2$ solution (400 g/l $MgCl_2 \cdot 6H_2O$) and 25 ml of NaCl solution (300 g/l NaCl). Pure cultures can be grown alternatively on soluble monocarboxylic acids. For instance, 5–10 ml of sodium formate solution (100 g/l) or a butyrate–caproate–caprylate mixture (containing per liter: 6.0, 2.5, and 1.0 g, respectively, of acids neutralized with NaOH) may be added per liter of medium.

sulfonema limicola, and *Desulfococcus multivorans* (Family *Desulfobacteraceae*). A signature sequence for *Desulfarculus baarsii* (also for *Desulfomonile tiedjei* and *Syntrophus* species) is 5′-GCCGGTGCTTCCTTTGGCGG-3′ (oligonucleotide probe DSMA 488), which binds to the 16S rDNA at *E. coli* position 5′-488–507-3′ (Manz et al., 1998).

List of species of the genus Desulfarculus

1. **Desulfarculus baarsii** *comb. nov.* (*Desulfovibrio baarsii* Widdel 1981, 382.)

   baar′ si.i. M.L. gen. n. *baarsii* named after J.K. Baars, a Dutch microbiologist, who did the first comprehensive studies on nutrition of sulfate-reducing bacteria.

   The morphology is as depicted in Fig. BXII.δ.39. Characteristics are as described for the genus. The type strain was isolated from ditch sediment near the University of Konstanz, Germany.

   The mol% G + C of the DNA is: 66 (T_m).

   Type strain: 2st14, ATCC 33931, DSM 2075.

   GenBank accession number (16S rRNA): M34403.

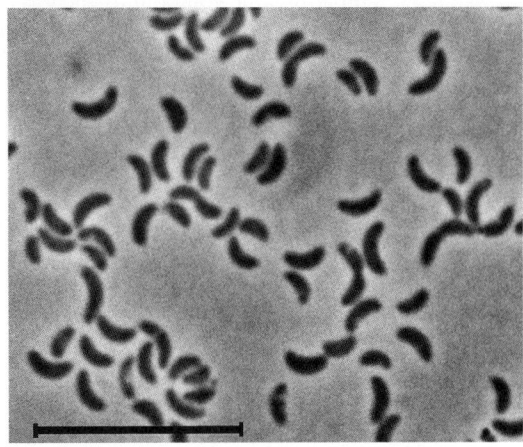

FIGURE BXII.δ.39. Phase-contrast micrograph of *Desulfarculus baarsii* (viable cells). Bar = 10 μm.

Order V. **Desulfuromonales** *ord. nov.*

Jan Kuever, Fred A. Rainey and Friedrich Widdel

De.sul.fu.ro.mo.na′les. M.L. fem. n. *Desulfuromonas* type genus of the type family of the order; suffix *ales* denotes order; M.L. fem. pl. n. *Desulfuromonales* the order that comprises *Desulfuromonaceae*.

The order *Desulfuromonales* has been established primarily based on 16S rRNA sequence analysis and combines three families of anaerobes, the *Desulfuromonaceae* and *Geobacteraceae* (see Fig. BXII.δ.40). Whereas the *Desulfuromonaceae* and *Geobacteraceae* also share several nutritional properties with each other, nutritional relationships to the family *Pelobacteraceae* are less obvious.

Cells are rod shaped and usually motile. **Strictly anaerobic chemolithoheterotrophs or chemoorganotrophs with respiratory or fermentative metabolism.** All members are moderate mesophiles with temperature optima for growth around 30°C. Menaquinones are present. Species have been isolated from anoxic freshwater and marine environments.

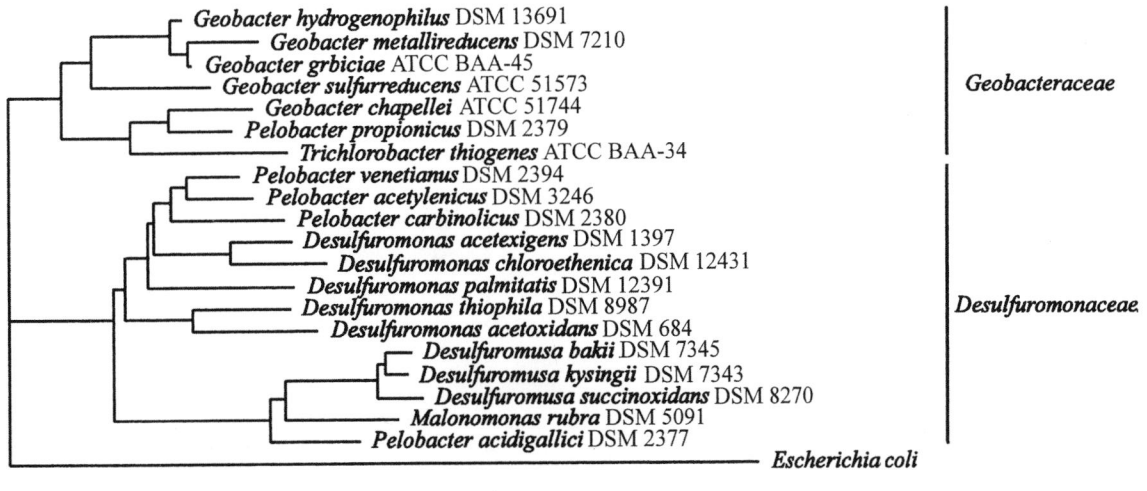

FIGURE BXII.δ.40. Phylogenetic relationships based on 16S rRNA gene sequence analyses of the taxa within the order *Desulfuromonales* for which 16S rRNA gene sequences are available. The scale bar represents 10 inferred nucleotide substitutions per 100 nucleotides. The dendrogram was reconstructed from distance matrices using the neighbor-joining method.

Most members of the *Desulfuromonaceae* and *Geobacteraceae* can use **ferric iron and elemental sulfur** (or polysulfide) as **electron acceptors** that are **reduced to ferrous iron and sulfide**; in addition, Mn(IV) and oxidized humic substances may be used by some species. Sulfate, thiosulfate, and sulfite do not serve as electron acceptors. Simple organic compounds such as acetate and other carboxylic acids serve as electron donors and carbon sources. Organic substrates are completely oxidized to CO_2; the mechanism of terminal oxidation is the citric acid cycle.

Members of the *Pelobacteraceae* show an exclusively fermentative metabolism; the main common fermentation product is acetate.

Type genus: **Desulfuromonas** Pfennig and Biebl 1977, 306.

Family I. **Desulfuromonaceae** *fam. nov.*

JAN KUEVER, FRED A. RAINEY AND FRIEDRICH WIDDEL

De.sul.fu.ro.mo.na.ce' ae. M.L. fem. n. *Desulfuromonas* type genus of the family; suffix *-ceae* denotes family; M.L. fem. pl. n. *Desulfuromonaceae* the family that comprises the genus *Desulfuromonas* and other genera.

The family comprises rod-shaped and curved cells. **Strictly anaerobic chemolithoheterotrophs or chemoorganotrophs with a respiratory or fermentative type of metabolism. Sulfur or ferric iron can act as electron acceptors for some members and are reduced to sulfide and ferrous iron**, respectively. Sulfate, sulfite, and thiosulfate are not used. Simple organic compounds such as long-chain fatty acids and alcohols serve as electron donors and carbon sources. Organic substrates are completely oxidized to CO_2. All members are mesophilic; thermophiles have not been isolated. Cells of most members contain various cytochromes and menaquinones. Mesophilic species have been isolated from various aquatic environments, including marine, brackish water, and freshwater habitats.

The family includes the genera *Desulfuromonas* and *Desulfuromusa*, *Malonomonas*, and *Pelobacter*. Differential characteristics for *Desulfuromonas* and *Desulfuromusa* are listed in Table BXII.δ.32.

Type genus: **Desulfuromonas** Pfennig and Biebl 1977, 306.

TABLE BXII.δ.32. Differential characteristics of the genera *Desulfuromonas* and *Desulfuromusa*[a]

Genus	*Desulfuromonas*	*Desulfuromusa*
Cell morphology	Rod-shaped	Rod-shaped
Motility	+	±
Mol% G + C content	54–62	46–47
Major menaquinone	MK-8	nr
Substrate oxidation	C	C
Typical electron donors and carbon sources:		
Acetate	+	+
Alanine	−[b]	+
Aspartate	−[b]	+
Fumarate	+[c]	+
Glutamate	−	+
Lactate	+[c]	+
Malate	+[c]	+
Maleate	−[b]	+
Oxalacetate	−	+
Propionate	−	+
Pyruvate	+[c]	+
Succinate	+[c]	+
Yeast extract	−	+
H_2 + acetate	+[c]	−
Long-chain fatty acids	+[c]	−
Chemolithoautotrophic growth	−	−
Fermentative growth	Some species	All species
Typical electron acceptors:		
Fumarate	+	+
Malate	+[c]	+
Sulfur	+	+
Tetrachloroethylene	+[c]	nr
Fe(III)	+[c]	+[c]
Mn(IV)	+[c]	nr
Humic substances	+[c]	nr
Optimal growth temperature (°C)	21–40	25–32
NaCl requirement	±	+
Habitat (anoxic):		
Brackish water	+	+
Freshwater	+	−
Marine	+	+

[a]Symbols: nr, not reported; +, present; −, not observed; ±, variable; C, complete.

[b]Not tested for all species.

[c]Some species.

Genus I. **Desulfuromonas** Pfennig and Biebl 1977, 306[AL]

JAN KUEVER, FRED A. RAINEY AND FRIEDRICH WIDDEL

De.sul.fu.ro.mo'nas. L. pref. *de* from; L. n. *sulfur* sulfur; Gr. n. *monas* a unit, monad; M.L. fem. n. *Desulfuromonas* a monad that reduces sulfur.

Straight or slightly curved rods and elongated ovoid rods, 0.4–0.9 × 1.0–4.0 µm in length. Special resting forms such as spores are not known to occur. Gram negative. **Motile, generally by means of a single flagellum located at a lateral or subpolar position**; cells exhibit a characteristic propeller-like movement. Some strains have polar flagella. Strictly anaerobic. **Possess mainly a respiratory type of metabolism with elemental sulfur serving as the terminal electron acceptor, and being reduced to H₂S** (dissimilatory sulfur reduction). L-Malate or fumarate may be fermented, giving succinate as the main product in the presence or absence of acetate. Betaine may be fermented to acetate, CO_2, and trimethylamine. Optimal growth temperature, 30°C. Colonies on anaerobic agar media are translucent to opaque and either **peach-colored or pink.** Chemoorganotrophic, using acetate and other simple organic compounds as carbon sources (usually together with CO_2) and electron donors; **acetate and other substrates are completely oxidized to CO_2** via the citric acid cycle as indicated by 2-oxoglutarate dehydrogenase activity. Occur regularly in anoxic sediments of salt lakes and in marine or brackish habitats; have also been enriched from freshwater sediments.

The mol% G + C of the DNA is: 54–62.

Type species: **Desulfuromonas acetoxidans** Pfennig and Biebl 1977, 306.

FURTHER DESCRIPTIVE INFORMATION

Strains of *Desulfuromonas* isolated from anoxic sediments of saline lakes or marine habitats are generally straight or slightly curved slender rods, which are highly motile by a single lateral to subpolar flagellum. Most strains from anoxic freshwater sediments are elongated ovoid rods, which are motile by means of a subpolar or polar flagellum. Cells of freshwater strains have a tendency to form nonmotile clumps, which stick to the bottom of the culture vessel.

The ability to use the following electron donors and carbon sources also varies among *Desulfuromonas* strains: ethanol, 1-propanol, pyruvate, lactate, propionate, higher fatty acids, and glutamate. The oxidation of acetyl-CoA proceeds via the TCA cycle (Gebhardt et al., 1985). The following substrates generally cannot be used as electron donors and carbon sources: polysaccharides, sugars, sugar alcohols, most amino acids, and H_2/CO_2.

Besides elemental sulfur, poly- or disulfide bond-containing compounds such as polysulfide, cystine, or oxidized glutathione can serve as electron acceptors. Sulfate, sulfite, thiosulfate, nitrate, or oxygen cannot act as electron acceptors.

Strains differ in their ability to ferment malate or fumarate. Many strains grow well on a 4:1 mixture (molar ratio) of fumarate (or malate) and acetate, which is converted to succinate (Gottschalk and Andreesen, 1979). With the strains isolated in the initial study (Pfennig and Biebl, 1976), no growth was observed within three weeks on malate or fumarate when acetate was omitted; however, later studies showed that malate or fumarate can be fermented in bicarbonate-free medium supplemented with yeast extract in the absence of additional electron donors (Pfennig and Widdel, 1981; Widdel and Pfennig, 1992). The type strain can ferment betaine to acetate, CO_2, and trimethylamine (Heijthuijsen and Hansen, 1989).

Most marine and salt lake strains require 20 g/l NaCl and 3 g/l MgCl₂·6H₂O for growth; some strains can grow in both seawater and freshwater media. Biotin may be required as a growth factor.

The pH range for growth is 6.5–8.5, with the optimal being 7.2–7.5. The temperature optimal for growth is around 30°C. Some species have an optimal at 40°C.

Colonies of *Desulfuromonas* on agar media, as well as accumulations of cells in sediments or centrifuged pellets, are pink- to peach-colored, due to *c*-type cytochromes with absorption maxima at 419, 523, and 553 nm. In redox difference spectra, the β and α peaks are at 522 and 551.5 nm, respectively. A major component of the cytochromes of *D. acetoxidans* strain 5071 is the low potential, tri-heme cytochrome $c_{551.5}$ (c_7) (Probst et al., 1977). All members of the genus *Desulfuromonas* investigated so far contain MK-8 as the major menaquinone (Collins and Widdel, 1986). Major cellular fatty acids of the *Desulfomonas* strains are $C_{16:0}$ and $C_{16:1\ \omega7c}$ (Kohring et al., 1994).

A characteristic of all *Desulfuromonas* strains is that they grow well in syntrophic cultures containing phototrophic green sulfur bacteria and acetate (Biebl and Pfennig, 1978). Marine or salt lake isolates show particularly efficient growth in combined cultures containing the green bacterium *Prosthecochloris aestuarii* at moderate light intensities (e.g., 25W tungsten lamp, 50 cm distance). This is readily attributed to the fact that the illuminated *Prosthecochloris* oxidizes H₂S almost exclusively to extracellular elemental sulfur, which in turn is immediately reduced back to H₂S by *Desulfuromonas* as long as acetate is present.

Dissimilatory sulfur-reducing bacteria resembling *Desulfuromonas* occur regularly in the anoxic sulfide-containing sediments of marine and brackish water habitats and salt lakes. They have also been isolated, although much less successfully, from sulfide-containing, anoxic, freshwater sediments.

All strains of *Desulfuromonas* can be cultured in a defined anaerobic medium[1] containing sulfide as a reductant. The electron donor and carbon source (usually acetate) is added aseptically from a sterile stock solution (final concentration around 0.5 g/l). The electron acceptor, elemental sulfur, is added aseptically from a sterile slurry of sulfur flower in distilled water;[2] a pea-size amount is used for 50 ml of medium. Fermentative growth in the absence of sulfur is most commonly achieved with a fumarate–acetate mixture (2 g/l and 0.5 g/l, respectively), but several strains may not grow fermentatively.

ENRICHMENT AND ISOLATION PROCEDURES

For the selective enrichment of *Desulfuromonas* strains, mineral medium supplemented with 0.5 g anhydrous sodium acetate and

---

1. Defined basal medium has the following composition (per liter of distilled water): KH₂PO₄, 0.5 g; NH₄Cl, 0.3 g; CaCl₂·2H₂O, 0.1 g; NaCl, 0.5 g (for salt water strains, 20.0 g); MgCl₂·6H₂O, 0.4 g (for salt water strains, 3.0 g); trace element solution (see chapter on *Desulfobacter*), 1.0 ml. After the medium has been autoclaved and cooled under anoxic conditions, the following components are added aseptically from sterile stock solutions: 20 ml of a solution of 84 g/l NaHCO₃ (saturated with CO_2 and autoclaved under a CO_2 atmosphere), 2–3 ml of a solution of 120 g/l Na₂S·9H₂O (autoclaved under an N₂ atmosphere), and 1 ml of a vitamin solution containing 4 mg of *p*-aminobenzoic acid and 1 mg of biotin per 100 ml. The pH of the medium is adjusted to 7.1–7.3 with sterile 1 M HCl or 0.5 M Na₂CO₃ solution (autoclaved in closed bottles with gas head phase). The medium is then distributed aseptically into sterile screw-capped bottles or test tubes and sealed against air.

2. Highly purified sulfur flower is ground thoroughly in a mortar together with distilled water. The slurry is autoclaved for 30 min at 112°C in a screw-capped bottle. The excess water is then decanted.

a spatula of ground, wetted sulfur flower is used. Bottles or tubes (100-, 50-, or 20-ml capacity) can serve as culture vessels and are kept anoxic with tight screw caps, or by using a headspace of N_2/CO_2 (90:10) and sealing with butyl rubber stoppers. The medium is inoculated with black anoxic mud from freshwater, brackish water, or marine habitats; 3–5% of the total culture volume should be added. *Desulfuromonas* species may be also enriched and isolated via direct serial dilution of anoxic environmental samples in liquid medium (according to most-probable-number method) or agar medium (see below) with acetate as organic substrate. Depending on the salinity of the anoxic mud sample that is used as the inoculum, salt- or freshwater medium is used. Sulfur flower may be kept in suspension by providing each bottle with a glass bead inside and incubation on a rotary shaker. Sulfur-reducing bacteria may become enriched under such conditions after 1 or 2 weeks of incubation at 28°C. Successful enrichment cultures are recognized by their strong odor of H_2S and by their lowered pH (6.5–6.8); microscopically, slender motile rods with a propeller-like movement are seen. Second and third transfers usually grow within a few days. Although the tolerance of *Desulfuromonas* to H_2S (some 10 mM) is considerable, growth is usually limited by the inhibitory effect of the H_2S.

The growth inhibition due to H_2S can be eliminated if the enrichment cultures for *Desulfuromonas* (in media with added sulfide as reductant) are inoculated with a phototrophic green sulfur bacterium and incubated in dim light from a tungsten lamp. In this case, no sulfur flower is added to the medium because the green sulfur bacterium continuously forms elemental sulfur and at the same time consumes the H_2S formed by the sulfur reducers being enriched. Repeated transfers may yield fast-growing syntrophic cultures, which can be used directly for isolation of a pure culture.

Pure cultures are obtained by repeated application of the agar dilution method. Repeatedly washed agar is used to prepare a sterile agar solution (33 g/l) in water, which is then dispensed in 3-ml amounts into test tubes. The molten agar is then combined with 6-ml amounts of prewarmed (42°C) complete culture medium, with 6–8 tubes being prepared for each dilution series. For use with agar, the complete culture medium is prepared with polysulfide solution instead of sulfur flower, in order to obtain a fine and homogeneous distribution of sulfur in the agar. A volume of 0.5 ml from an anoxically (under N_2) autoclaved polysulfide solution (10 g of $Na_2S \cdot 9H_2O$ and 3 g of sulfur flower dissolved in 45 ml of distilled water) are added per 50 ml of medium. The inoculated tubes are placed in a cold water bath and immediately sealed against air by overlaying an N_2/CO_2 mixture (9:1) and sealing with butyl rubber stoppers. The sealed tubes are incubated at 28–30°C. Clearing of the opaque yellowish sulfur and polysulfide and appearance of pink to peach-colored colonies indicate growth of *Desulfuromonas*.

In addition to microscopic examination, newly isolated strains are checked for purity by inoculating them into AC medium (Difco) and ethanol or lactate medium for sulfate-reducing bacteria. Absence of turbidity in both media supports absence of contamination.

MAINTENANCE PROCEDURES

Stock cultures are transferred every 3 months. Towards the end of growth or at the beginning of the stationary phase, cultures are refrigerated and stored at 2–5°C in the dark. Strains capable of using malate or fumarate are maintained on these substrates; strains lacking this ability are maintained with elemental sulfur as the electron acceptor. *Desulfuromonas* strains may be preserved indefinitely by suspension of cells in anoxic medium containing 5% dimethylsulfoxide and storage in liquid nitrogen.

DIFFERENTIATION OF THE GENUS *DESULFUROMONAS* FROM OTHER GENERA

The genus *Desulfuromonas* is distinguished from other anaerobic, dissimilatory sulfur-reducing bacteria primarily by physiological characteristics. The special feature of the genus is the capacity to grow on acetate as the sole organic substrate by an anaerobic respiration in which the oxidation of acetate to CO_2 is stoichiometrically linked to the reduction of elemental sulfur to sulfide. Sulfate, sulfite, or thiosulfate cannot be utilized as electron acceptors. Lacking the ability to reduce oxidized sulfur compounds distinguishes *Desulfuromonas* from acetate-oxidizing sulfate-reducing bacteria. The ability to reduce elemental sulfur to H_2S in a dissimilatory metabolism also occurs in *Wolinella succinogenes*, "*Sulfospirillum*" species (formerly designated as *Campylobacter* strains; Laanbroek et al., 1978), and some sulfate-reducing bacteria of the genus *Desulfomicrobium* (Biebl and Pfennig, 1977; Schumacher et al., 1992; see chapter for *Desulfomicrobium*); however, these organisms are unable to oxidize acetate to CO_2 and require H_2 or formate (or lactate in the case of *Desulfovibrio*) as electron donors for reduction of sulfur. The use of other electron acceptors (ferric iron, dimethyl sulfoxide, etc.) allows no clear differentiation from other genera. In contrast to the closely related *Desulfuromusa* spp., the number of substrates used for dissimilatory sulfur reduction by members of the genus *Desulfuromonas* is limited (see Table BXII.δ.33). The only exception is *D. palmitatis*, which can also use long-chain fatty acids as substrates. A clear differentiation between these genera and other closely related genera (*Pelobacter* and *Geobacter*) is possible by determination of the 16S rRNA gene sequence. Differentiation of *Geobacter*, *Desulfuromusa*, and *Desulfuromonas* from *Pelobacter* is possible due to the presence of cytochromes, which are missing in *Pelobacter* strains.

In contrast to other rod-shaped anaerobic bacteria exhibiting dissimilatory reduction of sulfur compounds, most *Desulfuromonas* strains are motile by means of a single flagellum located at a lateral or subpolar position.

TAXONOMIC COMMENTS

The closest relatives of the genus *Desulfuromonas* are *Geobacter*, *Pelobacter*, and *Desulfuromusa*.

FURTHER READING

Pfennig, N. 1978. *Rhodocyclus purpureus* gen. nov. and sp. nov. a ring-shaped, vitamin B_{12}-requiring member of the family *Rhodospirillaceae*. Int. J. Syst. Bacteriol. *28*: 283–288.

List of species of the genus Desulfuromonas

1. **Desulfuromonas acetoxidans** Pfennig and Biebl 1977, 306[AL]
   a.cet.o′xi.dans. L. n. *acetum* vinegar; M.L. n. *acidum aceticum* acetic acid; M.L. v. *oxido* make acid, oxidize; M.L. part. adj. *acetoxidans* oxidizing acetate.

Characteristics are summarized in Table BXII.δ.33. Straight and slightly curved slender rods, 0.4–0.8 × 1–4 µm. Motile by a single lateral to subpolar flagellum. Acetate, ethanol, 1-propanol, and other simple organic compounds

TABLE BXII.δ.33. Morphological and physiological characteristics of described species of the genus *Desulfuromonas*[a]

Species	*Desulfuromonas acetoxidans*[b]	*Desulfuromonas acetexigens*[c]	*Desulfuromonas chloroethenica*[d]	*Desulfuromonas palmitatis*[e]	*Desulfuromonas thiophila*[f]
Cell morphology	Rod-shaped	Rod-shaped	Rod-shaped	Rod-shaped	Oval- or rod-shaped
Cell size (μm)	0.4–0.8 × 1–4	0.9–1.2 × 1.3–3.5	0.6 × 1.0–1.7	0.3 × 1–2	0.7–0.8 × 1.5–2.0
Motility	+ (sl)	+ (sp)	+ (sp)	−	+ (sl)
Mol% G + C content	53.6	62.3	nr	54.7	61.6
Major menaquinone	MK-8	MK-8	nr	nr	nr
Optimal temperature (°C)	30	30	21–31	40	26–30
Oxidation of substrate	C	C	C	C	C
Compounds used as electron donors and carbon sources:					
H_2 + acetate	−	−	nr	+	−
Formate	−	−	nr	−	−
Acetate	+	+	+	+	+
Fatty acid: C atoms	−	−	−	14, 16, 18	−
Ethanol	+	−	−	−	−
Lactate	−	−	nr	+	−
Pyruvate	+	nr	+	nr	+
Fumarate	+	−	nr	+	+
Succinate	−	−	nr	+	+
Malate	+	−	−	nr	−
Benzoate	nr	−	nr	−	−
Others	1-Propanol	nr	nr	nr	nr
Fermentation of	Fumarate, malate, betaine	nr	nr	nr	−
Compounds used as electron acceptor:					
Sulfate	−	−	−	−	−
Sulfite	−	−	−	−	−
Sulfur	+	+	+	+	+
Thiosulfate	−	−	−	−	−
Fumarate	+	+	+	+	−
Nitrate	−	−	nr	−	−
Fe(III)	+	nr	+	+	+
Other electron acceptors	Malate	nr	Tetrachloroethylene, trichloroethylene	MnIV	nr
Growth factor requirement	bi	bi	nr	nr	bi
NaCl requirement (g/l)	20	−	−	20	−

[a]Symbols: sl, single lateral flagellum; sp, single polar flagellum; nr, not reported; C, complete; +, good growth; (+), poor growth; −, no growth; bi, biotin.

[b]Data from Pfennig and Biebl (1976) and Roden and Lovley (1993).

[c]Data from Finster et al. (1994).

[d]Data from Krumholz (1997).

[e]Date from Coates et al. (1995) and Finster et al. (1997b).

[f]Data from Finster et al. (1997b).

are oxidized to CO_2 and elemental sulfur is reduced to H_2S. L-Malate or fumarate may be used as electron acceptors instead of sulfur, and are reduced to succinate; malate, fumarate, and betaine may also be fermented. Colonies or densely packed cells are peach-colored or pink due to the presence of *c*-type cytochromes. Occur regularly in anoxic sediments from marine or brackish water habitats or salt lakes.

The mol% G + C of the DNA is: 51 (T_m); 53.6 (HPLC).
Type strain: 11070, DSM 684.
GenBank accession number (16S rRNA): M26634.

2. **Desulfuromonas acetexigens** Finster, Bak and Pfennig 1997a, 601[VP] (Effective publication: Finster, Bak and Pfennig 1994, 331.)

a.cet.ex' i.gens. L. n. *acetum* vinegar; M.L. n. *acidum aceticum* acetic acid; L. v. *exigo* to demand; M.L. part. adj. *acetexigens* demanding acetate for growth.

Characteristics are summarized in Table BXII.δ.33. Occur in freshwater habitats. The type strain was isolated from digester sludge of the treatment plant, Göttingen, Germany.

The mol% G + C of the DNA is: 62 (T_m); 62.3 (HPLC).
Type strain: 2873, DSM1397.
GenBank accession number (16S rRNA): U23140.

3. **Desulfuromonas chloroethenica** Krumholz 1997, 1263[VP]

chlo.ro.e.the' ni.ca. M.L. adj. *chloroethenicus, -a, -um*, pertaining to chloroethene, as the organism utilizes such compounds as electron acceptors.

Characteristics are summarized in Table BXII.δ.33. Occur in freshwater sediment contaminated with chlorinated ethylenes.

The mol% G + C of the DNA is: not known.
Type strain: TT4B, ATCC 700295.
GenBank accession number (16S rRNA): U49748.

4. **Desulfuromonas palmitatis** Coates, Lonergan, Philips, Jenter and Lovley 2000, 1699[VP] (Effective publication: Coates, Lonergan, Philips, Jenter and Lovley 1995, 411.)

pal.mi.ta' tis. N.L. n. *palmitas* palmitate (chemical); N.L. gen. n. *palmitatis* of palmitate, because it oxidizes palmitate.

Characteristics are summarized in Table BXII.δ.33. The type strain was enriched from marine sediment of Palleta Creek in San Diego Bay, San Diego, California, USA.

The mol% G + C of the DNA is: 54.7 (HPLC).
Type strain: SDBY1, ATCC 51701.
GenBank accession number (16S rRNA): U28172.

5. **Desulfuromonas thiophila** Finster, Coates, Liesack and Pfennig 1997b, 757[VP]

thi.o.phi′ la. Gr. n. *thios* sulfur; Gr. adj. *philos* loving; M.L. fem. adj. *thiophila* sulfur-loving.

Characteristics are summarized in Table BXII.δ.33. The type strain was enriched from anoxic freshwater mud from Ngawha Sulfur springs near Moerewa, New Zealand.

The mol% G + C of the DNA is: 62 (T_m); 62.3 (HPLC).
Type strain: NZ27, DSM 8987.
GenBank accession number (16S rRNA): Y11560.

Genus II. **Desulfuromusa** *Liesack and Finster 1994, 756^VP*

WERNER LIESACK AND KAI FINSTER

De.sul.fu.ro.mu′ sa. L. pref. *de* from; L. n. *sulfur* sulfur; M.L. n. *musa* banana; M.L. fem. n. *Desulfuromusa* a banana-shaped bacterium that reduces sulfur.

Slightly curved or rod-shaped cells 0.4–0.8 × 1–6 μm. Motile by means of a single, subpolarly inserted flagellum. Gram negative. Spore formation not observed. **Strictly anaerobic**, having mainly a respiratory type of metabolism; however, fermentative growth may occur with fumarate or malate as the sole substrate. **Chemoorganotrophic**; all species utilize acetate, propionate, succinate, lactate, pyruvate, oxaloacetate, malate, maleate, glutamate, alanine, and aspartate as electron donors and as carbon sources; these **substrates are completely oxidized to CO_2**. Alcohols and sugars are not utilized. **S^0 is used as terminal electron acceptor and is reduced to H_2S** (dissimilatory sulfur reduction). Sulfate, sulfite, thiosulfate, tetrathionate, trimethylammonium oxide, cystine, oxidized glutathione, or N_2O are not utilized as electron acceptors. Strains can be cultivated in a defined medium without addition of complex nutrients such as yeast extract or peptone. Occur regularly in **anoxic marine or estuarine mud**.

The mol% G + C of the DNA is: 47.

Type species: **Desulfuromusa kysingii** Liesack and Finster 1994, 757.

FURTHER DESCRIPTIVE INFORMATION

When grown in anaerobic agar media with S^0 as electron acceptor, the cells of all strains tested form whitish filamentous colonies. Microscopic examination reveals the presence of highly light-dispersive globular bodies, probably droplets of S^0, in the center of these colonies. When grown with fumarate as the sole substrate, the cells form peach-colored colonies which are lens-shaped and have entire edges. Grown with fumarate as the sole substrate, the generation times are generally shorter than with S^0.

When growing on S^0 as terminal electron acceptor, direct contact between cells and sulfur particles is not necessary. However, when cells are not in contact with the sulfur, H_2S is required to initiate growth, most likely due to the formation of polysulfide. The addition of a reducing agent such as H_2S or dithionite to the medium is not necessary when the culture is inoculated in oxygen-free medium. However, the lag phase for sulfur reduction is reduced from 2 d without H_2S to 12 h in its presence.

Cultures of *D. kysingii* are relatively tolerant towards oxygen. Culturable cells were present after three weeks of continuous exposure to air. Cultures of *Desulfuromusa bakii* and *Desulfuromusa succinoxidans* do not survive this treatment.

Strains of *D. kysingii* and *D. succinoxidans* can grow by disproportionation of aspartate and maleate. Growth with one of these two substrates without an additional electron acceptor is considerably slower than growth in the presence of S^0.

ENRICHMENT AND ISOLATION PROCEDURES

When samples of anoxic marine or estuarine muds were inoculated into media containing succinate, propionate, or lactate as the electron donor, a significant production of H_2S by sulfur reduction was observed after two weeks of anaerobic incubation. With valerate, sulfide production was noticeable after one month. Regardless of the electron donor used or the origin of the inoculum, after three transfers most enrichments were dominated by motile, rod-shaped, slightly curved cells (Fig. BXII.δ.41). Significantly larger cells developed only in an enrichment culture with lactate as electron donor. These cells were isolated and designated as strain Gylac, the type strain of the species *Desulfuromusa succinoxidans* (Table BXII.δ.34). In general, only a weak increase in medium turbidity was observed parallel to H_2S production. Most of the cells were attached to the sulfur precipitate. During growth, the pH of the medium decreased from 7.4 to 6.2–6.4. Pure cultures were obtained by repeated application of the deep-agar dilution method.

MAINTENANCE PROCEDURES

Pure cultures are maintained in liquid culture medium in screw-capped bottles or in anaerobically gassed bottles sealed with butyl rubber stoppers, with storage at 2–5°C. Stock cultures are transferred every three months.

DIFFERENTIATION OF THE GENUS *DESULFUROMUSA* FROM OTHER GENERA

Desulfuromusa species are obligately anaerobic, mesophilic sulfur reducers capable of complete oxidation of their substrates. Apart from one validly described species of the genus *Geobacter*, the dissimilatory Fe(III)-reducing *G. sulfurreducens* (Caccavo et al., 1994), these characteristics are shared only by members of the

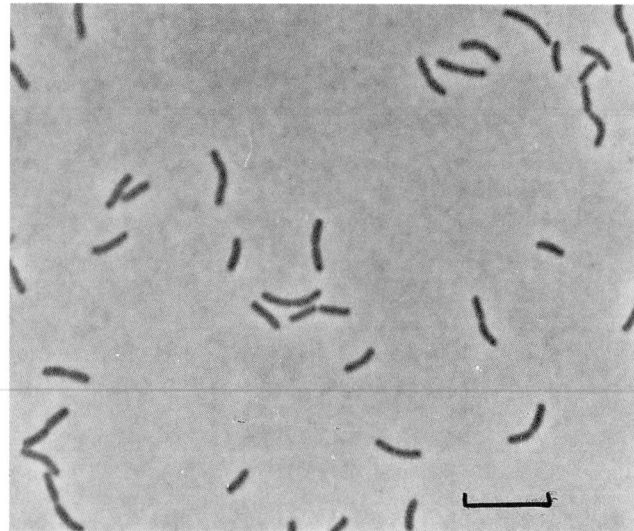

FIGURE BXII.δ.41. Phase-contrast photomicrograph showing cells of *Desulfuromusa kysingii* strain Kysw2 isolated from anoxic mud of the Kysing Fjord (Denmark). Bar = 10 μm.

TABLE BXII.δ.34. Differential characteristics of the species of the genus *Desulfuromusa*[a,b]

Characteristics	*D. kysingii*	*D. bakii*	*D. succinoxidans*
Cell dimensions, μm[c]	0.5–0.6 × 2–5	0.5–0.6 × 2–4	0.8–1.0 × 4–6
Electron donors used for sulfur reduction:			
Valerate	(+)	−	−
Citrate	+	+	−
Electron acceptors:[d]			
Nitrate	+	−	−
Dimethyl sulfoxide	+	−	−
Fe(III)-citrate	+	−	−
Disproportionation:			
Aspartate, maleate	+	−	+
Requirement for *p*-aminobenzoate	+	nd	nd
Diazotrophic growth	+	nd	nd
Oxygen tolerance[e]	+	−	−
Salt range[f]	Brackish	Marine	Brackish
pH optimal (range)	6.5–7.9 (5.8–.25)	6.7–7.4 (6.2–7.8)	6.5–7.9 (5.8– 8.25)
Optimal temperature (range), °C	~32 (24–35)	25–30 (8–32)	~32 (4–35)
Mol% G + C of DNA	47	46	46

[a]Symbols: +, utilized; (+), utilized by some strains; −, not utilized; nd, not determined.

[b]Data taken from Finster and Bak (1993).

[c]Cells grown on S^0 as electron acceptor.

[d]The alternative electron acceptors nitrate, dimethyl sulfoxide, and Fe(III)-citrate are reduced to ammonia, dimethyl sulfide, and dissolved Fe(II), respectively.

[e]Three weeks of continuous exposure to oxygen (2%) is tolerated, but no growth occurs in the presence of this compound.

[f]Brackish: Growth occurs equally well at concentrations between 114 and 340 mM for NaCl and between 3.3 and 14.7 mM for MgCl$_2$; no growth in the absence of NaCl or at a MgCl$_2$ concentration of 0.5 mM. Marine: Growth occurs equally well at concentrations between 230 and 340 mM for NaCl and between 9.8 and 14.7 mM for MgCl$_2$; no growth at a NaCl concentration of 114 mM and a MgCl$_2$ concentration of 3.3 mM.

genus *Desulfuromonas* (Widdel and Pfennig, 1992). Using S^0 as the sole electron acceptor, *G. sulfurreducens* is capable of growth with acetate but not with propionate, succinate, lactate, and several other substrates. In contrast, *Desulfuromusa* spp. are able to utilize these compounds as electron donors for growth (see generic description). Members of the genera *Desulfuromonas* and *Desulfuromusa* are phylogenetically related. (See Taxonomic Comments). The phylogenetically distinct *Desulfurella* species also grow by dissimilatory reduction of sulfur with complete oxidation of their substrates, but these microorganisms are moderate thermophiles with a temperature optimal around 55°C (Bonch-Osmolovskaya et al., 1990).

The *Desulfuromusa* species can be differentiated from the described members of the genus *Desulfuromonas* by their ability to utilize the amino acids alanine, aspartate, and glutamate as growth substrates. Apart from a very few exceptions listed in Table BXII.δ.34, the three described species of the genus *Desulfuromusa* share an identical set of electron donors which can be utilized for growth. With the exception of acetate, utilization patterns of electron donors are rather diverse between individual species of the genus *Desulfuromonas*. This heterogeneity might be a reflection of their phylogenetic diversity. The genus *Desulfuromonas* encompasses described species with up to 8% overall 16S rRNA gene sequence dissimilarity, while the three *Desulfuromusa* species form a phylogenetically tight cluster moderately related to, but clearly distinct from, members of the genus *Desulfuromonas* (see Taxonomic comments). The *Desulfuromonas* species tested so far, *D. acetoxidans* (Pfennig and Biebl, 1976), *D. acetexigens* (Finster et al., 1994), and *D. thiophila* (Finster et al., 1997b), require biotin as a growth factor, while members of the genus *Desulfuromusa* do not.

The mol% G + C values of the DNA are in a narrow range between 46 and 47 for members of the genus *Desulfuromusa*, while these values vary between 53.6 (*D. acetoxidans*) and 62.3 (*D. acetexigens*) for *Desulfuromonas* species.

TAXONOMIC COMMENTS

The three described species of the genus *Desulfuromusa*, i.e., *D. kysingii*, *D. bakii*, and *D. succinoxidans*, share overall 16S rDNA sequence similarities above 98%. However, the levels of DNA relatedness in DNA–DNA hybridization between strains of the three described *Desulfuromusa* species were less than 30%. This was well below the DNA reassociation level of 70% generally accepted as a threshold for assigning strains as identical or different genomic species (Wayne et al., 1987). The genus *Desulfuromusa* belongs to a sublineage within the *Deltaproteobacteria*. This sublineage contains, in addition to *Desulfuromusa*, members of the genera *Geobacter*, *Desulfuromonas*, and *Pelobacter*. This group of microorganisms has recently been unified in the family *Geobacteraceae* by Lonergan et al., (1996), which is phylogenetically divided into two subgroups, the *Geobacter* and *Desulfuromonas* clusters. The three species of the genus *Desulfuromusa*, together with *Pelobacter acidigallici*, form a distinct line of descent within the *Desulfuromonas* cluster. More recently, the microaerotolerant, fermenting *Malonomonas rubra* has been shown to be closely related to *Pelobacter acidigallici* and the *Desulfuromusa* species (Kolb et al., 1998).

The analysis presented in Fig.BXII.δ.40 of the description of the order *Desulfuromonales* suggests that *Desulfuromonas*, *Desulfuromusa*, *Malonomonas*, and *Pelobacter acidigallici* are related, while *Geobacter*, *Pelobacter propionicus*, and *Trichlorobacter thiogenes* are related. The family *Desulfuromonaceae* would then include the organisms in the first group and the family *Geobacteraceae*, the organisms in the second group. The order would become *Desulfuromonales* because *Desulfuromonas* was described first and would have precedence over *Geobacter*.

The ability to carry out anaerobic respiration with Fe(III) and/or S^0 as terminal electron acceptors (dissimilatory iron and/or sulfur reduction) has been proposed as a common physiological characteristic of members of the family *Geobacteraceae* (Lonergan et al., 1996). Indeed, the ability to grow with elemental sulfur has been demonstrated for *Geobacter sulfurreducens* (Caccavo et al., 1994), originally isolated and described as a Fe(III)-reducing microorganism, and *vice versa*, the ability to grow with Fe(III) as an electron acceptor has been shown for *Desulfuromonas acetoxidans* (Roden and Lovley, 1993) and *Desulfuromonas palmitatis* (Coates et al., 1995). Similarly, *D. kysingii* has been shown to utilize Fe(III) (added as Fe(III)-citrate) as an electron acceptor, while Fe(III)-citrate was not utilized by *D. bakii* and *D. succinoxidans* (Finster and Bak, 1993; Liesack and Finster, 1994; Lonergan et al., 1996). However, all three *Desulfuromusa* species were able to grow in medium with acetate as the electron donor and Fe(III)-NTA as the electron acceptor, with significant Fe(III) reduction over time (Lonergan et al., 1996). Finally, growth by dissimilatory reduction of Fe(III) and S^0 has also been demonstrated for *Pelobacter carbinolicus* (Lovley et al., 1995b), originally isolated and described as a strictly anaerobic, fermentative bacterium (Schink, 1984a, 1992). This mix-up of physiological traits originally considered as special characteristics of members of different genera is also reflected in the topology of the phylogenetic tree. The genus *Pelobacter* is phylogenetically intertwined with members of the genera *Geobacter* and *Desulfuromonas* (Evers et al., 1993; Lonergan et al., 1996). However, growth by dissimilatory reduction of Fe(III) or S^0 could not be demonstrated for *Malonomonas rubra* (Kolb et al., 1998). These data as a whole may indicate the need for a thorough study of the physiological features of members of the family *Geobacteraceae*, followed by a revision of the currently existing taxonomy within this family.

FURTHER READING

Finster, K. and F. Bak. 1993. Complete oxidation of propionate, valerate, succinate, and other organic compounds by newly isolated types of marine, anaerobic, mesophilic, Gram-negative, sulfur-reducing eubacteria. Appl. Environ. Microbiol. *59*: 1452–1640.

Liesack, W. and K. Finster. 1994. Phylogenetic analysis of five strains of Gram-negative, obligately anaerobic, sulfur-reducing bacteria and description of *Desulfuromusa* gen. nov., including *Desulfuromusa kysingii* sp. nov., *Desulfuromusa bakii* sp. nov. and *Desulfuromusa succinoxidans* sp. nov. Int. J. Syst. Bacteriol. *44*: 753–758.

DIFFERENTIATION OF THE SPECIES OF THE GENUS *DESULFUROMUSA*

Differential characteristics of the species *D. kysingii*, *D. bakii*, and *D. succinoxidans* are listed in Table BXII.δ.34.

List of species of the genus Desulfuromusa

1. **Desulfuromusa kysingii** Liesack and Finster 1994, 757[VP]

   ky.sin' gi.i. M.L. gen. n. *kysingii* of kysing; named after the Fjord south of Århus (Jutland, Denmark) from which the organism was isolated.

   Characteristics are as described for the genus and as shown in Table BXII.δ.34.

   Isolated from anoxic mud of the Kysing Fjord south of Århus, Denmark.

   The mol% G + C of the DNA is: 47 (HPLC).

   Type strain: Kysw2, DSM 7343.

   GenBank accession number (16S rRNA): X79414.

2. **Desulfuromusa bakii** Liesack and Finster 1994, 757[VP]

   ba' ki.i. M.L. gen. n. *bakii* of Bak, named in memory of F. Bak, a German microbiologist who has made important contributions to our knowledge of sulfur- and sulfate-reducing bacteria.

   Characteristics are as described for the genus and as shown in Table BXII.δ.34.

   Isolated from an anoxic mud sample collected from the Guayamas basin (Gulf of California) at a depth of about 2000 m.

   The mol% G + C of the DNA is: 46 (HPLC).

   Type strain: Gyprop, DSM 7345.

   GenBank accession number (16S rRNA): X79412.

3. **Desulfuromusa succinoxidans** Liesack and Finster 1994, 757[VP]

   suc.cin.ox' i.dans. L. n. *succinum* amber; M.L. n. *acidum succinicum* succinic acid (derived from amber); M.L. v. *oxido* make acid, oxidize; M.L. part. adj. *succinoxidans* oxidizing succinate.

   Characteristics are as described for the genus and as shown in Table BXII.δ.34.

   Isolated from an anoxic mud sample collected from the Guayamas basin (Gulf of California) at a depth of about 2000 m.

   The mol% G + C of the DNA is: 46 (HPLC).

   Type strain: Gylac, DSM 8270.

   GenBank accession number (16S rRNA): X79415.

Genus III. **Malonomonas** *Dehning and Schink 1990, 320[VP] (Effective publication: Dehning and Schink 1989a, 431)*

JAN KUEVER AND BERNHARD SCHINK

Ma.lo.no.mo' nas. M.L. n. *acidum malonicum* malonic acid; Gr. n. *monas* a unit, monad; M.L. fem. n. *Malonomonas* malonic-acid-utilizing monad.

Straight or slightly curved rods, 0.4 × 3.1–4.0 μm. Resting forms such as spores are not known to occur. Gram negative. **Motile, generally by means of a single polar flagellum. Microaerotolerant anaerobe. Chemoorganotroph. Possesses a fermentative type of metabolism. Malonate is decarboxylated to acetate. L-Malate or fumarate is fermented to succinate and CO_2.** Optimal growth temperature, 28–30°C. Colonies on or in anoxic agar media are **red-colored, because of high amounts of periplasmic cytochrome c of unknown function.** Occur in anoxic sediments in marine habitats.

The mol% G + C of the DNA is: 48.3.

Type species: **Malonomonas rubra** Dehning and Schink 1990, 320 (Effective publication: Dehning and Schink 1989a, 431.)

FURTHER DESCRIPTIVE INFORMATION

Strains of *Malonomonas* isolated from anoxic sediments of marine habitats are generally straight or slightly curved rods with rounded ends, 0.4 μm in diameter and 3.1–4.0 μm in length, which are motile by a single subpolar flagellum. Motility may occur only during early growth phases. Spore formation is absent. At the end of exponential growth, cells tend to form large aggregates, which settle as clumps to the bottom of the culture vessel.

The only substrates utilized are malonate, fumarate, and malate. Malonate is decarboxylated to acetate, whereas malate and fumarate are fermented to succinate and CO_2. Methylmalonate is not used, not even in the presence of 2 mM acetate. Sulfur, sulfate, sulfite, thiosulfate, nitrate, and N_2O cannot act as electron acceptors.

The pH range for growth is 6.0–8.5, with the optimal being 7.1–7.3. The temperature range for growth is 22–45°C with an optimal at 28–30°C.

Colonies of *Malonomonas* on or in agar media, as well as accumulations of sedimented or centrifuged cell pellets, are red-colored, due to *c*-type cytochromes with absorption maxima at 418, 522, and 551 nm. This cytochrome comprises up to 12% of the total cell protein. It has been assumed that this cytochrome does not fulfill a physiological function, but may be a remnant from sulfur- or ferric iron-reducing ancestors.

Malonomonas occurs in anoxic sulfide-containing sediments of marine habitats. Freshwater strains have not been isolated so far.

Malonomonas strains require at least 8–9 g/l NaCl for growth. All strains of *Malonomonas* can be cultured in a defined anoxic medium with cysteine as reducing agent under exclusion of air (Dehning and Schink, 1989a).

ENRICHMENT AND ISOLATION PROCEDURES

For selective enrichment of *Malonomonas* strains, basal mineral medium supplemented with 20 mM disodium malonate is used. Bottles or tubes (100-, 50-, or 20-ml capacity) can serve as culture vessels and are kept anoxic with tight screw caps, or by using a headspace of N_2/CO_2 (90:10) and sealing with butyl rubber stoppers. The medium is inoculated with anoxic mud from freshwater, brackish water, or marine habitats; 3–5% of the total cul-

ture volume should be added. *Malonomonas* species may also be enriched and isolated via direct serial dilution of anoxic environmental samples in liquid medium (according to most-probable-number method) or agar medium with malonate as the sole organic substrate. Depending on the salinity of the anaerobic mud sample that is used as the inoculum, salt- or freshwater medium is used. Gas production after 3–5 weeks incubation at 28–30°C indicates substrate degradation. Pure cultures are obtained by repeated application of the agar dilution method. Appearance of disk-shaped red-colored colonies indicates growth of *Malonomonas*.

MAINTENANCE PROCEDURES

Stock cultures are transferred every 3 months. Towards the end of growth or at the beginning of the stationary phase, cultures are refrigerated and stored at 2–5°C in the dark. *Malonomonas* strains may be preserved indefinitely by suspension of cells in anoxic medium containing 5% dimethylsulfoxide and storage in liquid nitrogen.

DIFFERENTIATION OF THE GENUS *MALONOMONAS* FROM OTHER GENERA

The genus *Malonomonas* is phenotypically distinguished from other fermentative bacteria by using only malonate, fumarate, and malate as substrates for growth. Inorganic electron acceptors are not used. However, a clear differentiation of this genus from other closely related genera (*Pelobacter*, *Desulfuromusa*, *Desulfuromonas*, and *Geobacter*) can be achieved by determination of the 16S rRNA gene sequence. Differentiation of *Malonomonas* from *Pelobacter* is possible due to the presence of cytochromes absent in *Pelobacter* strains.

TAXONOMIC COMMENTS

Malonomonas rubra is a member of the *Deltaproteobacteria*. The closest relatives are *Pelobacter acidigallici* and members of the genus *Desulfuromusa*.

FURTHER READING

Dehning, I. and B. Schink. 1989. *Malonomonas rubra* gen. nov. sp. nov., a microaerotolerant anaerobic bacterium growing by decarboxylation of malonate. Arch. Microbiol. *151*: 427–433.

List of species of the genus Malonomonas

1. **Malonomonas rubra** Dehning and Schink 1990, 320[VP] (Effective publication: Dehning and Schink 1989a, 431.)

   ru'bra. L. adj. *ruber* red, referring to the red color of the cells.

   The description is as for the genus. Isolated using mal-

onate as the sole substrate. Occur in anoxic sediments from marine or brackish water habitats.

   The mol% G + C of the DNA is: 48.3 (T_m).
   Type strain: GraMal1, DSM 5091.
   GenBank accession number (16S rRNA): Y17712.

Genus IV. **Pelobacter** *Schink and Pfennig 1983, 896*[VP] *(Effective publication: Schink and Pfennig 1982, 200)*

BERNHARD SCHINK

Pe.lo.bac'ter. Gr. masc. n. *pelos* mud; M.L. masc. n. *bacter* the equivalent of Gr. neut. n. *bakterion* a small rod; M.L. masc. n. *Pelobacter* a mud-inhabiting rod.

Rod-shaped cells, 0.5–1.0 × 1.5–6.0 μm, with rounded ends, single, in pairs, or in chains. Motile forms may occur in some species; other species are nonmotile. Gram negative. Endospores not formed. Anaerobic. **Have a chemoorganotrophic, fermentative type of metabolism, using few simple organic compounds as substrates. Carbohydrates not utilized.** Media containing a

reductant, e.g., sulfide, are necessary for growth. Catalase negative. Isolated from sediments of limnic or marine origin.

The mol% G + C of the DNA is: 50–58.

Type species: **Pelobacter acidigallici** Schink and Pfennig 1983, 896 (Effective publication: Schink and Pfennig 1982, 200.)

FURTHER DESCRIPTIVE INFORMATION

The restriction, in every instance, to only a few, structurally related substrates makes the pathways of substrate degradation simple. *P. acidigallici* and *P. massiliensis* are both restricted to the utilization of trihydroxybenzenes, which, after necessary modifications and reduction to dihydrofluoroglucinol, are degraded through β-thiolytic cleavage. The four other species described have been enriched with diols or acetylene, which all form acetaldehyde as a primary intermediate, which is subsequently fermented through various pathways to either acetate plus ethanol or acetate plus propionate. *P. venetianus*, *P. carbinolicus*, and *P. acetylenicus* can easily release electrons in the form of molecular hydrogen in syntrophic cooperation with, for example, methanogenic or acetogenic partner bacteria. They can also act as syntrophic ethanol oxidizers in the presence of such partners.

ENRICHMENT AND ISOLATION PROCEDURES

All *Pelobacter* species known so far have been enriched and isolated from sediments of limnic or marine origin in simple mineral media with specific organic enrichment substrates. The mineral media for enrichment must be low in phosphate (1–2 mM KPO_4) and rich in CO_2/HCO_3^- (>20 mM). Sulfide is usually used as a reductant. Since all *Pelobacter* species grow with substrates that yield only C_2 degradation intermediates, they must form pyruvate and sugars via reductive carboxylation of acetyl CoA and therefore need carbon dioxide for this reaction. The enrichment medium used successfully for these organisms is derived from that used for the cultivation of sulfate-reducing bacteria (Widdel and Pfennig, 1981), but without sulfate, and is described in detail in the original publications (e.g., Schink and Pfennig, 1982). The respective substrates are applied in a concentration range of 10–20 mM, and inocula of at least 2–5 ml are used to secure sufficient attachment surfaces, structure-bound sulfides, and trace cosubstrates in the initial growth steps.

MAINTENANCE PROCEDURES

Cultures are maintained either by repeated transfer at intervals of 2–3 months or by freezing in liquid nitrogen using techniques common for strictly anaerobic bacteria. No information exists about survival upon lyophilization.

DIFFERENTIATION OF THE GENUS *PELOBACTER* FROM OTHER GENERA

The genus *Pelobacter* differs from most other strictly anaerobic types of Gram-negative, rod-shaped bacteria by its lack of spore formation and lack of sugar utilization. All strains have been isolated in defined mineral media with single defined organic substrates and do not require complex, undefined medium additions. Differential characteristics of the genera *Pelobacter* and *Malonomonas* are listed in Table BXII.δ.35.

TAXONOMIC COMMENTS

The genus *Pelobacter* has been proposed as a taxonomic entity consisting of strictly anaerobic, Gram-negative, non-sporeforming, rod-shaped bacteria that use only a very limited number of substrates. All members of this genus are unable to ferment sugars and therefore cannot be grouped with any other genus in the family *Bacteroidaceae* (Holdeman et al., 1984). The genus so far comprises six different species: *P. acidigallici* (Schink and Pfennig, 1982), *P. venetianus* (Schink and Stieb, 1983), *P. carbinolicus* (Schink, 1984a), *P. propionicus* (Schink, 1984a), *P. acetylenicus* (Schink, 1985), and *P. massiliensis* (Schnell et al., 1991a), each of which has been proposed on the basis of 3–5 described strains.

Comparison of the various *Pelobacter* species by DNA–DNA hybridization experiments has revealed that the genus is heterogeneous (J.P. Touzel and B. Schink, unpublished observations). The species *P. venetianus*, *P. carbinolicus*, and *P. acetylenicus* form a rather homogeneous cluster within the family *Geobacteraceae* that is closely related to various *Desulfuromonas* and *Geobacter* repre-

TABLE BXII.δ.35. Differential characteristics of the genera *Pelobacter* and *Malonomonas*[a]

Genus	*Pelobacter*	*Malonomonas*
Cell morphology	Rod-shaped	Rod-shaped
Motility	+	+
Mol% G + C content	50–58	48
Major menaquinone	MK-8	nr
Typical substrates for fermentation:		
Acetoin	+[b]	nr
2,3-Butanediol	+[b]	nr
1-Butanol	+[b]	nr
Ethanol	+[b]	nr
Fumarate	−	+
Gallate	+[c]	nr
Malate	−	+
Malonate	nr	+
Phloroglucinol	+[c]	nr
1–Propanol	+[b]	nr
Pyrogallol	+[c]	nr
Chemolithoautotrophic growth	−	+
Fermentative growth	+	
Typical electron acceptors	Fermentative metabolism[d,e]	Fermentative metabolism
Optimal growth temperature (°C)	30	28–30
NaCl requirement	±	+
Habitat	Freshwater and marine	Marine

[a]Symbols: nr, not reported; +, present, observed in all species; −, not present, not observed; ±, variable.

[b]Used by all other described species of *Pelobacter*.

[c]Used by *P. acidigallici* and *P. massiliensis*.

[d]Some species use Fe(III) and sulfur as electron acceptors.

[e]Use of external electron acceptors is coupled to fermentation shift and incomplete oxidation.

sentatives. In addition, *Pelobacter propionicus* is directly related to *Geobacter sulfurreducens* based on 16S rRNA sequence similarity data. The tight relationship to sulfur- and ferric iron-respiring organisms has led to speculations that the genus *Pelobacter* represents a group of fermenting bacteria that developed a fermentative metabolism as a secondary evolutionary event, meaning that they are separate from primarily fermentative bacteria (Stackebrandt et al., 1989).

FURTHER READING

Schink, B. 1984. Fermentation of 2,3-butanediol by *Pelobacter carbinolicus*, new species and *Pelobacter propionicus*, new species and evidence for propionate formation from C-2 compounds. Arch. Microbiol. *137*: 33–41.

Schink, B. 1985. Fermentation of acetylene by an obligate anaerobe, *Pelobacter acetylenicus*, new species. Arch. Microbiol. *142*: 295–301.

Schink, B., D.R. Kremer and T.A. Hansen. 1987. Pathway of propionate formation from ethanol in *Pelobacter propionicus*. Arch. Microbiol. *147*: 321–327.

Schink, B. and N. Pfennig. 1982. Fermentation of trihydroxybenzenes by *Pelobacter acidigallici*, new genus new species: a new strictly anaerobic, non-spore-forming bacterium. Arch. Microbiol. *133*: 195–201.

Schink, B. and M. Stieb. 1983. Fermentative degradation of polyethylene glycol by a strictly anaerobic, gram-negative, non-spore-forming bacterium, *Pelobacter venetianus*, sp. nov. Appl. Environ. Microbiol. *45*: 1905–1913.

Stackebrandt, E., U. Wehmeyer and B. Schink. 1989. The phylogenetic status of *Pelobacter acidigallici*, *Pelobacter venetianus*, and *Pelobacter carbinolicus*. Syst. Appl. Microbiol. *11*: 257–260.

Widdel, F. and N. Pfennig. 1981. Studies on dissimilatory sulfate-reducing bacteria that decompose fatty acids. I. Isolation of new sulfate-reducing bacteria with acetate from saline environments. Description of *Desulfobacter postgatei* gen. nov., sp. nov. Arch. Microbiol. *129*: 395–400.

DIFFERENTIATION OF THE SPECIES OF THE GENUS *PELOBACTER*

The key differences in substrate utilization among the various species are listed in Table BXII.δ.36.

List of species of the genus Pelobacter

1. **Pelobacter acidigallici** Schink and Pfennig 1983, 896[VP] (Effective publication: Schink and Pfennig 1982, 200.)

a.ci.di.gal' li.ci. M.L. neut. n. *acidum gallicum* gallic acid; M.L. gen. n. *acidigallici* of gallic acid.

Rod-shaped cells, 0.5–0.8 × 1.5–3.5 μm, with rounded ends, single, in pairs, or in chains. Motile in young cultures; however, motility may be lost early. No spore formation. Gram negative.

Gallic acid, pyrogallol, 2,4,6-trihydroxybenzoic acid, and fluoroglucinol are the only fermentable substrates, and ace-tate and CO_2 are the only fermentation products. No other phenolic compounds, fatty acids, dicarboxylic acids, or alcohols are metabolized. Sulfate, sulfur, and nitrate are not reduced. Growth requires mineral media with a reductant. The marine type strain requires at least 10 g/l NaCl and 1.5 g/l $MgCl_2$, whereas freshwater isolates do not need enhanced salt concentrations. No growth factors or vitamins are needed. Indole is not formed; gelatin and urea are not hydrolyzed.

pH range: 5.3–8.2; optimal, 6.5–7.0. Temperature

TABLE BXII.δ.36. Differential characteristics of the species of the genus *Pelobacter*[a]

Characteristic	P. acidigallici	P. acetylenicus	P. carbinolicus	P. massiliensis	P. propionicus	P. venetianus
Width (μm)	0.5–0.8	0.6–0.8	0.5–0.7	0.4–0.5	0.5–0.7	0.5–1.0
Length (μm)	1.5–3.5	1.5–4.0	1.2–3.0	0.7–1.5	1.2–6.0	2.5
Mol% G + C of the DNA	51.8	57.1	52.3	59.0	57.4	52.2
Substrates metabolized:						
Gallic acid	+	−	−	+	−	−
Pyrogallol	+	−	−	+	−	−
Fluoroglucinol	+	−	−	+	−	−
Fluoroglucinol carboxylate	+	−	−	nd	−	−
Hydroxyhydroquinone	−	nd	nd	+	nd	nd
Acetoin	−	+	+	nd	+	+
2,3-Butanediol	−	+	+	nd	+	+
Ethylene glycol	−	−	+	−	−	+[b]
Polyethylene glycols	−	−	−	−	−	+
Ethanol	−	+[c]	+[c]	−	+	+[c]
n-Propanol	−	+[b,c]	+[b,c]	−	+[c]	+[b,c]
n-Butanol	−	+[c,d]	+[c,d]	−	+[d]	+[c,d]
1,2-Propanediol	−	+[d]	−	nd	−	+[d]
Acetylene	−	+	−	nd	−	−
Lactate	−	−	−	−	+	−
Pyruvate	−	−	−	−	+	−
Glycerol	−	+[d]	−	−	−	+[d]
Typical products:						
Acetate	+	+	+	+	+	+
Ethanol		+	+			+
Propionate					+	
CO$_2$	+					

[a]Symbols: +, growth; −, no growth (all strains tested); nd, not determined.

[b]Growth is possible only at very low concentration (<1 mM) or in continuous culture.

[c]Growth is possible only in the presence of a hydrogen-scavenging anaerobe, e.g., a methanogenic bacterium.

[d]Growth is possible only in the presence of small amounts of acetate for cell carbon synthesis.

range, 10–37°C; optimal, 35°C. No cytochromes detectable. Habitats: anaerobic muds of freshwater or marine origin.

The mol% G + C of the DNA is: 51.8 (T_m).

Type strain: Ma Gal 2, DSM 2377.

GenBank accession number (16S rRNA): X77216.

2. **Pelobacter acetylenicus** Schink 1986, 354[VP] (Effective publication: Schink 1985, 300.)

a.ce.ty.le′ ni.cus. M.L. adj. *acetylenicus* referring to acetylene utilization.

Rod-shaped cells, 0.6–0.8 × 1.5–4 µm, with slightly pointed ends, single, in pairs, or in chains. Motile in young cultures.

Acetylene, acetoin, ethanolamine, choline, 1,2-propanediol and glycerol are used as substrates, the latter two in the presence of acetate. Substrates are fermented to acetate and ethanol or the respective higher acids and alcohols. Ethanol is oxidized to acetate in the presence of H_2-scavenging anaerobes. Grows in freshwater medium, as well as in the presence of 2% (w/v) NaCl. Sulfate, sulfur, thiosulfate, sulfite, and nitrate are not reduced. Indole is not formed. Gelatin and urea are not hydrolyzed. Cytochromes are absent.

pH range, 6.5–7.5. Temperature range, 15–40°C; optimal, 28–34°C. Habitats: anoxic muds of freshwater or marine origin.

The mol% G + C of the DNA is: 57.1 ± 0.2 (T_m).

Type strain: WoAcy 1, DSM 3246.

GenBank accession number (16S rRNA): X70955.

3. **Pelobacter carbinolicus** Schink 1984b, 356[VP] (Effective publication: Schink 1984a, 39.)

car.bi.no′ li.cus. M.L. adj. *carbinolicus* referring to carbinols.

Rod-shaped cells, 0.5–0.7 × 1.2–3.0 µm, with rounded ends, single, in pairs, or in chains. Nonmotile.

2,3-Butanediol, acetoin, and ethylene glycol are utilized for growth and fermented to acetate and ethanol. No other organic substrates are utilized. In coculture with H_2-consuming bacteria, ethanol, propanol, and butanol are oxidized to the respective fatty acids plus hydrogen. Grows in freshwater medium, as well as in the presence of 2% NaCl. Sulfate, sulfur, thiosulfate, sulfite, and nitrate are not reduced. Indole is not formed. Gelatin and urea are not hydrolyzed. Cytochromes are absent.

pH range, 6.0–8.0; optimal 6.5–7.2. Temperature range, 15–40°C; optimal 35°C. Habitats: anoxic muds of marine origin.

The mol% G + C of the DNA is: 52.3 ± 1.0 (T_m).

Type strain: Gra Bd1, DSM 2380.

GenBank accession number (16S rRNA): X79413.

4. **Pelobacter massiliensis** Schnell, Brune and Schink 1991b, 580[VP] (Effective publication: Schnell, Brune and Schink 1991a, 515.)

mas.si.li.en′ sis. M.L. adj. *massiliensis* a citizen of Massilia (Latin), Marseille; referring to a strictly anaerobic fermenting bacterium enriched from sediments close to Marseille.

Rod-shaped cells, 0.4–0.5 × 0.7–1.5 µm. Motile in young cultures. Hydroxyhydroquinone, pyrogallol, fluoroglucinol, and gallic acid are the only fermentable substrates, and acetate, CO_2, and traces of resorcinol are the only fermen-

tation products. Sulfate, sulfite, thiosulfate, sulfur, nitrate, and fumarate are not reduced. Growth requires mineral medium with at least 10 g/l NaCl and 1.5 g/l $MgCl_2·6H_2O$ and sulfide as a reducing agent. Cytochromes not detectable.

pH optimal, 7.2. Temperature optimal, 30°C. *P. massiliensis* has been enriched and isolated so far only from marine sediments.

The mol% G + C of the DNA is: 59 (T_m).

Type strain: HHQ7, DSM 6233.

5. **Pelobacter propionicus** Schink 1984b, 356[VP] (Effective publication: Schink 1984a, 39.)

pro.pi.o′ ni.cus. M.L. n. *acidum propionicum* propionic acid; M.L. adj. *propionicus* forming propionic acid.

Rod-shaped cells, 0.5–0.7 × 1.2–6.0 µm, with rounded ends, single, in pairs, or in chains. Nonmotile.

2,3-Butanediol, acetoin, ethanol, pyruvate, and lactate are utilized for growth; propanol and butanol are used in the presence of acetate. Propionate and acetate are the main fermentation products; butanol and propanol are oxidized to the respective fatty acids, with concomitant reduction of acetate and bicarbonate to propionate. No other substrates are utilized. Sulfate, sulfite, thiosulfate, sulfur, and nitrate are not reduced. Growth requires mineral media with bicarbonate and a reductant. Indole is not formed. Gelatin and urea are not hydrolyzed. Catalase negative. A *b*-type cytochrome is present.

pH range: 6.5–8.4; optimal: 7.0–8.0. Temperature range: 4–45°C; optimal: 33°C. Habitats: anoxic muds of freshwater origin; anaerobic sewage sludge digesters.

The mol% G + C of the DNA is: 57.4 ± 1.0% (T_m).

Type strain: Ott Bd 1, DSM 2379.

GenBank accession number (16S rRNA): X70954.

6. **Pelobacter venetianus** Schink and Stieb 1984, 91[VP] (Effective publication: Schink and Stieb 1983, 1912.)

ve.ne.ti.a′ nus. L. n. *venetianus* from Venice, the origin of the type strain and the place where it was found in high numbers.

Cells rod-shaped, 0.5 × 1.0–2.5 µm, with rounded ends, occurring singly or in pairs. Motile in young cultures.

Polyethylene glycol, polyethylene glycol-containing compounds, and acetoin are the only substrates utilized. Substrates are fermented to nearly equal amounts of acetate and ethanol. 1,2-Propanediol, 1,2-butanediol, and glycerol are also fermented to the corresponding acids and alcohols, but for growth, acetate is necessary as a source of carbon. Sulfate, sulfur, thiosulfate, and nitrate are not reduced. Growth requires mineral media with a reductant. Growth yields and rates of the type strain are higher on media containing at least 10 g/l of NaCl and 1.5 g/l of $MgCl_2·6H_2O$. No growth factors or vitamins are required. Indole is not formed. Gelatin and urea are not hydrolyzed. In coculture with H_2-consuming methanogenic or homoacetogenic bacteria, growth on ethanol is achieved by oxidation of ethanol to acetate. No cytochromes are detectable.

pH range: 5.5–8.0; optimal: 7.0–7.5. Temperature range: 10–40°C; optimal: 33°C. Habitats: anaerobic mud of freshwater or marine origin.

The mol% G + C of the DNA is: 52.2 ± 1.0 (T_m).

Type strain: Gra PEG 1, DSM 2394.

Family II. **Geobacteraceae** fam. nov.

GEORGE M. GARRITY, JULIA A. BELL AND TIMOTHY LILBURN

Ge.o.bac.ter.a' ce.ae. M.L. masc. n. *Geobacter* type genus of the family; *-aceae* ending to denote family; M.L. fem. pl. n. *Geobacteraceae* the *Geobacter* family.

The family *Geobacteraceae* was circumscribed for this volume on the basis of phylogenetic analysis of 16S rRNA sequences; the family contains the genera *Geobacter* (type genus) and *Trichlorobacter*.

Nonmotile. Strict anaerobes. *Geobacter* utilizes Fe(III) as an electron acceptor; *Tricholorobacter* reduces trichloroacetic acid to dichloroacetic acid.

Type genus: **Geobacter** Lovley, Giovannoni, White, Champine, Phillips, Gorby and Goodwin 1995a, 619[VP] (Effective publication: Lovley, Giovannoni, White, Champine, Phillips, Gorby and Goodwin 1993a, 342.)

Genus I. **Geobacter** Lovley, Giovannoni, White, Champine, Phillips, Gorby and Goodwin 1995a, 619[VP] (Effective publication: Lovley, Giovannoni, White, Champine, Phillips, Gorby and Goodwin 1993a, 342)

JOHN D. COATES AND DEREK R. LOVLEY

Ge.o.bac' ter. Gr. n. *ge* the earth; masc. *bacter* equivalent of Gr. n. *bakterion* a small rod; M.L. masc. n. *Geobacter* a rod from the earth.

Cells are rod-shaped with rounded ends that grow as single cells, in pairs, or in chains. Cells are 1–4 µm × 0.6 µm, nonmotile, with no spore formation. They are Gram-negative **strict anaerobes** and belong to the *Deltaproteobacteria*. They are catalase and carotenoid negative. Cells contain menaquinone and *c*-type cytochromes, but lack ubiquinone. **Growth occurs chemoorganotrophically with Fe(III) serving as the sole electron acceptor. The organic electron donor is completely oxidized to carbon dioxide.** Vitamins are not required for growth, but growth is enhanced by their presence. Growth is also enhanced in media containing FeCl$_2$ as a reducing agent.

The mol% G + C of the DNA is: 50.2–60.6 (HPLC).

Type species: **Geobacter metallireducens** Lovley, Giovannoni, White, Champine, Phillips, Gorby and Goodwin 1995a, 619 (Effective publication: Lovley, Giovannoni, White, Champine, Phillips, Gorby and Goodwin 1993a, 342.)

FURTHER DESCRIPTIVE INFORMATION

Geobacter metallireducens was the first described dissimilatory Fe(III) reducer that is able to obtain energy to support growth from the complete oxidation of acetate and other multicarbon compounds coupled to the reduction of Fe(III) (Lovley et al., 1987; Lovley, and Phillips, 1988). Phylogenetic analysis of its 16S rRNA sequence placed it within the *Deltaproteobacteria* and closest to the acetate-oxidizing sulfur reducer *Desulfuromonas acetoxidans* (Lovley et al., 1993a). *D. acetoxidans* has subsequently been found to be able to grow with acetate as electron donor and Fe(III) as electron acceptor (Roden and Lovley, 1993). In addition to *G. metallireducens*, the genus *Geobacter* contains four other validly published species, *G. hydrogenophilus* (Coates et al., 2001), *G. grbiciae* (Coates et al., 2001), *G. chapellei* (Coates et al., 2001), and *G. sulfurreducens* (Caccavo et al., 1994), as well as "*G. humireducens*" (Coates et al., 1998, 2001).

The *Geobacter* species are the Fe(III)- and humics-reducing bacteria most readily isolated from a variety of freshwater, sedimentary, and soil environments (Coates et al., 1996, 1998). Analysis of 16S rDNA sequences in a broad diversity of environments including soils and sediments in which Fe(III) reduction was the terminal electron-accepting process indicated that *Geobacter* species were among the most numerous organisms present and orders of magnitude more numerous than other known Fe(III)-reducing microorganisms (Rooney-Varga et al., 1999; Snoeyenbos-West et al., 2000).

ENRICHMENT AND ISOLATION PROCEDURES

Geobacter species are mesophilic, complete-oxidizing, strict anaerobes. Amorphous Fe(III)-oxide (30 mM) (Lovley, 2000) is a suitable electron acceptor for *Geobacter* species with acetate (2 mM) as the sole electron donor. Enrichments are generally incubated at temperatures of 15–30°C. Positive enrichments can be visually identified by a color change in the amorphous Fe(III) from brown to black as the iron is reduced. If left for an extended period after complete reduction of the Fe(III) has taken place (1–2 weeks), the crystalline iron mineral magnetite (Fe$_3$O$_4$) will form, which can readily be identified by its magnetic properties. Initial enrichments usually take 1–2 weeks at 30°C. A suitable medium for *Geobacter* species is given below.[1] Initial positive enrichments should be transferred as soon as the iron precipitate in the medium has turned black. Inoculum transfers into fresh medium should be 10% of the culture volume. Sequential transfers should be done three times to remove residual particulates and biodegradable organics associated with the original sample. Enrichments should then be transferred into basal medium amended with a soluble form of Fe(III) such as Fe(III)-NTA (10 mM) (Roden and Lovley, 1993). Growth can be recognized by a color change in the medium from transparent orange to colorless and the formation of FeCO$_3$, which is seen as a white precipitate at the bottom of the culture vessel. Cell yields are generally poor, and an optical increase in cell density will not be apparent. Once growth has been achieved on Fe(III)-NTA, the enrichment should be transferred to solid medium with acetate (2 mM) and Fe(III)-NTA (10 mM) as the sole electron donor and acceptor, respectively, and amended with 2% by weight Noble agar. Colonies of Fe(III)-reducing bacteria can readily be recognized on the surface of agar medium as small white colonies

---

1. Freshwater Fe(III)-oxide basal medium contains (per liter): 30 ml amorphous Fe(III)-oxide, 0.25 g NH$_4$Cl, 0.60 g NaH$_2$PO$_4$, 1.36 g CH$_3$COONa·3H$_2$O, 2.5 g NaHCO$_3$ (primary buffer with CO$_2$), 0.1 g KCl, 10 ml each of vitamin and mineral solutions (Lovley and Phillips, 1988). The medium is dispensed into tubes before sparging with 80% N$_2$ and 20% CO$_2$ (at least 6 min, the last minute with the stopper in place). The amorphous Fe(III)-oxide can be replaced with alternative Fe(III) forms as needed.

surrounded by a colorless halo in tan-brown colored agar. Colonies should be picked with sterile glass Pasteur-pipettes in an anaerobic glove box and restreaked onto fresh solid media. After incubation, several of the restreaked colonies should be picked and used to inoculate a single tube of fresh anaerobic Fe(III)-NTA medium. Growth in these nascent cultures is significantly improved with the addition of $FeCl_2$ (2.5 mM) as a reducing agent. A viable Fe(III)-reducing culture should be obtained after 7–10 days.

MAINTENANCE PROCEDURES

Geobacter cultures can be maintained as frozen stocks at $-70°C$. Cultures should be grown to mid-exponential phase in medium amended with a suitable soluble electron donor and acceptor. Aliquots (1 ml) should be anaerobically transferred into small serum vials (10 ml) that have previously been thoroughly sparged with an N_2:CO_2 mixture (80:20; v/v) and heat sterilized. The vials should be amended with an anaerobic aqueous glycerol solution (100 μl) (25% v/v) mixed and frozen at $-70°C$ immediately. Frozen stocks should be checked regularly to ensure viability.

DIFFERENTIATION OF THE GENUS *GEOBACTER* FROM OTHER GENERA

The genus *Geobacter* differs from all other strictly anaerobic bacteria both phenotypically and genotypically. *Geobacter* species are the first complete-oxidizing dissimilatory Fe(III)-reducers described. Members of the genus *Geobacter* are also the only described organisms that can couple oxidation of monoaromatic compounds, including the hydrocarbon toluene, to Fe(III) reduction. All strains have been isolated in defined mineral medium and do not require complex undefined medium additions.

DIFFERENTIATION OF THE SPECIES OF THE GENUS *GEOBACTER*

The key differences in electron donor and acceptor utilization among the various *Geobacter* species are listed in Table BXII.δ.37.

List of species of the genus Geobacter

1. **Geobacter metallireducens** Lovley, Giovannoni, White, Champine, Phillips, Gorby and Goodwin 1995a, 619[VP] (Effective publication: Lovley, Giovannoni, White, Champine, Phillips, Gorby and Goodwin 1993a, 342.)
   me.tal.li.re.du' cens. L. n. *metallum* metal; L. part. adj. *reducens* converting to a different state; N.L. adj. *metallireducens* reducing metal.

   G. metallireducens is the type strain for the genus *Geobacter*. Its closest known relatives are *G. hydrogenophilus* (Lonergan et al., 1996) and *G. grbiciae* (Coates et al., 1996, 2001). It is

TABLE BXII.δ.37. Differentiation of the species of the genus *Geobacter*

Electron donors and acceptors	*Geobacter metallireducens*	*Geobacter chapellei*	*Geobacter grbiciae*	*Geobacter hydrogenophilus*	*Geobacter sulfurreducens*	"*Geobacter humireducens*"
Source	Aquatic sediments	Deep subsurface sediments	Aquatic sediments	Contaminated aquifer	Contaminated ditch	Contaminated wetland
Electron donors oxidized with Fe(III):						
Acetate	+	+	+	+	+	+
Benzoate	+	−	+	+	−	−
Formate	−	+	+	+	−	+
Ethanol	+	+	+	+	−	+
H_2	−	−	+	+	+	+
Lactate	−	+	−	−	−	+
Propionate	+	−	+	+	−	−
Other electron donors used:						
None		+			+	
Benzaldehyde	+					
Butyrate	+		+	+		
p-Cresol	+					
Isovalerate	+					
Phenol	+					
Pyruvate	+		+	+		+
Succinate				+		
Toluene	+		+			
Valerate	+					
Electron acceptors:						
AQDS[a]	+		+			+
Co(III)					+	
Fumarate		+		+	+	+
Humics	+					
Malate					+	
Nitrate	+					+
Mn(IV)	+	+				+
S⁰					+	+
U(VI)	+	+		+		

[a]2,6-anthraquinone disulfonate—a humic substances analog.

a strictly anaerobic, Gram-negative, nonmotile rod, 2–4 μm × 0.6 μm. *G. metallireducens* was first isolated from aquatic sediments that were enriched with acetate as the electron donor and amorphous Fe(III)-oxide as the electron acceptor. Cell walls display characteristics typical of a Gram-negative organism with an outer membrane separated from an inner membrane by a periplasmic region. Cellular inclusions or flagella are not detected. Cells divide by forming a central restriction between daughter cells. Colonies were also isolated on anaerobic agar slants with nitrate as the electron acceptor. Cells are catalase and ubiquinone negative but contain *c*-type cytochromes. In addition to Fe(III), *G. metallireducens* also utilizes nitrate, which is stoichiometrically reduced to ammonia. Although elemental sulfur is reduced by cell suspensions of *G. metallireducens*, energy for growth is not conserved during this metabolism.

The mol% G + C of the DNA is: 56.6 (T_m); 58.9 ± 0.3 (HPLC).

Type strain: GS-15, ATCC 53774, DSM 7210.

GenBank accession number (16S rRNA): L07834.

2. **Geobacter chapellei** Coates, Bhupathiraju, Achenbach, McInerney and Lovley 2001, 586[VP]

cha.pel' le.i. N.L. gen. masc. n. *chapellei* of Chapelle, named after Frank Chapelle, who contributed to our knowledge of subsurface biogeochemistry.

G. chapellei is a strictly anaerobic, Gram-negative, nonmotile rod, 1–2 × 0.5 μm and contains *c*-type cytochrome(s). Its closest known relative is *Pelobacter propionicus*. *G. chapellei* was isolated from an acetate-oxidizing, amorphous Fe(III)-reducing enrichment of pristine aquifer sediments from the deep subsurface. In contrast to the other *Geobacter* species, it has an optimal growth temperature of 25°C. It was isolated on Fe(III)-NTA agar plates with acetate as the electron donor, which it completely oxidizes to CO_2. It is relatively restricted in the range of electron donors it can use; in addition to acetate it can only use formate, ethanol, and lactate. In addition to Fe(III)-NTA, it can use amorphous Fe(III)-oxide, Mn(IV), or fumarate as electron acceptors. In contrast to other *Geobacter* species, it cannot use Fe(III) chelated with citrate as an electron acceptor, probably due to toxicity of citrate at elevated concentrations.

The mol% G + C of the DNA is: 50.2 ± 0.3 (HPLC).

Type strain: 172, DSM 13688.

GenBank accession number (16S rRNA): U41561.

3. **Geobacter grbiciae** Coates, Bhupathiraju, Achenbach, McInerney and Lovley 2001, 587[VP]

grb.i' ci.ae. N.L. gen. fem. n. *grbiciae* of Grbic, named in honor of Dunja Grbic-Galic, for her significant contributions to the field of anaerobic aromatic hydrocarbon oxidation.

G. grbiciae is a close relative (99% sequence identity based on 16S rRNA analysis) of *G. metallireducens* and was identified as a new species based on DNA–DNA hybridization (Coates et al., 2001). Morphologically *G. grbiciae* is a Gram-negative, strictly anaerobic, nonmotile, nonfermentative rod, 1–2 × 0.5 μm. It was isolated in agar shake tubes from a toluene-oxidizing, amorphous Fe(III)-oxide-reducing enrichment of aquatic sediments with benzoate as the electron donor and Fe(III)-NTA as the electron acceptor. Phenotyp-

ically, *G. grbiciae* is similar to *G. metallireducens*. It is a complete oxidizer and uses a diverse range of electron donors including toluene and H_2. With *G. metallireducens*, it is only the second Fe(III) reducer shown to utilize toluene as an electron donor; however, in contrast to *G. metallireducens*, it can also use H_2 as an electron donor. This ability to oxidize both aromatics and H_2 makes *G. grbiciae* unique among the known Fe(III)-reducers.

The mol% G + C of the DNA is: 57.4 ± 0.3 (HPLC).

Type strain: TACP-2, ATCC BAA-45, DSM 13689.

GenBank accession number (16S rRNA): AF335182.

4. **Geobacter hydrogenophilus** Coates, Bhupathiraju, Achenbach, McInerney and Lovley 2001, 586[VP]

hy.dro.ge.no' phi.lus. N.L. n. *hydrogenium* hydrogen; Gr. adj. *philos* friendly to; N.L. adj. *hydrogenophilus* liking hydrogen, referring to the ability of the organism to grow by oxidation of hydrogen.

G. hydrogenophilus is a strictly anaerobic, Gram-negative, nonfermentative, nonmotile rod, 1–2 × 0.5 μm. Its 16S rRNA sequence is 99% identical to that of *G. metallireducens*; however, DNA–DNA hybridization homology indicates that *G. hydrogenophilus* is a separate species (Coates et al., 2001). It exhibits phenotypic traits characteristic of *G. metallireducens*. It was isolated from a petroleum-contaminated aquifer with acetate as the electron donor and Fe(III) chelated with nitrilotriacetic acid (Fe(III)-NTA) (Roden and Lovley, 1993) as the electron acceptor. In addition to simple organics, *G. hydrogenophilus* can also obtain energy for growth from the oxidation of H_2. It cannot grow autotrophically and requires citrate or acetate as a carbon source. Growth on H_2 is significantly enhanced in cultures amended with 0.1 g/l yeast extract. When growing on H_2 and Fe(III) chelated with citrate (Fe(III)-citrate), *G. hydrogenophilus* can oxidize H_2 to a threshold value of 0.092 Pa, which is between H_2 threshold values reported for pure cultures of sulfate-reducing bacteria growing with sulfate and H_2 threshold values obtained with *Shewanella algae* growing with Fe(III)-citrate. In addition to the various forms of Fe(III), *G. hydrogenophilus* can also reduce elemental sulfur with acetate as the electron donor; however, energy to support growth is not obtained from this metabolism. In contrast to *G. metallireducens*, *G. hydrogenophilus* could not use Mn(IV) as an alternative electron acceptor in the absence of Fe(III).

The mol% G + C of the DNA is: 58.4 ± 0.2 (HPLC).

Type strain: H-2, ATCC 51590, DSM 13691.

GenBank accession number (16S rRNA): U28173.

5. **Geobacter sulfurreducens** Caccavo, Lonergan, Lovley, Davis, Stolz and McInerney 1995, 619[VP] (Effective publication: Caccavo, Lonergan, Lovley, Davis, Stolz and McInerney 1994, 3757.)

sul' fur.re.du' cens. L. n. *sulfur* sulfur; L. part. adj. *reducens* converting to a different state; N.L. adj. *sulfurreducens* reducing sulfur.

G. sulfurreducens is a nonfermentative, strictly anaerobic, Gram-negative, nonmotile rod, 1–2 × 0.5 μm. It was isolated from a contaminated drainage ditch with acetate as the electron donor and Fe(III)-pyrophosphate as the electron acceptor. *G. sulfurreducens* is very limited in the number of electron donors it can use and, of those tested, only acetate and H_2 supported growth and Fe(III) reduction. This is in contrast to *G. metallireducens*, which cannot oxidize H_2 but can use a broad range of organic compounds as electron donors. *G. sulfurreducens* was the first strictly anaerobic Fe(III)-reducer to be isolated that can obtain energy for growth by the oxidation of H_2. In addition to the various forms of Fe(III), *G. sulfurreducens* can also use sulfur, or cobalt chelated with EDTA (Co(III)-EDTA) as alternative electron acceptors. It is the first organism identified to be capable of sustaining growth with cobalt as an electron acceptor. Its ability to use elemental sulfur as well as its inability to use Mn(IV) or U(VI) as alternative electron acceptors is in contrast to *G. metallireducens*.

The mol% G + C of the DNA is: 60.6 ± 0.4 (HPLC).

Type strain: PCA, ATCC 51573, DSM 12127.

GenBank accession number (16S rRNA): U13928.

Other Organisms

1. *"Geobacter humireducens"*

hu.mi.re.du' cens. N.L. n. *humi* humic acids, derived from humus; N.L. n. *humus* soil; L. part. adj. *reducens* converting to a different state; N.L. adj. *humireducens* reducing humic substances.

"G. humireducens" was isolated from hydrocarbon-contaminated wetland sediments with acetate as the electron donor and the humic substances analog, 2,6-anthraquinone disulfonate (AQDS), as the electron acceptor. It is the first organism to be isolated as a dissimilatory humics-reducer (Coates et al., 1998, 2001). *"G. humireducens"* can use the broadest range of electron acceptors of any of the *Geobacter* species.

The mol% G + C of the DNA is: unknown.

GenBank accession number (16S rRNA): AF019932.

Genus II. **Trichlorobacter** *De Wever, Cole, Fettig, Hogan and Tiedje 2001, 1^{VP} (Effective publication: De Wever, Cole, Fettig, Hogan and Tiedje 2000, 2301)*

THE EDITORIAL BOARD

Tri.chlor.o.bac' ter. L. pref. *tri* three; Gr. n. *chloros* chlorine; M.L. masc. n. *bacter* equivalent of Gr. neut. n. *bactrum* a rod; M.L. masc. n. *Trichlorobacter* a TCA-dechlorinating rod.

The description of the genus *Trichlorobacter* is the same as that of *Trichlorobacter thiogenes*.

The mol% G + C of the DNA is: unknown.

Type species: **Trichlorobacter thiogenes** De Wever, Cole, Fettig, Hogan and Tiedje 2001, 1 (Effective publication: De Wever, Cole, Fettig, Hogan and Tiedje 2000, 2301.)

FURTHER DESCRIPTIVE INFORMATION

Trichlorobacter thiogenes grows anaerobically on acetate or acetoin and reduces trichloroacetic acid (TCA) to dichloroacetic acid. In some media and in some physiological states the organism appears to be able to couple TCA reduction to the oxidation of sulfide to sulfur with precipitation of sulfur in the medium. The precipitated sulfur is then reduced to sulfide with electrons derived from acetate. However, the redox cycle involving sulfur and sulfide is not required for TCA reduction (De Wever et al., 2000).

Analysis of 16S rDNA sequences placed *Trichlorobacter thiogenes* in the *Deltaproteobacteria* (De Wever et al., 2000).

ENRICHMENT AND ISOLATION PROCEDURES

The organism was obtained from contaminated soil incubated in liquid anaerobic enrichment medium containing acetate and trichloroacetic acid as described by De Wever et al. (2000).

MAINTENANCE PROCEDURES

The organism was maintained on the enrichment medium (De Wever et al., 2000).

List of species of the genus Trichlorobacter

1. **Trichlorobacter thiogenes** De Wever, Cole, Fettig, Hogan and Tiedje 2001, 1^{VP} (Effective publication: De Wever, Cole, Fettig, Hogan and Tiedje 2000, 2301.)

thi.o' gen.es. Gr. n. *theion* sulfur; M.L. n. *genes* production; M.L. n. *thiogenes* because it produces sulfur.

Gram-negative nonmotile curved rods. Strict anaerobe.

Oxidizes acetate with reduction of trichloroacetic acid to dichloroacetic acid. Sulfur and fumarate also used as electron acceptors.

The mol% G + C of the DNA is: unknown.

Type strain: K1, ATCC BAA-34.

GenBank accession number (16S rRNA): AF223382.

Order VI. **Syntrophobacterales** *ord. nov.*

JAN KUEVER, FRED A. RAINEY AND FRIEDRICH WIDDEL

Syn.troph.o.bac.te.ra' les. M.L. masc. n. *Syntrophobacter* type genus of the type family of the order; suffix *-ales* denotes an order; M.L. fem. pl. n. *Syntrophobacterales* the order that comprises the family *Syntrophobacteraceae* and others.

Cells are oval or rod shaped and often motile. **Strictly anaerobic chemoorganotrophic or chemolithoautotrophic growth by fermentative or respiratory metabolism. Several members use sulfate as an electron acceptor, which is reduced to sulfide; however, most described species of the genera** *Syntrophobacter, Syntrophus,* **and** *Smithella* **are unable to use sulfate as electron acceptor but rather reduce protons (or form formate) and thus require the presence of H$_2$-utilizing (or formate-utilizing) organisms** (methanogens or sulfate reducers) **as syntrophic partners.** In addition, sulfite or thiosulfate may be used; some members may also reduce elemental sulfur (or polysulfide) to sulfide. Simple organic compounds serve as electron donors and carbon sources that are oxidized incompletely to acetate or completely to CO$_2$. Most members are mesophilic; in addition, some moderate thermophiles with growth temperature optima at 60°C have been isolated. Cells contain various cytochromes and other redox proteins. Mesophilic species have been isolated from almost every type of anoxic aquatic environment, including marine and freshwater habitats, as well as sludge from sewage treatment plants. Thermophilic members have been isolated from geothermally heated marine environments. All members grow preferentially around neutral conditions.

The order comprises the families *Syntrophobacteraceae* and *Syntrophaceae.*

Type genus: **Syntrophobacter** Boone and Bryant 1984, 355 (Effective publication: Boone and Bryant 1980, 631).

Family I. **Syntrophobacteraceae** *fam. nov.*

JAN KUEVER, FRED A. RAINEY AND FRIEDRICH WIDDEL

Syn.troph.o.bac.te.ra.ce' ae. M.L. masc. n. *Syntrophobacter* type genus of the family; suffix *ceae* denotes family; M.L. fem. pl. n. *Syntrophobacteraceae* the family that comprises the genus *Syntrophobacter.*

The common electron acceptor for the genera *Desulfacinum, Thermodesulforhabdus, Desulforhabdus,* and *Desulfovirga* is sulfate, which is reduced to sulfide. In contrast, most members of the genus *Syntrophobacter* are unable to use sulfate as electron acceptor. Except for members of the genera *Thermodesulforhabdus* and *Desulfacinum,* all known members are mesophilic bacteria. Only members of the genus *Syntrophobacter* oxidize organic substrates incompletely to acetate. All other members show a complete substrate oxidation.

The family comprises the genera *Syntrophobacter, Desulfacinum, Thermodesulforhabdus, Desulforhabdus,* and *Desulfovirga.* Detailed properties of these genera are given in Table BXII.δ.38.

Type genus: **Syntrophobacter** Boone and Bryant 1984, 355 (Effective publication: Boone and Bryant 1980, 631).

Genus I. **Syntrophobacter** *Boone and Bryant 1984, 355^{VP} (Effective publication: Boone and Bryant 1980, 631)*

MICHAEL J. MCINERNEY, ALFONS J.M. STAMS AND DAVID R. BOONE

Syn.tro.pho.bac' ter. Gr. adj. *syn* together with; Gr. n. *trophos* one who feeds; M.L. n. *bacter* masc. equivalent of Gr. neut. n. *bacterion* small staff, rod; M.L. masc. n. *Syntrophobacter* rod which feeds together with (another species).

Straight rods with rounded ends, 0.6–1.2 × 1.0–4.5 μm, sometimes oval to ellipsoid, occur singly and in pairs. May form short chains and filamentous cells (up to 35 μm). **Gram negative. Endospores not formed. Strictly anaerobic chemoorganotroph. Mesophilic.** Growth occurs at neutral pH (6.2–8.0) in low salinity media. **Growth occurs by syntrophic metabolism and sulfate reduction.** Propionate oxidation to acetate and CO$_2$ requires the presence of either a H$_2$/formate-using organism (methanogen or sulfate reducer) or sulfate as the electron acceptor. **Acetate and other fatty acids not oxidized.** Some species grow fermentatively with pyruvate, malate, or fumarate. Lactate and propanol may serve as electron donors for syntrophic metabolism or sulfate reduction. **Other common bacterial substrates such as sugars and aromatic compounds are not used** either in coculture with a H$_2$/formate-using organism or in pure culture with sulfate as the electron acceptor. Sulfite or thiosulfate, but not nitrate, may be used as the electron acceptor. Habitat is sludge from anaerobic waste treatment facilities.

The mol% G + C of the DNA is: 57–61.

Type species: **Syntrophobacter wolinii** Boone and Bryant 1984, 355 (Effective publication: Boone and Bryant 1980, 631.)

FURTHER DESCRIPTIVE INFORMATION

Cell morphology differs markedly between the species in the genus *Syntrophobacter. S. wolinii* is a short rod, with the ends tapering to a blunt point (Fig. BXII.δ.42) (Liu et al., 1999) and some cells forming long filaments (Boone and Bryant, 1980). Both *S. pfennigii* and *S. fumaroxidans* are oval- to egg-shaped rods, and individual cells do not form filaments (Wallrabenstein et al., 1995b; Harmsen et al., 1998).

Cells of *S. wolinii* have a typical Gram-negative ultrastructure, and the cytoplasm of fumarate-grown cells contains numerous membrane-delimited inclusions (Fig. BXII.δ.42) (Liu et al., 1999). Details of the ultrastructure of *S. pfennigii* and *S. fumaroxidans* have not been elucidated.

Surface colonies of *S. wolinii* in coculture with a *Desulfovibrio*

TABLE BXII.δ.38. Differential characteristics of the genera of the family *Syntrophobacteraceae*[a]

Genus	*Syntrophobacter*	*Desulfacinum*	*Desulforhabdus*	*Desulfovirga*	*Thermodesulforhabdus*
Cell morphology	Rod-shaped	Oval to rod-shaped	Rod-shaped	Rod-shaped	Rod-shaped
Motility	±	±	−	+	+
Mol% G + C content	57–61	59.5–64	52.5	60	51
Sulfite reductase	nr	nr	nr	nr	nr
Major menaquinone	nr	nr	nr	nr	nr
Substrate oxidation	I	C	C	C	C
CO-DH[b]	nr	nr	+	nr	nr
Typical electron donors and carbon sources:					
H_2/CO_2	±[c]	+	+	−	−
Formate	±[c]	+	+	+	−
Acetate	−	+	+	+	+
Fatty acids (number of carbon atoms)	3	3–18	3–4	3–12	4–10, 13, 15–18
Isobutyrate	−	+	+	+	nr
Adipate	nr	nr	nr	+	nr
Ethanol	−	+	+	+	+
Fumarate	±	+	+	−	+
Succinate	±	+	−	−	+
Malate	±	+	−	−	+
Lactate	±	+	+	+	+
Pyruvate	+	+	+	+	+
Chemolithoautotrophic growth	−	+	+	−	−
Fermentative growth	±	+	−	−	−
Growth by disproportionation of reduced sulfur compounds	nr	nr	nr	nr	nr
Typical electron acceptors:					
Sulfur	nr	−	−	+	−
Sulfate	+	+	+	+	+
Sulfite	+[d]	+	+	+	+
Thiosulfate	+[d]	+	+	+	+
Fumarate	+[d]	−	−	−	
Optimal growth temperature (°C)	35–37	60	37	35	60
NaCl requirement	−	+	−	−	+
Habitat (anoxic):					
Freshwater	+	−	+	+	−
Marine	−	+	−	−	+

[a]Symbols: nr, not reported; +, present, observed in all species; −, not present, not observed; ±, variable; I, incomplete oxidation; C, complete oxidation.

[b]CO-DH, carbon monoxide dehydrogenase, a key enzyme indicative of the C_1-pathway for acetyl-CoA oxidation.

[c]Some species with fumarate as electron acceptor.

[d]Some species.

species are black-centered and convex, with circular, entire edges (Boone and Bryant, 1980). Colonies form in 3–5 weeks and may reach 2–3 mm in diameter. Subsurface colonies are lenticular. Colonies of a triculture of *S. wolinii*, *Methanospirillum hungatei*, and *Methanosarcina barkeri* are light yellow and take 6 months to develop in agar roll tubes (Liu et al., 1999). *S. pfennigii*, in association with *Methanospirillum hungatei*, forms large, yellow, lens-shaped, subsurface colonies after 10 weeks of incubation. Colony formation with *S. pfennigii* is observed only when agar medium is inoculated with a primary enrichment (Wallrabenstein et al., 1995b). *S. fumaroxidans* also forms brownish-yellow, lens-shaped colonies in agar medium with fumarate (Harmsen et al., 1998).

All three species of *Syntrophobacter* oxidize propionate to acetate, CO_2, and H_2 when grown in the presence of a hydrogen- and formate-using organism, such as a methanogen or a sulfate reducer. The presence of an organism that uses both hydrogen and formate is needed to maintain hydrogen and formate at low enough concentrations that the degradation of propionate is thermodynamically favorable (McInerney and Bryant, 1981; Schink, 1997). *S. fumaroxidans* is not able to degrade propionate in the presence of a methanogen that uses only hydrogen, suggesting that the use of both hydrogen and formate is required for syntrophic metabolism of propionate (Dong et al., 1994). In addition to propionate, *S. pfennigii* grows by the oxidation of propanol and lactate in the presence of a hydrogen/formate-

using organism. All three species grow in pure culture by propionate oxidation coupled to sulfate reduction. *S. wolinii* and *S. fumaroxidans* also grow in pure culture by the fermentation of pyruvate, fumarate, or malate.

Pure cultures of *S. wolinii* (Wallrabenstein et al., 1994), *S. pfennigii* (Wallrabenstein et al., 1995b), and *S. fumaroxidans* (Harmsen et al., 1998) grow by sulfate reduction in a mineral medium with a B-vitamin solution and propionate as the only organic compounds added to the medium. Whether the vitamin solution is required for growth has not been determined.

Syntrophobacter species degrade propionate by the reversal of the methylmalonyl-CoA pathway used by fermentative bacteria to form propionate. A key characteristic of this pathway is that uniformly labeled acetate is produced from either 2-^{13}C-propionate or 3-^{13}C-propionate (Houwen et al., 1987). ATP is likely formed by substrate-level phosphorylation reactions involving acetate kinase or succinate thiokinase. Hydrogen production from propionate requires energy in the form of a proton motive force, with succinate and malate oxidation steps being the thermodynamically most difficult steps to couple to hydrogen production (Schink, 1997; van Kuijk et al., 1998).

Syntrophic propionate oxidizers can be enriched from anaerobic environments where organic matter is completely mineralized to methane and carbon dioxide. An excellent source for enrichment is the biomass from methanogenic reactors that are

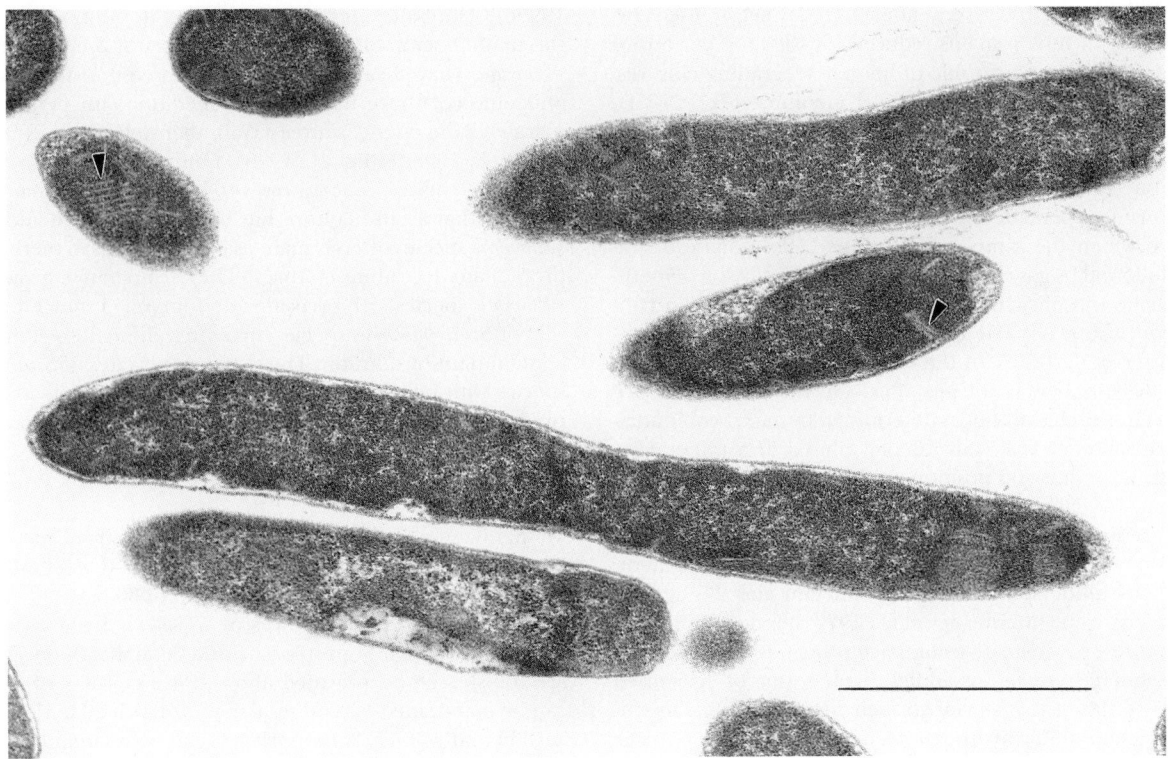

FIGURE BXII.δ.42. Transmission electron micrograph of *Syntrophobacter wolinii* strain DB. Cytoplasm contains flat parallel plates (*arrowheads*), which are invaginations of the plasma membrane. Bar = 1 μm. Micrograph by Henry C. Aldrich.

used for the stabilization of sewage sludge or the anaerobic treatment of industrial wastewater (Stams, 1994). Successful enrichment is also possible using other sources, such as freshwater sediments, rice paddy soils, and fermented manure. Syntrophic propionate oxidation is not restricted to only *Syntrophobacter* species; consequently, it is not yet clear in which methanogenic environments *Syntrophobacter* species are the main propionate oxidizers. In marine sediments, sulfate reduction is the main process for organic matter mineralization; therefore, syntrophic propionate oxidizers are not easily enriched from marine or brackish-water sediments. Although *Syntrophobacter* species are able to couple propionate to sulfate reduction, these bacteria are easily overgrown by *Desulfobulbus* species when sulfate is present. *Desulfobulbus* species grow much faster with propionate and sulfate than do *Syntrophobacter* species (Oude Elferink et al. 1994; Wallrabenstein et al., 1994; Van Kuijk and Stams 1995). Syntrophic propionate oxidation is a rather sensitive process and is easily inhibited by perturbations that lead to increases in the hydrogen, formate, or acetate concentrations (Boone and Xun, 1987; Schmidt and Ahring, 1993; Liu et al., 1999). Syntrophic propionate oxidizers live in close association with methanogens, allowing for efficient interspecies hydrogen and formate transfer. Growth conditions become less favorable upon disruption of the structure and subsequent cultivation in suspended culture.

ENRICHMENT AND ISOLATION PROCEDURES

Selective enrichment of bacteria that syntrophically oxidize propionate is accomplished by using an anaerobic medium that lacks electron acceptors—such as oxyanions of nitrogen, sulfur, S^0, or ferric iron—and by incubating the cultures in the dark (McInerney et al., 1979; Boone and Bryant, 1980; Wallrabenstein et

al., 1995b). Either a basal medium[1] with rumen fluid and B-vitamins or a mineral medium is used. The mineral medium has the same composition as the basal medium, except that it lacks organic components other than propionate and the B-vitamin solution. Anaerobic media and solutions are prepared using a modification of the Hungate technique (Bryant, 1972; Balch and Wolfe, 1976). Because of the slow growth rates of syntrophic associations, the medium must be prepared very carefully so that it remains reduced for a long period. The reducing solution is

---

1. The basal medium has the following composition (g/l distilled water): KH_2PO_4, 0.5 g; $MgCl_2 \cdot 6H_2O$, 0.3 g; NaCl, 0.4 g; NH_4Cl, 0.4 g; $CaCl_2 \cdot 2H_2O$, 0.5 g; resazurin, 0.001 g; clarified rumen fluid, 50 ml; B-vitamin solution, 5 ml; trace metal solution, 20 ml. The medium is boiled under a stream of N_2/CO_2 (80:20 v/v) gas mixture. Solid $NaHCO_3$ (3.5 g/l) is slowly added, and the medium is stoppered and brought inside an anaerobic chamber. The medium is dispensed into serum tubes, which are then stoppered and sealed. The gas phase is replaced by vacuuming and re-pressurizing each tube with an N_2/CO_2 (80:20 v/v) gas mixture using a gassing station (Balch and Wolfe, 1976). The medium is then autoclaved. Before use, the medium is reduced with a cysteine–sulfide reducing solution (0.2 ml/10 ml medium). Additions to the medium and inoculation are done using sterile needles and syringes that have been flushed with an anaerobic gas. The trace metal solution contains (g/l): nitrilotriacetic acid, 2.0; pH is then adjusted to 6 with KOH before addition of the other components (g/l): $MnSO_4 \cdot H_2O$, 1.0, $Fe(NH_4)_2(SO_4)_2 \cdot 6H_2O$, 0.8; $CoCl_2 \cdot 6H_2O$, 0.2; $ZnSO_4 \cdot 7H_2O$, 0.2; $CuCl_2 \cdot 6H_2O$, 0.02; $NiCl_2 \cdot 6H_2O$, 0.02; $Na_2MoO_4 \cdot 2H_2O$, 0.02; Na_2SeO_4, 0.02; Na_2WO_4, 0.02. The B-vitamin solution contains (mg/l): pyridoxine-HCl, 10; thiamin-HCl, 5; riboflavin, 5; calcium pantothenate, 5; *p*-aminobenzoic acid, 5; thiotic acid, 5; nicotinic acid, 5; vitamin B_{12}, 5; biotin, 2; folic acid, 2. Cysteine–sulfide solution contains (g/l): cysteine, 12.5, neutralized with NaOH before the addition of sodium sulfide; and NaS·9H_2O, 12.5. The solution is boiled and dispensed under a 100% nitrogen gas phase and then autoclaved. The sodium sulfide solution contains 25 g/l of NaS·9H_2O and is boiled and dispensed under a 100% nitrogen gas phase.

added to individual tubes several hours to a day before use. This ensures that each tube remains reduced for the long periods of time needed to grow syntrophic propionate-degrading cultures.

For primary enrichments, the basal medium (pH 7.2–7.4), with 20 mM sodium propionate and sodium sulfide as the reductant, is used. The inoculum is from a methanogenic environment, such as sewage sludge or a fresh water sediment. The enrichments are incubated at the temperature of the environment from which the sample was obtained. Propionate degradation is followed by gas chromatography or comparison of methane production in enrichments with propionate to that in controls without substrate. The enrichments are maintained by serially transferring 20–50% of the volume every 1–2 weeks.

When the numbers and types of cells in the enrichment are consistent after each transfer, as determined by microscopic analysis, the enrichment is serially 10-fold diluted in a mineral solution with the same composition as the medium used for the enrichment, but lacking the organic components. The highest dilutions are inoculated into tubes of molten agar medium having the same composition as the medium used for enrichment, plus cysteine–sulfide solution as the reductant and 2% agar as the solidifying agent (McInerney et al., 1979; Boone and Bryant, 1980). A hydrogen/formate-using bacterium (either a methanogen or a sulfate reducer) is added to each tube of agar or to the last few dilution tubes prior to their inoculation by the syntrophic enrichment. If a hydrogen/formate-using sulfate reducer is used, then 30 mM sodium sulfate is added to the agar medium. After inoculation, the agar tubes are spun quickly in ice water to solidify the agar along the side of the tube. The tubes are incubated in an upright position at the incubation temperature. The shake-tube method of Pfennig (1978), in which the agar medium is allowed to solidify in the base of the tube, can also be used. The agar concentration is about 0.5–1% for shake tubes.

Colonies of the propionate degrader will take weeks to months to appear. Colonies that appear within the first two weeks of incubation are probably not syntrophic associations and are marked so that late-forming colonies can be easily detected. Incubation of roll tubes inside anaerobic jars (or chambers) helps to maintain low redox conditions for the long periods of time needed for colonies to develop. Colonies with the sulfate reducer as the hydrogen/formate-using partner have black centers. Colonies with the methanogen as the hydrogen/formate-using partner are yellowish to greenish in color. Colonies are picked with bent Pasteur pipettes under a stream of anaerobic gas and are inoculated into agar slants containing the same medium as used for isolation. The addition of 0.1–0.5 ml of a mid-exponential phase culture of the hydrogen/formate-using partner to the medium in which the colony is transferred may increase the probability that the colonies will grow in new medium. Monoxenic cultures (cocultures) containing only the syntrophic bacterium and its partner may be obtained by repeating the above steps. Procedures for the growth of pure cultures of hydrogen/formate-using methanogens or sulfate reducers are described in Balch and Wolfe (1976) and McInerney et al. (1979).

Cocultures of a syntrophic propionate degrader with a hydrogen/formate-using methanogen have been obtained by serial dilution of the enrichment into propionate basal medium that has also been inoculated with a pure culture of the methanogen (Wallrabenstein et al., 1994, 1995b). After repeated attempts, a coculture containing the syntrophic propionate degrader and the methanogen may be obtained. Axenic cultures of *S. pfennigii* have been obtained by repeated transfer of the methanogenic coculture into mineral medium containing propionate sulfate and 5 mM bromoethanesulfonic acid (Wallrabenstein et al.,

1995b). The latter compound is added to inhibit the growth of the methanogen. Similarly, pure cultures of *S. wolinii* and *S. fumaroxidans* have been obtained by repeated transfer of the methanogenic coculture into mineral medium with pyruvate or fumarate as the energy source (Wallrabenstein et al., 1994; Harmsen et al., 1998; Liu et al., 1999). One must be cautioned about the use of this approach, however, because it is difficult to demonstrate that a pure culture has been obtained. In addition, *Desulfovibrio* species often remain as contaminants in methanogenic propionate-degrading cultures. The fermentative growth of *Desulfovibrio* species with pyruvate and fumarate is much faster than that of *Syntrophobacter* species, making it difficult to eliminate the contaminant by dilution. The slow growth rates of *Syntrophobacter* species with propionate and sulfate make it unlikely that syntrophobacters can be enriched and isolated with a propionate–sulfate medium.

MAINTENANCE PROCEDURES

Methanogenic cocultures are grown in the basal medium with 20 mM propionate and transferred every 2–4 weeks. Sulfate-reducing cocultures require the addition of 30 mM sodium sulfate to this medium. Pure cultures of *S. wolinii* are grown in basal medium with 20 mM propionate and 30 mM sodium sulfate and are transferred as indicated above. Pure cultures of *S. pfennigii* can be maintained in a mineral medium with 20 mM propionate and 30 mM sulfate. *S. fumaroxidans* can be maintained in a mineral medium with 20 mM fumarate. Addition of 50–100 μM dithionite to the above media helps to initiate growth after transfer. The optimal growth temperature for the species is 35–37°C. *S. wolinii* can be stored in sealed glass vessels in liquid nitrogen vapor.

PROCEDURES FOR TESTING SPECIAL CHARACTERS

The ability to metabolize propionate syntrophically is tested by inoculating the organism into tubes of basal medium containing 20 mM propionate and reduced with cysteine–sulfide solution. To each tube, 0.1–0.5 ml of a pure culture of a hydrogen/formate-using methanogens or a sulfate reducer is added. When the latter is used, 30 mM sodium sulfate is added to the medium. Growth, propionate degradation, and methane or sulfide production are monitored over time and compared to controls lacking propionate. In addition, media with and without propionate are inoculated with only the hydrogen/formate-using organism to show that this organism is not responsible for metabolism of propionate.

To confirm that methanogenic cocultures require hydrogen and/or formate removal for propionate degradation, the effect of bromoethanesulfonic acid—an inhibitor of methanogenesis—on propionate degradation and growth of the coculture is determined. For sulfate-reducing cocultures, the effect of high partial pressures of hydrogen on propionate degradation is determined.

DIFFERENTIATION OF THE GENUS *SYNTROPHOBACTER* FROM OTHER GENERA

The most closely related genera to *Syntrophobacter*, as determined by 16S rDNA sequence analysis, are shown in Fig. BXII.δ.43. *Syntrophobacter wolinii* differs from *Desulforhabdus amnigena* by 6.9% and from *Thermodesulforhabdus norvegica, Desulfacinum infernum, Desulfomonile tiedjei, Smithella propionica, Desulfarculus baarsii,* and *Desulfobulbus propionicus* by at least 9%. The genus *Syntrophobacter* can be differentiated from the above genera by morphological and physiological characteristics. *Syntrophobacter* species and *Smithella propionica* (Liu et al., 1999) syntrophically oxidize propionate. However, *Smithella propionica* forms small amounts of butyrate from propionate and syntrophically metabolizes butyrate, which *Syntrophobacter* species cannot use. *Syntrophobacter* species, *Desulfobulbus* species, and *Desulforhabdus amnigena* grow by using propionate and sulfate, but neither *Desulfobulbus* species nor *Desulforhabdus amnigena* metabolize propionate syntrophically. The growth rate for *Desulfobulbus* species (0.9–1.6 h^{-1}) with propionate and sulfate is much faster than the growth rates of *Syntrophobacter* species (0.02–0.06 h^{-1}) (Wallrabenstein et al., 1994; Van Kuijk and Stams, 1995). In addition, *Desulforhabdus amnigena* completely oxidizes its substrates and grows with acetate, butyrate, and isobutyrate, substrates that *Syntrophobacter* species cannot use. *Thermodesulforhabdus norvegica* and *Desulfacinum infernum* are both thermophiles, while *Syntrophobacter* species are mesophilic. *Thermodesulforhabdus norvegica* and *Desulfacinum infernum* use acetate and other fatty acids that *Syntrophobacter* species cannot use. Morphological characteristics distinguish *Syntrophobacter* species from *Desulfomonile tiedjei*, which has a distinctive collar morphology, and *Desulfarculus baarsii*, which is a vibrio. In addition, *Desulfomonile tiedjei* uses benzoate and *Desulfarculus baarsii* uses butyrate and longer chain fatty acids; these are substrates that *Syntrophobacter* species cannot use.

TAXONOMIC COMMENTS

Phylogenetic analysis of 16S rDNA sequences shows that the species of *Syntrophobacter* form a monophylogenetic group that is distinct from other genera in the *Deltaproteobacteria* (Harmsen et al., 1993, 1995). *Syntrophobacter* species are most closely related to group 7 of the sulfate-reducing bacteria as defined by Devereux et al. (1989). Distance-matrix-similarity coefficients between *Syntrophobacter* and other genera differ by at least 6.9% (Harmsen et al. 1993, 1995; Oude Elferink, et al., 1995). DNA–DNA hybridization studies show that there is little homology (<26%) between *Syntrophobacter wolinii*, *Syntrophobacter pfennigii*, *Syntrophobacter fumaroxidans*, and one phylogenetically related but as yet unnamed organism, strain HP1.1 (Harmsen et al., 1998). This supports the contention that all four organisms are distinct species. A signature sequence for *Syntrophobacter* species, *Desulforhabdus amnigena*, and other closely related sulfate-reducing bacteria is 5′-GCGGGUACUCAUUCCUG-3′, which hybridizes to the V5 region of the 16S RNA corresponding to positions 5′-835-850-3′ of the *Escherichia coli* 16S RNA sequence.

The three species of *Syntrophobacter* and one phylogenetically related but as yet unnamed organism, strain HP1.1, form a phylogenetically coherent group within the sulfate-reducing bacteria. All four organisms share the ability to metabolize propionate syntrophically, which suggests that this trait may be a suitable taxonomic marker for this group. However, there are several unnamed organisms that reduce sulfate or are thermophilic and that also have the ability to metabolize propionate syntrophically (Mucha et al., 1988; Heppner et al., 1992; Wu et al., 1992; Stams et al. 1993; Imachi et al., 2002; Plugge et al., 2002). Taxonomic characterization of these organisms may show that syntrophic propionate metabolism is not found exclusively in members of the genus *Syntrophobacter*.

Desulforhabdus amnigena is phylogenetically closely related to *S. wolinii* (93.1% sequence similarity) (Oude Elferink et al., 1995) and other *Syntrophobacter* species (Harmsen et al., 1998). However, the physiology of *D. amnigena* is significantly different from that of *Syntrophobacter* species in that *D. amnigena* can oxidize a wide range of organic compounds coupled to sulfate reduction and is unable to oxidize propionate syntrophically. Further work may show that some organisms that are unable to metabolize propionate syntrophically are also members of the genus *Syntrophobacter*.

ACKNOWLEDGMENTS

We thank Henry C. Aldrich for the electron photomicrograph.

FURTHER READING

Oude Elferink, S.H.J.W., A. Visser, L.W. Hulshoff Pol and A.J.M. Stams. 1994. Sulfate reduction in methanogenic bioreactors. FEMS Microbiol. Rev. *15*: 119–126.

Schink, B. 1997. Energetics of syntrophic cooperation in methanogenic degradation. Microbiol. Mol. Biol. Rev. *61*: 262–280.

Stams, A.M. 1994. Metabolic interactions between anaerobic bacteria in methanogenic environments. Antonie Leeuwenhoek *66*: 271–294.

DIFFERENTIATION OF THE SPECIES OF THE GENUS *SYNTROPHOBACTER*

Differential characteristics of the species of the genus *Syntrophobacter* are given in Table BXII.δ.39. Other characteristics are given in Table BXII.δ.40.

List of species of the genus Syntrophobacter

1. **Syntrophobacter wolinii** Boone and Bryant 1984, 355[VP] (Effective publication: Boone and Bryant 1980, 631.)
   wo.lin'i.i. M.L. gen. n. *wolinii* of Wolin; named after Meyer J. Wolin, an American microbiologist who contributed substantially to the understanding of interspecies hydrogen transfer.

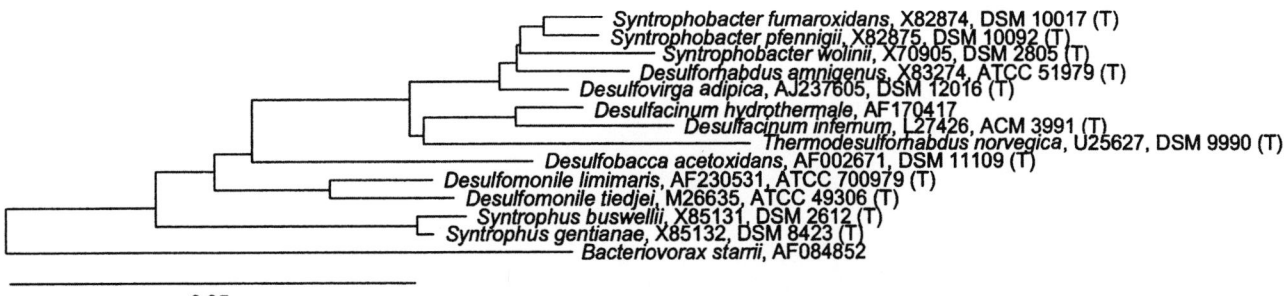

FIGURE BXII.δ.43. Phylogenetic tree showing the relationship of *Syntrophobacter* species to other members of the *Deltaproteobacteria*. The distances in the tree were calculated using 1101 positions (the least-squares method, Jukes-Cantor model). (Courtesy T. Lilburn of the Ribosomal Database Project.)

The characteristics are as described for the genus and as indicated in Tables BXII.δ.39 and BXII.δ.40. The width of the cells is sometimes irregular, especially in filaments. Cells are highly refractile.

No growth occurs in thioglycolate medium, suggesting that glucose and the materials present in yeast extract and a tryptic digest of casein do not support growth. Utilization of other carbohydrates or aromatic compounds has not been determined.

Habitat: Anaerobic sewage sludge

The mol% G + C of the DNA is: not determined.

Type strain: DB, DSM 2805.

GenBank accession number (16S rRNA): X70905.

Additional Remarks: S. wolinii was originally deposited as a coculture with Desulfovibrio sp. strain G11 under the designation DSM 2805 (Boone and Bryant, 1980). Subsequently, a pure culture of S. wolinii (Wallrabenstein et al., 1994) was obtained, which was deposited under the number given above.

2. **Syntrophobacter fumaroxidans** Harmsen, Van Kuijk, Plugge, Akkermans, De Vos and Stams 1998, 1386[VP]

fu.mar.ox'i.dans. M.L. neut. adj. fumaricum pertaining to fumaric acid (fumarate); L. adj. oxidans oxidizing; L. adj. fumaroxidans oxidizing fumarate. The ability to oxidize fu-

marate played an important role in the isolation and physiological characterization of this bacterium.

The characteristics are as described for the genus as indicated in Tables BXII.δ.39 and BXII.δ.40. Rod to eye-shaped cells with round ends; found singly or in pairs.

Succinate and formate also serve as electron donors for sulfate reduction, and aspartate also supports growth of the pure culture. Butyrate, isobutyrate, acetate, citrate, lactate, butanol, propanol, ethanol, methanol, glucose, fructose, xylose, and glutamate are not utilized. Thiosulfate and fumarate can serve as electron acceptors, but nitrate is not reduced. Growth is optimal in freshwater medium.

Habitat: granular sludge from an upflow anaerobic sludge bed (UASB) reactor treating sugar-beet processing wastewater.

The mol% G + C of the DNA is: 60.6 (HPLC).

Type strain: MPOB, DSM 10017.

GenBank accession number (16S rRNA): X82874.

3. **Syntrophobacter pfennigii** Wallrabenstein, Hauschild and Schink 1996, 836[VP] (Effective publication: Wallrabenstein, Hauschild and Schink 1995b, 351.)

pfen.ni'gi.i. M.L. gen. n. pfennigii of Pfennig; named after Norbert Pfennig, a German microbiologist who contributed

TABLE BXII.δ.39. Differential characteristics of the species of the genus Syntrophobacter[a]

Characteristic	S. wolinii	S. fumaroxidans	S. pfennigii
Morphology:			
Rods in pairs and long filaments	+		
Lemon-shaped rod		+	+
Motility	−	−	+
Gas vesicles	−	−	+
Substrates for axenic growth:			
Pyruvate	+	+	−
Fumarate	+	+	−
Malate	+	+	−
Products of pyruvate metabolism:			
Acetate	+	+	
Propionate	+		
Succinate		+	
CO₂	+	+	
Desulforubidin	−	nd	+
16S rRNA probe	5′-ACGCAGACTCATCCCCGTG-3′	5′-CGTCAGCCATGAAGCTTAT-3′	5′-TCAAGTCCCCAGTCTCTTCGA-3′
5′-start site for probe	223	464	460

[a]Symbols: see standard definitions; nd, not determined.

TABLE BXII.δ.40. Other characteristics of Syntrophobacter species[a]

Characteristic	S. wolinii	S. fumaroxidans	S. pfennigii
Dimensions, μm	0.6–1.0 × 1.0–4.5	1.1–1.6 × 1.8–2.5	1.0–1.2 × 2.2–3.0
Temperature range for growth, °C	27–37	20–40	20–37
Temperature optimal, °C	35	37	37
pH range	6.0–7.5	6.0–8.0	6.2–8.0
pH optimal	7.0	7.0	7.0–7.3
Electron donors (coculture):			
Acetate	−	−	−
Butyrate	−	−	−
Caproate	−	nd	−
Palmitate	−	nd	−
Lactate	nd	−	+
Propanol	nd	−	+
Glucose	−	−	−
Desulfoviridin	−	nd	−
Mol% G + C of DNA (HPLC)	nd	60.6	57.3

[a]Symbols: see standard definitions; nd, not determined.

substantially to the understanding of syntrophic relationships and of the physiology of sulfate-reducing bacteria.

The characteristics are as described for the genus and as indicated in Tables BXII.δ.39 and BXII.δ.40. Oval to egg-shaped rods, found singly, in pairs, or in chains. Motility is observed only in the early exponential phase of growth. Gas vesicles formed in the late exponential phase of growth.

Other substrates that do not support the growth of the pure culture include fructose, xylose, arabinose, ethanol, methanol, succinate, fumarate, oxaloacetate, malate, acry-late, valerate, H_2/CO_2/acetate, and formate/acetate. Propionate and lactate are metabolized to acetate and CO_2 in the presence of a H_2/formate-using organism (methanogen or sulfate reducer). Sulfite and thiosulfate, but not nitrate or fumarate, serve as electron acceptors. Growth requires a mineral medium with a strong reductant (dithionite).

Habitat: Anaerobic sewage sludge.

The mol% G + C of the DNA is: 57.3 (HPLC).

Type strain: KoProp1, DSM 10092.

GenBank accession number (16S rRNA): X82875.

Other Organisms

Strain HP1.1 is a mesophilic, lemon- to peanut-shaped, Gram-negative rod that oxidizes propionate to acetate and CO_2 in the presence of methanogens. 16S rDNA sequence analysis shows that strain HP1.1 is closely related to *S. wolinii* strain DB, *S. pfennigii* strain KoProp1, and *S. fumaroxidans* with 93.4, 95.7, and 95.4% sequence similarity, respectively (Zellner et al., 1996). This suggests that strain HP1.1 represents a new species of *Syntrophobacter*. Like other *Syntrophobacter* species, strain HP1.1 is able to couple propionate oxidation to sulfate reduction.

Genus II. **Desulfacinum** Rees, Grassia, Sheehy, Dwivedi and Patel 1995a, 88[VP]

GAVIN N. REES AND BHARAT K.C. PATEL

De.sul.fa.ci′num. L. pref. *de* from; L. n. *sulfur* sulfur; L. n. *acinum* a berry, especially a grape; L. n. *Desulfacinum* a berry-shaped, sulfate-reducing bacterium.

Oval cells, 1.5 × 2.5–3 μm, occurring singly or in pairs. **Gram negative. Nonmotile. Endospores not formed. Anaerobic, possessing a respiratory type of metabolism with sulfate, sulfite, and thiosulfate as terminal electron acceptors, which are reduced to sulfide.** Able to use a wide range of organic acids and alcohols as electron donors. **Autotrophic growth possible. Temperature range for growth, 40–65°C; optimum, 60°C.** Vitamins are required for growth and can be supplied as yeast extract. Isolated from the production fluids of high temperature petroleum reservoirs.

The mol% G + C of the DNA is: 64.

Type species: **Desulfacinum infernum** Rees, Grassia, Sheehy, Dwivedi and Patel 1995a, 88.

FURTHER DESCRIPTIVE INFORMATION

Thin sections examined by electron microscopy demonstrate a Gram-negative type of wall. Gram stain and lysis by KOH produce a Gram-negative reaction.

D. infernum can be grown in a phosphate-buffered, thioglycollate/ascorbic acid-reduced medium[1]. If a growth medium is required that does not have a high iron content, a bicarbonate-buffered, sulfide-reduced medium can be used[2]. Vitamins are required for growth and can be supplied either from a vitamin supplement or by the addition of 0.1 g/l yeast extract to the medium.

The pH range for growth is 6.6–8.4, with optimal growth occurring at 7.1–7.5.

The following compounds are used as electron donors and carbon sources: straight-chain monocarboxylic acids from C_1 to C_{18}, isobutyrate, isovalerate, lactate, pyruvate, fumarate, malate, succinate, glycerol, ethanol, propanol, butanol, hexanol, and alanine. Growth occurs with H_2 as an electron donor. Sulfate, thiosulfate, and sulfite are reduced to H_2S, with lactate, ethanol, or formate as the electron donor. Nitrate and sulfur are not used as terminal electron acceptors. Fermentative growth occurs on pyruvate in the absence of an exogenous electron acceptor. Fermentative growth does not occur on glycerol or fumarate.

Desulfoviridin is not present.

Current descriptions of *Desulfacinum* have been restricted to isolates obtained from the production fluids of high-temperature petroleum reservoirs. The type strain was obtained from the Beatrice Field, North Sea, United Kingdom, and a further isolate was obtained from an off-shore reservoir in Malaysia (Rees, unpublished data). Both reservoirs have approximately the same salinity as seawater. *Desulfacinum hydrothermale* was isolated from a submarine vent located in Palaeochori Bay, Milos, in the Aegean Sea.

ENRICHMENT AND ISOLATION PROCEDURES

No selective medium exists for the isolation of *Desulfacinum*. The greatest likelihood of success will occur if production waters from off-shore petroleum reservoirs (field temperature approximately 60–80°C) are used for isolation. *Desulfotomaculum* spp. may be present in enrichments with acetate and *Thermodesulfobacterium* spp. and *Desulfotomaculum* spp. may occur when lactate is added as the carbon and energy source. All cultures should be incubated at 60°C. Pure cultures are obtained by dilution in agar shake tubes with bicarbonate-buffered, sulfide-reduced medium containing 2 g/l of agar. Once tubes are inoculated and the agar is set, the headspace of the tubes is flushed with a mixture of N_2/CO_2 (80:20), and tubes are sealed with the appropriate butyl rubber stoppers. Tubes are inverted during incubation to prevent contamination of the surface of the agar plug with liquid formed due to incubation at 60°C. Further details on the agar shake dilution procedure are described in the section on *Desulfobacter*.

---

1. Phosphate-buffered, thioglycollate/ascorbic acid-reduced medium is prepared as described in the footnote of the section on *Desulfovibrio*, with the following additions (g/l): NaCl, 10; MgCl$_2$·6H$_2$O, 3.1. Sodium lactate (3.0–4.0 g/l) is added as the carbon and energy source.

2. The bicarbonate-buffered, sulfide-reduced medium is prepared as described in the footnote of the section on *Desulfobacter*, with the following addition (g/l): NaCl, 10; MgCl$_2$·6H$_2$O, 3.1. Sodium lactate (3.0–4.0 g/l) is added as the carbon and energy source.

MAINTENANCE PROCEDURES

Short- to medium-term storage is achieved by inoculating phosphate-buffered, iron-rich medium. Bottles are incubated until growth occurs and significant amounts of iron sulfide form in the medium. The resulting cultures are stored at 20°C and should be transferred to fresh medium every 2–3 months. Long term storage is achieved by lyophilization.

DIFFERENTIATION OF THE GENUS *DESULFACINUM* FROM OTHER GENERA

At present, species included in the genus are *Desulfacinum infernum* and *Desulfacinum hydrothermale*. The organisms most closely related to it are *Thermodesulforhabdus norvegica*, *Desulforhabdus amnigena*, and *Syntrophobacter wolinii* (with 16S rRNA similarities of 89.2, 90.3, and 90.6%, respectively). Characteristics differentiating *Desulfacinum* from these other species are presented in Table BXII.δ.41.

TAXONOMIC COMMENTS

The 16S rRNA gene of *Desulfacinum infernum* possesses the signature sequences that define the *Proteobacteria* (Stackebrandt et al., 1988b). Phylogenetically, *D. infernum* and *D. hydrothermale* are placed in the *Deltaproteobacteria*. The most closely related organisms are *Thermodesulforhabdus norvegica* (Beeder et al., 1995), *Desulforhabdus amnigena* (Oude Elferink et al., 1995), and *Syntrophobacter wolinii* (Boone and Bryant, 1980; Harmsen et al., 1993). *Desulfacinum infernum*, *Thermodesulforhabdus norvegica*, and *Desulforhabdus amnigena* are almost equidistant from one another, with similarities of approximately 90%. Although they are currently recognized as three separate genera, analysis of the 16S rRNA sequences seems to suggest that the three organisms may represent a single genus; however, further study is required before the taxonomic status of these organisms is re-evaluated.

A phylogenetic tree showing *Desulfacinum*, its closest relatives, and representative members of the *Deltaproteobacteria* are shown in Fig. BXII.δ.44.

TABLE BXII.δ.41. Differential characteristics of *Desulfacinum* and related species[a]

Characteristic	Desulfacinum infernum[b]	Desulfacinum hydrothermale[c]	Desulforhabdus amnigena[d]	Syntrophobacter wolinii[e]	Thermodesulforhabdus norvegica[f]
Morphology	Oval	Oval to short rod	Rod to ellipsoidal	Rod, in pairs or filaments	Rod
Cell width, μm	1.0	0.8–1.0	1.4–1.9	0.6–1.0	1.0
Cell length, μm	2.5	1.5–2.5	2.5–3.4	<35	2.5
Motility	−	+	−	−	+
Temperature optimal, °C	60	60	37	35	60
Salinity optimal, g/l	10	32–36	Freshwater[g]	0.4	16
Electron donors/carbon sources utilized:					
H$_2$/CO$_2$, formate	+	+	+	nd	−
Acetate, butyrate	+	+	+	−	−
Propionate	+	+	+	+	+
Ethanol	+	+	+	nd	+
Propanol	+	+	+	−	−
Lactate	+	+	+	−	+
Fumarate	+	−	−	nd	+
Malate	+	−	−	nd	+
Electron acceptors utilized:					
Sulfate	+	+	+	+	+
Thiosulfate	+	+	+	−	−
Sulfite	+	+	+	−	+
Fermentation of pyruvate	+	+	+	+	−
Source:					
Petroleum reservoir production fluids	+				+
Hydrothermal vent		+			
Granular sludge from an upflow anaerobic sludge bed digester			+		
Anaerobic sewage digester				+	
Mol% G + C of the DNA	64	59.5	52.5	nd	51

[a]For symbols see standard definitions; nd, not determined.

[b]Data from Rees et al. (1995a).

[c]Data from Sievert and Kuever (2000).

[d]Data from Oude Elferink et al. (1995).

[e]Data from Boone and Bryant (1980) and Wallrabenstein et al. (1994).

[f]Data from Beeder et al. (1995).

[g]Freshwater medium contains 1.0 g/l NaCl.

FURTHER READING

Rees, G., G.N. Grassia, A.J. Sheehy, P.P. Dwivedi and B.K.C. Patel. 1995. *Desulfacinum infernum* gen. nov., sp. nov., a thermophilic sulfate-reducing bacterium from a petroleum reservoir. Int. J. Syst. Bacteriol. *45*: 85–89.

Sievert, S.M. and J. Kuever. 2000. *Desulfacinum hydrothermale* sp. nov., a thermophilic sulfate-reducing bacterium from geothermally heated sediments near Milos Island (Greece). Int. J. Syst. Bacteriol. *50*: 1239–1246.

List of species of the genus Desulfacinum

1. **Desulfacinum infernum** Rees, Grassia, Sheehy, Dwivedi and Patel 1995a, 88[VP]

in.fer' num. L. adj. *infernum* that which is, or comes from below, especially the lower world.

The characteristics are as described for the genus and as listed in Table BXII.δ.41.

Isolated from production fluids of high petroleum reservoirs.

The mol% G + C of the DNA is: 64 (T_m).
Type strain: BaG1, ACM 3991, DSM 9756.
GenBank accession number (16S rRNA): L27426.

2. **Desulfacinum hydrothermale** Sievert and Kuever 2000, 1244[VP]

hy.dro.ther.ma' le. Gr. n. *hydro* water; Gr. n. *thermos* heat; M.L. neut. adj. *hydrothermale* pertaining to hot water, i.e., to hydrothermal vents.

Cells oval to short rods, 0.8–1.0 × 1.5–2.5 μm. Requires at least 15g/l NaCl, optimal between 32 and 36 g/l. Growth occurs on 3 to 5, 8, 10, 12, 16, and 18 carbon straight-chain carboxylic acids, alcohols, lactate, and pyruvate. Does not use fumarate or succinate.

Isolated from a shallow submarine hydrothermal vent.
The mol% G + C of the DNA is: 59.5 (HPLC).
Type strain: MT-96, DSM 13166.
GenBank accession number (16S rRNA): AF170417.

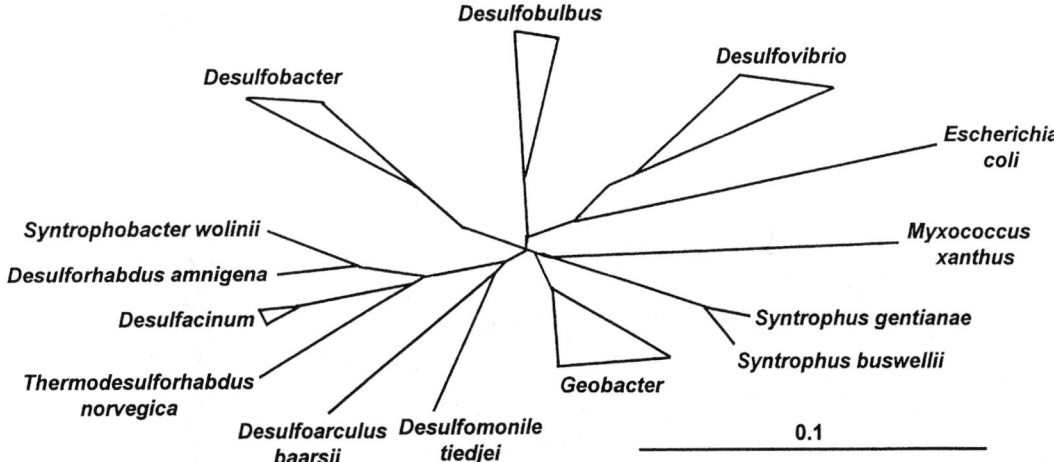

FIGURE BXII.δ.44. Phylogenetic tree of *Desulfacinum* spp. and related organisms. *Desulfovibrio, Desulfobulbus, Desulfobacter,* and *Geobacter* species are represented as clusters.

Genus III. **Desulforhabdus** Oude Elferink, Maas, Harmsen and Stams 1997, 1274[VP] (Effective publication: Oude Elferink, Maas, Harmsen and Stams 1995, 123)

JAN KUEVER, FRED A. RAINEY AND FRIEDRICH WIDDEL

De.sul.fo.rhab' dus. L. pref. *de* from; L. neut. n. *sulfur* sulfur; Gr. fem. n. *rhabdus* rod; M.L. fem. n. *Desulforhabdus* a rod-shaped sulfate reducer.

Cells are rod-like to ellipsoid, 1.4–1.9 × 2.5–3.4 μm. Occur singly, in pairs, or in long chains. Spore formation is not observed. Gram negative and nonmotile. Strictly anaerobic, having a respiratory type of metabolism. **Chemoautotrophic or chemoorganotrophic**, using H_2 (+ CO_2), formate, acetate, propionate, butyrate, isobutyrate, lactate, pyruvate, ethanol, 1-propanol, and 1-butanol **as electron donors and carbon sources**. Organic substrates are completely oxidized to CO_2 via the anaerobic C_1-pathway (carbon monoxide dehydrogenase pathway, Wood-pathway) as indicated by carbon monoxide dehydrogenase activity, but transient excretion of acetate, isobutyrate, and propionate can occur. For chemoautotrophic growth, **H_2 serves as the electron donor and** CO_2 **as the carbon source. Sulfate, sulfite, and thiosulfate serve as terminal electron acceptors and are reduced to H_2S.** Sulfur is not reduced. **No growth by fermentation in the absence of an external electron acceptor.**

The only described species of the genus is able to grow in simple defined media containing a reductant (usually sulfide); vitamins are not required. The optimal pH for growth is 7.2–7.6. The optimal temperature is 37°C. Thermophilic species are not known. NaCl is not required. Desulfoviridin is absent. Type strain was isolated from anoxic granular sludge from a reactor treating sulfate-rich wastewater.

The mol% G + C of the DNA is: 52.5.

Type species: **Desulforhabdus amnigena** Oude Elferink, Maas, Harmsen and Stams 1997, 1274 (Effective publication: *Desulforhabdus amnigenus* (sic) Oude Elferink, Maas, Harmsen and Stams 1995, 123.)

FURTHER DESCRIPTIVE INFORMATION

The type strain of *D. amnigena* has rod-shaped to ellipsoid cells under all growth conditions. The cells are nonmotile and stain Gram negative.

The temperature range for growth is 25–45°C. The type strain has an optimal growth temperature of 37°C. The pH range for growth is 6.6–8.5, with an optimal pH of 7.2–7.6.

D. amnigena reduces sulfate, sulfite, or thiosulfate to H_2S. Elemental sulfur, fumarate, and nitrate cannot serve as electron acceptors. The following compounds are utilized as electron donors and organic carbon sources: formate, acetate, propionate, butyrate, isobutyrate, lactate, pyruvate, ethanol, 1-propanol, and 1-butanol. Despite the capacity for complete oxidation of organic substrates, transient excretion of organic products may occur; these products are acetate during growth on propionate or 1-butanol, acetate and isobutyrate (presumably by an isomerizing side-reaction) during growth on butyrate, and acetate and propionate during growth on lactate. For chemoautotrophic growth, H_2 serves as electron donor and CO_2 as carbon source. No growth in the absence of an external electron acceptor.

Strains of *Desulforhabdus* may be cultured in a defined freshwater medium (e.g., as used for *Desulfobulbus*) with sulfate as electron acceptor and lactate or various other substrates as the electron donor and carbon source. Ammonium ions are used as the nitrogen source. Vitamins are not required as growth factors. Cells contain cytochromes of the *c*-type. Desulfoviridin is not present.

Desulforhabdus strains occur in sewage sludge and probably other freshwater habitats.

ENRICHMENT AND ISOLATION PROCEDURES

For the selective enrichment of *Desulforhabdus* strains, the acetate–sulfate medium for freshwater sulfate-reducing bacteria (e.g., as described for *Desulfobacter* without elevated salt concentration) should be used. The type strain was isolated from a serial dilution series with acetate as the only substrate. The inoculum was obtained from granular sludge of a reactor treating sulfate-rich wastewater. An enrichment culture with acetate, inoculated with 10% sewage sludge, may also be used. The cultures are incubated at 37°C. After intense formation of H_2S has occurred, transfers to fresh medium are made. Pure cultures of *Desulforhabdus* strains have been obtained by repeated application of the roll-tube-dilution method, but the agar dilution method (as described for *Desulfobacter* without elevated salt concentration) may also work.

MAINTENANCE PROCEDURES

Stock cultures are transferred every 3–5 months. Towards the end of growth or at the beginning of the stationary phase, cultures are refrigerated and stored at 2–5°C in the dark. *Desulforhabdus* strains may be preserved indefinitely by suspension of cells in anoxic medium containing 5% dimethylsulfoxide and storage in liquid nitrogen.

DIFFERENTIATION OF THE GENUS *DESULFORHABDUS* FROM OTHER GENERA

In contrast to morphologically similar *Desulfococcus* strains, which are also found in freshwater habitats, *Desulforhabdus* strains are restricted to fatty acids with shorter chains as substrates. Clear differentiation requires a comparative analysis of the 16S rDNA sequence.

TAXONOMIC COMMENTS

The closest relatives of *Desulforhabdus amnigena* are the syntrophic propionate oxidizer *Syntrophobacter wolinii* and the sulfate-reducing bacteria *Desulfomonile tiedjei* and *Desulfarculus baarsii*.

FURTHER READING

Oude Elferink, S.J.W.H., R.N. Maas, H.J.M. Harmsen and A.J.M. Stams. 1995. *Desulforhabdus amnigenus* gen. nov. sp. nov., a sulfate reducer isolated from anaerobic granular sludge. Arch. Microbiol. *164*: 119–124.

List of species of the genus Desulforhabdus

1. **Desulforhabdus amnigena** Oude Elferink, Maas, Harmsen and Stams 1997, 1274[VP] (Effective publication: *Desulforhabdus amnigenus* (sic) Oude Elferink, Maas, Harmsen and Stams 1995, 123.)

   am.ni' ge.na. L. masc. n. *amnis* water; L. fem. n. *genus* origin; L. adj. *amnigena* coming from water.

   Characteristics are as described for the genus. Type strain isolated from sludge of a reactor (Netherlands) supplemented with acetate.

   The mol% G + C of the DNA is: 52.5 (HPLC).
   Type strain: ASRB1, ATCC 51979, DSM 10338.
   GenBank accession number (16S rRNA): X83274.

Genus IV. **Desulfovirga** Tanaka, Stackebrandt, Tohyama and Eguchi 2000a, 643[VP]

JAN KUEVER, FRED A. RAINEY AND FRIEDRICH WIDDEL

De.sul.fo.vir' ga. L. pref. *de* from; L. n. *sulfur* sulfur; L. fem. n. *virga* twig, rod; M.L. fem. n. *Desulfovirga* a sulfate-reducing rod..

Rod-shaped cells, 0.8–2 × 2.2–4 μm. Occur singly or in pairs. Spore formation is not observed. Gram negative. Motile by a single polar flagellum. Strictly anaerobic, having a respiratory type of metabolism. **Chemoorganotrophic**, using C_1–C_{12} straight-chain fatty acids, other organic acids, and alcohols **as carbon sources and electron donors.** Electron donors are completely oxidized to CO_2. **Sulfate, sulfite, thiosulfate, and sulfur serve as terminal electron acceptors and are reduced to H_2S. Nitrate is not used as a terminal electron acceptor.** Optimal growth temperature, 35°C. Media containing a reductant and vitamins are necessary for growth. Cells contain cytochrome *c*, but no desulfoviridin. *Desulfovirga* species occur in freshwater habitats.

The mol% G + C of the DNA is: 60.

Type species: **Desulfovirga adipica** Tanaka, Stackebrandt, Tohyama and Eguchi 2000a, 643.

FURTHER DESCRIPTIVE INFORMATION

Cells are slightly curved short rods, 0.8–2 µm in diameter and 2.2–4.0 µm in length. Occur singly or in pairs. Spore formation is not observed. Gram negative. Cells are nonmotile or motile by subpolar flagella.

The temperature for growth is between 20 and 36°C with an optimal at 35°C. The pH range for growth is between 6.6 and 7.4 with an optimal at 7.0.

D. adipica uses sulfate, sulfite, thiosulfate, and sulfur as electron acceptors, whereas nitrate is not used. Adipate, (E)-2-hexendioate, (E)-hexenedioate, pyruvate, lactate, C_1–C_{12} straight-chain fatty acids, and C_2–C_{10} straight-chain primary alcohols serve as electron donors and also as carbon sources; these compounds are oxidized completely to CO_2. Adipate is degraded via (E)-2-hexenedioate by β-oxidation.

Media containing a reductant and vitamins are necessary for growth. Cells contain *c*-type cytochromes but no desulfoviridin. Ammonium ions are used as the nitrogen source. *Desulfovirga* strains occur in anoxic freshwater sediments rich in organic material.

ENRICHMENT AND ISOLATION PROCEDURES

For the selective enrichment of *Desulfovirga* spp., the same defined sulfate medium as described for the isolation of the type species (Tanaka et al., 2000a) may be used. Other freshwater media (e.g., as described for *Desulfobulbus*) may be used in the same way. Adipate as sole electron donor and carbon source favors the development of *Desulfovirga* strains. Bottles or tubes (100-, 50-, or 20-ml capacity) can serve as culture vessels and are kept anoxic with tight screw caps, or by using a headspace of N_2/CO_2 (90:10) and sealing with butyl rubber stoppers. The medium is inoculated with black, anoxic, freshwater sediment slurry; 3–5% of the total culture volume should be added. *Desulfovirga* species may also be enriched and isolated via direct serial dilution of anoxic environmental samples in liquid medium (according to most-probable-number method) or agar medium (see next paragraph) with adipate as organic substrate. Enrichment cultures are incubated at 30°C. Cultures should be briefly shaken once a week. After intense formation of H_2S has occurred, transfers to fresh medium are made. The enrichment cultures should be mixed well before subculturing.

Pure cultures of *Desulfovirga* strains are obtained by repeated application of the agar dilution method. The method is similar to that described for the genus *Desulfobacter*, except that other substrates (see above) are used.

MAINTENANCE PROCEDURES

Stock cultures are transferred every 3–5 months. Towards the end of growth or at the beginning of the stationary phase, cultures are refrigerated and stored at 2–5°C in the dark. *Desulfovirga* strains may be preserved indefinitely by suspension of cells in anoxic medium containing 5% dimethylsulfoxide and storage in liquid nitrogen.

DIFFERENTIATION OF THE GENUS *DESULFOVIRGA* FROM OTHER GENERA

Desulfovirga adipica can be distinguished from other mesophilic sulfate-reducing bacteria by its ability to utilize adipate. So far, this is the only known sulfate-reducing bacterium able to use adipate as sole electron donor and carbon source. However, reliable differentiation of adipate-utilizing sulfate-reducing bacteria from other types requires comparative 16S rDNA sequence analysis.

TAXONOMIC COMMENTS

The closest relatives of *Desulfovirga adipica* are members of the genera *Syntrophobacter* and *Desulforhabdus*.

FURTHER READING

Tanaka, K., E. Stackebrandt, S. Tohyama and T. Eguchi. 2000. *Desulfovirga adipica* gen. nov., sp. nov. an adipate-degrading, Gram-negative, sulfate-reducing bacterium. Int. J. Syst. Evol. Bacteriol. *50*: 639–644.

List of species of the genus Desulfovirga

1. **Desulfovirga adipica** Tanaka, Stackebrandt, Tohyama and Eguchi 2000a, 643[VP]

   a.di'pi.ca. M.L. n. *acidum adipinum* adipic acid (adipate); M.L. fem. adj. *adipica* pertaining to adipic acid and the organism's ability to degrade it.

   Characteristics are as described for the genus. The type strain was isolated from an anaerobic digestor of a municipal wastewater treatment center, (Tsuchiura, Ibaraki, Japan) with adipate as substrate.

   The mol% G + C of the DNA is: 60 (HPLC).
   Type strain: TsuAS1, DSM 12016.
   GenBank accession number (16S rRNA): AJ237605.

Genus V. **Thermodesulforhabdus** Beeder, Torsvik and Lien 1996, 625[VP] (Effective publication: Beeder, Torsvik and Lien 1995, 335)

JANICHE BEEDER

Ther.mo.de.sul.fo.rhab'dus. Gr. adj. *thermos* warm, hot; L. pref. *de* from; L. n. *sulfur* sulfur; Gr. fem. n. *rhabdos* rod; M.L. fem. n. *Thermodesulforhabdus* thermophilic, rod-shaped, sulfate reducer.

Rod-shaped cells 1 × 2.5 µm. Motile by means of a single polar flagellum. Gram-negative by staining and cell wall structure. **Anaerobic.** Chemoorganotrophic.

Nutritionally versatile, oxidizing various organic acids; these are oxidized completely to CO_2. Sulfate and sulfite are used as electron acceptors. Does not contain desulfoviridin, desulfofuscidin, desulforubidin, or P582.

Habitat: anaerobic warm oil field water.

The mol% G + C of the DNA is: 51.

Type species: **Thermodesulforhabdus norvegica** Beeder, Torsvik and Lien 1996, 625 (Effective publication: *Thermodesulforhabdus norvegicus* (sic) Beeder, Torsvik and Lien 1995, 335.)

FURTHER DESCRIPTIVE INFORMATION

Enrichment cultures were obtained from anoxic samples of hot oil field waters by the use of immunomagnetic beads (Christensen et al., 1992).

DIFFERENTIATION OF THE GENUS *THERMODESULFORHABDUS* FROM OTHER GENERA

Table BXII.δ.42 presents characteristics differentiating *Thermodesulforhabdus norvegica* from other phenotypically similar organisms.

TAXONOMIC COMMENTS

Only a single species has been described, *Thermodesulforhabdus norvegica*, which belongs to the *Deltaproteobacteria*.

Fig. BXII.δ.45 indicates the relationship of this species to other bacteria in this class, based on analysis of 16S rRNA sequences.

ACKNOWLEDGMENTS

The water samples from which *Thermodesulforhabdus* was isolated were contributed by the Norwegian oil company Statoil.

List of species of the genus Thermodesulforhabdus

1. **Thermodesulforhabdus norvegica** Beeder, Torsvik and Lien 1996, 625[VP] (Effective publication: *Thermodesulforhabdus norvegicus* (sic) Beeder, Torsvik and Lien 1995, 335.) nor.ve′gi.ca. M.L. fem. adj. *norvegica* Norwegian, describing the place of isolation.

   The characteristics are as described for the genus and listed in Table BXII.δ.42. Temperature range for growth, 44–74°C, pH range for growth, 6.1–7.7. With SO_4^{2-} as the electron acceptor, electron donors are acetate, lactate, pyruvate, butyrate, succinate, malate, fumarate, valerate, caproate, heptanoate, octanoate, nonadecanoate, decanoate, tridecanoate, pentadecanoate, palmitate, heptadecanoate,

   stearate, and ethanol. Compounds not used as electron acceptors include formate, propionate, benzoate, H_2/CO_2, acetate (1 mM) + H_2, butanol, 1-propanol, yeast extract, undecanoate, dodecanoate, tetradecanoate, and crude oil. With acetate as the electron donor, electron acceptors are SO_4^{2-} and SO_3^{2-}. Acetate is metabolized via the oxidative carbon monoxide dehydrogenase pathway. Nitrate, fumarate, sulfur, and thiosulfate are not used as electron acceptors. Isolated from hot North Sea oil field water from a Norwegian oil platform.

   The mol% G + C of the DNA is: 51 (HPLC).

   Type strain: A8444, DSM 9990.

   GenBank accession number (16S rRNA): U25627.

TABLE BXII.δ.42. Characteristics useful in differentiating *Thermodesulforhabdus norvegica* from *Desulfacinum infernum* and *Syntrophobacter wolinii*[a]

Characteristic	*T. norvegica*	*D. infernum*[b]	*S. wolinii*[c]
Shape:			
Rods	+	+	
Rods and filaments			+
Arrangement:			
Single	+	+	
Pairs			+
Dimensions, μm	1 × 2.5	1.5 × 2.5-3	0.6–1 × <35
Motility	+	−	−
Temperature optimal,°C	60	60	35
pH optimal	6.9	7.1–7.5	nd
Salinity optimal (g/l NaCl)	16	10	0.4
Pyruvate fermentation	−	+	−
Electron acceptors:			
Thiosulfate	−	+	nd
Sulfite	+	+	nd
Oxidation of organic substrates:			
Complete	+	+	
Incomplete			+
Electron donors:			
Acetate, lactate, C_{4-10} fatty acids, C_{13-18} fatty acids	+	+	−
Propionate	−	+	+
Formate, butanol, propanol	−	+	−
Autotrophic growth	−	+	nd
Vitamin requirement	−	+	nd
Mol% G + C	51	64	nd

[a]For symbols see standard definitions; nd, not determined.
[b]Data from Rees et al. (1995a).
[c]Data from Boone and Bryant (1980).

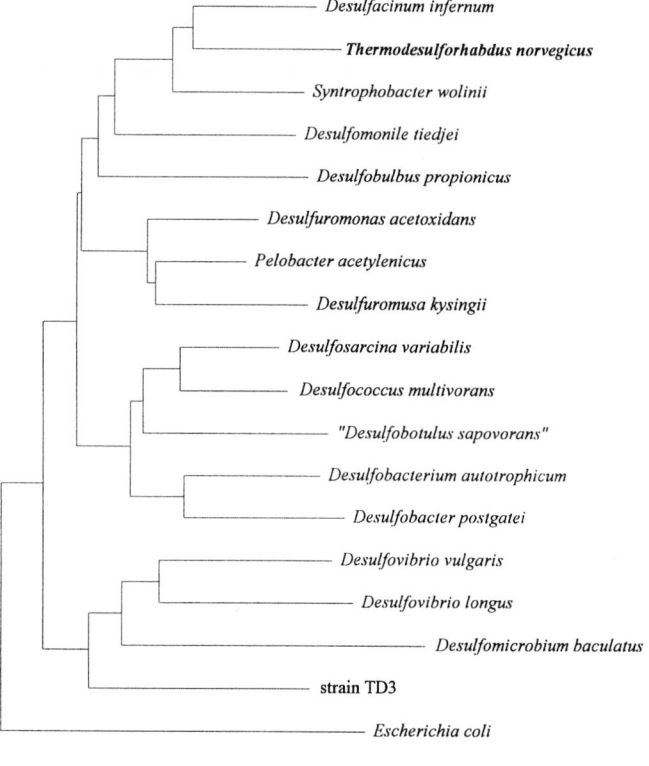

FIGURE BXII.δ.45. Phylogenetic dendrogram indicating the position of *Thermodesulforhabdus norvegica* within the sulfate-reducing bacteria belonging to the *Deltaproteobacteria*. Pairwise evolutionary distances were computed using the correction of Jukes and Cantor (1969). The least-squares distance method of De Soete (1983) was used in the construction of the phylogenetic dendrogram from the distance matrix. Bar = 5 inferred nucleotide substitutions per 100 nucleotides. (Reproduced by permission from J. Beeder et al., Archives of Microbiology *164*: 331–336, 1995, ©Springer-Verlag.)

Family II. **Syntrophaceae** *fam. nov.*

JAN KUEVER, FRED A. RAINEY AND FRIEDRICH WIDDEL

Syn.troph.a.ce' ae. Gr. masc. n. *Syntrophus* type genus of the family; suffix *-aceae* denotes family; M.L. fem. pl. n. *Syntrophaceae* the family that comprises the genus *Syntrophus.*

For the genera *Desulfomonile* and *Desulfobacca,* the common electron acceptor is sulfate, which is reduced to sulfide. Sulfate reduction is coupled to complete oxidation of organic substrates. Members of the genera *Syntrophus* and *Smithella* are unable to use sulfate as an electron acceptor; rather they grow in syntrophic association with H_2-utilizing microorganisms, and oxidize substrates incompletely to acetate. All known members are mesophilic bacteria. The family comprises the genera *Syntrophus, Desulfomonile, Desulfobacca,* and *Smithella.* Detailed properties of these genera are given in Table BXII.δ.43.

Type genus: **Syntrophus** Mountfort, Brulla, Krumholz and Bryant 1984, 216.

TABLE BXII.δ.43. Differential characteristics of the genera *Syntrophus, Desulfomonile, Desulfobacca,* and *Smithella* which constitute the family *Syntrophaceae*[a]

Genus	*Syntrophus*	*Desulfomonile*	*Desulfobacca*	*Smithella*
Cell morphology	Rod-shaped	Rod-shaped	Rod-shaped	Rod-shaped
Motility	±	−	−	+
Mol% G + C content	43–53	49	51	nr
Sulfite reductase	−[b]	DV	nr	−[b]
Major menaquinone	Not identified	nr	nr	nr
Substrate oxidation	I	C	C	I
CO–DH	nr	+	+	nr
Typical electron donors and carbon sources:				
Acetate	−	−	+	−
Benzoate	+	+	−	−
Butyrate	−	+	−	+
Crotonate	+	nr	−	+
Formate	−	+	−	−
Propionate	−	−	−	+
Pyruvate	−	+	−	−
H_2/CO_2	−	+	−	−
Methoxylated aromatic acids	nr	+	nr	nr
Chemolithoautotrophic growth	−	+	−	−
Fermentative growth	+	+	−	+
Growth by disproportionation of reduced sulfur compounds	nr	+	nr	nr
Typical electron acceptors:				
Crotonate	+[c]	nr	nr	+
Sulfate	−	+	+	−
Sulfite	−	+	+	−
Thiosulfate	−	+	+	−
m-Halogenated benzoates	nr	+	nr	nr
H_2/formate-using syntrophic partners	+	nr	nr	+
Optimal growth temperature (°C)	28–37	37	37	37
NaCl requirement	−	−	−	−
Habitat (anoxic):				
Freshwater	+	+	+	+
Waste water treatment plant	+	+	+	+

[a]Symbols: nr, not reported; +, present, observed in all species; −, not present, not observed; ±, variable; I, incomplete oxidation; C, complete oxidation; CO-DH, carbon monoxide dehydrogenase, a key enzyme indicative of the C_1-pathway for acetyl-CoA oxidation; DV, desulfoviriden.

[b]Not present, fermentative metabolism.

[c]The type species *S. buswellii* and *S. gentianae* can use crotonate as electron acceptor.

Genus I. **Syntrophus** *Mountfort, Brulla, Krumholz and Bryant 1984, 216*[VP]

JAN KUEVER AND BERNHARD SCHINK

Syn' tro.phus. Gr. masc. n. *syntrophos* foster brother; N.L. masc. n. *Syntrophus* one living syntrophically with another so that each produces a nutrient required by the other.

Cells are rod shaped with rounded ends, 0.5–0.8 × 1.0–2.0 μm. Occur singly and in pairs. Gram negative. Endospores not formed. The type species is motile by means of monotrichous polar flagella in the early stage of growth, other species show no motility. **Strictly anaerobic and chemoorganotrophic. Possesses a fermentative type of metabolism. Crotonate is fermented by all species, whereas aromatic compounds such as gentisate, hy-** droquinone, and cinnamate are only fermented by some species. Some substrates, such as benzoate or fatty acids, are oxidized in the presence of H_2/formate-utilizing methanogenic or sulfate-reducing partner bacteria. Some species can grow on some substrates in the presence of crotonate as electron acceptor. Substrate oxidation is incomplete and leads to acetate. Sulfate, sulfite, thiosulfate, sulfur, and nitrate are not used as electron ac-

ceptors. Cells contain cytochromes and menaquinones. Optimal growth temperature, 28–37°C. Occur in anoxic freshwater sediment or sludge from anaerobic wastewater treatment facilities.

The mol% G + C of the DNA is: 43–53.

Type species: **Syntrophus buswellii** Mountfort, Brulla, Krumholz and Bryant 1984, 216.

FURTHER DESCRIPTIVE INFORMATION

Strains of *Syntrophus* isolated from anoxic freshwater sediments or sewage sludge treatment plants may be motile only during early growth phases.

Crotonate is fermented by pure cultures of *Syntrophus* strains according to the following equation: 2 Crotonate$^-$ + 2H$_2$O → 2 Acetate$^-$ + Butyrate$^-$ + H$^+$. Gentisate and hydroquinone are fermented to benzoate, acetate, and CO$_2$ by a pure culture of *S. gentianae*. One mol benzoate is degraded by syntrophic co-cultures to 3 mol acetate, 1 mol CO$_2$, and 3 mol H$_2$, which is consumed by sulfate-reducing or methanogenic partner organisms.

Benzoate is activated by a benzoyl-CoA ligase and not a benzoyl-CoA transferase (Auburger and Winter, 1996; Schöcke and Schink, 1999). In *S. gentianae* the pyrophosphate formed by this reaction is cleaved by a membrane-bound proton-translocating pyrophosphatase. Benzoyl-CoA reductase is not present. The further breakdown is catalyzed by glutaryl-CoA dehydrogenase and enzymes of the β-oxidation pathway (Wallrabenstein and Schink, 1994; Auburger and Winter, 1996). Reversed electron transport coupled to cytochromes and menaquinones is involved in hydrogen release during benzoate oxidation in co-cultures (Wallrabenstein and Schink, 1994; Auburger and Winter, 1996;

Schöcke and Schink, 1999). Thus, these bacteria can synthesize no more than ⅛ to ⅔ mol ATP per mol benzoate oxidized (Warikoo et al., 1996; Schöcke and Schink, 1998, 1999).

The ability to use long-chain fatty acids and methyl esters of butyrate and hexanoate as electron donors and carbon sources is restricted to *S. aciditrophicus* (see Table BXII.δ.44).

The pH range for growth is 6.5–7.5, with the optimal being 7.1–7.4. The temperature range for growth is 10–42°C with an optimal at 28–37°C.

Cells of *Syntrophus* strains contain *b*- and *c*-type cytochromes and unidentified menaquinones. (Wallrabenstein and Schink, 1994; Auburger and Winter, 1996). Major cellular fatty acids of *Syntrophus aciditrophicus* are C$_{16:0}$, C$_{16:1\ \omega7c}$, and C$_{18:1\ \omega7c}$ (Jackson et al., 1999).

Syntrophus species occur in anoxic, sulfide-containing, freshwater sediments or sewage sludge treatment plants. Marine isolates have not been obtained so far.

Syntrophus strains can be cultured in a defined anaerobic medium with a reducing agent; *S. aciditrophicus* requires clarified rumen fluid as a supplement (Jackson et al., 1999).

ENRICHMENT AND ISOLATION PROCEDURES

For the selective enrichment of *Syntrophus* strains, an anoxic mineral medium (e.g., as prepared for *Desulfobacter*) supplemented with 5–10 mM sodium benzoate is used. Bottles or tubes (100-, 50-, or 20-ml capacity) can serve as culture vessels and are kept anoxic with tight screw caps, or by using a headspace of N$_2$/CO$_2$ (90:10) and sealing with butyl rubber stoppers. The medium is inoculated with anoxic mud from a freshwater habitat; 3–5% of the total culture volume should be added. *Syntrophus* species may also be enriched and isolated via direct serial dilution of anoxic

TABLE BXII.δ.44. Morphological and physiological characteristics of described species of the genus *Syntrophus*[a]

Species	Syntrophus buswellii	Syntrophus aciditrophicus	Syntrophus gentianae
Cell morphology	Rod-shaped	Rod-shaped	Rod-shaped
Cell size (μm)	0.8 × 1–2	0.5–0.7 × 1.0–1.6	0.8–1 × 1.3–1.6
Motility	± (sp)	−	−
Mol% G + C content	nr	43.1	53.2
Major menaquinone	nr	nr	nr
Optimal temperature (°C)	37	35	28
Oxidation of substrate	I	I	I
Fermentative growth by pure culture:			
Cinnamate	+[b]	nr	−
Crotonate	+	+	+
Gentisate	−	nr	+
Hydroquinone	−	nr	+
Electron donors used by pure culture with crotonate as electron acceptor:			
Benzoate	+[b]	−	+
3–Phenylpropionate	+[b]	nr	+
Electron donors used in co-culture with H$_2$/formate-using partners:			
Benzoate	+	+	+
Fatty acids (C4, 6, 7, 16, 18)	−	+	−
trans-2-Pentenoate	nr	+	nr
trans-2-Hexanoate	nr	+	nr
trans-3-Hexanoate	nr	+	nr
2-Octenoate	nr	+	nr
Methyl esters of butyrate and hexanoate	nr	+	nr
Electron acceptors used	Crotonate	−	Crotonate
Growth factor requirement	nr	0.5% clarified rumen fluid	nr
NaCl requirement (g/l)	−	−	−
Literature	Mountfort et al., 1984; Fardeau et al., 1993; Auburger and Winter, 1995; Wallrabenstein et al., 1995a	Jackson et al., 1999	Wallrabenstein et al., 1995a

[a]Symbols: sp, single polar flagellum; nr, not reported; I, incomplete; +, good growth; (+), poor growth; −, no growth.

[b]Might vary between strains.

environmental samples in liquid medium (according to most-probable-number method) or agar with benzoate as organic substrate. To prevent development of sulfate-reducing bacteria, the medium should contain either a low sulfate concentration (only for assimilatory purposes) or no sulfate (in which case the reductant, sulfide, serves as sulfur source). Depending on the salinity of the anaerobic mud sample that is used as the inoculum, salt- or freshwater medium is used. Gas (methane) production after 3–5 weeks incubation at 28–37°C indicates substrate degradation with the involvement of methanogens. Pure cultures are obtained by repeated application of the agar dilution method, using 20 mM crotonate as the sole electron donor and electron acceptor.

MAINTENANCE PROCEDURES

Stock cultures are transferred every 2–3 months. Towards the end of growth or at the beginning of the stationary phase, cultures are refrigerated and stored at 4–8°C in the dark. *Syntrophus* strains may be preserved indefinitely by suspension of cells in anoxic medium containing 5% dimethylsulfoxide and storage in liquid nitrogen.

DIFFERENTIATION OF THE GENUS *SYNTROPHUS* FROM OTHER GENERA

The genus *Syntrophus* is distinguished from other fermentative bacteria by its limitation to benzoate and crotonate as the only substrates for growth, although some species can also use other aromatic compounds, and one species uses various fatty acids in syntrophic co-cultures. Inorganic electron acceptors are not used. However, a clear differentiation of this genus and other physiologically related genera is only possible via determination of the 16S rDNA sequence.

TAXONOMIC COMMENTS

The closest relatives of the genus *Syntrophus* are *Smithella propionica*, *Desulfomonile tiedjei*, and *Desulfobacca acetoxidans*.

FURTHER READING

Mountfort, D.O., W.J. Brulla, L.R. Krumholz and M.P. Bryant. 1984. *Syntrophus buswellii* gen. nov., sp. nov.: a benzoate catabolizer from methanogenic ecosystems. Int. J. Syst. Bacteriol. *34*: 216–217.

Jackson, B.E., V.K. Bhupathiraju, R.S. Tanner, C.R. Woese and M.J. McInerney. 1999. *Syntrophus aciditrophicus* sp. nov., a new anaerobic bacterium that degrades fatty acids and benzoate in syntrophic association with hydrogen-using microorganisms. Arch. Microbiol. *171*: 107–114.

Wallrabenstein, C., N. Gorny, N. Springer, W. Ludwig and B. Schink. 1995. Pure culture of *Syntrophus buswellii*, definition of its phylogenetic status, and description of *Syntrophus gentianae* sp. nov. System. Appl. Microbiol. *18*: 62–66.

List of species of the genus Syntrophus

1. **Syntrophus buswellii** Mountfort, Brulla, Krumholz and Bryant 1984, 216[VP]

   bus.wel'li.i. N.L. masc. gen. n. *buswellii* in honor of A.M. Buswell, who first demonstrated degradation of benzoate in enrichments from methanogenic ecosystems.

   Characteristics are summarized in Table BXII.δ.44. The type strain was isolated with benzoate as the sole electron donor. Occurs in anoxic sediments from marine or brackish water habitats.

   The mol% G + C of the DNA is: 54.6 (T_m) (Fardeau et al., 1993); strain GA, 54.4 (T_m) (isolated by Auburger and Winter, 1995).

   Type strain: DM-2, DSM 2612.

   GenBank accession number (16S rRNA): X85831.

2. **Syntrophus aciditrophicus** Jackson, Bjupathiraju, Tanner, Woese and McInerney 2001, 793[VP] (Effective publication: Jackson, Bjupathiraju, Tanner, Woese and McInerney 1999, 112.)

   a.ci.di.tro'phi.cus. L. nom. n. *acidum* acid; Gr. v. *trephein* to feed; L. suff. *trophicus* relating to feeding; M.L. adj. *aciditrophicus* one that feed on acids, acid-feeding.

   Characteristics are summarized in Table BXII.δ.44. *S. aciditrophicus* is more versatile than the other described species (see Table BXII.δ.44). The type strain was isolated with benzoate as the sole electron donor in association with H_2/formate-using partners. Pure cultures are obtained with crotonate as the sole substrate. Occurs in sewage sludge of municipal sewage treatment plants.

   The mol% G + C of the DNA is: 53.2 (T_m).

   Type strain: SB, ATCC 700169.

   GenBank accession number (16S rRNA): U86447.

3. **Syntrophus gentianae** Wallrabenstein, Gorny, Springer, Ludwig and Schink 1996, 836[VP] (Effective publication: Wallrabenstein, Gorny, Springer, Ludwig and Schink 1995a, 65.)

   gen.ti.a'nae. M.L. fem. n. *gentiana, gentianae*, of gentian, referring to the substrate gentisic acid, which has been isolated from gentian plants.

   Characteristics are summarized in Table BXII.δ.44. The type strain was isolated with hydroquinone or gentisate as the sole electron donor. Occurs in anoxic freshwater sediments.

   The mol% G + C of the DNA is: 53.2 (T_m).

   Type strain: HQGö1, DSM 8423.

   GenBank accession number (16S rRNA): X85832.

Genus II. **Desulfobacca** Oude Elferink, Akkermans-van Vliet, Bogte and Stams 1999, 348[VP]

ALFONS J.M. STAMS AND STEFANIE J.W.H. OUDE ELFERINK

De.sul.fo.bac'ca. L. pref. *de* from; L. n. *sulfur* sulfur; M.L. pref. *Desulfo* desulfuricating, used to characterize a dissimilatory sulfate-reducing procaryote; L. fem. n. *baca, bacca* berry, especially olive; M.L. fem. n. *Desulfobacca* a sulfate-reducing olive-shaped bacterium.

Oval to rod-shaped cells, 1.3 × 1.9–2.2 μm, occurring singly or in pairs. Nonmotile. Gram negative. Endospores not formed. **Anaerobic, with respiratory type of metabolism. Sulfate and inorganic sulfur compounds including sulfite and thiosulfate, but**

not S⁰, act as terminal electron acceptors and are reduced to H_2S. Acetate is used as the sole electron donor and carbon source and is oxidized to CO_2 via the carbon monoxide dehydrogenase pathway. Optimal pH, 7.5–7.7. Optimal temperature, 37°C. Belongs to the class *Deltaproteobacteria*. Isolated from sludge from an anaerobic sludge bed reactor fed with acetate and sulfate.

The mol% G + C of the DNA is: 51.

Type species: **Desulfobacca acetoxidans** Oude Elferink, Akkermans-van Vliet, Bogte and Stams 1999, 349.

FURTHER DESCRIPTIVE INFORMATION

Refractile gas vacuoles may occur in cells during the late log phase or stationary phase of growth. Cytochromes of the *c* type are present. Desulfoviridin is absent.

Optimal growth occurs in freshwater media. Growth is retarded in brackish water media, and no growth occurs in marine media.

Common bacterial substrates such as glucose, volatile fatty acids, lactate, H_2/CO_2, alcohols (ethanol, propanol, butanol), pyruvate, fumarate, amino acids (aspartate, glutamate), and aromatic compounds (phenol, benzoate) are not oxidized.

ENRICHMENT AND ISOLATION PROCEDURES

Desulfobacca acetoxidans was isolated from granular sludge from a mesophilic upflow anaerobic sludge bed reactor fed with acetate and sulfate. In this reactor, it had outcompeted *Methanosaeta* species at low acetate concentrations. Granular sludge samples were crushed and diluted serially in liquid media containing 20 mM acetate and sulfate under a gas phase containing N_2 and CO_2 (Oude Elferink et al., 1995). After incubation at 30°C, the highest dilutions in which growth was detected were chosen for isolation of colonies using anaerobic roll-tubes. Purity of isolates may be verified microscopically and by failure to grow on media with substrates other than acetate.

DIFFERENTIATION OF THE GENUS *DESULFOBACCA* FROM OTHER GENERA

Desulfobacca differs from most other sulfate reducers in that it is restricted to acetate as its source of electrons and carbon. *Desulfobacca* is phenotypically similar to *Desulfobacter* in this respect, because both are restricted to acetate as an electron donor, although some *Desulfobacter* species can use H_2 or lactate. However, *Desulfobacter* species oxidize acetate by means of the citric acid cycle, whereas *Desulfobacca* uses the carbon monoxide dehydrogenase pathway. In addition, *Desulfobacter* is a typical marine sulfate reducer and does not grow in freshwater media, while *Desulfobacca* grows optimally in freshwater media. Moreover, some species of *Desulfobacter* are motile.

Desulfobacca differs from *Desulfotignum balticum* (originally described as "*Desulfarculus*") in its inability to degrade aromatic compounds, its inability to use lactate, pyruvate, malate, fumarate, crotonate, and butyrate as electron donors, its inability to ferment pyruvate in the absence of an exogenous electron acceptor, and its lack of an NaCl requirement. Unlike the anaerobic, dehalogenating, sulfate-reducing genus *Desulfomonile*, *Desulfobacca* lacks a collar-like morphological feature, does not grow on pyruvate, and does not ferment pyruvate to acetate and lactate in the absence of other electron acceptors. Unlike *Nitrospina*, *Desulfobacca* is not an aerobic nitrifier.

TAXONOMIC COMMENTS

In this edition of the *Manual*, the genus *Desulfobacca* has been placed in the class *Deltaproteobacteria*, the order *Syntrophobacterales*, and the family *Syntrophaceae*. Other genera in this family include *Desulfomonile*, *Syntrophus*, and *Smithella*.

In a study by Oude Elferink et al. (1999), *Desulfobacca* was only distantly related to the phenotypically similar genus *Desulfobacter*, and, in this edition of the *Manual*, *Desulfobacter* is placed in a different family, *Desulfobacteraceae*.

List of species of the genus Desulfobacca

1. **Desulfobacca acetoxidans** Oude Elferink, Akkermans-van Vliet, Bogte and Stams 1999, 349[VP]

   a.cet.o'xi.dans. L. n. *acetum* vinegar; M.L. part. pres. *oxidans* oxidizing; M.L. part. adj. *acetoxidans* acetate-oxidizing.

The characteristics are as described for the genus.
The mol% G + C of the DNA is: 51 (HPLC).
Type strain: ASRB2, DSM 11109.
GenBank accession number (16S rRNA): AF002671.

Genus III. **Desulfomonile** *DeWeerd, Mandelco, Tanner, Woese and Suflita 1991, 178*[VP]
(Effective publication: DeWeerd, Mandelco, Tanner, Woese and Suflita 1990, 28)

KIM A. DEWEERD, G. TODD TOWNSEND AND JOSEPH M. SUFLITA

De.sul.fo.mo.ni' le. L. pref. *de* from; L. n. *sulfur* sulfur; L. n. *monile* collar; M.L. neut. n. *Desulfomonile* a "collared" sulfate-reducer.

Single rod-shaped cells with rounded ends, 0.5–0.7 × 3–10 µm, each containing an invagination of the cell wall that resembles a collar. Cells may be slightly curved or bent at the collar region. Endospores not formed. Gram negative. Nonmotile. Strict anaerobe. A medium containing a reducing agent is required for growth. Possess a respiratory type of metabolism with sulfur oxyanions reduced to H_2S or *meta*-halobenzoates reduced to dehalogenated products (reductive dehalogenation). Pyruvate can be fermented to acetate and lactate. Optimal growth temperature 37°C. Cells grow poorly on solid agar medium, yielding small (0.5–1.0 mm diameter), white colonies after extended incubation (>3 weeks). Growth in a mineral medium requires the vitamins thiamine, nicotinamide, and 1,4-naphthoquinone. Isolated from a sewage sludge bacterial consortium enriched on 3-chlorobenzoate.

The mol% G + C of the DNA is: 49.0.

Type species: **Desulfomonile tiedjei** DeWeerd, Mandelco, Tanner, Woese and Suflita 1991, 178 (Effective publication: DeWeerd, Mandelco, Tanner, Woese and Suflita 1990, 28.)

FURTHER DESCRIPTIVE INFORMATION

Desulfomonile tiedjei, formerly known as strain DCB-1 ("dechlorinating bacterium"), was isolated from a sewage sludge bacterial consortium enriched on 3-chlorobenzoate as the sole carbon and source (Shelton and Tiedje, 1984). Its long rod-shaped cells display a unique collar structure located anywhere along the length of the cell, depending on the growth stage (see Fig. BXII.δ.46). Unidirectional polar growth of the cell originates at the collar and proceeds until the total cell length is approximately twice the size of the original cell. Mother and daughter cells separate at the central collar, with each newly divided cell containing a terminal collar (Mohn et al., 1990). Cells stain Gram-negative, and transmission electron micrographs show that *D. tiedjei* cell walls have a bimembrane structure typical of Gram-negative procaryotes. Cells also have internal membranes in a stacked configuration and electron dense, spherical bodies that likely contain glycogen occurring preferentially near the poles (Mohn et al., 1990). Cells have a tendency to settle and form a film on the bottom of culture vessels.

The temperature range for growth is 30–39°C with an optimal of 37°C. The pH range for growth is 6.5–7.8 with an optimal of 6.8–7.0.

D. tiedjei reductively dehalogenates *meta*-chlorobenzoates, *ortho*- and *meta*-bromobenzoates, and all isomers of iodobenzoate to benzoate as a form of anaerobic respiration. *D. tiedjei* also dechlorinates *meta*-chlorine substituents from pentachlorophenol and tri- and di-chlorophenols and dechlorinates tetrachloroethene and trichloroethene, but it requires a *meta*-halobenzoate or a number of benzenoid analogs to induce these activities (Mohn and Tiedje, 1992). *D. tiedjei* reduces sulfate, sulfite, thiosulfate, and dithionite to sulfide, nitrate to nitrite, and fumarate to succinate. The presence of sulfur oxyanions in growth medium inhibits reductive dehalogenation activity (Townsend and Suflita, 1997). *D. tiedjei* cannot reduce S^0, Fe(III), or U(VI). Pyruvate, 3-anisate, 4-anisate, vanillate, or isovanillate can support growth in the mineral medium buffered with bicarbonate. With thiosulfate as the electron acceptor, additional carbon substrates that can support growth are benzoate, butyrate, lactate, acetate, L- and D-malate, L- and D-arabinose, and glycerol. With sulfate as the electron acceptor, *D. tiedjei* can grow with pyruvate, lactate, butyrate,

and 3-anisate as sources of carbon and energy. In mineral salts medium containing the essential vitamins, *D. tiedjei* grows poorly as a chemolithoautotroph with hydrogen or formate as the electron donor and sulfur oxyanions or *meta*-chlorobenzoates as electron acceptors. The addition of acetate to the medium stimulates growth under these conditions. Growth of *D. tiedjei* on acetate and thiosulfate occurs by utilizing the disproportionation of the latter as an energy source (Mohn and Tiedje, 1990). *D. tiedjei* requires nicotinamide, thiamine, and 1,4-naphthoquinone as growth factors in defined medium. Although NH_4Cl is supplied as the nitrogen source in the medium, *D. tiedjei* can fix N_2 for growth. Initiation of growth is favored by the addition of dithionite as a reductant.

The bacterium is inhibited by H_2S, $CuCl_2$, Na_2MoO_4, or Na_2SeO_4. *D. tiedjei* is sensitive to the electron transport inhibitors carbonyl-cyanide-*m*-chlorophenylhydrazine (CCCP), 8-hydroxyquinoline, diphenylamine, dimercaprol, metronidazole, 2,4-dinitrophenol (DNP), *N,N'*-dicyclohexylcarbodiimide (DCCD), pentachlorophenol (PCP), gramicidin, and monensin. Antibiotics that inhibit *D. tiedjei* are tetracycline, chloramphenicol, kanamycin, and streptomycin.

D. tiedjei cells contain desulfoviridin, cytochrome c_3, and a unique membrane-associated c-type cytochrome that is coinduced with the dehalogenase enzyme (Louie et al., 1997). Cells contain polyhydroxybutyrate and glycogen and are esculin negative.

D. tiedjei was isolated from a consortium enriched on 3-chlorobenzoate as the sole carbon and energy source over a 2-year period. Inoculum for the enrichment culture was originally obtained from the sludge digester of the municipality of Adrian, Michigan. The enrichment was maintained by a 10% (v/v) transfer into fresh media every two months; 3-chlorobenzoate (Na^+ salt) (2–3 mM) was added to the medium and replenished when depleted. Recently, closely related strains have been isolated from brackish and marine environments.

ENRICHMENT AND ISOLATION PROCEDURES

For the selective enrichment of *Desulfomonile* strains, a defined anaerobic medium is used. Bottles or tubes sealed with gas-impermeable rubber stoppers can serve as culture vessels. The medium is inoculated with municipal digester sludge and amended with 3-chlorobenzoate as the sole carbon and energy source. Enrichments are incubated at 35–37°C in the dark without agitation. Continual replenishment of 3-chlorobenzoate to the medium and successive transfers (10%) into fresh media containing 3-chlorobenzoate should maximize the probability of obtaining higher numbers of *D. tiedjei*, despite the slow growth rate of *D. tiedjei*. The availability of benzoate as a product of the dehalogenation reaction precludes a numerically dominant enrichment. Pure cultures of *Desulfomonile* strains are obtained by serial dilution and repeated application of the agar roll tube method. Since *D. tiedjei* is a sulfate-reducing bacterium, several substrates, such as pyruvate, H_2 plus CO_2, formate plus CO_2, benzoate, or 4-anisate can be used. Using ferrous sulfate as the electron acceptor permits the detection of sulfide-producing bacteria through observation of black zones around colonies, which are due to an FeS precipitate. Alternately, H_2 plus acetate and CO_2 or formate plus acetate and CO_2 may be used as the carbon source with 3-chlorobenzoate as the electron acceptor. Regardless of substrate used, three weeks of incubation are typically required before the appearance of pinpoint colonies. Isolated colonies should be transferred to the defined medium containing pyru-

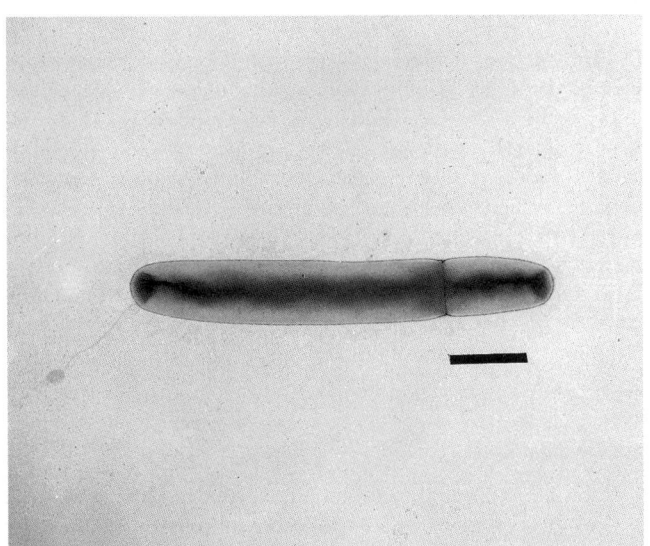

FIGURE BXII.δ.46. Electron micrograph of *Desulfomonile tiedjei*, displaying its distinctive collar-like morphology that serves as a site of cell elongation and probably cell division. Bar = 1 μm. (Reproduced with permission from K.A. DeWeerd et al., Archives of Microbiology, *154:* 23–30, 1990, ©Springer-Verlag, Berlin.) (Photograph courtesy of R.S. Tanner).

vate and 3-chlorobenzoate. Additions of yeast extract (0.1%) or clarified rumen fluid (10%) may be included in the liquid medium to stimulate growth. Once measurable growth occurs, identity can be presumptively verified by the unique collar structure and the dehalogenation of 3-chlorobenzoate.

MAINTENANCE PROCEDURES

Stock cultures may be maintained in the defined medium[1] in Balch tubes or serum bottles at 22–25°C with transfers every 1–3 months. *D. tiedjei* can be maintained on pyruvate alone or with 3-chlorobenzoate. Since H$_2$S is somewhat inhibitory to *D. tiedjei*, it is recommended that cultures not be maintained with sulfur oxyanions as electron acceptors.

For long-term preservation, an aliquot of culture can be mixed with an equal volume of sterile glycerol in a 5 ml serum bottle under a N$_2$ headspace and stored at −25°C. Cultures may also be lyophilized.

DIFFERENTIATION OF THE GENUS *DESULFOMONILE* FROM OTHER GENERA

The genus *Desulfomonile* is distinguished from other dissimilatory sulfate-reducing bacteria by its characteristic morphological and physiological attributes. Members of this genus contain an invagination of the cell wall, or "collar structure," which is visible under phase contrast microscopy. This feature is not present in other genera of sulfate-reducing bacteria. The ability of *Desulfomonile tiedjei* to reductively dehalogenate various halogenated benzoates distinguishes this bacterium from all of the other sulfate-reducing genera described. *Desulfitobacterium chlororespirans* can dechlorinate 3-chloro-4-hydroxybenzoate, but is unable to

dechlorinate 3-chlorobenzoate (Sanford et al., 1996). *Desulfomonile tiedjei* possesses the ability to dechlorinate the *meta*-chlorines of pentachlorophenol and several tri- and dichlorophenols. However, this activity must be induced by a *meta*-halobenzoate or a specific analog (Mohn and Kennedy, 1992). Other isolated bacteria that can dechlorinate chlorophenols are *Desulfitobacterium* species, which dechlorinate chlorophenols at the *ortho* position—including *Desulfitobacterium hafniense*, which dechlorinates at the *ortho* and *meta* positions, and *Desulfitobacterium frappieri*, which dechlorinates at all positions. *Desulfitobacterium* species can reduce thiosulfate, but not sulfate, and are placed in the *Desulfotomaculum–Clostridium* group. *Desulfomonile tiedjei* can also reductively dechlorinate tetra- and trichloroethylene, but this activity must be induced with 3-chlorobenzoate or a suitable analog. A large and diverse group of isolated anaerobic bacteria can also dechlorinate PCE and TCE. The anaerobic dechlorination activity of isolated bacteria has recently been summarized (El Fantroussi et al., 1998). The induction of the dehalogenase by *Desulfomonile tiedjei* is repressed by the presence of sulfur oxyanions, and sulfate and thiosulfate also directly inhibit the dehalogenation activity of the enzyme. The ability of *Desulfomonile tiedjei* to grow via the disproportionation of thiosulfate is a trait that is also found in *Desulfovibrio sulfodismutans*, *Desulfovibrio desulfuricans* CSN, *Desulfovibrio oxyclinae*, *Desulfocapsa thiozymogenes*, and several other strains of sulfate-reducing bacteria. The ability to fix nitrogen is found among members of the genera *Desulfovibrio*, *Desulfotomaculum*, and *Desulfobacter*. Two characteristic lipopolysaccharide hydroxy fatty acids (C$_{Br-19:0\ 3OH}$ and C$_{Br-21:0\ 3OH}$) have been used as signature molecules to detect *Desulfomonile tiedjei* inoculated in sediments (Ringelberg et al., 1994). *Desulfomonile tiedjei* was also detected in soil slurry microcosms using five specific oligonucleotide primers for PCR amplification of 16S rDNA genes of *D. tiedjei*. The five oligonucleotide primers used in a nested PCR analysis were located at 16S rRNA gene positions 59–79, 205–227, 476–495, 608–628, and 1032–1054 (El Fantroussi et al., 1997). Cell-specific polyclonal antibodies were also used to detect *D. tiedjei* in an upflow anaerobic granular-sludge blanket reactor (Ahring et al., 1992).

TAXONOMIC COMMENTS

The 16S rRNA of *Desulfomonile* shows two higher order structural features that link it strongly to the *Deltaproteobacteria*: (1) a lengthening of the stalk of the helix covering positions 184–193 from three pairs (characteristic of the *Alphaproteobacteria*, *Betaproteobacteria*, and *Gammaproteobacteria*) to about ten pairs and (2) a form of the helix covering positions 198–219 unique to the *Deltaproteobacteria* (Woese, 1987). 16S rRNA sequence analysis indicates that the genus belongs to the family *Syntrophaceae* family in the order *Syntrophobacterales* of the *Deltaproteobacteria*. The family *Syntrophaceae* contains two syntrophic genera (*Syntrophus* and *Smithella*) and one other sulfate-reducing genus (*Desulfobacca*).

---

1. Defined basal medium consists of (per liter of distilled water): PIPES (piperazine-*N*,*N*′-bis-[2-ethanesulfonic acid]) dipotassium salt, 1.5 g; NaCl, 0.8 g; NH$_4$Cl, 1.0 g; KCl, 0.1 g; KH$_2$PO$_4$, 0.1 g; MgCl$_2$·6H$_2$O, 0.2 g; CaCl$_2$·2H$_2$O, 0.02 g; NaHCO$_3$, 3.0 g; trace element solution, 10 ml; and resazurin solution (0.1%), 1.0 ml. The pH of the solution is adjusted to 7.8. Medium is prepared and dispensed using strict anaerobic technique and the gas phase is N$_2$/CO$_2$ (4:1) at 138 kPa positive pressure. After the medium has been autoclaved and cooled, the following components are added aseptically from filter sterilized, anoxic stock solutions: vitamin solution, 10 ml; sodium pyruvate (1 M), 40 ml; 3-chlorobenzoate (0.2 M in 0.2 M NaOH), 10 ml; and dithionite (10 mM), 10 ml. The pH of the autoclaved medium is 6.9–7.1. Trace element solution (per liter): nitrilotriacetic acid, 2.0 g; MnSO$_4$·H$_2$O, 1.0 g; Fe(NH$_4$)$_2$(SO$_4$)$_2$·6H$_2$O, 0.8 g; CoCl$_2$·6H$_2$O, 0.2 g; ZnSO$_4$·7H$_2$O, 0.2 g; CuCl$_2$·2H$_2$O, 0.02 g; NiCl$_2$·6H$_2$O, 0.02 g; Na$_2$MoO$_4$·2H$_2$O, 0.02 g; Na$_2$SeO$_4$, 0.02 g; Na$_2$WO$_4$, 0.02 g; pH adjusted to 6.0 with KOH. Vitamin solution (per liter): nicotinamide, 50.0 mg; 1,4-napthoquinone, 20.0 mg; thiamine, 5.0 mg; hemin, 5.0 mg; d-biotin, 5.0 mg; folic acid, 5.0 mg; pyridine-HCl, 5.0 mg; riboflavin, 5.0 mg, d-pantothenic acid, 5.0 mg; cyanocobalamin, 5.0 mg; *p*-aminobenzoate, 5.0 mg; lipoic acid (DL-6,8-thioctic acid), 5.0 mg. The 1,4-naphthoquinone and hemin are dissolved in 0.1 M NaOH to solubilize prior to addition to vitamin solution. The vitamin solution must be sterilized by filtration and not by autoclaving. Both the pyruvate stock solution and the dithionate stock solution should be freshly prepared and filter sterilized prior to use.

List of species of the genus Desulfomonile

1. **Desulfomonile tiedjei** DeWeerd, Mandelco, Tanner, Woese and Suflita 1991, 178 (Effective publication: DeWeerd, Mandelco, Tanner, Woese and Suflita 1990, 28.)

   tied′je.i. M.L. gen. n. *tiedjei* of Tiedje, named to recognize the contributions of James M. Tiedje and his laboratory to the field of microbial ecology.

   The description is as given for the genus, with the fol-

lowing additional information. Esculin negative. Can grow with H$_2$ or formate with CO$_2$ using sulfur oxyanions or *meta*-chlorobenzoates as electron acceptors, which are reduced to H$_2$S and benzoates, respectively. 3-Anisate, 4-anisate, vanillate, and isovanillate can serve as carbon sources, and benzoate is used as a growth substrate in the presence of sulfate or thiosulfate.

Occurs in municipal digester sludge enriched under anaerobic conditions to use 3-chlorobenzoate as sole carbon and energy source.

The mol% G + C of the DNA is: 49.0 (T_m).

Type strain: DCB-I, ATCC 49306, DSM 6799.

GenBank accession number (16S rRNA): M26635.

2. **Desulfomonile limimaris** Sun, Cole and Tiedje 2001, 370[VP]

li.mi.ma′ ris. M.L. gen. n. *limimaris* of mud from the sea, referring to its isolation from marine sediments.

Displays morphology and dehalogenation activities similar to those of *D. tiedjei*. 16S rRNA sequences show 93% similarity. Unlike *D. tiedjei*, growth and dehalogenation require 0.32 to 2.5% NaCl with an optimal of 1.25%. Other physiological traits that distinguish *D. limimaris* from *D. tiedjei* are its ability to use propionate and its inability to use acetate as electron donors and its inability to ferment pyruvate.

The mol% G + C of the DNA is: unknown.

Type strain: DCB-M, ATCC 700979.

GenBank accession number (16S rRNA): AF230531.

Genus IV. **Smithella** Liu, Balkwill, Aldrich, Drake and Boone 1999, 553[VP]

MARTIN SOBIERAJ AND DAVID R. BOONE

Smi.thel′ la. N.L. fem. n. *Smithella* named for Paul H. Smith in honor of his many contributions to the understanding of methanogenic propionate degradation.

Elongated, slightly sinuous rods, 0.5 μm in diameter, with round ends. **Weakly motile. Gram negative.** Endospores have not been observed. **Strictly anaerobic.** Mesophilic. **Grows on crotonate as energy source; also capable of using butyrate, propionate, malate, and fumarate in syntrophic association with H$_2$- or formate-utilizing microorganisms.** Catabolism of propionate involves its reductive carboxylation to butyrate (or butyryl-CoA).

Habitat: methanogenic environments in which propionate is degraded, such as anaerobic digestors.

Type species: **Smithella propionica** Liu, Balkwill, Aldrich, Drake and Boone 1999, 554.

FURTHER DESCRIPTIVE INFORMATION

Cells of *Smithella* are usually 3–5 μm long, with some longer cells up to 10 μm. They may contain granules of poly-β-hydroxyalkanoate that can be observed by staining with Sudan black. The cell wall has a typical Gram-negative structure.

The organism grows most rapidly at pH values near neutrality and at salt concentrations lower than 100 mM. *Smithella* grows on crotonate in pure culture. In coculture with *Methanospirillum hungatei*, it grows on a variety of substrates including crotonate, butyrate, propionate, malate, and fumarate, but not lactate, succinate, oxalate, valerate, caproate, or pyruvate. Yeast extract and peptones are required for growth.

Smithella (in coculture with *Methanospirillum hungatei*) degrades propionate to acetate, CO$_2$, and H$_2$ by an unknown catabolic pathway. The only other known syntrophic propionate-degrading bacterium, *Syntrophobacter wolinii*, uses the randomizing pathway, yielding 1 mol of acetate, 1 mol of CO$_2$, and 3 mol of H$_2$ per mol of propionate degraded (Houwen et al., 1990). In contrast, *Smithella* produced more acetate and less H$_2$. A ^{13}C-NMR study of *Smithella propionica* (de Bok et al., 2001) indicates a novel pathway of propionate oxidation that includes the condensation of two propionates into a 6-carbon intermediate (possibly bound to coenzyme A), with the 6-carbon intermediate subsequently cleaved to form acetate and butyrate.

ENRICHMENT AND ISOLATION PROCEDURES

Cultivation of *Smithella* requires stringent anaerobic techniques as described by Hungate (1969) or modifications of those techniques (Sowers and Noll, 1995). Cells can be enriched from methanogenic environments such as anaerobic digestors or methanogenic sediments. MS enrichment medium (Boone et al., 1989) at pH 7.2 with 20 mM propionate has been used to enrich *Smithella*, but any bicarbonate-buffered medium containing small amounts of yeast extract and peptones, plus propionate as catabolic substrate, would probably be satisfactory.

Although *Smithella* grows on crotonate in pure culture, it has not been directly isolated from natural environments or from enrichment cultures. Methanogenic cocultures are the first cultures obtained after enrichment. Such cocultures of *Smithella* with methanogens can be obtained by using a propionate-containing enrichment medium solidified in roll tubes. Dilutions of the enrichment culture are inoculated into the roll-tube media and co-inoculated with H$_2$-utilizing and acetate-utilizing methanogens. Colonies may take several months to develop, and they result in methanogenic cocultures. Axenic cultures are obtained from these cocultures by culturing in MS medium with 20 mM crotonate and then inoculating dilutions into medium with 20 mM crotonate as the catabolic substrate. The highest dilution that shows growth may contain only *Smithella*, but if methanogens are present, repetition of the extinction dilution technique will remove them. The complete removal of methanogens can be confirmed by inoculating the purified *Smithella* culture into media containing H$_2$ or acetate.

DIFFERENTIATION OF THE GENUS *SMITHELLA* FROM OTHER GENERA

The ability to grow on propionate in syntrophic association with methanogens is the most distinctive and perhaps the most important phenotypic feature of *Smithella*. The only other genus to contain such syntrophs is *Syntrophobacter*, and *Smithella* can be differentiated from *Syntrophobacter* by the pathway for propionate degradation. *Syntrophobacter* uses the randomizing pathway, resulting in 1 mol of acetate and 1 mol of CO$_2$ per mol of propionate degraded, whereas *Smithella* uses a different pathway that results in more than 1 mol of acetate, plus traces of butyrate, per mol of propionate degraded (Liu et al., 1999).

TAXONOMIC COMMENTS

Phylogenetic analysis of the 16S rDNA sequence of *Smithella propionica* LYP[T] indicates a relationship (88% sequence similarity) to the genus *Syntrophus*. However, this relationship is sufficiently distant to support the placement of *Smithella* as a separate genus.

The 16S rDNA sequence of *Smithella propionica* is very similar (99.4% sequence similarity) to the sequence obtained from a propionate enrichment culture named Syn7 (Harmsen et al., 1995). This culture was obtained from roll tubes by a colony pick, but a rod-shaped bacterium contaminates the culture in small numbers. To date, only the type strain of *Smithella* has been isolated in pure culture and characterized.

Phylogenetically, *Smithella* is most closely associated with *Syntrophus*. Table BXII.δ.45 compares some important phenotypic characteristics of *Smithella* with some of its phylogenetic relatives in the order *Syntrophobacterales*.

List of species of the genus Smithella

1. **Smithella propionica** Liu, Balkwill, Aldrich, Drake and Boone 1999, 554[VP]

   pro.pi.o' ni.ca. M.L. n. *acidum propionicum* propionic acid; M.L. adj. *propionica* pertaining to propionic acid, on which the bacterium grows.

   The characteristics are those given for the genus, with the following additional features. Cells are Gram-negative rods, elongated and slightly sinuous, 0.5 × 3–5 μm, with some cells as long as 10 μm. Contain granules of poly-β-hydroxybutyrate. Weakly motile. Strictly anaerobic. Fastest

growth at mesophilic temperatures, at near neutral pH, and in media with Na^+ and Cl^- concentrations less than 100 mM. Grows syntrophically on propionate with H_2/formate-utilizing methanogens. Grows slowly in pure culture by fermenting crotonate to butyrate and acetate. Habitat: methanogenic digestors.

   The mol% G + C of the DNA is: unknown.
   Type strain: LYP, OCM 661.
   GenBank accession number (16S rRNA): AF126282.

TABLE BXII.δ.45. Phenotypic characteristics of the genus *Smithella* and some of its phylogenetic relatives in the order *Syntrophobacterales*[a]

Characteristic	Smithella propionica	Syntrophobacter wolinii	Syntrophospora bryantii	Syntrophus buswellii	Syntrophus aciditrophicus	Syntrophus gentianae
Gram reaction	−	−	+[b]	−	−	−
Motility	W	nd	−	+	−	−
Mol% G + C of DNA	nd	nd	37.6	54.1	43.1	53.2
Optimal pH	6.5–7.5	~7	6.5–7.5	nd	nd	7.1–7.4
Optimal temperature, °C	33–35	34–37	28–34	nd	35	28
Endospore formation	−	−	+	−	−	−
Substrate utilization by pure cultures:						
Crotonate	+	+	+	+	+	+
Fumarate	−	+	nd	nd	−	nd
Substrate utilization in cocultures with H₂/formate-utilizing organisms:						
Propionate	+	+	−	−	−	−
Butyrate	+	+	+	−	+	−
C$_{5:0}$–C$_{8:0}$ (aliphatic) acids	nd	−	+	−	+	−
Acetate	−	−	nd	−	nd	−
Benzoate	nd	−	nd	+	+	+

[a]For symbols see standard definitions; W, weakly positive; nd, not determined.

[b]Ultrastructure is typical Gram positive, but Gram stain results are variable.

Order VII. **Bdellovibrionales** *ord. nov.*

GEORGE M. GARRITY, JULIA A. BELL AND TIMOTHY LILBURN

Bdel.lo.vib.ri.on.a' les. M.L. masc. n. *Bdellovibrio* type genus of the order; *-ales* ending to denote order; M.L. fem. n. *Bdellovibrionales* the *Bdellovibrio* order.

The order *Bdellovibrionales* was circumscribed for this volume on the basis of phylogenetic analysis of 16S rRNA sequences; the order contains the family *Bdellovibrionaceae*.

Description is the same as for the family *Bdellovibrionaceae*.
Type genus: **Bdellovibrio** Stolp and Starr 1963, 243.

Family I. **Bdellovibrionaceae** *fam. nov.*

GEORGE M. GARRITY, JULIA A. BELL AND TIMOTHY LILBURN

Bdel.lo.vib.ri.on.a' ce.ae. M.L. masc. n. *Bdellovibrio* type genus of the family; *-aceae* ending to denote family; M.L. fem. pl. n. *Bdellovibrionaceae* the *Bdellovibrio* family.

The family *Bdellovibrionaceae* was circumscribed for this volume on the basis of phylogenetic analysis of 16S rRNA sequences; the

family contains the genera *Bdellovibrio* (type genus), *Bacteriovorax*, *Micavibrio*, and *Vampirovibrio*.

Predatory bacteria with differing prey specificities and modes of attack. *Bdellovibrio*, *Bacteriovorax* and *Micavibrio* prey on other bacteria, while *Vampirovibrio* preys on the alga *Chlorella*. *Bdellovibrio* and *Bacteriovorax* invade the periplasmic space of the prey cell, while *Micavibrio* and *Vampirovibrio* do not.

Type genus: **Bdellovibrio** Stolp and Starr 1963, 243.

Genus I. **Bdellovibrio** *Stolp and Starr 1963, 243*[AL]*

HENRY N. WILLIAMS, MARCIE L. BAER AND JOHN J. TUDOR

Bdel.lo.vib'ri.o. Gr. n. *bdella* leech, sucker; M.L. masc. n. *Vibrio* a generic name; M.L. n. *Bdellovibrio* a leechlike vibrio.

Vibrios, 0.20–0.4 × 0.5–1.4 µm. Motile by a single, polar, sheathed flagellum. Gram negative. **Prey upon Gram-negative bacteria. Have a predatory lifestyle characterized by two distinct phases, a free-living attack phase and an intraperiplasmic growth phase** (Fig. BXII.δ.47). Prey-independent mutants that have lost their requirement for prey cells occur spontaneously and can be isolated in the laboratory. Some of these prey-independent mutants retain the ability to grow predaciously and are considered to be facultative with respect to their requirement for prey cells. **Aerobic**, having a strictly respiratory type of metabolism with oxygen as the terminal electron acceptor. Contain cytochromes *a*, *c*, and *b*. **Mesophilic. Oxidase positive.** Catalase variable. Occur in fresh and salt water, soil, and sewage.

The mol% G + C of the DNA is: 49.5–51 (Torrella et al., 1978).

Type species: **Bdellovibrio bacteriovorus** Stolp and Starr 1963, 243.

FURTHER DESCRIPTIVE INFORMATION

Morphology and structure Cells exhibit a typical Gram-negative morphology of an outer membrane surrounding a thin peptidoglycan layer. The highly motile, attack-phase bdellovibrios are propelled with a single, polar, sheathed flagellum (Fig. BXII.δ.48); the sheath is continuous with and similar in structure to the outer membrane of the cell envelope (Burnham et al., 1968; Seidler and Starr, 1968). Bdellovibrios represent some of the swiftest known bacteria, capable of exceeding 100 cell lengths/second.

Except for the presence of phosphonosphingolipids (Steiner et al., 1973) in their cell envelope, both the outer membrane and peptidoglycan have been shown to be chemically similar to most Gram-negative bacteria (Thomashow and Rittenberg, 1978a; Tudor et al., 1990). The function of these lipids is not known.

Cultivation Culture requirements for predacious *Bdellovibrio* strains include a dilute nutrient medium containing a prey bacterium; the medium is inoculated with bdellovibrios and incubated at 30°C, with shaking at 200 rpm overnight. Among the more cited media used to grow the organisms are NB/500[1] (Staples, 1973) and DNB[2] (Stolp, 1973). As the bdellovibrios grow, they lyse the prey bacteria in the culture fluid, producing a "lysate" that has a lower optical density than did the starting culture.

Bdellovibrios can also be cultured on agar plates as plaques in a lawn of prey cells. Molten agar medium (0.7% agar, cooled to 45°C) is inoculated with a suspension of prey bacteria and

---

Editorial Note: The previous edition of *Bergey's Manual of Systematic Bacteriology* described three species of *Bdellovibrio*: *B. bacteriovorus*, *B. starrii*, and *B. stolpii* (Burnham and Conti, 1984). However, the authors acknowledged that variability in physiological, biochemical, and genetic properties stressed the single-genus concept for these organisms. As a result of these and other differences revealed more recently by molecular studies, two of the species, *B. starrii* and *B. stolpii*, have been reclassified into a new genus, *Bacteriovorax*, to be described in the following chapter.

1. NB/500 is nutrient broth (Difco) at 1/500 normal strength (equals 0.016 g of dehydrated broth/liter).

2. DNB is a dilute nutrient broth, consisting of (per liter): nutrient broth (Difco), 0.8 g; yeast extract (Difco), 0.1 g; vitamin-free Casamino acids, 0.5 g; 2 mM CaCl₂; and 1 mM MgCl₂. The pH is adjusted to 7.2 with 1 M NaOH.

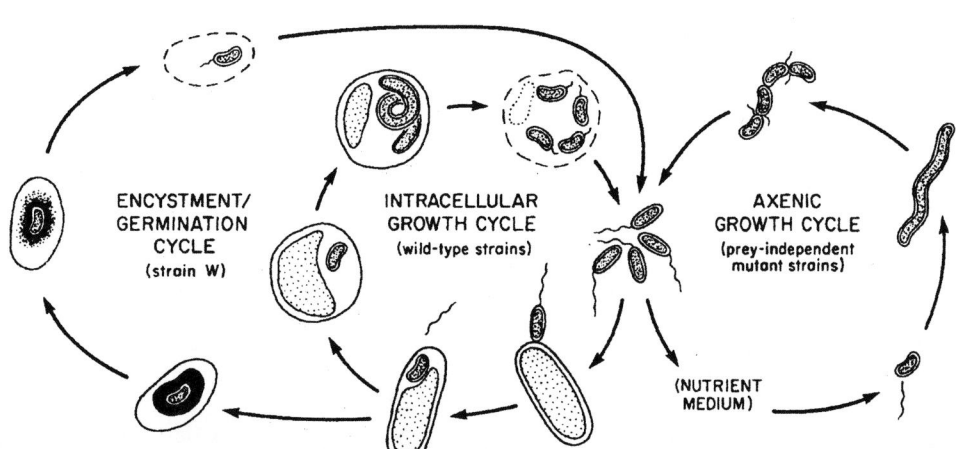

FIGURE BXII.δ.47. A schematic representation of the life cycles of prey-dependent (predatory), prey-independent, and facultative strains of the genus *Bdellovibrio*.

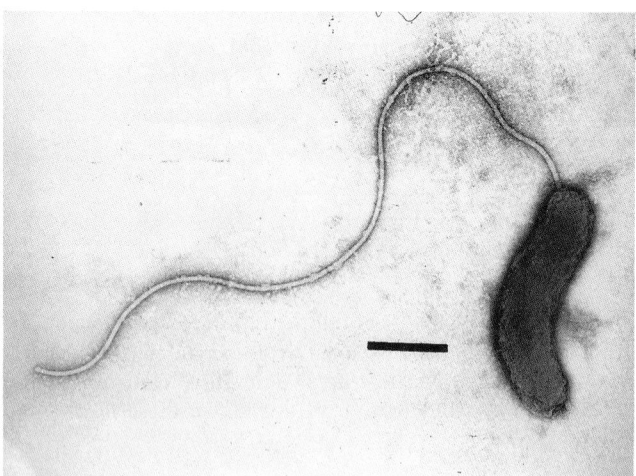

FIGURE BXII.δ.48. Negatively stained preparation of *Bdellovibrio bacteriovorus* ATCC 15143 showing the comma-shaped cell with its ensheathed polar flagellum. Bar = 0.2 μm.

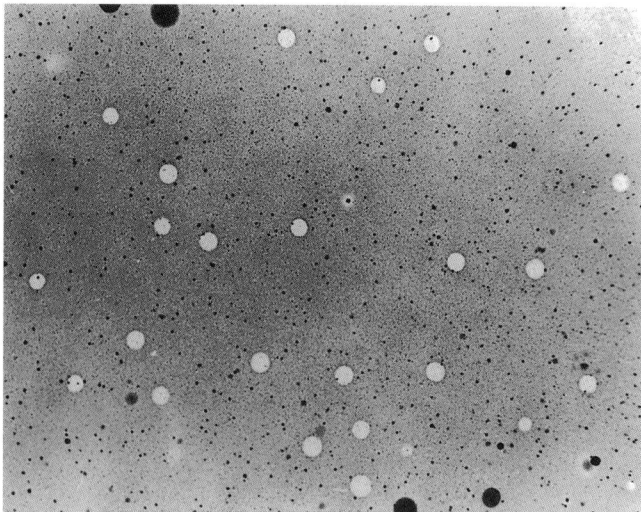

FIGURE BXII.δ.49. *Bdellovibrio* plaques demonstrating the typical clearing observed in a lawn of bacterial growth when bdellovibrios lyse their prey cells.

bdellovibrios. It is then poured over a layer of solidified agar (1.2% agar), allowed to cool and solidify, and incubated at the appropriate temperature. Plaques or clearings (Fig. BXII.δ.49) in the turbid lawn of prey cells will develop within 2–6 d. Plaques begin as small pinpoints and increase in size with continued incubation.

Prey-independent and facultative mutants of *Bdellovibrio* can be cultured in PYE[3] broth at the appropriate temperature with shaking, or on agar plates. Colonies will appear on plates after 3–4 d, and typically exhibit a yellow pigmentation.

Salt requirements Isolates from fresh water and from soil were observed to be inhibited by low concentrations (<0.85%) of salt (Varon and Shilo, 1968) and have been given the unofficial designation of fresh water or "terrestrial" bdellovibrios.

Life cycle These unusual bacteria were discovered by Stolp and Petzold (1962) while attempting to recover bacteriophage lytic against phytopathogenic pseudomonads (Stolp and Starr, 1963). In the initial description, bdellovibrios were reported to attach and lyse their prey from outside the cell wall. Subsequent studies by Starr and Baigent (1966) revealed that the predators breach the cell wall and penetrate into the periplasmic space.

Every *Bdellovibrio* strain that has been isolated from nature has been obligately predacious, requiring prey bacteria within the culture medium to initiate growth (Burnham and Conti, 1984; Ruby, 1992). The life cycle alternates between the motile, nongrowing attack phase, spent in a variety of macroenvironments, and the growth phase, spent within the periplasm of a susceptible bacterium. The mobility of the bdellovibrio enables it to collide with prey bacteria; however, collisions appear to be random and are not enhanced by chemotactic attraction toward prey (Straley and Conti, 1977). However, it has been shown that at least some strains of *Bdellovibrio* exhibit a chemoattraction to soluble nutrients such as amino acids and other organic compounds (Straley and Conti, 1974; Straley et al., 1979). This attraction may increase

the chances of a collision with prey since most of their prey would be attracted to the area by the same chemicals.

The attachment of the *Bdellovibrio* cell to its prey following the initial collision proceeds in two stages. Initial attachment is reversible and nonspecific. The second stage of attachment is specific and irreversible, and is followed immediately by penetration of the prey's outer envelope (Fig. BXII.δ.50) and the transformation of the prey cell into an osmotically stable growth chamber, termed a bdelloplast (Starr and Baigent, 1966; Abram et al., 1974) (Fig. BXII.δ.51). This irreversible attachment involves the interaction between the *Bdellovibrio* cell pole opposite the flagellum and the prey cell's outer membrane. The precise nature of this interaction is not understood, but there is evidence that it involves the core polysaccharide of the prey cell lipopolysaccharide (LPS) (Chemeris et al., 1984; Schelling and Conti, 1986). The specificity of this interaction is presumably responsible for the specific prey ranges of individual *Bdellovibrio* strains, although their prey ranges may be quite broad (Stolp, 1973). Most *Bdellovibrio* strains are capable of preying upon Gram-negative enteric bacteria and others with similar LPS structure. With the exception of a single report (Varon and Seijffers, 1975), attempts to isolate *Bdellovibrio* prey cells that have developed resistance to the predators have been unsuccessful. Gram-positive bacteria cannot serve as prey for the bdellovibrios, presumably due to the lack of outer membrane receptors.

Once irreversibly attached, the bdellovibrio flagellum continues to rotate, setting the entire invading cell spinning about the attachment axis. A few minutes following irreversible attachment, the bdellovibrio moves through the outer membrane and peptidoglycan of its prey, leaving its detached flagellum behind. It has been shown that a variety of enzymatic activities that are directed against the prey cell envelope are released at this time (Thomashow and Rittenberg, 1978a, b, c). Enzymes that have been detected include a glycanase, a peptidase, an *N*-deacetylase, an acylase directed against the prey peptidoglycan, a LPSase, and an enzyme that removes the Braun's lipoprotein. Presumably, the enzymatic activities permit the bdellovibrio to enter the prey cell periplasm through an entry pore and then to modify the penetrated cell envelope, producing a stabilized bdelloplast. Pep-

---

3. PYE is peptone-yeast extract medium (Seidler and Starr, 1969) and consists of (per liter): peptone (Bacto), 10.0 g; and yeast extract (Difco), 3.0 g. The pH is adjusted to 7.2 with 1 M NaOH.

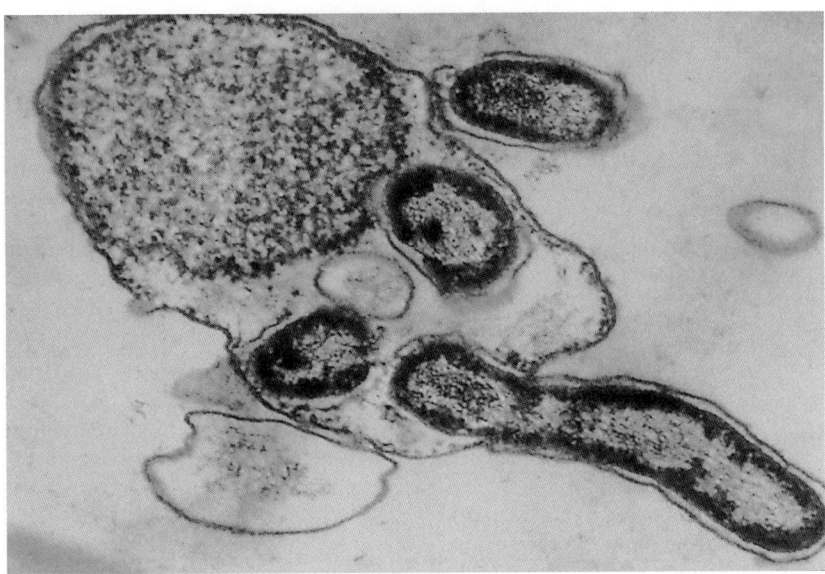

FIGURE BXII.δ.50. Electron micrograph of a thin section of *Bdellovibrio bacteriovorus* showing entrance of the predator through a pore. (Reprinted with permission from M.P. Starr and N.L. Baigent, Journal of Bacteriology *91:* 2006–2017, 1966 ©American Society for Microbiology.)

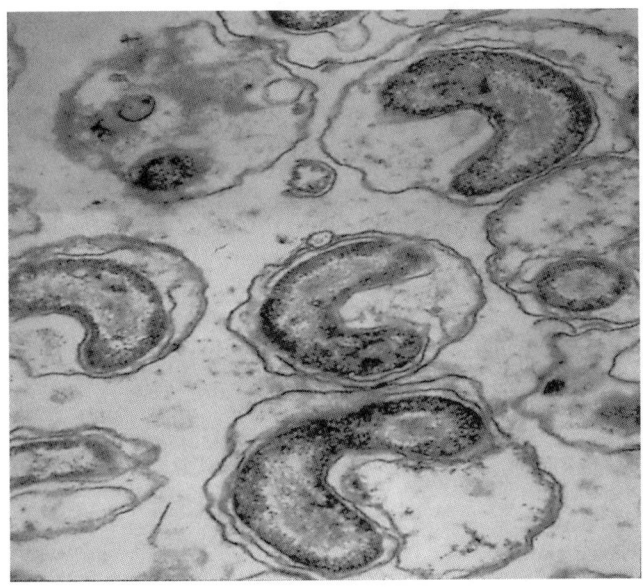

FIGURE BXII.δ.51. An electron micrograph of a thin section demonstrating spheroplasts and developing bdellovibrios. (Reprinted with permission from M.P. Starr and N.L. Baigent, Journal of Bacteriology *91:* 2006–2017, 1966 ©American Society for Microbiology.)

tidase activity is apparently required for penetration (Tudor et al., 1990), probably allowing the glycan chains of the peptidoglycan to spread enough to allow entry of the bdellovibrio. Penetration appears to be achieved by a combination of both mechanical and enzymatic means, and results in the almost immediate death of the invaded cell (Rittenberg and Shilo, 1970). The bdelloplast generally becomes spherical, providing the maximum area for growth of the bdellovibrio (Fig. BXII.δ.51). This rounding, the result of glycanase activity directed against the glycan chains of the peptidoglycan (Tudor et al., 1990), is abruptly stopped by the modification of the substrate by the

coordinate action of the *N*-deacetylase and acylase (Thomashow and Rittenberg, 1978a, b). Following penetration, the bdelloplast is typically immune to invasion by additional bdellovibrios, probably due to the extensive alterations of the outer membrane of the prey. The invading bdellovibrio is capable of modifying the prey cell envelope in other ways as well (Ruby, 1989). *Bdellovibrio* macromolecular synthesis and cellular growth appear to begin between 45 and 60 min following the initial attack.

The bdelloplast serves as a unique chamber specifically modified by the *Bdellovibrio* for its growth, which proceeds in a highly efficient manner (Rittenberg and Hespell, 1975). DNA replication is confined to the intraperiplasmic growth phase. It has been shown that access to the nutrients within the prey's protoplast is acquired through the translocation of outer membrane porins into the cytoplasmic membrane of the bdelloplast (Tudor and Karp, 1994). These porins either may be derived from the prey outer membrane during penetration, or may be *Bdellovibrio*-specific outer membrane proteins. It has been shown that the major *E. coli* outer membrane protein, OmpF, is translocated to the *Bdellovibrio* outer membrane during penetration (Diedrich et al., 1983, 1984). Degradative activity within the invaded prey's cytoplasm results from the methodical breakdown of the prey cell's macromolecules. This degradation of macromolecules produces potential biosynthetic building blocks for the growing bdellovibrios. These building blocks are typically small enough to move through the translocated porins or by other means of entry into the periplasmic space, but too large to escape the bdelloplast envelope. Macromolecular synthesis within the growing *Bdellovibrio* cell then results from the uptake of monomeric units derived from the degrading prey cell cytoplasm that have leaked into the periplasmic space (Matin and Rittenberg, 1972; Hespell et al., 1975; Kuenen and Rittenberg, 1975). The bdellovibrios are capable of taking up nucleotide and phospholipid precursors from their prey as the respective monophosphates (Rittenberg and Langley, 1975; Ruby et al., 1985; Ruby and McCabe, 1986). These unusual abilities provide the bdellovibrios with an energy efficiency that is almost twice that exhibited by most procaryotes.

Growth of the intraperiplasmic bdellovibrio proceeds with the elongation of the cell into a long aseptate spiral-shaped filament. The length of the filament is limited by the size of the invaded prey cell (Kessel and Shilo, 1976). A prey cell such as *E. coli* will give rise to an average of four to five progeny bdellovibrios (Thomashow and Rittenberg, 1979). When the bdelloplast is exhausted of nutrients, the elongated spiral receives a signal to fragment into individual attack-phase bdellovibrios, each with a newly synthesized polar sheathed flagellum. This signal may be a small peptide that accumulates and causes division to begin (Eksztejn and Varon, 1977). A lytic enzyme is then induced that breaks down the modified bdelloplast peptidoglycan from the inside, resulting in lysis of the bdelloplast and release of the progeny bdellovibrios (Thomashow and Rittenberg, 1978b). The length of the intraperiplasmic growth phase is approximately 3.5 h at 30°C for *B. bacteriovorus* using *E. coli* as prey. When grown axenically, prey-independent and facultative strains of *Bdellovibrio* follow the same pattern of growth elongation into nonmotile spiral-shaped filaments and fragmentation into individual motile progeny. These axenic mutants also appear to shed their flagella prior to growth (Burnham et al., 1970).

It has been proposed that at least two distinct signals are required for bdellovibrios to complete growth intraperiplasmically: one necessary to convert attack-phase cells into growth-phase cells, and the other needed for the initiation of DNA synthesis (Gray and Ruby, 1990, 1991). These potential signals could conceivably correspond to two major waves of protein synthesis that have been demonstrated to occur in the early stages of bdellovibrio development (Thomashow and Cotter, 1992). It is clear that the growth of intraperiplasmic bdellovibrios involves multiple cascades of protein synthesis indicative of tightly controlled differential gene expression (McCann et al., 1998).

Bdellovibrio strain W exhibits an additional variation in the life cycle in that the cells can produce encysted, resting stages, termed bdellocysts, within the bdelloplast (Fig. BXII.δ.52) (Burger et al., 1968; Tudor and Conti, 1978). Bdellocysts are characterized by a thick layer (25–30 nm) of peptidoglycan formed outside the original *Bdellovibrio* cell wall (Tudor and

Conti, 1977b; Tudor, 1980). This outer layer of cyst wall appears to be intimately associated with the bdelloplast wall and is synthesized *de novo*, not from preformed precursors from the prey wall (Tudor and Bende, 1986). Bdellocysts are much more resistant to physical stresses such as heat, starvation, desiccation, and disruption by nonionic detergents. Bdellocysts are produced at the expense of cellular reproduction since only one cyst is produced per prey cell, thus liberating only one attack-phase bdellovibrio upon germination. Bdellocysts contain more DNA, RNA, protein, and carbohydrate per cell than vegetative bdellovibrios (Tudor and Conti, 1977a). The increased carbohydrate represents large stores of glycogen within the bdellocyst that appear to be mobilized during germination (Tudor and Conti, 1978). *Bdellovibrio* strain W is a part of a unique group of procaryotic organisms that exhibits three different developmental stages: motile attack phase, intraperiplasmic growth phase, and dormant encystment phase.

Genetics   Genome sizes of *Bdellovibrio* strains have been determined chemically, and range from 1.32×10^9–1.7×10^9 Da (Torrella et al., 1978). These molecular weights correspond to genome sizes ranging from 2 to 2.6 Mb in length (Torrella et al., 1978; Baer, 1998), about half the size of the *E. coli* genome. The genome size for *B. bacteriovorus* (2.07×10^6 base pairs, estimated by pulsed-field gel electrophoresis of digested whole cell genomes) has been reported by Scognamiglio and Tudor (1996).

Prey-independent, or axenic, mutants have been reported to arise spontaneously at a frequency of about 1 in 10^6–10^7 cells, a rate that suggests that they result from a single mutation (Seidler and Starr, 1969). These mutants are either obligately prey-independent (i.e., have lost the ability to grow predaciously) or facultatively prey-independent, retaining the ability to grow on prey cells. It has been proposed that two independent mutational events may be involved: one (type I) relieves the negative regulation necessary to initiate the growth phase, and a second (type II) to initiate DNA synthesis (Thomashow and Cotter, 1992). Methods by which the basis of the obligately predacious and prey-independent phenotypes can be understood at the molecular

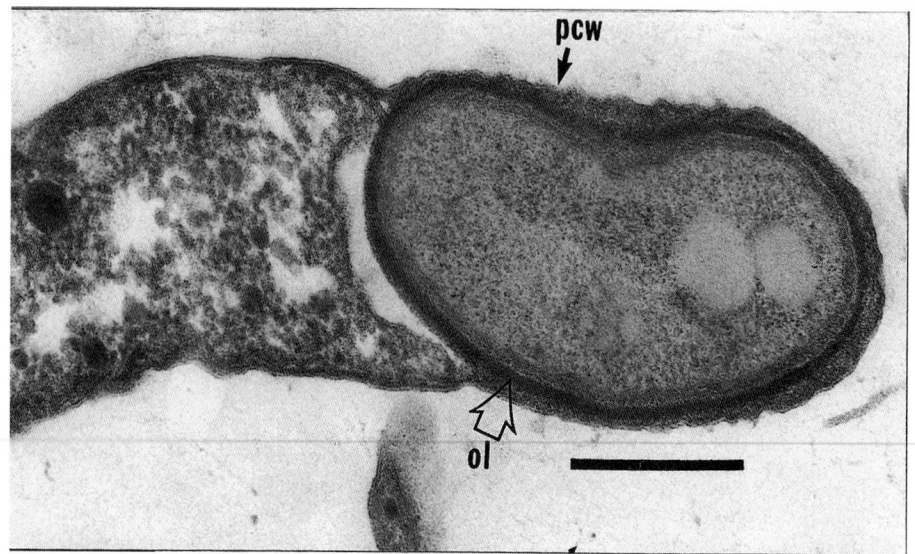

FIGURE BXII.δ.52.   Electron micrograph of *Bdellovibrio* sp. strain W during encystment in *Rhodospirillum rubrum*. The outer layer of the bdellovibrio has *(ol)* increased in thickness and is intimately associated with the prey cell wall *(pcw)*. Bar = 0.2 μm.

level have been devised. This was done through the development of a genetic system by which DNA can be transported via conjugation from *E. coli* to *B. bacteriovorus* (Cotter and Thomashow, 1992a; Jennings et al., 1998). To date, only a single locus has been identified that affects the ability of the prey-independent mutants to form plaques on susceptible prey cells. This locus codes for the production of a 10.5-kDa protein, termed the hit polypeptide, which relieves an apparent negative control on the ability of prey-independent mutants to produce large clear plaques (Cotter and Thomashow, 1992b). Recombination with the *Bdellovibrio* wild-type locus does not, however, abolish the ability of mutants to grow prey-independently. The *hit* locus appears to be present just in *B. bacteriovorus* and could not be detected in *B. stolpii* or *B. starrii* (Schwudke et al., 2001).

Phages and phage typing From 1970 to 1972, several laboratories reported the recovery of bacteriophage lytic for some strains of *Bdellovibrio*. Phage typing thus became an additional tool for assessing the diversity among the predators. Six phage-susceptibility groups have been reported thus far among the *Bdellovibrio* and *Bacteriovorax* groups (Hashimoto et al., 1970; Althauser et al., 1972; Varon and Levisohn, 1972). Groups I to IV contain the *B. bacteriovorus* strains. Brentlinger et al. (2002) described a novel single-stranded DNA, phi MH2K, of *B. bacteriovorus*. This phage is a member of the *Microviridae* family.

Antigenic structure Diversity among bdellovibrios has also been observed by the application of antigenic analysis. Serologic studies of terrestrial strains of *Bdellovibrio* have suggested the presence of serogroups as well as a common group antigen (Kramer and Westergaard, 1977; Schelling et al., 1977; Schelling and Conti, 1983). Serological studies of 21 strains of *B. bacteriovorus* revealed nine serovars (Burnham and Conti, 1984).

Antibiogram Patterns of sensitivity and resistance to a panel of antibiotics (antibiogram) may be a useful method for differentiating *Bdellovibrio* isolates. In one study (Guether and Williams, 1993), all *Bdellovibrio* strains tested were susceptible to antibiotic disks of ampicillin (20 µg), ampicillin (2 µg), carbenicillin (100 µg), gentamicin (10 µg), kanamycin (30 µg), and penicillin (10 U, BBL), and all were resistant to colistin (10 µg) (Table BXII.δ.46), metronidazole (80 µg), nalidixic acid (5 µg), and vancomycin (30 µg).

Pathogenicity As far as is known, the pathogenicity of the *Bdellovibrio* appears to be limited to the susceptible bacteria on which they prey. However, the ability to colonize animal tissue, an essential prerequisite for pathogenicity to animals, has been only superficially investigated and more extensive studies are warranted. Attempts to establish the organisms in the intestines of frogs, mice, and rabbits (Westergaard and Kramer, 1977) were unsuccessful. A report of bdellovibrio-like organisms from human fecal samples in China has been confirmed (Hu et al., 1990; Schwudke et al., 2001).

Habitat and distribution in nature Bdellovibrios are recovered from a wide variety of ecosystems. However, a common feature among the diverse habitats from which large numbers of the predators are recovered is an abundance of bacteria. Bdellovibrios have been recovered in relatively large numbers from sewage and other areas that are densely populated with bacteria. Presumably, many of the bacteria are prey to the bdellovibrios and account for the large numbers of the predators. However, bdellovibrios are not recovered from all habitats with an abundance of bacteria. This includes the intestinal tract of some animals (frogs, crabs) examined (Westergaard and Kramer, 1977; Kelley and Williams, 1992). *B. bacteriovorus* has been recovered from the gut of humans, horses, and chickens (Schwudke et al., 2001). Habitats with low abundance of bacteria such as pristine waters contain few, if any, culture-detectable bdellovibrios. Other specific habitats are described below.

Many studies have revealed that the distribution of bdellovibrios in nature is both ubiquitous and diverse. The organisms have been recovered from soil, sewage, and fresh waters (Klein and Casida, 1967; Taylor et al., 1974; Fry and Staples, 1976), including water distribution systems (Richardson, 1990).

The occurrence and distribution of bdellovibrios in nature is influenced by several known environmental factors and undoubtedly by factors that remain unknown. Among the known factors are salinity, temperature, habitat, prey population density, and pollution.

In regard to salinity, the presence of NaCl in as low a concentration as 0.5% can determine the distribution of bdellovibrios in the environment (Varon and Shilo, 1968). The halophilic bdellovibrio-like organisms (HBLO), for which the salt selects, occur in saltwater environments but not in fresh waters or soils where only the terrestrial strains are present. Shifts in the occurrence of HBLO and terrestrial bdellovibrio populations at the same site were observed by one of the authors (H. Williams) to correlate to salinity fluctuations between 0 and 1%.

Various types of pollutants may influence, positively or negatively, the density of *Bdellovibrio* populations in natural habitats. Fry and Staples (1976) reported the influence of sewage pollution on the occurrence of bdellovibrios. Numbers of the predators were greater in sewage-polluted rivers than in pristine waters. Moreover, *Bdellovibrio* numbers were greater at the point of entry of sewage effluent than either upstream or several hundred or more meters downstream. In this case, the source of the greater numbers of bdellovibrios appeared to be the sewage itself, rather than increased growth of autochthonous bdellovibrios promoted by prey bacteria or other organic pollutants in the sewage. Klein and Casida (1967) reported similar increased numbers of bdellovibrios in sewage-polluted soils versus normal soils.

The population of bdellovibrios also may be adversely affected by certain environmental chemical pollutants. Laboratory studies reported by Varon and Shilo (1981) have shown that the predatory activity of a *Bdellovibrio* isolate was inhibited by detergents, heavy metals, and pesticides. Undoubtedly, other factors such as acidity, cations (Bell and Latham, 1975), light (Friedberg, 1977), etc. influence the presence of bdellovibrios, but these effects have not been sufficiently investigated in nature.

The few attempts to establish that certain animals harbor *B. bacteriovorus* or other bdellovibrios have not met with much success, because the predators were either not recovered or were recovered in low, insignificant numbers. In one study, animals were forced-fed suspensions of *Bdellovibrio*, but the organisms failed to become established in the intestinal tract (Westergaard and Kramer, 1977). However, the isolation of *B. bacteriovorus* from the gut of humans and animals was recently reported (Schwudke et al., 2001).

Role in nature The early observations that bdellovibrio infection of a susceptible bacterium is a lethal event, and that in laboratory cultures the prey population is substantially reduced by the predators, provoked speculation on the potential role of bdellovibrios as biological control agents in the environment. Perhaps the expectation of total prey population eradication by bdellovibrios was unrealistically high, especially since the pred-

TABLE BXII.δ.46. Characteristics of *Bdellovibrio bacteriovorus* and of halophilic bdellovibrios[a]

Characteristic	*B. bacteriovorus*	Halophilic bdellovibrios (representative halophilic strain AQ)
Comma-shaped predacious cells, 0.35–1.2 µm	+	+
Motility	+	+
Single, polar, sheathed flagellum	+	+
Elongated to spiral-shaped forms develop during growth	+	+
Grow and divide within prey cell	+	+
Prey-dependent strains occur	+	+
Prey-independent variants occur	+	+
Gram-negative, but not Gram-positive, bacteria can serve as prey	+	+
Circular, plaque-like zones of clearing occur on agar-grown lawns of prey bacteria	+	+
Aerobic	+	+
Oxidase test	+	+
Catalase test[b]	+	+
Fermentative ability present	–	–
Gelatinase activity	+	+
Trypsin	+	+
Protease production		Low[c]
Susceptibility to antibiotics:[d]		
Carbenicillin, 100 µg	S	S
Gentamicin, 10 µg	S	S
Kanamycin, 30 µg	S	S
Penicillin, 10 U	S	S
Colistin, 10 µg	R	R

[a]Symbols: see standard definitions; S, sensitive; R, resistant.

[b]Cultures may be initially positive but become variable with subsequent transfer (Seidler and Starr, 1969).

[c]Low indicates <6% hydrolytic activity.

[d]Test results are the same for prey-dependent and prey-independent mutants tested.

ators do not completely eradicate the prey even in laboratory cultures under optimal conditions. Prey cultures initiated with 10^9 CFU/ml will be typically reduced by bdellovibrio predation to 10^2–10^3 CFU/ml. This is a logical strategy since complete elimination of the prey population would exhaust the food source of the predators, a condition that would bring about their own starvation and demise. Unfortunately, properly designed studies to determine the impact of bdellovibrios on the mortality of the bacterial population in nature have not been conducted. The few studies that have addressed this issue have yielded circumstantial evidence that is often conflicting with other reports. Several investigations have concluded that bdellovibrios were not a factor in reducing bacterial populations in water and sewage (Fry and Staples, 1974, 1976). If, in fact, bdellovibrio multiplication occurs exclusively by the predatory cycle described previously, then to some extent the organisms must contribute to the mortality of susceptible bacteria in nature. Perhaps the greatest role of bdellovibrios in nature is one of modulation of its prey population. In this scenario when the population becomes high, 10^6 or greater, predation becomes more efficient, subject to suitable temperature and other environmental parameters, and the rate of prey killing increases—resulting in lower numbers of prey. However, at numbers of 10^3 and below, the predators may have little, if any, effect on the prey population. This hypothesis, of course, remains speculative until properly designed studies to confirm it have been conducted.

ENRICHMENT AND ISOLATION PROCEDURES

Several procedures have been used for the enrichment and isolation of bdellovibrios. The procedure selected is dependent upon the nature of the source of the sample. Samples that potentially harbor large numbers of bacteria, protozoa, or fungi

that may overgrow and mask the presence and growth of small numbers of bdellovibrios may require special isolation procedures. In these cases, filtration, centrifugation, or other methods to reduce the number of the large heterotrophic bacterial population or other undesirable organisms may be necessary. These procedures respectively trap or sediment the larger, denser cells. In filtration, the filter size selected may vary; however, a sound rule is to use the largest pore size filter possible to reduce the cultivable bacteria, protozoa, and fungi to a level that does not mask bdellovibrio plaques on plates, and that minimally decreases the numbers of the bdellovibrios. A pore size of 1.2 µm has been used for some samples (Varon and Shilo, 1970). Centrifugation is done at low speed to sediment the larger bacteria while leaving the smaller bdellovibrios suspended in the supernatant fluid. Differential or gradient centrifugation may be used to achieve better separation of bdellovibrios from other microbial cells (Varon and Shilo, 1970).

Samples from environmental sources with relatively small numbers of cultivable bacteria such as pristine rivers or other bodies of water typically contain small numbers of bdellovibrios. In these samples, filtration or centrifugation to reduce the number of bacteria may reduce as well the bdellovibrio population, exacerbating the difficulty of their recovery. For such samples, and those with moderate numbers of cultivable bacteria, direct plating may be the most promising approach. Filtration and centrifugation procedures may also be used for samples with extremely small numbers of bdellovibrios to concentrate the organisms so as to yield numbers sufficient for recovery. In this case, the objective of filtration is to trap the bdellovibrios and other bacteria on the filter surface. For this, a small pore size filter (≤0.2 µm) is required. Concentration of bdellovibrios may also be accomplished by high speed centrifugation of samples to pellet all bacteria present. Alternatively, an initial low speed centrifugation (3–5 × g) may be done to pellet large cells, followed by high speed centrifugation to capture the smaller cells including the bdellovibrios. Gradient centrifugation may improve both the yield and separation of the bdellovibrios.

Plating environmental samples for the recovery, isolation, or direct quantitation of bdellovibrios is done using the double agar overlay procedure as described by Staples and Fry (1973). The volume of sample inoculum used by investigators has varied over a wide range from 0.1 to 10 ml (Taylor et al., 1974; Williams, 1988). Typically, samples are plated undiluted, although in some situations 10-fold dilutions are plated. Both the volume and dilution of inoculum plated is dependent upon the relative numbers of bdellovibrios and other cultivable bacteria expected to be in the sample. For ocean water samples where the total cultivable bacteria and bdellovibrio counts are typically low, Taylor et al. (1974) used 10 ml of sample as the inoculum. In several studies conducted in the Chesapeake Bay, the sample inoculum typically consisted of volumes of 0.1, 1.0, and 5.0 ml. In the colder months when bdellovibrio numbers decreased (Williams, 1988; Williams et al., 1995b) the 0.1-ml sample plating was eliminated. To accommodate the larger inoculum size for plating, Petri dishes of 150 × 25 mm or 150 × 15 mm are used. For plating of suspensions of sediment or surface biofilms from oyster shell, typically 0.1 and 1.0 ml of the undiluted sample and 1.0 ml of 10-fold serial dilutions up to 10^{-4} were plated. The sample is added to a tube of molten top (soft overlay) agar just previously inoculated with an appropriate prey suspension and held between 45 and 47°C. The agar concentration of the top agar is determined by the size of the sample inoculum to be added. The final agar concentration, following addition of sample and prey suspension, should range from 0.6 to 0.7%. The top agar suspension is mixed and overlaid onto prewarmed bottom agar (agar concentration 1.2 to 1.5%) plates. The plates are immediately swirled in a circular, figure-eight pattern to spread the molten agar evenly, before hardening, over the bottom agar surface.

The top agar is allowed to harden (~10 min) and the plates are inverted and incubated at 30°C. Plaques of bdellovibrios typically appear in 2–6 d and increase in size with prolonged incubation. *Bdellovibrio* plaques are usually round with a sharp, distinct boundary and clear in the center (Fig. BXII.δ.49). *Bdellovibrio* plaques must be differentiated from those caused by viruses, protozoa, myxobacteria, or other organisms. Plaques produced by bacteriophage will typically appear by 24 h incubation and do not increase in size. Plaque-like clearings with a central bacteria colony are not typical of those produced by bdellovibrios. *Bdellovibrio* plaques should be confirmed by microscopic examination. Phase contrast or dark-field microscopy is the most practical, but light microscopy of Gram-stained preparations is possible.

To obtain pure cultures from agar plates, plaque material may be removed with an inoculating loop or needle or the point of a scalpel blade and streaked for isolation directly on hardened top agar. Alternatively, the plaque material may be emulsified in a small volume (1–3 ml) of sterile medium that is subsequently serially diluted and plated to obtain a few well-isolated plaques on a plate. If desired, the plaque suspension may be filtered to remove extraneous organisms prior to plating. Material from a well-isolated plaque is again emulsified, serially diluted, and plated or transferred into broth and incubated. Several such passages of plaque material should be performed before the isolate is considered to be a pure culture.

Recovery of bdellovibrios is favored by use of a low-nutrient, minimal medium. Such a medium, when inoculated with a bacterial suspension to yield a final concentration of 10^7–10^9 prey per ml, retard the normal growth of the large numbers of heterotrophic cultivable bacteria typically found in environmental samples, and simultaneously promote rapid multiplication by the bdellovibrios. The medium most widely used for the recovery of *Bdellovibrio bacteriovorus* is NB/500 (Staples, 1973). The recommended prey is *Achromobacter* sp. (NCIB 8250) (Staples, 1973), although other genera and species have been used. The prey can be grown in an appropriate broth or agar medium and harvested to produce a concentrated suspension. Nutrient agar or other suitable medium can be used for this purpose. When grown in broth, the cells can be harvested by centrifugation, and if desired, washed by re-centrifugation in sterile medium. The pellet is then resuspended in the appropriate medium to yield a dense suspension. As an alternative to broth culture, prey cells can be grown on agar plates for 24–48 h. The colonies are harvested by flooding the plate with approximately 5 ml of sterile DNB. Following a period of about 5 min, the colonies are suspended in the overlying liquid by using a glass "hockey stick" in a spreading motion over the agar surface. The cell suspension is removed from the plate by pipetting and placed in a sterile tube and mixed. As a rinse, the agar plate may be flooded once again with 2–3 ml of sterile medium to harvest any residual bacteria from the plate and the rinse fluid added to the original tube. These procedures typically yield a suspension containing 10^8–10^{10} cells per ml. Just prior to inoculation into the top agar, the tube is mixed to prevent the settling of the cells.

Enrichment in broth may be preferred in situations where the primary objective is the recovery of bdellovibrios from environmental samples rather than quantitation. This procedure is also indicated in samples where the numbers of bdellovibrios are expected to be low, making their recovery by plaque assay difficult. Whereas the size of the sample inoculum in the double agar overlay plating procedure is limited to 10 ml, enrichment in broth has the advantage of accommodating a larger sample inoculum. For example, 100 or 1000 ml of an environmental sample could be collected in a flask and directly enriched with a concentrated prey suspension to yield a final prey cell density of 10^8–10^9 cells/ml. The larger sample size enhances recovery where bdellovibrios occur in very small numbers.

The typical enrichment procedure for recovery of the bdellovibrios is to add directly into the collected sample sufficient prey cells to yield a final concentration of approximately 10^9 cells/ml. At the discretion of the investigator, the sample may be prefiltered through a 0.3–1.2 μm pore size filter prior to adding to the enrichment vessel to reduce the number of heterotrophic bacteria. When working with soil or solid sample material, the sample may be added to a minimal broth medium. Natural water or a prepared salts solution is suggested. The lack of added chemical nutrients to the medium retards the growth of the large population of extraneous bacteria in the sample. The addition of a high concentration of prey cells to the minimal medium provides the nutrient source to promote the growth of bdellovibrios capable of predation on the prey selected. The enrichment culture should be incubated at 30°C. Enrichment cultures should be examined macroscopically for some degree of clearing in the broth, and microscopically, by phase contrast or dark-field microscopy, for the presence of small, highly motile, vibrio-like bacteria. Fluorescence microscopy using acridine orange or DAPI or the Gram stain may be used as alternatives but are not suitable for visualization of live cell preparations. With either clearing of broth cultures or microscopic observation of bdellovibrio-like organisms at 72 h incubation, samples of the enrichment culture should be serially diluted and plated for the recovery of bdellovibrio-type plaques. The enrichment sample may be pre-filtered prior to plating to reduce the number of persisting heterotrophic bacteria. If bdellovibrio-type plaques are not recovered, incubation should continue for at least 14 d with daily examination and periodic culturing for plaques. Plaques that are recovered may be subcultured to obtain pure cultures as described above.

MAINTENANCE PROCEDURES

Cultures of bdellovibrios can be maintained by either serial transfers of plaque material from isolated plaques or broth cultures as described in the first edition of *Bergey's Manual of Systematic Bacteriology* (Burnham and Conti, 1984). The use of a minimal medium and appropriate prey bacterium species is recommended. In addition to the use of NB/500, some investigators prefer to use dilute nutrient broth (DNB) or agar for the maintenance of isolates. The bacterial species selected for prey is typically the species on which the bdellovibrio isolate was initially recovered or other bacterium on which the predator grows well. Prey organisms reported to yield good recovery of bdellovibrios include *Achromobacter* sp. (NCIB 8250) (Staples, 1973) and *E. coli* (Klein and Casida, 1967).

When maintaining bdellovibrio cultures in broth, serial transfers should be done at least weekly. A small inoculum of 0.1 ml or less from a week-old bdellovibrio lysate into a dense prey suspension (approximately 10^9 per ml) will yield clearing of the suspension in several days typically and sustain the growing bdellovibrios for a week or more. A larger inoculum of the predators will exhaust the prey more rapidly and may require more frequent transfers. When maintaining bdellovibrio cultures in broth, it is recommended that periodically, and before using the culture for investigative purposes, a sample of the culture be plated to recover well-isolated plaques, which should be used to initiate a new broth culture. This procedure provides a safeguard for maintaining a single bdellovibrio culture and avoiding mixed cultures. This is important since bdellovibrios cannot be differentiated on the basis of cultural features and practical biochemical methods of distinguishing between isolates have not found wide use.

The preservation of bdellovibrio cultures remains as described in the first edition of *Bergey's Manual of Systematic Bacteriology* (Burnham and Conti, 1984). Cultures for preservation should be centrifuged at 10,000 × g for 10 min. The pellet should be suspended to the original volume in either skim milk or 12% sucrose (Burnham and Conti, 1984). This suspension can be distributed in ampoules for freeze-drying. Freeze-dried specimens can be revived by suspending in NB/500 broth or other medium in which they may be grown. The suspension may be inoculated either into broth or onto plates using the appropriate prey species.

PROCEDURES FOR TESTING SPECIAL CHARACTERS

Biochemical characterization The metabolic and biochemical characterization of bdellovibrios is handicapped, since the wild-type, predatory organisms have not been grown in pure culture and do not metabolize commonly used substrates. Special strategies must be employed to characterize and differentiate the bdellovibrios.

One such strategy employed is the use of tests that do not require growth of the organisms. The API-Zym identification system (BioMerieux) is a battery of miniaturized tests that characterize and differentiate among bacteria by the presence or absence of certain enzymes and does not require growth of the organism. Wild-type bdellovibrios have been characterized with this system (Guether and Williams, 1992; Deidrich, personal communication), and it proved useful in revealing some distinguishing traits among them. The important step prior to performing this test is to ensure that the bdellovibrio suspension used as inoculum is free from prey cells and is pure.

Another strategy for differentiation and characterization of bdellovibrios is to use prey-independent mutant strains from wild-type isolates. Mutants have been reported to arise at a relatively high rate among the bdellovibrios and are not too difficult to recover. However, this may not be true for all wild-type strains isolated. A protocol for recovering mutant strains has been described (Seidler and Starr, 1969). Evidence suggests that the mutant isolates result from a single mutation that relieves a negative repressor for multiplication. Preliminary reports suggest that in most cases, many properties of the mutant strains are very similar to, if not identical with, the wild type. Mutant strains may therefore be a useful research tool for studying the differential properties of the bdellovibrios for not only metabolic functions but also fatty acid analysis, genetic analysis by molecular "fingerprinting" techniques, and comparisons of 16S rRNA sequences. However, specific mutant strains used for such purposes should be given a specific designation and characterized since mutants having different properties may derive from a single wild-type clone (Barel and Jurkevitch, 2001).

Differentiation by molecular fingerprinting techniques Characterization of bdellovibrios using molecular techniques is a recent development (Scognamiglio and Tudor, 1996) and are useful for differentiating wild-type bdellovibrios, although some are beyond the means of most small laboratories. However, commercial companies offer as a service molecular bacterial identification and differential analysis. Molecular methods that have proven useful include pulsed field gel electrophoresis (PFGE), arbitrarily primed PCR (AP-PCR), and ribotyping.

Pulsed field gel electrophoresis (PFGE) PFGE was performed on five prey-independent strains of *Bdellovibrio bacteriovorus*, the two *Bacteriovorax* species, one environmental terrestrial isolate, and two halophilic *Bdellovibrio*-like organisms (Baer et al., 1998). It was observed that only the *Bdellovibrio bacteriovorus* strains (100, 109J, Ox9-2, E, and 2484Se2) were restricted by enzymes having an "8 base" recognition site (*Asc*I, *Fse*I, *Sfi*I, and *Not*I). Restriction patterns produced for the *B. bacteriovorus* strains were similar but strain differences (± 2–3 bands) were observed (Baer, 1998; Baer et al., 1998).

Arbitrarily primed PCR (AP-PCR) When amplification patterns produced by arbitrary primers were compared for 10 prey-independent isolates consisting of *B. bacteriovorus*, *Bacteriovorax*, and HBLO, each strain was observed to produce its own unique amplification banding pattern. However, the patterns produced by the *B. bacteriovorus* strains were the most similar (only 1–3 bands different) (Baer, 1998; Baer et al., 1998). Dendrograms based on the degree of similarity of the band patterns revealed that the *B. bacteriovorus* strains clustered as a group distinct from the HBLO and *Bacteriovorax* spp., which formed a second group.

Ribotyping *Bdellovibrio* and other predatory strains including 13 HBLO and the *Bacteriovorax* species were analyzed by ribotyping using two restriction enzymes, *Eco*RI and *Pvu*II. Similar results were observed as for PFGE and AP-PCR. *B. bacteriovorus* strains were found to be the most similar to each other, whereas the other strains tested produced their own distinct ribotyping profiles.

DIFFERENTIATION OF THE GENUS *BDELLOVIBRIO* FROM OTHER GENERA

Plaque-forming bacterial strains that are tentatively identified as members of the family *Bdellovibrionaceae* must be confirmed to grow in the periplasmic space of their respective prey bacteria. This characteristic serves as the primary factor that differentiates members of this family from other comma-shaped bacteria. Confirmation of the location of the bdellovibrios within the prey cell may be accomplished by examining plaque material, or dual predator-prey cultures, by phase-contrast or transmission electron microscopy.

Differentiation of the genus *Bdellovibrio* from the new genus *Bacteriovorax* may be accomplished through the combination of phenotypic and molecular analyses (see Table BXII.δ.48 in the *Bacteriovorax* chapter).

TAXONOMIC COMMENTS

In the previous edition of the *Manual*, all predatory, intraperiplasmic bacteria were placed into a single genus, *Bdellovibrio* (Burnham and Conti, 1984). However, the substantial diversity among the organisms was noted including a broad range of mol% G + C values (17%) and wide differences in DNA hybridization values. High levels of diversity among *Bdellovibrio* species have been observed in studies examining cultural, physiological, and biochemical characteristics such as fatty acid composition (Guether et al., 1993), antibiotic resistance patterns (Guether and Williams, 1993), enzymatic activities (Guether and Williams, 1992), serological properties (Schelling et al., 1977; Schoeffield et al., 1991), and phage typing (Hashimoto et al., 1970; Althauser et al., 1972; Varon and Levisohn, 1972). The level of diversity observed among these strains places considerable stress upon the concept of including these organisms within the same genus (Burnham and Conti, 1984). Recent analysis of complete 16S rDNA sequences and DNA–DNA hybridization (Baer et al., 2000) from *B. bacteriovorus* 100^T, *B. stolpii* Uki2^T, and *B. starrii* A3.12^T confirms the diversity among the predatory bacteria. The collective results from these studies demonstrate that it is no longer appropriate to use the presence of predatory behavior as the sole defining property for placing organisms within the *Bdellovibrio* genus. The bacteria formerly classified as *Bdellovibrio stolpii* and *Bdellovibrio starrii* have now been assigned to a new genus, *Bacteriovorax*, as described in the following chapter. The type species of the *Bdellovibrio* remains *Bdellovibrio bacteriovorus* strain 100 (ATCC 25622).

The genus *Bdellovibrio* is now better defined except for the halophilic strains, which are being further analyzed. The taxonomy of these strains remains in a primitive state with isolates being loosely designated as *Bdellovibrio* sp. since none of the organisms has been sufficiently characterized and defined to warrant designation as a taxonomic species. This lack of information has made it impossible to investigate the diversity of the organisms and the abundance and distribution of various species in nature. To avoid confusion in terminology and designations, the genus *Bacteriovorax* and the HBLO and other undefined isolates are referred to in this edition as bdellovibrio-like organisms (BLO). The halophilic bdellovibrios are referred to as the HBLO

as described above. All the intraperiplasmic predatory bacteria with the morphological and life cycle features as described for the bdellovibrios including *B. bacteriovorus*, *Bacteriovorax* sp., and the HBLO are referred to as the *Bdellovibrio* and like organisms.

From an evolutionary standpoint, it has been suggested that bacteria that are described as *Bdellovibrio* strains have originated from diverse organisms, which through converging patterns of evolution, have developed their predacious way of life (Varon and Shilo, 1980). The first phylogenetic tree for the *Bdellovibrio* was constructed by Donze et al. (1991). Results from this study revealed that the bdellovibrios clustered within the *Deltaproteobacteria*, clustering most closely with *Desulfovibrio desulfuricans* and *Myxococcus xanthus*. All *Bdellovibrio* strains were shown to be more closely related to each other than to other members of the *Deltaproteobacteria* or to organisms within the *Alphaproteobacteria*, *Betaproteobacteria*, and *Gammaproteobacteria*. The most interesting revelation was that the bdellovibrios formed two distinct branches. The first branch consisted of *B. bacteriovorus* and *Bdellovibrio* strains W, 6-5-S, 109, 109D, 114, Ox9-2, and Ox9–3. The second branch included *B. starrii*, *B. stolpii*, and their halophilic isolate BM4. The results reported by Donze et al. (1991) support the findings of several earlier independent studies by other investigators (Hespell et al., 1983; Stackebrandt and Woese, 1984; Oyaizu and Woese, 1985; Woese, 1987; Stackebrandt, 1988). The recent development of phylogenetic trees based on analysis of complete sequences of 16S rRNA genes has confirmed these findings (Fig. BXII.δ.53) (Baer et al., 1998). These data suggest a reevaluation of the assignment of the halophilic strains to the genus *Bdellovibrio*, as they are more closely related to members of the genus *Bacteriovorax* and may warrant a separate genus designation.

DNA–DNA hybridization A high degree of DNA–DNA hybridization values (ranging from 70 to 100%) has been observed among four of the different strains of *B. bacteriovorus* (100^T, 109J, Ox9-2, and E) (Baer et al., 2000). A lower degree of hybridization (ranging from 32 to 55%) has been noted between 2484Se2 and each of the other four *B. bacteriovorus* strains tested (Baer et al., 2000). A previous report (Seidler et al., 1972) demonstrated very little hybridization between *B. bacteriovorus* 100^T and related organisms (1% hybridization with *B. starrii* A3.12^T and no hybridization with *B. stolpii* Uki2^T, both of which are now classified as *Bacteriovorax*). Low hybridization values were observed between *B. bacteriovorus* and the halophilic isolates tested in this study (<5%).

FURTHER READING

Baer, M.L., J. Ravel, S.A. Pineiro, D. Guether-Borg and H.N. Williams. 2004. Reclassification of salt-water *Bdellovibrio* spp. as *Bacteriovorax marinus* sp. nov. and *Bacteriovorax litoralis* sp. nov. Int. J. Syst. Evol. Microbiol. *54*:: 1011–1016.

Guerrero, R., C. Pedros-Alio, I. Esteve, J. Mas, D. Chase and L. Margulis. 1986. Predatory prokaryotes: predation and primary consumption evolved in bacteria. Proc. Natl. Acad. Sci. U.S.A. *83*: 2138–2142.

Jurkevitch E. 2000. The genus *Bdellovibrio*. *In* Dworkin., Flakow, Rosenberg, Schleifer and Stackebrandt (Editors), The Prokaryotes: An Evolving Electronic Resource for the Microbiological Community, 3rd Ed., release 3.1, Springer-Verlag, New York.

Rittenberg, S.C. 1983. *Bdellovibrio*: attack, penetration, and growth on its prey. ASM News. *49*: 435–439.

Ruby, E.G. 1991. The genus *Bdellovibrio*. *In* Balows, Trüper, Dworkin, Harder and Schleifer (Editors), The Prokaryotes, 2nd Ed., Vol. 4, Springer-Verlag, New York. pp. 3400–3415.

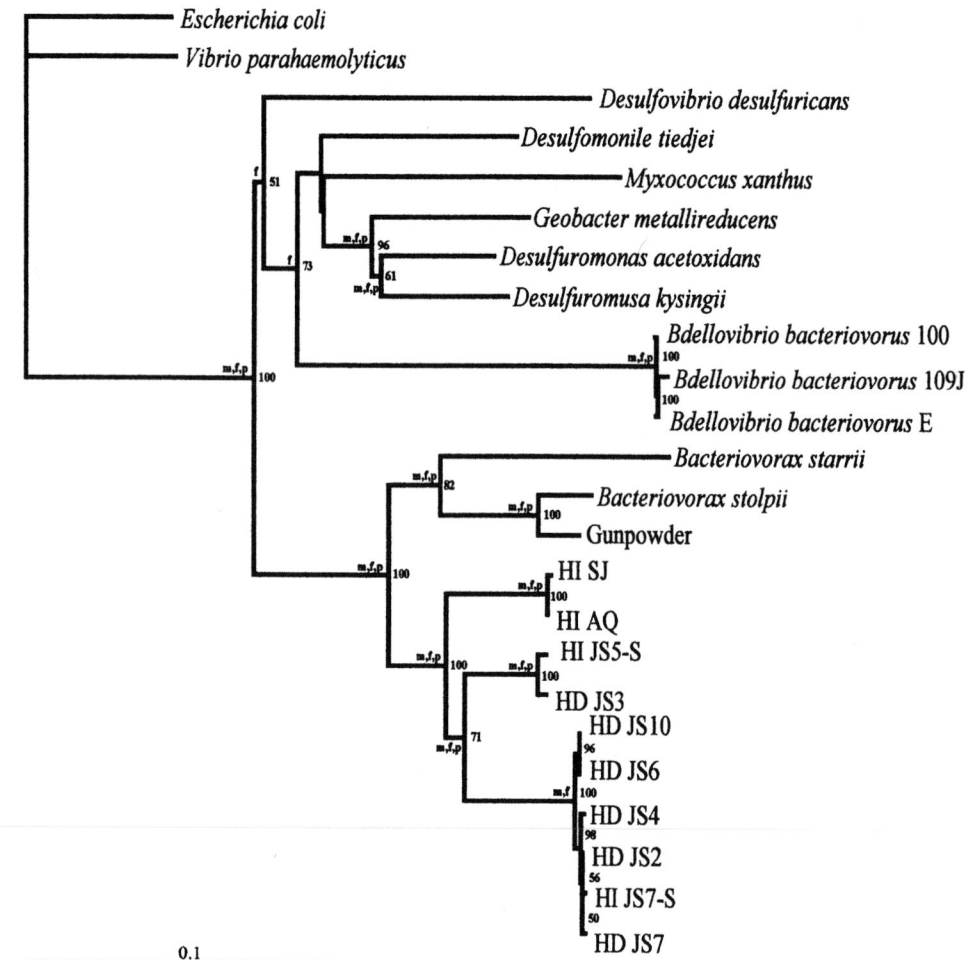

FIGURE BXII.δ.53. A neighbor-joining tree based on 16S rRNA sequences demonstrating the phylogenetic relationship between marine and terrestrial *Bdellovibrio* isolates. *m*, *f*, and *p* indicate branches that were also found using the FastDNAml, Fitch-Margoliash, and maximum parsimony methods, respectively.

List of species of the genus Bdellovibrio

1. **Bdellovibrio bacteriovorus** Stolp and Starr 1963, 243[AL]

   bac.te.ri.o'vo.rus. Gr. dim. n. *bacterium* a small rod; L. v. *voro* to devour; M.L. adj. *bacteriovorus* bacteria devouring.

   The characteristics are as described for the genus and as listed in Tables BXII.δ.46 and BXII.δ.47. Isolated from soil, sewage, and freshwater sources.

   The mol% G + C of the DNA is: 50.4 ± 0.9 (Bd).

   Type strain: HD 100, DSM 50701, ICPB 3268, NCIB 9529.

   GenBank accession number (16S rRNA): AJ292759.

Other Organisms

Isolates of the intraperiplasmic predatory bacteria (IPPB) from seas and oceans were observed to require NaCl for growth and were referred to as the "marine" bdellovibrios (Shilo, 1966). In the first edition of the *Manual*, these organisms were placed in the group *Bdellovibrio* species. It is now recognized that these salt-dependent organisms also occur in estuarine environments and hypersaline lakes. This makes the designation "halophilic" more appropriate. Conversely, isolates from fresh water and from soil were observed to be inhibited even by low concentrations (<0.85%) of salt (Varon and Shilo, 1968) and were given the common designation of fresh water or "terrestrial" bdellovibrios. Although initially named bdellovibrios, the halophilic isolates have long been recognized as being distinct from *Bdellovibrio*

bacteriovorus. Results from recent studies have revealed even greater differences and suggest that the variations between the organisms warrant consideration for a separate genus for the halophilic predators. Based on this, and to avoid confusion of the salt-tolerant IPPB with *Bdellovibrio*, they will be referred to in this volume as the halophilic bdellovibrio-like organisms (HBLO).

The HBLOs are like *Bdellovibrio bacteriovorus* and the *Bacteriovorax* sp. in major physical features and lifestyle. Many halophilic strains appear to be just slightly smaller in diameter and pass more readily through 0.3-μm pore size filters than the terrestrial *Bdellovibrio* and *Bacteriovorax*. Besides tolerance to sodium chloride, the most prominent distinctions between the halophilic

TABLE BXII.δ.47. Characteristics differentiating strains of *Bdellovibrio bacteriovorus* from those of halophilic bdellovibrios[a]

Characteristic	*B. bacteriovorus* strains				Halophilic strains			
	100[b]	109J[c]	0X9-2[b]	2484Se2[b]	AQ[c]	JS2[d]	JS4[d]	JS8[d]
Facultatively predatory[e]	−	−	−	−	−	−	−	−
Na$^+$ required for growth (75 mM or higher)	−	−	−	−	+	+	+	+
Source:								
Saltwater	−	−	−	−	+	+	+	+
Freshwater or terrestrial	+	+	+	+	−	−	−	−
Sensitive to vibriostatic agent 0/129[f]		+			−[g]			
Bacteriophage susceptibility[h]		+						
Mol% G + C of DNA	50.4	nd	nd	nd	strain values range from 33.4 to 38.6; 43.5[i]			
Optimal temperature, °C	28–30	20–30	20–30	20–30	20–25	20–25	20–25	20–25
Growth at 35°C	+	+[j]	+	+	−	−		+
Sensitivity to restriction enzymes (8 base cutters) AscI, FseI, SfiI and NotI	+	+[g]	+	+	−			
Enzyme activities detected:[k]								
C$_{14}$ lipase		−	−[j]	−		+[j]		
Valine amino peptidase	−	−	−	−	+[j]	+	+	+
Cystine amino peptidase	−	−	−	−	+[j]	+	+	+
Chymotrypsin	+	+	+	−	−			
Nitrate reduced to nitrite		−			+[j]			
Antibiotic sensitivity:								
Polymyxin B, 300 U	−	−[j]	−	−	+[j]	+	+	+
Streptomycin, 10 µg	+		−	−	−[j]			
Presence of fatty acids:								
C$_{16:1\ \omega 9c}$	−	−	−	−	+			
C$_{13:1\ iso\ 3OH}$	+	+	+	+	−			

[a]For symbols, see standard definitions; nd, no data.

[b]Prey-independent mutant strain derived from the wild-type strain.

[c]Both PD wild-type and PI mutant strains were used. Test results are from PD isolates unless otherwise noted.

[d]Test results from wild-type PD isolates.

[e]Defined to mean that some strains have been observed to complete their life cycle either prey-dependently or prey-independently.

[f]0/129, 24-diamino-6,7-diisopropylpteridine phosphate.

[g]Test result is from PI mutant.

[h]Data from Althauser et al. (1972) and Varon and Levisohn (1972).

[i]Most halophilic strains have a mol% G + C between 33.4–38.6 (Marbach et al., 1976); only one has been found to have a mol% G + C as high as 43.5 (Taylor et al.,1974).

[j]Test result is the same for both PD and PI mutants.

[k]Using API-Zym (BioMerieux) test system and prey-independent mutants of *Bdellovibrio*.

and terrestrial predators are G + C ratios, growth temperature, and nucleic acid molecular "fingerprint" as revealed by various techniques.

The properties of the DNA vary substantially between the HBLO and *Bdellovibrio bacteriovorus*. The mol% G + C content of the DNA of the HBLO ranges typically from 33.4 to 38.6, much lower than strains of *B. bacteriovorus*. The banding patterns produced by PFGE, ribotyping, and other molecular fingerprint techniques also revealed major differences between the two groups. Analysis of 16S rDNA sequence data from isolates of *Bdellovibrio*, *Bacteriovorax*, and HBLO revealed major differences (Baer, 1998). The level of similarity among *Bdellovibrio bacteriovorus* strains was >97%. However, when *B. bacteriovorus* 100 was compared to isolates of HBLO, lower similarity values, 80–82%, were observed. All HBLO isolates fell into three different groups within this same clade. The Chesapeake Bay isolates formed two groups; one contained isolates JS2, JS4, JS6, JS7, and JS10 and the other, JS5 and JS3. Ocean isolate SJ and aquarium salt-water isolate AQ were in the last group. These were more similar to each other than either was to the terrestrial strains (Fig. BXII.δ.53). Phylogenetic data suggest a reevaluation of the assignment of the HBLO to the genus *Bdellovibrio*.

The HBLOs, like *Bdellovibrio bacteriovorus* and the *Bacteriovorax* species, give rise spontaneously to mutants capable of growing without prey. The same procedure described to enrich the mutants of *Bdellovibrio* may be used to recover prey-independent HBLOs, using a medium with at least 1% salt. These mutants may be useful tools for studying the genetics and metabolism of the HBLOs and how they differ from the other members of the *Bdellovibrionaceae*.

Other differences that have been described between the HBLOs and the terrestrial predators include immunological properties, resistance to certain antibiotics, and metabolic properties (Table BXII.δ.47). Immunodiffusion analysis performed with antisera against eight isolates of HBLO and a single terrestrial isolate and antigen preparations from 45 HBLO environmental isolates yielded seven serogroups that were distinct from the terrestrial isolate (Falkler et al., 1979). Immunoelectrophoresis revealed a shared antigen between the halophilic and terrestrial isolates. Further study may reveal that this shared antigen is universal among the *Bdellovibrionaceae* (*Bdellovibrio bacteriovorus*, *Bacteriovorax* species, and HBLO). These results suggested the presence of various serotypes among the HBLO and inferred that these serologic differences would allow definition of subgroups of these organisms as well.

Differences in antibiotic susceptibilities among the *Bdellovibrionaceae* have been observed. These differences may be useful in differentiating the organisms within the family. The terrestrial and halophilic predators were differentiated based on their susceptibility to polymyxin B 300, being resistant and sensitive respectively (Guether and Williams, 1993).

The distribution of the *Bdellovibrionaceae* has been the subject of several reports. The HBLOs occur in oceans, seas, estuaries, man-made aquatic systems including a marine aquarium (Williams et al., 1987), and salt lakes, but not in fresh waters or soils. Strains of HBLO reportedly require ≥0.5% NaCl for growth. Although the HBLOs are not known to occur in fresh water, shifts in the occurrence of halophilic and fresh water bdellovibrio populations at the same site have been observed to correlate to salinity fluctuations between 0 (fresh water) and 1% (Kelley et al., 1997). Most distribution studies have focused on the occurrence of HBLO in the water column where the numbers recovered were typically low, i.e., less than 100 plaque-forming units (PFU) per ml of water (Marbach et al., 1976; Williams, 1988; Williams et al., 1995a, b). Earlier, Shilo (1969) had suggested the possibility that investigators had been examining the wrong habitats for greatest bdellovibrio abundance and activity. Subsequent studies in an estuary revealed that HBLOs were more abundant at the air–water interface, in sediments, and in microbial biofilms on surfaces (Williams et al., 1995a, b), all of which are densely populated with bacteria. These data confirm Shilo's hypothesis. The HBLOs, and perhaps the other *Bdellovibrionaceae*, are similar to many other aquatic bacteria in that they tend to associate with available surfaces (Bernheimer et al., 1993; Baer et al., 1994; Williams et al., 1995a, b). When sterile glass slides were submerged in the Patuxent River, bacteria, including HBLOs, associated rapidly with the surfaces (Williams et al., 1995b). Samples of biofilms from the surfaces of oyster and crab shells yielded the highest numbers of HBLO recorded from the aquatic ecosystem (Kelley and Williams, 1992; Williams et al., 1995a, b). Microbial biofilms appear to host the most enriched and stable population of the HBLO. The most obvious example of this is the stable occurrence of the predatory bacteria over an annual cycle in surface biofilms as opposed to the seasonal fluctuations observed in water and sediments (Kelley et al., 1997). Apparently, biofilms provide the nutritional (abundance of prey bacteria) and environmental conditions suitable for survival of the HBLO at relatively high levels under the harshest of conditions. This likely holds true for the terrestrial predatory organisms as well.

The HBLOs do show a seasonal distribution in the water column and sediment that has been correlated with temperature (Williams, 1988). At temperatures below 10°C, the numbers of the organisms were observed to decline rapidly in both the water column and sediments, although the numbers remaining in sediments were higher. Frequently at water temperatures below 5°C, the predatory HBLO could not be detected by culture. As the water temperature warmed in the spring, the numbers of the predators were first observed to increase in sediments and then in the water column. The effect of temperature on fluctuations in the numbers of HBLO was not nearly as dramatic in surface biofilms, where the organisms remained fairly constant, as that observed in water and sediments (Williams et al., 1995a, b). Rice et al. (1998) observed that a large majority of the bacteria recovered from the water, sediment, and biofilm in the Chesapeake Bay region was susceptible to predation by HBLO.

Clearly not all habitats are equal in their capacity to support the growth and maintenance of the HBLO and the other intraperiplasmic predatory bacteria (IPPB). In aquatic ecosystems, the predators do exhibit a preference for surface habitats over planktonic habitats.

The IPPB have been consistently isolated from animals and humans (Schwudke et al., 2001), including the blue crab, *Callinectus sapidus* (Kelley and Williams, 1992). HBLOs were recovered consistently from the gills of animals taken from estuarine waters and from seafood markets (Kelley and Williams, 1992; Schwudke et al., 2001).

For isolation of HBLO from environmental sources, samples are cultured in the same manner as their terrestrial counterparts except that a salt-containing (1–3%) medium such as Pp20[4] (Williams and Falkler, 1984) or MPY/10[5] (Marbach et al., 1976) is substituted for NB/500[6] (Staples, 1973) and DNB[7] (Stolp, 1973). The medium most often reported in the literature is Pp 20 (Williams et al., 1976; Schoeffield, 1990; Kelley et al., 1997) with *Vibrio parahaemolyticus* P-5 or other *Vibrio* sp. as prey. These species were observed to most consistently yield recovery of the greatest number of HBLO from the Chesapeake Bay (Schoeffield, 1990). However, when sampling other types of bodies of water, various species should be examined as prey. This may require using other media formulations such as MPY/10 that may be more suitable for production of confluent growth of the prey lawns required to maximize HBLO recovery (Marbach et al., 1976). The prey bacteria typically require salt, and they are grown on seawater yeast extract agar (SWYE[8]), or Marine broth (Difco). The colonies are harvested by flooding the plate with approximately 5 ml of sterile artificial seawater (ASW).

For initial recovery of HBLO, a solid agar or broth may be used. For samples in which HBLOs are expected to be abundant, recovery and isolation may be accomplished by direct plating on a solid medium using the double-agar overlay technique as described for the *Bdellovibrio*. This procedure has been used extensively for recovery from samples from rivers, estuaries, oceans, and marshes (Williams and Falkler, 1984; Williams, 1987; Williams et al., 1995a; Kelley et al., 1997; Rice et al., 1998). In addition, direct plating of samples should be considered when the water temperature is below 10°C, since it has been reported that the numbers of predatory bacteria are reduced at low temperatures (Williams, 1988). For the HBLO, plated samples are typically incubated at 22–25°C for up to 1 week, although plaques typically appear within 48–72 h. Individual plaque may be subcultured to obtain pure cultures. Samples thought to have numbers of HBLO too low to be readily detected by direct plating should be enriched. The typical enrichment procedure is to add a portion of the sample to a liquid medium containing prey cells. Good results have been observed for recovery of the HBLO using as culture medium filtered and sterilized natural seawater or estuarine water with no added nutrients or salts other than the

---

4. Pp20 is polypeptone 20 medium and consists of (g/l) polypeptone (Bacto), 1.0 g, and agar (Difco), 18.0 g. These ingredients are dissolved in 1 l of 70% artificial sea water. The 70% artificial sea water (ASW) consists of 26 g of Instant Ocean dissolved in 1 l of distilled water; the pH is adjusted to 8.0 with 1 M NaOH.

5. MPY/10 is a basal salt medium and consists of (per l basal salt solution) peptone (Bacto), 0.5 g; yeast extract, 0.3 g. The pH is adjusted to 7.4. Basal salt solution consists of (g/l) $MgSO_4·7H_2O$, 6.92; NaCl, 28.15; $MgCl_2·6H_2O$, 5.51; $CaCl_2·H_2O$, 1.45; KCl, 0.67.

6. NB/500 is nutrient broth (Difco) at 1/500 normal strength (equals 0.016 g of dehydrated broth/l).

7. DNB is a dilute nutrient broth, consisting of (per l) nutrient broth (Difco), 0.8 g; yeast extract (Difco), 0.1 g; vitamin-free casamino acids, 0.5 g; 2 mM $CaCl_2$; and 1 mM $MgCl_2$. The pH is adjusted to 7.2 with 1 M NaOH.

8. SWYE is salt water yeast extract medium and consists of (per l) peptone (Bacto), 10 g; yeast extract, 3.0 g dissolved in 1 l of 70% artificial seawater (ASW).

prey. In the previous edition of the *Manual*, the MPY medium (Taylor et al., 1974; Marbach et al., 1976) was cited for isolation, enrichment, and maintenance of HBLO. More recently, other media formulations have been used. For short-term maintenance in culture, satisfactory results have been obtained using prey–seawater medium consisting of prey cells suspended in either natural, filtered, or autoclaved seawater diluted to 70% or 70% artificial seawater (Instant Ocean, 26 g/l of distilled water) (personal observation). Estuarine water may be used undiluted. *V. parahaemolyticus* is a preferred prey for the growth and maintenance of the HBLO (Williams et al., 1976; Schoeffield, 1990). This method maintains cultures for 1–2 weeks only. For long-term maintenance, HBLO cultures may be pelleted as described for the terrestrial isolates, suspended in marine broth or other suitable medium containing 20% glycerol, quick-frozen in liquid nitrogen, and stored at −70°C. Storage for 6 months or less has been successful at −20°C.

Major phenotypic differences between *B. bacteriovorus* and the halophilic bdellovibrios are listed in Tables BXII.δ.46 and BXII.δ.47.

The taxonomy of the HBLO remains in a primitive state. In the first edition of the *Manual*, it was recommended that all halophilic and other low mol% G + C isolates be placed in the group designated *Bdellovibrio* sp. None of the HBLO organisms has been sufficiently characterized and defined to warrant designation as a taxonomic species. This lack of information has made it impossible to investigate the diversity of the organisms and the abundance and distribution of various species in nature. It is apparent that differences in the metabolic, genetic, and cultural properties between the HBLO and *Bdellovibrio* are sufficient to preclude them from belonging to the same genus (see Baer et al., 2004 in Further Reading section in this chapter).

Genus II. **Bacteriovorax** Baer, Ravel, Chun, Hill and Williams 2000, 222[VP]

HENRY N. WILLIAMS AND MARCIE L. BAER

Bac.te.ri.o.vo′rax. Gr. dim. n. *bacterion* small rod; L. n. *vorax* devourer; M.L. n. *Bacteriovorax* devourer of bacteria.

Members of the genus *Bacteriovorax* exhibit the same general morphological and life cycle features as described for the genus *Bdellovibrio*. However, genetically the two genera vary considerably, with *Bacteriovorax* sp. having a lower **mol% G + C content, little to no DNA–DNA hybridization (<4%), and <82% rDNA similarity with** *Bdellovibrio*.

The mol% G + C of the DNA is: 41–43.5.

Type species: **Bacteriovorax stolpii** (Seidler, Mandel and Baptist 1972) Baer, Ravel, Chun, Hill and Williams 2000, 223 (*Bdellovibrio stolpii* Seidler, Mandel and Baptist 1972, 216.)

FURTHER DESCRIPTIVE INFORMATION

To reflect the molecular genetic heterogeneity and biochemical and cultural diversity within the genus *Bdellovibrio*, a new genus, *Bacteriovorax*, was established and includes the organisms classified in the last edition of *Bergey's Manual of Systematic Bacteriology* (Burnham and Conti, 1984) as *Bdellovibrio stolpii* and *Bdellovibrio starrii*. Details of how the new genus came to be are discussed under Taxonomic Comments.

Like most *Bdellovibrio* strains, *Bacteriovorax* strains are capable of preying upon Gram-negative enteric bacteria and other bacteria with similar LPS structure. *Bacteriovorax stolpii* appears to be unaffected by changes in the LPS composition of bacteria but instead recognizes protein receptors in the outer membrane of the prey cell (Schelling and Conti, 1986). Gram-positive bacteria cannot serve as prey, presumably because of the lack of outer membrane receptors.

The physical and morphological features of *Bacteriovorax* have been confirmed to be as described for the *Bdellovibrio*. However, most of the physiological studies involving utilization of prey macromolecules, enzyme production at various stages of entry into prey cells, and development have been done on *Bdellovibrio bacteriovorus* 109 and other strains. It is presumed at the current time, but not yet confirmed experimentally, that *Bacteriovorax* species are similar or identical in these characteristics. However, further study on the new genus is warranted.

Genetics The genome sizes for the two species of *Bacteriovorax* has been reported to be in the range of 2.0–2.6 Mb (Torrella et al., 1978; Scognamiglio and Tudor, 1996; Baer, 1998), which is higher than that reported by Seidler et al. (1972).

Prey-independent mutants of *Bacteriovorax* have been isolated by the procedure described by Seidler and Starr (1969) and are available through the ATCC. These mutant strains were derived as described in the chapter on the genus *Bdellovibrio*.

Phages Several phage types among the predatory intraperiplasmic bacteria have been reported. Group V consists of *Bacteriovorax stolpii* and group VI, *Bacteriovorax starrii*. Groups V and VI are unrelated to *B. bacteriovorus* strains, which comprise groups I–IV. Further study involving additional phages may make phage typing a useful technique for distinguishing between *Bdellovibrio* and *Bacteriovorax* species.

Antigenic structure *Bacteriovorax stolpii* and *Bacteriovorax starrii* have been reported to be antigenically distinct from each other (Schoeffield, 1990). Both species were distinct from the *B. bacteriovorus* strains (Schelling et al., 1977).

Antibiotics The use of antibiotic sensitivity patterns has revealed some differences between the species of *Bacteriovorax* (Guether and Williams, 1993). Both species as well as the related organism *Bdellovibrio bacteriovorus* were susceptible by the disk diffusion method to ampicillin (20 μg and 2 μg), carbenicillin (100 μg), gentamicin (10 μg), kanamycin (30 μg), and penicillin (10 U, BBL) and were resistant to colistin (10 μg). Only *Bacteriovorax stolpii* was susceptible to metronidazole (80 μg), nalidixic acid (5 μg), and vancomycin (30 μg). *Bacteriovorax starrii* was resistant to nalidixic acid (5 μg) and vibriostatic agent 0/129.

Pathogenicity Other than the infection of susceptible bacteria, *Bacteriovorax* are not known to be pathogenic to other species including plants and animals. The *Bacteriovorax/Bdellovibrio* group of organisms has been recovered from the gut of humans and animals (Schwudke et al., 2001); whether these organisms are transient or permanent members of the intestinal flora is not known.

Ecology The ecology of the predatory, intraperiplasmic bacteria *Bacteriovorax/Bdellovibrio* has revealed the ubiquitous distribution of the group. However, past studies typically have not distinguished between the two genera or species within them. Currently the distribution and potential role of these predators in nature is presumed to be similar to that reported for the *Bdellovibrio* as described in the previous chapter of this volume since many of the isolates may have been of the *Bacteriovorax* species.

ENRICHMENT AND ISOLATION PROCEDURES

Conditions for the enrichment and isolation of *Bacteriovorax* are the same as described for the genus *Bdellovibrio*. The essential conditions consist of a heavy prey cell suspension (10^8–10^9 cells/ml) in a medium such as a buffered salts solution with few or no added nutrients. These conditions retard growth of many heterotrophic bacteria and typically enable the rapid proliferation of the *Bacteriovorax*. Attention should be given to the selection of the prey bacterium. Although the *Bacteriovorax* species were initially isolated on *P. aeruginosa* and are maintained on *E. coli* ML35, other species may be more efficient as prey, producing a greater yield of the predators and should be tested.

MAINTENANCE PROCEDURES

The procedures for maintaining *Bacteriovorax* in the laboratory are the same as those described in the chapter on *Bdellovibrio*.

PROCEDURES FOR TESTING SPECIAL CHARACTERS

Bacteriovorax species, like *Bdellovibrio* species, cannot be characterized by commonly used laboratory tests, because the organisms grow in dual culture and typically do not yield reactions in pure culture on artificial media. Tests that do not require growth of the organism may be useful, such as the API-Zym system (BioMereiux, France). These tests have revealed differential reactions between the *Bacteriovorax* and *Bdellovibrio* strains tested (Table BXII.δ.48). Prey-independent mutants—isolated from wild-type strains—have been useful in determining some biochemical reactions and other properties of the organisms. Based on limited data, it appears that the prey-independent mutants are identical in most properties to the wild type.

DIFFERENTIATION OF THE GENUS *BACTERIOVORAX* FROM OTHER GENERA

Bacteriovorax is most closely related to *Bdellovibrio*, the genus in which they were previously placed. Although the two groups of organisms share the same general distinct intraperiplasmic lifestyle, they may be differentiated by a number of properties, biochemical and molecular. Restriction analysis by pulsed field gel electrophoresis revealed differences in sensitivity to restriction enzymes and in restriction patterns. When treated with the enzymes *Asc*I, *Fse*I, *Sfi*I, and *Not*I only DNA of *Bdellovibrio bacteriovorus* strains was cut and produced banding patterns, whereas DNA of *Bacteriovorax* strains remained uncut. All the strains were cut with *Sma*I, *Sad*II, and *Xba*I. Other fingerprinting techniques such as AP-PCR and ribotyping yielded distinct separation of *Bacteriovorax* species from *Bdellovibrio bacteriovorus* strains. The halophilic strains previously considered to be a part of the *Bdellovibrio* genus clustered closer to *Bacteriovorax* species than to strains of *Bdellovibrio bacteriovorus*. *Bacteriovorax* is also differentiated from *Bdellovibrio* based on certain biochemical, physiological, genetic, and cultural properties presented in Table BXII.δ.49.

Key biochemical reactions of *Bacteriovorax* species on the API-Zym system is presented in Tables BXII.δ.48 and BXII.δ.49. This system allows differentiation between *Bacteriovorax stolpii* and *B. starrii* by their respective reactions to trypsin. The system is also useful for differentiating *Bacteriovorax* from *Bdellovibrio*. Fatty acid analysis has revealed that differentiation is possible among the *Bacteriovorax* species as well as between *Bacteriovorax* and *Bdellovibrio*. A few key biochemical reactions were also reported in the last edition of the *Manual*.

TAXONOMIC COMMENTS

When initially described as predatory bacteria in 1962 (Stolp and Petzold, 1962) and later as intraperiplasmic invaders (Starr and Baigent, 1966), bdellovibrios were the first bacteria reported to exhibit such features. For a time thereafter, all predatory bacteria isolated with similar morphology and a biphasic life cycle consisting of a free motile phase and an intraperiplasmic growth and reproductive phase were assigned to the genus *Bdellovibrio*, with a single species, *B. bacteriovorus*. Initially all isolates were considered to be a taxonomically homogeneous group, although important differences in their tolerance to sodium chloride and mol% G + C had been reported. In 1972, a report of molecular heterogeneity (based on DNA–DNA hybridization values among different isolates) resulted in the division of the genus into three species, *Bdellovibrio bacteriovorus*, *Bdellovibrio stolpii*, and *Bdellovibrio starrii* (Seidler et al., 1972). This was reflected in the previous edition of *Bergey's Manual of Systematic Bacteriology* (Burnham and Conti, 1984).

Based on the recent analysis of complete 16S rDNA sequences from *Bdellovibrio bacteriovorus* 100^T, *B. stolpii* Uki2^T, and *B. starrii* A3.12^T, and DNA–DNA hybridization studies (Baer et al., 2000), a single genus is no longer appropriate for this collection of molecularly diverse microorganisms. The 16S rDNA sequence similarities between *B. bacteriovorus* 100^T and *B. stolpii* Uki2^T and *B. starrii* A3.12^T were 81.7% and 81.2%, respectively. These similarity values are comparable to or lower than those between *B. bacteriovorus* 100^T and other organisms that phylogenetically group in the same clade but are considered to be members of separate genera (Fig. BXII.δ.54). The similarity between the rDNA sequences from *Bdellovibrio stolpii* Uki2^T and *Bdellovibrio starrii* A3.12^T is significantly higher (90.0%).

Comparison of the 16S rDNA sequences of *B. bacteriovorus* 100^T, *B. starrii* A3.12^T, and *B. stolpii* Uki2^T (M34125) revealed substantial sequence differences between all three organisms. *B. bacteriovorus* 100^T had 224 and 230 bp different from *B. stolpii* Uki2^T and *B. starrii* A3.12^T, respectively. Only 127 bp differences were found when *B. stolpii* Uki2^T and *B. starrii* A3.12^T sequences were compared to each other.

The distant relationship between the three *Bdellovibrio* species is also apparent from the rooted evolutionary tree (Fig. BXII.δ.54) based on the Jukes-Cantor distance model and the neighbor-joining methods. *B. bacteriovorus* 100^T clustered more closely with members of the *Desulfomonile*, *Desulfuromonas*, *Myxococcus*, *Geobacter*, and *Desulfuromusa* genera. *B. stolpii* Uki2^T and *B. starrii* A3.12^T clustered together but outside of the *B. bacteriovorus* 100^T clade. The nucleotide sequence data were also confirmed with three other tree-making algorithms: maximum parsimony, maximum likelihood, and Fitch-Margoliash methods. It is clear from the 16S rDNA sequence data that *B. bacteriovorus*

TABLE BXII.δ.48. Differentiation of the genus *Bacteriovorax* from other closely related taxa[a]

Characteristic	*Bacteriovorax*	*Bdellovibrio* (terrestrial strains)	*Bdellovibrio* (halophilic strains)
Typical mol% G + C of DNA[b]	41–43.5	Low 50s	33.4–38.6; 43.5[c]
Sensitivity to G 2.0	S	R	d[d]
Sensitivity to PB 300	R	R	S
Enzyme activities detected:[e]			
Valine aminopeptidase	+	−	+
Cystine aminopeptidase	+	−	+
Lipase (C − 14)	−	−	+
Chymotrypsin	−	+	−
Protease production	High to moderate	Low	nd
Presence of fatty acids:[f]			
$C_{16:1\ \omega 9c}$	+	−	+
$C_{13:1\ iso\ 3OH}$, $C_{17:1\ \omega 8c}$	−	+	−
$C_{17:1\ \omega 6c}$	+	−	−
Na$^+$ required for growth (75mM or higher)	−	−	+
Habitat:			
Freshwater or terrestrial	+	+	−
Sea water or brackish	−	−	+

[a]Symbols: see standard definitions; S, sensitive (based on zone size >10 mm using prey-independent strains); R, resistant (based on zone size <10 mm); nd, not determined.

[b]Based on the results from Burnham and Conti (1984).

[c]Mol% G + C range taken from Table 2.15 in Burnham and Conti (1984). Only one halophilic strain has been found to have a mol% G + C of 43.5.

[d]Strain 2484Se2 is positive.

[e]Result of API-Zym (BioMerieux) test system using prey-independent strains.

[f]Measured by the MIDI Sherlock Microbial Identification System, Newark, DE.

TABLE BXII.δ.49. Other characteristics of the species of the genus *Bacteriovorax*[a]

Characteristic	*B. stolpii* and *B. starrii*
Comma-shaped predacious cells, 0.35 × 1.2 μm	+
Motility	+
Single, polar, sheathed flagellum	+
Elongated to spiral-shaped forms develop during growth	
Grow and divide within prey cell	+
Prey-independent strains occur	+
Prey-independent variants occur	+
Gram-negative, but not Gram-positive, bacteria can serve as prey	+
Circular, plaques or zones of clearing occur on agar-grown lawns of prey bacteria	+
Aerobic	+
Bacteriophage susceptibility	+
Optimum growth temperature	28–30°C
Fermentative ability present	−
Susceptibility to antibiotics:[b]	
CB100	S
GM10	S
K30	S
P10	S
CL10	R
MET80	S,I[c]
VA30	S,I[c]
Presence of certain fatty acids:[d]	
$C_{16:1\ \omega 9}$	+
$C_{13:1\ 3OH}$	−

[a]Symbols: see standard definitions. S, susceptible; R, resistant; I, inconclusive result.

[b]Determined with prey-independent strains using a disk diffusion assay on agar plates.

[c]Results with *B. starrii* were inconclusive.

[d]As determined using the MIDI Sherlock Microbial Identification System, Newark, DE.

100^T belongs in a separate clade with only a distant relationship to *B. stolpii* and *B. starrii*.

Results from investigations into the degree of DNA–DNA hybridization between *Bdellovibrio* strains have been inconsistent. Baer et al. (2000) reported very low hybridization values between *B. stolpii* Uki2^T and *B. starrii* A3.12^T (<4%). A previous study (Seidler et al., 1972) reported a slightly higher degree of hybridization (16%) between these same strains. Hybridization values between Uki2^T and A3.A3.12^T and *Bdellovibrio bacteriovorus* were 1% or less, and the investigators of that study suggested that the low level of hybridization was sufficient to warrant the separation of the genus *Bdellovibrio* into three species, *B. bacteriovorus*, *Bacteriovorus stolpii*, and *Bacteriovorus starrii*. Low hybridization values were also observed between halophilic *Bdellovibrio* strains and their terrestrial counterparts (Baer, 1998).

This evidence, taken in conjunction with additional studies demonstrating the diversity among the organisms previously in the genus *Bdellovibrio*, has resulted in a taxonomic separation between *B. bacteriovorus* and the other two species, *B. stolpii* and *B. starrii*. *Bdellovibrio stolpii* Uki2^T and *Bdellovibrio starrii* A3.12^T have been reclassified into a new genus, *Bacteriovorax* gen. nov. as *Bacteriovorax stolpii* comb. nov. and *Bacteriovorax starrii* comb. nov. (Baer et al., 2000). *Bacteriovorax* also clusters within the *Deltaproteobacteria*. The two *Bacteriovorax* species, *B. starrii* and *B. stolpii*, are distinct from *Bdellovibrio bacteriovorus* based on 16S rDNA sequence comparison, pulsed field gel electrophoresis macrorestriction patterns, other genetic "fingerprint" techniques (arbitrarily primed PCR and ribotype patterns), fatty acid content (Guether et al., 1993), antibiotic sensitivity patterns, phage typing, antigenicity, and prey range (Althauser et al., 1972; Varon and Levisohn, 1972; Kramer and Westergaard, 1977; Nguyen et al., 1997; Schelling and Conti, 1983). Currently, the genus is composed of two species; however, molecular evidence reveals that the halophilic strains of the genus *Bdellovibrio* are more ge-

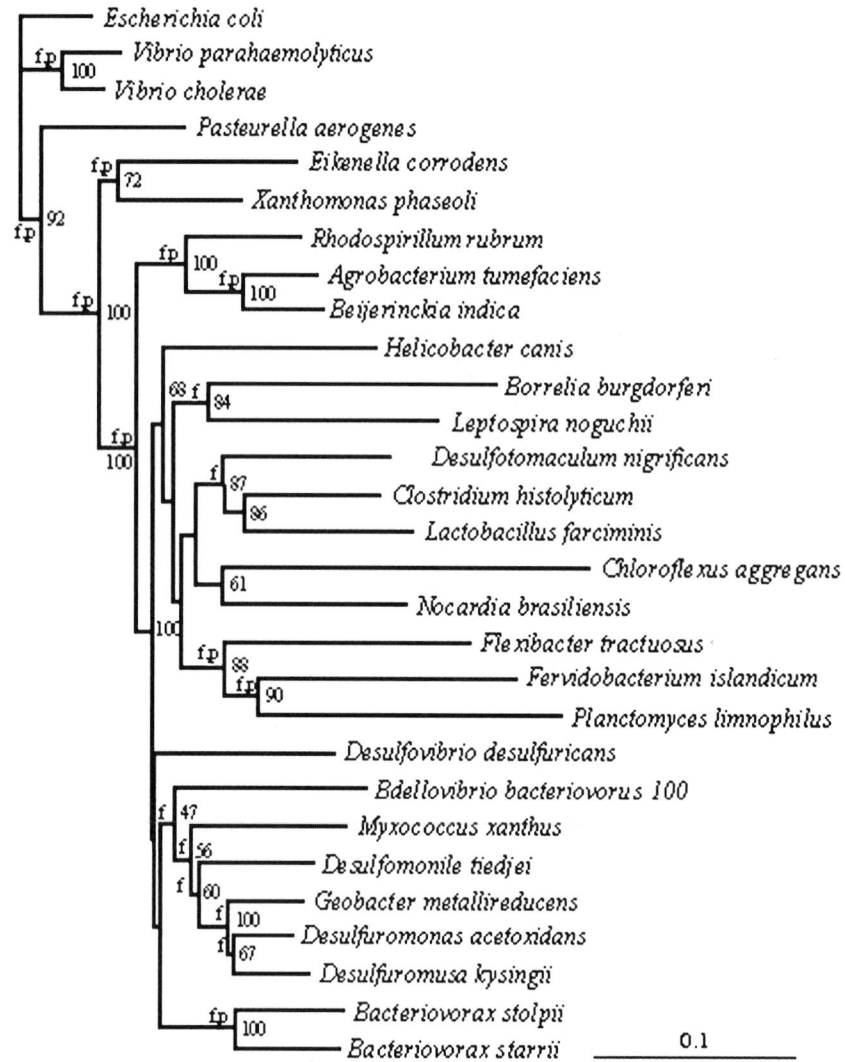

FIGURE BXII.δ.54. A neighbor-joining tree based on 16S rRNA sequences demonstrating the phylogenetic relationship between the *Bdellovibrio* and *Bacteriovorax* genera. *m*, *f*, and *p* indicate branches that were also found using the FastDNAml, Fitch-Margoliash, and maximum parsimony methods, respectively.

netically related to the genus *Bacteriovorax*. Fig. BXII.δ.53 in the section on the genus *Bdellovibrio* shows the phylogenetic relationships between different, unclassified, environmental isolates and known *Bdellovibrio* and *Bacteriovorax* sp. *B. bacteriovorus* strains 100, 109J, and E clustered together in one group. *Bacteriovorax stolpii* and *Bacteriovorax starrii* grouped into a separate clade, which included all the other environmental isolates. *B. stolpii* grouped most closely with the terrestrial environmental isolate Gunpowder, both of which formed a larger group with *B. starrii*. All marine isolates fell into three different groups within this same clade. The halophilic *Bdellovibrio* species were reclassified to the genus *Bacteriovorax* (see Baer et al., 2004 in Further Reading section below) after the cut-off date for inclusion in this edition of the *Manual*. It is expected, therefore, that the number

of species and strains of the genus will increase, as more information on the diversity of new isolates from nature is uncovered.

The establishment of the genus *Bacteriovorax* represents an advancement in the taxonomy and classification of the bacterial intraperiplasmic predatory bacteria and recognizes significant differences within this group of organisms that share a similar lifestyle.

FURTHER READING

Baer, M.L., J. Ravel, S.A. Pineiro, D. Guether-Borg and H.N. Williams. 2004. Reclassification of salt-water *Bdellovibrio* spp. as *Bacteriovorax marinus* sp. nov. and *Bacteriovorax litoralis* sp. nov. Int. J. Syst. Evol. Microbiol. *54*:: 1011–1016.

List of species of the genus Bacteriovorax

1. **Bacteriovorax stolpii** (Seidler, Mandel and Baptist 1972) Baer, Ravel, Chun, Hill and Williams 2000, 223[VP] (*Bdellovibrio stolpii* Seidler, Mandel and Baptist 1972, 216.)

stolp′i.i. M.L. gen. n. *stolpii* of Stolp, named after Heinz Stolp, discoverer of the bdellovibrios.

The characteristics are as described for the genus and as listed in Tables BXII.δ.48, BXII.δ.49, and BXII.δ.50.

The mol% G + C of the DNA is: 42 (Bd).

Type strain: Uki2, ATCC 27052, DSM 50722, ICPB 3291.

GenBank accession number (16S rRNA): M34125.

2. **Bacteriovorax starrii** (Seidler, Mandel and Baptist 1972) Baer, Ravel, Chun, Hill and Williams 2000, 223[VP] (*Bdellovibrio starrii* Seidler, Mandel and Baptist 1972, 216.)

starr' i.i. M.L. gen. n. *starrii* of Starr, named after M.P. Starr, an investigator of the bdellovibrios.

The characteristics are as described for the genus and as listed in Tables BXII.δ.48, BXII.δ.49, and BXII.δ.50.

The mol% G + C of the DNA is: 43.5 (Bd).

Type strain: A3.12, ATCC 15145.

GenBank accession number (16S rRNA): AF084852.

TABLE BXII.δ.50. Differential characteristics of the species of the genus *Bacteriovorax*[a]

Characteristic	B. stolpii	B. starrii
Facultatively predatory[b]	+	−
Sensitive to vibriostatic agent 0/129 150[c,d]	+	−
Mol% G + C of DNA	42	43.5
Growth at 35°C	+	−
Enzyme activities detected:[d]		
Trypsin	+	−
Protease production	High	Moderate
N-acetyl β-glucosaminidase	−	+
Catalase[e]	+	−
Nitrate reduced to nitrite	+	−

[a]Symbols: see standard definitions.

[b]Defined to mean that any individual comma-shaped cell may complete its life cycle either prey-dependently or prey-independently.

[c]O/129 = 24 diamino-6-7-disopropylpteridine phosphate. Test done on prey-independent mutants.

[d]Based on API Zym system using prey-independent mutants of *Bacteriovorax*.

[e]Cultures may be initially positive but become variable with subsequent transfer (Seidler and Starr, 1969).

Genus III. **Micavibrio** *Lambina, Afinogenova, Romai Penabad, Konovalova and Pushkareva 1989, 93[VP] (Effective publication: Lambina, Afinogenova, Romai Penabad, Konovalova and Pushkareva 1982, 105)*

MARCIE L. BAER AND HENRY N. WILLIAMS

Mi.ca.vib' ri.o. M.L. fem. n. *mica* a tiny thing; L. v. *vibrio* to move rapidly to and fro; M.L. masc. n. *vibrio* that which vibrates; M.L. masc. n. *Micavibrio* a tiny vibrio.

Curved rods 0.25–0.4 × 0.5–1.0 μm. Occur singly. **Motile by a single, sheathless, polar flagellum** approximately 13–15 nm in diameter. Gram negative. Electron microscopy demonstrates the cell wall infrastructure to be consistent with Gram-negative bacteria (Lambina et al., 1982). **Exoparasitic on other bacteria.** Attach themselves to their bacterial prey by the end opposite the flagellum and position themselves parallel to the longitudinal axis of the prey cell. **They remain along the outside of the prey cell** (unlike the genus *Bdellovibrio*, which penetrates its prey). **After attachment, the cells lose their motility, multiply by binary fission, and eventually cause lysis of their prey cell. The organisms have an obligate requirement for prey cells for multiplication** and lack the ability to propagate themselves on nutrient rich media. **Initial prey susceptibility appears to be specific for members of the genus *Pseudomonas* and *Xanthomonas*,** with different *Micavibrio* species having different prey requirements. Resistant to vibriostatic agent 0/129.

Initially isolated from wastewater in Pushchino.

The mol% G + C of the DNA is: 57.

Type species: **Micavibrio admirandus** Lambina, Afinogenova, Romai Penabad, Konovalova and Pushkareva 1989, 93 (Effective publication: Lambina, Afinogenova, Romai Penabad, Konovalova and Pushkareva 1982, 106.)

TAXONOMIC COMMENTS

Although *Micavibrio* has some characteristics that resemble those of the genus *Bdellovibrio*, it has several distinctive differences from this genus that warrant a separate taxonomic classification. Morphologically, bdellovibrios and micavibrios are similar in size and shape, but *Micavibrio* cells have a single unsheathed polar flagellum (Lambina et al., 1982). The flagellum of the *Bdellovibrio* sp. has a definite sheath, as shown by electron microscopy studies. Both genera are parasitic in nature; however, members of *Micavibrio* are exoparasites, while bdellovibrios are endoparasites. Prey specificity of *Micavibrio* species also appears to be much more narrow than the range observed for *Bdellovibrio* sp.

Micavibrio species also demonstrate morphological similarities to the *Chlorella* parasite, *Vampirovibrio chlorellavorus* (previously described as *Bdellovibrio chlorellavorus*). Both are exoparasites and are characterized by a high specificity with respect to their individual prey. However, *Micavibrio* spp. demonstrate an ability to lyse procaryotic bacterial cells while the chlorellavorus bacterium exclusively parasitizes eucaryotic algae (Lambina et al., 1982).

List of species of the genus Micavibrio

1. **Micavibrio admirandus** Lambina, Afinogenova, Romai Penabad, Konovalova and Pushkareva 1989, 93[VP] (Effective publication: Lambina, Afinogenova, Romai Penabad, Konovalova and Pushkareva 1982, 106.)

ad.mir.an' dus. M.L. adj. *admirandus* admirable.

The general morphological characteristics and parasitic nature are as described for the genus. Upon initial isolation, *M. admirandus* is characterized by a narrow species specificity for *Stenotrophomonas maltophilia*. However, studies have shown that upon repeated transfer and storage under laboratory conditions, *M. admirandus* loses its narrow prey specificity and becomes a parasite with a broad range of lytic ability against Gram-negative bacteria (Afinogenova et al., 1986).

The mol% G + C of the DNA is: 57.1 (analytical centrifugation in a CsCl density gradient).

Type strain: ARL-14, VKM B-1619.

Other Organisms

1. *"Micavibrio aeruginosavorus"* Lambina, Afinogenova, Romai Penabad, Konovalova and Andreev 1983, 777.

ae.ru.gi.no.sa'vo.rus. M.L. adj. *aeruginosus* pertaining to the bacterium *Pseudomonas aeruginosa*; L. v. *voro* to devour; M.L. adj. *aeruginosavorus* devouring *P. aeruginosa*.

The general morphological characteristics and parasitic nature are as described for the genus *Micavibrio*. "*M. aeruginosavorus*" exhibits a narrow prey range and specifically requires *P. aeruginosa* for growth and multiplication (Lambina et al., 1983).

An electrophoretically homogeneous lytic proteinase of molecular weight 39,000 ± 1500 Da was isolated from the *M. admirandus* by means of ion-exchange chromatography and gel chromatography (Severin et al., 1987). The proteinase lyses autoclaved *E. coli* cells.

The mol% G + C of the DNA is: 53.8 (analytical centrifugation in a CsCl density gradient).

Deposited strain: ARL-13.

Genus IV. **Vampirovibrio** *Gromov and Mamkaeva 1980a, 676*[VP] *(Effective publication:* *Gromov and Mamkaeva 1980b, 165)*

MARCIE L. BAER AND HENRY N. WILLIAMS

Vam.pi.ro.vib'ri.o. Fr. n. *vampire* vampire; L. v. *vibrio* to move rapidly to and fro; M.L. masc. n. *vibrio* that which vibrates; M.L. masc. n. *Vampirovibrio* a vampire-like vibrio.

Cell shapes range from vibrios to wider curved rods to cocci and range from 0.3–0.6 μm wide. Cell size and shape vary with life cycle form. **Gram negative.** Cell wall ultrastructure is consistent with that of Gram-negative bacteria. Electron micrographs indicate the presence of extracellular appendages **including a single, unsheathed polar flagellum** and fibrils. **Require viable cells of the algal genus** *Chlorella* for growth and development; reproduction occurs via binary fission. **Extracellular parasites**; penetration of the prey cell has never been observed, although the *Chlorella* cells are killed and the cell contents are digested (Coder and Starr, 1978).

The mol% G + C of the DNA is: 50 (Coder and Starr, 1978).

Type species: **Vampirovibrio chlorellavorus** Gromov and Mamkaeva 1980a, 676 (Effective publication: Gromov and Mamkaeva 1980b, 165.)

FURTHER DESCRIPTIVE INFORMATION

The cell wall ultrastructure is consistent with that of Gram-negative bacteria. Prey susceptibility appears to be limited to alga cells of the genus *Chlorella*, as no degradation was observed when lysates of the "chlorellavorus bacterium" were tested with suspensions of a variety of Gram-negative bacteria, other eucaryotic algae, or the cyanobacterium *Anacystis* (Mamkaeva, 1966; Coder and Starr, 1978).

Coder and Goff (1986) examined 76 algal strains to see whether they could serve as prey for *Vampirovibrio*. The bacterium attacked all 31 strains of the species *Chlorella vulgaris*, *Chlorella sorokiniana*, and *Chlorella kessleri* but only two of 39 strains of nine other *Chlorella* species. Neither of two strains of another algal genus, *Prototheca*, was susceptible to attack. It was suggested that this narrow host specificity may be related to cell surface properties.

TAXONOMIC COMMENTS

Although *Vampirovibrio* has similar characteristics to the genus *Bdellovibrio* including morphology, predatory ability, and DNA base composition, there are important differences between the two genera. *Vampirovibrio* demonstrates an obligate requirement for viable algal cells in order to grow and reproduce, whereas bdellovibrios are much more flexible in their developmental cycle, utilizing active prey cells, growing axenically in liquid/solid media, or growing saprophytically on heat-killed cellular extracts. The prey for *Vampirovibrio* is eucaryotic, not procaryotic, and the prey range is limited to one algal type (*Chlorella*), whereas bdellovibrios demonstrate the ability to parasitize a wide range of susceptible prey bacteria. The shape of the *Vampirovibrio* flagellum resembles that of *Bdellovibrio*, but it lacks a flagellar sheath. Finally, *Vampirovibrio* lacks the typical dimorphic life cycle observed for *Bdellovibrio* sp., i.e., there is an absence of long, spirillar forms, and the organisms never penetrate the selected prey cells. This type of parasitism is unlike its *Bdellovibrio* counterpart, in which the bacteria penetrate and lodge within the periplasmic space of the prey cell. These reasons support the creation of a new genus containing the "chlorellavorus bacterium".

Vampirovibrio, in fact, may be taxonomically closer to members of the genus *Micavibrio* (Lambina et al., 1982). Micavibrios have a similar morphology—small curved rods with a single non-sheathed polar flagellum—and reproductive cycle—multiplication by binary fission that is dependent upon the presence of prey cells. More importantly, micovibrios have an exoparasitic lifestyle like that of vampirovibrios: they never penetrate their prey, yet eventually destroy the prey organism. The major differences between these two genera are that micavibrios have a procaryotic prey (*Stenotrophomonas maltophilia*) and the mol% G + C of the DNA is higher (57% vs. 50%).

List of species of the genus Vampirovibrio

1. **Vampirovibrio chlorellavorus** Gromov and Mamkaeva 1980a, 676[VP] (Effective publication: Gromov and Mamkaeva 1980b, 165.)

chlo.rel.la'vo.rus. M.L. fem. n. *Chlorella* a genus of algae; L.

v. *voro* to devour; M.L. adj. *chlorellavorus* *Chlorella*-devouring.

The characteristics are as described for the genus.

The mol% G + C of the DNA is: 50 (method uncertain).

Type strain: ATCC 29753.

Order VIII. **Myxococcales** Tchan, Pochon and Prévot 1948, 398[AL]

HANS REICHENBACH

My.xo.coc.ca'les. M.L. fem. pl. n. *Myxococcaceae* type family of order; *-ales* ending to denote an order; M.L. fem. pl. n. *Myxococcales* the *Myxococcaceae* order.

Rod-shaped, Gram-negative bacteria that move by gliding, produce fruiting bodies, and desiccation-resistant myxospores. Although these organisms are extremely fascinating because of their intricate life cycles, which are unique among procaryotes, they have been studied by relatively few scientists. In consequence, almost nothing is known about most species other than *Myxococcus xanthus* and *Stigmatella aurantiaca*. The considerable interest that the organisms have gained recently because of their rich secondary metabolism may change that situation. Since the last edition of the *Manual,* a number of comprehensive reviews have appeared that summarize much of our present knowledge of the organism. These reviews can be consulted for further details (Shimkets, 1990b; Reichenbach and Dworkin, 1992; Dworkin and Kaiser, 1993; Dworkin, 1996; Dawid, 2000).

Type genus: **Myxococcus** Thaxter 1892, 403.

FURTHER DESCRIPTIVE INFORMATION

Cytology of the vegetative cell It was observed long ago that there are two different basic and species-specific shapes of myxobacterial cells: long slender cells with tapering ends and cylindrical cells with rounded ends (Krzemieniewska and Krzemieniewski, 1928). This difference in cell shape was an essential argument, among others, for removing *"Chondromyces aurantiacus"* from the genus *Chondromyces* and returning it to its original genus, *Stigmatella* (McCurdy and Khouw, 1969; Reichenbach and Dworkin, 1969; McCurdy, 1971a). The real taxonomic significance of the two basic cell shapes, however, is that they distinguish two suborders, the *Cystobacterineae* and the *Sorangineae,* which thus can be recognized by microscopic examination. This subdivision of the myxobacteria into suborders is clearly supported by 16S rRNA sequence comparisons (Ludwig et al., 1983; Shimkets and Woese, 1992; Spröer et al., 1999). In fact, there is still another cell type that is characteristic of a third suborder, the *Nannocystineae:* short, often almost cube-shaped cylindrical cells with rounded or truncated ends. In each case, the distinguishing cell type occurs together with other exclusive features. Thus, for example, the colony morphologies also differ in characteristic ways, and colonies of members of the *Cystobacterineae* adsorb Congo red, while those of the *Sorangineae* and *Nannocystineae* do not (McCurdy, 1969b).

Additional variation within each of the three cell types helps distinguish certain taxonomic groups at the genus or family level. Thus, all species of the family *Myxococcaceae* have cells of moderate length (3–6 µm), which are cigar-shaped or slightly boat-shaped; *Cystobacter* species have long (8–15 µm) needle-shaped cells; *Stigmatella* cells are boat shaped and of moderate length, resembling the cells of *Myxococcus fulvus*, etc.

Besides normal cells, aberrant cells are occasionally observed, e.g., branched cells (Krzemieniewska and Krzemieniewski, 1928; Reichenbach, 1965c) or screw-shaped cells (Reichenbach, 1965c). The significance of the latter will become apparent when gliding motility is discussed.

Most myxobacteria have relatively large, slender cells measuring typically 0.6–1.2 × 3–15 µm depending both on the species and, to a certain degree, on the medium and the age of the culture. Some genera, however, have much smaller cells; for example, *Nannocystis exedens* cells are 1.4–1.5 × 1.75–3.5 µm, and

some new myxobacteria have decidedly small, delicate cells, clearly a novel myxobacterial cell type.

Electron microscopic thin sections show a typical Gram-negative cell surface, with a thin peptidoglycan layer between outer and cytoplasmic membrane (Abadie, 1967, 1971; Voelz, 1967; Voelz and Reichenbach, 1969; MacRae and McCurdy, 1975; Lampky, 1976). The DNA region is clearly visible in electron micrographs. The early investigators were able to visualize several nucleoids within myxobacterial cells—variously called nucleus-like, nuclear, or chromatin bodies—under the light microscope after suitable staining. Growing vegetative cells may contain one, two, or four such bodies; myxospores apparently have only one (e.g., Thaxter, 1897; Vahle, 1910; Krzemieniewska and Krzemieniewski, 1928; Loebeck and Ordal, 1957; Reichenbach, 1965c).

Cell division is by septum formation (MacRae and McCurdy, 1975). Within the cells can be seen mesosomes (Voelz, 1965; Abadie, 1967, 1971; Lampky, 1976; Hirsch et al., 1978); polysomes, sometimes arranged in a helical pattern (Voelz 1966b; Abadie, 1971); polyphosphate bodies (Voelz, 1965; Voelz et al., 1966; Reichenbach et al., 1969; Abadie, 1971; Lampky, 1976); and various granules of reserve material, perhaps polysaccharide (Reichenbach et al., 1969; Schmidt-Lorenz and Kühlwein, 1969; Lampky, 1976) and certainly polyhydroxybutyric acid. In the myxospores, the polysaccharide granules may be surrounded by a dense monolayer of tiny electron dark particles (Voelz and Reichenbach, 1969; Galván et al., 1992). In addition, rather extended crystal-like structures may occur in the cytoplasm, which probably consist of protein (Voelz, 1968; Burchard et al., 1977a; Galván et al., 1992). Occasionally long filaments or tubules have been observed running lengthwise through the interior of the cell (Schmidt-Lorenz and Kühlwein, 1968; Abadie, 1971; MacRae and McCurdy, 1975; Burchard et al., 1977a; Galván et al., 1987). These tubules have been interpreted by some as the structural basis of gliding motility, but, in light of their rather irregular distribution, this hypothesis does not seem likely. There is, however, a regular structure just below the outer membrane that may be the still-unknown apparatus of gliding motility (Lünsdorf and Reichenbach, 1989; Freese et al., 1997). This structure will be discussed in more detail below.

The intracellular location of certain enzymes can be established by cytochemical methods. There is an ATPase in the cytoplasm and in the periplasmic space (Voelz, 1964; Abadie, 1971), acid phosphatase in the cytoplasm (Abadie, 1971), and alkaline phosphatase in the periplasm and, to a minor extent, in the cytoplasm (Voelz and Ortigoza, 1968; Abadie, 1971). In addition, the distribution of the stress protein SP21 was studied in indole-induced or heat-shocked cells of *Stigmatella aurantiaca* (Lünsdorf et al., 1995).

As shown by various electron microscopic techniques, the surface of myxobacterial cells is rather smooth. Sometimes small vesicles, or blebs, can be seen rising from the surface, perhaps connected with slime extrusion. There are also long filaments, probably composed of excreted slime or lipopolysaccharide (Abadie, 1967, 1971; McCurdy, 1969a; Schmidt-Lorenz and Kühlwein, 1969; MacRae and McCurdy, 1975, 1976; Lampky, 1976; MacRae et al., 1977; Galván et al., 1987, 1992). After fixation with glutaraldehyde, large polygonal holes have been observed in the surface and interpreted as collapsed slime vacuoles (Schmidt-Lorenz and Kühlwein, 1969).

Various appendages are produced by myxobacteria. Slime threads may be seen under the light microscope after treatment with India ink (Fluegel, 1963). Three types of fibers have been discovered in images obtained in the electron microscope. Long, thin fimbriae with a diameter of 7–8.5 nm are found at the cell poles of 12 species of myxobacteria; the only exception was *Nannocystis*. They appear to arise from holes with a collar surrounded by 12 spikes and have hypothetically been connected with gliding motility, because they are absent from nonmotile mutants (MacRae and McCurdy, 1975, 1976; MacRae et al., 1977). Even thinner fibrils have been seen in *Myxococcus* that have to do with slime (MacRae and McCurdy, 1976). Thick, flaccid fibrils of 30–50 nm diameter arise all over the cell and appear to be responsible for cell cohesion and thus for S-, or social, gliding. These fibrils are absent from nonmotile mutants and from cells made noncohesive by treatment with Congo red (Arnold and Shimkets, 1988); the fibrils form an extracellular network, or matrix, that bears a specific antigen, FA-1 (Behmlander and Dworkin, 1991). Their formation seems to be stimulated by cell–cell contact. In liquid cultures, such fibrils are responsible for keeping the cells together in tight spherules and help to ensnare cyanobacteria for lysis (Burnham et al., 1981). For a review of fibrils on myxobacteria, see Dworkin (1999).

Chemical composition Some data on the chemical composition of various myxobacteria are available. The main fatty acids are C_{15} and C_{17} isobranched and monounsaturated C_{16} fatty acids. There appear to be differences between the suborders with respect to hydroxy fatty acids. Bacteria of the *Cystobacterineae* (*Myxococcus, Corallococcus, Archangium, Stigmatella, Cystobacter*) contain substantial amounts of 2-hydroxy fatty acids (mainly iso $C_{17:1\ iso\ 2OH}$) and smaller quantities of 3-hydroxy fatty acids. Bacteria of the other two suborders (*Sorangium, Nannocystis*) appear to be free of such hydroxy fatty acids. The main fatty acid of *Sorangium* was $C_{16:0}$. *Sorangium* species contain appreciable amounts of isobranched fatty acids (Schröder and Reichenbach, 1970; Ware and Dworkin, 1973; Rosenfelder et al., 1974; Fautz et al., 1979, 1981; Yamanaka et al., 1988). While 3-OH fatty acids are well-known constituents of lipopolysaccharide, 2-hydroxy fatty acids may come from phospholipids and sphingolipids. Ceramides and cerebrosides have indeed been demonstrated in *Cystobacter fuscus* (Eckau et al., 1984; Dill et al., 1985). The respiratory quinone is always menaquinone MK-8. In *Myxococcus fulvus*, it occurs in rather large quantity (9.1 µmoles/g membrane protein) (Kleinig, 1972; Yamanaka et al., 1987; M.D. Collins, personal communication). Various phospho- and glycolipids have been found in myxobacteria. The main phospholipid is phosphatidylethanolamine. In membrane fractions of *Myxococcus fulvus*, it comprises 72% of total phospholipid; the rest consists of phosphatidylglycerol, phosphatidylinositol, phosphatidic acid, and unknown components (Kleinig, 1972; Yamanaka et al., 1988). The fatty acid patterns of the individual phospholipids differ in their composition. Inner and outer membranes of *Myxococcus xanthus* can be separated (Orndorff and Dworkin, 1980). In both, phosphoethanolamine constitutes 60–80% of total phospholipid, but the outer membrane contains three times as much phospholipid as the inner membrane and consequently has a lower buoyant density (1.166 vs. 1.221 for the inner membrane). Both features are unusual for Gram-negative bacteria, and, together with branched fatty acids, may give the outer membrane a high flexibility required for gliding motility. Practically all phos-

phatidylserine and cardiolipine is found in the inner membrane. About 40 proteins have been separated from the total membrane fraction of *Myxococcus xanthus*; about 65% originated from the inner membrane. Succinate dehydrogenase and CN-sensitive NADH oxidase activities were found in the inner membrane, while four times as much NADH dehydrogenase activity was found in the outer membrane. A plasmalogen, i.e., a monoalkylether monoacylester, has been identified in *Myxococcus stipitatus* (Stein and Budzikiewicz, 1987). In *Stigmatella aurantiaca* phosphatidylinositol is a major component and comprises 18–25% of total phospholipid. It occurs exclusively as dialkylether (Caillon et al., 1983). In *Cystobacter fuscus* two isomers of a glycolipid were found: 1- or 2-acyl 3-glucosyl glycerol (Scherer et al., 1992). The carotenoid glycosides, which are the main pigments of myxobacteria, are also glycolipids. They are mostly glucosides of the myxobacton and myxobactin type, which in addition bear an ester-bound fatty acid residue on the glucose, in contrast to the overall fatty acid pattern of mostly $C_{14:0}$, $C_{16:0}$, and $C_{18:0}$ (Kleinig et al., 1970; Reichenbach and Kleinig, 1971; Kleinig and Reichenbach, 1973). A nonesterified carotenoid rhamnoside was found in *Sorangium* (Kleinig et al., 1971).

Myxobacterial chromophores are always acyclic or monocyclic C_{40} carotenoids (Reichenbach and Kleinig, 1972). In *Myxococcus fulvus* carotenoids constitute 0.03% of the dry weight of acetone-extracted cells grown in the light, and 0.003% of cells grown in the dark (Reichenbach and Kleinig, 1971). The carotenoids are exclusively located in the membrane fraction, where they constitute up to 0.14% of the dry weight (Kleinig, 1972). They act as photoprotective agents. Some myxobacterial strains produce a deep violet pigment, e.g., *Cystobacter violaceus* (formerly "*Archangium violaceum*"), which probably consists of melanin-like compounds (Reichenbach, 1965c). A red pigment of *Cystobacter violaceus* is probably a dopachrome (Mayer, 1967). Liquid cultures of *Stigmatella aurantiaca* in peptone liquid media become pitch black within 1 to 2 h after reaching stationary phase, presumably because melanin-like pigments are produced by phenol oxidases set free from lysing cells (Reichenbach and Dworkin, 1969). Many myxobacteria produce brown, water-soluble, diffusing, melanin-like pigments on plates of peptone media. Several myxococci, especially *Myxococcus virescens*, excrete a greenish-yellow diffusing pigment that appears to be myxochromid (Trowitzsch-Kienast et al., 1993). Plate cultures of *Myxococcus stipitatus* show a bright yellow fluorescence under the UV lamp at 366 nm (Lampky and Brockman, 1977) that is caused by phenalamides or stipiamides (Kim et al., 1991; Trowitzsch-Kienast et al., 1992). This author has never observed fluorescence at that wavelength with any other myxobacterium (hundreds of strains tested).

Several but not all myxobacteria synthesize squalene (0.1–0.3% of dry cell mass) and, very unusually among procaryotes, true steroids. *Nannocystis* strains contain 1.5–2.2% (by dry weight) cholestenols and cholestadienols; *Polyangium* strains, 0.3% lanosterol; and *Cystobacter* strains, lanosterol-like compounds. Squalene is produced by *Corallococcus*, but neither squalene nor steroids are found in *Myxococcus, Chondromyces*, and *Sorangium*. In contrast to secondary metabolites, steroids are present in all strains of the species. Hopanoids have never been detected (Kohl et al., 1983; Zeggel, 1993).

Typical lipopolysaccharide (3–4% of cell dry weight) with lipid A, core region, and O-antigen side chains is found in the following myxobacteria: *Myxococcus fulvus, Myxococcus virescens, Myxococcus xanthus, Corallococcus exiguus, Cystobacter fuscus, Cystobacter ferrugineus, Cystobacter violaceus, Sorangium cellulosum* (Kleinig, 1972; Sutherland and Smith, 1973; Rosenfelder et al., 1974; Sutherland and Thomson, 1975; Sutherland, 1979b; Orndorff and Dworkin, 1980; Fink and Zissler, 1989). Lipid A contains glucosamine and 3-hydroxy fatty acids; the core region contains 2-keto-3-deoxyoctonate but no heptose. The main sugars are mannose, D-galactose, D-glucose, rhamnose, and galactosamine. Smaller amounts of ribose, D-xylose, and arabinose are found along with the unusual but characteristic sugar 3-O-methyl-D-xylose (Weckesser et al., 1971).

The peptidoglycan of *Myxococcus xanthus* (0.6% of cell dry weight) contains *N*-acetylglucosamine, *N*-acetylmuramic acid, glutamic acid, diaminopimelic acid, and alanine. It is unusual in that it does not form a continuous layer but is organized in patches interrupted by trypsin-sensitive areas (White et al., 1968). This arrangement may make the wall more flexible, which could be essential for gliding motility. In *Myxococcus fulvus* undecaprenol required for peptidoglycan synthesis is present in substantial amounts (Beyer and Kleinig, 1985).

In general, and in spite of being referred to as slime bacteria, myxobacteria do not excrete large quantities of slime. In liquid cultures in sugar-containing media, some myxobacteria produce a slightly viscous solution. The slime is difficult to analyze because it is normally contaminated by products of cell lysis and may consist of several macromolecules that are difficult to separate. All investigators agree, however, that it is composed of polysaccharide. The slime of *Cystobacter violaceus* can be removed from culture plates as a tough layer (Reichenbach, 1965c). If it is boiled for some time in a dilute NaOH solution, some material is removed, but the sheet from the older parts of the swarm colony remains. Under the electron microscope, this material shows a fibrillar structure, and it exhibits birefringence in polarized light. The material remains intact in 17.5% NaOH solution, but it is quickly hydrolyzed in concentrated HCl or 17% H_2SO_4. In addition, it dissolves in Schweizer's reagent (ammoniacal copper oxide used for dissolving cellulose), from which solution it precipitates in the form of loose flakes upon the addition of acid. The main components in the hydrolysate are glucose and an unidentified disaccharide, exactly as in hydrolysates of cotton wool. The slime hydrolysates contain two more components, one of which may be a uronic acid. Thus, the slime of this organism appears to be a cellulose-like polysaccharide. The extracellular slime of liquid cultures of *Myxococcus virescens* variants that grow in either a dispersed or a nondispersed manner can be precipitated with ethanol (Stahl, 1973). The yields are 120–260 mg lyophilized material per liter of medium. The composition of the slime differs between the dispersed and the nondispersed variants. While the crude precipitate mainly contains proteinaceous material, including 25% of the proteolytic activity of the culture, the purified high molecular weight slime is a polysaccharide of 20 million Da. The compositions of the exopolysaccharides of vegetative cultures, myxospores, and fruiting bodies of *Myxococcus* (Sutherland and Thomson, 1975) and of *Cystobacter, Archangium, Stigmatella*, and *Sorangium* species (Sutherland, 1979b) have been analyzed. *Cystobacter* and *Archangium* expolysaccharides contain glucose, rhamnose, and mannose as the main components (20–50% of the total); *Stigmatella aurantiaca*, glucose and galactose; and *Sorangium*, galactose and mannose but no rhamnose. *Myxococcus* exopolyaccharide contains mainly glucose, galactose, and mannose. The yields are 5–30% of the dry weight of the cell mass.

A lectin (MBHA) with hemagglutinin activity is found in *Myxococcus xanthus*. It is a protein located in the periplasmic space and on the cell surface in appreciable quantity; it increases 4-fold during development (Cumsky and Zusman, 1981a, b; Nelson et al., 1981).

A peculiar DNA–RNA hybrid, msDNA, was discovered first in *Myxococcus xanthus* and later in almost all other myxobacteria investigated (*Corallococcus coralloides, Cystobacter violaceus, Stigmatella aurantiaca, Nannocystis exedens*); the exceptions were *Cystobacter fuscus* and *Cystobacter ferrugineus* (Dhundale et al., 1985, 1987). The hybrid consists of a single-stranded DNA linked at its 5′ end by a 2′,5′-phosphodiester bond to a specific rG residue in a single-stranded RNA. The free end of the DNA hybridizes with the 3′ end of the RNA. In *Myxococcus xanthus* 500–700 copies of msDNA are found per chromosome. A second, smaller such hybrid also exists in this organism (Dhundale et al., 1988). The length of both nucleic acids varies somewhat with individual strains between 65 and 163 bases for the DNA and 49–77 bases for the RNA. The DNA is part of the chromosomal DNA. The gene is transcribed into an RNA approximately 375 bases long, from which msDNA is synthesized by a reverse transcriptase (Hsu et al., 1989; Inouye et al., 1989, 1990; Lampson et al., 1991). The msDNA gene can be inactivated by mutation without any adverse effect on growth and development of *Myxococcus xanthus* (Dhundale et al., 1988). Of course, there may be an adverse effect under natural conditions. (For a review of retroelements, see Lampson, 1993).

Motility Myxobacteria move by gliding, or creeping, along interfaces, e.g., along the surface of the culture vessel, the surface of agar plates, even the surface of liquids. They also are able to penetrate the substrate and move within it, e.g., agar plates at normal gel strength. Cells suspended in liquid are nonmotile, but when they attach themselves by one pole to the substrate they perform a gyrating movement (shown in films about myxobacteria: Reichenbach et al., 1965c, b). Gliding motility is also observed in many other, unrelated bacteria, e.g., the *Cytophaga–Flavobacterium* group, *Lysobacter, Herpetosiphon*, and many cyanobacteria, but its mechanism is not yet understood. There are many hypotheses about it. As there is an excellent review on gliding in myxobacteria (Spormann, 1999), this topic is not discussed here. Recently a complex apparatus has been discovered just below the outer membrane of *Myxococcus fulvus* and *Myxococcus xanthus* (Lünsdorf and Reichenbach, 1989; Freese et al., 1997). It consists of strands of tiny rings connected by two longitudinal elements and arranged in a densely packed helical belt running from pole to pole. The system occurs in two conformational states: either the rings stand on edge, or they are tilted, in which case the whole structure shrinks by 40% in width. Unfortunately, the apparatus is difficult to demonstrate *in situ* within the cell, but a helical structural element is clearly present in myxobacteria as well as in other gliding bacteria, because, first, as already mentioned, degenerating cells may develop a helical shape, and second, upon shock freezing a helical structure can be seen running from pole to pole (Lünsdorf and Schairer, 2001). Perhaps the conformational change starts at one cell pole and moves along the belt, thereby creating a contraction wave that propels the cell. The flexibility of the cell wall mentioned above would certainly be helpful. Some mechanism at the cell poles must be responsible for starting the movement and determining its direction. The gliding speed of myxobacteria is usually too slow to be readily seen under the microscope, but under optimal conditions it may reach 13–15 µm/min. Gliding cells always deposit a slime trail (Reichenbach et al., 1965a, b, c; Abadie, 1968), which may have a guiding effect on following cells (Reichenbach, 1965a, b). The role of the slime is probably simply to keep the cells on the surface, as has been shown for *Flexibacter*

(Humphrey et al., 1979). The movements may be directed by tactic behavior. Phototaxis has been demonstrated in myxobacteria, the mechanism of which is somewhat difficult to comprehend (Aschner and Chorin-Kirsch, 1970). Chemotaxis is clearly at work when sporangioles of *Chondromyces apiculatus* germinate, and the released small swarm colonies move like pseudoplasmodia and are attracted by each other (Reichenbach, 1965c). *Myxococcus xanthus* cells show directed movements toward latex or glass beads (Dworkin, 1983), but do not respond to chemical gradients (Dworkin and Eide, 1983). The movement toward the beads could in fact be due to elasticotaxis, the movement along stress lines in the agar (Stanier, 1942). If chemical gradients are steep enough, *Myxococcus xanthus* exhibits both positive (e.g., toward nutrient-rich areas) and negative chemotaxis (e.g., away from nutrient-poor regions or away from repellents like DMSO or short chain alcohols) (Shi et al., 1993). This response is under the control of *frz* genes, which are homologues of chemotaxis genes of *Escherichia* and *Salmonella*. The response is correlated with methylation (positive) or demethylation (negative response) of the FrzCD protein. Chemotaxis plays an essential role in fruiting, as can be shown with *frz* mutants. The compounds responsible for myxobacterial autochemotaxis are not known, but in *Stigmatella aurantiaca*, stigmolone, a pheromone of unknown function, is produced during fruiting body formation and has been characterized (Morikawa et al., 1998; Plaga et al., 1998).

Genetic studies have identified a large number of genes involved in gliding and reversal of movement (*frz* genes). There are two movement systems, A- (for adventurous) and S- (for social) gliding; the former controls movement of single cells, and the latter controls movement of groups of cells (Hodgkin and Kaiser, 1979). Pili (Kaiser, 1979) and surface agglutinins (Shimkets, 1986) appear to be required for S-gliding. The situation is rather complex and is not discussed here, as several reviews are available (e.g., Shimkets, 1990b; McBride et al., 1993).

Swarm colonies Because of gliding motility, the colonies of myxobacteria are film-like, spreading swarms. Swarms may cover an 8.5-cm plate within a week, which translates into a spreading speed of approximately 4 μm/min. The spreading is linear, indicating that expansion is essentially due to movement (Reichenbach, 1965c; Grimm and Kühlwein, 1973a). Growth and movement can be artificially separated by blocking the former with 200 μg/ml nalidixic acid (Shi et al., 1993). Swarming and gliding motility both depend on the composition of the medium. Rich media, especially media having high peptone concentrations, prevent spreading—if they allow growth at all—in which case round, compact, convex, slimy colonies are produced resembling those of nonmotile bacteria. A narrow fringe of swarming cells may still be present. The distribution of cells within the swarm is rarely uniform, and the swarms usually are structured by a more or less prominent pattern of veins. In addition, a concentration of cells along the edge in the form of a ridge or a band may develop. Flares emerge from the edge and sometimes become very long and filament-like. This behavior is typical of gliding bacteria in general.

Swarm morphology can be characteristic for certain taxonomic groups. Thus, swarms of the *Myxococcaceae*, including *Pyxicoccus*, usually are soft and slimy with flat knobs and snaking, undulating veins, or without any structure. Those of the *Cystobacteraceae* always have well-developed, often branching, radial veins, usually in a very tough slime sheet. Swarms of *Polyangium* and *Chondromyces* normally produce shallow depressions in the agar surface, tend to deeply penetrate the plate, and often etch

the surface of the plate in radial or irregular patterns. Often there is a dense, orange band of cells along the edge. Swarms of the genus *Sorangium* often develop prominent branched veins and may penetrate the culture plate. *Nannocystis* swarms corrode the agar plate and, on certain media, transform it into a spongy mass.

Swarm morphology may change dramatically when strains are transferred from plate to plate for some time or when variant strains adapted to dispersed growth in liquid medium are plated (Reichenbach, 1965c; Reichenbach and Dworkin, 1969; Grimm and Kühlwein, 1973a): The speed of spreading may be reduced substantially, the colonies may become soft and slimy, and structural features may be lost, so that the colony is no longer typical for the species. The cause of this phenomenon is not fully understood. In *Cystobacter violaceus*, for example, swarming "S" cells produce normal swarms, excrete less slime, and are up to four times more sensitive to NaCl than "K" cells that grow as compact colonies, are clearly impaired in motility, and grow more easily in suspension in liquid media. K and S cells convert into each other (Grimm and Kühlwein, 1973b, c). *Byssophaga cruenta*, which grows as pseudoplasmodia-like cell associations, forms homogeneous swarms after some months of transfers from plate to plate.

Because they are so thin, swarms usually are colorless in spite of striking pigmentation of the accumulated cell mass. Swarms of *Myxococcus virescens* and *Myxococcus xanthus* often become greenish yellow to orange; whereas those of *Myxococcus fulvus*, *Myxococcus stipitatus*, *Cystobacter*, and *Stigmatella* species may develop a deep brown color, particularly on peptone media. Swarms of *Cystobacter violaceus* and of some *Sorangium* strains are intensely violet. Light may stimulate carotenoid production, so that cultures kept in the light may become pink or red. Color variation has been observed in *Myxococcus xanthus* strain FB2 (Burchard et al., 1977b). When the strain is plated, tan and yellow colonies arise. While most yellow colonies yield only yellow colonies on repeated plating, tan colonies always produce both tan and yellow colonies. The conversion can be stimulated by UV irradiation, nalidixic acid, or cultivation at 36.5°C. The phase variation of pigmentation is paralleled by a variation in swarm morphology. Yellow colonies are rough and swarming; tan ones are mucoid with smooth edges. A genetic study showed that there is a specific promoter for yellow pigmentation, which apparently is controlled by a transcriptional regulator that in turn is regulated by a switch, the biochemistry of which is not known. A block in the switch leads to stable tan mutants with an extremely low conversion rate to yellow (Laue and Gill, 1994).

A peculiar feature of myxobacterial swarms is the occasional appearance of oscillating waves (Reichenbach, 1965b): Tiny, concentric ridges are arranged in a densely packed regular pattern, usually in the periphery of the swarm. In most myxobacteria, the waves occur in extended fields resembling ripples at the seashore. *Stigmatella* strains tend to produce isolated trains of waves, which is very typical for those organisms. When time-lapse movies of myxobacteria are made, the waves are seen to oscillate synchronously, and the whole pattern seems to vibrate. The vibration even may speed up and down, so that a second rhythm overlays the primary oscillation of the waves (Reichenbach et al., 1965b, c). The movements of the waves can be seen under the microscope by a patient observer. At high magnification, it becomes apparent that the waves consist of rows of cells arranged strictly parallel and in a radial direction. The movement of individual cells cannot be followed, because the cells are so densely packed. However, it appears that they really oscillate back and forth. This interpretation is suggested by a branching of the ridges at intervals and by the fact that the speed of the waves is faster than the spreading of the swarm. Neither the mechanism nor the biological importance of the phenomenon is understood. It may be an artifact in the unnatural environment of cultures. Perhaps pili play a role in this phenomenon. There is no experimental proof that the phenomenon is connected with fruiting body formation; indeed fruiting often takes place in total absence of oscillating waves. In *Myxococcus xanthus*, oscillating waves can be induced by peptidoglycan and certain components of it (Shimkets and Kaiser, 1982), but this is not a general rule. Lack of a reliable induction system and of optical resolution of the movements of individual cells still impedes the investigation of the phenomenon. Mathematical models have been developed to describe the behavior of moving myxobacterial cells in the swarm, e.g., during aggregation or rippling (e.g., Pfistner, 1990; Stevens, 1990; Deutsch, 1999). Of course, such analyses can only be schematic and cannot explain the biochemical mechanisms behind the phenomena.

When freshly isolated myxobacteria are inoculated into a liquid medium and the cultures shaken, the bacteria usually grow first along the glass wall of the flask and then as nodules and flakes in the medium. Until the 1960s, it was generally believed that myxobacteria cannot be made to grow dispersed in liquid medium. This belief was disproved first for a strain of *Myxococcus xanthus* (Mason and Powelson, 1958a). Later it was found that many strains of other myxobacteria of the suborder *Cystobacterineae* can also grow dispersed in liquid medium when several transfers are made from liquid medium to liquid medium, although occasionally a strain may refuse to do so even after many transfers. Strains that grow dispersed in liquid medium continue to do so after being frozen at −80°C and thawed. Suitable media are based on tryptic digests of casein, such as Difco Casitone. This transformation is more difficult to achieve in strains of the other two suborders. Several *Sorangium* strains have been made to grow dispersed in liquid medium, however, only after months of serial transfers (Irschik et al., 1987; Gerth et al., 1994). Fully dispersedly growing cultures of strains of *Chondromyces*, *Polyangium*, and *Nannocystis* have never been obtained to the author's knowledge. Various strategies have been proposed to prevent clumping and to stimulate dispersed growth, e.g., adding 0.04% agar to the medium (Schürmann, 1967), but no one trick leads to success in all cases. Homogenizing the inoculum mechanically only leads to more and smaller flakes and nodules. Grimm and Kühlwein (1973b) observed that variants of *Cystobacter violaceus* that produce more slime grow more readily in suspension, but those cultures were not agitated. It is not understood which factors determine clumping. A good guess is that the composition of the slime plays a role; the composition of the slime indeed differs between strains that grow dispersed in liquid medium and those that do not (Stahl, 1973). In addition, the presence of agglutinins on the cell surface (Arnold and Shimkets, 1988) may be crucial.

Plating for single-cell colonies as required for cloning and mutant selection depends, of course, on the availability of homogeneous cell suspensions. It is also a problem to get individual cells to grow into colonies (e.g., Grimm, 1967). Media that allow excellent growth when inoculated with copious amounts of cells often do not give high yields of single cell colonies, if the organisms grow at all. An efficient plating medium has to be developed for every strain by optimizing media components and their concentrations. The problem has been solved in many cases and a 100% plating efficiency obtained, although usually after months of experimentation, because colonies can be seen only

after 8–14 days. Sometimes it helps to add spent medium from old liquid cultures to the plating medium (Jaoua et al., 1992). The reasons for this failure of single cells to grow are not understood; perhaps some quorum-sensing mechanism is involved.

Nutritional requirements There are two nutritional types among myxobacteria. Whereas most species are proteolytic/bacteriolytic, the genera *Sorangium* and *Byssophaga* are cellulose decomposers. Both types can be cultivated on agar—and often also in liquid medium—containing autoclaved baker's yeast (e.g., VY/2-agar). The proteolytic species grow well on media containing a suitable peptone, e.g., tryptic digests of casein. Concentrations should not be too high, normally 0.3% to 1%. The inhibitory effect of higher concentrations is probably caused by an excess of phosphate, to which myxobacteria are rather sensitive. In addition, production of ammonia may become a limiting factor, especially with the genera *Myxococcus*, *Corallococcus*, and *Nannocystis* that do not utilize glucose or other mono- and disaccharides (Watson and Dworkin, 1968; Reichenbach, 1970). The situation with the latter genera may be more complex. *Myxococcus* and *Corallococcus* species always break down autoclaved yeast cells, and *Nannocystis* does the same with agar. It seems unlikely that the organisms do not make use of the degradation products. It has indeed been shown that a strain of *Corallococcus coralloides* decomposes starch into trisaccharides, and that the latter can be taken up and utilized, while mono- and disaccharides remain in the medium (Irschik and Reichenbach, 1985). A *Myxococcus fulvus* strain has been found to excrete a laminarinase that cleaves β-1,3-glycosidic bonds; this enzyme also produces trisaccharides (Borchers, 1982). However, the direct utilization of glucose and other sugars has been demonstrated for many other proteolytic myxobacteria, so that the addition of, for example, glucose or starch to the medium has a growth-stimulating effect. Yet those organisms usually do not grow faster with a carbohydrate in the medium, but only for a longer time and to a higher cell density. It appears that they utilize the sugars more as a source of building blocks than for energy. The cellulose degraders grow on very simple media, e.g., with nitrate (which they prefer to ammonium ion) as the nitrogen source and glucose as the carbon source. Many strains actually require a carbohydrate in the medium and do not grow on a pure peptone medium. In addition, those organisms grow better if an organic nitrogen source is available. Development of a defined amino acid medium for proteolytic myxobacteria is a tedious and time-consuming procedure, because not only the amino acid composition but also the concentrations of the individual amino acids must be carefully adjusted. Essential amino acids usually include the branched and the aromatic amino acids, glycine, and methionine. Defined media have been published for *Myxococcus xanthus* (Dworkin, 1962; Witkin and Rosenberg, 1970; Bretscher and Kaiser, 1978), *Cystobacter violaceus* (Mayer, 1967), *Cystobacter fuscus*, and *Cystobacter ferrugineus* (Reichenbach and Dworkin, 1992). In the latter case, the organisms become dependent on biotin and/or thiamine when the number of amino acids is reduced to three or four, or when amino acids are replaced by their respective keto acids. Every strain requires its own defined medium. Occasionally a myxobacterial strain depends on vitamin B_{12}. Many myxobacteria can be cultivated on technical substrates, such as soy meal or flour, skim milk, single-cell protein, or corn steep powder, a characteristic that may be useful for mass fermentations (Gerth et al., 1984). The mineral requirements are the usual ones (e.g., Coucke and Voets, 1967). Relatively high concentrations of Mg^{2+}

(0.15% MgSO$_4$·7H$_2$O) are often beneficial; some strains require Ca^{2+}. Sensitivity of many myxobacteria to high phosphate concentrations above 0.5 mM has already been mentioned. This concentration is reached in a 0.2% Casitone solution, so that usually no phosphate need be added to media with a peptone or a technical substrate. The phosphate problem may be mitigated by adding polyphosphate rather than salts of phosphoric acid. Many details on media recipes can be found in Reichenbach and Dworkin (1992).

Physiology The biochemistry and physiology of myxobacteria cannot be discussed here in detail; the review articles mentioned above provide that information. Myxobacteria are strictly aerobic organisms. Their temperature range for optimal growth is normally between 26° and 34°C, but some strains grow up to 40°C, and most probably also grow at much lower temperatures (e.g., Janssen et al., 1977). This author observed one strain of *Myxococcus fulvus* with a temperature optimum around 30°C still slowly growing at 6°C in the refrigerator. Psychrophilic myxobacteria were isolated from Antarctica that did not grow at 18°C (Dawid et al., 1988). The pH range for growth is usually 6.8–7.8. Cellulose degraders may grow at pH 5.8–6.4. There is rarely growth beyond pH 8.0. However, alkalophilic myxobacteria that grow at pH 9.5, but not at 7.2, have been found in an alkaline lake in East Africa (Dawid, 2000). Acidophilic myxobacteria have not been found so far. Salt tolerance of myxobacteria is usually low. The number of myxobacteria from terrestrial samples decreases dramatically when 0.8% NaCl is added to isolation media; usually only *Myxococcus virescens* (Rückert, 1978) and sometimes *Nannocystis* strains are obtained. At 1% NaCl, only a rare *Myxococcus virescens* strain may show up. While most myxobacteria grow well on yeast agar (VY/2) with 0.3% NaCl, only a few grow at 2% NaCl; these are mainly *Myxococcus* strains. Some *Myxococcus virescens* strains prefer a low salt concentration, while others grow well at 2% (Iizuka et al., 1998). Truly halotolerant and halophilic myxobacteria growing best at NaCl concentrations between 2% and 3% and not at all below 1% have been obtained from marine samples off Japan (Iizuka et al., 1998). In addition, 16S rRNA sequences typical of myxobacteria were discovered in samples from a deep-sea vent near Hawaii (Moyer et al., 1995). These two examples are both from the Pacific; efforts of several investigators to isolate marine myxobacteria from samples from the Atlantic were not successful. Myxospores seem to survive in a marine environment for some time, so that nonhalotolerant strains can be isolated from marine sediments (Rückert, 1984).

Myxobacteria efficiently decompose a number of biomacromolecules (Dworkin, 1966). All species degrade casein (e.g., in skim milk), gelatin, other proteins, starch, and dextrin, and all, with the exception of *Sorangium*, degrade RNA and DNA (e.g., Jahn, 1924; McCurdy, 1969b). Several myxobacteria hydrolyze chitin (always reprecipitated material and not natural chitin); this trait is typical of certain taxa and thus is of taxonomic relevance. Thus, all *Stigmatella* and several *Cystobacter* species, *Myxococcus stipitatus*, *Kofleria flava*, *Jahnia thaxteri*, and most *Sorangium* strains produce a complete clearing in a chitin layer. Most or all species appear to degrade xylan. Degradation of crystalline cellulose is probably restricted to the genera *Sorangium* and *Byssophaga*, and seems always to be found with the ability to utilize inorganic nitrogen compounds. In the older literature, cellulose decomposition is occasionally mentioned for *Myxococcus*, *Stigmatella* (*Podangium*), and *Angiococcus* species (Mishustin, 1938; Pronina, 1962). However, this author has not encountered cellulose-degrading *Myxococcus*, *Stigmatella*, or *Angiococcus* strains in

30 years of isolating cellulolytic myxobacteria and is inclined to believe that those "cellulose degraders" were myxobacterial contaminants of other cellulolytic bacteria. In fact, one always finds many noncellulolytic myxobacteria in enrichment cultures for cellulose-degrading organisms. Many myxobacteria decompose the β-glucans of yeast cell walls, and at least some excrete a laminarinase (Borchers, 1982). Agar degradation is rare among myxobacteria. Some *Sorangium* strains soften or even liquefy agar, and all *Nannocystis* strains digest holes and channels in an agar layer without liquefying it (Reichenbach, 1970). It is an open question whether *Nannocystis* can utilize agar as a substrate because in liquid media there seems to be no growth stimulation by agar. Perhaps the enzymes are directed against some other polysaccharide, e.g., some microbial products in the soil, where agar is absent.

The degrading enzymes are only partly known. They are mostly diffusing exoenzymes that thus can act far away from the swarm. The cellulases may be an exception, for it seems that cellulose fibers are only attacked when cells are in direct contact with them. While proteolytic activity has often been reported, relatively few studies deal with the enzymes responsible for this activity. *Myxococcus virescens* produces at least three extracellular proteases: a serine protease that has a molecular weight of about 26 kDa and shows activity on casein, hemoglobin, and, under alkaline conditions, on elastin. It cleaves preferentially near nonpolar amino acids (Gnosspelius, 1978). Three serine proteases are found in the culture supernatant of *Myxococcus xanthus*. Different strains show different protease patterns (Coletta and Miller, 1986). In addition, a 39 kDa extracellular metalloprotease with elastase activity can be isolated from *Myxococcus xanthus* (Dumont et al., 1994). A study of intracellular proteases in *Myxococcus xanthus* revealed mainly metalloproteases and traces of a serine protease. The intracellular enzymes of the strain are different from the extracellular ones (Orlowski and White, 1974). Initiation of the excretion of proteases when cells are growing on a protein substrate like casein requires a certain cell density (Rosenberg et al., 1977). An exocellulase with a pH optimum around 5.6 is released from "*Sorangium compositum*" by heating the culture at 100°C. This cellulase produces cellobiose, which is then cleaved into glucose by a heat-labile cellobiase. The cellulase also attacks carboxymethylcellulose with production of reducing sugars (Coucke and Voets, 1968). Addition of starch to the medium induces synthesis and excretion of amylase by *Corallococcus coralloides* (Farez-Vidal et al., 1990). A laminarinase (molecular weight 12 kDa, pH optimum 5.0–5.5) and an enzyme releasing trisaccharides from starch have already been mentioned. Nothing is known about myxobacterial xylanases and agarases. Restriction endonucleases have been demonstrated in various *Myxococcus* species, in *Cystobacter velatus*, and in *Archangium* (Morris and Parish, 1976; Mayer and Reichenbach, 1978). These enzymes are apparently located in the periplasm. They can have different specificities even if they come from the same species. Not all myxobacteria seem to contain endonucleases. Non-specific DNases (and RNases) have been observed in many myxobacteria (McCurdy, 1969b; Mayer and Reichenbach, 1978). A single cell-bound DNase can be isolated from several myxobacteria; it has an apparent molecular weight of about 200 kDa (Muñoz et al., 1989). A more careful study of a *Corallococcus coralloides* strain revealed three cellular DNases (Martinez-Cañamero et al., 1991). The purified enzymes have different temperature optima (between 25° and 37°C), require Mn^{2+} and Mg^{2+} for good activity, and have molecular weights of 39, 44, and 49 kDa. While DNases seem to be cell-bound in liquid cultures, they are excreted and diffuse through the agar in plate cultures. Several other degradative enzymes have been reported from myxobacteria. A coagulase (myxocoagulase) has been isolated in pure form from a strain of *Myxococcus fulvus* (Bojary and Dhala, 1989). It has a molecular weight of 57 kDa, is highly active on rabbit plasma, and is clearly different immunologically from *Staphylococcus aureus* coagulase. Esculin-cleaving glucosidases appear to be present in most myxobacteria, although not in *Myxococcus* (McCurdy, 1969b).

The potential of myxobacteria to lyse bacteria and yeasts was observed early, and the older literature is full of such reports (e.g., Beebe, 1941c; Singh, 1947; Nolte, 1957; Margalith, 1962; Bender, 1963; Raverdy, 1973; Yamanaka et al., 1993). The cocktail of hydrolases mentioned above, perhaps with the addition of antibiotics, is used for this activity. Among the bacteriolytic enzymes those that attack peptidoglycan play a special role. Six enzymes have been characterized in the culture supernatant of *Myxococcus xanthus* that have amidase, glucosaminidase, and peptidase activity and thus would be able to cleave the polysaccharide backbone as well as the cross-links (Sudo and Dworkin, 1972). An extracellular cell-wall-degrading enzyme has been isolated from *Cystobacter violaceus* in addition to three proteases (Hüttermann, 1969). Several *Myxococcus xanthus* and *Myxococcus fulvus* strains have been shown to control populations of the cyanobacterium, "*Phormidium luridus*", with predator-prey cycles of 9 days under experimental conditions. As few as 50 *Myxococcus* cells per 100 ml containing 10^7 cells of the prey were sufficient to start a lytic cycle (Burnham et al., 1984). The myxobacteria produced an enzyme resembling lysozyme (Daft et al., 1985). However, it appears that myxobacteria cannot control natural blooms of cyanobacteria, probably because of an insufficient supply of inorganic nutrients (Fraleigh and Burnham, 1988). It has been observed that myxobacteria digest holes in the cell wall of two fungi, *Cochleobolus* and *Rhizoctonia* (Homma, 1984), and of the green alga, *Cladophora* (Geitler, 1925), respectively. The myxobacteria enter the cells and produce fruiting bodies inside.

Most myxobacteria of the suborder *Cystobacterineae*, but not those of the other two suborders, autolyze soon after reaching the growth maximum, so that there is hardly a stationary phase. Five different autocides have been isolated from *Myxococcus xanthus*. AM I is a mixture of fatty acids (Varon et al., 1986); AM V consists of phosphoethanolamine containing unsaturated fatty acids (Gelvan et al., 1987). In the latter case, it appears that the cell population releases the fatty acids through the action of a phospholipase. In AM I unsaturated fatty acids are the most efficient autocides. While AM I and AM V act specifically on *Myxococcus* species, the other autocides act on a variety of myxobacteria. Mutants that are resistant to autocides can no longer produce fruiting bodies. In fact, during fruiting body formation in *Myxococcus xanthus* autolysis of part of the cell population takes place before sporulation begins (Wireman and Dworkin, 1977). During sporulation, cells reorganize the cell wall and become particularly prone to autolysis (Kottel and White, 1974). A unique feature of certain *Myxococcus xanthus* strains is their high sensitivity to monovalent cations; Na^+ (0.03 M) or NH_4^+ (0.03 M) induce rapid lysis at 45°C. Bulges arise at the sides or ends at the cell, gradually blow up, and transform the cell into an empty sphere (Mason and Powelson, 1958b). This course of cell degeneration is also seen in other species of the suborder *Cystobacterineae*.

The primary metabolism of myxobacteria resembles that of other aerobic Gram-negative bacteria. The catalase reaction is always positive. Oxidase is variable and often delayed or negative; oxidase production also depends on culture age and growth medium. The respiratory chain of myxobacteria appears to be the usual one. Cytochromes of the *a*, *b*, and *c* type, but not of the *o* type have been demonstrated in membrane preparations of *Myxococcus xanthus*, *Myxococcus fulvus*, *Sorangium cellulosum*, and *Stigmatella* (Dworkin and Niederpruem, 1964; McCurdy and Khouw, 1969; Kleinig, 1972; Sarao et al., 1985). NADH:cytochrome *c* reductase, cytochrome *c* oxidase, NADH oxidase and reductase activity, and flavoprotein are present. Cytochrome *c* reductase activity is 10 times higher in *Myxococcus xanthus* myxospores than in vegetative cells. The respiratory quinones are menaquinones, almost exclusively MK-8. A 27-kDa apoprotein of a FMN enzyme has been identified as the product of gene *fprA*, formerly *spoC* (Shimkets, 1990a). Acid and alkaline phosphatases have already been mentioned. Their activity in *Myxococcus xanthus* increases during sporulation (González et al., 1991). The enzymes are also present in *Corallococcus coralloides* (González et al., 1987; 1989). Two extracellular and two intracellular activities are produced constitutively. Carbohydrate metabolism has been studied in several myxobacteria. In *Myxococcus xanthus* a deficient Emden–Meyerhof pathway is present that lacks hexokinase and pyruvate kinase; this fact explains why the organism does not utilize glucose (Watson and Dworkin, 1968). The pathway may mainly be used for gluconeogenesis. A complete tricarboxylic acid cycle is present and is probably supplied with acetate from amino acids via pyruvate. The initial enzymes of the pentose phosphate pathway and of polysaccharide synthesis are also present. There is a glyoxylate shunt (Orlowski et al., 1972). During sporulation the key enzyme, isocitrate lyase, is downregulated. However, this pattern of sugar metabolism is not typical for myxobacteria in general. *Stigmatella aurantiaca* utilizes glucose with acid production (Reichenbach and Dworkin, 1969; Gerth and Reichenbach, 1978) as do *Cystobacter* strains (Reichenbach and Dworkin, 1992); *Sorangium cellulosum* (formerly *Polyangium cellulosum*) even grows on glucose and nitrate. Hexokinase is present in *Stigmatella erecta* (or "*Stigmatella brunnea*"), but, while growth is stimulated by sugars and starch, these substrates are not utilized (McCurdy and Khouw, 1969). *Corallococcus coralloides* contains both polyphosphate and ATP glucokinases with molecular weights of 61 kDa and 47 kDa, respectively (González et al., 1990). As mentioned before, *Corallococcus coralloides* does not utilize glucose or maltose, but degrades starch into trisaccharides, which it can take up and metabolize. This strain, too, has a hexokinase and other enzymes of glucose catabolism but seems to use glucose mainly to produce ribose and deoxyribose via the pentose phosphate pathway . The fact that starch does not speed up but only prolongs growth supports this interpretation (Irschik and Reichenbach, 1985). The enzymatic complement differs somewhat between individual *Sorangium cellulosum* strains, a situation that leads to different, strain-specific nutrient requirements. One strain carries out glycolysis, with a normal hexokinase, the TCA cycle (including the glyoxylate shunt), and the pentose phosphate pathway (Sarao et al., 1985). Strain So ce12 has the same pathways except for the glyoxylate shunt (Hofmann, 1989). Glucose is introduced into metabolism by glycolysis and the pentose phosphate pathway in a ratio of 95% to 5%. This strain can produce acetyl-CoA either from PEP with a pyruvate dehydrogenase or from external acetate with an acetyl-CoA synthetase; it can produce oxaloacetate from PEP with a PEP carboxylase. It is not capable of gluconeogenesis, or only carries out this process at a very low level, a fact that explains its dependence on a hexose, pentose, starch, or cellulose as a carbon source; glycerol or intermediates of the TCA cycle would not suffice. While enzyme activities in So ce12 are highest in early log phase, lactate dehydrogenase, glucose-6-phosphate dehydrogenase, and α-ketoglutarate dehydrogenase have their maxima in stationary phase. Growth of *Byssophaga cruenta*, the other myxobacterial cellulose degrader, is repressed by glucose at concentrations of more than 0.1%.

For most myxobacteria the only acceptable nitrogen sources are amino acids, which also supply carbon and energy. This fact clearly reflects their lifestyle as micropredators (Singh, 1947). The cellulose degraders, which grow on nitrate or ammonium, are an exception. They too can utilize amino acids, but usually cannot grow on them alone. Myxobacteria prefer peptones and proteins to free amino acids, which are acceptable only if their concentrations are carefully balanced in a narrow range. The enzymes involved in amino acid metabolism are as would be expected. A number of these enzymes have been characterized, e.g., aspartokinase in *Myxococcus xanthus* is regulated by feedback inhibition and feedback stimulation (Filer et al., 1973). Details may be found in the reviews mentioned above. A DOPA decarboxylase of unknown function was recently discovered in *Sorangium cellulosum* by gene sequence analysis (Müller et al., 2000); this enzyme was previously unknown in bacteria. Total dependence on amino acids, as is seen in *Myxococcus* species, has the consequence that substantial amounts of ammonia are produced during growth, which can become a serious problem in large-scale fermentations because ammonium ions have strong regulatory effects. Intracellular ammonia pools can be quite high: e.g., 80–140 mM in a *Myxococcus virescens* strain (Gerth and Reichenbach, 1986).

Other enzymatic activities of myxobacteria that have been studied include a DNA-dependent RNA polymerase, which in *Myxococcus xanthus* (Rudd and Zusman, 1982) and *Stigmatella aurantiaca* (Heidelbach et al., 1992) has the typical eubacterial α,β,β′ core enzyme (molecular weights 38, 140, and 145 kDa for *Myxococcus xanthus*; 40, 146, and 146 kDa for *Stigmatella aurantiaca*). The holoenzyme of *Myxococcus xanthus* may combine with two σ-factors (around 86 kDa); that of *Stigmatella aurantiaca* has one major σ-factor of 195 kDa. The major vegetative σ-factor of *Myxococcus xanthus*, σ-80, has high homology to σ-70 of *E. coli* and σ-43 of *Bacillus subtilis* (Inouye, 1990). Patterns of posttranslational phosphorylation and methylation of proteins and of DNA methylation change both with the growth stage and during conversion into myxospores (Komano et al., 1982; Yee and Inouye, 1982). Several kinds of kinases are also known from myxobacteria. For example, a serine/threonine protein kinase, the first example of this type of kinase in a procaryote, has been found in *Myxococcus xanthus* (Muñoz-Dorado et al., 1991). The enzyme is autophosphorylated and changes during development. A histidine protein kinase seems to be inserted in the membrane of *Myxococcus xanthus* and probably transduces starvation and cell density signals required for induction of development (Yang and Kaplan, 1997). A nucleoside diphosphate kinase from *Myxococcus xanthus* has been studied with its nucleotide in place by high-resolution X-ray crystallography (Williams et al., 1993), the first time that such an enzyme could be analyzed in this way.

Isoprenoids appear to be produced by the mevalonate pathway (Horbach et al., 1993). C_{15} to C_{60} polyprenols can be synthesized *in vitro* using cell free extracts of *Myxococcus fulvus* (Beyer and Kleinig, 1985). Polyketide synthases and peptide synthetases play an essential role in secondary metabolism; often both are combined in one enzyme complex (e.g., Paitan et al., 1999; Silakowski et al., 2001). Regulatory cyclic nucleotides are also known from myxobacteria. Adenylate and guanylate cyclases (Devi and McCurdy, 1984b), cyclic GMP- and cyclic AMP-binding

proteins (Devi and McCurdy, 1984a), and both cyclic AMP and cyclic GMP (Campos and Zusman, 1975; Passador and McCurdy, 1985) are found in *Myxococcus xanthus*. Concentrations of these components change during myxospore and fruiting body formation. A GTP-binding membrane G-protein—unusual for a procaryotic organism—exists in *Stigmatella aurantiaca* and may be involved in signal transduction (Dérijard et al., 1989). *Corallococcus coralloides* produces ppGpp beginning in stationary phase, but the observed kinetics do not support a role in a stringent response (Fernandez-Vivas et al., 1983). *Myxococcus xanthus* has an S-adenosylmethionine synthetase, which is gradually inhibited after glycerol induction of sporulation, so that the S-adenosylmethionine pool is reduced from 0.5 µmol/g dry weight—several times higher than the methionine pool—to 0.28 µmol/g (Jones and Wells, 1980). Myxobacterial ribosomes have the expected 30S/50S subunits, but conditions for their dissociation are not the same as in *E. coli* (Foster and Parish, 1973). Initiation factor IF2 of *Stigmatella aurantiaca* shows considerable homology to other bacterial IF2s, especially in the GTP-binding region, but is longer than the *E. coli* IF2 by 160 amino acids at the N-terminal end. The extension is rich in Ala, Pro, Glu, and Val residues and contains nine repetitive sequences. The function of the extension is unknown (Bremaud et al., 1997).

Relatively little is known about transport mechanisms in myxobacteria. In *Sorangium cellulosum* strain So ce12, glucose and fructose appear to be transported by a common uptake system that has a higher affinity for glucose and does not involve phosphorylation (Hofmann, 1989). Myxobacteria have been investigated as potential nonpathogenic expression hosts. Acid phosphatase and β-lactamase, which are typically located in the periplasm in Gram negative bacteria, have been cloned in *Myxococcus xanthus*. The enzymes are produced and excreted into the medium by *Myxococcus xanthus* (Breton and Guespin-Michel, 1987). Excretion is strictly regulated. This regulation can be ablated by Tn5 insertion mutagenesis. Mutants exhibit increased excretion (Nicaud et al., 1984). While native proteins are excreted directly into the medium, the foreign proteins appear to be actively transported first into the periplasmic space, from which they diffuse into the environment; the latter is the rate-limiting step (Letouvet-Pawlak et al., 1993). Iron transport may be mediated by iron chelators. Myxochelins were discovered in *Angiococcus disciformis*, *Myxococcus fulvus*, and *Stigmatella aurantiaca* (Kunze et al., 1989) and nannochelins in *Nannocystis exedens* (Kunze et al., 1992). The chelators are produced at high levels only when the iron concentration in the medium is low (10^{-7} M). The genes responsible for the biosynthesis of myxochelins in *Stigmatella aurantiaca* have been characterized (Silakowski et al., 2000b). Unexpectedly, a nonribosomal peptide synthetase appears to be involved.

Myxobacteria have an elaborate secondary metabolism. To date, 90 different basic structures and 350 structural variants of secondary metabolites have been characterized. Most are new compounds. Many are electron transport inhibitors (e.g., myxothiazol, stigmatellin). Epothilone efficiently stabilizes microtubuli, like paclitaxel (Taxol®). The ability to produce a certain compound is usually a strain and not a species characteristic. Many of these compounds are found in different species, genera, and even families, but rarely in different suborders. The genetic basis of myxobacterial secondary metabolism has been intensively studied (e.g., Silakowski et al., 2000a; Gaitatzis et al., 2001). Details may be found in the following reviews: Dworkin and Kaiser (1993); Reichenbach and Höfle (1993), Höfle and Reichenbach (1995), and Reichenbach and Höfle (1999).

Many myxobacterial cultures have an earthy odor that is due to geosmin, among other compounds Trowitzsch et al., 1981). *Chondromyces crocatus* produces a very characteristic odor that is unique among myxobacteria and that reminds one faintly of pyridine; it is produced by a mixture of pyrazine derivatives (Fuhlendorf and Schulz, personal communication).

It may be mentioned in passing that myxobacteria possess several features that are more characteristic of eucaryotes than of procaryotes, e.g., sterol, ceramide and cerebroside synthesis, a Ser/Thr protein kinase, DOPA decarboxylase, a G-protein, and calmodulin-like S-protein. This finding suggests that analysis of base sequences of the myxobacterial genome may contribute to a better understanding of sequences in genes of higher animals.

Antibiotic resistance varies with different myxobacteria (McCurdy, 1969b), but as a rule resistance varies among the strains of a species and not among species. *Sorangium cellulosum* is an exception. Almost all strains grow on yeast agar (VY/2-) containing 1000 µg/ml kanamycin sulfate; this characteristic can be used for isolation. Usually kanamycin resistance is found together with resistance to neomycin and gentamicin, but streptomycin resistance is rarely tolerated (50 µg/ml each). In contrast, the cellulose degrader *Byssophaga cruenta* does not grow in the presence of concentrations of kanamycin above 20 µg/ml. Not one of 81 tested *Myxococcus* strains belonging to all known species grows on yeast agar containing 250 µg/ml kanamycin sulfate, but about 50% of 95 tested *Corallococcus* strains do so. The opposite behavior was seen with streptomycin sulfate (50 µg/ml): 297 out of 306 *Corallococcus* strains do not grow, while 119 of 201 *Myxococcus* strains do. There is no antibiotic for which an occasional resistant isolate cannot be found. Practically all myxobacteria are resistant to β-lactam antibiotics, although not always to every one of the tested compounds (ampicillin, carbenicillin, cephalotin, cephalosporin C, 50 µg/ml each, all the following inhibitors are also at 50 µg/ml). An intracellular β-lactamase is found in *Myxococcus xanthus* DZ1 (Breton and Guespin-Michel, 1987). Many strains are insensitive to ribostamycin, kasugamycin, phosphomycin, polymyxin, cycloserine, and bacitracin, while resistance to paromomycin, spectinomycin, tobramycin, dibecacin, erythromycin, oxytetracycline, chloramphenicol, and rifampin is rare or exceptional. Kanamycin (and gentamicin) resistance of *Sorangium cellulosum* is apparently a property of ribosomal proteins, as the compounds are taken up and are not modified chemically, and the relevant portion of the 16S rRNA sequence is identical in *Myxococcus xanthus* and *E. coli*. Curiously, *in vitro* translation with *Sorangium cellulosum* ribosomes is only marginally inhibited by high concentrations of kanamycin, perhaps because it is inefficient. However, radioactive gentamicin clearly binds less strongly to *Sorangium* ribosomes than to those of *E. coli* (Pöhl, 1996). Myxobacteria are much more sensitive to actinomycin D (Dworkin, 1969) and acriflavin (Grimm, 1980) than ordinary Gram-negative bacteria. This sensitivity may mean that their outer membrane is more permeable to these compounds.

Intercellular communication Myxobacteria are decidedly social organisms. The reason may be found in their nutritional specialization. Because they depend on the degradation of various biomacromolecules, they must first produce exoenzymes, which diffuse into the environment and act on their substrates; the lysis products then diffuse in all directions. Consequently, only a small portion of the invested energy, if any, is returned to an individual cell. However, if a population is at work, the cells profit mutually from individual exertions, and the energy loss is minimized. As already mentioned, cell-to-cell communication occurs beginning at the swarm stage, and the mechanisms of communication are both mechanical and chemical in nature. Initially defined by genetic analyses, several signaling systems have been identified with various biochemical events; e.g., A-factor is a mixture of 8 amino acids, and C-factor is a 20-kDa surface-bound protein. Excellent reviews are available on this topic (Shimkets, 1990b; Kaiser and Kroos, 1993; Sager and Kaiser, 1994; Dworkin, 1996; Kaiser, 1998, 2000; Plaga and Schairer, 1999).

Morphogenesis Myxobacteria are the most sophisticated procaryotes with regard to morphogenesis. Under starvation conditions, hundreds of thousands to millions of cells aggregate, and if conditions are right, differentiate into fruiting bodies (cooperative morphogenesis). These fruiting bodies may be very simple structures, like the soft, slimy knobs produced by *Myxococcus* species. But many species surround the cell mass with a tough wall—often after a subdivision of the original mass into portions of approximately equal size—thus creating sporangioles, either singly or in clusters. The sporangiole wall is very thin, about 30 nm in *Chondromyces crocatus* (Abadie, 1971) and 200–300 nm in *Cystobacter* (Vahle, 1910; Jahn, 1924). It dissolves in potassium hypochlorite solution (eau de Javelle-Vahle, 1910) and is intensely colored by pigments that are not identical with those of the cells and that cannot be extracted by any of the normal solvents (Vahle, 1910). The chemical composition and the origin of the sporangiole wall are not known. It has been suggested that it is excreted by the cells lying at the surface (Jahn, 1924). Alternatively, it may be formed by fusion of the cells forming the top layer. In *Melittangium boletus* the cells in the developing sporangiole are arranged in a regular pattern in rows of parallel rods perpendicular to the surface. Thus, in this case, it seems that the wall would be excreted from one cell pole only (Jahn, 1924). How the subdivision of the sporangioles is determined within the undifferentiated cell mass is one of the many unsolved questions in myxobacterial morphogenesis. During germination the sporangiole wall is ruptured mechanically, and an empty husk remains after the cells leave.

The most elaborate fruiting bodies have white slime stalks, which are branched in *Chondromyces crocatus* and consist of fine tubules in *Stigmatella aurantiaca* and *Chondromyces* (McCurdy, 1969a; Voelz and Reichenbach, 1969; Abadie, 1971). The stalk bears at the end either a soft, slimy, globose head (*Myxococcus stipitatus*) or one or several sporangioles. Stalked fruiting bodies were apparently selected for independently in the suborders *Cystobacterineae* (*Myxococcus*, *Melittangium*, *Stigmatella*) and *Sorangineae* (*Chondromyces*). The stalks arise in a different way in the two suborders, however. In the *Cystobacterineae*, the piled-up and rotating cell mass differentiates morphologically into stalk and head, and the cells later simply leave the stalk region; in *Chondromyces* the stalk is excreted by the spherical cell mass, which sits initially on the substrate, is lifted upward as a whole by the developing stalk, and only differentiates into sporangioles at the end of the process (Thaxter, 1892; Jahn, 1924). The intricate designs of fruiting bodies made 19th century investigators believe that they were dealing with fungi.

The size of myxobacterial fruiting bodies varies between 10 and 1000 µm or more depending on the species, and many can be seen with the naked eye, especially since they are normally produced in large numbers and often are brightly colored in hues of yellow, orange, red, brown, or black. The distribution of fruiting bodies over the swarm is usually random, but sometimes fruiting bodies arise in concentric rings, along radial veins, or in various patterns controlled by elasticotaxis (Stanier, 1942). Fruiting proceeds in the swarm from center to periphery. The shape, structure, and color of fruiting bodies is species specific. In fact, myxobacterial species are mainly defined by their fruiting bodies, but similar shapes are developed by completely unrelated species, e.g., *Stigmatella* and *Chondromyces*, or *Cystobacter, Polyangium, Jahnia*, and *Sorangium*. In addition, shape and color of fruiting bodies may vary substantially within the species. During cultivation, fruiting bodies often degenerate quickly into shapes and structures that are no longer typical and resemble *Archangium* fruiting bodies, i.e., masses of myxospores and slime of ill-defined shape and size. This problem particularly affects *Cystobacter, Stigmatella*, and *Melittangium* species.

The fruiting bodies of myxobacteria may be interpreted as resting colonies. The function of fruiting bodies is probably to guarantee that a new life cycle is started with a population rather than a single cell, for the reasons discussed above in the section on intercellular communication. Fruiting bodies may also facilitate distribution into another habitat. Thus, some species produce fruiting bodies only in the light, a behavior that ensures that they arise on the surface of the substrate. This phenomenon has been demonstrated for *Chondromyces apiculatus* (Reichenbach et al., 1974) and *Stigmatella aurantiaca* (Qualls et al., 1978). No rearrangement of the genome appears to be connected with fruiting. The induction of fruiting body formation is a complex process and clearly more complicated than a mere response to starvation. Fruiting of *Myxococcus xanthus* (and other myxobacteria) takes place above and below dialysis membranes or cellophane sheets at exactly the same spot, suggesting a fruiting signal (McVittie and Zahler, 1962; Reichenbach., 1966). Fruiting bodies may arise in the middle of an active population of cells that do not participate in fruiting, which seems to indicate that the cells must be competent before they enter the fruiting process (Reichenbach et al., 1965b, c).

Within the maturing fruiting body, a cellular morphogenesis takes place: the production of myxospores. In the suborder *Cystobacterineae*, this process always involves a rather dramatic shortening and fattening of the vegetative rods into spheres or at least into short, fat rods. The myxospores surround themselves with a capsule that may be rather thin and recognizable only under the electron microscope. In addition, they become optically refractile. In the suborders *Sorangineae* and *Nannocystineae* the shape change is less extensive. The vegetative cells may become somewhat shorter and slightly constricted in the middle. Often they also become optically refractile and may (e.g., in *Chondromyces crocatus*; Abadie, 1971) or may not develop a thin capsule. The fine structure of myxospores, formerly also called microcysts, has been studied in six organisms: *Myxococcus xanthus* (Voelz and Dworkin, 1962; Voelz, 1966a; Bacon and Eiserling, 1967; Kottel et al., 1975), *Stigmatella aurantiaca* (Reichenbach et al., 1969; Voelz and Reichenbach, 1969), *Stigmatella erecta* (Galván et al., 1987), *Archangium gephyra* (Galván et al., 1992), *Sorangium cellulosum* (Lampky, 1976), and *Chondromyces crocatus* (McCurdy, 1969a; Abadie, 1971). *Myxococcus xanthus* has a very thick, layered capsule, which can be seen under the light microscope, especially after contrast staining with India ink. The capsule appears to be reduced in thickness during germination and is left behind as an empty shell by the emerging vegetative cell. The cell wall is retained intact during sporulation. An interesting but open question is what happens to the gliding apparatus during shape change. Polyphosphate and polysaccharide granules can be seen in the cytoplasm. Artificially induced myxospores fold their outer membranes into vesicular and tubular structures. About 75% of

the capsule is a polysaccharide that consists entirely of galactosamine and glucose. In *Stigmatella aurantiaca* fruiting body myxospores, the outer membrane becomes ruffled or folded; in artificially induced myxospores, these folds form densely packed membrane bodies between the cytoplasmic membrane and the capsule. It is assumed that the outer membrane cannot be melted in during cell conversion and therefore is conserved in this way. As artificially induced cells change their shape much faster than cells in developing fruiting bodies, the phenomenon is more pronounced in the former. Polyphosphate and polysaccharide granules can be seen in both types of myxospores; the surface of the latter is densely covered by small, dark, ribosome-like particles. Essentially the same type of myxospore is found in fruiting bodies of *Stigmatella erecta* and *Archangium gephyra*. The latter may contain large, fusiform bodies with a crystalline internal structure; these bodies are probably composed of protein crystals. Fruiting body myxospores of *Sorangium cellulosum* seem to be devoid of a capsule. They are full of large, electron translucent granules, probably composed of polysaccharide. Polysaccharide is also found near the poles of vegetative cells. In fact, phase bright pole regions are observed under the phase contrast microscope in older vegetative cells. Mesosome-like membrane stacks are seen deep in the interior of the myxospores. Fruiting body myxospores of *Chondromyces crocatus* are found with and without a capsule. In addition, mesosomes are present in *Chondromyces crocatus*. Furthermore, polysaccharide granules, slime vesicles, a ruffled outer membrane, and areas of vesicular or tubular outer membrane structures occur. The arrangement of myxospores within the sporangioles is normally random, but in *Melittangium boletus*, myxospores are oriented strictly perpendicularly to the outer wall (Jahn, 1924). In *Cystobacter fuscus* sporangioles, a pattern of surface-parallel myxospores can be seen impressed on the inner face of the wall (Jahn, 1924). In culture, myxospores are often also produced outside of fruiting bodies. A well-defined experimental system for the study of morphogenesis in myxobacteria became available with the discovery that the addition of glycerol to suspension cultures of *Myxococcus xanthus* induces the vegetative cells to convert synchronously into myxospores within 90 minutes (Dworkin and Gibson, 1964). This discovery made the organisms attractive as research subjects and led to a host of studies on the events during induction and conversion. However, no coherent picture of the involved biochemistry and genetics has been obtained thus far. Details can be found in the reviews mentioned in the beginning of this article and at the end of this section.

Many other myxobacteria can also be induced to sporulate by adding chemicals to suspension cultures. *Stigmatella aurantiaca* can be induced to sporulate by 40 different chemicals acting on at least three different receptors, as suggested by studies of reciprocal induction-resistant mutants (Gerth et al., 1993; Gerth and Reichenbach, 1994). Induction in *Stigmatella aurantiaca* is also triggered by high temperatures just below the temperature maximum, and induction by heat and by chemicals that complement each other. These observations could mean that stress receptors and heat shock proteins play a role in myxospore induction and possibly within fruiting bodies (Heidelbach et al., 1993). Myxospores are clearly designed for survival in unfavorable environmental conditions. Myxospores of *Myxococcus xanthus* are much more resistant to elevated temperatures (60°C), ultraviolet light, and sonic vibration than vegetative cells (Sudo and Dworkin, 1969). In contrast to vegetative cells, myxospores are completely desiccation resistant and can be lyophilized. Dry myxospores can survive for at least 20 years at room temperature.

The temperature resistance of dry myxospores is especially impressive. Myxospores of strains of *Myxococcus, Corallococcus, Angiococcus, Archangium, Cystobacter, Polyangium,* and *Sorangium* survive for 30 min at 140°C; myxospores of *Corallococcus, Archangium, Polyangium,* and *Sorangium* strains survive for 45 min; and myxospores of *Archangium* and *Polyangium* strains survive for 60 min. Exposure to 140°C for 240 min, or to 145°C for a short time kills all myxobacteria. Heat resistance of myxospores is often useful for the isolation of pure strains. Fruiting body suspensions can be heated at 58°C for 15 or 40 min to eliminate contaminants. In *Myxococcus xanthus* myxospore resistance to heat and desiccation appears to be effected by the accumulation of trehalose up to 1.100 µg/mg protein (McBride and Zusman, 1989). Details of numerous studies on myxobacterial morphogenesis can be found in several of the reviews mentioned at the beginning of this article (see also Sudo and Dworkin, 1973; Shimkets, 1987, 2000; Ward and Zusman, 2000; White and Schairer, 2000).

Genetics The DNA of myxobacteria has a mol% G + C content between 67 and 72; the species of the suborder *Cystobacterineae* tend to have lower values in this range (Mandel and Leadbetter, 1965; McCurdy and Wolf, 1967; Behrens et al., 1976). More than 90% of the codons have a G or a C in position 3 (Bremaud et al., 1997). The myxobacterial genomes are among the largest known in procaryotes, viz., 9.45 Mbp in *Myxococcus xanthus* (Chen et al., 1990b), 9.4 Mbp in *Stigmatella aurantiaca,* 9.7 Mbp in *Stigmatella erecta* (Neumann et al., 1992), and 12.2 Mbp in a strain of *Sorangium cellulosum* (Pradella et al., 2002). These sizes indicate a coding capacity of 8000–11,000 genes, or ¼ to ⅓ the number of human genes. Repetitive sequences have been identified in *Myxococcus xanthus* (Fujitani et al., 1991), *Stigmatella aurantiaca* (Bremaud et al., 1997), and *Nannocystis exedens* (Lampson and Rice, 1997). Physical maps of the chromosomes of *Myxococcus xanthus* (Chen et al., 1991a) and *Stigmatella aurantiaca* (Neumann et al., 1993) have been established. The chromosome is circular. Developmental and communication genes are found in all regions. The *Myxococcus xanthus* genome has been cloned in yeast artificial chromosomes (Kuspa et al., 1989). The complete base sequences of the genomes of three organisms, *Myxococcus xanthus, Stigmatella aurantiaca,* and *Sorangium cellulosum,* are about to be determined. *Myxococcus xanthus* DNA contains 5-methylcytosine, and the methylation pattern changes with culture development (Yee and Inouye, 1982).

Extrachromosomal DNA has been found in several *Myxococcus xanthus* and *Myxococcus fulvus* strains (Brown and Parish, 1976). Covalently closed plasmids were found in *Myxococcus xanthus* strain FBt after induction of chloramphenicol resistance. Chloramphenicol resistance could be transferred from *E. coli* via an R-plasmid, which also seemed to be retained free in the cytoplasm. However, plasmids normally seem to survive in myxobacterial cells only when inserted into the cellular DNA, perhaps because of efficient restriction systems.

Promoter regions (e.g., Inouye, 1984), genes for σ-factors (Inouye, 1990; Skladny et al., 1994), and for a number of enzymes (e.g., Muñoz-Dorado et al., 1991; Müller et al., 2000) were identified in several myxobacteria. A light-induced promoter controls carotenogenesis in *Myxococcus xanthus* (Balsalobre et al., 1987; Hodgson, 1993). In *Myxococcus xanthus,* induction of fruiting by starvation and of carotenogenesis by light are jointly controlled by gene *carD* (Nicolas et al., 1996). This gene appears to encode a transcription factor with high homology to nuclear proteins of plants and animals; it possesses four repeats of a DNA-binding domain.

Mutants of various myxobacteria can be obtained by UV irradiation and treatment with nitrosoguanidine (e.g., Grimm, 1978). Photorepair and perhaps other repair systems may become activated. In addition, insertional mutants have been obtained by introduction of transposons, mostly TN5, and plasmids and have been used in many studies (e.g., Starich et al., 1985; Jaoua et al., 1987; Glomp et al., 1988). Several gene transfer systems are available for myxobacteria. Chloramphenicol resistance, conferred by an R-factor, can be transferred from *E. coli* into several *Myxococcus* strains by ways not clearly established at the time of the initial studies (Parish, 1975). Later it was shown that gene transfer between *E. coli* and myxobacteria is possible through coliphage P1 as a suicide vector (Kaiser and Dworkin, 1975; Morris et al., 1978; Kuner and Kaiser, 1981; O'Connor and Zusman, 1983) and by conjugation (Breton et al., 1985; Glomp et al., 1988; Jaoua et al., 1992). Conjugal transfer also occurs from *Myxococcus xanthus* into *E. coli* and apparently also between *Myxococcus xanthus* strains. Gene transfer between *Myxococcus xanthus* strains is also possible using myxobacterial transducing phages (Campos et al., 1978; Geisselsoder et al., 1978), and with Mx alpha particles (probably a defective phage) (Starich and Zissler, 1989). More recently, electroporation has been used with good results (Hartzell and Kaiser, 1991; Silakowski et al., 2001). During the past 20 years many studies have been published making use of various gene transfer systems for the investigation of gliding motility, morphogenesis, carotenoid biosynthesis, and secondary metabolism. Methods and results are described in several reviews (see Zusman, 1980; Rosenberg and Dworkin, 1984; Kaiser, 1991; Dworkin and Kaiser, 1993; and reviews cited above).

Phages, Defective phages, Bacteriocins Phages are readily obtained for myxobacteria. Wild strains normally harbor prophage DNA inserted in the cellular genome and possess DNA restriction and modification systems, so that suitable indicator strains are required for the isolation of phages. *Myxococcus virescens* strain V2 is restrictive for *Myxococcus xanthus* phage MX-1 and contains two restriction endonucleases; this strain could be mutated to become phage-sensitive by loss of the restriction enzymes (Morris and Parish, 1976). The first myxophage to be isolated was *Myxococcus xanthus* phage MX-1 (Burchard and Dworkin, 1966), which came from a sample of cow dung and is a virulent phage resembling coliphage T2. MX-1 has a contractile tail sheath, a latent period of 2–2.5 h (30°C), a 2-h rise, and a burst size of about 100. MX-1 attacks *Myxococcus xanthus, Myxococcus virescens,* and *Myxococcus fulvus* strains. Two glycoproteins are found in its tail (Tsopanakis and Parish, 1976). The phage exhibits the classic pattern of infection and propagation known in *E. coli* phages (Voelz and Burchard, 1971) and survives during sporulation within myxospores; it induces lysis after germination (Burchard and Voelz, 1972). Many more phages have been isolated from *Myxococcus* species, mostly from *Myxococcus xanthus,* and from *Stigmatella aurantiaca* (e.g., Brown et al., 1976; Campos et al., 1978; Martin et al., 1978; Rodrigues et al., 1980). Many of them are temperate phages. In addition, several generalized transducing phages have been obtained (e.g., MX4, MX8). Myxophages do not differ in essence from other, comparable bacteriophages. All are DNA phages with double-stranded DNA of around 50 Kbp. Temperate MX8 of *Myxococcus xanthus* has DNA with terminal repetitions and circular permutation (Magrini et al., 1997). Its prophage DNA remains stably integrated in the cellular DNA for many generations and during development (Orndorff et al., 1983). It contains a gene of an adenine methylase that can be deleted, making room for cellular DNA to be transduced (Magrini et al., 1997). Several morphological types of myxophages are known, all with polyhedral heads and with either long contractile tails (e.g., MX1, MX4), or very short tails (e.g., MX8, MX9); λ-like myxophages have also been obtained (Rodrigues et al., 1980). Defective phages are also known from myxobacteria. The first one was found in *Cystobacter violaceus*. Particles resembling rhapidosomes, as described in *Saprospira grandis*, turned out to be extended and contracted phage tails, cores, and polysheaths (Reichenbach, 1965a, 1967). No complete phage has been discovered in the strain, and the defective phage seems not to interfere with growth and development of the organism. Phage tails have also been found in cultures of *Myxococcus xanthus* (Burchard and Dworkin, 1966; Brown et al., 1976). In addition, the transducing Mx alpha particles of *Myxococcus xanthus* could be a defective phage (Starich and Zissler, 1989).

Bacteriocins have been described in myxobacteria. Fulvocin C, one of three substances with bacteriocin-like activity from *Myxococcus fulvus* Mx f16 (Hirsch, 1977a), is a protein consisting of 45 amino acid residues (molecular weight 4672 Da) with no lipid or carbohydrate attached. Disulfide bridges (probably four) make it a very compact molecule (Tsai and Hirsch, 1981). It was the first bacteriocin for which the complete structure was elucidated. Fulvocin C killed 16 out of 17 tested *Myxococcus fulvus* strains, the exception being the strain that produces the molecule. Some strains of other myxobacteria, but no other eubacteria, were inhibited by fulvocin C. The minimal inhibitory concentration was 0.1 to 0.25 µM. Fulvocin C appears to act on the membrane system. Vesicles and stacks of membranes can be seen within sensitive cells treated with the bacteriocin, and the outer membranes of neighboring cells fuse (Hirsch et al., 1978).

A nonparticulate bacteriocin has been obtained from *Corallococcus coralloides* (Muñoz et al., 1984); it inhibits *Corallococcus*, *Myxococcus*, and *Stigmatella aurantiaca* strains. Xanthacin is produced by *Myxococcus xanthus* (McCurdy and McRae, 1974); production is induced by mitomycin C. Xanthacin is active exclusively on a few myxobacteria. It consists of spherical bodies of somewhat variable size, resembles membrane vesicles and the Mx alpha particles mentioned above, and may in fact be a defective phage. Two further "bacteriocins" reported from *Myxococcus* strains do not fit the definition of bacteriocins, as they also kill *Cytophaga* and *Salmonella* strains unrelated to the producer (Brown et al., 1976); they may be antibiotics.

Ecology As there are two recent reviews of this topic (Reichenbach, 1999; Dawid, 2000), only some general information will be presented here in addition to the facts that have already been mentioned. Myxobacteria are ubiquitous organisms living in soil, on decomposing plant material, on the bark of living and dead trees, on the droppings of wild rabbits, hares, sheep, and other herbivores, in freshwater habitats, and, perhaps, in the oceans. There are very rich populations of myxobacteria on *Zostera* leaves in the drift line on the shores of the German and Danish coast of the Baltic Sea. Like normal terrestrial myxobacteria, the latter isolates are only mildly halotolerant (0.3% NaCl). Myxobacteria are inhabitants of all climate zones and vegetation belts; particularly rich floras occur in warm, semi-arid regions such as the Southwest of the USA, Egypt, or India. They have been isolated from slightly acid to alkaline soils, but seem to prefer a more or less neutral pH. Acidophilic, thermophilic, and anaerobic myxobacteria have not been found so far. Myxospores and fruiting bodies are desiccation resistant and survive in air dried soil at environmental temperatures for long periods; this author isolated *Sorangium cellulosum*, *Corallococcus*, *Nannocystis*, *Ar-*

changium, and other myxobacteria from the same air-dried sample after 12 years of storage. Myxobacteria pathogenic for higher plants and animals have never been observed. (As a note of caution, it may be added that for some time *Cytophaga*-like bacteria were classified as myxobacteria and that among those organisms there are several serious fish pathogens and allergenic organisms. However, the two groups of organisms are not related and actually belong to different phyla.) In the older literature, there are several reports of symbiotic relationships between myxobacteria and other bacteria. While in some cases this conclusion may just have been a misinterpretation due to the inability to separate the bacteria with the then-available methods, there is indeed at least one case of a symbiosis, between *Chondromyces crocatus* strains and a *Sphingobacterium*-like organism, which always shows up with the *Chondromyces* regardless of where the latter has been isolated (Jacobi et al., 1997).

Isolation, Preservation, Cultivation Current methods and recipes can be found in *The Prokaryotes* (Reichenbach and Dworkin, 1992). Only some basic principles will be mentioned. Bacteriolytic myxobacteria can be enriched on spots or streaks of living *E. coli* on water agar. Swarms and fruiting bodies will be evident after 2–14 d at 30°C and can be transferred to new streaks for further enrichment. Cellulolytic myxobacteria are obtained on filter paper placed on a mineral salts agar with KNO_3 as the nitrogen source (ST21-agar). A selective cellodextrin-agarose medium is recommended for the isolation of cellulose degraders (Li et al., 1996). In both cases, it is advisable to suppress growth of fungi by adding a suitable antibiotic, e.g., cycloheximide, to which myxobacteria are completely resistant. The inoculum is taken from sources mentioned in the ecology section. Bark, rotting wood, and animal droppings, which are natural habitats or act as baits, can be incubated in a moist chamber. Fruiting bodies can be seen after a few days and can be used to inoculate streaks of *E. coli*. However, not all myxobacterial fruiting bodies are conspicuous enough to be easily seen on such substrates. Sterilized rabbit dung placed on top of collected soil in large dishes was a favored technique of the earlier investigators (Krzemieniewska and Krzemieniewski, 1926, 1927, 1930).

Some isolation methods are particularly useful for specific myxobacteria. *Myxococcus* species with their large and often colorful fruiting bodies are readily obtained if rabbit, sheep, or goat droppings are incubated in a moist chamber. *Stigmatella aurantiaca* seems to occur only on rotting wood. However, while the organism is very common in the Midwest of the USA, it seems to be exceedingly rare on the same substrate in Europe. Knowledge of exact habitats would certainly help to make the isolation of certain species more efficient. *Chondromyces crocatus* was isolated a hundred years ago from willow bark in the Prater in Vienna and was not obtained again from European samples until recently, when it was found in two samples from a compost heap near Basilea in Switzerland. Purification of strains is somewhat tedious and time consuming. Many strains are pure after a few transfers on living and autoclaved *E. coli*. If a series of transfers is not sufficient to purify a strain, antibiotics, treatment of suspended fruiting bodies at 58°C, or incubation of cultures at elevated temperature (up to 40°C) can be tried. Purity is tested on agar plates of a rich medium and in shaken liquid cultures, in which turbidity after 12 h usually indicates contamination.

Pure cultures can be propagated on yeast agar (VY/2-) or on peptone (CY-) agar. Transfers are necessary every 2–3 weeks at 30°C. Typical generation times at 30°C in liquid media range from 3 to 14 h. As mentioned above, *Chondromyces crocatus* requires the presence of a symbiont. For preservation, fruiting bodies and myxospores can be dried on filter paper or in skim milk. Fruiting bodies stored dry on filter paper at room temperature can be reactivated after as long as 22 years. Vegetative cells appear not to survive lyophilization but can be preserved frozen in liquid nitrogen or in a freezer at $-70°$ to $-80°C$. Frozen cultures remain viable for many years. In liquid nitrogen, they remain viable for decades.

Practical relevance   The only practical application myxobacteria have found thus far is as producers of secondary metabolites. About 650 chemical structures and structural variants have been obtained from myxobacteria, which have become one of the richest sources of such compounds among bacteria. Those compounds can either be a source of structural ideas or can be produced directly by fermentation (Reichenbach and Höfle, 1993, 1999; Höfle and Reichenbach, 1995; Reichenbach, 1999). Several myxobacterial metabolites are now useful tools in basic research, e.g., the electron transport inhibitors myxothiazol and stigmatellin and the inhibitors of eubacterial RNA polymerase sorangicin, myxo- and corallopyronin, and ripostatin. Ratjadon inactivates the CMR-1 shuttle that exports proteins from the nuclear into the cytoplasmic compartment specifically and with high mimetic epothilone (Gerth et al., 1996) is being investigated clinically as an antitumor drug. Other practical aspects are more hypothetical. Myxobacteria produce a host of enzymes and are a potential source of enzymes that has hardly been tapped. One interesting enzyme is a milk-clotting protease excreted by *Myxococcus xanthus* (Petit and Guespin-Michel, 1992). An extracellular glycopeptide, myxalin, which acts as a blood-anticoagulant, has been obtained from the same organism (Masson and Guespin-Michel, 1988; Akoum et al., 1992). Geosmin produced by myxobacteria may contribute to the unpleasant earthy smell of drinking water from reservoirs (the main culprits, however, are cyanobacteria and actinomycetes). Finally, micropredators like myxobacteria could suppress wanted organisms, such as *Azotobacter*, in soil or could help control water blooms of cyanobacteria. Quantitatively both effects are probably of minor importance.

Taxonomy, Phylogeny   The taxonomy and diagnostics of myxobacteria are based mainly on morphological characteristics: the shape and size of vegetative cells and myxospores; the morphology of the swarm colony; and especially the shape, size, color, and structure of the fruiting bodies. As mentioned above, the latter is a weak point in this endeavor, because the appearance of fruiting bodies may vary substantially from culture to culture, similar structures are produced by unrelated species, and fruiting body formation may degenerate or be lost during isolation or cultivation, so that atypical fruiting bodies or no fruiting bodies at all are found in pure cultures. All this led to much confusion in the past, and still is frustrating for everyone who tries to name new isolates. Lack of fruiting body formation also limits the usefulness of type strains. Sometimes altering laboratory conditions can help elicit fruiting. Most *Chondromyces* strains produce typical fruiting bodies when cultivated on filter paper with a crude mixture of cellulose-decomposing bacteria (no myxobacteria) and other organisms which can be obtained from enrichment cultures for *Sorangium cellulosum*. Other species of the suborder *Sorangineae* are sometimes stimulated to fruit under such conditions. Cultivation on pellets of (wild) rabbit feces may also be useful, probably because of the presence of a similar constellation of other organisms.

Efforts to find alternative characteristics for the classification of myxobacteria have not been particularly successful (e.g., McCurdy, 1969b). Cellulose degradation appears to be a reliable and stable feature and is probably always found together with the ability to use inorganic nitrogen sources, a character which is unusual for myxobacteria. This trait distinguishes the genera *Sorangium* and *Byssophaga* from all other myxobacteria. In addition, efficient chitin degradation appears to be rare and typical of certain species and genera, e.g., of all strains of the three *Stigmatella* species, of *Cystobacter fuscus*, *Cystobacter ferrugineus*, and *Cystobacter miniatus*, and of *Jahnia* and *Kofleria*. *Myxococcus stipitatus* cultures are supposed to show an intense yellow fluorescence under UV irradiation at 366 nm (Lampky and Brockman, 1977). This trait is found together with chitin degradation and poor swarming on chitin agar, perhaps due to inhibition by acetylglucosamine. Those characteristics are not found in typical *Myxococcus fulvus* strains, the closest relative of *Myxococcus stipitatus*. Yet there are many *Myxococcus fulvus* strains with small myxospores, which are typical of *Myxococcus stipitatus*, so that confirmation by the methods of molecular taxonomy would be desirable. There may be more phenotypic characteristics that would help distinguish between species, but they can be discovered only by a careful comparative study of many strains. In addition, the habitat may be helpful for diagnosis. For example, while *Stigmatella aurantiaca* appears to be restricted to rotting wood and bark, *Stigmatella erecta* and *Stigmatella hybrida* live in soil and on animal droppings. Analysis of 16S rRNA sequences largely supports classification by the procedures outlined above (Spröer et al., 1999). Groups of organisms remain that apparently represent several species—perhaps even several genera—that cannot be subdivided based on present knowledge. Such is the case with the genera *Corallococcus*, *Archangium*, *Sorangium*, *Polyangium*, and *Nannocystis*. DNA–DNA hybridization (Johnson and Ordal, 1969) and serological reactions (Grilione, 1968) have not been fully exploited for the classification of myxobacteria, but seem promising.

At the moment, about 50 species of myxobacteria are distinguishable and are classified in 16 genera, 5 families, and 3 suborders in the order *Myxococcales*. The order has changed its name three times during the author's scientific career, and one hopes that the present names will remain for some time to come. The suborders are easily recognizable by their characteristic vegetative cells, myxospores, and swarm patterns—the only problem arises with the genus *Kofleria* in the *Nannocystineae*—and are clearly separated in 16S rRNA similarity trees (Ludwig et al., 1983; Shimkets and Woese, 1992; Spröer et al., 1999). Still more so far undescribed species probably exist in nature, as is suggested by the many unusual organisms that have been isolated over the past 35 years, and the number of species may finally reach around 100.

Analysis of 16S rRNA sequences indicates that the myxobacteria are a phylogenetically coherent group and belong to the class *Deltaproteobacteria* (Stackebrandt et al., 1988b). Their nearest—if remote—relatives are, quite unexpectedly, *Bdellovibrio* and strictly anaerobic sulfate reducers (Oyaizu and Woese, 1985). The age of the group was estimated from 16S rRNA similarities to be 900–1000 million years (Shimkets, 1990b). Only a few strains of myxobacteria are found in most public collections. This author recently transferred a collection of roughly 7000 strains covering most of the described genera and species to the DSM (Deutsche Sammlung von Mikroorganismen und Zellkulturen, the German Culture Collection) in Braunschweig, so that ample material for all kinds of studies is now available.

Key to the suborders and families of the order Myxococcales

I. Vegetative cells with tapering ends. Myxospores much shorter and often fatter than vegetative cells, optically refractile when mature. Swarm colony adsorbs Congo red.

Suborder I. *Cystobacterineae*

    A. Myxospores spherical or ellipsoidal. Swarm colonies soft, slimy, often with undulating, snaking veins and flat humps.

Family I. *Myxococcaceae*

    B. Myxospores short, often relatively slender and curved rods. Swarm colonies with a tough slime sheet and long, often branching radial veins.

Family II. *Cystobacteraceae*

II. Vegetative cells cylindrical with rounded ends. Myxospores resemble vegetative cells but may be optically refractile. Swarm colony does not adsorb Congo red.

Suborder II. *Sorangineae*

    C. Swarm colonies sometimes without a coherent slime sheet but often with veins and a ridge or band of cells along the edge. Agar may be etched, and there may be shallow depressions in the agar surface. Often the swarm penetrates deeply into the agar plate, or, in several species, breaks up into small, independently migrating cell clumps.

Family III. *Polyangiaceae*

III. Vegetative cells either short, cylindrical or broadly ellipsoidal, with rounded or truncated ends, sometimes almost cube shaped; or slender, cylindrical rods with rounded ends resembling vegetative cells of the *Polyangiaceae*, but more delicate. Myxospores, as far as known (*Nannocystaceae*), are spherical or short and ellipsoidal and optically refractile. Swarm colonies do not adsorb Congo red.

Suborder III. *Nannocystineae*

    D. Swarm colonies without a slime sheet, tend to corrode the agar plate, on certain media very deeply so, transforming it into a spongy mass. Vegetative cells short rods. Fruiting bodies tiny to moderately sized sporangioles, solitary or aggregated.

Family IV. *Nannocystaceae*

    E. Swarm colonies consist of a slime sheet with more or less prominent radial veins, do not etch the agar, resemble the swarms of the *Cystobacteraceae* but do not adsorb Congo red. Usually produce many small to large globular knobs all over the swarm surface and dense, spherical cell aggregates within the agar resembling sporangioles, but real sporangioles and myxospores have yet to be found. Vegetative cells delicate, slender cylindrical rods with rounded ends. Besides terrestrial organisms, some marine, halophilic myxobacteria also appear to belong to this group.

Family V. *Kofleriaceae*

Comments In general, affiliation of an isolate with one of the suborders is easy to recognize by the shape of vegetative cells, myxospores, and swarm colonies (Fig. BXII.δ.55). The Congo red reaction is normally not required to distinguish the suborders; but sometimes an additional test may be welcome (e.g., with *Kofleria*). As first observed by McCurdy (1969b), the swarm colonies of some myxobacteria adsorb Congo red while those of others do not. The swarms are covered with an 0.01% aqueous Congo red solution. A positive reaction occurs when the swarm becomes purple-red within 5 min at room temperature; also, the swarm sheet then tends to float off the substrate. This result is typical for species of the suborder *Cystobacterineae*, and is especially striking in the family *Cystobacteraceae*. Swarms of species of the other two suborders do not give this reaction. Sometimes their fruiting bodies stain red with Congo red, e.g., those of *Sorangium cellulosum*. This reaction takes a much longer time, and the color is rather blood-red than purple. The Congo red reaction indicates a difference in slime chemistry. Incidentally, chitin flakes in chitin agar also stain intensely with Congo red, so that somewhat hazy lysis zones are made more distinct. Myxobacteria were originally regarded as strictly aerobic organisms; however, facultatively anaerobic, halorespiring myxobacteria have recently been reported and classified in a new genus, *Anaeromyxobacter* (Sanford et al., 2002).

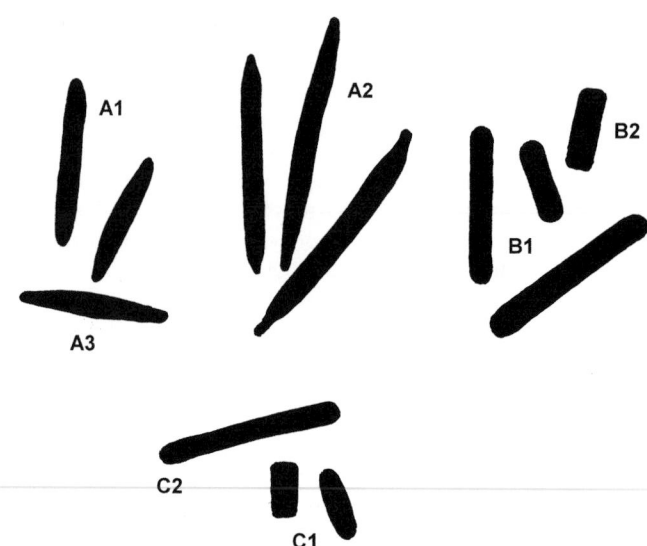

FIGURE BXII.δ.55. Many genera of myxobacteria may be recognized by their cell shape. A: Suborder *Cystobacterineae*. A1: *Myxococcus, Corallococcus, Pyxicoccus*; A2: *Archangium, Cystobacter, Melittangium*; A3: *Stigmatella*. B: Suborder *Sorangineae*. B1: *Polyangium, Byssophaga, Chondromyces, Jahnia*; B2: *Sorangium*. C. Suborder *Nannocystineae*. C1: *Nannocystis*; C2: *Kofleria*.

Family I. **Myxococcaceae** Jahn 1924, 84[AL]

HANS REICHENBACH

Myx.o.coc.ca' ce.ae. M.L. masc. n. *Myxococcus* type genus of the family; *-aceae* ending to denote a family; M.L. fem. pl. n. *Myxococcaceae* the *Myxococcus* family.

Vegetative cells slender rods with slightly tapering ends, cigar or boat-shaped. Myxospores spherical or ellipsoidal with a thick capsule, optically refractile. Fruiting bodies are naked masses of slime and myxospores or, in one genus, consisting of sporangioles with a wall. All species are of the proteolytic–bacteriolytic nutritional type.

Type genus: **Myxococcus** Thaxter 1892, 403.

Key to the genera of the family Myxococcaceae

I. Fruiting bodies are naked masses of slime and myxospores.
   A. Fruiting bodies are relatively large, soft and slimy, spherical knobs or heads with or without a stalk.

   Genus I. *Myxococcus*

   B. Fruiting bodies of cartilaginous consistency and very variable shape: pustules, ridges, often with horns or coralloid.

   Genus II. *Corallococcus*

II. Fruiting bodies consist of sporangioles with a wall. Sporangioles spherical or slightly elongated; arranged in packages and sheets.

   Genus III. *Pyxicoccus*

FURTHER DESCRIPTIVE INFORMATION

The family is well characterized by swarm morphology. The swarms are soft and slimy and have only rarely a thin, tough top layer. They are either without any structure, or develop snaking, undulating veins and flat humps, especially on peptone (CY-) agar. This normally allows for the recognition of members of the family at a glance. However, after prolonged cultivation, swarms of other myxobacteria may degenerate into unstructured, soft, slimy colonies, so that the initially clear difference from species of the *Myxococcaceae* becomes obscured. While the swarms of most species of the *Myxococcaceae* are unpigmented, swarms of two *Myxococcus* species may be intensely greenish yellow or orange. The Congo red reaction can be relatively weak and delayed but is usually clear. The second characteristic typical for all members of the family is the spherical or ellipsoidal myxospores. However, more or less spherical myxospores may also be produced by certain *Archangium* and *Cystobacter* species. With some experience, however, the two can be distinguished. Myxospores of the *Myxococcaceae* are more uniform and have a smoothly curved surface, probably because of their rather thick capsule, while those of the *Cystobacteraceae* are slightly angular and normally accompanied by more or less short, fat rods. Also, swarm morphology and shape of the vegetative cells differ. If for some reason a culture ceases development prematurely, cell conversion of the *Myxococcaceae* may stop at an intermediary stage of short, optically refractile rods. In this case, however, other cultures of the same strain usually have typical, spherical myxospores. Further, the vegetative cells of the *Myxococcaceae* are short to moderately long rods (3–6 µm), either cigar-shaped with slightly tapering ends or slender and boat-shaped, while vegetative cells of the *Cystobacteraceae* are often long, slender, needle-shaped rods (6–12 nm long).

In the last edition of the *Manual*, the species of the former genus *Chondrococcus* were incorporated in the genus *Myxococcus*. While the genus name *Chondrococcus* is not valid, because it was used earlier for a genus of red algae (Kützing, 1847; Papenfuss, 1940); this author wishes to propose a new genus, *Corallococcus*, for the organisms formerly known as *Chondrococcus* spp., because those species do indeed differ substantially and coherently in several respects from *Myxococcus* species. In general, fruiting cul-

tures of the two genera can be differentiated by eye, and if there is doubt, a touch with a needle will show whether the fruiting body is soft, as is typical of *Myxococcus*, or of cartilaginous consistency, as is characteristic of *Corallococcus*. In addition, the shape of the fruiting bodies normally differs. *Myxococcus* fruiting bodies are relatively large globular masses, knobs, or heads, often with a constriction at the base or even a stalk. They may degenerate in culture into deliquescent flat mounts or plump ridges, rarely into finger-like projections, but they always remain soft. *Corallococcus* fruiting bodies, in contrast, are smaller and very irregular in shape; they form pustules, branched or unbranched ridges, ribbon- or disk-like structures, often with finger-like projections that may or may not be branched; they can also develop into coralloid, much-branched masses. In culture, they are conspicuous because they are usually produced in tremendous numbers. The two genera show other differences as well. As mentioned before, *Myxococcus* strains do not grow in the presence of 250 µg/ml kanamycin sulfate, but 50% of *Corallococcus* strains do. On the other hand, while 60% of *Myxococcus* strains grow with 50 µg/ml streptomycin sulfate, only 3% of *Corallococcus* strains do. *Corallococcus* strains produce squalene, not so for *Myxococcus* strains. The number of secondary metabolites obtained from *Corallococcus* is much lower than it is with *Myxococcus*, and the array of compounds produced differs. In addition, there is a difference in salt tolerance. While *Myxococcus* strains still grow, if hesitantly, on yeast agar (VY/2-) with 2% NaCl, *Corallococcus* strains do not. Finally, the 16S rRNA sequences of *Corallococcus* strains cluster together and clearly separate from those of *Myxococcus* species (Spröer et al., 1999).

The third genus of the family, which is now called *Pyxicoccus*, is somewhat confusing. The roots of the genus go back to a bacterium described by Thaxter (1904) as *Myxococcus disciformis*. Because the spherical myxospores of the species are enclosed in sporangioles with a wall, Jahn (1924) created a new genus, *Angiococcus*, for it, apparently without ever having seen the organism. In spite of the fact that the same organism had been reported from Poland (Krzemieniewska and Krzemieniewski, 1926), it was later concluded that *Angiococcus* did not exist (Peterson and McDonald, 1966). Then the bacterium was rediscovered in the fossa of an alkaline bog (Hook et al., 1980). So the old species and genus names were revived, and a neotype strain of *Angiococcus disciformis* was deposited (ATCC 33172). The morphology of the organism suggests that it is really Thaxter's organism. However, it does not belong to the *Myxococcaceae*. In the last edition of this *Manual*, it was classified as *Cystobacter disciformis*. It has vegetative cells and a swarm morphology similar to those found in organisms of that genus. Its present classification is supported by 16S rRNA sequence data, which place the neotype strain close to *Cystobacter minus* (Spröer et al., 1999). There are, however, organisms with spherical myxospores, a swarm morphology characteristic of the *Myxococcaceae*, and 16S rRNA sequences that po-

sition them close to *Myxococcus* (Spröer et al., 1999). For organisms of this type, exemplified by strain An d1, a new genus name, *Pyxicoccus*, will be used here. This avoids confusion, especially as it is not yet established whether *Cystobacter disciformis* really belongs to the genus *Cystobacter*. If not, the genus name *Angiococcus* may have to be revived for it. It should be mentioned that *Pyxicoccus* fruiting bodies often degenerate into *Corallococcus*-like structures. The 16S rRNA study also proved that *Pyxicoccus* could be easily confused with other myxobacteria, as two of the three strains tentatively identified by the author as *Angiococcus disciformis* clustered in the *Archangium* group (Spröer et al., 1999).

Genus I. **Myxococcus** Thaxter 1892, 403[AL]

HANS REICHENBACH

Myx.o.coc'cus. Gr. fem. n. *myxa* mucus, slime; Gr. masc. n. *kokkos* berry; M.L. masc. n. *Myxococcus* slime coccus.

Vegetative cells are slender rods with tapering ends, either cigar-shaped or slender and boat-shaped, measuring about 0.8×3–6 μm. When typical, the fruiting bodies are soft, slimy globular masses, knobs, or heads, constricted at the base or on a slime stalk (but see comments on the family). Of the proteolytic–bacteriolytic nutritional type. Common organisms in soil, in freshwater, and on dung of herbivores. Readily obtained when rabbit, hare, or sheep droppings are incubated in a moist chamber at 30°C.

The mol% G + C of the DNA is: 68–71 (Mandel and Leadbetter, 1965), 67–68 (McCurdy and Wolf, 1967), 68–70 (Behrens et al., 1976), 65–67 (Yamanaka et al., 1987).

Type species: **Myxococcus fulvus** (Cohn 1875) Jahn 1911, 198.

FURTHER DESCRIPTIVE INFORMATION

Originally, the *Myxococcus* species were distinguished primarily by the color of the fruiting bodies. However, this color may vary quite considerably, depending on environmental factors such as light and growth media, and it may even be completely absent. A better feature appears to be the size of the myxospores, which seems to be fairly constant within each species. Yet questions remain, as the taxonomic significance of the observed variability between strains cannot be evaluated based on available data. The largest myxospores are those of *Myxococcus virescens*. *Myxococcus xanthus* has somewhat smaller myxospores; in addition, many strains produce a certain proportion of abortive, dark myxospores. *Myxococcus stipitatus* has small myxospores, which, in some strains, are decidedly ellipsoidal. In principle, the presence of a stalk should make it easy to recognize this species. However, well-developed stalks are normally only seen in crude cultures, when fruiting bodies develop on natural substrates. In culture, only rudimentary stalks or no stalks at all develop as a rule. Fruiting bodies often develop in groups or in lines on a common, cushion-like base; this behavior is not seen in *Myxococcus fulvus*, its nearest relative. Furthermore, it has been observed that cultures of *Myxococcus stipitatus* exhibit an intense yellow fluorescence under UV light at 366 nm (Lampky and Brockman, 1977). All such strains seem to degrade chitin and to spread very little on chitin agar; neither characteristic is seen in typical *Myxococcus fulvus* strains.

However, the size of the myxospores of *Myxococcus fulvus* varies from mid-sized to small in different strains, so that in this respect there is an overlap with *Myxococcus stipitatus*. DNA–DNA hybridization of a critical number of strains would be desirable to establish whether the above-mentioned distinguishing characteristics are reliable. The small differences in the 16S rRNA sequences are clearly not sufficient to classify the various *Myxococcus* species (Spröer et al., 1999).

Several *Myxococcus* species of questionable validity appear in the literature. In part, the confusion is due to variability of characteristics that could not be evaluated properly at the time the species were described, especially if the organisms were not cultivated. In addition, in the early days organisms were named *Myxococcus* that do not correspond with today's definition of the genus, as has just been shown with the example of *Angiococcus*. For the moment, it appears that all variants can be classified in the four species outlined below. This is probably also the case with the most recent addition, *Myxococcus flavescens* (Yamanaka et al., 1987). The size of its myxospores, its yellow color, and the myxovirescin type antibiotics produced by it (here called megovalicins; Miyashiro et al., 1988) are a perfect fit with the description of *Myxococcus virescens*. This identity is also supported by 16S rRNA sequence comparison (Spröer et al., 1999).

As happened with several myxobacteria, a *Myxococcus* species was observed and named before Thaxter (1892) discovered the peculiar life cycle of those bacteria, which he then described as myxobacteria. It appears that in 1875 Cohn gave a description of fruiting bodies of *Myxococcus fulvus*, which he named *"Micrococcus fulvus"*. Thaxter's *"Myxococcus rubescens"* was later regarded as being identical to *"Micrococcus fulvus"* and was consequently renamed (Jahn, 1911). Ferdinand Cohn, however, did not realize what an extraordinary developmental cycle was performed by his organism.

The relative abundance of the four *Myxococcus* species may be approximately estimated from the number of strains that have been isolated over 30 years: *Myxococcus fulvus* (572), *Myxococcus virescens* (264), *Myxococcus xanthus* (173), and *Myxococcus stipitatus* (94).

DIFFERENTIATION OF THE SPECIES OF THE GENUS MYXOCOCCUS

I. Myxospores middle-sized to small, 1.2–1.6 µm in diameter. Fruiting bodies pink, brownish, reddish, or white.

  A. Myxospores middle-sized to small, spherical. Fruiting bodies often with a constriction at the base and a pedicle, but never with a stalk. No yellow fluorescence of cultures.

M. fulvus

  B. Myxospores small, spherical, or ellipsoidal. Fruiting bodies with a long, white stalk. Cultures show a bright yellow fluorescence at 366 nm.

M. stipitatus

II. Myxospores large, 1.8–2.1 µm in diameter. Fruiting bodies often deep orange.

M. xanthus

III. Myxospores very large, 2.0–2.5 µm in diameter. Fruiting bodies often greenish yellow, or grayish and deliquescent.

M. virescens

List of species of the genus Myxococcus

1. **Myxococcus fulvus** (Cohn 1875) Jahn 1911, 198[AL]

ful' vus. L. masc. adj. *fulvus* reddish yellow, brown.

Vegetative cells are slender rods with tapering ends, often slightly boat-shaped, 0.6–0.8 × 4–8 µm (Fig. BXII.δ.56).

Fruiting bodies are spherical or slightly pear shaped, 50–250 µm and more in diameter (Fig. BXII.δ.57), often constricted at the base and on a pedicle or short stalk. In culture also plump ridges and deliquescent mounds may be found. Occur in a wide variety of colors: white, pink, red, brown. Myxospores spherical, optically refractile, 1.3–1.6 µm (range: 1.2—2.0 µm) in diameter, with a thick capsule (Fig. BXII.δ.58). Swarm colonies as are typical for the family (Fig. BXII.δ.59), rarely pigmented, sometimes greenish yellow when grown in the dark, pink to purplish red when cultivated in the light. A dark brown, diffusible pigment is often produced on peptone agar. Optimal pH 6.8–7.5. Temperature range 25–35°C (may be 6–38°C), optimal 26–32°C. Gelatin and casein are hydrolyzed. Does not utilize mono- and disaccharides.

Very common and readily obtained from dung of rabbits, hares, sheep, etc., but also from enrichment cultures with soil and rotting plant material on streaks of living *E. coli* or on plates for cellulose degraders. Also found in fresh water and samples from the marine littoral zone.

The wide range of myxospore diameters between individual strains may indicate that the species is not completely homogeneous.

The mol% G + C of the DNA is: 67–71 (Bd, T_m).

Type strain: ATCC 25199.

Additional Remarks: Reference strains include DSM 434 (Mx f2, isolated in 1967 from deer dung collected in Minnesota, USA).

2. **Myxococcus stipitatus** Thaxter 1897, 408[AL]

sti.pi.ta' tus. L. masc. n. *stipes,* gen. *stipitis* trunk, stalk; M.L. adj. *stipitatus* stalked.

Vegetative cells slender rods with tapering ends, 0.7–0.8 × 4–6 µm, normally shorter than those of *Myxococcus fulvus.* Fruiting bodies consist of a spherical mass of slime and myxospores, 40–140 µm and more in diameter, unpigmented or brownish, on a long, white slime stalk, 100–250 µm long and 30–60 µm wide (Fig. BXII.δ.60). While stalks usually are well developed in enrichment cultures with fruiting bodies sitting on the natural substrates, stalks may be mostly lacking in cultures on agar media, so that the fruiting bodies resemble those of *Myxococcus fulvus.* In contrast to *Myxococcus fulvus, Myxococcus stipitatus* fruiting bodies often arise in clusters or lines on a common, cushion-like base. In addition, in contrast to *Myxococcus fulvus, Myxococcus stipitatus* cultures show a bright yellow fluorescence under UV light at 366 nm, which is caused by phenalamide or stipiamide compounds. Further, *Myxococcus stipitatus* strains decompose chitin but spread very little on chitin agar plates. As mentioned above, it is not yet firmly established whether the latter two characteristics distinguish the two species reliably. Myxospores are spherical or sometimes ellipsoidal and rather small (1.3–1.5 × 1.5–1.6 µm in diameter) (Fig. BXII.δ.61). Swarm colonies as are typical for the family, unpigmented or yellowish. On peptone agar a brown, diffusible pigment is often produced. Nutritional requirements as for the other species of the genus.

Myxococcus stipitatus is common, especially on rotting plant material and in samples from the tropics, but may also be obtained from other sources. Strain Mx s8 was isolated at the Gesellschaft für Biotechnologische Forschung in 1982 from a soil sample collected near the Iguaçu waterfalls in Brazil, December 1981. Reference strain Mx s1 came from soil and rotting wood collected in Bangkok, Thailand, in August 1970 and was isolated at the University

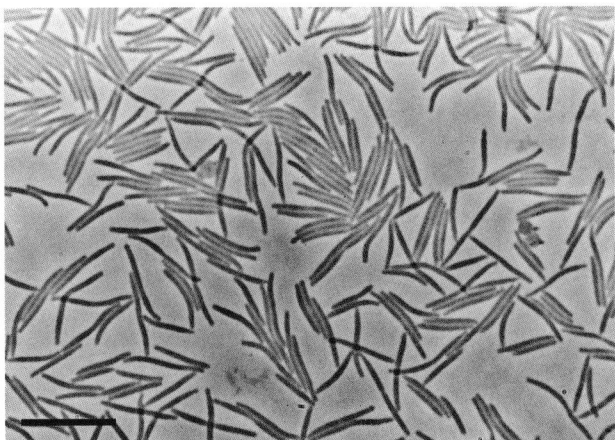

FIGURE BXII.δ.56. *Myxococcus fulvus:* vegetative cells from a liquid culture, in phase contrast. As the culture was already beyond its optimal, the cells are longer than usual. Bar = 10 µm.

FIGURE BXII.δ.57. *Myxococcus fulvus*: fruiting bodies: (*A*) on the natural substrate, in this case rotting wood; (*B*) on an agar surface. Typical fruiting bodies show a constriction at the base and a short pedicle. Sometimes the soft heads of neighboring fruiting bodies fuse. Typical fruiting bodies of *Myxococcus xanthus* and *Myxococcus virescens* would have the same shape, but a different color. (*A*) Bar = 165 μm; (*B*) Bar = 200 μm.

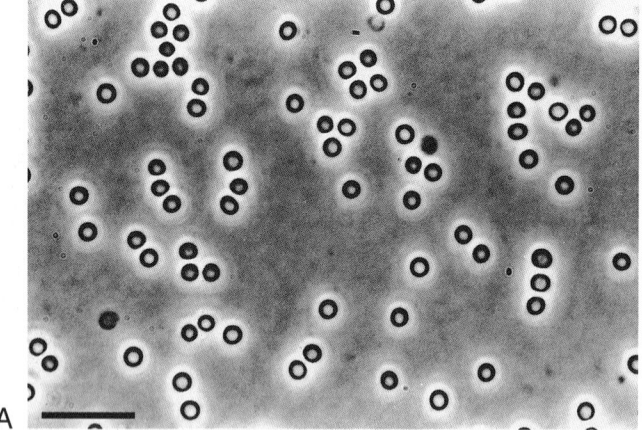
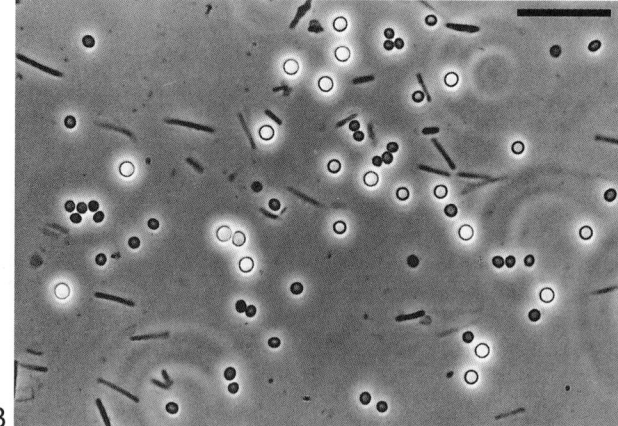

FIGURE BXII.δ.58. *Myxococcus fulvus*: myxospores from squashed fruiting bodies, in phase contrast; (*A*) a homogeneous suspension, (*B*) mixed with myxospores of *Myxococcus virescens* (the large, phase-bright cells): The size difference is recognizable at a glance through the microscope. (*A*) Bar = 8 μm; (*B*) Bar = 13 μm.

of Freiburg, Germany, in October 1970; reference strain Mx s2 was isolated in 1973, also in Freiburg, from a soil sample collected near the city.

The mol% G + C of the DNA is: 65 (T_m).

Type strain: Mx s8, DSM 14675, JCM 12634.

Additional Remarks: Reference strains include ATCC 29611, ATCC 29612, ATCC 29621, DSM 14676 (Mx s1), DSM 14700 (Mx s2).

3. **Myxococcus virescens** Thaxter 1892, 404[AL]

vi.res′ cens. L. part. adj. *virescens* becoming green.

Vegetative cells slender rods with shortly tapering ends, cigar-shaped, 0.7–0.8 × 4–8 μm. Fruiting bodies are globular or oval, rarely even finger-like (Fig. BXII.δ.62); on agar media also plump ridges and, very often, deliquescent mounds; greenish yellow when typical, but often yellow to orange, in cultures even unpigmented, gray. May become rather large, 150–500 μm. Myxospores are spherical, optically refractile, and very large, 1.9–2.4 μm in diameter (range: 1.7–2.6 μm) (Fig. BXII.δ.63), with a thick capsule. This large size is the most reliable distinguishing charac-

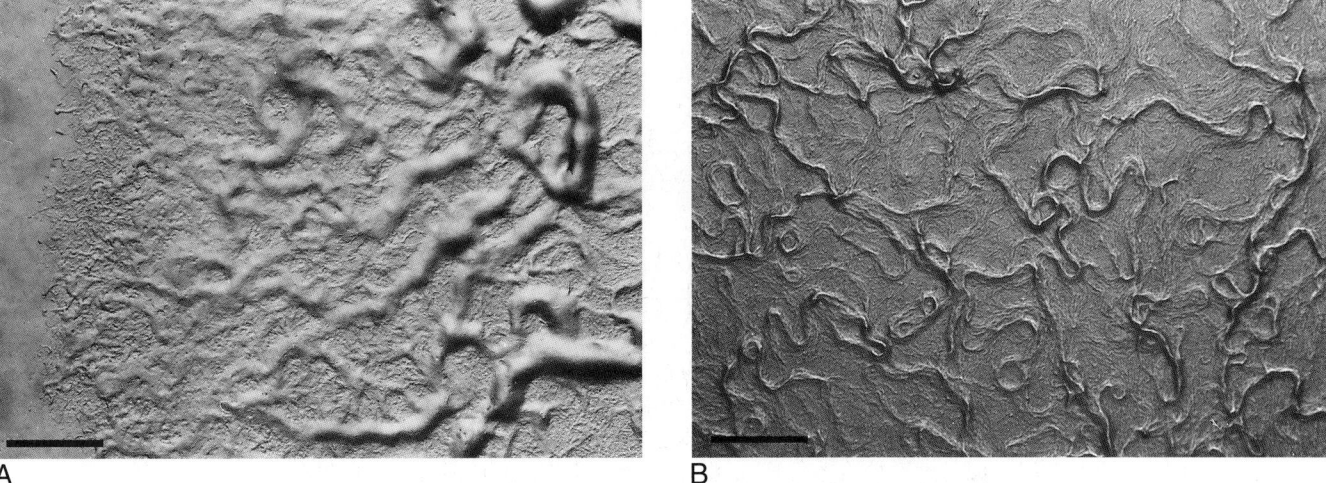

FIGURE BXII.δ.59. Swarm colonies of *Myxococcaceae*. The swarms of all species of the family look alike and are quite different from those of other myxobacteria. If swarms are structured at all, snaking, undulating veins are typical; they are particularly prominent on peptone-containing agar media (*A*). Both examples show swarms of *Corallococcus coralloides*. (*A*) Bar = 600 μm; (*B*) Bar = 800 μm.

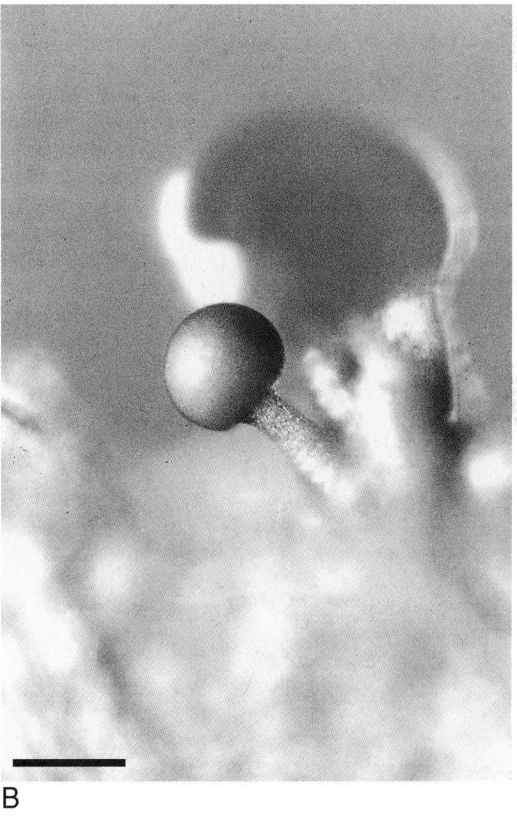

FIGURE BXII.δ.60. *Myxococcus stipitatus*: fruiting bodies. (*A*) Long stalks are normally produced on the natural substrate, in this case a crumb of soil, but (*B*) on agar plates stalks often are only rudimentary or absent. (*A*) Bar = 125 μm; (*B*) Bar = 75 μm.

teristic of the species. Swarm colonies as characteristic of the family, unpigmented or greenish yellow, sometimes orange. Temperature and pH range and nutritional requirements as for the other species of the genus.

Myxococcus virescens is a common organism in soil, in rotting plant material and in samples from the littoral; it is easily obtained when rabbit or sheep dung is incubated in a moist chamber. The species appears to be more halotolerant than most myxobacteria (Rükert, 1978).

The mol% G + C of the DNA is: 68–71 (Bd, T_m).

Type strain: M22, ATCC 25203, DSM 2260.

4. **Myxococcus xanthus** Beebe 1941a, 195[AL]

xan′thus. Gr. adj. *xanthos* yellow; M.L. masc. adj. *xanthus* yellow.

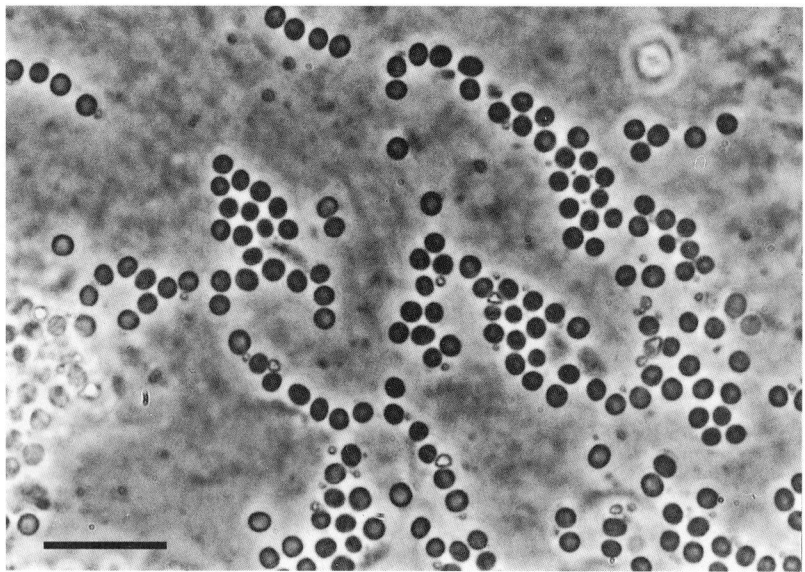

FIGURE BXII.δ.61. *Myxococcus stipitatus*: Myxospores from squashed fruiting body. At high magnification (100 × 2.0, phase contrast, oil immersion) the small myxospores do not show optical refractility which is readily seen using a 40× objective. Bar = 8 μm.

Vegetative cells slender rods with tapering ends, 0.7–0.8 × 4–8 μm. Fruiting bodies are spherical or oval; on agar plates plump ridges or deliquescent mounds are often formed, usually deep orange, but often yellow, greenish yellow, gray. or even unpigmented; 80–150 μm and up to 300–400 μm in diameter. Myxospores spherical, optically refractile, 1.7–2.2 μm in diameter. Often a subpopulation of dark, degenerate myxospores is seen among the typical ones. Swarm colonies as typical for the genus, unpigmented, orange or greenish yellow. Pigmentation of *Myxococcus xanthus* swarms overlaps with that of *Myxococcus virescens* and may change with cultivation. Temperature and pH range and nutritional requirements as for the other species of the genus.

Myxococcus xanthus appears to be somewhat less common than *Myxococcus fulvus* and *Myxococcus virescens* but it is still easily obtained from the usual sources, particularly when dung is incubated in a moist chamber.

The mol% G + C of the DNA is: 67–71. (Bd, T_m).

Type strain: FB, ATCC 25232, NCIB 9412, IFO 13542, DSM 6796.

Additional Remarks: Reference strains include DSM 435 (Mx x1, isolated in 1962 from a soil sample collected in 1961 in Liberia).

FIGURE BXII.δ.62. *Myxococcus virescens.* In agar cultures, the organism rarely produces atypical, finger-like fruiting bodies. Bar = 65 μm.

Other Organisms

A substantial number of additional *Myxococcus* species have been described, mainly in the early years of myxobacterial research. Most of them are synonymous with the above-listed species; others belong to other genera of myxobacteria, or are not myxobacteria at all. Almost none of the names are validly pub-lished and only in one case (*Myxococcus flavescens* DSM 4946, IAM 13189, JCM 6245) is a type strain still available.

The following species appear to be identical with *Myxococcus fulvus*: "*Myxococcus rubescens*" Thaxter 1892; "*Myxococcus pyriformis*" Smith 1901; "*Myxococcus ruber*" Baur 1905; "*Myxococcus ja-*

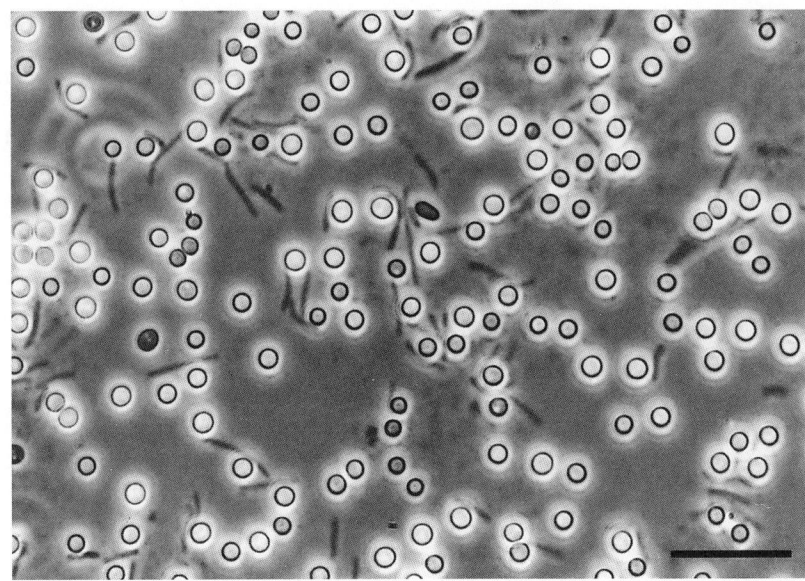

FIGURE BXII.δ.63. *Myxococcus virescens*: myxospores from squashed fruiting body, in phase contrast. Bar = 13 μm.

vanensis" de Kruyff 1908; perhaps also *"Myxococcus albus"* Finck 1950; *"Myxococcus brevipes"* Finck 1950; and most likely *"Myxococcus cellulosus"* Pronina 1962, which probably was not a cellulose degrader as stated (see Chapter on *Myxococcales*). *"Myxococcus ovalisporus"* Krzemieniewska and Krzemieniewski 1926 may have been *Myxococcus stipitatus*; the pale color of the fruiting bodies, the fact that they often arise in clusters on a common base, and oval myxospores would support this assumption. *"Myxococcus viperus"* Finck 1950 and *Myxococcus flavescens* Yamanaka et al., 1987 appear to be identical with *Myxococcus virescens* (see Further Descriptive Information). *"Myxococcus cirrhosus"* Thaxter 1897,

"Myxococcus digitatus" Quehl 1906, and *"Myxococcus clavatus"* Quehl 1906 may have been *Corallococcus coralloides* strains. *"Myxococcus polycystus"* Kofler 1913 was probably *C. coralloides*, too. *Myxococcus disciformis* Thaxter 1904 is *Angiococcus disciformis* Jahn 1924 and now *Cystobacter disciformis* McCurdy 1989. *"Myxococcus cruentus"* Thaxter 1897 is a cellulose-degrader belonging to the suborder *Sorangineae* and is now called *Byssophaga cruenta*. *"Myxococcus cerebriformis"* Kofler 1913 was probably *Archangium gephyra*. *"Myxococcus filiformis"* Solntseva 1940 is *Flexibacter filiformis*. *Myxococcus macrosporus* Zukal 1897b was obviously not a myxobacterium at all, but a fungus.

Genus II. **Corallococcus** gen. nov. (*Chondrococcus Jahn 1924, 85*)

HANS REICHENBACH

Co.ral.lo.coc' cus. Gr. neut. n. *korallion* coral; Gr. masc. n. *kokkos* berry; M.L. masc. n. *Corallococcus* coral-shaped coccus (i.e., with coral-shaped fruiting bodies).

Vegetative cells are slender rods with tapering ends, cigar-shaped, 0.6–0.8 × 3–8 μm. **Myxospores spherical or ellipsoidal**, often somewhat irregular, 1.3–2.4 μm in diameter. **Fruiting bodies very variable in shape and size** even within one culture: pustules, ridges, slabs, and irregular disks, often with finger-like projections, or raised and bizarre coralloid, **of cartilaginous consistency**. Swarm colonies as are typical for the family. The organisms do not utilize mono- and disaccharides, but may degrade polysaccharides, like starch, into tri- and oligosaccharides, which can be taken up and metabolized.

The mol% G + C of the DNA is: 66–68.

Type species: **Corallococcus coralloides** comb. nov. (*Myxococcus coralloides* Thaxter 1892, 404.)

FURTHER DESCRIPTIVE INFORMATION

The genus is readily distinguished from *Myxococcus* by the cartilaginous consistency of the fruiting bodies and their bizarre shape. In addition, *Corallococcus* fruiting bodies are usually much smaller and are produced in enormous numbers, even in agar cultures. There also are physiological differences, which are de-

scribed in the Further Descriptive Information section of the description of the family *Myxococcaceae*. *Corallococcus* fruiting bodies can be distinguished from *Archangium* fruiting bodies (which also lack a firm outer wall), by their small size, and by the absence of the large, extended aggregates of fruiting bodies typical of *Archangium*. The swarm structure is also different. *Archangium* swarms produce a tough slime sheet with radial veins, that are the shape and size of the vegetative cells (which are long needles in *Archangium*) and myxospores (which are more heterogeneous and rarely exactly spherical in *Archangium*).

An unsolved problem, however, is the subdivision of the genus *Corallococcus* into species, because of the extreme variability of the fruiting bodies. *Corallococcus coralloides* strains regarded here as typical have relatively large, bizarre coralloid fruiting bodies loosely scattered over the surface of the culture plate and myxospores around 1.6 μm in diameter. There are, however, many variant strains with all kinds of fruiting bodies and myxospores of very different size. Strains with large myxospores can be separated as the species *Corallococcus macrosporus*, and strains with many tiny, brownish fruiting bodies produced in very dense populations, as the species *Corallococcus exiguus*, but the remaining *Corallococcus coralloides* strains still must be regarded as forming a composite species. A further complication is that *Pyxicoccus* fruiting bodies may degenerate into *Corallococcus*-like structures, so that if typical *Pyxicoccus* fruiting bodies were not observed before degeneration occurred, *Pyxicoccus* strains would be classified in the *Corallococcus coralloides* complex.

Corallococcus strains are among the most common myxobacteria and are found in unfavorable habitats, such as high mountain or desert areas, where other myxobacteria are not found.

They can be isolated from soil, rotting plant material, dung of herbivores, and samples from the drift line at the sea shore. They are difficult to recognize on the surface of natural substrates, as are the myxococci, because *Corallococcus* fruiting bodies are much smaller and do not rise much from the surface. However, they may be conspicuous by their sheer number.

The number of strains in the author's collection may give a rough estimate of the relative abundance of the species: *Corallococcus coralloides* (1624), *Corallococcus exiguus* (172), and *Corallococcus macrosporus* (22).

Key to the species of the genus *Corallococcus*

I. Myxospores spherical or slightly irregular, small to moderately sized, 1.2–1.9 μm in diameter.

    A. Fruiting bodies are pustules or branched or unbranched ridges, often with long, finger-like projections or more or less coralloid; white, pale brown, pink, orange, or red; loosely to densely scattered over the agar plate.

          Corallococcus coralloides

    B. Fruiting bodies are small pustules or narrow ridges, only exceptionally with a short hump; brownish; in agar cultures very densely packed in extended fields. Myxospores large, 1.5–1.9 μm in diameter.

          Corallococcus exiguus

II. Myxospores spherical or short, fat rods, large, 1.8–2.2 μm in diameter. Fruiting bodies are ridges, often with finger-like projections; yellow to orange; loosely scattered over the agar surface. Swarm colonies often yellow to orange.

          Corallococcus macrosporus

List of species of the genus *Corallococcus*

1. **Corallococcus coralloides** *comb. nov.* (*Myxococcus coralloides* Thaxter 1892, 404.)
co.ral.lo.i' des. Gr. neut. n. *korallion* coral; Gr. neut. n. *eidos* shape, appearance; M.L. adj. *coralloides* coral-shaped.

Vegetative cells slender rods with tapering ends, 0.7–0.8 × 3–6 μm, tend to be shorter than *Myxococcus* cells.

Fruiting bodies 50–250 μm in size and extremely variable in shape (Fig. BXII.δ.64). May form pustules, ridges, or bands, with or without branches at the sides or at the ends, as well as star, staghorn, and anchor shapes that may end in a long, tapering tail of myxospores or in finger-like projections, which may themselves be branched or bizarrely coralloid and may extend root-like into the agar substrate. Fruiting bodies may be very different in color, depending at least partly on the culture conditions; occur singly or in dense clusters, loosely or densely scattered over the agar surface and normally produced in large numbers. Myxospores spherical, often somewhat deformed, optically refractile, with different strains of very different size, 1.5–1.7 μm in diameter (range: 1.2–1.8 μm); variability in size sometimes occurs within the same fruiting body.

Organisms of this kind are very common in nature, in fact, together with *Nannocystis*, the most common of all myxobacteria. However, the species as delineated here is probably a composite comprising several species. This conclusion is suggested by the very different sizes of myxospores among individual strains. In addition, differences in the specific shapes of fruiting bodies are seen among individual strains and could be a basis for a subdivision of the species. The problem, however, is that there is a considerable overlap of those characteristics, and that the shapes and patterns

are not reliably stable. Clearly, molecular taxonomy has to be applied here. The reference strains Cc c1613 and Cc c1616 were chosen because of their typical, coralloid fruiting bodies. Cc c1616 was isolated in 2001 from a sample of rotting *Zostera* leaves in the drift line on the shore of the Isle of Poel off the German Baltic Sea coast; Cc c1613 in 2001 from soil and decaying plant material collected in San Diego, CA, USA.

The mol% G + C of the DNA is: 68 (T_m).
Type strain: M2, ATCC 25202, DSM 2259.
GenBank accession number (16S rRNA): M94278.
Additional Remarks: Reference strains include DSM 14687 (Cc c1613), DSM 14688 (Cc c1616).

2. **Corallococcus exiguus** (ex Kofler 1913) *nom. rev., comb. nov.* ("*Myxococcus exiguus*" Kofler 1913, 367.)
ex.i' gu.us. L. masc. adj. *exigus* small.

Vegetative cells slender rods with tapering ends, 0.7–0.8 × 3–6 μm. Fruiting bodies small pustules or more or less long, narrow ridges, rarely with a low hump or short projection, 20–50 μm wide; brownish; single but in very densely packed, extended fields (Fig. BXII.δ.65). Myxospores spherical, optically refractile, tend to be larger than those of *Corallococcus coralloides*, 1.5–1.9 μm in diameter.

This organism is distinguishable from *Corallococcus coralloides* by its delicate, brownish fruiting bodies, which always appear in enormous numbers, by the lack of projections, and by its relatively large myxospores.

The mol% G + C of the DNA is: not determined.
Type strain: Cc e167, DSM 14696.
GenBank accession number (16S rRNA): DSM 14720 (Cc e163).

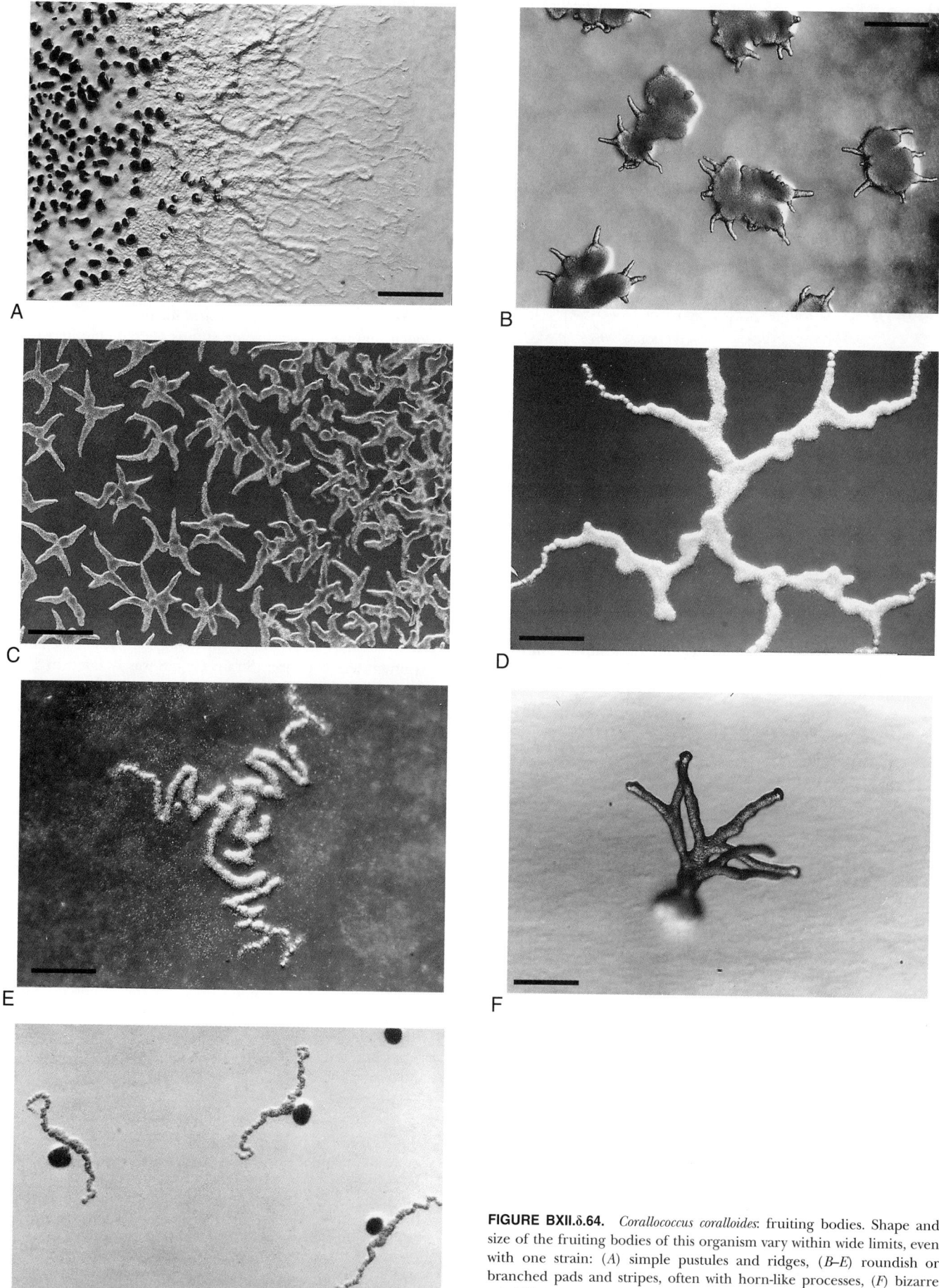

FIGURE BXII.δ.64. *Corallococcus coralloides*: fruiting bodies. Shape and size of the fruiting bodies of this organism vary within wide limits, even with one strain: (*A*) simple pustules and ridges, (*B–E*) roundish or branched pads and stripes, often with horn-like processes, (*F*) bizarre coral-like structures. (*G*) On nutrient-poor substrates, the fruiting bodies often develop long tails of myxospores. All specimens in agar cultures. Bars = 1100 μm (*A*), 330 μm (*B*), 380 μm (*C*), 260 μm (*D*), 160 μm (*E*), 200 μm (*F*), 130 μm (*G*).

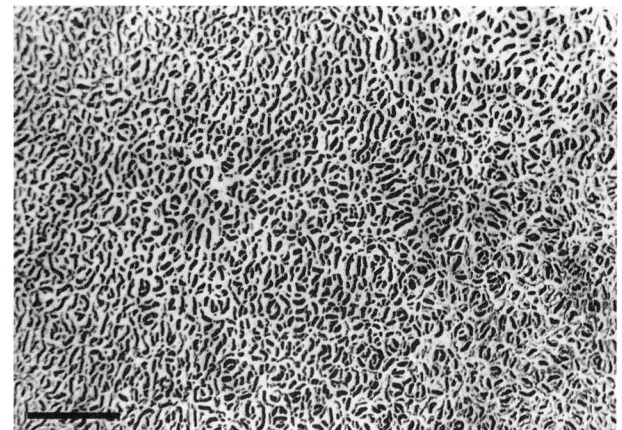

FIGURE BXII.δ.65. *Corallococcus exiguus:* fruiting bodies on agar surface, simple ridges in very dense populations. Bar = 940 μm.

Additional Remarks: While the general description of *Myxococcus exiguus* by Kofler (1913) fits the strains this text refers to quite well, the myxospores of this organism were much smaller, only 1–1.4 μm in diameter. However, measurements with microscopes of those days and of stained smears were perhaps not reliable. In any case, as no type material of Kofler's organism is available, there is no risk of confusion in defining the species anew. *Corallococcus exiguus* is found in the usual myxobacterial habitats and is moderately common. Strain Cc e167 comes from soil with decaying plant material collected in 1992 in Oranje Free State, Republic of South Africa, and was isolated in 1998; reference strain Cc c163 was isolated in 1998 from a soil sample collected in 1994 in Taiwan.

3. **Corallococcus macrosporus** *comb. nov.* (ex Krzemieniewska and Krzemieniewski 1926) *comb. nov.* (*Chondrococcus macrosporus* Krzemieniewska and Krzemieniewski 1926, 16.)
ma.cro.spor′us. Gr. adj. *makros* long, large; Gr. masc. n. *sporos* seed; M.L. masc. adj. *macrosporus* with large spores.

Vegetative cells slender rods with tapering ends, 0.7–0.8 × 4–8 μm, usually longer than those of the other *Corallococcus* species. Fruiting bodies short to moderately long ridges, often with long, finger-like projections; yellow to orange brown; loosely scattered over the agar surface. Myxospores spherical, optically refractile, 1.8–2.2 μm in diameter. The conversion of vegetative cells into myxospores is often incomplete, and the resulting myxospores are short, fat rods. Swarm colonies as typical for the family, often yellow to orange. At least some strains degrade chitin, which is exceptional among *Corallococcus*. May require vitamin B_{12}.

The mol% G + C of the DNA is: not determined.

Type strain: Cc m8, DSM 14697, JCM 12621.

Reference strains include ATCC 29619, ATCC 29620, ATCC 29039, DSM 14690 (Cc m15).

Additional Remarks: Corallococcus macrosporus is easy to distinguish from the other *Corallococcus* species by the yellow color of its swarm colonies and fruiting bodies, and particularly by its large myxospores. It may be confounded with *Myxococcus virescens*, which sometimes also produces finger-like fruiting bodies, that are, however, soft and not firm. Further, *Corallococcus macrosporus* may be mistaken for a degenerate *Myxococcus virescens* strain if typical fruiting bodies are lacking. *Corallococcus macrosporus* may be found in the same habitats as the other two *Corallococcus* species, but it is much less common. Strain Cc m8 was isolated in 1982 from soil collected in 1981 in Ceylon, Sri Lanka; reference strain Cc m15 was isolated in 1991 from bear feces collected in 1980 in Ontario, Canada.

Other Organisms

Many more *Corallococcus* (= *Chondrococcus*) species have been described. It appears that in most cases investigators were misled by the extreme variability of fruiting body shapes and sizes, and that most of those species belong to the *Corallococcus coralloides* complex. However, some of them may be stable varieties of the species, or may even turn out to be separate species when the *Corallococcus coralloides* complex can finally be broken down. No reference strains are available for any of them, and the names have not been validly published.

"*Chondrococcus coralloides*" Jahn 1924 is a synonym of *Myxococcus coralloides* Thaxter 1892. "*Chondrococcus cirrhosus*" Jahn 1924 (= "*Myxococcus cirrhosus*" Thaxter 1897), "*Chondrococcus coralloides*" var. clavatus Jahn 1924 (= "*Myxococcus clavatus*" Quehl 1906), "*Chondrococcus coralloides*" var. digitatus Jahn 1924 (= "*Myxococcus digitatus*" Quehl 1906), "*Chondrococcus blasticus*" Beebe 1941b probably belong to the same species. In addition, "*Chondrococcus coralloides*" var. maximus Lièvre 1927 and "*Chondrococcus coralloides*" var. polycistus Jahn 1924 (= "*Myxococcus polycistus*" Kofler) appear to have been *Corallococcus coralloides*, with the distinguishing characteristic that the fruiting bodies are subdivided into parcels, which were called cysts even though an outer wall was lacking; such composite fruiting bodies are indeed seen

with *Corallococcus coralloides* strains. "*Chondrococcus simplex*" Singh and Singh 1971 was most likely *Corallococcus coralloides*; the published figures would fit the species. "*Chondrococcus microsporus*" Singh and Singh 1971 had exceedingly small myxospores, 0.4–0.6 μm in diameter; the figure of the fruiting bodies seems to show walled sporangioles. The organism has been found only once, and its status must remain open for the moment. "*Chondrococcus cerebriformis*" Jahn 1924 (= "*Myxococcus cerebriformis*" Kofler 1913) looks more like an *Archangium*-type organism, as does "*Chondrococcus sorediatus*" Singh and Singh 1971 with its long vegetative cells and large, variably shaped myxospores. The same may be the case with "*Chondrococcus megalosporus*" Jahn 1924. "*Chondrococcus cruentus*" Krzemieniewska and Krzemieniewski 1927 and Krzemieniewska and Krzemieniewski 1930 was regarded as identical with "*Myxococcus cruentus*" Thaxter 1897; this is almost certainly not correct. Unfortunately, the description is very scanty, and the figures are not good enough to decide which species this was. "*Chondrococcus columnaris*" Ordal and Rucker 1944 (= "*Bacillus columnaris*" Davis 1921/1922) is not a myxobacterium but a fish-pathogenic *Cytophaga*-like bacterium, presently classified as *Flexibacter columnaris*.

Genus III. **Pyxicoccus** gen. nov.

HANS REICHENBACH

Py.xi.coc' cus. Gr. fem. n. *pyxis* box, case, container; Gr. masc. n. *kokkos* berry; M.L. masc. n. *Pyxicoccus* boxed coccus.

Vegetative cells are slender rods with tapering ends, 0.7–0.8 × 3–8 µm. **Myxospores spherical, often somewhat deformed, optically refractile**, 1.4–1.8 µm in diameter (sometimes up to 2.4 µm). **Fruiting bodies consist of more or less spherical sporangioles with a distinct outer wall, clustered in sori of variable size**. In culture, degenerate fruiting bodies resembling those of *Corallococcus coralloides* may be produced. Swarm colonies as are typical for the family: soft and slimy, with snaking, undulating veins, without striking pigmentation. Of the proteolytic-bacteriolytic nutritional type.

Type species: **Pyxicoccus fallax** *sp. nov.*

FURTHER DESCRIPTIVE INFORMATION

As already discussed, this organism is not identical with *Myxococcus disciformis* Thaxter (= *Angiococcus disciformis* Jahn = *Cystobacter disciformis* McCurdy). The sporangioles of the genus *Pyxicoccus* are smaller, not disk-shaped, and do not stand on edge. Its classification in the family *Myxococcaceae* is clearly supported by 16S rRNA sequence comparisons (Spröer et al., 1999): strain An d1 clusters with *Myxococcus* species. The same study shows, however, that strains belonging to the *Cystobacteraceae* can easily be mistaken for *Pyxicoccus fallax*, as strains An d4 and An d6, which were originally classified as *Pyxicoccus*, cluster with *Archangium*. In principle, the two species should be easy to distinguish, as *Archangium* has long, needle-shaped vegetative cells and swarm colonies with a tough slime sheet and conspicuous radial veins. However, strain degeneration during cultivation may obscure such clear characteristics. It is important to classify a strain as early as possible during isolation. The fruiting bodies of *Cystobacter disciformis* and *Cystobacter minus* could be mistaken for *Pyxicoccus*, as could those of *Hyalangium minus* gen. nov., sp. nov. This organism produces small, brownish sporangioles arranged in chains or sheets. However, they often are empty, translucent, glassy sporangioles among the mature ones, and vegetative cells, myxospores, and swarm structure are different.

List of species of the genus Pyxicoccus

1. **Pyxicoccus fallax** *sp. nov.*
   fal' lax. L. masc. adj. *fallax* deceptive.

   Vegetative cells, myxospores, and swarm colonies are as described for the genus. The fruiting bodies (Fig. BXII.δ.66) consist of golden brown, somewhat elongated or spherical sporangioles with a distinct wall containing the brown pigment, mostly 30–45 µm in diameter (range: 20–60 µm). The sporangioles are arranged as a monolayer or slightly piled up in sori of variable size, often around 450 × 200–250 µm on agar plates. The diameter of the myxospores (Fig. BXII.δ.67) averages 1.8 µm but varies from 1.5–2.1 µm. Some strains have larger (2.1–2.4 µm) myxospores, but otherwise look exactly like *Pyxicoccus fallax*. They could represent a different species.

   Pyxicoccus fallax is found in the usual myxobacterial habitats but is only moderately common. Strain Py f1 was isolated in 1992 from soil with decaying plant material collected a few months earlier on the island of Crete, Greece. Reference strain Py f4 was isolated in 1995 from a similar sample collected in 1994 on Crete, and reference strain Py f5 (= An d1) was isolated in 1980 from soil collected from a city dump near St. Thomé Madras, India.

   The mol% G + C of the DNA is: not determined.

   Type strain: Py f1, DSM 14698, JCM 12639.

   Additional Remarks: Reference strains include DSM 14689 (Py f4), DSM 14699 (Py f5 = An d1).

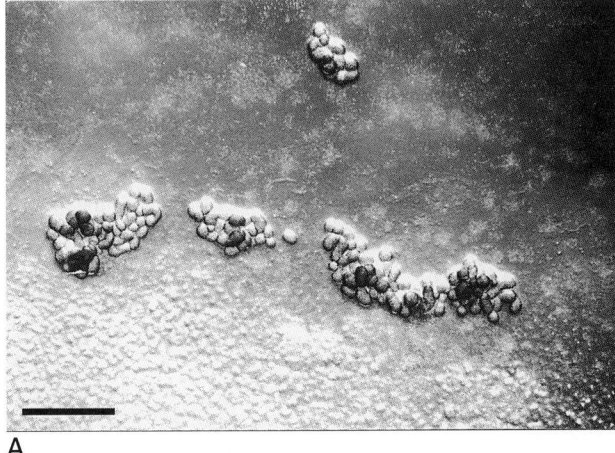

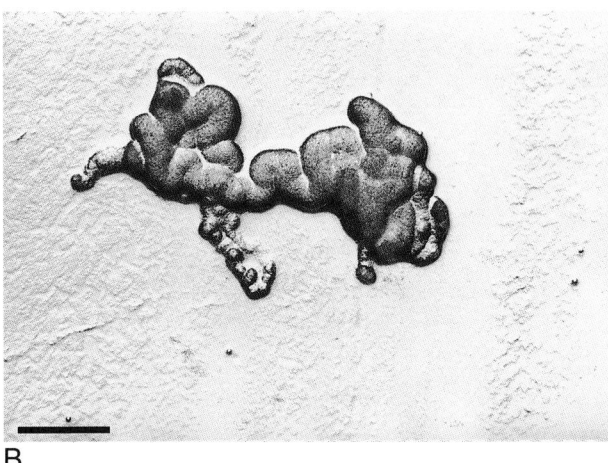

A   **B**

FIGURE BXII.δ.66. *Pyxicoccus fallax*: fruiting bodies on agar plates. *(A)* The fruiting bodies consist of sporangioles in piles or sori. In *B* an incompletely differentiated specimen is shown. The fruiting bodies in *D* and *E* are squeezed (see next page), in *E* more so than in *D*, so that the sporangiole walls can be seen more clearly; both in phase contrast. Bars = 320 µm *(A)*, 130 µm *(B,C)*, 80 µm *(D)*, 40 µm *(E)*.

(continued)

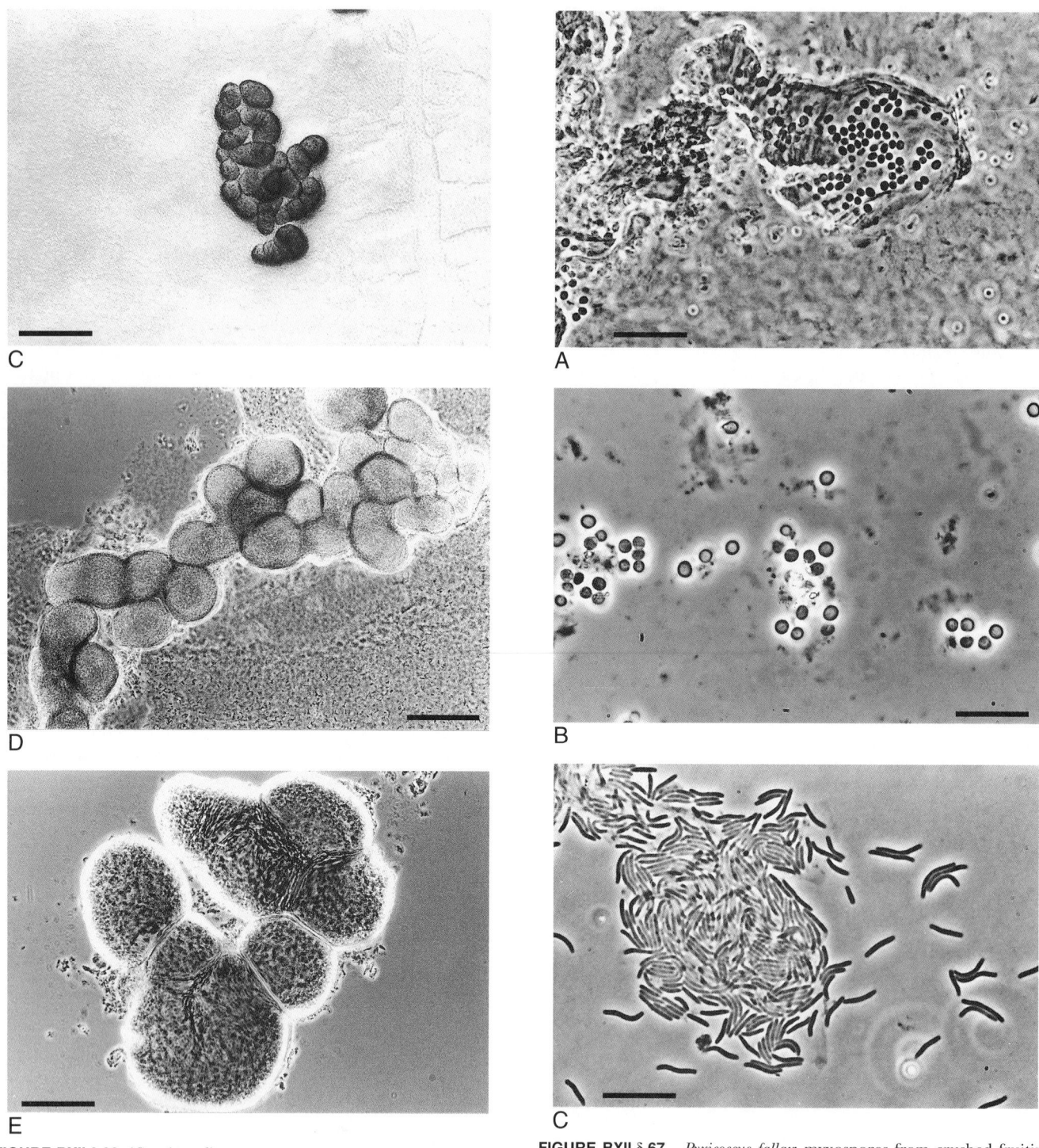

FIGURE BXII.δ.66 (Continued)
(for this figure legend see previous page)

FIGURE BXII.δ.67. *Pyxicoccus fallax*: myxospores from crushed fruiting bodies, and vegetative cells, in phase contrast. In *A*, parts of the sporangiole wall can be seen. *(C)* The vegetative cells of the organism are like those of *Corallococcus* and clearly different from those of *Cystobacter*. Bar = 16 μm *(A)*, 10 μm *(B)*, 13 μm *(C)*.

Family II. **Cystobacteraceae** McCurdy 1970, 286[AL]

HANS REICHENBACH

Cys.to.bac.ter.a' ce.ae. M.L. masc. n. *Cystobacter* type genus of the family; *-aceae* ending to denote a family; M.L. fem. pl. n. *Cystobacteraceae* the *Cystobacter* family.

Vegetative cells are slender, flexible rods with tapering ends, often long, needle shaped. **The fruiting bodies are either naked masses of slime and myxospores, of cartilaginous consistency,** and of very variable size and irregular shape, or, in most cases, **are made up of pale to deep brown sporangioles** in various arrangements, with or without a slime stalk or pedicle. **Myxospores are short, optically refractile rods to spheroids**, but in the latter case always slightly deformed and accompanied by short rods, **with rounded or tapering ends**.

The swarm colonies produce a firm slime layer with more or less conspicuous, straight, often branched radial veins. The slime sheet may become extremely tough, so that it can hardly be broken with the inoculation loop. Congo red is quickly adsorbed and gives the swarm a purple red color.

Of the proteolytic–bacteriolytic nutritional type. All species utilize mono-, di-, and polysaccharides; some vigorously degrade chitin.

Type genus: **Cystobacter** Schroeter 1886, 170.

Key to the genera of the family Cystobacteraceae

I. Fruiting bodies are naked masses of myxospores embedded in slime of cartilaginous consistency, more or less cushion-like, of very variable shape and size. Vegetative cells long, slender needles. Myxospores short, fat rods with rounded ends.

  Genus *Archangium*

II. Fruiting bodies composed of sporangioles with a distinct wall.

  A. Sporangioles sessile.

  1. Sporangioles relatively large, in more or less extended aggregates, often piled up and sometimes surrounded by a translucent slime layer.

  Genus *Cystobacter*

  2. Sporangioles rather small, often glassy, translucent and empty, in chains and sheets.

  Genus *Hyalangium*

  B. Sporangioles borne on (white) slime stalks.

  1. Sporangioles relatively small, spherical, bean- or calotte-shaped, pale brown, solitary on short, delicate stalks, do not degrade chitin.

  Genus *Melittangium*

  2. Sporangioles moderately large, spherical to club shaped, dark red brown, solitary or in clusters on conspicuous stalks. Vegetative cells of moderate length, slender, boat-shaped. Myxospores short, often slightly curved, fat rods with rounded or slightly tapering ends. Degrade chitin very efficiently.

  Genus *Stigmatella*

TAXONOMIC COMMENTS

While it is usually easy to place an isolate in this family, determination of its genus and species is frequently not as unequivocal as it may seem from the key above, because the organisms may not produce typical fruiting bodies. Rather, uncharacteristic structures are seen that are practically indistinguishable from *Archangium* fruiting bodies, i.e., structures that are very variable in shape and size, that have ridges or cushion-like masses, that sometimes pile up and form finger-like projections, and that are often very large due to fusion of adjacent fruiting bodies. For this reason, the family *Archangiaceae* has been omitted in this edition of the *Manual*: the borderline between *Archangium* and the genera with sporangioles is markedly fluid. One may even doubt whether a separate genus *Archangium* is justified. In problematic cases one has to rely on additional characteristics such as the shape and size of vegetative cells and myxospores (both of which appear to be relatively stable features), the swarm structure; or degradation of chitin. But the differences between genera and species with respect to these characteristics are often minimal, and their usefulness is limited by the degree of experience of the investigator. A further source of confusion lies in the fact that apparently not all species and even genera of the family have been isolated and studied, so that organisms that cannot readily be assigned to an existing genus may represent new taxa.

The nomenclature of the genera of the family is somewhat muddled, mainly because the taxonomic relevance of the shape of the vegetative cells was not understood until the 1960s. The type genus of the family, *Cystobacter*, was described by Schroeter (1886) with the two species *Cystobacter fuscus* and *"Cystobacter erectus"* (now *Stigmatella erecta*), although Schroeter did not recognize the peculiar life cycle of the organisms. In 1897 Thaxter described *Cystobacter fuscus* again, now as a myxobacterium. He acknowledged Schroeter's priority and used his name for the organism. However, in 1904, after having discovered the very old description of *Polyangium vitellinum* Link 1809, he renamed it *"Polyangium fuscum"*, being misled by similarities in fruiting body shape and structure. In the following years a series of further species were added to that genus. As is recognized today, they are actually quite different organisms and are separated in two suborders. This sequence of events has to be kept in mind when trying to understand the older literature. Only in 1970 were the real connections finally recognized, and the genus *Cystobacter* was reestablished (McCurdy, 1970).

The genus *Archangium* was established by Jahn (1924) to comprise several previous and two new species with fruiting bodies consisting of convoluted, braided strings of rod-shaped myxospores in hardened slime, without an outer wall. The genus is problematic, as will be discussed below.

The genus *Stigmatella* has a tortuous history as well. *Stigmatella aurantiaca* was first described by Berkeley and Curtis (in Berkeley, 1857) as a mold and was rediscovered by Thaxter in 1892, who named it "*Chondromyces aurantiacus*", again misled by a superficial similarity of fruiting body structure between *Chondromyces crocatus* and *Stigmatella aurantiaca*. In fact, the organism had in the meantime been described two more times as a mold, *Stilbum rhytidospora* Berkeley and Broome 1873 and "*Polycephalum aurantiacum*" Kalchbrenner and Cooke 1880. In 1969, workers in the field realized that the organism is not directly related to *Chondromyces*, and the genus *Stigmatella* was restored (McCurdy and Khouw, 1969; Reichenbach and Dworkin, 1969). The second of Schroe-ter's species, "*Cystobacter erectus*", was also found by Thaxter, who described it in 1897 as "*Chondromyces erectus*". Jahn (1924) created a new genus, *Podangium*, for it, because he felt that fruiting bodies with one single sporangiole on a stalk were sufficiently different from *Chondromyces* to justify that separation.

The genus *Melittangium* also has a complicated history. The first species was described by Thaxter in 1892 as "*Chondromyces lichenicolus*" (etymologically correct *lichenicola*). Jahn (1924) transferred it to the genus *Podangium*; at the same time he created the genus *Melittangium* for his new species, *Melittangium boletus*. Later the species "*C. lichenicola*" was transferred to the genus *Melittangium* (McCurdy, 1971b).

The new genus *Hyalangium* will be discussed below.

Genus I. Cystobacter Schroeter 1886, 170^AL

HANS REICHENBACH

Cys.to.bac'ter. Gr. fem. n. *kystis* bladder; M.L. masc. n. *bacter* from Gr. fem. n. *bakteria* rod, stick; M.L. masc. n. *Cystobacter* bladder-forming rod.

The history of the genus *Cystobacter* and the problems associated with its recognition are discussed above in the introduction to the family. The fruiting body shape and structure of *Cystobacter* is similar to fruiting bodies produced by *Polyangium*, *Sorangium*, *Byssophaga*, *Jahnia*, *Pyxicoccus*, and to a certain extent, *Stigmatella hybrida*. The genus as it is presented here still is heterogeneous and probably comprises at least three different genera. One group of species, with *Cystobacter fuscus* as the main representative, has slender, often slightly C- or S-shaped bent, rod-shaped myxospores with more or less tapering ends. Many strains of this group are efficient chitin degraders. The second group, represented by *Cystobacter disciformis*, has myxospores that are stout, short, fat rods, very much like myxospores of *Archangium gephyra*. These organisms rarely attack chitin. A third group, centered around *Cystobacter gracilis*, has delicate vegetative cells with tapering ends, and small myxospores that are straight or slightly bent rods with rounded ends. Analysis of the available 16S rRNA sequences separates at least the first two groups rather neatly (Spröer et al., 1999).

Type species: **Cystobacter fuscus** Schroeter 1886, 170.

Key to the species of the genus Cystobacter

I. Sporangioles brown; myxospores relatively slender rods of moderate length and with more or less tapering ends.

    A. Sporangioles large, spherical, chestnut brown, in sori with a thick, hyaline, and colorless slime capsule, glistening; efficient chitin degraders.

Cystobacter fuscus

    B. Sporangioles moderately large to large, of clearly different size within the sorus, spherical to elongated, dark brown and dull, in convoluted chains appressed to the surface, but ends often pointing upward, without a conspicuous slime matrix; efficient chitin degraders.

Cystobacter ferrugineus

    C. Sporangioles moderately large to large, spherical to elongated, in long chains that sometimes form three-dimensional networks, but also may be tightly packed and appressed to the surface, solitary sporangioles, usually on a short slime pedicle or stalk, always dark chestnut brown to almost black, dull, sorus without a slime matrix; efficient chitin degraders.

Cystobacter badius

    D. Sporangioles large, yellow-brown to red-brown, spherical, often in chains that may form a three-dimensional network, covered by a delicately plicated, thin slime sheet that gives many of the sporangioles a striking lengthwise striation; may or may not attack chitin.

Cystobacter velatus

II. Sporangioles brown; myxospores very short fat rods to almost spherical; most strains do not attack chitin.

    A. Sporangioles moderately large, in crude cultures disk-shaped, the disks often standing on edge, in pure cultures in sori appressed to the substrate; myxospores more or less spherical.

Cystobacter disciformis

    B. Sporangioles moderately large, spherical, in sori without a slime matrix; myxospores short, fat rods.

Cystobacter minus

   C. Sporangioles large, dark brown, dull, in large sori without a slime matrix; myxospores short rods with rounded ends, often slightly bent; some strains produce a dark violet pigment.
                              Cystobacter violaceus

   D. Sporangioles moderately large, in sori without a slime matrix; vegetative cells delicate, needle-shaped; myxospores small, short, slender rods with rounded ends; some strains degrade chitin.
                              Cystobacter gracilis

III. Sporangioles apricot colored or red.

   A. Sporangioles apricot colored, moderately large to large, in sori without a slime matrix; does not degrade chitin.
                              Cystobacter armeniaca

   B. Sporangioles cinnabar red, large, spherical to slightly elongated, in large sori that are often erect, board-like, with a thin, glassy slime envelope, efficient chitin degrader.
                              Cystobacter miniatus

List of species of the genus Cystobacter

1. **Cystobacter fuscus** Schroeter 1886, 170[AL]
   fus' cus. L. masc. adj. *fuscus* dark brown.

   Vegetative cells are long, slender rods with tapering ends, needle-shaped, i.e., with parallel sides and shortly pointed ends, 0.8 × 5–15 μm (Fig. BXII.δ.68). Sporangioles spherical or ellipsoidal, bright chestnut brown, glistening, 50–180 μm in diameter, in groups (sori) of up to 100 in a common gelatinous, hyaline, colorless slime matrix like a capsule (Fig. BXII.δ.69). Myxospores rather slender rods with pointed ends, often slightly C- or S-shaped, optically refractile, 0.8–1.5 × 2.5–5 μm. Swarm colonies produce a tough slime sheet with veins (Fig. BXII.δ.70), tenacious and difficult to break, often with extended fields of oscillating waves, on peptone agar producing a dark red-brown, diffusing pigment. Grows well on the standard media for myxobacteria. Relatively simple defined media have been found for the organism (Reichenbach and Dworkin, 1992).

   Cystobacter fuscus is found in soil and on dung of herbivorous animals; it is moderately common.

   The mol% G + C of the DNA is: 68 (T_m).

   Type strain: M29, ATCC 25194, DSM 2262.

   GenBank accession number (16S rRNA): M94726.

   Additional Remarks: Reference strains include ATCC 25195, ATCC 25529, ATCC 29623.

   The species is relatively easy to recognize. On yeast agar (VY/2-), it produces copious amounts of typical fruiting bodies for some time after isolation. Later, however, the usual degenerate ridges and padded masses arise. The glistening, chestnut brown fruiting bodies with a hyaline slime matrix, the slender myxospores, and efficient chitin degradation are the hallmarks of the species.

   Thaxter caused some confusion when he first renamed his *"Myxobacter aureus" "Cystobacter aureus"* (Thaxter 1897, 408) and then, in 1904, equated it with both *Polyangium*

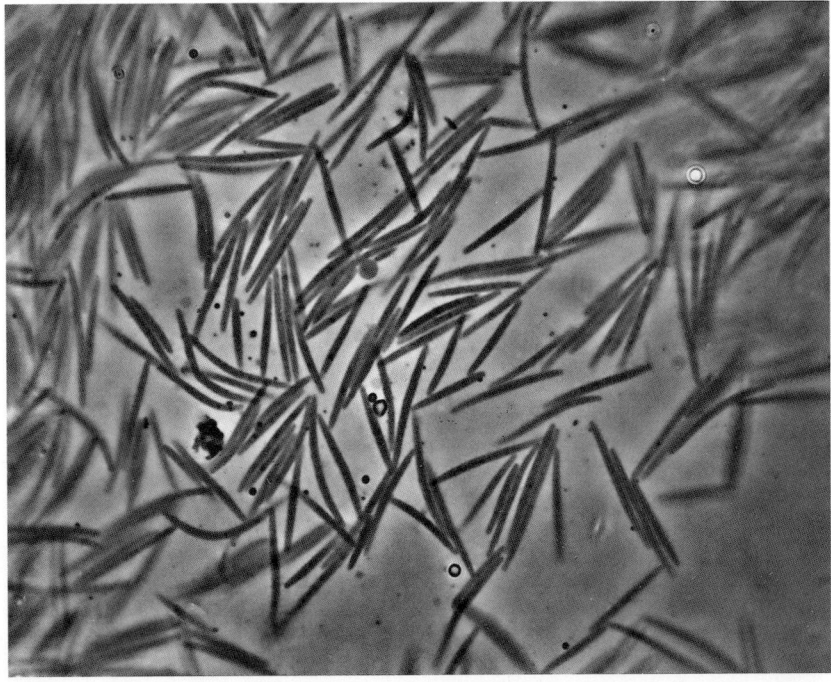

FIGURE BXII.δ.68. *Cystobacter fuscus* vegetative cells. Light micrograph (×1500).

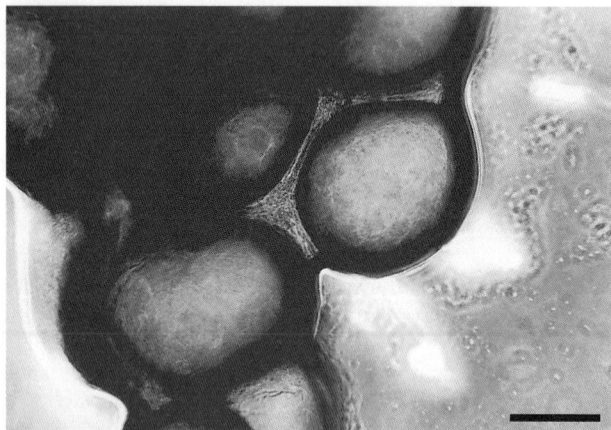

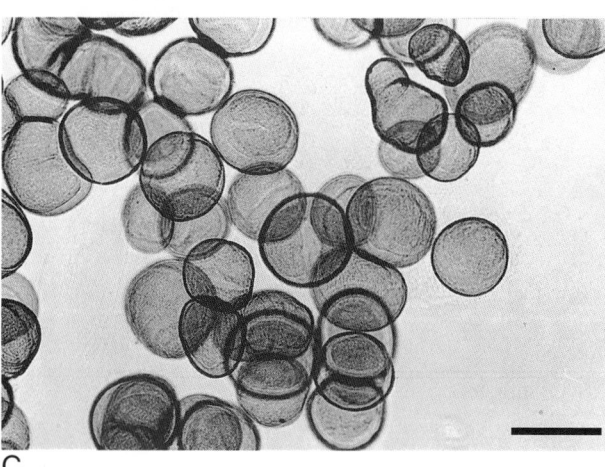

FIGURE BXII.δ.69. *Cystobacter fuscus*: fruiting bodies: (*A* and *B*) on agar surface, (*C*) in embedding medium. The spherical shape, and the rather uniform size of the shining sporangioles and the slime matrix around them are typical. Bar in *A* = 155 μm; bar in *B* = 50 μm; bar in *C* = 100 μm.

vitellinum (Thaxter, 1904), which is most likely correct, and with *Cystobacter fuscus*, which was obviously an error.

2. **Cystobacter armeniaca** *sp. nov.*

ar.me.nia′ ca. Gr. fem. n. *melea armeniaka* apricot tree; M.L. fem. n. *armeniaca* for apricot-colored.

Vegetative cells are rods with tapering ends, of moderate

length, 0.8 × 4–8 μm. Swarm colonies with a tough slime sheet and fine radial veins. Fruiting bodies consist of spherical, intensely apricot-colored, orange sporangioles, 60–150 μm in diameter, in sori of variable size, often piled up. Tend to degenerate into ridges and irregular masses of slime and myxospores. Myxospores moderately long and fat rods with rounded or slightly tapered ends, 1–1.2 × 2–4 μm, often C-or S-shaped.

Of the proteolytic–bacteriolytic nutritional type. Individual strains may degrade chitin.

Cystobacter armeniaca somewhat resembles *Cystobacter velatus*, but does not produce looping chains of striated sporangioles and has strikingly apricot-colored fruiting bodies.

Cystobacter armeniaca is found in the usual myxobacterial habitats; it is moderately common. Strain Cb a1 was isolated in 1997 from soil with decaying plant material collected in the Sonoran savanna west of Tucson, AZ, USA. Reference strain Cb a3 comes from a similar sample from the same region.

The mol% G + C of the DNA is: not determined.

Type strain: Cb a1, DSM 14710, JCM 12622.

Additional Remarks: Reference strains include DSM 14711 (Cb a3).

3. **Cystobacter badius** *sp. nov.*

ba′ di.us. L. masc. adj. *badius* chestnut brown.

Vegetative cells are long, slender rods with pointed ends, somewhat shorter than those of *Cystobacter fuscus*, 0.7–0.9 × 6–10 μm. The myxospores, too, are stouter and rod shaped with tapering or blunt ends, often slightly C- or S-shaped, 1.0–1.5 × 3–4 μm. The slime layer of the swarm colony is less tenacious and more frangible. The main distinguishing characteristic is the morphology of the fruiting bodies (Fig. BXII.δ.71). Sporangioles are spherical to elongated, of variable size, 30–180 μm in diameter; in chains that often loop upward but occasionally form three-dimensional networks; ends of sporangioles often point upward; surface of sori therefore rather turbulent; very dark brown to almost black; dull on the plate but glistening at higher magnification; only a trace of a slime matrix around the fruiting bodies, sometimes producing a fine striation; sporangioles often also solitary and then usually on a wrinkled slime cushion or pedicle. Due to the brittle slime sheet of the swarm, the fruiting bodies can easily be removed from the plate.

Growth pattern and nutritional requirements as for *C. fuscus* and *C. ferrugineus*.

Cystobacter badius differs from *Cystobacter ferrugineus* in the looping chains of sporangioles, the single sporangioles on slime cushions, and the stouter vegetative cells and myxospores; *Cystobacter badius* differs from *Cystobacter velatus* in the much darker color of the fruiting bodies, the absence or rare occurrence of three-dimensional networks of sporangioles, and the lack of a delicately plicated slime cover on the sporangioles, also in the occurrence of individual sporangioles on slime cushions. The two species are sometimes very similar. It seems, however, that, in contrast to *Cystobacter badius*, *Cystobacter velatus* normally does not degrade chitin. *Cystobacter badius* differs from *Stigmatella hybrida* in the lack of sporangioles with real stalks, in the production of large sori, in the shape of vegetative cells and myxospores, and in swarm morphology.

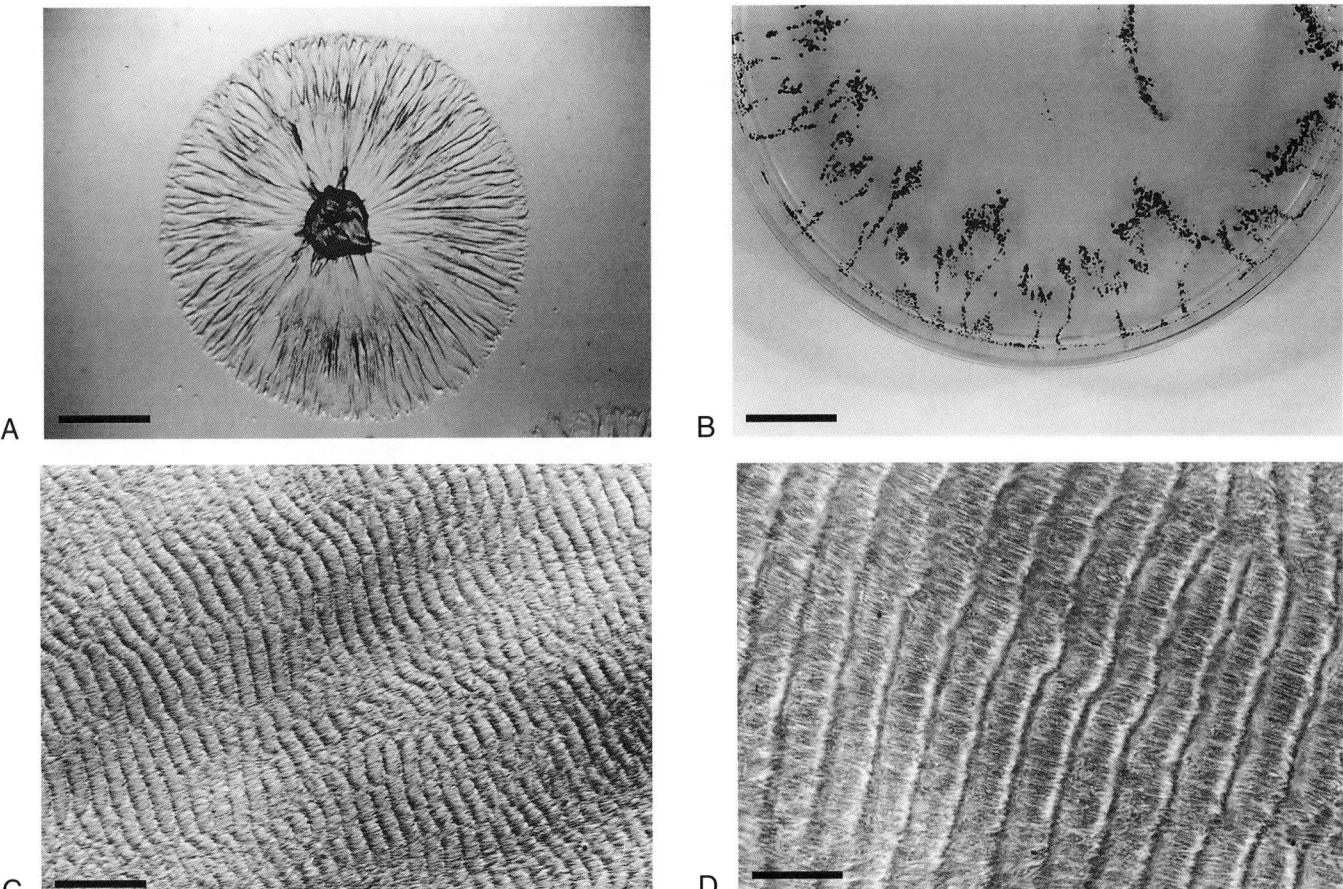

FIGURE BXII.δ.70. *Cystobacter fuscus*: swarm colonies on agar plates. (*A*) Young swarm with straight radial veins as is typical for the family. (*B*) Large swarm with many fruiting bodies. (*C*) The swarms often show extended fields of oscillating waves composed of rows of strictly aligned vegetative cells (*D*). Bar in *A* = 3 mm; bar in *B* = 13.5 mm; bar in *C* = 50 μm; bar in *D* = 20 μm.

Cystobacter badius may be obtained from soil and dung of herbivorous animals; it is moderately common, especially in habitats with a warm, semi-arid climate.

Strain Cb b2 was isolated in 1997 from a soil sample collected in the saguaro savanna west of Tucson, AZ, USA. Reference strain Sg a10 was isolated in 1998 from a similar sample from the island of Malta, and reference strain Cb b11 comes from a soil sample collected in the Dominican Republic and was isolated in 1998.

The mol% G + C of the DNA is: not determined.

Type strain: Cb b2, DSM 14723, JCM 12623.

Additional Remarks: Reference strains include DSM 14738 (Sg h10), DSM 14715 (Cb b11).

4. **Cystobacter disciformis** *comb. nov.* (Thaxter 1904) Brockman and McCurdy *in* McCurdy 1989, 2152 (*Myxococcus disciformis* Thaxter 1904, 89; *Angiococcus disciformis* (ex Jahn 1924, 89) Hook, Larkin and Brockman 1980, 142.)

dis.ci.for′ mis. Gr. masc. n. *diskos* disk; L. fem. n. *forma* form, shape; M.L. masc. adj. *disciformis* disk-shaped.

Vegetative cells are slender rods, needle-shaped, 0.7 × 5–10 μm. Fruiting bodies consist of spheroidal, disk-shaped sporangioles, 30–50 μm in diameter and 15 μm thick, in crude cultures often standing on edge, interconnected, forming heaped groups or a three-dimensional network, occasionally on a slime pedicle. In pure culture, the sporangioles tend to lie flat on the surface and are grouped in sori. At first yellowish, then later orange brown. Myxospores ellipsoidal or irregularly spherical, 1–1.5 μm in diameter.

Swarm colonies with a tough slime sheet and radial veins. Of the proteolytic–bacteriolytic nutritional type.

Originally obtained from muskrat and deer dung; appears to be common in slightly acid forest soils and in the fossa region of alkaline peat bogs in Michigan, USA.

The mol% G + C of the DNA is: 68 (Bd).

Deposited strain: ATCC 33172.

The species was placed in the genus *Cystobacter* by McCurdy (1989) in the last edition of this *Manual*. The problems associated with this species, the genus *Angiococcus*, and their relationships to the new genus *Pyxicoccus* are discussed in the Comments on the family *Myxococcaceae*. *Cystobacter disciformis* is clearly different from the above-described group of *Cystobacter* species and perhaps belongs in a genus of its own, which should be named *Angiococcus*. *Cystobacter disciformis* is similar to *Cystobacter minus*, but clearly differs from it in its disk-shaped sporangioles and small myxospores.

Three more species of *Angiococcus* have been described by Singh and Singh (1971), but their status is not clear. Most likely, the organisms are *Cystobacter* species. Type strains are not available. "*Angiococcus brunneus*" has large (90–185 μm), dark brown sporangioles, single or in heaps,

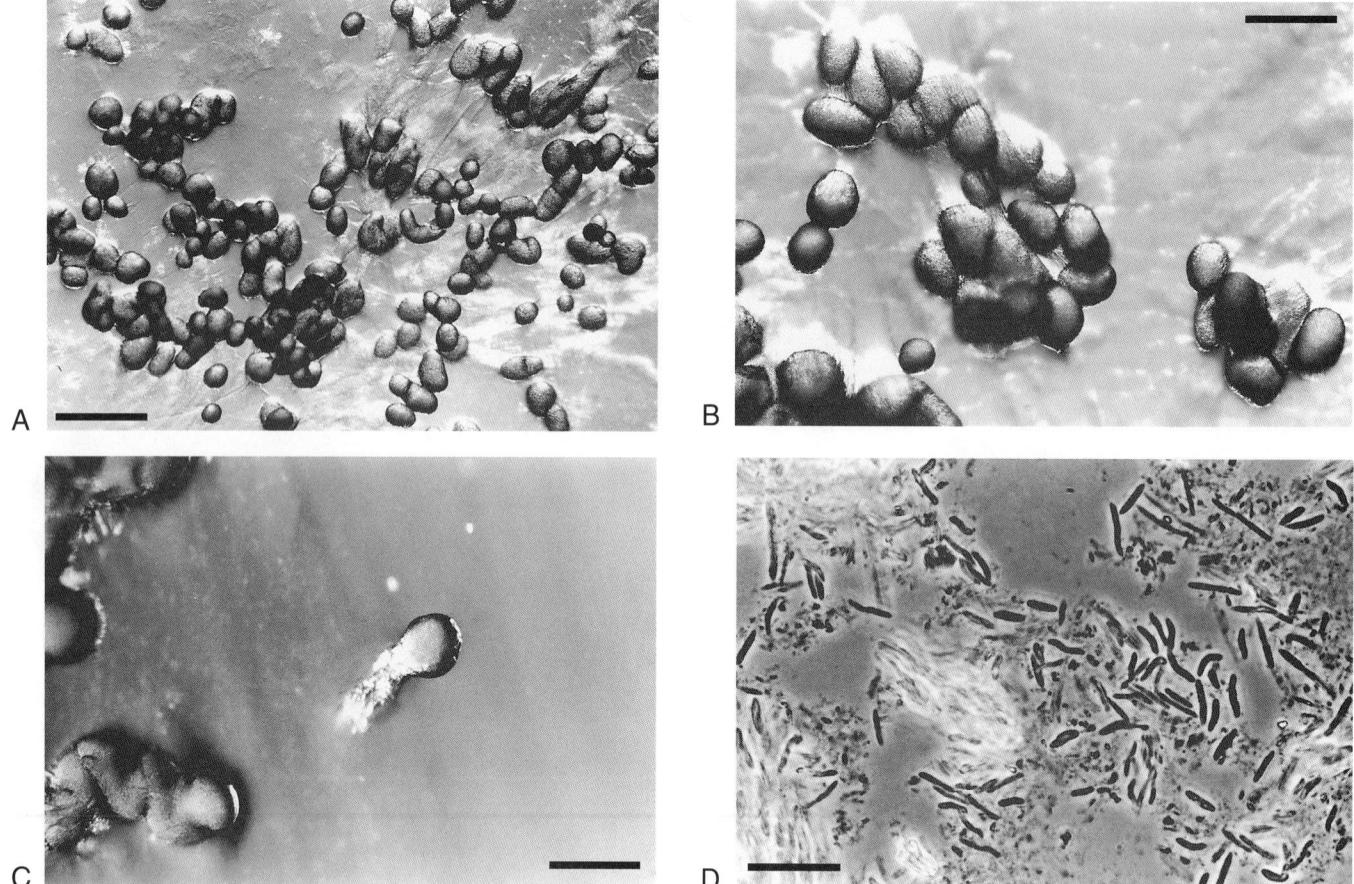

FIGURE BXII.δ.71. *Cystobacter badius*: fruiting bodies on agar plates, and myxospores (*D*) from crushed sporangioles, the latter in phase contrast. The sporangioles are typically either single, often on short slime pedicles (*C*), or in convoluted chains. Bar in *A* = 260 μm; bar in *B* = 125 μm; bar in *C* = 155 μm; bar in *D* = 13 μm.

and large, ellipsoidal to spherical myxospores. *"Angiococcus aureum"* (sic) produces fruiting bodies with very large (50–90 × 140–230 μm), ellipsoidal to elongated, golden colored sporangioles in loose groups. The myxospores are irregularly spherical to ellipsoidal and very large. The sporangioles of *"Angiococcus thaxteri"* are much smaller (30–60 μm) and are densely packed; the myxospores are spherical and medium sized. This species could actually fit *Cystobacter disciformis*. All three species have long, slender vegetative cells with tapering ends; those of *"Angiococcus aureum"* are very fat, measuring 1.0–1.4 μm in width. *"Angiococcus cellulosum"*, described by Mishustin (1938), turned out to be a fungus, *Rhizophlyctis rosea* (Singh and Singh, 1971), while tentatively classified *"Angiococcus moliroseus"* was an actinomycete, *Streptosporangium roseum* (Peterson and McDonald, 1966).

5. **Cystobacter ferrugineus** (Krzemieniewska and Krzemieniewski 1927) McCurdy 1970, 288[AL] (*"Polyangium ferrugineum"* Krzemieniewska and Krzemieniewski 1927, 89.)
fer.ru.gi′ ne.us. L. masc. adj. *ferrugineus* of the color of iron rust.

Vegetative cells, myxospores, swarm morphology, and growth characteristics as for *Cystobacter fuscus*. The species is distinguished by its fruiting bodies (Fig. BXII.δ.72). The sporangioles are moderately large to large, spherical and often elongated, of variable size even within the same fruit-

ing body, dark chestnut brown, and, in contrast to *Cystobacter fuscus*, dull. They are produced preferentially in long, convoluted chains, the ends of which often are pointing upward. The chains of sporangioles are usually tightly packed, somewhat piled up, and appressed to the substrate, without a prominent slime matrix. The diameter of the sporangioles is 30–180 μm. The usual degenerate fruiting bodies are formed after a strain has been cultivated for some time. The species closely resembles *Cystobacter badius*. It differs in the lighter color of the fruiting bodies and in the lack of solitary sporangioles on slime pedicles.

The organism is found in soil and on dung of herbivores, and is moderately common. Strain Cb fe18 was isolated in 1983 from a soil sample collected on the Chalkidike Peninsula, Greece; the reference strain, Cb fe37, comes from a soil sample with decaying plant material collected in 1996 near Braunschweig, Germany..

The mol% G + C of the DNA is: not determined.
Type strain: Cb fe18, DSM 14716, JCM 12624.
Additional Remarks: Reference strains include DSM 14717 (Cb fe37).

6. **Cystobacter gracilis** *sp. nov.*
gra′ ci.lis. L. adj. *gracilis* slim, slight.

Vegetative cells (Fig. BXII.δ.73) are rather short, delicate rods with tapering ends, 0.5–0.6 × 3–5 μm. Swarm colonies

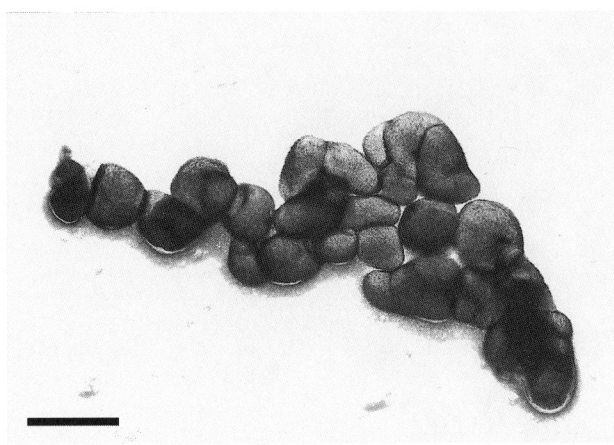

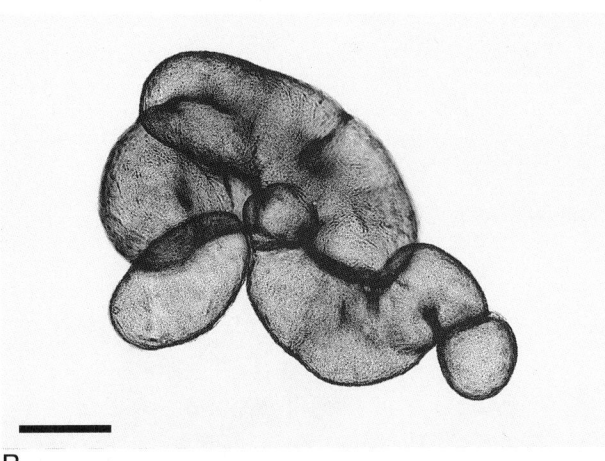

FIGURE BXII.δ.72. *Cystobacter ferrugineus*: fruiting bodies: (*A*) on agar surface, (*B*) in embedding medium. Dull, irregular sporangioles of very different size are typical. Bar in *A* = 125 μm; bar in *B* = 55 μm.

with a thin, tough, locally delicate plicated slime sheet and fine radial veins, more or less red-brown and even violet, sometimes intensely so. On peptone (CY-) agar the swarms spread only reluctantly and produce an intense, red-brown, diffusing pigment. Fruiting bodies (Fig. BXII.δ.74) orange-brown to dark red-brown; sporangioles more or less spherical, 30–80 μm in diameter. However, in cultures, degenerate fruiting bodies in the shape of short to moderately long, thin to somewhat broadened ridges and cushion-like, bulging masses are the rule. Myxospores are short, relatively slender, sometimes comma-shaped rods with rounded ends to almost spherical, small, 0.8–1.4 × 1–3 μm, optically refractile.

Of the proteolytic–bacteriolytic nutritional type. Does not attack chitin and cellulose.

The outstanding characteristics of this species are the short, delicate vegetative cells and rather small myxospores. Also typical is the thin but very tough slime sheet with very fine radial veins. *Cystobacter gracilis* probably represents a genus of its own in a complex of strains that may contain even more species.

Cystobacter gracilis is found in the usual myxobacterial habitats; it is moderately common. Strain Cb g1 was isolated in 1996 from a soil sample with decaying plant material

collected in the same year near Beni-Suef, Egypt, by Dr. Mohamed Khalil. The origins of the reference strains are as follows: Cb g2 was isolated in 1996 from a similar sample collected in 1996 near Giza, Egypt; Cb g13 was isolated in 1997 from soil with rabbit dung collected in 1996 near Benson, Arizona, USA; Cb g30 was isolated in 1998 from rotting wood collected in 1998 in the Alsace, France.

The mol% G + C of the DNA is: not determined.

Type strain: Cb g1, DSM 14753, JCM 12625.

Additional Remarks: Reference strains include DSM 14771 (Cb g2), DSM 14754 (Cb g13), DSM 14755 (Cb g30).

7. **Cystobacter miniatus** *sp. nov.*
mi.ni.a′ tus. L. masc. adj. *miniatus* cinnabar-red.

Vegetative cells are long, slender, needle-shaped rods, 0.8– 0.9 × 6–10 μm. Swarm colonies with fine radial veins in a brittle slime sheet. Sporangioles spherical to elongated, intensely cinnabar-red, 80–180 μm in diameter, densely packed in large sori with a common, glassy slime envelope, often as a slab strikingly board-like and erect (Fig. BXII.δ.75). Degenerate fruiting bodies in the form of jaggedly branched ridges, bands, knobs, and plaques; also intensely red Fig. BXII.δ.76. Myxospores relatively slender rods with tapering ends, 0.9–1.2 × 2–4 μm, optically refractile.

Of the proteolytic–bacteriolytic nutritional type. Does not attack cellulose but degrades chitin very efficiently.

Cystobacter miniatus is unique and easy to recognize by its cinnabar-red fruiting bodies in large, dense sori that stand in a board-like, upright position. Further, it is a strong chitin degrader, which distinguishes it from *Cystobacter armeniaca*. Its closest relatives are probably among the *Cystobacter fuscus* complex.

Cystobacter miniatus may be found in the usual myxobacterial habitats, but appears to be a rare species. Strain Cb a24 was isolated in 1999 from a sample of rotting sawdust and other decaying plant material collected on the island of Mallorca, Spain, reference strain Cb mi3 was isolated from soil with decaying plant material collected in 1998 in Botswana, South Africa.

The mol% G + C of the DNA is: not determined.

Type strain: Cb a24, DSM 14712, JCM 12626.

Additional Remarks: Reference strains include DSM 14756 (Cb mi3).

8. **Cystobacter minus** (Krzemieniewska and Krzemieniewski 1926) McCurdy 1970, 288[AL] (*Polyangium minus* Krzemieniewska and Krzemieniewski 1926, 33.)*
mi′ nus. L. neut. adj. *minus* smaller, less, comparative of *parvus* small.

Vegetative cells are slender, needle-shaped rods with tapered ends, 0.7 × 3–11 μm. Swarm colonies with fine radial veins in a tough slime sheet. Sporangioles spherical or ellipsoidal, relatively small, 20–70 × 20–50 μm, dull grayish to dark yellow-brown and red-brown, with a definite wall, 0.5–1 μm thick, arranged in flat sori up to 500 μm wide within a thin colorless slime layer, often also piled up, sometimes forming branched tree-like structures; when tightly

---

Editorial Note: The neuter form, *minus*, which was correct with *Polyangium*, has been used by mistake for *Cystobacter*, which is a masculine noun; however, according to the Bacteriological Code, the name cannot be corrected.

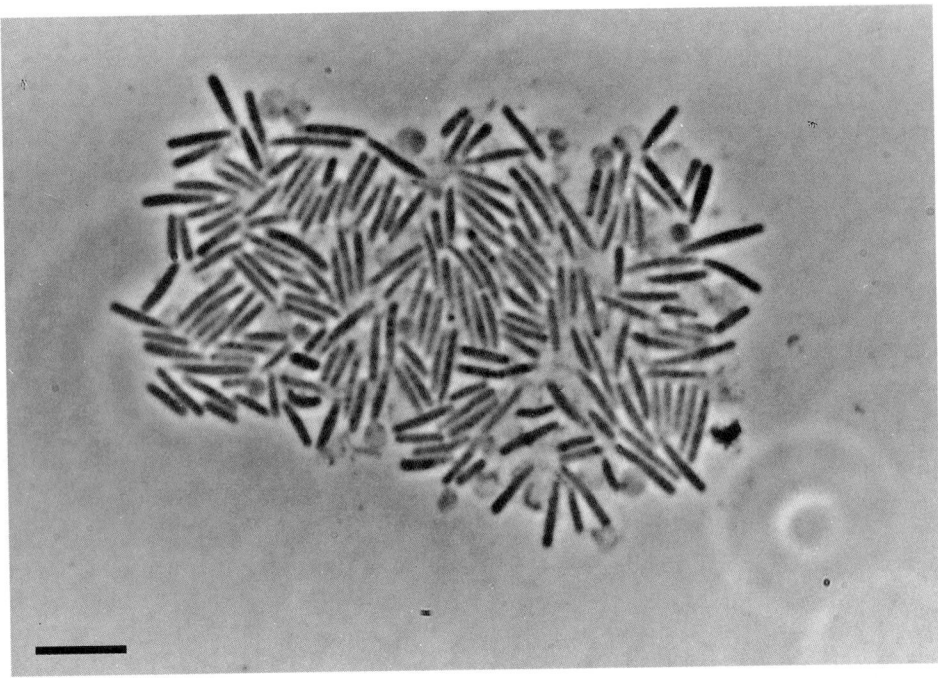

FIGURE BXII.δ.73. *Cystobacter gracilis*: vegetative cells, in phase contrast. Bar = 10 μm.

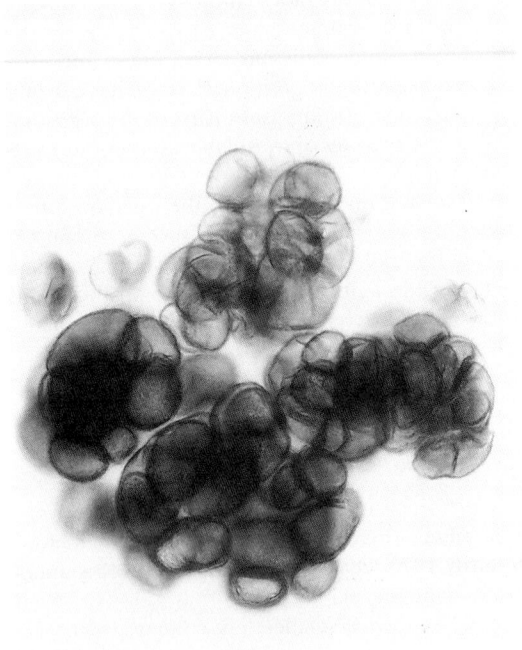

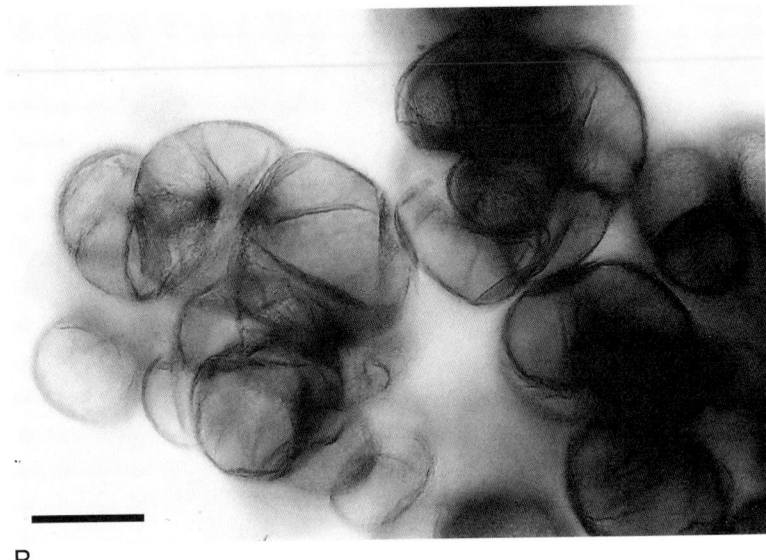

B

A

FIGURE BXII.δ.74. *Cystobacter gracilis*: fruiting bodies on agar surface under water and cover glass, in phase contrast. Bar in *A* = 85 μm; bar in *B* = 40 μm.

packed, the sporangioles may become polyhedral (Fig. BXII.δ.77). Myxospores stout, short rods with blunt rounded ends to almost spherical; optically refractile; 0.8–1.4 × 1.3–2.7 μm.

Of the proteolytic–bacteriolytic nutritional type. Some strains degrade chitin.

First isolated in Poland from sterilized rabbit dung placed on soil. Found in the usual myxobacterial habitats, moderately common. Strain Cb m2 was isolated in 1983 from soil collected in the same year in Arizona, USA, reference strain Cb m6 was isolated in 1990 from soil with decaying plant material collected near Jaipur, Rajasthan,

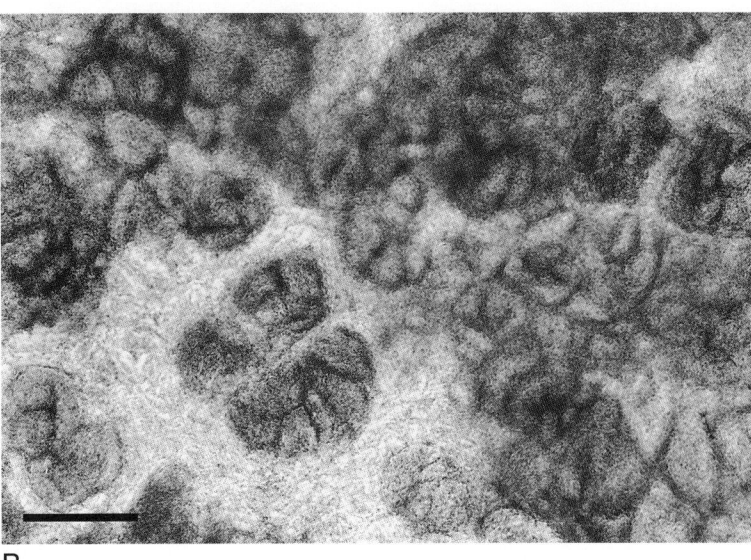

FIGURE BXII.δ.75. *Cystobacter miniatus*: fruiting bodies. (*A*) On agar surfaces the sporangioles often pile up in a peculiar slab-like fashion. (*B*) When squeezed in a slide mount, the sporangioles become distinguishable; in phase contrast. Bar in *A* = 260 µm; bar in *B* = 125 µm.

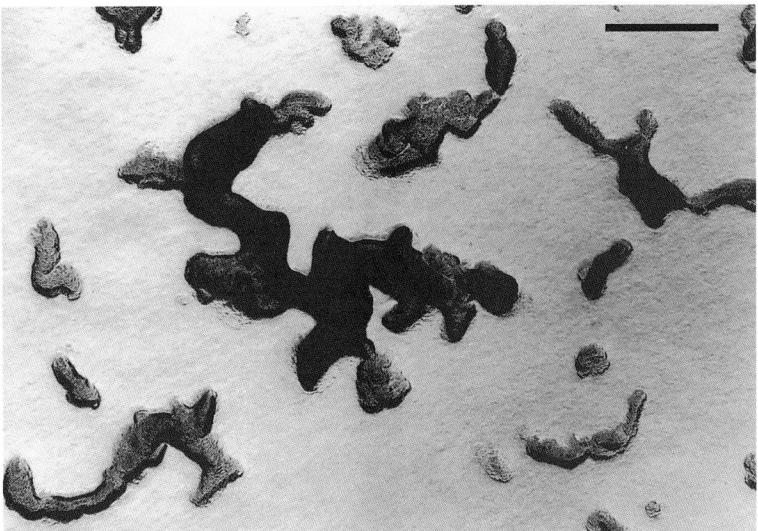

FIGURE BXII.δ.76. *Cystobacter miniatus*: degenerate fruiting bodies. Unlike those of other *Cystobacter* species, these are compact red ridges, often with short branches. Bar = 260 µm.

India, in 1990, and reference strain Cb m28 was isolated in 1998 from a similar sample collected in 1998 in Alsace, France.

The mol% G + C of the DNA is: not determined.

Type strain: Cb m2, DSM 14751, JCM 12627.

Additional Remarks: Reference strains include DSM 14772 (Cb m6), DSM 14752 (Cb m28).

This species is presently difficult to evaluate. It resembles *Cystobacter disciformis* but lacks disk-shaped sporangioles. The 16S rRNA sequences of the two species cluster closely together (Spröer et al., 1999). A comparative study using a number of strains of the two species would be desirable. After its original description, the species was observed only once by Singh and Singh (1971). However, the identity of the organisms of the two earlier descriptions with modern material must remain open. The photographs in the two

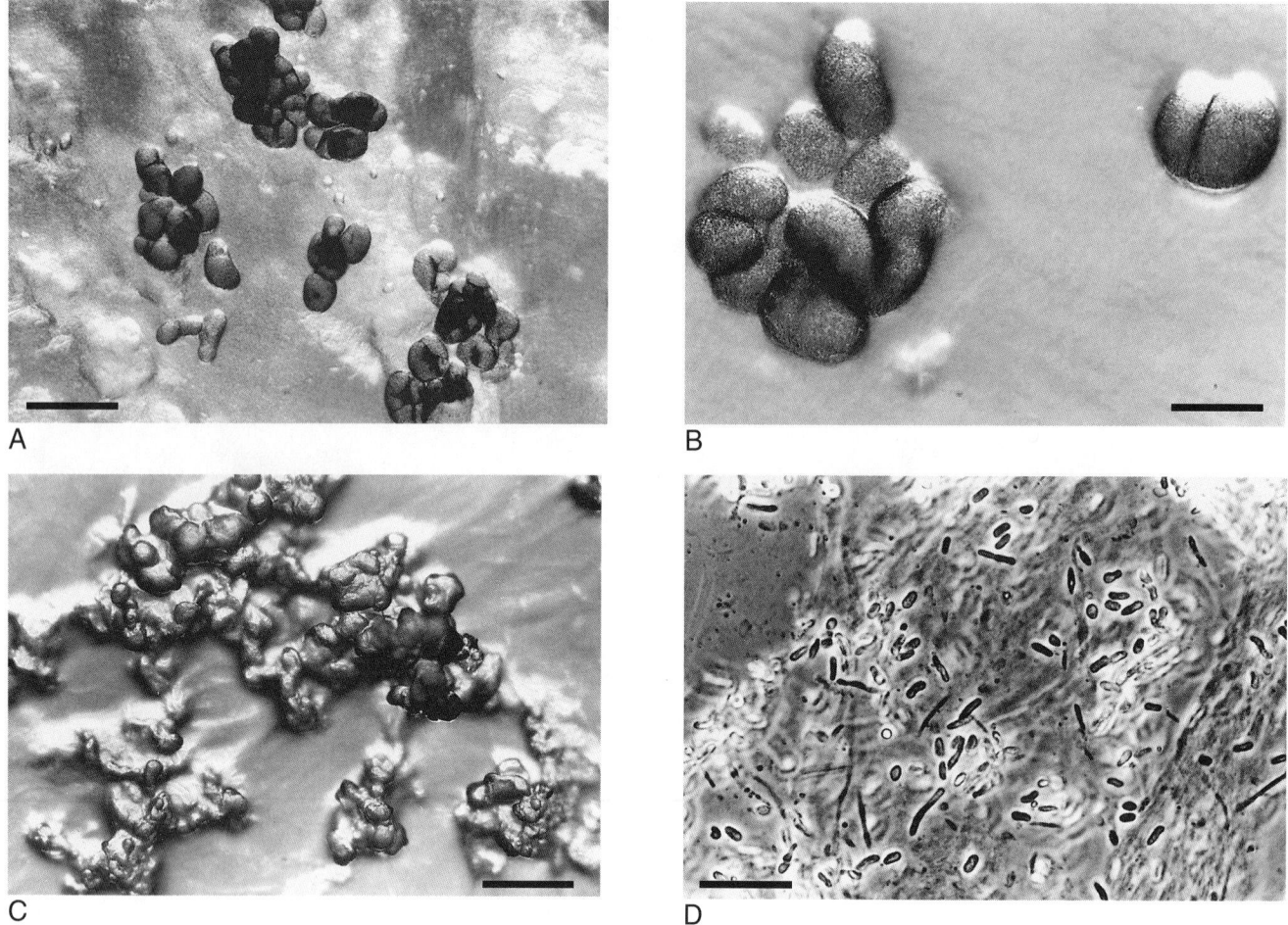

FIGURE BXII.δ.77. *Cystobacter minus*: fruiting bodies on an agar surface, and myxospores from crushed sporangioles (*D*). In *C, Archangium*-like degenerate fruiting bodies are seen, as are produced by all *Cystobacter* species. The sporangioles contain many degenerate cells besides typical myxospores (*D*). Bar in *A* = 200 μm; bar in *B* = 65 μm; bar in *C* = 260 μm; bar in *D* = 16 μm.

mentioned articles are not of good enough quality to allow a decision. In fact, some characteristics given in the two articles would even fit the new genus, *Hyalangium*, described below.

9. **Cystobacter velatus** *sp. nov.* (*Polyangium fuscum* biovar velatum Krzemieniewska and Krzemieniewki 1926.)
ve.la' tus. L. v. *velare* to veil, to cover; masc. part. perf. *velatus* veiled, covered.

   Cystobacter velatus closely resembles *Cystobacter badius* but differs in the following characteristics: sporangioles golden brown, preferentially in chains that tend to loop upward and to form three-dimensional networks, arise below a thin but tough, colorless, translucent slime layer that is pushed upward by the maturing sporangioles and becomes delicately plicated, giving the sporangioles a strikingly striated appearance (Fig. BXII.δ.78). Single sporangioles on pedicels are absent.

   Growth pattern and nutritional requirements as for *C. fuscus*, *C. badius*, and *C. ferrugineus*, with the exception that *Cystobacter velatus* may or may not attack chitin.

   Cystobacter velatus is found in the usual myxobacterial habitats and is moderately common.

The species was described by Krzemieniewska and Krzemieniewski (1926), who classified it as a variant of *Cystobacter fuscus* (then *"Polyangium fuscum"*). *Cystobacter velatus* differs substantially from the latter, so that it can be justifiably regarded as a separate species. The differentiation of *Cystobacter velatus* from *Cystobacter badius* can be very difficult, as the two overlap in most characteristics. The darker color of the sporangioles of the former, and the plicated slime cover on the fruiting bodies of the latter are not always prominent. Chitin degradation appears to be a distinguishing characteristic, but it remains to be established how reliable it is. Whether the two species are really separate is an open question at this time.

   Strain Cb v34 was isolated in 1996 from a soil sample with rotting wood collected near Jaipur, Rajasthan, India, in 1995, reference strain Cb v37 was isolated in the same year from a similar sample from the same site.

   The mol% G + C of the DNA is: not determined.
   Type strain: Cb v34, DSM 14718, JCM 14718.
   Additional Remarks: Reference strains include DSM 14719 (Cb v37).

10. **Cystobacter violaceus** (Kühlwein and Gallwitz 1958) *comb.*

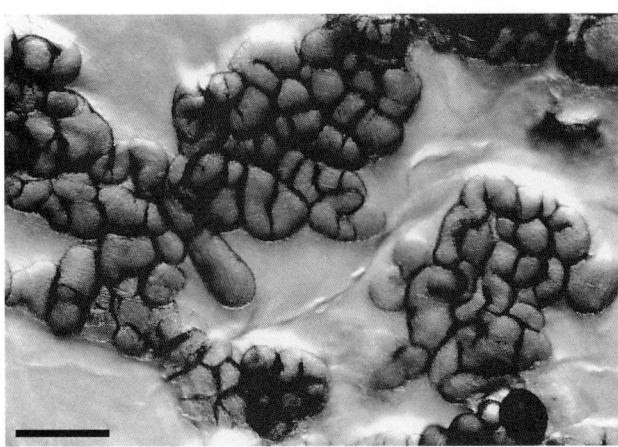

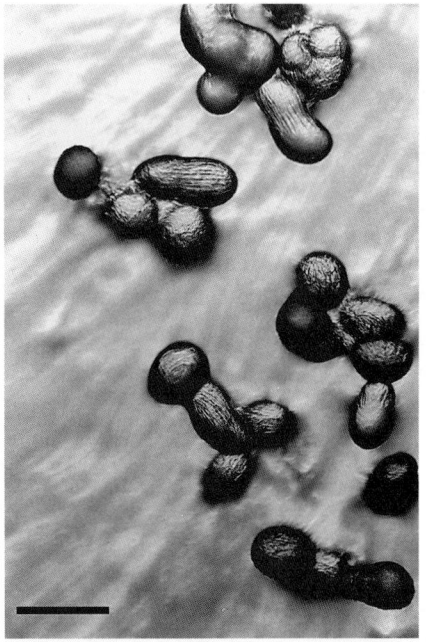

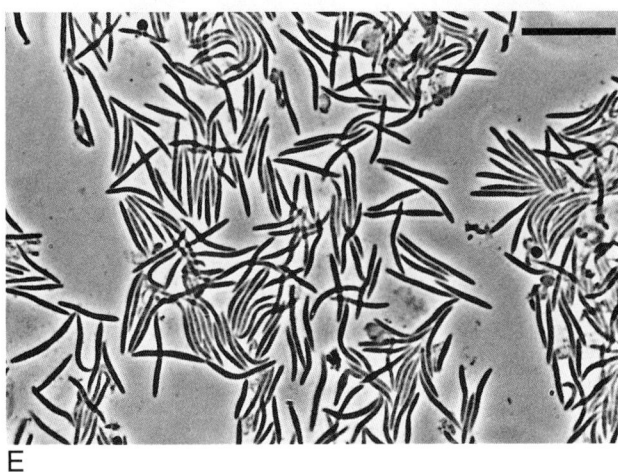

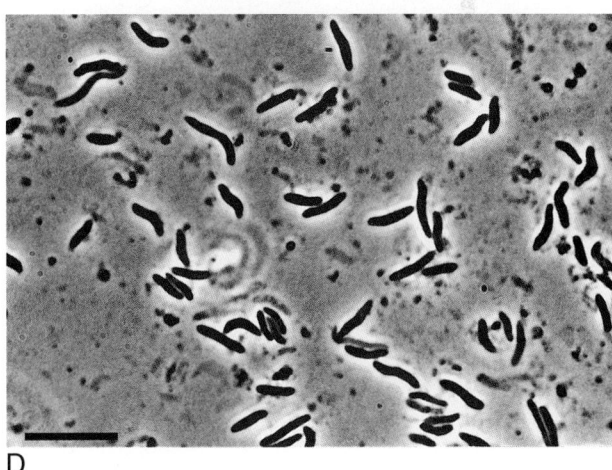

FIGURE BXII.δ.78. *Cystobacter velatus:* fruiting bodies on agar surface (*A–C*), myxospores from crushed sporangioles (*D*), and vegetative cells (*E*), the latter two in phase contrast. In *C*, degenerate fruiting bodies are seen. Bars in *A* and *C* = 200 μm; bar in *B* = 100 μm; bars in *D* and *E* = 13 μm.

nov. (*Polyangium violaceum* Kühlwein and Gallwitz 1958, 139; *Archangium violaceum* Kühlwein and Reichenbach 1964, 179.)

vi.o.la′ce.us. L. masc. adj. *violaceus* violet colored.

Vegetative cells are long, slender rods with tapered ends, 0.6–0.7 × 8–12 μm. Swarm colonies produce a tough slime sheet with fine radial veins; the slime apparently consists of a cellulose-like polysaccharide (Reichenbach, 1965c). Fruiting bodies (Fig. BXII.δ.79) consist of large, spherical to ellipsoidal sporangioles, 95–130 × 130–185 μm, single or in groups, dull brown to deep violet. After some time in

cultivation only degenerate fruiting bodies are produced in the form of meandering, twisting ridges, which may fuse into extended, dense, irregular masses. The diffusing violet pigment of the original isolate, which was once regarded as typical for the species, is not really a species-distinguishing characteristic. Violet pigments, apparently of a different chemical structure, occasionally occur in individual strains of other myxobacteria, e.g., other *Cystobacter* species and *Sorangium cellulosum.* Myxospores short, stout rods with rounded ends, often bean- or S-shaped, 0.8–1.8 × 1.8–4 μm.

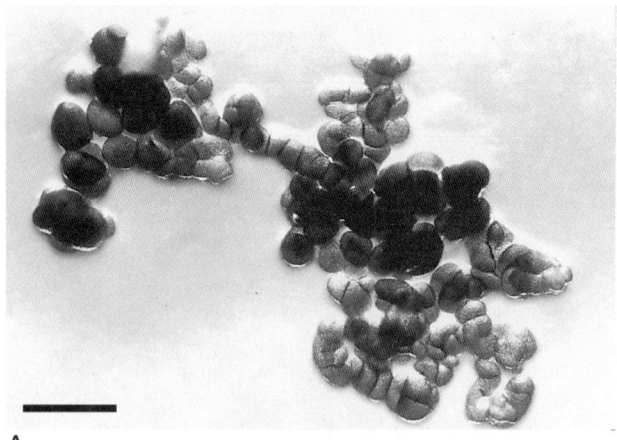

A

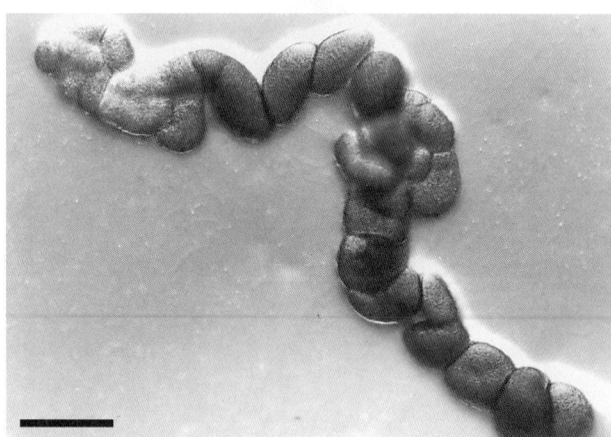

B

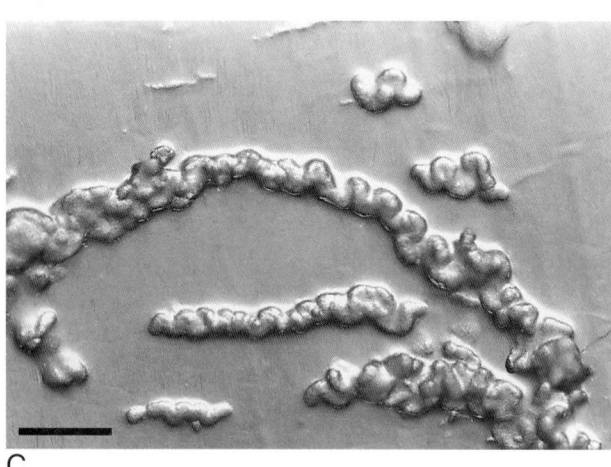

C

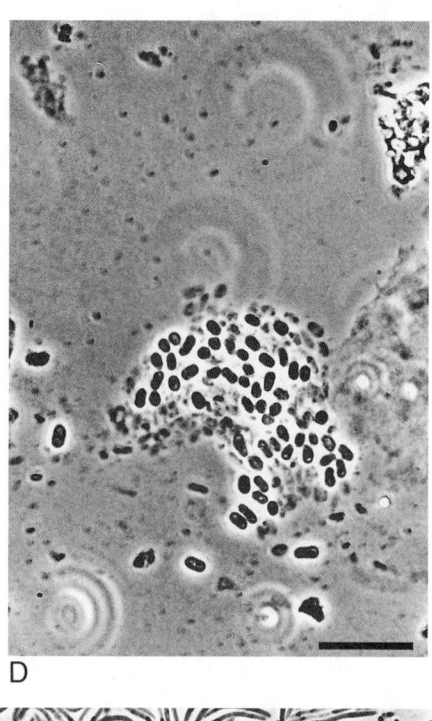

D

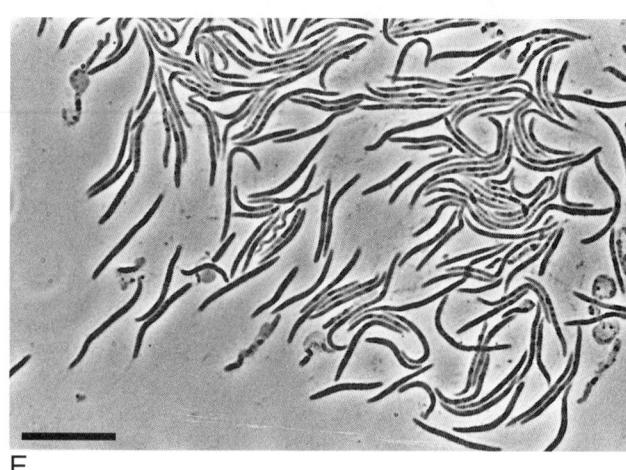

E

FIGURE BXII.δ.79. *Cystobacter violaceus*: fruiting bodies on agar surface (*A–C*), myxospores from crushed sporangioles (*D*), and vegetative cells (*E*), the latter two in phase contrast. Sporangioles are typically arranged in chains and three-dimensional networks and show a striated surface (*A, B*), but they may also occur densely packed in sori with the striation barely visible (*C*). Bar in *A* = 160 μm; bar in *B* = 135 μm; bar in *C* = 200 μm; bar in *D* = 10 μm; bar in *E* = 13 μm.

Of the proteolytic–bacteriolytic nutritional type. Efforts to demonstrate cellulose-degradation were negative. Most strains do not attack chitin.

The species bears some resemblance to *Cystobacter ferrugineus*, but has short to very short myxospores with rounded ends. The delineation of the species is presently somewhat blurred. Its history is discussed in the Comments on the genus *Archangium*.

The original isolate came from hare dung collected in Natal, South Africa, but is no longer available. May be found in the usual myxobacterial habitats; rather common. Strain

Cb vi61 was isolated in 1996 from goat dung, soil and decaying plant material collected the year before in Rajasthan, India. Reference strain Cb vi59 was isolated in 1996 from soil with rotting plant material collected in 1994 near Giza, Egypt; reference strain Cb vi6 was isolated in 1981 by Dr. Wolfgang Dawid from soil collected in the same year in Ceylon, Sri Lanka.

The mol% G + C of the DNA is: not determined.

Type strain: Cb vi61, DSM 14727, JCM 12629.

Additional Remarks: Reference strains include DSM 14758 (Cb vi6), DSM 14759 (Cb vi59).

Genus II. **Archangium** Jahn 1924, 66[AL]

HANS REICHENBACH

Ar.chan' gi.um. Gr. fem. n. *arche* beginning, origin, primitive; Gr. neut. n. *angion* vessel, container; *Archangium* primitive vessel.

Vegetative cells are long, slender, needle-shaped rods with tapering ends. **Fruiting bodies without sporangioles**, contorted strings of myxospores in hardened slime forming cushion-shaped masses, very variable in shape and size, may separate into packets when put under pressure. **Myxospores short, fat rods** with rounded ends to almost spherical, optically refractile. Swarm colonies with branched radial veins in a tough slime sheet. Does not degrade chitin.

Type species: **Archangium gephyra** Jahn 1924, 67.

FURTHER DESCRIPTIVE INFORMATION

The genus *Archangium* is problematic, because its definition rests on a fruiting body morphology that is produced by many other myxobacteria in place of their normal, typical fruiting bodies.

However, strains can be isolated from nature that never produce fruiting bodies with sporangioles, with specific types of vegetative cells and myxospores. These may be regarded as representatives of the genus, and they are placed here in just one species, *Archangium gephyra*. It must remain open for the moment whether this is correct and whether more species exist. Three strains that were classified as *Archangium gephyra* based on morphological characteristics, including strain ATCC 25201, possessed very similar 16S rRNA sequences (Spröer et al., 1999). They were also closely related to two strains tentatively identified as *Angiococcus disciformis* (*Cystobacter disciformis*). (The latter are probably *Cystobacter minus* strains.) The *Archangium* 16S rDNA cluster is closely associated with several *Cystobacter* species, which supports the position of the genus in the family *Cystobacteraceae*.

List of species of the genus Archangium

1. **Archangium gephyra** Jahn 1924, 67[AL]
   ge' phy.ra. Gr. fem. n. *gephyra* bridge.

Vegetative cells are long, slender, needle-like rods with tapering ends, 0.6–0.8 × 6–15 μm. The fruiting bodies (Fig. BXII.δ.80) are highly variable in size and shape, up to 1000 μm, but usually much smaller. They may be ridges, pads, or often mesenteric masses of contorted and convoluted strings of hardened slime and myxospores, 40–60 μm wide, subdivided by irregular constrictions, without an outer wall, with a knobby or brain-like surface, sometimes with finger-like projections, of firm consistency. When put under pressure, they usually separate into irregular parcels with diameters 15–30 μm and larger. Reddish, brownish, or bluish violet in color. Myxospores (Fig. BXII.δ.81) short, fat rods with rounded ends to almost spherical, often slightly bean-shaped, optically refractile, 1–2 × 1.5–2.8 μm in size. The swarm colonies have a tough slime sheet with radial veins. Of the proteolytic–bacteriolytic nutritional type. Does not degrade chitin. *Archangium gephyra* is found in the usual myxobacterial habitats and is a common organism.

The mol% G + C of the DNA is: 68 (T_m).

Type strain: M18, ATCC 25201, DSM 2261.

GenBank accession number (16S rRNA): M94273.

Other Organisms

Several more *Archangium* species have been described in the past; some may be identical with *Archangium gephyra*, while others are clearly different organisms, and still others cannot be recognized readily due to scanty descriptions. In addition, no reference strains are available. Part of the taxonomic problem arises from the fact that most species were never extensively cultivated or compared with a sufficiently large number of other strains.

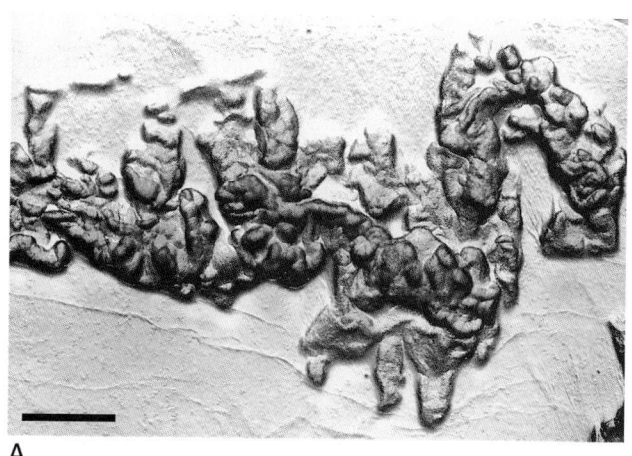
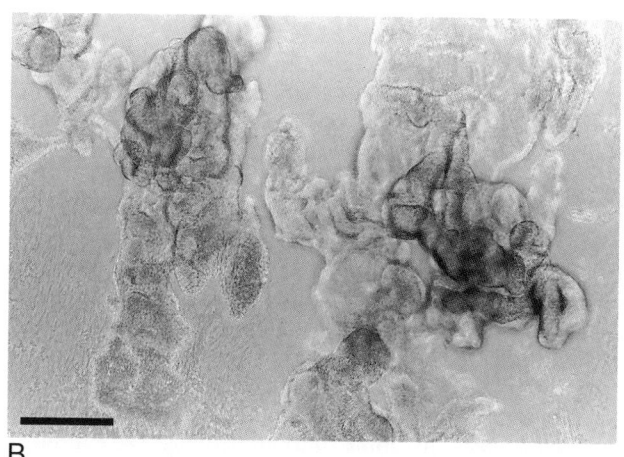

A   B

FIGURE BXII.δ.80. *Archangium gephyra*: fruiting bodies. (A) On agar plates: the mass of myxospores and hardened slime usually piles high and often produces horn-like processes. (*B*) Slide mount, phase contrast; the myxospores stick together in irregular parcels. Bar in *A* = 260 μm; bar in *B* = 95 μm.

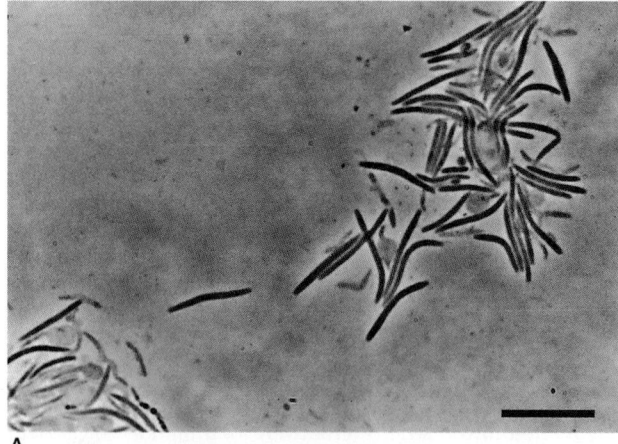

A

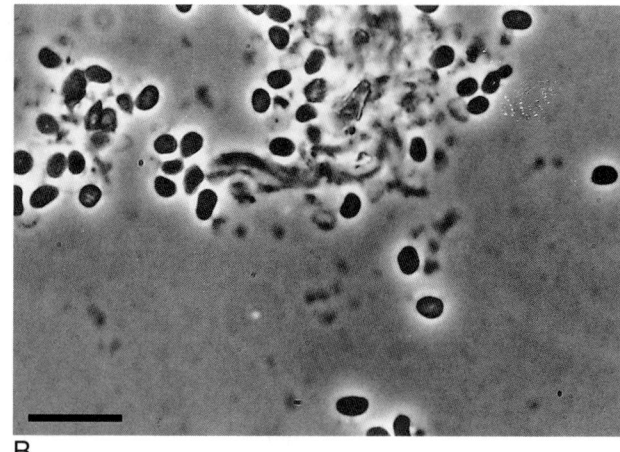

B

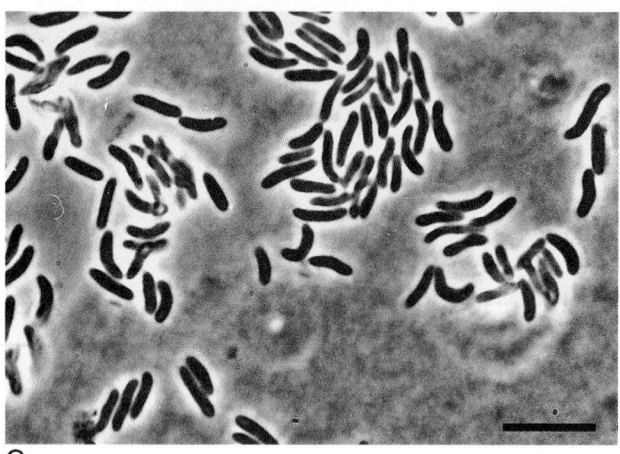

C

FIGURE BXII.δ.81. *Archangium* and *Cystobacter*: vegetative cells and myxospores. While the vegetative cells of all species are very similar in shape, like those of *Archangium gephyra* (*A*), the myxospores differ. Either they are short, stout rods, often almost spherical, as shown here for *Cystobacter minus* (*B*)—also typical of *A. gephyra*—or they are slender rods with tapering ends, like those of *Cystobacter ferrugineus* (*C*). These myxospores come from crushed sporangioles, but myxospores from degenerate fruiting bodies would look the same. All photographs in phase contrast. Bar in *A* = 10 μm; bars in *B* and *C* = 8 μm.

Archangium gephyra Jahn 1924 may be identical with *"Chondromyces serpens"* Thaxter 1892 (= *"Archangium serpens"* Jahn 1924) and *"Polyangium serpens"* Quehl 1906, as suggested by Krzemieniewska and Krzemieniewski (1926). While Thaxter's description would fit *Archangium gephyra*, his figure of the fruiting body does not; it shows rather loosely piled strings and tubes. In addition, the notion that the organism was found together with *Melittangium lichenicola* makes one suspect that *"Chondromyces serpens"* was an aberrant form of the latter. Therefore, Thaxter's priority is doubtful, and Jahn's name has been approved. Kofler's (1913) *"Myxococcus cerebriformis"* was probably the same organism, although the size of the myxospores was only 1.1–1.6 μm. Jahn's figure of the *Archangium gephyra* fruiting body reminds one of an immature *Cystobacter* fruiting body. *"Archangium primigenium"* Jahn 1924 (= *"Polyangium primigenium"* Quehl 1906) looks very much like *Archangium gephyra*, but Jahn gave the size of the myxospores as 0.8 × 4 μm. Such long, slender myxospores are typical of the *Cystobacter fuscus* complex, so that this organism may actually belong there. Jahn's variant, *assurgens*, is not very convincing, as finger-like, upward pointing projections are quite common with that type of fruiting body. Jahn's description of constrictions that make the tubes occasionally appear like a string of pearls supports the *Cystobacter* hypothesis. Quehl stressed that the fruiting bodies of *"Archangium primigenium"* are entirely homogeneous without tubular structures; this characteristic was also emphasized by Kofler (1913) and by Krzemieniewska and Krzemieniewski (1927) but may not really be a reliable distinguishing

characteristic. The latter authors described the myxospores as short rods with rounded ends, and they believed that Jahn's *"Archangium assurgens"* is a separate species. Quehl's observation that his strain produced fruiting bodies only on the glass wall of the dish, and not on the agar surface, suggests that his species is more likely to have been a degenerate *Cystobacter*. The question cannot be decided now, and the species is rightly dismissed. *"Polyangium flavum"* Kofler 1913 was renamed *"Archangium flavum"* by Jahn (1924); he felt that the species might be identical with his *"Archangium thaxteri"*. However, Kofler described the fruiting bodies of *"Polyangium flavum"* as globose or ellipsoidal, with a humpy surface, rather large, 600 × 400 μm, yellow, and easily confused with *Myxococcus virescens*, but containing rods rather than spheres. If such is the case, the organism may be identical with a new bacterium described below as *Kofleria flava* which belongs to the suborder *Nannocystineae*. Jahn gave the size of the rod-shaped cells inside those masses as 2–4 μm. *"Archangium flavum"* has been reported two other times. Krzemieniewska and Krzemieniewski (1930) described rod-shaped cells with rounded ends and fruiting bodies with a tubular structure; one variety had a gelatinous, hyaline slime matrix. Singh and Singh (1971) described fruiting bodies that were lemon yellow with a humped or padded surface; the myxospores were rods 1 × 2.5–5 μm in size. The reports may or may not refer to the same species. Jahn's (1924) *"Archangium thaxteri"* is almost certainly identical to an organism described below as *Jahnia thaxteri* in the *Chondromyces* group: convoluted tubules, 50 μm wide, in a dense mass on a stout, cushion-like slime stalk, with a yellow slime matrix. Kofler's (1913) "*Myxococcus polycystus*" was perhaps the same organism. "*Archangium violaceum*" Kühlwein and Reichenbach 1964 was an erroneous reclassification of "*Polyangium violaceum*" Kühlwein and Gallwitz 1958 and was in fact a *Cystobacter* species. The authors were misled by the fact that the type strain (which is no longer available) after years of cultivation under many different conditions did not produce fruiting bodies with sporangioles, which it did originally. At that time the degeneration of fruiting bodies of *Cystobacter* and other myxobacteria was not yet understood.

Jahn (1924) added a second new genus, *Stelangium*, to the new family *Archangiaceae*, with "*Stelangium muscorum*" (= "*Chondromyces muscorum*" Thaxter 1904) as the only species. The status of that species is obscure. Thaxter described the organism as similar to *Corallococcus coralloides*: bright yellow-orange, columnar or finger-like fruiting bodies, sometimes furcate, sessile without a stalk, erect, with a bluntly pointed apex, 90–300 × 20–50 μm,

apparently without an outer wall. Unfortunately, the myxospores were not described. The organism was reported again by Singh and Singh (1971), with similar fruiting body morphology and rod-shaped myxospores, 1–1.4 × 4–6 μm. In neither case was the organism cultivated. The description of the fruiting body reminds one of *Corallococcus macrosporus* or atypical fruiting bodies of *Myxococcus virescens* (Fig. BXII.δ.62 in the chapter on *Myxococcus*). In both cases, however, the myxospores were spherical. A second species, "*Stelangium vitreum*", was added by Peterson (1959). It had delicate crystalline orange fruiting bodies, either globose or columnar, 150–200 × 60–70 μm, without an outer wall. The myxospores were rods, 0.7 × 2 μm that occurred in small parcels that were arranged in the fruiting body in a basket-weave fashion. The vegetative cells (shape not mentioned) measured 0.7 × 4 μm. The organism was found on bark, and it, too, was not cultivated. Its identity is unclear.

Genus III. **Hyalangium** gen. nov.

HANS REICHENBACH

Hy.al.an'gi.um. Gr. fem. n. *hyalos* glass; Gr. neut. n. *angion* vessel, container; M.L. neut. n. *Hyalangium* glassy vessel.

Vegetative cells are delicate, slender rods with tapering ends. Fruiting bodies consist of small spherical sporangioles that are **often empty** and then look **glassy and transparent**, arranged in extended, dense sheets, or often, in chains. **Myxospores are short rods or irregularly spherical;** optically refractile. Swarm colonies with a thin but tough slime sheet with fine veins, adsorbs Congo red with a purple red color. Of the proteolytic–bacteriolytic nutritional type.

Type species: **Hyalangium minutum** *sp. nov.*

FURTHER DESCRIPTIVE INFORMATION

This organism differs substantially from the *Cystobacter* species discussed above and is therefore classified in a new genus. The

distinguishing characteristics are as follows. The vegetative cells are shorter and more delicate than those of *Cystobacter* species. The sporangioles are small and often empty and glassy. They are arranged in monolayer sheets or in short chains and do not pile up, or do so only slightly. The swarm colonies have a delicate slime sheet with very fine veins. The only species known so far somewhat resembles *Cystobacter disciformis*, although the latter has much different fruiting bodies. In addition, some traits are like those of *Polyangium minus* (now *Cystobacter*) in the original description of Krzemieniewska and Krzemieniewski (1926). The identity of *Hyalangium* with that organism cannot be ruled out with certainty, but even if the two organisms should be the same, a reclassification would be required.

List of species of the genus Hyalangium

1. **Hyalangium minutum** *sp. nov.*
   mi.nu' tum. L. neut. adj. *minutum* small, tiny.

   Vegetative cells are slender, delicate rods with tapering ends, 0.6–0.7 × 3–6 μm (Fig. BXII.δ.82). Fruiting bodies (Fig. BXII.δ.83) consist of spherical to slightly elongated, potato-like sporangioles, sometimes with unilateral constrictions by incomplete cross-walls, grayish brown or brown, but often empty, colorless, glassy, and transparent; sometimes one or two secondary sporangioles are produced within the empty husk. The sporangioles are surrounded by a definite, irregularly indented wall, 1.5–3.5 μm thick. Occasionally a fine, regular striation can be seen on the surface of the sporangioles, with a distance of the striae of 0.6 μm. The sporangioles measure between 20 × 30 and 40 × 80 μm, usually around 35–45 μm in diameter. They are arranged in extended monolayers, only rarely slightly piled up; sori 400–650 μm across, but often much larger. When fruiting starts to degenerate, the sporangioles arise in short chains. Degenerate fruiting bodies are narrow, delicate ridges with tapering ends, but never the cushion-like structures typical of *Cystobacter* spp. Myxospores (Fig.

BXII.δ.84) short rods with rounded ends, often slightly bent or bean-shaped, 1.5–1.8 × 2.1–4 μm, or irregularly spherical with a diameter between 1.8 and 2.6 μm, optically refractile.

   Swarm colonies with a delicate but tough slime sheet with very fine radial veins, often slightly wrinkled and tinted yellow brown by a diffusing pigment. Of the proteolytic–bacteriolytic nutritional type. Does not attack chitin.

   Hyalangium minutum is found in the usual myxobacterial habitats; it is rather common but easily overlooked because of its inconspicuous, translucent fruiting bodies and often complete degeneration of fruiting. The type strain, NOCB-2, was isolated in 1992 from soil with decaying plant material collected by Dr. S. Yamanaka in the mountains of Izu and Manazuru peninsula in Japan. Reference strain NOCB-4 comes from a sample collected in 1992 in Indianapolis, IN, USA, and was isolated in 1993; reference strain Hy m4 was isolated in 1998 from soil with rotting wood collected in 1997 in Iowa City, IA, USA.

   The mol% G + C of the DNA is: not determined.
   Type strain: NOCB-2, DSM 14724, JCM 12630.

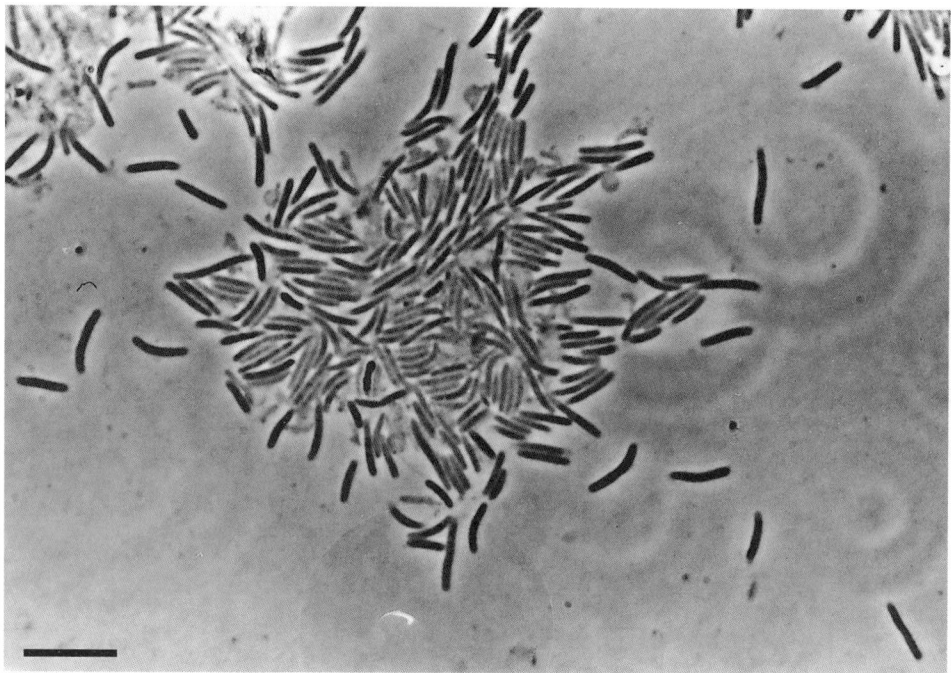

FIGURE BXII.δ.82. *Hyalangium minutum*: vegetative cells, in phase contrast. Bar = 13 μm.

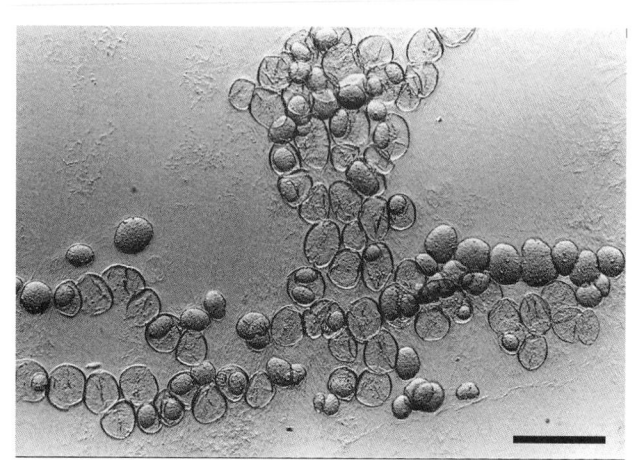

A

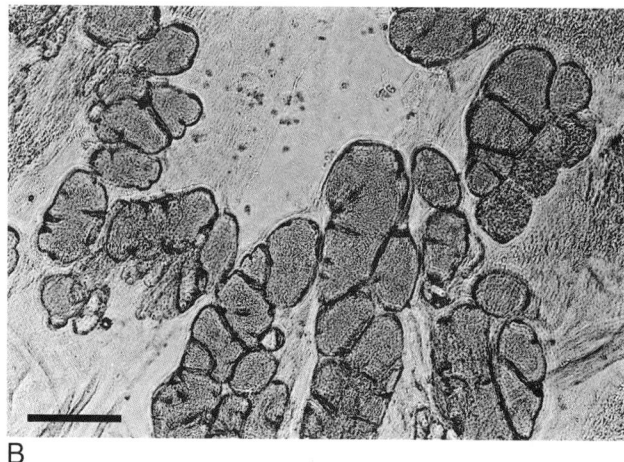

B

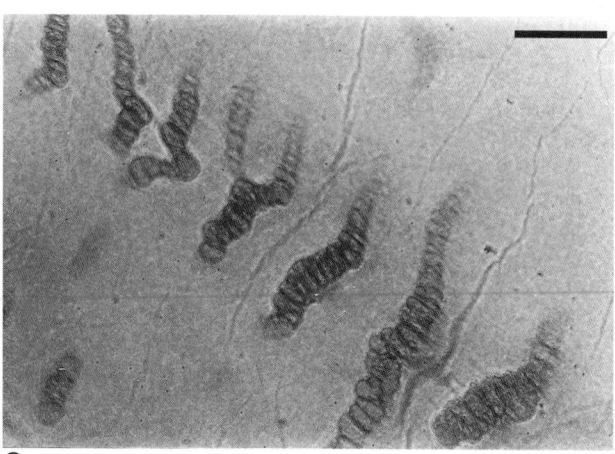

C

FIGURE BXII.δ.83. *Hyalangium minutum*: fruiting bodies. (*A*) Many empty, glassy sporangioles are seen on the agar surface . (*B*) Slide mount; indentations of the sporangiole wall are apparent. (*C*) When fruiting bodies start to degenerate, deformed sporangioles in short chains are produced, in phase contrast. Bar in *A* = 100 μm; bar in *B* = 55 μm; bar in *C* = 125 μm.

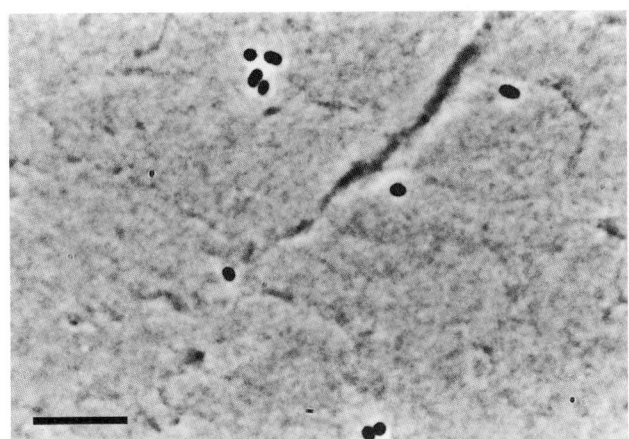

FIGURE BXII.δ.84. *Hyalangium minutum:* myxospores from crushed sporangioles, in phase contrast. Bar = 10 μm.

Additional Remarks: Reference strains include DSM 14725 (NOCB-4), DSM 14721 (Hy m4).

As mentioned above, it cannot be ruled out that the species is identical with *Polyangium minus* (now *Cystobacter*) Krzemieniewska and Krzemieniewski 1926. The 16S rRNA sequences of the type strain and of a second isolate, NOCB-4, cluster closely together and are well separated from *Cystobacter* species. The most closely related sequences are those of *Stigmatella* (Spröer et al., 1999).

Genus IV. **Melittangium** *Jahn 1924, 7*AL

HANS REICHENBACH

Me.lit.tan' gi.um. Gr. fem. n. *melitta* bee; Gr. neut. n. *angion* vessel; M.L. neut. n. *Melittangium* a vessel resembling a honeycomb.

Vegetative cells are slender rods with tapering ends. Fruiting bodies consist of solitary sporangioles on slime stalks. Myxospores rod shaped, optically refractile. Swarm colonies with a slime sheet and radial veins. Of the proteolytic–bacteriolytic nutritional type.

Type species: **Melittangium boletus** Jahn 1924, 78.

FURTHER DESCRIPTIVE INFORMATION

While the sporangioles of these organisms typically sit on stalks, sporangioles without stalks are also produced, and may be either solitary or arranged in looping chains. Degenerate fruiting bodies are narrow, delicate ridges, pustules, sometimes finger-like, erect, or irregular, gnarled, knobby masses. The fruiting bodies of *Melittangium* are smaller than those of the *Stigmatella* species and lighter in color. The 16S rRNA sequences of *Melittangium*

boletus and *Melittangium alboraceum* cluster together and are closely related to those of *Cystobacter fuscus* complex. The sequence of strain ATCC 25946 of *Melittangium lichenicola*, however, is most closely related to those of *Corallococcus* strains (Shimkets and Woese, 1992; Spröer et al., 1999). The GenBank sequence, M94277, does not, however, belong to strain ATCC 25946, as has recently been shown by a new sequence determination at the DSM using a fresh culture from the ATCC (Spröer, personal communication). *Melittangium boletus* and *Melittangium lichenicola* were placed for some time in the genus *Podangium* (Stanier, 1957). However, the type species of that genus, *"Podangium erectum"*, differs substantially from the *Melittangium* species and is now classified as *Stigmatella erecta.*

Key to the species of the genus *Melittangium*

I. Sporangiole like a mushroom cap, brown, on a short stalk; efficient chitin degrader.
Melittangium boletus
II. Sporangioles spherical, ellipsoidal or kidney-shaped, pale orange to light brown.
    A. Sporangiole on a short, delicate stalk, does not decompose chitin.
Melittangium lichenicola
    B. Sporangiole on a long, often bent stalk.
Melittangium alboraceum

List of species of the genus Melittangium

1. **Melittangium boletus** Jahn 1924, 78AL

bo.le' tus. L. masc. n. *boletus* a kind of mushroom.

Vegetative cells are slender, needle-shaped rods with tapering ends, 0.7 × 4–9 μm. The fruiting body (Fig. BXII.δ.85), when typical, consists of a single, colorless, grayish or yellowish brown to dark red brown semispherical sporangiole with a flat base like a mushroom cap, 50–100 μm wide and 40–50 μm thick. The sporangiole is borne on a delicate, short white stalk, 10–30 μm wide and up to 60 μm long, but usually much shorter. Sometimes a slime knob is seen at the base of the stalk. Fruiting bodies with

spherical, club-shaped, or kidney-shaped sporangioles may be found in addition to typical fruiting bodies. In addition, the stalk may be lacking, so that the sporangioles sit directly on the substrate, in which case they may be arranged singly, in chains, or in packets. After some time in culture, only degenerate fruiting bodies are produced. The fruiting bodies of *Melittangium boletus* arise singly or in loose groups. Myxospores short, often rather slender and slightly C- or S-shaped bent rods with rounded or slightly tapered ends, optically refractile, 0.7–1.0 × 1.5–5 μm.

Swarm colonies with radial veins in a tough slime sheet,

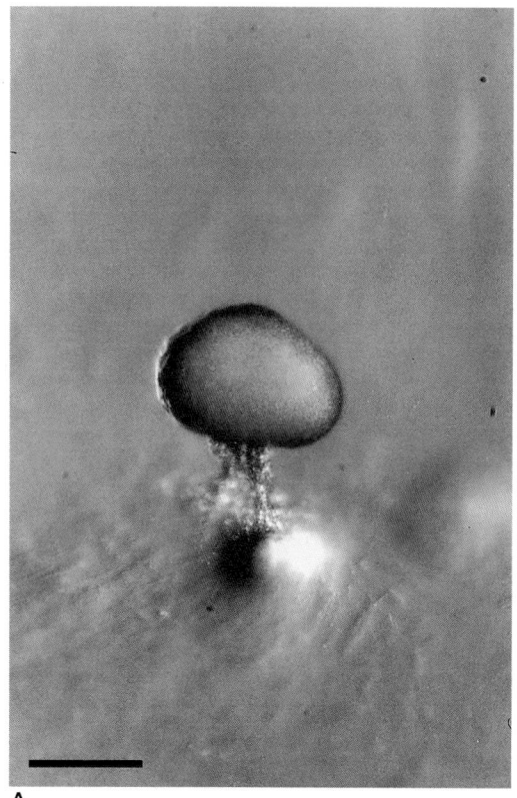

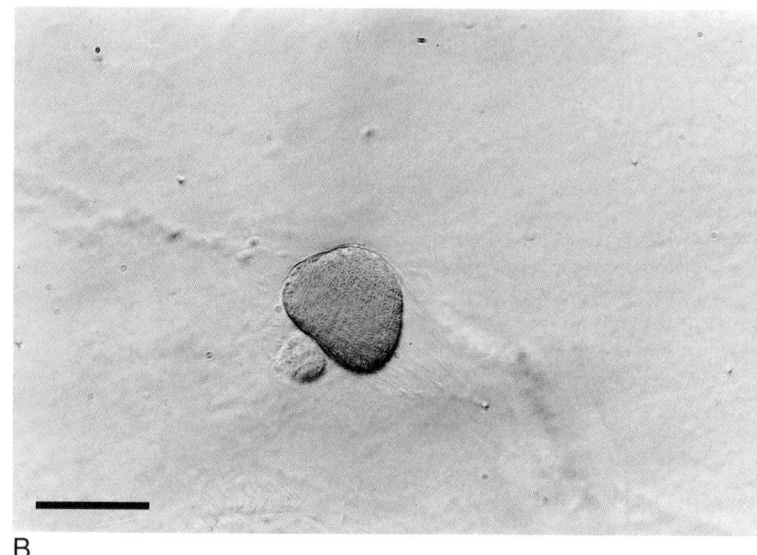

A

B

FIGURE BXII.δ.85. *Melittangium boletus*: fruiting bodies: (*A*) on agar surface, (*B*) in embedding medium, in phase contrast. Bar in *A* = 50 μm; bar in *B* = 30 μm.

on many media bright chrome yellow with a greenish tinge. Of the proteolytic-bacteriolytic nutritional type. Degrades starch and chitin very efficiently, but not filter paper cellulose.

The species is found in the usual myxobacterial habitats, but is not particularly common. Strain Me b8 was isolated in 1995 from a soil sample containing rotting wood and bark collected in the Nawalganj Bird Sanctuary southwest of Lucknow, Uttar Pradesh, India.

The mol% G + C of the DNA is: not determined.

Type strain: Me b8, DSM 14713, JCM 12633.

When typical fruiting bodies are lacking, the species resembles *Melittangium lichenicola*. However, it clearly differs in fruiting body shape and color, in the yellow pigmentation of the swarm colonies, and in efficient chitin degradation. Jahn emphasized that the myxospores on the surface of the spore mass stand parallel to one another and perpendicular on the sporangiole wall; the impressions of their tips in the wall generate a honeycomb-like pattern on the inner face (hence the name). This arrangement appeared unique enough to him to justify the creation of a new genus. However, while this pattern can indeed occasionally be observed, it is often absent. As the fruiting bodies of *Melittangium boletus* are so characteristic, the original description of the species was rather brief. It was soon complemented by other investigators (Krzemieniewska and Krzemieniewski, 1926; Solntseva, 1941). The figure of fruiting bodies in the article by Krzemieniewska and Krzemieniewski shows very long and relatively massive stalks, so that one is reminded more of *Melittangium alboraceum*.

2. **Melittangium alboraceum** (Peterson 1959) McCurdy 1971b, 54[AL]*

al.bo.ra′ ceum. L. masc. adj. *albus* white L. masc. n. *racemus* bunch of grapes, grape berry: was erroneously assumed to mean stalk; M.L. neut. adj. *alboraceum* with a white stalk.

Vegetative cells slender rods with square rather than tapering ends, 0.8–1.0 × 4.5–5.0 μm. Fruiting bodies consist of solitary, pale orange, crystalline, globose "cysts", around 35 μm in diameter, with an elastic, indiscernible membrane, on long, angular, distorted, corkscrew-like, snow-white stalks, 20 × 100–200 μm. Myxospores slightly curved rods, 0.8 × 2.5 μm. Found on bark of *Ulmus americana*.

The mol% G + C of the DNA is: not determined.

Type strain: none cultivated.

Additional Remarks: Reference material includes microscope slides Peterson 72 in the herbarium of the University of Missouri, Columbia, MO, USA.

The author reluctantly includes this species in the genus *Melittangium*. The square-ended vegetative cells do not even fit the description of the suborder, and the "cysts" are described to have a gelatinous, indiscernible wall and to be deliquescent, which does not resemble a sporangiole. The photographs in the original paper show long, wavy stalks, unlike those of other myxobacteria. The fact that the fruiting bodies disappeared completely and forever when the

---

*Editorial Note: The name given to the organism in the original description was *"Podangium alboracemum"*, which was later corrupted to *alboraceum* (McCurdy, 1971b); however, according to the Bacteriological Code, it cannot be corrected.

bark was dried suggests that these structures may have been immature fruiting bodies, perhaps of a *Chondromyces* species, which indeed often inhabit bark. *Chondromyces* fruiting bodies tend to degenerate into long white stalks with the gradual consumption of the bacterial cell mass at the tip. On the other hand, it should be noted that *Melittangium* strains with very long stalks have been described. They may represent a separate species, albeit not *Melittangium alboraceum*.

3. **Melittangium lichenicola** (Thaxter 1892) McCurdy 1971b, 53[AL]

li.che.ni' co.la. Gr. masc. n. *leichen* lichen; L. masc. n. *cola* inhabitant, dweller; *lichenicola* inhabitant of lichens.

Vegetative cells are slender, fusiform rods with tapering ends, $0.7–0.8 \times 5–8$ µm (Fig. BXII.δ.86). The fruiting bodies (Fig. BXII.δ.87) are solitary, small, pale yellowish brown to brown orange spherical, ellipsoidal or club- and kidney-shaped sporangioles, 25–40 µm across, on delicate short white stalks, $5–10 \times 10–40$ µm. Stalks may be lacking, and sporangioles may then arise in chains or clusters. Normally, the fruiting bodies are produced in large numbers in extended fields. In culture, soon only degenerate fruiting bodies arise as orange to orange brown padded masses, appressed convoluted tubules and ridges. Myxospores appear similar to those of *M. boletus* (see Fig. BXII.δ.88) and are

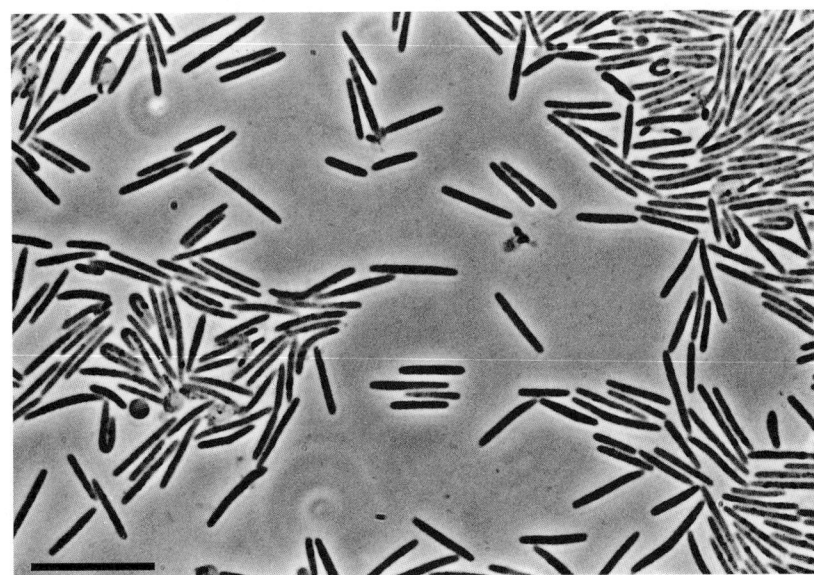

FIGURE BXII.δ.86. *Melittangium lichenicola*: vegetative cells, in phase contrast. Bar = 13 µm.

FIGURE BXII.δ.87. *Melittangium lichenicola*: fruiting bodies on agar ($\times$ 500).

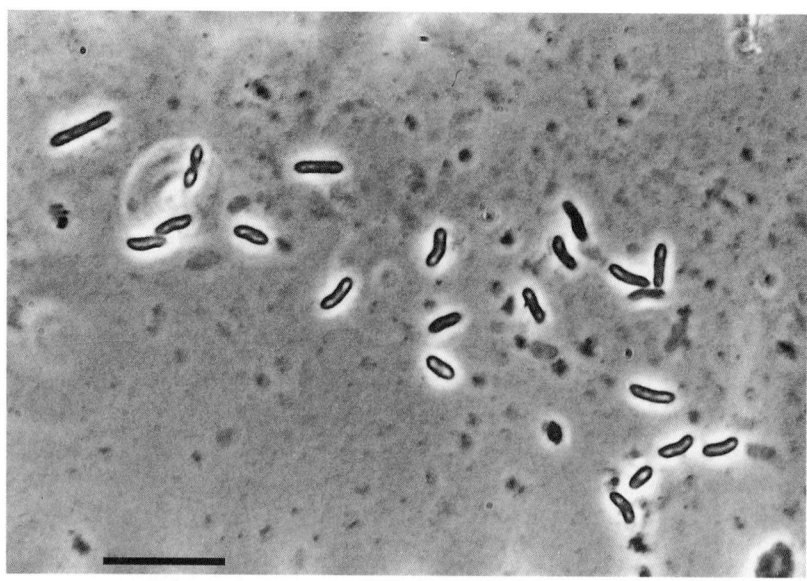

FIGURE BXII.δ.88. *Melittangium boletus:* myxospores from crushed sporangioles, in phase contrast. Bar = 13 μm.

relatively slender rods with rounded or pointed ends, often slightly C- and S-shaped bent and somewhat irregular, 0.8–1.2 × 2.5–4 μm.

Swarm colonies with fine radial veins. Of the proteolytic–bacteriolytic nutritional type. Degrades starch and xylan, but not cellulose, carboxymethyl cellulose, or chitin.

Originally discovered as a purported parasite of lichens, *Melittangium lichenicola* is found in the usual myxobacterial habitats; it is a rather common organism. Reference strain Me l 31 was isolated in 1990 from soil with decaying plant material collected in 1988 in Carinthia, Austria.

The mol% G + C of the DNA is: not determined.

Type strain: ATCC 25946, DSM 2275.

GenBank accession number (16S rRNA): M94277.

Additional Remarks: Reference strains include DSM 14740 (Me l 31).

As mentioned, the GenBank sequence published and used for the type strain of *Melittangium lichenicola* is probably that of a *Corallococcus* and does not belong to strain ATCC 25946.

Melittangium lichenicola can easily be confused with *Melittangium boletus.* The fruiting bodies of the latter, when typical, are somewhat larger, darker brown in color, and the sporangioles are shaped like a mushroom cap; also, the swarm colonies produce a yellow pigment and efficiently degrade chitin. It is not certain, however, whether the latter two traits are reliable distinguishing characteristics. Further, *Cystobacter disciformis* may be a source of error. However, the sporangioles of that organism are intensely orange-brown, disk-shaped, often piled up in opuntia-like clusters, and without stalks. *"Chondromyces gracilipes"* Thaxter 1897 (= *"Podangium lichenicolum"* Jahn 1924) is most likely identical with *Melittangium lichenicola.*

Genus V. **Stigmatella** *Berkeley and Curtis in Berkeley 1857, 313*

HANS REICHENBACH

Stig.ma.tel´la. L. neut. n. *stigma* gen. *stigmatis* brand, mark; L. fem. dim. *-ella* ending; M.L. fem. n. *Stigmatella* small dark spot.

Vegetative cells are moderately long, boat-shaped, fusiform cells with tapering ends, 0.7–0.8 × 4–8 μm. **Myxospores short, often bent rods with rounded ends,** optically refractile, 0.9–1.2 × 2–4 μm. **Fruiting bodies** consist of more or less spherical **sporangioles, either solitary or in clusters on stalks.** Swarm colonies with an extremely tough slime sheet with radial veins, often with oscillating waves that typically are arranged in long, narrow tracks rather than in extended fields. Adsorbs Congo red with a purple-red color. Of the proteolytic–bacteriolytic nutritional type. **Degrades chitin efficiently.**

Type species: **Stigmatella aurantiaca** Berkeley and Curtis *in* Berkeley 1857, 313.

FURTHER DESCRIPTIVE INFORMATION

Strains of the genus *Stigmatella* are usually easy to recognize by their relatively large, stalked, dark red brown fruiting bodies. In addition, the swarms with their tracks of oscillating waves are typical. The 16S rRNA sequences of the six strains studied cluster closely together as a group and are clearly separate from *Cystobacter, Archangium,* and *Melittangium* strains (Spröer et al., 1999). The differentiation of species is more difficult, because typical fruiting bodies are often not produced, and the degenerate forms are the same in all strains. In the case of *Stigmatella aurantiaca* and *Stigmatella erecta,* ecological observations may often help; while *Stigmatella aurantiaca* is only found on rotting wood,

Stigmatella erecta is obtained from soil and dung of herbivorous animals. *Stigmatella aurantiaca* was classified for many years as *"Chondromyces aurantiacus"*; *Stigmatella erecta* was classified as *"Podangium erectum"* (the genus was actually created for it by Jahn, 1924). The two look and behave very similarly and clearly belong in one genus. The genome sizes differ slightly between the two species (9.2–9.9 Mbp for *Stigmatella aurantiaca*, 9.7–10 Mbp for *Stigmatella erecta*; Neumann et al., 1992), but it is not known whether this is taxonomically relevant. Several more species, classified in the genus *Chondromyces* at the time but obviously belonging to *Stigmatella*, were described by Krzemieniewska and Krzemieniewski. Most have since been assigned to *Stigmatella aurantiaca*, and one to *Stigmatella erecta* (Reichenbach and Dworkin, 1969; McCurdy, 1971a). But the authors reported (Krzemieniewska and Krzemieniewski, 1946) that they could not isolate a real *"Chondromyces aurantiacus"*, i.e., *Stigmatella aurantiaca*, for 15 years; they finally succeeded in obtaining several strains from rotting beech wood and bark in Poland. While this is the proper habitat of *Stigmatella aurantiaca*, all earlier strains on which the various species descriptions were based appear to have been isolated from soil and rabbit dung, which are the typical habitats

of *Stigmatella erecta*. Yet at least some of their illustrations appear to show *Stigmatella aurantiaca* fruiting bodies. As strains are no longer available, the question of those "new" species cannot satisfactorily be decided. *"Chondromyces brunneus"* (Krzemieniewska and Krzemieniewski, 1946) was originally described as *"Chondromyces aurantiacus"* biovar frutescens (Krzemieniewska and Krzemieniewski, 1927) and later renamed *"Stigmatella brunneus"* (McCurdy, 1971a) and *"Stigmatella brunnea"* (McCurdy and Khouw, 1969). It may have been *Stigmatella erecta*; Krzemieniewska and Krzemieniewski recognized a similarity with that species. *"Chondromyces cylindricus"* (Krzemieniewska and Krzemieniewski, 1930) (= *"Stigmatella cylindrica"* McCurdy, 1971a), and *"Chondromyces medius"* (Krzemieniewska and Krzemieniewski, 1930) (= *"Stigmatella media"* McCurdy, 1971a) were most likely *Stigmatella aurantiaca*. *"Chondromyces minor"* (Krzemieniewska and Krzemieniewski, 1930) was later reinterpreted by Krzemieniewska and Krzemieniewski (1946) as a composite form of *"Podangium gracilipes"*, now *Melittangium lichenicola*, and withdrawn as a species. However, the organism may in fact have been *Stigmatella aurantiaca*. In this edition only three species of *Stigmatella* are listed; they can readily be distinguished from one another.

Key to the species of the genus *Stigmatella*

I. Clusters of sporangioles on an unbranched, common stalk.

Stigmatella aurantiaca

II. Sporangioles solitary on stalks, fruiting bodies often in tufts.

  A. Sporangioles usually stalked.

Stigmatella erecta

  B. Besides stalked sporangioles many sporangioles without a stalk, solitary or in chains and piles.

Stigmatella hybrida

List of species of the genus Stigmatella

1. **Stigmatella aurantiaca** Berkeley and Curtis *in* Berkeley 1857, 313[AL]

au.ran.ti'a.ca. M.L. fem. adj. *aurantiaca* orange colored.

Vegetative cells, myxospores, and swarm colony as described for the genus. Fruiting bodies (Fig. BXII.δ.89) consist of spherical to ovoid, bright orange to red brown sporangioles measuring 25–45 × 40–60 μm; 1–20, usually 5–15, sporangioles are borne on a common stalk, 60–140 μm high and 30–100 μm wide. The stalk is white but may become orange-brown near the upper end.

The species is easy to recognize when typical fruiting bodies are present, which is often not the case in culture. Stalks with a single sporangiole and completely degenerate fruiting bodies in the form of narrow ridges may be produced. In contrast to the other two *Stigmatella* species, *Stigmatella aurantiaca* seems to be restricted to rotting wood as a substrate, although it does not attack cellulose. It is a very common species in the Midwest of the United States, where it is found on bark and decaying wood. In Minneapolis, a soft-rotten willow trunk was found with large, bright orange-brown patches consisting of thousands and thousands of *Stigmatella aurantiaca* fruiting bodies. The species seems to be rather rare elsewhere, although it also has been found in other places, e.g., in Poland (Krzemieniewska and Krzemieniewski, 1946) and Germany (Dawid, 1978).

Stigmatella aurantiaca was classified for some time as

"Chondromyces aurantiacus". However, its fruiting bodies are much smaller than those of the *Chondromyces* species, which may occur in the same habitat. In addition, they arise in a different way. While the stalk of a *Chondromyces* fruiting body is excreted by a globular mass of cells, which is thus lifted upward and only differentiates into a cluster of bright yellow orange sporangioles when stalk formation is completed, the stalk and sporangioles of *Stigmatella aurantiaca* fruiting bodies emerge concomitantly from the aggregated cell mass, so that the stalk initially contains many cells. Most of those cells later emigrate from the stalk or die and lyse, while those in the sporangiole section convert into myxospores.

The mol% G + C of the DNA is: 67–68 (Bd, T_m).

Type strain: ATCC 25190.

GenBank accession number (16S rRNA): M94281.

Additional Remarks: Reference strains include ATCC 33878, DSM 1035 (Sg a8).

2. **Stigmatella erecta** (Schroeter 1886) McCurdy 1971a, 48[AL]
e.rec'ta. L. fem. adj. *erecta* erect, upright.

Vegetative cells, myxospores, and swarm colonies as described for the genus. Fruiting bodies (Fig. BXII.δ.90) consist of yellow-brown to red-brown spherical to club-shaped sporangioles, often slightly larger than those of *Stigmatella aurantiaca* and, in contrast to the latter, single on white stalks. The stalks often are much longer than those of *Stig-*

A

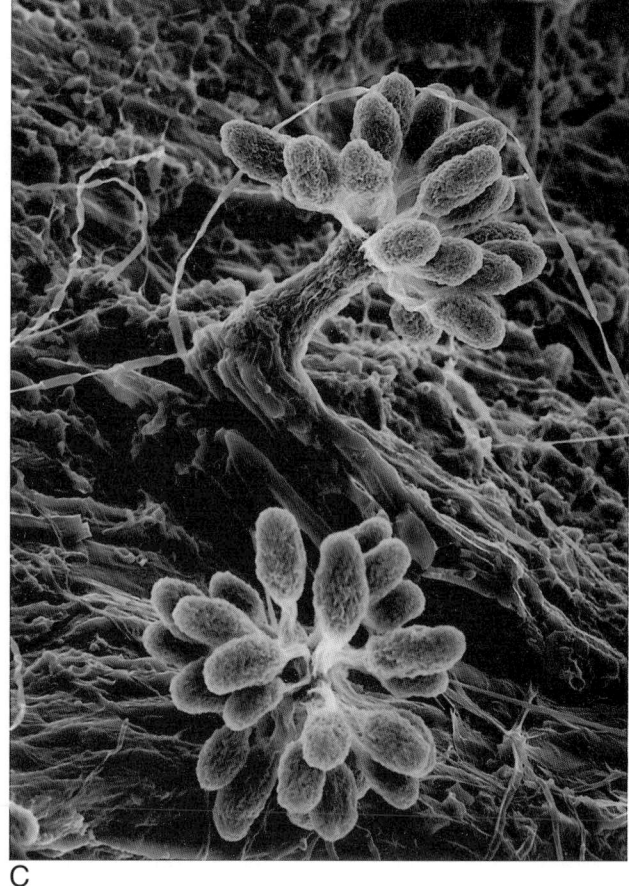

C

B

FIGURE BXII.δ.89. *Stigmatella aurantiaca*: fruiting bodies. (*A*) On agar surface, with sporangioles on long pedicles. (*B*) Young fruiting body at the edge of a piece of filter paper, under water and cover glass. (*C*) Scanning electron micrograph (× 315). Fruiting body on wood. Both a side and a top view are shown. Bar in *A* = 135 μm; bar in *B* = 65 μm.

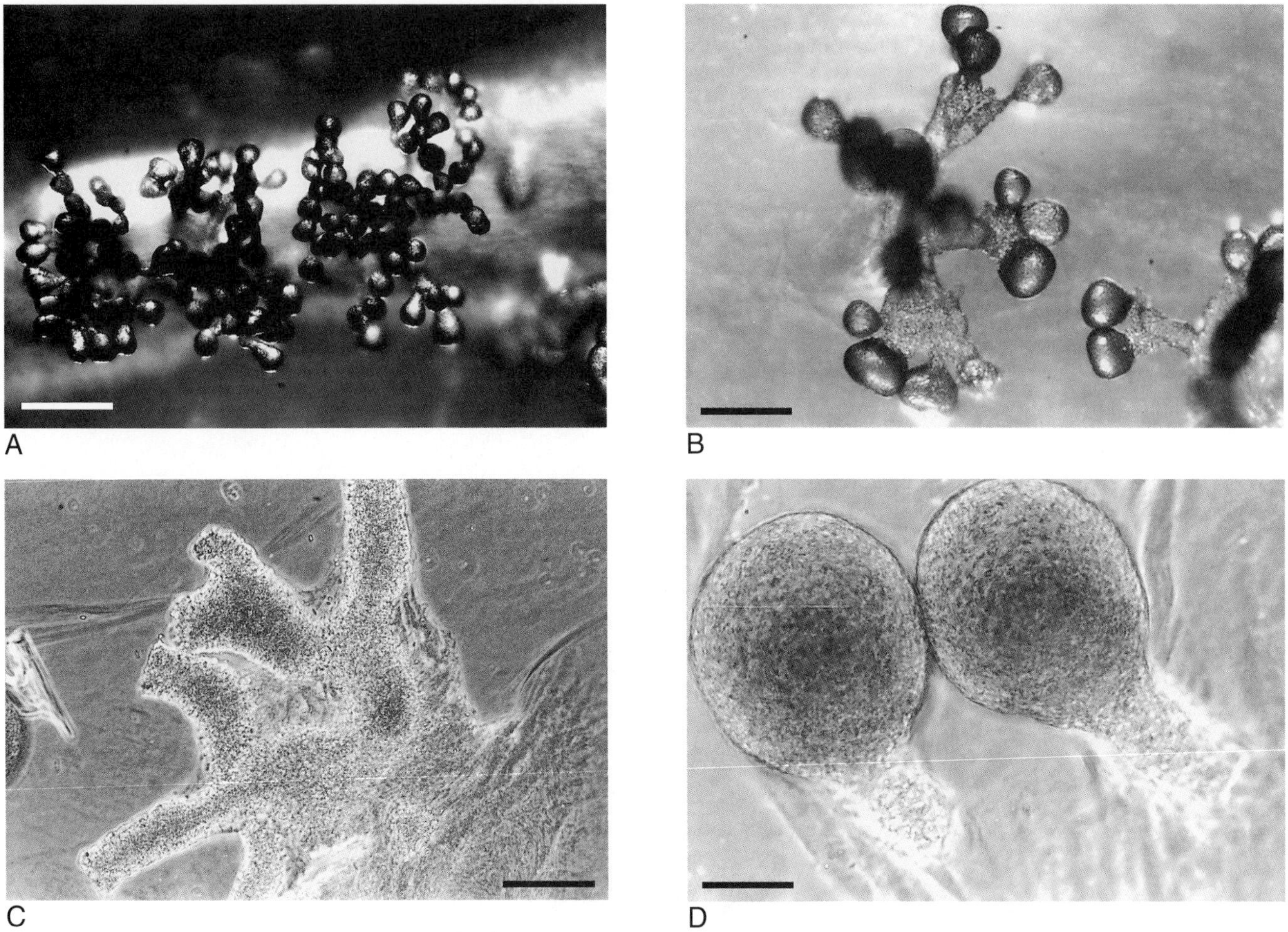

A

B

C

D

FIGURE BXII.δ.90. *Stigmatella erecta*: fruiting bodies. (*A* and *B*) On agar surface: The fruiting bodies tend to arise in dense clusters (*A*). As the stalks often fuse from the base upward, *S. erecta* fruiting bodies may resemble those of *Stigmatella aurantiaca* (*B*). (*C*) Cluster of stalks, with the sporangioles shed, in phase contrast. (*D*) Sporangioles, slide mount, in phase contrast. Bar in *A* = 160 μm; bar in *B* = 100 μm; bar in *C* = 40 μm; bar in *D* = 20 μm.

matella aurantiaca fruiting bodies, and sometimes are branched, usually near the base, but if the branching is higher up, the fruiting body can be mistaken for a degenerate *Stigmatella aurantiaca* fruiting body. The fruiting bodies of *Stigmatella erecta* often arise in large, dense tufts of 50–100 or more.

Stigmatella erecta can be found in soil and on dung of herbivorous animals and seems to be a more common and more widely distributed organism than *Stigmatella aurantiaca*.

The mol% G + C of the DNA is: 68 (T_m).

Type strain: ATCC 25191.

Additional Remarks: Reference strains include ATCC 25192, ATCC 29622.

3. **Stigmatella hybrida** *sp. nov.*

hy'bri.da. L. masc. fem. n. *hybrida, hibrida* half-breed, bastard.

Vegetative cells (Fig. BXII.δ.91), myxospores (Fig. BXII.δ.92), and swarm colonies as described for the genus. Fruiting bodies (Fig. BXII.δ.93) consist of moderately large to large dark red brown to almost black spherical to elongated sporangioles, 40–100 μm in diameter. The sporangioles can be either solitary on white, sometimes rather delicate, stalks of varying length, or they can be single on

a slime pad, or they can be arranged in long, often convoluted and looping chains; these chains sometimes condense into large piles. Stalked fruiting bodies may arise singly or in tufts. Thus, the fruiting bodies of *Stigmatella*

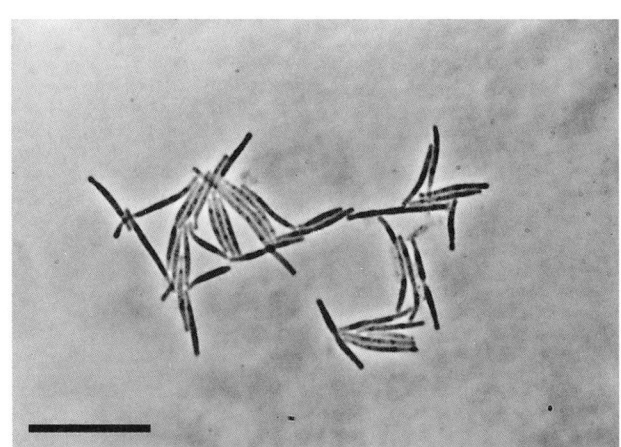

FIGURE BXII.δ.91. *Stigmatella aurantiaca*: vegetative cells from a liquid culture, in phase contrast. The vegetative cells of all *Stigmatella* species are identical in shape and size. Bar = 10 μm.

hybrida resemble those of *Stigmatella erecta*, but are darker in color and often sedentary or in chains. They resemble the fruiting bodies of *Cystobacter badius* and *Cystobacter velatus*, but differ in having their sporangioles on long stalks; further, *Stigmatella hybrida* has shorter, more delicate and fusiform vegetative cells and shorter, often bean-shaped myxospores. This mixture of morphological characteristics suggested the name of the species.

Stigmatella hybrida is relatively common in soil and on dung of herbivores and has a wide distribution. Strain Sg h20 was isolated in 2000 from soil with decaying plant material collected in 1999 on the island of Rhodos, Greece, and reference strain Sg h8 was isolated in 2000 from a similar sample collected in 1997 in the saguaro savanna west of Tucson and Sells, AZ, USA.

The mol% G + C of the DNA is: not determined.

Type strain: Sg h20, DSM 14722, JCM 12640.

Additional Remarks: Reference strains include DSM 14737 (Sg h8).

Additional support for the classification of the organism in the genus *Stigmatella* rather than *Cystobacter* rests on the observation that the pattern of secondary metabolites corresponds to that of the other two *Stigmatella* species and not to that of the cystobacters (Dr. Brigitte Kunze, personal communication).

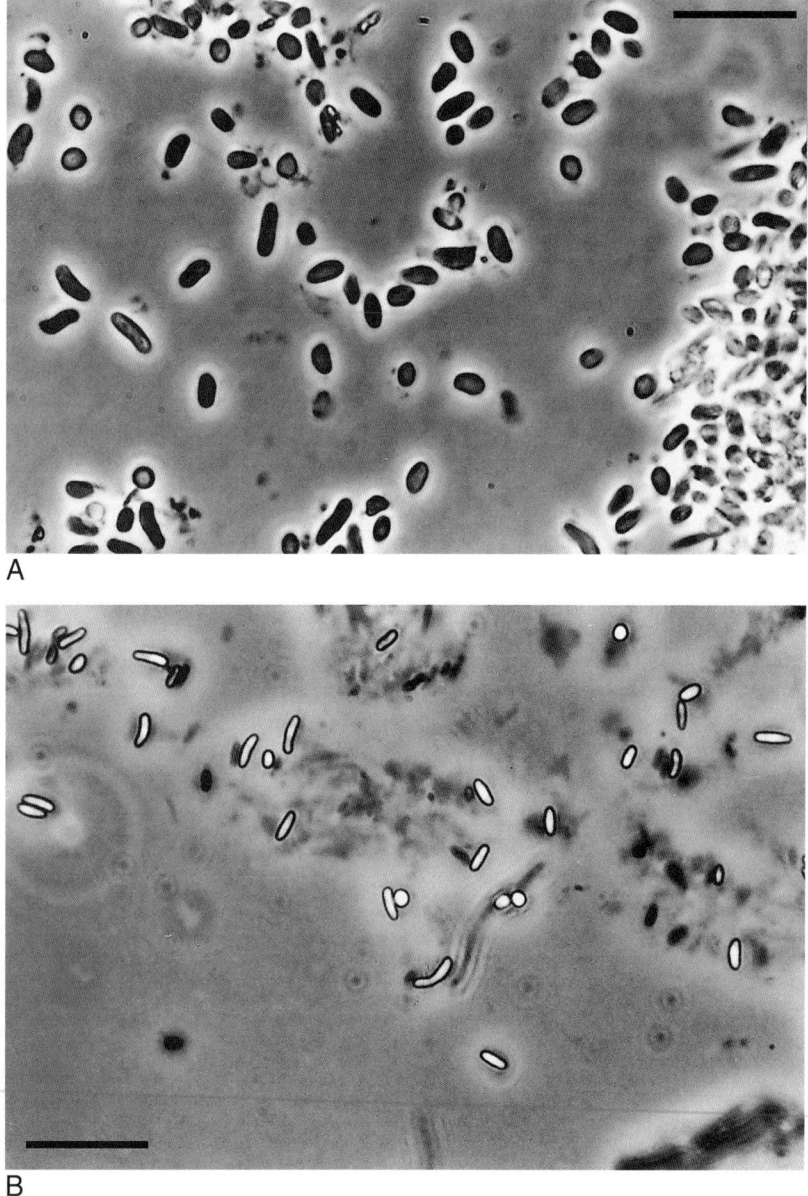

FIGURE BXII.δ.92. *Stigmatella*: myxospores. The myxospores of the *Stigmatella* species are the same in shape and size. (*A*) Myxospores of *Stigmatella erecta* from crushed sporangioles, in phase contrast. (*B*) Myxospores of *S. aurantiaca* produced outside of fruiting bodies on a thin agar layer in a chamber culture. Under such conditions, the optical refractility of the myxospores becomes particularly striking. Bar in *A* = 10 μm; bar in *B* = 25 μm.

Family III. **Polyangiaceae** Jahn 1924, 75[AL]

HANS REICHENBACH

Po.ly.an.gi.a' ce.ae. M.L. neut. n. *Polyangium* type genus of the family; *-aceae* ending to denote a family; M.L. fem. pl. n. *Polyangiaceae* the *Polyangium* family.

Vegetative cells slender, cylindrical rods with blunt, rounded ends. Often contain optically bright granules of reserve material. **The fruiting bodies always consist of sporangioles**, which may be single or, more often, **clustered in sedentary sori or borne on a slime stalk**. In cultures, strains often refuse to produce fruiting bodies. Yet degenerate fruiting bodies, as are so typical for the *Cystobacteraceae*, are only rarely formed. **The myxospores in the fruiting bodies are morphologically not much different from vegetative cells**, perhaps somewhat shorter and slightly constricted around the middle. However, they are true, optically refractile resting cells, which are as desiccation and heat resistant as the myxospores of the two previous families. **The swarm colonies** tend to etch the agar, producing radial tracks, holes, and pits in the surface of the plate, and often penetrate the substrate deeply. In the genera *Polyangium* and *Byssophaga* discontinuous swarm colonies with scattered cell clusters occur in addition to the more typical coherent swarms with slime sheets and veins. Congo red is not adsorbed to the slime of the swarms. Most species are of the proteolytic–bacteriolytic nutritional type, but some are cellulose degraders and, in contrast to all other myxobacteria, may be cultivated on very simple media with an inorganic nitrogen source and a sugar as the only carbon source.

Type genus: **Polyangium** Link 1809, 42.

Key to the genera of the family Polyangiaceae

I. Fruiting bodies consist of sporangioles without a stalk arranged singly or, more often, in sori, in which they may be so tightly packed as to become polyhedral.

    A. Of the proteolytic-bacteriolytic nutritional type, require an organic nitrogen source.

        1. Sporangioles in clusters (sori).

Genus *Polyangium*

        2. Sporangioles solitary.

Genus *Haploangium*

    B. Cellulose degraders; may be cultivated on inorganic nitrogen sources.

        1. Swarm colonies sheet-like, with more or less prominent veins. When grown on filter paper, sporangioles are produced in enormous numbers in the macerated paper; sori are yellow, orange, brown, or black, but never red.

Genus *Sorangium*

        2. Grows in the form of independently migrating cell companies, or pseudoplasmodia, that finally contract into intensely cinnabar- to carmine-red knob-like masses, resembling *Myxococcus* fruiting bodies. Sporangioles rarely produced; red.

Genus *Byssophaga*

II. Fruiting bodies consist of sporangioles on a slime stalk or cushion.

    A. Sporangioles bright yellow orange, borne on a long, sometimes branched white stalk.

Genus *Chondromyces*

    B. Sporangioles orange to red brown, in convoluted, brain-like chains, often on a prominent slime cushion.

Genus *Jahnia*

FURTHER DESCRIPTIVE INFORMATION

In most cases, it is easy to recognize that a strain belongs to this suborder, because the vegetative cells are so characteristic and clearly different from those of the *Cystobacterineae* and most organisms of the *Nannocystineae*. In addition, the swarm colonies with their pits and etched agar surface are typical. There are two confusing overlaps, however. Strains erroneously classified for some time as *Polyangium vitellinum* (Pl vt1) turned out to have a 16S rRNA sequence that connects them with the *Nannocystis* group (Spröer et al., 1999). Those organisms have the same cell type as *Polyangium*, i.e., slender, cylindrical rods with rounded ends, and fruiting bodies clearly different from those of *Nannocystis*. The swarms do not etch the agar and usually produce a very tough slime sheet. In fact, this bacterium was apparently found long ago by other investigators, e.g., by Kofler (1913), who regarded it as a new species, *"Polyangium flavum"* (*"Archangium flavum"* Jahn 1924). The organism is renamed here *Kofleria flava* in his honor and is transferred to the suborder *Nannocystineae*. The swarm colonies of *Nannocystis* are another possible source of confusion. Those organisms also etch the agar, sometimes transforming the plate into a spongy mass, but tracks and holes produced by *Nannocystis* and the cell clusters in them are usually much smaller than those of *Polyangium*. In addition, the vegetative cells, myxospores, and fruiting bodies of *Nannocystis* are different from those of *Polyangium*.

The type genus of the family, *Polyangium*, is the first myxobacterium ever to be described. Link (1809) did not recognize *Polyangium vitellinum* as a myxobacterium and thought it to be a tiny gasteromycete. But he published very clear drawings of it, so that there can be little doubt that Thaxter's (1892) *"Myxobacter aureus"* was the same organism. For a long time the genera *Polyangium* and *Cystobacter* were not distinguished, because their fruiting bodies look similar, and all those species were classified as *Polyangium*, until the fundamental difference in vegetative cells and myxospores of the two was recognized (McCurdy, 1970). Because of incomplete descriptions of some of those old species, it is impossible today to decide whether they belong to the one or the other genus. In addition, in the early days of myxobacterial research some species were labeled *Polyangium* even though sporangioles had never been observed. A similar conflation of the genera *Chondromyces* and *Stigmatella* occurred, as has been pointed out above. These cases show clearly that certain basic shapes of fruiting bodies have evolved independently in myxobacteria that are rather distantly related and that fruiting bodies alone are not a sufficient basis for classification.

The genus *Sorangium* was originally created by Jahn (1924) to comprise species with angular sporangioles. He even established a new family for those organisms. Angular sporangioles, however, are not a characteristic to build on, because the shape of the sporangioles depends very much on the culture conditions, which determine whether the sporangioles are densely or loosely packed. Spherical and polyhedral sporangioles may be found in the same culture, and polyhedral sporangioles are seen occasionally in *Cystobacter* and *Pyxicoccus* cultures. It was therefore justified to eliminate a genus based on such a definition (McCurdy, 1970). However, several of the first cellulose-degrading myxobacteria to be described (Krzemieniewska and Krzemieniewski, 1937a, b) were classified as *Sorangium* species, and as cellulose degradation appears to be a very stable and unique

characteristic, the genus *Sorangium* is revived here for those organisms. A new genus, *Byssophaga*, is established below for a second, newly discovered, and clearly different cellulose degrader. While *Sorangium* 16S rRNA sequences cluster closely with those of proteolytic *Polyangium*, the two differ in other characteristics besides cellulose degradation, e.g., in the ability to synthesize lanosterol (only *Polyangium*), in swarm morphology, and in sensitivity to certain antibiotics. Some *Sorangium* strains have the largest genomes known for myxobacteria (12 Mbp) (Pradella et al., 2002).

All seven *Chondromyces* strains, for which 16S rRNA sequences have been determined, form one cluster that is strongly separated from the *Polyangium*–*Sorangium* cluster. This finding could justify creating a separate family for the chondromycetes. For the moment, they are still included in the family *Polyangiaceae* until more information becomes available. *Chondromyces crocatus* is another one of those early myxobacteria that were first described as fungi, in this case a deuteromycete (Berkeley, 1857). The *Chondromyces* group received an unexpected addition through the 16S rRNA sequence of "*Polyangium thaxteri*" (Spröer et al., 1999). A new genus, *Jahnia*, will be created here for this organism.

Genus I. **Polyangium** *Link 1809, 42*[AL]

HANS REICHENBACH

Po.ly.an'gi.um. Gr. adj. *polys, poly* many, much; Gr. neut. n. *angion* vessel, container; M.L. neut. n. *Polyangium* many vessels.

Vegetative cells (Fig. BXII.δ.94) **slender, cylindrical rods with blunt, rounded ends. Fruiting bodies consist of spherical or polyhedral sporangioles,** usually in sedentary sori, on agar plates often deep within the substrate, sometimes surrounded by a prominent slime layer, or matrix. **Myxospores morphologically similar to vegetative cells** but optically refractile. **The swarm colonies** (Fig. BXII.δ.95) **often etch the agar** producing deep tracks and tunnels, holes, and pits; often penetrate the agar deeply with a curtain-like swarm edge; swarms often not coherent but broken up into many small, scattered groups of cells that migrate independently, like pseudoplasmodia.

Of the proteolytic–bacteriolytic nutritional type. All strains that have been investigated synthesize lanosterol (Zeggel, 1993).

Type species: **Polyangium vitellinum** Link 1809, 42.

FURTHER DESCRIPTIVE INFORMATION

While it is relatively easy to recognize a strain as a member of the genus *Polyangium*, the differentiation of species is not yet very reliable for several reasons. The first threshold is easy to cross. Beginning with Thaxter (1904) the *Cystobacter* species were lumped into the genus *Polyangium*, because the fruiting bodies looked so similar. The fundamental difference in cell shape be-

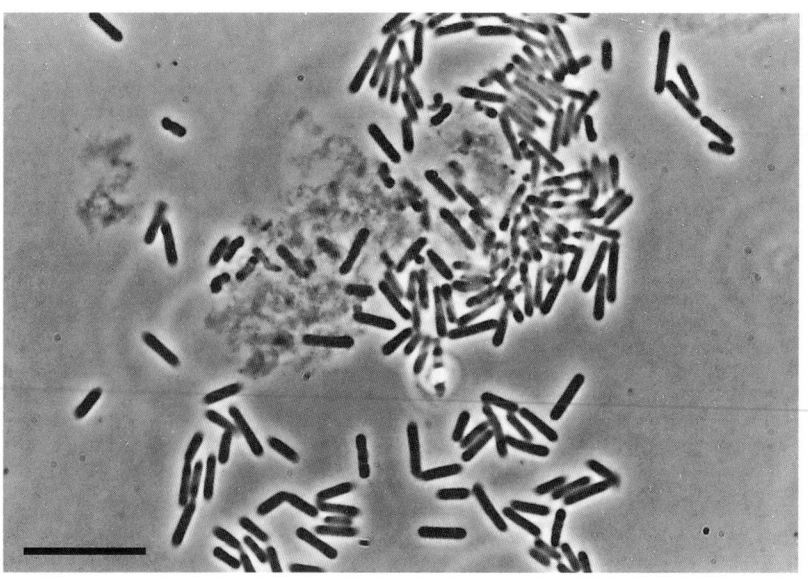

FIGURE BXII.δ.94. *Polyangium spumosum*: vegetative cells from yeast agar, phase contrast. The vegetative cells of all *Polyangium* species are very similar in shape and size. Bar = 13 μm.

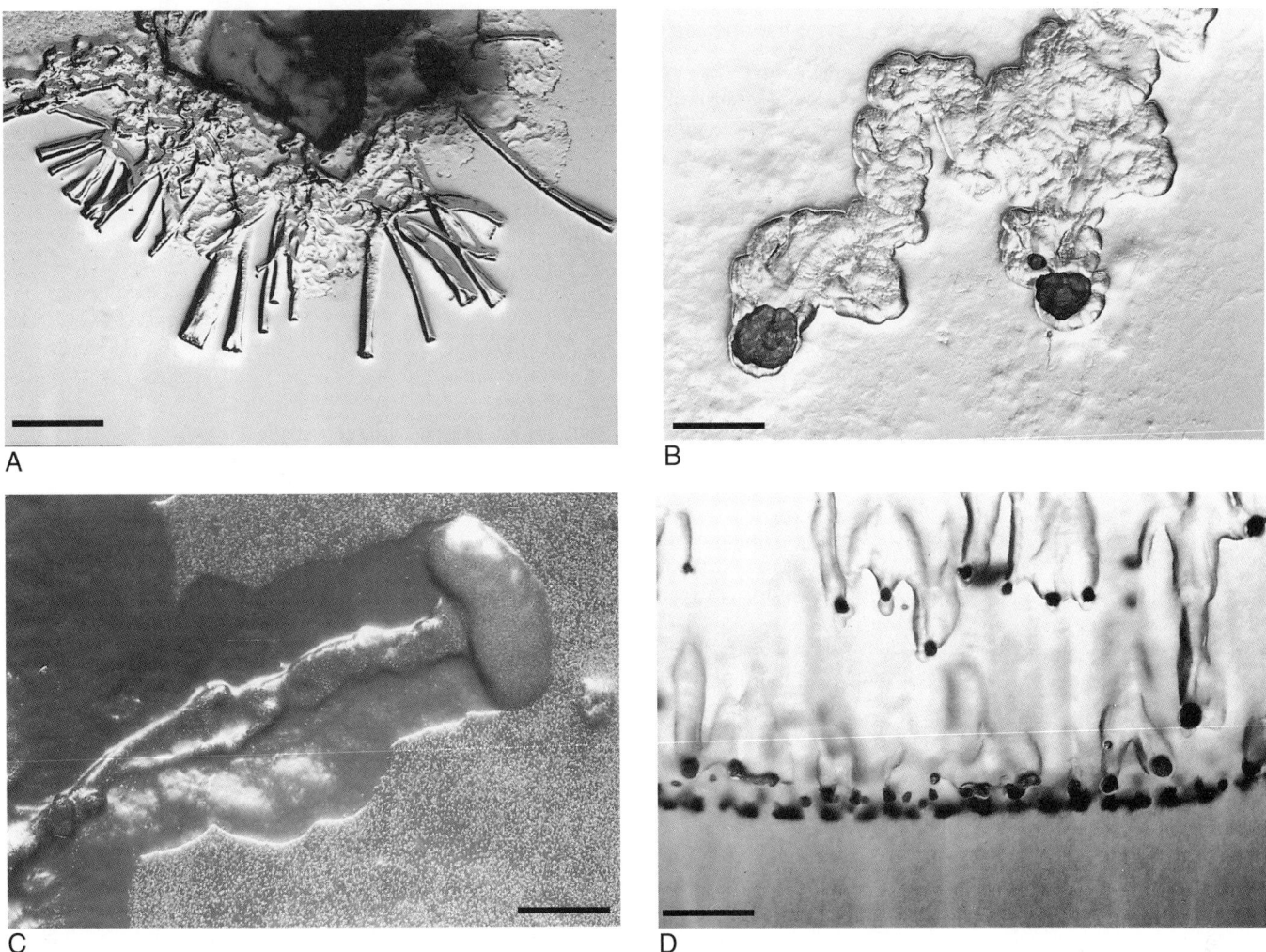

FIGURE BXII.δ.95. *Polyangium* species: swarm colonies on agar plates. (*A*) Growing on a streak of living *E. coli* on water agar, small packs of cells migrate radially away from the inoculum producing deep channels in the agar surface (bar = 1460 μm). (*B*) and (*C*) On streaks of autoclaved *E. coli* (bar in *B* = 960 μm; bar in *C* = 260 μm). (*C*) The cell mass eats a path into the *E. coli* streak and leaves a tail-like slime strand behind. (*D*) *Polyangium fumosum* on yeast agar: the swarm penetrates the agar plate deeply, the swarm edge is a curtain-like sheet of small cell clumps within the agar; note the lytic clearing of the yeast cells inside the swarm (bar = 50 μm).

tween *Polyangium* and *Cystobacter* was observed early (e.g., Krzemieniewska and Krzemieniewski, 1928), but the taxonomic relevance of this observation was not realized until much later (McCurdy, 1970). As mentioned above, the shape change during sporulation and, as a rule, the swarm morphology are clearly different in the two genera. Therefore, the separation of *Cystobacter* (and *Hyalangium*) and *Polyangium* is not problematic. However, over the years many organisms have been described that appear to be real *Polyangium* species, but often are difficult to distinguish and to evaluate based on present data. Most of those species are rare—or at least their preferred habitats have not yet been identified—and most of them have never been cultivated, so that many descriptions are incomplete and imprecise. There is often doubt as to whether later investigators were talking about the same organism when they used a name from the literature. Finally, no type strains are available. Consequently, many *Polyangium* species have been eliminated from the List of Approved Names and from the last edition of the *Manual*. Yet some of the descriptions sound realistic enough to make one suspect that those species may really exist as distinguishable taxa. Several of those organisms are included in the following key to facilitate orientation and to make it easier for future investigators to de-

termine whether a new isolate belongs to a previously described species. On the other hand, the author has isolated several strains that cannot be classified as any of the present species. Apparently, there are even more *Polyangium* species.

Polyangium strains are somewhat difficult to isolate and to handle, but with sufficient effort, all species should be culturable. They all grow on yeast agar (VY/2-), but for unexplained reasons the bacteria sometimes do not grow after a transfer to the same medium on which they grew well before. It seems that the age and perhaps the size of the inoculum is critical. However, fermentations at a relatively large scale (100 l) have successfully been performed with several strains (e.g., Kunze et al., 1993).

As already indicated, a reorganization of the genus is proposed below. The cellulose degraders are removed from the genus *Polyangium* and returned to the genus *Sorangium*. Further, species with solitary sporangioles are collected in the genus *Haploangium* as formerly defined. The remaining genus *Polyangium* may still not be completely homogeneous, as the species with very large sporangioles may not be closely related to the species with small, more or less angular sporangioles. This question can be answered only when more data become available.

Key to the species of the genus *Polyangium*

I. Not parasitic on freshwater algae.
  A. Sporangioles yellow to orange brown.
    1. Sporangioles spherical, bright yellow to orange, large (80–350 µm), only few in the sorus, sorus with a prominent slime envelope.
      a. Sorus in a hyaline, gelatinous, colorless slime matrix.
        Polyangium vitellinum
      b. Slime envelope of the sorus bright yellow.
        Polyangium luteum
    2. Sporangioles spherical or angular, colorless to golden brown, moderately large to small.

      a. Sporangioles golden brown, spherical, of moderate size (around 40 µm); sori with 2–12 sporangioles in a slime matrix that is often yellow.
        Polyangium aureum
      b. Sporangioles colorless to yellow-orange and reddish-brown, angular, small (5–15 µm), in large numbers densely packed, in sheet-like sori without a prominent slime envelope.
        Polyangium sorediatum
  B. Sporangioles brown or gray to almost black.
    1. Sporangioles in the sorus of very variable size, roundish, in three-dimensional translucent packets looking like a drop of foam, brownish.
      Polyangium spumosum
    2. Sporangioles in the sorus more or less of the same size, sorus often sheet-like or in the form of a ring, opaque, smoky-gray to blackish.
      Polyangium fumosum
II. Parasitizes freshwater algae (*Cladophora*).
      Polyangium parasiticum

List of species of the genus Polyangium

1. **Polyangium vitellinum** Link 1809, 42[AL]

vi.tel.li' num. L. masc. n. *vitellus* egg yolk; L. neut. adj. *vitellinum* in the color of egg yolk.

Vegetative cells cylindrical with rounded ends, 0.7–0.9 × 4–10 µm. Sporangioles spherical to ellipsoidal, golden yellow, 75–350 µm in diameter, with a thick, golden yellow wall; occur in groups of 4–6 (1–12 according to different authors) in a common, gelatinous, hyaline, and colorless slime envelope or matrix. Size of sori 0.7–1 mm (up to 4 mm) (the size of a large grain of sand; Link, 1809). Myxospores resemble vegetative cells in shape, but are shorter, 0.9 × 3.5 µm. Older sporangioles contain a yellow, oily material. Observed on decaying wood and bark, particularly under very wet conditions, e.g., in swamps. The organism appears to have never been cultivated.

The mol% G + C of the DNA is: not determined.

Type strain: Acc. no. 4564, Thaxter collection, Farlow Herbarium, Harvard University, Cambridge, MA, USA.

Additional Remarks: Polyangium vitellinum was well known to the botanists who used to roam forests and swamps in search of myxomycetes and fungi. They found the organism normally on wet wood and bark, and in pools, ditches, and swamps, a few sori at a time. Their descriptions of the organism are the most reliable ones (Thaxter, 1893; Zukal, 1897b; Quehl, 1906). There is some uncertainty whether later authors really were always dealing with the same organism when they wrote about *Polyangium vitellinum*, and how much they relied on earlier records for their descriptions. Jahn's description (Jahn, 1911, 1924) was essentially the same as that of Quehl (1906). Krzemieniewska and Krzemieniewski (1930) obtained an organism from forest

soil that resembled Thaxter's *Polyangium vitellinum*. In addition their photograph seems to fit the appearance of *Polyangium vitellinum*. The size of the sporangioles is at the lower end of the earlier measurements, and the habitat was perhaps not quite typical. Also McCurdy's description (McCurdy, 1970) may refer to *Polyangium vitellinum*. The color of the sporangioles was more reddish or brownish, the number of sporangioles in the sori (20) exceeded somewhat that given in the older reports, and the photograph does not show the prominent slime envelope (which may, however, have dried up). The organism described and cultivated by Dawid (1977) and illustrated in the former edition of the *Manual* differs from *Polyangium vitellinum* substantially in the size of the sporangioles, at least if the magnification of the photograph is given correctly. In addition, the ease with which it was cultivated is somewhat suspect. The bacterium identified by the author as *Polyangium vitellinum* and used for 16S rRNA sequencing (Spröer et al., 1999) was clearly not *Polyangium vitellinum*, and is now classified in the new genus *Kofleria*. In fact, this author has not yet seen a real *Polyangium vitellinum*. With the exception of Dawid's organism, the bacterium could never be cultivated in pure culture, but it was propagated on the natural substrate, wood, which was kept very wet for a long time in a dish with water (Quehl, 1906).

Polyangium vitellinum was first described by Link (1809) who classified it as a tiny gasteromycete. Incidentally, he also used first the term *sporangiole* for the encasement of the myxospores. The organism was again described and illustrated, this time in color and at its natural size, by Dittmar (1814), who in fact had collected the original sample on which Link's description was based. Zukal (1897b) recognized the identity of *Polyangium vitellinum* and *"Myxobacter aureus"* Thaxter 1892, which Thaxter (1904) readily accepted. The demise of the genus *Myxobacter* was already announced by Thaxter in 1897, when he recognized the similarity between *Myxobacter* and the older genus *Cystobacter*. *Polyangium vitellinum* became the type genus and species for the whole group of bacteria and even gave its name to the organisms, the polyangides (Die Polyangiden; Jahn, 1924) in the first monograph on myxobacteria.

Thaxter (1893) described the species *'Myxobacter simplex"*, which he later transferred first to *Cystobacter* (1897) and then to *Polyangium* (1904). It had very large, solitary, bright reddish yellow sporangioles, 250–400 μm in diameter. McCurdy (1970) regarded the species as a variant of *Polyangium vitellinum*; this conclusion was supported by his examination of Thaxter's specimens in the Farlow Herbarium at Harvard University. An organism later described as *"Haploangium simplex"* (Singh and Singh, 1971) and regarded as identical with *Polyangium simplex* Thaxter was evidently not the same bacterium but a member of the *Cystobacterineae*.

2. **Polyangium aureum** (ex Krzemieniewska and Krzemieniewski 1930) Brockman 1989b, 495[VP] (Effective publication: Brockman 1989a, 2162.)

au' re.um. L. neut. adj. *aureum* golden.

Vegetative cells cylindrical rods with rounded ends. Sporangioles spherical to oval, light brown to red brown with an orange-yellow wall, 15–50 × 20–60 μm, 32 × 37 μm on average. Older sporangioles may contain a colorless to yellow liquid besides a few myxospores. Sori contain 2–12 sporangioles in a transparent, sometimes yellow slime matrix. Myxospores resemble vegetative cells, 0.7–0.9 × 2.8–5 μm. Obtained from soil.

The mol% G + C of the DNA is: not determined.

Type strain: no culture.

Additional Remarks: The description seems to refer to a definite type, but the status of the species is not yet clear. Krzemieniewska and Krzemieniewski mentioned that their new species resembled *"Polyangium morula"* Jahn in some respects; the latter species differed mainly in the lack of a thick, yellow slime matrix around the sorus. Whether this is relevant is questionable, and the authors admitted that the two species could be identical. However this is of little help, as the status of *"Polyangium morula"* is also unclear.

3. **Polyangium fumosum** Krzemieniewska and Krzmieniewski 1930, 253[AL]

fu.mo' sum. L. neut. adj. *fumosum* smoke or soot colored.

Vegetative cells cylindrical with rounded ends, 0.6–0.9 × 4–7 μm. Sporangioles spherical or somewhat elongated, often slightly polyhedral, colorless to grayish, with transparent walls, 15–25 × 25–40 μm. Sori (Fig. BXII.δ.96) usually flat, sheet-like, roundish to elongated, often ring-shaped, consisting of 2–50 and more sporangioles; at first pale brown, later smoky gray to soot black. The pigment is found mainly in the slime envelope of the sorus, which, in older fruiting bodies, may transform into a wrinkled sac. Myxospores similar to vegetative cells in shape, 0.7–0.9 × 2.5–5.5 μm. In old sporangioles, the myxospores often degenerate and an oily liquid develops. The organism grows well on yeast agar (VY/2-) with strong lysis of the yeast cells, deeply penetrating the agar plate. Obtained from soil. Not uncommon. Strain Pl fu5 was isolated in 1994 from soil with decaying plant material collected in California, USA, the year before. Reference strain Pl fu11 was isolated in 1999 from a similar sample collected in the same year, also in California.

The mol% G + C of the DNA is: not determined.

Type strain: Pl fu5, DSM 14668, JCM 12636.

Additional Remarks: Reference strains include DSM 14669 (Pl fu 11) Windsor M257.

This *Polyangium* species is easy to distinguish. The moderately sized sporangioles in sheet-like, often ring-shaped, smoky gray to black sori are quite typical. The organism has also been reported and cultivated by other investigators (McCurdy, 1970, Dawid, 1977). It is found relatively often in soil with decaying plant material. It does not attack cellulose.

4. **Polyangium luteum** (ex Krzemieniewska and Krzemieniewski 1927) Brockman 1989b, 495[VP] (Effective publication: Brockman 1989a, 2160.)

lu' te.um. L. neut. adj. *luteum* saffron, golden yellow.

Vegetative cells cylindrical with rounded ends, 0.8–1.0 × 6–8 μm. Sori golden yellow with a thick, pale to bright yellow slime matrix containing 1–2, rarely 3–4 spherical or ellipsoidal sporangioles with thin, colorless walls, 40–120 × 40–90 μm in diameter. The yellow color of the sporangioles seems to be due to the cells within. Myxospores 0.7–0.8 × 4–6 μm. An organism described by Kühlwein and Schlicke (1971) as *Polyangium luteum* produced sori 100–180 × 85–170 μm, with a colorless slime matrix. This organism was isolated from rabbit dung and cultivated as a pure strain. Its swarm colony on agar with suspended *E. coli* resembled that of *Cystobacter fuscus*, which is not typical for a *Polyangium*. The organism hydrolyzed starch, but not cellulose or agar, and could be grown in liquid media in suspension culture, which again is unusual for a *Polyangium*. The author doubts whether this organism was really the *Polyangium luteum* of Krzemieniewska and Krzemieniewski; it is remotely similar to *Kofleria*. Found in soil and on rabbit dung, apparently very rare. New isolates for study would be desirable.

The mol% G + C of the DNA is: not determined.

Type strain: no type strain.

5. **Polyangium parasiticum** (ex Geitler 1925) Brockman 1989b, 495[VP] (Effective publication: Brockman 1989a, 2162.)

pa.ra.si' ti.cum. Gr. adj. *parasitikos;* L. neut. adj. *parasiticum* parasitic.

Vegetative cells cylindrical with rounded ends, 0.7 × 4–7 μm. Sporangioles spherical to somewhat elongated, sometimes slightly polygonal, with a sturdy red brown wall, 25–40 μm in diameter. Mature sporangioles may contain a large drop of an oily liquid. Sori contain 2–20 and more sporangioles within a colorless, hyaline slime matrix. Myxospores resemble vegetative cells, shorter by one-quarter of a cell length at the most. Found on the green alga, *Cladophora*, in a pond near Vienna. Cultivated only on the natural substrate. The organism first grew saprophytically on dead cells of the alga, but when the bacterial population became sufficiently large it attacked young, living cells by digesting a hole in the cell wall and entering the algal cells; the protoplasts, including storage granules of starch, were then completely lysed. Sporangioles were produced within

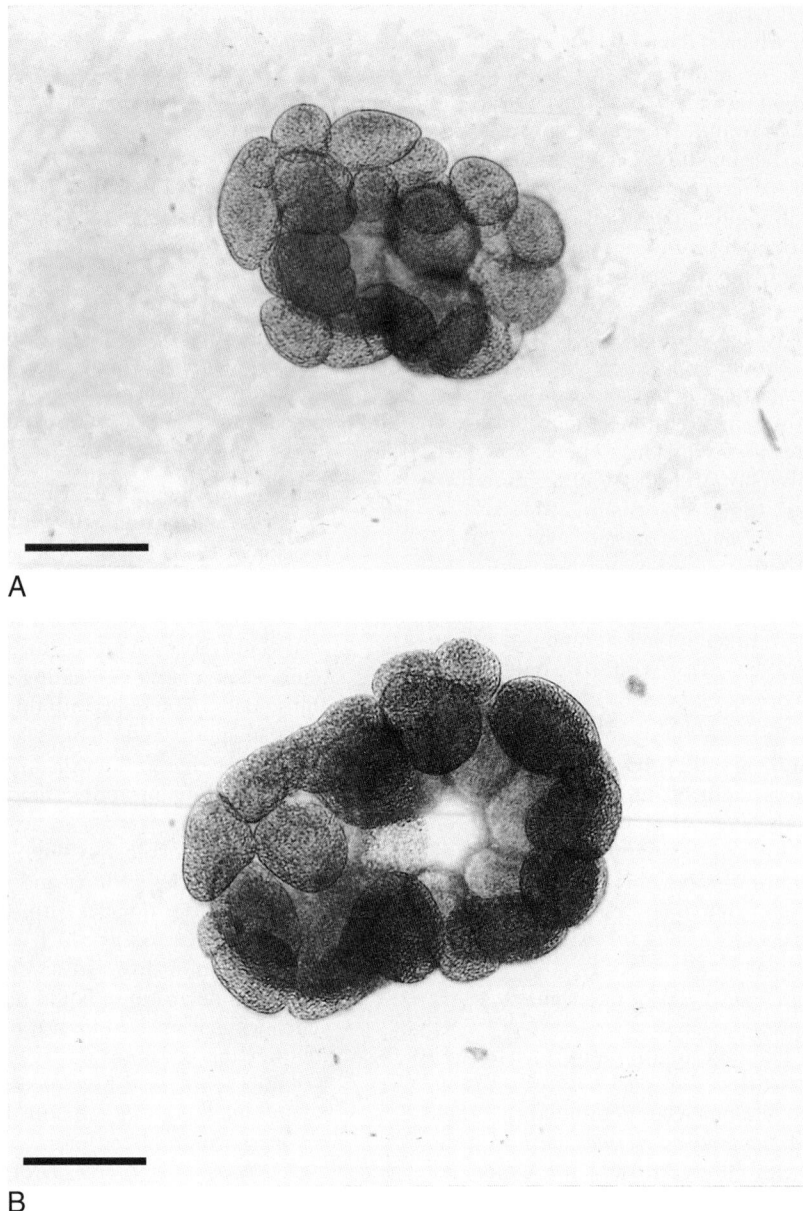

FIGURE BXII.δ.96. *Polyangium fumosum*: fruiting bodies on an agar plate under water and cover glass. The sporangioles are often arranged in rings. Bars = 50 μm.

and outside the *Cladophora* filaments. The myxobacterium did not colonize any other algae in the aquarium.

The mol% G + C of the DNA is: not determined.

Type strain: no type strain.

Additional Remarks: This organism was unquestionably a *Polyangium*. However, the status of the species is undecided. The bacterium also grew also saprophytically, as Geitler himself observed, and it very probably was a soil organism that facultatively parasitized *Cladophora* filaments. A good candidate may be *Polyangium aureum*, which also has red-brown sporangioles of about the same size. The reason why the myxobacterium was found exclusively on *Cladophora* is unknown; perhaps this alga was the only one in the culture with a surface suitable for the myxobacterium. The organism has not been seen since by any other investigator.

A similar case of a parasitic attack, this time on fungi, was reported many years later by Homma (1984). Again, the myxobacterium was a *Polyangium*. It had cylindrical vegetative cells with rounded ends and produced pale to red brown sporangioles that were, however, much smaller (12–22 μm in diameter) than those of *Polyangium parasiticum*. The myxobacterium made many small perforations in the fungal cell wall, entered the cells, and emptied them of their content. Finally, it produced sporangioles within and outside the fungal hyphae and conidia. The author did not give a name to his organism and left its status as a species open. It, too, may have been a facultatively parasitic soil *Polyangium*.

6. **Polyangium sorediatum** (ex Thaxter 1904) Brockman 1989b, 495[VP] (Effective publication: Brockman 1989a, 2161.)

so.re.di.a′ tum. Gr. masc. n. *soros* heap, pile; M.L. neut. adj. *sorediatum* piled up (i.e., the sporangioles).

Vegetative cells cylindrical rods with rounded ends, 0.8 × 3–6 μm. Sori (Fig. BXII.δ.97) colorless to yellow-orange and fox brown, with a very thin matrix, up to 0.5 mm large, flat, sheet-like or cushion-shaped, consisting of numerous roundish to polygonal sporangioles with definite walls, 5–15 μm in diameter, often in delimited clusters within the sorus. Myxospores resemble vegetative cells in shape, 0.8 × 3–4 μm. Readily cultivated on streaks of living or dead *E. coli* on water agar and on yeast agar (VY/2-); often continues to form fruiting bodies long after isolation. Swarm colonies with isolated, shallow tracks, cell clumps at the ends, often with a short slime tail. Does not degrade cellulose. Obtained from rabbit dung, soil, decaying plant material, bark. Moderately common organism, recognizable by its small sporangioles and sheet-like sori. Strain Pl s12

was isolated in 2001 from soil with decaying plant material collected in 2000 in San Diego, CA, USA. Reference strain Pl s4 was isolated in 1995 and came from a similar sample collected near Mumbay, India, in the same year.

The mol% G + C of the DNA is: not determined.

Type strain: Pl s12, DSM 14670, JCM 12637.

Additional Remarks: Reference strains include DSM 14693 (Pl ss4).

Polyangium sorediatum was reported repeatedly after Thaxter's description (Quehl, 1906; Krzemieniewska and Krzemieniewski, 1926, 1927; Dawid, 1977). Jahn (1924) felt that Quehl's strain was *"Polyangium compositum"* rather than *Polyangium sorediatum*, but this was refuted by Krzemieniewska and Krzemieniewski (1926). They also described (Krzemieniewska and Krzemieniewski, 1927) a variant: var. macrocystum, which had sporangioles of 6–10 × 7–16 μm, almost twice the size of Thaxter's organism. Whether this

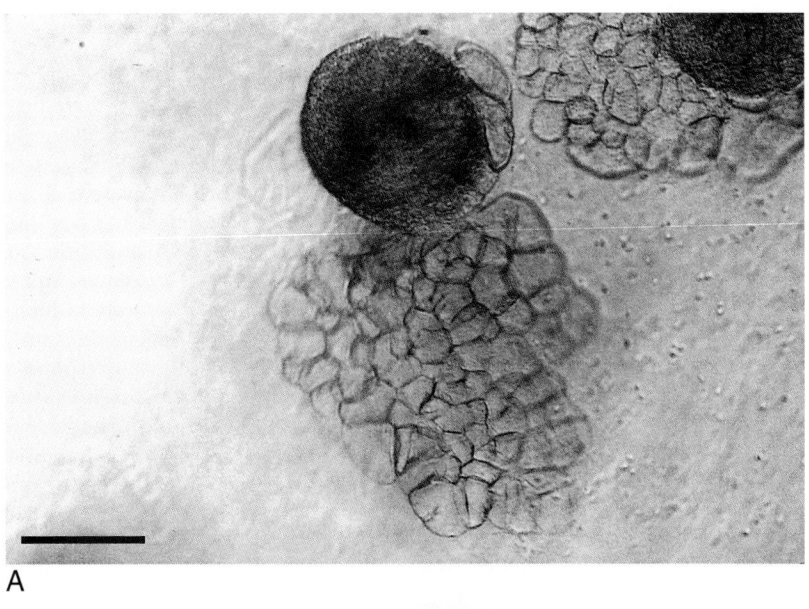

A

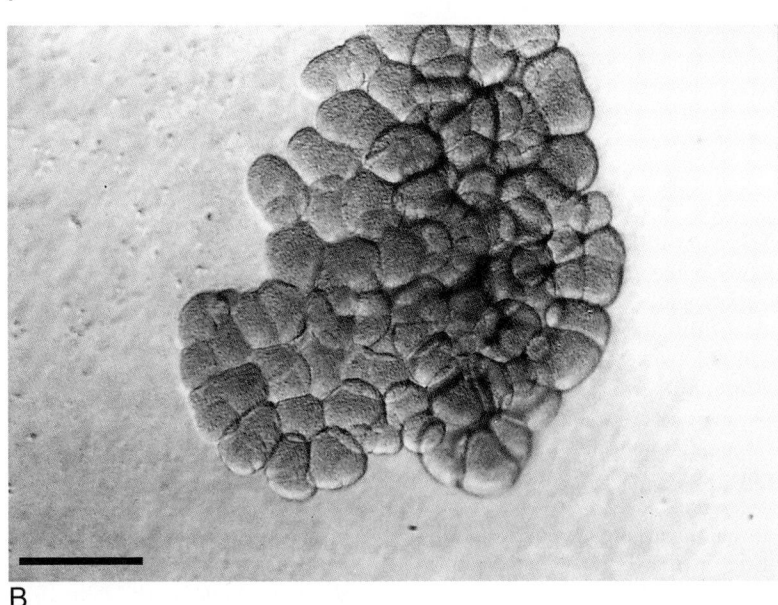

B

FIGURE BXII.δ.97. *Polyangium sorediatum*: fruiting bodies on agar surface. (*A*) The migrating cell mass has rounded up and is about to produce sporangioles. (*B*) Mature fruiting body. Bar in (*A*) = 65 μm; bar in (*B*) = 35 μm.

difference in size is a reliable characteristic, however, is disputable.

Several more species of *Polyangium* with small, more or less angular sporangioles were described in the early years of myxobacterial research. Jahn (1924) created the genus *Sorangium* for them. The sporangioles of some species were arranged in groups or packets within the sorus, with a well-defined slime envelope of their own. The species differed in the size of the sporangioles and packets and in the color of the sori, all highly variable characteristics that may change considerably even in subsequent cultures of a particular strain. In addition, none of those species has been cultivated, or at least not in pure form, and all were observed only once or rarely. Thus, while there may be fewer species than described by early authors, this author believes that some of them may be true species. Probably none of those species overlaps with *Polyangium sorediatum*, although that cannot be ruled out. For a quick survey, diagnoses of those species are given as far as one can deduce diagnostic characteristics from the old descriptions.

7. **Polyangium spumosum** (ex Krzemieniewska et Krzemieniewski 1930) Brockman 1989b, 495[VP] (Effective publication: Brockman 1989a, 2162.)

spu.mo′sum. L. neut. adj. *spumosum* foamy.

Vegetative cells cylindrical rods with rounded ends, 0.6–0.8 × 4–7 μm. Sporangioles spherical to ellipsoidal or slightly polyhedral with a colorless wall, 20–40 × 20–50 μm, 30 × 35 μm on average, about 20–60 per sorus. Sori (Fig. BXII.δ.98) spherical to somewhat elongated with an inconspicuous, hyaline slime envelope, colorless to light brown, 100–500 μm in size. The sporangioles in the sorus normally are of clearly different size. The sori are relatively transparent and usually three-dimensional resembling a bubble of foam. Myxospores similar to the vegetative cells in shape, only smaller. Older sporangioles may contain only a few myxospores in a granular mass, sometimes in addition to a colorless, oily liquid. Does not degrade cellulose. Grows rather well on yeast agar (VY/2-). Obtained originally from rabbit dung on soil. Not uncommon in soil and decaying plant material. Strain Pl sm5 was isolated in 1995 from soil with decaying plant material collected in 1994 near Braunschweig, Germany; the reference strains, Pl sm9 and Pl sm10, were both isolated in 1996 from similar samples collected at different places in Lucknow, Uttar Pradesh, India, in 1995.

The mol% G + C of the DNA is: not determined.

Type strain: Pl sm5, DSM 14734. JCM 12638.

Additional Remarks: Reference strains include DSM 14735 (Pl sm9), DSM 14736 (Pl sm10).

There is considerable confusion about this species. Krzemieniewska and Krzemieniewski (1927) described a new species, which they named "*Sorangium spumosum*". This bacterium clearly was not identical with the organism they reported in 1930 as *Polyangium spumosum*. The small size of the sporangioles, 8–26 × 7–20 μm, their colorless to brown walls, and their arrangement in single or double rows suggest that this was instead the bacterium that is listed in this edition as *Hyalangium minutum*. Therefore, the proposal of McCurdy (1970) to rename *Polyangium spumosum* as "*Polyangium paraspumosum*" in order to avoid confusion needs not to be followed; the older name is not valid anyway. Krzemieniewska and Krzemieniewski (1937a) described "*Sorangium spumosum*" again later, this time as a cellulose degrader. However, the vegetative cells were now much fatter, 1.1–1.5 μm wide, which would indeed be typical for *Sorangium cellulosum*. They distinguished two forms, one with smaller and one with somewhat larger sporangioles which, in both cases, were still smaller than those given for *Polyangium spumosum* but which would fit the description of *Sorangium cellulosum*. The author isolated about 20 *Polyangium* strains that correspond quite nicely with the original description of *Polyangium spumosum*. The fruiting bodies strikingly resemble a bubble of foam; this appearance which is caused by the three-dimensional arrangement of rather transparent sporangioles of different sizes. This organism grows very well on yeast agar (VY/2-) under lysis of the yeast cells, penetrating the plate deeply. It does not decompose cellulose. The author has deposited a strain of that organism.

Other Organisms

Apart from the *Cystobacter* species formerly named *Polyangium* and the species now classified as *Haploangium* and *Sorangium*, there are a few more *Polyangium* species in the literature that are now obsolete but may briefly be mentioned.

1. "*Polyangium compositum*" Thaxter 1904, 413. Sporangioles angular, small (former genus *Sorangium* Jahn 1924, 73). Sori completely resolved into sporangioles. Sorus dull yellowish orange; composed of one to six large packets with many sporangioles. Sori 100–170 μm; packets (often called "primary cysts", a confusing terminology now abandoned) 75 × 100 μm on average, sporangioles ("secondary cysts") 10–15 μm. Found on rabbit dung, not cultivated. Probably the same organism that has been reported repeatedly by later authors (Jahn, 1924; Krzemieniewska and Krzemieniewski, 1926, 1927) with slight variations in dimensions and color of the fruiting bodies. Even at that time there was some controversy about identity with Thaxter's species. The slime envelope of the packets was occasionally colored. The organisms were obtained from rabbit and deer dung and from soil, but apparently never cultivated. Jahn (1924) and Krzemieniewska and Krzemieniewski mention that sometimes the development of packets was inconspicuous or absent. Krzemieniewska and Krzemieniewski also described a variant with smaller packets, a further subdivision of the clusters, and sporangioles ("tertiary cysts") of about the size of the type. A similar morphology combined with a much wider variation in color is observed in the *Sorangium cellulosum* complex, so that the question arises whether at least some of the reported "*Polyangium compositum*" strains could have belonged to that species. *Sorangium cellulosum* may indeed be found on substrates like rabbit dung. Krzemieniewska and Krzemieniewski (1937a) reported later that "*Polyangium compositum*" indeed degraded cellulose. Whether or not that organism was identical with the earlier described strains must remain open.

2. "*Polyangium morula*" Jahn 1911, 202. Sporangioles spherical to slightly polygonal, moderately sized, 20–35 μm in diameter, with a thick wall, bright yellow, in mulberry-like sori

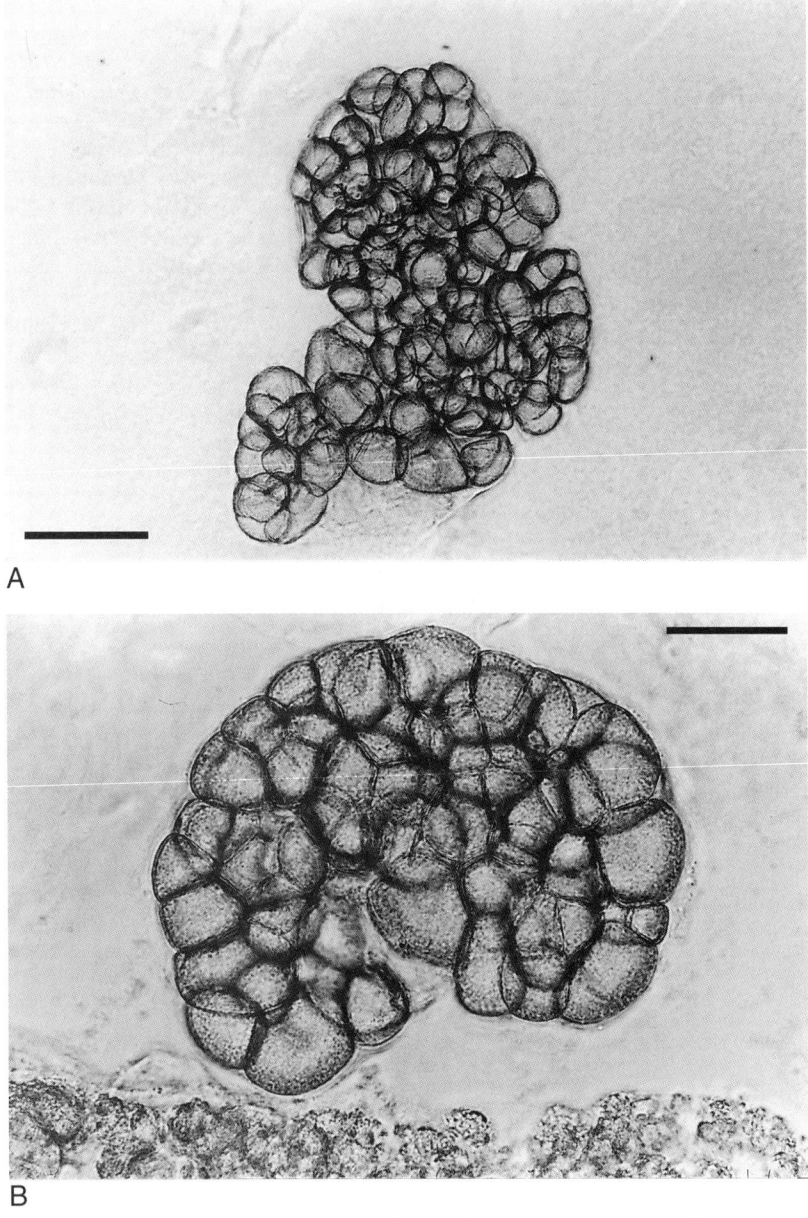

A

B

FIGURE BXII.δ.98. *Polyangium spumosum*: fruiting bodies, in slide mounts. Bar in *A* = 80 μm; bar in *B* = 40 μm.

100–200 μm large. Observed only once on rabbit dung and not cultivated. No other report on it seems to exist. Krzemieniewska and Krzemieniewski consider the possible identity of *"Polyangium morula"* and *Polyangium aureum*, but the sporangioles of the latter are said to have a deeper, golden brown color. The author once isolated a strain with striking bright yellow sori, which may correspond to Jahn's description. It was a cellulose decomposer, and thus would now be classified as a *Sorangium*.

3. *"Polyangium schroeteri"* Jahn 1924, 73. Sporangioles angular, small (former genus *Sorangium* Jahn 1924, 73). Sorus incompletely subdivided into sporangioles, orange red. Sporangioles angular, 12–14 μm in diameter. The sori often are kidney-shaped and may show brain-like convolutions. Found on rabbit dung; not cultivated. This "species" was probably an incompletely differentiated form of some other species; Krzemieniewska and Krzemieniewski suggested that

it was *"Polyangium compositum"*. The kidney shape of the sori and beginning septation would fit that hypothesis.

4. *"Polyangium septatum"* Thaxter 1904, 412. Sporangioles angular, small (former genus *Sorangium* Jahn 1924, 73). Sori completely resolved into sporangioles. Sorus yellowish-orange to dull orange when dry, 50 to several 100 μm, forming irregular, often confluent masses divided into small, subspherical to subcylindrical packets with a well-defined envelope, containing a few angular sporangioles each. Packets 18–22 × 12–22 μm; sporangioles 10–12 μm in diameter; myxospores 0.8–1 × 3–5 μm. Found in horse dung culture; not cultivated. The organism was seen again by Krzemieniewska and Krzemieniewski (1926, 1927), who obtained it from rabbit dung placed on soil. The packets contained two to seven sporangioles. The dimensions of the packets, sporangioles, and myxospores were essentially those given by Thaxter. In the 1927 paper they also described a variant

with smaller sporangioles (4–10 × 3–8 µm) (var. *microcystum*).

5. *"Polyangium stellatum"* Kofler 1913 had elongated sporangioles with red brown walls, 80–120 × 160–200 µm, arranged in a star-like pattern and sitting on a short, knob-like stalk, sometimes directly on the substrate. The shape of the vegetative cells was not mentioned. As already surmised by Jahn (1924), this may have been an organism belonging to the *Cystobacter–Stigmatella* group. *"Polyangium flavum"* Kofler 1913 (*"Archangium flavum"* Jahn 1924) is now classified as *Kofleria flava*. *"Polyangium indivisum"* Krzemieniewska and Krzemieniewski 1927 was described as similar to *"Polyangium ferrugineum"*; its sporangioles were much smaller and light orange-yellow within a colored slime matrix. It is stated to have had cylindrical cells with rounded ends, "like *Polyangium ferrugineum"*. This is contrary to the current concept of *Cystobacter ferrugineus*, and the standing of *"Polyangium indivisum"* is completely obscure. Peterson (1958) described *"Polyangium aurantiacum"* from a herbarium specimen in the Thaxter collection in the Farlow Herbarium at Harvard University and renamed it as *"Sorangium aurantiacum"*, with orange fruiting bodies consisting of angular sporangioles, mostly 7–8 µm wide, with bright yellow walls. The sori were 300–400 µm in diameter and 100 µm high. According to Thaxter the fruiting bodies were found only once on rat dung pellets. The description of the species is not sufficient to allow a reliable identification and evaluation. It may actually have been a *Polyangium sorediatum* strain.

Genus II. **Byssophaga** *gen. nov.*

HANS REICHENBACH

Bys.so.pha'ga. Gr. fem. n. *byssos* cotton fabric, fine linen (for cellulose); Gr. v. *phagein* to eat, devour; M.L. fem. n. *Byssophaga* devourer of cellulose.

Vegetative cells cylindrical rods with rounded ends. Swarm colonies consist of independently migrating, pseudoplasmodia-like cell assemblies. Intensely cinnabar- to blood-red. **Fruiting bodies are sori of roundish to polyhedral sporangioles. Myxospores resemble vegetative cells in shape but are optically refractile.** Swarm sheet does not adsorb Congo red. Cellulose degrader.

Type species: **Byssophaga cruenta** sp. nov., nom. rev. (*Myxococcus cruentus* Thaxter 1897, 409; non *Chondrococcus cruentus* Krzemieniewska et Krzemieniewski 1930, 250/268.)

FURTHER DESCRIPTIVE INFORMATION

The outstanding characteristic of the genus is the independently migrating flocks of cells. This colony type is unlike that of any other myxobacterium, so that a separate genus seems justified. The only species so far is *B. cruenta*, for which cellulose degradation and an extremely intense red color are typical. The author has occasionally isolated strains with a similar pseudoplasmodia-like pattern of swarm colonies. They were neither red nor cellulose degraders. It still is unclear whether they belong to this genus, and hence cellulose decomposition and color are not listed among the essential characteristics of the genus.

List of species of the genus Byssophaga

1. **Byssophaga cruenta** *sp. nov., nom. rev.* (*Myxococcus cruentus* Thaxter 1897, 409; non *Chondrococcus cruentus* Krzemieniewska et Krzemieniewski 1930, 250/268.)

cru.en'ta. L. fem. adj. *cruenta* blood-red.

Vegetative cells stout, cylindrical rods with rounded ends, 1.2–1.5 × 4–6 µm (Fig. BXII.δ.99). Swarm colonies on agar consist of scattered, independently migrating cell assemblies, in flocks, with a fan-like front and a tapering tail (Fig. BXII.δ.100), intensely cinnabar- to blood-red, becoming carmine-red when aging and dying. The migrating cell masses leave a parchment-like slime track behind, which often has a wrinkled surface. They also may deeply penetrate the agar plate. The pseudoplasmodia finally contract and become spherical knobs, not unlike *Myxococcus* fruiting bodies, 200–500 µm in diameter (Fig. BXII.δ.101). Often form massive rings, with an outer diameter of 330–440 µm and an inner diameter of 80–150 µm, and, rarely, a calyx-like structure with the slime track wrapped around it. Swarms with anastomosing veins, and eventually homogeneous swarms, may be produced upon subcultivation, but the organism may return to the original pattern after transfer to a fresh plate. Fruiting bodies (Fig. BXII.δ.102) are intensely red sori, 220–560 µm and more in diameter, with a thin outer slime envelope. They consist of densely packed roundish to polyhedral sporangioles with a delicate but firm wall, 60–180 µm wide, usually around 80–150 µm. Myxospores resemble vegetative cells in shape and size but are optically refractile.

The organism degrades crystalline cellulose completely and may be cultivated on filter paper as the only carbon source placed on a mineral salts agar with an inorganic nitrogen source, preferentially KNO_3. It grows better if an

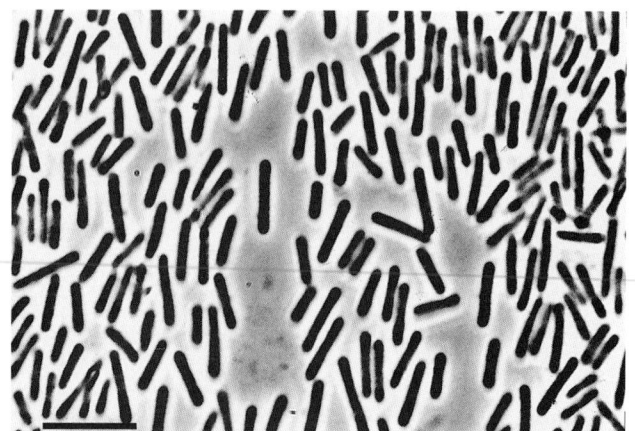

FIGURE BXII.δ.99. *Byssophaga cruenta*: vegetative cells from yeast-maltose agar, phase contrast. Bar = 10 µm.

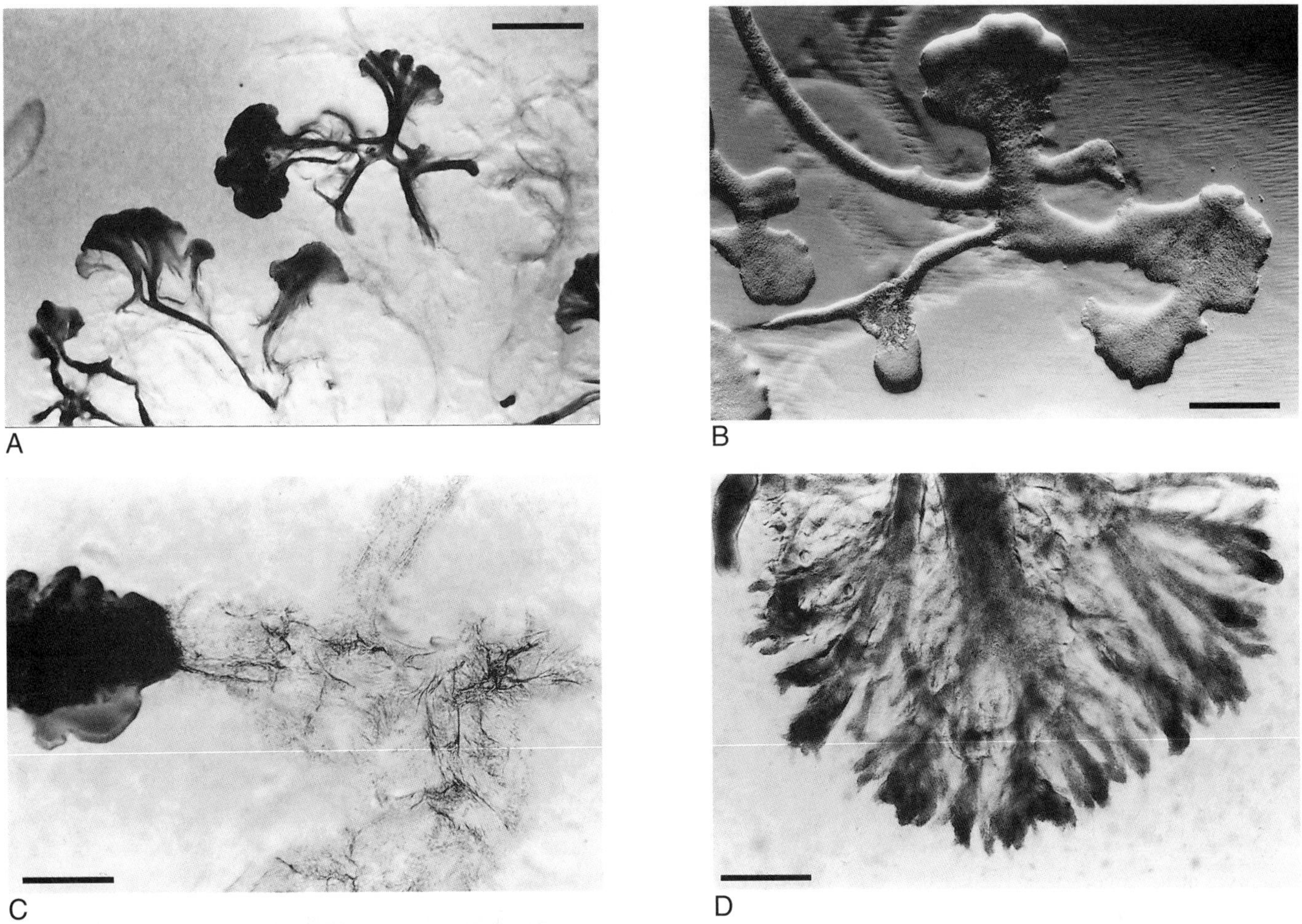

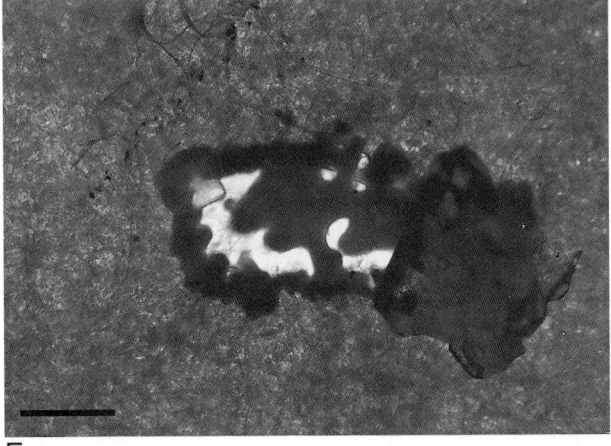

FIGURE BXII.δ.100. *Byssophaga cruenta*: swarm colonies. (*A* and *B*) The swarms of *Byssophaga cruenta* typically consist of independently moving, pseudoplasmodia-like cell packs (bar in *A* = 2100 μm; bar in *B* = 380 μm). (*C*) The migrating cell masses leave a parchment-like slime track behind (bar = 570 μm). (*D*) After many transfers, the organism may begin to grow in more or less coherent swarms (bar = 575 μm). (*E*) Filter paper on mineral salts agar is completely degraded (bar = 1100 μm).

organic nitrogen source is supplied, e.g., casein peptone at low concentrations (0.05–0.1%), but only when combined with a carbohydrate as a carbon source. While glucose is a strong growth repressor (above 0.05%), maltose (up to 4%) stimulates growth. Yeast agar (VY/2-) + 0.2% maltose is a good medium for stock cultures, especially if the inoculum is placed on a piece of filter paper on the agar surface. The yeast cells are degraded.

Starch is hydrolyzed but does not allow good growth, probably because of glucose repression. Growth on chitin agar and decomposition of chitin are slow but are somewhat better if the chitin top layer is on a water agar base (Chit7-

agar) and not on a peptone agar base (Chit6-agar; Reichenbach and Dworkin, 1992). The organism has also been cultivated successfully in liquid media. Grows well at 30°C and more slowly at 20°C.

Originally found on cow dung in Tennessee and in woods in the USA, the organism was not cultivated until recently. Strain By c2 was isolated by the author in 2001 from a soil sample with rotting plant material collected in June 1996 in a sagebrush steppe south of Holbrook, Arizona, USA. The sample was air dried and kept at room temperature for 5 years.

The mol% G + C of the DNA is: 69.9 (HPLC).

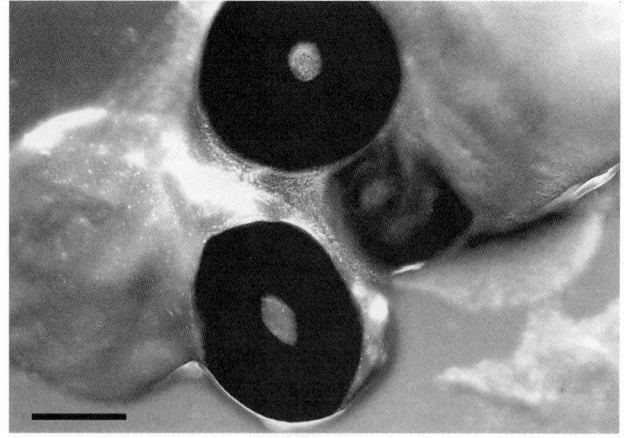

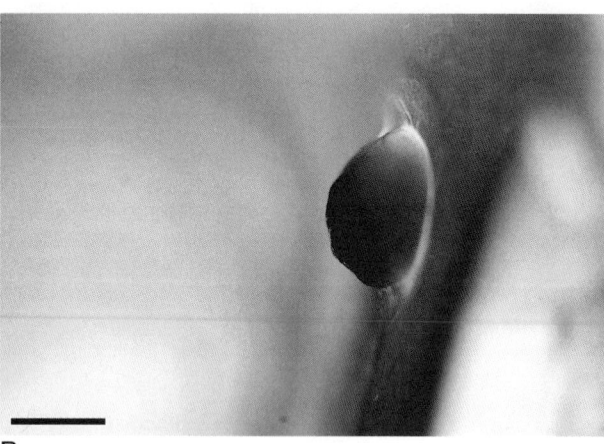

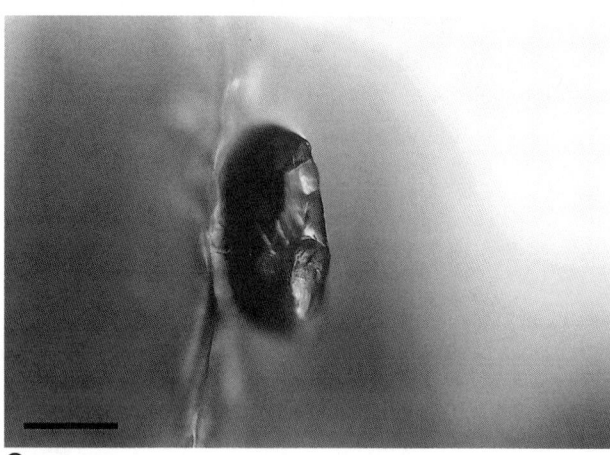

FIGURE BXII.δ.101. *Byssophaga cruenta*: cell aggregates. (*A*) Within the swarm the cells often assemble into massive rings (bar = 200 µm). (*B*) Normally, the pseudoplasmodia finally contract into bright red knobs resembling a *Myxococcus* fruiting body (bar = 260 µm). (*C*) Rarely they produce cup-like structures (bar = 380 µm).

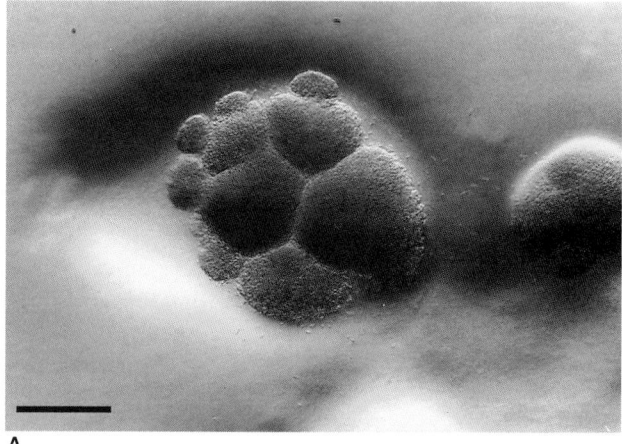

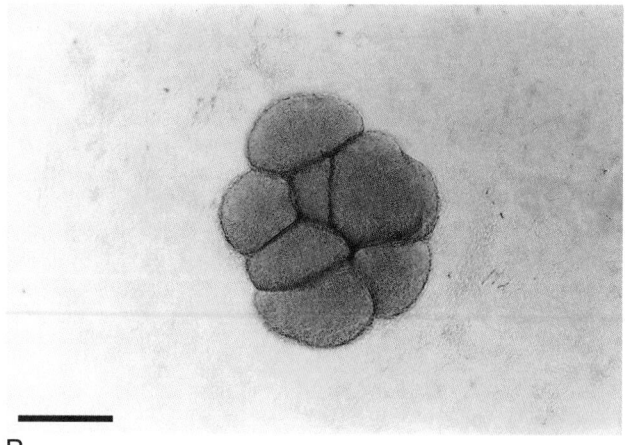

FIGURE BXII.δ.102. *Byssophaga cruenta*: fruiting bodies on surface of agar plates. (*A*) Early stage of fruiting: the developing sorus sits in a depression of the agar surface; sporangioles in the process of differentiation (bar = 80 µm). (*B*) Mature fruiting body, under water and cover glass: the sporangiole walls are clearly distinguishable (bar = 100 µm).

1.0 × 1.2–1.4 µm, i.e., as short rods. But he states that "the spores . . . are not as well developed as in the other species of *Myxococcus*, and seem to suggest a transitional form between *Chondromyces* and *Myxococcus*." As Thaxter could not cultivate the organism—another argument for identity, because cellulose-decomposing myxobacteria were not conceived of at the time—he could not describe the vegetative cells. It seems that no one since has seen, let alone cultivated, the organism. Krzemieniewska and Krzemieniewski (1927) mentioned that they obtained *"Myxococcus cruentus"* abundantly from beech forest soil in Poland, and gave a brief description of their organism (Krzemieniewska and Krzemieniewski, 1930). They curiously renamed it as *"Chondrococcus cruentus"*, although they seem to have seen cysts, i.e., sporangioles that are not produced by *Chondrococcus* (now *Corallococcus*). Their figure of vegetative cells seems to show rods with tapering ends, and the myxospores were essentially spherical. The author is quite sure that this organism was not the same as discussed above; perhaps it really was a red *Corallococcus*.

As a cellulose degrader, *B. cruenta* could qualify to be classified as a *Sorangium* species. There are, however, substantial differences that justify establishing a new genus. In the first place, the swarm pattern with its pseudoplasmodia

Type strain: DSM 14553, By c2, JCM 12614.

There is little doubt that Thaxter's organism is identical with the one described above. Cell aggregates resembling *Myxococcus* fruiting bodies, incredibly intense red pigmentation, and the structure of the fruiting bodies all are suggestive. The only difference is the shape of the myxospores, which Thaxter describes as oval or irregularly oblong, 0.9–

is specific to *B. cruenta*. Furthermore, while *S. cellulosum* always produces masses of fruiting bodies, especially on filter paper, *B. cruenta* is a very poor fruiter. In fact, the author has seen fruiting bodies only twice in 5 years. In addition, unlike *S. cellulosum*, *B. cruenta* is completely repressed by low concentrations of glucose and peptone and exhibits a rather low resistance to kanamycin sulfate (up to 10 µg/ml). Finally, the secondary metabolites produced by *Byssophaga* are not those typical of *Sorangium*.

Although this has not yet been proved, the shape of the vegetative cells and the structure of the fruiting bodies suggest that the organism belongs to the *Polyangiaceae*.

Genus III. **Chondromyces** Berkeley and Curtis in Berkeley 1874, 64[AL]

HANS REICHENBACH

Chon.dro.my′ces. Gr. masc. n. *chondros* cartilage; Gr. masc. n. *mykes* fungus; M.L. masc. n. *Chondromyces* cartilaginous fungus.

Vegetative cells are cylindrical rods with rounded ends. **Swarm colonies produce wide, shallow depressions in the agar surface,** often with an orange band of cells at the edge, agar sometimes slightly etched. Swarms often penetrate the agar deeply. **Clusters of sporangioles borne on long, branched or unbranched white slime stalks.** Myxospores resemble vegetative cells in shape but are optically refractile.

The mol% G + C of the DNA is: 69–70.

Type species: **Chondromyces crocatus** Berkeley and Curtis *in* Berkeley 1874, 64.

FURTHER DESCRIPTIVE INFORMATION

There are several other myxobacterial species with sporangioles on stalks, a fact which has caused some confusion, as already discussed in connection with the genus *Stigmatella*. Most of those species belong to the suborder *Cystobacterineae* and clearly differ from *Chondromyces* in the shape of their vegetative cells and myxospores. *Chondromyces* fruiting bodies are recognizable at once by the bright orange color of the sporangioles, the long, white stalk, and their sheer size. In cultures, older fruiting bodies often degenerate by "growing through", i.e., the bacterial mass on top of the stalk produces another stalk sitting on the original one, usually at an angle, and this process may repeat itself until only a tangle of stalks is left. In addition, species whose fruiting bodies normally have a simple stalk sometimes produce branched fruiting bodies and vice versa. Such aberrant specimens are infrequent and always occur among typical ones; however, a population of fruiting bodies should be available for a reliable diagnosis. In a further deviation from the type, all species sometimes produce stalks with a single sporangiole instead of a cluster.

None of the *Chondromyces* species listed below attacks chitin.

While most *Chondromyces* strains grow in pure culture, many *C. crocatus* strains require a specific symbiont for growth (Jacobi et al., 1996, 1997). There also may be strains of other *Chondromyces* species that need a companion. In pure culture, *Chondromyces* strains produce fruiting bodies only rarely and then under special induction conditions. For example, it was shown that a certain *C. apiculatus* strain requires light for differentiation (Reichenbach et al., 1974). All *Chondromyces* species, however, fruit reliably when grown on filter paper placed on a mineral salts agar, e.g., ST21-agar (Reichenbach and Dworkin, 1992) and inoculated with a mixture of cellulose degrading bacteria derived from nature (other than myxobacteria and cytophagas) and other organisms, such as may be obtained from enrichment cultures for cellulose decomposers. The cultures are kept at room temperature (around 22°C) in daylight.

Key to the species of the genus *Chondromyces*

I. Clusters of sporangioles on branched stalks.
  A. Clusters of tiny sporangioles; stalks normally branched.

C. crocatus

  B. Clusters of large, spherical, or bowl-shaped sporangioles with many tips, or tails, on the distal surface; stalks unbranched or, often, branched.

C. lanuginosus

II. Clusters of large sporangioles on simple, usually unbranched, stalks.
  A. Cluster composed of individual sporangioles.
    1. Sporangioles ending in tail-like tips.
      a. Sporangioles turnip-shaped with a single tip, fruiting bodies moderately large, 300–600 µm.

C. apiculatus

      b. Sporangioles bulbous to spherical, often with 2 or 3 tips, fruiting bodies large, 500–1000 µm.

C. robustus

    2. Sporangioles bell- or barrel-shaped with a flat end, often hanging down on long, delicate stalklets.

C. pediculatus

  B. Clusters composed of chains of sporangioles.

C. catenulatus

List of species of the genus Chondromyces

1. **Chondromyces crocatus** Berkeley and Curtis *in* Berkeley 1874, 64[AL]

cro.ca' tus. L. masc. adj. *crocatus* saffron-yellow.

Vegetative cells cylindrical rods with blunt, rounded ends, 1.1–1.4 × 3–12 µm (Fig. BXII.δ.103). Fruiting bodies (Fig. BXII.δ.104) are ≥1000 µm high and consist of white or orange branched slime stalks bearing dense, spherical clusters of small sporangioles at the tips of the branches. Often produced in large numbers, like a miniature forest. Sporangioles cylindrical to barrel-shaped with flat or rounded ends, straw-colored to orange, 20–30 × 30–50 µm, borne on very short pedicles or sitting directly on the stalk. Fruiting bodies with simple stalks may be found among those with branched stalks. Myxospores resemble vegetative cells in shape but are optically refractile, 1.0–1.3 × 3–6 µm.

Grows well on yeast agar (VY/2-) with lysis of the yeast cells. Good growth on casein peptone (e.g., CY-) agar. Most strains grow only in the presence of a companion bacterium, which has been found to be always the same regardless of where the *Chondromyces* strain was isolated (Jacobi et al., 1996, 1997). The companion, a *Sphingobacterium*-like organism, does not grow without the *Chondromyces*. However, there are *Chondromyces* strains that do not require a symbiont, e.g., strain Cm c5. *C. crocatus* cultures usually produce a very peculiar odor that is unique among myxobacteria and reminds one of pyridine or raw potatoes, so that new isolates may be identified as *C. crocatus* by this odor alone. Can be cultivated in liquid media together with the companion (Kunze et al., 1994, 1995). Produces at least seven different secondary metabolites, including crocacin, which blocks electron transport in the bc_1-complex of the respiratory chain (Kunze et al., 1994), and the chondramides, which stimulate actin polymerization (Sasse et al., 1998).

Found on decaying plant material, on bark, in the rhizosphere, and in soil.

The mol% G + C of the DNA is: 70 (T_m).

Type strain: Cm c5, DSM 14714, JCM 12616.

Additional Remarks: Reference strains include DSM 14606 (Cm c2) (Windsor M38) (ATCC 25193).

This is one of the most beautiful myxobacteria. Like *Polyangium vitellinum*, it was first seen by some early mycologists (Ravenel, Curtis, Berkeley) who took it, however, for a fungus. This history shows that the achievements of Roland Thaxter were by no means trivial when, in 1892, after careful and keen observations and experiments, he discovered the bacterial nature and most astonishing life cycle of those peculiar organisms, whereas other well-known and outstanding scientists failed to do so. The first description, essentially a figure, but an unmistakably characteristic one, was provided by Berkeley (1857). The type specimen, fruiting bodies on a decayed gourd or melon from South Carolina, came from the Curtis Herbarium and was probably originally collected by Ravenel. A brief species definition was later supplied (Berkeley, 1874). Thaxter (1892) realized at once the identity of his organism with the specimen in the Curtis collection and retained the name. Thaxter obtained his isolates from old straw in Cambridge, MA, and from Ceylon. A few years later, the same organism was described by Zukal (1896), this time as an unusual myxomycete, "*Myxobotrys variabilis*". He found the organism on the bark of a willow tree near Vienna. His figures leave no doubt—and he later realized his mistake and admitted (Zukal, 1897b), if reluctantly (Zukal, 1897a)—that *Myxobotrys* was a myxobacterium and identical with *C. crocatus*. It appears that *C. crocatus* has not been found in European samples for more than a hundred years. Only recently, it was isolated again from two samples from sites near Basle, Switzerland. *C. crocatus* has been found, together with other *Chondromyces* species, in the rhizosphere of plants in Assam, India (Agnihothrudu et al., 1959). The symbiont-free strain, Cm c5, was isolated by the author's laboratory in 1988 from a soil sample with decomposing plant material collected in the same year in the tropical rain forest near Iguaçu, Brazil; strain Cm c2, with its symbiont, was isolated in 1982 and comes from a similar sample from the island of Madeira, Portugal.

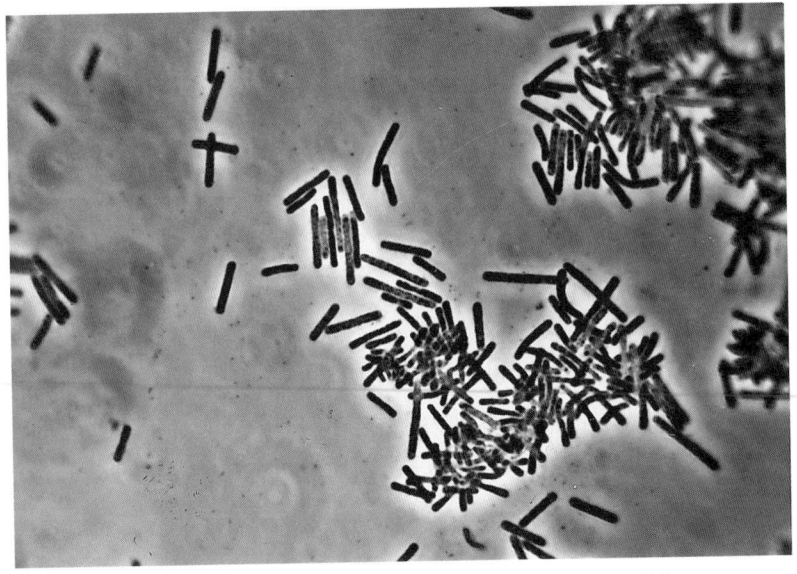

FIGURE BXII.δ.103. *Chondromyces crocatus* vegetative cells. Light micrograph (× 1000).

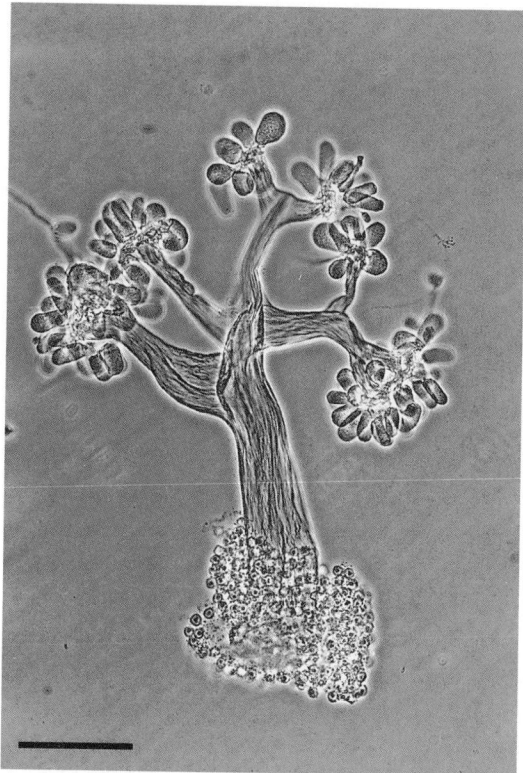

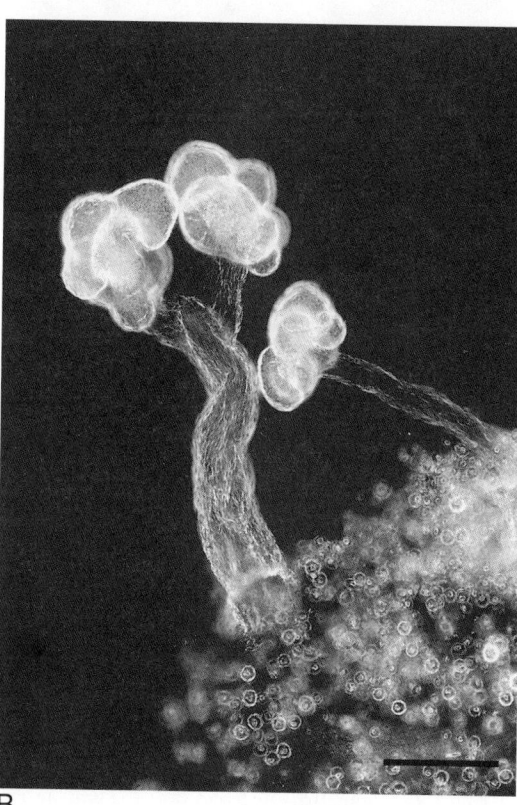

FIGURE BXII.δ.104. *Chondromyces crocatus*: fruiting bodies in embedding medium. (*A*) Fully differentiated fruiting body, in phase contrast; (*B*) immature stage, dark field. It should be noted that all photographs of *Chondromyces* fruiting bodies show small specimens, as depth of field is not sufficient for large ones. Bar in *A* = 125 μm; bar in *B* = 105 μm.

2. **Chondromyces apiculatus** Thaxter 1897, 405[AL]

a.pi.cu.la′tus. L. masc. n. *apex* genitive *apicis* tip, point; M.L. masc. adj. *apiculatus* pointed, with a tip or tail.

Vegetative cells cylindrical with rounded ends, 1.1–1.4 × 3–14 μm. Sporangioles bright orange, turnip-shaped when typical, but may also be either more bulbous or more slender; cylindrical, 25–40 × 35–50 μm; always with a conspicuous white tip or tail, up to 35 μm long. The sporangioles sit with or without a pedicle on an unbranched white stalk (Fig. BXII.δ.105). Stalks up to 700 μm long and 15–40 μm wide. Branched stalks are rarely seen. Shape and size of the fruiting bodies very variable and overlapping with those of the following species (see comments on *C. robustus*). Myxospores resemble vegetative cells in shape but are optically refractile, 1.0–1.3 × 3–6 μm.

Originally isolated from antelope dung from Liberia. The organism is very common on rotting wood and bark in the Midwest of the USA, but is also found in soil and similar habitats in other places. Strain Cm a14 was isolated in 1988 from rotting plant material collected in 1981 in Costa Rica, reference strain Cm a2 was isolated in 1966 from the surface of a rotting tree trunk collected in Minnesota, USA. The latter is the only *Chondromyces* strain of which the author is aware that can be induced to produce fruiting bodies under fully defined conditions (Reichenbach et al., 1974).

The mol% G + C of the DNA is: 69–70 (Bd, T_m).

Type strain: DSM 14605, Cm a14, JCM 12615.

Additional Remarks: Reference strains include DSM 14612 (Cm a2) (Windsor M6).

3. **Chondromyces catenulatus** Thaxter 1904, 410[AL]

ca.te.nu.la′tus. L. fem. n. *catena* chain; L. diminutive fem. n. *catenula* a small chain; M.L. masc. adj. *catenulatus* with small chains (of sporangioles).

Vegetative cells (in an earlier stage of fruiting) cylindrical rods with rounded ends, 1.0–1.3 × 4–6 μm. Sporangioles pale yellow to yellow orange, fusiform to elongated and ellipsoidal, 18 × 20–50 μm; 10 or 12 of them in sparingly (once or twice) branched chains; longer ones around 300 μm. The sporangioles are separated in the chains by shriveled, membranous links. The chains arise in dense clusters from the tips of unbranched, stout stalks, 180–360 μm high (Fig. BXII.δ.106).

Found on a rotten poplar log near Hanover, New Hampshire, USA, obtained only once by Thaxter. The organism could be kept on its natural substrate for some time, but was never cultivated in pure culture. There appears to be only one further report of *C. catenulatus*, from southeast Queensland, Australia, with scanning electron microscope images but without any details (McNeil and Skerman, 1972). Thus, the organism seems to be very rare. There are, however, beautiful specimens of Thaxter's in the Farlow Herbarium.

The mol% G + C of the DNA is: not determined.

Type specimen: Acc. no. 4517, Thaxter Collection, Farlow Herbarium, Harvard University, Cambridge, MA, USA.

4. **Chondromyces lanuginosus** Kofler 1913, 861[AL]

la.nu.gi.no′sus. L. fem. n. *lanugo* down, wool; M.L. masc. adj. *lanuginosus* downy, wooly.

Vegetative cells cylindrical rods with rounded ends, 0.9–1.0 × 3–8 μm. Fruiting bodies (Fig. BXII.δ.107) consist of

A

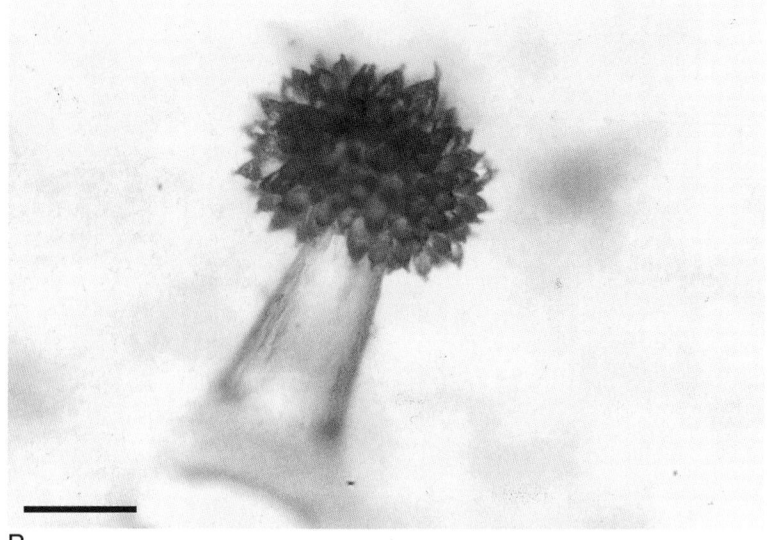

B

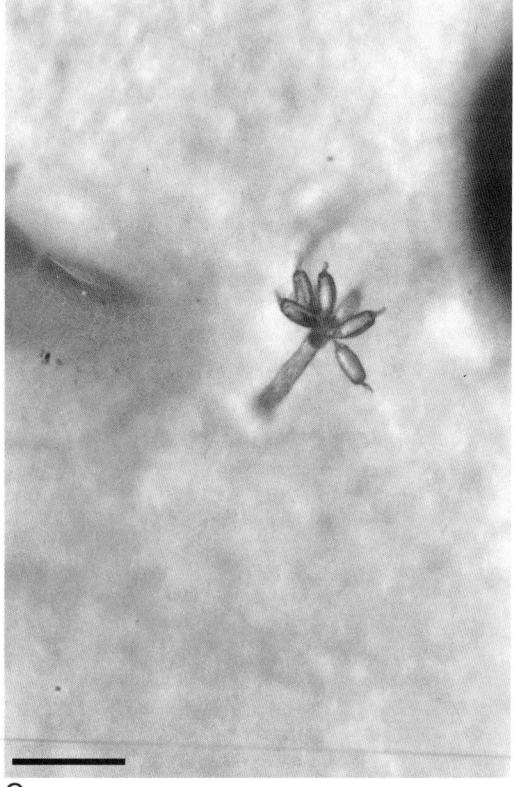

C

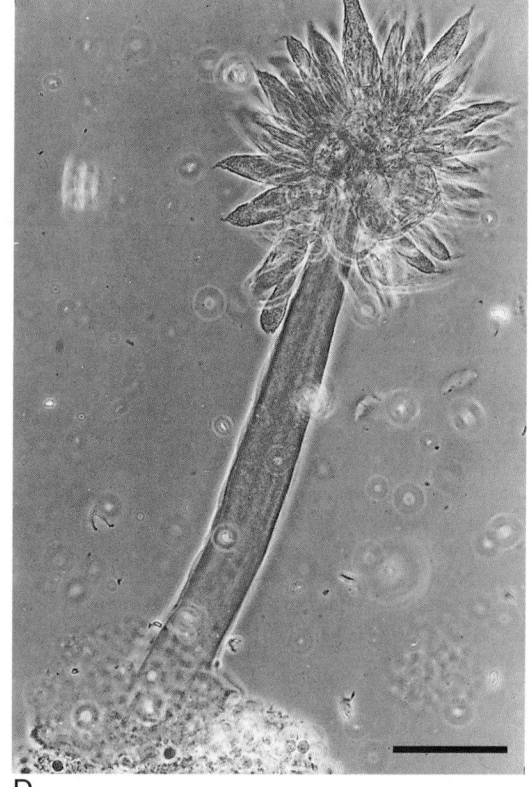

D

FIGURE BXII.δ.105. *Chondromyces apiculatus*: fruiting bodies on culture plates *(A–C)* and in embedding medium *(D)*; *(B* and *C)* show the enormous size differences seen in a single culture. Bar in *A* = 165 μm; bars in *B* and *C* = 100 μm; bar in *D* = 80 μm.

long, white or yellowish slime stalks, 700–1000 μm or more long, often rather slender, width 40–130 μm or more at the base, narrowing down to 20 μm at the tips; unbranched or, branched; sometimes repeatedly branched. At the tips of the stalks are clusters of 2–7, but sometimes as many as 20–30 large straw-colored to orange sporangioles, 65–110 μm

FIGURE BXII.δ.106. *Chondromyces catenulatus*: fruiting bodies. (*A*) Light micrograph (× 120). (*B*) Figures 1–4 from the original description by Roland Thaxter (1904). (*C* and *D*) Small, atypical fruiting bodies of *Chondromyces crocatus* with chains of sporangioles resembling *C. catenulatus*. Bar in *C* and *D* = 200 μm.

(range: 45–265 or more) wide. The sporangioles are spherical or cup-shaped, depressed spheres and bear many fine, white hair-like tails 10–50 μm long, giving the head a downy appearance. Myxospores resemble vegetative cells in shape but are slightly smaller, 0.7–1.1 × 2.5–3.5 μm, and optically refractile.

The organism grows well in pure culture, e.g., on yeast (VY/2-) and casein peptone (CY-) agar. Yeast cells are decomposed; chitin and cellulose are not degraded.

Originally obtained from hare dung found near Vienna and from deer dung in Canada; it also has been found in soil with decaying plant material and on bark. Appears to be a rare species. Strain Sy t2 was isolated in 1987 from soil with decaying plant material collected in 1986 in Upper Franconia, Germany; reference strain Sy t1 is a parallel isolate. Reference strain Sy t7 was isolated in 2000 from the bark collected from a diseased Ficus tree in Karnataka, India, in 1999.

The mol% G + C of the DNA is: not determined.

Type strain: Sy t2, DSM 14631, JCM 12617.

Additional Remarks: Reference strains include DSM 14630 (Sy t1) and DSM 14632 (Sy t7).

Additional Remarks: The fruiting bodies of *C. lanuginosus* are so characteristic and unique that the species cannot be mistaken. It was discovered at almost the same time in two laboratories and given two different names, as the authors did not know of one another, probably due to the disruption caused by World War I. Jahn (1924) later created a genus *Synangium* for those organisms. In addition, he listed *"Synangium lanuginosus"* and *"Synangium thaxteri"* as different species, but this seems not to be justified, as previously observed by Krzemieniewska and Krzemieniewski (1946),

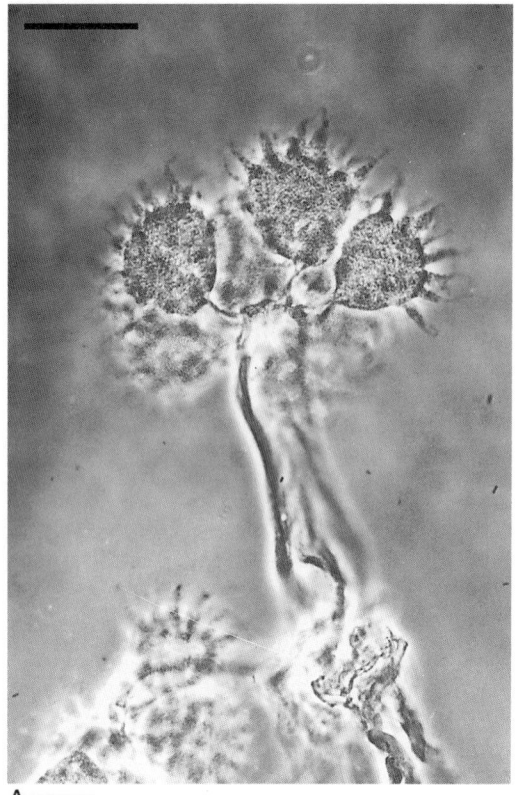

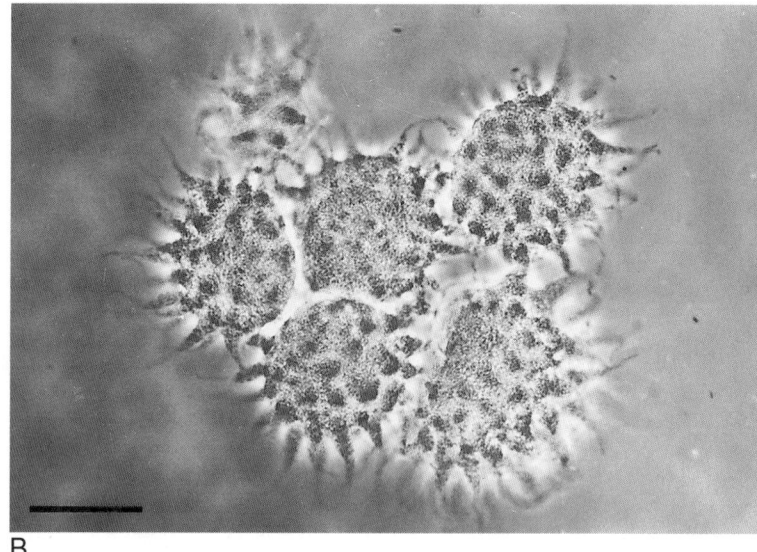

FIGURE BXII.δ.107. *Chondromyces lanuginosus*: fruiting bodies in embedding medium. The view from above shows that the end faces of the sporangioles are covered with tails (*B*). Bars = 65 μm.

for the two descriptions overlap virtually completely except for minor differences in dimensions. *C. lanuginosus* is in all respects comparable to other *Chondromyces* species, so that a genus of its own seems not to be meaningful. 16S rRNA sequence analysis positions *C. lanuginosus* strain Sy t2 close to one of the *C. apiculatus* strains, Cm a2, among the other *Chondromyces* species (Spröer et al., 1999).

This appears to be a less common *Chondromyces* species. Over a period of 25 years, the author's laboratory isolated seven strains from samples from Germany, Ceylon, India, and California. After duplicates were removed, the actual number was reduced to four. Three more strains have been reported in the literature—from Japan (Watanabe and Tanaka, 1929), Poland (Krzemieniewska and Krzemieniewski, 1946), and Tanzania (Dawid, 1980). Krzemieniewska and Krzemieniewski cultivated their strain on soaked hay.

Jahn (1924) added *"Chondromyces sessilis"* to his new genus *Synangium* Thaxter 1904. Krzemieniewska and Krzemieniewski (1946) later argued that Thaxter's description may have been founded on stalkless specimens of *C. lanuginosus* fruiting bodies, which may be correct, especially as Thaxter mentioned that "occasionally . . . there seems to have been an attempt to differentiate an irregular cystophore". The organism looks indeed like a degenerate *Chondromyces*—similar forms may occasionally be seen in cultures—but it more closely resembles *C. apiculatus*. At the moment the species cannot be regarded as valid.

5. **Chondromyces pediculatus** Thaxter 1904, 410[AL]

pe.di.cu.la' tus. L. masc. n. *pediculus* diminutive of *pes* foot, small foot; M.L. masc. adj. *pediculatus* with a small foot.

Vegetative cells cylindrical rods with rounded ends, 0.9–

1.2 × 3–6 μm. Sporangioles pale yellow to orange, cylindrical, barrel, or bell shaped, with a flat or slightly curved end face that sometimes has a very short tip; 25–40 × 35–60 μm. Clusters of a few to up to 60 sporangioles on long, slender, white slime stalks; often hanging down in an umbel on 20–40 μm long, fine, white pedicles (Fig. BXII.δ.108). A disk-shaped foot may be present at the base of the stalk (as is seen also with other *Chondromyces* species). The whole fruiting body can be up to 1000–1200 μm high. In culture, stouter fruiting bodies with sessile spherical sporangioles may also appear.

Obtained originally from goose dung from South Carolina, USA; the bacterium can also be isolated from soil containing decaying plant material, dung of herbivores, and similar substrates. Contrary to statements in the literature (Krzemieniewska and Krzemieniewski, 1946), the author finds that *C. pediculatus* is perhaps the most common of all *Chondromyces* species. Fifty-five strains were isolated over the years. Strain Cm p51 was isolated in 2000 from soil containing decaying plant material collected in Karnataka, India. In 1999; reference strain Cm p31 came from a similar sample from the island of Majorca, Spain, and was isolated in 1999.

The mol% G + C of the DNA is: not determined.

Type strain: Cm p51, DSM 14607, JCM 12618.

Additional Remarks: Reference strains include DSM 14660 (Cm p31).

C. pediculatus cannot be mistaken when typical fruiting bodies are produced as is usually the case in enrichment cultures, e.g., for cellulose degraders on filter paper over mineral salts agar. The umbels of hanging, bell-shaped sporangioles are quite typical. In addition, the stalks of *C. pedi-*

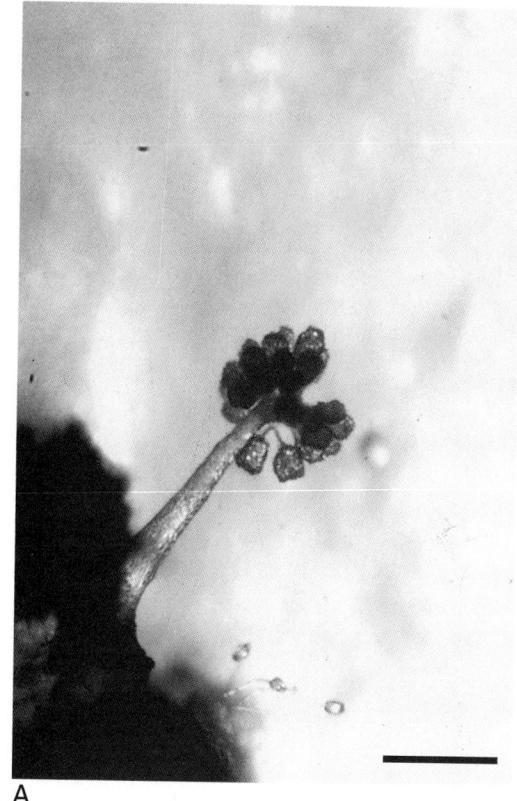

A

B

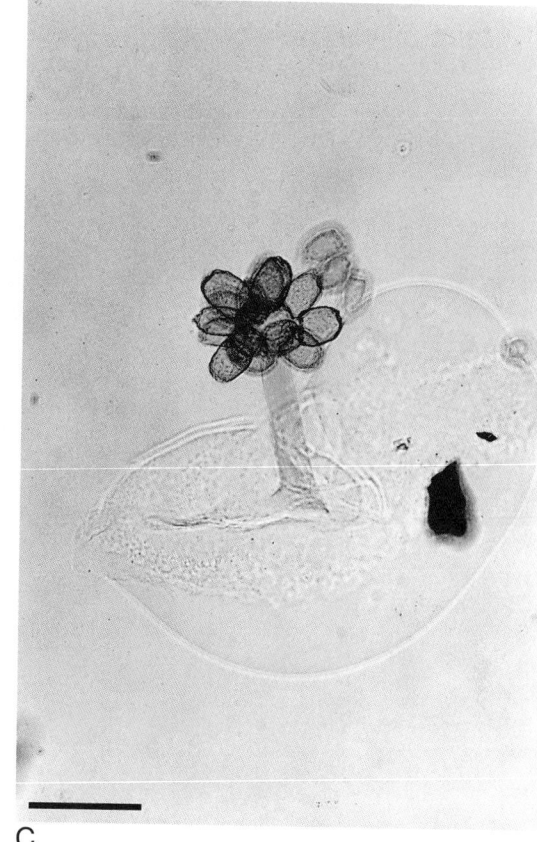

C

FIGURE BXII.δ.108. *Chondromyces pediculatus:* fruiting bodies. (*A*) In crude culture, sitting on a crumb of soil. (*B* and *C*) In embedding medium; (*C*) especially in cultures, the fruiting bodies do not always develop the characteristic overhanging, bell-shaped sporangioles. Bar in *A* = 100 μm; bar in *B* = 105 μm; bar in *C* = 125 μm.

culatus are normally more slender than those of most other *Chondromyces* species.

6. **Chondromyces robustus** *sp. nov.*

ro.bus′ tus. L. masc. adj. *robustus* made of oak wood, strong.

Vegetative cells cylindrical rods with rounded ends, 1.0–1.2 × 4–8 μm. Sporangioles bright orange, almost spherical to turnip-shaped, 70–120 μm long and 30–100 μm wide, with one to three 15–25 μm long, conspicuous white tips, or tails. The sporangioles are arranged in a cluster of a few

to 50 or more at the tip of a long, white, unbranched stalk, 200–1100 μm long and 50–400 μm wide (Fig. BXII.δ.109). The whole fruiting body may be up to 1300 μm high. Myxospores resemble vegetative cells in shape but are slightly smaller and optically refractile. Grows well in pure culture on yeast (VY/2-) and casein peptone (CY-) agar.

Obtained from decaying plant material, bark, and soil; moderately common and of worldwide distribution. Strain Cm a13 was isolated in 1980 from soil collected on Isla Mujeres, Yucatan, Mexico; reference strain Cm r9 was iso-

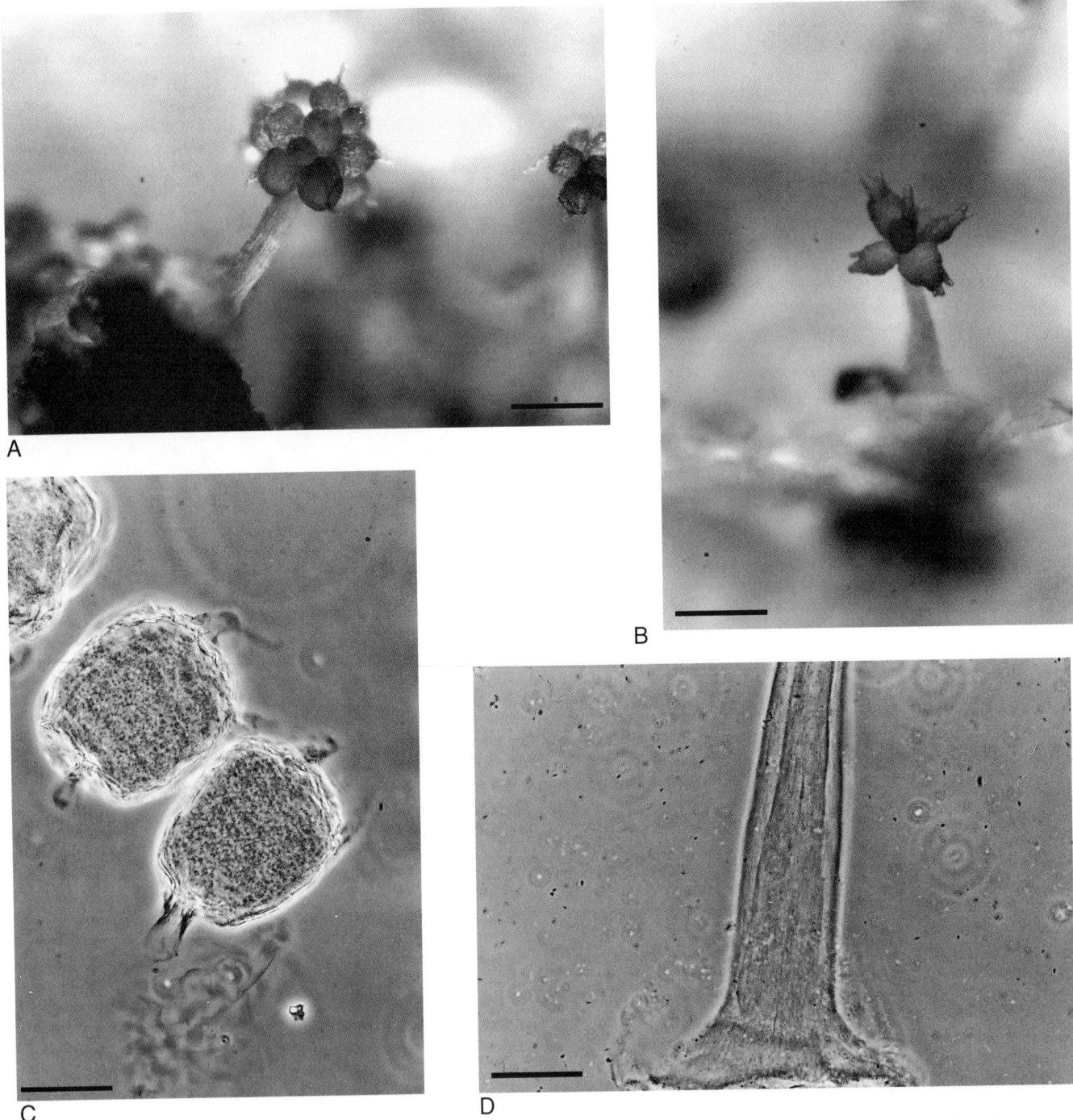

FIGURE BXII.δ.109. *Chondromyces robustus*: fruiting bodies. (*A* and *B*) On culture plates; the young, small fruiting body in *B*, still in the process of differentiation, shows sporangioles with several tips, which are characteristic for the species. (*C* and *D*) In embedding medium; (*C*) sporangioles, one with two tips; (*D*) base of a stalk with base plate. Bar in *A* = 130 μm; bar in *B* = 100 μm; bar in *C* = 30 μm; bar in *D* = 65 μm.

lated in 1996 from soil with decaying plant material collected in Karnataka, India, in 1995.

The mol% G + C of the DNA is: not determined.

Type strain: Cm a13, DSM 14608, JCM 12619.

Additional Remarks: Reference strains include DSM 14739 (Cm r9).

As observed by Thaxter (1897) when he described *C. apiculatus*, the fruiting bodies of that organism are highly variable. This is also the case with *C. robustus*, which displays considerable overlap with *C. apiculatus*. Hence, it is some-

what difficult to distinguish the two, and one should always have a population of fruiting bodies available for diagnosis. In fact, it is not certain whether Thaxter's organism was not what the author is now calling *C. robustus*. However, there is no question that two separate *Chondromyces* species with tailed sporangioles exist. The convincing evidence that Thaxter described a different organism from *C. apiculatus* was the small size of the sporangioles, which he gave as 35 × 28 μm, so that the whole cluster was only 200 μm in diameter on average. Strains with such characteristics are very common in the Midwest of the USA, although Thaxter's strain came from Liberia. The distinguishing characteristics of *C. robustus* are as follows: (1) sporangioles tend to be bulbous to spherical; (2) in most clusters there are sporangioles with two or three tips; (3) those sporangioles are not divided, so that the tails sit on branched sporangioles, as described by Thaxter for *C. apiculatus*; (4) the tails rise from one point; and (5) *C. robustus* fruiting bodies are much stouter and larger than those of *C. apiculatus*. When the author co-cultivated the type strains of the two species on mineral salts agar with filter paper inoculated with a mixture of cellulose degrading and other soil bacteria—a reliable method to induce fruiting of *Chondromyces* strains—he obtained large fruiting bodies intermingled with ones about one-half to two-thirds the size of the large ones. 16S rRNA sequence analysis showed a certain separation of *C. robustus* and *C. apiculatus*, but the 16S rRNA sequences of all *Chondromyces* strains are rather similar, and the resolution of the method may not allow distinction of species (Spröer et al., 1999). It appears that *C. robustus* is the more common species of the two.

Genus IV. **Haploangium** (ex Peterson 1959) gen.nov., nom. rev.*

HANS REICHENBACH

Ha.plo.an'gi.um. Gr. adj. *haplos* simple; Gr. neut. n. *angeion* vessel; M.L. neut. n. *Haploangium* simple vessel.

Vegetative cells cylindrical rods with rounded ends. Fruiting body consists of a single sporangiole. Myxospores similar to vegetative cells in shape.

Type species: **Haploangium rugiseptum** (ex McCurdy 1970) *comb. nov.* (*Haploangium rugiseptum* Peterson 1959, 5; *Polyangium rugiseptum* McCurdy 1970, 295.)

FURTHER DESCRIPTIVE INFORMATION

The genus *Haploangium* was established by Peterson (1959) to comprise species with *Polyangium*-type vegetative cells and isolated, solitary sporangioles. The name was not included on the Approved Lists. Peterson (1959) listed four species, two of which were new. *H. simplex* (= *Myxobacter simplex* Thaxter 1893 = *Polyangium simplex* Thaxter 1904) later turned out to be identical with *Polyangium vitellinum* (McCurdy, 1970). The position of "*Haploangium ochraceum*" (= *Polyangium ochraceum* Krzemieniewska and Krzemieniewski 1926) is obscure. The organism had spherical to ellipsoidal sporangioles that were orange to bright red, 30–60 × 50–130 µm, single, in a thick, yellow-brown slime envelope Krzemieniewska and Krzemieniewski usually found myxobacteria on rabbit dung pellets; the often-mentioned color of the slime envelopes may in fact have originated from diffusing brown pigments coming from the rabbit dung. The wall was occasionally seen to fold somewhat into the sporangiole, which, when observed from the side, gave the impression of a subdivision of the sporangiole. Myxospores 0.5 × 4–8 µm. Obtained from rabbit dung on soil. The organism was not cultivated. There is only one later report of "*H. ochraceum*" (Singh and Singh, 1971). The vegetative rods of that organism were thin with pointed ends, 0.5–0.7 × 7–12 µm, the myxospores were shortened rods, 1.4 × 2–3.5 µm, and the sporangioles were pale yellow to light orange. Therefore, this bacterium clearly does not belong to the genus *Haploangium* but to the *Cystobacterineae* and was most likely not the same bacterium as the one described by Krzemieniewska and Krzemieniewski. The Indian scientists also described another new *Haploangium* species, "*Haploangium krzemieniewskae*". This organism also does not fit the definition of *Haploangium*, as it had long, thin vegetative rods with somewhat tapering ends, 0.5–0.7 × 7–12 µm, and short myxospores, 0.5 × 2–3 µm. The solitary sporangioles were dull pink to gray, 15–35 × 15–50 µm. Found on bark. This species clearly also belongs to the *Cystobacterineae*.

The two new species listed below remain. None of the *Haploangium* species has ever been cultivated, so that all descriptions are incomplete. In the last edition of the *Manual*, following an earlier suggestion (McCurdy, 1970), these *Haploangium* species were included in the genus *Polyangium*. At least *H. rugiseptum* appears so unique in the structure of its fruiting bodies that a genus of its own seems justified. This species is also used as the type species, because Thaxter's *Polyangium simplex* formerly suggested as the type species belongs to a different genus.

While the genus is classified here in the family *Polyangiaceae*, this has still to be proved, e.g., by 16S rRNA sequence data.

Key to the species of the genus *Haploangium*

I. Sporangioles glistening orange-red with a flaky outer surface, heavily winkled when dry.
H. rugiseptum
II. Sporangioles brown to orange brown with a smooth surface.
H. minus

List of species of the genus Haploangium

1. **Haploangium rugiseptum** (ex McCurdy 1970) *comb. nov.* (*Haploangium rugiseptum* Peterson 1959, 5; *Polyangium rugiseptum* McCurdy 1970, 295.)
ru.gi.sep'tum. L. fem. n. *ruga* wrinkle; L. neut. n. *septum*, *saeptum* enclosure, fencing; M.L. neut. adj. *rugiseptum* with a wrinkled enclosure.

Vegetative cells not described. Sporangioles solitary, globose or oval; sessile or on a short stalk (Hu et al., 1985), up to 200 µm, 85 µm on average. Sporangiole wall with two indistinct layers: the inner one smooth and yellow, the outer one irregular, flaky and dark orange; heavily wrinkled when dry. The sporangiole sometimes contains fatty globules and amorphous material in addition to myxospores. Myxospores

---

Editorial Note: Readers are advised that the type material of *Polyangium rugiseptum* and *Polyangium minor* (sic) are not available as cited. As such, the proposals for new combinations presented here may result in illegitimate names.

cylindrical with blunt ends, 0.8×4–6 µm. Found on bark of various trees in Missouri, USA. Could not be cultivated.

The mol% G + C of the DNA is: not determined.

Deposited strain: Peterson 51, University of Missouri Herbarium, Columbia, MO, USA (type specimen).

The author found fruiting bodies that were identified as *H. rugiseptum* quite regularly on rotting wood and old stumps in the Black Forest in Germany, but also could not cultivate the organism. The sporangiole wall has a surface structure unlike that of any other myxobacterium. The species may be the only representative of the genus *Haploangium*.

2. **Haploangium minus** (ex McCurdy 1970) *comb. nov.* (*Haploangium minor* Peterson 1959, 4; *Polyangium minor* (sic) McCurdy 1970, 294.)

mi′ nus. L. neut. adj. comparative of *parvus* small; *minus* smaller, less.

Vegetative cells cylindrical with squarish ends, 0.7×3.5–4.5 µm. Sporangioles solitary but often in groups of 4–10,

globose, oval or bean-shaped, turgid, 60–140 µm, dull orange-brown when fresh, the wall becomes bright yellow-orange, collapses, and becomes wrinkled when drying. Myxospores cylindrical with blunt ends, 0.7×2.5–3.5 µm. Obtained on bark collected from various trees in Missouri. Could not be cultivated.

The mol% G + C of the DNA is: not determined.

Deposited strain: Peterson 41, University of Missouri Herbarium, Columbia, MO, USA (type specimen).

The description is reminiscent of *Melittangium lichenicola*, which, although living in the same habitat, has cells with tapering ends. "*H. minus*" was also reported from soil and bark in India by Singh and Singh (1971). Here, the vegetative cells were indeed described as having somewhat tapering ends, and the myxospores were described as being shortened rods. It is still possible that the Indian colleagues were not really describing the same organism. One always has to keep in mind that there are many examples of fruiting bodies of very similar shape and structure occurring in separate genera, families, and suborders of the myxobacteria.

Genus V. **Jahnia** gen. nov.

HANS REICHENBACH

Jah′ nia. named in honor of Eduard Adolf Wilhelm Jahn (1871–1942), who, in 1911, wrote the first synopsis, and, in 1924, the first monograph on myxobacteria (Claussen, 1942).

Vegetative cells are cylindrical rods with rounded ends. Fruiting bodies consist of coils of sporangioles sitting on a slime cushion or soft slime stalk. Myxospores are like vegetative cells in shape but are optically refractile.

Type species: **Jahnia thaxteri** (ex Jahn 1924) *sp. nov., nom. rev.* (*Archangium thaxteri* Jahn 1924, 71.)

List of species of the genus Jahnia

1. **Jahnia thaxteri** (ex Jahn 1924) *comb. nov., nom. rev.* (*Archangium thaxteri* Jahn 1924, 71.)

thax′ te.ri. M.L. gen. masc. n. *thaxteri* of Thaxter, in honor of Roland Thaxter (1858–1932), discoverer of myxobacteria (Pfister, 1993).

Vegetative cells are cylindrical rods with rounded ends, 0.7–0.9×3–8 µm. Sporangioles (Fig. BXII.δ.110) bright yellow orange to yellowish red brown, spherical to elongated, often incompletely separated, and rather large (60–90×80–120 µm). They usually form long, convoluted strings, but also occur in dense packages and then are sometimes slightly polygonal. They may be piled high, and arise either on the surface or within the agar layer. When typical, the sporangioles are positioned on a slime cushion or on a short yellowish to white stalk (Fig. BXII.δ.111). The whole fruiting body measures about 500–600 µm in height and about the same in width. At least in culture, stalks are usually absent. The first stage of fruiting often is a large, yellow, globular bacterial mass, like the initial stage of a *Chondromyces* fruiting body. Myxospores resemble vegetative cells in shape but are slightly smaller and optically refractile when mature.

Swarm colonies often etch the agar surface, and produce broad trails and wide depressions. A dense, orange band of vegetative cells is often seen along the edge of the swarm. Of the proteolytic–bacteriolytic nutritional type. May be isolated on streaks of living or dead *E. coli* on water agar. Grows

well in pure culture on yeast agar (VY/2-) with lysis of the yeast cells as well as on casein peptone (CY-) and skim milk agar. Casein, starch, and xylan are hydrolyzed, chitin is vigorously degraded, and cellulose is not attacked.

May be obtained from soil, dung of herbivores, and decaying plant material. Strains have been isolated from samples from Germany, Cyprus, Senegal, Arizona, and South Africa. Rather uncommon. Strain Pl t4 was isolated in 1988 from soil from South Africa. The reference strain, Pl t3, was originally isolated by W. Dawid from a soil sample from Arizona, USA.

The mol% G + C of the DNA is: not determined.

Type strain: Pl t4, DSM 14626, JCM 12631.

GenBank accession number (16S rRNA): DSM 14625 (Pl t3).

Additional Remarks: J. thaxteri is easy to recognize because of its large, bright orange brown, sporangioles in long strings. Stalks and slime cushions are often not produced, but when they are, the resulting fruiting body is unlike that of any other myxobacterium. (A good color photograph is found in the first edition of *The Prokaryotes* (Reichenbach and Dworkin, 1981). The cultures in the author's laboratory that have been deposited are most probably identical with *Archangium thaxteri* Jahn 1924. The main difference is that Jahn never observed sporangioles but only convoluted strands of tightly compacted cells, as in an *Archangium* fruiting body. In spite of this, he originally identified the organism as a *Polyangium* (Jahn, 1911), but this should not

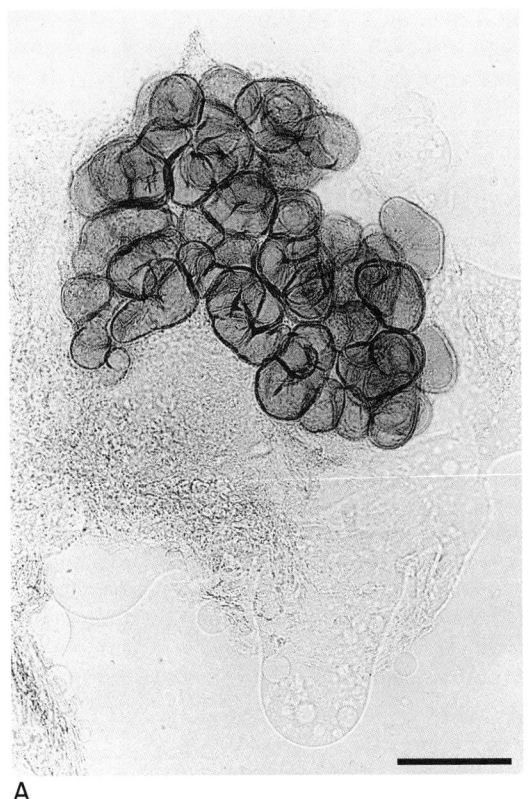

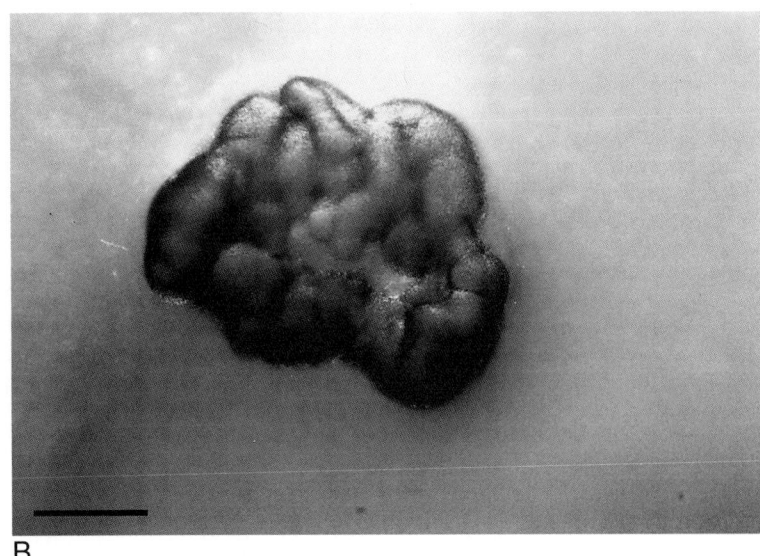

A

B

FIGURE BXII.δ.110. *Jahnia thaxteri*: fruiting bodies. (*A*) Fully differentiated sporangioles from a fruiting body with a slime pedicle (Fig. BXII.δ.111). (*B*) Partly differentiated cell mass resembling "*Archangium thaxteri*" Jahn. Bars = 200 μm.

FIGURE BXII.δ.111. *Jahnia thaxteri*: fruiting body with slime pedicle, on agar surface. Bar = 320 μm.

be taken too seriously because the definitions of genera were still rather hazy at that time. Jahn also mentioned that stalks were often lacking and that the bacterium was rare.

He found it only four times, always on rabbit dung from the Berlin area, Germany. In the author's opinion, Jahn also observed immature fruiting bodies. The strings of sporangioles arise from convoluted strands of cells that initially look somewhat brain-like; this stage exactly corresponds with Jahn's figures. Jahn was only able to propagate the organism on rabbit dung, but it is easily cultivated on various media developed more recently.

The isolates were originally classified as *Polyangium thaxteri*. This assignment to the genus *Polyangium* is consistent with the shape of the vegetative cells and myxospores, the etching of the agar by the swarm colony, and the structure of the fruiting bodies. It differs only in the production of stalks. But 16S rRNA sequence analysis (Spröer et al., 1999) clearly showed that the organism belongs in the *Chondromyces* group and not in the *Polyangium–Sorangium* cluster.

In the first stage of fruiting a large, yellow, globular mass is formed that clearly reminds one of a *Chondromyces*; however, the new bacterium differs substantially from *Chondromyces* species, which produce well-developed, long stalks with sporangioles, or chains of sporangioles, radiating in clusters from the tip of the stalk. In addition, in contrast to all *Chondromyces* species, this organism degrades chitin very efficiently and therefore should be transferred to a genus of its own.

There is a remote possibility that the present species is identical with one of the little known *Polyangium* species, e.g., *Polyangium aureum* or *Polyangium parasiticum*. However, the sporangioles of those species are considerably smaller, and a stalk has never been mentioned for them. The question can no longer be decided, because reference strains are not available.

Genus VI. **Sorangium** *(Jahn 1924) gen. nov., nom. rev.*

HANS REICHENBACH

So.ran' gi.um. Gr. masc. n. *soros* heap, pile; Gr. neut. n. *angion* vessel; M.L. neut. n. *Sorangium* piled up vessels.

Vegetative cells cylindrical with rounded or somewhat squarish ends. Sporangioles spherical to polyhedral, in sori of variable size, yellow, orange, brown, or black, but never bright red. **Myxospores like vegetative cells in shape**, optically refractile. **Cellulose decomposers**.

Type species: **Sorangium cellulosum** (Brockman 1989b) *comb. nov. (Polyangium cellulosum* (ex Imshenetski and Solntseva 1936) Brockman 1989b, 495.)

FURTHER DESCRIPTIVE INFORMATION

Jahn created the genus *Sorangium* to comprise species with sori composed of relatively small, angular sporangioles. He also established a family for the genus, the *"Sorangiaceae"*. However, this strict separation of the genera *Sorangium* and *Polyangium* was probably a misjudgment, and most of his *Sorangium* species are now classified in the older genus *Polyangium*. One subgroup among the old *Polyangium* and *Sorangium* species includes efficient cellulose degraders, and it is suggested here that the genus name *Sorangium* be used for them, and exclusively so. In the author's experience cellulose degradation is a unique and very stable characteristic, so that a separate genus *Sorangium* for such organisms may be justified. There are other unusual characteristics that distinguish *Sorangium* from *Polyangium*, such as growth on inorganic nitrogen compounds as the sole nitrogen source, absence of lanosterol synthesis, no etching of the agar surface, much elevated kanamycin resistance, and a different pattern of secondary metabolites. At the moment, only one species, *S. cellulosum* will be described although it appears to be a complex of several morphological and physiological variants that may represent different species, as will be discussed in more detail below.

Until recently, *S. cellulosum* was the only established cellulose degrader among myxobacteria. But now a new genus and species, *Byssophaga cruenta*, is being added to this group (see above). In their pioneering article on cellulose-decomposing myxobacteria, Krzemieniewska and Krzemieniewski (1937a) mentioned two forms of *Archangium* that decomposed cellulose. However, they were not sure whether the fruiting bodies of those strains were not just immature specimens of a *Sorangium*, and therefore did not name them. This seems to be the correct interpretation. Other myxobacteria stated to be cellulose degraders might have been contaminants in cellulose-decomposing cultures.

List of species of the genus Sorangium

1. **Sorangium cellulosum** (Brockman 1989b) *comb. nov. (Polyangium cellulosum* (ex Imshenetski and Solntseva 1936, 1115) Brockman 1989b, 495.) syn. *Sorangium compositum* Krzemieniewska and Krzemieniewski 1937a; *Polyangium compositum* Thaxter 1904, 413; *Sorangium compositum* Jahn 1924, 74; *Sorangium nigrescens* Krzemieniewska and Krzemieniewski 1937a, 21; *Sorangium nigrum* Krzemieniewska and Krzemieniewski 1937a, 22.

cel.lu.lo' sum. M.L. n. *cellulosum* cellulose.

Vegetative cells (Fig. BXII.δ.112) cylindrical with rounded or somewhat squarish ends, 0.8–1.2 × 3–8 µm. Sporangioles (Fig. BXII.δ.113) spherical to polyhedral, 15–40 µm in diameter, more or less densely packed in sori of variable size. When produced on filter paper, sporangioles often in long chains and stripes along decaying cellulose fibers, of many different colors: yellow, orange, brown, black, even colorless. When growing on an agar surface, e.g., on yeast agar (VY/2-), sporangioles may be single and scattered over the surface; fruiting bodies often arise also deep within the agar. Many strains show subdivision of the sori into clusters or packets of sporangioles. Fruiting bodies are produced on filter paper in enormous numbers, often forming a continuous mass in which individual sori can hardly be distinguished. Myxospores are similar to vegetative cells in shape, but are shorter (2–4 µm) and optically refractile. Swarm colonies on agar (Fig. BXII.δ.114) usually have a well-developed pattern of radiating veins. Agar surface normally not etched, but sometimes slightly depressed or ripped open by contraction of the firmly adhering slime sheet of the swarm. Congo red is not adsorbed, but may stain fruiting bodies upon longer exposure.

S. cellulosum readily decomposes crystalline cellulose and can be cultivated on filter paper as the only carbon source placed on a mineral salts agar (e.g., ST21-agar) with an inorganic nitrogen source (preferably KNO_3, but $(NH_4)_2SO_4$ can be used). Grows faster when an organic nitrogen source is supplied, e.g., tryptically digested casein, but many strains then require an additional carbon source, such as glucose, starch, or cellulose. All strains appear to degrade starch and xylan, and most of them degrade chitin very efficiently. Proteolytic activity, e.g., on skim milk agar, varies among individual strains from none to very strong. Yeast cells in yeast agar (VY/2) are attacked but only partly degraded, so that lysis zones remain rather turbid. Grows moderately on streaks of autoclaved *E. coli* on water agar but not on living bacteria. All strains are resistant to kanamycin sulfate, often up to exceedingly high concentrations (1000 µg/ml), a fact which can facilitate isolation. Tolerates a lower pH (5.8–6.2) than do most myxobacteria.

S. cellulosum grows well in liquid media (e.g., Gerth et al., 1994, 1996), first as small nodules and flakes, then—usually after many transfers—in homogeneous cell suspensions. Generation times range between 6 and 12–14 hours (30°C). Large-scale fermentations up to 100 m^3 have been performed successfully. These organisms produce a wide variety of bioactive secondary metabolites and are the source of many interesting compounds, such as epothilone, soraphen, sorangicin, and tartrolon.

Strains of *S. cellulosum* were found to have very large genomes, 10–12.2 Mbp.

Very common in soil with decaying plant material and on dung of herbivorous animals. Strain So ce1871 was isolated in 2001 from soil with decaying plant material collected in 2000 on the island of Rhodes, Greece; reference strain So ce1873 came from soil collected in Egypt in 2000 and was isolated in 2001. *"Sorangium nigrum"* strain So

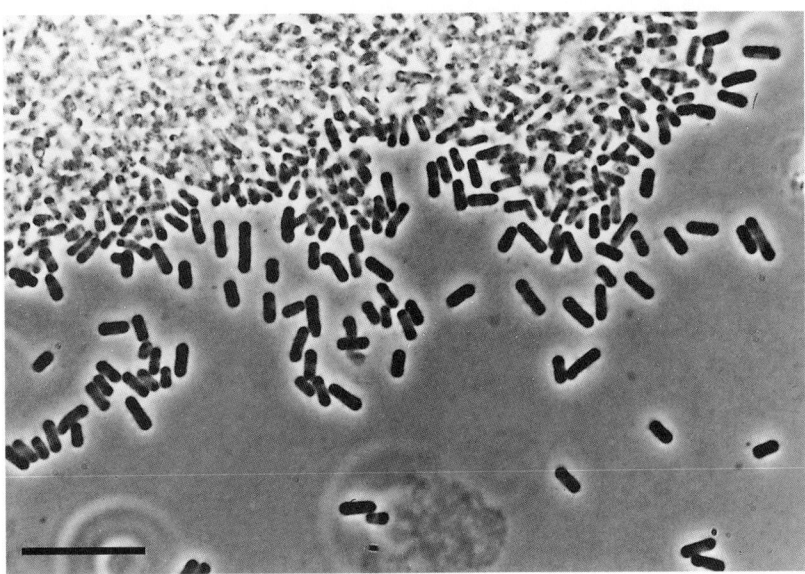

FIGURE BXII.δ.112. *Sorangium cellulosum*: vegetative cells from yeast agar, phase contrast. Bar = 13 μm.

ce1654 was isolated in 1999 from a sample from Bali, Indonesia, and reference strains So ce1695 and So ce1700 in 2000 from soil with rotting plant material collected in 1999 at different sites on the island of Rhodes, Greece.

The mol% G + C of the DNA is: 69 Bd (may be up to 70–72).

Type strain: So ce1871, DSM 14627, JCM 12641.

Additional Remarks: Reference strains include ATCC 15384, ATCC 25531, ATCC 25532, ATCC 25569, DSM 14671 (So ce1873).

Reference strains for *"Sorangium nigrum"* include DSM 14731 (So ce1654), DSM 14732 (So ce1695), DSM 14733 (So ce1700).

Cellulose-degrading myxobacteria are very common and have been isolated from all kinds of soils and environments from Russia and Sweden in the north to semiarid areas and tropical rain forests in the south. *S. cellulosum* clearly is a complex of several species. However, as most phenotypic characteristics of those organisms are extremely variable, all efforts at subdivision of this group of strains have failed so far. A comparative study making use of modern methods of molecular taxonomy may help to solve the problem.

The following phenotypic characteristics are potentially useful for species delimitation. (1) Size, shape, and color of fruiting bodies and sporangioles. However, fruiting bodies are always variable in size. They are sometimes sori composed of packets of sporangioles; in other cultures of the same strain they are homogeneous, and upon prolonged subcultivation may completely decay into scattered sporangioles, which sometimes form an uninterrupted, continuous mat of sporangioles when the organism is grown on filter paper. Shape and size of sporangioles is equally variable. Many strains initially produce small sporangioles that then increase substantially in size during cultivation. Polyhedral when densely packed, they become spherical when the sori are more loosely organized. However, there are fresh isolates that have tiny sporangioles and others that have large sporangioles. Such initial differences may indeed indicate the existence of different species. Color of fruiting

bodies tends to become ever paler with subsequent transfers. Many strains end with pale, brownish-orange fruiting bodies. Other strains retain an intense rust brown, orange, yellow, or black color for a long time, and pigmentation of fresh isolates may be a valuable taxonomic criterion. (2) Size and shape of the vegetative cells are similar for all strains. During cultivation, especially on agar media, cells tend to become slightly but recognizably fatter. There are, however, strains with relatively short, more delicate rod-shaped cells with squarish rather than rounded ends; these strains may really represent a separate species, *"Sorangium nigrum"*, which will be discussed below. (3) Other characteristics that may be taxonomically useful are chitin degradation, which varies from zero to very strong among different strains; growth on casein peptone as the only carbon and nitrogen source, which occurs in about 25% of strains; lysis of proteins, e.g., in skim milk agar; and the degree of kanamycin resistance. The unsatisfactory aspect of all those variable characteristics is that they vary independently, so that no stable patterns have become recognizable so far.

The variants and species discussed below are to be evaluated against this background.. Cellulose-decomposing myxobacteria were discovered independently in two laboratories in the 1930s. The earliest publication was by Imshenetski and Solntseva (1936), who named their organism *Polyangium cellulosum*. The bacterium produced red-brown fruiting bodies, sori were 40–45 × 110–160 μm and consisted of 12–40 roundish to polygonal orange sporangioles, 8–42, mostly 20–25 μm in diameter. The vegetative cells were cylindrical rods with rounded ends, 0.8–1 × 3.5–8.5 μm. The myxospores were similar in shape and measured 0.7–0.8 × 3-2–3.5 μm. Old sporangioles sometimes contained a drop of an oily liquid. The optimal growth temperature was 18–22°C, and development was much slower at 30°C. (The fact that the bacterium was isolated from soil in Russia exposed to a cold, continental climate may explain this unusually low optimal temperature.) The description of *Polyangium cellulosum* fits a large proportion

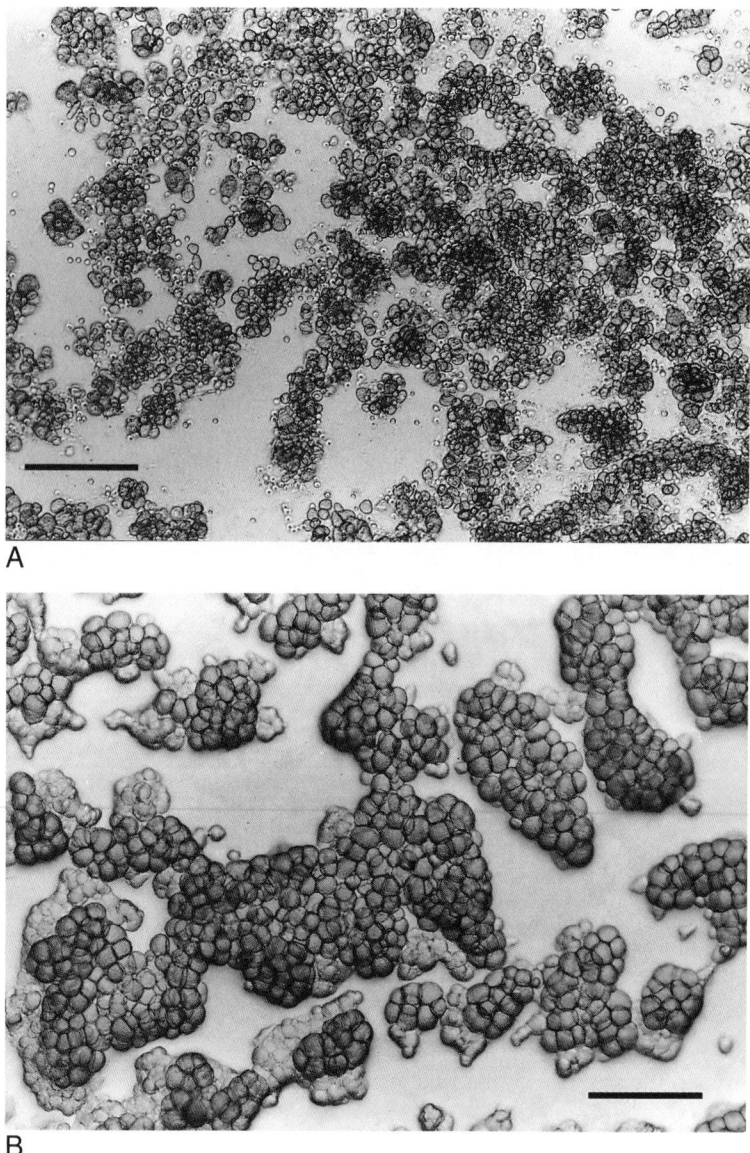

FIGURE BXII.δ.113. *Sorangium cellulosum*: fruiting bodies; (*A*) slide mount from filter paper on mineral salts agar, (*B*) on surface of yeast agar. Bar in *A* = 125 μm; bar in *B* = 100 μm.

of the author's 2000 *Sorangium* isolates. A strain has been deposited that corresponds to the original species definition in the color and structure of its fruiting bodies.

In the following year, Krzemieniewska and Krzemieniewski (1937a, b) published the results of their almost decade-long studies on cellulose decomposing myxobacteria. They recognized that *"P. compositum"*, which, following Jahn, they now named *"Sorangium compositum"*, was a cellulose decomposer (see comments on *Polyangium sorediatum*). The fruiting bodies of their strains were orange yellow. The sori were sometimes subdivided into packets of sporangioles; at other times, they were not. Krzemieniewska and Krzemieniewski distinguished two variants, one with smaller sporangioles, 8–12 (6–16) × 7–18 μm in size, and one with larger sporangioles, 10–15 (7–18) × 8–20 μm. A third, faster growing variant with small sporangioles, 6–13 × 7–14 μm, produced fruiting bodies that were scattered over the swarm from the beginning and did not arise gradually from the center outward. The vegetative cells of all three types were cylindrical rods, 0.6–0.9 × 2.6–6 μm. In young cultures on cellulose the rods were much wider, 1.1–1.4 μm; in young sporangioles, the rods were very short, 0.8 × 1.3 μm. The sporangioles sometimes contained globules of a fatty material. *"Sorangium compositum"* of Krzemieniewska and Krzemieniewski differs from *Polyangium cellulosum* in the color of its fruiting bodies and has smaller sporangioles, but it may still belong to the same species. Krzemieniewska and Krzemieniewski also discovered that their own *"S. spumosum"* was a cellulose decomposer. However, there is some uncertainty about that species (see comments on *Polyangium spumosum*). The cellulose degrading strains produced flesh red fruiting bodies in a colorless slime matrix, up to 400 μm, containing sporangioles of 16–24 μm diameter in one isolate and 9–15 μm in another. The sporangioles had a colorless wall. The vegetative cells measured 1.1–1.5 × 2.7–6 μm. This species was found by them to be the most com-

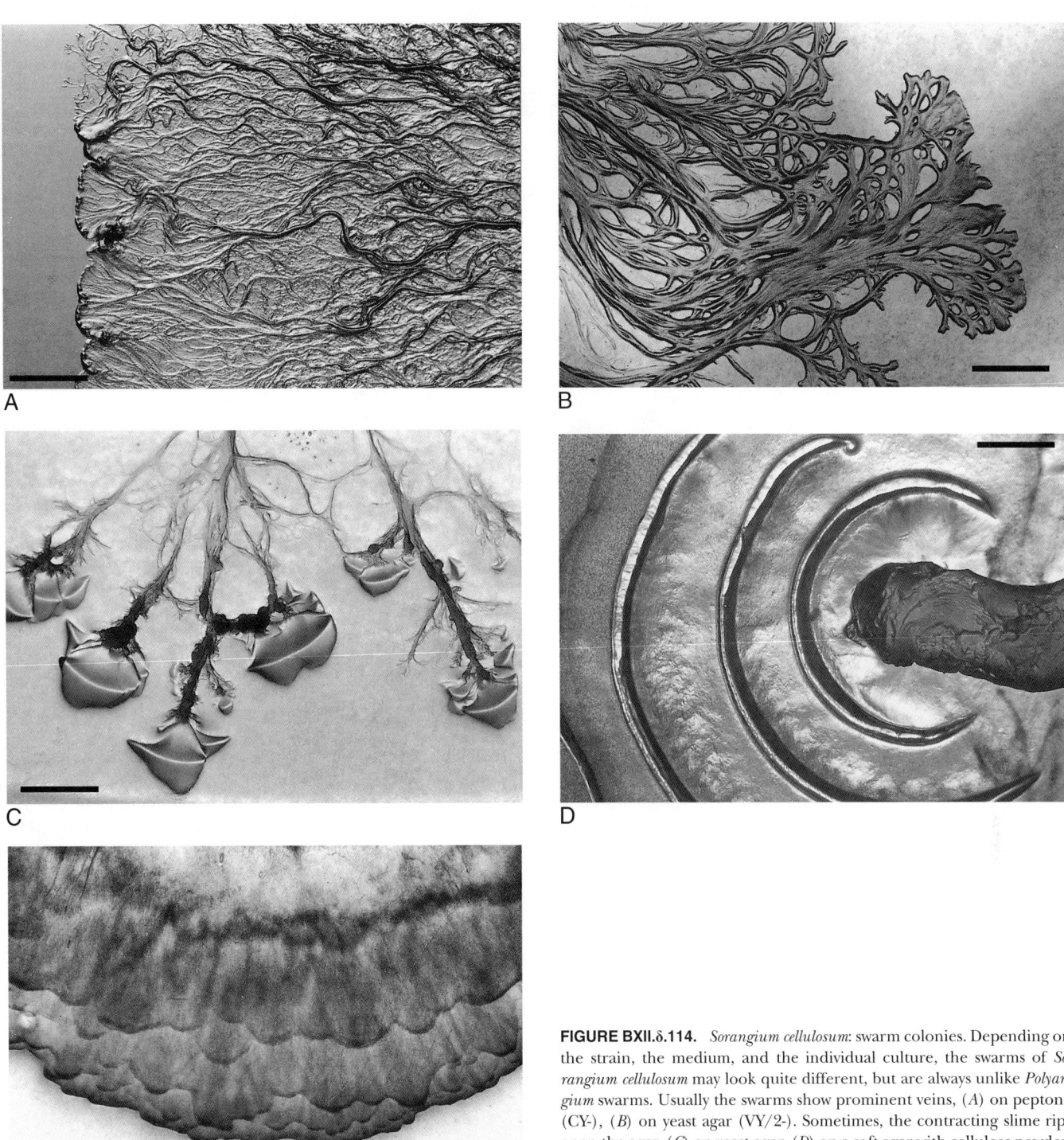

FIGURE BXII.δ.114. *Sorangium cellulosum*: swarm colonies. Depending on the strain, the medium, and the individual culture, the swarms of *Sorangium cellulosum* may look quite different, but are always unlike *Polyangium* swarms. Usually the swarms show prominent veins, (*A*) on peptone (CY-), (*B*) on yeast agar (VY/2-). Sometimes, the contracting slime rips open the agar, (*C*) on yeast agar, (*D*) on a soft agar with cellulose powder; note clearance of the cellulose in the swarm area. The swarm may also penetrate the agar deeply, producing a garland-like pattern at the swarm edge; on peptone (0.2% Casitone) soft agar (0.6%). Bar in *A* and *C* = 1100 μm; bar in *B* = 600 μm; bar in *D* and *E* = 3000 μm.

mon of all *Sorangium* species. Its status, however, is not clear. Krzemieniewska and Krzemieniewski described two other cellulose-decomposing *Sorangium* species. The first was *"Sorangium nigrescens"* Krzemieniewska and Krzemieniewski 1937a, 21. Mature sori were gray brown to blackish, with the color originating from the cell mass. Packets of sporangioles were about 200 μm wide, in a colorless slime matrix. Sporangioles were 5–12 × 5–15 μm, mostly 6–10 μm in diameter. Vegetative cells were 0.2–1.4 × 2.5–6.5 μm. While strains with fruiting bodies of that color do exist, their status as a species has still to be established. The second new species was *"Sorangium nigrum"* Krzemieniewska and Krzemieniewski 1937a, 22 (Fig. BXII.δ.115). The fruiting bodies were pitch black. Sporangioles were 9–16 × 9–

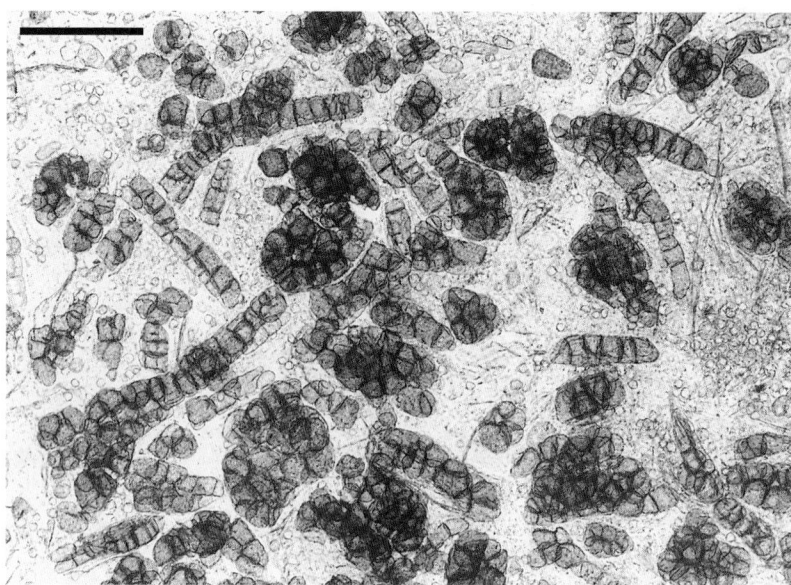

FIGURE BXII.δ.115. *"Sorangium nigrum"*: fruiting bodies, slide mount, from filter paper on mineral salts agar. Bar = 125 μm.

22 μm, mostly 16–18 μm, with rather stout walls, arranged in chains and strands along decaying cellulose fibers. Vegetative cells were 1.1–1.3 × 2.5–5.5 μm. Strains with such characteristics are often isolated. Swarm colonies on yeast agar (VY/2-) tend to become completely converted into fruiting bodies after some time. On filter paper, roundish sori are usually produced in addition to the strings and strands. The vegetative cells typically show somewhat squarish rather than rounded ends, and are often rather short. The author's impression is that this is a valid, distinguishable species, but molecular data should be presented before it is accepted. The author has deposited three characteristic strains.

Finally, a number of variants of *S. cellulosum* have been described. Mishustin (1938), who isolated cellulose degraders from Russian soils, distinguished four variants, mainly by the color of fruiting bodies. *Polyangium cellulosum* biovar fuscum—which he regarded as identical with the organism of Imshenetski and Solntseva (1936)—had brown fruiting bodies, biovar ferrugineum had dark red or red-brown fruiting bodies, biovar fulvum had pinkish brown to reddish yellow fruiting bodies, and biovar luteum had yellowish or-

ange fruiting bodies. There also were slight differences in the dimensions of sori, sporangioles, and vegetative cells, but all measurements overlapped. Strains with fruiting bodies of all those colors are abundant, but their taxonomic status is open. Pronina (1962) isolated cellulose degraders from spoiled material in a paper mill. She, too, described a number of variants (biovars) of *S. cellulosum*. Five of those had vegetative cells with slightly pointed ends and thus may not have been *Sorangium* strains at all (biovar flavovirens, biovar glaucum, biovar helvolum, biovar ochraceum, biovar album). *S. cellulosum* biovar umbrinum was brown on filter paper and on agar, but had very short rods of 1.0 × 2.7 μm; biovar aurescens was pale orange to pale brown and also had short vegetative cells of 0.7 × 2.7 μm; and biovar cremeum was pink on cellulose, and pale yellow to gray on agar; its vegetative cells measured 0.7 × 3.2 μm. The sporangioles were in all cases extremely small, 2.3–6 μm. In addition, the colors were in most instances quite atypical. It is not possible to decide now whether any of those organisms were really cellulose degrading members of the genus *Sorangium*.

Family IV. **Nannocystaceae** *fam. nov.*

HANS REICHENBACH

Nan.no.cys.ta' ce.ae. M.L. fem. n. *Nannocystis* type genus of the family; *-aceae* ending to denote a family; M.L. pl. fem. n. *Nannocystaceae* the *Nannocystis* family.

Vegetative cells are short, stout, cylindrical to somewhat bulging rods with rounded to squarish ends. Fruiting bodies consist of spherical to ellipsoidal, sometimes irregular sporangioles. Most of the sporangioles are solitary, scattered on top and within the agar plate; many strains also produce dense clusters, sheets, and packets of sporangioles. The size of the sporangioles usually varies very widely. **Myxospores are very short rods, ellipsoidal, or spherical. Swarm colonies etch and corrode the agar** to varying degrees, depending on the medium, from shallow depressions and pits (on yeast agar) to deep holes and channels, often transforming the agar plate into a spongy mass down to the bottom of the dish.

Of the proteolytic–bacteriolytic nutritional type. All strains grow well on yeast agar (VY/2-) and on streaks of living or autoclaved *E. coli* on water agar. Many strains also grow on casein peptone (CY-) agar. Agar corrosion is particularly dramatic on streaks of living and autoclaved *E. coli* and on CY-agar if the strain is grown on that medium. Sporangioles are usually produced abundantly on streaks of autoclaved *E. coli*. Because of agar corrosion, sporangioles are difficult to see under the dissecting microscope. However, when the swarm is covered with a drop of water, the sporangioles become clearly visible. The *Nannocystis* species appear not to utilize mono- and disaccharides but depend totally on peptones and proteins for growth. Do not degrade chitin. Synthesize cholestenols (Kohl et al., 1983; Zeggel, 1993), geosmin (Trowitzsch et al., 1981), and the iron chelator nannochelin (Kunze et al., 1992).

This family belongs to the third suborder of the *Myxococcales*, the *Nannocystineae*, together with the family *Kofleriaceae*. The suborder can currently be defined only by 16S rRNA sequence analysis, which clearly shows the separation of the organisms classified here from other myxobacteria (Spröer et al., 1999). The family, however, contains only one genus, *Nannocystis*, and is clearly defined by its phenotypic characteristics.

Some of the more unusual myxobacteria belong to this taxonomic group, in particular the psychrophilic myxobacteria (Dawid et al., 1988).

Type genus: **Nannocystis** Reichenbach 1970, 137.

Genus I. *Nannocystis* Reichenbach 1970, 137[AL]

HANS REICHENBACH

Nan.no.cys' tis. Gr. masc. n. *nannos* dwarf; Gr. fem. n. *kystis* bladder; M.L. fem. n. *Nannocystis* tiny bag.

Characteristics as given for the family. *Nannocystis* is perhaps the most common of all myxobacteria. It may be obtained from soil, decaying plant material, and similar substrates. It was overlooked for a long time because of its tiny sporangioles; it is best isolated on streaks of living *E. coli* on water agar. The swarm colonies of *Nannocystis* do not adsorb Congo red. The phenotypic charac-

teristics of *Nannocystis* strains are extremely variable, which makes distinguishing species difficult. Only two species are named in this edition. But a number of variants listed under *N. exedens* may qualify as species of their own. Representative strains have been deposited for further study.

Type species: **Nannocystis exedens** Reichenbach 1970, 137.

Key to the species of the genus *Nannocystis*

1. Sporangioles of very different sizes; spherical, ellipsoidal, or irregular in shape; solitary or in groups, sheets, and packets.

    N. exedens

2. Sporangioles rather uniform in size, very small, mostly spherical, always solitary.

    N. pusilla

List of species of the genus Nannocystis

1. **Nannocystis exedens** Reichenbach 1970, 137[AL]

    ex.e' dens. L. v. *exedere* to eat away; L. present part. *exedens* eating away, corroding (the agar).

    Vegetative cells (Fig. BXII.δ.116) are short, stout, cylindrical, or slightly bulging rods with rounded or squarish

ends, 1.1–2 × 1.5–5 µm. Move by gliding as is typical for myxobacteria (Fig. BXII.δ.117).

The fruiting bodies are solitary sporangioles (Fig. BXII.δ.118), spherical, ellipsoidal, or irregular, of very different size from very small, 3.5 × 6 µm containing just four or five myxospores, to rather large, 15 × 30 to 40 × 110 µm. Sometimes a fine, regular striation can be seen on the surface. Many strains produce, in addition to solitary sporangioles scattered on and within the agar plate, sporangioles

FIGURE BXII.δ.116. *Nannocystis exedens:* vegetative cells *in situ* on agar surface in a chamber culture. Phase-contrast micrograph. Bar = 13 µm.

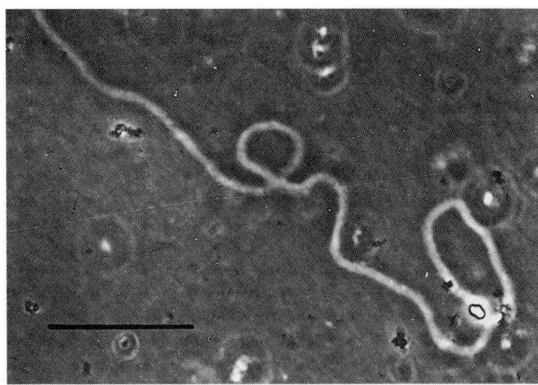

FIGURE BXII.δ.117. *Nannocystis exedens* slime track produced by a pack of a few cells migrating over a thin film of peptone agar in a chamber culture. Phase-contrast micrograph. Bar = 28 µm.

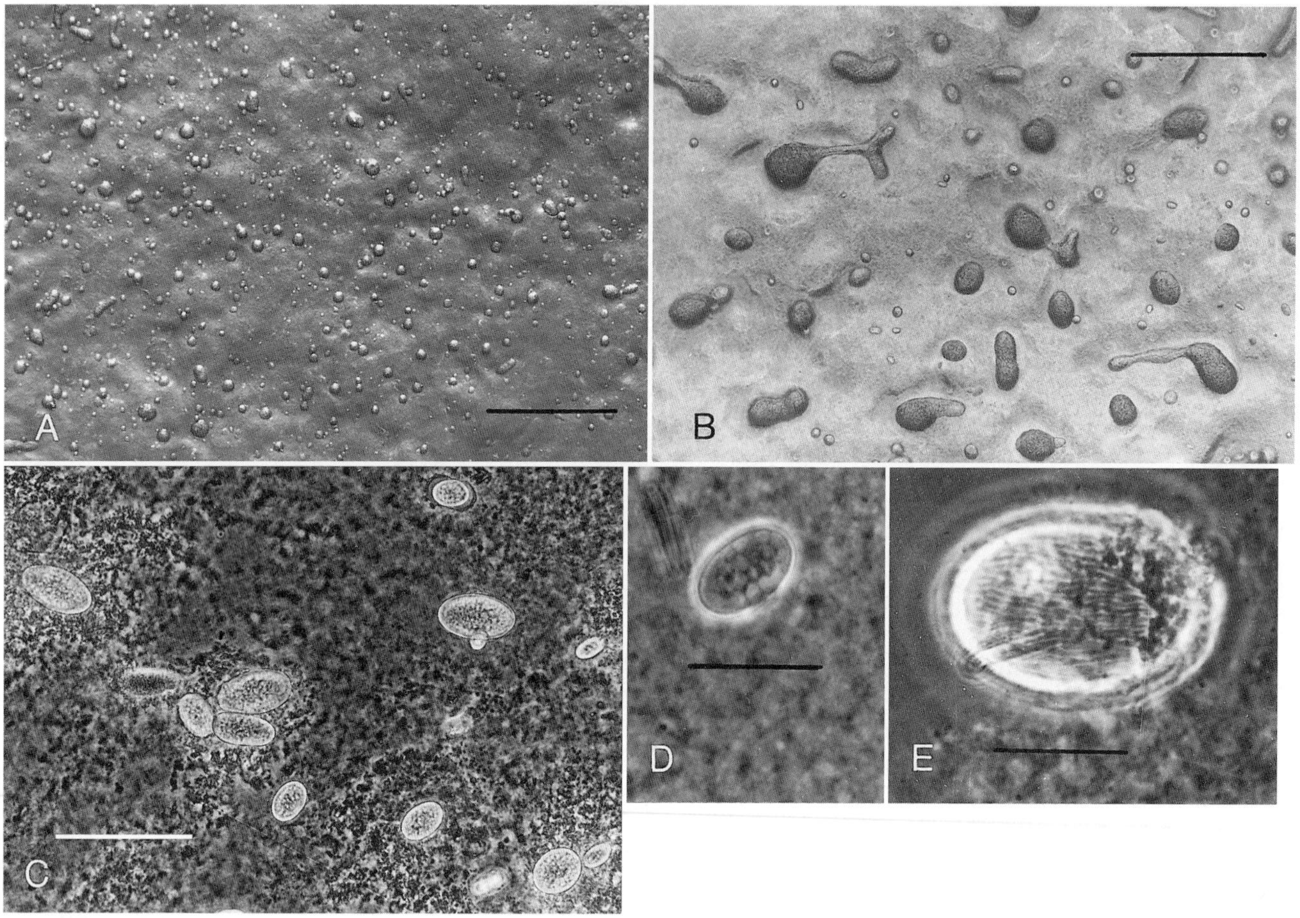

FIGURE BXII.δ.118. *Nannocystis exedens* fruiting bodies. (*A*), survey of the surface of a culture plate with numerous sporangia on and within the agar. Oblique illumination. Bar = 310 μm. (*B*), surface of a culture plate (at higher magnification) on which large irregularly shaped sporangia are seen on top of the agar and small ovoid ones are seen within it. Oblique illumination. Bar = 30 μm. (*C*), slide mount showing ovoid sporangia from within the agar. Phase-contrast micrograph. Bar = 50 μm. (*D*), small sporangium at high magnification. The myxospores within it are clearly recognizable. Phase-contrast micrograph. Bar = 13 μm. (*E*), ovoid sporangium from within the agar. A fine striation on its surface is shown. Phase-contrast micrograph. Bar = 12 μm.

that are arranged in loose or dense groups, sheets, and tightly packed sori. Such strains may represent different species and are described in more detail below. The sporangioles may be colorless, yellowish, or light to dark red-brown.

Myxospores (Fig. BXII.δ.119) are ellipsoidal to almost spherical, dark to optically refractile, and seem not to be surrounded by a capsule. Freeze-fracturing of sporangioles (Fig. BXII.δ.120) reveals a thick, compact wall in the electron microscope. A characteristic folding of the surface membranes on the myxospores gives them a walnut-like appearance.

On agar media, the organisms always penetrate the plate deeply, but the appearance of the swarm colonies depends on the medium used. On yeast agar (VY/2-) usually large, shallow depressions are produced without much agar corrosion. When the organism is grown on streaks of living or autoclaved *E. coli* on water agar, the agar is corroded in a highly variable and most striking pattern with holes, channels, and tunnels (Fig. BXII.δ.121). This pattern somewhat resembles the etching of the agar produced by *Polyangium* strains. However, the *Polyangium* pattern is much coarser,

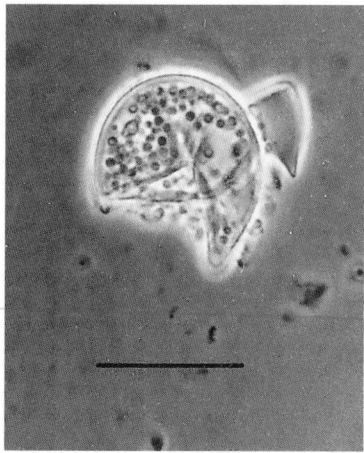

FIGURE BXII.δ.119. *Nannocystis exedens* crushed sporangium with spherical myxospores. Phase-contrast micrograph. Bar = 16 μm.

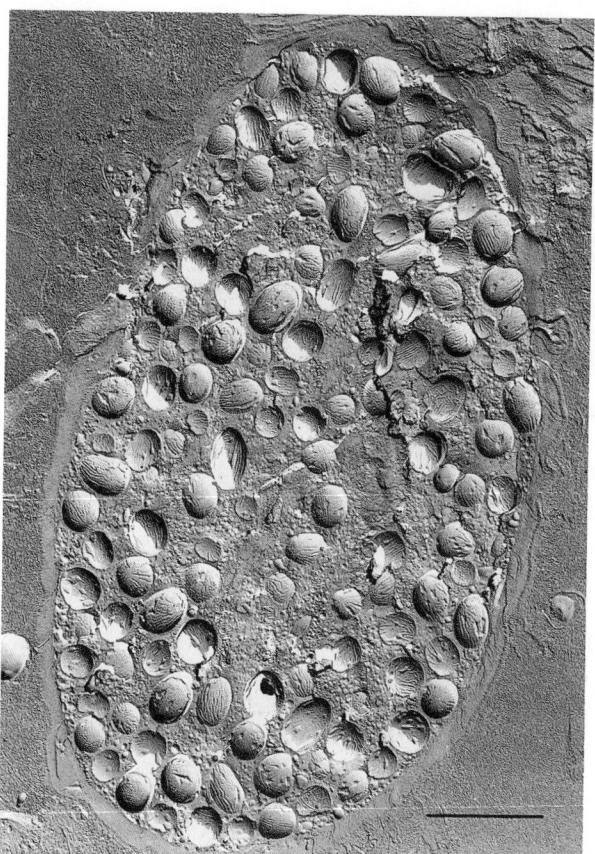

FIGURE BXII.δ.120. *Nannocystis exedens* sporangium with myxospores. The sporangium is surrounded by a thick wall. The surface membranes of the ovoid myxospores show a pattern of folds. Zeiss EM 10 B electron micrograph of a freeze fracture specimen. Bar = 3.2 μm. (Courtesy of I. Geffers.)

and the erosion of the agar plate is rarely as intense as with *Nannocystis*. Agar destruction is extreme on casein peptone (CY-) agar; however, not all strains grow on this medium. Development is normally slow, and the agar is transformed into a spongy mass; the swarms become bright red, orange, or orange brown.

Nannocystis can also be grown in liquid media, in which it forms tiny flakes or even homogeneous cell suspensions (e.g., Behrens et al., 1976; Kohl et al., 1983; Kunze et al., 1992).

An unusual retro element has been discovered in the genome of *N. exedens* (Lampson and Rice, 1997). There are two different satellite msDNAs, the genes of which are not flanked by the usual reverse transcriptase genes. Further, a set of somewhat variable repeated sequences related to msDNA were demonstrated throughout the chromosome.

The mol% G + C of the DNA is: 70–72 (Bd, T_m) (Behrens et al., 1976).

Type strain: Na e1, ATCC 25963, DSM 71.

GenBank accession number (16S rRNA): M94279.

Additional Remarks: Reference strains include ATCC 25965 (Na e17); ATCC 35989 (Na e465).

As mentioned, *Nannocystis* strains are extremely variable in their phenotypic appearance, so that the group cannot yet be classified into more than two species. However, in the author's collection of about 1900 strains, certain char-

acteristics appear to be connected with specific types of organisms. The following is an effort to bring some order into that assembly. The types are named as subspecies but may in fact represent species. It must be emphasized that the sporangioles regarded as typical for the variants always occur in a background of small, scattered sporangioles. Typical sporangioles are best observed on streaks of autoclaved *E. coli* on water agar.

a. **Nannocystis exedens** *subsp.* **exedens** *subsp. nov.*

Sporangioles small to moderately large, mostly spherical to ellipsoidal, colorless, yellowish brown or reddish, scattered.

b. **Nannocystis exedens** *subsp.* **aggregans** *subsp. nov.*
aggregans. L. present part. *aggregans* herding together.

Sporangioles small to moderately large, mostly spherical to ellipsoidal; brownish, in large, often branched, densely packed, sheet-like groups connected by a thin brown slime matrix (Fig. BXII.δ.122). The groups of sporangioles are usually only seen on streaks of autoclaved *E. coli*. 16S rRNA sequence comparison seems to support a separate position of this variant (Spröer et al., 1999). Strain Na a1 is from Franconia, Germany.

Type strain: Na a1, DSM 14639.

c. **Nannocystis exedens** *subsp.* **cinnabarina** *subsp. nov.*
cinnabarina. L. fem. adj. *cinnabarina* cinnabar red.

Sporangioles reddish, sometimes rather large and mostly spherical, may be produced in rather extended, sheet-like, dense groups, but often are scattered. Intensely red spherical globules of a fatty or oily material are seen between the sporangioles, but only on streaks of autoclaved *E. coli* (Fig. BXII.δ.123). Strain Na c1 is from Rhodos, Greece.

Type strain: Na c1, DSM 14641.

d. **Nannocystis exedens** *subsp.* **glomerata** *subsp. nov.*
glomerata. L. fem. part. perf. *glomerata* packed or pressed together.

Sporangioles small to moderately large, in small, more or less spherical, tightly packed sori (Fig. BXII.δ.124). Strain Na g1 is from Rajasthan, India.

Type strain: Na g1, DSM 14640.

e. **Nannocystis exedens** *subsp.* **pulla** *subsp. nov.*
pulla. L. fem. adj. *pulla* dark brown.

Sporangioles are small to moderately sized, narrow ellipsoidal to irregular, solitary or connected to one another in a net-like pattern, sooty-brown (Fig. BXII.δ.125). Strain Na a145 is from the Alsace, France.

Type strain: Na a145, DSM 14629.

2. **Nannocystis pusilla** *sp. nov.*
pu.sil'la. L. fem. adj. *pusilla* tiny.

Characteristics as described for the genus and *N. exedens* with the exception that the sporangioles are uniformly tiny, 8–15 μm large, scattered or in small, loose groups and short strings (Fig. BXII.δ.126). Strain Na p29 is from California, USA.

The mol% G + C of the DNA is: not determined.

Type strain: Na p29, DSM 14622, JCM 12635.

Additional Remarks: Reference strains include DSM 14621 (Na p23), from California, USA.

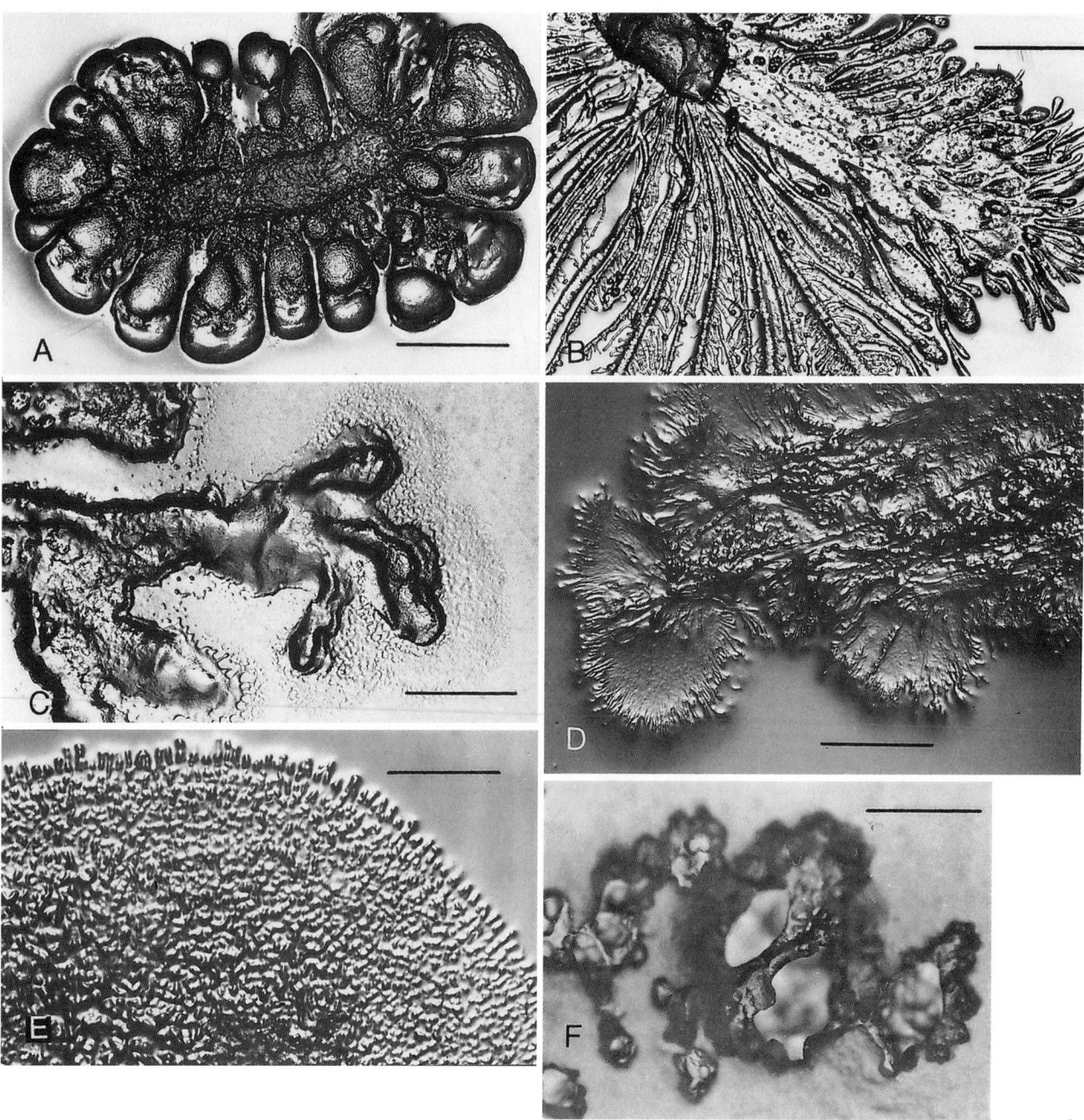

FIGURE BXII.δ.121. *Nannocystis exedens* swarm colonies. (*A*), small 4–week-old swarm colony on an agar medium containing low concentrations of casein peptone. The organism produced a series of deep bubblelike caverns within the agar plate. Bar = 6 mm. (*B*), swarm on a streak of autoclaved yeast on water agar. The organism etched a system of feathery branched trenches into the agar surface. Bar = 6 mm. (*C*), tip of a trench. At the very edge of the colony, the bacteria form a delicate veil-like swarm which later tends to contract somewhat and to sink into the agar. Bar = 1300 μm. (*D*), typical swarm pattern on water agar with streaks of living *E. coli*. The organism has decomposed the food bacteria, and the swarm is now expanding onto the agar surface. Bar = 2.5 mm. (*E*), on water agar with autoclaved *E. coli*, a delicate and uniform corrosion pattern may arise. Bar = 830 μm. (*F*), deep agar corrosion on yeast agar. Bar = 470 μm.

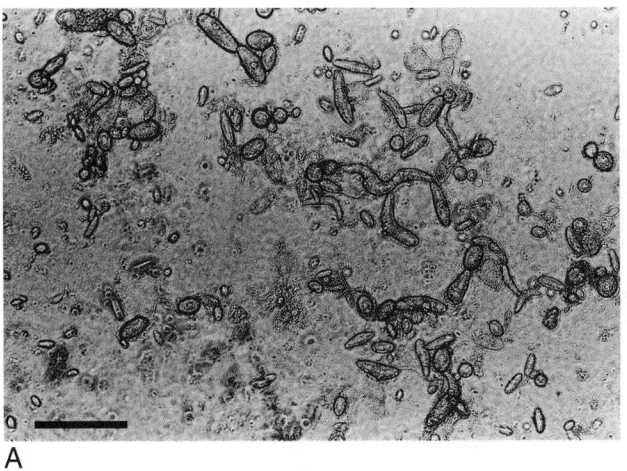

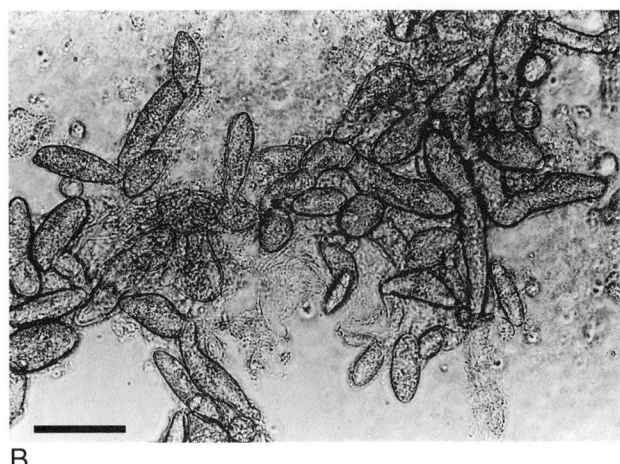

FIGURE BXII.δ.122. *"Nannocystis exedens* subsp. *pulla"*: fruiting bodies on agar plates. (*A*) Bar = 80 μm; (*B*) bar = 40 μm.

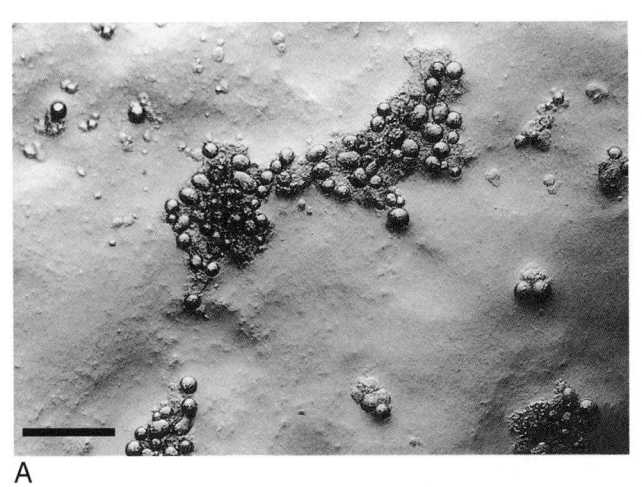

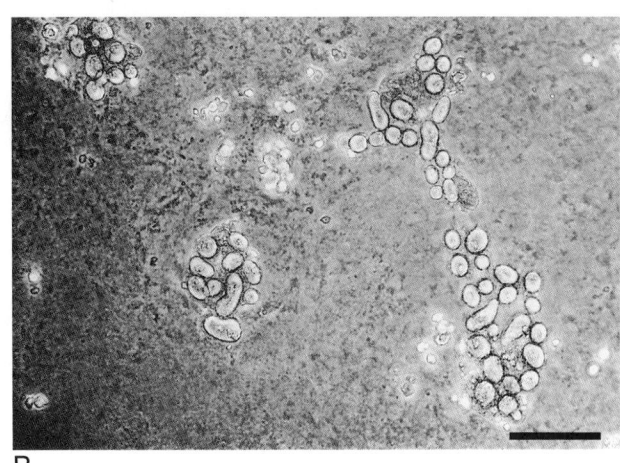

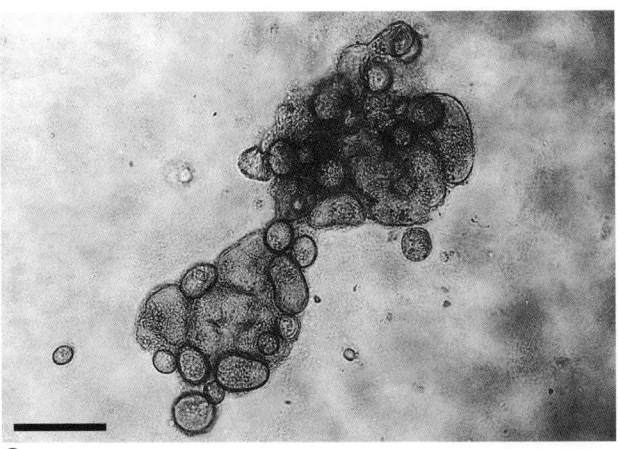

FIGURE BXII.δ.123. *" Nannocystis exedens* subsp. *aggregans"*: fruiting bodies on agar plates. (*A*) Bar = 100 μm; (*B*) bar = 85 μm; and (*C*) bar = 50 μm.

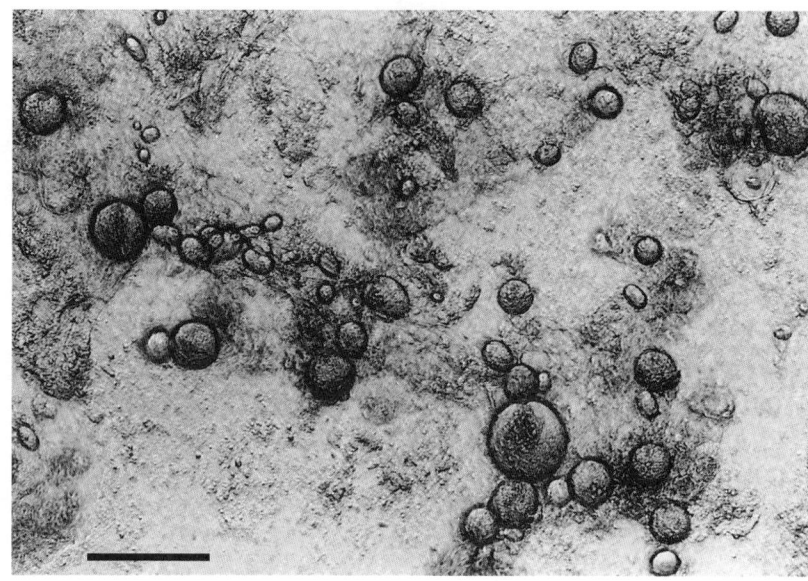

FIGURE BXII.δ.124. "*Nannocystis exedens* subsp. *glomerata*": fruiting bodies on agar plate. Bar = 65 µm.

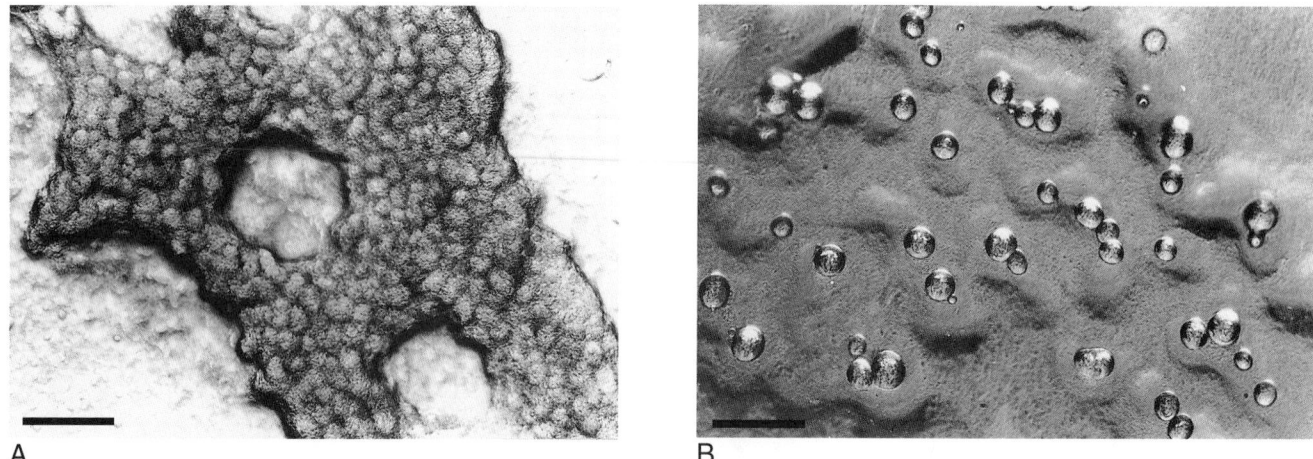

A   B

FIGURE BXII.δ.125. "*Nannocystis exedens* subsp. *cinnabarina*": fruiting bodies on agar plates. (*A*) Bar = 100 µm and (*B*) bar = 80 µm.

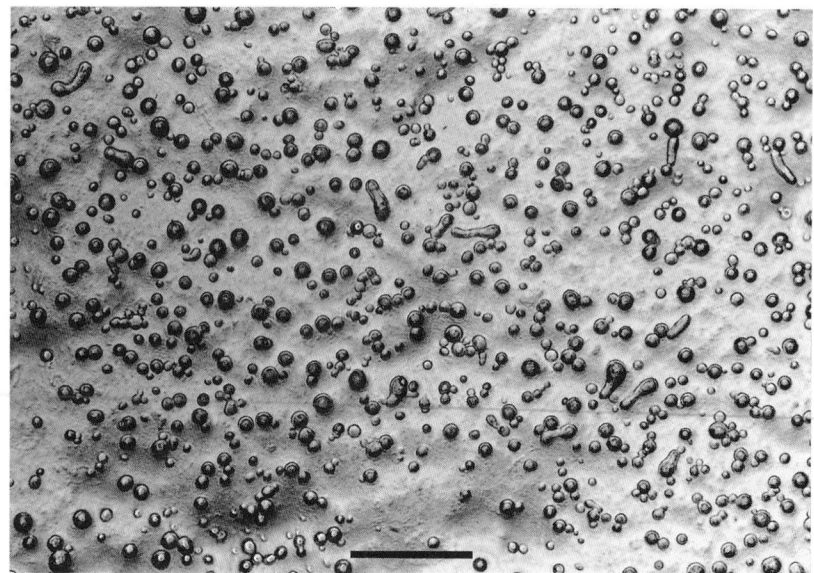

FIGURE BXII.δ.126. *Nannocystis pusilla*: fruiting bodies on agar plate. Bar = 50 µm.

Family V. **Kofleriaceae** *fam. nov.*

HANS REICHENBACH

Ko.fle.ri.a′ ce.ae. M.L. fem. n. *Kofleria* type genus of the family; *-aceae* ending to denote a family; M.L. pl. fem. n. *Kofleriaceae* the *Kofleria* family.

Vegetative cells are slender, cylindrical rods with rounded ends. Swarm colonies with radial veins that resemble those of the *Cystobacteraceae*. Does not adsorb Congo red. Does not etch the agar but may pull it apart. Fully developed fruiting bodies have not yet been seen, but probably consist of sporangioles. **Of the proteolytic–bacteriolytic nutritional type.**

Type genus: **Kofleria** *gen. nov.*

FURTHER DESCRIPTIVE INFORMATION

The vegetative cells of the *Kofleriaceae* resemble those of the *Polyangiaceae*, but the swarm colonies are different. These organisms were erroneously classified as *Polyangium vitellinum* strains (e.g.,

Pl vt1), but a 16S rRNA study clearly placed them in the suborder *Nannocystineae* (Spröer et al., 1999). This made it necessary to establish a new genus (*Kofleria*). As this genus is distant from *Nannocystis*, its classification in a new family seems justified.

It appears that some truly marine, halophilic myxobacteria also belong to this family (Moyer et al., 1995; Iizuka et al., 1998).*

Presently only one genus is distinguished in this family. The marine organisms may contribute further genera.

---

*Editorial Note: After this chapter was finished, a genus *Haliangium* with two new species *H. ochraceum* and *H. tepidum*, was proposed for the marine organisms belonging to this family (Fudou et al., 2002).

Genus I. **Kofleria** *gen. nov.*

HANS REICHENBACH

Ko.fle′ ria. M.L. fem. n. *Kofleria* in honor of Ludwig Kofler, the Austrian scientist who, in 1913, described the first species of the genus.

Vegetative cells are long, slender, cylindrical rods with rounded ends. Swarm colonies with radial veins in a tough slime sheet, with numerous small to very large globular masses, or knobs, all over the swarm. Congo red is not adsorbed; the agar is not etched. Mature fruiting bodies have not been observed but appear to consist of sporangioles. Cells within the knobs consist of long, cylindrical rods, optically refractile. Of the proteolytic–bacteriolytic nutritional type. Chitin degraders. Found in soil and similar substrates.

Type species: **Kofleria flava** (ex Kofler 1913) *sp. nov. nom. rev.*

List of species of the genus Kofleria

1. **Kofleria flava** (ex Kofler 1913) *sp. nov. nom. rev.*
fla′ va. L. fem. adj. *flava* yellow.

Vegetative cells are slender, cylindrical rods with rounded ends, often rather long, 0.6–0.7 × 4–6 µm (Fig. BXII.δ.127).

Swarm colonies produce a tough, more or less greenish-yellow slime sheet and prominent veins resembling a *Cystobacter* swarm. When cut, the slime layer tends to contract. The agar is not etched but is sometimes ripped open by the shrinking slime sheet.

Mature fruiting bodies have apparently not yet been observed. In the swarm colonies many large, greenish yellow spherical or slightly irregular globular masses, or knobs, arise (Fig. BXII.δ.128), with a smooth or bumpy surface, often constricted at the base or on a short stalk. The cells within those structures aggregate and stick together in spherical or convoluted masses; these cells are optically refractile, resemble vegetative cells in shape and size, but often are somewhat irregular in outline. Some of them may represent myxospores; others are clearly degenerate forms. The surface of the knobs often shows a boundary layer. The spherical cell masses within those knobs may have a thin envelope, like an immature sporangiole wall; their diameter varies between 40 and 120 µm. The size of the knobs is 200–600 µm. They may be interpreted as immature or degenerate fruiting bodies or may be the survival form of the organism.

Of the proteolytic-bacteriolytic nutritional type. Degrades chitin vigorously. Starch is hydrolyzed; cellulose is not attacked.

Rather uncommon. So far 11 strains of these myxobacteria have been isolated, four of which, however, come from the same site. The sources were mainly soil with decomposing plant material but include goat dung and the bark of a *Ficus* tree. The samples came from Germany, Spain, Greece, Siberia, North and South India, and Arizona and Pennsylvania in the USA. Obviously the organism has a worldwide distribution. Strain Pl vt1 was isolated in 1988 from a soil sample from Pennsylvania. The reference strains were obtained in 1998 from soil from Siberia (Pl vt4), in

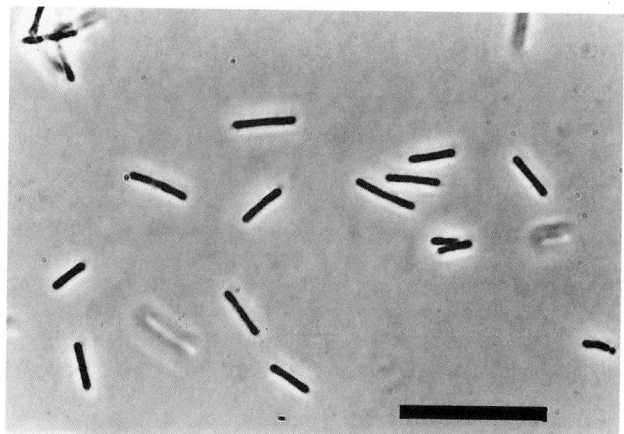

FIGURE BXII.δ.127. *Kofleria flava:* vegetative cells, in phase contrast. Bar = 10 µm.

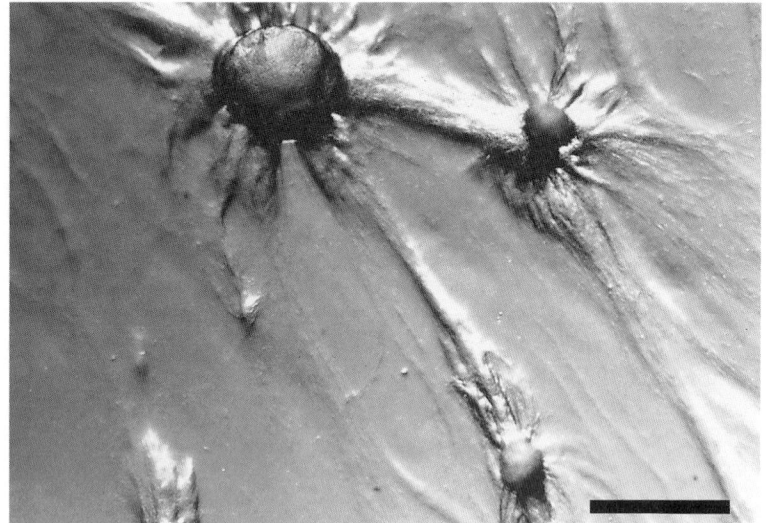

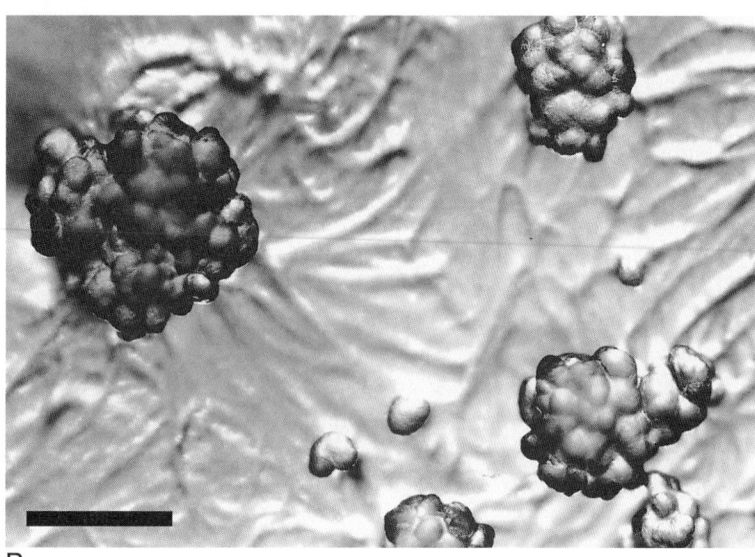

FIGURE BXII.δ.128. *Kofleria flava*: globular masses in swarm colonies, often elevated on a stout slime stalk (*C*). (*A*) Bar = 160 μm; (*B*) Bar = 260 μm; and (*C*) Bar = 320 μm.

2001 from soil from Arizona (Pl v10), and in 2001 from the bark of a *Ficus glomerata* from Karnataka, India (Pl vt11).

The mol% G + C of the DNA is: not determined.

Type strain: Pl vt1, DSM 14601, JCM 12632.

Additional Remarks: References strains include DSM 14659 (Pl vt4); DSM 14602 (Pl vt10); DSM 14620 (Pl vt11).

Initially these organisms were classified as *Polyangium vitellinum* because of the yellow color of the knobs. Further studies showed that these strains must be a different species. The absence of bright orange sporangioles, the greenish yellow color of the knobs and swarms, and the habitat did not fit the description of *Polyangium vitellinum*. Most likely the bacterium is identical to Kofler's (1913) *"Polyangium flavum"*, later renamed by Jahn (1924)—without ever having seen it—*"Archangium flavum"*. Kofler's note that *"Polyangium flavum"*, when observed by the naked eye, could easily be mistaken for *Myxococcus virescens* except that the surface of the "fruiting bodies" is not so smooth but rather bumpy, corresponds exactly to the author's observations. Furthermore, the yellow color of Kofler's organism, the size of the knobs (600 × 400 μm), the lack of "cysts", i.e., sporangioles, and the sticking together of the cells in clumps

when the "fruiting bodies" are squeezed is consistent with strains in the author's collection. Only the length of the cells (2–4 μm) given by Kofler is rather short but not beyond the usual variability of the population. Unfortunately, he did not describe their shape.

Kofler's organism was mentioned by Krzemieniewska and Krzemieniewski (1927). They obtained it from forest soil and described two new forms of the bacterium, one with a colorless, stiff envelope around the fruiting body. In addition, they pointed out that the vegetative cells were cylindrical rods with rounded ends. Krzemieniewska and Krzemieniewski were indeed probably describing *Kofleria flava*. A later report of *"Archangium flavum"* from India (Singh and Singh, 1971) is less likely to have been this organism, as the vegetative cells had slightly tapering ends. However, identity cannot be ruled out.

When the 16S rRNA sequence of strain Pl vt1 was determined and analyzed (Spröer et al., 1999), it became clear that the bacterium occupied a rather isolated position connected with *Nannocystis* at a deep level. This made the taxonomic adjustments necessary as presented here.

Class V. **Epsilonproteobacteria** *class. nov.*

GEORGE M. GARRITY, JULIA A. BELL AND TIMOTHY LILBURN

Ep.si.lon.pro.te.o.bac.te′ri.a. Gr. n. *epsilon* name of fifth letter of Greek alphabet; Gr. n. *Proteus* ocean god able to change shape; Gr. n. *bakterion* a small rod; M.L. fem. pl. n. *Epsilonproteobacteria* the class of bacteria having 16S rRNA gene sequences related to those of the members of the order *Campylobacterales*.

The class *Epsilonproteobacteria* was circumscribed for this volume on the basis of phylogenetic analysis of 16S rRNA sequences; the

class contains the order *Campylobacterales*.

Type order: **Campylobacterales** *ord. nov.*

Order I. **Campylobacterales** *ord. nov.*

GEORGE M. GARRITY, JULIA A. BELL AND TIMOTHY LILBURN

Cam.py.lo.bac.ter.a′ les. M.L. masc. n. *Campylobacter* type genus of the order; *-ales* ending to denote order; M.L. fem. n. *Campylobacterales* the *Campylobacter* order.

The order *Campylobacterales* was circumscribed for this volume on the basis of phylogenetic analysis of 16S rRNA sequences; the order contains the families *Campylobacteraceae*, *Helicobacteraceae*, and *"Nautiliaceae"*.

Actively growing cells generally curved or spiral-shaped, except for *Thiovulum*. Metabolically and ecologically diverse; in-

cludes human and animal pathogens. *Nautilia* and *Caminibacter* are anaerobic marine thermophiles.

Type genus: **Campylobacter** Sebald and Véron 1963, 907 emend. Vandamme, Falsen, Rossau, Hoste, Segers, Tytgat and De Ley 1991a, 98.

Family I. **Campylobacteraceae** Vandamme and De Ley 1991, 453[VP]

PETER VANDAMME, FLOYD E. DEWHIRST, BRUCE J. PASTER AND STEPHEN L.W. ON

Cam.py.lo.bac.ter.a′ ce.ae. M.L. masc. n. *Campylobacter* type genus of the family; suff. *-aceae* denoting family; M.L. masc. pl. n. *Campylobacteraceae*, the *Campylobacter* family.

Curved, S-shaped, or **spiral rods** that are 0.2–0.8 × 0.5–5 μm. Gram negative. Nonsporeforming. Cells in old cultures may form spherical or coccoid bodies. Mostly motile with a **characteristic corkscrew-like motion** by means of a single, **polar, unsheathed flagellum** at one or both ends of the cell.

Microaerophilic, with a respiratory type of metabolism. Some taxa grow also under aerobic or anaerobic conditions. Optimal temperature is 30–37°C. **Chemoorganotrophs. Carbohydrates are neither fermented nor oxidized.** Fumarate is reduced to succinate. Colonies are usually nonpigmented. Serum or blood enhances growth but is not necessary. **Energy is obtained from amino acids or tricarboxylic acid cycle intermediates, not carbohydrates.** Mostly oxidase positive. Methyl red and Voges–Proskauer negative, and no production of indole. Most species reduce nitrate and do not hydrolyze hippurate.

Menaquinones are the only respiratory quinones detected, with menaquinone-6 (three different structural types) and menaquinone-5 as major components. Internal transcribed spacers or intervening sequences occur in the 16S or 23S ribosomal RNA genes of strains of several species. Most species occur in man or animals, or both, primarily in the reproductive organs, intestinal tract, and oral cavity. Some species are considered true pathogens and are a significant cause of diarrheal disease; some species are associated with periodontal disease.

The mol% G + C of the DNA is: 27–47.

Type genus: **Campylobacter** Sebald and Véron 1963, 907 emend. Vandamme, Falsen, Rossau, Hoste, Segers, Tytgat and De Ley 1991a, 98.

FURTHER DESCRIPTIVE INFORMATION

The family *Campylobacteraceae* comprises the genera *Campylobacter*, *Arcobacter*, and *Sulfurospirillum*, and the generically misclassified species *Bacteroides ureolyticus* (Fig. BXII.ε.1). As described below, *B. ureolyticus* phylogenetically belongs to this family and shares most of its characteristics with other members of this family. It resembles *Campylobacter* species in its respiratory quinone content, its DNA base ratio, and most of its phenotypic characteristics but differs from campylobacters in its fatty acid composition and its proteolytic metabolism. *B. ureolyticus* was not formally reclassified pending the isolation and thorough taxonomic characterization of additional *B. ureolyticus*-like bacteria, and is thus considered species *incertae sedis*.

The delineation of this bacterial family was primarily based on phylogenetic criteria and is not supported by biochemical, chemotaxonomic, or ultrastructural characteristics. The genera *Helicobacter* and *Wolinella* are the closest phylogenetic neighbors of members of the *Campylobacteraceae*. Genus level identification is primarily achieved via the identification of strains to the species level.

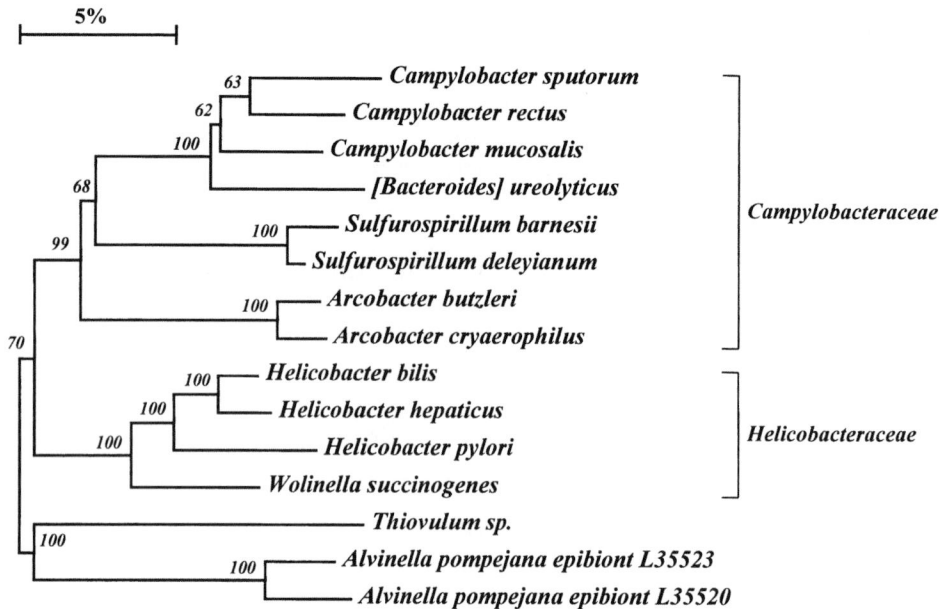

FIGURE BXII.ε.1. Phylogeny of the family *Campylobacteraceae*. The family *Campylobacteraceae* forms a coherent phylogenetic group in the *Epsilonproteobacteria* that is comprised of members of the genera *Campylobacter*, the misclassified *Bacteroides ureolyticus*, *Arcobacter*, and *Sulfurospirillum*. Bar = 5% difference in 16S rDNA nucleotide sequences. The Neighbor-Joining method was used for tree construction. One hundred bootstrap trees were generated, and bootstrap confidence levels (shown as percentages above the nodes) were determined.

Key to the genera of the family Campylobacteraceae

1. The mol% G + C content of the DNA is 29–47. Optimal growth is between 30 and 37°C, in microaerobic to anaerobic conditions. Menaquinone-6 and a methyl-substituted menaquinone-6 are the major respiratory quinones. No hydrolysis of casein or gelatin. Most strains do not hydrolyze urea. Isolated primarily from the reproductive organs, intestinal tract, and oral cavity of humans and animals. Internal transcribed spacers or intervening sequences occur in 16S and 23S rRNA genes of a variety of species.

    Genus *Campylobacter*

2. The mol% G + C content of the DNA is 27–31. Optimal growth between 15 and 30°C, in microaerobic to aerobic conditions. Menaquinone-6 and a second menaquinone-6, the detailed structure of which has not been determined, are the major respiratory quinones. No hydrolysis of casein or gelatin. Most strains do not hydrolyze urea. Isolated primarily from the reproductive organs and intestinal tract of man and animals. Internal transcribed spacers or intervening sequences have not been reported.

    Genus *Arcobacter*

3. The mol% G + C content of the DNA is 32–42. Optimal growth between 20 and 37°C, in microaerobic to anaerobic conditions. Menaquinone-6 and a methyl-substituted menaquinone-6 are the major respiratory quinones. Utilization of sulfur as electron acceptor. Free-living, found in freshwater and marine environments. Internal transcribed spacers or intervening sequences have not been reported.

    Genus *Sulfurospirillum*

4. The mol% G + C content of the DNA is 28–30. Optimal growth between 30 and 42°C, in microaerobic to anaerobic conditions. Menaquinone-6 and a methyl-substituted menaquinone-6 are the major respiratory quinones. Hydrolysis of casein, gelatin, and urea. Isolated primarily from superficial ulcers, soft-tissue infections, the urogenital tract, and oral cavity of humans.

    Species incertae sedis Bacteroides ureolyticus

Genus I. **Campylobacter** *Sebald and Véron 1963, 907,[AL] emend. Vandamme, Falsen, Rossau, Hoste, Segers, Tytgat and De Ley 1991a, 98*

PETER VANDAMME, FLOYD E. DEWHIRST, BRUCE J. PASTER AND STEPHEN L.W. ON

Cam.py' lo.bac.ter. Gr. adj. *campylo* curved; Gr. n. *bacter* rod; M.L. masc. n. *Campylobacter* a curved rod.

Cells of most species are **slender, spirally curved rods**, 0.2–0.8 × 0.5–5 µm; cells of some species are predominantly curved or straight rods. The rods may have one or more spirals and can be as long as 8 µm. They also appear S-shaped and gull-winged when two cells form short chains. Nonsporeforming. Cells in old cultures may form spherical or coccoid bodies. Cells have a multilaminar polar membrane at both ends of the cell that is located under the cytoplasmic membrane. Gram negative. Cells of most species are motile with a characteristic **corkscrewlike motion** by means of a single polar **unsheathed flagellum** at one or both ends of the cell. The flagella may be 2–3 times the length of the cells. Cells of other species are nonmotile (*Campylobacter gracilis*) or have multiple flagella (*Campylobacter showae*). Occasionally differences in the number of flagella shown by cells in a single culture are seen (*Campylobacter hyointestinalis*).

Microaerophilic, with a respiratory type of metabolism. Several species require anaerobiosis for optimal growth and grow only microaerobically in the presence of fumarate with formate or hydrogen. Require an oxygen concentration between 3% and 15% and a CO_2 concentration of 35%. Growth at 35–37°C, not at 4°C. **Chemoorganotrophs. Carbohydrates are neither fermented nor oxidized.** No acid or neutral end products produced. Serum or blood enhances, but is not required for, growth. **Energy is obtained from amino acids or tricarboxylic acid cycle intermediates, not carbohydrates.** Gelatin, casein, starch, and tyrosine are not hydrolyzed. Methyl red and Voges–Proskauer negative. **Oxidase activity** is present in all species except *C. gracilis* and sporadic isolates of *Campylobacter concisus* and *C. showae*. **Arylsulfatase activity is reported in some species**, but no lipase or lecithinase activity. Most species **reduce nitrate**. Pigments are not produced. Most species are pathogenic for man and animals. **Found in the reproductive organs, intestinal tract, and oral cavity of humans and animals.**

Internal transcribed spacers or intervening sequences (IVS) were first reported in rRNA genes of *Campylobacter* strains by Van Camp et al. (1993). Intervening sequences were present in the 16S rRNA genes of all *C. sputorum* strains (Van Camp et al., 1993; On et al., 1998) and some *C. hyointestinalis* subsp. *lawsonii* strains (Harrington and On, 1999) were examined. The IVS occurs in a 7-base stem-loop that is centered at position 210 (*E. coli* numbering). The length of the IVS replacing the 7-base stem-loop is 240 bases. Dewhirst and Paster (unpublished) have found IVS's in the type strain of *C. curvus* (147 bases), a different IVS in a *C. curvus*-like strain (SU C10; 208 bases), and in a *C. rectus*-like strain (CCUG 19168, 196 bases). Linton et al. (1994a) reported on the presence of an IVS (152 bases) in the 16S rRNA genes of some *C. helveticus* strains. 23S rRNA genes of all *C. fetus* strains examined (Van Camp et al., 1993), and the 23S rRNA genes of some of the *C. jejuni* and *C. upsaliensis* strains examined (Van Camp et al., 1993) have IVS elements. Hurtado and Owen (1997a) reported on intervening sequences in 23S rRNA genes of strains of *C. jejuni* (both subspecies), *C. coli*, *C. helveticus*, *C. fetus*, *C. sputorum*, and *C. upsaliensis*.

The mol% G + C of the DNA is: 29–47.

Type species: **Campylobacter fetus** (Smith and Taylor 1919) Sebald and Véron 1963, 907 (*Vibrio fetus* Smith and Taylor 1919, 301.)

FURTHER DESCRIPTIVE INFORMATION

The genera *Campylobacter*, *Arcobacter*, *Sulfurospirillum*, and the generically misclassified species *Bacteroides ureolyticus*, constitute the family *Campylobacteraceae* and belong to the *Epsilonproteobacteria*. Within the genus *Campylobacter*, the group of the thermophilic (or more accurately, thermotolerant) campylobacters (*C. jejuni*, *C. coli*, *C. lari*, *C. upsaliensis*, and *C. helveticus*) forms a distinct subcluster (Fig. BXII.ε.2). *C. fetus* and *C. hyointestinalis* are also close relatives, while the remaining species form a loose assemblage of predominantly hydrogen-requiring organisms. Low but significant DNA–DNA hybridization values (which vary with the hybridization techniques used) have been reported between (i) *C. fetus* and *C. hyointestinalis*; (ii) *C. jejuni* and *C. coli* (reviewed by Vandamme and Goossens, 1992); (iii) *C. upsaliensis* and *C. helveticus* (Stanley et al., 1992); and (iv) *C. showae* and *C. rectus* (Etoh et al., 1993); the level of DNA–DNA hybridization toward and between all other species were reported as not substantial.

Campylobacter cells are typically helically curved and have a very characteristic corkscrew-like darting type of motility, which is observed with phase contrast or darkfield microscopy. The growth medium becomes alkaline (pH 8.5–9.0), and coccoid forms occur under these unfavorable conditions. The coccoid forms are considered by many to be degenerative forms rather than a dormant stage of the organism and are difficult to detect using PCR methods. Increase of viable numbers of aged coccoid cultures is attributed to multiplication of residual viable cells (Hazeleger et al., 1994; Bovill and Mackey, 1997).

The outer cell membrane is double-layered, loosely fitted over the cell wall and has a wavy morphology. The cytoplasmic membrane is thickened at the polar region. This polar membrane can be seen at both ends of the cell and a similar structure has been reported in other bacteria (Smibert, 1978). In contrast, *C. rectus* exhibits a distinct cell wall structure in which the outer membrane is covered with a distinctive array of hexagonally packed macromolecular subunits (Lai et al., 1981).

Campylobacter species have a respiratory metabolism. *C. fetus* oxidizes citrate, *cis*-aconitate, isocitrate, α-ketoglutarate, succinate, fumarate, malate, and oxaloacetate. A complete tricarboxylic acid (TCA) cycle has been demonstrated. There is no oxidation or fermentation of carbohydrates. Energy for *C. fetus* is obtained from TCA intermediates and from amino acids such as glutamate and aspartate that can be deaminated to TCA intermediates. Mendz et al. (1997) reported on the role of pyruvate in the energy and biosynthesis metabolism of *Campylobacter* species. The important role of pyruvate was illustrated by the variety of products formed using pyruvate as the sole substrate and by the existence of anaplerotic sequences and anabolic pathways that employ pyruvate.

All campylobacters grow under microaerobic conditions. Several species (*C. concisus*, *C. curvus*, *C. rectus*, *C. mucosalis*, *C. showae*, *C. gracilis*, and, partly, *C. hyointestinalis*) require hydrogen or formate as an electron donor for microaerobic growth and some species grow preferentially under anaerobic conditions (Han et al., 1991). Hydrogen stimulates growth of the majority of species when inoculated on common agar bases. Cytochromes *b*, *c*, and carbon monoxide-binding cytochrome *c* have all been reported

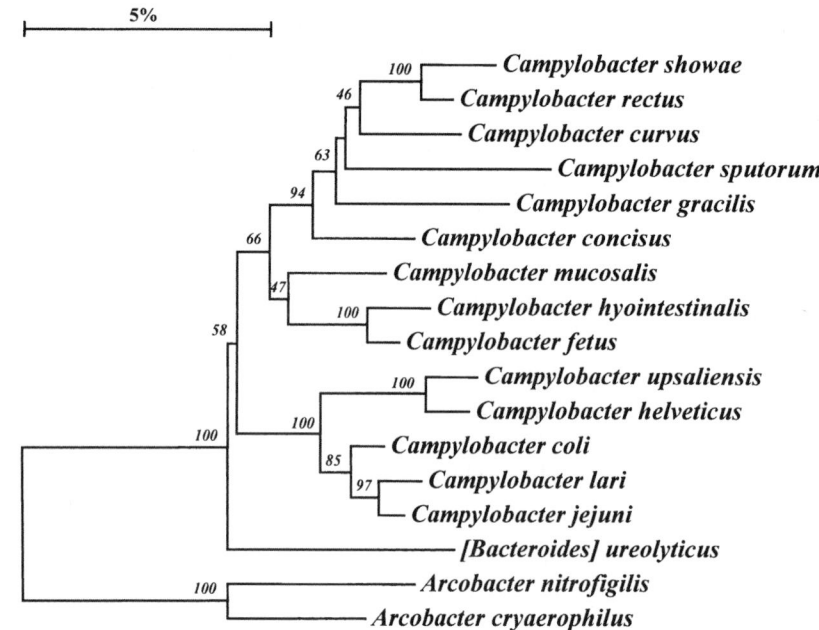

FIGURE BXII.ε.2. Phylogeny of *Campylobacter* species. The phylogenetic relationships of the named species of the genus *Campylobacter* and the misclassified *Bacteroides ureolyticus* are shown above. Bar = 5% difference in 16S rDNA nucleotide sequences. The Neighbor-Joining method was used for tree construction. One hundred bootstrap trees were generated, and bootstrap confidence levels (shown as percentages above the nodes) were determined.

in *Campylobacter* species, but not types *a* and *d*; the presence of cytochrome *o*, is controversial (Goodhew et al., 1988; Han et al., 1992). Many species also have catalase activity.

Menaquinone-6 (2-methyl-3-farnesyl-farnesyl-1,4-naphthoquinone) and a methyl-substituted menaquinone-6 (2, [5 or 8]-dimethyl-3-farnesyl-farnesyl-1,4-naphthoquinone) have been reported as major respiratory quinones in *Campylobacter* species (Moss et al., 1984, 1990b; Vandamme et al., 1995a).

Plasmids have been described in a variety of species including *C. jejuni, C. coli, C. upsaliensis, C. mucosalis, C. hyointestinalis,* and *C. fetus* (Taylor et al., 1981; Boosinger et al., 1990; Goossens et al., 1990a; Varga, 1991; Waterman et al., 1993). Tetracycline and kanamycin resistance were shown to be plasmid-mediated and transferable (Taylor et al., 1981; Cabrita et al., 1992; Velazquez et al., 1995).

Several serological typing systems have been developed but are focused primarily on *C. jejuni* and *C. coli* (reviewed by Patton and Wachsmuth, 1992). In addition, bacteriophage typing, biotyping, plasmid profiling, and a variety of DNA based typing methods such as pulsed field gel electrophoresis of large genomic fragments, ribotyping, and PCR based genomic fingerprinting, have all been used in epidemiological studies (Griffiths and Park, 1990; Mazurier et al., 1992; Patton and Wachsmuth, 1992; Owen et al., 1993).

Thermophilic campylobacters cause gastroenteritis in humans. In particular, *C. jejuni* is known worldwide as a major enteropathogen causing as much enteric disease in man as *Salmonella* and *Shigella*. *C. jejuni* infects people of all ages. It is more frequently diagnosed in children than adults and the seasonal incidence shows it to be higher in the summer and fall than in winter or spring (Butzler and Skirrow, 1979). The organism is found in the intestinal tract of a wide variety of animals. Enteric infection in animals has been reported. Transmission is most likely oral or fecal–oral. *Campylobacter* enteritis is a foodborne disease with meat, milk, and water as most important vehicles.

Other members of the thermophilic species group are *C. coli, C. lari,* and *C. upsaliensis,* all of which are known to cause enteritis in humans and to be carried in the intestinal tract of a variety of animals, in particular pigs, pets (cats and dogs), and poultry. Varying isolation rates—which may be due to genuine differences in prevalence, to inappropriate isolation or identification procedures, or to any combination of these factors—have been reported. *C. jejuni* infection is a known antecedent of Guillain–Barré syndrome and is associated with axonal degeneration, slow recovery, and severe residual disability (Rees et al., 1995b; Jacobs et al., 1996; Nachamkin et al., 1998). *C. rectus* and other oral species are associated with periodontal disease in humans, and may cause infections in other parts of the body. *C. fetus* and *C. hyointestinalis* are primarily important in veterinary medicine, causing sporadic abortion in cattle and abortion in sheep (*C. fetus* subsp. *fetus*), abortion and reproductive problems in cattle (*C. fetus* subsp. *venerealis*), and enteric disease in pigs (*C. hyointestinalis*). *C. fetus* subsp. *fetus* also causes bacteremia in humans. Thus far, *C. mucosalis* and *C. helveticus* are the only species not isolated from human infections.

The pathogenicity mechanisms by which campylobacters cause gastroenteritis are unclear, and a good animal model is not available. As with other enteropathogens, motility, chemotaxis, adherence, invasion, and toxin production have been recognized as virulence properties (Ketley, 1997). Flagella, lipopolysaccharides, outer membrane proteins, and a carbohydrate moiety may act as adhesins. An enterotoxin immunologically related to cholera toxin, a heat-labile trypsin-sensitive cytotoxin, a cytotoxin active on Vero and HeLa cells, a cytolethal distending toxin, a shiga-like toxin, a hemolytic toxin, and a hepatotoxin have all been documented. Their role in disease remains unclear (Griffiths and Park, 1990; Ketley, 1995, 1997, Wassenaar, 1997).

C. jejuni and *C. coli* are susceptible to a variety of antimicrobial agents including macrolides, fluoroquinolones, aminoglycosides, chloramphenicol, and tetracycline (Nachamkin, 1995). Erythromycin is the drug of choice for treating *C. jejuni* gastrointestinal infections, with ciprofloxacin as an alternative. However, considerable percentages of *C. coli* strains are resistant to erythromycin and emergence of fluoroquinolone resistance has been reported.

ENRICHMENT AND ISOLATION PROCEDURES

Isolation of *Campylobacter* species can be accomplished by two methods. The first involves filtration of the cells through membrane filters with a pore size of 0.45, 0.65, or 0.8 μm using a nonselective agar medium or broth medium. *Campylobacter* cells are small and highly motile and can penetrate the filter. All agar plates or broth cultures must be incubated in a microaerobic atmosphere preferably containing both CO_2 and H_2. Commercial products are available for establishing microaerobic environments. Microaerobic atmospheres without hydrogen will not allow growth of some *Campylobacter* species on commonly used agar bases.

The second method for isolating campylobacters is the use of selective agar media. A variety of different selective media has been described, some using blood agar, others a blood-free agar base as basal medium (Goossens et al., 1989; Griffiths and Park, 1990; Aspinall et al., 1993; Corry et al., 1995). However, none of these selective supplements supports growth of all of the *Campylobacter* species. Selective media reported to be suitable for the isolation of all thermophilic campylobacters including *C. upsaliensis* were reported by Burnens and Nicolet (1992) and Aspinall et al. (1996).

Incubation of inoculated media at 42–43°C will increase selectivity by the elimination or inhibition of many, but not all, other intestinal organisms and is particularly useful for the isolation of the thermotolerant campylobacters. It will, however, inhibit growth of some other *Campylobacter* species.

There is no gold standard for the routine isolation of all *Campylobacter* species. Simultaneous application of a microaerobic atmosphere containing hydrogen with a filtration method and a selective base is methodologically the optimal solution. However, *C. jejuni* and *C. coli*, the predominant species in human infection, can be readily grown in a microaerobic atmosphere on selective media without the necessity of using hydrogen. In order to evaluate the presence of other, less common species, appropriate culture conditions need to be applied.

MAINTENANCE PROCEDURES

Stock cultures of *Campylobacter* species can be maintained under microaerobic conditions by weekly transfer onto common blood agar bases. Addition of blood to media may increase survival. Cultures may be stored for many years by lyophilization, freezing at 80°C, or in liquid nitrogen. Cryoprotective agents such as 10% glycerol or DMSO should be added to cultures before freezing, and heavy cell concentrations should be used.

DIFFERENTIATION OF THE GENUS *CAMPYLOBACTER* FROM OTHER GENERA

Classical biochemical differentiation of the genus *Campylobacter* from related genera such as *Arcobacter* and *Helicobacter* is primarily achieved via the identification of the individual species. However, in general, campylobacters can be differentiated from arcobacters by the lower optimal growth temperatures (25–30°C compared to 30–42°C) and the aerotolerance of arcobacters. Analysis of respiratory quinones enables a clear differentiation of *Campylobacter* strains from *Arcobacter* or *Helicobacter* strains but is of limited value in a routine diagnostic laboratory.

TAXONOMIC COMMENTS

Since the creation of *Campylobacter* as a genus in 1963, a variety of Gram-negative, microaerobic to anaerobic, asaccharolytic, and oxidase-positive bacteria have been included in the genus. Phylogenetic work based on ribosomal RNA gene sequence analysis or hybridization experiments revealed within this ill-defined genus the presence of three independent subdivisions, each of which were subsequently given separate genus rank (Lau et al., 1987; Paster and Dewhirst, 1988; Thompson et al., 1988; Goodwin et al., 1989a; Vandamme et al., 1991a). The name *Campylobacter* was preserved for the lineage containing *C. fetus*, the type species. The genus *Arcobacter* was proposed to accommodate *C. nitrofigilis* and *C. cryaerophilus* and, later, *C. butzleri*, while the genus *Helicobacter* was proposed for *C. pylori* and *C. mustelae* and, subsequently, *C. cinaedi* and *C. fennelliae*. In addition, several so-called free-living campylobacters (Laanbroek et al., 1977; Wolfe and Pfennig, 1977) were found to constitute a separate fourth branch and were later classified into the genus *Sulfurospirillum* (Schumacher et al., 1992). Since the 1991 revision of its taxonomy and nomenclature, the classification of the genus *Campylobacter* is primarily phylogeny based. It should be emphasized that this phylogeny-driven taxonomy unified species with extremely different cellular morphologies and a range in DNA base ratio that exceeds that of most well-defined genera. Although the present classification has been generally accepted, these anomalies remain unexplained.

There are taxonomic problems at the species and infraspecific level. The name *C. sputorum* subsp. *mucosalis* was validly published but subsequent DNA–DNA hybridization and 16S rRNA gene sequence analysis demonstrated that the more recent classification of this organism as a separate species, *C. mucosalis*, is correct (Roop et al., 1985a; Thompson et al., 1988). Also, although the names *C. sputorum* subsp. *sputorum* and *C. sputorum* subsp. *bubulus* have standing in nomenclature, recent taxonomic analyses demonstrated that it is more appropriate to abandon this subspecies classification and to consider both taxa as a single biovar: *C. sputorum* biovar sputorum. The organism previously known as *"Campylobacter fecalis"*, and a newly described taxon, represent two additional biovars of *C. sputorum* (biovar faecalis and biovar paraureolyticus, respectively) (On et al., 1998).

Another problem concerns the taxonomic position of phenotypically unusual thermophilic campylobacters referred to as nalidixic acid-sensitive campylobacters, urease-positive thermophilic campylobacters, and others (Endtz et al., 1997). One dimensional whole-cell protein electrophoresis (Owen et al., 1988; Vandamme et al., 1991b), rRNA gene sequence analysis (Alderton et al., 1995), and semiquantitative DNA–DNA hybridization (Mégraud et al., 1988) indicated that these taxa are closely related to, or belong to, *C. lari*, and these strains have been referred to as *C. lari* variants. Quantitative DNA–DNA hybridization experiments between *C. lari* and a reference strain of the urease-positive thermophilic *Campylobacter* group proved that it indeed belonged to *C. lari* (Vandamme et al., 1997c). Subsequent work reported by Endtz et al. (1997) described a striking heterogeneity among and within the different groups of *C. lari* variants. The exact relationships between genuine *C. lari* strains and the various biochemical variants and protein electrophoretic subtypes should be explored further by means of DNA–DNA hybridization experiments.

In addition, the separation of *C. jejuni* subsp. *jejuni* and *C. coli* remains an important taxonomic problem. These taxa are remarkably similar by virtue of their overall phenotype and genotype and are often found in the same ecologic and pathologic niches. The most reliable and commonly used test is the hippurate hydrolysis test done according to the method of Hwang and Ederer (1975) as described by Harvey (1980), in which *C. coli* is negative. However, some strains of *C. jejuni* subsp. *jejuni* also give a negative result. Additional useful tests are hydrogen sulfide production in triple sugar iron agar and growth on a minimal medium (On et al., 1996). In contrast with *C. jejuni* subsp. *jejuni*, *C. coli* usually utilizes propionate when a commercial identification system (API Campy, bioMérieux, France) is used (Occhialini et al., 1996). Growth in 8% glucose and brilliant green media are not reliable traits for species identification (Skirrow and Benjamin, 1980). The complexity of this taxonomic area has been emphasized by data showing that strains first described as *C. hyoilei* (associated with porcine proliferative enteritis) are, in fact, *C. coli* (Vandamme et al., 1997c), despite a higher 16S rRNA gene sequence similarity to *C. jejuni* (Alderton et al., 1995). Moreover, the description of strains that closely resemble *C. coli*, yet are genotypically more divergent from the type strain (60% DNA–DNA relatedness) (Morris et al., 1985) than is usual, further emphasizes the problems associated with the taxonomy of the "thermotolerant campylobacters". There is a need for better genotypic and phenotypic markers to further investigate relationships among members of this group. Further studies are required to evaluate the taxonomic (and indeed clinical) significance of both genomically divergent *C. coli*-like strains (Morris et al., 1985), and pig isolates that may be referred to as a "*hyoilei*" variant of *C. coli* (Vandamme et al., 1997c).

The organisms listed in the "Other Organisms" section in the first edition of *Bergey's Manual of Systematic Bacteriology* have all been formally classified. "*Campylobacter fecalis*" is a distinct biovar of *C. sputorum*. The nitrogen-fixing organisms from salt marsh plants (McClung and Patriquin, 1980) and the aerotolerant campylobacters (Neill et al., 1985) are now classified as *Arcobacter* species (Vandamme et al., 1991a). "*Spirillum* 5175" and the free-living organism described by Laanbroek et al. (1977) are classified in the genus *Sulfurospirillum* (Schumacher et al., 1992).

ACKNOWLEDGMENTS

Floyd E. Dewhirst and Bruce J. Paster were supported by NIH grants DE-11443 and DE10374. We thank H.G. Trüper for clarifying the etymology of several *Campylobacter* taxa.

FURTHER READING

Griffiths, P.L. and R.W.A. Park. 1990. Campylobacters associated with human diarrhoeal disease. J. Appl. Bacteriol. *69*: 281–301.

On, S.L.W. 1996. Identification methods for campylobacters, helicobacters, and related organisms. Clin. Microbiol. Rev. *9*: 405–422.

Ursing, J.B., H. Lior and R.J. Owen. 1994. Proposal of minimal standards for describing new species of the family *Campylobacteraceae*. Int. J. Syst. Bacteriol. *44*: 842–845.

Vandamme, P., E. Falsen, R. Rossau, B. Hoste, P. Segers, R. Tytgat and J. De Ley. 1991. Revision of *Campylobacter*, *Helicobacter*, and *Wolinella* taxonomy: emendation of generic descriptions and proposal of *Arcobacter* gen. nov. Int. J. Syst. Bacteriol. *41*: 88–103.

Vandamme, P. and H. Goossens. 1992. Taxonomy of *Campylobacter*, *Arcobacter*, and *Helicobacter*: a review. Zentbl. Bakteriol. *276*: 447–472.

Wassenaar, T.M. 1997. Toxin production by *Campylobacter* spp. Clin. Microbiol. Rev. *10*: 466–476.

DIFFERENTIATION OF THE SPECIES OF THE GENUS *CAMPYLOBACTER*

Biochemical characteristics useful in distinguishing the various species of the genus *Campylobacter* are listed in Table BXII.ε.1. Additional descriptive characteristics are given in Table BXII.ε.2.

A variety of different methods has been applied for the identification of *Campylobacter* species; these methods have been recently reviewed by On (1996). Several authors studied the suitability of cellular fatty acid analysis for the differentiation and identification of campylobacters (reviewed by Vandamme and Goossens, 1992). Gas–liquid chromatography groups of *Campylobacter* species were defined by Lambert et al. (1987) and Goodwin et al. (1989b). Several of these gas–liquid chromatography groups consisted of more than one species, and several species have strains in different GLC groups. Additional phenotypic tests were often required for identification to the species level.

DNA probe- and PCR-based identification assays have been described for *C. fetus* (Ezaki et al., 1988; Chevrier et al., 1989;

Wesley et al., 1991; Bastyns et al., 1994; Blom et al., 1995; Eaglesome et al., 1995; Hum et al., 1997); *C. hyointestinalis* (Chevrier et al., 1989; Gebhart et al., 1989; Wesley et al., 1991; Bastyns et al., 1994); *C. mucosalis* (Gebhart et al., 1989; Bastyns et al., 1994); *C. concisus* (Bastyns et al., 1995b); *C. sputorum* (Bastyns et al., 1994); *C. jejuni* (Stucki et al. 1995; Occhialini et al., 1996; Day et al., 1997; Gonzalez et al., 1997a; Linton et al., 1997; Vandamme et al., 1997c; van Doorn et al., 1997); *C. coli* (Gonzalez et al., 1997a; Linton et al., 1997; Vandamme et al., 1997c; van Doorn et al., 1997); *C. lari* (Linton et al. 1996; Oyarzabal et al., 1997; van Doorn et al., 1997); *C. upsaliensis* (Lawson et al., 1997; van Doorn et al., 1997); and *C. helveticus* (Stanley et al., 1992; Lawson et al., 1997). Broad-spectrum molecular identification schemata based on restriction fragment analysis of PCR amplicons derived from 16S (Cardarelli-Leite et al., 1996) and 23S (Hurtado and Owen, 1997a) rRNA genes have also been described.

List of species of the genus Campylobacter

1. **Campylobacter fetus** (Smith and Taylor 1919) Sebald and Véron 1963, 907[AL] (*Vibrio fetus* Smith and Taylor 1919, 301.) *fe' tus*. L. masc. n. *fetus* fruit; L. gen. masc. n. *fetus* of a fetus.

   Slender curved rods that are 0.2–0.3 × 1.5–5 μm. They appear comma-, S-, and gull-shaped. The ends of the cells are pointed. Loosely wound spiral filaments up to 8 μm long appear in old cultures. Spherical or coccoid forms are also found in old cultures especially when grown on agar plates.

   Very actively motile with a characteristic darting and corkscrewlike motion. Motility and rotation of the cells are

so rapid the curvature of the cells may be overlooked. Best observed with a phase-contrast microscope.

Several types of colonies are found on agar on primary isolation (Bryner et al., 1962). Smooth colonies, the most frequently found, are small, 0.5 mm in diameter, round, slightly raised, smooth, colorless, and slightly translucent. "Cut-glass" colonies are 1 mm in diameter, round, raised, translucent, and granular with reflecting facets. Rough colonies are rare and similar to smooth colonies with the exception of being granular and more opaque. Mucoid colonies are similar to smooth and cut-glass colonies but are

TABLE BXII.ε.1. Differential characters for *Campylobacter* spp. and *Bacteroides ureolyticus*[a]

Characteristic	C. fetus subsp. fetus	C. fetus subsp. venerealis	C. coli	C. concisus	C. curvus	C. gracilis	C. helveticus	C. hominis	C. hyointestinalis subsp. hyointestinalis	C. hyointestinalis subsp. lawsonii	C. jejuni subsp. jejuni	C. jejuni subsp. doylei	C. lanienae	C. lari	C. mucosalis	C. rectus	C. showae	C. sputorum	C. upsaliensis	B. ureolyticus
Catalase	+	M	+	−	−	F	−	−	+	+	+	M	+	+	−	F	+	V	−	F
Urease	−	−	−	−	−	−	−	−	−	−	−	−	−	V	−	−	−	V	−	+
Indoxyl acetate hydrolysis	−	−	+	−	M	M	+	−	−	−	+	+	−	−	−	+	−	−	+	−
Hippurate hydrolysis	−	−	−	−	F	−	−	−	−	−	+	+	−	−	−	−	−	−	−	−
Nitrate reduction	+	M	+	F	+	M	+	−	+	+	+	−	+	+	+	+	+	+	+	+
Selenite reduction	M	−	+	F	−	−	−	−	+	+	+	−	+	+	−	+	+	+	+	
H$_2$S/TSI	−	−	F[b]	−[b]	F	−	−	−	M[b]	M[c]	−	−	−	−	+[c]	−	F	+[c]	−	−
25°C	+	M	−	−	−	−	−	−	−	−	−	−	−	−	−	−	−	−	−	−
42°C	M	−	+	M	M	M	+	F	+	+	+	−	+	+	+	F	F	+	M	M
1% glycine	+	−	+	F	+	+	F	+	+	F	M	F	−	+	F	+	F	+	+	+

[a]Symbols: +, 95–100% strains positive; −, 0–11% strains positive; V, test result varies between defined infrasubspecific taxa (see text for details); F, 14–50% strains positive; M, 60–93% strains positive. All data are based on reactions obtained by using recommended, standardized procedures (On and Holmes 1991, 1992, 1995).

[b]Trace quantities.

[c]Copious quantities.

TABLE BXII.ε.2. Additional diagnostic tests for *Campylobacter* spp. and *Bacteroides ureolyticus*[a]

Characteristic	C. fetus subsp. fetus	C. fetus subsp. venerealis	C. coli	C. concisus	C. curvus	C. gracilis	C. helveticus	C. hominis	C. hyointestinalis subsp. hyointestinalis	C. hyointestinalis subsp. lawsonii	C. jejuni subsp. jejuni	C. jejuni subsp. doylei	C. lanienae	C. lari	C. mucosalis	C. rectus	C. showae	C. sputorum	C. upsaliensis	B. ureolyticus
Oxidase	+	+	+	M	+	−	+	+	+	+	+	+	+	+	+	+	F	+	+	+
2.0% NaCl	−	−	−	F	F	F	−	M	−	−	−	−	−	M	M	M	+	+	−	+
2.0% ox-bile	+	M	M	F	−	−	F	F	+	−	M	−	−	+	M	−	−	I	+	+
Growth on minimal medium	F	M	+	−	M	F	−	−	M	M	−	−	−	−	−	−	F	I	−	F
Nalidixic acid	+	M	−	M	+	M	−	M	+	+	−	−	+	V	M	M	−	M	−	−
Metronidazole	M	F	+	F	−	−	F	−	−	I	M	F	+	+	M	−	+	F	M	−
Sodium fluoride	M	M	+	M	−	M	−	+	M	−	+	−	I	+	−	−	+	M	−	+
KMnO$_4$	+	F	+	−	M	+	−	−	F	−	+	M	−	F	−	−	−	−	F	−
0.02% safranin	+	M	+	F	+	+	−	−	+	+	+	−	F	+	+	−	−	I	+	F

[a]Symbols: +, 95–100% strains positive; −, 0–11% strains positive; V, test result varies between defined infrasubspecific taxa (see text for details); F, 14–50% strains positive; M, 60–93% strains positive; I, irreproducible. All data are based on reactions obtained by using recommended, standardized procedures (On and Holmes 1991, 1992, 1995).

viscid. On primary isolation, colonies sometimes occur as a thin veil of confluent growth that is translucent and a very light gray or tan color. Colonies on blood agar are nonhemolytic, round, 1 mm in diameter, smooth, raised, convex, and grayish white in appearance. Strains will grow on media containing 1.0–1.5% ox-bile and at 30°C, but not on media containing 0.04% triphenyl-tetrazolium chloride.

Most strains are tolerant to 0.032% methyl orange and 0.1% sodium deoxycholate.

Although the two subspecies of *C. fetus* are associated with distinct diseases in animals, their differentiation is not straightforward. Classical biochemical tests useful for the differentiation of these taxa are tolerance to glycine and the ability to produce hydrogen sulfide. Whole-cell protein electrophoresis does not separate the two taxa (Vandamme et al., 1991b). Salama et al. (1992a) reported the genomes of *C. fetus* subsp. *fetus* strains to be smaller (1.1 Mb) than those of *C. fetus* subsp. *venerealis* strains (1.3–1.5 MB). General differences between the *Sma*I-based macrorestriction profiles of the two subspecies have also been noted (Hum et al., 1997). A PCR-based assay designed for species and subspecies identification was 98% specific (Hum et al., 1997).

Erythromycin, ampicillin, or third-generation cephalosporins are usually effective against *C. fetus* infections (Neuzil et al., 1994). Aminoglycosides, ampicillin, and chloramphenicol have been successfully used to treat central nervous system or other serious infections caused by *C. fetus*.

The mol% G + C of the DNA is: 32–36 (T_m).

Type strain: CIP 5396, ATCC 27324, CCUG 6823, DSM 5361, LMG 6442, NCTC 10842.

GenBank accession number (16S rRNA): LO4314, M65012.

a. **Campylobacter fetus** subsp. **fetus** Véron and Chatelain 1973, 126[AL] (*Vibrio fetus* biovar intestinalis Florent 1959, 955; *Vibrio foetusovis* Buxton 1929, 47; *Campylobacter fetus* subsp. *intestinalis* Smibert 1974, 209.)

Morphology and characteristics as for species except as noted. Intermediate-sized spirals with an average wavelength of 1.8 μm and an average amplitude of 0.55 μm (Karmali et al., 1981a). Several types of colonies are found on agar on primary isolation (Bryner et al., 1962). Smooth colonies are 1 mm in diameter, colorless to slightly cream colored. Rough colonies are small, round, finely granular, opaque and white to cream or tan colored. They are 1–2 mm in diameter. "Cut-glass" colonies do not develop in primary cultures. Smooth colonies incubated for 6–8 days become mucoid. Upon subculture, smooth cut-glass and rough cut-glass colonies appear, as well as smooth colonies. On primary isolation, colonies are frequently low, flat, grayish to tan colored, and translucent with an irregular edge. They spread along the direction of the streak and coalesce. They may also form a thin veil of confluent growth on agar plates. Colonies on blood agar are nonhemolytic, round, 1–2 mm in diameter, smooth, convex, and grayish white or light tan colored. Grow on media containing 0.05% safranin and 64 mg/l cefoperazone.

Pathogenic. Cause of abortion in sheep and sporadic abortion in cattle, as well as a cause of human blood and, occasionally, gastrointestinal infections. Transmitted orally. Isolated from the placentas and stomach content of fetuses from aborted sheep and cattle and from the blood, intestinal content, and bile of infected ewes and cattle. Isolated from blood, spinal fluid, aborted fetuses, and abscesses from most parts of the body of humans. This organism will grow in the intestinal tract and gallbladder of man and animals (Bryner et al., 1964).

The mol% G + C of the DNA is: 33–36 (T_m).

Type strain: CIP 5396, ATCC 27324, CCUG 6823, DSM 5361, LMG 6442, NCTC 10842.

GenBank accession number (16S rRNA): LO4314, M65012.

b. **Campylobacter fetus** subsp. **venerealis** (Florent 1959) Véron and Chatelain 1973, 126[AL] (*Vibrio fetus* biovar vene-

realis Florent 1959, 955; *Campylobacter fetus* subsp. *fetus* Smibert 1974, 209.)

ve.ne′re.al.is. L. *venereus* from Venus goddess of love; L. adj. *veneral* L. gen. n. *venerealis* of Venus, goddess of love.

Morphology and characteristics are for species except as noted. Large spirals with an average wavelength of 2.43 μm and an average amplitude of 0.73 μm (Karmali et al., 1981a). Approximately 67% and 7% of strains will grow on 0.05% safranin- and 64 mg/l cefoperazone-containing media (On et al., 1996), respectively.

Pathogenic. A cause of abortion and infertility in cattle. Transmitted venereally. Found in the vaginal mucus of infected cows, the semen and prepuce of bulls, and in the placenta and tissues of aborted bovine fetuses. Pathogenic for cattle, guinea pigs, hamsters, and embryonated chicken eggs. Rarely isolated from human blood. Not pathogenic for rabbits, mice, or rats when injected intraperitoneally. Will not multiply in the intestinal tract of man and animals (Bryner et al., 1964).

The mol% G + C of the DNA is: 33–36 (T_m).

Type strain: CIP 68.29, ATCC 19483, CCUG 538, LMG 6443, NCTC 10354.

GenBank accession number (16S rRNA): L14633, M65011.

2. **Campylobacter coli** (Doyle 1948) Véron and Chatalain 1973, 127[AL] (*Vibrio coli* Doyle 1948, 50; *Campylobacter hyoilei* Alderton, Korolik, Coloe, Dewhirst and Paster 1995, 65.)

co′li. Gr. n. *colon* large intestine, colon; M.L. gen. n. *coli* of the colon.

Small, tightly coiled spiral, S-shaped or curved cells, 0.2–0.3 × 1.5–5.0 μm that transform rapidly to coccoid forms with age, or exposure to toxic concentrations of oxygen (Ng et al., 1985; Moran and Upton, 1987). Colonies are round, 1–2 mm in diameter, raised, convex, smooth, and glistening. On moist media, colonies are flat, grayish, and spread in the direction of the streak. Most, but not all strains are nonhemolytic. As with *C. jejuni*, the hemolytic activity (where noted) is usually cell-associated, with a possible secreted component also involved (Wassenaar, 1997). Blood enhances, but is not essential for, culture. Strains grow on solid media containing 1.0–1.5% ox-bile, 0.02% safranin, 32 mg/l cephalothin, and 0.04% triphenyl-tetrazolium chloride. Reduction of the latter substrate is also observed. Most (~76%) strains are resistant to 100 U/l 5-fluorouracil. As with *C. jejuni* subsp. *jejuni*, the proportion of strains that are resistant to the antibiotics nalidixic acid, tetracycline, chloramphenicol, kanamycin, and erythromycin (and related compounds) may significantly vary. However, in comparison with *C. jejuni*, a greater percentage of *C. coli* strains exhibit resistance to erythromycin. The genetic mechanisms determining these traits have been reviewed by Taylor (1992a, b) and Taylor and Courvalin (1988); several (tetracycline, chloramphenicol, kanamycin) are known to be plasmid borne. Aarestrup et al. (1997) showed that >90% of isolates from humans, pigs, and broiler chickens were sensitive to ampicillin (16 mg/l), apramycin (4 mg/l), carbadox (4 mg/l), chloramphenicol (8 mg/l), colistin (16 mg/l), enrofloxacin (16 mg/l), gentamicin (1 mg/l), nalidixic acid (128 mg/l), neomycin (4 mg/l), olaquindox (8 mg/l), spectinomycin (16 mg/l), spiramycin (128 mg/l), and tetracycline (2 mg/l). Aarestrup et al. (1997) determined that

13.8% and 61.7% of isolates showed resistance to streptomycin and tylosin, respectively, beyond the maximum minimal inhibitory concentration (MIC: 128 mg/l) used. *In vitro* activities of 47 antimicrobial agents toward *C. coli* were determined by Gebhart et al. (1985b).

As with *C. jejuni*, plasmid carriage in *C. coli* demonstrates considerable variation by virtue of plasmid presence in a given population, number, and size range (Tenover et al., 1985). It has been noted that conjugative plasmids conferring resistance to tetracycline, kanamycin, and chloramphenicol are more common in *C. coli* than in *C. jejuni* (Taylor, 1992b).

Differentiating *C. coli* from *C. jejuni* subsp. *jejuni* is difficult. The most common biochemical test used for this purpose is hippurate hydrolysis, for which *C. coli* is negative. However, some strains of *C. jejuni* subsp. *jejuni* also give a negative result. Additional tests that are of use were discussed above.

The taxonomic position of 11 isolates from lesions of proliferative enteritis in pigs was studied by Alderton et al. (1995), who considered the strains to represent a species closely related to both *C. coli* and *C. jejuni*. These authors subsequently proposed the name *Campylobacter hyoilei* for these strains. However, the taxonomic position of *C. hyoilei* was reevaluated by Vandamme et al. (1997c) using a range of phenotypic and genotypic methods, and, crucially, a classical quantitative DNA–DNA hybridization method. These data showed *C. hyoilei* to be indistinguishable from *C. coli*. Vandamme et al. (1997c) proposed that the two species be regarded as synonymous, with *C. coli* taking nomenclatural precedence. It was nonetheless noted that strains originally described as *C. hyoilei* may represent a variant of *C. coli* that is highly adapted for the porcine enteric tract, with pathologic consequences for the animal.

Pathogenic. Causes diarrhea, septicemia, and occasionally abortion in humans. May cause diarrhea in pigs and monkeys and abortion in rodents. Has been associated with hepatitis in certain bird species. Certain strains have been associated with proliferative enteritis in pigs.

The mol% G + C of the DNA is: 31–35 (T_m).

Type strain: CIP 7080, ATCC 33559, CCUG 11283, LMG 8847, NCTC 11366.

GenBank accession number (16S rRNA): L04312, M59073, L19738.

3. **Campylobacter concisus** Tanner, Badger, Lai, Listgarten, Visconti and Socransky 1981, 442^VP

con.ci'sus. L. part. adj. *concisus* brief, concise.

Cells are small and curved, 0.5×4 µm, with rounded ends. Rapid darting motility by means of a single polar flagellum. A membrane-like polar cap occurs at the ends of the cells. Colonies are convex, translucent, 1 mm in diameter, with entire edges. The agar is not pitted by the colonies. Does not grow microaerobically on common agar bases in an atmosphere without hydrogen. Will not grow in semisolid medium (0.16% agar), in air, or in an atmosphere containing $O_2/CO_2/N_2$ (5:10:85). Anaerobic growth occurs with formate and fumarate in the medium. Formate is oxidized to hydrogen and CO_2; and fumarate is reduced to succinate, which accumulates in the medium. Growth is stimulated by nitrate, formate, and fumarate. End products of metabolism when grown in a medium with formate and

fumarate are acetate, succinate, and H_2. Strains do not grow on MacConkey agar or on media containing 3.5% NaCl, 32 mg/l cephalothin, 64 mg/l cefoperazone, or 0.04% triphenyl-tetrazolium chloride. Most strains (70–80%) produce alkaline phosphatase and grow on media containing 0.032% methyl orange and 0.05% sodium fluoride. A few strains (14–29%) grow in the presence of 0.01% Janus green and 0.005% basic fuchsin.

Minimum inhibitory concentrations of antibiotics are (µg/ml): bacitracin, 128; chloramphenicol, 4.0; clindamycin, 24; colistin, 0.5–1.0; erythromycin, 4.0; gentamicin, 24; kanamycin, 12; metronidazole, 0.5–2.0; minocycline, 2; nalidixic acid, 64–128; neomycin, 16–32; penicillin, 0.5–4.0; polymyxin, 0.25–1.0; rifampin, 16–64; streptomycin, 12; tetracycline, 12; and vancomycin, 128. *C. concisus* strains were shown to be chemotactic toward formate (Paster and Gibbons, 1986). Whole-cell protein electrophoresis and DNA–DNA hybridization experiments (Vandamme et al., 1989; Van Etterijk et al., 1996) revealed that this is a heterogeneous species comprising many protein electrophoretic and several genotypic subgroups. The species is also phenotypically diverse (On et al., 1996), making definitive identification difficult. Found in the gingival crevices of humans with gingivitis, periodontitis, and periodontosis; in normal and diarrheic feces of humans, in human blood, and in human stomach and esophagus specimens.

The mol% G + C of the DNA is: 37–41 (T_m).

Type strain: FDC 484, ATCC 33237, CCUG 13144, LMG 7788, NCTC 11485.

GenBank accession number (16S rRNA): L04322.

4. **Campylobacter curvus** (Tanner, Listgarten and Ebersole 1984) Vandamme, Falsen, Rossau, Hoste, Segers, Tytgat and De Ley 1991a, 98^VP (*Wolinella curva* Tanner, Listgarten and Ebersole 1984, 279.)

curv'us. L. adj. *curvus* curved.

Cells are small and curved, $0.5–1 \times 2–6$ µm, with rounded or tapered ends. Helical or straight cells also occur. A membrane-like polar cap occurs at the ends of the cells. Rapid darting motility. Motile by means of a single polar flagellum or bipolar flagella. Translucent colonies are produced on blood agar bases. Different colony types are observed: small pinpoint colonies, 1 mm in diameter or spreading colonies up to 5 mm in diameter. Agar pitting is medium dependent; this trait was not seen in anaerobic, 3-d-old cultures on 5% blood agar (On et al., 1996).

Does not grow microaerobically on common agar bases in an atmosphere without hydrogen. Will not grow in air, in a CO_2-enriched atmosphere, or in an atmosphere containing $O_2/CO_2/N_2$ (5:10:85) on common agar bases. Anaerobic growth occurs with formate and fumarate in the medium. Hydrogen and formate are used as energy sources. Formate is oxidized to hydrogen and CO_2; and fumarate is reduced to succinate. Fumarate, nitrate, aspartate, asparagine, and malate serve as electron acceptors. Membrane-bound cytochrome *b*, cytochrome *c*, and CO-binding cytochrome *c*, and soluble cytochrome *c* and CO-binding cytochrome *c* are present. Strains grow in the presence of 0.005% basic fuchsin and 0.04% triphenyl-tetrazolium chloride; reduction of the latter is concurrently seen in many strains (60%). Most strains (80%) will grow on media containing 64 mg/l cefoperazone, 0.1% potassium permanganate, and 0.01% Janus green. Alkaline phosphatase activity has been detected in 40% of strains.

Minimum inhibitory concentrations of antibiotics are (μg/ml): bacitracin, >128; chloramphenicol, 2–4; clindamycin, 0.5–1; colistin, 4; erythromycin, 2; gentamicin, 2; kanamycin, 4–8; metronidazole, 1–2; minocycline, 2–4; nalidixic acid, 64–128; neomycin, 4–8; penicillin, 32; polymyxin B, 8; rifampin, 128 to >128; streptomycin, 2; tetracycline, 1; and vancomycin, >128. Strains were isolated from lesions in human oral cavities, from a blood culture, peritoneal fluid, and from normal and diarrheic feces of humans.

The mol% G + C of the DNA is: 43–47 (T_m).

Type strain: VPI 9584, ATCC 35224, CCUG 13146, DSM 6644, LMG 7609.

GenBank accession number (16S rRNA): L04313.

5. **Campylobacter gracilis** (Tanner, Badger, Lai, Listgarten, Visconti and Socransky 1981) Vandamme, Daneshvar, Dewhirst, Paster, Kersters, Goossens and Moss 1995a, 151^VP (*Bacteroides gracilis* Tanner, Badger, Lai, Listgarten, Visconti and Socransky 1981, 442.)

gra'cil.is. L. adj. *gracilis* slim, slender, thin.

Cells are small and straight, 0.4 × 4–6 μm, with rounded or tapered ends. Nonmotile. Intracytoplasmic, electron-dense inclusions, some membrane-bound, approximately 40 nm in diameter have been observed. Translucent colonies are produced on blood agar bases. Different colony types are observed: small pinpoint colonies, 1 mm in diameter or spreading colonies up to 5 mm in diameter. Agar pitting is medium dependent; this trait was not seen in anaerobic, 3-d-old cultures on 5% blood agar (On et al., 1996).

Optimal growth in anaerobic conditions. Does not grow microaerobically on common agar bases in an atmosphere without hydrogen. Will not grow in air, in a CO_2-enriched atmosphere, or in an atmosphere containing $O_2/CO_2/N_2$ (5:10:85) on common agar bases. Anaerobic growth occurs with formate and fumarate in the medium. Hydrogen and formate are used as energy sources. Formate is oxidized to hydrogen and CO_2, and fumarate is reduced to succinate. Fumarate, nitrate, nitrite, neutral red, benzyl viologen aspartate, asparagine, and malate serve as electron acceptors. Membrane-bound cytochrome *b*, cytochrome *c*, and CO-binding cytochrome *c*, and soluble cytochrome *c* and CO-binding cytochrome *c* are present. Strains will grow in the presence of 0.1% potassium permanganate and 0.05% basic fuchsin. Strains do not grow in the presence of 0.04% triphenyl-tetrazolium chloride, or 64 mg/l cefoperazone. A few strains (14%) grow on 0.01% Janus green medium. Alkaline phosphatase activity is not detected.

C. gracilis is the only oxidase-negative *Campylobacter* species. However, the pattern of cytochromes found in *C. gracilis* resembles that reported for other *Campylobacter* species in that it possesses cytochromes *b*, *c*, and CO-binding cytochrome *c* and does not possess detectable cytochromes *a* and *d*. Oxidase activity, as determined in the Kovacs test, is associated with cytochrome *c* and oxygen respiration, which is present in all *Campylobacter* species. Possible explanations for the failure to detect oxidase activity may be an incapability of the reagent to penetrate the cellular membranes, or the presence of a low-potential cytochrome *c* that cannot oxidize this reagent.

A group of bile-resistant *C. gracilis*-like organisms was recently reclassified in a novel genus *Sutterella* (Wexler et al., 1996a). This taxon is phylogenetically distinct from the campylobacters. Cellular fatty acid analysis, differences in dehydrogenase enzymes mobilities, and higher resistance to antimicrobial agents are useful to distinguish *Sutterella* strains from *C. gracilis*.

Minimum inhibitory concentrations of antibiotics are (μg/ml): bacitracin, >128; chloramphenicol, 2–8; clindamycin, 0.25–0.5; colistin, 0.5–1; erythromycin, 1–2; gentamicin, 2–4; kanamycin, <0.50–2; metronidazole, 0.12–1; minocycline, <0.5–2; nalidixic acid, 16–128; neomycin, 16–32; penicillin, 1–32; polymyxin B, <0.25–1; rifampin, 4–32; streptomycin, 1–2; tetracycline, 1–8; vancomycin, >128, amoxicillin/clavulanate, 0.06–0.5; cefoxitin, 1–16; ceftizoxime, 0.25–1; ceftriaxone, 0.016–0.25; meropenem, 0.03; piperacillin, 2–26; piperacillin/tazobactam, 0.016–4; and ticarcillin/clavulanate, 0.016. In general, *C. gracilis* is very susceptible to antimicrobial agents active against anaerobic bacteria.

Strains have been isolated from gingival crevices and from visceral, head, and neck infections; in soft tissue abscesses; pneumonia; empyema; and an ischial wound in humans. The association of *C. gracilis* with serious deep tissue infection, coupled with a high frequency of antibiotic resistance, suggests that its pathogenic role might be underestimated.

The mol% G + C of the DNA is: 44–46 (T_m).

Type strain: FDC 1084, ATCC 33236, CCUG 27720.

GenBank accession number (16S rRNA): L04320.

6. **Campylobacter helveticus** Stanley, Burnens, Linton, On, Costas and Owen 1993a, 398^VP (Effective publication: Stanley, Burnens, Linton, On, Costas and Owen 1992, 2302.)

hel.ve'ti.cus. L. adj. *helveticus* referring to Swiss, after the country of first isolation.

Cells are small curved, S-shaped, or helical rods, 0.2 × 1.5–3 μm, with rounded ends. Rapid darting motility. Motile by means of single bipolar flagella. Translucent, flat colonies, pinpoint to 0.5 mm in diameter are produced on blood agar bases after 48 h. Swarming may occur on moist agar surfaces. Grows microaerobically in the absence of hydrogen. Will not grow in air or in a CO_2-enriched atmosphere. The species is phenotypically, genotypically, and phylogenetically similar to *C. upsaliensis*, also a common inhabitant of domestic pets. *C. helveticus* grows on media containing 1.0–1.5% ox-bile and 100 U/15-fluorouracil, but not on potato starch or MacConkey agars. Sensitive to 32 mg/l cephalothin; some strains (22–44%) resistant to 64 mg/l cefoperazone.

Strains have been isolated from feces of diarrheic and asymptomatic domestic cats, more rarely from dogs. Pathogenicity is unknown.

The mol% G + C of the DNA is: 34 (T_m).

Type strain: ATCC 51209, CCUG 30682, LMG 12638, NCTC 12470.

GenBank accession number (16S rRNA): U03022.

7. **Campylobacter hominis** Lawson, On, Logan and Stanley 2001, 658^VP

hom.in'is. L. gen. n. *hominis* of man, from which the bacterium was first isolated.

Cells are small and straight, 0.25–0.5 × 0.5–1.8 μm (after 10 days of incubation) with blunt ends. Nonmotile. Some isolates produce irregular fimbria-like structures which are 4–8 nm wide and >1.0 μm long. Gray pinpoint colonies may be convex and entire or spreading and irregular on blood agar base media. No agar pitting is observed. Optimal growth in anaerobic conditions at 37°C. No growth in anaerobiosis at room temperature, 25°C, or 42°C. Poor growth, if any, under microaerobiosis with 2% hydrogen. Will not grow in air, in a CO_2-enriched atmosphere, or in an atmosphere containing 5% O_2, 10% CO_2 and 85% N_2 on common agar bases. Strains grow in the presence of 0.1% sodium fluoride, but not in the presence of 0.04% triphenyltetrazolium chloride. Alkaline phosphatase activity has not been detected. Strains have been isolated from asymptomatic human feces. Pathogenicity, if any, is unknown.

The mol% G + C of the DNA is: 32–33 (T_m).

Type strain: CH001A, LMG 19568, NCTC 13146.

GenBank accession number (16S rRNA): AJ251584.

8. **Campylobacter hyointestinalis** Gebhart, Edmonds, Ward, Kurtz and Brenner 1985a, 535[VP] (Effective publication: Gebhart, Edmonds, Ward, Kurtz and Brenner 1985b, 718.)

hy.o.in.tes′ tin.al.is. Gr. n. *hys, hyos* a hog; M.L. adj. *intestinalis* pertaining to the intestines; M.L. gen. n. *hyointestinalis* of a hog's intestine.

Cells are loosely spiraled, curved rods with characteristic darting motility due to a single polar flagellum. Occasionally, cells with two flagella at one pole have been observed. Cells are 0.2–0.5 × 1.2–2.5 μm. Filamentous forms, not coccoid bodies, are seen in old cultures. After 48 h of incubation, colonies are 1.5–2.0 mm in diameter, circular, and convex and do not swarm on moist media; they typically have a dirty yellowish color and are slightly mucoid.

When grown on common blood agar bases, only some strains grow microaerobically without hydrogen. All strains grow microaerobically in the presence of hydrogen. Many strains are weakly α-hemolytic; this is normally accompanied by a greenish hue around the bacterial growth. Strains grow in the presence of 1.0% ox-bile and 0.032% methyl orange. Most strains (~85%) are sensitive to cephalothin (32 mg/l).

Whole-cell protein electrophoretic analysis has revealed considerable protein electrophoretic diversity within this species and it has been suggested that this diversity offers potential for typing studies (Vandamme et al., 1990, 1991b; On et al., 1993). In addition, pulsed-field gel electrophoresis of large genomic fragments was shown to be a useful typing method (Salama et al., 1992b; On and Vandamme, 1997). Plasmids of about 38 mDa and 1.6 mDa in size were reported in some strains by Boosinger et al. (1990). Selective media were developed by different researchers (reviewed by Ohya et al., 1988). On and Vandamme (1997) reported additional groups of *C. hyointestinalis*-like bacteria; the exact taxonomic status of the "subgroup 3" and "subgroup 4" strains is at present unknown.

Minimum inhibitory concentrations of antibiotics are (μg/ml): bacitracin, >128; chloramphenicol, 2–8; clindamycin, 0.25–0.5; colistin, 0.5–1; erythromycin, 1–2; gentamicin, 2–4; kanamycin, <0.5–2; metronidazole, 0.12–1; minocycline, <0.5–2; nalidixic acid, 16 to >128; neomycin, 16–32; penicillin, 1–32; polymyxin B, <0.25–1; rifampin, 4–32; streptomycin, 1–2; tetracycline, 1–8; vancomycin, >128,

amoxicillin/clavulanate, 0.06–0.5; cefoxitin, 1–16; ceftizoxime, 0.25–1; ceftriaxone, 0.016–0.25; meropenem, 0.03; piperacillin, 2–26; piperacillin/tazobactam, 0.016–4; and ticarcillin/clavulanate, 0.016. *In vitro* activities of 47 antimicrobial agents toward *C. hyointestinalis* were determined by Gebhart et al. (1985b). Isolated from the intestines of pigs and hamsters, the stomach of pigs, and cattle, deer, and human feces. May be associated with porcine proliferative enteritis and diarrhea in animals and humans. Pathogenicity is not known.

The mol% G + C of the DNA is: 31–36 (T_m).

Type strain: 80-4577-4, ATCC 35217, CCUG 14169, LMG 7817, NCTC 11608.

GenBank accession number (16S rRNA): M65010, AF097689.

a. **Campylobacter hyointestinalis** *subsp.* **hyointestinalis** (Gebhart, Edmonds, Ward, Kurtz and Brenner 1985b) On, Bloch, Holmes, Hoste and Vandamme 1995, 773[VP] (*Campylobacter hyointestinalis* Gebhart, Edmonds, Ward, Kurtz and Brenner 1985b, 718.)

Morphology and characteristics as for species, except as noted. Strains grow in the presence of 0.01% Janus green. Most strains grow in the presence of 1.5% oxbile. Alkaline phosphatase activity has not been reported. Isolated from the intestines of pigs and hamsters, and cattle, deer, and human feces. May be associated with porcine proliferative enteritis and diarrhea in animals and humans. Pathogenicity is not known.

The mol% G + C of the DNA is: 33–36 (T_m).

Type strain: 80-4577-4, ATCC 35217, CCUG 14169, LMG 7817, NCTC 11608.

GenBank accession number (16S rRNA): M65010, AF097689.

b. **Campylobacter hyointestinalis** *subsp.* **lawsonii** On, Bloch, Holmes, Hoste and Vandamme 1995, 773[VP]

law.so′ ni.i. N.L. gen. n. *lawsonii* of Lawson, in honor of Gordon H.K. Lawson, a bacteriologist at Edinburgh University whose studies on enteric disease in pigs led to the delineation of *Campylobacter mucosalis* and the unculturable bacterium *Lawsonia intracellularis*.

Cells are loosely spiraled, curved rods. Cells are 0.2 × 1.42 μm. Morphology and characteristics as for species, except as noted. Few (11%) strains can grow on 1.5% bile media. Alkaline phosphatase activity has been seen in ~22% of strains. 44% of strains can grow on 0.01% Janus green medium. Isolated from the stomach of pigs. Pathogenicity is not known.

The mol% G + C of the DNA is: 31–33 (T_m).

Type strain: CHY 5, CCUG 34538, LMG 14432, NCTC 12901.

GenBank accession number (16S rRNA): AF097685.

9. **Campylobacter jejuni** (Jones, Orcutt and Little 1931) Véron and Chatelain 1973, 128[AL] (*Vibrio jejuni* Jones, Orcutt and Little 1931, 861; *Vibrio hepaticus* Mathey and Rissberger 1964 1339; *Campylobacter fetus* subsp. *jejuni* Smibert 1974, 209.)

je.ju′ ni. M.L. gen. neut. n. *jejuni* of the jejunum.

Small, tightly coiled spiral or S-shaped cells (average wavelength 1.12 μm and average amplitude of coils is 0.48 μm) which transform rapidly to coccoid forms with age, or exposure to toxic concentrations of oxygen (Ng et al., 1985; Moran and Upton, 1987). Electron microscopic studies reveal a ring-shaped cellular form that may represent an intermediate form between spiral and coccoid cells (Ng et al., 1985). Two types of colonies may be observed (Smibert, 1965, 1969). The first has a low, flat, grayish, finely granular, and translucent appearance with an irregular edge, and a tendency to spread along the direction of the streak, and to swarm and coalesce. The second is round (1–2 mm diameter), raised, convex, smooth, shiny, with an entire, translucent edge and a darker, opaque center. Most strains are weakly hemolytic on blood agar (Arimi et al., 1990; On et al., 1996), but this characteristic may be affected by composition and pH of the base medium used, the gas composition of the atmosphere, and the period and temperature of incubation (Arimi et al. 1990; Misawa et al., 1995). Hemolytic activity has been reported for rabbit, human, cattle, sheep, goat, horse, and chicken blood (Misawa et al., 1995) and appears to be principally cell associated, although a secreted component may also be involved (see Wassenaar, 1997 for a detailed overview). All strains will grow in the presence of 1.0% ox-bile. No growth is observed at 25°C. Strains are motile by means of a single polar flagellum (at one or both ends of the cell) which seems to be an important virulence factor, necessary for colonization of the intestinal tract (Ketley, 1997). In addition, phase and antigenic variation of the flagellar protein may serve as a means of evading the immunogenic response of the host. Variation of the flagellin gene loci (*flaA* and *flaB*) may be detected by PCR-based methods and used in molecular epidemiological studies (Nachamkin et al., 1993; Ayling et al., 1996; Meinersmann et al., 1997). Recombination may affect the stability of the *flaA*-based methods and possibly acts as a mechanism by which the immunogenic repertoire of a given strain is increased (Harrington et al., 1997).

The mol% G + C of the DNA is: 28–33 (T_m).

Type strain: CIP 702, ATCC 33560, CCUG 11284, LMG 8841, NCTC 11351.

GenBank accession number (16S rRNA): L04315, M59298.

a. **Campylobacter jejuni** *subsp.* **jejuni** (Jones, Orcutt and Little 1931) Véron and Chatelain 1973, 128[AL] (*Vibrio jejuni* Jones, Orcutt and Little 1931, 861; *Vibrio hepaticus* Mathey and Rissberger 1964, 1339; *Campylobacter fetus* subsp. *jejuni* Smibert 1974, 209.)

Cell walls contain D-galactose only, D-galactose and D-glucose, or D-galactose, D-glucose, and D-mannose (Smibert, 1970). Strains grow on solid media containing 1.0–1.5% ox-bile, and 0.02% safranin. Reduction and tolerance of 0.04% triphenyl-tetrazolium chloride is observed in 90% of strains. Most (90–95% respectively) strains grow in the presence of 100 mg/l 5-fluorouracil and 32 mg/l cephalothin. Older literature reports most strains as being sensitive to 16 mg/l nalidixic acid (Karmali et al., 1981b) but increasing fluoroquinolone resistance has been observed and the proportion of resistant strains now varies significantly (Reina et al. 1994; Koenraad et al., 1995; Aarestrup et al., 1997). Similar observations of resistance to other quinolones (ciprofloxacin, enrofloxacin, norfloxacin, etc.) have also been reported. Resistance may be conferred by mutations in the target gene, *gyrA* (Piddock and Guant, 1996). Strains may also differ in their susceptibility to other antibiotics

such as tetracycline, erythromycin, chloramphenicol, and kanamycin. The genetic mechanisms underlying these traits have been reviewed (Taylor and Courvalin, 1988; Taylor, 1992a, b). Aarestrup et al. (1997) showed that >90% of isolates from humans, cattle, chickens, and pigs were sensitive to ampicillin (16 mg/l), apramycin (2 mg/l), carbadox (0.5 mg/l), chloramphenicol (8 mg/l), colistin (16 mg/l), enrofloxacin (4 mg/l), erythromycin (4 mg/l), gentamicin (1 mg/l), nalidixic acid (32 mg/l), neomycin (1 mg/l), olaquindox (4 mg/l), spectinomycin (64 mg/l), spiramycin (8 mg/l), streptomycin (2 mg/l), tetracycline (1 mg/l), and tylosin (128 mg/l).

Two broad classes of antigens are recognized: heat-stable (HS) or somatic O-antigens, and heat-labile (HL) antigens. These form the basis of two recognized serotyping schemes, described by Penner et al. (1983) for HS antigens, and Lior et al. (1982) for HL antigens. The major component of the HS antigens is generally accepted as being lipopolysaccharide-based (Mills et al., 1992), although some workers have proposed that a capsule may be involved (Chart et al., 1996). The principal component of the HL antigen is traditionally believed to be the flagellar protein, although other somatic antigens appear to be involved in the recognition of HL-based serogroups (Taylor et al., 1988; Alm et al., 1991). However, *flaA* PCR-restriction fragment length polymorphism analysis showed limited correlation with HL serotyping suggesting that flagella may not be the major HL antigen (Mohran et al., 1996).

Considerable variation in the prevalence, number, and size of plasmids found in *C. jejuni* subsp. *jejuni* has been described. Between 19% (Austen and Trust, 1980) and 95% (Lee et al., 1994b) of strains may harbor 1–5 plasmids (Tenover et al., 1985), which range between 2.0 (Tenover et al., 1985) and 208 kb (Lee et al., 1994b) in size. Resistance to the antibiotics kanamycin, tetracycline, and chloramphenicol is often plasmid-mediated (Taylor, 1992b).

Pathogenic. Causes abortion in sheep; abortions in other animals, such as cattle and goats, have also been reported. May cause diarrhea in animals and has been associated with hepatitis in some bird species. In humans it is generally regarded as the most common bacterial cause of gastroenteritis worldwide; it also causes septicemia and abortion. Infection with certain strains of *C. jejuni* subsp. *jejuni* may be a predisposing factor to the development of the neurological disorders Guillain-Barré (GBS) (Kaldor and Speed, 1984; Nachamkin et al., 1998) and Miller-Fisher syndromes (MFS) (Roberts et al., 1987). Molecular mimicry is postulated as the mechanism for pathogenesis of these neuropathies; a trisaccharide moiety that resembles human gangliosides has been found in different HS serotypes recovered from GBS and MFS patients (Salloway et al., 1996). Strains are also found as normal intestinal flora of poultry and other bird species, cattle, sheep, pigs, goats, dogs, rabbits, and monkeys.

The mol% G + C of the DNA is: 30–33 (T_m).

Type strain: CIP 702, ATCC 33560, CCUG 11284, LMG 8841, NCTC 11351.

GenBank accession number (16S rRNA): L04315, M59298.

b. **Campylobacter jejuni** subsp. **doylei** Steele and Owen 1988, 316[VP]

doy.le′i. M.L. gen. n. doylei in honor of L.P. Doyle, an American veterinarian.

Cells may be spiral, S-shaped or, less frequently, straight rods. Cultures often demonstrate considerable pleomorphism that increases with age. Colonies are pinpoint to 1 mm diameter, grayish, smooth, glistening, and convex after 2–3 days growth on blood agar. Optimal growth temperature is 35–37°C; strains do not grow at 25°C and poorly, if at all, at 42°C. Reduction of, and tolerance to, 0.04% triphenyl-tetrazolium chloride observed in 40% of strains (On et al., 1996). Strains will not grow on solid media containing 0.02% safranin or 32 mg/l cephalothin. Principal tests differentiating this subspecies from C. jejuni subsp. jejuni are given in Table BXII.ε.1. These taxa can also be distinguished by numerical analysis of both whole-cell protein electrophoretograms (Vandamme et al., 1992a), and HaeIII restriction digest patterns (Owen et al., 1985).

In disk susceptibility tests, all strains are susceptible to penicillin (2 U), erythromycin (15 μg), and tetracycline (10 μg). Most strains are inhibited by nalidixic acid (30 μg). Pathogenicity unknown; has been isolated from ulcerated gastric tissue, diarrhea, and blood cultures of humans, notably infants (Steele and Owen, 1988; Lastovica, 1996).

The mol% G + C of the DNA is: 28–31 (T_m).

Type strain: 093, CCUG 24567, LMG 8843, NCTC 11951.

GenBank accession number (16S rRNA): L14630.

10. **Campylobacter lanienae** Logan, Burnens, Linton, Lawson and Stanley 2000, 870[VP]

lan.i.en′ae. L. n. laniena abattoir, after place of work of human carriers from whom the bacterium was first isolated.

Cells are slender, slightly spiral rods with rounded ends and are 1.2–2.4 μm long. Characteristic darting motility due to a single bipolar flagellum. Pinpoint colonies are visible on blood agar media after three days of incubation at 37°C. The colonies are smooth, entire, and translucent and cause some greening of blood agar through alpha-hemolytic activity. Optimal growth in microaerobic conditions. Growth is weak under anaerobic conditions. Microaerobic growth at 42°C, but not at 25°C. Catalase produced. Nitrate and selenite reduced. No hydrolysis of indoxyl acetate. The 16S rDNA sequence of C. lanienae, and thus its phylogenetic position, is remarkably similar to that of C. hyointestinalis subsp. lawsonii (Fig. BXII.ε.2). A 16S rDNA-based PCR test developed for the specific detection of C. lanienae cross-reacts with C. hyointestinalis subsp. lawsonii (S.L.W. On, unpublished observations). Extensive DNA–DNA hybridization experiments, however, confirmed that C. hyointestinalis subsp. hyointestinalis and C. hyointestinalis subsp. lawsonii strains represent a single species distinct from C. lanienae (P. Vandamme, unpublished observations). Strains have been isolated from asymptomatic human feces. Pathogenicity, if any, is unknown.

The mol% G + C of the DNA is: 36 (T_m).

Type strain: LMG21527, NCTC 13004.

GenBank accession number (16S rRNA): AF043425.

11. **Campylobacter lari** (corrig.) Benjamin, Leaper, Owen and Skirrow 1984, 270[VP] (Effective publication: Benjamin, Leaper, Owen and Skirrow 1983, 237 (Campylobacter laridis Benjamin, Leaper, Owen and Skirrow 1984, 270.)

la′ri. L. n. Larus gull; L. gen. n. lari of a gull.

Cells are small, curved, S-shaped, or helical rods, 0.3 × 1.7–2.4 μm, with rounded ends. Rapid transformation to coccoid forms in cultures exposed to air. Rapid darting motility. Motile by means of single bipolar flagella. Translucent, convex colonies, 1–1.5 mm in diameter are produced on blood agar bases after 48 h. Swarming may occur on very moist agar surfaces. Grows microaerobically in the absence of hydrogen. Anaerobic growth is observed in the presence of trimethylamine N-oxide hydrochloride. Will not grow in air or in a CO_2-enriched atmosphere. Strains grow in the presence of 1.0–1.5% ox-bile, 0.05% sodium fluoride, 32 mg/l cephalothin, and 64 mg/l cefoperazone. Most strains (93%) grow on, and reduce, 0.04% triphenyl-tetrazolium chloride medium. No growth on MacConkey agar.

C. lari was originally referred to as the group of nalidixic acid-resistant thermophilic campylobacters (NARTC group). These strains are primarily differentiated from other thermophilic Campylobacter species by their resistance to nalidixic acid, their anaerobic growth in the presence of trimethylamine N-oxide hydrochloride, and later also by the absence of indoxyl acetate hydrolysis. However, nalidixic acid-susceptible strains (NASC strains), urease-producing thermophilic strains (UPTC strains), and urease-producing, nalidixic acid-susceptible strains (UP-NASC strains) were later identified as C. lari variants by a variety of methods. Additional quantitative DNA–DNA hybridization experiments are required to establish the relationships between the different subgroups of strains presently classified as C. lari variants.

C. lari infections can be treated with aminoglycosides, erythromycin, clindamycin, and chloramphenicol; strains are generally resistant to third-generation cephalosporins, vancomycin, penicillin, and trimethoprim-sulfamethoxazole (Simor and Wilcox, 1987). Strains have been isolated from intestinal contents of seagulls and other animals, river water, and shellfish. Occasionally isolated from human diarrheic feces. Pathogenicity is unknown.

The mol% G + C of the DNA is: 31–33 (T_m).

Type strain: ATCC 35221, CCUG 23947, DSM 11375, LMG 8846, NCTC 11352.

GenBank accession number (16S rRNA): L04316.

12. **Campylobacter mucosalis** (Lawson, Leaver, Pettigrew and Rowland 1981) Roop, Smibert, Johnson and Krieg 1985a, 191[VP] (Campylobacter sputorum subsp. mucosalis Lawson, Leaver, Pettigrew and Rowland 1981, 385.)

mu.co.sa′lis. M.L. adj. mucosalis pertaining to the (tunica) mucosa mucous membrane.

Cells are short, irregularly curved, 0.25–0.30 × 1–3 μm. Motile by means of a single, polar flagellum. In old cultures, coccoid cells and filamentous forms 7–8 μm long are seen. Colonies are 1.5 mm in diameter, circular, raised with a flat surface, and have a dirty yellowish color. On moist agar, colonies tend to swarm along the line of inoculation.

Does not grow microaerobically on common agar bases in an atmosphere without hydrogen. Grows anaerobically in an atmosphere containing hydrogen and fumarate. Requires hydrogen as an electron donor. Formate may replace hydrogen for growth of most strains (Lawson et al., 1981). Anaerobic growth requires fumarate as an electron acceptor. Converts fumarate to succinate. Oxygen utilization by cell suspensions is greatly increased in the presence of hydrogen and formate. Utilization of oxygen with these substrates is unaffected by cyanide. Cells contain large amounts of cytochrome c_{553}, which is reduced when cell suspensions are incubated with hydrogen or formate. When exposed to air the reduced cytochrome c is reoxidized. Lactate, succinate, and NADH give slight reduction of cytochrome c while methanol, malate, glutamate, and serine are inactive. It is not affected by cyanide (Lawson et al., 1981). Strains grow on potato starch medium, and on media containing 1.0% ox-bile and 0.032% methyl orange. Most (70–90%) strains grow in the presence of 0.01% Janus green and 0.005% basic fuchsin.

Serological analysis shows that strains of this organism are closely related antigenically (Lawson et al., 1975). Antisera prepared against five strains agglutinated homologous and heterologous cells. When antisera were absorbed with one strain, the antisera reacted only with homologous cells. An antigenic analysis has been reported (Lawson et al., 1977) and three serovars, A, B, and C, have been described (Lawson et al., 1981). These three different serovars were shown to have strikingly distinct whole-cell protein patterns, but a high level of DNA–DNA hybridization was reported between representative strains (Costas et al., 1987; Vandamme et al., 1990). Selective media were developed by different researchers (reviewed by Ohya et al., 1988).

Minimum inhibitory concentrations of antibiotics are (μg/ml): bacitracin, >128; chloramphenicol, 2–8; clindamycin, 0.25–0.5; colistin, 0.5–1; erythromycin, 1–2; gentamicin, 2–4; kanamycin, <0.5–2; metronidazole, 0.12–1; minocycline, <0.5–2; nalidixic acid, 16 to >128; neomycin, 16–32; penicillin, 1–32; polymyxin B, <0.25–1; rifampin, 4–32; streptomycin, 1–2; tetracycline, 1–8; vancomycin, >128; amoxicillin/clavulanate, 0.06–0.5; cefoxitin, 1–16; ceftizoxime, 0.25–1; ceftriaxone, 0.016–0.25; meropenem, 0.03; piperacillin, 2–26; piperacillin/tazobactam, 0.016–4; and ticarcillin/clavulanate, 0.016. In vitro activities of 47 antimicrobial agents toward *C. mucosalis* were determined by Gebhart et al. (1985b).

Pathogenicity unknown. Originally believed to be a causal agent of proliferative enteritis in pigs but subsequent studies have identified *Lawsonia intracellularis* as the principal pathogen in this disease. Human infections with this organism have been reported but were shown to be caused by misidentified *C. concisus* strains (On, 1994; Anderson et al., 1996). Isolated from the intestinal mucosa of pigs with porcine intestinal adenamatosis, necrotic enteritis, regional ileitis, and proliferative hemorrhagic enteropathy; also isolated from the porcine oral cavity.

The mol% G + C of the DNA is: 36–38 (T_m).

Type strain: FS253/72, ATCC 43264, CCUG 6822, LMG 6448, NCTC 11000.

GenBank accession number (16S rRNA): L06978.

13. **Campylobacter rectus** (Tanner, Badger, Lai, Listgarten, Visconti and Socransky 1981) Vandamme, Falsen, Rossau, Hoste, Segers, Tytgat and De Ley 1991a, 98^{VP} (*Wolinella recta*

Tanner, Badger, Lai, Listgarten, Visconti and Socransky 1981, 441.)

rect'us. L. adj. *rectus* straight.

Cells are small and straight, 0.5 × 2–4 μm, with rounded ends. Rapid darting motility. Motile by means of a single polar flagellum. The outer surface is covered with a distinctive array of hexagonal, packed, macromolecular subunits, each about 17 nm in diameter. Translucent colonies are produced on blood agar bases. Different colony types are observed: small pinpoint colonies, 1 mm in diameter, or spreading colonies up to 5 mm in diameter. Agar pitting is medium dependent but most (80%) strains exhibit this trait after 3 d of anaerobic growth on 5% blood agar (On et al., 1996).

Optimal growth in anaerobic conditions. Does not grow microaerobically on common agar bases in an atmosphere without hydrogen. Will not grow in air, in a CO_2-enriched atmosphere, or in an atmosphere containing $O_2/CO_2/N_2$ (5:10:85) on common agar bases. Anaerobic growth occurs with formate and fumarate in the medium. Hydrogen and formate are used as energy sources. Formate is oxidized to hydrogen and CO_2; fumarate is reduced to succinate. Fumarate, nitrate, nitrite, neutral red, benzyl viologen aspartate, asparagine, and malate serve as electron acceptors. Membrane-bound cytochrome b, cytochrome c, and CO-binding cytochrome c, and soluble cytochrome c and CO-binding cytochrome c are present. Strains do not grow in the presence of 0.04% triphenyl-tetrazolium chloride, 0.01% Janus green or 64 mg/l cefoperazone. Alkaline phosphatase is not produced.

The mol% G + C of the DNA is: 42–46 (T_m).

Type strain: FDC 371, ATCC 33238, CCUG 20446, DSM 3260, LMG 18219.

GenBank accession number (16S rRNA): L04317, L06973.

14. **Campylobacter showae** Etoh, Dewhirst, Paster, Yamamoto and Goto 1993, 638^{VP}

sho'wae. L. n. *showae* referring to Showa University, Japan, where several of the first strains were isolated.

Cells are small and straight, 0.5–0.8 × 2–5 μm, with rounded ends. Motile by means of polar bundles of two to five flagella. Rapid darting motility. Translucent colonies are produced on blood agar base media.

Optimal growth in anaerobic conditions. Does not grow microaerobically on common agar bases in an atmosphere without hydrogen. Will not grow in air, in a CO_2-enriched atmosphere, or in an atmosphere containing $O_2/CO_2/N_2$ (5:10:85) on common agar bases. Anaerobic growth occurs with formate and fumarate in the medium. Fumarate is reduced to succinate. Fumarate, nitrate, and nitrite serve as electron acceptors. Strains grow in the presence of 0.05% sodium fluoride but not in the presence of 64 mg/l cefoperazone or 0.04% triphenyl-tetrazolium chloride. Alkaline phosphatase activity has not been detected. Strains have been isolated from human dental plaque and from infected root canals. Pathogenicity is unknown.

The mol% G + C of the DNA is: 44–46 (T_m).

Type strain: SU A4, ATCC 51146, CCUG 30254, LMG 12635.

GenBank accession number (16S rRNA): L06974.

15. **Campylobacter sputorum** (Prévot 1940) Véron and Chatelain 1973, 128^{AL} emend. Roop, Smibert, Johnson and

Krieg 1986, 348 emend. On, Atabay, Corry, Harrington and Vandamme 1998, 203 (*Vibrio sputorum* Prévot 1940, 85.) *spu.to' rum*. L. n. *sputum* spit, sputum; L. gen. pl. n. *sputorum* of sputa.

Slender, curved or spiral rods, 0.3–0.5 × 2–4 μm. Cells are usually S-shaped or gull-winged in appearance, but comma-shaped and unusually long (8 μm) cells are also observed. The ends of the cells are usually rounded. Motile with a characteristic darting and corkscrew-like movement by means of a single flagellum. Colonies of 3-d-old cultures on blood agar media are 1–2 mm in diameter, smooth, shiny, low convex, and round. Bacterial growth on blood agar usually shows a greenish hue, and it is accompanied by weak α-hemolytic activity. Growth in broth is light and easily dispersed. Strains may be cultured under microaerobic or anaerobic gaseous conditions. Anaerobic growth occurs in media containing fumarate only, formate and fumarate, or fumarate in the presence of hydrogen gas. Alkaline phosphatase is not produced. Neither growth nor reduction of triphenyl-tetrazolium chloride medium is detected. Strains will grow on media containing 0.032% methyl orange. Strains do not give reproducible results when tested for the ability to grow on media containing 3.5–4.0% NaCl, or 1.0–2.0% ox-bile.

Strains isolated from the human oral cavity and the genital tract of bulls were initially believed to be related at the subspecies level on the basis of their extensive biochemical similarities, and named *C. sputorum* subsp. *sputorum* and *C. sputorum* subsp. *bubulus* (Véron and Chatelain, 1973), respectively. Subsequent DNA homology studies (Tanner et al., 1981; Roop et al. 1985b) revealed a high level of DNA–DNA relatedness between these two taxa and between these taxa and "*C. fecalis*" (from sheep feces). Limited biochemical variation between these taxa had been noted (catalase production, and growth on 3.5% NaCl and 1.0% ox-bile media). Roop et al. (1985b) therefore proposed that the aforementioned taxa be referred to as source-specific biovars of *C. sputorum* (biovar sputorum, biovar bubulus, and biovar faecalis, respectively). The legitimacy of biovar bubulus as a distinct taxon was questioned when the tests used to distinguish biovar sputorum from biovar bubulus were found to be poorly reproducible, even when using highly standardized conditions for testing (On et al., 1994, 1998). The absolute validity of the "source-specific biovar" concept was also challenged since some *C. sputorum* strains from sheep and pigs cannot be distinguished from those of bovine and human origin (On et al., 1994, 1998). The recent identification of urease-positive isolates of *C. sputorum* from cattle feces resulted in the proposal of a new biovar structure, defined by reactions in two simple and reproducible tests (catalase and urease). The proposed biovar classification also agrees with the results of protein profile analysis (On et al., 1994, 1998).

Campylobacter sputorum biovar sputorum strains conform to the description of the species *C. sputorum*. Neither catalase nor urease is produced. Found in the oral cavity, feces (normal and diarrheic), and abscesses (and other skin lesions) of humans, the genital tract of bulls, aborted tissue of sheep, and the feces of sheep and pigs. Pathogenicity unknown. The reference strain is VPI S-17 (LMG 7795, NCTC 11528). GenBank accession numbers of the 16S rRNA gene sequences are X67775 (VPI S-17) and L04319 (ATCC 33491).

Campylobacter sputorum biovar faecalis strains conform to the description of the species *C. sputorum*. Catalase, but not urease, is produced. The biovar name refers to the fecal habitat. Isolated from the feces of sheep and cattle. Pathogenicity unknown. The recommended reference strain is LMG 8531 (CCUG 17761, NCTC 11415).

Campylobacter sputorum biovar paraureolyticus strains conform to the description of the species *C. sputorum*. Urease, but not catalase, is produced. The biovar name paraureolyticus indicates its resemblance to *Bacteroides ureolyticus*, a urease-producing bacterium closely related to other *Campylobacter* spp. Isolated from the feces of cattle and from human diarrhea. Pathogenicity unknown. Recommended reference strains are LMG 11764 (human) and LMG 17590 (CCUG 37579) (bovine). GenBank accession no. of the 16S rRNA gene sequence of strain LMG 17590 is AF022768.

The mol% G + C of the DNA is: 30–33 (T_m).

Type strain: VPI S 17, ATCC 35980, CCUG 9728, LMG 7795, NCTC 11528.

GenBank accession number (16S rRNA): X67775.

16. **Campylobacter upsaliensis** Sandstedt and Ursing 1991b, 331[VP] (Effective publication: Sandstedt and Ursing 1991a, 42.)
up.sa.li.en' sis. L. adj. *upsaliensis* referring to Uppsala, a Swedish city.

Cells are small, curved, S-shaped, or helical rods, 0.3–0.4 × 1.2–3 μm, with rounded ends. Rapid darting motility. Motile by means of a single polar flagellum or bipolar flagella. Transformation to coccoid forms in cultures exposed to air. Translucent, convex colonies, pinpoint to 1–2 mm in diameter are produced on blood agar bases after 48 h. Swarming may occur on very moist agar surfaces.

Grows microaerobically in the absence of hydrogen. Will not grow in air or in a CO_2-enriched atmosphere. Goossens et al. (1990a) reported that about 20% of 99 strains did not grow at 42°C; it therefore seems appropriate to consider this species as thermotolerant, not thermophilic. Strains grow on potato starch- and charcoal-based media, and on media containing 1.0–1.5% ox-bile and 100 U/l 5-fluorouracil. Strains do not grow on MacConkey agar. A few strains (11%) are resistant to cephalothin (32 mg/l).

A comprehensive overview of our present knowledge on this species was given by Bourke et al. (1998). A physical and genetic map of the genome and the complete sequence of the iron-uptake regulatory gene (*fur*) from the type strain were described by Bourke et al. (1995). Adherence to lipids and intestinal mucin were described by Sylvester et al. (1996). Selective media were reported to enable isolation of thermophilic campylobacters including *C. upsaliensis* strains (Burnens and Nicolet, 1992; Aspinall et al., 1996). A variety of plasmids has been reported in large percentages of strains (89%, Goossens et al., 1990a; 87%, Owen and Hernandez, 1990; 93%, Da Silva Tatley et al., 1992; 60%, Stanley et al., 1994a).

Minimum inhibitory concentrations of antibiotics are (μg/ml): bacitracin, >128; chloramphenicol, 2–8; clindamycin, 0.25– 0.5; colistin, 0.5–1; erythromycin, 1–2; gentamicin, 2–4; kanamycin, <0.5–2; metronidazole, 0.121; minocycline, <0.5–2; nalidixic acid, 16–128; neomycin, 16–32; penicillin, 1–32; polymyxin B, <0.25–1; rifampin, 4–32;

streptomycin, 1–2; tetracycline, 1–8; vancomycin, >128, amoxicillin/clavulanate, 0.06–0.5; cefoxitin, 1–16; ceftizoxime, 0.25–1; ceftriaxone, 0.016–0.25; meropenem, 0.03; piperacillin, 2–26; piperacillin/tazobactam, 0.016–4; and ticarcillin/clavulanate, 0.016.

Of 99 *C. upsaliensis* strains examined by Goossens et al. (1990a), all were generally susceptible to ampicillin, gentamicin, chloramphenicol, cefoperazone, colistin, vancomycin, rifampin, trimethoprim, and tetracycline; ten strains were resistant to erythromycin. Strains are isolated from blood specimens of humans, from feces from humans with gastrointestinal illness, and from asymptomatic humans, dogs, and cats; strains associated with a human abortion and a breast abscess were reported. Clinical data presented by Goossens et al. (1990a, b) suggest an enteropathogenic role in humans.

The mol% G + C of the DNA is: 32–36 (T_m).

Type strain: C 231, ATCC 43954, CCUG 14913, DSM 5365, LMG 8850, NCTC 11541.

GenBank accession number (16S rRNA): L14628.

Other Organisms

1. *Bacteroides ureolyticus* Jackson and Goodman 1978, 199[AL] (*Bacteroides corrodens* Eiken 1958, 415; *Ristella corrodens* (Eiken 1958) Prévot 1966, 118.)

ur′e.o.ly.ti.cus. M.L n. *urea* urea; Gr. adj. *lyticus* dissolving; M.L. adj. *ureolyticus* urea dissolving.

Cells are 0.5 × 1.5–4 μm. Nonmotile. Filaments exceeding 20 μm in length may occur. Cells of some strains have polar tufts of long pili in electron micrographs (Jackson et al., 1971) and exhibit "twitching" motility. The pili sometimes form a bundle and may be mistaken for flagella with light microscopy. Translucent colonies are produced on blood agar bases. Different colony types are observed: small pinpoint colonies, 1 mm in diameter or spreading colonies up to 5 mm in diameter. Agar pitting is medium dependent, but most strains (90%) exhibit this trait after 3 d of anaerobic growth on 5% blood agar (On et al., 1996).

Does not grow microaerobically on common agar bases in an atmosphere without hydrogen. Will not grow in air, in a CO_2-enriched atmosphere, or in an atmosphere containing $O_2/CO_2/N_2$ (5:10:85) on common agar bases. Anaerobic growth occurs with formate and fumarate in the medium. Fumarate is reduced to succinate; fumarate, nitrate, and nitrite serve as electron acceptors. Strains grow on media containing 3.5–4.0% NaCl and 0.032% methyl orange. As with other preferentially anaerobic *Campylobacter* species, strains are susceptible to a range of antibiotics including cephalothin (32 mg/l), carbenicillin (32 mg/l), cefoperazone (64 mg/l) and 5-fluorouracil (100 U/l). No growth is observed in the presence of 0.05% basic fuchsin. Contains cytochromes *b* and *c* (Jackson and Goodman, 1978).

Menaquinone-6 (2-methyl-3-farnesyl-farnesyl-1,4-naphthoquinone) and a methyl-substituted menaquinone-6 (2, [5 or 8]-dimethyl-3-farnesyl-farnesyl-1,4-naphthoquinone) have been reported as major respiratory quinones (Vandamme et al., 1995a). Strains have been isolated from superficial ulcers and soft tissue infections, nongonococcal, nonchlamydial urethritis, and periodontal disease. Its pathogenicity is difficult to assess because the strains are mostly recovered from mixed infections. Nevertheless, a potential pathogenic role is suggested by its predominance in mixed infections and its strong proteolytic activity, which may enable tissue destruction.

B. ureolyticus was included in a polyphasic taxonomic study to elucidate its taxonomic status (Vandamme et al., 1995a). This species resembles campylobacters in its respiratory quinone content, its DNA base ratio, and most of its phenotypic characteristics; it differs from campylobacters in its fatty acid composition and its proteolytic metabolism. Bootstrapping analysis of the 16S rRNA gene sequence-derived phylogeny separated *B. ureolyticus* from the *Campylobacter* clade in only 61% of trees generated. This organism was not formally reclassified pending the isolation and thorough taxonomic characterization of additional *B. ureolyticus*-like bacteria. The inclusion of *B. ureolyticus* in the genus *Campylobacter* would considerably extend the phenotypic heterogeneity of this genus. Exclusion from the genus *Campylobacter* would entail the creation of a monotypic taxon with the ability to digest casein and gelatin as differential features from the genus *Campylobacter*. *B. ureolyticus*-like oral isolates that were shown to be heterogeneous on the basis of phenotypic and biochemical criteria (Duerden et al., 1989) have been described. The taxonomic structure and position of these and other *B. ureolyticus*-like strains need further investigation.

The mol% G + C of the DNA is: 28–30 (T_m).

Deposited strain: ATCC 33387, DSM 20703, NCTC 10941.

GenBank accession number (16S rRNA): L04321.

The *Campylobacter mucosalis*-like strain CCUG 20705 was isolated from a porcine intestine in the United Kingdom and was originally identified as *C. mucosalis*. It occupied a distinct position in a dendrogram derived from DNA–rRNA hybridization experiments (Vandamme et al., 1991a). No significant DNA–DNA hybridization between CCUG 20705 and *C. mucosalis* reference strains (P. Vandamme, unpublished data) was detected although the 16S rRNA gene sequence of this strain (L14629) is about 98% similar to that of the type strain of *C. mucosalis* (Linton et al., 1994a).

Campylobacter sp. strain PGC 40-6AT, isolated from a pig stomach, was reported as one of several genetically similar isolates. A number of studies have remarked on its distinct phylogenetic position, suggesting it to represent a novel species. However, a study examining the phylogenetic position and diversity of the 16S rRNA gene from *C. hyointestinalis*, closely clustered PGC 40-6AT with six other sequences derived from reference strains of *C. hyointestinalis* subsp. *lawsonii* (Harrington and On, 1999). Since the latter subspecies is most commonly associated with porcine stomach tissue, these authors considered PGC 40-6AT identified as a strain of *C. hyointestinalis* subsp. *lawsonii*.

Genus II. **Arcobacter** Vandamme, Falsen, Rossau, Hoste, Segers, Tytgat and De Ley 1991a, 99[VP] emend. Vandamme, Vancanneyt, Pot, Mels, Hoste, Dewettinck, Vlaes, Van Den Borre, Higgins, Hommez, Kersters, Butzler and Goossens 1992b, 355

PETER VANDAMME, FLOYD E. DEWHIRST, BRUCE J. PASTER AND STEPHEN L.W. ON

Ar'co.bac.ter. L. n. arcus bow; Gr. n. bacter rod; M.L. masc. n. Arcobacter a curved rod.

Cells are slender, curved rods, 0.2–0.9 × 0.5–3 μm long; S-shaped or helical cells are often present. Nonsporeforming. Cells in old cultures may form spherical or coccoid bodies and loose spiral filaments up to 20 μm long. Gram negative. Motile with a **characteristic corkscrew-like motion** by means of a single **polar unsheathed flagellum** at one or both ends of the cell.

Optimal growth under microaerobic conditions, with a respiratory type of metabolism. **Able to grow in aerobic conditions.** Hydrogen is not required for microaerobic growth. Grows anaerobically. **Growth at 15 and 30°C, not at 4°C; growth is variable at 37°C. Chemoorganotrophs. Carbohydrates are neither fermented nor oxidized.** No acidic or neutral end products produced. Serum or blood enhances, but is not essential for, growth. **Energy is obtained from amino acids or tricarboxylic acid cycle intermediates, not carbohydrates. Oxidase activity is present; most strains produce catalase.** Most strains grow in the presence of 0.032% methyl orange. **Indoxyl acetate is hydrolyzed.** No hydrolysis of gelatin, casein, starch, hippurate, and tyrosine. Methyl red and Voges–Proskauer negative. No arylsulfatase or lecithinase activity. Pigments are not produced. **Habitat is extremely diverse.** Some species are pathogenic for humans and animals and are found in the reproductive organs and aborted fetuses of various animals and in the intestinal tract of man and animals. Detected in water reservoirs, sewage, oil field communities, and saline environments; one species is a plant-associated nitrogen fixer.

Internal transcribed spacers or intervening sequences in rRNA genes have not been reported.

The mol% G + C of the DNA is: 27–31.

Type species: **Arcobacter nitrofigilis** (McClung, Patriquin and Davis 1983) Vandamme, Falsen, Rossau, Hoste, Segers, Tytgat and De Ley 1991a, 100 (*Campylobacter nitrofigilis* McClung, Patriquin and Davis 1983, 610.)

FURTHER DESCRIPTIVE INFORMATION

Arcobacter, Campylobacter, the generically misclassified species *Bacteroides ureolyticus,* and *Sulfurospirillum* constitute the family *Campylobacteraceae* and belong to the *Epsilonproteobacteria* (Fig.

BXII.ε.3). *Arcobacter* is a most unusual genus, unifying species associated with disease or considered pathogenic in humans and animals, and a plant-associated nitrogen-fixing bacterium. This combination of plant- and animal-associated species within one genus is rarely seen among *Proteobacteria*. Although primarily known as human- and animal-associated bacteria that are of relevance in food microbiology, arcobacters were shown to be abundantly present in certain environmental niches including water reservoirs, sewage, oil field communities, and certain saline environments. Their role in the environment is not well documented, but some of these organisms were shown to be sulfide oxidizers (with the production of sulfur). It has been suggested that these organisms play a role in the sulfur cycle by reoxidizing sulfide formed by microbial sulfate or sulfur reduction (Voordouw et al., 1996).

Menaquinone-6 (2-methyl-3-farnesyl-farnesyl-1,4-naphthoquinone) and a second menaquinone-6, the detailed structure of which has not been determined, were reported as major respiratory quinones (Moss et al., 1990b).

Few typing studies have thus far been presented. A serotyping scheme based on heat-labile antigens using slide agglutination was developed by Lior and Woodward (1991). A biotyping scheme which involved four biochemical characteristics: the ability to produce urease, rapid hydrogen sulfide formation, DNase activity, and the utilization of sodium acetate was presented subsequently by the same authors (Lior and Woodward, 1993). In addition, genotypic methods such as ribotyping (Kiehlbauch et al., 1991b; Taylor et al., 1992b), pulsed field gel electrophoresis of macrorestriction fragments (Lior and Wang, 1993), and random and repetitive motif based polymorphic DNA analyses (Lior and Wang, 1993; Vandamme et al., 1993a) were performed on limited numbers of strains.

Little is known about virulence factors of *Arcobacter* strains. Musmanno et al. (1997) examined 18 isolates of *A. butzleri* from river water samples. Toxin profiles based on cytotonic, cytotoxic, and cytolethal distending factors were determined after analysis responses in Vero and CHO cells. Adhesivity and invasivity tests

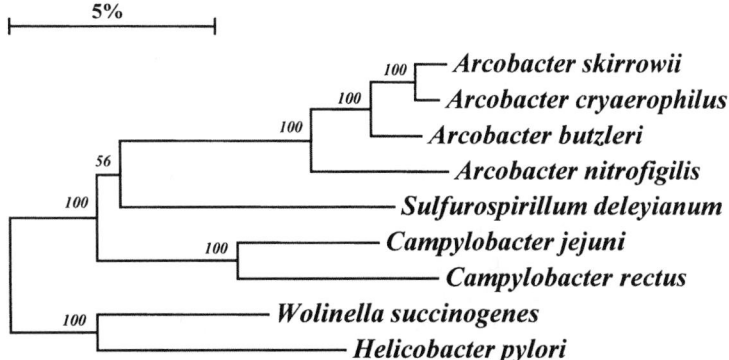

FIGURE BXII.ε.3. Phylogeny of *Arcobacter* species. The phylogenetic relationships of the named species of the genus *Arcobacter* compared to other selected members of the family *Campylobacteraceae* and family *Helicobacteraceae*. Bar = 5% difference in 16S rDNA nucleotide sequences. One hundred bootstrap trees were generated, and bootstrap confidence levels (shown as percentages above the nodes) were determined.

were performed on HeLa and Intestine 407 cells. All but one strain induced cytotoxic effects on cells in culture. The cytotoxic-negative strain caused elongation of CHO cells (a cytotonic-like effect). This strain was the only one that adhered to cells *in vitro*. Invasiveness was never observed.

In spite of accumulating evidence, little is known about the true distribution of arcobacters in human, animal, and environmental samples. Arcobacters may be readily overgrown by other bacteria, frequently by campylobacters, if inappropriate isolation methods are used. From an ecological perspective, this genus comprises two subgroups of species, those associated with human and animal disease, and those either associated with plants or free-living. Three species, *A. cryaerophilus*, *A. butzleri*, and *A. skirrowii*, have been isolated from feces of humans and various animals with diarrhea, from cattle with mastitis, and from a variety of meat products of animal origin. These species have been also associated with enteritis, recurrent cramps, bacteremia, endocarditis, peritonitis, appendicitis, abortion, and mastitis (Wesley, 1996). Evidence that arcobacters may be true pathogens is derived from a number of observations: (i) arcobacters are associated with clear clinical symptoms in the absence of other enteropathogens; (ii) in animals, arcobacters are more frequently recovered from aborted pig litters and from infertile sows with vaginal discharge than from healthy animals; (iii) experimental infection lowered conception rates in sows artificially inseminated with arcobacter-containing sperm and caused inflammation only in the inoculated udder of a dairy cow; (iv) *A. butzleri* in particular can colonize the intestine and other internal organs of neonatal pigs (Wesley et al., 1996). *A. nitrofigilis* and several unnamed arcobacters are free-living or plant-associated bacteria occurring in saline environments, sludge of wastewater plants, and oil field communities, and may play an important role in sulfur or nitrogen cycles (McClung et al., 1983; Teske et al., 1996; Voordouw et al., 1996; Snaidr et al., 1997).

ENRICHMENT AND ISOLATION PROCEDURES

Arcobacters may be isolated using selective media or filtration methods described for the isolation of *Campylobacter* species. However, these methods are suboptimal for growth of arcobacters, which may be overgrown by campylobacters present in the same specimens. Samples should be incubated at 24–30°C to enhance the selectivity of the procedure, and an enrichment step has been recommended (Atabay et al., 1998a).

Isolation of *Arcobacter* strains was originally described by Ellis et al. (1977) using a two-stage isolation method. *Leptospira* EMJH medium supplemented with 5-fluorouracil modified by the addition of 1% rabbit serum and 0.15% agar was used as the initial selective and enrichment medium. This semisolid medium was incubated at 30°C in air. Growth was seen as a distinct zone below the surface of the medium. Subsequently, an aliquot was withdrawn from the growth zone and inoculated onto a selective solid agar medium consisting of a blood agar base supplemented with 7% lysed horse blood and 125 µg/ml carbenicillin. The plates were then incubated at 30°C in a microaerobic atmosphere.

A selective medium developed for the isolation of arcobacters was recently described by de Boer et al. (1996) and is based on their swarming capacity in semisolid media. The enrichment broth contains *Brucella* broth (Difco) as a basal medium supplemented with lysed horse blood (5%), piperacillin (75 mg/l), cefoperazone (32 mg/l), trimethoprim (20 mg/l), and cycloheximide (100 mg/l). Mueller Hinton broth (Oxoid) supplemented with 0.25% agar and the same antibiotic supplement is used as the isolation medium. An aliquot of 40 µl of the enrichment culture is pipetted onto the center of the agar layer, and plates are incubated for two to three days at 24°C. Growth of arcobacters typically appears as gray, translucent zones of motility (de Boer et al., 1996).

Atabay and Corry used sensitive methods of isolation that allowed them to detect a wide range of *Campylobacter* and *Arcobacter* species from chicken carcasses (Atabay and Corry, 1997; Atabay et al., 1998a). These methods involved the use of different selective agents and/or lower concentrations of antibiotics in the isolation media. Hydrogen was also included in the microaerobic atmosphere, samples were plated onto nonselective blood agar, and incubation was carried out at various temperatures (most often 30°C and 37°C). Arcobacters were isolated only at 30°C, under aerobic conditions after enrichment. Modified CCDA (mCCDA) did not support the growth of arcobacters and strains of *A. skirrowii* were recovered only on nonselective blood agar. The reason for the failure to recover any of the *Arcobacter* strains on mCCDA could be sensitivity to the inhibitory substance(s) present in mCCDA and/or the synergistic inhibitory effect of the combination of inhibitors (cefoperazone and sodium deoxycholate). The *A. skirrowii* strains were sensitive to deoxycholate and this probably explains why they grew in CAT broth (which does not contain deoxycholate) but were isolated only from blood agar and not mCCDA or CAT agar (Atabay et al., 1998a).

MAINTENANCE PROCEDURES

Cultures may be stored for many years by lyophilization, by freezing at −80°C, or by storage in liquid nitrogen. Cryoprotective agents such as 10% glycerol or DMSO should be added to cultures before freezing, and heavy cell concentrations should be used. Grown cultures can be kept at room temperature in air but should be transferred weekly.

DIFFERENTIATION OF THE GENUS *ARCOBACTER* FROM OTHER GENERA

Classical biochemical differentiation of the genus *Arcobacter* from related genera such as *Campylobacter* and *Helicobacter*, is primarily achieved via the identification of the individual species. However, in general, arcobacters can be differentiated from campylobacters by their lower optimal growth temperatures (25–30°C compared to 30–42°C for campylobacters) and aerotolerance. Analysis of respiratory quinones enables a clear differentiation of *Arcobacter* strains from *Campylobacter* strains but is of limited value in a routine diagnostic laboratory.

TAXONOMIC COMMENTS

In 1985, Neill et al. (1985) performed an exhaustive phenotypic analysis of a large number of aerotolerant *Campylobacter* strains. Without insight into the phylogenetic and genotypic relationships of the organisms, these authors concluded that the group of aerotolerant campylobacters was more distantly related to any of the existing *Campylobacter* species than these species were to each other and emphasized that the aerotolerant campylobacters formed a heterogeneous group. These conclusions were confirmed by rRNA-based phylogenetic analyses and by integrated phenotypic, genotypic, and chemotaxonomic studies of, in part, the same isolates (Thompson et al., 1988; Kiehlbauch et al., 1991a; Vandamme et al., 1991a, 1992b). As a result, the genus *Arcobacter* was separated from the *Campylobacter* genus and three animal-associated species were described: *A. cryaerophilus*, *A. butz-*

leri, and *A. skirrowii* (Kiehlbauch et al., 1991a; Vandamme et al., 1991a, 1992b).

Few taxonomic problems are left. The phylogeny of these bacteria is well known and the relationships among the four named species have been examined by DNA–DNA hybridization experiments and by a variety of different chemotaxonomic and genotypic methods (see below). Two genotypic and chemotaxonomic subgroups were described within *A. cryaerophilus* (Kiehlbauch et al., 1991a; Vandamme et al., 1992b) but were not formally named pending the availability of simple diagnostic tests. Novel strains, some of which remained uncultured, that cluster phylogenetically among other *Arcobacter* strains have been described (Teske et al., 1996; Voordouw et al., 1996; Snaidr et al., 1997). Their relationships toward named *Arcobacter* species remain to be established.

ACKNOWLEDGMENTS

FED and BJP were supported by NIH Grants DE10374 and DE11443.

FURTHER READING

On, S.L. 1996. Identification methods for campylobacters, helicobacters, and related organisms. Clin. Microbiol. Rev. *9*: 405–422.

Ursing, J.B., H. Lior and R.J. Owen. 1994. Proposal of minimal standards for describing new species of the family *Campylobacteraceae*. Int. J. Syst. Bacteriol. *44*: 842–845.

Vandamme, P., E. Falsen, R. Rossau, B. Hoste, P. Segers, R. Tytgat and J. De Ley. 1991. Revision of *Campylobacter*, *Helicobacter*, and *Wolinella* taxonomy: emendation of generic descriptions and proposal of *Arcobacter* gen. nov. Int. J. Syst. Bacteriol. *41*: 88–103.

Vandamme, P., M. Vancanneyt, B. Pot, L. Mels, B. Hoste, D. Dewettinck, L. Vlaes, C. Van Den Borre, R. Higgins, J. Hommez, K. Kersters, J.-P. Butzler and H. Goossens. 1992. Polyphasic taxonomic study of the emended genus *Arcobacter* with *Arcobacter butzleri* comb. nov. and *Arcobacter skirrowii* sp. nov., an aerotolerant bacterium isolated from veterinary specimens. Int. J. Syst. Bacteriol. *42*: 344–356.

Wesley, I.V. 1996. *Helicobacter* and *Arcobacter* species: risks for foods and beverages. J. Food Prot. *59*: 1127–1132.

DIFFERENTIATION OF THE SPECIES OF THE GENUS *ARCOBACTER*

Biochemical characteristics useful in distinguishing the four named species of the genus *Arcobacter* are listed in Table BXII.ε.3.

Whole-cell protein electrophoresis (Vandamme et al., 1992b) and numerical analysis of 67 phenotypic characters (On and Holmes, 1995) correlated with percentage of DNA–DNA hybridization and were able to differentiate all species. Whole-cell fatty acid analysis was not able to distinguish *A. butzleri* from *A. cryaerophilus* subgroup 2, but differentiated all other *Arcobacter* taxa (Vandamme et al., 1992b).

16S rRNA-based DNA probes (Wesley et al., 1995) and PCR assays (Harmon and Wesley, 1997), ribotyping (Kiehlbauch et al.,

1991b), and restriction fragment length polymorphism analysis of a PCR-amplified fragment of the gene coding for 16S rRNA (Cardarelli-Leite et al., 1996) or 23S rRNA (Hurtado and Owen, 1997a) differentiated *A. butzleri* from other arcobacters but did not differentiate *A. cryaerophilus* from *A. skirrowii*. Using a variable 23S rRNA region, Bastyns et al. (1995a) developed genus- and species-specific PCR assays that differentiated *A. butzleri*, *A. cryaerophilus*, and *A. skirrowii*. Finally, Snaidr et al. (1997) described a set of genus-specific PCR primers that enabled detection of the four named species and an unnamed *A. nitrofigilis*-like taxon.

List of species of the genus Arcobacter

1. **Arcobacter nitrofigilis** (McClung, Patriquin and Davis 1983) Vandamme, Falsen, Rossau, Hoste, Segers, Tytgat and De Ley 1991a, 100[VP] (*Campylobacter nitrofigilis* McClung, Patriquin and Davis 1983, 610.)

   ni.tro.fig' i.lis. L. n. *nitrum* nitrate; L. v. *figo* to fix; L. adj. suff. *ilis* able to; M.L. adj. *nitrofigilis*, able to fix (nitrogen as) nitrate.

   Cells are small curved rods, 0.2–0.9 × 1–3 μm. Rapid corkscrew-like motility by means of a single polar flagellum. Occasionally longer, helical cells are observed. Coccoid forms develop in old cultures.

   Colonies of 48 h cultures are 2–3 mm in diameter, circular and convex, and have a beige to off-white color.

   Grows in media containing 2–70 g/l NaCl. Optimal growth with 10–40 g/l NaCl. Nitrogenase activity. Sulfide production from cysteine. A brown, water-soluble pigment is produced from tryptophan. Indole negative. Does not

   hydrolyze esculin. Incapable of growth in the presence of 1% glycine. Does not grow at 37°C under aerobic or microaerobic conditions. Grows on media containing 0.05% sodium fluoride, but not on charcoal-based media, or media containing 1.0– 2.0% ox bile. Sensitive to nalidixic acid (32 mg/l), cephalothin (32 mg/l), cefoperazone (64 mg/l), and carbenicillin (32 mg/l).

   Isolated from the roots and root-associated sediments of *Spartina alterniflora* growing in marshes with approximately oceanic salinity. Fixes nitrogen.

   The mol% G + C of the DNA is: 28–29 (T_m).
   Type strain: C1, CCUG 15893, LMG 7604.
   GenBank accession number (16S rRNA): L14627.

2. **Arcobacter butzleri** (Kiehlbauch, Brenner, Nicholson, Baker, Patton, Steigerwalt and Wachsmuth 1991a) Vandamme, Vancanneyt, Pot, Mels, Hoste, Dewettinck, Vlaes, Van Den Borre, Higgins, Hommez, Kersters, Butzler and

TABLE BXII.ε.3. Differential characteristics of the species of the genus *Arcobacter*[a]

Characteristic	Arcobacter nitrofigilis	Arcobacter butzleri	Arcobacter cryaerophilus	Arcobacter skirrowii
Urease	+	−	−	−
α-Hemolysis	−	F	−	+
4% NaCl	+	−	−	+
Cefoperazone, 64 mg/l	−	+	+	+
Growth on minimal medium	−	+	−	−
MacConkey agar	−	+	−	−

[a]Symbols: +, 95–100% strains positive; −, 0–11% strains positive; F, 14–50% strains positive; M, 60–93% strains positive. All data are based on reactions obtained by using recommended, standardized procedures (On and Holmes 1991, 1992, 1995).

Goossens 1992b, 355^VP (*Campylobacter butzleri* Kiehlbauch, Brenner, Nicholson, Baker, Patton, Steigerwalt and Wachsmuth 1991a, 384.)

butz'ler.i. N.L. gen. n. *butzleri* of Butzler, in honor of Jean-Paul Butzler, a Belgian microbiologist who was one of the first scientists to emphasize the importance of *Campylobacter* infections in humans.

Curved cells, 0.2–0.4 × 1–3 µm. Rapid darting motility. Motile by means of a single polar flagellum at one or both ends of the cell.

Colonies grown for 48 h under optimal conditions are 2–4 mm in diameter, convex with an entire edge. Bacterial growth has a whitish appearance. Strains will grow on charcoal-based and potato starch media, and on media containing 0.001% sodium arsenite, 32 mg/l carbenicillin, 64 mg/l cefoperazone, and 0.05% sodium fluoride. Most (91–96%) strains reduce nitrate and grow on, and reduce, 0.04% triphenyl-tetrazolium chloride. Indoxyl acetate is hydrolyzed.

Isolated from human blood and diarrheic feces; from feces of various animals with diarrhea including nonhuman primates, pigs, horses, cattle, an ostrich, and a tortoise; from bovine and porcine abortions; from various food products including ground pork, chicken, and turkey samples; from surface and drinking water reservoirs and canal waters. Associated with enteritis, abdominal cramps, bacteremia, and appendicitis in humans; enteritis and abortion in animals.

The mol% G + C of the DNA is: 28–31 (T_m).

Type strain: CDC D2686, CCUG 30485, LMG 10828.

3. **Arcobacter cryaerophilus** (Neill, Campbell, O'Brien, Weatherup and Ellis 1985) Vandamme, Falsen, Rossau, Hoste, Segers, Tytgat and De Ley 1991a, 100^VP (*Campylobacter cryaerophilus* Neill, Campbell, O'Brien, Weatherup and Ellis 1985, 354.)

cry.ae.ro'phi.lus. Gr. n. *cruos* cold; Gr. n. *aero* air; Gr. n. *philos* friend; L. adj. *cryaerophilus* friend of cold and air.

Curved cells, 0.2–0.4 × 1–3 µm. Rapid darting motility. Motile by means of a single polar flagellum at one or both ends of the cell.

Colonies grown for 48 h under optimal conditions are 1–3 mm in diameter, convex with an entire edge. Mostly with a dirty yellow pigment. Strains grow on nutrient agar; most (93–100%) strains grow on charcoal-based and potato starch media. Strains grow on media containing 0.001% sodium arsenite, 32 mg/l cephalothin, 100 U/l 5-fluorouracil, and 0.02% safranin. Strains vary (66%) in their ability to reduce nitrate.

Two subgroups, referred to as subgroup 1 or group 1A,

and subgroup 2 or group 1B, have been described (Kiehlbauch et al., 1991a; Vandamme et al., 1992b, respectively). These subgroups differ in their whole-cell protein and fatty acid patterns, and in the restriction fragment length polymorphisms of the ribosomal RNA genes. The DNA–DNA hybridization levels within these subgroups ranged from 56 to 100%, while values from 46 to 69% were found among strains of the different subgroups. These subgroups were not given formal names pending the availability of simple diagnostic tests. On et al. (1996) observed that most strains of subgroup 2 could grow under aerobic or microaerobic conditions at 37°C, in contrast with subgroup 1 strains, but this result seems to be highly method dependent.

Isolated from cases of human bacteremia and diarrhea; from chicken carcasses; from bovine, ovine, and porcine abortions; from porcine feces, and from mastitis in cattle. Pathogenicity is unknown.

The mol% G + C of the DNA is: 28–30 (T_m).

Type strain: Neill A 169/B, CCUG 17801, LMG 7536; CDC D2610.

GenBank accession number (16S rRNA): L14624.

Additional Remarks: LMG 10829 is a well-characterized reference strain of subgroup 2.

4. **Arcobacter skirrowii** Vandamme, Vancanneyt, Pot, Mels, Hoste, Dewettinck, Vlaes, Van Den Borre, Higgins, Hommez, Kersters, Butzler and Goossens 1992b, 355^VP

skir.ro'wi.i. N.L. gen. n. *skirrowii* of Skirrow, in honor of Martin B. Skirrow, a British microbiologist who was the first to describe a simple isolation method for *Campylobacter jejuni* obtained from stool specimens, which enabled most laboratories to routinely culture the organism.

Curved cells, 0.2–0.4 × 1–3 µm. Rapid darting motility. Motile by means of a single polar flagellum.

Colonies grown for 48 h under optimal conditions are 2–3 mm in diameter, and have a flat irregular shape. Bacterial growth has a grayish appearance and is not usually as profuse as in other *Arcobacter* spp. Strains grow on charcoal-based and potato starch media, and on media containing up to 2.0% ox-bile, and 32 mg/l cephalothin and 64 mg/l cefoperazone. Nitrate is reduced. Most strains (73%) are resistant to 0.001% sodium arsenite. No growth is observed in the presence of 0.05% basic fuchsin.

Isolated from preputial fluids of bulls; from bovine, ovine, and porcine abortions; from diarrheic feces of various animals with diarrhea including sheep, cattle; from chicken carcasses. Pathogenicity is unknown.

The mol% G + C of the DNA is: 29–30 (T_m).

Type strain: Skirrow 449/80, CCUG 10374, LMG 6621.

GenBank accession number (16S rRNA): L14625.

Other Organisms

Voordouw et al. (1996) examined oil field communities by cloning and sequencing of PCR-amplified 16S rRNA genes. Apart from a variety of Gram-negative, sulfate-reducing bacteria, a limited number of potential sulfide oxidizers and/or microaerophiles, including arcobacters, were detected. An organism most closely related to *A. nitrofigilis* was the source of the most frequently detected bacterial sequence from oil field DNA analyzed by the culture-independent approach.

Teske et al. (1996) examined the bacterial composition of a coculture capable of sulfate reduction after exposure to oxic and

micro-oxic conditions. This coculture was derived from cyanobacterial mats from Solar Lake (Sinai) that were introduced to an experimental hypersaline pond. PCR amplification of 16S ribosomal DNA fragments from the coculture, analyzed by denaturing gradient gel electrophoresis, resulted in two distinct 16S ribosomal DNA bands, one of which was derived from an *Arcobacter* strain. Molecular identification of the two components of this coculture allowed the design of specific culture conditions to separate and isolate both strains in pure culture. Curved to spiral rods, ~1–3 µm in length, characteristic for *Arcobacter* and related bacteria, were observed. The organism grew aerobically, microaerobically, and anaerobically on agar plates and used lactate, formate, malate, acetate, and glutamate as electron donors, but not succinate, fumarate, citrate, glycolate, methanol, ethanol, glucose, or fructose. Nitrate, fumarate, and oxygen, but not sulfate, sulfite, or thiosulfate, were used as electron acceptors. Its nearest relative was *A. nitrofigilis*. The GenBank accession number of the 16S rRNA gene sequence of this strain is L42994.

Snaidr et al. (1997) investigated the bacterial community structure of activated sludge of a large municipal wastewater treatment plant by cloning and sequencing of PCR-amplified 16S rRNA genes. Representatives of the genus *Arcobacter* were present in 4% of the activated sludge samples examined. These uncultured organisms were observed in sludge samples by *in situ* hybridization with fluorescently labeled oligonucleotide probes and had the typical *Arcobacter* morphology: slender, curved to spiral rods, 0.8×2–3 µm. Analysis of 16S rRNA gene fragments (some of which were nearly complete) revealed *A. cryaerophilus* as the closest neighbor.

Genus III. **Sulfurospirillum** Schumacher, Kroneck, and Pfennig 1993, 188[VP] (Effective publication: Schumacher, Kroneck and Pfennig 1992, 291) emend. Finster, Liesack and Tindall 1997d, 1216

JOHN F. STOLZ, RONALD S. OREMLAND, BRUCE J. PASTER, FLOYD E. DEWHIRST AND PETER VANDAMME

Sul.fu.ro.spi.ril' lum. L. n. *sulfur* sulfur; Gr. n. *spira* a spiral; M.L. neut. n. *Sulfurospirillum* a spirillum that reduces elemental sulfur.

Slender, vibrioid to spirally curved rods, 0.1–0.5 × 1.0–3 µm. May form helical chains of two or more cells. Motile by polar flagella. Gram negative. Nonsporeforming. **Microaerobic, with respiratory oxidation of succinate, fumarate, malate, lactate, or pyruvate, but not acetate.** Oxidase positive. Carbohydrates are not oxidized or fermented. **Reduces S⁰. Fumarate may be fermented under anoxic conditions.** Free living, found in freshwater and marine environments. Temperature growth range from 8 to 36°C. Based on 16S rRNA gene sequence comparisons, *Sulfurospirillum* falls in the family *Campylobacteraceae*, which belongs in the *Epsilonproteobacteria*.

Menaquinone-6, a methyl-substituted menaquinone-6 (referred to as thermoplasmaquinon-6), and **menaquinone-5** have been reported as respiratory quinones in *Sulfurospirillum deleyianum* (Collins and Widdel, 1986).

The mol% G + C of the DNA is: 32–42.

Type species: **Sulfurospirillum deleyianum** Schumacher, Kroneck, and Pfennig 1993, 188 (Effective publication: Schumacher, Kroneck and Pfennig 1992, 292.)

FURTHER DESCRIPTIVE INFORMATION

The members of the genus *Sulfurospirillum* are microaerobic sulfur reducers but exhibit metabolic versatility. *S. barnesii* has the distinction of being able to utilize selenate and, along with *S. arsenophilum*, arsenate as electron acceptors. This versatility is reflected in protein content. Zöphel et al. (1988, 1991) have identified NiFe hydrogenase, menaquinone (MK-6), a flexirubin-like pigment, fumarate reductase, and sulfur oxidoreductase in *S. deleyianum*. A pentaheme cytochrome *c* nitrate reducatase has also been purified from *S. deleyianum* (Schumacher and Kroneck 1991; Schumacher et al., 1992, 1994) and its crystal structure determined (Einsle et al., 1999). Finster et al. (1997d) detected the presence of *b* and *c* type cytochromes in *S. arcachonense*. Stolz et al. (1997) showed that protein composition, cytochrome content, and enzyme activity were affected by the electron acceptor on which cells of *S. barnesii* were grown. At least two types of cytochrome *c* (including nitrite reductase) and two types of cytochrome *b* were found associated with membrane fractions. Enzyme activities for nitrate, thiosulfate, fumarate, and selenate reduction were induced by their respective substrates (Stolz et al., 1997). Arsenate reductase activity was constitutive in washed cell suspensions (Laverman et al., 1995) and membrane fractions (D.K. Newman, personal communication). The respiratory arsenate reductase from *S. barnesii* has been purified and characterized (D.K. Newman personal communication).

ENRICHMENT AND ISOLATION PROCEDURES

There is no simple formulation for the isolation and culture of *Sulfurospirillum* species, since both freshwater and marine strains have been described and different enrichment strategies were used for their isolation. The type species, *S. deleyianum*, was isolated from an anaerobic enrichment culture for *Desulfuromonas* in which acetate was the electron donor and S⁰ was the acceptor (Wolfe and Pfennig, 1977). *S. arcachonense* was enriched in medium formulated for formate-utilizing sulfur reducers (Finster et al., 1997d). This species has an obligate requirement for sodium and magnesium chloride. *S. barnesii* was isolated from enrichments using agar plates containing a minimal salts medium with 20 mM acetate and 20 mM selenate, which were incubated in the presence of hydrogen (Oremland et al., 1994). The pure culture was subsequently grown with lactate as the electron donor. *S. arsenophilum* was enriched for by using arsenate as the terminal electron acceptor (Ahmann et al., 1994).

The medium for routine growth of *S. barnesii* as modified from Oremland et al. (1994) in Stolz et al. (1997), contains (per liter): K_2HPO_4, 0.225 g; KH_2PO_4, 0.225 g; NaCl, 0.46 g; $(NH_4)_2SO_4$, 0.225 g; $MgSO_4 \cdot 7H_2O$, 0.117 g; yeast extract, 1 g; sodium lactate, 2.24 g; $NaHCO_3$, 4.2 g; $Na_2S \cdot 9H_2O$, 0.1 g; trace element solution, 10 ml; vitamin solution, 10 ml; plus the addition of the electron acceptor such as Na_2SeO_4, 3.78 g. Cysteine-HCl (0.1 g/l) may be added as a reducing agent (Oremland et al., 1994). The pH is adjusted to 7.3 prior to autoclaving. The vitamin solution, lactate, reducing agent, and bicarbonate may be filter sterilized and added separately (Oremland et al., 1994). The trace element solution contains (per liter): nitrilotriacetic acid, 1.5 g; $MgSO_4 \cdot 7H_2O$, 3 g; $MnSO_4 \cdot H_2O$, 0.5 g; NaCl, 1.0 g; $FeSO_4 \cdot 7H_2O$, 0.1 g; $CaCl_2 \cdot 2H_2O$, 0.1 g; $CoCl_2 \cdot 6H_2O$, 0.1 g; ZnCl, 0.13 g; $CuSO_4 \cdot 2H_2O$, 0.01 g; $AlK(SO_4)_2 \cdot 12H_2O$, 0.01 g; H_3BO_3, 0.01 g, $Na_2MoO_4 \cdot 2H_2O$, 0.025 g; $NiCl_2 \cdot 6H_2O$, 0.024 g; and $Na_2WO_4 \cdot 2H_2O$, 0.025 g. The vitamin mixture contains (per liter): biotin, 2 mg; folic acid, 2 mg; pyridoxine-HCl, 10 mg; riboflavin, 5 mg; thiamine, 5 mg; nicotinic acid, 5 mg; pantothenic acid, 5 mg; B$_{12}$, 0.1 mg; *p*-aminobenzoic acid, 5 mg; and thioctic acid, 5 mg. The liquid medium is dispensed in crimped septum-sealed tubes or bottles (125 ml) and degassed with N$_2$/CO$_2$ (80:20) for 5 minutes. The headspace is then degassed for 1 minute and the vessels sealed, crimped, and autoclaved.

MAINTENANCE PROCEDURES

Cells can be preserved by lyophilization (e.g., ATCC stocks) or in glycerol stocks at $-20°C$.

DIFFERENTIATION OF THE GENUS *SULFUROSPIRILLUM* FROM OTHER GENERA

Members of the genera *Arcobacter* and *Campylobacter* are closely related to species of *Sulfurospirillum*. The members of the genus *Sulfurospirillum* are free-living, nonpathogenic, sulfur respirers. They are further distinguished from *Arcobacter* in that all four species grow at 25°C but don't grow at oxygen concentrations above 4%. Although they share many physiological characteristics with *Campylobacter*, it is clear from phylogenetic analysis that they form their own distinct clade (Fig. BXII.ε.4).

TAXONOMIC COMMENTS

Comparative 16S rRNA gene sequence analysis has shown that members of the genus *Sulfurospirillum* form a distinct clade separate from species of the genera *Campylobacter* and *Arcobacter* (Finster et al., 1997d; Stolz et al., 1999). The *Sulfurospirillum* species fall in the family *Campylobacteraceae*, which belongs in the *Epsilonproteobacteria* (see Fig. BXII.ε.4). DNA–DNA hybridization studies confirmed that *S. barnesii*, *S. arsenophilum*, and *S. deleyianum* are separate species with DNA relatedness levels between the species ranging from 31 to 53% (Stolz et al., 1999).

ACKNOWLEDGMENTS

J.F. Stolz was supported by EPA Grant G71A0007. P. Vandamme is indebted to the Fund for Scientific Research-Vlaanderen (Belgium) for a position as a postdoctoral research fellow. F.E. Dewhirst and B.J. Paster were supported by NIH Grants DE10374 and DE11443.

FURTHER READING

Finster, K., W. Liesack and B.J. Tindall. 1997. *Sulfurospirillum arcachonense* sp. nov., a new microaerophilic sulfur-reducing bacterium. Int. J. Syst. Bacteriol. *47*: 1212–1217.

Laanbroek, H.J., W. Kingma and H. Veldkamp. 1977. Isolation of an aspartate fermenting, free-living *Campylobacter* species. FEMS Microbiol. Lett. *1*: 99–102.

Oremland, R.S., J.S. Blum, C.W. Culbertson, P.T. Visscher, L.G. Miller, P. Dowdle and F.E. Strohmaier. 1994. Isolation, growth, and metabolism of an obligately anaerobic, selenate-respiring bacterium, strain SES-3. Appl. Environ. Microbiol. *60*: 3011–3019.

Stolz, J.F., D.J. Ellis, J.S. Blum, D. Ahmann, D.R. Lovley and R.S. Oremland. 1999. *Sulfurospirillum barnesii* sp. nov. and *Sulfurospirillum arsenophilum* sp. nov., new members of the *Sulfurospirillum* clade of the Epsilon *Proteobacteria*. Int. J. Syst. Bacteriol. *49*: 1177–1180.

Wolfe, R.S. and N. Pfennig. 1977. Reduction of sulfur by spirillum 5175 and syntrophism with *Chlorobium*. Appl. Environ. Microbiol. *33*: 427–433.

DIFFERENTIATION OF THE SPECIES OF THE GENUS *SULFUROSPIRILLUM*

Characteristics useful in distinguishing the four named species of the genus *Sulfurospirillum* are listed in Table BXII.ε.4. *S. barnesii* can be distinguished from *S. deleyianum* by its ability to respire arsenate and selenate, pH and salinity optima, use of lactate as an electron donor, and the inability to use sulfite as an electron acceptor. *S. arsenophilum* is the smallest of the species and couples the reduction of arsenate to the complete oxidation of lactate to CO$_2$. It does not, however, respire selenate. *S. arcachonense* is the sole marine strain, has a lower mol% G + C content of the DNA, and respires only oxygen and elemental sulfur.

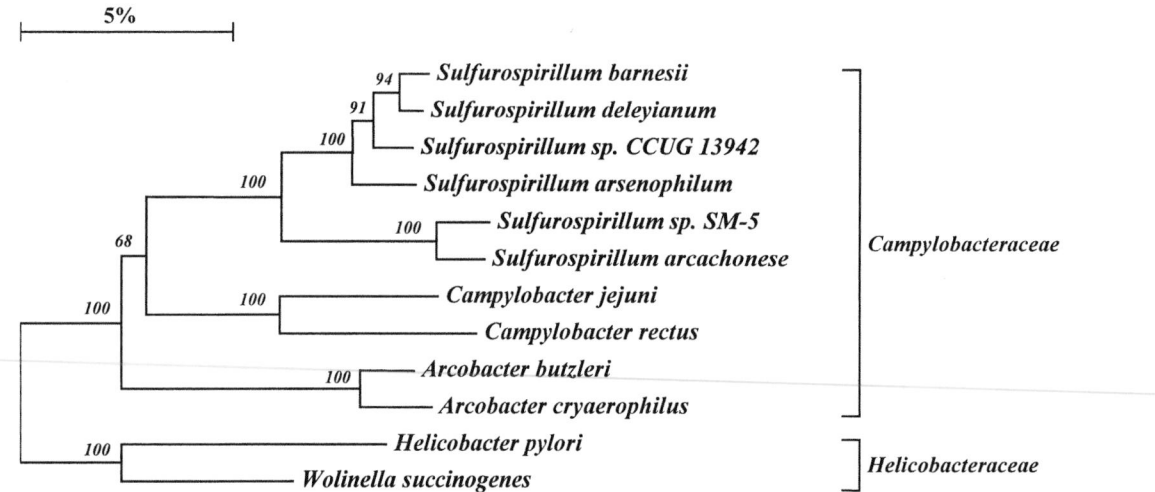

FIGURE BXII.ε.4. The phylogenetic relationships of the species of the genus *Sulfurospirillum* compared to other members of the family *Campylobacteraceae* and family *Helicobacteraceae*. Bar = 5% difference in 16S rDNA nucleotide sequences. The Neighbor-Joining method was used for tree construction. One hundred bootstrap trees were generated, and bootstrap confidence levels (shown as percentages above the nodes) were determined.

TABLE BXII.ε.4. Differentiation of known species of *Sulfurospirillum*[a]

Characteristics	*S. deleyianum*	*S. arcachonense*	*S. arsenophilum*	*S. barnesii*
Width, μm	0.3–0.5	0.3	0.1–0.3	0.3
Mol% G + C (HPLC)	40.6	32	40.9	40.8
Growth at:				
25°C	+	+	+	+
30°C	+	+	+	+
42°C	–	–	–	–
pH optimum	7.0–7.1	7–7.4	nd	7.5
NaCl, %	<0.2	1.2–2	0.1	0.8
Oxidase	+	+	nd	nd
Catalase	–	+	nd	nd
Electron acceptors:				
Nitrate	+	–	+	+
Nitrite	+	–	+	+
Thiosulfate	+	–	+	+
TMAO	+	–	nd	+
Fe(III)	nd	–	nd	+
Sulfite	+	–	nd	+
DMSO	+	–	nd	–
Arsenate	–	nd	nd	nd
Selenate	–	nd	+	+
Elemental sulfur	+	+	–	+
Oxygen (microaerobic)	+	+	+	+
Electron donors:				
Lactate	–	+	+	+
Pyruvate	+	nd	+	+
Fumarate	+	+	+	+
Formate	+	nd	+	+
Fermentation:				
Fumarate	+	nd	+	+
Malate	+	–	nd	nd
Growth with 1% glycerol	–	+	nd	nd

[a]For symbols see standard definitions; nd, not determined.

List of species of the genus Sulfurospirillum

1. **Sulfurospirillum deleyianum** Schumacher, Kroneck and Pfennig 1993, 188[VP] (Effective publication: Schumacher, Kroneck and Pfennig 1992, 292.)

   de.ley.ia' num. M.L. neut. n. *deleyianum* named after J. De Ley, for his contributions to genetic analysis in bacterial systematics.

   Slender, vibrioid to spirally curved rods, 0.3–0.5 × 1.0–3 μm. Colonies in deep agar cultures are lens-shaped with a yellowish color due to the presence of a flexirubin-type pigment. Anoxic growth occurs with hydrogen or formate as electron donor; acetate and hydrogen carbonate as the carbon source; and nitrate, nitrite, sulfite, thiosulfate, S[0], fumarate, malate, and aspartate as electron acceptors. Sulfate is not reduced. S[0] is reduced to sulfide. Nitrate and nitrite are reduced to ammonia. Fumarate and malate can be fermented.

   Microaerobic growth occurs with 1–4% oxygen, utilizing succinate, fumarate, malate, aspartate, pyruvate, oxoglutarate, and oxaloacetate as substrates. Glycerol and acetate are not oxidized.

   Exogenous source of reduced sulfur needed. No vitamins or amino acids required. Grows at temperatures from 20 to 36°C, but not at 42°C.

   Isolated from anoxic mud from a forest pond near Heiningen, Germany.

   The mol% G + C of the DNA is: 38.4 (T_m), 40.6 (HPLC).

   Type strain: Spirillum 5175, ATCC 51133, DSM 6946.

   GenBank accession number (16S rRNA): Y13671.

2. **Sulfurospirillum arcachonense** Finster, Liesack and Tindall 1997d, 1216[VP]

   ar.ca.cho.nen' se. M.L. adj. *arcachonense* pertaining to the city of Arcachon, France, from where the strain was isolated.

   Curved cells 0.3 × 1–2.5 μm. Colonies in deep agar are white, round, and filamentous when grown on S[0]. Colonies are lens-shaped and yellowish-brown when grown with oxygen or fumarate. Temperature range is 8–30°C with optimal growth at 26°C. Grows with 0.6–4% NaCl and at least 0.1% $MgCl_2 \cdot 6H_2O$ is required.

   Microaerobic growth occurs with up to 15% oxygen when agitated. Oxidase positive. Oxygen and S[0] can be used as an electron acceptor. Formate, acetate, proprionate, succinate, alanine, fumarate, lactate, α-ketoglutarate, citrate, malate, and yeast extract are oxidized with oxygen. Hydrogen, formate, lactate, pyruvate, α-ketoglutarate, glutarate, and yeast extract can be oxidized with S[0] as the electron acceptor.

   Fatty acid composition $C_{14:0}$ (2.7%), $C_{14:1}$ (3.2%), $C_{16:0}$ (31.0%), $C_{16:1}$ (40.8%), $C_{18:0}$ (20.8%), and $C_{18:1}$ (0.6%), with $C_{15:0}$, $C_{17:0}$, and $C_{14:0\ 3OH}$ in trace amounts. Polar lipids are phosphatidylglycerol, phophatidylethanolamine, and lysophosophatidylethanolamine.

   Isolated from oxidized surface sediment in an intertidal mud flat near Archachon, France.

   The mol% G + C of the DNA is: 32 (HPLC).

   Type strain: F1F6, DSM 9755.

   GenBank accession number (16S rRNA): Y11561.

3. **Sulfurospirillum arsenophilum** Stolz, Ellis, Switzer Blum, Ahmann, Lovley and Oremland 1999, 1179[VP]

ar.sen.o' phi.lum. L. v. *arsenicum* yellow pigment of gold; Gr. comb. form *philos* loving; N.L. adj. *arsenophilum* arsenic loving.

Vibrioid to spiral-shaped cells, 0.3 × 1–2 µm. Motile by single polar flagellum. Distinguished by the ability to respire arsenate. Grows at 20°C and as high as 30°C but not at 37°C or above. Grows in the presence of 0.1% NaCl but 1% NaCl is inhibitory. Hydrogen and formate can serve as electron donors when acetate is used as the carbon source. Lactate, pyruvate, and fumarate can be used as electron donors. Arsenate, nitrate, and fumarate can serve as terminal electron acceptors. Nitrate is reduced to ammonia. Can ferment fumarate. Isolated from arsenic-contaminated watershed sediments in eastern Massachusetts, USA.

The mol% G + C of the DNA is: 40.9 (HPLC).

Type strain: MIT-13, ATCC 700056, DSM 10659.

GenBank accession number (16S rRNA): U85964.

4. **Sulfurospirillum barnesii** Stolz, Ellis, Switzer Blum, Ahmann, Lovley and Oremland 1999, 1179[VP]

barnes' i.i. M.L. gen. n. *barnesii* of Barnes, named after I. Barnes, of the U.S. Geological Survey, who recognized the environmental significance of selenium.

Vibrioid to spiral-shaped cells, 0.3 × 1–2 µm. Motile by polar flagella. Growth optimal 33°C, pH 7.5, and 0.05% NaCl. Selenate, arsenate, thiosulfate, S^0, trimethylamine oxide, Fe(III), nitrate, fumarate, aspartate, and manganese dioxide can all be used as a terminal electron acceptor when grown under anoxic conditions with lactate as the electron donor. Capable of microaerobic growth. Selenate is reduced through selenite to elemental selenium. Nitrate is reduced to ammonium. Lactate and H_2 can serve as electron donors. When grown with hydrogen, acetate is required as the carbon source.

The periplasmic nitrate reductase (NapA) from *S. barnesii* shares a 72% identity and 91% similarity with the NapA homolog from *Campylobacter jejuni*.

SES-3 is sensitive to ampicillin (10 µg/disk), carbenicillin (100 µg/disk), tetracycline (30 µg/disk), and erythromycin (30 µg/disk), showing complete inhibition at those concentrations, but only moderately sensitive to chloramphenicol (30 µg/disk) and naladixic acid (30 µg/disk), with only a small zone of inhibition at the given concentrations.

Isolated from a selenium-contaminated Massie Slough, western Nevada, USA.

The mol% G + C of the DNA is: 40.8 (HPLC).

Type strain: SES-3, ATCC 700032, DSM 10660.

GenBank accession number (16S rRNA): AF038843 and U41564.

Other Organisms

A marine sulfur-reducing bacterium, designated strain SM-5, was isolated by Coleman et al. (1993) and implicated as the organism involved in the formation of ferrous nodules. The 16S rRNA sequence (U85965) of SM-5 indicated that it is the fifth member of the genus *Sulfurospirillum* (Lonergan et al., 1996). It has yet to be fully characterized.

A free-living strain, *Sulfurospirillum* sp. DSM 806, (CCUG 13942), that is capable of aspartate fermentation, was isolated from activated sludge (Laanbroek et al., 1977; Laanbroek and Veldkamp, 1979) and has been reassigned as a strain of *S. deleyianum*. However, phylogenetic analysis of the 16S rRNA gene sequence of this strain indicates that it may represent a separate species, most closely related to *S. deleyianum*. Growth between 15°C and 41°C, with an optimal growth temperature of 37°C.

The mol% G + C of the DNA is: 41.5 (T_m).

Family II. **Helicobacteraceae** fam. nov.

GEORGE M. GARRITY, JULIA A. BELL AND TIMOTHY LILBURN

He.li.co.bac.ter.a' ce.ae. M.L. masc. n. *Helicobacter* type genus of the family; -aceae ending to denote family; M.L. fem. pl. n. *Helicobacteraceae* the *Helicobacter* family.

The family *Helicobacteraceae* was circumscribed for this volume on the basis of phylogenetic analysis of 16S rRNA sequences; the family contains the genera *Helicobacter* (type genus), *Thiovulum*, and *Wolinella*.

Microaerophilic or anaerobic. Morphologically, metabolically, and ecologically diverse. Includes human and animal pathogens.

Type genus: **Helicobacter** Goodwin, Armstrong, Chilvers, Peters, Collins, Sly, McConnell, Harper 1989a, 403,[VP] emend. Vandamme, Falsen, Rossau, Hoste, Segers, Tytgat and De Ley 1991a, 100.

Genus I. **Helicobacter** Goodwin, Armstrong, Chilvers, Peters, Collins, Sly, McConnell, Harper 1989a, 403,[VP] emend. Vandamme, Falsen, Rossau, Hoste, Segers, Tytgat and De Ley 1991a, 100

STEPHEN L.W. ON, ADRIAN LEE, JANI L. O'ROURKE, FLOYD E. DEWHIRST, BRUCE J. PASTER, JAMES G. FOX AND PETER VANDAMME

He.li.co.bac'ter. Gr. n. *helix* a spiral; M.L. masc. n. *bacter* a staff; M.L. masc. n. *Helicobacter* a spiral rod.

Cells may be **curved, spiral, or fusiform rods**, 0.2–1.2 × 1.5–10.0 μm. **Spiral cells may be tightly or loosely wound** depending on the species, and on the age and condition of the culture examined. Cells in old cultures or those exposed to air become coccoid. Ultrastructural studies show that **periplasmic fibers may be observed on the cell surface** of a few taxa and an electron-dense glycocalyx- or capsule-like layer has been observed on the cell surface of some species. Nonsporeforming. Gram negative. **Motile with a rapid corkscrew- or slower wave-like motion** due to flagellar activity. **Multiple sheathed flagella are seen in most species**, frequently with a bipolar distribution. Typical cell morphologies of various *Helicobacter* species are shown in Figs. BXII.ε.5 and BXII.ε.6.

In the laboratory, **strains grow under microaerobic conditions at 37°C** and show a respiratory type of metabolism. **Chemoorganotrophs. Asaccharolytic when sugar catabolism is tested by standard method**; however, recent studies have indicated that **glucose oxidation occurs in at least one species (*H. pylori*)** (Chalk et al., 1994; Tomb et al., 1997). Gelatin, starch, casein, and tyrosine are not hydrolyzed. Methyl red- and Voges–Proskauer negative. **Oxidase activity in all species.** Strains of most species produce catalase. **Many species produce urease and/or alkaline phosphatase.** Pigments are not produced; bacterial growth of some species demonstrates a blue-gray hue when cultured on blood agar. **Many species are pathogenic for humans and animals. Found in the intestinal tract, oral cavity, and internal organs of man and animals.**

Internal transcribed spacers or intervening sequences have been described in the 16S ribosomal RNA genes of *H. canis* (Linton et al., 1994b), *H. typhlonius* (Franklin et al., 2001), and *H. bilis* (Fox et al., 1995). Intervening sequences have been described in the 23S rRNA gene of *H. canis*, *H. mustelae*, and *H. muridarum* (Hurtado et al., 1997). The length of intervening elements in the 16S rRNA gene ranges from 166 bp (*H. typhlonius*) to 235 bp (*H. canis*), with those found in the 23S rRNA gene ranging from 93 bp (23S rRNA spacer, *H. mustelae*) to 377 bp (*H. muridarum*).

The mol% G + C of the DNA is: 24–48.

Type species: **Helicobacter pylori** (Marshall, Royce, Annear, Goodwin, Pearman, Warren, Armstrong 1984) Goodwin, Armstrong, Chilvers, Peters, Collins, Sly, McConnell, Harper 1989a, 403 (*Campylobacter pylori* Marshall, Royce, Annear, Goodwin, Pearman, Warren, Armstrong 1984, 87.)

FURTHER DESCRIPTIVE INFORMATION

The genus *Helicobacter* currently comprises 23 validly published species. Two candidate (*candidatus*) species have also been proposed. At least 10 other taxa have been designated as potentially novel species. The closest phylogenetic relatives of the genus *Helicobacter* include *Wolinella* and *Thiovulum* spp., and members of the family *Campylobacteraceae* (encompassing the genera *Campylobacter, Arcobacter, Sulfurospirillum,* and the generically misnamed *Bacteroides ureolyticus*). *Helicobacter* is phylogenetically diverse in comparison with the aforementioned genera, although

most species associated with the gastric tract of different animals (*H. pylori, H. nemestrinae, H. acinonychis, H. felis, H. bizzozeronii,* and *H. salomonis*) form a distinct subcluster (Fig. BXII.ε.7). Significant levels of DNA–DNA hybridization (11–52%) have been reported between (i), *H. pylori* and *H. mustelae* (Fox et al., 1989); (ii), *H. cinaedi, H. fennelliae,* and *H. canis* (Stanley et al., 1993b); and (iii) *H. felis, H. bizzozeronii,* and *H. salomonis* (Jalava et al., 1997). The DNA–DNA relatedness of *H. acinonychis, H. pametensis, H. cholecystus, H. muridarum, H. hepaticus, H. trogontum, H. bilis, H. aurati, H. canadensis, H. ganmani, H. mesocricetorum, H. typhlonius,* and *H. rodentium* to other species has not been determined, but numerical analysis of whole-cell protein profiles show *H. acinonychis* and *H. pylori* to be closely related (Costas et al., 1993).

Electron micrographs of thin sections of *H. pylori, H. mustelae, H. felis, H. muridarum,* and *H. rodentium* cells show a distinct electron-dense glycocalyx-like or capsular structure external to the cell wall unit membrane (Goodwin et al., 1989a; Paster et al., 1991; Lee et al., 1992a; Shen et al., 1997). In these species, the glycocalyx is up to 40 nm thick and has a radial periodicity of approximately 14 nm. Periplasmic fibers may be coiled around the cells of *H. felis, H. muridarum, H. bilis, H. trogontum, H. aurati,* and "*H. rappini*". In *H. muridarum* the structure of the fibers has been determined as circular and hollow with a diameter of 28–32 nm. Electron-dense granular bodies have been observed in *H. pylori* (Bode et al., 1993) and *H. rodentium* (Shen et al., 1997). In *H. pylori* these bodies are known to be aggregates of polyphosphate and may serve as a reserve energy source.

Helicobacter species have a respiratory metabolism. By means of conventional bacteriological methods, neither oxidation nor fermentation of carbohydrates is observed. However, detailed studies of metabolic pathways in *H. pylori* have demonstrated catabolic pathways for glucose, the sole carbohydrate source. The glycolysis-gluconeogenesis metabolic axis probably comprises the principal means of energy production as well as the starting point for many biosynthetic pathways. In addition, glucose may also be oxidized by means of the Entner–Doudoroff pathway (Chalk et al., 1994; Tomb et al., 1997); enzymes for the pentose phosphate pathway are also present but may be involved principally in biosynthesis. Metabolic pathways for pyruvate and the amino acids serine, alanine, arginine, and glutamine (Chalk et al., 1994; Mendz and Hazell, 1995; Tomb et al., 1997) are also extant. Sequence analysis of the whole genome of *H. pylori* indicates that the tricarboxylic acid (TCA) cycle is incomplete, the glyoxylate shunt is absent, and that several substrates, including urea and ammonia, serve as important nitrogen sources for *H. pylori* (Tomb et al., 1997). The characteristically high degree of urease production in this species probably serves both as a mechanism for survival in the acid conditions of the stomach and as a means of providing nitrogen for essential cell functions (Chalk et al., 1994). Proton translocation is mediated by dehydrogenase NDH-1, and various cytochromes. Four respiratory electron-generating dehydrogenases have also been identified.

All validated *Helicobacter* spp. grow in a microaerobic (3–7% O_2) atmosphere. No growth is observed in aerobic conditions;

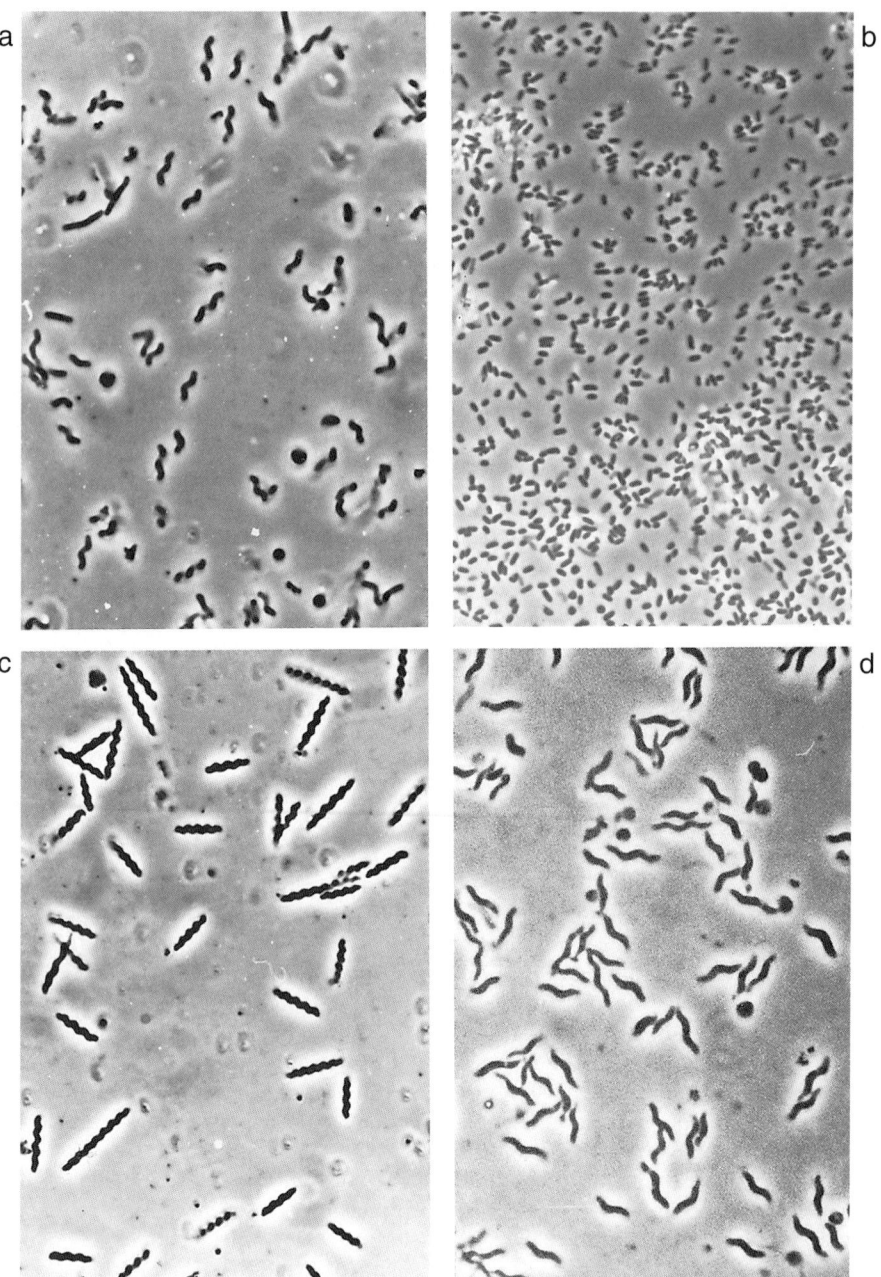

FIGURE BXII.ε.5. Phase contrast micrographs showing various morphologies of the *Helicobacter* spp. (*a*) Rod to S-shape of *H. pylori*; (*b*) short rod shape of *H. mustelae*; (*c*) tight helical shape of *H. felis*; and (*d*) S-shape of *H. muridarum*. (× 1000). (Micrographs courtesy of J. O'Rourke and A. Lee.)

some strains of certain species grow, usually poorly, in anaerobic conditions. *H. pylori*, *H. mustelae*, *H. nemestrinae*, and *H. acinonychis* do not require atmospheric hydrogen for growth in microaerobic conditions. The remaining species have all been cultured under microaerobic conditions containing hydrogen, and it is not known if this is an essential atmospheric component (as with certain *Campylobacter* species) or simply enhances growth on culture media. The metabolism of pyruvate by *H. pylori* reflects its microaerophilic nature since neither of the enzymes used in aerobic, or strictly anaerobic, processes are present. Instead, pyruvate is converted to acetyl CoA by pyruvate ferrodoxin oxidoreductase, previously found only in hyperthermophilic organisms. Oxidase activity is detected in all *Helicobacter* species; *H.*

pylori is known to contain cytochromes *b* and *c* but not *a* or *d* types. Most helicobacters produce catalase. Superoxide dismutase is present in *H. pylori*.

Major cellular fatty acids reported for *Helicobacter* species include tetradecanoic acid ($C_{14:0}$); most species contain moderate-to-high levels of hexadecanoic acid ($C_{16:0}$) and also octadecanoic acid ($C_{18:1}$) (Lambert et al., 1987; Goodwin et al., 1989a; Haque et al., 1996). The presence of 19-carbon cyclopropane fatty acid ($C_{19:0 \text{ cyclo}}$) is highly species dependent but appears to be associated principally with species for which the natural ecological niche is the gastric mucosa (Haque et al., 1996). The polar lipid profile of many species is characterized by the presence of relatively high levels of various cholesteryl glucosides (between 9.7

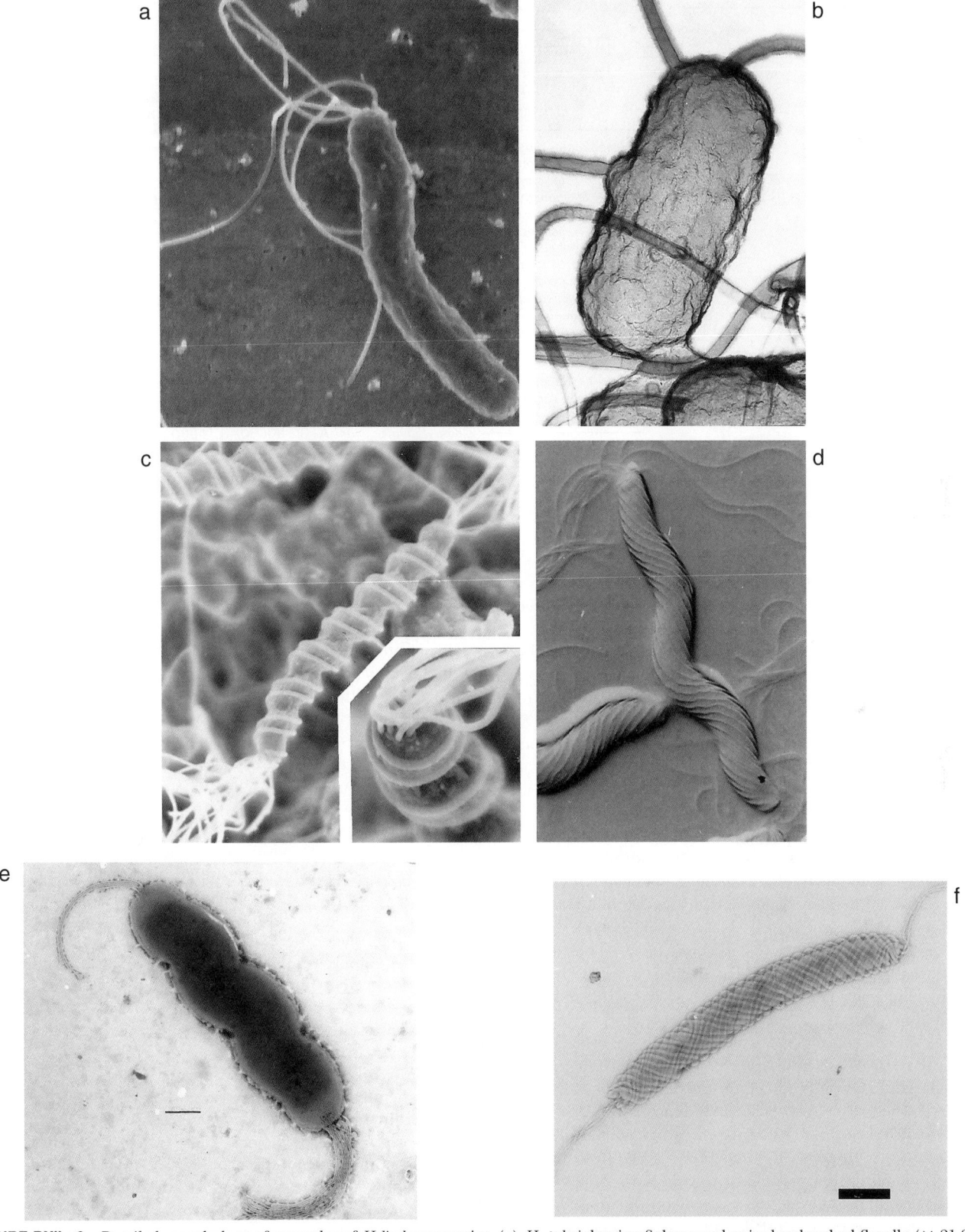

FIGURE BXII.ε.6. Detailed morphology of examples of *Helicobacter* species. (*a*) *H. pylori* showing S-shape and unipolar sheathed flagella (× 21,000). (*b*) *H. mustelae* showing rod shape with polar and lateral flagella (× 26,000) (Reprinted with permission from Lee and O'Rourke. *Helicobacter pylori* Biology and Clinical Practice, CRC Press, Boca Raton, 1993). (*c*) *H. felis* showing helical shape with tufts of bipolar flagella and periplasmic fibrils in pairs entwining the bacterium (*Inset*: terminal end showing detail of flagella insertion) (× 14,000 and × 38,000). (*d*) *H. muridarum* showing S-shape with bipolar flagella and periplasmic fibrils completely entwining the bacterium (× 14,000) (Reprinted with permission from Lee et al., International Journal of Systematic Bacteriology, *42*: 27–36, 1992, ©International Union of Microbiological Societies). (*e*) *H. salomonis* showing loosely helical, enlarged cell body with multiple bipolar flagella (× 6,000; bar = 1 μm). (*f*) *H. bilis* showing fusiform cell body encircled by tightly wound periplasmic fibers and multiple sheathed flagella (Bar = 0.5 μm). (*a* and *c*, critical point dried preparations viewed with a field emission SEM; *b* and *d*, freeze-dried preparations viewed by TEM; *e* and *f*, negatively stained cell preparations viewed by TEM). Micrographs *a–d* (Courtesy of J.L. O'Rourke, L. Thompson, M.W. Phillips, P. Marks, M. Dickson and A. Lee;) *e*, (Courtesy of K. Jalava and M.-L-Hänninen;) *f*, (Courtesy of B. Paster.)

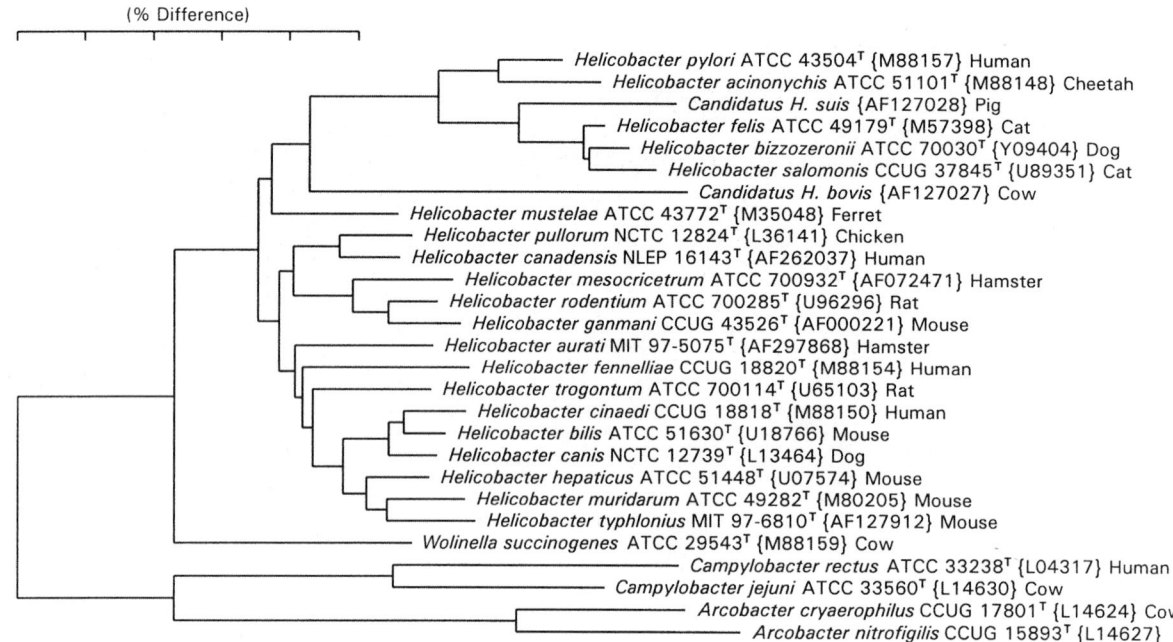

FIGURE BXII.ε.7. Phylogenetic relationships among *Helicobacter* species and close taxonomic relatives as evaluated by 16S rRNA gene sequence comparisons. Bar = 5% difference in nucleotide sequences as determined by measuring the lengths of the horizontal lines connecting two species.

and 33.1% total lipid content); these compounds have not been detected in *H. cinaedi* or *H. pametensis* (Haque et al., 1995, 1996). The isoprenoid quinone content of *H. pylori*, *H. cinaedi*, and *H. fennelliae* has been determined and includes menaquinone-6 (2-methyl-3-farnesyl-farnesyl-1,4-naphthoquinone) and an unidentified quinone (un-MK-6) as major respiratory quinones (Moss et al., 1990b).

Plasmids in the size range of 3.7 to >148 kb may be found in 35–58% of *H. pylori* strains (Tjia et al., 1987; Graham et al., 1988; Penfold et al., 1988). The plasmids do not contain any known virulence determinant. Kiehlbauch et al. (1995) demonstrated plasmids in the size range 1.8–45.0 kb in 14 of 42 strains of *H. cinaedi*, 1 of 5 strains of *H. fennelliae*, and one *H. cinaedi/H. fennelliae*-like strain from humans and animals. Of 14 *H. felis* strains tested, all contained more than one plasmid, and these ranged in size from 2 to >16 kb (Jalava et al., 1999a).

A wide range of phenotypic and genotypic methods for epidemiological typing of *H. pylori* has been described. These include biotyping; serotyping; whole-cell protein electrophoretic typing; plasmid profiling; restriction enzyme analysis of DNA with frequent- or rare-cutting (the latter resolved by pulsed-field gel electrophoresis, PFGE) restriction enzymes; ribotyping; restriction fragment length polymorphism (RFLP) analysis of DNA using species-specific probes; PCR-RFLP analysis of urease, vacuolating toxin (*vac*A), and adhesin genes; and other PCR-based fingerprinting methods using repetitive (REP-PCR), random (RAPD), or restriction halfsite-specific (AFLP) sequences as primers (Morgan and Owen, 1990; Owen and Desai, 1990; Simor et al., 1990; Costas et al., 1991; Foxall et al., 1992; Taylor et al., 1992a; Desai et al., 1993; Li et al., 1993; Evans et al., 1995; Forbes et al., 1995; Vandamme et al., 1995b; Simoons-Smit et al., 1996; Shortridge et al., 1997; Gibson et al., 1998; Salaun et al., 1998; van Doorn et al., 1998a, b). Application of the genetic methods in particular has demonstrated that *H. pylori* is a highly diverse species; the same genotype is usually only present within mem-

bers of a single family. The extensive genetic diversity in *H. pylori* is due to a range of genetic phenomena, including point mutations, genomic rearrangements, and transformation, and a panmitic population genetic structure which has been proposed for the organism (Salaun et al., 1998). By contrast, *H. mustelae* (the species closest to *H. pylori* in terms of DNA relatedness) is genetically highly conserved, with few types discernible by biotyping, ribotyping, or RFLP analysis of genomic DNA with frequent- or rare-cutting enzymes (Morgan and Owen, 1990; Owen and Desai, 1990; Taylor et al., 1994b). *H. hepaticus* strains can be differentiated by PFGE analysis, with most strains from the United States demonstrating less diversity than European strains (Saunders et al., 1997). In addition, PCR-RFLP of the urease gene identifies just three types, indicating this gene to be highly conserved in *H. hepaticus* (Shen et al., 1998). PFGE analysis of 24 *H. felis* strains distinguished up to 20 distinct types; strains found to be indistinguishable were isolated from different animals at distinct times. Similar results for *H. felis* were obtained with plasmid profiling and ribotyping (Jalava et al., 1999a). *H. cinaedi* and *H. fennelliae* strains from humans, hamsters, cats, dogs, and macaque monkeys are differentiated by ribotyping, with strains from different hosts clearly distinct (Kiehlbauch et al., 1995).

Helicobacter species may be associated with three types of disease, generally related to their natural ecological niche. Of the species found in the gastric mucosa, *H. pylori* is well established as the leading microbiological cause of gastric and duodenal ulcers in humans (Dunn et al., 1997). *H. pylori* infection is also associated with the development of atrophic gastritis, and subsequent progression to gastric cancer; MALT lymphoma and adenocarcinoma of the distal stomach; and Ménétrièr's disease, a rare disorder in which the gastric folds become hypertrophic. An association of *H. pylori* with nonulcer dyspepsia has been suggested but as yet remains unproven. The O-specific chain of the lipopolysaccharide in *H. pylori* mimics Lewis blood group antigens of the host and may serve as an immunoprotective "camouflage" mechanism or may contribute directly to the pathogenesis of autoimmune disease. The CagA protein is another major antigenic component in many strains. CagA is 120–140 kDa in size and serves as a marker for the *cag* pathogenicity island, which is ca. 35–40 kb in size. Although the function of the CagA protein and most of the *cag* pathogenicity island has not yet been determined, *cagA*-positive strains have been correlated with an increased risk of development of duodenal ulcers and adenocarcinoma of the distal stomach (Dunn et al., 1997).

Of the other gastric species, *H. mustelae*, *H. acinonychis*, *H. felis*, and *H. bizzozeronii* have all been observed in natural and/or experimental infections in host animals showing signs of gastric disease (Lee et al. 1993; Skirrow, 1994; Jalava et al., 1998). *H. felis* has been occasionally described as a possible cause of human gastritis (Germani et al., 1997; also see comments in *H. felis* species description), while a cultured strain of "*Helicobacter heilmannii*" (Andersen et al., 1996), a taxon associated with approximately 1% of human gastritis cases (Dunn et al., 1997), has been identified as *H. bizzozeronii* (Jalava et al., 2001). Experimental infection of mice with *H. muridarum* (a urease-producing species normally found in murine intestinal mucosa) also induces atrophic gastritis in aging animals (Lee et al., 1993).

Several species associated with the lower gastrointestinal tract (including *H. pullorum*, *H. fennelliae*, *H. cinaedi*, *H. canis*, and "*H. rappini*") have been isolated from cases of gastroenteritis in humans and/or animals (Skirrow, 1994; On, 1996); severe diarrhea in laboratory mice co-infected with *H. bilis* and *H. rodentium* has been noted (Shen et al., 1997). Inflammatory bowel disease in immunodeficient rats has been observed with natural and/or experimental infections using *H. bilis* and *H. typhlonius* (Haines et al., 1998; Fox et al., 1999). In immunocompromised human patients, complications such as septicemia and meningitis may occur. These enteric helicobacters are infrequently reported as agents of disease in humans, but their incidence could be significantly underestimated because of inadequate isolation and identification procedures (Skirrow, 1994; On, 1996).

Other enteric species (*H. hepaticus*, *H. cholecystus*, *H. bilis*, some strains of *H. pullorum*, *H. canis*, and "*H. rappini*") have been isolated from hepatitis, cholangiofibrosis, and pancreatitis lesions in host animals (Kirkbride et al., 1985; Fox et al., 1994b, 1996a; Stanley et al., 1994b; Franklin et al., 1996). PCR-based analysis of bile and gallbladder samples from humans afflicted with chronic cholecystitis indicated the presence of *H. bilis*, *H. pullorum*, and "*H. rappini*" (Fox et al., 1998a). A soluble factor toxic to liver cells has been demonstrated in strains of *H. hepaticus*, *H. felis*, *H. acinonychis*, *H. pylori*, and *H. mustelae*, but not in *H. muridarum* or "*H. rappini*" (Taylor et al., 1995). This hepatotoxic activity may represent different cytotoxic factors produced by these species with activity against liver, and other cells.

Enrichment and Isolation Procedures

There is no ideal method for the isolation of all *Helicobacter* species, irrespective of the sample under consideration. Recovery is optimized by the concurrent use of selective and nonselective methods to examine the freshest possible samples. In particular, for species colonizing the gastric or intestinal mucosae, it is recommended that specimens be processed within 1 hour of sampling (Hänninen et al., 1996; Jalava et al., 1998). Fresh mucosal scrapings, or tissue biopsy material crushed in a small volume of suitable medium (brain heart infusion (BHI) broth + serum is often used), have proven to be suitable for the isolation of species from gastric or intestinal mucosae or extragastrointestinal tissue such as liver. Enteric species may be recovered from fecal samples

and/or carcass washings by direct culture methods. Enrichment media commonly used to enhance the recovery of thermophilic *Campylobacter* species rarely have a similar effect on enteric helicobacters. *H. pametensis* was isolated by examining feces within 3 hours of collection (Seymour et al., 1994), and *H. pullorum* was only recovered from fresh, not chilled or stored, chicken carcass washings (Atabay et al., 1998b).

Selective media are frequently based upon the antibiotic supplement first described by Skirrow (1977) for the isolation of *C. jejuni* and *C. coli*, which includes vancomycin, trimethoprim, and polymixin B; amphotericin B may also be included to further restrict the growth of fungi. It is known that *H. pullorum* and some *H. pylori* strains cannot be cultured in the presence of polymixin B (Dent and McNulty, 1988; Atabay et al., 1998b). Cefsulodin has been used to replace polymixin B in some media for isolating *H. pylori* (Dent and McNulty, 1988; Tee et al., 1991). Most enteric species can be recovered on media incorporating cefoperazone, amphotericin B, and vancomycin (Burnens et al., 1993).

A range of solid media has been evaluated for isolating *H. pylori* (Hachem et al., 1995). BHI and *Brucella* agars supplemented with 5–10% blood are often used for isolating this, and other *Helicobacter* species. However, *Brucella* agar does not support the growth of *H. bizzozeronii* (Jalava et al., 1998), and certain formulations contain bisulfite compounds that inhibit the growth of *H. pylori* (Hawrylik et al., 1994). Nonselective growth media may be inoculated directly with the test sample, but this procedure normally results in an overgrowth of contaminating microflora. Instead, the sample may be passed through membrane filters with a pore size of 0.45, 0.65, or 0.8 µm onto the medium of choice. All plate cultures must be incubated for 3–10 d at 37°C under microaerobic conditions typically containing $CO_2/N_2/O_2/H_2$ (5:88:5:2). The latter enhances, and is sometimes essential for, helicobacterial growth. The isolation of several species, notably *H. bizzozeronii* and *H. salomonis*, has proven to be especially problematic. Special recommendations for the growth of these species, including the periodic addition of nutrient broth to plate cultures to inhibit surface dehydration (Jalava et al., 1998) have been made.

Maintenance Procedures

Stock cultures of *Helicobacter* species can be maintained under microaerobic conditions by transfer onto common blood agar bases every 4–7 d. Cultures may be stored for many years by freezing at −80°C, or in liquid nitrogen. Cryoprotective agents such as 10% glycerol or DMSO should be added to cultures before freezing, and heavy cell concentrations should be used. Many species can also be stored as lyophilized samples in 5% inositol serum, but more suitable methods for preserving fastidious species (*H. felis*, *H. muridarum*, *H. bizzozeronii*, *H. salomonis*) by this method require development.

Taxonomic Comments

The formation of the genus *Helicobacter* from the reclassification of *Campylobacter pylori* and *C. mustelae* by Goodwin et al. (1989a) reconciled substantial phenotypic and phylogenetic differences observed between these two species and members of the genera *Campylobacter* and *Wolinella*. The subsequent transfer of *C. cinaedi* and *C. fennelliae* to *Helicobacter* by Vandamme et al. (1991a) was also made on the basis of polyphasic taxonomic data and established the ecological divergence of the genus. Since this time, the classification of *Helicobacter* has been based principally on phylogenetic relationships as determined by analyses of the 16S rRNA gene. In phylogenetic terms, the genus is readily distinguished from other members of the *Epsilonproteobacteria*. Moreover, the natural ecological niche of helicobacters appears to be fairly conserved (the gastric tract or lower intestine), but considerable diversity in the morphological characteristics and DNA base ratios of different species is observed.

It is feasible to propose that *Helicobacter* and the monotypic genus *Wolinella* be included in a single bacterial family. Phylogenetic analyses based upon both $T_{m(e)}$ values of DNA–rRNA hybrids and sequence comparisons of the 16S rRNA gene consistently demonstrate that *Helicobacter* and *Wolinella* are at least as closely related as members of the family *Campylobacteraceae* (*Campylobacter, Arcobacter, Sulfurospirillum*, and the generically misnamed *Bacteroides ureolyticus*) (see Fig. BXII.ε.10; Vandamme et al., 1991a). The phenotype, morphology, natural habitat and DNA base ratio of *Wolinella* and *Helicobacter* appear similarly compatible, given the diversity in the latter genus.

Phylogenetic distinctions aside, there are presently no absolute defining features for the genus *Helicobacter* that distinguish it from the genus *Campylobacter*. Early studies indicated that flagellar sheathing and resistance to polymyxin B were consistent features and served to differentiate helicobacters from campylobacters (Han et al., 1989; Burnens and Nicolet, 1993). It is now clear that certain species (*H. pullorum, H. rodentium, H. canadensis, H. ganmani*, and *H. mesocricetorum*) give atypical results in at least one of these unifying characteristics. Other chemotaxonomic markers (cellular fatty acid composition, presence of steryl glucosides) similarly serve as useful distinguishing features for most, but not all, species, and there is a need for phenotypic characters that accurately reflect the overall phylogenetic coherence of the genus.

Elucidation of the whole genome sequence of *H. pylori*, combined with the intensive and wide-ranging studies made on this medically important organism, demonstrated its ability to metabolize glucose. This was a surprising discovery, since by all conventional methods no such activity has been detected. This finding may clarify, or further confound, the issue of generic consistency. It is not yet known whether similar metabolic pathways for glucose utilization are extant in other *Helicobacter* species. If present, then the characteristic would serve as an important taxonomic marker for the genus; if absent in certain, or all other, species, then the characteristics of the genus become even more divergent. It is however significant that comparison of the inferred metabolic pathways for *C. jejuni* from whole-genome sequence analysis (*C. jejuni* sequencing group at the Sanger Center, 1998) with those for *H. pylori* (Tomb et al., 1997) indicate that these phylogenetically related, but distinct bacteria share several important features, including elements of the TCA cycle and gluconeogenesis pathways. Thus, it is possible that all helicobacters, and not just *H. pylori*, have the ability to metabolize glucose. Furthermore, comparison of the whole-genome sequences of *C. jejuni* and *H. pylori* suggests salient differences in the metabolic pathways of the two species. For example, the Entner–Doudoroff pathway is absent in *C. jejuni*, and only *C. jejuni* appears able to operate a complete tricarboxylic acid cycle (Kelly, 2001). Such features may eventually prove useful characteristics for distinguishing these two distinct, but highly related genera.

There are a number of outstanding nomenclatural and taxonomic problems, principally at the species level. Strains described as *"Flexispira rappini"* are clearly *Helicobacter* species by phylogenetic analysis, and several workers already refer to these bacteria as *"H. rappini"*. However, the taxonomic status of *"H. rappini"* requires clarification. The name *"Flexispira rappini"* was first proposed by Bryner et al. (1986) for a group of ultrastructurally distinct, urease-producing strains isolated from aborted lambs, dog, and human feces, and subsequently from pig intestines (Bryner et al., 1986; Bryner, 1987). Archer et al. (1988) examined human, canine, and ovine strains using biochemical tests and whole-cell protein analysis and concluded all strains were similar, although some distinction between the ovine strain and fecal isolates was made. Since these first studies, strains have been identified as *"H. rappini"* by 16S rRNA gene sequence comparisons, despite notable morphologic and/or phenotypic differences (Tee et al., 1998). Such results may be due, in part, to errors made in the 16S rRNA gene sequence determination for a human strain often used as a study reference (ATCC 43879 CCUG 23435, GenBank no. M88137) as a consequence of strain mislabeling. This error has since been corrected by one of us (F.E. Dewhirst, personal communication). A subsequent study using the corrected sequence of M88137 (=M88138), and that of a different pig strain (ATCC 43968, GenBank no. U96300), shows that these strains are 95.6% similar by 16S rRNA gene sequence comparison and assigns these strains to radically different branches in a phylogenetic analysis (Shen et al., 1997). Considerable diversity in the 16S rRNA gene sequence of 36 strains exhibiting *"H. rappini"*-like morphology has since been observed (Dewhirst et al., 2000a). In addition, *"H. rappini"*-like strains from canine gastric mucosa resemble, but are phenotypically different from CCUG 23435, including their protein electrophoresis profiles (Jalava et al., 1998). Moreover, *H. bilis, H. aurati*, and *H. trogontum* bear a striking ultrastructural and phenotypic resemblance to *"H. rappini"* strains studied by Bryner et al. (1986), Bryner (1987), and Jalava et al. (1998). A polyphasic investigation of the relationships between all these strains is required to clarify the taxonomic position and status of this complex group. Available data strongly suggest that urease-positive bacteria with the complex, fusiform ultrastructure first associated with this taxon represent more than one species, of which three (*H. bilis, H aurati*, and *H. trogontum*) have standing in nomenclature. However, the phylogenetic classification of many *"H. rappini"* strains, along with the formal description of *"H. rappini"* for one of the major groups observed (Dewhirst et al., 2000a), will provide a useful basis for examining interstrain relationships in more detail.

Similar nomenclatural and taxonomic problems exist for the group of gastric bacteria often referred to as *"Gastrospirillum"* spp. and/or *"H. heilmannii"* (hereafter referred to as gastrospirilla). These organisms are principally characterized by their cell morphology (large, tightly coiled rods) and such cells have been observed in gastric biopsies of many hosts including humans, cats, dogs, pigs, monkeys, rats, and various captive exotic carnivores (Salomon, 1898; McNulty et al., 1989; Mendes et al., 1990; Jakob et al., 1997; Jalava et al., 1998). Gastrospirilla are also typified by their highly fastidious growth requirements, and much of the uncertainty concerning their taxonomic status has arisen because most investigators have failed to culture the strains *in vitro*. Phylogenetic analyses clearly identify all gastrospirilla as helicobacters, with names proposed for pig (*"Gastrospirillum suis"*, Candidatus Helicobacter suis), lemur (*"G. lemur"*), and human (either *"G. hominis"*, of which two phylogenetically distinct types have been described, or *"H. heilmannii"*) strains (McNulty et al. 1989; Mendes et al., 1990; Dewhirst et al., 1992; Solnick et al., 1993). The taxonomic status of these strains as distinct species has been largely assumed by the position of their 16S rDNA sequences in phylogenetic trees. However, studies of successfully cultured gastrospirilla-like strains from cats, dogs, and a human, indicate complex taxonomic relationships among these strains. *H. felis*, *H. bizzozeronii*, and *H. salomonis* represent a group of highly related species from domestic pets, with 16S rRNA gene sequences showing greater variation between strains of a single species than comparable differences between strains of different species (Jalava et al., 1998). Subsequent phylogenetic analysis clearly identifies these validly described species as being related to the two sequences representing "*G. hominis*", although the latter are somewhat distinct from the *H. felis*–*H. bizzozeronii*–*H. salomonis* group. Nevertheless, a study examining phenotypic characters, protein profiles, and DNA relatedness of a human strain initially described as "*H. heilmannii*" (Andersen et al., 1996) has identified the isolate as *H. bizzozeronii* (Jalava et al., 1999c). As the former name has no standing in nomenclature, this human isolate should be named *H. bizzozeronii*. It should not however be assumed that all human gastrospirilla are in fact *H. bizzozeronii*. The 16S rRNA gene sequence of "*H. heilmannii*" type 1 (Solnick et al., 1993) is 95.9–97.0% similar to those of the *H. felis*–*H. bizzozeronii*–*H. salomonis* group and "*H. heilmannii*" type 2 (Solnick et al., 1993), but 99.5% similar to "*G. suis*" from pigs (Mendes et al., 1994). The relatively low level of phylogenetic similarity between "*H. heilmannii*" type 1 and the *Helicobacter* spp. from cats and dogs suggests that *H. bizzozeronii* may represent only part of the human strains. The high degree of sequence relatedness between "*H. heilmannii*" type 1 and "*G. suis*" certainly indicates an additional animal reservoir. Nevertheless, although available data suggest that human gastrospirilla may represent more than one species, further polyphasic taxonomic studies on cultured gastrospirilla from each of the sources in which they have been observed, are required to clarify the taxonomic status of these bacteria.

In the case of many recently described or putative *Helicobacter* species, considerable weight has been placed on the position of 16S rRNA gene sequences of representative strains in phylogenetic analyses. Although this method is presently unsurpassed for determining broad evolutionary relationships, its efficacy in delineating taxa at or below the species level is now known to be limited (Stackebrandt and Goebel, 1994; Clayton et al., 1995). Recent studies on *H. bizzozeronii*, *H. salomonis*, and *H. felis* have shown that these taxa cannot be unequivocally separated by 16S

rRNA gene sequence analysis (Jalava et al., 1997) although their status as discrete, albeit closely related, species is substantiated by phenotypic and genotypic data (Jalava et al., 1997, 1998). The identification of *Helicobacter* sp. strain Mainz and "*H. westmeadii*" as *H. cinaedi* isolates (Vandamme et al., 2000), and of a "*H. heilmannii*" isolate as *H. bizzozeronii* (Jalava et al., 1999c), is similarly relevant, since these taxa have been postulated as potentially novel species on the basis of 16S rRNA gene sequence data. It is impossible to predict whether the taxonomic status of other helicobacters for which extensive polyphasic characterization has not been performed will be similarly amended. It is however clear that descriptions of putative new species should consider the present view that interstrain relationships should be circumscribed by means of a polyphasic taxonomic analysis. The provision of a set of minimal standards for describing new *Helicobacter* species will also prove valuable for future descriptions (Dewhirst et al., 2000b).

Since the initial revision of this chapter, *Candidatus* Helicobacter bovis and *Candidatus* H. suis have been described (De Groote et al., 1999a, b). These are noteworthy since they represent the first descriptions of *Helicobacter* species assigned to the provisional taxonomic status *Candidatus*. The designation *Candidatus* was first proposed by Murray and Schleifer (1994) to record the properties of putative new species of procaryotes that cannot be cultured at present under standard laboratory conditions. This proposal was recommended for implementation by the International Committee on Systematic Bacteriology (ICSB) and formal guidelines for its practice published (Murray and Stackebrandt, 1995). The provisional *Helicobacter* species listed were described in accordance with these guidelines.

ACKNOWLEDGMENTS

We gratefully thank all those who allowed the citation of data that were, at the time of writing, unpublished. K. Jalava (University of Helsinki, Finland) is also thanked for the electron micrograph of *H. salomonis*.

FURTHER READING

Dunn, B.E., H. Cohen and M.J. Blaser. 1997. *Helicobacter pylori*. Clin. Microbiol. Rev. *10*: 720–741.

On, S.L.W. 1996. Identification methods for campylobacters, helicobacters, and related organisms. Clin. Microbiol. Rev. *9*: 405–422.

Solnick, J.V. and D.B. Schauer. 2001. Emergence of diverse *Helicobacter* species in the pathogenesis of gastric and enterohepatic diseases. Clin. Microbiol. Rev. *14*: 59–97.

DIFFERENTIATION OF THE SPECIES OF THE GENUS *HELICOBACTER*

Phenotypic characteristics useful for differentiating the validly described *Helicobacter* species are given in Table BXII.ε.5. It should be noted that *H. pullorum* bears a close resemblance to certain *Campylobacter* species (notably *C. lari*), with which it also shows similarities in host range and disease associations. Several pairs of species, notably *H. pylori* and *H. acinonychis*, and *H. felis* and *H. bizzozeronii*, cannot be differentiated with conventional phenotypic tests.

A variety of different methods has been applied to identify *Helicobacter* species (reviewed by On, 1996). Of the phenotypic methods described thus far, standardized whole-cell protein profile analysis has proven to be consistently effective in species identification (Costas et al., 1993; Jalava et al., 1998). Of the

genetic methods described to date, restriction profile analysis of PCR amplicons derived from the 23S rRNA gene (Hurtado and Owen, 1997b) was shown to differentiate 13 species (including "*H. rappini*") but could not unequivocally discriminate *H. felis*, *H. bizzozeronii*, and *H. salomonis* from each other (Jalava et al., 1999b). Specific oligonucleotide probes and/or PCR assays for *H. pametensis* (and the related, but unnamed taxa *Helicobacter* bird-B and bird-C), *H. pylori*, *H. hepaticus*, *H. pullorum*, *H. bilis*, *H. trogontum*, and *H. canis* (Dewhirst et al., 1994; Fox et al., 1995; Mendes et al., 1996; On, 1996) have been described. It should be noted that, due to constant developments in the taxonomy of *Helicobacter*, none of the aforementioned genetic methods have been fully evaluated for all species thus far described.

TABLE BXII.ε.5. Characteristics which differentiate *Helicobacter* species[a]

Characteristic	H. pylori	H. acinonychis	H. aurati	H. bilis	H. bizzozeronii	H. canadensis	H. canis	H. cholecystus	H. cinaedi	H. felis	H. fennelliae	H. ganmani	H. hepaticus	H. mesocricetorum	H. muridarum	H. mustelae	H. nemestrinae	H. pametensis	H. pullorum	H. rodentium	H. salomonis	H. trogontum	H. typhlonius
Catalase production	+	+	+	+	+	+	−	+	+	+	+	−	+	+	+	+	+	+	+	+	+	+	−
Nitrate reduction	−	−	−	+	+	V	−	+	+	+	−	+	+	+	−	+	−	+	+	+	+	+	V
Alkaline phosphatase hydrolysis	+	+	−	−	V	−	+	+	D	V	D	−	−	+	+[b]	+	+	+	−	−	v	−	−
Urease	+	+	+	+	+					+			+		+	+	+				+	+	
Indoxyl acetate hydrolysis	−	D	+	−	D	+	+	−	−	D	+	−	+	nd	+[b]	+	−	−	−	−	−	−	−
γ-Glutamyl transpeptidase	+	+	+	+	+	−	+	−	−	+	−	nd	−	−	+	+	nd	−	nd	−	+	+	−
Growth at 42°C	D	D	+	+	V	+	+	+	D	V	D	−	−	+	−	V	+	+	+	+	−	+	+
Tolerance to 1.0% glycine	−	−	−	+	−	+	−	+	−	−	−	+	−	−	−	−	−	V	−	+	−	nd	+
Tolerance and reduction of 0.04% TTC	D	−	nd	nd	−	nd	V	+	+	−	+	+	+	nd	−	+	+	V	D	nd	−	+	nd
Periplasmic fibers	−	−	+	+	−	−	−	−	−	+	−	−	−	−	+	−	−	−	−	−	−	+	−
Sheathed flagella	+	+	+	+	+	−	+	+	+	+	+		+	+	+	+	+	+	−	−	+	+	+
Number of flagella	4–8	2–5	7–10	3–14	10–20	1–2	2	1	1–2	14–20	2	2	2	2	10–14	4–8	4–8	2	1	2	10–23	5–7	2
Distribution of flagella	B	B	B	B	B	M or B	B	M	M or B	B	B	B	B	B	B	P	B	B	M	B	B	B	B

[a]Symbols: +, 80–100% strains positive; V, 50–66% strains positive; D, 20–43% strains positive; −, 0–17% strains positive; nd, not determined. TTC, triphenyl-tetrazolium chloride. B, bipolar; M, monotrichous; P, peritrichous.

[b]Negative results when tested by different methods.

List of species of the genus Helicobacter

1. **Helicobacter pylori** (Marshall, Royce, Annear, Goodwin, Pearman, Warren, Armstrong 1984) Goodwin, Armstrong, Chilvers, Peters, Collins, Sly, McConnell, Harper 1989a, 403[VP] (*Campylobacter pylori* Marshall, Royce, Annear, Goodwin, Pearman, Warren, Armstrong 1984, 87.)

py.lo'ri. Gr. masc. n. *pyloros* gate keeper; L. masc. n. *pylorus* lower part of the stomach; L. gen. n. *pylori* of the pylorus.

Cells are helical or curved, 2.5–5 × 0.5 µm, with rounded ends and spiral periodicity. Polyphosphate granules seen in cells under certain conditions (Bode et al., 1993). Cells transform to coccoid forms with age (Jones et al., 1985). Motile by means of four to seven unipolar sheathed flagella with distinctive terminal bulbs. Nutritionally fastidious, translucent colonies 1–2 mm in diameter are produced on blood agar bases. Will grow in defined media (Reynolds and Penn, 1994) or liquid culture using brain–heart infusion or *Brucella* broth, growth enhanced by shaking the culture and the addition of serum or cyclodextrins (Olivieri et al., 1993). Grows microaerophilically and in the presence of air enriched with 10% CO_2. Some strains grow poorly at 30°C and no growth is obtained at 25°C. A few strains from rhesus monkeys grow at 42°C. Initial growth obtained after 4–5 d but will grow in 2 d on subsequent subculture. No growth in the presence of 1% glycine and 1.5% NaCl. Urease, catalase, and oxidase positive. Does not reduce nitrates or hydrolyze hippurate. Exhibits leucine arylamidase, alkaline phosphatase, and gamma-glutamyl-transpeptidase activities. Major fatty acids are tetradecanoic acid ($C_{14:0}$) and 19-carbon cyclopropane fatty acid ($C_{19:0 \text{ cyclo}}$) (Goodwin et al., 1989b). The major isoprenoid is MK-6; methylated MK-6 is absent (Moss et al., 1990b). Has a requirement for amino acids (arginine, histidine, isoleucine, leucine, methionine, phenylalanine, and valine). Some strains require alanine or serine (Nedenskov,

1994; Reynolds and Penn, 1994). The amino acids may be deaminated to provide carbon and energy to the cell. Amino acids may be the primary energy source of the species. Excess amine nitrogen may be removed from the cell via the urea cycle. Limited metabolism of carbohydrates. Able to take up and metabolize glucose via the pentose phosphate and the Entner–Doudoroff pathways. Glucose may be fermented to mixed acid products or provide metabolites to the TCA (Krebs) cycle (Mendz et al., 1994). Alcohol dehydrogenase activity reported (Kaihovaara et al., 1994). May use oxygen as the terminal receptor in aerobic respiration or fumarate in anaerobic respiration (Hazell and Mendz, 1997).

Resistant to nalidixic acid, trimethoprim, sulfonamides, and vancomycin, and sensitive to penicillin, ampicillin, cephalothin, kanamycin, gentamicin, rifampin, and tetracycline. Variable resistance to metronidazole and clarithromycin. FDA-approved treatment regimens include the use of metronidazole (250 mg) and tetracycline (500 mg), together with bismuth subsalicylate (262 mg; bismuth subcitrate often used outside of the USA), all taken four times daily for 14 d, or clarithromycin (500 mg three times daily) with either omeprazole (40 mg once daily) or ranitidine bismuth citrate (400 mg twice daily) for 14 d (Dunn et al., 1997).

Strains show a high level of genetic diversity with variable genome sizes, gene order, and nucleotide sequences (Langenberg et al., 1986; Akopyanz et al., 1992; Go et al., 1996). Evidence of gene mosaicism, e.g., six possible allelic versions of the *vacA* gene (Atherton et al., 1995). However, comparison of two complete genome sequences from geographically distinct locations indicated considerable conservation at the gene level (Alm et al., 1999).

Plasmids ranging in size from 3.7 to >148 kb found in 35–58% of strains (Tjia et al., 1987; Graham et al., 1988;

Penfold et al., 1988). Bacteriophages observed (Schmid et al., 1990).

Approximately 60% of strains possess a pathogenicity island (PAI), 40 kb in size with a mol% G + C content very different from the mean chromosomal average, associated with virulence factors and found in conjunction with the *cagA* gene (Censini et al., 1996). The 31 genes of *cag* encode for a new type IV secretion system (Salmond, 1996). The genome of strain 26695 has been completely sequenced (Tomb et al., 1997). This strain has a circular genome with 1590 predicted coding regions, 1091 of them matching identified database entries. The average mol% G + C content of strain 26695 from the sequence data was 39%, with five regions within the genome showing significantly different mol% G + C compositions.

The principal reservoir of *H. pylori* is the human, with infection occurring mainly in childhood (Mitchell et al., 1992). *H. pylori* has been found in rhesus monkeys, certain macaque species, and in a closed colony of barrier-maintained cats (Fox and Lee, 1997). The latter instance appears to be a unique report in domestic pets, and human-to-animal transmission has been suggested as the most likely explanation (El-Zaatari et al., 1997). Subsequent investigations have not detected *H. pylori* in either cats (Jalava et al., 1998) or dogs (Eaton et al., 1996a) from random sources. The presence of *H. pylori* in pigs is poorly documented. However, one strain of porcine origin is widely available (NCTC 11916). It is not known if this strain is derived from an experimental or natural infection. Costas et al. (1993) determined that this pig isolate, and strains from humans and rhesus monkeys, could be discriminated by whole-cell protein profiles and that strains from each source were also phenotypically distinguishable.

Pathogenic. Widespread infection causing gastritis in all infected humans and most naturally or experimentally infected animals (monkeys, pigs, cats, dogs), but is usually asymptomatic. Symptomatic diseases include peptic ulcers and gastric malignancy. Causes up to 95% of duodenal ulcers and 70–80% of gastric ulcers in humans (Dunn et al., 1997). Classified as a class 1 carcinogen by the International Agency for Research in Cancer of the World Health Organization (Anonymous, 1994). In association with other factors such as a high-salt diet and lack of vitamin C, *H. pylori* causes the majority of cases of gastric adenocarcinoma (Correa et al., 1998). The evidence is primarily epidemiological (Forman, 1998) although gastric cancer is induced in Mongolian gerbils (Watanabe et al., 1998). Causes low-grade B-cell gastric lymphomas (MALTomas), the majority of which regress on anti-*H. pylori* therapy (Isaacson, 1996). Urease enzyme activity is an essential colonizing factor allowing survival in gastric acid (Eaton et al., 1991a; Scott et al., 1998). Motility is also essential for colonization (Eaton et al., 1992). Severity of gastritis dependent on possession of the pathogenicity island via induction of proinflammatory cytokines such as IL-8 (Crabtree, 1998). No other specific virulence factors have been conclusively defined although many strains produce a vacuolating cytotoxin (Cover, 1996). Host factors most likely contribute to pathogenesis (Lee, 1997). Infection induces autoantibody to gastric tissue (Negrini et al., 1991). Molecular mimicry observed with the LPS structure (Appelmelk et al., 1996).

Lewis antigens also shown to be possible adhesins (Boren et al., 1993).

The mol% G + C of the DNA is: 36–39 (T_m).

Type strain: ATCC 43504, DSM 4867, JCM 7653, LMG 7539, NCTC 11637.

GenBank accession number (16S rRNA): Z25741, M88157, U01330; genome sequence, strain 26695: AE000511.

2. **Helicobacter acinonychis** Eaton, Dewhirst, Radin, Fox, Paster, Krakowa, Morgan 1993, 105[VP] (*Helicobacter acinonyx* (sic) Eaton, Dewhirst, Radin, Fox, Paster, Krakowa, Morgan 1993, 105.)

ac.i.non.y'chis. L. gen. n. *acinonychis* referring to the feline species *Acinonyx jubilatus*, from which the organism was isolated.

Cells are short spiral or curved rods, 0.3 × 1.5–2.0 μm. Motile by means of 2–5 multiple sheathed flagella, which exhibit a monopolar arrangement and possess terminal bulb-like structures. Colonies on blood agar are small, round (1–2 mm in diameter), colorless, and translucent. Swarming is not observed. Strains do not grow at 37°C in anaerobic conditions. Growth obtained on blood agar media containing 32 mg/l nalidixic acid, but not 32 mg/l cephalothin, after 3 d microaerobic incubation. Alpha-hemolytic after three days growth on 5% horse blood agar. Catalase, alkaline phosphatase, and urease produced. Nitrate is not reduced.

Numerical analysis of whole-cell protein profiles and sequence analysis of the 16S rRNA gene identify an especially close relationship between *H. acinonychis* and *H. pylori* (Costas et al., 1993; Fig. BXII.ε.7). These species are phenotypically indistinguishable (On et al., 1996). DNA–DNA hybridization studies between these species have not been performed, but differences in the mol% G + C content of their DNA have been noted (Eaton et al., 1993).

Pathogenicity undetermined. Isolated from mucosal lesions in cases of severe lymphoplasmacytic gastritis in a single colony of captive cheetahs (*Acinonyx jubilatus*). Restriction endonuclease analysis of genomic DNA indicated all strains to be an outbreak clone (Eaton et al., 1991b). However, protein profiles and extensive phenotypic testing identify one of the four strains available in international culture collections to differ from the others (Costas et al., 1993; On and Holmes, 1995).

Resistant to sulfamethoxazole, trimethoprim, and vancomycin. Sensitive to penicillin, ampicillin, nitrofurazone, erythromycin, gentamicin, chlortetracycline, and chloramphenicol.

The mol% G + C of the DNA is: 29.9 (T_m).

Type strain: 90-119, ATCC 51101, CCUG 29263.

GenBank accession number (16S rRNA): M88148.

Additional Remarks: Name corrected to *H. acinonychis* by Trüper and De' Clari, 1997.

3. **Helicobacter aurati** Patterson, Schrenzel, Feng, Dewhirst, Paster, Thibodeau, Versalovic and Fox 2002, 3[VP] (Effective publication: Patterson, Schrenzel, Feng, Dewhirst, Paster, Thibodeau, Versalovic and Fox 2000, 3727.)

au.ra'ti. L. gen. masc. n. of the golden one, named after the golden Syrian hamster, *Mesocricetus auratus*, from which it was isolated.

Cells are fusiform, 0.6 × 4–5 μm. Periplasmic fibers are coiled around the cell body, which gives a crisscross appearance to the bacterial surface. In older cultures, large coccoid forms are observed. Motile by means of tufts of 7–10 sheathed flagella in a bipolar arrangement. Culture on agar media containing 5% sheep blood or 5% fetal calf serum under microaerobic conditions does not result in single colonies; growth occurs as a thin spreading film. Catalase, urease, and γ-glutamyl transpeptidase produced. Indoxyl acetate hydrolyzed. No growth under anaerobic conditions. Sensitive to nalidixic acid (30 μg) but resistant to cephalothin (30 μg) in disk diffusion tests.

Pathogenicity unknown. Isolated from inflamed stomachs and ceca of the Syrian hamster *Mesocricetus auratus*.

Helicobacter aurati shares the same distinctive ultrastructure as *H. bilis* and *H. trogontum*, and thus these species represent distinct phylotypes of bacteria frequently referred to as "*Flexispira rappini*". As discussed in more detail above (see Taxonomic Comments), the genetic relationship of *H. aurati* to the wide range of "*Flexispira rappini*" phylotypes described from other sources (Dewhirst et al., 2000a) is presently unknown. It remains to be seen if *H. aurati* is host-specific, or if it possesses a broader host range.

The mol% G + C of the DNA is: not determined.

Type strain: Type strain: MIT 97-5075c.

GenBank accession number (16S rRNA): AF297868.

4. **Helicobacter bilis** Fox, Yan, Dewhirst, Paster, Shames, Murphy, Hayward, Belcher and Mendes 1997b, 601^VP (Effective publication: Fox, Yan, Dewhirst, Paster, Shames, Murphy, Hayward, Belcher and Mendes 1995, 453.)

bi' lis. L. n. *bilis* relating to the bile, referring to the bodily fluid from which it was isolated.

Cells are fusiform to slightly spiral, 0.5 × 4–5 μm. Periplasmic fibers are coiled around the cell body, which gives a crisscross appearance to the bacterial surface. Occasionally, loosening of the fibers results in the appearance of paired filaments. In older cultures, coccoid forms predominate; periplasmic fibers and flagella are usually seen. Motile by means of 3–14 sheathed flagella arranged in tufts at both ends of the cell body. Colonies are initially punctiform but often spread out as thin layers on the surface of agar media after 3–5 d microaerobic incubation at 37°C. Catalase and urease produced. Strains will grow on media containing up to 20% bile but not 1.5% NaCl. Sensitive to metronidazole (4 mg) but resistant to cephalothin (30 mg) and nalidixic acid (30 mg) in disk diffusion tests.

Pathogenicity unknown. First isolated from colonic crypts and ceca of mice, and the bile and livers of mice with hepatitis. Also causes a condition resembling inflammatory bowel disease (inflammation and hyperplasia of the lower gastrointestinal tract) in immunodeficient rats (Shomer et al., 1997; Haines et al., 1998).

Strains from human bile and gallbladder samples, and from a canine gastric biopsy, have also been identified as *H. bilis* by 16S rRNA gene sequence comparisons (Eaton et al., 1996a; Fox et al., 1998a). However, the close ultrastructural resemblance and taxonomic proximity of *H. bilis, H. trogontum, H. aurati*, and "*H. rappini*" (see Taxonomic Comments), and absence of other chemotaxonomic data that would clarify the relationships among these taxa, makes

it difficult to be certain that dog and human strains are indeed *H. bilis*.

The mol% G + C of the DNA is: not determined.

Type strain: Hb-1, ATCC 51630.

GenBank accession number (16S rRNA): U18766.

5. **Helicobacter bizzozeronii** Hänninen, Happonen, Saari and Jalava 1996, 165^VP

biz.zo' zer.o.ni.i. L. gen. n. *bizzozeronii* named in honor of Guilio Bizzozero, an Italian pathologist who was one of the first scientists to describe spiral organisms in the canine gastric tract.

Cells are tightly coiled helical rods, 0.3 × 5–10 μm. Coccoid forms are observed in aging cultures. Cells demonstrate a rapid, corkscrew-like motility achieved by means of multiple sheathed flagella in a bipolar arrangement. Culture of the bacteria is difficult and best achieved on freshly prepared, moist media. Consequently, individual colonies have not been observed and bacterial growth appears as a thin spreading film on agar media. All strains produce urease when tested *in situ* in mucosal tissue samples and when first isolated, but may spontaneously lose this activity after subcultivation. Most strains (80%) are resistant to 5-fluorouracil (100 U/l), but do not grow in the presence of 1% bile or metronidazole (4 mg/l). Variable results for alkaline phosphatase production and ability to hydrolyze indoxyl acetate have been reported.

H. bizzozeronii and *H. felis* share 2–30% DNA relatedness in optical DNA–DNA reassociation studies and 99.3–99.9% sequence homology in the 16S rRNA gene. Conventional phenotypic tests do not discriminate between these two species. Ultrastructural studies show that most *H. felis* strains exhibit periplasmic fibrils encircling the cell body, a feature absent in all *H. bizzozeronii* isolates examined to date. The two species can also be differentiated by whole-cell protein analysis and by dot-blot DNA–DNA hybridization assays.

H. bizzozeronii is morphologically indistinguishable from strains found in humans which are referred to as either "*H. heilmannii*" or "*Gastrospirillum hominis*". By 16S rRNA gene sequence comparisons, *H. bizzozeronii* and "*H. heilmannii*" appear closely related (96.7–98.7% similarity), but taxonomic studies to clarify the relationship between these taxa have been impaired due to the failure of most workers to culture the human strains. One successfully cultured strain of "*H. heilmannii*" (Andersen et al., 1996) has been identified as *H. bizzozeronii* by phenotypic, protein electrophoretic, and DNA–DNA hybridization experiments (Jalava et al., 2001). However, it is likely that *H. bizzozeronii* does not account for all such observations (see Taxonomic Comments and Other Organisms).

Pathogenicity unknown. Isolated from the canine gastric mucosa of healthy animals, and those showing signs of upper gastrointestinal dysfunction (vomiting or abdominal discomfort). One strain isolated from a gastric biopsy of a human with gastritis.

The mol% G + C of the DNA is: not determined.

Type strain: Storkis, CCUG 35545.

GenBank accession number (16S rRNA): Y09404.

6. *Candidatus* Helicobacter bovis De Groote, van Doorn, Ducatelle, Verschuuren, Tilmant, Quint, Haesebrouck and Vandamme 1999b, 1713^VP

bo' vis. L. n. *bos* the cow; L. gen. n. *bovis* of the cow.

Cells are helical rods showing 1–3 complete spiral turns with a wavelength of ~750 nm, 0.3 × 1.5–2.5 μm. Cells possess at least four flagella at one end of the cell, but further details of the flagellar arrangement and presence or absence of a flagellar sheath are not clear at this time. Urease activity was detected in all pyloric samples in which bacteria were seen. The species has not yet been cultured *in vitro*.

Pathogenicity unknown. Found in the gastric crypts of the pyloric part of the bovine abomasal stomach.

Comparison of 16S rRNA gene sequences shows that *Candidatus* H. bovis occupies a distinct branch in the genus. Its closest validly published phylogenetic neighbor is *H. bilis* (level of similarity, 92.8%), although clustering of sequences by the neighbor-joining method places *Candidatus* H. bovis with the majority of gastric *Helicobacter* species (De Groote et al., 1999b). A specific PCR assay and oligonucleotide probe is available for diagnostic purposes (De Groote et al., 1999b). Cross-reactivity with polyclonal *H. pylori*-derived antibodies is observed.

The mol% G + C of the DNA is: not determined.
Type strain: Not yet cultured.
GenBank accession number (16S rRNA): AF127027.
Additional Remarks: Candidate (*Candidatus*) species.

7. **Helicobacter canadensis** Fox, Chien, Dewhirst, Paster, Shen, Melito, Woodward and Rodgers 2002, 3[VP] (Effective publication: Fox, Chien, Dewhirst, Paster, Shen, Melito, Woodward and Rodgers 2000, 2549.)

can.ad.en.sis. L. n. from Canada, the country of original isolation.

Cells are curved to spiral (1–3 turns) rods, 0.3 × 1.5–4 μm. Motile by means of single unipolar or bipolar unsheathed flagella. Culture on agar media containing 5% sheep blood under microaerobic conditions does not result in single colonies; growth occurs as a thin spreading film. Catalase, but not urease, alkaline phosphatase, or γ-glutamyl transpeptidase produced. Indoxyl acetate hydrolyzed. Nitrate reduced. Resistant to nalidixic acid (30 μg) and cephalothin (30 μg) in disk diffusion tests.

Pathogenicity unknown. Isolated from human diarrhea.

In both phylogenetic and phenotypic analyses, *H. canadensis* is similar to *H. pullorum* although the inability of the latter species to hydrolyze indoxyl acetate is a useful distinguishing feature. The close resemblance of these two species led Fox et al. (2000) to ponder the prospect of *H. canadensis* sharing similar animal reservoirs. With the discovery of strains from wild birds sharing >99.1% similarity in their 16S rDNA sequences with that of *H. canadensis* and with a phenotype closely resembling the latter species (S.L.W. On and J. Waldenström, unpublished data), it seems *H. canadensis* and *H. pullorum* also share the ability to colonize the avian gut. The wider zoonotic potential of these enteric species should be a subject for careful consideration.

The mol% G + C of the DNA is: not determined.
Type strain: NLEP-16143, MIT 98-5491, ATCC 700968.
GenBank accession number (16S rRNA): AF262037.

8. **Helicobacter canis** Stanley, Linton, Burnens, Dewhirst, Owen, Porter, On and Costas 1994c, 370[VP] (Effective publication: Stanley, Linton, Burnens, Dewhirst, Owen, Porter, On and Costas 1993b, 2502.)

ca'nis. L. n. *canis* dog, after source of first isolation.

Curved or helical rods, typically 0.25 × 4 μm. Motile by means of two sheathed flagella showing a bipolar distribution. Colonies are pinpoint, translucent, and nonpigmented after 2 d microaerobic incubation on blood agar.

Neither catalase nor urease produced. Strains grow at 42°C and on media containing up to 1.5% ox-bile and 100 U/l 5-fluorouracil. No growth on media containing 0.032% methyl orange.

Insertion sequences ranging from 181 to 320 bp in size have been detected in the 16S and 23S rRNA genes of some strains (Linton et al., 1994b). The two 23S rRNA operons in *H. canis* may be polymorphic with respect to the presence of such insertion sequences.

In semi-quantitative DNA–DNA hybridization experiments, *H. canis* shares 30% and 52% DNA relatedness, respectively, with the species *H. fennelliae* and *H. cinaedi*. 16S rRNA gene sequence analyses identify the latter species, and *H. bilis* and "*H. rappini*" also, as reasonably close relatives (>97.8% sequence similarity). All these taxa are known to inhabit the lower intestinal tract of their hosts and have been associated with gastrointestinal disease; several (including *H. canis*, *H. bilis*, and "*H. rappini*") have been isolated from hepatic disorders in humans or animals.

Pathogenicity unknown. Isolated from diarrhea in dogs and humans, and from the liver of a dog suffering from multifocal necrotizing hepatitis (Stanley et al., 1993b; Fox et al., 1996a).

The mol% G + C of the DNA is: 48.2–48.8 (T_m).
Type strain: ATCC 51401, NCTC 12739.
GenBank accession number (16S rRNA): L13464.

9. **Helicobacter cholecystus** Franklin, Beckwith, Livingston, Riley, Gibson, Besch-Williford and Hook 1997, 601[VP] (Effective publication: Franklin, Beckwith, Livingston, Riley, Gibson, Besch-Williford and Hook 1996, 2957.)

cho.le' cyst.us. N.L. n. *cholecyst* gallbladder; L. gen. n. *cholecystus* relating to the gallbladder.

Cells are straight or curved rods, 0.5–0.65 × 2–5 μm. Motile by means of a single sheathed flagellum. Colonies are pinpoint and nonpigmented after 3 d microaerobic incubation on blood agar. Catalase, arginine aminopeptidase, and L-arginine arylamidase produced. Nitrate is reduced. Urease is not produced and indoxyl acetate is not hydrolyzed. Strains grow at 42°C and on media containing up to 5% bile salts. Growth observed under anaerobic conditions. Resistant to 30 mg/l cephalothin.

Syrian hamsters have been identified as a natural host of *H. cholecystus*. A number of helicobacters occur naturally in rodents and, of these, *H. bilis* and *H. hepaticus* are possibly closely related by 16S rRNA gene sequence comparisons (>96% similarity) to *H. cholecystus*. All of these species have been associated with various hepatic diseases in their respective hosts, possibly suggesting a common evolutionary origin.

Pathogenicity unknown. Isolated from the gall bladder of Syrian hamsters afflicted with cholangiofibrosis and pancreatitis.

The mol% G + C of the DNA is: not determined.
Type strain: Hkb-1, ATCC 700242.
GenBank accession number (16S rRNA): U46129.

10. **Helicobacter cinaedi** (Totten, Fennell, Tenover, Wezenberg, Perine, Stamm and Holmes 1985) Vandamme, Falsen, Ros-

sau, Hoste, Segers, Tytgat and De Ley 1991a, 100[VP] (*Campylobacter cinaedi* Totten, Fennell, Tenover, Wezenberg, Perine, Stamm and Holmes 1985, 138.)

ci.nae.di. L. gen. n. *cinaedi* of a homosexual.

Cells in 48 h old cultures are predominantly slender spiral rods, 0.3–0.5 × 1.5–5.0 μm. Rapid corkscrew-like motility by means of a single polar, sheathed flagellum. Single curved, S-shaped, and straight rods are occasionally seen. Coccoid forms develop in old cultures.

Colonies that are 48 h old are pinpoint and translucent. On freshly prepared or moist media, growth often occurs as a flat and spreading zone and discreet colonies may be rare. No distinct odor.

Grow in microaerobic conditions at 37°C, but not at 25°C or 42°C. Hydrogen stimulates or is essential for growth. No growth in aerobic or anaerobic conditions. Catalase is produced. Nitrate is reduced. No urease activity. Growth in 1% glycine, but not in 2% NaCl. Strains grow on media containing 0.04% triphenyl-tetrazolium chloride. No hydrolysis of indoxyl acetate.

Susceptible to ampicillin, gentamicin, doxycycline, tetracycline, ceftriaxone, rifampin, spectinomycin, nalidixic acid, and chloramphenicol. Most strains are susceptible to sulfamethoxazole (<4 μg/ml) and trimethoprim-sulfamethoxazole (<8 μg/ml), and resistant to 64 μg/ml of cefoperazone. Resistant to trimethoprim. Varying levels of resistance towards clindamycin, erythromycin, streptomycin, and metronidazole.

Pathogenicity unknown. Isolated from rectal swabs and blood of homosexual men. Associated with gastroenteritis, proctitis, proctocolitis, cellulitis, and arthritis with or without accompanying HIV infection. Rarely isolated from blood, cerebrospinal fluid, or feces of women and children without risk factors for HIV infection. Isolated from normal or diarrhetic feces of hamsters, cats, dogs, and foxes (Stanley et al., 1993b; Vandamme et al., 2000).

Campylobacter cinaedi was first described by Totten et al. (1985) in a taxonomic study of *Campylobacter*-like organisms from homosexual men with enteric disease. This species was proposed by these authors to encompass two genetic groups (CLO-1a and CLO-1b), although DNA hybridization values between CLO-1a and CLO-1b ranged from 42 to 51%. Comparable values between group CLO-2 strains (proposed as *C. fennelliae*) and the single (still unnamed: see below) CLO-3 isolate were, however, far lower (<7%). These results and the absence of phenotypic differences between CLO-1a and CLO-1b groups led Totten et al. (1985) to include both in a single species. *H. cinaedi* is thus known to be genotypically diverse, comprising at least two genomospecies. It is also noteworthy that *H. cinaedi* isolates from humans, dogs, and hamsters formed distinct ribotype pattern groups according to their host source, despite DNA–DNA hybridization values to the type strain exceeding 77% (Kiehlbauch et al., 1995). *H. cinaedi* is genetically most closely related to *H. canis* (52% DNA relatedness; Stanley et al., 1993b) and *H. fennelliae* (3–10% DNA relatedness; Totten et al., 1985). Phylogenetically, *H. cinaedi* forms a tight cluster with strains of *H. canis* and a human "*H. rappini*" strain. The three validly described species are phenotypically very similar and few tests unequivocally differentiate these species (see Table BXII.ε.5).

The mol% G + C of the DNA is: 37–38 (T_m).

Type strain: ATCC 35683, CCUG 18818, DSM 5359.

GenBank accession number (16S rRNA): M88150.

11. **Helicobacter felis** Paster, Lee, Fox, Dewhirst, Tordoff, Fraser, O'Rourke, Taylor and Ferrero 1991, 36[VP]

fe'lis. L. gen. n. *felis* of a cat.

Cells are tightly coiled helical cells, 0.4 × 5–7.5 μm. Cells transform to coccoid forms (diameter 2–4 μm) with age. Rapid corkscrew-like motility by means of 10–17 sheathed flagella (25 nm thick) with a bipolar distribution. The flagella are positioned slightly off center at each end of the cell. Cells of most strains are surrounded by periplasmic fibers, which appear as concentric helical ridges on the cell surface, either in pairs, threes, or singly. Highly fastidious. Grows as a thin, translucent film on enriched blood agar and does not readily form colonies. Colonies when observed are pinpoint and colorless after 3 d growth. Can be cultured using serum-enriched broths (brain-heart infusion and *Brucella* broths have been used), in which growth is enhanced by shaking the culture.

Some strains can grow on solid media under anaerobic conditions. Optimal growth is achieved at 37°C in a microaerobic atmosphere. Initial growth obtained after 4–5 days, will grow in 2 days on subsequent subculture. Some strains grow at 42°C; no growth obtained at 25°C. Urease produced, but spontaneous loss of this enzyme activity in a few strains after subcultivation may be observed (Jalava et al., 1998). Arginine aminopeptidase, leucine aminopeptidase, and gamma-glutamyl transpeptidase activities are detected. Most strains have histidine and leucine aminopeptidase activity. No production of *N*-acetylglucosaminidase, α-glucosidase, α-arabinosidase, β-glucosidase, α-fucosidase, α-galactosidase, β-galactosidase, proline aminopeptidase, pyroglutamic acid amylamidase, tyrosine aminopeptidase, alanine aminopeptidase, phenylalanine aminopeptidase, glycine aminopeptidase, or arginine dihydrolase. Nitrate is reduced to nitrite. Neither hippurate nor indoxyl acetate is hydrolyzed. In tests requiring growth on solid media, *H. felis* is typified by its unreactivity, a trait shared with its closest relatives *H. bizzozeronii* and *H. salomonis* (On et al., 1996; Jalava et al., 1998). Most strains are resistant to 5-fluorouracil (100 U/l).

Strains are genetically heterogeneous with no association to host species or country of origin. Most strains have multiple plasmids, 2 to >16 kb. Genome size 1.6 Mb (Jalava et al., 1999a). Pathogenicity island not detected.

Helical bacteria have been observed in canine and feline gastric mucosal samples for over a century (Bizzozero, 1893; Salomon, 1898). Such bacteria are known to be present in virtually all adult cats and dogs and three *Helicobacter* species (*H. felis*, *H. bizzozeronii*, *H. salomonis*) have been described from the gastric tract of domestic pets. These taxa are closely related, with DNA–DNA hybridization experiments indicating intraspecific DNA relatedness values of 11–39%. Phylogenetic analysis using 16S rRNA gene sequence comparisons shows strains of these species are so highly related that individual species cannot be clearly distinguished (Jalava et al., 1997). Similar results are inferred using restriction fragment length polymorphisms of the 23S rRNA gene, which cannot discriminate these species (Jalava et al., 1999a).

The relative prevalence and distribution of each species

in domestic pets is difficult to estimate since isolation and identification of the bacteria is extremely problematic (Eaton et al., 1996a; Jalava et al. 1998). Of 51 *Helicobacter* strains isolated from 95 dogs, 10 were *H. felis* (Jalava et al. 1998). The same study isolated only three strains from 22 cats examined; all were *H. felis*. This result may reflect the observation that this species was easiest to isolate.

Isolated from the gastric mucosa of cats and dogs (Lee et al., 1988). A human gastric biopsy strain identified as *H. felis* (99.5% similar to type strain sequence) by sequence comparison of a 967 bp segment of the 16S rRNA gene to similar data held in GenBank and EMBL databases, and by a PCR assay based on the urease gene sequence described as specific for *H. felis* (Germani et al., 1997). The result should however be treated with some caution due to the subsequent description of *H. salomonis* and *H. bizzozeronii*, which show up to 99.6% sequence homology in the 16S rRNA gene to other *H. felis* strains and which have not been examined with the urease gene assay applied by Germani et al. (1997).

Infection associated with lymphoid follicles in experimentally infected dogs. (Lee et al., 1992a). Colonizes the gastric mucosa of mice (Dick-Hegedus and Lee, 1991; Fox et al. 1991) and may cause a persistent chronic gastritis of approximately 1 year in duration in germ-free mice and their conventional non-germ-free counterparts (Fox et al., 1993). Widely used as an experimental model, including for vaccine studies (Lee, 1998).

The mol% G + C of the DNA is: 42.5 (T_m).

Type strain: CS1, ATCC 49179.

GenBank accession number (16S rRNA): M57398.

12. **Helicobacter fennelliae** (Totten, Fennell, Tenover, Wezenberg, Perine, Stamm and Holmes 1985) Vandamme, Falsen, Rossau, Hoste, Segers, Tytgat and De Ley 1991a, 100[VP] (*Campylobacter fennelliae* Totten, Fennell, Tenover, Wezenberg, Perine, Stamm and Holmes 1985, 138.)

fen.nel' li.ae. N.L. gen. n. *fennelliae* of C.L. Fennell, the person who first isolated this organism from rectal swabs of homosexual males.

Cells of 48 h old cultures are predominantly slender spiral rods, 0.3–0.5 × 1.5–5.0 μm. Rapid corkscrew-like motility by means of a single polar sheathed flagellum. Single curved, S-shaped, and straight rods are occasionally seen. Coccoid forms develop in old cultures.

Colonies of 48 h old cultures are pinpoint and translucent. On freshly prepared or moist media, growth often occurs as a flat and spreading zone, and discreet colonies may be rare. Strains may be characterized by an odor resembling that of hypochlorite cleaning powders.

Grow in microaerobic conditions at 37°C, but not at 25°C. Approximately one-third of strains grow at 42°C. Hydrogen stimulates or is essential for growth. No growth in aerobic or anaerobic conditions. Most strains produce catalase. No urease activity. Strains grow on media containing 0.04% triphenyl-tetrazolium chloride; reduction of this compound is concurrently observed and bacterial growth is of a metallic red sheen.

Susceptible to ampicillin, gentamicin, doxycycline, tetracycline, ceftriaxone, rifampin, spectinomycin, nalidixic acid, chloramphenicol, clindamycin, erythromycin, and metronidazole. Resistant to 5-fluorouracil, trimethoprim, and streptomycin. Most strains are susceptible to sulfamethoxazole (<4 μg/ml) and trimethoprim-sulfamethoxazole (<8 μg/ml).

DNA–DNA hybridization experiments indicate the closest taxonomic relatives to *H. fennelliae* to be *H. canis* (30% DNA relatedness) and *H. cinaedi* (3–10% DNA relatedness). By 16S rRNA gene sequence comparisons, *H. fennelliae* appears rather distinct (Fig. BXII.ε.7); Shen et al. (1997) showed the type strain of *H. pullorum* (NCTC 12824) was phylogenetically most closely related to *H. fennelliae* (95.9% sequence similarity).

Pathogenicity unknown. Isolated from rectal swabs and blood of homosexual men. Associated with proctitis and proctocolitis with or without accompanying HIV infection. Isolated from blood and feces of children with diarrhea.

The mol% G + C of the DNA is: 37–38 (T_m).

Type strain: 231, ATCC 35684, CCUG 18820, DSM 7491.

GenBank accession number (16S rRNA): M88154.

13. **Helicobacter ganmani** Robertson, O'Rourke, Vandamme, On and Lee 2001, 1888[VP]

gan.man' i. N.L. gen. n. *ganmani* an arbitrary name derived from *ganman*, which, in the language of the Gadigal people (indigenous Australians who live in the Sydney Harbour area), means "snake finder". Thus *ganmani* is intended to refer to both the spiral, snake-like morphology of the organism and to the area in Australia from which it was first described.

Cells are curved to spiral (2 turns per cell) rods 0.3 × 2.5 μm. Motile by means of single bipolar unsheathed flagellum. Culture on agar media containing 5% horse blood under anaerobic conditions occasionally results in translucent, irregular, and unpigmented colonies < 1 μm in diameter. However, growth is more frequently seen as a thin spreading film. Nitrate and triphenyl-tetrazolium chloride (TTC) reduced. Urease and alkaline phosphatase activity not detected. Weak catalase activity observed in a few strains. Indoxyl acetate not hydrolyzed. Resistant to cephalothin (32 μg/l), carbenicillin (32 μg/l), and cefoperazone in agar dilution tests.

Pathogenicity unknown. Isolated from the ceca, large and small bowels, and livers of laboratory mice.

Phenotypic, protein electrophoretic, and phylogenetic analyses indicated *Helicobacter ganmani* and *H. rodentium* were closely related, but distinct species. In addition, both species occupy the same ecological niche. All *H. ganmani* and 2/3 *H. rodentium* strains tested by Robertson et al. (2001) grew only under anaerobic conditions, a feature not observed with other *Helicobacter* species, although since *H. rodentium* had been isolated under microaerobic conditions (Shen et al., 1997), methodological differences may account for this anomaly. These species, along with *H. canadensis*, *H. mesocricetorum*, and *H. pullorum* are atypical from all other *Helicobacter* species in lacking flagellar sheathing. The significance of this is unknown. *H. ganmani* is the tenth validly described *Helicobacter* species isolated from the rodent intestine. This emphasizes the diversity of species that naturally occur in rodents and has wider implications for workers employing laboratory rodents for pathogenicity, immunological and other studies. The routine screening of laboratory rodents for indigenous *Helicobacter* species has been suggested (Shen et al., 1997; Robertson et al., 2001).

The mol% G + C of the DNA is: not determined.
Type strain: CMRI H02, CCUG 43526.
GenBank accession number (16S rRNA): AF000221.

14. **Helicobacter hepaticus** Fox, Dewhirst, Tully, Paster, Yan, Taylor, Collins, Gorelick and Ward 1994c, 595[VP] (Effective publication: Fox, Dewhirst, Tully, Paster, Yan, Taylor, Collins, Gorelick and Ward 1994c, 1243.)

he.pa' ti.cus Gr. adj. *hepatikos* relating to the liver.

Cells are slender, curved to spiral rods, 0.2–0.3 × 1.5–5 μm. Coccoid forms are found in older cultures. Motile by means of a sheathed, single bipolar flagellum. Colonies are pinpoint, but cultures often appear as a thin spreading layer on agar media after two to three days incubation at 37°C in microaerobic conditions. The spreading growth is either translucent or light blue-gray in color. Growth also obtained under anaerobic, but not aerobic conditions. Hemolytic. Catalase and urease produced. Strains grow on media containing 0.04% triphenyl-tetrazolium chloride; reduction of this compound is concurrently observed and bacterial growth is of a metallic red sheen. Strains also grow on media containing 0.1% sodium deoxycholate. Sensitive to metronidazole (4 mg/l) but resistant to 5-fluorouracil (100 U/l) and carbenicillin (32 mg/l).

H. hepaticus causes a chronic active hepatitis in certain strains of mice and, in A/JCr and B6C3F1 varieties, infection is linked to the development of hepatic adenoma and hepatocellular carcinoma (Fox et al., 1994b, 1996b, c; Ward et al., 1994; Whary et al., 1998). In some immunodysfunctional mice, the organism can cause severe inflammation of the colon and cecum, causing a condition closely resembling inflammatory bowel syndrome (Cahill et al., 1997). The organism colonizes the crypts of the cecum and colon, and, in susceptible strains of mice, the bile canaliculi of the liver. The prevalence of *H. hepaticus* in laboratory mice (including strains used in carcinogenesis bioassays) has been estimated from 47–100% among different vendors (Shames et al., 1995; Fox et al., 1998b).

The mol% G + C of the DNA is: not determined.
Type strain: Fox Hh-2, ATCC 51448.
GenBank accession number (16S rRNA): U07574.

15. **Helicobacter mesocricetorum** Simmons, Riley, Besch-Williford and Franklin 2000b, 1699[VP] (Effective publication: Simmons, Riley, Besch-Williford and Franklin 2000a, 1816.)

me' so.cric.et.or' um. L. n. from the Syrian hamster *Mesocricetus auratus*, the original source of isolation.

Cells are loosely spiral rods, 0.3–4 × 2–3 μm. Motile by means of single bipolar unsheathed flagella. Culture on agar media containing 5% sheep blood under microaerobic conditions results in either pinpoint colonies or a thin spreading film. Catalase and alkaline phosphatase, but not urease produced. No growth on media containing 1.0% glycine or 1.5% sodium chloride. Resistant to cephalothin (30 μg) in disk diffusion tests.

Pathogenicity unknown. Isolated from the feces of asymptomatic Syrian hamsters (*Mesocricetus auratus*).

The species *Helicobacter mesocricetorum*, *H. ganmani*, and *H. rodentium* bear a striking phenotypic resemblance to each other and appear closely related in 16S rRNA gene sequence comparisons. These species also share a cellular

morphotype uncommon among *Helicobacter* species, in that sheathing along the flagellar shaft is absent. Since these species appear to be indigenous to the rodent intestine, it is reasonable to presume they share a relatively recent common ancestor, although of these species only *H. mesocricetorum* has yet been recovered from Syrian hamsters. Nevertheless, *H. mesocricetorum* may be conveniently distinguished from the other taxa by its alkaline phosphatase activity and poor or no growth under anaerobic conditions.

The mol% G + C of the DNA is: not determined.
Type strain: MU 97-1514, ATCC 700932, CCUG 45420.
GenBank accession number (16S rRNA): AF072471.

16. **Helicobacter muridarum** Lee, Phillips, O'Rourke, Paster, Dewhirst, Fraser, Fox, Sly, Romaniuk, Trust and Kouprach 1992b, 35[VP]

mu.ri.da' rum. L. n. *Muridae* family name for Old World rats and mice; L. gen. n. *muridarum* of the *Muridae*.

Cells are helical rods, 3.5–5 × 0.5–0.6 μm with two to three spiral turns. Cells transform to coccoid forms with age. Rapid, darting motility by means of 10–14 bipolar sheathed flagella. Cells surrounded by 9–11 periplasmic fibers, which appear as concentric helical ridges on the cell surface. Highly fastidious. Grows as a fine, translucent spreading film on media enriched with blood or serum. Optimal growth occurs on a moist agar surface after 2–3 d incubation at 37°C in a microaerobic atmosphere (1–16% oxygen, with added 5–10% CO_2). Single colonies rarely obtained due to the preference of the bacterium for moist agar surfaces. Catalase, urease, and arginine aminopeptidase produced. No production of leucine aminopeptidase and γ-glutamyl transpeptidase. Nonhemolytic. No growth under aerobic or anaerobic conditions, or at 25°C or 42°C. No growth in the presence of 1.5% NaCl, >1% bile, or 0.1% sodium fluoride. Sensitive to cefoperazone and metronidazole. Indole and H_2S not produced. Gelatin, casein, starch, tributyrin, and hippurate are not hydrolyzed.

Isolated from the intestinal mucosa of rats and mice, mainly colonizing the mucus in the ileal crypts. In germ-free animals colonizes the cecal pits in large numbers (Phillips and Lee, 1983). Invades small intestinal tissue in immunologically compromised mice (Quastler and Hampton, 1962; Hampton, 1967). In older animals may colonize the gastric mucosa in large numbers (Lee et al., 1993).

The mol% G + C of the DNA is: 34 (T_m).
Type strain: ST1, ATCC 49282.
GenBank accession number (16S rRNA): M80205.

17. **Helicobacter mustelae** (Fox, Taylor, Edmonds and Brenner 1988) Goodwin, Armstrong, Chilvers, Peters, Collins, Sly, McConnell and Harper 1989a, 403[VP] (*Campylobacter mustelae* (Fox, Taylor, Edmonds and Brenner 1988) Fox, Chilvers, Goodwin, Taylor, Edmonds, Sly and Brenner 1989, 303; *Campylobacter pylori* subsp. *mustelae* Fox, Taylor, Edmonds and Brenner 1988, 370.)

mus.te' lae. L. gen. n. *mustelae* of a ferret.

Cells of 3–5 d old cultures are 0.5–1 × 3–5 μm. Coccoid forms predominate in older cultures. Colonies on blood agar are nonpigmented, translucent, and 1 mm in diameter. Straight or curved rods, 0.5–0.65 × 2–5 μm. Motile by means of multiple sheathed flagella with terminal bulb-like structures demonstrating a peritrichous arrangement. Catalase and urease produced. Urease activity can be spontaneously lost during subcultivation on artificial culture media (Costas et al., 1991). Grows in anaerobic conditions in the presence of 10% CO_2. Does not exhibit leucine arylamidase activity. Resistant to cephalothin (32 mg/l) and 5-fluorouracil (100 U/l), sodium fluoride (0.1%). Sensitive to metronidazole (4 mg/l).

Major fatty acids are $C_{16:0}$. Moderate amounts (given as % total fatty acid content) of $C_{14:0}$ (13–17%), $C_{18:1}$ (14–23%), and $C_{19:0\ cyclo}$ (8–15%) detected.

In a taxonomic investigation to determine the relationship between *Campylobacter pylori* and a bacterium isolated from ferret gastric mucosa, DNA–DNA hybridization values between these organisms were found to exceed 85%. Phenotypic differences between the ferret and human strains were noted, and these data used to propose that ferret isolates represented a subspecies of *C. pylori*, named by the authors *C. pylori* subsp. *mustelae* (Fox et al., 1988). Almost a year later, Fox et al. (1989) reported that subsequent DNA–DNA hybridization experiments using three different methods indicated a much lower level of DNA relatedness between the two taxa (1–49%) and the two subspecies were subsequently elevated to species status, the ferret isolates being assigned to *C. mustelae*. The subsequent issue of the *International Journal of Systematic Bacteriology* saw the transfer of both species to the new genus *Helicobacter* (Goodwin et al., 1989a). Although *H. mustelae* and *H. pylori* are closely related genetically and both are associated with gastric disease in their natural hosts, they occupy different clades in phylogenetic analyses based on 16S rRNA gene sequences (Fig. BXII.ε.7). In further contrast with *H. pylori*, *H. mustelae* appears to be genomically conserved (Taylor et al., 1994b).

H. mustelae is frequently used as an animal model of *H. pylori* infection and a number of important virulence traits have been characterized. Isogenic mutants lacking urease activity, as well as those without flagella, are unable to colonize gastric mucosa of ferrets (Andrutis et al., 1995, 1997).

Pathogenic in ferrets (*Mustela putorius furo*). Causes chronic gastritis and is clinically associated with gastric and duodenal ulcers as well as gastric adenocarcinoma and gastric MALT lymphoma (Fox et al., 1990, 1997a; Erdman et al., 1997). Isolated from the gastric mucosa and feces of infected ferrets and from the gastric mucosa of mink.

The mol% G + C of the DNA is: 36–41 (T_m).

Type strain: R85-13-6, ATCC 43772, CCUG 25715, NCTC 12198.

GenBank accession number (16S rRNA): M35048.

18. **Helicobacter nemestrinae** Bronsdon, Goodwin, Sly, Chilvers, Schoenknecht 1991, 150[VP]

ne.mes.tri.nae. N.L. gen. n. *nemestrinae* of the macaque species *M. nemestrina*, from which the organism was isolated.

Cells are helical or curved rods, 0.2–0.3 × 2.0–5 µm. Cells appear as loosely coiled wave-forms and small curved rods under light microscopy. Under certain culture conditions cells may demonstrate a high degree of pleomorphism and appear as bizarre twisted or knotted forms. Coccoid forms are observed in aging cultures. Motile by means of multiple sheathed flagella which exhibit a monopolar arrangement and possess terminal bulb-like structures.

Colonies on blood agar are small, round (0.5–1.0 mm in diameter), colorless, and translucent. Swarming is not observed but colonies may develop irregular edges after 3–5 d of incubation on moist growth media. Strains will grow at 42°C in a microaerobic atmosphere and at 37°C in anaerobic conditions. Weak growth obtained on blood agar media containing 32 mg/l nalidixic acid and 100 U 5-fluorouracil after 3 d microaerobic incubation. Weakly alpha-hemolytic after three days growth on 5% calf blood agar. Catalase, alkaline phosphatase, and urease produced. Indoxyl acetate not hydrolyzed.

Major cellular fatty acids are $C_{18:1}$ and $C_{16:0}$. Small amounts of $C_{14:0\ 3OH}$ and $C_{15:0}$ present but no $C_{19:0\ cyclo}$ or $C_{16:0\ 3OH}$.

At present, the type strain is the only known representative of the species. The DNA base composition is highly atypical for the genus, but phylogenetic analysis, overall phenotype, and the ecological niche of the organism are consistent with those recorded for other *Helicobacter* spp. Bronsdon et al. (1991) recorded very low levels of DNA relatedness between *H. nemestrinae* and the type strains of *H. pylori* (<10%) and *H. mustelae* (<1.0%). *H. nemestrinae* was found to be distinct from *H. felis* and *H. mustelae*, but not from all strains of *H. pylori*, when examined for restriction fragment length polymorphisms in the copper-transporting ATPase gene *copA* (Ge et al., 1996).

Sequence comparisons of nine housekeeping and flagellar genes of the type strain held at ATCC (ATCC 49396) suggest it to be a strain of *H. pylori* (Suerbaum et al., 2002). The status of the equivalent strains held at other international culture collections is at present unknown. Although current data suggest *H. nemestrinae* should be considered a junior synonym of *H. pylori*, DNA–DNA hybridization studies are required to confirm the genetic relationship between these two taxa.

Pathogenicity unknown. Isolated from the gastric mucosa of a pigtailed macaque monkey (*Macaca nemestrina*). No abnormal pathology was seen in naturally infected tissue samples, and infection was not associated with any morbidity or mortality of the animal.

Resistant to nalidixic acid, trimethoprim, sulfamethoxazole, and vancomycin. Sensitive to penicillin, ampicillin, erythromycin, gentamicin, cephalothin, cefoperazone, tetracycline, polymyxin, furoxone, rifampin, metronidazole, and chloromycetin.

The mol% G + C of the DNA is: 24 (T_m).

Type strain: T81213-NTB, ATCC 49396, CCUG 32350, DSM 7492.

GenBank accession number (16S rRNA): X67854.

19. **Helicobacter pametensis** Dewhirst, Seymour, Fraser, Paster and Fox 1994, 559[VP]

pa.me.ten' sis. N.L. of the Pamet River, Truro, Mass.

Cells are small curved rods, approximately 0.4 × 1.5 µm with rounded ends. Motile by means of two sheathed flagella, one at each end of the cell. Occasionally cells possess a third flagellum adjacent to one of the other flagella. Flagella are inserted subterminally. Colonies are pinpoint to 0.5 mm in diameter, translucent, colorless and weakly hemolytic after 3 d growth on 5% blood agar in microaerobic conditions at 37°C. Will also grow at 42°C and poorly under anaerobic conditions, but not at 25°C or in air. Catalase and arginine beta-naphthylamide aminopeptidase activity. Urease, gamma glutamyl transpeptidase, lysine, and ornithine decarboxylase, and proline, pyroglutamate, tyrosine, alanine, phenylalanine, and glycine beta-naphthylamide aminopeptidases are not detected. Nitrate and selenite reduced. Sensitive to nalidixic acid (32 mg/l) but resistant to cephalothin (32 mg/l) in disk diffusion and agar dilution assays.

Pathogenicity unknown. *H. pametensis* has been isolated from tern, gull, and swine feces and from a feline gastric biopsy (Seymour et al., 1994; Neiger et al., 1998). No disease manifestations have been recorded.

The mol% G + C of the DNA is: 38 (T_m).
Type strain: B9, ATCC 51478, CCUG 29255.
GenBank accession number (16S rRNA): M88147.

20. **Helicobacter pullorum** Stanley, Linton, Burnens, Dewhirst, On, Porter, Owen and Costas 1995, 418[VP] (Effective publication: Stanley, Linton, Burnens, Dewhirst, On, Porter, Owen and Costas 1994b, 3448.)

pul.lo′ rum. L. gen. n. *pullorum*, derived from *pullus*, chicken, after source of first isolation.

Cells are slightly curved rods, 0.3–0.5 × 3–4 μm. Motile by means of a single, unsheathed, polar flagellum. Colonies are pinpoint to 1 mm in diameter, nonpigmented, smooth, and translucent after 3 d incubation on blood agar at 37°C. Alpha-hemolytic. Most strains produce catalase (88%) and reduce nitrate (91%). Hydrogen sulfide production in triple sugar iron has been detected in many strains (76%); the use of freshly prepared media (≤3 days old) and a hydrogen-containing atmosphere may favor development of this trait. Urease is not produced and indoxyl acetate is not hydrolyzed. Most strains grow at 42°C (88%) and on 1% bile (79%) media under microaerobic conditions. Most strains (94%) are resistant to cephalothin (32 mg/l) and cefoperazone (64 mg/l) but not to nalidixic acid (18%).

H. pullorum is an unusual *Helicobacter* species in several respects. It is one of only four species (*H. rodentium, H. ganmani,* and *H. mesocricetorum* are the others) in which the flagellum is not sheathed. The fact that *H. pullorum* is also sensitive to polymyxin B (to which all other *Helicobacter* spp. tested thus far are resistant) makes the phenotype of *H. pullorum* more consistent with members of the genus *Campylobacter*. This is noteworthy since several important *Campylobacter* species (e.g., *C. lari, C. jejuni* subsp. *jejuni* and *C. coli*) are, like *H. pullorum*, of enteric origin and occur in poultry. There is thus a risk that *H. pullorum* could be misclassified as *Campylobacter* spp. in routine laboratories. However, the use of appropriate isolation and identification methods showed *H. pullorum* to be present in 60% of poultry samples (Atabay et al., 1998b). A species-specific PCR assay to facilitate its identification has been described (Stanley et al., 1994b), but cross-reacts with *H. canadensis* (Fox et al., 2000).

The avian origin of *H. pullorum* is also atypical of *Helicobacter* spp., most of which are found in mammalian hosts. Interestingly, the closest phylogenetic relative to *H. pullorum* is *H. pametensis*. At present, these species are the only validly described helicobacters from birds. Finally, *H. pullorum* has been associated with hepatitis in some host animals, an association shared by *H. hepaticus, H. cholecystus,* and "*H. rappini*". The natural ecological niche of all three of these taxa

is considered the lower gastrointestinal tract and thus an invasive capability, at least of some strains, is suggested.

Pathogenicity unknown. Isolated from the ceca of asymptomatic broiler chickens, the liver and intestinal contents of chickens suffering from vibrionic hepatitis, and from human diarrhea.

The mol% G + C of the DNA is: 33.8–35.1 (T_m).
Type strain: NCTC 12824.
GenBank accession number (16S rRNA): L36141.

21. **Helicobacter rodentium** Shen, Fox, Dewhirst, Paster, Foltz, Yan, Shames and Perry 1997, 633[VP]

ro.den′ ti.um. L. gen. n. *rodentium* of gnawing animals, referring to the first reported source (mice).

Cells are curved to spiral rods, 0.3 × 1.5–5 μm. Motile by means of two sheathed flagella showing a bipolar distribution. After 4–7 d of incubation under microaerobic conditions, colonies are 1–2 mm in diameter, but growth on solid media is frequently observed as a thin spreading film. Catalase, but not urease or γ-glutamyl transpeptidase, produced. Strains will grow under anaerobic conditions. Nonhemolytic.

Pathogenicity unknown. Isolated from the feces and intestinal tissue samples of healthy mice. Diarrhea was seen in *scid* mice coinfected with *H. bilis.*

The mol% G + C of the DNA is: not determined.
Type strain: MIT 95-1707, ATCC 700285.
GenBank accession number (16S rRNA): U96296.

22. **Helicobacter salomonis** Jalava, Kaartinen, Utriainen, Happonen and Hänninen 1997, 981[VP]

sa.lo.mo′ nis. L. gen. n. *salomonis* in honor of Hugo Salomon, a German scientist who was one of the first workers to describe three morphologically distinct spiral organisms in canine gastric mucosa.

Cells are loosely helical or curved rods 0.8–1.2 × 5–7 μm. Coccoid forms predominate in aging cultures. Cells exhibit a more pronounced corkscrew-like appearance when studied *in situ* in mucosal tissue samples. Cells exhibit a relatively slow, wave-like motility that is attained by means of multiple sheathed flagella arranged at one or both ends of the cell body. Culture of the bacteria is difficult and best achieved on freshly prepared, moist media. Consequently, individual colonies have not been observed and bacterial growth appears as a thin spreading film on agar media. All strains produce catalase and urease. Strains do not grow in the presence of 1% bile or cefoperazone (<60 mg/l). A few strains (17%) tolerate 5-fluorouracil (100 U/l). Variable results for alkaline phosphatase production and ability to hydrolyze indoxyl acetate have been reported.

H. salomonis shares 11–34% DNA relatedness with *H. bizzozeronii* and 26–39% DNA relatedness with *H. felis* in optical DNA–DNA reassociation studies. Similarly, *H. salomonis* shares 98.2–99.2% sequence similarity in the 16S rRNA gene with *H. bizzozeronii* and *H. felis*. 16S rRNA gene sequence similarity between two *H. salomonis* strains was found to be 98.4%. Unequivocal differentiation of *H. salomonis* from its two closest taxonomic relatives cannot be achieved by conventional phenotypic tests, but the cell morphology and motility of cultured strains, and inability to grow at 42°C or on 5-fluorouracil (100 U/l)-containing media serve as broadly distinguishing features. These species can also be differentiated by whole-cell protein analysis and by dot-blot DNA–DNA hybridization assays.

Pathogenicity unknown. Isolated from the gastric mucosa of healthy dogs.

The mol% G + C of the DNA is: not determined.

Type strain: Inkinen, CCUG 37845.

GenBank accession number (16S rRNA): U89351.

23. *Candidatus* Helicobacter suis De Groote, van Doorn, Ducatelle, Verschuuren, Haesebrouck, Quint, Jalava and Vandamme 1999a, 1775.

su′ is. L. n. *sus* the pig; L. gen. n. *suis* of the pig.

Cells are tightly coiled helical rods, 0.6 × 2.5–3.5 μm. Cells possess 1–5 flagella in a bipolar arrangement. Presence or absence of a flagellar sheath is not clear at this time. Urease activity was detected in all pig stomach samples in which bacteria were seen. The species has not yet been cultured *in vitro*.

Pathogenicity unknown. Found in the gastric and pyloric crypts of the pig stomach.

Comparison of 16S rRNA gene sequences shows *Candidatus* H. suis and "*H. heilmannii*" type 1 are 99.5% similar (De Groote et al., 1999a). This is in agreement with a similar analysis performed on sequences representing "*Gastrospirillum suis*" and "*H. heilmannii*" type 1 (Mendes et al., 1994). Since the ultrastructure and host source of *Candidatus* H. suis and "*G. suis*" are the same, these taxa are highly likely to represent the same species. The former name is validly published and should take nomenclatural precedence over the latter. Although present data suggest that *Candidatus* H. suis and "*H. heilmannii*" type 1 represent the same species with a potential for zoonotic transmission, the relationship between these taxa can be determined only when strains of both taxa are cultured (De Groote et al., 1999a). A specific PCR assay and oligonucleotide probe for diagnostic purposes is available (De Groote et al., 1999a). Cross-reactivity with polyclonal *H. pylori*-derived antibodies is observed.

The mol% G + C of the DNA is: not determined.

Type strain: Not yet cultured.

GenBank accession number (16S rRNA): AF1270278.

Additional Remarks: Candidate (*Candidatus*) species.

24. **Helicobacter trogontum** Mendes, Queiroz, Dewhirst, Paster, Moura and Fox 1996, 920[VP]

tro.gon′ tum. Gr. part. adj. *trogon* gnawing; M.L. gen. pl. n. *trogontum* of gnawing animals, referring to source of first isolation.

Cells are fusiform to slightly spiral with pointed ends that measure 0.6–0.7 × 4–6 μm. Cells also have periplasmic fibers coiled around the protoplasmic cylinder, which gives a crisscross appearance to the bacterial surface. In older cultures, coccoid forms with overlapping periplasmic fibers are common. Motile by means of tufts of 3–7 sheathed flagella at each end of the cell. Colonies are initially punctiform but often spread out as thin layers with a blue-gray hue on the surface of agar media after 3–5 d microaerobic incubation. Catalase, urease, gamma-glutamyl transpeptidase, and ornithine decarboxylase produced. Some strains produce alkaline phosphatase. Nitrate not reduced. No growth on agar media containing >1.5% NaCl. Sensitive to

metronidazole (4 mg/l) but resistant to cephalothin (32 mg/l).

The complex cellular ultrastructure of *H. trogontum* bears a striking resemblance to that exhibited by *H. bilis* and ovine, porcine, murine, and human isolates that have been referred to as "*Flexispira*" or "*Helicobacter rappini*". 16S rRNA gene sequence comparisons indicate the phylogenetic positions of all these taxa to be distinct, with similarity values between *H. trogontum* and *H. bilis* and human and porcine "*H. rappini*" strains ranging from 95.5–96.1% (Shen et al., 1997). Since the taxonomic status of "*H. rappini*" is unclear (see Taxonomic Comments) and the level of DNA–DNA relatedness among the aforementioned taxa has not been determined, it is likely that *H. trogontum* represents a murine subset of strains that at one time would have been referred to as "*H. rappini*". Indeed, strains from laboratory mice were identified by Schauer et al. (1993) as "*F. rappini*" by their distinctive cell ultrastructure and 16S rRNA gene sequence identity (98.0–99.3% similarity) to human and ovine "*F. rappini*" strains studied by Bryner et al. (1986, 1987) and Archer et al. (1988).

Pathogenicity unknown. Isolated from colonic mucosa of rats. No abnormal pathological changes in the colonized tissue have yet been noted.

The mol% G + C of the DNA is: not determined.

Type strain: LRB 8581, ATCC 700114.

GenBank accession number (16S rRNA): U65103.

25. **Helicobacter typhlonius** Franklin, Gorelick, Riley, Dewhirst, Livingston, Ward, Beckwith and Fox 2002, 686[VP] (Effective publication Franklin, Gorelick, Riley, Dewhirst, Livingston, Ward, Beckwith and Fox 2001, 3925.)

ty.phlo′ ni.us. Gr. n. *typhlon*, cecum; N.L. adj. *typhlonius* pertaining to the cecum, site of original isolation.

Cells are filamentous spiral or curved rods, 0.3 × 2–5 μm. Motile by means of single bipolar sheathed flagella. Culture on agar media containing 5% sheep blood under microaerobic conditions results in pinpoint colonies. No growth under anaerobic conditions. Catalase, but not alkaline phosphatase or urease produced. Indoxyl acetate not hydrolyzed. Weak growth on media containing 1.0% glycine. Resistant to cephalothin (30 μg) but sensitive to nalidixic acid (30 μg) in disk diffusion tests.

Pathogenicity uncertain. Causes typhlocolitis, hepatitis, and inflammatory bowel disease in immunodeficient mice, but can also be recovered from mice without abnormal gastrointestinal pathology. Isolated from feces and ceca of laboratory mice.

In 1999, two independent research groups described novel helicobacters associated with gastrointestinal lesions of various strains of immunodeficient mice (Fox et al., 1999; Franklin et al., 1999); the latter group proposed the name "*Helicobacter typhlonicus*" for their isolates. A subsequent collaboration between the two research groups determined that the strains studied by each respective group represented a distinct species, *Helicobacter typhlonius*. This specific epithet *typhlonius* was used in preference to the initial designation "*typhlonicus*" for nomenclatural reasons relating to use of Latin adjectives.

In 16S rRNA gene sequence comparisons, *Helicobacter typhlonius* forms a clade with other murine helicobacters *H. hepaticus* and *H. muridarum*. However, many features differentiate these species, including presence of periplasmic fibrils and multiple flagella on the cell body of *H. muridarum*, absence of urease in *H. typhlonius* and alkaline phosphatase activity in *H. muridarum* only. All strains of *H. typhlonius* examined have demonstrated an intervening sequence of 166 bp in their 16S rDNA amplicons (Franklin et al., 2001).

Thus, this species is one of only three species at present (including *H. bilis* and *H. canis*) in which such insertions have been detected in the 16S rRNA gene.

The mol% G + C of the DNA is: not determined.

Type strain: MIT 97-6810, ATCC BAA-367.

GenBank accession number (16S rRNA): AF127912.

Other Organisms

Helicobacter sp. strain CLO-3 was originally described by Fennell et al. (1984) in their study of *Campylobacter*-like organisms (CLO) isolated from homosexual males. The groups CLO-1 and CLO-2 were later formally classified as *Campylobacter* species by Totten et al. (1985), and reclassified as *Helicobacter cinaedi* and *Helicobacter fennelliae*, respectively, by Vandamme et al. (1991a). Only a single strain belonged to CLO-3, and the CLO-3 taxon was not formally named pending the isolation and characterization of additional strains belonging to the same species. This single isolate is well-characterized and has been included in a variety of taxonomic studies.

Cells of 48 h old cultures in microaerobic conditions at 37°C are predominantly slender spiral rods, 0.3–0.5 × 1.5–5.0 µm. Rapid corkscrew-like motility by means of a single, polar, sheathed flagellum. Single curved, S-shaped, and straight rods are occasionally seen. Coccoid forms develop in old cultures.

Colonies of 48 h old cultures are pinpoint and translucent. On freshly prepared or moist media, or in older cultures, growth usually occurs as a flat and spreading zone with a blue-gray hue and discreet colonies may be rare. No distinct odor.

No growth in aerobic or anaerobic conditions. Catalase is produced. Nitrate is not reduced. No urease activity. Growth in 1% glycine, but not in 2% NaCl. Grows on media containing 0.04% triphenyl-tetrazolium chloride; reduction of this compound is concurrently observed and bacterial growth is of a metallic red sheen. Hydrolysis of indoxyl acetate is weakly positive.

The mol% G + C of the DNA is: 45 (T_m).

Deposited strain: CLO-3, CCUG 14564.

GenBank accession number (16S rRNA): M88151.

In their study of *Helicobacter* strains from wild bird feces, Seymour et al. (1994) distinguished three biochemical groups of strains, referred to as "Bird-A", "Bird-B", and "Bird-C" strains. Strains belonging to the Bird-A taxon were formally classified as *Helicobacter pametensis* by Dewhirst et al. (1994); the Bird-B and Bird-C taxa were not formally classified.

Biochemically, Bird-C and Bird-B strains differed only in the degree of susceptibility towards nalidixic acid (the Bird-C strain was susceptible while Bird-B strains were only moderately susceptible); Bird-B and Bird-C strains were distinguished from Bird A strains by the absence of urease activity and susceptibility to cephalothin, which are characteristic for the latter. Bird-B and Bird-C strains produced urease, alkaline phosphatase, catalase, and oxidase, reduced nitrate, and hydrolyzed indoxyl acetate; they did not produce gamma-2-glutamyl transpeptidase. They were resistant to cephalothin but susceptible to nalidixic acid. They grew at 42°C and in the presence of 1% glycine, but not in the presence of 3.5% NaCl.

Old cultures of cells of Bird-B and -C strains 48 h old are predominantly slender spiral rods, 0.3–0.5 × 1.5–5.0 µm. Rapid corkscrew-like motility by means of a single, bipolar, sheathed flagellum. Coccoid forms develop in old cultures.

The 16S rRNA genes of *H. pametensis*, *H. mustelae*, and the Bird-B and Bird-C strains are more than 97% similar. The presence of several differential biochemical characteristics suggests that the Bird-B and Bird-C strains represent distinct species, but DNA–DNA hybridization experiments have not yet been performed.

Both Bird-B strains were isolated from tern feces; the Bird-C strains were isolated from feces of a house sparrow.

The mol% G + C of the DNA is: 31 and 30 (T_m) (Bird-B and -C strains, respectively).

Deposited strain: Bird-B, CCUG 29256, M88139; Bird-B, CCUG 29254, M88146; and Bird-C, CCUG 29261, M88144.

GenBank accession number (16S rRNA): Bird-B, M88139, M88146; Bird-C, M88144.

The name "*Flexispira rappini*" was proposed by Bryner (Bryner et al., 1986; Bryner, 1987) to denote a group of distinctive bacteria from enteric or genital tract samples of sheep, humans, and pigs. The organisms closely resembled strains observed by Rappin (1881) in dog feces and had a fusiform cell body encircled by spiral periplasmic fibers with bipolar tufts of sheathed flagella. In 1991, analysis of 16S rRNA gene sequences and DNA–rRNA hybridization experiments (Paster et al., 1991; Vandamme et al., 1991a) revealed that two human "*Flexispira rappini*" strains belonged to the genus *Helicobacter*. Since these studies, additional "*H. rappini*" strains have been identified, primarily based on their cell morphology and phylogenetic position. Moreover, there is increasing evidence indicating that strains named "*H. rappini*" represent more than one species. Indeed the murine species *H. aurati*, *H. bilis*, and *H. trogontum* closely resemble "*H. rappini*". A fuller description of the problems concerning the taxonomic status of these strains is given in the Taxonomic Comments section above. Validation of the name "*H. rappini*" (proposed for certain human, dog, and mouse strains that form a closely related phylogenetic group) is pending (Dewhirst et al., 2000a).

Biochemical characteristics of individual "*F. rappini*" strains have been described in various studies cited above. Bryner (1987) reported that seven "*F. rappini*" strains had an average DNA base ratio of 33 mol% G + C (thermal denaturation method). Vandamme et al. (1991a) reported that the DNA base composition of strains ATCC 43879 and ATCC 43880 was 33.7 and 34.3 mol% G + C (thermal denaturation method), respectively, confirming data reported by Bryner (1987).

Strains tentatively identified as "*F. rappini*" have been isolated from a variety of sources including the gastrointestinal mucosa and feces of humans and animals including dogs, mice, and pigs. Strains have also been isolated from human blood and from aborted ovine fetuses. It should however be emphasized that it is unclear whether these strains represent a single species. It is certainly possible that rodent strains represent either *H. bilis* or *H. trogontum*, which are ultrastructurally indistinguishable from "*F. rappini*".

Deposited strain: ATCC 43966, ATCC 43968, ATCC 43879.

GenBank accession number (16S rRNA): M88137, U96300, M88138.

The name "*H. westmeadii*" was proposed by Trivett-Moore et al. (1997) for two strains isolated from blood cultures of AIDS patients. The authors presented a formal description for this novel species but did not denote a type strain. Thus far, the name has not been validated and has no standing in nomenclature. Recently, detailed phenotypic, protein-electrophoretic, and DNA–DNA hybridization analyses have identified these strains as *H. cinaedi* (Vandamme et al., 2000).

Cells of 48 h old cultures are predominantly slender spiral rods, 0.5×1.5–2.0 µm. Rapid corkscrew-like motility by means of a single, unipolar, sheathed flagellum.

Colonies of 48 h old cultures are pinpoint and translucent. On freshly prepared or moist media, growth occurs as a fine, translucent, spreading zone, and discreet colonies are rare.

Although originally reported as an anaerobic organism, optimal growth is obtained in a microaerobic atmosphere containing hydrogen. Growth at 37°C, but not at 25°C or 42°C. No growth in aerobic conditions. Catalase, alkaline and acid phosphatase, C4 esterase, C8 ester lipase, leucine arylamidase, and naphthol-AS-BI-phosphohydrolase activities, but not lysine or ornithine decarboxylase, or arginine dihydrolase activity. Nitrate is reduced. No urease activity. Indole and hydrogen sulfide are not produced. Resistant to cephalothin but sensitive to nalidixic acid.

Isolated from blood cultures of HIV-positive males with AIDS.

Deposited strain: None designated.

GenBank accession number (16S rRNA): U44756.

Additional Remarks: Both strains were reported to have identical 16S rRNA gene sequences, but strain numbers were not specified.

Helicobacter sp. strain Mainz (CCUG 33804) was isolated from an effusion of a knee joint of an AIDS patient with septic arthritis (Husmann et al., 1994). The strain produces catalase but does not hydrolyze indoxyl acetate. Alkaline phosphatase and urease activity are absent. The strain was susceptible to cephalothin (30 µg), tetracycline (30 IU), and clindamycin (2 IU), and resistant to nalidixic acid (30 µg), ciprofloxacin (5 µg), and erythromycin (15 IU) (as determined by a disc diffusion assay). Analysis of the 16S rRNA gene sequence revealed a 97.7% similarity with *H. fennelliae*, its closest phylogenetic neighbor. The authors suggested that the strain may represent a novel species, but subsequent phenotypic, protein electrophoretypic, and DNA–DNA hybridization analyses identified the strain as *H. cinaedi* (Vandamme et al., 2000).

Deposited strain: CCUG 33804.

GenBank accession number (16S rRNA): X81028.

The name *"Gastrospirillum hominis"* was proposed by McNulty et al. (1989) to refer to tightly coiled spiral bacteria that were clearly distinct from *H. pylori* in human gastric biopsy samples. The organisms were not cultured. A subsequent study by Solnick et al. (1993) clearly identified *"G. hominis"* strains as helicobacters by 16S rRNA gene sequence comparisons, and also designated two phylogenetically distinct types. As a result of their findings, Solnick et al. (1993) proposed the name *"H. heilmannii"* to supercede *"G. hominis"* for reference to the human strains; some workers have referred to strains observed in domestic pets as *"H. heilmannii"* (Dieterich et al., 1998; Neiger et al. 1998). It should be noted that neither *"G. hominis"* nor *"H. heilmannii"* have been formally described (until recently, their cultivation has eluded most workers, with one exception) and neither at present has nomenclatural precedence over the other. The name *"Gastrospirillum"* is still used to designate large, tightly helical, uncultivable bacteria from the gastric mucosa of a wide variety of animals. Strains from pigs and lemurs have been referred to as *"G. suis"* and *"G. lemur"*, respectively (Mendes et al., 1990; Dewhirst et al., 1992).

The taxonomic and nomenclatural problems surrounding the gastrospirilla are discussed more completely in the Taxonomic Comments section above. The identification of a cultured human *"H. heilmannii"* strain (Andersen et al., 1996) as *H. bizzozeronii* (Jalava et al., 1999c) has significant implications from taxonomic, applied, and epidemiological perspectives. It is, however, probable that the gastrospirilla represent a number of distinct species, some of which may be novel. Nonetheless, it is important to understand that *H. felis, H. bizzozeronii*, and *H. salomonis* are inherently difficult to isolate (Eaton et al., 1996a; Jalava et al., 1998) and strains of these species that are not cultured are, by existing criteria, gastrospirilla. The close phylogenetic associations of all these strains make unequivocal differentiation of these taxa impossible by current methods. Results from PCR-based assays based on urease gene sequences (Germani et al., 1997; Dieterich et al., 1998; Neiger et al., 1998) require validation as to their species specificity. However, comparisons of urease gene sequences do differentiate *H. bizzozeronii, H. felis, H. salomonis*, and *Candidatus* H. suis (J.L. O'Rourke and A. Lee, personal communication); nevertheless, it is likely the gastrospirilla will remain a difficult and controversial group for some time.

Although there is no formal description, generally *"Gastrospirillum"* cells are large spiral rods, 0.4–0.9×4–10 µm with polar bundles of sheathed flagella. Cells have 3–8 spiral turns and some strains have periplasmic fibers, but these are not detectable using common light microscopy. The presence of gastrospirillum cells is always associated with urease activity in gastric biopsy samples.

Deposited strain: Not yet cultured.

GenBank accession number (16S rRNA): type 1, L10079, type 2, L10080.

The name *"Helicobacter suncus"* was proposed by Goto et al. (1998) to describe a group of 27 strains (of which five were characterized) recovered from the stomach of house musk shrews (*Suncus murinus*) suffering from chronic gastritis. Based on the biochemical and phylogenetic data presented, a close relationship with *H. mustelae* and the unnamed *Helicobacter* sp. Bird-B, described above, was indicated. *H. suncus* differs from the latter taxa in being resistant to nalidixic acid and failing to hydrolyze indoxyl acetate. A type strain was not designated, and as yet, the name has not been validated.

Cells are curved rods, typically 0.5×3.5–5 µm. Coccoid forms are predominant in 5–7 d old cultures. Motile by means of two sheathed flagella showing a bipolar distribution. Bacterial growth is described as transparent and mucoid by the authors after 5–7 d of incubation under microaerobic conditions. Catalase, urease, and alkaline phosphatase activity is observed. Nonhemolytic.

Pathogenicity unknown. Isolated from stomach homogenates of house musk shrews (*Suncus murinus*) suffering from chronic gastritis.

The mol% G + C of the DNA is: not determined.

Deposited strain: Kaz-1, Kaz-2.

GenBank accession number (16S rRNA): AB006147, AB006148.

In order to determine whether ulcerative colitis-like disease in cotton-top tamarins (*Saguinus oedipus*) was associated with infection with *Helicobacter* spp., eight strains representing a putative new species were isolated (Saunders et al., 1999). The ultrastructure and overall phenotype of the strains closely resembled that of *"H. rappini"*, *H. aurati, H. bilis*, and *H. trogontum*. However, in contrast to these four taxa, the cotton-top tamarin isolates were found to be urease negative. The 16S rRNA gene sequences of two representative tamarin strains were identical to one another, contained an intervening sequence of 350 bases, and were most similar to *H. canis*, although the analysis presented by the authors placed *H. fennelliae* as the closest phylogenetic relative.

Cells are fusiform rods exhibiting surface periplasmic fibers, 0.5×4–5 μm. Motile by means of two tufts of sheathed flagella (6–12 in total) showing a bipolar distribution. Bacterial growth appears as a thin, spreading film on sheep blood agar medium after 3–5 d of incubation under microaerobic conditions. Catalase positive. Urease, alkaline phosphatase, and gamma-glutamyl transpeptidase activity is not observed. Resistant to nalidixic acid, cephalothin, sulfamethoxazole, and trimethoprim.

Pathogenicity unknown. Isolated from feces and colonic biopsies of cotton-top tamarins (*Saguinus oedipus*) suffering from chronic colitis.

The mol% G + C of the DNA is: not determined.

Deposited strain: None.

GenBank accession number (16S rRNA): AF107494.

A spiral bacterium was observed in the feces of a cat with severe diarrhea (Foley et al., 1998). Attempts to isolate the strain were unsuccessful. A eubacterial PCR assay was applied directly to DNA from the fecal sample and sequence analysis of the amplicon revealed it to be 98.3% similar to *H. canis*. The ultrastructure of the strain differed from the latter species since bipolar tufts of multiple flagella were seen, in contrast with the 1–2 bipolar flagellar arrangement of *H. canis*. These distinctions led the authors to propose the name "*H. colifelis*" for the organism observed. However, absence of a cultured strain prohibits further studies.

Deposited strain: Not yet cultured.

GenBank accession number (16S rRNA): AF142062.

Melito et al., (2001) proposed the name "*Helicobacter winghamensis*" to designate a group of seven isolates from fecal samples of humans afflicted with gastroenteritis. The strains had been recovered from samples submitted to a National Reference Laboratory in Canada from 1997 to 1999. 16S rDNA sequence analysis revealed these strains were highly related and clearly distinct from other *Helicobacter* species, including *H. canadensis* and *H. pullorum*, enteric species also associated with human gastroenteritis to which "*Helicobacter winghamensis*" bears a close ultrastructural and biochemical resemblance. The specific epithet "*winghamensis*" refers to the area in Ontario, Canada, from which the first isolate came. Although the species description conforms with minimal standards for description of new *Helicobacter* species (Dewhirst et al., 2000a), the name has not yet been validly published.

Cells are slightly curved to spiral rods, 0.3–0.6×2 μm. Motile by means of 1–2 bipolar unsheathed flagella. Culture on agar media containing 10% sheep blood under microaerobic conditions may result in colony morphologies ranging from single and distinct to those with a spreading appearance, depending on the individual strain. Catalase, alkaline phosphatase, and urease activity not detected. Indoxyl acetate hydrolyzed. Nitrate not reduced. Growth at 37°C but not at 42°C under microaerobic conditions. Strains differ in their susceptibility to nalidixic acid (30 μg) and cephalothin (30 μg) in disk diffusion tests.

Pathogenicity unknown. Isolated from the feces of humans with gastroenteritis.

The mol% G + C of the DNA is: not determined.

Deposited strain: NLEP 97-1090 (proposed type strain).

GenBank accession number (16S rRNA): AF246984.

An isolate from the blood culture of a 14-month old child was identified as a *Helicobacter* species by 16S rRNA gene sequence comparisons (Tee et al., 2000); its sequence most closely resembled those of *H. fennelliae* and (to a lesser extent) *H. cinaedi* strains, although the presence of a large intervening sequence (IVS) of 353 bp was noted as a striking difference. Subsequent sequence analysis of the IVS indicated a significant similarity to an IVS previously recorded in the 23S rRNA gene of *H. canis*. These IVSs were however in opposite orientations to each other. Despite the unusual characteristics of the 16S rDNA sequence of the blood culture isolate, further work is required to determine its taxonomic status as a novel species. Comparison of the available recorded phenotypic characteristics of this strain (Tee et al., 2000) with those of other helicobacters (On et al., 1996) suggests a striking similarity to *H. fennelliae*.

Cells are spiral rods (dimensions not recorded). Colonies were pinpoint, grayish-white, and hemolytic after several days microaerobic incubation on 6% blood agar; growth was enhanced in an atmosphere containing 10% H_2. Catalase, but not alkaline phosphatase or urease produced. Resistant to nalidixic acid (30 μg) in disk diffusion tests.

The mol% G + C of the DNA is: not determined.

Deposited strain: VIDRL 6606, ATCC 700956.

GenBank accession number (16S rRNA): AF237612.

Microscopic examination of pus from an abdominal abscess revealed the presence of Gram-negative fusiform rods, thereby resembling the "*Flexispira rappini*" morphotype (Han et al., 2000). By use of direct PCR examination of DNA extracted from the pus, a 16S rDNA-derived amplicon of 1668 bp was obtained. Subsequent analysis showed the enlarged product was due to an intervening sequence of ca. 200 bp and that the amplicon shared 97% identity with strains of *H. bilis* and "*F. rappini*". Although the authors argued that their data suggested a potential new species, this assertion is weakened by current knowledge of natural 16S rRNA gene sequence diversity in helicobacters (see Taxonomic Comments above), and the microscopic resemblance of the strain to cells of extant species. Attempts to culture the strain from pus were unsuccessful.

Deposited strain: None deposited.

GenBank accession number (16S rRNA): AJ011431.

The name "*Helicobacter muricola*" was proposed by Won et al. (2002) to delineate two strains (of four isolates) recovered from the feces and ceca of the Korean wild mouse (*Mus musculus molossinus*). The specific epithet "*muricola*" refers to the murine origin of the strains. The 16S rRNA gene sequences of the two strains were identical and most similar to *H. muridarum* and *H. pullorum* (96.71–96.36% similarity respectively). Various ultrastructural and biochemical criteria distinguish "*Helicobacter muricola*" from these and other *Helicobacter* species (Won et al., 2002).

Cells are slightly curved to weakly spiral rods, 0.4–0.5×2–5.5 μm. Coccal forms are frequently observed. Motile by means of bipolar unsheathed flagella. Culture on agar media containing 5% sheep blood under microaerobic conditions may result in pinpoint colonies or a thin spreading film. Catalase, and urease, but not alkaline phosphatase detected. Indoxyl acetate not hydrolyzed. Nitrate reduced. Growth at 37°C but not at 42°C under microaerobic conditions. Resistant to cephalothin (30 μg) but not to nalidixic acid (30 μg) in disk diffusion tests.

Pathogenicity unknown. Isolated from the feces and ceca of the Korean wild mouse (*Mus musculus molossinus*).

The mol% G + C of the DNA is: not determined.

Deposited strain: w-06 (proposed type strain).

GenBank accession number (16S rRNA): AF264783.

In an investigation of a potential bacteriological cause of gastritis in the dolphin species *Lagenorhynchus acutus* and *Delphinus delphis*, three strains were isolated (Harper et al., 2000). All were shown to be *Helicobacter* species by use of 16S rRNA gene sequence analysis. Additional sequences resembling those derived from the above isolates were obtained by direct PCR examination of additional dolphin stomach tissue. All sequences formed a distinct, but somewhat diverse cluster with other gastric *Helicobacter* species. Although closely related, it cannot be discounted that certain sequences represent distinct species. Unlike all cultured gastric helicobacters, alkaline phosphatase activity was not detected in the dolphin isolates. Histological examination of dolphin gastric tissue revealed lesions resembling those associated with human *H. pylori* infection.

Cells are slightly spiral rods, 0.6 × 4.0 μm. Motile by means of bipolar sheathed flagella. The colonial morphology of the strains was not described. Catalase, urease, and γ-glutamyl transpeptidase, but not alkaline phosphatase produced. Indoxyl acetate not hydrolyzed. Nitrate not reduced. Growth at 37°C and at 42°C under microaerobic conditions. Susceptible to cephalothin (30 μg) but not to nalidixic acid (30 μg) in disk diffusion tests.

Pathogenicity unknown. Isolated from the glandular mucosa of the dolphin species *Lagenorhynchus acutus*. 16S rDNA sequences resembling, but distinct from these strains, were detected by PCR in gastritic stomach tissue from the dolphin species *Delphinus delphis*.

The mol% G + C of the DNA is: not determined.

Deposited strain: none.

GenBank accession number (16S rRNA): AF292377, AF292378, AF292379.

Rectal swab and gastric fluid samples were taken from each of three captive beluga whales, one of which exhibited signs of clinical illness, and assessed for the presence of helicobacters by use of culture and PCR-based methods (Harper et al., 2002). Examinations of the two healthy whales proved negative, but one isolate (MIT 00-7128) was obtained from the fecal sample of the symptomatic whale. Analysis of this isolate 16S rDNA sequence showed it to be most similar to a *Helicobacter* strain obtained from a dolphin species, *Lagenorhynchus acutus* (see Other Organisms above). Phenotypic testing of the strain revealed it could grow on 5% sheep blood agar at both 37°C and at 42°C under microaerobic conditions, produced catalase and urease, but not alkaline phosphatase, and was unable to reduce nitrate or hydrolyze indoxyl acetate. The strain proved susceptible to both cephalothin (30 μg) and nalidixic acid (30 μg) in disk diffusion tests. Culture of the gastric fluid sample of the symptomatic whale gave negative results, but PCR analysis using a *Helicobacter*-specific assay yielded an amplicon which, when sequenced, appeared distinct from that of the whale fecal strain and more similar to a distinct dolphin *Helicobacter* sequence. Restriction fragment length polymorphism analysis of *Helicobacter*-specific assay amplicons from the fecal isolate, and whale gastric fluid, indicated the respective strains were similar but not identical, since *Hha*I but not *Alu*I profiles were distinct. Both whale *Helicobacter* sequences clustered among those of gastric helicobacters.

Deposited strain: None deposited.

GenBank accession number (16S rRNA): AF455130, AF455131.

A sequence deposited in Genbank (AJ007931) refers to "*Helicobacter ulmiensis*", a species for which no description appears to be available in international peer-reviewed journals. Independent analyses of this strain by whole-cell protein profile comparisons and by PCR-RFLP analysis respectively has identified it as a *H. hepaticus* isolate (P. Vandamme and R J. Owen, unpublished data).

Deposited strain: none deposited.

GenBank accession number (16S rRNA): AF007931.

Genus II. *Thiovulum* Hinze 1913, 195[AL]

LESLEY A. ROBERTSON, J. GIJS KUENEN, BRUCE J. PASTER, FLOYD E. DEWHIRST AND PETER VANDAMME

Thi.o'vu.lum. Gr. n. *thios* sulfur; L. n. *ovum* egg; M.L. neut. dim. n. *Thiovulum* small sulfur egg.

Cells are round to ovoid, **5–25 μm in diameter. Cytoplasm is often concentrated at one end of the cell**, with the remaining space being occupied by a large vacuole. Cytoplasm normally contains **orthorhombic sulfur inclusions**, which are generally concentrated at one end and often almost filling cells completely (Fig. BXII.ε.8). The amount of sulfur present in the cells varies with the H$_2$S supply, and cells may be temporarily fully devoid of sulfur inclusions. The number of **greenish bodies (up to 5 μm in diameter)**, thought to be **storage material**, increases with decreasing numbers of sulfur globules (Hinze, 1913). **Cells are rapidly motile** by peritrichous flagella (Fig. BXII.ε.9; de Boer et al., 1961). Swimming **velocity of over 600 μm s⁻¹**, some of the highest recorded in the bacterial world (Garcia-Pichel, 1989). Forward movement is accompanied by rotation around the long axis. Unlike most motile bacteria, *Thiovulum* cells do swim backwards or tumble. Cells possess a **fibrillar, or antapical organelle** at the posterior end of the cell. Its function is to secrete a **slime stalk** or thread (Fig. BXII.ε.9) which is used by cells to attach to solid surfaces. No resting stages are known.

Organisms are found in freshwater and marine environments, e.g., marshes, where sulfide-containing water or mud layers are in contact with overlaying oxygen-containing water (Kuever et al., 1996).

Gram negative. Microaerophilic. No pure cultures are available. Chemolithotrophic, oxidize sulfide. Catalase negative. Multiplication is by constriction followed by fission. Cell division appeared to take place along the longitudinal axis and without any specific relation to the distribution of cell contents. The cell wall was reported to be of the normal double type comprising polysaccharides.

Type species: **Thiovulum majus** Hinze 1913, 195.

FURTHER DESCRIPTIVE INFORMATION

Strongly chemotactic with respect to O$_2$ (Fenchel and Glud, 1998). Chemotaxis with respect to H$_2$S may not occur (Fenchel, 1994). Cells concentrate in sharply defined, characteristic white veils or webs consisting of separate, individually moving cells held together to some extent by a loose slime matrix. These cell masses are found at the O$_2$/H$_2$S interface where both substrates occur at low concentrations (0–10 μM⁻¹) but are constantly replen-

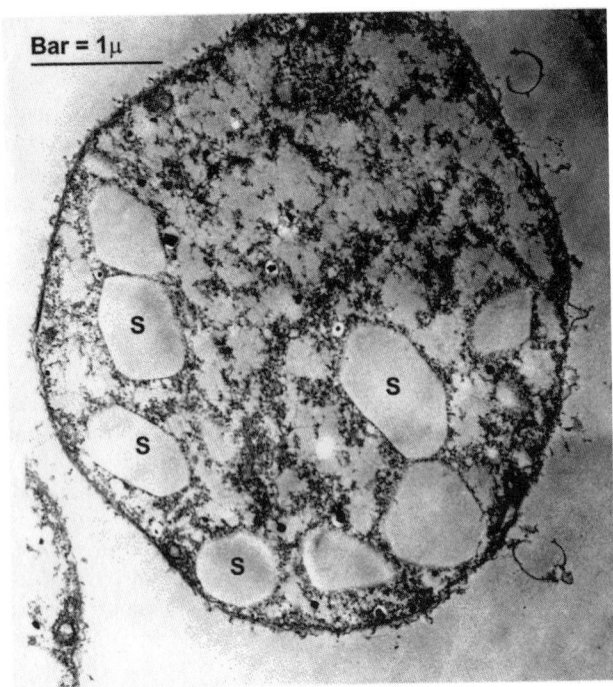

FIGURE BXII.ε.8. Ultrathin section of *Thiovulum* cell showing sulfur crystals (*S*) accumulating toward one end of the cell. (Photograph by the late Willemina E. de Boer.)

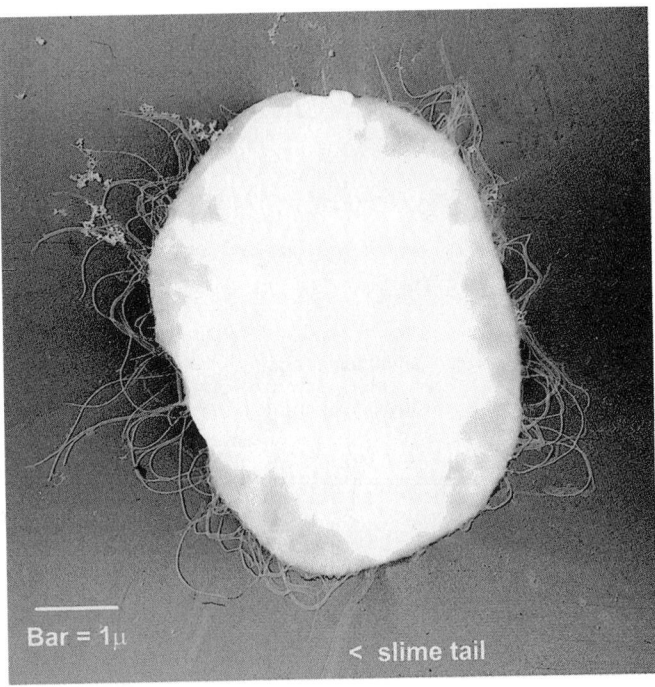

FIGURE BXII.ε.9. Platinum-shadowed *Thiovulum* cell showing flagella and the slime stalk or tail. (Photograph by the late Willemina E. de Boer.)

ished by diffusion (Jørgensen and Revsbech, 1983). *Thiovulum* veils may be as little as 200 μm thick (Bernard and Fenchel, 1995). New evidence has shown that *Thiovulum* veils also generate convective O_2 transport through the surrounding water. Oxygenated water passes down through the veil, and anoxic water is then recirculated through distinct holes in the veil, probably enhancing the rates of respiration and sulfide oxidation (Fenchel and Glud, 1998). Cells are killed by both anaerobic conditions and O_2 concentrations near air saturation values.

Although there are no pure cultures of *Thiovulum*, considerable progress has been made in the study of its physiology by means of crude laboratory enrichments (Wirsen and Jannasch, 1978; Jørgensen and Revsbech, 1983). Members of this genus may gain in importance when studies of sulfide oxidizers encountered at the deep-sea bottom around H_2S-emitting hydrothermal vents (Jannasch, 1984) have been made.

ENRICHMENT AND ISOLATION PROCEDURES

There are various methods for enrichment and maintenance of *Thiovulum* from seawater, all of which are based upon the provision of a H_2S-generating system coupled to a continuous and controlled supply of dissolved oxygen (Wirsen and Jannasch, 1978; la Rivière and Schmidt, 1981). In order to exclude photosynthetic sulfur bacteria, enrichment cultures must be kept in the dark; low temperatures (<15°C) appear to be favorable.

A good example is a system consisting of a 1–10 liter jar with a layer of decaying *Ulva* on the bottom and a continuous flow of seawater entering near the bottom. The sediment provides H_2S through reduction of sulfate supplied by the seawater, which is also a continuous O_2 source. *Thiovulum* cells maintain themselves at the optimal location in the gradient thus created and, in contrast to the nonchemotactic contaminants, are not flushed out. H_2S sources other than decaying *Ulva* can be employed, such as a pure culture of sulfate-reducing bacteria fed independently and kept separate from the *Thiovulum* culture by a dialysis membrane.

Vigorously growing laboratory enrichments can be maintained by such methods for many months and can provide ample material for morphological, physiological, and phylogenetic study. Unfortunately, pure cultures of *Thiovulum* have not been obtained.

TAXONOMIC COMMENTS

The names *Thiovulum majus*, *Thiovulum minus*, and *Thiovulum muelleri* have all been used to refer to *Thiovulum* cells that differed mainly in their cellular sizes and habitat (Starr and Skerman 1965; la Rivière, 1974). Currently only the name *T. majus* is used, although it is likely that there is more than one species within the genus *Thiovulum*. Comparative 16S rRNA gene sequence analysis of enriched *Thiovulum*-like cells has been achieved (Romaniuk et al., 1987; Lane et al., 1992) and phylogenetic analysis demonstrated that these unusual bacteria are closely related to the family *Helicobacteraceae*, which belongs in the *Epsilonproteobacteria* (see Fig. BXII.ε.10). To date, the closest relatives to *Thiovulum* are related only at the genus level and are species identified from 16S rRNA gene clonal analysis of bacteria from an active hydrothermal vent (Moyer et al., 1995) and epibiotic bacteria associated with the hydrothermal vent polychaete *Alvinella pompejana* (Haddad et al., 1995). In the absence of pure cultures, sequence analyses of cloned 16S rRNA genes amplified directly from environmental samples can establish the phylogenetic diversity within the genus.

ACKNOWLEDGMENTS

Bruce J. Paster and Floyd E. Dewhirst were supported by NIH Grants DE10374 and DE11443. Peter Vandamme is indebted to the Fund for Scientific Research Vlaanderen (Belgium) for a position as a postdoctoral fellow.

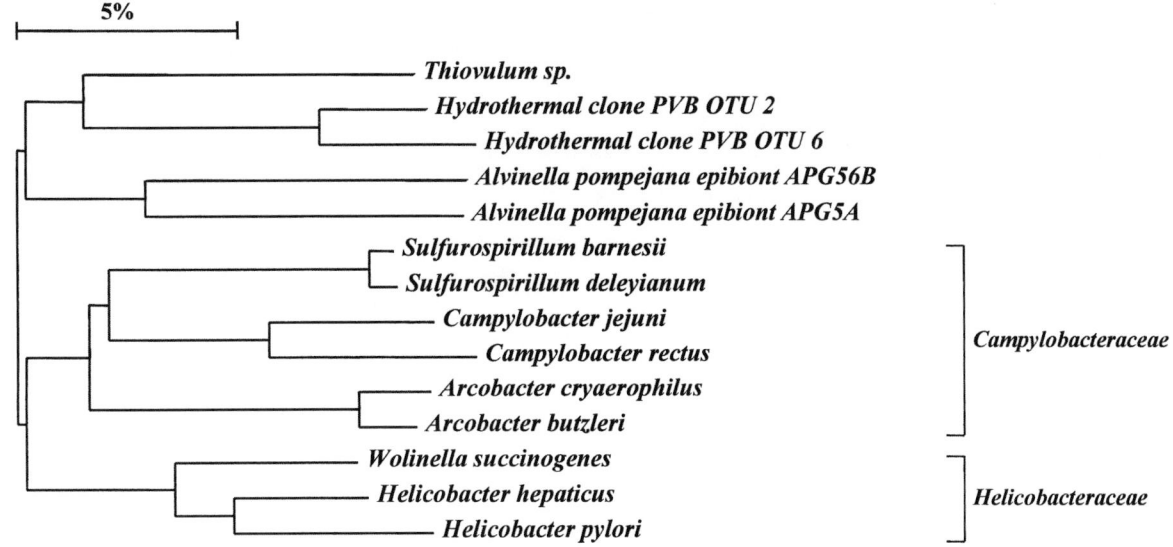

FIGURE BXII.ε.10. The phylogenetic relationships of species of the genus *Thiovulum* compared to other members of the *Epsilonproteobacteria*. *Thiovulum* is closely related to the family *Helicobacteraceae*. Bar = 5% difference in 16S rDNA nucleotide sequences. Due to the phylogenetic depth of *Thiovulum*, bootstrap confidence levels are not included.

List of species of the genus Thiovulum

1. **Thiovulum majus** Hinze 1913, 195[AL]

   ma'jus. L. comp. adj. *major* larger.

   For a description, see that of the genus.
   The mol% G + C of the DNA is: not known.
   Type strain: No culture isolated.

GenBank accession number (16S rRNA): M92323 (5′ end) and M92334 (3′ end).

Additional Remarks: GenBank accession numbers for operational taxonomic units (OTUs) from a hydrothermal vent: U15100–U15107.

Genus III. **Wolinella** Tanner, Badger, Lai, Listgarten, Visconti and Socransky 1981, 439[VP]

ACHIM KRÖGER, OLIVER KLIMMEK, PETER VANDAMME, FLOYD E. DEWHIRST AND BRUCE J. PASTER

Wo.li.nel'la. M.L. ending -*ella*; M.L. fem. n. *Wolinella* named after M.J. Wolin, American bacteriologist who first isolated the type species.

Helical, curved, or straight, unbranched cells, 0.5–1.0 × 2–6 μm, with rounded or tapered ends. Endospores are not produced. Gram negative. **Rapid, darting motility** is by means of a single polar flagellum. Colonies are pale yellow opaque to gray translucent with convex, pitting, and spreading variants.

Anaerobic. Hydrogen and formate are electron donors and are used as energy sources. **Fumarate, polysulfide, nitrate**, and other compounds are used as **electron acceptors. Formate is oxidized to CO₂**, while **fumarate is reduced to succinate. Carbohydrates are not fermented** and do not support growth. **Hydrogen sulfide** is produced by polysulfide reduction. Catalase negative. The type strain was isolated from the bovine rumen.

Menaquinone-6 (2-methyl-3-farnesyl-farnesyl-1,4-naphthoquinone) and **a methyl-substituted menaquinone-6** (2, [5 or 8]-dimethyl-3-farnesyl-farnesyl-1,4-naphthoquinone) have been reported as **major respiratory quinones** (Collins and Fernandez, 1984).

There is only one member of the genus, the type species *Wolinella succinogenes*.

The mol% G + C of the DNA is: 47.

Type species: **Wolinella succinogenes** (Wolin, Wolin and Jacobs 1961) Tanner, Badger, Lai, Listgarten, Visconti and Socransky 1981, 439 (*Vibrio succinogenes* Wolin, Wolin and Jacobs 1961, 917.)

FURTHER DESCRIPTIVE INFORMATION

L-Asparaginase has been isolated from *W. succinogenes* (Kafkewitz and Goodman, 1974; Albanese and Kafkewitz, 1978) and from similar organisms isolated from humans (Radcliffe et al., 1979). This enzyme has been of particular interest for its potential antitumor activity. For example, asparaginase from *W. succinogenes* was shown to inhibit the growth of cultured pancreatic carcinoma cells (Wu et al., 1978).

Growth of *W. succinogenes* *W. succinogenes* can grow by anaerobic respiration with fumarate, nitrate, nitrite, N₂O, polysulfide ([S]), or DMSO as terminal electron acceptor, and formate as electron donor (reaction a–f in Table BXII.ε.6). The stoichiometry of reactions (a), (b), (c), and (e) was confirmed experimentally with growing cultures, and cell formation was shown to be proportional to substrate consumption and product formation. Polysulfide is the actual terminal electron acceptor of sulfur respiration by *W. succinogenes*, and is formed abiotically from elemental sulfur in sulfide solutions. Sulfate, thiosulfate, organic disulfides (R-S-S-R), or TMAO are not used as terminal electron acceptors. Growth is observed with fumarate, nitrate, nitrite, polysulfide, or DMSO when formate is replaced by H₂.

Fumarate can be used as the sole source of carbon during

TABLE BXII.ε.6. Catabolic reactions sustaining growth of *Wolinella succinogenes*

Growth reaction	$t_d(h)$	Y (g dry cells/mol formate)	References
(a) $HCO_2^- + Fumarate + H^+ \rightarrow CO_2 + Succinate$	1.5	7	Bronder et al., 1982
(b) $HCO_2^- + NO_3^- + H^+ \rightarrow CO_2 + NO_2^- + H_2O$	1.4	5.2	Bokranz et al., 1983
(c) $3HCO_2^- + NO_2^- + 5H^+ \rightarrow 3CO_2 + NH_4^+ + 2H_2O$	1.4	5.2	Bokranz et al., 1983
(d) $HCO_2^- + N_2O + H^+ \rightarrow CO_2 + N_2 + H_2O$	$\leq 2.7^a$	5.7^a	Yoshinari, 1980
(e) $HCO_2^- + [S] \rightarrow CO_2 + HS^-$	2.3	3.2	Klimmek et al., 1991
(f) $HCO_2^- + (CH_3)_2SO + H \rightarrow CO_2 + (CH_3)_2S + H_2O$	14	6.7	Lorenzen et al., 1994
(g) $HS^- + Fumarate + H^+ \rightarrow [S] + Succinate$	3.6	2.4^b	Simon et al., 1998

[a]Evaluated from the data of Yoshinari (1980).

[b]g dry cells/mol fumarate.

growth by fumarate respiration. However, growth rate and growth yield are nearly doubled with small amounts of glutamate (1% of the fumarate amount) (Bronder et al., 1982). The growth parameters with yeast extract are slightly higher than those with glutamate. Glutamate is partially incorporated into the bacterial cell mass, specifically into the amino acid residues of the glutamate family. The residual glutamate is oxidized to succinate. Succinate (or fumarate) and glutamate serve as carbon sources during growth according to reactions b–f. Acetate and glutamate are used as carbon sources when polysulfide serves as terminal electron acceptor. While sulfate is used as the source of sulfur during growth with fumarate, either H_2S or cysteine has to be supplied for growth with nitrate.

ATP synthesis The ATP required for growth is synthesized according to a mechanism similar to that of oxidative phosphorylation. An electrochemical proton potential of 160 mV is generated across the cytoplasmic membrane of *W. succinogenes* by electron transport from formate or H_2 to fumarate or polysulfide (Mell et al., 1986; Wloczyk et al., 1989). In the second step, the free energy stored in the proton potential is used for ATP synthesis from ADP and phosphate, catalyzed by ATP synthase. The ATP synthase isolated from the membrane of *W. succinogenes* is similar to those of aerobic or phototrophic bacteria (Bokranz et al., 1985).

Electron transport with fumarate The electron transport chain catalyzing fumarate reduction by formate (reaction a) or H_2 consists of fumarate reductase, menaquinone, and formate dehydrogenase or hydrogenase (Fig. BXII.ε.11). Formate dehy-

drogenase or hydrogenase catalyzes menaquinone reduction by formate or H_2, and fumarate reductase catalyzes menaquinol reoxidation by fumarate. The three enzymes are built according to a common principle. Each enzyme consists of two hydrophilic and one hydrophobic subunit. The hydrophobic subunits are different diheme cytochromes *b* carrying the active sites for menaquinone or menaquinol. The substrate sites are located on the larger hydrophilic subunits. The smaller hydrophilic subunits represent iron–sulfur proteins that probably mediate electron transfer between the other two subunits. The catalytic subunits of hydrogenase and formate dehydrogenase protrude into the periplasmic space, while that of fumarate reductase faces the cytoplasm. It has been suggested that the electrochemical proton potential across the membrane is generated by transmembrane electron transport (Mell et al., 1986). Reconstitution of coupled fumarate respiration was carried out in liposomes by incorporating the electron transport enzymes isolated from *W. succinogenes* (Biel et al., 2002).

The genes encoding the subunits of hydrogenase (*hydABC*) and fumarate reductase (*frdCAB*) form separate operons on the genome of *W. succinogenes* (Lauterbach et al., 1990; Dross et al., 1992). A deletion mutant lacking the *hyd* genes did not grow with H_2 and fumarate (Gross et al., 1998). When grown with formate and fumarate, the mutant did not catalyze fumarate reduction by H_2, in contrast to the wild-type strain. Growth and electron transport activity with H_2 was restored upon integration of the *hyd* genes into the genome of the deletion mutant. A deletion mutant lacking the *frd* genes did not grow with fumarate and either H_2 or formate (Simon et al., 1998). The mutant grown

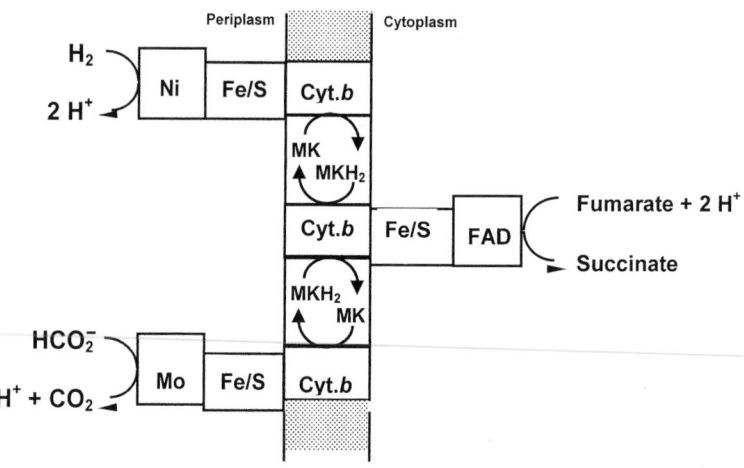

FIGURE BXII.ε.11. The electron transport system of *W. succinogenes* catalyzing fumarate reduction by H_2 or formate (Kröger et al., 1992). Hydrogenase contains nickel ion. Fumarate reductase contains covalently bound FAD. Formate dehydrogenase contains molybdenum ion. Fe/S, iron–sulfur centers; Cyt. *b*, diheme cytochrome *b*.

with nitrate as terminal electron acceptor did not catalyze fumarate reduction by H_2 or formate, in contrast to the wild-type strain. Integration of the *frd* genes into the genome of the deletion mutant caused recovery of growth and electron transport activity with fumarate. Two operons on the genome of *W. succinogenes* encode the subunits of formate dehydrogenase (Lenger et al., 1997). The operons differ in their promoter regions, but are nearly identical in their gene sequences. Deletion mutants lacking one of the operons still grow with formate and fumarate. The enzymes of the mutants appear to be identical.

Electron transport with polysulfide The electron transport chain catalyzing polysulfide reduction by H_2 or formate consists of polysulfide reductase, hydrogenase, or formate dehydrogenase (Schröder et al., 1988; Fauque et al., 1994). Hydrogenase and formate dehydrogenase are identical with the enzymes involved in fumarate reduction. The mutant lacking the *hyd* operon does not grow with H_2 and polysulfide, and growth with H_2 is restored upon insertion of the *hyd* operon into the genome of the deletion mutant (Gross et al., 1998). Polysulfide reductase is a molybdenum enzyme consisting of three different subunits (Fauque et al., 1994; Krafft et al., 1995). The catalytic subunit contains molybdenum coordinated by molybdopterin guanine dinucleotide (Jankielewicz et al., 1994). The second hydrophilic subunit is an iron–sulfur protein. The enzyme is anchored in the membrane by a hydrophobic subunit, which probably carries bound methylmenaquinone but no heme or other redox-active prosthetic groups (Dietrich and Klimmek, 2002). Electron transfer from hydrogenase to polysulfide reductase requires an intact hydrogenase cytochrome *b* (Gross et al., 1998). A deletion mutant lacking the polysulfide reductase genes (*psrABC*) does not catalyze polysulfide reduction by H_2 or formate when grown with fumarate, in contrast to the wild-type strain (Krafft et al., 1995). The deletion mutant grows with polysulfide. It forms a second polysulfide reductase that is different from the wild-type enzyme and is not present in the wild-type strain growing with polysulfide. The catalytic subunits of polysulfide reductase and formate dehydrogenase resemble those of other molybdooxidoreductases, including the formate dehydrogenase, nitrate reductase, and DMSO reductase of *Escherichia coli*.

ENRICHMENT AND ISOLATION PROCEDURES

W. succinogenes was isolated from an inoculum of bovine rumen fluid after serial transfer in an anaerobic methanogenic enrichment medium containing formate, sulfide, and inorganic salts. Secondary enrichment was made in a broth medium containing formate and fumarate (Wolin et al., 1961). Additional strains of *W. succinogenes* were recently isolated (unpublished data) using the same procedure, except that the primary medium contained fumarate. These strains were virtually indistinguishable from the original Wolin strain with respect to 16S rRNA gene sequence growth with fumarate, nitrate, or polysulfide, and the specific activities of electron transport from formate to fumarate or polysulfide. The genomes of the new isolates also contain two copies of the formate dehydrogenase operon. The media and buffer used for recent isolation of *W. succinogenes* were made anoxic by alternate evacuation and flushing with N_2. Bovine rumen fluid obtained from the local slaughterhouse was passed through filter paper, and the filtrate (250 ml) was centrifuged for 15 min at $10,000 \times g$. The sediment was resuspended in 0.2 liter Tris-buffer (50 mM, pH 8.0, 0°C). Centrifugation was repeated and the sediment was suspended in 0.2 l medium I (50 mM Tris-chloride, 20 mM sodium acetate, 0.1 M fumaric acid, 1 mM glutamic acid,

0.1 M sodium formate, 0.25 mM $MgCl_2$, 10 mM K_2HPO_4, 0.05 mM $CaCl_2$, 5 mM $(NH_4)_2SO_4$; trace element solution SL8 (Pfennig and Trüper, 1981), 0.2 ml/l culture; the pH was adjusted to 7.6–7.8 by KOH addition, and the medium was autoclaved). This suspension was used to inoculate (5%) culture medium I. The culture was kept for 24 h at 37°C. Using 5% inoculum and the same growth conditions, the bacteria were transferred to medium II (same as medium I, except it contains 10 mM NH_4Cl and 2 mM Na_2S instead of $(NH_4)_2SO_4$, and no acetate), then to medium I, and finally to agar (1%) plates containing medium I. Plating was repeated three times.

MAINTENANCE PROCEDURES

Wolinella succinogenes can be maintained in the laboratory by weekly transfer on the media described above or on commercially prepared Trypticase soy agar (BBL) supplemented with 5% sheep blood. Several broth media for the cultivation of *W. succinogenes* have been described (Wolin et al., 1961; Kafkewitz, 1975). *Wolinella succinogenes* will grow in a broth medium (MFF broth) consisting of *Mycoplasma* broth supplemented with hemin (5 mg/l), sodium formate (2 g/l), and sodium fumarate (3 g/l). The organism grows at 37°C in an atmosphere of $N_2/H_2/CO_2$ (80:10:10).

For preservation by lyophilization, a dense suspension of cells harvested from surface growth on blood agar media is prepared in broth containing 5% serum and 1% glucose. Organisms can also be preserved in liquid nitrogen by placing young colonies from blood agar plates into broth supplemented with 5% DMSO, for slow freezing before final storage.

DIFFERENTIATION OF THE GENUS *WOLINELLA* FROM OTHER GENERA

There are only a few phenotypic criteria that differentiate *W. succinogenes* from closely related taxa, namely species of the genera *Helicobacter*, *Campylobacter*, and *Arcobacter*. *W. succinogenes* is anaerobic, and not capable of microaerophilic growth as are members of the other three genera. *W. succinogenes* is catalase negative, whereas most species of *Helicobacter* and *Arcobacter* are catalase positive. Most species of *Campylobacter* are catalase positive. *W. succinogenes* is clearly distinguished from the other genera based on 16S rRNA gene sequence comparisons (Fig. BXII.ε.12).

TAXONOMIC COMMENTS

There is only one species of the genus, the type species *Wolinella succinogenes*. Tanner et al. (1981) had initially described two oral species of *Wolinella*, *W. recta*, and *W. curva*; however, 16S rRNA gene sequence analysis indicated that these two species were more closely related to campylobacters than to *W. succinogenes* (Paster and Dewhirst, 1988). These species are now described as members of the genus *Campylobacter* (Vandamme et al., 1991a). Cells are short, Gram-negative rods with a characteristic rapid darting motility. *W. succinogenes* is anaerobic and uses hydrogen and formate as electron donors. Fumarate, nitrate, and other compounds can be used as electron acceptors. Carbohydrates are not fermented and do not support growth. Hydrogen sulfide is produced. Comparative 16S rRNA gene sequence analysis has shown that *Wolinella succinogenes* falls in the family *Helicobacteraceae*, which belongs in the *Epsilonproteobacteria* (see Fig. BXII.ε.12). The closest relatives to *W. succinogenes* are species of the genus *Helicobacter*.

ACKNOWLEDGMENTS

Peter Vandamme is indebted to the Fund for Scientific Research - Vlaanderen (Belgium) for a position as a post-doctoral research fellow. Floyd

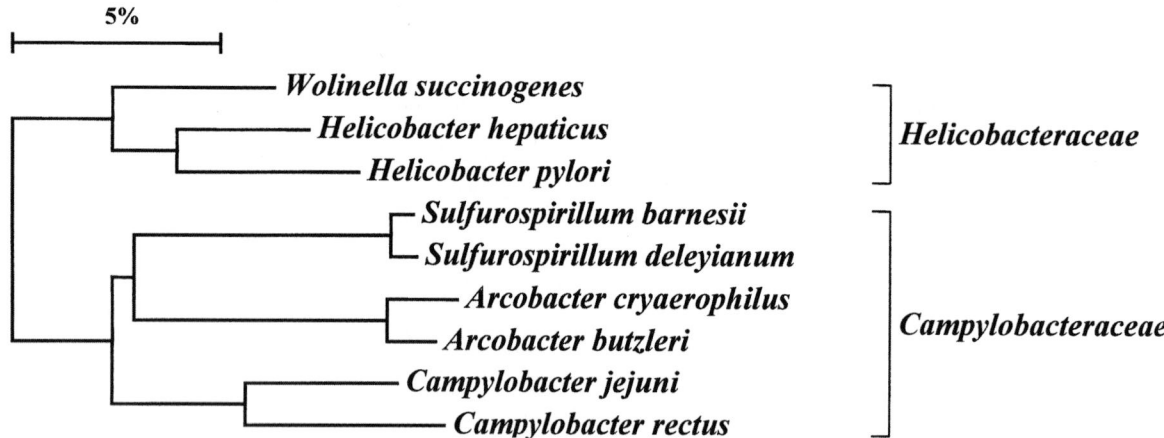

FIGURE BXII.ε.12. The phylogenetic relationships between *Wolinella succinogenes*, the only member of the genus, and other members of the family *Helicobacteraceae* and family *Campylobacteraceae*. Bar = 5% difference in nucleotide sequences.

E. Dewhirst and Bruce J. Paster were supported by NIH grants DE10374 and DE11443.

FURTHER READING

Smibert, R.M. and L.V. Holdeman. 1976. Clinical isolates of anaerobic Gram-negative rods with a formate-fumarate energy metabolism: *Bacteroides corrodens*, *Vibrio succinogenes*, and unidentified strains. J. Clin. Microbiol. *3*: 432–437.

Tanner, A.C.R., S. Badger, C.-H. Lai, M.A. Listgarten, R.A. Visconti and S.S. Socransky. 1981. *Wolinella* gen. nov., *Wolinella succinogenes* (*Vibrio succinogenes* Wolin et al.) comb. nov., and description of *Bacteroides gracilis* sp. nov., *Wolinella recta* sp. nov., *Campylobacter concisus* sp. nov., and *Eikenella corrodens* from humans with periodontal disease. Int. J. Syst. Bacteriol. *31*: 432–445.

Wolin, M.J., E.A. Wolin and N.J. Jacobs. 1961. Cytochrome-producing anaerobic vibrio, *Vibrio succinogenes*, sp. n. J. Bacteriol. *81*: 911–917.

List of species of the genus Wolinella

1. **Wolinella succinogenes** (Wolin, Wolin and Jacobs 1961) Tanner, Badger, Lai, Listgarten, Visconti and Socransky 1981, 439[VP] (*Vibrio succinogenes* Wolin, Wolin and Jacobs 1961, 917.)

suc.ci.no′ge.nes. M.L. n. *acidum succinicum* succinic acid; Gr. v. *gennaio* to produce; M.L. adj. *succinogenes* succinic acid-producing.

The characteristics are as described for the genus.

There is presently only one member of the genus, the type species *Wolinella succinogenes*. Description of species is as for the genus. Comparative 16S rRNA gene sequence analysis has shown that *W. succinogenes* falls in the family *Helicobacteraceae*, which belongs in the *Epsilonproteobacteria*. The closest relatives to *W. succinogenes* are species of the genus *Helicobacter*.

The mol% G + C of the DNA is: 47 (T_m).

Type strain: ATCC 29543, DSM 1740.

GenBank accession number (16S rRNA): M26636, M88159.

Other Organisms

Isolates resembling *W. succinogenes* have been isolated from sewage (Yoshinari, 1980) and from humans (Radcliffe et al., 1979), however no genetic information is available for these strains.

Bibliography

Aa, K. and R.A. Olsen. 1996. The use of various substrates and substrate concentrations by a *Hyphomicrobium* sp. isolated from soil: effect on growth rate and growth yield. Microb. Ecol. *31*: 67–76.

Aalen, R.B. and W.B. Gundersen. 1985. Polypeptides encoded by cryptic plasmids from *Neisseria gonorrhoeae*. Plasmid *14*: 209–216.

Aamand, J., T. Ahl and E. Spieck. 1996. Monoclonal antibodies recognizing nitrite oxidoreductase of *Nitrobacter hamburgensis*, *N. winogradskyi*, and *N. vulgaris*. Appl. Environ. Microbiol. *62*: 2352–2355.

Aarestrup, F.M., E.M. Nielsen, M. Madsen and J. Engberg. 1997. Antimicrobial susceptibility patterns of thermophilic *Campylobacter* spp. from humans, pigs, cattle, and broilers in Denmark. Antimicrob. Agents Chemother. *41*: 2244–2250.

Abadie, M. 1967. Formations intracytoplasmique du type "mésome" chez *Chondromyces crocatus* Berkeley et Curtis. C.R. Acad. Sci. Paris *265*: 2132–2134.

Abadie, M. 1968. Sur l'organisation des masses cellulaires végétatives chez les *Chondromyces*: Importance de la "matrice" initiale et de la trame muqueuse résiduelle. C.R. Acad. Sci. Paris *267*: 2037–2040.

Abadie, M. 1971. Contribution a la connaissance des myxobactéries supérieures. II. Données ultrastructurales et morphogénétiques sur le *Chondromyces crocatus*. Ann. Sci. Nat. Bot. Biol. Veg. *12*: 345–428.

Abalain, J.H., S. Di Stefano, M.L. Abalain-Colloc and H.H. Floch. 1995. Cloning, sequencing and expression of *Pseudomonas testosteroni* gene encoding 3-α-hydroxysteroid dehydrogenase. J. Steroid Biochem. Mol. Biol. *55*: 233–238.

Abdala, A.A., E. Pipano, D.H. Aguirre, A.B. Gaido, M.A. Zurbriggen, A.J. Mangold and A.A. Guglielmone. 1990. Frozen and fresh *Anaplasma centrale* vaccines in the protection of cattle against *Anaplasma marginale* infection. Rev. Elev. Med. Vet. Pays. Trop. *43*: 155–158.

Abe, M., R. Kawamura, S. Higashi, S. Mori, M. Shibata and T. Uchiumi. 1998. Transfer of the symbiotic plasmid from *Rhizobium leguminosarum* biovar trifolii to *Agrobacterium tumefaciens*. J. Gen. Appl. Microbiol. *44*: 65–74.

Abe, M. and T. Nakazawa. 1994. Characterization of hemolytic and antifungal substance, cepalycin, from *Pseudomonas cepacia*. Microbiol. Immunol. *38*: 1–9.

Abe, M., M. Tsuda, M. Kimoto, S. Inouye, A. Nakazawa and T. Nakazawa. 1996. A genetic analysis system of *Burkholderia cepacia*: construction of mobilizable transposons and a cloning vector. Gene *174*: 191–194.

Abraham, W.R., H. Meyer, S. Lindholst, M. Vancanneyt and J. Smit. 1997. Phospho- and sulfolipids as biomarkers of *Caulobacter sensu lato*, *Brevundimonas* and *Hyphomonas*. Syst. Appl. Microbiol. *20*: 522–539.

Abraham, W.R., C. Strömpl, H. Meyer, S. Lindholst, E.R. Moore, R. Christ, M. Vancanneyt, B.J. Tindall, A. Bennasar, J. Smit and M. Tesar. 1999. Phylogeny and polyphasic taxonomy of *Caulobacter* species. Proposal of *Maricaulis* gen. nov. with *Maricaulis maris* (Poindexter) comb. nov. as the type species, and emended description of the genera *Brevundimonas* and *Caulobacter*. Int. J. Syst. Bacteriol. *49*: 1053–1073.

Abraham, W.R., C. Strömpl, M. Vancanneyt, H. Lünsdorf and E.R. Moore. 2001. Determination of the systematic position of the genus *Asticca-caulis* Poindexter by a polyphasic analysis. Int. J. Syst. Evol. Microbiol. *51*: 27–34.

Abram, D., J. Castro e Melo and D. Chou. 1974. Penetration of *Bdellovibrio bacteriovorus* into host cells. J. Bacteriol. *118*: 663–680.

Abramochkina, F.N., L.V. Bezrukova, A.V. Koshelev, V.F. Gal'chenko and M.V. Ivanov. 1987. Microbial methane oxidation in a fresh-water reservoir. Mikrobiologiya *56*: 464–471.

Achenbach, L.A., U. Michaelidou, R.A. Bruce, J. Fryman and J.D. Coates. 2001. *Dechloromonas agitata* gen. nov., sp. nov. and *Dechlorosoma suillum* gen. nov., sp. nov., two novel environmentally dominant (per)chlorate-reducing bacteria and their phylogenetic position. Int. J. Syst. Evol. Microbiol. *51*: 527–533.

Achouak, W., R. Christen, M. Barakat, M.H. Martel and T. Heulin. 1999. *Burkholderia caribensis* sp. nov., an exopolysaccharide-producing bacterium isolated from vertisol microaggregates in Martinique. Int. J. Syst. Bacteriol. *49*: 787–794.

Adams, G.A. and A.S. Chaudhari. 1972. Galactosamine polymer isolated from the cell wall of *Neisseria sicca*. Can. J. Biochem. *50*: 345–351.

Adams, L.F. and W.C. Ghiorse. 1986. Physiology and ultrastructure of *Leptothrix discophora* SS-1. Arch. Microbiol. *145*: 126–135.

Adams, L.F. and W.C. Ghiorse. 1987. Characterization of extracellular Mn^{2+}-oxidizing activity and isolation of manganese-oxidizing protein from *Leptothrix discophora* SS-1. J. Bacteriol. *169*: 1279–1285.

Adler, O. 1904. Über Eisenbakterien in ihrer Beziehung zu den therapeutisch verwendteten natürlichen Eisenwässer. Zentbl. Bakteriol. *215*: 277.

Afinogenova, A.V., S.M. Konovalova and V.A. Lambina. 1986. The loss of monospecificity of exoparasitic bacteria of the *Micavibrio* genus. Mikrobiologiya *55*: 487–489.

Agathos, S.N., E. Hellin, H. Ali-Khodja, S. Deseveaux, F. Vandermesse and H. Naveau. 1997. Gas-phase methyl ethyl ketone biodegradation in a tubular biofilm reactor: microbiological and bioprocess aspects. Biodegradation *8*: 251–264.

Agnihothrudu, V., G.C.S. Barua and K.C. Barua. 1959. Occurrence of *Chondromyces* in the rhizosphere of plants. Indian Phytopathol. *12*: 158–160.

Aguero-Rosenfeld, M., H.W. Horowitz, G.P. Wormser, D.F. McKenna, J. Nowakowski, J. Munoz and J.S. Dumler. 1996. Human granulocytic ehrlichiosis (HGE): A series from a single medical center in New York State. Ann. Int. Med. *125*: 904–908.

Aguilar, O.M., H. Reilánder, W. Arnold and A. Pühler. 1987. *Rhizobium meliloti nifN* (*fixF*) gene is part of an operon regulated by a *nifA*-dependent promoter and codes for a polypeptide homologous to the *nifK* gene product. J. Bacteriol. *169*: 5393–5400.

Ahamed, N.M., H. Mayer, H. Biebl and J. Weckesser. 1982. Lipopolysaccharide with 2,3-diamino-2,3-dideoxyglucose containing lipid-A in *Rhodopseudomonas sulfoviridis*. FEMS Microbiol. Lett. *14*: 27–30.

Ahlers, B., W. König and E. Bock. 1990. Nitrite reductase activity in *Nitrobacter vulgaris*. FEMS Microbiol. Lett. *67*: 121–126.

Ahmad, D., J. Fraser, M. Sylvestre, A. Larose, A. Kahn, J. Bergeron, J.M. Juteau and M. Sondossi. 1995. Sequence of the *bphD* gene encoding

2-hydroxy-6-oxo-(phenyl/chlorophenyl)hexa-2,4 dienoic acid (HOP/cPDA) hydrolase involved in the biphenyl/polychlorinated biphenyl degradation pathway in *Comamonas testosteroni*: evidence suggesting involvement of Ser112 in catalytic activity. Gene *156*: 69–74.

Ahmann, D., A.L. Roberts, L.R. Krumholz and F.M. Morel. 1994. Microbe grows by reducing arsenic. Nature *371*: 750.

Aho, E.L. and J.G. Cannon. 1988. Characterization of a silent pilin gene locus from *Neisseria meningitidis* strain FAM18. Microb. Pathog. *5*: 391–398.

Aho, E.L., A.M. Keating and S.M. McGillivray. 2000. A comparative analysis of pilin genes from pathogenic and nonpathogenic Neisseria species. Microb. Pathog. *28*: 81–88.

Aho, E.L., G.L. Murphy and J.G. Cannon. 1987. Distribution of specific DNA sequences among pathogenic and commensal *Neisseria* species. Infect. Immun. *55*: 1009–1013.

Ahrens, A., A. Lipski, S. Klatte, H.J. Busse, G. Auling and K. Altendorf. 1997. Polyphasic classification of *Proteobacteria* isolated from biofilters. Syst. Appl. Microbiol. *20*: 255–267.

Ahrens, R. 1968. Taxonomische Untersuchungen an sternbildenden Agrobacterium-Arten aus der westlichen Ostsee. Kiel. Meeresforsch. *24*: 147–173.

Ahrens, R. and G. Rheinheimer. 1967. Über einige sternbildende Bakterien aus der Ostsee. Kiel. Meeresforsch. *23*: 127–136.

Ahring, B.K., N. Christiansen, I. Mathrani, H.V. Hendriksen, A.J.L. Macario and E. Conway De Macario. 1992. Introduction of a de novo bioremediation ability, aryl reductive dechlorination, into anaerobic granular sludge by inoculation of sludge with *Desulfomonile tiedjei*. Appl. Environ. Microbiol. *58*: 3677–3682.

Akagawa, M. and K. Yamasato. 1989. Synonymy of *Alcaligenes aquamarinus*, *Alcaligenes faecalis* subsp. *homari*, and *Deleya aesta*: *Deleya aquamarina* comb. nov. as the type species of the genus *Deleya*. Int. J. Syst. Bacteriol. *39*: 462–466.

Akagawa-Matsushita, M., T. Itoh, Y. Katayama, H. Kuraishi and K. Yamasato. 1992. Isoprenoid quinone composition of some marine *Alteromonas*, *Marinomonas*, *Deleya*, *Pseudomonas* and *Shewanella* species. J. Gen. Microbiol. *138*: 2275–2281.

Akiba, T., R. Usami and K. Horikoshi. 1983. *Rhodopseudomonas rutila*, a new species of nonsulfur purple photosynthetic bacteria. Int. J. Syst. Bacteriol. *33*: 551–556.

Akopyanz, N., N.O. Bukanov, T.U. Westblom and D.E. Berg. 1992. PCR-based RFLP analysis of DNA sequence diversity in the gastric pathogen *Helicobacter pylori*. Nucleic Acids Res. *20*: 6221–6225.

Akoum, A., R. Guidoin, M.W. King, Y. Marois, M. Sigot and M.F. Sigot-luizard. 1992. A new bioactive molecule for improving vascular graft patency—exploratory trials in dogs. Med. Clin. Exp. *15*: 318–330.

Ala'Aldeen, D.A.A. 1996. Transferrin receptors of *Neisseria meningitidis*: Promising candidates for a broadly cross-protective vaccine. J. Med. Microbiol. *44*: 237–243.

Ala'Aldeen, D.A.A. and S.P. Borriello. 1996. The meningococcal transferrin-binding proteins 1 and 2 are both surface exposed and generate bactericidal antibodies capable of killing homologous and heterologous strains. Vaccine *14*: 49–53.

Alarcón, B., M.M. Lopez, M. Cambra and J. Ortiz. 1987. Comparative study of *Agrobacterium* biotypes 1, 2, and 3 by electrophoresis and serological methods. J. Appl. Bacteriol. *62*: 295–308.

Alban, P.S. and N.R. Krieg. 1996. Improved method for colony counts of the microaerophile *Spirillum volutans*. Can. J. Microbiol. *42*: 701–704.

Alban, P.S. and N.R. Krieg. 1998. A hydrogen peroxide resistant mutant of *Spirillum volutans* has NADH peroxidase activity but no increased oxygen tolerance. Can. J. Microbiol. *44*: 87–91.

Alban, P.S., D.L. Popham, K.E. Rippere and N.R. Krieg. 1998. Identification of a gene for a rubrerythrin/nigerythrin-like protein in *Spirillum volutans* by using amino acid sequence data from mass spectrometry and NH₂-terminal sequencing. J. Appl. Microbiol. *85*: 875–882.

Albanese, E. and D. Kafkewitz. 1978. Effect of medium composition on

the growth and asparaginase production of *Vibrio succinogenes*. Appl. Environ. Microbiol. *36*: 25–30.

Albrecht, H. and A. Ghon. 1901. Über die Aetiologie und pathologische Anatomie der Meningitis cerebrospinalis epidemica. Wein. Klin. Wochenschr. *14*: 984–996.

Albritton, W.L., J.K. Setlow, M.L. Thomas and F.O. Sottnek. 1986. Relatedness within the family *Pasteurellaceae* as determined by genetic transformation. Int. J. Syst. Bacteriol. *36*: 103–106.

Albuquerque, L., J. Santos, P. Travassos, M.F. Nobre, F.A. Rainey, R. Wait, N. Empadinhas, M.T. Silva and M.S. da Costa. 2002. *Albidovulum inexpectatum* gen. nov., sp nov., a nonphotosynthetic and slightly thermophilic bacterium from a marine hot spring that is very closely related to members of the photosynthetic genus Rhodovulum. Appl. Environ. Microbiol. *68*: 4266–4273.

Alderton, M.R., V. Korolik, P.J. Coloe, F.E. Dewhirst and B.J. Paster. 1995. *Campylobacter hyoilei* sp. nov., associated with porcine proliferative enteritis. Int. J. Syst. Bacteriol. *45*: 61–66.

Aldon, D., B. Brito, C. Boucher and S. Genin. 2000. A bacterial sensor of plant cell contact controls the transcriptional induction of *Ralstonia solanacearum* pathogenicity genes. EMBO. *19*: 2304–2314.

Aldrich, H.C., L. McDowell, M.F. Barbosa, L.P. Yomano, R.K. Scopes and L.O. Ingram. 1992. Immunocytochemical localization of glycolytic and fermentative enzymes in *Zymomonas mobilis*. J. Bacteriol. *174*: 4504–4508.

Aldridge, K.E., G.T. Valainis and C.V. Sanders. 1988. Comparison of the *in vitro* activity of ciprofloxacin and 24 other antimicrobial agents against clinical strains of *Chromobacterium violaceum*. Diagn. Microbiol. Infect. Dis. *10*: 31–39.

Alkan, S., M.B. Morgan, R.L. Sandin, L.C. Moscinski and C.W. Ross. 1995. Dual role for *Afipia felis* and *Rochalimaea henselae* in cat-scratch disease. Lancet *345*: 385.

Allaker, R.P., K.A. Young and J.M. Hardie. 1994. Production of hydrolytic enzymes by oral isolates of *Eikenella corrodens*. FEMS Microbiol. Lett. *123*: 69-74.

Allardet-Servent, A., G. Bourg, M. Ramuz, M. Pages, M. Bellis and G. Roizes. 1988. DNA polymorphism in strains of the genus *Brucella*. J. Bacteriol. *170*: 4603–4607.

Allardet-Servent, A., M.J. Carles-Nurit, G. Bourg, S. Michaux and M. Ramuz. 1991. Physical map of the *Brucella melitensis* 16 M chromosome. J. Bacteriol. *173*: 2219–2224.

Allardet-Servent, A., S. Michaux Charachon, E. Jumas-Bilak, L. Karayan and M. Ramuz. 1993. Presence of one linear and one circular chromosome in the *Agrobacterium tumefaciens* C58 genome. J. Bacteriol. *175*: 7869–7874.

Alleman, A.R., S.M. Kamper, N. Viseshakul and A.F. Barbet. 1992. Analysis of the *Anaplasma marginale* genome by pulsed-field electrophoresis. J. Gen. Microbiol. *139*: 2439–2444.

Allen, A.G., R.M. Thomas, J.T. Cadisch and D.J. Maskell. 1998. Molecular and functional analysis of the lipopolysaccharide biosynthesis locus *wlb* from *Bordetella pertussis*, *Bordetella parapertussis* and *Bordetella bronchiseptica*. Mol. Microbiol. *29*: 27–38.

Allen, C., J. Gay and L. Simon-Buela. 1997. A regulatory locus, *pehSR*, controls polygalacturonase production and other virulence functions in *Ralstonia solanacearum*. Mol. Plant-Microbe Interact. *10*: 1054–1064.

Allen, E.K. and O.N. Allen. 1950. Biochemical and symbiotic properties of the rhizobia. Bacteriol. Rev. *14*: 273–330.

Allen, J.R., D.D. Clark, J.G. Krum and S.A. Ensign. 1999. A role for coenzyme M (2-mercaptoethanesulfonic acid) in a bacterial pathway of aliphatic epoxide carboxylation. Proc. Natl. Acad. Sci. U.S.A. *96*: 8432–8437.

Allen, O.N. and A.J. Holding. 1974. Genus II. *Agrobacterium* Conn. 1942, 359. *In* Buchanan and Gibbons (Editors), Bergey's Manual of Determinative Bacteriology, 8th Ed., The Williams & Wilkins Co., Baltimore. pp. 264–267.

Allison, M.J. and H.M. Cook. 1981. Oxalate degradation by microbes of the large bowel of herbivores: the effect of dietary oxalate. Science *212*: 675–676.

Allison, M.J., K.A. Dawson, W.R. Mayberry and J.G. Foss. 1985a. *Oxalo-*

bacter formigenes, gen. nov., sp. nov.: oxalate-degrading anaerobes that inhabit the gastrointestinal tract. Arch. Microbiol. *141*: 1–7.

Allison, M.J., K.A. Dawson, W.R. Mayberry and J.G. Foss. 1985b. *In* Validation of the publication of new names and new combinations previously effectively published outside the IJSB. List No. 18. Int. J. Syst. Bacteriol. *35*: 375–376.

Allison, N., Turner, J.E. and Wait, R.. 1995. Degradation of homovanillate by a strain of *Variovorax paradoxus* via ring hydroxylation. FEMS Microbiol. Lett. *134*: 213–219.

Allsopp, M., E.S. Visser, J.L. du Plessis, S.W. Vogel and B.A. Allsopp. 1997. Different organisms associated with heartwater as shown by analysis of 16S ribosomal RNA gene sequences. Vet. Parasitol. *71*: 283–300.

Allunans, J., M. Bjoras, E. Seeberg and K. Bøvre. 1998. Production, isolation and purification of bacteriocins expressed by two strains of *Neisseria meningitidis*. APMIS *106*: 1181–1187.

Allunans, J. and K. Bovre. 1996. Bacteriocins in *Neisseria meningitidis*—screening of systemic patient strains and pharyngeal isolates from healthy carriers. APMIS *104*: 206–212.

Alm, R.A., P. Guerry, M.E. Power, H. Lior and T.J. Trust. 1991. Analysis of the role of flagella in the heat-labile Lior serotyping scheme of thermophilic Campylobacters by mutant allele exchange. J. Clin. Microbiol. *29*: 2438–2445.

Alm, R.A., L.S. Ling, D.T. Moir, B.L. King, E.D. Brown, P.C. Doig, D.R. Smith, B. Noonan, B.C. Guild, B.L. deJonge, G. Carmel, P.J. Tummino, A. Caruso, M. Ruia-Nickelsen, D.M. Mills, C. Ives, R.J. Gibson, D. Merberg, S.D. Mills, Q. Jiang, E.W. Taylor, G.F. vovis and T.J. Trust. 1999. Genomic-sequence comparison of two unrelated isolates of the human gastric pathogen *Helicobacter pylori*. Nature (Lond.) *397*: 176–180.

Altenschmidt, U., B. Oswald, E. Steiner, H. Herrmann and G. Fuchs. 1993. New aerobic benzoate oxidation pathway via benzoyl-coenzyme A and 3-hydroxybenzoyl-coenzyme A in a denitrifying *Pseudomonas* sp. J. Bacteriol. *175*: 4851–4858.

Althauser, M., W.A. Samsonoff, C. Anderson and S.F. Conti. 1972. Isolation and preliminary characterization of bacteriophages for *Bdellovibrio bacteriovorus*. J. Virol. *10*: 516–523.

Alton, GG., L.M. Jones, R.D. Angus and J.M. Verger. 1988. Techniques for the Brucellosis Laboratory, Institut National de la Recherche Agronomique, Paris.

Altschul, S.F. 1989. Evolutionary trees for the genus *Bordetella*. J. Bacteriol. *171*: 1211–1213.

Altson, R.A. 1936. Studies on *Azotobacter* in Malayan soils. J. Agric. Sci. *26*: 268–280.

Alvarez, B. and G. Martínez-Drets. 1995. Metabolic characterization of *Acetobacter diazotrophicus*. Can. J. Microbiol. *41*: 918–924.

Amann, R.I., W. Ludwig, R. Schulze, S. Spring, E. Moore and K.H. Schleifer. 1996. rRNA-targeted oligonucleotide probes for the identification of genuine and former pseudomonads. Syst. Appl. Microbiol. *19*: 501–509.

Amann, R., N. Springer, W. Ludwig, H.D. Görtz and K.H. Schleifer. 1991. Identification in situ and phylogeny of uncultured bacterial endosymbionts. Nature (Lond.) *351*: 161–164.

Amano, K., M. Cedzynski, A.S. Swierzko, K. Kyohno and W. Kaca. 1996. Comparison of serological reactions of rickettsiae-infected patients and rabbit anti-*Proteus* OX antibodies with *Proteus* OX2, OX19 and OXK lipopolysaccharides. Arch. Immunol. Ther. Exp. *44*: 235–240.

Amano, K.I., J.C. Williams and G.A. Dasch. 1998. Structural properties of lipopolysaccharides from *Rickettsia typhi* and *Rickettsia prowazekii* and their chemical similarity to the lipopolysaccharide from *Proteus vulgaris* OX19 used in the Weil–Felix test. Infect. Immun. *66*: 923–926.

Amano, Y., J. Rumbea, J. Knobloch, J. Olson and M. Kron. 1997. Bartonellosis in Ecuador: serosurvey and current status of cutaneous verrucous disease. Am. J. Trop. Med. Hyg. *57*: 174–179.

Amaral, J.A., C. Archambault, S.R. Richards and R. Knowles. 1995. Denitrification associated with Groups I and II methanotrophs in a gradient enrichment system. FEMS Microbiol. Ecol. *18*: 289–298.

Amarger, N., V. Macheret and G. Laguerre. 1997. *Rhizobium gallicum* sp.

nov. and *Rhizobium giardinii* sp. nov., from *Phaseolus vulgaris* nodules. Int. J. Syst. Bacteriol. *47*: 996–1006.

Ambler, R.P. 1973. Bacterial cytochrome *c* and molecular evolution. Syst. Zool. *22*: 554–565.

Ambler, R.P., M. Daniel, J. Hermoso, T.E. Meyer, T.G. Bartsch and M.D. Kamen. 1979. Cytochrome c_2 sequence variation among the recognized species of purple nonsulfur photosynthetic bacteria. Nature *278*: 659–660.

Amerein, M.P., D. De Briel, B. Jaulhac, P. Meyer, H. Monteil and Y. Piemont. 1996. Diagnostic value of the indirect immunofluorescence assay in cat scratch disease with *Bartonella henselae* and *Afipia felis* antigens. Clin. Diagn. Lab. Immunol. *3*: 200–204.

Amils, R., N. Irazabal, D. Moreira, J.P. Abad and I. Marin. 1998. Genomic organization analysis of acidophilic chemolithotrophic bacteria using pulsed field gel electrophoretic techniques. Biochimie *80*: 911–921.

Amin, P.M. and S.V. Ganapati. 1967. Occurrence of *Zoogloea* colonies and protozoans at different stages of sewage purification. Appl. Microbiol. *15*: 17–21.

Amir, J. and P. Yagupsky. 1998. Invasive *Kingella kingae* infection associated with stomatitis in children. Pediatr. Infect. Dis. J. *17*: 757–758.

Anacker, R.L., R.E. Mann and C. Gonzales. 1987. Reactivity of monoclonal antibodies to *Rickettsia rickettsii* with spotted fever and typhus group rickettsiae. J. Clin. Microbiol. *25*: 167–171.

Anacker, R.L., T.F. McCaul, W. Burgdorfer and R.K. Gerloff. 1980. Properties of selected rickettsiae of the spotted fever group. Infect. Immun. *27*: 468–474.

Anantharam, V., M.J. Allison and P.C. Maloney. 1989. Oxalate:formate exchange: the basis for energy coupling in *Oxalobacter*. J. Biol. Chem. *264*: 7244–7250.

Anast, N. and J. Smit. 1988. Isolation and characterization of marine caulobacters and assessment of their potential for genetic experimentation. Appl. Environ. Microbiol. *54*: 809–817.

Anders, H.J., A. Kaetzke, P. Kämpfer, W. Ludwig and G. Fuchs. 1995. Taxonomic position of aromatic-degrading denitrifying pseudomonad strains K 172 and KB 740 and their description as new members of the genera *Thauera*, as *Thauera aromatica* sp. nov., and *Azoarcus*, as *Azoarcus evansii* sp. nov., respectively, members of the beta subclass of the *Proteobacteria*. Int. J. Syst. Bacteriol. *45*: 327–333.

Andersen, B.M., O. Solberg, K. Bryn, L.O. Froholm, P. Gaustad, E.A. Hoiby, B.E. Kristiansen and K. Bovre. 1987. Endotoxin liberation from *Neisseriam meningitidis* isolated from carriers and clinical cases. Scand. J. Infect. Dis. *19*: 409–419.

Andersen, B.M., A.G. Steigerwalt, S.P. O'Connor, D.G. Hollis, R.S. Weyant, R.E. Weaver and D.J. Brenner. 1993. *Neisseria weaveri* sp. nov., formerly CDC group M-5, a Gram-negative bacterium associated with dog-bite wounds. J. Clin. Microbiol. *31*: 2456–2466.

Andersen, L.P., A. Norgaard, S. Holck, J. Blom and L. Elsborg. 1996. Isolation of a "*Helicobacter heilmanii*"-like organism from the human stomach. Eur. J. Clin. Microbiol. Infect. Dis. *15*: 95–96.

Anderson, A.R. and L.W. Moore. 1979. Host specificity in the genus *Agrobacterium*. Phytopathology *69*: 320–323.

Anderson, B.E. 1990. The 17-kilodalton protein antigens of spotted fever and typhus group rickettsiae. Ann. N. Y. Acad. Sci. *590*: 326–333.

Anderson, B.E., J.E. Dawson, D.C. Jones and K.H. Wilson. 1991. *Ehrlichia chaffeensis*, a new species associated with human ehrlichiosis. J. Clin. Microbiol. *29*: 2838–2842.

Anderson, B.E., J.E. Dawson, D.C. Jones and K.H. Wilson. 1992a. *In* Validation of the publication of new names and new combinations previously effectively published outside the IJSB. List No. 41. Int. J. Syst. Bacteriol. *42*: 327–328.

Anderson, B.E., C.E. Greene, D.C. Jones and J.E. Dawson. 1992b. *Ehrlichia ewingii* sp. nov., the etiologic agent of canine granulocytic ehrlichiosis. Int. J. Syst. Bacteriol. *42*: 299–302.

Anderson, B.E., G.A. McDonald, D.C. Jones and R.L. Regnery. 1990. A protective protein antigen of *Rickettsia rickettsii* has tandemly repeated, near-identical sequences. Infect. Immun. *58*: 2760–2769.

Anderson, B., D. Scotchlas, D. Jones, A. Johnson, T. Tzianabos and B. Baumstark. 1997. Analysis of 36-kilodalton protein (PapA) associated with the bacteriophage particle of *Bartonella henselae*. DNA Cell Biol. *16*: 1223–1229.

Anderson, B.E., J.W. Sumner, J.E. Dawson, T. Tzianabos, C.R. Greene, J.G. Olson, D.B. Fishbein, O.-R. M., B.P. Holloway, E.H. George and A.F. Azad. 1992c. Detection of the etiologic agent of human ehrlichiosis by polymerase chain reaction. J. Clin. Microbiol. *30*: 775–780.

Anderson, G.W., Jr. and J.V. Osterman. 1980. Host defenses in experimental rickettsialpox: Resistance of C3H mouse sublines. Acta Virol. *24*: 294–296.

Anderson, I.C., M. Poth, J. Homstead and D. Burdige. 1993. A comparison of NO and N$_2$O production by the autotrophic nitrifier *Nitrosomonas europaea* and the heterotrophic nitrifier *Alcaligenes faecalis*. Appl. Environ. Microbiol. *59*: 3525–3533.

Anderson, J.D. and H. Smith. 1965. The metabolism of erythritol by *Brucella abortus*. J. Gen. Microbiol. *38*: 109–124.

Anderson, J.E., P.F. Sparling and C.N. Cornelissen. 1994. Gonococcal transferrin-binding protein-2 facilitates but is not essential for transferrin utilization. J. Bacteriol. *176*: 3162–3170.

Anderson, L.P., O. Andersen, S. Holck, J. Blom and T. Justesen. 1996. *Campylobacter mucosalis* in faeces from a child with severe haemorrhagic colitis. *In* Newell, Ketley and Feldman (Editors), Campylobacters, Helicobacters, and Related Organisms, Plenum Press, New York. pp. 503–506.

Anderson, L. and R.C. Fuller. 1967a. Photosynthesis in *Rhodospirillum rubrum*. I. Autotrophic carbon dioxide fixation. Plant. Physiol. *42*: 487–490.

Anderson, L. and R.C. Fuller. 1967b. Photosynthesis in *Rhodospirillum rubrum*. II. Photoheterotrophic carbon dioxide fixation. Plant. Physiol. *42*: 491–496.

Anderson, R.L., W.E. Bishop and R.L. Campbell. 1985. A review of the environmental and mammalian toxicology of nitrilotriacetic acid. Crit. Rev. Toxicol. *15*: 1–102.

Andersson, S.G., A. Zomorodipour, J.O. Andersson, T. Sicheritz-Ponten, U.C. Alsmark, R.M. Podowski, A.K. Naslund, A.S. Eriksson, H.H. Winkler and C.G. Kurland. 1998. The genome sequence of *Rickettsia prowazekii* and the origin of mitochondria. Nature *396*: 133–140.

Andreasen, J.R., Jr. and T. Sandhu. 1993. *Pasteurella anatipestifer*-like bacteria associated with respiratory disease in pigeons. Avian Dis. *37*: 908–911.

Andreesen, M. and H.G. Schlegel. 1974. A new coryneform bacterium: *Corynebacterium autotrophicum* strain 7c. II. Isolation of a slime-free mutant. Arch. Microbiol. *100*: 351–361.

Andreev, L.V. and V.F. Gal'chenko. 1983. Phospholipid composition and differentiation of methanotrophic bacteria. J. Liq. Chromatogr. *6*: 2699–2708.

Andrew, R., J.M. Bonnin and S. Williams. 1946. Tick typhus in North Queensland. Med. J. Aust. *2*: 253–258.

Andrews, K.T. and B.K.C. Patel. 1996. *Fervidobacterium gondwanense* sp. nov., a new thermophilic anaerobic bacterium isolated from nonvolcanically heated geothermal waters of the Great Artesian Basin of Australia. Int. J. Syst. Bacteriol. *46*: 265–269.

Andrutis, K.A., J.G. Fox, D.B. Schauer, R.P. Marini, X. Li, L. Yan, C. Josenhans and S. Suerbaum. 1997. Infection of the ferret stomach by isogenic flagellar mutant strains of *Helicobacter mustelae*. Infect. Immun. *65*: 1962–1966.

Andrutis, K.A., J.G. Fox, D.B. Schauer, R.P. Marini, J.C. Murphy, L. Yan and J.V. Solnick. 1995. Inability of an isogenic urease-negative mutant strain of *Helicobacter mustelae* to colonize the ferret stomach. Infect. Immun. *63*: 3722–3725.

Angus, B.J., S.T. Green, J.J. McKinley, D.J. Goldberg and M. Frischer. 1994. *Eikenella corrodens* septicemia among drug injectors: a possible association with licking wounds. J. Infect. *28*: 102–103.

Anonymous 1994. Schistosomes, liver flukes, and *Heliocobacter pylori*. IARC Monographs on the Evaluation of Carcinogenic Risks to Humans, Vol. 61, IARC Press, Lyon, France.

Antheunisse, J. 1972. Preservation of Microorganisms. J. Microbiol. Serol. *38*: 617–622.

Antheunisse, J. 1973. Viability of lyophilized microorganisms after storage. J. Microbiol. Serol. *39*: 243–248.

Anthony, C. 1982. The Biochemistry of Methylotrophs, Academic Press, Ltd., London.

Anthony, C. 1996. Quinoprotein-catalysed reactions. Biochem. J. *320*: 697–711.

Anzai, Y., H. Kim, J.Y. Park, H. Wakabayashi and H. Oyaizu. 2000. Phylogenetic affiliation of the pseudomonads based on 16S rRNA sequence. Int. J. Syst. Evol. Microbiol. *50*: 1563–1589.

Aoki, M., K. Uehara, K. Koseki, K. Tsuji, M. Iijima, K. Ono and T. Samejima. 1991. An antimicrobial substance produced by *Pseudomonas cepacia* B5 against the bacterial wilt disease pathogen, *Pseudomonas solanacearum*. Agric. Biol. Chem. *55*: 715–722.

Aoyama, T., Y. Murase, T. Iwata, A. Imaizumi, Y. Suzuki and Y. Sato. 1986. Comparison of blood-free medium (cyclodextrin solid medium) with Bordet-Gengou medium for clinical isolation of *Bordetella pertussis*. J. Clin. Microbiol. *23*: 1046–1048.

Apicella, M.A., M.A.J. Westerink, S.A. Morse, H. Schneider, P.A. Rice and J.M. Griffiss. 1986. Bactericidal antibody-response of normal human serum to the lipooligosaccharide of *Neisseria gonorrhoeae*. J. Infect. Dis. *153*: 520–526.

Appelmelk, B.J., I. Simoons-Smit, R. Negrini, A.P. Moran, G.O. Aspinall, J.G. Forte, T. De Vries, H. Quan, T. Verboom, J.J. Maaskant, P. Ghiara, E.J. Kuipers, E. Bloemena, T.M. Tadema, R.R. Townsend, K. Tyagarajan, J.M. Crothers, Jr., M.A. Monteiro, A. Savio and J. De Graaf. 1996. Potential role of molecular mimicry between *Helicobacter pylori* lipopolysaccharide and host Lewis blood group antigens in autoimmunity. Infect. Immun. *64*: 2031–2040.

Aragno, M. 1975. Mise en évidence d'hydrogénobactéries corynéformes auxohétérotrophes pour la biotine dans l'eau d'un lac eutrophe. Ann. Microbiol. (Inst. Pasteur) *126A*: 539–542.

Aragno, M. and H.G. Schlegel. 1977. *Alcaligenes ruhlandii* (Packer and Vishniac) comb. nov., a peritrichous hydrogen bacterium previously assigned to *Pseudomonas*. Int. J. Syst. Bacteriol. *27*: 279–281.

Aragno, M. and H.G. Schlegel. 1978. *Aquaspirillum autotrophicum*, a new species of hydrogen-oxidizing, facultatively autotrophic bacteria. Int. J. Syst. Bacteriol. *28*: 112–116.

Aragno, M. and H.G. Schlegel. 1992. The mesophilic hydrogen-oxidizing (Knallgas) bacteria. *In* Balows, Trüper, Dworkin, Harder and Schleifer (Editors), The Prokaryotes: A Handbook of Bacteria: Ecophysiology, Isolation, Identification, Applications, 2nd Ed., Vol. 1, Springer-Verlag, New York. pp. 344–384.

Aragno, M., A. Walther-Mauruschat, F. Mayer and H.G. Schlegel. 1977. Micromorphology of Gram-negative hydrogen bacteria. I. Cell morphology and flagellation. Arch. Microbiol. *114*: 93–100.

Arata, H., Y. Serikawa and K. Takamiya. 1988. Trimethylamine *N*-oxide respiration by aerobic photosynthetic bacterium, *Erythrobacter* sp. OCh 114. J. Biochem. (Tokyo) *103*: 1011–1015.

Arata, S., T. Hirayama, N. Kasai, T. Itoh and A. Ohsawa. 1989. Isolation of 9-hydroxy-delta-tetradecalactone from lipid-A of *Pseudomonas diminuta* and *Pseudomonas vesicularis*. FEMS Microbiol. Lett. *60*: 219–222.

Arata, S., N. Kasai, T.W. Klein and H. Friedman. 1994. *Legionella pneumophila* growth restriction and cytokine production by murine macrophages activated by a novel *Pseudomonas* lipid-A. Infect. Immun. *62*: 729–732.

Archer, J.R., S. Romero, A.E. Ritchie, M.E. Hamacher, B.M. Steiner, J.H. Bryner and R.F. Schell. 1988. Characterization of an unclassified microaerophilic bacterium associated with gastroenteritis. J. Clin. Microbiol. *26*: 101–105.

Arendsen, A.F., M.F.J.M. Verhagen, R.B.G. Wolbert, A.J. Pierik, A.J.M. Stams, M.S.M. Jetten and W.R. Hagen. 1993. The dissimilatory sulfite reductase from *Desulfosarcina variabilis* is a desulforubidin containing uncoupled metalated sirohemes and S /2 iron-sulfur clusters. Biochemistry *32*: 10323–10330.

Arensdorf, J.J. and D.D. Focht. 1995. A *meta* cleavage pathway for 4-chlorobenzoate, an intermediate in the metabolism of 4-chlorobiphenyl by *Pseudomonas cepacia* P166. Appl. Environ. Microbiol. *61*: 443–447.

Aretz, W., H. Kaspari and J.H. Klemme. 1978. Utilization of purines as nitrogen source by facultative phototropic bacteria. FEMS Microbiol. Lett. *4*: 249–253.

Arfmann, H.-A., K.N. Timmis and R.M. Wittich. 1997. Mineralization of 4-chlorodibenzofuran by a consortium consisting of *Sphingomonas* sp. strain RW1 and *Burkholderia* sp. strain JWS. Appl. Environ. Microbiol. *63*: 3458–3462.

Argall, M.E. and G.D. Smith. 1993. The use of trehalose-stabilized lyophilized methanol dehydrogenase from *Hyphomicrobium* X for the detection of methanol. Biochem. Mol. Biol. Int. *30*: 491–497.

Arico, B. and R. Rappuoli. 1987. *Bordetella parapertussis* and *Bordetella bronchiseptica* contain transcriptionally silent pertussis toxin genes. J. Bacteriol. *169*: 2847–2853.

Arimi, S.M., R.W.A. Park and C.R. Fricker. 1990. Study of haemolytic activity of some *Campylobacter* spp. on blood agar plates. J. Appl. Bacteriol. *69*: 384–389.

Arisoy, E.S., A.G. Correa, M.L. Wagner and S.L. Kaplan. 1999. Hepatosplenic cat-scratch disease in children: selected clinical features and treatment. Clin. Infect. Dis. *28*: 778–784.

Aristovskaya, T.V. 1961. Accumulation of iron in breakdown of organominereal humus complexes by microorganisms (in Russian). Dokl. Akad. Nauk. S.S.S.R. *136*: 954–957.

Aristovskaya, T.V. 1963. On the decomposition of organic minreal compounds in podzolic soils. Pochvoved. Akad. Nauk. S.S.S.R. *1*: 30–42.

Aristovskaya, T.V. 1964. The taxonomic position of the genus *Seliberia* Arist. et Parink. Mikrobiologiya *33*: 823–828.

Aristovskaya, T.V. 1974. *Seliberia. In* Buchanan and Gibbons (Editors), Bergey's Manual of Determinative Bacteriology, 8th Ed., The Williams & Wilkins Co., Baltimore. p. 160.

Aristovskaya, T.V. and V.V. Parinkina. 1961. New soil microorganism *Seliberia stellata* nov. gen. n. sp. Izvestiya Akad. Nauk SSSR, Ser. Biol. *28*: 49–56.

Aristovskaya, T.V. and V.V. Parinkina. 1963. New soil microorganism *Seliberia stellata* nov. gen., n. sp. Izv. Akad. Nauk S.S.S. R. Ser. Biol. *28*: 49–56.

Ark, P.A. and H.E. Thomas. 1946. Bacterial leaf spot and bud rot of orchids caused by *Phytomonas cattleyae*. Phytopathology *36*: 695–698.

Arko, R.J. and T. Odugbemi. 1984. Superoxol and amylase inhibition tests for distinguishing gonococcal and nongonococcal cultures growing on selective media. J. Clin. Microbiol. *20*: 1–4.

Armengaud, J., B. Happe and K.N. Timmis. 1998. Genetic analysis of dioxin dioxygenase of *Sphingomonas* sp. Strain RW1: catabolic genes dispersed on the genome. J. Bacteriol. *180*: 3954–3966.

Armengaud, J. and K.N. Timmis. 1997. Molecular characterization of Fdx1, a putidaredoxin-type [2Fe–2S] ferredoxin able to transfer electrons to the dioxin dioxygenase of *Sphingomonas* sp. RW1. Eur. J. Biochem. *247*: 833–842.

Armengaud, J. and K.N. Timmis. 1998. The reductase RedA2 of the multicomponent dioxin dioxygenase system of *Sphingomonas* sp. RW1 is related to class-I cytochrome P_{450}-type reductases. Eur. J. Biochem. *253*: 437–444.

Armengaud, J., K.N. Timmis and R.M. Wittich. 1999. A functional 4-hydroxysalicylate/hydroxyquinol degradative pathway gene cluster is linked to the initial dibenzo-*p*-dioxin pathway genes in *Sphingomonas* sp. strain RW1. J. Bacteriol. *181*: 3452–3461.

Arnold, J.W. and L.J. Shimkets. 1988. Cell-surface properties correlated with cohesion in *Myxococcus xanthus*. J. Bacteriol. *170*: 5771–5777.

Aronoff, S.C. 1988. Outer membrane permeability in *Pseudomonas cepacia* diminished porin content in a β-lactam resistant mutant and in resistant cystic fibrosis isolates. Antimicrob. Agents Chemother. *32*: 1636–1639.

Arp, L.H. and N.F. Cheville. 1984. Tracheal lesions in young turkeys infected with *Bordetella avium*. Am. J. Vet. Res. *45*: 2196–2200.

Arthur, L.O., J. Bulla, L.A., G. St. Julian and L.K. Nakamura. 1973. Carbohydrate metabolism in *Agrobacterium tumefaciens*. J. Bacteriol. *116*: 304–313.

Arthur, L.O., L.K. Nakamura, G. St. Julian and J. Bulla, L.A.. 1975. Car-

bohydrate catabolism of selected strains in the genus *Agrobacterium*. Appl. Microbiol. *30*: 731–737.

Artymiuk, P.J., E.R. Bauminger, P.M. Harrison, D.M. Lawson, I. Nowik, A. Treffry and S.J. Yewdall. 1991. Ferritin: A model system for iron biomineralization. *In* Frankel and Blakemore (Editors), Iron Biominerals, Plenum Press, New York. pp. 269–294.

Arzumanyan, V.G., Z.V. Sakharova, N.S. Panikov and V.S. Vshivtsev. 1997. Growth and nitrogen-fixing activity of the batch culture of *Xanthobacter autotrophicus* at various concentrations of dissolved oxygen. Mikrobiologiya *66*: 750–754.

Asai, T. 1935. Taxonomic studies on acetic acid bacteria and allied oxidative bacteria isolated from fruits. A new classification of the oxidative bacteria. J. Agr. Chem. Soc. Jpn. *11*: 674–708.

Asai, T. 1968. Acetic acid bacteria. Classification and biochemical activities, University of Tokyo Press, Tokyo.

Asanovich, K.M., J.S. Bakken, J.E. Madigan, M. Aguero-Rosenfeld, G.P. Wormser and J.S. Dumler. 1997. Antigenic diversity of granulocytic *Ehrlichia* isolates from humans in Wisconsin and New York and a horse in California. J. Infect. Dis. *176*: 1029–1034.

Aschner, M. and I. Chorin-Kirsh. 1970. Light-oriented locomotion in certain myxobacter species. Arch. Microbiol. *74*: 308–314.

Ashbolt, N.J. and P.A. Inkerman. 1990. Acetic acid bacterial biota of the pink sugar cane mealybug, *Saccharococcus sacchari*, and its environs. Appl. Environ. Microbiol. *56*: 707–712.

Ashdown, L.R. 1979a. Improved screening technique for isolation of *Pseudomonas pseudomallei* from clinical specimens. Pathology *11*: 293–297.

Ashdown, L.R. 1979b. Nosocomial infection due to *Pseudomonas pseudomallei*: two cases and an epidemiologic study. Rev. Infect. Dis. *1*: 891–894.

Aspinall, S.T. and R. Graham. 1989. Two sources of contamination of a hydrotherapy pool by environmental organisms. J. Hosp. Infect. *14*: 285–292.

Aspinall, S.T., D.R.A. Wareing, P.G. Hayward and D.N. Hutchinson. 1993. Selective medium for thermophilic campylobacters including *Campylobacter upsaliensis*. J. Clin. Pathol. *46*: 829–831.

Aspinall, S.T., D.R.A. Wareing, P.G. Hayward and D.N. Hutchinson. 1996. A comparison of a new campylobacter selective medium (CAT) with membrane filtration for the isolation of thermophilic campylobacters including *Campylobacter upsaliensis*. J. Appl. Bacteriol. *80*: 645–650.

Assinder, S.J. and P.A. Williams. 1990. The TOL plasmids: determinants of the catabolism of toluene and the xylenes. Adv. Microb. Physiol. *31*: 1–69.

Assmus, B., P. Hutzler, G. Kirchhof, R. Amann, J.R. Lawrence and A. Hartmann. 1995. In situ localization of *Azospirillum brasilense* in the rhizosphere of wheat with fluorescently labeled, rRNA-targeted oligonucleotide probes and scanning confocal laser microscopy. Appl. Environ. Microbiol. *61*: 1013–1019.

Assmus, B., M. Schloter, G. Kirchhof, P. Hutzler and A. Hartmann. 1997. Improved in situ tracking of rhizosphere bacteria using dual staining with fluorescence-labeled antibodies and rRNA-targeted oligonucleotides. Microb. Ecol. *33*: 32–40.

Atabay, H.I. and J.E. Corry. 1997. The prevalence of campylobacters and arcobacters in broiler chickens. J. Appl. Microbiol. *83*: 619–626.

Atabay, H.I., J.E.L. Corry and S.L. On. 1998a. Diversity and prevalence of *Arcobacter* spp. in broiler chickens. J. Appl. Microbiol. *84*: 1007–1016.

Atabay, H.I., J.E.L. Corry and S.L.W. On. 1998b. Identification of unusual *Campylobacter*-like isolates from poultry products as *Helicobacter pullorum*. J. Appl. Microbiol. *84*: 1017–1024.

Atherton, J.G. 1983. Evaluation of selective supplements used in media for the isolation of the causative organism of contagious equine metritis. Vet. Rec. *113*: 299–300.

Atherton, J.C., P. Cao, R.M. Peek, Jr., M.K.R. Tummuru, M.J. Blaser and T.L. Cover. 1995. Mosaicism in Vacuolating Cytotoxin Alleles of *Helicobacter pylori*: Association of specific *vacA* types with cytotoxin production and peptic ulceration. J. Biol. Chem. *270*: 17771–17777.

Atkey, P.T., T.R. Fermor and S.P. Lincoln. 1992. Electron microscopy of the infection process of rapid soft rot disease of the edible mushroom *Agaricus bitorquis.* Mycol. Res. *96:* 717–722.

Attwood, M.M. and W. Harder. 1972. A rapid and specific enrichment procedure for *Hyphomicrobium* spp. Antonie Leeuwenhoek *38:* 369–377.

Attwood, M.M. and W. Harder. 1974. The oxidation and assimilation of C₂ compounds by *Hyphomicrobium* sp. J. Gen. Microbiol. *84:* 350–356.

Attwood, M.M. and W. Harder. 1977. Isocitrate lyase activity in *Hyphomicrobium* spp. a critical reappraisal. FEMS Microbiol. Lett. *1:* 25–30.

Attwood, M.M., J.P. van Dijken and J.T. Pronk. 1991. Glucose metabolism and gluconic acid production by *Acetobacter diazotrophicus.* J. Ferment. Bioeng. *72:* 101–105.

Auburger, G. and J. Winter. 1995. Isolation and physiological characterization of *Syntrophus buswellii* strain GA from a syntrophic benzoate degrading strictly anaerobic coculture. Appl. Microbiol. Biotechnol. *44:* 241–248.

Auburger, G. and J. Winter. 1996. Activation and degradation of benzoate, 3-phenylpropianate and crotonate by *Syntrophus buswellii* strain GA. Evidence for electron transport phosphorylation during crotonate respiration. Appl. Microbiol. Biotechnol. *44:* 807–815.

Audy, J.R. 1968. Scrub-itch and the ecologist. *In* Audy (Editor), Red Mites and Typhus, Oxford University Press Inc., New York. pp. 1–27.

Auling, G., H.J. Busse, T. Egli, T. El-Banna and E. Stackebrandt. 1993a. Description of the Gram-negative, obligately aerobic, nitrilotriacetate (NTA)-utilizing bacteria as *Chelatobacter heintzii,* gen. nov., sp. nov., and *Chelatococcus asaccharovorans,* gen. nov., sp. nov. Syst. Appl. Microbiol. *16:* 104–112.

Auling, G., M. Dittbrenner, M. Maarzahl, T. Nokhal and M. Reh. 1980. Deoxyribonucleic acid relationships among hydrogen-oxidizing strains of the genera *Pseudomonas, Alcaligenes,* and *Paracoccus.* Int. J. Syst. Bacteriol. *30:* 123–128.

Auling, G., H.-J. Busse, T. Egli, T. El-Banna and E. Stackebrandt. 1993b. *In* Validation of the publication of new names and new combinations previously effectively published outside the IJSB. List No. 46. Int. J. Syst. Bacteriol. *43:* 624.

Auling, G., H.-J. Busse, M. Hahn, H. Hennecke, R.-M. Kroppenstedt, A. Probst and E. Stackebrandt. 1988. Phylogenetic heterogeneity and chemotaxonomic properties of certain Gram-negative aerobic carboxydobacteria. Syst. Appl. Microbiol. *10:* 264–272.

Auling, G., H.-J. Busse, F. Pilz, L. Webb, H. Kneifel and D. Claus. 1991. Rapid differentiation, by polyamine analysis, of *Xanthomonas* strains from phytopathogenic pseudomonads and other members of the class *Proteobacteria* interacting with plants. Int. J. Syst. Bacteriol. *41:* 223–228.

Auling, G., M. Reh, C.M. Lee and H.G. Schlegel. 1978. *Pseudomonas pseudoflava* a new species of hydrogen-oxidizing bacteria: its differentiation from *Pseudomonas flava* and other yellow-pigmented, Gram-negative, hydrogen oxidizing species. Int. J. Syst. Bacteriol. *28:* 82–95.

Auran, T.B. and E.L. Schmidt. 1972. Similarities between *Hyphomicrobium* and *Nitrobacter* with respect to fatty acids. J. Bacteriol. *109:* 450–451.

Austen, R.A. and T.J. Trust. 1980. Detection of plasmids in the related group of the genus *Campylobacter.* FEMS Microbiol. Lett. *8:* 201–204.

Austin, B. and M. Goodfellow. 1979. *Pseudomonas mesophilica,* a new species of pink bacteria isolated from leaf surfaces. Int. J. Syst. Bacteriol. *29:* 373–378.

Austin, B., C.J. Rodgers, J.M. Forns and R.R. Colwell. 1981. *Alcaligenes faecalis* subsp. *homari* subsp. nov., a new group of bacteria isolated from moribund lobsters. Int. J. Syst. Bacteriol. *31:* 72–76.

Austin, F.E., J. Turco and H.H. Winkler. 1987. *Rickettsia prowazekii* requires host cell serine and glycine for growth. Infect. Immun. *55:* 240–244.

Austin, F.E. and H.H. Winkler. 1988a. Proline incorporation into protein by *Rickettsia prowazekii* during growth in Chinese hamster ovary (CHO-K1) cells. Infect. Immun. *56:* 3167–3172.

Austin, F.E. and H.H. Winkler. 1988b. Relationship of rickettsial physiology and composition to the rickettsia-host cell interaction. *In* Walker (Editor), Biology of Rickettsial Diseases, CRC Press, Boca Raton. pp. 29–49.

Austin, J.W. and R.G.E. Murray. 1987. The perforate component of the regularly structured (RS) layer of *Lampropedia hyalina.* Can. J. Microbiol. *33:* 1039–1045.

Austin, J.W. and R.G.E. Murray. 1990. Isolation and *in vitro* assembly of the components of the outer S layer of *Lampropedia hyalina.* J. Bacteriol. *172:* 3681–3689.

Ayling, R.D., M.J. Woodward, S. Evans and D.G. Newell. 1996. Restriction fragment length polymorphism of polymerase chain reaction products applied to the differentiation of poultry campylobacters for epidemiological investigations. Res. Vet. Sci. *60:* 168–172.

Azegami, K., K. Nishiyama, Y. Watanabe, I. Kadota, A. Ohuchi and C. Fukazawa. 1987. *Pseudomonas plantarii* sp. nov., the causal agent of rice seedling blight. Int. J. Syst. Bacteriol. *37:* 144–152.

Aznar, R., R.J. Owen and J. Hernandez. 1992. DNA–DNA hybridization and ribotyping of *Acidovorax delafieldii* isolates from eels and aquatic environments. Lett. Appl. Microbiol. *14:* 185–188.

Baalsrud, K. and K.S. Baalsrud. 1954. Studies on *Thiobacillus denitrificans.* Arch. Mikrobiol. *20:* 34–62.

Babalis, T., Y. Tselentis, V. Roux, A. Psaroulaki and D. Raoult. 1994. Isolation and identification of a rickettsial strain related to *Rickettsia massiliae* in Greek ticks. Am. J. Trop. Med. Hyg. *50:* 365–372.

Babel, W. 1984. Assimilation of methanol by an acidophilic bacterium of the genus *Acetobacter.* Acta Biotechnol. *4:* 369–376.

Babinchak, J.A. and V.F. Gerencser. 1976. Bacteriophage typing of the "*Caulobacter* group". Int. J. Syst. Bacteriol. *26:* 82–84.

Babudieri, B. 1950. Natura delle cosidette "S-formen" delle leptospire. Loro identificazione con *Hyphomicrobium vulgare* Stutzer e Hartleb. Studio di quest. Ultimo germ. Estratto dai Rendiconti dell'Istituto Superiore di Sanita. *13:* 580-591.

Babudieri, B. 1973. Experimental infections by spirilla. *In* Eichler (Editor), Handbuch der Experimentellen Pharmakologie, Vol. 17 11B, Springer-Verlag, New York Berlin. pp. 43–49.

Bachhawat, A.K. and S. Ghosh. 1987a. Iron transport in *Azospirillum brasilense*: role of the siderophore spirilobactin. J. Gen. Microbiol. *133:* 1759–1765.

Bachhawat, A.K. and S. Ghosh. 1987b. Isolation and characterization of the outer membrane proteins of *Azospirillum brasilense.* J. Gen. Microbiol. *133:* 1751–1758.

Bachofen, R. and A. Schenk. 1998. Quorum sensing autoinducers: do they play a role in natural microbial habitats? Microbiol. Res. *153:* 61–63.

Bachrach, G., M. Banai, S. Bardenstein, G. Hoida, A. Genizi and H. Bercovier. 1994a. *Brucella* ribosomal protein L7/L12 is a major component in the antigenicity of brucellin INRA for delayed-type hypersensitivity in *Brucella* sensitized guinea pigs. Infect. Immun. *62:* 5361–5366.

Bachrach, G., D. Bar-Nir, M. Banai and H. Bercovier. 1994b. Identification and nucleotide sequence of *Brucella melitensis* L7/L12 ribosomal protein. FEMS Microbiol. Lett. *120:* 237–240.

Backman, A., P. Lantz, P. Radstrom and P. Olcen. 1999. Evaluation of an extended diagnostic PCR assay for detection and verification of the common causes of bacterial meningitis in CSF and other biological samples. Mol. Cell. Probes *13:* 49–60.

Bacon, K. and F.A. Eiserling. 1967. A unique structure in microcysts of *Myxococcus xanthus.* J. Ultrastruct. Res. *21:* 378–382.

Badger, S.J., T. Butler, C.K. Kim and K.H. Johnston. 1979. Experimental *Eikenella corrodens* endocarditis in rabbits. Infect. Immun. *23:* 751–757.

Badger, S.J. and A.C.R. Tanner. 1981. Serological studies of *Bacteroides gracilis, Campylobacter concisus, Wolinella recta,* and *Eikenella corrodens,* all from humans with periodontal disease. Int. J. Syst. Bacteriol. *31:* 446–451.

Baena, S., M.L. Fardeau, M. Labat, B. Ollivier, J.L. Garcia and B.K.C. Patel. 1998. *Desulfovibrio aminophilus* sp. nov., a novel amino acid degrading and sulfate reducing bacterium from an anaerobic dairy wastewater lagoon. Syst. Appl. Microbiol. *21:* 498–504.

Baena, S., M.L. Fardeau, M. Labat, B. Ollivier, J.L. Garcia and B.K.C. Patel. 1999. *In* Validation of the publication of new names and new combinations previously effectively published outside the IJSB. List no. 69. Int. J. Syst. Bacteriol. *49:* 341–342.

Baer, M.L. 1998. Molecular Characterization of the Bacterial Predator *Bdellovibrio*, Doctoral thesis, University of Maryland, Baltimore, Maryland.

Baer, M.L., J. Ravel, J. Chun, R.T. Hill and H.N. Williams. 2000. A proposal for the reclassification of *Bdellovibrio stolpii* and *Bdellovibrio starrii* into a new genus, *Bacteriovorax* gen. nov. as *Bacteriovorax stolpii* comb. nov. and *Bacteriovorax starrii* comb. nov., respectively. Int. J. Syst. Evol. Microbiol. 50: 219–224.

Baer, M.L., J. Ravel, A.J. Schoeffield, R.T. Hill and H.N. Williams. 1998. Analysis of *Bdellovibrio* spp. by arbitrarily primed PCR, pulsed field electrophoresis, ribotyping and 16S rDNA analysis. 98th Annual Meeting of the American Society for Microbiology, p. 481.

Baer, M.L., A.J. Schoeffield, D. Serio, C. Frederick, W. Buchanan, M. Challmes and H.N. Williams. 1994. Interaction of halophilic bdellovibrios with an attached *Vibrio vulnificus* community. 94th Annual Meeting of the American Society for Microbiology, p. 322.

Baev, N. and Á. Kondorosi. 1992. Nucleotide sequence of the *Rhizobium meliloti nodL* gene located in locus n5 of the *nod* regulon. Plant. Mol. Biol. 18: 843–846.

Baev, N., G. Endre, G. Petrovics, Z. Banfalvi and Á. Kondorosi. 1991. Six nodulation genes of *nod* box locus 4 in *Rhizobium meliloti* are involved in nodulation signal production: *nodM* codes for D-glucosamine synthetase. Mol. Gen. Genet. 228: 113–124.

Baev, N., M. Schultze, I. Barlier, D.C. Ha, H. Virelizier, E. Kondorosi and A. Kondorosi. 1992. *Rhizobium nodM* and *nodN* genes are common *nod* genes: *nodM* encodes functions for efficiency of nod signal production and bacteroid maturation. J. Bacteriol. 174: 7555–7565.

Baier, R., A. Meyer, V. DePalma, R. Krieg and M. Fornalik. 1983. Surface microfouling during the induction period. J. Heat Trans. 105: 618–624.

Bailie, W.E., E.C. Stowe and A.M. Schmitt. 1978. Aerobic bacterial flora of oral and nasal fluids of canines with reference to bacteria associated with bites. J. Clin. Microbiol. 7: 223–231.

Baird, R.W., M. Lloyd, J. Stenos, B.C. Ross, R.S. Stewart and B. Dwyer. 1992. Characterization and comparison of Australian human spotted fever group rickettsiae. J. Clin. Microbiol. 30: 2896–2902.

Bak, F. and N. Pfennig. 1987. Chemolithotrophic growth of *Desulfovibrio sulfodismutans*, new species by disproportionation of inorganic sulfur compounds. Arch. Microbiol. 147: 184–189.

Bak, F. and N. Pfennig. 1988. *In* Validation of the publication of new names and new combinations previously effectively published outside the IJSB. List No. 24. Int. J. Syst. Bacteriol. 38: 136–137.

Bak, F. and F. Widdel. 1986. Anaerobic degradation of indolic compounds by sulfate-reducing enrichment cultures, and description of *Desulfobacterium indolicum* gen. nov., sp. nov. Arch. Microbiol. 146: 170–176.

Bak, F. and F. Widdel. 1988. *In* Validation of the publication of new names and new combinations previously effectively published outside the IJSB. List No. 24. Int. J. Syst. Bacteriol. 38: 136–137.

Baker, C.A., G.W. Claus and P.A. Taylor. 1983. Predominant bacteria in an activated sludge reactor for the degradation of cutting fluids. Appl. Environ. Microbiol. 46: 1214–1223.

Baker, D.A. and R.W.A. Park. 1975. Changes in morphology and cell wall structure that occur during growth of *Vibrio* sp. NCTC 4716 in batch culture. J. Gen. Microbiol. 86: 12–28.

Baker, M.E. 1996. 3-α-hydroxysteroid dehydrogenase is homologous to a fusion of bacterial ribosomal L10 and L7/12 genes. J. Steroid Biochem. Mol. Biol. 59: 365–366.

Baker, P.J. and J.B. Wilson. 1965. Chemical composition and biological properties of endotoxin of *Brucella abortus*. J. Bacteriol. 90: 895–902.

Baker, S.C., S.J. Ferguson, B. Ludwig, M.D. Page, O.M.H. Richter and R.J.M. van Spanning. 1998. Molecular genetics of the genus *Paracoccus*: metabolically versatile bacteria with bioenergetic flexibility. Microbiol. Mol. Biol. Rev. 62: 1046–1078.

Baker, S.C., C.F. Goodhew, I.P. Thompson, A. Bramwell, G. Pettigrew and S.J. Ferguson. 1995. A study of *Paracoccus denitrificans* using fatty acid methyl ester analysis and cytochrome c_{550} amino acid sequence. *In* Schaffers and Dijken (Editors), Beijerinck Centennial. Microbial Phys-

iology and Gene Regulation: Emerging Principles and Applications, Delft University Press, The Netherlands. pp. 311–312.

Bakken, J.S., J.S. Dumler, S.M. Chen, M.R. Eckman, L.L. Van Etta and D.H. Walker. 1994. Human granulocytic ehrlichiosis in the upper Midwest United States. A new species emerging? JAMA. 272: 212–218.

Bakken, J.S., J. Krueth, C. Wilson-Nordskog, R.L. Tilden, K. Asanovich and J.S. Dumler. 1996. Clinical and laboratory characteristics of human granulocytic ehrlichiosis. JAMA. 275: 199–205.

Balashova, V.V. 1967a. Enrichment culture of *Gallionella filamenta* n. sp. Mikrobiologiya 36: 646–650.

Balashova, V.V. 1967b. Structure of the "stalk" fibers in a laboratory culture of *Gallionella filamenta*. Mikrobiologiya 36: 1050–1053.

Balashova, V.V. 1968. Taxonomy of the genus *Gallionella*. Mikrobiologiya 37: 715–723.

Balashova, V.V. and N.E. Cherni. 1970. Ultrastructure of *Gallionella filamenta*. Mikrobiologiya 39: 348–351.

Balasubramanian, A. and S.R. Prabhu. 1989. Occurrence of the new species *Azospirillum halopraeferens* in association with rice roots. Curr. Sci. (Bangalore) 58: 1391–1392.

Balch, W.E. and R.S. Wolfe. 1976. New approach to the cultivation of methanogenic bacteria: 2-mercaptoethanesulfonic acid (HS-CoM)-dependent growth of *Methanobacterium ruminantium* in a pressureized atmosphere. Appl. Environ. Microbiol. 32: 781–791.

Baldani, J.I., V.L.D. Baldani, M.J.A.M. Sampaio and J. Dobereiner. 1984. A fourth *Azospirillum* species from cereal roots. An. Acad. Bras. Cienc. 56: 365.

Baldani, J.I., L.D. Baldani, L. Seldin and J. Döbereiner. 1986a. Characterization of *Herbaspirillum seropedicae* gen. nov., sp. nov., a root-associated nitrogen-fixing bacterium. Int. J. Syst. Bacteriol. 36: 86–93.

Baldani, J.I., L. Caruso, V.L.D. Baldani, S.R. Goi and J. Doebereiner. 1997. Recent advances in BNF with non-legume plants. Soil Biol. Biochem. 29: 911–922.

Baldani, J.I., B. Pot, G. Kirchhof, E. Falsen, V.L.D. Baldani, F.L. Olivares, B. Hoste, K. Kersters, A. Hartmann, M. Gillis and J. Döbereiner. 1996. Emended description of *Herbaspirillum*: inclusion of "*Pseudomonas*" *rubrisubalbicans*, a mild plant pathogen, as *Herbaspirillum rubrisubalbicans* comb. nov.; and classification of a group of clinical isolates (EF group 1) as *Herbaspirillum* species 3. Int. J. Syst. Bacteriol. 46: 802–810.

Baldani, V.L.D., M.A. Alvarez, J.I. Baldani and J. Dobereiner. 1986b. Establishment of inoculated *Azospirillum* spp. in the rhizosphere and in roots of field grown wheat and sorghum. Plant Soil 90: 35–46.

Baldani, V.L.D., J.I. Baldani, F. Olivares and J. Dobereiner. 1992. Identification and ecology of *Herbaspirillum seropedicae* and the closely related *Pseudomonas rubrisubalbicans*. Symbiosis 13: 65–73.

Baldani, V.L.D., F.L. Olivares, S.R. Goi, J.I. Baldani and J. Döbereiner. 1994. Infection and colonization of rice and sugarcane plants by *Herbaspirillum* spp. Proceedings of the NATO Advanced Research Workshop on *Azospirillum* and related Microorganisms, Sárvar, Hungary. p. 5.

Bale, S.J., K. Goodman, P.A. Rochelle, J.R. Marchesi, J.C. Fry, A.J. Weightman and R.J. Parkes. 1997. *Desulfovibrio profundus* sp. nov., a novel barophilic sulfate-reducing bacterium from deep sediment layers in the Japan Sea. Int. J. Syst. Bacteriol. 47: 515–521.

Balke, E., A. Weber and B. Fronk. 1977. Untersuchungen des Aminosaurestoffwechsels mit der Dunnschichtchromatographie zur Differenzierung von Brucellen. Zentralbl. Bakteriol. Parasitenkd. Infektionskr. Hyg. Abt. I Orig.A. 237: 523–529.

Balkwill, D.L., G.R. Drake, R.H. Reeves, J.K. Fredrickson, D.C. White, D.B. Ringelberg, D.P. Chandler, M.F. Romine, D.W. Kennedy and C.M. Spadoni. 1997. Taxonomic study of aromatic-degrading bacteria from deep-terrestrial-subsurface sediments and description of *Sphingomonas aromaticivorans* sp. nov, *Sphingomonas subterranea* sp. nov., and *Sphingomonas stygia* sp. nov. Int. J. Syst. Bacteriol. 47: 191–201.

Balkwill, D.L., D. Maratea and R.P. Blakemore. 1980. Ultrastructure of a magnetotactic spirillum. J. Bacteriol. 141: 1399–1408.

Ballard, R.W., M. Doudoroff, R.Y. Stanier and M. Mandel. 1968. Taxonomy of the aerobic pseudomonads: *Pseudomonas diminuta* and *P.vesiculare*. J. Gen. Microbiol. *53*: 349–361.

Ballard, R.W., N.J. Palleroni, M. Doudoroff, R.Y. Stanier and M. Mandel. 1970. Taxonomy of the aerobic pseudomonads: *Pseudomonas cepacia, P. marginata, P. alliicola,* and *P. caryophylli*. J. Gen. Microbiol. *60*: 199–214.

Bally, M. 1994. Physiology and ecology of nitrilotriacetate-degrading bacteria in pure culture, activated sludge and surface waters, Thesis, Swiss Federal Institute of Technology, Zürich.

Bally, M. and T. Egli. 1996. Dynamics of substrate consumption and enzyme synthesis in *Chelatobacter heintzii* during growth in carbon-limited continuous culture with different mixtures of glucose and nitrilotriacetate. Appl. Environ. Microbiol. *62*: 133–140.

Bally, M., E. Wilberg, M. Kühni and T. Egli. 1994. Growth and regulation of enzyme synthesis in the nitrilotriacetic acid (NTA)-degrading bacterium *Chelatobacter heintzii* ATCC 29600. Microbiology *140*: 1927–1936.

Bally, R., D. Thomas-Bauzon, T. Heulin, J. Balandreau and C. Richard. 1983. Determination of the most frequent N_2-fixing bacteria in a rice rhizosphere. Can. J. Microbiol. *29*: 881–887.

Balsalobre, J.M., R.M. Ruizvazquez and F.J. Murillo. 1987. Light induction of gene expression in *Myxococcus xanthus*. Proc. Natl. Acad. Sci. U.S.A. *84*: 2359–2362.

Bambauer, A., F.A. Rainey, E. Stackebrandt and J. Winter. 1998a. Characterization of *Aquamicrobium defluvii* gen. nov. sp. nov., a thiophene-2-carboxylate-metabolizing bacterium from activated sludge. Arch. Microbiol. *169*: 293–302.

Bambauer, A., F.A. Rainey, E. Stackebrandt and J. Winter. 1998b. *In* Validation of the publication of new names and new combinations previously effectively published outside the IJSB. List No. 66. Int. J. Syst. Bacteriol. *48*: 631–632.

Banai, M., I. Mayer and A. Cohen. 1990. Isolation, identification, and characterization in Israel of *Brucella melitensis* biovar 1 atypical strains susceptible to dyes and penicillin, indicating the evolution of a new variant. J. Clin. Microbiol. *28*: 1057–1059.

Bandi, C., T.J. Anderson, C. Genchi and M.L. Blaxter. 1998. Phylogeny of *Wolbachia* in filarial nematodes. Proc. R. Soc. Lond. B Biol. Sci. *265*: 2407–2413.

Bandi, C., J.W. McCall, C. Genchi, S. Corona, L. Venco and L. Sacchi. 1999. Effects of tetracycline on the filarial worms *Brugia pahangi* and *Dirofilaria immitis* and their bacterial endosymbionts *Wolbachia*. Int. J. Parasitol. *29*: 357–364.

Bandi, C., M. Sironi, C.A. Nalepa, S. Corona and L. Sacchi. 1997. Phylogenetically distant intracellular symbionts in termites. Parassitologia. *39*: 71–75.

Bandi, C., A.J. Trees and N.W. Brattig. 2001. *Wolbachia* in filarial nematodes: evolutionary aspects and implications for the pathogenesis and treatment of filarial diseases. Vet. Parasitol. *98*: 215–238.

Banerjee, P.C., M.K. Ray, C. Koch, S. Bhattacharyya, S. Shivaji and E. Stackebrandt. 1996. Molecular characterization of two acidophilic heterotrophic bacteria isolated from a copper mine of India. Syst. Appl. Microbiol. *19*: 78–82.

Banfalvi, Z., A. Nieuwkoop, M. Schell, L. Besl and G. Stacey. 1988. Regulation of *nod* gene expression in *Bradyrhizobium japonicum*. Mol. Gen. Genet. *214*: 420–424.

Bang, B. 1897. Die aetiologie des seuchenhaften ("infektiosen") verwerfens. Z. Tiermed. *1*: 241–278.

Bano, N. and J.T. Hollibaugh. 2000. Diversity and distribution of DNA sequences with affinity to ammonia-oxidizing bacteria of the β subdivision of the class *Proteobacteria* in the Arctic Ocean. Appl. Environ. Microbiol. *66*: 1960–1969.

Bar, T. and Y. Okon. 1995. Conversion of tryptophan, indole-3-pyruvic acid, indole-3-lactic acid and indole- to indole-3-acetic acid by *Azospirillum brasilense* Sp7. NATO ASI Ser. Ser. G Ecol. Sci. *37*: 347–359.

Baraldes, M.A., P. Domingo, J.L. Barrio, R. Pericas, M. Gurgui and G. Vazquez. 2000. Meningitis due to *Neisseria subflava*: Case report and review. Clin. Infect. Dis. *30*: 615–617.

Baraoidan, M.R. 1981. Bacterial stripe of rice: occurrence, identification

and sources of inoculum, Thesis, University of the Philippines, Los Baños.

Barbet, A.F. 1995. Recent developments in the molecular biology of anaplasmosis. Vet. Parasitol. *57*: 43–49.

Barbosa, H.R. and F. Alterthum. 1992. The role of extracellular polysaccharide in cell viability and nitrogenase activity of *Beijerinckia derxii*. Can. J. Microbiol. *38*: 986–988.

Barbosa, H.R., M.F.A. Rodrigues, C.C. Campos, M.E. Chaves, I. Nunes, Y. Juliano and N.F. Novo. 1995. Counting of viable cluster forming and non cluster forming bacteria a comparison between the drop and the spread methods. J. Microbiol. Methods 22: 39–50.

Barbour, W.M., S.P. Wang and G. Stacey. 1992. Molecular genetics of *Bradyrhizobium* symbioses. *In* Stacey, Burris and Evans (Editors), Biological Nitrogen Fixation, Chapman and Hall, New York. 648–684.

Bardenstein, S., M. Mandelboim, T.A. Ficht, M. Baum and M. Banai. 2002. Identification of the *Brucella melitensis* vaccine strain Rev1 in animals and humans in Israel by PCR analysis of the PstI site polymorphism of its *omp2* gene. J. Clin. Microbiol. *40*: 1475–1480.

Barel, G. and E. Jurkevitch. 2001. Analysis of phenotypic diversity among host-independent mutants of *Bdellovibrio bacteriovorus* 109J. Arch. Microbiol. *176*: 211–216.

Barelmann, I., J.M. Meyer, K. Taraz and H. Budzikiewicz. 1996. Cepaciachelin, a new catecholate siderophore from *Burkholderia (Pseudomonas) cepacia*. Z. Naturforsch. Sect. C J. Biosci. *51*: 627–630.

Barker, B.T.P. 1948. Some recent studies on the nature and incidence of cider sickness. Annu. Rep. Agric. Hort. Res. Stn. Long Ashton Bristol, pp. 174–181.

Barker, B.T.P. and V.F. Hillier. 1912. Cider sickness. J. Agric. Sci. *5*: 67–85.

Barlough, J.E., G.H. Reubel, J.E. Madigan, L.K. Vredevoe, P.E. Miller and Y. Rikihisa. 1998. Detection of *Ehrlichia risticii*, the agent of Potomac horse fever, in freshwater stream snails (*Pleuroceridae: Juga* spp.) from northern California. Appl. Environ. Microbiol. *64*: 2888–2893.

Barnes, A., L. Galbraith and S.G. Wilkinson. 1989. The Presence of 11-methyloctadec-11-enoic acid in the extractable lipids of *Pseudomonas vesicularis*. FEMS Microbiol. Lett. *59*: 101–106.

Barnett, M.J., R.F. Fisher and 24 other authors. 2001. Nucleotide sequence and predicted functions of the entire *Sinorhizobium meliloti* pSymA megaplasmid. Proc. Natl Acad. Sci. U.S.A. *98*: 9883–9888.

Barnett, M.J. and S.R. Long. 1990. DNA sequence and translational product of a new nodulation-regulatory locus: *syrM* has sequence similarity to *nodD* proteins. J. Bacteriol. *172*: 3695–3700.

Barnewall, R.E., N. Ohashi and Y. Rikihisa. 1999. *Ehrlichia chaffeensis* and *E. sennetsu*, but not the human granulocytic ehrlichiosis agent, colocalize with transferrin receptor and up-regulate transferrin receptor mRNA by activating iron-responsive protein 1. Infect. Immun. *67*: 2258–2265.

Baron, E.J. 1997. B*ilophila wadsworthia*: a unique gram-negative anaerobic rod. Anaerobe *3*: 83–86.

Baron, E.J., R. Bennion, J. Thompson, C. Strong, P. Summanen, M. McTeague and S.M. Finegold. 1992a. A microbiological comparison between acute and complicated appendicitis. Clin. Infect. Dis. *14*: 227–231.

Baron, E.J., M. Curren, G. Henderson, H. Jousimies-Somer, K. Lee, K. Lechowitz, C.A. Strong, P. Summanen, K. Tuner and S.M. Finegold. 1992b. *Bilophila wadsworthia* isolates from clinical specimens. J. Clin. Microbiol. *30*: 1882–1884.

Baron, E.J., G. Ropers, P. Summanen and R.J. Courcol. 1993. Bactericidal activity of selected antimicrobial agents against *Bilophila wadsworthia* and *Bacteroides gracilis*. Clin. Infect. Dis. *16*: S339–S343.

Baron, E.J., P. Summanen, J. Downes, M.C. Roberts, H. Wexler and S.M. Finegold. 1989. *Bilophila wadsworthia*, gen. nov. and sp. nov., a unique Gram-negative anaerobic rod recovered from appendicitis specimens and human faeces. J. Gen. Microbiol. *135*: 3405–3411.

Baron, E.J., P. Summanen, J. Downes, M.C. Roberts, H. Wexler and S.M. Finegold. 1990. *In* Validation of the publication of new names and new combinations previously effectively published outside the IJSB. List No. 34. Int. J. Syst. Bacteriol. *40*: 320–321.

Barrera, L.L., M.E. Trujillo, M. Goodfellow, F.J. Garcia, I. Hernandez Lucas, G. Davila, P. van Berkum and E. Martinez Romero. 1997. Biodiversity of bradyrhizobia nodulating *Lupinus* spp. Int. J. Syst. Bacteriol. *47*: 1086–1091.

Barrett, P.A., E. Beveridge, P.L. Bradley, C.G.D. Brown, S.R.M. Bushby, M.L. Clarke, R.A. Neal, R. Smith and J.K.H. Wilde. 1965. Biological activities of some α-dithiosemicarbozones. Nature *206*: 1340–1341.

Barrett, S.J., L.K. Schlater, R.J. Montali and P.H.A. Sneath. 1994a. A New Species of *Neisseria* from iguanid Lizards, *Neisseria iguanae* sp. nov. Lett. Appl. Microbiol. *18*: 200–202.

Barrett, S.J., L.K. Schlater, R.J. Montali and P.H.A. Sneath. 1994b. *In* Validation of the publication of new names and new combinations previously effectively published outside the IJSB. Validation List 51. Int. J. Syst. Bacteriol. *44*: 852.

Barrett, S.J. and P.H. Sneath. 1994. A numerical phenotypic taxonomic study of the genus *Neisseria*. Microbiology *140*: 2867–2891.

Barrow, G.I. and R.K.A. Feltham. 1993. Appendix A: Preparation and control of culture media. *In* Barrow and Feltham (Editors), Cowan and Steele's Manual for the Identification of Medical bacteria, 3rd Ed., Cambridge University Press, Cambridge. pp. 188–213.

Barsomian, G. and T.G. Lessie. 1986. Replicon fusions promoted by insertion sequences on *Pseudomonas cepacia* plasmid pTGL6. Mol. Gen. Genet. *204*: 273–280.

Bart, A., J. Dankert and A. van der Ende. 1999. Antigenic variation of the class I outer membrane protein in hyperendemic *Neisseria meningitidis* strains in the Netherlands. Infect. Immun. *67*: 3842–3846.

Barta, J.R., Y. Boulard and S.S. Desser. 1989. Blood parasites of *Rana esculenta* from Corsica: comparison of its parasites with those of eastern North American ranids in the context of host phylogeny. Trans. Am. Microsc. Soc. *108*: 6–20.

Barthel, T., R. Jonas and H. Sahm. 1989. NADP$^+$-dependent acetaldehyde dehydrogenase from *Zymomonas mobilis*—isolation and partial characterization. Arch. Microbiol. *153*: 95–100.

Bartlett, J. and S.M. Finegold. 1978. Bacteriology of expectorated sputum with quantitative culture and wash technique compared to transtracheal aspirates. Am. Rev. Respir. Dis. *117*: 1019–1027.

Bartosch, S., I. Wolgast, E. Spieck and E. Bock. 1999. Identification of nitrite-oxidizing bacteria with monoclonal antibodies recognizing the nitrite oxidoreductase. Appl. Environ. Microbiol. *65*: 4126–4133.

Bashan, Y. and G. Holguin. 1997. *Azospirillum*-plant relationships: environmental and physiological advances (1990–1996). Can. J. Microbiol. *43*: 103–121.

Basnayake, W.V. and R.G. Birch. 1995. A gene from *Alcaligenes denitrificans* that confers albicidin resistance by reversible antibiotic binding. Microbiology (Read.) *141*: 551–560.

Bass, J.W., B.C. Freitas, A.D. Freitas, C.L. Sisler, D.S. Chan, J.M. Vincent, D.A. Person, J.R. Claybaugh, R.R. Wittler, M.E. Weisse, R.L. Regnery and L.N. Slater. 1998. Prospective randomized double blind placebo-controlled evaluation of azithromycin for treatment of cat-scratch disease. Pediatr. Infect. Dis. J. *17*: 447–452.

Bassalik, K. 1913. Über die verabeitung der oxalsäure durch *Bacillus extorquens* n. sp. Jahrb. Botan. *53*: 255–302.

Bastyns, K., D. Cartuyvels, S. Chapelle, P. Vandamme, H. Goossens and R. DeWachter. 1995a. A variable 23S rDNA region is a useful discriminating target for genus-specific and species-specific PCR amplification in *Arcobacter* species. Syst. Appl. Microbiol. *18*: 353–356.

Bastyns, K., S. Chapelle, P. Vandamme, H. Goossens and R. De Wachter. 1994. Species-specific detection of Campylobacters important in veterinary medicine by PCR amplification of 23S rDNA areas. Syst. Appl. Microbiol. *17*: 563–568.

Bastyns, K., S. Chapelle, P. Vandamme, H. Goossens and R. De Wachter. 1995b. Specific detection of *Campylobacter concisus* by PCR amplification of 23S rDNA areas. Mol. Cell. Probes *9*: 247–250.

Batie, C.J., E. Lahaie and D.P. Ballou. 1987. Purification and characterization of phthalate oxygenase and phthalate oxygenase reductase from *Pseudomonas cepacia*. J. Biol. Chem. *262*: 1510–1518.

Batrakov, S.G. and D.I. Nikitin. 1996. Lipid composition of the phos-phatidylcholine-producing bacterium *Hyphomicrobium vulgare* NP-160. Biochim. Biophys. Acta *1302*: 129–137.

Batrakov, S.G., D.I. Nikitin and I.A. Pitryuk. 1996. A novel glycolipid, 1,2-diacyl-3-α-D-glucuronopyranosyl-sn-glycerol taurineamide, from the budding seawater bacterium *Hyphomonas jannaschiana*. Biochim. Biophys. Acta *130*: 167–176.

Batrakov, S.G., D.I. Nikitin, V.I. Sheichenko and A.O. Ruzhitsky. 1997. Unusual lipid composition of the gram-negative, freshwater, stalked bacterium *Caulobacter bacteroides* NP-105. Biochim. Biophys. Acta *1347*: 127–139.

Battelli, C. 1947. Si di un piroplasma della Naia nigrocollis (*Aegyptianella carpani* n. sp.). Riv. Parassitol. *8*: 205–212.

Batterman, H.J., J.A. Peek, J.S. Loutit, S. Falkow and L.S. Tompkins. 1995. *Bartonella henselae* and *Bartonella quintana* adherence to and entry into cultured human epithelial cells. Infect. Immun. *63*: 4553–4556.

Bauld, J., R. Bigford and J.T. Staley. 1983. *Prosthecomicrobium litoralum*, a new species from marine habitats. Int. J. Syst. Bacteriol. *33*: 613–617.

Baumann, L. and P. Baumann. 1978. Studies of relationship among terrestrial *Pseudomonas*, *Alcaligenes*, and enterobacteria by an immunological comparison of glutamine synthetase. Arch. Microbiol. *119*: 25–30.

Baumann, L., P. Baumann, M. Mandel and R.D. Allen. 1972. Taxonomy of aerobic marine eubacteria. J. Bacteriol. *110*: 402–429.

Baumann, L., R.D. Bowditch and P. Baumann. 1983. Description of *Deleya* gen. nov. created to accommodate the marine species *Alcaligenes aestus*, *Alcaligenes pacificus*, *Alcaligenes cupidus*, *Alcaligenes venustus*, and *Pseudomonas marina*. Int. J. Syst. Bacteriol. *33*: 793–802.

Baumgarten, J., M. Reh and H.G. Schlegel. 1974. Taxonomic studies on some Gram-positive coryneform hydrogen bacteria. Arch. Microbiol. *100*: 207–217.

Baur, E. 1905. Myxobakterien studien. Arch. Protistenkd. *5*: 92–121.

Baxter, I.A. and P.A. Lambert. 1994. Isolation and partial purification of a carbapenem hydrolysing metallo-β-lactamase from *Pseudomonas cepacia*. FEMS Microbiol. Lett. *122*: 251–256.

Baxter, I.A., P.A. Lambert and I.N. Simpson. 1997. Isolation from clinical sources of *Burkholderia cepacia* possessing characteristics of *Burkholderia gladioli*. J. Antimicrob. Chemother. *39*: 169–175.

Bazlikova, M. and R. Brezina. 1978. Some biological properties of rickettsiae isolated in Armenian SSR. *In* Kazar, Ormsbee and Tarasevich (Editors), Rickettsiae and Rickettsial Diseases, VEDA, Bratislava. pp.155–159.

Bazylinski, D.A. and R.P. Blakemore. 1983a. Denitrification and assimilatory nitrate reduction in *Aquaspirillum magnetotacticum*. Appl. Environ. Microbiol. *46*: 1118–1124.

Bazylinski, D.A. and R.P. Blakemore. 1983b. Nitrogen fixing (acetylene reduction) in *Aquaspirillum magnetotacticum*. Curr. Microbiol. *9*: 305–308.

Bazylinski, D.A., L.K. Kimble, D. Schüler, E.P. Phillips and D.R. Lovley. 2000. N$_2$-dependent growth and nitrogenase activity in the metal-metabolizing bacteria *Geobacter metallireducens* and *Magnetospirillum* species. Env. Microbiol. *2*: 266–273.

Bazzicalupo, M. and E. Gallori. 1983. Genetic analysis in *Azospirillum*. *In* Klingmuller (Editor), *Azospirillum* II: Genetics, Physiology, Ecology, Birkhauser-Verlag, Basel. pp. 24–28.

Bazzocchi, C., F. Ceciliani, J.W. McCall, I. Ricci, C. Genchi and C. Bandi. 2000a. Antigenic role of the endosymbionts of filarial nematodes: IgG response against the *Wolbachia* surface protein in cats infected with *Dirofilaria immitis*. Proc. R. Soc. Lond. B Biol. Sci. *267*: 2511–2516.

Bazzocchi, C., W. Jamnongluk, S.L. O'Neill, T.J. Anderson, C. Genchi and C. Bandi. 2000b. *wsp* gene sequences from the *Wolbachia* of filarial nematodes. Curr. Microbiol. *41*: 96–100.

Beardsley, R.E. 1955. Phage production by crown-gall bacteria and the formation of plant tumors. Am. Natur. *89*: 175–176.

Beardsmore, A.J., P.N.G. Aperghis and J.R. Quayle. 1982. Characterization of the assimilatory and dissimilatory pathways of carbon metabolism during growth of *Methylophilus methylotrophus* on methanol. J. Gen. Microbiol. *128*: 1423–1440.

Beati, L., M. Meskini, B. Thiers and D. Raoult. 1997. *Rickettsia aeschlimannii* sp. nov., a new spotted fever group rickettsia associated with *Hyalomma marginatum* ticks. Int. J. Syst. Bacteriol. *47*: 548–554.

Beati, L., O. Péter, W. Burgdorfer, A. Aeschlimann and D. Raoult. 1993. Confirmation that *Rickettsia helvetica* sp. nov. is a distinct species of the spotted fever group of rickettsiae. Int. J. Syst. Bacteriol. *43*: 521–526.

Beati, L. and D. Raoult. 1993. *Rickettsia massiliae* sp. nov., a new spotted fever group rickettsia. Int. J. Syst. Bacteriol. *43*: 839–840.

Beati, L., V. Roux, A. Ortuno, J. Castella, F.S. Porta and D. Raoult. 1996. Phenotypic and genotypic characterization of spotted fever group rickettsiae isolated from Catalan *Rhipicephalus sanguineus* ticks. J. Clin. Microbiol. *34*: 2688–2694.

Beaulieu, C., L.J. Coulombe, R.L. Granger, B. Miki, C. Beauchamp, G. Rossignol and P. Dion. 1983. Characterization of opine-utilizing bacteria isolated from Quebec. Phytoprotection. *64*: 61–68.

Becking, J.H. 1959. Nitrogen-fixing bacteria of the genus *Beijerinckia* in South African soils. Plant Soil *11*: 193–206.

Becking, J.H. 1961. Studies on nitrogen-fixing bacteria of the genus *Beijerinckia*. I. Geographical and ecological distribution in soils. Plant Soil *14*: 49–81.

Becking, J.H. 1962. Species differences in molybdenum and vanadium requirements and combined nitrogen utilization by Azotobacteraceae. Plant Soil *16*: 171–201.

Becking, J.H. 1963. Fixation of molecular nitrogen by an aerobic *Vibrio* or *Spirillum*. Antonie van Leeuwenhoek J. Microbiol. Serol. *29*: 326.

Becking, J.H. 1974. Nitrogen-fixing bacteria of the genus *Beijerinckia*. Soil Sci. *118*: 196–212.

Becking, J.H. 1978. *Beijerinckia* in irrigated rice soils. *In* Environmental Role of Nitrogen-Fixing Blue-Green Algae and Asymbiotic Bacteria, Vol. Ecol. Bull. 26, Stockholm. pp. 116–129.

Becking, J.H. 1981. The family Azotobacteraceae. *In* Starr, Stolp, Trüper, Balows and Schlegel (Editors), The Prokaryotes: A Handbook on Habitats, Isolation, and Identification of Bacteria, 1st Ed., Vol. 1, Springer-Verlag, Berlin. pp. 795–817.

Becking, J.H. 1984a. Genus *Beijerinckia*. *In* Kreig and Holt (Editors), Bergey's Manual of Systematic Bacteriology, 1st Ed., Vol. 1, The Williams & Wilkins Co., Baltimore. pp. 311–321.

Becking, J.H. 1984b. Genus *Derxia*. *In* Kreig and Holt (Editors), Bergey's Manual of Systematic Bacteriology, 1st Ed., Vol. 1, The Williams & Wilkins Co., Baltimore. pp. 321–325.

Beebe, J.M. 1941a. The morphology and cytology of *Myxococcus xanthus*, n. sp. J. Bacteriol. *42*: 193–223.

Beebe, J.M. 1941b. Studies on the myxobacteria. Iowa State College J. Sci. *15*: 307–317.

Beebe, J.M. 1941c. Studies on the myxobacteria. 2. The role of myxobacteria as bacterial parasites. Iowa State College J. Sci. *15*: 319–337.

Beeder, J., T. Torsvik and T. Lien. 1995. *Thermodesulforhabdus norvegicus* gen. nov., sp. nov., a novel thermophilic sulfate-reducing bacterium from oil field water. Arch. Microbiol. *164*: 331–336.

Beeder, J., T. Torsvik and T. Lien. 1996. *In* Validation of the publication of new names and new combinations previously effectively published outside the IJSB. List No. 57. Int. J. Syst. Bacteriol. *46*: 625–626.

Beger, H. and G. Bringmann. 1953. Bisherige Anschauung über die Morphologie von Gallionella und neuere elektronenmikroskopische Befunde. Zentbl. Bakteriol. Parasitenkd. Infektkrankh. Hyg. Abt. II *107*: 305–318.

Behling, U.H., P. Phan and A. Nowotny. 1979. Biological activity of the slime and endotoxin of the periodontopathic organism *Eikenella corrodens*. Infect. Immunol. *26*: 580–584.

Behmlander, R.M. and M. Dworkin. 1991. Extracellular fibrils and contact-mediated cell interactions in *Myxococcus xanthus*. J. Bacteriol. *173*: 7810–7821.

Behrens, H., J. Flossdorf and H. Reichenbach. 1976. Base composition of deoxyribonucleic acid from *Nannocystis exedens* (*Myxobacterales*). Int. J. Syst. Bacteriol. *26*: 561–562.

Beier, C.L., M. Horn, R. Michel, M. Schweikert, H.D. Görtz and M. Wagner. 2002. The genus *Caedibacter* comprises endosymbionts of *Paramecium* spp. related to the *Rickettsiales* (*Alphaproteobacteria*) and to *Fran-*

cisella tularensis (*Gammaproteobacteria*). Appl. Biochem. Microbiol. *68*: 6043–6050.

Beijerinck, M.W. 1895. Über *Spirillum desulfuricans* als Ursache von Sulfatreduktion. Zentralbl. Bakteriol. Parasitenkd. Infektionskr. Hyg. Abt. I Orig. *1*: 1–9; 49–59; 104–114.

Beijerinck, M.W. 1898. Über die Arten der Essigbakterien. Zentralbl. Bakteriol. Parasitenkd. Infektionskr. Hyg.. Abt. II *4*: 209–216.

Beijerinck, M.W. 1904a. Phéromenes de réduction produits par les microbes. Arch. Neer. Sci. (Sect. 2) *9*: 131–157.

Beijerinck, M.W. 1904b. Über Bakterien welche sich im dunkeln mit Kohlensäure als Kohlenstoffquelle ernähren können. Zentbl. Bakteriol. Parasitenkd. Infectionskr. Hyg. Abt. II *11*: 593–599.

Beijerinck, M.W. 1916. Formation of pyruvic acid from malic acid by microbes. Verslag gewone Vergad. Akad. Amst. *18*: 1198–2000.

Beijerinck, M. 1925. Über ein *Spirillum*, welches freien Stickstoff binden kann? Zentralbl Bakteriol. Parasitenkd. Infektionskr. Hyg. Abt. 2 *63*: 353–359.

Beijerinck, M.W. and D.C. Minkman. 1910. Bildung und verbrauch von stickoxydul durch bakterien. Zentbl. Bakteriol. Parasitenik. Abt. II *25*: 30–63.

Beijerinck, M.W. and A. van Delden. 1902. Über die Assimilation des freien Stickstoffs durch Bakterien. Zentralbl. Bakteriol. Parasitenk. Infektionskr. Hyg. Abt. II *9*: 3–43.

Bejuk, D., J. Begovac, A. Bace, N. Kuzmanovic-Sterk and B. Aleraj. 1995. Culture of *Bordetella pertussis* from three upper respiratory tract specimens. Pediatr. Infect. Dis. J. *14*: 64–65.

Belaïch, J.P. and J.C. Senez. 1965. Influence of aeration and pantothenate on growth yields of *Zymomonas mobilis*. J. Bacteriol. *89*: 1195–1200.

Belbahri, L., C. Boucher, T. Candresse, M. Nicole, P. Ricci and H. Keller. 2001. A local accumulation of the *Ralstonia solanacearum* PopA protein in transgenic tobacco renders a compatible plant-pathogen interaction incompatible. Plant J. *28*: 419–430.

Bell, E.J., G.M. Kohls, H.G. Stoenner and D.B. Lackman. 1963. Nonpathogenic rickettsias related to the spotted fever group isolated from ticks, *Dermacentor variabilis* and *Dermacentor andersoni* from Eastern Montana. J.Immunol. *90*: 770–781.

Bell, E.J. and E.G. Pickens. 1953. A toxic substance associated with the rickettsias of the spotted fever group. J. Immunol. *70*: 461–472.

Bell, E.J. and H.G. Stoenner. 1960. Immunologic relationships among the spotted fever group of rickettsias determined by toxin neutralization tests in mice with convalescent animal serums. J. Immunol. *84*: 171–182.

Bell, R.G. and D.J. Latham. 1975. Influence of NaCl, Ca^{2+} and Mg^{2+} on the growth of marine *Bdellovibrio* spp. Estuar. Coast. Mar. Sci. *3*: 381–384.

Bell, S.C. and J.M. Turner. 1973. Iodinin biosynthesis by a pseudomonad. Biochem. Soc. Trans. *1*: 751–753.

Bella, F., B. Font, S. Uriz, T. Munoz, E. Espejo, J. Traveria, J.A. Serrano and F. Segura. 1990. Randomized trial of doxycycline versus josamycin for Mediterranean spotted fever. Antimicrob. Agents Chemother. *34*: 937–938.

Bellaire, B.H., P.H. Elzer, C.L. Baldwin and R.M. Roop. 1999. The siderophore 2,3-dihydroxybenzoic acid is not required for virulence of *Brucella abortus* in BALB/c mice. Infect. Immun. *67*: 2615–2618.

Beller, H.R. and A.M. Spormann. 1997. Anaerobic activation of toluene and *o*-xylene by addition to fumarate in denitrifying strain T. J. Bacteriol. *179*: 670–676.

Bellion, E. and L.B. Hersh. 1972. Methylamine metabolism in a *Pseudomonas* species. Arch. Biochem. Biophys. *153*: 368–374.

Bellion, E. and J.C. Spain. 1976. The distribution of the isocitrate lyase serine pathway amongst one-carbon utilizing organisms. Can. J. Microbiol. *22*: 404–408.

Bellmann, W. and F. Lingens. 1985. Demonstration of herbicide degrading bacteria by the detection of a bacteria specific fatty-acid in soil. Naturwissenschaften *72*: 599.

Belyaev, S.S. 1967. Distribution of the caulobacter group of bacteria in the Volga-Don reservoirs. Mikrobiologiya *36*: 157–162.

Belyaev, S.S. 1969. Certain aspects of the ecology of *Caulobacter*. Mikrobiologiya *38*: 499–504.

Bemis, D.A., L.E. Carmichael and M.J.G. Appel. 1977. Naturally occurring respiratory disease in a kennel caused by *Bordetella bronchiseptica*. Cornell Vet. *67*: 282–293.

Ben Dekhil, S., M. Cahill, E. Stackebrandt and L.I. Sly. 1997a. Transfer of *Conglomeromonas largomobilis* subsp. *largomobilis* to the genus *Azospirillum* as *Azospirillum largomobile* comb. nov., and elevation of *Conglomeromonas largomobilis* subsp. *parooensis* to the new type species of *Conglomeromonas*, *Conglomeromonas parooensis* sp. nov. Syst. Appl. Microbiol. *20*: 72–77.

Ben Dekhil, S., M. Cahill, E. Stackebrandt and L.I. Sly. 1997b. *In* Validation of the publication of new names and new combinations previously effectively published outside the IJSB. List No. 62. Int. J. Syst. Bacteriol. *47*: 915–916.

Ben Dekhil, S.M., M.M. Peel, V.A. Lennox, E. Stackebrandt and L.I. Sly. 1997c. Isolation of *Lautropia mirabilis* from sputa of a cystic fibrosis patient. J. Clin. Microbiol. *35*: 1024–1026.

Ben-Tovim, T., E. Eylan, A. Romano and R. Stein. 1974. Gram-negative bacteria isolated from external eye infections. Infection *2*: 162–165.

Bender, H. 1963. Untersuchungen an *Myxococcus xanthus*. II. Mitteilung. Partielle Lyse von Pullularia pullulans und einigen echten Hefen durch ein extracelluläres Enzymsystem. Arch. Microbiol. *45*: 407–422.

Bender, R.A., C.M. Refson and E.A. O'Neill. 1989. Role of the flagellum in cell-cycle-dependent expression of bacteriophage receptor activity in *Caulobacter crescentus*. J. Bacteriol. *171*: 1035–1040.

Bendis, I. and L. Shapiro. 1970. Properties of *Caulobacter* ribonucleic acid bacteriophage φCb5. J. Virol. *6*: 847–854.

Benjamin, J., S. Leaper, R.J. Owen and M.B. Skirrow. 1983. Description of *Campylobacter laridis*, a new speices comprising the nalidixic acid resistant thermophilic *Campylobacter* (NARTC) group. Curr. Microbiol. *8*: 231–238.

Benjamin, J., S. Leaper, R.J. Owen and M.B. Skirrow. 1984. *In* Validation of the publication of new names and new combinations previously effectively published outside the IJSB. List No. 14. Int. J. Syst. Bacteriol. *34*: 270–271.

Bennett, B.L., J.E. Smadel and R.L. Gauld. 1949. Studies on scrub typhus (tsutsugamushi disease). IV. Heterogeneity of strains of *R. tsutsugamushi* as demonstrated by cross-neutralization tests. J. Immunol. *62*: 453–461.

Bennion, R.S., E.J. Baron, J.E. Thompson, Jr., J. Downes, P. Summanen, D.A. Talan and S.M. Finegold. 1990a. The bacteriology of gangrenous and perforated appendicitis—revisited. Ann. Surg. *211*: 165–171.

Bennion, R.S., J.E. Thompson, E.J. Baron and S.M. Finegold. 1990b. Gangrenous and perforated appendicitis with peritonitis: treatment and bacteriology. Clin.Ther. *12 (Suppl. C)*: 31–44.

Beppu, T. 1993. Genetic organization of *Acetobacter* for acetic acid fermentation. Antonie Leeuwenhoek *64*: 121–135.

Berestetsky, O.A., L.F. Vasyuk, T.A. Elisashvili and A.V. Plyushch. 1985. The activity of nitrogen fixation and the effect of spirilla growing on plant roots. Mikrobiologiya *54*: 1002–1007.

Berg, R.H., M.E. Tyler, N.J. Novick, V. Vasil and I.K. Vasil. 1980. Biology of *Azospirillum*-sugarcane association: enhancement of nitrogenase activity. Appl. Environ. Microbiol. *39*: 642–649.

Berg, R.H., V. Vasil and I.K. Vasil. 1979. The biology of *Azospirillum*-sugarcane association. II. Ultrastructure. Protoplasma *101*: 143–163.

Bergan, T. 1981. Human- and animal-pathogenic members of the genus *Pseudomonas*. *In* Starr, Stolp, Trüper, Balows and Schlegel (Editors), The Prokaryotes: A Handbook on Habitats, Isolation, and Identification of Bacteria, Springer-Verlag, Berlin. 666–700.

Berger, U. 1960. *Neisseria animalis* nov. spec. Zeitschrift F. Hygiene. *147*: 158–161.

Berger, U. 1961a. Polysaccharidbildung durch saprophytische Neisserien. Zentralbl. Bakteriol. Parasitenkd. Infektionskr. Hyg. Abt. 1 Orig. *183*: 345–348.

Berger, U. 1961b. Untersuchungen ueber die Pigmentbildung durch Neisseria. Z. Hyg. *147*: 461–469.

Berger, U. 1962. Über das vorkommen von neisserien bei einigen tieren. Z. Hyg. Infektionskr. *148*: 445–457.

Berger, U. 1963. Reinzüchtung von *Simonsiella* spp. Z. Hyg. *149*: 336–340.

Berger, U. 1970. Untersuchungen yur Reduktionvon Nitrat und Nitrit durch *Neisseria gonorrhoeae* und *Neisseria meningitidis*. Z. Med. Mikrobiol. Immunol. *156*: 86–89.

Berger, U. 1971. *Neisseria mucosa* var. *heidelbergensis*. Z. Med. Mikrobiol. Immunol. *156*: 154–158.

Berger, U., I. Aboulkchair and W. Rottman. 1974. Sepsis und meningitis durch *Neisseria mucosa* var. *heidelbergensis*. Infection *2*: 108–110.

Berger, U. and H. Brunhoeber. 1961. *Neisseria flava* (Bergey et al. 1923). Art oder Varietät? Z. Hyg. Infektionskr. *148*: 39–44.

Berger, U. and E. Falsen. 1976. Über die Artenverteilung von *Moraxella* und *Moraxella*—ahnlichen Keimen im Nasopharynx gesunder Erwachsener. Med. Mikrobiol. Immunol. *162*: 239–249.

Berger, U. and R. Issi. 1971. Resistenz gegen Acetezolamid als taxonomisches Kriterium bei *Neisseria*. Arch. Hyg. *154*: 540–544.

Berger, U. and M. Miersch. 1970. The normal occurrence of *Neisseria mucosa* (Veron et al. 1959). Z Med Mikrobiol Immunol. *155*: 186–191.

Berger, U. and H.D. Piotrowski. 1974. Die biochemische Diagnose von *Neisseria elongata* (Bøvre und Holten, 1970). Med. Microbiol. Immunol. *159*: 309–316.

Berger, U. and B. Wulf. 1961. Üntersuchungen an saprophytischen Neisserien. Z. Hyg. Infektionskr. *147*: 257–268.

Berger, V. and B.W. Catlin. 1975. Biochemical differentiation between *N. sicca* and *N. perflava*. Zentralbl. Bakteriol. Parasitenkd. Infektionskr. Hyg. Abt. I. Orig. Reike A. *232*: 129–130.

Bergeron, H., D. Labbe, C. Turmel and P.C.K. Lau. 1998. Cloning, sequence and expression of a linear plasmid-based and a chromosomal homolog of chloroacetaldehyde dehydrogenase-encoding genes in *Xanthobacter autotrophicus* GJ10. Gene *207*: 9–18.

Bergeron, J., D. Ahmad, D. Barriault, A. Larose, M. Sylvestre and J. Powlowski. 1994. Identification and mapping of the gene translation products involved in the first steps of the *Comamonas testosteroni* B-356 biphenyl/chlorobiphenyl degradation pathway. Can. J. Microbiol. *40*: 743–753.

Bergey, D.H., F.C. Harrison, R.S. Breed, B.W. Hammer and F.M. Huntoon. 1923a. Bergey's Manual of Determinative Bacteriology, 1st Ed., The Williams & Wilkins Co., Baltimore. pp. 1–442.

Bergey, D.H., D.C. Harrison, R.S. Breed, B.W. Hammer and F.M. Huntoon. 1923b. Flavobacterium. *In* Bergey, Harrison, Breed, Hammer and Huntoon (Editors), Bergey's Manual of Determinative Bacteriology, 1st Ed., The Williams & Wilkins Co., Baltimore. p. 102.

Bergonzini, C. 1881. Sopra un nuovo bacterio colorato. Annuar. Soc. Nat. Modena, Ser. 2 *14*: 149–158.

Beringer, J.E. and D.A. Hopwood. 1976. Chromosomal recombination and mapping in *Rhizobium leguminosarum*. Nature (Lond.) *264*: 291–293.

Beringer, J.E., A.W. Johnston and Á. Kondorosi. 1987. Genetic maps of *Rhizobium meliloti* and *R. leguminosarum* biovar phaseoli, biovar trifolii and biovar viciae. *In* O'Brien (Editor), Genetic Maps, Vol. 4, Cold Spring Harbor Laboratories, Cold Spring Harbor, New York. 245–251.

Berkeley, M.J. 1857. Introduction to Cryptogamic Botany, H. Bailliere, London. 313, 315.

Berkeley, M.J. 1874. Notices of North American fungi. Grevillea *3*: 4–64.

Berkeley, M.J. and C.E. Broome. 1873. Enumeration of the fungi of Ceylon. J. Linn. Soc. Lond. Bot. *14*: 96.

Berkhoff, H.A. and G.D. Riddle. 1984. Differentiation of *Alcaligenes*-like bacteria of avian origin and comparison with *Alcaligenes* spp. reference strains. J. Clin. Microbiol. *19*: 477–481.

Berlier, Y.M. and P.A. Lespinat. 1980. Mass-spectrometric kinetic studies of the nitrogenase and hydrogenase activities in in vivo cultures of *Azospirillum brasilense*. Arch. Microbiol. *125*: 67–72.

Bermond, D., R. Heller, F. Barrat, G. Delacour, C. Dehio, A. Alliot, H. Monteil, B. Chomel, H.J. Boulouis and Y. Piémont. 2000. *Bartonella birtlesii* sp. nov., isolated from small mammals (*Apodemus* spp.). Int. J. Syst. Evol. Microbiol . *50*: 1973–1979.

Bernaerts, M.J. and J. De Ley. 1960a. Microbial formation and preparation of 3-ketoglycosides from disaccharides. J. Gen. Microbiol. *22*: 129–136.

Bernaerts, M.J. and J. De Ley. 1960b. The structure of 3-keto-glycosides formed from disaccharides by certain bacteria. J. Gen. Microbiol. *22*: 137–146.

Bernaerts, M.J. and J. De Ley. 1963. A biochemical test for crown gall bacteria. Nature *197*: 406–407.

Bernard, C. and T. Fenchel. 1995. Mats of colourless sulphur bacteria. II. Structure, composition of biota and successional patterns. Mar. Ecol. Prog. Ser. *128*: 171–179.

Bernard, D., G. Verschraegen, G. Claeys, S. Lauwers and P. Rosseel. 1994. *Bilophila wadsworthia* bacteremia in a patient with gangrenous appendicitis. Clin. Infect. Dis. *18*: 1023–1024.

Bernard, U., I. Probst and H.G. Schlegel. 1974. The cytochromes of some hydrogen bacteria. Arch. Microbiol. *95*: 29–37.

Bernardo, E.B., B.A. Neilan and I. Couperwhite. 1998. Characterization, differentiation and identification of wild-type cellulose-synthesizing *Acetobacter* strains involved in Nata de Coco production. Syst. Appl. Bacteriol. *21*: 599–608.

Berndt, H., D.J. Lowe and M.G. Yates. 1978. The nitrogen-fixing system of *Corynebacterium autotrophicum*. Purification and properties of the nitrogen- ase components and two ferredoxins. Eur. J. Biochem. *86*: 133–142.

Berndt, H., K.P. Ostwal, J. Lalucat, C. Schumann, F. Mayer and H.G. Schlegel. 1976. Identification and physiological characterization of the nitrogen fixing bacterium *Corynebacterium autotrophicum* GZ29. Arch. Microbiol. *108*: 17–26.

Berndt, H. and D. Wölfle. 1979. Hydrogenase: its role as electron generating in the nitrogen fixing bacterium *Xanthobacter autotrophicus*. *In* Schlegel and Schneider (Editors), Hydrogenases: Their Catalytic Activity, Structure and Function, Erich Goltze KG, Göttingen. 327–351.

Bernhard, M., T. Buhrke, B. Bleijlevens, A.L. De Lacey, V.M. Fernandez, S.P.J. Albracht and B. Friedrich. 2001. The H_2 sensor of *Ralstonia eutropha*. Biochemical characteristics, spectroscopic properties, and its interaction with a histidine protein kinase. J. Biol. Chem. *276*: 15592–15597.

Bernheimer, J., A.J. Schoeffield, M. Baer and B.D. Tall. 1993. Predation of a biofilm community by halophilic bdellovibrios. 93rd Annual Meeting of the American Society for Microbiology, p. 301.

Berry, A., D. Janssens, M. Hümbelin, J.P.M. Jore, I. Cleenwerck, M. Vancanneyt, W. Bretzel, A.F. Mayer, R. Lopez-Ulibarri, B. Shanmugam, J. Swings and L. Pasamontes. 2003. *Paracoccus zeaxanthinifaciens* sp. nov., a zeaxanthin-producing bacterium. Int. J. Syst. Evol. Microbiol. *53*: 231–238.

Bertini, I., F. Capozzi, A. Dikiy, B. Happe, C. Luchinat and K.N. Timmis. 1995. Evidence of histidine coordination to the catalytic ferrous ion in the ring-cleaving 2,2′,3-trihydroxybiphenyl dioxygenase from the dibenzofuran-degrading bacterium *Sphingomonas* sp. strain RW1. Biochem. Biophys. Res. Commun. *215*: 855–860.

Bertolla, F., R. Pepin, E. Passelegue-Robe, E. Paget, A. Simkin, X. Nesme and P. Simonet. 2000. Plant genome complexity may be a factor limiting in situ the transfer of transgenic plant genes to the phytopathogen *Ralstonia solanacearum*. Appl. Environ. Microbiol. *66*: 4161–4167.

Bertrand, J.L., B.A. Ramsay, J.A. Ramsay and C. Chavarie. 1990. Biosynthesis of poly-β-hydroxyalkanoates from pentoses by *Pseudomonas pseudoflava*. Appl. Environ. Microbiol. *56*: 3133–3138.

Best, D.J. and I.J. Higgins. 1981. Methane-oxidizing activity and membrane morphology in a methanol grown obligate methanotroph, *Methylosinus trichosporium* OB3b. J. Gen. Microbiol. *125*: 73–84.

Beuscher, N., F. Mayer and G. Gottschalk. 1974. Citrate lyase from *Rhodopseudomonas gelatinosa*, electron microscopy and subunit structure. Arch. Microbiol. *100*: 307–328.

Bevan, L.G.W. 1930. Blood culture in undulant fever. Br. Med. J. *2*: 267.

Beveridge, T.J. and R.G.E. Murray. 1976. Dependence of the superficial layers of *Spirillum putridiconchylium* on Ca^{2+} or Sr^{2+}. Can. J. Microbiol. *22*: 1233–1244.

Bevivino, A., S. Tabacchioni, L. Chiarini, M.V. Carusi, M. Del Gallo and

P. Visca. 1994. Phenotypic comparison between rhizosphere and clinical isolates of *Burkholderia cepacia*. Microbiology *140*: 1069–1077.

Beyer, P. and H. Kleinig. 1985. *In vitro* synthesis of C_{15}–C_{60} polyprenols in a cell-free system of *Myxococcus fulvus* and determination of chain length by high performance liquid chromatography. Methods Enzymol. *110*: 299–303.

Beynon, L.M., A.D. Cox, C.J. Taylor, S.G. Wilkinson and M.B. Perry. 1995. Characterization of a lipopolysaccharide O antigen containing two different trisaccharide repeating units from *Burkholderia cepacia* serotype E (O2). Carbohydr. Res. *272*: 231–239.

Bezrukova, L.V., N. Y.I., A.I. Nesterov, V.F. Gal'chenko and M.V. Ivanov. 1983. Comparative serological analysis of methanotrophic bacteria. Mikrobiologiya *52*: 800–805.

Bhat, M.A., T. Ishida, K. Horiike, C.S. Vaidyanathan and M. Nozaki. 1993. Purification of 3,5-dichlorocatechol 1,2-dioxygenase, a nonheme iron dioxygenase and a key enzyme in the biodegradation of a herbicide, 2,4- dichlorophenoxyacetic acid (2,4-D), from *Pseudomonas cepacia* CSV90. Arch. Biochem. Biophys. *300*: 738–746.

Bhat, M.A., M. Tsuda, K. Horiike, M. Nozaki, C.S. Vaidyanathan and T. Nakazawa. 1994. Identification and characterization of a new plasmid carrying genes for degradation of 2,4-dichlorophenoxyacetate from *Pseudomonas cepacia* CSV90. Appl. Environ. Microbiol. *60*: 307–312.

Bianco, N., S. Neshat and K. Poole. 1997. Conservation of the multidrug resistance efflux gene *oprM* in *Pseudomonas aeruginosa*. Antimicrob. Agents Chemother. *41*: 853–856.

Biebl, H. 1973. Die Verbreitung der schwefelfreien Purpurbakterien im Plubsee und anderen Seen Ostholsteins, Ph.D. thesis, University of Freiburg, F.R.G.

Biebl, H. and G. Drews. 1969. Das in-vivo-Spektrum als taxonomisches Merkmal bei Untersuchungen zur Bergreitung von *Athiorhodoaceae*. Zentbl. Bakteriol. Parasitenkd. Infektkrankh. Hyg. Abt. II Orig. *123*: 425–452.

Biebl, H. and N. Pfennig. 1977. Growth of sulfate-reducing bacteria with sulfur as electron acceptor. Arch. Microbiol. *112*: 115–117.

Biebl, H. and N. Pfennig. 1978. Growth yields of green sulfur bacteria in mixed cultures with sulfur and sulfate reducing bacteria. Arch. Microbiol. *117*: 9–16.

Biebl, H. and N. Pfennig. 1981. Isoation of members of *Rhodospirillaceae*. *In* Starr, Stolp, Trüper, Balows and Schlegel (Editors), The Prokaryotes: A Handbook on Habitats, Isolation, and Identification of Bacteria, Springer-Verlag, Berlin. pp. 267–273.

Biel, S., J. Simon, R. Gross, M. Ruitenberg and A. Kroger. 2002. Reconstitution of coupled fumarate respiration in liposomes by incorporating the electron transport enzymes isolated from *Wolinella succinogenes*. Eur. J. Biochem. *269*: 1974–1983.

Biggins, D.R. and J.R. Postgate. 1969. Nitrogen fixation by cultures and cell-free extracts of *Mycobacterium flavum* 301. J. Gen. Microbiol. *56*: 181–193.

Biggins, D.R. and J.R. Postgate. 1971. Nitrogen fixation by extracts of *Mycobacterium flavum* 301. Use of natural electron donors and oxygensensitivity of cell-free preparations. Eur. J. Biochem. *19*: 408–415.

Billings, A.N., G.J. Teltow, S.C. Weaver and D.H. Walker. 1998. Molecular characterization of a novel *Rickettsia* species from *Ixodes scapularis* in Texas. Emerg. Infect. Dis. *4*: 305–309.

Bilos, Z.J., A. Kucharchuk and W. Metzger. 1978. *Eikenella corrodens* in human bites. Clin. Orthop. *134*: 320–324.

Binns, A.N. and P. Costantino. 1998. The *Agrobacterium* oncogenes. *In* Spaink, Kondorosi and Hooykaas (Editors), The *Rhizobiaceae*: Molecular Biology of Model Plant Associated Bacteria, Kluwer Academic, Dordrecht; Boston. 251–266.

Birch, L. and H. Brandl. 1996. A rapid method for the determination of metal toxicity to the biodegradation of water insoluble polymers. Fresenius' J. Anal. Chem. *354*: 760–762.

Bird, R.G. and P.C. Garnham. 1969. *Aegyptianella pullorum* Carpano 1928--fine structure and taxonomy. Parasitology. *59*: 745–752.

Birgisson, H., O. Steingrimsson and T. Gudnason. 1997. *Kingella kingae* infections in paediatric patients: 5 cases of septic arthritis, osteomyelitis and bacteraemia. Scand. J. Infect. Dis. *29*: 495–498.

Birkness, K.A., V.G. George, E.H. White, D.S. Stephens and F.D. Quinn. 1992. Intracellular growth of *Afipia felis*, a putative etiologic agent of cat scratch disease. Infect. Immun. *60*: 2281–2287.

Birtles, R.J. 1995. Differentiation of *Bartonella* species using restriction endonuclease analysis of PCR-amplified 16S rRNA genes. FEMS Microbiol. Lett. *129*: 261–265.

Birtles, R.J., T.G. Harrison, N.A. Saunders and D.H. Molyneux. 1995. Proposals to unify the genera *Grahamella* and *Bartonella*, with descriptions of *Bartonella talpae* comb. nov., *Bartonella peromysci* comb. nov., and three new species, *Bartonella grahamii* sp. nov., *Bartonella taylorii* sp. nov., and *Bartonella doshiae* sp. nov. Int. J. Syst. Bacteriol. *45*: 1–8.

Birtles, R.J. and D. Raoult. 1996. Comparison of partial citrate synthase gene (*gltA*) sequences for phylogenetic analysis of *Bartonella* species. Int. J. Syst. Bacteriol. *46*: 891–897.

Birtles, R.J., T.J. Rowbotham, R. Michel, D.G. Pitcher, B. Lascola, S. Alexiou-Daniels and D. Raoult. 2000. *Candidatus* Odyssella thessalonicensis gen. nov., sp nov., an obligate intracellular parasite of *Acanthamoeba* species. Int. J. Syst. Evol. Microbiol. *50*: 63–72.

Birtles, R.J., T.J. Rowbotham, D. Raoult and T.G. Harrison. 1996. Phylogenetic diversity of intra-amoebal legionellae as revealed by 16S rRNA gene sequence comparison. Microbiology (Reading) *142*: 3525–3530.

Bisacchi, G.S., D.R. Hockstein, W.H. Koster, W.L. Parker, M.L. Rathnum and S.E. Unger. 1987. Xylocandin: a new complex of antifungal peptides II. Structural studies and chemical modifications. J. Antibiot. *40*: 1520–1529.

Biswas, B., R. Vemulapalli and S.K. Dutta. 1998. Molecular basis for antigenic variation of a protective strain-specific antigen of *Ehrlichia risticii*. Infect. Immun. *66*: 3682–3628.

Biswas, G.D. and P.F. Sparling. 1995. Characterization of Lbpa, the structural gene for a lactoferrin receptor in *Neisseria gonorrhoeae*. Infect. Immun. *63*: 2958–2967.

Bizzozero, G. 1893. Über die schlauchförmigen Drusen des magendarmkanals und die Beziehungen ihres Epithels zu dem Oberflächenepithel der Schleimhaut. Arch. Mikrosk. Anat. *42*: 82.

Black, C.G., J.A. Fyfe and J.K. Davies. 1998. Absence of an SOS-like system in *Neisseria gonorrhoeae*. Gene *208*: 61–66.

Blackall, L.L., A.C. Hayward and L.I. Sly. 1985. Cellulolytic and dextranolytic Gram-negative bacteria - revival of the genus *Cellvibrio*. J. Appl. Bacteriol. *59*: 81–97.

Blackall, L.L., S. Rossetti, C. Christensson, M. Cunningham, P. Hartman, P. Hugenholtz and V. Tandoi. 1997. The characterization and description of representatives of "G" bacteria from activated sludge plants. Lett. Appl. Microbiol. *25*: 63–69.

Blackall, P.J. and C.M. Doheny. 1987. Isolation and characterisation of *Bordetella avium* and related species and an evaluation of their role in respiratory disease in poultry. Aust. Vet. J. *64*: 235–239.

Blackall, P.J., L.E. Eaves and M. Fegan. 1995. Antimicrobial sensitivity testing of Australian isolates of *Bordetella avium* and the *Bordetella avium*-like organism. Aust. Vet. J. *72*: 97–100.

Blake, C.K. and G.D. Hegeman. 1987. Plasmid pCBI carries genes for anaerobic benzoate catabolism in *Alcaligenes xylosoxidans* subsp. *denitrificans* PN-1. J. Bacteriol. *169*: 4878–4883.

Blake, M.S., C.M. Blake, M.A. Apicella and R.E. Mandrell. 1995. Gonococcal opacity — lectin-like interactions between Opa proteins and lipooligosaccharide. Infect. Immun. *63*: 1434–1439.

Blake, M.S., C.M. Macdonald and K.P. Klugman. 1989. Colony morphology of piliated *Neisseria meningitidis*. J. Exp. Med. *170*: 1727–1736.

Blakebrough, I.S., B.M. Greenwood, H.C. Whittle, A.K. Bradley and H.M. Gilles. 1982. The epidemiology of infections due to *Neisseria meningitidis* and *Neisseria lactamica* in a northern Nigerian community. J. Infect. Dis. *146*: 626–637.

Blakemore, R.P., N.A. Blakemore, D.A. Bazylinski and T.T. Moench. 1989. Magnetotactic bacteria. *In* Staley, Bryant, Pfennig and Holt (Editors), Bergey's Manual of Systematic Bacteriology, 1st ed., Vol. 3, The Williams & Wilkins Co., Baltimore.

Blakemore, R.P. and R.B. Frankel. 1981. Magnetic navigation in bacteria. Sci. Am. *245*: 42–49.

Blakemore, R.P., D. Maratea and R.S. Wolfe. 1979. Isolation and pure culture of a freshwater magnetic spirillum in chemically defined medium. J. Bacteriol. *140*: 720–729.

Blakemore, R.P., K.A. Short, D.A. Bazylinski, C. Rosenblatt and R.B. Frankel. 1985. Microaerobic conditions are required for magnetite formation within *Aquaspirillum magnetotacticum*. Geomicrobiol. J. *4*: 53–71.

Blatny, J.M., ., T. Brautaset, H.C. Winther Larsen, K. Haugan and S. Valla. 1997. Construction and use of a versatile set of broad-host-range cloning and expression vectors based on the RK2 replicon. Appl. Environ. Microbiol. *63*: 370–379.

Bleumink-Pluym, N.M.C. 1995. *Taylorella equigenitalis*: epidemiology and pathogenicity, Thesis, Utrecht University.

Bleumink-Pluym, N.M.C., E.A. ter Laak, D.J. Houwers and B.A.M. van der Zeijst. 1996. Differences between *Taylorella equigenitalis* strains in their invasion of and replication in cultured cells. Clin. Diagn. Lab. Immunol. *3*: 47–50.

Bleumink-Pluym, N.M.C., E.A. ter Laak and B.A.M. van der Zeijst. 1990. Epidemiologic study of *Taylorella equigenitalis* strains by field inversion gel electrophoresis of genomic restriction endonuclease fragments. J. Clin. Microbiol. *28*: 2012–2016.

Bleumink-Pluym, N.M.C., L. van Dijk, A.H.M. van Vliet, J.W.B. van der Giessen and B.A.M. van der Zeijst. 1993. Phylogenetic position of *Taylorella equigenitalis* determined by analysis of amplified 16S ribosomal DNA sequences. Int. J. Syst. Bacteriol. *43*: 618–621.

Bleumink-Pluym, N.M.C., M.E. Werdler, D.J. Houwers, J.M. Parlevliet, B. Colenbrander and B.A.M. van der Zeijst. 1994. Development and evaluation of PCR test for detection of *Taylorella equigenitalis*. J. Clin. Microbiol. *32*: 893–896.

Bloch, M. 1918. Beiträg zür Untersuchungen über die *Zoogloca ramigera* (Itzigsohn) auf Grund von Reinkulturen. Zentbl. Bakteriol. Parasitenkd. Infektkrankh. Hyg. Abt. II *48*: 44–62.

Blom, K., C.M. Patton, M.A. Nicholson and B. Swaminathan. 1995. Identification of *Campylobacter fetus* by PCR-DNA probe method. J. Clin. Microbiol. *33*: 1360–1362.

Blümel, S., H.-J. Busse, A. Stolze and P. Kämpfer. 2001a. *Xenophilus azovorans* gen. nov., sp. nov., a soil bacterium that is able to degrade azo dyes of the Orange II type. Int. J. Syst. Evol. Microbiol. *51*: 1831–1837.

Blümel, S., B. Mark, H.J. Busse, P. Kämpfer and A. Stolz. 2001b. *Pigmentiphaga kullae* gen. nov., sp. nov., a novel member of the family *Alcaligenaceae* with the ability to decolorize azo dyes aerobically. Int. J. Syst. Evol. Microbiol. *51*: 1867–1871.

Blundell, J.K. and H.R. Perkins. 1981. Effects of beta-lactam antibiotics on peptidoglycan synthesis in growing *Neisseria gonorrhoeae*, including changes in the degree of *O*-acetylation. J. Bacteriol. *147*: 633–641.

Bock, E., H.-P. Koops, H. Harms and B. Ahlers. 1991. The biochemistry of nitrifying organisms. *In* Shively and Barton (Editors), Variations in Autotrophic Life, Academic Press, Ltd., London. pp. 171–199.

Bock, E. and H.-P., Koops. 1992. The genus *Nitrobacter* and related genera. *In* Balows, Trüper, Dworkin, Harder and Schleifer (Editors), The Prokaryotes: A Handbook of Bacteria: Ecophysiology, Isolation, Identification, Applications, 2nd ed., Vol. 3, Springer-Verlag, New York. pp. 2302–2309.

Bock, E., H.P. Koops, U.C. Möller and M. Rudert. 1990. A new facultatively nitrite oxidizing bacterium, *Nitrobacter vulgaris*, new species. Arch. Microbiol. *153*: 105–110.

Bock, E., H. Sundermeyer-Klinger and E. Stackebrandt. 1983. New facultative lithoautotrophic nitrite-oxidizing bacteria. Arch. Microbiol. *136*: 281–284.

Bock, E., H. Sundermeyer-Klinger and E. Stackebrandt. 2001. *In* Validation of the publication of new names and new combinations previously effectively published outside the IJSEM. List No. 78. Int. J.Syst. Bacteriol. *51*: 1–2.

Bock, E., P.A. Wilderer and A. Freitag. 1988. Growth of *Nitrobacter* in the absence of dissolved oxygen. Water Res. *22*: 245–250.

Bode, G., F. Mauch, H. Ditschuneit and P. Malfertheiner. 1993. Identification of structures containing polyphosphate in *Helicobacter pylori*. J. Gen. Microbiol. *139*: 3029–3033.

Bodrossy, L., E.M. Holmes, A.J. Holmes, K.L. Kovacs and J.C. Murrell. 1997. Analysis of 16S rRNA and methane monooxygenase gene sequences reveals a novel group of thermotolerant and thermophilic methanotrophs, *Methylocaldum* gen. nov. Arch. Microbiol. *168*: 493–503.

Boesch, C. 1998. Nucleotide sequences as a basis to genetic and taxonomic investigations of *Acetobacter* species isolated from industrial fermentations, Thesis, Swiss Federal Institute of Technology, Zurich (ETHZ), Zurich.

Boesch, C., J. Trcek, M. Sievers and M. Teuber. 1998. *Acetobacter intermedius*, sp. nov. Syst. Appl. Microbiol. *21*: 220–229.

Boivin, C., I. Ndoye, G. Lortet, A. Ndiaye, P. deLajudie and B. Dreyfus. 1997. The Sesbania root symbionts *Sinorhizobium saheli* and *S. teranga* biovar sesbanie can form stem nodules on *Sesbania rostrata*, although they are less adapted to stem nodulation than *Azorhizobium caulinodans*. Appl. Environ. Microbiol. *63*: 1040–1047.

Boivin, M.F., V.L. Morris, E.C.M. Lee-Chan and R.G.E. Murray. 1985. Deoxyribonucleic acid relatedness between selected members of the genus *Aquaspirillum* by slot blot hydridization: *Aquaspirillum serpens* (Mueller 1786) Hylemon, Wells, Krieg, and Jannasch 1973 emended to include *Aquaspirillum bengal* as a subjective synonym. Int. J. Syst. Bacteriol. *35*: 512–517.

Boivin-Jahns, V., A. Bianchi, R. Ruimy, J. Garcin, S. Daumus and R. Christen. 1995. Comparison of phenotypical and molecular methods for the identification of bacterial strains isolated from a deep subsurface environment. Appl. Environ. Microbiol. *61*: 3400–3406.

Bojary, M.R. and S.A. Dhala. 1989. Coagulase of *Myxococcus fulvus* NK-35. 1. Purification and partial characterization. Zentbl. Mikrobiol. *144*: 347–354.

Bokranz, M., J. Katz, I. Schröder, A.M. Roberton and A. Kröger. 1983. Energy metabolism and biosynthesis of *Vibrio succinogenes* growing with nitrate or nitrite as terminal electron acceptor. Arch. Microbiol. *135*: 36–41.

Bokranz, M., E. Mörschel and A. Kröger. 1985. Phosphorylation and phosphate-ATP exchange catalyzed by the ATP synthase isolated from *Wolinella succinogenes*. Biochim. Biophys. Acta *810*: 332–339.

Boll, M., S.S.P. Albracht and G. Fuchs. 1997. Benzoyl-CoA reductase (dearomatizing), a key enzyme of anaerobic aromatic metabolism: A study of adenosinetriphosphatase activity, ATP stoichiometry of the reaction and EPR properties of the enzyme. Eur. J. Biochem. *244*: 840–851.

Bolton, H.J., D.C. Girvin, A.E. Plymale, S.D. Harvey and D.J. Workman. 1996. Degradation of metal-nitrilotriacetate complexes by *Chelatobacter heintzii*. Environ. Sci. Technol. *30*: 931–938.

Bonam, D., L. Lehman, G.P. Roberts and P.W. Ludden. 1989. Regulation of carbon monoxide dehydrogenase and hydrogenase in *Rhodospirillum rubrum* effects of carbon monoxide and oxygen on synthesis and activity. J. Bacteriol. *171*: 3102–3107.

Bonch-Osmolovskaya, E.A., T.G. Sokolova, N.A. Kostrikina and G.A. Zavarzin. 1990. *Desulfurella acetivorans* gen. nov. and sp. nov.—a new thermophilic sulfur-reducing eubacterium. Arch. Microbiol. *153*: 151–155.

Bonch-Osmolovskaya, E.A., T.G. Sokolova, N.A. Kostrikina and G.A. Zavarzin. 1993. *In* Validation of the publication of new names and new combinations previously effectively published outside the IJSB, List No. 46. Int. J. Syst. Bacteriol. *43*: 624–625.

Bonnet, D., I. Artraud, C. Moali, D. Petre and D. Mansuy. 1997. Highly efficient control of iron-containing nitrile hydratase by stoichiometric amounts of nitric oxide and light. FEBS Lett. *409*: 216–220.

Bonora, P., I. Principi, A. Hochkoeppler, R. Borgheses and D. Zannoni. 1998. The respiratory chain of the halophilic anoxygenic purple bacterium *Rhodospirillum sodomense*. Arch. Microbiol. *170*: 435–441.

Boogerd, F.C., H.W. van Verseveld and A.H. Stouthamer. 1981. Respiration-driven proton translocation with nitrite and nitrous oxide in *Paracoccus denitrificans*. Biochim. Biophys. Acta *638*: 181–191.

Boogerd, F.C., H.W. van Verseveld and A.H. Stouthamer. 1983. Dissimi-

latory nitrate uptake in *Paracoccus denitrificans* via a proton motive force-dependent system and a nitrate-nitrite antiport system. Biochim. Biophys. Acta *723*: 415–427.

Boon, N., J. Goris, P. De Vos, W. Verstraete and E.M. Top. 2001. Genetic diversity among 3-chloroaniline and aniline-degrading strains of the *Comamonadaceae*. Appl. Environ. Microbiol. *67*: 1107–1115.

Boon, P.I., P. Virtue and P.D. Nichols. 1996. Microbial consortia in wetland sediments—a biomarker analysis of the effects of hydrological regime, vegetation and season on benthic microbes. Mar. Freshw. Res. *47*: 27–41.

Boone, D.R. and M.P. Bryant. 1980. Propionate-degrading bacterium, *Syntrophobacter wolinii* sp. nov. gen. nov., from methanogenic ecosystems. Appl. Environ. Microbiol. *40*: 626–632.

Boone, D.R. and M.P. Bryant. 1984. *In* Validation of the publication of new names and new combinations previously effectively published outside the IJSB. List No. 15. Int. J. Syst. Bacteriol. *34*: 355–357.

Boone, D.R., R.L. Johnson and Y. Liu. 1989. Diffusion of the interspecies electron carriers H_2 and formate in methanogenic ecosystems and its implications in the measurement of K_m for H_2 or formate uptake. Appl. Environ. Microbiol. *55*: 1735–1741.

Boone, D.R. and L. Xun. 1987. Effects of pH, temperature, and nutrients on propionate degradation by a methanogenic enrichment culture. Appl. Environ. Microbiol. *53*: 1589–1592.

Boosinger, T.R., W.T. Blevins, J.V. Heron and J.L. Sunter. 1990. Plasmid profiles of six species of *Campylobacter* from human beings, swine, and sheep. Am. J. Vet. Res. *51*: 718–722.

Booth, A.J., L. Stogdale and J.A. Grigor. 1984. Salmon poisoning disease in dogs on Southern Vancouver Island. Can. Vet. J. *25*: 2–6.

Bøvre, K. 1984. Family VIII. *Neisseriaceae*. *In* Krieg and Holt (Editors), Bergey's Manual of Systematic Bacteriology, Vol. 1, The Williams & Wilkins Co., Baltimore. pp. 288–290.

Borchers, M. 1982. Isolierung und charakterisierung hefelytischer Enzyme aus dem gleitenden Bakterium *Myxococcus fulvus* MX f80 (*Myxobacterales*), Thesis, Technical University Braunschweig, Germany, 90 pp..

Borchsenius, O.N., I.I. Skoblo and D.V. Ossipov. 1983. *Holospora curviuscula* - a new species of macronuclear symbiotic bacteria of *Paramecium bursaria*. Tsitologiya *25*: 91–97.

Bordet, J. and O. Gengou. 1906. Le microbe de al coqueluche. Ann. Inst. Pasteur (Paris) *20*: 731–741.

Boren, T., P. Falk, K.A. Roth, G. Larson and S. Normark. 1993. Attachment of *Helicobacter pylori* to human gastric epithelium mediated by blood group antigens. Science *262*: 1892–1895.

Borodina, E., D.P. Kelly, F.A. Rainey, N.L. Ward-Rainey and A.P. Wood. 2000. Dimethylsulfone as a growth substrate for novel methylotrophic species of *Hyphomicrobium* and *Arthrobacter*. Arch. Microbiol. *173*: 425–437.

Borremans, B., J.L. Hobman, A. Provoost, N.L. Brown and D. van der Lelie. 2001. Cloning and functional analysis of the pbr lead resistance determinant of *Ralstonia metallidurans* CH34. J. Bacteriol. *183*: 5651–5658.

Borrow, R., H. Claus, M. Guiver, L. Smart, D.M. Jones, E.B. Kaczmarski, M. Frosch and A.J. Fox. 1997. Non-culture diagnosis and serogroup determination of meningococcal B and C infection by a sialyltransferase (*siaD*) PCR ELISA. Epidemiol. Infect. *118*: 111–117.

Borsodi, A. K., A. Micsinai, G. Kovacs, E. Toth, P. Schumann, A. L. Kovacs, B. Boddi and K. Marialigeti. 2003. *Pannonibacter phragmitetus* gen. nov., sp nov., a novel alkalitolerant bacterium isolated from decomposing reed rhizomes in a Hungarian soda lake. Int. J. Syst. Evol. Microbiol. *53*: 555–561.

Boschker, H.T.S., S.C. Nold, P. Wellsbury, D. Bos, W. De Graaf, R. Pel, R.J. Parkees and T.E. Cappenberg. 1998. Direct linking of microbial populations to specific biogeochemical processes by ^{13}C-labelling of biomarkers. Nature *392*: 801–805.

Boss, A.O.L., O.N. Borchsenius and D.V. Ossipov. 1987. *Pseudolyticum multiflagellatum* Ng, n. sp. - a new symbiotic bacterium in the cytoplasm of *Paramecium caudatum* (Ciliata, Protozoa). Tsitologiya *29*: 94–99.

Bossier, P. and W. Verstraete. 1996. *Comamonas testosteroni* colony phenotype influences exopolysaccharide production and coaggregation with yeast cells. Appl. Environ. Microbiol. *62*: 2687–2691.

Bothe, H., B. Klein, M.P. Stephan and J. Dobereiner. 1981. Transformations of inorganic nitrogen by *Azospirillum* spp. Arch. Microbiol. *130*: 96–100.

Bothe, H. and M.G. Yates. 1976. The electron transport to nitrogenase in *Mycobacterium flavum*. Arch. Microbiol. *107*: 25–31.

Bottini, R., M. Fulchieri, D. Pearce and R.P. Pharis. 1989. Identification of gibberellins A1, A3, and iso-A3 in cultures of *Azospirillum lipoferum*. Plant Physiol. *90*: 45–47.

Bottomley, P.J., H.H. Cheng and S.R. Strain. 1994. Genetic structure and symbiotic characteristics of a *Bradyrhizobium* population recovered from a pasture soil. Appl. Environ. Microbiol. *60*: 1754–1761.

Bottone, E.J., J. Kittick and S.S. Schneierson. 1973. Isolation of *Bacillus* HB-1 from human clinical sources. Am. J. Clin. Pathol. *59*: 560–566.

Boucher, C., A. Martinel, P. Barberis, G. Alloing and C. Zischek. 1986. Virulence genes are carried by a megaplasmid of the plant pathogen *Pseudomonas solanacearum*. Mol. Gen. Genet. *205*: 270–275.

Bouchon, D., T. Rigaud and P. Juchault. 1998. Evidence for widespread *Wolbachia* infection in isopod crustaceans: Molecular identification and host feminization. Proc. R. Soc. Lond. B Biol. Sci. *265*: 1081–1090.

Boulygina, E.S., K.M. Chumakov and A.I. Netrusov. 1993. Systematics of Gram-negative methylotrophic bacteria based on 5S rRNA sequences. *In* Murrell and Kelly (Editors), Microbial growth on C₁ compounds: Proceedings of the 7th International Symposium, Intercept Ltd., Andover, UK. 275–284.

Bourbeau, P., V. Holla and S. Piemontese. 1990. Ophthalmia neonatorum caused by *Neisseria cinerea*. J. Clin. Microbiol. *28*: 1640–1641.

Bourke, B., V.L. Chan and P. Sherman. 1998. *Campylobacter upsaliensis*: waiting in the wings. Clin. Microbiol. Rev. *11*: 440–449.

Bourke, B., P. Sherman, H. Louie, E. Hani, P. Islur and V.L. Chan. 1995. Physical and genetic map of the genome of *Campylobacter upsaliensis*. Microbiology *141*: 2417–2424.

Bousfield, I.J. and P.N. Green. 1985. Reclassification of bacteria of the genus *Protomonas* Urakami and Komagata 1984 in the genus *Methylobacterium* (Patt, Cole, and Hanson) emend. Green and Bousfield 1983. Int. J. Syst. Bacteriol. *35*: 209.

Bouyer, D.H., J. Stenos, P. Crocquet-Valdes, C.G. Moron, V.L. Popov, J.E. Zavala-Velazquez, L.D. Foil, D.R. Stothard, A.F. Azad and D.H. Walker. 2001. *Rickettsia felis*: Molecular characterization of a new member of the spotted fever group. Int. J. Syst. Evol. Microbiol. *51*: 339–347.

Bouzar, H. 1994. Request for a judicial opinion concerning the type species of *Agrobacterium*. Int. J. Syst. Bacteriol. *44*: 373–374.

Bouzar, H., W.S. Chilton, X. Nesme, Y. Dessaux, V. Vaudequin, A. Petit, J.B. Jones and N.C. Hodge. 1995. A new *Agrobacterium* strain isolated from aerial tumors on *Ficus benjamina* L. Appl. Environ. Microbiol. *61*: 65–73.

Bouzar, H., D. Ouadah, Z. Krimi, J.B. Jones, M. Trovato, A. Petit and Y. Dessaux. 1993. Correlative association between resident plasmids and the host chromosome in a diverse *Agrobacterium* soil population. Appl. Environ. Microbiol. *59*: 1310–1317.

Bovarnick, M.R., J.C. Miller and J.C. Snyder. 1950. The influence of certain salts, amino acids, sugars and proteins on the stability of rickettsiae. J. Bacteriol. *59*: 509–522.

Bovarnick, M.R. and J.C. Snyder. 1949. Respiration of typhus rickettsiae. J. Exp. Med. *89*: 561–565.

Bové, J.M., E.C. Calavan, S.P. Capoor, R.E. Cortez and R.E. Schwarz. 1974. Influence of temperature on symptoms of California stubborn, South African greening, Indian citrus decline and Philippines leaf mottling diseases. Proc. 6th Conf. Org. Citrus Virol. 12–15.

Bové, J.M. and M. Garnier. 1992. Citrus greening and its bacterial agent. Proc. Int. Soc. Citriculture *3*: 1283–1289.

Bovill, R.A. and B.M. Mackey. 1997. Resuscitation of 'non-culturable' cells from aged cultures of *Campylobacter jejuni*. Microbiology *143*: 1575–1581.

Bøvre, K. 1964. Studies on transformation in *Moraxella* and organisms assumed to be related to *Moraxella*. 2. Quantitative transformation

reactions between *Moraxella nonliquefaciens* strains, with streptomycin resistance marked DNA. Acta Pathol. Microbiol. Scand. *62*: 239–248.

Bøvre, K. 1965. Studies on transformation in *Moraxella* and organisms assumed to be related to *Moraxella*. 4. Streptomycin resistance transformation between asaccharolytic *Neisseria* strains. Acta Pathol. Microbiol. Scand. *64*: 229–242.

Bøvre, K. 1969. Identification of an asaccharolytic *Neisseria* strain causing meningitis. Acta Pathol. Microbiol. Scand. *76*: 148–149.

Bøvre, K. 1979. Proposal to divide the genus *Moraxella* Lwoff 1939 emend. Henriksen and Bøvre 1968 into two subgenera — subgenus *Moraxella* (Lwoff 1939) Bøvre 1979 and subgenus *Branhamella* (Catlin 1970) Bøvre 1979. Int. J. Syst. Bacteriol. *29*: 403–406.

Bøvre, K. 1980. Progress in classification and identification of *Neisseriaceae* based on genetic affinity. *In* Goodfellow and Board (Editors), Microbiological Classification and Identification, The Society for Applied Bacteriology Symposium Series No. 8, Academic Press, London. pp. 55–72.

Bøvre, K. 1984. Genus II. *Moraxella*. *In* Krieg and Holt (Editors), Bergey's Manual of Systematic Bacteriology, 1st Ed., Vol. 1, The Williams & Wilkins Co., Baltimore. pp. 296–303.

Bøvre, K., T. Bergan and L.O. Frøholm. 1970. Electron microscopical and serological characteristics associated with colony type in *Moraxella nonliquefaciens*. Acta Pathol. Microbiol. Scand. B Microbiol. Immunol. *78*: 765–779.

Bøvre, K., K. Bryn, O. Closs, N. Hagen and L.O. Frøholm. 1983. Surface polysaccharide of *Moraxella nonliquefaciens* identical to *Neisseria meningitidis* group B capsular polysaccharide: a chemical and immunological investigation. NIPH (Natl. Inst. Public Health) Ann. *6*: 65–74.

Bøvre, K. and L.O. Frøholm. 1971. Competence of genetic transformation correlated with the occurrence of fimbriae in three bacterial species. Nat. New Biol. *234*: 151–152.

Bøvre, K. and L.O. Frøholm. 1972. Competence in genetic transformation related to colony type and fimbriation in three species of *Moraxella*. Acta Pathol. Microbiol. Scand. B Microbiol. Immunol. *80*: 649–659.

Bøvre, K., L.O. Froholm, S.D. Henriksen and E. Holten. 1977. Relationship of *Neisseria elongata* subsp. *glycolytica* to other members of family *Neisseriaceae*. Acta Pathol. Microbiol. Scand. Sect. B Microbiol. *85*: 18–26.

Bøvre, K., J.E. Fuglesang, N. Hagen, E. Jantzen and L.O. Frøholm. 1976. *Moraxella atlantae* sp. nov. and its distinction from *Moraxella phenylpyruvica*. Int. J. Syst. Bacteriol. *26*: 511–521.

Bøvre, K. and N. Hagen. 1981. The family *Neisseriaceae*: rod-shaped species of the genera *Moraxella*, *Acinetobacter*, *Kingella*, and *Neisseria*, and the *Branhamella* group of cocci. *In* Starr, Stolp, Trüper, Balows and Schlegel (Editors), The Prokaryotes: A Handbook on Habitats, Isolation and Identification of Bacteria, 1st Ed., Vol. 2, Springer-Verlag, New York. pp. 1506–1529.

Bøvre, K. and S.D. Henriksen. 1962. An approach to transformation studies in *Moraxella*. Acta Pathol. Microbiol. Scand. *56*: 223–228.

Bøvre, K., S.D. Henriksen and V. Jonsson. 1974. Correction of specific epithet *kingii* in the combinations *Moraxella kingii* Henriksen and Bøvre 1968 and *Pseudomonas kingii* Jonsson 1970 to *kingae*. Int. J. Syst. Bacteriol. *24*: 307.

Bøvre, K. and E. Holten. 1970. *Neisseria elongata* sp.nov., a rod-shaped member of the genus *Neisseria*. Re-evaluation of cell shape as a criterion in classification. J. Gen. Microbiol. *60*: 67–75..

Bøvre, K., R. Hytta, E. Jantzen and L.O. Frøholm. 1972. Gas chromatography of bacterial whole cell methanolysates. 3. Group relations of Neisseriae and Moraxellae. Acta Pathol. Microbiol. Scand. [B] Microbiol. Immunol. *80*: 683–689.

Bowdre, J.H. and N.R. Krieg. 1974. Water quality monitoring: bacteria as indicators. Va. Polytech. Inst. State Univ. Water Resour. Res. Cent. Bull. *69*:

Bowdre, J.H., N.R. Krieg, P.S. Hoffman and R.M. Smibert. 1976. Stimulatory effect of dihydroxyphenyl compounds on the aerotolerance of *Spirillum volutans* and *Campylobacter fetus* subsp. *jejuni*. Appl. Environ. Microbiol. *31*: 127–133.

Bowers, T.J., D. Sweger, D. Jue and B. Anderson. 1998. Isolation, sequencing and expression of the gene encoding a major protein from the backteriophage associated with *Bartonella henselae*. Gene *206*: 49–52.

Bowien, B. and H.G. Schlegel. 1981. Physiology and biochemistry of aerobic hydrogen-oxidizing bacteria. Annu. Rev. Microbiol. *35*: 405–452.

Bowman, J.P. 1992. The systematics of methane-utilizing bacteria., Doctoral thesis, University of Queensland, Brisbane, Australia.

Bowman, J.P., L. Jimenez, I. Rosario, T.C. Hazen and G.S. Sayler. 1993a. Characterization of the methanotrophic bacterial community present in a trichloroethylene-contaminated subsurface groundwater site. Appl. Environ. Microbiol. *59*: 2380–2387.

Bowman, J.P., S.A. McCammon, M.V. Brown, D.S. Nichols and T.A. McMeekin. 1997a. Diversity and association of psychrophilic bacteria in Antarctic sea ice. Appl. Environ. Microbiol. *63*: 3068–3078.

Bowman, J.P., S.A. McCammon and J.H. Skerratt. 1997b. *Methylosphaera hansonii* gen. nov., sp. nov., a psychrophilic, group I methanotroph from antarctic marine-salinity, meromictic lakes. Microbiology *143*: 1451–1459.

Bowman, J.P. and G.S. Sayler. 1994. Optimization and maintenance of soluble methane monooxygenase activity in *Methylosinus trichosporium* OB3b. Biodegradation *5*: 1–11.

Bowman, J.P., J.H. Skerratt, P.D. Nichols and L.I. Sly. 1991a. Phospholipid fatty acid and lipopolysaccharide fatty acid signature lipids in methane-utilising bacteria. FEMS Microbiol. Ecol. *85*: 15–22.

Bowman, J.P., L.I. Sly and A.C. Hayward. 1988. *Pseudomonas mixta* sp. nov., a bacterium from soil with degradative activity on a variety of complex polysaccharides. Syst. Appl. Microbiol. *11*: 53–59.

Bowman, J.P., L.I. Sly and A.C. Hayward. 1989. *In* Validation of the publication of new names and new combinations previously effectively published outside the IJSB. List No. 29. Int. J. Syst. Bacteriol. *39*: 205–206.

Bowman, J.P., L.I. Sly and A.C. Hayward. 1991b. Contribution of genome characteristics to assessment of taxonomy of obligate methanotrophs. Int. J. Syst. Bacteriol. *41*: 301–305.

Bowman, J.P., L.I. Sly, A.C. Hayward, Y. Spiegel and E. Stackebrandt. 1993b. *Telluria mixta* (*Pseudomonas mixta* Bowman, Sly, and Hayward 1988) gen. nov., comb. nov., and *Telluria chitinolytica* sp. nov., soil-dwelling organisms which actively degrade polysaccharides. Int. J. Syst. Bacteriol. *43*: 120–124.

Bowman, J.P., L.I. Sly, P.D. Nichols and A.C. Hayward. 1993c. Revised taxonomy of the methanotrophs: description of *Methylobacter* gen. nov., emendation of *Methylococcus*, validation of *Methylosinus* and *Methylocystis* species, and a proposal that the family *Methylococcaceae* includes only the group I methanotrophs. Int. J. Syst. Bacteriol. *43*: 735–753.

Bowser, D.V., R.W. Wheat, J.W. Foster and D. Leong. 1974. Occurrence of quinovosamine in lipopolysaccharides of *Brucella* species. Infect. Immun. *9*: 772–774.

Box, S.J., A.G. Brown, M.L. Gilpin, M.N. Gwynn and S.R. Spear. 1988. Mm-42842, a new member of the monobactam family produced by *Pseudomonas cocovenenans* II. Production, isolation and properties of Mm-42842. J. Antibiot. *41*: 7–12.

Boyce, K.J. and E. A.W.. 1966. Production of freeze dried *Brucella abortus* strain 19 vaccine using cells produced by continuous culture. J. Appl. Bacteriol. *29*: 401–408.

Boyd, R.J., A.C. Hildebrandt and O.N. Allen. 1970a. Electron microscopy of phages for *Agrobacterium tumefaciens*. Arch. Mikrobiol. *73*: 47–54.

Boyd, R.J., A.C. Hildebrandt and Allen, O.N.. 1970b. Specificity patterns of *Agrobacterium tumefaciens* phages. Arch. Mikrobiol. *73*: 324–330.

Boyle, L., S.L. O'Neill, H.M. Robertson and T.L. Karr. 1993. Interspecific and intraspecific horizontal transfer of *Wolbachia* in *Drosophila*. Science *260*: 1796–1799.

Bozeman, F.M., B.L. Elisberg, J.W. Humphries, K. Runcik and D.B.J. Palmer. 1970. Serologic evidence of *Rickettsia canada* infection of man. J. Infect. Dis. *121*: 367–371.

Bozeman, F.M., J.M. Humphries, J.M. Campbell and P.L. O'Hara. 1960. Laboratory studies of the spotted fever group of rickettsiae. *In* Wiseman (Editor), Symposium on the Spotted Fever Group of Rickettsiae, Med Sci. Publ. 7, Walter Reed Army Inst. Res., Washington. pp. 7–11.

Bozeman, F.M., S.A. Masiello, M.S. Williams and B.L. Elisberg. 1975. Epidemic typhus rickettsiae isolated from flying squirrels. Nature (Lond.) *255*: 545–547.

Bozeman, F.M., A. Shirai, J.W. Humphries and F. H.S.. 1967. Ecology of Rocky Mountain spotted fever. II. Natural infection of wild mammals and birds in Virginia and Maryland. Am. J. Trop. Med. Hyg. *16*: 48–59.

Bozeman, F.M., D.E. Sonenshine, M.S. Williams, D.P. Chadwick, D.M. Lauer and B.L. Elisberg. 1981. Experimental infection of ectoparasitic arthropods with *Rickettsia prowazekii* (GvF-16 strain) and transmission to flying squirrels. Am. J. Trop. Med. Hyg. *30*: 253–263.

Bradbury, J.F. 1973. *Xanthomonas ampelina*. Commonwealth Mycological Institute Descriptions of Pathogenic Fungi and Bacteria. No. 378, The Eastern Press Ltd., London.

Bradbury, J.F. 1984. Genus II. *Xanthomonas*. *In* Krieg and Holt (Editors), Bergey's Manual of Systematic Bacteriology, 1st Ed., Vol. 1, The Williams & Wilkins Co., Baltimore. pp. 199–210.

Bradbury, J.F. 1986. Guide to Plant Pathogenic Bacteria, CAB International Mycological Institute, Kew.

Bradley, D.E. 1967. Ultrastructure of bacteriophages and bacteriocins. Bacteriol. Rev. *31*: 230–314.

Bradley, D.E. 1980. Function of *Pseudomonas aeruginosa* PAO polar pili - twitching motility. Can. J. Microbiol. *26*: 146–154.

Braig, H.R., H. Guzman, R.B. Tesh and S.L. O'Neill. 1994. Replacement of the natural *Wolbachia* symbiont of *Drosophila simulans* with a mosquito counterpart. Nature *367*: 453–455.

Braker, G., A. Fesefeldt and K.P. Witzel. 1998. Development of PCR primer systems for amplification of nitrite reductase genes (*nirK* and *nirS*) to detect denitrifying bacteria in environmental samples. Appl. Environ. Microbiol. *64*: 3769–3775.

Brambilla, E., H. Hippe, A. Hagelstein, B.J. Tindall and E. Stackebrandt. 2001. 16S rDNA diversity of cultured and uncultured prokaryotes of a mat sample from Lake Fryxell, McMurdo Dry Valleys, Antarctica. Extremophiles *5*: 23–33.

Bramer, C.O. and A. Steinbuchel. 2001. The methylcitric acid pathway in *Ralstonia eutropha*: new genes identified involved in propionate metabolism. Microbiology *147*: 2203–2214.

Brämer, C.O., P. Vandamme, L.F. da Silva, J.G.C. Gomez and A. Steinbüchel. 2001. *Burkholderia sacchari* sp. nov., a polyhydroxyalkanoate-accumulating bacterium isolated from soil of a sugar-cane plantation in Brazil. Int. J. Syst. Evol. Microbiol. *51*: 1709–1713.

Brandis-Heep, A., N.A. Gebhardt, R.K. Thauer, F. Widdel and N. Pfennig. 1983. Anaerobic acetate oxidation to carbon dioxide by *Desulfobacter postgatei*. 1. Demonstration of all enzymes required for the operation of the citric acid cycle. Arch. Microbiol. *136*: 222–229.

Brandsma, A.R., S.E. Little, J.M. Lockhart, W.R. Davidson, D.E. Stallknecht and J.E. Dawson. 1999. Novel *Ehrlichia* organism (*Rickettsiales*: *Ehrlichieae*) in white-tailed deer associated with lone star tick (Acari: Ixodidae) parasitism. J. Med. Entomol. *36*: 190–194.

Brandt, K.K. and K. Ingvorsen. 1997. *Desulfobacter halotolerans* sp. nov., a halotolerant acetate-oxidizing sulfate-reducing bacterium isolated from sediments of Great Salt Lake, Utah. Syst. Appl. Microbiol. *20*: 366–373.

Brandt, K.K. and K. Ingvorsen. 1998. *In* Validation of the publication of new names and new combinations previously effectively published outside the IJSB. List No. 64. Int. J. Syst. Bacteriol. *48*: 327–328.

Brandt, K.K., F. Vester, A.N. Jensen and K. Ingvorsen. 2001. Sulfate reduction dynamics and enumeration of sulfate-reducing bacteria in hypersaline sediments of the Great Salt Lake (Utah, USA). Microb. Ecol. *41*: 1–11.

Brandt K., K., B.K.C. Patel and K. Ingvorsen. 1999. *Desulfocella halophila* gen. nov., sp. nov., a halophilic, fatty-acid-oxidizing, sulfate-reducing bacterium isolated from sediments of the Great Salt Lake. Int. J. Syst. Bacteriol. *49*: 193–200.

Brandtzæg, P. 1995. Pathogenesis of meningococcal infections. *In* Cartwright (Editor), Meningococcal Disease, John Wiley and Sons Ltd., Chichester. pp. 71–114.

Branham, S. 1930. A new meningococcus-like organism (*Neisseria flavescens* n. sp.) from epidemic meningitis. U.S. Public Health Serv. Rep. *45*: 845–846.

Brannan, D.K. and D.E. Caldwell. 1980. *Thermothrix thiopara*: growth and metabolism of a newly isolated thermophile capable of oxidizing sulfur and sulfur compounds. Appl. Environ. Microbiol. *40*: 211–216.

Brannan, D.K. and D.E. Caldwell. 1982. Evaluation of a proposed surface colonization equation using *Thermothrix thiopara* as a model organism. Microb. Ecol. *8*: 15–21.

Brannan, D.K. and D.E. Caldwell. 1983. Growth kinetics and yield coefficients of the extreme thermophile *Thermothrix thiopara* in continuous culture. Appl. Environ. Microbiol. *45*: 169–173.

Bratina, B.J., G.A. Brusseau and R.S. Hanson. 1992. Use of 16S rRNA analysis to investigate phylogeny of methylotrophic bacteria. Int. J. Syst. Bacteriol. *42*: 645–648.

Brattig, N.W., U. Rathjens, M. Ernst, F. Geisinger, A. Renz and F.W. Tischendorf. 2000. Lipopolysaccharide-like molecules derived from *Wolbachia* endobacteria of the filaria *Onchocerca volvulus* are candidate mediators in the sequence of inflammatory and antiinflammatory responses of human monocytes. Microbes Infect. *2*: 1147–1157.

Braude, A.I. 1951. Studies in the pathology and pathogenesis of experimental brucellosis. I. A comparison of the pathogenicity of *Brucella abortus*, *Brucella melitensis*, and *Brucella suis* for guinea pigs. J. Infect. Dis. *89*: 76–86.

Brauman, A., J.F. Köenig, J. Dutreix and J.L. Garcia. 1990. Characterization of two sulfate-reducing bacteria from the gut of the soil-feeding termite, *Cubitermes speciosus*. Antonie Leeuwenhoek *58*: 271–275.

Braun, K. and D.T. Gibson. 1984. Anaerobic degradation of 2-aminobenzoate (anthranilic acid) by denitrifying bacteria. Appl. Environ. Microbiol. *48*: 102–107.

Breed, R.S. 1957. Genus X. *Mycoplana*. Gray and Thornton, 1928. *In* Breed, Murray and Smith (Editors), Bergey's Manual of Determinative Bacteriology, 7th ed., The Williams & Wilkins Co., Baltimore. pp. 204–206.

Breed, R.S., E.G.D. Murray and N.R. Smith (Editors). 1957. Bergey's Manual of Determinative Bacteriology, 7th Ed., The Williams & Wilkins Co., Baltimore.

Breedveld, M.W. and K.J. Miller. 1994. Cyclic β-glucans of members of the family *Rhizobiaceae*. Microbiol. Rev. *58*: 145–161.

Breil, B., J. Borneman and E.W. Triplett. 1996. A newly discovered gene, *tfuA*, involved in the production of the ribosomally synthesized peptide antibiotic trifolitoxin. J. Bacteriol. *178*: 4150–4156.

Breil, B.T., P.W. Ludden and E.W. Triplett. 1993. DNA sequence and mutational analysis of genes involved in the production and resistance of the antibiotic peptide trifolitoxin. J. Bacteriol. *175*: 3693–3702.

Breitschwerdt, E., B.C. Hegarty, M.G. Davidson and N.S. Szabados. 1995. Evaluation of the pathogenic potential of *Rickettsia canada* and *Rickettsia prowazekii* organisms in dogs. J. Am. Vet. Med. Assoc. *207*: 58–63.

Bremaud, L., S. Laalami, B. Derijard and Y. Cenatiempo. 1997. Translation initiation factor IF2 of the myxobacterium *Stigmatella aurantiaca*: Presence of a single species with an unusual N-terminal sequence. J. Bacteriol. *179*: 2348–2355.

Brenner, D.J., D.G. Hollis, C.W. Moss, C.K. English, G.S. Hall, J. Vincent, J. Radosevic, K.A. Birkness, W.F. Bibb, F.D. Quinn, B. Swaminathan, R.E. Weaver, M.W. Reeves, S. O'Connor, P. Hayes, F. Tenover, A.G. Steigerwalt, B. Perkins, M.I. Daneshvar, B.C. Hill, J.A. Washington, T. Woods, S. Hunter, D.J. Wear and J. Wenger. 1991a. Proposal of *Afipia*, gen. nov., with *Afipia felis*, sp. nov. (formerly the cat scratch disease bacillus), *Afipia clevelandensis*, sp. nov. (formerly the Cleveland Clinic Foundation strain), *Afipia broomeae*, sp. nov.,and three unnamed genospecies. J. Clin. Microbiol. *29*: 2450–2460.

Brenner, D.J., D.G. Hollis, C.W. Moss, C.K. English, G.S. Hall, J. Vincent, J. Radosevic, K.A. Birkness, W.F. Bibb, F.D. Quinn, B. Swaminathan, R.E. Weaver, M.W. Reeves, S. O'connor, P. Hayes, F. Tenover, A.G. Steigerwalt, B. Perkins, M.I. Daneshvar, B.C. Hill, J.A. Washington, T. Woods, S. Hunter, D.J. Wear and J. Wenger. 1992. *In* Validation of the publication of new names and new combinations previously effectively published outside the IJSB. List No. 41. Int. J. Syst. Bacteriol. *42*: 327–328.

Brenner, D.J., S.P. O'Connor, D.G. Hollis, R.E. Weaver and A.G. Steigerwalt. 1991b. Molecular characterization and proposal of a neotype strain for *Bartonella bacilliformis*. J. Clin. Microbiol. *29*: 1299–1302.

Brenner, D.J., S.P. O'Connor, H.H. Winkler and A.G. Steigerwalt. 1993. Proposals to unify the genera *Bartonella* and *Rochalimaea*, with descriptions of *Bartonella quintana* comb. nov., *Bartonella vinsonii* comb. nov., *Bartonella henselae* comb. nov., and *Bartonella elizabethae* comb. nov., and to remove the family *Bartonellaceae* from the order *Rickettsiales*. Int. J. Syst. Bacteriol. *43*: 777–786.

Brenner, S.A., J.A. Rooney, P. Manzewitsch and R.L. Regnery. 1997. Isolation of *Bartonella* (*Rochalimaea*) *henselae*: effects of methods of blood collection and handling. J. Clin. Microbiol. *35*: 544–547.

Brentlinger, K.L., S. Hafenstein, C.R. Novak, B.A. Fane, R. Borgon, R. McKenna and M. Agbandje-McKenna. 2002. *Microviradae*, a family divided: Isolation, characterization and genome sequence of *phiMH2K*, a bacteriophage of the obligate intracellular parasite bacterium *Bdellovibrio bacteriovorus*. J. Bacteriol. *184*: 1089–1094.

Breton, A.M. and J.F. Guespin-Michel. 1987. *Escherichia coli* pH 2.5 acid phosphate and β-lactamase TEM2 are secreted into the medium by *Myxococcus xanthus*. FEMS Microbiol. Lett. *40*: 183–188.

Breton, A.M., S. Jaoua and J. Guespin-Michel. 1985. Transfer of plasmid RP4 to *Myxococcus xanthus* and evidence for its integration into the chromosome. J. Bacteriol. *161*: 523–528.

Bretscher, A.P. and D. Kaiser. 1978. Nutrition of *Myxococcus xanthus*, a fruiting myxobacterium. J. Bacteriol. *133*: 763–768.

Brett, P.J., D. Deshazer and D.E. Woods. 1997. Characterization of *Burkholderia pseudomallei* and *Burkholderia pseudomallei*-like strains. Epidemiol. Infect. *118*: 137–148.

Brett, P.J., D. DeShazer and D.E. Woods. 1998. *Burkholderia thailandensis* sp. nov., a *Burkholderia pseudomallei*-like species. Int. J. Syst. Bacteriol. *48*: 317–320.

Brett, P.J., D.C. Mah and D.E. Woods. 1994. Isolation and characterization of *Pseudomonas pseudomallei* flagellin proteins. Infect. Immun. *62*: 1914–1919.

Brett, P.J. and D.E. Woods. 1996. Structural and immunological characterization of *Burkholderia pseudomallei* O-polysaccharide flagellin protein conjugates. Infect. Immun. *64*: 2824–2828.

Brew, S.D., L.L. Perrett, J.A. Stack, A.P. MacMillan and N.J. Staunton. 1999. Human exposure to *Brucella* recovered from a sea mammal. Vet. Rec. *144*: 483.

Breznak, J.A. and R.N. Costilow. 1994. Physiochemical factors in growth. *In* Gerhardt, Murray, Wood and Krieg (Editors), Methods for General and Molecular Bacteriology, American Society for Microbiology, Washington, D.C. pp. 137–154.

Bricker, B.J., L.B. Tabatabai, B.A. Judge, B.L. Deyoe and J.E. Mayfield. 1990. Cloning, expression, and occurrence of the *Brucella* Cu-Zn superoxide dismutase. Infect. Immun. *58*: 2935–2939.

Brilon, C., W. Beckmann and H.-J. Knackmuss. 1981. Catabolism of naphthalenesulfonic acids by *Pseudomonas* sp. A3 and *Pseudomonas* sp. C22. Appl. Environ. Microbiol. *42*: 44–55.

Bringer, S., R.K. Finn and H. Sahm. 1984. Effect of oxygen on the metabolism of *Zymomonas mobilis*. Arch. Microbiol. *139*: 376–381.

Bringer, S., T. Hartner, K. Poralla and H. Sahm. 1985. Influence of ethanol on the hopanoid content and the fatty acid pattern in bath and continuous cultures of *Zymomonas mobilis*. Arch. Microbiol. *140*: 312–316.

Bringer-Meyer, S. and H. Sahm. 1988. Metabolic shifts in *Zymomonas mobilis* in response to growth conditions. FEMS Microbiol. Rev. *54*: 131–142.

Bringer-Meyer, S. and H. Sahm. 1989. Junctions of catabolic and anabolic pathways in *Zymomonas mobilis*: phosphoenolpyruvate carboxylase and malic enzyme. Appl. Microbiol. Biotechnol. *31*: 529–536.

Bringer-Meyer, S. and H. Sahm. 1993. Formation of acetyl-CoA in *Zy*-

momonas mobilis by a pyruvate dehydrogenase complex. Arch. Microbiol. *159*: 197–199.

Brinkmann, U. and W. Reineke. 1992. Degradation of chlorotoluenes by in vivo constructed hybrid strains: problems of enzyme specificity, induction and prevention of *meta*-pathway. FEMS Microbiol. Lett. *75*: 81–87.

Brinton, C.C., J. Bryan, J.-A. Dillon, N. Guerina, L.J. Jacobson, A. Labik, S. Lee, A. Levine, S. Lim, J. McMichael, S. Polen, K. Rogers, A.C.-C. To and S.C.-M. To. 1978. Uses of pili in gonorrhea control: Role of bacterial pili in disease, purification and properties of gonococcal pili, and progress in the development of a gonococcal pilus vaccine for gonorrhea. *In* Brooks, Gotschlich, Homes, Sawyer and Young (Editors), Immunobiology of *Neisseria gonorrhoeae*, American Society for Microbiology Press, Washington, D.C. pp. 155–178.

Brisbane, P.G. and A. Kerr. 1983. Selective media for 3 biovars of *Agrobacterium*. J. Appl. Bacteriol. *54*: 425–432.

Britschgi, T.B. and S.J. Giovannoni. 1991. Phylogenetic analysis of a natural marine bacterioplankton population by rRNA gene cloning and sequencing. Appl. Environ. Microbiol. *57*: 1707–1713.

Brock, F.M. and R.G. Murray. 1988. The ultrastructure and ATPase nature of polar membrane in *Campylobacter jejuni*. Can. J. Microbiol. *34*: 594–604.

Brock, T.D. 1978. The genus *Sulfolobus*. *In* Starr (Editor), Thermophilic Microorganisms and Life at High Temperatures, Springer-Verlag, Heidelberg. pp. 117–179.

Brockman, E.R. 1989a. *Polyangium*. *In* Staley, Bryant, Pfennig and Holt (Editors), Bergey's Manual of Systematic Bacteriology, Vol. 3, The Williams & Wilkins Co., Baltimore. pp. 2159–2162.

Brockman, E.R. 1989b. *In* Validation of the publication of new names and new combinations previously effectively published outside the IJSB. List No. 31. Int. J. Syst. Bacteriol. *39*: 495–497.

Brockmann, H., Jr. and G. Knobloch. 1972. Ein neues bacteriochlorophyll aus *Rhodospirillum rubrum*. Arch. Mikrobiol. *85*: 123–126.

Brokamp, A., B. Happe and F.R. Schmidt. 1997. Cloning and nucleotide sequence of a D,L-haloalkanoic acid dehalogenase encoding gene from *Alcaligenes xylosoxidans* subsp. *denitrificans* ABIV. Biodegradation *7*: 383–396.

Brokamp, A. and F.R.J. Schmidt. 1991. Survival of *Alcaligenes xylosoxidans* degrading 2,2-dichloropropionate and horizontal transfer of its halidohydrolase in a soil microcosm. Curr. Microbiol. *22*: 299–306.

Brom, S., E. Martínez, G. Dávila and R. Palacios. 1988. Narrow-host-range and broad-host-range symbiotic plasmids of *Rhizobium* spp. strains that nodulate *Phaseolus vulgaris*. Appl. Environ. Microbiol. *54*: 1280–1283.

Bronder, M., H. Mell, E. Stupperich and A. Kröger. 1982. Biosynthetic pathways of *Vibrio succinogenes* growing with fumarate as terminal electron acceptor and sole carbon source. Arch. Microbiol. *131*: 216–223.

Bronsdon, M.A., C.S. Goodwin, L.I. Sly, T. Chilvers and F.D. Schoenknecht. 1991. *Helicobacter nemestrinae*, new species, a spiral bacterium found in the stomach of a pigtailed macaque (*Macaca nemestrina*). Int. J. Syst. Bacteriol. *41*: 148–153.

Brooke, A.G. and M.M. Attwood. 1983. Regulation of enzyme synthesis during the growth of *Hyphomicrobium* X on mixtures of methylamine and ethanol. J. Gen. Microbiol. *129*: 2399–2404.

Brooke, A.G. and M.M. Attwood. 1984. Methylamine uptake by the facultative methylotroph *Hyphomicrobium* X. J. Gen. Microbiol. *130*: 459–464.

Brooke, A.G. and M.M. Attwood. 1985. Regulation of enzyme-synthesis in *Hyphomicrobium* x: growth on mixtures of methylamine and ethanol in continuous cultures. FEMS Microbiol. Lett. *29*: 251–256.

Brooke, A.G., M.G. Duchars and M.M. Attwood. 1987. Nitrogen assimilation in the facultative methylotroph *Hyphomicrobium* x. FEMS Microbiol. Lett. *41*: 41–45.

Brooks, G.F., J.M. O'Donoghue and J.P. Rissing. 1974. *Eikenella corrodens*, a recently recognized pathogen: infections in medical-surgical patients and in association with methylphenylate abuse. Medicine (Baltimore) *53*: 325–342.

Brosius, J., T.J. Dull, D.D. Sleeter and H.F. Noller. 1981. Gene organi-

zation and primary structure of a ribosomal RNA operon from *Escherichia coli*. J. Mol. Biol. *148*: 107–128.

Brosius, J., M.L. Palmer, P.J. Kennedy and H.F. Noller. 1978. Complete nucleotide sequence of a 16S ribosomal RNA gene from *Escherichia coli*. Proc. Natl. Acad. Sci. U.S.A. *75*: 4801–4805.

Brothers, P.N., G. Blotny, L. Qi and R.M. Pollack. 1995. An active site phenylalanine of 3-oxo-delta 5-steroid isomerase is catalytically important for proton transfer. Biochemistry *34*: 15453–15458.

Broughton, E.S. and K.L. Jahans. 1997. The differentiation of *Brucella* species by substrate specific tetrazolium reduction. Vet. Microbiol. *57*: 253–271.

Brouqui, P., M.L. Birg and D. Raoult. 1994. Cytopathic effect, plaque formation, and lysis of *Ehrlichia chaffeensis* grown on continuous cell lines. Infect. Immun. *62*: 405–411.

Brouqui, P. and D. Raoult. 1993. Proteinase K-sensitive and filterable phagosome-lysosome fusion inhibiting factor in *Afipia felis*. Microb. Pathog. *15*: 187–195.

Brouqui, P. and D. Raoult. 1996. *Bartonella quintana* invades and multiplies within endothelial cells *in vitro* and *in vivo* and forms intracellular blebs. Res. Microbiol. *147*: 719–731.

Brown, A.J. 1886. On an acetic ferment which forms cellulose. J. Chem. Soc. (London) *49*: 432–439.

Brown, C.M. and R.A. Herbert. 1977. Ammonia assimilation in members of *Rhodospirillaceae*. FEMS Microbiol. Lett. *1*: 43–45.

Brown, G.M., C.R. Ranger and D.J. Kelley. 1971. Selective media for the isolation of *Brucella ovis*. Cornell Vet. *61*: 265–280.

Brown, N.L., R.P. Burchard, D.W. Morris, J.H. Parish, N.D. Stow and C. Tsopanakis. 1976. Phage and defective phage of strains of *Myxococcus*. Arch. Microbiol. *108*: 271–279.

Brown, N.L. and J.H. Parish. 1976. Extrachromosomal DNA in chloramphenicol resistant *Myxococcus* strains. J. Gen. Microbiol. *93*: 63–68.

Brown, W.C., V. Shkap, D. Zhu, T.C. McGuire, W. Tuo, T.F. McElwain and G.H. Palmer. 1998. CD4(+) T-lymphocyte and immunoglobulin G2 responses in calves immunized with *Anaplasma marginale* outer membranes and protected against homologous challenge. Infect. Immun. *66*: 5406–5413.

Bruce, D. 1893. Sur une nouvelle forme de fièvre rencontrée sur les bords de la Mediterranée. Ann. Inst. Pasteur (Paris) *7*: 289–304.

Bruce, R.A., L.A. Achenbach and J.D. Coates. 1999. Reduction of (per)chlorate by a novel organism isolated from paper mill waste. Environ. Microbiol. *1*: 319–329.

Bruckner, R.J. and S.H. Fahey. 1969. A giant bacterial form (*Simonsiella*) seen in oral exfoliative cytology preparations. Oral Surg. Oral Med. Oral Pathol. *28*: 197–201.

Brumbley, S.M., B.F. Carney and T.P. Denny. 1993. Phenotype conversion in *Pseudomonas solanacearum* due to spontaneous inactivation of PhcA, a putative LysR transcriptional regulator. J. Bacteriol. *175*: 5477–5487.

Brumbley, S.M. and T.P. Denny. 1990. Cloning of wild-type *Pseudomonas solanacearum phcA*, a gene that when mutated alters expression of multiple traits that contribute to virulence. J. Bacteriol. *172*: 5677–5685.

Brumpt, E. 1911. Note sur le parasite des Hematies de la Taupe: *Grahamella talpae* n. g. n. sp. Bulletin de la Societe de Pathologie Exotique. *4*: 514–517.

Brumpt, E. 1922. Précis de parasitologie, 3rd Ed., Masson and Co., Paris.

Brumpt, E. 1932. Longévité de virus de la fièvre boutonneuse (*Rickettsia conori* n. sp.) chez la tique, *Rhipicephalus sanguineus*. C.R. Séances Soc. Biol. Filiales. *110*: 1199–1209.

Brumpt, E. and G. Lavier. 1935. Sur un piroplasmide nouveau, parasite de tortue *Tunetella emydis* N.G., N. Sp. Ann. Parasitol. Hum. Comp. *13*: 544–550.

Brune, A., W. Ludwig and B. Schink. 2002. *Propionivibrio limicola* sp. nov., a fermentative bacterium specialized in the degradation of hydroaromatic compounds, reclassification of *Propionibacter pelophilus* and *Propionivibrio pelophilus* comb. nov. and amended description of the genus *Propionivibrio*. Int. J. Syst. Evol. Microbiol. *52*: 441–444.

Brunen, M. and H. Engelhardt. 1995. Significance of positively charged amino acids for the function of the *Acidovorax delafieldii* porin Omp34. FEMS Microbiol. Lett. *126*: 127–132.

Brunen, M., H. Engelhardt, A. Schmid and R. Benz. 1991. The major outer membrane protein of *Acidovorax delafieldii* is an anion-selective porin. J. Bacteriol. *173*: 4182–4187.

Brunham, R.C., F. Plummer, L. Slaney, F. Rand and W. Dewitt. 1985. Correlation of auxotype and protein I-type with expression of disease due to *Neisseria gonorrhoeae*. J. Infect. Dis. *152*: 339–343.

Brunker, P., W. Minas, P.T. Kallio and J.E. Bailey. 1998. Genetic engineering of an industrial strain of *Saccharopolyspora erythraea* for stable expression of the *Vitreoscilla* haemoglobin gene (*vhb*). Microbiology *144*: 2441–2448.

Brusseau, G.A., E.S. Bulygina and R.S. Hanson. 1994. Phylogenetic analysis and development of probes for differentiating methylotrophic bacteria. Appl. Environ. Microbiol. *60*: 626–636.

Brusseau, G.A., H.-C. Tsien, R.S. Hanson and L.P. Wackett. 1990. Optimization of trichloroethylene oxidation by methanotrophs and the use of a colorimetric assay to detect soluble methane monooxygenase activity. Biodegradation *1*: 19–29.

Bruun, B. 1982. Studies on a collection of strains of the genus *Flavobacterium*. 1. Biochemical studies. Acta Pathol. Microbiol. Immunol. Scand. [B]. *90*: 415–421.

Bryant, M.P. 1972. Commentary on the Hungate technique for culture of anaerobic bacteria. Am. J. Clin. Nutr. *25*: 1324–1328.

Bryant, M.P. 1973. Nutritional requirements of the predominant rumen cellulolytic bacteria. Fed. Proc. *32*: 1809–1813.

Bryant, R.D., K.M. McGroarty, J.W. Costerton and E.J. Laishley. 1983. Isolation and characterization of a new acidophilic *Thiobacillus* species (*T. albertis*). Can. J. Microbiol. *29*: 1159–1170.

Bryn, K., E. Jantzen and K. Bøvre. 1977. Occurrence and patterns of waxes in *Neisseriaceae*. J. Gen. Microbiol. *102*: 33–43.

Bryner, J.H. 1987. *Flexispira rappini*, gen. nov., sp. nov., a motile, urease-producing rod similar to *Campylobacter pyloridis*. Proceedings of the Fourth International Workshop on *Campylobacter* Infections, University of Göteborg, Sweden. pp. 440–442.

Bryner, J.H., A.H. Frank and P.A. O'Berry. 1962. Dissociation studies of *Vibrio* from the bovine genital tract. Am. J. Vet. Res. *23*: 32–41.

Bryner, J.H., J. Littleton, C. Gates, C.A. Kirkbride, A.E. Richie and J.R. Archer. 1986. *Flexispira rappini* gen. nov., sp. nov., a Gram-negative rod from mammalian fetus and feces. XIV International Congress of Microbiology, Manchester, England. p. 307.

Bryner, J.H., P.A. O'Berry and A.H. Frank. 1964. *Vibrio* infection of the digestive organs of cattle. Am. J. Vet. Res. *25*: 1048–1050.

Brysch, K., C. Schneider, G. Fuchs and F. Widdel. 1987. Lithoautotrophic growth of sulfate-reducing bacteria, and description of *Desulfobacterium autotrophicum* gen. nov., sp. nov. Arch. Microbiol. *148*: 264–274.

Brysch, K., C. Schneider, G. Fuchs and F. Widdel. 1988. *In* Validation of the publication of new names and new combinations previously effectively published outside the IJSB. List No. 26. Int. J. Syst. Bacteriol. *38*: 328–329.

Brzin, B. 1966a. Dependence of the cell morphology of *Vitreoscilla* on the temperature of incubation. Experientia (Basel) *22*: 804–805.

Brzin, B. 1966b. Morphology of *Vitreoscilla* grown at different incubation temperatures. Zentbl. Bakteriol. Parasitenkd. Infektkrankh. Hyg. Abt. II *120*: 611–615.

Buchanan, G.E. and D.A. Kuhn. 1978. Patterns of growth and gliding motility in *Simonsiella*. Curr. Microbiol. *1*: 257–262.

Buchanan, R.E. 1926. What names should be used for the organisms producing nodules on the roots of leguminous plants? Proc. Iowa Acad. Sci. *33*: 81–90.

Buchanan, R.E. and W.E. Gibbons (Editors). 1974. Bergey's Manual of Determinative Bacteriology, 8th Ed., The Williams & Wilkins Co., Baltimore.

Buchanan, T.M. 1978. Antigen-specific serotyping of *Neisseria gonorrhoeae*. 1. Use of an enzyme-linked immunosorbent assay to quantitate pilus antigens on gonococci. J. Infect. Dis. *138*: 319–325.

Buchanan, T.M., D.A. Eschenbach, J.S. Knapp and K.K. Holmes. 1980. Gonococcal salpingitis is less likely to recur with *Neisseria gonorrhoeae*

of the same principal outer-membrane protein antigenic type. Am. J. Obstet. Gynecol. *138*: 978–980.

Bucheli-Witschel, M. and T. Egli. 2001. Environmental fate and microbial degradation of aminopolycarboxylic acids. FEMS Microbiol Rev. *25*: 69-106.

Buchholz, S.E., M.M. Dooley and D.E. Eveleigh. 1987. *Zymomonas* - an alcoholic enigma. Trends Biotechnol. *5*: 199–204.

Buchholz, S.E. and D.E. Eveleigh. 1989. Effect of dilution buffers on cell viability of *Zymomonas mobilis*. J. Microbiol. Meth. *10*: 65–69.

Buchholz, S.E., P. O'Mullan and D.E. Eveleigh. 1988. Growth of *Zymomonas mobilis* CP4 on mannitol. Appl. Microbiol. Biotechnol. *29*: 275–281.

Buckle, K.A. and E. Kartadarma. 1990. Inhibition of bongkrek acid and toxoflavin production in tempe bongkrek containing *Pseudomonas cocovenenans*. J. Appl. Bacteriol. *68*: 571–576.

Buckmire, F.L.A. 1971. A protective role for a cell wall protein layer of *Spirillum serpens* against infection by *Bdellovibrio bacteriorovorus*, Abstract G122. *In* Bacteriological Proceedings, American Society for Microbiology, Washington D.C. p. 43.

Buckmire, F.L.A. and R.G.E. Murray. 1970. Studies on the cell wall of *Spirillum serpens*. 1. Isolation and partial purification of the outermost cell wall layer. Can. J. Microbiol. *16*: 1011–1022.

Buddle, M.B. 1956. Studies on *Brucella ovis* (n. sp.), a cause of genital disease of sheep in New Zealand and Australia. J. Hyg. Camb. *54*: 351–364.

Buhrke, T., B. Bleijlevens, S.P.J. Albracht and B. Friedrich. 2001. Involvement of *hyp* gene products in maturation of the H_2-sensing (NiFe) hydrogenase of *Ralstonia eutropha*. J. Bacteriol. *183*: 7087–7093.

Bulygina, E.S., V.F. Galchenko, N.I. Govorukhina, A.I. Netrusov, D.I. Nikitin, Y.A. Trotsenko and K.M. Chumakov. 1990. Taxonomic studies on methylotrophic bacteria by 5S RNA sequencing. J. Gen. Microbiol. *136*: 441–446.

Bulygina, E.S., N.I. Govorukhina, A.I. Netrusov, Y.A. Trotsenko and K.M. Chumakov. 1993. Comparative studies on 5S RNA sequences and DNA–DNA hybridization of obligately and restricted facultatively methylotrophic bacteria. Syst. Appl. Microbiol. *16*: 85–91.

Bulygina, E.S., O.M. Gulikova, E.M. Dikanskaya, A.I. Netrusov, T.P. Tourova and K.M. Chumakov. 1992. Taxonomic studies of the genera *Acidomonas*, *Acetobacter* and *Gluconobacter* by 5S ribosomal RNA sequencing. J. Gen. Microbiol. *138*: 2283–2286.

Bundle, D.R., J.W. Cherwonogrodzky, M. Caroff and M.B. Perry. 1987. The lipopolysaccharides of *Brucella abortus* and *B. melitensis*. Ann. Inst. Pasteur Microbiol. (Paris) *138*: 92–98.

Bünz, P.V. and A.M. Cook. 1993. Dibenzofuran 4,4a-dioxygenase from *Sphingomonas* sp. strain RW1: angular dioxygenation by a three-component enzyme system. J. Bacteriol. *175*: 6467–6475.

Bünz, P.V., R. Falchetto and A.M. Cook. 1993. Purification of two isofunctional hydrolases (EC 3.7.1.8) in the degradative pathway for dibenzofuran in *Sphingomonas* sp. strain RW1. Biodegradation *4*: 171–178.

Buonaurio, R., V.M. Stravato and C. C. Cappelli. 2001. Brown spot caused by *Sphingomonas* sp. on yellow Spanish melon fruits in Spain. Plant Pathol. (Oxf.) *50*: 397–401.

Buonaurio, R., V.M. Stravato, Y. Kosako, N. Fujiwara, T. Naka, K. Kobayashi, C. Cappelli and E. Yabuuchi. 2002. *Sphingomonas melonis* sp. nov., a novel pathogen that causes brown spots on yellow Spanish melon fruits. Int. J. Syst. Evol. Microbiol. *52*: 2081–2087.

Burbage, D.A. and M. Sasser. 1982. A medium selective for *Pseudomonas cepacia*. Phytopathology *72*: 706–706.

Burchard, A.C., R.P. Burchard and J.A. Kloetzel. 1977a. Intracellular, periodic structures in the gliding bacterium *Myxococcus xanthus*. J. Bacteriol. *132*: 666–672.

Burchard, R.P., A.C. Burchard and J.H. Parish. 1977b. Pigmentation phenotype instability in *Myxococcus xanthus*. Can. J. Microbiol. *23*: 1657–1662.

Burchard, R.P. and M. Dworkin. 1966. A bacteriophage for *Myxococcus xanthus*: isolation, characterization and relation of infectivity to host morphogenesis. J. Bacteriol. *91*: 1305–1313.

Burchard, R.P. and H. Voelz. 1972. Bacteriophage infection of *Myxococcus xanthus* during cellular differentiation and vegetative growth. Virology *48*: 555–566..

Burdman, S., E. Jurkevitch, B. Schwartsburd, M. Hampel and Y. Okon. 1998. Aggregation in *Azospirillum brasilense*: Effects of chemical and physical factors and involvement of extracelullar components. Microbiology (Read.) *144*: 1989–1999.

Burgdorfer, W. 1988. Ecological and epidemiological considerations of Rocky Mountain spotted fever and scrub typhus. *In* Walker (Editor), Biology of Rickettsial Diseases, CRC Press, Boca Raton. pp. 35–50.

Burgdorfer, W., A. Aeschlimann, O. Peter, S.F. Hayes and R.N. Philip. 1979. *Ixodes ricinus*: Vector of a hitherto undescribed spotted fever group agent in Switzerland. Acta Trop. *36*: 357–367.

Burgdorfer, W. and L.P. Brinton. 1975. Mechanisms of transovarial infection of spotted fever rickettsiae in ticks. Ann. N.Y. Acad. Sci. *266*: 61–72.

Burgdorfer, W., L.P. Brinton, W.L. Krynski and R.N. Philip. 1978. *Rickettsia rhipicephali*, a new spotted fever group rickettsia from the brown dog tick, *Rhipicephalus sanguineus*. *In* Kazár, Ormsbee and Tarasevich (Editors), Rickettsiae and Rickettsial Diseases, VEDA, Bratislava. pp. 307–316.

Burgdorfer, W., J.C. Cooney and L.A. Thomas. 1974. Zoonotic potential (Rocky Mountain spotted fever and tularemia) in the Tennessee Valley region. II. Prevalence of *Rickettsia rickettsii* and *Francisella tularensis* in mammals and ticks from Land Between the Lakes. Am. J. Trop. Med. Hyg. *23*: 109–117.

Burgdorfer, W., S.F. Hayes, T. L.A. and J.L. Lancaster Jr.. 1981. A new spotted fever group *Rickettsia* from the lone star tick, *Amblyomma americanum*. *In* Burgdorfer and Anacker (Editors), Rickettsiae and Rickettsial Diseases, Academic Press, New York. pp. 595–625.

Burgdorfer, W., V.F. Newhouse, E.G. Pickens and D.B. Lackman. 1962. Ecology of Rocky Mountain spotted fever in Western Montana. I. Isolation of *Rickettsia rickettsii* from wild mammals. Am. J. Hyg. *76*: 293–301.

Burgdorfer, W., D.J. Sexton, R.K. Gerloff, R.L. Anacker, R.N. Philip and L.A. Thomas. 1975. *Rhipicephalus sanguineus*: Vector of a new spotted fever group rickettsia in the United States. Infect. Immun. *12*: 205–210.

Burger, A., G. Drews and R. Ladwig. 1968. Host range and infection cycle of a newly isolated strain of *Bdellovibrio bacteriovorus*. Arch. Mikrobiol. *61*: 261–279.

Burgess, A.W. and B.E. Anderson. 1998. Outer membrane proteins of *Bartonella henselae* and their interaction with human endothelial cells. Microb. Pathog. *25*: 157–164.

Burgess, J.G., R. Kawaguchi, T. Sakaguchi, R.H. Thornhill and T. Matsunaga. 1993. Evolutionary relationships among *Magnetospirillum* strains inferred from phylogenetic analysis of 16S rDNA sequences. J. Bacteriol. *175*: 6689–6694.

Burkhead, K.D., D.A. Schisler and P.J. Slininger. 1994. Pyrrolnitrin production by biological control agent *Pseudomonas cepacia* B37w in culture and in colonized wounds of potatoes. Appl. Environ. Microbiol. *60*: 2031–2039.

Burkholder, W.H. 1942. Three bacterial plant pathogens: *Phytomonas caryophylli* sp. n. *Phytomonas alllicola* sp. n., and *Phytomonas manihotis* (Arthaud-Berthet and Bondar) Viegas. Phytopathology *32*: 141–149.

Burkholder, W.H. and M.P. Starr. 1948. The generic and specific characters of phytopathogenic species of *Pseudomonas* and *Xanthomonas*. Phytopathology *38*: 494–502.

Burnens, A.P. and J. Nicolet. 1992. Detection of *Campylobacter upsaliensis* in diarrheic dogs and cats, using a selective medium with cefoperazone. Am. J. Vet. Res. *53*: 48–51.

Burnens, A.P. and J. Nicolet. 1993. Three supplementary diagnostic tests for *Campylobacter* species and related organisms. J. Clin. Microbiol. *31*: 708–710.

Burnens, A.P., J. Stanley, U.B. Schaad and J. Nicolet. 1993. Novel *Campylobacter*-like organism resembling *Helicobacter fennelliae* isolated from a boy with gastroenteritis and from dogs. J. Clin. Microbiol. *31*: 1916–1917.

Burnham, J.C., S.A. Collart and M.J. Daft. 1984. Myxococcal predation of the cyanobacterium *Phormidium luridum* in aqueous environments. Arch. Microbiol. *137*: 220–225.

Burnham, J.C., S.A. Collart and B.W. Highison. 1981. Entrapment and lysis of the cyanobacterium *Phormidium luridum* by aqueous colonies of *Myxococcus xanthus* PCO₂. Arch. Microbiol. *129*: 285–294.

Burnham, J.C. and S.F. Conti. 1984. Genus *Bdellovibrio*. *In* Krieg and Holt (Editors), Bergey's Manual of Systematic Bacteriology, 1st Ed., Vol. 1, The Williams & Wilkins Co., Baltimore. pp. 118–124.

Burnham, J.C., T. Hashimoto and S.F. Conti. 1968. Electron microscopic observations on the penetration of *Bdellovibrio bacteriovorus* into Gramnegative bacterial hosts. J. Bacteriol. *96*: 1366–1381.

Burnham, J.C., T. Hashimoto and S.F. Conti. 1970. Ultrastructure and cell division of a facultatively parasitic strain of *Bdellovibrio bacteriovorus*. J. Bacteriol. *101*: 997–1004.

Burns, J.L. and D.K. Clark. 1992. Salicylate inducible antibiotic resistance in *Pseudomonas cepacia* associated with absence of a pore forming outer membrane protein. Antimicrob. Agents Chemother. *36*: 2280–2285.

Burns, J.L., C.D. Wadsworth, J.J. Barry and C.P. Goodall. 1996. Nucleotide sequence analysis of a gene from *Burkholderia (Pseudomonas) cepacia* encoding an outer membrane lipoprotein involved in multiple antibiotic resistance. Antimicrob. Agents Chemother. *40*: 307–213.

Burr, T.J., A.L. Bishop, B.H. Katz, L.M. Blanchard and C. Bazzi. 1987. A Root-specific decay of grapevine caused by *Agrobacterium tumefaciens* and *Agrobacterium radiobacter* biovar 3. Phytopathology *77*: 1424–1427.

Burris, R.H. 1994. Comparative study of the response of *Azotobacter vinelandii* and *Acetobacter diazotrophicus* to changes in pH. Protoplasma. *183*: 62–66.

Burrows, K.J., A. Cornish, D. Scott and I.J. Higgins. 1984. Substrate specificities of the soluble and particulate methane monooxygenases of *Methylosinus trichosporium* OB3b. J. Gen. Microbiol. *130*: 3327–3333.

Büsing, K.H., W. Döll and K. Freytag. 1953. Die bakterienflora der medizinische blutegel. Arch. Mikrobiol. *19*: 52–86.

Busse, H.J. and G. Auling. 1988. Polyamine pattern as a chemotaxonomic marker within the *Proteobacteria*. Syst. Appl. Microbiol. *11*: 1–8.

Busse, H.J. and G. Auling. 1992. The genera *Alcaligenes* and "*Achromobacter*". *In* Balows, Trüper, Dworkin, Harder and Schleifer (Editors), The Prokaryotes: A Handbook of Bacteria: Ecophysiology, Isolation, Identification, Applications, 2nd Ed., Vol. 3, Springer-Verlag, New York. pp. 2544–2555.

Busse, H.J., T. el Banna, H. Oyaizu and G. Auling. 1992. Identification of xenobiotic-degrading isolates from the beta subclass of the *Proteobacteria* by a polyphasic approach including 16S rRNA partial sequencing. Int. J. Syst. Bacteriol. *42*: 19–26.

Busse, H.J., P. Kämpfer and E.B.M. Denner. 1999. Chemotaxonomic characterisation of *Sphingomonas*. J. Ind. Microbiol. Biotechnol. *23*: 242–251.

Buswell, C.M., Y.M. Herlihy, P.D. Marsh, C.W. Keevil and S.A. Leach. 1997. Coaggregation amongst aquatic biofilm bacteria. J. Appl. Microbiol. *83*: 477–484.

Butler, B.J., K.L. McCallum and W.E. Inniss. 1989. Characterization of *Aquaspirillum arcticum* sp. nov., a new psychrophilic bacterium. Syst. Appl. Microbiol. *12*: 263–266.

Butler, B.J., K. McCallum and W.E. Inniss. 1990. *In* Validation of the publication of new names and combinations previously effectively published outside the IJSB. List No. 34. Int. J. Syst. Bacteriol. *40*: 320–321.

Butterfield, C.T. 1935. Studies of sewage purification. II. A zoogloeaforming organism found in activated sludge. Pub. Health Rep. *50*: 671–684.

Butzler, J.P. and M.B. Skirrow. 1979. *Campylobacter enteritis*. Clinics in Gastroenterology *8*: 737–765.

Buxton, J.B. 1929. A note on *Vibrio foetus* ovis in the ram, . Univ. of Cambridge. Report nr First Report of Director, Inst. Anim. Pathol., 1929–1930

Byng, G.S., J.L. Johnson, R.J. Whitaker, R.L. Gherna and R.A. Jensen. 1983. The evolutionary pattern of aromatic amino acid biosynthesis and the emerging phylogeny of pseudomonad bacteria. J. Mol. Evol. *19*: 272–282.

Byng, G.S., R.J. Whitaker, R.L. Gherna and R.A. Jensen. 1980. Variable enzymological patterning in tyrosine biosynthesis as a means of determining natural relatedness among the *Pseudomonadaceae*. J. Bacteriol. *144*: 247–257.

Byrd, J.J., L.R. Zeph and L.E. Casida, Jr.. 1985. Bacterial control of *Agromyces ramosus* in soil. Can. J. Microbiol. *31*: 1157–1163.

Byron, C.M., M.T. Stankovich, M. Hussain and V.L. Davidson. 1989. Unusual redox properties of electron-transfer flavoprotein from *Methylophilus methylotrophus*. Biochemistry *28*: 8582–8587.

Caballero-Mellado, J., L.E. Fuentes-Ramírez, V.M. Reis and E. Martínez-Romero. 1995. Genetic structure of *Acetobacter diazotrophicus* populations and identification of a new genetically distant group. Appl. Environ. Microbiol. *61*: 3008–3013.

Caballero-Mellado, J., L. López-Reyes and R. Bustillos-Cristales. 1999. Presence of 16S rRNA genes in multiple replicons in *Azospirillum brasilense*. FEMS Microbiol. Biol. Lett. *178*: 283–288.

Caballero-Mellado, J. and E. Martínez-Romero. 1994. Limited genetic diversity in the endophytic sugarcane bacterium *Acetobacter diazotrophicus*. Appl. Environ. Microbiol. *60*: 1532–1537.

Caballero-Mellado, J. and E. Martínez-Romero. 1999. Soil fertilization limits the genetic diversity of *Rhizobium* in bean nodules. Symbiosis *26*: 111–121.

Cabrita, J., J. Rodrigues, F. Braganca, C. Morgado, I. Pires and A.P. Goncalves. 1992. Prevalence, biotypes, plasmid profile and antimicrobial resistance of *Campylobacter* isolated from wild and domestic animals from northeast Portugal. J. Appl. Bacteriol. *73*: 279–285.

Caccavo, F., Jr., D.J. Lonergan, D.R. Lovley, M. Davis, J.F. Stolz and M.J. McInerney. 1994. *Geobacter sulfurreducens* sp. nov., a hydrogen- and acetate-oxidizing dissimilatory metal-reducing microorganism. Appl. Environ. Microbiol. *60*: 3752–3759.

Caccavo, F., D.J. Lonergan, D.R. Lovley, M. Davis, J.F. Stolz and M.J. McInerney. 1995. *In* Validation of the publication of new names and new combinations previously effectively published outside the IJSB. List No. 54. Int. J. Syst. Bacteriol. *45*: 619–620.

Cadwallader, K.R., R.J. Braddock and M.E. Parish. 1992. Isolation of alpha-terpineol dehydratase from *Pseudomonas gladioli*. J. Food. Sci. *57*: 241–244.

Cahill, R.J., C.J. Foltz, J.G. Fox, C.A. Dangler, F. Powrie and D.B. Schauer. 1997. Inflammatory bowel disease: an immunity-mediated condition triggered by bacterial infection with *Helicobacter hepaticus*. Infect. Immun. *65*: 3126–3131.

Caillon, E., B. Lubochinsky and D. Rigomier. 1983. Occurrence of dialkyl ether phospholipids in *Stigmatella aurantiaca* DW4. J. Bacteriol. *153*: 1348–1351.

Caldwell, D.E., S.J. Caldwell and J.P. Laycock. 1981. *In* Validation of the publication of new names and new combinations previously effectively published outside the IJSB. List No. 6. Int. J. Syst. Bacteriol. *31*: 215–218.

Caldwell, D.E., S.J. Caldwell and J.P. Laylock. 1976. *Thermothrix thioparus* gen. et sp. nov. a facultatively anaerobic facultative chemolithotroph living at neutral pH and high temperature. Can. J. Microbiol. *22*: 1509–1517.

Caldwell, S.R., J.R. Newcomb, K.A. Schlecht and F.M. Raushel. 1991. Limits of diffusion in the hydrolysis of substrates by the phosphotriesterase from *Pseudomonas diminuta*. Biochemistry *30*: 7438–7444.

Calhoun, A. and G.M. King. 1998. Characterization of root-associated methanotrophs from three freshwater macrophytes: *Pontederia cordata*, *Sparganium eurycarpum*, and *Sagittaria latifolia*. Appl. Environ. Microbiol. *64*: 1099–1105.

Calubiran, O.V., P.E. Schoch and B.A. Cunha. 1990. *Pseudomonas paucimobilis* bacteraemia associated with haemodialysis. J. Hosp. Infect. *15*: 383–388.

Camer, A., J. Masangkay, H. Satoh, T. Okabayashi, S. Norizuki, Y. Motoi, H. Ueno and C. Morita. 2000. Prevalence of spotted fever rickettsial antibodies in dogs and rodents in the Philippines. Jpn. J. Infect. Dis. *53*: 162–163.

Cammack, R., G. Fauque, J.J.G. Moura and J. LeGall. 1984. ESR studies of cytochrome c_3 from *Desulfovibrio desulfuricans* strain Norway 4: midpoint potentials of the 4 hemes and interactions with ferredoxin and colloidal sulfur. Biochim. Biophys. Acta *784*: 68–74.

Campbell, L.L., M.A. Kasprzycki and J.R. Postgate. 1966. *Desulfovibrio africanus* sp. n., a new dissimilatory sulfate-reducing bacterium. J. Bacteriol. *92*: 1122–1127.

Campbell, P.W., III, J.A. Phillips, III, G.J. Heidecker, M.R. Krishnamani, R. Zahorchak and T.L. Stull. 1995. Detection of *Pseudomonas (Burkholderia) cepacia* using PCR. Pediatr. Pulmonol. *20*: 44–49.

Campbell, R.W. and R. Domrow. 1974. Rickettsioses in Australia: Isolation of *Rickettsia tsutsugamushi* and *R. australis* from naturally infected arthropods. Roy. Soc. Trop. Med. Hyg. *68*: 397–402.

Campêlo, A.B. and J. Döbereiner. 1970.. Ocorrêntia de *Derxia* sp. em solos de alguns Estados Brasileiro. Pesqui. Agropecuária Bras. *5*: 327–332.

Campos, J.M., J. Geisselsoder and D.R. Zusman. 1978. Isolation of bacteriophage MX4, a generalized transducing phage for *Myxococcus xanthus*. J. Mol. Biol. *119*: 167–178.

Campos, J.M. and D.R. Zusman. 1975. Regulation of development in *Myxococcus xanthus*: Effect of $3':5'$-cyclic AMP, ADP, and nutrition. Proc. Natl. Acad. Sci. U.S.A. *72*: 518–522.

Canale-Parola, E., S.L. Rosenthal and D.G. Kupfer. 1966. Morphological and physiological characteristics of *Spirillum gracile* sp. n. Antonie van Leeuwenhoek J. Microbiol. Serol. *32*: 113–124.

Cannon, J.G., T.M. Buchanan and P.F. Sparling. 1983. Confirmation of association of protein-i serotype of *Neisseria gonorrhoeae* with ability to cause disseminated infection. Infect. Immun. *40*: 816–819.

Cannon, R.E. and S.M. Anderson. 1991. Biogenesis of bacterial cellulose. Crit. Rev. Microbiol. *17*: 435–447.

Cao, Y., J. Zhou and H. Chen. 1984. Differentiation and viability of nodule bacteria in host cells. Sci. Sin. Ser. B (Chem. Biol. Agric. Med. Earth Sci.) *27*: 593–600.

Capoor, S.P., D.G. Rao and S.M. Viswanath. 1967. *Diaphorina citri* Kuway., a vector of greening disease of citrus in India. India J. Agric. Sci. *37*: 572–576.

Carandina, G., M. Bacchelli, A. Virgili and R. Strumia. 1984. *Simonsiella* filaments isolated from erosive lesions of the human oral cavity. J. Clin. Microbiol. *19*: 931–933.

Caraway, B.H. and N.R. Krieg. 1974. Aerotaxis in *Spirillum volutans*. Can. J. Microbiol. *20*: 1367–1377.

Carbonetti, N.H., V.I. Simnad, H.S. Seifert, M. So and P.F. Sparling. 1988. Genetics of Protein-I of *Neisseria gonorrhoeae* — Construction of hybrid porins. Proc. Natl. Acad. Sci. U.S.A. *85*: 6841–6845.

Cardarelli-Leite, P., K. Blom, C.M. Patton, M.A. Nicholson, A.G. Steigerwalt, S.B. Hunter, D.J. Brenner, T.J. Barrett and B. Swaminathan. 1996. Rapid identification of *Campylobacter* species by restriction fragment length polymorphism analysis of a PCR-amplified fragment of the gene coding for 16S rRNA. J. Clin. Microbiol. *34*: 62–67.

Carey, V.C. and L.O. Ingram. 1983. Lipid composition of *Zymomonas mobilis*: effects of ethanol and glucose. J. Bacteriol. *154*: 1291–1300.

Carifo, K. and B.W. Catlin. 1973. *Neisseria gonorrhoeae* auxotyping: differentiation of clinical isolates based on growth responses on chemically defined media. Appl. Microbiol. *26*: 223–230.

Carlson, R.W. 1982. Surface chemistry. *In* Broughton (Editor), Nitrogen Fixation II: Biology of the nitrogen-fixing organisms, Vol. 2, Oxford University Press, Oxford. 199–234.

Carlson, R.W., J. Sanjuan, U.R. Bhat, J. Glushka, H.P. Spaink, A.H.M. Wijfjes, A.A.N. van Brussel, T.J.W. Stokkermans, N.K. Peters and G. Stacey. 1993. The structures and biological activities of the lipo-oligosaccharide nodulation signals produced by type-I and type-II strains of *Bradyrhizobium japonicum*. J. Biol. Chem. *268*: 18372–18381.

Carmichael, L.E. and D.W. Bruner. 1968. Characteristics of a newly recognized species of *Brucella* responsible for infectious canine abortion. Cornell Vet. *58*: 579–592.

Carmichael, L.E., R. Flores-Castro and S. Zoha. 1980. Brucellosis caused by *Brucella canis* (*Br. canis*): an update of infection in animals and man, World Health Organization Brucellosis document. WHO/BRUC/80.361.

Carney, J.F., L. Wan, T.E. Lovelace and R.R. Colwell. 1975. Numerical taxonomy study of *Vibrio* and *Spirillum* spp. Int. J. Syst. Bacteriol. *25*: 38–46.

Caroff, M., D.R. Bundle and M.B. Perry. 1984a. Structure of the O-chain of the phenol phase soluble cellular lipopolysaccharide of *Yersinia enterocolitica* serotype O:9. Eur. J. Biochem. *139*: 195–200.

Caroff, M., D.R. Bundle, M.B. Perry, J.W. Cherwonogrodzky and J.R. Duncan. 1984b. Antigenic S-type lipopolysaccharide of *Brucella abortus* 1119-3. Infect. Immun. *46*: 384–388.

Carpano, M. 1929. Su di un piroplasma osservato nei polli egitto (*Aegyptianella pullorum*). Bull. Minist. Agric. Egypt. *86*: 1–12.

Carr, J.G. 1958. *Acetobacter estunense* nov. spec. An addition to Frateur's ten basic species. Antonie Leeuwenhoek *24*: 157–160.

Carr, J.G. 1974. Genus *Zymomonas* Kluyver and van Niel. *In* Buchanan and Gibbons (Editors), Bergey's Manual of Determinative Bacteriology, 8th Ed., The Williams & Wilkins Co., Baltimore. 352–353.

Carr, J.G. and S.M. Passmore. 1971. Discovery of the "cider sickness" bacterium *Zymomonas anaerobia* in apple pulp. J. Inst. Brew. London. *77*: 462–466.

Carr, J.G. and S.M. Passmore. 1979. Methods for identifying acetic acid bacteria. *In* Skinner and Lovelock (Editors), Identification Methods for Microbiologists, 2nd Ed., Academic Press, London. pp. 33–47.

Carr, R.T., S. Balasubramanian, P.C.D. Hawkins and S.J. Benkovic. 1995. Mechanism of metal-independent hydroxylation by *Chromobacterium violaceum* phenylalanine hydroxylase. Biochemistry *34*: 7525–7532.

Carr, R.T. and S.J. Benkovic. 1993. An examination of the copper requirement of phenylalanine hydroxylase from *Chromobacterium violaceum*. Biochemistry *32*: 14132–14138.

Carrick, C.S., J.A. Fyfe and J.K. Davies. 1998. *Neisseria gonorrhoeae* contains multiple copies of a gene that may encode a site-specific recombinase and is associated with DNA rearrangements. Gene *220*: 21–29.

Carruthers, M.M. and H.M. Sommers. 1973. *Eikenella corrodens* osteomyelitis. Ann. Intern. Med. *79*: 900.

Carter, H.V. 1888. Note on the occurrence of a minute blood-spirillum in an Indian rat. Sci. Mem. Offrs. Army India *3*: 45–48.

Carter, R.N. and J.M. Schmidt. 1976. Fatty acid composition of selected prosthecate bacteria. Arch. Microbiol. *110*: 91–94.

Cartwright, K.A.V. 1995. Meningococcal Disease, John Wiley and Sons Ltd., Chichester.

Casadevall, A., L.F. Freundlich and L. Pirofski. 1992. Septic shock caused by *Pseudomonas paucimobilis*. Clin. Infect. Dis. *14*: 784.

Casalta, J.P., Y. Peloux, D. Raoult, P. Brunet and H. Gallais. 1989. Pneumonia and meningitis caused by a new nonfermentative unknown gram-negative bacterium. J. Clin. Microbiol. *27*: 1446–1448.

Casao, M.A., J. Leiva, R. Diaz and C. Gamazo. 1998. Anti-phosphatidylcholine antibodies in patients with brucellosis. J. Med. Microbiol. *47*: 49–54.

Cascio, A., C. Colomba, D. Di Rosa, L. Salsa, L. di Martino and L. Titone. 2001. Efficacy and safety of clarithromycin as treatment for Mediterranean spotted fever in children: A randomized controlled trial. Clin. Infect. Dis. *33*: 409–411.

Casida, L.E., Jr. 1980. Bacterial predators of *Micrococcus luteus* in soil. Appl. Environ. Microbiol. *39*: 1035–1041.

Casida, L.E., Jr. 1982. *Ensifer adhaerens* gen. nov., sp. nov.—a bacterial predator of bacteria in soil. Int. J. Syst. Bacteriol. *32*: 339–345.

Casida, L.E., Jr. 1984. A growth initiation factor involved in magnesium utilization by certain soil bacteria. Can. J. Microbiol. *30*: 824–829.

Casida, L.E., Jr 1987. Relation to copper of N-1, a nonobligate bacterial predator. Appl. Environ. Microbiol. *53*: 1515–1518.

Casida, L.E., Jr. 1989a. *Arthrobacter* species as a prey cell reservoir for nonobligate bacterial predators in soil. Can. J. Microbiol. *35*: 559–564.

Casida, L.E., Jr 1989b. Protozoan response to the addition of bacterial predators and other bacteria to soil. Appl. Environ. Microbiol. *55*: 1857–1859.

Casiraghi, M., T.J. Anderson, C. Bandi, C. Bazzocchi and C. Genchi. 2001a. A phylogenetic analysis of filarial nematodes: Comparison with the phylogeny of *Wolbachia* endosymbionts. Parasitology. *122*: 93–103.

Casiraghi, M., G. Favia, G. Cancrini, A. Bartoloni and C. Bandi. 2001b. Molecular identification of *Wolbachia* from the filarial nematode *Mansonella ozzardi*. Parasitol. Res. *87*: 417–420.

Casiraghi, M., J.W. McCall, L. Simoncini, L.H. Kramer, L. Sacchi, C. Genchi, J.H. Werren and C. Bandi. 2002. Tetracycline treatment and sex-ratio distortion: a role for *Wolbachia* in the moulting of filarial nematodes? Int. J. Parasitol. *32*: 1457–1468.

Castañeda, M.R. 1947. A practical method for routine blood cultures in brucellosis. Proc. Soc. Exp. Biol. Med. *64*: 114–115.

Castellani, A. and A.J. Chalmers. 1919. Manual of Tropical Medicine, 3rd Ed., William Wood and Company, New York.

Castle, M.D. and B.M. Christensen. 1985. Isolation and identification of *Aegyptianella pullorum* (*Rickettsiales, Anaplasmataceae*) in wild turkeys form North America. Avian Dis. *29*: 437–445.

Castro, C.E., S.K. O'Shea, W. Wang and E.W. Bartnicki. 1996. Biodehalogenation: oxidative and hydrolytic pathways in the transformations of acetonitrile, chloroacetonitrile, chloroacetic acid, and chloroacetamide by *Methylosinus trichosporium* OB-3b. Environ. Sci. Technol. *30*: 1180–1184.

Cataldi, M.S. 1939. Estudio fisiólogico y sistemático de algunas *Chlamydobacteriales*, Thesis, University of Buenos Aires

Catlin, B.W. 1960. Transformation of *Neisseria meningitidis* by deoxyribonucleates from cells and from slime. J. Bacteriol. *79*: 579–590.

Catlin, B.W. 1970. Transfer of the organism named *Neisseria catarrhalis* to *Branhamella* gen. nov. Int. J. Syst. Bacteriol. *20*: 155–159.

Catlin, B.W. 1973. Nutritional profiles of *Neisseria gonorrhoeae*, *Neisseria meningitidis*, *Neisseria lactamica* in chemically defined media and use of growth requirements for gonococcal typing. J. Infect. Dis. *128*: 178–194.

Catlin, B.W. 1977. Nutritional requirements and auxotyping. *In* Roberts (Editor), The Gonococcus, John Wiley and Sons, New York. pp. 92–109.

Catlin, B.W. 1978. Characterization and auxotyping of *Neisseria gonorrhoeae*. *In* Bergan and Norris (Editors), Methods in Microbiology, Vol. 10, Academic Press, New York. pp. 345–380.

Catlin, B.W. 1991. *Branhamaceae* fam. nov., a proposed family to accommodate the genera *Branhamella* and *Moraxella*. Int. J. Syst. Bacteriol. *41*: 320–323.

Catlin, B.W. and L.S. Cunningham. 1961. Transforming activities and base contents of deoxyribonucleate preparations from various neisseriae. J. Gen. Microbiol. *26*: 303–312.

Caugant, D.A., L.O. Froholm, R.K. Selander and K. Bovre. 1989. Sulfonamide resistance in *Neisseria meningitidis* isolates of clones of the Et-5 complex. APMIS *97*: 425–428.

Caumette, P. 1993. Ecology and physiology of phototrophic bacteria and sulfate-reducing bacteria in marine salterns. Experientia (Basel) *49*: 473–481.

Caumette, P., Y. Cohen and R. Matheron. 1991a. Isolation and characterization of *Desulfovibrio halophilus* sp. nov., isolated from Solar Lake (Sinai). Syst. Appl. Microbiol. *14*: 33–38.

Caumette, P., Y. Cohen and R. Matheron. 1991b. *In* Validation of the publication of new names and new combinations previously effectively published outside the IJSB. List No. 37. Int. J. Syst. Bacteriol. *41*: 331.

Cavalcante, V.A. and J. Döbereiner. 1988. A new acid-tolerant nitrogen-fixing bacterium associated with sugarcane. Plant Soil *108*: 23–31.

Cech, J.S. and P. Hartman. 1990. Glucose-induced breakdown of enhanced biological phosphate removal. Environ. Technol. *11*: 651–656.

Cech, J.S. and P. Hartman. 1993. Competition between polyphosphate and polysaccharide accumulating bacteria in enhanced biological phosphate removal systems. Water Res. *27*: 1219–1225.

Censini, S., C. Lange, Z. Xiang, J.E. Crabtree, P. Ghiara, M. Borodovsky, R. Rappuoli and A. Covacci. 1996. Cag, a pathogenicity island of *Helicobacter pylori*, encodes type I-specific and disease-associated virulence factors. Proc. Natl. Acad. Sci. U.S.A. *93*: 14648–14653.

Centers for Disease Control and Prevention 1998. Nosocomial *Ralstonia pickettii* colonization associated with intrinsically contaminated saline solution—Los Angeles, California, 1998. Morb. Mortal. Wkly. Rep. *47*: 285–286.

Centers for Disease Control and Prevention 2001. Recommended childhood immunization schedule - United States. Morb. Mortal. Wkly. Rep. *50*: 7–10.

Cerantola, S. and H. Montrozier. 1997. Structural elucidation of two polysaccharides present in the lipopolysaccharide of a clinical isolate of *Burkholderia cepacia*. Eur. J. Biochem. *246*: 360–366.

Cervantes, E., S.B. Sharma, F. Maillet, J. Vasse, G. Truchet and C. Rosenberg. 1989. The *Rhizobium meliloti* host range *nodQ* gene encodes a protein which shares homology with translation elongation and initiation factors. Mol. Microbiol. *3*: 745–755.

Cha, J.M., W.S. Cha and J.H. Lee. 1999. Removal of organo-sulphur odour compounds by *Thiobacillus novellus* SRM, sulphur-oxidizing microorganisms. Process Biochem. *34*: 659–665.

Cha, J.-S., C. Pujol and C.I. Kado. 1997. Identification and characterization of a *Pantoea citrea* gene encoding glucose dehydrogenase that is essential for causing pink disease of pineapple. Appl. Environ. Microbiol. *63*: 71–76.

Chadwick, P.R., H. Malnick and A.O. Ebizie. 1995. *Haemophilus paraphrophilus* infection: a pitfall in laboratory diagnosis. J. Infect. *30*: 67–69.

Chae, J.S., N. Pusterla, E. Johnson, E. Derock, S.P. Lawler and J.E. Madigan. 2000. Infection of aquatic insects with trematode metacercariae carrying *Ehrlichia risticii*, the cause of Potomac horse fever. J. Med. Entomol. *37*: 619–625.

Chaichanasiriwithaya, W., Y. Rikihisa, S. Yamamoto, S. Reed, T.B. Crawford, L.E. Perryman and G.H. Palmer. 1994. Antigenic, morphologic, and molecular characterization of new *Ehrlichia risticii* isolates. J. Clin. Microbiol. *32*: 3026–3033.

Chakrabarty, A.M. 1976. Plasmids in *Pseudomonas*. Annu. Rev. Genet. *10*: 7–30.

Chakraborty, B. and K.R. Samaddar. 1995. Evidence for the occurrence of an alternative nitrogenase system in *Azospirillum brasilense*. FEMS Microbiol. Lett. *127*: 127–131.

Chalcroft, J.P., H. Engelhardt and W. Baumeister. 1986. 3-Dimensional structure of a regular surface-layer from *Pseudomonas acidovorans*. Arch. Microbiol. *144*: 196–200.

Chalk, P.A., A.D. Roberts and W.M. Blows. 1994. Metabolism of pyruvate and glucose by intact cells of *Helicobacter pylori* studied by ^{13}C NMR spectroscopy. Microbiology (Reading) *140*: 2085–2092.

Chan, C.L., T.A. Lumpkin and C.S. Root. 1988. Characterization of *Bradyrhizobium* sp. (*Astragalus sinicus* L.) using serological agglutination, intrinsic antibiotic resistance, plasmid visualization, and field performance. Plant Soil *109*: 85–92.

Chan, H.T.C. and C. Anthony. 1991. The o-type oxidase of the acidophilic methylotroph *Acetobacter methanolicus*. J. Gen. Microbiol. *137*: 693–704.

Chan, K.K., R. Bakhtiar and C. Jiang. 1997. Depsipeptide (FR901228, NSC-630176) pharmacokinetics in the rat by LC/MS/MS. Investig. New Drugs. *15*: 195–206.

Chan, Y.K., L.M. Nelson and R. Knowles. 1980. Hydrogen metabolism of *Azospirillum brasilense* in nitrogen-free medium. Can. J. Microbiol. *26*: 1126–1131.

Chandrasekar, P.H., E. Arathoon and D.P. Levine. 1986. Infections due to *Achromobacter xylosoxidans*. Case report and review of the literature. Infection *14*: 279–282.

Chang, C.C., R.K. Jayaswal, C.M. Chen and S.B. Gelvin. 1989. Altered imino diacid synthesis and transcription in crown gall tumors with transposon Tn5 insertions in the 3' end of the octopine synthase gene. J. Bacteriol. *171*: 5922–5927.

Chang, J.P. and J.G. Morris. 1962. Studies of the utilization of nitrate by *Micrococcus denitrificans*. J. Gen. Microbiol. *29*: 301–310.

Chang, W.L. and M.J. Pan. 1996. Specific amplification of *Ehrlichia platys* DNA from blood specimens by two-step PCR. J. Clin. Microbiol. *34*: 3142–3146.

Chang, Y.F. and D.S. Feingold. 1970. D-Glucaric acid and galactaric acid catabolism by *Agrobacterium tumefaciens*. J. Bacteriol. *102*: 85–96.

Chang, Y.-H., J. Han, J. Chun, K.C. Lee, M.S. Rhee, Y.B. Kim and K.S.

Bae. 2002. *Comamonas koreensis* sp. nov., a non-motile species from wetland in Woopo, Korea. Int. J. Syst. Evol. Microbiol. *52*: 377–381.

Chanway, C.P. and F.B. Holl. 1993. First year field performance of spruce seedlings inoculated with plant growth promoting rhizobacteria. Can. J. Microbiol. *39*: 1084–1088.

Chapman, J.A., R.G.E. Murray and M.R.J. Salton. 1963. The surface anatomy of *Lampropedia hyalina*. Proc. Roy. Soc. B. *158*: 498–513.

Chapman, S.J. and H.R. Perkins. 1983. Peptidoglycan-degrading enzymes in ether-treated cells of *Neisseria gonorrhoeae*. J. Gen. Microbiol. *129*: 877–883.

Charest, P.J. and P. Dion. 1985. The influence of temperature on tumorigenesis induced by various strains of *Agrobacterium tumefaciens*. Can. J. Bot.-Rev. Can. Bot. *63*: 1160–1167.

Charles, I.G., G. Dougan, D. Pickard, S. Chatfield, M. Smith, P. Novotny, P. Morrissey and N.F. Fairweather. 1989. Molecular cloning and characterization of protective outer membrane protein P.69 from *Bordetella pertussis*. Proc. Natl. Acad. Sci. U.S.A. *86*: 3554–3558.

Charles, I., N. Fairweather, D. Pickard, J. Beesley, R. Anderson, G. Dougan and M. Roberts. 1994. Expression of the *Bordetella pertussis* P.69 pertactin adhesin in *Escherichia coli*: fate of the carboxy-terminal domain. Microbiology *140*: 3301–3308.

Charlet, E. and W. Schwartz. 1954. Beiträge zur Biologie der Eisenmikroben. I. Untersuchungen über die Lebensweise von *Leptothrix ochracea* und einigen begleitenden Eisenmikroben. Schwiez. Z. Hydrol. *16*: 318–341.

Chart, H., J.A. Frost, A. Oza, R. Thwaites, S. Gillanders and B. Rowe. 1996. Heat-stable serotyping antigens expressed by strains of *Campylobacter jejuni* are probably capsular and not long-chain lipopolysaccharide. J. Appl. Bacteriol. *81*: 635–640.

Chase, A.R., J.W. Miller and J.B. Jones. 1984. Leaf spot and blight of *Asplenium nidus* caused by *Pseudomonas gladioli*. Plant Dis. *68*: 344–347.

Cheah, E., K. Macpherson, D. Quiggin, P. Keese and D.L. Ollis. 1998. Crystallization and preliminary X-ray analysis of IND, an enzyme with indole oxygenase activity from *Chromobacterium violaceum*. Acta Crystallographica Section D Biological Crystallography. *54*:

Chee-Sanford, J.C., J.W. Frost, M.R. Fries, J. Zhou and J.M. Tiedje. 1996. Evidence for acetyl coenzyme A and cinnamoyl coenzyme A in the anaerobic toluene mineralization pathway in *Azoarcus tolulyticus* Tol-4. Appl. Environ. Microbiol. *62*: 964–973.

Chekanova, Y.A. and G.A. Dubinina. 1990. Cytochemical localization of hydrogen peroxide and superoxide radicals in the cells of the colorless sulfur bacterium *Macromonas bipunctata*. Mikrobiologiya *59*: 856–862.

Chemeris, N.A., A.V. Afinogenova and T.S. Tsarikaeva. 1984. Role of carbohydrate-protein recognition in the process of *Bdellovibrio* attachment to host bacterial cells. Mikrobiologiya *53*: 556–558.

Chen, C. 1996. Distribution of a newly described species, *Kingella oralis*, in the human oral cavity. Oral Microbiol. Immunol. *11*: 425–427.

Chen, C., E.M. Bauske, G. Musson, R. Rodriguez-Kabana and J.W. Kloepper. 1995a. Biological control of *Fusarium* wilt on cotton by use of endophytic bacteria. Biol. Control *5*: 83–91.

Chen, C.-K.C., T.V. Potts and M.E. Wilson. 1990a. DNA homologies shared among *E. corrodens* isolates and other corroding bacilli from the oral cavity. J. Peridont. *25*: 106–112.

Chen, D.-Q., B.C. Campbell and A.H. Purcell. 1996. A new rickettsia from a herbivorous insect, the pea aphid *Acyrthosiphon pisum* (Harris). Curr. Microbiol. *33*: 123–128.

Chen, H.W., I.M. Keseler and L.J. Shimkets. 1990b. Genome size of *Myxococcus xanthus* determined by pulsed-field gel electrophoresis. J. Bacteriol. *172*: 4206–4213.

Chen, H.W., A. Kuspa, I.M. Keseler and L.J. Shimkets. 1991a. Physical map of the *Myxococcus xanthus* chromosome. J. Bacteriol. *173*: 2109–2115.

Chen, H.K., F.D. Li and Y.Z. Cao. 1992. Characteristics, distribution, ecology, and utilization of *Astrogalus sinicus*-rhizobia symbiosis. *In* Hong (Editor), Nitrogen Fixation in China, Shanghai Scientific and Technical Publishers, Shanghai. pp. 439–445.

Chen, H.K. and M.K. Shu. 1944. Notes on the root nodule bacteria of *Astrogalus sinicus* L. Soil Sci. *58*: 291–293.

Chen, S.M., J.S. Dumler, J.S. Bakken and D.H. Walker. 1994. Identification of a granulocytotropic *Ehrlichia* species as the etiologic agent of human disease. J. Clin. Microbiol. *32*: 589–595.

Chen, S.M., V.L. Popov, H.M. Feng, J. Wen and D.H. Walker. 1995b. Cultivation of *Ehrlichia chaffeensis* in mouse embryo, Vero, BGM, and L929 cells and study of *Ehrlichia*-induced cytopathic effect and plaque formation. Infect. Immun. *63*: 647–655.

Chen, T., F. Grunert, A. MedinaMarino and E.C. Gotschlich. 1997a. Several carcinoembryonic antigens (CD66) serve as receptors for gonococcal opacity proteins. J. Exp. Med. *185*: 1557–1564.

Chen, W., M.R. Alley, B.W. Manktelow, D. Hopcroft and R. Bennett. 1988a. Pneumonia in lambs inoculated with *Bordetella parapertussis* - bronchoalveolar lavage and ultrastructural studies. Vet. Pathol. *25*: 297–303.

Chen, W.M., S. Laevens, T.M. Lee, T. Coenye, M. De Vos, M. Mergeay and P. Vandamme. 2001. *Ralstonia taiwanensis* sp. nov., isolated from root nodules of *Mimosa* species and sputum of a cystic fibrosis patient. Int. J. of Syst. Evol. Microbiol. *51*: 1729–1735.

Chen, W.X., G.S. Li, Y.L. Qi, E.T. Wang, H.L. Yuan and J.L. Li. 1991b. *Rhizobium huakuii*, sp. nov. isolated from the root nodules of *Astragalus sinicus*. Int. J. Syst. Bacteriol. *41*: 275–280.

Chen, W., E. Wang, S. Wang, Y. Li, X. Chen and Y. Li. 1995c. Characteristics of *Rhizobium tianshanense* sp. nov., a moderately and slowly growing root nodule bacterium isolated from an arid saline environment in Xinjiang, People's Republic of China. Int. J. Syst. Bacteriol. *45*: 153–159.

Chen, W.X., G.H. Yan and J.L. Li. 1988b. Numerical taxonomic study of fast-growing soybean rhizobia and a proposal that *Rhizobium fredii* be assigned to *Sinorhizbium* gen. nov. Int. J. Syst. Bacteriol. *38*: 392–397.

Chen, W.X., Z.-Y. Tan, J.-L. Gao, Y. Li and E.-T. Wang. 1997b. *Rhizobium hainanense* sp. nov., isolated from tropical legumes. Int. J. Syst. Bacteriol. *47*: 870–873.

Chen, X.-Y. and W.-N. Xiang. 1986. A strain of *Agrobacterium radiobacter* inhibits growth and gall formation by biotype III strains of *A. tumefaciens* from grapvine (English translation). Acta Microbiol. Sin. *26*: 193–199.

Chen, Y.P., G. Lopezdevictoria and C.R. Lovell. 1993. Utilization of aromatic compounds as carbon and energy sources during growth and N_2 fixation by free living nitrogen fixing bacteria. Arch. Microbiol. *159*: 207–212.

Cheng, H.P. and T.G. Lessie. 1994. Multiple replicons constituting the genome of *Pseudomonas cepacia* 17616. J. Bacteriol. *176*: 4034–4042.

Chern, C.-K., A. Ando, I. Kusaka and S. Fukni. 1976a. A succinate dehydrogenase-deficient mutant of *Agrobacterium tumefaciens*. Agr. Biol. Chem. *40*: 144–149.

Chern, C.-K., I. Kusaka and S. Fukni. 1976b. Significance of pyruvate carboxylase in sugar metabolism of *Agrobacterium tumefaciens*. Agr. Biol. Chem. *40*: 136–143.

Chernin, L.S., M.K. Winson, J.M. Thompson, S. Haran, B.W. Bycroft, I. Chet, P. Williams and G.S.A.B. Stewart. 1998. Chitinolytic activity in *Chromobacterium violaceum*: substrate analysis and regulation by quorum sensing. J. Bacteriol. *180*: 4435–4441.

Cheron, M., E. Abachin, E. Guerot, M. el Bez and M. Simonet. 1994. Investigation of hospital-acquired infections due to *Alcaligenes denitrificans* subsp. *xylosoxydans* by DNA restriction fragment length polymorphism. J. Clin. Microbiol. *32*: 1023–1026.

Cherry, J.D. 1996. Historical review of pertussis and the classical vaccine. J. Infect. Dis. *174 (Suppl 3)*: S259–263.

Cherwonogrodzky, J., G. Dubray, E. Moreno and H. Mayer. 1990. Antigens of *Brucella*. *In* Nielsen and Duncan (Editors), Animal Brucellosis, CRC Press, Boca Raton. pp. 19–64.

Chester, I.R. and R.G.E. Murray. 1975. Analysis of the cell wall and lipopolysaccharide of *Spirillum serpens*. J. Bacteriol. *124*: 1168–1176.

Chester, I.R. and R.G.E. Murray. 1978. Protein-lipid-lipopolysaccharide association in the superficial layer of *Spirillum serpens* cell walls. J. Bacteriol. *133*: 932–941.

Chevrier, D., D. Larzul, F. Mégraud and J.L. Guesdon. 1989. Identification and classification of *Campylobacter* strains by using nonradioactive DNA probes. J. Clin. Microbiol. *27*: 321–326.

Chilton, M.-D., R.K. Saiki, N. Yadav, M.P. Gordon and Q. F.. 1980. T-DNA from *Agrobacterium* Ti plasmid is in the nuclear DNA fraction of crown gall tumor cells. Proc. Natl. Acad. Sci. U. S. A. *77*: 4060–4064.

Chilton, W.S., A.M. Stomp, V. Beringue, H. Bouzar, V. Vaudequindransart, A. Petit and Y. Dessaux. 1995. The chrysopine family of amadori-type crown gall opines. Phytochemistry (Oxf) *40*: 619–628.

Ching, W.M., G.A. Dasch, M. Carl and M.E. Dobson. 1990. Structural analyses of the 120-kDa serotype protein antigens of typhus group rickettsiae. Comparison with other S-layer proteins. Ann. N. Y. Acad. Sci. *590*: 334–351.

Ching, W.M., H. Wang, J. Davis and G.A. Dasch. 1993. Amino acid analysis and multiple methylation of lysine residues in the surface protein antigen of *Rickettsia prowazeki*. *In* Angeletti (Editor), Techniques in Protein Chemistry IV, Academic Press Inc., San Diego. pp. 307–314.

Cho, K.S., M. Hirai and M. Shoda. 1992. Enhanced removability of odorous sulfur-containing gases by mixed cultures of purified bacterial from peat biofilters. J. Ferment. Bioeng. *73*: 219–224.

Choi, E.-S., E.-H. Lee and S.-K. Rhee. 1995a. Purification of a membrane-bound sorbitol dehydrogenase from *Gluconobacter suboxydans*. FEMS Microbiol. Lett. *125*: 45–49.

Choi, M.H., J.J. Song and S.C. Yoon. 1995b. Biosynthesis of copolyesters by *Hydrogenophaga pseudoflava* from various lactones. Can. J. Microbiol. *41 (Suppl. 1)*: 60–67.

Cholodny, N. 1924. Zur Morphologie der Eisenbakterien *Gallionella* und *Spirophyllum*. Ber. Dtsch. Bot. Ges. *42*: 35–44.

Cholodny, N. 1926. *In* Kolkwitz (Editor), Die Eisenbakterien. Beiträge zur einer Monographie. Pflanzenforsch. Heft 4, G. Fischer, Jena. 1–162.

Cholodny, N. 1929. Zur kenntnis der Eisenbakterien aus der Gattung *Gallionella*. Planta. *8*: 252–268.

Chomel, B.B., R.W. Kasten, K. Floyd Hawkins, B. Chi, K. Yamamoto, J. Roberts Wilson, A.N. Gurfield, R.C. Abbott, N.C. Pedersen and J.E. Koehler. 1996. Experimental transmission of *Bartonella henselae* by the cat flea. J. Clin. Microbiol. *34*: 1952–1956.

Chong, C.Y. and M.S. Lam. 1997. Case report and review of chromobacterium sepsis: a gram-negative sepsis mimicking melioidosis. SMJ. *38*: 263–265.

Chorpenning, F.W., D.H. Schmidt, H.B. Stamper and P.R. Dugan. 1978. Antigenic relationships among floc-forming *Pseudomonadaceae*. Ohio J. Sci. *78*: 29–33.

Chow, T.C. and J.M. Schmidt. 1974. Fatty acid composition of *Caulobacter crescentus*. J. Gen. Microbiol. *83*: 369–373.

Christensen, B., T. Torsvik and T. Lien. 1992. Immunomagnetically captured thermophilic sulfate-reducing bacteria from North Sea oil field waters. Appl. Environ. Microbiol. *58*: 1244–1248.

Christenson, J.C., D.F. Welch, G. Mukwaya, M.J. Muszynski, R.E. Weaver and D.J. Brenner. 1989. Recovery of *Pseudomonas gladioli* from respiratory tract specimens of patients with cystic fibrosis. J. Clin. Microbiol. *27*: 270–273.

Christiansen-Weniger, C. and J.A. Van Veen. 1991. NH_4^+-excreting *Azospirillum brasilense* mutants enhance the nitrogen supply of a wheat host. Appl. Environ. Microbiol. *56*: 3006–3012.

Christofferson, F.A. and H.E. Ottosen. 1941. Recent staining methods. Skand. Vettidskr. *31*: 599–607.

Christopher, W.N. and C.W. Edgerton. 1930. Bacterial stripe diseases of sugarcane in Louisiana. J. Agr. Res. *41*: 259–267.

Chyi, Y.S., R.A. Jorgensen, D. Goldstein, S.D. Tanksley and F. Loaizafigueroa. 1986. Locations and stability of *Agrobacterium*-mediated transfer DNA insertions in the *Lycopersicon* genome. Mol. Gen. Genet. *204*: 64–69.

Cihlar, R.L., T.G. Lessie and S.C. Holt. 1978. Characterization of bacteriophage Cp1, an organic solvent sensitive phage associated with *Pseudomonas cepacia*. Can. J. Microbiol. *24*: 1404–1412.

Claesson, B., E. Falsen and B. Kjellman. 1985. *Kingella kingae* infections:

a review and a presentation of data from 10 Swedish cases. Scand. J. Infect. Dis. *17*: 233–243.

Claflin, L.E., B.A. Ramundo, J.E. Leach and I.D. Erinle. 1989. *Pseudomonas avenae*, causal agent of bacterial leaf stripe on pearl millet. Plant Dis. *73*: 1010–1014.

Clark, A.G. 1969. A selective medium for the isolation of *Agrobacterium* species. J. Appl. Bacteriol. *32*: 348–351.

Clark, V.L., L.A. Campbell, D.A. Palermo, T.M. Evans and K.W. Klimpel. 1987. Induction and repression of outer-membrane proteins by anaerobic growth of *Neisseria gonorrhoeae*. Infect. Immun. *55*: 1359–1364.

Clark, V.L., J.S. Knapp, S. Thompson and K.W. Klimpel. 1988. Presence of antibodies to the major anaerobically induced gonococcal outer-membrane protein in sera from patients with gonococcal infections. Microb. Pathog. *5*: 381–390.

Clark, W.A., D.G. Hollis, R.E. Weaver and P. Riley. 1984. Identification of unusual pathogenic gram-negative aerobic and facultatively anaerobic bacteria, U.S. Dept. of Health and Human Services, Public Health Service, Centers for Disease Control, Atlanta.

Clark-Walker, G.D. 1969. Association of microcyst formation in *Spirillum itersonii* with the spontaneous induction of a defective bacteriophage. J. Bacteriol. *97*: 885–892.

Clark-Walker, G.D. and J. Lascelles. 1970. Cytochrome c_{550} from *Spirillum itersonii*: purification and some properties. Arch. Biochem. Biophys. *136*: 153–159.

Clark-Walker, G.D. and S.B. Primrose. 1971. Isolation and characterization of a bacteriophage Si 1 for *Spirillum itersonii*. J. Gen. Virol. *11*: 139–145.

Clark-Walker, G.D., B. Rittenberg and J. Lascelles. 1967. Cytochrome synthesis and its regulation in *Spirillum itersonii*. J. Bacteriol. *94*: 1648–1655.

Clarke, R.T.J. 1979. Niche in pasture-fed ruminants for the large rumen bacteria *Oscillospira*, *Lampropedia*, and Quinn's and Eadie's ovals. Appl. Environ. Microbiol. *37*: 654–657.

Claros, M.C., U. Schumacher, M. Jacob, S. Hunt Gerardo, N. Kleinkauf, E.J.C. Goldstein, S.M. Finegold and A.C. Rodloff. 1999. Characterization of *Bilophila wadsworthia* isolates using PCR fingerprinting. Anaerobe *5*: 589–593.

Claus, G. and H.J. Kutzner. 1985. Denitrification of nitrate and nitric acid with methanol as carbon source. Appl. Microbiol. Biotechnol. *22*: 378–381.

Clausen, V., J.G. Jones and E. Stackebrandt. 1985. 16S ribosomal RNA analysis of *Filibacter limicola* indicates a close relationship to the genus *Bacillus*. J. Gen. Microbiol. *131*: 2659–2663.

Claussen, P. 1942. Eduard Jahn. Ber. Dtsch. Bot. Gesellsch. *60*: (164)–(176).

Clavareau, C., V. Wellemans, K. Walravens, M. Tryland, J.M. Verger, M. Grayon, A. Cloeckaert, J.J. Letesson and J. Godfroid. 1998. Phenotypic and molecular characterization of a *Brucella* strain isolated from a minke whale (*Balaenoptera acutorostrata*). Microbiology *144*: 3267–3273.

Clayton, R.A., G. Sutton, P.S. Hinkle, Jr., C.J. Bult and C. Field. 1995. Intraspecific variation in small-subunit rRNA sequences in GenBank: why single sequences may not adequately represent prokaryotic taxa. Int. J. Syst. Bacteriol. *45*: 595–599.

Cleasby, A., E. Garman, M.R. Egmond and M. Batenburg. 1992. Crystallization and preliminary X-ray study of a lipase from *Pseudomonas glumae*. J. Mol. Biol. *224*: 281–282.

Cleenwerck, I., M. DeWachter, B. Hoste, D. Janssens and J. Swings. 2003. *Aquaspirillum dispar* Hylemon et al. 1973 and *Microvirgula aerodenitrificans* Patureau et al. 1998 are subjective synonyms; the name *Microvirgula dispar* comb. nov. is proposed for this taxon. Int. J. Syst. Evol. Microbiol. *53*: 1457–1459.

Cleton-Jansen, A.M., S. Dekker, P. van de Putte and N. Goosen. 1991. A single amino acid substitution changes the substrate specificity of quinoprotein glucose dehydrogenase in *Gluconobacter oxydans*. Mol. Gen. Genet. *229*: 206–212.

Cloeckaert, A., A. Tibor and M.S. Zygmunt. 1999. *Brucella* outer mem-

brane lipoproteins share antigenic determinants with bacteria of the family *Rhizobiaceae*. Clin. Diagn. Lab. Immunol. *6*: 627–629.

Cloeckaert, A., J.M. Verger, M. Grayon and O. Grepinet. 1995. Restriction site polymorphism of the genes encoding the major 25 kDa and 36 kDa outer-membrane proteins of *Brucella*. Microbiology *141*: 2111–2121.

Cloeckaert, A., V. Weynants, J. Godfroid, J.M. Verger, M. Grayon and M.S. Zygmunt. 1998. O-Polysaccharide epitopic heterogeneity at the surface of *Brucella* spp. studied by enzyme linked immunosorbent assay and flow cytometry. Clin. Diagn. Lab. Immunol. *5*: 862–870.

Cloeckaert, A., M.S. Zygmunt, J.C. Nicolle, G. Dubray and J.N. Limet. 1992. O-chain expression in the rough *Brucella melitensis* strain B115: induction of O-polysaccharide specific monoclonal antibodies and intracellular localization demonstrated by immunoelectron microscopy. J. Gen. Microbiol. *138*: 1211–1219.

Close, T.J., R.C. Tait and C.I. Kado. 1985. Regulation of Ti plasmid virulence genes by a chromosomal locus of *Agrobacterium tumefaciens*. J. Bacteriol. *164*: 774–781.

Clough, S.J., K.E. Lee, M.A. Schell and T.P. Denny. 1997. A two-component system in *Ralstonia (Pseudomonas) solanacearum* modulates production of PhcA-regulated virulence factors in response to 3-hydroxypalmitic acid methyl ester. J. Bacteriol. *179*: 3639–3648.

Clough, S.J., M.A. Schell and T.P. Denny. 1994. Evidence for involvement of a volatile extracellular factor in *Pseudomonas solanacearum* virulence gene expression. Mol. Plant-Microbe Interact. *7*: 621–630.

Coates, J.D., V.K. Bhupathiraju, L.A. Achenbach, M.J. McInerney and D.R. Lovley. 2001. *Geobacter hydrogenophilus*, *Geobacter chapellei* and *Geobacter grbiciae*—three new, strictly anaerobic, dissimilatory Fe(III)-reducers. Int. J. Syst. Evol. Microbiol. *51*: 581–588.

Coates, J.D., D.J. Ellis, E.L. Blunt-Harris, C.V. Gaw, E.E. Roden and D.R. Lovley. 1998. Recovery of humic-reducing bacteria from a diversity of environments. Appl. Environ. Microbiol. *64*: 1504–1509.

Coates, J.D., D.J. Lonergan, E.J. Philips, H. Jenter and D.R. Lovley. 1995. *Desulfuromonas palmitatis* sp. nov., a marine dissimilatory Fe(III) reducer that can oxidize long-chain fatty acids. Arch. Microbiol. *164*: 406–413.

Coates, J.D., D.J. Lonergan, E.J.P. Philips, H. Jenter and D.R. Lovley. 2000. *In* Validation of the publication of new names and new combinations previously effectively published outside the IJSB. List No. 76. Int. J. Syst. Evol. Microbiol. *50*: 1699–1700.

Coates, J.D., U. Michaelidou, R.A. Bruce, S.M. O'Connor, J.N. Crespi and L.A. Achenbach. 1999. Ubiquity and diversity of dissimilatory (per)chlorate-reducing bacteria. Appl. Environ. Microbiol. *65*: 5234–5241.

Coates, J.D., E.J. Phillips, D.J. Lonergan, H. Jenter and D.R. Lovley. 1996. Isolation of *Geobacter* species from diverse sedimentary environments. Appl. Environ. Microbiol. *62*: 1531–1536.

Coder, D.M. and L.J. Goff. 1986. The host range of the chlorellavorous bacterium ("*Vampirovibrio chlorellavorus*"). J. Phycol. *22*: 543–546.

Coder, D.M. and M.P. Starr. 1978. Antagonistic association of the chlorellavorus bacterium ("*Bdellovibrio*" *chlorellavorus*) with *Chlorella vulgaris*. Curr. Microbiol. *1*: 59–64.

Coenye, T., E. Falsen, B. Hoste, M. Ohlen, J. Goris, J.R.W. Govan, M. Gillis and P. Vandamme. 2000. Description of *Pandoraea* gen. nov. with *Pandoraea apista* sp. nov., *Pandoraea pulmonicola* sp. nov., *Pandoraea pnomenusa* sp. nov., *Pandoraea sputorum* sp. nov. and *Pandoraea norimbergensis* comb. nov. Int. J. Syst. Evol. Microbiol. *50*: 887–899.

Coenye, T., E. Falsen, M. Vancanneyt, B. Hoste, J.R.W. Govan, K. Kersters and P. Vandamme. 1999a. Classification of *Alcaligenes faecalis*-like isolates from the environment and human clinical samples as *Ralstonia gilardii* sp. nov. Int. J. Syst. Bacteriol. *49*: 405–413.

Coenye, T., J. Goris, T. Spilker, P. Vandamme and J.J. LiPuma. 2002. Characterization of unusual bacteria isolated from respiratory secretions of cystic fibrosis patients and description of *Inquilinus limosus* gen. nov., sp nov. J. Clin. Microbiol. *40*: 2062–2069.

Coenye, T., B. Holmes, K. Kersters, J.R. Govan and P. Vandamme. 1999b. *Burkholderia cocovenenans* (van Damme et al. 1960) Gillis et al. 1995 and *Burkholderia vandii* Urakami et al. 1994 are junior synonyms of

Burkholderia gladioli (Severini 1913) Yabuuchi et al. 1993 and *Burkholderia plantarii* (Azegami et al. 1987) Urakami et al. 1994, respectively. Int J Syst Bacteriol. *49*: 37–42.

Coenye, T., S. Laevens, A. Willems, M. Ohlen, W. Hannant, J.R.W. Govan, M. Gillis, E. Falsen and P. Vandamme. 2001a. *Burkholderia fungorum* sp. nov., and *Burkholderia caledonica* sp. nov., two new species isolated from the environment, animals and human clinical samples. Int. J. Syst. Evol. Microbiol. *51*: 1099–1107.

Coenye, T. and J.J. LiPuma. 2002. Use of the *gyrB* gene for the identification of *Pandoraea* species. FEMS Microbiol. Lett. *208*: 15–19.

Coenye, T., L.X. Liu, P. Vandamme and J.J. LiPuma. 2001b. Identification of *Pandoraea* species by 16S ribosomal DNA-based PCR assays. J. Clin. Microbiol. *39*: 4452–4455.

Coenye, T., E. Mahenthiralingam, D. Henry, J.J. LiPuma, S. Laevens, M. Gillis, D.P. Speert and P. Vandamme. 2001c. *Burkholderia ambifaria* sp. nov., a novel member of the *Burkholderia cepacia* complex including biocontrol and cystic fibrosis-related isolates. Int. J. of Syst. Evol. Microbiol. *51*: 1481–1490.

Cohen-Bazire, G., W.R. Sistrom and R.Y. Stanier. 1957. Kinetic studies of pigment synthesis by non-sulfur purple bacteria. J. Cell. Comp. Physiol. *49*: 25–68.

Cohn, F. 1875. Untersuchungen uber Bakterien II. Beitr. Biol. Pflanz. *1875 1 (Heft 3)*: 141–207.

Cojho, E.H., V.M. Reis, A.C.G. Schenberg and J. Döbereiner. 1993. Interactions of *Acetobacter diazotrophicus* with an amylolytic yeast in nitrogen-free batch culture. FEMS Microbiol. Lett. *106*: 341–346.

Colby, J., H. Dalton and R. Whittenbury. 1979. Biological and biochemical aspects of microbial growth on C1 compounds. Annu. Rev. Microbiol. *33*: 481–517.

Colby, J. and L.J. Zatman. 1973. Trimethylamine metabolism in obligate and facultative methylotrophs. Biochem. J. *132*: 101–112.

Cole, J.A. and S.C. Rittenberg. 1971. A comparison of respiratory processes in *Spirillum volutans*, *Spirillum itersonii*, and *Spirillum serpens*. J. Gen. Microbiol. *69*: 375–383.

Cole, M.A. and G.H. Elkan. 1973. Transmissible resistance to penicillin G, neomycin, and chloramphenicol in *Rhizobium japonicum*. Antimicrob. Agents Chemother. *4*: 248–253.

Coleman, M.L., D.B. Hedrick, D.R. Lovley, D.C. White and K. Pye. 1993. Reduction of Fe(III) in sediments by sulfate-reducing bacteria. Nature (Lond.) *361*: 436–438.

Coletta, P.L. and P.G.G. Miller. 1986. The extracellular proteases of *Myxococcus xanthus*. FEMS Microbiol. Lett. *37*: 203–207.

Collin, B. 1913. Sur en ensemble de protistes parasites des bactraciens (note préliminaire). Arch. Zool. Exp. Gen. Notes Rev. *51*: 59–76.

Collins, M.D. and F. Fernandez. 1984. Menaquinone-6 and thermoplasmaquinone-6 in *Wolinella succinogenes*. FEMS Microbiol. Lett. *22*: 273–276.

Collins, M.D. and P.N. Green. 1985. Isolation and characterization of a novel coenzyme Q from some methane-oxidizing bacteria. Biochem. Biophys. Res. Commun. *133*: 1125–1131.

Collins, M.D. and D. Jones. 1981. Distribution of isoprenoid quinone structural types in bacteria and their taxonomic implication. Microbiol. Rev. *45*: 316–354.

Collins, M.D., D. Jones, R.M. Keddie and P.H.A. Sneath. 1980. Reclassification of *Chromobacterium iodinum* in a redefined genus *Brevibacterium* as *Brevibacterium iodinum*, nom. rev., comb. nov. J. Gen. Microbiol. *120*: 1–10.

Collins, M.D. and F. Widdel. 1986. Respiratory quinones of sulphate-reducing and sulphur-reducing bacteria: a systematic investigation. Syst. Appl. Microbiol. *8*: 8–18.

Collins, R.F., L. Davidsen, J.P. Derrick, R.C. Ford and T. Tonjum. 2001. Analysis of the PilQ secretin from *Neisseria meningitidis* by transmission electron microscopy reveals a dodecameric quaternary structure. J. Bacteriol. *183*: 3825–3832.

Collins, R.F., R.C. Ford, A. Kitmitto, R.O. Olsen, T. Tønjum and J.P. Derrick. 2003. Three-dimensional structure of the *Neisseria meningitidis* secretin PilQ determined from negative-stain transmission electron microscopy. J. Bacteriol. *185*: 2611–2617.

Colloc, M.L., O. Masure, Y. Perramant, B. Lejeune and C. Chastel. 1980. Actualités des infections à *Eikenella corrodens*. Medicine et Maladies Infectieuses. *10*: 387–390.

Colores, G.M., P.M. Radehaus and S.K. Schmidt. 1995. Use of a pentachlorophenol degrading bacterium to bioremediate highly contaminated soil. Appl. Biochem. Biotechnol. *54*: 271–275.

Condon, C.R., R.J. Fitzgerald and F. O'Gara. 1991. Conjugation and heterologous gene expression in *Gluconobacter oxydans* subsp. *suboxydans*. FEMS Microbiol. Lett. *80*: 173–178.

Condon, C., C. Squires and C.L. Squires. 1995. Control of rRNA transcription in *Escherichia coli*. Microbiol. Rev. *59*: 623–645.

Conn, H.J. 1938. Taxonomic relationships of certain non-sporeforming rods in soil. J. Bacteriol. *36*: 320-321.

Conn, H.J. 1942. Validity of the genus *Alcaligenes*. J. Bacteriol. *44*: 353–360.

Connell, T.D., W.J. Black, T.H. Kawula, D.S. Barritt, J.A. Dempsey, K. Kverneland, A. Stephenson, B.S. Schepart, G.L. Murphy and J.G. Cannon. 1988. Recombination among protein-ii genes of *Neisseria gonorrhoeae* generates new coding sequences and increases structural variability in the protein-ii family. Mol. Microbiol. *2*: 227–236.

Conti, S.F. and P. Hirsch. 1965. Biology of budding bacteria: III. Fine structure of *Rhodomicrobium* and *Hyphomicrobium* spp. J. Bacteriol. *89*: 503–512.

Contzen, M., E.R.B. Moore, S. Blümel, A. Stolz and P. Kämpfer. 2000. *Hydrogenophaga intermedia* sp. nov., a 4-aminobenzene-sulfonate degrading organism. Syst. Appl. Microbiol. *23*: 487–493.

Contzen, M., E.R.B. Moore, S. Blümel, A. Stolz and P. Kämpfer. 2001. *In* Validation of the publication of new names and new combinations previously effectively published outside the IJSEM. List No. 80. Int. J. Syst. Evol. Microbiol. *51*: 793–794.

Conway, T. 1992. The Entner-Doudoroff pathway: history, physiology and molecular biology. FEMS Microbiol. Rev. *9*: 1–27.

Conway, T., M.O.K. Byun and L.O. Ingram. 1987. Expression vector for *Zymomonas mobilis*. Appl. Environ. Microbiol. *53*: 235–241.

Cook, A.M., H. Laue and F. Junker. 1999. Review — microbial desulfonation. FEMS Microbiol. Rev. *22*: 399–419.

Cook, G.T. 1950. A plate test for nitrate reduction. J. Clin. Pathol. *3*: 359–362.

Cook, J.M. and R.D.J. Butcher. 1999. The transmission and effects of *Wolbachia* bacteria in parasitoids. Res. Popul. Ecol. *41*: 15–28.

Cooke, V.M., M.N. Hughes and R.K. Poole. 1995. Reduction of chromate by bacteria isolated from the cooling water of an electricity generating station. J. Ind. Microbiol. Biotechnol. *14*: 323–328.

Cookson, B.T., P. Vandamme, L.C. Carlson, A.M. Larson, J.V. Sheffield, K. Kersters and D.H. Spach. 1994. Bacteremia caused by a novel *Bordetella* species, *Candidatus* B. hinzii. J. Clin. Microbiol. *32*: 2569–2571.

Cooper, D.M. and C.J. Gebhart. 1998. Comparative aspects of proliferative enteritis. J. Am. Vet. Med. Assoc. *212*: 1446–1451.

Cooper, D.M., D.L. Swanson, S.M. Barns and C.J. Gebhart. 1997. Comparison of the 16S ribosomal DNA sequences from the interacellular agents of proliferative enteritis in a hamster, deer, and ostrich with the sequence of a porcine isolate of *Lawsonia intracellularis*. Int. J. Syst. Bacteriol. *47*: 635–639.

Coote, J.G. and H. Hassal. 1973. The degradation of L-histidine, imidazolyl-L-lactate and imidazolyl-propionate by *Pseudomonas testosteroni*. Biochem. J. *132*: 409–422.

Corbel, M.J. 1977a. Isolation and partial characterization of a phage receptor from *Brucella neotomae* 5K33. Ann. Sclavo. *19*: 131–142.

Corbel, M.J. 1977b. Production of a phage variant lytic for non smooth *Brucella* strains. Ann. Sclavo. *19*: 99–108.

Corbel, M.J. 1979. Isolation and properties of a phage lytic for non-smooth *Brucella* organisms. J. Biol. Stand. *7*: 349–360.

Corbel, M.J. 1982. International committee on nomenclature of bacteria subcommittee on the taxonomy of *Brucella*. Minutes of meeting, 4 and 5 September, 1978. Int. J. Syst. Bacteriol. *32*: 260–261.

Corbel, M.J. 1985. Recent advances in the study of *Brucella* antigens and their serological cross reactions. Vet. Bull. *54*: 927–942.

Corbel, M.J. 1987. *Brucella* phages: advances in the development of a reliable phage typing system for smooth and non-smooth *Brucella* isolates. Ann. Inst. Pasteur. Microbiol. *138*: 70–75.

Corbel, M.J. 1988. International committee on systematic bacteriology subcommittee on the taxonomy of *Brucella* report of the meeting, 5 September 1986, Manchester, England. Int. J. Syst. Bacteriol. *38*: 450–452.

Corbel, M.J. 1989a. Brucellosis: epidemiology and prevalence worldwide. *In* Young and Corbel (Editors), Brucellosis: Clinical and Laboratory Aspects, CRC Press, Boca Raton. pp. 25–40.

Corbel, M.J. 1989b. Microbiology of the genus *Brucella*. *In* Young and Corbel (Editors), Brucellosis: Clinical and Laboratory Aspects, CRC Press, Boca Raton. pp. 53–72.

Corbel, M.J. 1991. Identification of dye sensitive strains of *Brucella melitensis*. J. Clin. Microbiol. *29*: 1066–1068.

Corbel, M.J., C.D. Bracewell, E.L. Thomas and K.P.W. Gill. 1979. Techniques in the identification and classification of *Brucella* species. *In* Skinner and Lovelock (Editors), Identification Methods for Microbiologists, Technical Series No. 14, Academic Press, London. pp. 71–122.

Corbel, M.J. and W.J. Brinley-Morgan. 1984. Genus *Brucella* Meyer and Shaw. *In* Krieg, N.R. and Holt, J.G. (Editors), Bergey's Manual of Systematic Bacteriology, Vol. 1, The Williams & Wilkins Co., Baltimore. pp. 377–388.

Corbel, M.J., K.P.W. Gill and E.L. Thomas. 1978. Methods for the identification of *Brucella*, Ministry of Agriculture, Fisheries and Food, Pinner, Middlesex.

Corbel, M.J. and J.A. Morris. 1974. Studies on a smooth phage resistant variant of *Brucella abortus*. I. Immunological properties. Br. J. Exp. Pathol. *55*: 78–87.

Corbel, M.J. and J.A. Morris. 1975. Studies on a smooth phage resistant variant of *Brucella abortus*. 2. Mechanism of phage resistance. Br. J. Exp. Pathol. *56*: 1–7.

Corbel, M.J., A.C. Scott and H.M. Ross. 1980. Properties of a cell wall defective variant of *Brucella abortus* of bovine origin. J. Hyg. Camb. *85*: 103–113.

Corbel, M.J. and E.L. Thomas. 1976. Properties of some new *Brucella* phage isolates: evidence for lysogeny within the genus. Develop. Biol. Stand. *31*: 38–45.

Corbel, M.J. and E.L. Thomas. 1980. The *Brucella* phages: their properties characterization and applications, Ministry of Agriculture, Fisheries and Food, Pinner, Middlesex.

Corbel, M.J., F. Tolari and V.K. Yadava. 1988. Characterisation of a new phage lytic for both smooth and non-smooth *Brucella* species. Res. Vet. Sci. *44*: 45–49.

Corbett, M.J., R.J. Black and C.E.I. Wilde. 1986. Antibodies to outer membrane protein–macromolecular complex (OMP-MC) are bactericidal for serum resistant gonococci. *In* Poolman, Zanen, Meyer, Heckels, Mäkelä, Smith and Beuvery (Editors), Gonococci and Meningococci, Klüwer Academic Publishers, Dordrecht. pp. 685–691.

Cord-Ruwisch, R., B. Ollivier and J.L. Garcia. 1986. Fructose degradation by *Desulfovibrio* sp. in pure culture and in coculture with *Methanospirillum hungatei*. Curr. Microbiol. *13*: 285–289.

Cornelissen, C.N., M. Kelley, M.M. Hobbs, J.E. Anderson, J.G. Cannon, M.S. Cohen and P.F. Sparling. 1998. The transferrin receptor expressed by gonococcal strain FA1090 is required for the experimental infection of human male volunteers. Mol. Microbiol. *27*: 611–616.

Cornick, N.A. and M.J. Allison. 1996a. Anabolic incorporation of oxalate by *Oxalobacter formigenes*. Appl. Environ. Microbiol. *62*: 3011–3013.

Cornick, N.A. and M.J. Allison. 1996b. Assimilation of oxalate, acetate, and CO_2 by *Oxalobacter formigenes*. Can. J. Microbiol. *42*: 1081–1086.

Corpe, W.A. 1951. A study of the wide spread distribution of *Chromobacterium* species in soil by a simple technique. J. Bacteriol. *62*: 515–517.

Correa, P., G. Malcom, B. Schmidt, E. Fontham, B. Ruiz, J.C. Bravo, L.E. Bravo, G. Zarama and J.L. Realpe. 1998. Review article: Antioxidant micronutrients and gastric cancer. Aliment Pharmacol. Ther. *12* (Suppl. 1): 73–82.

Corry, J.E.L., D.E. Post, P. Colin and M.J. Laisney. 1995. Culture media for the isolation of campylobacters. Int. J. Food Microbiol. *26*: 43–76.

Corstjens, P.L.A.M., J.P.M. De Vrind, P. Westbroek and E.W. De Vrind-De Jong. 1992. Enzymatic iron oxidation by *Leptothrix discophora*: identification of an iron-oxidizing protein. Appl. Environ. Microbiol. *58*: 450–454.

Corstjens, P. and G. Muyzer. 1993. Phylogenetic analysis of the metal-oxidizing bacteria *Leptothrix discophora* and *Sphaerotilus natans* using 16S-rDNA sequencing data. Syst. Appl. Microbiol. *16*: 219–223.

Cortes, G., A. Mendoza and D. Munoz. 1996. Toxicity evaluation using bioassays in Rural Developing District 063 Hidalgo, Mexico. Environ. Toxicol. Water Qual. *11*: 137–143.

Costas, M., S.L.W. On, R.J. Owen, B. Lopez-Urquijo and A.J. Lastovica. 1993. Differentiation of *Helicobacter* species by numerical analysis of their one-dimensional electrophoretic protein patterns. Syst. Appl. Microbiol. *16*: 396–404.

Costas, M., R.J. Owen, J. Bickley and D.R. Morgan. 1991. Molecular techniques for studying the epidemiology of infection by *Helicobacter pylori*. Scand. J. Gastroenterol. Suppl. *26*: 20–32.

Costas, M., R.J. Owen and P.J.H. Jackman. 1987. Classification of *Campylobacter sputorum* and allied campylobacters based on numerical analysis of electrophoretic protein patterns. Syst. Appl. Microbiol. *9*: 125–131.

Costerton, J.W., R.T. Irvin and K.J. Cheng. 1981. The bacterial glycocalyx in nature and disease. Annu. Rev. Microbiol. *35*: 299–324.

Costerton, J.W.F., R.G.E. Murray and C.F. Robinow. 1961. Observations on the motility and the structure of *Vitreoscilla*. Can. J. Microbiol. *7*: 329–339.

Cotter, P.A., M.H. Yuk, S. Mattoo, B.J. Akerley, J. Boschwitz, D.A. Relman and J.F. Miller. 1998. Filamentous hemagglutinin of *Bordetella bronchiseptica* is required for efficient establishment of tracheal colonization. Infect. Immun. *66*: 5921–5929.

Cotter, T.W. and M.F. Thomashow. 1992a. A conjugation procedure for *Bdellovibrio bacteriovorus* and its use to identify DNA sequences that enhance the plaque-forming ability of a spontaneous host-independent mutant. J. Bacteriol. *174*: 6011–6017.

Cotter, T.W. and M.F. Thomashow. 1992b. Identification of a *Bdellovibrio bacteriovorus* genetic locus, hit, associated with the host-independent phenotype. J. Bacteriol. *174*: 6018–6024.

Coty, V.F. 1967. Atmospheric nitrogen fixation by hydrocarbon-oxidizing bacteria. Biotechnol. Bioengin. *9*: 25–32.

Coucheron, D.H. 1991. An *Acetobacter xylinum* insertion sequence element associated with inactivation of cellulose production. J. Bacteriol. *173*: 5723–5731.

Coucheron, D.H. 1993. A family of IS1031 elements in the genome of *Acetobacter xylinum*: nucleotide sequences and strain distribution. Mol. Microbiol. *9*: 211–218.

Coucheron, D.H. 1997. *Acetobacter* strains contain DNA modified at GAATTC and GANTC. Can. J. Microbiol. *43*: 456–460.

Coucke, P. and J.P. Voets. 1967. The mineral requirements of *Polyangium cellulosum*. Z. Allg. Mikrobiol. *7*: 175–182.

Coucke, P.L. and J.P. Voets. 1968. Etude de la cellulolyse enzymatique par *Sorangium compositum*. Ann. Inst. Pasteur *115*: 549–560.

Coulton, J.W. and R.G.E. Murray. 1978. Cell envelope associations of *Aquaspirillum serpens* flagella. J. Bacteriol. *136*: 1037–1049.

Couso, R.O., L. Ielpi and M.A. Dankert. 1987. A xantham-gum-like polysaccharide from *Acetobacter xylinum*. J. Gen. Microbiol. *133*: 2123–2135.

Covacevich, M.T. and G.N. Richards. 1978. Studies on dextranases. 7. Purification of intracellular dextranases and deuterium-glucosidases from *Pseudomonas* UQM 733. Carbohydr. Res. *64*: 169–180.

Covacevich, M.T. and G.N. Richards. 1979. Studies on dextranases. 8. Modes of action of intracellular dextranase and 3 oligoglucanases from *Pseudomonas* UQM 733. Carbohydr. Res. *70*: 283–293.

Cover, T.L. 1996. The vacuolating cytotoxin of *Helicobacter pylori*. Mol. Microbiol. *20*: 241–246.

Cover, W.H. 1978. Studies of the microaerophilic nature of *Spirillum volutans*, Thesis, Virginia Polytechnic Institute and State University, Blacksburg, VA.

Cowan, S.T. 1974. Cowan and Steel's manual for the identification of medical bacteria, 2nd Ed., Cambridge University Press, London.

Cowdry, E.V. 1925. Studies on the etiology of heartwater. I. Observation of a rickettsia, *Rickittsia ruminantium* (n. sp.) in the tissues of infected animals. J. Exp. Med. *42*: 231–252.

Cox, A.D. and S.G. Wilkinson. 1989a. Polar lipids and fatty acids of *Pseudomonas cepacia*. Biochim. Biophys. Acta *1001*: 60–67.

Cox, A.D. and S.G. Wilkinson. 1989b. Structures of the O-Specific polymers from the lipopolysaccharides of the reference strains for *Pseudomonas cepacia* serogroup-O3 and serogroup-O5. Carbohydr. Res. *195*: 123–129.

Cox, A.D. and S.G. Wilkinson. 1990a. Structure of the O-specific polymer for *Pseudomonas cepacia* serogroup O7. Carbohydr. Res. *198*: 153–156.

Cox, A.D. and S.G. Wilkinson. 1990b. Structure of the putative O-antigen containing 2-amino-2-deoxy-L-glucose in the reference strain for *Pseudomonas cepacia* serogroup-O1. Carbohydr. Res. *195*: 295–301.

Cox, A.D. and S.G. Wilkinson. 1991. Ionizing groups in lipopolysaccharides of *Pseudomonas cepacia* in relation to antibiotic resistance. Mol. Microbiol. *5*: 641–646.

Cox, C.D., K.L. Rinehart, M.L. Moore and J.C. Cook. 1981. Pyochelin: novel ntructure of an iron chelating growth promoter for *Pseudomonas aeruginosa*. Proc. Nat. Acad. Sci. U.S.A. *78*: 4256–4260.

Cox, H.R. 1941. Cultivation of rickettsia of Rocky Mountain spotted fever, typhus and Q-fever groups in the embryonic tissues of developing chicks. Science *94*: 399–403.

Cox, T.L. and L.I. Sly. 1997. Phylogenetic relationships and uncertain taxonomy of *Pedomicrobium* species. Int. J. Syst. Bacteriol. *47*: 377–380.

Coykendall, A.L. and K.S. Kaczmarek. 1980. DNA homologies among *Eikenella corrodens* strains. J. Periodontal Res. *15*: 615–620.

Crabtree, J.E. 1998. Role of cytokines in pathogenesis of *Helicobacter pylori*-induced mucosal damage. Dig. Dis. Sci. *43*: 46S–55S.

Crabtree, K., W. Boyle, E. McCoy and G.A. Rohlich. 1966. A mechanism of floc formation by *Zoogloea ramigera*. J. Water Pollut. Control Fed. *38*: 1968–1980.

Crabtree, K. and E. McCoy. 1967. *Zoogloea ramigera* Itzigsohn, identification and description. Request for an opinion as to the status of the generic name *Zoogloea*. Int. J. Syst. Bacteriol. *17*: 1–10.

Craig, A.S., R.M. Greenwood and K.I. Williamson. 1973. Ultrastructural inclusions of rhizobial bacteroids of Lotus nodules and their taxonomic significance. Arch Mikrnhinl. *89*: 22.

Cramm, R., A. Pohlmann and B. Friedrich. 1999. Purification and characterization of the single-component nitric oxide reductase from *Ralstonia eutropha* H16. FEBS Lett. *460*: 6–10.

Crane, L.R., L.C. Tagle and W.A. Palutke. 1981. Outbreak of *Pseudomonas paucimobilis* in an intensive care facility. JAMA. *246*: 985–987.

Creaven, M., R.J. Fitzgerald and F. O'Gara. 1994. Transformation of *Gluconobacter oxydans* subsp. *suboxydans* by electroporation. Can. J. Microbiol. *40*: 491–494.

Cremers, H.C.J.C., C.A. Wijffelman, E. Pees, B.G. Rolfe, M.A. Djordievic and B.J.J. Lugtenberg. 1988. Host specific nodulation of plants of the pea cross-inoculation group is influenced by genes in fast growing *Rhizobium* downstream *nodC*. J. Plant Physiol. *132*: 398–404.

Cripps, R.E. and A.S. Noble. 1973. The metabolism of nitrilotriacetate by a pseudomonad. Biochem. J. *136*: 1059–1068.

Cronin, D., Y. Moenne-Loccoz, C. Dunne and F. O'Gara. 1997. Inhibition of egg hatch of the potato cyst nematode *Globodera rostochiensis* by chitinase-producing bacteria. Eur. J. Plant Pathol. *103*: 433–440.

Cross, T. and M. Goodfellow. 1973. Taxonomy and classification of the actinomycetes. *In* Sykes and Skinner (Editors), *Actinomycetales*: characteristics and practical importance, Academic Press, London, New York,. pp. 11–112.

Crossman, L.C., J.W.B. Moir, J.M. Wehrfritz, A. Keech, A. Thomson, S. Spiro and D.J. Richardson. 1995. Heterotrophic nitrification in *Paracoccus denitrificans*. *In* Scheffers and Dijken (Editors), Proceedings Beijerinck Centennial Symposium. Microbial physiology and gene regulation. Emerging principles and applications, Delft University Press, Delft, The Netherlands. pp. 44–45..

Crotchfelt, K.A., L.E. Welsh, D. DeBonville, M. Rosenstraus and T.C.

Quinn. 1997. Detection of *Neisseria gonorrhoeae* and *Chlamydia trachomatis* in genitourinary specimens from men and women by a coamplification PCR assay. J. Clin. Microbiol. *35*: 1536–1540.

Croucher, S.C. and E.M. Barnes. 1983. The occurrence and properties of *Gemmiger formicilis* and related anaerobic budding bacteria in the avium caecum. J. Appl. Bacteriol. *54*: 7–22.

Cruden, D.L. and A.J. Markovetz. 1981. Relative numbers of selected bacterial forms in different regions of the cockroach *Eublaberus posticus* hind gut. Arch. Microbiol. *129*: 129–134.

Crutzen, P.J. 1991. Methane's sinks and sources. Nature *350*: 380–381.

Cuadra, M. and J. Takano. 1969. The relationship of *Bartonella bacilliformis* to the red blood cell as revealed by electron microscopy. Blood. *33*: 708–716.

Cullinane, L.C., M.R. Alley, R.B. Marshall and B.W. Manktelow. 1987. *Bordetella parapertussis* from lambs. N. Z. Vet. J. *35*: 175–175.

Cummings, D.E., F. Caccavo, S. Spring and R.F. Rosenzweig. 1999. *Ferribacterium limneticum*, gen. nov., sp. nov., an Fe(III)- reducing microorganism isolated from mining-impacted freshwater lake sediments. Arch. Microbiol. *171*: 183–188.

Cummings, D.E., F. Caccavo, S. Spring and R.F. Rosenzweig. 2000. *In* Validation of publication of new names and new combinations previously effectively published outside the IJSEM. List No.77. Int. J. Syst. Evol. Microbiol. *50*: 1953.

Cumsky, M.G. and D.R. Zusman. 1981a. Binding properties of Myxobacterial hemagglutinin. J. Biol. Chem. *256*: 2596–2599.

Cumsky, M.G. and D.R. Zusman. 1981b. Purification and characterization of myxobacterial hemagglutinin, a development-specific lectin of *Myxococcus xanthus*. J. Biol. Chem. *256*: 2581–2588.

Curasson, G. 1938. Notes sur la piroplasmose aviaire en E.O.F. Bull. Serv. Zootech. Epiz. A.O.F. *1*: 33–35.

Curasson, G. and P. Andrjesky. 1929. Sur les "corps de Balfour" du sang de la poule. Bulletin de la Societe de Pathologie Exotique. *22*: 316–317.

Currier, T.C. and E.W. Nester. 1976. Evidence for diverse types of large plasmids in tumor-inducing strains of *Agrobacterium*. J. Bacteriol. *126*: 157–165.

Cypionka, H. and O. Meyer. 1983a. Carbon monoxide insensitive respiratory chain of *Pseudomonas carboxydovorans*. J. Bacteriol. *156*: 1178–1187.

Cypionka, H. and O. Meyer. 1983b. The cytochrome composition of carboxydotrophic bacteria. Arch. Microbiol. *135*: 293–298.

Cypionka, H., O. Meyer and H.G. Schlegel. 1980. Physiological characteristics of various species of strains of carboxydobacteria. Arch. Microbiol. *127*: 301–307.

Czurda, V. and E. Maresch. 1937. Beitrag zur Kenntnis der Atiorhodobakterien-Gesellschaften. Archiv Mikrobiol. *8*: 99–124.

da Rocha-Lima, H. 1916. Zür Aetiologie des Fieck-fiebers. Berlin Klin. Wochenschr. *53*: 567–569.

Da Silva Tatley, F.M., A.J. Lastovica and L.M. Steyn. 1992. Plasmid profiles of "*Campylobacter upsaliensis*" isolated from blood cultures and stools of paediatric patients. J. Med. Microbiol. *37*: 8–14.

Daane, L.L., J.A. Molina, E.C. Berry and M.J. Sadowsky. 1996. Influence of earthworm activity on gene transfer from *Pseudomonas fluorescens* to indigenous soil bacteria. Appl. Environ. Microbiol. *62*: 515–521.

Dadds, M.J.S. 1971. The detection of *Zymomonas anaerobia*. *In* Shapton and Board (Editors), Isolation of Anaerobes, Academic Press, Inc., New York. 219–222.

Dadds, M.J.S., P.A. Martin and J.G. Carr. 1973. The doubtful status of the species *Zymomonas anaerobia* and *Z. mobilis*. J. Appl. Bacteriol. *36*: 531–539.

Daft, M.J., J.C. Burnham and Y. Yamamoto. 1985. Lysis of *Phormidium luridum* by *Myxococcus fulvus* in continuous flow cultures. J. Appl. Bacteriol. *59*: 73–80.

Dagasan, L. and R.M. Weiner. 1986. Contribution of the electrophoretic pattern of cell envelope protein to the taxonomy of *Hyphomonas* spp. Int. J. Syst. Bacteriol. *36*: 192–196.

Dagher, F., E. Deziel, P. Lirette, G. Paquette, J.G. Bisaillon and R. Villemur. 1997. Comparative study of five polycyclic aromatic hydrocarbon degrading bacterial strains isolated from contaminated soils. Can. J. Microbiol. *43*: 368–377.

Dagley, S. and M.D. Patel. 1957. Oxidation of *p*-cresol and related compounds by a *Pseudomonas*. Biochem. J. *66*: 227–233.

Dahm, H., H. Rozycki, E. Strzelczyk and C.Y. Li. 1993. Production of B-group vitamins by *Azospirillum* spp. grown in media of different pH at different temperatures. Zentbl. Mikrobiol. *148*: 195–203.

Dailey, H.A., Jr. 1976. Membrane-bound respiratory chain of *Spirillum itersonii*. J. Bacteriol. *127*: 1286–1291.

Dailey, H.A., Jr. and J. Lascelles. 1974. Ferrochelatase activity in wild-type and mutant strains of *Spirillum itersonii*. Arch. Biochem. Biophys. *160*: 523–529.

Dainty, R.H., D.J. Etherington, B.G. Shaw, J. Barlow and G.T. Banks. 1978. Studies on the production of extracellular proteinases by a non-pigmented strain of *Chromobacterium lividum* isolated from abbatoir effluent. J. Appl. Bacteriol. *45*: 111–124.

Dale, C.J.H., E.K. Moses, C.C. Ong, C.J. Morrow, M.B. Reed, D. Hasse and R.A. Strugnell. 1998. Identification and sequencing of the groE operon and flanking genes of *Lawsonia intracellularis*: use in phylogeny. Microbiology *144*: 2073–2084.

Dalton, H. 1992. Methane oxidation by methanotrophs: physiological and mechanistic implications. *In* Murrell and Dalton (Editors), Methane and Methanol Utilizers, Plenum Press, New York. pp. 85–114.

Daly, J.S., M.G. Worthington, D.J. Brenner, C.W. Moss, D.G. Hollis, R.S. Weyant, A.G. Steigerwalt, R.E. Weaver, M.I. Daneshvar and S.P. O'Connor. 1993. *Rochalimaea elizabethae* sp. nov. isolated from a patient with endocarditis. J. Clin. Microbiol. *31*: 872–881.

D'Amato, R.F., L.A. Eriquez, K.M. Tomfahrde and E. Singerman. 1978. Rapid identification of *Neisseria gonorrhoeae* and *Neisseria meningitidis* by using enzymatic profiles. J. Clin. Microbiol. *7*: 77–81.

Daneshvar, M.I., D.G. Hollis, A.G. Steigerwalt, A.M. Whitney, L. Spangler, M.P. Douglas, J.G. Jordan, J.P. MacGregor, B.C. Hill, F.C. Tenover, D.J. Brenner and R.S. Weyant. 2001. Assignment of CDC weak oxidizer group 2 (WO-2) to the genus *Pandoraea* and characterization of three new *Pandoraea* genomospecies. J. Clin. Microbiol. *39*: 1819–1826.

Danganan, C.E., S. Shankar, R.W. Ye and A.M. Chakrabarty. 1995. Substrate diversity and expression of the 2,4,5-trichlorophenoxyacetic acid oxygenase from *Burkholderia cepacia* AC1100. Appl. Environ. Microbiol. *61*: 4500–4504.

Danganan, C.E., R.W. Ye, D.L. Daubaras, L. Xun and A.M. Chakrabarty. 1994. Nucleotide sequence and functional analysis of the genes encoding 2,4,5- trichlorophenoxyacetic acid oxygenase in *Pseudomonas cepacia* AC1100. Appl. Environ. Microbiol. *60*: 4100–4106.

Dangeard, P.A. 1926. Recherches sur les tubercles radicaux des Légumineuses. Botaniste (Paris) *16*: 1–275.

Dangmann, E., A. Stolz, A.E. Kuhm, A. Hammer, B. Feigel, R.N. Noisommit, M. Rizzi, M. Reuss and H.J. Knackmuss. 1996. Degradation of 4-aminobenzenesulfonate by a two-species bacterial coculture. Biodegradation *7*: 223–229.

Daniel, R.M., A.W. Limmer, K.W. Steele and I.M. Smith. 1982. Anaerobic growth, nitrate reduction and denitrification in 46 *Rhizobium* strains. J. Gen. Microbiol. *128*: 1811–1815.

Daniel, S.L., H.M. Cook, P.A. Hartman and M.J. Allison. 1989. Enumeration of anaerobic oxalate-degrading bacteria in the ruminal contents of sheep. FEMS Microbiol. Ecol. *62*: 329–334.

Daniel, S.L., P.A. Hartman and M.J. Allison. 1987a. Intestinal colonization of laboratory rats with *Oxalobacter formigenes*. Appl. Environ. Microbiol. *53*: 2767–2770.

Daniel, S.L., P.A. Hartman and M.J. Allison. 1987b. Microbial degradation of oxalate in the gastrointestinal tracts of rats. Appl. Environ. Microbiol. *53*: 1793–1797.

Das, S.K. and A.K. Mishra. 1996. Transposon mutagenesis affecting thiosulfate oxidation in *Bosea thiooxidans*, a new chemolithoheterotrophic bacterium. J. Bacteriol. *178*: 3628–3633.

Das, S.K., A.K. Mishra, B.J. Tindall, F.A. Rainey and E. Stackebrandt. 1996. Oxidation of thiosulfate by a new bacterium, *Bosea thiooxidans* (strain BI-42) gen. nov., sp. nov.: Analysis of phylogeny based on chemotax-

onomy and 16S ribosomal DNA sequencing. Int. J. Syst. Bacteriol. *46*: 981–987.

Dasch, G.A. 1981. Isolation of species-specific protein antigens of *Rickettsia typhi* and *Rickettsia prowazekii* for immunodiagnosis and immunoprophylaxis. J. Clin. Microbiol. *14*: 333–341.

Dasch, G.A., A.L. Bourgeois and F.M. Rollwagen. 1999. The surface protein antigen of *Rickettsia typhi*: *In vitro* and *in vivo* immunogenicity and protective efficacy in mice. *In* Raoult and Brouqui (Editors), Rickettsiae and Rickettsial Diseases at the Turn of the Third Millenium, Elsevier, Paris. pp.116–122.

Dasch, G.A., J.R. Samms and E. Weiss. 1978. Biochemical characteristics of typhus group rickettsiae with special attention to the *Rickettsia prowazekii* strains isolated from flying squirrels. Infect. Immun. *19*: 676–685.

Dasen, S.E., J.J. LiPuma, J.R. Kostman and T.L. Stull. 1994. Characterization of PCR-ribotyping for *Burkholderia (Pseudomonas) cepacia*. J. Clin. Microbiol. *32*: 2422–2424.

Daubaras, D.L., C.E. Danganan, A. Hubner, R.W. Ye, W. Hendrickson and A.M. Chakrabarty. 1996a. Biodegradation of 2,4,5-trichlorophenoxyacetic acid by *Burkholderia cepacia* strain AC1100: evolutionary insight. Gene *179*: 1–8.

Daubaras, D.L., K. Saido and A.M. Chakrabarty. 1996b. Purification of hydroxyquinol 1,2-dioxygenase and maleylacetate reductase: the lower pathway of 2,4,5-trichlorophenoxyacetic acid metabolism by *Burkholderia cepacia* AC1100. Appl. Environ. Microbiol. *62*: 4276–4279.

Davidson, D., B. Beheshti and M.W. Mittelman. 1996. Effects of *Arthrobacter* sp., *Acidovorax delafieldii*, and *Bacillus megaterium* colonisation on copper solvency in a laboratory reactor. Biofouling *9*: 279–292.

Davies, G., C.N. Hébert and A.D. Casey. 1973. Preservation of *Brucella abortus* (Strain 544) in liquid nitrogen and its virulence when subsequently used as a challenge. J. Biol. Stand. *1*: 165–170.

Davis, D.H., M. Doudoroff, R.Y. Stanier and M. Mandel. 1969. Proposal to reject the genus *Hydrogenomonas*: taxonomic implications. Int. J. Syst. Bacteriol. *19*: 375–390.

Davis, D.H., R.Y. Stanier, M. Doudoroff and M. Mandel. 1970. Taxonomic studies on some Gram negative polarly flagellated "hydrogen bacteria" and related species. Arch. Mikrobiol. *70*: 1–13.

Davis, E.P., R. Boopathy and J. Manning. 1997. Use of trinitrobenzene as a nitrogen source by *Pseudomonas vesicularis* isolated from soil. Curr. Microbiol. *34*: 192–197.

Davis, E.O., I.J. Evans and A.W. Johnston. 1988. Identification of *nodX*, a gene that allows *Rhizobium leguminosarum* biovar viciae strain TOM to nodulate Afghanistan peas. Mol. Gen. Genet. *212*: 531–535.

Davis, G.H.G. and R.W.A. Park. 1962. A taxonomic study of certain bacteria currently classified as *Vibrio* species. J. Gen. Microbiol. *27*: 101–119.

Davis, H.S. 1921/1922. A new bacterial disease of fresh water fishes. Bull. U.S. Bureau Fish. *38*: 261–280.

Davis, M.J., Z. Ying, B.R. Brunner, A. Pantoja and F.H. Ferwerda. 1998. Rickettsial relative associated with papaya bunchy top disease. Curr. Microbiol. *36*: 80–84.

Davis, R.J. 1962. The resistance of rhizobia to antimicrobial agents. J. Bacteriol. *84*: 187–188.

Davis, R.H. and M.R. Salton. 1975. Some properties of a D-alanine carboxypeptidase in envelope fractions of *Neisseria gonorrhoeae*. Infect. Immun. *12*: 1065–1069.

Davison, A.D., M.R. Gillings, D.R. Jardine, P. Karuso, A.S. Nouwens, J.J. French, D.A. Veal and N. Altavilla. 1999. *Sphingomonas paucimobilis* BPSI-3 mutant AN2 produces a red catabolite during biphenyl degradation. J. Ind. Microbiol. Biotechnol. *23*: 314–319.

Davison, A.D., P. Karuso, D.R. Jardine and D.A. Veal. 1996. Halopicolinic acids, novel products arising through the degradation of chloro- and bromo-biphenyl by *Sphingomonas paucimobilis* BPSI-3. Can. J. Microbiol. *42*: 66–71.

Davydov, N.N. 1961. Properties of *Brucella* isolated from reindeer (In Russian). Trudy Vsyesoyuz Inst. Eksp. Vet. *27*: 24–31.

Dawes, E.A., M. Midgley and M. Ishaq. 1970. The endogenous metabolism of anaerobic bacteria, European Research Office, U.S. Army.

Dawid, W. 1977. Fruchtkörperbildende Myxobakterien. V. Die *Polyangium*-Arten: *P. cellulosum, P. fumosum, P. sorediatum, P. vitellinum.* Mikrokosmos *12*: 364–373.

Dawid, W. 1978. Fruchtkorperbildende Myxobacterien VI. Die *Stigmatella*-Arten: *S. erecta, S. aurantiaca.* Mikrokosmos *67*: 43–50.

Dawid, W. 1980. Fruchtkörperbildende Myxobakterien. VII. Die *Chondromyces*-Arten: *Ch. apiculatus* und *Ch. lanuginosus.* Mikrokosmos *69*: 73–79.

Dawid, W. 2000. Biology and global distribution of myxobacteria in soils. FEMS Microbiol. Rev. *24*: 403–427.

Dawid, W., C.A. Gallikowski and P. Hirsch. 1988. Psychrophilic myxobacteria from Antarctic soils. Polarforschung. *58*: 271–278.

Dawson, J.E., B.E. Anderson, D.B. Fishbein, J.L. Sanchez, C.S. Goldsmith, K.H. Wilson and C.W. Duntley. 1991. Isolation and characterization of an *Ehrlichia* sp. from a patient diagnosed with human ehrlichiosis. J. Clin. Microbiol. *29*: 2741–2745.

Dawson, J.E., F.J. Candal, V.G. George and E.W. Ades. 1993. Human endothelial cells as an alternative to DH82 cells for isolation of *Ehrlichia chaffeensis, E. canis,* and *Rickettsia rickettsii.* Pathobiology *61*: 293–296.

Dawson, J.E., M. Ristic, C.J. Holland, R.H. Whitlock and J. Sessions. 1987. Isolation of *Ehrlichia risticii,* the causative agent of Potomac horse fever, from the fetus of an experimentally infected mare. Vet. Rec. *121*: 232.

Dawson, J.E., C.K. Warner, V. Baker, S.A. Ewing, D.E. Stallknecht, W.R. Davidson, A.A. Kocan, J.M. Lockhart and J.G. Olson. 1996. *Ehrlichia*-like 16S rDNA sequence from wild white-tailed deer (*Odocoileus virginianus*). J. Parasitol. *82*: 52–58.

Dawson, K.A. 1979. Enrichment, isolation and characterization of anaerobic oxalate-degrading bacteria from the rumen, Doctoral thesis, Iowa State University. 92 pp.

Dawson, K.A., M.J. Allison and P.A. Hartman. 1980a. Characteristics of anaerobic oxalate-degrading enrichment cultures from the rumen. Appl. Environ. Microbiol. *40*: 840–846.

Dawson, K.A., M.J. Allison and P.A. Hartman. 1980b. Isolation and some characteristics of anaerobic oxalate-degrading bacteria from the rumen. Appl. Environ. Microbiol. *40*: 833–839.

Day, W.A., Jr., I.L. Pepper and L.A. Joens. 1997. Use of an arbitrarily primed PCR product in the development of a *Campylobacter jejuni*-specific PCR. Appl. Environ. Microbiol. *63*: 1019–1023.

De Baere, T., S. Steyaert, G. Wauters, P. De Vos, J. Goris, T. Coenye, T. Suyama, G. Verschraegen and M. Vaneechoutte. 2001. Classification of *Ralstonia pickettii* biovar 3/'thomasii' strains (Pickett 1994) and of new isolates related to nosocomial recurrent meningitis as *Ralstonia mannitolytica* sp. nov. Int. J. Syst. Evol. Microbiol. *51*: 547–558.

de Beer, R., J.A. Duine, J. Frank and J. Westerling. 1983. The role of pyrrolo-quinoline semiquinone forms in the mechanism of action of methanol dehydrogenase. Eur. J. Biochem. *130*: 105–109.

de Boer, E., J.J. Tilburg, D.L. Woodward, H. Lior and W.M. Johnson. 1996. A selective medium for the isolation of *Arcobacter* from meats. Lett. Appl. Microbiol. *23*: 64–66.

de Boer, W.E., J.W.M. La Rivière and A.L. Houwink. 1961. Observations on the morphology of *Thiovulum majus* Hinze. Antonie Leeuwenhoek *27*: 447–456.

de Bok, F.A.M., A.J.M. Stams, C. Dijkema and D.R. Boone. 2001. Pathway of propionate oxidation by a syntrophic culture of *Smithella propionica* and *Methanospirillum hungatei.* Appl. Environ. Microbiol. *67*: 1800–1804.

De Bont, J.A.M. and M.W.M. Leijten. 1976. Nitrogen fixation by hydrogen-utilizing bacteria. Arch. Microbiol. *107*: 235–240.

de Bont, J.A., A. Scholten and T.A. Hansen. 1981a. DNA-DNA hybridization of *Rhodopseudomonas capsulata, Rhodopseudomonas sphaeroides* and *Rhodopseudomonas sulfidophila* strains. Arch. Microbiol. *128*: 271–274.

De Bont, J.A.M., J.P. Van Dijken and W. Harder. 1981b. Dimethyl sulfoxide and dimethyl sulfide as a carbon, sulfur and energy source for growth of *Hyphomicrobium* strain S. J. Gen. Microbiol. *127*: 315–324.

De Cleene, M. 1979. Crown gall: economic importance and control. Zentralbl. Bakteriol. Parasitenk. Infektionskr. Hyg. II Abt. *134*: 551–554.

De Cleene, M. and J. De Ley. 1976. The host range of crown gall. Bot. Rev. *42*: 389–466.

De Cleene, M. and J. De Ley. 1981. The host range of infectious hairy-root. Bot. Rev. *47*: 147–194.

de Gier, J.-W., M. Lübben, W.N. Reijnders, C.A. Tipker, D.J. Slotboom, R.J.M. van Spanning, A.H. Stouthamer and J. van der Oost. 1994. The terminal oxidases of *Paracoccus denitrificans.* Mol. Microbiol. *13*: 183–196.

De Groot, M.J.A., P. Bundock, P.J.J. Hooykaas and A.G.M. Beijersbergen. 1998. *Agrobacterium tumefaciens* mediated transformation of filamentous fungi. Nat. Biotechnol. *16*: 839–842.

De Groote, D., L.J. van Doorn, R. Ducatelle, A. Verschuuren, F. Haesebrouck, Q. W.G.V., K. Jalava and P. Vandamme. 1999a. "*Candidatus* Helicobacter suis", a gastric helicobacter from pigs, and its phylogenetic relatedness to other gastrospirilla. Int. J. Syst. Bacteriol. *49*: 1769–1777.

De Groote, D., L.J. van Doorn, R. Ducatelle, A. Verschuuren, K. Tilmant, W.G.V. Quint, F. Haesebrouck and P. Vandamme. 1999b. Phylogenetic characterization of "*Candidatus* Helicobacter bovis", a new gastric helicobacter in cattle. Int. J. Syst. Bacteriol. *49*: 1707–1715.

De Iannino, N.I., G. Briones, M. Tolmasky and R.A. Ugalde. 1998. Molecular cloning and characterization of *cgs,* the *Brucella abortus* cyclic beta(1-2) glucan synthetase gene: genetic complementation of *Rhizobium meliloti ndvB* and *Agrobacterium tumefaciens chvB* mutants. J. Bacteriol. *180*: 4392–4400.

de Jong, G.A., A. Geerlof, J. Stoorvogel, J.A. Jongejan, S. de Vries and J.A. Duine. 1995. Quinohaemoprotein ethanol dehydrogenase from *Comamonas testosteroni.* Purification, characterization and reconstitution of the apoenzyme with pyrroloquinoline quinone analogues. Eur. J. Biochem. *230*: 899–905.

de Kruyff, E. 1908. Die lebensgeschichte von *Myxococcus javanensis* sp.n. Zentbl. Bakteriol. *21*: 385–386.

De Lajudie, P., A. Laurent-Fulele, U. Willems, R. Torck, R. Coopman, M.D. Collins, K. Kersters, B. Dreyfus and M. Gillis. 1998a. Description of *Allorhizobium undicola* gen. nov., sp. nov., for nitrogen-fixing bacteria efficiently nodulating *Neptunia natans* in Senegal. Int. J. Syst. Bacteriol. *48*: 1277–1290.

de Lajudie, P., A. Willems, G. Nick, F. Moreira, F. Molouba, B. Hoste, U. Torck, M. Neyra, M.D. Collins, K. Lindstrom, B. Dreyfus and M. Gillis. 1998b. Characterization of tropical tree rhizobia and description of *Mesorhizobium plurifarium* sp. nov. Int. J. Syst. Bacteriol. *48*: 369–382.

de Lajudie, P., A. Willems, B. Pot, D. Dewettinck, G. Maestrojuan, M. Neyra, M.D. Collins, B. Dreyfus, K. Kersters and M. Gillis. 1994. Polyphasic taxonomy of rhizobia: emendation of the genus *Sinorhizobium* and description of *Sinorhizobium meliloti* comb. nov., *Sinorhizobium saheli* sp. nov., and *Sinorhizobium teranga* sp. nov. Int. J. Syst. Bacteriol. *44*: 715–733.

De Ley, J. 1961. Comparative carbohydrate metabolism and a proposal for a phylogenetic relationship of the acetic acid bacteria. J. Gen. Microbiol. *24*: 31–50.

De Ley, J. 1968. DNA base composition and hybridization in the taxonomy of the phytopathogenic bacteria. Annu. Rev. Phytopathol. *6*: 63–90.

De Ley, J. 1969. Compositional nucleotide distribution and the theoretical prediction of homology in bacterial DNA. J. Theoret. Biol. *22*: 89–116.

De Ley, J. 1972. *Agrobacterium*: intrageneric relationships and evolution. Proc 3rd Int. Con. Plant Path. Bact, Wageningen, The Netherlands: Pudoc. pp. 251–259.

De Ley, J. 1974. Phylogeny of procaryotes. Taxon. *23*: 291–300.

De Ley, J. 1978. Modern molecular methods in bacterial taxonomy: evaluation, application, prospects. Proc. 4th Int. Conf. Plant Path. Bact., Tours, France.1978 Gilbert-Clarey. pp. 347–357.

De Ley, J. 1992. The *Proteobacteria* ribosomal RNA cistron similarities and bacterial taxonomy. *In* Balows, Trüper, Dworkin, Harder and Schleifer (Editors), The Prokaryotes: A Handbook of Bacteria: Ecophysiology, Isolation, Identification, Applications, 2nd Ed., Vol. 2, Springer-Verlag, New York. pp. 2111–2140.

De Ley, J., M. Bernaerts, A. Rassel and J. Guilmot. 1966. Approach to an improved taxonomy of the genus *Agrobacterium*. J. Gen. Microbiol. *43*: 7–17.

De Ley, J., H. Cattoir and A. Reynaerts. 1970a. The quantitative measurement of DNA hybridization from renaturation rates. Eur. J. Biochem. *12*: 133–142.

De Ley, J. and J. Frateur. 1974. *Acetobacter. In* Buchanan and Gibbons (Editors), Bergey's Manual of Determinative Bacteriology, 8th Ed., The Williams & Wilkins Co., Baltimore. pp. 251–253.

De Ley, J., M. Gillis, C.F. Pootjes, K. Kersters, R. Tytgat and M. Van Braekel. 1972. Relationship among temperate *Agrobacterium* phage genomes and coat proteins. J. Gen. Virol. *16*: 199–214.

De Ley, J., K. Kersters, J. Khan-Matsubara and J.M. Shewan. 1970b. Comparative D-gluconate metabolism and DNA base composition in *Achromobacter* and *Alcaligenes*. Antonie Leeuwenhoek J. Microbiol. Serol. *36*: 193–207.

De Ley, J., W. Mannheim, P. Segers, A. Lievens, M. Denijn, M. Vanhoucke and M. Gillis. 1987. Ribosomal ribonucleic acid cistron similarities and taxonomic neighborhood of *Brucella* and CDC group Vd. Int. J. Syst. Bacteriol. *37*: 35–42.

De Ley, J. and J.V. Muylem. 1963. Some applications of deoxyribonucleic acid base composition in bacterial taxonomy. Anton Leeuwenhoek J. Microbiol. Serol. *29*: 344–358.

De Ley, J. and I.W. Park. 1966. Molecular biological taxonomy of some free living nitrogen-fixing bacteria. Antonie Leeuwenhoek J. Microbiol. Serol. *32*: 6–16.

De Ley, J. and A. Rassel. 1965. DNA base composition, flagellation and taxonomy of the genus *Rhizobium*. J. Gen. Microbiol. *41*: 85–91.

De Ley, J., P. Segers and M. Gillis. 1978. Intra- and intergeneric similarities of *Chromobacterium* and *Janthinobacterium* ribosomal ribonucleic acid cistrons. Int. J. Syst. Bacteriol. *28*: 154–168.

De Ley, J., P. Segers, K. Kersters, W. Mannheim and A. Lievens. 1986. Intrageneric and intergeneric similarities of the *Bordetella* ribosomal ribonucleic acid cistrons: proposal for a new family, *Alcaligenaceae*. Int. J. Syst. Bacteriol. *36*: 405–414.

De Ley, J. and J. Swings. 1976. Phenotypic description, numerical analysis and a proposal for an improved taxonomy and nomenclature of the genus *Zymomonas* Kluyver and van Niel 1936. Int. J. Syst. Bacteriol. *26*: 146–157.

De Ley, J., J. Swings and F. Gosselé. 1984. Genus I. *Acetobacter* Beijerinck. *In* Krieg and Holt (Editors), Bergey's Manual of Systematic Bacteriology, 1st Ed., Vol. 1, The Williams & Wilkins Co., Baltimore. pp. 268–274.

De Ley, J., R. Tijtgat, J. De Smedt and M. Michiels. 1973. Thermal stability of DNA:DNA hybrids within the genus *Agrobacterium*. J. Gen. Microbiol. *78*: 241–252.

De Maagd, R.A., H.P. Spaink, E. Pees, I.H. Mulders, A. Wijfjes, C.A. Wijffelman, R.J. Okker and B.J. Lugtenberg. 1989. Localization and symbiotic function of a region on the *Rhizobium leguminosarum* Sym plasmid pRL1JI responsible for a secreted, flavonoid- inducible 50-kilodalton protein. J. Bacteriol. *171*: 1151–1157.

de Man, J.C., M. Rogosa and M.E. Sharpe. 1960. A medium for the cultivation of lactobacilli. J. Appl. Bacteriol. *23*: 130–135.

De Marco, P., J.C. Murrell, A.A. Bordalo and P. Moradas-Ferreira. 2000. Isolation and characterization of two new methanesulfonic acid-degrading bacterial isolates from a Portuguese soil sample. Arch. Microbiol. *173*: 146–153.

De Petris, S., K. G. and R.W.I. Kessel. 1964. Ultra structure of S and R variants of *Brucella abortus* grown on a lifeless medium. J. Gen. Microbiol. *35*: 373–382.

De Polli, H., B.B. Bohlool and J. Doebereiner. 1980. Serological differentiation of *Azospirillum* spp. belonging to different host-plant specificity groups. Arch. Microbiol. *126*: 217–222.

De Rore, H., K. Demolder, K. De Wilde, E. Top, F. Houwen and W. Verstraete. 1994. Transfer of the catabolic plasmid RP4::Tn4371 to indigenous soil bacteria and its effect on respiration and biphenyl breakdown. FEMS Microbiol. Ecol. *15*: 71–77.

De Smedt, J., M. Banwens, R. Tijtgat and J. De Ley. 1980. Intra- and

intergeneric similarities of ribosomal ribonucleic acid cistrons of free living, nitrogen-fixing bacteria. Int. J. Syst. Bacteriol. *30*: 106–122.

De Smedt, J. and J. D.Ley. 1977. Intra- and inter-generic similarities of *Agrobacterium* ribosomal ribonucleic acid cistrons. Int. J. Syst. Bacteriol. *27*: 222–240.

De Smedt, J. and De Ley, J. 1977. Intra- and inter-genic similarities of *Agrobacterium* ribosomal ribonucleic acid cistrons. Int. J. Syst. Bacteriol. *27*: 222-240.

De Soete, G. 1983. A least squares alogorithm for fitting additive trees to proximity data. Psychometrika. *48*: 621–626.

de Souza, M.P. and D.C. Yoch. 1995. Purification and characterization of dimethylsulfoniopropionate lyase from an *Alcaligenes* like dimethyl sulfide-producing marine isolate. Appl. Environ. Microbiol. *61*: 21–26.

de Toni, J.B. and V. Trevisan. 1889. *Schizomycetaceae* Naeg. *In* Saccardo (Editor), Sylloge fungorum omnium hujusque cognitorum, Vol. 8, pp. 923–1087.

De Vos, P. and J. De Ley. 1983. Intra- and intergeneric similarities of *Pseudomonas* and *Xanthomonas* ribosomal ribonucleic acid cistrons. Int. J. Syst. Bacteriol. *33*: 487–509.

De Vos, P., M. Goor, M. Gillis and J. De Ley. 1985a. Ribosomal ribonucleic acid cistron similarities of phytopathogenic *Pseudomonas* species. Int. J. Syst. Bacteriol. *35*: 169–184.

De Vos, P., K. Kersters, E. Falsen, B. Pot, M. Gillis, P. Segers and J. De Ley. 1985b. *Comamonas* Davis and Park 1962 gen. nov., nom. rev. emend., and *Comamonas terrigena* Hugh 1962 sp. nov., nom. rev. Int. J. Syst. Bacteriol. *35*: 443–453.

De Vos, P., A. Van Landschoot, P. Segers, R. Tytgat, M. Gillis, M. Bauwens, R. Rossau, M. Goor, B. Pot, K. Kersters, P. Lizzaraga and J. De Ley. 1989. Genotypic relationships and taxonomic localization of unclassified *Pseudomonas* and *Pseudomonas*-like strains by deoxyribonucleic acid: ribonucleic acid hybridizations. Int. J. Syst. Bacteriol. *39*: 35–49.

de Vries, F.P., R. Cole, J. Dankert, M. Frosch and J.P.M. van Putten. 1998. *Neisseria meningitidis* producing the Opc adhesin binds epithelial cell proteoglycan receptors. Mol. Microbiol. *27*: 1203–1212.

De Wever, H., J.R. Cole, M.R. Fettig, D.A. Hogan and J.M. Tiedje. 2000. Reductive dehalogenation of trichloroacetic acid by *Trichlorobacter thiogenes* gen. nov., sp. nov. Appl. Environ. Microbiol. *66*: 2297–2301.

De Wever, H., J.R. Cole, M.R. Fettig, D.A. Hogan and J.M. Tiedje. 2001. *In* Validation of the publication of new names and new combinations previously effectively published oustide the IJSEM. List No. 78. Int. J. Syst. Evol. Microbiol. *51*: 1–2.

Deanda, K., M. Zhang, C. Eddy and S. Picataggio. 1996. Development of an arabinose-fermenting *Zymomonas mobilis* strain by metabolic pathway engineering. Appl. Environ. Microbiol. *62*: 4465–4470.

Debellé, F., C. Rosenberg, J. Vasse, F. Maillet, E. Martinez, J. Dénarié and G. Truchet. 1986. Assignment of symbiotic developmental phenotypes to common and specific nodulation (*nod*) genetic loci of *Rhizobium meliloti*. J. Bacteriol. *168*: 1075–1086.

Decker, C.F., R.E. Hawkins and G.L. Simon. 1992. Infections with *Pseudomonas paucimobilis*. Clin. Infect. Dis. *14*: 783–784.

Dedeine, F., F. Vavre, F. Fleury, B. Loppin, M.E. Hochberg and M. Bouletreau. 2001. Removing symbiotic *Wolbachia* bacteria specifically inhibits oogenesis in a parasitic wasp. Proc. Natl. Acad. Sci. U.S.A. *98*: 6247–6252.

Dedysh, S.N., M. Derakshani and W. Liesack. 2001. Detection and enumeration of methanotrophs in acidic sphagnum peat by 16S rRNA fluorescence *in situ* hybridization, including the use of newly developed oligonucleotide probes for *Methylocella palustris*. Appl. Environ. Microbiol. *67*: 4850–4857.

Dedysh, S.N. , V.N. Khmelenina, N.E. Suzina, Y.A. Trotsenko, J.D. Semrau, W. Liesack and J.M. Tiedje. 2002. *Methylocapsa acidiphila* gen. nov., sp. nov., a novel methane-oxidizing and dinitrogen-fixing acidophilic bacterium from Sphagnum bog. Int. J. Syst. Evol. Microbiol. *52*: 251–261.

Dedysh, S.N., W. Liesack, V.N. Khmelenina, N.E. Suzina, Y.A. Trotsenko, J.D. Semrau, A.M. Bares, N.S. Panikov and J.M. Tiedje. 2000. *Methylocella palustris* gen. nov., sp. nov., a new methane oxidizing acidophilic bacterium from peat bags, representing a novel subtype of serine pathway methanotrophs. Int. J. Syst. Evol. Microbiol. *50*: 955–969.

Dedysh, S.N., N.S. Panikov, W. Liesack, R. Grosskopf, J.Z. Zhou and J.M. Tiedje. 1998a. Isolation of acidophilic methane oxidizing bacteria from northern peat wetlands. Science *282*: 281–284.

Dedysh, S.N., N.S. Panikov and J.M. Tiedje. 1998b. Acidophilic methanotrophic communities from Sphagnum peat bogs. Appl. Environ. Microbiol. *64*: 922–929.

Defives, C., D. Ochin, J.P. Hornez and M. Werquin. 1990. Accidents de fabrication du vinaigre causes par un bacteriophage. Microbiol. Alim. Nutrit. *8*: 77–79.

Degli-Innocenti, F., E. Ferdani, B. Pesenti-Barili, M. Dani, L. Giovanetti and S. Ventura. 1990. Identification of microbial isolates by DNA fingerprinting: analysis of ATCC *Zymomonas* strains. J. Biotechnol. *13*: 335–346.

Degryse, E., N. Glansdorff and A. Pierard. 1978. A comparative analysis of extreme thermophilic bacteria belonging to the genus *Thermus*. Arch. Microbiol. *117*: 189–196.

Dehning, I. and B. Schink. 1989a. *Malonomonas rubra* gen. nov. sp. nov., a microaerotolerant anaerobic bacterium growing by decarboxylation of malonate. Arch. Microbiol. *151*: 427–433.

Dehning, I. and B. Schink. 1989b. Two new species of anaerobic oxalate-fermenting bacteria, *Oxalobacter vibrioformis* sp. nov. and *Clostridium oxalicum* sp. nov., from sediment samples. Arch. Microbiol. *153*: 79–84.

Dehning, I. and B. Schink. 1990. *In* Validation of the publication of new names and new combinations previously effectively published outside the IJSB. List No. 34. Int. J. Syst. Bacteriol. *40*: 320–321.

Dejsirilert, S., R. Butraporn, D. Chiewsilp, E. Kondo and K. Kanai. 1989. High activity of acid phosphatase of *Pseudomonas pseudomallei* as a possible attribute relating to its pathogenicity. Jpn. J. Med. Sci. Biol. *42*: 39–49.

Del Gallo, M. and I. Fendrick. 1994. The rhizosphere and *Azospirillum*. *In* Okon (Editor), *Azospirillum*-plant associations, CRC Press, Boca Raton. pp. 57–75.

Delafield, F.P., M. Doudoroff, N.J. Palleroni, C.J. Lusty and R. Contopoulos. 1965. Decomposition of poly-β-hydroxybutyrate by pseudomonads. J. Bacteriol. *90*: 1455–1466.

Delaporte, B. 1964. Etude comparee de grands spirilles format des spores: *Sporospirillum (Spirillum) praeclarum* (Collie) n. g., *Sporospirillum gyrini* n sp. et *Sporospirillum bisporum* n. sp. Ann. Inst. Pasteur (Paris) *107*: 246–262.

Delmer, D.P. and Y. Amor. 1995. Cellulose biosynthesis. Plant Cell 7: 987–1000.

DelVecchio, V.G., V. Kapatral, R.J. Redkar, G. Patra, C. Mujer, T. Los, N. Ivanova, I. Anderson, A. Bhattacharyya, A. Lykidis, G. Reznik, L. Jablonski, N. Larsen, M. D'Souza, A. Bernal, M. Mazur, E. Goltsman, E. Selkov, P.H. Elzer, S. Hagius, D. O'Callaghan, J.J. Letesson, R. Haselkorn, N. Kyrpides and R. Overbeek. 2002. The genome sequence of the facultative intracellular pathogen *Brucella melitensis*. Proc. Natl. Acad. Sci. U.S.A. *99*: 443–448.

DeMello, F.J. and M.S. Leonard. 1979. *Eikenella corrodens*, a new pathogen. Oral Surg. *45*: 401–404.

Dempsey, J.A., A.B. Wallace and J.G. Cannon. 1995. The physical map of the chromosome of a serogroup A strain of *Neisseria meningitidis* shows complex rearrangements relative to the chromosomes of the two mapped strains of the closely related species *N. gonorrhoeae*. J. Bacteriol. *177*: 6390–6400.

Den Dooren de Jong, L.E. 1926. Bijdrage tot de kennis van het mineralisatieproces, Nijgh and van Ditmar Uitgevers-Mij, Rotterdam. pp. 1–200.

Dénarié, J., F. Debellé and J.C. Promé. 1996. *Rhizobium* lipo-chitooligosaccharide nodulation factors: signaling molecules mediating recognition and morphogenesis. Annu. Rev. Biochem. *65*: 503–535.

Dénarié, J., F. Debellé and C. Rosenberg. 1992. Signaling and host range variation in nodulation. Annu. Rev. Microbiol. *46*: 497–531.

Dénarié, J., G. Truchet and B. Bergeron. 1976. Effects of some mutations on symbiotic properties of *Rhizobium*. *In* Nutman (Editor), Symbiotic

nitrogen fixation in plants: International Biological Programme 7, Cambridge University Press, Cambridge [Eng.]; New York. 47–61.

Denisov, I.I. and V.I. Kapliev. 1991. The level of spontaneous phage production and sensitivity to melioidosis phages of museum cultures of *Pseudomonas pseudomallei*. Mikrobiol. Zh. *53*: 66–70.

Denisov, I.I. and V.I. Kapliev. 1995. The isolation and characteristics of cloned strains of *Pseudomonas pseudomallei* phages. Mikrobiol. Zh. *57*: 53–56.

Denner, E.B., P. Kämpfer, H.J. Busse and E.R. Moore. 1999. Reclassification of *Pseudomonas echinoides* Heumann 1962, 343[AL], in the genus *Sphingomonas* as *Sphingomonas echinoides* comb. nov. Int. J. Syst. Bacteriol. *49*: 1103–1109.

Denner, E.B., S. Paukner, P. Kämpfer, E.R. Moore, W.R. Abraham, H.J. Busse, G. Wanner and W. Lubitz. 2001. *Sphingomonas pituitosa* sp. nov., an exopolysaccharide-producing bacterium that secretes an unusual type of sphingan. Int. J. Syst. Evol. Microbiol. *51*: 827–841.

Dennis, R.T. and T.W. Young. 1982. A simple, rapid method for the detection of subspecies of *Zymomonas mobilis*. J. Inst. Brew. *88*: 25–29.

Denoel, P.A., M.S. Zygmunt, V. Weynants, A. Tibor, B. Lichtfouse, P. Briffeuil, J.N. Limet and J.J. Letesson. 1995. Cloning and sequencing of the bacterioferritin gene of *Brucella melitensis* 16M strain. FEBS Lett. *361*: 238–242.

Dent, J.C. and C.A.M. McNulty. 1988. Evaluation of a new selective medium for *Campylobacter pylori*. Eur. J. Clin. Microbiol. Infect. Dis. *7*: 555–558.

Dérijard, B., M. Benaissa, B. Lubochinsky and Y. Cenatiempo. 1989. Evidence for a membrane-associated GTP-binding protein in *Stigmatella aurantiaca*, a prokaryotic cell. Biochem. Biophys. Res. Commun. *158*: 562–568.

Derrick, E.H. and H.E. Brown. 1949. Isolation of the Karp strain of *R. tsutsugamushi*. Lancet *2*: 150–151.

Derrick, J.P., R. Urwin, J. Suker, I.M. Feavers and M.C.J. Maiden. 1999. Structural and evolutionary inference from molecular variation in *Neisseria* porins. Infect. Immun. *67*: 2406–2413.

DerVartanian, D.V. 1994. Desulforubidin: dissimilatory, high-spin sulfite reductase of *Desulfomicrobium* species. Methods Enzymol. *243*: 270–276.

Derx, H.G. 1950a. *Beijerinckia*, a new genus of nitrogen-fixing bacteria occurring in tropical soils. Proc. Kon. Ned. Akad. Wet. Ser. *C53*: 140–147.

Derx, H.G. 1950b. Further researches on *Beijerinckia*. Ann. Bogor. *1*: 1–12.

Desai, M., D. Linton, R.J. Owen, H. Cameron and J. Stanley. 1993. Genetic diversity of *Helicobacter pylori* indexed with respect to clinical symptomatology, using a 16S rRNA and a species-specific DNA probe. J. Appl. Bacteriol. *75*: 574–582.

Desiervo, A.J. 1985. High-levels of glycolipid and low-levels of phospholipid in a marine caulobacter. J. Bacteriol. *164*: 684–688.

Desiervo, A.J. and A.D. Homola. 1980. Analysis of *Caulobacter crescentus* lipids. J. Bacteriol. *143*: 1215–1222.

Desjardins, M., C. Fenlon and D. Madison. 1999. Non-chromogenic *Chromobacterium violaceum* bacteremia. Clin. Microbiol. Newsl. *21*: 14–16.

Dessaux, Y., A. Petit, S.K. Ferrand and P.J. Murphy. 1998. Opines and opine like compounds in plant-*Rhizoiaceae* interactions. *In* Spaink, Kondorosi and Hooykaas (Editors), The *Rhizobiaceae*: Molecular Biology of Model Plant-Associated Bacteria, Kluwer Academic, Dordrecht ; Boston. 173–197.

Dessaux, Y., A. Petit and J. Tempé. 1992. Opines in *Agrobacterium* biology. *In* Verma (Editor), Molecular Signals in Plant-Microbe Communications, CRC Press, Boca Raton, Fla. 109–136.

Desser, S.S. 1987. *Aegyptianella ranarum* sp. n. (*Rickettsiales, Anaplasmataceae*): ultrastructure and prevalence in frogs from Ontario. J. Wildlife Dis. *23*: 52–59.

Desser, S.S. and J.R. Barta. 1989. The morphological features of *Aegyptianella bacterifera*: an intraerythrocytic rickettsia of frogs from Corsica. J. Wildlife Dis. *25*: 313–318.

Deutsch, A. 1999. Principles of biological pattern formation: swarming and aggregation viewed as self-organization phenomena. J. Biosci. *24*: 115–120.

Deveer, A.M.T.J., R. Dijkman, M. Leuvelingtjeenk, L. Vandenberg, S. Ransac, M. Batenburg, M. Egmond, H.M. Verheij and G.H. Dehaas. 1991. A monolayer and bulk study on the kinetic behavior of *Pseudomonas glumae* lipase using synthetic pseudoglycerides. Biochemistry *30*: 10034–10042.

Devereux, J., P. Haeberli and O. Smithies. 1984. A comprehensive set of sequence analysis programs for the VAX. Nucleic Acids Res. *12*: 387–395.

Devereux, R., M. Delaney, F. Widdel and D.A. Stahl. 1989. Natural relationships among sulfate-reducing eubacteria. J. Bacteriol. *171*: 6689–6695.

Devereux, R., S.H. He, C.L. Doyle, S. Orkland, D.A. Stahl, J. Le Gall and W.B. Whitman. 1990. Diversity and origin of *Desulfovibrio* species: phylogenetic definition of a family. J. Bacteriol. *172*: 3609–3619.

Devereux, R., M.D. Kane, J. Winfrey and D.A. Stahl. 1992. Genus- and group-specific hybridization probes for determinative and environmental studies of sulfate-reducing bacteria. Syst. Appl. Microbiol. *15*: 601–609.

Devi, A.L. and H.D. McCurdy. 1984a. Adenylate cyclase and guanylate cyclase in *Myxococcus xanthus*. J. Gen. Microbiol. *130*: 1851–1856.

Devi, A.L. and H.D. McCurdy. 1984b. Cyclic GMP and cyclic-AMP binding proteins in *Myxococcus xanthus*. J. Gen. Microbiol. *130*: 1845–1849.

Devine, T.E.,, L.D. Kuykendall and J.J. O'Neill. 1988. DNA homology group and the identity of bradyrhizobial strains producing rhizobitoxine-induced foliar chlorosis on soybean. Crop Sci. *28*: 939–941.

Devine, T.E., L.D. Kuykendall and J.J. O'Neill. 1990. The Rj4 allele in soybean represses nodulation by chlorosis-inducing bradyrhizobia classified as DNA homology group II by antibiotic-resistance profiles. Theor. Appl. Genet. *80*: 33–37.

Devine, T.E. and L.D. Kuykendall. 1996. Host genetic control of symbiosis in soybean (Glycine max L). Plant Soil *186*: 173–187.

DeVoe, I.W. and J.E. Gilchrist. 1975. Pili on meningococci from primary cultures of nasopharyngeal carriers and cerebrospinal fluid of patients with acute disease. J. Exp. Med. *141*: 297–305.

Devries, J.T. and H.G. Derx. 1953. On the occurance of *Mycoplana rubra* and its identity with *Protaminobacter rubrum*. Ann. Bogor. *1*: 53–60.

DeWeerd, K.A., L. Mandelco, R.S. Tanner, C.R. Woese and J.M. Suflita. 1990. *Desulfomonile tiedjei* gen. nov. and sp. nov., a novel anaerobic, dehalogenating, sulfate-reducing bacterium. Arch. Microbiol. *154*: 23–30.

DeWeerd, K.A., L. Mandelco, R.S. Tanner, C.R. Woese and J.M. Suflita. 1991. *In* Validation of the publication of new names and new combinations previously effectively published outside the IJSB. List No. 41. Int. J. Syst. Bacteriol. *41*: 178–179.

Dewhirst, F.E., C.K.C. Chen, B.J. Paster and J.J. Zambon. 1993. Phylogeny of species in the family *Neisseriaceae* isolated from human dental plaque and description of *Kingella orale*, sp. nov. Int. J. Syst. Bacteriol. *43*: 490–499.

Dewhirst, F.E., J.G. Fox, E.N. Mendes, B.J. Paster, C.E. Gates, C.A. Kirkbride and K.A. Eaton. 2000a. *"Flexispira rappini"* strains represent at least 10 *Helicobacter* taxa. Int. J. Syst. Evol. Microbiol. *50*: 1781–1787.

Dewhirst, F.E., J.G. Fox and S.L.W. On. 2000b. Recommended minimal standards for describing new species of the genus *Helicobacter*. Int. J. Syst. Evol. Microbiol. *50*: 2231–2237.

Dewhirst, F.E., B.J. Paster and P.L. Bright. 1989. *Chromobacterium, Eikenella, Kingella, Neisseria, Simonsiella*, and *Vitreoscilla* species comprise a major branch of the beta group *Proteobacteria* by 16S ribosomal ribonucleic acid sequence comparison: transfer of *Eikenella* and *Simonsiella* to the family *Neisseriaceae* (emend). Int. J. Syst. Bacteriol. *39*: 258–266.

Dewhirst, F.E., B.J. Paster, G.J. Fraser and I. Olsen. 1992. Phylogeny of 9 major groups of eubacteria based on 16S rRNA sequence comparisons. Conference on Taxonomy and Automated Identification of Bacteria, Prague, Czechoslovakia. p. 6.

Dewhirst, F.E., B.J. Paster, S. La Fontaine and J.I. Rood. 1990. Transfer of *Kingella indologenes* (Snell and Lapage 1976) to the genus *Suttonella* gen. nov. as *Suttonella indologenes* comb. nov.; transfer of *Bacteroides*

nodosus (Beveridge 1941) to the genus *Dichelobacter* gen. nov. as *D. nodosus* comb. nov.; and assignment of the genera *Cardiobacterium, Dichelobacter*, and *Suttonella* to *Cardiobacteriaceae* fam. nov. in the gamma division of *Proteobacteria* on the basis of 16S rRNA sequence comparisons. Int. J. Syst. Bacteriol. *40*: 426–433.

Dewhirst, F.E., C. Seymour, G.J. Fraser, B.J. Paster and J.G. Fox. 1994. Phylogeny of *Helicobacter* isolates from bird and swine feces and description of *Helicobacter pametensis* sp. nov. Int. J. Syst. Bacteriol. *44*: 553–560.

Dhundale, A.R., T. Furuichi, S. Inouye and M. Inouye. 1985. Distribution of multicopy single-stranded-DNA among *Myxobacteria* and related species. J. Bacteriol. *164*: 914–917.

Dhundale, A., M. Inouye and S. Inouye. 1988. A new species of multicopy single-stranded-DNA from *Myxococcus xanthus* with conserved structural features. J. Biol. Chem. *263*: 9055–9058.

Dhundale, A., B. Lampson, T. Furuichi, M. Inouye and S. Inouye. 1987. Structure of msDNA from *Myxococcus xanthus*—Evidence for a long, self-annealing RNA precursor for the covalently linked, branched RNA. Cell *51*: 1105–1112.

Di Giovanni, G.D., J.W. Neilson, I.L. Pepper and N.A. Sinclair. 1996. Plasmid diversity within a 2, 4-dichlorophenoxyacetic acid-degrading *Variovorax paradoxus* population isolated from a contaminated soil. J. Environ. Sci. Health Part A Environ. Sci.Eng. Toxic Hazard. Subst. Control *31*: 963–976.

Dianese, J.C., J. Dobereiner and L.T. Dos Santos. 1989. Membrane protein patterns of three *Azospirillum* species and *Herbaspirillum seropedicae*. Ann. Acad. Bras. Cienc. *61*: 223–230.

Dias, F.F. and J.V. Bhat. 1964. Microbial ecology of activated sludge. Appl. Microbiol. *12*: 412–417.

Diaz, R., L.M. Jones and J.B. Wilson. 1968. Antigenic relationship of the Gram negative organism causing canine abortion to smooth and rough brucellae. J. Bacteriol. *95*: 618–624.

Dick-Hegedus, E. and A. Lee. 1991. Use of a mouse model to examine anti-*Helicobacter pylori* agents. Scand. J. Gastroenterol. *26*: 909–915.

Dickerson, R.E. 1980a. Cytochrome *c* and the evolution of energy metabolism. Sci. Am. *242*: 137–153.

Dickerson, R.E. 1980b. Evolution and gene transfer in purple photosynthetic bacteria. Nature *283*: 210–212.

Dieckmann, J. 1977. An infectious bacterium in the cytoplasm of *Paramecium caudatum*. *In* Abstract Volume. International Congress on Protozoology, Warszawa. p. 77.

Diedrich, D.L., C.P. Duran and S.F. Conti. 1984. Acquisition of *Escherichia coli* outer membrane proteins by *Bdellovibrio* sp. strain 109D. J. Bacteriol. *159*: 329–334.

Diedrich, D.L., C.A. Portnoy and S.F. Conti. 1983. *Bdellovibrio* possesses a prey-derived OmpF protein in its outer membrane. Curr. Micriobiol. *8*: 51–56.

Dieterich, C., P. Wiesel, R. Neiger, A. Blum and T.I. Corthesy. 1998. Presence of multiple *"Helicobacter heilmannii"* strains in an individual suffering from ulcers and in his two cats. J. Clin. Microbiol. *36*: 1366–1370.

Dietrich, W. and O. Klimmek. 2002. The function of methyl-menaquinone-6 and solysulfide reductase membrane anchor (PsrC) in polysulfide respiration of *Wolinella succinogenes*. Eur. J. Biochem. *269*: 1085–1095.

Dijkhuizen, L., W. Harder, L. De Boer, A. Van Boven, W. Clement, S. Bron and G. Venema. 1984. Genetic manipulation of the restricted facultative methylotroph *Hyphomicrobium* X by the R-plasmid-mediated introduction of the *Escherichia coli pdh* genes. Arch. Microbiol. *139*: 311–318.

Dijkstra, M., J. Frank and J.A. Duine. 1988a. Methanol oxidation under physiological conditions using methanol dehydrogenase and a factor isolated from *Hyphomicrobium* X. FEBS Lett. *227*: 198–202.

Dijkstra, M., J.J. Frank and A. Duine. 1989. Studies on electron transfer from methanol dehydrogenase to cytochrome c_L, both purified from *Hyphomicrobium* X. Biochem. J. *257*: 87–94.

Dijkstra, M., J. Frank, J.F. Van Wielink and J.A. Duine. 1988b. The soluble cytochrome *c* of methanol-grown *Hyphomicrobium* X: evidence against the involvement of autoreduction in electron-acceptor functioning of cytochrome c_1. Biochem. J. *251*: 467–474.

Diks, R.M.M., S.P.P. Ottengraf and A.H.C. Van Der Oever. 1994a. The influence of NaCl on the degradation rate of dichloromethane by *Hyphomicrobium* sp. Biodegradation *5*: 129–141.

Diks, R.M.M., S.P.P. Ottengraf and S. Vrijland. 1994b. The existence of a biological equilibrium in a trickling filter for waste gas purification. Biotechnol. Bioeng. *44*: 1279–1287.

Dill, D., H. Eckau and H. Budzikiewicz. 1985. Neuartige cerebroside aus *Cystobacter fuscus* (*Myxobacterales*). Z. Naturforsch. *40b*: 1738–1746.

Dilling, W. and H. Cypionka. 1990. Aerobic respiration in sulfate-reducing bacteria. FEMS Microbiol. Lett. *71*: 123–127.

Dillon, J.R. and M. Pauze. 1981. Appearance in Canada of *Neisseria gonorrhoeae* strains with a 3.2 megadalton penicillinase-producing plasmid and a 24.5 megadalton transfer plasmid. Lancet *2*: 700.

Dillon, J.R., M. Pauze and K.H. Yeung. 1983. Spread of penicillinase-producing and transfer plasmids from the gonococcus to *Neisseria meningitidis*. Lancet *1*: 779–781.

Dillon, J.A. and K.H. Young. 1989. Beta-lactamase plasmids and chromosomally mediated antibiotic resistance in pathogenic *Neisseria* species. Clin. Microbiol. Rev. *2 Suppl*: S125–S133.

Dilts, J.A. 1976. Covalently closed, circular DNA in kappa endosymbionts of *Paramecium*. Genet. Res. *27*: 161–170.

Dimarco, A.A. and A.H. Romano. 1985. D-Glucose transport system of *Zymomonas mobilis*. Appl. Environ. Microbiol. *49*: 151–157.

Dippell, R.V. 1950. Mutation of the killer cytoplasmic factor in *Paramecium aurelia*. Heredity *4*: 165–187.

Dittmar, L.P.F. 1814. Die Pilze Deutschlands. *In* Sturm (Editor), Deutschlands Flora, III. Abt. Heft 2, Kosten des Herausgebers, Nürnberg. pp. 55–56.

Ditzelmüller, G., M. Loidl and F. Streichsbier. 1989. Isolation and characterization of a 2,4-dichlorophenoxyacetic acid degrading soil bacterium. Appl. Microbiol. Biotechnol. *31*: 93–96.

Dobbek, H., L. Gremer, R. Kiefersauer, R. Juber and O. Meyer. 2002. Catalysis at a dinuclear [CuSMoOOH] cluster in a CO dehydrogenase resolved at 1.1 Å resolution. Proc. Natl. Acad. Sci. U.S.A. *99*: 15971–15976.

Döbereiner, J. 1991. The genera *Azospirillum* and *Herbaspirillum*. *In* Balows, Trüper, Dworkin, Harder and Schleifer (Editors), The Prokaryotes: A Handbook of Bacteria: Ecophysiology, Isolation, Identification, Applications, 2nd ed., Vol. 3, Springer-Verlag, New York. 2236–2253.

Döbereiner, J. 1992a. The genera *Azospirillum* and *Herbaspirillum*. *In* Balows, Trüper, Dworkin, Harder and Schleifer (Editors), The Prokaryotes: A Handbook of Bacteria: Ecophysiology, Isolation, Identification, Applications, 2nd Ed., Vol. 3, Springer-Verlag, New York. pp. 2236–2253.

Döbereiner, J. 1992b. History and new perspective of diazotrophs in association with non-leguminous plants. Symbiosis *13*: 1–13.

Döbereiner, J. and V.L. Baldani. 1979. Selective infection of maize roots by streptomycin-resistant *Azospirillum lipoferum* and other bacteria. Can. J. Microbiol. *25*: 1264–1269.

Döbereiner, J. and V.L.D. Baldani. 1998. Biological nitrogen fixation by endophytic diazotrophs in non-leguminous crops in the tropics. *In* Malik, Mirza and Ladha (Editors), Nitrogen Fixation with Non-Legumes, Kluwer Academic Publishers, Dordecht. pp. 3–7.

Döbereiner, J., V.L.D. Baldani and J.I. Baldani. 1995. Como isolar e identificar bacterias diazotroficas de plants nao-leguminosas, Embrapa-SPI: Itaguai, RJ: Embrapa-CNPAB, Brasilia.

Döbereiner, J. and J.M. Day. 1976. Associative symbioses in tropical grasses: Characterization of microorganisms and dinitrogen fixing sites. *In* Newton and Nymans (Editors), Symposium on Nitrogen Fixation, Washington State University Press, Pullman, Washington. 518–538.

Döbereiner, J. and F.O. Pedrosa. 1987. Nitrogen-Fixing Bacteria in Non-Leguminous Crop Plants. Brock/Springer Series in Contemporary Bioscience, Springer-Verlag, New York.

Döbereiner, J. and A.P. Ruschel. 1958. Uma nova espécie de *Beijerinckia*. Rev. Biol. *1*: 261–272.

Dobson, S.J. and P.D. Franzmann. 1996. Unification of the genera *Deleya* (Baumann et al. 1983), *Halomonas* (Vreeland et al. 1980), and *Halovibrio* (Fendrich 1988) and the species *Paracoccus halodenitrificans* (Robinson and Gibbons 1952) into a single genus, *Halomonas*, and placement of the genus *Zymobacter* in the family *Halomonadaceae*. Int. J. Syst. Bacteriol. *46*: 550–558.

Dobson, S.J., T.A. McMeekin and P.D. Franzmann. 1993. Phylogenetic relationships between some members of the genera *Deleya*, *Halomonas*, and *Halovibrio*. Int. J. Syst. Bacteriol. *43*: 665–673.

Dockendorff, T.C., A.J. Sharma and G. Stacey. 1994. Identification and characterization of the *nolYZ* genes of *Bradyrhizobium japonicum*. Mol. Plant-Microbe Interact. *7*: 173–180.

Dodatko, T.A., E.A. Kiprianova and V.V. Smirnov. 1989a. Bacteriocin typing of *Pseudomonas cepacia* strains isolated from clinical sources and the rhizosphere of plants. Zh. Mikrobiol. Epidemiol. Immunobiol. *1*: 21–26.

Dodatko, T.A., E.A. Kiprianova and V.V. Smirnov. 1989b. The biological activity and physicochemical properties of a new bacteriocin from a strain of *Pseudomonas cepacia* 5779. Mikrobiol. Zh. *51*: 68–74.

Doelle, H.W., L. Kirk, R. Crittenden, H. Toh and M.B. Doelle. 1993. *Zymomonas mobilis*—science and industrial application. Crit. Rev. Biotechnol. *13*: 57–98.

Doetsch, R.N. 1981. Determinative methods of light microscopy. *In* Gerhardt (Editor), Manual of Methods for General Microbiology, American Society for Microbiology, Washington, D.C. pp. 21–33.

Dohra, H. and M. Fujishima. 1999. Cell structure of the infectious form of *Holospora*, an endonuclear symbiotic bacterium of the ciliate *Paramecium*. Zool. Sci. *16*: 93–98.

Dohra, H., M. Fujishima and K. Hoshide. 1994. Monoclonal antibodies specific for periplasmic materials of the macronuclear specific bacterium *Holospora obtusa* of the ciliate *Paramecium caudatum*. Eur. J. Protistol. *30*: 288–294.

Dohra, H., M. Fujishima and H. Ishikawa. 1998. Structure and expression of a *GroE*-homologous operon of a macronucleus-specific symbiont *Holospora obtusa* of the ciliate *Paramecium caudatum*. J. Eukaryot. Microbiol. *45*: 71–79.

Dohra, H., K. Yamamoto, M. Fujishima and H. Ishikawa. 1997. Cloning and sequencing of gene coding for a periplasmic 5.4 kDa peptide of the macronucleus-specific symbiont *Holospora obtusa* of the ciliate *Paramecium caudatum*. Zool. Sci. *14*: 69–75.

Don, R.H. and J.M. Pemberton. 1981. Properties of six pesticide degradation plasmids isolated from *Alcaligenes paradoxus* and *Alcaligenes eutrophus*. J. Bacteriol. *145*: 681–686.

Donatien, A. and F. Lestaquard. 1936. *Rickettsia bovis* nouvelle espèce pathogène pour le boeuf. Bull. Soc. Pathol. Exot. *29*: 1057–1061.

Donatien, A. and F. Lestoquard. 1935. Existence en Algérie d' une *Rickettsia* du chien. Bull. Soc. Pathol. Exot. *28*: 418–419.

Dondero, N.C. 1975. The *Sphaerotilus-Leptothrix* group. Annu. Rev. Microbiol. *29*: 407–428.

Dong, X., C.M. Plugge and A.J.M. Stams. 1994. Anaerobic degradation of propionate by a mesophilic acetogenic bacterium in coculture and triculture with different methanogens. Appl. Environ. Microbiol. *60*: 2834–2838.

Donze, D., J.A. Mayo and D.L. Diedrich. 1991. Relationships among the Bdellovibrios revealed by partial sequences of 16S ribosomal RNA. Current Microbiology *23*: 115-120.

Dorfer, J., G. Layh, J. Eberspächer and F. Lingens. 1985. Relationships of *Phenylobacterium immobile* and purple nonsulfur bacteria on the basis of surface antigens. FEMS Microbiol. Lett. *28*: 151–155.

Dorff, G.F., L.J. Jackson and M.W. Rytel. 1974. Infections with *Eikenella corrodens*, a newly recognized human pathogen. Ann. Intern. Med. *80*: 305–309.

Dorff, P. 1934. Die Eisenorganismen. Pflanzenforsch. Heft 16. Hrsg. von Kolkwitz, F. Fischer, Jena.

Dorittke, C., P. Vandamme, K.H. Hinz, E.M. Schemken-Birk and C.H. Wirsing von Konig. 1995. Isolation of a *Bordetella avium*-like organism from a human specimen. Eur. J. Clin. Microbiol. Infect. Dis. *14*: 451–454.

Doronina, N.V. 1985. The properties of a new *Hyphomicrobium vulgare* strain. Mikrobiologiya *54*: 538–544.

Doronina, N.V., S.A. Braus-Stromeyer, T. Leisinger and Y.A. Trotsenko. 1995. Isolation and characterization of a new facultatively methylotrophic bacterium: Description of *Methylorhabdus multivorans*, gen. nov., sp. nov. Syst. Appl. Microbiol. *18*: 92–98.

Doronina, N.V., S.A. Braus-Stromeyer, T. Leisinger and Y.A. Trotsenko. 1996a. *In* Validation of the publication of new names and new combinations previously effectively published outside the IJSB. List No. 56. Int. J. Syst. Bacteriol. *46*: 362–363.

Doronina, N.V., N.I. Govorukhina and Y.A. Trotsenko. 1983. *Blastobacter aminooxidans*, a new species of bacteria growing autotrophically on methylated amines. Microbiology *52*: 547–553.

Doronina, N.V., A.P. Sokolov and Y.A. Trotsenko. 1996b. Isolation and initial characterization of aerobic chloromethane-utilizing bacteria. FEMS Microbiol. Lett. *142*: 179–183.

Doronina, N.V. and Y.A. Trotsenko. 1992. Method for storage of methylotrophic and heterotrophic microorganisms. Appl. Biochem. Microbiol. *28*: 476–479.

Doronina, N.V. and Y.A. Trotsenko. 1994. *Methylophilus leisingerii* sp. nov., a new species of restricted facultatively methylotrophic bacteria. Mikrobiologiya *63*: 529–536.

Doronina, N.V. and Y.A. Trotsenko. 2000. A novel plant-associated thermotolerant alkaliphilic methylotroph of the genus *Paracoccus*. Microbiology *69*: 593–598.

Doronina, N.V. and Y.A. Trotsenko. 2001a. *In* Validation of the publication of new names and new combinations previously effectively published outside the IJSB. List No. 82. Int. J. Syst. Evol. Microbiol. *51*: 1619–1620.

Doronina, N.V. and Y.A. Trotsenko. 2001b. *In* Validation of the publication of new names and new combinations previously effectively published outside the IJSEM. List No. 78. Int. J. Syst. Evol. Bacteriol. *51*: 1.

Doronina, N.V. and Y.A. Trotsenko. 2003. Reclassification of 'Blastobacter viscosus' 7d and 'Blastobacter aminooxidans' 14a as *Xanthobacter viscosus* sp. nov. and *Xanthobacter aminoxidans* sp. nov. Int. J. Syst. Evol. Microbiol. *53*: 179–182.

Doronina, N.V., Y.A. Trotsenko, V.I. Krausova, E.S. Boulygina and T.P. Tourova. 1998a. *Methylopila capsulata* gen. nov., sp. nov., a novel non-pigmented aerobic facultatively methylotrophic bacterium. Int. J. Syst. Bacteriol. *48*: 1313–1321.

Doronina, N.V., Y.A. Trotsenko, V.I. Krausova and N.E. Suzina. 1998b. *Paracoccus methylutens* sp. nov. — a new aerobic facultatively methylotrophic bacterium utilizing dichloromethane. Syst. Appl. Microbiol. *21*: 230–236.

Doronina, N.V., Y.A. Trotsenko, V.I. Krausova and N.E. Suzina. 1998c. *In* Validation of the publication of new names and new combinations previously effectively published outside the IJSB. List No. 67. Int. J. Syst. Bacteriol. *48*: 1083–1084.

Doronina, N.V., Y.A. Trotsenko, V.I. Krauzova and N.E. Suzina. 1996c. New methylotrophic isolates of the genus *Xanthobacter*. Mikrobiologiya *65*: 245–253.

Doronina, N.V., Y.A. Trotsenko and T.P. Tourova. 2000a. *Methylarcula marina* gen. nov., sp. nov. and *Methylarcula terricola* sp. nov.: novel aerobic, moderately halophilic, facultatively methylotrophic bacteria from coastal saline environments. Int. J. Syst. Evol. Microbiol. *50*: 1849–1859.

Doronina, N.V., Y.A. Trotsenko, T.P. Tourova, B.B. Kuznetsov and T. Leisinger. 2000b. *Methylopila helvetica* sp. nov. and *Methylobacterium dichloromethanicum* sp. nov.—novel aerobic facultatively methylotrophic bacteria utilizing dichloromethane. Syst. Appl. Microbiol. *23*: 210–218.

Doronina, N.V., Y.A. Trotsenko, T.P. Tourova, B.B. Kuznetsov and T. Leisinger. 2000c. *In* Validation of the publication of new names and new combinations previously effectively published outside the IJSB. List No. 77. Int. J. Syst. Evol. Microbiol. *50*: 1953.

Doronina, N.V., Y.A. Trotsenko, T.P. Tourova, B.B. Kuznetsov and T. Leisinger. 2001. *Albibacter methylovorans* gen. nov., sp. nov., a novel aerobic,

facultatively autotrophic and methylotrophic bacterium that utilizes dichloromethane. Int. J. Syst. Evol. Microbiol. *51*: 1051–1058.

Dörr, J., T. Hurek and B. Reinhold-Hurek. 1998. Type IV pili are involved in plant-microbe and fungus-microbe interactions. Mol. Microbiol. *30*: 7–17.

Dorsch, M., E. Moreno and E. Stackebrandt. 1989. Nucleotide sequence of the 16S rRNA from *Brucella abortus*. Nucleic Acids Res. *17*: 1765.

Doudoroff, M. and N.J. Palleroni. 1974. Genus I. *Pseudomonas. In* Buchanan and Gibbons (Editors), Bergey's Manual of Determinative Bacteriology, 8th Ed., The Williams & Wilkins Co., Baltimore. pp. 217–243.

Doudoroff, M. and R.Y. Stanier. 1959. Role of poly-β-hydroxybutyric acid in the assimilation of organic carbon by bacteria. Nature *183*: 1440–1442.

Douglas, J.T. and S.S. Elberg. 1976. Isolation of *Brucella melitensis* phage of broad biotype and species specificity. Infect. Immun. *14*: 306–308.

Douglas, J.T. and S.S. Elberg. 1978. Properties of the berkeley phage Iytic for *Brucella melitensis* and other species. Ann. Sclavo. *20*: 681–691.

Douglas, J.T., E.Y. Rosenberg, H. Nikaido, D.R. Verstreate and A.J. Winter. 1984. Porins of *Brucella* species. Infect. Immun. *44*: 16–21.

Douglas, S.L., M.K. Lee and H. Nikaido. 1981. Protein I of *Neisseria gonorrhoeae* is a porin. FEMS Microbiol. Lett. *12*: 305–309.

Dowdle, S.F. and B.B. Bohlool. 1985. Predominance of fast-growing *Rhizobium japonicum* in a soybean field in the People's Republic of China. Appl. Environ. Microbiol. *50*: 1171–1176.

Dowling, N.J.E., F. Widdel and D.C. White. 1986. Phospholipid ester-linked fatty acid biomarkers of acetate-oxidizing sulfate-reducers and other sulfide-forming bacteria. J. Gen. Microbiol. *132*: 1815–1825.

Downes, J., J.I. Mangels, J. Holden, M.J. Ferraro and E.J. Baron. 1990. Evaluation of two single-plate incubation systems and the anaerobic chamber for the cultivation of anaerobic bacteria. J. Clin. Microbiol. *28*: 246–248.

Downs, J. and D.E.F. Harrison. 1974. Studies on the production of pink pigment in *Pseudomonas extorquens* NCIB 9399 growing in continuous culture. J. Appl. Bacteriol. *37*: 65–74.

Doyle, L.P. 1948. The etiology of swine dysentery. Am. J. Vet. Res. *9*: 50–51.

Drake, S.L. and M. Koomey. 1995. The product of the pilQ gene is essential for the biogenesis of type IV pili in *Neisseria gonorrhoeae*. Mol. Microbiol. *18*: 975–986.

Drake, S.L., S.A. Sandstedt and M. Koomey. 1997. PilP, a pilus biogenesis lipoprotein in *Neisseria gonorrhoeae*, affects expression of PilQ as a high-molecular-mass multimer. Mol. Microbiol. *23*: 657–668.

Drescher, N. and S. Otto. 1969. Über den abbau von 1-phenyl-4-amino-5-chloro-pyridazon-6 (pyrazon) im boden. Z Pflanzenkr. Pflanzenschutz. *76*: 27–33.

Drewlo, S., C.O. Bramer, M. Madkour, F. Mayer and A. Steinbuchel. 2001. Cloning and expression of a *Ralstonia eutropha* HF39 gene mediating indigo formation in *Escherichia coli*. Appl. Environ. Microbiol. *67*: 1964–1969.

Drews, G. 1981. *Rhodospirillum salexigens* sp. nov., an obligatory halophilic phototrophic bacterium. Arch. Microbiol. *130*: 325–327.

Drews, G. 1982. *In* Validation of the publication of new names and new combinations previously effectively published outside the IJSB. List No. 9. Int. J. Syst. Bacteriol. *32*: 384–385.

Drews, G. 1983. Mikrobiologisches Praktikum (4. Aufl.), Springer-Verlag, Heidelberg.

Drews, G. and P. Giesbrecht. 1966. *Rhodopseudomonas viridis*, n. sp., a newly isolated, obligate phototrophic bacterium. Arch. Mikrobiol. *53*: 255–262.

Dreyfus, B.L. and Y.R. Dommergues. 1981. Nodulation of Acacia species by fast- and slow-growing tropical strains of *Rhizobium*. Appl. Environ. Microbiol. *41*: 97–99.

Dreyfus, B.L., C. Elmerich and Y.R. Dommergues. 1983. Free-living *Rhizobium* strains able to grow on N₂ as the sole nitrogen source. Appl. Environ. Microbiol. *45*: 711–713.

Dreyfus, B., J.L. Garcia and M. Gillis. 1988. Characterization of *Azorhizobium caulinodans* gen. nov., sp. nov., a stem-nodulating nitrogen-fixing bacterium isolated from *Sesbania rostrata*. Int. J. Syst. Bacteriol. *38*: 89–98.

Drobner, E., H. Huber, R. Rachel and K.O. Stetter. 1992. *Thiobacillus plumbophilus*, sp. nov., a novel galena and hydrogen oxidizer. Arch. Microbiol. *157*: 213–217.

Dross, F., V. Geisler, R. Lenger, F. Theis, T. Krafft, F. Fahrenholz, E. Kojro, A. Duchêne, D. Tripier, K. Juvenal and A. Kröger. 1992. The quinone-reactive Ni/Fe-hydrogenase of *Wolinella succinogenes*. Eur. J. Biochem. *206*: 93–102.

Droz, S., B. Chi, E. Horn, A.G. Steigerwalt, A.M. Whitney and D.J. Brenner. 1999. *Bartonella koehlerae* sp. nov., isolated from cats. J. Clin. Microbiol. *37*: 1117–1122.

Droz, S., B. Chi, E. Horn, A.G. Steigerwalt, A.M. Whitney and D.J. Brenner. 2000. *In* Validation of the publication of new names and new combinations previously effectively published outside the IJSB. List No. 73. Int. J. Syst. Evol. Microbiol. *50*: 423–424.

Drzyzga, O., A. Schmidt and K.-H. Blotevogel. 1996. Cometabolic transformation and cleavage of nitrodiphenylamines by three newly isolated sulfate-reducing bacterial strains. Appl. Environ. Microbiol. *62*: 1710–1716.

D'Souza-Ault, M.R., M.B. Cooley and C.I. Kado. 1993. Analysis of the Ros repressor of *Agrobacterium virC* and *virD* operons - molecular intercommunication between plasmid and chromosomal genes. J. Bacteriol. *175*: 3486–3490.

Du Plessis, L., F. Reyers and K. Stevens. 1997. Morphological evidence for infection of impala, *Aepyceros melampus*, platelets by a rickettsia-like organism. Onderstepoort J. Vet. Res. *64*: 317–318.

Dubinina, G.A. 1978a. Functional significance of bivalent iron and manganese oxidation in *Leptothrix pseudoochraceae*. Mikrobiologiya *47*: 783–789.

Dubinina, G.A. 1978b. Mechanism of oxidation of bivalent iron and manganese by iron bacteria growing at neutral acidity of the medium. Mikrobiologiya *47*: 591–599.

Dubinina, G.A., V.V. Churikova, M. Yu Grabovich, N.A. Chernykh, M.V. Raichenstein and N.E. Petukhova. 1989. Isolation and characteristics of *Aquaspirillum denitrificans* sp. nov. Mikrobiologiya *58*: 824–829.

Dubinina, G.A. and M.Y. Grabovich. 1984. Isolation, cultivation and characteristics of *Macromonas bipunctata*. Mikrobiologiya *53*: 748–755.

Dubinina, G.A. and M.Y. Grabovich. 1989. *In* Validation of the publication of new names and new combinations previously effectively published outside the IJSB. List No. 31. Int. J. Syst. Bacteriol. *39*: 495–497.

Dubinina, G.A., M.Y. Grabovich, A.M. Lysenko, N.A. Chernykh and V.V. Churikova. 1993. Revision of taxonomic position of colorless sulfur spirilla of the genus *Thiospira* and description of a new species *Aquaspirillum bipunctata* comb. nov. Microbiology *62*: 368–644.

DuBow, M.S. and T. Ryan. 1977. Host Factor for coliphage QbRNA replication as an aid in elucidating phylogenetic relationships: the genus *Pseudomonas*. J. Gen. Microbiol. *102*: 263–268.

Dubray, G. 1972. Étude ultrastructurale des bactéries de colonies lisses (S) et rugueuses (R) du genre *Brucella*. Ann. Inst. Pasteur (Paris) *123*: 171–193.

Dubray, G. 1976. Localization cellulaire des polyosides des bactéries des genres *Brucella* et *Escherichia* en phase lisse (S) ou rugueuse (R). Ann. Microbiol. (Paris) *1276*: 133–140.

Dubray, G. and J.N. Limet. 1987. Evidence of heterogeneity of lipopolysaccharides among *Brucella* biovars in relation to A and M specificities. Ann. Inst. Pasteur (Paris) *138*: 84–87.

Dubray, G. and M. Plommet. 1976. Structure et constituents des *Brucella*. Characterization des fractions et propriétés biologiques. Develop. Biol. Stand. *31*: 68–91.

Duchars, M.G. and M.M. Attwood. 1987. NADP$^+$ dependent glutamate-dehydrogenase from the facultative methylotroph *Hyphomicrobium* X. FEMS Microbiol. Lett. *48*: 133–137.

Duchars, M.G. and M.M. Attwood. 1989. The influence of carbon:nitrogen ratio in the growth medium on the cellular composition and regulation of enzyme activity in *Hyphomicrobium* X. J. Gen. Microbiol. *135*: 787–794.

Duchars, M.G. and M.M. Attwood. 1991. Purification, localization, prop-

erties and regulation of glutamine synthetase from *Hyphomicrobium* X. J. Gen. Microbiol. *137*: 1345–1354.

Duchow, E. and H.C. Douglas. 1949. *Rhodomicrobium vannielii*, a new photoheterotrophic bacterium. J. Bacteriol. *58*: 409–416.

Dudley, J.P., E.J.C. Goldstein, W.L. George, B.V. Bock, B.D. Kirby and S.M. Finegold. 1978. Sinus infection due to *Eikenella corrodens*. Arch. Otolaryngol. *104*: 462–463.

Dudman, W.F. 1976. The extracellular polysaccharides of *Rhizobium japonicum*: compositional studies. Carbohydr. Res. *46*: 97–110.

Dudman, W.F. 1978. Structural studies of the extracellular polysaccharides of *Rhizobium japonicum* strain 71A, CC708 and CB1795. Carbohydr. Res. *66*: 9–23.

Duerden, B.I., A. Eley, L. Goodwin, J.T. Magee, J.M. Hindmarch and K.W. Bennett. 1989. A comparison of *Bacteroides ureolyticus* isolates from different clinical sources. J. Med. Microbiol. *29*: 63–73.

Duetz, W.A., C. Dejong, P.A. Williams and J.G. Vanandel. 1994. Competition in chemostat culture between *Pseudomonas* strains that use different pathways for the degradation of toluene. Appl. Environ. Microbiol. *60*: 2858–2863.

Dufresne, C. and E. Farnworth. 2000. Tea, Kombucha, and health: a review. Food Res. Int. *33*: 409–421.

Dugan, P.R. and D.G. Lundgren. 1960. Isolation of the floc-forming organism *Zoogloea ramigera* and its culture in complex and synthetic media. Appl. Microbiol. *8*: 357–361.

Duine, J.A. 1990. NAD-linked, factor-independent, and glutathione-independent aldehyde dehydrogenase from *Hyphomicrobium* X. Method Enzymol. *188*: 327–330.

Duine, J.A. and J. Frank. 1979. Purification and properties of methanol dehydrogenase from *Hyphomicrobium* X. Antonie Van Leeuwenhoek *45*: 151–152.

Duine, J.A. and J. Frank, Jr.. 1980a. The prosthetic group of methanol dehydrogenase: purification and some of its properties. Biochem. J. *187*: 221–226.

Duine, J.A. and J. Frank, Jr.. 1980b. Studies on methanol dehydrogenase from *Hyphomicrobium* X: Isolation of an oxidized form of the enzyme. Biochem. J. *187*: 213–220.

Duine, J.A. and J. Frank. 1990. The role of PQQ and quinoproteins in methylotrophic bacteria. FEMS Microbiol. Rev. *87*: 221–225.

Duine, J.A., J. Frank and P.E.J. Verwiel. 1980. Structure and activity of the prosthetic group of methanol dehydrogenase. Eur. J. Biochem. *108*: 187–192.

Duine, J.A., J.J. Frank and P.E.J. Verwiel. 1981. Characterization of 2nd prosthetic group in methanol dehydrogenase from *Hyphomicrobium* strain X. Eur. J. Biochem. *118*: 395–400.

Duine, J.A., J. Frank and J. Westerling. 1978. Purification and properties of methanol dehydrogenase from *Hyphomicrobium* X. Biochim. Biophys. Acta *524*: 277–287.

Duma, R.J., D.E. Sonenshine, F.M. Bozeman, J. Veazey, M. Jr., B.L. Elisberg, D.P. Chadwick, N.I. Stocks, T.M. McGill, G.B. Miller and J.N. MacCormack. 1981. Epidemic typhus in the United States associated with flying squirrels. J. Amer. Med. Assoc. *245*: 2318–2323.

Dumler, J.S., K.M. Asanovich, J.S. Bakken, P. Richter, R. Kimsey and J.E. Madigan. 1995. Serologic cross-reactions among *Ehrlichia equi*, *Ehrlichia phagocytophila*, and human granulocytic *Ehrlichia*. J. Clin. Microbiol. *33*: 1098–1103.

Dumler, J.S., A.F. Barbet, C.P.J. Bekker, G.A. Dasch, G.H. Palmer, S.C. Ray, Y. Rikihisa and F.R. Rurangirwa. 2001. Reorganization of genera in the families *Rickettsiaceae* and *Anaplasmataceae* in the order *Rickettsiales*: Unification of some species of *Ehrlichia* with *Anaplasma*, *Cowdria* with *Ehrlichia* and *Ehrlichia* with *Neorickettsia*, descriptions of six new species combinations and designation of *Ehrlichia equi* and 'HGE agent' as subjective synonyms of *Ehrlichia phagocytophila*. Int. J. Syst. Evol. Microbiol. *51*: 2145–2165.

Dumont, L., B. Verneuil, R. Julien and J. Wallach. 1994. Elastolytic activity of Map1, a protease from *Myxococcus xanthus*. Biochem. Mol. Biol. Int. *33*: 535–542.

Dunbar, J.M., I. Zlatkin and L.J. Forney. 1995. Variation in genome organization among *Variovorax paradoxus* clones isolated from soil. *In*

Center for Microbial Ecology, an NSF Science and Technology Center, Research Findings, Center for Microbial Ecology, Michigan State University, East Lansing.

Duncan, M.J. 1979. L-arabinose metabolism in *Rhizobium*. J. Gen. Microbiol. *113*: 177–179.

Dunn, B.E., H. Cohen and M.J. Blaser. 1997. *Helicobacter pylori*. Clin. Microbiol. Rev. *10*: 720–741.

Dunne, W.M., Jr. and S. Maisch. 1995. Epidemiological investigation of infections due to *Alcaligenes* species in children and patients with cystic fibrosis: use of repetitive-element-sequence polymerase chain reaction. Clin. Infect. Dis. *20*: 836–841.

Dupuy, N., A. Willems, B. Pot, D. Dewettinck, I. Vandenbruaene, G. Maestrojuan, B. Dreyfus, K. Kersters, M.D. Collins and M. Gillis. 1994. Phenotypic and genotypic characterization of bradyrhizobia nodulating the leguminous tree *Acacia albida*. Int. J. Syst. Bacteriol. *44*: 461–473.

Duran, N., R.V. Antonio, M. Haun and R.A. Pilli. 1994. Biosynthesis of a trypanocide by *Chromobacterium violaceum*. World J. Microbiol. Biotechnol. *10*: 686–690.

Duran, N. and C.F. Menck. 2001. *Chromobacterium violaceum*: a review of pharmacological and industiral perspectives. Crit. Rev. Microbiol. *27*: 201–222.

Dutka, B.J., N. Nyholm and J. Petersen. 1983. Comparison of several microbiological toxicity screening tests. Water Res. *17*: 1363–1368.

Dutta, S.K., A.C. Myrup, R.M. Rice, M.G. Robl and R.C. Hammond. 1985. Experimental reproduction of Potomac horse fever in horses with a newly isolated *Ehrlichia* organism. J. Clin. Microbiol. *22*: 265–269.

Dutta, S.K., R. Vemulapalli and B. Biswas. 1998. Association of deficiency in antibody response to vaccine and heterogeneity of *Ehrlichia risticii* strains with Potomac horse fever vaccine failure in horses. J. Clin. Microbiol. *36*: 506–512.

Dutton, P.L. and W.C. Evans. 1969. The metabolism of aromatic compounds by *Rhodopseudomonas palustris*. Biochem. J. *113*: 525–536.

Dutton, P.L. and W.C. Evans. 1978. The metabolism of aromatic compounds by *Rhodospirillaceae*. *In* Clayton and Sistrom (Editors), The Photosynthetic Bacteria, Plenum Press, New York. 719–726.

Dworkin, M. 1962. Nutritional requirements for vegetative growth of *Myxococcus xanthus*. J. Bacteriol. *84*: 250–257.

Dworkin, M. 1966. Biology of the myxobacteria. Annu. Rev. Microbiol. *20*: 75–106.

Dworkin, M. 1969. Sensitivity of gliding bacteria to actinomycin D. J. Bacteriol. *98*: 851–852.

Dworkin, M. 1983. Tactic behavior of *Myxococcus xanthus*. J. Bacteriol. *154*: 452–459.

Dworkin, M. 1996. Recent advances in the social and developmental biology of the myxobacteria. Microbiol. Rev. *60*: 70–102.

Dworkin, M. 1999. Fibrils as extracellular appendages of bacteria: their role in contact mediated cell–cell interactions in *Myxococcus xanthus*. Bioessays *21*: 590–595.

Dworkin, M. and D. Eide. 1983. *Myxococcus xanthus* does not respond chemotactically to moderate concentration gradients. J. Bacteriol. *154*: 437–442.

Dworkin, M. and S.M. Gibson. 1964. A system for studying microbial morphogenesis: rapid formation of microcysts in *Myxococcus xanthus*. Science (Wash. D. C.) *146*: 243–244.

Dworkin, M. and D. Kaiser. 1993. Myxobacteria II, American Society for Microbiology, Washington, D.C.

Dworkin, M. and D.J. Niederpruem. 1964. Electron transport system in vegetative cells and microcysts of *Myxococcus xanthus*. J. Bacteriol. *87*: 316–322.

Dye, D.W., J.F. Bradbury, M. Goto, A.C. Hayward, R.A. Lelliott and M.N. Schroth. 1980. International standards for naming pathovars of phytopathogenic bacteria and a list of pathovar names and pathotype strains. Rev. Plant Pathol. *59*: 153–168.

Eadie, J.M. 1962. The development of rumen microbial populations in lambs and calves under various conditions of management. J. Gen. Microbiol. *29*: 563–578.

Eaglesome, M.D., M.I. Sampath and M.M. Garcia. 1995. A detection assay for *Campylobacter fetus* in bovine semen by restriction analysis of PCR amplified DNA. Vet. Res. Commun. *19*: 253–263.

Eardly, B.D., L.A. Materon, N.H. Smith, D.A. Johnson, M.D. Rumbaugh and R.K. Selander. 1990. Genetic structure of natural populations of the nitrogen-fixing bacterium *Rhizobium meliloti*. Appl. Environ. Microbiol. *56*: 187–194.

Eardly, B.D., F.S. Wang and P. vanBerkum. 1996. Corresponding 16S rRNA gene segments in *Rhizobiaceae* and *Aeromonas* yield discordant phylogenies. Plant Soil *186*: 69–74.

Eaton, K.A., C.L. Brooks, D.R. Morgan and S. Krakowka. 1991a. Essential role of urease in pathogenesis of gastritis induced by *Helicobacter pylori* in gnotobiotic piglets. Infect. Immun. *59*: 2470–2475.

Eaton, K.A., F.E. Dewhirst, B.J. Paster, N. Tzellas, B.E. Coleman, J. Paola and R. Sherding. 1996a. Prevalence and varieties of *Helicobacter* species in dogs from random sources and pet dogs, animal and public health implications. J. Clin. Microbiol. *34*: 3165–3170.

Eaton, K.A., F.E. Dewhirst, M.J. Radin, J.G. Fox, B.J. Paster, S. Krakowka and D.R. Morgan. 1993. *Helicobacter acinonyx* sp. nov. isolated from cheetahs with gastritis. Int. J. Syst. Bacteriol. *43*: 99–106.

Eaton, K.A., D.R. Morgan and S. Krakowka. 1992. Motility as a factor in the colonisation of gnotobiotic piglets by *Helicobacter pylori*. J. Med. Microbiol. *37*: 123–127.

Eaton, K.A., M.J. Radin, L. Kramer, R. Wack, R. Sherding, S. Krakowka and D.R. Morgan. 1991b. Gastric spiral bacilli in captive cheetahs. Scand. J. Gastroenterol. Suppl. *26*: 38–42.

Eaton, S.L., S.M. Resnick and D.T. Gibson. 1996b. Initial reactions in the oxidation of 1,2-dihydronaphthalene by *Sphingomonas yanoikuyae* strains. Appl. Environ. Microbiol. *62*: 4388–4394.

Eberhardt, U. 1971. The cell wall as the site of carotenoid in the "Knallgas" bacterium 12/60/x. Arch. Mikrobiol. *80*: 32–37.

Ebisu, S., H. Nakae and H. Okada. 1988. Coaggregation of *Eikenella corrodens* with oral bacteria mediated by bacterial lectin-like substance. Adv. Dent. Res. *2*: 323–327.

Ebisu, S. and H. Okada. 1983. Agglutination of human erythrocytes by *Eikenella corrodens*. FEMS Microbiol. Lett. *18*: 153–156.

Eckau, H., D. Dill and H. Budzikiewicz. 1984. Neuartige ceramide aus Cystobacter fuscus (Myxobacterales). Z. Naturforsch. *39c*: 1–9.

Eckersley, K. and C.S. Dow. 1980. *Rhodopseudomonas blastica* sp. nov.—a member of the *Rhodospirillaceae*. J. Gen. Microbiol. *119*: 465–473.

Eckersley, K. and C.S. Dow. 1981. *In* Validation of the publication of new names and new combinations previously effectively published outside the IJSB. List No. 6. Int. J. Syst. Bacteriol. *31*: 216.

Eckert, B., O.B. Weber, G. Kirchhof, A. Halbritter, M. Stoffels and A. Hartmann. 2001. *Azospirillum doebereinerae* sp. nov., a nitrogen-fixing bacterium associated with the C4-grass *Miscanthus*. Int. J. Syst. Evol. Bacteriol. *51*: 17–26.

Eckhardt, F.E.W., P. Roggentin and P. Hirsch. 1979. Fatty-acid composition of various hyphal budding bacteria. Arch. Microbiol. *120*: 81–85.

Eckhardt, M.M., I.R. Baldwin and E.B. Fred. 1931. Studies on the root nodule bacteria of *Lupinus*. J. Bacteriol. *21*: 273–285.

Economou, A., W.D. Hamilton, A.W. Johnston and J.A. Downie. 1990. The *Rhizobium* nodulation gene *nodO* encodes a Ca^{2+}-binding protein that is exported without N-terminal cleavage and is homologous to haemolysin and related proteins. Eur. Mol. Biol. Organ. J. *9*: 349–354.

Edelman, D.C. and J.S. Dumler. 1996. Evaluation of an improved PCR diagnostic assay for human granulocytic ehrlichiosis. Mol. Diagn. *1*: 41–49.

Eden, P.A. and R.P. Blakemore. 1991. Electroporation and conjugal plasmid transfer to members of the genus *Aquaspirillum*. Arch. Microbiol. *155*: 449–452.

Ederer, M.M., R.L. Crawford, R.P. Herwig and C.S. Orser. 1997. PCP degradation is mediated by closely related strains of the genus *Sphingomonas*. Mol. Ecol. *6*: 39–49.

Edwards, C.E. and R. Kraus. 1960. *Spirillum serpens* meningitis: Report of a case. N. Engl. J. Med. *262*: 458–460.

Egelhoff, T.T., R.F. Fisher, T.W. Jacobs, J.T. Mulligan and S.R. Long. 1985.

Nucleotide sequence of *Rhizobium meliloti* 1021 nodulation genes: *nodD* is read divergently from *nodABC*. DNA. *4*: 241–248.

Egelhoff, T.T. and S.R. Long. 1985. *Rhizobium meliloti* nodulation genes: identification of *nodDABC* gene products, purification of *nodA* protein, and expression of *nodA* in *Rhizobium meliloti*. J. Bacteriol. *164*: 591–599.

Egener, T., T. Hurek and B. Reinhold-Hurek. 1999. Endophytic expression of *nif* genes of *Azoarcus* sp. strain BH72 in rice roots. Mol. Plant-Microbe Interact. *12*: 813–819.

Egener, T., D.E. Martin, A. Sarkar and B. Reinhold-Hurek. 2001. Role of a ferredoxin gene cotranscribed with the *nifHDK* operon in N₂ fixation and nitrogenase "switch-off" of *Azoarcus* sp. strain BH72. J. Bacteriol. *183*: 3752–3760.

Egert, M., A. Hamann, R. Kömen and C.G. Friedrich. 1993. Methanol and methylamine utilization result from mutational events in *Thiosphaera pantotropha*. Arch. Microbiol. *159*: 364–371.

Egli, T. 1988. (An)aerobic breakdown of chelating agents used in household detergents. Microbiol. Sci. *5*: 36–41.

Egli, T. 1994. Biochemistry and physiology of the degradation of nitrilotriacetic acid and other metal complexing agents. *In* Ratledge (Editor), Biochemistry of Microbial Degradation, Kluwer Academic Publishers, Dordrecht. pp. 179–195.

Egli, T. 1995. The ecological and physiological significance of the growth of heterotrophic microorganisms with mixtures of substrates. Adv. Microb. Ecol. *14*: 305–386.

Egli, T., M. Bally and T. Uetz. 1990. Microbial degradation of chelating agents used in detergents with special reference to nitrilotriacetic acid (NTA). Biodegradation *1*: 121–132.

Egli, T. and H.-U. Weilenmann. 1989. Isolation, characterization, and physiology of bacteria able to degrade nitrilotriacetate. Toxic. Assess. *4*: 23–34.

Egli, T., H.-U. Weilenmann, T. El-Banna and G. Auling. 1988. Gram-negative, aerobic, nitrilotriacetate-utilizing bacteria from wastewater and soil. Syst. Appl. Microbiol. *10*: 297–305.

Egyed, Z., T. Sreter, Z. Szell, G. Nyiro, K. Marialigeti and I. Varga. 2002. Molecular phylogenetic analysis of *Onchocerca lupi* and its *Wolbachia* endosymbiont. Vet. Parasitol. *108*: 153–161.

Ehrenberg, C.G. 1832. Beiträge zür Kenntnis der Organization der Infusorien und ihrer geographischen Verbreitung, besonders in Sibirien, Abh. Konig Akad. Wiss., 1830, Berlin. pp. 88.

Ehrenberg, C.G. 1836. Vorlaufige Mittheilung uber das wirkliche Vorkommen fossiler Infusorien und ihre grosse Verbreitung. Ann. Phys. Chem. *38*: 213–227.

Ehrenberg, C.G. 1838. Die Infusionthierchen als vollkommene Organismen: ein Blick in das tiefere organische Leben der Natur, L. Voss, Leipzig. pp. i–xvii; 1–547.

Ehrenreich, P. 1996. Anaerobes Wachstum neuartiger sulfatreduzierender und nitarreduzierender Bakterien auf n-Alkanen und Erdöl, University of Bremen. Shaker Verlag, Aachen.

Ehrich, S., D. Behrens, E. Lebedeva, W. Ludwig and E. Bock. 1995. A new obligately chemolithoautotrophic, nitrite-oxidizing bacterium, *Nitrospira moscoviensis* sp. nov. and its phylogenetic relationship. Arch. Microbiol. *164*: 16–23.

Eidels, L., P.L. Edelmann and J. Preiss. 1970. Biosynthesis of bacterial glycogen. VIII. Activation and inhibition of the adenosine diphosphoglucose pyrophosphorylases of *Rhodopseudomonas capsulata* and of *Agrobacterium tumefaciens*. Arch. Biochem. Biophys. *140*: 60–74.

Eikelboom, D.H. 1975. Filamentous organisms observed in activated sludge. Water Res. *9*: 365–388.

Eiken, M. 1958. Studies on an anaerobic rod-shaped Gram-negative microorganism: *Bacteroides corrodens* N. sp. Acta Pathol. Microbiol. Scand. *43*: 404–416.

Einsle, O., A. Messerschmidt, P. Stach, G.P. Bourenkov, H.D. Bartunik, R. Huber and P.M.H. Kroneck. 1999. Structure of cytochrome *c* nitrite reductase. Nature *400*: 476–480.

Eisenberg, J. 1891. Bacteriologische Diagnostik Hiflstabellen zum Gebrauche beim Praktischen Arbeiten 3 Aufl, Leopold Voss, Hamburg.

Eisenstein, B.I., T. Sox, G. Biswas, E. Blackman and P.F. Sparling. 1977.

Conjugal transfer of the gonococcal penicillinase plasmid. Science *195*: 998–1000.

Ekendahl, S., J. Arlinger, F. Ståhl and K. Pedersen. 1994. Characterization of attached bacterial populations in deep granitic groundwater from the Stripa research mine by 16S ribosomal RNA gene sequencing and scanning electron microscopy. Microbiology (Reading) *140*: 1575–1583.

Eksztejn, J. and M. Varon. 1977. Elongation and cell division in *Bdellovibrio bacteriovorus*. Arch. Microbiol. *114*: 175–181.

El Fantroussi, S., J. Mahillon, H. Naveau and S.N. Agathos. 1997. Introduction of anaerobic dechlorinating bacteria into soil slurry microcosms and nested PCR monitoring. Appl. Environ. Microbiol. *63*: 806–811.

El Fantroussi, S., H. Naveau and S.N. Agathos. 1998. Anaerobic dechlorinating bacteria. Biotechnol. Prog. *14*: 167–188.

El-Banna, T. 1989. Characterization of some unclassified *Pseudomonas* species, Thesis, University of Hannover, Germany University of Tanta, Egypt, Hannover Tanta.

El-Zaatari, F.A.K., J.S. Woo, A. Badr, M.S. Osato, H. Serna, L.M. Lichtenberger, R.M. Genta and D.Y. Graham. 1997. Failure to isolate *Helicobacter pylori* from stray cats indicates that *H. pylori* in cats may be an anthroponosis: an animal infection with a human pathogen. J. Med. Microbiol. *46*: 372–376.

Elander, R.P., J.A. Mabe, R.H. Hamill and M. Gorman. 1968. Metabolism of tryptophans by *Pseudomonas aureofaciens*. VI. Production of pyrrolnitrin by selected *Pseudomonas* species. Appl. Microbiol. *16*: 753–758.

Eldering, G. and P.L. Kendrick. 1938. *Bacillus parapertussis*: A species resembling both *Bacillus pertussis* and *Bacillus bronchisepticus* but identical to neither. J. Bacteriol. *35*: 561–572.

Elkan, G.H. and J.R. Kuykendall. 1982. Carbohydrate metabolism in *Rhizobium*. *In* Broughton (Editor), Nitrogen Fixation II: Biology of the Nitrogen Fixing Organisms, Oxford University Press, Oxford, UK. 147–167.

Elkan, G.H. and I. Kwik. 1968. Nitrogen, energy and vitamin nutrition of *Rhizobium japonicum*. J. Appl. Bacteriol. *31*: 399–404.

Ellis, J.G., A. Kerr, M. Van Montagu and J. Schell. 1979. *Agrobacterium*: genetic studies on agrocin 84 production and the biological control of crown gall. Physiol. Plant Pathol. *15*: 311–319.

Ellis, W.A., S.D. Neill, J.J. O'Brien, H.W. Ferguson and J. Hanna. 1977. Isolation of *Spirillum/Vibrio*-like organisms from bovine fetuses. Vet. Rec. *100*: 451–452.

Elmerich, C. 1983. *Azospirillum* genetics. *In* Puhler (Editor), Molecular Genetics of the Bacteria–Plant Interaction, Springer-Verlag, Berlin. 367–372.

Elmerich, C., M. De Zamaroczy, F. Arsene, L. Pereg, A. Paquelin and A. Kaminski. 1997. Regulation of *nif* gene expression and nitrogen metabolism in *Azospirillum*. Soil Biol. Biochem. *29*: 847–852.

Elmerich, C., B. Quiviger, C. Rosenberg, C. Franche, P. Laurent and J. Dobereiner. 1982. Characterization of a temperate bacteriophage for *Azospirillum*. Virology *122*: 29–37.

Elmerich, C., W. Zimmer and C. Vieille. 1991. Associative nitrogen fixing bcteria. *In* Stacey, Evans and Burris (Editors), Biological Nitrogen Fixation, Chapman and Hill, New York. 212–258.

ElRayes, E.G., I.M. Banat and I.Y. Hamdan. 1991. Methanol metabolism and ammonia assimilation in four *Methylophilus* strains. Acta Biotechnol. *11*: 87–93.

Ely, B., A.B.C. Amarasinghe and R.A. Bender. 1978. Ammonia assimilation and glutamate formation in *Caulobacter crescentus*. J. Bacteriol. *133*: 225–230.

Embuscado, M.E., J.N. BeMiller and J.S. Marks. 1996. Isolation and partial characterization of cellulose produced by *Acetobacter xylinum*. Food Hydrocolloid. *10*: 75–82.

Embuscado, M.E., J.S. Marks and J.N. Bemiller. 1994. Bacterial cellulose. I. Factors affecting the production of cellulose by *Acetobacter xylinum*. Food Hydrocolloid. *8*: 407–418.

Emerson, D. and W.C. Ghiorse. 1993a. Role of disulfide bonds in maintaining the structural integrity of the sheath of *Leptothrix discophora* SP-6. J. Bacteriol. *175*: 7819–7827.

Emerson, D. and W.C. Ghiorse. 1993b. Ultrastructure and chemical composition of the sheath of *Leptothrix discophora* SP-6. J. Bacteriol. *175*: 7808–7818.

Enatsu, T., H. Urakami and A. Tamura. 1999. Phylogenetic analysis of *Orientia tsutsugamushi* strains based on the sequence homologies of 56-kDa type-specific antigen genes. FEMS Microbiol. Lett. *180*: 163–169.

Endtz, H.P., J.S. Vliegenthart, P. Vandamme, H.W. Weverink, N.P. van den Braak, H.A. Verbrugh and A. van Belkum. 1997. Genotypic diversity of *Campylobacter lari* isolated from mussels and oysters in The Netherlands. Int. J. Food Microbiol. *34*: 79–88.

Enfors, S.-O. and N. Molin. 1973a. Biodegradation of nitrilotriacetate (NTA) by bacteria. I. Isolation of bacteria able to grow anaerobically with NTA as a sole carbon source. Water Res. *7*: 881–888.

Enfors, S.-O. and N. Molin. 1973b. Biodegradation of nitrilotriacetate (NTA) by bacteria. II. Cultivation of an NTA-degrading bacterium in anaerobic medium. Water Res. *7*: 889–893.

Engelhard, M., T. Hurek and B. Reinhold-Hurek. 2000. Preferential occurrence of diazotrophic endophytes, *Azoarcus* spp., in wild rice species and land races of *Oryza sativa* in comparison with modern races. Environ. Microbiol. *2*: 131–141.

Engelhardt, H. and J.H. Klemme. 1978. Characterization of an allosteric, nucleotide-unspecific glutamate dehydrogenase from *Rhodopseudomonas sphaeroides*. FEMS Microbiol. Lett. *3*: 287–290.

English, C.K., D.J. Wear, A.M. Margileth, C.R. Lissner and G.P. Walsh. 1988. Cat-scratch disease: isolation and culture of the bacterial agent. J. Med. Am. Soc. *259*: 1347–1352.

Engval, A., E. Olsson, N. Bleumink-Pluym, E.A. ter Laak and B.A.M. van der Zeijst. 1991. Epidemiology and control of contagious equine metritis in Sweden. *In* Equine Infectious Diseases VI: Proceedings of the Sixth International Conference, University Press of Kentucky, Lexington, KY. pp. 89–93.

Engvild, K.C. and H.L. Jensen. 1969. Microbial decomposition of the herbicide Pyrazon. Soil Biol. Biochem. *1*: 295–300.

Enright, M.C., P.E. Carter, I.A. MacLean and H. McKenzie. 1994. Phylogenetic relationships between some members of the genera *Neisseria*, *Acinetobacter*, *Moraxella*, and *Kingella* based on partial 16S ribosomal DNA sequence analysis. Int. J. Syst. Bacteriol. *44*: 387–391.

Enright, M.C. and B.G. Spratt. 1999. Multilocus sequence typing. Trends Microbiol. *7*: 482–487.

Ensminger, P.W. 1953. Pigment production by *Haemophilus parapertussis*. J. Bacteriol. *63*: 509–510.

Entani, E., S. Ohmori, H. Masai and K.I. Suzuki. 1985. *Acetobacter polyoxogenes* sp. nov., a new species of an acetic acid bacterium useful for producing vinegar with high acidity. J. Gen. Appl. Microbiol. *31*: 475–490.

Erasmus, H.D., F.N. Matthee and H.A. Louw. 1974. A comparison between plant pathogenic species of *Pseudomonas*, *Xanthomonas*, and *Erwinia* with special reference to the bacterium responsible for bacterial blight of vines. Phtyophylactica. *6*: 11–18.

Erdman, S.E., P. Correa, L.A. Coleman, M.D. Schrenzel, X. Li and J.G. Fox. 1997. *Helicobacter mustelae*-associated gastric MALT lymphoma in ferrets. Am. J. Pathol. *151*: 273–280.

Eremeeva, M.E., G.A. Dasch and D.J. Silverman. 2001. Quantitative analyses of variations in the injury of endothelial cells elicited by 11 isolates of *Rickettsia rickettsii*. Clin. Diagn. Lab. Immunol. *8*: 788–796.

Eremeeva, M.E., A. Madan, J. Malek, J. Wierzbowski and G.A. Dasch. 2002. The genome sequences of *Rickettsia akari*, *R. sibirica*, and *R. rickettsii*: Analysis and implications for functional proteomics. International Conference on Rickettsiae and Rickettsial Diseases Joint Meeting with ASR Conference 2002, Ljubljana, Slovenia. p. 64.

Eremeeva, M.E., V. Roux and D. Raoult. 1993. Determination of genome size and restriction pattern polymorphism of *Rickettsia prowazekii* and *Rickettsia typhi* by pulsed field gel electrophoresis. FEMS Microbiol. Lett. *112*: 105–112.

Escalante-Semerena, J.C., R.P. Blakemore and R.S. Wolfe. 1980. Nitrate dissimilation under microaerophilic conditions by a magnetic spirillum. Appl. Environ. Microbiol. *40*: 429–430.

Eskew, D.L., D.D. Focht and I.P. Ting. 1977. Nitrogen fixation, denitrification, and pleomorphic growth in a highly pigmented *Spirillum lipoferum*. Appl. Environ. Microbiol. *34*: 582–585.

Esmarch, E. 1887. Über die Reinkultur eines Spirillum. Zentralbl. Bakteriol. Parasitenkd. Infektionskr. Hyg. Abt. I Orig. *1*: 225–230.

Esnault, G., P. Caumette and J.L. Garcia. 1988a. Characterization of *Desulfovibrio giganteus* sp. nov., a sulfate-reducing bacterium isolated from a brackish coastal lagoon. Syst. Appl. Microbiol. *10*: 147–151.

Esnault, G., P. Caumette and J.L. Garcia. 1988b. *In* Validation of the publication of new names and new combinations previously effectively published outside the IJSB. List No. 26. Int. J. Syst. Bacteriol. *38*: 328–329.

Estéve, J.C. 1978. Wild strain of *Paramecium caudatum* showing a killer trait (Ehrenberg). Protistologica *14*: 201–207.

Etchebehere, C., I. Errazquin, E. Barrandeguy, P. Dabert, R. Moletta and L. Muxi. 2001a. Evaluation of the denitrifying microbiota of anoxic reactors. FEMS Microbiol. Ecol. *35*: 259–265.

Etchebehere, C., M.I. Errazquin, R. Dabert, R. Moletta and L. Muxí. 2001b. *Comamonas nitrativorans* sp. nov., a novel denitrifier isolated from a denitifying reactor treating landfill leachate. Int. J. Syst. Evol. Microbiol. *51*: 977–983.

Etoh, Y., F.E. Dewhirst, B.J. Paster, A. Yamamoto and N. Goto. 1993. *Campylobacter showae* sp. nov., isolated from the human oral cavity. Int. J. Syst. Bacteriol. *43*: 631–639.

Euzéby, J.P. 1997. Revised nomenclature of specific or subspecific epithets that do not agree in gender with generic names that end in -bacter. Int. J. Syst. Bacteriol. *47*: 585.

Euzéby, J.P. 1998a. Proposal to amend Rule 61 of the International Code of Nomenclature of Bacteria (1990 revision). Int. J. Syst. Bacteriol. *48*: 611–612.

Euzéby, J.P. 1998b. Taxonomic note: necessary correction of specific and subspecific epithets according to Rules 12c and 13b of the International Code of Nomenclature of Bacteria (1990 revision). Int. J. Syst. Bacteriol. *48*: 1073–1075.

Evans, D.G., D.J. Evans, Jr., H.C. Lampert and D.Y. Graham. 1995. Restriction fragment length polymorphism in the adhesin gene *hpaA* of *Helicobacter pylori*. Am. J. Gastroenterol. *90*: 1282–1288.

Evans, I.J. and J.A. Downie. 1986. The *nodI* gene product of *Rhizobium leguminosarum* is closely related to ATP-binding bacterial transport proteins; nucleotide sequence analysis of the *nodI* and *nodJ* genes. Gene *43*: 95–101.

Evers, S., M. Weizenegger, W. Ludwig, B. Schink and K.H. Schleifer. 1993. The phylogenetic positions of *Pelobacter acetylenicus* and *Pelobacter propionicus*. Syst. Appl. Microbiol. *16*: 216–218.

Ewanowich, C.A., L.W. Chui, M.G. Paranchych, M.S. Peppler, R.G. Marusyk and W.L. Albritton. 1993. Major outbreak of pertussis in northern Alberta, Canada: analysis of discrepant direct fluorescent-antibody and culture results by using polymerase chain reaction methodology. J. Clin. Microbiol. *31*: 1715–1725.

Ewers, J., D. Freier Schroder and H.J. Knackmuss. 1990. Selection of trichloroethene (TCE) degrading bacteria that resist inactivation by TCE. Arch. Microbiol. *154*: 410–413.

Ewing, E.P., A. Takeuchi, A. Shirai and J.V. Osterman. 1978. Experimental infection of mouse peritoneal mesothelium with scrub typhus rickettsiae: An ultrastructural study. Infect. Immun. *19*: 1068–1075.

Ewing, S.A., U.G. Munderloh, E.F. Blouin, K.M. Kocan and T.J. Kurtti. 1995. *Ehrlichia canis* in tick cell culture. Proceedings of the 76th Conference of Research Workers in Animal Diseases, Chicago. Iowa State University Press.

Ezaki, T., N. Takeuchi, S.L. Liu, A. Kai, H. Yamamoto and E. Yabuuchi. 1988. Small-scale DNA preparation for rapid genetic identification of *Campylobacter* species without radioisotope. Microbiol. Immunol. *32*: 141–150.

Faast, R., M.A. Ogierman, U.H. Stroeher and P.A. Manning. 1989. Nucleotide sequence of the structural gene, *tcpA*, for a major pilin subunit of *Vibrio cholerae*. Gene *85*: 227-231.

Facinelli, B. and P.E. Varaldo. 1987. Plasmid-mediated sulfonamide resistance in *Neisseria meningitidis*. Antimicrob. Agents Chemother. *31*: 1642–1643.

Fåhraeus, G. 1957. The infection of clover root hairs by nodule bacteria studied by a simple glass slide technique. J. Gen. Microbiol. *16*: 374–381.

Falcao de Morais, J.O., E.M. Rios, G.M. Calazans and C.E. Lopes. 1993. *Zymomonas mobilis* research in the Pernambuco Federal University. J. Biotechnol. *31*: 75–91.

Falk, E.C., J. Dobreiner, J.L. Johnson and N.R. Krieg. 1985. Deoxyribonucleic acid homology of *Azospirillum amazonense* Magalhaes et. al., 1984 and emendation of the description of the genus *Azospirillum*. Int. J. Syst. Bacteriol. *35*: 117–118.

Falk, E.C, J.L. Johnson, V.L.D. Baldani, J. Döbereiner and N.R. Krieg. 1986. Deoxyribonucleic and ribonucleic acid homology studies of the genera *Azospirillum* and *Conglomeromonas*. Int. J. Syst. Bacteriol. *36*: 80–85.

Falkler, W.A.J., H.N. Williams and C.N. Smoot. 1979. Antigenicity of marine and terrestrial bdellovibrios. 79th Annual Meeting of the American Society for Microbiology, p. 188.

Fallik, E. and Y. Okon. 1996. Inoculants of *Azospirillum brasilense*: biomass production, survival and growth promotion of *Setaria italica* and *Zea mays*. Soil Biol. Biochem. *21*: 123–126.

Fallik, E., Y. Okon, E. Epstein, A. Goldman and M. Fischer. 1989. Identification and quantification of IAA and IBA in *Azospirillum brasilense* inoculated maize roots. Soil Biol. Biochem. *21*: 147–154.

Falsen, E. 1996. Catalogue of strains, CCUG Culture Collection, University of Göteborg, Sweden.

Famureva, O., H.G. Sonntag and P. Hirsch. 1983. Avirulence of 27 bacteria that are budding, prosthecate or both. Int. J. Syst. Bacteriol. *33*: 565–572.

Fang, J.S., M.J. Barcelona and J.D. Semrau. 2000. Characterization of methanotrophic bacteria on the basis of intact phospholipid profiles. FEMS Microbiol. Lett. *189*: 67–72.

Fani, R., C. Bandi, M. Bazzicalupo, M.T. Ceccherini, S. Fancelli, E. Gallori, L. Gerace, A. Grifoni, N. Miclaus and G. Damiani. 1995a. Phylogeny of the genus *Azospirillum* based on 16S rDNA sequence. FEMS Microbiol. Lett. *129*: 195–200.

Fani, R., C. Bandi, M. Bazzicalupo, G. Damiani, F. Di Cello, S. Fancelli, E. Gallori, L. Gerace, A. Grifoni and P. Lio. 1995b. Phylogenetic studies of the genus *Azospirillum*. NATO ASI ser, Ser G: Ecol. Sci. *37*: 59–75.

Fani, R., M. Bazzicalupo, P. Coianiz and M. Polsinelli. 1986. Plasmid transformation of *Azospirillum brasilense*. FEMS Microbiol. Lett. *35*: 23–27.

Fantino, M.G. and C. Bazzi. 1982. Azione antagonista di *Pseudomonas cepacia* verso *Fusarium oxysporum* f.sp. cepae. Inf. Fitopol. *32*: 55–58.

Fardeau, M.L., B. Ollivier, A. Soubrane, P. Sauve, G. Prensier, J.L. Garcia and J.P. Belaich. 1993. Determination of the G + C content of 2 *Syntrophus buswellii* strains by ultracentrifugation techniques. Curr. Microbiol. *26*: 185–189.

Farez-Vidal, M.E., A. Fernandez-Vivas and J.M. Arias. 1990. Amylase programming during the life cycle of *Myxococcus coralloides*. J. Appl. Bacteriol. *69*: 119–124.

Farrah , S.R. and R.F. Unz. 1975. Fluorescent antibody study of natural finger-like zoogloeae. Appl. Microbiol. *30*: 132–139.

Farrah, S.R. and R.F. Unz. 1976. Isolation of exocellular polymer from *Zoogloea* strains MP6 and 106 and from activated sludge. Appl. Environ. Microbiol. *32*: 33–37.

Farrar, J.A., A.J. Thomson, M.R. Cheesman, D.M. Dooley and W.G. Zumft. 1994. A model of the copper centres of nitrous oxide reductase (*Pseudomonas stutzeri*). Evidence from optical, EPR and MCD spectroscopy. FEBS Lett. *294*: 11–15.

Farrell, I.D. 1974. The development of a new selective medium for the isolation of *Brucella abortus* from contaminated sources. Res. Vet. Sci. *16*: 280–286.

Fass, R.J. and J. Barnishan. 1976. Acute meningitis due to a *Pseudomonas*-like Group Va-1 bacillus. Ann. Intern. Med. *84*: 51–52.

Fauque, G., D. Herve and J. LeGall. 1979. Structure-function relationship in hemoproteins: the role of cytochrome c_3 in the reduction of colloidal sulfur by sulfate-reducing bacteria. Arch. Microbiol. *121*: 261–264.

Fauque, G.D., O. Klimmek and A. Kröger. 1994. Sulfur reductases from spirilloid mesophilic sulfur-reducing eubacteria. Methods Enzymol. *243*: 367–383.

Faur, Y.C., M.H. Weisburd and M.E. Wilson. 1973. A new medium for the isolation of pathogenic *Neisseria* (NYC nmedium). 3. Performance as a culture and transport medium without addition of ambient carbon dioxide. Health Lab. Sci. *10*: 61–74.

Faure, D., M.L. Bouillant, C. Jacoud and R. Bally. 1996. Phenolic derivatives related to lignin metabolism as substrates for *Azospirillum* laccase activity. Phytochemistry. *42*: 357–359.

Fauré-Fremiet, E. 1952. Symbiontes bactériens des ciliés du genre *Euplotes*. C.R Acad. Sci. *235*: 402–403.

Fautz, E., L. Grotjahn and H. Reichenbach. 1981. Hydroxy fatty acids as valuable chemosystematic markers in gliding bacteria and flavobacteria. *In* Reichenbach and Weeks (Editors), The *Flavobacterium–Cytophaga* Group, Verlag Chemie, Weinheim. pp. 127–133.

Fautz, E., G. Rosenfelder and L. Grotjahn. 1979. *Iso*-branched 2-hydroxy and 3-hydroxy fatty acids as characteristic lipid constituents of some gliding bacteria. J. Bacteriol. *140*: 852–858.

Favinger, J., R. Stadtwald and H. Gest. 1989. *Rhodospirillum centenum*, sp. nov., a thermotolerant cyst-forming anoxygenic photosynthetic bacterium. Antonie Leeuwenhoek *55*: 291–296.

Favinger, J., R. Stadtwald and H. Gest. 1994. *In* Validation of the publication of new names and new combinations previously effectively published outside the IJSB. List No. 48. Int. J. Syst. Bacteriol. *44*: 182–183.

Feavers, I.M. and M.C.J. Maiden. 1998. A gonococcal *porA* pseudogene: implications for understanding the evolution and pathogenicity of *Neisseria gonorrhoeae*. Mol. Microbiol. *30*: 647–656.

Federov, M.V. and T.A. Kalininskaya. 1961. A new species of a nitrogen-fixing mycobacterium and its physiological peculiarities. Mikrobiologiya *30*: 9–14.

Feil, E.J., M.C. Enright and B.G. Spratt. 2000. Estimating the relative contributions of mutation and recombination to clonal diversification: a comparison between *Neisseria meningitidis* and *Streptococcus pneumoniae*. Res. Microbiol. *151*: 465–469.

Feil, E.J., E.C. Holmes, D.E. Bessen, M.S. Chan, N.P. Day, M.C. Enright, R. Goldstein, D.W. Hood, A. Kalia, C.E. Moore, J. Zhou and B.G. Spratt. 2001. Recombination within natural populations of pathogenic bacteria: short-term empirical estimates and long-term phylogenetic consequences. Proc. Natl. Acad. Sci. U.S.A. *98*: 182–187.

Feil, E., J. Zhou, J. Maynard Smith and B.G. Spratt. 1996. A comparison of the nucleotide sequences of the *adk* and *recA* genes of pathogenic and commensal *Neisseria* species: evidence for extensive interspecies recombination in *adk*. J. Mol. Evol. *43*: 631-640.

Fein, J.E., R.C. Charley, K.A. Hopkins, B. Lavers and H.G. Lawford. 1983. Development of a simple defined medium for continuous ethanol production by *Zymomonas mobilis*. Biotechnol. Lett. *5*: 1–6.

Feio, M.J., I.B. Beech, M. Carepo, J.M. Lopes, C.W.S. Cheung, R. Franco, J. Guezennec, J.R. Smith, J.I. Mitchell, J.J.G. Moura and A.R. Lino. 1998. Isolation and characterization of a novel sulphate-reducing bacterium of the *Desulfovibrio* genus. Anaerobe *4*: 117–130.

Feio, M.J., I.B. Beech, M. Carepo, J.M. Lopes, C.W.S. Cheung, R. Franco, J. Guezennec, J.R. Smith, J.I. Mitchell, J.J.G. Moura and A.R. Lino. 2000. *In* Validation of the publication of new names and new combinations previously effectively published outside the IJSB. List No. 75. Int. J. Syst. Evol. Microbiol. *50*: 1415–1417.

Feldmann, S., H. Sahm and G.A. Sprenger. 1992. Pentose metabolism in *Zymomonas mobilis* wild-type and recombinant strains. Appl. Microbiol. Biotechnol. *38*: 354–361.

Fell, J.W. 1966. *Sterigmatomyces*, a new fungal genus from marine areas. Antonie Van Leeuwenhoek J. Microbiol. Serol. *32*: 99–104.

Fellinger, B.E. 1924. Untersuchungen über die Mundoscillarien des Menschen. Zentralbl. Bakteriol. Abt. 1 Orig. *91*: 398–401.

Felsenstein, J. 1981. Evolutionary trees from DNA sequences: a maximum likelihood approach. J. Mol. Evol. *17*: 368–376.

Felsenstein, J. 1982. Numerical methods for inferring evolutionary trees. Q. Rev. Biol. *57*: 379–404.

Felsenstein, J. 1985. Confidence limits on phylogenies: an approach using the bootstrap. Evolution. *39*: 783–791.

Felsenstein, J. 1989. PHYLIP-Phylogeny Inference Package. Cladistics *5*: 164–166.

Felsenstein, J. 1993. PHYLIP (Phylogeny Inference Package), Version 3.5c. Department of Genetics, University of Washington, Seattle.

Felter, R.A., R.R. Colwell and G.B. Chapman. 1969. Morphology and round body formation in *Vibrio marinus*. J. Bacteriol. *99*: 326–335.

Felzenberg, E.R., G.A. Yang, J.G. Hagenzieker and J.S. Poindexter. 1996. Physiologic, morphologic and behavioral responses of perpetual cultures of *Caulobacter crescentus* to carbon, nitrogen and phosphorus limitations. J. Indust. Microbiol. *17*: 235–252.

Fenchel, T. 1994. Motility and chemosensory behaviour of the sulphur bacterium *Thiovulum majus*. Microbiology (Reading) *140*: 3109–3116.

Fenchel, T. and R.N. Glud. 1998. Veil architecture in a sulphide-oxidizing bacterium enhances countercurrent flux. Nature *394*: 367–369.

Feng, H.M., T.S. Chen, B.H. Lin, Y.Z. Lin, P.F. Wang, Q.H. Su, H.B. Xia, K. Kumano and T. Uchida. 1991. Serologic survey of spotted fever group rickettsiosis on Hainan Island of China. Microbiol. Immunol. *35*: 687–694.

Fennell, C.L., P.A. Totten, T.C. Quinn, D.L. Patton, K.K. Holmes and W.E. Stamm. 1984. Characterization of *Campylobacter*-like organisms isolated from homosexual men. J. Infect. Dis. *149*: 58–66.

Fensom, A.H., W.M. Kurowski and S.J. Pirt. 1974. The use of ferricyanide for the production of 3-ketosugars by non-growing suspensions of *Agrobacterium tumefaciens*. J. Appl. Chem. Biotechnol. *24*: 457–467.

Fernandez, R.C. and A.A. Weiss. 1994. Cloning and sequencing of a *Bordetella pertussis* serum resistance locus. Infect. Immun. *62*: 4727–4738.

Fernandez-Vivas, A., C. de Haro, J.M. ARias and E. Montoya. 1983. Detection of guanosine 5′-diphosphate 3′-diphosphate in *Myxococcus coralloides* D vegetative cells. FEMS Microbiol. Lett. *20*: 17–22.

Ferreira, M.C.B., M.S. Fernandes and J. Dobereiner. 1987. Role of *Azospirillum brasilense* nitrate reductase in nitrate assimilation by wheat plants. Biol. Fertil. Soils *4*: 47–54.

Ferreiros, C.M., L. Ferron and M.T. Criado. 1994. *In vivo* human immune-response to transferrin-binding protein-2 and other iron-regulated proteins of *Neisseria meningitidis*. FEMS Immunol. Med. Microbiol. *8*: 63–68.

Ferry, N.S. 1911. Etiology of canine distemper. J. Infect. Dis. *8*: 399–420.

Ferry, N.S. 1912. *Bacillus bronchisepticus* (bronchicanis): the cause of distemper in dogs and a similar disease in other animals. Vet. J. *68*: 376–391.

Ferschl, A., M. Loidl, G. Ditzelmuller, C. Hinteregger and F. Streichsbier. 1991. Continuous degradation of 3-chloroaniline by calcium-alginate entrapped cells of *Pseudomonas acidovorans* Ca28 - influence of additional substrates. Appl. Microbiol. Biotechnol. *35*: 544–550.

Fesefeldt, A. 1998. Molekularbiologische charakterisieruing einer methylotrophen bakteriengattung und untersuchung der spezifischen genexpression der methanol-dehydrogenase, University of Kiel. Kiel, Germany. . p. 164.

Fesefeldt, A. and C.G. Gliesche. 1997. Identification of *Hyphomicrobium* spp. using PCR-amplified fragments of the *mxaF* gene as a molecular marker. Syst. Appl. Microbiol. *20*: 387–396.

Fesefeldt, A., N.C. Holm and C.G. Gliesche. 1998a. Genetic diversity and population dynamics of *Hyphomicrobium* spp. in a sewage treatment plant and its receiving lake. Water Sci. Technol. *37*: 113–116.

Fesefeldt, A., K. Kloos, H. Bothe, H. Lemmer and C.G. Gliesche. 1998b. Distribution of denitrification and nitrogen fixation genes in *Hyphomicrobium* spp. and other budding bacteria. Can. J. Microbiol. *44*: 181–186.

Fesefeldt, A., M. Poetsch and C.G. Gliesche. 1997. Development of a species-specific gene probe for *Hyphomicrobium facilis* with the inverse PCR. Appl. Environ. Microbiol. *63*: 335–337.

Fetzner, S., R. Muller and F. Lingens. 1989. Degradation of 2-chlorobenzoate by *Pseudomonas cepacia* 2CBS. Bio. Chem. Hoppe-Seyler *370*: 1173–1182.

Fetzner, S., R. Muller and F. Lingens. 1992. Purification and some properties of 2-halobenzoate 1,2-dioxygenase, a two-component enzyme system from *Pseudomonas cepacia* 2CBS. J. Bacteriol. *174*: 279–290.

Ficht, T.A., S.W. Bearden, B.A. Sowa and H. Marquis. 1990. Genetic variation at the *omp2* porin locus of the brucellae: species-specific markers. Mol. Microbiol. *4*: 1135–1142.

Ficht, T.A., H.S. Husseinen, J. Derr and S.W. Bearden. 1996. Species-specific sequences at the *omp2* locus of *Brucella* type strains. Int. J. Syst. Bacteriol. *46*: 329–331.

Filer, D., E. Rosenberg and S.H. Kindler. 1973. Aspartokinase of *Myxococcus xanthus*: "Feedback stimulation" by required amino acids. J. Bacteriol. *115*: 23–28.

Finan, T.M., S. Weidner, K. Wong, J. Buhrmester, P. Chain, F.J. Vorhoelter, I. Hernandez-Lucas, A. Becker, A. Cowie, J. Gouzy, B. Golding and A. Puehler. 2001. The complete sequence of the 1,683-kb pSymB megaplasmid from the N$_2$-fixing endosymbiont *Sinorhizobium meliloti*. Proc. Natl. Acad. Sci. U.S.A. *98*: 9889–9894.

Finck, G. 1950. Biologische und stoffwechselphysiologische Studien an Myxococcaceen. Arch. Microbiol. *15*: 358–388.

Finegold, S.M. and H. Jousimies-Somer. 1997. Recently described clinically important anaerobic bacteria: medical aspects. Clin. Infect. Dis. *25 (Suppl. 2)*: S88–S93.

Fink, J.M. and J.F. Zissler. 1989. Characterization of lipopolysaccharide from *Myxococcus xanthus* by use of monoclonal antibodies. J. Bacteriol. *171*: 2028–2032.

Finn, T.M. and L.A. Stevens. 1995. Tracheal colonization factor: a *Bordetella pertussis* secreted virulence determinant. Mol. Microbiol. *16*: 625–634.

Finne, J.M., M. Leinonen and P.H. Mäkelä. 1983. Antigenic similarities between brain components and bacteria causing meningitis. Lancet *ii*: 7175–7179.

Finstein, M.S. 1967. Growth and flocculation in a *Zoogloea* culture. Appl. Microbiol. *15*: 962–963.

Finster, K. and F. Bak. 1993. Complete oxidation of propionate, valerate, succinate, and other organic compounds by newly isolated types of marine, anaerobic, mesophilic, gram-negative, sulfur-reducing eubacteria. Appl. Environ. Microbiol. *59*: 1452–1460.

Finster, K., F. Bak and N. Pfennig. 1994. *Desulfuromonas acetexigens* sp. nov., a dissimilatory sulfur-reducing eubacterium from anoxic freshwater sediments. Arch. Microbiol. *161*: 328–332.

Finster, K., F. Bak and N. Pfennig. 1997a. *In* Validation of the publication of new names and new combinations previously effectively published outside the IJSB. List No. 61. Int. J. Syst. Bacteriol. *47*: 601–602.

Finster, K., J.D. Coates, W. Liesack and N. Pfennig. 1997b. *Desulfuromonas thiophila* sp. nov., a new obligately sulfur-reducing bacterium from anoxic freshwater sediments. Int. J. Syst. Bacteriol. *47*: 754–758.

Finster, K., W. Liesack and B. Thamdrup. 1998. Elemental sulfur and thiosulfate disproportionation by *Desulfocapsa sulfoexigens* sp. nov., a new anaerobic bacterium isolated from marine surface sediment. Appl. Environ. Microbiol. *64*: 119–125.

Finster, K., W. Liesack and B. Thamdrup. 2000. *In* Validation of the publication of new names and new combinations previously published outside the IJSB. List No. 76. Int. J. Syst. Bacteriol. *50*: 1699–1700.

Finster, K., W. Liesack and B.J. Tindall. 1997c. *Desulfospira joergensenii*, gen. nov., sp. nov., a new sulfate-reducing bacterium isolated from marine surface sediment. Syst. Appl. Microbiol. *20*: 201–208.

Finster, K., W. Liesack and B.J. Tindall. 1997d. *Sulfurospirillum arcachonense* sp. nov., a new microaerophilic sulfur-reducing bacterium. Int. J. Syst. Bacteriol. *47*: 1212–1217.

Finster, K., W. Liesack and B.J. Tindall. 1997e. *In* Validation of the publication of new names and new combinations previously effectively published outside the IJSB. List No. 63. Int. J. Syst. Bacteriol. *47*: 1274.

Fischer, A., T. Roggentin, H. Schlesner and E. Stackebrandt. 1985. 16S ribosomal RNA oligonucleotide cataloging and the phylogenetic position of *Stella humosa*. Syst. Appl. Microbiol. *6*: 43–47.

Fisher, R.F., J.A. Swanson, J.T. Mulligan and S.R. Long. 1987. Extended region of nodulation genes in *Rhizobium meliloti* 1021 II. nucleotide-sequence, transcription start sites and protein products. Genetics. *117*: 191–201.

Fitch, M.W., D.W. Graham, R.G. Arnold, S.K. Agarwal, P. Phelps, G.E. Speitel, Jr. and G. Georgiou. 1993. Phenotypic characterization of copper-resistant mutants of *Methylosinus trichosporium* OB3b. Appl. Environ. Microbiol. *59*: 2771–2776.

Flad, H.D. and A.J. Ulmer. 1995. Glycosphingolipids from *Sphingomonas paucimobilis* induce monokine production in human mononuclear cells. Infect. Immun. *63*: 2899–2905.

Flavier, A.B., S.J. Clough, M.A. Schell and T.P. Denny. 1997a. Identification of 3-hydroxypalmitic acid methyl ester as a novel autoregulator controlling virulence in *Ralstonia solanacearum*. Mol. Microbiol. *26*: 251–259.

Flavier, A.B., L.M. Ganova-Raeva, M.A. Schell and T.P. Denny. 1997b. Hierarchical autoinduction in *Ralstonia solanacearum*: Control of acyl-homoserine lactone production by a novel autoregulatory system responsive to 3-hydroxypalmitic acid methyl ester. J. Bacteriol. *179*: 7089–7097.

Flesch, G. and M. Rohmer. 1989. Prokaryotic triterpenoids. A novel hopanoid from the ethanol-producing bacterium *Zymomonas mobilis*. Biochem. J. *262*: 673–675.

Flesher, S.A. and E.J. Bottone. 1989. *Eikenella corrodens* cellulitis and arthritis of the knee. J. Clin. Microbiol. *27*: 2606–2608.

Fletcher, L.M. and M.E. Rhodes-Roberts. 1976. The bacterial leaf nodule association in *Psychotria*. Soc. Appl. Bacteriol. Tech. Ser. *12*: 99–118.

Fletcher, M.T., P.J. Blackall and C.M. Doheny. 1987. A note on the isoprenoid quinone content of *Bordetella avium* and related species. J. Appl. Bacteriol. *62*: 275–277.

Florent, A. 1959. Les deux vibrioses genitales de la bete bovine: La vibriose venerienne, due a *Vibrio foetus* venerealis, et la vibriose d'origine intestinale due a *V. foetus* intestinales. Proc. 10th Int. Vet. Cong, Madrid. *2*: 953–957.

Flores-Encarnación, M., M. Contreras-Zentella, L. Soto-Urzua, G.R. Aguilar, B.E. Baca and J.E. Escamilla. 1999. The respiratory system and diazotrophic activity of *Acetobacter diazotrophicus* PAL5. J. Bacteriol. *181*: 6987–6995.

Florin, C., T. Kohler, M. Grandguillot and P. Plesiat. 1996. *Comamonas testosteroni* 3-ketosteroid-d4(5-a)-dehydrogenase: gene and protein characterization. J. Bacteriol. *178*: 3322–3330.

Fluegel, W. 1963. Simple method for demonstrating myxobacterial slime. J. Bacteriol. *85*: 1173–1174.

Flügge, C. 1886. Die Microorganismen, F. C. W. Vogel, Leipzig.

Flynn, J. and M.J. McEntegart. 1972. Bacteriocins from *Neisseria gonorrhoeae* and their possible role in epidemiological studies. J. Clin. Pathol. *25*: 60–61.

Focht, D.D. and H.A. Joseph. 1971. Bacterial degradation of nitrilotriacetic acid (NTA). Can. J. Microbiol. *17*: 1553–1556.

Foggie, A. 1951. Studies on the infectious agent of tick-borne fever in sheep. J. Pathol. Bacteriol. *63*: 1–15.

Foissner, W. 1977. *Euplotes moebiusi* f. quadricirratus (Ciliophora, Hypotrichida). II. Die Feinstruktur einiger cytoplasmatischer Organellen. Naturkd. Jahrb. Stadt Linz. *23*: 17–24.

Fokin, S.I. 1989. Bacterial endobionts of the ciliate *Paramecium woodruffi*: 1. Endobionts of the macronucleus. Tsitologiya *31*: 839–844.

Fokin, S.I. 1991. *Holospora recta* sp. nov., - a micronucleus-specific endobiont of the ciliate *Paramecium caudatum*. Tsitologiya *33*: 135–141.

Fokin, S.I., T. Brigge, J. Brenner and H.D. Görtz. 1996. *Holospora* species infected the nuclei of *Paramecium* appear to belong in two groups of bacteria. Europ. J. Protistol. *32(suppl.1)*: 19–24.

Fokin, S. and H.D. Görtz. 1993. *Caedibacter macronucleorum* sp. nov., a bacterium inhabiting the macronucleus of *Paramecium duboscqui*. Arch. Protistenkd. *143*: 319–324.

Fokin, S.I. and D.V. Ossipov. 1986. *Pseudocaedibacter glomeratus* sp. n. - a

cytoplasmic symbiont of the ciliate *Paramecium pentaurelia*. Tsitologiya *28*: 1000–1004.

Fokin, S. and E. Sabaneyeva. 1993. Bacterial endocytobionts of the ciliate *Paramecium calkinsi*. Eur. J. Protistol. *29*: 390–395.

Fokin, S.I. and E. Sabaneyeva. 1997. Release of endonucleobiotic bacteria *Holospora bacillata* and *Holospora curvata* from the macronucleus of their host cells *Paramecium woodruffi* and *Paramecium calkinsi*. Endocytobiosis Cell Res. *12*: 49–55.

Fokin, S.I. and I.N. Skovorodkin. 1991. *Holospora obtusa*-endonucleobiont of the ciliate *Paramecium caudatum* in search of the macronucleus. Tsitologiya *33*: 101–115.

Fokin, S.I. and I.N. Skovorodkin. 1997. Experimental analysis of the resistance of *Paramecium caudatum* (Ciliophora) against infection by bacterium *Holospora undulata*. Eur. J. Protistol. *33*: 214–218.

Foley, J.E., J.V. Solnick, J.M. Lapointe, S.S. Jang and N.C. Pedersen. 1998. Identification of a novel enteric *Helicobacter* species in a kitten with severe diarrhea. J. Clin. Microbiol. *36*: 908–912.

Folkerts, M., U. Ney, H. Kneifel, E. Stackebrandt, E.G. Witte, H. Förstel, S.M. Schoberth and H. Sahm. 1989a. *Desulfovibrio furfuralis* sp. nov., a furfural degrading strictly anaerobic bacterium. Syst. Appl. Microbiol. *11*: 161–169.

Folkerts, M., U. Ney, H. Kneifel, E. Stackebrandt, E.G. Witte, H. Förstel, S. Schoberth and H. Sahm. 1989b. *In* Validation of the publication of new names and new combinations previously effectively published outside the IJSB. List No. 31. Int. J. Syst. Bacteriol. *39*: 495–497.

Fomenkov, A., J.P. Xiao and S.Y. Xu. 1995. Nucleotide sequence of a small plasmid isolated from *Acetobacter pasteurianus*. Gene *158*: 143–144.

Fontaine, T., M.P. Stephan, L. Debarbieux, J.O. Previato and L. Mendonca-Previato. 1995. Lipopolysaccharides from six strains of *Acetobacter diazotrophicus*. FEMS Microbiol. Lett. *132*: 45–50.

Fontana, J.D., V.C. Franco, S.J. de Souza, I.N. Lyra and A.M. de Souza. 1991. Nature of plant stimulators in the production of *Acetobacter xylinum* ("tea fungus") biofilm used in skin therapy. Appl. Biochem. Biotechnol. *28-29*: 341–351.

Forbes, K.J., Z. Fang and T.H. Pennington. 1995. Allelic variation in the *Helicobacter pylori* flagellin genes *flaA* and *flaB*: its consequences for strain typing schemes and population structure. Epidemiol. Infect. *114*: 257–266.

Forget, P. and P. F.. 1965. Le cycle tricarboxylique chez une bactérie denitrifiante obligatoire. Ann. Inst. Pasteur (Paris) *108*: 364–377.

Forman, D. 1998. *Helicobacter pylori* infection and cancer. Br. Med. Bull. *54*: 71–78.

Forsman, M., G. Sandström and A. Sjöstedt. 1994. Analysis of 16S ribosomal DNA sequences of *Francisella* strains and utilization for determination of the phylogeny of the genus and for identification of strains by PCR. Int. J. Syst. Bacteriol. *44*: 38–46.

Fortnagel, P., H. Harms, R.M. Wittich, S. Krohn, H. Meyer, V. Sinnwell, H. Wilkes and W. Francke. 1990. Metabolism of dibenzofuran by *Pseudomonas* sp. strain HH69 and the mixed culture HH27. Appl. Environ. Microbiol. *56*: 1148–1156.

Foss, S. and J. Harder. 1998. *Thauera linaloolentis* sp. nov. and *Thauera terpenica* sp. nov., isolated on oxygen-containing monoterpenes (linalool, menthol, and eucalyptol) and nitrate. Syst. Appl. Microbiol. *21*: 365–373.

Foss, S. and J. Harder. 1999. *In* Validation of the publication of new names and new combinations previously effectively published outside the IJSB. List No. 68. Int. J. Syst. Bacteriol. *49*: 1–3.

Foss, S., U. Heyen and J. Harder. 1998a. *Alcaligenes defragrans* sp. nov., description of four strains isolated on alkenoic monoterpenes ((+)-menthene, α-pinene, 2-carene, and α-phellandrene) and nitrate. Syst. Appl. Microbiol. *21*: 237–244.

Foss, S., U. Heyen and J. Harder. 1998b. *In* Validation of the publication of new names and new combinations previously effectively published outside the IJSB. List No. 67. Int. J. Syst. Bacteriol. *48*: 1083–1084.

Foster, H.A. and J.H. Parish. 1973. Ribosomes, ribosomal subunits and ribosomal proteins from *Myxococcus xanthus*. J. Gen. Microbiol. *75*: 391–400.

Foster, J.W. 1944. Microbiological aspects of riboflavin. I. Introduction.

II. Bacterial oxidation of riboflavin to lumichrome. J. Bacteriol. *47*: 27–41.

Fothergill, J.C. and J.R. Guest. 1977. Catabolism of L-lysine by *Pseudomonas aeruginosa*. J. Gen. Microbiol. *99*: 139–155.

Fournier, P.E., H. Fujita, N. Takada and D. Raoult. 2002. Genetic identification of rickettsiae isolated from ticks in Japan. J. Clin. Microbiol. *40*: 2176–2181.

Fournier, P.E., F. Grunnenberger, B. Jaulhac, G. Gastinger and D. Raoult. 2000. Evidence of *Rickettsia helvetica* infection in humans, eastern France. Emerg. Infect. Dis. *6*: 389–392.

Fox, G.E., E. Stackebrandt, R.B. Hespell, J. Gibson, J. Maniloff, T.A. Dyer, R.S. Wolfe, W.E. Balch, R.S. Tanner, L.J. Magrum, L.B. Zablen, R. Blakemore, R. Gupta, L. Bonen, B.J. Lewis, D.A. Stahl, K.R. Luehrsen, K.N. Chen and C.R. Woese. 1980. The phylogeny of prokaryotes. Science *209*: 457–463.

Fox, J.P. 1956. Immunization against epidemic typhus. Am. J. Trop. Med. Hyg. *5*: 464–479.

Fox, J.G., M. Blanco, J.C. Murphy, N.S. Taylor, A. Lee, Z. Kabok and J. Pappo. 1993. Local and systemic immune responses in murine *Helicobacter felis* active chronic gastritis. Infect. Immun. *61*: 2309–2315.

Fox, J.G., C.C. Chien, F.E. Dewhirst, B.J. Paster, Z. Shen, P.L. Melito, D.L. Woodward and F.G. Rodgers. 2000. *Helicobacter canadensis* sp. nov. isolated from humans with diarrhea as an example of an emerging pathogen. J. Clin. Microbiol. *38*: 2546–2549.

Fox, J.G., C.C. Chien, F.E. Dewhirst, B.J. Paster, Z. Shen, P.L. Melito, D.L. Woodward and F.G. Rodgers. 2002. *In* Validation of new names and new combinations previously effectively published outside the IJSEM. List No. 84. Int. J. Syst. Evol. Microbiol. *52*: 3–4.

Fox, J.G., T. Chilvers, C.S. Goodwin, N.S. Taylor, P. Edmonds, L.I. Sly and D.J. Brenner. 1989. *Campylobacter mustelae*, new species resulting from the elevation of *Campylobacter pylori* subsp. *mustelae* to species status. Int. J. Syst. Bacteriol. *39*: 301–303.

Fox, J.G., P. Correa, N.S. Taylor, A. Lee, G. Otto, J.C. Murphy and R. Rose. 1990. *Helicobacter mustelae* associated gastritis in ferrets: an animal model of *Helicobacter pylori* gastritis in humans. Gastroenterology *99*: 352–361.

Fox, J.G., C.A. Dangler, W. Sager, R. Borkowski and J.M. Gliatto. 1997a. *Helicobacter mustelae*-associated gastric adenocarcinoma in ferrets (*Mustela putorius furo*). Vet. Pathol. *34*: 225–229.

Fox, J.G., F.E. Dewhirst, G.J. Fraser, B.J. Paster, B. Shames and J.C. Murphy. 1994a. Intracellular *Campylobacter*-like organism from ferrets and hamsters with proliferative bowel disease is a *Desulfovibrio* sp. J. Clin. Microbiol. *32*: 1229–1237.

Fox, J.G., F.E. Dewhirst, Z. Shen, Y. Feng, N.S. Taylor, B.J. Paster, R.L. Ericson, C.N. Lau, P. Correa, J. Araya, C. and I. Roa. 1998a. Hepatic *Helicobacter* species identified in bile and gallbladder tissue from Chileans with chronic cholecystitis. Gastroenterology *114*: 755–763.

Fox, J.G., F.E. Dewhirst, J.G. Tully, B.J. Paster, L. Yan, N.S. Taylor, M.J. Collins, Jr., P.L. Gorelick and J.M. Ward. 1994b. *Helicobacter hepaticus* sp. nov., a microaerophilic bacterium isolated from livers and intestinal mucosal scrapings from mice. J. Clin. Microbiol. *32*: 1238–1245.

Fox, J.G., F.E. Dewhirst, J.G. Tully, B.J. Paster, L. Yan, N.S. Taylor, M.J. Collins, Jr., P.L. Gorelick and J.M. Ward. 1994c. *In* Validation of the publication of new names and new combinations previously effectively published outside the IJSB. List No. 50. Int. J. Syst. Bacteriol. *44*: 595.

Fox, J.G., R. Drolet, R. Higgins, S. Messier, L. Yan, B.E. Coleman, B.J. Paster and F.E. Dewhirst. 1996a. *Helicobacter canis* isolated from a dog liver with multifocal necrotizing hepatitis. J. Clin. Microbiol. *34*: 2479–2482.

Fox, J.G., P.L. Gorelick, M.C. Kullberg, Z. Ge, F.E. Dewhirst and J.M. Ward. 1999. A novel urease-negative *Helicobacter* species associated with colitis and typhlitis in IL-10-deficient mice. Infect. Immun. *67*: 1757–1762.

Fox, J.G. and A. Lee. 1997. The role of *Helicobacter* species in newly recognized gastrointestinal tract diseases of animals. Lab. Anim. Sci. *47*: 222–255.

Fox, J.G., A. Lee, G. Otto, N.S. Taylor and J.C. Murphy. 1991. *Helicobacter*

felis gastritis in gnotobiotic rats: an animal model of *Helicobacter pylori* gastritis. Infect. Immun. *59*: 785–791.

Fox, J.G., X. Li, L. Yan, R.J. Cahill, R. Hurley, R. Lewis and J.C. Murphy. 1996b. Chronic proliferative hepatitis in A/JCr mice associated with persistent *Helicobacter hepaticus* infection: A model of *Helicobacter*-induced carcinogenesis. Infect. Immun. *64*: 1548–1558.

Fox, J.G., J.A. Macgregor, Z. Shen, X. Li, R. Lewis and C.A. Dangler. 1998b. Comparison of methods of identifying *Helicobacter hepaticus* in B6C3F1 mice used in a carcinogenesis bioassay. J. Clin. Microbiol. *36*: 1382–1387.

Fox, J.G., N.S. Taylor, P. Edmonds and D.J. Brenner. 1988. *Campylobacter pylori* subsp. *mustelae*, new subspecies isolated from the gastric mucosa of ferrets (*Mustela putorius furo*), and an emended description of *Campylobacter pylori*. Int. J. Syst. Bacteriol. *38*: 367–370.

Fox, J.G., L.L. Yan, F.E. Dewhirst, B.J. Paster, B. Shames, J.C. Murphy, A. Hayward, J.C. Belcher and E.N. Mendes. 1995. *Helicobacter bilis* sp. nov., a novel *Helicobacter* species isolated from bile, livers, and intestines of aged, inbred mice. J. Clin. Microbiol. *33*: 445–454.

Fox, J.G., L.L. Yan, F.E. Dewhirst, B.J. Paster, B. Shames, J.C. Murphy, A. Hayward, J.C. Belcher and E.N. Mendes. 1997b. *In* Validation of the publication of new names and new combinations previously effectively published outside the IJSB. List No. 61. Int. J. Syst. Bacteriol. *47*: 601–602.

Fox, J.G., L. Yan, B. Shames, J. Campbell, J.C. Murphy and X. Li. 1996c. Persistent hepatitis and enterocolitis in germfree mice infected with *Helicobacter hepaticus*. Infect. Immun. *64*: 3673–3681.

Fox, K.F., A. Fox, M. Nagpal, P. Steinberg and K. Heroux. 1998c. Identification of *Brucella* by ribosomal spacer region PCR and differentiation of *Brucella canis* from other *Brucella* spp. pathogenic for humans by carbohydrate profiles. J. Clin. Microbiol. *36*: 3217–3222.

Fox, K.K. and J.S. Knapp. 1999. Antimicrobial resistance in *Neisseria gonorrhoeae*. Curr. Opin. Urol. *9*: 65–70.

Foxall, P.A., L.T. Hu and H.L.T. Mobley. 1992. Use of polymerase chain reaction-amplified *Helicobacter pylori* urease structural genes for differentiation of isolates. J. Clin. Microbiol. *30*: 739–741.

Fraenkel, D.G. 1992. Genetics and intermediary metabolism. Annu. Rev. Gen. *26*: 159–177.

Fraleigh, P.C. and J.C. Burnham. 1988. Myxococcal predation on cyanobacterial populations—Nutrient effects. Limnol. Oceanogr. *33*: 476–483.

Franche, C., E. Canelo, D. Gauthier and C. Elmerich. 1981. Mobilization of the chromosome of *Azospirillum brasilense* by plasmid R68-45. FEMS Microbiol. Lett. *10*: 199–202.

Franche, C. and C. Elmerich. 1981. Physiological properties and plasmid content of several strains of *Azospirillum brasilense* and *Azospirillum lipoferum*. Ann. Microbiol. *132*: 3–18.

Francis, E. and A.C. Evans. 1926. Agglutination, cross-agglutination and agglutinin absorption in tularaemia. Public Health Rep. *41*: 1273–1295.

Frank, B. 1879. Ueber die Parasiten in den Wurzelanschwillungen der Papilionaceen. Ber. Dtsch. Bot. Ges. *37*: 376–387 394–399.

Frank, B. 1889. Ueber die Pilzsymbiose der Leguminosen. Ber. Dtsch. Bot. Ges. *7*: 332–346.

Frank, J., M. Dijkstra, J.A. Duine and C. Balny. 1988. Kinetic and spectral studies on the redox forms of methanol dehydrogenase from *Hyphomicrobium* X. Eur. J. Biochem. *174*: 331–338.

Frank, J. and J.A. Duine. 1990a. Cytochrome c_L and cytochrome c_H from *Hyphomicrobium*-X. Method Enzymol. *188*: 303–308.

Frank, J. and J.A. Duine. 1990b. Methanol dehydrogenase from *Hyphomicrobium* X. Method Enzymol. *188*: 202–209.

Frank, R. and C.M. Switzer. 1969. Behavior of Pyrazon in soil. Weed Sci. *17*: 323–326.

Franke, I.H., M. Fegan, C. Hayward, G. Leonard and L.I. Sly. 2000. Molecular detection of *Gluconacetobacter sacchari* associated with the pink sugarcane mealybug *Saccharicoccus sacchari* (Cockerell) and the sugarcane leaf sheath microenvironment by FISH and PCR. FEMS Microbiol. Ecol. *31*: 61–71.

Franke, I.H., M. Fegan, C. Hayward, G. Leonard, E. Stackebrandt and

L.I. Sly. 1999. Description of *Gluconacetobacter sacchari* sp. nov., a new species of acetic acid bacterium isolated from the leaf sheath of sugar cane and from the pink sugar-cane mealy bug. Int. J. Syst. Bacteriol. *49*: 1681–1693.

Franke, I.H., M. Fegan, A.C. Hayward and L.I. Sly. 1998. Nucleotide sequence of the *nifH* gene coding for nitrogen reductase in the acetic acid bacterium *Acetobacter diazotrophicus*. Lett. Appl. Microbiol. *26*: 12–16.

Frankel, R.B., D.A. Bazylinski, M.S. Johnson and B.L. Taylor. 1997. Magneto-aerotaxis in marine coccoid bacteria. Biophys. J. *73*: 994–1000.

Franklin, C.L., C.S. Beckwith, R.S. Livingston, L.K. Riley, S.V. Gibson, W.C.L. Besch and R.R. Hook, Jr. 1996. Isolation of a novel *Helicobacter* species, *Helicobacter cholecystus* sp. nov., from the gallbladders of Syrian hamsters with cholangiofibrosis and centrilobular pancreatitis. J. Clin. Microbiol. *34*: 2952–2958.

Franklin, C.L., C.S. Beckwith, R.S. Livingston, L.K. Riley, S.V. Gibson, W.C.L. Besch and R.R. Hook, Jr. 1997. *In* Validation of the publication of new names and new combinations previously effectively published outside the IJSB. List No. 61. Int. J. Syst. Bacteriol. *47*: 601–602.

Franklin, C.L., P.L. Gorelick, L.K. Riley, F.E. Dewhirst, R.S. Livingston, J.M. Ward, C.S. Beckwith and J.G. Fox. 2001. *Helicobacter typhlonius* sp. nov., a novel urease-negative *Helicobacter* species. J. Clin. Microbiol. *39*: 3920–3926.

Franklin, C.L., P.L. Gorelick, L.K. Riley, F.E. Dewhirst, R.S. Livingston, J.M. Ward, C.S. Beckwith and J.G. Fox. 2002. *In* Validation of new names and new combinations previously effectively published outside the IJSEM. List No. 85. Int. J. Syst. Evol. Microbiol. *52*: 685–690.

Franklin, C.L., L.K. Riley, R.S. Livingston, C.S. Beckwith, R.R. Hook Jr., C.L. Besch-Williford, R. Hunziker and P.L. Gorelick. 1999. Enteric lesions in SCID mice infected with "*Helicobacter typhlonicus*", a novel urease-negative *Helicobacter* species. Lab. Anim. Sci. *49*: 496–505.

Franzmann, P.D. and B.J. Tindall. 1990. A chemotaxonomic study of members of the family *Halomonadaceae*. Syst. Appl. Microbiol. *13*: 142–147.

Frasch, C.E. 1980. Role of lipopolysaccharide in wheat germ agglutinin-mediated agglutination of *Neisseria meningitidis* and *Neisseria gonorrhoeae*. J. Clin. Microbiol. *12*: 498–501.

Frasch, C.E. 1989. Vaccines for prevention of meningococcal disease. Clin. Microbiol. Rev. *2*: S134–S138.

Frateur, J. 1950. Essai sur la systématique des Acetobacters. La Cellule. *53*: 287–392.

Frazier, W.C. 1926. A method for the detection of changes in gelatin due to bacteria. J. Infect. Dis. *39*: 302–309.

Frébortová, J., K. Matsushita and O. Adachi. 1997. Effect of growth substrates on formation of alcohol dehydrogenase in *Acetobacter methanolicus* and *Acetobacter aceti*. J. Ferment. Bioeng. *83*: 21–25.

Fredrickson, J.K., D.L. Balkwill, G.R. Drake, M.F. Romine, D.B. Ringelberg and D.C. White. 1995. Aromatic-degrading *Sphingomonas* isolates from the deep subsurface. Appl. Environ. Microbiol. *61*: 1917–1922.

Freeman, B.A., J.R. McGhee and R.E. Baughn. 1970. Some physical, chemical and taxonomic features of the soluble antigens of the Brucellae. J. Infect. Dis. *121*: 522–527.

Freese, A., H. Reichenbach and H. Lünsdorf. 1997. Further characterization and *in situ* localization of chain-like aggregates of the gliding bacteria *Myxococcus fulvus* and *Myxococcus xanthus*. J. Bacteriol. *179*: 1246–1252.

Freiberg, C., R. Fellay, A. Bairoch, W.J. Broughton, A. Rosenthal and X. Perret,. 1997. Molecular basis of symbiosis between *Rhizobium* and legumes. Nature *387*: 394–401.

Freitag, A., M. Rudert and E. Bock. 1987. Growth of *Nitrobacter* by dissimilatoric nitrate reduction. FEMS Microbiol. Lett. *48*: 105–109.

Freitag, N.E., H.S. Seifert and M. Koomey. 1995. Characterization of the *pilF-pilD* pilus-assembly locus of *Neisseria gonorrhoeae*. Mol. Microbiol. *16*: 575–586.

Freitas, M., F.A. Rainey, M.F. Nobre, A.J. Silvestre and M.S. da Costa. 2003. *Tepidimonas aquatica* sp. nov., a new slightly thermophilic β-proteobacterium isolated from a hot water tank. Syst. Appl. Microbiol. *26*: 376-381.

French, D.M., T.F. McElwain, T.C. McGuire and G.H. Palmer. 1998. Expression of *Anaplasma marginale* major surface protein 2 variants during persistent cyclic rickettsemia. Infect. Immun. *66*: 1200–1207.

French, T.W. and J.W. Harvey. 1983. Serologic diagnosis of infectious cyclic thrombocytopenia in dogs using an indirect fluorescent antibody test. Am. J. Vet. Res. *44*: 2407–2411.

Frenkiel-Krispin, D. and A. Minsky. 2002. Biocrystallization: A last-resort survival strategy in bacteria. ASM News. *68*: 277–283.

Friedberg, D. 1977. Effect of light on *Bdellovibrio bacteriovorus*. J. Bacteriol. *131*: 399–404.

Friedman, B.A. and P.R. Dugan. 1968. Identification of *Zoogloea* species and the relationship to zoogloeal matrix and floc formation. J. Bacteriol. *95*: 1903–1909.

Friedman, B.A., P.R. Dugan, R.M. Pfister and C.C. Remsen. 1968. Fine structure and composition of the zoogloeal matrix surrounding *Zoogloea ramigera*. J. Bacteriol. *96*: 2144–2153.

Friedman, B.A., P.R. Dugan, R.M. Pfister and C.C. Remsen. 1969. Structure of exocellular polymers and their relationship to bacterial flocculation. J. Bacteriol. *98*: 1328–1334.

Friedman, R.L., R.L. Fiederlein, L. Glasser and J.N. Galgiani. 1987. *Bordetella pertussis* adenylate cyclase: effects of affinity-purified adenylate cyclase on human polymorphonuclear leukocyte functions. Infect. Immun. *55*: 135–140.

Friedrich, C.G. 1998. Physiology and genetics of sulfur-oxidizing bacteria. Adv. Microb. Physiol. *39*: 235–289.

Friedrich, C.G. and G. Mitrenga. 1981. Oxidation of thiosulfate by *Paracoccus denitrificans* and other hydrogen bacteria. FEMS Microbiol. Lett. *10*: 209–212.

Friedrich, M. and B. Schink. 1995. Isolation and characterization of a desulforubidin-containing sulfate-reducing bacterium growing with glycolate. Arch. Microbiol. *164*: 271–279.

Friedrich, M., N. Springer, W. Ludwig and B. Schink. 1996. Phylogentic positions of *Desulfofustis glycolicus* gen. nov., sp. nov., and *Syntrophobotulus glycolicus* gen. nov., sp. nov., two new strict anaerobes growing with glycolic acid. Int. J. Syst. Bacteriol. *46*: 1065–1069.

Fries, M.R., L.J. Forney and J.M. Tiedje. 1997. Phenol- and toluene-degrading microbial populations from an aquifer in which successful trichloroethene cometabolism occurred. Appl. Environ. Microbiol. *63*: 1523–1530.

Fries, M.R., J. Zhou, J. Chee-Sanford and J.M. Tiedje. 1994. Isolation, characterization, and distribution of denitrifying toluene degraders from a variety of habitats. Appl. Environ. Microbiol. *60*: 2802–2810.

Frisk, C.S. and J.E. Wagner. 1977. Experimental hamster enteritis: an electron microscopic study. Am. J. Vet. Res. *38*: 1861–1868.

Fritsche, K., G. Auling, J.R. Andreesen and U. Lechner. 1999a. *Defluvibacter lusatiae* gen. nov., sp. nov., a new chlorophenol-degrading member of the alpha-2 subgroup of *Proteobacteria*. Syst. Appl. Microbiol. *22*: 197–204.

Fritsche, K., G. Auling, J.R. Andreesen and U. Lechner. 1999b. *In* Validation of the publication of new names and new combinations previously effectively published outside the IJSB. List No. 71. Int. J. Syst. Bacteriol. *49*: 1325–1326.

Fröhlich, J., H. Sass, H.D. Babenzien, T. Kuhnigk, A. Varma, S. Saxena, C. Nalepa, P. Pfeiffer and H. König. 1999a. Isolation of *Desulfovibrio intestinalis* sp. nov. from the hindgut of the lower termite *Mastotermes darwiniensis*. Can. J. Microbiol. *45*: 145–152.

Fröhlich, J., H. Sass, H.D. Babenzien, T. Kuhnigk, A. Varma, S. Saxena, C. Nalepa, P. Pfeiffer and H. König. 1999b. *In* Validation of the publication of new names and new combinations previously effectively published outside the IJSB. List No. 71. Int. J. Syst. Bacteriol. *49*: 1325–1326.

Fröhner, C., O. Oltmanns and F. Lingens. 1970. Isolation and characterization of bacteria, growing on pyrazon. Arch. Mikrobiol. *74*: 82–89.

Frøholm, L.O. and K. Bøvre. 1972. Fimbriation associated with the spreading-corroding colony type of *Moraxella kingii*. Acta Pathol. Microbiol. Scand Sect. B. *80*: 641–648.

Frøholm, L.O., K. Jyssum and K. Bovre. 1973. Electron microscopical and cultural features of *Neisseria meningitidis* competence variants. Acta Pathol. Microbiol. Scand. [B] Microbiol. Immunol. *81*: 525–537.

Frøholm, L.O. and K. Sletten. 1977. Purification and *N*-terminal sequence of a fimbrial protein from *Moraxella nonliquefaciens*. FEBS Lett. *73*: 29–32.

Frolund, B., P.A. Suci, S. Langille, R.M. Weiner and G.G. Geesey. 1996. Influence of protein conditioning films on binding of bacterial polysaccharide adhesin from *Hyphomonas* MHS-3. Biofouling *10*: 17–30.

Froman, B.E., R.C. Tait and L.D. Gottlieb. 1989. Isolation and characterization of the phosphoglucose isomerase gene from *Escherichia coli*. Mol. Gen. Genet. *217*: 126–131.

Frosch, M. and A. Muller. 1993. Phospholipid substitution of capsular polysaccharides and mechanisms of capsule formation in *Neisseria meningitidis*. Mol. Microbiol. *8*: 483–493.

Frosch, M., C. Weisgerber and T.F. Meyer. 1989. Molecular characterization and expression in *Escherichia coli* of the gene complex encoding the polysaccharide capsule of *Neisseria meningitidis* group B. Proc. Natl. Acad. Sci. U.S.A. *86*: 1669–1673.

Fründ, C. and Y. Cohen. 1992. Diurnal cycles of sulfate reduction under oxic conditions in cyanobacterial mats. Appl. Environ. Microbiol. *58*: 70–77.

Frunzke, K. and O. Meyer. 1990. Nitrate respiration, denitrification, and utilization of nitrogen sources by aerobic carbon monoxide-oxidizing bacteria. Arch. Microbiol. *154*: 168–174.

Fry, J.C. and D.G. Staples. 1974. The occurrence and role of *Bdellovibrio bacteriovorus* in polluted river water. Water Res. *8*: 1029–1035.

Fry, J.C. and D.G. Staples. 1976. Distribution of *Bdellovibrio bacteriovorus* in sewage works, river waters, and sediments. Appl. Environ. Microbiol. *31*: 469–474.

Fu, C. and R. Knowles. 1988. H_2 supports nitrogenase activity in carbon-starved *Azospirillum lipoferum* and *A. amazonense*. Can. J. Microbiol. *34*: 825–829.

Fu, C. and R. Knowles. 1989. Intracellular location and O_2 sensitivity of uptake hydrogenase in *Azospirillum* sp. Appl. Environ. Microbiol. *55*: 2315–2319.

Fu, H., A. Hartmann, R.G. Lowery, W.P. Fitzmaurice, G.P. Roberts and R.H. Burris. 1989. Posttranslational regulatory system for nitrogenase activity in *Azospirillum* spp. J Bacteriol. *171*: 4679–4685.

Fudou, R., Y. Jojima, T. Iizuka and S. Yamanaka. 2002. *Haliangium ochraceum* gen. nov., sp. nov. and *Haliangium tepidum* sp. nov.: novel moderately halophilic myxobacteria isolated from coastal saline environments. J Gen Appl Microbiol. *48*: 109-116.

Fuentes, L. 1979. Primer caso de fiebre de las Montanas Rocosas en Costa Rica, America Central. Rev. Latinoam. Microbiol. *58*: 227–237.

Fuentes-Ramírez, L.E., R. Bustillos-Cristales, A. Tapia-Hernández, T. Jiménez-Salgado, E.T. Wang, E. Martínez-Romero and J. Caballero-Mellado. 2001. Novel nitrogen-fixing acetic acid bacteria, *Gluconacetobacter johannae* sp. nov. and *Gluconacetobacter azotocaptans* sp. nov., associated with coffee plants. Int. J. Syst. Evol. Microbiol. *51*: 1305–1314.

Fuentes-Ramírez, L.E., T. Jimenez-Salgado, I.R. Abarca-Ocampo and J. Caballero-Mellado. 1993. *Acetobacter diazotrophicus*, an indoleacetic acid producing bacterium isolated from sugarcane cultivars of Mexico. Plant Soil *154*: 145–150.

Fuerst, J.A. 1995. The planctomycetes: emerging models for microbial ecology, evolution and cell biology. Microbiology *141*: 1493–1506.

Fuerst, J.A., J.A. Hawkins, A. Holmes, L.I. Sly, C.J. Moore and E. Stackebrandt. 1993. *Porphyrobacter neustonensis* gen. nov., sp. nov., an aerobic bacteriochlorophyll-synthesizing budding bacterium from fresh water. Int. J. Syst. Bacteriol. *43*: 125–134.

Fuerst, J.A. and A.C. Hayward. 1969a. The sheathed flagellum of *Pseudomonas stizolobii*. J. Gen. Microbiol. *58*: 239–245.

Fuerst, J.A. and A.C. Hayward. 1969b. Surface appendages similar to fimbriae (pili) on *Pseudomonas* species. J. Gen. Microbiol. *58*: 227–237.

Fuerst, J.A. and R.I. Webb. 1991. Membrane-bounded nucleoid in the eubacterium *Gemmata obscuriglobus*. Proc. Natl. Acad. Sci. U.S.A. *88*: 8184–8188.

Fuhrman, J.A., K. McCallum and A.A. Davis. 1993. Phylogenetic diversity of subsurface marine microbial communities from the Atlantic and Pacific Oceans. Appl. Environ. Microbiol. *59*: 1294–1302.

Fuhrmann, S., M. Ferner, T. Jeffke, A. Henne, G. Gottschalk and O. Meyer. 2003. Complete nucleotide sequence of the circular megaplasmid pHCG3 of *Oligotropha carboxidovorans*: function in the chemolithoautotrophic utilization of CO, H_2 and CO_2. Gene *322*: 67-75.

Fujii, K., N. Urano, H. Ushio, M. Satomi and S. Kimura. 2001. *Sphingomonas cloacae* sp. nov., a nonylphenol-degrading bacterium isolated from wastewater of a sewage-treatment plant in Tokyo. Int. J. Syst. Evol. Microbiol. *51*: 603–610.

Fujishima, M. 1993. Control of morphological changes of the endonuclear symbiont *Holospora* of the ciliate *Paramecium*. *In* Sato, Ishida and Ishikawa (Editors), Endocytobiology V: 5th International Colloquim on Endocytobiology and Symbiosis, Tübingen University Press, Tübingen. 505–508.

Fujishima, M., H. Dohra and M. Kawai. 1997. Quantitative changes in periplasmic proteins of the macronucleus specific bacterium *Holospora obtusa* in the infection process of the ciliate *Paramecium caudatum*. J. Eukaryot. Microbiol. *44*: 636–642.

Fujishima, M. and K. Heckmann. 1984. Intraspecies and interspecies transfer of endosymbionts in *Euplotes*. J. Exp. Zool. *230*: 339–345.

Fujishima, M., K. Nagahara and Y. Kojima. 1990. Changes in morphology, buoyant density and protein composition in differentiation from the reproductive short form to the infectious long form of *Holospora obtusa*, a macronucleus - specific symbiont of the ciliate *Paramecium caudatum*. Zool. Sci. *7*: 849–860.

Fujita, S., T. Yoshida and F. Matsubara. 1981. *Pseudomonas pickettii* bacteremia. J. Clin. Microbiol. *13*: 781–782.

Fujitani, S., T. Komano and S. Inouye. 1991. A unique repetitive DNA sequence in the *Myxococcus xanthus* genome. J. Bacteriol. *173*: 2125–2127.

Fukaya, M., H. Takemura, H. Okumura, Y. Kawamura, S. Horinouchi and T. Beppu. 1990. Cloning of genes responsible for acetic acid resistance in *Acetobacter aceti*. J. Bacteriol. *172*: 2096–2104.

Fukaya, M., H. Takemura, K. Tayama, H. Okumura, Y. Kawamura, S. Horinouchi and T. Beppu. 1993. The *aarC* gene responsible for acetic acid assimilation confers acetic acid resistace on *Acetobacter acetii*. J. Ferment. Bioeng. *76*: 270–275.

Fukaya, M., K. Tayama, T. Tamaki, H. Tagami, H. Okumura, Y. Kawamura and T. Beppu. 1989. Cloning of the membrane-bound aldehyde dehydrogenase gene of *Acetobacter polyoxogenes* and improvement of acetic acid production by use of the cloned gene. Appl. Environ. Microbiol. *55*: 171–176.

Fukuda, T., T. Kitao and Y. Keida. 1954. Studies on the causative agent of "Hyuganetsu" disease. I. Isolation of the agent and its inoculation trial in human beings. Med. Biol. *32*: 200–209.

Fukuda, T., T. Sasahara and T. Kitao. 1973. Causative agent of "Hyuganetsu" disease. 11. Characteristics of rickettsia-like organisms isolated from metacercaria of *Stellantchasmus falcatus*. Kansenshogaku Zasshi. *47*: 474–482.

Fukui, M., A. Teske, B. Assmus, G. Muyzer and F. Widdel. 1999. Physiology, phylogenetic relationships, and ecology of filamentous sulfate-reducing bacteria (genus *Desulfonema*). Arch. Microbiol. *172*: 193–203.

Fukui, M., A. Teske, B. Assmus, G. Muyzer and F. Widdel. 2000. *In* Validation of the publication of new names and new combinations previously effectively published outside the IJSEM. List No. 75. Int. J. Syst. Evol. Microbiol. *50*: 1415–1417.

Fukui, S., R.M. Hochster, R. Durbin, E.E. Grebner and D.S. Feingold. 1963. The conversion of sucrose to α-D-ribohexopyranosyl-3-ulose-β-3-D-fructofuranoside by cultures of *Agrobacterium tumefaciens*. Bull. Res. Counc. Israel. *11A4*: 314–320.

Fukumori, Y., K. Watanabe and T. Yamanaka. 1987. Cytochrome-aa_3 from the aerobic photoheterotroph *Erythrobacter longus* - purification, and enzymatic and molecular-features. J. Biochem. (Tokyo) *102*: 777–784.

Fukumoto, F., M. Sato and Y. Minobe. 1997. Transformation of pBR322-derived plasmids in phytopathogenic *Pseudomonas avenae* and enhanced transformation in its proline-auxotrophic mutant. Curr. Microbiol. *34*: 138 –143.

Fuller, H.S., E.S. Murray, J.C. Ayres, J.C. Snyder and L. Potash. 1951. Studies of rickettsialpox. I. Recovery of the causative agent from house mice in Boston, Massachusetts. Am. J. Hyg. *54*: 82–100.

Fulop, V., J.W. Moir, S.J. Ferguson and J. Hajdu. 1993. Crystallization and preliminary crystallographic study of cytochrome *cd*₁ nitrite reductase from *Thiosphaera pantotropha*. J. Mol. Biol. *232*: 1211–1212.

Fulthorpe, R.R. and D.G. Allen. 1995. A comparison of organochlorine removal from bleached kraft pulp and paper-mill effluents by dehalogenating *Pseudomonas, Ancylobacter* and *Methylobacterium* strains. Appl. Microbiol. Biotechnol. *42*: 782–789.

Fulthorpe, R.R., S.N. Liss and D.G. Allen. 1993. Characterization of bacteria isolated from a bleached kraft pulp mill wastewater treatment system. Can. J. Microbiol. *39*: 13–24.

Fulton, H.R., P.O. Sikorowski and B.R. Morment. 1974. A survey of North Mississippi mosquitoes for pathogenic microorganisms. Mosq. News. *34*: 86–90.

Fung-Tomc, J., K. Bush, B. Minassian, B. Kolek, R. Flamm, E. Gradelski and D. Bonner. 1997. Antibacterial activity of BMS-180680, a new catechol-containing monobactam. Antimicrob. Agents Chemother. *41*: 1010–1016.

Funke, G., T. Hess, A. von Graevenitz and P. Vandamme. 1996. Characteristics of *Bordetella hinzii* strains isolated from a cystic fibrosis patient over a 3-year period. J. Clin. Microbiol. *34*: 966–969.

Gaffney, T.D. and T.G. Lessie. 1987. Insertion sequence dependent rearrangements of *Pseudomonas cepacia* plasmid Ptgl1. J. Bacteriol. *169*: 224–230.

Gagnon, H. and R.K. Ibrahim. 1998. Aldonic acids: A novel family of *nod* gene inducers of *Mesorhizobium loti, Rhizobium lupini,* and *Sinorhizobium meliloti*. Mol. Plant-Microbe Interact. *11*: 988–998.

Gahrn-Hansen, B., P. Alstrup, R. Dessau, K. Fuursted, A. Knudsen, H. Olsen, H. Oxhoj, A.R. Petersen, A. Siboni and K. Siboni. 1988. Outbreak of infection with *Achromobacter xylosoxidans* from contaminated intravascular pressure transducers. J. Hosp. Infect. *12*: 1–6.

Gaitatzis, N., A. Hans, R. Müller and S. Beyer. 2001. The *mtaA* gene of the myxothiazol biosynthetic gene cluster from *Stigmatella aurantiaca* DW4/3-1 encodes a phosphopantetheinyl transferase that activates polyketide synthases and polypeptide synthetases. J. Biochem. (Tokyo) *129*: 119–124.

Galarneault, T.P. and E. Leifson. 1956. Taxonomy of *Lophomonas* N. Gen. Can. J. Microbiol. *2*: 102–110.

Galarneault, T.P. and E. Leifson. 1964. *Pseudomonas vesiculare* (Büsing et al.) nov. comb. Int. Bull. Bacteriol. Nomencl. Taxon. *14*: 165–168.

Gal'chenko, V.F. 1977. New species of methanotrophic bacteria. *In* Skryabin, Kondratjeva, Zavarzin, Trotsenko, and Nesterov (Editors), 2nd International Symposium, Microbial Growth on C1 Compounds, Vol. 1, USSR Academy of Sciences, Pushchino. 10–12.

Gal'chenko, V.F. 1994. Sulfate reduction, methane generation and methane oxidation in different Bunger Hills reservoirs (Antarctica). Mikrobiologiya *63*: 388–396.

Gal'chenko, V.F., F.N. Abramochkina, L.V. Bezrukova, E.N. Sokolova and M.V. Ivanov. 1988. The species composition of aerobic methanotrophic microflora in the Black Sea. Mikrobiologiya *57*: 248–253.

Gal'chenko, V.F. and A.I. Nesterov. 1981. Numerical analysis of protein electrophoretograms for obligate methanotrophic bacteria. Mikrobiologiya *50*: 725–730.

Gal'chenko, V.F., V.N. Shishkina, N.E. Suzina and I.A. Trotsenko. 1977. Isolation and properties of new strains of obligate methanotrophs. Mikrobiologiya *46*: 890–897.

Gal'chenko, V.F., V.N. Shishkina, V.S. Tiurin and I.A. Trotsenko. 1975. Isolation of pure methanotrophic cultures and their properties. Mikrobiologiya *44*: 844–850.

Galibert, F., T.M. Finan, S.R. Long, A. Puhler, P. Abola, F. Ampe, F. Barloy-Hubler, M.J. Barnett, A. Becker, P. Boistard, G. Bothe, M. Boutry, L. Bowser, J. Buhrmester, E. Cadieu, D. Capela, P. Chain, A. Cowie, R.W. Davis, S. Dreano, N.A. Federspiel, R.F. Fisher, S. Gloux, T. Godrie, A. Goffeau, B. Golding, J. Gouzy, M. Gurjal, I. Hernandez-Lucas, A. Hong, L. Huizar, R.W. Hyman, T. Jones, D. Kahn, M.L. Kahn, S. Kalman, D.H. Keating, E. Kiss, C. Komp, V. Lelaure, D. Masuy, C.

Palm, M.C. Peck, T.M. Pohl, D. Portetelle, B. Purnelle, U. Ramsperger, R. Surzycki, P. Thebault, M. Vandenbol, F.J. Vorholter, S. Weidner, D.H. Wells, K. Wong, K.C. Yeh and J. Batut. 2001. The composite genome of the legume symbiont *Sinorhizobium meliloti*. Science *293*: 668–672.

Galimand, M. 1999. High-level chloramphenicol resistance in *Neisseria meningitidis*. New Engl. J. Med. *340*: 824–824.

Galimand, M., G. Gerbaud and P. Courvalin. 2000. Spectinomycin resistance in *Neisseria* spp. due to mutations in 16S rRNA. Antimicrob. Agents Chemother. *44*: 1365–1366.

Gälli, R. and T. Leisinger. 1985. Specialized bacterial strains for the removal of dichloromethane from industrial waste. Conserv. Recycl. *8*: 91–100.

Gallus, C., N. Gorny, W. Ludwig and B. Schink. 1997. Anaerobic degradation of alpha-resorcylate by a nitrate-reducing bacterium, *Thauera aromatica* strain AR-1. Syst. Appl. Microbiol. *20*: 540–544.

Gallus, C. and B. Schink. 1998. Anaerobic degradation of alpha-resorcylate by *Thauera aromatica* strain AR-1 proceeds via oxidation and decarboxylation to hydroxyhydroquinone. Arch. Microbiol. *169*: 333–338.

Galushko, A.S. and E.P. Rosanova. 1991. *Desulfobacterium cetonicum* spec. nov., a sulfate-reducing bacterium oxidizing fatty acids and ketones. Mikrobiologiya *60*: 102–107.

Galushko, A.S. and E.P. Rozanova. 1994. *In* Validation of the publication of new names and new combinations previously effectively published outside the IJSB. List No. 49. Int. J. Syst. Bacteriol. *44*: 370–371.

Galván, A., F. de Castro and D. Fernandez-Galiano. 1987. Ultrastructure of the fruiting body of *Stigmatella erecta* (*Myxobacterales*). Trans Am. Microsc. Soc. *106*: 89–93.

Galván, A., M.A. Marcotegui and F. Decastro. 1992. Ultrastructure of natural and induced myxospores of *Archangium gephyra*. Can. J. Microbiol. *38*: 130–134.

Gamble, T.N., M.R. Betlach and J.M. Tiedje. 1977. Numerically dominant denitrifying bacteria from world soils. Appl. Environ. Microbiol. *33*: 926–939.

Gambrill, M.R. and C.L. Wisseman, Jr.. 1973. Mechanisms of immunity in typhus infections. II. Multiplication of typhus richettsiae in human macrophage cell cultures in the nonimmune system: Influence of virulence of rickettsial strains and of chloramphenicol. Infect. Immun. *8*: 519–527.

Gandy, D.E. 1968. A technique for screen bacteria causing Brown blotch of cultivated mushrooms, Rep. Glasshouse Crops Res. Inst. . 150–154.

Gao, S., M. Garnier and J.M. Bové. 1993. Production of monoclonal antibodies recognizing most strains of the greening BLO by *in vitro* immunization with an antigenic protein purified from the BLO. Proc. 12th Conf. Intern. Org. Citrus Virol, 244–249.

Gao, W.M. and S.S. Yang. 1995. A *Rhizobium* strain that nodulates and fixes nitrogen in association with alfalfa and soybean plants. Microbiology *141*: 1957–1962.

Garcia, J.L., S. Roussos, D. Gauthier, G. Rinaudo and M. Mandel. 1983. Taxonomical study of free-living N₂-fixing bacteria isolated from the endorhizosphere of rice. Ann. Microbiol. *B134*: 329–346.

García-de los Santos, A. and S. Brom. 1997. Characterization of two plasmid-borne lpsβ loci of *Rhizobium etli* required for lipopolysaccharide synthesis and for optimal interaction with plants. Mol. Plant-Microbe Interact. *10*: 891–902.

Garcia-Pichel, F. 1989. Rapid bacterial swimming measured in swarming cells of *Thiovulum majus*. J. Bacteriol. *171*: 3560–3563.

García-Valdés, E., E. Cozar, R. Rotger, J. Lalucat and J. Ursing. 1988. New naphthalene-degrading marine *Pseudomonas* strains. Appl. Environ. Microbiol. *54*: 2478–2485.

Gardan, L., C. Dauga, P. Prior, M. Gillis and G.S. Saddler. 2000. *Acidovoarx anthurii* sp. nov., a new phytopathogenic bacterium which causes bacterial leaf-spot of anthurium. Int. J. Syst. Evol. Microbiol. *50*: 235–246.

Gardener, B.B.M. and F.J. de Bruijn. 1998. Detection and isolation of novel rhizopine-catabolizing bacteria from the environment. Appl. Environ. Microbiol. *64*: 4944–4949.

Garg, R.P., J. Huang, W. Yindeeyoungyeon, T.P. Denny and M.A. Schell. 2000. Multicomponent transcriptional regulation at the complex promoter of the exopolysaccharide I biosynthetic operon of *Ralstonia solanacearum*. J. Bacteriol. *182*: 6659–6666.

Garnier, M. and J.M. Bové. 1983. Transmission of the organism associated with citrus greening disease from sweet orange (*Citrus sinensis* cultivar *Madame Vinous*) to periwinkle (*Vinca rosea*) by dodder (*Cuscuta campestris*). Phytopathology *73*: 1358–1363.

Garnier, M. and J.M. Bové. 1993. Citrus greening disease and the greening bacterium. Proc. 12th Conf. Intern. Org. Citrus Virol, 212–219.

Garnier, M. and J.M. Bové. 1996. Distribution of the huanglongbing (greening) liberobacter species in fifteen african and asian countries. Proc. 13th Conf. Intern. Org. Citrus Viro., 388–391.

Garnier, M., N. Danel and J.M. Bové. 1984. Aetiology of citrus greening disease. Ann. Inst. Pasteur *135A*: 169–179.

Garnier, M., S. Jagoueix-Eveillard, C.P.R. Cronje, H.P. Le Roux and J.M. Bové. 2000. Genomic characterization of a liberibacter present in an ornamental rutaceous tree, *Calodendrum capense*, in the Western Cape province of South Africa. Proposal of "*Candidatus* Liberibacter africanus subsp. *capensis*". Int. J. Syst. Evol. Microbiol. *50*: 2119–2125.

Garrard, W.T. 1971. Selective release of proteins from *Spirillum itersonii* by tris(hydroxymethyl)aminomethane and ethylenediaminetetraacetate. J. Bacteriol. *105*: 93–100.

Garrard, W.T. 1972. Synthesis, assembly, and localization of periplasmic cytochrome *c*. J. Biol. Chem. *247*: 5935–5943.

Gaubier, P., D. Vega and R. Cooke. 1992. Nucleotide sequence of a 2 kb plasmid from *Pseudomonas cepacia* implicated in the degradation of phenylcarbamate herbicides. DNA Seq. *2*: 269–271.

Gaudy, E. and R.S. Wolfe. 1961. Factors affecting filamentous growth of *Sphaerotilus natans*. Can. J. Microbiol. *9*: 580–584.

Gaudy, E. and R.S. Wolfe. 1962. Compostition of an extracellular polysaccharaide produced by *Sphaerotilus natans*. Appl. Microbiol. *10*: 200–205.

Gaunt, M.W., S.L. Turner, L. Rogottier-Gois, S.A. Lloyd-Macgilp and J.P.W. Young. 2001. Phylogenies of *atpD* and *recA* support the small subunit rRNA-based classification of rhizobia. Int. J. Syst. Evol. Microbiol. *51*: 2037–2048.

Gaur, D. and S.G. Wilkinson. 1996. Structure of the O-specific polysaccharide from *Burkholderia vietnamiensis* strain LMG 6998. Carbohydr. Res. *295*: 179–184.

Gaur, Y.D. and A.N. Sen. 1979. Cross inoculation group specificity in Cicer *Rhizobium* symbiosis. New Phytol. *83*: 745–754.

Gauthier, D.K., G.D. Clark-Walker, W.T. Garrard, Jr. and J. Lascelles. 1970. Nitrate reductase and soluble cytochrome *c* in *Spirillum itersonii*. J. Bacteriol. *102*: 797–803.

Gauthier, M.J. 1976. Morphological, physiological, and biochemical characteristics of some violet-pigmented bacteria isolated from seawater. Can. J. Microbiol. *22*: 138–149.

Gauthier, V., S. Redercher and J.C. Block. 1999. Chlorine inactivation of *Sphingomonas* cells attached to goethite particles in drinking water. Appl. Environ. Microbiol. *65*: 355–357.

Ge, Z., Q. Jiang and D.E. Taylor. 1996. Conservation and diversity of the *Helicobacter pylori* copper-transporting ATPase gene (*copA*) sequence among *Helicobacter* species and *Campylobacter* species detected by PCR and RFLP. Helicobacter *1*: 112–117.

Gebers, R. 1981. Enrichment, isolation, and emended description of *Pedomicrobium ferrugineum* Aristovskaya and *Pedomicrobium manganicum* Aristovskaya. Int. J. Syst. Bacteriol. *31*: 302–316.

Gebers, R. 1989. Genus *Pedomicrobium* Aristoskaya 1961, 957[AL], emend. Gebers 1981, 313. *In* Staley, Bryant, Pfennig and Holt (Editors), Bergey's Manual of Systematic Bacteriology, 1st ed., Vol. 3, The Williams & Wilkins Co., Baltimore. 1910–1919.

Gebers, R. and M. Beese. 1988. *Pedomicrobium americanum* sp. nov. and *Pedomicrobium australicum* sp. nov. from aquatic habitats, *Pedomicrobium* gen. emend., and *Pedomicrobium ferrugineum* sp. emend. Int. J. Syst. Bacteriol. *38*: 303–315.

Gebers, R. and P. Hirsch. 1978. Isolation and investigation of *Pedomicrobium* spp., heavy metal-depositing bacteria from soil habitats. *In* Krum-

bein (Editor), Environmental Biogeochemistry and Geomicrobiology, Vol. 3: Methods, Metals, and Assessment, Ann Arbor Science Publ. Inc., Ann Arbor. 911–922.

Gebers, R. and P. Hirsch. 1979. Isolierung und Eigenchaften neuer eisen und mangan-ablagernder bakterien der gattung *Pedomicrobium* aus boden. Mitt. Deutsch. Bodenkundl. Gesellsch. *29*: 473–478.

Gebers, R., M. Mandel and P. Hirsch. 1981a. Deoxyribonucleic-acid base composition and nucleotide distribution of *Pedomicrobium* spp. Zbl. Bakt. Mikrobio. Hyg. C-Allg. *2*: 332–338.

Gebers, R., R.L. Moore and P. Hirsch. 1981b. DNA/DNA reassociation studies on the genus *Pedomicrobium*. FEMS Microbiol. Lett. *11*: 283–286.

Gebers, R., R.L. Moore and P. Hirsch. 1984. Physiological properties and DNA–DNA homologies of *Hyphomonas polymorpha* and *Hyphomonas neptunium*. Syst. Appl. Microbiol. *5*: 510–517.

Gebers, R., U. Wehmeyer, T. Roggentin, H. Schlesner, J. Kölbel-Boelke and P. Hirsch. 1985. Deoxyribonucleic acid base compositions and nucleotide distributions of 65 strains of budding bacteria. Int. J. Syst. Bacteriol. *35*: 260–269.

Gebers, R., U. Wehmeyer, T. Roggentin, H. Schlesner, J. Kölbel-Boelke and P. Hirsch. 1986. Deoxyribonucleic acid homologies of *Hyphomicrobium* spp., *Hyphomonas* spp., and other hyphal, budding bacteria. Int. J. Syst Bacteriol. *36*: 241-245.

Gebhardt, N.A., D. Linder and R.K. Thauer. 1983. Anaerobic acetate oxidation to carbon dioxide by *Desulfobacter postgatei*. 2. Evidence from ^{14}C-labeling studies of the operation of the citric acid cycle. Arch. Microbiol. *136*: 230–233.

Gebhardt, N.A., R.K. Thauer, D. Linder, P.M. Kaulfers and N. Pfennig. 1985. Mechanism of acetate oxidation to CO_2 with elemental sulfur in *Desulfuromonas acetoxidans*. Arch. Microbiol. *141*: 392–398.

Gebhart, C.J., S.M. Barns, S. McOrist, G.-F. Lin and G.H.K. Lawson. 1993. Ileal symbiont intracellularis, an obligate intracellular bacterium of porcine intestines showing a relationship to *Desulfovibrio* species. Int. J. Syst. Bacteriol. *43*: 533–538.

Gebhart, C.J., P. Edmonds, G.E. Ward, H.J. Kurtz and B. D.J.. 1985a. In Validation of the publication of new names and new combinations previously effectively published outside the IJSB. List No. 19. Int. J. Syst. Bacteriol. *35*: 535.

Gebhart, C.J., G.F. Lin, S.M. McOrist, G.H. Lawson and M.P. Murtaugh. 1991. Cloned DNA probes specific for the intracellular *Campylobacter*-like organism of porcine proliferative enteritis. J. Clin. Microbiol. *29*: 1011–1015.

Gebhart, C.J., S. McOrist, G.H. Lawson, J.E. Collins and G.E. Ward. 1994. Specific *in situ* hybridization of the intracellular organism of porcine proliferative enteropathy. Vet. Pathol. *31*: 462–467.

Gebhart, C.J., G.E. Ward and H.J. Kurtz. 1985b. In vitro activities of 47 antimicrobial agents against three *Campylobacter* spp. from pigs. Antimicrob. Agents Chemother. *27*: 55–59.

Gebhart, C.J., G.E. Ward and M.P. Murtaugh. 1989. Species-specific cloned DNA probes for the identification of *Campylobacter hyointestinalis*. J. Clin. Microbiol. *27*: 2717–2723.

Geelen, D., P. Mergaert, R.A. Geremia, S. Goormachtig, M. Van Montagu and M. Holsters. 1993. Identification of *nodSUIJ* genes in Nod locus 1 of *Azorhizobium caulinodans*: evidence that *nodS* encodes a methyltransferase involved in Nod factor modification. Mol. Microbiol. *9*: 145–154.

Geerlof, A., J. Stoorvogel, J.A. Jongejan, E.J.T.M. Leenen, T.J.G.M. Van Dooren, W.J.J. Van den Tweel and J.A. Dunie. 1994a. Studies on the production of (S)-(+)-solketal (2,2-dimethyl-1, 3-dioxylane-4-methanol) by enantioselective oxidation of racemic solketal with *Comamonas testosteroni*. Appl. Microbiol. Biotechnol. *42*: 8–15.

Geerlof, A., J.B.A. Van Tol, J.A. Jongejan and J.A. Duine. 1994b. Enantioselective conversions of the racemic C_3-alcohol synthons, glycidol (2,3-epoxy-1-propanol), and solketal (2,2-dimethyl-4-(hydroxymethyl)-1,3-dioxolane) by quinohaemoprotein alcohol dehydrogenases and bacteria containing such enzymes. Biosci. Biotechnol. Biochem. *58*: 1028–1036.

Gehring, F. 1962. Untersuchungen über den Infektionsverlauf einer durch *Pectobacterium parthenii* (Starr) Hellmers var. dianthicola Hellmers verursachten Nelkenbakteriose sowie über enymatische Eigenschaften dieses Bakteriums in Verleich mit *Pseudomonas caryophylli* (Burkholder) Starr et Burkholder und einigen typyschen Nassfäuleerregern. Phytopathol. Zeitschr. *43*: 383-407.

Geisselsoder, J., J.M. Campos and D.R. Zusman. 1978. Physical characterization of bacteriophage MX4, a generalized transducing phage for *Myxococcus xanthus*. J. Mol. Biol. *119*: 179-189.

Geitler, L. 1925. Über *Polyangium parasiticum* n.sp., eine submerse, parasitische Myxobacteriacee. Arch. Protistenkd. *50*: 67-88.

Gelvan, I., M. Varon and E. Rosenberg. 1987. Cell-density-dependent killing of *Myxococcus xanthus* by Autocide Amv. J. Bacteriol. *169*: 844-849.

Gelvin, S.B. 1992. Chemical signaling between *Agrobacterium* and its host plant. *In* Verma (Editor), Molecular Signals in Plant-Microbe Communications, CRC Press, Boca Raton, Fla. 137-167.

Genet, R., P.H. Benetti, A. Hammadi and A. Menez. 1995. L-Tryptophan 2',3'-oxidase from *Chromobacterium violaceum*: substrate specificity and mechanistic implications. J. Biol. Chem. *270*: 23540-23545.

Gerardo, S.H., M. Marina, D.M. Citron, M.C. Claros, M.K. Hudspeth and E.J. Goldstein. 1997. *Bilophila wadsworthia* clinical isolates compared by polymerase chain reaction fingerprinting. Clin. Infect. Dis. *25*: S291-S294.

Germani, Y., C. Dauga, P. Duval, M. Huerre, M. Levy, G. Pialoux, P. Sansonetti and P.A.D. Grimont. 1997. Strategy for the detection of *Helicobacter* species by amplification of 16S rRNA genes and identification of *H. felis* in a human gastric biopsy. Res. Microbiol. *148*: 315-326.

Germida, J.J. 1984. Spontaneous induction of bacteriophage during growth of *Azospirillum brasilense* in complex media. Can. J. Microbiol. *30*: 805-808.

Germida, J.J. and L.E. Casida, Jr.. 1983. *Ensifer adhaerens* predatory activity against other bacteria in soil, as monitored by indirect phage analysis. Appl. Environ. Microbiol. *45*: 1380-1388.

Germida, J.J., R.E. Karamanos and J.W.B. Stewart. 1985. A simple microbial bioassay for plant available manganese. Soil Sci. Soc. Am. J. *49*: 1411-1415.

Gerner-Smidt, P., H. Keiser-Nielsen, M. Dorsch, E. Stackebrandt, J. Ursing, J. Blom, A.C. Christensen, J.J. Christensen, W. Frederiksen, S. Hoffmann, W. Holten-Andersen and Y.T. Ying. 1994. *Lautropia mirabilis* gen. nov., sp. nov., a Gram-negative motile coccus with unusual morphology isolated from the human mouth. Microbiology *140*: 1787-1797.

Gerner-Smidt, P., H. Keiser-Nielsen, M. Dorsch, E. Stackebrandt, J. Ursing, J. Blom, A.C. Christensen, J.J. Christensen, W. Frederiksen, S. Hoffmann, W. Holten-Andersen and Y.T. Ying. 1995. *In* Validation of the publication of new names and new combinations previously effectively published outside the IJSB. List No. 53. Int. J. Syst. Bacteriol. *45*: 418-419.

Gerstenberg, C., B. Friedrich and H.G. Schlegel. 1982. Physical evidence for plasmids in autotrophic, especially hydrogen-oxidizing bacteria. Arch. Microbiol. *133*: 90-96.

Gerth, K., N. Bedorf, G. Hofle, H. Irschik and H. Reichenbach. 1996. Epothilons A and B: antifungal and cytotoxic compounds from *Sorangium cellulosum* (Myxobacteria). Production, physico-chemical and biological properties. J. Antibiot. *49*: 560-563.

Gerth, K., N. Bedorf, H. Irschik, G. Hofle and H. Reichenbach. 1994. The soraphens—a family of novel antifungal compounds from *Sorangium cellulosum* (Myxobacteria). 1. Soraphen a(1-Alpha) - fermentation, isolation, biological properties. J. Antibiot. *47*: 23-31.

Gerth, K., R. Metzger and H. Reichenbach. 1993. Induction of myxospores in *Stigmatella aurantiaca* (Myxobacteria)—Inducers and inhibitors of myxospore formation, and mutants with a changed sporulation behavior. J. Gen. Microbiol. *139*: 865-871.

Gerth, K. and H. Reichenbach. 1978. Induction of myxospore formation in *Stigmatella aurantiaca* (*Myxobacterales*). 1. General characterization of system. Arch. Microbiol. *117*: 173-182.

Gerth, K. and H. Reichenbach. 1986. Determination of bacterial am-

monia pools using *Myxococcus virescens* as an example. Anal. Biochem. *152*: 78-82.

Gerth, K. and H. Reichenbach. 1994. Induction of myxospores in *Stigmatella aurantiaca* (Myxobacteria)—Analysis of inducer-inducer and inducer-inhibitor interactions by dose-response curves. Microbiology *140*: 3241-3247.

Gerth, K., W. Trowitzsch, G. Piehl, R. Schultze and J. Lehmann. 1984. Inexpensive media for mass cultivation of myxobacteria. Appl. Microbiol. Biotechnol. *19*: 23-28.

Gessard, C. 1981. Variété mélanogène du bacillu pyocyanique. Ann. Inst. Pasteur *15*: 817-831.

Gessner, A.R. and J.E. Mortensen. 1990. Pathogenic factors of *Pseudomonas cepacia* isolates from patients with cystic fibrosis. J. Med. Microbiol. *33*: 115-120.

Gest, H., M.W. Dits and J.L. Favinger. 1983. Characterization of *Rhodopseudomonas sphaeroides* strain "cordata/81-1". FEMS Microbiol. Lett. *17*: 321-325.

Ghadi, S.C. and U.M. Sangodkar. 1994. Identification of a *meta* cleavage pathway for metabolism of phenoxyacetic acid and phenol in *Pseudomonas cepacia* AC1100. Biochem. Biophys. Res. Commun. *204*: 983-993.

Ghai, S.K., M. Hisamatsu, A. Amemura and T. Harada. 1981. Production and chemical composition of extracellular polysaccharides of *Rhizobium*. J. Gen. Microbiol. *122*: 33-40.

Gherna, R.L. 1994. Culture preservation. *In* Gerhardt, Murray, Wood and Krieg (Editors), Methods for General and Molecular Bacteriology, 2nd Ed., American Society for Microbiology, Washington, D.C. pp. 278-292.

Ghiorse, W.C. and S.D. Chapnick. 1983. Metal-depositing bacteria and the distribution of manganese and iron in swamp waters. *In* Hallberg (Editor), Environmental Biogeochemistry. Ecol. Bull. 35, FRN, Stockholm. pp. 367-376.

Ghiorse, W.C. and P. Hirsch. 1978a. Association of iron and manganese oxides with extracellular polymer of metal depositing, *Pedomicrobium* - like budding bacteria. *In* Abstr. Annu. Meet. Am. Soc. Microbiol, Abstract N7, p.163.

Ghiorse, W.C. and P. Hirsch. 1978b. Iron and manganese deposition by budding bacteria. *In* Krumbein (Editor), Environmental Biogeochemistry and Geomicrobiology, Vol. 3: Methods, Metals, and Assessment, Ann Arbor Science Publ. Inc., Ann Arbor. 897-909.

Ghiorse, W.C. and P. Hirsch. 1979. Ultrastructural-study of iron and manganese deposition associated with extracellular polymers of *Pedomicrobium*-like budding bacteria. Arch. Microbiol. *123*: 213-226.

Ghisalba, O., P. Cevey, M. Kuenzi and H.P. Schär. 1985. Biodegradation of chemical waste by specialized methylotrophs, an alternative to physical methods of waste-disposal. Conserv. Recycl. *8*: 47-71.

Ghisalba, O. and M. Küenzi. 1983. Biodegradation and utilization of monomethyl sulfate by specialized methylotrophs. Experientia (Basel) *39*: 1257-1263.

Ghisalba, O., M. Kuenzi, G.M.R. Tombo and H.P. Schär. 1987. Microbial-degradation and utilization of selected organophosphorus compounds - strategies and applications. Chimia. *41*: 206-215.

Ghisalba, O., H.P. Schär and G.M. Ramos Tombo. 1986. Applications of microbes and microbial enzymes in environmental control and organic synthesis. *In* Schneider (Editor), Enzymes as Catalysts in Organic Synthesis, Kluwer Academic Publishers, Dordrecht; Boston. 233-250.

Ghosh, S.K., P.B. Doctor and P.K. Kulkarni. 1996. Toxicity of zinc in three microbial test systems. Environ. Toxicol. Water Qual. *11*: 13-19.

Ghosh, S., N.R. Mahapatra and P.C. Banerjee. 1997. Metal resistance in *Acidocella* strains and plasmid-mediated transfer of this characteristic to *Acidiphilium multivorum* and *Escherichia coli*. Appl. Environ. Microbiol. *63*: 4523-4527.

Giangiordano, R.A. and D.A. Klein. 1994. Silver ion effects on *Hyphomicrobium* species growth initiation and apparent minimum growth temperatures. Lett. Appl. Microbiol. *18*: 181-183.

Giardini, F., S. Caramello and C. Attisano. 1997. Study of uncommon bacterial flora responsible for conjunctival infection. Igene Moderna *107*: 89–96.

Gibbins, A.M. and K.F. Gregory. 1972. Relatedness among *Rhizobium* and *Agrobacterium* species determined by three methods of nucleic acid hybridization. J. Bacteriol. *111*: 129–141.

Gibson, J. and C.S. Harwood. 1995. Degradation of aromatic compounds by nonsulfer purple bacteria. *In* Blankenship, Madigan and Bauer (Editors), Anoxygenic Photsynthetic Bacteria, Kluwer Academic Publ., Netherlands. 991–1003.

Gibson, J.R., E. Slater, J. Xerry, D.S. Tompkins and R.J. Owen. 1998. Use of an amplified-fragment length polymorphism technique to fingerprint and differentiate isolates of *Helicobacter pylori*. J. Clin. Microbiol. *36*: 2580–2585.

Gibson, J., E. Stackebrandt, L.B. Zablen, R. Gupta and C.R. Woese. 1979. A phylogenetic analysis of the purple photosynthetic bacteria. Curr. Microbiol. *3*: 59–64.

Giesberger, G. 1936. Beitrage zur Kenntnis der Gattung Spirillum, Utrecht. pp. 1–136.

Giesberger, G. 1947. Some observations on the culture, physiology and morphology of some brown-red *Rhodosprillum* species. Antonie Leeuwenhoek J. Microbiol. Serol. *13*: 135–148.

Gieszczykiewicz, M. 1939. Zagadniene systematihki w bakteriologii — Zür Frage der Bakterien-Systematic. Bull. Acad. Polon. Sci., Ser. Sci. Biol. *1*: 9–27.

Giffhorn, F., N. Beuscher and G. Gottschalk. 1972. Regulation of citrate lyase activity in *Rhodopseudomonas gelatinosa*. Biochem. Biophys. Res. Commun. *49*: 467–471.

Gilad, J., A. Borer, N. Peled, K. Riesenberg, S. Tager, A. Appelbaum and F. Schlaeffer. 2000. Hospital-acquired *Brevundimonas vesicularis* septicaemia following open-heart surgery: Case report and literature review. Scand. J. Infect. Dis. *32*: 90–91.

Giladi, M., B. Avidor, Y. Kletter, S. Abulafia, L.N. Slater, D.F. Welch, D.J. Brenner, A.G. Steigerwalt, A.M. Whitney and M. Ephros. 1998. Cat scratch disease: the rare role of *Afipia felis*. J. Clin. Microbiol. *36*: 2499–2502.

Gilardi, G.L. 1971. Characterization of nonfermentative nonfastidious Gram negative bacteria encountered in medical bacteriology. J. Appl. Bacteriol. *34*: 623–644.

Gilardi, G.L. 1976. *Pseudomonas* species in clinical microbiology. Mt. Sinai J. Med. *43*: 710–726.

Gilardi, G.L. 1978a. Identification of miscellaneous glucose nonfermenting Gram-negative bacteria. *In* Gilardi (Editor), Glucose nonfermenting Gram-negative Bacteria in Clinical Microbiology, CRC Press, West Palm Beach. pp. 45–65.

Gilardi, G.L. 1978b. Identification of *Pseudomonas* and related bacteria. *In* Gilardi (Editor), Glucose Nonfermenting Gram-Negative Bacteria in Clinical Microbiology, CRC Press, West Palm Beach, Florida. 15–44.

Gilardi, G.L. 1985. *Pseudomonas*. *In* Lennette, Balows, Hausler and Shadomy (Editors), Manual of Clinical Microbiology, 4th ed, American society for Microbiology, Washington, D.C. pp. 350–372.

Gilardi, G.L. 1991. *Pseudomonas* and related genera. *In* Balows (Editor), Manual of Clinical Microbiology, Vol. 5, American Society for Microbiology, Washington DC. 429–441.

Gilardi, G.L. and Y.C. Faur. 1984. *Pseudomonas mesophilica* and an unnamed taxon, clinical isolates of pink-pigmented oxidative bacteria. J. Clin. Microbiol. *20*: 626–629.

Gilligan, P. 1995. *Pseudomonas* and *Burkholderia*. *In* Murray, Baron, Pfaller, Tenover and Yolken (Editors), Manual of Clinical Microbiolgy, American Society for Microbiology, Washington DC. 509–519.

Gilligan, P.H. and M.C. Fisher. 1984. Importance of culture in laboratory diagnosis of *Bordetella pertussis* infections. J. Clin. Microbiol. *20*: 891–893.

Gillis, M., J. Döbereiner, B. Pot, M. Goor, E. Falsen, B. Hoste, B. Reinhold and K. Kersters. 1991. Taxonomic relationships between *[Pseudomonas] rubrisubalbicans*, some clinical isolates (EF group 1), *Herbaspirillum seropedicae* and *[Aquaspirillum] autotrophicum*. *In* Polsinelli, Materassi

and Vincenzini (Editors), Nitrogen Fixation, Kluwer Academic Publishers, Dordrecht. pp. 293–294.

Gillis, M. and J. De Ley. 1980. Intra- and intergeneric similarities of the ribosomal ribonucleic acid cistrons of *Acetobacter* and *Gluconobacter*. Int. J. Syst Bacteriol. *30*: 7–27.

Gillis, M. and J. De Ley. 1992. The genera *Chromobacterium* and *Janthinobacterium*. *In* Balows, Trüper, Dworkin, Harder and Schleifer (Editors), The Prokaryotes: A Handbook of Bacteria: Ecophysiology, Isolation, Identification, Applications, 2nd ed., Vol. 4, Springer-Verlag, New York. pp. 2591–2600.

Gillis, M., J. Dejonghe, A. Smet, G. Onghenae and J. De Ley. 1982. Intra- and intergeneric similarities of the ribosomal ribonucleic acid cistrons in the *Rhodospirillaceae*. Abstract A-16. IV. Intern. Symp. Photosynthetic Procaryotes. Bombannes-Bordeaux :

Gillis, M., K. Kersters, B. Hoste, D. Janssens, R.M. Kroppenstedt, M.P. Stephan, K.R.S. Teixeira, J. Dobereiner and J. De Ley. 1989. *Acetobacter diazotrophicus* sp. nov., a nitrogen-fixing acetic acid bacterium associated with sugarcane. Int. J. Syst. Bacteriol. *39*: 361–364.

Gillis, M. and B. Reinhold-Hurek. 1994. Taxonomy of *Azospirillum*. *In* Okon (Editor), *Azospirillum*/Plant Associations, CRC Press, Boca Raton. 1–14.

Gillis, M., T. Van Van, R. Bardin, M. Goor, P. Hebbar, A. Willems, P. Segers, K. Kersters, T. Heulin and M.P. Fernandez. 1995. Polyphasic taxonomy in the genus *Burkholderia* leading to an emended description of the genus and proposition of *Burkholderia vietnamiensis* sp. nov. for N_2-fixing isolates from rice in Vietnam. Int. J. Syst. Bacteriol. *45*: 274–289.

Gilmore, R.D., Jr. 1993. Comparison of the *rompA* gene repeat regions of *Rickettsiae* reveals species-specific arrangements of individual repeating units. Gene *125*: 97–102.

Giménez, D.F. 1964. Staining rickettsiae in yolk-sac cultures. Stain Technol. *39*: 135–140.

Giron, J.A., A.S. Ho and G.K. Schoolnik. 1991. An inducible bundle-forming pilus of enteropathogenic *Escherichia coli*. Science *254*: 710–713.

Giron, J.A., M.M. Levine and J.B. Kaper. 1994. Longus: a long pilus ultrastructure produced by human enterotoxigenic *Escherichia coli*. Mol. Microbiol. *12*: 71–82.

Girsch, P. and S. de Vries. 1997. Purification and initial kinetic and spectroscopic characterization of NO reductase from *Paracoccus denitrificans*. Biochim. Biophys. Acta *1318*: 202–216.

Gitahy, P.M., J.F. Salles, K.R.S. Teixeira, L. Skot and J.I. Baldani. 1997. Expression of *Bacillus thuringiensis cry3A* gene in the endophytic diazotrophic bacteria of the genus *Herbaspirillum*. Proceedings of the XXI Reunião de Genética de Microoganismos, Universidade Estadual de Londrina.p. 83.

Givaudan, A., A. Effosse and R. Bally. 1991. Melanin production by *Azospirillum lipoferum* strains. *In* Polsinelli, Meterassi and Vincenzini (Editors), Nitrogen fixation, Kluwer Academic Publishers, Dordrecht. pp. 311–312.

Glaeser, J. and J. Overmann. 1999. Selective enrichment and characterization of *Roseospirillum parvum*, gen. nov. and sp. nov., a new purple nonsulfur bacterium with unusual light absorption properties. Arch. Microbiol. *171*: 405–416.

Glaeser, J. and J. Overmann. 2001. *In* Validation of the publication of new names and new combinations previously effectively published outside the IJSB. List No. 80. Int. J. Syst. Evol. Bacteriol. *51*: 793–794.

Glaser, P., H. Sakamoto, J. Bellalou, A. Ullmann and A. Danchin. 1988. Secretion of cyclolysin, the calmodulin-sensitive adenylate cyclase-haemolysin bifunctional protein of *Bordetella pertussis*. Embo J. *7*: 3997–4004.

Gliesche, C.G. 1997. Transformation of methylotrophic bacteria by electroporation. Can. J. Microbiol. *43*: 197–201.

Gliesche, C.G. and F.E.W. Eckhardt. 1991. Isolation of nonmotile and morphologically altered mutants of *Hyphomicrobium facilis* B-522. *In* Abstracts of the Annual Meeting of the American Society for Microbiology, American Society for Microbiology., Washington D.C. 159, H–127.

1244   BIBLIOGRAPHY

Gliesche, C.G. and A. Fesefeldt. 1998. Monitoring the denitrifying *Hyphomicrobium* DNA/DNA hybridization group HR 27 in activated sludge and lake water using MPN cultivation and subsequent screening with the gene probe *Hvu*-1. Syst. Appl. Microbiol. *21*: 315–320.

Gliesche, C.G. and P. Hirsch. 1992. Mutagenesis and chromosome mobilization in *Hyphomicrobium facilis* B-522. Can. J. Microbiol. *38*: 1167–1174.

Gliesche, C.G., N.C. Holm, M. Beese, M. Neumann, H. Voelker, R. Gebers and P. Hirsch. 1988. New bacteriophages active on strains of *Hyphomicrobium*. J. Gen. Microbiol. *134*: 1339–1354.

Gliesche, C.G., M. Menzel and A. Fesefeldt. 1997. A rapid method for creating species-specific gene probes for methylotrophic bacteria. J. Microbiol. Methods *28*: 25–34.

Glomp, I., P. Saulnier, J. Guespinmichel and H.U. Schairer. 1988. Transfer of IncP plasmids into *Stigmatella aurantiaca* leading to insertional mutants affected in spore development. Mol. Gen. Genet. *214*: 213–217.

Glupczynski, Y., W. Hansen, M. Dratwa, C. Tielemans, R. Wens, F. Collart and E. Yourassowsky. 1984. *Pseudomonas paucimobilis* peritonitis in patients treated by peritoneal dialysis. J. Clin. Microbiol. *20*: 1225–1226.

Glupczynski, Y., W. Hansen, J. Freney and E. Yourassowsky. 1988. *In vitro* susceptibility of *Alcaligenes denitrificans* pathovar *xylosoxidans* to 24 antimicrobial agents. Antimicrob. Agents Chemother. *32*: 276–278.

Gnida, M., R. Ferner, L. Gremer, O. Meyer and W. Meyer-Klaucke. 2003. A novel binuclear [CuSMo] cluster at the active site of carbon monoxide dehydrogenase: characterization by X-ray absorption spectroscopy. Biochemistry *42*: 222-230.

Gnosspelius, G. 1978. Purification and properties of an extracellular protease from *Myxococcus virescens*. J. Bacteriol. *133*: 17–25.

Go, M.F., V. Kapur, D.Y. Graham and J.M. Musser. 1996. Population genetic analysis of *Helicobacter pylori* by multilocus enzyme electrophoresis: extensive allelic diversity and recombinational population structure. J. Bacteriol. *178*: 3934–3938.

Goatcher, L.J., A.A. Qureshi and I.D. Gaudet. 1984. Evaluation and refinement of *Spirillum volutans* test for use in toxicity screening. *In* Dickson and Dutka (Editors), Toxicity Screening Procedures Using Bacterial Systems, Marcel Dekker, New York. pp. 89–108.

Gober, J.W. and M.V. Marques. 1995. Regulation of cellular differentiation in *Caulobacter crescentus*. Microbiol. Rev. *59*: 31–47.

Godfrey, C.A. 1972. The carotenoid pigment and deoxyribonucleic base ratio of a *Rhizobium* which nodulated *Lotononis bainesii* Baker. J. Gen. Microbiol. *72*: 399–402.

Godfroid, F., B. Taminiau, I. Danese, P. Denoel, A. Tibor, V. Weynants, A. Cloeckaert, J. Godfroid and J.J. Letesson. 1998. Identification of the perosamine synthetase gene of *Brucella melitensis* 16M and involvement of lipopolysaccharide O side chain in *Brucella* survival in mice and in macrophages. Infect. Immun. *66*: 5485–5493.

Goebel, B.M., P.R. Norris and N.P. Burton. 2000. Acidophiles in biomining. *In* Priest and Goodfellow (Editors), Applied Microbial Systematics, Kluwer, Dordrecht. pp. 293–314.

Goebel, E.M. and N.R. Krieg. 1984. Fructose catabolism in *Azospirillum brasilense* and *Azospirillum lipoferum*. J. Bacteriol. *159*: 86–92.

Goedert, M. 1973. *Agrobacterium tumefaciens* (Smith et Town) Conn: Détermination des concentrations minimales inhibitrices de différents antibiotiqes et sulfamides. Ann. Microbiol. (Inst. Pasteur) *124A*: 237–241.

Goethals, K., M. Gao, K. Tomekpe, M. Van Montagu and M. Holsters. 1989. Common *nodABC* genes in nod locus 1 of *Azorhizobium caulinodans*: nucleotide sequence and plant-inducible expression. Mol. Gen. Genet. *219*: 289–298.

Goethals, K., P. Mergaert, M. Gao, D. Geelen, M. Van Montagu and M. Holsters. 1992a. Identification of a new inducible nodulation gene in *Azorhizobium caulinodans*. Mol. Plant-Microbe Interact. *5*: 405–411.

Goethals, K., G. Van den Eeede, M. Van Montagu and M. Holsters. 1990. Identification and characterization of a functional *nodD* gene in *Azorhizobium caulinodans* ORS571. J. Bacteriol. *172*: 2658–2666.

Goethals, K., M. Van Montagu and M. Holsters. 1992b. Conserved motifs in a divergent nod box of *Azorhizobium caulinodans* ORS571 reveal a common structure in promoters regulated by LysR-type proteins. Proc. Natl. Acad. Sci. U.S.A. *89*: 1646–1650.

Goethert, H.K. and S.R. Telford. 2000. Enzootic transmission of *Ehrlichia bovis* in cottontail rabbits on Nantucket Island. 15th Meeting of the American Society for, Captiva Island, Florida. American Society for Rickettsiology. :

Gogotov, I.N. and V.M. Gorlenko. 1995. Influence of cultivation conditions on the composition of quinones in purple bacteria and freshwater erythrobacteria. Microbiology *64*: 654–656.

Gogotov, J.N. and H.F. Schlegel. 1974. N₂-fixation by chemoautotrophic hydrogen bacteria. Arch. Microbiol. *97*: 359–362.

Gogotova, G.I. and M.B. Vainshtein. 1989. Description of sulfate reducing bacterium *Desulfobacterium macestii* sp. nov., which is capable of autotrophic growth. Mikrobiologiya *58*: 64–68.

Goldbaum, F.A., C.A. Velikovsky, P.C. Baldi, S. Mortl, A. Bacher and C.A. Fossati. 1999. The 18 kDa cytoplasmic protein of *Brucella* species an antigen useful for diagnosis is a lumazine synthase. J. Med. Microbiol. *48*: 833–839.

Goldberg, J.D., P. Brick, T. Yoshida, T. Mitsunaga, T. Oshiro, M. Shimao and Y. Izumi. 1992. Crystallization and preliminary diffraction studies of hydroxypyruvate reductase (D-glycerate dehydrogenase) from *Hyphomicrobium methylovorum*. J. Mol. Biol. *225*: 909–911.

Goldberg, J.D., T. Yoshida and P. Brick. 1994. Crystal structure of a NAD-dependent D-glycerate dehydrogenase at 2–4 Angstrom resolution. J. Mol. Biol. *236*: 1123–1140.

Goldfine, H. and P. Hagen. 1968. N-methyl groups in bacterial lipids. III. Phospholipids of hyphomicrobia. J. Bacteriol. *95*: 367–375.

Goldman, W.E., D.G. Klapper and J.B. Baseman. 1982. Detection, isolation, and analysis of a released *Bordetella pertussis* product toxic to cultured tracheal cells. Infect. Immun. *36*: 782–794.

Goldmann, D.A. and J.D. Klinger. 1986. *Pseudomonas cepacia*: biology, mechanisms of virulence, epidemiology. J. Pediatr. *108*: 806–812.

Goldstein, E.J.C., E.O. Agyare and R. Silletti. 1981. Comparative growth of *Eikenella corrodens* on fifteen media in three atmospheres of incubation. J. Clin. Microbiol. *13*: 951–953.

Goldstein, E.J.C., V.L. Sutter and S.M. Finegold. 1978. The susceptibility of *Eikenella corrodens* to 10 cephalosporins. Antimicrob. Agents Chemother. *14*: 404.

Goldstein, R., L. Sun, R.Z. Jiang, U. Sajjan, J.F. Forstner and C. Campanelli. 1995. Structurally variant classes of pilus appendage fibers coexpressed from *Burkholderia* (*Pseudomonas*) *cepacia*. J. Bacteriol. *177*: 1039–1052.

Golecki, J.R. and G. Drews. 1980. Cellular organization of the halophilic phototrophic bacterium strain WS 68. Eur. J. Cell Biol. *22*: 654–660.

Gomez-Miguel, M.J., I. Moriyon and J. Lopez. 1987. *Brucella* outer membrane lipoprotein shares antigenic determinants with *Escherichia coli* Braun lipoprotein and is exposed on the cell surface. Infect. Immun. *55*: 258–262.

Gommers, P.J.F. and J.G. Kuenen. 1988. *Thiobacillus* strain Q, a chemolithoheterotrophic sulfur bacterium. Arch. of Microbiol. *150*: 117–125.

Gonçalves, A.F.S. and R. Oliveira. 1998. Cyanide production by brazilian strains of *Azospirillum*. Rev. Microbiol. *29*: 36–39.

Gonçalves de Lima, O., J.M. De Araújo, I.E. Schumacher and E. Cavalcanti Da Silva. 1970. Estudos de microorganismos antagonistas presentes nas bebidas fermentadas usadas pelo povo do Recife. I. Sôbre uma variedade de *Zymomonas mobilis* (Lindner) (1928) Kluyver e van Niel (1936) *Zymomonas mobilis* biovar recifensis (Gonçalves de Lima, Araújo, Schumacher and Cavalcanti) (1970), isolada de bebida popular denominada "caldo-de-cana picado". Rev. Inst. Antibtio. Univ. Recife. *10*: 3–15.

González, C.F. and A.K. Vidaver. 1979. Bacteriocin, plasmid and pectolytic diversity in *Pseudomonas cepacia* of clinical and plant origin. J. Gen. Microbiol. *110*: 161–170.

González, F., J.M. Arias and E. Montoya. 1987. Phosphatase activities in the life cycle of *Myxococcus coralloides* D. J. Gen. Microbiol. *133*: 2327–2332.

González, F., A. Fernández-Vivas, J.M. Arias and E. Montoya. 1990. Polyphosphate glucokinase and ATP glucokinase activities in *Myxococcus coralloides* D. Arch. Microbiol. *154*: 438–442.

González, F., J. Munoz, J.M. Arias and E. Montoya. 1989. Production of acid and alkaline phosphatases by *Myxococcus coralloides*. Folia Microbiol. *34*: 185–194.

González, F., A. Vargas, J.M. Arias and E. Montoya. 1991. Phosphatase activity during development cycle of *Myxococcus xanthus*. Can. J. Microbiol. *37*: 74–77.

Gonzalez, I., K.A. Grant, P.T. Richardson, S.F. Park and M.D. Collins. 1997a. Specific identification of the enteropathogens *Campylobacter jejuni* and *Campylobacter coli* by using a PCR test based on the *ceuE* gene encoding a putative virulence determinant. J. Clin. Microbiol. *35*: 759–763.

González, J.M., R.P. Kiene and M.A. Moran. 1999. Transformation of sulfur compounds by an abundant lineage of marine bacteria in the alpha-subclass of the class *Proteobacteria*. Appl. Environ. Microbiol. *65*: 3810–3819.

González, J.M., F. Mayer, M.A. Moran, R.E. Hodson and W.B. Whitman. 1997b. *Sagittula stellata* gen. nov., sp. nov., a lignin-transforming bacterium from a coastal environment. Int. J. Syst. Bacteriol. *47*: 773–780.

González, J.M. and M.A. Moran. 1997. Numerical dominance of a group of marine bacteria in the alpha-subclass of the class *Proteobacteria* in coastal seawater. Appl. Environ. Microbiol. *63*: 4237–4242.

Goodfellow, M., G.P. Manfio and J. Chun. 1997. Towards a practical species concept for cultivable bacteria. *In* Claridge (Editor), Species: the Units of Biodiversity, 1st ed., Chapman & Hall, London ; New York. 25–59.

Goodfellow, M. and A.G. O'Donnell. 1993. Roots of bacterial systematics. *In* Goodfellow and O'Donnell (Editors), Handbook of New Bacterial Systematics, Academic Press Limited, London. 3–56.

Goodhew, C.F., A.B. Elkurdi and G.W. Pettigrew. 1988. The microaerophilic respiration of *Campylobacter mucosalis*. Biochim. Biophys. Acta *933*: 114–123.

Goodhew, C.F., G.W. Pettigrew, B. Devreese, J. VanBeeumen, R.J.M. van Spanning, S.C. Baker, N. Saunders, S.J. Ferguson and I.P. Thompson. 1996. The cytochromes c_{550} of *Paracoccus denitrificans* and *Thiosphaera pantotropha*: a need for re-evaluation of the history of *Paracoccus* cultures. FEMS Microbiol. Lett. *137*: 95–101.

Goodman, A.D. 1977. *Eikenella corrodens* isolated in oral infections of dental origin. Oral Surg. *44*: 128–134.

Goodman, J.L., C. Nelson, B. Vitale, J.E. Madigan, J.S. Dumler, T.J. Kurtti and U.G. Munderloh. 1996. Direct cultivation of the causative agent of human granulocytic ehrlichiosis. N. Engl. J. Med. *334*: 209–215.

Goodman, S.D. and J.J. Scocca. 1988. Identification and arrangement of the DNA-sequence recognized in specific transformation of *Neisseria gonorrhoeae*. Proc. Natl. Acad. Sci. U.S.A. *85*: 6982–6986.

Goodner, B., G. Hinkle and 29 other authors. 2001. Genome sequence of the plant pathogen and biotechnology agent *Agrobacterium tumefaciens* C58. Science *294*: 2323–2328.

Goodnow, R.A. 1980. Biology of *Bordetella bronchiseptica*. Microbiol. Rev. *44*: 722–738.

Goodwin, C.S., J.A. Armstrong, T. Chilvers, M. Peters, M.D. Collins, L. Sly, W. McConnell and W.E.S. Harper. 1989a. Transfer of *Campylobacter pylori* and *Campylobacter mustelae* to *Helicobacter* gen. nov. as *Helicobacter pylori* comb. nov. and *Helicobacter mustelae* comb. nov., respectively. Int. J. Syst. Bacteriol. *39*: 397–405.

Goodwin, C.S., W. McConnell, R.K. McCulloch, C. McCullough, R. Hill, M.A. Bronsdon and G. Kasper. 1989b. Cellular fatty acid composition of *Campylobacter pylori* from primates and ferrets compared with those of other campylobacters. J. Clin. Microbiol. *27*: 938–943.

Goodwin, P.M. and C. Anthony. 1998. The biochemistry, physiology and genetics of PQQ and PQQ-containing enzymes. Adv. Microb. Physiol. *40*: 1–80.

Goossens, H., B. Pot, L. Vlaes, C. Van den Borre, R. Van den Abbeele, C. Van Naelten, J. Levy, H. Cogniau, P. Marbehant, J. Verhoef, K. Kersters, J.-P. Butzler and P. Vandamme. 1990a. Characterization and

description of "*Campylobacter upsaliensis*" isolated from human feces. J. Clin. Microbiol. *28*: 1039–1046.

Goossens, H., L. Vlaes, M. De Boeck, B. Pot, K. Kersters, J. Levy, P. De Mol, J.-P. Butzler and P. Vandamme. 1990b. Is "*Campylobacter upsaliensis*" an unrecognised cause of human diarrhoea? Lancet *335*: 584–586.

Goossens, H., L. Vlaes, I. Galand, C. Van den Borre and J.-P. Butzler. 1989. Semisolid blood-free selective-motility medium for the isolation of campylobacters from stool specimens. J. Clin. Microbiol. *27*: 1077–1080.

Gorby, Y.A., T.J. Beveridge and R.P. Blakemore. 1988. Characterization of the bacterial magnetosome membrane. J. Bacteriol. *170*: 834–841.

Gordon, D.M., M.H. Ryder, K. Heinrich and P.J. Murphy. 1996. An experimental test of the rhizopine concept *Rhizobium meliloti*. Appl. Environ. Microbiol. *62*: 3991–3996.

Gordon, W.S., A. Brownlee, D.R. Wilson and J. MacLeod. 1932. "Tickborne fever" (A hitherto undescribed disease of sheep). J. Comp. Pathol. Therap. *65*: 301–307.

Goris, J., P. De Vos, T. Coenye, B. Hoste, D. Janssens, H. Brim, L. Diels, M. Mergeay, K. Kersters and P. Vandamme. 2001. Classification of metal-resistant bacteria from industrial biotopes as *Ralstonia campinensis* sp. nov., *Ralstonia metallidurans* sp. nov. and *Ralstonia basilensis* Steinle et al. 1998 emend. Int. J. Syst. Evol. Microbiol. *51*: 1773–1782.

Gorrell, T.E. and R.L. Uffen. 1977. Fermentative metabolism of pyruvate by *Rhodospirillum rubrum* after anaerobic growth in darkness. J. Bacteriol. *131*: 533–543.

Görtz, H.D. 1983. Endonuclear symbionts in ciliates. *In* Jeon (Editor), Intracellular Symbiosis, Academic Press, New York. 145–176.

Görtz, H.D. 1986. Endonucleobiosis in Ciliates. Int. Rev. Cytol. *102*: 169–213.

Görtz, H.D. and J. Dieckmann. 1980. Life cycle and infectivity of *Holospora elegans* Hafflkine a micronucleus-specific symbiont of *Paramecium caudatum* (Ehrenberg). Protistologica *16*: 591–603.

Görtz, H.D. and M. Fujishima. 1983. Conjugation and meiosis of *Paramecium caudatum* infected with the micronucleus specific bacterium *Holospora elegans*. Eur. J. Cell Biol. *32*: 86–91.

Görtz, H.D., S. Lellig, O. Miosga and M. Wiemann. 1990. Changes in fine structure and polypeptide pattern during development of *Holospora obtusa*, a bacterium infecting the macronucleus of *Paramecium caudatum*. J. Bacteriol. *172*: 5664–5669.

Görtz, H.D. and M. Wiemann. 1989. Route of infection of the bacteria *Holospora elegans* and *Holospora obtusa* into the nuclei of *Paramecium caudatum*. Eur. J. Protistol. *24*: 101–109.

Gosink, J.J., R.P. Herwig and J.T. Staley. 1997. *Octadecabacter arcticus* gen. nov., sp. nov., nonpigmented, psychrophilic gas vacuolate bacteria from polar sea ice and water. Syst. Appl. Microbiol. *20*: 356–365.

Gosink, J.J., R.P. Herwig and J.T. Staley. 1998. *In* Validation of the publication of new names and new combinations previously effectively published outside the IJSB. List No. 64. Int. J. Syst. Bacteriol. *48*: 327–328.

Gosink, J.J. and J.T. Staley. 1995. Biodiversity of gas vacuolate bacteria from Antarctic sea ice and water. Appl. Environ. Microbiol. *61*: 3486–3489.

Gosselé, F. and J. Swings. 1986. Identification of *Acetobacter liquefaciens* as causal agent of pink-disease of pineapple fruit. J. Phytopathol. (Berl.) *116*: 167–175.

Gosselé, F., J. Swings and J. De Ley. 1980. A rapid, simple and simultaneous detection of 2-keto, 5-keto- and 2,5-diketogluconic acids by thin-layer chromatography in culture media of acetic acid bacteria. Zentralbl. Bakteriol. Hyg. I Abt. Orig. *1*: 178–181.

Gosselé, F., J. Swings, K. Kersters and J. De Ley. 1983a. Numerical analysis of phenotypic features and protein gel electropherograms of *Gluconobacter* Asai 1935 emend. mut. char. Asai, Iizuka, and Komagata 1964. Int. J. Syst. Bacteriol. *33*: 65–81.

Gosselé, F., J. Swings, K. Kersters, P. Pauwels and J. De Ley. 1983b. Numerical analysis of phenotypic features and protein gel electropherograms of a wide variety of *Acetobacter* strains: proposal for the improvement of the taxonomy of the genus *Acetobacter* Beijerinck 1898, 215. Syst. Appl. Microbiol. *4*: 338–368.

Gosselé, F., J. Swings, K. Kersters, P. Pauwels and J. De Ley. 1983c. *In* Validation of the publication of new names and new combinations previously effectively published outside the IJSB. List No. 12. Int. J. Syst. Bacteriol. *33*: 896–897.

Gossling, J. and W.E.C. Moore. 1975. *Gemmiger formicilis*, n. gen., n. sp., an anaerobic budding bacterium from intestines. Int. J. Syst. Bacteriol. *25*: 202–207.

Gothe, R. 1967a. Ein Beiträg zür systematischen Stellung von *Aegyptianella pullorum* Carpano 1928. Z. Parasitenkd. *29*: 119–129.

Gothe, R. 1967b. Untersuchungen über die Entwicklung und den Infektionsverlauf von *Aegyptianella pullorum* Carpano 1928, im Huhn. Z. Parasitenkd. *29*: 149–158.

Gothe, R. 1967c. Zür Entwicklung von *Aegyptianella pullorum* Carpano 1928, in der Lederzecke *Argas (Persicargas) persicus* (Oken, 1818) und Ubertragung. Z. Parasitenkd. *29*: 103–118.

Gothe, R. 1971. Wirt-Parasit-Verhaltnis von *Aegyptianella pullorum* Carpano 1928, im biologischen Übertrager *Argas (Persicargas) persicus* (Oken, 1818) und im Wirbeltierwirt Gallus gallus domesticus L. Fortschr. Veterinaermed. *Heft 16*:

Gothe, R. 1978. New aspects of the epizootiology of aegyptianellosis in poultry. *In* Wilde (Editor), Tick-borne Diseases and their Vectors, Centre for Tropical Veterinary Medicine, University of Edinburgh, Lewis Reprints Ltd., Tonbridge. pp. 201–204.

Gothe, R. and E. Burkhardt. 1979. The erythrocytic entry- and exit-mechanism of *Aegyptianella pullorum* Carpano, 1928. Z. Parasitenkd. *60*: 221–227.

Gothe, R. and S. Hartmann. 1979. The viability of cryopreserved *Aegyptianella pullorum* Carpano, 1928 in the vector *Argas (Persicargas) walkerae* Kaiser and Hoogstraal, 1969. Z. Parasitenkd. *58*: 189–190.

Gothe, R. and J.P. Kreier. 1984. Genus II. *Aegyptianella* Carpano 1929, 12AL. *In* Krieg and Holt (Editors), Bergey's Manual of Systematic Bacteriology, 1st Ed., Vol. 1, The Williams & Wilkins Co., Baltimore. pp. 722–723.

Gothe, R. and H. Mieth. 1979. Zür Wirksamkeit von Pleuromutilinen bei *Aegyptianella pullorum*-Infectionen der Kuken. Tropenmed. Parasitol. *30*: 323–327.

Goto, E., T. Kodama and Y. Minoda. 1977. Isolation and culture conditions of thermophilic hydrogen bacteria. Agric. Biol. Chem. *41*: 685–690.

Goto, E., T. Kodama and Y. Minoda. 1978. Growth and taxonomy of thermophilic hydrogen bacteria. Agric. Biol. Chem. *42*: 1305–1308.

Goto, K., H. Ohashi, S. Ebukuro, K. Itoh, Y. Tohma, A. Takakura, S. Wakana, M. Ito and T. Itoh. 1998. Isolation and characterization of *Helicobacter* species from the stomach of the house musk shrew (*Suncus murinus*) with chronic gastritis. Curr. Microbiol. *37*: 44–51.

Goto, M. 1983. *Pseudomonas pseudoalcaligenes* subsp. *konjaci* subsp. nov., the causal agent of bacterial leaf blight of konjac (*Amorphopallus konjac* Koch). Int. J. Syst. Bacteriol. *33*: 539–545.

Goto, M. and M.P. Starr. 1971. A comparative study of *Pseudomonas andropogonis*, *P stizolobii* and *P. alboprecipitans*. Ann. Phytopathol. Soc. Japan *37*: 233–241.

Gotoh, N., K. Nagino, K. Wada, H. Tsujimoto and T. Nishino. 1994a. *Burkholderia* (formerly *Pseudomonas*) *cepacia* porin is an oligomer composed of two component proteins. Microbiology *140*: 3285–3291.

Gotoh, N., N.J. White, W. Chaowagul and D.E. Woods. 1994b. Isolation and characterization of the outer-membrane proteins of *Burkholderia* (*Pseudomonas*) *pseudomallei*. Microbiology *140*: 797–805.

Gotschlich, E.C., B.A. Fraser, O. Nishimura, J.B. Robbins and T.Y. Liu. 1981. Lipid on capsular polysaccharides of gram-negative bacteria. J. Biol. Chem. *256*: 8915–8921.

Gotschlich, E.C., M. Seiff and M.S. Blake. 1987a. The DNA-sequence of the structural gene of gonococcal protein-iii and the flanking region containing a repetitive sequence — homology of protein-iii with enterobacterial Ompa proteins. J. Exp. Med. *165*: 471–482.

Gotschlich, E.C., M.E. Seiff, M.S. Blake and M. Koomey. 1987b. Porin

protein of *Neisseria gonorrhoeae* - cloning and gene structure. Proc. Natl. Acad. Sci. U.S.A. *84*: 8135–8139.

Gottfert, M., P. Grob and H. Hennecke. 1990a. Proposed regulatory pathway encoded by the *nodV* and *nodW* genes, determinants of host specificity in *Bradyrhizobium japonicum*. Proc. Natl. Acad.Sci. U.S.A. *87*: 2680–2684.

Gottfert, M., S. Hitz and H. Hennecke. 1990b. Identification of *nodS* and *nodU*, two inducible genes inserted between the *Bradyrhizobium japonicum nodYABC* and *nodIJ* genes. Mol. Plant-Microbe Interact. *3*: 308–316.

Gottfert, M., D. Holzhauser, D. Bani and H. Hennecke. 1992. Structural and functional analysis of two different *nodD* genes in *Bradyrhizobium japonicum* USDA110. Mol. Plant-Microbe Interact. *5*: 257–265.

Göttfert, M., B. Horvath, E. Kondorosi, P. Putnoky, F. Rodriguez-Quiñones and Á. Kondorosi. 1986. At least two *nodD* genes are necessary for efficient nodulation of alfalfa by *Rhizobium meliloti*. J. Mol. Biol. *191*: 411–420.

Gottschal, J.C. and J.G. Kuenen. 1980. Selective enrichment of facultatively chemolithotrophic thiobacilli and related organisms in continuous culture. FEMS Microbiol. Lett. *7*: 2.

Gottschalk, G. and J.R. Andreesen. 1979. Chapter 3, Energy metabolism in anaerobes. *In* Quayle (Editor), International Review of Biochemistry, Microbial Biochemistry, Vol. 21, University Park Press, Baltimore. p. 86–108.

Gould, D.J. and M.L. Miesse. 1954. Recovery of a rickettsia of the spotted fever group from *Microtus pennsylvanicus* from Virginia. Proc. Soc. Exp. Biol. Med. *85*: 558–561.

Gounot, A.M. 1991. Bacterial life at low-temperature - physiological aspects and biotechnological implications. J. Appl. Bacteriol. *71*: 386–397.

Goutzmanis, J.J., G. Gonis and G.L. Gilbert. 1991. *Kingella kingae* infection in children: ten cases and a review of the literature. Pediatr. Infect. Dis. J. *10*: 677–683.

Govan, J.R.W. and G. Harris. 1985. Typing of *Pseudomonas cepacia* by bacteriocin susceptibility and production. J. Clin. Microbiol. *22*: 490–494.

Govindaraj, S., E. Eisenstein, L.H. Jones, J. Sanders Loehr, A.Y. Chistoserdov, V.L. Davidson and S.L. Edwards. 1994. Aromatic amine dehydrogenase, a second tryptophan tryptophylquinone enzyme. J. Bacteriol. *176*: 2922–2929.

Govorukhina, N.I., L.V. Kletsova, Y.D. Tsygankov, Y.A. Trotsenko and A.I. Netrusov. 1987. Characteristics of a new obligate methylotroph. Mikrobiologiya *56*: 849–854.

Govorukhina, N.I., L.V. Kletsova, Y.D. tsygankov, Y.A. Trotsenko and A.I. Netrusov. 1998. *In* Validation of the publication of new names and new combinations previously effectively published outside the IJSB. List No. 66. Int. J. Syst. Bacteriol. *48*: 631–632.

Govorukhina, N.I. and Y.A. Trotsenko. 1991. *Methylovorus*, a new genus of restricted facultatively methylotrophic bacteria. Int. J. Syst. Bacteriol. *41*: 158–162.

Goyal, A.K. and G.J. Zylstra. 1996. Molecular cloning of novel genes for polycyclic aromatic hydrocarbon degradation from *Comamonas testosteroni* GZ29. Appl. Environ. Microbiol. *62*: 230–236.

Grabovich, M.Y., V.V. Churikova, N.A. Chernykh, I.O. Kononykhina and I.P. Popravko. 1987. Isolation and characteristics of strains belonging to *Aquaspirillum voronezhense*, new species. Mikrobiologiya *56*: 666–672.

Grabovich, M.Y., V.V. Churikova, N.A. Chernykh, N.V. Leshcheva, O.E. Pushkina, L.I. Shipilova and E.E. Panteleyeva. 1990. Isolation and description of the new species *Aquaspirillum elegans* sp. nov. Mikrobiologiya *59*: 205–209.

Grabovich, M.Y., G.A. Dubinina, V.V. Churikova, S.N. Churikov and T.I. Korovina. 1995. Mechanisms of synthesis and utilization of oxalate inclusions in the colorless sulfur bacterium *Macromonas bipunctata*. Mikrobiologiya *64*: 630–636.

Grabovich, M.Y., G.A. Dubinina, V.V. Churikova and A.E. Glushkov. 1993. Peculiarities of carbon metabolism in the colorless sulfur bacterium *Macromonas bipunctata*. Mikrobiologiya *62*: 421–428.

Graham, D.W., J.A. Chaudhary, R.S. Hanson and R.G. Arnold. 1993.

Factors affecting competition between type I and type II methanotrophs in 2-organism, continuius-flow reactors. Microb. Ecol. *25*: 1–17.

Graham, D.Y., P.D. Klein, A.R. Opekun and T.W. Boutton. 1988. Effect of age on the frequency of active *Campylobacter pylori* infection diagnosed by the ^{13}C urea breath test in normal subjects and patients with peptic ulcer disease. J. Infect. Dis. *157*: 777–780.

Graham, D.W., D.G. Korich, R.P. Leblanc, N.A. Sinclair and R.G. Arnold. 1992. Applications of a colorimetric plate assay for soluble methane monooxygenase activity. Appl. Environ. Microbiol. *58*: 2231–2236.

Graham, P.H. 1963. Antigenic affinities of the root-nodule bacteria of legumes. Antonie Leeuwenhoek J. Microbiol. Serol. *29*: 281–291.

Graham, P.H. 1964. The application of computer techniques to the taxonomy of the root nodule bacteria of legumes. J. Gen. Microbiol. *35*: 511–517.

Graham, P.H. 1971. Serological studies with *Agrobacterium radiobacter, A. tumefaciens,* and *Rhizobium* strains. Arch. Mikrobiol. *78*: 70–75.

Graham, P.H. 1976. Identification and classification of root nodule bacteria. *In* Nutman (Editor), Symbiotic Nitrogen Fixation in Plants: International Biological Programme 7, Cambridge University Press, Cambridge [Eng.] ; New York. 99–112.

Graham, P.H. and C.A. Parker. 1964. Diagnostic features in the characterization of the root-nodule bacteria of legumes. Plant Soil *20*: 383–396.

Graham, P.H., M.J. Sadowsky, H.H. Keyser, Y.M. Barnet, R.S. Bradley, J.E. Cooper, D.J. Deley, B.D.W. Jarvis, E.B. Roslycky, B.W. Strijdom and J.P.W. Young. 1991. Proposed minimal standards for the description of new genera and species of root-nodulating and stem-nodulating bacteria. Int. J. Syst. Bacteriol. *41*: 582–587.

Graham, P.H., M.J. Sadowsky, S.W. Tighe, J.A. Thompson, R.A. Date, J.G. Howieson and R. Thomas. 1995. Differences among strains of *Bradyrhizobium* in fatty acid-methyl ester (FAME) analysis. Can. J. Microbiol. *41*: 1038–1042.

Granada, G.A. and L. Sequeira. 1983. Survival of *Pseudomonas solanacearum* in soil rhizosphere and plant roots. Can. J. Microbiol. *29*: 433–440.

Granato, P.A. and M.R. Franz. 1990. Use of the Gen-Probe PACE system for the detection of *Neisseria gonorrhoeae* in urogenital samples. Diagn. Microbiol. Infect. Dis. *13*: 217–221.

Granstrom, M., A.M. Olindernielsen, P. Holmblad, A. Mark and K. Hanngren. 1991. Specific immunoglobulin for treatment of whooping-cough. Lancet *338*: 1230–1233.

Grant, P.E., D.J. Brenner, A.G. Steigerwalt, D.G. Hollis and R.E. Weaver. 1990. *Neisseria elongata* subsp. *nitroreducens* subsp. nov., formerly CDC group M-6, a gram-negative bacterium associated with endocarditis. J. Clin. Microbiol. *28*: 2591–2596.

Grassè, P.P. 1924. Notes protistologiques I. La sporulation des Oscillospiracèes. II Le genre Alysiella Langeron 1923. Archives de Zoologie Expérimentale et Gènèrale. *62*: 25–34.

Grasso, S., W.J. Moller, E. Refatti, G.M. Di San Lio and G. Granata. 1979. *Xanthomonas ampelina* as causal agent of a grape (*Vitis vinifera*) decline in Sicily, Italy. Rivista di Patologia Vegetale. *15*: 91–106.

Gray, K.M. and E.G. Ruby. 1990. Prey-derived signals regulating duration of the developmental growth phase of *Bdellovibrio bacteriovorus.* J. Bacteriol. *172*: 4002–4007.

Gray, K.M. and E.G. Ruby. 1991. Intercellular signalling in the *Bdellovibrio* developmental cycle. *In* Dworkin (Editor), Microbial Cell–Cell Interactions, American Society for Microbiology, Washington D.C. pp. 333–366.

Gray, P.H.H. and H.G. Thornton. 1928. Soil bacteria that decompose certain aromatic compounds. Centralbl. Bakteriol. Parasitenkd. Infektionskr. 2. Abt. *73*: 74–96.

Gräzer-Lampart, S.D., T. Egli and G. Hamer. 1986. Growth of *Hyphomicrobium* ZV620 in the chemostat: regulation of ammonium-assimilating enzymes and cellular composition. J. Gen. Microbiol. *132*: 3337–3348.

Green, P.N. 1992. The Genus *Methylobacterium. In* Balows, Trüper, Harder and Schleifer (Editors), The Prokaryotes: A Handbook of Bacteria:

Ecophysiology, Isolation, Identification, Applications, 2nd Ed., Vol. 3, Springer-Verlag, Berlin. pp. 2342–2349.

Green, P.N. and I.J. Bousfield. 1981. The taxonomy of pink-pigmented facultatively methylotrophic bacteria. *In* Dalton (Editor), Microbial Growth on C1-Compounds, Heyden and Son, London. 285–293.

Green, P.N. and I.J. Bousfield. 1982. A taxonomic study of some gram-negative facultatively methylotrophic bacteria. J. Gen. Microbiol. *128*: 623–638.

Green, P.N. and I.J. Bousfield. 1983. Emendation of *Methylobacterium* Patt, Cole, and Hanson 1976; *Methylobacterium rhodinum* (Heumann 1962) comb. nov. corrig.; *Methylobacterium radiotolerans* (Ito and Iizuka 1971) comb. nov., corrig.; and *Methylobacterium mesophilicum* (Austin and Goodfellow 1979) comb. nov. Int. J. Syst. Bacteriol. *33*: 875–877.

Green, P.N., I.J. Bousfield and D. Hood. 1988. Three new *Methylobacterium* species: *Methylobacterium rhodesianum*, sp. nov., *Methylobacterium zatmanii*, sp. nov., and *Methylobacterium fujisawaense*, sp. nov. Int. J. Syst. Bacteriol. *38*: 124–127.

Green, P.N. and M. Gillis. 1989. Classification of *Pseudomonas aminovorans* and some related methylated amine utilizing bacteria. J. Gen. Microbiol. *135*: 2071–2076.

Greenfield, S. and G.W. Claus. 1972. Nonfunctional tricarboxylic acid cycle and the mechanism of glutamate biosynthesis in *Acetobacter suboxydans.* J. Bacteriol. *112*: 1295–1301.

Greenwood, J.A., J. Mills, P.D. Tyler and C.W. Jones. 1998. Physiological regulation, purification and properties of urease from *Methylophilus methylotrophus.* FEMS Microbiol. Lett. *160*: 131–135.

Gregory, E. and J.T. Staley. 1982. Widespread distribution of ability to oxidize manganese among fresh-water bacteria. Appl. Environ. Microbiol. *44*: 509–511.

Greig, B., K.M. Asanovich, P.J. Armstrong and J.S. Dumler. 1996. Geographic, clinical, serologic, and molecular evidence of granulocytic ehrlichiosis, a likely zoonotic disease, in Minnesota and Wisconsin dogs. J. Clin. Microbiol. *34*: 44–48.

Gremer, L., S. Kellner, H. Dobbek, R. Huber and O. Meyer. 2000. Binding of flavin adenine dinucleotide to molybdenum-containing carbon monoxide dehydrogenase from *Oligotropha carboxidovorans* — structural and functional analysis of a carbon monoxide dehydrogenase species in which the native flavoprotein has been replaced by its recombinant counterpart produced in *Escherichia coli.* J. Biol. Chem. *275*: 1864–1872.

Gresshoff, P.M., M.L. Skotnicki and B.G. Rolfe. 1979. Crown gall teratoma formation is plasmid and plant controlled. J. Bacteriol. *137*: 1020–1021.

Grey, B.E. and T.R. Steck. 2001. The viable but nonculturable state of *Ralstonia solanaceurum* may be involved in long-term survival and plant infection. Appl. Environ. Microbiol. *67*: 3866–3872.

Gribble, D.H. 1969. Equine ehrlichiosis. J. Am. Vet. Med. Assoc. *155*: 462–469.

Griffin, A.M., K.J. Edwards, V.J. Morris and M.J. Gasson. 1997. Genetic analysis of acetan biosynthesis in *Acetobacter xylinum*: DNA sequence analysis of the *aceM* gene encoding an UDP-gluco dehydrogenase. Biotechnol. Lett. *19*: 469–474.

Griffin, A.M., V.J. Morris and M.J. Gasson. 1994. Genetic analysis of the acetan biosynthetic pathway in *Acetobacter xylinum.* Int. J. Biol. Macromol. *16*: 287–289.

Griffin, A.M., V.J. Morris and M.J. Gasson. 1996. Identification, cloning and sequencing the *aceA* gene involved in acetan biosynthesis in *Acetobacter xylinum.* FEMS Microbiol. Lett. *137*: 115–121.

Griffin, P.J. and E. Racker. 1956. The carbon dioxide requirement of *Neisseria gonorrhoeae.* J. Bacteriol. *71*: 717–721.

Griffiss, J.M., J.P. Obrien, R. Yamasaki, G.D. Williams, P.A. Rice and H. Schneider. 1987. Physical heterogeneity of neisserial lipooligosaccharides reflects oligosaccharides that differ in apparent molecular weight, chemical composition, and antigenic expression. Infect. Immun. *55*: 1792–1800.

Griffiss, J.M., H. Schneider, R.E. Mandrell, R. Yamasaki, G.A. Jarvis, J.J. Kim, B.W. Gibson, R. Hamadeh and M.A. Apicella. 1988. Lipooligosaccharides—The principal glycolipids of the neisserial outer membrane. Rev. Infect. Dis. *10*: S287–S295.

Griffith, B.M. 1853. *Gallionella ferruginea.* Ehr. Ann. Mag. Nat. Hist. II Ser. *12*: 438.

Griffiths, P.L. and R.W.A. Park. 1990. Campylobacters associated with human diarrhoeal disease. J. Appl. Bacteriol. *69*: 281–301.

Grifoll, M., S.A. Selifonov, C.V. Gatlin and P.J. Chapman. 1995. Actions of a versatile fluorene degrading bacterial isolate on polycyclic aromatic compounds. Appl. Environ. Microbiol. *61*: 3711–3723.

Grigioni, S., R. Boucher-Rodoni, A. Demarta, M. Tonolla and R. Peduzzi. 2000. Phylogenetic characterisation of bacterial symbionts in the accessory nidamental glands of the sepioid *Sepia officinalis* (Cephalopoda: Decapoda). Mar. Biol. *136*: 217–222.

Grilione, P. 1968. Serological reactions of some higher myxobacteria. J. Bacteriol. *95*: 1202–1204.

Grimes, D.J., C.R. Woese, M.T. MacDonell and R.R. Colwell. 1997. Systematic study of the genus *Vogesella* gen. nov. and its type species, *Vogesella indigofera* comb. nov. Int. J. Syst. Bacteriol. *47*: 19–27.

Grimm, K. 1967. Clone cultures of *Archangium violaceum.* Arch. Mikrobiol. *57*: 283–284.

Grimm, K. 1978. Comparison of spontaneous, UV-induced, and nitrosoguanidine-induced mutability to drug resistance in myxobacteria. J. Bacteriol. *135*: 748–753.

Grimm, K. 1980. Mutation to acriflavine resistance in some myxobacteria. Microbios. *27*: 193–206.

Grimm, K. and H. Kühlwein. 1973. Untersuchungen an spontanen Mutanten von *Archangium violaceum* (*Myxobacterales*). III. Über weitere Eigenschaften der K- und S-Zellen. Arch. Microbiol. *89*: 133–146.

Grimm, K. and H. Kuhlwein. 1973a. Untersuchungen an spontanen Mutanten von *Archangium violaceum* (*Myxobacterales*). I. Bewegliche und unbewegliche Zellen von *A. violaceum.* Arch. Microbiol. *89*: 105–119.

Grimm, K. and H. Kuhlwein. 1973b. Untersuchungen an spontanen Mutanten von *Archangium violaceum* (*Myxobacterales*). II. Über den Einfluss des Schleims auf die Bewegung der Zellen und die entstehung stabiler Suspensionskulturen. Arch. Mikrobiol. *89*: 121–132.

Grimont, P.A.D., M. Vancanneyt, M. Lefèvre, K. Vandemeulebroecke, L. Vauterin, R. Brosch, K. Kersters and F. Grimont. 1996. Ability of Biolog and Biotype-100 systems to reveal the taxonomic diversity of the pseudomonads. Syst. Appl. Microbiol. *19*: 510–527.

Grishanin, R.N., I.I. Chalmina and I.B. Zhulin. 1991. Behavoir of *Azospirillum brasilense* in a spatial gradient of oxygen an in a "redox" gradient of an artificial electron acceptor. J Gen Microbiol. *137*: 2781–2785.

Gromet-Elhanan, Z. 1995. The proton-translocating F_0F_1 ATP synthase-ATPase complex. *In* Blankenship, Madigan and Bauer (Editors), Anoxygenic Photosynthetic Bacteria, Kluwer Academic Publishers, Netherlands. 807–830.

Gromov, B.V. and K.A. Mamkaeva. 1980a. Proposal of a new genus *Vampirovibrio* for chorellavorus bacteria previously assigned to *Bdellovibrio.* Mikrobiologiya *49*: 165–167.

Gromov, B.V. and K.A. Mamkaeva. 1980b. *In* Validation of the publication of new names and new combinations previously effectively published outside the IJSB. List. No. 5. Int. J. Syst. Bacteriol. *30*: 676–677.

Gromov, B.V. and D.V. Ossipov. 1981. *Holospora* (Ex Hafkine 1890) nom. rev., a genus of bacteria inhabiting the nuclei of paramecia. Int. J. Syst. Bacteriol. *31*: 348–352.

Grones, J. and J. Turna. 1995. Transformation of microorganisms with the plasmid vector with the replicon from pAC1 from *Acetobacter pasteurianus.* Biochem. Biophys. Res. Commun. *206*: 942–947.

Gross, M.J. and B.E. Logan. 1995. Influence of different chemical treatments on transport of *Alcaligenes paradoxus* in porous media. Appl. Environ. Microbiol. *61*: 1750–1756.

Gross, R., J. Simon, F. Theis and A. Kröger. 1998. Two membrane anchors of *Wolinella succinogenes* hydrogenase and their function in fumarate and polysulfide respiration. Arch. Microbiol. *170*: 50–58.

Grothues, D. and B. Tümmler. 1991. New approaches in genome analysis by pulsed-field gel- electrophoresis - application to the analysis of *Pseudomonas* species. Mol. Microbiol. *5*: 2763–2776.

Groves, M.G., D.L. Rosenstreich, B.A. Taylor and J.V. Osterman. 1980. Host defenses in experimental scrub typhus: Mapping the gene that controls natural resistance in mice. J. Immunol. *125*: 1395–1399.

Grundmann, G.L., M. Neyra and P. Normand. 2000. High-resolution phylogenetic analysis of NO_2-oxidizing *Nitrobacter* species using the rrs-rrl IGS sequence and rrl genes. Int. J. Syst. Evol. Microbiol. *50*: 1893–1898.

Grzeszik, C., T. Jeffke, J. Schaeferjohann, B. Kusian and B. Bowien. 2000. Phosphoenolpyruvate is a signal metabolite in transcriptional control of the *cbb* CO_2 fixation operons in *Ralstonia eutropha.* J. Mol. Microbiol. Biotechnol. *2*: 311–320.

Guay, R. and M. Silver. 1975. *Thiobacillus acidophilus* sp. nov.: isolation and some physiological characteristics. Can. J. Microbiol. *21*: 281–288.

Gubish, E.R.J., K.C. Chen and T.M. Buchanan. 1982. Detection of a gonococcal endo-β-N-acetyl-D-glucosaminidase and its peptidoglycan cleavage site. J. Bacteriol. *151*: 172–176.

Guckert, J.B., D.B. Ringelberg, D.C. White, R.S. Hanson and B.J. Bratina. 1991. Membrane fatty acids as phenotypic markers in the polyphasic taxonomy of methylotrophs within the *Proteobacteria.* J. Gen. Microbiol. *137*: 2631–2641.

Güde, H., B. Haibel and H. Müller. 1985. Development of planktonic bacterial populations in a water column of Lake Constance (Bodensee-Obersee). Arch. Hydrobiol. *104*: 59–78.

Gueirard, P., A. Druilhe, M. Pretolani and N. Guiso. 1998. Role of adenylate cyclase-hemolysin in alveolar macrophage apoptosis during *Bordetella pertussis* infection in vivo. Infect. Immun. *66*: 1718–1725.

Guerin, W.F. and R.P. Blakemore. 1992. Redox cycling of iron supports growth and magnetite synthesis by *Aquaspirillum magnetotacticum.* Appl. Environ. Microbiol. *58*: 1102–1109.

Guerin, W.F. and S.A. Boyd. 1995. Maintenance and induction of naphthalene degradation activity in *Pseudomonas putida* and an *Alcaligenes* sp. under different culture conditions. Appl. Environ. Microbiol. *61*: 4061–4068.

Guether, D.L., G.J. Osterhout, J.D. Dick and H.N. Williams. 1993. Analysis of fatty acid composition of *Bdellovibrio* isolates. 93rd Annual Meeting of the American Society for Microbiology, p. 391.

Guether, D.L. and H.N. Williams. 1992. Reactions of *Bdellovibrio* isolates in miniature rapid test systems. 92nd Annual Meeting of the American Society for Microbiology, 372.

Guether, D.L. and H.N. Williams. 1993. Antibiogram characterization of aquatic and terrestrial *Bdellovibrio* isolates. 93rd Annual Meeting of the American Society for Microbiology, 391.

Guezennec, J. and A. Fiala-Medioni. 1996. Bacterial abundance and diversity in the Barbados Trench determined by phospholipid analysis. FEMS Microbiol. Ecol. *19*: 83–93.

Guglielmone, A.A., O.S. Anziani, A.J. Mangold, M.M. Volpogni and A. Vogel. 1996. Enrofloxacin to control *Anaplasma marginale* infections. Ann. N. Y. Acad. Sci. *791*: 471–472.

Gully, N.J. and A.H. Rogers. 1996. Energy production and peptidase activity in *Eikenella corrodens.* FEMS Microbiol. Lett. *139*: 209–213.

Gumaelius, L., G. Magnusson, B. Pettersson and G. Dalhammar. 2001. *Comamonas denitrificans* sp. nov., an efficient denitrifying bacterium isolated from activated sludge. Int. J. Syst. Evol. Microbiol. *51*: 999–1006.

Gündisch, C., G. Kirchhof, M. Baur, W. Bode and A. Hartmann. 1993. Identification of *Azospirillum* species by RFLP and pulsed-field gel electrophoresis. Microb. Releases. *2*: 41–45.

Günther, K.A. 1894. Über einen neuen, im Erdboden gefundenen Kommabacillus. Centralbl. Bakteriol. Parasitenkd. *16*: 746–747.

Gupta, A., V. Verma and G.N. Qazi. 1997. Transposon induced mutation in *Gluconobacter oxydans* with special reference to its direct-glucose oxidation metabolism. FEMS Microbiol. Lett. *147*: 181–188.

Guptill, L., L.N. Slater, C.C. Wu, T.L. Lin, L.T. Glickman, D.F. Welch, J. Tobolski and H. HogenEsch. 1998. Evidence of reproductive failure and lack of perinatal transmission of *Bartonella henselae* in experimentally infected cats. Vet. Immunol. Immunopathol. *65*: 177–189.

Gürgün, V., G. Kirchner and N. Pfennig. 1976. Vergärung von pyuvat durch sieben arten phototropher purpurbacterien. Z. Allg. Mikrobiol. *16*: 573–586.

Guris, D., P.M. Strebel, B. Bardenheier, M. Brennan, R. Tachdjian, E. Finch, M. Wharton and J.R. Livengood. 1999. Changing epidemiology of pertussis in the United States: Increasing reported incidence among adolescents and adults, 1990-1996. Clin. Infect. Dis. *28*: 1230–1237.

Gutiérrez-Zamora, M.L. and E. Martínez-Romero. 2001. Natural endophytic association between *Rhizobium etli* and maize. J. Biotechnol. *91*: 117–126.

Guyon, P., A. Petit, J. Tempe and Y. Dessaux. 1993. Transformed plants producing opines specifically promote growth of opine-degrading agrobacteria. Mol. Plant-Microbe Interact. *6*: 92–98.

Gwynn, M.N., S.J. Box, A.G. Brown and M.L. Gilpin. 1988. MM 42842, a new member of the monobactam family produced by *Pseudomonas cocovenenans*. I. Identification of the producing organism. J. Antibiot. *41*: 1–6.

Haak, B., S. Fetzner and F. Lingens. 1995. Cloning, nucleotide sequence, and expression of the plasmid-encoded genes for the two component 2-halobenzoate 1,2-dioxygenase from *Pseudomonas cepacia* 2CBS. J. Bacteriol. *177*: 667–675.

Haake, D.A., T.A. Summers, A.M. McCoy and W. Schwartzman. 1997. Heat shock response and groEL sequence of *Bartonella henselae* and *Bartonella quintana*. Microbiology *143*: 2807–2815.

Haarbrink, M., G.K. Abadi, W.A. Buurman, M.A. Dentener, A.J. Terhell and M. Yazdanbakhsh. 2000. Strong association of interleukin-6 and lipopolysaccharide-binding protein with severity of adverse reactions after diethylcarbamazine treatment of microfilaremic patients. J. Infect. Dis. *182*: 564–569.

Haars, E.G. and J.M. Schmidt. 1974. Stalk formation and its inhibition in *Caulobacter crescentus*. J. Bacteriol. *120*: 1409–1416.

Haas, R. and T.F. Meyer. 1986. The repertoire of silent pilus genes in *Neisseria gonorrhoeae*—evidence for gene conversion. Cell *44*: 107–115.

Haber, C.L., L.N. Allen and R.S. Hanson. 1984. Methylotrophic bacteria: biochemical diversity and genetics. Science *221*: 1147–1153.

Hachem, C.Y., J.E. Clarridge, D.G. Evans and D.Y. Graham. 1995. Comparison of agar based media for primary isolation of *Helicobacter pylori*. J. Clin. Pathol. *48*: 714–716.

Hackstadt, T., R. Messer, W. Cieplak and M.G. Peacock. 1992. Evidence for proteolytic cleavage of the 120-kilodalton outer membrane protein of rickettsiae: Identification of an avirulent mutant deficient in processing. Infect. Immun. *60*: 159–165.

Hadani, A. and Y. Dinur. 1968. Studies on the transmission of *Aegyptianella pullorum* by the tick *Argas persicus*. J. Protozool. *15*: 45.

Haddad, A., F. Camacho, P. Durand and S.C. Cary. 1995. Phylogenetic characterization of the epibiotic bacteria associated with the hydrothermal vent polychaete *Alvinella pompejana*. Appl. Environ. Microbiol. *61*: 1679–1687.

Haeckel, E. 1866. Generelle Morphologie der Organismen: allgemeine Grundzüge der organischen Formen-Wissenschaft: mechanisch begründet durch die von Charles Darwin reformirte Descendenz-Theorie, G. Reimer, Berlin.

Hafkine, M.W. 1890. Maladies infectieuses des Paramécies. Ann. Inst. Pasteur *4*: 148–162.

Häfliger, M., H. Spillmann and Z. Puhan. 1991a. Einfluss des Ruhrens auf die Entwicklung der Mikroflora des Kefirs, unter besonderer Berucksichtigung der Essigsaurebakterien. Schewiz. Milchw. Forschung. *20*: 55–62.

Häfliger, M., H. Spillmann and Z. Puhan. 1991b. Selektiver Nachweis und Identifizierung von Essigsaurebakterien in Kefirkornern und Kefir. Schweiz. Milchw. Forschung. *20*: 3–8.

Hagedorn, C., W.D. Gould, T.R. Bardinelli and D.R. Gustavson. 1987. A selective medium for enumeration and recovery of *Pseudomonas cepacia* biotypes from soil. Appl. Environ. Microbiol. *53*: 2265–2268.

Hagishita, T., T. Yoshida, Y. Izumi and T. Mitsunaga. 1996a. Cloning and expression of the gene for serine-glyoxylate aminotransferase from an obligate methylotroph *Hyphomicrobium methylovorum* GM2. Eur. J. Biochem. *241*: 1–5.

Hagishita, T., T. Yoshida, Y. Izumi and T. Mitsunaga. 1996b. Immunological characterization of serine-glyoxylate aminotransferase and hydroxypyruvate reductase from a methylotrophic bacterium, *Hyphomicrobium methylovorum* GM2. FEMS Microbiol. Lett. *142*: 49–52.

Hagman, M. and D. Danielsson. 1989. Increased adherence to vaginal epithelial cells and phagocytic killing of gonococci and urogenital meningococci associated with heat modifiable proteins. APMIS *97*: 839–844.

Haines, D.C., P.L. Gorelick, J.K. Battles, K.M. Pike, R.J. Anderson, J.G. Fox, N.S. Taylor, Z. Shen, F.E. Dewhirst, M.R. Anver and J.M. Ward. 1998. Inflammatory large bowel disease in immunodeficient rats naturally and experimentally infected with *Helicobacter bilis*. Vet. Pathol. *35*: 202–208.

Haines, K.A., L. Yeh, M.S. Blake, P. Cristello, H. Korchak and G. Weissmann. 1988. Protein-I, a translocatable ion channel from *Neisseria gonorrhoeae*, selectively inhibits exocytosis from human neutrophils without inhibiting O$_2$ generation. J. Biol. Chem. *263*: 945–951.

Hajiroussou, V., B. Holmes, J. Bullas and C.A. Pinning. 1979. Meningitis caused by *Pseudomonas paucimobilis*. J. Clin. Pathol. *32*: 953–955.

Hall, G.S., K. Pratt-Rippin and J.A. Washington. 1991. Isolation of agent associated with cat scratch disease bacillus from pretibial biopsy. Diagn. Microbiol. Infect. Dis. *14*: 511–514.

Hall, P.E., S.M. Anderson, D.M. Johnston and R.E. Cannon. 1992. Transformation of *Acetobacter xylinum* with plasmid DNA by electroporation. Plasmid *28*: 194–200.

Hallander, H.O., E. Reizenstein, B. Renemar, G. Rasmuson, L. Mardin and P. Olin. 1993. Comparison of nasopharyngeal aspirates with swabs for culture of *Bordetella pertussis*. J. Clin. Microbiol. *31*: 50–52.

Hallbeck, L. and K. Pedersen. 1990. Culture parameters regulating stalk formation and growth rate of *Gallionella ferruginea*. J. Gen. Microbiol. *136*: 1675–1680.

Hallbeck, L. and K. Pedersen. 1991. Autotrophic and mixotrophic growth of *Gallionella ferruginea*. J. Gen. Microbiol. *137*: 2657–2661.

Hallbeck, L. and K. Pedersen. 1995. Benefits associated with the stalk of *Gallionella ferruginea*, evaluated by comparison of a stalk-forming and a non-stalk-forming strain and biofilm studies *in situ*. Microb. Ecol. *30*: 257–268.

Hallbeck, L., F. Stahl and K. Pedersen. 1993. Phylogeny and phenotypic characterization of the stalk-forming and iron-oxidizing bacterium *Gallionella ferruginea*. J. Gen. Microbiol. *139*: 1531–1535.

Halling, S.M. and E.S. Zehr. 1990. Polymorphism in *Brucella* spp. due to highly repeated DNA. J. Bacteriol. *172*: 6637–6640.

Hallmann, J., J.W. Kloepper and R. Rodriguez-Kabana. 1997. Application of the Scholander pressure bomb to studies on endophytic bacteria of plants. Can. J. Microbiol. *43*: 411–416.

Halperin, S.A., R. Bortolussi and A.J. Wort. 1989. Evaluation of culture, immunofluorescence, and serology for the diagnosis of pertussis. J. Clin. Microbiol. *27*: 752–757.

Hamamura, N., C. Page, T. Long, L. Semprini and D.J. Arp. 1997. Chloroform cometabolism by butane-grown CF8, *Pseudomonas butanovora*, and *Mycobacterium vaccae* JOB5 and methane-grown *Methylosinus trichosporium* OB3b. Appl. Environ. Microbiol. *63*: 3607–3613.

Hamana, K. 1997. Polyamine distribution patterns within the families *Aeromonadaceae*, *Vibrionaceae*, *Pasteurellaceae*, and *Halomonadaceae*, and related genera of the gamma subclass of the *Proteobacteria*. J. Gen. Appl. Microbiol. *43*: 49–59.

Hamana, K., M. Kamekura, H. Onishi, T. Akazawa and S. Matsuzaki. 1985. Polyamines in photosynthetic eubacteria and extreme-halophilic archaebacteria. J. Biochem. (Tokyo) *97*: 1653–1658.

Hamana, K. and S. Matsuzaki. 1993. Polyamine distribution patterns serve as a phenotypic marker in the chemotaxonomy of the *Proteobacteria*. Can. J. Microbiol. *39*: 304–310.

Hamana, K., T. Sakane and A. Yokota. 1994. Polyamine analysis of the genera, *Aquaspirillum*, *Magnetospirillum*, *Oceanospirillum* and *Spirillum*. J. Gen. Appl. Microbiol. *40*: 75–82.

Hamana, K. and M. Takeuchi. 1998. Polyamine profiles as chemotaxonomic markers within alpha, beta, gamma, delta, and epsilon subclasses of class *Proteobacteria*: Distribution of 2-hydroxyputrescine and homospermidine. Microbiol. Cult. Coll. *14*: 1–14.

Hamilton, R.D. and K.E. Austin. 1967. Physiological and cultural characteristics of *Chromobacterium marinum* sp. n. Antonie Leeuwenhoek J. Microbiol. Serol. *33*: 257–264.

Hammadi, A., A. Menez and R. Genet. 1997. Asymmetric deuteration of *N*-acetyl-(Z)-alpha,beta-dehydrotryptophan-(L)-phenylalanine methyl ester produced by (L)-tryptophan 2′,3′-oxidase from *Chromobacterium violaceum*. A new route for stereospecific labelling of peptides. Tetrahedron. *53*: 16115–16122.

Hammerschmidt, S., A. Muller, H. Sillmann, M. Muhlenhoff, R. Borrow, A. Fox, J. vanPutten, W.D. Zollinger, R. Gerardy-Schahn and M. Frosch. 1996. Capsule phase variation in *Neisseria meningitidis* serogroup B by slipped-strand mispairing in the polysialyltransferase gene (*siaD*): Correlation with bacterial invasion and the outbreak of meningococcal disease. Mol. Microbiol. *20*: 1211–1220.

Hampton, J.C. 1967. The effects of nitrogen mustard on the intestinal epithelium of the mouse. Radiat. Res. *30*: 576–589.

Han, S.R., C. Schindel, R. Genitsariotis, E. Märker-Hermann, S. Bhakdi and M.J. Maeurer. 2000. Identification of a unique *Helicobacter* species by 16S rRNA gene analysis in an abdominal abscess from a patient with X-linked hypogammaglobulinemia. J. Clin. Microbiol. *38*: 2740–2742.

Han, Y.-H., R.M. Smibert and N.R. Krieg. 1989. Occurrence of sheathed flagella in *Campylobacter cinaedi* and *Campylobacter fennelliae*. Int. J. Syst. Bacteriol. *39*: 488–490.

Han, Y.-H., R.M. Smibert and N.R. Krieg. 1991. *Wolinella recta*, *Wolinella curva*, *Bacteroides ureolyticus*, and *Bacteroides gracilis* are microaerophiles, not anaerobes. Int. J. Syst. Bacteriol. *41*: 218–222.

Han, Y.-H., R.M. Smibert and N.R. Krieg. 1992. Cytochrome composition and oxygen-dependent respiration-driven proton translocation in *Wolinella curva*, *Wolinella recta*, *Bacteroides ureolyticus*, and *Bacteroides gracilis*. Can. J. Microbiol. *38*: 104–110.

Hanada, S., A. Hiraishi, K. Shimada and K. Matsuura. 1995. Isolation of *Chloroflexus aurantiacus* and related thermophilic phototrophic bacteria from Japanese hot springs using an improved isolation procedure. J. Gen. Appl. Microbiol. *41*: 119–130.

Hanada, S., Y. Kawase, A. Hiraishi, S. Takaichi, K. Matsuura, K. Shimada and K.V.P. Nagashima. 1997. *Porphyrobacter tepidarius* sp. nov., a moderately thermophilic aerobic photosynthetic bacteium isolated from a hot spring. Int. J. Syst. Bacteriol. *47*: 408–413.

Handrick, R., S. Reinhardt and D. Jendrossek. 2000. Mobilization of poly(3-hydroxybutyrate) in *Ralstonia eutropha*. J. Bacteriol. *182*: 5916–5918.

Hanert, H.H. 1989. Genus *Gallionella*. *In* Staley, Bryant, Pfenning and Holt (Editors), Bergey's Manual of Systematic Bacteriology, 1st Ed., Vol. 3, The Williams & Wilkins Co., Baltimore. pp. 1974–1979.

Hänninen, M.L., I. Happonen, S. Saari and K. Jalava. 1996. Culture and characteristics of *Helicobacter bizzozeronii*, a new canine gastric *Helicobacter* sp. Int. J. Syst. Bacteriol. *46*: 160–166.

Hansen, E.C. 1879. Bidrag til kundskab om hvilke organismer der kunne forekomme og leve i øl og ølurt. Medd. Carlsberg Lab. *1*: 185–234.

Hansen, M.V. and C.E. Wilde. 1984. Conservation of peptide structure of outer-membrane protein- macromolecular complex from *Neisseria gonorrhoeae*. Infect. Immun. *43*: 839–845.

Hansen, T.A. 1974. Sulfide als electronendonor voor *Rhodospirillaceae*, University of Groningen, The Netherlands.

Hansen, T.A. and J.F. Imhoff. 1985. *Rhodobacter veldkampii* sp. nov., a new species of phototrophic purple nonsulfur bacteria. Int. J. Syst. Bacteriol. *35*: 115–116.

Hansen, T.A., H.E. Nienhuis-Kuiper and A.J.M. Stams. 1990. A rod-shaped, Gram-negative, propionigenic bacterium with a wide substrate range and the ability to fix molecular nitrogen. Arch. Microbiol. *155*: 42–45.

Hansen, T.A., A.B.J. Sepers and H. van Gemerden. 1975. A new purple bacterium that oxidizes sulfide to extracellular sulfur and sulfate. Plant Soil *43*: 17–27.

Hansen, T.A. and H. van Gemerden. 1972. Sulfide utilization by purple nonsulfur bacteria. Arch. Mikrobiol. *86*: 49–56.

Hansen, T.A. and H. Veldkamp. 1973. *Rhodopseudomonas sulfidophila*, nov. spec., a new species of the purple nonsulfur bacteria. Arch. Mikrobiol. *92*: 45–58.

Hanski, E. and Z. Farfel. 1985. *Bordetella pertussis* invasive adenylate cyclase, partial resolution and properties of its cellular penetration. J. Biol. Chem. *260*: 5526–5532.

Hanson, B.A., C.L. Wisseman, Jr., A. Waddell and D.J. Silverman. 1981. Some characteristics of heavy and light bands of *Rickettsia prowazekii* on Renografin gradients. Infect. Immun. *34*: 596–604.

Hanson, R.S., B.J. Bratina and G.A. Brusseau. 1993. Phylogeny and ecology of methylotrophic bacteria. *In* Kelley (Editor), Microbial Growth on C1 Compounds, Intercept Press, Ltd., Andover, UK. 285–302.

Hanson, R.S. and T.E. Hanson. 1996. Methanotrophic bacteria. Microbiol. Rev. *60*: 439–471.

Hanson, R.S., A.I. Netrusov and K. Tsuji. 1992. The obligate methanotrophic bacteria *Methylococcus*, *Methylomonas*, and *Methylosinus*. *In* Balows, Trüper, Dworkin, Harder and Schleifer (Editors), The Prokaryotes: A Handbook of Bacteria: Ecophysiology, Isolation, Identification, Applications, 2nd Ed., Vol. 3, Springer-Verlag, New York. pp. 2350–2364.

Hanus, F.J., R.J. Maier and H.J. Evans. 1979. Autotrophic growth of hydrogen uptake-positive strains of *Rhizobium japonicum* in an atmosphere supplied with hydrogen gas. Proc. Nat. Acad. Sci. U.S.A. *76*: 1788–1792.

Happe, B., L.D. Eltis, H. Poth, R. Hedderich and K.N. Timmis. 1993. Characterization of 2,2′,3-trihydroxybiphenyl dioxygenase, an extradiol dioxygenase from the dibenzofuran- and dibenzo-*p*-dioxin-degrading bacterium *Sphingomonas* sp. strain RW1. J. Bacteriol. *175*: 7313–7320.

Happe, R.P., W. Roseboom, G. Egert, C.G. Friedrich, C. Massanz, B. Friedrich and S.P.J. Albracht. 2000. Unusual FTIR and EPR properties of the H_2-activating site of the cytoplasmic NAD-reducing hydrogenase from *Ralstonia eutropha*. FEBS Lett. *466*: 259–263.

Haque, M., Y. Hirai, K. Yokota, N. Mori, I. Jahan, H. Ito, H. Hotta, I. Yano, Y. Kanemasa and K. Oguma. 1996. Lipid profile of *Helicobacter* spp.: presence of cholesteryl glucoside as a characteristic feature. J. Bacteriol. *178*: 2065–2070.

Haque, M., Y. Hirai, K. Yokota and K. Oguma. 1995. Steryl glycosides: a characteristic feature of the *Helicobacter* spp.? J. Bacteriol. *177*: 5334–5337.

Harashima, K., J.I. Hayashi, T. Ikari and T. Shiba. 1980. O_2-Stimulated sythesis of bacteriochlorophyll and carotenoids in marine bacteria. Plant Cell Physiol. *21*: 1283–1294.

Harashima, K., K. Kawazoe, I. Yoshida and H. Kamata. 1987. Light-stimultated aerobic growth of *Erythrobacter* species Och-114. Plant Cell Physiol. *28*: 365–374.

Harashima, K. and H. Nakada. 1983. Carotenoids and ubiquinone in aerobically grown cells of aerobic photosynthetic bacterium *Erythrobacter* species (OCh114). Agric. Biol. Chem. *47*: 1057–1063.

Harashima, K., M. Nakagawa and N. Murata. 1982. Photochemical activities of bacteriochlorophyll in aerobically grown cells of aerobic heterotrophs, *Erythrobacter* species (Och 114) and *Erythrobacter longus* (Och 101). Plant Cell Physiol. *23*: 185–193.

Harashima, K., T. Shiba, T. Totsuka, U. Simidu and N. Taga. 1978. Occurrence of bacteriochlorophyll-*a* in a strain of an aerobic heterotrophic bacterium. Agr. Biol. Chem. Tokyo *42*: 1627–1628.

Harayama, S. 1997. Polycyclic aromatic hydrocarbon bioremediation design. Curr. Opin. Biotechnol. *8*: 268–273.

Harder, J. 1997. Anaerobic degradation of cyclohexane-1,2-diol by a new *Azoarcus* species. Arch. Microbiol. *168*: 199–204.

Harder, W., A. Matin and M.M. Attwood. 1975. Studies on the physiological significance of the lack of a pyruvate dehydrogenase complex in *Hyphomicrobium* sp. J. Gen. Microbiol. *86*: 319–326.

Hardy, S.J., M. Christodoulides, R.O. Weller and J.E. Heckels. 2000. Interactions of *Neisseria meningitidis* with cells of the human meninges. Mol. Microbiol. *36*: 817–829.

Harker, M., J. Hirschberg and A. Oren. 1998. *Paracoccus marcusii* sp. nov., an orange Gram-negative coccus. Int. J. Syst. Bacteriol. *48*: 543–548.

Harkness, J.E. and J.E. Wagner. 1995. *Bordetella bronchiseptica* infections. *In* Biology and Medicine of Rabbits and Rodents, Williams & Wilkins Co., Baltimore. 182–185.

Harmon, K.M. and I.V. Wesley. 1997. Multiplex PCR for the identification of *Arcobacter* and differentiation of *Arcobacter butzleri* from other arcobacters. Vet. Microbiol. *58*: 215–227.

Harms, G., R. Rabus and F. Widdel. 1999a. Anaerobic oxidation of the aromatic plant hydrocarbon *p*-cymene by newly isolated denitrifying bacteria. Arch. Microbiol. *172*: 303–312.

Harms, G., K. Zengler, R. Rabus, F. Aeckersberg, D. Minz, R. Rosselló-Mora and F. Widdel. 1999b. Anaerobic oxidation of *o*-xylene, *m*-xylene, and homologous alkylbenzenes by new types of sulfate-reducing bacteria. Appl. Environ. Microbiol. *65*: 999–1004.

Harms, H., H.P. Koops and H. Wehrmann. 1976. An ammonia-oxidizing bacterium, *Nitrosovibrio tenuis* nov. gen. nov. sp. Arch. Microbiol. *108*: 105–111.

Harms, H., H. Wilkes, R.M. Wittich and P. Fortnagel. 1995. Metabolism of hydroxydibenzofurans, methoxydibenzofurans, acetoxydibenzofurans, and mitrodibenzofurans by *Sphingomonas* sp. strain HH69. Appl. Environ. Microbiol. *61*: 2499–2505.

Harms, H. and A.J. Zehnder. 1994. Influence of substrate diffusion on degradation of dibenzofuran and 3-chlorodibenzofuran by attached and suspended bacteria. Appl. Environ. Microbiol. *60*: 2736–2745.

Harms, N., J. Ras, S. Koning, W.N.M. Reijnders, A.H. Stouthamer and R.J.M. Van Spanning. 1996. Genetics of C1 metabolism regulation in *Paracoccus denitrificans*. *In* Lidstrom and Tabita (Editors), Microbial Growth on C1 Compounds, Kluwer Academic Publishers, The Netherlands. pp. 126–132.

Harmsen, D., J. Rothganger, C. Singer, J. Albert and M. Frosch. 1999. Intuitive hypertext-based molecular identification of microorganisms. Lancet *353*: 291–291.

Harmsen, D., C. Singer, J. Rothganger, T. Tonjum, G.S. de Hoog, H. Shah, J. Albert and M. Frosch. 2001. Diagnostics of *Neisseriaceae* and *Moraxellaceae* by ribosomal DNA sequencing: ribosomal differentiation of medical microorganisms. J. Clin. Microbiol. *39*: 936–942.

Harmsen, H.J.M., H.M.P. Kengen, A.D.L. Akkermans and A.J.M. Stams. 1995. Phylogenetic analysis of two syntrophic propionate oxidizing bacteria in enrichments cultures. Syst. Appl. Microbiol. *18*: 67-73.

Harmsen, H.J.M., D. Prieur and C. Jeanthon. 1997. Distribution of microorganisms in deep-sea hyrothermal vent chimneys investigated by whole-cell hybridization and enrichment culture of thermophilic subpopulations. Appl. Environ. Microbiol. *63*: 2876–2883.

Harmsen, H.J.M., B.L.M. van Kuijk, C.M. Plugge, A.D.L. Akkermans, W.M. De Vos and A.J.M. Stams. 1998. *Syntrophobacter fumaroxidans* sp. nov., a syntrophic propionate-degrading sulfate-reducing bacterium. Int. J. Syst. Bacteriol. *48*: 1383–1387.

Harmsen, H.J.M., B. Wullings, A.D.L. Akkermans, W. Ludwig and A.J.M. Stams. 1993. Phylogenetic analysis of *Syntrophobacter wolinii* reveals a relationship with sulfate-reducing bacteria. Arch. Microbiol. *160*: 238–240.

Harper, C.M.G., C.A. Dangler, S. Xu, Y. Feng, Z. Shen, B. Sheppard, A. Stamper, F.E. Dewhirst, B.J. Paster and J.G. Fox. 2000. Isolation and characterisation of a *Helicobacter* sp. from the gastric mucosa of dolphins, *Lagenorhynchus acutus* and *Delphinus delphus*. Appl. Environ. Microbiol. *66*: 4751–5757.

Harper, C.M.G., S. Xu, Y. Feng, L. Dunn, N.S. Taylor, F.E. Dewhirst and J.G. Fox. 2002. Isolation of novel *Helicobacter* spp. from a beluga whale. Appl. Environ. Microbiol. *68*: 2040–2043.

Harrington, C.S. and S.L. On. 1999. Extensive 16S rRNA gene sequence diversity in *Campylobacter hyointestinalis* strains: taxonomic and applied implications. Int. J. Syst. Bacteriol. *49*: 1171–1175.

Harrington, C.S., F.M. Thomson Carter and P.E. Carter. 1997. Evidence for recombination in the flagellin locus of *Campylobacter jejuni*: implications for the flagellin gene typing scheme. J. Clin. Microbiol. *35*: 2386–2392.

Harrington, R., Jr., D.R. Bond and G.M. Brown. 1977. Smooth phage resistant *Brucella abortus* from bovine tissue. J. Clin. Microbiol. *5*: 663–664.

Harrison, A.P. 1981. *Acidiphilium cryptum* gen. nov., sp. nov., heterotrophic bacterium from acidic minteral environments. Int. J. Syst. Bacteriol. *31*: 327–332.

Harrison, A.P. 1982. Genomic and physiological diversity amongst strains of thiobacillus ferrooxidans and genomic comparison with *Thiobacillus thiooxidans*. Arch. Microbiol. *131*: 68–76.

Harrison, A.P., Jr. 1983. Genomic and physiological comparisons between heterotrophic thiobacilli and *Acidiphilum cryptum, Thiobacillus versutus* sp. nov., and *Thiobacillus acidophilus* nom. rev. Int. J. Syst. Bacteriol. *33*: 211–217.

Harrison, A.P. 1989. Genus *Acidiphilium*. *In* Staley, Bryant, Pfennig and Holt (Editors), Bergey's Manual of Systematic Bacteriology, Vol. 3, The Williams & Wilkins Co., Baltimore. pp. 1863–1868.

Harrison, C.P., B.W. Jarvis and J.L. Johnson. 1980. Heterotrophic bacteria from cultures of autotrophic *Thiobacillus ferrooxidans*: relationship as studied by means of deoxyribonucleic acid homology. J. Bacteriol. *143*: 448–445.

Hartmann, A. 1988. Ecophysiological aspects of growth and nitrogen fixation in *Azospirillum* spp. Plant Soil *110*: 225–238.

Hartmann, A. and R.H. Burris. 1987. Regulation of nitrogenase activity by oxygen in *Azospirillum brasilense* and *Azospirillum lipoferum*. J. Bacteriol. *169*: 944–948.

Hartmann, A., H. Fu and R.H. Burris. 1986. Regulation of nitrogenase activity of ammonium chloride in *Azospirillum* spp. J. Bacteriol. *165*: 864–870.

Hartmann, A., H.A. Fu and R.H. Burris. 1988. Influence of amino acids on nitrogen fixation activity and growth of *Azospirillum* spp. Appl. Environ. Microbiol. *54*: 87–93.

Hartmann, A., C. Gündisch and W. Bode. 1992. *Azospirillum* mutants improved in iron acquisition and osmotolerance as tools for the investigation of environmental fitness traits. Symbiosis *13*: 271–279.

Hartmann, A., H.-A. Fu, S.-D. Song and R.H. Burris. 1985. Comparison of nitrogenase regulation in *A. brasilense, A. lipoferum* and *A. amazonense*. *In* Klingmuller (Editor), *Azospirillum* III: Genetics, Physiology and Ecology, Springer-Verlag, Berlin. pp. 116–126.

Hartmann, A. and T. Hurek. 1988. Effect of carotenoid overproduction on oxygen tolerance of nitrogen fixation in *Azospirillum brasilense* Sp7. J Gen. Microbiol. *143*: 2449–2455.

Hartmann, A., S.R. Prabhu and E.A. Galinski. 1991. Osmotolerance of diazotrophic rhizosphere bacteria. Plant Soil *48*: 155–159.

Hartmann, A., M. Singh and W. Klingmüller. 1983. Isolation and characterization of *Azospirillum* mutants excreting high amounts of indoleacetic acid. Can. J. Microbiol. *29*: 916–923.

Hartmann, A. and W. Zimmer. 1994. Physiology of *Azospirillum*. *In* Okon (Editor), Azospirillum-Plant Associations, CRC Press, Boca Raton. pp. 15–39.

Hartmans, S., A. Schmuckle, A.M. Cook and T. Leisinger. 1986. Methyl chloride: Naturally occurring toxicant and C-1 growth substrate. J. Gen. Microbiol. *132*: 1139–1142.

Hartmans, S., J.P. Smits, M.J. Vanderwerf, F. Volkering and J.A. De Bont. 1989. Metabolism of styrene oxide and 2-phenylethanol in the styrene-degrading *Xanthobacter* strain 124x. Appl. Environ. Microbiol. *55*: 2850–2855.

Hartzell, P. and D. Kaiser. 1991. Upstream gene of the Mgl operon controls the level of MglA protein in *Myxococcus xanthus*. J. Bacteriol. *173*: 7625–7635.

Harvey, S.M. 1980. Hippurate hydrolysis by *Campylobacter fetus*. J. Clin. Microbiol. *11*: 435–437.

Harvill, E.T., P.A. Cotter, M.H. Yuk and J.F. Miller. 1999. Probing the function of *Bordetella bronchiseptica* adenylate cyclase toxin by manipulating host immunity. Infect. Immun. *67*: 1493–1500.

Harwood, C.S., G. Burchhardt, H. Herrmann and G. Fuchs. 1998. Anaerobic metabolism of aromatic compounds via the benzoyl-CoA pathway. FEMS Microbiol. Rev. *22*: 439–458.

Harwood, C.S. and R.E. Parales. 1996. The beta-ketoadipate pathway and the biology of self identity. Annu. Rev. Microbiol. *50*: 553–590.

Hashem, F.M., J.S. Angle and P.A. Ristiano. 1986. Isolation and characterization of rhizobiophages specific for *Bradyrhizobium japonicum* USDA 117. Can. J. Microbiol. *32*: 326–329.

Hashem, F.M., L.D. Kuykendall, G. ElFadly and T.E. Devine. 1997. Strains of *Rhizobium fredii* effectively nodulate and efficiently fix nitrogen with *Medicago sativa* and *Glycine max*. Symbiosis *22*: 255–264.

Hashem, F.M., L.D. Kuykendall, S.E. Udell and P.M. Thomas. 1996. Phage susceptibility and plasmid profile analysis of *Sinorhizobium fredii*. Plant Soil *186*: 127–134.

Hashidoko, Y., M. Urashima and T. Yoshida. 1994. Predominant epiphytic bacteria on damaged *Polymnia sonchifolia* leaves, and their metabolic properties on phenolics of plant origin. Biosci. Biotechnol. Biochem. *58*: 1894–1896.

Hashimoto, T., D.L. Diedrich and S.F. Conti. 1970. Isolation of a bacteriophage for *Bdellovibrio bacteriovorus*. J. Virol. *5*: 97–98.

Hashimoto, W. and K. Murato. 1998. α-Rahamnosidase of *Sphingomonas* sp. R1 producing an unusual exopolysaccharide of sphingan. Biosci. Biotechnol. Biochem. *62*: 1068–1074.

Hassan, H., S. Susntharalingam and K.S. Dhillon. 1993. Fatal *Chromobacterium violaceum* septicaemia. Singapore Med. J. *34*: 456–458.

Hassan, I.J. and L. Hayek. 1993. Endocarditis caused by *Kingella denitrificans*. J. Infect. *27*: 291–295.

Hatten, B.A. 1973. Growth characteristics of microorganisms occurring in penicillin treated *Brucella abortus* cultures. Proc. Soc. Exp. Biol. Med. *142*: 909–914.

Hatton, A.D., G. Malin and A.G. McEwan. 1994. Identification of a periplasmic dimethylsulphoxide reductase in *Hyphomicrobium* EG grown under chemolithoheterotrophic conditions with dimethylsulphoxide as carbon source. Arch. Microbiol. *162*: 148–150.

Hattori, R., H. Watanabe, A. Tonosaki and T. Hattori. 1995. Unusual morphology of *Agromonas oligotrophica* and the effect of NaCl and organic nutrient on its fine-structure. J. Gen. Appl. Microbiol. *41*: 23–30.

Hattori, T. 1981. Enrichment of oligotrophic bacteria at microsites of soil. J. Gen. Appl. Microbiol. *27*: 43–55.

Haubold, R. 1978. Two different types of surface structures of methane utilizing bacteria. Z. Allg. Mikrobiol. *18*: 511–515.

Haugland, R.A., U.M.X. Sangodkar, P.R. Sferra and A.M. Chakrabarty. 1991. Cloning and characterization of a chromosomal DNA region required for growth on 2,4,5-T by *Pseudomonas cepacia* Ac1100. Gene *100*: 65–73.

Haugland, R.A., D.J. Schlemm, R.P. Lyons, III, P.R. Sferra and A.M. Chakrabarty. 1990. Degradation of the chlorinated phenoxyacetate herbicides 2,4- dichlorophenoxyacetic acid and 2,4,5-trichlorophenoxyacetic acid by pure and mixed bacterial cultures. Appl. Environ. Microbiol. *56*: 1357–1362.

Haukka, K., K. Lindström and J.P. Young. 1998. Three phylogenetic groups of *nodA* and *nifH* genes in *Sinorhizobium* and *Mesorhizobium* isolates from leguminous trees growing in Africa and Latin America. Appl. Environ. Microbiol. *64*: 419–426.

Havel, J. and W. Reineke. 1991. Total degradation of various chlorobiphenyls by cocultures and *in vivo* constructed hybrid pseudomonads. FEMS Microbiol. Lett. *78*: 163–170.

Havenner, J.A., B.A. McCardell and R.M. Weiner. 1979. Development of defined, minimal, and complete media for growth of *Hyphomicrobium neptunium*. Appl. Environ. Microbiol. *38*: 18–23.

Hawrylik, S.J., D.J. Wasilko, S.L. Haskell, T.D. Gootz and S.E. Lee. 1994. Bisulfite or sulfite inhibits growth of *Helicobacter pylori*. J. Clin. Microbiol. *32*: 790–792.

Hayano, K. and S. Fukni. 1967. Purification and properties of 3-keto-sucrose-forming enzyme from the cells of *Agrobacterium tumefaciens*. J. Biol. Chem. *242*: 3665–3672.

Hayano, K. and S. Fukui. 1970. α-3-Ketoglucosidase of *Agrobacterium tumefaciens*. J. Bacteriol. *101*: 692–697.

Hayano, K., Y. Tsubouchi and S. Fukni. 1973. 3-Ketoglucose reductase of *Agrobacterium tumefaciens*. J. Bacteriol. *113*: 652–657.

Hayashi, N. 1920. Etiology of tsutsugamushi disease. J. Parasitol. *7*: 53–68.

Hayashi, N.R., T. Ishida, A. Yokota, T. Kodama and Y. Igarashi. 1999. *Hydrogenophilus thermoluteolus* gen. nov., sp. nov., a thermophilic, facultatively chemolithoautotrophic, hydrogen-oxidizing bacterium. Int. J. Syst. Bacteriol. *49*: 783–786.

Hayes, S.F. and W. Burgdorfer. 1982. Reactivation of *Rickettsia rickettsii* in *Dermacentor andersoni* ticks: An ultrastructural analysis. Infect. Immun. *37*: 779–785.

Haynes, W.C. and W.H. Burkholder. 1957. Genus I *Pseudomonas*. *In* Breed, Murray and Smith (Editors), Bergey's Manual of Determinative Bacteriology, 7th Ed., The Williams & Wilkins Co., Baltimore. pp. 89–152.

Hayward, A.C. 1962. Studies on bacterial pathogens of sugar cane. II. Differentiation, taxonomy and nomenclature of the bacteria causing red stripe and mottled stripe diseases. Mauritius Sugar Ind. Res. Inst. Occas. Pap. *13*: 13–27.

Hayward, A.C. 1964. Characteristics of *Pseudomonas solanacearum*. J. Appl. Bacteriol. *27*: 265–277.

Hayward, A.C. 1972. A bacterial disease of clover in Hawaii. Plant Dis. Rep. *56*: 446–450.

Hazeleger, W., C. Arkesteijn, A. Toorop Bouma and R. Beumer. 1994. Detection of the coccoid form of *Campylobacter jejuni* in chicken products with the use of the polymerase chain reaction. Int. J. Food Microbiol. *24*: 273–281.

Hazell, S.L. and G.L. Mendz. 1997. How *Helicobacter pylori* works: an overview of the metabolism of *Helicobacter pylori*. Helicobacter *2*: 1–12.

Hazeu, W. and P.J. Steenis. 1970. Isolation and characterization of two vibrio-shaped methane-oxidizing bacteria. Antonie Leeuwenhoek *36*: 67–72.

Hebbar, K.P., A.G. Davey, J. Merrin, T.J. McLoughlin and P.J. Dart. 1992. *Pseudomonas cepacia*, a potential suppressor of maize soil borne diseases seed inoculation and maize root colonization. Soil Biol. Biochem. *24*: 999–1007.

Hebeler, B.H. and F.E. Young. 1976. Chemical composition and turnover of peptidoglycan in *Neisseria gonorrhoeae*. J. Bacteriol. *126*: 1180–1185.

Heberlein, G.T., J. De Ley and R. Tijtgat. 1967. Deoxyribonucleic acid homology and taxonomy of *Agrobacterium*, *Rhizobium* and *Chromobacterium*. J. Bacteriol. *94*: 116–124.

Heckels, J.E. 1989. Structure and function of pili of pathogenic *Neisseria* species. Clin. Microbiol. Rev. *2*: S66–S73.

Heckly, R. 1961. Preservation of bacteria by lyophilization. Adv. Appl. Microbiol. *3*: 1–76.

Heckmann, K. 1975. Omikron, ein essentieller Endosymbiont von *Euplotes aediculatus*. J. Protozool. *22*: 97–104.

Heckmann, K. 1983. Endosymbionts of *Euplotes*. Int. Rev. Cytol. Suppl. *14*: 111–114.

Heckmann, K., J.R. Preer, Jr. and W.H. Straetling. 1967. Cytoplasmic particles in the killers of *Euplotes minuta* and their relationship to the killer substance. J. Protozool. *14*: 360–363.

Heckmann, K. and H.J. Schmidt. 1987. *Polynucleobacter necessarius* gen. nov., sp. nov., an obligately endosymbiotic bacterium living in the cytoplasm of *Euplotes aediculatus*. Int. J. Syst. Bacteriol. *37*: 456–457.

Heckmann, K., R. Tenhagen and H.D. Görtz. 1983. Fresh water *Euplotes* species with a 9-type-1 cirrus pattern depend upon endosymbionts. J. Protozool. *30*: 284–289.

Hedlund, B.P. and J.T. Staley. 2002. Phylogeny of the genus *Simonsiella* and other members of the *Neisseriaceae*. Int. J. Syst. Evol. Microbiol. *52*: 1377–1382.

Heidelbach, M., H. Skladny and H.U. Schairer. 1992. Purification of the DNA-dependent RNA polymerase from the myxobacterium *Stigmatella aurantiaca*. J. Bacteriol. *174*: 2733–2735.

Heidelbach, M., H. Skladny and H.U. Schairer. 1993. Heat-shock and development induce synthesis of a low molecular weight stress-responsive protein in the myxobacterium *Stigmatella aurantiaca*. J. Bacteriol. *175*: 7479–7482.

Heider, J., M. Boll, K. Breese, S. Breinig, J.C. Ebenau, U. Feil, N. Gad'on, D. Laempe, B. Leuthner, M.E. Mohamed, S. Schneider, G. Burchhardt and G. Fuchs. 1998. Differential induction of enzymes involved in anaerobic metabolism of aromatic compounds in the denitrifying bacterium *Thauera aromatica*. Arch. Microbiol. *170*: 120–131.

Heider, J. and G. Fuchs. 1997a. Anaerobic metabolism of aromatic compounds. Eur. J. Biochem. *243*: 577–596.

Heider, J. and G. Fuchs. 1997b. Microbial anaerobic aromatic metabolism. Anaerobe *3*: 1–22.

Heidt, A., H. Monteil and C. Richard. 1983. O and H serotyping of *Pseudomonas cepacia*. J. Clin. Microbiol. *18*: 738–740.

Heijthuijsen, J.H.F.G. and T.A. Hansen. 1989. Betaine fermentation and oxidation by marine *Desulfuromonas* strains. Appl. Environ. Microbiol. *55*: 965–969.

Heimbrook, M.E., W.L.L. Wang and G. Campbell. 1989. Staining bacterial flagella easily. J. Clin. Microbiol. *27*: 2612–2615.

Heintz, W. 1862. Über dem Ammoniaktypus angehörige organische Säuren. Annalen der Chemie *122*: 257–294.

Heintz, W. 1865. Beiträge zür Kenntnifs der Glycolamidsäuren. Annalen der Chemie *136*: 213–223.

Heinzen, R.A., S.S. Grieshaber, L.S. Van Kirk and C.J. Devin. 1999. Dynamics of actin-based movement by *Rickettsia rickettsii* in Vero cells. Infect. Immun. *67*: 4201–4207.

Heinzen, R.A., S.F. Hayes, M.G. Peacock and T. Hackstadt. 1993. Directional actin polymerization associated with spotted fever group *Rickettsia* infection of Vero cells. Infect. Immun. *61*: 1926–1935.

Heising, S. and B. Schink. 1998. Phototrophic oxidation of ferrous iron by a *Rhodomicrobium vannielii* strain. Microbiology *144*: 2263–2269.

Heiske, A. and R. Mutters. 1994. Differentiation of selected members of the family *Neisseriaceae* (*Alysiella, Eikenella, Kingella, Simonsiella* and CDC groups EF-4 and M-5) by carbohydrate fingerprints and selected phenotypic features. Zentbl. Bakteriol. *281*: 67–79.

Heller, R., M. Artois, V. Xemar, D. De Briel, H. Gehin, B. Jaulhac, H. Monteil and Y. Piemont. 1997. Prevalence of *Bartonella henselae* and *Bartonella clarridgeiae* in stray cats. J. Clin. Microbiol. *35*: 1327-1331.

Heller, R., M. Kubina, P. Mariet, P. Riegel, G. Delacour, C. Dehio, F. Lamarque, R. Kasten, H.J. Boulouis, H. Monteil, B. Chomel and Y. Piemont. 1999. *Bartonella alsatica* sp. nov., a new *Bartonella* species isolated from the blood of wild rabbits. Int. J. Syst. Bacteriol. *49 Pt 1*: 283–288.

Heller, R., P. Riegel, Y. Hansmann, G. Delacour, D. Bermond, C. Dehio, F. Lamarque, H. Monteil, B. Chomel and Y. Piemont. 1998. *Bartonella tribocorum* sp. nov., a new *Bartonella* species isolated from the blood of wild rats. Int. J. Syst. Bacteriol. *48 Pt 4*: 1333–1339.

Hemelt, I.E., G.E. Lewis, D.L. Huxsoll and E.H. Stephenson. 1980. Serial propagation of *Ehrlichia canis* in primary canine peripheral blood monocyte cultures. Cornell Vet. *70*: 36–42.

Hemmingsen, S.M., C. Woolford, S.M. Vandervies, K. Tilly, D.T. Dennis, C.P. Georgopoulos, R.W. Hendrix and R.J. Ellis. 1988. Homologous plant and bacterial proteins chaperone oligomeric protein assembly. Nature *333*: 330–334.

Henderson, I.R., F. Navarro-Garcia and J.P. Nataro. 1998. The great escape: Structure and function of the autotransporter proteins. Trends Microbiol. *6*: 370–378.

Hendrie, M.S., A.J. Holding and J.M. Shewan. 1974. Emended descriptions of the genus *Alcaligenes* and of *Alcaligenes faecalis* and proposal that the generic name *Achromobacter* be rejected; status of the named species of *Alcaligenes* and *Achromobacter*. Int. J. Syst. Bacteriol. *24*: 534–550.

Henkle-Dührsen, K., V.H. Eckelt, G. Wildenburg, M. Blaxter and R.D. Walter. 1998. Gene structure, activity and localization of a catalase from intracellular bacteria in *Onchocerca volvulus*. Mol. Biochem. Parasitol. *96*: 69–81.

Henneberg, W. 1897. Beiträge zür Kenntnis der Essigbakterien. Zentbl. Bakteriol. Parasitenkd. Infektkrankh. Hyg. Abt. II *3*: 223–231.

Henneberg, W. 1898. Weitere Untersuchungen ueber Essigbakteriën. Zentbl. Bakteriol. Parasitenkd. Infektkrankh. Hyg. Abt. II *4*: 14–20 67–73 138–147.

Henneberg, W. 1906. Zür Kenntnis der Schnellessig und Weinessigbakteriën. Deut. Essigindustrie. *10*: 89–93, 98–99, 106–108, 113–116, 121–124, 129–132, 137–140, 146–148.

Henrichsen, J. 1972. Bacterial surface translocation: a survey and a classification. Bacteriol. Rev. *36*: 478–503.

Henrichsen, J. 1975a. The occurrence of twitching motility among Gram-negative bacteria. Acta Pathol. Microbiol. Scand. Sect. B Microbiol. *83*: 171–178.

Henrichsen, J. 1975b. On twitching motility and its mechanism. Acta Pathol. Microbiol. Scand. [B]. *83*: 187–190.

Henrichsen, J. and J. Blom. 1975. Examination of fimbriation of some Gram-negative rods with and without twitching and gliding motility. Acta Pathol. Microbiol. Scand. Sect. B Microbiol. *83*: 161–170.

Henrichsen, J., L.O. Frøhlm and K. Bøvre. 1972. Studies on bacterial surface translocation 2. Correlation of twitching motility and fimbriation in colony variants of *Moraxella nonliquefaciens, M. bovis* and *M. kingii*. Acta Pathol. Microbiol. Scand. Sect. B Microbiol. *80*: 445–452.

Henrici, A.T. and D. Johnson. 1935a. Stalked bacteria, a new order of *Schizomycetes*. J. Bacteriol. *29*: 3-4.

Henrici, A.T. and D.E. Johnson. 1935b. Studies on freshwater bacteria. II. Stalked bacteria, a new order of schizomycetes. J. Bacteriol. *30*: 61–93.

Henriksen, S.D. 1948. Studies on Gram-negative anaerobes II. Gram-negative anaerobic rods with spreading colonies. Acta Pathol. Microbiol. Scand. *25*: 368.

Henriksen, S.D. 1952. *Moraxella*: classification and taxonomy. J. Gen. Microbiol. *6*: 318–328.

Henriksen, S.D. 1969a. Corroding bacteria from the respiratory tract. I. *Moraxella kingii*. Acta Pathol. Microbiol. Scand. *75*: 85–90.

Henriksen, S.D. 1969b. Designation of the type strain of *Bacteroides corrodens* Eiken 1958. Int. J. Syst. Bacteriol. *19*: 165–166.

Henriksen, S.D. 1976. *Moraxella, Neisseria, Branhamella*, and *Acinetobacter*. Annu. Rev. Microbiol. *30*: 63–83.

Henriksen, S.D. and K. Bøvre. 1968a. *Moraxella kingii* spec. nov., a haemolytic, saccharolytic species of the genus *Moraxella*. J. Gen. Microbiol. *51*: 377–385.

Henriksen, S.D. and K. Bøvre. 1968b. The taxonomy of the genera *Moraxella* and *Neisseria*. J. Gen. Microbiol. *51*: 387–392.

Henriksen, S.D. and K. Bøvre. 1976. Transfer of *Moraxella kingae* Henriksen and Bøvre to the genus *Kingella* gen. nov. in the family *Neisseriaceae*. Int. J. Syst. Bacteriol. *26*: 447-450.

Henriksen, S.D. and E. Holten. 1976. *Neisseria elongata* subsp. *glycolytica* subsp. nov. Int. J. Syst. Bacteriol. *26*: 478–481.

Henry, B.S. 1933. Dissociation in the genus *Brucella*. J. Infect. Dis. *52*: 374–402.

Henry, D.A., E. Mahenthiralingam, P. Vandamme, T. Coenye and D.P. Speert. 2001. Phenotypic methods for determining genomovar status of the *Burkholderia cepacia* complex. J. Clin. Microbiol. *39*: 1073–1078.

Henrysson, T. and B. Mattiasson. 1993. A microbial biosensor system for dihalomethanes. Biodegradation *4*: 101–105.

Hensel, G. and H.G. Trüper. 1976. Cysteine and S-sulfocysteine biosynthesis in phototrophic bacteria. Arch. Microbiol. *109*: 101–103.

Hensgens, C.M.H., M.E. Nienhuis-Kuiper and T.A. Hansen. 1994. Effects of tungstate on the growth of *Desulfovibrio gigas* NCIMB 9332 and other sulfate-reducing bacteria with ethanol as a substrate. Arch. Microbiol. *162*: 143–147.

Hentschel, U., J. Hopke, M. Horn, A.B. Friedrich, M. Wagner, J. Hacker and B.S. Moore. 2002. Molecular evidence for a uniform microbial community in sponges from different oceans. Appl. Environ. Microbiol. *68*: 4431–4440.

Heppner, B., G. Zellner and H. Diekmann. 1992. Start-up and operation of a propionate-degrading fluidized-bed reactor. Appl. Microbiol. Biotechnol. *36*: 810–816.

Herbert, R.A., E. Siefert and N. Pfennig. 1978. Nitrogen assimilation in *Rhodopseudomonas acidophila*. Arch. Microbiol. *119*: 1–5.

Hermans, M.A., B. Neuss and H. Sahm. 1991. Content and composition of hopanoids in *Zymomonas mobilis* under various growth conditions. J. Bacteriol. *173*: 5592–5595.

Hermans, P.G., C.A. Hart and A.J. Trees. 2001. *In vitro* activity of antimicrobial agents against the endosymbiont *Wolbachia pipientis*. J. Antimicrob. Chemother. *47*: 659–663.

Hermodson, M.A., K.C.S. Chen and T.M. Buchanan. 1978. *Neisseria* pili proteins—Amino-terminal amino acid sequences and identification of an unusual amino acid. Biochemistry *17*: 442–445.

Hernández, L., J. Arrieta, C. Menéndez, R. Vazquez, A. Coego, V. Suárez, G. Selman, M.F. Petit-Glatron and R. Chambert. 1995. Isolation and enzymic properties of levansucrase secreted by *Acetobacter diazotrophicus* SRT4, a bacterium associated with sugar cane. Biochem. J. *309*: 113–118.

Hernández, L., R. Ramírez, J.V. Hormaza, J. Madrazo and J. Arrieta. 1999. Increased levansucrase production by a genetically modified *Acetobacter diazotrophicus* strain in shaking batch cultures. Lett. Appl. Microbiol. *28*: 41–44.

Herron, M.J., C.M. Nelson, J. Larson, K.R. Snapp, G.S. Kansas and J.L. Goodman. 2000. Intracellular parasitism by the human granulocytic ehrlichiosis bacterium through the P-selectin ligand, PSGL-1. Science *288*: 1653–1656.

Hertig, M. 1936. The rickettsia, *Wolbachia pipientis* (gen. et sp. n.) and associated inclusions of the mosquito, *Culex pipiens.* Parasitol. *28*: 453–486.

Hertig, M. and S.B. Wolbach. 1924. Studies on rickettsia-like microorganisms in insects. J. Med. Res. *44*: 329–374.

Hertzberg, S., G. Borch and S. Liaaen-Jensen. 1976. Bacterial carotenoids. L. Absolute configuration of zeaxanthin dirhamnoside. Arch. Microbiol. *110*: 95–99.

Hespell, R.B., G.F. Miozzari and S.C. Rittenberg. 1975. Ribonucleic acid destruction and synthesis during intraperiplasmic growth of *Bdellovibrio bacteriovorus.* J. Bacteriol. *123*: 481–491.

Hespell, R.B., B.J. Paster, T.J. Macke and C.R. Woese. 1983. The origin and phylogeny of the bdellovibrios. Syst. Appl. Microbiol. *5*: 196–203.

Hess, A., B. Zarda, D. Hahn, A. Haner, D. Stax, P. Hohener and J. Zeyer. 1997. *In situ* analysis of denitrifying toluene- and *m*-xylene-degrading bacteria in a diesel fuel-contaminated laboratory aquifer column. Appl. Environ. Microbiol. *63*: 2136–2141.

Heumann, W. 1962. Die mtodik der keuzung sternbildender bkterien. Biol. Zentbl. *81*: 341–354.

Heumann, W. and R. Marx. 1964. Feinstruktur und Funktion der Fimbrien bei den sternbildenden Bakterium *Pseudomonas echinoides.* Arch. Mikrobiol. *47*: 325–337.

Hewlett, E.L. 1990. *Bordetella* species. *In* Mandell, Douglas and Bennett (Editors), Principles and Practice of Infectious Diseases, Churchhill Livingstone Inc., New York. 1756–1762.

Hewlett, E.L. 1997. Pertussis: current concepts of pathogenesis and prevention. Pediatr. Infect. Dis. J. *16*: S78–84.

Heyer, J., Y.R. Malashenko, U. Berger and E. Budkova. 1984. Vergreitung methanotropher bakterien. Z. Allg. Mikrobiol. *24*: 725–744.

Hickman, D.D. and A.W. Frenkel. 1965a. Observations on the structure of *Rhodospirillum molischianum.* J. Cell Biol. *25*: 261–278.

Hickman, D.D. and A.W. Frenkel. 1965b. Observations on the structure of *Rhodospirillum rubrum.* J. Cell Biol. *25*: 279–291.

Higgins, T.P., M.J.E. Hewlins and G.F. White. 1996. A ^{13}C-NMR study of the mechanism of bacterial metabolism of monomethyl sulfate. Eur. J. Biochem. *236*: 620–625.

Hildebrand, D.C., N.J. Palleroni and M. Doudoroff. 1973. Synonymy of *Pseudomonas gladioli* Severini 1913 and *Pseudomonas marginata* (McCulloch 1921) Stapp 1928. Int. J. Syst. Bacteriol. *23*: 433–437.

Hildebrand, E.M. 1940. Cane gall of brambles caused by *Phytomonas rubi* n. sp. J. Agr. Res. *61*: 685–696.

Hildebrandt, P.K., D.L. Huxsoll, J.S. Walker, R.M. Nims, R. Taylor and M. Andrews. 1973. Pathology of canine ehrlichiosis (tropical canine pancytopenia). Am. J. Vet. Res. *34*: 1309–1320.

Hilger, F. 1965. Études sur la systematique du genre *Beijerinckia* Derx. Ann. Inst. Pasteur (Paris) *109*: 406–423.

Hill, B. and M.M. Attwood. 1974. The purification of glycerate kinase from *Hyphomicrobium* sp. and *Pseudomonas* AM1: product identification. J. Gen. Microbiol. *83*: 187–190.

Hill, B. and M.M. Attwood. 1976a. The effect of adenosine triphosphate on phosphoglycerate mutase activity from *Hyphomicrobium* X and *Pseu-*

domonas AM1 grown on reduced one-carbon compounds. J. Gen. Microbiol. *97*: 335–338.

Hill, B. and M.M. Attwood. 1976b. Purification and characterization of phosphoglycerate mutase from methanol-grown *Hyphomicrobium* X and *Pseudomonas* AM1. J. Gen. Microbiol. *96*: 185–193.

Hill, L.R., J.J.S. Snell and S.P. Lapage. 1970. Identification and characteristics of *Bacteroides corrodens.* J. Med. Microbiol. *3*: 483–491.

Hill, S. 1971. Influence of oxygen concentration on the colony type of *Derxia gummosa* grown on nitrogen-free media. J. Gen. Microbiol. *67*: 77–83.

Hill, S. and J.R. Postgate. 1969. Failure of putative nitrogen-fixing bacteria to fix nitrogen. J. Gen. Microbiol. *58*: 277–285.

Hines, W.D., B.A. Freeman and G.A. Pearson. 1964. Production and characterization of *Brucella* spheroplasts. J. Bacteriol. *87*: 438–445.

Hinteregger, C., M. Loidl and F. Streichsbier. 1994. *Pseudomonas acidovorans* - a bacterium capable of mineralizing 2-chloroaniline. J. Basic Microbiol. *34*: 77–85.

Hinz, K.H., G. Glünder and H. Lüders. 1978. Acute respiratory disease in turkey poults caused by *Bordetella bronchiseptica*-like bacteria. Vet. Rec. *103*: 262–263.

Hinze, G.. 1913. Beiträge zur Kenntnis der farblosen Schweferlbakterien. Ber. Dtsch. Bot. Ges. *31*: 189–202.

Hippe, H., A. Hagelstein, I. Kramer, J. Swiderski and E. Stackebrandt. 1999. Phylogenetic analysis of *Formivibrio citricus, Propionivibrio dicarboxylicus, Anaerobiospirillum thomasii, Succinimonas amylolytica* and *Succinivibrio dextrinosolvens* and proposal for *Succinivibrionaceae* fam. nov. Int. J. Syst. Bacteriol. *49*: 779-782.

Hiraishi, A. 1988a. Bicarbonate-stimulated dark fermentative growth of a phototrophic purple nonsulfur bacterium. FEMS Microbiol. Lett. *56*: 199–202.

Hiraishi, A. 1988b. Fumarate reduction systems in members of the family *Rhodospirillaceae* with different quinone types. Arch. Microbiol. *150*: 56–60.

Hiraishi, A. 1994. Phylogenetic affiliations of *Rhodoferax fermentans* and related species of phototrophic bacteria as determined by automated 16S DNA sequencing. Curr. Microbiol. *28*: 25–28.

Hiraishi, A. 1997. Transfer of the bacteriochlorophyl *b* containing phototrophic bacteria *Rhodopseudomonas viridis* and *Rhodopseudomonas sulfoviridis* to the genus *Blastochloris* gen. nov. Int. J. Syst. Bacteriol. *47*: 217–219.

Hiraishi, A. and Y. Hoshino. 1984. Distribution of rhodoquinone in *Rhodospirillaceae* and its taxonomic implications. J. Gen. Appl. Microbiol. *30*: 435–448.

Hiraishi, A., Y. Hoshino and T. Satoh. 1991a. *Rhodoferax fermentans* gen. nov., sp. nov., a phototrophic purple nonsulfur bacterium previously referred to as the "*Rhodocyclus gelatinosus*-like" group. Arch. Microbiol. *155*: 330–336.

Hiraishi, A., Y. Hoshino and T. Satoh. 1992a. *In* Validation of the publication of new names and new combinations previously effectively published outside the IJSB. List No. 40. Int. J.Syst. Bacteriol. *42*: 191–192.

Hiraishi, A. and H. Kitamura. 1984. Distribution of phototrophic purple nonsulfur bacteria in activated sludge systems and other aquatic environments. Bull. Jpn. Soc. Sci. Fish. *50*: 1929–1938.

Hiraishi, A. and K. Komagata. 1989a. Effects of the growth medium composition on menaquinone homolog formation in *Micrococcus luteus.* J. Gen. Appl. Microbiol. *35*: 311–318.

Hiraishi, A. and K. Komagata. 1989b. Isolation of rhodoquinone-containing chemoorganotrophic bacteria from activated sludge. FEMS Microbiol. Lett. *58*: 55–58.

Hiraishi, A., H. Kuraishi and K. Kawahara. 2000a. Emendation of the description of *Blastomonas natatoria* (Sly 1985) Sly and Cahill 1997 as an aerobic photosynthetic bacterium and reclassification of *Erythromonas ursincola* Yurkov et al. 1997 as *Blastomonas ursincola* comb. nov. Int. J. Syst. Evol. Microbiol. *50*: 1113–1118.

Hiraishi, A., Y. Matsuzawa, T. Kanbe and N. Wakao. 2000b. *Acidisphaera rubrifaciens* gen. nov., sp. nov., an aerobic bacteriochlorophyll-containing bacterium isolated from acidic environments. Int. J. Syst. Bacteriol. *50*: 1539–1546.

Hiraishi, A., Y. Morishima and H. Kitamura. 1991b. Use of isoprenoid quinone profiles to study the bacterial community structure and population dynamics in the photosynthetic sludge system. Water Sci. Technol. *23*: 937–945.

Hiraishi, A., K. Muramatsu and Y. Ueda. 1996. Molecular genetic analyses of *Rhodobacter azotoformans* sp. nov. and related species of phototrophic bacteria. Syst. Appl. Microbiol. *19*: 168–177.

Hiraishi, A., K. Muramatsu and Y. Ueda. 1997a. *In* Validation of the publication of new names and new combinations previously effectively published outside the IJSB. List No. 61. Int. J. Syst. Bacteriol. *47*: 601–602.

Hiraishi, A., K. Muramatsu and K. Urata. 1995a. Characterization of new denitrifying *Rhodobacter* strains isolated from photosynthetic sludge for waste-water treatment. J. Ferment. Bioeng. *79*: 39–44.

Hiraishi, A., K.V.P. Nagashima, K. Matsuura, K. Shimada, S. Takaichi, N. Wakao and Y. Katayama. 1998. Phylogeny and photosynthetic features of *Thiobacillus acidophilus* and related acidophilic bacteria: its transfer to the genus *Acidiphilium* as *Acidiphilium acidophilum* comb. nov. Int. J. Syst. Bacteriol. *48*: 1389–1398.

Hiraishi, A., T.S. Santos, J. Sugiyama and K. Komagata. 1992b. *Rhodopseudomonas rutila* is a later subjective synonym of *Rhodopseudomonas palustris*. Int. J. Syst. Bacteriol. *42*: 186–188.

Hiraishi, A., Y.K. Shin and J. Sugiyama. 1995b. *Brachymonas denitrificans* gen. nov., sp. nov., an aerobic chemoorganotrophic bacterium which cantains rhodoquinones, and evolutionary relationships of rhodoquinone producers to bacterial species with various quinone classes. J. Gen. Appl. Microbiol. *41*: 99–117.

Hiraishi, A., Y.K. Shin and J. Sugiyama. 1995c. *In* Validation of the publication of new names and new combinations previously effectively published outside the IJSB. List No. 55. Int. J. Syst. Bacteriol. *45*: 879–880.

Hiraishi, A., Y.K. Shin and J. Sugiyama. 1997b. Proposal to reclassify *Zoogloea ramigera* IAM, 12670 (P.R. Dugan 115) as *Duganella zoogloeoides* gen. nov., sp. nov. Int. J. Syst. Bacteriol. *47*: 1249–1252.

Hiraishi, A., Y.K. Shin, J. Sugiyama and K. Komagata. 1992c. Isoprenoid quinones and fatty acids of *Zoogloea*. Antonie Leeuwenhoek *61*: 231–236.

Hiraishi, A. and Y. Ueda. 1994a. Intrageneric structure of the genus *Rhodobacter*: transfer of *Rhodobacter sulfidophilus* and related marine species to the genus *Rhodovulum* gen. nov. Int. J. Syst. Bacteriol. *44*: 15–23.

Hiraishi, A. and Y. Ueda. 1994b. *Rhodoplanes* gen. nov., a new genus of phototrophic bacteria including *Rhodopseudomonas rosea* as *Rhodoplanes roseus* comb. nov. and *Rhodoplanes elegans* sp. nov. Int. J. Syst. Bacteriol. *44*: 665–673.

Hiraishi, A. and Y. Ueda. 1995. Isolation and characterization of *Rhodovulum strictum* sp. nov., and some other purple nonsulfur bacteria from colored blooms in tidal and seawater pools. Int. J. Syst. Bacteriol. *45*: 319–326.

Hiraishi, A., K. Urata and T. Satoh. 1995d. A new genus of marine budding phototrophic bacteria, *Rhodobium* gen. nov., which includes *Rhodobium orientis* sp. nov. and *Rhodobium marinum* comb. nov. Int. J. Syst. Bacteriol. *45*: 226–234.

Hirsch, H.J. 1977a. Bacteriocins from *Myxococcus fulvus* (*Myxobacterales*). Arch. Microbiol. *115*: 45–49.

Hirsch, H.J., H. Tsai and I. Geffers. 1978. Purification and effects of fulvocin-C, a bacteriocin from *Myxococcus fulvus* MX-F16. Arch. Microbiol. *119*: 279–286.

Hirsch, P. 1968. Biology of budding bacteria. IV. Epicellular deposition of iron by aquatic budding bacteria. Arch. Mikrobiol. *60*: 201–216.

Hirsch, P. 1970. Budding, nitrifying bacteria: the nomenclatural status of *Nitromicrobium germinans* Stutzer und Hartlieb 1899 and *Nitrobacter*. Int. J. Syst. Bacteriol. *20*: 317–320.

Hirsch, P. 1974a. Budding bacteria. Ann. Rev. Microbiol. *28*: 391–444.

Hirsch, P. 1974b. Genus *Hyphomicrobium* Stutzer and Hartleb 1898, 76[AL]. *In* Buchanan and Gibbons (Editors), Bergey's Manual of Determi-

native Bacteriology, 8th ed., The Williams & Wilkins Co., Baltimore. 148–150.

Hirsch, P. 1977b. Ecology and morphogenesis of *Thiopedia* spp. in ponds, lakes, and laboratory cultures. Proceedings of the Second International Symposium on Photosynthetic Prokaryotes, Dundee, Scotland. pp. 13–15.

Hirsch, P. 1980. Distribution and pure culture studies of morphologically distinct Solar Lake microoganisms. *In* Nissenbaum (Editor), Hypersaline Brines and Evaporitic Environments, Elsevier, Amsterdam. 41–60.

Hirsch, P. 1984. Genus *Hyphomicrobium*. *In* Staley, Bryant, Pfennig and Holt (Editors), Bergey's Manual of Systematic Bacteriology, 1st Ed., Vol. 3, The Williams & Wilkins Co., Baltimore. pp. 1895–1904.

Hirsch, P. 1989. Genus *Hyphomicrobium*. *In* Staley, Bryant, Pfennig and Holt (Editors), Bergey's Manual of Systematic Bacteriology, 1st Ed., Vol. 3, The Williams & Wilkins Co., Baltimore. pp. 1895–1904.

Hirsch, P. and S.F. Conti. 1964a. Biology of budding bacteria: I. Enrichment, isolation, and morphology of *Hyphomicrobium* spp. Arch. Mikrobiol. *48*: 339–357.

Hirsch, P. and S.F. Conti. 1964b. Biology of budding bacteria: II. Growth and nutrition of *Hyphomicrobium* spp. Arch. Mikrobiol. *48*: 358–367.

Hirsch, P. and S.F. Conti. 1965. Enrichment and isolation of stalked and budding bacteria (*Hyphomicrobium*, *Rhodomicrobium*, and *Caulobacter*). Sympos. Anreicherungskultur und Mutantenauslese, Göttingen. Zentralbl. Bakteriol. Parasitenkd. Infektionskr. Hyg. I Suppl. *1*: 253–255.

Hirsch, P. and B. Hoffmann. 1989a. *Dichotomicrobium thermohalophilum*, gen. nov., spec. nov., budding prosthecate bacteria from the Solar Lake (Sinai) and some related strains. Syst. Appl. Microbiol. *11*: 291–301.

Hirsch, P. and B. Hoffmann. 1989b. *In* Validation of the publication of new names and new combinations previously effectively published outside the IJSB. List No. 31. Int. J. Syst. Bacteriol. *39*: 495–497.

Hirsch, P. and M. Müller. 1985. *Blastobacter aggregatus* sp. nov., *Blastobacter capsulatus* sp. nov., and *Blastobacter denitrificans* sp. nov., new budding bacteria from fresh water habitats. Syst. Appl. Microbiol. *6*: 281–286.

Hirsch, P. and M. Müller. 1986. *In* Validation of the publication of new names and new combinations previously effectively published outside the IJSB. List No. 20. Int. J. Syst. Bacteriol. *36*: 354–356.

Hirsch, P., M. Müller and H. Schlesner. 1977. New aquatic budding and prosthecate bacteria and their taxonomic position. *In* Skinner and Shewan (Editors), Aquatic Microbiology, Academic Press, London. pp. 107–133.

Hirsch, P. and G. Rheinheimer. 1968. Biology of budding bacteria. V. Budding bacteria in aquatic habitats: occurrence, enrichment and isolation. Arch. Mikrobiol. *62*: 289–306.

Hirsch, P. and H. Schlesner. 1981. The genus *Stella*. *In* Starr, , Stolp, Trüper, Balows and Schlegel (Editors), The Prokaryotes: A Handbook on Habitats, Isolation, and Identification of Bacteria, Springer-Verlag, Berlin. pp. 461–465.

Hitchcock, P.J. 1989. Unified nomenclature for pathogenic *Neisseria* species. Clin. Microbiol. Rev. *2*: S64–S65.

Hitchcock, P.J., T.M. Brown, D. Corwin, S.F. Hayes, A. Olszewski and W.J. Todd. 1985. Morphology of three strains of contagious equine metritis organism. Infect. Immun. *48*: 94–108.

Hitschmann, A. and H. Stockinger. 1985. Oxygen deficiency and its effect on the adenylate system in *Acetobacter* in the submerse acetic fermentation. Appl. Microbiol. Biotechnol. *22*: 46–49.

Hitzig, W.M. and A. Liebesman. 1944. Subacute endocarditis associated with infection by a spirillum. Arch. Intern. Med. *73*: 415–424.

Ho, Y.K. and J. Lascelles. 1971. δ-Aminolevulinic acid dehydratase of *Spirillum itersonii* and the regulation of tetrapyrrole synthesis. Arch. Biochem. Biophys. *144*: 734–740.

Hobbs, M.M., B. Malorny, P. Prasad, G. Morelli, B. Kusecek, J.E. Heckels, J.G. Cannon and M. Achtman. 1998. Recombinational reassortment among *opa* genes from ET-37 complex *Neisseria meningitidis* isolates of diverse geographical origins. Microbiology-Uk. *144*: 157–166.

Hochkoeppler, A., S. Ciurli, G. Venturoli and D. Zannoni. 1995a. The high potential iron-sulfur protein (HiPIP) from *Rhodoferax fermentans* is competent in photosynthetic electron transfer. FEBS Lett. *357*: 70–74.

Hochkoeppler, A., G. Moschettini and D. Zannoni. 1995b. The electron transport system of the facultative phototroph *Rhodoferax fermentans*. I. A functional, thermodynamic and spectroscopic study of the respiratory chain of dark- and light-grown cells. Biochim. Biophys. Acta *1229*: 73–80.

Hochkoeppler, A., D. Zannoni, S. Ciurli, T.E. Meyer, M.A. Cusanovich and G. Tollin. 1996. Kinetics of photo-induced electron transfer from high-potential iron-sulfur protein to the photosynthetic reaction center of the purple phototroph *Rhodoferax fermentans*. Proc. Natl. Acad. Sci. U.S.A. *93*: 6998–7002.

Hocquellet, A., J.M. Bové and M. Garnier. 1997. Production and evaluation of non-radioactive probes for the detection of the two "*Candidatus* Liberobacter" species associated with citrus huanglongbing (greening). Mol. Cell. Probes *11*: 433–438.

Hocquellet, A., P. Toorawa, J.M. Bové and M. Garnier. 1999. Detection and identification of the two "*Candidatus* Liberobacter" species associated with citrus huanglongbing by PCR amplification of ribosomal protein genes of the beta operon. Mol. Cell. Probes *13*: 373–379.

Hodgkin, J. and D. Kaiser. 1979. Genetics of gliding motility in *Myxococcus xanthus* (*Myxobacterales*). 2 gene systems control movement. Mol. Gen. Genet. *171*: 177–191.

Hodgson, D.A. 1993. Light-induced carotenogenesis in *Myxococcus xanthus*: genetic analysis of the *carR* region. Mol. Microbiol. *7*: 471–488.

Hoeniger, J.F., H.D. Tauschel and J.L. Stokes. 1973. The fine structure of *Sphaerotilus natans*. Can. J. Microbiol. *19*: 309–313.

Hoerauf, A., K. Nissen-Pahle, C. Schmetz, K. Henkle-Duhrsen, M.L. Blaxter, D.W. Buttner, M.Y. Gallin, K.M. Al-Qaoud, R. Lucius and B. Fleischer. 1999. Tetracycline therapy targets intracellular bacteria in the filarial nematode *Litomosoides sigmodontis* and results in filarial infertility. J. Clin. Invest. *103*: 11–18.

Hoerauf, A., L. Volkmann, C. Hamelmann, O. Adjei, I.B. Autenrieth, B. Fleischer and D.W. Buttner. 2000a. Endosymbiotic bacteria in worms as targets for a novel chemotherapy in filariasis. Lancet *355*: 1242–1243.

Hoerauf, A., L. Volkmann, K. Nissen-Paehle, C. Schmetz, I. Autenrieth, D.W. Buttner and B. Fleischer. 2000b. Targeting of *Wolbachia* endobacteria in *Litomosoides sigmodontis*: Comparison of tetracyclines with chloramphenicol, macrolides and ciprofloxacin. Trop. Med. Int. Health *5*: 275–279.

Hofer, A.W. 1941. A characterization of *Bacterium radiobacter* (Beijerinck and van Delden) Löhnis. J. Bacteriol. *41*: 193–224.

Höfle, G. and H. Reichenbach. 1995. The biosynthetic potential of the myxobacteria. *In* Kuhn, W. and H.P. Fiedler (Editors), Sekundärmetabolismus bei Mikroorganismen. Beiträge zur Forschung, Attempto Verlag, Tübingen, pp. 61–78.

Höfle, M.G. 1990. Transfer RNA as genotypic fingerprints of eubacteria. Arch. Microbiol. *153*: 299–304.

Hofmann, D.U. 1989. Physiologisch Studien an *Sorangium cellulosum*, So ce12, Thesis, Technical University Braunschweig, Germany . 160 pp.

Hofmeister, E.K., C.P. Kolbert, A.S. Abdulkarim, J.M. Magera, M.K. Hopkins, J.R. Uhl, A. Ambyaye, S.R. Telford, III, F.R. Cockerill, III and D.H. Persing. 1998. Cosegregation of a novel *Bartonella* species with *Borrelia burgdorferi* and *Babesia microti* in *Peromyscus leucopus*. J. Infect. Dis. *177*: 409–416.

Hofstad, T., O. Hope and E. Falsen. 1998. Septicaemia with *Neisseria elongata* ssp. nitroreducens in a patient with hypertrophic obstructive cardiomyopathia. Scand. J. Infect. Dis. *30*: 200–201.

Höhnl, G. 1955. Ein Beitrag zur Physiologie der Eisenbakterien. Vom Wasser *22*: 176–193.

Hoke, C. and N.A. Vedros. 1982a. Characterization of atypical aerobic Gram-negative cocci isolated from humans. J. Clin. Microbiol. *15*: 906–914.

Hoke, C. and N.A. Vedros. 1982b. Taxonomy of the *Neisseriae*. Fatty acid analysis, aminopeptidase activity, and pigment extraction. Int. J. Syst. Bacteriol. *32*: 51–56.

Hoke, C. and N.A. Vedros. 1982c. Taxonomy of the Neisseriae—DNA base composition, interspecific transformation, and DNA hybridization. Int. J. Syst. Bacteriol. *32*: 57–66.

Holdeman, L.V., I.J. Good and W.E.C. Moore. 1976. Human fecal flora: variation in bacterial composition within individuals and a possible effect of emotional stress. Appl. Environ. Microbiol. *31*: 359–375.

Holdeman, L., R.W. Kelly and W.E.C. Moore. 1984. Family I. *Bacteroidaceae*. *In* Krieg and Holt (Editors), Bergey's Manual of Systematic Bacteriology, 1st Ed., Vol. 1, The Williams & Wilkins Co., Baltimore. pp. 602–631.

Holding, A.J. and J.M. Shewan. 1974. Genus *Alcaligenes*. *In* Buchanan and Gibbons (Editors), Bergey's Manual of Determinative Bacteriology, 8th Ed., The Williams & Wilkins Co., Baltimore. pp. 273–275.

Holland, C.J., M. Ristic, A.I. Cole, P. Johnson, G. Baker and T.E. Goetz. 1985. Isolation, experimental transmission, and characterization of causative agent of Potomac horse fever. Science *227*: 522–524.

Holländer, R. and W. Mannheim. 1975. Characterization of hemophilic and related bacteria by their respiratory quinones and cytochromes. Int. J. Syst. Bacteriol. *25*: 102–107.

Hollender, J., J. Hopp and W. Dott. 1997. Degradation of 4-cholorophenol via the meta cleavage pathway by *Comamonas testosteroni* JH5. Appl. Environ. Microbiol. *63*: 4567–4572.

Hollis, A.B., W.E. Kloss and C.H. Elkan. 1981. DNA:DNA hybridization studies of *Rhizobium japonicum* and related *Rhizobiaceae*. J. Gen. Microbiol. *123*: 215–222.

Hollis, D.G., R.E. Weaver and P.S. Riley. 1983. Emended description of *Kingella denitrificans* (Snell and Lapage 1976): correction of the maltose reaction. J. Clin. Microbiol. *18*: 1174–1176.

Hollis, D.G., G.L. Wiggins and R.E. Weaver. 1969. *Neisseria lactamicus* sp. n., a lactose-fermenting species resembling *Neisseria meningitidis*. Appl. Microbiol. *17*: 71–77..

Hollis, D.G., G.L. Wiggins and R.E. Weaver. 1972. An unclassified Gram-negative rod isolated from the pharynx on Thayer-Martin medium (selective agar). Appl. Microbiol. *24*: 772–777.

Holm, N.C. 1991. Diversität und struktur von *Hyphomicrobium* populationen im plöner klärwerk und im kleinen plöner see, University of Kiel. Kiel, Germany. . p. 120.

Holm, N.C., C.G. Gliesche and P. Hirsch. 1996. Diversity and structure of *Hyphomicrobium* populations in a sewage treatment plant and its adjacent receiving lake. Appl. Environ. Microbiol. *62*: 522–528.

Holm, P. 1950. Studies on the etiology of human actinomycosis. I. The "other microbes" and their importance. Acta Pathol. Microbiol. Scand. *27*: 736–751.

Holman, R.C., C.D. Paddock, A.T. Curns, J.W. Krebs, J.H. McQuiston and J.E. Childs. 2001. Analysis of risk factors for fatal Rocky Mountain spotted fever: Evidence for superiority of tetracyclines for therapy. J. Infect. Dis. *184*: 1437–1444.

Holmes, A.J., A. Costello, M.E. Lidstrom and J.C. Murrell. 1995a. Evidence that particulate methane monooxygenase and ammonia monooxygenase may be evolutionarily related. FEMS Microbiol. Lett. *132*: 203–208.

Holmes, A.J., D.P. Kelly, S.C. Baker, A.S. Thompson, P. De Marco, E.M. Kenna and J.C. Murrell. 1997. *Methylosulfonomonas methylovora* gen. nov., sp. nov., and *Marinosulfonomonas methylotropha* gen. nov., sp. nov.: novel methylotrophs able to grow on methanesulfonic acid. Arch. Microbiol. *167*: 46–53.

Holmes, A.J., N.J. Owens and J.C. Murrell. 1995b. Detection of novel marine methanotrophs using phylogenetic and functional gene probes after methane enrichment. Microbiology *141*: 1947–1955.

Holmes, A.J., N.P.J. Owens and J.C. Murrell. 1996. Molecular analysis of enrichment cultures of marine methen oxidising bacteria. J. Exp. Mar. Biol. Ecol. *203*: 27–38.

Holmes, B. 1988. The taxonomy of *Agrobacterium*. Acta Hort. *225*: 47–52.

Holmes, B., M. Costas, S.L.W. On, P. Vandamme, E. Falsen and K. Kersters. 1993. *Neisseria weaveri* sp. nov. (formerly CDC group M-5), from dog-bite wounds of humans. Int. J. Syst. Bacteriol. *43*: 687–693.

Holmes, B. and C.A. Dawson. 1983. Numerical taxonomic studies on *Achromobacter* isolates from clinical material. *In* Leclerc (Editor), Gram Negative Bacteria of Medical and Public Health Importance: Taxonomy-Identification-Applications, Les Colloques de l' INSERM. Vol. 114, L' Institut National de la Sante et de la Recherche Medicale, Paris. pp. 331–341.

Holmes, B., R.J. Owen, A. Evans, H. Malnick and W.R. Willcox. 1977a. *Pseudomonas paucimobilis*, a new species isolated from human clinical specimens, the hospital environment, and other sources. Int. J. Syst. Bacteriol. *27*: 133–146.

Holmes, B., M. Popoff, M. Kiredjian and K. Kersters. 1988. *Ochrobactrum anthropi* gen. nov., sp. nov. from human clinical specimens and previously known as groupVd. Int. J. Syst. Bacteriol. *38*: 406–416.

Holmes, B. and P. Roberts. 1981. The classification, identification and nomenclature of agrobacteria: incorporating revised descriptions for each of *Agrobacterium tumefaciens*, *Agrobacterium rhizogenes* and *Agrobacterium rubi*. J. Appl. Bacteriol. *50*: 443–468.

Holmes, B., J.J. Snell and S.P. Lapage. 1977b. Strains of *Achromobacter xylosoxidans* from clinical material. J. Clin. Pathol. (Lond.) *30*: 595–601.

Holmes, B., A.G. Steigerwalt, R.E. Weaver and D.J. Brenner. 1987. *Chryseomonas luteola*, comb. nov. and *Flavimonas oryzihabitans*, gen.nov., comb. nov., *Pseudomonas*-like species from human clinical specimens and formerly known, respectively, as groups Ve-1 and Ve-2. Int. J. Syst. Bacteriol. *37*: 245–250.

Holmes, E.C., R. Urwin and M.C. Maiden. 1999. The influence of recombination on the population structure and evolution of the human pathogen *Neisseria meningitidis*. Mol. Biol. Evol. *16*: 741–749.

Holmes, P.A. 1985. Application of PHB—a microbially produced biodegradable thermoplastic. Phys. Technol. *16*: 32–36.

Holt, J.G., N.R. Krieg, P.H.A. Sneath, J.T. Staley and S.T. Williams (Editors). 1994. Bergey's Manual of Determinative Bacteriology, 9th Ed., The Williams & Wilkins Co., Baltimore.

Holten, E. 1973. Glutamate dehydrogenases in genus *Neisseria*. Acta Pathol. Microbiol. Scand. [B] Microbiol. Immunol. *81*: 49–58.

Holten, E. 1975. Radiorespirometric studies in genus *Neisseria*. I. The catabolism of glucose. Acta Pathol. Microbiol. Scand. [B]. *83*: 353–366.

Holten, E. 1976a. Pyridine nucleotide independent oxidation of L-malate in genus *Neisseria*. Acta Pathol. Microbiol. Scand. [B]. *84*: 17–21.

Holten, E. 1976b. Radiorespirometric studies in genus Neisseria. 2. The catabolism of glutamate and fumarate. Acta Pathol. Microbiol. Scand. [B]. *84*: 1–8.

Holten, E. 1977. The catabolism of glucose, glutamate pyruvate and acetate in *Neisseria elongata* subsp. *glycolytica*. Acta Pathol Microbiol Scand [B]. *85*: 117–124.

Holten, E., D. Bratlid and K. Bovre. 1978. Carriage of *Neisseria meningitidis* in a semi-isolated arctic community. Scand. J. Infect. Dis. *10*: 36–40.

Holten, E. and K. Jyssum. 1973. Glutamate dehydrogenases in *Neisseria meningitidis*. Acta Pathol. Microbiol. Scand. [B] Microbiol. Immunol. *81*: 43–48.

Holten, E. and K. Jyssum. 1974. Activities of some enzymes concerning pyruvate metabolism in *Neisseria*. Acta Pathol. Microbiol. Scand. [B] Microbiol. Immunol. *82*: 843–848.

Holtzman, H.E. 1959. A kappa-like particle in a non-killer stock of *Paramecium aurelia*, syngen 5. J. Protozool. *6 (suppl.)*: 26.

Homma, Y. 1984. Perforation and lysis of hyphae of *Rhizoctonia solani* and conidia of *Cochliobolus miyabeanus* by soil myxobacteria. Phytopathology *74*: 1234–1239.

Homma, Y., Z. Sato, F. Hirayama, K. Konno, H. Shirahama and T. Suzui. 1989. Production of antibiotics by *Pseudomonas cepacia* as an agent for biological control of soilborne plant pathogens. Soil Biol. Biochem. *21*: 723–728.

Hong, S.B. and F.M. Raushel. 1996. Metal-substrate interactions facilitate the catalytic activity of the bacterial phosphotriesterase. Biochemistry *35*: 10904–10912.

Honma, M.A., M. Asomaning and F.M. Ausubel. 1990. *Rhizobium meliloti nodD* genes mediate host-specific activation of *nodABC*. J. Bacteriol. *172*: 901–911.

Honma, M.A. and F.M. Ausubel. 1987. *Rhizobium meliloti* has three functional copies of the *nodD* symbiotic regulatory gene. Proc. Natl. Acad. Sci. U. S. A. *84*: 8558–8562.

Hood, B.L. and R. Hirschberg. 1995. Purification and characterization of *Eikenella corrodens* type IV pilin. Infect. Immun. *63*: 3693–3696.

Hood, D.W., C.S. Dow and P.N. Green. 1987. DNA:DNA hybridization studies on the pink-pigmented facultative methylotrophs. J. Gen. Microbiol. *133*: 709–720.

Hood, D.W., C.S. Dow and P.N. Green. 1988. Electrophoretic comparison of total soluble proteins in the pink-pigmented facultative methylotrophs. J. Gen. Microbiol. *134*: 2375–2384.

Hood, M.A. and J.M. Schmidt. 1996. The examination of *Seliberia stellata* exopolymers using lectin assays. Microb. Ecol. *31*: 281–290.

Hoogstraal, H. 1967. Ticks in relation to human diseases caused by rickettsia species. Annu.Rev. Entomol. *12*: 377–420.

Hoogstraal, H. 1981. Changing patterns of tickborne diseases in modern society. Annu. Rev. Entomol. *26*: 75–99.

Hoogstraal, H., M.N. Kaiser, R.A. Ormsbee, D.J. Osborn, I. Hemly and S. Gaber. 1967. *Hyalomma (Hyalommina) rhipicephaloides* Neumann (Ixodoidea: Ixodidae): Its identity, hosts, and ecology, and *Rickettsia conori*, *R. prowazeki*, and *Coxiella burneti* infections in rodent hosts in Egypt. J. Med. Entomol. *4*: 391–400.

Hook, E.W., S.F. Ching, J. Stephens, K.F. Hardy, K.R. Smith and H.H. Lee. 1997. Diagnosis of *Neisseria gonorrhoeae* infections in women by using the ligase chain reaction on patient-obtained vaginal swabs. J. Clin. Microbiol. *35*: 2129–2132.

Hook, L.A., J.M. Larkin and E.R. Brockman. 1980. Isolation, characterization, and emendation of description of *Angiococcus disciformis* (Thaxter 1904) Jahn 1924 and proposal of a neotype strain. Int. J. Syst. Bacteriol. *30*: 135–142.

Hooykaas, P.J.J. and A.G.M. Beijersbergen. 1994. The virulence system of *Agrobacterium tumefaciens*. Annu. Rev. Phytopathol. *32*: 157–179.

Hooykaas, P.J.J., P.M. Klapwijk, M.P. Nuti, R.A. Schilperoort and A. Rörsch. 1977. Transfer of *Agrobacterium tumefaciens* Ti plasmid to avirulent agrobacteria and to *Rhizobium* ex-planta. J. Gen. Microbiol. *98*: 477–484.

Höpfl, P., W. Ludwig, K.H. Schleifer and N. Larsen. 1989. The 23S ribosomal RNA higher order structure of *Pseudomonas cepacia* and other prokaryotes. Eur. J. Biochem. *185*: 355–364.

Hoppe, J.E. 1988. Methods for isolation of *Bordetella pertussis* from patients with whooping cough. Eur. J. Clin. Microbiol. Infect. Dis. *7*: 616–620.

Hoppe, J.E. 1999. *Bordetella. In* Murray, Baron, Phaller, Tenover and Yolken (Editors), Manual of Clinical Microbiology, American Society for Microbiology, Washington DC. 614–624.

Hoppe, J.E. and T. Paulus. 1998. Comparison of three media for agar dilution susceptibility testing of *Bordetella pertussis* using six antibiotics. Eur. J. Clin. Microbiol. Infect. Dis. *17*: 391–393.

Hoppe, J.E. and C.G. Simon. 1990. In vitro susceptibilities of *Bordetella pertussis* and *Bordetella parapertussis* to seven fluoroquinolones. Antimicrob. Agents Chemother. *34*: 2287–2288.

Hoppe, J.E. and the Erythromycin Study Group. 1992. Comparison of erythromycin estolate and erythromycin ethylsuccinate for treatment of pertussis. Pediatr. Infect. Dis. J. *11*: 189–193.

Hoppe, J.E. and T. Tschirner. 1997. Comparison of Etest and agar dilution for testing the activity of three macrolides against *Bordetella parapertussis*. Diagn. Microbiol. Infect. Dis. *28*: 49–51.

Hoppe, J.E. and A. Weiss. 1987. Recovery of *Bordetella pertussis* from four kinds of swabs. Eur. J. Clin. Microbiol. *6*: 203–205.

Horbach, S., H. Sahm and R. Welle. 1993. Isoprenoid biosynthesis in bacteria—2 different pathways. FEMS Microbiol. Lett. *111*: 135–140.

Horbach, S., J. Strohhacker, R. Welle, A. de Graaf and H. Sahm. 1994. Enzymes involved in the formation of glycerol 3-phosphate and the by-products dihydroxyacetone and glycerol in *Zymomonas mobilis*. FEMS Microbiol. Lett. *120*: 37–44.

Horn, M., T.R. Fritsche, R.K. Gautom, K.H. Schleifer and M. Wagner. 1999. Novel bacterial endosymbionts of *Acanthamoeba* spp. related to the *Paramecium caudatum* symbiont *Caedibacter caryophilus*. Environ. Microbiol. *1*: 357–367.

Horvath, B., E. Kondorosi, M. John, J. Schmidt, I. Török, Z. Györgypal,

I. Barabas, U. Wieneke, J. Schell and A. Kondorosi. 1986. Organization, structure and symbiotic function of *Rhizobium meliloti* nodulation genes determining host specificity for alfalfa. Cell *46*: 335–343.

Hoshino, T., T. Hayashi and T. Uchiyama. 1994. Pseudodeoxyviolacein, a new red pigment produced by the tryptophan metabolism of *Chromobacterium violaceum*. Biosci. Biotechnol. Biochem. *58*: 279–282.

Hoshino, T., Y. Kojima, T. Hayashi, T. Uchiyama and K. Kaneko. 1993a. A new metabolite of tryptophan, chromopyrrolic acid, produced by *Chromobacterium violaceum*. Biosci. Biotechnol. Biochem. *57*: 775–781.

Hoshino, T., T. Sugisawa, M. Tazoe, M. Shinjoh and A. Fujiwara. 1990. Metabolic pathway for 2-keto-L-gluconic acid formation in *Gluconobacter melanogenus* IFO 3293. Agric. Biol. Chem. *54*: 1211–1218.

Hoshino, T., M. Yamamoto and T. Uchiyama. 1993b. Formations of (5-hydroxy)indole S-(−)-lactic acid, N-acetyl-5-hydroxy-L-tryptophan, and (5-hydroxy)indole carboxylic acid in the metabolism of tryptophan and 5-hydroxytryptophan by *Chromobacterium violaceum*. Biosci. Biotechnol. Biochem. *57*: 1609–1610.

Hoskin, F.C.G., J.E. Walker, W.D. Dettbarn and J.R. Wild. 1995. Hydrolysis of tetriso by an enzyme derived from *Pseudomonas diminuta* as a model for the detoxication of o-ethyl S-(2- diisopropylaminoethyl) methylphosphonothiolate(Vx). Biochem. Pharmacol. *49*: 711–715.

Hou, C.T. (Editor). 1984. Methylotrophs: microbiology, biochemistry, and genetics, CRC Press, Boca Raton, Florida.

Houck, D.R., J.L. Hanners, C.J. Unkefer, M.A. van Kleef and J.A. Duine. 1989. PQQ: biosynthetic studies in *Methylobacterium* AM1 and *Hyphomicrobium* X using specific ^{13}C labeling and NMR. Antonie Leeuwenhoek *56*: 93–101.

Hougardy, A. and J.-H. Klemme. 1995. Nitrate reduction in a new strain of *Rhodoferax fermentans*. Arch. Microbiol. *164*: 358–362.

Hougardy, A., B.J. Tindall and J.H. Klemme. 2000. *Rhodopseudomonas rhenobacensis* sp. nov., a new nitrate-reducing purple non-sulfur bacterium. Int. J. Syst. Evol. Microbiol. *50*: 985–992.

Houston, L.S., R.G. Cook and S.J. Norris. 1990. Isolation and characterization of a *Treponema pallidum* major 60 kilodalton protein resembling the Groel protein of *Escherichia coli*. J. Bacteriol. *172*: 2862–2870.

Houwen, F.P., C. Dijkema, C.H.H. Schoenmakers, A.J.M. Stams and A.J.B. Zehnder. 1987. ^{13}C-NMR study of propionate degradation by a methanogenic coculture. FEMS Microbiol. Lett. *41*: 269–274.

Houwen, F.P., J. Plokker, C. Kijkema and A.J.M. Stams. 1990. Syntrophic propionate oxidation. *In* Bélaich, Bruschi and Garcia (Editors), Microbiology and Biochemistry of Strict Anaerobes Involved in Interspecies Hydrogen Transfer, Plenum Press, New York.

Houwink, A.L. 1955. *Caulobacter*: its morphogenesis, taxonomy, and parasitism. Antonie Leeuwenhoek J. Microbiol. Serol. *21*: 49–64.

Hoyer, B.H. and N.B. McCullough. 1968a. Homologies of deoxyribonucleic acids from *Brucella ovis*, canine abortion organisms and other *Brucella* species. J. Bacteriol. *96*: 1783–1790.

Hoyer, B.H. and N.B. McCullough. 1968b. Polynucleotide homologies of *Brucella* nucleic acids. J. Bacteriol. *95*: 444–448.

Hsiao, C. and T. Hsiao. 1985. *Rickettsia* as the cause of cytoplasmic incompatibility in the alfalfa weevil, *Hypera postica*. J. Invert. Pathol. *45*: 244–246.

Hsu, M.Y., S. Inouye and M. Inouye. 1989. Structural requirements of the RNA precursor for the biosynthesis of the branched RNA-linked multicopy single-stranded DNA of *Myxococcus xanthus*. J. Biol. Chem. *264*: 6214–6219.

Hu, F.P. and J.M. Young. 1998. Biocidal activity in plant pathogenic *Acidovorax*, *Burkholderia*, *Herbaspirillum*, *Ralstonia* and *Xanthomonas* spp. J. Appl. Microbiol. *84*: 263–271.

Hu, F.P., J.M. Young and C.M. Triggs. 1991. Numerical analysis and determinative tests for nonfluorescent plant-pathogenic *Pseudomonas* spp. and genomic analysis and reclassification of species related to *Pseudmonas avenae* Manns 1909. Int. J. Syst. Bacteriol. *41*: 516–525.

Hu, H.L., J.E. Peterson and E.R. Brockman. 1985. Stalked sporangia of *Polyangium rugiseptum*. Int. J. Syst. Bacteriol. *35*: 362–363.

Hu, L.T., L. Binyang and S. Qin. 1990. Isolation of *Bdellovibrio bacteriovorus* from human stool. J. Chinese Microbiol. Epidemiol. *10*: 95–98.

Hu, W.J., X.M. Chen, H.D. Meng and Z.H. Meng. 1989. Fermented corn flour poisoning in rural areas of China. III. Isolation and identification of main toxin produced by causal microorganisms. Biomed. Environ. Sci. *2*: 65–71.

Huang, J., W. Yindeeyoungyeon, R.P. George, T.P. Denny and M.A. Schell. 1998. Joint transcriptional control of *xpsR*, the unusual signal integrator of the *Ralstonia solanacearum* virulence gene regulatory network, by a response regulator and a LysR-type transcriptional activator. J. Bacteriol. *180*: 2736–2743.

Huang, Q. and C. Allen. 1997. An exo-poly-α-D-galacturonosidase, PehB, is required for wild-type virulence of *Ralstonia solanacearum*. J. Bacteriol. *179*: 7369–7378.

Huber, H. and K.O. Stetter. 1989. *Thiobacillus prosperus*, sp. nov., represents a new group of halotolerant metal-mobilizing bacteria isolated from a marine geothermal field. Arch. Microbiol. *151*: 479–485.

Huber, H. and K.O. Stetter. 1990. *Thiobacillus cuprinus* sp. nov., a novel facultatively organotrophic metal mobilizing bacterium. Appl. Environ. Microbiol. *56*: 315–322.

Huber, R., T. Wilharm, D. Huber, A. Trincone, S. Burggraf, H. König, R. Rachel, I. Rockinger, H. Fricke and K.O. Stetter. 1992. *Aquifex pyrophilus* gen. nov. sp. nov., represents a novel group of marine hyperthermophilic hydrogen-oxidizing bacteria. Syst. Appl. Microbiol. *15*: 340–351.

Hubner, A. and W. Hendrickson. 1997. A fusion promoter created by a new insertion sequence, IS1490, activates transcription of 2,4,5-trichlorophenoxyacetic acid catabolic genes in *Burkholderia cepacia* AC1100. J. Bacteriol. *179*: 2717–2723.

Huchzermeyer, F.W., I.G. Horak, J.F. Putterill and R.A. Earle. 1992. Description of *Aegyptianella botuliformis* n. sp. (*Rickettsiales: Anaplasmataceae*) from the helmeted guineafowl, *Numida meleagris*. Onderstepoort J. Vet. Res. *59*: 97–101.

Huddleson, I.F. 1929. The differentiation of the species of the genus *Brucella*. Bull. Mich. Agric. Exp. Sta. *100*: 1–6.

Huddleson, I.F. 1957. Genus III. *Brucella* Meyer and Shaw 1920. *In* Breed, Murray and Smith (Editors), Bergey's Manual of Determinative Bacteriology, 7th Ed., The Williams & Wilkins Co., Baltimore. pp. 404–406.

Hudson, J.R. 1950. The recognition of tick-borne fever as a disease of cattle. Brit. Vet. J. *106*: 3–17.

Hudson, R.A., D.E. Thompson and M.D. Collins. 1993. Genetic interrelationships of saccharolytic *Clostridium botulinum* types B, E and F and related clostridia by small subunit rRNA gene sequences. FEMS Microbiol. Lett. *108*: 103–110.

Huebner, R.J., W.L. Jellison and C. Pomerantz. 1946. Rickettsialpox, newly recognized rickettsial disease. IV. Isolation of a rickettsia apparently identical with the causative agent of rickettsialpox from *Allodermanyssus sanguineus*, a rodent mite. Pub. Health Rep. *61*: 1677–1682.

Hugendieck, I. and O. Meyer. 1992. The structural genes encoding CO dehydrogenase subunits (*coxL*, *M* and *S*) in *Pseudomonas carboxydovorans* OM5 reside on plasmid pHCG3 and are, with the exception of *Streptomyces thermoautotrophicus*, conserved in carboxydotrophic bacteria. Arch. Microbiol. *157*: 301–304.

Hugenholtz, P., E. Stackebrandt and J.A. Fuerst. 1994. A phylogenetic analysis of the genus *Blastobacter* with a view to its future reclassification. System. Appl. Microbiol. *17*: 51–57.

Hugh, R. 1962. *Comamonas terrigena* comb. nov. with proposal of a neotype and request for an opinion. Int. Bull. Bacteriol. Nomencl. Taxon. *12*: 33–35.

Hugh, R. 1965. A comparison of *Pseudomonas testosteroni* and *Comamonas terrigena*. Int. Bull. Bacteriol. Nomencl. Taxon. *15*: 125–132.

Hugh, R. and E. Leifson. 1953. The taxonomic significance of fermentative versus oxidative metabolism of carbohydrates by various Gram-negative bacteria. J. Bacteriol. *66*: 24–26.

Hughes, M.L. 1893. The natural history of certain fevers occurring in the Mediterranean. Mediterranean Nat. *2*: 299–300; 325–327; 332–334.

Hum, S., K. Quinn, J. Brunner and S.L.W. On. 1997. Evaluation of a PCR assay for identification and differentiation of *Campylobacter fetus* subspecies. Aust. Vet. J. *75*: 827–831.

Humphrey, B.A., M.R. Dickson and K.C. Marshall. 1979. Physicochemical and *in situ* observations on the adhesion of gliding bacteria to surfaces. Arch. Microbiol. *120*: 231–238.

Humphrey, B., J.M. Vincent and V. Skerdleta. 1973. Group antigens in slow-growing *Rhizobium*. Arch. Mikrobiol. *89*: 79–82.

Hungate, R.E. 1966. The rumen and its microbes, Academic Press, New York.

Hungate, R.E. 1969. A roll tube method for the cultivation of strict anaerobes. Methods Microbiol. *3B*: 117–132.

Hungerer, C., B. Troup, U. Romling and D. Jahn. 1995. Regulation of the *hemA* gene during 5-aminolevulinic acid formation in *Pseudomonas aeruginosa*. J. Bacteriol. *177*: 1435–1443.

Hunter, W.J. and L.D. Kuykendall. 1990. Enhanced nodulation and nitrogen fixation by a revertant of a nodulation-defective *Bradyrhizobium japonicum* tryptophan auxotroph. Appl. Environ. Microbiol. *56*: 2399–2403.

Hurek, T., S. Burggraf, C.R. Woese and B. Reinhold-Hurek. 1993. 16S rRNA-targeted polymerase chain reaction and oligonucleotide hybridization to screen for *Azoarcus* spp., grass-associated diazotrophs. Appl. Environ. Microbiol. *59*: 3816–3824.

Hurek, T., T. Egener and B. Reinhold-Hurek. 1997a. Divergence in nitrogenases of *Azoarcus* spp., *Proteobacteria* of the beta subclass. J. Bacteriol. *179*: 4172–4178.

Hurek, T., B. Reinhold, I. Fendrik and E.G. Niemann. 1987. Root-zone-specific oxygen tolerance of *Azospirillum* spp. and diazotrophic rods closely associated with Kallar grass. Appl. Environ. Microbiol. *53*: 163–169.

Hurek, T. and B. Reinhold-Hurek. 1995. Identification of grass-associated and toluene-degrading diazotrophs, *Azoarcus* spp., by analyses of partial 16S ribosomal DNA sequences. Appl. Environ. Microbiol. *61*: 2257–2261.

Hurek, T. and B. Reinhold-Hurek. 1999. Interactions of *Azoarcus* sp. with rhizosphere fungi. *In* Varma, A. and B. Hock (Editors), Mycorrhiza: Structure, Function, Molecular Biology and Biotechnology, 2nd Ed., Springer, Berlin. 595–614.

Hurek, T., B. Reinhold-Hurek, G.L. Turner and F.J. Bergersen. 1994a. Augmented rates of respiration and efficient nitrogen fixation at nanomolar concentrations of dissolved O_2 in hyperinduced *Azoarcus* sp. strain BH72. J. Bacteriol. *176*: 4726–4733.

Hurek, T., B. Reinhold-Hurek, M. Van Montagu and E. Kellenberger. 1994b. Root colonization and systemic spreading of *Azoarcus* sp. strain BH72 in grasses. J. Bacteriol. *176*: 1913–1923.

Hurek, T., M. Van Montagu, E. Kellenberger and B. Reinhold-Hurek. 1995. Induction of complex intracytoplasmic membranes related to nitrogen fixation in *Azoarcus* sp. BH72. Mol. Microbiol. *18*: 225–236.

Hurek, T., B. Wagner and B. Reinhold-Hurek. 1997b. Identification of N_2-fixing plant and fungus associated *Azoarcus* species by PCR-based genomic fingerprints. Appl. Environ. Microbiol. *63*: 4331–4339.

Hurst, G.D.D., F.M. Jiggins, J.H. von der Schulenburg, D. Bertrand, S.A. West, I.I. Goriacheva, I.A. Zakharov, J.H. Werren, R. Stouthamer and M.E. Majerus. 1999. Male-killing *Wolbachia* in two insect species. Proc. R. Soc. Lond. B Sci. *266*: 735–740.

Hurtado, A., J.P. Clewley, D. Linton, R.J. Owen and J. Stanley. 1997. Sequence similarities between large subunit ribosomal RNA gene intervening sequences from different *Helicobacter* species. Gene *194*: 69–75.

Hurtado, A. and R.J. Owen. 1997a. A molecular scheme based on 23S rRNA gene polymorphisms for rapid identification of *Campylobacter* and *Arcobacter* species. J. Clin. Microbiol. *35*: 2401–2404.

Hurtado, A. and R.J. Owen. 1997b. A rapid identification scheme for *Helicobacter pylori* and other species of *Helicobacter* based on 23S rRNA gene polymorphisms. Syst. Appl. Microbiol. *20*: 222–231.

Hurtubise, Y., D. Barriault, J. Powlowski and M. Sylvestre. 1995. Purification and characterization of the *Comamonas testosteroni* B-356 biphenyl dioxygenase components. J. Bacteriol. *177*: 6610–6618.

Hurtubise, Y., D. Barriault and M. Sylvestre. 1996. Characterization of active recombinant his-tagged oxygenase component of *Comamonas testosteroni* B-356 biphenyl dioxygenase. J. Biol. Chem. *271*: 8152–8156.

Huska, J., I. Zavdska, D. Toth, M. Dobrotova and P. Gemeiner. 1996. Immobilization of surfactant degrading bacteria in alginate gel. Biologia *51*: 279–283.

Husmann, M., C. Gries, P. Jehnichen, T. Woelfel, G. Gerken, W. Ludwig and S. Bhakdi. 1994. *Helicobacter* sp. strain Mainz isolated from an AIDS patient with septic arthritis: case report and nonradioactive analysis of 16S rRNA sequence. J. Clin. Microbiol. *32*: 3037–3039.

Hüttermann, A. 1969. Studies on a bacteriolytic enzyme of *Archangium violaceum* (*Myxobacteriales*). II. Partial purification and properties of the enzyme. Arch. Mikrobiol. *67*: 306–317.

Hwang, M.N. and G.M. Ederer. 1975. Rapid hippurate hydrolysis method for presumptive identification of Group B Streptococci. J. Clin. Microbiol. *1*: 114–115.

Hylemon, P.B., N.R. Krieg and P.V. Phibbs, Jr.. 1974. Transport and catabolism of D-fructose by *Spirillum itersonii*. J. Bacteriol. *117*: 144–150.

Hylemon, P.B., J.S. Wells, Jr., J.H. Bowdre, T.O. MacAdoo and N.R. Krieg. 1973a. Designation of *Spirillum volutans* Ehrenberg 1832 as type species of the genus *Spirillum* Ehrenberg 1832 and designation of the neotype strain of *S. volutans*. Request for an opinion. Int. J. Syst. Bacteriol. *23*: 20–27.

Hylemon, P.B., J.S. Wells, Jr., N.R. Krieg and H.W. Jannasch. 1973b. The genus *Spirillum*: a taxonomic study. Int. J. Syst. Bacteriol. *23*: 340–380.

Hynes, M.F. and N.F. McGregor. 1990. Two plasmids other than the nodulation plasmid are necessary for formation of nitrogen-fixing nodules by *Rhizobium leguminosarum*. Mol. Microbiol. *4*: 567–574.

Hynes, M.F., R. Simon, P. Müller, K. Niehaus, M. Labes and A. Pühler. 1986. The 2 megaplasmids of *Rhizobium meliloti* are involved in the effective nodulation of alfalfa. Mol. Gen. Genet. *202*: 356–362.

Hynes, R.K., A.L. Ding and L.M. Nelson. 1985. Denitrification by *Rhizobium fredii*. FEMS Microbiol. Lett. *30*: 183–186.

Iba, K., K. Takamiya, Y. Toh and N. M.. 1988. Roles of bacteriochlorophyll and carotenoid synthesis in formation of intracytoplasmic membrane systems and pigment-protein complexes in an aerobic photosynthetic bacterium, *Erythrobacter* sp. strain OCh114. J. Bacteriol. *170*: 1843–1847.

Ihn, K.S., S.H. Han, H.R. Kim, M.S. Huh, S.Y. Seong, J.S. Kang, T.H. Han, I.S. Kim and M.S. Choi. 2000. Cellular invasion of *Orientia tsutsugamushi* requires initial interaction with cell surface heparan sulfate. Microb. Pathog. *28*: 227–233.

Iida, T., Y. Haishima, A. Tanaka, K. Nishiyama, S. Saito and K. Tanamoto. 1996. Chemical structure of lipid A isolated from *Comamonas testosteroni* lipopolysaccharide. Eur. J. Biochem. *237*: 468–475.

Iizuka, T., Y. Jojima, R. Fudou and S. Yamanaka. 1998. Isolation of myxobacteria from the marine environment. FEMS Microbiol. Lett. *169*: 317–322.

Ikemoto, S., K. Katoh and K. Komagata. 1978a. Cellular fatty-acid composition in methanol-utilizing bacteria. J. Gen. Appl. Microbiol. *24*: 41–49.

Ikemoto, S., H. Kuraishi, K. Komagata, R. Azuma, T. Suto and H. Murooka. 1978b. Cellular fatty-acid composition in *Pseudomonas* species. J. Gen. Appl. Microbiol. *24*: 199–213.

Ikushiro, H., H. Hayashi and H. Kagamiyama. 2001. A water-soluble homodimeric serine palmitoyltransferase from *Sphingomonas paucimobilis* EY2395^T strain. Purification, characterization, cloning, and overproduction. J. Biol. Chem. *276*: 18249–18256.

Imachi, H., Y. Sekiguchi, Y. Kamagata, S. Hanada, A. Ohashi and H. Harada. 2002. *Pelotomaculum thermopropionicum* gen. nov., sp. nov., an anaerobic, thermophilic, syntrophic propionate-oxidizing bacterium. Int. J. Syst. Evol. Microbiol. *52*: 1729–1735.

Imaizumi, A., Y. Suzuki, S. Ono, H. Sato and Y. Sato. 1983. Heptakis(2,6-O-dimethyl)β-cyclodextrin: a novel growth stimulant for *Bordetella pertussis* phase I. J. Clin. Microbiol. *17*: 781–786.

Imanaka, H., M. Kousaka, G. Tamura and K. Arima. 1965. Studies on pyrrolnitrin, a new antibiotic. 3. Structure of pyrrolnitrin. J. Antibiot. *18*: 207–210.

Imhoff, J.F. 1982. Occurrence and evolutionary significance of two sulfate assimilation pathways in the *Rhodospirillaceae*. Arch. Microbiol. *132*: 197–203.

Imhoff, J.F. 1983. *Rhodopseudomonas marina* sp. nov., a new marine phototropic purple bacterium. Syst. Appl. Microbiol. *4*: 512–521.

Imhoff, J.F. 1984a. Quinones of phototrophic purple bacteria. FEMS Microbiol. Lett. *256*: 85–89.

Imhoff, J.F. 1984b. *In* Validation of the publication of new names and new combinations previously effectively published outside the IJSB. List No. 14. Int. J. Syst. Bacteriol. *34*: 270–271.

Imhoff, J.F. 1988. Anoxygenic phototrophic bacteria. *In* Austin (Editor), Methods in Aquatic Bacteriology, John Wiley & Sons Ltd., Chichester. pp. 207–240.

Imhoff, J.F. 1989. Genus *Rhodobacter*. *In* Staley, Bryant, Pfennig and Holt (Editors), Bergey's Manual of Systematic Bacteriology, 1st ed., Vol. 3, The Williams & Wilkins Co., Baltimore. 1668–1672.

Imhoff, J.F. 1991. Polar lipids and fatty acids in the genus *Rhodobacter*. Syst. Appl. Microbiol. *14*: 228–234.

Imhoff, J.F. 1992. The family *Ectothiorhodospiraceae*. *In* Balows, Trüper, Dworkin, Harder and Schleifer (Editors), The Prokaryotes: A Handbook of Bacteria: Ecophysiology, Isolation, Identification, Applications, 2nd Ed., Springer Verlag, New York. 3222–3229.

Imhoff, J. F. 2001. Transfer of *Rhodopseudomonas acidophila* to the new genus *Rhodoblastus* as *Rhodoblastus acidophilus* gen. nov., comb. nov. Int. J. Syst. Evol. Microbiol. *51*: 1863–1866.

Imhoff, J.F. and U. Bias-Imhoff. 1995. Lipids, quinones and fatty acids of anoxygenic phototrophic bacteria. *In* Blankenship, Madigan and Bauer (Editors), Anoxygenic Photosynthetic Bacteria, Kluwer Academic Publishing, The Netherlands. pp. 179–205.

Imhoff, J.F., D.J. Kushner, S.C. Kushwaha and M. Kates. 1982. Polar lipids in phototrophic bacteria of the *Rhodospirillaceae* and *Chromatiaceae* families. J. Bacteriol. *150*: 1192–1201.

Imhoff, J.F., R. Petri and J. Süling. 1998. Reclassification of species of the spiral-shaped phototrophic purple non-sulfur bacteria of the α-Proteobacteria: description of the new genera *Phaeospirillum* gen. nov., *Rhodovibrio* gen. nov., *Rhodothalassium* gen. nov. and *Roseospira* gen. nov. as well as transfer of *Rhodospirillum fulvum* to *Phaeospirillum fulvum* comb. nov., of *Rhodospirillum molischianum* to *Phaeospirillum molischianum* comb. nov., of *Rhodospirillum salinarum* to *Rhodovibrio salinarum* comb. nov., of *Rhodospirillum sodomense* to *Rhodovibrio sodomensis* comb. nov., of *Rhodospirillum salexigens* to *Rhodothalassium salexigens* comb. nov. and of *Rhodospirillum mediosalinum* to *Roseospira mediosalina* comb. nov. Int. J. Syst. Bacteriol. *48*: 793–798.

Imhoff, J.F., J. Then, F. Hashwa and H.G. Trüper. 1981. Sulfate assimilation in *Rhodopseudomonas globiformis*. Arch. Microbiol. *130*: 234–237.

Imhoff, J.F. and H.G. Trüper. 1976. Marine sponges as habitats of anaerobic phototrophic bacteria. Microb. Ecol. *3*: 1–9.

Imhoff, J.F. and H.G. Trüper. 1977. *Ectothiorhodospira halochloris* sp. nov. new extremely halophilic phototropic bacterium containing bacteriochlorophyll *b*. Arch. Microbiol. *114*: 115–121.

Imhoff, J.F. and H.G. Trüper. 1992. The genus *Rhodospirillum* and related genera. *In* Balows, Trüper, Dworkin, Harder and Schleifer (Editors), The Prokaryotes: A Handbook of Bacteria: Ecophysiology, Isolation, Identification, Applications, 2nd ed., Springer-Verlag, New York. 2141–2155.

Imhoff, J.F., H.G. Trüper and N. Pfennig. 1984. Rearrangement of the species and genera of the phototrophic "purple nonsulfur bacteria". Int. J. Syst. Bacteriol. *34*: 340–343.

Imhoff, J.F. and H.G. Trüper. 1989. Genus *Rhodocyclus*. *In* Staley, Bryant, Pfennig and Holt (Editors), Bergey's Manual of Systematic Bacteriology, Vol. 3, The Williams & Wilkins Co., Baltimore. pp. 1678–1682.

Imhoff-Stuckle, D. and N. Pfennig. 1983. Isolation and characterization of a nicotinic acid-degrading sulfate-reducing bacterium, *Desulfococcus niacini*, sp. nov. Arch. Microbiol. *136*: 194–198.

Impey, C.S. and B.A. Phillips. 1991. Maintenance of anaerobic bacteria. *In* Kirsop and Doyle (Editors), Maintenance of Microorganisms and Cultured Cells, A Manual of Laboratory Methods, 2nd Ed., Academic Press Ltd., London. pp. 71–80.

Imshenetski, A.A. and L. Solntseva. 1936. On aerobic cellulose decomposing bacteria. Izv. Akad. Nauk S.S.S.R. Ser. Viol. *6*: 1115–1172.

Inagaki, K., J. Tomono, N. Kishimoto, T. Tano and H. Tanaka. 1993.

Cloning and sequence of the *recA* gene of *Acidiphilium facilis*. Nucleic Acids Res. *21*: 4149.

Inglis, T.J.J., D. Chiang, G.S.H. Lee and K.L. Chor. 1998. Potential misidentification of *Burkholderia pseudomallei* by API 20NE. Pathology *30*: 62–64.

Ingram, L.O., C.K. Eddy, K.F. Mackenzie, T. Conway and F. Alterhum. 1989. Genetics of *Zymomonas mobilis* and ethanol production. Dev. Ind. Microbiol. *30*: 53-69.

Inguva, S. and G.S. Shreve. 1999. Biodegradation kinetics of trichloroethylene and 1,2-dichloroethane by *Burkholderia (Pseudomonas) cepacia* PR131 and *Xanthobacter autotrophicus* GJ10. Int. Biodeter. Biodegrad. *43*: 57–61.

Ingvorsen, K., B. Hojer-Pedersen and S.E. Godtfredsen. 1991. Novel cyanide-hydrolyzing enzyme from *Alcaligenes xylosoxidans* subsp. *denitrificans*. Appl. Environ. Microbiol. *57*: 1783–1789.

Inoue, K. and K. Komagata. 1976. Taxonomic study on obligately psychrophilic bacteria isolated from Antarctica. J. Gen. Appl. Microbiol. *22*: 165–176.

Inoue, T., M. Sunagawa, A. Mori, C. Imai, M. Fukuda, M. Takagi and K. Yano. 1989. Cloning and sequencing of the gene encoding the 72-kilodalton dehydrogenase subunit of alcohol dehydrogenase from *Acetobacter aceti*. J. Bacteriol. *171*: 3115–3122.

Inoue, T., M. Sunagawa, A. Mori, C. Imai, M. Fukuda, M. Takagi and K. Yano. 1992. Nucleotide sequence of the gene encoding the 45-kilodalton subunit of alcohol dehydrogenase from *Acetobacter aceti*. J. Ferment. Bioeng. *73*: 419–424.

Inouye, S. 1984. Identification of a development specific promoter of *Myxococcus xanthus*. J. Mol. Biol. *174*: 113–120.

Inouye, S. 1990. Cloning and DNA sequence of the gene coding for the major sigma factor from *Myxococcus xanthus*. J. Bacteriol. *172*: 80–85.

Inouye, S., P.J. Herzer and M. Inouye. 1990. Two independent retrons with highly diverse reverse transcriptases in *Myxococcus xanthus*. Proc. Natl. Acad. Sci. U.S.A. *87*: 942–945.

Inouye, S., M.Y. Hsu, S. Eagle and M. Inouye. 1989. Reverse transcriptase associated with the biosynthesis of the branched RNA Linked msDNA in *Myxococcus xanthus*. Cell *56*: 709–717.

Iosipenko, A. and V. Ignatov. 1995. Physiological aspects of phytohormone production by *Azospirillum brasilense* Sp7. NATO ASI Ser. Ser. G Ecol. Sci. *37*: 271–278.

Irgens, R.L. 1977. *Meniscus*, a new genus of aerotolerant, gas-vacuolated bacteria. Int. J. Syst. Bacteriol. *27*: 38–43.

Irgens, R.L., J.J. Gosink and J.T. Staley. 1996. *Polaromonas vacuolata* gen. nov., sp. nov., a psychrophilic, marine, gas vacuolate bacterium from Antarctica. Int. J. Syst. Bacteriol. *46*: 822–826.

Irgens, R.L., K. Kersters, P. Segers, M. Gillis and J.T. Staley. 1991. *Aquabacter spiritensis*, gen. nov., sp. nov. an aerobic, gas-vacuolate aquatic bacterium. Arch. Microbiol. *155*: 137–142.

Irgens, R.L., K. Kersters, P. Segers, M. Gillis and J.T. Staley. 1993. *In* Validation of the publication of new names and new combinations previously effectively published outside the IJSB. List No. 47. Int. J. Syst. Bacteriol. *43*: 864–865.

Irgens, R.L., I. Suzuki and J.T. Staley. 1989. Gas vacuolate bacteria obtained from marine waters of Antarctica. Curr. Microbiol. *18*: 261–265.

Irschik, H., R. Jansen, K. Gerth, G. Hofle and H. Reichenbach. 1987. Antibiotics from gliding bacteria. 32. The sorangicins, novel and powerful inhibitors of eubacterial RNA polymerase isolated from myxobacteria. J. Antibiot. *40*: 7–13.

Irschik, H. and H. Reichenbach. 1985. An unusual pattern of carbohydrate utilization in *Corallococcus (Myxococcus) corralloides* (*Myxobacterales*). Arch. Microbiol. *142*: 40–44.

Isaac, L. and G.C. Ware. 1974. The flexibility of bacterial cell walls. J. Appl. Bacteriol. *37*: 335–339.

Isaacson, P.G. 1996. Recent developments in our understanding of gastric lymphomas. Am. J. Surg. Pathol. *20*: (Suppl. 1) S1–S7.

Isaksen, M.F. and A. Teske. 1996. *Desulforhopalus vacuolatus* gen. nov., sp. nov., a new moderately psychrophilic sulfate-reducing bacterium with gas vacuoles isolated from a temperate estuary. Arch. Microbiol. *166*: 160–168.

Isaksen, M.F. and A. Teske. 1999. *In* Validation of the publication of new names and new combinations previously published outside the IJSB. List No. 70. Int. J. Syst. Bacteriol. *49*: 935–936.

Isayama, Y., R. Azuma, S. Tanaka and T. Suto. 1977. The pathogenicity and antigenicity of *Brucella canis* QE13 for experimental animals. Ann. Sclavo. *19*: 89–98.

Ison, C.A., C.M. Bellinger and J. Walker. 1986. Homology of cryptic plasmid of *Neisseria gonorrhoeae* with plasmids from *Neisseria meningitidis* and *Neisseria lactamica*. J. Clin. Pathol. *39*: 1119–1123.

Ito, H. and H. Iizuka. 1971. Taxonomic studies on radio-resistant *Pseudomonas*. Part XII. Studies on the microorganisms of cereal grains. Agric. Biol. Chem. *35*: 1566–1571.

Itoh, J., S. Miyadoh, S. Takahasi, S. Amano, N. Ezaki and Y. Yamada. 1979. Studies on antibiotic Bn-227 and antibiotic Bn227-F, new antibiotics I. Taxonomy, isolation and characterization. J. Antibiot. *32*: 1089–1095.

Itoh, S., M. Iwaki, N. Wakao, K. Yoshizu, A. Aoki and K. Tazaki. 1998. Accumulation of Fe, Cr, Ni metals inside cells of acidophilic bacterium *Acidiphilium rubrum* that produces Zn-containing bacteriochlorophyll *a*. Plant Cell Physiol. *39*: 740–744.

Itoh, Y. and D. Haas. 1985. Cloning vectors derived from the *Pseudomonas* plasmid Pvs1. Gene *36*: 27–36.

Itoh, Y., J.M. Watson, D. Haas and T. Leisinger. 1984. Genetic and molecular characterization of the *Pseudomonas* plasmid Pvs1. Plasmid *11*: 206–220.

Itzigsohn, H. 1868. Entwicklungsvorgange von *Zoogloea, Oscillaria, Synedra, Staurastrum, Spirotaenia* und *Chroolepus*. *In* S. B. Ges. Natur Fr. (19 Nov. 1967), Berlin. pp. 30–31.

Ivanova, T.L., T.P. Turova and A.S. Antonov. 1988. DNA-DNA hybridization studies on some purple nonsulfur bacteria. Syst. Appl. Microbiol. *10*: 259–263.

Iversen, T., R. Standal, T. Pedersen and D.H. Coucheron. 1994. IS1032 from *Acetobacter xylinum*, a new mobile insertion sequence. Plasmid *32*: 46–54.

Ives, T.J., P. Manzewitsch, R.L. Regnery, J.D. Butts and M. Kebede. 1997. In vitro susceptibilities of *Bartonella henselae, B. quintana, B. elizabethae, Rickettsia rickettsii, R. conorii, R. akari,* and *R. prowazekii* to macrolide antibiotics as determined by immunofluorescent-antibody analysis of infected Vero cell monolayers. Antimicrob. Agents Chemother. *41*: 578–582.

Izumi, Y., M. Takizawa, Y. Tani and H. Yamada. 1982. An obligate methylotrophic *Hyphomicrobium* strain identification, growth characteristics and cell composition. J. Ferment. Technol. *60*: 371–375.

Izumi, Y., M. Takizawa, Y. Tani and H. Yamada. 1983. *In* Validation of the publication of new names and new combinations previously effectively published outside the IJSB. List No. 10. Int. J. Syst. Bacteriol. *33*: 438–440.

Izumi, Y., T. Yoshida, T. Hagishita, Y. Tanaka, T. Mitsunaga, T. Ohshiro, T. Tanabe, A. Miyata, C. Yokoyama, J.D. Goldberg and P. Brick. 1996. Structure and function of the serine pathway enzymes in *Hyphomicrobium*. *In* Lidstrom and Tabita (Editors), Microbial growth on C₁ compounds: Proceedings of the 8th International Symposium, Kluwer Academic, Dordrecht; Boston. 25–32.

Izumi, Y., T. Yoshida, S.S. Miyazaki, T. Mitsunaga, T. Ohshiro, M. Shimao, A. Miyata and T. Tanabe. 1993. L-Serine production by a methylotroph and its related enzymes. Appl. Microbiol. Biotechnol. *39*: 427–432.

Jackson, B.E., V.K. Bhupathiraju, R.S. Tanner, C.R. Woese and M.J. McInerney. 1999. *Syntrophus aciditrophicus* sp. nov., a new anaerobic bacterium that degrades fatty acids and benzoate in syntrophic association with hydrogen-using microorganisms. Arch. Microbiol. *171*: 107–114.

Jackson, B.E., V.K. Bhupathiraju, A.C. Tanner, C.R. Woese and B.V. McInerney. 2001. *In* Validation of the publication of new names and new combinations previously effectively published outside the IJSB, List No. 80. Int. J. Syst. Bacteriol. *51*: 793–794.

Jackson, E.B., J.X. Danauskas, M.C. Coale and J.E. Smadel. 1957. Recovery

of *Rickettsia akari* from the Korean vole *Microtus fortis pelliceus*. Am. J. Hyg. *66*: 301–308.

Jackson, F.L. and Y.E. Goodman. 1972. Transfer of the facultatively anaerobic organism *Bacteroides corrodens* Eiken to a new genus, *Eikenella*. Int. J. Syst. Bacteriol. *22*: 73–77.

Jackson, F.L. and Y.E. Goodman. 1978. *Bacteroides ureolyticus,* a new species to accommodate strains previously identified as *Bacteroides corrodens,* anaerobic. Int. J. Syst. Bacteriol. *28*: 197–200.

Jackson, F.L., Y.E. Goodman, F.R. Bel, P.C. Wong and R.L.S. Whitehouse. 1971. Taxonomic status of facultative and strictly anaerobic corroding bacilli that have been classified as *Bacteroides corrodens*. J. Med. Microbiol. *4*: 171–184.

Jackwood, M.W., D.A. Hilt and P.A. Dunn. 1991. Observations on colonial phenotypic variation in *Bordetella avium*. Avian Dis. *35*: 496–504.

Jackwood, M.W. and Y.M. Saif. 1987. Pili of *Bordetella avium*: expression, characterization, and role in in vitro adherence. Avian Dis. *31*: 277–286.

Jackwood, M.W., Y.M. Saif, P.D. Moorhead and R.N. Dearth. 1985. Further characterization of the agent causing coryza in turkeys. Avian Dis. *29*: 690–705.

Jacobi, C.A., B. Assmus, H. Reichenbach and E. Stackebrandt. 1997. Molecular evidence for association between the sphingobacterium-like organism "*Candidatus* comitans" and the myxobacterium *Chondromyces crocatus*. Appl. Environ. Microbiol. *63*: 719–723.

Jacobi, C.A., H. Reichenbach, B.J. Tindall and E. Stackebrandt. 1996. "*Candidatus* comitans" a bacterium living coculture with *Chondromyces crocatus* (myxobacteria). Int. J. Syst. Bacteriol. *46*: 119–122.

Jacobs, B.C., P.A. van Doorn, P.I. Schmitz, A.P. Tio Gillen, P. Herbrink, L.H. Visser, H. Hooijkass and F.G. van der Meche. 1996. *Campylobacter jejuni* infections and anti-GM1 antibodies in Guillain-Barre syndrome. Ann. Neurol. *40*: 181–187.

Jacobs, T.W., T.T. Egelhoff and S.R. Long. 1985. Physical and genetic map of a *Rhizobium meliloti* nodulation gene region and nucleotide sequence of nodC. J. Bacteriol. *162*: 469–476.

Jacobsen, G. 1975. Untersuchungen zur rolle der poly-β-hydroxybuttersäure-granula bei *Hyphomicrobium* spp, University of Kiel. Kiel, Germany . p. 70.

Jagoueix, S., J.M. Bové and M. Garnier. 1994. The phloem-limited bacterium of greening disease of citrus is a member of the alpha subdivision of the *Proteobacteria*. Int. J. Syst. Bacteriol. *44*: 379–386.

Jagoueix, S., J.M. Bové and M. Garnier. 1996. PCR detection of the two *Candidatus* Liberobacter species associated with greening disease of citrus. Mol. Cell. Probes *10*: 43–50.

Jahn, E. 1911. *Myxobacteriales*. Kryptogamenflora der Mark Brandenburg *5*: 187–206.

Jahn, E. 1924. Beitrage zur botanischen Protistologie I. Die Polyangiden, Verlag Gebruder Borntraeger, Leipzig. 107 pp. + 102 plates.

Jahng, D., C.S. Kim, R.S. Hanson and T.K. Wood. 1996. Optimization of trichloroethylene degradation using soluble methane monooxygenase of *Methylosinus trichosporium* OB3d expressed in recombinant bacteria. Biotechnol. Bioeng. *51*: 349–359.

Jahng, D. and T.K. Wood. 1994. Trichloroethylene and chloroform degradation by a recombinant pseudomonad expressing soluble methane monooxygenase from *Methylosinus trichosporium* OB3b. Appl. Environ. Microbiol. *60*: 2473–2482.

Jahnke, M., T. El-Banna, R. Klintworth and G. Auling. 1990. Mineralization of orthanilic acid is a plasmid-associated trait in *Alcaligenes* sp. O-1. J. Gen. Microbiol. *136*: 2241–2249.

Jahnke, M., F. Lehmann, A. Schoebel and G. Auling. 1993. Transposition of the TOL catabolic genes (Tn4651) into the degradative plasmid pSAH of *Alcaligenes* sp. O-1 ensures simultaneous mineralization of sulpho- and methyl-substituted aromatics. J. Gen. Microbiol. *139*: 1959–1966.

Jakob, W., M. Stolte, A. Valentin and H.D. Schröder. 1997. Demonstration of *Helicobacter pylori*-like organisms in the gastric mucosa of captive exotic carnivores. J. Comp. Pathol. *116*: 21–33.

Jalava, K., M.C. De Ungria, J. O'Rourke, A. Lee, U. Hirvi and M.L. Hänninen. 1999a. Characterization of *Helicobacter felis* by pulsed-field gel electrophoresis, plasmid profiling and ribotyping. Helicobacter *4*: 17–27.

Jalava, K., S. Hielm, U. Hirvi and M.L. Hänninen. 1999b. Evaluation of a molecular identification scheme based on 23S rRNA gene polymorphisms for differentiating canine and feline gastric *Helicobacter* spp. Lett. Appl. Microbiol. *28*: 269–274.

Jalava, K., M. Kaartinen, M. Utriainen, I. Happonen and M.L. Hänninen. 1997. *Helicobacter salomonis* sp. nov., a canine gastric *Helicobacter* sp. related to *Helicobacter felis* and *Helicobacter bizzozeronii*. Int. J. Syst. Bacteriol. *47*: 975–982.

Jalava, K., S.L.W. On, C.S. Harrington, L.P. Andersen, M.L. Hänninen and P.A.R. Vandamme. 1999c. *Helicobacter heilmannii*, a human gastric pathogen, and *H. bizzozeronii*, a frequent canine gastric coloniser, represent the same species. Abstracts of the 10th International Workshop on *Campylobacter, Helicobacter* and Related Organisms, Baltimore, MD. p. 116.

Jalava, K., S.L.W. On, C.S. Harrington, L.P. Anderson, M.L. Hänninen and P. Vandamme. 2001. A cultured strain of "*Helicobacter heilmannii*", a human gastric pathogen, identified as *Helicobacter bizzozeronii*: evidence for zoonotic potential of *Helicobacter*. Emerg. Inf. Dis. *7*: 1036–1038.

Jalava, K., S.L.W. On, P.A.R. Vandamme, I. Happonen, A. Sukura and M.L. Hänninen. 1998. Isolation and identification of *Helicobacter* spp. from canine and feline gastric mucosa. Appl. Environ. Microbiol. *64*: 3998–4006.

James, E.K. and F.L. Olivares. 1997. Infection and colonization of sugar cane and other graminaceous plants by endophytic diazotrophs. Crit. Rev. Plant Sci. *17*: 77–119.

James, J.F. and J. Swanson. 1978. Piliation of gonococci *in vivo*. J. Infect. Dis. *137*: 94–96.

Jamnongluk, W., P. Kittayapong, V. Baimai and S.L. O'Neill. 2002. *Wolbachia* infections of tephritid fruit flies: molecular evidence for five distinct strains in a single host species. Curr. Microbiol. *45*: 255–260.

Jang, S.S., J.M. Donahue, A.B. Arata, J. Goris, L.M. Hansen, D.L. Earley, P.A.R. Vandamme, P.J. Timoney and D.C. Hirsh. 2001. *Taylorella asinigenitalis* sp. nov., a bacterium isolated from the genital tract of male donkeys (*Equus asinus*). Int. J. Syst. Evol. Microbiol. *51*: 971–976.

Janik, A., E. Juni and G.A. Heym. 1976. Genetic transformation as a tool for detection of *Neisseria gonorrhoeae*. J. Clin. Microbiol. *4*: 71–81.

Janisiewicz, W.J. and J. Roitman. 1988. Biological control of blue mold and gray mold on apple and pear with *Pseudomonas cepacia*. Phytopathology *78*: 1697–1700.

Jankielewicz, A., R.A. Schmitz, O. Klimmek and A. Kröger. 1994. Polysulfide reductase and formate dehydrogenase from *Wolinella succinogenes* contain molybdopterin guanine dinucleotide. Arch. Microbiol. *162*: 238–242.

Jannasch, H.W. 1965. Die Isolierung heterotropher aquatischer Spirillen. *In* Schlegel (Editor), Anreicherungskultur und Mutantenauslese, Gustav Fischer Verlag, Stuttgart. 198–203.

Jannasch, H.W. 1984. Microbial processes at deep sea hydrothermal vents. *In* Rona, Bostrom, Laubier and Smith (Editors), Hydrothermal Processes at Seafloor Spreading Centers, Plenum Publishing, New York. pp. 667–709.

Jannasch, H.W. and C.O. Wirsen. 1981. Morphological survey of microbial mats near deep-sea thermal vents. Appl. Environ. Microibol. *41*: 528–538.

Janssen, D.B., F. Pries, J.V. Ploeg, B. Kazemier, P. Terpstra and B. Witholt. 1989. Cloning of 1,2-dichloroethane degradation genes of *Xanthobacter autotrophicus* Gj10 and expression and sequencing of the *Dhla* gene. J. Bacteriol. *171*: 6791–6799.

Janssen, D.B., A. Scheper, L. Dijkhuizen and B. Witholt. 1985. Degradation of halogenated aliphatic compounds by *Xanthobacter autotrophicus* Gj-10. Appl. Environ. Microbiol. *49*: 673–677.

Janssen, G.R., J.W. Wireman and M. Dworkin. 1977. Effect of temperature on the growth of *Myxococcus xanthus*. J. Bacteriol. *130*: 561–562.

Janssen, P.H. and C.G. Harfoot. 1987. Phototrophic growth on *n*-fatty acids by members of the family *Rhodospirillaceae*. Syst. Appl. Microbiol. *9*: 9–11.

Janssen, P.H. and C.G. Harfoot. 1991. *Rhodopseudomonas rosea* sp. nov., a new purple nonsulfur bacterium. Int. J. Syst. Bacteriol. *41*: 26–30.

Janssen, P.H. and B. Schink. 1995a. Metabolic pathways and energetics of the acetone-oxidizing, sulfate-reducing bacterium, *Desulfobacterium cetonicum*. Arch. Microbiol. *163*: 188–194.

Janssen, P.H. and B. Schink. 1995b. Pathway of butyrate catabolism by *Desulfobacterium cetonicum*. J. Bacteriol. *177*: 3870–3872.

Janssen, P.H., A. Schuhmann, F. Bak and W. Liesack. 1996. Disproportionation of inorganic sulfur compounds by the sulfate-reducing bacterium *Desulfocapsa thiozymogenes* gen. nov., sp. nov. Arch. Microbiol. *166*: 184–192.

Janssen, P.H., A. Schuhmann, F. Bak and W. Liesack. 1997. *In* Validation of the publication of new names and new combinations previously published outside the IJSB. List No. 61. Int. J. Syst. Bacteriol. *47*: 601–602.

Jansson, P.E., B. Lendberg and P.A. Sandford. 1983. Structural studies of gellan gum, an extracellular polysaccharide elaborated by *Pseudomonas elodea*. Carbohydr. Res. *124*: 135–139.

Jantzen, E., K. Bryn and K. Bøvre. 1976. Cellular monosaccharide patterns of *Neisseriaceae*. Acta Pathol. Microbiol. Scand. [B]. *84*: 177–188.

Jantzen, E., K. Bryn, T. Bergan and K. Bøvre. 1975. Gas chromatography of bacterial whole cell methanolysates. VII. Fatty acid composition of *Acinetobacter* in relation to the taxonomy of *Neisseriaceae*. Acta Pathol Microbiol Scand Suppl. *83*: 569-580.

Jantzen, E., K. Bryn, T. Bergan and K. Bøvre. 1974. Gas chromatography of bacterial whole cell methanolysates. V. Fatty acid composition of neisseriae and moraxellae. Acta Pathol. Microbiol. Scand. Sect. B Microbiol. *82*: 767–779.

Jaoua, S., J.F. Guespinmichel and A.M. Breton. 1987. Mode of insertion of the broad host range plasmid RP4 and Its derivatives into the chromosome of *Myxococcus xanthus*. Plasmid *18*: 111–119.

Jaoua, S., S. Neff and T. Schupp. 1992. Transfer of mobilizable plasmids to *Sorangium cellosum* and evidence for their integration into the chromosome. Plasmid *28*: 157–165.

Jarvis, B.D., H.L. Downer and J.P. Young. 1992. Phylogeny of fast-growing soybean-nodulating rhizobia support synonymy of *Sinorhizobium* and *Rhizobium* and assignment to *Rhizobium fredii*. Int. J. Syst. Bacteriol. *42*: 93–96.

Jarvis, B.D.W., M. Gillis and J. De Ley. 1986. Intra- and intergeneric similarities between the ribosomal ribonucleic acid cistrons of *Rhizobium* and *Bradyrhizobium* species and some related bacteria. Int. J. Syst. Bacteriol. *36*: 129–138.

Jarvis, B.D.W., C.E. Pankhurst and J.J. Patel. 1982. *Rhizobium loti*, a new species of legume root nodule bacteria. Int. J. Syst. Bacteriol. *32*: 378–380.

Jarvis, B.D.W., S.W. Sivakumaran, S.W. Tighe and M. Gillis. 1996. Identification of *Agrobacterium* and *Rhizobium* species based on cellular fatty acid composition. Plant Soil *184*: 143–158.

Jarvis, B.D.W., P. van Berkum, W.X. Chen, S.M. Nour, M.P. Fernandez, J.C. Cleyet-Marel and M. Gillis. 1997. Transfer of *Rhizobium loti, Rhizobium huakuii, Rhizobium ciceri, Rhizobium mediterraneum*, and *Rhizobium tianshanense* to *Mesorhizobium* gen. nov. Int. J. Syst. Bacteriol. *47*: 895–898.

Jarvis, G.A. and N.A. Vedros. 1987. Sialic acid of group B *Neisseria meningitidis* regulates alternative complement pathway activation. Infect. Immun. *55*: 174–180.

Jayasekara, N.Y., G.M. Heard, J.M. Cox and G.H. Fleet. 1999. Association of micro-organisms with the inner surfaces of bottles of non-carbonated mineral waters. Food Microbiol. *16*: 115–128.

Jayaswal, R.K., M.A. Fernandez and R.G. Schroeder. 1990. Isolation and characterization of a *Pseudomonas* strain that restricts growth of various phytopathogenic fungi. Appl. Environ. Microbiol. *56*: 1053–1058.

Jayaswal, R.K., M. Fernandez, R.S. Upadhyay, L. Visintin, M. Kurz, J. Webb and K. Rinehart. 1993. Antagonism of *Pseudomonas cepacia* against phytopathogenic fungi. Curr. Microbiol. *26*: 17–22.

Jayaswal, R.K., M.A. Fernandez, L. Visintin and R.S. Upadhyay. 1992. Transposon Tn5-259 mutagenesis of *Pseudomonas cepacia* to isolate mutants deficient in antifungal activity. Can. J. Microbiol. *38*: 309–312.

Jendrossek, D. 2001. Transfer of *[Pseudomonas] lemoignei*, a Gram-negative rod with restricted catabolic capacity, to *Paucimonas* gen. nov. with one species, *Paucimonas lemoignei* comb. nov. Int. J. Syst. Evol. Microbiol. *51*: 905–908.

Jendrossek, D., M. Backhaus and M. Andermann. 1995. Characterization of the extracellular poly(3-hydroxybutyrate) depolymerase of *Comamonas* sp. and of its structural gene. Can. J. Microbiol. *41*: 160–169.

Jendrossek, D., I. Knoke, R.B. Habibian, A. Steinbüchel and H.G. Schlegel. 1993. Degradation of poly(3-hydroxybutyrate), PHB, by bacteria and purification of a novel PHB depolymerse from *Comamonas* sp. J. Environ. Polym. Degr. *1*: 53–63.

Jenkins, C.L., A.G. A.G. Andrewes, T.J. T.J. McQuade and M.P. M.P. Starr. 1979. The pigment of *Pseudomonas paucimobilis* is a carotenoid (Nostoxanthin), rather than a brominated aryl-polyene (Xanthomonadin). Curr. Microbiol. *3*: 1–4.

Jenkins, C.L., D.A. Kuhn and K.R. Daly. 1977. Fatty acid composition of *Simonsiella* strains. Arch. Microbiol. *113*: 209–213.

Jenkins, C.L. and M.P. Starr. 1985. Formation of halogenated aryl-polyene (xanthomonadin) pigments by the type and other yellow-pigmented strains of *Xanthomonas maltophilia*. Ann. Inst. Pasteur. Microbiol. *136B*: 257–264.

Jenkins, D., M.G. Richard and G.T. Daigger. 1986. Manual on the causes and control of activated sludge bulking and foaming, Water Research Commission, Rep. of South Africa, and U.S: EPA, Ohio, U.S.A. 165 pp.

Jenkins, O., D. Byrom and D. Jones. 1987. *Methylophilus*: a new genus of methanol-utilizing bacteria. Int. J. Syst. Bacteriol. *37*: 446–448.

Jenni, B. and M. Aragno. 1987. *Xanthobacter agilis* sp. nov., a motile, dinitrogen-fixing, hydrogen-oxidizing bacterium. Syst. Appl. Microbiol. *9*: 254–257.

Jenni, B. and M. Aragno. 1988. *In* Validation of the publication of new names and new combinations previously effectively published outside the IJSB. List No. 24. Int. J. Syst. Bacteriol. *38*: 136–137.

Jenni, B., C. Isch and M. Aragno. 1989. Nitrogen fixation by new strains of *Pseudomonas pseudoflava* and related bacteria. J. Gen. Microbiol. *135*: 461–468.

Jennings, D., P. Volpe and J.J. Tudor. 1998. Characterization and complementation of temperature-sensitive mutants of *Bdellovibrio bacteriovorus* 109J. 98th Annual Meeting of the American Society for Microbiology, p. 321.

Jensen, H.L., E.J. Petersen, P.K. De and R. Bhattacharya. 1960. A new nitrogen-fixing bacterium: *Derxia gummosa* nov. gen. nov. spec. Arch. Mikrobiol. *36*: 182– 195.

Jensen, K.T., H. Schønheyder and V.F. Thomsen. 1994. *In vitro* activity of β-lactam and other antimicrobial agents against *Kingella kingae*. J. Antimicrob. Chemother. *33*: 635–640.

Jensen, N.S. and M.G. Allison. 1994. Studies on the diversity among amaerobic oxalate-degrading bacteria now in the species *Oxalobacter formigenes*. 94th Annual Meeting of the American Society for Microbiology, American Society for Microbiology. p. 255.

Jensen, P., A. Fomsgaard, N. Hoiby and P. Hindersson. 1995. Cloning and nucleotide sequence comparison of the *groE* operon of *Pseudomonas aeruginosa* and *Burkholderia cepacia*. APMIS *103*: 113–123.

Jensen, V. and E. Holm. 1975. Associative growth of nitrogen-fixing bacteria with other microorganisms. *In* Stewart (Editor), Nitrogen fixation by Free-Living Microorganisms, Cambridge University Press, Cambridge. pp. 101–119.

Jephcott, A.E., A. Reyn and A. Birch-Andersen. 1971. *Neisseria gonorrhoeae* 3. Demonstration of presumed appendages to cells from different colony types. Acta Pathol. Microbiol. Scand. [B] Microbiol. Immunol. *79*: 437–439.

Jessen, J. 1934. Studien über gramnegative Kokken. Zentralbl. Bakteriol. Parasitenkd. Infektionskr. Hyg. Abt. I. Orig. *133*: 73–88.

Jeyaprakash, A. and M.A. Hoy. 2000. Long PCR improves *Wolbachia* DNA amplification: *wsp* sequences found in 76% of sixty-three arthropod species. Insect Mol. Biol. *9*: 393–405.

Jeyaretnam, B., J. Glushka, V.S. Kumar Kolli and R.W. Carlson. 2002. Characterization of a novel lipid A from *Rhizobium* species Sin-1. A unique lipid A structure that is devoid of phosphate and has a glycosyl backbone consisting of glucosamine and 2-aminogluconic acid. J. Biol. Chem. *277*: 41802–41801.

Jiggins, F.M., J.H. von der Schulenburg, G.D. Hurst and M.E. Majerus. 2001. Recombination confounds interpretations of *Wolbachia* evolution. Proc. R. Soc. Lond. B Biol. Sci. *268*: 1423–1427.

Jímenez-Salgado, T., L.E. Fuentes-Ramírez, A. Tapia Hernández, M.A. Mascarúa Esparza, E. Martínez-Romero and J. Caballero Mellado. 1997. *Coffea arabica* L., a new host plant for *Acetobacter diazotrophicus* and isolation of other nitrogen-fixing acetobacteria. Appl. Environ. Microbiol. *63*: 3676–3683.

Johannson, B.C. and H. Gest. 1976. Inorganic nitrogen assimilation by the photosynthetic bacterium *Rhodopseudomonas capsulata*. J. Bacteriol. *128*: 683–688.

Johns, M.R., P.F. Greenfield and H.W. Doelle. 1992. Byproducts from *Zymomonas mobilis*. Adv. Biochem. Eng. Biotechnol. *44*: 97–121.

Johnson, D.B. 1998. Biodiversity and ecology of acidophilic microorganisms. FEMS Microbiol. Ecol. *27*: 307–317.

Johnson, D.A., U.H. Behling, C.H. Lain, M. Listgarten, S. Socransky and A. Nowotny. 1978. Role of bacterial products in periodontitis: immune response in gnotobiotic rats monoinfected with *Eikenella corrodens*. Infect. Immun. *19*: 246–253.

Johnson, D.B. and W.I. Kelso. 1983. Detection of heterotrophic contaminants in cultures of *Thiobacillus ferrooxidans* and their elimination by subculturing in media containing copper sulphate. J. Gen. Microbiol. *129*: 2969–2972.

Johnson, D.B. and S. McGinness. 1991. Ferric iron reduction by acidophilic heterotrophic bacteria. Appl. Environ. Microbiol. *57*: 207–211.

Johnson, E.A. and R.O. Jacoby. 1978. Transmissible ileal hyperplasia. II. Ultrastructure. Am. J. Pathol. *91*: 451–468.

Johnson, G.V., H.J. Evans and C. T.M.. 1966. Enzymes of the glyoxylate cycle in rhizobia and nodules of legumes. Plant Physiol. *41*: 1330–1336.

Johnson, J.L. and E.J. Ordal. 1969. Deoxyribonucleic acid homology among the fruiting myxobacteria. J. Bacteriol. *3*: 319–320.

Johnson, K.G., I.J. McDonald and M.B. Perry. 1976. Studies on the cellular and free lipopolysaccharides from *Branhamella catarrhalis*. Can. J. Microbiol. *22*: 460–467.

Johnson, R. and P.H.A. Sneath. 1973. Taxonomy of *Bordetella* and related organisms of the families *Achromobacteraceae*, *Brucellaceae* and *Neisseriaceae*. Int. J. Syst. Bacteriol. *22*: 381–404.

Johnson, S.M. and G.A. Pankey. 1976. *Eikenella corrodens* osteomyelitis, arthritis and cellulitis of the hand. South. Med. J. *69*: 535–539.

Joklik, W.K., H.P. Willet and D.B. Amos (Editors). 1980. Zinsser Microbiology, 17th Ed., Appleton-Century Crofts, New York.

Jones, C.W., J.M. Brice and C. Edwards. 1977. The effect of respiratory chain composition on the growth efficiencies of aerobic bacteria. Arch. Microbiol. *115*: 85–93.

Jones, D.M., A. Curry and A.J. Fox. 1985. An ultrastructural study of the gastric campylobacter-like organism *Campylobacter pyloridis*. J. Gen. Microbiol. *131*: 2335–2342.

Jones, D. and N.R. Krieg. 1984. Bacterial classification V. Serology and chemotaxonomy. *In* Krieg and Holt (Editors), Bergey's Manual of Systematic Bacteriology, 1st Ed., Vol. 1, The Williams & Wilkins Co., Baltimore. 15–18.

Jones, F.S., F. Orcutt and R.B. Little. 1931. Vibrios (*Vibrio jejuni* n. sp.) associated with intestinal disorders of cows and calves. J. Exp. Med. *53*: 853–864.

Jones, G.F., G.E. Ward, M.P. Murtaugh, G. Lin and C.J. Gebhart. 1993. Enhanced detection of intracellular organism of swine proliferative enteritis, ileal symbiont intracellularis, in feces by polymerase chain reaction. J. Clin. Microbiol. *31*: 2611–2615.

Jones, H.E. and P. Hirsch. 1968. Cell wall composition of *Hyphomicrobium* spp. J. Bacteriol. *96*: 1037–1041.

Jones, H.C. and J.M. Schmidt. 1973. Ultrastructural study of crossbands occurring in the stalks of *Caulobacter crescentus*. J. Bacteriol. *116*: 466–470.

Jones, J.L. and D.A. Romig. 1979. *Eikenella corrodens*: a pathogen in head and neck infections. Oral Surg. *47*: 501–505.

Jones, L.M. 1967. Report to the international committee on nomenclature of bacteria by the sub committee on taxonomy of Brucellae. Minutes of meeting, 22–23 July 1966. Int. J. Syst. Bacteriol. *17*: 371–375.

Jones, L.M. and W.J.B. Morgan. 1958. A preliminary report on a selective medium for the culture of *Brucella*, including fastidious types. Bull. W.H.O. *19*: 200–203.

Jones, L.M. and W. Wundt. 1971. International committee on nomenclature of bacteria subcommittee on the taxonomy of *Brucella*. Minutes of meeting 7 August 1970. Int. J. Syst. Bacteriol. *21*: 126–128.

Jones, M.V. and V.E. Wells. 1980. S-Adenosylmethionine biosynthesis in *Myxococcus xanthus*. FEBS Lett. *117*: 103–106.

Jones, R.D., R.Y. Morita, H.P. Koops and S.W. Watson. 1988. A new marine ammonium oxidizing bacterium, *Nitrosomonas cryotolerans* sp. nov. Can. J. Microbiol. *34*: 1122–1128.

Jones, W.J., J.P. Guyot and R.S. Wolfe. 1984. Methanogenesis from sucrose by defined immobilized consortia. Appl. Environ. Microbiol. *47*: 1–6.

Jönsson, A.B., G. Nyberg and S. Normark. 1991. Phase variation of gonococcal pili by frameshift mutation in *pilC*, a novel gene for pilus assembly. EMBO J. *10*: 477–488.

Jordan, D.C. 1982. Transfer of *Rhizobium japonicum* Buchanan 1980 to *Bradyrhizobium*, gen. nov. a slow-growing, root nodule bacterium from leguminous plants. Int. J. Syst. Bacteriol. *32*: 136–139.

Jordan, D.C. 1984a. Family III. *Rhizobiaceae*. *In* Krieg and Holt (Editors), Bergey's Manual of Systematic Bacteriology, 1st Ed., Vol. 1, The Williams & Wilkins Co., Baltimore.

Jordan, D.C. 1984b. Genus I. *Rhizobium*. *In* Krieg and Holt (Editors), Bergey's Manual of Systematic Bacteriology, 1st Ed. , Vol. 1, The Williams & Wilkins Co., Baltimore. pp. 235–242.

Jordan, D.C. 1984c. Genus II. *Bradyrhizobium*. *In* Krieg and Holt (Editors), Bergey's Manual of Systematic Bacteriology, 1st Ed., Vol. 1, The Williams & Wilkins Co., Baltimore.

Jordan, D.C. and O.N. Allen. 1974. Family III. *Rhizobiaceae*. *In* Buchanan and Gibbons (Editors), Bergey's Manual of Determinative Bacteriology, 8th Ed., The Williams & Wilkins Co., Baltimore. pp. 261–264.

Jordan, S.L., I.R. McDonald, A.J. Kraczkiewicz-Dowjat, D.P. Kelly, F.A. Rainey, J.C. Murrell and A.P. Wood. 1997. Autotrophic growth on carbon disulfide is a property of novel strains of *Paracoccus denitrificans*. Arch. Microbiol. *168*: 225–236.

Jørgensen, B.B. Fenchel, T. and . 1974. The sulfur cycle of a marine sediment model system. Mar. Biol. *14*: 189–201.

Jørgensen, B.B. and N.P. Revsbech. 1983. Colorless sulfur bacteria, *Beggiatoa* spp. and *Thiovulum* spp. in oxygen and hydrogen sulfide microgradients. Appl. Environ. Microbiol. *45*: 1261–1270.

Jorgensen, S., K.W. Skov and B. Diderichsen. 1991. Cloning, sequence, and expression of a lipase gene from *Pseudomonas cepacia* lipase production in heterologous hosts requires two *Pseudomonas* genes. J. Bacteriol. *173*: 559–567.

Jose, J., J. Kramer, T. Klauser, J. Pohlner and T.F. Meyer. 1996. Absence of periplasmic DsbA oxidoreductase facilitates export of cysteine-containing passenger proteins to the *Escherichia coli* cell surface via the Iga beta autotransporter pathway. Gene *178*: 107–110.

Joshi, B. and S. Walia. 1995. Characterization by arbitrary primer polymerase chain reaction of polychlorinated biphenyl (PCB)-degrading strains of *Comamonas testosteroni* isolated from PCB-contaminated soil. Can. J. Microbiol. *41*: 612–619.

Joshi, M., S. Mande and K.L. Dikshit. 1998. Hemoglobin biosynthesis in *Vitreoscilla stercoraria* DW: cloning, expression, and characterization of a new homolog of a bacterial globin gene. Appl. Environ. Microbiol. *64*: 2220–2228.

Jousimies-Somer, H. 1997. Recently described clinically important anaerobic bacteria: taxonomic aspects and update. Clin. Infect. Dis. *25* (Suppl. 2): S78–S87.

Joyce, G.H. and P.R. Dugan. 1970. The role of floc-forming bacteria in BOD removal from waste water. Develop. Ind. Microbiol. *11*: 377–386.

Jucker, W. and L. Ettlinger. 1981. Host range of a bacteriophage of acetic acid bacteria. Int. J. Syst. Bacteriol. *31*: 245–246.

Judd, R.C. and S.F. Porcella. 1993. Isolation of the periplasm of *Neisseria gonorrhoeae*. Mol. Microbiol. *10*: 567–574.

Judicial Commission. 1958a. Opinion 19. Conservation of the generic name *Rickettsia* da Rocha-Lima and of the species name *Rickettsia prowazekii* da Rocha-Lima. Int. Bull. Bacteriol. Nomencl. Taxon. *8*: 158–159.

Judicial Commission. 1958b. Rejection of the generic names *Nitromonas* Winogradsky 1890 and *Nitromonas* Orla-Jensen 1909, conservation of the generic names *Nitrosomonas* Winogradsky 1892, *Nitrosococcus* Winogradsky 1892, and *Nitrobacter* Winogradsky 1892, and the designation of the type species of these genera. Int. Bull. Bacteriol. Nomencl. Taxon. *8*: 169–170.

Judicial Commission. 1970. Opinion 33. Conservation of the generic name *Agrobacterium* Conn 1942. Int. J. Syst. Bacteriol. *20*: 10.

Judicial Commission. 1979. Minutes of the meeting, 3 September 1978, Munich, West Germany. Int. J. Syst. Bacteriol. *29*: 267–269.

Jukes, T.H. and R.R. Cantor. 1969. Evolution of protein molecules. *In* Munzo (Editor), Mammalian Protein Metabolism, Academic Press, New York. pp. 21–132.

Jumas-Bilak, E., C. Maugard, S. Michaux-Charachon, A. Allardet-Servent, A. Perrin, D. O'Callaghan and M. Ramuz. 1995. Study of the organization of the genomes of *Escherichia coli*, *Brucella melitensis* and *Agrobacterium tumefaciens* by insertion of a unique restriction site. Microbiology *141*: 2425–2432.

Jumas-Bilak, E., S. Michaux-Charachon, G. Bourg, D. O'Callaghan and M. Ramuz. 1998. Differences in chromosome number and genome rearrangements in the genus *Brucella*. Mol. Microbiol. *27*: 99–106.

Jungermann, K. and G. Schön. 1974. Pyruvate formate lyase in *Rhodospirillum rubrum* Ha adapted to anaerobic dark conditions. Arch. Microbiol. *99*: 109–116.

Juni, E. 1972. Interspecies transformation of *Acinetobacter*: genetic evidence for a ubiquitous genus. J. Bacteriol. *112*: 917–931.

Juni, E. 1974. Simple genetic transformation assay for rapid diagnosis of *Moraxella osloensis*. Appl. Microbiol. *27*: 16–24.

Juni, E. 1990. Application of genetic transformation in identification of Gram-negative bacteria. *In* Olsvik and Bukholm (Editors), Application of Molecular Biology in Diagnosis of Infectious Diseases, Norwegian College of Veterinary Medicine, Oslo. pp. 61–68.

Juni, E. and G.A. Heym. 1977. Simple method for distinguishing gonococcal colony types. J. Clin. Microbiol. *6*: 511–517.

Juni, E. and G.A. Heym. 1980. Transformation assay for identification of psychrotrophic achromobacters. Appl. Environ. Microbiol. *40*: 1106–1114.

Juni, E. and A. Janik. 1969. Transformation of *Acinetobacter calco-aceticus* (*Bacterium anitratum*). J. Bacteriol. *98*: 281–288.

Junker, F. and A.M. Cook. 1997. Conjugative plasmids and the degradation of arylsulfonates in *Comamonas testosteroni*. Appl. Environ. Microbiol. *63*: 2403–2410.

Junker, F., R. Kiewitz and A.M. Cook. 1997. Characterization of the *p*-toluenesulfonate operon *tsaMBCD* and *tsaR* in *Comamonas testosteroni* T-2. J. Bacteriol. *179*: 919–927.

Junker, F., E. Saller, H.R. Schlafi Oppenberg, P.M.H. Kroneck, T. Leisinger and A.M. Cook. 1996. Degradative pathways for *p*-toluenecarboxylate and *p*-toluenesulfonate and their multicomponent oxygenases in *Comamonas testosteroni* strains PSB-4 and T-2. Microbiology *14*: 2419–2427.

Jurtshuk, P. and T.W. Milligan. 1974. Quantitation of the tetramethyl-*p*-phenylenediamine oxidase reaction in *Neisseria* species. Appl. Microbiol. *28*: 1079–1081.

Justin, P. and D.P. Kelly. 1978. Growth kinetics of *Thiobacillus denitrificans* in anaerobic and aerobic chemostat culture. J. Gen. Bacteriol. *107*: 123–130.

Juteau, P., R. Larocque, D. Rho and A. LeDuy. 1999. Analysis of the relative abundance of different types of bacteria capable of toluene degradation in a compost biofilter. Appl. Microbiol. Biotechnol. *52*: 863–868.

Jüttner, R.R., R.M. Lafferty and H.J. Knackmuss. 1975. A simple method for the determination of poly-β-hydroxybutyric acid in microbial biomass. Eur. J. Appl. Microbiol. *1*: 233–237.

Jyssum, K. 1959. Assimilation of nitrogen in meningococci grown with the ammonium ion as sole nitrogen source. Acta Pathol. Microbiol. Scand. *46*: 320–332.

Jyssum, K. 1960. Intermediate reactions of the tricarboxylic acid cycle in meningococci. Acta Pathol. Microbiol. Scand. *48*: 121–132.

Jyssum, K. and S. Jyssum. 1962. Phosphoenolpyruvic carboxylase activity in extracts from *Neisseria meningitidis* isolated from patients in Norway. Acta Pathol. Microbiol. Scand. *74*: 93–100.

Jyssum, K. and S. Jyssum. 1968. Isolation of variants with increased mutability from *Neisseria meningitidis*. Acta Pathol. Microbiol. Scand. *74*: 93–100.

Jyssum, K. and S. Lie. 1965. Genetic factors determining competence in transformation of *Neisseria meningitidis* I. A permanent loss of competence. Acta Pathol. Microbiol. Scand. [B] Microbiol. Immunol. *63*: 306–316.

Jyssum, S. and K. Bøvre. 1974. Search for thymidine phosphorylase, nucleoside deoxyribosyltransferase and thymidine kinase in *Moraxella*, *Acinetobacter*, and allied bacteria. Acta Pathol. Microbiol. Scand. *82B*: 57–66.

Ka, J.O., W.E. Holben and J.M. Tiedje. 1994a. Analysis of competition in soil among 2,4-dichlorophenoxyacetic acid degrading bacteria. Appl. Environ. Microbiol. *60*: 1121–1128.

Ka, J.O., W.E. Holben and J.M. Tiedje. 1994b. Genetic and phenotypic diversity of 2,4-dichlorophenoxyacetic acid (2,4-D)-degrading bacteria isolated from 2,4-D-treated field soils. Appl. Environ. Microbiol. *60*: 1106–1115.

Ka, J.O., W.E. Holben and J.M. Tiedje. 1994c. Use of gene probes to aid in recovery and identification of functionally dominant 2,4-dichlorophenoxyacetic acid-degrading populations in soil. Appl. Environ. Microbiol. *60*: 1116–1120.

Ka, J.O. and J.M. Tiedje. 1994. Integration and excision of a 2,4-dichlorophenoxyacetic acid-degradative plasmid in *Alcaligenes paradoxus* and evidence of its natural intergeneric transfer. J. Bacteriol. *176*: 5284–5289.

Kabir, M.M., D. Faure, J. Haurat, P. Normand, C. Jacoud, P. Wadoux and R. Bally. 1995. Oligonucleotide probes based on 16S rRNA sequences for the identification of four *Azospirillum* species. Can. J. Microbiol. *41*: 1081–1087.

Kado, C.I. 1991. Molecular mechanisms of crown gall tumorigenesis. Crit. Rev. Plant Sci. *10*: 1–32.

Kado, C.I. and M.G. Heskett. 1970. Selective media for isolation of *Agrobacterium*, *Corynebacterium*, *Erwinia*, *Pseudomonas*, and *Xanthomonas*. Phytopathology 60: 969–976.

Kadota, I., A. Mizuno and K. Nishiyama. 1996. Detection of a protein specific to the strain of *Pseudomonas avenae* Manns 1909 pathogenic to rice. Ann. Phytopathol. Soc. Jpn. *62*: 425–428.

Kafkewitz, D. 1975. Improved growth media for *Vibrio succinogenes*. Appl. Microbiol. *29*: 121–122.

Kafkewitz, D. and D. Goodman. 1974. L-Asparaginase production by the rumen anaerobe *Vibrio succinogenes*. Appl. Microbiol. *27*: 206–209.

Kahan, A., A. Philippon, G. Paul, S. Weber, C. Richard, G. Hazebroucq and M. Degeorges. 1983. Nosocomial infections by chlorhexidine solution contaminated with *Pseudomonas pickettii* (Biovar VA-I). J. Infect. 7: 256–263.

Kahl, A. 1930. Die Tierwelt Deutschlands und der angrenzenden Meeresteile. *In* Teil (Editor), Urtiere oder Protozoa I: Wimpertiere oder Ciliata (Infusoria), Gustav Fischer Verlag, Jena.

Kaihovaara, P., K.S. Salmela, R.P. Roine, T.U. Kosunen and M. Salaspuro. 1994. Purification and characterization of *Helicobacter pylori* alcohol dehydrogenase. Alcohol. Clin. Exp. Res. *18*: 1220–1225.

Kaijser, B. 1975. Immunological studies of an antigen common to many gram-negative bacteria with special reference to *E. coli*. Characteri-

zation and biological significance. Int. Arch. Allergy Appl. Immunol. *48*: 72–81.

Kaiser, A., H.G. Classen, J. Eberspächer and F. Lingens. 1981. Acute toxicity testing of some herbicides, alkaloids, and antibiotic-metabolizing soil bacteria in the rat. Zentralbl. Bakteriol. Mikrobiol. Hyg. B *173*: 173–179.

Kaiser, D. 1979. Social gliding Is correlated with the presence of pili in *Myxococcus xanthus*. Proc. Natl. Acad. Sci. U.S.A. *76*: 5952–5956.

Kaiser, D. 1991. Genetic systems in myxobacteria. Method Enzymol. *204*: 357–372.

Kaiser, D. 1998. How and why myxobacteria talk to each other. Curr. Opin. Microbiol. *1*: 663–668.

Kaiser, D. 2000. Cell-interactive sensing of the environment. *In* Brun and Shimkets (Editors), Prokaryotic Development, American Society for Microbiology, Washington, D.C. pp. 263–275.

Kaiser, D. and M. Dworkin. 1975. Gene transfer to a myxobacterium by *Escherichia coli* phage P1. Science *187*: 653–654.

Kaiser, D. and L. Kroos. 1993. Intercellular signalling. *In* Dworkin and Kaiser (Editors), Myxobacteria II, American Society for Microbiology, Washington, D.C. pp. 257–283.

Kaiser, G.E. and M.J. Starzyk. 1973. Ultrastructure and cell division of an oral bacterium resembling *Alysiella filiformis*. Can. J. Microbiol. *19*: 325–327.

Kakii, K., H. Yamaguchi, Y. Iguchi, M. Teshima, T. Shirakashi and M. Kuriyama. 1986. Isolation and growth characteristics of nitrilotriacetate-degrading bacteria. J. Ferment. Technol. *64*: 103–108.

Kalchbrenner, C. and M.C. Cooke. 1880. South African fungi. Grevillea *9*: 2.

Kaldor, J. and B.R. Speed. 1984. Guillain-Barré syndrome and *Campylobacter jejuni*: a serological study. Br. Med. J. Clin. Res. *288*: 1867–1870.

Kalmbach, S., W. Manz, B. Bendinger and U. Szewzyk. 2000. *In situ* probing reveals *Aquabacterium commune* as a widespread and highly abundant bacterial species in drinking water biofilms. Water Res. *34*: 575–581.

Kalmbach, S., W. Manz and U. Szewzyk. 1997. Isolation of new bacterial species from drinking water biofilms and proof of their in situ dominance with highly specific 16S rRNA probes. Appl. Environ. Microbiol. *63*: 4164–4170.

Kalmbach, S., W. Manz, J. Wecke and U. Szewzyk. 1999. *Aquabacterium* gen. nov., with description of *Aquabacterium citratiphilum* sp. nov., *Aquabacterium parvum* sp. nov. and *Aquabacterium commune* sp. nov., three in situ dominant bacterial species from the Berlin drinking water system. Int. J. Syst. Bacteriol. *49*: 769–777.

Kalnenieks, U., A.A. de Graaf, S. Bringer-Meyer and H. Sahm. 1993. Oxidative phosphorylation in *Zymomonas mobilis*. Arch. Microbiol. *160*: 74–79.

Kaluza, K., M. Hahn and H. Hennecke. 1985. Repeated sequences similar to insertion elements clustered around the *nif* region of the *Rhizobium japonicum* genome. J. Bacteriol. *162*: 535–542.

Kaluza, K. and H. Hennecke. 1984. Fine structure analysis of the *nifDK* operon encoding the α subunits and β subunits of dinitrogenase from *Rhizobium japonicum*. Mol. Gen. Genet. *196*: 35–42.

Kamagata, Y., R.R. Fulthorpe, K. Tamura, H. Takami, L.J. Forney and J.M. Tiedje. 1997. Pristine environments harbor a new group of oligotrophic 2,4-dichlorophenoxyacetic acid-degrading bacteria. Appl. Environ. Microbiol. *63*: 2266–2272.

Kämpfer, P. 1995. Physiological and chemotaxonomic characterization of filamentous bacteria belonging to the genus *Haliscomenobacter*. Syst. Appl. Microbiol. *18*: 363–367.

Kämpfer, P. 1997. Detection and cultivation of filamentous bacteria from activated sludge. FEMS Microbiol. Ecol. *23*: 169–181.

Kämpfer, P. 1998. Some chemotaxonomic and physiological properties of the genus *Sphaerotilus*. Syst. Appl. Microbiol. *21*: 156–162.

Kämpfer, P., E.B. Denner, S. Meyer, E.R. Moore and H.J. Busse. 1997. Classification of "*Pseudomonas azotocolligans*" Anderson 1955, 132, in the genus *Sphingomonas* as *Sphingomonas trueperi* sp. nov. Int. J. Syst. Bacteriol. *47*: 577–583.

Kämpfer, P., R. Erhart, C. Beimfohr, J. Böhringer, M. Wagner and R.

Amann. 1996. Characterization of bacterial communities from activated sludge: culture-dependent numerical identification versus in situ identification using group- and genus-specific rRNA-targeted oligonucleotide probes. Microb. Ecol. *32*: 101–121.

Kämpfer, P., C. Müller, M. Mau, A. Neef, G. Auling, H.J. Busse, A.M. Osborn and A. Stolz. 1999. Description of *Pseudaminobacter* gen. nov. with two new species, *Pseudaminobacter salicylatoxidans* sp. nov. and *Pseudaminobacter defluvii* sp. nov. Int. J. Syst. Bacteriol. *49*: 887–897.

Kämpfer, P., A. Neef, M.S. Salkinoja-Salonen and H.J. Busse. 2002. *Chelatobacter heintzii* (Auling et al. 1993) is a later subjective synonym of *Aminobacter aminovorans* (Urakami et al. 1992). Int. J. Syst. Evol. Microbiol. *52*: 835–839.

Kämpfer, P., M. Steiof, P.M.L. Becker and W. Dott. 1993. Characterization of chemoheterotrophic bacteria associated with the in situ bioremediation of a waste-oil contaminated site. Microb. Ecol. *26*: 161–188.

Kämpfer, P., D. Weltin, D. Hoffmeister and W. Dott. 1995. Growth requirements of filamentous bacteria isolated from bulking and scumming sludge. Water Res. *29*: 1585–1588.

Kanabrocki, J.A., J. Lalucat, B.J. Cox and R.L. Quackenbush. 1986. Comparative study of refractile (R) bodies and their genetic determinants: relationship of type 51 R bodies to R bodies produced by *Pseudomonas taeniospiralis*. J. Bacteriol. *168*: 1019–1022.

Kanamaru, K., T. Hieda, Y. Iwamuro, Y. Mikami, Y. Obi and T. Kisaki. 1982a. Isolation and characterization of a *Hyphomicrobium* species and its polysaccharide formation from methanol. Agric. Biol. Chem. *46*: 2411–2417.

Kanamaru, K., Y. Iwamuro, Y. Mikami, Y. Obi and T. Kisaki. 1982b. 2-O-methyl-D-mannose in an extracellular polysaccharide from *Hyphomicrobium* sp. Agric. Biol. Chem. *46*: 2419–2424.

Kanazawa, S. and K. Mori. 1996. Isolation of cadmium-resistant bacteria and their resistance mechanisms. I. Isolation of Cd-resistant bacteria from soils contaminated with heavy metals. Soil Sci. Plant Nutr. *42*: 725–730.

Kandler, O., H. Koenig, J. Wiegel and D. Claus. 1983. Occurrence of poly-γ-D-glutamic acid and poly-α-L-glutamine in the genera *Xanthobacter, Flexithrix, Sporosarcina* and *Planococcus*. Syst. Appl. Microbiol. *4*: 34–41.

Kaneko, T., Y. Nakamura, S. Sato, K. Minamisawa, T. Uchiumi, S. Sasamoto, A. Watanabe, K. Idesawa, M. Iriguchi, K. Kawashima, M. Kohara, M. Matsumoto, S. Shimpo, H. Tsuruoka, T. Wada, M. Yamada and S. Tabata. 2002. Complete genomic sequence of nitrogen-fixing symbiotic bacterium *Bradyrhizobium japonicum* USDA110. DNA Res. *9*: 189–197.

Kanemaru, T., M. Kamada, R. Wada, T. Anzai, T. Kumanomido, H. Yoshikawa and T. Yoshikawa. 1992. Electron microscopic observation of *Taylorella equigenitalis* with pili in vivo. J. Vet. Med. Sci. *54*: 345–347.

Kang, H.Y., T.J. Brickman, F.C. Beaumont and S.K. Armstrong. 1996. Identification and characterization of iron-regulated *Bordetella pertussis* alcaligin siderophore biosynthesis genes. J. Bacteriol. *178*: 4877–4884.

Kang, H.-L. and H.-S. Kang. 1998. A physical map of the genome of ethanol fermentative bacterium *Zymomonas mobilis* ZM4 and localization of genes on the map. Gene *206*: 223–228.

Kang, K.S. and G.T. Veeder. 1982a. Heteropolysaccharide S-130. United States Patent 4342866,

Kang, K.S. and G.T. Veeder. 1982b. Polysaccharide S-60 and bacterial fermentation process for its preparation. United States Patent No. 4326053,

Kang, K.S. and G.T. Veeder. 1985. Heteropolysaccharide S-88. United States Patent 4535153,

Kang, K.S., G.T. Veeder, P.J. Mirrasoul, T. Kaneto and I.W. Cottrell. 1982. Agar-like polysaccharide produced by a *Pseudomonas* sp.: production and base properties. Appl. Environ. Microbiol. *43*: 1086–1091.

Kanso, S. and B.K.C. Patel. 2003. *Microvirga subterranea* gen. nov., sp nov., a moderate thermophile from a deep subsurface Australian thermal aquifer. Int. J. Syst. Evol. Microbiol. *53*: 401–406.

Kanter, M., J. Mott, N. Ohashi, B. Fried, S. .Reed and Y.C. Lin. 2000. Analysis of 16S rRNA and 51-kilodalton antigen gene and transmission

in mice of *Ehrlichia risticii* in virgulate trematodes from *Elimia livescens* snails in Ohio. J. Clin. Microbiol. *38*: 3349–3358.

Kaplan, J.M., G.H. McCracken and J.D. Nelson. 1973. Infections in children caused by the HB group of bacteria. J. Pediatr. *82*: 398–403.

Kaplan, R.L., D.B. Yelton and V.F. Gerencser. 1976. Biochemical and biophysical properties of *Hyphomicrobium* bacteriophage Hyø30. J. Virol. *19*: 899–902.

Kappler, U., B. Bennett, J. Rethmeier, G. Schwarz, R. Deutzmann, A.G. McEwan and C. Dahl. 2000. Sulfite:cytochrome *c* oxidoreductase from *Thiobacillus novellus*: purification, characterization, and molecular biology of a heterodimeric member of the sulfite oxidase family. J. Biol. Chem. *275*: 13202–13212.

Kappler, U., C.G. Friedrich, H.G. Trüper and C. Dahl. 2001. Evidence for two pathways of thiosulfate oxidation in *Starkeya novella* (formerly *Thiobacillus novellus*). Arch. Microbiol. *175*: 102–111.

Karlson, U., F. Rojo, J.D. Van Elsas and E. Moore. 1995. Genetic and serological evidence for the recognition of four pentachlorophenol-degrading bacterial strains as a species of the genus *Sphingomonas*. Syst. Appl. Microbiol. *18*: 539–548.

Karmali, M.A., A.K. Allen and P.C. Fleming. 1981a. Differentiation of catalase-positive campylobacters with special reference to morphology. Int. J. Syst. Bacteriol. *31*: 64–71.

Karmali, M.A., S. DeGrandis and P.C. Fleming. 1981b. Antimicrobial susceptibility of *Campylobacter jejuni* with special reference to resistance patterns of Canadian isolates. Antimicrob. Agents Chemother. *19*: 593–597.

Karpati, F. and J. Jonasson. 1996. Polymerase chain reaction for the detection of *Pseudomonas aeruginosa, Stenotrophomonas maltophilia* and *Burkholderia cepacia* in sputum of patients with cystic fibrosis. Mol. Cell. Probes *10*: 397–403.

Kashima, Y., Y. Nakajima, T. Nakano, K. Tayama, Y. Koizumi, S. Udaka and F. Yanagida. 1999. Cloning and characterization of ethanol-regulated esterase genes in *Acetobacter pasteurianus*. J. Biosci. Bioeng. *87*: 19–27.

Kaspari, H. 1979. Reductive pyrimidine catabolism in *Rhodopseudomonas capsulata*. Microbiologica. *2*: 231–241.

Kasprzak, A.A., E.J. Papas and D.J. Steenkamp. 1983. Identity of the subunits and the stoichiometry of prosthetic groups in trimethylamine dehydrogenase and dimethylamine dehydrogenase. Biochem. J. *211*: 535–542.

Kasprzak, A.A. and D.J. Steenkamp. 1983. Localization of the major dehydrogenases in 2 methylotrophs by radiochemical labeling. J. Bacteriol. *156*: 348–353.

Kasprzak, A.A. and D.J. Steenkamp. 1984. Differential labeling with imidoesters of cytoplasmic and extra cytoplasmic proteins in two methylotrophs. *In* Crawford and Hanson (Editors), Microbial Growth on C₁ Compounds: Proceedings of the 4th International Symposium, American Society for Microbiology, Washington, DC. 147–154.

Kasten, M.J., J.E. Rosenblatt and D.R. Gustafson. 1992. *Bilophila wadsworthia* bacteremia in two patients with hepatic abscesses. J. Clin. Microbiol. *30*: 2502–2503.

Katada, T. and M. Ui. 1982. ADP ribosylation of the specific membrane-protein of C6 cells by islet-activating protein associated with modification of adenylate-cyclase activity. J. Biol. Chem. *257*: 7210–7216.

Katashima, R. 1965. Mate-killing in *Euplotes patella*, syngen 1. Annot. Zool. Jpn. *38*: 207–215.

Katayama, Y., A. Hiraishi and H. Kuraishi. 1995. *Paracoccus thiocyanatus* sp. nov., a new species of thiocyanate-utilizing facultative chemolithotroph, and transfer of *Thiobacillus versutus* to the genus *Paracoccus* as *Paracoccus versutus* comb. nov. with emendation of the genus. Microbiology *141*: 1469–1477.

Katayama, Y., A. Hiraishi and H. Kuraishi. 1996. *In* Validation of the publication of new names and new combinations previously effectively published outside the IJSB. List. No. 57. Int. J. Syst. Bacteriol. *46*: 625–626.

Katayama-Fujimura, Y., Y. Enokizono, T. Kaneko and H. Kuraishi. 1983. DNA homologies among species of the genus *Thiobacillus*. J. Gen. Appl. Microbiol. *29*: 287–296.

Katayama-Fujimura, Y., I. Kawashima, N. Tsuzaki and H. Kuraishi. 1984a. Physiological characteristics of the facultatively chemolithotrophic *Thiobacillus* species *Thiobacillus delicatus* nov. rev., emended, *Thiobacillus perometabolis*, and *Thiobacillus intermedius*. Int. J. Syst. Bacteriol. *34*: 139–144.

Katayama-Fujimura, Y. and H. Kuraishi. 1983. Emendation of *Thiobacillus perometablis* London and Rittenberg 1967. Int. J. Syst. Bacteriol. *33*: 650–651.

Katayama-Fujimura, Y., N. Tsuzaki, A. Hirata and H. Kuraishi. 1984b. Polyhedral inclusion bodies (carboxysomes) in *Thiobacillus* species with reference to the taxonomy of the genus *Thiobacillus*. J. Gen. Appl. Microbiol. *30*: 211–222.

Katayama-Fujimura, Y., N. Tsuzaki and H. Kuraishi. 1982. Ubiquinone, fatty acid and DNA base composition determination as a guide to the taxonomy of the genus *Thiobacillus*. J. Gen. Microbiol. *128*: 1599–1612.

Katoh, T. 1963. Nitrate reductase in the photosynthetic bacterium *Rhodospirillum rubrum*. Purification and properties of nitrate reductase in nitrate-adapted cells. Plant Cell Physiol. *4*: 13–28.

Katsura, K., H. Kawasaki, W. Potacharoen, S. Saono, T. Seki, Y. Yamada, T. Uchimura and K. Komagata. 2001. *Asaia siamensis* sp. nov., an acetic acid bacterium in the alpha-*Proteobacteria*. Int. J. Syst. Evol. Microbiol. *51*: 559–563.

Katsura, K., Y. Yamada, T. Uchimura and K. Komagata. 2002. *Gluconobacter asaii* Mason and Claus 1989 is a junior subjective synonym of *Gluconobacter cerinus* Yamada and Akita 1984. Int. J. Syst. Evol. Microbiol. *52*: 1635–1640.

Kattar, M.M., J.F. Chavez, A.P. Limaye, S.L. Rassoulian-Barrett, S.L. Yarfitz, L.C. Carlson, Y. Houze, S. Swanzy, B.L. Wood and B.T. Cookson. 2000. Application of 16S rRNA gene sequencing to identify *Bordetella hinzii* as the causative agent of fatal septicemia. J. Clin. Microbiol. *38*: 789–794.

Katznelson, H. and A.C. Zagallo. 1957. Metabolism of rhizobia in relation to effectiveness. Can. J. Microbiol. *3*: 879–884.

Katzy, E., L. Petrova, I. Borisov and V. Panasenko. 1995. Genetical aspects of indole acetate production in *Azospirillum brasilense* Sp245. NATO ASI Ser. Ser. G Ecol. Sci. *37*: 113–119.

Kauffmann, J. and P. Toussaint. 1951. Un nouveau germe fixateur de l'azote atmosphérique: *Azotobacter lacticogenes*. C.R. Acad. Sci. (Paris) *223*: 710–711.

Kawagoshi, Y. and M. Fujita. 1997. Purification and properties of polyvinyl alcohol oxidase with broad substrate range obtained from *Pseudomonas vesicularis* var. povalolyticus PH. World J. Microbiol. Biotechnol. *13*: 273–277.

Kawagoshi, Y., Y. Mitihiro, M. Fujita and S. Hashimoto. 1997. Production and recovery of an enzyme from *Pseudomonas vesicularis* var. povalolyticus PH that degrades polyvinyl alcohol. World J. Microbiol. Biotechnol. *13*: 63–67.

Kawahara, K., S. Dejsirilert, H. Danbara and T. Ezaki. 1992. Extraction and characterization of lipopolysaccharide from *Pseudomonas pseudomallei*. FEMS Microbiol. Lett. *75*: 129–133.

Kawahara, K., H. Kuraishi and U. Zahringer. 1999. Chemical structure and function of glycosphingolipids of *Sphingomonas* spp. and their distribution among members of the alpha-4 subclass of *Proteobacteria*. J. Ind. Microbiol. Biotechnol. *23*: 408–413.

Kawahara, K., M. Matsuura and H. Danbara. 1990. Chemical structure and biological activity of lipooligosaccharide isolated from *Sphingomonas paucimobilis*, a gram-negative bacterium lacking usual lipopolysaccharide. Jpn. J. Med. Sci. Biol. *43*: 250.

Kawahara, K., H. Moll, Y.A. Knirel, U. Seydel and U. Zahringer. 2000. Structural analysis of two glycosphingolipids from the lipopolysaccharide-lacking bacterium *Sphingomonas capsulata*. Eur. J. Biochem. *267*: 1837–1846.

Kawahara, K., U. Seydel, M. Matsuura, H. Danbara, E.T. Rietschel and U. Zahringer. 1991. Chemical structure of glycosphingolipids isolated from *Sphingomonas paucimobilis*. FEBS Lett. *292*: 107–110.

Kawai, F. 1995. Bacterial degradation of glycol ethers. Appl. Microbiol. Biotechnol. *44*: 532–538.

Kawai, M. and M. Fujishima. 2000. Invasion of the macronucleus of *Paramecium caudatum* by the bacterium *Holospora obtusa*: Fates of the bacteria and timings of invasion steps. Eur. J. Protistol. *36*: 46–52.

Kawai, Y., I. Yano, K. Kaneda and E. Yabuuchi. 1988. Ornithine-containing lipids of some *Pseudomonas* species. Eur. J. Biochem. *175*: 633–641.

Kawamoto, S.O. and J.W. Lorbeer. 1976. Protection of onion seedlings from *Fusarium oxysporum* f.sp. cepae by seed and soil infestation with *Pseudomonas cepacia*. Plant Dis. Rep. *60*: 189–191.

Kawasaki, H., Y. Hoshino, A. Hirata and K. Yamasato. 1993a. Is intracytoplasmic membrane structure a generic criterion? It does not coincide with phylogenetic interrelationships among phototrophic purple nonsulfur bacteria. Arch. Microbiol. *160*: 358–362.

Kawasaki, H., Y. Hoshino, A. Hirata and K. Yamasato. 1994a. *In* Validation of the publication of new names and new combinations previously effectively published outside the IJSB. List No. 51. Int. J. Syst. Bacteriol. *44*: 852.

Kawasaki, H., Y. Hoshino, H. Kuraishi and K. Yamasato. 1992. *Rhodocista centenaria* gen. nov., sp. nov., a cyst-forming anoxygenic photosynthetic bacterium and its phylogenetic postion in the *Proteobacteria* alpha group. J. Gen. Appl. Microbiol. *38*: 541–551.

Kawasaki, H., Y. Hoshino, Y. Kuraishi and K. Yamasato. 1994b. *In* Validation of the publication of new names and new combinations previously effectively published outside the IJSB. List No. 48. Int. J. Syst. Bacteriol. *44*: 182–183.

Kawasaki, H., Y. Hoshino and K. Yamasato. 1993b. Phylogenetic diversity of phototrophic purple non-sulfur bacteria in the α*Proteobacteria* group. FEMS Microbiol. Lett. *112*: 61–66.

Kawasaki, H., K. Yamasato and J. Sugiyama. 1997. Phylogenetic relationships of the helical-shaped bacteria in the α *Proteobacteria* inferred from 16S rDNA sequences. J. Gen. Appl. Microbiol. *43*: 89–95.

Kawasaki, S., R. Moriguchi, K. Sekiya, T. Nakai, E. Ono, K. Kume and K. Kawahara. 1994c. The cell envelope structure of the lipopolysaccharide-lacking gram-negative bacterium *Sphingomonas paucimobilis*. J. Bacteriol. *176*: 284–290.

Kawasumi, T., Y. Igarashi, T. Kodama and Y. Minoda. 1984. *Hydrogenobacter thermophilus* gen. nov. sp. nov., an extremely thermophilic, aerobic, hydrogen-oxidizing bacterium. Int. J. Syst. Bacteriol. *34*: 5–10.

Kawula, T.H., S.M. Spinola, D.G. Klapper and J.G. Cannon. 1987. Localization of a conserved epitope and an azurin-like domain in the H.8 protein of pathogenic *Neisseria*. Mol. Microbiol. *1*: 179–185.

Kazunga, C. and M.D. Aitken. 2000. Products from the incomplete metabolism of pyrene by polycyclic aromatic hydrocarbon-degrading bacteria. Appl. Environ. Microbiol. *66*: 1917–1922.

Keane, P.J., A. Kerr and P.B. New. 1970. Crown gall of stone fruit. II. Identification and nomenclature of *Agrobacterium* isolates. Aust. J. Biol. Sci. *23*: 585–595.

Kee, S.H., K.A. Cho, M.K. Kim, B.U. Lim, W.H. Chang and J.S. Kang. 1999. Disassembly of focal adhesions during apoptosis of endothelial cell line ECV304 infected with *Orientia tsutsugamushi*. Microb. Pathog. *27*: 265–271.

Keeble, J.R. and T. Cross. 1977. An improved medium for the enumeration of *Chromobacterium* in soil and water. J. Appl. Bacteriol. *43*: 325–327.

Keele, B.B., Jr., P.B. Hamilton and G.H. Elkan. 1969. Glucose catabolism in *Rhizobium japonicum*. J. Bacteriol. *97*: 1184–1191.

Keele, B.B., Jr., P.B. Hamilton and G.H. Elkan. 1970. Gluconate catabolism in *Rhizobium japonicum*. J. Bacteriol. *101*: 698–704.

Keeler, R.F., A.E. Ritchie, J.H. Bryner and J. Elmore. 1966. The preparation and characterization of cell walls and the preparation of flagella of *Vibrio fetus*. J. Gen. Microbiol. *43*: 439–462.

Keil, D.J. and B. Fenwick. 1998. Role of *Bordetella bronchiseptica* in infectious tracheobronchitis in dogs. J. Am. Vet. Med. Assoc. *212*: 200–207.

Keim, T., W. Francke, S. Schmidt and P. Fortnagel. 1999. Catabolism of 2,7-dichloro- and 2,4,8-trichlorodibenzofuran by *Sphingomonas* sp strain RW1. J. Ind. Microbiol. Biotechnol. *23*: 359–363.

Keiser, P.B., S.M. Reynolds, K. Awadzi, E.A. Ottesen, M.J. Taylor and T.B.

Nutman. 2002. Bacterial endosymbionts of *Onchocerca volvulus* in the pathogenesis of posttreatment reactions. J. Infect. Dis. *185*: 805–811.

Keister, D.L. 1975. Acetylene reduction by pure cultures of Rhizobia. J. Bacteriol. *123*: 1265–1268.

Kekcheeva, N.G., I.N. Kokorin and E.D. Miskarova. 1978. Study of rickettsial infection in inbred mice. *In* Kazar, Ormsbee and Tarasevich (Editors), Rickettsiae and Rickettsial Diseases, VEDA, Bratislava. pp. 189–196.

Kellenberger, E., A. Ryter and J. Sechaud. 1958. Electron microscope study of DNA-containing plasmids. II. Vegetative and mature phage DNA as compared with normal bacterial nucleoids in different physiological states. J. Biophys. Biochem. Cytol. *4*: 671–678.

Keller, B., E. Keller and F. Lingens. 1982. Arogenate (pretyrosine) as an obligatory intermediate of the biosynthesis of L-tyrosine in chloridazon degrading bacteria. FEMS Microbiol. Lett. *13*: 121–123.

Kelley, B.C., R.H. Dunstan and D.J.D. Nicholas. 1982. Respiration-dependent nitrogenase activity in the dark in a denitrifying bacterium *Rhodopseudomonas sphaeroides* f. sp. denitrificans. FEMS Microbiol. Lett. *13*: 253–258.

Kelley, J.I., B.F. Turng, H.N. Williams and M.L. Baer. 1997. Effects of temperature, salinity, and substrate on the colonization of surfaces *in situ* by aquatic Bdellovibrios. Appl. Environ. Microbiol. *63*: 84–90.

Kelley, J.I. and H.N. Williams. 1992. Bdellovibrios in *Callinectes sapidus*, the blue crab. Appl. Environ. Microbiol. *58*: 1408–1410.

Kelln, R.A. and R.A. Warren. 1971. Isolation and properties of a bacteriophage lytic for a wide range of pseudomonads. Can. J. Microbiol. *17*: 677–682.

Kellogg, D.S., I.R. Cohen, L.C. Norins, A.L. Schroeter and G. Reising. 1986. *Neisseria gonorrhoeae*. II. Colonial variation and pathogenicity during 35 months *in vitro*. J. Bacteriol. *96*: 596–605.

Kellogg, D.S., J.A. Crawford and C.S. Callaway. 1983. Cultivation of *Neisseria gonorrhoeae* under low-oxygen conditions. J. Clin. Microbiol. *18*: 178–184.

Kellogg, D.S., Jr., W.L. Peacock, Jr., W.E. Deacon, L. Brown and C.I. Pirkle. 1963. *Neisseria gonorrhoeae*. I. Virulence genetically linked to clonal variation. J. Bacteriol. *85*: 1274–1279.

Kelly, D.J. 2001. The physiology and metabolism of *Camplyobacter jejuni* and *Helicobacter pylori*. Symp. Ser. Soc. Appl. Microbiol. *30*: 16S–24S.

Kelly, D.P. and A.P. Harrison. 1989. Genus *Thiobacillus*. *In* Staley, Bryant, Pfennig and Holt (Editors), Bergey's Manual of Systematic Bacteriology, 1st Ed., Vol. 3, The Williams & Williams Co., Baltimore. pp. 1842–1858.

Kelly, D.P., I.R. McDonald and A.P. Wood. 2000. Proposal for the reclassification of *Thiobacillus novellus* as *Starkeya novella* gen. nov., comb. nov., in the alpha-subclass of the *Proteobacteria*. Int. J. Syst. Evol. Microbiol. *50*: 1797–1802.

Kelly, D.P., J.K. Shergill, W.P. Lu and A.P. Wood. 1997. Oxidative metabolism of inorganic sulfur compounds by bacteria. Antonie Leeuwenhoek *71*: 95–107.

Kelly, D.P., E. Stackebrandt, J. Burghardt and A.P. Wood. 1998a. Confirmation that *Thiobacillus halophilus* and *Thiobacillus hydrothermalis* are distinct species within the gamma-sublass of the *Proteobacteria*. Arch. Microbiol. *170*: 138–140.

Kelly, D.P. and A.P. Wood. 1998. Microbes of the sulfur cycle. *In* Burlage, Atlas, Stahl, Geesey and Sayler (Editors), Techniques in Microbial Ecology, Oxford University Press, New York. pp. 31–57.

Kelly, D.P. and A.P. Wood. 2000a. Confirmation of *Thiobacillus denitrificans* as a species of the genus *Thiobacillus*, in the β-subclass of the *Proteobacteria*, with strain NCIMB 9548 as the type strain. Int. J. Syst. Evol. Microbiol. *50*: 547–550.

Kelly, D.P. and A.P. Wood. 2000b. Reclassification of some species of *Thiobacillus* to the newly designated genera *Acidithiobacillus* gen. nov., *Halothiobacillus* gen. nov. and *Thermithiobacillus* gen. nov. Int. J. Syst. Evol. Microbiol. *50*: 511–516.

Kelly, P.J., L. Beati, P.R. Mason, L.A. Matthewman, V. Roux and D. Raoult. 1996. *Rickettsia africae* sp. nov., the etiological agent of African tick bite fever. Int. J. Syst. Bacteriol. *46*: 611–614.

Kelly, T.M., I. Padmalayam and B.R. Baumstark. 1998b. Use of the cell division protein FtsZ as a means of differentiating among *Bartonella* species. Clin. Diagn. Lab. Immunol. *5*: 766–772.

Kendrick, P.L. and G. Eldering. 1969. Microbiology of whooping cough. *In* Ocklitz (Editor), Der Keuchusten VI. Infectionskrankheiten und ihre Erreger, Vol. 8, Veb Gustav Fischer, Jena. 259–282.

Kenne, L., B. Lindberg, P. Unger, B. Gustafsson and T. Holme. 1982. Structural studies of the *Vibrio cholerae* O-antigen. Carbohydr. Res. *100*: 341–349.

Kennedy, L.D. 1976. Isolation of 3-0-methyl-D-ribose from *Rhizobium polysaccharide*. Carbohydr. Res. *52*: 259–261.

Kennedy, L.D. and R.W. Bailey. 1976. Monomethyl sugars in extracellular polysaccharides from slow-growing rhizobia. Carbohydr. Res. *49*: 451–454.

Keppen, O.I. and V.M. Gorlenko. 1975. Characteristics of a new species of budding purple bacteria containing bacteriochlorophyll *b*. Mikrobiologiia *44*: 258–264.

Kerby, R.L., P.W. Ludden and G.P. Roberts. 1995. Carbon monoxide-dependent growth of *Rhodospirillum rubrum*. J. Bacteriol. *177*: 2241–2244.

Kerlikowske, K. and H.F. Chambers. 1989. *Kingella kingae* endocarditis in a patient with the acquired immunodeficiency syndrome. West. J. Med. *151*: 558–560.

Kern, R. 1985. Production of *draB* reducing polymers by hydrogen bacteria. AIP Conference Proceedings, Polymer-Flow Interaction American Institute of Physics. 135–142.

Kern, W.V., M. Oethinger, A. Kaufhold, E. Rozdzinski and R. Marre. 1993. *Ochrobactrum anthropi* bacteremia: report of four cases and short review. Infection *21*: 306–310.

Kerr, A. 1980. Biological control of crown gall through production of agrocin 84. Plant Dis. *64*: 25–30.

Kerr, A. 1992. The genus *Agrobacterium*. *In* Balows, Trüper, Dworkin, Harder and Schleifer (Editors), The Prokaryotes: A Handbook of Bacteria: Ecophysiology, Isolation, Identification, Applications, 2nd Ed., Springer-Verlag, New York. pp. 2214–2235.

Kerr, A. and K. Htay. 1974. Biological control of crown gall through bacteriocin production. Physiol. Plant Pathol. *4*: 37–44.

Kerr, A. and C.G. Panagopoulos. 1977. Biotypes of *Agrobacterium radiobacter* var. *tumefaciens* and their biological control. Phytopath. Z. *90*: 172–179.

Kersters, K. and J. De Ley. 1975. Identification and grouping of bacteria by numerical analysis of their electrophoretic protein patterns. J. Gen. Microbiol. *87*: 333–342.

Kersters, K. and J. De Ley. 1984a. Genus *Agrobacterium*. *In* Krieg and Holt (Editors), Bergey's Manual of Systematic Bacteriology, 1st Ed., Vol. 1, The Williams & Wilkins Co., Baltimore. pp. 244–254.

Kersters, K. and J. De Ley. 1984b. Genus *Alcaligenes*. *In* Krieg and Holt (Editors), Bergey's Manual of Systematic Bacteriology, 1st Ed., Vol. 1, The Williams & Wilkins Co., Baltimore. pp. 361–373.

Kersters, K., J. De Ley, P.H.A. Sneath and M. Sackin. 1973. Numerical taxonomic analysis of *Agrobacterium*. J. Gen. Microbiol. *78*: 227–239.

Kersters, K., K.H. Hinz, A. Hertle, P. Segers, A. Lievens, O. Siegmann and J. De Ley. 1984. *Bordetella avium* sp. nov., isolated from the respiratory tracts of turkeys and other birds. Int. J. Syst. Bacteriol. *34*: 56–70.

Kessel, M. and M. Shilo. 1976. Relationship of *Bdellovibrio* elongation and fission to host cell size. J. Bacteriol. *128*: 477–480.

Kesseler, F.P. and A.C. Schwartz. 1995. Dye-linked aldehyde dehydrogenase from methanol-grown *Hyphomicrobium* contains a binuclear iron-sulfer center. *In* Abstracts of the 8th International Symposium on Microbial Growth on C_1 Compounds, San Diego, California. 63, B–31.

Ketchum, P.A. and C.L. Sevilla. 1973. *In vitro* formation of nitrate reductase using extracts of the nitrate reductase mutant of *Neurospora crassa*, nit-1, and *Rhodospirillum rubrum*. J. Bacteriol. *116*: 600–609.

Ketkar, S.D. 1967. MS Thesis, University of Bombay

Ketkar, S.D. and S.A. Dhala. 1978. Annual Conference, Association of Microbiologists of India, Baroda, India.

Ketley, J.M. 1995. Virulence of *Campylobacter* species: a molecular genetic approach. J. Med. Microbiol. *42*: 312–327.

Ketley, J.M. 1997. Pathogenesis of enteric infection by *Campylobacter*. Microbiology *143*: 5–21.

Khairat, O. 1967. *Bacteroides corrodens* isolated from bacteremias. J. Pathol. Bacteriol. *94*: 29–40.

Khakmun, T. 1967. Iron- and manganese-oxidizing microorganisms in soils of South Sakhalin (in Russian). Mikrobiologiya *36*: 337–344.

Khambata, S.R. and J.V. Bhat. 1953. Studies on a new oxalate-decomposing bacterium, *Pseudomonas oxalaticus*. J. Bacteriol. *66*: 505–507.

Khammas, K.M., E. Ageron, P.A.D. Grimont and P. Kaiser. 1989. *Azospirillum irakense*, sp. nov.: a nitrogen-fixing bacterium associated with rice roots and rhizosphere soil. Res. Microbiol. *140*: 679–694.

Khammas, K.M., E. Ageron, P.A.D. Grimont and P. Kaiser. 1991. *In* Validation of the publication of new names and new combinations previously effectively published outside the IJSB. List No. 39. Int. J. Syst. Bacteriol. *41*: 580–581.

Khammas, K.M. and P. Kaiser. 1991. Characterization of a pectinolytic activity in *Azospirillum irakense*. Plant Soil *137*: 75–80.

Khan, A.A., R.F. Wang, W.W. Cao, W. Franklin and C.E. Cerniglia. 1996a. Reclassification of a polycyclic aromatic hydrocarbon-metabolizing bacterium, *Beijerinckia* sp. strain B1, as *Sphingomonas yanoikuyae* by fatty acid analysis, protein pattern analysis, DNA–DNA hybridization, and 16S ribosomal DNA sequencing. Int. J. Syst. Bacteriol. *46*: 466–469.

Khan, A.A., R.F. Wang, M.S. Nawaz, W.W. Cao and C.E. Cerniglia. 1996b. Purification of 2,3-dihydroxybiphenyl 1,2-dioxygenase from *Pseudomonas putida* OU83 and characterization of the gene (*bphC*). Appl. Environ. Microbiol. *62*: 1825–1830.

Khan, A.A., R.F. Wang, M.S. Nawaz and C.E. Cerniglia. 1997. Nucleotide sequence of the gene encoding *cis*-biphenyl dihydrodiol dehydrogenase (*bphB*) and the expression of an active recombinant His-tagged *bphB* gene product from a PCB degrading bacterium, *Pseudomonas putida* OU83. FEMS Microbiol. Lett. *154*: 317–324.

Khan, S.U., S.M. Gordon, P.C. Stillwell, T.J. Kirby and A.C. Arroliga. 1996c. Empyema and bloodstream infection caused by *Burkholderia gladioli* in a patient with cystic fibrosis after lung transplantation. Pediatr. Infect. Dis. J. *15*: 637–639.

Khattak, M.N. and R.C. Matthews. 1993. Genetic relatedness of *Bordetella* species as determined by macrorestriction digests resolved by pulsed-field gel electrophoresis. Int. J. Syst. Bacteriol. *43*: 659–664.

Khlebnikov, A. and P. Peringer. 1996. Biodegradation of *p*-toluenesulphonic acid by *Comamonas testosteroni* in an aerobic counter-current structured packing biofilm reactor. Water Sci. Technol. *34*: 257–266.

Khlebnikov, A., V. Zhoukov and P. Peringer. 1997. Comparison of *p*-toluenesulphonic acid degradation by two *Comamonas* strains. Biotechnol. Lett. *19*: 389–393.

Khosla, C. and J.E. Bailey. 1988a. Heterologous expression of a bacterial haemoglobin improves the growth properties of recombinant *Escherichia coli*. Nature *331*: 633–635.

Khosla, C. and J.E. Bailey. 1988b. The *Vitreoscilla* hemoglobin gene: molecular cloning, nucleotide sequence and genetic expression in *Escherichia coli*. Mol. Gen. Genet. *214*: 158–161.

Kiehlbauch, J.A., D.J. Brenner, D.N. Cameron, A.G. Steigerwalt, J.M. Makowski, C.N. Baker, C.M. Patton and I.K. Wachsmuth. 1995. Genotypic and phenotypic characterization of *Helicobacter cinaedi* and Candidatus Helicobacter fennelliae strains isolated from humans and animals. J. Clin. Microbiol. *33*: 2940–2947.

Kiehlbauch, J.A., D.J. Brenner, M.A. Nicholson, C.N. Baker, C.M. Patton, A.G. Steigerwalt and I.K. Wachsmuth. 1991a. *Campylobacter butzleri* sp. nov. isolated from humans and animals with diarrheal illness. J. Clin. Microbiol. *29*: 376–385.

Kiehlbauch, J.A., B.D. Plikaytis, B. Swaminathan, D.N. Cameron and I.K. Wachsmuth. 1991b. Restriction fragment length polymorphisms in the ribosomal genes for species identification and subtyping of aerotolerant *Campylobacter* species. J. Clin. Microbiol. *29*: 1670–1676.

Kiessling, M. and O. Meyer. 1982. Profitable oxidation of carbon monoxide or hydrogen during heterotrophic growth of *Pseudomonas carboxydoflava*. FEMS Microbiol. Lett. *13*: 333–338.

Kim, E. and G.J. Zylstra. 1999. Functional analysis of genes involved in biphenyl, naphthalene, phenanthrene, and m-xylene degradation by *Sphingomonas yanoikuyae* B1. J. Ind. Microbiol. Biotechnol. *23*: 294–302.

Kim, E., G.J. Zylstra, J.P. Freeman, T.M. Heinze, J. Deck and C.E. Cerniglia. 1997. Evidence for the role of 2-hydroxychromene-2-carboxylate isomerase in the degradation of anthracene by *Sphingomonas yanoikuyae* B1. FEMS Microbiol. Lett. *153*: 479–484.

Kim, S.J., J. Chun, K.S. Bae and Y.C. Kim. 2000a. Polyphasic assignment of an aromatic-degrading *Pseudomonas* sp., strain DJ77, in the genus *Sphingomonas* as *Sphingomonas chungbukensis* sp. nov. Int. J. Syst. Evol. Microbiol. *50*: 1641–1647.

Kim, S.W., K.S. Ihn, S.H. Han, S.Y. Seong, I.S. Kim and M.S. Choi. 2001. Microtubule- and dynein-mediated movement of *Orientia tsutsugamushi* to the microtubule organizing center. Infect. Immun. *69*: 494–500.

Kim, T.H. and T.S. Yu. 1998. Chemical modification of extracellular cytosine deaminase from *Chromobacterium violaceum* YK 391. J. Microbiol. Biotechnol. *8*: 581–587.

Kim, Y.J., K. Furihata, S. Yamanaka, R. Fudo and H. Seto. 1991. Isolation and structural elucidation of stipiamide, a new antibiotic effective to multidrug-resistant cancer cells. J. Antibiot. *44*: 553–555.

Kim, Y.J., K.B. Song and S.K. Rhee. 1995. A novel aerobic respiratory chain-linked NADH oxidase system in *Zymomonas mobilis*. J. Bacteriol. *177*: 5176–5178.

Kim, Y.S. and S.C. Tu. 1989. Molecular cloning of salicylate hydroxylase genes from *Pseudomonas cepacia* and *Pseudomonas putida*. Arch. Biochem. Biophys. *269*: 295–304.

Kim, Y., K.H. Yoon, Y. Khang, S. Turley and W.G.J. Hol. 2000b. The 2.0 angstrom crystal structure of cephalosporin acylase. Structure (Lond.) *8*: 1059–1068.

Kimura, M. 1980. A simple method for estimating evolutionary rates of base substitutions through comparative studies of nucleotide sequences. J. Mol. Evol. *16*: 111–120.

Kimura, N., A. Nishi, M. Goto and K. Furukawa. 1997. Functional analyses of a variety of chimeric dioxygenases constructed from two biphenyl dioxygenases that are similar structurally but different functionally. J. Bacteriol. *179*: 3936–3943.

King, E.O. 1964. The identification of unusual pathogenic Gram-negative bacteria, Communicable Disease Center, Atlanta.

King, E.O., W.K. Ward and D.E. Raney. 1954. Two simple media for the demonstration of pyocyanin and fluorescein. J. Lab. Clin. Med. *44*: 301–307.

Kingma-Boltjes, T.Y. 1936. Über Hyphomicrobium vulgare Stutzer et Hartleb. Arch. Mikrobiol. *7*: 188–205.

Kingsbury, D.T. 1966. Bacteriocin production by strains of *Neisseria meningitidis*. J. Bacteriol. *91*: 1696–1699.

Kingsbury, D.T. 1967. Deoxyribonucleic acid homologies among species of the genus *Neisseria*. J. Bacteriol. *94*: 870–874.

Kingsbury, D.T., G.R. Fanning, K.E. Johnson and D.J. Brenner. 1969. Thermal stability of interspecies *Neisseria* DNA duplexes. J. Gen. Microbiol. *55*: 201–208..

Kingsley, M.T. and B.B. Bohlool. 1983. Characterization of *Rhizobium* sp. (*Cicer arietinum* L.) by immunofluorescence, immunodiffusion and intrinsic antibiotic resistance. Can. J. Microbiol. *29*: 518–526.

Kirchhof, G., B. Eckert, M. Stoffels, J.I. Baldani, V.M. Reis and A. Hartmann. 2001. *Herbaspirillum frisingense* sp. nov., a new nitrogen-fixing bacterial species that occurs in C4 fibre plants. Int. J. Syst. Evol. Microbiol. *51*: 157–168.

Kirchhof, G., V.M. Reis, J.I. Baldani, B. Eckert, J. Dobereiner and A. Hartmann. 1997a. Occurrence, physiological and molecular analysis of endophytic diazotrophic bacteria in gramineous energy plants. Plant Soil *194*: 45–55.

Kirchhof, G., M. Schloter, B. Assmus and A. Hartmann. 1997b. Molecular microbial ecology approaches applied to diazotrophs associated with non-legumes. Soil Biol. Biochem. *29*: 853–862.

Kirchner, O. 1896. Die Wurzelknöllchen der Sojabohne. Beitr. Biol. Pflanz. *7*: 213–224.

Kiredjian, M., B. Holmes, K. Kersters, I. Guilvout and J. De Ley. 1986. *Alcaligenes piechaudii*, a new species from human clinical specimens and the environment. Int. J. Syst. Bacteriol. *36*: 282–287.

Kiredjian, M., M. Popoff, C. Coynault, M. Lefevre and M. Lemelin. 1981. Taxonomie du genre *Alcaligenes*. Ann. Microbiol. (Paris) *B132*: 337–374.

Kirk, L.A., R.I. Webb and H.W. Doelle. 1994. Capsule formation in *Zymomonas mobilis* grown on sucrose. World J. Microbiol. Biotechnol. *10*: 481–482.

Kirkbride, C.A., C.E. Gates, J.E. Collins and A.E. Ritchie. 1985. Ovine abortion associated with an anaerobic bacterium. J. Am. Vet. Med. Assoc. *186*: 789–791.

Kirstein, K. and E. Bock. 1993. Close genetic relationship between *Nitrobacter hamburgensis* nitrite oxidoreductase and *Escherichia coli* nitrate reductases. Arch. Microbiol. *160*: 447–453.

Kishimoto, N., K. Adachi, S. Tamura, M. Nishihara, K. Inagaki, T. Sugio and T. Tano. 1993a. Lipoamino acids isolated from *Acidiphilium organovorum*. Syst. Appl. Microbiol. *16*: 17–21.

Kishimoto, N., F. Fukaya, K. Inagaki, T. Sugio, H. Tanaka and T. Tano. 1995a. Distribution of bacteriochlorophyll *a* among aerobic and acidophilic bacteria and light-enhanced CO_2 incorporation in *Acidiphilium rubrum*. FEMS Microbiol. Ecol. *16*: 291–296.

Kishimoto, N., K. Inagaki, T. Sugio and T. Tano. 1990. Growth inhibition of *Acidiphilium* species by organic acids contained in yeast extract. J. Ferment. Bioeng. *70*: 7–10.

Kishimoto, N., Y. Kosako and T. Tano. 1993b. *Acidiphilium aminolytica*, new species: an acidophilic chemoorganotrophic bacterium isolated from acidic mineral environment. Curr. Microbiol. *27*: 131–136.

Kishimoto, N., Y. Kosako, N. Wakao, T. Tano and A. Hiraishi. 1995b. Transfer of *Acidiphilium facilis* and *Acidiphilium aminolytica* to the genus *Acidocella* gen. nov., and emendation of the genus *Acidiphilium*. Syst. Appl. Microbiol. *18*: 85–91.

Kishimoto, N., Y. Kosako, N. Wakao, T. Tano and A. Hiraishi. 1996. *In* Validation of the publication of new names and new combinations previously effectively published outside the IJSB. List No. 56. Int. J. Syst. Bacteriol. *46*: 362–363.

Kishimoto, N. and T. Tano. 1987. Acidophilic heterotrophic bacteria isolated from acidic mine drainage, sewage, and soils. J. Gen. Appl. Microbiol. *33*: 11–26.

Kishino, H. and Hasegawa, M. 1989. Evaluation of the maximum likelihood estimate of the evolutionary tree topologies from DNA sequence data, and the branching order in hominoidea. J. Mol. Evol. *29*: 170–179.

Kiss, E., P. Mergaert, B. Olàh, A. Kereszt, C. Staehelin, A.E. Davies, J.A. Downie, Á. Kondorosi and E. Kondorosi. 1998. Conservation of *nolR* in the *Sinorhizobium* and *Rhizobium* genera of the *Rhizobiaceae* family. Mol. Plant-Microbe Interact. *11*: 1186–1195.

Kistner, A. 1953. On a bacterium oxidizing carbon monoxide. Proc. K. Ned. Akad. Wet., Ser. C. *56*: 443–450.

Kistner, A. 1954. Conditions determining the oxidation of carbon monoxide by *Hydrogenomonas carboxydovorans*. Proc. K. ned. Akad. Wet., Ser. C. *57*: 186–195.

Kita, K., K. Ishimaru, M. Teraoka, H. Yanase and N. Kato. 1995. Properties of poly(3-hydroxybutyrate) depolymerase from a marine bacterium, *Alcaligenes faecalis* AE122. Appl. Environ. Microbiol. *61*: 1727–1730.

Klassen, G., F.O. Pedrosa, E.M. Souza, S. Funayama and L.U. Rigo. 1997. Effect of nitrogen compounds on nitrogenase activity in *Herbaspirillum seropedicae* SMR1. Can. J. Microbiol. *43*: 887–891.

Klaveness, D. 1982. The *Cryptomonas–Caulobacter* consortium: facultative ectocommensalism with possible taxonomic consequences? Nord. J. Bot. *2*: 183–188.

Klecka, G.M. and D.T. Gibson. 1980. Metabolism of dibenzo-*p*-dioxin and chlorinated dibenzo-*p*-dioxins by a *Beijerrinckia* species. Appl. Environ. Microbiol. *39*: 288–296.

Kleczkowska, J., P.S. Nutman, F.A. Skinner and J.M. Vincent. 1968. The identification and classification of *Rhizobium*. *In* Gibbs and Shapton (Editors), Identification Methods for Microbiologists Part B, Academic Press, New York. 51–65.

Klee, S.R., X. Nassif, B. Kusecek, P. Merker, J.L. Beretti, M. Achtman and C.R. Tinsley. 2000. Molecular and biological analysis of eight genetic islands that distinguish *Neisseria meningitidis* from the closely related pathogen *Neisseria gonorrhoeae*. Infect. Immun. *68*: 2082–2095.

Kleihues, L., O. Lenz, M. Bernhard, T. Buhrke and B. Friedrich. 2000. The H_2 sensor of *Ralstonia eutropha* is a member of the subclass of regulatory (NiFe) hydrogenases. J. Bacteriol. *182*: 2716–2724.

Klein, C.R., F.P. Kesseler, C. Perrei, J. Frank, J.A. Duine and A.C. Schwartz. 1994. A novel dye-linked formaldehyde dehydrogenase with some properties indicating the presence of a protein-bound redox-active quinone cofactor. Biochem. J. *301*: 289–295.

Klein, D.A. and L.E. Casida. 1967. Occurrence and enumeration of *Bdellovibrio bacteriovorus* in soil capable of parasitizing *Escherichia coli* and indigenous soil bacteria. Can. J. Microbiol. *13*: 1235–1241.

Klein, M.B., S. Hu, C.C. Chao and J.L. Goodman. 2000. The agent of human granulocytic ehrlichiosis induces the production of myelosuppressing chemokines without induction of proinflammatory cytokines. J. Infect. Dis. *182*: 200–205.

Klein, M.B., C.M. Nelson and J.L. Goodman. 1997. Antibiotic susceptibility of the newly cultivated agent of human granulocytic ehrlichiosis: promising activity of quinolones and rifamycins. Antimicrob. Agents Chemother. *41*: 76–79.

Klein, R.M. and I.L. Tenebaum. 1955. A quantitative bioassay for crown-gall tumor formation. Am. J. Bot. *42*: 709–712.

Kleinig, H. 1972. Membranes from *Myxococcus fulvus* (*Myxobacterales*) containing carotenoid glucosides. I. Isolation and composition. Biochim. Biophys. Acta *274*: 489–498.

Kleinig, H. and H. Reichenbach. 1973. A new carotenoid glucoside ester from *Chondromyces apiculatus*. Phytochemistry (Oxf) *12*: 2483–2485.

Kleinig, H., H. Reichenbach and H. Achenbach. 1970. Carotenoid pigments of *Stigmatella aurantiaca* (*Myxobacterales*). II. Acylated carotenoid glucosides. Arch. Mikrobiol. *74*: 223–234.

Kleinig, H., H. Reichenbach, H. Achenbach and J. Stadler. 1971. Carotenoid pigments of *Sorangium compositum* (*Myxobacterales*) including two new carotenoid glucoside esters and two new carotenoid rhamnosides. Arch. Mikrobiol. *78*: 224–233.

Klemme, J.H. 1968. Untersuchungen zur Photoautotrophie mit molekularem Wasserstoff bei neuisolierten schwefelfreien Purpurbakterien. Arch. Mikrobiol. *64*: 29–42.

Klemme, J.H. 1979. Occurrence of assimilatory nitrate reduction in phototropic bacteria of the genera *Rhodospirillum* and *Rhodopseudomonas*. Microbiologica. 2: 415–420.

Klemme, J.H., I. Chyla and M. Preuss. 1980. Dissimilatory nitrate reduction by strains of the facultative phototropic bacterium *Rhodopseudomonas palustris*. FEMS Microbiol. Lett. *9*: 137–140.

Klemme, J.-H. and C. Pfleiderer. 1977. production of extracellular proteolytic enzymes by phototropic bacteria. FEMS Lett. *1*: 297–299.

Klimmek, O., A. Kröger, R. Steudel and G. Holdt. 1991. Growth of *Wolinella succinogenes* with polysulfide as terminal acceptor of phosphorylative electron transport. Arch. Microbiol. *155*: 177–182.

Kloos, K., A. Fesefeldt, C.G. Gliesche and H. Bothe. 1995. DNA-probing indicates the occurrence of denitrification and nitrogen fixation genes in *Hyphomicrobium*. Distribution of denitrifying and nitrogen fixing isolates of *Hyphomicrobium* in a sewage treatment plant. FEMS Microbiol. Ecol. *18*: 205–213.

Kloos, W.E., W.J. Dobrogosz, J.W. Ezzell, B.R. Kimbro and C.R. Manclark. 1979. DNA-DNA hybridization, plasmids, and genetic exchange in *Bordetella*. *In* Manclark and Hill (Editors), International Symposium on Pertussis, U.S. Government Printing Office, Washington, D.C. 70–80.

Kloos, W.E., N. Mohapatra, W.J. Dobrogosz, J.W. Ezzell and C.R. Manclark. 1981. Deoxyribonucleotide sequence relationships among *Bordetella* species. Int. J. Syst. Bacteriol. *31*: 173–176.

Kluyver, A.J. 1957. *Zymomonas* Kluyver and van Viel 1936. *In* Breed, Murray and Smith (Editors), Bergey's Manual of Determinative Bacteriology, 7th Ed., The Williams & Wilkins Co., Baltimore. pp. 199–200.

Kluyver, A.J. and J.H. Becking. 1955. Some observations on the nitrogen-fixing bacteria of the genus *Beijerinckia* Derx. Ann. Acad. Sci. Fennicae A II *60*: 367–380.

Kluyver, A.J. and W.J. Hoppenbrouwers. 1931. Ein merkwürdiges Gärungsbakterium: Lindner's *Termobacterium mobile*. Arch. Mikrobiol. *2*: 245–260.

Kluyver, A.J. and A. Manten. 1942. Some observations on the metabolism of bacteria oxidizing molecular hydrogen. Antonie Leeuwenhoek J. Microbiol. Serol. *8*: 71–85.

Kluyver, A.J. and C.B. van Niel. 1936. Prospects for a natural system of classification of bacteria. Zentbl. Bakteriol. Parasitenkd. Infektkrankh. Hyg. Abt. II *94*: 369–403.

Knapp, J.S. 1988. Historical perspectives and identification of *Neisseria* and related species. Clin Microbiol Rev. *1*: 415–431.

Knapp, J.S. and V.L. Clark. 1984. Anaerobic growth of *Neisseria gonorrhoeae* coupled to nitrite reduction. Infect. Immun. *46*: 176–181.

Knapp, J.S., S.R. Johnson, J.M. Zenilman, M.C. Roberts and S.A. Morse. 1988. High-level tetracycline resistance resulting from TetM in strains of *Neisseria* spp., *Kingella denitrificans*, and *Eikenella corrodens*. Antimicrob. Agents Chemother. *32*: 765–767.

Knapp, J.S. and E.H. Koumanis. 1999. *Neisseria* and *Branhamella*. *In* Murray, Baron, Pfaller, Tenover and Yolken (Editors), Manual of Clinical Microbiology, 7th Ed., American Society for Microbiology, Washington, DC. pp. 586–603.

Knippschild, M. and R. Ansorg. 1998. Epidemiological typing of *Alcaligenes xylosoxidans* subsp. *xylosoxidans* by antibacterial susceptibility testing, fatty acid analysis, PAGE of whole-cell protein and pulsed-field gel electrophoresis. Zentbl. Bakteriol. *288*: 145–157.

Knippschild, M., E.N. Schmid, M. Uppenkamp, E. Konig, P. Meusers, G. Brittinger and H.G. Hoffkes. 1996. Infection by *Alcaligenes xylosoxidans* subsp. *xylosoxidans* in neutropenic patients. Oncology (Basel) *53*: 258–262.

Knittel, J.P., D.I. Larson, D.L. Harris, M.B. Roof and S. McOrist. 1996. United States isolates of *Lawsonia intracellularis* from porcine proliferative enteropathy resemble European isolates. Swine Health Prod. *3*: 118–122.

Knobel, H.-R. 1997. Genetic study of bacterial nitrilotriacetate-degrading enzymes, Thesis, Swiss Federal Institute of Technology, Zürich. p. 106.

Knobel, H.-R., T. Egli and J.R. van der Meer. 1996. Cloning and characterization of the genes encoding nitrilotriacetate monooxygenase of *Chelatobacter heintzii* ATCC 29600. J. Bacteriol. *178*: 6123–6132.

Knoblauch, C., K. Sahm and B.B. Jorgensen. 1999. Psychrophilic sulfate-reducing bacteria isolated from permanently cold Arctic marine sediments: description of *Desulfofrigus oceanense* gen. nov., sp. nov., *Desulfofrigus fragile* sp. nov., *Desulfofaba gelida* gen. nov., sp. nov., *Desulfotalea psychrophila* gen. nov., sp. nov and *Desulfotalea arctica* sp. nov. Int. J. Syst. Bacteriol. *49*: 1631–1643.

Knösel, D. 1962. Prufung von Bakterien auf Fahigkeit zur Sternbildung. Zentbl. Bakteriol. Parasitenkd. Infektkrankh. Hyg. Abt II *116*: 79–100.

Knösel, D. 1984a. Genus IV. *Phyllobacterium* (ex Knosel 1962) nom. rev. (*Phyllobacterium* Knosel 1962, 96). *In* Krieg and Holt (Editors), Bergey's Manual of Systematic Bacteriology, 1st Ed., Vol. 1, The Williams & Wilkins Co., Baltimore. pp. 254–256.

Knösel, D.H. 1984b. *In* Validation of the publication of new names and new combinations previously effectively published outside the IJSB. List No. 15. Int. J. Syst. Bacteriol. *34*: 355–357.

Knudsen, T.D. and E.J. Simko. 1995. *Eikenella corrodens*: an unexpected pathogen causing a persistent peritonsillar abscess. Ear Nose Throat J. *74*: 114–117.

Kobayashi, D.Y., A.W. Stretch and P.V. Oudemans. 1995. A bacterial leaf spot of highbush blueberry hardwood cuttings caused by *Pseudomonas andropogonis*. Plant Dis. *79*: 839–842.

Kobayashi, H., H. Saito and T. Kakegawa. 2000. Bacterial strategies to inhabit acidic environments. J. Gen. Appl. Microbiol. *46*: 235–243.

Kobayashi, M., M. Akiyama, M. Yamamura, H. Kise, S. Takaichi, T. Watanabe, K. Shimada, M. Iwaki, S. Itoh, N. Ishida, M. Koizumi, H. Kano, N. Wakao and A. Hiraishi. 1998. Structural determination of the novel

Zn-containing bacteriochlorophyll in *Acidiphilium rubrum*. Photomed. Photobiol. *20*: 75–80.

Kobayashi, T., A. Sugiyama, Y. Kawase, T. Saito, J. Mergaert and J. Swings. 1999. Biochemical and genetic characterization of an extracellular poly(3-hydroxybutyrate) depolymerase from *Acidovorax* sp. strain TP4. J. Environ. Polym. Degrad. *7*: 9–18.

Koburger, J.A. and S.O. May. 1982. Isolation of *Chromobacterium* spp. from foods, soil, and water. Appl. Environ. Microbiol. *44*: 1463–1465.

Kocan, K.M., J.H. Venable, K.C. Hsu and W.E. Brock. 1978. Ultrastructural localization of anaplasmal antigens (Pawhuska isolate) with ferritin-conjugated antibody. Am. J. Vet. Res. *39*: 1131–1135.

Koch, R. 1877. Untersuchungen über Bacterien. VI. Verfahren zůr Untersuchung zum Conserviren und Photographiren der Bacterien. Beitr. Biol. Pflanz. *2*: 399–440.

Koch, R.H. 1883. Bericht über die Thätigkeit der deutschen Cholerakommisionen in Aegypten und Ostindien. Wien. Med. Wochenschr. *33*: 1548–1551.

Koch, W. 1964. Verzeichnis der Sammlung von Algenkulturen am Pflanzenphysiologischen Institut der Universität Göttingen. Arch. Mikrobiol. *47*: 402–432.

Kock, N.D., A.H. van Vliet, K. Charlton and F. Jongejan. 1995. Detection of *Cowdria ruminantium* in blood and bone marrow samples from clinically normal, free-ranging Zimbabwean wild ungulates. J. Clin. Microbiol. *33*: 2501–2504.

Kocur, M., T. Martinec and K. Mazanec. 1968. Fine structure of *Micrococcus denitrificans* and *M. halodenitrificans* in relation to their taxonomy. Antonie Leeuwenhoek J. Microbiol. Serol. *34*: 19–26.

Kodaka, H., A.Y. Armfield, G.L. Lombard and V.R. Dowell, Jr.. 1982. Practical procedure for demonstrating bacterial flagella. J. Clin. Microbiol. *16*: 948–952.

Koechlein, D.J. and N.R. Krieg. 1998. Viable but nonculturable coccoid forms of *Prolinoborus fasciculus* (*Aquaspirillum fasciculus*). Can. J. Microbiol. *44*: 910–912.

Koehler, J.E., F.D. Quinn, T.G. Berger, P.E. LeBoit and J.W. Tappero. 1992. Isolation of *Rochalimaea* species from cutaneous and osseous lesions of bacillary angiomatosis. New Engl. J. Med. *327*: 1625–1631.

Koenraad, P.M., W.F. Jacobs Reitsma, T. Van der Laan, R.R. Beumer and F.M. Rombouts. 1995. Antibiotic susceptibility of *Campylobacter* isolates from sewage and poultry abattoir drain water. Epidemiol. Infect. *115*: 475–483.

Kofler, L. 1913. Die Myxobakterien der Umgebung von Wien. Sitzungsber. Akad. Wiss. Math. Naturwiss. Kl. Abt. I. *122*: 845–876.

Koga, T., K. Ishimoto and S. Lory. 1993. Genetic and functional characterization of the gene cluster specifying expression of *Pseudomonas aeruginosa* pili. Infect. Immun. *61*: 1371–1377.

Kohl, W., A. Gloe and H. Reichenbach. 1983. Steroids from the myxobacterium *Nannocystis exedens*. J. Gen. Microbiol. *129*: 1629–1635.

Köhler, J., F.P. Kesseler and A.C. Schwartz. 1985. Aldehyde oxidation and electron transport in the methylotrophic *Hyphomicrobium* ZV 580. *In* Schäfer (Editor), Third European Bioenergetics Conference: Proceedings of the 3rd EBEC meeting, Hannover, F.R.G., September 2-7, 1984. International Council of Scientific Unions Short Reports, Vol. 3, Cambridge University Press, Cambridge [Cambridgeshire], New York. 220–221.

Köhler, J. and A.C. Schwartz. 1981. Respiratory ubiquinone-9 from *Hyphomicrobium* spec. strain ZV 580. Z. Allg. Mikrobiol. *21*: 117–123.

Köhler, J. and A.C. Schwartz. 1982. Oxidation of aromatic-aldehydes and aliphatic secondary alcohols by *Hyphomicrobium* spp. Can. J. Microbiol. *28*: 65–72.

Köhler, J. and A.C. Schwartz. 1983. Multiple cytochromes *c* in the electron-transport system of the methylotrophic bacterium *Hyphomicrobium* Zv-580. Hoppe-Seyler.s Z. Physiol. Chem. *364*: 1162–1162.

Kohler-Staub, D., S. Frank and T. Leisinger. 1995. Dichloromethane as the sole carbon source for *Hyphomicrobium* sp. strain DM2 under denitrification conditions. Biodegradation *6*: 229–235.

Kohler-Staub, D., S. Hartmans, R. Gälli, F. Suter and T. Leisinger. 1986. Evidence for identical dichloromethane dehalogenases in different methylotrophic bacteria. J. Gen. Microbiol. *132*: 2837–2844.

Kohler-Staub, D. and T. Leisinger. 1985. Dichloromethane dehalogenase of *Hyphomicrobium* sp. strain DM2. J. Bacteriol. *162*: 676–681.

Kohlmiller, E.F., Jr. and H. Gest. 1951. A comparative study of the light and dark fermentations of organic acids by *Rhodospirillum rubrum*. J. Bacteriol. *61*: 269–282.

Kohring, L.L., D.B. Ringelberg, R. Devereux, D.A. Stahl, M.W. Mittelman and D.C. White. 1994. Comparison of phylogenetic-relationships based on phospholipid fatty-acid profiles and ribosomal-RNA sequence similarities among dissimilatory sulfate-reducing bacteria. FEMS Microbiol. Lett. *119*: 303–308.

Kokorin, I.N., E.D. Miskarova, O.S. Gudima, E.A. Kabanova and T. Kiet. 1978. Intracellular development of rickettsiae. *In* Kazar, Ormsbee and Tarasevich (Editors), Rickettsiae and Rickettsial Diseases, VEDA, Bratislava. pp. 107–203.

Kolar, G.W. 1974. Description of *Pseudomonas aenacaerulea* sp. nov., a blue pigmented fluorescent Pseudomonad, University of Wisconsin La Crosse.

Kolb, S., S. Seeliger, N. Springer, W. Ludwig and B. Schink. 1998. The fermenting bacterium *Malonomonas rubra* is phylogenetically related to sulfur-reducing bacteria and contains a *c*-type cytochrome similar to those of sulfur and sulfate reducers. Syst. Appl. Microbiol. *21*: 340–345.

Kölbel-Boelke, J., R. Gebers and P. Hirsch. 1985. Genome size determinations for 33 strains of budding bacteria. Int. J. Syst. Bacteriol. *35*: 270–273.

Kollars, T.M., B. Tippayachai and D. Bodhidatta. 2001. Short report: Thai tick typhus, *Rickettsia honei*, and a unique *Rickettsia* detected in *Ixodes granulatus* (Ixodidae: acari) from Thailand. Am. J. Trop. Med. Hyg. *65*: 535–537.

Komagata, Y., R.R. Fulthorpe, K. Tamura, H. Takami, L.J. Forney and J. Tiedje. 1997. Pristine environments harbor a new group of oligotrophic 2,4-dichlorophenoxy acetic acid-degrading bacteria. Appl. Environ. Microbiol. *63*: 2266–2272.

Komano, T., N. Brown, S. Inouye and M. Inouye. 1982. Phosphorylation and methylation of proteins during *Myxococcus xanthus* spore formation. J. Bacteriol. *151*: 114–118.

Kompantseva, E.J. 1981. Utilization of sulfide by nonsulfur purple bacteria *Rhodopseudomonas capsulata*. Mikrobiologiya *50*: 429–436.

Kompantseva, E.J. 1985. *Rhodobacter euryhalinus* sp. nov., a new halophilic purple bacterial species. Mikrobiologiya *54*: 974–982.

Kompantseva, E.I. 1989a. A new species of budding purple bacterium: *Rhodopseudomonas julia* sp. nov. Mikrobiologiya *58*: 319–325.

Kompantseva, E.J. 1989b. *In* Validation of the publication of new names and new combinations previously effectively published outside the IJSB. List No. 29. Int. J. Syst. Bacteriol. *39*: 205–206.

Kompantseva, E.I. 1993. *In* Validation of the publication of new names and new combinations previously effectively published outside the IJSB. List No. 44. Int. J. Syst. Bacteriol. *43*: 188–189.

Kompantseva, E.I. and V.M. Gorlenko. 1984. A new species of moderately halophilic purple bacterium, *Rhodospirillum mediosalinum*. Mikrobiologiya *53*: 954–961.

Kondo, K., T. Beppu and S. Horinouchi. 1995. Cloning, sequencing, and characterization of the gene encoding the smallest subunit of the three-component membrane-bound alcohol dehydrogenase from *Acetobacter pasteurianus*. J. Bacteriol. *177*: 5048–5055.

Kondo, K. and S. Horinouchi. 1997a. Characterization of an insertion sequence, IS12528, from *Gluconobacter suboxydans*. Appl. Environ. Microbiol. *63*: 1139–1142.

Kondo, K. and S. Horinouchi. 1997b. Characterization of the genes encoding the three-component membrane-bound alcohol dehydrogenase from *Gluconobacter suboxydans* and their expression in *Acetobacter pasteurianus*. Appl. Environ. Microbiol. *63*: 1131–1138.

Kondo, K. and S. Horinouchi. 1997c. A new insertion sequence IS1452 from *Acetobacter pasteurianus*. Microbiology *143*: 539–546.

Kondorosi, A., G.B. Kiss, T. Forrai, E. Vincze and Z. Banfalvi. 1977. Circular linkage map of the *Rhizobium meliloti* chromosome. Nature (Lond.) *268*: 525–527.

Kondorosi, E., M. Buiré, M. Cren, N. Iyer, B. Hoffmann and Á. Kondorosi.

1991a. Involvement of the *syrM* and *nodD3* genes of *Rhizobium meliloti* in *nod* gene activation and in optimal nodulation of the plant host. Mol. Microbiol. *5*: 3035–3048.

Kondorosi, E., M. Pierre, M. Cren, U. Haumann, M. Buiré, B. Hoffmann, J. Schell and Á. Kondorosi. 1991b. Identification of NolR, a negative transacting factor controlling the *nod* regulon in *Rhizobium meliloti*. J. Mol. Biol. *222*: 885–896.

Kondratieva, E.N., N. Pfenning and H.G. Trüper. 1992. The phototrophic prokaryotes. *In* Balows, Trüper, Dworkin, Harder and Schlefer (Editors), The Prokaryotes: A Handbook of Bacteria: Ecophsiology, Isolation, Identification, Applications, 2nd Ed., Vol. 1, Springer-Verlag, New York. pp.312–330.

Konishi, H. and Z. Yoshii. 1986. Determination of the spiral conformation of *Aquaspirillum* spp. by scanning electron-microscopy of elongated cells induced by cephalexin treatment. J. Gen. Microbiol. *132*: 877–881.

Konopka, A.E., J.C. Lara and J.T. Staley. 1977. Isolation and characterization of gas vesicles from *Microcyclus aquaticus*. Arch. Microbiol. *112*: 133–140.

Konopka, A.E., R.L. Moore and J.T. Staley. 1976. Taxonomy of *Microcyclus* and other nonmotile ring-forming bacteria. Int. J. Syst. Bacteriol. *26*: 505–510.

Koomey, J.M. and S. Falkow. 1984. Nucleotide sequence homology between the immunoglobulin A1 protease genes of *Neisseria gonorrhoeae*, *Neisseria meningitidis*, and *Haemophilus influenzae*. Infect. Immun. *43*: 101–107.

Koomey, J.M. and S. Falkow. 1987. Cloning of the *recA* gene of *Neisseria gonorrhoeae* and construction of gonococcal *recA* mutants. J. Bacteriol. *169*: 790–795.

Koomey, J.M., R.E. Gill and S. Falkow. 1982. Genetic and biochemical-analysis of gonococcal iga1 protease—cloning in *Escherichia coli* and construction of mutants of gonococci that fail to produce the activity. Proc. Nat. Acad. Sci. U.S.A. *79*: 7881–7885.

Koomey, M. 1998. Competence for natural transformation in *Neisseria gonorrhoeae*: a model system for studies of horizontal gene transfer. APMIS *84*: 56–61.

Koomey, M., S. Bergstrom, M. Blake and J. Swanson. 1991. Pilin expression and processing in pilus mutants of *Neisseria gonorrhoeae*: critical role of Gly-1 in assembly. Mol. Microbiol. *5*: 279–287.

Koomey, M., E.C. Gotschlich, K. Robbins, S. Bergstrom and J. Swanson. 1987. Effects of *recA* mutations on pilus antigenic variation and phase-transitions in *Neisseria gonorrhoeae*. Genetics. *117*: 391–398.

Koops, H.P., B. Böttcher, U.C. Möller, A. Pommerening-Röser and G. Stehr. 1991. Classification of 8 new species of ammonia-oxidizing bacteria *Nitrosomonas communis* sp. nov., *Nitrosomonas ureae* sp. nov., *Nitrosomonas aestuarii* sp. nov., *Nitrosomonas marina* sp. nov., *Nitrosomonas nitrosa* sp. nov., *Nitrosomonas eutropha* sp. nov., *Nitrosomonas oligotropha* sp. nov., and *Nitrosomonas halophila* sp. nov. J. Gen. Microbiol. *137*: 1689–1699.

Koops, H.P. and H. Harms. 1985. Deoxyribonucleic acid homologies among 96 strains of ammonia-oxidizing bacteria. Arch. Microbiol. *141*: 214–218.

Koops, H.P., H. Harms and H. Wehrmann. 1976. Isolation of a moderate halophilic ammonia-oxidizing bacterium, *Nitrosococcus mobilis* nov. sp. Arch. Microbiol. *107*: 277–282.

Kopmans-Gargantiel, A.I. and C.L. Wisseman, Jr.. 1981. Differential requirements for enriched atmospheric carbon dioxide content for intracellular growth in cell culture among selected members of the genus *Rickettsia*. Infect. Immun. *31*: 1277–1280.

Koppe, F. 1924. Die Schlammflora der ostholsteinischen Seen und des Bodensees. Arch. Hydrobiol. *14*: 619–672.

Kordick, D.L., T.T. Brown, K. Shin and E.B. Breitschwerdt. 1999. Clinical and pathologic evaluation of chronic *Bartonella henselae* or *Bartonella clarridgeiae* infection in cats. J. Clin. Microbiol. *37*: 1536–1547.

Kordick, D.L., E.J. Hilyard, T.L. Hadfield, K.H. Wilson, A.G. Steigerwalt, D.J. Brenner and E.B. Breitschwerdt. 1997. *Bartonella clarridgeiae*, a newly recognized zoonotic pathogen causing inoculation papules, fever, and lymphadenopathy (cat scratch disease). J. Clin. Microbiol. *35*: 1813–1818.

Kordick, D.L., B. Swaminathan, C.E. Greene, K.H. Wilson, A.M. Whitney, S. O'Connor, D.G. Hollis, G.M. Matar, A.G. Steigerwalt, G.B. Malcolm, P.S. Hayes, T.L. Hadfield, E.B. Breitschwerdt and D.J. Brenner. 1996. *Bartonella vinsonii* berkhoffii subsp. nov., isolated from dogs; *Bartonella vinsonii* subsp. *vinsonii*; and emended description of *Bartonella vinsonii*. Int. J. Syst. Bacteriol. *46*: 704–709.

Korgenski, E.K. and J.A. Daly. 1997. Surveillance and detection of erythromycin resistance in *Bordetella pertussis* isolates recovered from a pediatric population in the intermountain west region of the United States. J. Clin. Microbiol. *35*: 2989–2991.

Kornberg, H.L., J.F. Collins and D. Bigley. 1960. The influence of growth substrates on metabolic pathways in *Micrococcus denitrificans*. Biochim. Biophys. Acta *39*: 9–24.

Kortlüke, C., K. Breese, N. Gad'on, A. Labahn and G. Drews. 1997. Structure of the *puf* operon of the obligately aerobic, bacteriochlorophyll alpha-containing bacterium *Roseobacter denitrificans* OCh114 and its expression in a *Rhodobacter capsulatus puf puc* deletion mutant. J. Bacteriol. *179*: 5247–5258.

Korvick, J.A., J.D. Rihs, G.L. Gilardi and V.L. Yu. 1989. A pink-pigmented, oxidative, nonmotile bacterium as a cause of opportunistic infections. Arch. Intern. Med. *149*: 1449–1451.

Kosako, Y., E. Yabuuchi, T. Naka, N. Fujiwara and K. Kobayashi. 2000a. Proposal of *Sphingomonadaceae* fam. nov., consisting of *Sphingomonas* Yabuuchi et al. 1990, *Erythrobacter* Shiba and Shimidu 1982, *Erythromicrobium* Yurkov et al. 1994, *Porphyrobacter* Fuerst et al. 1993, *Zymomonas* Kluyver and van Niel 1936, and *Sandaracinobacter* Yurkov et al. 1997, with the type genus *Sphingomonas* Yabuuchi et al. 1990. Microbiol. Immunol. *44*: 563–575.

Kosako, Y., E. Yabuuchi, T. Naka, N. Fujiwara and K. Kobayashi. 2000b. *In* Validation of the publication of new names and new combinations previously effectively published outside the IJSEM. List No. 77. Int. J. Syst. Evol. Bacteriol. *50*: 1953.

Kosoy, M.Y., R.L. Regnery, O.I. Kosaya, D.C. Jones, E.L. Marston and J.E. Childs. 1998. Isolation of *Bartonella* spp. from embryos and neonates of naturally infected rodents. J. Wildlife Dis. *34*: 305–309.

Kosoy, M.Y., R.L. Regnery, T. Tzianabos, E.L. Marston, D.C. Jones, D. Green, G.O. Maupin, J.G. Olson and J.E. Childs. 1997. Distribution, diversity, and host specificity of *Bartonella* in rodents from the Southeastern United States. Am. J. Trop. Med. Hyg. *57*: 578–588.

Kotob, S.I., S.L. Coon, E.J. Quintero and R.M. Weiner. 1995. Homogentisic acid is the primary precursor of melanin synthesis in *Vibrio cholerae*, a *Hyphomonas* strain, and *Shewanella colwelliana*. Appl. Environ. Microbiol. *61*: 1620–1622.

Kottel, R.H., K. Bacon, D. Clutter and D. White. 1975. Coats from *Myxococcus xanthus*: characterization and synthesis during myxospore differentiation. J. Bacteriol. *124*: 550–557.

Kottel, R.H. and H.D. Raj. 1973. Pathways of carbohydrate metabolism in *Microcyclus* species. J. Bacteriol. *113*: 341–349.

Kottel, R. and D. White. 1974. Autolytic activity associated with myxospore formation in *Myxococcus xanthus*. Arch. Microbiol. *95*: 91–95.

Koukkou, A.-I., E. Douka, C. Drainas, C. Pale-Grosdemange and M. Rohmer. 1998. A dialkylcyclohexadienecarbinol from the bacterium *Zymomonas mobilis*, a novel type of potential membrane lipid. Tetrahedron Lett. *39*: 5193–5194.

Kouno, K. and A. Ozaki. 1975. Distribution and identification of methanol-utilizing bacteria. *In* Committee (Editor), Microbial Growth on C_1-Compounds, Society of Fermentation Technology, Osaka, Japan. pp. 11–21.

Kovach, M.E., R.W. Phillips, P.H. Elzer, R.M. Roop and K.M. Peterson. 1994. pBBR1MCS: a broad host range cloning vector. Biotechniques *16*: 800–802.

Kovács, N. 1956. Identification of *Pseudomonas pyocyanea* by the oxidase reaction. Nature *178*: 703.

Koval, S.F. and S.H. Hynes. 1991. Effect of paracrystalline protein surface layers on predation by *Bdellovibrio bacteriovorus*. J. Bacteriol. *173*: 2244–2249.

Kowal, J. 1961. *Spirillum* fever: Report of a case and review of the literature. N. Engl. J. Med. *264*: 123–128.

Kowallik, U. and E.G. Pringsheim. 1966. The oxidation of hydrogen sulfide by *Beggiatoa*. Am. J. Bot. *53*: 801–806.

Kozek, W.J. 1977. Transovarially-transmitted intracellular microorganisms in adult and larval stages of *Brugia malayi*. J. Parasitol. *63*: 992–1000.

Kozek, W.J. and H.F. Marroquin. 1977. Intracytoplasmic bacteria in *Onchocerca volvulus*. Am. J. Trop. Med. Hyg. *26*: 663–678.

Kozinska, A. and J. Antychowicz. 1996. Isolation of a new bacteria pathogenic to carp. Med. Weter. *52*: 657–660.

Krafft, T., A. Bowen, F. Theis and J.M. Macy. 2000. Cloning and sequencing of the genes encoding the periplasmic-cytochrome B-containing selenate reductase of *Thauera selenatis*. DNA Seq. *10*: 365–377.

Krafft, T., R. Gross and A. Kröger. 1995. The function of *Wolinella succinogenes* psr genes in electron transport with polysulphide as the terminal electron acceptor. Eur. J. Biochem. *230*: 601–606.

Krambovitis, E., M.B. McIllmurray, P.A. Lock, H. Holzel, M.R. Lifely and C. Moreno. 1987. Murine monoclonal-antibodies for detection of antigens and culture identification of *Neisseria meningitidis* group b and *Escherichia coli* K-1. J. Clin. Microbiol. *25*: 1641–1644.

Kramer, T.T. and J.M. Westergaard. 1977. Antigenicity of bdellovibrios. Appl. Environ. Microbiol. *33*: 967–970.

Krasil'nikov, N.A. 1949. Guide to the bacteria and actinomycetes, Akad. Nauk SSSR, Moscow. 1–830.

Krasil'nikov, N.A. and S.S. Belyaev. 1973. Systematics and classification of the genus *Caulobacter*. Izv. Akad. Nauk. SSR Ser. Biol. *3*: 313–323.

Kraut, M., I. Hugendieck, S. Herwig and O. Meyer. 1989. Homology and distribution of CO dehydrogenase structural genes in carboxydotrophic bacteria. Arch. Microbiol. *152*: 335–341.

Kraut, M. and O. Meyer. 1988. Plasmids in carboxydotrophic bacteria: physical and restriction analysis. Arch. Microbiol. *149*: 540–546.

Kreier, J.P. and M. Ristic. 1963. Anaplasmosis. XII. The growth and survival in deer and sheep of the parasites present in the blood of calves infected with the Oregon strain of *Anaplasma marginale*. Am. J. Vet. Res. *24*: 697–702.

Kreis, M., J. Eberspächer and F. Lingens. 1981. Detection and characterization of plasmids in chloridazon and antipyrin degrading bacteria. Zentbl. Bakteriol. Mikrobiol. Hyg. 1 Abt. Orig. C *2*: 45–60.

Kreke, B. and H. Cypionka. 1995. Energetics of sulfate transport in *Desulfomicrobium baculatum*. Arch. Microbiol. *163*: 307–309.

Krekeler, D., P. Sigalevich, A. Teske, H. Cypionka and Y. Cohen. 1997. A sulfate-reducing bacterium from the oxic layer of a microbial mat from Solar Lake (Sinai), *Desulfovibrio oxyclinae* sp. nov. Arch. Microbiol. *167*: 369–375.

Krekeler, D., P. Sigalevich, A. Teske, H. Cypionka and Y. Cohen. 2000. *In* Validation of the publication of new names and new combinations previously effectively published outside the IJSB. List No. 76. Int. J. Syst. Evol. Microbiol. *50*: 1699–1700.

Kreutzer, D.L. and D.C. Robertson. 1979. Surface macromolecules and virulence in intracellular parasitism: comparison of cell envelope components of smooth and rough strains of *Brucella abortus*. Infect. Immun. *23*: 819–828.

Kreutzer, D.L., J.W. Scheffel, L.R. Draper and D.C. Robertson. 1977. Mitogenic activity of cell wall components from smooth and rough strains of *Brucella abortus*. Infect. Immun. *15*: 842–845.

Krieg, A. 1961. Grundlagen der Insektenpathologie; viren-, rickettsien- und bakterien-Infektionen, Vol. 69: Wissenschaftliche Forschungsberichte. Naturwissenschaftliche, D. Steinkopef, Darmstadt.

Krieg, N.R. 1974. The genus *Spirillum*. *In* Buchanan and Gibbons (Editors), Bergey's Manual of Determinative Bacteriology, 8th ed., The Williams & Wilkins Co., Baltimore. pp. 196–207.

Krieg, N.R. 1976. Biology of the chemoheterotrophic spirilla. Bacteriol. Rev. *40*: 55–115.

Krieg, N.R. 1984a. Aerobic/microaerophilic, motile, helical/vibroid gram-negative bacteria. *In* Krieg and Holt (Editors), Bergey's Manual of Systematic Bacteriology, Vol. 1, The Williams & Wilkins Co., Baltimore. pp. 71–93.

Krieg, N.R. 1984b. Genus *Oceanospirillum*. *In* Krieg and Holt (Editors),

Bergey's Manual of Systematic Bacteriology, 1st Ed., Vol. 1, The Williams & Wilkins Co., Baltimore. pp. 104–110.

Krieg, N.R. and J. Döbereiner. 1984. Genus *Azospirillum*. *In* Krieg and Holt (Editors), Bergey's Manual of Systematic Bacteriology, Vol. 1, The Williams & Wilkins Co., Baltimore. pp. 94–104.

Krieger, C.J., W. Roseboom, S.P. Albracht and A.M. Spormann. 2001. A stable organic free radical in anaerobic benzylsuccinate synthase of *Azoarcus* sp. strain T. J. Biol. Chem. *276*: 12924–12927.

Kröger, A., V. Geisler, E. Lemma, F. Theis and R. Lenger. 1992. Bacterial fumarate respiration. Arch. Microbiol. *158*: 311–314.

Krooneman, J., E.B. Wieringa, E.R. Moore, J. Gerritse, R.A. Prins and J.C. Gottschal. 1996. Isolation of *Alcaligenes* sp. strain L6 at low oxygen concentrations and degradation of 3-chlorobenzoate via a pathway not involving (chloro)catechols. Appl. Environ. Microbiol. *62*: 2427–2434.

Kropinski, A.M. 1975. A chemically defined medium for *Aquaspirillum aquaticum* ATCC 11330. Can. J. Microbiol. *21*: 1886–1889.

Krum, J.G. and S.A. Ensign. 2000. Heterologous expression of bacterial epoxyalkane: Coenzyme M transferase and inducible coenzyme M biosynthesis in *Xanthobacter* strain Py2 and *Rhodococcus rhodochrous* B276. J. Bacteriol. *182*: 2629–2634.

Krumbein, W.E. and H.J. Altmann. 1973. A new method for the detection and enumeration of manganese oxidizing and reducing microorganisms. Helgol. Wiss. Meersunters *25*: 347–356.

Krumholz, L.R. 1997. *Desulfuromonas chloroethenica* sp. nov. uses tetrachloroethylene and trichloroethylene as electron acceptors. Int. J. Syst. Bacteriol. *47*: 1262–1263.

Krumholz, L.R., S.H. Harris, S.T. Tay and J.M. Suflita. 1999. Characterization of two subsurface hydrogen utilizing bacteria: *Desulfomicrobium hypogeium* sp. nov., *Acetobacterium psammolithicum* sp. nov. and their ecological roles. Appl. Environ. Microbiol. *65*: 2300–2306.

Kryukov, V.R., N.D. Savelyeva and M.A. Pusheva. 1983. *Calderobacterium hydrogenophilum*, nov. gen. nov. sp., an extreme thermophilic hydrogen bacterium, and its hydrogenase activity. Mikrobiologiya *52*: 781–788.

Krzemieniewska, H. and S. Krzemieniewski. 1926. Miksobakterje Polski (Die Myxobakterien von Polen). Acta Soc. Bot. Pol. *4*: 1–54.

Krzemieniewska, H. and S. Krzemieniewski. 1927. Miksobakterje Polski. Uzupelnienie. Acta Soc. Bot. Pol. *5*: 79–98.

Krzemieniewska, H. and S. Krzemieniewski. 1928. Morfologja komórki miksobakteryi (Zur Morphologie der Myxobakterienzelle). Acta Soc. Bot. Pol. *5*: 46–90.

Krzemieniewska, H. and S. Krzemieniewski. 1930. Miksobakterje Polski. Crtrzecia. Acta Soc. Bot. Pol. *7*: 250–173.

Krzemieniewska, H. and S. Krzemieniewski. 1937a. Die zellulosezersetzenden Myxobakterien. Bull. Int. Acad. Cracovie (Acad. Pol. Sci.) Ser. B. Sci. Nat. *1*: 11–31.

Krzemieniewska, H. and S. Krzemieniewski. 1937b. Über die Zersetzung der Zellulose durch Myxobakterien. Bull. Int. Acad. Cracovie (Acad. Pol. Sci.) Ser. B. Sci. Nat. *1*: 33–59.

Krzemieniewska, H. and S. Krzemieniewski. 1946. Myxobacteria of the species *Chondromyces* Berkeley and Curtis. Bull. Int. Acad. Cracovie (Acad. Pol. Sci.) Ser. B. Sci. Nat. *1*: 31–48.

Krziwon, C., U. Zahringer, K. Kawahara, B. Weidemann, S. Kusumoto, E.T. Rietschel, H.D. Flad and A.J. Ulmer. 1995. Glycosphingolipids from *Sphingomonas paucimobilis* induce monokine production in human mononuclear cells. Infect. Immun. *63*: 2899–2905.

Kucera, S. and R.S. Wolfe. 1957. A selective enrichment method for *Gallionella ferruginea*. J. Bacteriol. *74*: 344–349.

Kuehn, M., K. Lent, J. Haas, J. Hagenzieker, M. Cervin and A.L. Smith. 1992. Fimbriation of *Pseudomonas cepacia*. Infect. Immun. *60*: 2002–2007.

Kuenen, J.G. and S.C. Rittenberg. 1975. Incorporation of long-chain fatty acids of the substrate organism by *Bdellovibrio bacteriovorus* during intraperiplasmic growth. J. Bacteriol. *121*: 1145–1157.

Kuenen, J.G., L.A. Robertson and O.H. Tuovinen. 1992. The genera *Thiobacillus*, *Thiomicrospira*, and *Thiosphaera*. *In* Balows, A., H.G. Trüper, M. Dworkin, W. Harder and K.-H. Schleifer (Editors), The Prokaryotes: A Handbook of Bacteria: Ecophysiology, Isolation, Identi-

fication, Applications, 2nd Ed., Vol. 3, Springer-Verlag, New York. pp. 2638–2657.

Kuever, J., M. Könneke, A. Galushko and O. Drzyzga. 2001. Reclassification of *Desulfobacterium phenolicum* as *Desulfobacula phenolica* comb. nov. and description of strain SaxT as *Desulfotignum balticum* gen. nov., sp. nov. Int. J. Syst. Evol. Microbiol. *51*: 171–177.

Kuever, J., C. Wawer and R. Lillebaek. 1996. Microbiological observations in the anoxic basin Golfo Dulce, Costa Rica. Rev. Biol. Trop. *44*: 49–57.

Kühlwein, H. and E. Gallwitz. 1958. *Polyangium violaceum* nov. spec. ein Beitrag zur Kenntnis der Myxobakterien. Arch. Mikrobiol. *31*: 139–145.

Kühlwein, H. and H. Reichenbach. 1964. Ein neuer Vertreter der Myxobakteriengattung *Archangium* Jahn. Arch. Mikrobiol. *48*: 179–184.

Kühlwein, H. and B. Schlicke. 1971. *Polyangium luteum* Krzemieniewski in pure culture. J. Appl. Bact. *34*: 515–519.

Kuhn, D.A. 1981. The genera *Simonsiella* and *Alysiella*. *In* Starr, Stolp, Trüper, Balows and Schlegel (Editors), The Prokaryotes: A Handbook on Habitats, Isolation and Identification of Bacteria, Springer-Verlag, New York. 390–399.

Kuhn, D.A. and D.A. Gregory. 1978. Emendation of *Simonsiella muelleri* Schmid and description of *Simonsiella steedae* sp. nov., with designations of the respective proposed neotype and holotype strains. Curr. Microbiol. *1*: 11–14.

Kuhn, D.A., D.A. Gregory, G.E.J. Buchanan, M.D. Nyby and K.R. Daly. 1978. Isolation, characterization, and numerical taxonomy of *Simonsiella* strains from the oral cavities of cats, dogs, sheep, and humans. Arch. Microbiol. *118*: 235–241.

Kuhn, D.A., D.A. Gregory, M.D. Nyby and M. Mandel. 1977. Deoxyribonucleic acid base composition of *Simonsiellaceae*. Arch. Microbiol. *113*: 205–297.

Kuhn, D.A., D.A. Gregory, J. Pangborn and M. Mandel. 1974. *Simonsiella* strains from the human oral cavity. J. Dent. Res. *53*: 108.

Kuhn, D.A. and M.P. Starr. 1965. Clonal morphogenesis of *Lampropedia hyalina*. Arch. Microbiol. *52*: 360–375.

Kuhner, C.H., P.A. Hartman and M.J. Allison. 1996. Generation of a proton motive force by the anaerobic oxalate-degrading bacterium *Oxalobacter formigenes*. Appl. Environ. Microbiol. *62*: 2494–2500.

Kulla, H.G., R. Krieg, T. Zimmermann and T. Leisinger. 1984. Experimental evolution of azo dye-degrading bacteria. *In* Klug and Reddy (Editors), Current Perspectives in Microbial Ecology, American Society for Microbiology, Washington D.C. pp. 663–667.

Kumar, R., A.K. Banerjee, J.H. Bowdre, L.J. McElroy and N.R. Krieg. 1974. Isolation, characterization, and taxonomy of *Aquaspirillum bengal* sp. nov. Int. J. Syst. Bacteriol. *24*: 453–458.

Kummer, R.M. and L.D. Kuykendall. 1989. Symbiotic properties of amino acid auxotrophs of *Bradyrhizobium japonicum*. Soil Biol. Biochem. *21*: 779–782.

Kundig, C., C. Beck, H. Hennecke and M. Gottfert. 1995. A single rRNA gene region in *Bradyrhizobium japonicum*. J. Bacteriol. *177*: 5151–5154.

Kundig, C., H. Hennecke and M. Gottfert. 1993. Correlated physical and genetic map of the *Bradyrhizobium japonicum* 110 genome. J. Bacteriol. *175*: 613–622.

Kuner, J.M. and D. Kaiser. 1981. Introduction of transposon Tn5 into *Myxococcus* for analysis of developmental and other nonselectable mutants. Proc. Natl. Acad. Sci. U.S.A. *78*: 425–429.

Kunze, B., N. Bedorf, W. Kohl, G. Höfle and H. Reichenbach. 1989. Myxochelin-a, a new iron-chelating compound from *Angiococcus disciformis* (*Myxobacterales*)—Production, isolation, physicochemical and biological properties. J. Antibiot. *42*: 14–17.

Kunze, B., R. Jansen, G. Höfle and H. Reichenbach. 1994. Crocacin, a new electron transport inhibitor from *Chondromyces crocatus* (Myxobacteria)—Production, isolation, physicochemical and biological properties. J. Antibiot. *47*: 881–886.

Kunze, B., R. Jansen, L. Pridzun, E. Jurkiewicz, G. Hunsmann, G. Höfle and H. Reichenbach. 1993. Thiangazole, a new thiazoline antibiotic from *Polyangium* sp. (Myxobacteria)—Production, antimicrobial activity and mechanism of action. J. Antibiot. *46*: 1752–1755.

Kunze, B., R. Jansen, F. Sasse, G. Höfle and H. Reichenbach. 1995. Chondramides A-D, new antifungal and cytostatic depsipeptides from *Chondromyces crocatus* (Myxobacteria)—Production, physicochemical and biological properties. J. Antibiot. *48*: 1262–1266.

Kunze, B., W. Trowitzsch-Kienast, G. Höfle and H. Reichenbach. 1992. Nannochelins A, B and C, new iron-chelating compounds from *Nannocystis exedens* (myxobacteria). Production, isolation, physico-chemical and biological properties. J. Antibiot. *45*: 147–150.

Künzler, A. and N. Pfennig. 1973. Das Vorkommen von Bacteriochlorophyll a_P und a_{Gg} in Stammen aller Arten der *Rhodospirillaceae*. Arch. Mikrobiol. *91*: 83–86.

Kupsch, E.M., D. Aubel, C.P. Gibbs, A.F. Kahrs, T. Rudel and T.F. Meyer. 1996. Construction of Hermes shuttle vectors: A versatile system useful for genetic complementation of transformable and non-transformable *Neisseria* mutants. Mol. Gen. Genet. *250*: 558–569.

Kurita, T. and H. Tabei. 1967. On the pathogenic bacterium of bacterial grain rot of rice. Ann. Phytopathol. Soc. Japan *33*: 111.

Kurowski, W.M., A.H. Fensom and S.J. Pirt. 1975. Factors influencing the formation and stability of D-glucoside 3-dehydrogenase activity in cultures of *Agrobacterium tumefaciens*. J. Gen. Microbiol. *90*: 191–202.

Kurowski, W.M. and S.J. Pirt. 1971. The iron requirement of *Agrobacterium tumefaciens* for growth and 3-ketosucrose production. The removal of iron from solutions by Seitz filters. J. Gen. Microbiol. *68*: 65–69.

Kurtz, W.G.W. and T.A. La Rue. 1975. Nitrogenase activity in rhizobia in absence of host plant. Nature *256*: 407–409.

Kurzynski, T.A., D.M. Boehm, J.A. Rott-Petri, R.F. Schell and P.E. Allison. 1988. Antimicrobial susceptibilities of *Bordetella* species isolated in a multicenter pertussis surveillance project. Antimicrob. Agents Chemother. *32*: 137–140.

Küsel, K., T. Dorsch, G. Acker and E. Stackebrandt. 1999. Microbial reduction of Fe(III) in acidic sediments: isolation of *Acidiphilium cryptum* JF-5 capable of coupling the reduction of Fe(III) to the oxidation of glucose. Appl. Environ. Microbiol. *65*: 3633–3640.

Kuspa, A., D. Vollrath, Y. Cheng and D. Kaiser. 1989. Physical mapping of the *Myxococcus xanthus* genome by random cloning in yeast artificial chromosomes. Proc. Natl. Acad. Sci. U.S.A. *86*: 8917–8921.

Kuttler, K.L. 1984. *Anaplasma* infections in wild and domestic ruminants: a review. J. Wildl. Dis. *20*: 12–20.

Kutuzova, R.S., D.R. Gabe and I.M. Kravkina. 1972. Electron microscopic study of the growth of ooze iron-manganese microorganisms. Mikrobiologiia *41*: 1099–1102.

Kützing, F.T. 1833. Beitrag zur Kenntnis über die Entstehung und Metamorphose der niederen vegetalischen Organismen, nebst einer systematischen Zusammensetzung der hierher gehörigen niederen Algenformen. Linnaea *8*: 335–387.

Kützing, F.T. 1843. Phycologia Generales, Leipzig.

Kützing, F.T. 1847. Diagnosen und Bemerkungen zu oder kritischen Algen. Bot. Ztg. *5*: 22–25.

Kuykendall, L.D. 1981. Mutants of *Rhizobium* that are altered in legume interaction and nitrogen fixation. *In* Giles and Atherly (Editors), Biology of the *Rhizobiaceae*: International Review of Cytology, Supplement 13, Academic Press, New York. pp. 299–309.

Kuykendall, L.D. 1987. Isolation and identification of genetically marked strains of nitrogen-fixing microsymbionts of soybeans. *In* Elkan (Editor), Symbiotic Nitrogen Fixation Technology, Marcel Dekker, New York. pp. 205–220.

Kuykendall, L.D. and G.H. Elkan. 1976. *Rhizobium japonicum* derivatives differing in nitrogen-fixing efficiency and carbohydrate utilization. Appl. Environ. Microbiol. *32*: 511–519.

Kuykendall, L.D., Y.D. Gaur and S.K. Dutta. 1993a. Genetic diversity among *Rhizobium* strains from *Cicer arietinum* L. Lett. Appl. Microbiol. *17*: 259–263.

Kuykendall, L.D., F.M. Hashem, G.R. Bauchan, T.E. Devine and R.B. Dadson. 1999. Symbiotic competence of *Sinorhizobium fredii* on twenty alfalfa cultivars of diverse dormancy. Symbiosis *27*: 1–16.

Kuykendall, L.D. and W.J. Hunter. 1991. Enhancement of nitrogen fixation with *Bradyrhizobium japonicum* mutants, United States Patent No. 5,021,076

Kuykendall, L.D., M.A. Roy, J.J. O'Neill and T.E. Devine. 1988. Fatty acids, antibiotic resistance, and deoxyribonucleic acid homology groups of *Bradyrhizobium japonicum*. Int. J. Syst. Bacteriol. *38*: 358–361.

Kuykendall, L.D., B. Saxena, T.E. Devine and S.E. Udell. 1992. Genetic diversity in *Bradyrhizobium japonicum* Jordan 1982 and a proposal for *Bradyrhizobium elkanii* sp.nov. Can. J. Microbiol. *38*: 501–505.

Kuykendall, L.D., B. Saxena, T.E. Devine and S.E. Udell. 1993b. *In* Validation of the publication of new names and new combinations previously effectively published outside the IJSB. List No. 45. Int. J. Syst. Bacteriol. *43*: 398–399.

la Rivière, J.W.M. 1974. Genus *Thiovulum*. *In* Buchanan and Gibbons (Editors), Bergey's Manual of Determinative Bacteriology, 8th Ed., The Williams & Wilkins Co., Baltimore. p. 463.

la Rivière, J.W.M. and K. Schmidt. 1981. Morphologically conspicuous sulfur-oxidizing bacteria. *In* Starr, Stolp, Trüper, Balows and Schlegel (Editors), The Prokaryotes: A Handbook on Habitats, Isolation and Identification of Bacteria, Springer-Verlag, Berlin. pp. 1037–1048.

La Scola, B., R.J. Birtles, M.N. Mallet and D. Raoult. 1998a. *Massilia timonae* gen. nov., sp. nov., isolated from blood of an immunocompromised patient with cerebellar lesions. J. Clin. Microbiol. *36*: 2847–2852.

La Scola, B., R.J. Birtles, M.N. Mallet and D. Raoult. 2000. *In* Validation of publication of new names and new combinations previously effectively published outside the IJSEM. List No. 73. Int. J. Syst. Evol. Microbiol. *50*: 423–424.

La Scola, B., I. Iorgulescu and C. Bollini. 1998b. Five cases of *Kingella kingae* skeletal infection in a French hospital. Eur. J. Clin. Microbiol. Infect. Dis. *17*: 512–515.

La Scola, B. and D. Raoult. 1996a. Diagnosis of Mediterranean spotted fever by cultivation of *Rickettsia conorii* from blood and skin samples using the centrifugation-shell vial technique and by detection of *R. conorii* in circulating endothelial cells: A 6-year follow-up. J. Clin. Microbiol. *34*: 2722–2727.

La Scola, B. and D. Raoult. 1996b. Serological cross-reactions between *Bartonella quintana*, *Bartonella henselae*, and *Coxiella burnetii*. J. Clin. Microbiol. *34*: 2270—2274.

La Scola, B. and D. Raoult. 1999a. *Afipia felis* in hospital water supply in association with free-living amoebae. Lancet *353*: 1330.

La Scola, B. and D. Raoult. 1999b. Culture of *Bartonella quintana* and *Bartonella henselae* from human samples: a 5-year experience (1993 to 1998). J. Clin. Microbiol. *37*: 1899–1905.

La Scola, B. and D. Raoult. 1999c. Third human isolate of a *Desulfovibrio* sp. identical to the provisionally named *Desulfovibrio fairfieldensis*. J. Clin. Microbiol. *37*: 3076–3077.

Laanbroek, H.J., T. Abee and I.L. Voogd. 1982. Alcohol conversions by *Desulfobulbus propionicus* Lindhorst in the presence and absence of sulfate and hydrogen. Arch. Microbiol. *133*: 178–184.

Laanbroek, H.J. and H.J. Geerligs. 1983. Influence of clay particles (illite) on substrate utilization by sulfate-reducing bacteria. Arch. Microbiol. *134*: 161–163.

Laanbroek, H.J., W. Kingma and H. Veldkamp. 1977. Isolation of an aspartate fermenting, free-living *Campylobacter* species. FEMS Microbiol. Lett. *1*: 99–102.

Laanbroek, H.J. and N. Pfennig. 1981. Oxidation of short-chain fatty acids by sulfate-reducing bacteria in freshwater and in marine sediments. Arch. Microbiol. *128*: 330–335.

Laanbroek, H.J., L.J. Stal and H. Veldkamap. 1978. Utilization of hydrogen and formate by *Campylobacter* spec. under aerobic and anaerobic conditions. Arch. Microbiol. *119*: 99–102.

Laanbroek, H.J. and H. Veldkamp. 1979. Growth yield and energy generation in anaerobically-grown *Campylobacter* species. Arch. Microbiol. *120*: 47–51.

Labarca, J.A., W.E. Trick, C.L. Peterson, L.A. Carson, S.C. Holt, M.J. Arduino, M. Meylan, L. Mascola and W.R. Jarvis. 1999. A multistate nosocomial outbreak of *Ralstonia pickettii* colonization associated with an intrinsically contaminated respiratory care solution. Clin. Infect. Dis. *29*: 1281–1286.

Labaw, L.W. and V.M. Mosley. 1955. Periodic structure of the flagella of *Brucella bronchiseptica*. Biochim. Biophys. Acta *17*: 322–324.

Labbé, A. 1894. Recherches zoologiques et biologiques sur les parasites endoglobulaires du sang des vertébrés. Arch. Zool. Exp. Gén. (3e Série, Tome II) *22*: 55–258.

Labbé, M., W. Hansen, E. Schoutens and E. Yourassowsky. 1977. Isolation of *Bacteroides corrodens* and *Eikenella corrodens* from human clinical specimens. Comparative study of incidence and methods of identification. Infection *5*: 159–162.

Labrenz, M., M.D. Collins, P.A. Lawson, B.J. Tindall, G. Braker and P. Hirsch. 1998. *Antarctobacter heliothermus* gen. nov., sp. nov., a budding bacterium from hypersaline and heliothermal Ekho Lake. Int. J. Syst. Bacteriol. *48*: 1363–1372.

Labrenz, M., M.D. Collins, P.A. Lawson, B.J. Tindall, P. Schumann and P. Hirsch. 1999. *Roseovarius tolerans* gen. nov., sp. nov., a budding bacterium with variable bacteriochlorophyll *a* production from hypersaline Ekho Lake. Int. J. Syst. Bacteriol. *49*: 137–147.

Labrenz, M., B.J. Tindall, P.A. Lawson, M.D. Collins, P. Schumann and P. Hirsch. 2000. *Staleya guttiformis* gen. nov., sp. nov. and *Sulfitobacter brevis* sp. nov., α-3-*Proteobacteria* from hypersaline, heliothermal and meromictic antarctic Ekho Lake. Int. J. Syst. Evol. Microbiol. *50*: 303–313.

Labuzek, S., A. Mrozik, J. Pajak and J. Kasiak. 1994. Transformation of *E. coli* with plasmids coding for degradation of aromatic structure of phenols. Acta Biochim. Pol. *41*: 127–128.

Lacava, P.M. and M.R. Ortolono. 1997. Utilization of *Spirillum volutans* for monitoring the toxicity of effluents of a cellulose and paper industry. Rev. Microbiol. *28*: 23–24.

Lackman, D.B., E.J. Bell, H.G. Stoenner and E.G. Pickens. 1965. The Rocky Mountain spotted fever group of rickettsias. Health Lab. Sci. *2*: 135–141.

Lackman, D.B., R.R. Parker and R.K. Gerloff. 1949. Serological characteristics of a pathogenic rickettsia occurring in *Amblyomma maculatum*. Pub. Health Rep. *64*: 1342–1349.

Lacy, B.W. 1960. Antigenic modulation of *Bordetella pertussis*. J. Hyg. *58*: 57–93.

Ladha, J.K. and R.B. So. 1994. Numerical taxonomy of photosynthetic rhizobia nodulating *Aeschynomene* species. Int. J. Syst. Bacteriol. *44*: 62–73.

Laempe, D., W. Eisenreich, A. Bacher and G. Fuchs. 1998. Cyclohexa-1,5-diene-1-carboxyl-CoA hydratase, an enzyme involved in anaerobic metabolism of benzoyl-CoA in the denitrifying bacterium *Thauera aromatica*. Eur. J. Biochem. *255*: 618–627.

Laempe, D., M. Jahn and G. Fuchs. 1999. 6-Hydroxycyclohex-1-ene-1-carbonyl-CoA dehydrogenase and 6-oxocyclohex-1-ene-1-carbonyl-CoA hydrolase, enzymes of the benzoyl-CoA pathway of anaerobic aromatic metabolism in the denitrifying bacterium *Thauera aromatica*. Eur. J. Biochem. *263*: 420–429.

Lafay, B. and J.J. Burdon. 1998. Molecular diversity of Rhizobia occuring on native shrubby legumes in south eastern Austrailia. Appl. Environ. Microbiol. *64*: 3989–3997.

Lafay, B., R. Ruimy, C. Rausch de Traubenberg, V. Breittmayer, M.J. Gauthier and R. Christen. 1995. *Roseobacter algicola* sp. nov., a new marine bacterium isolated from the phycosphere of the toxin-producing dinoflagellate *Prorocentrum lima*. Int. J. Syst. Bacteriol. *45*: 290–296.

Lafitskaya, T.N. and L.V. Vasilyeva. 1976. A new triangle bacterium. Mikrobiologiya *45*: 812–816.

Lafitskaya, T.N., L.V. Vasilyeva, E.N. Krasilnikova and N.I. Alexandruskina. 1976. Physiology of *Prosthecomicrobium polyspheroidum*. Izv. Akad. Nauk S.S.S.R. Biol. *6*: 849–857.

Lagenaur, C., S. Farmer and N. Agabian. 1977. Adsorption properties of stage-specific *Caulobacter* phage φCbK. Virology *77*: 401–407.

Laguerre, G., M.R. Allard, F. Revoy and N. Amarger. 1994. Rapid identification of rhizobia by restriction fragment length polymorphism analysis of PCR-amplified 16S rRNA genes. Appl. Environ. Microbiol. *60*: 56–63.

Laguerre, G., B. Bossand and R. Bardin. 1987. Free-living dinitrogen-fixing bacteria isolated from petroleum refinery oily sludge. Appl. Environ. Microbiol. *53*: 1674–1678.

Laguerre, G., P. Van Berkum, N. Amarger and D. Prevost. 1997. Genetic diversity of rhizobial symbionts isolated from legume species within the genera *Astragalus*, *Oxytropis*, and *Onobrychis*. Appl. Environ. Microbiol. *63*: 4748–4758.

Lai, C.-H., M.A. Listgarten, A.C.R. Tanner and S.S. Socransky. 1981. Ultrastructures of *Bacteroides gracilis*, *Campylobacter concisus*, *Wolinella recta*, and *Eikenella corrodens*, all from human periodontal disease. Int. J. Syst. Bacteriol. *31*: 465–475.

Laird, M. and F.A. Lari. 1957. The avian blood parasite *Babesia moshkovskii* (Schurenkova, 1938), with a record from *Corvus splejdens* Vieillot in Pakistan. Can. J. Zool. *35*: 783–795.

Lajoie, C.A., A.C. Layton, I.R. Gregory, G.S. Sayler, D.E. Taylor and A.J. Meyers. 2000. Zoogleal clusters and sludge dewatering potential in an industrial activated-sludge wastewater treatment plant. Water Environ. Res. *72*: 56–64.

Lalucat, J., O. Meyer, F. Mayer, R. Pares and H.G. Schlegel. 1979. R-bodies in newly isolated free-living hydrogen-oxidizing bacteria. Arch. Microbiol. *121*: 9–15.

Lalucat, J., R. Parés and H.G. Schlegel. 1982. *Pseudomonas taeniospiralis* sp. nov., an R-body-containing hydrogen bacterium. Int. J. Syst. Bacteriol. *32*: 332–338.

Lamb, J.W. and H. Hennecke. 1986. *In Bradyrhizobium japonicum* the common nodulation genes, *nodABC*, are linked to *nifA* and *fixA*. Mol. Gen. Genet. *202*: 512–517.

Lambert, B., H. Joos, S. Dierickx, R. Vantomme, J. Swings, K. Kersters and M. Van Montagu. 1990. Identification and plant interaction of a *Phyllobacterium* sp., a predominant rhizobacterium of young sugar beet plants. Appl. Environ. Microbiol. *56*: 1093–1102.

Lambert, M.A., C.M. Patton, T.J. Barrett and C.W. Moss. 1987. Differentiation of *Campylobacter* and *Campylobacter*-like organisms by cellular fatty acid composition. J. Clin. Microbiol. *25*: 706–713.

Lambina, V.A., A.V. Afinogenova, S.R. Penabad, S.M. Konovalova and L.V. Andreev. 1983. A new species of the exoparasitic bacterium, *Micavibrio*, destroying gram-negative bacteria. Mikrobiologiya *52*: 777–780.

Lambina, V.A., A.V. Afinogenova, S. Romai Penabad, S.M. Konovalova and A.P. Pushkareva. 1982. *Micavibrio admirandus* gen. et sp. nov. Mikrobiologiya *51*: 114–117.

Lambina, V.A., A.V. Afinogenova, S. Romai Penabad, S.M. Konovalova and A.P. Pushkareva. 1989. *In* Validation of the publication of new names and new combinations previously effectively published outside the IJSB. List No. 28. Int. J. Syst. Bacteriol. *39*: 93–94.

Lamm, R.B. and C.A. Neyra. 1981. Characterization and cyst production of azospirilla isolated from selected grasses growing in New Jersey and New York, U.S.A. Can. J. Microbiol. *27*: 1320–1325.

Lampky, J.R. 1976. Ultrastructure of *Polyangium cellulosum*. J. Bacteriol. *126*: 1278–1284.

Lampky, J.R. and E.R. Brockman. 1977. Fluorescence of *Myxococcus stipitatus*. Int. J. Syst. Bacteriol. *27*: 161–161.

Lampson, B.C. 1993. Retroelements of the myxobacteria. *In* Dworkin and Kaiser (Editors), Myxobacteria II, American Society for Microbiology, Washington. D.C. pp. 109–128.

Lampson, B.C., M. Inouye and S. Inouye. 1991. Survey of multicopy single-stranded DNAs and reverse transcriptase genes among natural isolates of *Myxococcus xanthus*. J. Bacteriol. *173*: 5363–5370.

Lampson, B.C. and S.A. Rice. 1997. Repetitive sequences found in the chromosome of the myxobacterium *Nannocystis exedens* are similar to msDNA: A possible retrotransposition event in bacteria. Mol. Microbiol. *23*: 813–823.

Landa, A.S., E.M. Sipkema, J. Weijma, A.A. Beenackers, J. Dolfing and D.B. Janssen. 1994. Cometabolic degradation of trichloroethylene by *Pseudomonas cepacia* G4 in a chemostat with toluene as the primary substrate. Appl. Environ. Microbiol. *60*: 3368–3374.

Lane, D.J., A.P. Harrison, Jr., D. Stahl, B. Pace, S.J. Giovannoni, G.J. Olsen and N.R. Pace. 1992. Evolutionary relationships among sulfur- and iron-oxidizing eubacteria. J. Bacteriol. *174*: 269–278.

Lane, D.J., D.A. Stahl, G.J. Olsen, D.J. Heller and N.R. Pace. 1985. Phylogenetic analysis of the genera *Thiobacillus* and *Thiomicrospira* by 5S ribosomal RNA sequences. J. Bacteriol. *163*: 75–81.

Lane, R.S., R.W. Emmons, D.V. Dondero and B.C. Nelson. 1981. Ecology of tick-borne agents in California. I. Spotted fever group rickettsiae. Am. J. Trop. Med. Hyg. *30*: 239–252.

Lang, E. and K.A. Malik. 1996. Maintenance of biodegradation capacities of aerobic bacteria during long-term preservation. Biodegradation 7: 65–71.

Langenberg, W., E.A.J. Rauws, A. Widjojokusumo, G.N.J. Tytgat and H.C. Zanen. 1986. Identification of *Campylobacter pyloridis* isolates by restriction endonuclease DNA analysis. J. Clin. Microbiol. *24*: 414–417.

Langendijk, P.S., E.M. Kulik, H. Sandmeier, J. Meyer and J.S. van der Hoeven. 2001. Isolation of *Desulfomicrobium orale* sp. nov. and *Desulfovibrio* strain NY682, oral sulfate-reducing bacteria involved in human periodontal disease. Int. J. Syst. Evol. Bacteriol. *51*: 1035–1044.

Langeron, M. 1923. Les oscillariées parasites du tube digestif de l' homme et des animaux. Ann. Parisitol. Hum. Comp. *1*: 113–123.

Langille, S.E. and R.M. Weiner. 1998. Spatial and temporal deposition of *Hyphomonas* strain VP-6 capsules involved in biofilm formation. Appl. Environ. Microbiol. *64*: 2906–2913.

Langworthy, N.G., A. Renz, U. Mackenstedt, K. Henkle-Duhrsen, M.B. de Bronsvoort, V.N. Tanya, M.J. Donnelly and A.J. Trees. 2000. Macrofilaricidal activity of tetracycline against the filarial nematode *Onchocerca ochengi*: Elimination of *Wolbachia* precedes worm death and suggests a dependent relationship. Proc. R. Soc. Lond. B Biol. Sci. *267*: 1063–1069.

Lanys, S.G. 1972. Morphological studies of cell envelope differences among colony variants of *Lampropedia hyalina*, Thesis, University of Western Ontario, London, Ontario.

Lapage, S.P., L.R. Hill and J.D. Reeve. 1968. *Pseudomonas stutzeri* in pathological material. J. Med. Microbiol. *1*: 195–202.

Lapage, S.P., J.E. Shelton, T.G. Mitchell and A.R. MacKenzie. 1970. Culture collections in the preservation of bacteria. *In* Norris and Ribbons (Editors), Methods Microbiol, Vol. 3A, Academic Press, London. pp. 136–228.

Lapage, S.P., P.H.A. Sneath, E.F. Lessel Jr., V.B.D. Skerman, H.P.R. Seeliger and W.A. Clark (Editors). 1992. International Code of Nomenclature of Bacteria (1990) Revision. Bacteriological Code, American Society for Microbiology, Washington, DC.

Lapchine, L. 1979. Regularly arranged structures on the surface of some *Pseudomonas* sp. FEMS Microbiol. Lett. *5*: 223–225.

Lara, J.C. and A. Konopka. 1987. Isolation of motile variants from gas-vacuolate strains of *Ancylobacter aquaticus*. J. Gen. Microbiol. *133*: 1489–1494.

Large, P.J. and H. MacDougall. 1975. An enzymic method for the microestimation of trimethylamine. Anal. Biochem. *64*: 304–310.

Large, P.J., J.B.M. Meiberg and W. Harder. 1979. Cytochrome c_{co} is not a primary electron-acceptor for the amine dehydrogenases of *Hyphomicrobium* X. FEMS Microbiol. Lett. *5*: 281–286.

Large, P.J. and J.R. Quayle. 1963. Microbial growth on C_1 compounds. 5. Enzyme activities of extracts of *Pseudomonas* AM1. Biochem. J. *87*: 386–395.

Larkin, J.M. and R. Borrall. 1978. *Spirosomaceae*, a new family to contain the genera *Spirosoma* Migula 1894, *Flectobacillus* Larkin et al, 1977, and *Runella* Larkin and Williams 1978. Int. J. Syst. Bacteriol. *28*: 595–596.

Larkin, J.M. and R. Borrall. 1979. Proposal of ATCC 25396 as the neotype strain of *Microcyclus aquaticus* Ørskov 1928. Int. J. Syst. Bacteriol. *29*: 414–415.

Larkin, J.M., P.M. Williams and R. Taylor. 1977. Taxonomy of the genus *Microcyclus* Ørskov 1928: reintroduction and emendation of the genus *Spirosoma* Migula 11894 and proposal of a new genus, *Flectobacillus*. Int. J. Syst. Bacteriol. *27*: 147–156.

Larsen, N., G.J. Olsen, B.L. Maidak, R. McCaughey, R. Overbeek, T.J. Macke, T.L. Marsh and C.R. Woese. 1993. RNA database project. Nucleic Acids Res. *21*: 191–198.

Larson, R.J. and J.L. Pate. 1975. Growth and morphology of *Asticcacaulis biprosthecum* in defined media. Arch. Microbiol. *106*: 147–157.

LaScolea, L.J., Jr. and F.E. Young. 1974. Development of defined minimal medium for the growth of *Neisseria gonorrhoeae*. Appl. Microbiol. *28*: 70–76.

Lasko, D.R., C. Schwerdel, J.E. Bailey and U. Sauer. 1997. Acetate-specific stress response in acetate-resistant bacteria: an analysis of protein patterns. Biotechnol. Prog. *13*: 519–523.

Lasserre, E., F. Godard, T. Bouquin, J.A. Hernandez, J.C. Pech, D. Roby and C. Balague. 1997. Differential activation of two ACC oxidase gene promoters from melon during plant development and in response to pathogen attack. Mol. Gen. Genet. *256*: 211–222.

Lastovica, A.J. 1996. *Campylobacter/Helicobacter* bacteraemia in Cape Town, South Africa, 1977–1995. *In* Newell, Ketley and Feldman (Editors), Campylobacters, Helicobacters, and Related Organisms, Plenum Press, New York. pp. 475–479.

Latin, R.X. and D.L. Hopkins. 1995. Bacterial fruit blotch of watermelon—the hypothetical exam question becomes reality. Plant Dis. *79*: 761–765.

Lau, P.P., B. DeBrunner-Vossbrinck, B. Dunn, K. Miotto, M.T. MacDonell, D.M. Rollins, C.J. Pillidge, R.B. Hespell, R.R. Colwell, M.L. Sogin and G.E. Fox. 1987. Phylogenetic diversity and position of the genus *Campylobacter*. Syst. Appl. Microbiol. *9*: 231–238.

Laue, B.E. and R.E. Gill. 1994. Use of a phase variation-specific promoter of *Myxococcus xanthus* in a strategy for isolating a phase-locked mutant. J. Bacteriol. *176*: 5341–5349.

Laue, H., K. Denger and A.M. Cook. 1997. Taurine reduction in anaerobic respiration of *Bilophila wadsworthia* RZATAU. Appl. Environ. Microbiol. *63*: 2016–2021.

Laue, H., M. Friedrich, J. Ruff and A.M. Cook. 2001. Dissimilatory sulfite reductase (desulfoviridin) of the taurine-degrading, non-sulfate-reducing bacterium *Bilophila wadsworthia* RZATAU contains a fused DsrB–DsrD subunit. J. Bacteriol. *183*: 1727–1733.

Laughon, B.E. 1973. MS Thesis, Virginia Polytechnic Institute and State University

Laughon, B.E. and N.R. Krieg. 1974. Sugar catabolism in *Aquaspirillum gracile*. J. Bacteriol. *119*: 691–697.

Lauquin, G.J., A.M. Duplaa, G. Klein, A. Rousseau and P.V. Vignais. 1976. Isobongkrekic acid, a new inhibitor of mitochondrial ADP–ATP transport: radioactive labeling and chemical and biological properties. Biochemistry 15: 2323–2327.

Lauritis, J.A. 1967. Fine structure of an unusual photosynthetic bacterium, Thesis, Iowa State University, Ames, Iowa.

Lauterbach, F., C. Körtner, S.P.J. Albracht, G. Unden and A. Kröger. 1990. The fumarate reductase operon of *Wolinella succinogenes*: sequence and expression of the *frdA* and *frdB* genes. Arch. Microbiol. *154*: 386–393.

Lauterborn, R. 1915. Die sapropelische Lebewelt. Verh. Natu rh.-Med. Ver. Heidelb. *13*: 437–438.

Lautrop, H. 1960. Laboratory diagnosis of whooping-cough or *Bordetella* infections. Bull. W.H.O. *23*: 15–31.

Lautrop, H. 1967. *Agrobacterium* spp. isolated from clinical specimens. Acta Path. Microbiol. Scand., Suppl. *187*: 63–64.

Lautrop, H., K. Bovre and W. Frederiksen. 1970. A *Moraxella*-like microorganism isolated from the genito-urinary tract of man. Acta Pathol. Microbiol. Scand. Sect. B Microbiol. Immunol. *78*: 255–256.

Laverman, A.M., J.S. Blum, J.K. Schaefer, E.J.P. Phillips, D.R. Lovley and R.S. Oremland. 1995. Growth of strain SES-3 with arsenate and other diverse electron acceptors. Appl. Environ. Microbiol. *61*: 3556–3561.

Law, I.J. 1979. Resistance of *Rhizobium* specific for *Lotononis bainesii* to ultra-violet radiation. Soil Biol. Biochem. *11*: 87–88.

Lawson, A.J., D. Linton, J. Stanley and R.J. Owen. 1997. Polymerase chain reaction detection and speciation of *Campylobacter upsaliensis* and *C. helveticus* in human faeces and comparison with culture techniques. J. Appl. Microbiol. *83*: 375–380.

Lawson, A.J., S.L.W. On, J.M.J. Logan and J. Stanley. 2001. *Campylobacter hominis* sp. nov., from the human gastrointestinal tract. Int. J. Syst. Evol. Microbiol. *51*: 651–660.

Lawson, G.M. 1940. Modified technique for staining capsules of *Haemophilus pertussis*. J. Lab. Clin. Med. *25*: 435–438.

Lawson, G.H.K., J.L. Leaver, G.W. Pettigrew and A.C. Rowland. 1981. Some features of *Campylobacter sputorum* subsp. *mucosali* subsp. nov., nom. rev. and their taxonomic significance. Int. J. Syst. Bacteriol. *31*: 385–391.

Lawson, G.H., R.A. Mackie, D.G. Smith and S. McOrist. 1995. Infection of cultured rat enterocytes by ileal symbiont intracellularis depends on host cell function and actin polymerisation. Vet. Microbiol. *45*: 339–350.

Lawson, G.H., S. McOrist, S. Jasni and R.A. Mackie. 1993. Intracellular bacteria of porcine proliferative enteropathy: cultivation and maintenance *in vitro*. J. Clin. Microbiol. *31*: 1136–1142.

Lawson, G.H.K., A.C. Rowland and L. Roberts. 1977. The surface antigens of *Campylobacter sputorum* subsp. *mucosalis*. Res. Vet. Sci. *23*: 378–382.

Lawson, G.H.K., A.C. Rowland and P. Wooding. 1975. The characterization of *Campylobacter sputorum* subsp. *mucosalis* isolated from pigs. Res. Vet. Sci. *18*: 121–126.

Lawson, P.A. and M.D. Collins. 1996a. Description of *Bartonella clarridgeiae* sp nov isolated from the cat of a patient with *Bartonella henselae* septicemia. Med. Microbiol. Lett. *5*: 64–73.

Lawson, P.A. and M.D. Collins. 1996b. *In* Validation of the publication of new names and new combinations previously effectively published outside the IJSB. List No. 58. Int. J. Syst. Bacteriol. *46*: 836–837.

Lawton, W.D., M.A. Bellinger and M.G. Schling, H.A.. 1976. Bacteriocin production by *Neisseria gonorrhoeae*. Antimicrob. Agents Chemother. *10*: 417–420.

Layh, G., R. Böhm, J. Eberspächer and F. Lingens. 1983. Serological studies on chloridazon degrading bacteria. Syst. Appl. Microbiol. *4*: 459–469.

Layton, A.C., P.N. Karanth, C.A. Lajoie, A.J. Meyers, I.R. Gregory, R.D. Stapleton, D.E. Taylor and G.S. Sayler. 2000. Quantification of *Hyphomicrobium* populations in activated sludge from an industrial wastewater treatment system as determined by 16S rRNA analysis. Appl. Environ. Microbiol. *66*: 1167–1174.

Le Gall, J. 1963. A new species of *Desulfovibrio*. J. Bacteriol. *86*: 1120.

Leahy, J.G., A.M. Byrne and R.H. Olsen. 1996. Comparison of factors influencing trichloroethylene degradation by toluene oxidizing bacteria. Appl. Environ. Microbiol. *62*: 825–833.

Leahy, J.G. and R.R. Colwell. 1990. Microbial degradation of hydrocarbons in the environment. Microbiol. Rev. *54*: 305–315.

Lebedinskii, A.V. 1981. *Hyphomicrobium vulgare* growth with various electron acceptors. Mikrobiologiya *50*: 665–669.

Lebedinskii, A.V. and J.Y. Vedenina. 1981. Production of nitrous and nitric oxides by methylotrophic denitrifying organisms. Mikrobiologiya *50*: 757–762.

Lebedinskii, A.V. and I.Y. Vedenina. 1987. Electron-transport from methanol to nitrate in *Hyphomicrobium* Z-3. Microbiology *56*: 563–567.

Lebuhn, M., W. Achouak, M. Schloter, O. Berge, H. Meier, M. Barakat, A. Hartmann and T. Heulin. 2000. Taxonomic characterization of *Ochrobactrum* sp. isolates from soil samples and wheat roots, and description of *Ochrobactrum tritici* sp. nov. and *Ochrobactrum grignonense* sp. nov. Int. J. Syst. Evol. Microbiol. *50*: 2207–2223.

Lechevalier, H.A. and M.P. Lechevalier. 1981a. Actinomycete genera "in search of a family". *In* Starr, Stolp, Trüper, Balows and Schlegel (Editors), The Prokaryotes: A Handbook on Habitats, Isolation, and Identification of Bacteria, Vol. 2, Springer -Verlag, Berlin. pp. 2118–2119.

Lechevalier, H.A. and M.P. Lechevalier. 1981b. Genus *Oerskovia* Prauser, Lechevalier and Lechevalier 1970, 534; emended Lechevalier 1972, 263^{AL}. *In* Starr, Stolp, Trüper, Balows and Schlegel (Editors), The Prokaryotes: A Handbook on Habitats, Isolation, and Identification of Bacteria, Vol. 2, Springer -Verlag, Berlin. pp. 1489–1491.

Lechevalier, H.A. and M.P. Lechevalier. 1981c. Introduction to the order *Actinomycetales*. *In* Starr, Stolp, Trüper, Balows and Schlegel (Editors), The Prokaryotes: A Handbook on Habitats, Isolation, and Identification of Bacteria, Vol. 2, Springer -Verlag, Berlin. pp. 1915–1922.

Lechner, S. and R. Conrad. 1997. Detection in soil of aerobic hydrogen-oxidizing bacteria related to *Alcaligenes eutrophus* by PCR and hybridization assays targeting the gene of the membrane-bound (NiFe) hydrogenase. FEMS Microbiol. Ecol. *22*: 193–206.

Lechner, U., R. Baumbach, D. Becker, V. Kitunen, G. Auling and M. Salkinoja-Salonen. 1995. Degradation of 4-chloro-2-methylphenol by an activated sludge isolate and its taxonomic description. Biodegradation *6*: 83–92.

Lee, A. 1997. The pathogenesis of *Helicobacter pylori* infection. Baillieres Clin. Infect. Dis. *4*: 341–365.

Lee, A. 1998. Animal models for host-pathogen interaction studies. Br. Med. Bull. *54*: 163–173.

Lee, A., M. Chen, N. Coltro, J. O'Rourke, S.L. Hazell, P. Hu and Y. Li. 1993. Long term infection of the gastric mucosa with *Helicobacter* species does induced atrophic gastritis in an animal model of *Helicobacter pylori* infection. Zentbl. Bakteriol. *280*: 38–50.

Lee, A.., S.L. Hazell, J. O'Rourke and S. Kouprach. 1988. Isolation of a spiral-shaped bacterium from the cat stomach. Infect. Immun. *56*: 2843–2850.

Lee, A., S. Krakowka, J.G. Fox, G. Otto, K.A. Eaton and J.C. Murphy. 1992a. Role of *Helicobacter felis* in chronic canine gastritis. Vet. Pathol. *29*: 487–494.

Lee, A., M.W. Phillips, J.L. O'Rourke, B.J. Paster, F.E. Dewhirst, G.J. Fraser, J.G. Fox, L.I. Sly, P.J. Romaniuk, T.J. Trust and S. Kouprach. 1992b. *Helicobacter muridarum* sp. nov., a microaerophilic helical bacterium with a novel ultrastructure isolated from the intestinal mucosa of rodents. Int. J. Syst. Bacteriol. *42*: 27–36.

Lee, C.H., S. Kim, B. Hyun, J.W. Suh, C. Yon, C. Kim and Y. Lim. 1994a. Cepacidine A, a novel antifungal antibiotic produced by *Pseudomonas cepacia*. I. Taxonomy, production, isolation and biological activity. J. Antibiot. *47*: 1402–1405.

Lee, C.M. and H.G. Schlegel. 1981. Physiological characterization of *Pseudomonas pseudoflava* GA3. Curr. Microbiol. *5*: 333–337.

Lee, C.Y., C.L. Tai, S.C. Lin and Y.T. Chen. 1994b. Occurrence of plasmids and tetracycline resistance among *Campylobacter jejuni* and *Campylobacter coli* isolated from whole market chickens and clinical samples. Int. J. Food Microbiol. *24*: 161–170.

Lee, H.A., H.A. Purdy, C.C. Barnum and J.P. Martin. 1925. A comparison of red-stripe of sugar cane and other grasses. *In* Red-stripe Disease Studies, Bull. Expt. Sta. Hawaiian Sugar Planters' Assoc., pp. 1–99.

Lee, J.-H., T. Omori and T. Kodama. 1994c. Identification of the metabolic intermediates of phthalate by Tn5 mutants of *Pseudomonas testosteroni* and analysis of the 4,5-dihydroxyphthalate decarboxylase gene. J. Ferment. Bioeng. *77*: 583–590.

Lee, J.S., Y.K. Shin, J.H. Yoon, M. Takeuchi, Y.R. Pyun and Y.H. Park. 2001. *Sphingomonas aquatilis* sp. nov., *Sphingomonas koreensis* sp. nov., and *Sphingomonas taejonensis* sp. nov., yellow-pigmented bacteria isolated from natural mineral water. Int. J. Syst. Evol. Microbiol. *51*: 1491–1498.

Lee, J.P., C.S. Yi, J. LeGall and H.D. Peck, Jr. 1973. Isolation of a new pigment, desulforubidin, from *Desulfovibrio desulfuricans* (Norway strain) and its role in sulfite reduction. J. Bacteriol. *115*: 453–455.

Lee, K.J., D.E. Tribe and P.L. Rogers. 1979. Ethanol production by *Zymomonas mobilis* in continuous culture at high glucose concentrations. Biotechnol. Lett. *1*: 421–426.

Lee, N.M. and T. Welander. 1996. The effect of different carbon sources on respiratory denitrification in biological wastewater treatment. J. Ferment. Bioeng. *82*: 277–285.

Lee, S., A. Reth, D. Meletzus, M. Sevilla and C. Kennedy. 2000. Characterization of a major cluster of *nif*, *fix*, and associated genes in a sugarcane endophyte, *Acetobacter diazotrophicus*. J. Bacteriol. *182*: 7088–7091.

Legrain, M., B. Rokbi, D. Villeval and E. Jacobs. 1998. Characterization of genetic exchanges between various highly divergent *tbpBs*, having occurred in *Neisseria meningitidis*. Gene *208*: 51–59.

Legrand, C. and E. Anaissie. 1992. Bacteremia due to *Achromobacter xylosoxidans* in patients with cancer. Clin. Infect. Dis. *14*: 479–484.

Lehmann, A.K., A.R. Gorringe, K.M. Reddin, K. West, I. Smith and A. Halstensen. 1999. Human opsonins induced during meningococcal disease recognize transferrin binding protein complexes. Infect. Immun. *67*: 6526–6532.

Lehmann, H. 1976. Wachstumsphysiologische Untersuchungen an fakultativ aeroben und mikroaeroben *Rhodospirillaceae* und Charakterisierung ihrer Elektronentransportsysteme, Ph.D. Thesis, University of Göttingen, F.R.G.

Lehmann, K.B. and R.O. Neumann. 1927. Bakteriologie: insbesondere bakteriologische Diagnostik, 7th ed. ed., Lehmann, Munchen.

Lehmicke, L.G. and M.E. Lidstrom. 1985. Organization of genes necessary for growth of the hydrogen-methanol autotroph *Xanthobacter* sp. strain H4-14 on hydrogen and carbon dioxide. J. Bacteriol. *162*: 1244–1249.

Leidy, G., E. Hahn and H.E. Alexander. 1956. On the specificity of the desoxyribonucleic acid which induces streptomycin resistance in *Haemophilus*. J. Exp. Med. *104*: 305–320.

Leifson, E. 1956. Morphological and physiological characters of the genus *Chromobacterium*. J.Bacteriol. *71*: 393–400.

Leifson, E. 1962. The bacterial flora of distilled and stored water. III. New species of the genera *Corynebacterium, Flavobacterium, Spirillum* and *Pseudomonas*. Int. Bull. Bacteriol. Nomencl. Taxon. *12*: 161–170.

Leifson, E. 1964. *Hyphomicrobium neptunium* sp. n. Antonie Leeuwenhoek J. Microbiol. Serol. *30*: 249–256.

Leifson, E. and R. Hugh. 1954a. *Alcaligenes denitrificans* n. sp. J. Gen. Microbiol. *11*: 512–513.

Leifson, E. and R. Hugh. 1954b. A new type of polar monotrichous flagellation. J. Gen. Microbiol. *10*: 68–70.

Leigh, D., R.K. Scopes and P.L. Rogers. 1984. A proposed pathway for sorbitol production by *Zymomonas mobilis*. Appl. Microbiol. Biotechnol. *20*: 413–415.

Leisinger, T. 1965. Untersuchungen zu Systematik und Stoffwechsel der Essigsäurebakteriën. Zentbl. Bakteriol. Parasitenkd. Infektkrankh. Hyg. Abt. II *119*: 329–376.

Leisinger, T. and D. Kohlerstaub. 1990. Dichloromethane dehalogenase from *Hyphomicrobium* Dm2. Method Enzymol. *188*: 355–361.

Lejbkowicz, F., L. Cohn, N. Hashman and I. Kassis. 1999. Revovery of *Kingella kingae* from blood and synovial fluid of two pediatric patients by using the BacT/Alert system. J. Clin. Microbiol. *37*: 878.

Lemaitre, D., A. Elaichouni, M. Hundhausen, G. Claeys, P. Vanhaesebrouck, M. Vaneechoutte and G. Verschraegen. 1996. Tracheal colonization with *Sphingomonas paucimobilis* in mechanically ventilated neonates due to contaminated ventilator temperature probes. J. Hosp. Infect. *32*: 199–206.

Lemke, M.J., C.J. McNamara and L.G. Leff. 1997. Comparison of methods for the concentration of bacterioplankton for *in situ* hybridization. J. Microbiol. Methods *29*: 23–29.

Lemmer, H., A. Zaglauer, A. Neef, H. Meier and R. Amann. 1997. Denitrification in a methanol-fed fixed-bed reactor. 2. Composition and ecology of the bacterial community in the biofilms. Water Res. *31*: 1903–1908.

Lemmers, M., M. De Beuckeleer, M. Holsters, P. Zambryski, A. Depicker, J.P. Hernalsteens, M. Van Montagu and J. Schell. 1980. Internal organization boundaries and integration of Ti-plasmid DNA in nopaline crown gall tumours. J. Mol. Biol. *144*: 355–378.

Lenger, R., U. Herrmann, R. Gross, J. Simon and A. Kröger. 1997. Structure and function of a second gene cluster encoding the formate dehydrogenase of *Wolinella succinogenes*. Eur. J. Biochem. *246*: 646–651.

Lennon, E. and B.T. DeCicco. 1991. Plasmids of *Pseudomonas cepacia* strains of diverse origins. Appl. Environ. Microbiol. *57*: 2345–2350.

Lepo, J.E., F.J. Hanus and H.J. Evans. 1980. Chemoautotrophic growth of hydrogen-uptake positive strains of *Rhizobium japonicum*. J. Bacteriol. *141*: 664–670.

Lerner, S.A., E.L. Friedman, E.J. Dudek, G. Kominski, M. Bohnhoff and J.A. Morello. 1980. Absence of acetohydroxy acid synthetase in a clinical isolate of *Neisseria gonorrhoeae* requiring isoleucine and valine. J. Bacteriol. *142*: 344–346.

Lerouge, P., P. Roche, C. Faucher, F. Maillet, G. Truchet, J.C. Promé and J. Dénarié. 1990. Symbiotic host-specificity of *Rhizobium meliloti* is determined by a sulphated and acylated glucosamine oligosaccharide signal. Nature (Lond.) *344*: 781–784.

Lersten, N.R. and H.T. Horner, Jr. 1976. Bacterial leaf nodule symbiosis

in angiosperms with emphasis on Rubiaceae and Myrsinaceae. Bot. Rev. *42*: 145–214.

Leslie, P.H. and A.D. Gardner. 1931. The phases of *Hemophilus pertussis*. J. Hyg. *31*: 423–434.

Lessie, T.G. and T. Gaffney. 1986. Catabolic potential of *Pseudomonas cepacia*. *In* Sokatch (Editor), The Biology of *Pseudomonas*: A Treatise on Structure and Function, Vol. 10, Academic Press, Orlando. pp. 439–481.

Lessie, T.G., W. Hendrickson, B.D. Manning and R. Devereux. 1996. Genomic complexity and plasticity of *Burkholderia cepacia*. FEMS Microbiol. Lett. *144*: 117–128.

Lestoquard, F. 1924. Seuscieme note sue les piraplasmoses du mouton en Algerie. L' anaplasmose. *Anaplasma ovis* nov. sp. Bull. Soc. Pathol. Exot. *17*: 784–787.

Letouvet-Pawlak, B., S. Barray, K. Lavalfavre and J.F. Guespin-Michel. 1993. Kinetics of secretion of recombinant acid phosphatase by *Myxococcus xanthus*—a sensitive probe for the assay of protein translocation through the envelopes. J. Gen. Microbiol. *139*: 3243–3252.

Leung, K.T., S. Campbell, Y. Gan, D.C. White, H. Lee and J.T. Trevors. 1999. The role of the *Sphingomonas* species UG30 pentachlorophenol-4-monooxygenase in *p*-nitrophenol degradation. FEMS Microbiol. Lett. *173*: 247–253.

Levin, R.E. and R.H. Vaughn. 1968. Spontaneous spheroplast formation by *Desulfovibrio aestuarii*. Can. J. Microbiol. *14*: 1271–1276.

Levine, M. and F.C. Miller. 1996. An *Eikenella corrodens* toxin detected by plaque toxin-neutralizing monoclonal antibodies. Infect. Immun. *64*: 1672–1678.

Levy-Schil, S., F. Soubrier, A.M. Crutz-Le Coq, D. Faucher, J. Crouzet and D. Petre. 1995. Aliphatic nitrilase from soil-isolated *Comamonas testosteroni* sp.: gene cloning and overexpression, purification and primary structure. Gene *161*: 15–20.

Lewin, R.A. and W.T. Hughes. 1966. *Neisseria subflava* as a cause of meningitis and septicemia in children. Report of five cases. JAMA. *195*: 821–823.

Lewis, K., M.A. Saubolle, F.C. Tenover, M.F. Rudinsky, S.D. Barbour and J.D. Cherry. 1995. Pertussis caused by an erythromycin-resistant strain of *Bordetella pertussis*. Pediatr. Infect. Dis. J. *14*: 388–391.

Lewis, L., F. Stock, D. Williams, S. Weir and V.J. Gill. 1997. Infections with *Roseomonas gilardii* and review of characteristics used for biochemical identification and molecular typing. Am. J. Clin. Pathol. *108*: 210–216.

Li, C.Y., D.A. Ferguson, Jr., T. Ha, D.S. Chi and E. Thomas. 1993. A highly specific and sensitive DNA probe derived from chromosomal DNA of *Helicobacter pylori* is useful for typing *Helicobacter pylori* isolates. J. Clin. Microbiol. *31*: 2157–2162.

Li, H. and D.H. Walker. 1998. rOmpA is a critical protein for the adhesion of *Rickettsia rickettsii* to host cells. Microb. Pathog. *24*: 289–298.

Li, T.S.C. and P.L. Sholberg. 1992. *Pseudomonas*-like early blight on sweet cherries. Can. Plant Dis. Surv. *72*: 120–121.

Li, X. 1993. Studies on *Pseudomonas andropogonis* and related pseudomonads, Doctoral thesis, University of Queensland, St. Lucia, Queensland, Australia. .

Li, X. and A.C. Hayward. 1994. Bacterial whole cell protein profiles of the ribosomal RNA group II pseudomonads. J. Appl. Bacteriol. *77*: 308–318.

Li, X.Z., J. Liu and P.J. Gao. 1996. A simple method for the isolation of cellulolytic myxobacteria and cytophagales. J. Microbiol. Methods *25*: 43–47.

Lidstrom, M.E. 1988. Isolation and characterization of marine methanotrophs. Antonie Leeuwenhoek *54*: 189–200.

Lidstrom, M.E. and D.I. Stirling. 1990. Methylotrophs: genetics and commercial applications. Annu. Rev. Microbiol. *44*: 27–58.

Lidstrom-O'Connor, M.E., G.L. Fulton and A.E. Wopat. 1983. "*Methylobacterium ethanolicum*": a syntrophic association of 2 methylotrophic bacteria. J. Gen. Microbiol. *129*: 3139–3148.

Lie, T.J., M.L. Clawson, W. Godchaux and E.R. Leadbetter. 1999. Sulfidogenesis from 2-aminoethanesulfonate (taurine) fermentation by a morphologically unusual sulfate-reducing bacterium, *Desulforhopalus singaporensis* sp. nov. Appl. Environ. Microbiol. *65*: 3328–3334.

Lie, T.J., M.L. Clawson, W. Godchaux and E.R. Leadbetter. 2000. *In* Validation of the publication of new names and new combinations previously published outside the IJSB. List No. 76. Int. J. Syst. Bacteriol. *50*: 1699–1700.

Lieberman, J.M. 1996. Safety and immunogenicity of a serogroups A/C *Neisseria meningitidis* oligosaccharide-protein conjugate vaccine in young children. A randomized controlled trial. JAMA. *275*: 1499–1503.

Lien, T. and J. Beeder. 1997. *Desulfobacter vibrioformis* sp. nov., a sulfate reducer from a water-oil separation system. Int. J. Syst. Bacteriol. *47*: 1124–1128.

Lien, T., M. Nadsen, I.H. Steen and K. Gjerdevik. 1998. *Desulfobulbus rhabdoformis* sp. nov., a sulfate reducer from a water-oil separation system. Int. J. Syst. Bacteriol. *48*: 469–474.

Liesack, W. and K. Finster. 1994. Phylogenetic analysis of five strains of gram-negative, obligately anaerobic, sulfur-reducing bacteria and description of *Desulfuromusa* gen. nov., including *Desulfuromusa kysingii* sp. nov., *Desulfuromusa bakii* sp. nov., and *Desulfuromusa succinoxidans* sp. nov. Int. J. Syst. Bacteriol. *44*: 753–758.

Lieske, R. 1911. Beiträge zur Kenntnis der Physiologie von *Spirophyllum ferrugineum* Ellis, einem typischen Eisenbakterium. Jahrbuch für Wissenschaftliche Botanik. *49*: 91–127.

Lieske, R. 1912. Untersuchungen über die Physiologie die denitrifizierenden Schwefelbakterien. Ber. Dtsch. Bot. Ges. *30*: 12–22.

Lieske, R. 1919. Zur Ernährungsphysiologie der Eisenbakterien. Zentralbl. Bakteriol. Parasitenk. Infektionskr. Hyg. Abt. II *49*: 413–425.

Liessens, J., R. Germonpre, I. Kersters, S. Beernaert and W. Verstraete. 1993. Removing nitrate with a methylotrophic fluidized bed: Microbiological water quality. Am. Water Works Assoc. J. *85*: 155–161.

Lièvre, H. 1927. Les myxobactéries de l'Afrique du Nord. Bull. Soc. Hist. Nat. Afrique Nord. *18*: 186–189.

Lim, S.T., K. Andersen, R. Tait and R.C. Valentine. 1980. Genetic engineering in agriculture: hydrogen uptake (*hup*) genes. Trends Biochem. Sci. *5*: 167–170.

Lim, Y., J.W. Suh, S. Kim, B. Hyun, C. Kim and C.H. Lee. 1994. Cepacidine A, a novel antifungal antibiotic produced by *Pseudomonas cepacia*. II. Physico-chemical properties and structure elucidation. J. Antibiot. *47*: 1406–1416.

Lin, M. and Y. Rikihisa. 2003. *Ehrlichia chaffeensis* and *Anaplasma phagocytophilum* lack genes for lipid A biosynthesis and incorporate cholesterol for their survival. Infect. Immun. *71*: 5324–5331.

Lin, Y.-F. and K.-C. Chen. 1995. Denitrification and methanogenesis in a co-immobilized mixed culture system. Water Res. *29*: 35–43.

Lincoln, S.P., T.R. Fermor, D.E. Stead and J.E. Sellwood. 1991. Bacterial soft rot of *Agaricus bitorquis*. Plant Pathol. (Oxf.) *40*: 136–144.

Lincoln, S.P., T.R. Fermor and B.J. Tindall. 1999. *Janthinobacterium agaricidamnosum* sp. nov., a soft rot pathogen of *Agaricus bisporus*. Int. J. Syst. Bacteriol. *49*: 1577–1589.

Lind, I. 1990. Epidemiology of antibiotic resistant *Neisseria gonorrhoeae* in industrialized and developing countries. Scand. J. Infect. Dis. *69S*: 77–82.

Lindberg, G.D. 1981. An antibiotic lethal to fungi. Plant Dis. *65*: 680–683.

Lindner, P. 1928a. Atlas der mikroskopischen Grundlagen der Gärungskunde, 3rd Ed., Tafel 68, Berlin.

Lindner, P. 1928b. Gärungsstudien über pulque in Mexiko. Ber. Westpreuss Bot. Zool. Ver. *50*: 253–255.

Lindner, P. 1929. Allgemeine Betrachtungen über Gärung und Fäulnis und die Anwendung von Gärungsmikroben in der Milchwirtschaft. Süddeutsche Molk. *50*: 889–891.

Lindner, P. 1931. *Termobacterium mobile*, ein mexikanisches Bakterium als neues Einsäuerungsbakterium für Rübenschnitzel. Z. Ver. Dsch. Zuckerind. *81*: 25–36.

Lindquist, S.W., D.J. Weber, M.E. Mangum, D.G. Hollis and J. Jordan. 1995. *Bordetella holmesii* sepsis in an asplenic adolescent. Pediatr. Infect. Dis. J. *14*: 813–815.

Lindsay, M.R., R.I. Webb and J.A. Fuerst. 1997. Pirellulosomes: a new type of membrane-bounded cell compartment in planctomycete bacteria of the genus *Pirellula*. Microbiology *143*: 739–748.

Lindström, K. 1989. *Rhizobium galegae*, a new species of legume root nodule bacteria. Int. J. Syst. Bacteriol. *39*: 365–367.

Lindström, K., G. Laguerre, P. Normand, R. T., T. Heulin, B.W.D. Jarvis, P. De Lajudie, E. Martínez–Romero and W.X. Chen. 1998. Taxonomy and phylogeny of diazotrophs. *In* Elmerich, Kondorosi and William (Editors), Biological Nitrogen Fixation for the 21st Century: Proceedings of the 11th International Congress on Nitrogen Fixation, Institut Pasteur, Paris, France, July 20-25, 1997, Kluwer Academic Publishers, Dordrecht; Boston. pp. 559-570.

Lindström, K. and S. Lehtomäki. 1988. Metabolic properties, maximum growth temperature and phage sensitivity of *Rhizobium* sp. (Galega) compared with other fast-growing rhizobia. FEMS Microbiol. Lett. *50*: 277–287.

Lindström, K., P. Van Berkum, M. Gillis, E. Martinez, N. Novikova and B.D.W. Jarvis. 1995. Report from the roundtable on *Rhizobium* taxonomy. *In* Tikhonovich, Provorov, Romanov and Newton (Editors), Nitrogen Fixation: Fundamental and aApplication, Kluwer Academic Publishers, Dordrecht. pp. 807–810.

Line, M.A. 1997. A nitrogen-fixing consortia associated with the bacterial decay of a wooden pipeline. Lett. Appl. Microbiol. *25*: 220–224.

Lingens, F., R. Blecher, H. Blecher, F. Blobel, J. Eberspächer, C. Fröhner, H. Görisch and G. Layh. 1985. *Phenylobacterium immobile* gen. nov., sp. nov., a Gram-negative bacterium that degrades the herbicide chloridazon. Int. J. Syst. Bacteriol. *35*: 26–39.

Link, H.F. 1809. Observationes in Ordines plantarum naturales. Dissertatio Ima complectens Amandarum ordines Epiphytas, Mucedines, Gastromycos et Fungos. Magaz. Ges. Nat. Freunde Berlin *3*: 3–42.

Linnemann, C.C.J., A.E. Schaeffer, W. Burgdorfer, L. Hutchinson and R.N. Philip. 1980. Rocky Mountain spotted fever in Clermont County, Ohio. II. Distribution of population and infected ticks in an endemic area. Am. J. Epidemiol. *111*: 31–36.

Linton, D., F.E. Dewhirst, J.P. Clewley, R.J. Owen, A.P. Burnens and J. Stanley. 1994a. Two types of 16S rRNA gene are found in *Campylobacter helveticus*: analysis, applications and characterization of the intervening sequence found in some strains. Microbiology *140*: 847–855.

Linton, D., A.J. Lawson, R.J. Owen and J. Stanley. 1997. PCR detection, identification to species level, and fingerprinting of *Campylobacter jejuni* and *Campylobacter coli* direct from diarrheic samples. J. Clin. Microbiol. *35*: 2568–2572.

Linton, D., R.J. Owen and J. Stanley. 1996. Rapid identification by PCR of the genus *Campylobacter* and of five *Campylobacter* species enteropathogenic for man and animals. Res. Microbiol. *147*: 707–718.

Linton, J., P. Clewley, A. Burnens, R.J. Owen and J. Stanley. 1994b. An intervening sequence (IVS) in the 16S rRNA gene of the eubacterium *Helicobacter canis*. Nucleic Acids Res. *22*: 1954–1958.

Lior, H. and G. Wang. 1993. Differentiation of *Arcobacter butzleri* by pulsefield gel electrophoresis (PFGE) and random amplified polymorphic DNA (RAPD). Acta Gastro-Enterol. Belg. *56*: 29.

Lior, H. and D. Woodward. 1991. A serotyping scheme for *Campylobacter butzleri*. Microb. Ecol. Health. Dis. *4*: S93.

Lior, H. and D.L. Woodward. 1993. *Arcobacter butzleri*: a biotyping scheme. Acta Gastro-Enterol. Belg. *56*: 28.

Lior, H., D.L. Woodward, J.A. Edgar, L.J. Laroche and P. Gill. 1982. Serotyping of *Campylobacter jejuni* by slide agglutination based on heatlabile antigenic factors. J. Clin. Microbiol. *15*: 761–768.

Lippincott, J.A. and B.B. Lippincott. 1969. Tumor initiating ability and nutrition in the genus *Agrobacterium*. J. Gen. Microbiol. *59*: 57–75.

Lipscomb, J.D. 1994. Biochemistry of the soluble methane monooxygenase. Annu. Rev. Microbiol. *48*: 371–399.

Lipski, A., S. Klatte, B. Bendinger and K. Altendorf. 1992. Differentiation of gram-negative, nonfermentative bacteria isolated from biofilters on the basis of fatty acid composition, quinone system, and physiological reaction profiles. Appl. Environ. Microbiol. *58*: 2053–2065.

Lipski, A., K. Reichert, B. Reuter, C. Spröer and K. Altendorf. 1998. Identification of bacterial isolates from biofilters as *Paracoccus alkenifer* sp. nov. and *Paracoccus solventivorans* with emended description of *Paracoccus solventivorans*. Int. J. Syst. Bacteriol. *48*: 529–536.

Lipski, A., E. Spieck, A. Makolla and K. Altendorf. 2001. Fatty acid profiles of nitrite-oxidizing bacteria reflect their phylogenetic heterogeneity. Syst. Appl. Microbiol. *24*: 377–384.

Lisdiyanti, P., H. Kawasaki, T. Seki, Y. Yamada, T. Uchimura and K.J. Komagata. 2000. Systematic study of the genus *Acetobacter* with descriptions of *Acetobacter indonesiensis* sp. nov., *Acetobacter tropicalis* sp. nov., *Acetobacter orleanensis* (Henneberg 1906) comb. nov., *Acetobacter lovaniensis* (Frateur 1950) comb. nov., and *Acetobacter estunensis* (Carr 1958) comb. nov. Gen. Appl. Microbiol. *46*: 147–165.

Lisdiyanti, P., H. Kawasaki, T. Seki, Y. Yamada, T. Uchimura and K.J. Komagata. 2001a. Identification of *Acetobacter* strains isolated from Indonesian sources, and proposals of *Acetobacter syzygii* sp. nov., *Acetobacter cibinongensis* sp. nov., and *Acetobacter orientalis* sp. nov. Gen. Appl. Microbiol. *47*: 119–131.

Lisdiyanti, P., H. Kawasaki, T. Seki, Y. Yamada, T. Uchimura and K. Komagata. 2001b. *In* Validation of the publication of new names and new combinations previously effectively published outside the IJSEM. List No. 51. Int. J. Syst. Evol. Microbiol. *51*: 263–265.

Lisdiyanti, P., H. Kawasaki, T. Seki, Y. Yamada, T. Uchimura and K. Komagata. 2002. *In* Validation of the publication of new names and new combinations previously effectively published outside the IJSEM. List No. 84. Int. J. Syst. Evol. Microbiol. *52*: 3–4.

Lissman, B.A. and J.L. Benach. 1980. Rocky Mountain spotted fever in dogs. J. Am. Vet. Med. Assoc. *176*: 994–995.

Lissolo, L., G. Maitrewilmotte, P. Dumas, P. Mignon, B. Danve and M.J. Quentinmillet. 1995. Evaluation of transferrin-binding protein-2 within the transferrin-binding protein complex as a potential antigen for future meningococcal vaccines. Infect. Immun. *63*: 884–890.

Listgarten, M.A., D. Johnson, A. Nowotny, A.C.R. Tanner and S.S. Socransky. 1978. Histopathology of periodontal disease in gnotobiotic rats monoinfected with *Eikenella corrodens*. J. Periodontol. Res. *13*: 134–148.

Little, S.E., J.E. Dawson, J.M. Lockhart, D.E. Stallknecht, C.K. Warner and W.R. Davidson. 1997. Development and use of specific polymerase reaction for the detection of an organism resembling *Ehrlichia* sp. in white-tailed deer. J. Wildl. Dis. *33*: 246–253.

Litwin, C.M., T.B. Martins and H.R. Hill. 1997. Immunologic response to *Bartonella henselae* as determined by enzyme immunoassay and Western blot analysis. Am. J. Clin. Pathol. *108*: 202–209.

Liu, D.F., E. Phillips, T.M. Wizemann, M.M. Siegel, K. Tabei, J.L. Cowell and E. Tuomanen. 1997. Characterization of a recombinant fragment that contains a carbohydrate recognition domain of the filamentous hemagglutinin. Infect. Immun. *65*: 3465–3468.

Liu, H., Y. Kang, S. Genin, M.A. Schell and T.P. Denny. 2001. Twitching motility of *Ralstonia solanacearum* requires a type IV pilus system. Microbiology (Reading) *147*: 3215–3229.

Liu, J.W., W.H. Yap, T. Thanabalu and A.G. Porter. 1996. Efficient synthesis of mosquitocidal toxins in *Asticcacaulis excentricus* demonstrates potential of gram-negative bacteria in mosquito control. Nat. Biotechnol. *14*: 343–347.

Liu, K.-C. and L.E. Casida, Jr.. 1983. Survival of *Myxobacter* strain 8 in natural soil in the presence and absence of host cells. Soil Biol. Biochem. *15*: 551–555.

Liu, Y., D.L. Balkwill, H.C. Aldrich, G.R. Drake and D.R. Boone. 1999. Characterization of the anaerobic propionate-degrading syntrophs *Smithella propionica* gen. nov., sp. nov. and *Syntrophobacter wolinii*. Int. J. Syst. Bacteriol. *49*: 545–556.

Livey, I. and A.C. Wardlaw. 1984. Production and properties of *Bordetella pertussis* heat-labile toxin. J. Med. Microbiol. *17*: 91–103.

Ljungdahl, L.G. and J. Wiegel. 1986. Working with anaerobic bacteria. *In* Demain and Solomon (Editors), Manual of Industrial Microbiology and Biotechnology, American Society for Microbiology, Washington, D.C. pp. 84–96.

Lloyd-Jones, G. and P.C. Lau. 1997. Glutathione S-transferase-encoding gene as a potential probe for environmental bacterial isolates capable

of degrading polycyclic aromatic hydrocarbons. Appl. Environ. Microbiol. *63*: 3286–3290.

Lo, N., M. Casiraghi, E. Salati, C. Bazzocchi and C. Bandi. 2002. How many wolbachia supergroups exist? Mol. Biol. Evol. *19*: 341–346.

Lobas, D., M. Nimtz, V. Wray, A. Schumpe, C. Proppe and W.D. Deckwer. 1994. Structure and physical properties of the extracellular polysaccharide PS-P4 produced by *Sphingomonas paucimobilis* P4 (DSM 6418). Carbohydr. Res. *251*: 303–313.

Lobos, J.H., T.E. Chisolm, L.H. Bopp and D.S. Holmes. 1986. *Acidiphilium organovorum* sp. nov., an acidophilic heterotroph isolated from a *Thiobacillus ferrooxidans* culture. Int. J. Syst. Bacteriol. *36*: 139–144.

Locher, H.H., B. Poolman, A.M. Cook and W.N. Konings. 1993. Uptake of 4-toluene sulfonate by *Comamonas testosteroni* T-2. J. Bacteriol. *175*: 1075–1080.

Loebeck, M.E. and E.J. Ordal. 1957. The nuclear cycle of *Myxococcus fulvus*. J. Gen. Microbiol. *16*: 76–85.

Logan, J.M.J., A. Burnens, D. Linton, A.J. Lawson and J. Stanley. 2000. *Campylobacter lanienae* sp. nov., a new species isolated from workers in an abattoir. Int. J. Syst. Evol. Microbiol. *50*: 865–872.

Logan, N.A. 1989. Numerical taxonomy of violet-pigmented, Gram-negative bacteria and description of *Iodobacter fluviatile* gen. nov., comb. nov. Int. J. Syst. Bacteriol. *39*: 450–456.

Logan, N.A. and M.O. Moss. 1992. Identification of *Chromobacterium, Janthinobacterium* and *Iodobacter* species. *In* Board, Jones and Skinner (Editors), Identification Methods in Applied and Environmental Microbiology, The Society for Applied Bacteriology Technical Series Blackwell Scientific Publications, Oxford. pp. 183–192.

Loganathan, P., R. Sunita, A.K. Parida and S. Nair. 1999. Isolation and characterization of two genetically distant groups of *Acetobacter diazotrophicus* from a new host plant *Eleusine coracana* L. J. Appl. Microbiol. *87*: 167–172.

Loginova, N.V., B.B. Namsaraev and Y.A. Trotsenko. 1978. Autotrophic metabolism of methanol in *Microcyclus aquaticus*. Microbiology *47*: 134–135.

Loginova, N.V., V.N. Shishkina and I.U. Trotsenko. 1976. Primary metabolic pathways of methylated amines in *Hyphomicrobium vulgare*. Mikrobiologiya *45*: 41–47.

Loginova, N.V. and Y.A. Trotsenko. 1979. *Blastobacter viscosus* - a new species of autotrophic bacteria utilizing methanol. Microbiology *48*: 644–651.

London, J. 1963. *Thiobacillus intermedius* nov. sp. a novel type of facultative autotroph. Arch. Mikrobiol. *46*: 329–337.

London, J. and S.C. Rittenberg. 1967. *Thiobacillus perometabolis* nov. sp., a non-autotrophic *Thiobacillus*. Arch. Mikrobiol. *59*: 218–225.

Lonergan, D.J., H.L. Jenter, J.D. Coates, E.J.P. Phillips, T.M. Schmidt and D.R. Lovely. 1996. Phylogenetic analysis of dissimilatory Fe(III)-reducing bacteria. J. Bacteriol. *178*: 2402–2408.

Long, M.T., T.E. Goetz, H.E. Whiteley, I. Kakoma and T.E. Lock. 1995. Identification of *Ehrlichia risticii* as the causative agent of two equine abortions following natural maternal infection. J. Vet. Diagn. Invest. *7*: 201–205.

Long, P.A., L.I. Sly, A.V. Pham and G.H.G. Davis. 1981. Characterization of *Morococcus cerebrosus* gen. nov., sp. nov. and comparison with *Neisseria mucosa*. Int. J. Syst. Bacteriol. *31*: 294–301.

Lontoh, S. and J.D. Semrau. 1998. Methane and trichloroethylene degredation by *Methylosinus trichosporium* OB3d expressing particulate methane monooxygenase. Appl. Environ. Microbiol. *64*: 1106–1114.

Loos, H., R. Kramer, H. Sahm and G.A. Sprenger. 1994. Sorbitol promotes growth of *Zymomonas mobilis* in environments with high concentrations of sugar: evidence for a physiological function of glucose-fructose oxidoreductase in osmoprotection. J. Bacteriol. *176*: 7688–7693.

Loos, H., M. Voller, B. Rehr, Y-D. Stierhof, H. Sahm and G.A. Sprenger. 1991. Localisation of the glucose-fructose oxidoreductase in wild type and overproducing strains of *Zymomonas mobilis*. FEMS Microbiol. Lett. *84*: 211–216.

Loper, J.E. and C.I. Kado. 1979. Host range conferred by the virulence-specifying plasmid of *Agrobacterium tumefaciens*. J. Bacteriol. *130*: 591–596.

Lopez, M. 1978. Characteristics of French isolates of *Agrobacterium*. Proc. 4th Int. Conf. Plant Path. Bact., Angers 1978. 233–237.

Lopez, M.M., M.T. Gorris and A.M. Montojo. 1988. Opine utilization by Spanish isolates of *Agrobacterium tumefaciens*. Plant Pathol. (Oxf.) *37*: 565–572.

López, M.M., M. Gracia and M. Sampayo. 1987. Current status of *Xanthomonas ampelina* in Spain and susceptibility of Spanish cultivars to bacterial necrosis. Bull. OEPP *17*: 231–236.

López-López, A. , Pujalte, M.J. , Benlloch, S. , Mata-Roig, M., Rosselló-Mora, R., Garay, E. and Rodríguez-Valera, F.. 2002. *Thalassospira lucentensis* gen. nov., sp nov., a new marine member of the α-*Proteobacteria*. Int. J. Syst. Evol. Microbiol. *52*: 1277–1283.

Lorenzen, D.R., F. Dux, U. Wolk, A. Tsirpouchtsidis, G. Haas and T.F. Meyer. 1999. Immunoglobulin A1 protease, an exoenzyme of pathogenic Neisseriae, is a potent inducer of proinflammatory cytokines. J. Exp. Med. *190*: 1049–1058.

Lorenzen, J., S. Steinwachs and G. Unden. 1994. DMSO respiration by the anaerobic rumen bacterium *Wolinella succinogenes*. Arch. Microbiol. *162*: 277–281.

Lorite, M.J., J. Tachil, J. Sanjuan, O. Meyer and E.J. Bedmar. 2000. Carbon monoxide dehydrogenase activity in *Bradyrhizobium japonicum*. Appl. Environ. Microbiol. *66*: 1871–1876.

Lorquin, J., G. Lortet, M. Ferro, N. Mear, J.C. Promé and C. Boivin. 1997. *Sinorhizobium teranga* biovar acaciae ORS1073 and *Rhizobium* sp. strain ORS1001, two distantly related *Acacia*-nodulating strains, produce similar Nod factors that are *O*-carbamoylated, *N*-methylated, and mainly sulfated. J. Bacteriol. *179*: 3079–3083.

Lortet, G., N. Mear, J. Lorquin, B. Dreyfus, P. De Lajudie, C. Rosenberg and C. Boivin. 1996. Nod factor thin-layer chromatography profiling as a tool to characterize symbiotic specificity of rhizobial strains: application to *Sinorhizobium saheli*, *S. teranga*, and *Rhizobium* sp. strains isolated from Acacia and Sesbania. Mol. Plant-Microbe Interact. *9*: 736–747.

Loubinoux, J., F. Mory, I.A.C. Pereira and A.E. Le Faou. 2000. Bacteremia caused by a strain of *Desulfovibrio* related to the provisionally named *Desulfovibrio fairfieldensis*. J. Clin. Microbiol. *38*: 931–934.

Loubinoux, J., F.M.A. Valente, I.A.C. Pereira, A. Costa, P.A.D. Grimont and A.E. Le Faou. 2002. Reclassification of the only species of the genus *Desulfomonas*, *Desulfomonas pigra*, as *Desulfovibrio piger* comb. nov. Int. J. Syst. Evol. Microbiol. *52*: 1305–1308.

Louie, T.M., S. Ni, L. Xun and W.W. Mohn. 1997. Purification, characterization and gene sequence analysis of a novel cytochrome *c* co-induced with reductive dechlorination activity in *Desulfomonile tiedjei* DCB-1. Arch. Microbiol. *168*: 520–527.

Louis, C. and L. Nigro. 1989. Ultrastructural evidence of *Wolbachia Rickettsiales* in *Drosophila simulans* and their relationships with unidirectional cross-incompatibility. J. Ivert. Pathol. *54*: 39–44.

Lovley, D.R. 2000. Dissimilatory Fe(III)- and Mn(IV)-reducing prokaryotes. *In* Dworkin, Falkow, Rosenberg, Schleifer and Stackebrandt (Editors), The Prokaryotes: An Evolving Electronic Resource for the Microbiological Community, 3 Ed., Vol. release 3.4, Springer-Verlag, New York. www.prokaryotes.com.

Lovley, D.R., S.J. Giovannoni, D.C. White, J.E. Champine, E.J.P. Phillips, Y.A. Gorby and S. Goodwin. 1993a. *Geobacter metallireducens* gen. nov. sp. nov., a microorganism capable of coupling the complete oxidation of organic compounds to the reduction of iron and other metals. Arch. Microbiol. *159*: 336–344.

Lovley, D.R., S.J. Giovannoni, D.C. White, J.E. Champine, E.J. Phillips, Y.A. Gorby and S. Goodwin. 1995a. *In* Validation of the publication of new names and new combinations previously effectively published outside the IJSB. List No. 54. Int. J. Syst. Bacteriol. *45*: 619–620.

Lovley, D.R. and E.J.P. Phillips. 1986. Organic matter mineralization with reduction of ferric iron in anaerobic sediments. Appl. Environ. Microbiol. *51*: 683–689.

Lovley, D.R. and E.J.P. Phillips. 1988. Novel mode of microbial energy metabolism: organic carbon oxidation coupled to dissimilatory reduction of iron or manganese. Appl. Environ. Microbiol. *54*: 1472–1480.

Lovley, D.R. and E.J.P. Phillips. 1992. Reduction of uranium by *Desulfovibrio desulfuricans*. Appl. Environ. Microbiol. *58*: 850–856.

Lovley, D.R. and E.J.P. Phillips. 1994a. Novel processes for anaerobic sulfate production from elemental sulfur by sulfate-reducing bacteria. Appl. Environ. Microbiol. *60*: 2394–2399.

Lovley, D.R. and E.J.P. Phillips. 1994b. Reduction of chromate by *Desulfovibrio vulgaris* and its c_3 cytochrome. Appl. Environ. Microbiol. *60*: 726–728.

Lovley, D.R., E.J.P. Phillips, D.J. Lonergan and P.K. Widman. 1995b. Fe(III) and S^0 reduction by *Pelobacter carbinolicus*. Appl. Environ. Microbiol. *61*: 2132–2138.

Lovley, D.R., E.E. Roden, E.J.P. Phillips and J.C. Woodward. 1993b. Enzymatic iron and uranium reduction by sulfate-reducing bacteria. Mar. Geol. *113*: 41–53.

Lovley, D.R., J.F. Stolz, G.L. Nord and E.J.P. Phillips. 1987. Anaerobic production of magnetite by a dissimilatory iron- reducing microorganism. Nature *330*: 252–254.

Loy, J.K., F.E. Dewhirst, W. Weber, P.F. Frelier, T.L. Garbar, S.I. Tasca and J.W. Templeton. 1996. Molecular phylogeny and in situ detection of the etiologic agent of necrotizing hepatopancreatitis in shrimp. Appl. Environ. Microbiol. *62*: 3439–3445.

Lu, S.F., F.L. Lee and H.K. Chen. 1999. A thermotolerant and high acetic acid-producing bacterium *Acetobacter* sp. I14-2. J. Appl. Microbiol. *86*: 55–62.

Lucey, D., M.J. Dolan, C.W. Moss, M. Garcia, D.G. Hollis, S. Wegner, G. Morgan, R. Almeida, D. Leong, K.S. Greisen, D.F. Welch and L.N. Slater. 1992. Relapsing illness due to *Rochalimaea henselae* in immunocompetent hosts: implication for therapy and new epidemiological associations. Clin. Infect. Dis. *14*: 683–688.

Ludden, P.W., Y. Okon and R.H. Burris. 1978. The nitrogenase system of *Sprillum lipoferum*. Biochem. J. *173*: 1001–1003.

Ludden, P.W. and G.P. Roberts. 1995. The biochemistry and genetics of nitrogen fixation by photosynthetic bacteria. *In* Blankenship, Madigan and Bauer (Editors), Anoxygenic Photosynthetic Bacteria, Kluwer Academic Publishers, Netherlands. pp. 929–947.

Ludwig, W., J. Eberspächer, F. Lingens and E. Stackebrandt. 1984. 16S Ribosomal RNA studies on the relationship of a chloridazon-degrading Gram negative eubacterium. Syst. Appl. Microbiol. 5: 241–246.

Ludwig, W., G. Mittenhuber and C.G. Friedrich. 1993. Transfer of *Thiosphaera pantotropha* to *Paracoccus denitrificans*. Int. J. Syst. Bacteriol. *43*: 363–367.

Ludwig, W., R. Rosselló-Mora, R. Aznar, S. Klugbauer, S. Spring, K. Reetz, C. Beimfohr, E. Brockmann, G. Kirchhof, S. Dorn, M. Bachleitner, N. Klugbauer, N. Springer, D. Lane, R. Nietupsky, M. Weiznegger and K.H. Schleifer. 1995. Comparative sequence analysis of 23S rRNA from *Proteobacteria*. Syst. Appl. Microbiol. *18*: 164–188.

Ludwig, W., K.H. Schleifer, H. Reichenbach and E. Stackebrandt. 1983. A phylogenetic analysis of the myxobacteria *Myxococcus fulvus*, *Stigmatella aurantiaca*, *Cystobacter fuscus*, *Sorangium cellulosum*, and *Nannocystis exedens*. Arch. Microbiol. *135*: 58–62.

Luker, K.E., J.L. Collier, E.W. Kolodziej, G.R. Marshall and W.E. Goldman. 1993. *Bordetella pertussis* tracheal cytotoxin and other muramyl peptides: distinct structure-activity relationships for respiratory epithelial cytopathology. Proc. Natl. Acad. Sci. U. S. A. *90*: 2365–2369.

Lumsden, R.D. and M. Sasser. 1986. Medium for the isolation of *Pseudomonas cepacia* biotype from soil and the isolated biotype. U.S. Patent #4,588,584.

Lundy, D.W. and D.K. Kehl. 1998. Increasing prevalence of *Kingella kingae* in osteoarticular infections in young children. J. Pediatr. Orthop. *18*: 262–267.

Lünsdorf, H. and H. Reichenbach. 1989. Ultrastructural details of the apparatus of gliding motility of *Myxococcus fulvus* (*Myxobacterales*). J. Gen. Microbiol. *135*: 1633–1641.

Lünsdorf, H. and H.U. Schairer. 2001. Frozen motion of gliding bacteria outlines inherent features of the motility apparatus. Microbiology *147*: 939–947.

Lünsdorf, H., H.U. Schairer and M. Heidelbach. 1995. Localization of the stress protein SP21 in indole-induced spores, fruiting bodies, and heat-shocked cells of *Stigmatella aurantiaca*. J. Bacteriol. *177*: 7092–7099.

Lütke-Eversloh, T. and A. Steinbuchel. 1999. Biochemical and molecular characterization of a succinate semialdehyde dehydrogenase involved in the catabolism of 4-hydroxybutyric acid in *Ralstonia eutropha*. FEMS Microbiol. Lett. *181*: 63–71.

Lütters-Czekalla, S. 1990. Lithoautotrophic growth of the iron bacterium *Gallionella ferruginea* with thiosulfate or sulfide as energy source. Arch. Microbiol. *154*: 417–421.

Lyman, J. and R.H. Fleming. 1940. Composition of sea water. J. Mar. Res. *3*: 134–146.

Lyons, N.F. and J.D. Taylor. 1990. Serological detection and identification of bacteria from plants by the conjugated *Staphylococcus aureus* slide agglutination test. Plant Pathol. (Oxf.) *39*: 584–590.

Lysenko, A.M., V.F. Gal'chenko and N.A. Chernykh. 1988. Taxonomic study of obligate methanotrophic bacteria using the DNA–DNA hybridization technique. Mikrobiologiya *57*: 653–658.

Lysenko, A.M., A.M. Semenov and L.V. Vasilyeva. 1984. DNA nucleotide composition of prosthecate bacteria with a radial cell symmetry. Mikrobiologiya *53*: 859–861.

Lysko, P.G. and S.A. Morse. 1981. *Neisseria gonorrhoeae* cell-envelope permeability to hydrophobic molecules. J. Bacteriol. *145*: 946–952.

Ma, D.Q., Y.G. Lin, J. Zhou, W.N. Xiang, J.F. You, X.M. Xie and P.M. Chen. 1985. Biotype and plasmid type of *Agrobacterium tumefaciens* isolated from the crown gall of grapevine in North China. Acta Microbiol. Sin. *25*: 45–53.

Mack, E.E., L. Mandelco, C.R. Woese and M.T. Madigan. 1993. *Rhodospirillum sodomense*, sp. nov., a Dead Sea *Rhodospirillum* species. Arch. Microbiol. *160*: 363–371.

Mack, E.E., L. Mandelco, C.R. Woese and M.T. Madigan. 1996. *In* Validation of the publication of new names and new combinations previously effectively published outside the IJSB. List No. 59. Int. J. Syst. Bacteriol. *46*: 1189–1190.

MacKenzie, C.R., K.G. Johnson and I.J. McDonald. 1977. Glycogen synthesis by amylosucrase from *Neisseria perflava*. Can. J. Microbiol. *23*: 1303–1307.

MacKenzie, C.R., I.J. McDonald and K.G. Johnson. 1978a. Glycogen metabolism in the genus *Neisseria*: synthesis from sucrose by amylosucrase. Can. J. Microbiol. *24*: 357–362.

MacKenzie, C.R., I.J. McDonald and K.G. Johnston. 1978b. Sucrose uptake by *Neisseria denitrificans*. Can. J. Microbiol. *24*: 569–573.

MacKenzie, C.R., M.B. Perry, I.J. McDonald and K.G. Johnson. 1978c. Structure of the D-glucans produced by *Neisseria perflava*. Can. J. Microbiol. *24*: 1419–1422.

MacLeod, J.R. and W.S. Gordon. 1933. Studies in tick-borne fever of sheep. I. Transmission by the tick, *Ixodes ricinus*, with a description of the disease produced. Parsitol. *25*: 273–285.

MacLeod, M.N. and I.W. DeVoe. 1981. Localization of carbonic anhydrase in the cytoplasmic membrane of *Neisseria sicca* (strain 19). Can. J. Microbiol. *27*: 87–92.

MacRae, J.D. and J. Smit. 1991. Characterization of caulobacters isolated from wastewater treatment systems. Appl. Environ. Microbiol. *57*: 751–758.

Macrae, T.H., W.J. Dobson and H.D. McCurdy. 1977. Fimbriation in gliding bacteria. Can. J. Microbiol. *23*: 1096–1108.

Macrae, T.H. and H.D. McCurdy. 1975. Ultrastructural studies of *Chondromyces crocatus* vegetative cells. Can. J. Microbiol. *21*: 1815–1826.

Macrae, T.H. and H.D. McCurdy. 1976. Evidence for motility-related fimbriae in the gliding microorganism *Myxococcus xanthus*. Can. J. Microbiol. *22*: 1589–1593.

Macy, J.M., S. Rech, G. Auling, M. Dorsch, E. Stackebrandt and L.I. Sly. 1993. *Thauera selenatis* gen. nov. sp. nov., a member of the beta subclass of *Proteobacteria* with a novel type of anaerobic respiration. Int. J. Syst. Bacteriol. *43*: 135–142.

Maddison, W.P. and Maddison, D.R.. 1992. MacClade: Analysis of Phylogeny and Character Evolution v3.0, Sinauer Associates, Sunderland, Massachusetts.

Madigan, J.E. and D.H. Gribble. 1987. Equine ehrlichiosis in northern California: 49 cases (1968–1981). J. Am. Vet. Med. Assoc. *190*: 445–448.

Madigan, J.E., N. Pusterla, E. Johnson, J.S. Chae, J.B. Pusterla, E. Derock and S.P. Lawler. 2000a. Transmission of *Ehrlichia risticii*, the agent of Potomac horse fever, using naturally infected aquatic insects and helminth vectors: Preliminary report. Equine Vet. J. *32*: 275–279.

Madigan, M.T. and S.S. Cox. 1982. Nitrogen fixation in *Rhodopseudomonas globiformis*. Arch. Microbiol. *133*: 6–10.

Madigan, M., S.S. Cox and R.A. Stegeman. 1984. Nitrogen-fixation and nitrogenase activities in members of the Family *Rhodospirillaceae*. J. Bacteriol. *157*: 73–78.

Madigan, M.T. and H. Gest. 1978. Growth of a photosynthetic bacterium anaerobically in darkness, supported by "oxidant-dependent" sugar fermentation. Arch. Microbiol. *117*: 119–122.

Madigan, M.T. and H. Gest. 1979. Growth of the photosynthetic bacterium *Rhodopseudomonas capsulata* chemoautotrophically in darkness with H_2 as the energy source. J. Bacteriol. *137*: 524–530.

Madigan, M.T., D.O. Jung, C.R. Woese and L.A. Achenbach. 2000b. *Rhodoferax antarcticus* sp. nov., a moderately psychrophilic purple nonsulfur bacterium isolated from an Antarctic microbial mat. Arch. Microbiol. *173*: 269–277.

Madigan, M.T., J.D. Wall and H. Gest. 1979. Dark anaerobic dinitrogen fixation by a photosynthetic microorganism. Science *240*: 1429–1430.

Madkour, M.M. 2000. Madkour's Brucellosis, 2nd Ed., Springer, Berlin.

Magalhães, F.M., J.I. Baldani, S.M. Souto, J.R. Kuykendall and J. Döbreiner. 1984. *In* Validation of the publication of new names and new combinations previously effectively published outside the IJSB. List No. 15. Int. J. Syst. Bacteriol. *34*: 355–357.

Magalhães, F.M., J.I. Baldani, S.M. Souto, J.R. Kuykendall and J. Dobreiner. 1983. A new acid-tolerant *Azospirillum* species. Nac. Acad. Bras. Cienc. *55*: 417–430.

Magalhães, L.M.S., C.A. Neyra and J. Dobereiner. 1978. Nitrate and nitrite reductase negative mutants of N_2-fixing *Azospirillum* spp. Arch. Microbiol. *117*: 247–252.

Maggs, A.F., J.M.J. Logan, P.E. Carter and T.H. Pennington. 1998. The detection of penicillin insensitivity in *Neisseria meningitidis* by polymerase chain reaction. J. Antimicrob. Chemother. *42*: 303–307.

Magot, M., P. Caumette, J.M. Desperrier, R. Matheron, C. Dauga, F. Grimont and L. Carreau. 1992. *Desulfovibrio longus* sp. nov., a sulfate-reducing bacterium isolated from an oil-producing well. Int. J. Syst. Bacteriol. *42*: 398–403.

Magrini, V., D. Salmi, D. Thomas, S.K. Herbert, P.L. Hartzell and P. Youderian. 1997. Temperate *Myxococcus xanthus* phage Mx8 encodes DNA adenine methylase, Mox. J. Bacteriol. *179*: 4254–4263.

Mahapatra, N.R. and P.C. Banerjee. 1996. Extreme tolerance to cadmium and high resistance to copper, nickel and zinc in different *Acidiphilium* strains. Lett. Appl. Microbiol. *23*: 393–397.

Mahara, F. 1997. Japanese spotted fever: Report of 31 cases and review of the literature. Emerg. Infect. Dis. *3*: 105–111.

Mahenthiralingam, E., D.A. Simpson and D.P. Speert. 1997. Identification and characterization of a novel DNA marker associated with epidemic *Burkholderia cepacia* strains recovered from patients with cystic fibrosis. J. Clin. Microbiol. *35*: 808–816.

Mahoney, R.P. and M.R. Edwards. 1966. Fine structure of *Thiobacillus thiooxidans*. J. Bacteriol. *92*: 487–495.

Maidak, B.L., J.R. Cole, C.T. Parker, Jr., G.M. Garrity, N. Larsen, B. Li, T.G. Lilburn, M.J. McCaughey, G.J. Olsen, R. Overbeek, S. Pramanik, T.M. Schmidt, J.M. Tiedje and C.R. Woese. 1999. A new version of the RDP (Ribosomal Database Project). Nucleic Acids Res. *27*: 171–173.

Maidak, B.L., G.J. Olsen, N. Larsen, R. Overbeek, M.J. McCaughey and C.R. Woese. 1996. The ribosomal database project (RDP). Nucleic Acids Res. *24*: 82–85.

Maidak, B.L., G.J. Olsen, N. Larsen, R. Overbeek, M.J. McCaughey and C.R. Woese. 1997. The RDP (Ribosomal Database Project). Nucleic Acids Res. *25*: 109–111.

Maiden, M.C.J., J.A. Bygraves, E. Feil, G. Morelli, J.E. Russell, R. Urwin, Q. Zhang, J.J. Zhou, K. Zurth, D.A. Caugant, I.M. Feavers, M. Achtman and B.G. Spratt. 1998. Multilocus sequence typing: A portable approach to the identification of clones within populations of pathogenic microorganisms. Proc. Natl. Acad. Sci. U. S. A. *95*: 3140–3145.

Maier, R.J. 1981. *Rhizobium japonicum* mutant strains unable to grow chemoautotrophically with H₂. J. Bacteriol. *145*: 533–540.

Majewski, D.M. 1986. Molekularbiologische charakterisierung von bakteriophagen knospender bakterien, Dipl. Arb. Univ. Kiel, Germany. 135.

Makkar, N.S. and L.E. Casida, Jr.. 1987a. *Cupriavidus necator* gen. nov., sp. nov.: a nonobligate bacterial predator of bacteria in soil. Int. J. Syst. Bacteriol. *37*: 323–326.

Makkar, N.S. and L.E. Casida, Jr.. 1987b. Technique for estimating low numbers of a bacterial strain(s) in soil. Appl. Environ. Microbiol. *53*: 887–888.

Makula, R.A. 1978. Phospholipid composition of methane-utilizing bacteria. J. Bacteriol. *134*: 771–777.

Malashenko, Y.R., Y.U. Khaier, E.N. Budkova, Y.U. Isagulova, U. Berger, T.P. Krishtab, D.V. Chernyshenko and V.A. Romanovskaya. 1987. Methane-oxidizing microflora in fresh and saline water reservoirs. Mikrobiologiya *56*: 115–120.

Malashenko, Y.R., V.A. Romanovskaya and V.N. Bogachenko. 1975. Method of isolating pure cultures of mesophilic, thermotolerant and thermophilic methane-utilizing bacteria. Mikrobiologiya *44*: 707–713.

Málek, I. and Kazdová-Kozisková. 1946. *Pseudomonas odorans* n. sp., a new microbe discovered from diagnostic material. Sb. Lek. *47*: 189–194.

Málek, I., M. Radochová and O. Lysenko. 1963. Taxonomy of the species *Pseudomonas odorans*. J. Gen. Microbiol. *33*: 349–355.

Málek, K.A. and H.G. Schlegel. 1981. Chemolithoautotrophic growth of bacteria able to grow under N₂-fixing conditions. FEMS Microbiol. Lett. *11*: 63–67.

Malik, K.A. 1975. Preservation of Knallgas bacteria. Fifth International Fermentation Symposium, Berlin. Westkreuz Druckerei und Verlag, Berlin-Bonn. p. 180.

Malik, K.A. 1990a. A simplified liquid-drying method for the preservation of microorganisms sensitive to freezing and freeze-drying. J. Microbiol. Methods *12*: 125–132.

Malik, K.A. 1990b. Use of activated charcoal for the preservation of anaerobic phototrophic and other sensitive bacteria by freeze-drying. J. Microbiol. Methods *12*: 117–124.

Malik, K.A. and D. Claus. 1979. *Xanthobacter flavus*, a new species of nitrogen-fixing hydrogen bacteria. Int. J. Syst. Bacteriol. *29*: 283–287.

Maliszewski, C.R. and S.J. Badger. 1980. Group and type antigens of *Eikenella corrodens*. Abstracts of the 80th General Meeting of the American Society for Microbiology, American Society for Microbiology. p. 94.

Maliszewski, C.R., C.W. Shuster and S.J. Badger. 1983. A type-specific antigen of *Eikenella corrodens* is the major outer membrane protein. Infect. Immun. *42*: 208–213.

Malmqvist, Å., T. Welander and L. Gunnarsson. 1991. Anaerobic growth of microorganisms with chlorate as an electron-acceptor. Appl. Environ. Microbiol. *57*: 2229–2232.

Malmqvist, A., T. Welander, E. Moore, A. Ternström, G. Molin and I.-M. Stenström. 1994a. *Ideonella dechloratans* gen. nov., sp. nov., a new bacterium capable of growing anaerobically with chlorate as an electron acceptor. Syst. Appl. Microbiol. *17*: 58–64.

Malmqvist, A., T. Welander, E. Moore, A. Ternstrom, G. Molin and I.M. Stenstrom. 1994b. *In* Validation of the publication of new names and new combinations previously effectively published outside the IJSB. List No. 50. Int. J. Syst. Bacteriol. *44*: 595.

Malofeeva, I.V. and D. Laush. 1976. Utilization of various nitrogen-compounds by phototropic bacteria. Microbiology *45*: 441–443.

Mamat, U., E.T. Rietschel and G. Schmidt. 1995. Repression of lipopolysaccharide biosynthesis in *Escherichia coli* by an antisense RNA of *Acetobacter methanolicus* phage Acm1. Mol. Microbiol. *15*: 1115–1125.

Mamkaeva, K.A. 1966. Observations on the lysis of cultures of the genus *Chlorella*. Mikrobiologiya *35*: 853–859.

Manasse, R.J., R.C. Staples, R.R. Granados and E.G. Barnes. 1972. Mor-

phological, biological, and physical properties of *Agrobacterium tumefaciens* bacteriophages. Virology *47*: 375–384.

Mandel, M., P. Hirsch and S.F. Conti. 1972. Deoxyribonucleic acid base compositions of hyphomicrobia. Arch. Mikrobiol. *81*: 289–294.

Mandel, M. and E.R. Leadbetter. 1965. Deoxyribonucleic acid base composition of myxobacteria. J. Bacteriol. *90*: 1795–1796.

Mandrell, R.E., J.M. Griffiss and B.A. Macher. 1988. Lipooligosaccharides (Los) of *Neisseria gonorrhoeae* and *Neisseria meningitidis* have components that are immunochemically similar to precursors of human-blood group antigens—carbohydrate sequence specificity of the mouse monoclonal-antibodies that recognize crossreacting antigens on Los and human-erythrocytes. J. Exp. Med. *168*: 107–126.

Mangels, L.A., J.L. Favinger, M.T. Madigan and H. Gest. 1986. Isolation and characterization of the N₂-fixing marine photosynthetic bacterium *Rhodopseudomonas marina*, variety agilis. FEMS Microbiol. Lett. *36*: 99–104.

Mann, S. 1969. Über melaninbildende Stämme von *Pseudomonas aeruginosa*. Arch. Microbiol. *65*: 359–379.

Manna, A. and G.C. Sadhukhan. 1991. Lethality induced to embryos by the bacterium *Xanthobacter autotrophicus* to normally pregnant female mice at various days of their gestation. Chromosome Information Service. *51*: 10–12.

Manna, G.K. and G.C. Sadhukhan. 1992. *Xanthobacter flavus* induced chromosome aberrations in human leucocyte culture. Nucleus. *35*: 70–73.

Manna, G.K. and G.C. Sadhukhan. 1993. Cytogenetic assays of mice treated with various samples of the bacterium *Xanthobacter flavus*. Kromosomo. *2*: 2385–2394.

Manniello, J.M., H. Heymann and F.W. Adair. 1979. Isolation of atypical lipopolysaccharides from purified cell walls of *Pseudomonas cepacia*. J. Gen. Microbiol. *112*: 397–400.

Manning, H.L. 1975. New medium for isolating iron-oxidizing and heterotrophic acidophilic bacteria from acid mine drainage. Appl. Microbiol. *30*: 1010–1016.

Manns, T.F. 1909. The blade blight of oats. A bacterial disease. Bull. Ohio Agr. Expt. Sta. *210*: 91–167.

Manor, E., N.H. Carbonetti and D.J. Silverman. 1994. *Rickettsia rickettsii* has proteins with cross-reacting epitopes to eukaryotic phospholipase A2 and phospholipase C. Microb. Pathog. *17*: 99–109.

Manz, W., M. Eisenbrecher, T.R. Neu and U. Szewzyk. 1998. Abundance and spatial organization of Gram-negative sulfate- reducing bacteria in activated sludge investigated by in situ probing with specific 16S rRNA targeted oligonucleotides. FEMS Microbiol. Ecol. *25*: 43–61.

Maratea, D. and R.P. Blakemore. 1981. *Aquaspirillum magnetotacticum*, sp. nov., a magnetic spirillum. Int. J. Syst. Bacteriol. *31*: 452–455.

Marbach, A., M. Varon and M. Shilo. 1976. Properties of marine bdellovibrios. Microb. Ecol. *2*: 284–295.

Marchessault, R.H., E.M. Debzi, J.F. Revol and A. Steinbüchel. 1995. Single crystals of bacterial and synthetic poly(3-hydroxyvalerate). Can. J. Microbiol. *41*: 297–302.

Marchitto, K.S., S.G. Smith, C. Locht and J.M. Keith. 1987. Nucleotide sequence homology to pertussis toxin gene in *Bordetella bronchiseptica* and *Bordetella parapertussis*. Infect. Immun. *55*: 497–501.

Marcus, P.I. and P. Talalay. 1956. Induction and purification of alpha- and beta- hydroxysteroid dehydrogenases. J. Biol. Chem. *218*: 661–674.

Marczynski, G.T. and L. Shapiro. 1995. The control of asymmetric gene expression during *Caulobacter* cell differentiation. Arch. Microbiol. *163*: 313–321.

Mardh, P.A., L. Westrom and M. Akerlund. 1975. *In vitro* experiments on adherence of bacteria to vaginal epithelial cells. Acta Obstet. Gynecol. Scand. *54*: 193–194.

Margalith, P. 1962. Bacteriolytic principles of *Myxococcus fulvus*. Nature *196*: 1335–1336.

Mariette, I., E. Schwarz, R.F. Vogel and W.P. Hammes. 1991. Characterization by plasmid profile analysis of acetic acid bacteria from wine, spirit and cider acetators for industrial vinegar production. J. Appl. Bacteriol. *71*: 134–138.

Marín, I., R. Amils and J.P. Abad. 1995. Linear extrachromosomal DNA of *Thiobacillus cuprinus* DSM 5495 in relation to its chemolithotrophic growth. FEMS Microbiol. Lett. *134*: 75–78.

Marini, R.P., C.J. Foltz, D. Kersten, M. Batchelder, W. Kaser and X. Li. 1996. Microbiologic, radiographic, and anatomic study of the nasolacrimal duct apparatus in the rabbit (*Oryctolagus cuniculus*). Lab. Anim. Sci. *46*: 656–662.

Marison, I.W. and M.M. Attwood. 1980. Partial purification and characterization of a dye-linked formaldehyde dehydrogenase from *Hyphomicrobium* X. J. Gen. Microbiol. *117*: 305–314.

Marison, I.W. and M. Attwood. 1982. A possible alternative mechanism for the oxidation of formaldehyde to formate. J. Gen. Microbiol. *128*: 1441–1446.

Markiewicz, Z., B. Glauner and U. Schwarz. 1983. Murein structure and lack of DD- and LD-carboxypeptidase activities in *Caulobacter crescentus*. J. Bacteriol. *156*: 649–655.

Marmur, J. 1961. A procedure for the isolation of deoxyribonucleic acid from microorganisms. J. Mol. Biol. *3*: 208–218.

Marmur, J. and P. Doty. 1962. Determination of the base composition of deoxyriboniucleic acid from its thermal denaturation temperature. J. Mol. Biol. *5*: 109–118.

Maroye, P., H.P. Doermann, A.M. Rogues, J.P. Gachie and F. Megraud. 2000. Investigation of an outbreak of *Ralstonia pickettii* in a paediatric hospital by RAPD. J. Hosp. Infect. *44*: 267–272.

Marsden, H.B. and W.A. Hyde. 1971. Isolation of *Bacteroides corrodens* from infections in children. J. Clin. Pathol. *24*: 117–119.

Marshall, B.J., H. Royce, D.I. Annear, C.S. Goodwin, J.W. Pearman, J.R. Warren and J.A. Armstrong. 1984. Original isolation of *Campylobacter pyloridis* from human gastric mucosa. Microbios Lett. *25*: 83–88.

Martin, D., N. Cadieux, J. Hamel and B.R. Brodeur. 1997. Highly conserved *Neisseria meningitidis* surface protein confers protection against experimental infection. J. Exp. Med. *185*: 1173–1183.

Martin, H. and J.C. Murrell. 1995. Methane monooxygenase mutants of *Methylosinus trichosporium* constructed by marker-exchange mutagenesis. FEMS Microbiol. Lett. *127*: 243–248.

Martin, J.E., J.H. Armstrong and P.B. Smith. 1974. New system for cultivation of *Neisseria gonorrhoeae*. Appl. Microbiol. *27*: 802–805.

Martin, J. and J. Brimacombe. 1992. *Chromobacterium violaceum* septicaemia: the intensive care management of two cases. Anaesth. Intensive Care. *20*: 88–90.

Martin, J.P., J. Fleck, M. Mock and J.M. Ghuysen. 1973. The wall peptidoglycans of *Neisseria perflava*, *Moraxella glucidolytica*, *Pseudomonas alcaligenes* and *Proteus vulgaris* strain P18. Eur. J. Biochem. *38*: 301–306.

Martin, S., E. Sodergren, T. Masuda and D. Kaiser. 1978. Systematic isolation of transducing phages for *Myxococcus xanthus*. Virology *88*: 44–53.

Martin, W.B. 1996. Respiratory infections of sheep. Comp. Immunol. Microbiol. Infect. Dis. *19*: 171–179.

Martin-Didonet, C.C.G., L.S. Chubatsu, E.M. Souza, M. Kleina, F.G.M. Rego, L.U. Rigo, M.G. Yates and F.O. Pedrosa. 2000. Genome structure of the genus *Azospirillum*. J. Bacteriol. *182*: 4113–4116.

Martinez, D., C. Sheikboudou, P.O. Couraud and A. Bensaid. 1993. *In vitro* infection of bovine brain endothelial cells by *Cowdria ruminantium*. Res. Vet. Sci. *55*: 258–260.

Martínez, E., R. Palacios and F. Sánchez. 1987. Nitrogen-fixing nodules induced by *Agrobacterium tumefaciens* harboring *Rhizobium phaseoli* plasmids. J. Bacteriol. *169*: 2828–2834.

Martinez, R.J. 1963. On the nature of the granules of the genus *Spirillum*. Arch. Microbiol. *44*: 334–343.

Martínez De Drets, G. and A. Arias. 1970. Metabolism of some polyols by *Rhizobium meliloti*. J. Bacteriol. *103*: 97–103.

Martinez-Cañamero, M.M., J. Muñoz, A.L. Extremera and J.M. Arias. 1991. Deoxyribonuclease activities in *Myxococcus coralloides* D. J. Appl. Bacteriol. *71*: 170–175.

Martínez-De Drets, G. and A. Arias. 1972. Enzymatic basis for differentiation of *Rhizobium* into fast- and slow-growing groups. J. Bacteriol. *109*: 467–470.

Martinez-Drets, G., E. Fabiano and A. Cardona. 1985. Carbohydrate catabolism in Azospirillum amazonense. Appl. Environ. Microbiol. *50*: 183–185.

Martínez-Romero, E. 1994. Recent developments in *Rhizobium* taxonomy. Plant Soil *161*: 11–20.

Martínez-Romero, E. and J. CaballeroMellado. 1996. *Rhizobium* phylogenies and bacterial genetic diversity. Crit. Rev. Plant Sci. *15*: 113–140.

Martínez-Romero, E., L. Segovia, F.M. Mercante, A.A. Franco, P. Graham and M.A. Pardo. 1991. *Rhizobium tropici*, a novel species nodulating *Phaseolus vulgaris* L. beans and *Leucaena* sp. trees. Int. J. Syst. Bacteriol. *41*: 417–426.

Mason, D.J. and D. Powelson. 1958a. The cell wall of *Myxococcus xanthus*. Biochim. Biophys. Acta *29*: 1–7.

Mason, D.J. and D. Powelson. 1958b. Lysis of *Myxococcus xanthus*. J. Gen. Microbiol. *19*: 65–70.

Mason, L.M. and G.W. Claus. 1989. Phenotypic characteristics correlated with deoxyribonucleic acid sequence similarities for three species of *Gluconobacter*: *Gluconbacter oxydans* (Henneberg 1897) De Ley 1961, *Gluconbacter frateurii* sp. nov., and *Gluconobacter asaii* sp. nov. Int. J. Syst. Bacteriol. *39*: 174–184.

Masson, L., Y. Comeau, R. Brousseau, R. Samson and C. Greer. 1993. Construction and application of chromosomally integrated *lac-lux* gene markers to monitor the fate of a 2,4-dichlorophenoxyacetic acid degrading bacterium in contaminated soils. Microb. Releases. *1*: 209–216.

Masson, L. and B.E. Holbein. 1983. Physiology of sialic acid capsular polysaccharide synthesis in serogroup B *Neisseria meningitidis*. J. Bacteriol. *154*: 728–736.

Masson, P.J. and J.F. Guespin-Michel. 1988. An extracellular blood-anticoagulant glycopeptide produced exclusively during vegetative growth by *Myxococcus xanthus* and other myxobacteria is not co-regulated with other extracellular macromolecules. J. Gen. Microbiol. *134*: 801–806.

Masters, R.A. and M.T. Madigan. 1983. Nitrogen metabolism in the phototrophic bacteria *Rhodocyclus purpureus* and *Rhodospirillum tenue*. J. Bacteriol. *155*: 222–227.

Masuda, M., H. Odake, K. Miura and K. Oba. 1995. Biodegradation of 2-sulfonato fatty-acid-methyl-ester (α-Sfme) identification of microorganisms isolated from activated-sludge and their capability for degrading α-Sfme. Appl. Microbiol. Biotechnol. *43*: 379–382.

Masuda, T., K. Inoue, M. Masuda, M. Nagayama, A. Tamaki, H. Ohta, H. Shimada and K.I. Takamiya. 1999. Magnesium insertion by magnesium chelatase in the biosynthesis of zinc bacteriochlorophyll *a* in an aerobic acidophilic bacterium *Acidiphilium rubrum*. J. Biol. Chem. *274*: 33594–33600.

Masui, S., S. Kamoda, T. Sasaki and H. Ishikawa. 2000. Distribution and evolution of bacteriophage WO in *Wolbachia*, the endosymbiont causing sexual alterations in arthropods. J. Mol. Evol. *51*: 491–497.

Masui, S., T. Sasaki and H. Ishikawa. 1997. *groE*-homologous operon of *Wolbachia*, an intracellular symbiont of arthropods: A new approach for their phylogeny. Zool. Sci. *14*: 701–706.

Maszenan, A.M., R.J. Seviour, B.K.C. Patel, G.N. Rees and B.M. McDougall. 1997. *Amaricoccus* gen. nov., a gram-negative coccus occurring in regular packages or tetrads, isolated from activated sludge biomass, and descriptions of *Amaricoccus veronensis* sp. nov., *Amaricoccus tamworthensis* sp. nov., *Amaricoccus macauensis* sp. nov., and *Amaricoccus kaplicensis* sp. nov. Int. J. Syst. Bacteriol. *47*: 727–734.

Maszenan, A.M., R.J. Seviour, B. Patel, K.C., G.N. Rees and B. McDougall. 1998. The hunt for the G-bacteria in activated sludge biomass. Water Sci. Technol. *37*: 65–69.

Maszenan, A.M., R.J. Seviour, B.K.C. Patel and P. Schumann. 2002. *Quadricoccus australiensis* gen. nov., sp nov., a β-proteobacterium from activated sludge biomass. Int. J. Syst. Evol. Microbiol. *52*: 223–228.

Maszenan, A.M., R.J. Seviour, B.K.C. Patel and J. Wanner. 2000. A fluorescently-labelled rRNA targeted oligonucleotide probe for the *in situ* detection of G-bacteria of the genus *Amaricoccus* in activated sludge. J. Appl. Microbiol. *88*: 826–835.

Materassi, R., G. Florenzano, W. Balloni and F. Flavilli. 1966. Su una nuova specie di *Beijerinckia* (*Beijerinckia venezuelae* nov. sp.) isolate da terreni venezuelani. Ann. Microbiol. Enzimol. *16*: 201–215.

Matevosyan, S.R., I.B. Utkin and A.M. Bezborodov. 1989. Physico-chemical properties of catechol 1.2-dioxygenase from *Alcaligenes paradoxus* SM 6226. Biokhimiya *54*: 1394–1399.

Matheson, V.G., J. Munakata-Marr, G.D. Hopkins, P.L. McCarty, J.M. Tiedje and L.J. Forney. 1997. A novel means to develop strain specific DNA probes for detecting bacteria in the environment. Appl. Environ. Microbiol. *63*: 2863–2869.

Mathew, J.S., S.A. Ewing, G.L. Murphy, K.M. Kocan, R.E. Corstvet and J.C. Fox. 1997. Characterization of a new isolate of *Ehrlichia platys* (Order *Rickettsiales*) using electron microscopy and polymerase chain reaction. Vet. Parasitol. *68*: 1–10.

Mathey, W.J. 1956. A diphtheroid stomatitis of chickens apparently due to a spirillum, *Spirillum pulli*, species nova. Am. J. Vet. Res. *17*: 742–746.

Mathey, W.J. and A.C. Rissberger. 1964. A turkey sinus vibrio (*Vibrio maleagridis* n. sp.) compared with the avian hepatitis vibrio (*Vibrio hepaticus*, n. sp.). Poultry Sci. *43*: 1339.

Matin, A. and S.C. Rittenberg. 1972. Kinetics of deoxyribonucleic acid destruction and synthesis during growth of *Bdellovibrio bacteriovorus* strain 109D on *Pseudomonas putida* and *Escherichia coli*. J. Bacteriol. *111*: 664–673.

Matson, B.A. 1967. Theileriosis in Rhodesia: I. A study of diagnostic specimens over two seasons. J. S. African Vet. Med. Assoc. *38*: 93–102.

Matsuda, M., Y. Asami, T. Miyazawa, T. Samata, Y. Isayama, M. Honda and Y. Ide. 1993. Analysis of chromosome-sized DNA and genome typing of isolated strains of *Taylorella equigenitalis*. Vet. Res. Commun. *18*: 93–98.

Matsumoto, H., Y. Itoh, S. Ohta and Y. Terawaki. 1986. A generalized transducing phage of *Pseudomonas cepacia*. J. Gen. Microbiol. *132*: 2583–2586.

Matsunaga, I., M. Yamada, E. Kusunose, T. Miki and K. Ichihara. 1998. Further characterization of hydrogen peroxide-dependent fatty acid alpha-hydroxylase from *Sphingomonas paucimobilis*. J. Biochem. (Tokyo) *124*: 105–110.

Matsunaga, I., M. Yamada, E. Kusunose, Y. Nishiuchi, I. Yano and K. Ichihara. 1996. Direct involvement of hydrogen peroxide in bacterial alpha-hydroxylation of fatty acid. FEBS Lett. *386*: 252–254.

Matsunaga, I., N. Yokotani, O. Gotoh, E. Kusunose, M. Yamada and K. Ichihara. 1997. Molecular cloning and expression of fatty acid alpha-hydroxylase from *Sphingomonas paucimobilis*. J. Biol. Chem. *272*: 23592–23596.

Matsushita, K., H. Ebisuya, M. Ameyama and O. Adachi. 1992a. Change of the terminal oxidase from cytochrome a_1 in shaking cultures to cytochrome *o* in static cultures of *Acetobacter aceti*. J. Bacteriol. *174*: 122–129.

Matsushita, K., H. Ebisuya, M. Ameyama and O. Adachi. 1992b. Homology in the structure and the prosthetic groups between two different terminal ubiquinol oxidases, cytochrome a_1 and cytochrome *o*, of *Acetobacter aceti*. J. Biol. Chem. *267*: 24748–24753.

Matsushita, K., Y. Nagatani, E. Shinagawa, O. Adachi and M. Ameyama. 1989. Effect of extracellular pH on the respiratory chain and energetics of *Gluconobacter suboxydans*. Agric. Biol. Chem. *53*: 2895–2902.

Matsushita, K., K. Takahashi and O. Adachi. 1993. A novel quinoprotein methanol dehydrogenase containing an additional 32-kilodalton peptide purified from *Acetobacter methanolicus*: identification of the peptide as a MoxJ product. Biochemistry *32*: 5576–5582.

Matsushita, K., K. Takahashi, M. Takahashi, M. Ameyama and O. Adachi. 1992c. Methanol and ethanol oxidase respiratory chains of the methylotrophic acetic acid bacterium, *Acetobacter methanolicus*. J. Biochem. (Tokyo) *111*: 739–747.

Matsushita, K., Y. Takaki, E. Shinagawa, M. Ameyama and O. Adachi. 1992d. Ethanol oxidase respiratory chain of acetic acid bacteria. Reactivity with ubiquinone of pyrroloquinone quinone-dependent alcohol dehydrogenases purified from *Acetobacter aceti* and *Gluconobacter suboxydans*. Biosci. Biotechnol. Biochem. *56*: 304–310.

Matsushita, K., H. Toyama and O. Adachi. 1994. Respiratory chains and bioenergetics of acetic acid bacteria. Adv. Microb. Physiol. *36*: 247–301.

Matsuura, M., K. Kawahara, T. Ezaki and M. Nakano. 1996. Biological activities of lipopolysaccharide of *Burkholderia* (*Pseudomonas*) *pseudomallei*. FEMS Microbiol. Lett. *137*: 79–83.

Matsuyama, N. 1995. Application of the direct colony TLC for identification of phytopathogenic bacteria III. Distinction of the pseudomonads in the rRNA homology group II (*Burkholderia* spp.). J. Fac. Agric. Kyushu Univ. *40*: 189–196.

Matsuzawa, Y., T. Kanbe, J. Suzuki and A. Hiraishi. 2000. Ultrastructure of the acidophilic aerobic photosynthetic bacterium *Acidiphilium rubrum*. Curr. Microbiol. *40*: 398–401.

Matthysse, A.G., K.V. Holmes and R.H.G. Gurlitz. 1981. Elaboration of cellulose fibrils by *Agrobacterium tumefaciens* during attachment to carrot cells. J. Bacteriol. *145*: 583–595.

Matzen, N. and P. Hirsch. 1982a. Continuous culture and synchronization of *Hyphomicrobium* sp. B-522. Arch. Microbiol. *132*: 96–99.

Matzen, N. and P. Hirsch. 1982b. Improved growth conditions for *Hyphomicrobium* sp. B-522 and 2 additional strains. Arch. Microbiol. *131*: 32–35.

Maurer, J., J. Jose and T.F. Meyer. 1997. Autodisplay: one-component system for efficient surface display and release of soluble recombinant proteins from *Escherichia coli*. J. Bacteriol. *179*: 794–804.

Maurin, M., F. Eb, J. Etienne and D. Raoult. 1997. Serological cross-reactions between *Bartonella* and *Chlamydia* species: implications for diagnosis. J. Clin. Microbiol. *35*: 2283–2287.

Maurin, M., S. Gasquet, C. Ducco and D. Raoult. 1995. MICs of 28 antibiotic compounds for 14 *Bartonella* (formerly *Rochalimaea*) isolates. Antimicrob. Agents Chemother. *39*: 2387–2391.

Maurin, M., H. Lepocher, D. Mallet and D. Raoult. 1993. Antibiotic susceptibilities of *Afipia felis* in axenic medium and in cells. Antimicrob. Agents Chemother. *37*: 1410–1413.

Maxcy, K.F. 1929. Endemic typhus fever of the Southeastern United States: Reaction of the guinea pig. Publ. Health Rep. *44*: 589–600.

Maxwell-Lyons, F. and A. C.R.. 1949. The epidemiology and prevention of the acute ophthalmias of Egypt. Ophthamological Society of Egypt. *42*: 116–134.

Mayer, D. 1967. Ernährungsphysiologische untersuchungen an *Archangium violaceum*. Arch. Mikrobiol. *58*: 186–200.

Mayer, H., E. Bock and J. Weckesser. 1983. 2,3-Diamino-2,3-dideoxyglucose containing lipid A in the *Nitrobacter* strain X14. FEMS Microbiol. Lett. *17*: 93–96.

Mayer, H. and H. Reichenbach. 1978. Restriction endonucleases—general survey procedure and survey of gliding bacteria. J. Bacteriol. *136*: 708–713.

Mayer, H., P.V. Salimath, O. Holst and J. Weckesser. 1984. Unusual lipid-A types in phototrophic bacteria and related species. Rev. Infect. Dis. *6*: 542-545.

Mayfield, C.I. and W.E. Inniss. 1977. A rapid, simple method for staining bacterial flagella. Can. J. Microbiol. *23*: 1311–1313.

Mayfield, D.C. and A.S. Kester. 1972. Physiological studies on *Vitreoscilla stercoraria*. J. Bacteriol. *112*: 1052–1056.

Mayfield, D.C. and A.S. Kester. 1975. Nutrition of *Vitreoscilla stercoraria*. Can. J. Microbiol. *21*: 1947–1951.

Mazengia, E., E.A. Silva, J.A. Peppe, R. Timperi and H. George. 2000. Recovery of *Bordetella holmesii* from patients with pertussis-like symptoms: use of pulsed-field gel electrophoresis to characterize circulating strains. J. Clin. Microbiol. *38*: 2330–2333.

Mazloum, H., P.A. Totten, G.F. Brooks, C.R. Dawson, S. Falkow, J.F. James, J.S. Knapp, J.M. Koomey, C.J. Lammel and D. Peters. 1986. An unusual *Neisseria* isolated from conjunctival cultures in rural Egypt. J. Infect. Dis. *154*: 212–224.

Mazurier, S., A. van de Giessen, K. Heuvelman and K. Wernars. 1992. RAPD analysis of *Campylobacter* isolates: DNA fingerprinting without the need to purify DNA. Lett. Appl. Microbiol. *14*: 260–262.

McAnulla, C., C.A. Woodall, I.R. McDonald, A. Studer, S. Vuilleumier, T. Leisinger and J.C. Murrell. 2001. Chloromethane utilization gene cluster from *Hyphomicrobium chloromethanicum* strain CM2[T] and development of functional gene probes to detect halomethane-degrading bacteria. Appl. Environ. Microbiol. *67*: 307–316.

McBride, J.W., X.J. Yu and D.H. Walker. 2000. A conserved, transcriptionally active p28 multigene locus of *Ehrlichia canis*. Gene *254*: 245–252.

McBride, M.J., P. Hartzell and D.R. Zusman. 1993. Motility and tactic behavior of *Myxococcus xanthus. In* Dworkin and Kaiser (Editors), Myxobacteria II, American Society for Microbiology, Washington, D.C. pp. 285–305.

McBride, M.J. and D.R. Zusman. 1989. Trehalose accumulation in vegetative cells and spores of *Myxococcus xanthus.* J. Bacteriol. *171*: 6383–6386.

McCall, J.W., J.J. Jun and C. Bandi. 1999. *Wolbachia* and the antifilarial properties of tetracycline. Ital. J. Zool. *66*: 7–10.

McCallum, K.L. and W.E. Inniss. 1990. Thermotolerance, cell filamentation, and induced protein-synthesis in psychrophilic and psychrotrophic bacteria. Arch. Microbiol. *153*: 585–590.

McCann, M.P., H.T. Solimeo, F. Cusick, Jr., B. Panunti and C. McCullen. 1998. Developmentally regulated protein synthesis during intraperiplasmic growth of *Bdellovibrio bacteriovorus* 109J. Can. J. Microbiol. *44*: 50–55.

McCarthy, D.L., A.A. Claude and S.D. Copley. 1997. *In vivo* levels of chlorinated hydroquinones in a pentachlorophenol-degrading bacterium. Appl. Environ. Microbiol. *63*: 1883–1888.

McClean, K.H., M.K. Winson, L. Fish, A. Taylor, S.R. Chhabra, M. Camara, M. Daykin, J.H. Lamb, S. Swift, B.W. Bycroft, G.S.A.B. Stewart and P. Williams. 1997. Quorum sensing and *Chromobacterium violaceum*: Exploitation of violacein production and inhibition for the detection of *N*-acylhomoserine lactones. Microbiology (Reading) *143*: 3703–3711.

McClung, C.R. and D.G. Patriquin. 1980. Isolation of a nitrogen-fixing *Campylobacter* species from the roots of *Spartina alterniflora* Loisel. Can. J. Microbiol. *26*: 881–886.

McClung, C.R., D.G. Patriquin and R.E. Davis. 1983. *Campylobacter nitrofigilis* sp. nov., a nitrogen-fixing bacterium associated with roots of *Spartina alterniflora* Loisel. Int. J. Syst. Bacteriol. *33*: 605–612.

McComb, J.A., J. Elliot and M.J. Dilworth. 1975. Acetylene reduction by *Rhizobium* in pure culture. Nature (Lond.) *256*: 409–410.

McCowan, R.P., K.J. Cheng and J.W. Costerton. 1979. Colonization of a portion of the bovine tongue by unusual filamentous bacteria. Appl. Environ. Microbiol. *37*: 1224–1229.

McCrary, A.L. and W.L. Howard. 1979. Chelating agents. *In* Grayson and Eckroth (Editors), Kirk-Othmer Encyclopedia of Chemical Technology, 3rd Ed., Vol. 5, John Wiley & Sons, New York. pp. 339–368.

McCullough, N.B. and G.A. Beal. 1951. Growth and manometric studies on carbohydrate utilisation of *Brucella.* J. Infect Dis. *89*: 266–271.

McCullough, N.B. and L.A. Dick. 1943. Growth of *Brucella* in a simple chemically defined medium. Proc. Soc. Exp. Biol. Med. *52*: 310–311.

McCurdy, H.D. 1969a. Light and electron microscope studies on the fruiting bodies of *Chondromyces crocatus.* Arch. Microbiol. *65*: 380–390.

McCurdy, H.D. 1969b. Studies on the taxonomy of the *Myxobacterales.* I. Record of Canadian isolates and survey of methods. Can. J. Microbiol. *15*: 1453–1461.

McCurdy, H.D. 1970. Studies on the taxonomy of the *Myxobacterales.* II. *Polyangium* and the demise of the *Sorangiaceae.* Int. J. Syst. Bacteriol. *20*: 283–296.

McCurdy, H.D. 1971a. Studies on the taxonomy of the *Myxobacterales.* III. *Chondromyces* and *Stigmatella.* Int. J. Syst. Bacteriol. *58*: 40–49.

McCurdy, H.D. 1971b. Studies on the taxonomy of the *Myxobacterales.* IV. *Melittangium.* Int. J. Syst. Bacteriol. *21*: 50–54.

McCurdy, H.D. 1989. Genus I. *Cystobacter. In* Staley, Bryant, Pfennig and Holt (Editors), Bergey's Manual of Systematic Bacteriology, 1st Ed., Vol. 3, The Williams & Wilkins Co., Baltimore. pp. 2150–2153.

McCurdy, H.D. and B.T. Khouw. 1969. Studies on *Stigmatella brunnea.* Can. J. Microbiol. *15*: 731–738.

McCurdy, H.D. and T.H. MacRae. 1974. Xanthacin. A bacteriocin of *Myxococcus xanthus* fb. Can. J. Microbiol. *20*: 131–135.

McCurdy, H.D. and S. Wolf. 1967. Deoxyribonucleic acid base compositions of fruiting *Myxobacterales.* Can. J. Microbiol. *13*: 1707–1708.

McDade, J.E., C.C. Shepard, M.A. Redus, V.F. Newhouse and J.D. Smith.

1980. Evidence of *Rickettsia prowazekii* infections in the United States. Am. J. Trop. Med. Hyg. *29*: 277–284.

McDermott, E.N. 1928. Rat-bite fever: a study of the experimental disease, with a critical review of the literature. Q. J. Med. *21*: 433–438.

McDonald, I.R., N.V. Doronina, Y.A. Trotsenko, C. McAnulla and J.C. Marrell. 2001. *Hyphomicrobium chloromethanicum* sp. nov. and *Methylobacterium chloromethanicum* sp. nov., chloromethane-utilizing bacteria isolated from a polluted environment. Int. J. Syst. Evol. Microbiol. *51*: 119–122.

McDonald, I.R., G.H. Hall, R.W. Pickup and J.C. Murrell. 1996. Methane oxidation potential and preliminary analysis of methanotrophs in blanket bog peat using molecular ecology techniques. FEMS Microbiol. Ecol. *21*: 197–211.

McDonald, I.J. and K.G. Johnson. 1975. Nutritional requirements of some nonpathogenic *Neisseria* grown in simple synthetic media. Can. J. Microbiol. *21*: 1198–1204.

McDonald, I.R., D.P. Kelly, J.C. Murrell and A.P. Wood. 1997. Taxonomic relationships of *Thiobacillus halophilus*, *Thiobacillus aquaesulis*, and other species of *Thiobacillus*, as determined using 16S rRNA sequencing. Arch. Microbiol. *166*: 394–398.

McDonald, I.R., E.M. Kenna and J.C. Murrell. 1995. Detection of methanotrophic bacteria in environmental samples with the PCR. Appl. Environ. Microbiol. *61*: 116–121.

McDonald, I.R. and J.C. Murrell. 1997a. The methanol dehydrogenase structural gene *mxaF* and its use as a functional gene probe for methanotrophs and methylotrophs. Appl. Environ. Microbiol. *63*: 3218–3224.

McDonald, I.R. and J.C. Murrell. 1997b. The particulate methane mono-oxygenase gene *pmoA* and its use as a functional gene probe for methanogens. FEMS Microbiol. Lett. *156*: 205–210.

McDougall, R., J. Robson, D. Paterson and W. Tee. 1997. Bacteremia caused by a recently described novel *Desulfovibrio* species. J. Clin. Microbiol. *35*: 1805–1808.

McElroy, L.J. and N.R. Krieg. 1972. A serological method for the identification of spirilla. Can. J. Microbiol. *18*: 57–64.

McEwan, A.G., S.J. Ferguson and J.B. Jackson. 1983. Electron flow to dimethylsulphoxide or trimethylamine-*N*-oxide generates a membrane potential in *Rhodopseudomonas capsulata.* Arch. Microbiol. *136*: 300–305.

McGarey, D.J. and D.R. Allred. 1994. Characterization of hemagglutinating components on the *Anaplasma marginale* initial body surface and identification of possible adhesins. Infect. Immun. *62*: 4587–4593.

McGee, Z.A., R.R. Dourmashkin, J.G. Gross, J.B. Clark and D. Taylor-robinson. 1977. Relationship of pili to colonial morphology among pathogenic and nonpathogenic species of *Neisseria.* Infect. Immun. *15*: 594–600.

McGee, Z.A., C.H. Street, C.L. Chappell, E.S. Cousar, F. Morris and R.G. Horn. 1979. Pili of *Neisseria meningitidis*: effect of media on maintenance of piliation, characteristics of pili, and colonial morphology. Infect. Immun. *24*: 194–201.

McGhee, J.R. and B.A. Freeman. 1970a. Osmotically sensitive *Brucella* in infected, normal and immune macrophages. Infect. Immun. *1*: 146–150.

McGhee, J.R. and B.A. Freeman. 1970b. Separation of soluble *Brucella* antigens by gel filtration chromatography. Infect. Immun. *2*: 48–53.

McGuckin, M.B., R.J. Thorpe, K.M. Koch, A. Alavi, M. Staum and E. Abrutyn. 1982. An outbreak of *Achromobacter xylosoxidans* related to diagnostic tracer procedures. Am. J. Epidemiol. *115*: 785–793.

McGuire, T.C., G.H. Palmer, W.L. Goff, M.I. Johnson and W.C. Davis. 1984. Common and isolate-restricted antigens of *Anaplasma marginale* detected with monoclonal antibodies. Infect. Immun. *45*: 697–700.

McInerney, M.J. and M.P. Bryant. 1981. Review of nethane fermentation fundamentals. *In* Wise (Editor), Fuel Gas Production from Biomass, CRC Press, Boca Raton. pp. 19–46.

McInerney, M.J., M.P. Bryant and N. Pfennig. 1979. Anaerobic bacterium that degrades fatty acids in syntrophic association with methanogens. Arch. Microbiol. *122*: 129–136.

McKiel, J.A., E.J. Bell and D.B. Lackman. 1967. *Rickettsia canada*: A new member of the typhus group of rickettsiae isolated from *Haemophysalis leporispalustris* ticks in Canada. Can. J. Microbiol. *13*: 503–510.

McLaren, D.J., M.J. Worms, B.R. Laurence and M.G. Simpson. 1975. Micro-organisms in filarial larvae (Nematoda). Trans. R. Soc. Trop. Med. Hyg. *69*: 509–514.

McLean, A.P.D. and P.C.J. Oberholzer. 1965. *Citrus psylla*, a vector of the greening disease of sweet orange. South Africa J. Agric. Sci. *24*: 89–95.

McLoughlin, T.J., J.P. Quinn, A. Bettermann and R. Bookland. 1992. *Pseudomonas cepacia* suppression of sunflower wilt fungus and role of antifungal compounds in controlling the disease. Appl. Environ. Microbiol. *58*: 1760–1763.

McNeil, K.E. and V.B.D. Skerman. 1972. Examination of myxobacteria by scanning electron microscopy. Int. J. Syst. Bacteriol. *22*: 243–250.

McNeil, M.M., S.L. Solomon, R.L. Anderson, B.J. Davis, R.F. Spengler, B.E. Reisberg, C. Thornsberry and W.J. Martone. 1985. Nosocomial *Pseudomonas pickettii* colonization associated with a contaminated respiratory therapy solution in a special care nursery. J. Clin. Microbiol. *22*: 903–907.

McNicol, P., S.M. Giercke, M. Gray, D. Martin, B. Brodeur, M.S. Peppler, T. Williams and G. Hammond. 1995. Evaluation and validation of a monoclonal immunofluorescent reagent for direct detection of *Bordetella pertussis*. J. Clin. Microbiol. *33*: 2868–2871.

McNulty, C.A.M., J.C. Dent, A. Curry, J.S. Uff, G.A. Ford, M.W.L. Gear and S.P. Wilkinson. 1989. New spiral bacterium in gastric mucosa. J. Clin. Pathol. *42*: 585–591.

McOrist, S., R. Boid and G.H. Lawson. 1989a. Antigenic analysis of *Campylobacter* species and an intracellular *Campylobacter*-like organism associated with porcine proliferative enteropathies. Infect. Immun. *57*: 957–962.

McOrist, S., R. Boid, G.H. Lawson and I. McConnell. 1987. Monoclonal antibodies to intracellular *Campylobacter*-like organisms of the porcine proliferative enteropathies. Vet. Rec. *121*: 421–422.

McOrist, S., C.J. Gebhart, R. Boid and S. Barns. 1995a. Characterization of *Lawsonia intracellularis* gen. nov., sp. nov., the obligately intracellular bacterium of porcine proliferative enteropathy. Int. J. Syst. Bacteriol. *45*: 820–825.

McOrist, S., C.J. Gebhart and G.H. Lawson. 1994a. Polymerase chain reaction for diagnosis of porcine proliferative enteropathy. Vet. Microbiol. *41*: 205–212.

McOrist, S., S. Jasni, R.A. Mackie, H.M. Berschneider, A.C. Rowland and G.H. Lawson. 1995b. Entry of the bacterium ileal symbiont intracellularis into cultured enterocytes and its subsequent release. Res. Vet. Sci. *59*: 255–260.

McOrist, S., S. Jasni, R.A. Mackie, N. MacIntyre, N. Neef and G.H. Lawson. 1993. Reproduction of porcine proliferative enteropathy with pure cultures of ileal symbiont intracellularis. Infect. Immun. *61*: 4286–4292.

McOrist, S., G.H. Lawson, A.C. Rowland and N. MacIntyre. 1989b. Early lesions of proliferative enteritis in pigs and hamsters. Vet. Pathol. *26*: 260–264.

McOrist, S., G.H. Lawson, D.J. Roy and R. Boid. 1990. DNA analysis of intracellular *Campylobacter*-like organisms associated with the porcine proliferative enteropathies: novel organism proposed. FEMS Microbiol. Lett. *69*: 189–193.

McOrist, S., R.A. Mackie and G.H. Lawson. 1995c. Antimicrobial susceptibility of ileal symbiont intracellularis isolated from pigs with proliferative enteropathy. J. Clin. Microbiol. *33*: 1314–1317.

McOrist, S., R.A. Mackie, G.H. Lawson and D.G. Smith. 1997. *In vitro* interactions of *Lawsonia intracellularis* with cultured enterocytes. Vet. Microbiol. *54*: 385–392.

McOrist, S., R.A. Mackie, N. Neef, I. Aitken and G.H. Lawson. 1994b. Synergism of ileal symbiont intracellularis and gut bacteria in the reproduction of porcine proliferative enteropathy. Vet. Rec. *134*: 331–332.

McOrist, S., L. Roberts, S. Jasni, A.C. Rowland, G.H. Lawson, C.J. Gebhart and B. Bosworth. 1996. Developed and resolving lesions in porcine

proliferative enteropathy: possible pathogenic mechanisms. J. Comp. Pathol. *115*: 35–45.

McVittie, A. and S.A. Zahler. 1962. Chemotaxis in *Myxococcus*. Nature *194*: 1299–1300.

Meade, B.D. and A. Bollen. 1994. Recommendations for use of the polymerase chain reaction in the diagnosis of *Bordetella pertussis* infections. J. Med. Microbiol. *41*: 51–55.

Meade, H. and E. Singer. 1977. Genetic mapping of *Rhizobium meliloti*. Proc. Natl. Acad. Sci. U.S.A. *74*: 2076–2078.

Mechichi, T., E. Stackebrandt, N. Gad'on and G. Fuchs. 2002. Phylogenetic and metabolic diversity of bacteria degrading aromatic compounds under denitrifying conditions, and description of *Thauera phenylacetica* sp. nov., *Thauera aminoaromatica* sp. nov., and *Azoarcus buckelii* sp. nov. Arch. Microbiol. *178*: 26–35.

Meckenstock, R.U., K. Krusche, L.A. Staehelin, M. Cyrklaff and H. Zuber. 1994. The six fold symmetry of the B880 light-harvesting complex and the structure of the photosynthetic membranes of *Rhodopseudomonas marina*. Biol. Chem. Hoppe Seyler *375*: 429–438.

Megharaj, M., R.M. Wittich, R. Blasco and D.H. Pieper. 1997. Superior survival and degradation of dibenzo-*p*-dioxin and dibenzofuran in soil by soil-adapted *Sphingomonas* sp. Strain RW1. Appl. Microbiol. Biotechnol. *48*: 109–114.

Mégraud, F., D. Chevrier, N. Desplaces, A. Sedallian and J.L. Guesdon. 1988. Urease-positive thermophilic *Campylobacter* (*Campylobacter laridis* variant) isolated from an appendix and from human feces. J. Clin. Microbiol. *26*: 1050–1051.

Mégraud, F., J.L. Traissac and J. Latrille. 1981. Abscès à *Eikenella corrodens*: à propos d'un cas grave. Medicine et Maladies Infectieuses. *11*: 39–43.

Mehock, J.R., C.E. Greene, F.C. Gherardini, T.W. Hahn and D.C. Krause. 1998. *Bartonella henselae* invasion of feline erythrocytes in vitro. Infect. Immun. *66*: 3462–3466.

Meiberg, J.B.M., P.M. Bruinenberg and W. Harder. 1980. Effect of dissolved oxygen tension on the metabolism of methylated amines in *Hyphomicrobium* X in the absence and presence of nitrate: evidence for aerobic denitrification. J. Gen. Microbiol. *120*: 453–464.

Meiberg, J.B.M. and W. Harder. 1978. Aerobic and anaerobic metabolism of trimethylamine, dimethylamine and methylamine in *Hyphomicrobium* X. J. Gen. Microbiol. *106*: 265–276.

Meiberg, J.B.M. and W. Harder. 1979. Dimethylamine dehydrogenase from *Hyphomicrobium* X: purification and some properties of a new enzyme that oxidizes secondary amines. J. Gen. Microbiol. *115*: 49–58.

Meijer, E.M., H.W. van Verseveld, E.G. van der Beek and A.H. Stouthamer. 1977. Energy conservation during aerobic growth in *Paracoccus denitrificans*. Arch. Microbiol. *112*: 25–34.

Meijer, W.G. 1994. The Calvin cycle enzyme phosphoglycerate kinase of *Xanthobacter flavus* required for autotrophic CO_2 fixation is not encoded by the *cbb* operon. J. Bacteriol. *176*: 6120–6126.

Meijer, W.G., A.C. Arnberg, H.G. Enequist, P. Terpstra, M.E. Lidstrom and L. Dijkhuizen. 1991. Identification and organization of carbon dioxide fixation genes in *Xanthobacter flavus* H4-14. Mol. Gen. Genet. *225*: 320–330.

Meijer, W.G., L.M. Croes, B. Jenni, L.G. Lehmicke, M.E. Lidstrom and L. Dijkhuizen. 1990. Characterization of *Xanthobacter* strains H4-14 and 25a and enzyme profiles after growth under autotrophic and heterotrophic conditions. Arch. Microbiol. *153*: 360–367.

Meijer, W.G., P. De Boer and G. van Keulen. 1997. *Xanthobacter flavus* employs a single triosephosphate isomerase for heterotrophic and autotrophic metabolism. Microbiology (Read.) *143*: 1925–1931.

Meijer, W.G., M.E. Nienhuis-Kuiper and T.A. Hansen. 1999. Fermentative bacteria from estuarine mud: phylogenetic position of *Acidaminobacter hydrogenoformans* and description of a new type of Gram-negative, propionigenic bacterium as *Propionibacter pelophilus* gen. nov., sp. nov. Int. J. Syst. Bacteriol. *49*: 1039–1044.

Meincke, M., E. Bock, D. Kastrau and P.M.H. Kroneck. 1992. Nitrite oxidoreductase from *Nitrobacter hamburgensis*: redox centers and their catalytic role. Arch. Microbiol. *158*: 127–131.

Meinersmann, R.J., L.O. Helsel, P.I. Fields and K.L. Hiett. 1997. Discrimination of *Campylobacter jejuni* isolates by *fla* gene sequencing. J. Clin. Microbiol. *35*: 2810–2814.

Melito, P.L., C. Munro, P.R. Chipman, D.L. Woodward, T.F. Booth and F.G. Rodgers. 2001. *Helicobacter winghamensis* sp. nov., a novel *Helicobacter* sp. isolated from patients with gastroenteritis. J. Clin. Microbiol. *39*: 2412–2417.

Mell, H., C. Wellnitz and A. Kröger. 1986. The electrochemical proton potential and the proton/electron ratio of the electron transport with fumarate in *Wolinella succinogenes*. Biochim. Biophys. Acta *852*: 212–221.

Melly, M.A., Z.A. McGee, R.G. Horn, F. Morris and A.D. Glick. 1979. An electron microscopic India ink technique for demonstrating capsules on microorganisms: Studies with *Streptococcus pneumoniae*, *Staphylococcus aureus*, and *Neisseria gonorrhoeae*. J. Infect. Dis. *140*: 605–609.

Mendelman, P.M., J. Campos, D.O. Chaffin, D.A. Serfass, A.L. Smith and J.A. Saez-Nieto. 1988. Relative penicillin G resistance in *Neisseria meningitidis* and reduced affinity of penicillin-binding protein 3. Antimicrob. Agents Chemother. *32*: 706–709.

Mendes, E.N., D.M.M. Queiroz, F.E. Dewhirst, B.J. Paster, S.B. Moura and J.G. Fox. 1996. *Helicobacter trogontum* sp. nov., isolated from the rat intestine. Int. J. Syst. Bacteriol. *46*: 916–921.

Mendes, E.N., D.M.M. Queroz, G.A. Rocha, S.B. Moura, V.H.R. Leite and M.E.F. Fonseca. 1990. Ultrastructure of a spiral microorganism from pig gastric mucosa ("*Gastrospirillum suis*"). J. Med. Microbiol. *33*: 61–66.

Mendes, E.N., D.M.M. Queroz, F.E. Dewhirst, B.J. Paster, G.A. Rocha and J.G. Fox. 1994. Are pigs a resevoir for human *Helicobacter infections*? Am. J. Gastroenterol. *89*: 1296.

Mendoza, Y.A., F.O. Gulacar, Z.L. Hu and A. Buchs. 1987. Unsubstituted and hydroxy substituted fatty acids in recent lacustrine sediment. Int. J. Environ. Anal. Chem. *31*: 107–127.

Mendz, G.L., G.E. Ball and D.J. Meek. 1997. Pyruvate metabolism in *Campylobacter* spp. Biochim. Biophys. Acta *1334*: 291–302.

Mendz, G.L. and S.L. Hazell. 1995. Amino acid utilization by *Helicobacter pylori*. Int. J. Biochem. Cell Biol. *27*: 1085–1093.

Mendz, G.L., S.L. Hazell and B.P. Burn. 1994. The Entner-Doudoroff pathway in *Helicobacter pylori*. Arch. Biochem. Biophys. *312*: 349–356.

Mensah, K., A. Philippon, C. Richard and P.A.D. Grimont. 1989. Infections nosocomiales a *Alcaligenes denitrificans* xylosoxidans: sensibilite de 41 souches a 38 antibiotiques. Med. Mal. Infect. *19*: 167–172.

Menzel, U. and G. Gottschalk. 1985. The internal pH of *Acetobacterium wieringae* and *Acetobacter aceti* during growth and production of acetic acid. Arch. Microbiol. *143*: 47–51.

Mercier, E., E. Jumas-Bilak, A. Allardet-Servent, D. O'Callaghan and M. Ramuz. 1996. Polymorphism in *Brucella* strains detected by studying distribution of two short repetitive DNA elements. J. Clin. Microbiol. *34*: 1299–1302.

Mergaert, J., M.C. Cnockaert and J. Swings. 2002. *Phyllobacterium myrsinacearum* (subjective synonym *Phyllobacterium rubiacearum*) emend. Int. J. Syst. Evol. Microbiol. *52*: 1821-1823.

Mergaert, J., A. Schirmer, L. Hauben, M. Mau, B. Hoste, K. Kersters, D. Jendrossek and J. Swings. 1996. Isolation and identification of poly(3-hydroxyvalerate) degrading strains of *Pseudomonas lemoignei*. Int. J. Syst. Bacteriol. *46*: 769–773.

Mergaert, J. and J. Swings. 1996. Biodiversity of microorganisms that degrade bacterial and synthetic polyesters. J. Ind. Microbiol. Biotechnol. *17*: 463–469.

Mergaert, P., M. Van Montagu, J.-C. Prome and M. Holsters. 1993. Three unusual modifications, a D-arabinosyl, a *N*-methyl, and a carbamoyl group, are present on the *Nod* factors of *Azorhizobium caulinodans* ORS571. Proc. Natl. Acad. Sci. U.S.A. *90*: 1551–1555.

Merker, P., J. Tommassen and B. Kusecek. 1997. Two-dimensional structure of the Opc invasin from *Neisseria meningitidis*. Mol. Microbiol. *23*: 281–293.

Merker, R.I. and J. Smit. 1988. Characterization of the adhesive holdfast of marine and fresh- water Caulobacters. Appl. Environ. Microbiol. *54*: 2078–2085.

Merlo, D.J. and E.W. Nester. 1977. Plasmids in avirulent strains of *Agrobacterium*. J. Bacteriol. *129*: 76–80.

Mermod, N., J.L. Ramos, P.R. Lehrbach and K.N. Timmis. 1986. Vector for regulated expression of cloned genes in a wide range of Gram-negative bacteria. J. Bacteriol. *167*: 447–454.

Merz, A.J., M. So and M.P. Sheetz. 2000. Pilus retraction powers bacterial twitching motility. Nature *407*: 98–102.

Mesbah, M., U. Premachandran and W.B. Whitman. 1989. Precise measurement of the G + C content of deoxyribonucleic acid by high-performance liquid chromatography. Int. J. Syst. Bacteriol. *39*: 159–167.

Mesnard, R., J.M. Sire, P.Y. Donnio, J.Y. Riou and J.L. Avril. 1992. Septic arthritis due to *Oligella urethralis*. Eur. J. Clin. Microbiol. Infect. Dis. *11*: 195–196.

Messick, J.B., L.M. Berent and S.K. Cooper. 1998. Development and evaluation of a PCR-based assay for detection of *Haemobartonella felis* in cats and differentiation of *H. felis* from related bacteria by restriction fragment length polymorphism analysis. J. Clin. Microbiol. *36*: 462–466.

Messick, J.B. and Y. Rikihisa. 1992a. Presence of parasite antigen on the surface of P388D1 cells infected with *Ehrlichia risticii*. Infect. Immun. *60*: 3079–3086.

Messick, J.B. and Y. Rikihisa. 1992b. Suppression of Ia antigen expression on gamma interferon treated macrophages infected with *Ehrlichia risticii*. Vet. Immunol. Immunopathol. *32*: 225–241.

Messick, J.B. and Y. Rikihisa. 1993. Characterization of *Ehrlichia risticii* binding, internalization, and proliferation in host cells by flow cytometry. Infect. Immun. *61*: 3803–3810.

Messick, J.B. and Y. Rikihisa. 1994. Inhibition of binding, entry, or intracellular proliferation of *Ehrlichia risticii* in P388D1 cells by anti-*E. risticii* serum, immunoglobulin G, or Fab fragment. Infect. Immun. *62*: 3156–3161.

Messner, P. and U.B. Sleytr. 1992. Crystalline bacterial cell-surface layers. Adv. Microb. Physiol. *33*: 213–275.

Metzner, P. 1920. Die Bewegung und Reizbeantwortung der bipolar gegeisselten Spirillen. Jahr. Wiss. Bot. *59*: 325–412.

Meulenberg, R., J.T. Pronk, W. Hazeu, P. Bos and J.G. Kuenen. 1992. Oxidation of reduced sulphur compounds by intact cells of *Thiobacillus acidophilus*. Arch. Microbiol. *157*: 161–168.

Mevius, W.J. 1953. Beiträge zur Kenntnis von *Hyphomicrobium vulgare* Stutzer et Hartleb. Arch. Mikrobiol. *19*: 1–29.

Meyer, J. 1977. Contributions to the taxonomy of methane-utilizing bacteria. *In* Skryabin, Kondratjeva, Zavarzin, Trotsenko and Nesterov (Editors), 2nd International Symposium, Microbial Growth on C1 Compounds, Vol. 1, USSR Academy of Sciences, Pushchino. 17–18.

Meyer, J., R. Haubold, J. Heyer and W. Bockel. 1986. Contribution to the taxonomy of methanotrophic bacteria: correlation between membrane type and GC-value. Z. Allg. Mikrobiol. *26*: 155–160.

Meyer, J.M., D. Hohnadel and F. Halle. 1989. Cepabactin from *Pseudomonas cepacia*, a new type of siderophore. J. Gen. Microbiol. *135*: 1479–1487.

Meyer, J.M., V.T. Van, A. Stintzi, O. Berge and G. Winkelmann. 1995. Ornibactin production and transport properties in strains of *Burkholderia vietnamiensis* and *Burkholderia cepacia* (formerly *Pseudomonas cepacia*). Biometals *8*: 309–317.

Meyer, K.F. and E.B. Shawl. 1920. A comparison of the morphological, cultural and biochemical characteristics of *B. abortus* and *B. melitensis*. Studies on genus *Brucella* nov. gen. J. Infect. Dis. *27*: 173–184.

Meyer, M.E. 1961. Metabolic characterization of the genus *Brucella*. III. Oxidative metabolism of strains that show anomalous characteristics by conventional methods. J. Bacteriol. *82*: 401–410.

Meyer, M.E. 1969. *Brucella* organisms isolated from dogs: comparison of characteristics of members of the genus *Brucella*. Amer. J. Vet. Res. *30*: 1751–1756.

Meyer, M.E. 1976. Evolution and taxonomy in the genus *Brucella*: steroid hormone induction of filterable forms with altered characteristics after reversion. Am. J. Vet. Res. *37*: 207–210.

Meyer, M.E. and H.S. Cameron. 1961a. Metabolic characterization of the

Brucella. II.Oxidative metabolic patterns of the described species. J. Bacteriol. *82*: 396–400.

Meyer, M.E. and H.S. Cameron. 1961b. Metabolic characterization of the genus *Brucella.* I. Statistical evaluation of the oxidative rates by which type I of each species can be identified. J. Bacteriol. *82*: 387–395.

Meyer, M.E. and W.J.B. Morgan. 1962. Metabolic charcterization of strains that show conflicting identity by biochemical and serological methods. Bull. WHO *26*: 823–827.

Meyer, O. 1997. Köhlerei im Fichtelgebirge, Frankenwald und Bayerischen Wald, Goltze, Göttingen. 168 pp.

Meyer, O., L. Gremer, R. Ferner, M. Ferner, H. Dobbek, M. Gnida, W. Meyer-Klaucke and R. Huber. 2000. The role of Se, Mo and Fe in the structure and function of carbon monoxide dehydrogenase. Biol. Chem. *381*: 865–876.

Meyer, O., J. Lalucat and H.G. Schlegel. 1980. *Pseudomonas carboxydohydrogena* (Sanjieva and Zavarzin) comb. nov., a monotrichous, non budding, strictly aerobic, carbon monoxide-utilizing hydrogen bacterium previously assigned to *Seliberia.* Int. J. Syst. Bacteriol. *30*: 189–195.

Meyer, O., M. Meyer, D. Gadkari, H. Zellmann, K. Schricker and M. Schmitt. 1991. Microorganisms and their activities in the covering soil of a burning charcoal pile. Proceedings of the Third Symposium on Biotechnology of Coal and Coal-derived substances, Essen, Germany. pp. 111-121.

Meyer, O. and M. Rohde. 1984. Enzymology and bioenergetics of carbon monoxide-oxidizing bacteria. *In* Crawford, R. and R. Hanson (Editors), Microbial growth on C1 compounds, American Society for Microbiology, American Society for Microbiology, Washington. pp. 26–33.

Meyer, O. and H.G. Schlegel. 1978. Reisolation of the carbon monoxide utilizing hydrogen bacterium *Pseudomonas carboxydovorans* (Kistner) comb. nov. Arch. Microbiol. *118*: 35–43.

Meyer, O. and H.G. Schlegel. 1983. Biology of aerobic carbon monoxide-oxidizing bacteria. Annu. Rev. Microbiol. *37*: 277–310.

Meyer, O., E. Stackebrandt and G. Auling. 1993. Reclassification of ubiquinone Q-10 containing carboxidotrophic bacteria: transfer of "*[Pseudomonas] carboxydovorans*" OM5 ^{T} to *Oligotropha*, gen. nov., as *Oligotropha carboxidovorans*, comb. nov., transfer of "*[Alcaligenes] carboxydus*" DSM 1086^T to *Carbophilus*, gen. nov., as *Carbophilus carboxidus*, comb. nov., transfer of "*[Pseudomonas] comparansoris*" DSM 1231^T to *Zavarzinia*, gen., nov., as *Zavarzinia compransoris*, comb. nov., and amended descriptions of the new genera. Syst. Appl. Microbiol. *16*: 390–395.

Meyer, O., E. Stackebrandt and G. Auling. 1994. *In* Validation of the publication of new names and new combinations previously effectively published outside the IJSB. List No. 48. Int. J. Syst. Bacteriol. *44*: 182–183.

Meyer, T.F., E. Billyard, R. Haas, S. Storzbach and M. So. 1984. Pilus genes of *Neisseria gonorrheae*—chromosomal organization and DNA sequence. Proc. Nat. Acad. Sci. U.S.A. *81*: 6110–6114.

Meyer, T.F. and J.P.M. Vanputten. 1989. Genetic mechanisms and biological implications of phase variation in pathogenic neisseriae. Clin. Microbiol. Rev. *2*: S139–S145.

Meyers, A.J., Jr. 1982. Obligate methylotrophy: evaluation of methyl formate as a C1 compound. Can. J. Microbiol. *28*: 1401–1404.

Meyers, E., G.S. Bisacchi, L. Dean, W.C. Liu, B. Minassian, D.S. Slusarchyk, R.B. Sykes, S.K. Tanaka and W. Trejo. 1987. Xylocandin, a new complex of antifungal peptides. I. Taxonomy, isolation and biological activity. J. Antibiot. *40*: 1515–1519.

Micales, B.K., J.L. Johnson and G.W. Claus. 1985. Deoxyribonucleic acid homologies among organisms in the genus *Gluconobacter.* Int. J. Syst. Bacteriol. *35*: 79–85.

Michaux, S., J. Paillisson, M.J. Carles-Nurit, G. Bourg, A. Allardet-Servent and M. Ramuz. 1993. Presence of two independent chromosomes in the *Brucella melitensis* 16M genome. J. Bacteriol. *175*: 701–705.

Michaux-Charachon, S., G. Bourg, E. Jumas-Bilak, P. Guigue-Talet, A. Allardet-Servent, D. O'Callaghan and M. Ramuz. 1997. Genome struc-

ture and phylogeny in the genus *Brucella.* J. Bacteriol. *179*: 3244–3249.

Michel, G.P. and J. Starka. 1986. Effect of ethanol and heat stresses on the protein pattern of *Zymomonas mobilis.* J. Bacteriol. *165*: 1040–1042.

Michiels, K.W., C.L. Croes and J. Vanderleyden. 1991. Two different modes of attachment of *Azospirillum brasilense* Sp7 to wheat roots. J. Gen. Microbiol. *137*: 2241–2246.

Michiels, K., J. Vanderleyden and C. Elmerich. 1994. Genetics and molecular biology of *Azospirillum. In* Okon (Editor), *Azospirillum*-plant associations, CRC Press, Boca Raton. pp. 41–56.

Midani, S. and M. Rathore. 1998. *Chromobacterium violaceum* infection. South. Med. J. *91*: 464–466.

Midgley, J., S.P. LaPage, B.A.G. Jenkins, G.I. Barrow, M.E. Roberts and A.G. Buck. 1970. *Cardiobacterium hominis* endocarditis. J. Med.Microbiol. *3*: 91–98.

Miguez, C.B., D. Bourque, J.A. Sealy, C.W. Greer and D. Groleau. 1997. Detection and isolation of methanotrophic bacteria possessing soluble methane monooxygenase (sMMO) genes using the polymerase chain reaction (PCR). Microb. Ecol. *33*: 21–31.

Miguez, C.B., C.W. Greer, J.M. Ingram and R. MacLeod. 1995. Uptake of benzoic acid and chloro-substituted benzoic acids by *Alcaligenes denitrificans* BRI 3010 and BRI 6011. Appl. Environ. Microbiol. *61*: 4152–4159.

Migula, W. 1894. Über ein neues System der Bakterien. Arb. Bakteriol. Inst. Karlsruhe. *1*: 235–238.

Migula, W. 1900. System der Bakterien, Vol. 2, Gustav Fischer, Jena.

Mikes, V., H. Chvalova and L. Matlova. 1991. Assimilation of ammonia in *Paracoccus denitrificans.* Folia Microbiol. *36*: 35–41.

Milde, K. and E. Bock. 1985. Comparative studies on membrane proteins of *Nitrobacter hamburgensis* and *Nitrobacter winogradskyi.* FEMS Microbiol. Lett. *26*: 135–139.

Milford, A.D., L.A. Achenbach, D.O. Jung and M.T. Madigan. 2000. *Rhodobaca bogoriensis* gen. nov. and sp. nov., an alkaliphilic purple nonsulfur bacterium from African Rift Valley soda lakes. Arch. Microbiol. *174*: 18–27.

Milford, A.D., L.A. Achenbach, D.O. Jung and M.T. Madigan. 2001. *In* Validation of the publication of new names and new combinations previously effectively published outside the IJSB. List No. 80. Int. J. Syst. Bacteriol. *51*: 793–794.

Millemann, R.E. and S.E. Knapp. 1970. Biology of *Nanophyetus salmincola* and "salmon poisoning" disease. Adv. Parasitol. *8*: 1–41.

Miller, C.E., S. Wuertz and J.J. Cooney. 1995. Plasmids in tributyltin resistant bacteria from fresh and estuarine waters. J. Indust. Microbiol. *14*: 337–342.

Miller, J.D.A., P.M. Neumann, L. Elford and D.S. Wakerley. 1970. Malate dismutation by *Desulfovibrio.* Arch. Mikrobiol. *71*: 214–219.

Miller, J.F., S. R.J., T.M. Saito and J.B. Humber. 1943. An agglutinative reaction for *Haemophilus pertussis* II. Its relation to clinical immunity. J. Pediat. *22*: 644–651.

Miller, J.D.A. and A.M. Saleh. 1964. A sulphate-reducing bacterium containing cytochrome c_3 but lacking desulfoviridin. J. Gen. Microbiol. *37*: 419–423.

Miller, K. and M.E. Neville. 1976. Evaluation of alternate coupling reagents to replace α-naphthyl amine for the detection of nitrate reduction. Microbios. *17*: 207–212.

Miller, R.D., K.E. Brown and S.A. Morse. 1977. Inhibitory action of fattyacids on growth of *Neisseria gonorrhoeae.* Infect. Immun. *17*: 303–312.

Miller, W.G., L.G. Adams, T.A. Ficht, N.F. Cheville, J.P. Payeur, D.R. Harley, C. House and S.H. Ridgway. 1999. *Brucella* induced abortions and infection in bottlenose dolphins (*Tursiops truncatus*). J. Zoo. Wildl. Med. *30*: 100–110.

Millis, N.F. 1951. Some Bacterial Fermentations of Cider, Thesis, University of Bristol, Bristol, U.K.

Millis, N.F. 1956. A study of cider-sickness bacillus — a new variety of *Zymomonas anaerobia.* J. Gen. Microbiol. *15*: 521–528.

Mills, S.D., G.O. Aspinall, A.G. McDonald, T.S. Raju, L.A. Kurjanczyk and J.L. Penner. 1992. Lipopolysaccharide antigens of *Campylobacter jejuni. In* Nachamkin, Blaser and Tompkins (Editors), *Campylobacter jejuni.*

Current Status and Future Trends, American Society for Microbiology, Washington, D.C. pp. 223–229.

Min, K.T. and S. Benzer. 1997. *Wolbachia*, normally a symbiont of *Drosophila*, can be virulent, causing degeneration and early death. Proc. Natl. Acad. Sci. U.S.A. *94*: 10792–10796.

Minamisawa, K., T. Isawa, Y. Nakatsuka and N. Ichikawa. 1998. New *Bradyrhizobium japonicum* strains that possess high copy numbers of the repeated sequence RS alpha. Appl. Environ. Microbiol. *64*: 1845–1851.

Minamoto, G.Y. and E.M. Sordillo. 1992. *Kingella denitrificans* as a cause of granulomatous disease in a patient with AIDS [letter]. Clin. Infect. Dis. *15*: 1052–1053.

Minges, C.G., J.A. Titus and W.R. Strohl. 1983. Plasmid DNA in colorless filamentous gliding bacteria. Arch. Microbiol. *134*: 38–44.

Minnick, M.F. and K.D. Barbian. 1997. Identification of *Bartonella* using PCR; genus- and species-specific primer sets. J. Microbiol. Methods *31*: 51–57.

Mintz, C.S., M.A. Apicella and S.A. Morse. 1984. Electrophoretic and serological characterization of the lipopolysaccharide produced by *Neisseria gonorrhoeae*. J. Infect. Dis. *149*: 544–552.

Miroshnichenko, M.L., G.A. Gongadze, A.M. Lysenko and E.A. Bonch-Osmolovskaya. 1994. *Desulfurella multipotens* sp. nov., a new sulfur-respiring thermophilic eubacterium from Raoul Island (Kermadec Archipelago, New Zealand). Arch. Microbiol. *161*: 88–93.

Miroshnichenko, M.L., G.A. Gongadze, A.M. Lysenko and E.A. Bonch-Osmolovskaya. 1996. *In* Validation of the publication of new names and new combinations previously effectively published outside the IJSB, List No. 57. Int. J. Syst. Bacteriol. *46*: 625–626.

Miroshnichenko, M.L., F.A. Rainey, H. Hippe, N.A. Chernyh, N.A. Kostrikina and E.A. Bonch-Osmolovskaya. 1998. *Desulfurella kamchatkensis* sp. nov. and *Desulfurella propionica* sp. nov., new sulfur-respiring thermophilic bacteria from Kamchatka thermal environments. Int. J. Syst. Bacteriol. *48*: 475–479.

Miroshnichenko, M.L., F.A. Rainey, M. Rhode and E.A. Bonch-Osmolovskaya. 1999. *Hippea maritima* gen. nov., sp. nov., a new genus of thermophilic, sulfur-reducing bacterium from submarine hot vents. Int. J. Syst. Bacteriol. *49*: 1033–1038.

Misao, R. and Y. Kobayashi. 1956. Infectious mononucleosis (glandular fever). J. Jpn. Assoc. Infect. Dis. *30*: 453–465.

Misao, T. and Y. Kobayashi. 1954. Studies on infectious mononucleosis I. Isolation of etiologic agent from blood, bone marrow, and lymph node of a patient with infectious mononucleosis by using mice. Tokyo Iji Shinshi. *71*: 683–686.

Misawa, N., K. Hirayama, K. Itoh and E. Takahashi. 1995. Detection of α- and β-hemolytic-like activity from *Campylobacter jejuni*. J. Clin. Microbiol. *33*: 729–731.

Misawa, N. and K. Nakamura. 1989. Nucleotide sequence of the 2.7 kb plasmid of *Zymomonas mobilis* ATCC 10988. J. Biotechnol. *12*: 63–70.

Misawa, N., S. Yamano and H. Ikenaga. 1991. Production of beta-carotene in *Zymomonas mobilis* and *Agrobacterium tumefaciens* by introduction of the biosynthesis genes from *Erwinia uredovora*. Appl. Environ. Microbiol. *57*: 1847–1849.

Mishra, A.K., P. Roy and S. Bhattacharya. 1979. Deoxyribonucleic acid-mediated transformation of *Sprillum lipoferum*. J. Bacteriol. *137*: 1425–1427.

Mishustin, E.N. 1938. Cellulose decomposing myxobacteria. Mikrobiologiya *7*: 427–444.

Mishustin, E., T. Kalininskaja and T. Redkina. 1984. New forms of nitrogen fixing aerobic bacteria from USSR soils. *In* Veeger and Newton (Editors), Advances in Nitrogen Fixation research: Proceedings of the 5th International Symposium on Nitrogen Fixation, Nijhoff/Junk Publishers, The Hague. p. 352.

Mitchell, D. and J. Smit. 1990. Identification of genes affecting production of the adhesion organelle of *Caulobacter crescentus* CB2. J. Bacteriol. *172*: 5425–5431.

Mitchell, H.M., Y.Y. Li, P.J. Hu, Q. Liu, M. Chen, G.G. Du, Z.J. Wang, A. Lee and S.L. Hazell. 1992. Epidemiology of *Helicobacter pylori* in southern China: identification of early childhood as the critical period for acquisition. J. Infect. Dis. *166*: 149–153.

Mitchell, R.G. and S.K. Clarke. 1965. An *Alcaligenes* species with distinctive properties isolated from human sources. J. Gen. Microbiol. *40*: 343–348.

Mitsuoka, T. 1980. The World of Intestinal Bacteria — The Isolation and Identification of Anaerobic Bacteria; A Color Atlas of Anaerobic Bacteria, Sobunsha (Sobun Press), Tokyo.

Miyashiro, S., S. Yamanaka, S. Takayama and H. Shibai. 1988. Novel macrocyclic antibiotics: megovalicins A, B, C, D, G and H. I. Screening of antibiotics-producing myxobacteria and production of megovalicins. J. Antibiot. (Tokyo) *41*: 433–438.

Miyata, A., T. Yoshida, K. Yamaguchi, C. Yokoyama, T. Tanabe, H. Toh, T. Mitsunaga and Y. Izumi. 1993. Molecular cloning and expression of the gene for serine hydroxymethyltransferase from an obligate methylotroph *Hyphomicrobium methylovorum* GM2. Eur. J. Biochem. *212*: 745–750.

Miyauchi, K., S.K. Suh, Y. Nagata and M. Takagi. 1998. Cloning and sequencing of a 2,5-dichlorohydroquinone reductive dehalogenase gene whose product is involved in degradation of gamma-hexachlorocyclohexane by *Sphingomonas paucimobilis*. J. Bacteriol. *180*: 1354–1359.

Miyazaki, S.S., S.I. Toki, Y. Izumi and H. Yamada. 1986. Crystalline serine hydroxymethyltransferase from an obligate methylotroph, *Hyphomicrobium methylovorum*. Biochem. Biophys. Res. Commun. *139*: 71–78.

Miyazaki, S.S., S.I. Toki, Y. Izumi and H. Yamada. 1987a. Further characterization of serine hydroxymethyltransferase from a serine-producing methylotroph, *Hyphomicrobium methylovorum*. Agric. Biol. Chem. *51*: 2587–2589.

Miyazaki, S.S., S.I. Toki, Y. Izumi and H. Yamada. 1987b. Immunological characterization of serine hydroxymethyltransferase of methylotrophic *Hyphomicrobium* strains. Arch. Microbiol. *147*: 328–333.

Miyazaki, S.S., S.I. Toki, Y. Izumi and H. Yamada. 1987c. Purification and characterization of methanol dehydrogenase of a serine-producing methylotroph, *Hyphomicrobium methylovorum*. J. Ferment. Technol. *65*: 371–378.

Miyazaki, S.S., S.I. Toki, Y. Izumi and H. Yamada. 1987d. Purification and characterization of serine hydroxymethyltransferase from an obligate methylotroph, *Hyphomicrobium methylovorum* GM2. Eur. J. Biochem. *162*: 533–540.

Miyazaki, Y., S. Oka, S. Yamaguchi, S. Mizuno and I. Yano. 1995. Stimulation of phagocytosis and phagosome-lysosome fusion by glycosphingolipids from *Sphingomonas paucimobilis*. J. Biochem. (Tokyo) *118*: 271–277.

Mizoguchi, T., T. Sato and T. Okabe. 1976. New sulfur-oxidizing bacteria capable of growing heterotrophically, *Thiobacillus rubellus* nov. sp. and *Thiobacillus delicatus* nov. sp. J. Ferment. Technol. *54*: 181–191.

Möbus, E., M. Jahn, R. Schmid, D. Jahn and E. Maser. 1997. Testosterone-regulated expression of enzymes involved in steroid and aromatic hydrocarbon catabolism in *Comamonas testosteroni*. J. Bacteriol. *179*: 5951–5955.

Mochida, C., Y. Hirakata, J. Matsuda, F. Iori, Y. Ozaki, M. Nakano, K. Hamaguchi, K. Izumikawa, T. Yamaguchi, K. Tomono, S. Maesaki, Y. Yamada, S. Kohno and S. Kamihira. 1998. Antimicrobial susceptibility testing of *Bilophila wadsworthia* isolates submitted for routine laboratory examination. J. Clin. Microbiol. *36*: 1790–1792.

Moffett, M.L. and R.R. Colwell. 1968. Adansonian analysis of the *Rhizobiaceae*. J. Gen. Microbiol. *51*: 245–266.

Mohamed, M.E. 2000. Biochemical and molecular characterization of phenylacetate-coenzyme A ligase, an enzyme catalyzing the first step in aerobic metabolism of phenylacetic acid in *Azoarcus evansii*. J. Bacteriol. *182*: 286–294.

Mohamed, M.E., A. Zaar, C. Ebenau-Jehle and G. Fuchs. 2001. Reinvestigation of a new type of aerobic benzoate metabolism in the proteobacterium *Azoarcus evansii*. J. Bacteriol. *183*: 1899–1908.

Mohn, W.W. 1995. Bacteria obtained from a sequencing batch reactor that are capable of growth on dehydroabietic acid. Appl. Environ. Microbiol. *61*: 2145–2150.

Mohn, W.W. and K.J. Kennedy. 1992. Reductive dehalogenation of chlorophenols by *Desulfomonile tiedjei* DCB-1. Appl. Environ. Microbiol. *58*: 1367–1370.

Mohn, W.W., T.G. Linkfield, H.S. Pankratz and J.M. Tiedje. 1990. Involvement of a collar structure in polar growth and cell division of strain DCB-1. Appl. Environ. Microbiol. *56*: 1206–1211.

Mohn, W.W. and J.M. Tiedje. 1990. Catabolic thiosulfate disproportionation and carbon dioxide reduction in strain DCB-1, a reductively dechlorinating anaerobe. J. Bacteriol. *172*: 2065–2070.

Mohn, W.W. and J.M. Tiedje. 1992. Microbial reductive dehalogenation. Microbiol. Rev. *56*: 482–507.

Mohn, W.W., A.E. Wilson, P. Bicho and E.R.B. Moore. 1999a. Physiological and phylogenetic diversity of bacteria growing on resin acids. Syst. Appl. Microbiol. *22*: 68–78.

Mohn, W.W., A.E. Wilson, P. Bicho and E.R.B. Moore. 1999b. *In* Validation of publication of new names and new combinations previously effectively published outside the IJSB. List No. 70. Int. J. Syst. Bacteriol. *49*: 935–936.

Mohran, Z.S., P. Guerry, H. Lior, J.R. Murphy, A.M. el Gendy, M.M. Mikhail and B.A. Oyofo. 1996. Restriction fragment length polymorphism of flagellin genes of *Campylobacter jejuni* and/or *C. coli* isolates from Egypt. J. Clin. Microbiol. *34*: 1216–1219.

Moir, J., W. , D. Baratta, D.J. Richardson and S.J. Ferguson. 1993. The purification of a cd1-type nitrite reductase from, and the absence of a copper-type nitrite reductase from, the aerobic denitrifier *Thiosphaera pantotropha*: the role of pseudoazurin as an electron donor. Eur. J. Biochem. *212*: 377–385.

Moir, J.W., L.C. Crossman, S. Spiro and D.J. Richardson. 1996a. The purification of ammonia monooxygenase from *Paracoccus denitrificans*. FEBS Lett. *387*: 71–74.

Moir, J.W. and S.J. Ferguson. 1993. Spontaneous mutation of *Thiosphaera pantotropha* enabling growth on methanol correlates with synthesis of a 26 kDa *c*-type cytochrome. FEMS Microbiol. Lett. *113*: 321–326.

Moir, J.W., J.M. Wehrfritz, S. Spiro and D.J. Richardson. 1996b. The biochemical characterization of a novel non-haem-iron hydroxylamine oxidase from *Paracoccus denitrificans* GB17. Biochem. J. *319*: 823–827.

Molisch, H. 1907. Die Purpurbakterien Nach Neuen Untersuchungen, G. Fischer, Jena.

Molisch, H. 1910. Die Eisenbakterien, Gustav Fischer Verlag, Jena.

Molitoris, E., H.M. Wexler and S.M. Finegold. 1997. Sources and antimicrobial susceptibilities of *Campylobacter gracilis* and *Sutterella wadsworthensis*. Clin. Infect. Dis. *25 (Suppl. 2)*: S264–S265.

Mollee, T., P. Kelly and M. Tilse. 1992. Isolation of *Kingella kingae* from a corneal ulcer. J. Clin. Microbiol. *30*: 2516–2517.

Møller, V. 1955. Simplified test for some amino acid decarboxylases and arginine dihydrolase system. Acta Pathol. Microbiol. Scand. *36*: 158–172.

Monteiro, J.L. 1931. Estudos sobre o typho exanthematico de São Paulo. Mem. Inst. Butantan São Paulo. *6*: 3–135.

Montenegro-James, S., M.A. James, M.T. Benitez, E. Leon, B.K. Baek and A.T. Guillen. 1991. Efficacy of purified *Anaplasma marginale* initial bodies as a vaccine against anaplasmosis. Parasitol. Res. *77*: 93–101.

Montie, T.C. and G.B. Stover. 1983. Isolation and characterization of flagellar preparations from *Pseudomonas* species. J. Clin. Microbiol. *18*: 452–456.

Montoya, A.L., M.D. Chilton, M.P. Gordon, D. Sciaky and E.W. Nester. 1977. Octopine and nopaline metabalism in *Agrobacterium tumefaciens* and crown gall tumor cells: role of plasmld genes. J. Bacteriol. *129*: 101–107.

Mooi, F.R., W.H. Jansen, H. Brunings, H. Gielen, H.G. van der Heide, H.C. Walvoort and P.A. Guinee. 1992. Construction and analysis of *Bordetella pertussis* mutants defective in the production of fimbriae. Microb. Pathog. *12*: 127–135.

Mooi, F.R., H.G. van der Heide, A.R. ter Avest, K.G. Welinder, I. Livey, B.A. van der Zeijst and W. Gaastra. 1987. Characterization of fimbrial subunits from *Bordetella* species. Microb. Pathog. *2*: 473–484.

Moore, C.J., H. Mawhinney and P.J. Blackall. 1987. Differentiation of *Bordetella avium* and related species by cellular fatty acid analysis. J. Clin. Microbiol. *25*: 1059–1062.

Moore, E.R.B., R.M. Wittich, P. Fortnagel and K.N. Timmis. 1993. 16S ribosomal RNA sequence characterization and phylogenetic analysis of a dibenzo-*p*-dioxin-degrading isolate within the new genus *Sphingomonas*. Lett. Appl. Microbiol. *17*: 115–118.

Moore, J.E., T. Coenye, P. Vandamme and J.S. Elborn. 2001. First report of *Pandoraea norimbergensis* isolated from food-potential clinical significance. Food Microbiol. *18*: 113–114.

Moore, L.W., A. Anderson and C.I. Kado. 1980. *Agrobacterium. In* Schaad (Editor), Laboratory Guide for Identification of Plant Pathogenic Bacteria, American Phytopathological Society, St Paul, Minnesota. pp. 17–25.

Moore, L.W., W.S. Chilton and M.L. Canfield. 1997. Diversity of opines and opine-catabolizing bacteria isolated from naturally occurring crown gall tumors. Appl. Environ. Microbiol. *63*: 201–207.

Moore, L.W. and D.A. Cooksey. 1981. Biology of *Agrobacterium tumefaciens*: plant interactions. *In* Giles and Atherly (Editors), Biology of the *Rhizobiaceae*: International Review of Cytology, Supplement 13, Academic Press, New York. pp. 15–46.

Moore, L.W., C.I. Kado and H. Bouzar. 1988. *Agrobacterium. In* Schaad (Editor), Laboratory Guide for Identification of Plant Pathogenic Bacteria, 2nd Ed., APS Press, St. Paul, Minnesota. pp. 16–36.

Moore, L.W. and G. Warren. 1979. *Agrobacterium radiobacter* strain 84 and biological control of crown gall. Annu. Rev. Phytopathol. *17*: 163–179.

Moore, L., G. Warren and G. Strobel. 1979. Involvement of a plasmid in the hairy root disease of plants caused by *Agrobacterium rhizogenes*. Plasmid *2*: 617–626.

Moore, R.L. 1977. Ribosomal ribonucleic acid cistron homologies among *Hyphomicrobium* and various other bacteria. Can. J. Microbiol. *23*: 478–481.

Moore, R.L. 1981a. The biology of *Hyphomicrobium* and other prosthecate, budding bacteria. Annu. Rev. Microbiol. *35*: 567–594.

Moore, R.L. 1981b. The genera *Hyphomicrobium*, *Pedomicrobium*, and *Hyphomonas*. *In* Starr, Stolp, Trüper, Balows and Schlegel (Editors), The Prokaryotes: A Handbook on Habitats, Isolation, and Identification of Bacteria, Springer-Verlag, Berlin. pp. 480–487.

Moore, R.L. 1984. Methods for increasing the usefulness of the *Spirillum volutans* motility test. *In* Dickson and Dutka (Editors), Toxicity Screening Procedures Among Bacterial Systems, Marcel Dekker, New York. pp. 109–124.

Moore, R.A. and R.E.W. Hancock. 1986. Involvement of outer membrane of *Pseudomonas cepacia* in aminoglycoside and polymyxin resistance. Antimicrob. Agents Chemother. *30*: 923–926.

Moore, R.L. and P. Hirsch. 1972. Deoxyribonucleic acid base sequence homologies of some budding and prosthecate bacteria. J. Bacteriol. *110*: 256–261.

Moore, R.L. and P. Hirsch. 1973a. First generation synchrony of isolated *Hyphomicrobium* swarmer populations. J. Bacteriol. *116*: 418–423.

Moore, R.L. and P. Hirsch. 1973b. Nuclear apparatus of *Hyphomicrobium*. J. Bacteriol. *116*: 1447–1455.

Moore, R.L., J. Schmidt, J. Poindexter and J.T. Staley. 1978. Deoxyribonucleic-acid homology among Caulobacters. Int. J. Syst. Bacteriol. *28*: 349–353.

Moore, R.L. and J.T. Staley. 1976. Deoxyribonucleid acid homology of *Prosthecomicrobium* and *Ancalomicrobium*. Int. J. Syst. Bacteriol. *26*: 283–285.

Moore, R.L., R.M. Weiner and R. Gebers. 1984. Genus *Hyphomonas* Pongratz 1957 nom. rev. emend., *Hyphomonas polymorpha* Pongratz 1957 nom. rev. emend., and *Hyphomonas neptunium* (Leifson 1964) comb. nov. emend., (*Hyphomicrobium neptunium*). Int. J. Syst. Bacteriol. *34*: 71–73.

Moore, W.E.C. and L.V. Holdeman. 1974. Human fecal flora: the normal flora of 20 Japanese-Hawaiians. Appl. Microbiol. *27*: 961–979.

Moore, W.E.C., J.L. Johnson and L.V. Holdeman. 1976. Emendation of *Bacteroides* and *Butyrivibrio* and descriptions of *Desulfomonas* gen. nov. and ten new species in the genera *Desulfomonas, Butyrivibrio, Eubac-*

terium, Clostridium, and *Ruminococcus.* Int. J. Syst. Bacteriol. *26:* 238–252.

Moore, W.E.C. and L.V.H. Moore. 1992. Index of the bacterial and yeast nomenclatural changes: Published in the International Journal of Systematic Bacteriology since the 1980 Approved lists of bacterial names: (1st January 1980 to 1st January 1992), American Society for Microbiology, Washington, D.C.

Moorhouse, R. 1987. Structure/property relationships of a family of microbial polysaccharides. *In* Yalpani (Editor), Industrial Polysaccharides: Genetic Engineering, Structure/Property Relatedness and Applications, Elsevier, Amsterdam. pp. 187–206.

Mooser, H. 1928. Experiments relating to the pathology and the etiology of Mexican typhus (tabardillo). J. Infect. Dis. *43:* 241–272.

Morais, P.V. and M.S. Dacosta. 1990. Alterations in the major heterotrophic bacterial-populations isolated from a still bottled mineral water. J. Appl. Bacteriol. *69:* 750–757.

Moran, A.P., M.M. Prendergast and B.J. Appelmelk. 1996. Molecular mimicry of host structures by bacterial lipopolysaccharides and its contribution to disease. FEMS Immunol. Med. Microbiol. *16:* 105–115.

Moran, A.P. and M.E. Upton. 1987. Factors affecting production of coccoid forms by *Campylobacter jejuni* on solid media during incubation. J. Appl. Bacteriol. *62:* 527–537.

Moran, B.N. and W.J. Hickey. 1997. Trichloroethylene biodegradation by mesophilic and psychrophilic ammonia oxidizers and methanotrophs in groundwater microcosms. Appl. Environ. Microbiol. *63:* 3866–3871.

Moreira, C., F.A. Rainey, M.F. Nobre, M.T. da Silva and M.S. da Costa. 2000. *Tepidimonas ignava* gen. nov., sp nov., a new chemolithoheterotrophic and slightly thermophilic member of the β-*Proteobacteria.* Int. J. Syst. Evol. Microbiol. *50:* 735–742.

Moreira, D. and R. Amils. 1997. Phylogeny of *Thiobacillus cuprinus* and other mixotrophic thiobacilli: proposal for *Thiomonas* gen. nov. Int. J. Syst. Bacteriol. *47:* 522–528.

Moreira, J.A. and O. de Magalhaes. 1937. Typho exanthematico de Minas Geraes. Brasil-Med. *51:* 583–584.

Moreira-Jacob, M. 1968. New group of virulent bacteriophages showing differential affinity for *Brucella* species. Nature *219:* 752–753.

Morello, J.A. and M. Bonhoff. 1980. *Neisseria* and *Branhamella.* *In* Lennette, Balows, Hausler and Truant (Editors), Manual of Clinical Microbiology, 3rd Ed., American Society for Microbiology, Washington, D.C. pp. 111–130.

Moreno, E. 1998. Genome evolution within the alpha *Proteobacteria*: why do some bacteria not possess plasmids and others exhibit more than one different chromosome? FEMS Microbiol. Rev. *22:* 255–275.

Moreno, E., L.M. Jones and D.T. Berman. 1984. Immunochemical characterization of rough *Brucella* lipopolysaccharides. Infect. Immun. *43:* 779–782.

Moreno, E., M.W. Pitt, L.M. Jones, G.G. Schurig and D.T. Berman. 1979. Purification and characterization of smooth and rough lipopolysaccharides from *Brucella abortus.* J. Bacteriol. *138:* 361–369.

Moreno, E., S.L. Speth, L.M. Jones and D.T. Berman. 1981. Immunochemical characterization of *Brucella* lipopolysaccharides and polysaccharides. Infect. Immun. *31:* 214–222.

Moreno-López, M. 1952. El genero Bordetella. Microbiol. Esp. *5:* 177–181.

Morgan, C.A. and R.C. Wyndham. 1996. Isolation and characterization of resin acid degrading bacteria found in effluent from a bleached kraft pulp mill. Can. J. Microbiol. *42:* 423–430.

Morgan, D.D. and R.J. Owen. 1990. Use of DNA restriction endonuclease digest and ribosomal RNA gene probe patterns to fingerprint *Helicobacter pylori* and *Helicobacter mustelae* isolated from human and animal hosts. Mol. Cell. Probes *4:* 321–334.

Morgan, W.J.B. and S.G.M. Gower. 1966. Techniques in the identification and classification of *Brucella.* *In* Gibbs and Skinner (Editors), Identification Methods of Microbiologists, Academic Press, London. pp. 35–50.

Moriguchi, M. and K. Ideta. 1988. Production of D-aminoacylase from

Alcaligenes denitrificans subsp. *xylosoxydans* MI-4. Appl. Environ. Microbiol. *54:* 2767–2770.

Morikawa, Y., S. Takayama, R. Fudo, S. Yamanaka, K. Mori and A. Isogai. 1998. Absolute chemical structure of the myxobacterial pheromone of *Stigmatella aurantiaca* that induces the formation of its fruiting body. FEMS Microbiol. Lett. *165:* 29–34.

Morita, R.Y. 1975. Psychrophilic bacteria. Bacteriol. Rev. *39:* 144–167.

Mormak, D.A. and L.E. Casida, Jr.. 1985. Study of *Bacillus subtilis* endospores in soil by use of a modified endospore stain. Appl. Environ. Microbiol. *49:* 1356–1360.

Morrill, W.E., J.M. Barbaree, B.S. Fields, G.N. Sanden and W.T. Martin. 1988. Effects of transport temperature and medium on recovery of *Bordetella pertussis* from nasopharyngeal swabs. J. Clin. Microbiol. *26:* 1814–1817.

Morris, D.W., S.R. Ogden-Swift, V. Virrankoski-Castrodeza, K. Ainley and J.H. Parish. 1978. Transduction of *Myxococcus virescens* by coliphage P1cm—Generation of plasmids containing both phage and myxococcus genes. J. Gen. Microbiol. *107:* 73–83.

Morris, D.W. and J.H. Parish. 1976. Restriction in *Myxococcus virescens.* Arch. Microbiol. *108:* 227–230.

Morris, G.K., M.R. el Sherbeeny, C.M. Patton, H. Kodaka, G.L. Lombard, P. Edmonds, D.G. Hollis and D.J. Brenner. 1985. Comparison of four hippurate hydrolysis methods for identification of thermophilic *Campylobacter* spp. J. Clin. Microbiol. *22:* 714–718.

Morris, J.A. and M.J. Corbel. 1973. Properties of a new phage lytic for *Brucella suis.* J. Gen. Virol. *21:* 539–544.

Morris, J.T. and M. Myers. 1998. Bacteremia due to *Bordetella holmesii.* Clin. Infect. Dis. *27:* 912–913.

Morris, M.B. and J.B. Roberts. 1959. A group of pseudomonads able to synthesize poly-β-hydroxybutyric acid. Nature (Lond.) *183:* 1538–1539.

Morse, J.H. and S.I. Morse. 1970. Studies on the ultrastructure of *Bordetella pertussis.* I. Morphology, origin, and biological activity of structures present in the extracellular fluid of liquid cultures of *Bordetella pertussis.* J. Exp. Med. *131:* 1342–1357.

Morse, S.A. and L. Bartenstein. 1974. Factors affecting autolysis of *Neisseria gonorrhoeae.* Proc. Soc. Exp. Biol. Med. *145:* 1418–1421.

Morse, S.A., S.R. Johnson, J.W. Biddle and M.C. Roberts. 1986. High-level tetracycline resistance in *Neisseria gonorrhoeae* is result of acquisition of streptococcal tetM determinant. Antimicrob. Agents Chemother. *30:* 664–670.

Morse, S.A. and J.S. Knapp. 1987. Neisserial Infections. *In* Wentworth (Editor), Diagnostic Procedures for Bacterial Infections, American Public Health Association, Washington, D.C. pp. 407–429.

Morse, S.A. and J.S. Knapp. 1989. The genus *Neisseria.* *In* Starr, Stolp, Trüper, Balows and Schlegel (Editors), The Prokaryotes: A Handbook on Habitats, Isolation, and Identification of Bacteria, Springer-Verlag, Berlin. pp. 2495–2529.

Mortensen, J.E., A. Brumbach and T.R. Shryock. 1989. Antimicrobial susceptibility of *Bordetella avium* and *Bordetella bronchiseptica* isolates. Antimicrob. Agents Chemother. *33:* 771–772.

Mortensen, J.E., D.V. Schidlow and E.M. Stahl. 1988. *Pseudomonas gladioli* (marginata) isolated from a patient with cystic fibrosis. Clin. Microbiol. Newslett. *10:* 29–39.

Mosca, A., M. D'Alagni, R. Del Prete, G.P. De Michele, P.H. Summanen, S.M. Fingold and G. Miragliotta. 1995. Preliminary evidence of endotoxic activity of *Bilophila wadsworthia.* Anaerobe *1:* 21–24.

Moschettini, G., A. Hochkoeppler, B. Monti, B. Benelli and D. Zannoni. 1997. The electron transport system of the halophilic purple nonsulfur bacterium *Rhodospirillum salinarum.* 1. A functional and thermodynamic analysis of the respiratory chain in aerobically and photosynthetically grown cells. Arch. Microbiol. *168:* 302–309.

Moshkovski, S.D. 1945. Cytotropic inducers of infection and the classification of the *Rickettsiae* with *Chlamydozoa* (in Russian, English summary). Adv. Mod. Biol. (Moscow) *19:* 1–44.

Moshkovski, S.D. 1947. Comments by readers. Science *106:* 62.

Moss, C.W., M.I. Daneshvar, D.G. Hollis and K.A. Birkness. 1991. Isoprenoid quinones of "*Afipia*" spp. J. Clin. Microbiol. *29:* 2904–2905.

Moss, C.W., S.B. Dees and G.O. Guerrant. 1980. Gas-liquid chromatography of bacterial fatty acids with a fused-silica capillary column. J. Clin. Microbiol. 12: 127–130.

Moss, C.W., G. Holzer, P.L. Wallace and D.G. Hollis. 1990a. Cellular fatty acid compositions of an unidentified organism and a bacterium associated with cat scratch disease. J. Clin. Microbiol. 28: 1071–1074.

Moss, C.W., A. Kai, M.A. Lambert and C. Patton. 1984. Isoprenoid quinone content and cellular fatty acid composition of Campylobacter species. J. Clin. Microbiol. 19: 772–776.

Moss, C.W., M.A. Lambert-Fair, M.A. Nicholson and G.O. Guerrant. 1990b. Isoprenoid quinones of Campylobacter cryaerophila, C. cinaedi, C. fennelliae, C. hyointestinalis, C. pylori, and "C. upsaliensis". J. Clin. Microbiol. 28: 395–397.

Moss, C.W., P.L. Wallace, D.G. Hollis and R.E. Weaver. 1988. Cultural and chemical characterization of CDC groups EO-2, M-5, and M-6, Moraxella (Moraxella) species, Oligella urethralis, Acinetobacter species, and Psychrobacter immobilis. J. Clin. Microbiol. 26: 484–492.

Moss, M.O. and T.N. Bryant. 1982. DNA:ribosomal RNA hybridization studies of Chromobacterium fluviatile. J. Gen. Microbiol. 128: 829–834.

Moss, M.O. and C. Ryall. 1981. The genus Chromobacterium. In Starr, Stolp, Trüper, Balows and Schlegel (Editors), The Prokaryotes: A Handbook on Habitats, Isolation, and Identification of Bacteria, Springer-Verlag, Berlin. pp. 1355–1364.

Moss, M.O., C. Ryall and N.A. Logan. 1978. The classification and characterization of chromobacteria from a lowland river. J. Gen. Microbiol. 105: 11–21.

Moss, M.O., C. Ryall and N.A. Logan. 1981. In Validation of the publication of new names and new combinations previously effectively published in the IJSB. List No. 6. Int. J. Syst. Bacteriol. 31: 215–218.

Motamedi, M. and K. Pedersen. 1998. Desulfovibrio aespoeensis sp. nov., a mesophilic sulfate-reducing bacterium from deep groundwater at Äspö hard rock laboratory, Sweden. Int. J. Syst. Bacteriol. 48: 311–315.

Mott, J., R.E. Barnewall and Y. Rikihisa. 1999. Human granulocytic ehrlichiosis agent and Ehrlichia chaffeensis reside in different cytoplasmic compartments in HL-60 cells. Infect. Immun. 67: 1368–1378.

Mott, J., Y. Muramatsu, E. Seaton, C. Martin, S. Reed and Y. Rikihisa. 2002. Molecular analysis of Neorickettsia risticii in adult aquatic insects in Pennsylvania, in horses infected by ingestion of insects, and isolated in cell culture. J. Clin. Microbiol. 40: 690–693.

Mott, J., Y. Rikihisa, Y. Zhang, S.M. Reed and C.Y. Yu. 1997. Comparison of PCR and culture to the indirect fluorescent-antibody test for diagnosis of Potomac horse fever. J. Clin. Microbiol. 35: 2215–2219.

Mottola, H.A. 1974. Nitrilotriacetic acid as a chelating agent: applications, toxicology and bio-environmental impact. Toxicol. Environ. Chem. Rev. 2: 99–161.

Mountfort, D.O., W.J. Brulla, L.R. Krumholz and M.P. Bryant. 1984. Syntrophus buswellii gen. nov., sp. nov.: a benzoate catabolizer from methanogenic ecosystems. Int. J. Syst. Bacteriol. 34: 216–217.

Moyer, C.L., F.C. Dobbs and D.M. Karl. 1995. Phylogenetic diversity of the bacterial community from a microbial mat at an active, hydrothermal vent system, Loihi Seamount, Hawaii. Appl. Environ. Microbiol. 61: 1555–1562.

Mucha, H., F. Lingens and W. Trösch. 1988. Conversion of propionate to acetate and methane by syntrophic consortia. Appl. Microbiol. Biotechnol. 27: 581–586.

Mudd, S. and S. Warren. 1923. A readily cultivable vibrio, filterable through Berkefeld "V" candles, Vibrio percolans (new species). J. Bacteriol. 8: 447–454.

Muddaris, M. and B. Austin. 1988. Quantitative and qualitative studies of the bacterial microflora of turbot Scophthalmus maximus L. gills. J. Fish. Biol. 32: 223–229.

Mueller, J.G., R. Devereux, D.L. Santavy, S.E. Lantz, S.G. Willis and P.H. Pritchard. 1997. Phylogenetic and physiological comparisons of PAH-degrading bacteria from geographically diverse soils. Antonie Leeuwenhoek 71: 329–343.

Mueller, J.A., A.S. Galushko, A. Kappler and B. Schink. 1999. Anaerobic degradation of m-cresol by Desulfobacterium cetonicum is initiated by formation of 3-hydroxybenzylsuccinate. Arch. Microbiol. 172: 287–294.

Mukwaya, G.M. and D.F. Welch. 1989. Subgrouping of Pseudomonas cepacia by cellular fatty acid composition. J. Clin. Microbiol. 27: 2640–2646.

Mulder, E.G. 1989a. Genus Leptothrix. In Staley, Bryant, Pfennig and Holt (Editors), Bergey's Manual of Systematic Bacteriology, Vol. 3, The Williams & Wilkins Co., Baltimore. pp. 1998–2003.

Mulder, E.G. 1989b. Genus Sphaerotilus Kützing 1833, 386^AL. In Staley, Bryant, Pfennig and Holt (Editors), Bergey's Manual of Systematic Bacteriology, Vol. 3, Williams & Wilkins Co., Baltimore. pp. 1994–1998.

Mulder, E.G. and M.H. Deinema. 1981. The sheathed bacteria. In Starr, Stolp, Trüper, Balows and Schlegel (Editors), The Prokaryotes: A Handbook on Habitats, Isolation and Identification of Bacteria, 1st ed., Springer-Verlag, Berlin. pp. 425–440.

Mulder, E.G. and W.L. van Veen. 1963. Investigations on the Sphaerotilus-Leptothrix group. Antonie Leeuwenhoek J. Microbiol. Serol. 29: 121–153.

Mulder, E.G. and W.L. van Veen. 1965. Anreicherundskultur von Organismen der Sphaerotilus-Leptothrix-Gruppe Aus: Anreicherungskultur und Mutantelauslese, Symp. Göttingen, 1964. Zentralbl. Bakteriol. Parasitenkd. Infektionskr. Hyg. Abt. I Suppl. 1: 28–46.

Mulks, M.H. and J.S. Knapp. 1987. Immunoglobulin A1 protease types of Neisseria gonorrhoeae and their relationship to auxotype and serovar. Infect. Immun. 55: 931–936.

Mulks, M.H., A.G. Plaut, H.A. Feldman and B. Frangione. 1980. IgA proteases of two distinct specificities are released by Neisseria meningitidis. J. Exp. Med. 152: 1442–1447.

Müller, A., D. Gunther, F. Dux, M. Naumann, T.F. Meyer and T. Rudel. 1999. Neisserial porin (PorB) causes rapid calcium influx in target cells and induces apoptosis by the activation of cysteine proteases. EMBO J. 18: 339–352.

Muller, F.M., J.E. Hoppe and C.H. Wirsing von König. 1997a. Laboratory diagnosis of pertussis: state of the art in 1997. J. Clin. Microbiol. 35: 2435–2443.

Müller, H.E. 1995. Investigations of culture and properties of Afipia spp. Zentbl. Bakteriol. 282: 18–23.

Müller, H.P., A. Heinecke, M. Borneff, A. Knopf, C. Kiencke and S. Pohl. 1997b. Microbial ecology of Actinobacillus actinomycetemcomitans, Eikenella corrodens and Capnocytophaga spp. in adult periodontitis. J. Periodontal Res. 32: 530–542.

Muller, M. and A. Hildebrandt. 1993. Nucleotide sequences of the 23S rRNA genes from Bordetella pertussis, B.parapertussis, B.bronchiseptica and B.avium, and their implications for phylogenetic analysis. Nucleic Acids Res. 21: 3320.

Müller, O.F. 1786. Animalcula Infusoria Fluviatilia et Marina, Quae Detexit, Systematice Descripsit et Ad Vivum Delineari Curavit, pp. 1–367.

Müller, R., K. Gerth, P. Brandt, H. Blöcker and S. Beyer. 2000. Identification of an L-dopa decarboxylase gene from Sorangium cellulosum So ce90. Arch. Microbiol. 173: 303–306.

Mulongoy, K. and G.H. Elkan. 1977. Glucose catabolism in two derivatives of a Rhizobium japonicum strain differing in nitrogen-fixing efficiency. J. Bacteriol. 131: 179–187.

Munderloh, U.G., E.F. Blouin, K.M. Kocan, N.L. Ge, W.L. Edwards and T.J. Kurtti. 1996a. Establishment of the tick (Acari:Ixodidae)-borne cattle pathogen Anaplasma marginale (Rickettsiales: Anaplasmataceae) in tick cell culture. J. Med. Entomol. 33: 656–664.

Munderloh, U.G., J.E. Madigan, J.S. Dumler, J.L. Goodman, S.F. Hayes, J.E. Barlough, C.M. Nelson and T.J. Kurtti. 1996b. Isolation of the equine granulocytic ehrlichiosis agent, Ehrlichia equi, in tick cell culture. J. Clin. Microbiol. 34: 664–670.

Munkley, A., C.R. Tinsley, M. Virji and J.E. Heckels. 1991. Blocking of bactericidal killing of Neisseria meningitidis by antibodies directed against class-4 outer-membrane protein. Microb. Pathog. 11: 447–452.

Muñoz, J., J.M. Arias and E. Montoya. 1984. Production and properties of a bacteriocin from Myxococcus coralloides-D. J. Appl. Bacteriol. 57: 69–74.

Muñoz, J., F. González, M.M. Martinez-Cañamero, M.A. Goicoechea, A.L. Extremera and J.M. Arias. 1989. Deoxyribonuclease and phosphatase activities in myxobacteria. Microbios. *58*: 43–47.

Muñoz-Dorado, J., S. Inouye and M. Inouye. 1991. A gene encoding a protein serine/threonine kinase is required for normal development of *M. xanthus*, a Gram-negative bacterium. Cell *67*: 995–1006.

Murai, F., M. Fukuda and K. Yano. 1990. Indigenous plasmids of *Beijerinckia* strains. Agric. Biol. Chem. *54*: 545–546.

Murphy, C.I., J.R. Storey, J. Recchia, L.A. Doros-Richert, C. Gingrich-Baker, K. Munroe, J.S. Bakken, R.T. Coughlin and G.A. Beltz. 1998. Major antigenic proteins of the agent of human granulocytic ehrlichiosis are encoded by members of a multigene family. Infect. Immun. *66*: 3711–3718.

Murphy, G.L., T.D. Connell, D.S. Barritt, M. Koomey and J.G. Cannon. 1989. Phase variation of gonococcal protein-ii—regulation of gene expression by slipped-strand mispairing of a repetitive DNA sequence. Cell *56*: 539–547.

Murphy, P.J., N. Heycke, Z. Banfalvi, M.E. Tate, F. Debruijn, A. Kondorosi, J. Tempé and J. Schell. 1987. Genes for the catabolism and synthesis of an opine-like compound in *Rhizobium meliloti* are closely linked and on the Sym plasmid. Proc. Natl. Acad. Sci. U.S.A. *84*: 493–497.

Murphy, P.J., N. Heycke, S.P. Trenz, P. Ratet, F.J. de Bruijn and J. Schell. 1988. Synthesis of an opine-like compound, a rhizopine, in alfalfa nodules is symbiotically regulated. Proc. Natl. Acad. Sci. U.S.A. *85*: 9133–9137.

Murphy, P.J., W. Wexler, W. Grzemski, J.P. Rao and D. Gordon. 1995. Rhizopines - their role in symbiosis and competition. Soil Biol. Biochem. *27*: 525–529.

Murray, E.G.D. 1929. Medical Research Council (Great Britain). Special Report Series. *124*: 7–142.

Murray, E.G.D. 1939. Family *Neisseriaceae*. *In* Bergey, Breed, Murray and Hitchens (Editors), Bergey's Manual of Determinative Bacteriology, 5th Ed., The Williams & Wilkins Co., Baltimore. p. 309.

Murray, E.S. and S.B. Torrey. 1975. Virulence of *Rickettsia prowazeki* for head lice. Ann. N. Y. Acad. Sci. *266*: 25–34.

Murray, R.G.E. 1963. Role of superficial structures in the characteristic morphology of *Lampropedia hyalina*. Can. J. Microbiol. *9*: 593–600.

Murray, R.G.E. and A. Birch-Andersen. 1963. Specialized structure in the region of the flagella tuft in *Spirillum serpens*. Can. J. Microbiol. *9*: 393–401.

Murray, R.G.E., D.J. Brenner, R.R. Colwell, P. DeVos, M. Goodfellow, P.A.D. Grimont, N. Pfennig, E. Stackebrandt and G.A. Zavarzin. 1990. Report of the ad hoc committee on approaches to taxonomy within the proteobacteria. Int. J. Syst. Bacteriol. *40*: 213–215.

Murray, R.G.E. and K.H. Schleifer. 1994. Taxonomic notes: a proposal for recording the properties of putative taxa of procaryotes. Int. J. Syst. Bacteriol. *44*: 174–176.

Murray, R.G.E. and E. Stackebrandt. 1995. Taxonomic note: implementation of the provisional status *Candidatus* for incompletely described procaryotes. Int. J. Syst. Bacteriol. *45*: 186–187.

Murray, R.G.E., P. Steed and H.E. Elson. 1965. The location of the mucopeptide in sections of the cell wall of *Escherichia coli* and other Gram-negative bacteria. Can. J. Microbiol. *11*: 547–560.

Murrell, J.C. 1992. Genetics and molecular biology of methanotrophs. FEMS Microbiol. Rev. *8*: 233-248.

Murrell, J.C. and H. Dalton. 1983a. Ammonia assimilation in *Methylococcus capsulatus* (Bath) and other obligate methanotrophs. J. Gen. Microbiol. *129*: 1197–1206.

Murrell, J.C. and H. Dalton. 1983b. Nitrogen fixation in obligate methanotrophs. J. Gen. Microbiol. *129*: 3481–3486.

Murty, M.G. 1984. Phyllosphere of cotton as a habitat for diazotrophic microorganisms. Appl. Environ. Microbiol. *48*: 713–718.

Musmanno, R.A., M. Russi, H. Lior and N. Figura. 1997. In vitro virulence factors of *Arcobacter butzleri* strains isolated from superficial water samples. New Microbiol. *20*: 63–68.

Musser, J.M., D.A. Bemis, H. Ishikawa and R.K. Selander. 1987. Clonal diversity and host distribution in *Bordetella bronchiseptica*. J. Bacteriol. *169*: 2793–2803.

Musser, J.M., E.L. Hewlett, M.S. Peppler and R.K. Selander. 1986. Genetic diversity and relationships in populations of *Bordetella* spp. J. Bacteriol. *166*: 230–237.

Mutinga, M.J. and O.O. Dipeolu. 1989. Saurian Malaria in Kenya: description of new species of haemoproteid and haemogregarine parasites, *Anaplasma*-like and *Pirhemocyton*-like organisms in the blood of lizards in West Pokot district Kenya. Insect Sci. Appl. *10*: 401–412.

Myerhoff, M. 1911. Sur la conjonctivite gonococcique epidemique d'Egypt. Et ses rapports avec la trachoma. Arch. d'Ophthalmol. (Paris) *21*: 1–34.

Myers, W.F. and C.L. Wisseman, Jr.. 1980. Genetic relatedness among the typhus group of rickettsiae. Int. J. Syst. Bacteriol. *30*: 143–150.

Myers, W.F. and C.L. Wisseman, Jr.. 1981. The taxonomic relationship of *Rickettsia canada* to the typhus and spotted fever groups of the genus *Rickettsia*. *In* Burgdorfer and Anacker (Editors), Rickettsiae and Rickettsial Diseases, Academic Press, New York. pp. 313–325.

Nachamkin, I. 1995. *Campylobacter* and *Arcobacter*. *In* Murray, Baron, Pfaller, Tenover and Yolken (Editors), Manual of Clinical Microbiology, 6th Ed., ASM Press, Washington D.C. pp. 483–491.

Nachamkin, I., B.M. Allos and T. Ho. 1998. *Campylobacter* species and Guillain-Barre syndrome. Clin. Microbiol. Rev. *11*: 555–567.

Nachamkin, I., K. Bohachick and C.M. Patton. 1993. Flagellin gene typing of *Campylobacter jejuni* by restriction fragment length polymorphism analysis. J. Clin. Microbiol. *31*: 1531–1536.

Nagashima, K.V.P., A. Hiraishi, K. Shimada and K. Matsuura. 1997a. Horizontal transfer of genes coding for the photosynthetic reaction centers of purple bacteria. J. Mol. Evol. *45*: 131–136.

Nagashima, K.V.P., K. Matsuura, N. Wakao, A. Hiraishi and K. Shimada. 1997b. Nucleotide sequences of genes coding for photosynthetic reaction centers and light-harvesting proteins of *Acidiphilium rubrum* and related aerobic acidophilic bacteria. Plant Cell Physiol. *38*: 1249–1258.

Naik, G.A., L.N. Bhat, B.A. Chopade and J.M. Lynch. 1994. Transfer of broad host range antibiotic resistance plasmids in soil microcosms. Curr. Microbiol. *28*: 209–215.

Nair, S.K., P. Jara, B. Quiviger and C. Elmerich. 1983. Recent developments in the genetics of nitrogen fixation in *Azospirillum*. *In* Klingmuller (Editor), *Azospirillum* II: Genetics, Physiology, Ecology, Birkhauser-Verlag, Basel. pp. 29–38.

Nairn, C.A., J.A. Cole, P.V. Patel, N.J. Parsons, J.E. Fox and H. Smith. 1988. Cytidine 5′-monophospho-*N*-acetylneuraminic acid or a related compound is the low Mr factor from human red blood cells which induces gonococcal resistance to killing by human serum. J. Gen. Microbiol. *134*: 3295–3306.

Naka, T., N. Fujiwara, E. Yabuuchi, M. Doe, K. Kobayashi, Y. Kato and I. Yano. 2000. A novel sphingoglycolipid containing galacturonic acid and 2-hydroxy fatty acid in cellular lipids of *Sphingomonas yanoikuyae*. J. Bacteriol. *182*: 2660–2663.

Nakagawa, Y., T. Sakane and A. Yokota. 1996. Transfer of *"Pseudomonas riboflavina"* (Foster 1944), a gram-negative, motile rod with long-chain 3-hydroxy fatty acids, to *Devosia ribiflavina* gen. nov., sp. nov., nom. rev. Int. J. Syst. Bacteriol. *46*: 16–22.

Nakagawa, Y. and K. Yamasato. 1993. Phylogenetic diversity of the genus *Cytophaga* revealed by 16S rRNA sequencing and menaquinone analysis. J. Gen. Microbiol. *139*: 1155–1161.

Nakai, T., A. Moriya, N. Tonouchi, T. Tsuchida, F. Yoshinaga, S. Horinouchi, Y. Sone, H. Mori, F. Sakai and T. Hayashi. 1998. Control of expression by the cellulose synthase (*bcsA*) promotor region from *Acetobacter xylinum* BPR 2001. Gene *213*: 93–100.

Nakamura, L.K. and D.D. Tyler. 1977. Induction of D-aldohexoside: cytochrome *c* oxidoreductase in *Agrobacterium tumefaciens*. J. Bacteriol. *129*: 830–835.

Nakamura, Y., S. Hyodo, E. Chonan, S. Shigeta and E. Yabuuchi. 1986. Serological classification of *Pseudomonas cepacia* by somatic antigen. J. Clin. Microbiol. *24*: 152–154.

Nakamura, Y., J.I. Someya and T. Suzuki. 1985. Nickel requirement of oxygen-resistant hydrogen bacterium *Xanthobacter autotrophicus* strain Y-38. Agric. Biol. Chem. *49*: 1711–1718.

Nakanishi, I., K. Kimura, T. Suzuki, M. Ishikawa, I. Banno, T. Sakane and T. Harada. 1976. Demonstration of curdlan-type polysaccharide and some other β-1,3-glucan in microorganisms with aniline blue. J. Gen. Appl. Microbiol. *22*: 1–11.

Nakase, Y. 1957. Studies on *Hemophilus bronchisepticus*. IV. Serological relation of *H. bronchisepticus* from guinea pig, dog and human. Kitasato Arch. Exp. Med. *30*: 85–94.

Nakashita, H. and H. Seto. 1991. A microorganism with both abilities to form and cleave C-P bonds. Agric. Biol. Chem. *55*: 2913–2915.

Nakashita, H., A. Shimazu and H. Seto. 1991. A new screening method for C-P compound producing organisms by the use of phosphoenolpyruvate phosphomutase. Agric. Biol. Chem. *55*: 2825–2829.

Nakazawa, T., M. Kimoto and M. Abe. 1990. Cloning, sequencing, and transcriptional analysis of the *RecA* gene of *Pseudomonas cepacia*. Gene *94*: 83–88.

Namsaraev, B.B. 1973. Growth of *Microcyclus* on methanol. Microbiology *42*: 986–987.

Namsaraev, B.B. and A.N. Nozhevnikova. 1978. Autotrophic growth of *Microcyclus aquaticus* in an atmosphere of hydrogen. Microbiology *47*: 315–318.

Namsaraev, B.B. and G.A. Zavarzin. 1972. Trophic links in cultures oxidizing methane. Mikrobiologiya *41*: 999–1006.

Namsaraev, B.B. and G.A. Zavarzin. 1974. Growth of the "tetraedron" budding bacterium on monocarbon compounds. Mikrobiologiya *43*: 406–409.

Nanninga, H.J. and J.C. Gottschal. 1986. Isolation of a sulfate-reducing bacterium growing with methanol. FEMS Microbiol. Ecol. *38*: 125–130.

Nanninga, H.J. and J.C. Gottschal. 1987. Properties of *Desulfovibrio carbinolicus* sp. nov. and other sulfate-reducing bacteria isolated from an anaerobic-purification plant. Appl. Environ. Microbiol. *53*: 802–809.

Nanninga, H.J. and J.C. Gottschal. 1995. *In* Validation of the publication of new names and new combinations previously effectively published outside the IJSB. List No. 55. Int. J. Syst. Bacteriol. *45*: 879–880.

Napoli, C., R. Sanders, R. Carlson and P. Albersheim. 1980. Host-symbiont interactions: recognizing *Rhizobium*. *In* Newton and Orme-Johnson (Editors), Nitrogen Fixation, Vol. II, University Park Press, Baltimore. pp. 189–203.

Nassif, X. 1999. Interaction mechanisms of encapsulated meningococci with eucaryotic cells: what does this tell us about the crossing of the blood-brain barrier by *Neisseria meningitidis*? Curr. Opin. Microbiol. *2*: 71–77.

Nassif, X., J.L. Beretti, J. Lowy, P. Stenberg, P. Ogaora, J. Pfeifer, S. Normark and M. So. 1994. Roles of pilin and Pilc in adhesion of *Neisseria meningitidis* to human epithelial and endothelial cells. Proc. Natl. Acad. Sci. U.S.A. *91*: 3769–3773.

Navani, N.K., M.A. Joshi and K.L. Dikshit. 1996. Genetic transformation of *Vitreoscilla* sp. Gene *177*: 265–266.

Navarro, E., M.P. Fernandez, F. Grimont, A. Clays-Josserand and R. Bardin. 1992. Genomic heterogeneity of the genus *Nitrobacter*. Int. J. Syst. Bacteriol. *42*: 554–560.

Nedenskov, P. 1994. Nutritional requirements for growth of *Helicobacter pylori*. Appl. Environ. Microbiol. *60*: 3450–3453.

Nedwell, D.B. and M. Rutter. 1994. Influence of temperature on growth rate and competition between two psychrotolerant antarctic bacteria: low temperature diminishes affinity for substrate uptake. Appl. Environ. Microbiol. *60*: 1984–1992.

Neef, A. 1997. Anwendung der *In situ* Einzelzell Identifizierung von Bakterien zur Populationsanalyse in Komplexen Mikrobiellen Biozönosen., Thesis, Technical University of Munich, Munich.

Neef, A., A. Zaglauer, H. Meier, R. Amann, H. Lemmer and K.H. Schleifer. 1996. Population analysis in a denitrifying sand filter: conventional and *in situ* identification of *Paracoccus* spp. in methanol-fed biofilms. Appl. Environ. Microbiol. *62*: 4329–4339.

Negrini, R., L. Lisato, I. Zanella, L. Cavazzini, S. Gullini, V. Villanacci, C. Poiesi, A. Albertini and S. Ghielmi. 1991. *Helicobacter pylori* infection induces antibodies cross-reacting with human gastric mucosa. Gastroenterology *101*: 437–445.

Neiger, R., C. Dieterich, A. Burnens, A. Waldvogel, I. Corthesy-Theulaz, F. Halter, B. Lauterburg and A. Schmassmann. 1998. Detection and prevalence of *Helicobacter* infection in pet cats. J. Clin. Microbiol. *36*: 634–637.

Neill, S.C., J.N. Campbell, J.J. O'Brien, S.T.C. Weatherup and W.A. Ellis. 1985. Taxonomic position of *Campylobacter cryaerophila* sp. nov. Int. J. Syst. Bacteriol. *35*: 342–356.

Neilson, J.W., K.L. Josephson, I.L. Pepper, R.B. Arnold, G.D. Di Giovanni and N.A. Sinclair. 1994. Frequency of horizontal gene transfer of a large catabolic plasmid (pJP4) in soil. Appl. Environ. Microbiol. *60*: 4053–4058.

Nelson, D.C. and R.W. Castenholz. 1981. Use of reduced sulfur compounds by *Beggiatoa* sp. J. Bacteriol. *147*: 140–154.

Nelson, D.R., M.G. Cumsky and D.R. Zusman. 1981. Localization of myxobacterial hemagglutinin in the periplasmic space and on the cell surface of *Myxococcus xanthus* during developmental aggregation. J. Biol. Chem. *256*: 2589–2595.

Nelson, E.L. and M.J. Pickett. 1951. The recovery of L-forms of *Brucella* and their relation to *Brucella* phage. J. Infect. Dis. *89*: 226–232.

Nelson, L.M. and R. Knowles. 1978. Effect of oxygen and nitrate on nitrogen fixation and denitrification by *Azospirillum brasilense* grown in continuous culture. Can. J. Microbiol. *24*: 1395–1403.

Nesme, X., M.F. Michel and B. Digat. 1987. Population heterogeneity of *Agrobacterium tumefaciens* in galls of *Populus L* from a single nursery. Appl. Environ. Microbiol. *53*: 655–659.

Nester, E.W., M.P. Gordon, R.M. Amasino and M.F. Yanofsky. 1984. Crown gall - a molecular and physiological analysis. Annu. Rev. Plant Physiol. Plant Molec. Biol. *35*: 387–413.

Neuer, G., A. Kronenberg and H. Bothe. 1985. Denitrification and nitrogen fixation by *Azospirillum*:3. Properties of a wheat and *Azospirillum* association. Arch. Microbiol. *141*: 364–370.

Neumann, B., A. Pospiech and H.U. Schairer. 1992. Size and stability of the genomes of the myxobacteria *Stigmatella aurantiaca* and *Stigmatella erecta*. J. Bacteriol. *174*: 6307–6310.

Neumann, B., A. Pospiech and H.U. Schairer. 1993. A physical and genetic map of the *Stigmatella aurantiaca* DW4/3.1 chromosome. Mol. Microbiol. *10*: 1087–1099.

Neumann, U., H. Mayer, E. Schiltz, R. Benz and J. Weckesser. 1995. Lipopolysaccharide and porin of *Roseobacter denitrificans*, confirming its phylogenetic relationship to the alpha-3 subgroup of *Proteobacteria*. Microbiology *141*: 2013–2017.

Neunlist, S. and M. Rohmer. 1985. The hopanoids of *Methylosinus trichosporium* are aminobacteriohopanetriol and aminobacteriohopanetetrol. J. Gen. Microbiol. *131*: 1363–1367.

Neutzling, O., J.F. Imhoff and H.G. Trüper. 1984. *In* Validation of the publication of new names and new combinations previously effectively published outside the IJSB. List No. 16. Int. J. Syst. Bacteriol. *34*: 503–504.

Neutzling, O. and H.G. Trüper. 1982. Assimilatory sulfur metabolism in *Rhodopseudomonas sulfoviridis*. Arch. Microbiol. *133*: 145–148.

Neuzil, K.M., E. Wang, D.W. Haas and M.J. Blaser. 1994. Persistence of *Campylobacter fetus* bacteremia associated with absence of opsonizing antibodies. J. Clin. Microbiol. *32*: 1718–1720.

Neveling, U., R. Klasen, S. Bringer Meyer and H. Sahm. 1998. Purification of the pyruvate dehydrogenase multienzyme complex of *Zymomonas mobilis* and identification and sequence analysis of the corresponding genes. J. Bacteriol. *180*: 1540–1548.

New, P.B. and A. Kerr. 1972. Biological control of crown gall: field measurements and glass-house experiments. J. Appl. Bacteriol. *35*: 279–287.

Newcombe, J., K. Cartwright, S. Dyer and J. McFadden. 1998. Naturally occurring insertional inactivation of the *porA* gene of *Neisseria meningitidis* by integration of IS 1301. Mol. Microbiol. *30*: 453–454.

Newhall, W.J., W.D. Sawyer and R.A. Haak. 1980a. Cross-linking analysis of the outer membrane proteins of *Neisseria gonorrhoeae*. Infect. Immun. *28*: 785–791.

Newhall, W.J., C.E. Wilde, W.D. Sawyer and R.A. Haak. 1980b. High-molecular-weight antigenic protein complex in the outer membrane of *Neisseria gonorrhoeae*. Infect. Immun. *27*: 475–482.

Newman, L.M. and L.P. Wackett. 1995. Purification and characterization of toluene 2-monooxygenase from *Burkholderia cepacia* G4. Biochemistry *34*: 14066–14076.

Newman, L.M. and L.P. Wackett. 1997. Trichloroethylene oxidation by purified toluene 2-monooxygenase: products, kinetics, and turnover dependent inactivation. J. Bacteriol. *179*: 90–96.

Newton, J.W., A.G. Marr and J.B. Wilson. 1954. Fixation of $^{14}CO_2$ into nucleic acid constituents by *Brucella abortus*. J. Bacteriol. *67*: 233–236.

Neyra, C.A., A. Atkinson and O. Olubayi. 1995. Coaggregation of *Azospirillum* with other bacteria: basis for functional diversity. NATO ASI Ser. Ser. G Ecol. Sci. *37*: 429–439.

Neyra, C.A., J. Dobereiner, R. Lalande and R. Knowles. 1977. Denitrification by N_2-fixing *Sprillum lipoferum*. Can. J. Microbiol. *23*: 300–305.

Neyra, C.A. and P. Van Berkum. 1977. Nitrate reduction and nitrogenase activity in *Spirillum lipoferum*. Can. J. Microbiol. *23*: 306–310.

Ng, L.-K., R. Sherburne, D.E. Taylor and M.E. Stiles. 1985. Morphological forms and viability of *Campylobacter* species studied by electron microscopy. J. Bacteriol. *164*: 338–343.

Nga, D.P., D.T.C. Ha, L.T. Hien and H. Stan-Lotter. 1996. *Desulfovibrio vietnamensis* sp. nov., a halophilic sulfate-reducing bacterium from Vietnamese oil fields. Anaerobe *2*: 385–392.

Nga, D.P., D.T.C. Ha, T.H. Lai and H. Stan-Lotter. 2002. *In* Validation of the publication of new names and new combinations previously effectively published outside the IJSEM. List No. 86. Int. J. Syst. Evol. Microbiol. *52*: 1075–1076.

Ngoc Dung, N., M. Kraut and W. Klingmuller. 1995. Isolation and characterization of *Azospirillum* strains from soil and rice plants in North Vietnam. NATO ASI Ser. Ser. G Ecol. Sci. *37*: 559–566.

Nguyen, B., T. Scognamiglio and J.J. Tudor. 1997. RAPD and macrorestriction polymorphism analysis of *Bdellovibrio* strains. 97th Annual Meeting of the American Society for Microbiology, Miami Beach, Florida.

Ni, H., A.I. Knight, K. Cartwright, W.H. Palmer and J. McFadden. 1992. Polymerase chain reaction for diagnosis of meningococcal meningitis. Lancet *340*: 1432–1434.

Nicaud, J.M., A. Breton, G. Younes and J. Guespin-Michel. 1984. Mutants of *Myxococcus xanthus* impaired in protein secretion—an approach to study of a secretory mechanism. Mol. Microbiol. *20*: 344–350.

Nichols, P.D., J.M. Henson, C.P. Antworth, J. Parsons, J.T. Wilson and D.C. White. 1987. Detection of a microbial consortium, including type II methanotrophs, by use of phospholipid fatty acids in an aerobic halogenated hydrocarbon-degrading soil column enriched with natural gas. Environ. Toxicol. Chem. *6*: 89–98.

Nichols, P.D., G.A. Smith, C.P. Antworth, R.S. Hanson and D.C. White. 1985. Phospholipid and lipopolysaccharide normal and hydroxy fatty acids as potential signature for methane-utilizing bacteria. FEMS Microbiol. Ecol. *32*: 327–335.

Nichols, P., B.K. Stulp, J.G. Jones and D.C. White. 1986. Comparison of fatty acid content and DNA homology of the filamentous gliding bacteria *Vitroscilla, Flexibacter, Filibacter*. Arch. Microbiol. *146*: 1–6.

Nick, G., P. De Lajudie, B.D. Eardly, S. Suomalainen, L. Paulin, X. Zhang, M. Gillis and K. Lindström. 1999. *Sinorhizobium arboris* sp. nov. and *Sinorhizobium kostiense* sp. nov., isolated from leguminous trees in Sudan and Kenya. Int. J. Syst. Bacteriol. *49*: 1359–1368.

Nickens, D., C.J. Fry, L. Ragatz, C.E. Bauer and H. Gest. 1996. Biotype of the purple nonsulfur photosynthetic bacterium, *Rhodospirillum centenum*. Arch. Microbiol. *165*: 91–96.

Nicolas, F.J., M.L. Cayuela, I.M. Martinez-Argudo, R.M. Ruiz-Vazquez and F.J. Murillo. 1996. High mobility group I(Y)-like DNA-binding domains on a bacterial transcription factor. Proc. Natl. Acad. Sci. U.S.A. *93*: 6881–6885.

Niebylski, M.L., M.G. Peacock, E.R. Fischer, S.F. Porcella and T.G. Schwan. 1997a. Characterization of an endosymbiont infecting wood ticks, *Dermacentor andersoni*, as a member of the genus *Francisella*. Appl. Environ. Microbiol. *63*: 3933–3940.

Niebylski, M.L., M.G. Peacock, M.E. Schrumpf, W. Burgdorfer, E.R.

Fischer, K.L. Gage and T.G. Schwan. 1996. Characterization of the east side agent, a Spotted Fever group rickettsia infecting wood ticks, *Dermacentor andersoni*, in western Montana. Rickettsiae and Rickettsial Diseases, Proceedings of the Vth International Symposium, Bratislavia. pp. 227–232.

Niebylski, M.L., M.G. Peacock and T.G. Schwan. 1999. Lethal effect of *Rickettsia rickettsii* on its tick vector (*Dermacentor andersoni*). Appl. Environ. Microbiol. *65*: 773–778.

Niebylski, M.L., M.E. Schrumpf, W. Burgdorfer, E.R. Fischer, K.L. Gage and T.G. Schwan. 1997b. *Rickettsia peacockii* sp. nov., a new species infecting wood ticks, *Dermacentor andersoni*, in western Montana. Int. J. Syst. Bacteriol. *47*: 446–452.

Nielsen, A.K., K. Gerdes and J.C. Murrell. 1997. Copper-dependent reciprocal transcriptional regulation of methane monooxygenase genes in *Methylococcus capsulatus* and *Methylosinus trichosporium*. Mol. Microbiol. *25*: 399–409.

Nielsen, J.T., W. Liesack and K. Finster. 1999. *Desulfovibrio zosterae* sp. nov., a new sulfate reducer isolated from surface-sterilized roots of the seagrass *Zostera marina*. Int. J. Syst. Bacteriol. *49*: 859–865.

Nierman, W.C., T.V. Feldblyum, M.T. Laub, I.T. Paulsen, K.E. Nelson, J. Eisen, J.F. Heidelberg, M.R. Alley, N. Ohta, J.R. Maddock, I. Potocka, W.C. Nelson, A. Newton, C. Stephens, N.D. Phadke, B. Ely, R.T. DeBoy, R.J. Dodson, A.S. Durkin, M.L. Gwinn, D.H. Haft, J.F. Kolonay, J. Smit, M.B. Craven, H. Khouri, J. Shetty, K. Berry, T. Utterback, K. Tran, A. Wolf, J. Vamathevan, M. Ermolaeva, O. White, S.L. Salzberg, J.C. Venter, L. Shapiro and C.M. Fraser. 2001. Complete genome sequence of *Caulobacter crescentus*. Proc. Natl. Acad. Sci. U.S.A. *98*: 4136–4141.

Nieuwkoop, A.J., Z. Banfalvi, N. Deshmane, D. Gerhold, M.G. Schell, K.M. Sirotkin and G. Stacey. 1987. A locus encoding host range is linked to the common nodulation genes of *Bradyrhizobium japonicum*. J. Bacteriol. *169*: 2631–2638.

Nikaido, H. 1996. Multidrug efflux pumps of Gram negative bacteria. J. Bacteriol. *178*: 5853–5859.

Nikitin, D.I. 1971. A new soil microorganism—*Renobacter vaculatum* gen. et sp. n. Dokl. Akad. Nauk SSSR *198*: 447–448.

Nikitin, D.I. and L.V. Vasil'eva. 1968. A new soil microorganism *Agrobacterium polyspheroidum*, n. sp. Izv. Akad. Nauk S.S.S.R. Biol. *3*: 443–444.

Nikitin, D.I., L.V. Vasilyeva and R.A. Lokhmacheva. 1966. New and rare forms of soil microorganisms, Nauka, Moscow.

Nikitin, D.I., O.Y. Vishnewetskaya, K.M. Chumakov and I.V. Zlatkin. 1990. Evolutionary relationship of some stalked and budding bacteria (genera *Caulobacter*, "*Hyphobacter*", *Hyphomonas* and *Hyphomicrobium*) as studied by the new integral taxonomical method. Arch. Microbiol. *153*: 123–128.

Niklewski, B. 1910. Über die wasserstoffoxydation durch mikroorganismen. Jahrb. Wiss. Bot. *48*: 113–142.

Nilsson, K., O. Lindquist and C. Pahlson. 1999. Association of *Rickettsia helvetica* with chronic perimyocarditis in sudden cardiac death. Lancet *354*: 1169–1173.

Nishimura, K., H. Shimada, H. Ohta, T. Masuda, Y. Shioi and K. Takamiya. 1996. Expression of the *puf* operon in an aerobic photosynthetic bacterium, *Roseobacter denitrificans*. Plant Cell Physiol. *37*: 153–159.

Nishimura, Y., Y. Muroga, S. Saito, T. Shiba, K. Takamiya and Y. Shioi. 1994. DNA relatedness and chemotaxonomic feature of aerobic bacteriochlorophyll containing bacteria isolated from coasts of Australia. J. Gen. Appl. Microbiol. *40*: 287–296.

Nishimura, Y., M. Shimadzu and H. Iizuka. 1981. Bacteriochlorophyll formation in radiation-resistant *Pseudomonas radiodora*. J. Gen. Appl. Microbiol. *27*: 427–430.

Nishio, T. and Y. Ishida. 1990. Production of dihydroxamate siderophore alcaligin by *Alcaligenes xylosoxidans* subsp. *xylosoxidans*. Agric. Biol. Chem. *54*: 1837–1839.

Nishio, T., N. Tanaka, J. Hiratake, Y. Katsube, Y. Ishida and J. Oda. 1988. Isolation and structure of the novel dihydroxamate siderophore alcaligin. J. Am. Chem. Soc. *110*: 8733–8734.

Nishiyama, K., T. Kusaba, K. Ohta, K. Nahata and A. Ezuka. 1979. Bacterial black rot of tulip caused by *Pseudomonas andropogonis*. Ann. Phytopathol. Soc. Japan *45*: 668–674.

Nissen, H. and I.D. Dundas. 1984. *Rhodospirillum salinarum* sp. nov., a halophilic photosynthetic bacterium isolated from a Portuguese saltern. Arch. Microbiol. *138*: 251–256.

Nissen, H. and I.D. Dundas. 1985. *In* Validation of the publication of new names and new combinations previously effectively published outside the IJSB. List No. 17. Int. J. Syst. Bacteriol. *35*: 223–225.

Noegel, A. and E.C. Gotschlich. 1983. Isolation of a high molecular weight polyphosphate from *Neisseria gonorrhoeae*. J. Exp. Med. *157*: 2049–2060.

Noguchi, T., H. Hayashi, K. Shimada, S. Takaichi and M. Tasumi. 1992. In vivo states and functions of carotenoids in an aerobic photosynthetic bacterium, *Erythrobacter longus*. Photosynth. Res. *31*: 21–30.

Noguchi, T.T., R. Nachum and C.A. Lawrence. 1963. Acute purulent meningitis caused by chromogenic *Neisseria*. A case report and literature review. Med. Art. Sci. *17*: 11–18.

Noguchi, Y., T. Fujiwara, K. Yoshimatsu and Y. Fukumori. 1999. Iron reductase for magnetite synthesis in the magnetotactic bacterium *Magnetospirillum magnetotacticum*. J. Bacteriol. *181*: 2142–2147.

Nohynek, L.J., E.L. Nurmiaho-Lassila, E.L. Suhonen, H.J. Busse, M. Mohammadi, J. Hantula, F. Rainey and M.S. Salkinoja-Salonen. 1996a. Description of chlorophenol-degrading *Pseudomonas* sp. strains KF1^T, KF3, and NKF1 as a new species of the genus *Sphingomonas*, *Sphingomonas subarctica* sp. nov. Int. J. Syst. Bacteriol. *46*: 1042–1055.

Nohynek, L.J., E.L. Suhonen, E.L. Nurmiaho-Lassila, J. Hantula and M. Salkinoja-Salonen. 1995. Description of four pentachlorophenol-degrading bacterial strains as *Sphingomonas chlorophenolica* sp. nov. Syst. Appl. Microbiol. *18*: 527–538.

Nohynek, L.J., E.L. Suhonen, E.L. Nurmiaho-Lassila, J. Hantula and M. Salkinoja-Salonen. 1996b. *In* Validation of the publication of new names and new combinations previously effectively published outside the IJSB. List No. 57. Int. J. Syst. Bacteriol. *46*: 625–626.

Nojiri, M. and T. Saito. 1997. Structure and function of poly(3-hydroxybutyrate) depolymerase from *Alcaligenes faecalis* T1. J. Bacteriol. *179*: 6965–6970.

Nokhal, T.H. and F. Mayer. 1979. Structural analysis of four strains of *Paracoccus denitrificans*. Antonie Leeuwenhoek J. Microbiol. Serol. *45*: 185–197.

Nokhal, T.H. and H.G. Schlegel. 1980. The regulation of hydrogenase formation as a differentiating character of strains of *Paracoccus denitrificans*. Antonie Leeuwenhoek *46*: 143–155.

Nokhal, T.H. and H.G. Schlegel. 1983. Taxonomic study of *Paracoccus denitrificans*. Int. J. Syst. Bacteriol. *33*: 26–37.

Nold, S.C., E.D. Kopczynski and D.M. Ward. 1996. Cultivation of aerobic chemoorganotrophic proteobacteria and Gram-positive bacteria from a hot spring microbial mat. Appl. Environ. Microbiol. *62*: 3917–3921.

Nöller, W. 1917. Blut- und Insektenflagellaten Zuchtung auf Platten. Arch. Schiffs- u. Tropen-Hyg. *21*: 53–94.

Nolte, E.M. 1957. Untersuchungen über Ernährung und Fruchtkörperbildung von Myxobakterien. Arch. Microbiol. *28*: 191–218.

Nomura, Y., N. Takada and Y. Oshima. 1989. Isolation and identification of phthalate-utilizing bacteria. J. Ferment. Bioeng. *67*: 297–299.

Nomura, Y., M. Yamamoto and K. Sugisawa. 1995a. Prevention by acetic acid bacteria of coloration by amino-carbonyl reaction during food storage. Biosci. Biotechnol. Biochem. *59*: 21–25.

Nomura, Y., M. Yamamoto, K. Sugisawa and H. Kumagai. 1995b. Reducing the stale flavor of cooked rice by treating with cells of acetic acid bacteria. Biosci. Biotechnol. Biochem. *59*: 1402–1406.

Norqvist, A., J. Davies, L. Norlander and S. Normark. 1978. Effect of iron starvation on outer membrane-protein composition of *Neisseria gonorrhoeae*. FEMS Microbiol. Lett. *4*: 71–75.

Norris, D.O. 1958. A red strain of *Rhizobium* from *Lotononis bainesii* Baker. Aust. J. Agric. Res. *9*: 629–632.

Norris, D.O. 1963. A porcelain bead method for storing *Rhizobium*. J. Exp. Agric. *31*: 255–258.

Norris, D.O. 1965. Acid production by *Rhizobium*. A unifying concept. Plant Soil *22*: 143–166.

Nörtemann, B. 1987. Bakterieller Abbau von Amino- und Hydroxy-naphthalinsulfonsäuren, Ph.D. thesis, University of Stuttgart, Germany.

Norton, R., B. Roberts, M. Freeman, M. Wilson, C. Ashhurst-Smith, W. Lock, D. Brookes and J. La Brooy. 1998. Characterisation and molecular typing of *Burkholderia pseudomallei*: are disease presentations of melioidosis clonally related? FEMS Immunol. Med. Microbiol. *20*: 37–44.

Nosko, P., L.C. Bliss and F.D. Cook. 1994. The association of free-living nitrogen-fixing bacteria with the roots of high arctic graminoids. Arct. Alp. Res. *26*: 180–186.

Nour, S.M., M.P. Fernandez, P. Normand and J.-C. Cleyet-Merel. 1994a. *Rhizobium ciceri* sp. nov., consisting of strains that nodulate chickpeas (*Cicer arietinum* L.). Int. J. Syst. Bacteriol. *44*: 511–522.

Nour, S.M., J.-C. Cleyet-Marel, D. Beck, A. Effosse and M.P. Fernandez. 1994b. Genotypic and phenotypic diversity of *Rhizobium* isolated from chickpea (*Cicer arietinum* L.). Can. J. Microbiol. *40*: 345–354.

Nour, S.M., J.-C. Cleyet-Marel, P. Normand and M.P. Fernandez. 1995. Genomic heterogeneity of strains nodulating chickpeas (*Cicer arietinum* L.) and description of *Rhizobium mediterraneum* sp. nov. Int. J. Syst. Bacteriol. *45*: 640–648.

Novikova, N. and V. Safronova. 1992. Transconjugants of *Agrobacterium radiobacter* harbouring *sym* genes of *Rhizobium galegae* can form an effective symbiosis with *Medicago sativa*. FEMS Microbiol. Lett. *72*: 261–268.

Novotny, P. and J.E. Brookes. 1975. The use of *Bordetella pertussis* preserved in liquid nitrogen as a challenge suspension in the Kendrick mouse protection test. J. Biol. Stand. *3*: 11–29.

Nozhevnikova, A.N. and G.A. Zavarzin. 1974. Taxonomy of CO-oxidizing Gram-negative bacteria. Izv. Akad. Nauk Uzb. SSR Ser. Biol. *3*: 436–440.

Nunley, J.W. and N.R. Krieg. 1968. Isolation of *Gallionella ferruginea* by use of formalin. Can. J. Microbiol. *14*: 385–389.

Nunn, D.N. and S. Lory. 1991. Product of the *Pseudomonas aeruginosa* gene *pilD* is a prepilin leader peptidase. Proc. Natl. Acad. Sci. U.S.A. *88*: 3281–3285.

Nur, I., Y. Okon and Y. Henis. 1980. Comparative studies of nitrogen-fixing bacteria associated with grasses in Israel with *Azospirillum brasilense*. Can. J. Microbiol. *26*: 714–718.

Nur, I., Y.L. Steinitz, Y. Okon and Y. Henis. 1981. Carotenoid composition and function in nitrogen-fixing bacteria of the genus *Azospirillum*. J. Gen. Microbiol. *122*: 27–32.

Nwankwo, D.O., R.E. Maunus and S. Xu. 1997. Cloning and expression of *Aat*II restriction-modification system in *Escherichia coli*. Gene *185*: 105–109.

Nyberg, P.A., S.E. Knapp and R.E. Milleman. 1967. "Salmon poisoning" disease. IV. Transmission of the disease to dogs by *Nanophyetus salmincola* eggs. J. Parasitol. *53*: 694–699.

Nyberg, U., H. Aspegren, B. Andersson, J.L. Jansen and I.S. Villadsen. 1992. Full-scale application of nitrogen removal with methanol as carbon source. Water Sci. Technol. *26*: 1077–1086.

Nyby, M.D., D.A. Gregory, D.A. Kuhn and J. Pangborn. 1977. Incidence of *Simonsiella* in the oral cavity of dogs. J. Clin. Microbiol. *6*: 87–88.

Nyindo, M.B.A., M. Ristic, D.L. Huxsoll and A.R. Smith. 1971. Tropical canine pancytopenia: *in vitro* cultivation of the causative agent *Ehrlichia canis*. Am. J. Vet. Res. *32*: 1651–1658.

Oakley, C.J. and J.C. Murrell. 1993. *nifH* genes in the obligate methane oxidizing bacteria. FEMS Microbiol. Lett. *49*: 53–57.

Oaks, S.C., Jr. and J.V. Osterman. 1979. The influence of temperature and pH on the growth of *Rickettsia conorii* in irradiated mammalian cells. Acta Virol. *23*: 67–72.

Obijeski, J.F., E.L. Palmer and T. Tzianabos. 1974. Proteins of purified rickettsiae. Microbios. *11*: 61–76.

O'Callaghan, D., C. Cazevieille, A. Allardet-Servent, M.L. Boschiroli, G. Bourg, V. Foulongne, P. Frutos, Y. Kulakov and M. Ramuz. 1999. A homologue of the *Agrobacterium tumefaciens* VirB and *Bordetella pertussis* Ptl type IV secretion systems is essential for intracellular survival of *Brucella suis*. Mol. Microbiol. *33*: 1210–1220.

Occhialini, A., V. Stonnet, J. Hua, C. Camou, J.L. Guesdon and F. Megraud. 1996. Identification of strains of *Campylobacter jejuni* and *Campylobacter coli* by PCR and correlation with phenotypic characterisitics.

In Newell, Ketley and Feldman (Editors), Campylobacters, Helicobacters, and Related Organisms, Plenum Press, New York. pp. 217–219.

O'Connor, K.A. and D.R. Zusman. 1983. Coliphage P1-mediated transduction of cloned DNA from *Escherichia coli* to *Myxococcus xanthus*—Use for complementation and recombinational analyses. J. Bacteriol. *155*: 317–329.

Odintsova, E.V., H.W. Jannasch, J.A. Mamone and T.A. Langworthy. 1996. *Thermothrix azorensis* sp. nov., an obligately chemolithoautotrophic, sulfur-oxidizing, thermophilic bacterium. Int. J. Syst. Bacteriol. *46*: 422–428.

Odom, J.M. and R. Singleton. 1993. The Sulfate-Reducing bacteria: Contemporary Perspectives, Springer-Verlag, New York.

Odugbemi, T., C. Nwofor and K.T. Joiner. 1988. Isolation of an unidentified pink-pigmented bacterium in a clinical specimen. J. Clin. Microbiol. *26*: 1072–1073.

Ogata, H., S. Audic, P. Renesto-Audiffren, P.E. Fournier, V. Barbe, D. Samson, V. Roux, P. Cossart, J. Weissenbach, J.M. Claverie and D. Raoult. 2001. Mechanisms of evolution in *Rickettsia conorii* and *R. prowazekii*. Science *293*: 2093–2098.

Ogata, N. 1931. Aetiologie der Tsutsugamuchi-Krankheit: *Rickettsia tsutsugamushi*. Zentralbl. Bakteriol. Parasitenkd. Infektionskr. Hyg. Abt. I Orig. *122*: 249–253.

Ogg, J.E. 1962. Studies on the coccoid form of ovine *Vibrio fetus*. I. Cultural and serologic investigations. Am. J. Vet. Res. *23*: 354–358.

Ogram, A.V., Y. Duana, S.L. Trabuea, X. Feng, H. Castroa and L. Ou. 2000. Carbofuran degradation mediated by three related plasmid systems. FEMS Microbiol. Ecol. *32*: 197–203.

Oh, J.K. and I. Suzuki. 1977. Isolation and characterization of a membrane-associated thiosulfate-oxidizing system of *Thiobacillus novellus*. J. Gen. Microbiol. *99*: 397–412.

Ohad, I., D. Danon and S. Hestrin. 1962. Synthesis of cellulose by *Acetobacter xylinum*. V. Ultrastructure of polymer. J. Cell. Biol. *12*: 31–46.

Ohara, M., Y. Katayama, M. Tsuzaki, S. Nakamoto and H. Kuraishi. 1990. *Paracoccus kocurii* sp. nov., a tetramethylammonium-assimilating bacterium. Int. J. Syst. Bacteriol. *40*: 292–293.

Ohara, S., T. Sato and M. Homma. 1974. Serological studies on *Francisella tularensis*, *Francisella novicida*, *Yersinia philomiragia*, and *Brucella abortus*. Int. J. Syst. Bacteriol. *24*: 191–196.

Ohashi, N., H. Nashimoto, H. Ikeda and A. Tamura. 1992. Diversity of immunodominant 56-kDa type-specific antigen (TSA) of *Rickettsia tsutsugamushi*. Sequence and comparative analyses of the genes encoding TSA homologues from four antigenic variants. J. Biol. Chem. *267*: 12728–12735.

Ohashi, N., A. Tamura and T. Suto. 1988. Immunoblotting analysis of anti-rickettsial antibodies produced in patients of Tsutsugamushi disease. Microbiol. Immunol. *32*: 1085–1092.

Ohashi, N., A. Unver, N. Zhi and Y. Rikihisa. 1998a. Cloning and characterization of multigenes encoding the immunodominant 30-kilodalton major outer membrane proteins of *Ehrlichia canis* and application of the recombinant protein for serodiagnosis. J. Clin. Microbiol. *36*: 2671–2680.

Ohashi, N., N. Zhi, Y. Zhang and Y. Rikihisa. 1998b. Immunodominant major outer membrane proteins of *Ehrlichia chaffeensis* are encoded by a polymorphic multigene family. Infect. Immun. *66*: 132–139.

Ohmori, S., H. Masai, K. Arima and T. Beppu. 1980. Isolation and identification of acetic acid bacteria for submerged acetic acid fermentation at high temperature. Agric. Biol. Chem. *44*: 2901–2906.

Ohta, H. 2000. Growth characteristics of *Agromonas oligotropha* on ferulic acid. Microbes and Environments. *15*: 133–142.

Ohta, H. and T. Hattori. 1980. Bacteria sensitive to nutrient broth medium in terrestrial environments. Soil Sci. Plant Nutr. *26*: 99–107.

Ohta, H. and T. Hattori. 1983. *Agromonas oligotrophica* gen. nov., sp. nov., a nitrogen-fixing oligotrophic bacterium. Antonie Leeuwenhoek J. Microbiol. *49*: 429–446.

Ohta, H. and T. Hattori. 1985. *In* Validation of the publication of new names and new combinations previously effectively published outside the IJSB. List No. 17. Int. J. Syst. Bacteriol. *35*: 223–225.

Ohta, H. and S. Taniguchi. 1988a. Growth-characteristics of the soil oligotrophic bacterium - *Agromonas oligotrophica* JCM-1494 on diluted nutrient broth. J. Gen. Appl. Microbiol. *34*: 349–353.

Ohta, H. and S. Taniguchi. 1988b. Respiratory characteristics of two oligotrophic bacteria - *Agromonas oligotrophica* JCM-1494 and *Aeromonas hydrophila* 315. J. Gen. Appl. Microbiol. *34*: 355–365.

Ohtsubo, Y., K. Miyauchi, K. Kanda, T. Hatta, H. Kiyohara, T. Senda, Y. Nagata, Y. Mitsui and M. Takagi. 1999. PcpA, which is involved in the degradation of pentachlorophenol in *Sphingomonas chlorophenolica* ATCC39723, is a novel type of ring-cleavage dioxygenase. FEBS Lett. *459*: 395–398.

Ohya, T., M. Kubo and H. Watase. 1988. New selective media for the isolation of *Campylobacter mucosalis* and *Campylobacter hyointestinalis*. Jpn. J. Vet. Sci. *50*: 1103–1106.

Oikawa, T., T. Ohtori and M. Ameyama. 1995. Production of cellulose from D-mannitol by *Acetobacter xylinum* KU-1. Biosci. Biotechnol. Biochem. *59*: 331–332.

Ojcius, D.M., C. Thibon, C. Mounier and A. Dautry-Varsat. 1995. pH and calcium dependence of hemolysis due to *Rickettsia prowazekii*: Comparison with phospholipase activity. Infect. Immun. *63*: 3069–3072.

Okabe, N. 1934. Bacterial diseases of plants occurring in Formosa. IV. J. Soc. Trop. Agric. Taiwan *6*: 54–63.

Okada, G. and E.J. Hehre. 1974. New studies on amylosucrase, a bacterial alpha-D-glucosylase that directly converts sucrose to a glycogen-like alpha-glucan. J. Biol. Chem. *249*: 126–135.

Okamura, K., K. Takamiya and M. Nishimura. 1985. Photosynthetic electron transfer system is inoperative in anaerobic cells of *Erythrobacter* species strain OCh114. Arch. Microbiol. *142*: 12–17.

Okon, Y., S.L. Albrecht and R.H. Burris. 1976. Carbon and ammonia metabolism of *Spirillum lipoferum*. J. Bacteriol. *128*: 592–597.

Okon, Y., L. Cakmakci, I. Nur and I. Chet. 1980. Aerotaxis and chemotaxis of *Azospirillum brasilense*: a note. Microb. Ecol. *6*: 277–280.

Okon, Y. and R. Itzigsohn. 1992. Poly-β-hydroxybutyrate metabolism in *Azospirillum brasilense* and the ecological role of PHB in the rhizosphere. FEMS Microbiol. Lett. *103*: 131–139.

Okon, Y. and C.A. Labandera-Gonzalez. 1994. Agronomic applications of *Azospirillum*: An evaluation of 20 years worldwide field inoculation. Soil Biol. Biochem. *26*: 1591–1601.

Olafson, R.W., P.J. McCarthy, A.R. Bhatti, J.S. Dooley, J.E. Heckels and T.J. Trust. 1985. Structural and antigenic analysis of meningococcal piliation. Infect. Immun. *48*: 336–342.

Oldenhuis, R., R.L.J.M. Vink, D.B. Janssen and B. Witholt. 1989. Degradation of chlorinated aliphatic hydrocarbons by *Methylosinus trichosporium* OB3b expressing soluble methane monooxygenase. Appl. Environ. Microbiol. *55*: 2819–2826.

Olivares, F.L. 1997. Taxonomia, ecologia e mecanismos envolvidos na infecção e colonização de plantas de cana-de-açúcar (*Saccharum* sp., hibrido) por bactérias diazotróficas endofíticas do genero *Herbaspirillum*, Tese de Doutorado, UFRRJ . 358 pp.

Olivares, F.L., V.L.D. Baldani, V.M. Reis, J.I. Baldani and J. Dobereiner. 1996. Occurrence of the endophytic diazotrophs *Herbaspirillum* spp in roots, stems, and leaves, predominantly of Gramineae. Biol. Fertil. Soils *21*: 197–200.

Olivares, F.L., E.K. James, J.I. Baldani and J. Dobereiner. 1997. Infection of mottled stripe disease-susceptible and resistant sugar cane varieties by the endophytic diazotroph *Herbaspirillum*. New Phtyologist. *135*: 723–737.

Oliveira, R.G.B. and A. Drozdowicz. 1981. Bacteriocins in the genus *Azospirillum*. Rev. Microbiol. *12*: 42–47.

Oliveira, R.G.B. and A. Drozdowicz. 1988. Are *Azospirillum* bacteriocins produced and active in soil? *In* Klingmuller (Editor), *Azospirillum* IV: Genetics, Physiology, Ecology, Springer-Verlag, Berlin. pp. 101–108.

Oliveira, S.C. and G.A. Splitter. 1996. Immunization of mice with recombinant L7/L12 ribosomal protein confers protection against *Brucella abortus* infection. Vaccine *14*: 959–962.

Olivieri, R., M. Bugnoli, D. Armellini, S. Bianciardi, R. Rappuoli, P.F. Bayeli, L. Abate, E. Esposito, L. De Gregorio, J. Aziz, C. Basagni and

N. Figura. 1993. Growth of *Helicobacter pylori* in media containing cyclodextrins. J. Clin. Microbiol. *31*: 160–162.

Ollivier, B., P. Caumette, J.L. Garcia and R.A. Mah. 1994. Anaerobic bacteria from hypersaline environments. Microbiol. Rev. *58*: 27–38.

Ollivier, B., R. Cord-Ruwisch, E.C. Hatchikian and J.L. Garcia. 1988. Characterization of *Desulfovibrio fructosovorans* sp. nov. Arch. Microbiol. *149*: 447–450.

Ollivier, B., R. Cord-Ruwisch, E.C. Hatchikian and J.L. Garcia. 1990. *In* Validation of the publication of new names and new combinations previously effectively published outside the IJSB. List No. 32. Int. J. Syst. Bacteriol. *40*: 105–106.

Ollivier, B., C.E. Hatchikian, G. Prensier, J. Guezennec and J.-L. Garcia. 1991. *Desulfohalobium retbaense* gen. nov., sp. nov., a halophilic sulfate-reducing bacterium from sediments of a hypersaline lake in Senegal. Int. J. Syst. Bacteriol. *41*: 74–81.

Olopoenia, L.A., V. Mody and M. Reynolds. 1994. *Eikenella corrodens* endocarditis in an intravenous drug user: case report and literature review. J. Nat. Med. Assoc. *86*: 313–315.

Olsen, G.J., H. Matsuda, R. Hagstrom and R. Overbeek. 1994. FastDNAml: A tool for construction of phylogenetic trees of DNA sequences using maximum likelihood. Comput. Appl. Biosci. *10*: 41–48.

Olubayi, O., R. Caudales, A. Atkinson and C.A. Neyra. 1998. Differences in chemical composition between nonflocculated and flocculated *Azospirillum brasilense* Cd. Can. J. Microbiol. *44*: 386–390.

Olyhoek, A.J.M., J. Sarkari, M. Bopp, G. Morelli and M. Achtman. 1991. Cloning and expression in *Escherichia coli* of *opc*, the gene for an unusual class-5 outer-membrane protein from *Neisseria meningitidis* (meningococci surface antigen). Microb. Pathog. *11*: 249–257.

On, S.L.W. 1994. Confirmation of human *Campylobacter concisus* isolates misidentified as *Campylobacter mucosalis* and suggestions for improved differentiation between the two species. J. Clin. Microbiol. *32*: 2305–2306.

On, S.L.W. 1996. Identification methods for campylobacters, helicobacters, and related organisms. Clin. Microbiol. Rev. *9*: 405–422.

On, S.L.W., H.I. Atabay, J.E.L. Corry, C.S. Harrington and P. Vandamme. 1998. Emended description of *Campylobacter sputorum* and revision of its infrasubspecific (biovar) divisions, including *C. sputorum* biovar paraureolyticus bv. nov., a urease-producing variant from cattle and humans. Int. J. Syst. Bacteriol. *48*: 195–206.

On, S.L.W., B. Bloch, B. Holmes, B. Hoste and P. Vandamme. 1995. *Campylobacter hyointestinalis* subsp. *lawsonii* subsp. nov., isolated from the porcine stomach, and an emended description of *Campylobacter hyointestinalis*. Int. J. Syst. Bacteriol. *45*: 767–774.

On, S.L.W., M. Costas and B. Holmes. 1993. Identification and intraspecific heterogeneity of *Campylobacter hyointestinalis* based on numerical analysis of electrophoretic protein profiles. Syst. Appl. Microbiol. *16*: 37–46.

On, S.L.W., M. Costas and B. Holmes. 1994. Classification and identification of *Campylobacter sputorum* using numerical analyses of phenotypic tests and of one-dimensional electrophoretic protein profiles. Syst. Appl. Microbiol. *17*: 543–553.

On, S.L.W. and B. Holmes. 1991. Reproducibility of tolerance tests that are useful in the identification of campylobacteria. J. Clin. Microbiol. *29*: 1785–1788.

On, S.L.W. and B. Holmes. 1992. Assessment of enzyme detection tests useful in identification of campylobacteria. J. Clin. Microbiol. *30*: 746–749.

On, S.L.W. and B. Holmes. 1995. Classification and identification of campylobacters, helicobacters and allied taxa by numerical analysis of phenotypic characters. Syst. Appl. Microbiol. *18*: 374–390.

On, S.L.W., B. Holmes and M.J. Sackin. 1996. A probability matrix for the identification of campylobacters, helicobacters and allied taxa. J. Appl. Bacteriol. *81*: 425–432.

On, S.L.W. and P. Vandamme. 1997. Identification and epidemiological typing of *Campylobacter hyointestinalis* subspecies by phenotypic and genotypic methods and description of novel subgroups. Syst. Appl. Microbiol. *20*: 238–247.

O'Neill, S.L., R. Giordano, A.M. Colbert, T.L. Karr and H.M. Robertson.

1992. 16S rRNA phylogenetic analysis of the bacterial endosymbionts associated with cytoplasmic incompatibility in insects. Proc. Natl. Acad. Sci. U.S.A. *89*: 2699–2702.

O'Neill, S.L., M.M. Pettigrew, S.P. Sinkins, H.R. Braig, T.G. Andreadis and R.B. Tesh. 1997. *In vitro* cultivation of *Wolbachia pipientis* in an *Aedes albopictus* cell line. Insect Mol. Biol. *6*: 33–39.

Onyeocha, I., C. Vielle, W. Zimmer, B.E. Baca, M. Flores, R. Palacios and C. Elmerich. 1990. Physical map and properties of a 90-MDa plasmid of *Azospirillum brasilense* Sp7. Plasmid *23*: 169–182.

Ophel, K. and A. Kerr. 1990. *Agrobacterium vitis* sp. nov. for strains of *Agrobacterium* biovar 3 from grapevines. Int. J. Syst. Bacteriol. *40*: 236–241.

Opitz, R. 1977. Die Verwertung organischer Substrate and die Regulation der Abbauwege bei *Corynebacterium autotrophicum* Stamm 19/-/x, Dissertation, University of Göttingen.

Opitz, R. and H.G. Schlegel. 1978. Allosteric inhibition by phosphoenolpyruvate of glucose-6-phosphate dehydrogenase from bacteria and its taxonomic importance. Biochem. Syst. Ecol. *6*: 149–155.

Oppenberg, H.R.S., G. Chen, T. Leisinger and A.M. Cook. 1995. Regulation of the degradative pathways from 4-toluenesulphonate and 4-toluenecarboxylate to protocatechuate in *Comamonas testosteroni* T-2. Microbiology *141*: 1891–1899.

Oppermann, U.C., I. Belai and E. Maser. 1996. Antibiotic resistance and enhanced insecticide catabolism as consequence of steroid induction in the gram-negative bacterium *Comamonas testosteroni*. J. Steroid Biochem. Mol. Biol. *58*: 217–223.

Oppermann, U.C., C. Filling, K.D. Berndt, B. Persson, J. Benach, R. Ladenstein and H. Jornvall. 1997. Active site directed mutagenesis of 3ß/17ß-hydroxysteroid dehydrogenase establishes differential effects on short-chain dehydrogenase/reductase reactions. Biochemistry *36*: 34–40.

Oppermann, U.C. and E. Maser. 1996. Characterization of a 3-α-hydroxysteroid dehydrogenase/carbonyl reductase from the gram-negative bacterium *Comamonas testosteroni*. Eur. J. Biochem. *241*: 744–749.

Ordal, E.J. and R.R. Rucker. 1944. Pathogenic myxobacteria. Proc. Soc. Exp. Biol. Med. *56*: 15–18.

Oremland, R.S., J.S. Blum, C.W. Culbertson, P.T. Visscher, L.G. Miller, P. Dowdle and F.E. Strohmaier. 1994. Isolation, growth, and metabolism of an obligately anaerobic, selenate-respiring bacterium, strain SES-3. Appl. Environ. Microbiol. *60*: 3011–3019.

Orla-Jensen, S. 1909. Die Hauptlinien des natürlichen Bakterien-systems. Zentralbl. Bakteriol. Parasitenkd. Infektionskr. Hyg. Abt. II *22*: 97–98 305–346.

Orlowski, M., P. Martin, D. White and M.C. Wong. 1972. Changes in activity of glyoxylate cycle enzymes during myxospore development in *Myxococcus xanthus*. J. Bacteriol. *111*: 784–790.

Orlowski, M. and D. White. 1974. Intracellular proteolytic activity in developing myxospores of *Myxococcus xanthus*. Arch. Microbiol. *97*: 347–357.

Ormsbee, R., M. Peacock, R. Gerloff, G. Tallent and D. Wike. 1978. Limits of rickettsial infectivity. Infect. Immun. *19*: 239–245.

Orndorff, P.E. and M. Dworkin. 1980. Separation and properties of the cytoplasmic and outer membranes of vegetative cells of *Myxococcus xanthus*. J. Bacteriol. *141*: 914–927.

Orndorff, P., E. Stellwag, T. Starich, M. Dworkin and J. Zissler. 1983. Genetic and physical characterization of lysogeny by bacteriophage-Mx8 in *Myxococcus xanthus*. J. Bacteriol. *154*: 772–779.

Ørskov, J. 1928. Beschreibung eines neuen Mikroben, *Microcyclus aquaticus*, mit eigentumlicher Morphologie. Zentralbl. Bakteriol. Parasitenk. Infektionskr. Hyg. Abt. I Orig. *107*: 180–184.

Ørskov, J. 1930. Untersuchungen über einen in Mundhöhle und oberen Luftwegen häufig vorkommenden, zur Sarcina gruppe gehörigen Microben, der eigentümlische Verhältnisse aufweist. Acta Pathol. Microbiol. Immunol. Scand. Suppl. III : 519–541.

Ørskov, J. 1953. Microcyclus. Riassunti d. communicazioni. Vl. Cong. Int. Microbiol. Roma. *1*: 24–25.

Orso, S., M. Gouy, E. Navarro and P. Normand. 1994. Molecular phylogenetic analysis of *Nitrobacter* spp. Int. J. Syst. Bacteriol. *44*: 83–86.

Ossipov, D.V. and S.A. Podlipaev. 1977. Electron microscope examination of early stages of infection of *Paramecium caudatum* by bacterial symbionts of the macronucleus (iota-bacteria). Acta Protozool. *16*: 289–308.

Ossipov, D.V., I.I. Skoblo, O.N. Borchsenius and N.A. Lebedeva. 1993. Interactions between *Paramecium bursaria* (Protozoa, Ciliophora, Hymenostomatida) and their nuclear symbionts .1. Phenomenon of symbiogenic lysis of the bacterium *Holospora acuminata*. Eur. J. Protistol. *29*: 61–71.

Ossipov, D.V., Skoblo, II, O.N. Borkhsenius, M.S. Rautian and S.A. Podlipaev. 1980. *Holospora acuminata* - a new species of symbiotic bacterium from the micronucleus of the ciliate *Paramecium bursaria* Focke. Tsitologiya *22*: 922–929.

Ostle, A.G. and J.G. Holt. 1982. Nile blue A as a fluorescent stain for poly-β-hydroxybutyrate. Appl. Environ. Microbiol. *44*: 238–241.

Otte, S., N.G. Grobben, L.A. Robertson, M.S. Jetten and J.G. Kuenen. 1996. Nitrous oxide production by *Alcaligenes faecalis* under transient and dynamic aerobic and anaerobic conditions. Appl. Environ. Microbiol. *62*: 2421–2426.

Otten, L., P. De Ruffray, E.A. Momol, M.T. Momol and T.J. Burr. 1996. Phylogenetic relationships between *Agrobacterium vitis* isolates and their Ti plasmids. Mol. Plant-Microbe Interact. *9*: 782–786.

Ottow, J.C. 1975. Ecology, physiology, and genetics of fimbriae and pili. Annu. Rev. Microbiol. *29*: 79–108.

Ouahrani, S., S. Michaux, J. Sri Widada, G. Bourg, R. Tournebize, M. Ramuz and J.P. Liautard. 1993. Identification and sequence analysis of IS6501, an insertion sequence in *Brucella* spp.: relationship between genomic structure and the number of IS6501 copies. J. Gen. Microbiol. *139*: 3265–3273.

Ouattara, A.S., B.K.C. Patel, J.L. Cayol, N. Cuzin, A.S. Traore and J.L. Garcia. 1999. Isolation and characterization of *Desulfovibrio burkinensis* sp. nov. from an African ricefield, and phylogeny of *Desulfovibrio alcoholivorans*. Int. J. Syst. Bacteriol. *49*: 639–643.

Ouchi, K., M. Abe, M. Karita, T. Oguri, J. Igari and T. Nakazawa. 1995. Analysis of strains of *Burkholderia (Pseudomonas) cepacia* isolated in a nosocomial outbreak by biochemical and genomic typing. J. Clin. Microbiol. *33*: 2353–2357.

Oude Elferink, S.J., W.M. Akkermans-van Vliet, J.J. Bogte and A.J. Stams. 1999. *Desulfobacca acetoxidans* gen. nov., sp. nov., a novel acetate-degrading sulfate reducer isolated from sulfidogenic granular sludge. Int. J. Syst. Bacteriol. *49*: 345–350.

Oude Elferink, S.J.W.H., R.N. Maas, H.J.M. Harmsen and A.J.M. Stams. 1995. *Desulforhabdus amnigenus* gen. nov. sp. nov., a sulfate reducer isolated from anaerobic granular sludge. Arch. Microbiol. *164*: 119–124.

Oude Elferink, S.J.W.H., R.N. Maas, H.J.M. Harmsen and A.J.M. Stams. 1997. *In* Validation of the publication of new names and new combinations previously effectively published outside the IJSB. List No. 63. Int. J. Syst. Bacteriol. *47*: 1274.

Oude Elferink, S.H.J.W., A. Visser, L.W. Hulshoff Pol and A.J.M. Stams. 1994. Sulfate reduction in methanogenic bioreactors. FEMS Microbiol. Rev. *15*: 119–126.

Ourisson, G., M. Rohmer and K. Poralla. 1987. Prokaryotic hopanoids and other polyterpenoid sterol surrogates. Annu. Rev. Microbiol. *41*: 301–333.

Owen, R.J., A. Beck and P. Borman. 1985. Restriction endonuclease digest patterns of chromosomal DNA from nitrate-negative *Campylobacter jejuni*-like organisms. Eur. J. Epidemiol. *1*: 281–287.

Owen, R.J., M. Costas, L.L. Sloss and F.J. Bolton. 1988. Numerical analysis of electrophoretic protein patterns of *Campylobacter laridis* and allied thermophilic campylobacters from the natural environment. J. Appl. Bacteriol. *65*: 69–78.

Owen, R.J. and M. Desai. 1990. Preformed enzyme profiling of *Helicobacter pylori* and *Helicobacter mustelae* from human and animal sources. Lett. Appl. Microbiol. *2*: 103–105.

Owen, R.J., M. Desai and S. Garcia. 1993. Molecular typing of thermotolerant species of *Campylobacter* with ribosomal RNA gene patterns. Res. Microbiol. *144*: 709–720.

Owen, R.J. and J. Hernandez. 1990. Occurrence of plasmids in *"Campylobacter upsaliensis"* (catalase negative or weak group) from geographically diverse patients with gastroenteritis or bacteraemia. Eur. J. Epidemiol. *6*: 111–117.

Owen, R.J. and P.J. Jackman. 1982. The similarities between *Pseudomonas paucimobilis* and allied bacteria derived from analysis of deoxyribonucleic acids and electrophoretic protein patterns. J. Gen. Microbiol. *128*: 2945–2954.

Oyaizu, H. and K. Komagata. 1983. Grouping of *Pseudomonas* species on the basis of cellular fatty acid composition and the quinone system with special reference to the existence of 3-hydroxy fatty acids. J. Gen. Appl. Microbiol. *29*: 17–40.

Oyaizu, H. and C.R. Woese. 1985. Phylogenetic relationships among the sulfate respiring bacteria, myxobacteria and purple bacteria. Syst. Appl. Microbiol. *6*: 257–263.

Oyaizu-Masuchi, Y. and K. Komagata. 1988. Isolation of free-living nitrogen-fixing bacteria from the rhizosphere of rice. J. Gen. Appl. Microbiol. *34*: 127–164.

Oyarzabal, O.A., I.V. Wesley, J.M. Barbaree, L.H. Lauerman and D.E. Conner. 1997. Specific detection of *Campylobcter lari* by PCR. J. Microbiol. Methods *29*: 97–102.

Ozawa, H., H. Tanaka, Y. Ichinose, T. Shiraishi and T. Yamada. 2001. Bacteriophage P4282, a parasite of *Ralstonia solanacearum*, encodes a bacteriolytic protein important for lytic infection of its host. Mol. Genet. Genom. *265*: 95–101.

Pacios-Bras, C., Y.E.M. van der Burgt, A.M. Deedler, P. Vinuesa, D. Werner and H.P. Spaink. 2002. Novel lipochitin oligosaccharide structures produced by *Rhizobium elti* KIM5s. Carbobydr. Res. *337*: 1191–1200.

Packer, L. and W. Vishniac. 1955. Chemosynthetic fixation of carbon dioxide and characteristics of hydrogenase in resting cell suspensions of *Hydrogenomonas ruhlandii* nov. spec. J. Bacteriol. *70*: 216–223.

Padden, A.N., D.P. Kelly and A.P. Wood. 1998. Chemolithoautotrophy and mixotropy in the thiophene-2-carboxylic acid-utilizing *Xanthobacter tagetidis*. Arch. Microbiol. *169*: 249–256.

Padden, A.N., F.A. Rainey, D.P. Kelly and A.P. Wood. 1997. *Xanthobacter tagetidis* sp. nov., an organism associated with *Tagetes* species and able to grow on substituted thiophenes. Int. J. Syst. Bacteriol. *47*: 394–401.

Padgett, P.J. 1981. Factors relating to the growth and aerotolerance of *Spirillum volutans*, Thesis, Virginia Polytechnic Institute and State University, Blacksburg, VA.

Padgett, P.J., W.H. Cover and N.R. Krieg. 1982. The microaerophile *Spirillum volutans*: cultivation on complex liquid and solid media. Appl. Environ. Microbiol. *43*: 469–477.

Padgett, P.J., N.W. Friedman and N.R. Krieg. 1983. Straight mutants of *Spirillum volutans* can swim. J. Bacteriol. *153*: 1543–1544.

Padilla, L., V. Matus, P. Zenteno and B. Gonzalez. 2000. Degradation of 2,4,6-trichlorophenol via chlorohydroxyquinol in *Ralstonia eutropha* JMP134 and JMP222. J. Basic Microbiol. *40*: 243–249.

Padmalayam, I., B. Anderson, M. Kron, T. Kelly and B. Baumstark. 1997. The 75-kilodalton antigen of *Bartonella bacilliformis* is a structural homolog of the cell division protein FtsZ. J. Bacteriol. *179*: 4545–4552.

Pagan, J.D., J.J. Child, W.R. Scowcroft and A.H. Gibson. 1975. Nitrogen fixation by *Rhizobium* cultured in a defined medium. Nature *256*: 406–407.

Pagotto, F., A.T. Aman, L.K. Ng, K.H. Yeung, M. Brett and J.A. Dillon. 2000. Sequence analysis of the family of penicillinase-producing plasmids of *Neisseria gonorrhoeae*. Plasmid *43*: 24–34.

Pagotto, F. and J.A. Dillon. 2001. Multiple origins and replication proteins influence biological properties of β-lactamase-producing plasmids from *Neisseria gonorrhoeae*. J. Bacteriol. *183*: 5472–5481.

Paitan, Y., G. Alon, E. Orr, E.Z. Ron and E. Rosenberg. 1999. The first gene in the biosynthesis of the polyketide antibiotic TA of *Myxococcus xanthus* codes for a unique PKS module coupled to a peptide synthetase. J. Mol. Biol. *286*: 465–474.

Palacios, R., P. Boistard, G. Dvila, M. Fonstein, M. Gttfert, X. Perret, C. Ronson and B. Sobral. 1998. Genome structure in nitrogen-fixing organisms. *In* Elmerich, Kondoroski and Newton (Editors), Biological

Nitrogen Fixation for the 21st Century, Kluwer Academic Publishers, Dordrecht. pp. 541–547.

Palleroni, N.J. 1984. Genus I *Pseudomonas*. *In* Krieg and Holt (Editors), Bergey's Manual of Systematic Bacteriology, 1st Ed., Vol. 1, The Williams & Wilkins Co., Baltimore. pp. 141–199.

Palleroni, N.J. and M. Doudoroff. 1972. Some properties and taxonomic subdivisions of the genus *Pseudomonas*. Annu. Rev. Phytopathol. *10*: 73–100.

Palleroni, N.J. and B. Holmes. 1981. *Pseudomonas cepacia* sp. nov., nom. rev. Int. J. Syst. Bacteriol. *31*: 479–481.

Palleroni, N.J., R. Kunisawa, R. Contopoulou and M. Doudoroff. 1973. Nucleic acid homologies in the genus *Pseudomonas*. Int. J. Syst. Bacteriol. *23*: 333–339.

Palleroni, N.J. and A.V. Palleroni. 1978. *Alcaligenes latus*, a new species of hydrogen-utilizing bacteria. Int. J. Syst. Bacteriol. *28*: 416–424.

Palmer, G.H., J.R. Abbott, D.M. French and T.F. McElwain. 1998. Persistence of *Anaplasma ovis* infection and conservation of the *msp*-2 and *msp*-3 multigene families within the genus *Anaplasma*. Infect. Immun. *66*: 6035–6039.

Palmer, G.H., A.F. Barbet, A.J. Musoke, J.M. Katende, F. Rurangirwa, V. Shkap, E. Pipano, W.C. Davis and T.C. McGuire. 1988a. Recognition of conserved surface protein epitopes on *Anaplasma centrale* and *Anaplasma marginale* isolates from Israel, Kenya and the United States. Int. J. Parasitol. *18*: 33–38.

Palmer, G.H., S.M. Oberle, A.F. Barbet, W.L. Goff, W.C. Davis and T.C. McGuire. 1988b. Immunization of cattle with a 36-kilodalton surface protein induces protection against homologous and heterologous *Anaplasma marginale* challenge. Infect. Immun. *56*: 1526–1531.

Palmer, H.M., H. Mallinson, R.L. Wood and A.J. Herring. 2003. Evaluation of the specificities of five DNA amplification methods for the detection of *Neisseria gonorrhoeae*. J. Clin. Microbiol. *41*: 835–837.

Palmer, J.E., C.E. Benson and R.H. Whitlock. 1990. Resistance to development of equine ehrlichial colitis in experimentally inoculated horses and ponies. Am. J. Vet. Res. *51*: 763–765.

Palumbo, J.D., D.A. Phillips and C.I. Kado. 1998. Characterization of a new *Agrobacterium tumefaciens* strain from alfalfa (*Medicago sativa* L.). Arch. Microbiol. *169*: 381–386.

Panagopoulos, C.G. 1969. The disease "Tsilik Marasi" of grapevine, its description and identification of tbe causal agent (*Xanthomonas ampelina* sp. nov.). Ann. Inst. Phytopathol. Benaki. *9*: 59–81.

Panagopoulos, C.G. and P.G. Psallidas. 1973. Characteristics of Greek isolates of *Agrobacterium tumefaciens* (E. F. Smith and Townsend) Conn. J. Appl. Bacteriol. *36*: 233–240.

Panagopoulos, C.G., P.G. Psallidas and A.S. Alivizatos. 1978. Studies on biotype 3 of *Agrobacterium radiobacter* var. *tumefaciens*. Proc. 4th Int. Conf. Plant Path. Bact, Angers. 221–228.

Pang, H. and H.H. Winkler. 1994. Analysis of the peptidoglycan of *Rickettsia prowazekii*. J. Bacteriol. *176*: 923–926.

Pangborn, J., D.A. Kuhn and J.R. Woods. 1977. Dorsal-ventral differentiation in *Simonsiella* and other aspects of its morphology and ultrastructure. Arch. Microbiol. *113*: 197–204.

Pangborn, J. and M.P. Starr. 1966. Ultrastructure of *Lampropedia hyalina*. J. Bacteriol. *91*: 2025–2030.

Pankova, L.M., Y.E. Shvinka, M.E. Beker and E.E. Slava. 1985. Effect of aeration on *Zymomonas mobilis* metabolism. Microbiology *54*: 120–124.

Paoletti, L.C. and R.P. Blakemore. 1986. Hydroxamate production by *Aquaspirillum magnetotacticum*. J. Bacteriol. *167*: 153–163.

Papen, H., R. von Berg, I. Hinkel, B. Thoene and H. Rennenberg. 1989. Heterotrophic nitrification by *Alcaligenes faecalis*: NO_2^-, NO_3^-, N_2O, and NO production in exponentially growing cultures. Appl. Environ. Microbiol. *55*: 2068–2072.

Papenfuss, G.F. 1940. Notes on South African marine algae. Bot Not. *93*: 200–226.

Parales, R.E., T.A. Ontl and D.T. Gibson. 1997. Cloning and sequence analysis of a catechol 2,3-dioxygenase gene from the nitrobenzene-degrading strain *Comamonas* sp. JS765. J. Ind. Microbiol. Biotechnol. *19*: 385–391.

Parish, J.H. 1975. Transfer of drug resistance to *Myxococcus* from bacteria carrying drug resistance factors. J. Gen. Microbiol. *87*: 198–210.

Park, C.E. and L.R. Berger. 1967. Fatty acids of extractable and bound lipids of *Rhodomicrobium vannielii*. J. Bacteriol. *93*: 230–236.

Park, J. and Y. Rikihisa. 1991. Inhibition of *Ehrlichia risticii* infection in murine peritoneal macrophages by gamma interferon, a calcium ionophore, and concanavalin A. Infect. Immun. *59*: 3418–3423.

Park, J. and Y. Rikihisa. 1992. L-arginine-dependent killing of intracellular *Ehrlichia risticii* by macrophages treated with gamma interferon. Infect. Immun. *60*: 3504–3508.

Parke, J.L., R.E. Rand, A.E. Joy and E.B. King. 1991. Biological control of *Pythium* damping off and *Aphanomyces* root rot of peas by application of *Pseudomonas cepacia* or *P. fluorescens* to seed. Plant Dis. *75*: 987–992.

Parker, C., W.O. Barnell, J.L. Snoep, L.O. Ingram and T. Conway. 1995. Characterization of the *Zymomonas mobilis* glucose facilitator gene product (glf) in recombinant *Escherichia coli*: examination of transport mechanism, kinetics and the role of glucokinase in glucose transport. Mol. Microbiol. *15*: 795–802.

Parker, R.R., G.M. Kohls, G.W. Cox and G.E. Davis. 1939. Observations on an infectious agent from *Amblyomma maculatum*. Pub. Health Rep. *54*: 1482–1484.

Parker, W.L., M.L. Rathnum, V. Seiner, W.H. Trejo, P.A. Principe and R.B. Sykes. 1984. Cepacin A and Cepacin B, two new antibiotics produced by *Pseudomonas cepacia*. J. Antibiot. *37*: 431–440.

Parkhill, J., M. Achtman, K.D. James, S.D. Bentley, C. Churcher, S.R. Klee, G. Morelli, D. Basham, D. Brown, T. Chillingworth, R.M. Davies, P. Davis, K. Devlin, T. Feltwell, N. Hamlin, S. Holroyd, K. Jagels, S. Leather, S. Moule, K. Mungall, M.A. Quail, M.A. Rajandream, K.M. Rutherford, M. Simmonds, J. Skelton, S. Whitehead, B.G. Spratt and B.G. Barrell. 2000. Complete DNA sequence of a serogroup A strain of *Neisseria meningitidis* Z2491. Nature *404*: 502–506.

Parola, P., L. Beati, M. Cambon and D. Raoult. 1998. First isolation of *Rickettsia helvetica* from *Ixodes ricinus* ticks in France. Eur. J. Clin. Microbiol. Infect. Dis. *17*: 95–100.

Parr, T.R., R.A. Moore, L.V. Moore and R.E.W. Hancock. 1987. Role of porins in intrinsic antibiotic resistance of *Pseudomonas cepacia*. Antimicrob. Agents Chemother. *31*: 121–123.

Parsons, A.B. and P.R. Dugan. 1971. Production of extracellular polysaccharide matrix by *Zoogloea ramigera*. Appl. Microbiol. *21*: 657–661.

Parsons, J.R., D. Sijm, A. Vanlaar and O. Hutzinger. 1988. Biodegradation of chlorinated biphenyls and benzoic acids by a *Pseudomonas* strain. Appl. Microbiol. Biotechnol. *29*: 81–84.

Parsons, N.J., J.R.C. Andrade, P.V. Patel, J.A. Cole and H. Smith. 1989. Sialylation of lipopolysaccharide and loss of absorption of bactericidal antibody during conversion of gonococci to serum resistance by cytidine 5'-monophospho-*n*-acetyl neuraminic acid. Microb. Pathog. *7*: 63–72.

Parton, R. 1996. New perspectives on *Bordetella* pathogenicity. J. Med. Microbiol. *44*: 233–235.

Pashkova, N.I., N.G. Starostina and A.B. Tsiomenko. 1997. A secretory protein involved in the antagonistic interactions between methanotrophic bacteria. Biochemistry *62*: 386–390.

Passador, L. and H.D. McCurdy. 1985. Cyclic nucleotides and development of *Myxococcus xanthus*—analysis of mutants. Curr. Microbiol. *12*: 289–294.

Paster, B.J. and F.E. Dewhirst. 1988. Phylogeny of campylobacters, wolinellas, *Bacteroides gracilis*, and *Bacteroides ureolyticus* by 16S ribosomal ribonucleic acid sequencing. Int. J. Syst. Bacteriol. *38*: 56–62.

Paster, B.J. and R.J. Gibbons. 1986. Chemotactic response to formate by *Campylobacter concisus* and its potential role in gingival colonization. Infect. Immun. *52*: 378–383.

Paster, B.J., A. Lee, J.G. Fox, F.E. Dewhirst, L.A. Tordoff, G.J. Fraser, J.L. O'Rourke, N.S. Taylor and R. Ferrero. 1991. Phylogeny of *Helicobacter felis*, sp. nov., *Helicobacter mustelae*, and related bacteria. Int. J. Syst. Bacteriol. *41*: 31–38.

Pasteur, L. 1864. Mémoire sur la fermentation acétique. Ann. Sci. Ec. norm. sup., Paris *1*: 113–158.

Pate, J.L., S.J. Petzold and T.H. Umbreit. 1979. Two flagellotropic phages and one pilus-specific phage active against *Asticcacaulis biprosthecum*. Virology *94*: 24–37.

Pate, J.L., J.S. Porter and T.L. Jordan. 1973. *Asticcacaulis biprosthecum* sp. nov. life cycle, morphology and cultural characteristics. Antonie Leeuwenhoek *39*: 569–583.

Patel, R.N. and L.N. Orston. 1976. Immunological comparison of enzymes of the beta-ketoadipate pathway. Arch. Microbiol. *110*: 27–36.

Patriquin, D.G. and J. Dobereiner. 1978. Light microscopy observations of tetrazolium-reducing bacteria in the endorhizosphere of maize and other grasses in Brazil. Can. J. Microbiol. *24*: 734–742.

Patt, T.E., G.C. Cole, J. Bland and R.S. Hanson. 1974. Isolation and characterization of bacteria that grow on methane and organic compounds as sole sources of carbon and energy. J. Bacteriol. *120*: 955–964.

Patt, T.E., G.E. Cole and R.S. Hanson. 1976. *Methylobacterium*, a new genus of facultatively methylotrophic bacteria. Int. J. Syst. Bacteriol. *26*: 226–229.

Patterson, M.M., M.D. Schrenzel, Y. Feng, S. Xu, F.E. Dewhirst, B.J. Paster, S.A. Thibodeau, J. Versalovic and J.G. Fox. 2000. *Helicobacter aurati* sp. nov., a urease-positive *Helicobacter* species cultured from gastointestinal tissues of Syrian hamsters. J. Clin. Microbiol. *38*: 3722–3728.

Patterson, M.M., M.D. Schrenzel, Y. Feng, S. Xu, F.E. Dewhirst, B.J. Paster, S.A. Thibodeau, J. Versalovic and J.G. Fox. 2002. *In* Validation of new names and new combinations previously effectively published outside the IJSEM. List No. 84. Int. J. Syst. Evol. Microbiol. *52*: 3–4.

Patton, C.M. and I.K. Wachsmuth. 1992. Typing Schemes: are current methods useful? *In* Nachamkin, Blaser and Tompkins (Editors), *Campylobacter jejuni*. Current Status and Future Trends, American Society for Microbiology, Washington, D.C. pp. 110–128.

Patureau, D., N. Bernet and R. Moletta. 1997. Combined nitrification and denitrification in a single aerated reactor using the aerobic denitrifier *Comamonas* sp. strain SGLY2. Water Res. *31*: 1363–1370.

Patureau, D., J.J. Godon, P. Dabert, T. Bouchez, N. Bernet, J.P. Delgenes and R. Moletta. 1998. *Microvirgula aerodenitrificans* gen. nov., sp. nov., a new gram-negative bacterium exhibiting co-respiration of oxygen and nitrogen oxides up to oxygen-saturated conditions. Int. J. Syst. Bacteriol. *48*: 775–782.

Pauley, E.H. and N.R. Krieg. 1974. Long-term preservation of *Spirillum volutans*. Int. J. Syst. Bacteriol. *24*: 292–293.

Paulus, F., J. Canaday and L. Otten. 1991a. Limited host range Ti plasmids - recent origin from wide host range Ti plasmids and involvement of a novel IS element, IS868. Mol. Plant-Microbe Interact. *4*: 190–197.

Paulus, F., B. Huss, B. Tinland, A. Herrmann, J. Canaday and L. Otten. 1991b. Role of T-region borders in *Agrobacterium* host range. Mol. Plant-Microbe Interact. *4*: 163–172.

Pavarino, G.L. 1911. Malattie causate da bacteri nelle orchidee. Atti Accad. Lincei. *20*: 233–237.

Pawluk, A., R.K. Scopes and K. Griffiths-Smith. 1986. Isolation and properties of the glycolytic enzymes from *Zymomonas mobilis*. The five enzymes from glyceraldehyde-3-phosphate dehydrogenase through to pyruvate kinase. Biochem. J. *238*: 275–281.

Payne, S.M. and R.A. Finkelstein. 1975. Pathogenesis and immunology of experimental gonococcal infection: Role of iron in virulence. Infect. Immun. *12*: 1313–1318.

Peace, T.A., K.V. Brock and H.F. Stills, Jr.. 1994. Comparative analysis of the 16S rDNA gene sequence of the putative agent of proliferative ileitis of hamsters. Int. J. Syst. Bacteriol. *44*: 832–835.

Pear, J.R., Y. Kawagoe, W.E. Schreckengost, D.P. Delmer and D.M. Stalker. 1996. Higher plants contain homologs of the bacterial *celA* genes encoding the catalytic subunit of cellulose synthase. Proc. Natl. Acad. Sci. U.S.A. *93*: 12637–12642.

Pearson, H.W., R. Howsley and S.T. Williams. 1982. A study of nitrogenase activity in *Mycoplana* species and free-living actinomycetes. J. Gen. Microbiol. *128*: 2073–2080.

Pedersen, C.E., Jr. and V.D. Walters. 1978. Comparative electrophoresis of spotted fever group rickettsial proteins. Life Sci. *22*: 583–587.

Pedersen, M.M., E. Marso and M.J. Pickett. 1970. Nonfermentative bacilli

associated with man. III. Pathogenicity and antibiotic susceptibility. Am. J. Clin. Pathol. *54*: 178–192.

Pedrosa, F.O., J. Döbereiner and M.G. Yates. 1980. Hydrogen-dependent growth and autotrophic carbon dioxide fixation in *Derxia*. J. Gen. Microbiol. *119*: 547–551.

Pedrosa, F.O., K.R.S. Teixeira, I.M.P. Machado, M.B.R. Steffens, G. Klassen, E.M. Benelli, H.B. Machado, S. Funayama, L.U. Rigo, M.L. Ishida, M.G. Yates and E.M. Souza. 1997. Structural organization and regulation of the *nif* genes of *Herbaspirillum seropedicae*. Soil Biol. Biochem. *29*: 843–846.

Pedrosa, F.O. and M.G. Yates. 1984. Regulation of nitrogen fixation (*nif*) genes of *Azospirillum brasilense* by *nifA* and *ntr* (gln) type gene products. FEMS Microbiol. Lett. *23*: 95–101.

Peik, J.A., S.M. S.M. Steenbergen and H.R. H.R. Hayden. 1985. Heteropolysaccharide S-198. United States Patent 4529797,

Peik, J.A., S.M. Steenbergen and H.R. Hayden. 1983. Heteropolysaccharides-194. United States Patent 4401760,

Pellegrin, Y., Juretschko, S., Wagner, M. and Cottenceau, G.. 1999. Morphological and biochemical properties of *Sphaerotilus* sp. isolated from paper mill slimes. Appl. Environ. Microbiol. *65*: 156–162.

Pellerin, N.B. and H. Gest. 1983. Diagnostic features of the photosynthetic bacterium *Rhodopseudomonas sphaeroides*. Curr. Microbiol. *9*: 339–344.

Peñaloza-Vazquez, A., G.L. Mena, L. Herreraestrella and A.M. Bailey. 1995. Cloning and sequencing of the genes involved in glyphosate utilization by *Pseudomonas pseudomallei*. Appl. Environ. Microbiol. *61*: 538–543.

Penfold, S.S., A.J. Lastovica and B.G. Elisha. 1988. Demonstration of plasmids in *Campylobacter pylori*. J. Infect. Dis. *157*: 850–851.

Penner, J.L., J.N. Hennessy and R.V. Congi. 1983. Serotyping of *Campylobacter jejuni* and *Campylobacter coli* on the basis of thermostable antigens. Eur. J. Clin. Microbiol. *2*: 378–383.

Pereg-Gerk, L., F. Arsene, P. Gounon, A. Paquelin, R. Carreno-Lopez, I.R. Kennedy and C. Elmerich. 1998. Colonization of wheat roots by *Azospirillum brasilense* wild type and by mutant strains impaired in flocculation and motility. *In* Elmerich, Kondorosi and Newton (Editors), Biological Nitrogen Fixation for he 21st Century, Kluwer Academic Publishers, Dordrecht. p. 395.

Pereira, A.S., R. Franco, M.J. Feio, C. Pinto, J. Lampreia, M.A. Reis, J. Calvete, I. Moura, I. Beech, A.R. Lino and J.J. Moura. 1996. Characterization of representative enzymes from a sulfate reducing bacterium implicated in the corrosion of steel. Biochem. Biophys. Res. Commun. *221*: 414–421.

Pereira, J.A.R., V.A. Cavalcante, J.I. Baldani and J. Dobereiner. 1988. Field inoculation of Sorghum and rice with *Azospirillum* spp. and *Herbaspirillum seropedicae*. Plant Soil *110*: 269–274.

Perez, J.M., D. Martinez, A. Debus, C. Sheikboudou and A. Bensaid. 1997. Development of an *in vitro* cloning method for *Cowdria ruminantium*. Clin. Diagn. Lab. Immunol. *4*: 620–623.

Pérez-Ramírez, N.O., M.A. Rogel, E. Wang, J.Z. Castellanos and E. Martínez-Romero. 1998. Seeds of *Phaseolus vulgaris* bean carry *Rhizobium etli*. FEMS Microbiol. Ecol. *26*: 289–296.

Peros, J.P., G. Berger and M. Ridé. 1995. Effect of grapevine cultivar, strain of *Xylophilus ampelinus* and culture medium on *in vitro* development of bacterial necrosis. Vitis. *34*: 189–190.

Perry, A.C., I.J. Nicolson and J.R. Saunders. 1988. *Neisseria meningitidis* C114 contains silent, truncated pilin genes that are homologous to *Neisseria gonorrhoeae* pil sequences. J. Bacteriol. *170*: 1691–1697.

Perry, M.B. and D.R. Bundle. 1990. Lipopolysaccharide antigens and carbohydrates of *Brucella*. *In* Adams (Editor), Advances in Brucellosis Research, 1st Ed., Texas A&M University Press, College Station. pp. 76–88.

Perschmann, G. and W. Gräf. 1970. Über eine neus Spezies von *Vitreoscilla* (*Vitreoscilla proteolytica*) im Bodensee. Arch. Hyg. Bakteriol. *154*: 128–137.

Peschkov, J.J. and V. Feodorov. 1978. Comparative study on the ultrastructure of L-forms obtained from S and R variants of *Brucella suis*

1330. Zentbl. Bakteriol. Parasitenkd. Infektionskr. Hyg. Abt. I Orig. A. *240*: 94–105.

Peters, D. and R. Wigand. 1955. *Bartonellaceae*. Bacteriol. Rev. *19*: 150–155.

Peterson, J.E. 1958. Two new fifty-year-old species of myxobacteria. Mycologia. *50*: 628–633.

Peterson, J.E. 1959. A monocystic genus of the *Myxobacterales* (Schizomycetes). Mycologia. *51*: 1–8.

Peterson, J.E. and J.C. McDonald. 1966. The demise of the myxobacterial genus *Angiococcus*. Mycologia. *58*: 962–965.

Petit, A., C. David, G.A. Dahl, J.G. Ellis, P. Guyon, F. Cassedelbart and J. Tempé. 1983. Further extension of the opine concept - plasmids in *Agrobacterium rhizogenes* cooperate for opine degradation. Mol. Gen. Genet. *190*: 204–214.

Petit, F. and J.F. Guespin-Michel. 1992. Production of an extracellular milk-clotting activity during development in *Myxococcus xanthus*. J. Bacteriol. *174*: 5136–5140.

Petitprez, M., A. Petitprez, H. Leclerc and E. Vivier. 1969. Some structural aspects of *Sphaerotilus natans*. Ann. Inst. Pasteur Lille *20*: 103–113.

Petroni, E.A. and L. Ielpi. 1996. Isolation and nucleotide sequence of the GDP-mannose:cellobiosyl-diphosphopolyprenol alpha-mannosyltransferase gene from *Acetobacter xylinum*. J. Bacteriol. *178*: 4814–4821.

Petruschky, J. 1896. *Bacillus faecalis alcaligenes* (n. sp.). Zentbl. Bakteriol. Parasitenk. Infektionskr. Hyg. Abt. I. *19*: 187–191.

Pettersson, A., V. Klarenbeek, J. Vandeurzen, J.T. Poolman and J. Tommassen. 1994. Molecular characterization of the structural gene for the lactoferrin receptor of the meningococcal strain H44/76. Microb. Pathog. *17*: 395–408.

Pettersson, A., T. Prinz, A. Umar, J. van der Biezen and J. Tommassen. 1998a. Molecular characterization of LbpB, the second lactoferrin-binding protein of *Neisseria meningitidis*. Mol. Microbiol. *27*: 599–610.

Pettersson, B., A. Kodjo, M. Ronaghi, M. Uhlen and T. Tønjum. 1998b. Phylogeny of the family *Moraxellaceae* by 16S rDNA sequence analysis, with special emphasis on differentiation of *Moraxella* species. Int. J. Syst. Bacteriol. *48*: 75–89.

Pfaller, S.L., S.D. Sutton and B.K. Kinkle. 1999. *Sphingomonas* sp. strain Lep1: an aerobic degrader of 4-methylquinoline. Can. J. Microbiol. *45*: 623–626.

Pfennig, N. 1965. Anreicherungskulturen fur rote und grune Schwefelbakterien. Zentbl. Bakteriol. Parasitenkd. Infektkrankh. Hyg. Abt. I Orig. *Suppl. I*: 179–189, 503–505.

Pfennig, N. 1969a. *Rhodopseudomonas acidophila*, sp. n., a new species of the budding purple nonsulfur bacteria. J. Bacteriol. *99*: 597–602.

Pfennig, N. 1969b. *Rhodospirillum tenue* sp. n., a new species of the purple nonsulfur bacteria. J. Bacteriol. *99*: 619–620.

Pfennig, N. 1974. Rhodopseudomonas globiformis, sp. n., a new species of the Rhodospirillaceae. Arch. Microbiol. *100*: 197–206.

Pfennig, N. 1978. *Rhodocyclus purpureus* gen. nov. and sp. nov. a ring-shaped, vitamin B_{12}-requiring member of the family *Rhodospirillaceae*. Int. J. Syst. Bacteriol. *28*: 283–288.

Pfennig, N. and H. Beibl. 1981. The dissimilatory sulfur-reducing bacteria. *In* Starr, M.P., H. Stolp, H.G. Trüper, A. Balows and H.G. Schlegel (Editors), The Prokaryotes: A Handbook on Habitats, Isolation, and Identification of Bacteria, 1st Ed., Springer-Verlag, Berlin, New York. pp. 941-947.

Pfennig, N. and H. Biebl. 1976. *Desulfuromonas acetoxidans* gen. nov. and sp. nov., a new anaerobic, sulfur-reducing, acetate-oxidizing bacterium. Arch. Microbiol. *110*: 3–12.

Pfennig, N. and H. Biebl. 1977. Announcement of the valid publication of new names and new combinations previously effectively published outside the IJSB. List No. 1. Int. J. Syst. Bacteriol. *27*: 306.

Pfennig, N., K.E. Eimhjellen and S. Liaaen-Jensen. 1965. A new isolate of the *Rhodospirillum fulvum* group and its photosynthetic pigments. Arch. Mikrobiol. *51*: 258–266.

Pfennig, N., H. Lünsdorf, J. Süling and J.F. Imhoff. 1998. *In* Validation of the publication of new names and new combinations previously effectively published outside the IJSB. List No. 64. Int. J. Syst. Bacteriol. *48*: 327–328.

Pfennig, N. and K.D. Lippert. 1966. Über das Vitamin B_{12}-Bedürfnis phototropher Schwefelbakterien. Arch. Microbiol. *55*: 245–256.

Pfennig, N., H. Lunsdorf, J. Suling and J.F. Imhoff. 1997. *Rhodospira trueperi* gen. nov., sp. nov., a new phototrophic proteobacterium of the alpha group. Arch. Microbiol. *168*: 39–45.

Pfennig, N. and H.G. Trüper. 1971. Higher taxa of the phototrophic bacteria. Int. J. Syst. Bacteriol. *21*: 17–18.

Pfennig, N. and H.G. Trüper. 1974. The phototrophic bacteria. *In* Buchanan and Gibbons (Editors), Bergey's Manual of Determinative Bacteriology, 8th Ed., The Williams & Wilkins Co., Baltimore. pp. 24–60.

Pfennig, N. and H.G. Trüper. 1981. Isolation of the members of the families Chromatiaceae and Chlorobiaceae. *In* Starr, Stolp, Truper, Balows and Schlegel (Editors), The Prokaryotes: A Handbook on Habitats, Isolation and Identification of Bacteria, 1st Ed., Vol. 1, Springer-Verlag, New York. pp. 279–289.

Pfennig, N. and H.G. Trüper. 1992. The family *Chromatiaceae*. *In* Balows, Trüper, Dworkin, Harder and Schleifer (Editors), The Prokaryotes: A Handbook of Bacteria: Ecophysiology, Isolation, Identification, Applications, 2nd Ed., Vol. 4, Springer-Verlag, New York. pp. 3200–3221.

Pfennig, N. and S. Wagener. 1986. An improved method of preparing wet mounts for photomicrographs of microorganisms. J. Microbiol. Methods *4*: 303–306.

Pfennig, N. and F. Widdel. 1981. Ecology and physiology of some anaerobic bacteria from the microbial sulfur cycle. *In* Bothe and Trebst (Editors), Biology of Inorganic Nitrogen and Sulfur, Springer-Verlag, Heidelberg. pp. 169–177.

Pfennig, N., F. Widdel and H.G. Trüper. 1981. The dissimilatory sulfate-reducing bacteria. *In* Starr, Stolp, Truper, Balows and H.G. Schlegel (Editors), The Prokaryotes: A Handbook on Habitats, Isolation and Identification of Bacteria, 1st Ed., Vol. 1, Springer-Verlag, Berlin. pp. 926–940.

Pfister, D.H. 1993. Roland Thaxter and the myxobacteria. *In* Dworkin and Kaiser (Editors), Myxobacteria II, American Society for Microbiology, Washington, D.C. pp. 1–11.

Pfistner, B. 1990. A one dimensional model for the swarming behaviour of myxobacteria. *In* Alt and Hoffmann (Editors), Biological Motion, Springer-Verlag, Heidelberg. pp. 556–563.

Phaup, J.D. 1968. The biology of *Sphaerotilus* species. Water Res. *2*: 597–614.

Phelps, L.N. 1967. Isolation and characterization of bacteriophages for *Neisseria*. J. Gen. Virol. *1*: 529–532.

Phelps, P.A., S.K. Agarwal, G.E. Speitel, Jr. and G. Georgiou. 1992. *Methylosinus trichosporium* OB3b mutants having constitutive expression of soluble methane monooxygenase in the presence of high levels of copper. Appl. Environ. Microbiol. *58*: 3701–3708.

Philip, C.B. 1943. Nomenclature of the pathogenic rickettsiae. Am. J. Hyg. *37*: 301–309.

Philip, C.B. 1950. Miscellaneous human rickettsioses. *In* Pullen (Editor), Communicable Diseases, Lea and Febiger Co., Philadelphia. pp. 781–788.

Philip, C.B. 1956. Comments on the classification of the order *Rickettsiales*. Can. J. Microbiol. *2*: 261–270.

Philip, C.B. 1957. *Anaplasmataceae*. *In* Breed, Murray and Smith (Editors), Bergey's Manual of Determinative Bacteriology, 7th Ed., The Williams & Wilkins Co., Baltimore. pp. 980–984.

Philip, C.B. 1962. Appendix G. Summary of tick-borne rickettsioses, Rome. wholeFAO/OIE Expert Panel Tick-borne Dis. Livestock, Cairo United Nations. Report nr Rep., 2nd Mtg., pp. 41-43

Philip, C.B., W.J. Hadlow and L.E. Hughes. 1953. *Neorickettsia helmintheca*, a new rickettsia-like disease agent in dogs in western United States transmitted by a helminth. Riass. Comun. VI Congr. Int. Microbiol., Roma. *2*: 256–257.

Philip, R.N. and E.A. Casper. 1981. Serotypes of spotted fever group rickettsiae isolated from *Dermacentor andersoni* (Stiles) ticks in Western Montana. Amer. J. Trop. Med. Hyg. *30*: 230–238.

Philip, R.N., E.A. Casper, R.L. Anacker, J. Cory, S.F. Hayes, W. Burgdorfer and C.E. Yunker. 1983. *Rickettsia bellii* sp. nov.: a tick-borne rickettsia, widely distributed in the United States, that is from the spotted fever and typhus biogroup. Int. J. Syst. Bacteriol. *33*: 94–106.

Philip, R.N., E.A. Casper, W. Burgdorfer, R.K. Gerloff, L.E. Hughes and E.J. Bell. 1978. Serologic typing of rickettsiae of the spotted fever group by microimmunofluorescence. J. Immunol. *121*: 1961–1968.

Philipp, B. and B. Schink. 2000. Two distinct pathways for anaerobic degradation of aromatic compounds in the denitrifying bacterium *Thauera aromatica* strain AR-1. Arch. Microbiol. *173*: 91–96.

Phillips, C.J., Z. Smith, T.M. Embley and J.I. Prosser. 1999. Phylogenetic differences between particle associated and planktonic ammonia-oxidizing bacteria of the beta subdivision of the class *Proteobacteria* in the northwestern Mediterranean Sea. Appl. Environ. Microbiol. *65*: 779–786.

Phillips, I., S. Eykyn and M. Laker. 1972. Outbreak of hospital infection caused by contaminated autoclaved fluids. Lancet *1*: 1258–1260.

Phillips, M.W. and A. Lee. 1983. Isolation and characterization of a spiral bacterium from the crypts of rodent gastrointestinal tracts. Appl. Environ. Microbiol. *45*: 675–683.

Pichinoty, F., M. Mandel, B. Greenway and J.-L. Garcia. 1977a. Isolation and properties of a denitrifying bacterium related to *Pseudomonas lemoignei*. Int. J. Syst. Bacteriol. *27*: 346–348.

Pichinoty, F., M. Mandel, B. Greenway and J.-L. Garcia. 1977b. Study of 14 denitrifying soil bacteria of *Pseudomonas stutzeri* group isolated by enrichment culture in presence of nitrous oxide. Ann. Microbiol. (Paris) *A128*: 75–89.

Pichinoty, F., M. Mandel and J.-L. Garcia. 1977c. Étude physiologique et taxonomique de *Paracoccus denitrificans*. Ann. Inst. Pasteur Microbiol. *128B*: 243–251.

Pichinoty, F., M. Veron, M. Mandel, M. Durand, C. Job and J.L. Garcia. 1978. Etude physiologique et taxonomique du genre *Alcaligenes*: *A. denitrificans*, *A. odorans* et *A.faecalis*. Can. J. Microbiol. *24*: 743–753.

Pickett, M.J. and J.R. Greenwood. 1980. A study of the Va-1 group of pseudomonads and its relationship to *Pseudomonas pickettii*. J. Gen. Microbiol. *120*: 439–446.

Pickett, M.J. and E.L. Nelson. 1950. *Brucella* bacteriophage. J. Hyg. Camb. *48*: 500–503.

Pickett, M.J. and E.L. Nelson. 1955. Speciation within the genus *Brucella*. IV. Fermentation of carbohydrates. J. Bacteriol. *69*: 333–336.

Piddock, L.J.V. and N. Guant. 1996. Epidmiology, and mechanism, of ciprofloxacin resistant campylobacters isolated in the UK. *In* Newell, Ketley and Feldman (Editors), Campylobacters, Helicobacters, and Related Organisms, Plenum Press, New York. pp. 257–258.

Piéron, R. and Y. Mafart. 1977. Infection d' un kyste branchial à *Eikenella corrodens*. Sem. Hop. *53*: 1087–1091.

Pikuta, E.V., A.M. Lysenko and T.N. Zhilina. 1997. Distribution of *Desulfonatronovibrio hydrogenovorans* in soda lakes of Tuva. Mikrobiologiya *66*: 216–221.

Pikuta, E.V., T.N. Zhilina and F.A. Rainey. 1998a. *In* Validation of the publication of new names and new combinations previously effectively published outside the IJSB. List No. 66. Int. J.Syst. Bacteriol. *48*: 631–632.

Pikuta, E.V., T.N. Zhilina, G.A. Zavarzin, N.A. Kostrikina, G.A. Osipov and F.A. Rainey. 1998b. *Desulfonatronum lacustre* gen. nov., sp. nov.: a new alkaliphilic sulfate-reducing bacterium utilizing ethanol. Mikrobiologiya *67*: 105–113.

Pillon, L., M. Chan, J. Franczyk and M. Goldner. 1982. Comparative use of amino acids by three auxotypes of *Neisseria gonorrhoeae*. Antonie Leeuwenhoek *534*: 139–148.

Pimentel, J.P., F.L. Olivares, M. Pitard, S. Urquiaga, F. Akiba and J. Döbereiner. 1991. Dinitrogen fixation and infection of grass leaves by *Pseudomonas rubrisubalbicans* and *Herbaspirillum serpedicae*. Plant Soil *137*: 61–65.

Pinkerton, H. 1936. Criteria for the accurate classification of the rickettsial diseases (rickettsioses) and of their etiological agents. Parasitol. *28*: 172–189.

Pinkwart, M., H. Bahl, M. Reimer, D. Wölfle and H. Berndt. 1979. Activity of the H₂-oxidizing hydrogenase in different N₂-fixing bacteria. FEMS Microbiol. Lett. *6*: 177–181.

Pinkwart, M., K. Schneider and H.G. Schlegel. 1982. The hydrogenase systems of recently isolated aerobic hydrogen bacteria. Zentbl. Bakteriol. Mikrobiol. Hyg. I Abt. Orig. C. *3*: 542.

Pintado, C., C. Salvador, R. Rotger and C. Nombela. 1985. Multiresistance plasmid from commensal *Neisseria* strains. Antimicrob. Agents Chemother. *27*: 120–124.

Pitt, T.L. 1998. *Pseudomonas, Burkholderia,* and related genera. *In* Balows and Duerden (Editors), Systematic Bacteriology, Vol. 2, Arnold, London. pp. 1109–1138.

Pitt, T.L., H. Aucken and D.A.B. Dance. 1992. Homogeneity of lipopolysaccharide antigens in *Pseudomonas pseudomallei*. J. Infect. *25*: 139–146.

Pittman, M. 1984a. The concept of pertussis as a toxin-mediated disease. Pediatr. Infect. Dis. *3*: 467–486.

Pittman, M. 1984b. Genus *Bordetella*. *In* Krieg and Holt (Editors), Bergey's Manual of Systematic Bacteriology, 1st Ed., Vol. 1, The Williams & Wilkins Co., Baltimore. pp. 388–393.

Pizza, M., V. Scarlato, V. Masignani, M.M. Giuliani, B. Arico, M. Comanducci, G.T. Jennings, L. Baldi, E. Bartolini, B. Capecchi, C.L. Galeotti, E. Luzzi, R. Manetti, E. Marchetti, M. Mora, S. Nuti, G. Ratti, L. Santini, S. Savino, M. Scarselli, E. Storni, P.J. Zuo, M. Broeker, E. Hundt, B. Knapp, E. Blair, T. Mason, H. Tettelin, D.W. Hood, A.C. Jeffries, N.J. Saunders, D.M. Granoff, J.C. Venter, E.R. Moxon, G. Grandi and R. Rappuoli. 2000. Identification of vaccine candidates against serogroup B meningococcus by whole-genome sequencing. Science *287*: 1816–1820.

Plaga, W. and H.U. Schairer. 1999. Intercellular signalling in *Stigmatella aurantiaca*. Curr. Opin. Microbiol. *2*: 593–597.

Plaga, W., I. Stamm and H.U. Schairer. 1998. Intercellular signaling in *Stigmatella aurantiaca*: Purification and characterization of stigmolone, a myxobacterial pheromone. Proc. Natl. Acad. Sci. U.S.A. *95*: 11263–11267.

Planes, A.M., A. Ramirez, F. Fernandez, J.A. Capdevila and C. Tolosa. 1992. *Pseudomonas vesicularis* bacteremia. Infection *20*: 367–368.

Plantard, O., J.Y. Rasplus, G. Mondor, I. Le Clainche and M. Solignac. 1999. Distribution and phylogeny of *Wolbachia* inducing thelytoky in *Rhoditini* and '*Aylacini*' (Hymenoptera: Cynipidae). Insect Mol. Biol. *8*: 185–191.

Plante, M., N. Cadieux, C.R. Rioux, J. Hamel, B.R. Brodeur and D. Martin. 1999. Antigenic and molecular conservation of the gonococcal NspA protein. Infect. Immun. *67*: 2855–2861.

Platen, H., A. Temmes and B. Schink. 1990. Anaerobic degradation of acetone by *Desulfococcus biacutus* spec. nov. Arch. Microbiol. *154*: 355–361.

Platen, H., A. Temmes and B. Schink. 1991. *In* Validation of the publication of new names and new combinations previously effectively published outside the IJSB. List No. 39. Int. J. Syst. Bacteriol. *41*: 580–581.

Plaut, A.G., J.V. Gilbert, M.S. Artenstein and J.D. Capra. 1975. *Neisseria gonorrhoeae* and *Neisseria meningitidis*: extracellular enzyme cleaves human immunoglobulin A. Science *190*: 1103–1105.

Plazinski, J., P.J. Dart and B.G. Rolfe. 1983. Plasmid visualization and *nif* gene location in nitrogen-fixing *Azospirillus* strains. J. Bacteriol. *155*: 1429–1433.

Plenge-Bönig, A., M. Kromer and D.W. Buttner. 1995. Light and electron microscopy studies on *Onchocerca jakutensis* and *O. flexuosa* of red deer show different host-parasite interactions. Parasitol. Res. *81*: 66–73.

Plugge, C.M., M. Balk and A.J.M. Stams. 2002. *Desulfotomaculum thermobenzoicum* subsp. *thermosyntrophicum* subsp. nov., a thermophilic, syntrophic, propionate-oxidizing, spore-forming bacterium. Int. J. Syst. Evol. Microbiol. *52*: 391–399.

Plummer, F.A., H. Chubb, J.N. Simonsen, M. Bosire, L. Slaney, I. Maclean, J.O. Ndinyaachola, P. Waiyaki and R.C. Brunham. 1993. Antibody to Rmp (outer-membrane protein 3) increases susceptibility to gonococcal infection. J. Clin. Investig. *91*: 339–343.

Poels, P.A. and J.A. Duine. 1989. NAD-linked, GSH- and factor-independent aldehyde dehydrogenase of the methylotrophic bacterium, *Hyphomicrobium* X. Arch. Biochem. Biophys. *271*: 240–245.

Poget, C., B. Dubuis and U. von Stockar. 1994. Acetate production from lactate and glucose fermentation by *Gluconobacter oxydans*. Biotechnol. Lett. *16*: 1293–1298.

Pöhl, H. 1996. Untersuchungen zum Mechanismus der Aminoglycosid-resistenz bei *Sorangium cellulosum*, Thesis, Technical University, Braunschweig, Germany, 97 pp..

Pohlmann, A., R. Cramm, K. Schmelz and B. Friedrich. 2000. A novel NO-responding regulator controls the reduction of nitric oxide in *Ralstonia eutropha*. Mol. Microbiol. *38*: 626–638.

Pohlner, J., R. Halter, K. Beyreuther and T.F. Meyer. 1987. Gene structure and extracellular secretion of *Neisseria gonorrhoeae* IgA protease. Nature *325*: 458–462.

Poindexter, J.S. 1964. Biological properties and classification of the *Caulobacter* group. Bacteriol. Rev. *28*: 231–295.

Poindexter, J.S. 1974. Genus *Caulobacter* Henrici and Johnson 1935 and Genus *Asticcaulis* Poindexter 1964. *In* Buchanan and Gibbons (Editors), Bergey's Manual of Determinative Bacteriology, 8th Ed., The Williams & Wilkins Co., Baltimore. pp. 153–156.

Poindexter, J.S. 1981. The Caulobacters - ubiquitous unusual bacteria. Microbiol. Rev. *45*: 123–179.

Poindexter, J.S. 1984. The role of calcium in stalk development and in phosphate acquisition in *Caulobacter crescentus*. Arch. Microbiol. *138*: 140–152.

Poindexter, J.S. 1989. The genus *Caulobacter* and the genus *Asticcacaulis*. *In* Staley, Byrant, Pfenning and Holt (Editors), Bergey's Manual of Systematic Bacteriology, 1st Ed., Vol. 3, The Williams & Wilkins Co., Baltimore. pp. 1924–1942.

Poindexter, J.S. 1992. Dimorphic prosthecate bacteria: the genera *Caulobacter, Asticcaulis, Hyphomicrobium, Pedomicrobium, Hyphomonas*, and *Thiodendron*. *In* Balows, Truper, Dworkin, Harder and Schlieffer (Editors), The Prokaryotes, 2nd Ed., Vol. 3, Springer-Verlag, New York. pp. 2176–2196.

Poindexter, J.S. and J.G. Hagenzieker. 1982. Novel peptidoglycans in *Caulobacter* and *Asticcacaulis* spp. J. Bacteriol. *150*: 332–347.

Poindexter, J.S., P.R. Hornack and P.A. Armstrong. 1967. Intracellular development of a large DNA bacteriophage lytic for *Caulobacter crescentus*. Arch. Mikrobiol. *59*: 237–246.

Poindexter, J.S. and R.F. Lewis. 1966. Recommendations for revision of the taxonomic treatment of stalked bacteria. Int. J. Syst. Bacteriol. *16*: 377–382.

Poindexter, J.S., K.P. Pujara and J.T. Staley. 2000. In situ reproductive rate of freshwater *Caulobacter* spp. Appl. Environ. Microbiol. *66*: 4105–4111.

Poindexter, J.S. and J.T. Staley. 1996. *Caulobacter* and *Asticcaulis* stalk bands as indicators of stalk age. J. Bacteriol. *178*: 3939–3948.

Pol, A., H.J.M. Op Den Camp, S.G.M. Mees, M.A.S.H. Kersten and C. Van Der Drift. 1994. Isolation of a dimethylsulfide-utilizing *Hyphomicrobium* species and its application in biofiltration of polluted air. Biodegradation *5*: 105–112.

Polderman-Tijmes, J.J., P.A. Jekel, E.J. de Vries, A.E.J. van Merode, R. Floris, J.M. van der Laan, T. Sonke and D.B. Janssen. 2002. Cloning, sequence analysis, and expression in *Escherichia coli* of the gene encoding an α-amino acid ester hydrolase from *Acetobacter turbidans*. Appl. Environ. Microbiol. *68*: 211–218.

Policastro, P.F., U.G. Munderloh, E.R. Fischer and T. Hackstadt. 1997. *Rickettsia rickettsii* growth and temperature-inducible protein expression in embryonic tick cell lines. J. Med. Microbiol. *46*: 839–845.

Policastro, P.F., M.G. Peacock and T. Hackstadt. 1996. Improved plaque assays for *Rickettsia prowazekii* in Vero 76 cells. J. Clin. Microbiol. *34*: 1944–1948.

Polisson, C. and R.D. Morgan. 1988. *Ase*I, a restriction endonuclease from *Aquaspirillum serpens* which recognizes 5' AT–TAAT 3'. Nucleic Acids Res. *16*: 10365.

Pollock, T. 1993. Gellan-related polysaccharide secreted by *Sphingomonas*. J. Gen. Microbiol. *139*: 1939–1945.

Pollock, T.J., L. Thorne, M. Yamazaki, M.J. Mikolajczak and R.W. Armentrout. 1994. Mechanism of bacitracin resistance in gram-negative bacteria that synthesize exopolysaccharides. J. Bacteriol. *176*: 6229–6237.

Pollock, T.J., W.A. van Workum, L. Thorne, M.J. Mikolajczak, M. Yamazaki, J.W. Kijne and R.W. Armentrout. 1998. Assignment of biochemical functions to glycosyl transferase genes which are essential for biosynthesis of exopolysaccharides in *Sphingomonas* strain S88 and *Rhizobium leguminosarum*. J. Bacteriol. *180*: 586–593.

Polman, J.K. and J.M. Larkin. 1990. Nitrogen fixation in *Vitreoscilla* and *Thiothrix*. Microbios Lett. *44*: 65–68.

Pond, F.R., I. Gibson, J. Lalucat and R.L. Quackenbush. 1989. R-body producing bacteria. Microbiol. Rev. *53*: 25–67.

Pongratz, E. 1957. D'une bacterie pediculee isolee d'un pus de sinus. Schweiz. Z. Allg. Pathol. Bakteriol. *20*: 593–608.

Pongsunk, S., P. Ekpo and T. Dharakul. 1996. Production of specific monoclonal antibodies to *Burkholderia pseudomallei* and their diagnostic application. Asian Pac. J. Allergy Immunol. *14*: 43–47.

Ponomareva, W.W. 1964. Theory of Podalization. *In* Academy of Science, Moscow, pp. 71–73.

Poole, K. and R.E.W. Hancock. 1986. Phosphate starvation induced outer membrane proteins of members of the families *Enterobacteriaceae* and *Pseudomonodaceae* demonstration of immunological cross reactivity with an antiserum specific for porin protein P of *Pseudomonas aeruginosa*. J. Bacteriol. *165*: 987–993.

Poolman, J., P.A. van der Ley and J. Tommassen. 1995. Surface structures and secreted products of meningococci. *In* Cartwright (Editor), Meningococcal Disease, John Wiley and Sons Ltd., Chichester. pp. 21–34.

Pootjes, C.F. 1977. Evidence for plasmid coding of the ability to utilize hydrogen gas by *Pseudomonas facilis*. Biochem. Biophys. Res. Commun. *76*: 1002–1006.

Popkhadze, N.Z. and T.G. Abashidze. 1957. Characteristics of the *Brucella* bacteriophage isolated at the Tbilisi NIIVS. (In Russian). Bakteriofagiya *5*: 321–325.

Popov, V.L., V.C. Han, S.M. Chen, J.S. Dumler, H.M. Feng, T.G. Andreadis, R.B. Tesh and D.H. Walker. 1998. Ultrastructural differentiation of the genogroups in the genus *Ehrlichia*. J. Med. Microbiol. *47*: 235–251.

Porat, N., M.A. Apicella and M.S. Blake. 1995. A Lipooligosaccharide binding site on Hepg2 cells similar to the gonococcal opacity-associated surface protein Opa. Infect. Immun. *63*: 2164–2172.

Porra, R.J., W. Schäfer, N. Gad'on, I. Katheder, G. Drews and H. Scheer. 1996. Origin of the two carbonyl oxygens of bacteriochlorophyll a. Demonstration of two different pathways for the formation of ring E in *Rhodobacter sphaeroides* and *Roseobacter denitrificans*, and a common hydratase mechanism for 3-acetyl group formation. Eur. J. Biochem. *239*: 85–92.

Porritt, R.J., J.L. Mercer and R. Munro. 2000. Detection and serogroup determination of *Neisseria meningitidis* in CSF by polymerase chain reaction (PCR). Pathology *32*: 42–45.

Porter, J.F., K. Connor and W. Donachie. 1994. Isolation and characterization of *Bordetella parapertussis* like bacteria from ovine lungs. Microbiology *140*: 255–261.

Porter, J.F., K. Connor and W. Donachie. 1996. Differentiation between human and ovine isolates of *Bordetella parapertussis* using pulsed-field gel electrophoresis. FEMS Microbiol. Lett. *135*: 131–135.

Porter, J.F., K. Connor, N. Krueger, J.C. Hodgson and W. Donachie. 1995a. Predisposition of specific pathogen-free lambs to *Pasteurella haemolytica* pneumonia by *Bordetella parapertussis* infection. J. Comp. Pathol. *112*: 381–389.

Porter, J.F., K. Connor, A. van der Zee, F. Reubsaet, P. Ibsen, I. Heron, R. Chaby, K. Le Blay and W. Donachie. 1995b. Characterisation of ovine *Bordetella parapertussis* isolates by analysis of specific endotoxin (lipopolysaccharide) epitopes, filamentous haemagglutinin production, cellular fatty acid composition and antibiotic sensitivity. FEMS Microbiol. Lett. *132*: 195–201.

Porter, J.F., C.S. Mason, N. Krueger, K. Connor and W. Donachie. 1995c. Bronchopneumonia in mice caused by *Pasteurella haemolytica* A2 after predisposition by ovine *Bordetella parapertussis*. Vet. Microbiol. *46*: 393–400.

Postgate, J.R. 1959. A diagnostic reaction of *Desulphovibrio desulphuricans*. Nature *183*: 481–482.

Postgate, J.R. 1984a. Genus *Desulfovibrio*. *In* Krieg and Holt (Editors), Bergey's Manual for Systematic Bacteriology, Vol. 1, Williams & Wilkins Co., Baltimore. pp. 666–672.

Postgate, J.R. 1984b. The Sulfate-Reducing Bacteria, 2nd ed., Cambridge University Press, Cambridge, New York.

Postgate, J.R. and L.L. Campbell. 1966. Classification of *Desulfovibrio* species the nonsporulating sulfate-reducing bacteria. Bacteriol. Rev. *30*: 732–738.

Pot, B. 1996. De fylogenie van chemo-organotrofe spirillen. (The phylogeny of chemoorganotrophic spirilla) Proefschrift ingediend tot het behalen van de graad van Doctor in de wetenschappen, University of Gent, Gent, Belgium.

Pot, B., M. Gillis and J. De Ley. 1992a. The genus *Aquaspirillum*. *In* Balows, Trüper, Dworkin, Harder and Schleifer (Editors), The Prokaryotes: A Handbook of Bacteria: Ecophysiology, Isolation, Identification, Applications, 2nd Ed., Vol. 4, Springer-Verlag, New York. pp. 2569–2582.

Pot, B., A. Willems, M. Gillis and J. De Ley. 1992b. Intra- and intergeneric relationships of the genus *Aquaspirillum*: *Prolinoborus*, a new genus for *Aquaspirillum fasciculus*, with the species *Prolinoborus fasciculus* comb. nov. Int. J. Syst. Bacteriol. *42*: 44–57.

Potts, W.J. and J.R. Saunders. 1988. Nucleotide sequence of the structural gene for class I pilin from *Neisseria meningitidis*: homologies with the *pilE* locus of *Neisseria gonorrhoeae*. Mol. Microbiol. *2*: 647–653.

Poupot, R., E. Martínez-Romero, N. Gautier and J.C. Promé. 1995. Wild type *Rhizobium etli*, a bean symbiont, produces acetyl-fucosylated, *N*-methylated, and carbamoylated nodulated factors. J. Biol. Chem. *270*: 6050–6055.

Poupot, R., E. Martínez-Romero and J.C. Promé. 1993. Nodulation factors from *Rhizobium tropici* are sulfated or nonsulfated chitopentasaccharides containing an *N*-methyl-*N*-acylglucosaminyl terminus. Biochem. *32*: 10430–10435.

Powell, D.A. 1979. Structure, solution properties and biological interactions of some microbial extracellular polysaccharides. *In* Berkeley, Gooday and Ellwood (Editors), Microbial Polysaccharides and Polysaccharases, Academic Press, New York. pp. 117–160.

Powell, D.M., B.S. Roberson and R.M. Weiner. 1980. Serological relationships among budding, prosthecate bacteria. Can. J. Microbiol. *26*: 209–207.

Pradella, S., A. Hans, C. Sproer, H. Reichenbach, K. Gerth and S. Beyer. 2002. Characterisation, genome size and genetic manipulation of the myxobacterium *Sorangium cellulosum* So ce56. Arch Microbiol. *178*: 484–492.

Prakash, D., A. Chauhan and R.K. Jain. 1996. Plasmid encoded degradation of *p*-nitrophenol by *Pseudomonas cepacia*. Biochem. Biophys. Res. Commun. *224*: 375–381.

Präve, P. 1957. Untersuchungen über die Stoffwechselphysiologie des Eisenbakteriums *Leptothrix ochracea*. Arch. Mikrobiol. *27*: 33–62.

Preer, J.R., Jr. and L.B. Preer. 1982. Revival of names of protozoan endosymbionts and proposal of *Holospora caryophila* nom. nov. Int. J. Syst. Bacteriol. *32*: 140–141.

Preer, J.R., Jr., L.B. Preer and A. Jurand. 1974. Kappa and other endosymbionts in *Paramecium aurelia*. Bacteriol. Rev. *38*: 113–163.

Preer, L.B. 1969. Alpha, an infectious macronuclear symbiont of *Paramecium aurelia*. J. Protozool. *16*: 570–578.

Prefontaine, G. and F.L. Jackson. 1972. Cellular fatty acid profiles as an aid to the classification of "corroding bacilli" and certain other bacteria. Int. J. Syst. Bacteriol. *22*: 210–217.

Preissner, W.C., S. Maier, H. Voelker and P. Hirsch. 1988. Isolation and partial characterization of a bacteriophage active on *Hyphomicrobium* sp. WI-926. Can. J. Microbiol. *34*: 101–106.

Prere, M.F., O. Fayet, C. Delmas, M.B. Lareng and H. Dabernat. 1985. Presence of plasmids in *Neisseria meningitidis*. Ann Inst Pasteur Microbiol. *136A*: 271–276.

Preston, N.W. 1963. Type-specific immunity against whooping cough. Br. Med. J. *2*: 724–726.

Preston, N.W. 1970. Technical problems in the laboratory diagnosis and prevention of whooping-cough. Lab. Pract. *19*: 482–486.

Preston, N.W. and R.C. Matthews. 1998. Acellular pertussis vaccines: progress but deja vu. Lancet *351*: 1811–1812.

Preston, N.W. and T.N. Stanbridge. 1972. Efficacy of pertussis vaccines: a brighter horizon. Br. Med. J. *3*: 448–451.

Preston, N.W., N. Surapatana and E.J. Carter. 1982. A reappraisal of serotype factors 4, 5, and 6 of *Bordetella pertussis*. J. Hyg. (Lond) *88*: 39–46.

Preston, N.W., A.A. Zorgani and E.J. Carter. 1990. Location of the three major agglutinogens of *Bordetella pertussis* by immuno-electronmicroscopy. J. Med. Microbiol. *32*: 63–68.

Pretorius, A.M. and R.J. Birtles. 2002. *Rickettsia aeschlimannii*: A new pathogenic spotted fever group rickettsia, South Africa. Emerg. Infect. Dis. *8*: 874.

Pretorius, W.A. 1963. A systematic study of genus *Spirillum* which occurs in oxidation ponds, with a description of a new species. J. Gen. Microbiol. *32*: 403–408.

Pretzman, C., D. Ralph, D.R. Stothard, P.A. Fuerst and Y. Rikihisa. 1995. 16S rRNA gene sequence of *Neorickettsia helminthoeca* and its phylogenetic alignment with members of the genus *Ehrlichia*. Int. J. Syst. Bacteriol. *45*: 207–211.

Prévot, A.R. 1933. Études de systématique bactérienne. I. Lois générales. II. Cocci anaérobius. Ann. Sci. Natur. Zool. Biol. Anim. *15*: 23–260.

Prévot, A.R. 1940. Études de systématique bactérienne. Annales de l' Institute Pasteur *64*: 117–125.

Prévot, A.R. 1966. Manual for the Classification and Determination of the Anaerobic Bacteria, 1st Amer. Ed. ed., Lea and Febiger, Philadelphia.

Preziosi, L., G.P.F. Michel and J. Baratti. 1990. Sucrose metabolism in *Zymomonas mobilis*: production and localization sucrase and levansucrase activities. Can. J. Microbiol. *36*: 159–163.

Price, N.P., B. Relic, F. Talmont, A. Lewin, D. Promé, S.G. Pueppke, F. Maillet, J. Dénarié, D. Promé and W.J. Broughton. 1992. Broad-host-range *Rhizobium* species strain NGR234 secretes a family of carbamoylated, and fucosylated, nodulation signals that are o-acetylated or sulphated. Mol. Microbiol. *6*: 3575–3584.

Prince, R.C. 1994. Haloalkane dehalogenase caught in the act. Trends Biochem. Sci. *19*: 3–4.

Pringsheim, E.G. 1949a. The filamentous bacteria *Sphaerotilus*, *Leptothrix*, *Cladothrix*, and their relation to iron and manganese. Philos. Trans. R. Soc. London Ser. B. *223*: 453–482.

Pringsheim, E.G. 1949b. Iron Bacteria. Biol. Rev. Camb. Philos. Soc. *24*: 200–245.

Pringsheim, E.G. 1949c. The relationship between bacteria and the *Myxophyceae*. Bacteriol. Rev. *13*: 47–91.

Pringsheim, E.G. 1951. The *Vitreoscillaceae*: a family of colourless, gliding, filamentous organisms. J. Gen. Microbiol. *5*: 124–149.

Pringsheim, E.G. 1955. *Lampropedia hyalina* Schroeter 1886 and *Vannielia aggregata* n.g., n.sp., with remarks on natural and on organized colonies in bacteria. J. Gen. Microbiol. *13*: 285–291.

Pringsheim, E.G. 1964. Heterotrophism and species concepts in *Beggiatoa*. Am. J. Bot. *51*: 898–913.

Pringsheim, E.G. 1966. *Lampropedia hyalina* Schroeter, eine apochlorotische *Merismopedia* (Cyanophyceae). Kleine Mitteilungen über Flagellation und Algen. XII. Arch. Mikrobiol. *55*: 200–208.

Probst, J., M. Bruschi, N. Pfennig and J. LeGall. 1977. Cytochrome $c_{551.5}$ (c_7) from *Desulfuromonas acetoxidans*. Biochim. Biophys. Acta *460*: 58–64.

Proenca, R., W.W. Niu, G. Cacalano and A. Prince. 1993. The *Pseudomonas cepacia* 249 chromosomal penicillinase is a member of the AmpC family of chromosomal β-lactamases. Antimicrob. Agents Chemother. *37*: 667–674.

Progulske, A. and S.C. Holt. 1980. Transmission-scanning electron microscopic observations of selected *Eikenella corrodens* strains. J. Bacteriol. *143*: 1003–1018.

Progulske, A., R. Mishell, C. Trummel and S.C. Holt. 1984. Biological activities of *Eikenella corrodens* outer membrane and lipopolysaccharide. Infect. Immun. *43*: 178–182.

Pronina, N.I. 1962. Description of new species and varieties of cellulose-decomposing myxobacteria. Microbiology *31*: 384–390.

Pronk, J.T. and D.B. Johnson. 1992. Oxidation and reduction of iron by acidophilic bacteria. Geomicrobiology J. *10*: 153–171.

Pronk, J.T., P.R. Levering, W. Olijve and J.P. van Dijken. 1989. Role of NADP-dependent and quinoprotein glucose dehydrogenases in gluconic acid production by *Gluconobacter oxydans*. Enzyme Microb. Technol. *11*: 160–164.

Pronk, J.T., R. Meulenberg, W. Hazeu, P. Bos and J.G. Kuenen. 1990. Oxidation of reduced inorganic sulfur compounds by acidophilic thiobacilli. FEMS Microbiol. Lett. *75*: 293–306.

Psallidas, P.G. 1988. Large-scale application of biological control of crown gall in Greece. EPPO Bulletin/Bulletin OEPP *18*: 61–66.

Pueppke, S.G. and W.J. Broughton. 1999. *Rhizobium* sp. strain NGR234 and *R. fredii* USDA257 share exceptionally broad host ranges. Mol. Plant Microbe Interact. *12*: 293–318.

Pugliese, A., B. Pacris, P.E. Schoch and B.A. Cunha. 1993. *Oligella urethralis* urosepsis. Clin. Infect. Dis. *17*: 1069-1070.

Pugsley, A.P. 1993. The complete general secretory pathway in gram-negative bacteria. Microbiol. Rev. *57*: 50–108.

Pukall, R., D. Buntefuß, A. Frühling, M. Rohde, R.M. Kroppenstedt, J. Burghardt, P. Lebaron, L. Bernard and E. Stackebrandt. 1999. *Sulfitobacter mediterraneus* sp. nov., a new sulfite-oxidizing member of the α-*Proteobacteria*. Int. J. Syst. Bacteriol. *49*: 513–519.

Pukall, R., M. Laroche, R.M. Kroppenstedt, P. Schumann, E. Stackebrandt and R. Ulber. 2003. *Paracoccus seriniphilus* sp. nov., an L-serine-dehydratase-producing coccus isolated from the marine bryozoan *Bugula plumosa*. Int. J. Syst. Evol. Microbiol. *53*: 443–447.

Punkosdy, G.A., V.A. Dennis, B.L. Lasater, G. Tzertzinis, J.M. Foster and P.J. Lammie. 2001. Detection of serum IgG antibodies specific for *Wolbachia* surface protein in rhesus monkeys infected with *Brugia malayi*. J. Infect. Dis. *184*: 385–389.

Punsalang, A.P. and W.D. Sawyer. 1973. Role of pili in the virulence of *Neisseria gonorrhoeae*. Infect. Immun. *8*: 255–263.

Pusterla, N., J. Huder, C. Wolfensberger, B. Litschi, A. Parvis and H. Lutz. 1997. Granulocytic ehrlichiosis in two dogs in Switzerland. J. Clin. Microbiol. *35*: 2307–2309.

Pusterla, N., J.E. Madigan, K.M. Asanovich, J.S. Chae, E. Derock, C.M. Leutenegger, J.B. Pusterla, H. Lutz and J.S. Dumler. 2000a. Experimental inoculation with human granulocytic *Ehrlichia* agent derived from high- and low-passage cell culture in horses. J. Clin. Microbiol. *38*: 1276–1278.

Pusterla, N., J.E. Madigan, J.S. Chae, E. DeRock, E. Johnson and J.B. Pusterla. 2000b. Helminthic transmission and isolation of *Ehrlichia risticii*, the causative agent of Potomac horse fever, by using trematode stages from freshwater stream snails. J. Clin. Microbiol. *38*: 1293–1297.

Puttlitz, D.H. and H.W. Seeley, Jr.. 1968. Physiology and nutrition of *Lampropedia hyalina*. J. Bacteriol. *96*: 931–938.

Puustinen, A., M. Finel, M. Virkki and M. Wikström. 1989. Cytochrome *o* (*bo*) is a proton pump in *Paracoccus denitrificans* and *Escherichia coli*. FEBS Lett. *249*: 163–167.

Puustinen, A. and M. Wikström. 1991. The heme groups of cytochrome *o* from *Escherichia coli*. Proc. Natl. Acad. Sci. U.S.A. *88*: 6122–6126.

Qatibi, A.I., V. Niviere and J.L. Garcia. 1991. *Desulfovibrio alcoholovorans* sp. nov., a sulfate-reducing bacterium able to grow on glycerol, 1,2- and 1,3-propanediol. Arch. Microbiol. *155*: 143–148.

Qatibi, A.I., V. Niviere and J.L. Garcia. 1995. *In* Validation of the publication of new names and new combinations previously effectively published outside the IJSB. List No. 55. Int. J. Syst. Bacteriol. *45*: 879–880.

Qazi, G.N., R. Parshad, V. Verma, C.L. Chopra, R. Buse, M. Trager and U. Onken. 1991. Diketo-gluconate fermentation by *Gluconobacter oxydans*. Enzyme Microb. Technol. *13*: 504–507.

Quackenbush, R.L. 1978. Genetic relationships among bacterial endosymbionts of *Paramecium aurelia*: deoxyribonucleotide sequence re-lationships among members of *Caedobacter*. J. Gen. Microbiol. *108*: 181–187.

Quackenbush, R.L. 1982. Validation of the publication if new names and new combinations previously effectively published outside the IJSB, List No. 8. Int. J. Syst. Bacteriol. *32*: 266–268.

Quackenbush, R.L. 1988. Endosymbionts of killer paramecia. *In* Görtz (Editor), Paramecium, Springer Verlag, Heidelberg, New York. pp. 406–418.

Qualls, G.T., K. Stephens and D. White. 1978. Light-stimulated morphogenesis in fruiting myxobacterium *Stigmatella aurantiaca*. Science *201*: 443–444.

Quaroni, A. and R.J. May. 1980. Establishment and characterization of intestinal epithelial cell cultures. Methods Cell Biol. *21B*: 403–427.

Quastler, H. and J.C. Hampton. 1962. Effect of ionizing radiation on the fine structure and function of the intestinal epithelium of the mouse. I. Villue epithelium. Radiat. Res. *17*: 914–931.

Quayle, J.R. 1972. The metabolism of one-carbon compounds by microorganisms. *In* Rose and Tempest (Editors), Advances in Microbial Physiology, Vol. 7, Academic Press, London. pp. 119–203.

Quehl, A. 1906. Untersuchungen über Myxobakterien. Zentbl. Bakteriol. Parasitenkd. Infektionskr. Hyg. Abt. II *16*: 9–34.

Quinlivan, D., T.M. Davis, F.J. Daly and H. Darragh. 1996. Hepatic abscess due to *Eikenella corrodens* and *Streptococcus milleri*: implications for antibiotic therapy. J. Infect. *33*: 47–48.

Quintero, E.J., K. Busch and R.M. Weiner. 1998. Spatial and temporal deposition of adhesive extracellular polysaccharide capsule and fimbriae by *Hyphomonas* strain MHS-3. Appl. Environ. Microbiol. *64*: 1246–1255.

Quintero, E.J. and R.M. Weiner. 1995. Evidence for the adhesive function of the exopolysaccharide of *Hyphomonas* strain MHS-3 in its attachment to surfaces. Appl. Environ. Microbiol. *61*: 1897–1903.

Rabenhorst, E.G. 1854. *Gloeosphaera ferruginea* RABENHORST. Algen Mitteleuropas N. 387. Hedwigia. *1*: 43–45.

Rabus, R., T.A. Hansen and F. Widdel. 2000a. Dissimilatory sulfate- and sulfer-reducing prokaryotes. *In* Dworkin, Falkow, Rosenberg, Schleifer and Stackebrandt (Editors), The Prokaryotes, electronic edition, Springer-Verlag, New York.

Rabus, R., R. Nordhaus, W. Ludwig and F. Widdel. 1993. Complete oxidation of toluene under strictly anoxic conditions by a new sulfate-reducing bacterium. Appl. Environ. Microbiol. *59*: 1444–1451.

Rabus, R., R. Nordhaus, W. Ludwig and F. Widdel. 2000b. *In* Validation of the publication of new names and new combinations previously effectively published outside the IJSEM. List No. 75. Int. J. Syst. Evol. Microbiol. *50*: 1415–1417.

Rabus, R. and F. Widdel. 1995. Anaerobic degradation of ethylbenzene and other aromatic hydrocarbons by new denitrifying bacteria. Arch. Microbiol. *163*: 96–103.

Radcliffe, C.W., D. Kaflkewitz and A. Abuchowski. 1979. Asparaginase production by human clinical isolates of *Vibrio succinogenes*. Appl. Environ. Microbiol. *38*: 761–762.

Rådstrøm, P., C. Fermer, B.E. Kristiansen, A. Jenkins, O. Skold and G. Swedberg. 1992. Transformational exchanges in the dihydropteroate synthase gene of *Neisseria meningitidis*–a novel mechanism for acquisition of sulfonamide resistance. J. Bacteriol. *174*: 6386–6393.

Radulovic, S., H.M. Feng, M. Morovic, B. Djelalija, V. Popov, P. Crocquet-Valdes and D.H. Walker. 1996. Isolation of *Rickettsia akari* from a patient in a region where Mediterranean spotted fever is endemic. Clin. Infect. Dis. *22*: 216–220.

Raether, W. and H. Seidenath. 1977. Survival of *Aegyptianella pullorum*, *Anaplasma marginale* and various parasitic protozoa following prolonged storage in liquid nitrogen. Z. Parasitenkd. *53*: 41–46.

Ragatz, L., Z.Y. Jiang, C.E. Bauer and H. Gest. 1995. Macroscopic phototactic behavior of the purple photosynthetic bacterium *Rhodospirillum centenum*. Arch. Microbiol. *163*: 1–6.

Rahmatullah, S.M. and M.C.M. Beveridge. 1993. Ingestion of bacteria in suspension Indian major carps (*Catla catla*, *Labeo rohita*) and Chinese carps (*Hypophthalmichthys molitrix*, *Aristichthys nobilis*). Hydrobiologia *264*: 79–84.

Rainey, F.A., D.P. Kelly, E. Stackebrandt, J. Burghardt, A. Hiraishi, Y. Katayama and A.P. Wood. 1999. A re-evaluation of the taxonomy of *Paracoccus denitrificans* and a proposal for the combination *Paracoccus pantotrophus* comb. nov. Int. J. Syst. Bacteriol. *49*: 645–651.

Rainey, F.A., R. Toalster and E. Stackebrandt. 1993. *Desulfurella acetivorans*, a thermophilic, acetate-oxidizing and sulfur-reducing organism, represents a distinct lineage within the *Proteobacteria*. Syst. Appl. Microbiol. *16*: 373–379.

Rainey, F.A., N. Ward-Rainey, C.G. Gliesche and E. Stackebrandt. 1998. Phylogenetic analysis and intrageneric structure of the genus *Hyphomicrobium* and the related genus *Filomicrobium*. Int. J. Syst. Bacteriol. *48*: 635–639.

Rainey, F.A. and J. Wiegel. 1996. 16S ribosomal DNA sequence analysis confirms the close relationship between the genera *Xanthobacter*, *Azorhizobium*, and *Aquabacter* and reveals a lack of phyloganetic coherence among *Xanthobacter* species. Int. J. Syst. Bacteriol. *46*: 607–610.

Raitio, M. and M. Wikström. 1994. An alternative cytochrome oxidase of *Paracoccus denitrificans* functions as a proton pump. Biochim. Biophys. Acta *1186*: 100–106.

Raj, H.D. 1970. A new species—*Microcyclus flavus*. Int. J. Syst. Bacteriol. *20*: 61–81.

Raj, H.D. 1977. *Microcyclus* and related ring-forming bacteria. Crit. Rev. Microbiol. *5*: 243–269.

Raj., H.D. 1983. Proposal of *Ancylobacter* gen. nov. as a substitute for the bacterial genus *Microcyclus* Ørskov 1928. Int. J. Syst. Bacteriol. *33*: 397–398.

Ralston, E., N.J. Palleroni and M. Doudoroff. 1973. *Pseudomonas pickettii*, a new species of clinical origin related to *Pseudomonas solanacearum*. Int. J. Syst. Bacteriol. *23*: 15–19.

Ramia, M. and M. Swan. 1994. The swimming of unipolar cells of *Spirillum volutans*: theory and observations. J. Exp. Biol. *187*: 75–100.

Ramm, L.E. and H.H. Winkler. 1976. Identification of cholesterol in the receptor site for rickettsiae on sheep erythrocyte membranes. Infect. Immun. *13*: 120–126.

Ramsay, B.A., K. Lomaliza, C. Chavarie, B. Dube, P. Bataille and J.A. Ramsay. 1990. Production of poly-(β-hydroxybutyric-co-β-hydroxyvaleric) acids. Appl. Environ. Microbiol. *56*: 2093–2098.

Ramundo, B.A. and L.E. Claflin. 1990. Demonstration of synonymy between the plant pathogens *Pseudomonas avenae* and *Pseudomonas rubrilineans*. J. Gen. Microbiol. *136*: 2029–2033.

Raoult, D., M. Drancourt and G. Vestris. 1990. Bactericidal effect of doxycycline associated with lysosomotropic agents on *Coxiella burnetii* in P388D1 cells. Antimicrob. Agents Chemother. *34*: 1512–1514.

Raoult, D., P.E. Fournier, P. Abboud and F. Caron. 2002. First documented human *Rickettsia aeschlimannii* infection. Emerg. Infect. Dis. *8*: 748–749.

Raoult, D., P.E. Fournier, M. Drancourt, T.J. Marrie, J. Etienne, J. Cosserat, P. Cacoub, Y. Poinsignon, P. Leclercq and A.M. Sefton. 1996. Diagnosis of 22 new cases of *Bartonella* endocarditis. Ann. Intern. Med. *125*: 646–652.

Raoult, D., B. La Scola, M. Enea, P.E. Fournier, V. Roux, F. Fenollar, M.A. Galvao and X. de Lamballerie. 2001. A flea-associated *Rickettsia* pathogenic for humans. Emerg. Infect. Dis. *7*: 73–81.

Rappin, G. 1881. Contribution à l'étude de bacteries de la bouche à l'état normal et dans la fièvre typhoide, Doctoral thesis, Collège de France, Nantes.

Rarick, H.R., P.S. Riley and R. Martin. 1978. Carbon substrate utilization studies of some cultures of *Alcaligenes denitrificans*, *Alcaligenes faecalis*, and *Alcaligenes odorans* isolated from clinical specimens. J. Clin. Microbiol. *8*: 313–319.

Ras, J., M.J. Hazelaar, L.A. Robertson, J.G. Kuenen, R.J.M. van Spanning, A.H. Stouthamer and N. Harms. 1995. Methanol oxidation in a spontaneous mutant of *Thiosphaera pantotropha* with a methanol-positive phenotype is catalyzed by a dye-linked ethanol dehydrogenase. FEMS Microbiol. Lett. *127*: 159–164.

Rasul, G., M.S. Mirza, F. Latif and K.A. Malik. 1998. Identification of plant hormones produced by bacterial isolates from rice, wheat and kallar grass. *In* Malik, Mirza and Ladha (Editors), Nitrogen Fixation with Non-legumes, Kluwer Academic Publishers, Dordrecht. pp. 25–37.

Rauch, H.C. and M.J. Pickett. 1961. *Bordetella bronchiseptica* bacteriophage. Can. J. Microbiol. *7*: 125–133.

Rault-Leonardon, M., M.A. Atkinson, C.A. Slaughter, C.R. Moomaw and P.A. Srere. 1995. *Azotobacter vinelandii* citrate synthase. Biochemistry *34*: 257–263.

Ravaz, L. 1895. La maladie d'Oléron. Ann. Ecole Natl. Agric. Montpellier. *9*: 299–317.

Raveh, D., A. Simhon, Z. Gimmon, T. Sacks and M. Shapiro. 1993. Infections caused by *Pseudomonas pickettii* in association with permanent indwelling intravenous devices: Four cases and a review. Clin. Infect. Dis. *17*: 877–880.

Ravenschlag, K., K. Sahm, C. Knoblauch, B.B. Jorgensen and R. Amann. 2000. Community structure, cellular rRNA content, and activity of sulfate-reducing bacteria in marine Arctic sediments. Appl. Environ. Microbiol. *66*: 3592–3602.

Raverdy, J. 1973. Sur l'isolement et l'activité bactériolytique de quelques myxobactéries isolées de l'eau. Water Res. *7*: 687–693.

Reasoner, D.J. and E.E. Geldreich. 1985. A new medium for the enumeration and subculture of bacteria from potable water. Appl. Environ. Microbiol. *49*: 1–7.

Redburn, A.C. and B.K.C. Patel. 1994. *Desulfovibrio longreachii* sp. nov., a sulfate-reducing bacterium isolated from the Great Artesian Basin of Australia. FEMS Microbiol. Lett. *115*: 33–38.

Redburn, A.C. and B.K.C. Patel. 1995. *In* Validation of the publication of new names and new combinations previously effectively published outside the IJSB. List No. 55. Int. J. Syst. Bacteriol. *45*: 879–880.

Reddy, G.R., C.R. Sulsona, A.F. Barbet, S.M. Mahan, M.J. Burridge and A.R. Alleman. 1998. Molecular characterization of a 28 kDa surface antigen gene family of the tribe *Ehrlichiae*. Biochem. Biophys. Res. Commun. *247*: 636–643.

Redfearn, M.S. and N.J. Palleroni. 1975. Glanders and meliodosis. *In* Hubbert, McCulloch, Schnurrenberger and Hull (Editors), Diseases Transmitted from Animals to Man, 6th Ed., Vol. 6, C. C. Thomas, Springfield. pp. 110–128.

Redfearn, M.S., N.J. Palleroni and R.Y. Stanier. 1966. A comparative study of *Pseudomonas pseudomallei* and *Bacillus mallei*. J. Gen. Microbiol. *43*: 293–313.

Reding, H.K. 1991. Ecological, Physiological, and Taxonomical Studies of Xanthobacter Strains Isolated from the Roots of Wetland Rice, Dissertation, University of Georgia, Athens, Georgia.

Reding, H.K., G.L.M. Croes, L. Dijkhuizen and J. Wiegel. 1992. Emendation of *Xanthobacter flavus* as a motile species. Int. J. Syst. Bacteriol. *42*: 309–311.

Reding, H.K., P.G. Hartel and J. Wiegel. 1991. Effect of *Xanthobacter* isolated and characterized from rice roots on growth of wetland rice. Plant Soil *138*: 221–230.

Reding, H.K. and J. Wiegel. 1993. Motility and chemotaxis of a *Xanthobacter* sp. isolated from roots of rice. J. Gen. Microbiol. *139*: 815–820.

Redmond, J. 1979. The structure of the O-antigenic side chain of the lipopolysaccharide of *Vibrio cholerae* 569B (Inaba). Biochem. Biophys. Acta *584*: 346–352.

Redway, K.F. and S.P. Lapage. 1974. Effect of carbohydrates and related compounds on the long-term preservation of freeze-dried bacteria. Cryobiology. *11*: 73–79.

Reeburgh, W.S., S.C. Whalen and M.L. Alpern. 1993. The role of methylotrophy in the global methane budget. *In* Murrell and Kelly (Editors), Microbial growth on C_1 compounds, Intercept Press Ltd., Andover. pp. 1–14.

Reed, W.M. and P.R. Dugan. 1978. Distribution of *Methylomonas methanica* and *Methylosinus trichosporium* in Cleveland Harbor as determined by an indirect fluorescent antibody-membrane filter technique. Appl. Environ. Microbiol. *35*: 422–430.

Reed, W.M., J.A. Titus, P.R. Dugan and R.M. Pfister. 1980. Structure of *Methylosinus trichosporium* exospores. J. Bacteriol. *141*: 908–913.

Rees, G., G.N. Grassia, A.J. Sheehy, P.P. Dwivedi and B.K.C. Patel. 1995a. *Desulfacinum infernum* gen. nov., sp. nov., a thermophilic sulfate-reducing bacterium from a petroleum reservoir. Int. J. Syst. Bacteriol. *45*: 85–89.

Rees, G.N., C.G. Harfoot and A.J. Sheehy. 1998. Amino acid degradation by the mesophilic sulfate-reducing bacterium *Desulfobacterium vacuolatum*. Arch. Microbiol. *169*: 76–80.

Rees, G.N. and B.K.C. Patel. 2001. *Desulforegula conservatrix* gen. nov., sp. nov., a long-chain fatty acid-oxidizing, sulfate-reducing bacterium isolated from sediments of a freshwater lake. Int. J. Syst. Evol. Bacteriol. *51*: 1911–1916.

Rees, J.H., S.E. Soudain, N.A. Gregson and R.A. Hughes. 1995b. *Campylobacter jejuni* infection and Guillain-Barré syndrome. N. Engl. J. Med. *333*: 1374–1379.

Regan, J. and F. Lowe. 1977. Enrichment medium for the isolation of *Bordetella*. J. Clin. Microbiol. *6*: 303–309.

Regnery, R.L., B.E. Anderson, J.E. Clarridge, III, M.C. Rodriguez Barradas, D.C. Jones and J.H. Carr. 1992. Characterization of a novel *Rochalimaea* species, *R. henselae* sp. nov., isolated from blood of a febrile, human immunodeficiency virus-positive patient. J. Clin. Microbiol. *30*: 265–274.

Reichenbach, H. 1965a. Rhapidosomen bei Myxobakterien. Arch. Microbiol. *50*: 246–255.

Reichenbach, H. 1965b. Rhythmische vorgänge bei der schwarmentfaltung von myxobakterien. Ber. Deutsch. Bot. Ges. *78*: 102–105.

Reichenbach, H. 1965c. Untersuchungen an *Archangium violaceum*. Arch. Microbiol. *52*: 376–403.

Reichenbach, H. 1966. *Myxococcus* spp. (*Myxobacterales*): Schwarmentwicklung und Bildung von Protocysten. Text with Film E 778, Encyclopaedia Cinematographica, Göttingen. Institut fur den Wissenschaftlichen Filmen.

Reichenbach, H. 1967. Die wahre natur der Myxobakterien-"Rhapidosomen". Arch. Microbiol. *56*: 371–383.

Reichenbach, H. 1970. *Nannocystis exedens* gen. nov., spec. nov., a new myxobacterium of the family *Sorangiaceae*. Arch. Microbiol. *70*: 119–138.

Reichenbach, H. 1999. The ecology of the myxobacteria. Environ. Microbiol. *1*: 15–21.

Reichenbach, H. and M. Dworkin. 1969. Studies on *Stigmatella aurantiaca* (*Myxobacterales*). J. Gen. Microbiol. *58*: 3–14.

Reichenbach, H. and M. Dworkin. 1981. The order *Myxobacterales*. *In* Starr, Stolp, Trüper, Balows and Schlegel (Editors), The Prokaryotes: A Handbook on Habitats, Isolation and Identification of Bacteria, Vol. I, Springer-Verlag, Berlin. pp. 328–355.

Reichenbach, H. and M. Dworkin. 1992. The Myxobacteria. *In* Balows, Trüper, Dworkin, Harder and Schleifer (Editors), The Prokaryotes: A Handbook of Bacteria: Ecophysiology, Isolation, Identification, Applications, 2 ed., Vol. 4, Springer-Verlag, New York. pp. 3417–3487.

Reichenbach, H. and G. Höfle. 1993. Biologically active secondary metabolites from myxobacteria. Biotechnol. Adv. *11*: 219–277.

Reichenbach, H. and G. Höfle. 1999. Myxobacteria as producers of secondary metabolites. *In* Grabley and Thiericke (Editors), Drug Discovery from Nature, Springer-Verlag, Berlin, Berlin. pp. 149–179.

Reichenbach, H., H.H. Heunert and H. Kuczka. 1965a. *Chondromyces apiculatus* (*Myxobacterales*): Schwarmentwicklung und Morphogenese. Film E 779, Encyclopaedia Cinematographica, Göttingen. Institut fur den Wissenschaftlichen Filmen.

Reichenbach, H., H.H. Heunert and H. Kuczka. 1965b. *Myxococcus* spp. (*Myxobacterales*): Schwarmentwicklung und Bildung von Protocysten. Film E 778, Encyclopaedia Cinematographica, Göttingen. Institut fur den Wissenschaftlichen Filmen.

Reichenbach, H., H.H. Heunert and H. Kuczka. 1965c. Schwarmentwicklung und Morphogenese bei Myxobakterien - *Archangium, Myxococcus, Chondrococcus, Chondromyces*. Film C 893, Göttingen. Institut fur den Wissenschaftlichen Filmen.

Reichenbach, H., H.H. Heunert and H. Kuczka. 1974. *Chondromyces apiculatus* (*Myxobacterales*): Schwarmentwicklung und Morphogenese, Encyclopaedia Cinematographica Sekt. Biologie, Göttingen. Institut fur den Wissenschaftlichen Filmen.

Reichenbach, H. and H. Kleinig. 1971. The carotenoids of *Myxococcus fulvus* (*Myxobacterales*). Arch. Microbiol. *76*: 364–380.

Reichenbach, H. and H. Kleinig. 1972. Die Carotinoide der Myxobakterien. Zent Bakteriol Hyg I Abt. Orig. A *220*: 458–463.

Reichenbach, H., W. Ludwig and E. Stackebrandt. 1986. Lack of relationship between gliding cyanobacteria and filamentous gliding heterotrophic eubacteria: comparison of 16S ribosomal RNA catalogues of *Spirulina, Saprospira, Vitreoscilla, Leucothrix*, and *Herpetosiphon*. Arch. Microbiol. *145*: 391–395.

Reichenbach, H., H. Voelz and M. Dworkin. 1969. Structural changes in *Stigmatella aurantiaca* during myxospore induction. J. Bacteriol. *97*: 905–911.

Reichenbecher, W. and B. Schink. 1997. *Desulfovibrio inopinatus*, sp. nov., a new sulfate-reducing bacterium that degrades hydroxyhydroquinone (1,2,4-trihydroxybenzene). Arch. Microbiol. *168*: 338–344.

Reichenbecher, W. and B. Schink. 1999. *In* Validation of the publication of new names and new combinations previously effectively published outside the IJSB. List No. 68. Int. J. Syst. Bacteriol. *49*: 1–3.

Reij, M.W. and S. Hartmans. 1996. Propene removal from synthetic waste gas using a hollow-fibre membrane bioreactor. Appl. Microbiol. Biotechnol. *45*: 730–736.

Reij, M.W., J. Kieboom, J.A.M. De Bont and S. Hartmans. 1995. Continuous degradation of trichloroethylene by *Xanthobacter* sp. strain Py2 during growth on propene. Appl. Environ. Microbiol. *61*: 2936–2942.

Reimer, B. and H. Schlesner. 1989. Isolation of 11 strains of star-shaped bacteria from aquatic habitats and investigation of their taxonomic position. Syst. Appl. Microbiol. *12*: 156–158.

Reina, J., M. Antich, B. Siquier and P. Alomar. 1988. Nosocomial outbreak of *Achromobacter xylosoxidans* associated with a diagnostic contrast solution. J. Clin. Pathol. (Lond.) *41*: 920–921.

Reina, J., M.J. Ros and A. Serra. 1994. Susceptibilities to 10 antimicrobial agents of 1,220 *Campylobacter* strains isolated from 1987 to 1993 from feces of pediatric patients. Antimicrob. Agents. Chemother. *38*: 2917–2920.

Reinhold, B., T. Hurek, I. Fendrik, B. Pot, M. Gillis, K. Kersters, S. Thielemans and J. De Ley. 1987. *Azospirillum halopraeferens*, sp. nov., a nitrogen-fixing organism associated with roots of Kallar grass (*Leptochloa fusca* (L.) Kunth). Int. J. Syst. Bacteriol. *37*: 43–51.

Reinhold, B., T. Hurek, E.G. Niemann and I. Fendrik. 1986. Close association of *Azospirillum* and diazotrophic rods with different root zones of Kallar grass. Appl. Environ. Microbiol. *52*: 520–526.

Reinhold, L. 1966. Untersuchungen an *Bacteroides corrodens* (Eileen, 1958). Zentbl. Bakteriol. Parasitenkd. Infektionskr. Hyg. Abt. I. *201*: 49–57.

Reinhold-Hurek, B. and T. Hurek. 1998. Life in grasses: diazotrophic endophytes. Trends Microbiol. *6*: 139–144.

Reinhold-Hurek, B. and T. Hurek. 2000. Reassessment of the taxonomic structure of the diazotrophic genus *Azoarcus sensu lato* and description of three new genera and new species, *Azovibrio restrictus* gen. nov., sp. nov., *Azospira oryzae* gen. nov., sp. nov. and *Azonexus fungiphilus* gen. nov., sp. nov. Int. J. Syst. Evol. Microbiol. *50*: 649–659.

Reinhold-Hurek, B., T. Hurek, M. Claeyssens and M. van Montagu. 1993a. Cloning, expression in *Escherichia coli*, and characterization of cellulolytic enzymes of *Azoarcus* sp., a root-invading diazotroph. J. Bacteriol. *175*: 7056–7065.

Reinhold-Hurek, B., T. Hurek, M. Gillis, B. Hoste, M. Vancanneyt, K. Kersters and J. De Ley. 1993b. *Azoarcus* gen. nov., nitrogen-fixing proteobacteria associated with roots of kallar grass (*Leptochloa fusca* (L.) Kunth), and description of two species, *Azoarcus indigens* sp. nov. and *Azoarcus communis* sp. nov. Int. J. Syst. Bacteriol. *43*: 574–584.

Reis, J., F.B., V.M. Reis, M. Schloter and J. Döbereiner. 1998. Utlzação da metodologia de imunocaptura para o isolamento de bactérias diazotróficas endofíticas em plantas de cana-de-açúcar. Proceedings of the Reunião Brasileira de Fertilidade do Solo e Mutrição de Plantas, p. 606.

Reis, V.M. and J. Döbereiner. 1998. Effect of high sugar concentration on nitrogenase activity of *Acetobacter diazotrophicus*. Arch. Microbiol. *171*: 13–18.

Relman, D.A., P.W. Lepp, K.N. Sadler and T.M. Schmidt. 1992. Phylogenetic relationships among the agent of bacillary angiomatosis, *Bartonella bacilliformis*, and other alpha-proteobacteria. Mol. Microbiol. *6*: 1801–1807.

Relman, D.A., J.S. Loutit, T.M. Schmidt, S. Falkow and L.S. Tompkins. 1990a. The agent of bacillary angiomatosis. An approach to the identification of uncultured pathogens. New Engl. J. Med. *323*: 1573–1580.

Relman, D., E. Tuomanen, S. Falkow, D.T. Golenbock, K. Saukkonen and S.D. Wright. 1990b. Recognition of a bacterial adhesion by an integrin: macrophage CR3 (α M β 2, CD11b/CD18) binds filamentous hemagglutinin of *Bordetella pertussis*. Cell *61*: 1375–1382.

Remsen, C.C. and S.W. Watson. 1972. Freeze-etching of bacteria. Int. Rev. Cytol. *33*: 253–296.

Ren, T., J.A. Amaral and R. Knowles. 1997. The response of methane consumption by pure cultures of methanotrophic bacteria to oxygen. Can. J. Microbiol. *43*: 925–928.

Rest, R.F. and D.C. Robertson. 1974. Glucose transport in *Brucella abortus*. J. Bacteriol. *118*: 250–259.

Rest, R.F. and D.C. Robertson. 1975. Characterization of the electron transport system in *Brucella abortus*. J. Bacteriol. *122*: 139–144.

Reubel, G.H., J.E. Barlough and J.E. Madigan. 1998. Production and characterization of *Ehrlichia risticii*, the agent of Potomac horse fever, from snails (*Pleuroceridae: Juga* spp.) in aquarium culture and genetic comparison to equine strains. J. Clin. Microbiol. *36*: 1501–1511.

Reverdy, M.E., J. Freney, J. Fleurette, M. Coulet, M. Surgot, D. Marmet and C. Ploton. 1984. Nosocomial colonization and infection by *Achromobacter xylosoxidans*. J. Clin. Microbiol. *19*: 140–143.

Reyes, L. and R.K. Scopes. 1991. Membrane-associated ATPase from *Zymomonas mobilis*: purification and characterization. Biochim. Biophys. Acta *1068*: 174–178.

Reyn, A. 1974. The genus *Neisseria*. *In* Buchanan and Gibbons (Editors), Bergey's Manual of Determinative Baeteriology, 8th Ed., The Williams & Wilkins Co., Baltimore. pp. 428–432.

Reyn, A., A.E. Jephcott and H. Raun. 1971. *Neisseria gonorrhoeae*. Colony variation II. Acta Pathol. Microbiol. Scand. Sect. B Microbiol. *79*: 435–436.

Reynders, L. and K. Vlassak. 1979. Conversion of tryptophan to indolacetic acid by *Azospirillum brasilense*. Soil Biol. Biochem. *11*: 547–548.

Reynolds, D.M., E.J. Laishley and J.W. Costerton. 1981. Physiological and ultrastructural characterization of a new acidophilic *Thiobacillus* species (*Thiobacillus kabobis*, sp. nov.). Can. J. Microbiol. *27*: 151–161.

Reynolds, D.J. and C.W. Penn. 1994. Characteristics of *Helicobacter pylori* growth in a defined medium and determination of its amino acids requirements. Microbiology (Reading) *140*: 2649–2656.

Rezanka, T., I.V. Zlatkin, I. Viden, O.I. Slabova and D.I. Nikitin. 1991. Capillary gas-chromatography mass-spectrometry of unusual and very long-chain fatty-acids from soil oligotrophic bacteria. J. Chromatogr. *558*: 215–221.

Rezsöhazy, R., B. Hallet, J. Delcour and J. Mahillon. 1993. The IS4 family of insertion sequences: evidence for a conserved transposase motif. Mol. Microbiol. *9*: 1283–1295.

Rhee, S.K., G.M. Lee, J.H. Yoon, Y.H. Park, H.S. Bae and S.T. Lee. 1997. Anaerobic and aerobic degradation of pyridine by a newly isolated denitrifying bacterium. Appl. Environ. Microbiol. *63*: 2578–2585.

Rice, P.A. and D.L. Kasper. 1977. Characterization of gonococcal antigens responsible for induction of bactericidal antibody in disseminated infection. J. Clin. Invest. *60*: 1149–1158.

Rice, R.J. and J.S. Knapp. 1994. Antimicrobial susceptibilities of *Neisseria gonorrhoeae* strains representing five distinct resistance phenotypes. Antimicrob. Agents Chemother. *38*: 155–158.

Rice, T.D., H.N. Williams and B.F. Turng. 1998. Susceptibility of bacteria in estuarine environments to autochthonous bdellovibrios. Microb. Ecol. *35*: 256–264.

Richard, C. and M. Kiredjian. 1995. Méthodes de Laboratoire pour l' Identification des Bacilles à Gram-Négative Aerbies Stricts, Institut Pasteur, Paris, France.

Richard, C., H. Monteil, F. Megraud, R. Chatelain and B. Laurent. 1981. Phenotype characteristics of 100 strains of *Pseudomonas cepacia* proposition of a biovars classification. Ann. Biol. Clin. *39*: 9–15.

Richards, G.N. and M. Streamer. 1972. Studies on dextranases. 1. Isolation of extracellular, bacterial dextranases. Carbohydr. Res. *25*: 323–332.

Richards, M. and D.A. Corbey. 1974. Isolation of *Zymomonas* from primed beer. J. Inst. Brew. *80*: 241–244.

Richardson, I.R. 1990. The incidence of *Bdellovibrio* spp. in man-made water systems: coexistence with legionellas. J. Appl. Bacteriol. *69*: 134–140.

Richardson, R.L., C. Hansen and J. Schmidt. 1966. Isolation, morphology, and cultural characteristics of *Simonsiella*. J. Dent. Res. *53*: 78.

Richter, O.M.H., J.S. Tao, A. Turba and B. Ludwig. 1994. A cytochrome *ba*(3) functions as a quinol oxidase in *Paracoccus denitrificans*: purification, cloning and sequence comparison. J. Biol. Chem. *269*: 23079–23086.

Richter, P.J., R.B. Kimsey, J.E. Madigan, J.E. Barlough, J.S. Dumler and D.L. Brooks. 1996. *Ixodes pacificus* (Acari: Ixodidae) as a vector of *Ehrlichia equi* (*Rickettsiales: Ehrlichieae*). J. Med. Entomol. *33*: 1–5.

Rickard, A.H., S.A. Leach, C.M. Buswell, N.J. High and P.S. Handley. 2000. Coaggregation between aquatic bacteria is mediated by specific-growth-phase-dependent lectin-saccharide interactions. Appl. Environ. Microbiol. *66*: 431–434.

Ricketts, H.T. 1911. Contributions to medical science by Howard Taylor Ricketts, 1870–1910, University of Chicago Press, Chicago.

Ridé, M. 1984. La nécrose bactérienne. Phytoma. *36*: 33–36.

Ridé, M. 1996. La nécrose bactérienne de la vigne: données biologiques et épidémiologiques, base d'une stratégie de lutte. Comptes rendus de l'Académie d'Agriculture de France. *82*: 31–50.

Rigaud, T., C. Souty-Grosset, R. Raimond, J. Mocquard and P. Juchault. 1991. Feminizing endocytobiosis in the terrestrial crustacean *Armadillidum vulgare* LATR. (Isopoda): Recent acquisitions. Endocytobiosis Cell. Res. *7*: 259–273.

Rigby, C.E. 1990. The *Brucella* phages. *In* Nielsen and Duncan (Editors), Animal brucellosis, CRC Press, Boca Raton. pp. 121–130.

Rigby, C.E. and A.D. Fraser. 1989. Plasmid transfer and plasmid-mediated genetic exchange in *Brucella abortus*. Can. J. Vet. Res. *53*: 326–330.

Rihs, J.D., D.J. Brenner, R.E. Weaver, A.G. Steigerwalt, D.G. Hollis and V.L. Yu. 1993. *Roseomonas*, a new genus associated with bacteremia and other human infections. J. Clin. Microbiol. *31*: 3275–3283.

Rihs, J.D., D.J. Brenner, R.E. Weaver, A.G. Steigerwalt, D.G. Hollis and V.L. Yu. 1998. *In* Validation of the publication of new names and new combinations previously effectively published outside the IJSB. List No. 65. Int. J. Syst. Bacteriol. *48*: 627.

Riker, A.J., W.M. Banfield, W.H. Wright, G.W. Keitt and H.E. Sagen. 1930. Studies on infectious hairy root of nursery apple trees. J. Agr. Res. *41*: 507–540.

Rikihisa, Y. 1990a. Growth of *Ehrlichia risticii* in human colonic epithelial cells. *In* Hechemy (Editor), Rickettsiology: Current Issues and Perspectives, New York Academy of Science, New York. pp. 104–110.

Rikihisa, Y. 1990b. Ultrastructure of rickettsiae with special emphasis on ehrlichia. *In* Williams and Kakoma (Editors), Ehrlichiosis: A vector-borne Disease of Animals and Humans, Current Topics in Veterinary Medicine and Animal Sciences Vol. 54, Kluwer Publishing Co, Norwell. pp. 22–31.

Rikihisa, Y. 1991a. Protection against murine potomac horse fever by an inactivated *Ehrlichia risticii* vaccine. Vet. Microbiol. *27*: 339–350.

Rikihisa, Y. 1991b. The tribe Ehrlichieae and ehrlichial diseases. Clin. Microbiol. Rev. *4*: 286–308.

Rikihisa, Y. 1996. Ehrlichiae. Proceedings of the 5th International Symposium on Rickettsiae and Rickettsial Diseases, Slovak Academy of Sciences, Bratislava, Slovak Republic: International Society of Rickettsiae and Rickettsial Diseases. pp. 272–286.

Rikihisa, Y. 1997. Rickettsial diseases in horses. *In* Bayly and Reed (Editors), Equine Internal Medicine, W.B. Saunders, Philadelphia. pp.112–123.

Rikihisa, Y. and S. Ito. 1982. Entry of *Rickettsia tsutsugamushi* into polymorphonuclear leukocytes. Infect. Immun. *38*: 343–350.

Rikihisa, Y. and B.M. Jiang. 1988. *In vitro* susceptibilities of *Ehrlichia risticii* to eight antibiotics. Antimicrob. Agents Chemother. *32*: 986–991.

Rikihisa, Y. and B.M. Jiang. 1989. Effect of antibiotics on clinical, pathologic and immunologic responses in murine Potomac horse fever: protective effects of doxycycline. Vet. Microbiol. *19*: 253–262.

Rikihisa, Y., G.C. Johnson and C.J. Burger. 1987. Reduced immune responsiveness and lymphoid depletion in mice infected with *Ehrlichia risticii*. Infect. Immun. *55*: 2215–2222.

Rikihisa, Y., G.C. Johnson, Y.Z. Wang, S.M. Reed, R. Fertel and H.J. Cooke. 1992. Loss of absorptive capacity for sodium and chloride in the colon causes diarrhoea in Potomac horse fever. Res. Vet. Sci. *52*: 353–362.

Rikihisa, Y. and B.D. Perry. 1984. Causative agent of Potomac Horse Fever. Vet. Rec. *115*: 554.

Rikihisa, Y. and B.D. Perry. 1985. Causative ehrlichial organisms in Potomac horse fever. Infect. Immun. *49*: 513–517.

Rikihisa, Y., B. Perry and D. Cordes. 1984. Rickettsial link with acute equine diarrhoea. Vet. Rec. *115*: 390.

Rikihisa, Y., B.D. Perry and D.O. Cordes. 1985. Ultrastructural study of ehrlichial organisms in the large colons of ponies infected with Potomac horse fever. Infect. Immun. *49*: 505–512.

Rikihisa, Y., C.I. Pretzman, G.C. Johnson, S.M. Reed, S. Yamamoto and F. Andrews. 1988. Clinical, histopathological, and immunological responses of ponies to *Ehrlichia sennetsu* and subsequent *Ehrlichia risticii* challenge. Infect. Immun. *56*: 2960–2966.

Rikihisa, Y., S.M. Reed, R.A. Sams, J.C. Gordon and C.I. Pretzman. 1990. Serosurvey of horses with evidence of equine monocytic ehrlichiosis. J. Am. Vet. Med. Assoc. *197*: 1327–1332.

Rikihisa, Y., H. Stills and G. Zimmerman. 1991. Isolation and continuous culture of *Neorickettsia helminthoeca* in a macrophage cell line. J. Clin. Microbiol. *29*: 1928–1933.

Rikihisa, Y., R. Wada, S.M. Reed and S. Yamamoto. 1993. Development of neutralizing antibody in horses infected with *Ehrlichia risticii*. Vet. Microbiol. *36*: 139–147.

Rikihisa, Y., C. Zhang and B.M. Christensen. 1200. DMolecular characterization of *Aegyptianella pullorum* (Rickettsiales, Anaplasmataceae). J Clin. Microbiol. *41*: 5294–5297.

Rikihisa, Y., Y. Zhang and J. Park. 1994. Inhibition of infection of macrophages with *Ehrlichia risticii* by cytochalasins, monodansylcadaverine, and taxol. Infect. Immun. *62*: 5126–5132.

Rikihisa, Y., Y. Zhang and J. Park. 1995. Role of Ca^{2+} and calmodulin in ehrlichial infection in macrophages. Infect. Immun. *63*: 2310–2316.

Rikihisa, Y. and G. Zimmerman. 1995. Salmon poisoning disease. *In* Kirk and Bonagura (Editors), Current Veterinary Therapy XII Small Animal Practice,, W.B. Saunders Co., Philadelphia. pp. 297–300.

Riley, P.S., H.W. Tatum and P.E. Weaver. 1973. Identity of HB-1 of King and *Eikenella corrodens* (Eiken) Jackson and Goodman. Int. J. Syst. Bacteriol. *23*: 75–76.

Riley, P.S. and R.E. Weaver. 1977. Comparison of thirty-seven strains of Vd-3 bacteria with *Agrobacterium radiobacter*, morphological and physiological observations. J. Clin. Microbiol. *5*: 172–177.

Riley, U.B., G. Bignardi, L. Goldberg, A.P. Johnson and B. Holmes. 1996. Quinolone resistance in *Oligella urethralis*—associated chronic ambulatory peritoneal dialysis peritonitis. J. Infect. *32*: 155–156.

Rimler, R.B. and D.G. Simmons. 1983. Differentiation among bacteria isolated from turkeys with coryza (rhinotracheitis). Avian Dis. *27*: 491–500.

Ringel, M., R. Gross, T. Krafft, A. Kröger and R. Schauder. 1996. Growth of *Wolinella succinogenes* with elemental sulfur in the absence of polysulfide. Arch. Microbiol. *165*: 62–64.

Ringelberg, D.B., G.T. Townsend, K.A. DeWeerd, J.M. Suflita and D.C. White. 1994. Detection of the anaerobic dechlorinating microorganism *Desulfomonile tiedjei* in environmental matrixes by its signature lipopolysaccharide branced-long-chain hydroxy fatty acids. FEMS Microbiol. Ecol. *14*: 9–18.

Rioche, M. 1966. La rickettsiose générale bovine au Sénégal. Rev. Élev. Méd. Véterin. Pays. Trop. *19*: 485–494.

Riou, J.Y. 1977. Diagnostic bactériologiques des espèces des genres *Neisseria* et *Branhamella*. Ann. Biol. Clin. *35*: 73–87.

Riou, J.Y., J. Buissière, C. Richard and M. Guibourdenche. 1982. Gamma-Glutamyl-transferase activity in the family "*Neisseriaceae*". Ann. Microbiol. (Paris) *133*: 387–392.

Riou, J.Y. and M. Guibourdenche. 1987. *Neisseria polysaccharea* sp.nov. Int. J. Syst. Bacteriol. *37*: 163–165.

Riou, J.-Y., M. Guibourdenche, M.B. Perry, L.L. MacLean and D.W. Griffith. 1986. Structure of the exocellular D-glucan produced by Neisseria polysaccharea. Can. J. Microbiol. *32*: 909–911.

Riou, N. and D.L. Rudulier. 1990. Osmoregulation in *Azospirillum brasilense* glycine betaine transport enhances growth and nitrogen fixation under salt stress. NATO ASI ser, Ser G. *37*: 271–278.

Rippey, S.R. and V.J. Cabelli. 1979. Membrane filter procedure for enumeration of *Aeromonas hydrophila* in fresh waters. Appl. Environ. Microbiol. *38*: 106–113.

Ristic, M. 1981. Anaplasmosis. *In* Ristic and McIntyre (Editors), Diseases of Cattle in the Tropics, Martinus Nijhoff Publishers, The Hague. pp. 327–344.

Ristic, M. 1990. Current strategies in research on ehrlichiosis. *In* Williams and Kakoma (Editors), Erhlichiosis: A Vector-borne Disease of Animals and Humans, Kluwer Academic Publishers, Boston. pp. 136–153.

Ristic, M. and D.L. Huxsoll. 1984. *In* Validation of the publication of new names and new combinations previously effectively published outside the IJSB. List No. 15. Int. J. Syst. Bacteriol. *34*: 355–357.

Ristic, M. and J.P. Kreier. 1984a. Genus *Anaplasma*. *In* Krieg and Holt (Editors), Bergey's Manual of Systematic Bacteriology, Vol. 1, The Williams & Wilkins Co., Baltimore. pp. 720–722.

Ristic, M. and J.P. Kreier. 1984b. Genus II. *Grahamella*. *In* Krieg and Holt (Editors), Bergey's Manual of Systematic Bacteriology, 1st Ed., Vol. 1, The Williams & Wilkins Co., Baltimore. pp. 718–719.

Ristic, M. and J.P. Kreier. 1984c. *In* Validation of the publication of new names and new combinations previously effectively published outside the IJSB. List No. 15. Int. J. Syst. Bacteriol. *34*: 355–357.

Ritchie, A.E., R.F. Keeler and J.H. Bryner. 1966. Anatomical features of *Vibrio fetus*: An electron microscope survey. J. Gen. Microbiol. *43*: 427–438.

Rittenberg, B.T. and S.C. Rittenberg. 1962. The growth of *Spirillum volutans* Ehrenberg in mixed and pure cultures. Arch. Mikrobiol. *42*: 138–153.

Rittenberg, S.C. and R.B. Hespell. 1975. Energy efficiency of intraperiplasmic growth of *Bdellovibrio bacteriovorus*. J. Bacteriol. *121*: 1158–1165.

Rittenberg, S.C. and D. Langley. 1975. Utilization of nucleoside monophosphates per se for intraperiplasmic growth of *Bdellovibrio bacteriovorus*. J. Bacteriol. *121*: 1137–1144.

Rittenberg, S.C. and M. Shilo. 1970. Early host damage in the infection cycle of *Bdellovibrio bacteriovorus*. J. Bacteriol. *102*: 149–160.

Robakis, N.K., N.J. Palleroni, M. Boublik and C.W. Despreaux. 1985a. Construction of a restriction map of the *Gluconobacter bacteriophage* A-1 genome. J. Gen. Microbiol. *131*: 2475–2477.

Robakis, N.K., N.J. Palleroni, C.W. Despreaux, M. Boublik, C.A. Baker, P.J. Churn and G.W. Claus. 1985b. Isolation and characterization of two phages for *Gluconobacter oxydans*. J. Gen. Microbiol. *131*: 2467–2473.

Roberts, G.P., W.T. Leps, L.E. Silver and W.J. Brill. 1980. Use of two-dimensional polyacrylamide gel electrophoresis to identify and classify *Rhizobium* strains. Appl. Environ. Microbiol. *39*: 414–422.

Roberts, L.A., P.J. Collignon, V.B. Cramp, S. Alexander, A.E. McFarlane, E. Graham, A. Fuller, V. Sinickas and A. Hellyar. 1990. An Australia-wide epidemic of *Pseudomonas pickettii* bacteremia due to contaminated sterile water for injection. Med. J. Aust. *152*: 652–655.

Roberts, M.C. 1989. Plasmids of *Neisseria gonorrhoeae* and other *Neisseria* species. Clin. Microbiol. Rev. *2*: S18–23.

Roberts, M.C. and J.S. Knapp. 1988a. Host range of the conjugative 25.2-megadalton tetracycline resistance plasmid from *Neisseria gonorrhoeae* and related species. Antimicrob. Agents Chemother. *32*: 488–491.

Roberts, M.C. and J.S. Knapp. 1988b. Transfer of β-lactamase plasmids from *Neisseria gonorrhoeae* to *Neisseria meningitidis* and commensal *Neis-*

seria species by the 25.2-megadalton conjugative plasmid. Antimicrob. Agents Chemother. *32*: 1430–1432.

Roberts, S.A., M. A.J., N. McIvor and R. Ellis-Pegler. 1997. *Chromobacterium violaceum* infection of the deep neck tissues in a traveler to Thailand. Clin. Infect. Dis. *25*: 334–335.

Roberts, T., A. Shah, J.G. Graham and I.N. McQueen. 1987. The Miller Fischer syndrome following campylobacter enteritis: a report of two cases. J. Neurol. Neurosurg. Psychiatry *50*: 1557–1558.

Roberts, W.P., M.E. Tate and A. Kerr. 1977. Agrocin 84 is a 6-N-phosphoramidate of an adenine nucleotide analogue. Nature (Lond.) *265*: 379–381.

Robertson, A. 1924. Observation of the causal organism of rat-bite fever in man. Ann. Trop. Med. Parasitol. *18*: 157–175.

Robertson, B.R., J. O'Rourke, P. Vandamme, S.L.W. On and A. Lee. 2001. *Helicobacter ganmani* sp. nov., a urease-negative anaerobe isolated from the intestines of laboratory mice. Int. J. Syst. Bacteriol. *51*: 1881–1889.

Robertson, D.C. and W.G. McCullough. 1968a. The glucose catabolism of the genus *Brucella*. I. Evaluation of pathways. Arch. Biochem. Biophys. *127*: 263–273.

Robertson, D.C. and W.G. McCullough. 1968b. The glucose catabolism of the genus *Brucella*. II. Cell-free studies with *B. abortus* (S-19). Arch. Biochem. Biophys. *127*: 445–456.

Robertson, G.T. and R.M. Roop, Jr.. 1999. The *Brucella abortus* host factor I (HF-I) protein contributes to stress resistance during stationary phase and is a major determinant of virulence in mice. Mol. Microbiol. *34*: 690–700.

Robertson, J.N., P. Vincent and M.E. Ward. 1977. The preparation and properties of gonococcal pili. J. Gen. Microbiol. *102*: 169–177.

Robertson, L.A. and J.G. Kuenen. 1984. *In* Validation of the publication of new names and new combinations previously effectively published outside the IJSB. List No. 13. Int. J. Syst. Bacteriol. *34*: 91–92.

Robertson, R.G. and C.L. Wisseman, Jr.. 1973. Tick-borne rickettsiae of the spotted fever group in West Pakistan. II. Serological classification of isolates from West Pakistan and Thailand: Evidence for two new species. Am. J. Epidemiol. *97*: 55–64.

Robinson, J.V.A. and A.L. James. 1973. Some serological studies on *Bacteroides corrodens*. J. Gen. Microbiol. *78*: 193–197.

Robinson, J.V.A. and A.L. James. 1974. *In vitro* susceptibility of *Bacteroides corrodens* to ten chemotherapeutic agents. Antimicrob. Agents Chemother. *6*: 545–548.

Robinson, P.D. and A.J. Stipanovic. 1989. Method for oil recovery using a modified heteropolysaccharide, United States Patent 4874044.

Rockhill, R.C. and L.I. Lutwick. 1978. Group IVe-like gram-negative bacillemia in a patient with obstructive uropathy. J. Clin. Microbiol. *8*: 108–109.

Rodelas, B., V. Salmeron, M.V. Martinez-Toledo and J. Gonzalez-Lopez. 1993. Production of vitamins by *Azospirillum brasilense* in chemically-defined media. Plant Soil *153*: 97–101.

Roden, E.E. and D.R. Lovley. 1993. Dissimilatory Fe(III) reduction by the marine microorganism *Desulfuromonas acetoxidans*. Appl. Environ. Microbiol. *59*: 734–742.

Rodley, P.D., U. Romling and B. Tummler. 1995. A physical genome map of the *Burkholderia cepacia* type strain. Mol. Microbiol. *17*: 57–67.

Rodrigues, F.K., V. Virrankoski-Castrodeza, J.H. Parish and K. Grimm. 1980. Isolation and characterization of new bacteriophages for *Myxococcus xanthus*. Arch. Microbiol. *126*: 175–180.

Rodrigues-Pereira, A.S., P.J.W. Houwen, H.W.J. Deurenberg-Vos and E.B.F. Pey. 1972. Cytokinins and the bacterial symbiosis of *Ardisia* species. Z. Pflanzenphysiol. *68*: 170–177.

Rodriguez-Valera, F., A. Ventosa, G. Juez and J.F. Imhoff. 1985. Variation of environmental features and microbial populations with salt concentrations in a multi-pond saltern. Microb. Ecol. *11*: 107–116.

Roessler, W.G., T.H. Sanders, J. Dulberg and C.R. Brewer. 1952. Anaerobic glycolysis by enzyme preparations of *Brucella suis*. J. Biol. Chem. *194*: 207–213.

Rogers, P.L., K.J. Lee, M.L. Skotnicki and D.E. Tribe. 1982. Ethanol production by *Zymomonas mobilis*. Adv. Biochem. Eng. Biotechnol. *23*: 27–84.

Rogers, S.R. and J.J. Anderson. 1976a. Measurement of growth and iron deposition in *Sphaerotilus discophorus*. J. Bacteriol. *126*: 257–263.

Rogers, S.R. and J.J. Anderson. 1976b. Role of iron deposition in *Sphaerotilus discophorus*. J. Bacteriol. *126*: 264–271.

Roggentin, P. 1980. Untersuchung der zellwand knospender und prosthekater bakterien unter besonderer berücksichtigung und morphogenetischer aspekte, University of Kiel. Kiel, Germany. . p. 73.

Roggentin, P. and P. Hirsch. 1982. Peptidoglycan composition of hyphal and budding bacteria. *In* Abstr. Annu. Meet. Am. Soc. Microbiol, CC II 10, p.172.

Roggentin, T. and P. Hirsch. 1989. Ribosomal RNA cistron similarities among *Hyphomicrobium* species and several other hyphal, budding bacteria. Syst. Appl. Microbiol. *11*: 140–147.

Rogosa, M. 1980. *Streptobacillus moniliformis* and Spirillum minor. *In* Lennette, Balows, Hausler and Truant (Editors), Manual of Clinical Microbiology, 3rd Ed.., American Society for Microbiology, Washington, D.C. pp. 350–356.

Rohde, M., F. Mayer and O. Meyer. 1984. Immunocytochemical localization of carbon monoxide oxidase in *Pseudomonas carboxydovorans*. J. Biol. Chem. *259*: 14788–14792.

Rohmer, M., P. Bouvier and G. Ourisson. 1979. Molecular evolution of biomembranes: structural equivalents and phylogenetic precursors of sterols. Proc. Natl. Acad. Sci. U.S.A. *76*: 847–851.

Rohmer, M., P. Bouvier-Nave and G. Ourisson. 1984. Distribution of hopanoid triterpenes in prokaryotes. J. Gen. Microbiol. *130*: 1137–1150.

Rohmer, M., M. Knani, P. Simonin, B. Sutter and H. Sahm. 1993. Isoprenoid biosynthesis in bacteria: a novel pathway for the early steps leading to isopentenyl diphosphate. Biochem. J. *295*: 517–524.

Rokbi, B., M. Mignon, D.A. Caugant and M.J. QuentinMillet. 1997. Heterogeneity of *tbpB*, the transferrin-binding protein B gene, among serogroup B *Neisseria meningitidis* strains of the ET-5 complex. Clin. Diagn. Lab. Immunol. *4*: 522–529.

Rolls, J.P. and E.S. Lindstrom. 1967. Effect of thiosulfate on the photosynthetic growth of *Rhodopseudomonas palustris*. J. Bacteriol. *94*: 860–869.

Romagnoli, S., A. Hochkoeppler, L. Damgaard and D. Zannoni. 1997. The effect of respiration on the phototactic behavior of the purple nonsulfur bacterium *Rhodospirillum centenum*. Arch. Microbiol. *167*: 99–105.

Romaniuk, P.J., B. Zoltowska, T.J. Trust, D.J. Lane, G.J. Olsen, N.R. Pace and D.A. Stahl. 1987. *Campylobacter pylori*, the spiral bacterium associated with human gastritis, is not a true *Campylobacter* sp. J. Bacteriol. *169*: 2137–2141.

Romano, A.H. and J.P. Peloquin. 1963. Composition of the sheath of *Sphaerotilus natans*. J. Bacteriol. *86*: 252–258.

Romanovskaya, V.A., Y.R. Malashenko and V.N. Bogachenko. 1978. Corrected diagnoses of the genera and species of methane-utilizing bacteria. Mikrobiologiya *47*: 96–103.

Rome, S., M.P. Fernandez, B. Brunel, P. Normand and J.-C. Cleyet-Marel. 1996. *Sinorhizobium medicae* sp. nov., isolated from annual *Medicago* spp. Int. J. Syst. Bacteriol. *46*: 972–980.

Ronson, C.W. and S.B. Primrose. 1979. Carbohydrate metabolism in *Rhizobium trifolii*. Identification and symbiotic properties of mutants. J. Gen. Microbiol. *112*: 77–88.

Rooney-Varga, J.N., R.T. Anderson, J.L. Fraga, D. Ringelberg and D.R. Lovley. 1999. Microbial communities associated with anaerobic benzene degradation in a petroleum-contaminated aquifer. Appl. Environ. Microbiol. *65*: 3056–3063.

Roop, R.M. 1984. *Bordetella* and *Alcaligenes*. *In* Carter and Cole (Editors), Diagnostic Procedures in Veterinary Bacteriology and Mycology, 5th ed., Academic Press Inc., San Diego, CA. pp. 87–93.

Roop, R.M., R.M. Smibert, J.L. Johnson and N.R. Krieg. 1985a. *Campylobacter mucosalis* (Lawson, Leaver, Pettigrew, and Rowland 1981) comb. nov.: emended description. Int. J. Syst. Bacteriol. *35*: 189–192.

Roop, R.M., R.M. Smibert, J.L. Johnson and N.R. Krieg. 1985b. DNA homology studies of the catalase-negative campylobacters and "*Campylobacter fecalis*", an emended description of *Campylobacter sputorum*, and proposal of the neotype strain of *Campylobacter sputorum*. Can. J. Microbiol. *31*: 823–831.

Roop, R.M., R.M. Smibert, J.L. Johnson and N.R. Krieg. 1986. Designation of the neotype strain for *Campylobacter sputorum* (Prévot) Véron adn Chatelain 1973. Int.J. Syst. Bacteriol. *36*: 348.

Rosati, G. and F. Verni. 1975. Macronuclear symbionts in *Euplotes crassus* (Ciliata, Hypotrichida). Boll. Zool. *42*: 231–232.

Rosati, G. and F. Verni. 1977. Bacteria-like endosymbionts in *Euplotes crassus*, Abstract. *In* 5th International Congress on Protozoology, Allen Press, Kansas. p. 443.

Rosati, G., F. Verni and P. Luporini. 1976. Cytoplasmic bacteria-like endosymbionts in *Euplotes crassus* (Dujardin) (Ciliata, Hypotrichida). Monitore Zool. Ital. (N.S.) *10*: 449–460.

Rosen, H.R. 1922. A bacterial disease of foxtail (*Chaetochloa lutescens*). Ann. Mo. Bot. Gard. *9*: 333–402.

Rosen, M.R., L. Coshell, J.V. Turner and R.J. Woodbury. 1996. Hydrochemistry and nutrient cycling in Yalgorup National Park, Western Australia. J. Hydrol. *185*: 241–274.

Rosenberg, C., F. Casse-Delbart, I. Dusha, M. David and C. Boucher. 1982. Megaplasmids in the plant-associated bacteria *Rhizobium meliloti* and *Pseudomonas solanacearum*. J. Bacteriol. *150*: 402–406.

Rosenberg, E. and M. Dworkin (Editors). 1984. Myxobacteria. Development and Cell Interactions, Springer-Verlag, New York.

Rosenberg, E., K.H. Keller and M. Dworkin. 1977. Cell density-dependent growth of *Myxococcus xanthus* on casein. J. Bacteriol. *129*: 770–777.

Rosenfelder, G., O. Lüderitz and O. Westphal. 1974. Composition of lipopolysaccharides from *Myxococcus fulvus* and other fruiting and non-fruiting myxobacteria. Eur. J. Biochem. *44*: 411–420.

Rosenthal, R.S. 1979. Release of soluble peptidoglycan from growing gonococci: Hexaminidase and amidase activities. Infect. Immun. *24*: 869–878.

Rosenthal, R.S., R.M.S. Wright and R.K. Sinha. 1980. Extent of peptide cross-linking in the peptidoglycan of *Neisseria gonorrhoeae*. Infect. Immun. *28*: 867–875.

Roslev, P. and G.M. King. 1994. Survival and recovery of methanotrophic bacteria starved under oxic and anoxic conditions. Appl. Environ. Microbiol. *60*: 2602–2608.

Roslev, P. and G.M. King. 1995. Aerobic and anaerobic starvation metabolism in methanotrophic bacteria. Appl. Environ. Microbiol. *61*: 1563–1570.

Roslycky, E.B. 1967. Bacteriocin production in the rhizobia bacteria. Can. J. Microbiol. *13*: 431–432.

Roslycky, E.B., O.N. Allen and E. McCoy. 1963. Serological properties of phages of *Agrobacterium radiobacter*. Can. J. Microbiol. *9*: 709–717.

Ross, H.M. and M.J. Corbel. 1980. Isolation of a cell wall defective strain of *Brucella abortus* from bovine tissue. Vet. Rec. *106*: 242.

Ross, H.M., K.L. Jahans, A.P. MacMillan, R.J. Reid, P.M. Thompson and G. Foster. 1996. *Brucella* species infection in North Sea seal and cetacean populations. Vet. Rec. *138*: 647–648.

Ross, J.L., P.I. Boon, P. Ford and B.T. Hart. 1997. Detection and quantification with 16S rRNA probes of planktonic methylotrophic bacteria in a floodplain lake. Microb. Ecol. *34*: 97–108.

Ross, J.P., S.M. Holland, V.J. Gill, E.S. DeCarlo and J.I. Gallin. 1995. Severe *Burkholderia (Pseudomonas) gladioli* infection in chronic granulomatous disease: report of two successfully treated cases. Clin. Infect. Dis. *21*: 1291–1293.

Ross, P., R. Mayer and M. Benziman. 1991. Cellulose biosynthesis and function in bacteria. Microbiol. Rev. *55*: 35–58.

Ross, P., H. Weinhouse, Y. Aloni, D. Michaeli, P. Weinberger-Ohana, R. Mayer, S. Braun, E. de Vroom, G.A. van der Marel, J.H. van Boom and M. Benziman. 1987. Regulation of cellulose synthesis in *Acetobacter xylinum* by cyclic diguanylic acid. Nature (Lond.) *325*: 279–281.

Rossau, R., K. Kersters, E. Falsen, E. Jantzen, P. Segers, A. Union, L. Nehls and J. De Ley. 1987. *Oligella*, a new genus including *Oligella urethralis* comb. nov. (formerly *Moraxella urethralis*) and *Oligella ureolytica* sp. nov. (formerly CDC group IVe): relationship to *Taylorella equigenitalis* and related taxa. Int. J. Syst. Bacteriol. *37*: 198–210.

Rossau, R., A. Van Landschoot, M. Gillis and J. De Ley. 1991. Taxonomy of *Moraxellaceae* fam. nov., a new bacterial family to accomodate the genera *Moraxella*, *Acinetobacter*, and *Psychrobacter* and related organisms. Int. J. Syst. Bacteriol. *41*: 310–319.

Rossau, R., A. Van Landschoot, W. Mannheim and J. De Ley. 1986. Intergeneric and intrageneric similarities of ribosomal RNA cistrons of the *Neisseriaceae*. Int. J. Syst. Bacteriol. *36*: 323–332.

Rossau, R., G. Vandenbusche, S. Thielemans, P. Segers, H. Grosch, E. Goethe, W. Mannheim and J. De Ley. 1989. Ribosomal RNA cistron similarities and DNA homologies of *Neisseria*, *Kingella*, *Eikenella*, *Simonsiella*, *Alysiella*, and Centers for Disease Control groups EF-4 and M-5 in the emended family *Neisseriaceae*. Int. J. Syst. Bacteriol. *39*: 185–198.

Rosselló-Mora, R.A., W. Ludwig and K.H. Schleifer. 1993. *Zoogloea ramigera*: A phylogenetically diverse species. FEMS Micriobiol. Lett. *114*: 129–134.

Rosselló-Mora, R.A., M. Wagner, R. Amann and K.H. Schleifer. 1995. The abundance of *Zoogloea ramigera* in sewage treatment plants. Appl. Environ. Microbiol. *61*: 702–707.

Rossen, L., A.W. Johnston and J.A. Downie. 1984. DNA sequence of the *Rhizobium leguminosarum* nodulation genes *nodAB* and *C* required for root hair curling. Nucleic Acids Res. *12*: 9497–9508.

Rossmann, S.N., P.H. Wilson, J. Hicks, B. Carter, S.G. Cron, C. Simon, C.M. Flaitz, G.J. Demmler, W.T. Shearer and M.W. Kline. 1998. Isolation of *Lautropia mirabilis* from oral cavities of human immunodeficiency virus-infected children. J. Clin. Microbiol. *36*: 1756–1760.

Rossolini, G.M., M.A. Condemi, F. Pantanella, J.D. Docquier, G. Amicosante and M.C. Thaller. 2001. Metallo-beta-lactamase producers in environmental microbiota: new molecular class B enzyme in *Janthinobacterium lividum*. Antimicrob. Agents Chemother. *45*: 837–844.

Rostas, K., E. Kondorosi, B. Horvath, A. Simoncsits and Á. Kondorosi. 1986. Conservation of extended promoter regions of nodulation genes in *Rhizobium*. Proc. Natl. Acad. Sci. U. S. A. *83*: 1757–1761.

Rotger, R., F. Rubio and C. Nombela. 1986. A multi-resistance plasmid isolated from commensal *Neisseria* species is closely related to the enterobacterial plasmid RSF1010. J. Gen. Microbiol. *132*: 2491–2496.

Roth, A.W. 1797. Catalecta botanica quibus plantae novae et minus cognitae describuntur atque illistrantur. *In* Lipsiae in Bibliopolio I.G. Mulleriano, fasc. 1.

Rothe, B., A. Fischer, P. Hirsch, M. Sittig and E. Stackebrandt. 1987. The phylogenetic position of the budding bacteria *Blastobacter aggregatus* and *Gemmobacter aquatilis* gen. nov., sp. nov. Arch. Microbiol. *147*: 92–99.

Rothe, B., A. Fischer, P. Hirsch, M. Sittig and E. Stackebrandt. 1988. *In* Validation of the publication of new names and new combinations previously effectively published outside the IJSB. List No. 26. Int. J. Syst. Bacteriol. *38*: 328–329.

Rouf, M.A. and J.L. Stokes. 1964. Morphology, nutrition and physiology of *Sphaerotilus discophorus*. Arch. Mikrobiol. *49*: 132–149.

Rouppe van der Voort, E., M. Schuller, J. Holst, P. de Vries, P. van der Ley, G. van den Dobbelsteen and J. Poolman. 2000. Immunogenicity studies with a genetically engineered hexavalent PorA and a wild-type meningococcal group B outer membrane vesicle vaccine in infant *Cynomolgus* monkeys. Vaccine *18*: 1334–1343.

Rousset, F., D. Bouchon, B. Pintureau, P. Juchault and M. Solignac. 1992. *Wolbachia* endosymbionts responsible for various alterations of sexuality in arthropods. Proc. R. Soc. Lond. B Biol. Sci. *250*: 91–98.

Roux, J. and J. Sassine. 1971. Étude d' une souche fixée de sphéroplastes (formes L) de *Brucella melitensis*. Ann. Inst. Pasteur (Paris) *120*: 174–185.

Roux, V., M. Drancourt and D. Raoult. 1992. Determination of genome sizes of *Rickettsia* spp. within the spotted fever group, using pulsed-field gel electrophoresis. J. Bacteriol. *174*: 7455–7457.

Roux, V. and D. Raoult. 2000. Phylogenetic analysis of members of the genus *Rickettsia* using the gene encoding the outer-membrane protein rOmpB (*ompB*). Int. J. Syst. Evol. Microbiol. *50 Pt 4*: 1449–1455.

Rowatt, E. 1957. The growth of *Bordetella pertussis*: a review. J. Gen. Microbiol. *17*: 297–326.

Rowland, A.C. and G.H. Lawson. 1974. Intestinal adenomatosis in the pig: immunofluorescent and electron microscopic studies. Res. Vet. Sci. *17*: 323–330.

Roy, A.B. and S. Sen. 1962. A new species of *Derxia*. Nature *194*: 604–605.

Roy, M.A. and M. Sasser. 1983. A medium selective for *Agrobacterium tumefaciens* biotype 3. Phytopathology *73*: 810–810.

Rozanova, E.P. and T.N. Nazina. 1976. A mesophilic, sulfate-reducing, rod-shaped nonspore-forming bacterium. Mikrobiologiya *45*: 825–830.

Rozanova, E.P., T.N. Nazina and A.S. Galushko. 1988. A new genus of sulfate-reducing bacteria and the description of its new species, *Desulfomicrobium apsheronum*, new genus new species. Mikrobiologiya *57*: 514–520.

Rozanova, E.P., T.N. Nazina and A.S. Galushko. 1994. *In* Validation of the publication of new names and new combinations previously effectively published outside the IJSB. List No. 49. Int. J. Syst. Bacteriol. *44*: 370–371.

Ruan, Z.S., V. Anantharam, I.T. Crawford, S.V. Ambudkar, S.Y. Rhee, M.J. Allison and P.C. Maloney. 1992. Identification, purification, and reconstitution of OxlT, the oxalate-formate antiport protein of *Oxalobacter formigenes*. J. Biol. Chem. *267*: 10537–10543.

Rubenstein, J.E., M.F. Lieberman and N. Gadoth. 1976. Central nervous system infection with *Eikenella corrodens*: report of two cases. Pediatrics. *57*: 264–265.

Rubin, S.J., P.A. Granato and B.L. Wasilauskas. 1980. Glucose-nonfermenting Gram-negative bacteria. *In* Lennette, Balows, J. Hausler and Truant (Editors), Manual of Clinical Microbiology, 3rd Ed., American Society for Microbiology, Washington, D.C. pp. 263–287.

Rubin, S.J., P.A. Granato and B.L. Wasilauskas. 1985. Glucose-nonfermenting Gram-negative bacteria. *In* Lennette, Balows, Hausler and Shadomy (Editors), Manual of Clinical Microbiology, 4th ed., American Society for Microbiology, Washington, D.C. pp. 330–349.

Ruby, E.G. 1989. Cell-envelope modifications accompanying intracellular growth of *Bdellovibrio bacteriovorus*. *In* Moulder (Editor), Intracellular Parasitism, CRC Press, Boca Raton. pp. 17–34.

Ruby, E.G. 1992. The genus *Bdellovibrio*. *In* Balows, Trüper, Dworkin, Harder and Schleifer (Editors), The Prokaryotes: A Handbook of Bacteria: Ecophysiology, Isolation, Identification, Applications, 2nd Ed, Vol. 4, Springer-Verlag, New York. pp. 3400–3415.

Ruby, E.G. and J.B. McCabe. 1986. An ATP transport system in the intracellular bacterium, *Bdellovibrio bacteriovorus* 109J. J. Bacteriol. *167*: 1066–1070.

Ruby, E.G., J.B. McCabe and J.I. Barke. 1985. Uptake of intact nucleoside monophosphates by *Bdellovibrio bacteriovorus* 109J. J. Bacteriol. *163*: 1087–1094.

Rückert, G. 1978. Förderung der Fruchtkörper-bildung von *Myxococcus virescens* Thaxter (*Myxobacterales*) in Rohkulturen durch Salzzusatz. Z. Allg. Mikrobiol. *18*: 69–71.

Rückert, G. 1984. Untersuchungen zum Vorkommen von Myxobakterien in von Meerwasser beeinflussten Substraten unter besonderer Berücksichtigung der Insel Helgoland. Helgol. Meeresunters. *38*: 179–184.

Rudd, K.E. and D.R. Zusman. 1982. RNA polymerase of *Myxococcus xanthus*: Purification and selective transcription in vitro with bacteriophage templates. J. Bacteriol. *151*: 89–105.

Rudel, T., H.J. Boxberger and T.F. Meyer. 1995a. Pilus biogenesis and epithelial-cell adherence of *Neisseria gonorrhoeae pilC* double knockout mutants. Mol. Microbiol. *17*: 1057–1071.

Rudel, T., D. Facius, R. Barten, I. Scheuerpflug, E. Nonnenmacher and T.F. Meyer. 1995b. Role of pili and the phase-variable PilC protein in natural competence for transformation of *Neisseria gonorrhoeae*. Proc. Natl. Acad. Sci. U.S.A. *92*: 7986–7990.

Rueter, P., R. Rabus, H. Wilkes, F. Aeckersberg, F.A. Rainey, H.W. Jannasch and F. Widdel. 1994. Anaerobic oxidation of hydrocarbons in crude oil by new types of sulphate-reducing bacteria. Nature *372*: 455–458.

Rüger, H.J. and M.G. Höfle. 1992. Marine star-shaped-aggregate-forming bacteria: *Agrobacterium atlanticum*, sp. nov.; *Agrobacterium meteori*, sp. nov.; *Agrobacterium ferrugineum*, sp. nov., nom. rev.; *Agrobacterium ge-*

latinovorum, sp. nov., nom. rev., and *Agrobacterium stellulatum*, sp. nov., nom. rev. Int. J. Syst. Bacteriol. *42*: 133–143.

Rüger, H.-J. and T.L. Tan. 1983. Separation of *Alcaligenes denitrificans* sp. nov., nom. rev. from *Alcaligenes faecalis* on the basis of DNA base composition, DNA homology, and nitrate reduction. Int. J. Syst. Bacteriol. *33*: 85–89.

Ruinen, J. 1956. Occurrence of *Beijerinckia* species in the "phyllosphere". Nature *177*: 220–221.

Ruinen, J. 1961. The phyllosphere. I. An ecologically neglected milieu. Plant Soil *15*: 81–109.

Ruiz, A., M. Poblet, A. Mas and J.M. Guillamon. 2000. Identification of acetic acid bacteria by RFLP of PCR-amplified 16S rDNA and 16S–23S rDNA intergenic spacer. Int. J. Syst. Evol. Microbiol. *50*: 1981–1987.

Ruiz-Ponte, C., V. Cilia, C. Lambert and J.L. Nicolas. 1998. *Roseobacter gallaeciensis* sp. nov., a new marine bacterium isolated from rearings and collectors of the scallop *Pecten maximus*. Int. J. Syst. Bacteriol. *48*: 537–542.

Rurangirwa, F.R., D. Stiller, D.M. French and G.H. Palmer. 1999. Restriction of major surface protein 2 (MSP2) variants during tick transmission of the ehrlichia *Anaplasma marginale*. Proc. Natl. Acad. Sci. U.S.A. *96*: 3171–3176.

Rurangirwa, F.R., D. Stiller and G.H. Palmer. 2000. Strain diversity in major surface protein 2 expression during tick transmission of *Anaplasma marginale*. Infect. Immun. *68*: 3023–3027.

Rushing, B.G., M.M. Yelton and S.R. Long. 1991. Genetic and physical analysis of the *nodD3* region of *Rhizobium meliloti*. Nucleic Acids Res. *19*: 921–927.

Russel, M. 1998. Macromolecular assembly and secretion across the bacterial cell envelope: type II protein secretion systems. J. Mol. Biol. *279*: 485–499.

Russel, S. and S. Muszynski. 1995. Reduction of 4-chloronitrobenzene by *Azospirillum lipoferum*. NATO ASI Ser. Ser. G Ecol. Sci. *37*: 549–554.

Russell, R.R.B., K.G. Johnson and I.J. McDonald. 1975. Envelope proteins in *Neisseria*. Can. J. Microbiol. *21*: 1519–1534.

Rutter, J.M. 1985. Atrophic rhinitis in swine. Adv. Vet. Sci. Comp. Med. *29*: 239–279.

Ryall, C. and M.O. Moss. 1975. Selective media for the enumeration of *Chromobacterium* spp. in soil and water. J. Appl. Bacteriol. *38*: 53–59.

Rydkina, E., V. Roux, N. Balayeva and D. Raoult. 1996. Determination of the genome size of ehrlichiae by pulsed-field gel electrophoresis. Proceedings of the 5th International Symposium on Rickettsiae and Rickettsial Diseases, Bratislava, Slovak Republic: Slovak Academy of Sciences, International Society of Rickettsiae and Rickettsial Diseases. pp. 318–323.

Rydkina, E., V. Roux and D. Raoult. 1999. Determination of the genome size of *Ehrlichia* spp., using pulsed field gel electrophoresis. FEMS Microbiol. Lett. *176*: 73–78.

Ryu, E. 1937. A simple method of staining bacterial flagella. Kitisato Arch. Exp. Med. *14*: 218–219.

Sacks, V., I. Eshkenazi, T. Neufeld, C. Dosoretz and J. Rishpon. 2000. Immobilized parathion hydrolase: An amperometric sensor for parathion. Anal. Chem. *72*: 2055–2058.

Sadowsky, M.J., P.B. Cregan, M. Gottfert, A. Sharma, D. Gerhold, F. Rodriguez Quinones, H.H. Keyser, H. Hennecke and G. Stacey. 1991. The *Bradyrhizobium japonicum nolA* gene and its involvement in the genotype-specific nodulation of soybeans. Proc. Natl. Acad. Sci. U.S.A. *88*: 637–641.

Saegusa, H., M. Shiraki, C. Kanai and T. Saito. 2001. Cloning of an intracellular poly(D(−)-3-hydroxybutyrate) depolymerase gene from *Ralstonia eutropha* H16 and characterization of the gene product. J. Bacteriol. *183*: 94–100.

Saeki, A., K. Matsushita, S. Takeno, M. Taniguchi, H. Toyama, G. Theeragool, N. Lotong and O. Adachi. 1999. Enzymes responsible for acetate oxidation by acetic acid bacteria. Biosci. Biotechnol. Biochem. *63*: 2102–2109.

Saeki, A., G. Theeragool, K. Matsushita, H. Toyama, N. Lotong and O. Adachi. 1997. Development of thermotolerant acetic acid bacteria useful for vinegar fermentation at higher temperatures. Biosci. Biotechnol. Biochem. *61*: 138–145.

Sager, B. and D. Kaiser. 1994. Intercellular C-aignaling and the traveling waves of myxococcus. Genes Dev. *8*: 2793–2804.

Sahin, N., K. Isik, A.U. Tamer and M. Goodfellow. 2000a. Taxonomic position of "*Pseudomonas oxalaticus*" strain Ox1ᵀ (DSM 1105ᵀ) (Khambata and Bhat, 1953) and its description in the genus *Ralstonia* as *Ralstonia oxalatica* comb. nov. Syst. Appl. Microbiol. *23*: 206–209.

Sahin, N., K. Isik, A.U. Tamer and M. Goodfellow. 2000b. *In* Validation of the publication of new names and new combinations previously effectively published outside the IJSEM. List No. 77. Int. J. Syst. Evol. Microbiol. *50*: 1953.

Sahm, H., S. Bringer-Meyer and G. Sprenger. 1992. The Genus *Zymomonas. In* Balows, Trüper, Dworkin, Harder and Schleifer (Editors), The Prokaryotes: A Handbook of Bacteria: Ecophysiology, Isolation, Identification, Applications, 2nd ed., Vol. 3, Springer-Verlag, New York. pp. 2287–2301.

Sahm, H., R.B. Cox and J.R. Quayle. 1976. Metabolism of methanol by *Rhodopseudomonas acidophila.* J. Gen. Microbiol. *94*: 313–322.

Sahm, H., M. Rohmer, S. Bringer Meyer, G.A. Sprenger and R. Welle. 1993. Biochemistry and physiology of hopanoids in bacteria. Adv. Microb. Physiol. *35*: 247–273.

Sahm, K., C. Knoblauch and R. Amann. 1999. Phylogenetic affiliation and quantification of psychrophilic sulfate-reducing isolates in marine arctic sediments. Appl. Environ. Microbiol. *65*: 3976–3981.

Sahni, S.K., D.J. Van Antwerp, M.E. Eremeeva, D.J. Silverman, MarderV.J. and L.A. Sporn. 1998. Proteasome-independent activation of nuclear factor κB in cytoplasmic extracts from human endothelial cells by *Rickettsia rickettsii.* Infect. Immun. *66*: 1827–1833.

Saier, M.H. and J.T. Staley. 1977. Phosphoenolpyruvate: sugar phosphotransferase system in *Ancalomicrobium adetum.* J. Bacteriol. *131*: 716–718.

Saif, Y.M., P.D. Moorhead, R.N. Dearth and D.J. Jackwood. 1980. Observations on *Alcaligenes faecalis* infection in turkeys. Avian Dis. *24*: 665–684.

Saile, E., J.A. McGarvey, M.A. Schell and T.P. Denny. 1997. Role of extracellular polysaccharide and endoglucanase in root invasion and colonization of tomato plants by *Ralstonia solanacearum.* Phytopathology *87*: 1264–1271.

Saiman, L., G. Cacalano and A. Prince. 1990. *Pseudomonas cepacia* adherence to respiratory epithelial cells is enhanced by *Pseudomonas aeruginosa.* Infect. Immun. *58*: 2578–2584.

Saiman, L., J. Sadoff and A. Prince. 1989. Cross reactivity of *Pseudomonas aeruginosa* antipilin monoclonal antibodies with heterogeneous strains of *P. aeruginosa* and *Pseudomonas cepacia.* Infect. Immun. *57*: 2764–2770.

Saint, C.P. and D.W. Ribbons. 1990. A catabolic plasmid involved in 4-methyl-ortho-phthalate and 4-hydroxy-iso-phthalate degradation in *Pseudomonas cepacia.* FEMS Microbiol. Lett. *69*: 323–328.

Saint, C.P. and P. Romas. 1996. 4-Methylphthalate catabolism in *Burkholderia (Pseudomonas) cepacia* Pc701: a gene encoding a phthalate specific permease forms part of a novel gene cluster. Microbiology *142*: 2407–2418.

Saint André, A., N.M. Blackwell, L.R. Hall, A. Hoerauf, N.W. Brattig, L. Volkmann, M.J. Taylor, L. Ford, A.G. Hise, J.H. Lass, E. Diaconu and E. Pearlman. 2002. The role of endosymbiotic *Wolbachia* bacteria in the pathogenesis of river blindness. Science *295*: 1892–1895.

Saito, A., H. Mitsui, R. Hattori, K. Minamisawa and T. Hattori. 1998. Slow-growing and oligotrophic soil bacteria phylogenetically close to *Bradyrhizobium japonicum.* FEMS Microbiol. Ecol. *25*: 277–286.

Saito, T., K. Suzuki, J. Yamamoto, T. Fukui, K. Miwa, K. Tomita, S. Nakanishi, S. Odani, J. Suzuki and K. Ishikawa. 1989. Cloning, nucleotide sequence, and expression in *Escherichia coli* of the gene for poly(3-hydroxybutyrate) depolymerase from *Alcaligenes faecalis.* J. Bacteriol. *171*: 184–189.

Saito, Y., Y. Ishii, H. Hayashi, Y. Imao, T. Akashi, K. Yoshikawa, Y. Noguchi, S. Soeda, M. Yoshida, M. Niwa, J. Hosoda and K. Shimomura. 1997. Cloning of genes coding for ʟ-sorbose and ʟ-sorbosone dehydrogenases from *Gluconobacter oxydans* and microbial production of 2-keto-ʟ-gulonate, a precursor of ʟ-ascorbic acid, in a recombinant *G. oxydans* strain. Appl. Environ. Microbiol. *63*: 454–460.

Saitoh, S. and Y. Nishimura. 1996. Taxonomic characterization of novel bacteriochlorophyll-containing bacteria isolated from soil. J. Gen. Microbiol. *42*: 121–140.

Saitoh, S., T. Suziki and Y. Nishimura. 1998. Proposal of *Craurococcus roseus* gen. nov., sp. nov. and *Paracraurococcus ruber* gen. nov., sp. nov., novel aerobic bacteriochlorophyll *a*-containing bacteria from soil. Int. J. Syst. Bacteriol. *48*: 1043–1047.

Saitou, N. and M. Nei. 1987. The neighbor-joining method: a new method for reconstructing phylogenetic trees. Mol. Biol. Evol. *4*: 406–425.

Sajjan, U.S. and J.F. Forstner. 1993. Role of a 22-kilodalton pilin protein in binding of *Pseudomonas cepacia* to buccal epithelial cells. Infect. Immun. *61*: 3157–3163.

Sajjan, U.S., L. Sun, R. Goldstein and J.F. Forstner. 1995. Cable (cbl) type II pili of cystic fibrosis associated *Burkholderia (Pseudomonas) cepacia*: nucleotide sequence of the *cblA* major subunit pilin gene and novel morphology of the assembled appendage fibers. J. Bacteriol. *177*: 1030–1038.

Sakai, S.N., A. Numata, A. Kawai, M. Katano and E. Yabuuchi. 1978. *Pseudomonas paucimobilis* isolated from a patient with postoperative meningitis. Jpn. J. Clin. Microbiol. *26*: 450.

Sakane, T. and K. Kuroshima. 1997. Viabilities of dried cultures of various bacteria after preservation for over 20 years and their prediction by the accelerated storage test. Microbiol. Cult. Coll. *13*: 1–7.

Sakane, T., T. Nishii, T. Itoh and K. Mikata. 1996. Protocols for long-term preservation of microorganisms by ʟ-drying (in Japanese). Microbiol. Cult. Coll. *12*: 91–97.

Sakane, T. and A. Yokota. 1994. Chemotaxonomic investigation of heterotrophic, aerobic and microaerophilic spirilla, the genera *Aquaspirillum, Magnetospirillum* and *Oceanospirillum.* Syst. Appl. Microbiol. *17*: 128–134.

Sako, W. 1947. Studies on pertussis immunization. J. Pediat. *30*: 29–40.

Salama, S.M., M.M. Garcia and D.E. Taylor. 1992a. Differentiation of the subspecies of *Campylobacter fetus* by genomic sizing. Int. J. Syst. Bacteriol. *42*: 446–450.

Salama, S.M., H. Tabor, M. Richter and D.E. Taylor. 1992b. Pulsed-field gel electrophoresis for epidemiologic studies of *Campylobacter hyointestinalis* isolates. J. Clin. Microbiol. *30*: 1982–1984.

Salanitro, J.P., P.A. Muirhead and J.R. Goodman. 1976. Morphological and physiological characteristics of *Gemmiger formicilis* isolated from chicken ceca. Appl. Environ. Microbiol. *32*: 623–632.

Salanoubat, M., S. Genin, F. Artiguenave, J. Gouzy, S. Mangenot, M. Arlat, A. Billault, P. Brottier, J.C. Camus, L. Cattolico, M. Chandler, N. Choisne, C. Claudel Renard, S. Cunnac, N. Demange, C. Gaspin, M. Lavie, A. Moisan, C. Robert, W. Saurin, T. Schiex, P. Siguier, P. Thebault, M. Whalen, P. Wincker, M. Levy, J. Weissenbach and C.A. Boucher. 2002. Genome sequence of the plant pathogen *Ralstonia solanacearum.* Nature *415*: 497–502.

Salaun, L., C. Audibert, G. Le Lay, C. Burucoa, J.L. Fauchere and B. Picard. 1998. Panmictic structure of *Helicobacter pylori* demonstrated by the comparative study of six geneic markers. FEMS Microbiol. Lett. *16*: 231–239.

Salloway, S., L.A. Mermel, M. Seamans, G.O. Aspinall, J.E. Nam Shin, L.A. Kurjanczyk and J.L. Penner. 1996. Miller-Fisher syndrome associated with *Campylobacter jejuni* bearing lipopolysaccharide molecules that mimic human ganglioside GD3. Infect. Immun. *64*: 2945–2949.

Salmond, G.P. 1996. Pili, peptidases and protein secretion: curious connections. Trends Microbiol. *4*: 474–476.

Salomon, H. 1898. Über das *Spirillum* des Säugetiermagens und sein Verhalten zu den Belegzellen. Zentralbl. Bakteriol. (Orig. A) *19*: 422–441.

Salvatore, P., C. Bucci, C. Pagliarulo, M. Tredici, R. Colicchio, G. Cantalupo, M. Bardaro, L. Del Giudice, D.R. Massardo, A. Lavitola, C.B. Bruni and P. Alifano. 2002. Phenotypes of a naturally defective *recB* allele in *Neisseria meningitidis* clinical isolates. Infect. Immun. *70*: 4185–4195.

Samain, E., H.C. Dubourguier and G. Albagnac. 1984. Isolation and Characterization of *Desulfobulbus elongatus* sp. nov. from a mesophilic industrial digester. Syst. Appl. Microbiol. *5*: 391–401.

Samain, E., H.C. Dubourguier and G. Albagnac. 1985. *In* Validation of the publication of new names and new combinations previously effectively published outside the IJSB. List No. 17. Int. J. Syst. Bacteriol. *35*: 223–225.

Sampaio, M.J.A.M., F.O. Pedrosa and J. Dobreiner. 1982. Growth of *Derxia gummosa* and *Azospirillum* spp. on C₁-compuonds. An. Acad. Bras. Cienc. *54*: 457–458.

Sampaio, M.J.A.M., E.M.R. Silva, J. Döobreiner, M.G. Yates and F.O. Pedrosa. 1981. Autotrophy and methylotrophy in *Derxia gummosa*, *Azospirillum brasilense* and *A. lipoferum*. Current Perspectives in Nitrogen Fixation, Proc. 4th Int. Symp. of Nitrogen Fixation, Australian Academy of Science. p. 444.

Samuels, S.B., C.W. Moss and R.E. Weaver. 1973. The fatty acids of *Pseudomonas multivorans* (*Pseudomonas cepacia*) and *Pseudomonas kingii*. J. Gen. Microbiol. *74*: 275–279.

San Miguel, V.V., J.P. Lavery, J.C. York and J.R. Lisse. 1991. *Achromobacter xylosoxidans* septic arthritis in a patient with systemic lupus erythematosus. Arthritis Rheum. *34*: 1484–1485.

Sander, A., M. Ruess, S. Bereswill, M. Schuppler and B. Steinbrueckner. 1998. Comparison of different DNA fingerprinting techniques for molecular typing of *Bartonella henselae* isolates. J. Clin. Microbiol. *36*: 2973–2981.

Sandkvist, M., M. Bagdasarian, S.P. Howard and V.J. DiRita. 1995. Interaction between the autokinase EpsE and EpsL in the cytoplasmic membrane is required for extracellular secretion in *Vibrio cholerae*. EMBO (Eur. Mol. Biol. Organ.) J. *14*: 1664–1673.

Sandoe, J.A.T., H. Malnick and K.W. Loudon. 1997. A case of peritonitis caused by *Roseomonas gilardii* in a patient undergoing continuous ambulatory peritoneal dialysis. J. Clin. Microbiol. *35*: 2150–2152.

Sandros, J. and E. Tuomanen. 1993. Attachment factors of *Bordetella pertussis*: mimicry of eukaryotic cell recognition molecules. Trends Microbiol. *1*: 192–196.

Sandstedt, K. and J. Ursing. 1991a. Description of *Campylobacter upsaliensis* sp. nov. previously known as the CNW group. Syst. Appl. Microbiol. *14*: 39–45.

Sandstedt, K. and J. Ursing. 1991b. *In* Validation of the publication of new names and new combinations previously effectively published outside the IJSB. List No. 37. Int. J. Syst. Bacteriol. *41*: 331.

Sandstrøm, E.G., K.C. Chen and T.M. Buchanan. 1982. Serology of *Neisseria gonorrhoeae*: coagglutination serogroups WI and WII/III correspond to different outer membrane protein I molecules. Infect Immun. *38*: 462-470.

Sandström, E.G., J.S. Knapp, L.B. Reller, S.E. Thompson, E.W. Hook and K.K. Holmes. 1984. Serogrouping of *Neisseria gonorrhoeae*—Correlation of serogroup with disseminated gonococcal infection. Sex. Transm. Dis. *11*: 77–80.

Sanford, R.A., J.R. Cole, F.E. Löffler and J.M. Tiedje. 1996. Characterization of *Desulfitobacterium chlororespirans* sp. nov., which grows by coupling the oxidation of lactate to the reductive dechlorination of 3-chloro-4-hydroxybenzoate. Appl. Environ. Microbiol. *62*: 3800–3808.

Sanford, R.A., J.R. Cole and J.M. Tiedje. 2002. Characterization and description of *Anaeromyxobacter dehalogenans* gen. nov., sp. nov., an aryl-halorespiring facultative anaerobic myxobacterium. Appl. Environ. Microbiol. *68*: 893-900.

Sangodkar, U.M.X., P.J. Chapman and A.M. Chakrabarty. 1988. Cloning, physical mapping and expression of chromosomal genes specifying degradation of the herbicide 2,4,5-T by *Pseudomonas cepacia* Ac1100. Gene *71*: 267–277.

Santiago, B. and O. Meyer. 1997. Purification and molecular characterization of the H-2 uptake membrane-bound NiFe-hydrogenase from the carboxidotrophic bacterium *Oligotropha carboxidovorans*. J. Bacteriol. *179*: 6053–6060.

Santiago, B., U. Schubel, C. Egelseer and O. Meyer. 1999. Sequence analysis, characterization and CO-specific transcription of the *cox* gene cluster on the megaplasmid pHCG3 of *Oligotropha carboxidovorans*. Gene *236*: 115–124.

Santos, J.M., D.R. Verstreate, V.Y. Perera and A.J. Winter. 1984. Outer membrane proteins from rough strains of four *Brucella* species. Infect. Immun. *46*: 188–194.

Santos Ferreira, M.O., M.M. Canica and M.J. Bacelar. 1985. *Pseudomonas cepacia*: the sensitivity of nosocomial strains to new antibiotics. J. Int. Med. Res. *13*: 270–275.

Saphir, D.A. and G.R. Carter. 1976. Gingival flora of the dog with special reference to bacteria associated with bites. J. Clin. Microbiol. *3*: 344–349.

Sapico, F.L., D. Reeves, H.M. Wexler, J. Duncan, K.H. Wilson and S.M. Finegold. 1994. Preliminary study using species-specific oligonucleotide probe for rRNA of *Bilophila wadsworthia*. J. Clin. Microbiol. *32*: 2510–2513.

Saralov, A.I. and T.R. Babanazarov. 1982. The microflora and molecular nitrogen fixation in takyr-like soils of rice fields in Karakalpak ASSR (Uzbek SSR, USSR). Mikrobiologiya *51*: 847–853.

Saralov, A.I., I.N. Krylova, E.E. Saralova and S.I. Kuznetsov. 1984. The distribution and species composition of methane-oxidizing bacteria in lake water. Mikrobiologiya *53*: 695–700.

Sarao, R., H.D. McCurdy and L. Passador. 1985. Enzymes of the intermediary carbohydrate metabolism of *Polyangium cellulosum*. Can. J. Microbiol. *31*: 1142–1146.

Saraste, M. and J. Castresana. 1994. Cytochrome oxidase evolved by tinkering with denitrification enzymes. FEBS Lett. *341*: 1–4.

Sarkar, P.K. and A.K. Banerjee. 1980. Nicotinic acid as an essential growth factor of *Rhodospirillum photometricum* and a new *Rhodosprillum* isolate. Naturwissenschaften *67*: 41–42.

Sarkari, J., N. Pandit, E.R. Moxon and M. Achtman. 1994. Variable expression of the Opc outer-membrane protein in *Neisseria meningitidis* is caused by size variation of a promoter containing poly-cytidine. Mol. Microbiol. *13*: 207–217.

Sarles, L.S. and F.R. Tabita. 1983. Derepression of the synthesis of D-ribulose 1,5-bisphosphate carboxylase oxygenase from *Rhodospirillum rubrum*. J. Bacteriol. *153*: 458–464.

Sasajima, K.I. and A.J. Sinskey. 1979. Oxidation of L-glucose by a pseudomonad. Biochim. Biophys. Acta *571*: 120–126.

Sass, H., M. Berchtold, J. Branke, H. König, H. Cypionka and H.D. Babenzien. 1998a. Psychrotolerant sulfate-reducing bacteria from an oxic freshwater sediment, description of *Desulfovibrio cuneatus* sp. nov. and *Desulfovibrio litoralis* sp. nov. Syst. Appl. Microbiol. *21*: 212–219.

Sass, H., M. Berchtold, J. Branke, H. König, H. Cypionka and H.D. Babenzien. 1998b. *In* Validation of the publication of new names and new combinations previously effectively published outside the IJSB. List No. 67. Int. J. Syst. Bacteriol. *48*: 1083–1084.

Sasse, F., B. Kunze, T.M.A. Gronewold and H. Reichenbach. 1998. The chondramides: Cytostatic agents from myxobacteria acting on the actin cytoskeleton. J. Natl. Cancer Inst. *90*: 1559–1563.

Sasser, M. 1990a. Identification of bacteria by gas chromatography of cellular fatty acids. MIDI technical note 101, MIDI, Newark, DE.

Sasser, M. 1990b. "Tracking" a strain using the microbial identification system. Technical Note # 102, MIDI, Newark, DE.

Sato, H. and Y. Sato. 1990. Protective activities in mice of monoclonal antibodies against pertussis toxin. Infect. Immun. *58*: 3369–3374.

Sato, K. 1978. Bacteriochlorophyll formation by facultative methylotrophs, *Protaminobacter ruber* and *Pseudomonas* AM 1. FEBS Lett. *85*: 207–210.

Sato, K., T. Shiba and Y. Shioi. 1989. Regulation of the biosynthesis of bacteriochlorophyll. *In* Harashima, Shima and Murata (Editors), Aerobic Photosynthetic Bacteria, Japan Scientific Societies Press, Springer-Verlag, Tokyo, Berlin, Heidelberg, New York, London, Paris. pp. 95–124.

Sato, K. and S. Shimizu. 1979. The conditions for bacteriochlorophyll formation and the ultrastructure of a methanol-utilizing bacterium, *Protaminobacter ruber*, classified as a nonphotosynthetic bacterium. Agric. Biol. Chem. *43*: 1669–1676.

Satoh, T., Y. Hoshina and H. Kitamura. 1974. Isolation of denitrifying photosynthetic bacteria. Agric. Biol. Chem. *38*: 1749–1751.

Satoh, T., Y. Hoshina and H. Kitamura. 1976. *Rhodopseudomonas sphaeroides* f. sp. denitrificans, a denitrifying strain as a subspecies of *Rhodopseudomonas sphaeroides*. Arch. Microbiol. *108*: 265–269.

Sauer, K. and R.K. Thauer. 2000. Methyl-coenzyme M formation in methanogenic archaea. Involvement of zinc in coenzyme M activation. Eur. J. Biochem. *267*: 2498–2504.

Saunders, K.E., K.J. McGovern and J.G. Fox. 1997. Use of pulsed-field gel electrophoresis to determine genomic diversity in strains of *Helicobacter hepaticus* from geographically distant locations. J. Clin. Microbiol. *35*: 2859–2863.

Saunders, K.E., Z. Shen, F.E. Dewhirst, B.J. Paster, C.A. Dangler and J.G. Fox. 1999. Novel intestinal *Helicobacter* species isolated from cotton-top tamarins (*Saguinus oedipus*) with chronic colitis. J. Clin. Microbiol. *37*: 146–151.

Săvulescu, T. 1947. Contribution à la classification des bactériacées phytopathogénes. Anal. Acad. Romane Ser. III *22*: 1–26.

Sawada, H. and H. Ieki. 1992a. Crown gall of kiwifruit caused by *Agrobacterium tumefaciens* in Japan. Plant. Dis. *76*: 212.

Sawada, H. and H. Ieki. 1992b. Phenotypic characteristics of the genus *Agrobacterium*. Ann. Phytopathol. Soc. Jpn. *58*: 37–45.

Sawada, H., H. Ieki, S. Kobayashi and I. Oiyama. 1992a. Grouping of tumorigenic *Agrobacterium* spp. based on Ti plasmid-related phenotypes. Ann. Phytopathol. Soc. Jpn. *58*: 244–252.

Sawada, H., H. Ieki and I. Matsuda. 1995. Pcr detection of Ti and Ri plasmids from phytopathogenic *Agrobacterium* strains. Appl. Environ. Microbiol. *61*: 828–831.

Sawada, H., H. Ieki, H. Oyaizu and S. Matsumoto. 1993. Proposal for rejection of *Agrobacterium tumefaciens* and revised descriptions for the genus *Agrobacterium* and for *Agrobacterium radiobacter* and *Agrobacterium rhizogenes*. Int. J. Syst. Bacteriol. *43*: 694–702.

Sawada, H., H. Ieki and Y. Takikawa. 1990. Identification of grapevine crown gall bacteria isolated in Japan. Ann. Phytopathol. Soc. Jpn. *56*: 199–206.

Sawada, H., J. Imada and H. Ieki. 1992b. Evaluation of serodiagnosis for differentiating serogroups of *Agrobacterium tumefacians* biovar 3. Ann. Phytopathol. Soc. Jpn. *58*: 91–94.

Sawada, H., J. Imada and H. Ieki. 1992c. Serogroups of *Agrobacterium tumefaciens* biovar 3 determined using somatic antigens. Ann. Phytopathol. Soc. Jpn. *58*: 52–57.

Sawada, H., Y. Takikawa and H. Ieki. 1992d. Fatty acid methyl ester profiles of the genus *Agrobacterium*. Ann. Phytopathol. Soc. Japan *58*: 46–51.

Saxena, B., M. Modi and V.V. Modi. 1986. Isolation and characterization of siderophores from *Azospirillum lipoferum* D-2. J. Gen. Microbiol. *132*: 2219–2224.

Saxena, I.M. and J. Brown, R.M.. 1995. Identification of a second cellulose synthase gene (*acsAII*) in *Acetobacter xylinum*. J. Bacteriol. *177*: 5276–5283.

Saxena, I.M., K. Kudlicka, K. Okuda and R.M. Brown, Jr.. 1994. Characterization of genes in the cellulose-synthesizing operon (*acs* operon) of *Acetobacter xylinum*: implications for cellulose crystallization. J. Bacteriol. *176*: 5735–5752.

Saxena, I.M., F.C. Lin and R.M. Brown, Jr.. 1990. Cloning and sequencing of the cellulose synthase catalytic subunit gene of *Acetobacter xylinum*. Plant Mol. Biol. *15*: 673–683.

Scandola, M., L. Finelli, B. Sarti, J. Mergaert, J. Swings, K. Ruffieux, E. Wintermantel, J. Boelens, B. de Wilde, W.-R. Müller, A. Schäfer, A.-B. Fink and H.G. Bader. 1998. Biodegradation of a starch containing thermoplastic in standardized test systems. J.M.S.-Pure Appl. Chem. *A35*: 589–608.

Schaab, C., F. Giffhorn, S. Schobert, N. Pfennig and G. Gottschalk. 1972. Phototrophic growth of *Rhodopseudomonas gelatinosa* on citrate: accumulation and subsequent utilization of cleavage products. Z. Naturforsch. *27b*: 962–967.

Schaad, N.W., C.I. Kado and D.R. Sumner. 1975. Synonymy of *Pseudomonas avenae* Manns 1905 and *Pseudomonas alboprecipitans* Rosen 1922. Int. J. Syst. Bacteriol. *25*: 133–137.

Schaad, N.W., G. Sowell, Jr., R.W. Goth, R.R. Colwell and R.E. Webb. 1978. *Pseudomonas pseudoalcaligenes* subsp. *citrulli* subsp. nov. Int. J. Syst. Bacteriol. *28*: 117–125.

Schach, S., B. Tshisuaka, S. Fetzner and F. Lingens. 1995. Quinoline 2-oxidoreductase and 2-oxo-1,2-dihydroquinoline 5,6-dioxygenase from *Comamonas testosteroni* 63. The first two enzymes in quinoline and 3-methylquinoline degradation. Eur. J. Biochem. *232*: 536–544.

Schad, G.A., R. Knowles and E. Meerovitch. 1964. The occurrence of *Lampropedia* in the intestines of some reptiles and nematodes. Can. J. Microbiol. *10*: 801–804.

Schaechter, M., F.M. Bozeman and J.E. Smadel. 1957. Study on the growth of rickettsiae. II. Morphologic observations of living rickettsiae in tissue culture cells. Virol. *3*: 160–172.

Schaefer, J.K., K.D. Goodwin, I.R. McDonald, J.C. Murrell and R.S. Oremland. 2002. *Leisingera methylohatidivorans* gen. nov., sp nov., a marine methylotroph that grows on methyl bromide. Int. J. Syst. Evol. Microbiol. *52*: 851–859.

Schank, S.C., R.L. Smith, G.C. Weiser, D.A. Zuberer, J.H. Bouton, K.H. Quesenberry, M.E. Tyler, J.R. Milam and R.C. Littell. 1979. Fluorescent antibody technique to identify *Azospirillum brasilense* associated with roots of grasses. Soil Biol. Biochem. *11*: 287–295.

Schär, H.P., P. Chemla and O. Ghisalba. 1985. Methanol dehydrogenase from *Hyphomicrobium* Ms-223. FEMS Microbiol. Lett. *26*: 117–122.

Schär, H.P. and O. Ghisalba. 1985. *Hyphomicrobium* bacterial electrode for determination of monomethyl sulfate. Biotechnol. Bioeng. *27*: 897–901.

Schatz, A. and C. Bovell, Jr.. 1952. Growth and hydrogenase activity of a new bacterium, *Hydrogenomonas facilis*. J. Bacteriol. *63*: 87–98.

Schauder, R., B. Eikmanns, R.K. Thauer, F. Widdel and G. Fuchs. 1986. Acetate oxidation of carbon dioxide in anaerobic bacteria via a novel pathway not involving reactions of the citric acid cycle. Arch. Microbiol. *145*: 162–172.

Schauer, D.B., N. Ghori and S. Falkow. 1993. Isolation and characterization of "*Flexispira rappini*" from laboratory mice. J. Clin. Microbiol. *31*: 2709–2714.

Scheinert, P., R. Krausse, U. Ullmann, R. Söller and G. Krupp. 1996. Molecular differentiation of bacteria by PCR amplification of the 16S–23S rRNA spacer. J. Microbiol. Methods *26*: 103–117.

Schell, J., M. Van Montagu, M. De Beuckeleer, M. De Block, A. Depicker, M. De Wilde, G. Engler, C. Genetello, J.P. Hernalsteens, M. Holsters, J. Seurinck, A. Silva, F. Van Vliet and R. Villaroel. 1979. Interactions and DNA transfer between *Agrobacterium tumefaciens*, the Ti-plasmid and the plant host. Proc. R. Soc. Lond. B. *204*: 251–266.

Schell, M.A. 2000. Control of virulence and pathogenicity genes of Ralstonia solanacearum by an elaborate sensory. Annu. Rev. Phytopathol. *38*: 263–292.

Schelling, J.E., C. Anderson and S.F. Conti. 1977. Serotyping of bdellovibrios by agglutination and indirect immunofluorescence. 77th Annual Meeting of the American Society for Microbiology, p. 179.

Schelling, M.E. and S.F. Conti. 1983. Serotyping of bdellovibrios by agglutination and indirect immunofluorescence. Int. J. Syst. Bacteriol. *33*: 816–821.

Schelling, M. and S.F. Conti. 1986. Host receptor sites involved in the attachment of *Bdellovibrio bacteriovorus* and *Bdellovibrio stolpii*. FEMS Microbiol. Lett. *36*: 319–323.

Scherer, O.W., H. Budzikiewicz, R. Hartmann, R.A. Klein and H. Egge. 1992. The structural elucidation of the two positional isomers of a mono-glucopyranosyl mono-acyl glycerol derivative from *Cystobacter fuscus* (*Myxobacterales*). Biochim. Biophys. Acta *1117*: 42–46.

Schink, B. 1984a. Fermentation of 2,3-butanediol by *Pelobacter carbinolicus*, new species and *Pelobacter propionicus*, new species and evidence for propionate formation from C-2 compounds. Arch. Microbiol. *137*: 33–41.

Schink, B. 1984b. *In* Validation of the publication of new names and new combinations previously effectively published outside the IJSB. List No. 15. Int. J. Syst. Bacteriol. *34*: 355–357.

Schink, B. 1985. Fermentation of acetylene by an obligate anaerobe, *Pelobacter acetylenicus*, new species. Arch. Microbiol. *142*: 295–301.

Schink, B. 1986. *In* Validation of the publication of new names and new combinations previously effectively published outside the IJSB. List No. 20. Int. J. Syst. Bacteriol. *36*: 354–356.

Schink, B. 1992. The genus *Pelobacter*. *In* Balows, Trüper, Dworkin, Harder and Schleifer (Editors), The Prokaryotes: A Handbook of Bacteria: Ecophysiology, Isolation, Identification, Applications, 2nd ed., Vol. 4, Springer-Verlag, New York. pp. 3393–3399.

Schink, B. 1997. Energetics of syntrophic cooperation in methanogenic degradation. Microbiol. Mol. Biol. Rev. *61*: 262–280.

Schink, B. and N. Pfennig. 1982. Fermentation of trihydroxybenzenes by *Pelobacter acidigallici*, new genus new species: a new strictly anaerobic, non-spore-forming bacterium. Arch. Microbiol. *133*: 195–201.

Schink, B. and N. Pfennig. 1983. *In* Validation of the publication of new names and new combinations previously effectively published outside the IJSB. List No. 12. Int. J. Syst. Bacteriol. *33*: 896–897.

Schink, B. and H.G. Schlegel. 1980. The membrane-bound hydrogenase of *Alcaligenes eutrophus*. II. Localization and immunological comparison with other hydrogenase systems. Antonie Leeuwenhoek J. Microbiol. Serol. *46*: 1–14.

Schink, B. and M. Stieb. 1983. Fermentative degradation of polyethylene glycol by a strictly anaerobic, gram-negative, non-spore-forming bacterium, *Pelobacter venetianus*, sp. nov. Appl. Environ. Microbiol. *45*: 1905–1913.

Schink, B. and M. Stieb. 1984. *In* Validation of the publication of new names and new combinations previously effectively published outside the IJSB. List No. 13. Int. J. Syst. Bacteriol. *34*: 91–92.

Schläfi, H.R., M.A. Weiss, T. Leisinger and A.M. Cook. 1994. Terephthalate 1,2-dioxygenase system from *Comamonas testosteroni* T-2: purification and some properties of the oxygenase component. J. Bacteriol. *176*: 6644–6652.

Schleifer, K.H. and O. Kandler. 1972. Peptidoglycan types of bacterial cell walls and their taxonomic implications. Bacteriol. Rev. *36*: 407–477.

Schleifer, K.H., D. Schüler, S. Spring, M. Weizenegger, R. Amann, W. Lduwig and M. Köhler. 1991. The genus *Magnetospirillum*, gen. nov., description of *Magnetospirillum gryphiswaldense*, sp. nov. and transfer of *Aquaspirillum magnetotacticum* to *Magnetospirillum magnetotacticum*, comb. nov. Syst. Appl. Microbiol. *14*: 379–385.

Schleifer, K.H., D. Schüler, S. Spring, M. Weizenegger, R. Amann, W. Lduwig and M. Köhler. 1992. *In* Validation of the publication of new names and new combinations previously effectively published outside the IJSB. List No. 40. Int. J. Syst. Bacteriol. *42*: 191–192.

Schlesner, H. 1983. Isolierug und Beschreibung knospender und prosthekater Bakterien aus der Kieler Foerde, Thesis, Christian-Albrechts-Universitat zu Kiel, Kiel.

Schlesner, H. 1986. *Pirella marina* sp. nov., a budding, peptidoglycan-less bacterium from brackish water. Syst. Appl. Microbiol. *8*: 177–180.

Schlesner, H. 1987. *Filomicrobium fusiforme* gen. nov., sp. nov., a slender budding, hyphal bacterium from brackish water. Syst. Appl. Microbiol. *10*: 63–67.

Schlesner, H. 1988. *In* Validation of the publication of new names and new combinations previously effectively published outside the IJSB. List No. 25. Int. J. Syst. Bacteriol. *38*: 220–222.

Schlesner, H. 1992. The genus *Stella*. *In* Balows, Trüper, Dworkin, Harder and Schleifer (Editors), The Prokaryotes: A Handbook of Bacteria: Ecophysiology, Isolation, Identification, Applications, 2nd. Ed., Vol. 3, Springer-Verlag, New York. pp. 2167–2170.

Schlesner, H. 1994. The development of media suitable for the microorganisms morphologically resembling *Planctomyces* spp., *Pirellula* spp., and other *Planctomycetales* from various aquatic habitats using dilute media. Syst. Appl. Microbiol. *17*: 135–145.

Schlesner, H., C. Bartels, M. Sittig, M. Dorsch and E. Stackebrandt. 1990. Taxonomic and phylogenetic studies on a new taxon of budding, hyphal *Proteobacteria*, *Hirschia baltica* gen. nov., sp. nov. Int. J. Syst. Bacteriol. *40*: 443–451.

Schlesner, H., T. Kath, A. Fischer and E. Stackebrandt. 1989. Studies on the phylogentic position of *Prosthecomicrobium pneumaticum*, *P. enhydrum*, *Ancalomicrobium adetum*, and various *Prosthecomicrobium*-like bacteria. Syst. Appl. Microbiol. *12*: 150–155.

Schlomann, M. 1994. Evolution of chlorocatechol catabolic pathways. Conclusions to be drawn from comparisons of lactone hydrolases. Biodegradation *5*: 301–321.

Schlomann, M., K.L. Ngai, L.N. Ornston and H.J. Knackmuss. 1993. Dienelactone hydrolase from *Pseudomonas cepacia*. J. Bacteriol. *175*: 2994–3001.

Schloter, M., W. Bode and A. Hartmann. 1997. Characterization and application of a strain-specific monoclonal antibody against the rhizosphere bacterium *Azospirillum brasilense* Wa5. Hybridoma. *16*: 183–187.

Schloter, M. and A. Hartmann. 1996. Production and characterization of strain-specific monoclonal antibodies against outer membrane components of *Azospirillum brasilense* Sp245. Hybridoma. *15*: 225–232.

Schloter, M. and A. Hartmann. 1998. Endophytic and surface colonization of wheat roots (*Triticum aestivum*) by different *Azospirillum brasilense* strains studied with strain-specific monoclonal antibodies. Symbiosis *25*: 159–179.

Schloter, M., S. Moens, C. Croes, G. Reidel, M. Esquenet, R. De Mot, A. Hartmann and K. Michiels. 1994. Characterization of cell surface components of *Azospirillum brasilense* Sp7 as antigen determinants for strain-specific monoclonal antibodies. Microbiology *140*: 823–828.

Schmid, E.N., G. von Recklinghausen and R. Ansorg. 1990. Bacteriophages in *Helicobacter pylori*. J. Med. Microbiol. *32*: 101–104.

Schmid-Appert, M., K. Zoller, H. Traber, S. Vuilleumier and T. Leisinger. 1997. Association of newly discovered IS elements with the dichloromethane utilization genes of methylotrophic bacteria. Microbiology (Reading) *143*: 2557–2567.

Schmider, F. and J.C.G. Ottow. 1986. Characterization of denitrifying bacteria in the various compartments of a biological sewage plant. Arch. Hydrobiol. *106*: 497–512.

Schmidt, H.J. 1982. Isolation of Omikron endosymbionts from mass-cultures of *Euplotes aediculatus* and characterization of their DNA. Exp. Cell Res. *140*: 417–425.

Schmidt, H.J., H.D. Görtz and R.L. Quackenbush. 1987a. *Caedibacter caryophila* sp. nov., a killer symbiont inhabiting the macronucleus of *Paramecium caudatum*. Int. J. Syst. Bacteriol. *37*: 459–462.

Schmidt, J. 1901. *In* Schmidt and Weis Bakteriene. Naturhistorisk Grundlag for det bakteriologiske Studium, Morten Porsild, København 1899-1901. p. 266.

Schmidt, J.M. 1966. Observations on the adsorption of Caulobacter bacteriophages containing ribonucleic acid. J. Gen. Microbiol. *45*: 347–353.

Schmidt, J.E. and B.K. Ahring. 1993. Effects of hydrogen and formate on the degradation of propionate and butyrate in thermophilic granules from an upflow anaerobic sludge blanket reactor. Appl. Environ. Microbiol. *59*: 2546–2551.

Schmidt, J.M. and R.Y. Stanier. 1965. Isolation and characterization of bacteriophages active against stalked bacteria. J. Gen. Microbiol. *39*: 95–107.

Schmidt, J.M. and R.Y. Stanier. 1966. The development of cellular stalks in bacteria. J. Cell. Biol. *28*: 423–436.

Schmidt, J.M. and M.P. Starr. 1984. Unidirectional polar growth of cells of *Seliberia stellata* and aquatic *Seliberia*-like bacteria revealed by immunoferritin labeling. Arch. Microbiol. *138*: 89–95.

Schmidt, J.M. and J.R. Swafford. 1975. Ultrastructure of crossbands in prosthecae of *Asticcacaulis* species. J. Bacteriol. *124*: 1601–1603.

Schmidt, J.M. and J.R. Swafford. 1979. Isolation and morphology of helically sculptured, rosette-forming, freshwater bacteria resembling *Seliberia*. Curr. Microbiol. *3*: 65–70.

Schmidt, J.M. and J.R. Swafford. 1981. The genus *Seliberia*. *In* Starr, Stolp, Trüper, Balows and Schlegel (Editors), The Prokaryotes: A Handbook on Habitats, Isolation, and Identification of Bacteria, Springer-Verlag, Berlin. pp.516–519.

Schmidt, J.M. and J.R. Swafford. 1992. The genus *Seliberia*. *In* Balows, Trüper, Dworkin, Harder and Schleifer (Editors), The Prokaryotes:

A Handbook of Bacteria: Ecophysiology, Isolation, Identification, Applications, 2nd Ed., Springer-Verlag, New York. pp. 2490–2494.

Schmidt, K. 1978. Biosynthesis of carotenoids. *In* Clayton and Sistrom (Editors), The Photosynthetic Bacteria, Plenum Press, New York. pp. 729–750.

Schmidt, K. and B. Bowien. 1983. Notes on the description of *Rhodopseudomonas blastica*. Arch. Microbiol. *136*: 242.

Schmidt, K. and S. Liaasen-Jensen. 1973. Bacterial carotenoids XLII. New ketocarotenoids from *Rhodopseudomonas globiformis*. Acta Chem. Scand. *27*: 3040–3052.

Schmidt, S., P. Fortnagel and R.M. Wittich. 1993. Biodegradation and transformation of 4,4′- and 2,4-dihalodiphenyl ethers by *Sphingomonas* sp. strain SS33. Appl. Environ. Microbiol. *59*: 3931–3933.

Schmidt, S., R.M. Wittich, D. Erdmann, H. Wilkes, W. Francke and P. Fortnagel. 1992a. Biodegradation of diphenyl ether and its monohalogenated derivatives by *Sphingomonas* sp. strain SS3. Appl. Environ. Microbiol. *58*: 2744–2750.

Schmidt, S., R.M. Wittich, P. Fortnagel, D. Erdmann and W. Francke. 1992b. Metabolism of 3-methyldiphenyl ether by *Sphingomonas* sp. SS31. FEMS Microbiol. Lett. *75*: 253–258.

Schmidt, T.M., B. Arieli, Y. Cohen, E. Padan and W.R. Strohl. 1987b. Sulfur metabolism in *Beggiatoa alba*. J. Bacteriol. *169*: 5466–5472.

Schmidt, T. and H.G. Schlegel. 1994. Combined nickel-cobalt-cadmium resistance encoded by the *ncc* locus of *Alcaligenes xylosoxidans* 31A. J. Bacteriol. *176*: 7045–7054.

Schmidt, T., R.D. Stoppel and H.G. Schlegel. 1991. High-level nickel resistance in *Alcaligenes xylosoxydans* 31A and *Alcaligenes eutrophus* KTO2. Appl. Environ. Microbiol. *57*: 3301–3309.

Schmidt-Lorenz, W. and H. Kühlwein. 1968. Intracelluläre Bewegungsorganellen der Myxobakterien. Arch. Microbiol. *60*: 95–98.

Schmidt-Lorenz, W. and H. Kühlwein. 1969. Beiträge zur kenntnis der Myxobakterienzelle. 2. Mitteilung. oberflächenstrukturen der schwarmzellen. Arch. Microbiol. *68*: 405–426.

Schmincke, A. 1917. Histopathologischer Befund in Roseolen der Haut bei wolhynischem Fieber. Muench. Med. Wochenschr. *64*: 961.

Schmitt, S., R. Müller, W. Wegst and F. Lingens. 1984. Chloridazon catechol dioxygenases, a distinct group of meta-cleaving enzymes. Hoppe-Seyler′s Z Physiol. Chem. *365*: 143–150.

Schneider, H., A.S. Cross, R.A. Kuschner, D.N. Taylor, J.C. Sadoff, J.W. Boslego and C.D. Deal. 1995. Experimental human gonococcal urethritis—250 *Neisseria gonorrhoeae* Ms11mkc are infective. J. Infect. Dis. *172*: 180–185.

Schneider, H., J.M. Griffiss, J.W. Boslego, P.J. Hitchcock, K.M. Zahos and M.A. Apicella. 1991. Expression of paragloboside-like lipooligosaccharides may be a necessary component of gonococcal pathogenesis in men. J. Exp. Med. *174*: 1601–1605.

Schneider, H., J.M. Griffiss, R.E. Mandrell and G.A. Jarvis. 1985. Elaboration of a 3.6-kilodalton lipooligosaccharide, antibody against which is absent from human sera, is associated with serum resistance of *Neisseria gonorrhoeae*. Infect. Immun. *50*: 672–677.

Schneider, H., K.A. Schmidt, D.R. Skillman, L. VandeVerg, R.L. Warren, H.J. Wylie, J.C. Sadoff, C.D. Deal and A.S. Cross. 1996. Sialylation lessens the infectivity of *Neisseria gonorrhoeae* MS11mkC. J. Infect. Dis. *173*: 1422–1427.

Schneider, K., V. Rudolph and H.G. Schlegel. 1973. Description and physiological characterization of a coryneform hydrogen bacterium, strain 14g. Arch. Mikrobiol. *93*: 179–193.

Schneider, K. and H.G. Schlegel. 1977. Localization and stability of hydrogenases from aerobic hydrogen bacteria. Arch. Microbiol. *112*: 229–238.

Schneider, R.P., F. Zürcher, T. Egli and G. Hamer. 1988. Determination of nitrilotriacetate in biological matrices using ion exclusion chromatography. Anal. Biochem. *173*: 278–284.

Schnell, S., A. Brune and B. Schink. 1991a. Degradation of hydroxyhydroquinone by the strictly anaerobic fermenting bacterium *Pelobacter massiliensis*, sp. nov. Arch. Microbiol. *155*: 511–516.

Schnell, S., A. Brune and B. Schink. 1991b. *In* Validation of the publication of new names and new combinations previously effectively published outside the IJSB. List No. 39. Int. J. Syst. Bacteriol. *41*: 580–581.

Schnell, S. and H.M. Steinman. 1995. Function and stationary-phase induction of periplasmic copper-zinc superoxide dismutase and catalase/peroxidase in *Caulobacter crescentus*. J. Bacteriol. *177*: 5924–5929.

Schoberth, S. 1973. A new strain of *Desulfovibrio gigas* isolated from a sewage plant. Arch. Mikrobiol. *92*: 365–368.

Schoch, P.E. and B.A. Cunha. 1988. Nosocomial *Achromobacter xylosoxidans* infections. Infect. Control Hosp. Epidemiol. *9*: 84–87.

Schocher, A.J., H. Kuhn, B. Schlindler, N.J. Palleroni, C.W. Despreaux, M. Boublik and P.A. Miller. 1979. *Acetobacter* bacteriophage A-1. Arch. Microbiol. *121*: 193–197.

Schöcke, L. and B. Schink. 1998. Membrane-bound proton-translocating pyrophosphatase of *Syntrophus gentianae*, a syntrophically benzoate-degrading fermenting bacterium. Eur. J. Biochem. *256*: 589–594.

Schöcke, L. and B. Schink. 1999. Energetics and biochemistry of fermentative benzoate degradation by *Syntrophus gentianae*. Arch. Microbiol. *171*: 331–337.

Schoeffield, A.J. 1990. Ecological, Serological and Molecular Characterization of Halophilic Bdellovibrios, Doctoral thesis, University of Maryland at Baltimore. Baltimore, Maryland.

Schoeffield, A.J., W.A.J. Falkler, D. Desai and H.N. Williams. 1991. Serogrouping of halophilic bdellovibrios from Chesapeake Bay and environs by immunodiffusion and immunoelectrophoresis. Appl. Environ. Microbiol. *57*: 3471–3475.

Schoenlein, P.V. and B. Ely. 1983. Plasmids and bacteriocins in *Caulobacter* species. J. Bacteriol. *153*: 1092–1094.

Scholla, M.H. and G.H. Elkan. 1984. *Rhizobium fredii* sp. nov., a fast-growing species that effectively nodulates soybeans. Int. J. Syst. Bacteriol. *34*: 484–486.

Scholten, E., T. Lukow, G. Auling, R.M. Kroppenstedt, F.A. Rainey and H. Diekmann. 1999. *Thauera mechernichensis* sp. nov., an aerobic denitrifier from a leachate treatment plant. Int. J. Syst. Bacteriol. *49*: 1045–1051.

Schouls, L.M., I. Van De Pol, S.G. Rijpkema and C.S. Schot. 1999. Detection and identification of *Ehrlichia*, *Borrelia burgdorferi* sensu lato, and *Bartonella* species in Dutch *Ixodes ricinus* ticks. J. Clin. Microbiol. *37*: 2215–2222.

Schräder, T., G. Zarnt and J.R. Andreesen. 2001. NAD(P)-dependent aldehyde dehydrogenases induced during growth of Ralstonia eutropha strain Bo on tetrahydrofurfuryl alcohol. J. Bacteriol. *183*: 7408–7411.

Schrautemeier, B. 1981. The role of ferredoxin in the nitrogen-fixing hydrogen bacterium *Xanthobacter autotrophicus*. FEMS Microbiol. Lett. *12*: 153–157.

Schreckenberger, P.C., J.M. Janda, J.D. Wong and E.J. Baron. 1999. Algorithms for identification of aerobic Gram-negative bacteria. *In* Murray, Baron, Pfaller, Tenover and Yolken (Editors), Manual of Clinical Microbiology, 7th Ed., American Society for Microbiology, Washington, D.C. pp. 438–441.

Schreckenberger, P.C. and A. von Graevenitz. 1999. *Acinetobacter*, *Achromobacter*, *Alcaligenes*, *Moraxella*, *Methylobacterium*, and other nonfermtative Gram-negative rods. *In* Murray, Baron, Pfaller, Tenover and Yolken (Editors), Manual of Clinical Microbiology, 7th Ed., American Society for Microbiology, Washington, D.C. pp. 539–560.

Schröder, I., A. Kröger and J.M. Macy. 1988. Isolation of the sulfur reductase and reconstitution of the sulfur respiration of *Wolinella succinogenes*. Arch. Microbiol. *149*: 572–579.

Schröder, I., S. Rech, T. Krafft and J.M. Macy. 1997. Purification and characterization of the selenate reductase from *Thauera selenatis*. J. Biol. Chem. *272*: 23765–23768.

Schröder, J. and H. Reichenbach. 1970. The fatty acid composition of vegetative cells and myxospores of *Stigmatella aurantiaca* (*Myxobacterales*). Arch. Microbiol. *71*: 384–390.

Schröder, M. 1932. Die assimilation des luftstickstoffs durch einige Bakterien. Zentbl. Bakteriol. Infektionskr. Hyg. Abst. 2 *85*: 177–212.

Schroeter, J. 1885-1889. *In* Cohn, F. (Editor), Kryptogamenflora von Schlesien. Bd.3, Heft.3, Pilze J.U, Kern′s Verlag, Breslau. 1–814.

Schroeter, J. 1886. Schizomycetes. *In* Cohn (Editor), Kryptogamenflora von Schlesien, Bd. 3, Heft 3, Pilze, J.U. Kern's Verlag, Breslau. pp. 1–814.

Schroll, G., H.J. Busse, G. Parrer, S. Rölleke, W. Lubitz and E.B.M. Denner. 2001a. *Alcaligenes faecalis* subsp. *parafaecalis* subsp. nov., a bacterium accumulating poly-beta-hydroxybutyrate from acetone-butanol bioprocess residues. Syst. Appl. Microbiol. *24*: 37–43.

Schroll, G., H.J. Busse, G. Parrer, S. Rölleke, W. Lubitz and E.B.M. Denner. 2001b. *In* Validation of the publication of new names and new combinations previously effectively published outside the IJSEM. List No. 82. Int. J. Syst. Evol. Microbiol. *51*: 1619–1620.

Schröter, G. 1975. The detection of twitching motility in *Eikenella corrodens*. Z. Med. Microbiol. Immunol. (Berl.) *161*: 41–46.

Schröter, G. and J. Stawru. 1970. Die bedeutung von *Bacteroides corrodens* Eiken, 1958, in Rhamen der Tonsillenflora. Z. Med. Mikrobiol. Immunol. *155*: 241–247.

Schroth, M.N., J.P. Thompson and D.C. Hildebrand. 1965. Isolation of *Agrobacterium tumefaciens–A. radiobacter* group from soil. Phytopathology *55*: 645–647.

Schryvers, A.B. and I. Stojiljkovic. 1999. Iron acquisition systems in the pathogenic *Neisseria*. Mol. Microbiol. *32*: 1117–1123.

Schübel, U., M. Kraut, G. Morsdorf and O. Meyer. 1995. Molecular characterization of the gene cluster *coxMSL* encoding the molybdenum-containing carbon monoxide dehydrogenase of *Oligotropha carboxidovorans*. J. Bacteriol. *177*: 2197–2203.

Schubert, P., A. Steinbuchel and H.G. Schlegel. 1988. Cloning of the *Alcaligenes eutrophus* genes for synthesis of poly-β-hydroxybutyric acid (PHB) and synthesis of PHB in *Escherichia coli*. J. Bacteriol. *170*: 5837–5847.

Schubert, W.W., G.A. Nelson and G.E. Petersen. 1986. Extracellular polysaccharides from diverse genera exhibiting the rheological property of drag reduction. Abstract of the 86th General Meeting of the American Society for Microbiology. *86*: 171.

Schüler, D. and E. Baeuerlein. 1996. Iron-limited growth and kinetics of iron uptake in *Magnetospirillum gryphiswaldense*. Arch. Microbiol. *166*: 301–307.

Schüler, D. and E. Baeuerlein. 1997. Iron transport and magnetite crystal formation of the magnetic bacterium *Magnetospirillum gryphiswaldense*. J. Phys. IV. *7*: 647–650.

Schüler, D. and E. Baeuerlein. 1998. Dynamics of iron uptake and Fe_3O_4 biomineralization during aerobic and microaerobic growth of *Magnetospirillum gryphiswaldense*. J. Bacteriol. *180*: 159–162.

Schüler, D. and M. Köhler. 1992. The isolation of a new magnetic spirillum. Zentbl. Mikrobiol. *147*: 150–151.

Schüler, D., S. Spring and D.A. Bazylinski. 1999. Improved technique for the isolation of magnetotactic spirilla from a freshwater sediment and their phylogenetic characterization. Syst. Appl. Microbiol. *22*: 466–471.

Schüler, D., R. Uhl and E. Baeuerlein. 1995. A simple light scattering method to assay magnetism in *Magnetospirillum gryphiswaldense*. FEMS Microbiol. Lett. *132*: 139–145.

Schüller, G., C. Hertel and W.P. Hammes. 2000. *Gluconacetobacter entanii* sp. nov., isolated from submerged high-acid industrial vinegar fermentations. Int. J. Syst. Evol. Microbiol. *50*: 2013–2020.

Schultheiss, D. and D. Schüler. 2003. Development of a genetic system for *Magnetospirillum gryphiswaldense*. Arch. Microbiol. *179*: 89–94.

Schultz, E. and P. Hirsch. 1973. Morphologically unusual bacteria in acid bog water habitats. Abstr. Ann. Meet. Am. Soc. Microbiol. *73*: 60.

Schultz, G.A., G. Chaconas and R.L. Moore. 1978. Polyadenylic acid sequences in the RNA of *Hyphomicrobium*. J. Bacteriol. *133*: 569–575.

Schultz, J.E. and P.F. Weaver. 1982. Fermentation and anearobic respiration by *Rhodospirillum rubrum* and *Rhodopseudomonas capsulata*. J. Bacteriol. *149*: 181–190.

Schultze, M., E. Kondorosi, P. Ratet, M. Buire and A. Kondorosi. 1994. Cell and molecular biology of *Rhizobium*-plant interactions. Int. Rev. Cytol. *156*: 1–75.

Schulz, S., W.B. Dong, U. Groth and A.M. Cook. 2000. Enantiomeric degradation of 2-(4-sulfophenyl) butyrate via 4-sulfocatechol in *Defltia acidovorans* SPB1. Appl. Environ. Microbiol. *66*: 1905–1910.

Schulze, R., S. Spring, R. Amann, I. Huber, W. Ludwig, K.H. Schleifer and P. Kämpfer. 1999a. Genotypic diversity of *Acidovorax* strains isolated from activated sludge and description of *Acidovorax defluvii* sp. nov. Syst. Appl. Microbiol. *22*: 205–214.

Schulze, R., S. Spring, R. Amann, I. Huber, W. Ludwig, K.H. Schleifer and P. Kämpfer. 1999b. *In* Validation of the publication of new names and new combinations previously effectively published outside the IJSB. List No. 71. Int. J. Syst. Evol. Microbiol. *49*: 1325–1326.

Schumacher, U.K. 1997. Adherence of *Bilophila wadsworthia* to laminin and fibronectin. Clin. Infect. Dis. *25*: S180.

Schumacher, U. and M. Bucheler. 1997. First isolation of *Bilophila wadsworthia* in otitis externa. HNO. *45*: 567–569.

Schumacher, U.K., M. Maennel and H. Werner. 2000. Adherence of *Bacteroides* species and *Bilophila wadsworthia* to phospholipids and glycolipids. Anaerobe *6*: 61–63.

Schumacher, W., U. Hole and P.M.H. Kroneck. 1994. Ammonia-forming cytochrome *c* nitrite reductase from *Sulfurospirillum deleyianum* is a tetraheme protein: new aspects of the molecular composition and spectroscopic properties. Biochem. Biophys. Res. Commun. *205*: 911–916.

Schumacher, W. and P.M.H. Kroneck. 1991. Dissimilatory hexaheme-C nitrite reductase of *Spirillum* strain 5175: purification and properties. Arch. Microbiol. *156*: 70–74.

Schumacher, W., P.M.H. Kroneck and N. Pfennig. 1992. Comparative systematic study on "*Spirillum*" 5175, *Campylobacter* and *Wolinella* species. Description of "*Spirillum*" 5175 as *Sulfurospirillum deleyianum* gen. nov., spec. nov. Arch. Microbiol. *158*: 287–293.

Schumacher, W., P.M.H. Kroneck and N. Pfennig. 1993. *In* Validation of the publication of new names and new combinations previously effectively published outside the IJSB. List No. 44. Int. J. Syst. Bacteriol. *43*: 188–189.

Schurenkova, A. 1938. *Sogdianella moshkowskii* gen. nov.-a parasite belonging to the *Piroplasmidea* in a raptororial bird-*Gypaetus barbatus*. L. Med. Parazitol. Parazit. Bolezni. *7*: 932–937.

Schürmann, C. 1967. Growth of myxococci in suspension in liquid media. Appl. Microbiol. *15*: 971–974.

Schwartz, E. and B. Friedrich. 2001. A physical map of the megaplasmid pHG1, one of three genomic replicons in *Ralstonia eutropha* H16. FEMS Microbiol. Lett. *201*: 213–219.

Schwedock, J. and S.R. Long. 1989. Nucleotide sequence and protein products of two new nodulation genes of *Rhizobium meliloti*, *nodP* and *nodQ*. Mol. Plant Microbe Interact. *2*: 181–194.

Schwedock, J. and S.R. Long. 1994. An open reading frame downstream of *Rhizobium meliloti nodQ1* shows nucleotide sequence similarity to an *Agrobacterium tumefaciens* insertion sequence. Mol. Plant Microbe Interact. *7*: 151–153.

Schwudke, D., E. Strauch, M. Kruger and B. Appel. 2001. Taxonomic studies of predatory Bdellovibrios based on 16S rRNA analysis, ribotyping and the hit locus and characterization of isolates from the gut of animals. Syst. Appl. Microbiol. *24*: 385–394.

Scognamiglio, T. and J.J. tudor. 1996. Macrorestriction patterns and partial genomic map of *Bdellovibrio*. 96th Annual Meeting of the American Society for Microbiology, p. 492.

Scopes, R.K., V. Testolin, A. Stoter, K. Griffiths Smith and E.M. Algar. 1985. Simultaneous purification and characterization of glucokinase, fructokinase and glucose-6-phosphate dehydrogenase from *Zymomonas mobilis*. Biochem. J. *228*: 627–634.

Scordaki, A. and C. Drainas. 1987. Analysis of natural plasmids of *Zymomonas mobilis* ATCC 10988. J. Gen. Microbiol. *133*: 2547–2556.

Scordilis, G.E., H. Ree and T.G. Lessie. 1987. Identification of transposable elements which activate gene expression in *Pseudomonas cepacia*. J. Bacteriol. *169*: 8–13.

Scott, D., J. Brannan and I.J. Higgins. 1981. The effect of growth conditions on intracytoplasmic membranes and methane monooxygenase activities in *Methylosinus trichosporium* OB3b. J. Gen. Microbiol. *125*: 63–72.

Scott, D.B., C.A. Scott and J. Dobereiner. 1979. Nitrogenase activity and nitrate respiration in *Azospirillum* spp. Arch. Microbiol. *121*: 141–145.

Scott, D.R., D.P. Weeks, C. Hong, S. Postius, K. Melchers and G. Sachs. 1998. The role of internal urease in acid resistance of *Helicobacter pylori*. Gastroenterology *114*: 58–70.

Scott, D.B., C.A. Young, J.M. Collins-Emerson, E.A. Terzaghi, E.S. Rockman, P.E. Lewis and C.E. Pankhurst. 1996. Novel and complex chromosomal arrangement of *Rhizobium loti* nodulation genes. Mol. Plant-Microbe Interact. *9*: 187–197.

Scully, D.A. and N.C. Dondero. 1973. Estimation with several culture media of spirilla of 11 natural sources. Can. J. Microbiol. *19*: 983–989.

Sears, H.J., S. Spiro and D.J. Richardson. 1997. Effect of carbon substrate and aeration on nitrate reduction and expression of the periplasmic and membrane-bound nitrate reductases in carbon-limited continuous cultures of *Paracoccus denitrificans* Pd1222. Microbiology *143*: 3767–3774.

Sebald, M. and M. Véron. 1963. Teneur en bases de l' ADN et classification des vibrions. Ann. Inst. Pasteur (Paris) *105*: 897–910.

Sedina, S.A. and V.N. Ivanov. 1991. Composition and metabolic activity of a bacterial association that decomposes diethylene glycol. Mikrobiol. Zh. *53*: 30–38.

Seeley, H.W., Jr. 1974. Genus *Lampropedia* Schroeter 1886, 151. *In* Buchanan and Gibbons (Editors), Bergey's Manual of Determinative Bacteriology, 8th Ed., The Williams & Wilkins Co., Baltimore. pp. 440–441.

Seewaldt, E., K.H. Schleifer, E. Bock and E. Stackebrandt. 1982. The close phylogenetic relationship of *Nitrobacter* and *Rhodopseudomonas palustris*. Acta Microbiol. *131*: 287–290.

Segers, P., M. Vancanneyt, B. Pot, U. Torck, B. Hoste, D. Dewettinck, E. Falsen, K. Kersters and P. De Vos. 1994. Classification of *Pseudomonas diminuta* Leifson and Hugh 1954 and *Pseudomonas vesicularis* Busing, Doll, and Freytag 1953 in *Brevundimonas* gen. nov. as *Brevundimonas diminuta* comb. nov. and *Brevundimonas vesicularis* comb. nov., respectively. Int. J. Syst. Bacteriol. *44*: 499–510.

Segonds, C., E. Bingen, G. Couetdic, S. Mathy, N. Brahimi, N. Marty, P. Plesiat, Y. Michel-Briand and G. Chabanon. 1997. Genotypic analysis of *Burkholderia cepacia* isolates from 13 French cystic fibrosis centers. J. Clin. Microbiol. *35*: 2055–2060.

Segovia, L., D. Piñero, R. Palacios and E. Martínez-Romero. 1991. Genetic structure of a soil population of nonsymbiotic *Rhizobium leguminosarum*. Appl. Environ. Microbiol. *57*: 426–433.

Segovia, L., J.P.W. Young and E. Martínez-Romero. 1993. Reclassification of American *Rhizobium legminosarum* biovar phaseoli type I strains as *Rhizobium etli* sp. nov. Int. J. Syst. Bacteriol. *43*: 374–377.

Seidler, R.J., M. Mandel and J.N. Baptist. 1972. Molecular heterogeneity of the bdellovibrios: evidence of two new species. J. Bacteriol. *109*: 209–217.

Seidler, R.J. and M.P. Starr. 1968. Structure of the flagellum of *Bdellovibrio bacteriovorus*. J. Bacteriol. *95*: 1952–1955.

Seidler, R.J. and M.P. Starr. 1969. Isolation and characterization of host-independent bdellovibrios. J. Bacteriol. *100*: 769–785.

Seifert, H.S., R.S. Ajioka, D. Paruchuri, F. Heffron and M. So. 1990. Shuttle mutagenesis of *Neisseria gonorrhoeae*: Pilin null mutations lower DNA transformation competence. J. Bacteriol. *172*: 40–46.

Seitz, H.J. and H. Cypionka. 1986. Chemolithotrophic growth of *Desulfovibrio desulfuricans* with hydrogen coupled to ammonification of nitrate or nitrite. Arch. Microbiol. *146*: 63–67.

Sekeyova, Z., V. Roux, W. Xu, J. Reháček and D. Raoult. 1998. *Rickettsia slovaca* sp. nov., a member of the spotted fever group rickettsiae. Int. J. Syst. Bacteriol. *4*: 1455–1462.

Selander, R.K., D.A. Caugant, H. Ochman, J.M. Musser, M.N. Gilmour and T.S. Whittam. 1986. Methods of multilocus enzyme electrophoresis for bacterial population genetics and systematics. Appl. Environ. Microbiol. *51*: 873–884.

Selenska-Pobell, S. 2002. Diversity and activity of bacterium in uranium waste piles. *In* Keith-Roach and Livens (Editors), Interactions of Microorganisms with Radionuclides, Elsevier, Amsterdam. pp. 225–253.

Sellmer, S., M. Sievers and M. Teuber. 1992. Morphology, virulence and epidemiology of bacteriophage particles isolated from industrial vinegar fermentations. Syst. Appl. Microbiol. *15*: 610–616.

Semrau, J.D., A. Chistoserdov, J. Lebron, A. Costello, J. Davagnino, E. Kenna, A.J. Holmes, R. Finch, J.C. Murrell and M.E. Lidstrom. 1995. Particulate methane monooxygenase genes in methanotrophs. J. Bacteriol. *177*: 3071–3079.

Senff, L.M., W.S. Wegener, G.F. Brooke, W.R. Finnerty and R.A. Makula. 1976. Phospholipid composition and phospholipase activity of *Neisseria gonorrhoeae*. J. Bacteriol. *127*: 874–880.

Seong, S.Y., M.S. Choi and I.S. Kim. 2001. *Orientia tsutsugamushi* infection: Overview and immune responses. Microbes Infect. *3*: 11–21.

Serfontein, S., J.J. Serfontein, W.J. Botha and J.L. Staphorst. 1997. The isolation and characterisation of *Xylophilus ampelinus*. Vitis. *36*: 209–210.

Severin, A.I., N.Y. Markelova, A.V. Afinogenova and I.S. Kulaev. 1987. Isolation and some physicochemical properties of lytic proteinase of the parasitic bacterium *Micavibrio admirandus*. Biokhimiya *52*: 1594–1599.

Severin, J., A. Wohlfarth and E.A. Galinski. 1992. The predominant role of recently discovered tetrahydropyrimidines for the osmoadaptation of halophilic eubacteria. J. Gen. Microbiol. *138*: 1629–1638.

Severini, G. 1913. Una bactériosi dell' Ixia maculata e del *Gladiolus coluilli*. Ann. Bot. (Rome) *11*: 413–424.

Sevilla, M., A. de Oliveira, I. Baldani and C. Kennedy. 1998. Contributions of the bacterial endophyte *Acetobacter diazotrophicus* to sugarcane nutrition: a preliminary study. Symbiosis *25*: 181–191.

Seviour, E.M., C. Williams, B. Degrey, J.A. Soddell, R.J. Seviour and K.C. Lindrea. 1994. Studies on filamentous bacteria from Australian activated sludge plants. Water Res. *28*: 2335–2342.

Seviour, R.J., A.M. Maszenan, J.A. Soddell, V. Tandoi, B.K. Patel, Y. Kong and P. Schumann. 2000. Microbiology of the "G-bacteria" in activated sludge. Environ. Microbiol. *2*: 581–593.

Seyfried, B., A. Tschech and G. Fuchs. 1991. Anaerobic degradation of phenylacetate and 4-hydroxyphenylacetate by denitrifying bacteria. Arch. Microbiol. *155*: 249–255.

Seymour, C., R.G. Lewis, M. Kim, D.F. Gagnon, J.G. Fox, F.E. Dewhirst and B.J. Paster. 1994. Isolation of *Helicobacter* strains from wild bird and swine feces. Appl. Environ. Microbiol. *60*: 1025–1028.

Shah, H.N. and M.D. Collins. 1989. Proposal to restrict the genus *Bacteroides* (Castellani and Chalmers) to *Bacteroides fragilis* and closely related species. Int. J. Syst. Bacteriol. *39*: 85-87.

Shah, S., V. Karkhanis and A. Desai. 1992. Isolation and characterization of siderophore, with antimicrobial activity, from *Azospirillum lipoferum* M. Curr. Microbiol. *25*: 347–351.

Shames, B., J.G. Fox, F.E. Dewhirst, L. Yan, Z. Shen and N.S. Taylor. 1995. Identification of widespread *Helicobacter hepaticus* infection in feces in commercial mouse colonies by culture and PCR assay. J. Clin. Microbiol. *33*: 2968–2972.

Shankar, H.N.R., I.R. Kennedy and P.B. New. 1986. Autotrophic growth and nitrogen fixation in *Derxia gummosa*. J. Gen. Microbiol. *132*: 1797–1804.

Shanks, J.C. and C.N. Hale. 1984. New plant disease record in New Zealand bacterial leaf spot of carnation. N. Z. J. Agric. Res. *27*: 437–439.

Sharak Genthner, B.R., S.D. Friedman and R. Devereux. 1997a. Reclassification of *Desulfovibrio desulfuricans* Norway 4 as *Desulfomicrobium norvegicum* comb. nov. and confirmation of *Desulfomicrobium escambiense* (corrig., formerly "escambium") as a new species in the genus *Desulfomicrobium*. Int. J. Syst. Bacteriol. *47*: 889–892.

Sharak Genthner, B.R., G. Mundfrom and R. Devereux. 1994. Characterization of *Desulfomicrobium escambium* sp. nov. and proposal to assign *Desulfovibrio desulfuricans* strain Norway 4 to the genus *Desulfomicrobium*. Arch. Microbiol. *161*: 215–219.

Sharak Genthner, B.R., G. Mundfrom and R. Devereux. 1996. *In* Validation of the publication of new names and new combinations previously effectively published outside the IJSB. List No. 59. Int. J. Syst. Bacteriol. *46*: 1189–1190.

Sharak Genthner, B.R., G.T. Townsend and B.O. Blattmann. 1997b. Reduction of 3-chlorobenzoate, 3-bromobenzoate, and benzoate to corresponding alcohols by *Desulfomicrobium escambiense*, isolated from a 3-chlorobenzoate-dechlorinating coculture. Appl. Environ. Microbiol. *63*: 4698–4703.

Sharp, J.T. 1968. Isolation of L-forms of *Bartonella bacilliformis*. Proc. Soc. Exp. Biol. Med. *128*: 1072–1075.

Sharypova, L.A. , K. Niehaus, H. Scheidle, O. Holst and A. Becker. 2003. *Sinorhizobium meliloti* acpXL mutant lacks the C28 hydroxylated fatty acid moiety of lipid A and does not express a slow migrating form of lipopolysaccharide. J. Biol. Chem. *278*: 12946–12954.

Shaw, D., I.R. Poxton and J.R. Govan. 1995. Biological activity of *Burkholderia (Pseudomonas) cepacia* lipopolysaccharide. FEMS Immunol. Med. Microbiol. *11*: 99–106.

Shearman, C.A., L. Rossen, A.W.B. Johnston and J.A. Downie. 1986. The *Rhizobium leguminosarum* nodulation gene *Nodf* encodes a polypeptide similar to acyl-carrier protein and is regulated by NodD plus a factor in pea root exudate. Embo J. *5*: 647–652.

Sheikholeslam, S., B.C. Lin and C.I. Kado. 1979. Multiple-size plasmids in *Agrobacterium radiobacter* and *A. tumefaciens*. Phytopathology *69*: 54–58.

Shelton, D.R. and J.M. Tiedje. 1984. Isolation and partial characterization of bacteria in an anaerobic consortium that mineralizes 3-chlorobenzoic acid. Appl. Environ. Microbiol. *48*: 840–848.

Shen, N., L. Dagasan, D. Sledjeski and R.M. Weiner. 1989. Major outer membrane proteins unique to reproductive cells of *Hyphomonas jannaschiana*. J. Bacteriol. *171*: 2226–2228.

Shen, N. and R. Weiner. 1998. Isolation and characterization proteins with S-layer properties from the prothescate bacteria, *Hyphomonas jannaschiana*. Microbios. *93*: 7–16.

Shen, Z., J.G. Fox, F.E. Dewhirst, B.J. Paster, C.J. Foltz, L. Yan, B. Shames and L. Perry. 1997. *Helicobacter rodentium* sp. nov., a urease-negative *Helicobacter* species isolated from laboratory mice. Int. J. Syst. Bacteriol. *47*: 627–634.

Shen, Z., D.B. Schauer, H.L.T. Mobley and J.G. Fox. 1998. Development of a PCR-restriction fragment length polymorphism assay using the nucleotide sequence of the *Helicobacter hepaticus* urease structural genes *ureAB*. J. Clin. Microbiol. *36*: 2447–2453.

Shepard, M.C. and C.D. Lunceford. 1976. Differential agar medium (A7) for identification of *Ureaplasma urealyticum* (human T mycoplasmas) in primary cultures of clinical material. J. Clin. Microbiol. *3*: 613–625.

Sheu, D.S., Y.T. Wang and C.Y. Lee. 2000. Rapid detection of polyhydroxyalkanoate-accumulating bacteria isolated from the environment by colony PCR. Microbiology (Read.) *146*: 2019–2025.

Shi, B.H. , V. Arunpairojana, S. Palakawong and A. Yokota. 2002. *Tistrella mobilis* gen. nov., sp nov., a novel polyhydroxyalkanoate-producing bacterium belonging to α-*Proteobacteria*. J. Gen. Appl. Microbiol. *48*: 335–343.

Shi, J., V. Coyne and R. Weiner. 1997. Identification of an alkaline metalloprotease produced by the hydrothermal vent bacterium, *Hyphomonas jannachiana* VP3. Microbios. *91*: 15–26.

Shi, W.Y., T. Kohler and D.R. Zusman. 1993. Chemotaxis plays a role in the social behavior of *Myxococcus xanthus*. Mol. Microbiol. *9*: 601–611.

Shiba, T. 1984. Utilization of light energy by the strictly aerobic bacterium *Erythrobacter* sp. och 114. J. Gen. Appl. Microbiol. *30*: 239–244.

Shiba, T. 1987. O₂ Regulation of bacteriochlorophyll synthesis in the aerobic bacterium *Erythrobacter*. Plant Cell Physiol. *28*: 1313–1320.

Shiba, T. 1989. Taxonomy and ecology of marine bacteria. *In* Harashima, Shima and Murata (Editors), Aerobic Photosynthetic Bacteria, Japan Scientific Societies Press, Springer-Verlag, Tokyo, Berlin, Heidelberg, New York, London, Paris. pp. 9–23.

Shiba, T. 1991a. *Roseobacter litoralis* gen. nov., sp. nov., and *Roseobacter denitrificans* sp. nov., aerobic pink-pigmented bacteria which contain bacteriochlorophyll *a*. Syst. Appl. Microbiol. *14*: 140–145.

Shiba, T. 1991b. *In* Validation of the publication of new names and new combinations previously effectively published outside the IJSB. List No. 37. Int. J. Syst. Bacteriol. *41*: 331.

Shiba, T. 1995. Distribution of aerobic bacteriochlorophyll-containing bacteria in Otsuchi Bay, Iwate. Fish. Sci. *61*: 245–248.

Shiba, T. and K. Harashima. 1986. Aerobic photosynthetic bacteria. Microbiol. Sci. *3*: 376–378.

Shiba, T., Y. Shioi, K. Takamiya, D.C. Sutton and C.R. Wilkinson. 1991. Distribution and physiology of aerobic-bacteria containing bacteriochlorophyll *a* on the east and west coasts of Australia. Appl. Environ. Microbiol. *57*: 295–300.

Shiba, T. and U. Simidu. 1982. *Erythrobacter longus*, gen. nov., sp. nov., an aerobic bacterium which contains bacteriochlorophyll *a*. Int. J. Syst. Bacteriol. *32*: 211–217.

Shiba, T., U. Simidu and N. Taga. 1979. Distribution of aerobic bacteria which contain bacteriochlorophyll *a*. Appl. Environ. Microbiol. *38*: 43–45.

Shibata, S., Y. Isayama and T. Shimizu. 1962. A possibility of variation in *Brucella abortus* from type II to type I. Natl. Inst. Anim. Health Q. (Tokyo) *2*: 10–14.

Shields, M.S., S.O. Montgomery, S.M. Cuskey, P.J. Chapman and P.H. Pritchard. 1991. Mutants of *Pseudomonas cepacia* G4 defective in catabolism of aromatic compounds and trichloroethylene. Appl. Environ. Microbiol. *57*: 1935–1941.

Shields, M.S. and M.J. Reagin. 1992. Selection of a *Pseudomonas cepacia* strain constitutive for the degradation of trichloroethylene. Appl. Environ. Microbiol. *58*: 3977–3983.

Shields, M.S., M.J. Reagin, R.R. Gerger, R. Campbell and C. Somerville. 1995. TOM, a new aromatic degradative plasmid from *Burkholderia (Pseudomonas) cepacia* G4. Appl. Environ. Microbiol. *61*: 1352–1356.

Shigeta, S., Y. Yasunaga, K. Honzumi, H. Okamura, R. Kumata and S. Endo. 1978. Cerebral ventriculitis associated with *Achromobacter xylosoxidans*. J. Clin. Pathol. *31*: 156–161.

Shilo, M. 1966. Predatory bacteria. Science *2*: 33–37.

Shilo, M. 1969. Morphological and physiological aspects of the interaction of *Bdellovibrio* with host bacteria. Curr. Top. Microbiol. Immunol. *50*: 174–204.

Shimada, K. 1995. Aerobic anoxygenic phototrophs. *In* Blankenship, Madigan and Bauer (Editors), Anoxygenic Photosynthetic Bacteria, Kluwer Academic Publishers, Dordrecht. 105–122.

Shimada, K., H. Hayashi and M. Tasumi. 1985. Bacteriochlorophyll-protein complexes of aerobic-bacteria, *Erythrobacter longus* and *Erythrobacter* species Och-114. Arch. Microbiol. *143*: 244–247.

Shimkets, L.J. 1986. Correlation of energy-dependent cell cohesion with social motility in *Myxococcus xanthus*. J. Bacteriol. *166*: 837–841.

Shimkets, L.J. 1987. Control of morphogenesis in Myxobacteria. CRC Crit. Rev. Microbiol. *14*: 195–227.

Shimkets, L.J. 1990a. The *Myxococcus xanthus* Fpra protein causes increased flavin biosynthesis in *Escherichia coli*. J. Bacteriol. *172*: 24–30.

Shimkets, L.J. 1990b. Social and developmental biology of the myxobacteria. Microbiol. Rev. *54*: 473–501.

Shimkets, L.J. 2000. Growth, sporulation, and other tough decisions. *In* Brun and Shimkets (Editors), Prokaryotic Development, American Society for Microbiology, Washington, D.C. pp. 277–284.

Shimkets, L.J. and D. Kaiser. 1982. Induction of coordinated movement of *Myxococcus xanthus* cells. J. Bacteriol. *152*: 451–461.

Shimkets, L.J. and C.R. Woese. 1992. A phylogenetic analysis of the myxobacteria: Basis for their classification. Proc. Natl. Acad. Sci. U.S.A. *89*: 9459–9463.

Shimwell, J.L. 1937. Study of a new type of beer disease bacterium (*Achromobacter anaerobium* sp. nov.) producing alcoholic fermentation of glucose. J. Inst. Brew. *43*: 507–509.

Shimwell, J.L. 1950. *Saccharomonas*, a proposed new genus for bacteria producing a quantitative alcoholic fermentation of glucose. J. Inst. Brew. London. *56*: 179–182.

Shin, Y.K., A. Hiraishi and J. Sugiyama. 1993. Molecular systematics of the genus *Zoogloea* and emendation of the genus. Int. J. Syst. Bacteriol. *43*: 826–831.

Shinhar, E., J. Silver, R. Yeivin and M. Shapiro. 1980. Polymicrobial endocarditis due to *Eikenella corrodens* and group B β-hemolytic streptococcus. Isr. J. Med. Sci. *16*: 458–459.

Shinjoh, M. and T. Hoshino. 1995. Development of a stable shuttle vector and a conjugative transfer system for *Gluconobacter oxydans.* J. Ferment. Bioeng. *79:* 95–99.

Shinjoh, M., N. Tomiyama, A. Asakura and T. Hoshino. 1995. Cloning and nucleotide sequencing of the membrane-bound L-sorbosone dehydrogenase gene of *Acetobacter liquefaciens* IFO 12258 and its expression in *Gluconobacter oxydans.* Appl. Environ. Microbiol. *61:* 413–420.

Shinners, E.N. and B.W. Catlin. 1982. Arginine and pyrimidine biosynthetic defects in *Neisseria gonorrhoeae* strains isolated from patients. J. Bacteriol. *151:* 295–302.

Shinoda, Y., Y. Sakai, M. Ue, A. Hiraishi and N. Kato. 2000. Isolation and characterization of a new denitrifying spirillum capable of anaerobic degradation of phenol. Appl. Environ. Microbiol. *66:* 1286–1291.

Shinomiya, M., T. Iwata, K. Kasuya and Y. Doi. 1997. Cloning of the gene for poly(3-hydroxybutyric acid) depolymerase of *Comamonas testosteroni* and functional analysis of its substrate-binding domain. FEMS Microbiol. Lett. *154:* 89–94.

Shioi, Y. 1986. Growth characteristics and substrate specificity of aerobic photosynthetic bacterium *Erythrobacter* sp. Och-114. Plant Cell Physiol. *27:* 567–572.

Shioi, Y. and M. Doi. 1988. Control of bacteriochlorophyll accumulation by light in an aerobic photosynthetic bacterium *Erythrobacter* sp. Och 114. Arch. Biochem. Biophys. *266:* 470–477.

Shiraki, M., T. shimada, M. Tatsumichi and Y. Saito. 1995. Purification and characterization of extracellular poly(3-hydroxybutyrate) depolymerases. J. Environ. Polym. Degrad. *7:* 12–21.

Shirata, A., T. Tsukamoto, H. Yasui, T. Hata, S. Hayasaka, A. Koijma and H. Kato. 2000. Isolation of bacteria producing bluish-purple pigment and use for dyeing. Japan Agricultural Research Quarterly *34:* 131–140.

Shirazi-Beechey, S.P. and C.J. Knowles. 1984. Serine hydroxymethyltransferase (EC 2.1.2.1) and glycine cleavage enzyme (EC 2.1.2.10) from the cyanogenic bacterium *Chromobacterium violaceum.* J. Gen. Microbiol. *130:* 521–526.

Shirley, E.K. and K. Schmidt-Nielsen. 1967. Oxalate metabolism in the pack rat, sand rat, hamster and white rat. J. Nutr. *91:* 496–502.

Shishido, A., M. Ohtawara, S. Tateno, S. Mizuno, M. Ogura and M. Kitaoka. 1958. The nature of immunity against scrub typhus in mice. I. The resistance of mice, surviving subcutaneous infection of scrub typhus rickettsia, to intraperitoneal reinfection of the same agent. Jpn. J. Med. Sci. Biol. *11:* 383–399.

Shishkina, V.N. and Y.A. Trotsenko. 1974. Properties of a new strain of *Hyphomicrobium,* utilizing 1-carbon compounds. Mikrobiologiia *43:* 765–770.

Shishkina, V.N. and Y.A. Trotsenko. 1979. Pathways of ammonia assimilation in obligate methane utilizers. FEMS Microbiol. Lett. *5:* 187–191.

Shively, J.M., G.L. Decker and J.W. Greenawalt. 1970. Comparative ultrastructure of the thiobacilli. J. Bacteriol. *101:* 618–627.

Shively, J.M., G. van Keulen and W.G. Meijer. 1998. Something from almost nothing: carbon dioxide fixation in chemoautotrophs. Annu. Rev. Microbiol. *52:* 191–230.

Shomer, N.H., C.A. Dangler, M.D. Schrenzel and J.G. Fox. 1997. *Helicobacter bilis*-induced inflammatory bowel disease in *scid* mice with defined flora. Infect. Immun. *65:* 4858–4864.

Shooner, F., J. Bousquet and R.D. Tyagi. 1996. Isolation, phenotypic characterization, and phylogenetic position of a novel, facultatively autotrophic, moderately thermophilic bacterium, *Thiobacillus thermosulfatus* sp. nov. Int. J. Syst. Bacteriol. *46:* 409–415.

Shortridge, V.D., G.G. Stone, R.K. Flamm, J. Beyer, J. Versalovic, D.W. Graham and S.K. Tanaka. 1997. Molecular typing of *Helicobacter pylori* isolates from a multicenter U.S. clinical trial by *ure*C restriction fragment length polymorphism. J. Clin. Microbiol. *35:* 471–473.

Shuttleworth, K.L. 1996. Isolation and characterization of *Hyphomicrobium* capable of growth on N, N-dimethylformamide. *In* Abstracts of the Annual Meeting of the American Society for Microbiology, American Society for Microbiology, Washington, D.C. 435, Q–284.

Shuttleworth, K.L. and C.E. Cerniglia. 1996. Bacterial degradation of low concentrations of phenanthrene and inhibition by naphthalene. Microb. Ecol. *31:* 305–317.

Shuttleworth, K.L., J. Sung, E. Kim and C.E. Cerniglia. 2000. Physiological and genetic comparison of two aromatic hydrocarbon-degrading *Sphingomonas* strains. Mol. Cell *10:* 199–205.

Shuttleworth, K.L., R.F. Unz and P.L. Wichlacz. 1985. Glucose catabolism in strains of acidophilic, heterotrophic bacteria. Appl. Environ. Microbiol. *50:* 573–579.

Shvinka, J.E., L.M. Pankova, I.N. Mezbarde and L.J. Licis. 1989. Hydrogen peroxide production by *Zymomonas mobilis.* Appl. Microbiol. Biotechnol. *31:* 240–245.

Siddiqui, A. and I.D. Goldberg. 1975. Intergenic transformation of *Neisseria gonorrhoeae* and *Neisseria perflava* to streptomycin and nutritional independence. J. Bacteriol. *124:* 1359–1365.

Sidhu, H., M. Allison and A.B. Peck. 1997a. Identification and classification of *Oxalobacter formigenes* strains by using oligonucleotide probes and primers. J. Clin. Microbiol. *35:* 350–353.

Sidhu, H., L. Enatska, S. Ogden, W.N. Williams, M.J. Allison and A.B. Peck. 1997b. Evaluating children in the Ukraine for colonization with the intestinal bacterium *Oxalobacterium formigenes,* using a polymerase chain reaction-based system. Mol. Diagn. *2:* 89–97.

Sidhu, H., R.P. Holmes, M.J. Allison and A.B. Peck. 1999. Direct quantification of the enteric bacterium *Oxalobacter formigenes* in human fecal samples by quantitative competitive-template PCR. J. Clin. Microbiol. *37:* 1503–1509.

Sidhu, H., B. Hoppe, A. Hesse, K. Tenbrock, S. Bromme, E.T. Rietschel and A.B. Peck. 1998. Absence of *Oxalobacter formigenes* in cystic fibrosis patients: a risk factor for hyperoxaluria. Lancet *352:* 1026–1029.

Siebert, D. 1969. Über Propan verwertende, Wasserstoff oxidierende Bakterien und die Charakterisierung eines Förderungsfaktors, Dissertation, University of Göttingen.

Siefert, E. 1976. Die Fixierung von molekularem Stickstoff bei phototrophen Bakterien am Beispiel von *Rhodopseudomonas acidophila,* Ph.D. thesis, University of Göttingen, Germany

Siefert, E., R.L. Irgens and N. Pfennig. 1978. Phototrophic purple and green bacteria in a sewage treatment plant. Appl. Environ Microbiol. *35:* 38–44.

Siefert, E. and V.B. Koppenhagen. 1982. Studies on the vitamin B$_{12}$ auxotrophy of *Rhodocyclus purpureus* and two other vitamin B$_{12}$-requiring purple nonsulfur bacteria. Arch. Microbiol. *132:* 173–178.

Siefert, E. and N. Pfennig. 1979. Chemoautotrophic growth of *Rhodopseudomonas* species with hydrogen and chemotrophic utilization of methanol and formate. Arch. Microbiol. *122:* 177–182.

Siefert, E. and N. Pfennig. 1980. Diazotrophic growth of *Rhodopseudomonas acidophila* and *Rhodopseudomonas capsulata* under microaerobic conditions in the dark. Arch. Microbiol. *125:* 73–77.

Siering, P.L. and W.C. Ghiorse. 1996. Phylogeny of the *Sphaerotilus leptothrix* group inferred from morphological comparisons, genomic fingerprinting, and 16S ribosomal DNA sequence analyses. Int. J. Syst. Bacteriol. *46:* 173–182.

Siering, P.L. and W.C. Ghiorse. 1997. Development and application of 16S rRNA-targeted probes for detection of iron- and manganese-oxidizing sheathed bacteria in environmental samples. Appl. Environ. Microbiol. *63:* 644–651.

Sievers, M., L. Alonso, S. Gianotti, C. Boesch and M. Teuber. 1996. 16S-23S ribosomal RNA spacer regions of *Acetobacter europaeus* and *A. xylinum,* tRNA genes and antitermination sequences. FEMS Microbiol. Lett. *142:* 43–48.

Sievers, M., C. Gaberthüel, C. Boesch, W. Ludwig and M. Teuber. 1995a. Phylogenetic position of *Gluconobacter* species as a coherent cluster separated from all *Acetobacter* species on the basis of 16S ribosomal RNA sequences. FEMS Microbiol. Lett. *126:* 123–126.

Sievers, M., C. Lanini, A. Weber, U. Schuler-Schmid and M. Teuber. 1995b. Microbiology and fermentation balance in a kombucha beverage obtained from a tea fungus fermentation. Syst. Appl. Microbiol. *18:* 590–594.

Sievers, M., W. Ludwig and M. Teuber. 1994a. Phylogentic positioning of *Acetobacter, Gluconobacter, Rhodopila* and *Acidiphilium* species as a branch of acidophilic bacteria in the alpha-subclass of *Proteobacteria* based on 16S ribosomal DNA sequences. Syst. Appl. Microbiol. *17*: 189–196.

Sievers, M., W. Ludwig and M. Teuber. 1994b. Revival of the species *Acetobacter methanolicus* (ex Uhlig et al. 1986) nom. rev. Syst. Appl. Microbiol. *17*: 352–354.

Sievers, M., H.G. Schlegel, J. Caballero Mellado, J. Dobereiner and W. Ludwig. 1998. Phylogenetic identification of two major nitrogen-fixing bacteria associated with sugarcane. Syst. Appl. Microbiol. *21*: 505–508.

Sievers, M., S. Sellmer and M. Teuber. 1992. *Acetobacter europaeus* sp. nov, a main component of industrial vinegar fermenters in Central Europe. Syst. Appl. Microbiol. *15*: 386–392.

Sievers, M., M. Stöckli and M. Teuber. 1997. Purification and properties of citrate synthase from *Acetobacter europaeus*. FEMS Microbiol. Lett. *146*: 53–58.

Sievers, M. and M. Teuber. 1995. The microbiology and taxonomy of *Acetobacter europaeus* in commercial vinegar production. J. Appl. Bacteriol. (Suppl.) *79*: S84–S95.

Sievert, S.M. and J. Kuever. 2000. *Desulfacinum hydrothermale* sp. nov., a thermophilic, sulfate-reducing bacterium from geothermally heated sediments near Milos Island (Greece). Int. J. Syst. Evol. Microbiol. *50*: 1239–1246.

Silakowski, B., B. Kunze and R. Müller. 2000a. *Stigmatella aurantiaca* Sg a15 carries genes encoding type I and type II 3-deoxy-D-arabino-heptulosonate-7-phosphate synthases: Involvement of a type II synthase in aurachin biosynthesis. Arch. Microbiol. *173*: 403–411.

Silakowski, B., B. Kunze, G. Nordsiek, H. Blöcker, G. Höfle and R. Müller. 2000b. The myxochelin iron transport regulon of the myxobacterium *Stigmatella aurantiaca* Sg a15. Eur. J. Biochem. *267*: 6476–6485.

Silakowski, B., G. Nordsiek, B. Kunze, H. Blöcker and R. Müller. 2001. Novel features in a combined polyketide synthase/non-ribosomal peptide synthetase: the myxalamid biosynthetic gene cluster of the myxobacterium *Stigmatella aurantiaca* Sga15. Chem. Biol. *8*: 59–69.

Siller, H., F.A. Rainey, E. Stackebrandt and J. Winter. 1996. Isolation and characterization of a new Gram-negative, acetone-degrading, nitrate-reducing bacterium from soil, *Paracoccus solventivorans* sp. nov. Int. J. Syst. Bacteriol. *46*: 1125–1130.

Sillman, C.E. and L.E. Casida, Jr.. 1986. Isolation of nonobligate bacterial predators of bacteria from soil. Can. J. Microbiol. *32*: 760–762.

Silman, N.J., M.A. Carver and C.W. Jones. 1991. Directed evolution of amidase in *Methylophilus methylotrophus*: purification and properties of amidases from wild-type and mutant strains. J. Gen. Microbiol. *137*: 169–178.

Silverman, D.J. and L.A. Santucci. 1988. Potential for free radical-induced lipid peroxidation as a cause of endothelial cell injury in Rocky Mountain spotted fever. Infect. Immun. *56*: 3110–3115.

Silverman, D.J. and L.A. Santucci. 1990. A potential protective role for thiols against cell injury caused by *Rickettsia rickettsii*. Ann. N.Y. Acad. Sci. *590*: 111–117.

Silverman, D.J., L.A. Santucci, N. Meyers and Z. Sekeyova. 1992. Penetration of host cells by *Rickettsia rickettsii* appears to be mediated by a phospholipase of rickettsial origin. Infect. Immun. *60*: 2733–2740.

Silverman, D.J. and C.L. Wisseman, Jr.. 1978. Comparative ultrastructural study on the cell envelopes of *Rickettsia prowazeki, Rickettsia rickettsii*, and *Rickettsia tsutsugamushi*. Infect. Immun. *21*: 1020–1023.

Silverman, D.J. and C.L. Wisseman. 1979. *In vitro* studies of rickettsia-host cell-interactions - ultrastructural changes induced by *Rickettsia rickettsii* infection of chicken embryo fibroblasts. Infect. Immun. *26*: 714-727.

Silverman, D.J., C.L. Wisseman, Jr. and A. Waddell. 1980. *In vitro* studies of rickettsia-host cell interactions: Ultrastructural study of *Rickettsia prowazekii*-infected chicken embryo fibroblasts. Infect. Immun. *29*: 778–790.

Simmons, D.G., D.E. Davis, L.P. Rose, J.G. Gray and G.H. Luginbuhl. 1981. *Alcaligenes faecalis*-associated respiratory-disease of chickens. Avian Dis. *25*: 610–613.

Simmons, D.G. and J.G. Gray. 1979. Transmission of acute respiratory disease (rhinotracheitis) of turkeys. Avian Dis. *23*: 131–138.

Simmons, D.G., J.G. Gray, L.P. Rose, R.C. Dillman and S.E. Miller. 1979. Isolation of an etiologic agent of acute respiratory-disease (rhinotracheitis) of turkey poults. Avian Dis. *23*: 194–203.

Simmons, J.H., L.K. Riley, C.L. Besch-Williford and C.L. Franklin. 2000a. *Helicobacter mesocricetorum* sp. nov., a novel *Helicobacter* isolated from the feces of Syrian hamsters. J. Clin. Microbiol. *38*: 1811–1817.

Simmons, J.H., L.K. Riley, C.L. Besch-Williford and C.L. Franklin. 2000b. *In* Validation of the publication of new names and new combinations previously effectively published outside the IJSB. List No. 76. Int. J. Syst. Evol. Bacteriol. *50*: 1699–1700.

Simon, J., R. Gross, M. Ringel, E. Schmidt and A. Kröger. 1998. Deletion and site-directed mutagenesis of the *Wolinella succinogenes* fumarate reductase operon. Eur. J. Biochem. *251*: 418–426.

Simonin, P., B. Tindall and M. Rohmer. 1994. Structure elucidation and biosynthesis of 31-methylhopanoids from *Acetobacter europaeus*. Studies on a new series of bacterial triterpenoids. Eur. J. Biochem. *225*: 765–771.

Simons, H. 1922. Saprophytische Oscillarien des Menschen und der Tiere. Zentralbl Bakteriol. Abt. 1 Orig. *88*: 501–510.

Simoons-Smit, I.M., B.J. Appelmelk, T. Verboom, R. Negrini, J.L. Penner, G.O. Aspinall, A.P. Moran, S.F. Fei, S. Bi-Shan, W. Rudnica, A. Savio and J. de Graaff. 1996. Typing of *Helicobacter pylori* with monoclonal antibodies against Lewis antigens in lipopolysaccharide. J. Clin. Microbiol. *34*: 2196–2200.

Simor, A.E., B. Shames, B. Drumm, P. Sherman, D.E. Low and J.L. Penner. 1990. Typing of *Campylobacter pylori* by bacterial DNA restriction endonuclease analysis and determination of plasmid profile. J. Clin. Microbiol. *28*: 83–86.

Simor, A.E. and L. Wilcox. 1987. Enteritis associated with *Campylobacter laridis*. J. Clin. Microbiol. *25*: 10–12.

Simpson, C.F. 1972. Structure of *Ehrlichia canis* in blood monocytes of a dog. Am. J. Vet. Res. *33*: 2451–2454.

Simpson, C.F. 1974. Relationship of *Ehrlichia canis*-infected mononuclear cells to blood vessels of lungs. Infect. Immun. *10*: 590–596.

Simpson, R.M. and S.D. Gaunt. 1991. Immunocytochemical detection of *Ehrlichia platys* antigens in canine blood platelets. J. Vet. Diagn. Invest. *3*: 228–231.

Simser, J.A., A.T. Palmer, V. Fingerle, B. Wilske, T.J. Kurtti and U.G. Munderloh. 2002. *Rickettsia monacensis* sp. nov., a spotted fever group *Rickettsia*, from ticks (*Ixodes ricinus*) collected in a European city park. Appl. Environ. Microbiol. *68*: 4559–4566.

Simser, J.A., A.T. Palmer, U.G. Munderloh and T.J. Kurtti. 2001. Isolation of a spotted fever group *Rickettsia, Rickettsia peacockii*, in a Rocky Mountain wood tick, *Dermacentor andersoni*, cell line. Appl. Environ. Microbiol. *67*: 546–552.

Singh, B.N. 1947. Myxobacteria in soils and composts; their distribution, number and lytic action on bacteria. J. Gen. Microbiol. *1*: 1–10.

Singh, B.N. and N.B. Singh. 1971. Distribution of fruiting myxobacteria in Indian soils, bark of trees an dung of herbivorous animals. Indian J. Microbiol. *11*: 47–92.

Singleton, R., L.L. Campbell and F.M. Hawkridge. 1979. Cytochrome c_3 from the sulfate-reducing anaerobe *Desulfovibrio africanus* Benghazi - purification and properties. J. Bacteriol. *140*: 893–901.

Sinha, R.K. and R. R.S.. 1980. Release of soluble peptidoglycan from growing gonococci: Demonstration of anhydro-muramyl-containing fragments. Infect. Immun. *29*: 914–924.

Sironi, M., C. Bandi, L. Sacchi, B. Di Sacco, G. Damiani and C. Genchi. 1995. Molecular evidence for a close relative of the arthropod endosymbiont *Wolbachia* in a filarial worm. Mol. Biochem. Parasitol. *74*: 223–227.

Sistrom, W.R. 1960. A requirement for sodium in the growth of *Rhodopseudomonas sphaeroides*. J. Gen. Microbiol. *22*: 778–782.

Sittig, M. and P. Hirsch. 1992. Chemotaxonomic investigation of budding and/or hyphal bacteria. Syst. Appl. Bacteriol. *15*: 209–222.

Sittig, M. and H. Schlesner. 1993. Chemotaxonomic investigation of various prosthecate and/or budding bacteria. Syst. Appl. Microbiol. *16*: 92–103.

Sivakumaran, S., P.J. Lockhart and B.D. Jarvis. 1997. Identification of soil bacteria expressing a symbiotic plasmid from *Rhizobium leguminosarum* biovar trofolii. Can. J. Microbiol. *43*: 164–177.

Sivendra, R. 1976. Unusual *Chromobacterium violaceum*: aerogenic strains. J. Clin. Microbiol. *3*: 70–71.

Sivendra, R. and H.S. Lo. 1975. Identification of *Chromobacterium violaceum*: pigmented and non-pigmented strains. J. Gen. Microbiol. *90*: 21–31.

Skaar, E.P., M.P. Lazio and H.S. Seifert. 2002. Roles of the *recJ* and *recN* genes in homologous recombination and DNA repair pathways of *Neisseria gonorrhoeae*. J. Bacteriol. *184*: 919–927.

Skeeles, J.K. and L.H. Arp. 1997. Bordetellosis (turkey coryza). *In* Calnek, Barnes, Beard, McDougald and Saif (Editors), Diseases of Poultry, Iowa State University Press, Ames, IA. 275–287.

Skerman, V.B.D., V. McGowan and P.H.A. Sneath. 1980. Approved lists of bacterial names. Int. J. Syst. Bacteriol. *30*: 225–420.

Skerman, V.B.D., V. McGowan and P.H.A. Sneath. 1989. Approved lists of bacterial names, amended edition, American Society for Microbiology, Washington, D.C.

Skerman, V.B.D., L.I. Sly and M.L. Williamson. 1983. *Conglomeromonas largomobilis*, gen. nov., sp. nov., a sodium-sensitive, mixed-flagellated organism from fresh waters. Int. J. Syst. Bacteriol. *33*: 300–308.

Skerratt, J.H., P.D. Nichols, J.P. Bowman and L.I. Sly. 1992. Occurrence and significance of long chain (ω-1)-hydroxy fatty acids in methane utilizing bacteria. Org. Geochem. *18*: 189–194.

Skirrow, M.B. 1977. *Campylobacter* enteritis: a "new" disease. Br. Med. J. *2*: 9–11.

Skirrow, M.B. 1994. Diseases due to *Campylobacter*, *Helicobacter* and related bacteria. J. Comp. Pathol. *111*: 113–149.

Skirrow, M.B. and J. Benjamin. 1980. '1001' campylobacters: cultural characteristics of intestinal campylobacters from man and animals. J. Hyg. Camb. *85*: 427–442.

Skladny, H., M. Heidelbach and H.U. Schairer. 1994. Cloning and characterization of the gene encoding the major sigma-factor of *Stigmatella aurantiaca*. Gene *143*: 123–127.

Skoblo, I.I., O.N. Borchsenius, N.A. Lebedeva and D.V. Ossipov. 1990. Symbiogenic lysis of bacteria *Holospora acuminata* in the generative nucleus of the ciliate *Paramecium bursaria*. Tsitologiya *32*: 515–519.

Skotnicki, M.L., A.E. Goodman, R.G. Warr and P.L. Rogers. 1984. Isolation and characterization of *Zymomonas mobilis* plasmids. Microbios. *40*: 53–61.

Skuja, H. 1964. Grundzüge der Algenflora und Algenvegetation der Fjeldgegenden um Abisko in Schwedisch-Lappland. Nova Acta Reg. Soc. Sci. Upsal. Ser. IV. *18*: 1–465.

Skyring, G.W. and H.E. Jones. 1972. Guanine plus cytosine contents of the deoxyribonucleic acids of some sulfate-reducing bacteria: a reassessment. J. Bacteriol. *109*: 1298–1300.

Slabas, A.R. and L.R. Whatley. 1977. Metabolic regulation of pyruvate kinase isolated from autotrophically and heterotrophically grown *Paracoccus denitrificans*. Arch. Microbiol. *115*: 67–71.

Slee, A.M. and J.M. Tanzer. 1978. Selective medium for isolation of *Eikenella corrodens* from periodontal lesions. J. Clin. Microbiol. *8*: 459–462.

Sleytr, U. and M. Kocur. 1973. Structure of *Micrococcus denitrificans* and *M. halodenitrificans* revealed by freeze-etching. J. Appl. Bacteriol. *36*: 19–22.

Sleytr, U.B., P. Messner, D. Pum and M. Sara. 1988. Crystalline bacterial cell surface layers, Springer-Verlag, Berlin, New York.

Sluis, M.K., F.J. Small, J.R. Allen and S.A. Ensign. 1996. Involvement of an ATP-dependent carboxylase in a CO_2-dependent pathway of acetone metabolism by *Xanthobacter* strain Py2. J. Bacteriol. *178*: 4020–4026.

Sly, L.I. 1985. Emendation of the genus *Blastobacter* Zavarzin 1961 and description of *Blastobacter natatorius* sp. nov. Int. J. Syst. Bacteriol. *35*: 40–45.

Sly, L.I. and V. Arunpairojana. 1987. Isolation of manganese-oxidizing

Pedomicrobium cultures from water by micromanipulation. J. Microbiol. Methods *6*: 177–182.

Sly, L.I., V. Arunpairojana and M.C. Hodgkinson. 1988. *Pedomicrobium manganicum* from drinking-water distribution systems with manganese-related dirty water problems. Syst. Appl. Microbiol. *11*: 75–84.

Sly, L.I. and M.M. Cahill. 1997. Transfer of *Blastobacter natatorius* (Sly 1985) to the genus *Blastomonas* gen. nov. as *Blastomonas natatoria* comb. nov. Int. J. Syst. Bacteriol. *47*: 566–568.

Sly, L.I., M.M. Cahill, K. Majeed and G. Jones. 1997. Reassessment of the phylogenetic position of *Caulobacter subvibrioides*. Int. J. Syst. Bacteriol. *47*: 211–213.

Sly, L.I., T.L. Cox and T.B. Beckenham. 1999. The phylogenetic relationships of *Caulobacter*, *Asticcacaulis* and *Brevundimonas* species and their taxonomic implications. Int. J. Syst. Bacteriol. *49*: 483–488.

Sly, L.I. and M.H. Hargreaves. 1984. Two unusual budding bacteria isolated from a swimming pool. J. Appl. Bacteriol. *56*: 479–486.

Sly, L.I. and E. Stackebrandt. 1999. Description of *Skermanella parooensis* gen. nov., sp. nov. to accommodate *Conglomeromonas largomobilis* subsp. *parooensis* following the transfer of *Conglomeromonas largomobilis* subsp. *largomobilis* to the genus *Azospirillum*. Int. J. Syst. Bacteriol. *49*: 541–544.

Smadel, J.E., revised by B.L. Elisberg 1965. Scrub typhus rickettsia. *In* Horsfall and Tamm (Editors), Viral and Rickettsial Infections of Man, 4th Ed., J.B. Lippincott, Philadelphia. pp. 1130–1143.

Smarda, J. 1985. A new epiphytic bacterium of a cyanophyte. Syst. Appl. Microbiol. *6*: 298–301.

Smet, E., G. Chasaya, H. Van Langenhove and W. Verstraete. 1996. The effect of inoculation and the type of carrier material used on the biofiltration of methyl sulphides. Appl. Microbiol. Biotechnol. *45*: 293–298.

Smibert, R.M. 1965. *Vibrio fetus* biovar intestinalis isolated from fecal and intestinal contents of clinically normal sheep: Biochemical and cultural characteristics of microaerophilic vibrios isolated from the intestinal contents of sheep. Am. J. Vet. Res. *26*: 315–319.

Smibert, R.M. 1969. *Vibrio fetus* biovar intestinalis isolated from the intestinal content of birds. Am. J. Vet. Res. *30*: 1437–1442.

Smibert, R.M. 1970. Cell wall composition in the classification of *Vibrio fetus*. Int. J. Syst. Bacteriol. *20*: 407–412.

Smibert, R. 1974. *Campylobacter*. *In* Buchanan and Gibbons (Editors), Bergey's Manual of Determinative Bacteriology, 8th Ed., The Williams & Wilkins Co., Baltimore. pp. 207–212.

Smibert, R.M. 1978. The genus *Campylobacter*. Annu. Rev. Microbiol. *32*: 673–709.

Smibert, R.M. and L.V. Holdeman. 1976. Clinical isolates of anaerobic Gram-negative rods with a formate-fumarate energy metabolism: *Bacteroides corrodens*, *Vibrio succinogenes*, and unidentified strains. J. Clin. Microbiol. *3*: 432–437.

Smiles, J. and M.J. Dobson. 1956. Direct ultra-violet and ultra-violet negative phase contrast micrography of bacteria from the stomachs of sheep. J. R. Microsc. Soc. *75*: 244–253.

Smirnova, Z.S. and V.R. Arkhipova. 1981. Use of the fluorescent antibody method for control over the species composition of mixed cultures of methanotrophic bacteria during continuous cultivation. Prikl. Biokhim. Mikrobiol. *17*: 159–165.

Smit, G., T.J. Logman, M.E. Boerrigter, J.W. Kijne and B.J. Lugtenberg. 1989. Purification and partial characterization of the *Rhizobium leguminosarum* biovar viciae Ca^{2+} dependent adhesin, which mediates the first step in attachment of cells of the family *Rhizobiaceae* to plant root hair tips. J. Bacteriol. *171*: 4054–4062.

Smith, A.L. 1901. Myxobacteria. J. Bot. *39*: 69–72.

Smith, E.F. 1896. A bacterial disease of the tomato, egg plant and Irish potato (*Bacillus solanacearum* n. sp.). U.S. Dept. Div. Veg. Phys. Pathol. Bull. *12*: 1–28.

Smith, E.F. 1911. Bacteria in Relation to Plant Diseases, Vol. 2, Carnegie Inst. Wash. Publ. pp. 1–368.

Smith, E.F. 1914. Bacteria in relation to plant diseases. Carnegie Institute, Washington *3*: 1–309.

Smith, E.F. and C.O. Townsend. 1907. A plant-tumor of bacterial origin. Science *25*: 671–673.

Smith, H.O., M.L. Gwinn and S.L. Salzberg. 1999a. DNA uptake signal sequences in naturally transformable bacteria. Res. Microbiol. *150*: 603–616.

Smith, L.D. and F. Heffron. 1987. Transposon Tn5 mutagenesis of *Brucella abortus*. Infect. Immun. *55*: 2774–2776.

Smith, M.R., W.J.J. Van den Tweel and J.A.M. De Bont. 1991. Degradation of 3-chloro-2-methylpropionic acid by *Xanthobacter* sp. Cimw 99. Appl. Microbiol. Biotechnol. *36*: 246–251.

Smith, M.D., V. Wuthiekanun, A.L. Walsh and N.J. White. 1996. *In vitro* activity of carbapenem antibiotics against beta lactam susceptible and resistant strains of *Burkholderia pseudomallei*. J. Antimicrob. Chemother. *37*: 611–615.

Smith, N.H., E.C. Holmes, G.M. Donovan, G.A. Carpenter and B.G. Spratt. 1999b. Networks and groups within the genus *Neisseria*: Analysis of *argF*, *recA*, *rho*, and 16S rRNA sequences from human *Neisseria* species. Mol. Biol. Evol. *16*: 773–783.

Smith, R.L., F.E. Strohmaier and R.S. Oremland. 1985. Isolation of anaerobic oxalate-degrading bacteria from freshwater lake sediments. Arch. Microbiol. *141*: 8–13.

Smith, T. and M.S. Taylor. 1919. Some morphological and biochemical characteristics of the spirilla (*Vibrio fetus*, n. sp.) associated with disease of the fetal membranes in cattle. J. Exp. Med. *30*: 299–311.

Smith-Greenier, L.L. and A. Adkins. 1996. Degradation of diclofop-methyl by pure cultures of bacteria isolated from Manitoban soils. Can. J. Microbiol. *42*: 227–233.

Snaidr, J., R. Amann, I. Huber, W. Ludwig and K.H. Schleifer. 1997. Phylogenetic analysis and *in situ* identification of bacteria in activated sludge. Appl. Environ. Microbiol. *63*: 2884–2896.

Sneath, P.H.A. 1956. Cultural and biochemical characteristics of the genus *Chromobacterium*. J. Gen. Microbiol. *15*: 70–98.

Sneath, P.H.A. 1960. A study of the bacterial genus *Chromobacterium*. Iowa St. J. Sc. *34*: 243–500.

Sneath, P.H.A. 1979. Identification methods applied to *Chromobacterium*. *In* Skinner and Lovelock (Editors), Identification Methods for Microbiologists, 2nd Ed., Academic Press, London, New York. pp. 167–175.

Sneath, P.H.A. 1984a. Genus *Chromobacterium*. *In* Krieg and Holt (Editors), Bergey's Manual of Systematic Bacteriology, 1st Ed., Vol. 1, The Williams & Wilkins Co., Baltimore. pp. 580–582.

Sneath, P.H.A. 1984b. Genus *Janthinobacterium*. *In* Krieg and Holt (Editors), Bergey's Manual of Systematic Bacteriology, Vol. 1, The Williams & Wilkins Co., Baltimore. 376–377.

Sneath, P.H.A. and S.J. Barrett. 1996. A new species of *Neisseria* from the dental plaque of the domestic cow, *Neisseria dentiae* sp. nov. Lett. Appl. Microbiol. *23*: 355–358.

Sneath, P.H.A. and S.J. Barrett. 1997. *In* Validation of the publication of new names and new combinations previously effectively published outside the IJSB. List No. 62. Int. J. Syst. Bacteriol. *47*: 915–916.

Sneath, P.H.A. and R.R. Sokal. 1973. Numerical taxonomy. The principles and practice of numerical classification, W. H. Freeman, San Francisco.

Snell, J.J.S. 1984. Genus IV. *Kingella*. *In* Krieg and Holt (Editors), Bergey's Manual of Systematic Bacteriology, 1st Ed., Vol. 1, The Williams & Wilkins Co., Baltimore. pp. 307–309.

Snell, J.J.S., L.R. Hill and S.P. Lapage. 1972. Identification and characterization of *Moraxella phenylpyruvica*. J. Clin. Pathol. *25*: 959–965.

Snell, J.J.S. and S.P. LaPage. 1976. Transfer of some saccharolytic *Moraxella* species to *Kingella* Henriksen and Bøvre 1976, with descriptions of *Kingella indologenes* sp. nov. and *Kingella denitrificans* sp. nov. Int. J. Syst. Bacteriol. *26*: 451–458.

Snoeck, C., E. Luyten, V. Poinsot, A. Savagnac, J. Vanderleyden and J.C. Promé. 2001. *Rhizobium* sp. BR816 produced a complex mixture of known and novel lipochitooligosaccharide molecules. Mol. Plant-Microbe Interact. *14*: 678–684.

Snoeyenbos-West, O.L., K.P. Nevin, R.T. Anderson and D.R. Lovley. 2000. Enrichment of *Geobacter* species in response to stimulation of Fe(III) reduction in sandy aquifer sediments. Microb. Ecol. *39*: 153–167.

Snyder, J.C. 1965. Typhus fever rickettsiae. *In* Horsfall and Tamm (Editors), Viral and rickettsial infections of man, 4th Ed., J.B. Lippincott Co., Philadelphia. pp. 1059–1143.

Snyder, S.W. and T.C. Hollocher. 1987. Purification and some characteristics of nitrous oxide reductase from *Paracoccus denitrificans*. J. Biol. Chem. *262*: 6515–6525.

So, R.B., J.K. Ladha and J.P. Young. 1994. Photosynthetic symbionts of *Aeschynomene* spp. form a cluster with bradyrhizobia on the basis of fatty acid and rRNA analyses. Int. J. Syst. Bacteriol. *44*: 392–403.

Soberón-Chávez, G. and R. Nájera. 1989. Isolation from soil of *Rhizobium leguminosarum* lacking symbiotic information. Can. J. Microbiol. *35*: 464–468.

Socransky, S.S. 1977. Microbiology of periodontal disease—present status and future considerations. J. Periodontol. *48*: 497–504.

Sokol, P.A. 1986. Production and utilization of pyochelin by clinical isolates of *Pseudomonas cepacia*. J. Clin. Microbiol. *23*: 560–562.

Sokol, P.A., C.J. Lewis and J.J. Dennis. 1992. Isolation of a novel siderophore from *Pseudomonas cepacia*. J. Med. Microbiol. *36*: 184–189.

Sokollek, S.J. and W.P. Hammes. 1997. Description of a starter culture preparation for vinegar fermentation. Syst. Appl. Microbiol. *20*: 481–491.

Sokollek, S.J., C. Hertel and W.P. Hammes. 1998a. Cultivation and preservation of vinegar bacteria. J. Biotechnol. *60*: 195–206.

Sokollek, S.J., C. Hertel and W.P. Hammes. 1998b. Description of *Acetobacter oboediens* sp. nov. and *Acetobacter pomorum* sp. nov., two new species isolated from industrial vinegar fermentations. Int. J. Syst. Bacteriol. *48*: 935–940.

Sola-Landa, A., J. Pizarro-Cerda, M.J. Grillo, E. Moreno, I. Moriyon, J.M. Blasco, J.P. Gorvel and I. Lopez-Goni. 1998. A two component regulatory system playing a critical role in plant pathogens and endosymbionts is present in *Brucella abortus* and controls cell invasion and virulence. Mol. Microbiol. *29*: 125–138.

Soldo, A.T. and G.A. Godoy. 1973. Observations on the production of folic acid by symbiont lambda particles of *Paramecium aurelia* stock 299. J. Protozool. *20*: 502.

Solnick, J.V., J. O'Rourke, A. Lee, B.J. Paster, F.E. Dewhirst and L.S. Tompkins. 1993. An uncultured gastric spiral organism is a newly identified *Helicobacter* in humans. J. Infect. Dis. *168*: 379–385.

Solntseva, L. 1940. Biology of myxobacteria I. *Myxococcus*. Mikrobiologiya *9*: 217–232.

Solntseva, L. 1941. The biology of the myxobacteria. II. The genera *Melittangium* and *Chondromyces*. Mikrobiologiya *10*: 505–525.

Somasegaran, P. and H.J. Hoben. 1994. Handbook for Rhizobia : Methods in Legume-Rhizobium Technology, Springer-Verlag, New York. p. 450.

Sommer, C. and H. Gorisch. 1997. Enzymology of the degradation of (di)chlorobenzenes by *Xanthobacter flavus* 14p1. Arch. Microbiol. *167*: 384–391.

Song, B., M.M. Haggblom, J. Zhou, J.M. Tiedje and N.J. Palleroni. 1999. Taxonomic characterization of denitrifying bacteria that degrade aromatic compounds and description of *Azoarcus toluvorans* sp. nov. and *Azoarcus toluclasticus* sp. nov. Int. J. Syst. Bacteriol. *49 Pt 3*: 1129–1140.

Song, B., N.J. Palleroni and M.M. Häggblom. 2000a. Description of strain 3CB-1, a genomovar of *Thauera aromatica*, capable of degrading 3-chlorobenzoate coupled to nitrate reduction. Int. J. Syst. Evol. Microbiol. *50*: 551–558.

Song, B., N.J. Palleroni and M.M. Häggblom. 2000b. Isolation and characterization of diverse halobenzoate-degrading denitrifying bacteria from soils and sediments. Appl. Environ. Microbiol. *66*: 3446–3453.

Song, B., N.J. Palleroni, L.J. Kerkhof and M.M. Häggblom. 2001. Characterization of halobenzoate-degrading, denitrifying *Azoarcus* and *Thauera* isolates and description of *Thauera chlorobenzoica* sp. nov. Int. J. Syst. Evol. Microbiol. *51*: 589–602.

Song, B., L.Y. Young and N.J. Palleroni. 1998. Identification of denitrifier strain T1 as *Thauera aromatica* and proposal for emendation of the genus *Thauera* definition. Int. J. Syst. Bacteriol. *48*: 889–894.

Song, J.J., S. Zhang, R.W. Lenz and S. Goodwin. 2000c. *In vitro* polymerization and copolymerization of 3-hydroxypropionyl-CoA with the PHB synthase from *Ralstonia eutropha*. Biomacromolecules. *1*: 433–439.

Song, S.D., A. Hartmann and R.H. Burris. 1985. Purification and properties of the nitrogenase of *Azospirillum amazonense*. J. Bacteriol. *164*: 1271–1277.

Song, W.X., L. Ma, R.W. Chen and D.C. Stein. 2000d. Role of lipooligosaccharide in Opa-independent invasion of *Neisseria gonorrhoeae* into human epithelial cells. J. Exp. Med. *191*: 949–959.

Sordillo, E.M., M. Rendel, R. Sood, J. Belinfanti, O. Murray and D. Brook. 1993. Septicemia due to β-lactamase-positive *Kingella kingae* [letter]. Clin. Infect. Dis. *17*: 818–819.

Sorokin, D.Y. 1992. *Catenococcus thiocyclus*, gen. nov. sp. nov. a new facultatively anaerobic bacterium from a near-shore sulphidic hydrothermal area. J. Gen. Microbiol. *138*: 2287–2292.

Sorokin, D.Y. 1994. *In* Validation of the publication of new names and new combinations previously effectively published outside the IJSB. List No. 51. Int. J. Syst. Bacteriol. *44*: 852.

Sorokin, D.Y. 1995. *Sulfitobacter pontiacus* gen. nov., sp. nov., a new heterotrophic bacterium from the Black Sea, specialized on sulfite oxidation. Microbiology *64*: 295–305.

Sorokin, D.Y. 1996. *In* Validation of the publication of new names and new combinations previously effectively published outside the IJSB. List No. 56. Int. J. Syst. Bacteriol. *46*: 362–363.

Sorokin, D.Y. and A.M. Lysenko. 1993. Heterotrophic bacteria from the Black Sea oxidizing reduced sulfur compounds to sulfate. Microbiology *62*: 594–602.

Sorokin, D.Y., G. Muyzer, T. Brinkhoff, J.G. Kuenen and M.S.M. Jetten. 1998. Isolation and characterization of a novel facultatively alkaliphilic *Nitrobacter* species, *N. alkalicus* sp. nov. Arch. Microbiol. *170*: 345–352.

Sorokin, D.Y., G. Muyzer, T. Brinkhoff, J.G. Kuenen and M.S.M. Jetten. 1999. *In* Validation of the publication of new names and new combinations previously effectively published outside the IJSEM. List No. 71. Int. J. Syst. Bacteriol. *49*: 1325–1326.

Sorokin, D.Y., T.P. Tourova, B.B. Kuznetsov, I.A. Bryantseva and V.M. Gorlenko. 2000a. *Roseinatronobacter thiooxidans* gen. nov., sp. nov., a new alkaliphilic aerobic bacteriochlorophyll *a*-containing bacterium isolated from a soda lake. Mikrobiologiya *69*: 89–97.

Sorokin, D.Y., T.P. Tourova, B.B. Kuznetsov, I.A. Bryantseva and V.M. Gorlenko. 2000b. *In* Validation of the publication of new names and new combinations previously effectively published outside the IJSEM. List No. 75. Int. J. Syst. Evol. Microbiol. *50*: 1415.

Southerland, W.M. and F. Toghrol. 1983. Sulfite oxidase activity in *Thiobacillus novellus*. J. Bacteriol. *156*: 941–944.

Sowers, K.R. and K.M. Noll. 1995. Techniques for anaerobic growth. *In* Robb, Sowers, DasSarma, Place, Schreier and Fleischmann (Editors), Archaea: A Laboratory Manual, Cold Spring Harbor Laboratory Press, Plainview, NY. pp. 15–48.

Spach, D.H., A.S. Kanter, N.A. Daniels, D.J. Nowowiejski, A.M. Larson, R.A. Schmidt, B. Swaminathan and D.J. Brenner. 1995. *Bartonella (Rochalimaea)* species as a cause of apparent culture-negative endocarditis. Clin. Infect. Dis. *20*: 1044–1047.

Sparling, P.F. 1966. Genetic transformation of *Neisseria gonorrhoeae* to streptomycin resistance. J. Bacteriol. *92*: 1364–1371.

Spaulding, E.H. 1974. Introduction. *In* Prier, J.E. and H. Friedman (Editors), Opportunistic Pathogens, University Park Press, Baltimore. pp. xiii–xv.

Speakman, A.J., S. Dawson, S.H. Binns, C.J. Gaskell, C.A. Hart and R.M. Gaskell. 1999. *Bordetella bronchiseptica* infection in the cat. J. Small Anim. Pract. *40*: 252–256.

Spencer, R.R. and R.R. Parker. 1923. Rocky Mountain spotted fever: Infectivity of fasting and recently fed ticks. Public Health Rep. *38*: 333–339.

Sperl, G.T. and D.S. Hoare. 1971. Denitrification with methanol: a selective enrichment for *Hyphomicrobium* species. J. Bacteriol. *108*: 733–736.

Spieck, E., J. Aamand, S. Bartosch and E. Bock. 1996a. Immunocytochemical detection and location of the membrane-bound nitrite ox-

idoreductase in cells of *Nitrobacter* and *Nitrospira*. FEMS Microbiol. Lett. *139*: 71–76.

Spieck, E., S. Muller, A. Engel, E. Mandelkow, H. Patel and E. Bock. 1996b. Two-dimensional structure of membrane-bound nitrite oxidoreductase from *Nitrobacter hamburgensis*. J. Struct. Biol. *117*: 117–123.

Spiegel, Y., E. Cohn, S. Galper, E. Sharon and I. Chet. 1991. Evaluation of a newly isolated bacterium, *Pseudomonas chitinolytica* sp. nov., for controlling the root-knot nematode *Meloidogyne javanica*. Biocontrol Sci. Technol. *1*: 115–125.

Spiess, E. and H. Gorisch. 1996. Purification and characterization of chlorobenzene *cis*-dihydrodiol dehydrogenase from *Xanthobacter flavus* 14p1. Arch. Microbiol. *165*: 201–205.

Spiess, E., C. Sommer and H. Gorisch. 1995. Degradation of 1,4-dichlorobenzene by *Xanthobacter flavus* 14p1. Appl. Environ. Microbiol. *61*: 3884–3888.

Spiff, E.D. and C.T. Odu. 1973. Acetylene reduction by *Beijerinckia* under various partial pressures of oxygen and acetylene. J. Gen. Microbiol. *78*: 207–209.

Spormann, A.M. 1999. Gliding motility in bacteria: Insights from studies of *Myxococcus xanthus*. Microbiol. Mol. Biol. Rev. *63*: 621–641.

Spormann, A.M. and F. Widdel. 2000. Metabolism of alkylbenzenes, alkanes, and other hydrocarbons in anaerobic bacteria. Biodegradation *11*: 85–105.

Sporn, L.A., S.K. Sahni, N.B. Lerner, V.J. Marder, D.J. Silverman, L.C. Turpin and A.L. Schwab. 1997. *Rickettsia rickettsii* infection in cultured human endothelial cells induces NK-kB activation. Infect. Immun. *65*: 2786–2791.

Spratt, B.G. 1988. Hybrid penicillin-binding proteins in penicillin-resistant strains of *Neisseria gonorrhoeae*. Nature *332*: 173–176.

Spratt, B.G., L.D. Bowler, Q.Y. Zhang, J.J. Zhou and J.M. Smith. 1992. Role of interspecies transfer of chromosomal genes in the evolution of penicillin resistance in pathogenic and commensal *Neisseria* species. J. Mol. Evol. *34*: 115–125.

Spratt, B.G., Q.Y. Zhang, D.M. Jones, A. Hutchison, J.A. Brannigan and C.G. Dowson. 1989. Recruitment of a penicillin-binding protein gene from *Neisseria flavescens* during the emergence of penicillin resistance in *Neisseria meningitidis*. Proc. Natl. Acad. Sci. U.S.A. *86*: 8988–8992.

Sprenger, G.A. 1993. Approaches to broaden the substrate and product range of the ethanologenic bacterium *Zymomonas mobilis* by genetic engineering. J. Biotechnol. *27*: 225–237.

Sprenger, G.A. 1996. Carbohydrate metabolism in *Zymomonas mobilis*: a catabolic highway with some scenic routes. FEMS Microbiol. Lett. *145*: 301–307.

Sprenger, G.A., M.A. Typas and C. Drainas. 1993. Genetics and genetic engineering on *Zymomonas mobilis*. World J. Microbiol. Biotechnol. *9*: 17–24.

Spring, S., P. Kämpfer and K.-H. Schleifer. 2001. *Limnobacter thiooxidans* gen. nov., sp nov., a novel thiosulfate-oxidizing bacterium isolated from freshwater lake sediment. Int. J. Syst. Evol. Microbiol. *51*: 1463–1470.

Spring, S., P. Kämpfer, W. Ludwig and K.-H. Schleifer. 1996. Polyphasic characterization of the genus *Leptothrix*: new descriptions of *Leptothrix mobilis* sp. nov. and *Leptothrix discophora* sp. nov. nom. rev. and emended description of *Leptothrix cholodnii* emend. Syst. Appl. Microbiol. *19*: 634–643.

Spring, S., P. Kämpfer, W. Ludwig and K.-H. Schleifer. 1997. *In* Validation of the publication of new names and new combinations previously effectively published outside the IJSB. List No. 64. Int. J. Syst. Bacteriol. *47*: 601–602.

Spring, S. and K.H. Schleifer. 1995. Diversity of magnetotactic bacteria. Syst. Appl. Microbiol. *18*: 147–153.

Springael, D., J. van Thor, H. Goorissen, A. Ryngaert, R. De Baere, P. Van Hauwe, L.C. Commandeur, J.R. Parsons, R. De Wachter and M. Mergeay. 1996. RP4::Mu3A-mediated *in vivo* cloning and transfer of a chlorobiphenyl catabolic pathway. Microbiology *142*: 3283–3293.

Springer, N., R. Amann, W. Ludwig, K.H. Schleifer and H. Schmidt. 1996. *Polynucleobacter necessarius*, an obligate bacterial endosymbiont of the hypotrichous ciliate *Euplotes aediculatus*, is a member of the beta-subclass of *Proteobacteria*. FEMS Microbiol. Lett. *135*: 333–336.

Springer, N., W. Ludwig, R. Amann, H.J. Schmidt, H.D. Görtz and K.H. Schleifer. 1993. Occurrence of fragmented 16S ribosomal-RNA in an obligate bacterial endosymbiont of *Paramecium caudatum*. Proc. Natl. Acad. Sci. U.S.A. *90*: 9892–9895.

Springer, N., W. Ludwig, B. Philipp and B. Schink. 1998. *Azoarcus anaerobius* sp. nov., a resorcinol-degrading, strictly anaerobic, denitrifying bacterium. Int. J. Syst. Bacteriol. *48*: 953–956.

Spröer, C., H. Reichenbach and E. Stackebrandt. 1999. The correlation between morphological and phylogenetic classification of myxobacteria. Int. J. Syst. Bacteriol. *49*: 1255–1262.

Stableforth, A.W. 1959. Diseases due to bacteria. *In* Stableforth and Galloway (Editors), Infectious Diseases of Animals, Vol. 1, Butterworths, London. p. 60.

Stableforth, A.W. and L.M. Jones. 1963. Report of the sub committee on Taxonomy of the genus *Brucella*. Speciation in the genus *Brucella*. Int. Bull. Bacteriol. Nomencl. Taxon. *13*: 145–158.

Stacey, G., S. Luka, J. Sanjuan, Z. Banfalvi, A.J. Nieuwkoop, J.Y. Chun, L.S. Forsberg and R. Carlson. 1994. *nodZ*, a unique host-specific nodulation gene, is involved in the fucosylation of the lipooligosaccharide nodulation signal of *Bradyrhizobium japonicum*. J. Bacteriol. *176*: 620–633.

Stacey, G., J. Sanjuan, S. Luka, T. Dockendorff and R.W. Carlson. 1995. Signal exchange in the *Bradyrhiobium*-soybean symbiosis. Soil Biol. Biochem. *27*: 473–483.

Stackebrandt, E. 1988. Phylogenetic relationships vs. phenotypic diversity: how to achieve a phylogenetic classification system of the eubacteria. Can. J. Microbiol. *34*: 552–556.

Stackebrandt, E. 1992. Unifying phylogeny and phenotypic diversity. *In* Balows, Trüper, Dworkin, Harder and Schleifer (Editors), The Prokaryotes: A Handbook of Bacteria: Ecophysiology, Isolation, Identification, Applications, 2nd Ed., Springer-Verlag, New York. pp. 19–47.

Stackebrandt, E., A. Fischer, T. Roggentin, U. Wehmeyer, D. Bomar and J. Smida. 1988a. A phylogenetic survey of budding, and/or prosthecate, non-phototrophic eubacteria: membership of *Hyphomicrobium*, *Hyphomonas*, Pedomicrobiumu, *Filomicrobium*, *Caulobacter* and "*Dichotomicrobium*" to the alpha-subdivision of purple non-sulfur bacteria. Arch. Microbiol. *149*: 547–556.

Stackebrandt, E. and B.M. Goebel. 1994. Taxonomic note: A place for DNA–DNA reassociation and 16S rRNA sequence analysis in the present species definition in bacteriology. Int. J. Syst. Bacteriol. *44*: 846–849.

Stackebrandt, E., R.G.E. Murray and H.G. Trüper. 1988b. *Proteobacteria* classis nov., a name for the phylogenetic taxon that includes the "purple bacteria and their relatives". Int. J. Syst. Bacteriol. *38*: 321–325.

Stackebrandt, E., U. Wehmeyer and B. Schink. 1989. The phylogenetic status of *Pelobacter acidigallici*, *Pelobacter venetianus*, and *Pelobacter carbinolicus*. Syst. Appl. Microbiol. *11*: 257–260.

Stackebrandt, E. and C.R. Woese. 1984. The phylogeny of prokaryotes. Microbiol. Sci. *1*: 117–122.

Stadtwalddemchick, R., F.R. Turner and H. Gest. 1990. *Rhodopseudomonas cryptolactis*, sp. nov, a new thermotolerant species of budding phototrophic purple bacteria. FEMS Microbiol. Lett. *71*: 117–121.

Stahl, D.A., R. Key, B. Flesher and J. Smit. 1992. The phylogeny of marine and freshwater caulobacters reflects their habitat. J. Bacteriol. *174*: 2193–2198.

Stahl, D.A., D.J. Lane, G.J. Olsen, D.J. Heller, T.M. Schmidt and N.R. Pace. 1987. Phylogenetic analysis of certian sulfide-oxidizing and related morphologically conspicuous bacteria by 5S ribosomal ribonucleic acid sequences. Int. J. Syst. Bacteriol. *37*: 116–122.

Stahl, S. 1973. Slime of *Myxococcus virescens*. Physiol. Plant. *28*: 523–529.

Stainthorpe, A.C., V. Lees, G.P. Salmond, H. Dalton and J.C. Murrell. 1991. Screening of obligate methanotrophs for soluble methane monooxygenase genes. FEMS Microbiol. Lett. *70*: 211–216.

Staley, J.T. 1968. *Prosthecomicrobium* and *Ancalomicrobium*: new prosthecate freshwater bacteria. J. Bacteriol. *95*: 1921–1942.

Staley, J.T. 1971. Incidence of prosthecate bacteria in a polluted stream. Appl. Microbiol. *22*: 496–502.

Staley, J.T. 1973a. Budding and prosthecate bacteria. *In* Laskin and Lechevalier (Editors), Organismic Microbiology: Handbook of Microbiology, Vol. 1, CRC Press, Cleveland. pp. 29–49.

Staley, J.T. 1973b. Budding bacteria of the *Pasteuria-Blastobacter* group. Can. J. Microbiol. *19*: 609–614.

Staley, J.T. 1981a. The genera *Prosthecomicrobium* and *Ancalomicrobium*. *In* Starr, Stolp, Trüper, Balows and Schlegel (Editors), The Prokaryotes: A Handbook on Habitats, Isolation and Identification of Bacteria, Springer-Verlag, Berlin. pp. 456–460.

Staley, J.T. 1981b. The genus *Pasteuria*. *In* Starr, Stolp, Trüper, Balows and Schlegel (Editors), The Prokaryotes: A Handbook on Habitats, Isolation and Identification of Bacteria, Vol. 1, Springer-Verlag, New York. pp. 490–492.

Staley, J.T. 1984. *Prosthecomicrobium hirschii* a new species in a redefined genus. Int. J. Syst. Bacteriol. *34*: 304–308.

Staley, J.T. 1989. Genus *Ancalomicrobium*. *In* Staley, Bryant, Pfennig and Holt (Editors), Bergey's Manual of Systematic Bacteriology, 1st Ed., Vol. 3rd, The Williams & Wilkins Co., Baltimore. pp. 1914–1916.

Staley, J.T., P. Hirsch and J.M. Schmidt. 1981. Introduction to the budding and/or appendaged bacteria. *In* Starr, Stolp, Trüper, Balows and Schlegel (Editors), The Prokaryotes: A Handbook on Habitats, Isolation and Identification of Bacteria, 1st Ed., Vol. 1, Springer-Verlag, Berlin. pp. 448–455.

Staley, J.T., R.L. Irgens and R.P. Herwig. 1989. Gas vacuolate bacteria from the sea ice of Antarctica. Appl. Environ. Microbiol. *55*: 1033–1036.

Staley, J.T. and T.L. Jordan. 1973. Crossbands of *Caulobacter crescentus* stalks serve as indicators of cell age. Nature (Lond.) *246*: 155–156.

Staley, J.T. and M. Mandel. 1973. Deoxyribonucleic acid base composition of *Prosthecomicrobium* and *Ancalomicrobium* strains. Int. J. Syst. Bacteriol. *23*: 271–273.

Staley, J.T., K.C. Marshall and V.B.D. Skerman. 1980. Budding and prosthecate bacteria from freshwater habitats of various trophic states. Microb. Ecol. *5*: 245–251.

Stalon, V. and A. Mercenier. 1984. L-Arginine utilization by *Pseudomonas* species. J. Gen. Microbiol. *130*: 69–76.

Stamm, W.W., M. Kittelmann, H. Follmann and H.G. Trüper. 1989. The occurrence of bacteriophages in spirit vinegar fermentation. Appl. Microbiol. Biotechnol. *30*: 41–46.

Stamp, J.T., A.D. McEwen, I.A.A. Watt and D.I. Nisbet. 1950. Enzootic abortion in ewes. Transmission of the disease. Vet. Rec. *62*: 251–254.

Stams, A.M. 1994. Metabolic interactions between anaerobic bacteria in methanogenic environments. Antonie Leeuwenhoek *66*: 271–294.

Stams, A.M., J.B. Van Dijk, C. Dijkema and C.M. Plugge. 1993. Growth of syntrophic propionate-oxidizing bacteria with fumarate in the absence of methanogenic bacteria. Appl. Environ. Microbiol. *59*: 1114–1119.

Stanbridge, T.N. and N.W. Preston. 1974. Experimental pertussis infection in the marmoset: type specificity of active immunity. J. Hyg. (Lond) *72*: 213–228.

Standal, R., T-G. Iversen, D.H. Coucheron, E. Fjaervik, J.M. Blatny and S. Valla. 1994. A new gene required for cellulose production and a gene encoding cellulolytic activity in *Acetobacter xylinum* are colocalized with the *bcs* operon. J. Bacteriol. *176*: 665–672.

Stanier, R.Y. 1942. A note on elasticotaxis in myxobacteria. J. Bacteriol. *44*: 405–412.

Stanier, R.Y. 1957. Order VIII. *Myxobacterales* Jahn. *In* Breed, Murray and Smith (Editors), Bergey's Manual of Determinative Bacteriology, 7th Ed., The Williams & Wilkins Co., Baltimore. pp. 854–891.

Stanier, R.Y., N.J. Palleroni and M. Doudorof. 1966. The aerobic pseudomonads: a taxonomic study. J. Gen. Microbiol. *43*: 159–271.

Stanley, J., A.P. Burnens, D. Linton, S.L.W. On, M. Costas and R.J. Owen. 1992. *Campylobacter helveticus* sp. nov., a new thermophilic species from domestic animals: characterization and cloning of a species-specific DNA probe. J. Gen. Microbiol. *138*: 2293–2303.

Stanley, J., A.P. Burnens, D. Linton, S.L.W. On, M. Costas and R.J. Owen.

1993a. *In* Validation of the publication of new names and new combinations previously effectively published outside the IJSB. List No. 45. Int. J. Syst. Bacteriol. *43*: 398–399.

Stanley, J., C. Jones, A. Burnens and R.J. Owen. 1994a. Distinct genotypes of human and canine isolates of *Campylobacter upsaliensis* determined by 16S rRNA gene typing and plasmid profiling. J. Clin. Microbiol. *32*: 1788–1794.

Stanley, J., D. Linton, A.P. Burnens, F.E. Dewhirst, S.L.W. On, A. Porter, R.J. Owen and M. Costas. 1994b. *Helicobacter pullorum* sp. nov.: genotype and phenotype of a new species isolated from poultry and from human patients with gastroenteritis. Microbiology (Reading) *140*: 3441–3449.

Stanley, J., D. Linton, A.P. Burnens, F.E. Dewhirst, S.L.W. On, A. Porter, R.J. Owen and M. Costas. 1995. *In* Validation of the publication of new names and new combinations previously effectively published outside the IJSB. List No. 53. Int. J. Syst. Bacteriol. *45*: 418–419.

Stanley, J., D. Linton, A.P. Burnens, F.E. Dewhirst, O. R.J., A. Porter, S.L.W. On and M. Costas. 1993b. *Helicobacter canis* sp. nov., a new species from dogs: an integrated study of phenotype and genotype. J. Gen. Microbiol. *139*: 2495–2504.

Stanley, J., D. Linton, A.P. Burnens, F.E. Dewhirst, O. R.J., A. Porter, S.L.W. On and M. Costas. 1994c. *In* Validation of the publication of new names and new combinations previously effectively published outside the IJSB. List No. 49. Int. J. Syst. Bacteriol. *44*: 370–371.

Stanley, P.M., R.L. Moore and J.T. Staley. 1976. Characterization of two new isolates of mushroom-shaped budding bacteria. Int. J. Syst. Bacteriol. *26*: 522–527.

Stanley, P.M., E.J. Ordal and J.T. Staley. 1979. High numbers of prosthecate bacteria in pulp mill waste aeration lagoons. Appl. Environ. Microbiol. *37*: 1007–1011.

Stannard, A.A., D.H. Gribble and R.S. Smith. 1969. Equine ehrlichiosis: A disease with similarities to tick-borne fever and bovine petechial fever. Vet. Rec. *84*: 149–150.

Staphorst, J.L., F.G.H. Van Zyl, B.W. Strijdom and Z.E. Groenewold. 1985. Agrocin producing pathogenic and nonpathogenic biotype-3 strains of *Agrobacterium tumefaciens* active against biotype-3 pathogens. Curr. Microbiol. *12*: 45–52.

Staples, D.G. 1973. The ecology and physiology of *Bdellovibrio bacteriovorus*, Doctoral thesis, University of Wales, Institute of Science and Technology, Cardiff.

Staples, D.G. and J.C. Fry. 1973. Factors which influence the enumeration of *Bdellovibrio bacteriovorus* in sewage and river water. J. Appl. Bacteriol. *36*: 1–11.

Stapp, C. 1928. *Schizomycetes* (Spaltpilze oder Bakterien). *In* Sorauer (Editor), Handbuch der Pflanzenkrankheiten, 5th Ed., Vol. 2, Paul Parey, Berlin. pp. 1–295.

Stapp, C. and D. Knösel. 1954. Zur genetik sternbildender Bakterien. Zentralbl. Bakteriol. Parasitenk. Infektionskr. Hyg. Abt. II *108*: 243–259.

Stapp, C. and D. Knösel. 1956. Fortgeführte Untersuchungen über den Entwicklungscyclus und die Karyologie sternbildender Bakterien. Zentralbl. Bakteriol. Parasitenkd. Infektionskr. Hyg. Abt. II *109*: 416–428.

Starich, T., P. Cordes and J. Zissler. 1985. Transposon tagging to detect a latent virus in *Myxococcus xanthus*. Science *230*: 541–543.

Starich, T. and J. Zissler. 1989. Movement of multiple DNA units between *Myxococcus xanthus* cells. J. Bacteriol. *171*: 2323–2336.

Starkey, R.L. 1934. Cultivation of organisms concerned in the oxidation of thiosulfate. J. Bacteriol. *28*: 365–386.

Starkey, R.L. 1935a. Isolation of some bacteria which oxidize thiosulfate. Soil Sci. *39*: 197–219.

Starkey, R.L. 1935b. Products of the oxidation of thiosulfate by bacteria in mineral media. J. Gen. Physiol. *18*: 325–349.

Starkey, R.L. and P.K. De. 1939. A new species of *Azotobacter*. Soil Sci. *47*: 329–343.

Starostina, N.G. and N.I. Pashkova. 1993. Interactions between populations in a 3-component mixed culture of methanotrophic and lytic bacteria. Mikrobiologiya *62*: 213–218.

Starostina, N.G., K.D. Wendlandt, G. Rogge and M. Jechorek. 1995. Interactions between populations of methanotrophic bacteria in a two-component culture. Mikrobiologiya *64*: 722–724.

Starr, D.J. and T.W. Cline. 2002. A host parasite interaction rescues *Drosophila* oogenesis defects. Nature (Lond.) *418*: 76–79.

Starr, M.P. 1946. The nutrition of phytopathogenic bacteria. II. The genus *Agrobacterium*. J. Bacteriol. *52*: 187–194.

Starr, M.P. 1981. The Genus *Lampropedia*. *In* Starr, Stolp, Trüper, Balows and Schlegel (Editors), The Prokaryotes: A Handbook on Habitats, Isolation and Identification of Bacteria, Springer-Verlag, Berlin. pp. 1530–1536.

Starr, M.P. and N.L. Baigent. 1966. Parasitic interaction of *Bdellovibrio bacteriovorus* with other bacteria. J. Bacteriol. *91*: 2006–2017.

Starr, M.P., W. Blau and G. Cosens. 1960. The blue pigment of Pseudomonas lemonnieri. Biochem. Z. *333*: 328–334.

Starr, M.P. and W.H. Burkholder. 1942. Lipolytic activity of phytopathogenic bacteria determined by means of spirit blue agar and its taxonomic significance. Phytopathology *32*: 598–604.

Starr, M.P., C.L. Jenkins, L.B. Bussey and A.G. Andrewes. 1977. Chemotaxonomic significance of the xanthomonadins, novel brominated aryl polyene pigments produced by bacteria of the genus *Xanthomonas*. Arch. Microbiol. *113*: 1–9.

Starr, M.P. and V.B. Skerman. 1965. Bacterial diversity: the natural history of selected morphologically unusual bacteria. Annu. Rev. Microbiol. *19*: 407–454.

Starr, M.P. and J.E. Weiss. 1943. Growth of phytopathogenic bacteria in a synthetic asparagin medium. Phytopathology *33*: 314–318.

Staskawicz, B.J., M.B. Mudgett, J.L. Dangl and J.E. Galan. 2001. Common and contrasting themes of plant and animal diseases. Science *292*: 2285–2289.

Stead, D.E. 1992. Grouping of plant-pathogenic and some other *Pseudomonas* spp. by using cellular fatty acid profiles. Int. J. Syst. Bacteriol. *42*: 281–295.

Stechmann, A. and U. Berger. 1964. On age-related changes in the oral flora. Z. Hyg. Infektionskr. *150*: 18–28.

Steed, P.D.M. 1962. *Simonsiellaceae* fam. nov. with characterization of *Simonsiella crassa* and *Alysiella filiformis*. J. Gen. Microbiol. *29*: 615–624.

Steeghs, L., R. den Hartog, A. den Boer, B. Zomer, P. Roholl and P. van der Ley. 1998. Meningitis bacterium is viable without endotoxin. Nature *392*: 449–450.

Steele, K.E., Y. Rikihisa and A.M. Walton. 1986. *Ehrlichia* of Potomac horse fever identified with a silver stain. Vet. Pathol. *23*: 531–533.

Steele, T.W. and R.J. Owen. 1988. *Campylobacter jejuni* subsp. *doylei* subsp. nov., a subspecies of nitrate-negative campylobacters isolated from human clinical specimans. Int. J. Syst. Bacteriol. *38*: 316–318.

Steenkamp, D.J. 1979. Identification of the prosthetic groups of dimethylamine dehydrogenase from *Hyphomicrobium* X. Biochem. Biophys. Res. Commun. *88*: 244–250.

Steenkamp, D.J. and H. Beinert. 1982a. Mechanistic studies on the dehydrogenases of methylotrophic bacteria. 1. The influence of substrate binding to reduced trimethylamine dehydrogenase on the intramolecular electron transfer between its prosthetic groups. Biochem. J. *207*: 233–240.

Steenkamp, D.J. and H. Beinert. 1982b. Mechanistic studies on the dehydrogenases of methylotrophic bacteria. 2. Kinetic studies on the intramolecular electron transfer in trimethylamine dehydrogenase and dimethylamine dehydrogenase. Biochem. J. *207*: 241–252.

Steffan, R.J. and R.M. Atlas. 1988. DNA amplification to enhance detection of genetically engineered bacteria in environmental samples. Appl. Environ. Microbiol. *54*: 2185–2191.

Stein, J. and H. Budzikiewicz. 1987. 1-*O*-(13-Methyl-1-Z-tetradecenyl)-2-*O*-(13-methyltetradecanoyl)-glycero-3-phospho-ethanolamin, ein plasmalogen aus *Myxococcus stipitatus*. Z. Naturforsch. *42b*: 1017–1020.

Stein, L.Y., M.T. La Duc, T.J. Grundl and K.H. Nealson. 2001. Bacterial and archaeal populations associated with freshwater ferromanganous micronodules and sediments. Environ. Microbiol. *3*: 10–18.

Steinbüchel, A., E.M. Debzi, R.H. Marchessault and A. Timm. 1993. Synthesis and production of poly(3-hydroxyvaleric acid) homopolyester by *Chromobacterium violaceum.* Appl. Microbiol. Biotechnol. *39*: 443–449.

Steinbüchel, A. and G. Schmack. 1995. Large-scale production of poly(3-hydroxyvaleric acid) by fermentation of *Chromobacterium violaceum*, processing, and characterization of the homopolyester. J. Environ. Polym. Degr. *3*: 243–258.

Steiner, S., S.F. Conti and R.L. Lester. 1973. Occurrence of phosphonosphingolipids in *Bdellovibrio bacteriovorus* strain UKi2. J. Bacteriol. *116*: 1199–1211.

Steinle, P., G. Stucki, R. Stettler and K.W. Hanselmann. 1998. Aerobic mineralization of 2,6-dichlorophenol by *Ralstonia* sp. strain RK1. Appl. Environ. Microbiol. *64*: 2566–2571.

Steinle, P., G. Stucki, R. Stettler and K.W. Hanselmann. 1999. *In* Validation of the publication of new names and new combinations previously effectively published outside the IJSB. List No. 71. Int. J. Syst. Bacteriol. *49*: 1325–1326.

Steinman, H.M. and B. Ely. 1990. Copper-zinc superoxide dismutase of *Caulobacter crescentus*: cloning, sequencing, and mapping of the gene and periplasmic location of the enzyme. J. Bacteriol. *172*: 2901–2910.

Steinmetz, I., M. Rohde and B. Brenneke. 1995. Purification and characterization of an exopolysaccharide of *Burkholderia (Pseudomonas) pseudomallei.* Infect. Immun. *63*: 3959–3965.

Stenos, J., V. Roux, D. Walker and D. Raoult. 1998. *Rickettsia honei* sp. nov., the aetiological agent of Flinders Island spotted fever in Australia. Int. J. Syst. Bacteriol. *4*: 1399–1404.

Stenos, J. and D.H. Walker. 2000. The rickettsial outer-membrane protein A and B genes of *Rickettsia australis*, the most divergent rickettsia of the spotted fever group. Int. J. Syst. Evol. Microbiol. *50*: 1775–1779.

Stephan, H., S. Freund, W. Beck, G. Jung, J.M. Meyer and G. Winkelmann. 1993a. Ornibactins: a new family of siderophores from *Pseudomonas*. Biometals *6*: 93–100.

Stephan, H., S. Freund, J.M. Meyer, G. Winkelmann and G. Jung. 1993b. Structure elucidation of the gallium ornibactin complex by 2d NMR spectroscopy. Liebigs Annalen Der Chemie *1*: 43–48.

Stephan, M.P., M. Oliveira, K.R.S. Teixeira, G. Martinez-Drets and J. Dobereiner. 1991. Physiology and dinitrogen fixation of *Acetobacter diazotrophicus*. FEMS Microbiol. Lett. *77*: 67–72.

Stephen, J.R., A.E. McCaig, Z. Smith, J.I. Prosser and T.M. Embley. 1996. Molecular diversity of soil and marine 16S rRNA gene sequences related to β-subgroup ammonia-oxidizing bacteria. Appl. Environ. Microbiol. *62*: 4147–4154.

Stephens, D.S. and Z.A. McGee. 1981. Attachment of *Neisseria meningitidis* to human mucosal surfaces—influence of pili and type of receptor cell. J. Infect. Dis. *143*: 525–532.

Stephenson, E.H. and J.V. Osterman. 1977. Canine peritoneal macrophages: Cultivation and infection with *Ehrlichia canis*. Am. J. Vet. Res. *38*: 1815–1819.

Stern, A. and T.F. Meyer. 1987. Common mechanism controlling phase and antigenic variation in pathogenic neisseriae. Mol. Microbiol. *1*: 5–12.

Steudel, A., D. Miethe and W. Babel. 1980. Bacterium MB 58, a methylotrophic "acetic acid bacterium". Z. Allg. Mikrobiol. *20*: 663–672.

Stevens, A. 1990. Simulations of the gliding behaviour and aggregation of myxobacteria. *In* Alt and Hoffmann (Editors), Biological Motion, Springer-Verlag, Heidelberg. pp. 548–555.

Stevens, L.A., R. Giordano and R.F. Fialho. 2001. Male-killing, nematode infections, bacteriophage infection and virulence of cytoplasmic bacteria in the genus *Wolbachia*. Annu. Rev. Ecol. Syst. *32*: 519–545.

Stewart, M., T.J. Beveridge and R.G.E. Murray. 1980. Structure of the regular surface layer of *Spirillum putridiconchylium*. J. Mol. Biol. *137*: 1–8.

Stich, R.W., J.A. Bantle, K.M. Kocan and A. Fekete. 1993. Detection of *Anaplasma marginale* (*Rickettsiales: Anaplasmataceae*) in hemolymph of *Dermacentor andersoni* (Acari: Ixodidae) with the polymerase chain reaction. J. Med. Entomol. *30*: 781–788.

Stich, R.W., K.M. Kocan, R.T. Damian and M. Fechheimer. 1997. Inclusion appendages associated with the intraerythrocytic rickettsial parasite

Anaplasma marginale are composed of bundled actin filaments. Protoplasma. *199*: 93–98.

Stich, R.W., K.M. Kocan, G.H. Palmer, S.A. Ewing, J.A. Hair and S.J. Barron. 1989. Transstadial and attempted transovarial transmission of *Anaplasma marginale* by *Dermacentor variabilis*. Am. J. Vet. Res. *50*: 1377–1380.

Stigter, E.C.A., G.A.H. de Jong, J.A. Jongejan, J.A. Duine, J.P. van der Lugt and W.A.C. Somers. 1997. Electron transfer and stability of a quinohaemoprotein alcohol dehydrogenase electrode. J. Chem. Technol. Biotechnol. *68*: 110–116.

Stills, H.F., Jr. 1991. Isolation of an intracellular bacterium from hamsters (*Mesocricetus auratus*) with proliferative ileitis and reproduction of the disease with a pure culture. Infect. Immun. *59*: 3227–3236.

Stoenner, H.G. and D.B. Lackman. 1957. A new species of *Brucella* isolated from the desert wood rat, *Neotoma lepida* Thomas. Amer. J. Vet. Res. *18*: 942–951.

Stoenner, H.G., D.B. Lackman and E.J. Bell. 1962. Factors affecting the growth of rickettsias of the spotted fever group in fertile hens' eggs. J. Infect. Dis. *110*: 121–128.

Stoesser, G., W. Baker, A. van den Broek, E. Camon, M. Garcia-Pastor, C. Kanz, T. Kulikova, V. Lombard, R. Lopez, H. Parkinson, N. Redaschi, P. Sterk, P. Stoehr and M.A. Tuli. 2001. The EMBL nucleotide sequence database. Nucleic Acids Res. *29*: 17–21.

Stoffels, M., T. Castellanos and A. Hartmann. 2001. Design and application of a new 16S rRNA-target oligonucleotide probe for the *Azospirillum–Skermanella–Rhodocista* cluster. Syst. Appl. Microbiol. *24*: 83–97.

Stöhr, R., A. Waberski, W. Liesack, H. Völker, U. Wehmeyer and M. Thomm. 2001. *Hydrogenophilus hirschii* sp. nov., a novel thermophilic hydrogen-oxidizing bega-proteobacterium isolated from Yellowstone National Park. Int. J. Syst. Evol. Microbiol. *51*: 481–488.

Stokes, H.W., E.L. Dally, M.D. Yablonsky and D.E. Eveleigh. 1983. Comparison of plasmids in strains of *Zymomonas mobilis*. Plasmid *9*: 138–146.

Stokes, J.L. 1954. Studies on the filamentous sheathed iron bacterium *Sphaerotilus natans*. J. Bacteriol. *67*: 278–291.

Stokes, J.L. and A.H. Johnson. 1965. Growth factor requirements of two strains of *Sphaerotilus discophorus*. Antonie Leeuwenhoek J. Microbiol. Serol. *31*: 175–180.

Stokes, J.L. and M.T. Powers. 1965. Formation of rough and smooth strains of *Sphaerotilus discophorus*. Antonie Leeuwenhoek J. Microbiol. Serol. *31*: 157–164.

Stolp, H. 1973. The bdellovibrios: bacterial parasites of bacteria. Annu. Rev. Phytopath. *11*: 53–76.

Stolp, H. and H. Petzhold. 1962. Untersuchugen über einen obligat poarsitischen mikroorganismus mit lytischer activitat für *Pseudomonas*-bakterien. Phytopath. Z. *45*: 364–390.

Stolp, H. and M.P. Starr. 1963. *Bdellovibrio bacteriovorus* gen. et sp. n., a predatory, ectoparasitic, and bacteriolytic microorganism. Antonie Leeuwenhoek J. Microbiol. Serol. *29*: 217–248.

Stolz, A., C. Schmidt-Maag, E.B.M. Denner, H.-J. Busse, T. Egli and P. Kämpfer. 2000. Description of *Sphingomonas xenophaga* sp. nov. for strains BN6^T and N,N which degrade xenobiotic aromatic compounds. Int. J. Syst. Evol. Microbiol. *50*: 35–41.

Stolz, J.F., D.J. Ellis, J.S. Blum, D. Ahmann, D.R. Lovley and R.S. Oremland. 1999. *Sulfurospirillum barnesii* sp. nov. and *Sulfurospirillum arsenophilum* sp. nov., new members of the *Sulfurospirillum* clade of the Epsilon *Proteobacteria*. Int. J. Syst. Bacteriol. *49*: 1177–1180.

Stolz, J.F., T. Gugliuzza, J.S. Blum, R. Oremland and F.M. Murillo. 1997. Differential cytochrome content and reductase activity in *Geospirillum barnesii* strain SeS-3. Arch. Microbiol. *167*: 1–5.

Stone, R.L., C.G. Gulbertson and H.M. Powell. 1956. Studies of a bacteriophage active against a chromogenic *Neisseria*. J. Bacteriol. *71*: 516–520.

Stonier, T., J. McSharry and T. Speitel. 1967. *A. tumefaciens* Conn. IV. bacteriophage PB2 and its inhibitory effect on tumor induction. J. Virol. *1*: 268–273.

Stoorvogel, J., D.E. Kraayveld, C.A. Van Sluis, J.A. Jongejan, S. De Vries and J.A. Duine. 1996. Characterization of the gene encoding quinohaemoprotein ethanol dehydrogenase of *Comamonas testosteroni*. Eur. J. Biochem. *235*: 690–698.

Stoppel, R.D. and H.G. Schlegel. 1995. Nickel-resistant bacteria from anthropogenically nickel-polluted and naturally nickel-percolated ecosystems. Appl. Environ. Microbiol. *61*: 2276–2285.

Stothard, D.R., J.B. Clark and P.A. Fuerst. 1994. Ancestral divergence of *Rickettsia bellii* from the spotted fever and typhus groups of *Rickettsia* and antiquity of the genus *Rickettsia*. Int. J. Syst. Bacteriol. *44*: 798–804.

Stouthamer, A.H. 1980. Bioenergetic studies on *Paracoccus denitrificans*. Trends Biochem. Sci. *5*: 164–166.

Stouthamer, A.H. 1992. Metabolic pathways in *Paracoccus denitrificans* and closely related bacteria in relation to the phylogeny of prokaryotes. Antonie Leeuwenhoek *61*: 1–33.

Stouthamer, A.H., A.P. de Boer, J. van der Oost and R.J. van Spanning. 1997. Emerging principles of inorganic nitrogen metabolism in *Paracoccus denitrificans* and related bacteria. Antonie Leeuwenhoek *71*: 33–41.

Stouthamer, R., J.A.J. Breeuwer and G.D.D. Hurst. 1999. *Wolbachia pipientis*: Microbial manipulator of arthropod reproduction. Annu. Rev. Microbiol. *53*: 71–102.

Stouthamer, R., J.A.J. Breeuwer, R.F. Luck and J.H. Werren. 1993. Molecular identification of microorganisms associated with parthenogenesis. Nature *361*: 66–68.

Stovall, I. and M. Cole. 1978. Organic acid metabolism by isolated *Rhizobium japonicum* bacteroids. Plant Physiol. (Rockv.) *61*: 787–790.

Stover, C.K., D.P. Marana, G.A. Dasch and E.V. Oaks. 1990. Molecular cloning and sequence analysis of the Sta58 major antigen gene of *Rickettsia tsutsugamushi*: Sequence homology and antigenic comparison of Sta58 to the 60-kilodalton family of stress proteins. Infect. Immun. *58*: 1360–1368.

Straley, S.C. and S.F. Conti. 1974. Chemotaxis in *Bdellovibrio bacteriovorus*. J. Bacteriol. *120*: 549–551.

Straley, S.C. and S.F. Conti. 1977. Chemotaxis by *Bdellovibrio bacterionorus* toward prey. J. Bacteriol. *132*: 628–640.

Straley, S.C., A.G. Lamarre, L.J. Lawrence and S.F. Conti. 1979. Chemotaxis of *Bdellovibrio bacteriovorus* toward pure compounds. J. Bacteriol. *140*: 634–642.

Straub, K.L., F.A. Rainey and F. Widdel. 1999. *Rhodovulum iodosum* sp. nov. and *Rhodovulum robiginosum* sp. nov., two new marine phototrophic ferrous-iron-oxidizing purple bacteria. Int. J. Syst. Bacteriol. *49*: 729–735.

Straus, D.C., M.K. Lonon, D.E. Woods and C.W. Garner. 1990. 3-Deoxy-D-manno-2-octulosonic acid in the lipopolysaccharide of various strains of *Pseudomonas cepacia*. J. Med. Microbiol. *33*: 265–269.

Strauss, D.G. and U. Berge. 1983. Methylosin A and B pigments from *Methylosinus trichosporium*. Z. Allg. Mikrobiol. *23*: 661–668.

Strebel, P.M., S.L. Cochi, K.M. Farizo, B.J. Payne, S.D. Hanauer and A.L. Baughman. 1993. Pertussis in Missouri: evaluation of nasopharyngeal culture, direct fluorescent antibody testing, and clinical case definitions in the diagnosis of pertussis. Clin. Infect. Dis. *16*: 276–285.

Streichan, M. and B. Schink. 1986. Microbial populations in wetwood of European white fir (*Abies alba* Mill). FEMS Microbiol. Ecol. *38*: 141–150.

Strength, W.J., B. Isani, D.M. Linn, F.D. Williams, G.E. Vandermolen, B.E. Laughon and N.R. Krieg. 1976. Isolation and characterization of *Aquaspirillum fasciculus* sp. nov., a rod-shaped nitrogen-fixing bacterium having unusual flagella. Int. J. Syst. Bacteriol. *26*: 253–268.

Strohdeicher, M., B. Neuss, S. Bringer-Meyer and H. Sahm. 1990. Electron transport chain of *Zymomonas mobilis*. Interaction with the membrane-bound glucose dehydrogenase and identification of ubiquinone 10. Arch. Microbiol. *154*: 536–543.

Strohdeicher, M., B. Schmitz, S. Bringer-Meyer and H. Sahm. 1988. Formation and degradation of gluconate by *Zymomonas mobilis*. Appl. Microbiol. Biotechnol. *27*: 378–382.

Strohhäcker, J., A.A. de Graaf, S.M. Schoberth, R.M. Wittig and H. Sahm.

1993. ^{31}P nuclear magnetic resonance studies of ethanol inhibition in *Zymomonas mobilis*. Arch. Microbiol. *159*: 484–490.

Strohl, W.R. and J.M. Larkin. 1978a. Cell division and trichome breakage in *Beggiatoa*. Curr. Microbiol. *1*: 151–155.

Strohl, W.R. and J.M. Larkin. 1978b. Enumeration, isolation and characterization of *Beggiatoa* from freshwater sediments. Appl. Environ. Microbiol. *36*: 755–770.

Strohl, W.R. and T.M. Schmidt. 1984. Mixotrophy in *Beggiatoa* and *Thiothrix*. *In* Strohl and Tuovinen (Editors), Microbial Chemoautotrophy, The Ohio State University Press, Columbus. pp. 79–95.

Strohl, W.R., T.M. Schmidt, N.H. Lawry, M.J. Mezzino and J.M. Larkin. 1986a. Characterization of *Vitreoscilla beggiatoides* and *Vitreoscilla filiforms* sp. nov., nom. rev., and comparison with *Vitreoscilla stercoraria* and *Beggiatoa alba*. Int. J. Syst. Bacteriol. *36*: 302–313.

Strohl, W.R., T.M. Schmidt, V.A. Vinci and J.M. Larkin. 1986b. Electron transport and respiration in *Beggiatoa* and *Vitreoscilla*. Arch. Microbiol. *145*: 71–75.

Strom, M.S. and S. Lory. 1986. Cloning and expression of the pilin gene of *Pseudomonas aeruginosa* PAK in *Escherichia coli*. J. Bacteriol. *165*: 367–372.

Strom, M.S., D. Nunn and S. Lory. 1991. Multiple roles of the pilus biogenesis protein PilD: involvement of PilD in excretion of enzymes from *Pseudomonas aeruginosa*. J. Bacteriol. *173*: 1175–1180.

Strom, T., T. Ferenci and J.R. Quayle. 1974. The carbon assimilation pathways of *Methylococcus capsulatus*, *Pseudomonas methanica* and *Methylosinus trichosporium* (OB3b) during growth on methane. Biochem. J. *144*: 465–476.

Strong, R.P., E.E. Tyzzer, C.T. Brues, A.W. Sellards and J.A. Gastiaburu. 1913. Verruga peruviana, Oroya fever and uta. J. Am. Med. Assoc. *61*: 1713–1716.

Strong, R.P., E.E. Tyzzer and A.W. Sellards. 1915. Oroya fever, second report. J. Am. Med. Assoc. *64*: 806–808.

Strunk, O. and W. Ludwig. 1998. Arb: a software environment for sequence data, Technical University of Munich, Munich, Germany.

Struthers, M., J. Wong and J.M. Janda. 1996. An initial appraisal of the clinical significance of *Roseomonas* species associated with human infections. Clin. Infect. Dis. *23*: 729–733.

Strzelcowa, A. 1968. The use of the technique of van Schreven for the taxonomy of *Rhizobium* strains. Acta Microbiol. Polon. *17*: 263–268.

Stubner, S., T. Wind and R. Conrad. 1998. Sulfur oxidation in rice field soil: activity, enumeration, isolation and characterization of thiosulfate-oxidizing bacteria. Syst. Appl. Microbiol. *21*: 569–578.

Stucki, G., R. Gälli, H.R. Ebersold and T. Leisinger. 1981. Dehalogenation of dichloromethane by cell-extracts of *Hyphomicrobium* Dm2. Arch. Microbiol. *130*: 366–371.

Stucki, U., J. Frey, J. Nicolet and A.P. Burnens. 1995. Identification of *Campylobacter jejuni* on the basis of a species-specific gene that encodes a membrane protein. J. Clin. Microbiol. *33*: 855–859.

Stumm, W. and J.J. Morgan. 1996. Aquatic Chemistry. Chemical Equilibria and Rates in Natural Waters, 3rd Ed., John Wiley & Sons, Inc., New York.

Stutzer, A. and R. Hartleb. 1898. Untersuchungen über die bei der bildung von salpeter beobachteten mikroorganismen. Abhandlungen und Mitteilungen des Landwirtschaftlichen Instituts der Königlichen Universität Breslau. *1*: 75–100.

Subbao-Rao, N.S., P.F. Mateos, D. Baker, H.S. Pankratz, J. Palma, F.B. Dazzo and J.I. Sprent. 1995. The unique root-nodule symbiosis between *Rhizobium* and the aquatic legume *Neptunia natans* (L.f.) Druce. Planta. *196*: 311–320.

Sudo, S.Z. and M. Dworkin. 1969. Resistance of vegetative cells and microcysts of *Myxococcus xanthus*. J. Bacteriol. *98*: 883–887.

Sudo, S. and M. Dworkin. 1972. Bacteriolytic enzymes produced by *Myxococcus xanthus*. J. Bacteriol. *110*: 236–245.

Sudo, S.Z. and M. Dworkin. 1973. Comparative biology of prokaryotic resting cells. Adv. Microbial Physiol. *9*: 153–223.

Suerbaum, S., C. Kraft, F.E. Dewhirst and J.G. Fox. 2002. *Helicobacter nemestrinae* ATCC 49396^T is a strain of *Helicobacter pylori* (Marshall et al. 1985) Goodwin et al. 1989, and *Helicobacter nemestrinae* Bronsdon et al. 1991 is therefore a junior heterotypic synonym of *Helicobacter pylori.* Int. J. Syst. Evol. Bacteriol. *52:* 437–439.

Sugimoto, C., M. Eguchi, Y. Haritani, Y. Isayama and M. Kashiwazaki. 1988. Isolation and characterization of the outer membrane of *T. equigenitalis. In* Powell (Editor), Equine Infectious Diseases V: Proceedings of the Fifth International Conference, University Press of Kentucky, Lexington, KY. pp. 164–167.

Sugimoto, C., Y. Isayama, R. Sakazaki and S. Kuramochi. 1983. Transfer of *Haemophilus equigenitalis* Taylor et al. 1978 to the genus *Taylorella*, gen. nov. as *Taylorella equigenitalis*, comb. nov. Curr. Microbiol. *9:* 155–162.

Sugimoto, C., Y. Isayama, R. Sakazaki and S. Kuramochi. 1984. *In* Validation of the publication of new names and new combinations previously effectively published outside the IJSB. List No. 16. Int. J. Syst. Bacteriol. *34:* 503–504.

Sugisawa, T., T. Hoshino, S. Masuda, S. Nomura, Y. Setoguchi, M. Tazoe, M. Shinjoh, S. Someha and A. Fujiwara. 1990. Microbial production of 2-keto-L-gulonic acid from L-sorbose and D-sorbitol by *Gluconobacter melanogenus*. Agric. Biol. Chem. *54:* 1201–1209.

Sugisawa, T., S. Ojima, P.K. Matzinger and T. Hoshino. 1995. Isolation and characterization of a new vitamin C producing enzyme (L-gulono-gamma-lactone dehydrogenase) of bacterial origin. Biosci. Biotechnol. Biochem. *59:* 190–196.

Sukapure, R.S., M.P. Lechevalier, H. Reber, M.L. Higgins, H.A. Lechevalier and H. Prauser. 1970. Motile nocardoid *Actinomycetales*. Appl. Microbiol. *19:* 527–533.

Süle, S. 1978. Biotypes of *Agrobacterium tumefaciens* in Hungary. J. Appl. Bacteriol. *44:* 207–213.

Sullivan, J.T., B.D. Eardly, P. Van Berkum and C.W. Ronson. 1996. Four unnamed species of nonsymbiotic rhizobia isolated from the rhizosphere of *Lotus corniculatus*. Appl. Environ. Microbiol. *62:* 2818–2825.

Sullivan, J.T., H.N. Patrick, W.L. Lowther, D.B. Scott and C.W. Ronson. 1995. Nodulating strains of *Rhizobium loti* arise through chromosomal symbiotic gene transfer in the environment. Proc. Natl. Acad. Sci. U.S.A. *92:* 8985–8989.

Sullivan, J.T. and C.W. Ronson. 1998. Evolution of rhizobia by acquisition of a 500-kb symbiosis island that integrates into a phe-tRNA gene. Proc. Natl. Acad. Sci. U. S. A. *95:* 5145–5149.

Summanen, P., J. Downes, M. Karl, K. Lounaimaa, E. Baron, H. Jousimies-Somer and S. Finegold. 1989. Characteristics and ultrastructure of an unusual Gram-negative bacillus isolated from inflamed and noninflamed appendices. Abstr. Eur. Soc. Clin. Microbiol. *5:* p. 15.

Summanen, P.H., H. Jousimies-Somer, S. Manley, D. Bruckner, M. Marina, E.J. Goldstein and S.M. Finegold. 1995. *Bilophila wadsworthia* isolates from clinical specimens. Clin. Infect. Dis. *20:* S210–S211.

Summanen, P., H.M. Wexler and S.M. Finegold. 1992. Antimicrobial susceptibility testing of *Bilophila wadsworthia* by using triphenyltetrazolium chloride to facilitate endpoint determination. Antimicrob. Agents Chemother. *36:* 1658–1664.

Summanen, P., H.M. Wexler, K. Lee, S.A. Becker, M.M. Garcia and S.M. Finegold. 1993. Morphological response of *Bilophila wadsworthia* to imipenem: correlation with properties of penicillin-binding proteins. Antimicrob. Agents Chemother. *37:* 2638–2644.

Sumner, J.W., W.L. Nicholson and R.F. Massung. 1997. PCR amplification and comparison of nucleotide sequences from the *groESL* heat shock operon of *Ehrlichia* species. J. Clin. Microbiol. *35:* 2087–2092.

Sumner, J.W., K.G. Sims, D.C. Jones and B.E. Anderson. 1995. Protection of guinea-pigs from experimental Rocky Mountain spotted fever by immunization with baculovirus-expressed *Rickettsia rickettsii* rOmpA protein. Vaccine *13:* 29–35.

Sun, B., J.R. Cole and J.M. Tiedje. 2001. *Desulfomonile limimaris* sp. nov., an anaerobic dehalogenating bacterium from marine sediments. Int. J. Syst. Evol. Microbiol. *51:* 365–371.

Sundermeyer, H. and E. Bock. 1981. Characterization of the nitrite-oxidizing system in *Nitrobacter. In* Bothe and Trebst (Editors), Biology of Inorganic Nitrogen and Sulfur, Springer-Verlag, Berlin. pp. 317–324.

Sundermeyer-Klinger, H., W. Meyer, B. Warninghoff and E. Bock. 1984.

Membrane-bound nitrite oxidoreductase of *Nitrobacter*: evidence for a nitrate reductase system. Arch. Microbiol. *140:* 153–158.

Sundh, I., M. Nilsson and P. Borga. 1997. Variation in microbial community structure in two boreal peatlands as determined by analysis of phospholipid fatty acid profiles. Appl. Environ. Microbiol. *63:* 1476–1482.

Suputtamongkol, Y., D.A.B. Dance, W. Chaowagul, Y. Wattanagoon, V. Wuthiekanun and N.J. White. 1991. Amoxicillin clavulanic acid treatment of melioidosis. Trans. Roy. Soc. Trop. Med. Hyg. *85:* 672–675.

Surin, B.P. and J.A. Downie. 1988. Characterization of the *Rhizobium leguminosarum* genes *nodLMN* involved in efficient host-specific nodulation. Mol. Microbiol. *2:* 173–183.

Surin, B.P., J.M. Watson, W.D. Hamilton, A. Economou and J.A. Downie. 1990. Molecular characterization of the nodulation gene, *nodT*, from two biovars of *Rhizobium leguminosarum*. Mol. Microbiol. *4:* 245–252.

Surovtseva, E.G., V.S. Ivoilov, G.K. Vasil'eva and S.S. Belyaev. 1996. Degradation of chlorinated anilines by certain representatives of the genera *Aquaspirillum* and *Paracoccus*. Mikrobiologiya *65:* 553–559.

Sutherland, I.W. 1979a. Microbial exopolysaccharides: control of synthesis and acylation. *In* Berkeley, Gooday and Ellwood (Editors), Microbial Polysaccharides and Polysaccharases, Academic Press, New York. pp. 1–34.

Sutherland, I.W. 1979b. Polysaccharides produced by *Cystobacter, Archangium, Sorangium* and *Stigmatella* species. J. Gen. Microbiol. *111:* 211–216.

Sutherland, I.W. 1994. Structure function relationships in microbial exopolysaccharides. Biotechnol. Advances. *12:* 393–448.

Sutherland, I.W. and A.F.D. Kennedy. 1986. Comparison of bacterial lipopolysaccharides by high-performance liquid chromatography. Appl. Environ. Microbiol. *52:* 948–950.

Sutherland, I.W. and L. Kennedy. 1996. Polysaccharide lyases from gellan-producing *Sphingomonas* spp. Microbiology *142:* 867–872.

Sutherland, I.W. and M.L. Smith. 1973. The lipopolysaccharides of fruiting and non-fruiting myxobacteria. J. Gen. Microbiol. *74:* 259–266.

Sutherland, I.W. and S. Thomson. 1975. Comparison of polysaccharides produced by *Myxococcus* strains. J. Gen. Microbiol. *89:* 124–132.

Sutter, B. 1991. Biosynthese isoprenique chez les procaryotes: un model innovateur *Zymomonas mobilis*, Université de Haute-Alsace, Mulhouse, France.

Suyama, T., H. Hosoya and Y. Tokiwa. 1998a. Bacterial isolates degrading aliphatic polycarbonates. FEMS Microbiol. Lett. *16:* 255–261.

Suyama, T., T. Shigematsu, S. Takaichi, Y. Nodasaka, S. Fujikawa, H. Hosoya, Y. Tokowa, T. Kanagawa and S. Hanada. 1999. *Roseateles depolymerans* gen. nov., sp. nov. a new bacteriochlorophyll *a*-containing obligate aerobe belonging to the β-subclass of the *Proteobacteria*. Int. J. Syst. Bacteriol. *49:* 449–457.

Suyama, T., Y. Tokiwa, P. Ouichanpagdee, T. Kanagawa and Y. Kamagata. 1998b. Phylogenetic affiliation of soil bacteria that degrade aliphatic polyesters available commercially as biodegradable plastics. Appl. Environ. Microbiol. *64:* 5008–5011.

Suylen, G.M.H. and J.G. Kuenen. 1986. Chemostat enrichment and isolation of *Hyphomicrobium* EG: a dimethyl sulfide oxidizing methylotroph and reevaluation of *Thiobacillus* MS1. Antonie Leeuwenhoek *52:* 281–294.

Suylen, G.M.H., P.J. Large, J.P. Vandijken and J.G. Kuenen. 1987. Methyl mercaptan oxidase, a key enzyme in the metabolism of methylated sulfur-compounds by *Hyphomicrobium* EG. J. Gen. Microbiol. *133:* 2989–2997.

Suylen, G.M.H., G.C. Stefess and J.G. Kuenen. 1986. Chemolithotrophic potential of a *Hyphomicrobium* species, capable of growth on methylated sulfur compounds. Arch. Microbiol. *146:* 192–198.

Suzuki, K., N. Wakao, T. Kimura, K. Sakka and K. Ohmiya. 1998. Expression and regulation of the arsenic resistance operon of *Acidiphilium multivorum* AIU 301 plasmid pKW301 in *Escherichia coli*. Appl. Environ. Microbiol. *64:* 411–418.

Suzuki, M., M.S. Rappe, Z.W. Haimberger, H. Winfield, N. Adair, J. Strobel and S.J. Giovannoni. 1997. Bacterial diversity among small-subunit rRNA gene clones and cellular isolates from the same seawater sample. Appl. Environ. Microbiol. *63*: 983–989.

Suzuki, S., T. Kohzuma, S. Shidara, K. Ohki and T. Aida. 1993. Novel spectroscopic aspects of type-I copper in *Hyphomicrobium* nitrite reductase. Inorg. Chim. Acta *208*: 107–109.

Suzuki, T., Y. Muroga, M. Takahama and Y. Nishimura. 1999a. *Roseivivax halodurans* gen. nov., sp. nov. and *Roseivivax halotolerans* sp. nov., aerobic bacteriochlorophyll-containing bacteria isolated from a saline lake. Int. J. Syst. Bacteriol. *49*: 629–634.

Suzuki, T., Y. Muroga, M. Takahama and Y. Nishimura. 2000. *Roseibium denhamense* gen. nov., sp. nov. and *Roseibium hamelinense* sp. nov., aerobic bacteriochlorophyll-containing bacteria isolated from the east and west coasts of Australia. Int. J. Syst. Evol. Microbiol. *50*: 2151–2156.

Suzuki, T., Y. Muroga, M. Takahama, T. Shiba and Y. Nishimura. 1999b. *Rubrimonas cliftonensis* gen. nov., sp. nov., an aerobic bacteriochlorophyll-containing bacterium isolated from a saline lake. Int. J. Syst. Bacteriol. *49*: 201–205.

Suzuki, T., E. Sugimoto, Y. Tahara and Y. Yamada. 1996. Cloning and nucleotide sequence of *Apa*LI restriction-modification system from *Acetobacter pasteurianus* IFO 13753. Biosci. Biotechnol. Biochem. *60*: 1401–1405.

Švitel, J., O. Curilla and J. Tkac. 1998. Microbial cell-based biosensor for sensing glucose, sucrose or lactose. Biotechnol. Appl. Biochem. *27*: 153–158.

Swann, A.I., C.L. Garby, P.R. Schnurrenberger and R.R. Brown. 1981. Safety aspects in preparing suspensions of field strains of *Brucella abortus* for serological identification. Vet. Rec. *109*: 254–255.

Swanson, J. 1973. Studies on gonococcus infection. IV. Pili: their role in attachment of gonococci to tissue culture cells. J. Exp. Med. *137*: 571–589..

Swanson, J. 1978. Studies on gonococcus infection. XII. Colony color and opacity variations of gonococci. Infect. Immun. *19*: 320–331.

Swanson, J., O. Barrera, J. Sola and J. Boslego. 1988. Expression of outer membrane protein II by gonococci in experimental gonorrhea. J. Exp. Med. *168*: 2121–2129.

Swanson, J., S. Bergstrom, K. Robbins, O. Barrera, D. Corwin and J.M. Koomey. 1986. Gene conversion involving the pilin structural gene correlates with pilus⁺ in equilibrium with pilus⁻ changes in *Neisseria gonorrhoeae*. Cell *47*: 267–276.

Swanson, J. and J.M. Koomey. 1989. Mechanism for variation of pili and outer membrane protein II in *Neisseria gonorrheae*. *In* Berg and Howe (Editors), Mobile DNA, American Society for Microbiology, Washington, D.C. pp. 743–761.

Swanson, J., S.J. Kraus and E.C. Gotschlich. 1971. Studies on gonococcus infection. I. Pili and zones of adhesion: their relation to gonococcal growth patterns. J. Exp. Med. *134*: 886–906..

Swanson, J., K. Robbins, O. Barrera, D. Corwin, J. Boslego, J. Ciak, M. Blake and J.M. Koomey. 1987. Gonococcal pilin variants in experimental gonorrhea. J. Exp. Med. *165*: 1344–1357.

Swaving, J. and J.A.M. De Bont. 1998. Microbial transformation of epoxides. Enzyme Microb. Technol. *22*: 19–26 .

Swaving, J., W. van Leest, A.J.J. Vanooyen and J.A.M. De Bont. 1996. Electrotransformation of *Xanthobacter autotrophicus* GJ10 and other *Xanthobacter* strains. J. Microbiol. Methods *25*: 343–348.

Swaving, J., C.A.G.M. Weijers, A.J.J. Vanooyen and J.A.M. De Bont. 1995. Complementation of *Xanthobacter* Py2 mutants defective in epoxyalkane degradation, and expression and nucleotide sequence of the complementing DNA fragment. Microbiology (Read.) *141*: 477–484.

Swings, J. 1992. The genera *Acetobacter* and *Gluconobacter*. *In* Balows, Trüper, Dworkin, Harder and Schleifer (Editors), The Prokaryotes: A Handbook of Bacteria: Ecophysiology, Isolation, Identification, Applications, 2nd Ed., Vol. 3, Springer-Verlag, New York. pp. 2268–2286.

Swings, J. and J. De Ley. 1975. Genome deoxyribonucleic acid of the genus *Zymomonas* Kluyver and van Niel 1936: base composition, size, and similarities. Int. J. Syst. Bacteriol. *25*: 324–328.

Swings, J. and J. De Ley. 1977. The biology of *Zymomonas*. Bacteriol. Rev. *41*: 1–46.

Swings, J., K. Kersters and J. De Ley. 1976. Numerical analysis of electrophoretic protein patterns of *Zymomonas* strains. J. Gen. Microbiol. *93*: 266–271.

Swings, J., K. Kersters and J. De Ley. 1977. Taxonomic position of additional *Zymomonas mobilis* strains. Int. J. Syst. Bacteriol. *27*: 271–273.

Swings, J., B. Lambert, K. Kersters and B. Holmes. 1992. The genera *Phyllobacterium* and *Ochrobactrum*. *In* Balows, Trüper, Harder and Schleifer (Editors), The Prokaryotes: A Handbook of Bacteria: Ecophysiology, Isolation, Identification, Applications, 2nd Ed., Vol. 3, Springer Verlag, New York. pp. 2601–2604.

Swings, J. and W. Van Pee. 1977. Infra-red spectroscopy of *Zymomonas* cells. J. Gen. Appl. Microbiol. *23*: 297–301.

Swisher, L.A., J.R. Roberts and M.J. Glynn. 1994. Needle licker's osteomyelitis. Am. J. Emerg. Med. *12*: 343–346.

Swofford, D.L. 1991. PAUP: Phylogenetic Analysis Using Parsimony v3.1.1, Illinois Natural History Survey, Champaign, IL.

Sylvester, F.A., D. Philpott, B. Gold, A. Lastovica and J.F. Forstner. 1996. Adherence to lipids and intestinal mucin by a recently recognized human pathogen, *Campylobacter upsaliensis*. Infect. Immun. *64*: 4060–4066.

Sylvestre, M., Y. Hurtubise, D. Barriault, J. Bergeron and D. Ahmad. 1996a. Characterization of active recombinant 2,3-dihydro-2,3-dihydroxybiphenyl dehydrogenase from *Comamonas testosteroni* B-356 and sequence of the encoding gene (*bphB*). Appl. Environ. Microbiol. *62*: 2710–2715.

Sylvestre, M., M. Sirois, Y. Hurtubise, J. Bergeron, D. Ahmad, F. Shareck, D. Barriault, I. Guillemette and J.M. Juteau. 1996b. Sequencing of *Comamonas testosteroni* strain B-356 biphenyl/chlorobiphenyl dioxygenase genes: evolutionary relationships among Gram-negative bacterial biphenyl dioxygenases. Gene *174*: 195–202.

Szegedi, E., M. Czakó, L. Otten and C.S. Koncz. 1988. Opines in crown gall tumors induced by biotype-3 isolates of *Agrobacterium tumefaciens*. Physiol. Mol. Plant Pathol. *32*: 237–247.

Tabacchioni, S., P. Visca, L. Chiarini, A. Bevivino, C. Di Serio, S. Fancelli and R. Fani. 1995. Molecular characterization of rhizosphere and clinical isolates of *Burkholderia cepacia*. Res. Microbiol. *146*: 531–542.

Tabe, Y. and J. Igari. 1994. Susceptibilities of glucose non-fermentative gram-negative bacilli to antibiotics. Jpn. J. Antibiot. *47*: 1030–1040.

Tabita, F.R. 1995. The biochemistry and metabolic regulation of carbon metabolism and CO_2 fixation in purple bacteria. *In* Blankenship, Madigan and Bauer (Editors), Anoxygenic Photosynthetic Bacteria, Kluwer Academic Publishers, Netherlands. pp. 885–914.

Tabita, F.R. and B.A. McFadden. 1974a. D-Ribulose 1,5-diphosphate carboxylase from *Rhodospirillum rubrum*. I. Levels, purification, and effects of metallic ions. J. Biol. Chem. *249*: 3453–3458.

Tabita, F.R. and B.A. McFadden. 1974b. D-Ribulose 1,5-diphosphate carboxylase from *Rhodospirillum rubrum*. II. Quaternary structure, composition, catalytic, and immunological properties. J. Biol. Chem. *249*: 3459–3464.

Tachil, J. and O. Meyer. 1997. Redox state and activity of molybdopterin cytosine dinucleotide (MCD) of CO dehydrogenase from *Hydrogenophaga pseudoflava*. FEMS Microbiol. Lett. *148*: 203–208.

Tagami, H., K. Tayama, T. Tohyama, M. Fukaya, H. Okumura, Y. Kawamura, S. Horinouchi and T. Beppu. 1988. Purification and properties of a site-specific restriction endonuclease AaaI from *Acetobacter aceti* subsp. *aceti* no. 1023. FEMS Microbiol. Lett. *56*: 161–165.

Taguchi, S., A. Maehara, K. Takase, M. Nakahara, H. Nakamura and Y. Doi. 2001. Analysis of mutational effects of a polyhydroxybutyrate (PHB) polymerase on bacterial PHB accumulation using an *in vivo* assay system. FEMS Microbiol. Lett. *198*: 65–71.

Tahara, Y. and M. Kawazu. 1990. Isolation of free ceramide from *Zymomonas mobilis*. Agric. Biol. Chem. *54*: 1581–1582.

Tahara, Y. and M. Kawazu. 1994. Isolation of glucuronic acid-containing glycosphingolipid from *Zymomonas mobilis*. Biosci. Biotechnol. Biochem. *58*: 586–587.

Takada, N. 1975. A new species of *Hyphomicrobium*. *In* The Organization Committee. (Editor), Microbial Growth on C_1 Compounds: Proceedings of the International Symposium on Microbial Growth on C_1

Compounds, September 5, 1974, Tokyo, Japan, The Society of Fermentation Technology, Printed by Nakanishi Printing Co. LTD, Kyoto, Japan. 29–33.

Takahashi, E., M. Furui and T. Shibatani. 1997. Scale-up of D-lysine production from L-lysine by successive chemical racemization and microbial asymmetric degradation. Biotechnol. Lett. *19*: 245–249.

Takahashi, T., Y. Yamazaki and K. Kato. 1974. Substrate specificity of an α-amino acid ester hydrolase produced by *Acetobacter turbidans* ATCC 9325. Biochem. J. *137*: 497–503.

Takahashi, T., Y. Yamazaki, K. Kato and M. Isona. 1972. Enzymatic synthesis of cephalosporins. J. Am. Chem. Soc. *94*: 4035–4037.

Takaichi, S., K. Furihata, J. Ishidsu and K. Shimada. 1991. Carotenoid sulfates from the aerobic photosynthetic bacterium, *Erythrobacter longus*. Phytochemistry (Oxf) *30*: 3411–3415.

Takaichi, S., K. Shimada and J. Ishidsu. 1988. Monocyclic cross-conjugated carotenal from an aerobic photosynthetic bacterium, *Erythrobacter longus*. Phytochemistry (Oxf) *27*: 3605–3609.

Takaichi, S., K. Shimada and J. Ishidsu. 1990. Carotenoids from the aerobic photosynthetic bacterium, *Erythrobacter longus* - Beta-carotene and its hydroxyl derivatives. Arch. Microbiol. *153*: 118–122.

Takaichi, S., N. Wakao, A. Hiraishi, S. Itoh and K. Shimada. 1999. Nomenclature of metal-substituted (bacterio)chlorophylls in natural photosynthesis: Metal-(bacterio)chlorophyll and M-(B)Chl. Photosynth. Res. *59*: 255–256.

Takamiya, K., H. Arata, Y. Shioi and M. Doi. 1988. Restoration of the optimal redox state for the photosynthetic electron transfer system by auxiliary oxidants in an aerobic photosynthetic bacterium, *Erythrobacter* sp. OCh114. Biohcim. Biophys. Acta *935*: 26–33.

Takamiya, K.I., Y. Shioi, H. Shimada and H. Arata. 1992. Inhibition of accumulation of bacteriochlorophyll and carotenoids by blue light in an aerobic photosynthetic bacterium, *Roseobacter denitrificans*, during anaerobic respiration. Plant Cell Physiol. *33*: 1171–1174.

Takeda, K. 1988. Characteristics of a nitrogen-fixing methanotroph, *Methylocystis* T-1. Antonie Leeuwenhoek *54*: 521–534.

Takeda, M., K. Iohara, S. Shinmaru, I. Suzuki and J. Koizumi. 2000. Purification and properties of an enzyme capable of degrading the sheath of *Sphaerotilus natans*. Appl. Environ. Microbiol. *66*: 4998–5004.

Takeda, M., F. Nakano, T. Nagase, K. Iohara and J. Koizumi. 1998. Isolation and chemical composition of the sheath of *Sphaerotilus natans*. Biosci. Biotechnol. Biochem. *62*: 1138–1143.

Takemura, H., S. Horinouchi and T. Beppu. 1991. Novel insertion sequence IS1380 from *Acetobacter pasteurianus* is involved in loss of ethanol-oxidizing ability. J. Bacteriol. *173*: 7070–7076.

Takemura, H., K. Kondo, S. Horinouchi and T. Beppu. 1993. Induction by ethanol of alcohol dehydrogenase activity in *Acetobacter pasteurianus*. J. Bacteriol. *175*: 6857–6866.

Takeuchi, M., K. Hamana and A. Hiraishi. 2001. Proposal of the genus *Sphingomonas sensu stricto* and three new genera, *Sphingobium*, *Novosphingobium* and *Sphingopyxis*, on the basis of phylogenetic and chemotaxonomic analyses. Int. J. Syst. Evol. Bacteriol. *51*: 1405–1417.

Takeuchi, M. and K. Hatano. 1998. Proposal of six new species in the genus *Microbacterium* and transfer of *Flavobacterium marinotypicum* ZoBell and Upham to the genus *Microbacterium* as *Microbacterium maritypicum* comb. nov. Int. J. Syst. Bacteriol. *48*: 973–982.

Takeuchi, M., F. Kawai, Y. Shimada and A. Yokota. 1993a. Taxonomic study of polyethylene glycol-utilizing bacteria: emended description of the genus *Sphingomonas* and new descriptions of *Sphingomonas macrogoltabidus* sp. nov., *Sphingomonas sanguis* sp. nov. and *Sphingomonas terrae* sp. nov. Syst. Appl. Microbiol. *16*: 227–238.

Takeuchi, M., F. Kawai, Y. Shimada and A. Yokota. 1993b. *In* Validation of the publication of new names and new combinations previously effectively published outside the IJSB. List No. 47. Int. J. Syst. Bacteriol. *43*: 864–865.

Takeuchi, M., T. Sakane, M. Yanagi, K. Yamasato, K. Hamana and A. Yokota. 1995. Taxonomic study of bacteria isolated from plants: Proposal of *Sphingomonas rosa* sp. nov., *Sphingomonas pruni* sp. nov., *Sphingomonas asaccharolytica* sp. nov., and *Sphingomonas mali* sp. nov. Int. J. Syst. Bacteriol. *45*: 334–341.

Takeuchi, M., H. Sawada, H. Oyaizu and A. Yokota. 1994. Phylogenetic evidence for *Sphingomonas* and *Rhizomonas* as nonphotosynthetic members of the alpha-4 subclass of the *Proteobacteria*. Int. J. Syst. Bacteriol. *44*: 308–314.

Tal, R., H.C. Wong, R. Calhoon, D. Gelfand, A.L. Fear, G. Volman, R. Mayer, P. Ross, D. Amikam, H. Weinhouse, A. Cohen, S. Sapir, P. Ohana and M. Benziman. 1998. Three *cdg* operons control cellular turnover of cyclic di-GMP in *Acetobacter xylinum*: genetic organization and occurrence of conserved domains in isoenzymes. J. Bacteriol. *180*: 4416–4425.

Talalay, P., M.M. Dobson and D.F. Tapley. 1952. Oxidative degradation of testosterone by adaptive enzymes. Nature *170*: 620–621.

Talley, R.S. and C.L. Baugh. 1975. Effects of bicarbonate on growth of *Neisseria gonorrhoeae*: replacement of gaseous CO_2 atmosphere. Appl. Microbiol. *29*: 469–471.

Tamaki, T., M. Fukaya, H. Takemura, K. Tayama, H. Okumura, Y. Kawamura, M. Nishiyama, S. Horinouchi and T. Beppu. 1991. Cloning and sequencing of the gene cluster encoding two subunits of membrane-bound alcohol dehydrogenase from *Acetobacter polyoxogenes*. Biochim. Biophys. Acta *1088*: 292–300.

Tamaki, T., S. Horinouchi, M. Fukaya, H. Okumura, Y. Kawamura and T. Beppu. 1989. Nucleotide sequence of the membrane-bound aldehyde dehydrogenase gene from *Acetobacter polyoxogenes*. J. Biochem. (Tokyo) *106*: 541–544.

Tamaoka, J., D.-M. Ha and K. Komagata. 1987. Reclassification of *Pseudomonas acidovorans* den Dooren de Jong 1926 and *Pseudomonas testosteroni* Marcus and Talalay 1956 as *Comamonas acidovorans* comb. nov. and *Comamonas testosteroni* comb. nov., with an emended description of the genus *Comamonas*. Int. J. Syst. Bacteriol. *37*: 52–59.

Tamaoka, J., Y. Katayama-Fujimura and H. Kuraishi. 1983. Analysis of bacterial menaquinone mixtures by high performance liquid chromatography. J. Appl. Bacteriol. *54*: 31–36.

Tamegai, H. and Y. Fukumori. 1994. Purification, and some molecular and enzymatic features of a novel *ccb*-type cytochrome *c* oxidase from a microaerobic denitrifier, *Magnetospirillum magnetotacticum*. FEBS Lett. *347*: 22–26.

Tamegai, H., T. Yamanaka and Y. Fukumori. 1993. Purification and properties of a cytochrome a_1-like hemoprotein from a magnetotactic bacterium, *Aquaspirillum magnetotacticum*. Biochim. Biophys. Acta *1158*: 237–243.

Tamer, A.U. , M. Aragno and N. Sahin. 2002. Isolation and characterization of a new type of aerobic, oxalic acid utilizing bacteria, and proposal of *Oxalicibacterium flavum* gen. nov., sp nov. Syst. Appl. Microbiol. *25*: 513–519.

Tamura, A., N. Ohashi, H. Urakami and S. Miyamura. 1995. Classification of *Rickettsia tsutsugamushi* in a new genus, *Orientia* gen. nov., as *Orientia tsutsugamushi* comb. nov. Int. J. Syst. Bacteriol. *45*: 589–591.

Tamura, A., N. Ohashi, H. Urakami, K. Takahashi and M. Oyanagi. 1985. Analysis of polypeptide composition and antigenic components of *Rickettsia tsutsugamushi* by polyacrylamide gel electrophoresis and immunoblotting. Infect. Immun. *48*: 671–675.

Tamura, A., H. Urakami and T. Tsuruhara. 1982. Purification of *Rickettsia tsutsugamushi* by Percoll density gradient centrifugation. Microbiol. Immunol. *26*: 321–328.

Tan, Z.-Y., X.-D. Xu, E.-T. Wang, J.-L. Gao, E. Martinez-Romero and W.-X. Chen. 1997. Phylogentic and genetic relationships of *Mesorhizobium tianhanense* and related rhizobia. Int. J. Syst. Bacteriol. *47*: 874–879.

Tanaka, K., K. Nakamura and E. Mikami. 1990. Fermentation of maleate by a gram-negative strictly anaerobic non-spore-former, *Propionivibrio dicarboxylicus* gen. nov., sp. nov. Arch. Microbiol. *154*: 323–328.

Tanaka, K., K. Nakamura and E. Mikami. 1991a. Fermentation of S-citramalate, citrate, mesaconate, and pyruvate by a gram-negative strictly anaerobic non-spore-former, *Formivibrio citricus* gen. nov., sp. nov. Arch. Microbiol. *155*: 491–495.

Tanaka, K., K. Nakamura and E. Mikami. 1991b. *In* Validation of the publication of new names and new combinations previously effectively published outside the IJSB. List No. 37. Int. J. Syst. Bacteriol. *41*: 331.

Tanaka, K., K. Nakamura and E. Mikami. 1991c. *In* Validation of the publication of new names and new combinations previously effectively published outside the IJSB. List No. 39. Int. J. Syst. Bacteriol. *41*: 580–581.

Tanaka, K., E. Stackebrandt, S. Tohyama and T. Eguchi. 2000a. *Desulfovirga adipica* gen. nov., sp. nov. an adipate-degrading, Gram-negative, sulfate-reducing bacterium. Int. J. Syst. Evol. Bacteriol. *50*: 639–644.

Tanaka, M., S. Murakami, R. Shinke and K. Aoki. 1999. Reclassification of the strains with low G + C contents of DNA belonging to the genus *Gluconobacter* Asai 1935 (*Acetobacteraceae*). Biosci. Biotechnol. Biochem. *63*: 989–992.

Tanaka, M., S. Murakami, R. Shinke and K. Aoki. 2000b. Genetic characteristics of cellulose-forming acetic acid bacteria identified phenotypically as *Gluconacetobacter xylinus*. Biosci. Biotechnol. Biochem. *64*: 757–760.

Tanaka, S., T. Suto, Y. Isayama, R. Azuma and H. Hatakeyama. 1977. Chemotaxonomical studies on fatty acids of *Brucella* species. Ann. Sclavo. *19*: 67–82.

Tanaka, Y., Y. Fukumori and T. Yamanaka. 1983. Purification of cytochrome $a_1 c_1$ from Nitrobacter agilis and characterization of nitrite oxidation system of the bacterium. Arch. Microbiol. *135*: 265–271.

Tanaka, Y., T. Yoshida, K. Watanabe, Y. Izumi and T. Mitsunaga. 1997a. Characterization, gene cloning and expression of isocitrate lyase involved in the assimilation of one-carbon compounds in *Hyphomicrobium methylovorum* GM2. Eur. J. Biochem. *249*: 820–825.

Tanaka, Y., T. Yoshida, K. Watanabe, Y. Izumi and T. Mitsunaga. 1997b. Cloning and analysis of methanol oxidation genes in the methylotroph *Hyphomicrobium methylovorum* GM2. FEMS Microbiol. Lett. *154*: 397–401.

Tang, Y.W., M.K. Hopkins, C.P. Kolbert, P.A. Hartley, P.J. Severance and D.H. Persing. 1998. *Bordetella holmesii* like organisms associated with septicemia, endocarditis, and respiratory failure. Clin. Infect. Dis. *26*: 389–392.

Taniguchi, S. and M.D. Kamen. 1963. On the nitrate metabolism of facultative photoheterotrophs. *In* Japanese Society of Plant Physiologists (Editor), Microalgae and Photosynthetic Bacteria, University of Tokyo Press, Tokyo. pp. 465–484.

Tanner, A.C.R., S. Badger, C.-H. Lai, M.A. Listgarten, R.A. Visconti and S.S. Socransky. 1981. *Wolinella* gen. nov., *Wolinella succinogenes* (*Vibrio succinogenes* Wolin et al.) comb. nov., and description of *Bacteroides gracilis* sp. nov., *Wolinella recta* sp. nov., *Campylobacter concisus* sp. nov., and *Eikenella corrodens* from humans with periodontal disease. Int. J. Syst. Bacteriol. *31*: 432–445.

Tanner, A.C.R., M.A. Listgarten and J.L. Ebersole. 1984. *Wolinella curva* sp. nov.: *Vibrio succinogenes* of human origin. Int. J. Syst. Bacteriol. *34*: 275–282.

Tans-Kersten, J., H. Huang and C. Allen. 2001. *Ralstonia solanacearum* needs motility for invasive virulence on tomato. J. Bacteriol. *183*: 3597–3605.

Tardy-Jacquenod, C., M. Magot, F. Laigret, M. Kaghad, B.K.C. Patel, J. Guezennec, R. Matheron and P. Caumette. 1996. *Desulfovibrio gabonensis* sp. nov., a new moderately halophilic sulfate-reducing bacterium isolated from an oil pipeline. Int. J. Syst. Bacteriol. *46*: 710–715.

Tarlera, S. and E.B.M. Denner. 2003. *Sterolibacterium denitrificans* gen. nov., sp nov., a novel cholesterol-oxidizing, denitrifying member of the β-*Proteobacteria*. Int. J. Syst. Evol. Microbiol. *53*: 1085–1091.

Tarrand, J.J., N.R. Krieg and J. Dobereiner. 1978. A taxonomic study of the *Spirillum lipoferum* group, with descriptions of a new genus, *Azospirillum* gen. nov. and two species, *Azospirillum lipoferum* (Beijerinck) comb. nov. and *Azospirillum brasilense* sp. nov. Can. J. Microbiol. *24*: 967–980.

Tarrand, J.J., N.R. Krieg and J. Dobreiner. 1979. *In* Validation of the publication of new names and new combinations previously effectively published outside the IJSB. List No.2. Int. J. Syst. Bacteriol. *29*: 79–80.

Tartakovsky, B., C.B. Miguez, L. Petti, D. Bourque, D. Groleau and S.R. Guiot. 1998. Tetrachloroethylene dechlorination using a consortium of coimmobilized methanogenic and methanotrophic bacteria. Enzyme Microb. Technol. *22*: 255–260.

Tasaki, M., Y. Kamagata, K. Nakamura and E. Mikami. 1990. Isolation of a propionate using sulfate-reducing bacterium. *In* Bélaich, Bruschi and Garcia (Editors), Microbiology and Biochemistry of Strict Anaerobes Involved in Interspecies Hydrogen Transfer. FEMS symposium; No. 54, Plenum Press, New York. pp. 477–479.

Tayama, K., M. Fukaya, H. Okumura, Y. Kawamura and T. Beppu. 1989. Purification and characterization of membrane-bound alcohol dehydrogenase from A *cetobacter polyoxogenes* sp. nov. Appl. Microbiol. Biotechnol. *32*: 181–185.

Tayama, K., M. Fukaya, H. Okumura, Y. Kawamura, S. Horinouchi and T. Beppu. 1994. Transformation of *Acetobacter polyoxogenes* with plasmid DNA by electroporation. Biosci. Biotechnol. Biochem. *58*: 974–975.

Taylor, B.F., W.L. Campbell and I. Chinoy. 1970. Anaerobic degradation of the benzene nucleus by a facultatively anaerobic microorganism. J. Bacteriol. *102*: 430–437.

Taylor, B.F., D.S. Hoare and S.L. Hoare. 1971. *Thiobacillus denitrificans* as an obligate chemolithotroph. I. Isolation and growth studies. Arch. Mikrobiol. *80*: 193–204.

Taylor, C.J., A.J. Anderson and S.G. Wilkinson. 1994a. Structure of the O9 antigen from *Burkholderia* (*Pseudomonas*) *cepacia*. FEMS Microbiol. Lett. *115*: 201–204.

Taylor, C.J., A.J. Anderson and S.G. Wilkinson. 1998. Phenotypic variation of lipid composition in *Burkholderia cepacia*: a response to increased growth temperature is a greater content of 2-hydroxy acids in phosphatidylethanolamine and ornithine amide lipid. Microbiology *144*: 1737–1745.

Taylor, C.E.D., R.O. Rosenthal, D.F.J. Brown, S.P. Lapage, L.R. Hill and R.M. Legros. 1978. The causative organism of contagious equine metritis 1977: proposal for a new species to be known as *Haemophilus equigenitalis*. Equine Vet. J. *10*: 136–144.

Taylor, D.E. 1992a. Antimicrobial resistance of *Campylobacter jejuni* and *Campylobacter coli* to tetracycline, chloramphenicol, and erythromycin. *In* Nachamkin, Blaser and Tompkins (Editors), *Campylobacter jejuni*. Current Status and Future Trends, American Society for Microbiology, Washington, D.C. pp. 74–86.

Taylor, D.E. 1992b. Genetics of *Campylobacter* and *Helicobacter*. Annu. Rev. Microbiol. *46*: 35–64.

Taylor, D.E., N. Chang, N.S. Taylor and J.G. Fox. 1994b. Genome conservation in *Helicobacter mustelae* as determined by pulsed-field gel electrophoresis. FEMS Microbiol. Lett. *118*: 31–36.

Taylor, D.E. and P. Courvalin. 1988. Mechanisms of antibiotic resistance in *Campylobacter* species. Antimicrob. Agents Chemother. *32*: 1107–1112.

Taylor, D.E., S.A. DeGrandis, M.A. Karmali and P.C. Fleming. 1981. Transmissible plasmids from *Campylobacter jejuni*. Antimicrob. Agents Chemother. *19*: 831–835.

Taylor, D.E., M. Eaton, N. Chang and S.M. Salama. 1992a. Construction of a *Helicobacter pylori* genome map and demonstration of diversity at the genome level. J. Bacteriol. *174*: 6800–6806.

Taylor, D.E., H. Eng, H. Lior and W.M. Wenman. 1988. Role of flagellar protein and other proteins in *Campylobacter jejuni* heat-labile serotyping system. Serodiagn. Immunother. Infect. Dis. *2*: 27–32.

Taylor, D.N., J.A. Kiehlbauch, W. Tee, C. Pitarangsi and P. Echeverria. 1992b. Isolation of group 2 aerotolerant *Campylobacter* species from Thai children with diarrhea. J. Infect. Dis. *163*: 1062–1067.

Taylor, J. and R.J. Parkes. 1983. The cellular fatty acids of the sulfate-reducing bacteria, *Desulfobacter* sp., *Desulfobulbus* sp. and *Desulfovibrio desulfuricans*. J. Gen. Microbiol. *129*: 3303–3309.

Taylor, M.J., K. Bilo, H.F. Cross, J.P. Archer and A.P. Underwood. 1999. 16S rDNA phylogeny and ultrastructural characterization of *Wolbachia* intracellular bacteria of the filarial nematodes *Brugia malayi*, *B. pahangi*, and *Wuchereria bancrofti*. Exp. Parasitol. *91*: 356–361.

Taylor, M.J., H.F. Cross and K. Bilo. 2000. Inflammatory responses induced by the filarial nematode *Brugia malayi* are mediated by lipopolysaccharide-like activity from endosymbiotic *Wolbachia* bacteria. J. Exp. Med. *191*: 1429–1436.

Taylor, M.J. and A. Hoerauf. 1999. *Wolbachia* bacteria of filarial nematodes. Parasitol. Today. *15*: 437–442.

Taylor, N.S., J.G. Fox and L. Yan. 1995. *In vitro* hepatotoxic factor in *Helicobacter hepaticus, H. pylori* and other *Helicobacter* species. J. Med. Microbiol. *42*: 48–52.

Taylor, R.M., M. Lisbonne and G. Roman. 1932. Recherches sur l'identification des Brucella isolées en France par l'action bactériostatic des matières colorantes et la production d'hydrogène sulfuré (Huddleson). Ann. Inst. Pasteur (Paris) *49*: 284–302.

Taylor, V.I., P. Baumann, J.L. Reichelt and R.D. Allen. 1974. Isolation, enumeration, and host range of marine bdellovibrios. Arch. Microbiol. *98*: 101–114.

Tazuke, Y., K. Matsuda, K. Adachi and Y. Tsukada. 1994. Purification and properties of bile acid sulfate sulfatase from *Pseudomonas testosteroni.* Biosci. Biotechnol. Biochem. *58*: 889–894.

Tchan, Y.T. 1957. Studies of nitrogen-fixing bacteria. VI. A new species of nitrogen-fixing bacteria. Proc. Linn. Soc. N.S.W. *82*: 314–316.

Tchan, Y.T., J. Pochon and A.R. Prévot. 1948. Études de systématique bactérienne. VIII. Essai de classification des *Cytophaga.* Ann. Inst. Pasteur *74*: 394–400.

Tebele, N., T.C. McGuire and G.H. Palmer. 1991. Induction of protective immunity by using *Anaplasma marginale* initial body membranes. Infect. Immun. *59*: 3199–3204.

Tee, W., M. DyallSmith, W. Woods and D. Eisen. 1996. Probable new species of *Desulfovibrio* isolated from a pyogenic liver abscess. J. Clin. Microbiol. *34*: 1760–1764.

Tee, W., S. Fairley, R. Smallwood and B. Dwyer. 1991. Comparative evaluation of three selective media and a nonselective medium for the culture of *Helicobacter pylori* from gastric biopsies. J. Clin. Microbiol. *29*: 2587–2589.

Tee, W., S. Hinds, J. Montgomery and M.L. Dyall-Smith. 2000. A probable new *Helicobacter* species isolated from a patient with bacteremia. J. Clin. Microbiol. *38*: 3846–3848.

Tee, W., K. Leder, E. Karroum and M. Dyall-Smith. 1998. *Flexispira rappini* bacteremia in a child with pneumonia. J. Clin. Microbiol. *36*: 1679–1682.

Tegtmeyer, B., J. Weckesser, H. Mayer and J.F. Imhoff. 1985. Chemical composition of the lipopolysaccharides of *Rhodobacter sulfidophilus, Rhodopseudomonas acidophila,* and *Rhodopseudomonas blastica.* Arch. Microbiol. *143*: 32–36.

Teichmann, E. 1935. Vergleichende Untersuchungen über die Kultur und Morphologie einiger Eisenorganismen., Thesis, Prague.

Teixeira, K.R.S., M. Wulling, T. Morgan, R. Galler, E.M. Zellermann, J.I. Baldani, C. Kennedy and D. Meletzus. 1999. Molecular analysis of the chromosomal region encoding the *nifA* and *nifB* genes of *Acetobacter diazotrophicus.* FEMS Microbiol. Lett. *176*: 301–309.

Telford, S.R., J.E. Dawson, P. Katavolos, C.K. Warner, C.P. Kolbert and D.H. Persing. 1996. Perpetuation of the agent of human granulocytic ehrlichiosis in a deer tick-rodent cycle. Proc. Natl. Acad. Sci. U.S.A. *93*: 6209–6214.

Tellez, C.M., K.P. Gaus, D.W. Graham, R.G. Arnold and R.Z. Guzman. 1998. Isolation of copper biochelates from *Methylosinus trichosporium* OB3b and soluble methane monooxygenase mutants. Appl. Environ. Microbiol,. *64*: 1115–1122.

Tenover, F.C., S. Williams, K.P. Gordon, C. Nolan and J.J. Plorde. 1985. Survey of plasmids and resistance factors in *Campylobacter jejuni* and *Campylobacter coli.* Antimicrob. Agents Chemother. *27*: 37–41.

Terasaki, Y. 1961a. On *Spirillum putridiconchylium* nov. sp. Bot. Mag. *74*: 79–85.

Terasaki, Y. 1961b. On two new species of *Spirillum.* Bot. Mag. *74*: 220–227.

Terasaki, Y. 1963. On the isolation of *Spirillum.* Bull. Suzugamine Women's Coll. Nat. Sci. *10*: 1–10.

Terasaki, Y. 1970. Über die Anhäufung von in Süsswasser und Meerwasser vorkommenden *Spirillum.* Bull. Suzagamine Women's Coll. Nat. Sci. *15*: 1–7.

Terasaki, Y. 1972. Studies on the genus *Spirillum* Ehrenberg. I. Morpho-

logical, physiological, and biochemical characteristics of water spirilla. Bull. Suzugamine Women's Coll. Nat. Sci. *16*: 1–146.

Terasaki, Y. 1973. Studies on the genus *Spirillum* Ehrenberg. II. Comments on type and reference strains of *Spirillum* and descriptions of new species and new subspecies. Bull. Suzugamine Women's Coll. Nat. Sci. *17*: 1–71.

Terasaki, Y. 1975. Freeze-dried cultures of water spirilla made on experimental basis. Bull. Suzugamine Women's Coll., Nat. Sci. *19*: 1–10.

Terasaki, Y. 1979. Transfer of five species and two subspecies of *Spirillum* to other genera (*Aquaspirillum* and *Oceanospirillum*), with emended descriptions of the species and subspecies. Int. J. Syst. Bacteriol. *29*: 130–144.

Terasaki, Y. 1980. Enrichment and isolation of aerobic chemoheterotrophic spirilla from mud and sand samples. J. Gen. Appl. Microbiol. *26*: 395–402.

Terefework, Z., G. Nick, S. Suomalainen, L. Paulin and K. Lindström. 1998. Phylogeny of *Rhizobium galegae* with respect to other rhizobia and agrobacteria. Int. J. Syst. Bacteriol. *48*: 349–356.

Terekhova, L.P., A.V. Laiko, P.A. Lokhmacheva, T.N. Lafitskaya and L.V. Vasil'eva. 1981. The susceptibility of some genera of prosthecobacteria and budding bacteria to antibiotics. Izv. Akad. Nauk Uzb. SSSR Ser. Biol. *1*: 143–147.

Tesh, R.B. and G.B. Modi. 1983. Development of a continuous cell line from the sand fly *Lutzomyia longipalpis* (Diptera: Psychodidae), and its susceptibility to infection with arboviruses. J. Med. Entomol. *20*: 199–202.

Teske, A., E. Alm, J.M. Regan, S. Toze, B.E. Rittmann and D.A. Stahl. 1994. Evolutionary relationships among ammonia- and nitrite-oxidizing bacteria. J. Bacteriol. *176*: 6623–6630.

Teske, A., P. Sigalevich, Y. Cohen and G. Muyzer. 1996. Molecular identification of bacteria from a coculture by denaturing gradient gel electrophoresis of 16S ribosomal DNA fragments as a tool for isolation in pure cultures. Appl. Environ. Microbiol. *62*: 4210–4215.

Tesoriero, L.A., P.C. Fahy and L.V. Gunn. 1982. First record of bacterial rot of onion in Australia caused by *Pseududomonas gladioli* pv. *allicola* and association with internal browning caused by *Pseudomonas aeruginosa.* Australas. Plant Pathol. *11*: 56–57.

Tett, V.A., A.J. Willets and H.M. LappinScott. 1994. Enantioselective degradation of the herbicide mecoprop [2-(2-methyl-4-chlorophenoxy) propionic acid] by mixed and pure bacterial cultures. FEMS Microbiol. Ecol. *14*: 191–199.

Tett, V.A., A.J. Willets and H.M. LappinScott. 1997. Biodegradation of the chlorophenoxy herbicide (R)-(+)-mecoprop by *Alcaligenes denitrificans.* Biodegradation 8: 43–52.

Tettelin, H., N.J. Saunders, J. Heidelberg, A.C. Jeffries, K.E. Nelson, J.A. Eisen, K.A. Ketchum, D.W. Hood, J.F. Peden, R.J. Dodson, W.C. Nelson, M.L. Gwinn, R. DeBoy, J.D. Peterson, E.K. Hickey, D.H. Haft, S.L. Salzberg, O. White, R.D. Fleischmann, B.A. Dougherty, T. Mason, A. Ciecko, D.S. Parksey, E. Blair, H. Cittone, E.B. Clark, M.D. Cotton, T.R. Utterback, H. Khouri, H.Y. Qin, J. Vamathevan, J. Gill, V. Scarlato, V. Masignani, M. Pizza, G. Grandi, L. Sun, H.O. Smith, C.M. Fraser, E.R. Moxon, R. Rappuoli and J.C. Venter. 2000. Complete genome sequence of *Neisseria meningitidis* serogroup B strain MC58. Science *287*: 1809–1815.

Teuber, M., A. Andresen and M. Sievers. 1987a. Bacteriophage problems in vinegar fermentations. Biotechnol. Lett. *9*: 37–38.

Teuber, M., M. Sievers and A. Andresen. 1987b. Characterization of the microflora of high acid submerged vinegar fermenters by distinct plasmid profiles. Biotechnol. Lett. *9*: 265–268.

Teysseire, N., C. Chiche-Portiche and D. Raoult. 1992. Intracellular movements of *Rickettsia conorii* and *R. typhi* based on actin polymerization. Res. Microbiol. *143*: 821–829.

Tezuka, Y. 1973. A zoogloea bacterium with gelatinous mucopolysaccharide matrix. J. Water Pollut. Control Fed. *45*: 531–536.

Thakur, I.S. 1996. Use of monoclonal antibodies against dibenzo-*p*-dioxin degrading *Sphingomonas* sp. strain RW1. Lett. Appl. Microbiol. *22*: 141–144.

Thanabalu, T., J. Hindley, S. Brenner, C. Oei and C. Berry. 1992. Expression of the mosquitocidal toxins of *Bacillus sphaericus* and *Bacillus thuringiensis* subsp. israelensis by recombinant *Caulobacter crescentus*, a vehicle for biological control of aquatic insect larvae. Appl. Environ. Microbiol. *58*: 905–910.

Tharanathan, R.N., A. Yokota, H. Rau and H. Mayer. 1993. Isolation and chemical characterization of lipopolysaccharides from 4 *Mycoplana* species (*M. bullata*, *M. segnis*, *M. ramosa* and *M. dimorpha*). Arch. Microbiol. *159*: 445–452.

Thaxter, R. 1892. On the *Myxobacteriaceae*, a new order of Schizomycetes. Bot. Gaz. *17*: 389–406.

Thaxter, R. 1893. A new order of Schizomycetes. Bot. Gaz. *18*: 29–30.

Thaxter, R. 1897. Further observations on the *Myxobacteriaceae*. Bot. Gaz. *23*: 395–411.

Thaxter, R. 1904. Notes on the *Myxobacteriaceae*. Bot. Gaz. *37*: 405–416.

Theiler, A. 1910. *Anaplasma marginale* (gen. and spec. nov.). The marginal points in the blood of cattle suffering from specific disease. Transvaal S. Afr. Rep. Vet. Bacteriol. Dept. Agr. *1908-9*: 7–64.

Theiler, A. 1911. Further investigations into anaplasmosis of South African cattle. 1st Rep. Dir. Vet. Res. August. *1911*: 7–46.

Then, J. and H.G. Trüper. 1981. The role of thiosulfate in sulfur metabolism of *Rhodopseudomonas globiformis*. Arch. Microbiol. *130*: 143–146.

Thiele, O.W. and W. Kehr. 1969. The "free" lipids of *Brucella abortus* Bang. Concerning the neutral lipids. Eur. J. Biochem. *9*: 167–175.

Thiele, O.W. and G. Schwinn. 1973. The free lipids of *Brucella melitensis* and *Bordetella pertussis*. Eur. J. Biochem. *34*: 333–344.

Thies, K.L., D.E. Griffin, C.H. Graves and C.P. Hegwood. 1991. Characterization of *Agrobacterium* isolates from muscadine grape. Plant Dis. *75*: 634–637.

Thiesen, B., B. Greenwood, N. Brieske and M. Achtman. 1997. Persistence of antibodies to meningococcal IgA1 protease versus decay of antibodies to group A polysaccharide and Opc protein. Vaccine *15*: 209–219.

Thomas, C.E. and P.F. Sparling. 1996. Isolation and analysis of a fur mutant of *Neisseria gonorrhoeae*. J. Bacteriol. *178*: 4224–4232.

Thomas, J.M. and M. Alexander. 1987. Colonization and mineralization of palmitic acid by *Pseudomonas pseudoflava*. Microb. Ecol. *14*: 75–80.

Thomas, J.C., Y. St. Pierre, R. Beaudet and R. Villemur. 2000. Monitoring by laser-flow-cytometry of the polycyclic aromatic hydrocarbon-degrading *Sphingomonas* sp. strain 107 during biotreatment of a contaminated soil. Can. J. Microbiol. *46*: 433–440.

Thomas, R.A.P., K. Lawlor, M. Bailey and L.E. Macaskie. 1998. Biodegradation of metal-EDTA complexes by an enriched microbial population. Appl. Environ. Microbiol. *64*: 1319–1322.

Thomashow, M.F. and T.W. Cotter. 1992. *Bdellovibrio* host dependence: the search for signal molecules and genes that regulate the intraperiplasmic growth cycle. J. Bacteriol. *174*: 5767–5771.

Thomashow, M.F. and S.C. Rittenberg. 1978a. Intraperiplasmic growth of *Bdellovibrio bacteriovorus* 109J: attachment of long chain fatty acids to *Escherichia coli* peptidoglycan. J. Bacteriol. *135*: 1015–1023.

Thomashow, M.F. and S.C. Rittenberg. 1978b. Intraperiplasmic growth of *Bdellovibrio bacteriovorus* 109J: N-deacylation of *Escherichia coli* peptidoglycan amino sugars. J. Bacteriol. *135*: 1008–1014.

Thomashow, M.F. and S.C. Rittenberg. 1978c. Intraperiplasmic growth of *Bdellovibrio bacteriovorus* 109J: solubilization of *Escherichia coli* peptidoglycan. J. Bacteriol. *135*: 998–1007.

Thomashow, M.F. and S.C. Rittenberg. 1979. the intraperiplasmic growth cycle — the lifestyle of the bdellovibrios. *In* Parish (Editor), Developmental Biology of Prokaryotes, University of California Press, Berkeley. pp. 115–138.

Thompson, J.P. 1987. Survival of *Azotobacteraceae* dessicated over silica gel. J. Appl. Microbiol. Biotechnol. *3*: 185–195.

Thompson, J.D., D.G. Higgins and T.J. Gibson. 1994. CLUSTAL W: improving the sensitivity of progressive multiple sequence alignment through sequence weighting, position-specific gap penalties and weight matrix choice. Nucleic Acids Res. *22*: 4673–4680.

Thompson, J.P. and V.B.D. Skerman. 1979. *Azotobacteraceae*: The Taxonomy and Ecology of the Aerobic Nitrogen-Fixing Bacteria, Academic Press, London.

Thompson, J.P. and V.B.D. Skerman. 1981. *In* Validation of the publication of new names and new combinations previously effectively published outside the IJSB. List No. 6. Int. J.Syst. Bacteriol. *31*: 215–218.

Thompson, L.M., R.M. Smibert, J.L. Johnson and N.R. Krieg. 1988. Phylogenetic study of the genus *Campylobacter*. Int. J. Syst. Bacteriol. *38*: 190–200.

Thomsen, A. 1929. Smitsom kasthingsenzooti (Bang-lnfektion) blandt veer i Midtjylland. Mskr. Dyrläg. *41*: 386.

Thurner, C.A.K. 1997. Ethanol metabolism in *Acetobacter europaeus*: purification, characterization and genetic analysis of the membrane-associated alcohol and aldehyde dehydrogenase complexes, Thesis, Swiss Federal Institute of Technology, Zurich (ETHZ), Zurich.

Thurner, C., C. Vela, L. Thöny Meyer, L. Meile and M. Teuber. 1997. Biochemical and genetic characterization of the acetaldehyde dehydrogenase complex from *Acetobacter europaeus*. Arch. Microbiol. *168*: 81–91.

Thurnheer, T., T. Kohler, A.M. Cook and T. Leisinger. 1986. Orthanilic acid and analogues as carbon sources for bacteria: growth physiology and enzymic desulphonation. J. Gen. Microbiol. *132*: 1215–1220.

Ti, T.Y., W.C. Tan, A.P.Y. Chong and E.H. Lee. 1993. Nonfatal and fatal infectious caused by *Chromobacterium violaceum*. Clinical Infectious Diseases. *17*: 505–507.

Tibazarwa, C., S. Wuertz, M. Mergeay, L. Wyns and D. van der Lelie. 2000. Regulation of the *cnr* cobalt and nickel resistance determinant of *Ralstonia eutropha* (*Alcaligenes eutrophus*) CH34. J. Bacteriol. *182*: 1399–1409.

Tibor, A., B. Decelle and J.J. Letesson. 1999. Outer membrane proteins Omp10, Omp16, and Omp19 of *Brucella* spp. are lipoproteins. Infect. Immun. *67*: 4960–4962.

Tiedje, J.M. 1980. Nitrilotriacetate: hindsights and gunsights. *In* Maki, Dickson and Cairns (Editors), Biotransformation and Fate of Chemicals in the Aquatic Environment, American Society for Microbiology, Washington, D.C. pp. 114–119.

Tiedje, J.M., B.B. Mason, C.B. Warren and E.J. Malec. 1973. Metabolism of nitrilotriacetate by cells of *Pseudomonas* species. Appl. Microbiol. *25*: 811–818.

Tien, T.M., H.G. Diem, M.H. Gaskins and D.H. Hubbell. 1981. Polygalacturonic acid transeliminase production by *Azospirillum* species. Can. J. Microbiol. *27*: 426–431.

Tien, T.M., M.H. Gaskins and D.H. Hubbell. 1979. Plant growth substances produced by *Azospirillum brasilense* and their effect on the growth of pearl millet. Appl. Environ. Microbiol. *37*: 1016–1024.

Tighe, S.W., P. de Lajudie, K. Dipietro, K. Lindstrom, G. Nick and B.D. Jarvis. 2000. Analysis of cellular fatty acids and phenotypic relationships of *Agrobacterium*, *Bradyrhizobium*, *Mesorhizobium*, *Rhizobium*, and *Sinorhizobium* species using the Sherlock Microbial Identification System. Int. J. Syst. Evol. Microbiol. *50*: 787–801.

Tilak, K.V.B.R., M. Lakshmi-Kumari and C.S. Nautiyal. 1979. Survival of *Azospirillum brasilense* in different carriers. Curr. Sci. (Bangalore) *48*: 412–413.

Tilden, A.R., M.A. Becker, L.L. Amma, J. Arciniega and A.K. McGaw. 1997. Melatonin production in an aerobic photosynthetic bacterium: An evolutionarily early association with darkness. J. Pineal Res. *22*: 102–106.

Timkovich, R., R. Dhesi, K.J. Martinkus, M.K. Robinson and T.M. Rea. 1982. Isolation of *Paracoccus denitrificans* cytochrome cd_1: comparative kinetics with other nitrite reductases. Arch. Biochem. Biophys. *215*: 47–58.

Timmermans, P. and A. Van Haute. 1983. Denitrification with methanol: fundamental study of the growth and denitrification capacity of *Hyphomicrobium* sp. Water Res. *17*: 1249–1256.

Tinsley, C.R. and J.E. Heckels. 1986. Variation in the expression of pili and outer-membrane protein by *Neisseria meningitidis* during the course of meningococcal infection. J. Gen. Microbiol. *132*: 2483–2490.

Tinsley, C.R. and X. Nassif. 1996. Analysis of the genetic differences between *Neisseria meningitidis* and *Neisseria gonorrhoeae*: Two closely related bacteria expressing two different pathogenicities. Proc. Natl. Acad. Sci. U.S.A. *93*: 11109–11114.

Titus, J.A., W.M. Reed, R.M. Pfister and P.R. Dugan. 1982. Exospore formation in *Methylosinus trichosporium*. J. Bacteriol. *149*: 354–360.

Tiunov, A.V., T.G. Dobovol'skaya and L.M. Polyanskaya. 1997. Microbial community of the *Lumbricus terrestris* L. earthworm burrow walls. Microbiology *66*: 349–353.

Tjia, T.N., W.E.S. Harper, C.S. Goodwin and W.B. Grubb. 1987. Plasmids in *Campylobacter pyloridis*. Microbios Lett. *36*: 7–12.

Todorov, T. and P. Koleva-Todorova. 1971. Brucellacins and their formation (In Bulgarian), Second Congress in Microbiology, Sofia. p. 219.

Toghrol, F. and W.M. Southerland. 1983. Purification of *Thiobacillus novellus* sulfite oxidase: evidence for the presence of heme and molybdenum. J. Biol. Chem. *258*: 6762–6765.

Toledo, I., L. Lloret and E. Martínez-Romero. 2003. *Sinorhizobium americanum* sp. nov., a new *Sinorhizobium* species nodulating native *Acacia* spp. in Mexico. Syst. Appl. Microbiol. *26*: 54–64.

Tolxdorff-Neutzling, R. and J.H. Klemme. 1982. Metabolic role and regulation of L-alanine dehydrogenase in *Rhodopseudomonas capsulata*. FEMS Microbiol. Lett. *13*: 155–159.

Tomasek, P.H., B. Frantz, U.M.X. Sangodkar, R.A. Haugland and A.M. Chakrabarty. 1989. Characterization and nucleotide sequence determination of a repeat element isolated from a 2,4,5-T degrading strain of *Pseudomonas cepacia*. Gene *76*: 227–238.

Tomasi, I., I. Artaud, Y. Bertheau and D. Mansuy. 1995. Metabolism of polychlorinated phenols by *Pseudomonas cepacia* AC1100: determination of the first two steps and specific inhibitory effect of methimazole. J. Bacteriol. *177*: 307–311.

Tomb, J.F., O. White, A.R. Kerlavage, R.A. Clayton, G.G. Sutton, R.D. Fleischmann, K.A. Ketchum, H.P. Klenk, S. Gill, B.A. Dougherty, K. Nelson, J. Quackenbush, L. Zhou, E.F. Kirkness, S.N. Peterson, B. Loftus, D. Richardson, R. Dodson, H.G. Khalak, A. Glodek, K. McKenney, L.M. Fitzgerald, N.A. Lee, M.D. Adams, E.K. Hickey, D.E. Berg, J.D. Gocayne, T.R. Utterback, J.D. Peterson, J.M. Kelley, M.D. Cotton, J.M. Weidman, C. Fujii, C. Bowman, L. Watthey, E. Wallin, W.S. Hayes, M. Borodovsky, P.D. Karp, H.O. Smith, C.M. Fraser and J.C. Venter. 1997. The complete genome sequence of the gastric pathogen *Helicobacter pylori*. Nature *388*: 539–547.

Tomita, B., H. Inoue, K. Chaya, A. Nakamura, N. Hamamura, K. Ueno, K. Watanabe and Y. Ose. 1987. Identification of dimethyl disulfide-forming bacteria isolated from activated sludge. Appl. Environ. Microbiol. *53*: 1541–1547.

Tommassen, J., P. Vermeij, M. Struyve, R. Benz and J.T. Poolman. 1990. Isolation of *Neisseria meningitidis* mutants deficient in class-1 (PorA) and Class-3 (PorB) outer membrane proteins. Infect. Immun. *58*: 1355–1359.

Tønjum, T., K. Bøvre and E. Juni. 1995a. Fastidious gram-negative bacteria: meeting the diagnostic challenge with nucleic acid analysis. APMIS *103*: 609–627.

Tønjum, T., G. Bukholm and K. Bøvre. 1989. Differentiation of some species of *Neisseriaceae* and other bacterial groups by DNA—DNA hybridization. APMIS *97*: 395–405.

Tønjum, T., D.A. Caugant, S.A. Dunham and M. Koomey. 1998. Structure and function of repetitive sequence elements associated with a highly polymorphic domain of the *Neisseria meningitidis* PilQ protein. Mol. Microbiol. *29*: 111–124.

Tønjum, T., D.A. Caugant, E. Namork and J.M. Koomey. 1994. Characterization of the phosphoglucose isomerase of *Neisseria gonorrhoeae* and *Neisseria meningitidis*. Proceedings of the 9th International Pathogenic Neisseria Conference, Winchester, U.K.. p. 208.

Tønjum, T., N.E. Freitag, E. Namork and M. Koomey. 1995b. Identification and characterization of *pilG*, a highly conserved pilus assembly gene in pathogenic *Neisseria*. Mol. Microbiol. *16*: 451–464.

Tønjum, T., N. Hagen and K. Bøvre. 1985. Identification of *Eikenella corrodens* and *Cardiobacterium hominis* by genetic transformation. Acta Pathol. Microbiol. Immunol. Scand. B Microbiol. *93*: 389–394.

Tønjum, T. and M. Koomey. 1997. The pilus colonization factor of pathogenic neisserial species: Organelle biogenesis and structure/function relationships—A review. Gene *192*: 155–163.

Tønjum, T., C.F. Marrs, F. Rozsa and K. Bøvre. 1991. The type 4 pilin of *Moraxella nonliquefaciens* exhibits unique similarities with the pilins of *Neisseria gonorrhoeae* and *Dichelobacter nodosus*. J. Gen. Microbiol. *137*: 2483–2490.

Tønjum, T., S. Weir, K. Bøvre, A. Progulske-Fox and C.F. Marrs. 1993. Sequence divergence in 2 tandemly located pilin genes of *Eikenella corrodens*. Infect. Immun. *61*: 1909–1916.

Tonomura, K., N. Kurose, S. Konishi and H. Kawasaki. 1982. Occurence of plasmids in *Zymomonas mobilis*. Agric. Biol. Chem. *46*: 2851–2853.

Tonouchi, N., N. Thara, T. Tsuchida, F. Yoshinaga, T. Beppu and S. Horinouchi. 1995. Addition of a small amount of an endoglucanase enhances cellulose production by *Acetobacter xylinum*. Biosci. Biotechnol. Biochem. *59*: 805–808.

Tonouchi, N., T. Tsuchida, F. Yoshinaga, S. Horinouchi and T. Beppu. 1994. A host-vector system for a cellulose-producing *Acetobacter* strain. Biosci. Biotechnol. Biochem. *58*: 1899–1901.

Torioni de Echaide, S., D.P. Knowles, T.C. McGuire, G.H. Palmer, C.E. Suarez and T.F. McElwain. 1998. Detection of cattle naturally infected with *Anaplasma marginale* in a region of endemicity by nested PCR and a competitive enzyme-linked immunosorbent assay using recombinant major surface protein 5. J. Clin. Microbiol. *36*: 777–782.

Tornabene, T.G., G. Holzer, A.S. Bittner and K. Grohmann. 1982. Characterization of the total extractable lipids of *Zymomonas mobilis* biovar mobilis. Can. J. Microbiol. *28*: 1107–1118.

Török, I., E. Kondorosi, T. Stepkowski, J. Pósfai and A. Kondorosi. 1984. Nucleotide sequence of *Rhizobium meliloti* nodulation genes. Nucleic Acids Res. *12*: 9509–9524.

Torrella, F., R. Guerrero and R.J. Seidler. 1978. Further taxonomic characterization of the genus *Bdellovibrio*. Can. J. Microbiol. *24*: 1387–1394.

Toth, D., J. Huska, I. Zavadska and M. Dobrotova. 1996. Effect of bacterial starvation on surfactant biotransformation. Folia Microbiol. *41*: 477–479.

Totten, P.A., C.L. Fennell, F.C. Tenover, J.M. Wezenberg, P.L. Perine, W.E. Stamm and K.K. Holmes. 1985. *Campylobacter cinaedi* (sp. nov.) and *Campylobacter fennelliae* (sp. nov.): two new *Campylobacter* species associated with enteric disease in homosexual men. J. Infect. Dis. *151*: 131–139.

Toukdarian, A.E. and M.E. Lidstrom. 1984a. DNA hybridization analysis of the *nif* region of two methylotrophs and molecular cloning of *nif*-specific DNA. J. Bacteriol. *157*: 925–930.

Toukdarian, A.E. and M.E. Lidstrom. 1984b. Nitrogen metabolism in a new obligate methanotroph, *Methylosinus* strain 6. J. Gen. Microbiol. *130*: 1827–1837.

Townsend, G.T. and J.M. Suflita. 1997. Influence of sulfur oxyanions on reductive dehalogenation activities in *Desulfomonile tiedjei*. Appl. Environ. Microbiol. *63*: 3594–3599.

Towson, S., D. Hutton, J. Siemienska, L. Hollick, T. Scanlon and S.K. Tagboto. 2000. Antibiotics and *Wolbachia* in filarial nematodes: antifilarial activity of rifampicin, oxytetracycline and chloramphenicol against *Onchocerca gutturosa*, *Onchocerca lienalis* and *Brugia pahangi*. Ann. Trop. Med. Parasitol. *94*: 801–816.

Toyosaki, H., Y. Kojima, T. Tsuchida, K.-I. Hoshino, Y. Yamada and F. Yoshinaga. 1995. The characterization of an acetic acid bacterium useful for producing bacterial cellulose in agitation cultures: the proposal of *Acetobacter xylinum* subsp. *sucrofermentans* subsp. nov. J. Gen. Appl. Microbiol. *41*: 307–314.

Trallero, E.P., J.M.G. Arenzana, G.C. Eguiluz and J.T. Larrucea. 1986. Beta-lactamase-producing *Eikenella corrodens* in an intraabdominal abscess [letter]. J. Infect. Dis. *153*: 379–380.

Tram, U. and W. Sullivan. 2002. Role of delayed nuclear envelope breakdown and mitosis in *Wolbachia*-induced cytoplasmic incompatibility. Science *296*: 1124–1126.

Tran Van, V., S. Ngôkê, O. Berge, D. Faure, R. Bally, P. Hebbar and T. Heulin. 1997. Isolation of *Azospirillum lipoferum* from the rhizosphere of rice by a new, simple method. Can. J. Microbiol. *43*: 486–490.

Traub, R. and C.L. Wisseman, Jr.. 1974. The ecology of chigger-borne rickettsiosis (scrub typhus). J. Med. Entomol. *11*: 237–303.

Traub, R., C.L. Wisseman, Jr. and A. Farhang-Azad. 1978. The ecology of murine typhus: A critical review. Trop. Dis. Bull. *75*: 237–317.

Trček, J., J. Ramus and P. Raspor. 1997. Phenotypic characterization and RAPD-PCR profiling of *Acetobacter* sp. isolated from spirit vinegar production. Food Technol. Biotechnol. *35*: 63–67.

Trček, J., P. Raspor and M. Teuber. 2000. Molecular identification of *Acetobacter* isolates from submerged vinegar production, sequence analysis of plasmid pJK2-1 and application in the development of a cloning vector. Appl. Microbiol. Biotechnol. *53*: 289–295.

Tremblay, G., R. Gagliardo, W.S. Chilton and P. Dion. 1987a. Diversity among opine-utilizing bacteria - identification of coryneform isolates. Appl. Environ. Microbiol. *53*: 1519–1524.

Tremblay, G., R. Lambert, H. Lebeuf and P. Dion. 1987b. Isolation of bacteria from soil and crown-gall tumors on the basis of their capacity for opine utilization. Phytoprotection. *68*: 35–42.

Trevisan, V. 1885. Caratteri di alcuni nuovi generi di Batteriacee. Atti Accad. Fis-Med-Stat. Milano (Ser.4) *3*: 92–107.

Trevisan, V. 1889. I generi e le specie delle batteriacee, Zanaboni and Gabuzzi, Milan. 1–35.

Trinick, M.J. 1973. Symbiosis between *Rhizobium* and the non-legume *Trema aspera*. Nature (Lond.) *244*: 459–460.

Trinick, M.J. 1976. *Rhizobium* symbiosis with a non-legume. Proceedings of the First International Symposium on Nitrogen Fixation, Washington State University Press, Pullman. pp. 507–517.

Trinkerl, M., A. Breunig, R. Schauder and H. König. 1990. *Desulfovibrio termitidis* sp. nov., a carbohydrate-degrading sulfate-reducing bacterium from the hindgut of a termite, *Heterotermes indicola* (Wasman). Syst. Appl. Microbiol. *13*: 372–377.

Trinkerl, M., A. Breunig, R. Schauder and H. König. 1991. *In* Validation of the publication of new names and new combinations previously effectively published outside the IJSB. List No. 36. Int. J. Syst. Bacteriol. *41*: 178–179.

Triplett, E.W., B.T. Breil and G.A. Splitter. 1994. Expression of *tfx* and sensitivity to the rhizobial peptide antibiotic trifolitoxin in a taxonomically distinct group of alpha-proteobacteria including the animal pathogen *Brucella abortus*. Appl. Environ. Microbiol. *60*: 4163–4166.

Triplett, E.W. and M.J. Sadowsky. 1992. Genetics of competition for nodulation of legumes. Annu. Rev. Microbiol. *46*: 399–428.

Trivett-Moore, N.L., W.D. Rawlinson, M. Yuen and G.L. Gilbert. 1997. *Helicobacter westmeadii* sp. nov., a new species isolated from blood cultures of two AIDS patients. J. Clin. Microbiol. *35*: 1144–1150.

Trollfors, B., J. Taranger, T. Lagergard, L. Lind, V. Sundh, G. Zackrisson, C.U. Lowe, W. Blackwelder and J.B. Robbins. 1995. A placebo-controlled trial of a pertussis toxoid vaccine. N. Engl. J. Med. *333*: 1045–1050.

Trotsenko, Y.A., N.V. Doronina and P. Hirsch. 1989. Genus *Blastobacter*. *In* Staley, Bryant, Pfennig and Holt (Editors), Bergey's Manual of Systematic Bacteriology, Vol. 3, The Williams & Wilkins Co., Baltimore. pp. 1963–1968.

Trotter, J.A., T.L. Kuhls, D.A. Pickett, S.R. Delarocha and D.F. Welch. 1990. Pneumonia caused by a newly recognized pseudomonad in a child with chronic granulomatous disease. J. Clin. Microbiol. *28*: 1120–1124.

Trowitzsch, W., L. Witte and H. Reichenbach. 1981. Geosmin from earthy-smelling cultures of *Nannocystis exedens* (*Myxobacterales*). FEMS Microbiol. Lett. *12*: 257–260.

Trowitzsch-Kienast, W., E. Forche, V. Wray, H. Reichenbach, E. Jurkiewicz, G. Hunsmann and G. Höfle. 1992. Phenalamide, neue HIV-1-Inhibitoren aus *Myxococcus stipitatus* Mx s40. Liebigs Ann. Chem. *1992*: 659–664.

Trowitzsch-Kienast, W., K. Gerth, V. Wray, H. Reichenbach and G. Höfle. 1993. Myxochromid A: ein hochungesättigtes ipopeptidlacton aus *Myxococcus virescens*. Liebigs Ann. Chem. *1993*: 1233–1237.

Trüper, H.G. and L. De' Clari. 1997. Taxonomic note: necessary correction of specific epithets formed as substantives (nouns) "in apposition". Int. J. Syst. Bacteriol. *47*: 908–909.

Trüper, H.G. and J.F. Imhoff. 1992. The genera *Rhodocyclus* and *Rubrivivax*. *In* Balows, Trüper, Dworkin, Harder and Schleifer (Editors), The Prokaryotes: A Handbook of Bacteria: Ecophysiology, Isolation, Identification, Applications, 2nd Ed., Vol. 3, Springer-Verlag, New York. pp. 2556–2561.

Tsai, C.M., W.H. Chen and P.A. Balakonis. 1998. Characterization of terminal NeuNAc-alpha2-3Galbeta1-4GlcNAc sequence in lipooligosaccharides of *Neisseria meningitidis*. Glycobiology. *8*: 359–365.

Tsai, H. and H.J. Hirsch. 1981. The primary structure of fulvocin-C from *Myxococcus fulvus*. Biochim. Biophys. Acta *667*: 213–217.

Tsai, W.M., S.H. Larsen and C.E. Wilde. 1989. Cloning and DNA sequence of the *omc* gene encoding the outer membrane protein macromolecular complex from *Neisseria gonorrhoeae*. Infect. Immun. *57*: 2653–2659.

Tsai, Y.-C., C.-P. Tseng, K.-M. Hsiao and L.-Y. Chen. 1988. Production and purification of D-aminoacylase from *Alcaligenes denitrificans* and taxonomic study of the strain. Appl. Environ. Microbiol. *54*: 984–989.

Tschantz, M.F., J.P. Bowman, P.R. Bienkowski, T.L. Donaldson, J.M. Strong-Gunderson, A.V. Palumbo and G.S. Sayler. 1995. Methanotrophic TCE biodegradation in a multi-stage bioreactor. Environ. Sci. Technol. *29*: 2073–2082.

Tschech, A. and G. Fuchs. 1987. Anaerobic degradation of phenol by pure cultures of newly isolated denitrifying pseudomonads. Arch. Microbiol. *148*: 213–217.

Tsien, H.C., G.A. Brusseau, R.S. Hanson and L.P. Wackett. 1989. Biodegradation of trichloroethylene by *Methylosinus trichosporium* OB3b. Appl. Environ. Microbiol. *55*: 3155–3161.

Tsien, H.C. and R.S. Hanson. 1992. Soluble methane monooxygenase component B gene probe for identification of methanotrophs that rapidly degrade trichloroethylene. Appl. Environ. Microbiol. *58*: 953–960.

Tsien, H.C., R. Lambert and H. Laudelout. 1968. Fine structure and the localization of the nitrite oxidizing system in *Nitrobacter winogradskyi*. Antonie Leeuwenhoek *34*: 483–494.

Tsopanakis, C. and J.H. Parish. 1976. Bacteriophage MX-1: properties of the phage and its structural proteins. J. Gen. Virol. *30*: 99–112.

Tsu, I.H., C.Y. Huang, J.L. Garcia, B.K.C. Patel, J.L. Cayol, L. Baresi and R.A. Mah. 1998. Isolation and characterization of *Desulfovibrio senezii* sp. nov., a halotolerant sulfate reducer from a solar saltern and phylogenetic confirmation of *Desulfovibrio fructosovorans* as a new species. Arch. Microbiol. *170*: 313–317.

Tsu, I.H., C.Y. Huang, J.L. Garcia, B.K.C. Patel, J.L. Cayol, L. Baresi and R.A. Mah. 1999. *In* Validation of the publication of new names and new combinations previously effectively published outside the IJSB. List No. 69. Int. J. Syst. Bacteriol. *49*: 341–342.

Tsubokura, A., H. Yoneda and H. Mizuta. 1999. *Paracoccus carotinifaciens* sp. nov., a new aerobic Gram-negative astaxanthin-producing bacterium. Int. J. Syst. Bacteriol. *49*: 277–282.

Tsuji, K., H.C. Tsien, R.S. Hanson, S.R. DePalma, R. Scholtz and S. LaRoche. 1990. 16S ribosomal RNA sequence analysis for determination of phylogenetic relationship among methylotrophs. J. Gen. Microbiol. *136*: 1–10.

Tsujimoto, H., N. Gotoh, J. Yamagishi, Y. Oyamada and T. Nishino. 1997. Cloning and expression of the major porin protein gene opcP of *Burkholderia* (formerly *Pseudomonas*) *cepacia* in *Escherichia coli*. Gene *186*: 113–118.

Tsuruhara, T., H. Urakami and A. Tamura. 1982. Surface morphology of *Rickettsia tsutsugamushi*-infected mouse fibroblasts. Acta Virol. *26*: 506–511.

Tudor, J.J. 1980. Chemical analysis of the outer cyst wall and inclusion material of *Bdellovibrio* bdellocysts. Curr. Microbiol. *4*: 251–256.

Tudor, J.J. and S.M. Bende. 1986. The outer cyst wall of *Bdellovibrio* bdellocysts is made *de novo* and not from preformed units from the prey wall. Curr. Microbiol. *13*: 185–190.

Tudor, J.J. and S.F. Conti. 1977a. Characterization of bdellocysts of *Bdellovibrio* sp. J. Bacteriol. *131*: 314–322.

Tudor, J.J. and S.F. Conti. 1977b. Ultrastructural changes during encystment and germination of *Bdellovibrio*. J. Bacteriol. *131*: 323–330.

Tudor, J.J. and S.F. Conti. 1978. Characterization of germination and activation of *Bdellovibrio* bdellocysts. J. Bacteriol. *133*: 130–138.

Tudor, J.J. and M.A. Karp. 1994. Translocation of an outer membrane protein into prey cytoplasmic membranes by bdellovibrios. J. Bacteriol. *176*: 948–952.

Tudor, J.J., M.P. McCann and I.A. Acrich. 1990. A new model for the penetration of prey cells by Bdellovibrios. J. Bacteriol. *172*: 2421–2426.

Tuhela, L., J.B. Robinson, S. Fishbain, D.A. Stahl and O.H. Tuovinen. 1997. Phylogenetic and *narG* analysis of a *Hyphomicrobium* isolate. Curr. Microbiol. *35*: 244–248.

Tuhela, L., J.B. Robinson and O.H. Tuovinen. 1998. Characterization of chemotactic responses and flagella of *Hyphomicrobium* strain W1-1B. J. Bacteriol. *180*: 3003–3006.

Tunail, N. and H.G. Schlegel. 1972. Phosphoenolpyruvate, a new inhibitor of glucose-6-phosphate dehydrogenase. Biochem. Biophys. Res. Commun. *49*: 1554–1560.

Tunail, N. and H.G. Schlegel. 1974. A new coryneform hydrogen bacterium: *Corynebacterium autotrophicum* strain 7C. I. Characterization of the wild type strain. Arch. Microbiol. *100*: 341–350.

Tung, S.Y. and T.H. Kuo. 1999. Requirement for phosphoglucose isomerase of *Xanthomonas campestris* in pathogenesis of citrus canker. Appl. Environ. Microbiol. *65*: 5564–5570.

Tuovinen, O.H. and D.P. Kelly. 1973. Studies on the growth of *Thiobacillus ferrooxidans*. I. Use of membrane filters and ferrous iron agar to determine viable numbers, and comparison with $^{14}CO_2$-fixation and iron oxidation as measures of growth. Arch. Mikrobiol. *88*: 285–298.

Turner, A.W. 1954. Bacterial oxidation of arsenite. I. Description of bacteria isolated from arsenical cattle-dipping fluids. Aust. J. Biol. Sci. *7*: 452–478.

Turner, L.R., J.C. Lara, D.N. Nunn and S. Lory. 1993. Mutations in the consensus ATP-binding sites of XcpR and PilB eliminate extracellular protein secretion and pilus biogenesis in *Pseudomonas aeruginosa*. J. Bacteriol. *175*: 4962–4969.

Turova, T.P., M.V. Burkal'tseva, E.S. Bulygina and V.M. Gorlenko. 1995. Phylogenetic position of freshwater erythrobacteria studied by 5S rRNA analysis. Mikrobiologiya *64*: 782–787.

Turton, J.A., T.C. Katsande, M.B. Matingo, W.K. Jorgensen, U. Ushewokunze-Obatolu and R.J. Dalgliesh. 1998. Observations on the use of *Anaplasma centrale* for immunization of cattle against anaplasmosis in Zimbabwe. Onderstepoort J. Vet. Res. *65*: 81–86.

Tuttle, D.M. and H.W. Scherp. 1952. Studies on carbon dioxide requirement of *Neisseria meningitidis*. J. Bacteriol. *64*: 171–182.

Tuttle, J.H., C.I. Randles and P.R. Dugan. 1968. Activity of microorganisms in acid mine water. I. Influence of acid water on aerobic heterotrophs of a normal stream. J. Bacteriol. *95*: 1495–1503.

Tyler, D.D. and L.K. Nakamura. 1971. Conditions for production of 3-ketomaltose from *Agrobacterium tumefaciens*. Appl. Microbiol. *21*: 175–180.

Tyler, M.E., J.R. Milam, R.L. Smith, S.C. Schank and D.A. Zuberer. 1979. Isolation of *Azospirillum* from diverse geographic regions. Can. J. Microbiol. *25*: 693–697.

Tyler, P.A. and K.C. Marshall. 1967. Pleomorphy in stalked, budding bacteria. J. Bacteriol. *93*: 1132–1136.

Tyler, S.D., K.R. Rozee and W.M. Johnson. 1996. Identification of IS1356, a new insertion sequence, and its association with IS402 in epidemic strains of *Burkholderia cepacia* infecting cystic fibrosis patients. J. Clin. Microbiol. *34*: 1610–1616.

Tyler, S.D., C.A. Strathdee, K.R. Rozee and W.M. Johnson. 1995. Oligonucleotide primers designed to differentiate pathogenic pseudomonads on the basis of the sequencing of genes coding for 16S–23S rRNA internal transcribed spacers. Clin. Diagn. Lab. Immunol. *2*: 448–453.

Tyzzer, E.E. 1938. *Cytoectes microti* N.G., (n. sp.) a parasite developing in granulocytes and infective for small rodents. Parasitology. *30*: 242–257.

Tyzzer, E.E. 1942. A comparison study of Grahamellae, Haemobartonellae

and Eperythrozoa in small mammals. Proc. Am. Phil. Soc. *85*: 359–398.

Uchida, T., F. Mahara, Y. Tsuboi and A. Oya. 1985. Spotted fever group rickettsiosis in Japan. Jpn. J. Med. Sci. Biol. *38*: 151–153.

Uchida, T., T. Uchiyama, K. Kumano and D.H. Walker. 1992. *Rickettsia japonica* sp. nov., the etiological agent of spotted fever group rickettsiosis in Japan. Int. J. Syst. Bacteriol. *42*: 303–305.

Uchino, Y., T. Hamada and A. Yokota. 2002. Proposal of *Pseudorhodobacter ferrugineus* gen. nov., comb. nov., for a non-photosynthetic marine bacterium, *Agrobacterium ferrugineum*, related to the genus *Rhodobacter*. J. Gen. Appl. Microbiol. *48*: 309–319.

Uchino, Y., A. Hirata, A. Yokota and J. Sugiyama. 1998. Reclassification of marine *Agrobacterium* species: proposals of *Stappia stellulata* gen. nov., comb. nov., *Stappia aggregata* sp. nov., nom. rev., *Ruegeria atlantica* gen. nov., comb. nov., *Ruegeria gelatinovora* comb. nov., *Ruegeria algicola* comb. nov., and *Ahrensia kieliense* gen. nov., sp. nov., nom. rev. J. Gen. Appl. Microbiol. *44*: 201–210.

Uchino, Y., A. Hirata, A. Yokota and J. Sugiyama. 1999. *In* Validation of the publication of new names and new combinations previously effectively published outside the IJSB. List No. 68. Int. J. Syst. Bacteriol. *49*: 1–3.

Uchino, Y., A. Yokota and J. Sugiyama. 1997. Phylogenetic position of the marine subdivision of *Agrobacterium* species based on 16S rRNA sequence analysis. J. Gen. Appl. Microbiol. *43*: 243–247.

Uchiyama, T. 1999. Role of major surface antigens of *Rickettsia japonica* in the attachment to host cells. *In* Brouqui and Raoult (Editors), Rickettsiae and Rickettsial Diseases at the Turn of the Third Millennium, Elsevier, Paris. pp. 182–188.

Uebayasi, M., S. Kawamura, N. Tomizuka and A. Kamibayashi. 1985. Comparison of the key enzymes of *Hyphomicrobium* sp. 53-49 under various growth conditions: aerobic, denitrifying and autotrophic. Agric. Biol. Chem. *49*: 1799–1808.

Uebayasi, M., S. Kawamura, N. Tomizuka, A. Kamibayashi and K. Tonomura. 1984. Incorporation of [^{14}C]methanol and [^{14}C]bicarbonate by *Hyphomicrobium* sp. 53-49 under various growth conditions: aerobic, denitrifying and autotrophic. Agric. Biol. Chem. *48*: 1395–1404.

Uebayasi, M., N. Tomizuka, A. Kamibayashi and K. Tonomura. 1981. Autotrophic growth of a *Hyphomicrobium* sp. and its hydrogenase activity. Agric. Biol. Chem. *45*: 1783–1790.

Ueda, H., H. Nakajima, Y. Hiro, T. Goto and M. Okuhara. 1994. Action of FR901228, a novel antitumor bicyclic depsipeptide produced by *Chromobacterium violaceum* No. 968, on Ha-ras transformed NIH3T3 cells. Biosci. Biotechnol. Biochem. *58*: 1579–1583.

Ueda, T., Y. Suga, N. Yahiro and T. Matsuguchi. 1995. Remarkable N_2-fixing bacterial diversity detected in rice roots by molecular evolutionary analysis of *nifH* gene sequences. J. Bacteriol. *177*: 1414–1417.

Uetz, T.A. 1992. Biochemistry of nitrilotriacetate degradation in obligately aerobic, Gram-negative bacteria, Swiss Federal Institute of Technology Zurich. pp. 98.

Uetz, T. and T. Egli. 1993. Characterization of an inducible, membrane-bound iminodiacetate dehydrogenase from *Chelatobacter heintzii* ATCC29600. Biodegradation *3*: 423–434.

Uetz, T., R. Schneider, M. Snozzi and T. Egli. 1992. Purification and characterization of a two-component monooxygenase that hydroxylates nitrilotriacetate from Chelatobacter strain ATCC 29600. J. Bacteriol. *174*: 1179–1188.

Uffen, R.L. 1973. Growth properties of *Rhodospirillum rubrum* mutants and fermentation of pyruvate in anaerobic, dart conditions. J. Bacteriol. *116*: 874–884.

Uffen, R.L. 1976. Anaerobic growth of a *Rhodopseudomonas* species in the dark with carbon monoxide as sole carbon and energy substrate. Proc. Natl. Acad. Sci. U.S.A. *73*: 3298–3302.

Uffen, R.L. 1983. Metabolism of carbon monoxide by *Rhodopseudomonas gelatinosa*: cell growth and properties of the oxidation system. J. Bacteriol. *155*: 956–965.

Uhlenbusch, I., H. Sahm and G.A. Sprenger. 1991. Expression of an L-alanine dehydrogenase gene in *Zymomonas mobilis* and excretion of L-alanine. Appl. Environ. Microbiol. *57*: 1360–1366.

Uhlig, H., K. Karbaum and A. Steudel. 1986. *Acetobacter methanolicus* sp. nov., an acidophilic facultatively methylotrophic bacterium. Int. J. Syst. Bacteriol. *36*: 317–322.

Ui, M. 1998. The multiple biological activities of pertussis toxin. *In* Wardlay and Parton (Editors), Pathogenesis and Immunity in Pertussis, Wiley, New York. 121–146.

Uilenberg, G., C.J. van Vorstenbosch and N.M. Perie. 1979. Blood parasites of sheep in the Netherlands. I. *Anaplasma mesaeterum* sp.n. (*Rickettsiales, Anaplasmataceae*). Tijdschr. Diergeneeskd. *104*: 14–22.

Umali-Garcia, M., D.H. Hubbell, M.H. Gaskins and F.B. Dazzo. 1980. Association of *Azospirillum brasilense* with grass roots. Appl. Environ. Microbiol. *39*: 219–226.

Umbreit, W.W., R.H. Burriss and J.F. Stauffer. 1972. Manometric techniques, 5th Ed., Burgess Publishing Co., Minneapolis.

Unger, L., S.F. Ziegler, G.A. Huffman, V.C. Knauf, R. Peet, L.W. Moore, M.P. Gordon and E.W. Nester. 1985. New class of limited-host-range *Agrobacterium* mega-tumor-inducing plasmids lacking homology to the transferred DNA of a wide-host-range, tumor-inducing plasmid. J. Bacteriol. *164*: 723–730.

Unz, F.R. 1984. Genus IV. *Zoogloea. In* Krieg and Holt (Editors), Bergey's Manual of Systematic Bacteriology, 1st Ed., Vol. 1, The Williams & Wilkins Co., Baltimore. pp. 214–219.

Unz, R.F. 1971. Neotype strain of *Zoogloea ramigera* Itzigsohn. Request for an opinion. Int. J. Syst. Bacteriol. *21*: 91–99.

Unz, R.F. and N.C. Dondero. 1967a. The predominant bacteria in natural zoogloeal colonies. I. Isolation and identification. Can. J. Microbiol. *13*: 1671–1682.

Unz, R.F. and N.C. Dondero. 1967b. The predominant bacteria in natural zoogloeal colonies. II. Physiology and nutrition. Can. J. Microbiol. *13*: 1683–1694.

Unz, R.F. and S.R. Farrah. 1972. Use of aromatic compounds for growth and isolation of *Zoogloea*. Appl. Microbiol. *23*: 524–530.

Unz, R.F. and S.R. Farrah. 1976a. Exopolymer production and flocculation by *Zoogloea* MP6. Appl. Environ. Microbiol. *31*: 623–626.

Unz, R.F. and S.R. Farrah. 1976b. Observations on the formation of wastewater zoogloeae. Water Res. *10*: 665–671.

Upadhyay, R.S. and R.K. Jayaswal. 1992. *Pseudomonas cepacia* causes mycelial deformities and inhibition of conidiation in phytopathogenic fungi. Curr. Microbiol. *24*: 181–187.

Upadhyay, R.S., L. Visintin and R.K. Jayaswal. 1991. Environmental factors affecting the antagonism of *Pseudomonas cepacia* against *Trichoderma viride*. Can. J. Microbiol. *37*: 880–884.

Urakami, H., N. Ohashi, T. Tsuruhara and A. Tamura. 1986. Characterization of polypeptides in *Rickettsia tsutsugamushi*: Effect of preparative conditions on migration of polypeptides in polyacrylamide gel electrophoresis. Infect. Immun. *51*: 948–952.

Urakami, T., H. Araki and H. Kobayashi. 1990a. Isolation and identification of tetramethylammonium-biodegrading bacteria. J. Ferment. Bioeng. *70*: 41–44.

Urakami, T., H. Araki and K. Komagata. 1995a. Characteristics of newly isolated *Xanthobacter* strains and fatty acid compositions and quinone systems in yellow-pigmented hydrogen-oxidizing bacteria. Int. J. Syst. Bacteriol. *45*: 863–867.

Urakami, T., H. Araki, H. Oyanagi, K.-I. Suzuki and K. Komagata. 1992. Transfer of *Pseudomonas aminovorans* (den Dooren de Jong 1926) to *Aminobacter* gen. nov. as *Aminobacter aminovorans* comb. nov. and description of *Aminobacter aganoensis* sp. nov. and *Aminobacter niigataensis* sp. nov. Int. J. Syst. Bacteriol. *42*: 84–92.

Urakami, T., H. Araki, H. Oyanagi, K. Suzuki and K. Komagata. 1990b. *Paracoccus aminophilus* sp. nov. and *Paracoccus aminovorans* sp. nov., which utilize N,N-dimethylformamide. Int. J. Syst. Bacteriol. *40*: 287–291.

Urakami, T., H. Araki, K.I. Suzuki and K. Komagata. 1993. Further studies of the genus *Methylobacterium* and description of *Methylobacterium aminovorans*, sp. nov. Int. J. Syst. Bacteriol. *43*: 504–513.

Urakami, T., C. Ito-Yoshida, H. Araki, T. Kijima, K.-I. Suzuki and K. Komagata. 1994. Transfer of *Pseudomonas plantarii* and *Pseudomonas glumae*

to *Burkholderia* as *Burkholderia* spp. and description of *Burkholderia vandii* sp. nov. Int. J. Syst. Bacteriol. *44*: 235–245.

Urakami, T., H. Kobayashi and H. Araki. 1990c. Isolation and identification of N, N-dimethylformamide-biodegrading bacteria. J. Ferment. Bioeng. *70*: 45–47.

Urakami, T. and K. Komagata. 1979. Cellular fatty acid composition and coenzyme Q system in gram-negative methanol-utilizing bacteria. J. Gen. Appl. Microbiol. *25*: 343–360.

Urakami, T. and K. Komagata. 1981. Electrophoretic comparison of enzymes in the Gram-negative methanol-utilizing bacteria. J. Gen. Appl. Microbiol. *27*: 381–404.

Urakami, T. and K. Komagata. 1984. *Protomonas*, new genus of facultatively methylotrophic bacteria. Int. J. Syst. Bacteriol. *34*: 188–201.

Urakami, T. and K. Komagata. 1986a. Emendation of *Methylobacillus* Yordy and Weaver 1977, a genus for methanol-utilizing bacteria. Int. J. Syst. Bacteriol. *36*: 502–511.

Urakami, T. and K. Komagata. 1986b. Occurrence of isoprenoid compounds in gram-negative methanol-utilizing, methane-utilizing, and methylamine-utilizing bacteria. J. Gen. Appl. Microbiol. *32*: 317–341.

Urakami, T. and K. Komagata. 1987a. Cellular fatty acid composition with special reference to the existence of hydroxy fatty acids in gram-negative methanol-utilizing, methane-utilizing, and methylamine-utilizing bacteria. J. Gen. Appl. Microbiol. *33*: 135–165.

Urakami, T. and K. Komagata. 1987b. Characterization and identification of methanol-utilizing Hyphomicrobium strains and a comparison with species of *Hyphomonas* and *Rhodomicrobium*. J. Gen. Appl. Microbiol. *33*: 521–542.

Urakami, T. and K. Komagata. 1988. Cellular fatty-acid composition with special reference to the existence of hydroxy fatty-acids, and the occurrence of squalene and sterols in species of *Rhodospirillaceae* genera and *Erythrobacter longus*. J. Gen. Appl. Microbiol. *34*: 67–84.

Urakami, T., H. Oyanagi, H. Araki, K.I. Suzuki and K. Komagata. 1990d. Recharacterization and emended description of the genus *Mycoplana* and description of two new species, *Mycoplana ramosa* and *Mycoplana segnis*. Int. J. Syst. Bacteriol. *40*: 434–442.

Urakami, T., J. Sasaki, K.I. Suzuki and K. Komagata. 1995b. Characterization and description of *Hyphomicrobium denitrificans* sp. nov. Int. J. Syst. Bacteriol. *45*: 528–532.

Urakami, T., J. Tamaoka, K.-I. Suzuki and K. Komagata. 1989a. *Acidomonas* gen. nov., incorporating *Acetobacter methanolicus* as *Acidomonas methanolica* comb. nov. Int. J. Syst. Bacteriol. *39*: 50–55.

Urakami, T., J. Tamaoka, K.-I. Suzuki and K. Komagata. 1989b. *Paracoccus alcaliphilus* sp. nov., an alkaliphilic and facultatively methylotrophic bacterium. Int. J. Syst. Bacteriol. *39*: 116–121.

Urakami, T., J. Tamaoka and K. Komagata. 1985. DNA-base composition and DNA-DNA homologies of methanol-utilizing bacteria. J. Gen. Appl. Microbiol. *31*: 243–253.

Urbance, J.W., B.J. Bratina, S.F. Stoddard and T.M. Schmidt. 2001. Taxonomic characterization of *Ketogulonigenium vulgare* gen. nov., sp. nov. and *Ketogulonigenium robustum* sp. nov., which oxidize L-sorbose to 2-keto-L-gulonic acid. Int. J. Syst. Evol. Microbiol. *51*: 1059–1070.

Ureta, A., B. Alvarez, A. Ramon, M.A. Vera and G. Martinez-Drets. 1995. Identification of *Acetobacter diazotrophicus*, *Herbaspirillum seropedicae* and *Herbaspirillum rubrisubalbicans* using biochemical and genetic criteria. Plant Soil *172*: 271–277.

Ursing, J.B., R.A. Rosselló-Mora, E. Garcia-Valdes and J. Lalucat. 1995. Taxonomic note-a pragmatic approach to the nomenclature of phenotypically similar genomic groups. Int. J. Syst. Bacteriol. *45*: 604.

Utsumi, R., T. Yagi, S. Katayama, K. Katsuragi, K. Tachibana, H. Toyoda, S. Ouchi, K. Obata, Y. Shibano and M. Noda. 1991. Molecular cloning and characterization of the fusaric acid resistance gene from *Pseudomonas cepacia*. Agric. Biol. Chem. *55*: 1913–1918.

Vahle, C. 1910. Vergleichende ntersuchungen über die Myxobakteriazeen und Bakteriazeen. Zentbl. Bakteriol. Abt 2 *25*: 178–260.

Vainshtein, M., H. Hippe and R.M. Kroppenstedt. 1992. Cellular fatty acid composition of *Desulfovibrio* species and its use in classification of sulfate-reducing bacteria. Syst. Appl. Microbiol. *15*: 554–566.

Väisänen, O.M., E.L. Nurmiaho-Lassila, S.A. Marmo and M.S. Salkinoja-

Salonen. 1994. Structure and composition of biological slimes on paper and board machines. Appl. Environ. Microbiol. *60*: 641–653.

Valla, S., D.H. Coucheron and J. Kjosbakken. 1985. Conjugative transfer of the naturally occurring plasmids of *Acetobacter xylinum* by IncP-plasmid-mediated mobilization. J. Bacteriol. *165*: 336–339.

Vallis, A.J., V. Finck-Barbancon, T.L. Yahr and D.W. Frank. 1999. Biological effects of *Pseudomonas aeruginosa* type III-secreted proteins on CHO cells. Infect. Immun. *67*: 2040–2044.

Valls, M., S. Atrian, V. de Lorenzo and L.A. Fernandez. 2000. Engineering a mouse metallothionein on the cell surface of *Ralstonia eutropha* CH34 for immobilization of heavy metals in soil. Nat. Biotechnol. *18*: 661–665.

Van Beeumen, J. and J. De Ley. 1968. Hexopyranoside: cytochrome *c* oxidoreductase from *Agrobacterium tumefaciens*. Eur. J. Biochem. *6*: 331–343.

Van Beeumen, J., P. Tempst, P. Stevens, D. Bral, J. Van Damme and J. De Ley. 1980. Cytochromes *c* of two different sequence classes in *Agrobacterium tumefaciens*. *In* Peeters (Editor), Protides of the Biological Fluids 28, Pergamon Press, Oxford. 69–74.

van Berkum, P., D. Beyene, G. Bao, T.A. Campbell and B.D. Eardly. 1998. *Rhizobium mongolense* sp. nov. is one of three rhizobial genotypes identified which nodulate and form nitrogen-fixing symbioses with *Medicago ruthenica* [(L.) Ledebour]. Int. J. Syst. Bacteriol. *48*: 13–22.

van Bruggen, A.H.C., K.N. Jochimsen and P.R. Brown. 1990. *Rhizomonas suberifaciens* gen. nov., sp. nov., the causal agent of corky root of lettuce. Int. J. Syst. Bacteriol . *40*: 175–188.

Van Camp, G., Y. Van de Peer, S. Nicolai, J.-M. Neefs, P. Vandamme and R. De Wachter. 1993. Structure of 16S and 23S ribosomal RNA genes in *Campylobacter* species: phylogenetic analysis of the genus *Campylobacter* and presence of internal transcribed spacers. Syst. Appl. Microbiol. *16*: 361–368.

van Damme, P.A., A.G. Johannes, H.C. Cox and W. Berends. 1960. On toxoflavin, the yellow poison of *Pseudomonas cocovenenans*. Recl. Trav. Chim. Pays-Bas Belg. *79*: 255–267.

Van de Peer, Y. and R. de Wachter. 1994. TREECON for windows: A software package for the construction and drawing of evolutionary trees for the microsoft windows environment. Comp. Appl. Biosci. *10*: 569–570.

Van Den Akker, W.M. 1998. Lipopolysaccharide expression within the genus *Bordetella*: influence of temperature and phase variation. Microbiology *144*: 1527–1535.

Van den Mooter, M. and J. Swings. 1990. Numerical analysis of 295 phenotypic features of 266 *Xanthomonas* strains and related strains and an improved taxonomy of the genus. Int. J. Syst. Bacteriol. *40*: 348–369.

Van den Tweel, W.J.J., R.J.J. Janssens and J.A.M. De Bont. 1986. Degradation of 4-hydroxyphenylacetate by *Xanthobacter* 124x physiological resemblance with other Gram-negative bacteria. Journal of Microbiology *52*: 309–318.

Van den Tweel, W.J.J., J.B. Kok and J.A.M. de Bont. 1987. Reductive dechlorination of 2,4-dichlorobenzoate to 4- chlorobenzoate and hydrolytic dehalogenation of 4-chlorobenzoate, 4-bromobenzoate, and 4-iodobenzoate by *Alcaligenes denitrificans* Ntb-1. Appl. Environ. Microbiol. *53*: 810–815.

Van den Wijngaard, A.J., K.W. van der Kamp, J. van der Ploeg, F. Pries, B. Kazemier and D.B. Janssen. 1992. Degradation of 1,2-dichloroethane by *Ancylobacter aquaticus* and other facultative methylotrophs. Appl. Environ. Microbiol. *58*: 976–983.

Van der Drift, C., F.E. de Windt and H.J. Doddema. 1981. Metabolism of allantoin in *Hyphomicrobium* species. Antonie Van Leeuwenhoek *47*: 565–570.

Van der Ende, A., C.T.P. Hopman and J. Dankert. 1999. Deletion of *porA* by recombination between clusters of repetitive extragenic palindromic sequences in *Neisseria meningitidis*. Infect. Immun. *67*: 2928–2934.

van der Ende, A., C.T. Hopman, S. Zaat, B.B. Essink, B. Berkhout and J. Dankert. 1995. Variable expression of class 1 outer membrane protein in *Neisseria meningitidis* is caused by variation in the spacing between the −10 and −35 regions of the promoter. J. Bacteriol. *177*: 2475–2480.

Van der Heyden, J.H., M.A. Catchpole, W.J. Paget and A. Stroobant. 2000. Trends in gonorrhoea in nine western European countries, 1991–6. European Study Group. Sex. Transm. Infect. *76*: 110–116.

van der Maarel, M.J.E.C., S. van Bergeijk, A.F. van Werkhoven, A.M. Laverman, W.G. Meijer, W.T. Stam and T.A. Hansen. 1996. Cleavage of dimethylsulfoniopropionate and reduction of acrylate by *Desulfovibrio acrylicus* sp. nov. Arch. Microbiol. *166*: 109–115.

van der Maarel, M.J.E.C., S. van Bergeijk, A.F. van Werkhoven, A.M. Laverman, W.G. Meijer, W.T. Stam and T.A. Hansen. 1997. *In* Validation of the publication of new names and new combinations previously effectively published outside the IJSB. List No. 60. Int. J. Syst. Bacteriol. *47*: 242.

van der Oost, J., A.P. de Boer, J.W. de Gier, W.G. Zumft, A.H. Stouthamer and R.J. van Spanning. 1994. The heme-copper oxidase family consists of three distinct types of terminal oxidases and is related to nitric oxide reductase. FEMS Microbiol. Lett. *121*: 1–9.

Van der Schaaf, A. and M. Rosa. 1940. Brucellosis oncho-cerciasis in verband met een chronisch gewrichtslijden bij runderen. Ned.-lnd. Blad. Diergeneesk. *52*: 1–20.

Van Der Voort, E.R., P. Van Der Ley, J. Van Der Biezen, S. George, O. Tunnela, H. Van Dijken, B. Kuipers and J. Poolman. 1996. Specificity of human bactericidal antibodies against PorA p1.7,16 induced with a hexavalent meningococcal outer membrane vesicle vaccine. Infect. Immun. *64*: 2745–2751.

van Der Zee, A., H. Groenendijk, M. Peeters and F.R. Mooi. 1996. The differentiation of *Bordetella parapertussis* and *Bordetella bronchiseptica* from humans and animals as determined by DNA polymorphism mediated by two different insertion sequence elements suggests their phylogenetic relationship. Int. J. Syst. Bacteriol. *46*: 640–647.

van Der Zee, A., F. Mooi, J. Van Embden and J. Musser. 1997. Molecular evolution and host adaptation of *Bordetella* spp.: phylogenetic analysis using multilocus enzyme electrophoresis and typing with three insertion sequences. J. Bacteriol. *179*: 6609–6617.

van Dijck, P., M. Delmee, H. Ezzedine, A. Deplano and M.J. Struelens. 1995. Evaluation of pulsed-field gel electrophoresis and rep-PCR for the epidemiological analysis of *Ochrobactrum anthropi* strains. Eur. J. Clin. Microbiol. Infect. Dis. *14*: 1099–1102.

van Doorn, L.J., C. Figueiredo, R. Sanna, S. Pena, P. Midolo, E.K.W. Ng, J.C. Atherton, M.J. Blaser and W.G.V. Quint. 1998a. Expanding allelic diversity of *Helicobacter pylori vacA*. J. Clin. Microbiol. *36*: 2597–2603.

van Doorn, L.J., B.A.J. Giesendorf, R. Bax, B.A.M. van der Zeijst, P. Vandamme and W.G. Quint. 1997. Molecular discrimination between *Campylobacter jejuni, Campylobacter coli, Campylobacter lari* and *Campylobacter upsaliensis* by polymerase chain reaction based on a novel putative GTPase gene. Mol. Cell. Probes *11*: 177–185.

van Doorn, N.E.M., F. Namavar, J.G. Kusters, E.P. van Rees, E.J. Kuipers and J. de Graaff. 1998b. Genomic DNA fingerprinting of clinical isolates of *Helicobacter pylori* by REP-PCR and restriction fragment end-labelling. FEMS Microbiol. Lett. *160*: 145–150.

van Elsas, J.D., P. Kastelein, P.M. de Vries and L.S. van Overbeek. 2001. Effects of ecological factors on the survival and physiology of *Ralstonia solanacearum* bv. 2 in irrigation water. Can. J. Microbiol. *47*: 842–854.

Van Ert, M. and J.T. Staley. 1971. Gas-vacuolated strains of *Microcyclus aquaticus*. J. Bacteriol. *108*: 236–240.

Van Etterijck, R., J. Breynaert, H. Revets, T. Devreker, Y. Vandenplas, P. Vandamme and S. Lauwers. 1996. Isolation of *Campylobacter concisus* from feces of children with and without diarrhea. J. Clin. Microbiol. *34*: 2304–2306.

Van Ginkel, C.G. and J.A.M. De Bont. 1986. Isolation and characterization of alkene-utilizing *Xanthobacter* spp. Arch. Microbiol. *145*: 403–407.

Van Ginkel, C.G., H.G.J. Welten and J.A.M. De Bont. 1986. Epoxidation of alkenes by alkene-grown *Xanthobacter* spp. Appl. Microbiol. Biotechnol. *24*: 334–337.

van Heeckeren, A.M., Y. Rikihisa, J. Park and R. Fertel. 1993. Tumor necrosis factor alpha, interleukin-1 alpha, interleukin-6, and prostaglandin E2 production in murine peritoneal macrophages infected with *Ehrlichia risticii*. Infect. Immun. *61*: 4333–4337.

Van Hove, C. 1976. Bacterial leaf symbiosis and nitrogen fixation. *In* Nutman (Editor), Symbiotic Nitrogen Fixation in Plants, Cambridge University Press, Cambridge. pp. 551–560.

van Iterson, W. 1958. *Gallionella ferruginea* Ehrenberg in a different light. Verh. K. Ned. Akad. Wet. Afd. Natuurkd. Tweede Reeks. *52*: 1–185.

Van Keer, C., K. Kersters and J. De Ley. 1976. L-Sorbose metabolism in *Agrobacterium tumefaciens*. Antonie van Leeuwenhoek J. Microbiol. Serol. *42*: 13–24.

Van Kuijk, B.M., E. Schlösser and A.M. Stams. 1998. Investigation of the fumarate metabolism of the syntrophic propionate-oxidizing bacterium strain MPOB. Arch. Microbiol. *169*: 346–352.

Van Kuijk, B.M. and A.J.M. Stams. 1995. Sulfate reduction by a syntrophic propionate-oxidizing bacterium. Antonie Leeuwenhoek *68*: 293–296.

Van Neerven, A. and J.T. Staley. 1988. Mixed acid fermentation by a budding prosthecate bacterium. Arch. Microbiol. *149*: 335–338.

van Niel, C.B. 1944. The culture, general physiology, morphology and classification of the nonsulfur purple and brown bacteria. Bacteriol. Rev. *8*: 1–118.

van Niel, E.W., K.J. Braber, L.A. Robertson and J.G. Kuenen. 1992. Heterotrophic nitrification and aerobic denitrification in *Alcaligenes faecalis* strain TUD. Antonie Leeuwenhoek *62*: 231–237.

Van Outryve, M.F., F. Gosselé, K. Kersters and J. Swings. 1988. The composition of the rhizosphere of chicory (*Cichorium intybus* var. *foliosum* Hegi). Can. J. Microbiol. *34*: 1203–1208.

van Overbeek, L.S., M. Cassidy, J. Kozdroj, J.T. Trevors and J.D. van Elsas. 2002. A polyphasic approach for studying the interaction between *Ralstonia solanacearum* and potential control agents in the tomato phytosphere. J. Microbiol. Methods *48*: 69–86.

Van Pee, W., M. Vanlaar and J. Swings. 1974. The nutrition of *Zymomonas*. Acad. R. Sci. Outre-Mer (Brussels) Bull. Séances. *2*: 206–211.

van Putten, J.P.M., T.D. Duensing and J. Carlson. 1998. Gonococcal invasion of epithelial cells driven by P.IA, a bacterial ion channel with GTP binding properties. J. Exp. Med. *188*: 941–952.

van Putten, J.P.M. and S.M. Paul. 1995. Binding of syndecan-like cell surface proteoglycan receptors is required for *Neisseria gonorrhoeae* entry into human mucosal cells. EMBO J. *14*: 2144–2154.

van Rhijn, P. and J. Vanderleyden. 1995. The *Rhizobium*-plant symbiosis. Microbiol. Rev. *59*: 124–142.

Van Rossum, D., F.P. Schuurmans, M. Gillis, A. Muyotcha, H.W. Van Verseveld, A.H. Stouthamer and F.C. Boogerd. 1995. Genetic and phenetic analyses of *Bradyrhizobium* strains nodulating peanut (*Arachis hypogaea* L.) roots. Appl. Environ. Microbiol. *61*: 1599–1609.

van Schie, P.M. and L.Y. Young. 1998. Isolation and characterization of phenol-degrading denitrifying bacteria. Appl. Environ. Microbiol. *64*: 2432–2438.

Van Spanning, R.J., W.N. Reijnders and A.H. Stouthamer. 1995. Integration of heterologous DNA into the genome of *Paracoccus denitrificans* is mediated by a family of IS1248-related elements and a second type of integrative recombination event. J. Bacteriol. *177*: 4772–4778.

Van Spanning, R.J., C.W. Wansell, W.N. Reijnders, N. Harms, J. Ras, L.F. Oltmann and A.H. Stouthamer. 1991. A method for introduction of unmarked mutations in the genome of *Paracoccus denitrificans*: construction of strains with multiple mutations in the genes encoding periplasmic cytochromes c_{550}, c_{551i}, and c_{553i}. J. Bacteriol. *173*: 6962–6970.

van Veen, R.J.M., H. den Dulk-Ras, T. Bisseling, R.A. Schilperoort and P.J.J. Hooykaas. 1988. Crown gall tumor and root nodule formation by the bacterium *Phylobacterium myrsinacearum* after the introduction of an *Agrobacterium* Ti plasmid or a *Rhizobium* Sym plasmid. Mol. Plant Microbe Interact. *1*: 231–234.

van Veen, W.L., E.G. Mulder and M.H. Deinema. 1978. The *Sphaerotilus-Leptothrix* group of bacteria. Microbiol. Rev. *42*: 329–356.

van Verseveld, H.W., J.P. Boon and A.H. Stouthamer. 1979. Growth yields and the efficiency of oxidative phosphorylation of *Paracoccus denitrificans* during two-(carbon) substrate-limited growth: evidence for the induction of site 3-phosphorylation by methanol in heterotrophically grown cells. Arch. Microbiol. *121*: 213–223.

van Verseveld, H.W., J.A. de Hollander, J. Frankena, M. Braster, F.J. Leeuwerik and A.H. Stouthamer. 1986. Modeling of microbial substrate conversion, growth and product formation in a recycling fermentor. Antonie Leeuwenhoek *52*: 325–342.

van Verseveld, H.W., K. Krab and A.H. Stouthamer. 1981. Proton pump coupled to cytochrome *c* oxidase in *Paracoccus denitrificans*. Biochim. Biophys. Acta *635*: 525–534.

van Verseveld, H.W. and A.H. Stouthamer. 1978. Growth yields and the efficiency of oxidative phosphorylation during autotrophic growth of *Paracoccus denitrificans* on methanol and formate. Arch. Microbiol. *118*: 21–26.

van Verseveld, H.W. and A.H. Stouthamer. 1992. The Genus *Paracoccus*. *In* Balows, Trüper, Dworkin, Harder and Schleifer (Editors), The Prokaryotes: A Handbook of Bacteria: Ecophysiology, Isolation, Identification, Applications, 2nd Ed., Vol. 3, Springer-Verlag, New York. pp 2321–2334.

van Wegen, R.J., S.Y. Lee and A.P.J. Middelberg. 2001. Metabolic and kinetic analysis of poly(3-hydroxybutyrate) production by recombinant *Escherichia coli*. Biotechnol. Bioeng. *74*: 70–80.

Van Zyl, F.G.H., B.W. Strijdom and J.L. Staphorst. 1986. Susceptibility of *Agrobacterium tumefaciens* strains to 2 agrocin producing *Agrobacterium* strains. Appl. Environ. Microbiol. *52*: 234–238.

Vanbrabant, J., P. de Vos, M. Vancanneyt, J. Liessens, W. Verstraete and K. Kersters. 1993. Isolation and identification of autotrophic and heterotrophic bacteria from an autohydrogenotrophic pilot-plant for denitrification of drinking water. Syst. Appl. Microbiol. *16*: 471–482.

Vancanneyt, M., F. Schut, C. Snauwaert, J. Goris, J. Swings and J.C. Gottschal. 2001. *Sphingomonas alaskensis* sp. nov., a dominant bacterium from a marine oligotrophic environment. Int. J. Syst. Evol. Microbiol. *51*: 73–79.

Vancanneyt, M., P. Vandamme and K. Kersters. 1995. Differentiation of *Bordetella pertussis*, *B. parapertussis*, and *B. bronchiseptica* by whole-cell protein electrophoresis and fatty acid analysis. Int. J. Syst. Bacteriol. *45*: 843–847.

Vancanneyt, M., S. Witt, W.R. Abraham, K. Kersters and H.L. Fredrickson. 1996. Fatty acid content in whole cell hydrolysates and phospholipid fractions of pseudomonads: a taxonomic evaluation. Syst. Appl. Microbiol. *19*: 528–540.

Vance, C.P. 1998. Legume symbiotic nitrogen fixation: agronomic aspects. *In* Spaink, Kondorosi and Hooykaas (Editors), The *Rhizobiaceae*: Molecular Biology of Model Plant-Associated Bacteria, Kluwer Academic, Dordrecht ; Boston. 509–530.

Vandamme, P., M.I. Daneshvar, F.E. Dewhirst, B.J. Paster, K. Kersters, H. Goossens and C.W. Moss. 1995a. Chemotaxonomic analyses of *Bacteroides gracilis* and B*acteroides ureolyticus* and reclassification of *B.gracilis* as *Campylobacter gracilis* comb. nov. Int. J. Syst. Bacteriol. *45*: 145–152.

Vandamme, P. and J. De Ley. 1991. Proposal for a new family, *Campylobacteraceae*. Int. J. Syst. Bacteriol. *41*: 451–455.

Vandamme, P., D. Dewettinck and K. Kersters. 1992a. Application of numerical analysis of electrophoretic protein profiles for the identification of thermophilic campylobacters. Syst. Appl. Microbiol. *15*: 402–408.

Vandamme, P., E. Falsen, B. Pot, B. Hoste, K. Kersters and J. De Ley. 1989. Identification of EF group 22 campylobacters from gastroenteritis cases as *Campylobacter concisus*. J. Clin. Microbiol. *27*: 1775–1781.

Vandamme, P., E. Falsen, R. Rossau, B. Hoste, P. Segers, R. Tytgat and J. De Ley. 1991a. Revision of *Campylobacter*, *Helicobacter*, and *Wolinella* taxonomy: emendation of generic descriptions and proposal of *Arcobacter* gen. nov. Int. J. Syst. Bacteriol. *41*: 88–103.

Vandamme, P., B.A. Giesendorf, A. van Belkum, D. Pierard, S. Lauwers, K. Kersters, J.P. Butzler, H. Goossens and W.G. Quint. 1993a. Discrimination of epidemic and sporadic isolates of *Arcobacter butzleri* by polymerase chain reaction-mediated DNA fingerprinting. J. Clin. Microbiol. *31*: 3317–3319.

Vandamme, P., M. Gillis, M. Vancanneyt, B. Hoste, K. Kersters and E. Falsen. 1993b. *Moraxella lincolnii* sp. nov., isolated from the human respiratory tract, and reevaluation of the taxonomic position of *Moraxella osloensis*. Int. J. Syst. Bacteriol. *43*: 474–481.

Vandamme, P., Y. Glupczynski, A.P. Lage, C. Lammens, W.G.V. Quint and H. Goossens. 1995b. Evaluation of random and repetitive motif primed polymerase chain reaction typing of *Helicobacter pylori*. Syst. Appl. Microbiol. *18*: 357–362.

Vandamme, P. and H. Goossens. 1992. Taxonomy of *Campylobacter, Arcobacter,* and *Helicobacter*: a review. Zentbl. Bakteriol. *276*: 447–472.

Vandamme, P., J. Goris, T. Coenye, B. Hoste, D. Janssens, K. Kersters, P. De Vos and E. Falsen. 1999. Assignment of Centers for Disease Control group IVc-2 to the genus Ralstonia as Ralstonia paucula sp. nov. Int. J. Syst. Bacteriol. *49*: 663-669.

Vandamme, P.A.R., C.S. Harrington, K. Jalava and S.L.W. On. 2000. Misidentifying helicobacters: the *Helicobacter cinaedi* example. J. Clin. Microbiol. *38*: 2261–2266.

Vandamme, P., M. Heyndrickx, I. DeRoose, C. Lammens, P. DeVos and K. Kersters. 1997a. Characterization of *Bordetella* strains and related bacteria by amplified ribosomal DNA restriction analysis and randomly and repetitive element-primed PCR. Int. J. Syst. Bacteriol. *47*: 802–807.

Vandamme, P., M. Heyndrickx, M. Vancanneyt, B. Hoste, P. De Vos, E. Falsen, K. Kersters and K.H. Hinz. 1996a. *Bordetella trematum* sp. nov., isolated from wounds and ear infections in humans, and reassessment of *Alcaligenes denitrificans* Ruger and Tan 1983. Int. J. Syst. Bacteriol. *46*: 849–858.

Vandamme, P., B. Holmes, M. Vancanneyt, T. Coenye, B. Hoste, R. Coopman, H. Revets, S. Lauwers, M. Gillis, K. Kersters and J.R. Govan. 1997b. Occurrence of multiple genomovars of *Burkholderia cepacia* in cystic fibrosis patients and proposal of *Burkholderia multivorans* sp. nov. Int. J. Syst. Bacteriol. *47*: 1188–1200.

Vandamme, P., J. Hommez, M. Vancanneyt, M. Monsieurs, B. Hoste, B. Cookson, C.H. Wirsing von Konig, K. Kersters and P.J. Blackall. 1995c. *Bordetella hinzii* sp. nov., isolated from poultry and humans. Int. J. Syst. Bacteriol. *45*: 37–45.

Vandamme, P., B. Pot, E. Falsen, K. Kersters and J. De Ley. 1990. Intraspecific and interspecific relationships of veterinary campylobacters revealed by numerical analysis of electrophoretic protein profiles and DNA:DNA hybridizations. Syst. Appl. Microbiol. *13*: 295–303.

Vandamme, P., B. Pot, M. Gillis, P. De Vos, K. Kersters and J. Swings. 1996b. Polyphasic taxonomy, a consensus approach to bacterial systematics. Microbiol. Rev. *60*: 407–438.

Vandamme, P., B. Pot and K. Kersters. 1991b. Differentiation of Campylobacters and *Campylobacter*-like organisms by numerical analysis of one-dimensional electrophoretic protein patterns. Syst. Appl. Microbiol. *14*: 57–66.

Vandamme, P., P. Segers, M. Ryll, J. Hommez, M. Vancanneyt, R. Coopman, R. de Baere, Y. van de Peer, K. Kersters, R. de Wachter and K.H. Hinz. 1998. *Pelistega europaea* gen. nov., sp. nov., a bacterium associated with respiratory disease in pigeons: taxonomic structure and phylogenetic allocation. Int. J. Syst. Bacteriol. *48*: 431–440.

Vandamme, P., L.-J. Van Doorn, S.T.A. Rashid, W.G.V. Quint, J. Van Der Plas, V.L. Chan and S.L.W. On. 1997c. *Campylobacter hyoilei* Alderton et al. 1995 and *Campylobacter coli* Veron and Chatelain 1973 are subjective synonyms. Int. J. Syst. Bacteriol. *47*: 1055–1060.

Vandamme, P., M. Vancanneyt, B. Pot, L. Mels, B. Hoste, D. Dewettinck, L. Vlaes, C. Van Den Borre, R. Higgins, J. Hommez, K. Kersters, J.-P. Butzler and H. Goossens. 1992b. Polyphasic taxonomic study of the emended genus *Arcobacter* with *Arcobacter butzleri* comb. nov. and *Arcobacter skirrowii* sp. nov., an aerotolerant bacterium isolated from veterinary specimens. Int. J. Syst. Bacteriol. *42*: 344–356.

Vande Broek, A., A.M. Bekri, F. Dosselaere, D. Fuaure, M. Lambrecht, Y. Okon, A. Costacurta, E. Prinsen, P. De Troch, J. Desair, V. Keijers and J. Vanderleyden. 1998. *Azospirillum*–plant root association: genetics of IAA biosynthesis and plant cell wall degradation. *In* Elmerich, Kondorosi and Newton (Editors), Biological Nitrogen Fixation for the 21st Century, Kluwer Academic Publishers, Dordrecht. p. 375.

Vande Broek, A., V. Keijers and J. Vanderleyden. 1996. Effect of oxygen on the free-living nitrogen fixation activity and expression of the *Azospirillum brasilense nifH* gene in various plant-associated diazotrophs. Symbiosis *21*: 25–40.

Vande Broek, A., J. Michiels, A. Van Gool and J. Vanderleyden. 1993. Spatial-temporal colonization patterns of *Azospirillum brasilense* on the wheat root surface and expression of the bacterial *nifH* gene during association. Mol. Plant-Microbe Interact. *6*: 592–600.

Vande Broek, A., A. van Gool and J. Vanderleyden. 1989. Electroporation of *Azospirillum brasilense* with plasmid DNA. FEMS Microbiol. Lett. *61*: 177–182.

Vande Broek, A. and J. Vanderleyden. 1995. The genetics of the *Azospirillum*–plant root association. Crit. Rev. Plant Sci. *14*: 445–466.

Vandekerckhove, T.T., S. Watteyne, A. Willems, J.G. Swings, J. Mertens and M. Gillis. 1999. Phylogenetic analysis of the 16S rDNA of the cytoplasmic bacterium *Wolbachia* from the novel host *Folsomia candida* (Hexapoda, Collembola) and its implications for wolbachial taxonomy. FEMS Microbiol. Lett. *180*: 279–286.

Vandenberghe, A., A. Wassink, P. Raeymaekers, R. De Baere, E. Huysmans and R. De Wachter. 1985. Nucleotide sequence, secondary structure and evolution of the 5S ribosomal RNA from five bacterial species. Eur. J. Biochem. *149*: 537–542.

Vander Wauven, C. and V. Stalon. 1985. Occurrence of succinyl derivatives in the catabolism of arginine in *Pseudomonas cepacia*. J. Bacteriol. *164*: 882–886.

Vanderberg-Twary, L., K. Steenhoudt, B.J. Travis, J.L. Hanners, T.M. Foreman and J.R. Brainard. 1997. Biodegradation of paint stripper solvents in a modified gas lift loop bioreactor. Biotechnol. Bioeng. *55*: 163–169.

Vaneechoutte, M., T. de Baere, G. Wauters, S. Steyaert, G. Claeys, D. Vogelaers and G. Verschraegen. 2001. One case each of recurrent meningitis and hemoperitoneum infection with *Ralstonia mannitolilytica*. J. Clin. Microbiol. *39*: 4588–4590.

Vaneechoutte, M., R. Rossau, P. De Vos, M. Gillis, D. Janssens, N. Paepe, A. De Rouck, T. Fiers, G. Claeys and K. Kersters. 1992. Rapid identification of bacteria of the *Comamonadaceae* with amplified ribosomal DNA-restriction analysis (ARDRA). FEMS Microbiol. Lett. *72*: 227–233.

Van't Wout, J., W.N. Burnette, V.L. Mar, E. Rozdzinski, S.D. Wright and E.I. Tuomanen. 1992. Role of carbohydrate recognition domains of pertussis toxin in adherence of *Bordetella pertussis* to human macrophages. Infect. Immun. *60*: 3303–3308.

Vardanis, A. and R.M. Hochster. 1961. On the mechanism of glucose metabolism in the plant tumor-inducing organism *Agrobacterium tumefaciens*. Can. J. Biochem. Physiol. *39*: 1165–1182.

Varga, J. 1991. Comparison of surface antigens of some *Campylobacter fetus* subsp. *fetus* strains of ovine origin by polyacrylamide gel electrophoresis and immunoblotting. Zentbl. Vetmed. Reihe B. *38*: 497–504.

Varon, M. and R. Levisohn. 1972. Three-membered parasitic system: a bacteriophage, *Bdellovibrio bacteriovorus*, and *Escherichia coli*. J. Virol. *9*: 519–525.

Varon, M. and J. Seijffers. 1975. Symbiosis-independent and symbiosis-incompetent mutants of *Bdellovibrio bacteriovorus* 109J. J. Bacteriol. *124*: 1191–1197.

Varon, M. and M. Shilo. 1968. Interaction of *Bdellovibrio bacteriovorus* and host bacteria. I. Kinetic studies of attachment and invasion of *Escherichia coli* B by *Bdellovibrio bacteriovorus*. J. Bacteriol. *95*: 744–753.

Varon, M. and M. Shilo. 1970. Methods for separation of *Bdellovibrio* from mixed bacterial population by filtration through Millipore filters or by gradient differential centrifugation. Rev. Int. Oceanogr. Med. *18-19*: 145–152.

Varon, M. and M. Shilo. 1980. Ecology of aquatic bdellovibrios. Adv. Aquat. Microbiol. *2*: 1–48.

Varon, M. and M. Shilo. 1981. Inhibition of the predatory activity of *Bdellovibrio* by various environmental pollutants. Microb. Ecol. *7*: 107–112.

Varon, M., A. Tietz and E. Rosenberg. 1986. *Myxococcus xanthus* autocide ami. J. Bacteriol. *167*: 356–361.

Vasilyeva, G.K. 1972a. On the cycle of development and cytological properties of new soil microorganisms possessing prosthecae. Izv. Akad. Nauk S.S.S.R. Ser. Biol. *6*: 860–864.

Vasilyeva, G.K. and T.N. Lafitskaya. 1976. Assignment of *Agrobacterium polyspheroidum* to the genus *Prosthecomicrobium polyspheroidum* comb. nov. Izv. Akad. Nauk S.S.S.R. Ser. Biol. *6*: 768–772.

Vasilyeva, G.K., E.G. Surovtseva and L.P. Bakhaeva. 1996. Use of microorganisms mixed with a sorbent for bioremediation of soil polluted with chloroanilines, Abstract Q-421, Session 260. *In* Abstracts of the 96th general meeting of the American Society for Microbiology. p. 459.

Vasilyeva, L.V. 1970. A star-shaped soil microorganism. Izv. Akad. Nauk S.S.S.R. Ser. Biol. *2*: 308–310.

Vasilyeva, L.V. 1972b. The peculiarities of ultrastructure and the cycle of development of the bacterium *Stella humosa*. Izv. Akad. Nauk S.S.S.R. Ser. Biol. *5*: 782–788.

Vasilyeva, L.V. 1980. Morphological grouping of prosthecobacteria. Izv. Akad. Nauk S.S.S.R. Ser. Biol. *5*: 719–737.

Vasilyeva, L.V. 1984. Oligotrophic microorganisms as components of biogeocoenosis. *In* Mishustin (Editor), Soil Microorganisms as Components of Biogeocoenosis, Nauka, Moscow. pp. 232–241.

Vasilyeva, L.V. 1985. *Stella*, new genus of soil prosthecobacteria, with proposals for *Stella humosa*, sp. nov. and *Stella vacuolata*, sp. nov. Int. J. Syst. Bacteriol. *35*: 518–521.

Vasilyeva, L.V. 1986. New unusual bacteria with radial symmetry. Proceedings of the International Symposium on Microbial Ecology-4: 147–152.

Vasilyeva, L.V. 1989. Genus *Angulomicrobium* (Vasilyeva, Lafitskaya and Namsaraev, 1979, 1037[VP]). *In* Holt, Staley, Bryant and Pfennig (Editors), Bergey's Manual of Systematic Bacteriology, 9th Ed., Vol. 3, The Williams & Wilkins Co., Baltimore. pp. 1969–1971.

Vasilyeva, L.V., T.N. Lafitskaya, N.I. Aleksandruskina and E.N. Krasil'nikova. 1974. Physiologo-biochemical peculiarities of prosthecobacteria *Stella humosa* and *Prosthecomicrobium* sp. Izv. Akad. Nauk S.S.S.R. Ser. Biol. *5*: 699–714.

Vasilyeva, L.V., T.N. Lafitskaya and B.B. Namsaraev. 1979. A new genus of budding bacteria, *Angulomicrobium tetraedale*. Mikrobiologiya *48*: 1033–1039.

Vasilyeva, L.V., T.N. Lafitskaya and B.B. Namsaraev. 1986. *In* Validation of the publication of new names and new combinations previously effectively published outside the IJSB. List No. 20. Int. J. Syst. Bacteriol. *36*: 354–356.

Vasilyeva, L.V. and A.M. Semenov. 1984. *Labrys monahos*, a new budding prosthecate bacterium with radial symmetry. Mikrobiologiya *53*: 85–92.

Vasilyeva, L.V. and A.M. Semenov. 1985. *In* Validation of the publication of new names and new combinations previously effectively published outside the IJSB. List No. 18. Int. J. Syst. Bacteriol. *35*: 375–376.

Vasilyeva, L.V., A.M. Semenov and A.I. Giniyatullina. 1991. New soil bacterial species belonging to the *Prosthecomicrobium* genus. Mikrobiologiya *60*: 350–359.

Vasilyeva, L.V. and G.A. Zavarzin. 1995. Dissipotrophs in the microbiol community. Mikrobiologiya *64*: 239–244.

Vasse, J., S. Genin, P. Frey, C. Boucher and B. Brito. 2000. The *hrpB* and *hrpG* regulatory genes of *Ralstonia solanacearum* are required for different stages of the tomato root infection process. Mol. Plant-Microbe Interact. *13*: 259–267.

Vatter, A.E. and R.S. Wolfe. 1956. Electron microscopy of *Gallionella ferruginea*. J. Bacteriol. *72*: 248–252.

Vazquez, J.A., S. Berron, M. Orourke, G. Carpenter, E. Feil, N.H. Smith and B.G. Spratt. 1995. Interspecies recombination in nature—a meningococcus that has acquired a gonococcal *pib* porin. Mol. Microbiol. *15*: 1001–1007.

Vázquez, M., A. Dávalos, A. de las Peñas, F. Sánchez and C. Quinto. 1991. Novel organization of the common nodulation genes in *Rhizobium leguminosarum* biovar phaseoli strains. J. Bacteriol. *173*: 1250–1258.

Vázquez, M., O. Santana and C. Quinto. 1993. The NodI and NodJ proteins from *Rhizobium* and *Bradyrhizobium* strains are similar to capsular

polysaccharide secretion proteins from gram-negative bacteria. Mol. Microbiol. *8*: 369–377.

Vecherskaya, M.S., V.F. Galchenko, E.N. Sokolova and V.A. Samarkin. 1993. Activity and species composition of aerobic methanotrophic communities in tundra soils. Curr. Microbiol. *27*: 181–184.

Vedenina, I.Y. and D.Y. Sorokin. 1992. ATP Synthesis by heterotrophic bacteria on oxidation of thiosulfate to tetrathionate. Microbiology *61*: 530–534.

Vedenina, J.Y., N.A. Chernykh, A.V. Lebedinskii and L.V. Andreev. 1991. The correlation of properties in the genus *Hyphomicrobium*. Mikrobiologiya *59*: 895–902.

Vedenina, J.Y. and N.I. Govorukhina. 1988. Formation of a methylotrophous denitrifying coenosis in a system of sewage purification from nitrates. Mikrobiologiya *57*: 320–328.

Vedros, N.A. 1981. The Genus *Neisseria*. *In* Starr, Stolp, Trüper, Balows and Schlegel (Editors), The Prokaryotes: A Handbook of Habitats, Isolation, and Identification of Bacteria, Springer-Verlag, Berlin. pp. 1497–1505.

Vedros, N.A. 1984. Genus I. *Neisseria*. *In* Krieg and Holt (Editors), Bergey's Manual of Systematic Bacteriology, Vol. 1, The Williams & Wilkins Co., Baltimore. pp. 290–296.

Vedros, N.A., C. Hoke and P. Chun. 1983. *Neisseria macacae* sp. nov., a new neisseria species isolated from the oropharynges of rhesus monkeys (*Macaca mulatta*). Int. J. Syst. Bacteriol. *33*: 515–520.

Vedros, N.A., D.G. Johnston and P.I. Warren. 1973. *Neisseria* species isolated from dolphins. J. Wild. Dis. *9*: 241–244.

Velasco, J., C. Romero, I. Lopez-Goni, J. Leiva, R. Diaz and I. Moriyon. 1998. Evaluation of the relatedness of *Brucella* spp. and *Ochrobactrum anthropi* and description of *Ochrobactrum intermedium* sp. nov., a new species with a closer relationship to *Brucella* spp. Int. J. Syst. Bacteriol. *48*: 759–768.

Velázquez, E., J.M. Igual, A. Willems, M.P. Fernández, E. Muñoz, P.F. Mateos, A. Abril, N. Toro, P. Normand, E. Cervantes, M. Gillis and E. Martínez-Molina. 2001. *Mesorhizobium chacoense* sp. nov., a novel species that nodulates *Prosopis alba* in the Chaco Arido region (Argentina). Int. J. Syst. Evol. Bacteriol. *51*: 1011–1021.

Velazquez, J.B., A. Jimenez, B. Chomon and T.G. Villa. 1995. Incidence and transmission of antibiotic resistance in *Campylobacter jejuni* and *Campylobacter coli*. J. Antimicrob. Chemother. *35*: 173–178.

Vemulapalli, R., B. Biswas and S.K. Dutta. 1998. Cloning and molecular analysis of genes encoding two immunodominant antigens of *Ehrlichia risticii*. Microb. Pathog. *24*: 361–372.

Venter, A.P., S. Twelker, I.J. Oresnik and M.F. Hynes. 2001. Analysis of the genetic region encoding a novel rhizobiocin from *Rhizobium leguminosarum* biovar viciae strain 306. Can. J. Microbiol. *47*: 495–502.

Ventosa, A., M.C. Gutierrez, M.T. Garcia and F. Ruiz-Berraquero. 1989. Classification of "*Chromobacterium marismortui*" in a new genus, *Chromohalobacter* gen. nov. as *Chromohalobacter marismortui*, comb. nov., nov. rev. Int. J. Syst. Bacteriol. *39*: 382–386.

Verger, J.M. and M. Grayon. 1977. Oxidative metabolic profiles of *Brucella* species. Ann. Sclavo. *19*: 45–60.

Verger, J.M., M. Grayon, E. Chaslus-Dancla, M. Meurisse and J.P. Lafont. 1993. Conjugative transfer and *in vitro/in vivo* stability of the broad host range IncP R751 plasmid in Brucella spp. Plasmid *29*: 142–146.

Verger, J.M., F. Grimont, P.A.D. Grimont and M. Grayon. 1985. *Brucella*, a monospecific genus as shown by deoxyribonucleic Acid hybridization. Int. J. Syst. Bacteriol. *35*: 292–295.

Verma, V., P. Qazi, J. Cullum and G.N. Qazi. 1997. Genetic heterogeneity among keto-acid-producing strains of *Gluconobacter oxydans*. World J. Microbiol. Biotechnol. *13*: 289–294.

Véron, M. and R. Chatelain. 1973. Taxonomic study of the genus *Campylobacter* Sebald and Véron and designation of the neotype strain for the type species, *Campylobacter fetus* (Smith and Taylor) Sebald and Véron. Int. J. Syst. Bacteriol. *23*: 122–134.

Véron, M., P. Thibault and L. Second. 1959. *Neisseria mucosa* (*Diplococcus mucosus* Lingelsheim). Ann. Inst. Pasteur *97*: 497–510.

Verschraegen, G., G. Claeys, G. Meeus and M. Delanghe. 1985. *Pseudo-*

monas pickettii as a cause of pseudobacteremia. J. Clin. Microbiol. *21*: 278–279.

Verstreate, D.R. and A.J. Winter. 1984. Comparison of sodium dodecyl sulfate polyacrylamide gel electrophoresis profiles and antigenic relatedness among outer membrane proteins of 49 *Brucella abortus* strains. Infect. Immun. *46*: 182–187.

Vervliet, G., M. Holsters, H. Teuchy, M. Van Montagu and J. Schell. 1975. Characterization of different plaque-forming and defective temperate phages in *Agrobacterium* strains. J. Gen. Virol. *26*: 33–48.

Viallard, V., I. Poirier, B. Cournoyer, J. Haurat, S. Wiebkin, K. Ophel-Keller and J. Balandreau. 1998. *Burkholderia graminis* sp. nov., a rhizospheric *Burkholderia* species, and reassessment of *[Pseudomonas] phenazinium, [Pseudomonas] pyrrocinia* and *[Pseudomonas] glathei* as *Burkholderia*. Int. J. Syst. Bacteriol. *48*: 549–563.

Vidaver, A.K. and R.R. Carlson. 1978. Leaf spot of field corn caused by Pseudomonas andropogonis. Plant Disease Reporter *62*: 213–216.

Videira, P.A., L.L. Cortes, A.M. Fialho and I. Sa-Correia. 2000. Identification of the *pgmG* gene, encoding a bifunctional protein with phosphoglucomutase and phosphomannomutase activities, in the gellan gum-producing strain *Sphingomonas paucimobilis* ATCC 31461. Appl. Environ. Microbiol. *66*: 2252–2258.

Viikari, L. 1984. Formation of sorbitol by *Zymomonas mobilis*. Appl. Microbiol. Biotechnol. *20*: 118–123.

Vilcheze, C., P. Llopiz, S. Neunlist, K. Poralla and M. Rohmer. 1994. Prokaryotic triterpenoids: new hopanoids from the nitrogen-fixing bacteria *Azotobacter vinelandii, Beijerinckia indica* and *Beijerinckia mobilis*. Microbiology (Read.) *140*: 2749–2753.

Villarreal, D.T., R.F. Turco and A. Konopka. 1991. Propachlor degradation by a soil bacterial community. Appl. Environ. Microbiol. *57*: 2135–2140.

Villechanoux, S., M. Garnier, J. Renaudin and J.M. Bové. 1992. Detection of several strains of the bacterium-like organism of citrus greening disease by DNA probes. Curr. Microbiol. *24*: 89–96.

Vincent, J.M. 1970. A Manual for the Practical Study of Root Nodule Bacteria, Blackwell Scientific Publishers, Oxford.

Vincent, J.M. 1977. *Rhizobium* - general microbiology. *In* Hardy and Silver (Editors), A Treatise in Dinitrogen Fixation, Section III, J. Wiley and Sons, New York. pp. 277–366.

Vincent, J.M. 1982. The basic serology of rhizobia. *In* Vincent (Editor), Nitrogen Fixation in Legumes, Academic Press, New York. pp. 13–26.

Vincent, J.M. and B.A. Humphrey. 1970. Taxonomically significant group antigens in *Rhizobium*. J. Gen. Microbiol. *63*: 379–382.

Vincent, J.M., P.S. Nutman and F.A. Skinner. 1979. The identification and classification of *Rhizobium. In* Skinner and Lovelock (Editors), Identification Methods for Microbiologists. Soc. Appl. Bacteriol. Tech. Ser. 14, 2nd Ed., Academic Press, New York, London. pp. 49–69.

Vinuesa, P., B.L. Reuhs, C. Breton and D. Werner. 1999. Identification of a plasmid-borne locus in *Rhizobium etli* KIM5s involved in lipopolysaccharide *O*-chain biosynthesis and nodulation of *Phaseolus vulgaris*. J. Bacteriol. *181*: 5606–5614.

Virji, M., J.N. Fletcher, K. Zak and J.E. Heckels. 1987. The potential protective effect of monoclonal antibodies to gonococcal outer membrane protein-Ia. J. Gen. Microbiol. *133*: 2639–2646.

Virji, M., K. Zak and J.E. Heckels. 1986. Monoclonal antibodies to gonococcal outer membrane protein-ib—use in investigation of the potential protective effect of antibodies directed against conserved and type-specific epitopes. J. Gen. Microbiol. *132*: 1621–1629.

Visalli, M.A., S. Bajaksouzian, M.R. Jacobs and P.C. Appelbaum. 1997. Comparative activity of trovafloxacin, alone and in combination with other agents, against gram-negative nonfermentative rods. Antimicrob. Agents Chemother. *41*: 1475–1481.

Visca, P., A. Ciervo, V. Sanfilippo and N. Orsi. 1993. Iron regulated salicylate synthesis by *Pseudomonas* spp. J. Gen. Microbiol. *139*: 1995–2001.

Viseshakul, N., S. Kamper, M.V. Bowie and A.F. Barbet. 2000. Sequence and expression analysis of a surface antigen gene family of the rickettsia *Anaplasma marginale*. Gene *253*: 45–53.

vishniac, W. 1974. Genus *Thiobacillus. In* Buchanan and Gibbons (Editors), Bergey's Manual of Determinative Bacteriology, 8th Ed., The Williams & Wilkins Co., Baltimore. pp. 456–461.

Vishniac, W. and M. Santer. 1957. The thiobacilli. Bacteriol. Rev. *21*: 195–213.

Vishwanath, S. 1991. Antigenic relationships among the rickettsiae of the spotted fever and typhus groups. FEMS Microbiol. Lett. *65*: 341–344.

Visser, E.S., T.C. McGuire, G.H. Palmer, W.C. Davis, V. Shkap, E. Pipano and D.P. Knowles. 1992. The *Anaplasma marginale* msp5 gene encodes a 19-kilodalton protein conserved in all recognized *Anaplasma* species. Infect. Immun. *60*: 5139–5144.

Visser't Hooft, F. 1925. Biochemische Onderzoekingen over het geslacht *Acetobacter*, Thesis, Techn. Univ., Meinema, Delft, . pp. 1–129.

Voelz, H. 1964. Sites of adenosine triphosphatase activity in bacteria. J. Bacteriol. *88*: 1196–1198.

Voelz, H. 1965. Formation and structure of mesosomes in *Myxococcus xanthus*. Arch. Mikrobiol. *51*: 60–70.

Voelz, H. 1966a. The fate of the cell envelopes of *Myxococcus xanthus* during microcyst germination. Arch. Mikrobiol. *55*: 110–115.

Voelz, H. 1966b. *In vivo* induction of "polysomes" by limiting phosphate and the structural consequences in *Myxococcus xanthus*. Sixth International Congress for Electron Microscopy, Kyoto. pp. 255–256.

Voelz, H. 1967. The physical organization of the cytoplasm in *Myxococcus xanthus* and the fine structure of its components. Arch. Microbiol. *57*: 181–195.

Voelz, H. 1968. Structural comparison between intramitochondrial and bacterial crystalloids. J. Ultrastruct. Res. *25*: 29–36.

Voelz, H. and R.P. Burchard. 1971. Fine structure of bacteriophage infected *Myxococcus xanthus*. I. The lytic cycle in vegetative cells. Virology *43*: 243–250.

Voelz, H. and M. Dworkin. 1962. Fine structure of *Myxococcus xanthus* during morphogenesis. J. Bacteriol. *84*: 943–952.

Voelz, H., V.F. Gerencser and R. Kaplan. 1971. Bacteriophage replication in *Hyphomicrobium*. Virology *44*: 622–630.

Voelz, H. and R.O. Ortigoza. 1968. Cytochemistry of phosphatases in *Myxococcus xanthus*. J. Bacteriol. *96*: 1357–1365.

Voelz, H. and H. Reichenbach. 1969. Fine structure of fruiting bodies of *Stigmatella aurantiaca* (*Myxobacterales*). J. Bacteriol. *99*: 856-866.

Voelz, H., U. Voelz and R.O. Ortigoza. 1966. The "polyphosphate overplus" phenomenon in *Myxococcus xanthus* and its influence on the architecture of the cell. Arch. Microbiol. *53*: 371–388.

Voges, O. 1893. Ueber einige im Wasser vorkommende Pigmentbakterien. Zentralbl. Bakteriol. Parasitenk. Infektionskr. Hyg. Abt. I. *14*: 301–314.

Vogt, M. 1965. Wachstumsphysiologische Untersuchungen an *Micrococcus denitrificans*. Arch. Mikrobiol. *50*: 256–281.

Volk, M., O. Meyer and K. Frunzke. 1994. Metabolic relationship between the CO dehydrogenase molybdenum cofactor and the excretion of urothione by *Hydrogenophaga pseudoflava*. Eur. J. Biochem. *225*: 1063–1071.

Volpon, A.G.T., J. Dobereiner and H. De Polli. 1981. Physiology of nitrogen fixation in *Azospirillum lipoferum* Br 17 (ATCC 29709). Arch. Microbiol. *128*: 371–375.

Von Lichtenberg, F. 1984. Viral, Chlamydial, Rickettsial, and Bacterial Diseases. *In* Robbins, Cotran and Kumar (Editors), Pathologic Basis of Disease, 3rd Ed., W.B. Saunders Company, Philadelphia. pp. 273–350.

von Lingelsheim, W. 1906. Die bakteriologischen Arbeiten der Kgl. Hygienischen Station zu Beuthen O.-Schl. wahrend der Genickstarreepedemie in Oberschlesien in Winter 1904/05. Klin. Jahrb. *15*: 373–489.

von Lingelsheim, W. 1908. Beiträge zŭr Atiologie der epidemischen Genickstarre nach Ergebnissen der letzten Jahre. Z. Hyg. Infektionskr. *59*: 457–476.

Voordouw, G. 1995. The genus *Desulfovibrio* - the centennial. Appl. Environ. Microbiol. *61*: 2813–2819.

Voordouw, G., S.M. Armstrong, M.F. Reimer, B. Fouts, A.J. Telang, Y. Shen and D. Gevertz. 1996. Characterization of 16S rRNA genes from oil field microbial communities indicates the presence of a variety of sulfate-reducing, fermentative, and sulfide-oxidizing bacteria. Appl. Environ. Microbiol. *62*: 1623–629.

Vorholt, J.A., L. Chistoserdova, S.M. Stolyar, R.K. Thauer and M.E. Lidstrom. 1999. Distribution of tetrahydromethanopterin-dependent enzymes in methylotrophic bacteria and phylogeny of methenyl tetrahydromethanopterin cyclohydrolases. J. Bacteriol. *181*: 5750–5757.

Vuilleumier, S., H. Sorribas and T. Leisinger. 1997. Identification of a novel determinant of glutathione affinity in dichloromethane dehalogenases/glutathione S-transferases. Biochem. Biophys. Res. Commun. *238*: 452–456.

Vuilleumier, S., Z. Ucurum, S. Oelhafen, T. Leisinger, J. Armengaud, R.M. Wittich and K.N. Timmis. 2001. The glutathion S-transferase OrfE3 of the dioxin-degrading bacterium *Sphingomonas* sp. RW1 has maleylpyruvate isomerase activity. Chem. Biol. Interact. *133*: 265–267.

Wada, R., M. Kamada, Y. Fukunaga and T. Kumanomido. 1983. Studies on contagious equine metritis. IV. Pathology in horses experimentally infected with *Haemophilus equigenitalis*. Bull. Equine Res. Inst. *20*: 133–143.

Wagenbreth, D. 1961. Ein Beiträg zür Systematischen Einordnung der Knollchenbakterien durch Bestimmung des relativen Basengehaltes ihrer Desoxyribonucleinsäuren. Flora. *151*: 219–230.

Waghela, S.D., D. Cruz, R.E. Droleskey, J.R. DeLoach and G.G. Wagner. 1997. *In vitro* cultivation of *Anaplasma marginale* in bovine erythrocytes co-cultured with endothelial cells. Vet. Parasitol. *73*: 43–52.

Waghorn, D.J. and C.H. Cheetham. 1997. *Kingella kingae* endocarditis following chickenpox in infancy. Eur. J. Clin. Microbiol. Infect. Dis. *16*: 944–946.

Wagner, C. and M.E. Levitch. 1975. Enzymes involved in the assimilation of one-carbon units by *Pseudomonas* MS. J. Bacteriol. *122*: 905–910.

Wagner, G.H., F.P. Cooper and C.T. Bishop. 1973. Extracellular carbohydrate antigens from some non-pathogenic *Neisseria* species. Can. J. Microbiol. *19*: 703–708.

Wagner-Döbler, I. , H. Rheims, A. Felske, R. Pukall and B.J. Tindall. 2003. *Jannaschia helgolandensis* gen. nov., sp nov., a novel abundant member of the marine *Roseobacter* clade from the North Sea. Int. J. Syst. Evol. Microbiol. *53*: 731–738.

Wakabayashi, S., H. Matsubara and D.A. Webster. 1986. Primary sequence of a dimeric bacterial haemoglobin from *Vitreoscilla*. Nature *322*: 481–483.

Wakao, N., N. Nagasawa, T. Matsuura, H. Matsukura, T. Matsumoto, A. Hiraishi, Y. Sakurai and H. Shiota. 1994. *Acidiphilium multivorum* sp. nov., an acidophilic chemoorganotrophic bacterium from pyritic acid mine drainage. J. Gen. Appl. Microbiol. *40*: 143–159.

Wakao, N., N. Nagasawa, T. Matsuura, H. Matsukura, T. Matsumoto, A. Hiraishi, Y. Sakurai and H. Shiota. 1995. *In* Validation of the publication of new names and new combinations previously effectively published outside the IJSB. List No. 52. Int. J. Syst. Bacteriol. *45*: 197–198.

Wakao, N., T. Shiba, A. Hiraishi, M. Ito and Y. Sakurai. 1993. Distribution of bacteriochlorophyll *a* in species of the genus *Acidiphilium*. Curr. Microbiol. *27*: 277–279.

Wakao, N., N. Yokoi, N. Isoyama, A. Hiraishi, K. Shimada, M. Kobayashi, H. Kise, M. Iwaki, S. Itoh, S. Takaichi and Y. Sakurai. 1996. Discovery of natural photosynthesis using Zn-containing bacteriochlorophyll in an aerobic bacterium *Acidiphilium rubrum*. Plant Cell Physiol. *37*: 889–893.

Wali, T.W., G.R. Hudson, D.A. Danald and R.M. Weiner. 1980. Timing of swarmer cell cycle morphogenesis and macromolecular synthesis by *Hyphomicrobium neptunium* in synchronous culture. J. Bacteriol. *144*: 406–412.

Walia, S.K., V.C. Carey, B.P.d. All and L.O. Ingram. 1984. Self-transmissible plasmid in *Zymomonas mobilis* carrying antibiotic resistance. Appl. Environ. Microbiol. *47*: 198–200.

Walker, D.H. 1996. Rickettsiae. *In* Baron (Editor), Medical Microbiology, University of Texas Medical Branch, Galveston. pp. 487–501.

Walker, D.H. and J.S. Dumler. 1996. Emergence of the ehrlichioses as human health problems. Emerg. Infect. Dis. *2*: 18–29.

Walker, D.H., H.M. Feng and V.L. Popov. 2001. Rickettsial phospholipase A2 as a pathogenic mechanism in a model of cell injury by typhus and spotted fever group rickettsiae. Am. J. Trop. Med. Hyg. *65*: 936–942.

Walker, D.H., W.T. Firth, J.G. Ballard and B.C. Hegarty. 1983. Role of phospholipase-associated penetration mechanism in cell injury by *Rickettsia rickettsii*. Infect. Immun. *40*: 840–842.

Walker, D.H. and D.J. Sexton. 1999. *Rickettsia rickettsii. In* Yu, Merigan and Barriere (Editors), Antimicrobial Therapy and Vaccines, The Williams & Wilkins Co., Baltimore. pp. 562–568.

Walker, D.H., R.R. Tidwell, T.M. Rector and J.D. Geratz. 1984. Effect of synthetic protease inhibitors of the amidine type on cell injury by *Rickettsia rickettsii*. Antimicrob. Agents Chemother. *25*: 582–585.

Walker, S.G., R.E. Hancock and J. Smit. 1991. Expression in *Caulobacter crescentus* of the phosphate-starvation- inducible porin *OprP* of *Pseudomonas aeruginosa*. FEMS Microbiol. Lett. *61*: 217–222.

Walker, S.G., S.H. Smith and J. Smit. 1992. Isolation and comparison of the paracrystalline surface layer proteins of freshwater caulobacters. J. Bacteriol. *174*: 1783–1792.

Walker, T.S. 1984. Rickettsial interactions with human endothelial cells *in vitro*: Adherence and entry. Infect. Immun. *44*: 205–210.

Walker, T.S. and H.H. Winkler. 1978. Penetration of cultured mouse fibroblasts (L cells) by *Rickettsia prowazeki*. Infect. Immun. *22*: 200–208.

Wall, J.D., P.F. Weaver and H. Gest. 1975. Gene transfer agents, bacteriophages, and bacteriocins of *Rhodopseudomonas capsulata*. Arch. Microbiol. *105*: 217–224.

Wallace, P.L., D.G. Hollis, R.E. Weaver and C.W. Moss. 1990. Biochemical and chemical characterization of pink-pigmented oxidative bacteria. J. Clin. Microbiol. *28*: 689–693.

Wallen, L.L. and E.N. Davis. 1972. Biopolymers of activated sludge. Environ. Sci. Technol. *6*: 161–164.

Wallrabenstein, C., N. Gorny, N. Springer, W. Ludwig and B. Schink. 1995a. Pure culture of *Syntrophus buswellii*, definition of its phylogenetic status, and description of *Syntrophus gentianae* sp. nov. System. Appl. Microbiol. *18*: 62–66.

Wallrabenstein, C., E. Hauschild and B. Schink. 1994. Pure culture and cytological properties of '*Syntrophobacter wolinii*'. FEMS Microbiol. Lett. *123*: 249–254.

Wallrabenstein, C., E. Hauschild and B. Schink. 1995b. *Syntrophobacter pfennigii* sp. nov., new syntrophically propionate-oxidizing anaerobe growing in pure culture with propionate and sulfate. Arch. Microbiol. *164*: 346–352.

Wallrabenstein, C., E. Hauschild and B. Schink. 1996. *In* Validation of the publication of new names and new combinations previously effectively published outside the IJSB. List No. 58. Int. J. Syst. Bacteriol. *46*: 836–837.

Wallrabenstein, C. and B. Schink. 1994. Evidence of reversed electron-transport in syntrophic butyrate or benzoate oxidation by *Syntrophomonas wolfei* and *Syntrophus buswellii*. Arch. Microbiol. *162*: 136–142.

Walsh, R.D., N.C. Klein and B.A. Cunha. 1993. *Achromobacter xylosoxidans* osteomyelitis. Clin. Infect. Dis. *16*: 176–178.

Walther-Mauruschat, A., M. Aragno, F. Mayer and H.G. Schlegel. 1977. Micromorphology of Gram-negative hydrogen bacteria. II. Cell envelope, membranes and cytoplasmic inclusions. Arch. Microbiol. *114*: 101–110.

Wan, C.C., M.S. Kablaoui and W.C. Gates. 1988. Novel polysaccharide by *Xanthobacter* sp. useful as viscosifier in oil field applications. Abstracts of Papers of the American Chemical Society, p. 118.

Wang, E.T., J. Martinez-Romero and E. Martinez-Romero. 1999a. Genetic diversity of rhizobia from *Leucaena leucocephala* nodules in Mexican soils. Mol. Ecol. *8*: 711–724.

Wang, E.T., P. van Berkum, D. Beyene, X.H. Sui, O. Dorado, W.X. Chen and E. Martinez-Romero. 1998. *Rhizobium huautlense* sp. nov., a symbiont of *Sesbania herbacea* that has a close phylogenetic relationship with *Rhizobium galegae*. Int. J. Syst. Bacteriol. *48*: 687–699.

Wang, E.T., P. Van Berkum, X.H. Sui, D. Beyene, W.X. Chen and E. Martinez-Romero. 1999b. Diversity of rhizobia associated with *Amor-*

pha fruticosa isolated from Chinese soils and description of *Mesorhizobium amorphae* sp. nov. Int. J. Syst. Bacteriol. *49*: 51–65.

Wang, J.F., J. Olivier, P. Thoquet, B. Mangin, L. Sauviac and N.H. Grimsley. 2000. Resistance of tomato line Hawaii7996 to *Ralstonia solanacearum* Pss4 in Taiwan is controlled mainly by a major strain-specific locus. Mol. Plant-Microbe Interact. *13*: 6–13.

Wang, L.H., R.Y. Hamzah, Y. Yu and S.C. Tu. 1987. *Pseudomonas cepacia* 3-hydroxybenzoate 6-hydroxylase induction, purification, and characterization. Biochemistry *26*: 1099–1104.

Wang, R.C. and D.J.D. Nicholas. 1985. Some properties of glutamine synthetase and glutamate synthase from *Derxia gummosa*. Phytochemistry. *24*: 1133–1139.

Wang, R.C. and D.J.D. Nicholas. 1986a. Derepression of nitrate reductase from *Derxia gummosa* and some properties of the purified enzyme. Arch. Microbiol. *145*: 20–26.

Wang, R.C. and D.J.D. Nicholas. 1986b. Electron transfer during hydrogen oxidation in cell membranes of *Derxia gummosa*. Biochem Int. *13*: 633–640.

Wang, R.C. and D.J.D. Nicholas. 1986c. Regulation of nitrogen fixation by nitrite and glutamine in *Derxia gummosa*. FEMS Microbiol. Lett. *35*: 147–150.

Wang, R.C. and D.J.D. Nicholas. 1986d. Some properties of nitrite and hydroxylamine reductases from *Derxia gummosa*. Phytochemistry. *25*: 2463–2469.

Wang, S.P. and G. Stacey. 1991. Studies of the *Bradyrhizobium japonicum nodD1* promoter: a repeated structure for the *nod* box. J. Bacteriol. *173*: 3356–3365.

Wanner, J. and P. Grau. 1989. Identification of filamentous microorganisms from activated- sludge - a compromise between wishes, needs and possibilities. Water Res. *23*: 883–891.

Ward, J.M., M.R. Anver, D.C. Haines and R.E. Benveniste. 1994. Chronic active hepatitis in mice caused by *Helicobacter hepaticus*. Am. J. Pathol. *145*: 959–968.

Ward, M.E., P.J. Watt and A.A. Glynn. 1970. Gonococci in urethral exudates possess a virulence factor lost on subculture. Nature *227*: 382–384.

Ward, M.E., P.J. Watt and J.N. Rbertson. 1974. The human fallopian tube: A model for gonococcal infection. J. Infect. Dis. *129*: 650–659.

Ward, M.J. and D.R. Zusman. 2000. Developmental aggregation and fruiting body formation in the gliding bacterium *Myxococcus xanthus*. *In* Brun and Shimkets (Editors), Prokaryotic Development, American Society of Microbiology, Washington, D.C. pp. 243–262.

Ward, N.R., R.L. Wolfe, C.A. Justice and B.H. Olson. 1986. The identification of gram-negative, nonfermentative bacteria from water: problems and alternative approaches to identification. Adv. Appl. Microbiol. *31*: 293–365.

Ward-Rainey, N., F.A. Rainey and E. Stackebrandt. 1996. A study of the bacterial flora associated with *Holothuria atra*. J. Exp. Mar. Biol. Ecol. *203*: 11–26.

Wardlaw, A.C. and R. Parton. 1988. The host-parasite relationship in pertussis. *In* Wardlaw and Parton (Editors), Pathogenesis and Immunity in Pertussis, John Wiley & Sons, Chichester. pp. 327–352.

Ware, J.C. and M. Dworkin. 1973. Fatty acids of *Myxococcus xanthus*. J. Bacteriol. *115*: 253–261.

Warikoo, V., M.J. McInerney, J.A. Robinson and J.M. Suflita. 1996. Interspecies acetate transfer influences the extent of anaerobic benzoate degradation by syntrophic consortia. Appl. Environ. Microbiol. *62*: 26–32.

Warner, P.J., J.W. Drozd and I.J. Higgins. 1983. The effect of amino acids and amino acid analogues on growth of an obligate methanotroph, *Methylosinus trichosporium* OB3b. J. Chem. Technol. Biotechnol. *33B*: 29–34.

Warr, R.G., A.E. Goodman, P.L. Rogers and M.L. Skotnicki. 1984. Isolation and characterization of highly productive strains of *Zymomonas mobilis*. Microbios. *40*: 71–78.

Warrelmann, J. and B. Friedrich. 1986. Mutants of *Pseudomonas facilis* defective in lithoautotrophy. J.Gen. Microbiol. *132*: 91–96.

Wassenaar, T.M. 1997. Toxin production by *Campylobacter* spp. Clin. Microbiol. Rev. *10*: 466–476.

Watanabe, A. and I. Tanaka. 1929. Notiz über eine Myxobakterie. Botany Magazine, Tokyo *43*: 227–228.

Watanabe, T., M. Tada, H. Nagai, S. Sasaki and M. Nakao. 1998. *Helicobacter pylori* infection induces gastric cancer in Mongolian gerbils. Gastroenterology *115*: 642–648.

Waterman, S.R., J. Hackett and P.A. Manning. 1993. Characterization of the replication region of the small cryptic plasmid of *Campylobacter hyointestinalis*. Gene *125*: 11–17.

Watson, B.F. and M. Dworkin. 1968. Comparative intermediary metabolism of vegetative cells and microcysts of *Myxococcus xanthus*. J. Bacteriol. *96*: 1465–1473.

Watson, S.W. 1971a. Reisolation of *Nitrosospira briensis* S. Winogradsky and H. Winogradsky 1933. Arch. Mikrobiol. *75*: 179–188.

Watson, S.W. 1971b. Taxonomic considerations of the family *Nitrobacteraceae* Buchanan. Request for opinions. Int. J. Syst. Bacteriol. *21*: 254–270.

Watson, S.W., E. Bock, H. Harms, H.P. Koops and A.B. Hooper. 1989. Nitrifying bacteria. *In* Staley, Bryant, Pfennig and Holt (Editors), Bergey's Manual of Systematic Bacteriology, 1st Ed., Vol. 3, The Williams & Wilkins Co., Baltimore. pp. 1808–1833.

Watson, S.W. and C.C. Remsen. 1969. Macromolecular subunits in the walls of marine nitrifying bacteria. Science *163*: 685–686.

Watson, S.W. and J.B. Waterbury. 1971. Characteristics of two marine nitrite oxidizing bacteria, *Nitrospina gracilis* nov. gen. nov. sp. and *Nitrococcus mobilis* nov. gen. nov. sp. Arch. Microbiol. *77*: 203–230.

Wauters, G., G. Claeys, G. Verschraegen, T. de Baere, E. Vandecruys, L. Van Simaey, C. De Ganck and M. Vaneechoutte. 2001. Case of catheter sepsis with *Ralstonia gilardii* in a child with acute lymphoblastic leukemia. J. Clin. Microbiol. *39*: 4583–4584.

Wayne, L.G., D.J. Brenner, R.R. Colwell, P.A.D. Grimont, O. Kandler, M.I. Krichevsky, L.H. Moore, W.E.C. Moore, R.G.E. Murray, E. Stackebrandt, M.P. Starr and H.G. Trüper. 1987. Report of the ad hoc committee on reconciliation of approaches to bacterial systematics. Int. J. Syst. Bacteriol. *37*: 463–464.

Weaver, P.F., J.D. Wall and H. Gest. 1975a. Characterization of *Rhodopseudomonas capsulata*. Arch. Microbiol. *105*: 207–216.

Weaver, T.L., M.A. Patrick and P.R. Dugan. 1975b. Whole-cell and membrane lipids of the methylotrophic bacterium *Methylosinus trichosporium*. J. Bacteriol. *124*: 602–605.

Weber, A. 1976. Untersuchugen zur mikrobiologischen Diagnose and Epidemiologie der Brucella canis-Infection des Hundes, University of Giessen

Weber, O.B., V.L.D. Baldani, K.R.S. Teixeira, G. Kirchhof, J.I. Baldani and J. Dobereiner. 1999. Isolation and characterization of diazotrophic bacteria from banana and pineapple plants. Plant Soil *210*: 103–113.

Weber, O.B., B. Eckert, J.I. Baldani and J. Döbereiner. 1997. Caracterização de bactérias diazotróficas do tipo *Herbaspirillum* isoladas de abacaxizeiros e bananeiras. Proceedings of the XIX Congresso Brasileiro de Microbiolgoia, Rio de Janeiro. Sociedade Brasileira de Microbiologia. p. 225.

Webster, J., M. Dossantos and J.A. Thomson. 1986. Agrocin-producing *Agrobacterium tumefaciens* strain active against grapevine isolates. Appl. Environ. Microbiol. *52*: 217–219.

Webster, J. and J. Thomson. 1988. Genetic-analysis of an *Agrobacterium tumefaciens* strain producing an agrocin active against biotype-3 pathogens. Mol. Gen. Genet. *214*: 142–147.

Webster, P., J.W. IJdo, L.M. Chicoine and E. Fikrig. 1998. The agent of Human Granulocytic Ehrlichiosis resides in an endosomal compartment. J. Clin. Invest. *101*: 1932–1941.

Wecker, M.S.A. and R.R. Zall. 1987. Production of acetaldehyde by *Zymomonas mobilis*. Appl. Environ. Microbiol. *53*: 2815–2820.

Weckesser, J., J.G. Drews and H.-D. Tauschel. 1969. Zur Feinstruktur von *Rhodopseudomonas gelatinosa*. Arch. Mikrobiol. *65*: 346–358.

Weckesser, J., J.G. Drews and H. Mayer. 1979. Lipopolysaccharides of photosyntetic procaryotes. Annu. Rev. Microbiol. *33*: 215–239.

Weckesser, J. and H. Mayer. 1987. Lipopolysaccharide aus phototrophen bakterien. Forum Mikrobiol. *10*: 242–248.

Weckesser, J. and H. Mayer. 1988. Different lipid A types in lipopolysaccharides of phototrophic and related non-phototrophic bacteria. FEMS Microbiol. Rev. *4*: 143–153.

Weckesser, J., H. Mayer, G. Drews and I. Fromme. 1975. Lipophilic O-antigens containing D-glycero-D-mannoheptose as the sole neutral sugar in *Rhodopseudomonas gelatinosa*. J. Bacteriol. *123*: 449–455.

Weckesser, J., H. Mayer and G. Shulz. 1995. Anoxygenic phototrophic bacteria: model organisms for studies on cell wall macromolecules. *In* Blankenship, Madigan and Bauer (Editors), Anoxygenic Photosynthetic Bacteria, Kluwer Academic Publishing, The Netherlands. pp. 207–230.

Weckesser, J., G. Rosenfelder, H. Mayer and O. Luderitz. 1971. The identification of 3-*O*-methyl-D-xylose and 3-*O*-methyl-L-xylose as constituents of the lipopolysaccharides of *Myxococcus fulvus* and *Rhodopseudomonas viridis*, respectively. Eur. J. Biochem. *24*: 112–115.

Wehrfritz, J.M., A. Reilly, S. Spiro and D.J. Richardson. 1993. Purification of hydroxylamine oxidase from *Thiosphaera pantotropha*. Identification of electron acceptors that couple heterotrophic nitrification to aerobic denitrification. FEBS Lett. *335*: 246–250.

Wehrli, E. and T. Egli. 1988. Morphology of nitrilotriacetate-utilizing bacteria. Syst. Appl. Microbiol. *10*: 306–312.

Weidner, S., W. Arnold and A. Puhler. 1996. Diversity of uncultured microorganisms associated with the seagrass *Halophila stipulacea* estimated by restriction fragment length polymorphism analysis of PCR-amplified 16S rRNA genes. Appl. Environ. Microbiol. *62*: 766–771.

Weiner, R.M. 1998. Plasticity of bacteria. Survival mechanisms. *In* Rosenberg (Editor), Recent Advances in Microbial Ecology and Infectious Disease, ASM Press, Washington, D.C. pp. 17–29.

Weiner, R.M., R.A. Devine, D.M. Powell, L. Dagasan and R.L. Moore. 1985. *Hyphomonas oceanitis* sp. nov., *Hyphomonas hirschiana* sp. nov., and *Hyphomonas jannaschiana* sp. nov. Int. J. Syst. Bacteriol. *35*: 237–243.

Weiner, R.M., M. Melick, K. O'Neill and E.J. Quintero. 2000. *Hyphomonas adhaerens* sp. nov., *Hyphomonas johnsonii* sp. nov. and *Hyphomonas rosenbergii* sp. nov., marine budding and prosthecate bacteria. Int. J. Syst. Evol. Microbiol. *50*: 459–469.

Weir, S., L.W. Lee and C.F. Marrs. 1996. Identification of four complete type 4 pilin genes in a single *Kingella denitrificans* genome. Infect. Immun. *64*: 4993–4999.

Weir, S. and C.F. Marrs. 1992. Identification of type 4 pili in *Kingella denitrificans*. Infect. Immun. *60*: 3436–3441.

Weisburg, W.G., M.E. Dobson, J.E. Samuel, G.A. Dasch, L.P. Mallavia, O. Baca, L. Mandelco, J.E. Sechrest, E. Weiss and C.R. Woese. 1989. Phylogenetic diversity of the *Rickettsiae*. J. Bacteriol. *171*: 4202–4206.

Weisrock, W.P. and R.M. Johnson. 1966. Marine species of *Hyphomicrobium*. *In* Bacteriol. Proc, 22, G–36.

Weiss, A.A. 1992. The genus *Bordetella*. *In* Balows, Truper, Dworkin, Harder and Schleifer (Editors), The Prokaryotes: A Handbook of Bacteria: Ecophysiology, Isolation, Identification, Applications, 2nd ed., Vol. 3, Springer-Verlag, Berlin, Germany. pp. 2530–2543.

Weiss, A.A. and S. Falkow. 1984. Genetic analysis of phase change in *Bordetella pertussis*. Infect. Immun. *43*: 263–269.

Weiss, A.A. and E.L. Hewlett. 1986. Virulence factors of *Bordetella pertussis*. Annu. Rev. Microbiol. *40*: 661–686.

Weiss, A.A., E.L. Hewlett, G.A. Myers and S. Falkow. 1983. Tn5-induced mutations affecting virulence factors of *Bordetella pertussis*. Infect. Immun. *42*: 33–41.

Weiss, E. 1960. Some aspects of variation in rickettsial virulence. Ann. N.Y. Acad. Sci. *88*: 1287–1297.

Weiss, E. 1973. Growth and physiology of rickettsiae. Bacteriol. Rev. *37*: 259–283.

Weiss, E., J.C. Coolbaugh and J.C. Williams. 1975. Separation of viable *Rickettsia typhi* from yolk sac and L cell host components by Renografin density gradient centrifugation. Appl. Microbiol. *30*: 456–463.

Weiss, E. and G.A. Dasch. 1982. Differential characteristics of strains of *Rochalimaea*: *Rochalimaea vinsonii* sp. nov., the Canadian vole agent. Int. J. Syst. Bacteriol. *32*: 305–314.

Weiss, E., G.A. Dasch and K.P. Chang. 1984. *Wolbachieae*. *In* Kreig and Holt (Editors), Bergey's Manual of Systematic Bacteriology, The Williams & Wilkins Co., Baltimore. pp. 711–713.

Weiss, E., G.A. Dasch, Y.H. Kang and H.N. Westfall. 1988. Substrate utilization by *Ehrlichia sennetsu* and *Ehrlichia risticii* separated from host constituents by renografin gradient centrifugation. J. Bacteriol. *170*: 5012–5017.

Weiss, E. and H.R. Dressler. 1958. Growth of *Rickettsia prowozeki* in irradiated monolayer cultures of chick embryo entodermal cells. J. Bacteriol. *75*: 544–552.

Weiss, E. and H.R. Dressler. 1962. Increased resistance to chloramphenicol in *Rickettsia prowozekii* with a note on failure to demonstrate genetic interaction among strains. J. Bacteriol. *83*: 409–414.

Weiss, E., A.E. Green, R. Grays and L.M. Newman. 1973. Metabolism of *Richettsia tsutsugamushi* and *Rickettsia rickettsi* in irradiated host cells. Infect. Immun. *8*: 4–7.

Weiss, E. and J.W. Moulder. 1984a. Genus *Rickettsia*. *In* Krieg and Holt (Editors), Bergey's Manual of Systematic Bacteriology, Vol. 1, The Williams & Wilkins Co., Baltimore. pp. 688–698.

Weiss, E. and J.W. Moulder. 1984b. *In* Validation of the publication of new names and new combinations previously effectively published outside the IJSB. List No. 15. Int. J. Syst. Bacteriol. *34*: 355–357.

Weiss, E. and J.W. Moulder. 1988. *In* Validation of the publication of new names and new combinations previously effectively published outside the IJSB. List No. 25. Int. J. Syst. Bacteriol. *38*: 220–222.

Weiss, E., L.W. Newman, R. Grays and A.E. Green. 1972. Metabolism of *Rickettsia typhi* and *Rickettsia akari* in irradiated L cells. Infect. Immun. *6*: 50–57.

Weiss, E., J.C. Williams, G.A. Dasch and Y.H. Kang. 1989. Energy metabolism of monocytic *Ehrlichia*. Proc. Natl. Acad. Sci. U.S.A. *86*: 1674–1678.

Weissenfels, W.D., M. Beyer and J. Klein. 1990. Degradation of phenanthrene, fluorene and fluoranthene by pure bacterial cultures. Appl. Microbiol. Biotechnol. *32*: 479–484.

Weisser, P., R. Kramer, H. Sahm and G.A. Sprenger. 1995. Functional expression of the glucose transporter of *Zymomonas mobilis* leads to restoration of glucose and fructose uptake in *Escherichia coli* mutants and provides evidence for its facilitator action. J. Bacteriol. *177*: 3351–3354.

Weisser, P., R. Kramer and G.A. Sprenger. 1996. Expression of the *Escherichia coli pmi* gene, encoding phosphomannose isomerase in *Zymomonas mobilis*, leads to utilization of mannose as a novel growth substrate, which can be used as a selective marker. Appl. Environ. Microbiol. *62*: 4155–4161.

Weisshaar, R. and F. Lingens. 1983. The lipopolysaccharide of a chloridazon–degrading bacterium. Eur. J. Biochem. *137*: 155–161.

Weitzman, P.D. 1987. Patterns of diversity of citric acid cycle enzymes. Biochemical Society Symposium, pp. 33–43.

Weitzman, P.D. and D. Jones. 1975. The mode of regulation of bacterial citrate synthase as a taxonomic tool. J. Gen. Microbiol. *89*: 187–189.

Welch, D.F., K.C. Carroll, E.K. Hofmeister, D.H. Persing, D.A. Robison, A.G. Steigerwalt and D.J. Brenner. 1999. Isolation of a new subspecies, *Bartonella vinsonii* arupensis, from a cattle rancher: identity with isolates found in conjunction with *Borrelia burgdorferi* and *Babesia microti* among naturally infected mice. J. Clin. Microbiol. *37*: 2598–2601.

Welch, D.F., K.C. Carroll, E.K. Hofmeister, D.H. Persing, D.A. Robison, A.G. Steigerwalt and D.J. Brenner. 2000. *In* Validation of the publication of new names and new combinations previously effectively published outside the IJSB. List No. 72. Int. J. Syst. Evol. Microbiol. *50*: 3–4.

Welch, D.F., D.A. Pickett, L.N. Slater, A.G. Steigerwalt and D.J. Brenner. 1992. *Rochalimaea henselae* sp. nov., a cause of septicemia, bacillary angiomatosis, and parenchymal bacillary peliosis. J. Clin. Microbiol. *30*: 275–280.

Welch, W.D., R.K. Porschen and B. Luttrell. 1983. Minimal inhibitory concentrations of 19 antimicrobial agents for 96 clinical isolates of group IVe bacteria. Antimicrob. Agents Chemother. *24*: 432–433.

Weller, S.J., G.D. Baldridge, U.G. Munderloh, H. Noda, J. Simser and T.J. Kurtti. 1998. Phylogenetic placement of rickettsiae from the ticks *Amblyomma americanum* and *Ixodes scapularis*. J. Clin. Microbiol. *36*: 1305–1317.

Wells, B. and R.W. Horne. 1983. The ultrastructure of *Pseudomonas avenae*. 2. Intracellular refractile (R-body) structure. Micron Microsc. Acta *14*: 329–344.

Wells, J.M., J.E. Butterfield and L.G. Revear. 1993. Identification of bacteria associated with postharvest diseases of fruits and vegetables by cellular fatty acid composition an expert system for personal computers. Phytopathology *83*: 445–455.

Wells, J.S., Jr. and N.R. Krieg. 1965. Cultivation of *Spirillum volutans* in a bacteria-free environment. J. Bacteriol. *90*: 817–818.

Wells, M.Y. and Y. Rikihisa. 1988. Lack of lysosomal fusion with phagosomes containing *Ehrlichia risticii* in P388D1 cells: abrogation of inhibition with oxytetracycline. Infect. Immun. *56*: 3209–3215.

Wells, S.E. and L.D. Kuykendall. 1983. Tryptophan auxotrophs of *Rhizobium japonicum*. J. Bacteriol. *156*: 1356–1358.

Wen, A., M. Fegan, C. Hayward, S. Chakraborty and L.I. Sly. 1999. Phylogenetic relationships among members of the *Comamonadaceae*, and description of *Delftia acidovorans* (den Dooren de Jong 1926 and Tamaoka et al. 1987) gen. nov., comb. nov. Int. J. Syst. Bacteriol. *49*: 567–576.

Wen, B., Y. Rikihisa, P.A. Fuerst and W. Chaichanasiriwithaya. 1995a. Diversity of 16S rRNA genes of new *Ehrlichia* strains isolated from horses with clinical signs of Potomac horse fever. Int. J. Syst. Bacteriol. *45*: 315–318.

Wen, B., Y. Rikihisa, J. Mott, P.A. Fuerst, M. Kawahara and C. Suto. 1995b. *Ehrlichia muris* sp. nov., identified on the basis of 16S rRNA base sequences and serological, morphological, and biological characteristics. Int. J. Syst. Bacteriol. *45*: 250–254.

Wen, B., Y. Rikihisa, S. Yamamoto, N. Kawabata and P.A. Fuerst. 1996. Characterization of the SF agent, an *Ehrlichia* sp. isolated from the fluke *Stellantchasmus falcatus*, by 16S rRNA base sequence, serological, and morphological analyses. Int. J. Syst. Bacteriol. *46*: 149–154.

Wenseleers, T., F. Ito, S. Van Borm, R. Huybrechts, F. Volckaert and J. Billen. 1998. Widespread occurrence of the micro-organism *Wolbachia* in ants. Proc. R. Soc. Lond. B Biol. Sci. *265*: 1447–1452.

Werneburg, B. and H. Monteil. 1989. New serotypes of *Pseudomonas cepacia*. Res. Microbiol. *140*: 17–20.

Wernegreen, J.J. and M.A. Riley. 1999. Comparison of the evolutionary dynamics of symbiotic and housekeeping loci: a case for the genetic coherence of rhizobial lineages. Mol. Biol. Evol. *16*: 98–113.

Werner, J.K. 1993. Blood parasites of amphibians from Sichuan Province, People's Republic of China. J. Parasitol. *79*: 356–363.

Werren, J.H. 1997. Biology of *Wolbachia*. Annu. Rev. Entomol. *42*: 587–609.

Werren, J.H. and J.D. Bartos. 2001. Recombination in *Wolbachia*. Curr. Biol. *11*: 431–435.

Werren, J.H., L.R. Guo and D.W. Windsor. 1995a. Distribution of *Wolbachia* in neotropical arthropods. Proc. R. Soc. London Ser. B. *262*: 197–204.

Werren, J.H. and D.M. Windsor. 2000. *Wolbachia* infection frequencies in insects: evidence of a global equilibrium. Proc. R. Soc. Lond. B Biol. Sci. *267*: 1277–1285.

Werren, J.H., W. Zhang and L.R. Guo. 1995b. Evolution and phylogeny of *Wolbachia*: Reproductive parasites of arthropods. Proc. R. Soc. Lond. B Biol. Sci. *261*: 55–63.

Wertheim, W.A. and D.M. Markovitz. 1992. Osteomyelitis and intervertebral discitis caused by *Pseudomonas pickettii*. J. Clin. Microbiol. *30*: 2506–2508.

Wertlake, P.T. and T.W. Williams, Jr.. 1968. Septicaemia caused by *Neisseria flavescens*. J Clin Pathol. *21*: 437–439..

Werwath, J., H.A. Arfmann, D.H. Pieper, K.N. Timmis and R.M. Wittich. 1998. Biochemical and genetic characterization of a gentisate 1, 2-dioxygenase from *Sphingomonas* sp. strain RW5. J. Bacteriol. *180*: 4171–4176.

Wesley, I.V. 1996. *Helicobacter* and *Arcobacter* species: risks for foods and beverages. J. Food Prot. *59*: 1127–1132.

Wesley, I.V., A.L. Baetz and D.J. Larson. 1996. Infection of cesarean-derived colostrum-deprived 1-day-old piglets with *Arcobacter butzleri*, *Arcobacter cryaerophilus*, and *Arcobacter skirrowii*. Infect. Immun. *64*: 2295–2299.

Wesley, I.V., L. Schroeder-Tucker, A.L. Baetz, F.E. Dewhirst and B.J. Paster. 1995. *Arcobacter*-specific and *Arcobacter butzleri*-specific 16S rRNA-based DNA probes. J. Clin. Microbiol. *33*: 1691–1698.

Wesley, I.V., R.D. Wesley, M. Cardella, F.E. Dewhirst and B.J. Paster. 1991. Oligodeoxynucleotide probes for *Campylobacter fetus* and *Campylobacter hyointestinalis* based on 16S rRNA sequences. J. Clin. Microbiol. *29*: 1812–1817.

West, S.A., J.M. Cook, J.H. Werren and H.C. Godfray. 1998. *Wolbachia* in two insect host-parasitoid communities. Mol. Ecol. *7*: 1457–1465.

West, T.P. 1992. Pyrimidine base and ribonucleoside catabolic enzyme-activities of the *Pseudomonas diminuta* group. FEMS Microbiol. Lett. *99*: 305–310.

Westergaard, J.M. and T.T. Kramer. 1977. *Bdellovibrio* and the intestinal flora of vertebrates. Appl. Environ. Microbiol. *34*: 506–511.

Wetzler, L.M., K. Barry, M.S. Blake and E.C. Gotschlich. 1992a. Gonococcal lipooligosaccharide sialylation prevents complement-dependent killing by immune sera. Infect. Immun. *60*: 39–43.

Wetzler, L.M., M.S. Blake, K. Barry and E.C. Gotschlich. 1992b. Gonococcal porin vaccine evaluation—Comparison of Por proteosomes, liposomes, and blebs isolated from *rmp* deletion mutants. J. Infect. Dis. *166*: 551–555.

Wexler, H.M., D. Reeves, P.H. Summanen, E. Molitoris, M. McTeague, J. Duncan, K.H. Wilson and S.M. Finegold. 1996a. *Sutterella wadsworthensis* gen. nov., sp. nov., bile-resistant microaerophilic *Campylobacter gracilis*-like clinical isolates. Int. J. Syst. Bacteriol. *46*: 252–258.

Wexler, M., D. Gordon and P.J. Murphy. 1995. The distribution of inositol rhizopine genes in *Rhizobium* populations. Soil Biol. Biochem. *27*: 531–537.

Wexler, M., D.M. Gordon and P.J. Murphy. 1996b. Genetic relationships among rhizopine-producing *Rhizobium* strains. Microbiology-Uk. *142*: 1059–1066.

Weyant, R.S., D.G. Hollis, R.E. Weaver, M.F. Amin, A.G. Steigerwalt, S.P. O'Connor, A.M. Whitney, M.I. Daneshvar, C.W. Moss and D.J. Brenner. 1995a. *Bordetella holmesii* sp. nov., a new gram-negative species associated with septicemia. J. Clin. Microbiol. *33*: 1–7.

Weyant, R.S., C.W. Moss, R.E. Weaver, D.G. Hollis, J.G. Jordan, E.C. Cook and M.I. Daneshvar. 1995b. *In* Validation of the publication of new names and new combinations previously effectively published outside the IJSB. List No. 54. Int. J. Syst. Bacteriol. *45*: 619–620.

Weyant, R.S., C.W. Moss, R.E. Weaver, D.G. Hollis, J.G. Jordan, E.C. Cook and M.I. Daneshvar. 1996. Identification of Unusual Pathogenic Gram-negative Aerobic and Facultatively Anaerobic Bacteria, 2nd Ed., The Williams & Wilkins Co., Baltimore.

Weyant R.S., C.W. Moss, R.E. Weaver, D.G. Hollis, J.G. Jordan, E.C. Cook and M.I. Daneshvar. 1995. Identification of Unusual Pathogenic Gram-Negative Aerobic and Facultatively Anaerobic Bacteria, 2nd Ed., Williams & Wilkins, Baltimore.

Weyer, F. and R.J. Reiss-Gutfreund. 1973. Verhalten von *Rickettsia montana* und *R. canada* in Kleiderlausen. Acta Trop. *30*: 177–192.

Whary, M.T., T.J. Morgan, C.A. Dangler, K.J. Gaudes, N.S. Taylor and J.G. Fox. 1998. Chronic active hepatitis induced by *Helicobacter hepaticus* in the A/JCr mouse is associated with a Th1 cell-mediated immune response. Infect. Immun. *66*: 3142–3148.

Whatley, M.H., J.S. Bodwin, B.B. Lippincott and J.A. Lippincott. 1976. Role for *Agrobacterium* cell envelope lipopolysaccharide in infection site attachment. Infect. Immunol. *13*: 1080–1083.

Wheelis, M.L., N.J. Palleroni and R.Y. Stanier. 1967. The metabolism of aromatic acids by *Pseudomonas testosteroni* and *P. acidovorans*. Arch. Mikrobiol. *59*: 302–314.

Whiley, R.A. and D. Beighton. 1991. Emended descriptions and recognition of *Streptococcus constellatus*, *Streptococcus intermedius*, and *Streptococcus anginosus* as distinct species. Int. J. Syst. Bacteriol. *41*: 1–5.

Whitaker, R.J., G.S. Byng, R.L. Gherna and R.A. Jensen. 1981. Comparative allostery of 3-deoxy-D-arabino-heptulosonate 7-phosphate synthetase as an indicator of taxonomic relatedness in pseudomonad genera. J. Bacteriol. *145*: 752–759.

Whitby, G.E. and R.G.E. Murray. 1980. Defined medium for *Aquaspirillum serpens* VHL effective in batch and continuous culture. Appl. Environ. Microbiol. *39*: 20–24.

Whitchurch, C.B., M. Hobbs, S.P. Livingston, V. Krishnapillai and J.S. Mattick. 1991. Characterisation of a *Pseudomonas aeruginosa* twitching motility gene and evidence for a specialised protein export system widespread in eubacteria. Gene *101*: 33–44.

White, D., M. Dworkin and D.J. Tipper. 1968. Peptidoglycan of *Myxococcus xanthus*: Structure and relation to morphogenesis. J. Bacteriol. *95*: 2186–2197.

White, D. and H.U. Schairer. 2000. Development of *Stigmatella*. *In* Brun and Shimkets (Editors), Prokaryotic Development, ASM Press, Washington, D.C. pp. 285–294.

White, F.F. and E.W. Nester. 1980a. Hairy root: plasmid encodes virulence traits in *Agrobacterium rhizogenes*. J. Bacteriol. *141*: 1134–1141.

White, F.F. and E.W. Nester. 1980b. Relationship of plasmids responsible for hairy root and crown gall tumorigenicity. J. Bacteriol. *144*: 710–720.

White, G.F., K.S. Dodgson, I. Davies, P.J. Matts, J.P. Shapleigh and W.J. Payne. 1987. Bacterial utilization of short-chain primary alkyl sulfate esters. FEMS Microbiol. Lett. *40*: 173–177.

White, G.A. and C.H. Wang. 1964a. The dissimilation of glucose and gluconate by *Acetobacter xylinum*. 1. The origin and the fate of triose phosphate. Biochem. J. *90*: 408–423.

White, G.A. and C.H. Wang. 1964b. The dissimilation of glucose and gluconate by *Acetobacter xylinum*. 2. Pathway evaluation. Biochem. J. *90*: 424–433.

White, L.O. 1972. The taxonomy of the crown-gall organism *Agrobacterium tumefaciens* and its relationship to rhizobia and other agrobacteria. J. Gen. Microbiol. *72*: 565–574.

White, P.G. and J.B. Wilson. 1951. Differentiation of smooth and non-smooth colonies of Brucellae. J. Bacteriol. *61*: 239–240.

Whitehouse, R.L., H. Jackson, M.C. Jackson and M.M. Ramji. 1987. Isolation of *Simonsiella* sp. from a neonate. J. Clin. Microbiol. *25*: 522–525.

Whitman, W.B., E. Ankwanda and R.S. Wolfe. 1982. Nutrition and carbon metabolism of *Methanococcus voltae*. J. Bacteriol. *149*: 852–863.

Whitmore, A. 1913. An account of a "landers like" disease occurring in Rangoon. J. Hyg. *13*: 1–34.

Whittenbury, R.A., S.L. Davies and J.F. Davey. 1970a. Exospores and cysts formed by methane-utilizing bacteria. J. Gen. Microbiol. *61*: 219–226.

Whittenbury, R. and C.S. Dow. 1977. Morphogenesis and differentiation in *Rhodomicrobium vannielii* and other budding and prosthecate bacteria. Bacteriol. Rev. *41*: 754–808.

Whittenbury, R.A. and N.R. Krieg. 1984. Family IV. *Methylococcaceae*. *In* Krieg and Holt (Editors), Bergey's Manual of Systematic Bacteriology, 1st Ed., Vol. 1, The Williams & Wilkins Co., Baltimore. pp. 256–261.

Whittenbury, R.A. and J.M. Nicoll. 1971. A new, mushroom-shaped budding bacterium. J. Gen. Microbiol. *66*: 123–126.

Whittenbury, R.A., K.C. Phillips and J.F. Wilkinson. 1970b. Enrichment, isolation and some properties of methane-utilizing bacteria. J. Gen. Microbiol. *61*: 205–218.

Whitworth, T., V. Popov, V. Han, D. Bouyer, J. Stenos, S. Graves, L. Ndip and D. Walker. 2003. Ultrastructural and genetic evidence of a reptilian tick, *Aponomma hydrosauri*, as a host of *Rickettsia honei* in Australia: possible transovarial transmission. Ann. N.Y. Acad. Sci. *990*: 67–74.

Wichlacz, P.L. and R.F. Unz. 1981. Acidophilic heterotrophic bacteria of acid mine waters. Appl. Environ. Microbiol. *41*: 1254–1261.

Wichlacz, P.L., R.F. Unz and T.A. Langworthy. 1986. *Acidiphilium angustum*, sp. nov., *Acidiphilium facilis*, sp. nov. and *Acidiphilium rubrum*, sp. nov.: acidophilic heterotrophic bacteria isolated from acidic coal mine drainage. Int. J. Syst. Bacteriol. *36*: 197–201.

Widdel, F. 1980. Anaerober Abbau von Fettsäuren und Benzoesäure durch neu Isolierte Arten Sulfat-reduzierender Bakterien, Universität zu Göttingen, Lindhorst/Schaumburg-Lippe Göttingen.

Widdel, F. 1981. *In* Validation of the publication of new names and new combinations previously published outside the IJSB. List No. 7. Int. J. Syst. Bacteriol. *31*: 382–383.

Widdel, F. 1983. Methods for enrichment and pure culture isolation of filamentous gliding sulfate-reducing bacteria. Arch. Microbiol. *134*: 282–285.

Widdel, F. 1987. New types of acetate-oxidizing sulfate-reducing *Desulfobacter* species, *D. hydrogenophilus* sp. nov., *D. latus* sp. nov. and *D. curvatus* sp. nov. Arch. Microbiol. *148*: 286–291.

Widdel, F. 1988a. Microbiology and ecology of sulfate and sulfur-reducing bacteria. *In* Zehnder (Editor), Environmental Microbiology of Anaerobic Bacteria, Wiley and Sons, New York.

Widdel, F. 1988b. *In* Validation of the publication of new names and new combinations previously effectively published outside the IJSB. List No. 26. Int. J. Syst. Bacteriol. *38*: 328–329.

Widdel, F. and F. Bak. 1992. Gram negative mesophilic sulfate-reducing bacteria. *In* Balows, Trüper, Dworkin, Harder and Schleifer (Editors), The Prokaryotes: A Handbook of Bacteria: Ecophysiology, Isolation, Identification, Applications, 2nd Ed., Vol. 4, Springer Verlag, New York. pp. 3352-3378.

Widdel, F., G.-W. Kohring and F. Mayer. 1983. Studies on dissimilatory sulfate-reducing bacteria that decompose fatty acids III. Characterization of the filamentous gliding *Desulfonema limnicola* gen. nov. sp. nov., and *Desulfonema magnum* sp. nov. Arch. Microbiol. *134*: 286–294.

Widdel, F. and T.A. Hansen. 1992. The dissimilatory sulfate- and sulfur-reducing bacteria. *In* Balows, Trüper, Dworkin, Harder and Schleifer (Editors), The Prokaryotes: A Handbook of Bacteria: Ecophysiology, Isolation, Identification, Applications, 2nd Ed., Vol. 1, Springer-Verlag, New York. pp. 583–624.

Widdel, F. and N. Pfennig. 1981. Studies on dissimilatory sulfate-reducing bacteria that decompose fatty acids. I. Isolation of new sulfate-reducing bacteria with acetate from saline environments. Description of *Desulfobacter postgatei* gen. nov., sp. nov. Arch. Microbiol. *129*: 395–400.

Widdel, F. and N. Pfennig. 1982. Studies on dissimilatory sulfate-reducing bacteria that decompose fatty-acids. II. Incomplete oxidation of propionate by *Desulfobulbus propionicus* gen. nov., sp. nov. Arch. Microbiol. *131*: 360–365.

Widdel, F. and N. Pfennig. 1992. The genus *Desulfuromonas* and other Gram-negative sulfur-reducing eubacteria. *In* Balows, Trüper, Dworkin, Harder and Schleifer (Editors), The Prokaryotes: A Handbook of Bacteria: Ecophysiology, Isolation, Identification, Applications, 2nd Ed., Vol. 4, Springer-Verlag, New York. pp. 3379–3389.

Widdel, F., S. Schnell, S. Heising, A. Ehrenreich, B. Assmus and B. Schink. 1993. Ferrous iron oxidation by anoxygenic phototrophic bacteria. Nature (Lond.) *362*: 834–836.

Wieczorek, L. and P. Hirsch. 1979. Survival and growth of wildtype mutant bacteria in the original aquatic habitat. *In* Abstracts of the Annual Meeting of the American Society for Microbiology, American Society for Microbiology., Washington, D.C. p. 189.

Wieczorek, R., A. Steinbuchel and B. Schmidt. 1996. Occurrence of polyhydroxyalkanoic acid granule-associated proteins related to the *Alcaligenes eutrophus* H16 GA24 protein in other bacteria. FEMS Microbiol. Lett. *135*: 23–30.

Wiegel, J. 1981. Distinction between the Gram-reaction and the Gram-type of bacteria. Int. J. Syst. Bacteriol. *31*: 88.

Wiegel, J. 1986. The genus *Xanthobacter*. *In* Balows, Trüper, Dworkin, Harder and Schleifer (Editors), The Prokaryotes: A Handbook of Bacteria: Ecophysiology, Isolation, Identification, Applications, Springer-Verlag, New York. pp. 2363–2383.

Wiegel, J. and D. Kleiner. 1982. Survey of ammonium (methylammonium) transport by aerobic N_2-fixing bacteria-the specieal case of *Rhizobium*. FEMS Microbiol. Lett. *15*: 61–63.

Wiegel, J. and F. Mayer. 1978. Isolation of lipopolysaccharides and the effect of polymyxin B on the outer membrane of *Corynebacterium autotrophicum*. Arch. Microbiol. *118*: 67–69.

Wiegel, J. and L. Quandt. 1982. Determination of the Gram type using the reaction between polymyxin B and lipopolysaccharides of the outer cell wall of whole bacteria. J. Gen. Microbiol. *128*: 2261–2270.

Wiegel, J. and H.G. Schlegel. 1976. Enrichment and isolation of nitrogen-fixing hydrogen bacteria. Arch. Microbiol. *107*: 139–142.

Wiegel, J., D. Wilke, J. Baumgarten, R. Opitz and H.G. Schlegel. 1978. Transfer of the nitrogen-fixing hydrogen bacterium *Corynebacterium autotrophicum* Baumgarten et al. to *Xanthobacter* gen. nov. Int. J. Syst. Bacteriol. *28*: 573–581.

Wiemann, M. and H.D. Gortz. 1991. Identification and localization of major stage-specific polypeptides of infectious *Holospora obtusa* with monoclonal antibodies. J. Bacteriol. *173*: 4842–4850.

Wike, D.A. and W. Burgdorfer. 1972. Plaque formation in tissue cultures by *Rickettsia rickettsi* isolated directly from whole blood and tick hemolymph. Infect. Immun. *6*: 736–738.

Wilberg, E., T. El-Banna, G. Auling and T. Egli. 1993. Serological studies on nitrilotriacetic acid (NTA)-utilizing bacteria: Distribution of *Chelatobacter heintzii* and *Chelatococcus asaccharovorans* in sewage treatment plants and aquatic ecosystems. Syst. Appl. Microbiol. *16*: 147–152.

Wilde, E. and H.G. Schlegel. 1982. Oxygen tolerance of strictly aerobic hydrogen-oxidizing bacteria. Antonie Leeuwenhoek *48*: 131–143.

Wilke, D. 1980. Conjugational gene transfer in *Xanthobacter autotrophicus* GZ29. J. Gen. Microbiol. *117*: 431–436.

Wilke, D. and H.G. Schlegel. 1979. A defective generalized transducing bacteriophage in *Xanthobacter autotrophicus* GZ29. J. Gen. Microbiol. *115*: 403–410.

Wilkinson, B.J., M.R. Morman and D.C. White. 1972. Phospholipid composition and metabolism of *Micrococcus denitrificans*. J. Bacteriol. *112*: 1288–1294.

Wilkinson, S.G. 1968. Studies on the cell walls of *Pseudomonas* spp. resistant to ethylene diaminetetra-acetic acid. J. Gen. Microbiol. *54*: 195–213.

Wilkinson, S.G. 1981a. Isolation of D-threo-pent-2-ulose from the lipopolysaccharide of *Pseudomonas diminuta* Nctc 8545. Carbohydr. Res. *98*: 247–252.

Wilkinson, S.G. 1981b. Structural studies of an acetylated mannan from *Pseudomonas diminuta* Nctc-8545. Carbohydr. Res. *93*: 269–278.

Wilkinson, S.G. and M.E. Bell. 1971. The phosphoglucolipid from *Pseudomonas diminuta*. Biochim. Biophys. Acta *248*: 293–299.

Wilkinson, S.G. and L. Galbraith. 1979. Polar lipids of *Pseudomonas vesicularis* presence of a heptosyldiacylglycerol. Biochim. Biophys. Acta *575*: 244–254.

Wilkinson, S.G. and D.P. Taylor. 1978. Occurrence of 2,3-diamino-2,3-dideoxy-D-glucose in lipid A from lipopolysaccharide of *Pseudomonas diminuta*. J. Gen. Microbiol. *109*: 367–370.

Wilkinson, T.G., H.H. Topiwala and G. Hamer. 1974. Interactions in a mixed bacterial population growing on methane in continuous culture. Biotechnol. Bioeng. *16*: 41–59.

Wilks, K.E., K.L. Dunn, J.L. Farrant, K.M. Reddin, A.R. Gorringe, P.R. Langford and J.S. Kroll. 1998. Periplasmic superoxide dismutase in meningococcal pathogenicity. Infect. Immun. *66*: 213–217.

Willems, A., J. Busse, M. Goor, B. Pot, E. Falsen, E. Jantzen, B. Hoste, M. Gillis, K. Kersters, G. Auling and J. De Ley. 1989. *Hydrogenophaga*, a new genus of hydrogen-oxidizing bacteria that includes *Hydrogenophaga flava* comb. nov. (formerly *Pseudomonas flava*), *Hydrogenophaga palleronii* (formerly *Pseudomonas palleronii*), *Hydrogenophaga pseudflava* (formerly *Pseudomonas pseudoflava* and *Pseudomonas carboxydoflava*), and *Hydrogenophaga taeniospiralis* (formerly *Pseudomonas taeniospiralis*). Int. J. Syst. Bacteriol. *39*: 319–333.

Willems, A. and M.D. Collins. 1992. Evidence for a close genealogical relationship between *Afipia* (the causal organism of cat scratch disease), *Bradyrhizobium japonicum* and *Blastobacter denitrificans*. FEMS Microbiol. Lett. *96*: 241–246.

Willems, A. and M.D. Collins. 1993. Phylogenetic analysis of rhizobia and agrobacteria based on 16S rRNA gene sequences. Int. J. Syst. Bacteriol. *43*: 305–313.

Willems, A., R. Coopman and M. Gillis. 2001. Phylogenetic and DNA-DNA hybridization analyses of *Bradyrhizobium* species. Int. J. Syst. Evol. Microbiol. *51*: 111–117.

Willems, A., J. De Ley, M. Gillis and K. Kersters. 1991a. *Comamonadaceae*, a new family encompassing the acidovorans ribosomal RNA complex, including *Variovorax paradoxus* gen.nov., comb. nov., for *Alcaligenes paradoxus* (Davis 1969). Int. J. Syst. Bacteriol. *41*: 445–450.

Willems, A., E. Falsen, B. Pot, E. Jantzen, B. Hoste, P. Vandamme, M. Gillis, K. Kersters and J. De Ley. 1990. *Acidovorax*, a new genus for *Pseudomonas facilis*, *Pseudomonas delafieldii*, E. falsen (EF) group 13, EF group 16, and several clinical isolates, with the species *Acidovorax facilis* comb. nov., *Acidovorax delafieldii* comb. nov., and *Acidovorax temperans* sp. nov. Int. J. Syst. Bacteriol. *40*: 384–398.

Willems, A., M. Gillis and J. De Ley. 1991b. Transfer of *Rhodocyclus gelatinosus* to *Rubrivivax gelatinosus* gen. nov., comb. nov., and phylogenetic relationships with *Leptothrix*, *Sphaerotilus natans*, *Pseudomonas saccharophila* and *Alcaligenes latus*. Int. J. Syst. Bacteriol. *41*: 65–73.

Willems, A., M. Gillis, K. Kersters, L. Van den Broecke and J. De Ley. 1987. Transfer of *Xanthomonas ampelina* Panagopoulos 1969 to a new genus, *Xylophilus* gen. nov., as *Xylophilus ampelinus* (Panagopoulos 1969) comb. nov. Int. J. Syst. Bacteriol. *37*: 422–430.

Willems, A., M. Goor, S. Thielemans, M. Gillis, K. Kersters and J. De Ley. 1992a. Transfer of several phytopathogenic u species to *Acidovorax* as *Acidovorax avenae* subsp. *avenae* subsp. nov., comb. nov., *Acidovorax avenae* subsp. *citrulli*, *Acidovorax avenae* subsp. *cattleyae*, and *Acidovorax konjaci*. Int. J. Syst. Bacteriol. *42*: 107–119.

Willems, A., B. Pot, E. Falsen, P. Vandamme, M. Gillis, K. Kersters and J. De Ley. 1991. Polyphasic taxonomic study of the emended genus *Comamonas*: Relationship to *Aquaspirillum aquaticum*, E. Falsen group 10, and other clinical isolates. Int. J. Syst. Bacteriol. *41*: 427–444.

Willems, R.J., H.G. Van Der Heide and F.R. Mooi. 1992b. Characterization of a *Bordetella pertussis* fimbrial gene cluster which is located directly downstream of the filamentous haemagglutinin gene. Mol. Microbiol. *6*: 2661–2671.

Williams, H.N. 1987. The recovery of high numbers of bdellovibrios from the surface water microlayer. Can. J. Microbiol. *33*: 572–575.

Williams, H.N. 1988. A study of the occurrence and distribution of bdellovibrios in estuarine sediment over an annual cycle. Microb. Ecol. *15*: 9–20.

Williams, H.N. and W.A. Falkler, Jr.. 1984. Distribution of bdellovibrios in the water column of an estuary. Can. J. Microbiol. *30*: 971–974.

Williams, H.N., W.A. Falkler, Jr. and D.E. Shay. 1976. Incidence of marine bdellovibrios lytic against *Vibrio parahaemolyticus* in Chesapeake Bay, U.S.A. Appl. Environ. Microbiol. *40*: 970–972.

Williams, H.N., J.I. Kelley, M.L. Baer and B.F. Turng. 1995a. The association of bdellovibrios with surfaces in the aquatic environment. Can. J. Microbiol. *41*: 1142–1147.

Williams, H.N., A.J. Schoeffield, D. Guether, J. Kelley, D. Shah and W.A. Falkler, Jr.. 1995b. Recovery of bdellovibrios from submerged surfaces and other aquatic habitats. Microb. Ecol. *29*: 39–48.

Williams, H.N., S. Toon, E. Faulk and W.A.J. Falkler. 1987. The incidence of bdellovibrios in an artificial environment: The National Aquarium in Baltimore (Maryland, U.S.A.). Can. J. Microbiol. *33*: 483–488.

Williams, J. 1971. The growth in vitro of killer particles from *Paramecium aurelia* and the axenic culture of this protozoan. J. Gen. Microbiol. *68*: 253–262.

Williams, J.C. and J.C. Peterson. 1976. Enzymatic activities leading to pyrimidine nucleotide biosynthesis from cell-free extracts of *Rickettsia typhi*. Infect. Immun. *14*: 439–448.

Williams, M.A. 1959a. Some Problems in the identification and classification of species of *Spirillum* II. Later taxonomy of the genus *Spirillum*. Int. Bull. Bacteriol. Nomencl. Taxon. *9*: 137–157.

Williams, M.A. 1959b. Some problems in the identification and classification of *Spirillum*. Int. Bull. Bacteriol. Nomencl. Taxon. *9*: 35–55.

Williams, M.A. 1960. Flagellation in six species of *Spirillum* - a correction. Int. Bull. Bacteriol. Nomencl. Taxon. *10*: 193–196.

Williams, M.A. and S.C. Rittenberg. 1957. A taxonomic study of the genus *Spirillum* Ehrenberg. Int. Bull. Bacteriol. Nomencl. Taxon. *7*: 49–111.

Williams, R.A.D. and M.S. Da Costa. 1992. The genus *Thermus* and related microorganisms. *In* Balows, Trüper, Dworkin, Harder and Schleifer (Editors), The Prokaryotes: A Handbook of Bacteria: Ecophysiology, Isolation, Identification, Applications, 2nd Ed., Springer-Verlag, New York. pp. 3745–3753.

Williams, R.L., D.A. Oren, J. Munoz-Dorado, S. Inouye, M. Inouye and E. Arnold. 1993. Crystal structure of *Myxococcus xanthus* nucleoside diphosphate kinase and its interaction with a nucleotide substrate at 2.0 Å resolution. J. Mol. Biol. *234*: 1230–1247.

Williams, T.M. and R.F. Unz. 1985. Isolation and characterization of filamentous bacteria present in bulking activated-sludge. Appl. Microbiol. Biotechnol. *22*: 273–282.

Williamson, P. and R. Matthews. 1999. Development of neutralising human recombinant antibodies to pertussis toxin. FEMS Immunol. Med. Microbiol. *23*: 313–319.

Willmitzer, L., M. De Beuckeleer, M. Lemmers, M. Van Montagu and J. Schell. 1980. DNA from Ti plasmid present in nucleus and absent from plastics of crown gall plant cells. Nature (Lond.) *287*: 359–361.

Wilson, G.S. 1933. The classification of the *Brucella* group: a systematic study. J. Hyg. Camb. *33*: 516–541.

Wilson, G.S. and A.A. Miles. 1932. The serological differentiation of smooth strains of the *Brucella* group. Br. J. Exp. Pathol. *13*: 1–13.

Windham, R.C., A.E. Cashore, C.H. Nakutsu and M.C. Peel. 1994. Catabolic transposons. Biodegradation *5*: 323–342.

Winet, H. and S.R. Keller. 1976. *Spirillum* swimming: theory and observations of propulsion by the flagellar bundle. J. Exp. Biol. *65*: 577–602.

Winkelmann, G., B. Busch, A. Hartmann, G. Kirchhof, R. Suessmuth and G. Jung. 1999. Degradation of desferrioxamines by *Azospirillum irakense*: assignment of metabolites by HPLC/electrospray mass spectrometry. Biometals *12*: 255–264.

Winkelmann, G., K. Schmidtkunz and F.A. Rainey. 1996. Characterization of a novel *Spirillum*-like bacterium that degrades ferrioxamine-type siderophores. BioMetals. *9*: 78–83.

Winkler, H.H. 1990. *Rickettsia* species (as organisms). Ann. Rev. Microbiol. *44*: 131-153.

Winkler, H.H. and R.M. Daugherty. 1986. Acquisition of glucose by *Rickettsia prowazekii* through the nucleotide intermediate uridine 5′-diphosphoglucose. J. Bacteriol. *167*: 805–808.

Winkler, H.H. and R.M. Daugherty. 1989. Phospholipase A activity associated with the growth of *Rickettsia prowazekii* in L929 cells. Infect. Immun. *57*: 36–40.

Winkler, H.H., L. Day and R. Daugherty. 1994. Analysis of hydrolytic products from choline-labeled host cell phospholipids during growth of *Rickettsia prowazekii*. Infect. Immun. *62*: 1457–1459.

Winkler, H.H. and E.T. Miller. 1981. Immediate cytotoxicity and phospholipase A: The role of phospholipase A in the interaction of *R. prowazeki* and L-cells. *In* Burgdorfer and Anacker (Editors), Rickettsiae and Rickettsial Diseases, Academic Press, New York. pp. 327–333.

Winkler, H.H. and E.T. Miller. 1982. Phospholipase A and the interaction of *Rickettsia prowazekii* and mouse fibroblasts (L-929 cells). Infect. Immun. *38*: 109–113.

Winkler, S., W. Ockels, H. Budzikiewicz, H. Korth and G. Pulverer. 1986. 2-Hydroxy-4-methoxy-5-methyl pyridine N-oxide, an Al3 + complexing metabolite from *Pseudomonas cepacia*. Z. Naturforsch. C J. Biosci. *41*: 807–808.

Winogradsky, S. 1888. Uber Eisenbakterien. Bot. Ztg. *46*: 261–270.

Winogradsky, S. 1892. Contributions a la morphologie des organismes de la nitrification. Arch. Sci. Biol. (St. Petersb.) *1*: 86–137.

Winogradsky, S. 1922. Eisenbakterien als Anorgoxydanten. Zentbl. Bakteriol. Parasitenkd. Infektkrankh. Hyg. Abt. II *57*: 1–21.

Winogradsky, S. 1932. Études sur la microbiologie du sol. 5ᵉ mémoire. Analyse microbiologique du sol. Principes d'une nouvelle méthode. Ann. Inst. Pasteur *48*: 89–134.

Winogradsky, S. and H. Winogradsky. 1933. Études sur la microbiologie du sol. VII. Nouvelles recherches sur les organismes de la nitrification. Ann. Inst. Pasteur (Paris) *50*: 350–432.

Winslow, C.-E.A., J. Broadhurst, R.E. Buchanan, C.J. Krumwiede, L.A. Rogers and G.H. Smith. 1917. The families and genera of the bacteria. Preliminary report of the Committee of the Society of American Bacteriologists on characterization and classification of bacterial types. J. Bacteriol. *2*: 506–566.

Winterstein, C. and B. Ludwig. 1998. Genes coding for respiratory complexes map on all three chromosomes of the *Paracoccus denitrificans* genome. Arch. Microbiol. *169*: 275–281.

Wireman, J.W. and M. Dworkin. 1977. Developmentally induced autolysis during fruiting body formation by *Myxococcus xanthus*. J. Bacteriol. *129*: 796–802.

Wirsen, C.O. and H.W. Jannasch. 1978. Physiological and morphological observations on *Thiovulum* sp. J. Bacteriol. *136*: 765–774.

Wise, M.G., L.J. Shimkets and J.V. McArthur. 1995. Genetic structure of a lotic population of *Burkolderia (Pseudomonas) cepacia*. Appl. Environ. Microbiol. *61*: 1791–1798.

Wisseman, C.L., Jr. and A.D. Waddell. 1975. *In vitro* studies on *Rickettsia*-host cell interactions: Intracellular growth cycle of virulent and attenuated *Rickettsia prowazeki* in chicken embryo cells in slide chamber cultures. Infect. Immun. *11*: 1391–1401.

Wisseman, C.L., Jr., A.D. Waddell and D.J. Silverman. 1976. *In vitro* studies on *Rickettsia*-host cell interactions: Lag phase in intracellular growth cycle as a function of stage of growth of infecting *Rickettsia prowazeki*, with preliminary observations on inhibition of rickettsial uptake by host cell fragments. Infect. Immun. *13*: 1749–1760.

Wistreich, G.A. and R.F. Baker. 1971. The presence of fimbriae (pili) in three species of *Neisseria*. J. Gen. Microbiol. *65*: 167–173.

Witkin, S.S. and E. Rosenberg. 1970. Induction of morphogenesis by methionine starvation in *Myxococcus xanthus*: polyamine control. J. Bacteriol. *103*: 641–649.

Witschel, M., S. Nagel and T. Egli. 1997. Identification and characterization of the two-enzyme system catalyzing oxidation of EDTA in the EDTA-degrading bacterial strain DSM 9103. J. Bacteriol. *179*: 6937–6943.

Wittich, R.M., C. Strömpl, E.R. Moore, R. Blasco and K.N. Timmis. 1999. Interaction of *Sphingomonas* and *Pseudomonas* strains in the degradation of chlorinated dibenzofurans. J. Ind. Microbiol. Biotechnol. *23*: 353–358.

Wittich, R.M., H. Wilkes, V. Sinnwell, W. Francke and P. Fortnagel. 1992. Metabolism of dibenzo-*p*-dioxin by *Sphingomonas* sp. strain RW1. Appl. Environ. Microbiol. *58*: 1005–1010.

Wittke, R., W. Ludwig, S. Peiffer and D. Kleiner. 1997. Isolation and characterization of *Burkholderia norimbergensis* sp. nov., a mildly alkaliphilic sulfur oxidizer. Syst. Appl. Microbiol. *20*: 549–553.

Wittke, R., W. Ludwig, S. Peiffer and D. Kleiner. 1998. *In* Validation of the publication of new names and new combinations previously effectively published outside the IJSB. List No. 66. Int. J. Syst. Bacteriol. *48*: 631–632.

Wloczyk, C., A. Kröger, T. Göbel, G. Holdt and R. Steudel. 1989. The electrochemical proton potential generated by sulfur respiration of *Wolinella succinogenes*. Arch. Microbiol. *152*: 600–605.

Wober, W., O.W. Thiele and D. Urbaschek. 1964. Die "Freien Lipide" aus *Brucella abortus* Bang. I. Untersuchung der Phosphatide. Biochem. Biophys. Acta *84*: 376–390.

Wodara, C., F. Bardischewsky and C.G. Friedrich. 1997. Cloning and characterization of sulfite dehydrogenase, two *c*-type cytochromes, and a flavoprotein of *Paracoccus denitrificans* GB17: essential role of sulfite dehydrogenase in lithotrophic sulfur oxidation. J. Bacteriol. *179*: 5014–5023.

Woese, C.R. 1987. Bacterial evolution. Microbiol. Rev. *51*: 221–271.

Woese, C.R., P. Blanz and C.M. Hahn. 1984a. What isn't a pseudomonad: the importance of nomenclature in bacterial classification. Syst. Appl. Microbiol. *5*: 179–195.

Woese, C.R., P. Blanz, R.B. Hespell and C.M. Hahn. 1982. Phylogenetic relationships among various helical bacteria. Curr. Microbiol. *7*: 119–124.

Woese, C.R., S. Maloy, L. Mandelco and H.D. Raj. 1990. Phylogenetic placement of the *Spirosomaceae*. Syst. Appl. Microbiol. *13*: 19–23.

Woese, C.R., E. Stackebrandt, W. Weisburg, B.J. Paster, M.T. Madigan,

V.J. Fowler, C.M. Hahn, P. Blanz, R. Gupta, K.H. Nealson and G.E. Fox. 1984b. The phylogeny of purple bacteria: the alpha subdivision. Syst. Appl. Microbiol. *5*: 315–326.

Woese, C.R., W.G. Weisburg, C.M. Hahn, B.J. Paster, L.B. Zablen, B.J. Lewis, T.J. Macke, W. Ludwig and E. Stackebrandt. 1985. The phylogeny of purple bacteria: the gamma subdivision. Syst. Appl. Microbiol. *6*: 25–33.

Woese, C.R., W.G. Weisburg, B.J. Paster, C.M. Hahn, R.S. Tanner, N.R. Krieg, H.P. Koops, H. Harms and E. Stackebrandt. 1984. The phylogeny of purple bacteria: the beta subdivision. Syst. Appl. Microbiol. *5*: 327–336.

Wolbach, S.B. 1919. Studies on Rocky Mountain spotted fever. J. Med. Res. *41*: 1–197.

Wolbach, S.B. and J.L. Todd. 1920. Note sur l'étiologie et l'anatomie pathologique du typhus exanthématique au Mexique. Ann. Inst. Pasteur (Paris) *34*: 153–158.

Wolf, K. and P.A. Gilbert. 1992. EDTA-Ethylenediaminetetraacetic Acid. *In* Hutzinger and de Oude (Editors), The Handbook of Environmental Chemistry, Vol. 3 Part F, Anthropogenic Compounds, Detergents, Spinger-Verlag, Berlin. pp. 243–259.

Wolf, Y.I., L. Aravind and E.V. Koonin. 1999. Rickettsiae and Chlamydiae: Evidence of horizontal gene transfer and gene exchange. Trends Genet. *15*: 173–175.

Wolfe, R.S. 1964. Iron and manganese bacteria. *In* Heukelekian and Dondero (Editors), Principles and Applications in Aquatic Microbiology, John Wiley & Sons, Inc., New York. pp. 82–97.

Wolfe, R.S. and I.J. Higgins. 1979. Microbial biochemistry of methane — a study of contrasts. *In* Quayle (Editor), Microbial Biochemistry, University Park Press, Baltimore. pp. 267–353.

Wolfe, R.S. and N. Pfennig. 1977. Reduction of sulfur by spirillum 5175 and syntrophism with *Chlorobium*. Appl. Environ. Microbiol. *33*: 427–433.

Wolfe, R.S., R.K. Thauer and N. Pfennig. 1987. A "capillary racetrack" method for isolation of magnetotactic bacteria. FEMS Microbiol. Lett. *45*: 31–36.

Wolff, J., G.H. Cook, A.R. Goldhammer and S.A. Berkowitz. 1980. Calmodulin activates prokaryotic adenylate cyclase. Proc. Natl. Acad. Sci. U. S. A. *77*: 3841–3844.

Wolff, K. and A. Stern. 1995. Identification and characterization of specific sequences encoding pathogenicity associated proteins in the genome of commensal *Neisseria* species. FEMS Microbiol. Lett. *125*: 255–263.

Wolff, M. 1907. *Pedioplana haeckeli* n.g., n.sp. und *Planosarcina schaudinni* n.sp. zwei neue bewegliche Coccaceen. Zentbl. Bakteriol. Parasitenkd. Infektkrankh. Hyg. Abt. II *18*: 9–26.

Wolfgang, M., P. Lauer, H.S. Park, L. Brossay, J. Hebert and M. Koomey. 1998a. *PilT* mutations lead to simultaneous defects in competence for natural transformation and twitching motility in piliated *Neisseria gonorrhoeae*. Mol. Microbiol. *29*: 321–330.

Wolfgang, M., H.S. Park, S.F. Hayes, J.P.M. van Putten and M. Koomey. 1998b. Suppression of an absolute defect in Type IV pilus biogenesis by loss-of-function mutations in *pilT*, a twitching motility gene in *Neisseria gonorrhoeae*. Proc. Natl. Acad. Sci. U.S.A. *95*: 14973–14978.

Wolfgang, M., J.P. van Putten, S.F. Hayes, D. Dorward and M. Koomey. 2000. Components and dynamics of fiber formation define a ubiquitous biogenesis pathway for bacterial pili. Embo J. *19*: 6408–6418.

Wolfrum, T., G. Gruner and H. Stolp. 1986. Nucleic acid hybridization of pink-pigmented facultative methylotrophs and pseudomonads. Int. J. Syst. Bacteriol. *36*: 24–28.

Wolfrum, T. and H. Stolp. 1987. Comparative studies on 5S RNA sequences of ribulose monophosphate-type methylotrophic bacteria. Syst. Appl. Microbiol. *9*: 273–276.

Wolin, E.A., M.J. Wolin and R.S. Wolfe. 1963. Formation of methane by bacterial extracts. J. Biol. Chem. *238*: 2882–2886.

Wolin, M.J., E.A. Wolin and N.J. Jacobs. 1961. Cytochrome-producing anaerobic vibrio, *Vibrio succinogenes*, sp. n. J. Bacteriol. *81*: 911–917.

Wollenweber, H.W. and E.T. Rietschel. 1990. Analysis of lipopolysaccharide (lipid A) fatty acids. J. Microbiol. Methods *11*: 195–211.

Won, Y.S., J.H. Yoon, C.H. Lee, B.H. Kim, B.H. Hyun and Y.K. Choi. 2002. *Helicobacter muricola* sp. nov., a novel *Helicobacter* species isolated from the ceca and feces of Korean wild mouse (*Mus musculus molossinus*). FEMS Microbiol. Lett. *209*: 45–51.

Wong, B., C. Singer, D. Armstrong and S.J. Millian. 1979. Rickettsialpox - Case report and epidemiologic review. J. Am. Med. Assoc. *242*: 1998–1999.

Wong, F.Y.K., E. Stackebrandt, J.K. Ladha, D.E. Fleischman, R.A. Date and J.A. Fuerst. 1994. Phylogenetic analysis of *Bradyrhizobium japonicum* and photosynthetic stem-nodulating bacteria from *Aeschynomene* species grown in separated geographical regions. Appl. Environ. Microbiol. *60*: 940–946.

Wong, H.C., A.L. Fear, R.D. Calhoon, G.H. Eichinger, R. Mayer, D. Amikam, M. Benziman, D.H. Gelfand, J.H. Meade, A.W. Emerick, R. Bruner, A. Ben-Bassat and R. Tal. 1990. Genetic organization of the cellulose synthase operon in *Acetobacter xylinum*. Proc. Natl. Acad. Sci. U.S.A. *87*: 8130–8134.

Wong, K. and T.A. Gill. 1987. Enzymatic determination of trimethylamine and its relationship to fish quality. J. Food. Sci. *52*: 1–3.

Wong, M.T., M.J. Dolan, C.P. Lattuada, R.L. Regnery, M.L. Garcia, E.C. Mokulis, R.C. Labarre, D.P. Ascher, J.A. Delmar, J.W. Kelly, D.R. Leigh, A.C. McRae, J.B. Reed, R.E. Smith and G.P. Melcher. 1995a. Neuroretinitis, aseptic-meningitis, and lymphadenitis associated with *Bartonella (Rochalimaea) henselae* infection in immunocompetent patients and patients infected with human-immunodeficiency-virus type-1. Clin. Infect. Dis. *21*: 352–360.

Wong, M.T., D.C. Thornton, R.C. Kennedy and M.J. Dolan. 1995b. A chemically defined liquid medium that supports primary isolation of *Rochalimaea (Bartonella) henselae* from blood and tissue specimens. J. Clin. Microbiol. *33*: 742–744.

Wong, P.P., N.E. Stenberg and L. Edgar. 1980a. Characterization of a bacterium of the genus *Azospirillum* from cellulolytic nitrogen-fixing mixed cultures. Can. J. Microbiol. *26*: 291–296.

Wong, T.P., R.K. Shockley and K.H. Johnston. 1980b. WSJM, a simple chemically defined medium for growth of *Neisseria gonorrhoeae*. J. Clin. Microbiol. *11*: 363–369.

Wong, W.C. and T.F. Preece. 1979. Identification of *Pseudomonas tolaasi* - white line in agar and mushroom tissue block rapid pitting tests. J. Appl. Bacteriol. *47*: 401–407.

Wood, A.P. and D.P. Kelly. 1988. Isolation and physiological characterization of *Thiobacillus aquaesulis*, sp. nov., a novel facultatively autotrophic moderate thermophile. Arch. Microbiol. *149*: 339–343.

Wood, A.P. and D.P. Kelly. 1995. *In* Validation of the publication of new names and new combinations previously effectively published outside the IJSB. List No. 53. Int. J. Syst. Bacteriol. *45*: 418–419.

Wood, A.P., D.P. Kelly, I.R. McDonald, S.L. Jordan, T.D. Morgan, S. Khan, J.C. Murrell and E. Borodina. 1998. A novel pink-pigmented facultative methylotroph, *Methylobacterium thiocyanatum* sp. nov., capable of growth on thiocyanate or cyanate as sole nitrogen sources. Arch. Microbiol. *169*: 148–158.

Wood, A.P., D.P. Kelly, I.R. McDonald, S.L. Jordan, T.D. Morgan, S. Khan, J.C. Murrell and E. Borodina. 1999. *In* Validation of the publication of new names and new combinations previously effectively published outside the IJSB. List No. 69. Int. J. Syst. Bacteriol. *49*: 341–342.

Wood, A.G., E.M. Menezes, C. Dykstra and D.E. Duggan. 1982. Methods to demonstrate the megaloplasmids (or minichromosomes) in *Azospirillum*. *In* Klingmuller (Editor), *Azospirillum*: Genetics, Physiology, Ecology, Birkhauser, Basel. pp. 18–34.

Woodman, D.R., E. Weiss, G.A. Dasch and F.M. Bozeman. 1977. Biological properties of *Rickettsia prowazekii* strains isolated from flying squirrels. Infect. Immun. *16*: 853–860.

Woodward, T.E. and E.R. Jackson. 1965. Spotted fever rickettsiae. *In* Horsfall and Tamm (Editors), Viral and Rickettsial Infections of Man, 4th Ed., J.B. Lippincott, Philadelphia. pp. 1095–1129.

Woolfrey, B.F. and J.A. Moody. 1991. Human infections associated with *Bordetella bronchiseptica*. Clin. Microbiol. Rev. *4*: 243–255.

Woolley, K.J. 1987. The soluble *c*-type cytochromes from the bacterium

Aquaspirillum itersonii - the complete amino acid sequence of the cytochrome c_{550}. Eur. J. Biochem. *166*: 131–137.

Wright, J.D. and A.R. Barr. 1980. The ultrastructure and symbiotic relationships of *Wolbachia* of mosquitoes of the *Aedes scutellaris* group. J. Ultrastruct. Res. *72*: 52–64.

Wright, J.D., F.S. Sjostrand, J.K. Portaro and A.R. Barr. 1978. The ultrastructure of the rickettsia-like microorganism *Wolbachia pipientis* and associated virus-like bodies in the mosquito *Culex pipiens*. J. Ultrastruct. Res. *63*: 79–85.

Wright, P.J., R.G. Clark and C.N. Hale. 1993. A storage soft rot of New Zealand onions caused by *Pseudomonas gladioli* pv. *alliicola*. N. Z. J. Crop Hortic. Sci. *21*: 225–227.

Wu, B.J. and S.T. Thompson. 1984. Selective medium for *Pseudomonas cepacia* containing 9-chloro-9-(4-diethylaminophenyl)10-phenylacridan and polymyxin B sulfate. Appl. Environ. Microbiol. *48*: 743–746.

Wu, J. and A. Newton. 1997. Regulation of the *Caulobacter* flagellar gene hierarchy; not just for motility. Mol. Microbiol. *24*: 233–239.

Wu, M.C., G.K. Arimura and A.A. Yunis. 1978. Mechanism of sensitivity of cultured pancreatic carcinoma to asparaginase. Int. J. Cancer *22*: 728–733.

Wu, W. and H. Chen. 1991. A new species of genus *Derxia*. Acta Microbiol. Sinica. *30*: 243–248.

Wu, W.M., M.K. Jain, E. Conway De Macario, J.H. Thiele and J.G. Zeikus. 1992. Microbial composition and characterization of prevalent methanogens and acetogens isolated from syntrophic methanogenic granules. Appl. Microbiol. Biotechnol. *38*: 282–290.

Wundt, W. and W.J.B. Morgan. 1975. International committee on systematic bacteriology subcommittee on taxonomy of *Brucella*. Minutes of meeting 3 September 1974. Int. J. Syst. Bacteriol. *25*: 235–236.

Wünsche, L., H. Fischer and B. Kiesel. 1983. Lysogeny and lysogenic conversion in methylotropic bacteria. I. Demonstration of the lysogenic state of the facultative methanol-assimilating strain of *Acetobacter* MB 58/1 and characterization of its temperate phage MO 1. Z. Allg. Mikrobiol. *23*: 81–94.

Wuthiekanun, V., M.D. Smith, D.A. Dance, A.L. Walsh, T.L. Pitt and N.J. White. 1996. Biochemical characteristics of clinical and environmental isolates of *Burkholderia pseudomallei*. J. Med. Microbiol. *45*: 408–412.

Wynn-Williams, D.D. 1983. Distribution and characteristics of *Chromobacterium* in the maritime and sub-Antarctic. Polar Biology. *2*: 101–108.

Xia, Y., T.M. Embley and A.G. O'Donnell. 1994. Phylogenetic analysis of *Azospirillum* by direct sequencing of PCR amplified 16S rDNA. Syst. Appl. Microbiol. *17*: 197–201.

Xie, X.-M., J.-F. You, P.-M. Chen and J.-M. Guo. 1993. On a strain MI 15 of *Agrobacterium radiobacter* for the biological control of grapevine crown gall (English Translation). Acta Microbiol. Sinica. *23*: 137–141.

Xu, L.M., C. Ge, Z. Cui, J. Li and H. Fan. 1995. *Bradyrhizobium liaoningense* sp. nov., isolated from the root nodules of soybeans. Int. J. Syst. Bacteriol. *45*: 706–711.

Xu, L., K. Resing, S.L. Lawson, P.C. Babbitt and S.D. Copley. 1999. Evidence that *pcpA* encodes 2,6-dichlorohydroquinone dioxygenase, the ring cleavage enzyme required for pentachlorophenol degradation in *Sphingomonas chlorophenolica* strain ATCC 39723. Biochemistry *38*: 7659–7669.

Xu, Y., M.W. Mortimer, T.S. Fisher, M.L. Kahn, F.J. Brockman and L. Xun. 1997. Cloning, sequencing, and analysis of a gene cluster from *Chelatobacter heintzii* ATCC 29600 encoding nitrilotriacetate monooxygenase and NADH:flavin mononucleotide oxidoreductase. J. Bacteriol. *179*: 1112–1116.

Xu, Y. and Y. Murooka. 1995. A large plasmid isolated from *Rhizobium huakuii* biovar renge that includes genes for both nodulation of *Astragalus sinicus* cv. Japan and nitrogen fixation. J. Ferment. Bioeng. *80*: 276–279.

Xun, L. 1996. Purification and characterization of chlorophenol 4-monooxygenase from Burkholderia cepacia AC1100. J. Bacteriol. *178*: 2645–2649.

Xun, L., J. Bohuslavek and M. Cai. 1999. Characterization of 2,6-dichloro-

p-hydroquinone 1,2-dioxygenase (PcpA) of *Sphingomonas chlorophenolica* ATCC 39723. Biochem. Biophys. Res. Commun. *266*: 322–325.

Yablonsky, M.D., A.E. Goodman, N. Stevnsborg, O. Goncalves de Lima, J.O. Falcao de Morais, H.G. Lawford, P.L. Rogers and D.E. Eveleigh. 1988. *Zymomonas mobilis* CP4: a clarification of strains via plasmid profiles. J. Biotechnol. *9*: 71–80.

Yabuuchi, E., Y. Kawamura, Y. Kosako and T. Ezaki. 1998a. Emendation of genus *Achromobacter* and *Achromobacter xylosoxidans* (Yabuuchi and Yano) and proposal of *Achromobacter ruhlandii* (Packer and Vishniac) comb. nov., *Achromobacter piechaudii* (Kiredjian et al.) comb. nov., and *Achromobacter xylosoxidans* subsp. *dentrificans* (Ruger and Tan) comb. nov. Microbiol. Immunol. *42*: 429–438.

Yabuuchi, E., Y. Kawamura, Y. Kosako and T. Ezaki. 1998b. *In* Validation of the publication of new names and new combinations previously effectively published outside the IJSB. List No. 67. Int. J. Syst. Bacteriol. *48*: 1083–1084.

Yabuuchi, E., Y. Kosako, N. Fujiwara, T. Naka, I. Matsunaga, H. Ogura and K. Kobayashi. 2002. Emendation of the genus *Sphingomonas* Yabuuchi et al. 1990 and junior objective synonymy of the species of three genera, *Sphingobium*, *Novosphingobium* and *Sphingopyxis*, in conjunction with *Blastomonas ursincola*. Int. J. Syst. Evol. Microbiol. *52*: 1485–1496.

Yabuuchi, E., Y. Kosako, T. Naka, S. Suzuki and I. Yano. 1999a. Proposal of *Sphingomonas suberifaciens* (van Bruggen, Jochimsen and Brown 1990) comb. nov., *Sphingomonas natatoria* (Sly 1985) comb. nov., *Sphingomonas ursincola* (Yurkov et al. 1997) comb. nov., and emendation of the genus *Sphingomonas*. Microbiol. Immunol. *43*: 339–349.

Yabuuchi, E., Y. Kosako, T. Naka, S. Suzuki and I. Yano. 1999b. *In* Validation of the publication of new names and new combinations previously effectively published outside the IJSB. List No. 70. Int. J. Syst. Bacteriol. *49*: 935–936.

Yabuuchi, E., Y. Kosako, H. Oyaizu, I. Yano, H. Hotta, Y. Hashimoto, T. Ezaki and M. Arakawa. 1992. Proposal of *Burkholderia* gen. nov. and transfer of seven species of the genus *Pseudomonas* homology group II to the new genus, with the type species *Burkholderia cepacia* (Palleroni and Holmes 1981) comb. nov. Microbiol. Immunol. *36*: 1251–1275.

Yabuuchi, E., Y. Kosako, H. Oyaizu, I. Yano, H. Hotta, Y. Hashimoto, T. Ezaki and M. Arakawa. 1993. *In* Validation of the publication of new names and new combinations previously effectively published outside the IJSB. List No. 45. Int. J. Syst. Bacteriol. *43*: 398–399.

Yabuuchi, E., Y. Kosako, I. Yano, H. Hotta and Y. Nishiuchi. 1995. Transfer of two *Burkholderia* and an *Alcaligenes* species to *Ralstonia* gen. nov.: Proposal of *Ralstonia pickettii* (Ralston, Palleroni and Doudoroff 1973) comb. nov., *Ralstonia solanacearum* (Smith 1896) comb. nov. and *Ralstonia eutropha* (Davis 1969) comb. nov. Microbiol. Immunol. *39*: 897–904.

Yabuuchi, E., Y. Kosako, I. Yano, H. Hotta and Y. Nishiuchi. 1996. *In* Validation of the publication of new names and new combinations previously effectively published outside the IJSB. List No. 57. Int. J. Syst. Bacteriol. *46*: 625–626.

Yabuuchi, E. and A. Ohyama. 1971. *Achromobacter xylosoxidans* n. sp. from human ear discharge. Jpn. J. Microbiol. *15*: 477–481.

Yabuuchi, E. and A. Ohyama. 1972. Characterization of "pyomelanin"-producing strains of *Pseudomonas aeruginosa*. Int. J. Syst. Bacteriol. *22*: 53–64.

Yabuuchi, E., E. Tanimura, A. Ohyama, I. Yano and A. Yamamoto. 1979. *Flavobacterium devorans* ATCC 10829: A strain of *Pseudomonas paucimobilis*. J. Gen. Appl. Microbiol. *25*: 95–107.

Yabuuchi, E., H. Yamamoto, S. Terakubo, N. Okamura, T. Naka, N. Fujiwara, K. Kobayashi, Y. Kosako and A. Hiraishi. 2001. Proposal of *Sphingomonas wittichii* sp. nov. for strain RW1^{T}, known as a dibenzo-*p*-dioxin metabolizer. Int. J. Syst. Evol. Microbiol. *51*: 281–292.

Yabuuchi, E. and I. Yano. 1981. *Achromobacter* gen. nov. and *Achromobacter xylosoxidans* (ex Yabuuchi and Ohyama 1971) nom. rev. Int. J. Syst. Bacteriol. *31*: 477–478.

Yabuuchi, E., I. Yano, S. Goto, E. Tanimura, T. Ito and A. Ohyama. 1974.

Description of *Achromobacter xylosoxidans* Yabuuchi and Ohyama 1971. Int. J. Syst. Bacteriol. *24*: 470–477.

Yabuuchi, E., I. Yano, H. Oyaizu, Y. Hashimoto, T. Ezaki and H. Yamamoto. 1990a. Proposals of *Sphingomonas paucimobilis* gen. nov. and comb. nov., *Sphingomonas parapaucimobilis* sp. nov., *Sphingomonas yanoikuyae* sp. nov., *Sphingomonas adhaesiva* sp. nov., *Sphingomonas capsulata* comb. nov., and two genospecies of the genus *Sphingomonas*. Microbiol. Immunol. *34*: 99–119.

Yabuuchi, E., I. Yano, H. Oyaizu, Y. Hashimoto, T. Ezaki and H. Yamamoto. 1990b. *In* Validation of the publication of new names and new combinations previously effectively published outside the IJSB. List No. 34. Int. J. Syst. Bacteriol. *40*: 320–321.

Yagupsky, P. and R. Dagan. 1997. *Kingella kingae*: an emerging cause of invasive infections in young children. Clin. Infect. Dis. *24*: 860–866.

Yagupsky, P., R. Dagan, C.W. Howard, M. Einhorn, I. Kassis and A. Simu. 1992. High prevalence of *Kingella kingae* in joint fluid from children with septic arthritis revealed by the BACTEC blood culture system. J. Clin. Microbiol. *30*: 1278–1281.

Yagupsky, P., M. Merires, J. Bahar and R. Dagan. 1995. Evaluation of novel vancomycin-containing medium for primary isolation of *Kingella kingae* from upper respiratory tract specimens. J. Clin. Microbiol. *33*: 1426–1427.

Yagupsky, P. and J. Press. 1997. Use of the Isolator 1.5 microbial tube for culture of synovial fluid from patients with septic arthritis. J. Clin. Microbiol. *35*: 2410–2412.

Yakubu, D.E., F.J. Abadi and T.H. Pennington. 1999. Molecular typing methods for *Neisseria meningitidis*. J. Med. Microbiol. *48*: 1055–1064.

Yamada, H., S.S. Miyazaki and Y. Izumi. 1986. L-Serine production by a glycine-resistant mutant of methylotrophic *Hyphomicrobium methylovorum*. Agric. Biol. Chem. *50*: 17–22.

Yamada, Y. 1984. *In* Validation of the publication of new names and new combinations previously effectively published outside the IJSB. List No. 14. Int. J. Syst. Bacteriol. *34*: 270–271.

Yamada, Y. 2000. Transfer of *Acetobacter oboediens* Sokollek et al. 1998 and *Acetobacter intermedius* Boesch et al. 1998 to the genus *Gluconacetobacter* as *Gluconacetobacter oboediens* comb. nov. and *Gluconacetobacter intermedius* comb. nov. Int. J. Syst. Evol. Microbiol. *50*: 2225–2227.

Yamada, Y., K. Aida and T. Uemura. 1969. Enzymatic studies on the oxidation of sugar and sugar alcohol. V. Ubiquinone of acetic acid bacteria and its relation to classification of genera *Gluconobacter* and *Acetobacter*, especially of the so-called intermediate strains. J. Gen. Appl. Microbiol. *15*: 181–196.

Yamada, Y. and M. Akita. 1984a. An electrophoretic comparison of enzymes in strains of *Gluconobacter* species. J. Gen. Appl. Microbiol. *30*: 115–126.

Yamada, Y. and M. Akita. 1984b. *In* Validation of the publication of new names and new combinations previously effectively published outside the IJSB. List No. 16. Int. J. Syst. Bacteriol. *34*: 503–504.

Yamada, Y., K. Hoshino and T. Ishikawa. 1997. The phylogeny of acetic acid bacteria based on the partial sequences of 16S ribosomal RNA: the elevation of the subgenus *Gluconoacetobacter* to generic level. Biosci. Biotechnol. Biochem. *61*: 1244–1251.

Yamada, Y., K. Hoshino and T. Ishikawa. 1998a. *In* Validation of the publication of new names and new combinations previously effectively published outside the IJSB. List No. 64. Int. J. Syst. Bacteriol. *48*: 327–328.

Yamada, Y., K. Hoshino and T. Ishikawa. 1998b. *In* Validation of the publication of new names and new combinations previously effectively published outside the IJSB. List No. 64. Int. J. Syst. Bacteriol. *48*: 327–328.

Yamada, Y., R. Hosono, P. Lisdiyanti, Y. Widyastuti, S. Saono, T. Uchimura and K. Komagata. 1999. Identification of acetic acid bacteria isolated from Indonesian sources, especially of isolates classified in the genus *Gluconobacter*. J. Gen. Appl. Microbiol. *45*: 23–28.

Yamada, Y., K. Katsura, H. Kawasaki, Y. Widyastuti, S. Saono, T. Seki, T. Uchimura and K. Komagata. 2000. *Asaia bogorensis* gen. nov., sp. nov., an unusual acetic acid bacterium in the alpha-*Proteobacteria*. Int. J. Syst. Evol. Microbiol. *50*: 823–829.

Yamada, Y. and K. Kondo. 1984. *Gluconoacetobacter*, a new subgenus comprising the acetate-oxidizing acetic acid bacteria with ubiquinone-10 in the genus *Acetobacter*. J. Gen. Appl. Microbiol. *30*: 297–303.

Yamada, Y., M. Nunoda, T. Ishikawa and Y. Tahara. 1981. Taxonomic studies on acetic acid bacteria and allied organisms. 3. The cellular fatty acid composition in acetic acid bacteria. J. Gen. Appl. Microbiol. *27*: 405–417.

Yamada, Y., H. Takinaminakamura, Y. Tahara, H. Oyaizu and K. Komagata. 1982. Significance of the ubiquinone and menaquinone systems in the classification of Gram-negative and Gram-positive bacteria VIII. The ubiquinone systems in the strains of *Pseudomonas* species. J. Gen. Appl. Microbiol. *28*: 7–12.

Yamagishi, Y., J. Fujita, K. Takigawa, H. Miyawaki, Y. Yamaji, J. Takahara, K. Negayama, K. Kawanishi, M. Abe and T. Nakazawa. 1993. Epidemiological study on nosocomial infection of *Pseudomonas cepacia* by plasmid analyses. J. Jpn. Assoc. Infect. Dis. *67*: 1160–1166.

Yamamoto, A., I. Yano, M. Masui and E. Yabuuchi. 1978. Isolation of a novel sphingoglycolipid containing glucuronic acid and 2-hydroxy fatty acid from *Flavobacterium devorans* ATCC 10829. J. Biochem. (Tokyo) *83*: 1213–1216.

Yamamoto, K., B.B. Chomel, R.W. Kasten, C.C. Chang, T. Tseggai, P.R. Decker, M. Mackowiak, K.A. Floyd Hawkins and N.C. Pedersen. 1998. Homologous protection but lack of heterologous-protection by various species and types of *Bartonella* in specific pathogen-free cats. Vet. Immunol. Immunopathol. *65*: 191–204.

Yamamoto, K., K. Oishi, I. Fujimatsu and K. Komatsu. 1991. Production of R-(−)-mandelic acid from mandelonitrile by *Alcaligenes faecalis* ATCC 8750. Appl. Environ. Microbiol. *57*: 3028–3032.

Yamamoto, S. 1978. Studies on the causative agent of Hyuga-feveri: Cultivation of rickettsia-like organisms isolated from metacercariae of *Stellantchasumus falcatus* in tissue culture cell and their antigenic relation to *Rickettsia sennetsu*. Kansenshogaku Zasshi. *52*: 240–245.

Yamamoto, T., P. Naigowit, S. Dejsirilert, D. Chiewsilp, E. Kondo, T. Yokota and K. Kanai. 1990. *In vitro* susceptibilities of *Pseudomonas pseudomallei* to 27 antimicrobial agents. Antimicrob. Agents Chemother. *34*: 2027–2029.

Yamanaka, S., R. Fudo, A. Kawaguchi and K. Komagata. 1988. Taxonomic significance of hydroxy fatty acids in myxobacteria with special reference to 2-hydroxy fatty acids in phospholipids. J. Gen. Appl. Microbiol. *34*: 57–66.

Yamanaka, S., S. Kanbe and R. Fudo. 1993. Lysis of basidiomycetous yeast, *Rhodotorula glutinis* caused by myxobacteria. J. Gen. Appl. Microbiol. *39*: 419–427.

Yamanaka, S., A. Kawaguchi and K. Komagata. 1987. Isolation and identification of myxobacteria from soils and plant materials, with special reference to DNA base composition, quinone system, and cellular fatty acid composition, and with a description of a new species, *Myxococcus flavescens*. J. Gen. Appl. Microbiol. *33*: 247–265.

Yamanaka, T. and Y. Fukumori. 1988. The nitrite oxidizing system of *Nitrobacter winogradskyi*. FEMS Microbiol. Rev. *4*: 259–270.

Yamano, S., J. Tanaka and T. Inoue. 1994. Cloning and expression of the gene encoding alpha-acetolactate decarboxylase from *Acetobacter aceti* ssp. *xylinum* in brewer's yeast. J. Biotechnol. *32*: 165–171.

Yamasaki, R., D.E. Kerwood, H. Schneider, K.P. Quinn, J.M. Griffiss and R.E. Mandrell. 1994. The structure of lipooligosaccharide produced by *Neisseria gonorrhoeae*, strain-15253, isolated from a patient with disseminated infection—evidence for a new glycosylation pathway of the gonococcal lipooligosaccharide. J. Biol. Chem. *269*: 30345–30351.

Yamasato, K., M. Akagawa, N. Oishi and H. Kuraishi. 1982. Carbon substrate assimilation profiles and other taxonomic features of *Alcaligenes faecalis*, *Alcaligenes ruhlandii* and *Achromobacter xylosoxidans*. J. Gen. Appl. Microbiol. *28*: 195–213.

Yamazaki, T., H. Oyanagi, T. Fujiwara and Y. Fukumori. 1995. Nitrite reductase from the magnetotactic bacterium *Magnetospirillum magnetotacticum*: a novel cytochrome cd-1 with Fe(II):nitrite oxidoreductase activity. Eur. J. Biochem. *233*: 665–671.

Yanagi, M. and K. Yamasato. 1993. Phylogenetic analysis of the family

Rhizobiaceae and related bacteria by sequencing of 16S rRNA gene using PCR and DNA sequencer. FEMS Microbiol. Lett. *107*: 115–120.

Yang, C. and H.B. Kaplan. 1997. *Myxococcus xanthus sasS* encodes a sensor histidine kinase required for early developmental gene expression. J. Bacteriol. *179*: 7759–7767.

Yang, C.C., L.C. Packman and N.S. Scrutton. 1995a. The primary structure of *Hyphomicrobium* X dimethylamine dehydrogenase: relationship to trimethylamine dehydrogenase and implications for substrate recognition. Eur. J. Biochem. *232*: 264–271.

Yang, G.-P., F. Debellé, A. Savagnac, M. Ferro, O. Schiltz, F. Maillet, D. Promé, M. Treilhou, C. Vialas, K. Lindstrom, J. Dénarié and J.C. Promé. 1999. Structure of the *Mesorhizobium huakuii* and *Rhizobium galegae* Nod factors: a cluster of phylogenetically related legumes are nodulated by rhizobia producing Nod factors with α,β-unsaturated *N*-acyl substitutions. Mol. Microbiol. *34*: 227–237.

Yang, H.M., W. Chaowagul and P.A. Sokol. 1991a. Siderophore production by *Pseudomonas pseudomallei*. Infect. Immun. *59*: 776–780.

Yang, H., C.D. Kooi and P.A. Sokol. 1993. Ability of *Pseudomonas pseudomallei* malleobactin to acquire transferrin bound, lactoferrin bound, and cell derived iron. Infect. Immun. *61*: 656–662.

Yang, Q.L. and E.C. Gotschlich. 1996. Variation of gonococcal lipooligosaccharide structure is due to alterations in poly-G tracts in *lgt* genes encoding glycosyl transferases. J. Exp. Med. *183*: 323–327.

Yang, S., S. Tong and Z. Lu. 1995b. Geographical distribution of *Pseudomonas pseudomallei* in China. Southeast Asian J. Trop. Med. Public Health *26*: 636–638.

Yang, Y.B., C.S. Lin, C.P. Tseng, Y.J. Wang and Y.C. Tsai. 1991b. Purification and characterization of D-aminoacylase from *Alcaligenes faecalis* strain DA1. Appl. Environ. Microbiol. *57*: 1259–1260.

Yanni, Y.G., R.Y. Rizk, V. Corich, A. Squartini, K. Ninke, S. Philip-Hollingsworth, G. Orgambide, F. deBruijn, J. Stoltzfus, D. Buckley, T.M. Schmidt, P.F. Mateos, J.K. Ladha and F.B. Dazzo. 1997. Natural endophytic association between *Rhizobium leguminosarum* biovar trifolii and rice roots and assessment of its potential to promote rice growth. Plant Soil *194*: 99–114.

Yasuda, M., T. Sakamoto, R. Sashida, M. Ueda, Y. Morimoto and T. Nagasawa. 1995. Microbial hydroxylation of 3-cyanopyridine to 3-cyano-6-hydroxypyridine. Biosci. Biotechnol. Biochem. *59*: 572–575.

Yates, M.G., J. O'Donnell, D.J. Lowe and H. Bothe. 1978. Ferredoxins from nitrogen-fixing bacteria. Physical and chemical characterisation of two ferredoxins from *Mycobacterium flavam* 301. Eur. J. Biochem. *85*: 291–299.

Yee, T. and M. Inouye. 1982. Two-dimensional DNA electrophoresis applied to the study of DNA methylation and the analysis of genome size in *Myxococcus xanthus*. J. Mol. Biol. *154*: 181–196.

Yelton, D.B., V.F. Gerencser and H.G. Voelz. 1979. Isolation and preliminary characterization of 3 bacteriophages which adsorb specifically to the developing daughter cells of *Hyphomicrobium*. J. Gen. Virol. *43*: 29–38.

Yen, H.C. and B. Marrs. 1977. Growth of *Rhodopseudomonas capsulata* under anaerobic dark conditions with dimethyl sulfoxide. Arch. Biochem. Biophys. *181*: 411–418.

Yen, J.H. and A.R. Barr. 1974. Incompatability in *Culex pipiens*. *In* Pal and Whitten (Editors), The Use of Genetics in Insect Control, Elsevier, Amsterdam. pp. 97–118.

Yih, W.K., E.A. Silva, J. Ida, N. Harrington, S.M. Lett and H. George. 1999. *Bordetella holmesii*-like organisms isolated from Massachusetts patients with pertussis-like symptoms. Emerg. Infect. Dis. *5*: 441–443.

Yildiz, F.H., H. Gest and C.E. Bauer. 1991. Attenuated effect of oxygen on photopigment synthesis in *Rhodospirillum centenum*. J. Bacteriol. *173*: 5502–5506.

Yohalem, D.S. and J.W. Lorbeer. 1994. Intraspecific metabolic diversity among strains of *Burkholderia cepacia* isolated from decayed onions, soils, and the clinical environment. Antonie Leeuwenhoek *65*: 111–131.

Yokota, A. 1989. Taxonomic significance of cellular fatty acid composition in Rhizobium, Bradyrhizobium and Agrobacterium species. IFO Res. Comm. *14*: 25–39.

Yoneyama, A., H. Yano, S. Hitomi, K. Okuzumi, R. Suzuki and S. Kimura. 2000. *Ralstonia pickettii* colonization of patients in an obstetric ward caused by a contaminated irrigation system. J. Hosp. Infect. *46*: 79–80.

Yoon, H.J., W. Hashimoto, Y. Katsuya, Y. Mezaki, K. Murata and B. Mikami. 2000a. Crystallization and preliminary X-ray crystallographic analysis of alginate lyase A1-II from *Sphingomonas* species A1. Biochim. Biophys. Acta *1476*: 382–385.

Yoon, H.J., W. Hashimoto, O. Miyake, M. Okamoto, B. Mikami and K. Murata. 2000b. Overexpression in *Escherichia coli*, purification, and characterization of *Sphingomonas* sp. A1 alginate lyases. Protein Expr. Purif. *19*: 84–90.

Yordy, J.R. and T.L. Weaver. 1977. *Methylobacillus*: a new genus of obligately methylotrophic bacteria. Int. J. Syst. Bacteriol. *27*: 247–255.

York, G.M., B.H. Junker, J.A. Stubbe and A.J. Sinskey. 2001. Accumulation of the PhaP phasin of *Ralstonia eutropha* is dependent on production of polyhydroxybutyrate in cells. J. Bacteriol. *183*: 4217–4226.

York, W.S., M. McNeil, A.G. Darvill and P. Albersheim. 1980. Beta-2-linked glucans secreted by fast-growing species of Rhizobium. J. Bacteriol. *142*: 243–248.

Yoshida, T., K. Fukuta, T. Mitsunaga, H. Yamada and Y. Izumi. 1992. Purification and characterization of glycerate kinase from a serine-producing methylotroph, *Hyphomicrobium methylovorum* GM2. Eur. J. Biochem. *210*: 849–854.

Yoshida, T., T. Mitsunaga, H. Yamada and Y. Izumi. 1993. Enzymatic assay for L-serine and glyoxylate involving the enzymes in the serine pathway of a methylotroph. Anal. Biochem. *208*: 296–299.

Yoshida, T., Y. Tanaka, T. Hagishita, T. Mitsunaga and Y. Izumi. 1995a. Evidence for the existence of isocitrate lyase activity in methylotrophic *Hyphomicrobium* strains. FEMS Microbiol. Lett. *126*: 221–225.

Yoshida, T., Y. Tanaka, T. Mitsunaga and Y. Izumi. 1995b. CoASAc-independent phosphoenolpyruvate carboxylase from an obligate methylotroph *Hyphomicrobium methylovorum* GM2-purification and characterization. Biosci. Biotechnol. Biochem. *59*: 140–142.

Yoshida, T., K. Yamaguchi, T. Hagishita, T. Mitsunaga, A. Miyata, T. Tanabe, H. Toh, T. Ohshiro, M. Shimao and Y. Izumi. 1994. Cloning and expression of the gene for hydroxypyruvate reductase (D-glycerate dehydrogenase) from an obligate methylotroph *Hyphomicrobium methylovorum* GM2. Eur. J. Biochem. *223*: 727–732.

Yoshida, Y., R. Suzuki and Y. Yagi. 1990. Production of levan by a *Zymomonas* sp. J. Ferment. Bioeng. *70*: 269–271.

Yoshiie, K., H.Y. Kim, J. Mott and Y. Rikihisa. 2000. Intracellular infection by the human granulocytic ehrlichiosis agent inhibits human neutrophil apoptosis. Infect. Immun. *68*: 1125–1133.

Yoshimatsu, K., T. Fujiwara and Y. Fukumori. 1995. Purification, primary structure, and evolution of cytochrome *c*-550 from the magnetic bacterium, *Magnetospirillum magnetotacticum*. Arch. Microbiol. *163*: 400–406.

Yoshinari, T. 1980. N_2O reduction by *Vibrio succinogenes*. Appl. Environ. Microbiol. *39*: 81–84.

Yoshinari, T. 1985. Nitrite and nitrous oxide production by *Methylosinus trichosporium*. Can. J. Microbiol. *31*: 139–144.

You, C.B., W. Song, H.X. Wang, J.P. Li, M. Lin and W.L. Hai. 1991. Association of *Alcaligenes faecalis* with wetland rice. *In* Polsinelli, Materassi and Vincenzini (Editors), Nitrogen Fixation, Kluwer Academic Publishers, Dordrecht. pp. 195–199.

Young, E.J. and M.J. Corbel (Editors). 1989. Brucellosis: Clinical and Laboratory Aspects, CRC Press Inc., Boca Raton.

Young, J.P.W. 1992. Phylogenetic classification of nitrogen-fixing organisms. *In* Stacey, Roberts and Evans (Editors), Biological Nitrogen Fixation, Chapman & Hall, New York. pp. 43–86.

Young, J.P.W. 1994. All those new names an overview of the molecular phylogeny of plant-associated bacteria. *In* Daniels, Downie and Osbourn (Editors), Advances in Molecular Genetics of Plant-Microbe Interactions, Vol. 3, Kluwer Academic, Dordrecht. Boston. pp. 73–80.

Young, J.M., D.W. Dye, J.F. Bradbury, C.G. Panagopoulos and C.F. Robbs. 1978. A proposed nomenclature and classification for plant pathogenic bacteria. N. Z. J. Agric. Res. *21*: 153–177.

Young, J.P.W. and K.E. Haukka. 1996. Diversity and phylogeny of rhizobia. New Phytol. *133*: 87–94.

Young, J.P.W. and A.W.B. Johnston. 1989. The evolution of specificity in the legume *Rhizobium* symbiosis. Trends Ecol. Evol. *4*: 341–349.

Young, J.M., L.D. Kuykendall, E. Martínez-Romero, A. Kerr and H. Sawada. 2001. A revision of *Rhizobium* Frank 1889, with an emended description of the genus, and the inclusion of all species of *Agrobacterium* Conn 1942 and *Allorhizobium undicola* de Lajudie et al. 1998 as new combinations: *Rhizobium radiobacter, R. rhizogenes, R. rubi, R. undicola* and *R. vitis*. Int. J. Syst. Evol. Microbiol. *51*: 89–103.

Young, J.M., Y. Takikawa, L. Gardan and D.E. Stead. 1992. Changing concepts in the taxonomy of plant pathogenic bacteria. Annu. Rev. Phytopathol. *30*: 67–105.

Young, K.A., R.P. Allaker, J.M. Hardie and R.A. Whiley. 1996. Interactions between *Eikenella corrodens* and "*Streptococcus milleri* group" organisms: possible mechanisms of pathogenicity in mixed infections. Antonie Leeuwenhoek *69*: 371–373.

Yrjala, K., S. Suomalainen, E.L. Suhonen, S. Kilpi, L. Paulin and M. Romantschuk. 1998. Characterization and reclassification of an aromatic- and chloroaromatic-degrading *Pseudomonas* sp., strain HV3, as *Sphingomonas* sp. HV3. Int. J. Syst. Bacteriol. *48*: 1057–1062.

Yu, X., J.W. McBride, X. Zhang and D.H. Walker. 2000a. Characterization of the complete transcriptionally active *Ehrlichia chaffeensis* 28 kDa outer membrane protein multigene family. Gene *248*: 59–68.

Yu, X.-J., X.-F. F. Zhang, V.L. Popov and I. Aravind. 2000b. Characterization of a surface protein of *Rickettsia prowazekii* containing sequence similar to the catalytic domain of phospholipase A2. 49th Annual Meeting of the American Society of Tropical Medicine and Hygiene, Houston, Texas. p. 339.

Yu, Y., L.H. Wang and S.C. Tu. 1987. *Pseudomonas cepacia* 3-hydroxybenzoate 6-hydroxylase stereochemistry, isotope effects, and kinetic mechanism. Biochemistry *26*: 1105–1110.

Yu, Z.T. and W.W. Mohn. 1999. Isolation and characterization of thermophilic bacteria capable of degrading dehydroabietic acid. Can. J. Microbiol. *45*: 513–519.

Yuen, K.Y., P.C. Woo, J.L. Teng, K.W. Leung, M.K. Wong and S.K. Lau. 2001. *Laribacter hongkongensis* gen. nov., sp. nov., a novel gram-negative bacterium isolated from a cirrhotic patient with bacteremia and empyema. J. Clin. Microbiol. *39*: 4227–4232.

Yumoto, H., H. Azakami, H. Nakae, T. Matsuo and S. Ebisu. 1996. Cloning, sequencing and expression of an *Eikenella corrodens* gene encoding a component protein of the lectin-like adhesin complex. Gene *183*: 115–121.

Yun, N.R., Y.K. Shin, S.Y. Hwang, H. Kuraishi, J. Sugiyama and K. Kawahara. 2000a. Chemotaxonomic and phylogenetic analyses of *Sphingomonas* strains isolated from ears of plants in the family Gramineae and a proposal of *Sphingomonas roseoflava* sp. nov. J. Gen. Appl. Microbiol. *46*: 9–18.

Yun, N.R., Y.K. Shin, S.Y. Hwang, H. Kuraishi, J. Sugiyama and K. Kawahara. 2000b. *In* Validation of the publication of new names and new combinations previously effectively published outside the IJSEM. List No. 75. Int. J. Syst. Evol. Microbiol. *50*: 1415–1417.

Yunker, C.E. 1995. Current status of *in vitro* cultivation of *Cowdria ruminantium*. Vet. Parasitol. *57*: 205–211.

Yurkov, V.V. and J.T. Beatty. 1998a. Aerobic anoxygenic phototrophic bacteria. Microbiol. Mol. Biol. Rev. *62*: 695–724.

Yurkov, V.V. and J.T. Beatty. 1998b. Isolation of aerobic anoxygenic photosynthetic bacteria from black smoker plume wates of the Juan de Fuca Ridge in the Pacific Ocean. Appl. Environ. Microbiol. *64*: 337–341.

Yurkov, V., N. Gad'on, A. Angerhofer and G. Drews. 1994a. Light-harvesting complexes of aerobic bacteriochlorophyll-containing bacteria *Roseococcus thiosulfatophilus* RB2 and *Erythromicrobium ramosum* E5 and the transfer of excitation energy from carotenoids to bacteriochlorophyll. Zeitschrift fur Naturforsch. *49*: 579–586.

Yurkov, V., N. Gad'on and G. Drews. 1993. The major part of polar carotenoids of the aerobic bacteria *Roseococcus thiosulfatophilus* RB3 and *Erythromicrobium ramosum* E5 is not bound to the bacteriochloro-

phyll *a*-complexes of the photosynthetic apparatus. Arch. Microbiol. *160*: 372–376.

Yurkov, V.V. and V.M. Gorlenko. 1990. *Erythrobacter sibiricus*, sp. nov., a new fresh-water aerobic bacterial species containing bacteriochlorophyll *a*. Mikrobiologiya *59*: 120–126.

Yurkov, V.V. and V.M. Gorlenko. 1991. Freshwater aerobic bacteria containing bacteriochlorophyll *a*, *Roseococcus* gen.nov. Mikrobiologiya *60*: 902–907.

Yurkov, V.V. and V.M. Gorlenko. 1992. New species of aerobic bacteria from the genus *Erythromicrobium* containing bacteriochlorophyll *a*. Mikrobiologiya *61*: 163–168.

Yurkov, V.V., V.M. Gorlenko and E.I. Kompantseva. 1992a. A new genus of orange-coloured bacteria containing bacteriochlorophyll *a*: *Erythromicrobium* gen. nov. Mikrobiologiya *61*: 256–260.

Yurkov, V., J. Jappe and A. Vermeglio. 1996. Tellurite resistance and reduction by obligately aerobic photosynthetic bacteria. Appl. Environ. Microbiol. *62*: 4195–4198.

Yurkov, V., E.N. Krasil'nikov and V.M. Gorlenko. 1992b. Enzymes involved in heterotrophic carbon metabolism of aerobic *Erythrobacter sibericus* and *Erythrobacter longus*, bacteria containing bacteriochlorophyll *a*. Microbiology *60*: 401–403.

Yurkov, V.V., E.N. Krasilnikova and V.M. Gorlenko. 1994b. Thiosulfate metabolism by aerobic bacteriochlorophyll *a*-containing bacteria *Erythromicrobium hydrolyticum* and *Roseococcus thiosulfatophilus*. Mikrobiologiya *63*: 181–187.

Yurkov, V.V., S. Krieger, E. Stackebrandt and T. Beatty. 1999. *Citromicrobium bathyomarinum*, a novel aerobic bacterium isolated from deep-sea hydrothermal vent plume waters that contains photosynthetic pigment-protein complexes. J. Bacteriol. *181*: 4517–4525.

Yurkov, V.V., A.M. Lysenko and V.M. Gorlenko. 1991a. Hybridization analysis of the classification of bacteriochlorophyll *a*-containing freshwater aerobic bacteria. Microbiology *60*: 362–366.

Yurkov, V., V. Menin, B. Schoepp and A. Vermeglio. 1998a. Purification and characterization of reaction centers from the obligate aerobic phototrophic bacteria *Erythrobacter litoralis*, *Erythromonas ursincola* and *Sandaracinobacter sibiricus*. Photosynth. Res. *57*: 129–138.

Yurkov, V.V., L.L. Mityushina and V.M. Gorlenko. 1991b. The fine structure of the aerobic bacterium *Erythrobacter sibiricus* containing bacteriochlorophyll *a*. Mikrobiologiya *60*: 339–344.

Yurkov, V., B. Schoepp and A. Vermeglio. 1995. Electron transfer carriers in obligately aerobic photosynthetic bacteria from genera *Roseococcus* and *Erythromicrobium*. *In* Matthis (Editor), Photosynthesis: From Light to Biosphere, Kluwer Academic Publishers, Dordrecht. pp. 545–546.

Yurkov, V., B. Schoepp and A. Vermeglio. 1998b. Photoinduced electron transfer and cytochrome content in obligate aerobic phototrophic bacteria from genera *Erythromicrobium*, *Sandaracinobacter, Erythromonas, Roseococcus* and *Erythrobacter*. Photosyn. Res. *57*: 117–128.

Yurkov, V., E. Stackebrandt, O. Buss, A. Vermeglio, V. Gorlenko and J.T. Beatty. 1997. Reorganization of the genus *Erythromicrobium*: description of "*Erythromicrobium sibiricum*" as *Sandaracinobacter sibiricus* gen. nov., sp. nov., and of "*Erythromicrobium ursincola*" as *Erthyromonas ursincola* gen. nov., sp. nov. Int. J. Syst. Bacteriol. *47*: 1172–1178.

Yurkov, V., E. Stackebrandt, A. Holmes, J.A. Fuerst, P. Hugenholtz, J. Golecki, N. Gad'on, V.M. Gorlenko, E.I. Kompantseva and G. Drews. 1994c. Phylogenetic positions of novel aerobic, bacteriochlorophyll *a*-containing bacteria and description of *Roseococcus thiosulfatophilus* gen. nov., sp. nov., *Erythromicrobium ramosum* gen. nov., sp. nov., and *Erythrobacter litoralis* sp. nov. Int. J. Syst. Bacteriol. *44*: 427–434.

Zachariou, M. and R.K. Scopes. 1986. Glucose-fructose oxidoreductase, a new enzyme isolated from *Zymomonas mobilis* that is responsible for sorbitol production. J. Bacteriol. *167*: 863–869.

Zafar, Y., K.A. Malik and E. Niemann. 1987. Studies on nitrogen-fixing bacteria associated with salt-tolerant grass, *Leptochloa fusca* (L.) Kunth. J. Appl. Microbiol. Biotechnol. *3*: 45–56.

Zaitsev, G.M., T.V. Tsoi, V.G. Grishenkov, E.G. Plotnikova and A.M. Boronin. 1991. Genetic control of degradation of chlorinated benzoic acids in *Arthrobacter globiformis*, *Corynebacterium sepedonicum* and *Pseudomonas cepacia* strains. FEMS Microbiol. Lett. *81*: 171–176.

Zak, K., J.L. Diaz, D. Jackson and J.E. Heckels. 1984. Antigenic variation during infection with *Neisseria gonorrhoeae*—Detection of antibodies to surface proteins in sera of patients with gonorrhea. J. Infect. Dis. *149*: 166–174.

Zambryski, P.C. 1992. Chronicles from the *Agrobacterium*-plant cell DNA transfer story. Annu. Rev. Plant Physiol. Plant Molec. Biol. *43*: 465–490.

Zambryski, P., M. Holsters, K. Kruger, A. Depicker, J. Schell, M. Van Montagu and H.M. Goodman. 1980. Tumor DNA structure in plant cells transformed by *A. tumefaciens*. Science *209*: 1385–1391.

Zarnt, G., T. Schrader and J.R. Andreesen. 2001. Catalytic and molecular properties of the quinohemoprotein tetrahydrofurfuryl alcohol dehydrogenase from *Ralstonia eutropha* strain Bo. J. Bacteriol. *183*: 1954–1960.

Zavarzin, G.A. 1960. The life cycle and nuclear apparatus in *Hyphomicrobium vulgare* Stutzer and Hartleb. Mikrobiol. *29*: 38–42.

Zavarzin, G.A. 1961. Budding bacteria. Mikrobiologiya *30*: 952–975.

Zavarzin, G.A. 1970. The notion of microflora of dispersion in the carbon cycle. J. Gen. Biol. Akad. Nauk. S.S.S.R. *31*: 386–393.

Zavarzin, G.A. 1978. Hydrogen Bacteria and Carboxydobacteria, Nauka Publishing, Moscow.

Zavarzin, G. and A.N. Nozhevnikova. 1976. CO oxidizing bacteria. *In* Schlegel, Gottschalk and Pfenning (Editors), Microbial Production and Utilization of Gases, E. Goltze KG, Gottingen. pp. 207–213.

Zavarzin, G.A. and A.N. Nozhevnikova. 1977. Aerobic carboxydobacteria. Microb. Ecol. *3*: 305–326.

Zavarzin, G.A., T.N. Zhilina and E.V. Pikuta. 1996. Secondary anaerobes in haloalkaliphilic communities in lakes of Tuva. Mikrobiologiya *65*: 480–486.

Zdrodovskii, P.F. 1949. Systematics and comparative characterization of endemic rickettsioses. Zhur. Mikrobiol. Epidemiol. *10*: 19–28.

Zeggel, B. 1993. Steroide bei Myxobakterien, Technical University Braunschweig, Germany, . p. 134.

Zellner, G., K. Bleicher, E. Braun, H. Kneifel, B.J. Tindall, E. Conway de Macario and J. Winter. 1990. *In* Validation of the publication of new names and new combinations previously effectively published outside the IJSB. List No. 35. Int. J. Syst. Bacteriol. *40*: 470–471.

Zellner, G., A. Busmann, F.A. Rainey and H. Diekmann. 1996. A syntrophic propionate-oxidizing, sulfate-reducing bacterium from a fluidized bed reactor. Syst. Appl. Microbiol. *19*: 414–420.

Zellner, G., E. Feuerhake, H.J. Jördening, A.J.L. Macario and E. Conway De Macario. 1995. Denitrifying and methanogenic bacteria in the biofilm of a fixed-film reactor operated with methanol/nitrate demonstrated by immunofluorescence and microscopy. Appl. Microbiol. Biotechnol. *43*: 566–571.

Zellner, G., P. Messner, H. Kneifel and J. Winter. 1989. *Desulfovibrio simplex* spec. nov., a new sulfate-reducing bacterium from a sour whey digester. Arch. Microbiol. *152*: 329–334.

Zeph, L.R. 1986. Indirect phage analysis studies of the activity of non-obligate predator bacteria in soil, Thesis, Pennsylvania State University, University Park, Pennsylvania.

Zeph, L.R., . and L.E. Casida, Jr. 1986. Gram-negative vs. gram-positive (actinomycete) nonobligate bacterial predators of bacteria in soil. Appl. Environ. Microbiol. *52*: 819–823.

Zerfas, P.M., M. Kessel, E.J. Quintero and R.M. Weiner. 1997. Fine-structure evidence for cell membrane partitioning of the nucleoid and cytoplasm during bud formation in *Hyphomonas* species. J. Bacteriol. *179*: 148–156.

Zhang, H., S. Hanada, T. Shigematsu, K. Shibuya, Y. Kamagata, T. Kanagawa and R. Kurane. 2000a. *Burkholderia kururiensis* sp. nov., a trichloroethylene (TCE)- degrading bacterium isolated from an aquifer polluted with TCE. Int. J. Syst. Evol. Microbiol. *50*: 743–749.

Zhang, J.Z., M.Y. Fan, Y.M. Wu, P.E. Fournier, V. Roux and D. Raoult. 2000b. Genetic classification of "*Rickettsia heilongjiangii*" and "*Rickettsia hulinii*", two Chinese spotted fever group rickettsiae. J. Clin. Microbiol. *38*: 3498–3501.

Zhang, L., M. Hirai and M. Shoda. 1991. Removal characteristics of di-

methyl sulfide, methanethiol and hydrogen sulfide by *Hyphomicrobium* sp. I55 isolated from peat biofilter. J. Ferment. Bioeng. *72*: 392–396.

Zhang, M., C. Eddy, K. Deanda, M. Finkelstein and S. Picataggio. 1995. Metabolic engineering of a pentose metabolism pathway in ethanologenic *Zymomonas mobilis*. Science *267*: 240–243.

Zhang, Q.Y., D. Deryckere, P. Lauer and M. Koomey. 1992. Gene conversion in *Neisseria gonorrhoeae*—evidence for its role in pilus antigenic variation. Proc. Natl. Acad. Sci. U.S.A. *89*: 5366–5370.

Zhang, S., T. Yasuo, R.W. Lenz and S. Goodwin. 2000c. Kinetic and mechanistic characterization of the polyhydroxybutyrate synthase from *Ralstonia eutropha*. Biomacromolecules. *1*: 244–251.

Zhang, Y., N. Ohashi, E.H. Lee, A. Tamura and Y. Rikihisa. 1997. *Ehrlichia sennetsu groE* operon and antigenic properties of the *GroEL* homolog. FEMS Immunol. Med. Microbiol. *18*: 39–46.

Zhang, Y., N. Ohashi and Y. Rikihisa. 1998. Cloning of the heat shock protein 70 (HSP70) gene of *Ehrlichia sennetsu* and differential expression of HSP70 and HSP60 mRNA after temperature upshift. Infect. Immun. *66*: 3106–3112.

Zhang, Y. and Y. Rikihisa. 1997. Tyrosine phosphorylation is required for ehrlichial internalization and replication in P388D1 cells. Infect. Immun. *65*: 2959–2964.

Zhao, N., C. Qu, E. Wang and W. Chen. 1995. Phylogenetic evidence for the transfer of *Pseudomonas cocovenenans* (van Damme et al. 1960) to the genus *Burkholderia* as *Burkholderia cocovenenans* (van Damme et al. 1960) comb. nov. Int. J. Syst. Bacteriol. *45*: 600–603.

Zhao, Q., C. Abeygunawardana and A.S. Mildvan. 1997. NMR studies of the secondary structure in solution and the steroid binding site of d5-3-ketosteroid isomerase in complexes with diamagnetic and paramagnetic steroids. Biochemistry *36*: 3458–3472.

Zhao, Q., C. Abeygunawardana, P. Talalay and A.S. Mildvan. 1996. NMR evidence for the participation of a low-barrier hydrogen bond in the mechanism of d5-3-ketosteroid isomerase. Proc. Natl. Acad. Sci. U.S.A. *93*: 8220–8224.

Zhdanov, V. and R.S. Korenblit. 1950. Systematics and nomenclature of viruses. Zhur. Mikrobiol. Epidemiol. Immunobiol. *9*: 40–44.

Zhi, N., N. Ohashi and Y. Rikihisa. 1999. Multiple p44 genes encoding major outer membrane proteins are expressed in the human granulocytic ehrlichiosis agent. J. Biol. Chem. *274*: 17828–17836.

Zhi, N., N. Ohashi, Y. Rikihisa, H.W. Horowitz, G.P. Wormser and K. Hechemy. 1998. Cloning and expression of the 44-kilodalton major outer membrane protein gene of the human granulocytic ehrlichiosis agent and application of the recombinant protein to serodiagnosis. J. Clin. Microbiol. *36*: 1666–1673.

Zhi, N., Y. Rikihisa, H.Y. Kim, G.P. Wormser and H.W. Horowitz. 1997. Comparison of major antigenic proteins of six strains of the human granulocytic ehrlichiosis agent by Western immunoblot analysis. J. Clin. Microbiol. *35*: 2606–2611.

Zhilina, T.N. and G.A. Zavarzin. 1994. Alkaliphilic anaerobic community at pH 10. Curr. Microbiol. *29*: 109–112.

Zhilina, T.N., G.A. Zavarzin, F.A. Rainey, E.N. Pikuta, G.A. Osipov and N.A. Kostrikina. 1997. *Desulfonatronovibrio hydrogenovorans* gen. nov., sp. nov., an alkaliphilic, sulfate-reducing bacterium. Int. J. Syst. Bacteriol. *47*: 144–149.

Zhou, J., L.D. Bowler and B.G. Spratt. 1997. Interspecies recombination, and phylogenetic distortions, within the glutamine synthetase and shikimate dehydrogenase genes of *Neisseria meningitidis* and commensal *Neisseria* species. Mol. Microbiol. *23*: 799–812.

Zhou, J., M.R. Fries, J. Chee-Sanford and J.M. Tiedje. 1995. Phylogenetic analyses of a new group of denitrifiers capable of anaerobic growth on toluene and description of *Azoarcus tolulyticus* sp. nov. Int. J. Syst. Bacteriol. *45*: 500–506.

Zhou, W., F. Rousset and S.L. O'Neill. 1998. Phylogeny and PCR-based classification of *Wolbachia* strains using *wsp* gene sequences. Proc. R. Soc. London Ser B. *265*: 509-515.

Zhu, P.X., G. Morelli and M. Achtman. 1999. The *opcA* and *(psi)opcB* regions in *Neisseria*: Genes, pseudogenes, deletions, insertion elements and DNA islands. Mol. Microbiol. *33*: 635–650.

Ziegler, K., R. Buder, J. Winter and G. Fuchs. 1989. Activation of aromatic acids and aerobic 2-aminobenzoate metabolism in a denitrifying *Pseudomonas* strain. Arch. Microbiol. *151*: 171–176.

Ziegler, M., M. Lange and W. Dott. 1990. Isolation and morphological and cytological characterization of filamentous bacteria from bulking-sludge. Water Res. *24*: 1437–1451.

Zimmer, D., C. Aparicio and C. Elmerich. 1991. Relationship between tryptophan biosynthesis and indole-3-acetic acid production in *Azospirillum*: identification and sequencing of a *trpGDC* cluster. Mol. Gen. Genet. *229*: 41–51.

Zimmer, W., K. Kloos, B. Hundeshagen, E. Niederau and H. Bothe. 1995. Auxin biosynthesis and denitrification in plant growth promoting bacteria. *In* Fendrik, Del Gallo, Vanderleyden and de Zamaroczy (Editors), *Azospirillum* VI and Plant Related Microorganisms, Vol. G. 37, Springer-Verlag, Berlin. pp. 121–128.

Zimmer, W., M.P. Stephan and H. Bothe. 1984. Denitrification by *Azospirillum brasilense* Sp 7. 1. Growth with nitrite as respiratory electron acceptor. Arch. Microbiol. *138*: 206–211.

Zimmerer, R.P., R.H. Hamilton and C. Pootjes. 1966. Isolation and morphology of temperate *Agrobacterium tumejaciens* bacteriophage. J. Bacteriol. *92*: 746–750.

Zimmermann, O.E.R.. 1890. Die Bakterien unserer Trink- und Nutzwässer insbesondere des Wassers der Chemnitzer Wasserleitung. Elfter Bericht. Naturwiss. Ges. Chemnitz, pp. 53–154.

Zinner, S.H., A.K. Daly and W.M. McCormakc. 1973. Isolation of *Eikenella corrodans* in a general hospital. Appl. Microbiol. *25*: 705–708.

Zinsser, H. 1935. Rats, lice, and history, Little, Brown and Co., Boston.

Zipper, C., K. Nickel, W. Angst and H.-P.E. Kohler. 1996. Complete microbial degradation of both enantiomers of the chiral herbicide mecoprop [(RS)-2-(4-chloro-2-methylphenoxy)propionic acid] in an enantioselective manner by *Sphingomonas herbicidovorans* sp. nov. Appl. Environ. Microbiol. *62*: 4318–4322.

Zipper, C., K. Nickel, W. Angst and H.-P.E. Kohler. 1997. *In* Validation of the publication of new names and new combinations previously effectively published outside the IJSB. List No. 61. Int. J. Syst. Bacteriol. *47*: 601–602.

ZoBell, C.E. 1941. Studies on marine bacteria. I. The cultural requirements of heterotrophic aerobes. J. Mar. Res. *4*: 99–106.

ZoBell, C.E. and K.F. Meyer. 1932. Metabolism studies on the *Brucella* group. VI. Nitrates and nitrite reduction. J. Infect. Dis. *51*: 99–108.

ZoBell, C.E. and H.C. Upham. 1944. A list of marine bacteria including descriptions of sixty new species. Bull. Scripps Inst. Oceanogr. Univ. Calif. *5*: 239–292.

Zolg, W. and J.C.G. Ottow. 1975. *Pseudomonas glathei* sp. nov., a new nitrogen scavenging rod isolated from acid lateritic relicts in Germany. Z. Allg. Mikrobiol. *15*: 287–299.

Zopf, W. 1879. Entwicklungsgeschichtliche Untersuchung über Crenothrix polyspora, die Ursache der Berliner Wässercalamitat. Österr. Bot. Z. *29*: 372–373.

Zopf, W. 1883. Die Spaltpilze, Edward Trewendt, Breslau.

Zopf, W. 1885. Die Spaltpilze, 3rd Ed. ed., Edward Trewendt, Breslau. pp. 1–127.

Zöphel, A., M.C. Kennedy, H. Beinert and P.M.H. Kroneck. 1988. Investigations on microbial sulfur respiration. 1. Activation and reduction of elemental sulfur in several strains of eubacteria. Arch. Microbiol. *150*: 72–77.

Zöphel, A., M.C. Kennedy, H. Beinert and P.M.H. Kroneck. 1991. Investigations on microbial sulfur respiration. Isolation, purification, and characterization of cellular components from *Spirillum* 5175. Eur. J. Biochem. *195*: 849–856.

Zou, X.H., F.D. Li and H.K. Chen. 1997. Characteristics of plasmids in *Rhizobium huakuii*. Curr. Microbiol. *35*: 215–220.

Zukal, H. 1896. Myxobotrys variabilis Zuk., als repräsentant einer neuen Myxomyceten-ordnung. Ber. Deutsch. Bot. Ges. *14*: 340–347.

Zukal, H. 1897a. Notiz zu meiner Mittheilung über Myxobotrys variabilis Zuk. im 9. Hefte des jahrganges 1896. Ber. Deutsch. Bot. Ges. *15*: 17–18.

Zukal, H. 1897b. Über die Myxobacterien. Ber. Deutsch. Bot. Ges. *15*: 542–552.

Zumft, W.G. and F. Castillo. 1978. Regulatory properties of nitrogenase from *Rhodopseudomonas palustris*. Arch. Microbiol. *117*: 53–60.

Zusman, D.R. 1980. Genetic approaches to the study of development in the myxobacteria. *In* Leighton and Loomis (Editors), The Molecular Genetics of Development, Academic Press, New York. pp. 41–77.

Zylstra, G.J., R.H. Olsen and D.P. Ballou. 1989a. Cloning, expression, and regulation of the *Pseudomonas cepacia* protocatechuate 3,4-dioxygenase genes. J. Bacteriol. *171*: 5907–5914.

Zylstra, G.J., R.H. Olsen and D.P. Ballou. 1989b. Genetic organization and sequence of the *Pseudomonas cepacia* genes for the alpha and beta subunits of protocatechuate 3,4-dioxygenase. J. Bacteriol. *171*: 5915–5921.

Index of Scientific Names of *Archaea* and *Bacteria*

Key to the fonts and symbols used in this index:

Nomenclature Lower case, Roman	Genera, species, and subspecies of bacteria. Every bacterial name mentioned in the *Manual* is listed in the index. Specific epithets are listed individually and also under the genus.*
CAPITALS, ROMAN:	Names of taxa higher than genus (tribes, families, orders, classes, divisions, kingdoms).
Pagination Roman:	Pages on which taxa are mentioned.
Boldface:	Indicates page on which the description of a taxon is given.†

---

* Infrasubspecific names, such as serovars, biovars, and pathovars, are not listed in the index.

† A description may not necessarily be given in the *Manual* for a taxon that is considered as *incertae sedis* or that is listed in an addendum or note added in proof; however, the page on which the complete citation of such a taxon is given is indicated in boldface type.

Index of Scientific Names of *Archaea* and *Bacteria*